2014 IEEE 64th Electronic Components and Technology Conference

(ECTC 2014)

Lake Buena Vista, Florida, USA
27-30 May 2014

Pages 1-758

IEEE Catalog Number:	**CFP14ECT-POD**
ISBN:	**978-1-4799-2408-0**

Copyright © 2014 by the Institute of Electrical and Electronic Engineers, Inc
All Rights Reserved

Copyright and Reprint Permissions: Abstracting is permitted with credit to the source. Libraries are permitted to photocopy beyond the limit of U.S. copyright law for private use of patrons those articles in this volume that carry a code at the bottom of the first page, provided the per-copy fee indicated in the code is paid through Copyright Clearance Center, 222 Rosewood Drive, Danvers, MA 01923.

For other copying, reprint or republication permission, write to IEEE Copyrights Manager, IEEE Service Center, 445 Hoes Lane, Piscataway, NJ 08854. All rights reserved.

***This publication is a representation of what appears in the IEEE Digital Libraries. Some format issues inherent in the e-media version may also appear in this print version.

IEEE Catalog Number: CFP14ECT-POD
ISBN 13:   978-1-4799-2408-0

Additional Copies of This Publication Are Available From:

Curran Associates, Inc
57 Morehouse Lane
Red Hook, NY 12571 USA
Phone: (845) 758-0400
Fax: (845) 758-2633
E-mail: curran@proceedings.com
Web: www.proceedings.com

TABLE OF CONTENTS

1: Interposer Technologies
Chairs: Subhash L. Shinde, *Sandia National Laboratory*
John Knickerbocker, *IBM Corporation*

Integration Study of Die Strength and Various Bumping Volume and Reliability Performance on 2.5D Silicon Interposer Assembly 1
Shih-Liang Peng, *Siliconware Precision Industries Co., Ltd.*; Chen-Yu Huang, *Siliconware Precision Industries Co., Ltd.*; Ming-Hsien Yang, *Siliconware Precision Industries Co., Ltd.*; Stephen Tseng, *Siliconware Precision Industries Co., Ltd.*; J.Y. Lai, *Siliconware Precision Industries Co., Ltd.*; Terren Lu, *Siliconware Precision Industries Co., Ltd.*; Hsien-Wen Chen, *Siliconware Precision Industries Co., Ltd.*; Steve Chiu, *Siliconware Precision Industries Co., Ltd.*; Stephen Chen, *Siliconware Precision Industries Co., Ltd.*

Process Integration, Improvements, and Testing of Si Interposers for Embedded Computing Applications 8
S. Goodwin, *RTI International*; J. Lannon, Jr., *RTI International*; A. Hilton, *RTI International*; A. Huffman, *RTI International*; M. Lueck, *RTI International*; E. Vick, *RTI International*; G. Cunningham, *RTI International*; D. Malta, *RTI International*; C. Gregory, *RTI International*; D. Temple, *RTI International*

Mechanically Flexible Interconnects with Highly Scalable Pitch and Large Stand-off Height for Silicon Interposer Tile and Bridge Interconnection 13
Chaoqi Zhang, *Georgia Institute of Technology*; Hyung Suk Yang, *Georgia Institute of Technology*; Muhannad S. Bakir, *Georgia Institute of Technology*

Advancements in Fabrication of Glass Interposers 20
Aric Shorey, *Corning Incorporated*; Philippe Cochet, *Rudolph Technologies*; Alan Huffman, *RTI International*; John Keech, *Corning Incorporated*; Matt Lueck, *RTI International*; Scott Pollard, *Corning Incorporated*; Klaus Ruhmer, *Rudolph Technologies*

Large Area Interposer Lithography 26
Warren Flack, *Ultratech, Inc.*; Robert Hsieh, *Ultratech, Inc.*; Gareth Kenyon, *Ultratech, Inc.*; Manish Ranjan, *Ultratech, Inc.*; John Slabbekoorn, *IMEC*; Andy Miller, *IMEC*; Eric Beyne, *IMEC*; Medhat Toukhy, *AZ Electronics Materials USA Corporation*; PingHung Lu, *AZ Electronics Materials USA Corporation*; Chunwei Chen, *AZ Electronics Materials USA Corporation*

Minimizing Interposer Warpage by Process Control and Design Optimization 33
Mikael Detalle, *IMEC*; B. Vandevelde, *IMEC*; P. Nolmans, *IMEC*; J. De Messemaeker, *IMEC*; M. Gonzalez, *IMEC*; A. Miller, *IMEC*; A. La Manna, *IMEC*; G. Beyer, *IMEC*; E. Beyne, *IMEC*

High Performance IPDs (Integrated Passive Devices) and TGV (Through Glass Via) Interposer Technology Using the Photosensitive Glass 41
Jong-Min Yook, *Korea Electronics Technology Institute*; Dongsu Kim, *Korea Electronics Technology Institute*; Jun Chul Kim, *Korea Electronics Technology Institute*

2: Advances in Copper Pillar & Solder Based Flip Chip Technologies

Chairs: Tom Gregorich, *Micron*
Bernd Ebersberger, *Intel Corporation*

Challenges and Opportunities of Chip Package Interaction with Fine Pitch Cu Pillar for 28nm 47
Andy Bao, *Qualcomm, Inc.*; Lily Zhao, *Qualcomm, Inc.*; Yangyang Sun, *Qualcomm, Inc.*; Michael Han, *Qualcomm, Inc.*; Geoffrey Yeap, *Qualcomm, Inc.*; Steve Bezuk, *Qualcomm, Inc.*; Pat Holmes, *Qualcomm, Inc.*; Cecille Alcira, *Qualcomm, Inc.*; Xuefeng Zhang, *Qualcomm, Inc.*; Kenny Lee, *Qualcomm, Inc.*

Electromigration for Advanced Cu Interconnect and the Challenges with Reduced Pitch Bumps 50
Nokibul Islam, *STATS ChipPAC, Inc.*; Gwang Kim, *STATS ChipPAC, Inc.*;
KyungOe Kim, *STATS ChipPAC, Inc.*

Electromigration Performance of Cu Pillar Bump for Flip Chip Packaging with Bump on Trace by Using Thermal Compression Bonding 56
Kuei Hsiao (Frank) Kuo, *Siliconware Precision Industries Co., Ltd.*; Jason Lee, *Siliconware Precision Industries Co., Ltd.*; F.L. Chien, *Siliconware Precision Industries Co., Ltd.*; Rick Lee, *Siliconware Precision Industries Co., Ltd.*; Cindy Mao, *Siliconware Precision Industries Co., Ltd.*; John Lau, *ITRI*

Flip-Chip Bonding Alignment Accuracy Enhancement Using Self-Aligned Interconnection Elements to Realize Low-Temperature Construction of Ultrafine-Pitch Copper Bump Interconnections 62
Bui Thanh Tung, *Nanoelectronics Research Institute; Institute for Photonics-Electronics Convergence System Technology*; Naoya Watanabe, *Nanoelectronics Research Institute*; Fumiki Kato, *Nanoelectronics Research Institute*; Katsuya Kikuchi, *Nanoelectronics Research Institute*; Masahiro Aoyagi, *Nanoelectronics Research Institute; Institute for Photonics-Electronics Convergence System Technology*

Development of Second-Level Connection Method for Large-Size CPU Package 68
Shunji Baba, *Fujitsu Advanced Technologies, Ltd.*; Masateru Koide, *Fujitsu Advanced Technologies, Ltd.*; Manabu Watanabe, *Fujitsu Advanced Technologies, Ltd.*; Kenji Fukuzono, *Fujitsu Advanced Technologies, Ltd.*; Tsuyoshi Yamamoto, *Fujitsu Advanced Technologies, Ltd.*; Seiki Sakuyama, *Fujitsu Laboratories, Ltd.*; Kozo Shimizu, *Fujitsu Laboratories, Ltd.*; Keishiro Okamoto, *Fujitsu Laboratories, Ltd.*; Daisuke Mizutani, *Fujitsu Laboratories, Ltd.*

Development of Fine Pitch Area Array Cu Pillar/Lead Free Solder Bumps for Large 28nm Die in Large Organic Flip Chip Packages 74
John Osenbach, *LSI Corporation*; Sue Emerich, *LSI Corporation*; S. Cate, *LSI Corporation*; D. Brady, *Amkor Technology, Inc.*; Seung Min Hwang, *Amkor Technology, Inc.*; J. Dang, *Kyocera America Inc.*; D. Crouthamel, *LSI Corporation*

ELK Delaminate Improvement Methodology on Cu Pillar Interconnect BOP Structure 81
Nistec Chang, *Siliconware Precision Industries Co., Ltd.*; Albert Lan, *Siliconware Precision Industries Co., Ltd.*; Mark Liao, *Siliconware Precision Industries Co., Ltd.*; Eason Chen, *Siliconware Precision Industries Co., Ltd.*

3: Dynamic Mechanical Characterization

Chairs: Darvin R. Edwards, *Edwards Enterprises*
Tim Chaudhry, *Broadcom Corporation*

Transient Dynamics Model and 3D-DIC Analysis of New-Candidate for JEDEC JESD22-B111 Test Board 85
Pradeep Lall, *Auburn University*; Kalyan Dornala, *Auburn University*; Di Zhang, *Auburn University*; Dongji Xie, *Nvidia Corporation*; Andy Zhang, *Texas Instruments, Inc.*

Interconnect Reliability Prediction for Wafer Level Packages (WLP) for Temperature Cycle and Drop Load Conditions 100
Tong Cui, *Qualcomm Technologies, Inc.*; Ahmer Syed, *Qualcomm Technologies, Inc.*; Beth Keser, *Qualcomm Technologies, Inc.*; Rey Alvarado, *Qualcomm Technologies, Inc.*; Steven Xu, *Qualcomm Technologies, Inc.*; Mark Schwarz, *Qualcomm Technologies, Inc.*

A Novel Drop Test Methodology for Highly Stressed Interconnects in Automotive Electronic Control Units 108

M.H. Shirangi, *Robert Bosch GmbH*; Simo G. Tsebo, *Robert Bosch GmbH*; Z. Wang, *Bosch Automotive Products (Suzhou) Co., Ltd.*; R. Unnikrishnan, *RWTH Aachen University*; T. Heinrich, *Robert Bosch GmbH*

Early-State Crack Detection Method for Heel-Cracks in Wire Bond Interconnects 114

Michael Krüger, *Technical University Berlin*; Stefan Trampert, *Fraunhofer IZM*; Andreas Middendorf, *Technical University Berlin*; Stefan Schmitz, *Fraunhofer IZM*; Klaus-Dieter Lang, *Technical University Berlin*

Accelerated Vibration Reliability Testing of Electronic Assemblies Using Sine Dwell with Resonance Tracking 119

Quang Su, *Binghamton University*; James Pitarresi, *Binghamton University*; Mohammad Gharaibeh, *Binghamton University*; Aaron Stewart, *Binghamton University*; Gaurang Joshi, *Binghamton University*; Martin Anselm, *Universal Instruments Corporation*

Crack Monitoring and Life Modeling Technique Towards High Thermal Cyclic and Mechanical Reliability of fcBGA Solder Joint 126

Dongji Xie, *Nvidia Corporation*; Zhongming Wu, *Nvidia Corporation*; Min Woo, *Nvidia Corporation*; Tom McMullen, *Nvidia Corporation*

Fatigue Properties of Lead-Free Solder Joints in Electronic Packaging Assembly Investigated by Isothermal Cyclic Shear Fatigue 133

Huili Xu, *University of Texas, Arlington*; *Intel Corporation*; Tae-Kyu Lee, *Cisco Systems, Inc.*; Choong-Un Kim, *University of Texas, Arlington*

4: Bio & Flexible Electronics

Chairs: Joana Maria, *IBM Corporation*
C.S. Premachandran, *GLOBALFOUNDRIES*

MEMS-Based Implantable Heart Monitoring System with Integrated Pacing Function 139

Fjodors Tjulkins, *Buskerud and Vestfold University College*; Anh Tuan Thai Nguyen, *Buskerud and Vestfold University College*; Erik Andreassen, *Buskerud and Vestfold University College*; *SINTEF Materials and Chemistry*; Nils Hoivik, *Buskerud and Vestfold University College*; Knut Aasmundtveit, *Buskerud and Vestfold University College*; Lars Hoff, *Buskerud and Vestfold University College*; Ole Johannes Grymyr, *Oslo University Hospital Intervention Centre*; Per Steinar Halvorsen, *Oslo University Hospital Intervention Centre*; Kristin Imenes, *Buskerud and Vestfold University College*

Archipelago Platform for Skin-Mounted Wearable and Stretchable Electronics 145

Yung-Yu Hsu, *MC10, Inc.*; Cole Papakyrikos, *MC10, Inc.*; Milan Raj, *MC10, Inc.*; Mitul Dalal, *MC10, Inc.*; Pinghung Wei, *MC10, Inc.*; Xianyan Wang, *MC10, Inc.*; Gil Huppert, *MC10, Inc.*; Briana Morey, *MC10, Inc.*; Roozbeh Ghaffari, *MC10, Inc.*

Inkjet Printing in Manufacturing of Stretchable Interconnects 151

Toni Liimatta, *Tampere University of Technology*; Eerik Halonen, *Tampere University of Technology*; Hannu Sillanpää, *Tampere University of Technology*; Juha Niittynen, *Tampere University of Technology*; Matti Mäntysalo, *Tampere University of Technology*

Ultra Small Hearing Aid Electronic Packaging Enabled by Chip-in-Flex 157

John Dzarnoski, *Starkey Hearing Technologies*; Susie Johansson, *Starkey Hearing Technologies*

Fabrication of Silicon Based Microfluidics Device for Cell Sorting Application 165

Bivragh Majeed, *IMEC*; Chengxun Liu, *IMEC*; Lut Van Acker, *IMEC*; Robert Daily, *IMEC*; Tomokazu Miyazaki, *JSR Micro NV*; Deniz Sabuncuoglu, *IMEC*; Liesbet Lagae, *IMEC*

A Novel 3D Neural Probe with Integrated Channel and Its Package 170

Xingming Fu, *Wuhan University*; Yong Xu, *Wayne State University*; Yuefa Li, *Wayne State University*; Jinshen Zhang, *Wayne State University*; Xiaobing Luo, *Huazhong University of Science & Technology*; Sheng Liu, *Wuhan University*

CMOS Multiplexer for Portable Biosensing System with Integrated Microfluidic Interface 173

Tetiana Voitsekhivska, *Technical University, Dresden*; Eike Suthau, *Technical University, Dresden*; Klaus-Juergen Wolter, *Technical University, Dresden*

5: Silicon Photonics & LEDs

Chairs: Fuad Doany, *IBM Corporation*
Stefan Weiss, *II-VI Laser Enterprise GmbH*

Assembly of Mechanically Compliant Interfaces between Optical Fibers and Nanophotonic Chips 179

Tymon Barwicz, *IBM Corporation*; Yoichi Taira, *IBM Corporation*; Hidetoshi Numata, *IBM Corporation*; Nicolas Boyer, *IBM Corporation*; Stephane Harel, *IBM Corporation*; Swetha Kamlapurkar, *IBM Corporation*; Shotaro Takenobu, *Asahi Glass Corporation*; Simon Laflamme, *IBM Corporation*; Sebastian Engelmann, *IBM Corporation*; Yurii Vlasov, *IBM Corporation*; Paul Fortier, *IBM Corporation*

Proposal of Integrated-Optic Wavelength-Selective Modulator Based on Coupling-Efficiency Control of Distributed Bragg Reflector in Straight Waveguide 186

Shogo Ura, *Kyoto Institute of Technology*; Testunosuke Miura, *Kyoto Institute of Technology*; Satoshi Kawanami, *Kyoto Institute of Technology*; Kenji Kintaka, *National Institute of Advanced Industrial Science and Technology*; Kosuke Asai, *Kyoto Institute of Technology*; Kenzo Nishio, *Kyoto Institute of Technology*; Yasuhiro Awatsuji, *Kyoto Institute of Technology*

Porous Silicon Technology, a Breakthrough for Silicon Photonics: From Packaging to Monolithic Integration 194

M. Balucani, *Sapienza University of Rome*; A. Klyshko, *Sapienza University of Rome*; K. Kholostov, *Sapienza University of Rome*; A. Benedetti, *Sapienza University of Rome*; A. Belardini, *Sapienza University of Rome*; C. Sibilia, *Sapienza University of Rome*; M. Izzi, *Enea Casaccia Research Centre Rome*; M. Tucci, *Enea Casaccia Research Centre Rome*; H. Bandarenka, *Belarusian State University of Informatics and Radioelectronics*; V. Bondarenko, *Belarusian State University of Informatics and Radioelectronics*

High Power Density LED Modules with Silver Sintering Die Attach on Aluminum Nitride Substrates 203

Marc Schneider, *Karlsruhe Institute of Technology*; Benjamin Leyrer, *Karlsruhe Institute of Technology*; Christian Herbold, *Karlsruhe Institute of Technology*; Stefan Maikowske, *Karlsruhe Institute of Technology*

Effect of Optical Design on the Thermal Management for the Smart TV LED Backlight Systems 209

Kivanc Karsli, *Vestel AS*; Mehmet Arik, *Ozyegin University*

Wafer Level LED Packaging with Optimal Light Output and Thermal Dissipation for High-Brightness Lighting 215

Liang Wang, *Invensas Corporation*; Gabe Guevara, *Invensas Corporation*; Hala Shaba, *Invensas Corporation*; Roseann Alatorre, *Invensas Corporation*; Rey Co, *Invensas Corporation*; Ron Zhang, *Invensas Corporation*

High Power Laser Packaging Challenges and Standardization 221

Eric Zhou, *LDX Optronics, Inc.*; Jeffrey Morris, *LDX Optronics, Inc.*; Hanguo Wang, *University of California, Los Angeles*

6: Adhesives, Underfills, and Thermal Interface Materials
Chairs: Don Frye, *ATMI*
C. Robert Kao, *National Taiwan University*

Novel Highly Moisture Resistant Optical Adhesives and Their High Power Resistivity 230
Seiko Mitachi, *Tokyo University of Technology*; Kazushi Kimura, *Yokohama Rubber Co. Ltd.*

Engineered Thermal Interface Material 236
Lyndon Larson, *Dow Corning Corporation*; Yin Tang, *Dow Corning Corporation*; Loren Durfee, *Dow Corning Corporation*; Cassandra Hale, *Dow Corning Corporation*; David Plante, *Dow Corning Corporation*; Sushumna Iruvanti, *IBM Corporation*; Rebecca Wagner, *IBM Corporation*; Taryn Davis, *IBM Corporation*; Hai Longworth, *IBM Corporation*; Annique LaVoie, *IBM Corporation*; Richard Langlois, *IBM Corporation*

Degradation Mechanisms in Electronic Mold Compounds Subjected to High Temperature in Neighborhood of 200°C 242
Pradeep Lall, *Auburn University*; Shantanu Deshpande, *Auburn University*; Yihua Luo, *Auburn University*; Mike Bozack, *Auburn University*; Luu Nguyen, *Texas Instruments*; Masood Murtuza, *Texas Instruments*

Time, Temperature, and Mechanical Fatigue Dependence on Underfill Adhesion 255
Joseph Cremaldi, *Tulane University*; Michael Gaynes, *IBM Corporation*; Peter Brofman, *IBM Corporation*; Noshir Pesika, *Tulane University*; Eric Lewandowski, *IBM Corporation*

Study on Isotropic Electrically Conductive Adhesive for Medical Device Applications 263
Shawn Shi, *Medtronic, Inc.*; Scott Sleeper, *Medtronic, Inc.*; Chunho Kim, *Medtronic, Inc.*

Effect of Aligned Nanofiber in Nanofiber Solder Anisotropic Conductive Films (ACFs) on the Solder Ball Movement for Flex-on-Flex (FOF) Assembly 271
Tae-Wan Kim, *Korea Advanced Institute of Science and Technology (KAIST)*; Sang-Hoon Lee, *Korea Advanced Institute of Science and Technology (KAIST)*; Kyung-Wook Paik, *Korea Advanced Institute of Science and Technology (KAIST)*

Adhesive Enabling Technology for Directly Plating Metal on Molding Resin 279
Kwonil Kim, *Atotech USA, Inc.*; Kenichiroh Mukai, *Atotech USA, Inc.*; Brian Eastep, *Atotech USA, Inc.*; Lee Gaherty, *Atotech USA, Inc.*; Anirudh Kashyap, *Atotech USA, Inc.*; Lutz Brandt, *Atotech USA, Inc.*

7: Interposers & 3D Integration
Chairs: Katsuyuki Sakuma, *IBM Corporation*
Lou Nicholls, *Amkor Technology, Inc*

Modeling, Design, and Demonstration of Low-Temperature Cu Interconnections to Ultra-Thin Glass Interposers at 20 μm Pitch 284
Tao Wang, *Georgia Institute of Technology*; Vanessa Smet, *Georgia Institute of Technology*; Makoto Kobayashi, *Namics Corporation*; Venky Sundaram, *Georgia Institute of Technology*; P. Mardkondeya Raj, *Georgia Institute of Technology*; Rao Tummala, *Georgia Institute of Technology*

Low-Cost TSH (Through-Silicon Hole) Interposers for 3D IC Integration 290

John H. Lau, *Industrial Technology Research Institute (ITRI)*; Ching-Kuan Lee, *Industrial Technology Research Institute (ITRI)*; Chau-Jie Zhan, *Industrial Technology Research Institute (ITRI)*; Sheng-Tsai Wu, *Industrial Technology Research Institute (ITRI)*; Yu-Lin Chao, *Industrial Technology Research Institute (ITRI)*; Ming-Ji Dai, *Industrial Technology Research Institute (ITRI)*; Ra-Min Tain, *Industrial Technology Research Institute (ITRI)*; Heng-Chieh Chien, *Industrial Technology Research Institute (ITRI)*; Chun-Hsien Chien, *Industrial Technology Research Institute (ITRI)*; Ren-Shin Cheng, *Industrial Technology Research Institute (ITRI)*; Yu-Wei Huang, *Industrial Technology Research Institute (ITRI)*; Yuan-Chang Lee, *Industrial Technology Research Institute (ITRI)*; Zhi-Cheng Hsiao, *Industrial Technology Research Institute (ITRI)*; Wen-Li Tsai, *Industrial Technology Research Institute (ITRI)*; Pai-Cheng Chang, *Industrial Technology Research Institute (ITRI)*; Huan-Chun Fu, *Industrial Technology Research Institute (ITRI)*; Yu-Mei Cheng, *Industrial Technology Research Institute (ITRI)*; Li-Ling Liao, *Industrial Technology Research Institute (ITRI)*; Wei-Chung Lo, Industrial Technology Research Institute (ITRI); Ming-Jer Kao, *Industrial Technology Research Institute (ITRI)*

Cu Pattern Density Impacts on 2.5D TSI Warpage Using Experimental and FEM Analysis 297

C.T. Yeh, *United Microelectronics Corporation*; C.Y. Wu, *United Microelectronics Corporation*; C.F. Lin, *United Microelectronics Corporation*; K.M. Chen, *United Microelectronics Corporation*; M.J. Lin, *United Microelectronics Corporation*; Y.C. Lin, *United Microelectronics Corporation*; C.L. Kuo, *United Microelectronics Corporation*

A Resilient 3-D Stacked Multicore Processor Fabricated Using Die-Level 3-D Integration and Backside TSV Technologies 304

K.W. Lee, *Tohoku University*; H. Hashimoto, *Tohoku University*; M. Onishi, *Tohoku University*; Y. Sato, *Tohoku University*; M. Murugesan, *Tohoku University*; J.-C. Bae, *Tohoku University*; T. Fukushima, *Tohoku University*; T. Tanaka, *Tohoku University*; M. Koyanagi, *Tohoku University*

3D Stacking Induced Mechanical Stress Effects 309

V. Cherman, *IMEC*; G. Van der Plas, *IMEC*; J. De Vos, *IMEC*; A. Ivankovic, *IMEC; KU Leuven*; M. Lofrano, *IMEC*; V. Simons, *IMEC*; M. Gonzalez, *IMEC*; K. Vanstreels, *IMEC*; T. Wang, *IMEC*; R. Daily, *IMEC*; W. Guo, *IMEC*; G. Beyer, *IMEC*; A. La Manna, *IMEC*; I. De Wolf, *IMEC*; E. Beyne, *IMEC*

Six-Die Stacking: Three-Dimensional Interconnects Using Au and Pillar Bumps 316

Fei-Jain Wu, *Chipbond Technology Corporation*; Lung-Hua Ho, *Chipbond Technology Corporation*; Chih-Ming Kuo, *Chipbond Technology Corporation*; Chia-Jung Tu, *Chipbond Technology Corporation*; Chih-Hsien Ni, *Chipbond Technology Corporation*; Shih-Chieh Chang, *Chipbond Technology Corporation*; Chuan-Yu Wu, *Chipbond Technology Corporation*; Kung-An Lin, *Chipbond Technology Corporation*; Wei-Hsin Wu, *Chipbond Technology Corporation*; Yung Shen Wu, *Chipbond Technology Corporation*

TSV-Less 3D Stacking of MEMS and CMOS via Low Temperature Al-Au Direct Bonding with Simultaneous Formation of Hermetic Seal 324

S.L. Chua, *Nanyang Technological University*; A. Razzaq, *Nanyang Technological University*; K.H. Wee, *DSO National Laboratory*; K.H. Li, *Nanyang Technological University*; H. Yu, *Nanyang Technological University*; C.S. Tan, *Nanyang Technological University*

8: Flip Chip Packaging & Advanced Substrate
Chairs: Young-Gon Kim, *IDT*
Omar Bchir, *Qualcomm, Inc.*

Chip Package Interaction: An Experiment Study on White Bump Mitigation Using Flat Laminates 332

Yi Pan, *IBM Corporation*; Jeffrey A. Zitz, *IBM Corporation*; David L. Questad, *IBM Corporation*; Kamal K. Sikka, *IBM Corporation*

Design and Package Technology Development of Face-to-Face Die Stacking as a Low Cost Alternative for 3D IC Integration 338

Zhe Li, *Altera Corporation*; Yuan Li, *Altera Corporation*; John Xie, *Altera Corporation*

From C4 to Micro-Bump: Adapting Lead Free Solder Electroplating Processes to Next-Gen Advanced Packaging Applications 342

Julia Woertink, *Dow Electronic Materials*; Yi Qin, *Dow Electronic Materials*; Jonathan Prange, *Dow Electronic Materials*; Pedro Lopez-Montesinos, *Dow Electronic Materials*; Inho Lee, *Dow Electronic Materials*; Yil-Hak Lee, *Dow Electronic Materials*; Masaaki Imanari, *Dow Electronic Materials*; Jianwei Dong, *Dow Electronic Materials*; Jeffrey Calvert, *Dow Electronic Materials*

Development of New 2.5D Package with Novel Integrated Organic Interposer Substrate with Ultra-Fine Wiring and High Density Bumps 348

Kiyoshi Oi, *Shinko Electric Industries Company, Ltd.*; Satoshi Otake, *Shinko Electric Industries Company, Ltd.*; Noriyoshi Shimizu, *Shinko Electric Industries Company, Ltd.*; Shoji Watanabe, *Shinko Electric Industries Company, Ltd.*; Yuji Kunimoto, *Shinko Electric Industries Company, Ltd.*; Takashi Kurihara, *Shinko Electric Industries Company, Ltd.*; Toshinori Koyama, *Shinko Electric Industries Company, Ltd.*; Masato Tanaka, *Shinko Electric Industries Company, Ltd.*; Lavanya Aryasomayajula, *GLOBALFOUNDRIES, Inc.*; Zafer Kutlu, *GLOBALFOUNDRIES, Inc.*

Package Embedded Decoupling Capacitor Impact on Core Power Delivery Network for ARM SoC Application 354

Ga Won Kim, *Samsung Semiconductor Inc.*; Max (Sungwan) Min, *Samsung Semiconductor Inc.*; Melinda (Ling) Yang, *Samsung Semiconductor Inc.*; Anil Gundurao, *Samsung Semiconductor Inc.*; Eileen You, *Samsung Semiconductor Inc.*; Harpreet Gill, *Samsung Semiconductor Inc.*; Seungyong Cha, *Samsung Electronics Corporation*; Younghoon Kim, *Samsung Electronics Corporation*; Se-Ho You, *Samsung Electronics Corporation*; Seungbae Lee, *Samsung Electronics Corporation*; Woonghwan Ryu, *Samsung Electronics Corporation*

Embed Glass Interposer to Substrate for High Density Interconnection 360

Dyi-Chung Hu, *Unimicron Technology Corporation*; Yin-Po Hung, *Unimicron Technology Corporation*; Yu-Hua Chen, *Unimicron Technology Corporation*; Ra-Min Tain, *Unimicron Technology Corporation*; Wei-Chung Lo, *Industrial Technology Research Institute (ITRI)*

First Demonstration of a Surface Mountable, Ultra-Thin Glass BGA Package for Smart Mobile Logic Devices 365

Venky Sundaram, *Georgia Institute of Technology*; Yoichiro Sato, *Asahi Glass Company*; Toshitake Seki, *NGK Spark Plug Co., Ltd.*; Yutaka Takagi, *NGK Spark Plug Co., Ltd.*; Vanessa Smet, *Georgia Institute of Technology*; Makoto Kobayashi, *Namics Corporation*; Rao Tummala, *Georgia Institute of Technology*

9: Interconnect Reliability

Chairs: Tz-Cheng Chiu, *National Cheng Kung University*
Vikas Gupta, *Texas Instruments*

Towards a Quantitative Mechanistic Understanding of the Thermal Cycling of SnAgCu Solder Joints 371

D. Schmitz, *Binghamton University*; S. Shirazi, *Binghamton University*; L. Wentlent, *Binghamton University*; S. Hamasha, *Binghamton University*; L. Yin, *GE Global Research*; A. Qasaimeh, *Tennessee Tech University*; P. Borgesen, *Binghamton University*

Exploration of Aging Induced Evolution of Solder Joints Using Nanoindentation and Microdiffraction 379

Mohammad Hasnine, *Auburn University*; Jeffrey C. Suhling, *Auburn University*; Barton C. Prorok, *Auburn University*; Michael J. Bozack, *Auburn University*; Pradeep Lall, *Auburn University*

Accessing Adhesive Induced Risk for BGAs in Temperature Cycling 395

Guruprasad Arakere, *Intel Corporation*; Milena Vujosevic, *Intel Corporation*; Min Pei, *Intel Corporation*

Characteristics of Ceramic BGA Using Polymer Core Solder Balls 404

Hiroya Ishida, *Sekisui Chemical Co., Ltd.*; Kiyoto Matsushita, *Sekisui Chemical Co., Ltd.*

Lifetime Prediction of Cu-Al Wire Bonded Contacts for Different Mould Compounds 411

René Rongen, *NXP Semiconductors*; G.M. O'Halloran, *NXP Semiconductors*; Amar Mavinkurve, *NXP Semiconductors*; Leon Goumans, *NXP Semiconductors*; Mark-Luke Farrugia, *NXP Semiconductors*

The Corrosion Performance of Cu Alloy Wire Bond on Al Pad in Molding Compounds of Various Chlorine Contents under Biased-HAST 419

Ying-Ta Chiu, *ASE Group*; Tzu-Hsing Chiang, *ASE Group*; Yin-Fa Chen, *ASE Group*; Ping-Feng Yang, *ASE Group*; Louie Huang, *ASE Group*; Kwang-Lung Lin, *National Cheng Kung University*

The Effect of Nickel Microalloying on Thermal Fatigue Reliability and Microstructure of SAC105 and SAC205 Solders 425

Richard Coyle, *Alcatel-Lucent*; Richard Parker, *iNEMI*; Babak Arfaei, *Universal Instruments*; Francis Mutuku, *Binghamton University*; Keith Sweatman, *Nihon Superior Co., Ltd.*; Keith Howell, *Nihon Superior Co., Ltd.*; Stuart Longgood, *Delphi*; Elizabeth Benedetto, *Hewlett Packard Company*

10: Novel Materials & Processes
Chairs: Ivan Shubin, *Oracle*
Bing Dang, *IBM Corporation*

Flexible Non-Volatile Cu/CuxO/Ag ReRAM Memory Devices Fabricated Using Ink-Jet Printing Technology 441

Simin Zou, *Auburn University*; Michael C. Hamilton, *Auburn University*

Ultra-High Refractive Index LED Encapsulant 447

Chia-Chi Tuan, *Georgia Institute of Technology*; Ziyin Lin, *Georgia Institute of Technology*; Yan Liu, *Georgia Institute of Technology*; Kyoung-Sik Moon, *Georgia Institute of Technology*; Sehoon Yoo, *Korea Institute of Industrial Technology*; Myong-Gi Jang, *El Lighting Co. Ltd.*; Ching-Ping Wong, *Georgia Institute of Technology; Chinese University of Hong Kong*

A Novel Methodology for Wafer-Specific Feed-Forward Management of Backside Silicon Removal by Wafer Grinding for Optimized Through Silicon Via Reveal 452

Tyson Alvanos, *Disco Hi Tec America, Inc.*; John Garant, *IBM Corporation*; Yu Iijima, *Disco Hi Tec America, Inc.*; Richard Indyk, *IBM Corporation*; Christopher Rosenthal, *Lasertec USA, Inc.*; Osamu Sato, *Lasertec Corporation*; Naoki Sugase, *Lasertec Corporation*; Hideo Takizawa, *Lasertec Corporation*; Frank Wei, *Disco Hi Tec America, Inc.*

Thermal Characterization of Power Devices Using Graphene-Based Film 459

Pengtu Zhang, *Chalmers University of Technology; East China University of Science and Technology*; Nan Wang, *Chalmers University of Technology*; Carl Zandén, *Chalmers University of Technology*; Lilei Ye, *Smart High Tech AB*; Yifeng Fu, *Smart High Tech AB*; Johan Liu, *Chalmers University of Technology*

High Performance Phase Change Thermal Interface Materials Based on Porous Graphitic Carbon Spheres-Paraffin Wax Composite 464

Zhihua Cao, *Shenzhen Institutes of Advanced Technology, Chinese Academy of Sciences; University of Science and Technology of China*; Kai Zhang, *Hong Kong University of Science and Technology*; Gaugping Zhang, *Shenzhen Institutes of Advanced Technology, Chinese Academy of Sciences*; Matthew M.F. Yuen, *Hong Kong University of Science and Technology*; Ping Gu, *University of Science and Technology of China*; Xianzhu Fu, *Shenzhen Institutes of Advanced Technology, Chinese Academy of Sciences*; Rong Sun, *Shenzhen Institutes of Advanced Technology, Chinese Academy of Sciences*; C.P. Wong, *Chinese University of Hong Kong*

High Sensitivity In-Plane Strain Measurement Using a Laser Scanning Technique 470

Hanshuang Liang, *Arizona State University*; Teng Ma, *Arizona State University*; Cheng Lv, *Arizona State University*; Hoa Nguyen, *Arizona State University*; George Chen, *Arizona State University*; Hao Wu, *Arizona State University*; Rui Tang, *Arizona State University*; Hanqing Jiang, *Arizona State University*; Hongbin Yu, *Arizona State University*

Biophysicochemical Evaluation of Passivation Layers for the Packaging of Silicon Microsystems in Medical Devices 478

Jorge Mario Herrera Morales, *CEA-LETI*; Jean-Charles Souriau, *CEA-LETI*; David Ratel, *CEA-LETI*; François Berger, *CEA-LETI*; Gilles Simon, *CEA-LETI*

11: Innovative Packaging Technologies

Chairs: Paul Tiner, *Texas Instruments*
Shichun Qu, *Fairchild Semiconductor*

A New Era in Manufacturable, Low-Temperature and Ultra-Fine Pitch Cu Interconnections and Assembly without Solders 484

Vanessa Smet, *Georgia Institute of Technology*; Makoto Kobayashi, *Namics Corporation*; Tao Wang, *Georgia Institute of Technology*; Pulugurtha Markondeya Raj, *Georgia Institute of Technology*; Rao Tummala, *Georgia Institute of Technology*

Enabling Fine Pitch Cu & Ag Alloy Wire Bond Assessment for 28nm Ultra Low-k Structure 490

John D. Beleran, *United Test and Assembly Center, Ltd.*; Ninoy Milanes II, *United Test and Assembly Center, Ltd.*; Gaurav Mehta, *United Test and Assembly Center, Ltd.*; Nathapong Suthiwongshunthorn, *United Test and Assembly Center, Ltd.*; Ranjan Rajoo, *GLOBALFOUNDRIES, Inc.*; Chan Kai Chong, *GLOBALFOUNDRIES, Inc.*

Assembly of Multiple Chips on Flexible Substrate Using Anisotropic Conductive Film for Medical Imaging Applications 498

Hoang-Vu Nguyen, *Buskerud and Vestfold University College*; Trym Eggen, *GE Vingmed Ultrasound AS*; Bjørnar Sten-Nilsen, *GE Vingmed Ultrasound AS*; Kristin Imenes, *Buskerud and Vestfold University College*; Knut E. Aasmundtveit, *Buskerud and Vestfold University College*

High Frequency High Current Point of Load Modules with Integrated Planar Inductors 504

Wenli Zhang, *Virginia Polytechnic Institute and State University*; Yipeng Su, *Virginia Polytechnic Institute and State University*; David Gilham, *Virginia Polytechnic Institute and State University*; Mingkai Mu, *Virginia Polytechnic Institute and State University*; Qiang Li, *Virginia Polytechnic Institute and State University*; Fred C. Lee, *Virginia Polytechnic Institute and State University*

Integrated Microprobe Array and CMOS MEMS by TSV Technology for Bio-Signal Recording Application 512

Lei-Chun Chou, *National Chiao Tung University*; Shih-Wei Lee, *National Chiao Tung University*; Po-Tsang Huang, *National Chiao Tung University*; Chih-Wei Chang, *University of California, Los Angeles*; Shang-Lin Wu, *National Chiao Tung University*; Jin-Chern Chiou, *National Chiao Tung University; China Medical University*; Ching-Te Chuang, *National Chiao Tung University*; Wei Hwang, *National Chiao Tung University; Advanced Semiconductor Engineering, Inc.*; Chung-Hsi Wu, *Advanced Semiconductor Engineering, Inc.*; Kuo-Hua Chen, *Advanced Semiconductor Engineering, Inc.*; Chi-Tsung Chiu, *Advanced Semiconductor Engineering, Inc.*; Ho-Ming Tong, *Advanced Semiconductor Engineering, Inc.*; Kuan-Neng Chen, *National Chiao Tung University*

Material Characterization of a Novel Lead-Free Solder Material - SACQ 518

Tak-Sang Yeung, *Broadcom Corporation*; Henry Sze, *Broadcom Corporation*; Keith Tan, *Broadcom Corporation*; Javed Sandhu, *Broadcom Corporation*; Chong-Wei Neo, *Broadcom Corporation*; Edward Law, *Broadcom Corporation*

Lithography Challenges for 2.5D Interposer Manufacturing 523

Klaus Ruhmer, *Rudolph Technologies, Inc.*; Philippe Cochet, *Rudolph Technologies, Inc.*; Roger McCleary, *Rudolph Technologies, Inc.*; Rich Rogoff, *Rudolph Technologies, Inc.*; Rajiv Roy, *Rudolph Technologies, Inc.*

12: Power Integrity & Passive Component Modeling
Chairs: Wendem Beyene, *Rambus Inc.*
Daniel de Araujo, *Nimbic, Inc.*

Package Embedded Inductors for Integrated Voltage Regulators 528
William J. Lambert, *Intel Corporation*; Michael J. Hill, *Intel Corporation*; Kaladhar Radhakrishnan, *Intel Corporation*; Leigh Wojewoda, *Intel Corporation*; Anne E. Augustine, *Intel Corporation*

Power Supply Filter for PLL Circuit in Digital Systems 535
Nam Pham, *IBM Corporation*; Faraydon Pakbaz, *IBM Corporation*; Zhenrong Jin, *IBM Corporation*; Lloyd Walls, *IBM Corporation*

Coaxial Through-Package-Vias (TPVs) for Enhancing Power Integrity in 3D Double-Side Glass Interposers 541
Gokul Kumar, *Georgia Institute of Technology*; P. Markondeya Raj, *Georgia Institute of Technology*; Jounghyun Cho, *Korea Advanced Institute of Science and Technology (KAIST)*; Saumya Gandhi, *Georgia Institute of Technology*; Parthasarathi Chakraborti, *Georgia Institute of Technology*; Venky Sundaram, *Georgia Institute of Technology*; Joungho Kim, *Korea Advanced Institute of Science and Technology (KAIST)*; Rao Tummala, *Georgia Institute of Technology*

Modeling of Switching Noise and Coupling in Multiple Chips of 3D TSV-Based Systems 548
Huanyu He, *Rensselaer Polytechnic Institute*; Xiaoxiong Gu, *IBM Corporation*; Jian-Qiang Lu, *Rensselaer Polytechnic Institute*

Characterization of On-Die Power Supply Noise in FCBGA (Flip-Chip Ball Grid Array) Packages 554
Hyunho Baek, *University of Florida*; William R. Eisenstadt, *University of Florida*

An Enhanced Power Integrity Analysis Flow Based on the Interdependence between Simultaneous Switching Output Noise and Static IR Drop 560
Minghui Han, *Samsung Display*; Amir Amirkhany, *Samsung Display*; Wei Xiong, *Samsung Display*

Improving the Target Impedance Method for PCB Decoupling of Core Power 566
Guang Chen, *Altera Corporation*; Dan Oh, *Altera Corporation*

13: 3D Process Integration & Die Stacking
Chairs: Rozalia Beica, *Yole Developpement*
Jianwei Dong, *Dow Electronic Materials*

Process Development to Enable 3D IC Multi-Tier Die Bond for 20µm Pitch and Beyond 572
Y.H. Hu, *TSMC*; C.S. Liu, *TSMC*; M.T. Chen, *TSMC*; M.D. Cheng, *TSMC*; H.J. Kuo, *TSMC*; M.J. Lii, *TSMC*; A. LaManna, *IMEC*; K.J. Rebibis, *IMEC*; T. Wang, *IMEC*; S.V. Huylenbroeck, *IMEC*; R. Daily, *IMEC*; G. Capuz, *IMEC*; D. Velenis, *IMEC*; G. Beyer, *IMEC*; E. Beyne, *IMEC*; Doug C.H. Yu, *TSMC*

Factors in the Selection of Temporary Wafer Handlers for 3D/2.5D Integration 576
Bing Dang, *IBM Corporation*; Bucknell Webb, *IBM Corporation*; Cornelia Tsang, *IBM Corporation*; Paul Andry, *IBM Corporation*; John Knickerbocker, *IBM Corporation*

Optimization and Challenges on TSV MEOL Integration 582
DoHyeong Kim, *Amkor Technology Korea, Inc.*; DongHun Lee, *Amkor Technology Korea, Inc.*; YoungChul Seo, *Amkor Technology Korea, Inc.*; JungSoo Park, *Amkor Technology Korea, Inc.*; SeungChul Han, *Amkor Technology Korea, Inc.*; BoRa Jang, *Amkor Technology Korea, Inc.*; JooHyun Khim, *Amkor Technology Korea, Inc.*; YoungSuk Chung, *Amkor Technology Korea, Inc.*; SeongMin Seo, *Amkor Technology Korea, Inc.*; ChoonHeung Lee, *Amkor Technology Korea, Inc.*

TSV Integration on 20nm Logic Si: 3D Assembly and Reliability Results 590

Rahul Agarwal, *GLOBALFOUNDRIES, Inc.*; Dave Hiner, *Amkor Technology, Inc.*; Sukeshwar Kannan, *GLOBALFOUNDRIES, Inc.*; KiWook Lee, *Amkor Technology, Inc.*; DoHyeong Kim, *Amkor Technology, Inc.*; JongSik Paek, *Amkor Technology, Inc.*; SungGeun Kang, *Amkor Technology, Inc.*; Yong Son, *Amkor Technology, Inc.*; Sebastian Dej, *GLOBALFOUNDRIES, Inc.*; Dan Smith, *GLOBALFOUNDRIES, Inc.*; Sara Thangaraju, *GLOBALFOUNDRIES, Inc.*; Jens Paul, *GLOBALFOUNDRIES, Inc.*

TSV MEOL (Mid End of Line) and Packaging Technology of Mobile 3D-IC Stacking 596

Duk Ju Na, *STATS ChipPAC, Ltd.*; Kyaw Oo Aung, *STATS ChipPAC, Ltd.*; Won Kyung Choi, *STATS ChipPAC, Ltd.*; Tsuyoshi Kida, *Renesas Electronics Company*; Toshihiko Ochiai, *Renesas Electronics Company*; Tomoaki Hashimoto, *Renesas Electronics Company*; Michitaka Kimura, *Renesas Electronics Company*; Keiichirou Kata, *Renesas Electronics Company*; Seung Wook Yoon, *STATS ChipPAC, Ltd.*; Andy Chang Bum Yong, *STATS ChipPAC, Ltd.*

Thermally Enhanced 3 Dimensional Integrated Circuit (TE3DIC) Packaging 601

S. Snyder, *Harris Corporation GCSD*; J. Thompson, *Harris Corporation GCSD*; A. King, *Harris Corporation GCSD*; E. Walters, *Harris Corporation GCSD*; P. Tyler, *Harris Corporation GCSD*; M.R. Weatherspoon, *Harris Corporation GCSD*

Filler Trap and Solder Extrusion in 3D IC Thermo-Compression Bonded Microbumps 609

Yingxia Liu, *University of California, Los Angeles*; Menglu Li, *University of California, Los Angeles*; Dong Wook Kim, *Qualcomm, Inc.*; Sam Gu, *Qualcomm, Inc.*; Dilworth Y. Parkinson, *Lawrence Berkeley National Laboratory*; Justin Blair, *Lawrence Berkeley National Laboratory*; K.N. Tu, *University of California, Los Angeles*

14: TSV Fabrication & Its Reliability Impact

Chairs: Li Li, *Cisco Systems, Inc.*
Wei-Chung Lo, *ITRI*

Correlation between Cu Microstructure and TSV Cu Pumping 613

Joke De Messemaeker, *IMEC*; Olalla Varela Pedreira, *IMEC*; Harold Philipsen, *IMEC*; Eric Beyne, *IMEC*; Ingrid De Wolf, *IMEC*; Tom Van der Donck, *KU Leuven*; Kristof Croes, *IMEC*

TSV Reliability Model under Various Stress Tests 620

Ben-Je Lwo, *National Defense University*; Frank M.-S. Lin, *National Defense University*; Kuo-Hsin Huang, *National Defense University*

Development of Process and Design Criteria for Stress Management in Through Silicon Vias 625

O. Hölck, *Fraunhofer IZM*; M. Nuss, *Fraunhofer IZM*; A. Grams, *Fraunhofer IZM*; T. Prewitz, *Fraunhofer IZM*; P. John, *Fraunhofer IZM*; C. Fiedler, *Fraunhofer IZM*; M. Böttcher, *Fraunhofer IZM*; H. Walter, *Fraunhofer IZM*; M.J. Wolf, *Fraunhofer IZM*; O. Wittler, *Fraunhofer IZM*; K.-D. Lang, *Technical University Berlin*

High-Speed Wet Etching of Through Silicon Vias (TSVs) in Micro- and Nanoscale 631

Liyi Li, *Georgia Institute of Technology*; Ching-Ping Wong, *Georgia Institute of Technology; Chinese University of Hong Kong*

Replacing the PECVD-SiO$_2$ in the Through-Silicon Via of High-Density 3D LSIs with Highly Scalable Low Cost Organic Liner: Merits and Demerits 636

Murugesan Mariappan, *NICHe, Tohoku University*; Takafumi Fukushima, *NICHe, Tohoku University*; JiChel Beatrix, *NICHe, Tohoku University*; Hiroyuki Hashimoto, *NICHe, Tohoku University*; Yutaka Sato, *NICHe, Tohoku University*; Kangwook Lee, *NICHe, Tohoku University*; Tetsu Tanaka, *NICHe, Tohoku University*; Mitsumasa Koyanagi, *NICHe, Tohoku University*

Investigation of a TSV-RDL In-line Fault-Diagnosis System and Test Methodology for Wafer-level Commercial Production 641

Runiu Fang, *Peking University*; Min Miao, *Beijing Information Science and Technology University*; Xin Sun, *Peking University*; Yunhui Zhu, *Peking University*; Guanjiang Wang, *Peking University Shenzhen Graduate School*; Yichao Xu, *Peking University Shenzhen Graduate School*; Minggang Sun, *Beijing Information Science and Technology University*; Yufeng Jin, *Peking University*

Bonding Technologies for Chip Level and Wafer Level 3D Integration 647

Katsuyuki Sakuma, *IBM Corporation*; Spyridon Skordas, *IBM Corporation*; Jeffrey Zitz, *IBM Corporation*; Eric Perfecto, *IBM Corporation*; William Guthrie, *IBM Corporation*; Luc Guerin, *IBM Corporation*; Richard Langlois, *IBM Corporation*; Hsichang Liu, *IBM Corporation*; Koushik Ramachandran, *IBM Corporation*; Wei Lin, *IBM Corporation*; Kevin Winstel, *IBM Corporation*; Sayuri Kohara, *IBM Corporation*; Kuniaki Sueoka, *IBM Corporation*; Matthew Angyal, *IBM Corporation*; Troy Graves-Abe, *IBM Corporation*; Daniel Berger, *IBM Corporation*; John Knickerbocker, *IBM Corporation*; Subramanian Iyer, *IBM Corporation*

15: Solder Joint Reliability
Chairs: Keith Newman, *Hewlett-Packard Company*
Toni Mattila, *Aalto University*

Dependence of Solder Joint Reliability on Solder Volume, Composition and Printed Circuit Board Surface Finish 655

Babak Arfaei, *Universal Instruments Corporation*; Francis Mutuku, *Binghamton University*; Keith Sweatman, *Nihon-Superior*; Ning-Cheng Lee, *Indium Corporation*; Eric Cotts, *Binghamton University*; Richard Coyle, *Alcatel-Lucent*

The Effects of Aging on the Fatigue Life of Lead Free Solders 666

Muhannad Mustafa, *Auburn University*; Jordan C. Roberts, *Auburn University*; Jeffrey C. Suhling, *Auburn University*; Pradeep Lall, *Auburn University*

Solder Joint Height Impact on Temperature Cycle Reliability of BGA Components with Thermal Enabling Load 684

Yun Ge, *Intel Corporation*; Jeffery Cook, *Intel Corporation*; Min Pei, *Intel Corporation*; Milena Vujosevic, *Intel Corporation*; Bite Zhou, *Intel Corporation*; Suddhasattwa Nad, *Intel Corporation*

Controlling the Sn Grain Morphology of SnAg C4 Solder Bumps 690

Gregory Parks, *Binghamton University*; Minhua Lu, *IBM Corporation*; Eric Perfecto, *IBM Corporation*; Eric Cotts, *Binghamton University*

The Impact of Microstructure Evolution, Localized Recrystallization and Board Thickness on Sn-Ag-Cu Interconnect Board Level Shock Performance 697

Tae-Kyu Lee, *Cisco Systems, Inc.*; Weidong Xie, *Cisco Systems, Inc.*; Thomas R. Bieler, *Michigan State University*; Choong-Un Kim, *University of Texas, Arlington*

Thermal Cycle Fatigue Life Prediction for Flip Chip Solder Joints 703

Robert Darveaux, *Skyworks Solutions, Inc.*

High Thermo-Mechanical Fatigue and Drop Impact Resistant Ni-Bi Doped Lead Free Solder 712

Jae Hong Lee, *MK Electron, Ltd.*; Santosh Kumar, *MK Electron, Ltd.*; Hui Joong Kim, *MK Electron, Ltd.*; Young Woo Lee, *MK Electron, Ltd.*; Jeong Tak Moon, *MK Electron, Ltd.*

16: Advances in Signal Integrity & High-Speed System Design
Chairs: Xiaoxiong (Kevin) Gu, *IBM Corporation*
Kemal Aygun, *Intel Corporation*

Optimal Relaxation of I/O Electrical Requirements under Packaging Uncertainty by Stochastic Methods 717

Xu Chen, *University of Illinois, Urbana-Champaign*; Juan S. Ochoa, *University of Illinois, Urbana-Champaign*; José E. Schutt-Ainé, *University of Illinois, Urbana-Champaign*; Andreas C. Cangellaris, *University of Illinois, Urbana-Champaign*

An Accurate and Convenient Lumped/Discrete Port De-Embedding Method for the 3D Integration and Packaging Full-Wave Modeling by Splitting and Absorbing the Error-Cancelling Network 723

Zhaoqing Chen, *IBM Corporation*

Design, Modeling, and Characterization of Passive Channels for Data Rates of 50 Gbps and Beyond 730

Wendemagegnehu Beyene, *Rambus, Inc.*; Yeon-Chang Hahm, *Rambus, Inc.*; Dave Secker, *Rambus, Inc.*; Don Mullen, *Rambus, Inc.*; Yuriy Shlepnev, *Simberian Inc.*

Low Loss Conductors for CMOS and Through Glass/Silicon Via (TGV/TSV) Structures Using Eddy Current Cancelling Superlattice Structure 736

Arian Rahimi, *University of Florida*; Yong-Kyu Yoon, *University of Florida*

Modeling, Design, Fabrication and Characterization of First Large 2.5D Glass Interposer as a Superior Alternative to Silicon and Organic Interposers at 50 Micron Bump Pitch 742

Brett Sawyer, *Georgia Institute of Technology*; Hao Lu, *Georgia Institute of Technology*; Yuya Suzuki, *Zeon Corporation*; Yutaka Takagi, *NGK Spark Plug Co. Ltd.*; Makoto Kobayashi, *Namics Corporation*; Vanessa Smet, *Georgia Institute of Technology*; Taiji Sakai, *Fujitsu Laboratories Ltd.*; Venky Sundaram, *Georgia Institute of Technology*; Rao Tummala, *Georgia Institute of Technology*

Coupling Impact of Single Ended Signals to LVDS Interface 748

June Feng, *Altera Corporation*; Chooi Ian Loh, *Altera Corporation*; Edward Lin, *Altera Corporation*; Ellen Du, *Altera Corporation*; Guang Chen, *Altera Corporation*; Dan Oh, *Altera Corporation*

Analysis on Interference between Multi-Giga Bit Display Serial Link and RF Components in Smart Mobile Device 753

Youchul Jeong, *Silicon Image Inc.*; Jaemin Kim, *Silicon Image Inc.*; Baegin Sung, *Silicon Image Inc.*

17: Emerging Wireless Technologies & Design
Chairs: Amit P. Agrawal, *Cisco Systems, Inc.*
Lih-Tyng Hwang, *National Sun Yat-Sen University*

Novel Highly-Efficient and Misalignment Insensitive Wireless Power Transfer Systems Utilizing Strongly Coupled Magnetic Resonance Principles 759

Daerhan Daerhan, *Florida International University*; Olutola Jonah, *Florida International University*; Hao Hu, *Florida International University*; Stavros V. Georgakopoulos, *Florida International University*; Manos M. Tentzeris, *Georgia Institute of Technology*

A Wireless Charging and Near-field Communication Combination Module for Mobile Applications 763

Hiroki Shibuya, *Renesas Electronics Corporation*; Tatsuaki Tsukuda, *Renesas Electronics Corporation*; Hiroko Suzuki, *Renesas Electronics Corporation*; Tadashi Shimizu, *Renesas Electronics Corporation*; Masahiro Dobashi, *Renesas Electronics Corporation*; Shinji Nishizono, *Renesas Electronics Corporation*; Mikio Baba, *Renesas Electronics Corporation*; Hideki Sasaki, *Renesas Electronics Corporation*; Katsushi Terajima, *Renesas Electronics Corporation*

Enhanced-Performance Wireless Conformal "Smart Skins" Utilizing Inkjet-Printed Carbon-Nanostructures 769

Taoran Le, *Georgia Institute of Technology*; Ziyin Lin, *Georgia Institute of Technology*; C.P. Wong, *Georgia Institute of Technology*; M.M. Tentzeris, *Georgia Institute of Technology*

Novel THz Imaging Array Using High Resistivity Metasurfaces 775

Kyoung Youl Park, *Michigan State University*; Premjeet Chahal, *Michigan State University*

Magneto-Dielectric Characterization and Antenna Design 782

Kyu Han, *Georgia Institute of Technology*; Madhavan Swaminathan, *Georgia Institute of Technology*; P. Markondeya Raj, *Georgia Institute of Technology*; Himani Sharma, *Georgia Institute of Technology*; Rao Tummala, *Georgia Institute of Technology*; Vijay Nair, *Intel Corporation*

Flexible Liquid Crystal Polymer Based Complementary Split Ring Resonator Loaded Quarter Mode Substrate Integrated Waveguide Filters for Compact and Wearable Broadband RF Applications 789

David E. Senior, *Universidad Tecnológica de Bolívar; University of Florida*; Arian Rahimi, *University of Florida*; Pitfee Jao, *University of Florida*; Yong-Kyu Yoon, *University of Florida*

A Dual-Band Power Amplifier Based on Composite Right/Left-Handed Matching Networks 796

Kyriaki Niotaki, *Centre Tecnologic de Telecomunicacions de Catalunya*; Ana Collado, *Centre Tecnologic de Telecomunicacions de Catalunya*; Apostolos Georgiadis, *Centre Tecnologic de Telecomunicacions de Catalunya*; John Vardakas, *Iquadrat S. L.*

18: WLCSP, Flip Chip, and PoP
Chairs: Valerie Oberson, *IBM Corporation*
Sa Huang, *Medtronic Corporation*

Wafer-Level Non Conductive Films for Exascale Servers 803

A. Horibe, *IBM Corporation*; S. Kohara, *IBM Corporation*; H. Mori, *IBM Corporation*; Y. Orii, *IBM Corporation*; S. Kawamoto, *Namics Corporation*; H. Sone, *Namics Corporation*; M. Hoshiyama, *Namics Corporation*

Bump Geometric Deviation on the Reliability of BOR WLCSP 808

Yumin Liu, *Fairchild Semiconductor Corporation*; Yong Liu, *Fairchild Semiconductor Corporation*; Shichun Qu, *Fairchild Semiconductor Corporation*

Experimental Identification of Warpage Origination During the Wafer Level Packaging Process 815

Chunsheng Zhu, *Chinese Academy of Sciences*; Wenguo Ning, *Chinese Academy of Sciences*; Heng Lee, *Chinese Academy of Sciences*; Jiaotuo Ye, *Chinese Academy of Sciences*; Gaowei Xu, *Chinese Academy of Sciences*; Le Luo, *Chinese Academy of Sciences*

A Stress-Based Effective Film Technique for Wafer Warpage Prediction of Arbitrarily Patterned Films 821

Gregory T. Ostrowicki, *Texas Instruments, Inc.*; Siva P. Gurrum, *Texas Instruments, Inc.*

Drop Test and TCT Reliability of Buffer Coating Material for WLCSP 829

Nobuhiro Anzai, *Asahi Kasei E-Materials Corporation*; Mitsuru Fujita, *Asahi Kasei E-Materials Corporation*; Atsushi Fujii, *Asahi Kasei E-Materials Corporation*

Optimization of Compression Bonding Processing Temperature for Fine Pitch Cu-Column Flip Chip Devices 836

Yonghyuk Jeong, *STATS ChipPAC, Inc.*; Joonyoung Choi, *STATS ChipPAC, Inc.*; Youjoung Choi, *STATS ChipPAC, Inc.*; Nokibul Islam, *STATS ChipPAC, Inc.*; Eric Ouyang, *STATS ChipPAC, Inc.*

Reliability Improvement Methods of Solder Anisotropic Conductive Film (ACF) Joints Using Morphology Control of Solder ACF Joints 841

Yoo-Sun Kim, *Korea Advanced Institute of Science and Technology (KAIST)*; Seung-Ho Kim, *Korea Advanced Institute of Science and Technology (KAIST)*; Jiwon Shin, *Korea Advanced Institute of Science and Technology (KAIST)*; Kyung-Wook Paik, *Korea Advanced Institute of Science and Technology (KAIST)*

19: Progress in 3D Integration
Chairs: Shawn Shi, *Medtronic Corporation*
Mark Gerber, *Texas Instruments*

Development of the Technology to Control the Spatial Distribution of Plasma Using Double ICP Coil 846

T. Sakuishi, *ULVAC, Inc.; NMEMS Technology Research Organization*; T. Murayama, *ULVAC, Inc.; NMEMS Technology Research Organization*; Y. Morikawa, *ULVAC, Inc.; NMEMS Technology Research Organization*; K. Suu, *ULVAC, Inc.; NMEMS Technology Research Organization*

Defect Detection in Through Silicon Vias by GHz Scanning Acoustic Microscopy: Key Ultrasonic Characteristics 850

Alain Phommahaxay, *IMEC*; Ingrid De Wolf, *IMEC; KU Leuven*; Tatjana Djuric, *PVA TePla Analytical Systems GmbH*; Peter Hoffrogge, *PVA TePla Analytical Systems GmbH*; Sebastian Brand, *Fraunhofer IWM*; Peter Czurratis, *PVA TePla Analytical Systems GmbH*; Harold Philipsen, *IMEC*; Gerald Beyer, *IMEC*; Herbert Struyf, *IMEC*; Eric Beyne, *IMEC*

Temporary Spin-on Glass Bonding Technologies for Via-Last/Backside-Via 3D Integration Using Multichip Self-Assembly 856

H. Hashiguchi, *Tohoku University*; T. Fukushima, *Tohoku University*; A. Noriki, *Tohoku University*; H. Kino, *Tohoku University*; K.-W. Lee, *Tohoku University*; T. Tanaka, *Tohoku University*; M. Koyanagi, *Tohoku University*

TSV Module Optimization for High Performance Silicon Interposer 862

Andrew Cao, *Invensas Corporation*; Thomas Dinan, *Invensas Corporation*; Zhuowen Sun, *Invensas Corporation*; Guilian Gao, *Invensas Corporation*; Cyprian Uzoh, *Invensas Corporation*; Bong-Sub Lee, *Invensas Corporation*; Liang Wang, *Invensas Corporation*; Hong Shen, *Invensas Corporation*; Sitaram Arkalgud, *Invensas Corporation*

Study of TSV Thinning Wafer Strength Enhancement for 3DIC Package 868

Jyun-Ling Tsai, *Siliconware Precision Industries Co., Ltd.*; Chun-Chieh Chao, *Siliconware Precision Industries Co., Ltd.*; Hsiao-Chun Huang, *Siliconware Precision Industries Co., Ltd.*; Cheng-Hsiang Liu, *Siliconware Precision Industries Co., Ltd.*; Hung-Hsein Chang, *Siliconware Precision Industries Co., Ltd.*; Chang-Lun Lu, *Siliconware Precision Industries Co., Ltd.*; Shi-Ching Chen, *Siliconware Precision Industries Co., Ltd.*

Challenges in 3D Die Stacking 873

Juergen Grafe, *Fraunhofer IZM*; Wieland Wahrmund, *Fraunhofer IZM*; Stephan Dobritz, *Fraunhofer IZM*; Juergen Wolf, *Fraunhofer IZM*; Klaus-Dieter Lang, *Fraunhofer IZM*

Wet Silicon Etch Process for TSV Reveal 878

Laura B. Mauer, *Solid State Equipment, LLC*; John Taddei, *Solid State Equipment, LLC*; Ramey Youssef, *Solid State Equipment, LLC*; Yongqiang Lu, *SACHEM, Inc.*; Sian Collins, *SACHEM, Inc.*; Kevin Mclaughlin, *SACHEM, Inc.*; Craig Allen, *SACHEM, Inc.*

20: 3D Materials & Processing

Chairs: Myung Jin Yim, *Intel Corporation*
Daniel D. Lu, *Henkel Corporation*

Advanced Wafer Bonding and Laser Debonding 883

P. Andry, *IBM Corporation*; R. Budd, *IBM Corporation*; R. Polastre, *IBM Corporation*; C. Tsang, *IBM Corporation*; B. Dang, *IBM Corporation*; J. Knickerbocker, *IBM Corporation*; M. Glodde, *IBM Corporation*

Versatile Thin Wafer Stacking Technology for Monolithic Integration of Temporary Bonded Thin Wafers 888

Thomas Uhrmann, *EV Group*; Jürgen Burggraf, *EV Group*; Julian Bravin, *EV Group*; Viorel Dragoi, *EV Group*; Markus Wimplinger, *EV Group*; Thorsten Matthias, *EV Group*; Paul Lindner, *EV Group*

Temporary Bonding for High-Topography Applications: Spin-on Material versus Dry Film 894

Anne Jourdain, *IMEC*; Alain Phommahaxay, *IMEC*; Greet Verbinnen, *IMEC*; Alice Guerrero, *Brewer Science, Inc.*; Susan Bailey, *Brewer Science, Inc.*; Mark Privett, *Brewer Science, Inc.*; Kim Arnold, *Brewer Science, Inc.*; Andy Miller, *IMEC*; Kenneth Rebibis, *IMEC*; Gerald Beyer, *IMEC*; Eric Beyne, *IMEC*

Development of New Concept Thermoplastic Temporary Adhesive for 3D-IC Integration 899

A. Kubo, *Tokyo Ohka Kogyo Co., Ltd.*; K. Tamura, *Tokyo Ohka Kogyo Co., Ltd.*; H. Imai, *Tokyo Ohka Kogyo Co., Ltd.*; T. Yoshioka, *Tokyo Ohka Kogyo Co., Ltd.*; S. Oya, *Tokyo Ohka Kogyo Co., Ltd.*; S. Otaka, *Tokyo Ohka Kogyo Co., Ltd.*

Underfilling Techniques Comparison in 3D CtW Stacking Approach 906
A. Garnier, *CEA-LETI*; A. Jouve, *CEA-LETI*; R. Franiatte, *CEA-LETI*; S. Chéramy, *CEA-LETI*

High Throughput Thermal Compression NCF Bonding 913
Toshihisa Nonaka, *Toray Industries, Inc.*; Yuta Kobayashi, *Toray Industries, Inc.*; Noboru Asahi, *Toray Industries, Inc.*; Shoichi Niizeki, *Toray Industries, Inc.*; Koichi Fujimaru, *Toray Industries, Inc.*; Yoshiyuki Arai, *Toray Engineering Co., Ltd.*; Toshifumi Takegami, *Toray Engineering Co., Ltd.*; Yoshinori Miyamoto, *Toray Engineering Co., Ltd.*; Masatsugu Nimura, *Toray Engineering Co., Ltd.*; Hiroyuki Niwa, *Toray International America Inc.*

Through Silicon Underfill Dispensing for 3D Die/Interposer Stacking 919
Fuliang Le, *Hong Kong University of Science and Technology*; S.W. Ricky Lee, *Hong Kong University of Science and Technology*; Kei May Lau, *Hong Kong University of Science and Technology*; C. Patrick Yue, *Hong Kong University of Science and Technology*; Johnny K.O. Sin, *Hong Kong University of Science and Technology*; Philip K.T. Mok, *Hong Kong University of Science and Technology*; Wing-Hung Ki, *Hong Kong University of Science and Technology*; Hoi Wai Choi, *University of Hong Kong*

21: Wafer-Level & Fan-Out Packages
Chairs: Christopher Bower, *X-Celeprint Ltd.*
E. Jan Vardaman, *TechSearch International, Inc.*

Board Level Reliability and Surface Mount Assembly of 0.35mm and 0.3mm Pitch Wafer Level Packages 925
Beth Keser, *Qualcomm Technologies, Inc.*; Rey Alvarado, *Qualcomm Technologies, Inc.*; Alan Choi, *Qualcomm Technologies, Inc.*; Mark Schwarz, *Qualcomm Technologies, Inc.*; Steve Bezuk, *Qualcomm Technologies, Inc.*

Encapsulated Wafer Level Package Technology (eWLCS) 931
Tom Strothmann, *STATS ChipPAC, Inc.*; Seung Wook Yoon, *STATS ChipPAC, Ltd.*; Yaojian Lin, *STATS ChipPAC, Ltd.*

Enabling of Fan-Out WLP for More Demanding Applications by Introduction of Enhanced Dielectric Material for Higher Reliability 935
Rodrigo Almeida, *Namium, S.A.*; Isabel Barros, *Namium, S.A.*; José Campos, *Namium, S.A.*; Paulo Cardoso, *Namium, S.A.*; José Castro, *Namium, S.A.*; Vitor Henriques, *Namium, S.A.*; Eoin O'Toole, *Namium, S.A.*; Nelson Pinho, *Namium, S.A.*

24" x 18" Fan-Out Panel Level Packaging 940
T. Braun, *Fraunhofer IZM*; K.-F. Becker, *Fraunhofer IZM*; S. Voges, *Technical University Berlin*; J. Bauer, *Fraunhofer IZM*; R. Kahle, *Technical University Berlin*; V. Bader, *Fraunhofer IZM*; T. Thomas, *Technical University Berlin*; R. Aschenbrenner, *Fraunhofer IZM*; K.-D. Lang, *Technical University Berlin*

Development and Characterization of New Generation Panel Fan-Out (P-FO) Packaging Technology 947
Hong-Da Chang, *Siliconware Precision Industries Co., Ltd.*; David Chang, *Siliconware Precision Industries Co., Ltd.*; Kenny Liu, *Siliconware Precision Industries Co., Ltd.*; H.S. Hsu, *Siliconware Precision Industries Co., Ltd.*; Rui-Feng Tai, *Siliconware Precision Industries Co., Ltd.*; Hsiao-Chun Huang, *Siliconware Precision Industries Co., Ltd.*; Yi-Che Lai, *Siliconware Precision Industries Co., Ltd.*; Chang-Lun Lu, *Siliconware Precision Industries Co., Ltd.*; Chun-Tang Lin, *Siliconware Precision Industries Co., Ltd.*; Steve Chiu, *Siliconware Precision Industries Co., Ltd.*

Development of Exposed Die Large Body to Die Size Ratio Wafer Level Package Technology 952
J. Osenbach, *LSI Corporation*; S. Emerich, *LSI Corporation*; L. Golick, *LSI Corporation*; S. Cate, *LSI Corporation*; M. Chan, *STATS ChipPAC, Ltd.*; S.W. Yoon, *STATS ChipPAC, Ltd.*; Y.J. Lin, *STATS ChipPAC, Ltd.*; K. Wong, *STATS ChipPAC, Inc.*

3D Rectangular Waveguide Integrated in Embedded Wafer Level Ball Grid Array (eWLB) Package 956

E. Seler, *Friedrich-Alexander University Erlangen-Nuremberg*; M. Wojnowski, *Infineon Technologies AG*; W. Hartner, *Infineon Technologies AG*; J. Böck, *Infineon Technologies AG*; R. Lachner, *Infineon Technologies AG*; R. Weigel, *University of Erlangen-Nuremberg*; A. Hagelauer, *Friedrich-Alexander University Erlangen-Nuremberg*

22: System-Level Thermal & Mechanical Models I

Chairs: Yong Liu, *Fairchild Semiconductor Corporation*
Sandeep Sane, *Intel Corporation*

Interplay and Influence of Thermomechanical Stress in Copper-Filled TSV Interposers 963

Sheng-Tsai Wu, *Industrial Technology Research Institute (ITRI)*; Cheng-Fu Chen, *University of Alaska, Fairbanks*; Heng-Chieh Chien, *Industrial Technology Research Institute (ITRI)*

Does Current Crowding Induce Vacancy Concentration Singularity in Electromigration? 967

Ozgur Taner, *Lamar University*; Kasemsak Kijkanjanapaiboon, *Lamar University*; Xuejun Fan, *Lamar University*

Hygro-Thermo-Mechanical Analysis and Failure Prediction in Electronic Packages by Using Peridynamics 973

Selda Oterkus, *University of Arizona*; Erdogan Madenci, *University of Arizona*; Erkan Oterkus, *University of Strathclyde*; Yuchul Hwang, *Samsung Electronics Company, Ltd.*; Jangyong Bae, *Samsung Electronics Company, Ltd.*; Sungwon Han, *Samsung Electronics Company, Ltd.*

Cohesive Zone Experiments for Copper/Mold Compound Delamination 983

William E.R. Krieger, *Georgia Institute of Technology*; Sathyanarayanan Raghavan, *Georgia Institute of Technology*; Abhishek Kwatra, *Georgia Institute of Technology*; Suresh K. Sitaraman, *Georgia Institute of Technology*

Damage Pre-Cursor Based Life Prediction of the Effect of Mean Temperature of Thermal Cycle on the SnAgCu Solder Joint Reliability 990

Pradeep Lall, *Auburn University*; Kazi Mirza, *Auburn University*; Jeff Suhling, *Auburn University*

Methodology Development of Warpage Analysis of Polymer Based Packaging Substrate 1004

Cheolgyu Kim, *Korea Advanced Institute of Science and Technology (KAIST)*; Taeik Lee, *Korea Advanced Institute of Science and Technology (KAIST)*; Hyeseon Choi, *Korea Advanced Institute of Science and Technology (KAIST)*; Min Sung Kim, *Samsung Electro-Mechanics*; Taek-Soo Kim, *Korea Advanced Institute of Science and Technology (KAIST)*

Simulations for the Impact of Warpage on the Accuracy of Attitude and Heading Reference System 1010

Shengzhi Zhang, *Huazhong University of Science & Technology*; *Wuhan National Laboratory for Optoelectronics*; Qiang Dan, *Huazhong University of Science & Technology*; *Wuhan National Laboratory for Optoelectronics*; Chaojun Liu, *Huazhong University of Science & Technology*; *Wuhan National Laboratory for Optoelectronics*; Yong Xu, *Wayne State University*; Xin Wu, *Wayne State University*; Sheng Liu, *Wuhan University*; Xing Guo, *Huazhong University of Science & Technology*; *Wuhan National Laboratory for Optoelectronics*; Ming Wen, *Huazhong University of Science & Technology*; *Wuhan National Laboratory for Optoelectronics*

23: Optical Interconnects

Chairs: Hiren Thacker, *Oracle*
Ping Zhou, *LDX Optronics, Inc.*

Multicore Fiber 4 TX + 4 RX Optical Transceiver Based on Holey SiGe IC 1016

Fuad E. Doany, *IBM Corporation*; Daniel M. Kuchta, *IBM Corporation*; Alexander V. Rylyakov, *IBM Corporation*; Christian Baks, *IBM Corporation*; Shurong Tian, *IBM Corporation*; Mark Schultz, *IBM Corporation*; Frank Libsch, *IBM Corporation*; Clint L. Schow, *IBM Corporation*

336-Channel Electro-Optical Interconnect: Underfill Process Improvement, Fiber Bundle and Reliability Results 1021

Shuki Benjamin, *Compass-EOS*; Kobi Hasharoni, *Compass-EOS*; Avi Maman, *Compass-EOS*; Stanislav Stepanov, *Compass-EOS*; Michael Mesh, *Compass-EOS*; Helge Luesebrink, *PVA TePla AG*; Roland Steffek, *PVA TePla AG*; Wolfgang Pleyer, *PVA TePla AG*; Christian Stömmer, *PVA TePla AG*

Development of Optical Multi-Channel Connector for Rigid Waveguide – Fiber Optical Interconnection 1028

Kazumi Nakazuru, *Kyocera Corporation*; Satoshi Asai, *Kyocera Corporation*; Masatoshi Tsunoda, *Kyocera Corporation*; Naoki Takahashi, *Kyocera Corporation*; Takahiro Matsubara, *Kyocera Corporation*

Electro-Optical Backplane Demonstrator with Gradient-Index Multimode Glass Waveguides for Board-to-Board Interconnection 1033

Lars Brusberg, *Fraunhofer Institute IZM*; Henning Schröder, *Fraunhofer Institute IZM*; Richard Pitwon, *Xyratex Technology Ltd.*; Simon Whalley, *ILFA Feinstleitertechnik GmbH*; Allen Miller, *Xyratex Technology Ltd.*; Christian Herbst, *Technical University of Berlin*; Julia Röder, *Fraunhofer Institute IZM*; Daniel Weber, *Fraunhofer Institute IZM*; Klaus-Dieter Lang, *Technical University of Berlin*

Three-Dimensional High-Density Channel Integration of Polymer Optical Waveguide Using the Mosquito Method 1042

Takaaki Ishigure, *Keio University*; Daisuke Suganuma, *Keio University*; Kazutomo Soma, *Keio University*

Novel Trace Design for High Data-Rate, Multi-Channel Optical Transceiver Assembled Using Flip-Chip Bonding 1048

Takatoshi Yagisawa, *Fujitsu Laboratories, Ltd.*; Takashi Shiraishi, *Fujitsu Laboratories, Ltd.*; Mariko Sugawara, *Fujitsu Laboratories, Ltd.*; Kazuhiro Tanaka, *Fujitsu Laboratories, Ltd.*

Modeling, Design, and Demonstration of Ultra-Miniaturized and High Efficiency 3D Glass Photonics Modules 1054

Bruce C. Chou, *Georgia Institute of Technology*; Sandeep Razdan, *TE Connectivity*; Haipeng Zhang, *TE Connectivity*; Jibin Sun, *TE Connectivity*; Terry Bowen, *TE Connectivity*; Vanessa Smet, *Georgia Institute of Technology*; Gee-Kung Chang, *Georgia Institute of Technology*; Venky Sundaram, *Georgia Institute of Technology*; Rao Tummala, *Georgia Institute of Technology*

24: Innovative Interconnections

Chairs: James E. Morris, *Portland State University*
Nathan Lower, *Rockwell Collins, Inc.*

A Study on Nanofiber Anisotropic Conductive Films (ACFs) for Fine Pitch Chip-on-Glass (COG) Interconnections 1060

Sang Hoon Lee, *Korea Advanced Institute of Science and Technology (KAIST)*; Tae Wan Kim, *Korea Advanced Institute of Science and Technology (KAIST)*; Kyung-Wook Paik, *Korea Advanced Institute of Science and Technology (KAIST)*

Study of Fine Pitch Micro-Interconnections Formed by Low Temperature Bonded Copper Nanowires Based Anisotropic Conductive Film 1064

Jing Tao, *University College Cork*; Alan Mathewson, *University College Cork*; Kafil M. Razeeb, *University College Cork*

Carbon Nanofibers (CNF) for Enhanced Solder-Based Nano-Scale Integration and On-Chip Interconnect Solutions 1071

V. Desmaris, *Smoltek AB*; A.M. Saleem, *Smoltek AB*; S. Shafiee, *Smoltek AB*; J. Berg, *Smoltek AB*; M.S. Kabir, *Smoltek AB*; A. Johansson, *Smoltek AB*; Phil Marcoux, *PPM Associates*

Pressure-Less Plasma Sintering of Cu Paste for SiC Die-Attach of High-Temperature Power Device Manufacturing 1077

S. Nagao, *Osaka University*; K. Kodani, *Nissin, Inc.*; S. Sakamoto, *Osaka University*; S.-W. Park, *Osaka University*; T. Sugahara, *Osaka University*; K. Suganuma, *Osaka University*

Bonding 1200 V, 150 A IGBT Chips (13.5 mm x 13.5 mm) with DBC Substrate by Pressureless Sintering Nanosilver Paste for Power Electronic Packaging .... N/A

Shancan Fu, *Tianjin University*; Yunhui Mei, *Tianjin University*; Guo-Quan Lu, *Tianjin University, Virginia Tech*; Xin Li, *Tianjin University*; Gang Chen, *Tianjin University*; Xu Chen, *Tianjin University*

Flip Chip Based on Compliant Double Helix Interconnect for High Frequency Applications 1086

Pingye Xu, *Auburn University*; George A. Hernandez, *Auburn University*; Shiqiang Wang, *Auburn University*; Jie Zhong, *Auburn University*; Charles D. Ellis, *Auburn University*; Michael C. Hamilton, *Auburn University*

Modeling of Crosstalk Effects in Coupled MLGNR Interconnects Based on FDTD Method 1091

Vobulapuram Ramesh Kumar, *Indian Institute of Technology Roorkee*; Brajesh Kumar Kaushik, *Indian Institute of Technology Roorkee*; Amalendu Patnaik, *Indian Institute of Technology Roorkee*

25: Recent Advances in 3D Package Reliability
Chairs: Deepak Goyal, *Intel Corporation*
Jeffrey Suhling, *Auburn University*

First Demonstration of Reliable Copper-Plated 30µm Diameter Through-Package-Vias in Ultra-Thin Bare Glass Interposers 1098

Kaya Demir, *Georgia Institute of Technology*; Andac Armutlulu, *Georgia Institute of Technology*; Jialing Tong, *Georgia Institute of Technology*; Raghuram Pucha, *Georgia Institute of Technology*; Venkatesh Sundaram, *Georgia Institute of Technology*; Rao Tummala, *Georgia Institute of Technology*

Through-Glass Interposer Integrated High Quality RF Components 1103

Cheolbok Kim, *University of Florida*; David E. Senior, *University of Florida*; *Universidad Tecnológica de Bolívar*; Aric Shorey, *Corning, Inc.*; Hyup Jong Kim, *University of Florida*; Windsor Thomas, *Corning, Inc.*; Yong-Kyu Yoon, *University of Florida*

Minimization of Keep-Out Zone (KOZ) in 3D IC by Local Bending Stress Suppression with Low Temperature Curing Adhesive 1110

Hisashi Kino, *Tohoku University*; Hideto Hashiguchi, *Tohoku University*; Yohei Sugawara, *Tohoku University*; Seiya Tanikawa, *Tohoku University*; Takafumi Fukushima, *Tohoku University*; Kangwook Lee, *Tohoku University*; Mitsumasa Koyanagi, *Tohoku University*; Tetsu Tanaka, *Tohoku University*

Effect of Thermal Annealing on TSV Cu Protrusion and Local Stress 1116

Xiangmeng Jing, *National Center for Advanced Packaging; Chinese Academy of Sciences*; Hongwen He, *National Center for Advanced Packaging; Chinese Academy of Sciences*; Liang Ji, *National Center for Advanced Packaging*; Cheng Xu, *National Center for Advanced Packaging*; Kai Xue, *National Center for Advanced Packaging*; Meiying Su, *National Center for Advanced Packaging; Chinese Academy of Sciences*; Chongshen Song, *National Center for Advanced Packaging; Chinese Academy of Sciences*; Daquan Yu, *National Center for Advanced Packaging; Chinese Academy of Sciences*; Liqiang Cao, *National Center for Advanced Packaging; Chinese Academy of Sciences*; Wenqi Zhang, *National Center for Advanced Packaging*; Dongkai Shangguan, *National Center for Advanced Packaging; Chinese Academy of Sciences*

Effect of High Temperature Storage on the Stress and Reliability of 3D Stacked Chips 1122

Tengfei Jiang, *University of Texas, Austin*; Chenglin Wu, *University of Texas, Austin*; Peng Su, *Cisco Systems, Inc.*; Pierre Chia, *Cisco Systems, Inc.*; Li Li, *Cisco Systems, Inc.*; Ho-Young Son, *SK Hynix, Inc.*; Min-Suk Suh, *SK Hynix, Inc.*; Nam-Seog Kim, *SK Hynix, Inc.*; Jay Im, *University of Texas, Austin*; Rui Huang, *University of Texas, Austin*; Paul S. Ho, *University of Texas, Austin*

A Novel Fine Pitch TSV Interconnection Method Using NCF with Zn Nano-Particles 1128

Ji-Won Shin, *Korea Advanced Institute of Science and Technology (KAIST)*; Yong-Won Choi, *Korea Advanced Institute of Science and Technology (KAIST)*; Young Soon Kim, *Korea Advanced Institute of Science and Technology (KAIST)*; Un Byung Kang, *Samsung Electronics Company, Ltd.*; Sun Kyung Seo, *Samsung Electronics Company, Ltd.*; Kyung-Wook Paik, *Korea Advanced Institute of Science and Technology (KAIST)*

Residual Stress Investigations at TSVs in 3D Micro Structures by HR-XRD, Raman Spectroscopy and fibDAC 1134

U. Zschenderlein, *Technical University Chemnitz*; D. Vogel, *Fraunhofer ENAS*; E. Auerswald, *Fraunhofer ENAS*; O. Hölck, *Technical University Chemnitz*; H. Rajendran, *Technical University Chemnitz*; P. Ramm, *Fraunhofer EMFT*; R. Pufall, *Infineon Technologies*; B. Wunderle, *Technical University Chemnitz*; *Fraunhofer ENAS*

26: 3D Microbumps
Chairs: Kathy Cook, *Ziptronix*
Lei Shan, *IBM Corporation*

Formic Acid Treatment with Pt Catalyst for Cu Direct and Hybrid Bonding at Low Temperature 1143

Tadatomo Suga, *University of Tokyo*; Masakate Akaike, *University of Tokyo*; Wenhua Yang, *University of Tokyo*

Direct Multichip-to-Wafer 3D Integration Technology Using Flip-Chip Self-Assembly of NCF-Covered Known Good Dies 1148

Yuka Ito, *Tohoku University; Sumitomo Bakelite Co., Ltd.*; Mariappan Murugesan, *Tohoku University*; Takafumi Fukushima, *Tohoku University*; Kang-Wook Lee, *Tohoku University*; Koji Choki, *Sumitomo Bakelite Co., Ltd.*; Tetsu Tanaka, *Tohoku University*; Mitsumasa Koyanagi, *Tohoku University*

Maskless Screen Printing Technology for 20μm-Pitch, 52InSn Solder Interconnections in Display Applications 1154

Kwang-Seong Choi, *ETRI*; Haksun Lee, *ETRI*; Hyun-Cheol Bae, *ETRI*; Yong-Sung Eom, *ETRI*

Accelerated SLID Bonding Using Thin Multi-Layer Copper-Solder Stack for Fine-Pitch Interconnections 1160

Chinmay Honrao, *Georgia Institute of Technology*; Ting-Chia Huang, *Georgia Institute of Technology*; Makoto Kobayashi, *Namics Corporation*; Vanessa Smet, *Georgia Institute of Technology*; P. Markondeya Raj, *Georgia Institute of Technology*; Rao Tummala, *Georgia Institute of Technology*

Study of Electro-Migration Resistivity of Micro Bump Using SnBi Solder 1166

Kei Murayama, *Shinko Electric Industries Company, Ltd.*; Mitsuhiro Aizawa, *Shinko Electric Industries Company, Ltd.*; Mitsutoshi Higashi, *Shinko Electric Industries Company, Ltd.*

The Impact of Different Under Bump Metallurgies and Redistribution Layers on the Electromigration of Solder Balls for Wafer-Level Packaging 1173

Christine Hau-Riege, *Qualcomm Technologies, Inc.*; Beth Keser, *Qualcomm Technologies, Inc.*; Rey Alvarado, *Qualcomm Technologies, Inc.*; Ahmer Syed, *Qualcomm Technologies, Inc.*; YouWen Yau, *Qualcomm Technologies, Inc.*; Steve Bezuk, *Qualcomm Technologies, Inc.*; Kevin Caffey, *Qualcomm Technologies, Inc.*

Low-Pressure Sintering Bonding with Cu and CuO Flake Paste for Power Devices 1179

S.W. Park, *Osaka University*; R. Uwataki, *Osaka University*; S. Nagao, *Osaka University*; T. Sugahara, *Osaka University*; Y. Katoh, *Denso Corporation*; H. Ishino, *Denso Corporation*; K. Sugiura, *Denso Corporation*; K. Tsuruta, *Denso Corporation*; K. Suganuma, *Osaka University*

27: Sensors & MEMS Technologies
Chairs: Joseph W. Soucy, *Draper Laboratory*
Daniel Baldwin, *Engent, Inc.*

A Novel 3D Packaging Concept for RF Powered Sensor Grains 1183

Walther Pachler, *Graz University of Technology*; Klaus Pressel, *Infineon Technologies AG*; Jasmin Grosinger, *Graz University of Technology*; Gottfried Beer, *Infineon Technologies AG*; Wolfgang Bösch, *Graz University of Technology*; Gerald Holweg, *Infineon Technologies AG*; Christian Zilch, *Magna Diagnostics GmbH*; Manfred Meindl, *Danube Mobile Communications Engineering GmbH & Co. KG*

A Novel Sound Sensor and Its Package Used in Lung Sound Diagnosis 1189

Xingming Fu, *Wuhan University*; Chaojun Liu, *Wuhan University*; Yong Xu, *Wuhan University*; *Wayne State University*; Yating Hu, *Wayne State University*; Xiaobing Luo, *Huazhong University of Science & Technology*; Xin Wu, *Wayne State University*; Sheng Liu, *Wuhan University*

Novel System-in-Package Design and Packaging Solution for Solid State Lighting Systems 1192

Mingzhi Dong, *Delft University of Technology*; *State Key Laboratory of Solid State Lighting*; Fabio Santagata, *Delft University of Technology*; *State Key Laboratory of Solid State Lighting*; Jia Wei, *Delft University of Technology*; *State Key Laboratory of Solid State Lighting*; Cadmus Yuan, *Chinese Academy of Sciences*; *State Key Laboratory of Solid State Lighting*; Guoqi Zhang, *Chinese Academy of Sciences*; *Delft University of Technology*

Implantable Device Including a MEMS Accelerometer and an ASIC Chip Encapsulated in a Hermetic Silicon Box for Measurement of Cardiac Physiological Parameter 1198

Jean-Charles Souriau, *CEA-LETI*; Laetitia Castagné, *CEA-LETI*; Guy Parat, *CEA-LETI*; Gilles Simon, *CEA-LETI*; Karima Amara, *Sorin CRM SAS*; Philippe D'hiver, *Sorin CRM SAS*; Renzo Dal Molin, *Sorin CRM SAS*

Capping Technologies for Wafer Level MEMS Packaging Based on Permanent and Temporary Wafer Bonding 1204

K. Zoschke, *Fraunhofer IZM*; M. Wilke, *Fraunhofer IZM*; M. Wegner, *Fraunhofer IZM*; K. Kaletta, *Fraunhofer IZM*; C.-A. Manier, *Fraunhofer IZM*; H. Oppermann, *Fraunhofer IZM*; M. Wietstruck, *IHP GmbH*; B. Tillack, *IHP GmbH*; M. Kaynak, *IHP GmbH*; K.-D. Lang, *Technical University Berlin*

The Novel Assembly Method of a Field Deployable Biosensor Unit 1212

P. Xu, *East China Normal University*; F.M. Guo, *East China Normal University*; X.Y. Liu, *East China Normal University*; J.H. Shen, *East China Normal University*; L. Ding, *East China Normal University*; W. Wang, *East China Normal University*; Y.Q. Li, *East China Normal University*; Y.P. Ge, *East China Normal University*; S.H. Zhang, *East China Normal University*; M.J. Wang, *East China Normal University*; H.Z. Zheng, *East China Normal University*; J.T. Ye, *Chinese Academy of Sciences*; L.; Luo Chinese Academy of Sciences

SIMEIT-Project: High Precision Inertial Sensor Integration on a Modular 3D-Interposer Platform 1218

Wolfram Steller, *Fraunhofer IZM*; Christoph Meinecke, *Technical University Chemnitz*; Knut Gottfried, *Fraunhofer ENAS*; Gregor Woldt, *Microelectronic Packaging Dresden GmbH*; Wolfgang Günther, *GEMAC*; M. Juergen Wolf, *Fraunhofer IZM*; K. Dieter Lang, *Fraunhofer IZM*

28: System-Level Thermal & Mechanical Models II

Chairs: Pradeep Lall, *Auburn University*
Xuejun Fan, *Lamar University*

Mechanical Stress Management for Electrical Chip-Package Interaction (e-CPI) 1226

Wei Zhao, *Qualcomm Technologies, Inc.*; Mark Nakamoto, *Qualcomm Technologies, Inc.*; Vidhya Ramachandran, *Qualcomm Technologies, Inc.*; Riko Radojcic, *Qualcomm Technologies, Inc.*

Cu Pillar Flip Chip Assembly: Chip Attach Process Failure Mode Study 1231

Shengmin Wen, *Amkor Technology*; Bora Baloglu, *Amkor Technology*; Guangfeng Li, *Amkor Assembly and Test (Shanghai) Co., Ltd.*

Mechanical and Thermo-Mechanical Stress Considerations in Applying 3D ICs to a Design 1235

Jia-Shen Lan, *National Sun Yat-Sen University*; Mei-Ling Wu, *National Sun Yat-Sen University*

Modeling Microstructure Effects on Electromigration in Lead-Free Solder Joints 1241

Jiamin Ni, *Rensselaer Polytechnic Institute*; Yong Liu, *Fairchild Semiconductor*; Jifa Hao, *Fairchild Semiconductor*; Antoinette Maniatty, *Rensselaer Polytechnic Institute*; Barry O'Connell, *Fairchild Semiconductor*

Experimental Demonstration of the Effect of Copper TPVs (Through Package Vias) on Thermal Performance of Glass Interposers 1247

Sangbeom Cho, *Georgia Institute of Technology*; Yoichiro Sato, *Asahi Glass*; Venky Sundaram, *Georgia Institute of Technology*; Yogendra Joshi, *Georgia Institute of Technology*; Rao Tummala, *Georgia Institute of Technology*

Failure Mechanism Investigation of Stacked Via Cracking in Organic Chip Carrier 1253

Shidong Li, *IBM Corporation*; Yi Pan, *IBM Corporation*; Sushumna Iruvanti, *IBM Corporation*; David L. Questad, *IBM Corporation*; Randall J. Werner, *IBM Corporation*

A Novel Method to Predict Fluid/Structure Interaction in IC Packaging 1258

Chih-Chung Hsu, *National Tsing Hua University*; Tzu-Chang Wang, *CoreTech System (Moldex3D) Co., Ltd.*; Yen-Chi Chen, *CoreTech System (Moldex3D) Co., Ltd.*; Yang-Kai Lin, *CoreTech System (Moldex3D) Co., Ltd.*

29: Integrated RF & Power Modules

Chairs: Rockwell Hsu, *Cisco Systems, Inc.*
              P. Markondeya Raj, *Georgia Institute of Technology*

Modeling, Design and Demonstration of Multi-Die Embedded WLAN RF Front-End Module with Ultra-Miniaturized and High-Performance Passives 1264

Srikrishna Sitaraman, *Georgia Institute of Technology*; Yuya Suzuki, *Zeon Corporation*; Christopher White, *Georgia Institute of Technology*; Vijay Nair, *Intel Corporation*; Telesphor Kamgaing, *Intel Corporation*; Frank Juskey, *TriQuint Semiconductor*; Sung Jin Kim, *Georgia Institute of Technology*; P. Markondeya Raj, *Georgia Institute of Technology*; Venky Sundaram, *Georgia Institute of Technology*; Rao Tummala, *Georgia Institute of Technology*

A Compact 4-Chip Package with 64 Embedded Dual-Polarization Antennas for W-Band Phased-Array Transceivers 1272

Xiaoxiong Gu, *IBM Corporation*; Duixian Liu, *IBM Corporation*; Christian Baks, *IBM Corporation*; Alberto Valdes-Garcia, *IBM Corporation*; Ben Parker, *IBM Corporation*; Md. R. Islam, *IBM Corporation*; Arun Natarajan, *IBM Corporation; Oregon State University*; Scott K. Reynolds, *IBM Corporation*

Active Die Embedded Small Form Factor RF Packages for Ultrabooks and Smartphones 1278

Vijay K. Nair, *Intel Corporation*; Carlton Hanna, *Intel Corporation*; Ronald Spreitzer, *Intel Corporation*; Johanna Swan, *Intel Corporation*

Design and Material Contributions to Second-Harmonic Nonlinearities in RF Silicon Integrated Passive Devices 1284

Robert Frye, *RF Design Consulting, LLC*; Robert Melville, *Emecon, LLC*; Kai Liu, *STATS ChipPAC, Inc.*

Integration of Magnetic Materials into Package RF and Power Inductors on Organic Substrates for System in Package (SiP) Applications 1290

Hao Wu, *Arizona State University*; Donald S. Gardner, *Intel Corporation*; Cheng Lv, *Arizona State University*; Zhihua Zou, *Intel Corporation*; Hongbin Yu, *Arizona State University*

Through Silicon Capacitor Co-Integrated with TSV as an Efficient 3D Decoupling Capacitor Solution for Power Management on Silicon Interposer 1296

O. Guiller, *STMicroelectronics*; S. Joblot, *STMicroelectronics*; Y. Lamy, *CEA-LETI*; A. Farcy, *STMicroelectronics*; E. Defay, *CEA-LETI*; K. Dieng, *Université de Savoie*

Design of RF and Thermal Pads of CMOS PAs Using Copper to Copper Bonding Technology 1303

Lih-Tyng Hwang, *National Sun Yat-Sen University*; An-Yu Kuo, *Cadence Design Systems, Inc.*

30: Solders & Bonding
Chairs: Mikel Miller, *Draper Laboratory*
Grace Yi Li, *Intel Corporation*

Wafer IMS (Injection Molded Solder) – A New Fine Pitch Solder Bumping Technology on Wafers with Solder Alloy Composition Flexibility 1308
Jae-Woong Nah, *IBM Corporation*; Jeffrey Gelorme, *IBM Corporation*; Peter Sorce, *IBM Corporation*; Paul Lauro, *IBM Corporation*; Eric Perfecto, *IBM Corporation*; Mark McLeod, *IBM Corporation*; Kazushige Toriyama, *IBM Corporation*; Yasumitsu Orii, *IBM Corporation*; Peter Brofman, *IBM Corporation*; Takashi Nauchi, *Senju Metal Industry Co., Ltd.*; Akira Takaguchi, *Senju System Technology Co., Ltd.*; Kazuya Ishiguro, *Senju System Technology Co., Ltd.*; Tomoyasu Yoshikawa, *Senju Comtek Corporation*; Derek Daily, *Senju Comtek Corporation*; Ryoichi Suzuki, *Senju Metal Industry Co., Ltd.*

Reliability of Paste Based Transient Liquid Phase Sintered Interconnects 1314
Hannes Greve, *University of Maryland*; S. Ali Moeini, *University of Maryland*; F. Patrick McCluskey, *University of Maryland*

A Lead Free Joining Technology for High Temperature Interconnects Using Transient Liquid Phase Soldering (TLPS) 1321
Christian Ehrhardt, *Technical University Berlin*; Matthias Hutter, *Fraunhofer IZM*; Hermann Oppermann, *Fraunhofer IZM*; Klaus-Dieter Lang, *Technical University Berlin*

Developments of High-Bi Alloys as a High Temperature Pb-Free Solder 1328
Sandeep Mallampati, *Binghamton University*; Harry Schoeller, *Universal Instruments Corporation*; Liang Yin, *GE Global Research*; David Shaddock, *GE Global Research*; Junghyun Cho, *Binghamton University*

The Quantum Theory of Solid-State Atomic Bonding 1335
Chin C. Lee, *University of California, Irvine*; Lianxi Cheng, *University of California, Irvine*

Effective Method to Disperse and Incorporate Carbon Nanotubes in Electroless Ni-P Deposits 1342
Sha Xu, *City University of Hong Kong*; Yan Cheong Chan, *City University of Hong Kong*; Xiaoxin Zhu, *University of Greenwich*; Hua Lu, *University of Greenwich*; Chris Bailey, *University of Greenwich*

Electroless Ni-W-P Alloy as a Barrier Layer between Zn-Based High Temperature Solders and Cu Substrates 1348
Li Liu, *Loughborough University*; Longzao Zhou, *Huazhong University of Science & Technology*; Changqing Liu, *Loughborough University*

31: PoP, SiP, and Die Stacking
Chairs: Raj N. Master, *Microsoft Corporation*
Deborah Patterson, *Amkor Technology, Inc.*

Fabrication and Reliability Evaluation of a Novel Package-on-Package (PoP) Structure Based on Organic Substrate 1354
Xiaofeng Sun, *National Center for Advanced Packaging*; *Chinese Academy of Sciences*; Lixi Wan, *Chinese Academy of Sciences*; Yuan Lu, *National Center for Advanced Packaging*; *Chinese Academy of Sciences*

Strip Grinding Introduction for Thin PoP 1361
Jinseong Kim, *Amkor Technology Korea, Inc.*; Yesul Ahn, *Amkor Technology Korea, Inc.*; Gyuwan Han, *Amkor Technology Korea, Inc.*; Byoungwoo Cho, *Amkor Technology Korea, Inc.*; Dongjoo Park, *Amkor Technology Korea, Inc.*; Juhoon Yoon, *Amkor Technology Korea, Inc.*; Choonheung Lee, *Amkor Technology Korea, Inc.*; Lou Nicholls, *Amkor Technology Inc.*; Shengmin Wen, *Amkor Technology Inc.*

Cost and Performance Effective Silicon Interposer and Vertical Interconnect for 3D ASIC and Memory Integration 1366
Li Li, *Cisco Systems, Inc.*; Mitsutoshi Higashi, *Shinko Electric Industries Company, Ltd.*; Akihito Takano, *Shinko Electric Industries Company, Ltd.*; Jie Xue, *Cisco Systems, Inc.*; Gary Ikari, *Shinko Electric Industries Company, Ltd.*

Assembly and Packaging of Non-Bumped 3D Chip Stacks on Bumped Substrates 1372

Bing Dang, *IBM Corporation*; Joana Maria, *IBM Corporation*; Qianwen Chen, *IBM Corporation*; Jae-Woong Nah, *IBM Corporation*; Paul Andry, *IBM Corporation*; Cornelia Tsang, *IBM Corporation*; Katsuyuki Sakuma, *IBM Corporation*; Christy Tyberg, *IBM Corporation*; Raphael Robertazzi, *IBM Corporation*; Michael Scheuermann, *IBM Corporation*; Michael Gaynes, *IBM Corporation*; John Knickerbocker, *IBM Corporation*

The Miniaturization of a Micro-Ball Endoscope by SiP Approach 1378

Xunxun Zhu, *Tsinghua University*; Jian Cai, *Tsinghua University*; Yu Chen, *Tsinghua University*; Yingke Gu, *Tsinghua University*; Xiang Xie, *Tsinghua University*; Qian Wang, *Tsinghua University*; Zhihua Wang, *Tsinghua University*; Xiaofeng Sun, *Chinese Academy of Sciences*; Lixi Wan, *Chinese Academy of Sciences*

Design and Demonstration of Paper-Thin and Low-Warpage Single and 3D Organic Packages with Chip-Last Embedding Technology for Smart Mobile Applications 1384

Sung Jin Kim, *Georgia Institute of Technology*; Zihan Wu, *Georgia Institute of Technology*; Makoto Kobayashi, *Namics Corporation*; Fuhan Liu, *Georgia Institute of Technology*; Vanessa Smet, *Georgia Institute of Technology*; P. Markondeya Raj, *Georgia Institute of Technology*; Venky Sundaram, *Georgia Institute of Technology*; Rao Tummala, *Georgia Institute of Technology*

Manufacturing Readiness of BVA Technology for Ultra-High Bandwidth Package-on-Package 1389

Rajesh Katkar, *Invensas Corporation*; Rey Co, *Invensas Corporation*; Wael Zohni, *Invensas Corporation*

32: Substrates
Chairs: Yu-Hua Chen, *Unimicron*
Dong Wook Kim, *Qualcomm, Inc.*

Improvement of Substrate and Package Warpage by Copper Plating Process Optimization 1396

Omar Bchir, *Qualcomm Technologies, Inc.*; Houssam Jomaa, *Qualcomm Technologies, Inc.*; Chin Kwan Kim, *Qualcomm Technologies, Inc.*; Layal Rouhana, *Qualcomm Technologies, Inc.*; Kuiwon Kang, *Qualcomm Technologies, Inc.*; Milind Shah, *Qualcomm Technologies, Inc.*; Steve Bezuk, *Qualcomm Technologies, Inc.*

Coreless Substrate with Asymmetric Design to Improve Package Warpage 1401

Wei Lin, *Amkor Technology*; Bora Baloglu, *Amkor Technology*; Ken Stratton, *Amkor Technology*

Ultra Low CTE (1.8 ppm/°C) Core Material for Next Generation Thin CSP 1407

Tomohiko Kotake, *Hitachi Chemical Co., Ltd.*; Hikari Murai, *Hitachi Chemical Co., Ltd.*; Shin Takanezawa, *Hitachi Chemical Co., Ltd.*; Masato Miyatake, *Hitachi Chemical Co., Ltd.*; Masaaki Takekoshi, *Hitachi Chemical Co., Ltd.*; Masahisa Ose, *Hitachi Chemical Co., Ltd.*

A Novel Redistribution Layer Tailored by Nanotwinned Copper Decreases Warpage in Wafer Level Packaging 1411

Heng Li, *Shanghai Institute of Microsystem and Information Technology, Chinese Academy of Sciences*; Wenguo Ning, *Shanghai Institute of Microsystem and Information Technology, Chinese Academy of Sciences*; Chunsheng Zhu, *Shanghai Institute of Microsystem and Information Technology, Chinese Academy of Sciences*; Gaowei Xu, *Shanghai Institute of Microsystem and Information Technology, Chinese Academy of Sciences*; Le Luo, *Shanghai Institute of Microsystem and Information Technology, Chinese Academy of Sciences*

Demonstration of 3–5 µm RDL Line Lithography on Panel-Based Glass Interposers 1416

Hao Lu, *Georgia Institute of Technology*; Yutaka Takagi, *NGK Spark Plug Co., Ltd.*; Yuya Suzuki, *Georgia Institute of Technology*; Brett Sawyer, *Georgia Institute of Technology*; Robin Taylor, *Atotech GmbH*; Venky Sundaram, *Georgia Institute of Technology*; Rao Tummala, *Georgia Institute of Technology*

Characterization of Thin Polymer Films with the Focus on Lateral Stress and Mechanical Properties and Their Relevance to Microelectronics 1421

Markus Woehrmann, *Technical University Berlin*; Thorsten Fischer, *Fraunhofer IZM*; Hans Walter, *Fraunhofer IZM*; Michael Toepper, *Fraunhofer IZM*; Klaus-Dieter Lang, *Technical University Berlin*

Thin Polymer Dry-Film Dielectric Material and a Process for 10 μm Interlayer Vias in High Density Organic and Glass Interposers 1427

Yuya Suzuki, *Zeon Corporation; Georgia Institute of Technology*; Yutaka Takagi, *NGK Spark Plug Co., Ltd.*; Venky Sundaram, *Georgia Institute of Technology*; Rao Tummala, *Georgia Institute of Technology*

33: Novel Test Methods

Chairs: Lakshmi N. Ramanathan, *Microsoft Corporation*
Sridhar Canumalla, *Microsoft Corporation*

Pad Crater Detection Using Acoustic Waveform Analysis 1433

W. Carter Ralph, *Southern Research Institute*; Elizabeth E. Benedetto, *Hewlett Packard*; Aileen M. Allen, *Hewlett Packard*; Keith Newman, *Hewlett Packard*

High Acceleration Board Level Reliability Drop Test Using Dual Mass Shock Amplifier 1441

Andy Zhang, *Texas Instruments, Inc.*

Non-Destructive Crack and Defect Detection in SAC Solder Interconnects Using Cross-Sectioning and X-Ray Micro-CT Using Cross-Sectioning and X-Ray Micro-CT 1449

Pradeep Lall, *Auburn University*; Shantanu Deshpande, *Auburn University*; Junchao Wei, *Auburn University*; Jeff Suhling, *Auburn University*

High Resolution and Fast Throughput-Time X-Ray Computed Tomography for Semiconductor Packaging Applications 1457

Yan Li, *Intel Corporation*; Mario Pacheco, *Intel Corporation*; Deepak Goyal, *Intel Corporation*; John W. Elmer, *Lawrence Livermore National Laboratory*; Holly D. Barth, *Lawrence Livermore National Laboratory*; Dula Parkinson, *Lawrence Berkeley National Laboratory*

In-Situ Measurements of the Relative Thermal Resistance: Highly Sensitive Method to Detect Crack Propagation in Solder Joints 1464

Gordon Elger, *Technische Hochschule Ingolstadt*; Shri Vishnu Kandaswamy, *Technische Hochschule Ingolstadt*; Maarten von Kouwen, *Philips Technology GmbH*; Robert Derix, *Philips Technology GmbH*; Fosca Conti, *University of Padova*

Reliability Testing of Wire Bonds Using Pad Resistance with van der Pauw Method 1471

Michael Mayer, *University of Waterloo*; Samuel Kim, *University of Waterloo*

Colour Shift in Remote Phosphor Based LED Products 1477

M. Yazdan Mehr, *Materials Innovation Institute; Delft University of Technology*; W.D. Van Driel, *Philips Lighting; Delft University of Technology*; G.Q. Zhang, *Delft University of Technology*

34: Novel Packaging

Chairs: Vasudeva P. Atluri, *Renavitas Technologies*
Jai Agrawal, *Purdue University*

Multifunctional System Integration in Flexible Substrates 1482

K. Bock, *Fraunhofer EMFT*; E. Yacoub-George, *Fraunhofer EMFT*; W. Hell, *Fraunhofer EMFT*; A. Drost, *Fraunhofer EMFT*; H. Wolf, *Fraunhofer EMFT*; D. Bollmann, *Fraunhofer EMFT*; C. Landesberger, *Fraunhofer EMFT*; G. Klink, *Fraunhofer EMFT*; H. Gieser, *Fraunhofer EMFT*; C. Kutter, *Fraunhofer EMFT*

Preparation of a Micro Rubidium Vapor Cell and Its Integration in a Chip-Scale Atomic Magnetometer 1488

Yu Ji, *Southeast University*; Jintang Shang, *Southeast University*; Youpeng Chen, *Southeast University*; Ching-Ping Wong, *Chinese University of Hong Kong*

Nanowires-Based High-Density Capacitors and Thinfilm Power Sources in Ultra-Thin 3D Glass Modules 1492

Saumya Gandhi, *Georgia Institute of Technology*; Liyi Li, *Georgia Institute of Technology*; Ho-Yee Hui, *Georgia Institute of Technology*; Parthasarathi Chakraborti, *Georgia Institute of Technology*; Himani Sharma, *Georgia Institute of Technology*; P. Markondeya Raj, *Georgia Institute of Technology*; C.P. Wong, *Georgia Institute of Technology*; Rao Tummala, *Georgia Institute of Technology*

Development of a High Density Glass Interposer Based on Wafer Level Packaging Technologies 1498

Michael Töpper, *Fraunhofer IZM*; Markus Wöhrmann, *Technical University Berlin*; Lars Brusberg, *Fraunhofer IZM*; Nils Jürgensen, *Fraunhofer IZM*; Ivan Ndip, *Fraunhofer IZM*; Klaus-Dieter Lang, *Technical University Berlin*

Novel Sealing Technology for Organic EL Display and Lighting by Means of Modified Surface Activated Bonding Method 1504

Takashi Matsumae, *University of Tokyo*; Masahisa Fujino, *University of Tokyo*; Tadatomo Suga, *University of Tokyo*

Solder Joint Inspection with Induction Thermography 1509

Johannes Bohm, *Technical University Dresden*; Klaus-Juergen Wolter, *Technical University Dresden*; Henning Heuer, *Technical University Dresden*

Development of B-Spline X-Ray Diffraction Imaging Techniques for Die Warpage and Stress Monitoring inside Fully Encapsulated Packaged Chips 1517

C.S. Wong, *Dublin City University*; A. Ivankovic, *IMEC; KU Leuven*; A. Cowley, *Dublin City University*; N.S. Bennett, *Dublin City University*; A.N. Danilewsky, *Albert-Ludwigs-Universität*; M. Gonzalez, *IMEC*; V. Cherman, *IMEC*; B. Vandevelde, *IMEC*; I. De Wolf, *IMEC; KU Leuven*; P.J. McNally, *Dublin City University*

35: Innovations in Wirebond Technology

Chairs: William Chen, *Advanced Semiconductor Engineering, Inc.*
Gilles Poupon, *CEA-LETI*

Process Optimization and Reliability Study for Cu Wire Bonding Advanced Nodes 1523

Ivy Qin, *Kulicke and Soffa Industries, Inc.*; Hui Xu, *Kulicke and Soffa Industries, Inc.*; Basil Milton, *Kulicke and Soffa Industries, Inc.*; Nestor Mendoza, *Kulicke and Soffa Industries, Inc.*; Horst Clauberg, *Kulicke and Soffa Industries, Inc.*; Bob Chylak, *Kulicke and Soffa Industries, Inc.*; Hidenori Abe, *Hitachi Chemical Co., Ltd.*; Dongchul Kang, *Hitachi Chemical Co., Ltd.*; Yoshinori Endo, *Hitachi Chemical Co., Ltd.*; Masahiko Osaka, *Hitachi Chemical Co., Ltd.*; Shinya Nakamura, *Hitachi Chemical Co., Ltd.*

Silver-Assisted Copper Wire Bonding Using Solid-State Processes 1529

Yi-Ling Chen, *University of California, Irvine*; Yuan-Yun Wu, *University of California, Irvine*; Chin C. Lee, *University of California, Irvine*

Ag Alloy Wire Characteristic and Benefits 1533

Jensen Tsai, *Siliconware Precision Industries Co., Ltd.*; Albert Lan, *Siliconware Precision Industries Co., Ltd.*; D.S. Jiang, *Siliconware Precision Industries Co., Ltd.*; Li Wei Wu, *Siliconware Precision Industries Co., Ltd.*; Joseph Huang, *Siliconware Precision Industries Co., Ltd.*; J.B. Hong, *Siliconware Precision Industries Co., Ltd.*

Copper versus Palladium Coated Copper Wire Process and Reliability Differences 1539

Chu-Chung (Stephen) Lee, *Freescale Semiconductor, Inc.*; TuAnh Tran, *Freescale Semiconductor, Inc.*; Dan Boyne, *Freescale Semiconductor, Inc.*; Leo Higgins, *Freescale Semiconductor, Inc.*; Andrew Mawer, *Freescale Semiconductor, Inc.*

Improving the Bond Quality of Copper Wire Bonds Using a Friction Model Approach 1549

Simon Althoff, *University of Paderborn*; Jan Neuhaus, *University of Paderborn*; Tobias Hemsel, *University of Paderborn*; Walter Sextro, *University of Paderborn*

High Aspect Ratio Lithography for Litho-Defined Wire Bonding 1556

Zahra Kolahdouz Esfahani, *Delft University of Technology*; Henk van Zeijl, *Delft University of Technology*; G.Q. Zhang, *Delft University of Technology*

Comprehensive Intermetallic Compound Phase Analysis and Its Thermal Evolution at Cu Wirebond Interface 1562

In-Tae Bae, *Binghamton University*; Dae Young Jung, *Binghamton University*; Jenny Chang, *Advanced Semiconductor Engineering, Inc.*; Scott Chen, *Advanced Semiconductor Engineering, Inc.*

36: Recent Advancement in Manufacturing Technology
Chairs: Paul Houston, *Engent*
Hirofumi Nakajima, *Consultant*

High Uniformity and High Speed Copper Pillar Plating Technique 1571

Konstantin Kholostov, *Sapienza University of Rome*; Aliaksei Klyshko, *Sapienza University of Rome*; Danilo Ciarniello, *Rise Technology S.r.l.*; Paolo Nenzi, *Rise Technology S.r.l.*; Roberto Pagliucci, *Rise Technology S.r.l.*; Rocco Crescenzi, *Sapienza University of Rome*; Dario Bernardi, *2BG*; Marco Balucani, *Sapienza University of Rome, Rise Technology S.r.l.*

Plasma-Based Die Singulation Processing Technology 1577

Kenneth D. Mackenzie, *Plasma-Therm LLC*; David Pays-Volard, *Plasma-Therm LLC*; Linnell Martinez, *Plasma-Therm LLC*; Christopher Johnson, *Plasma-Therm LLC*; Thierry Lazerand, *Plasma-Therm LLC*; Russell Westerman, *Plasma-Therm LLC*

Removed Organic Solderability Preservative (OSP) by Ar/O2 Microwave Plasma to Improve Solder Joint in Thermal Compression Flip Chip Bonding 1584

Jr-Wei Peng, *ASE Group*; Yan-Siang Chen, *ASE Group*; Yi Chen, *ASE Group*; Jiang-Long Liang, *National Cheng Kung University*; Kwang-Lung Lin, *National Cheng Kung University*; Yuh-Lang Lee, *National Cheng Kung University*

A PoP Structure to Support I/O over 2000 1590

Dyi-Chung Hu, *Unimicron Technology Corporation*; Puru Lin, *Unimicron Technology Corporation*; Yu Hua Chen, *Unimicron Technology Corporation*; Chun-Ting Lin, *Unimicron Technology Corporation*

Enabling Eutectic Soldering of 3D Opto-Electronics onto Low Tg Flexible Interposers 1595

Meriem Ben-Salah Akin, *Leibniz University of Hanover*; Lutz Rissing, *Leibniz University of Hanover*; Wolfgang Heumann, *Leibniz University of Hanover*

Parameter Optimization in Assembly Manufacturing Process for a Power Module 1601

Yumin Liu, *Fairchild Semiconductor Corporation*; Yong Liu, *Fairchild Semiconductor Corporation*

Automated Inspection and Metrology for 2.5D and 3D/TSV Process Assurance 1606

James Wood, *IBM Corporation*; Vilmarie Soler, *IBM Corporation*; Eric Perfecto, *IBM Corporation*; Thomas Luckenbach, *Camtek USA*; Aki Shoukrun, *Camtek Ltd.*

37: Interactive Presentations 1
Chairs: Mark Poliks, *i3 Electronics, Inc.*
Ibrahim Guven, *University of Arizona*

Investigation of a Photodefinable Glass Substrate for Millimeter-Wave Radios on Package 1610

Telesphor Kamgaing, *Intel Corporation*; Adel A. Elsherbini, *Intel Corporation*; Torrey W. Frank, *Intel Corporation*; Sasha N. Oster, *Intel Corporation*; Valluri R. Rao, *Intel Corporation*

Design and Fabrication of Low-Pressure Piezoresistive MEMS Sensor for Fuel Cell Electric Vehicles 1616

Minkyu Lee, *Hyundai Motor Company*; Kiyoung Nam, *Hyundai Motor Company*; Seungyong Lee, *Hyundai Motor Company*; Hakgu Kim, *Hyundai Motor Company*; Chimyung Kim, *Hyundai Motor Company*; Yongsun Park, *Hyundai Motor Company*; Byungki Ahn, *Hyundai Motor Company*; Taewan Kim, *Sejong Industrial Company, Ltd.*; Hochul Seo, *Sejong Industrial Company, Ltd.*

Demonstration of TCNCP Flip Chip Reliability with 30μm Pitch Cu Bump and Substrate with Thin Ni and Thick Au Surface Finish 1622

Weihong Zhang, *Nantong Fujitsu Microelectronics Co., Ltd.*; Shengping Hong, *Nantong Fujitsu Microelectronics Co., Ltd.*; Xiaolong Yan, *Nantong Fujitsu Microelectronics Co., Ltd.*; Feng Zhou, *Nantong Fujitsu Microelectronics Co., Ltd.*; Tonglong Zhang, *Nantong Fujitsu Microelectronics Co., Ltd.*

Integrated Process Characterization and Fabrication Challenges for 2.5D IC Packaging Utilizing Silicon Interposer with Backside Via Reveal Process 1628

Cheng-Hsiang Liu, *Siliconware Precision Industries Co., Ltd.*; Jyun-Ling Tsai, *Siliconware Precision Industries Co., Ltd.*; Hung-Hsien Chang, *Siliconware Precision Industries Co., Ltd.*; Chang-Lun Lu, *Siliconware Precision Industries Co., Ltd.*; Shih-Ching Chen, *Siliconware Precision Industries Co., Ltd.*

Structure Effects on the Electrical Reliability of Fine-Pitch Cu Micro-Bumps for 3D Integration 1635

Byeong-Rok Lee, *Andong National University*; June-Bum Kim, *Andong National University*; Seung-Hyun Kim, *Andong National University*; Byeong-Hyun Bae, *Andong National University*; Ho-Young Son, *SK Hynix Inc.*; Tac-Keun Oh, *SK Hynix Inc.*; Min-Suk Suh, *SK Hynix Inc.*; Nam-Seog Kim, *SK Hynix Inc.*; Young-Bae Park, *Andong National University*

Demonstration of Low Cost TSV Fabrication in Thick Silicon Wafers 1641

E. Vick, *RTI International*; D.S. Temple, *RTI International*; R. Anderson, *RTI International*; J. Lannon, *RTI International*; C. Li, *DRS RSTA, Inc.*; K. Peterson, *DRS RSTA, Inc.*; G. Skidmore, *DRS RSTA, Inc.*; C.J. Han, *DRS RSTA, Inc.*

X-Ray Micro-Beam Diffraction Measurement of the Effect of Thermal Cycling on Stress in Cu TSV: A Comparative Study 1648

Chukwudi Okoro, *NIST*; Lyle E. Levine, *NIST*; Ruqing Xu, *Argonne National Laboratory*; Klaus Hummler, *SEMATECH*; Yaw Obeng, *NIST*

Adhesive Enabling Technology for Directly Plating Copper onto Glass/Ceramic Substrates 1652

Hailuo Fu, *Atotech USA Inc.*; Sara Hunegnaw, *Atotech USA Inc.*; Zhiming Liu, *Atotech USA Inc.*; Lutz Brandt, *Atotech USA Inc.*; Tafadzwa Magaya, *Atotech USA Inc.*

Very Thin POP and SIP Packaging Approaches to Achieve Functionality Integration Prior to TSV Implementation 1656

Fernando Roa, *Amkor Technology, Inc.*

A Study on the Fine Pitch Chip Interconnection Using Cu/SnAg Bumps and B-Stage Non-Conductive Films (NCFs) for 3D-TSV Vertical Interconnection 1661

Yongwon Choi, *Korea Advanced Institute of Science and Technology (KAIST)*; Jiwon Shin, *Korea Advanced Institute of Science and Technology (KAIST)*; Young Soon Kim, *Korea Advanced Institute of Science and Technology (KAIST)*; Kyung-Lim Suk, *Korea Advanced Institute of Science and Technology (KAIST)*; Il Kim, *Korea Advanced Institute of Science and Technology (KAIST)*; Kyung-Wook Paik, *Korea Advanced Institute of Science and Technology (KAIST)*

Pathfinding Methodology for Optimal Design and Integration of 2.5D/3D Interconnects 1667

Farhang Yazdani, *BroadPak Corporation*; John Park, *Mentor Graphics Corporation*

Cost Effective Interposer for Advanced Electronic Packages 1673

Satoru Kuramochi, *Dai Nippon Printing Co., Ltd.*; Sumio Koiwa, *Dai Nippon Printing Co., Ltd.*; Kousuke Suzuki, *Dai Nippon Printing Co., Ltd.*; Yoshitaka Fukuoka, *WEISTI*

Thermal Management for Wafer Level Packaging (WLP) 1679

Tiao Zhou, *Maxim Integrated*; Arkadii Samoilov, *Maxim Integrated*

Inkjet Printed Nano-Particle Cu Process for Fabrication of Re-Distribution Layers on Silicon Wafer 1685
Ayat Soltani, *Tampere University of Technology*; Tero Kumpulainen, *Tampere University of Technology*; Matti Mäntysalo, *Tampere University of Technology*

Design of Multi-Sensor for Safety Monitoring of Heavy Machinery 1690
Long Li, *Huazhong University of Science & Technology*; Fei Hou, *Dongfeng Automobile Electronics Co., Ltd.*; Jinghao Qiu, *Nanjing University of Aeronautics and Astronautics*; Zhang Luo, *Huazhong University of Science & Technology*; Shengzhi Zhang, *Huazhong University of Science & Technology*; Qiang Dan, *Huazhong University of Science & Technology*; Sheng Liu, *Wuhan University*

Novel TSV Process Technologies for 2.5D/3D Packaging 1697
Y. Morikawa, *ULVAC, Inc.; NMEMS Technology Research Organization*; T. Murayama, *ULVAC, Inc.; NMEMS Technology Research Organization*; T. Sakuishi, *ULVAC, Inc.; NMEMS Technology Research Organization*; A. Suzuki, *ULVAC, Inc.*; Y. Nakamuta, *ULVAC, Inc.*; K. Suu, *ULVAC, Inc.; NMEMS Technology Research Organization*

Increasing the Lifetime of Electronic Packaging by Higher Temperatures: Solders vs. Silver Sintering 1700
Aaron Hutzler, *Fraunhofer IISB*; Adam Tokarski, *Fraunhofer IISB*; Silke Kraft, *Fraunhofer IISB*; Sigrid Zischler, *Fraunhofer IISB*; Andreas Schletz, *Fraunhofer IISB*

Comparison of New Die-Attachment Technologies for Power Electronic Assemblies 1707
Eike Möller, *University of Freiburg*; Adeel Ahmad Bajwa, *University of Freiburg*; Eugen Rastjagaev, *Infineon Technologies AG*; Jürgen Wilde, *University of Freiburg*

High Vacuum Wafer Level Packaging for High-Value MEMS Applications 1714
S. Nicolas, *CEA-LETI*; F. Greco, *CEA-LETI*; S. Caplet, *CEA-LETI*; C. Coutier, *CEA-LETI*; C. Dressler, *CEA-LETI*; M. Audoin, *CEA-LETI*; X. Baillin, *CEA-LETI*; G. Dehag, *CEA-LETI*; F. Souchon, *CEA-LETI*; S. Fanget, *CEA-LETI*

Thermal and Electrical Tests of Air-Gap TSV 1722
Cui Huang, *Tsinghua University*; Dong Wu, *Tsinghua University*; Zheyao Wang, *Tsinghua University*

Heterogeneous System Integration Pseudo-SoC Technology for Smart-Health-Care Intelligent Life Monitor Engine & Eco-System (SILMEE) 1729
Hiroshi Yamada, *Toshiba Corporation*; Yasuhiro Sato, *Toshiba Corporation*; Nobuhiro Ooshima, *Toshiba Corporation*; Hiroyuki Hirai, *Toshiba Corporation*; Takuji Suzuki, *Toshiba Corporation*; Shigenobu Minami, *Toshiba Corporation*

Effects of Various Environmental Conditions on the Electrical Properties and Interfacial Reliability of Printed Ag/Polyimide System 1735
Byung-Hyun Bae, *Andong National University*; Min-Su Jeong, *Andong National University*; Byeong Rok Lee, *Andong National University*; Joung-Hoon Choo, *HICEL*; Eun-Kuk Choi, *HICEL*; Jong-Sun Yoon, *HICEL*; Young-Bae Park, *Andong National University*

Wafer Level Warpage Characterization for Backside Manufacturing Processes of TSV Interposers 1740
Feng Jiang, *National Center for Advanced Packaging*; Qibin Wang, *National Center for Advanced Packaging; Chinese Academy of Sciences*; Kai Xue, *National Center for Advanced Packaging; Chinese Academy of Sciences*; Xiangmeng Jing, *National Center for Advanced Packaging; Chinese Academy of Sciences*; Daquan Yu, *National Center for Advanced Packaging; Chinese Academy of Sciences*; Dongkai Shangguan, *National Center for Advanced Packaging; Chinese Academy of Sciences*

Stretchable and Transparent Silicone/Zinc Oxide Nanocomposite for Advanced LED Packaging 1745
Xueying Zhao, *Georgia Institute of Technology*; Liyi Li, *Georgia Institute of Technology*; Zhuo Li, *Georgia Institute of Technology*; Ching-Ping Wong, *Georgia Institute of Technology; Chinese University of Hong Kong*

Warpage Characterization of Panel Fan-Out (P-FO) Package 1750

Hung-Wen Liu, *Siliconware Precision Industries Co., Ltd.*; Yi-Wei Liu, *Siliconware Precision Industries Co., Ltd.*; Jason Ji, *Siliconware Precision Industries Co., Ltd.*; Jash Liao, *Siliconware Precision Industries Co., Ltd.*; Agassi Chen, *Siliconware Precision Industries Co., Ltd.*; Yan-Heng Chen, *Siliconware Precision Industries Co., Ltd.*; Nicholas Kao, *Siliconware Precision Industries Co., Ltd.*; Yi-Che Lai, *Siliconware Precision Industries Co., Ltd.*

38: Interactive Presentations 2

Chairs: Mark Eblen, *Kyocera America, Inc.*
Michael Mayer, *University of Waterloo*

A Novel Double Layer NCF for Highly Reliable Micro-Bump Interconnection 1755

Ji-Won Shin, *Korea Advanced Institute of Science and Technology (KAIST)*; Yong-Won Choi, *Korea Advanced Institute of Science and Technology (KAIST)*; Young Soon Kim, *Korea Advanced Institute of Science and Technology (KAIST)*; Un Byung Kang, *Samsung Electronics Company, Ltd.*; Sun Kyung Seo, *Samsung Electronics Company, Ltd.*; Kyung-Wook Paik, *Korea Advanced Institute of Science and Technology (KAIST)*

CO2-Laser Drilling of TGVs for Glass Interposer Applications 1759

Lars Brusberg, *Fraunhofer IZM*; Marco Queisser, *Technical University Berlin*; Marcel Neitz, *Technical University Berlin*; Henning Schröder, *Fraunhofer IZM*; Klaus-Dieter Lang, *Technical University Berlin*

Effects of Pad Surface Finish on Interfacial Reliabilities of Cu-Pillar/Sn-Ag Bumps of 2.5D TSV-Interposer on PCB Applications 1765

Youngsoon Kim, *Samsung Electro-Mechanics Company, Ltd.*; Ji-Won Shin, *Korea Advanced Institute of Science and Technology (KAIST)*; Young Won Choi, *Korea Advanced Institute of Science and Technology (KAIST)*; Kyung-Wook Paik, *Korea Advanced Institute of Science and Technology (KAIST)*

Effect of Variation in the Reflow Profile on the Microstructure of Near Eutectic SnAgCu Alloys 1769

Francis Mutuku, *Binghamton University*; Babak Arfaei, *Binghamton University; Universal Instruments Corporation*; Eric J. Cotts, *Binghamton University*

Development of the Thin Film with High Thermal Conductivity for Power Devices 1776

Hiroshi Takasugi, *Namics Corporation*; Shin Teraki, *Namics Corporation*; Tsuyoshi Kurokawa, *Namics Corporation*; Issei Aoki, *Namics Corporation*

Development of Electroless Nickel-Iron Plating Process for Microelectronic Applications 1782

Yu Luo, *IBM Corporation*; Sung K. Kang, *IBM Corporation*; Oblesh Jinka, *IBM Corporation*; Maurice Mason, *IBM Corporation*; Steven A. Cordes, *IBM Corporation*; Lubomyr T. Romankiw, *IBM Corporation*

Novel Conductive Paste Using Hybrid Silver Sintering Technology for High Reliability Power Semiconductor Packaging 1790

Howard (Hwa Il) Jin, *Alpha Advanced Materials*; Senthil Kanagavel, *Alpha Advanced Materials*; Wai Foo Chin, *Alpha Advanced Materials*

Novel Low Temperature Curable Photo-Sensitive Insulator 1796

Kenji Okamoto, *JSR Corporation*; Hikaru Mizuno, *JSR Corporation*; Tomohiko Sakurai, *JSR Corporation*; Katsumi Inomata, *JSR Corporation*

3D and 2.5D Packaging Assembly with Highly Silica Filled One Step Chip Attach Materials for Both Thermal Compression Bonding and Mass Reflow Processes 1803

Christopher Breach, *Kester Inc.*; Daniel Duffy, *Kester Inc.*; David Eichstadt, *Kester Inc.*

Process Compatibility of Conventional and Low-Temperature Curable Organic Insulation Materials for 2.5D and 3D IC Packaging – A User's Perspective 1810

Guilian Gao, *Invensas Corporation*; Bong-Sub Lee, *Invensas Corporation*; Andrew Cao, *Invensas Corporation*; Ellis Chau, *Invensas Corporation*

Optimization of CMP Process for TSV Reveal in Consideration of Critical Defect 1816

DongHoon Lee, *Amkor Technology Korea, Inc.; Sungkyunkwan University*; DoHyeong Kim, *Amkor Technology Korea, Inc.*; SeungChul Han, *Amkor Technology Korea, Inc.*; JooHyun Kim, *Amkor Technology Korea, Inc.*; JungSoo Park, *Amkor Technology Korea, Inc.*; BoRa Jang, *Amkor Technology Korea, Inc.*; YoungSuk Chung, *Amkor Technology Korea, Inc.*; SeongMin Seo, *Amkor Technology Korea, Inc.*; YongSang Kim, *Sungkyunkwan University*; ChoonHeung Lee, *Amkor Technology Korea, Inc.*

High Throughput Roller Type Nano-Pattern Transfer Technique on Both Rigid Flexible Substrates and Mold Deformation Analysis under Atmospheric Imprint Environment 1822

Yinsheng Zhong, *Hong Kong University of Science and Technology*; Matthew M.F. Yuen, *Hong Kong University of Science and Technology*

Capacitive Deionization of Water Coolant Using Hybrid Carbon Electrodes for High Power Electronic Applications 1828

Ziyin Lin, *Georgia Institute of Technology*; Zhuo Li, *Georgia Institute of Technology*; Kyoung-Sik Moon, *Georgia Institute of Technology*; Ching-Ping Wong, *Georgia Institute of Technology; Chinese University of Hong Kong*

A Microfluidic Chip Integrated with a Sono-Transducer Using Combined Resonance between Oscillations of Hemispherical Micro Glass Shell and Enclosed Microfluid .... N/A

Jiafeng Xu, *Southeast University*; Jintang Shang, *Southeast University*; Ching-Ping Wong, *Chinese University of Hong Kong*

RF Energy Harvesting 1838

Parvizso Aminov, *Purdue University*; Jai P. Agrawal, *Purdue University*

Localized Metal Plating on Aluminum Back Side PV Cells 1842

M. Balucani, *Sapienza University of Rome; Rise Technology S.r.l.*; K. Kholostov, *Sapienza University of Rome*; L. Serenelli, *ENEA Casaccia Research Centre*; M. Izzi, *ENEA Casaccia Research Centre*; D. Bernardi, *2BG S.r.l.*; M. Tucci, *ENEA Casaccia Research Centre*

Wet Etching of Deep Trenches on Silicon with Three-Dimensional (3D) Controllability 1848

Liyi Li, *Georgia Institute of Technology*; Ching-Ping Wong, *Georgia Institute of Technology; Chinese University of Hong Kong*

An Innovative Bumpless Stacking with Through Silicon Via for 3D Wafer-On-Wafer (WOW) Integration 1853

Sue-Chen Liao, *Industrial Technology Research Institute (ITRI)*; Erh-Hao Chen, *Industrial Technology Research Institute (ITRI)*; Chien-Chou Chen, *Industrial Technology Research Institute (ITRI)*; Shang-Chun Chen, *Industrial Technology Research Institute (ITRI)*; Jui-Chin Chen, *Industrial Technology Research Institute (ITRI)*; Po-Chih Chang, *Industrial Technology Research Institute (ITRI)*; Yiu-Hsiang Chang, *Industrial Technology Research Institute (ITRI)*; Cha-Hsin Lin, *Industrial Technology Research Institute (ITRI)*; Tzu-Kun Ku, *Industrial Technology Research Institute (ITRI)*; Ming-Jer Kao, *Industrial Technology Research Institute (ITRI)*; Young Suk Kim, *Tokyo Institute of Technology*; Nobuhide Maeda, *Tokyo Institute of Technology*; Shoichi Kodama, *Tokyo Institute of Technology*; Hideki Kitada, *Tokyo Institute of Technology*; Koji Fujimoto, *Tokyo Institute of Technology*; Takayuki Ohba, *Tokyo Institute of Technology*

3D Integration and Assembly of Wireless Sensor Nodes for 'Green' Sensor Networks 1857

Jian Lu, *National Institute of AIST; NMEMS Technology Research Organization*; Hironao Okada, *National Institute of AIST; NMEMS Technology Research Organization*; Toshihiro Itoh, *National Institute of AIST; NMEMS Technology Research Organization*; Takeshi Harada, *NMEMS Technology Research Organization*; Ryutaro Maeda, *National Institute of AIST; NMEMS Technology Research Organization*

New Demultiplexer Component for Optical Polymer Fiber Communication Systems 1862

S. Höll, *Harz University of Applied Sciences*; M. Haupt, *Harz University of Applied Sciences*; U.H.P. Fischer, *Harz University of Applied Sciences*

Nanofiller Based Spin-on Materials for Negligible Reflection of Silicon Photonic External Coupling 1870

Yoichi Taira, *IBM Corporation*; Ryuma Mizusawa, *Tokyo Ohka Kogyo Co., Ltd.*; Rie Matsumoto, *Tokyo Ohka Kogyo Co., Ltd.*; Kuniaki Suaoke, *IBM Corporation*; Hidetoshi Numata, *IBM Corporation*

Effect of Patterned Substrate on Light Extraction Efficiency of Chip-on-Board Packaging LEDs 1876

Huai Zheng, *Huazhong University of Science & Technology*; Zhili Zhao, *Huazhong University of Science & Technology*; Yiman Wang, *Huazhong University of Science & Technology*; Lang Li, *Huazhong University of Science & Technology*; Sheng Liu, *Huazhong University of Science & Technology*; Xiaobing Luo, *Huazhong University of Science & Technology*

39: Interactive Presentations 3

Chairs: Patrick Thompson, *Texas Instruments, Inc.*
Rao Bonda, *Amkor Technology, Inc.*

Transferrable Fine Pitch Probe Technology 1880

Y. Liu, *IBM Corporation*; S.L. Wright, *IBM Corporation*; B. Dang, *IBM Corporation*; P. Andry, *IBM Corporation*; R. Polastre, *IBM Corporation*; J. Knickerbocker, *IBM Corporation*

Improvement of the Crystallinity of Electroplated Copper Thin Films for Highly Reliable 3D Interconnections 1885

Chuanhong Fan, *Tohoku University*; Osamu Asai, *Tohoku University*; Ryosuke Furuya, *Tohoku University*; Ken Suzuki, *Tohoku University*; Hideo Miura, *Tohoku University*

Process, Assembly and Electromigration Characteristics of Glass Interposer for 3D Integration 1891

Chun-Hsien Chien, *Industrial Technology Research Institute (ITRI)*; Ching-Kuan Lee, *Industrial Technology Research Institute (ITRI)*; Chun-Te Lin, *Industrial Technology Research Institute (ITRI)*; Yu-Min Lin, *Industrial Technology Research Institute (ITRI)*; Chau-Jie Zhan, *Industrial Technology Research Institute (ITRI)*; Hsiang-Hung Chang, *Industrial Technology Research Institute (ITRI)*; Chao-Kai Hsu, *Industrial Technology Research Institute (ITRI)*; Huan-Chun Fu, *Industrial Technology Research Institute (ITRI)*; Wen-Wei Shen, *Industrial Technology Research Institute (ITRI)*; Yu-Wei Huang, *Industrial Technology Research Institute (ITRI)*; Cheng-Ta Ko, *Industrial Technology Research Institute (ITRI)*; Wei-Chung Lo, *Industrial Technology Research Institute (ITRI)*; Yung Jean (Rachel) Lu, *Corning Inc.*

Improved PCB Via Pattern to Reduce Crosstalk at Package BGA Region for High Speed Serial Interface 1896

Yujeong Shim, *Altera Corporation*; Dan Oh, *Altera Corporation*

A Wafer Level Through-Stack-Via Integration Process with One-Time Bottom-up Copper Filling 1902

Yunhui Zhu, *Peking University*; Shenglin Ma, *Xiamen University, Peking University*; Xin Sun, *Peking University*; Runiu Fang, *Peking University*; Xiao Zhong, *Peking University*; Yuan Bian, *Peking University*; Yong Guan, *Peking University*; Jing Chen, *Peking University*; Min Miao, *Peking University, Beijing Information Science and Technology University*; Yufeng Jin, *Peking University*

Effect of Joint Shape Controlled by Thermocompression Bonding on the Reliability Performance of 60μm-Pitch Solder Micro Bump Interconnections 1908

Yu-Wei Huang, *Industrial Technology Research Institute (ITRI)*; Chau-Jie Zhan, *Industrial Technology Research Institute (ITRI)*; Jing-Ye Juang, *Industrial Technology Research Institute (ITRI)*; Yu-Min Lin, *Industrial Technology Research Institute (ITRI)*; Shin-Yi Huang, *Industrial Technology Research Institute (ITRI)*; Su-Mei Chen, *Industrial Technology Research Institute (ITRI)*; Chia-Wen Fan, *Industrial Technology Research Institute (ITRI)*; Ren-Shin Cheng, *Industrial Technology Research Institute (ITRI)*; Shu-Han Chao, *National Chiao Tung University*; Wan-Lin Hsieh, *National Chiao Tung University*; Chih Chen, *National Chiao Tung University*; John H. Lau, *Industrial Technology Research Institute (ITRI)*

Development of Micro Bump Joints Fabrication Process Using Cone Shape Au Bumps for 3D LSI Chip Stacking 1915

Fumito Imura, *National Institute of AIST*; Naoya Watanabe, *National Institute of AIST*; Shunsuke Nemoto, *National Institute of AIST*; Wei Feng, *National Institute of AIST*; Katsuya Kikuchi, *National Institute of AIST*; Hiroshi Nakagawa, *National Institute of AIST*; Masahiro Aoyagi, *National Institute of AIST*

Effect of Polymer Liners in CNT Based Through Silicon Vias 1921
    Archana Kumari, *Indian Institute of Technology Roorkee*; M.K. Majumder, *Indian Institute of Technology Roorkee*; B.K. Kaushik, *Indian Institute of Technology Roorkee*; S.K. Manhas, *Indian Institute of Technology Roorkee*

Investigation of Low-Temperature Deposition High-Uniformity Coverage Parylene-HT as a Dielectric Layer for 3D Interconnection 1926
    Bui Thanh Tung, *National Institute of AIST*; Xiaojin Cheng, *National Institute of AIST; Loughborough University*; Naoya Watanabe, *National Institute of AIST*; Fumiki Kato, *National Institute of AIST*; Katsuya Kikuchi, *National Institute of AIST*; Masahiro Aoyagi, *National Institute of AIST*

Arrays of Millimeter-Wave Silicon Waveguides for Interchip Communication on Glass Interposer 1932
    Qidong Wang, *Chinese Academy of Sciences; National Center for Advanced Packaging*; Daniel Guidotti, *Chinese Academy of Sciences; National Center for Advanced Packaging*; Liqiang Cao, *Chinese Academy of Sciences; National Center for Advanced Packaging*; Delong Qiu, *National Center for Advanced Packaging*; Daquan Yu, *Chinese Academy of Sciences; National Center for Advanced Packaging*; Shuling Wang, *Chinese Academy of Sciences; National Center for Advanced Packaging*; Xugang Wang, *Chinese Academy of Sciences; National Center for Advanced Packaging*; Tiachun Ye, *Chinese Academy of Sciences; National Center for Advanced Packaging*; Lixi Wan, *National Center for Advanced Packaging*

Effect of Ag and Cu Content in Sn Based Pb-Free Solder on Electromigration 1940
    Minhua Lu, *IBM Corporation*; Charles Goldsmith, *IBM Corporation*; Thomas Wassick, *IBM Corporation*; Eric Perfecto, *IBM Corporation*; Charles Arvin, *IBM Corporation*

Low Loss Transmission Lines on Flexible COP Substrate by Standard Lamination Process 1944
    Chang-Ho Liou, *Industrial Technology Research Institute (ITRI)*; Hsin-Chia Lu, *National Taiwan University*; Yi-Fan Lin, *National Taiwan University*; Shih-Keng Chuang, *National Taiwan University*; Wen-Ching Ko, *Industrial Technology Research Institute (ITRI)*; Je-Ping Hu, *Industrial Technology Research Institute (ITRI)*

FBEOL No-Aluminum Pad Integration in Pb-Free C4 Products for Environmental, Cost and Reliability Benefits 1949
    E. Misra, *IBM Corporation*; T. Daubenspeck, *IBM Corporation*; T. Wassick, *IBM Corporation*; K. Tunga, *IBM Corporation*; D. Questad, *IBM Corporation*

Preparing 25Gbps Electrical I/O for Exascale Computing Systems 1955
    Lei Shan, *IBM Corporation*; Young Kwark, *IBM Corporation*; Renato Rimolo-Donadio, *IBM Corporation*; Christian Baks, *IBM Corporation*; Michael Gaynes, *IBM Corporation*; Timothy Chainer, *IBM Corporation*; Manabu Hoshino, *Zeon Corporation*; Masakazu Hashimoto, *Zeon Corporation*; Toshihiko Jimbo, *Zeon Corporation*; Junji Kodemura, *Zeon Corporation*; Ikkei Matsuura, *Zeon Corporation*

Large Low-CTE Glass Package-to-PCB Interconnections with Solder Strain-Relief Using Polymer Collars 1959
    Gary Menezes, *Georgia Institute of Technology*; Vanessa Smet, *Georgia Institute of Technology*; Makoto Kobayashi, *Namics Corporation*; Venky Sundaram, *Georgia Institute of Technology*; Pulugurtha Markondeya Raj, *Georgia Institute of Technology*; Rao Tummala, *Georgia Institute of Technology*

The Study of Bare-Die FCBGA Die Damage in Response to Applied Mechanical Stress During Heat Sink Assembly 1965
    Heidi S.Y. Ho, *Broadcom Corporation*; Daijiao Wang, *Broadcom Corporation*; Michael Johnson, *Amkor Technology, Inc.*; C.J. Berry, *Amkor Technology, Inc.*

Prognostication of Copper-Aluminum Wirebond Reliability under High Temperature Storage and Temperature-Humidity 1973
    Pradeep Lall, *Auburn University*; Shantanu Deshpande, *Auburn University*; Luu Nguyen, *Texas Instruments*; Masood Murtuza, *Texas Instruments*

Low-Frequency Testing of Through Silicon Vias for Defect Diagnosis in Three-Dimensional Integration Circuit Stacking Technology 1986

Yichao Xu, *Peking University*; Min Miao, *Beijing Information Science & Technology University*; *Peking University*; Runiu Fang, *Peking University*; Xin Sun, *Peking University*; Yunhui Zhu, *Peking University*; Minggang Sun, *Beijing Information Science & Technology University*; Guanjiang Wang, *Peking University*; Yufeng Jin, *Peking University*

Fast Estimation of LED's Accelerated Lifetime by Online Test Method 1992

Qi Chen, *Huazhong University of Science & Technology*; Quan Chen, *Huazhong University of Science & Technology*; Xiaobing Luo, *Huazhong University of Science & Technology*

Methodology and Apparatus for Rapid Power Cycle Accumulation and In-Situ Incipient Failure Monitoring for Power Electronic Modules 1996

Roy I. Davis, *Fairchild Semiconductor Corporation*; Daniel J. Sprenger, *Fairchild Semiconductor Corporation*

Fine-Pitch Probing on TSVs and Microbumps Using a Chip Prober Having a Transparent Membrane Probe Card 2003

Naoya Watanabe, *National Institute of AIST*; Michiyuki Eto, *STK Technology Co., Ltd.*; Kenji Kawano, *STK Technology Co., Ltd.*; Masahiro Aoyagi, *National Institute of AIST*

40: Interactive Presentations 4

Chairs: Nam Pham, *IBM Corporation*
Rabindra N. Das, *MIT Lincoln Labs*

Thermal Management of 3D RF PoP Based on Ceramic Substrate 2008

Fengze Hou, *National Center for Advanced Packaging*; *Chinese Academy of Sciences*; Fengman Liu, *National Center for Advanced Packaging*; *Chinese Academy of Sciences*; Yi He, *Chinese Academy of Sciences*; Xiaomeng Wu, *Chinese Academy of Sciences*; Xia Zhang, *National Center for Advanced Packaging*; *Chinese Academy of Sciences*; Liqiang Cao, *National Center for Advanced Packaging*; *Chinese Academy of Sciences*; Yuan Lu, *National Center for Advanced Packaging*; *Chinese Academy of Sciences*; Dongkai Shangguan, *National Center for Advanced Packaging*; *Chinese Academy of Sciences*

Bump Pattern Optimization and Stress Comparison Study for DCA Packages 2014

Akash Agrawal, *Micron Technology Inc.*; Owen Fay, *Micron Technology Inc.*; Mark Johnson, *Micron Technology Inc.*

Characterization of In-Plane Stress in TSV Array – A Unit Model Approach 2020

Cheng-Fu Chen, *University of Alaska, Fairbanks*

Electrical-Thermal Characterization of Wires in Packages 2027

Kai Liu, *STATS ChipPAC, Inc.*; Robert Frye, *STATS ChipPAC, Inc.*; HyunTai Kim, *STATS ChipPAC, Inc.*; YongTaek Lee, *STATS ChipPAC, Inc.*; Gwang Kim, *STATS ChipPAC, Inc.*; Susan Park, *STATS ChipPAC, Inc.*; Billy Ahn, *STATS ChipPAC, Inc.*

Computational Investigation of Failure in Anodized Aluminum 2035

Sabrina Ball, *University of Arizona*; Ibrahim Guven, *University of Arizona*; Pankaj Sinha, *Intel Corporation*; Rajiv Rastogi, *Intel Corporation*; Brian McCarson, *Intel Corporation*

Study on Prediction about Residual Position of Void Generated by Resin Flow 2042

Masayuki Mino, *Hitachi, Ltd.*; Naoya Suzuki, *Hitachi Chemical Co., Ltd.*; Hiroshi Takahashi, *Hitachi Chemical Co., Ltd.*; Tsutomu Kono, *Hitachi, Ltd.*

Modeling and Analysis of Temperature Effect on MEMS Gyroscope 2048

Ming Wen, *Huazhong University of Science & Technology*; Weihui Wang, *Huazhong University of Science & Technology*; Zhang Luo, *Huazhong University of Science & Technology*; Yong Xu, *Wuhan University*; *Wayne State University*; Xin Wu, *Wayne State University*; Fei Hou, *Dongfeng Automobile Electronics Co., Ltd.*; Sheng Liu, *Wuhan University*

Life Prediction and Classification of Failure Modes in Solid State Luminaires Using Bayesian Probabilistic Models 2053
Pradeep Lall, *Auburn University*; Junchao Wei, *Auburn University*; Peter Sakalaukus, *Auburn University*

Modeling for Reliability of Ultra Thin Chips in a System in Package 2063
Richard Qian, *Fairchild Semiconductor Corporation*; Yong Liu, *Fairchild Semiconductor Corporation*

Development of Effective Thermal Characterization on Handheld Devices by Matrix Method 2069
Tai-Yu Chen, *MediaTek Inc.*; Chung-Fa Lee, *MediaTek Inc.*

Comprehensive Design Optimization for 2.133 Gbps LPDDR3 Extension for Mobile Platform System 2075
Chanmin Jo, *Samsung Electronics*; Jaemin Shin, *Samsung Electronics*; BaekKyu Choi, *Samsung Electronics*; Sangmin Lee, *Samsung Electronics*; Seongjae Moon, *Samsung Electronics*; Sungjoo Kim, *Samsung Electronics*; Woong Hwan Ryu, *Samsung Electronics*

Estimation of Mode Conversion and Crosstalk Impact from a Single-Ended Aggressor to a Differential Victim Using Statistical BER Analysis 2081
Arun Reddy Chada, *Missouri S&T EMC Laboratory*; Jun Fan, *Missouri S&T EMC Laboratory*; James L. Drewniak, *Missouri S&T EMC Laboratory*; Bhyrav Mutnury, *Dell, Inc.*

Power Distribution Network Worst-Case Power Noise and an Efficient Estimation Method 2088
Jiangyuan Qian, *Broadcom Corporation*; Shiji Pan, *University of California, Irvine*

Fast Calculation of Electromagnetic Interference by Through-Silicon Vias 2094
Aosheng Rong, *University of Illinois, Urbana-Champaign*; Andreas C. Cangellaris, *University of Illinois, Urbana-Champaign*; Feng Ling, *Nanjing University of Science and Technology*

Electrical Simulation and Analysis of Si Interposer for 3D IC Integration 2099
Xin Sun, *Peking University*; Min Miao, *Beijing Information Science & Technology University*; Yunhui Zhu, *Peking University*; Runiu Fang, *Peking University*; Guanjiang Wang, *Peking University*; Wengao Lu, *Peking University*; Jing Chen, *Peking University*; Yufeng Jin, *Peking University*

A SPICE Model of Multi-Mode Optical Fiber in Mid-Channel Link for Package System SI Transient Simulations 2104
Zhaoqing Chen, *IBM Corporation*

Next Generation Package-on-Package Solution to Support Wide IO and High Bandwidth Interface 2112
Hung-Hsiang Cheng, *Advanced Semiconductor Engineering, Inc.*; Chang-Chi Lee, *Advanced Semiconductor Engineering, Inc.*; Ming-Feng Chung, *Advanced Semiconductor Engineering, Inc.*; Po-Chih Pan, *Advanced Semiconductor Engineering, Inc.*; Ping-Feng Yang, *Advanced Semiconductor Engineering, Inc.*; Chi-Tsung Chiu, *Advanced Semiconductor Engineering, Inc.*; Chih-Pin Hung, *Advanced Semiconductor Engineering, Inc.*; Chen-Chao Wang, *Advanced Semiconductor Engineering, Inc.*

Package-Level Electromagnetic Interference Analysis 2119
Namhoon Kim, *Broadcom Corporation*; Leo Hongyu Li, *Broadcom Corporation*; Sam Karikalan, *Broadcom Corporation*; Reza Sharifi, *Broadcom Corporation*; Henry Kim, *Broadcom Corporation*

A Path Finding Based SI Design Methodology for 3D Integration 2124
Bill Martin, *E-System Design*; KiJin Han, *UNIST*; Madhavan Swaminathan, *Georgia Institute of Technology*

Design and Implementation of a 700-2600 MHz RF SiP for Micro Base Station .... N/A
Yi He, *National Center for Advanced Packaging; Chinese Academy of Sciences*; Fengman Liu, *National Center for Advanced Packaging; Chinese Academy of Sciences*; Anmou Liao, *National Center for Advanced Packaging; Chinese Academy of Sciences*; Jun Li, *National Center for Advanced Packaging; Chinese Academy of Sciences*; Xiaomeng Wu, *National Center for Advanced Packaging*; Peng Wu, *National Center for Advanced Packaging; Chinese Academy of Sciences*; Liqiang Cao, *National Center for Advanced Packaging; Chinese Academy of Sciences*; Dongkai Shangguan, *National Center for Advanced Packaging*

Dielectric Lens Optimization for Conical Helix THz Antennas 2137

Paolo Nenzi, *ENEA Frascati Research Center*; Volha Varlamava, *Sapienza University of Rome*; Frank Silvio Marzano, *Sapienza University of Rome*; Fabrizio Palma, *Sapienza University of Rome*; Marco Balucani, *Sapienza University of Rome*

Embedded Diodes for Microwave and Millimeter Wave Circuits 2144

Xianbo Yang, *Michigan State University*; Amanpreet Kaur, *Michigan State University*; Premjeet Chahal, *Michigan State University*

PCIe Gen3 Link Design and Tuning in Server Systems with End Devices from Multiple IP Suppliers 2151

Si T. Win, *IBM Corporation*; Daniel Rodriguez, *IBM Corporation*; Nanju Na, *IBM Corporation*

A Low-Cost PCB Fabrication Process 2159

Jack Ou, *Sonoma State University*; Alberto Maldonado, *Sonoma State University*; Chio Saephan, *Sonoma State University*; Farid Farahmand, *Sonoma State University*; Michael Caggiano, *Rutgers University*

Novel Band-Pass Filters on Thin Glass Substrate with Through Glass Vias (TGVs) .... N/A

Cheng Pang, *National Center for Advanced Packaging*; *Chinese Academy of Sciences*; Wenya Shang, *National Center for Advanced Packaging*; *Chinese Academy of Sciences*; Mingchuan Zhang, *Chinese Academy of Sciences*; Zheng Qin, *National Center for Advanced Packaging*; *Chinese Academy of Sciences*; Huijuan Wang, *National Center for Advanced Packaging*; *Chinese Academy of Sciences*; Xiaoli Ren, *National Center for Advanced Packaging*; Jie Pan, *Chinese Academy of Sciences*; Daquan Yu, *National Center for Advanced Packaging*; *Chinese Academy of Sciences*; Dongkai Shangguan, *National Center for Advanced Packaging*; *Chinese Academy of Sciences*

Study of Microwave Circuits Based on Metal-Insulator-Metal (MIM) Diodes on Flex Substrates 2168

Amanpreet Kaur, *Michigan State University*; Xianbo Yang, *Michigan State University*; Premjeet Chahal, *Michigan State University*

41: Student Interactive Presentations

Chairs: Mark Poliks, *i3 Electronics, Inc.*
Ibrahim Guven, *University of Arizona*

Nanocomposite Pastes for Thermal and Mechanical Bonding 2175

Tingting Zhang, *Binghamton University*; Bahgat Sammakia, *Binghamton University*; Howard Wang, *Binghamton University*

Assembly and Packaging Technologies for High-Temperature and High-Power GaN HEMTs 2181

A.A. Bajwa, *University of Freiburg*; Y. Qin, *University of Freiburg*; J. Wilde, *University of Freiburg*; R. Reiner, *Fraunhofer Institute IAF*; P. Waltereit, *Fraunhofer Institute IAF*; R. Quay, *Fraunhofer Institute IAF*

Flip-Chip on Glass (FCOG) Package for Low Warpage 2189

Scott R. McCann, *Georgia Institute of Technology*; Venkatesh Sundaram, *Georgia Institute of Technology*; Rao R. Tummala, *Georgia Institute of Technology*; Suresh K. Sitaraman, *Georgia Institute of Technology*

Laser-Based Conductive Film Forming with Gold Nanoparticles for Electrical Contacts 2194

Mitsugu Yamaguchi, *Ibaraki University*; Shinji Araga, *Ibaraki Giken Ltd.*; Mamoru Mita, *M&M Research Laboratory*; Kazuhiko Yamasaki, *Ibaraki University*; Katsuhiro Maekawa, *Ibaraki University*

Analysis of Modes Effect on Signal/Power Integrity in Finite Cavity for Chip and Die Level Packaging Based on a Hybrid Full Wave Method 2200

Xin Chang, *University of Washington*; Leung Tsang, *University of Washington*

Directed Self-Assembly of Mesoscopic Dies Using Magnetic Force and Shape Recognition 2207

Anton Tkachenko, *Rensselaer Polytechnic Institute*; Robert F. Karlicek, Jr., *Rensselaer Polytechnic Institute*; James J.-Q. Lu, *Rensselaer Polytechnic Institute*

Controlled Silicon IC Thinning on Individual Die Level for Active Implant Integration Using a Purely Mechanical Process 2213

Vasiliki Giagka, *University College London*; Nooshin Saeidi, *University College London*; Andreas Demosthenous, *University College London*; Nick Donaldson, *University College London*

Connectors and Vibrations – Damages in Different Electrical Environments .... N/A

A. Berghuvud, *Blekinge Institute of Technology*; T. Björnängen, *Blekinge Institute of Technology*; T. Gissila, *Blekinge Institute of Technology*

Study of Extreme Low Temperature and Load Solid-Phase Sn-Ag System Bonding Mechanism for 3D ICs 2227

Kiyoto Yoneta, *Osaka University*; Ryohei Sato, *Osaka University*; Yoshiharu Iwata, *Osaka University*; Koichiro Atsumi, *Osaka University*; Kazuya Okamoto, *Osaka University*; Yukihiro Satio, *Osaka University*; Takumi Shigemoto, *Osaka University*

Self-Patterning, Pre-Applied Underfilling Technology for Stack-Die Packaging 2231

Chia-Chi Tuan, *Georgia Institute of Technology*; Ziyin Lin, *Georgia Institute of Technology*; Yan Liu, *Georgia Institute of Technology*; Kyoung-Sik Moon, *Georgia Institute of Technology*; Ching-Ping Wong, *Georgia Institute of Technology*; *Chinese University of Hong Kong*

Study of High CRI White Light-Emitting Diode Devices with Multi-Chromatic Phosphor 2236

Min Zheng, *Xi'an Jiaotong University*; Wen Ding, *Xi'an Jiaotong University*; Feng Yun, *Xi'an Jiaotong University*; Deyang Xia, *Xi'an Jiaotong University*; Yaping Huang, *Xi'an Jiaotong University*; Yukun Zhao, *Xi'an Jiaotong University*; Weihan Zhang, *Xi'an Jiaotong University*; Minyan Zhang, *Xi'an Jiaotong University*; Maofeng Guo, *Xi'an Jiaotong University*; Ye Zhang, *Xi'an Jiaotong University*

The Effects of Self-Fluxing Additives in Solder Anisotropic Conductive Films (ACFs) on Solder Wettability and Joint Reliability of Flex-on-Board (FOB) Assemblies 2241

Seung-Ho Kim, *Korea Advanced Institute of Science and Technology (KAIST)*; Yongwon Choi, *Korea Advanced Institute of Science and Technology (KAIST)*; Yoosun Kim, *Korea Advanced Institute of Science and Technology (KAIST)*; Kyung-Wook Paik, *Korea Advanced Institute of Science and Technology (KAIST)*

Modeling and Analysis of Frequency Shift of MEMS Gyroscope Subjected to Temperature Change 2245

Weihui Wang, *Huazhong University of Science & Technology*; Sheng Liu, *Wuhan University*; Zhang Luo, *Huazhong University of Science & Technology*; Ming Wen, *Huazhong University of Science & Technology*; Qiang Dan, *Huazhong University of Science & Technology*; Man Yu, *Huazhong University of Science & Technology*; Yong Xu, *Wuhan University, Wayne State University*; Xin Wu, *Wayne State University*

Interaction Effect between Electromigration and Microstructure Evolution in Cu/Sn-58Bi/Cu Solder Interconnect 2249

Hong-Bo Qin, *South China University of Technology*; Bin Li, *Southern Methodist University*; Wu Yue, *South China University of Technology*; Chang-Bo Ke, *South China University of Technology*; Min-Bo Zhou, *South China University of Technology*; Xin-Ping Zhang, *South China University of Technology*

Effects of Alignment of Graphene Flakes on Water Permeability of Graphene-Epoxy Composite Film 2255

Seong-Yoon Jung, *Korea Advanced Institute of Science and Technology (KAIST)*; Kyung-Wook Paik, *Korea Advanced Institute of Science and Technology (KAIST)*

Characterization of Alternate Power Distribution Methods for 3D Integration 2260

David C. Zhang, *Georgia Institute of Technology*; Madhavan Swaminathan, *Georgia Institute of Technology*; David Keezer, *Georgia Institute of Technology*; Satyanarayana Telikepalli, *Georgia Institute of Technology*

Adhesion and Reliability of Direct Cu Metallization of Through-Package Vias in Glass Interposers 2266

Timothy Huang, *Georgia Institute of Technology*; Venky Sundaram, *Georgia Institute of Technology*; P. Markondeya Raj, *Georgia Institute of Technology*; Himani Sharma, *Georgia Institute of Technology*; Rao Tummala, *Georgia Institute of Technology*

High-Frequency Characterization of Through-Package Vias Formed by Focused Electrical-Discharge in Thin Glass Interposers 2271

Jialing Tong, *Georgia Institute of Technology*; Yoichiro Sato, *Asahi Glass Company*; Shintaro Takahashi, *Asahi Glass Company*; Nobuhiko Imajyo, *Asahi Glass Company*; Andrew F. Peterson, *Georgia Institute of Technology*; Venky Sundaram, *Georgia Institute of Technology*; Rao Tummala, *Georgia Institute of Technology*

Interfacial Reactions between Cu and Sn, Sn-Ag, Sn-Bi, Sn-Zn Solder under Space Confinement for 3D IC Micro Joint Applications 2277

T.L. Yang, *National Taiwan University*; W.L. Shih, *National Taiwan University*; J.J. Yu, *National Taiwan University*; C.R. Kao, *National Taiwan University*

Simulation and Optimization of a Micro Flow Sensor 2283

Xing Guo, *Huazhong University of Science & Technology*; Chunlin Xu, *Huazhong University of Science & Technology*; Shengzhi Zhang, *Huazhong University of Science & Technology*; Yong Xu, *Wayne State University*; Xin Wu, *Wayne State University*; Sheng Liu, *Wuhan University*

Minimizing Coupling of Power Supply Noise between Digital and RF Circuit Blocks in Mixed Signal Systems 2287

Satyanarayana Telikepalli, *Georgia Institute of Technology*; Madhavan Swaminathan, *Georgia Institute of Technology*; David Keezer, *Georgia Institute of Technology*

A Feasibility Study of Flip-Chip Packaged Gallium Nitride HEMTs on Organic Substrates for Wideband RF Amplifier Applications 2293

Spyridon Pavlidis, *Georgia Institute of Technology*; A. Cagri Ulusoy, *Georgia Institute of Technology*; Wasif T. Khan, *Georgia Institute of Technology*; Outmane Lemtiri Chlieh, *Georgia Institute of Technology*; Edward Gebara, *I2R Nanowave Inc.*; John Papapolymerou, *Georgia Institute of Technology*

A Novel Molding Process for Wafer Level LED Packaging Using Uniform Micro Glass Bubble Arrays 2299

Yu Zou, *Southeast University*; Jintang Shang, *Southeast University*; Yu Ji, *Southeast University*; Li Zhang, *Jiangyin Changdian Advanced Packaging Co. Ltd.*; Chiming Lai, *Jiangyin Changdian Advanced Packaging Co. Ltd.*; Dong Chen, *Jiangyin Changdian Advanced Packaging Co. Ltd.*; Kim-Hui Chen, *Jiangyin Changdian Advanced Packaging Co. Ltd.*; Ching-Ping Wong, *Chinese University of Hong Kong*

Analysis of Room-Temperature Bonded Compliant Bump with Ultrasonic Bonding 2303

Keiichiro Iwanabe, *Kyushu University*; Takanori Shuto, *Kyushu University*; Tanemasa Asano, *Kyushu University*

FOREWORD

On behalf of the Program Committee and Executive Committee, it is our pleasure to welcome you to the 64th Electronic Components and Technology Conference (ECTC) which will be held at the Walt Disney World Swan and Dolphin Resort in Lake Buena Vista, Florida, USA from May 27-30, 2014. This premier international conference is sponsored by the IEEE Components, Packaging, and Manufacturing Technology Society (CPMT).

The ECTC Program Committee has selected over 350 papers which will be presented in 36 oral sessions and 5 interactive presentation sessions including 1 student interactive presentation session. The oral sessions will feature papers on 3D and TSV technologies, wafer level packaging, electrical and mechanical modeling, RF packaging, system design, and optical interconnects. Session topics will cover advanced packaging technologies, material development and characterization, reliability, assembly and manufacturing, and interposers. Four Interactive presentation sessions showcase papers in a format that encourages more in-depth discussion and interaction with authors about their work. Similarly, the student Interactive Presentation session will focus on the research being done at academic institutions around the world by emerging scientists and engineers. The Program Committee has created sessions which cover the ongoing technological challenges of established disciplines as well as addressing emerging topics of interest to the industry. Authors from companies, research institutions, and universities from over 25 countries will present their work at ECTC, illustrating the conference's global focus.

ECTC will also feature panel and special sessions with industry experts covering a number of important and emerging topic areas. On Tuesday, May 27 at 10 a.m., Karlheinz Bock will chair a session entitled "Flexible Electronics - Packaging Technology and Application Trends" where a panel of experts will discuss the recent advancements and market perspectives for this emerging technology area. Tuesday at 2 p.m., Manos Tentzeris will chair a special session sponsored by the Electronic Components & RF technical subcommittee on "Wireless Power Transfer Systems". The ECTC Panel Discussion, "Emerging Technologies and Market Trends of Silicon Photonics" will be chaired by Ricky Lee and Jie Xue and features panelists from various segments of the industry to discuss the growing influence of photonics technologies on microelectronic systems. Nancy Stoffel will chair the ECTC Plenary Session titled "Influence of Packaging on System Integration and Performance" on Wednesday evening at 7:00 p.m. where a panel of experts will discuss system integration and the role that component packaging technologies play. On Thursday evening at 8:00 p.m. the CPMT Seminar titled "Latest Advances in Organic Interposers" will be moderated by Kishio Yokouchi and Venky Sundaram.

Supplementing the technical program, ECTC will also offer several Professional Development Courses (PDCs) and the Technology Corner Exhibits. ECTC will offer 18 Professional Development Courses for 2014, covering a wide array of technical areas and organized by the PDC Committee chaired by Kitty Pearsall. The PDCs will take place on Tuesday, May 27 (8:00 a.m. – 5:30 p.m.) and are taught by distinguished experts in their respective fields. The Technology Corner Exhibits will showcase the latest products and technologies offered by leading companies in the electronic components, materials, packaging, and services fields. Technology Exhibits will be open Wednesday and Thursday starting at 9 a.m. ECTC also offers attendees numerous opportunities for networking and discussion with colleagues during coffee breaks, daily luncheons, and nightly receptions. We are pleased to announce that Dr. Peter Bocko of Corning Glass Technologies will give the invited keynote talk at the ECTC Luncheon on Wednesday.

Whether you are an engineer, a manager, a student, or an executive, ECTC offers something for everyone in the packaging industry. We would like to take this opportunity to thank our sponsors, exhibitors, authors, speakers, PDC instructors, session chairs, program committee members, as well as all the volunteers who have helped to make the 64th ECTC another resounding success. Once again, thank you for being a part of the 64th ECTC.

Wolfgang Sauter
General Chair
IBM Corporation

Alan Huffman
Program Chair
RTI International

Professional Development Courses for the 64th ECTC

May 27, 2014

Kitty Pearsall
Chair
Boss Precision, Inc.
kitty.pearsall@gmail.com
Phone: +1-512-845-3287

Jeffrey Suhling
Assistant Chair
Auburn University
jsuhling@eng.auburn.edu
Phone: +1-334-844-3332

Eddie Kobeda
IBM Corporation
kobeda@us.ibm.com
Phone: +1-919-543-2946

Albert F. Puttlitz
Independent Consultant
alputtlitz@yahoo.com
Phone: +1-802-899-4692

MORNING COURSES 8:00 AM - 12:00 PM

1. **Lead-Free Solder Joint Reliability—Material Consideration**
   Course Leader: Ning-Cheng Lee – Indium Corporation

2. **Wafer Level Packaging**
   Course Leader: Luu Nguyen – Texas Instruments

3. **Package Failure Analysis—Failure Mechanisms and Analytical Tools**
   Course Leaders: Rajen Dias and Deepak Goyal – Intel Corporation

4. **Next Frontier in Electronics: Systems Scaling for Smart Mobile Systems**
   Course Leader: Rao R. Tummala – Georgia Institute of Technology

5. **Polymers and Nano-Composites for Electronic and Photonic Packaging: Recent Advances on Materials and Processes**
   Course Leaders: C.P. Wong – Georgia Institute of Technology; Daniel Lu – Henkel Corporation

6. **Power and Electronics Packaging, Reliability, and Thermal Management**
   Course Leader: Patrick McCluskey and Avram Bar-Cohen - University of Maryland

7. **Fundamental Concepts of Reliability and Mechanics in Electronic Packaging**
   Course Leaders: Shubhada Sahasrabudhe and Sandeep Sane – Intel Corporation

8. **Product Qualification and Supply Chain Responsibilities**
   Course Leader: Michael Pecht – University of Maryland

AFTERNOON COURSES 1:15 P.M. - 5:15 PM

9. **High-Frequency Modeling and Optimization of Interconnections in Electronic Packaging**
   Course Leaders: Ivan Ndip and Michael Töpper -- Fraunhofer IZM

10. **Shock - Impact Reliability of Portable Electronics**
    Course Leader: Pradeep Lall – Auburn University

11. **Moisture and Media Influence on Microelectronic Package Reliability**
    Course Leaders: Tanja Braun and Hans Walter – Fraunhofer IZM

12. **3D IC Packaging and 3D Si Integration**
    Course Leader: John Lau – Industrial Technology Research Institute

13. **Polymers in Semiconductor Packaging Including 2.5D and 3D Integration**
    Course Leader: Jeffrey Gotro – InnoCentrix, LLC

14. **Flip Chip Technology**
    Course Leaders: Eric Perfecto – IBM Corporation; Shengmin Wen – Amkor Technology

15. **Package Failure Mechanisms, Reliability, and Solutions**
    Course Leader: Darvin Edwards – Edwards Enterprises

16. **Statistics for Engineering**
    Course Leader: Patrick Thompson – Texas Instruments

17. **Thermal Materials and Metrology for Advanced Packaging**
    Course Leaders: Ravi Mahajan - Intel Corporation; Kenneth Goodson – Stanford University

18. **Thermo-Electrical Co-Design of 3D Chip Stacks**
    Course Leaders: Avram Bar-Cohen and Ankur Srivastava – University of Maryland

Continuing Education Units

The IEEE Components, Packaging and Manufacturing Technology Society (CPMT) has been authorized to offer Continuing Education Units (CEUs) by the International Association for Continuing Education and Training (IACET) for all Professional Development Courses that will be presented at the 64th ECTC. CEUs are recognized by employers for continuing professional development as a formal measure of participation and attendance in "non-credit" self-study courses, tutorials, symposia and workshops. Complete details, including voluntary enrollment forms, will be available at the conference. All costs associated with ECTC Professional Development Courses CEUs will be underwritten by the conference, i.e. there are no additional costs for Professional Development Courses' attendees to obtain CEU credit.

Plan to propose a Professional Development Course at ECTC in San Diego, CA in 2015!

To propose a professional development course for the next ECTC, see the ECTC website (**www.ectc.net**) for submission and selection guidelines. Or send the proposed course title and outline to the 2015 Professional Development Courses Chair:

Kitty Pearsall, *Chair* Boss Precision, Inc., Phone: +1-512-845-3287, Email: kitty.pearsall@gmail.com

Best of Conference Papers — 2013

The Electronic Components and Technology Conference is proud to announce the "Best of Conference" papers selected from the 63rd ECTC proceedings. The authors of the Best Session Paper share a check for US $2,500 and the authors of the Best Interactive Presentation share a check for US $1,500. The winning authors also receive a personalized plaque commemorating their achievement.

Best Session Paper

Grain Structure Evolution and Its Impact on the Fatigue Reliability of Lead-Free Solder Joints in BGA Packaging Assembly
*Huili Xu, Choong-Un Kim–University of Texas, Arlington;
Tae-Kyu Lee, Cisco Systems, Inc.*

Best Interactive Presentation

Bath Chemistry and Copper Overburden as Influencing Factors of the TSV Annealing
P. Saettler, M. Böttcher, Catharina Rudolph, and K.-J. Wolter–Technische Universität Dresden

Outstanding Papers – 2013

The winning authors for Conference Outstanding Session Paper and Interactive Presentation receive a personalized plaque commemorating their achievement and will share a check for US $1,000.

Outstanding Session Paper

Effects of Varying Amplitudes on the Fatigue Life of Lead Free Solder Joints
M. Obaidat, S. Hamasha, Y. Jaradat, A. Qasaimeh, P. Borgesen–State University of New York, Binghamton; B. Arfaei, M. Anselm–Universal Instruments Corporation

Outstanding Interactive Presentation

Fabrication and Characterization of Novel Photodefined Polymer-Enhanced Through-Silicon Vias for Silicon Interposers
Paragkumar A. Thadesar and Muhannad S. Bakir–Georgia Institute of Technology

Intel Best Student Paper – 2013

The winning student receives a personalized plaque and a check for $2,500. The following paper was selected based on the Intel Best Student Paper competition conducted at the 63rd ECTC:

Low Temperature Fine Pitch Flex-on-Flex (FOF) Assembly Using Nanofiber Sn58Bi Solder Anisotropic Conductive Films (ACFs) and Ultrasonic Bonding Method
Tae Wan Kim, Kyung-Lim Suk and Kyung-Wook Paik–KAIST

Committee Members

64th Electronic Components and Technology Conference

May 27 – May 30, 2014   Orlando, FL USA

Executive Committee

General Chair, 64th ECTC	Wolfgang Sauter	IBM Corporation
Vice-General Chair, 64th ECTC	Beth Keser	Qualcomm, Inc.
Jr. Past General Chair, 62nd ECTC	David McCann	GLOBALFOUNDRIES
Sr. Past General Chair, 61st ECTC	Rajen Dias	Intel Corporation
Program Chair	Alan Huffman	RTI International
Assistant Program Chair	Henning Braunisch	Intel Corporation
Arrangements Chair	Lisa Renzi	Renzi & Company, Inc.
Short Course Chair	Kitty Pearsall	Boss Precision, Inc.
Web Administrator	Sam Karikalan	Broadcom Corporation
Finance Chair	Patrick Thompson	Texas Instruments, Inc.
Treasurer	Tom Reynolds	T3 Group LLC
Publications Chair	Steve Bezuk	Qualcomm CDMA Technologies
Publicity Chair	Eric Perfecto	IBM Corporation
Sponsorship Chair	David McCann	GLOBALFOUNDRIES
Exhibits Chair	Joe Gisler	Vector Associates
CPMT Representative	C. P. Wong	Georgia Institute of Technology

****

Advanced Packaging

Chair

John Knickerbocker .... IBM Corporation

Assistant Chair

Christopher Bower .... X-Celeprint Ltd.

Muhannad Bakir .... Georgia Institute of Technology
Daniel Baldwin .... Engent, Inc.
Omar Bchir .... Qualcomm, Inc.
Rozalia Beica .... Yole Developpement
Jianwei Dong .... Dow Electronic Materials
Paul M. Harvey .... IBM Corporation
Altaf Hasan .... Intel Corporation
Eric Jung .... Fraunhofer IZM
Sam Karikalan .... Broadcom Corporation

Beth Keser .... Qualcomm, Inc.
Young-Gon Kim .... IDT
Jeffrey A. Knight .... Elliott Manufacturing
John H. Lau .... ITRI
Raj N. Master .... Microsoft Corporation
Luu T. Nguyen .... Texas Instruments
Deborah Patterson .... Amkor Technology, Inc.
Raj Pendse .... STATS ChipPAC, Inc.
Peter Ramm .... Fraunhofer EMFT
Subhash L. Shinde .... Sandia National Laboratory
Joseph W. Soucy .... Draper Laboratory
E. Jan Vardaman .... TechSearch International, Inc.
James Jian Zhang .... Micron Technology, Inc.

Applied Reliability

Chair
Scott Savage .... Medtronic Microelectronics Center

Assistant Chair
Vikas Gupta .... Texas Instruments
Jo Caers .... Royal Philips
Sridhar Canumalla .... Microsoft Corporation
Tim Chaudhry .... Broadcom Corporation
Tz-Cheng Chiu .... National Cheng Kung University
Darvin R. Edwards .... Edwards Enterprises
Deepak Goyal .... Intel Corporation
Dongming He .... Qualcomm, Inc.
Toni Mattila .... Aalto University
Keith Newman .... Hewlett-Packard
Donna M. Noctor .... Siemens Industry, Inc.
John H. L. Pang .... Nanyang Technological University
S. B. Park .... Binghamton University
Lakshmi N. Ramanathan .... Microsoft Corporation
Ephraim Suhir .... University of California, Santa Cruz
Jeffrey Suhling .... Auburn University
Dongji Xie .... NVIDIA Corporation

Assembly & Manufacturing Technology

Chair
Shichun Qu .... Fairchild Semiconductor

Assistant Chair
Shawn Shi .... Medtronic Corporation
Sai Ankireddi .... Soraa, Inc.
Sharad Bhatt .... Shanta Systems, Inc
Claudius Feger .... IBM Corporation
Mark Gerber .... Texas Instruments
Paul Houston .... Engent
Sa Huang .... Medtronic Corporation
Vijay Khanna .... IBM Corporation
Chunho Kim .... Medtronic Corporation
Wei Koh .... Pacrim Technology
Choon Heung Lee .... Amkor Technology
Mali Mahalingam .... Freescale Semiconductor, Inc.
Debendra Mallik .... Intel Corporation
Jae-Woong Nah .... IBM Corporation
Hirofumi Nakajima .... Consultant
Valerie Oberson .... IBM Corporation
Tom Poulin .... Aerie Engineering
Tom Swirbel .... Motorola, Inc.
Paul Tiner .... Texas Instruments
Sean Too .... Microsoft Corporation
Andy Tseng .... Taiwan Semiconductor Manufacturing
Shaw Fong Wong .... Intel Corporation
Jie Xue .... Cisco Systems, Inc.
Jin Yang .... Intel Corporation
Tonglong Zhang .... Nantong Fujitsu Microelectronics, Ltd.

Electronic Components & RF

Chair
Manos M. Tentzeris .... Georgia Institute of Technology

Assistant Chair
Nanju Na .... IBM Corporation
Amit P. Agrawal .... Cisco Systems, Inc.
Eric Beyne .... IMEC
Prem Chahal .... Michigan State University

Craig Gaw .... Freescale Semiconductor, Inc.
Apostolos Georgiadis .... CTTC
Abhilash Goyal .... Oracle
T. S. Horng .... National Sun Yat-Sen University
Rockwell Hsu .... Cisco Systems, Inc.
C. P. Hung .... Advanced Semiconductor Engineering, Inc.
Lih-Tyng Hwang .... National Sun Yat-Sen University
Mahadevan K. Iyer .... Texas Instruments, Inc.
Timothy G. Lenihan .... TGL Consulting
Li Li .... Freescale Semiconductor, Inc.
Sebastian Liau .... ITRI
Lianjun Liu .... Freescale Semiconductor, Inc.
Albert Lu .... Singapore Inst. of Manufacturing Technology
Andrea Paganini .... IBM Corporation
Albert F. Puttlitz .... Mechanical Eng. Consultant
P. Markondeya Raj .... Georgia Institute of Technology
Tom Reynolds .... T3 Group LLC
Luca Roselli .... University of Perugia
Clemens Ruppel .... TDK
Hideki Sasaki .... Renesas Electronics Corporation
Grit Sommer .... Infineon Technologies AG
Frank Theunis .... Qualcomm Technologies
Leena Ukkonen .... Tampere University of Technology

Interactive Presentations

Chair
Mark Poliks .... i3 Electronics, Inc.

Assistant Chair
Ibrahim Guven .... University of Arizona
Swapan Bhattacharya .... Georgia Institute of Technology
Rao Bonda .... Amkor Technology
Mark Eblen .... Kyocera America, Inc.
Michael Mayer .... University of Waterloo
Nam Pham .... IBM Corporation
Patrick Thompson .... Texas Instruments, Inc.

Interconnections

Chair
Matthew Yao .... GE Energy Management

Assistant Chair
Li Li .... Cisco Systems, Inc.
William Chen .... Advanced Semiconductor Engineering, Inc.
Kathy Cook .... Ziptronix
Rajen Dias .... Intel Corporation
Bernd Ebersberger .... Intel Mobile Communications
Takafumi Fukushima .... Tohoku University
Tom Gregorich .... Micron
Changqing Liu .... Loughborough University
Wei-Chung Lo .... ITRI
Nathan Lower .... Rockwell Collins, Inc.
James Lu .... Rensselaer Polytechnic Institute
Voya Markovich .... Microelectronic Advanced Hardware
Consulting, LLC
James E. Morris .... Portland State University
Lou Nicholls .... Amkor Technology, Inc.
Gilles Poupon .... CEA-LETI
Katsuyuki Sakuma .... IBM Corporation
Lei Shan .... IBM Corporation
Katsuaki Suganuma .... Osaka University
Chuan Seng Tan .... Nanyang Technological University

Materials & Processing

Chair

Diptarka Majumdar.................................. Superior Graphite

Assistant Chair

Stephanie Potisek...Dow Chemical

Choong Kooi Chee ....KBU International College

Tim Chen............................. Darbond Technology Co., Ltd.

Yu-Hua Chen...Unimicron

Bing Dang.. IBM Corporation

Don Frye...ATMI

C. Robert Kao ....National Taiwan University

Dong Wook Kim.................................... Qualcomm, Inc.

Chin C. Lee............................ University of California, Irvine

Grace Yi Li.................................... Intel Corporation

Kwang-Lung Lin .... National Cheng Kung University

Daniel D. Lu .... Henkel Corporation

Hongtao Ma...................................... Lightera Corporation

Mikel Miller....................................... Draper Laboratory

Kyung-Wook Paik .... KAIST

Mark Poliks...i3 Electronics, Inc.

Dwayne Shirley ....Qualcomm Technologies, Inc.

Ivan Shubin... Oracle

Yoichi Taira... IBM Japan

Lejun Wang ....Qualcomm Technologies, Inc.

Myung Jin Yim............................... Intel Corporation

Tieyu Zheng............................... Microsoft Corporation

Modeling & Simulation

Chair

Zhaoqing Chen.. IBM Corporation

Assistant Chair

Yong Liu ....Fairchild Semiconductor Corporation

Ramachandra Achar................................. Carleton University

Kemal Aygun.. Intel Corporation

Wendem Beyene...Rambus, Inc.

Daniel de Araujo...Nimbic, Inc.

L .J. Ernst...Ernst Consultants

Xuejun Fan ....Lamar University

Xiaoxiong (Kevin) Gu .... IBM Corporation

Bruce Kim.......................... City University of New York

Pradeep Lall.. Auburn University

Michael Lamson ....Consultant

En-Xiao Liu .... Institute of High Performance

Computing, A*STAR

Sheng Liu........ Huazhong University of Science & Technology

Erdogan Madenci.................................... University of Arizona

Tony Mak ....Middlesex Community College

Dan Oh..Altera Corporation

Gamal Refai-Ahmed...........PreQual Technologies Corporation

Sandeep Sane... Intel Corporation

Jaemin Shin............................. Samsung Electronics Co., Ltd.

Suresh K. Sitaraman ....Georgia Institute of Technology

G. Q. (Kouchi) Zhang...............Delft University of Technology

Opto-Electronics

Chair

Kannan Raj...Oracle

Assistant Chair

Stefan Weiss...........................II-VI Laser Enterprise GmbH

Fuad Doany ....IBM Corporation

Gordon Elger........................... Technical University Ingolstadt

Z. Rena Huang ....Rensselaer Polytechnic Institute

Takaaki Ishigure.. Keio University

Soon Jang ....ficonTEC USA

Harry G. Kellzi............. Teledyne Microelectronic Technologies

Michael Leers..Fraunhofer ILT

Masanobu Okayasu .... Oclaro Japan, Inc.

Alex Rosiewicz ....Gooch & Housego

Henning Schroeder...Fraunhofer IZM

Andrew Shapiro .... JPL

Hiren Thacker ....Oracle

Masao Tokunari ....IBM Corporation

Jean Trewhella ....IBM Corporation

Shogo Ura .... Kyoto Institute of Technology

Ping Zhou...LDX Optronics, Inc.

Special Topics (Emerging Technologies)

Chair

Nancy Stoffel ....GE Global Research

Assistant Chair

C. S. Premachandran .... GLOBALFOUNDRIES

Isaac Robin Abothu........ Siemens Medical Solutions USA, Inc.

Jai Agrawal..Purdue University

Vasudeva P. Atluri ....Renavitas Technologies

Mark Bachman......................... University of California, Irvine

Karlheinz Bock .... Fraunhofer EMFT; University of Berlin

John Cunningham ....Oracle

Rabindra N. Das...................................... MIT Lincoln Labs

Steve Greathouse...Plexus Corporation

Joana Maria...IBM Corporation

Goran Matijasevic ....University of California, Irvine

Dave Peard .... Henkel Corporation

Koneru Ramakrishna.......................Hewlett-Packard Company

Jintang Shang...................................... Southeast University

Klaus Jürgen Wolter ....Technical University Dresden

Allison Xiao..Henkel Corporation

Yimin Yao... Intel Corporation

Professional Development Courses

Chair

Kitty Pearsall...Boss Precision, Inc.

Assistant Chair

Jeffrey Suhling ....Auburn University

Eddie Kobeda...IBM Corporation

Al Puttlitz.....................................Mechanical Eng. Consultant

Integration Study of Die Strength and Various Bumping Volume and Reliability Performance on 2.5D Silicon Interposer Assembly

Shih-Liang Peng, Chen-Yu Huang, Ming-Hsien Yang, Stephen Tseng, J.Y. Lai, Terren Lu, Hsien-Wen Chen,
Steve Chiu, Stephen Chen
Siliconware Precision Industries Co., Ltd.
No. 153, Sec.3, Chung-Shan Fong Rd, Tantzu, Taichung, Taiwan, R.O.C.
Tel: 886-4-25341525 ext 7927, Fax: 886-4-25325030
sampeng@spil.com.tw

Abstract

The applications of 2.5D/3D IC integration have been increasing in recent years, which enable high-density and heterogeneous ICs that can be assembled in one package using through silicon interposer (TSI). In this paper, the integration study of interposer in backside via reveal (BVR) process, various bumping volume and stacking reliability were introduced.

Different sizes of solder bump forms on the same side of interposer may become essential in order to reduce not only package parasitic and shorten the interconnection length but also can be developed to cost down for equal IC packaging performance in any kind of 2.5D system packaging. The proposed electrical plating solder bump with different size for top functional dies attachment. This technique shortens overall process time in manufacturing because it provides simpler and shorter total amount of process steps. We studies high volume and low volume solder bump with logic circuit and memory integrated on silicon interposer. The result shows electrical function is feasible.

The interposer BVR process makes die strength reduction while wafer thickness reduced and nitride film covered, It is found that grinding degrades strength significantly if surface is rougher; the chemical vapor deposition(CVD) passive film makes interposer structure as a composite material and dropped die strength significantly from dozen kilogram to hundreds gram level; in further processing, the chemical mechanical planarization(CMP) and C4 bumping will reduce strength a lot about 10 times from several kilogram to hundreds gram level. The strength enhancement were experimentally studied in terms of grinding optimization to control roughness under 1nm; the various kinds of CVD film were deposited; then CMP experiments added. Finally, we gain 2~3 times of die strength totally.

The stacking process scenario: Chip on Chip first (CoC first) was evaluated by various 2.5D assembly structure. The process flow that active dies are mounted on TSI firstly and then the CoC module is mounted on organic substrate is more practicable and reliable. The evaluations were experimentally studied in terms of warpage and thermo-mechanical stress and finally demonstrated the CoC first process flow is more reliable to assure high assembling yield. The robustness of 3D integrated package was also evaluated by QTC (Quick Temperature Cycle) test and the result showed low stress of passive film of TSI promised superior solder joint performance during tests. The package reliability tests were further conducted in terms of preconditioning (MSL-4), THT (Temperature Humidity Testing), un-biased HAST (Highly Accelerated Stress Testing), HTST (High Temperature Storage Test) and TCT (Thermal Cycling Test). The O/S (Open/Short) test and C-SAM imaging were applied for these reliability tests to monitor package integrity. The ASIC-die 2.5D integrated packages have passed O/S test after preconditioning and 2000X TC-B test, no significant issues were found. Furthermore, most integrated samples have passed the other reliability tests such as THT 1000hrs, and HTST 1000hrs.

Introduction

Electroplating of solder wafer developed as more flexible method than evaporation and generally is less cost than evaporated bumping. Electroplating has become more popular for high bump count chips due to its smaller feature size and precision. Traditional flip-chip packaging with electroplating, which uses solder bumps on wafer with under-bump-metallurgy (UBM) to attach to substrate along with under-fill. The UBM is typically an adhesion layer of Titanium-Copper in a thickness of 4K angstroms with an extending copper wetting layer and diffusion barrier layer to limit the diffusion of solder into the underlying material. The UBM is well adhesive to the passive film that is fully preventing oxidization of the underlying layers.

In silicon interposer BVR process, the wafer strength will be degraded along with thickness reduction. As shown in Figure 1, the BVR process includes major step as carrier bond; wafer grinding; silicon reaction ion etch (RIE); CVD nitride film deposition ; CMP through silicon via (TSV) opening ; C4 bumping and carrier de-bond step.

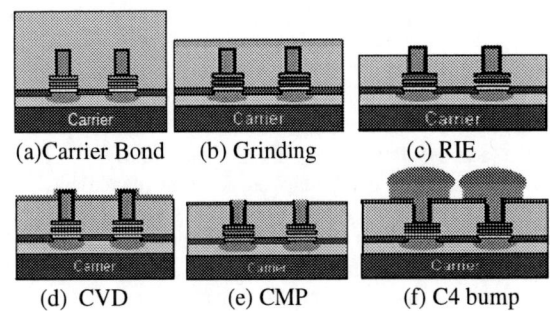

(a)Carrier Bond   (b) Grinding   (c) RIE

(d) CVD   (e) CMP   (f) C4 bump

Figure 1. General Interposer BVR Process

After BVR process, the thinned interposer thickness will be reduced to 100um. It is a high risk for die crack / broken in rear 2.5D package assemble as shown in Figure 2. In order to understand their strength degrading mechanism and further enhance their strength; we monitored die strength variation

2014 Electronic Components & Technology Conference

978-1-4799-2408-0/14 $31.00 © 2014 IEEE

after BVR process step by step to split out key factors for die strength reduction, and further to enhance them.

Figure 2. 2.5D package schematic diagram

The silicon die failure strength via their effects of die thickness, die size and backside grinding patterns on the die stress has been investigated [1], the results showed that the die strength is largely dependent on its geometry and damages due to wafer processing .The die strength enhancement has been under extensive studies [2-5], the die strength degraded major come from backside grinding and dicing mechanical defect and micro crack. Therefore, to eliminate or minimize these kinds of residue damage is very critical for thinned die with enough strength.

The strength properties of silicon substrate containing dense oxide and nitride films were investigated [6], the specified film thickness containing nano-indentations, whereas the oxide film enhances strength, the nitride film degrades it. The impact of the die backside defects on the stress is also investigated [7], it is found that coating a thin layer of nitride on the die backside can increase the die breaking strength.

2.5D and 3D IC stacking technologies using Si interposer are gaining position due to the continuously demand on homo- and heterogeneous IC integration solutions that enable ultrahigh I/O counts, low power consumption, and short interconnect length [9-11]. In design of 2.5D IC integration, two or more chips are placed side by side and assembled face down onto a Si interposer. The copper pillar micro-bumps manufactured on the active side of the chips are jointed with micro-pads on the front side of the interposer. The Si interposer acting as a pitch adapter which redistributes the electrical lines from the active chips down to the organic substrate through micro-joints, redistributed layers (RDLs), TSVs and C4 solder bumps in the assembled packages. The quality of bump joints including micro- and C4 bumps is governed by the warpage behavior of Si interposer which is mainly induced by large CTE mismatch between Si chip and organic substrate. Severe warpage could lead to passivation cracks and solder joints failure under thermal loading.

To address the large warpage issue during assembling process, the optimized chip stacking sequence have been proposed. The "CoC First" where chips are mounted on the interposer first shows better warpage performance than "CoS First" where the interposer is mounted on the substrate first and chips are mounted onto the interposer last. Furthermore, the interposer warpage is significantly affected by the film stress of metallization and surface passivation on the front and back side of a wafer [12]. Some warpage control solutions and simulated warpage results for 2.5D and 3D assembling scenario are also investigated [13-15]. At the recent papers, more 3D-IC relevant research released the long-term reliability test results. Most of the studies were focusing on thermal-mechanical reliability and electrical performance of copper pillar micro-bumps [16-19]. Various reliability failures occurring at package micro-bump level were found after different test condition and to be further improved using specific bump design [11].

Various Bumping Volume

Micro-bump typically grows by Cu pillar bump with Ni and Au of total height 6-8 micrometer or another choice is to grow reflowable small bumps using Sn rich pre-solder which have an advantage of wider range of bump height. Due to its advantage, various sizes of Sn pre-solder bump can be applied to front side of silicon interposer to die bonding with different logical top dies. In Table 1 provides the features of central pre-solder and its U% along with SEM inspection. Second different logical dies can be die bonding with higher pre-solder bumps as shown in Figure 3.

Check Item	Location			Avg.	U%
	Center	Middle	Edge		
Diameter	53.61	52.88	53.11	53.2	0.69
Height (8 +/- 1um)	8.31	8.54	8.13	8.39	3.28
	8.26	8.68	8.42		

Table 1. Central pre-solder profile and its picture by SEM

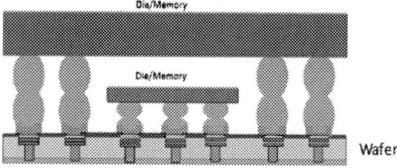

Figure 3. Cross section view of double sizes of micro-bump attaches to different logical top dies.

Die Strength Experiments

In our investigation, we monitor die strength change in BVR process step by step and try to enhance die strength to adjust different process parameters and different film types. And also check the processing surface roughness if any relevant to die strength migration.

In grinding process, there are three types of grinding wheel (Z1, Z2, Z3), we design three legs in Z1+Z2 only (Leg1) and Z1+Z2+Z3 with different wheel feed rate in low (Leg2) and high (Leg3). We also try different CVD film deposition experiments, which are combined by T-SiN (Tensile silicon nitride); C-SiN (Compressive silicon nitride); H-C-SiN (High hardness compressive silicon nitride); Bulk C-SiN (Bulk compressive silicon nitride); SiO2 (Silicon oxide); SiON (Oxy-nitride) in single layer/ bi-layer structure. Finally, we try different pressure and process time in CMP step.

The roughness observation is carried out by AFM (Atomic Force Microscope), the film hardness and young's modulus is carried by Nano-indenter. The die strength be measured by modified ball on ring tester as shown in Figure 4. It is similar with ball on ring test method, which be studied in [8].

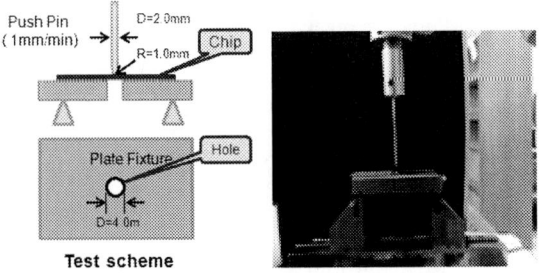

Figure 4. Modified Ball on Ring Die Strength Test Scheme

Die Strength Results and Discussions

From the experiment results, we found the die strength is degrading step by step in BVR process as shown in Figure 5. The 775um bare silicon wafer thickness be reduced to around 115um still with strong die strength after grinding process. The next RIE process to reveal TSV around 5um, the wafer thickness to be reduced to 100um and die strength was degraded about 45% in this stage. The following CVD process deposited the specific thickness silicon nitride film to strengthen the TSV structure to against CMP which possible induced in-process mechanical force, this deposited nitride film made wafer structure as composite material and changed their strength properties, the die strength was huge degraded about 85% to several kilogram level. And next CMP cut down nitride film thickness about 20%; therefore, the die strength also was degraded about 20% proportionally. Finally, the C4 bumping, which combined UBM, electric plating, etching, and reflow also degraded strength about 85% to hundreds of gram level.

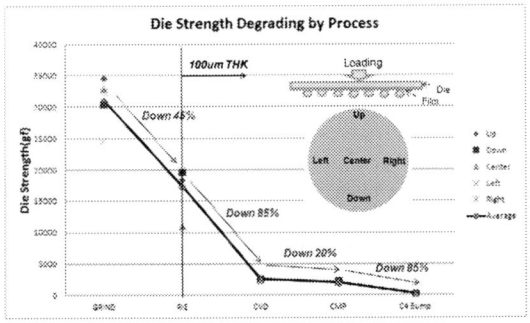

Figure 5. Interposer strength degrading by BVR process

In grinding Z1+Z2 (Leg1) only and Z1+Z2+Z3 with different wheel feed rate in low (Leg2) and high (Leg3) experiments, the results as shown in Figure 6, the Z1+Z2 only with obvious saw mark residue with Ra 22.3nm and die strength results in only hundreds gram. But after RIE dry etching, the wafer surface becomes more smooth with Ra 9.17nm and the sharp grinding saw marks were eliminated, the die strength returns to dozen of kilogram level. The corresponding 3D surface morphology measured by AFM as shown in Figure 7.

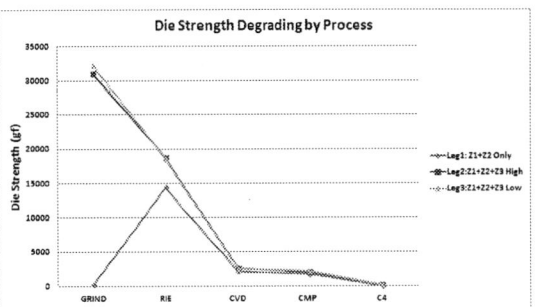

Figure 6. Grinding experiment die strength chart

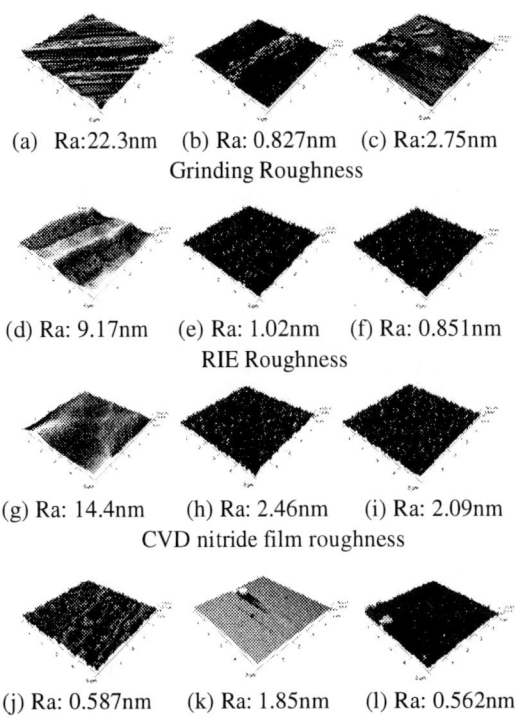

(a) Ra:22.3nm   (b) Ra: 0.827nm   (c) Ra:2.75nm
Grinding Roughness

(d) Ra: 9.17nm   (e) Ra: 1.02nm   (f) Ra: 0.851nm
RIE Roughness

(g) Ra: 14.4nm   (h) Ra: 2.46nm   (i) Ra: 2.09nm
CVD nitride film roughness

(j) Ra: 0.587nm   (k) Ra: 1.85nm   (l) Ra: 0.562nm
CMP roughness

Figure7. Surface roughness after BVR:
Leg1(a)(d)(g)(j), Leg2(b)(e)(h)(k), Leg3(c)(f)(i)(l)

The CVD film deposition experiments results as shown in Figure 8. The H-C-SiN (high hardness compressive nitride) film result in higher die strength not only in single layer structure but also in bi-layer structure. The H-C-SiN film with CVD process parameters controlled to decreased SiH4 and NH3 flow rate and increased RF power to get film characteristics in high hardness (>20GPa) and high young's modulus (E>180GPa), the film hardness and young's module (E) comparison table as shown in Table 2.

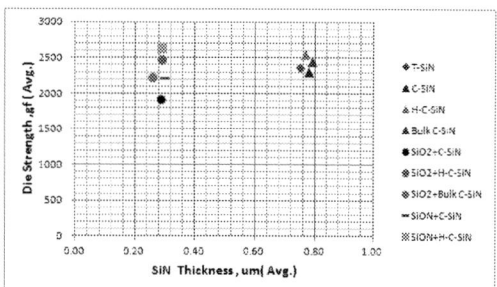

Fig. 8. Strength distribution of CVD film

Film Type	Hardness(GPa)	E(Gpa)
TiSiN	13.32	129.85
C-SiN	16.9	153.7
H-C-SiN	21.76	186.32
SiO2+C-SiN	17.88	122.49
SiO2+H-C-SiN	14.11	130.21
SiON+C-SiN	12.15	121.01
SiON+H-C-SiN	13.89	126.96

Table 2. CVD film hardness and young's modulus

Furthermore, several kind of film's die strength after C4 bumping tested results as shown in Figure 9, the H-C-SiN film results in better die strength with 2~3 times of C-SiN film not only in eutectic bump; but also in lead free bump. It shows that the harder film gains higher strength. From experiment results also discloses that the different metallization in C4 bumping process affects die strength a lot due to C4 metallization induced stress reaction in surface of nitride film.

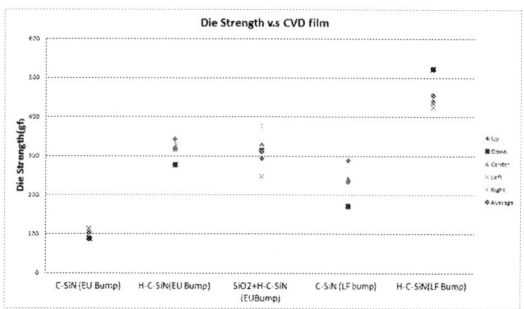

Figure 9. Strength distribution after C4 bumping

This dielectric layers are used to passivate wafers and to redistribute the I/O pattern. Silicon nitride film serves as an important metal protection layer. There are some of possible issues of reliability failure needed to be discussed. Adhesion of silicon nitride to the adjacent metal is often an anxious problem if it is not well controlled and thickness of silicon nitride can vary the warpage of die and wafer level. This may cause serious difficulty to stacking.

In order to improve the quality of silicon nitride film, indeed, high temperature vapor deposition (CVD) defines high density and good quality. Additionally, the film adhesion also be studied in the same time, the experiment in single layer nitride and bi-layer oxide plus nitride film with surface descum treatment or over Q-time conditions. The results told us a good way to improve adhesion is to use bi-layer. To add a

second layer SiO2 to the silicon nitride film finding the adhesion is 2 times higher silicon nitride only as show in Figure 10.

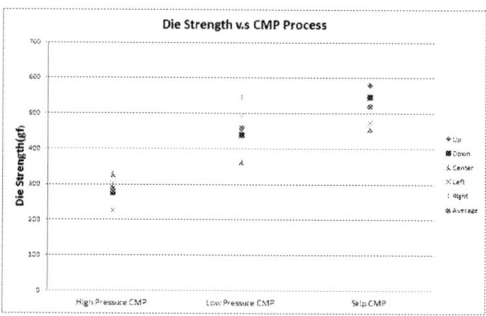

Figure 10. Adhesion distribution of CVD film

With H-C-SiN film samples were continue produced in CMP, the die strength results as shown in Figure 11. Lower CMP pressure and increase process time, the strength result is about 1.5 times of higher pressure case.

Figure 11. CMP experiment result char

Assembling Process and Test Vehicle

As shown in Fig. 12, the CoC first process flow was selected to build the 2.5D packages regarding to the CTE mismatch between Si interposer and organic substrate. At first, the Si interposer was temporarily bonded on a carrier which can suppress the interposer warpage variation during reflow process. Then the 45um pitch Cu pillar micro-bumps on top dies were mounted with micro-pad at interposer front side followed by reflow, underfill dispensing and resin curing. Next, the individual CoC modules were simultaneously de-bonded from carrier wafer using specific tape. Finally, the conventional FCBGA bonding technique was implemented for CoC module mounted on organic substrate. The relevant information of the packages investigated in this paper is summarized in Table 3 below. Sample A is homogeneous die integration whereas sample B is heterogeneous one.

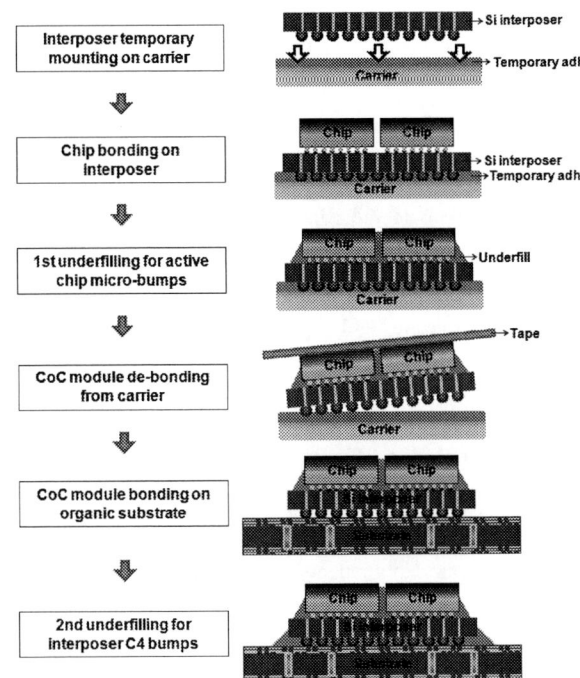

Figure 12. "Chip on Chip First" 2.5D assembly process flow

Overall package	2.5D IC Assembly		
	Item	Sample A	Sample B
Top Die	Micro-bump type	Cu pillar bump	Cu pillar bump
	Bump count	> 50000	> 40000
	Bump pitch	45 um	45 um
Interposer	C4 bump type	Eutectic	Eutectic
	Bump count	> 18000	> 9000
	Bump pitch	180 um	180 um
Substrate	Thickness	1.5 mm	1.6 mm
	BGA pitch	1 mm	0.8 mm

Table 3. 2.5D IC Packages information

Assembly Experimental Analysis

For 2.5D IC integration, the 100um thick Si interposer usually suffers many package level reliability challenges. The interposer crack and UBM delamination are the major concerns for a thin/large size Si interposer. In this paper, some examining methods were implemented to determine the qualified and failed bump condition. The QTC (Quick Thermal Cycling) test (-40°C ~ 60°C) was implemented to quickly evaluate C4 bump integrity and robustness of the Si interposer using the CoC module mounted on substrate without underfilling. Secondly, a non-destructive method called C-Mode Scanning Acoustic Microscopy (C-SAM)technology was employed to in-situ inspect the bump crack or other failure modes for the sample lots post specified cycle times (0X, 5X, 10X, 20X, 30X....). The cross-sectional analysis was further conducted to confirm crack interface once the bump failure location was found. Figure 13 shows the typical graphs and failure mode of white bump location screened by C-SAM inspection.

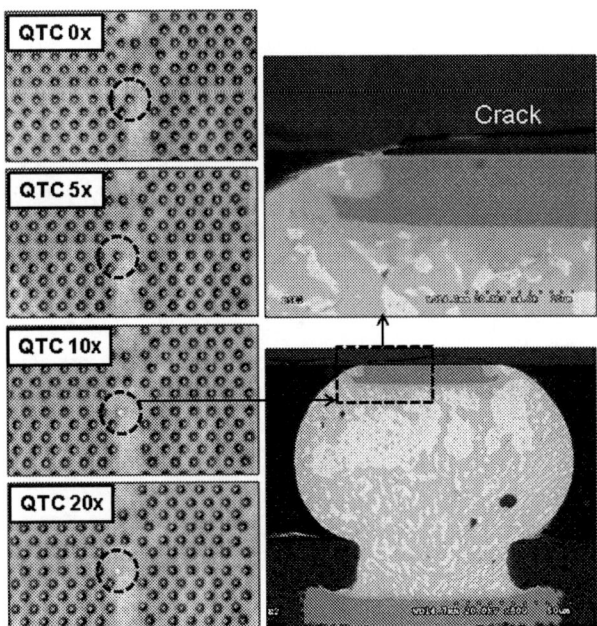

Figure 13. The typical graphs of C-SAM inspection and the failure mode at white bump location.

Assembly Results and Discussion

CoC Module Warpage Measurement Results

In this paper, "CoC first" process was adopting for the 2.5D FCBGA packages building. The process of CoC module mounting on substrate was highly affected by the warpage behavior of CoC module. The die strength of interposer used for Sample A has been enhanced and the CoC warpage under thermal loading was also improved as shown in Figure 14.

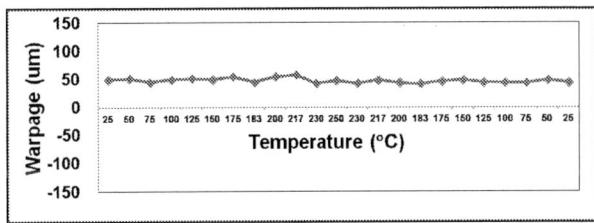

Figure 14. CoC warpage (Sample A) variation under thermal loading

Test Vehicle Reliability Tests

The full assemblies were successfully processed through an improved stacking process flow. The reliability results of different packages including sample A (5 lots prepared) and sample B (1 lot only) are presented in Table 4 & Table 5, and the tested conditions followed JEDEC standard package level covering preconditioning with moisture sensitivity level 4 (MSL-4), TH (85°C/85%RH) 1000hr, uHAST (110°C/85%RH) 168hr, HTS (150°C) 1000hr and TC-B (-55°C~125°C) 2000X. In addition, the quick temperature cycling (QTC, -40°C~60°C) was also conducted to further evaluate the integrity of interposer C4 bumps, especially focus on UMB structure.

978-1-4799-2408-0/14 $31.00 © 2014 IEEE

Sample ID	Lot ID	QTC (-40°C~60°C)	Precondition (MSL4)	TH (85°C/85%RH)	
				500hr	1000hr
Sample A	Lot 1	Failed (10X)	Passed	Passed	Passed
	Lot 2	Failed (10X)	Passed	Failed	N/A
	Lot 3	N/A	Passed	Passed	Passed
	Lot 4	N/A	Passed	Failed	N/A
	Lot 5	Passed (30X)	Passed	Passed	N/A

Table 4. A summary of reliability test results (QTC, Precondition and TH) of sample A

Sample ID	Lot ID	uHAST (110°C/85%RH)	HTS (150°C)			TC-B (-55°C~125°C)		
		168hr	250hr	500hr	1000hr	1000X	1500X	2000X
Sample A	Lot 1	N/A	Passed	Passed	Passed	Passed	Passed	Passed
	Lot 2	N/A	Passed	Passed	Passed	Passed	Passed	Passed
	Lot 3	N/A	Passed	Passed	Passed	N/A		
	Lot 4	N/A	Passed	Passed	Passed	N/A		
	Lot 5	Passed	Passed	Passed	Passed	N/A		

Table 5. A summary of reliability test results (HTS and TC-B) of sample A

As shown in Table 4 and Table 5, all lots have passed preconditioning, HTS 1000hrs and TC-B 2000X test conditions according to latest update. The QTC test is more critical for C4 bumps which connecting CoC module onto substrate without underfilling than normal thermal cycling test. Lot 1 and Lot 2 suffered early failure at C4 bumps after 10X QTC test. Stress induced silicon crack and UBM delamination occurred because the C4 bumps were build using previous design. To address this issue, a well-designed UBM structure for interposer C4 bumps was successfully implemented and the QTC tested result was presented as Lot 5 in Table 4. The UBM interface showed no significant issue after 30X QTC cycles and the packages can endure high-temperature storage at 150°C for 1000hr.

Sample ID	Lot ID	Precon (MSL4)	TC-C (-65°C~150°C)			
			300X	600X	900X	1200X
Sample B	Lot 1	Pass	Pass	Pass	Pass	Pass

Table 6. A summary of reliability test (TC-C) results of sample B

Table 6 shows the reliability results of sample B after MSL4 preconditioning and temperature cycle-C (TC-C, -65°C~150°C) that is more stringent test condition than TC-B (-55°C~125°C). Main focus of this lot is to quickly evaluate the impact of temperature cycling on the micro-bumps, underfill adhesion, and interposer C4 bump integrity. All tested packages passed 1200 cycles of TC-C without any failure.

Conclusion

In this paper, the BVR wafer thinning process induces die strength degrading and die strength improvement has been investigated. The grinding saw mark defects and deposited CVD film reduced die strength obviously from dozen of kilogram to several kilogram level. The CMP degrades die strength 20% off with nitride film thickness 20% off; the final backside C4 bumping process degrades strength huge around

10 times from several kilogram level to hundreds gram level. The enhancement experiments improved die strength 2~3times of baseline strength totally.

Then, a comprehensive study on interposer structure design, stacking process, interconnection integrity and reliability evaluation of 2.5D IC assembly has been outlined. The QTC test combining with C-SAM inspection has been used to evaluate the robustness of interposer C4 bump. A new-designed UBM structure for interposer C4 bump has been proved to have better solder bump integrity. Furthermore, the long-term reliability results have been presented to prove process capability for 2.5 IC assemblies.

References

1. Desmond Y.R. Chong, "Mechnical Characterization In Failure Strength Of Silicon Dice," 2004 Inter Society Conference on Thermal Phenomena, June 2004, pp. 203–210.
2. Shinya Takyu, "Novel Wafer Dicing and Chip Thinning Technologies Realizing High Chip Strength," 56th Electronic Components and Technology Conference, San Diego, CA, May 2006, pp. 1623–1627.
3. Tsai, M.Y. et al., "Determination of Silicon Die Strength" 55th Electronic Components and Technology Conference, Orlando, FL, May 2005, pp.1155-1162.
4. Chen, S. et al., "The Evaluation of Wafer Thinning and Singulating Processes to Enhance Chip Strength" 55th Electronic Components and Technology Conference, Orlando, FL, May 2005, pp.1526-1530.
5. Betty H. Yeung, "Assessment of Backside Processes Through Die Strength Evaluation" IEEE Transaction On Advanced Packaging, VOL.23, NO.3, August 2000, pp.582-587.
6. Yeon-Gil Jung, "Effect of oxide and nitride films on strength of silicon: A study using controlled small-scale flaws," J. Mater. Res., Vol. 19, No. 12, Dec 2004, pp. 3569–3575.
7. J. Liao, S.H Liu, "Study of the effect on die backside stress from coating of a nitride layer," 2012 13th International Conference on Electronic Packaging Technology & High Density Packaging, Aug 2012, pp. 817–821.
8. De-Shin Liu, "Evaluate Breaking Strenght Silicon Dei by Ball-on-ring Microforce Tests and Finite Element Analysis," Microsystem, Packaging, Assembly and Circuits Technology Conference (IMPACT), 2011 6th InternationElecton Devices Meeting (IEDM), 201 IEEE International, Oct 2011, pp. 188–190.
9. Patrick Dorsey, "Xilinx Stacked Silicon Interconnect Technology Delivers Breakthrough FPGA Capacity, Bandwidth, and Power Efficiency", Xilinx White Paper: Virtex-7 FPGAs, WP380, October 27, 2010, pp. 1-10.
10. Suresh Ramalingam, "Assembly Process Qualification and Reliability Evaluations for Heterogeneous 2.5D FPGA with HiCTE Ceramic", in Proc. IEEE Electronic Components and Technol. Conf. (ECTC), Las Vegas, Nevada, May 28-31, 2013, pp. 904-908.
11. C. J. Zhan et al., "Assembly Process and Reliability Assessment of TSV/RDL/IPD Interposer with Multi-Chip-Stacking for 3D IC Integration SiP", in Proc. IEEE

Electronic Components and Technol. Conf. (ECTC), San Diego, CA, May 29, 2012, pp. 548-554.

12. Robert L. Hubbard and Bong-Sub Lee, "Low Warpage and Improved 2.5/3DIC Process Capability with a Low Stress Polyimide Dielectric", *International Wafer-Level Packaging Conference, IWLPC 2014*, San Jose, CA, Nov. 2013.

13. K. Murayama et al., "Warpage Control of Silicon Interposer for 2.5D Package Application", in *Proc. IEEE Electronic Components and Technol. Conf. (ECTC)*, Las Vegas, Nevada, May 28-31, 2013, pp. 879-884.

14. Takashi Hisada et al., "Study of Warpage and Mechanical Stress of 2.5D Package Interposers during Chip and Interposer Mount Process", in *Proceedings of IMAPS 2012*, San Diego, CA,USA, Sept. 2012, pp. 1209-1213.

15. Takashi Hisada et al., "FEM Analysis on Warpage and Stress at the Micro Joint of Multiple Chip Stacking", *Transactions of The Japan Institute of Electronics Packaging*, Vol.6, No.1, Dec. 2013.

16. Larry Lin et al., "Reliability Characterization of Chip-on-Wafer-on-Substrate (CoWoS) 3D IC Integration Technology", in *Proc. IEEE Electronic Components and Technol. Conf. (ECTC)*, Las Vegas, Nevada, May 28-31, 2013, pp. 366-371.

17. Dong-Wook Kim et al., "Development of 3D Through Silicon Stack (TTS) Assembly for Wide IO Memory to Logic Device Integration", in *Proc. IEEE Electronic Components and Technol. Conf. (ECTC)*, Las Vegas, Nevada, May 28-31, 2013, pp. 77-80.

18. Woon-Seong Kwon et al., "Enabling a Manufacturable 3D Technologies and Ecosystem using 28nm FPGA with Stack Silicon Interconnect Technology", in *Proceedings of IMAPS*, Orlando, FL, Sept 30, 2013.

19. Ho-Young Son et al., "Reliability Studies on Micro-Bumps for 3-D TSV Integration", in *Proc. IEEE Electronic Components and Technol. Conf. (ECTC)*, Las Vegas, Nevada, May 28-31, 2013, pp. 29-34.

Process Integration, Improvements, and Testing of Si Interposers for Embedded Computing Applications

S. Goodwin, J. Lannon Jr., A. Hilton, A. Huffman, M. Lueck, E. Vick, G. Cunningham, D. Malta,
C. Gregory, and D. Temple
RTI International, Research Triangle Park, NC

Abstract

A high performance embedded computing module was enabled and demonstrated with the implementation of a 3D Si interposer. The interposer contained front and backside multi-level metallization (MLM) with through-Si vias (TSVs) on 150mm wafers. The front-side MLM (5 levels) was fabricated with a dual damascene process. Four 2 μm thick Cu routing layers with 2 μm oxide dielectric layers and one pad layer were used in the front-side MLM. The TSVs were fabricated using a vias-last, unfilled via process. Due to improved process modules, contact chain test structures between the front-side MLM layers with 20,064 vias had electrical yields as high as 100%. Etching process conditions for the TSV process flow were also optimized to result in 100% yield on contact chains that contain up to 540 TSVs. These optimized etching conditions produced low TSV resistances (<30 mΩ) and high TSV isolation resistance (>100MΩ/via at 3.3V) for the embedded computing module (ECM). Two die from the 1st generation interposer (3.97 cm x 3.67 cm die size) showed good continuity and isolation for 99% of the functional circuit path nets. A second generation design was recently fabricated that, through a combination of design changes and process optimizations, resulted in improved test capacitor performance, higher via chain yields, and increased power plane yields. Design changes were also implemented to enhance the high speed signal propagation properties of the TSVs. Specifically, the selection of 80 μm TSV diameters in 500 μm thick 100 Ω-cm substrates was made to improve the S_{11} and S_{21} properties of the TSVs over the frequency range of 1-4 GHz. Details of the design changes and process improvements implemented on the completed second generation ECM die and test die are discussed, along with test results from each type of die.

Introduction

Si interposers have been the recipient of more attention in the industry recently, as higher performance needs and reduced size, power consumption, and weight have driven microsystems to greater integration density levels than what have traditionally been achieved with chip to PCB packages [1]. A Si interposer can provide up to approximately $10^6/cm^2$ I/O densities and wiring pitches as low as 5μm, substantially higher than the $10^3/cm^2$ wiring densities and 50μm wiring pitches of ceramic or organic packages [1]. The design and fabrication of advanced interposers require the careful integration of three primary process modules: through-Si vias (TSVs), front-side multi-level metallization (MLM), and backside metallization. The interposer requirements and specifications determine which materials are used and in what order the modules are implemented in the process flow. For the metal routing layers, the specification of the line/space density and the impedance of the lines will dictate whether a wafer-level packaging technology, with lower line density and fabrication cost, or a dual-damascene technology, with higher line density and fabrication cost, is the correct approach [2,3] In the same manner, the choice of the technology used in the formation of the TSVs is determined by the required via size and density. A higher density of TSVs may result in the use of a filled, vias-first technology in a thinned interposer, while a lower density of TSVs may allow for the use of un-filled, vias-last technology in a full thickness substrate. The high frequency performance requirements of the TSVs will also contribute to the design and technology approach selected for the TSV process module [4,5].

In this paper, the choices of the process technology modules for the embedded computing module are described. In addition, the improvements in those process technology modules chosen for the 3D Si interposer, are described and validated with electrical test results. These improvements were implemented in a second generation interposer, which included both design and process integration elements.

TSV Si Interposer Test Vehicle

For the embedded computing module application, the primary components to be mounted on the interposer are a Xilinx FPGA (V5LX220T) and two DDR2 (1Gb each) memories. The 1st generation interposer was based on the Sixis SXM100 design which had a 3.97 cm x 3.67cm die size. The 2nd generation interposer was based on the Sixis SXM110 design, which led to a smaller 3.24cm x 2.29cm die size. The reduction in interposer die size was due to the removal of

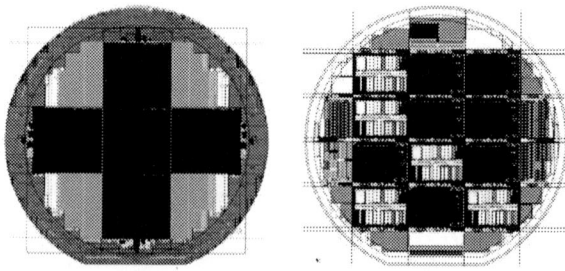

Figure 1: Plots of the first generation design with five active die (left), and the second generation design with seven active die and five test die (right). The active interposer dies are black in both plots. The five test die in the second generation layout are the die with vertical features that fill out the remainder of the 3 x 4 die layout on the wafer.

2014 Electronic Components & Technology Conference

some of the discrete termination components from the interposer and a more careful design methodology. Both generations of interposers were fabricated on 150mm diameter wafers. The 1st generation design contained five active interposer die per wafer. The 2nd generation design had space for 12 interposer die per wafer; seven die were active interposers and five die were dedicated test die. Figure 1 shows wafer plots of the two generation designs with the distribution of die.

Multi-level Metallization Process Module

For the embedded computing module application, high wiring densities are needed to adequately route signals between the FPGA and memory chips, thus requiring the dual damascene based metallization approach for the front-side metal layers. The front-side MLM had five metal levels, with metal layers 2 – 4 being self-aligned through the dual damascene process. The first metal layer (primarily a ground plane) used a subtractive wet etch to define the metal features. The first four metal layers were constructed from 2μm thick plated Cu films with a 2μm thick interlayer dielectric made from SiO_2 films deposited by plasma enhanced chemical vapor deposition (PECVD). These layer thicknesses, coupled with the design produced signal lines with characteristic impedances (Z_o) of 50Ω. Metal layer 5 was a pad layer constructed from a plated Ni/Au film. Due to the planarizing characteristic of the dual damascene process at each metal level, the vias from metal level 1 to metal level 5 could be stacked on top of each other as shown in Figure 2 in metal levels 3-5. However, the front side Ni/Au pads of level 5 could not be placed over the TSVs. Metal levels 1-4 also utilized dummy fill features to maintain a consistent amount of metal coverage across the interposer for good chemical mechanical planarization (CMP) results. Large metal features were perforated as well in order to assist in consistent metal coverage and to improve the overall integrity of the metal stack. Metal level 2 was primarily the power plane, while metal levels 3 and 4 contained combinations of power levels and signal lines. A more detailed description of the MLM process flow can be found in the literature [2].

For good via resistance between the dual damascene Cu levels, a thin cap layer of metal was deposited and patterned after the CMP step. We saw some instances of delamination in the 1st generation of interposer test vehicles, as shown in the x-SEM image in Figure 3. Through failure analysis, we determined the delamination was caused by a temperature

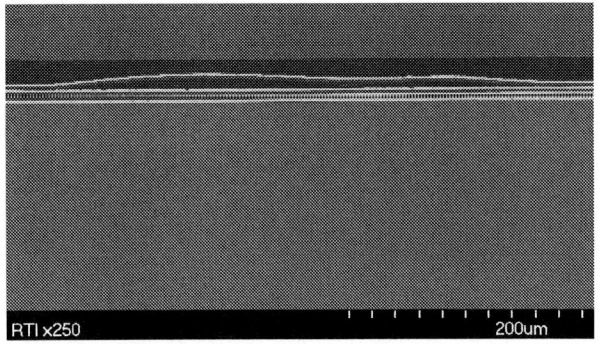

Figure 3: x-SEM images of metal interconnections levels showing delamination of the top metal layer.

induced interaction during the PECVD oxide deposition over the cap layer. At higher temperatures, the cap layer allowed the underlying Cu to interact with the SiO_2 and form copper oxides (see top image in Figure 4). By depositing the first portion of the SiO_2 at a lower temperature with the bulk of the insulating layer then deposited at a higher temperature, the migration of the Cu was stopped and the lower temperature SiO_2 film held the cap layer in place during subsequent high temperature depositions (bottom image of Figure 4). This hybrid temperature deposition process was subsequently used for the 2nd generation interposers.

To validate the process improvements in the front-side

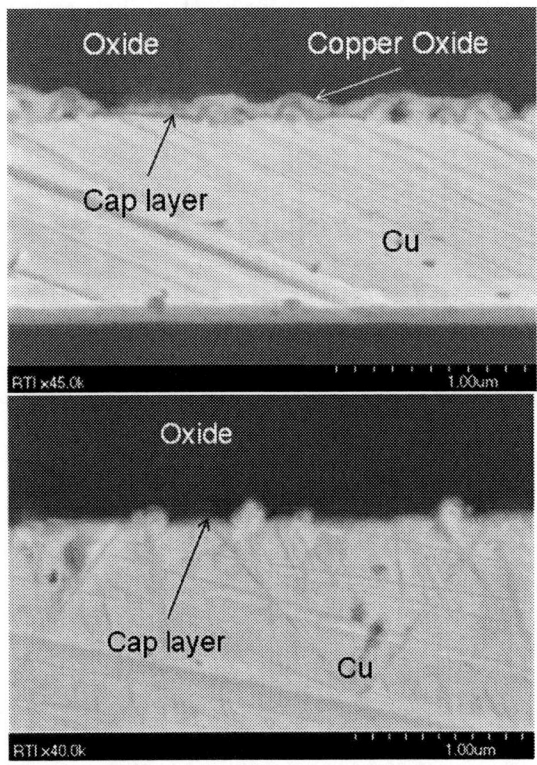

Figure 4: x-SEM images of high temperature PECVD oxide over a cap layer/Cu stack (top), and a hybrid low/high temperature PECVD oxide (bottom). Formation of surface nodules was independent of PECVD deposition temperature.

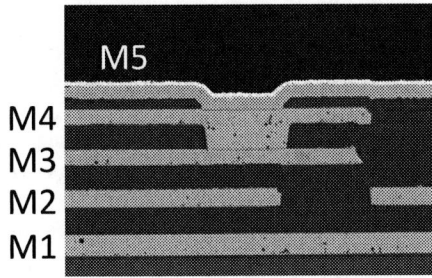

Figure 2: x-SEM image of the front-side MLM showing all five levels, including the dual damascene based process for M2-M4.

978-1-4799-2408-0/14 $31.00 © 2014 IEEE

MLM, test structures were included in the test die. These test structures included via chains and capacitors that enable the evaluation of the contact resistance and the connectivity between metal layers. The electrical results for each test vehicle were previously reported [6]. In short, the modifications in depositions on metal cap layers, die design, and fill structures for improved CMP uniformity were directly responsible for the improved via chain and area capacitor yields obtained from the 2nd generation test vehicle.

Through-Si Via Process Module

Based on the requirements of the embedded computing module, the vias-last (unfilled) approach was selected. Figure 5 shows some examples of unfilled/barrel coated vias created with this approach. The density of the required TSVs did not require the higher density available with a vias-first technology. As a result, front-side MLM was possible on a full thickness substrate, without the need for wafer thinning [7]. To minimize the real estate consumed by the TSVs, a TSV diameter of 80μm was used in a 500μm thick substrate. This is smaller than the 100μm diameter used in the 1st generation interposer design [2]. While interposers with smaller diameter TSVs and thinner substrates have been modeled and predicted to exhibit improved high frequency electrical performance, the additional processing risk required to fabricate thin interposers was not chosen for this demonstration. However, the use of a thinner substrate maintained the 6:1 via aspect ratio used on the 1st generation lot, which mitigated potential limitations on subsequent deposition and etch processes. The fabrication sequence for a vias-last process flow has been described elsewhere [2, 8]. For the second generation interposer design, an optimized

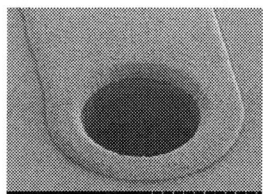 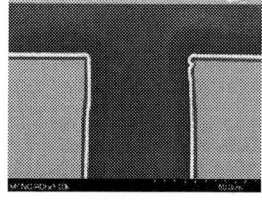

Figure 5: Example SEM images of unfilled/barrel coated vias.

TSV process flow that improves their electrical performance was used [6].

Since the resistivity of the substrates used for the interposer has large effects on the RF transmission characteristics of the TSVs, TSV structures were modeled with Computer Simulation Technology's Microwave Studio tool to determine how much performance gain is possible with a simple change in substrate resistivity. The TSV geometries were based on the 2nd generation dimensions of 80μm diameter through a 500μm substrate. The simulations showed that increases in the interposer substrate resistivity from 20Ω-cm (1st generation substrates) to 100Ω-cm to 1000Ω-cm substantially decreases the input voltage reflection coefficient (S_{11}) over the modeled frequency range of 1-4GHz and substantially increases the forward voltage gain (S_{21}), which should result in substantially improved performance. The simulations show that at 4GHz the S_{11} decreases from -13.5dB

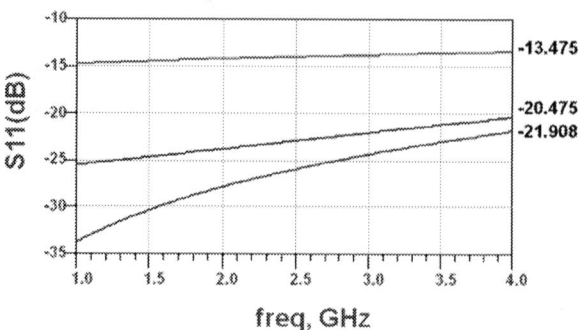

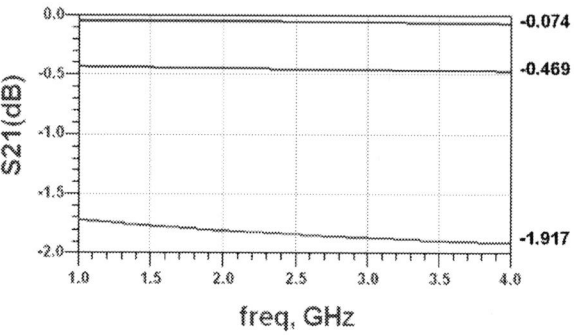

Figure 6: Plots of simulated voltage reflection coefficient S_{11} and voltage forward gain S_{21} for interposer substrate resistivities of 20Ω-cm (pink), 100Ω-cm (blue), and 1000Ω-cm (red).

for a 20Ω-cm substrate to -21.9dB for a 1000Ω-cm substrate, and the S21 increases from -1.92dB for a 20Ω-cm substrate to -0.07dB for a 1000Ω-cm substrate. The results of the simulations are graphically illustrated in Figure 6. Although 1000Ω-cm substrates are predicted to have better performance, 100Ω-cm substrates are more readily available and still provide significant increases in performance. For this reason, 100Ω-cm Si wafers were used as the 2nd generation interposer substrates.

Backside MLM Process Module

The backside MLM of the embedded computing module consisted of only two metal layers – a routing layer and a pad layer. The metal organic chemical vapor deposition (MOCVD) Cu that was used to line the barrel TSVs also served as the backside routing layer. A TiW film was deposited over the MOCVD Cu to prevent solder from wetting down the Cu lines. A dry film resist was used as a plating mold for the Ni/Au pads and then a second dry film resist was used to mask the etch of the TiW/Cu plating seed layers. The dry film resist easily bridged over the TSV openings, protecting the TSVs from the seed layer etch.

TSV Si Interposers Results

A second generation embedded computing module was created using a Si interposer based on the sequence of process modules described above (cross-section schematic shown in Figure 7). The front side of the interposer connects to a Xilinix FPGA and two memories using Ni/Au pads and four levels of Cu routing to distribute power and signals. Figure 8 shows a photograph of a populated first generation Si

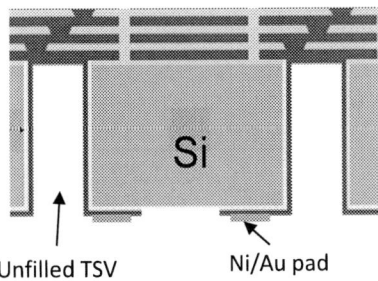

Figure 7: A cross-section schematic of the 3D Si interposer for the embedded computing module.

interposer showing the FPGA, memory die, and discrete components. The interposer includes TSVs that enable power and signals to be brought from the backside of the interposer to the front. The backside Cu layer is used as a redistribution layer (RDL) with the Ni/Au pad layer providing connections to the underlying package. Figure 9 shows a photograph of the front surface of the completed interposer. The gold circles are the Ni/Au pads, and the green level is the dual damascene M4 layer which shows both dummy fill regions as well as perforations in the metal planes to maintain a consistent metal coverage. The grey features are dual damascene M3 layer features. Invisible in this view is a TSV that connects to M1 and then up vertically through stacked vias to the M4 square feature that is roughly centered in between the four Ni/Au pads. A cross section SEM image of a completed Si interposer from the first generation is shown in Figure 10 The inset in Figure 10 shows the cross-section stacked metal and via layers over a TSV such as is shown in the plan view of

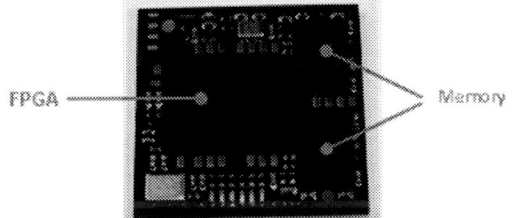

Figure 8: A photograph of a populated first generation embedded computing module.

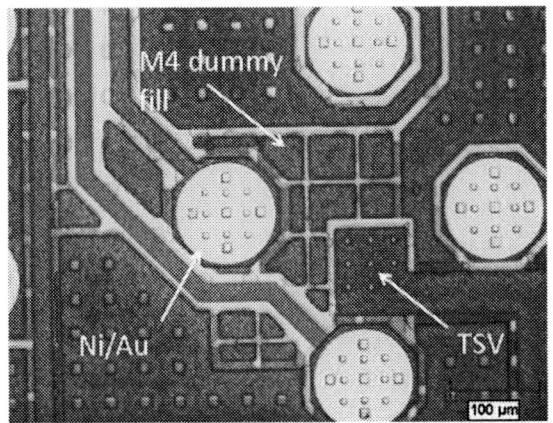

Figure 9: Optical micrograph of the front surface of the completed interposer showing M3, M4, and M5.

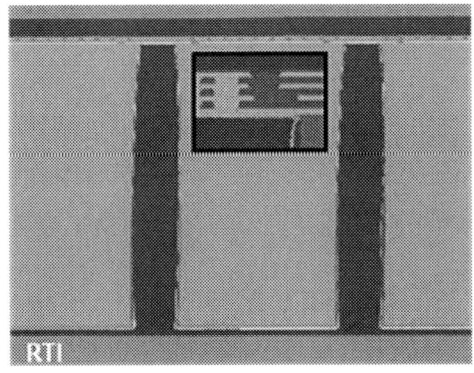

Figure 10: x-SEM images of fully fabricated dual damascene based, vias-last embedded computing module Si interposer.

Figure 9.

Testing of the isolation of the power planes from themselves and from ground was performed after the completion of the front side metallization and after the completion of the TSVs and backside metallization. With the large areas of the power and ground planes in the interposer, and the limited temperature of the PECVD oxide deposition, achieving good isolation and high yields is difficult. In the 1st generation interposer, after front side metallization, 4% of the die had excellent power plane isolation with impedances of over 1MΩ and 20% of the die had acceptable power plane isolation of over 200Ω. Since the FPGA has such a large current consumption, leakage corresponding to 200Ω is relatively small. However, once the backside metallization and TSVs were completed, the power plane yield for the 1st generation interposer dropped to zero.

Testing of signal lines was performed on two unpackaged die from the 1st generation lot. Simultaneous front-side and backside testing was done with a flying probe setup. The testing consisted of 9523 measurements taken on 2067 nets. Each net consists of an interconnect path connecting different parts of the circuit. For example, a net could be the node of an interconnect between the input of an FPGA to an external pad, where the net included TSVs connecting the front and backside metal lines. For the two die tested, the yields of passing nets which had good isolation and continuity were 99.1% and 99.6%. The failing nets were primarily shorts to ground, with one of the die also exhiting open circuits. Failure analysis revealed that MLM delamination in the front-side MLM and erosion of TSV sidewall passivation were the primary defects affecting yield.

The process improvements implemented in the 2nd generation interposer were thus focused on addressing these sources of known yield losses from the 1st lot. As a result of these process improvements, higher yields were observed during preliminary lot testing. The power plane yields of the 2nd generation (tested after front side metallization) were 27% for impedances greater than 1MΩ, and 38% for impedances over 200Ω. After the backside processing, the yield dropped to 5%. While this is a significant improvement over the 1st generation results, there is still considerable room for improvement. Additional testing and analysis is ongoing to better understand the failure mechanisms as well as to determine approaches to further improve the yield.

Conclusions

A full thickness Si interposer test vehicle was designed and fabricated for an embedded computing module. Processes were selected for the MLM and through-Si via (TSV) based on the design and performance requirements of the application. The process modules for dual damascene MLM and vias–last TSVs were developed, improved, and subsequently integrated to fabricate a second generation TSV Si interposer. The process and design improvements included cap metal layers for improved adhesion, higher resistivity substrates for improved high frequency performance, and smaller TSVs (80 µm diameter x 500 µm deep) for higher design densities. The resulting interposer had contact chain structures containing 20,064 MLM vias with yields as high as 100%, and functional nets in the embedded computing module with yields of over 99%. The average contact resistance between MLM layers and of vias-last TSVs was ≤ 4 mΩ and 25 mΩ, respectively. In short, the improvements implemented on the 2nd generation interposer produced the desired increases in both performance and yield. Testing and analysis for further embedded compute module optimization continues.

Acknowledgments

The authors wish to thank staff members of the RTI microfabrication lab for support of wafer processing, and the RTI analytical lab for sample preparation and characterization. We would also like to acknowledge Bill Batchelor for his role in the interposer design, and Michael Steer and Zhiping Feng of NCSU for their assistance with the RF modeling.

References

1. Knickerbocker, J.U. *et al*, "Development of Next-Generation System-on-Package (SOP) Technology Based on Silicon Carriers with Fine Pitch Interconnection," IBM Journal of Research and Development, Vol. 49, No. 4/5 (2005), pp. 725-753.

2. Vick, E. *et al*, "Electrical Demonstration of TSV Interconnects and Multilevel Metallization for 3D Si Interposer Applications," *Proc. of IMAPS 43rd International Symposium on Microelectronics*, Research Triangle Park, NC, Oct 31-Nov 4, 2010.

3. Huffman, A. et al, "On the Origins, Status, and Future of Flip Chip and Wafer Level Packaging," Huffman, A. & Garrou, P., *Advancing Microelectronics*, Vol. 38, No. 3, Washington, DC: IMAPS, May 2011, pp. 10–17.

4. Lu, K. and Horng, T., "Comparative Modeling of Single-ended Through-Silicon Vias in GS and GSG Configurations up to V-Band Frequencies", Progress in Electromagnetics Research, Vol. 143, (2013), pp. 559-574.

5. Yao, W. *et al*, "Power-Bandwidth Trade-off on TSV Array in 3D IC and TSV-RDL Junction Design Challenges", *Proceedings of 21st IEEE Electrical Performance of Electronic Packaging and Systems*, Tempe AZ, Oct. 21-24, 2012.

6. Lannon, J. *et al*, "Process Integration and Testing of TSV Si Interposers for 3D Integration Applications," *Proceedings of the 62nd IEEE Electronic Component and Technology Conference*, San Diego, CA, May 2012.

7. Dang, B., "3D Chip Stacking with C4 Technology," IBM Journal of Research and Development, Vol. 52, No. 6 (2008), pp.599-609.

8. Vick, E. *et al*, "Vias-last Process Technology for Thick 2.5D Si Interposers," *Proceedings of the IEEE 3D SIC Conference,* January, Osaka, Japan., Jan. 30 – Feb. 3, 2012.

Mechanically Flexible Interconnects with Highly Scalable Pitch and Large Stand-off Height for Silicon Interposer Tile and Bridge Interconnection

Chaoqi Zhang, Hyung Suk Yang, and Muhannad S. Bakir
School of Electrical and Computer Engineering
Georgia Institute of Technology, Atlanta, Georgia, 30332
Email: chqzhang@gatech.edu

Abstract

This paper reports novel interconnect technologies to enable a large scale 'interposer tile' and 'silicon bridge' interconnection platform. Microfabricated self-alignment structures enable high alignment accuracy between the components. Mechanically flexible interconnects (MFIs) are utilized to enable rematable electrical interconnects. Moreover, a proof of concept demonstration with interposer tiles directly mounted on FR4 board and interconnected by silicon bridges is reported.

I. Introduction

Silicon interposer based 2.5D integration has received significant interest because it can provide a high-bandwidth and low-energy interconnect platform for heterogeneous systems [1-6]. However, for state-of-the-art 2.5D integrated systems, as shown in Figure 1 (a), the high performance interposer interconnections are only available for chips mounted on a single interposer. Given that the size of interposers is limited by the reticle size as well as cost, there exists a limit on the number of chips that can be integrated. Therefore, an innovative interconnection platform between interposers is needed to extend interconnect benefits over a large-scale system.

As shown in Figure 1 (b), we propose a novel vision to realize large scale multi-interposer systems using positive self-alignment structures (PSAS) and mechanically flexible interconnects (MFIs). 'Interposer tiles', which are essentially silicon interposers with alignment structures, can possibly be directly mounted on the motherboard (or package for some applications). The adjacent interposer tiles are interconnected by 'silicon bridges', which are silicon chips with MFIs and corresponding routing designs. The tile-to-motherboard and bridge-to-tile electrical interconnects are enabled by MFIs with various pitches and heights. Interposer tiles, silicon bridges and the motherboard (or package) are self-aligned with each other using PSAS and inverted pyramid pit pairs. Our proposed concept is an extension of the macro-chip concept demonstrated in [4-6]. Key features of our large scale silicon platform include the following:

1) Low-cost and high-accuracy self-aligned assembly of the components (i.e. motherboard, interposer tiles and silicon bridges) is obtained using PSAS and pyramid pit pairs [7]. Besides electrical interconnects shown in Figure 1 (b), the proposed interposer tile-silicon bridge platform can utilize silicon nanophotonic interconnects as well. Using a similar self-alignment capability, a 15Gpbs silicon photonic link with 300 fJ/bit has been demonstrated [4].

2) Fine pitch MFIs are used to enable high density and robust I/Os between interposer tiles and silicon bridges.

3) Highly flexible interconnects are used to mitigate the stress induced by the coefficient of thermal expansion (CTE) mismatch between the silicon interposer tile and the substrate, which could be an organic or ceramic package. Moreover, the silicon interposer tile may be directly mounted on the motherboard, which not only shortens the interconnect length, increases interconnect density, and minimizes impedance discontinuities, but also lowers the system package thickness.

4) The system level electrical interconnects (i.e. tile-to-bridge and tile-to-board) enabled by Au-NiW MFIs [9] and PSAS are rematable, therefore a system-level test can be accomplished before the permanent integration of all silicon interposer tiles and bridges. In state-of-the art 2.5D systems, as the number and diversity of chips increases, the system yield and cost suffer. However, in our proposed system, non-functional chips detected during testing can be replaced, which increases the system yield and lowers cost.

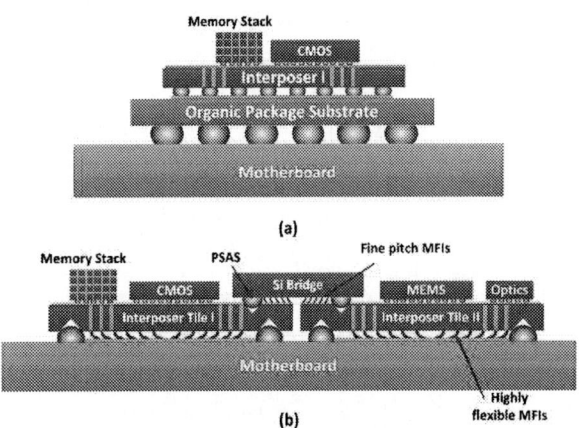

Figure 1: (a) State-of-the art 2.5D integration, (b) interposer tile and silicon bridge interconnection platform

PSAS and MFIs are the two key technologies to enable the envisioned system in Figure 1 (b). Since self-alignment assembly using PSAS was previously demonstrated in [7, 8], in this paper we focus on the MFIs, which are a key technology to enable our envisioned system shown in Figure 1 (b). Various flexible interconnect technologies have been investigated over the past decades [9-17]. However, it is very challenging for the traditional technologies to enable the above envisioned platform. In this paper, we report an MFI technology and process designed for silicon bridge-to-interposer tile and interposer tile-to-motherboard interconnections. Specifically, the reported MFI process can realize MFIs with a wide range of pitches (from 150 μm to 50 μm). Moreover, the reported MFIs feature a large vertical gap

2014 Electronic Components & Technology Conference

978-1-4799-2408-0/14 $31.00 © 2014 IEEE

and truncated-cone tip for improved temporary electrical interconnection. We also report PSAS fabrication on FR4 and the assembly of interposer tiles and bridges as a proof of concept demonstration of the envisioned large-scale platform in Figure 1 (b).

II. Advanced MFI Technology

In this section, a spray coating based fabrication process is demonstrated to enable MFI fabrication with a wide range of pitches, a large vertical gap and a truncated-cone tip. Four-point resistance measurements and mechanical indentation tests are also reported for the fabricated MFIs.

A. Spray Coating Based Fabrication

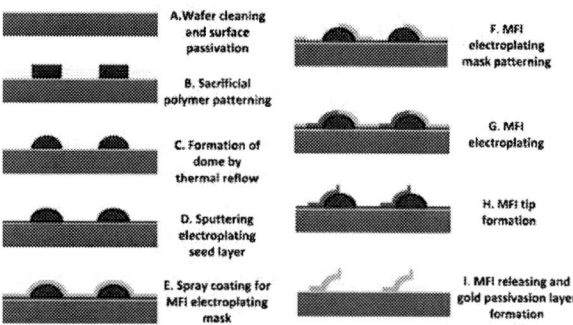

Figure 2: Spray coating based MFI fabrication process

The fabrication process of the MFIs is shown in Figure 2. The process begins with the formation of sacrificial polymer domes, which is accomplished by patterning and thermally reflowing a spin coated polymer layer on a nitride passivated silicon wafer [9,16,17]. Next, a Ti/Cu/Ti film is sputter coated on top of the 65 µm tall polymer domes as an electroplating seed layer. A thick conformal negative photoresist layer is next spray coated and patterned on top of the seed layer as the electroplating mold. Following the electroplating of MFIs, the photoresist plating mold is removed and followed by the patterning of another photoresist layer for the subsequent tip electroplating. Once the tips with a truncated-cone profile are formed on top of the MFIs, the tip electroplating mold, the seed layer and the polymer domes are stripped leaving behind MFIs with truncated-cone tips and a 65 µm vertical gap above the substrate. Finally, the free-standing NiW MFIs on the wafer are passivated by an electroless gold finish [9].

As shown in Figure 2, a key step to enable MFI fabrication is the spray coating of photoresist on the polymer domes. Because of the large height of the domes, a spin-coated photoresist process, similar to that used in prior flexible interconnects [10-15], would not work in this application.

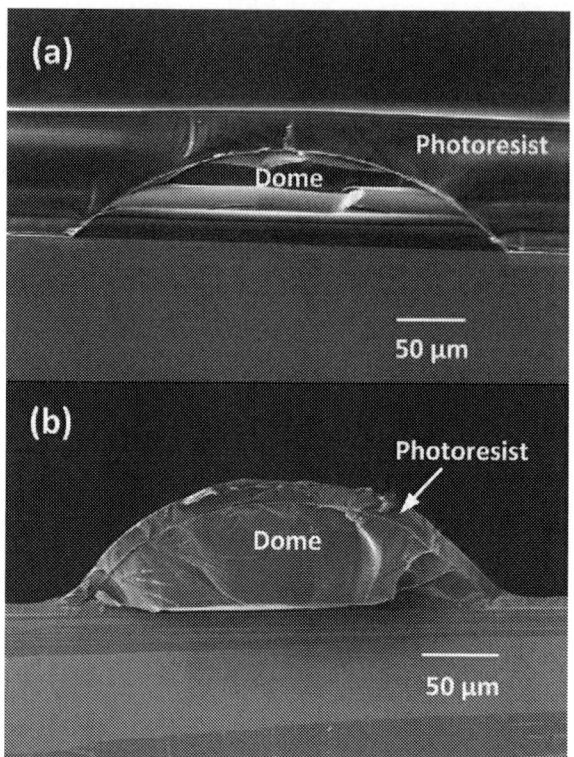

Figure 3: Compared to spin coating (a), photoresist spray coating (b) can form a uniform photoresist layer on top of sacrificial domes

Figure 3 (a) illustrates the photoresist layer coated over the substrate with domes by a traditional spin coating process. The photoresist layer is thin on top of the domes and thick in the valley between the domes. The non-uniform photoresist thickness makes the spin coating approach not feasible to fabricate the MFIs for the following reasons: 1) it is not possible to obtain a proper exposure and development simultaneously on top of the domes and in the valley between the domes; 2) a high soft bake temperature (> 110 °C) is needed for the thick photoresist in the valleys, but such high temperature softens the sacrificial polymer domes, which leads to breaking of the plated MFIs. The photoresist spray coating process solves these challenges. As shown in Figure 3 (b), a conformal photoresist layer can be formed over the surface of the domes by spray coating. Therefore, the exposure dose can be optimized since the thickness of the photoresist is identical both on the top of the domes and in the valleys. In addition, since the solvent evaporates quickly during the spray coating process, this precludes the need for a high temperature soft bake process. A possible drawback of spray coating is that surface roughness becomes worse as the thickness of photoresist layer increases. However, after a low temperature (<80 °C) oven bake step, the surface variation can be controlled to within 0.5 µm, as shown in Figure 4.

978-1-4799-2408-0/14 $31.00 © 2014 IEEE   14

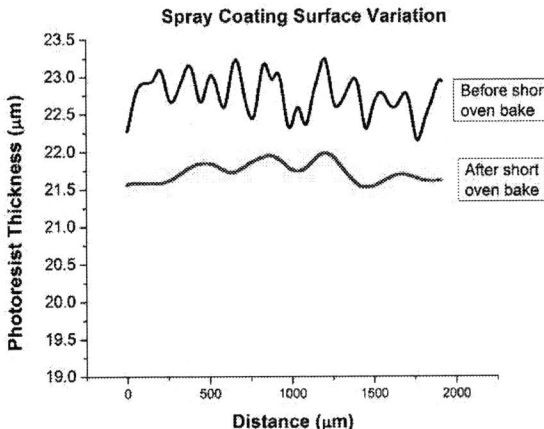

Figure 4: Surface roughness of the spray coated photoresist can be improved by a short oven bake step

B. MFIs with Highly Scalable Pitch

With the spray coating based process, wafer-level batch fabricated MFIs with highly scalable pitch can be attained. Such dense MFIs would be used for interposer tile-to-silicon bridge interconnection. Figure 5 illustrates 7.2 µm thick MFI array with a vertical gap of 65 µm and on 150µm, 75µm and 50 µm pitches.

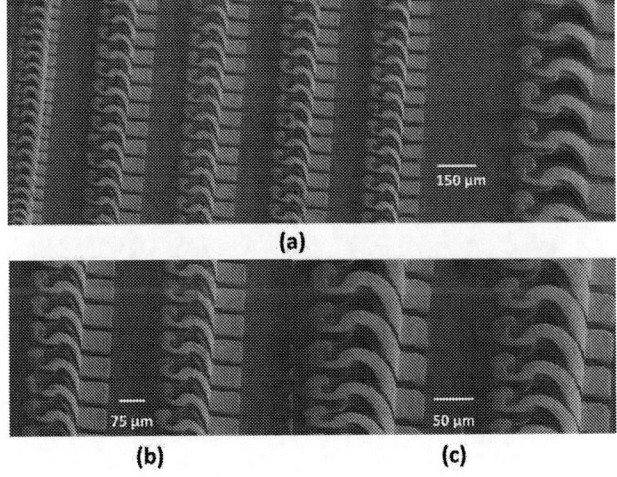

Figure 5: MFIs with highly scalable pitch (150 µm, 75 µm, and 50 µm), and a vertical gap of 65 µm

Four-point resistance measurement is performed using a probe station. During the measurements, the tested MFIs are partially bent to attain a stable resistance reading. The measured data is summarized in Table I and indicates that the resistance of the MFIs on different pitches does not vary appreciably. The MFIs with pitches of 150 µm to 50 µm are designed using the rules developed in [9,16,17] and scaled down in all dimensions except thickness with the same factor.

TABLE I
MEASURED RESISTANCE AND COMPLIANCE OF MFIS

Pitch (µm)	Average Resistance (mΩ)	Standard Deviation (mΩ)	Compliance (mm/N)
150	133.2	3.79	5.32
75	134.3	4.02	2.72
50	119.1	3.87	1.20

The mechanical deformation properties of MFIs are verified by indentation tests. Each indentation cycle includes a forward and a backward step. In the forward step, a predefined forced is applied on top of a free standing MFI by a piezo-driven indentation head, which bends the MFI to a specific depth. In the backward step, the MFI is released to recover its pre-indentation profile. The real-time position and reaction force of the tip are recorded in Figure 6 and the corresponding compliance of the MFIs is calculated and summarized in Table I. As shown in Figure 6, for a given thickness (7.2 µm), the compliance of MFIs decreases as the pitch scales. The compliance can be increased (to avoid rigid MFIs) by reducing the MFI thickness.

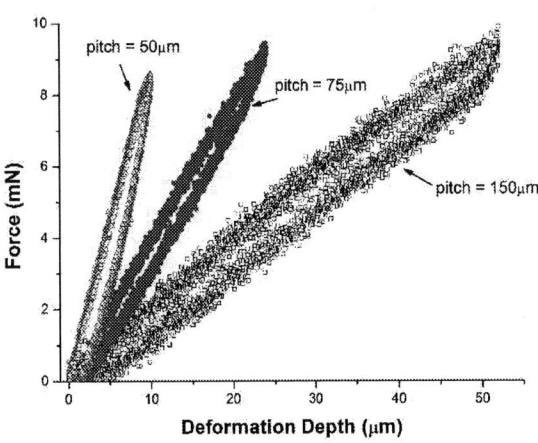

Figure 6: Indentation results of MFIs with pitches of 150 µm, 75 µm, and 50 µm

C. MFIs with truncated-cone tip

In order to overcome substrate surface variation and warpage, which is critical for interposer tiles on motherboard (or package) assembly, MFIs with truncated-cone tips are fabricated, as shown in Figure 7. The aggregate height of the MFI and tip is approximately 95 µm (65 µm tall vertical gap and 30 µm tall tip). The blunt tip can enhance the scratching capability while maintaining long tip life-time by avoiding tip plastic deformation, which occurs to sharp tips.

978-1-4799-2408-0/14 $31.00 © 2014 IEEE

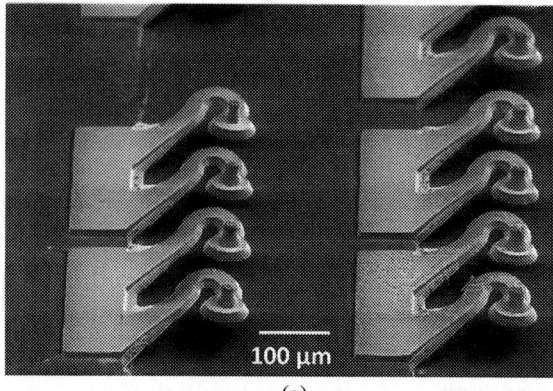

(a)

(b)

Figure 7: Overview (a) and front view (b) of Au-NiW MFIs with truncated-cone tip

Figure 8 illustrates gold-coated MFIs with truncated-cone tip. Compared with other passivation approaches, such as polymer coating, electroless gold passivation is a simple and low cost approach to enhance the life time of NiW MFIs as reported in [9]. In addition, the gold layer can lower the MFI resistance and MFI/pad contact resistance as well.

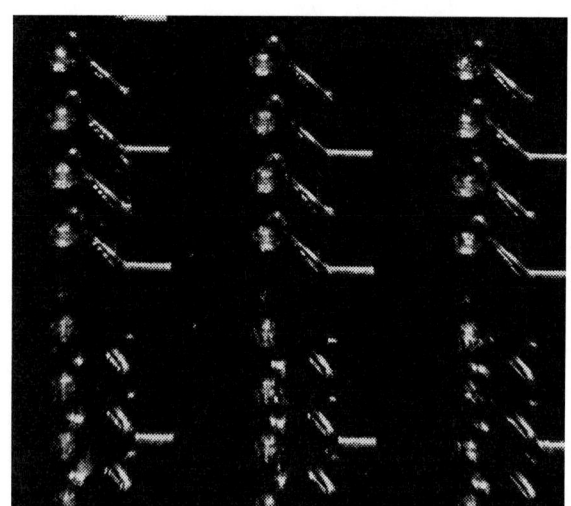

Figure 8: NiW MFIs after being passivated by electroless gold

The experimental results of interposer tile assembly with MFIs are discussed next:

1) Test bed fabrication

To demonstrate the mechanical robustness of the MFIs, a special substrate was designed and fabricated: bonding pads with different heights were formed across the substrate. In this manner, the only way to attain electrical contact across the entire chip is for the MFIs to locally deform to mate with the variable height pads. In this test bed, the pad height variation was 45 µm. For comparison, a substrate with all uniform height pads was also fabricated. The test chip with MFIs is shown in Figure 9 (a). The substrate with uniform and non-uniform pad heights are shown in Figure 9 (b) and (c), respectively. A profilometer measurement of the pads with differing height is shown in Figure 10.

2) Assembly

The assembly demonstrations are performed using a Finetech flip-chip bonder to align and place the interposer tile with MFIs onto the substrate. After assembly, an X-ray imaging tool, Dage X-Ray XD7600NT, was used for assembly verification. As shown in Figure 11, the interposer tile with MFIs is well aligned with the test substrate.

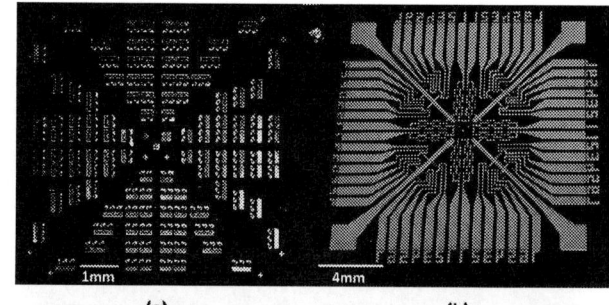

(a)   (b)

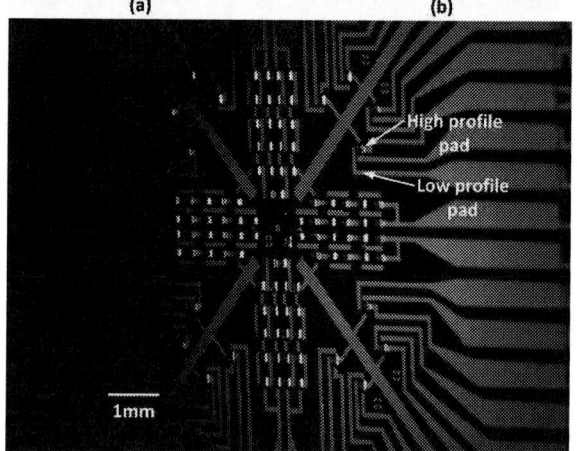

Figure 9: Interposer tiles with Au-NiW MFIs (a) and corresponding substrate with uniform pads (b) and non-uniform pads (c) for four-point resistance measurement.

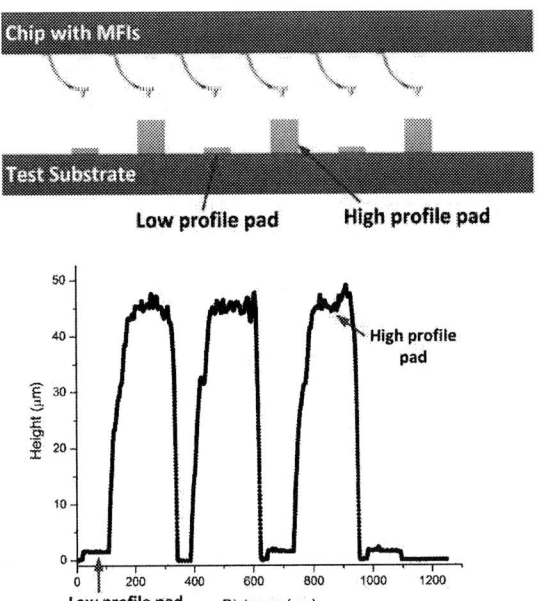

Figure 10: Assembly test bed cross-section and profilometer data of the non-uniform height pads on the substrate

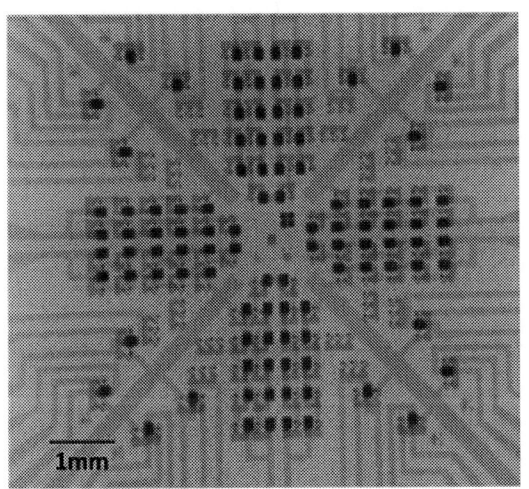

Figure 11: X-ray image of the assembled interposer tile and substrate with non-uniform pads

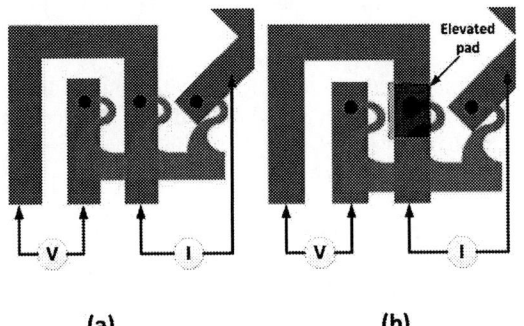

Figure 12: Four-point measurement design for the assembled interposer tile with MFIs

The four-point test designs shown in Figure 12 were used to characterize the resistance of assembled MFIs. The measured results are plotted in Figure 13. The average resistance and the corresponding standard deviation are summarized in Table II. The results indicate that the resistance of MFIs assembled on the substrate with non-uniform height pads is slightly larger than those assembled on the substrate with uniform pads. The resistance difference is believed to be mainly from the larger contact resistance of MFIs assembled to elevated pads, which have worse contact interface than MFIs assembled to low-profile pads. As shown in Figure 10, the surface roughness of elevated pads (5 μm) is much larger than the low-profile pads (1μm).

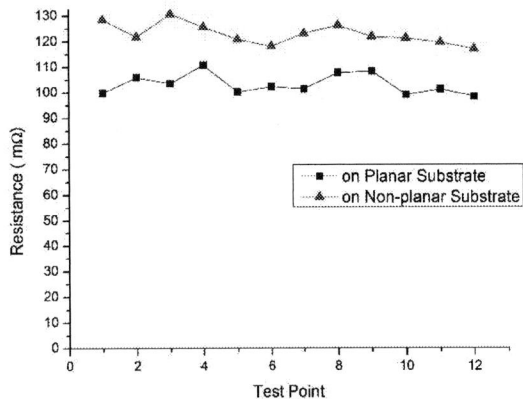

Figure 13: Four-point resistance measurements of MFIs assembled on substrates with uniform and non-uniform pads

TABLE II
RESISTANCE CHARACTERIZATION FOR REMATABLE ASSEMBLY

	Average Resistance (mΩ)	Standard Deviation (mΩ)
Assembly on substrate with uniform height pads	103.21	4.06
Assembly on substrate with non-uniform pads	122.81	4.16

III. Self-Aligned Interposer Tiles and Bridges Assembly on FR4

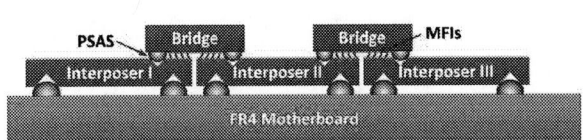

Figure 14: Self-aligned interposer tiles and bridges assembly on FR4 board using PSAS and MFIs

In this section, we report PSAS assisted assembly of interposer tiles and bridges on FR4 board, as shown in Figure 14. The demonstrated test bed contains: 1) three silicon interposer tiles, which are essentially 20x20 mm^2 silicon interposers with inverted pyramid pits on both sides; 2) two 20x6 mm^2 silicon bridges with PSAS on the side facing the interposer tiles; 3) one large FR4 board with PSAS on the top side. The details of PSAS/pit pair and assembly process are discussed next.

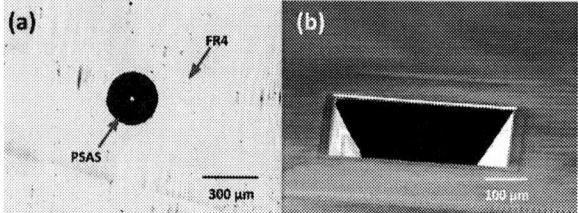

Figure 15: Key technologies enabling the proposed multi-interposer system: (a) PSAS, and (b) inverted pyramid pit

Figure 15 illustrates the key structures enabling the test bed: 1) the PSAS, as shown in Figure 15 (a), is a truncated polymer sphere with perfectly smooth surface and fabricated using a thermal reflow process; 2) the inverted pyramid pit structure, as shown in Figure 15 (b), is fabricated by an anisotropic silicon wet etch. The fabrication and characterization details of the PSAS and pits are reported in [7].

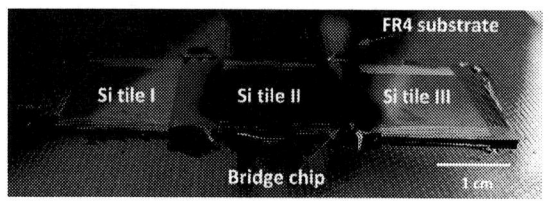

Figure 16: Self-aligned test bed includes three interposer tiles and two silicon bridge chips on FR4 board

As shown in Figure 16, three interposer tiles are mounted on top of the FR4 board. The position of each tile is determined by the PSAS on the FR4 board since they are designed to mate with the pits on the back side of the silicon tiles. Next, the silicon bridges are assembled across the adjacent tiles with PSAS side facing downward. Finally, all components of the test bed are glued by applying epoxy around the edge (this was used to simplify the mounting).

The alignment accuracy of the assembled test bed is measured by observing vernier patterns via infrared microscopy. As summarized in Table III and Table IV, the maximum misalignment between silicon interposer and FR4 is 4.4 μm; the maximum misalignment between silicon bridges and interposer tiles is at the top left corner of the bridge 2, which is about 7.6 μm. This alignment system, with further optimization, can be used to support silicon nanophotonic interconnection between silicon tiles using the silicon bridge.

TABLE III
PSAS ASSISTED SELF-ALIGNMENT ACCURACY OF SI/FR4

Regions	Si/FR4	
	Horizontal	Vertical
Bottom Left	4.4	2.0
Bottom Right	3.2	-3.2
Top Right	-1.6	-3.2
Top Left	-2.8	2.4

TABLE IV
PSAS ASSISTED SELF-ALIGNMENT ACCURACY OF BRIDGE/TILE

Regions	Si Bridge I		Si Bridge II	
	Horizontal	Vertical	Horizontal	Vertical
Bottom Left	-4.0	4.6	-5.2	-5.0
Bottom Right	-5.4	-4.8	-5.0	-5.0
Top Right	5.8	3.2	-5.8	-5.2
Top Left	6.0	-5.0	-7.6	-5.0

IV. Conclusion

Mechanically flexible interconnects featuring 1) a wide range of pitches (150 μm to 50 μm), 2) a vertical gap of 65 μm, and 3) a truncated-cone tip (30 μm tall) are reported in this paper. Assembly test beds to electrically and mechanically characterize the MFIs are also reported in this paper. Finally, an interposer tile and silicon bridge based system is demonstrated using PSAS assisted self-assembly.

Acknowledgments

We gratefully thank Drs. James Mitchell, Hiren Thacker, John Cunningham and Ivan Shubin at Oracle Labs for very valuable discussions, guidance, and support.

References

[1] B. Banijamali, *et al.*, "Advanced reliability study of TSV interposers and interconnects for the 28nm technology FPGA," in *Proc. Electronic Components and Technology Conference*, 2011, pp. 285-290.

[2] R. Chaware, *et al.*, "Assembly and reliability challenges in 3D integration of 28nm FPGA die on a large high density 65nm passive interposer," in *Proc. Electronic Components and Technology Conference*, 2012, pp. 279-283.

[3] http://www.eetimes.com/document.asp?doc_id=1262783

[4] A. V. Krishnamoorthy, *et al.*, "Computer systems based on silicon photonic interconnects," *Proc. IEEE*, vol. 97, no. 7, pp. 1337–1361, 2009.

[5] H. Thacker, *et al.*, "Hybrid integration of silicon nanophotonics with 40nm-cmos vlsi drivers and

receivers," in *Proc. Electronic Components and Technology Conference*, 2011, pp. 829-835.

[6] J. E. Cunningham, *et al.*, "Integration and packaging of a macrochip with silicon nanophotonic links," *J. Sel. Topics Quantum Electron,*2011,vol. 17, no. 3, pp.546–558.

[7] H.S. Yang, *et al.*, "A low-cost self-alignment structures for heterogeneous 3D integration," in *Proc. Electronic Components and Technology Conference*, 2013, pp. 232-239.

[8] A. Krishnamoorthy, *et al.*, "Alignment features for proximity communication." U.S. Patent No. 7,923,845. 12 Apr. 2011.

[9] C. Zhang, *et al.*, "Highly Elastic Gold Passivated Mechanically Flexible Interconnects," *IEEE Trans. Compon., Packag. Manuf. Technol.* vol. 3, no. 10, pp.1632-1639, Oct. 2013.

[10] M. Bakir, *et al.*, "SoL Ultra high density wafer level chip input/output interconnections for gigascale integration (GSI)," *IEEE Trans Electron Devices,* 2003, pp. 2039-2048.

[11] Q. Zhu, *et al.*, "β-Helix: a lithography-based compliant off-chip interconnect," *IEEE Trans. Components and Packaging Technology,* 2003, pp. 582-590.

[12] Q. Zhu, *et al.*, "Design and optimization of a novel compliant off-chip interconnect one-turn helix," in *Proc. Electronic Components and Technology Conference*, 2002, pp. 910-914.

[13] K. Kacker, *et al.*, "FlexConnects: a cost-effective implementation of compliant chip-to-substrate interconnects," in *Proc. Electronic Components and Technology Conference*, 2007, pp. 1678-1684.

[14] S. Muthukumar, *et al.*, "High-density compliant die-package interconnects," in *Proc. Electronic Components and Technology Conference*, 2006, pp. 1233-1238.

[15] G. Spanier, *et al.*, "Platform for temporary testing of hybrid microsystems at high frequencies," *J. MEMS* 2007, pp. 1367-1377.

[16] H. S. Yang and M. S. Bakir, "3D integration of CMOS and MEMS using mechanically flexible interconnects (MFI) and through silicon vias(TSV)," in *Proc. Electronic Components and Technology Conference*, 2010, pp. 822–828.

[17] H. S. Yang and M. S. Bakir, "Design, fabrication, and characterization of freestanding mechanically flexible interconnects using curved sacrificial layer," *IEEE Trans. Compon., Packag. Manuf. Technol.*, vol. 2, no. 4, pp. 561–568, Apr. 2012.

Advancements in Fabrication of Glass Interposers

Aric Shorey[1], Philippe Cochet[3], Alan Huffman[2], John Keech[1], Matt Lueck[2], Scott Pollard[1], and Klaus Ruhmer[3]

[1]Corning Incorporated

[2]RTI International

[3]Rudolph Technologies

Abstract

There is growing interest in applying glass as an interposer substrate for 2.5D/3D as well as component substrates for radio frequency (RF) applications. The list of important advantages provided by glass in these applications include material properties (e.g. electrical performance, ability to adjust coefficient of thermal expansion (CTE) to improve reliability) as well as the significant opportunities for cost advantages that glass based solutions provide over other approaches.

The feasibility of fabricating high quality holes in glass substrates has been demonstrated. While work in hole fabrication continues, additional efforts to demonstrate and mature downstream processing of glass substrates has accelerated. These include hole metallization and redistribution layers (RDL) in both wafer and panel formats, as well as initial characterization and demonstration of reliability. Significant progress in these areas is reported here.

1. Introduction

A lot of work is being done to validate the value of glass as an interposer substrate. [1,2] Because glass is an insulator, it is expected to have better electrical performance than silicon, particularly at higher frequencies. Electrical characterization and electrical models demonstrate the advantages of the insulating properties of glass, and its positive impact on functional performance.[3] Further advantages have been measured in reliability performance, because of the ability to adjust thermal properties such as coefficient of thermal expansion (CTE) of glass. [4]

Additionally, significant progress has been made in the demonstration of glass interposer fabrication. Fully patterned wafers and panels with through holes and blind holes are being fabricated today. An important step in the commercialization of glass based interposers and electrical components is to demonstrate the ability to leverage existing wafer based tools for metallization of holes as well as redistribution layers. Progress in hole fabrication and initial metallization has been reported. [5-8] Below we present new developments in these areas.

One challenge to adopting silicon interposer technology into high volume applications is the ability to achieve low cost targets required. An important way to achieve cost effectiveness for these high volume applications is to leverage glass forming techniques that enable high quality substrates in panel format. The ability to apply existing downstream panel-based processes for metallization and electrical distribution on these substrates is also important for cost effectiveness and ease of transition into production. This paper specifically discusses a suitable lithography technology which enables cost effective patterning of high-density interposer panels. Although leading edge front-end lithography is pushing into the single-digit nanometer scale already, the feature sizes needed for interposer panels (2.5/3D, advanced packaging) are much larger. Back-End lithography is driven by a whole different set of requirements related to thick resist, depth of focus, sidewall angle, panel handling capability and exposure field size only to name a few. The work summarizes the status as well as initial demonstrator results for applying existing panel lithography technology for glass interposer fabrication.

2. Glass Substrates

For 2.5D and 3D-IC interposer technology application, the substrate surface flatness and TTV are critical factors, as they will impact yields in device fabrication. The advantages given by Corning Incorporated's fusion forming process for supplying substrates for electronics applications, particularly interposers, has been previously reported. [5, 6] Supplies of 300 mm diameter wafers with TTV < 2 μm are readily available today. Because the fusion process is capable of delivering sheets several meters in size, high volume scaling of quality wafers and panels for the semiconductor industry is also easily achievable. In addition to scaling glass substrate size, it is possible to scale the process to deliver ultra-slim flexible glass to thicknesses down to ~100 μm (see Fig. 1). Providing large substrates in wafer or panel format at 100 μm thickness gives significant opportunities to reduce manufacturing costs of interposers because there are opportunities to significantly reduce manufacturing costs by eliminating process steps such as backgrinding.

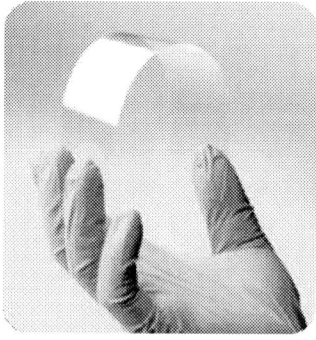

Fig. 1. Manufacture of high quality ultra-slim flexible glass provides substantial opportunities to deliver substrates for through glass via (TGV) that do not require post processing.

It is important to be able to leverage both wafer and panel based tools to best utilize the industry infrastructure for adoption. Corning's process for fabricating high quality vias has been developed to enable fabrication of both through and blind vias in thin (e.g. 100 um) and thick (e.g. 700 um) glass substrates to provide opportunities to leverage existing tools. Figure 2 shows recent demonstration of the ability to provide

978-1-4799-2408-0/14 $31.00 © 2014 IEEE

2014 Electronic Components & Technology Conference

through and blind holes in glass substrates as well as the ability to leverage existing tools and processes to metalize these holes. Another essential aspect of the hole fabrication is to make sure that high strength of the substrate is maintained. We have previously reported high strength of glass substrates after hole fabrication. [6] Below we report on progress of continued characterization of reliability after downstream processing.

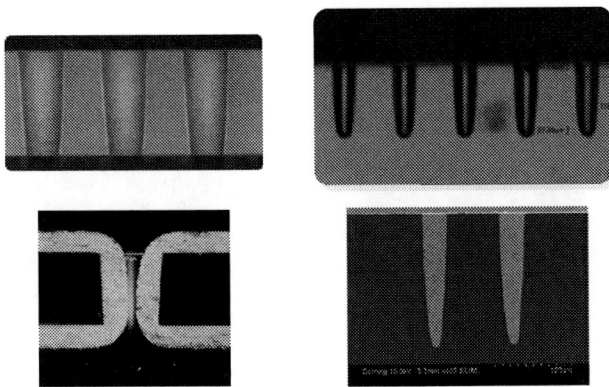

Fig. 2a Through Holes   **Fig. 2b** Blind Holes

Fig. 2. Demonstration of formation and metallization of high quality through and blind holes in glass.

3. Fabrication and Test of Glass Interposers in Wafer Format

The fabrication of thin glass interposers with Cu filled through glass vias (TGV) was done using standard back end of line (BEOL) fabrication tools with no significant modification of any of the equipment wafer handling to accommodate glass wafers. In order to test the effect of the glass CTE on the long term reliability of the glass interposers, 150 mm glass wafers formulated with two different CTEs, 3 ppm/°C and 8 ppm/°C, were used in the fabrication process.

Full thickness 150 mm glass wafers with 35 μm x 135 μm blind TGVs were sputtered with a thin adhesion layer of Ti and Cu. Highly conformal copper seed layers were deposited using metal-organic chemical vapor deposition (MOCVD), in preparation for TGV plating. The seed layers were nominally 0.75μm in thickness, which was uniform throughout the TGVs.

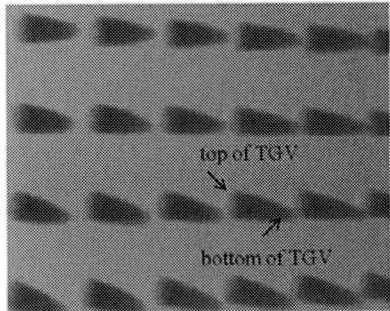

Fig. 3. High resolution X-ray images of void-free Cu-filled 35 μm x 135 μm TGVs

Copper electroplating of the TGVs was done with Enthone's MICROFAB DVF200 chemistry, a copper methanesulfonic acid bath which has been formulated for deep via fill applications. The chemical components included a make-up solution, an accelerator (B-component), and a primary suppressor (C-component). The plating was done in a Nexx Stratus 100 electro-deposition system. The plating additives and current density were adjusted to obtain the desired plating profiles for void-free filling. Measurement of the organic additive concentrations was done by cyclic voltammetric stripping (CVS). Further detail on the plating development for such TGVs has been published previously. [9,10]

After the Cu-plating step, TGVs were characterized using high resolution X-ray imaging (see Fig. 3), to verify Cu fill quality and look for hidden void defects. A high-rate Cu polish process was utilized to remove the Cu overburden and planarize the front surface [11]. The high rate Cu slurry was found to preferentially etch the Cu around the outside of the TGV leaving a small moat between the glass and Cu. Depending on the first metal level process applied, this feature could be an area of concern for continuity of thin metal layers. A two-step CMP process was introduced to limit the time of exposure of the high rate Cu slurry. SEM images of the TGVs before and after the process optimization are shown in Fig. 4.

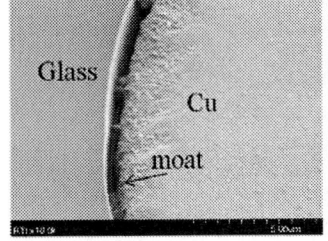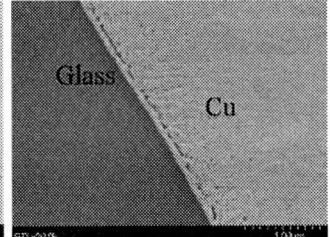

Fig. 4. SEM images of the edge of Cu filled TGVs showing (a) the moat feature left after non-optimized Cu overburden removal CMP and (b) a similar TGV after CMP optimization.

Annealing of Cu filled vias in silicon wafers has previously been found to stabilize the vias and prevent extrusion of the Cu that could happen during high temperature processing steps later in the buildup of routing metal and dielectric layers on the surface of interposers [9]. A standard stabilization anneal, 400 °C for 1 hour, was applied to the glass wafers. A second Cu CMP step, typically necessary on silicon wafers to remove extruded Cu, was not done here because the extrusion of the Cu in the TGVs was found to be less than 1 μm.

After completion of the TGVs, frontside routing metal was patterned on the wafers using a sputtered Ti / Cu seed metal layer, a photoresist plating template, and 3 μm thick electroplated Cu.

Thinning and backside reveal of the TGVs was done after attaching a temporary carrier to the frontside of the glass wafers. 3M's Wafer Support System (WSS) was utilized to provide mechanical support to the glass wafers as they were thinned to 125 μm and during the processing of the wafer

978-1-4799-2408-0/14 $31.00 © 2014 IEEE   21

backside through a routing metal layer fabrication. The adhesive thickness was nominally 50 μm and the TTV of the wafer stacks, measured after bonding using infrared interferometry, was 3 μm or less.

Backgrinding of the glass wafers was done, stopping 10 – 15 μm short of the bottom of the TGV's. The TGV reveal process was done through glass lapping. The endpoint of the lapping was chosen to be 10 – 15 μm past the initial exposure of the TGV to ensure complete removal of the rounded portion of the TGV bottom. This was followed by a brief chemical mechanical polish (CMP) step to remove any damaged layers introduced by mechanical grinding. Reveal of the TGVs without lapping was found to produce uneven surface quality in the low CTE wafers in addition to requiring longer CMP polish times. The TTV of the wafer and carrier stack was measured to be 4 μm or less after lapping.

An image of the 150 mm diameter after thinning and de-bond is shown in Fig. 5. An optical microscope image of the backside of the TGVs after reveal is shown in Fig. 6.

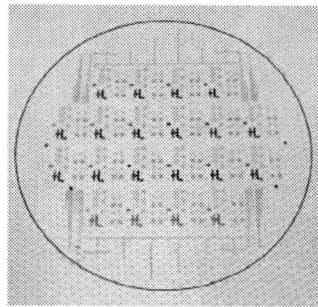

Fig. 5. An image of a 150 mm, 120 μm thick glass interposer test wafer, released from the temporary carrier onto dicing tape. A drawn outline has been added for clarity.

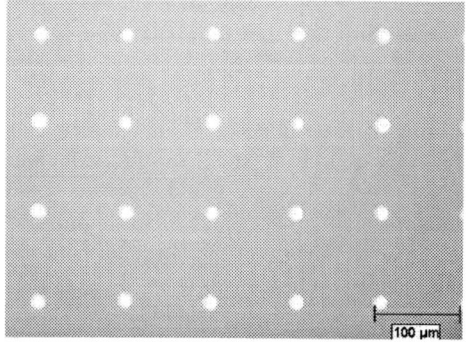

Fig. 6. Optical microscope image of TGVs after backside reveal

After wafer thinning, some of the wafers were diced for thermal cycle testing (TCT) while other wafers continued through additional process steps to complete the interposer substrate fabrication.

Reliability testing consisted of 500 cycles from -40°C to 125 °C with 1 hour cycle time. A subset of samples was pulled after 100 cycles to provide an intermediate observation. SEM images of the TGVs and routing metal before and after TCT is shown in Fig. 7.

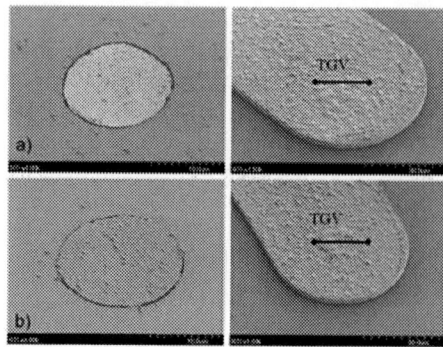

Fig. 7. SEM images of TGVs and first level metal routing layers (a) before and (b) after 100 thermal cycles. No evidence of cracking or lifting was seen in any of the samples after cycling. The black line indicates the position and diameter of the TGV under the Cu routing line

Table 1. The results of 2-wire electrical continuity tests on 20 x 20 arrays of TGVs on 100 μm pitch

Wafer	CTE (ppm/°C)	No. of 20x20 arrays tested	Yield of TGVs & routing metal (%)
SGW3 - Wafer 1	3	8	99.97
SGW3 - Wafer 2	3	8	99.97
SGW8 - Wafer 1	8.5	8	99.72
SGW8 - Wafer 2	8.5	8	100.00

After wafer thinning and TGV reveal, a backside metal routing layer was applied to complete daisy chain and RF test structures. Electrical continuity testing was done on eight test arrays, consisting of 20 x 20 TGVs on 100 μm pitch, across the diameter of four wafers. The results are shown in Table 1.

4. Glass Panel Fabrication

An important aspect of realizing cost effective glass interposers is to leverage the ability to manufacture in panel format and associated economies of scale. Many manufacturing tools exist today that enable fabrication of panel-based glass interposers and electronic components including steps for via fill and lithography.

One example of this is shown by Figs. 8 and 9. Just like with wafers, there are two ways one could envision metallizing the vias of a large panel. In the case of blind vias for panels, however, it would be preferable to not develop backgrinding operations for large panel. Instead, Corning is developing techniques to temporarily bond thin glass (e.g. 100-200 um thick) with through holes to thicker (e.g. 0.5-0.7 mm) carrier panels. This will allow processing of hole and front-side metallization prior to de-bond and backside processing. Figure 8 shows an example of a thin (~130um) glass sheet 370x470 mm in size bonded to a carrier. The partially patterned panel (~100,000 holes) was then processed with a sputtered Ti/Cu adhesion and seed layers as shown in the figure.

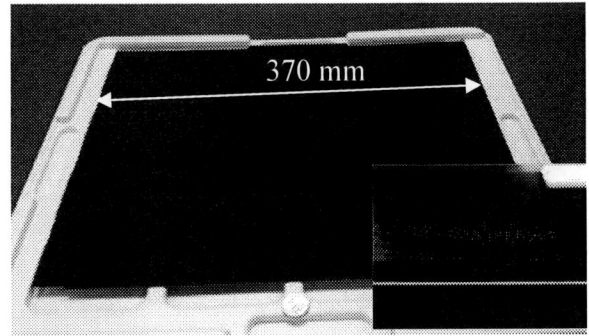

Fig. 8 Sputter coated 370x470 mm panel (~130 um thick) with ~100,000 holes on a panel carrier.

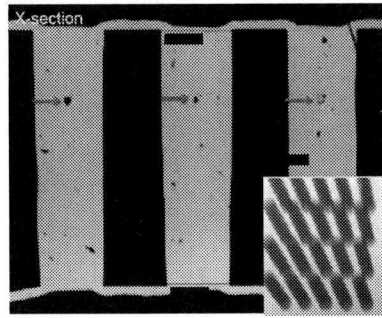

Fig. 9. Results from double side plating using Atotech's Process.

Another, particularly cost effective, method to fill through holes in thin glass panels is to process them using a double side semi additive plating (SAP) process. There are challenges in applying standard SAP processes to glass, but Atotech has recently reported significant progress in the ability to metallize glass vias using new adhesion promoters. [12] Figure 9 shows the ability to fill ~80 um through holes in 300 um thick glass. As the cross-section and inset x-ray images show, this is done nearly void free today with very little overburden remaining. The advantage of this approach is that it has the opportunity to leverage existing plating tools and techniques in the industry and provide excellent throughput for panel based interposer manufacture.

5. Panel Level Lithography

Once the holes are filled, it is necessary to have panel level tools to achieve electrical redistribution with fine line and spacing. The stability of glass relative to organic substrates provides the opportunity to approach line/space that is achievable in silicon, but in a panel format. Couple this with advancements in the ability to fill holes in a thin glass panel format, significant cost advantages will be realized without compromising L/S requirements. There is an opportunity to leverage tools designed for the flat panel display (FPD) industry to achieve < 3 um L/S.

Reduction lithography technology suitable to manufacture panel level glass interposers for advanced packaging has been developed by Rudolph Technologies based on Azores Flat Panel Display PanelPrinter™ product. Although leading edge front-end lithography is pushing into the single-digit nanometer scale, the feature sizes needed for interposer panels (2.5/3D, advanced packaging) are much larger. Back-End lithography is driven by a whole different set of requirements related to thick resist, depth of focus, sidewall angle, panel handling capability, exposure field size and more.

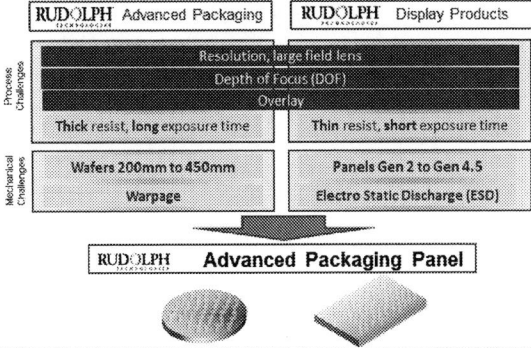

Fig. 10. Lithography challenges and solutions for panel-based interposers.

Figure 10 summarizes how Rudolph is able to address the challenges of panel level lithography from two directions to achieve the attributes required for advanced packaging. The production lithography challenges and solutions are categorized in three different parts:

- Process (resolution, exposure wavelength, resist thickness, process complexity in terms of number of layers, overlay, Depth of Focus and Dose/exposure time)
- Mechanical (substrate handling, substrate size, Electro Static Discharge (ESD), edge protection and/or edge exposure depending on the type of resist used –Positive or negative) and
- Cost. Throughput and a large field lens, in this case an 84mm diameter, is necessary to minimize the number of exposures and therefore the cost. The penalization of the back-end of the Advanced Packaging semiconductor industry becomes inevitable as the economics of panel lithography become more and more cost effective.
  - In Advanced Packaging, more chips on the substrate means lower cost per chip. In this case, we replace the word chip with chip interposer (200mm wafers, 300mm wafers, 450mm wafers, Gen 2 size panels)
  - As proven in Flat Panel Display, the larger the glass panels are the lower Cost Of Ownership (COO) when compared to wafers.

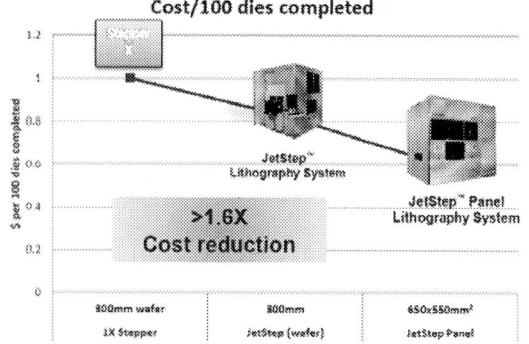

Fig. 11. Example of real reduced COO as economies of scale leveraged in FPD.

Figure 11 indicates how Rudolph applied technologies/features developed for back-end Advanced Packaging lithography and for Flat Panel Display can be used to develop a production lithography tool for glass interposers (JetStep™ Panel Lithography System.)

Initial work has begun to demonstrate the achievable L/S on glass substrates. Figure 12 shows examples of this work on 150x150 mm substrates using technology that is scalable to large format (e.g. 500 mm and greater) panel substrates. The scanning electron microscope (SEM) images show that the process is capable to provide sufficient exposure of the resist layer to 3.5 um today and work is ongoing to demonstrate this, and subsequent metallization to this level and targeting < 2 um.

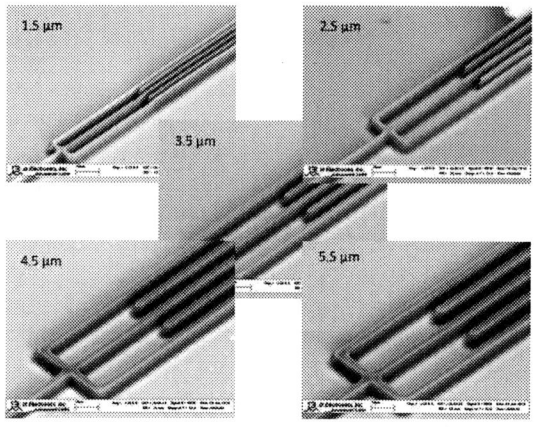

Fig. 12. SEM images of developed photoresist structures.

6. Conclusions

The electrical performance of glass, and tunability of material properties such as CTE and ability to form in thin large sheets of high quality allowing cost effective processes, generates tremendous incentive for using glass as a TGV substrate for 2.5D and 3D applications.

The ability to generate well-formed through and blind vias has been demonstrated, and fully populated test vehicles using glass interposers have been created. This work shows that existing metallization technology can be leveraged to generate very good Cu filling performance in glass in both wafer and panel formats.

Wafer level is ahead of panel level fabrication for TGV substrates. As reported here, significant process learning and development has been realized in wafer formats with very good results. The RDL results also show that metal can be smoothly plated on the glass with good results. Test vehicles indicate that good electrical, thermal and reliability performance can be achieved using glass vias and the expected advantages in electrical performance of glass relative to silicon have been demonstrated in previous work. This work continues to further demonstrate reliability of these structures on glass substrates. We report here initial progress in the ability to fabricate and handle large and thin glass panels. With glass, opportunities for truly cost-effective interposer solutions with improved reliability exist, and sustained demonstration of downstream processing continues to move these solutions closer to commercialization.

Acknowledgments

We acknowledge the strong support from Industrial Technology Research Institute (ITRI), Georgia Tech's Packaging Research Center, Atotech Deutschland GmbH and i3 Electronics, Incorporated for this work.

References

[1] V. Sukumaran et al., "Through-Package-Via Formation and Metallization of Glass Interposers", Electronic Components and Technology Conference (ECTC), IEEE 60th (2010).

[2] I. Ndip et al., "Characterization of Interconnects and RF Components on Glass Interposers", IMAPs 45th International Symposium on Microelectronics, (2012).

[3] C. H. Chien et al., "Performance and Process Comparison between Glass and Si Interposer for 3D-IC Integration", IMAPs 46th International Symposium on Microelectronics, (2013).

[4] X. Qin, N. Kumbhat, V. Sundaram, R. Tummala, "Highly-Reliable Silicon and Glass Interposers-to-Printed Wiring Board SMT Interonnections: Modeling, Design, Fabrication and Reliability", Electronic Components and Technology Conference (ECTC), IEEE 62nd (2012).

[5] Shorey, A; Pollard, S.; Streltsov, A.; Piech, G.; Wagner, R., Electronic Components and Technology Conference (ECTC), IEEE 62nd (2012).

[6] Keech, J.; Piech, G.; Pollard, S.; Shorey, A., "Development and Demonstration of 3D-IC Glass Interposers", Electronic Components and Technology Conference (ECTC), IEEE 63rd (2013).

[7] Shorey, A; Chaparala, S.; Pollard, S; Piech G.; and Keech, J.; "Glass Interposer Substrates: Fabrication, Characterization and Modeling", IMAPs 46th International Symposium on Microelectronics, (2013).

[8] A. Shorey, S. Pollard, J. Keech, "Progress in Fabrication and Test of Glass Interposer Substrates", IMAPs 10th Device Packaging Conference (2014).

[9] D. Malta, C. Gregory, M. Lueck, J. Lannon, D. Temple, P. DiFonzo, F. Naumann, and M. Petzold, "Characterization and Modeling of Copper TSVs for Silicon Interposers," *Proc. of 63rd ECTC*, May 2013.

[10] M. Lueck, D. Malta, A. Huffman, C. Gregory, M. Butler, J. Lannon and D. S. Temple, "Fabrication and Testing of Thin Silicon Interposers with Multilevel Frontside and Backside Metallization and Cu-filled TSVs," *Proc. of 63rd ECTC*, May 2013.

[11] R. Rhoades and D. Malta, "Advances in CMP for TSV Reveal," *Proc. of Int. Conf. on Planarization/CMP Tech.*, Oct. 2012.

[12] S. Bamberg et al., "Challenges of Adhesion Promotion for the Metallization of Glass Interposers", IMAPs 46th International Symposium on Microelectronics, (2013).

Large Area Interposer Lithography

Warren Flack, Robert Hsieh, Gareth Kenyon, Manish Ranjan
Ultratech, Inc
3050 Zanker Road, San Jose. CA. 95124
wflack@ultratech.com +1 408-577-3443

John Slabbekoorn, Andy Miller, Eric Beyne
IMEC
Kapeldreef 75 B-3001 Leuven, Belgium
millera@imec.be +32 16-28-76-15

Medhat Toukhy, PingHung Lu, Yi Cao, Chunwei Chen
AZ Electronic Materials USA Corp.
70 Meister Ave, Somerville, NJ 08876 USA
Yi.Cao@AZEM.com +32 16-28-31-33

Abstract

Large area silicon or glass interposers may exceed the maximum imaging field of step and repeat lithography tools. This paper discusses the lithographic process used to create a large area interposer on a stepper by the combination of multiple subfield exposures. Overlay metrology structures are used to confirm the relative placement of the subfields to construct the interposer. Routing lines from 1.5 to 4.0 µm in width are evaluated to measure critical dimension (CD) control where the lines cross the subfield boundaries. CD metrology at the bottom and top of the photoresist is performed using a top down CD-SEM. Finally large area test interposers are patterned using two subfields on a 1X stepper and processed through a Cu electroplating module for detailed characterization.

The CD control of routing lines as they cross the subfield boundary can be optimized by using a shaped or tapered line end design. Lithography simulation using Prolith modeling software by KLA-Tencor is matched to experimental results and then used to evaluate performance of various line end designs. Larger latitude for overlap error was observed for the tapered line end compared to the standard square line end.

The experimental and modeled results in this study show the capability of using stepper lithography to produce large area interposers with 1.5 µm I/O routing line dimensions.

Introduction

Over the last few decades, IC technology has used shrinking gate dimensions to increase gate switching speed and decreased operating voltage to reduce power consumption. As the demand for improved form factor and superior battery life accelerates, the semiconductor manufacturing supply chain is taking a closer look at back end of line (BEOL) manufacturing technology. Innovative IC packaging solutions are being developed to meet the needs in consumer electronics. Furthermore, IC packaging now is widely seen as a method to prolong Moore's law [1]. Many companies are evaluating the use of silicon interposers with through silicon vias (TSV) to address requirements for higher performance and smaller form factor packages. Interposer technology is less complex than full 3D stacking and therefore offers an advantage in time to market. It is currently viewed as the next major step in cost effective advanced packaging technology.

Key advantages of silicon interposers include high routing line density, excellent electrical and thermal performance, lower power requirements than equivalent single-chip packages through combination of multiple chips on one substrate, and the possibility of integrating passives into the substrate [2,3]. The individual device die can provide numerous functions including memory, logic, analog and MEMS (micro-electromechanical systems). Achieving high bandwidth between individual die on the interposer requires fine pitch routing lines. For advanced wide I/O applications, it is anticipated that interconnect line widths of less than 2 µm will be needed.

A step and repeat (stepper) lithography system provides the necessary patterning capability for high resolution devices with zero printable defects. However, for some designs the interposer area can exceed the maximum stepper field size.

Large Interposer Fabrication

The requirement for patterning large area devices with stepper lithography is not new. Superchips from the VHSIC (Very High Speed Integrated Circuit) program were constructed using macrocells [4]. Each macrocell was a self-contained integrated circuit placed in a single stepper field and only interconnect layout rules were used at crossing field boundaries [5, 6]. A more recent demand for large area devices is for infrared focal plane arrays used for aerospace applications [7]. The image sensor pixels are contained within a stepper field and only interconnect routing is allowed to cross stepper field boundaries similar to superchips.

Since interposers are designed to interconnect single chip devices, the field stitching considerations are similar to focal plane arrays and superchips. A large area interposer can be fabricated by splitting the interposer design into multiple sections where each section is smaller than the maximum field size of the step-and-repeat lithography system.

Figure 1 shows a 50 by 50 mm interposer split into a top half (purple) and a bottom half (pink). This two subfield approach would work for a lithography system with a field size greater than 50 by 25 mm.

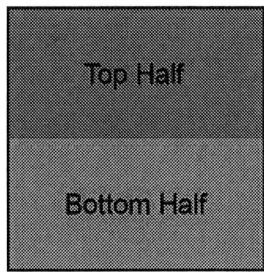

Figure 1. The large area interposer design is split into top and bottom halves for stepper lithography layout.

A reticle for the lithography system is then fabricated for each section of the interposer. Figure 2 shows the layout for a 1x reticle that supports placing multiple fields on one plate. Here the top half of the interposer (purple) is field 1 on the reticle and the bottom half of the interposer (pink) is field 2 on the reticle.

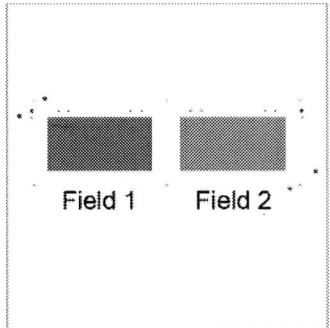

Figure 2. 1X reticle layout for an interposer split into two fields. The reticle size is 150 by 150mm.

This reticle can then be used on the lithography stepper to image the full interposer on the wafer by alternating the patterning of rows of field 1 (purple) and field 2 (pink) as shown in figure 3. The interposer routing lines that cross the boundary of the top half and bottom half of the interposer are stitched by allowing a small amount of Y overlap between the two fields. The same approach could be used to fabricate even larger area interposers by having additional reticle fields stitched in both the X and Y direction. This paper evaluates using this field stitching process to meet the requirements for advanced silicon interposer manufacturing.

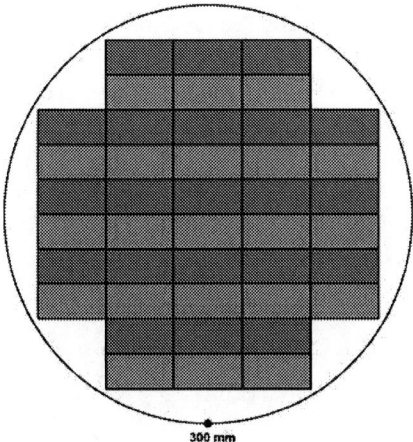

Figure 3. 300mm wafer layout with 21 interposers. The interposer size for this case is 50 by 50 mm.

Experimental Methods

Exposures are performed on an Ultratech AP300 advanced packaging stepper with a 0.16 NA Wynne-Dyson lens [8]. This catadioptric optical system design permits the use of broadband illumination from a mercury arc lamp, and the system used in this study has a capability to select i-line, gh-line or ghi-line wavelengths. The tool is equipped with a WEE (Wafer Edge Exposure) unit for exposing the edge of the wafer and a WEP (Wafer Edge Protection) unit for protecting a predefined outer edge of the wafer. The WEE enables precise removal of photoresist from the wafer edge creating an electrical contact that is required for electroplating. The WEP blocks exposure light and can be used to retain photoresist on a thin ring inside the WEE creating a protective seal ring to prevent leakage of solution during the electroplating step.

The large area interposer evaluated for this study is 44.0 by 44.0 mm with interconnect lines covering the whole chip area. Since the exposure field of the 1X stepper used in this study is 44x26.7mm, the interposer design is split in two fields each with a size of 44 by 22mm similar to figures 1 and 2. The reticle set was designed to include test structures that provide evaluation of overlay and CD performance at the field boundaries. For this case only a Y stitch is required.

Multiple test structures were created to evaluate reticle field stitching performance. Figure 4 shows a line integrity structure with six sets of line and space patterns with varying CD and pitch. The patterns above the red stitch line are on reticle field one and the patterns below the red stitch line are on reticle field two. The CD of the test structures vary from 1.5 μm line and space on the left side to as large as 4.0 μm line and 2.0 μm space on the right. The 0.5 in blue above the pattern indicates that the top and bottom half have a stitch overlap in Y of +0.5 μm between the fields. Additional line integrity structures were created with field stitch overlaps varying from as small as -0.5 μm to as large as +10.0 μm.

Figure 4. CD performance features with varying pitches. The red line indicates the stitch between field 1 and field 2. The 0.5 above the structure indicates the Y overlap in microns.

An electrical test structure was also created to evaluate the field stitch performance of electroplated Cu lines. Figure 5 shows a serpentine/comb four point probe structure. The pattern above the red stitch line is on reticle field one and the pattern below the red stitch line is on reticle field two. The test structures range from 1.5 μm line and space to 3.0 μm line and space. The field overlap in Y was +0.5 μm.

Initial test of the interposer stitching was performed on bare silicon wafers using JSR IX845 positive tone photo resist. The resultant photoresist structures are evaluated for overlay performance and CD behavior at the stitching

978-1-4799-2408-0/14 $31.00 © 2014 IEEE

boundary and are used as the reference for the simulation modeling discussed in the "Simulation of Field Stitching" section of this paper.

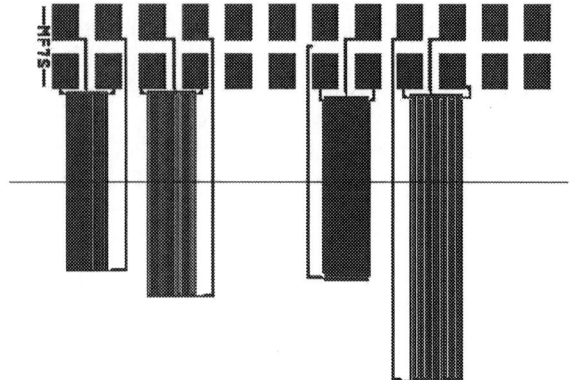

Figure 5. Serpentine/Comb structure. The red line indicates the stitch between field 1 and field 2.

Next the lithographic process was evaluated on Cu seed 200mm wafers using actual device process conditions. The Cu interconnect lines are fabricated using a semi-additive electroplating technique as shown in figure 6. In this technique a Cu seed layer consisting of 30 nm TiW and 50 nm of Cu is deposited on the wafer which acts as the current distributing layer during the electroplating process (figure 6.1). Next a 3.5 μm thick positive photoresist is coated on the wafer and the area to be electroplated is opened to the Cu seed via the lithography process (figure 6.2). The resist is descummed and then 2.5 μm of Cu is electroplated on the wafer (figure 6.3). The photoresist is then stripped off of the wafer (figure 6.4). The Cu seed is wet etched followed by wet etch of the TiW to create the final structure (figure 6.5).

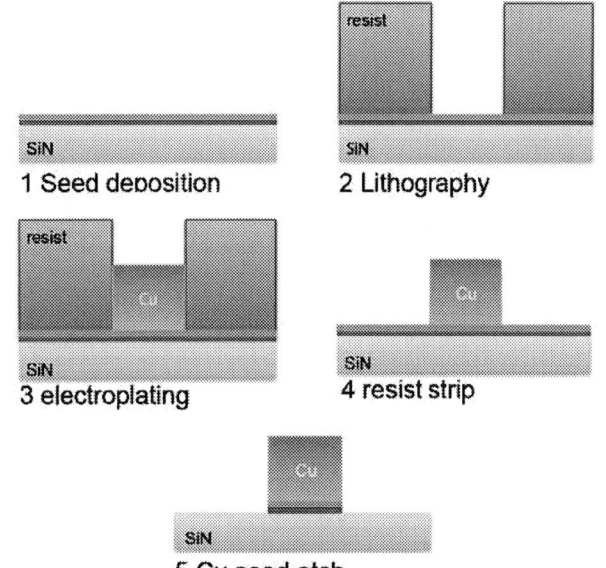

Figure 6. Wafer Process Flow to fabricate Cu interconnect lines using of a semi-additive electroplating technique.

The novolak/diazonaphthoquinone (DNQ) positive tone resist platform is approaching its process limits to adequately meet the demands of advanced packaging applications. DNQ high absorption and low sensitivity impose significant limitations on its resolution, pattern profiles, and photospeed over a wide range of film thickness. Therefore, chemically amplified (CA) resist platforms are believed to be more suitable than (DNQ) platform to fulfill the future needs for advanced packaging applications such as interposers. The higher photospeed of CA resist reduces exposure times and considerably improves the cost of ownership of the lithography tool.

This study employed AZ EXP CN-3 positive resist, which is based on a phenolic polymer, CA platform. This resist can produce vertical sidewalls with minimal footing on Cu substrates, and is capable of resolving submicron patterns in 3.5μm thick resist using i-line lithography. The exposure latitude of 1.5 μm lines and spaces measured on a 0.16 NA stepper was 18% with a ±10% CD criterion.

Processing conditions for AZ EXP CN-3 are summarized in table 1. All resist processing was performed on a TEL ACT12 Clean Track which is equipped with high viscosity pumps for thick resist processing. CD metrology was performed on a KLA 8250XR CD-SEM.

Process Step	Conditions
Soft Bake	120 seconds at 110°C
Post Exposure Bake	60 seconds at 90°C
Development	30 seconds x 2 (double puddle) in 0.26N TMAH developer with surfactant

Table 1. Process conditions used for AZ EXP CN-3 resist.

The 1X stepper offers multiple alignment options, and for this study both blindstep and zero layer alignment were evaluated. Blindstep uses the XY stage encoder along with a previously calibrated transform to accurately print a multiple field array. The zero layer technique requires a dedicated array of field alignment targets to be printed on the wafer. These targets are used to align and stitch two adjacent reticle fields together to optimize overlay. Overlay metrology was performed using microscope measurements and using the self-metrology feature on the 1X stepper.

Results

For this study the AZ EXP CN-3 resist was exposed at i-line using a nominal exposure dose of 140 mJ/cm^2 . The resist was optimized to produce a 1.5 μm line and space pitch on Cu seed wafers. A cross section of 3.5 μm thick AZ EXP CN3 photoresist is shown in figure 7(a). The resist exhibits an excellent profile with minimal footing. Cross sections of Cu plated lines are shown in figure 7(b).

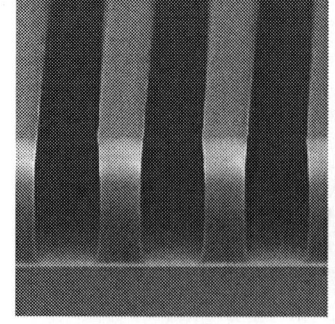 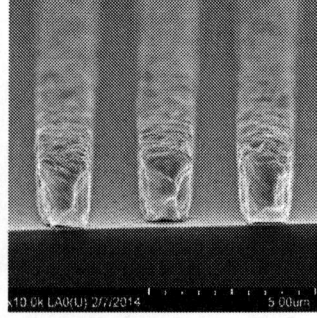

(a) AZ CN3 Resist   (b) Cu Plated Lines

Figure 7. (a) Cross section of 3.5 μm thick AZ EXP CN3 photoresist. The exposure dose is 140 mJ/cm^2 and the focus

978-1-4799-2408-0/14 $31.00 © 2014 IEEE   28

offset is 0 μm. (b) Cross section of Cu plated metal lines before Cu seed etch. Both cases are for line and space pattern with 3 μm pitch.

Sample interposer structures were stitched together using two lithography fields. The resist line at the stitch area was evaluated for three different Y axial offsets (overlap). Figure 8(a) shows a positive overlap of 1.0 μm and the positive resist line at the stitch decreases in CD. Figure 8(c) shows a negative overlap of 0.5 μm and the resist line bulges out and merges with adjacent lines. Note that these images are taken before the descum step which would further open the spaces between resist lines.

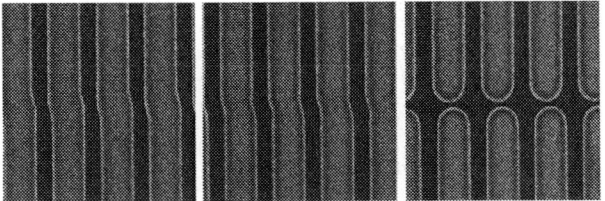

(a) Overlap of +1.0 μm (b) Overlap of +0.5 μm (c) Overlap of -0.5 μm

Figure 8. Effect of Y overlap at the stitch of resist spaces for a 3 μm pitch structure. The resist feature is dark in these images. The lateral offset is 0.25 μm. These images are taken before the resist descum process.

The effect of the Y overlap was also evaluated after Cu electroplating. Figure 9 shows three cases: (a) 1.0 μm overlap, (b) 0.5 μm overlap, (c) -0.5 μm overlap (a gap). A large overlap creates a bulge in the line at the stitch whereas a gap creates a line constriction. Taken to the extreme, a large overlap can create an electrical short and a large gap can create and electrical open. Therefore to preserve line integrity the Y overlap error must be controlled. Based on these experimental results a 0.5 μm overlap is used in the rest of this study.

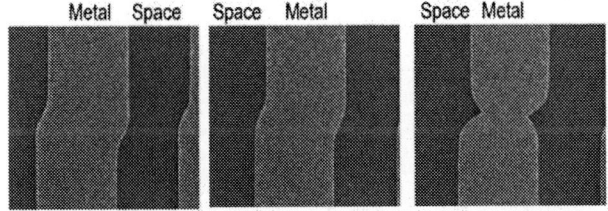

(a) +1.0 μm overlap (b) +0.5 μm overlap (c) -0.5 μm overlap (gap)

Figure 9. Effect of Y axial offset at the stitch for plated metal lines for a 3 μm pitch structure.

Figure 10(a) shows a top down SEM of stitched dense plated lines with 3 μm pitch. The plated lines form within the spaces of the resist pattern. This view shows an optimized stitch for which the actual stitch location is difficult to discern. A red horizontal line is added to the stitch location. For instructive purposes it is useful to include moderate amounts of lateral offset in the stitch in order to clearly denote its position. Figure 10(b) shows stitched plated line with a large lateral offset of 0.5 μm. With large lateral offset the line is well defined and maintains adequate width, however the spaces between the lines become constricted.

Tilted SEM images of the plated lines are shown in figure 11(a) and 11(b). The images illustrate ability to fabricate line/space metal lines down to the 1.5 μm and to stitch these

lines across a field boundary. In this case the lateral offset was set at 0.25 μm.

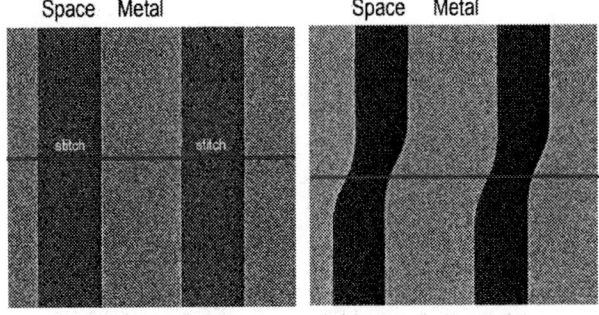

(a) No lateral offset   (b) Lateral offset of 0.5 μm

Figure 10. Stitched plated metal lines at 3 μm pitch with (a) no lateral offset and (b) a lateral offset of 0.5 μm. The overlap is 0.5 μm. The red line is the field stitch location.

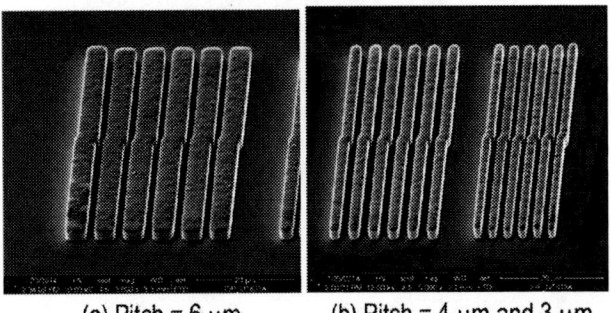

(a) Pitch = 6 μm   (b) Pitch = 4 μm and 3 μm

Figure 11. Electroplated metal lines at stitch with (a) 4 μm line and 2 μm space and (b) 4 μm and 3 μm pitches equal line and space pitches. Both cases have a lateral offset of 0.25 μm.

The steps following resist development also influence the shape and size of the Cu lines. Both descum and the seed removal steps need to be optimized for effective control of CD. Descum is essential for uniform plating results because it reduces the surface tension of the photoresist to allow proper wetting to the Cu seed. However, this process consumes photoresist and the CD of the resist opening increases as a result. Figure 11(a) shows one line exhibiting some non-uniformity which indicates that the descum is on the limit of being too soft.

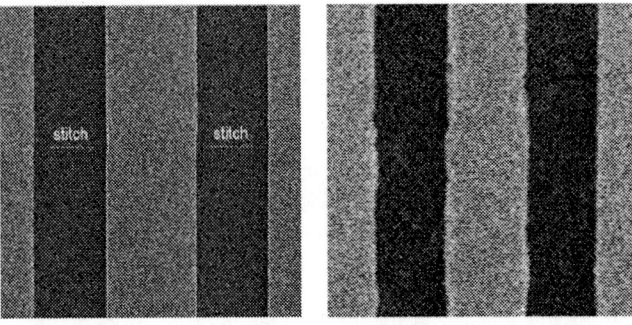

(a) Before Cu seed etch   (b) After Cu seed etch

Figure 12. Top down view of Cu plated metal lines (a) before Cu seed etch and (b) after seed etch. Both cases are for 3 μm pitch, line and space pattern.

The seed etch also can have a large impact on the shape and size of the Cu lines. During wet etching of the Cu seed, the electroplated structures are also etched with a reduction of

CD as a result as shown in Figure 12. To reduce this effect the Cu seed thickness needs to be as thin as possible in order to minimize the etching time. For example, using a standard Cu seed thickness of 150nm results in more than 400 nm CD loss, which would be unacceptable for creating 1.5 μm Cu lines. This is the reason that a 50nm Cu seed was used in this study.

Figure 13 shows top down view of a Cu electroplated serpentine/comb structure with a 3μm pitch. Visual inspection reveals no line breaks or shorts in any of the serpentine/comb structures. Future work includes electrical characterization of these four point probe structures.

Figure 13. Cu electroplated serpentine/comb structure with a 3 μm pitch. Visual inspection reveals no line breaks or shorts in the structure.

Simulation of Field Stitching

A lithography simulation program (Prolith version 14.1.1.1 from KLA-Tencor) was used to study the effect of stitching overlap and lateral offset on the resist pattern for stitched vertical lines. The pattern consists of 1.5 μm dense vertical lines. Two photomask passes are used to accurately simulate the separate top and bottom field exposures. Figure 14 illustrates the construction of the Prolith model for a standard square corner line end.

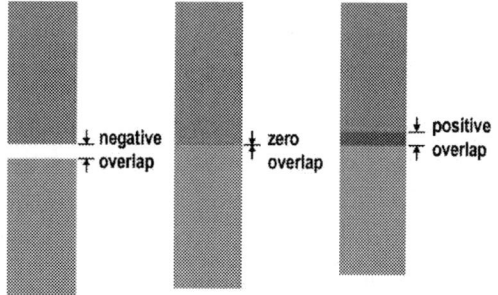

Figure 14. Simulation conditions for stitching line with square ends with lateral offset and overlap. The top and bottom exposures are independently simulated.

The construction of a tapered line end is shown in figure 15. The objective is to determine whether changing the shape of the line ends from square ends can improve process margin for misalignment. For the square line ends the zero overlap condition is defined where the two line ends just meet. For the tapered line ends the zero overlap is defined where the full width shoulders of the tapers meet. In both cases, a positive overlap moves the bottom feature up in Y relative to the top feature. For lateral offset a positive shift moves the bottom feature to the right in X relative to the top feature.

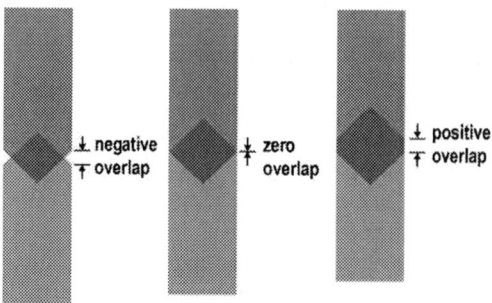

Figure 15. Simulation conditions for stitching line with 45 degree tapered ends with lateral offset and overlap.

Experimental test patterns were exposed in 2.7 μm thick JSR IX845 resist on a 200 mm wafer, and SEM photos were taken to document resist performance at the stitch for different offsets. Figure 16(a) shows top down SEM of a stitch with measured X and Y offset of 0.2 and 0.75 μm respectively. The corresponding figure 16(b) shows a Prolith 3-dimensional resist simulation of the same conditions. Note the resist feature is grey in this figure. Experimental and simulated results show very good agreement, which gives confidence in using modeling to investigate the effect of stitching overlap and lateral offset on line quality.

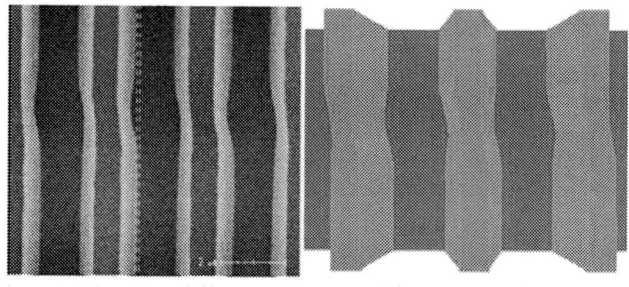

(a) Top down SEM   (b) Resist simulation

Figure 16. 1.5 μm stitched lines with lateral offset of 0.2 μm and overlap of 0.75 μm. (a) Top down SEM of resist pattern. (b) Prolith simulation of the resist pattern with square ends.

To minimize the CD variation across the stitch a reasonable starting goal is to hold the lateral offset to 20% of the line width. For 1.5 μm line/space this translates into a ±0.3 μm offset range. A similar number can be used for the overlap range. Prolith simulation was used to determine the effectiveness of these criteria for stitched lines.

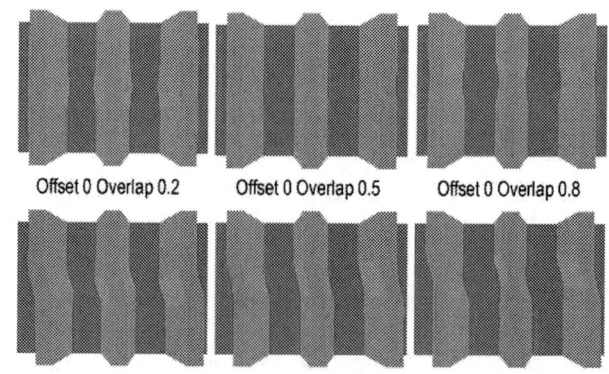

Offset 0 Overlap 0.2   Offset 0 Overlap 0.5   Offset 0 Overlap 0.8

Offset 0.3 Overlap 0.2   Offset 0.3 Overlap 0.5   Offset 0.3 Overlap 0.8

Figure 17. Square end stitch simulation of 1.5 μm lines with overlaps varying from 0.2 μm to 0.8 μm and lateral offsets of 0.0 (top row) and 0.3 μm (bottom row).

978-1-4799-2408-0/14 $31.00 © 2014 IEEE

Figure 17 shows a sequence of simulations for different values of overlap and lateral offset for square line ends. Overlap range shown in figure 17 is 0.2 to 0.8 µm. Note that resist feature is grey in this figure. CD at the stitch is well controlled, however one can see that as overlap decreases, the resist line starts to bulge at the stitch; and with increasing overlap, the resist line constricts. The effect of a lateral offset of 0.3 µm is shown in the bottom row. For large lateral offsets the rotation of the dense lines at the stitch reduces the spacing between adjacent lines. Both overlap and lateral offset need to be controlled in order to maintain line integrity across the stitch. However the overlap parameter has the larger effect on CD at the stitch.

The effect of tapering the line ends was studied using simulation. A 45 degree taper ending in a point was investigated. For this end shape, the definition of zero overlap is where the full width bases of the tapered lines meet. Prolith simulations show a range of overlap from -0.5 to 0.1 µm where linewidth is moderately well controlled as shown in figure 18. The effect of a lateral offset of 0.3 µm is shown in the bottom row. The stitched lines with tapered line ends have smoother transitions than the stitched lines with square line ends.

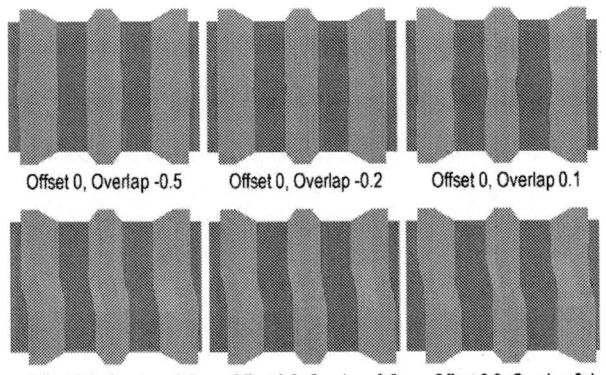

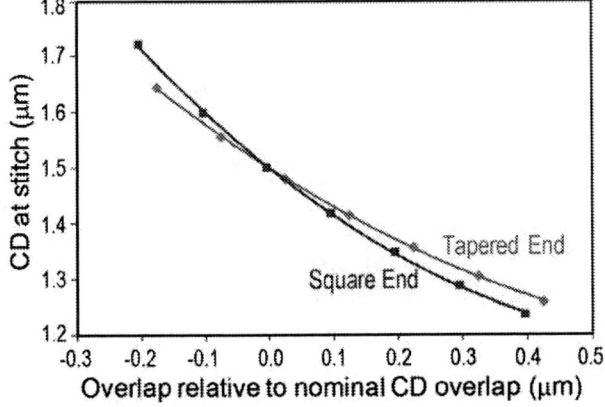

Figure 18. Tapered ends stitch simulation of 1.5 µm lines with overlaps varying from -0.5 µm to +0.1 µm and lateral offsets of 0.0 (top row) and 0.3 µm (bottom row).

Figure 19. Comparison of resist line CD versus overlap for square and tapered line ends. Data from Prolith modeling of JSR IX845 resist. Overlap scale is shifted so that zero overlap corresponds to the condition where CD at the stitch equals the nominal linewidth of 1.5 µm.

The resist linewidth at the simulated stitch was measured for various overlaps, with zero lateral offset as shown in figure 19. For both square and tapered line end types the CD versus overlap behavior can be well described by a quadratic fit. For comparison between square and tapered line ends the overlap is shifted such that zero overlap corresponds to the condition where the CD at the stitch equals the nominal linewidth of 1.5 µm. The nominal CD is obtained at an overlap of 0.41 µm for the square ends and -0.32 µm for the tapered ends. It is apparent that the slope of the tapered end curve is less than the square end curve, indicating larger latitude for overlap error for the tapered end. If a ±10% CD tolerance is allowed the overlap range is 25% larger for the tapered line end relative to the square line end.

The resist model based on JSR IX845 resist characteristics can be compared to actual measured CD data using AZ EXP CN-3 resist. Note that these measurements are of the space between resist lines, since the space defines the size of the plated lines. Figure 20 shows the 1.5 µm resist features on a 3.0 µm pitch with 0.3 µm and 0.8 µm overlap. Here the total overlap is calculated by adding the design overlap and the measured registration at the stitch site. The CD at the stitching line is measured and compared to the "nominal" CD just above and below the stitching line. At 0.3 µm overlap the CD of the space is 342 nm smaller than nominal and at 0.8 µm the CD of the space is 256 nm larger than nominal. The estimated optimum CD at the stitch occurs at 0.59 µm overlap based on linear interpolation, or 0.55 µm based on fitting to the curvature in figure 19 for square end lines. The experimental data for AZ EXP CN-3 resist shows a reasonably good fit to the model. Further experiments would be needed to fine tune the model specifically for AZ EXP CN-3 resist.

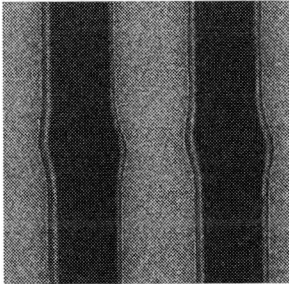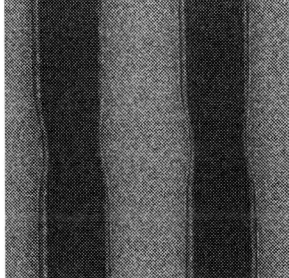

Figure 20. AZ EXP CN-3 resist lines (dark features) at stitch. 1.5 µm CD lines and spaces across the stitching line for +0.3 and +0.8 µm overlap. Lateral offset is 0.1 µm.

Conclusions

Extending device performance increasingly relies on advancements in back end technologies such as the use of very large interposer designs with aggressive interconnect density requirements. A stepper system provides the necessary patterning capability for high resolution devices with zero printable defects. However to produce large area interposers requires stitching of stepper subfields.

This study experimentally investigated patterning copper lines with lateral dimensions as small as 1.5 µm line/space in a vertically stitched 44 by 44 mm device. To achieve this an experimental AZ EXP CN-3 resist was employed. The resist must meet a combination of requirements for this application including high resolution, steep sidewalls and chemical resistance to the Cu electroplating process.

Lithography simulation was used to further study the line fidelity across the subfield boundary, including the effect of stitching error and effect of changing the line end shape at the stitch. Larger latitude for overlap error was observed for the tapered line end compared to the standard square line end. This characterization is useful for improving process control of the lithography step.

This work demonstrates that stitching of subfields for interposer interconnects can be achieved by leveraging existing stepper lithography and processes technologies.

Acknowledgments

We would like to thank Patrick Jaenen and Inge de Preter of IMEC for the cross-sectional SEM images. We would also like to thank Geraldine Jamieson, Sofie Robert, Joris Tuinstra and Sander van Gompel of IMEC for their help setting up the wafer processes.

References

1. Flemming, J., et al.; "Cost Effective 3D Glass Microfabrication for Advanced Electronic Packages", 2013 IPC APEX Conference & Exhibition, San Diego, California, February 2013.

2. Vardaman, J. et. al., *TechSearch International: Developments in 2.5D: The Role of Silicon Interposers*, August 2013.

3. Hogan, M., "Silicon interposers: building blocks for 3D-ICs", *Solid State Technology*, June, 2011.

4. Stroll, Z. "VHSIC Submicron technology at TRW", *GOMAC Digest*, 1989.

5. Flack, W. et. al., "Lithographic Manufacturing Techniques for Wafer Scale Integration", *International Conference on Wafer Scale Integration*, San Francisco, 1992.

6. Flores, G. et. al., "Lithography Strategies for Wafer Scale Integration", *KTI Microlithography Seminar Proceedings*, San Diego, 1990.

7. Rogalski, A., "Semiconductor Detectors and Focal Plane Arrays for Far-Infrared Imaging", *Opto-Electronics Review*, **21**(4) 2013.

8. Flores, G. et. al, "Lithographic Performance of a New Generation i-line Optical System," *Optical/Laser Lithography VI Proceedings*, SPIE 1927 (1993).

Minimizing Interposer Warpage by Process Control and Design Optimization

Mikael Detalle, B. Vandevelde, P. Nolmans, J. De Messemaeker, M. Gonzalez, A. Miller, A. La Manna, G. Beyer and E. Beyne

imec vzw, Kapeldreef 75, B-3001 Leuven, Belgium
Tel: +32 16 287986 Fax: +32 16 281501 Email: detalle@imec.be

Abstract

An analytical model simulating the bowing at wafer or thin die level was applied to imec's 3D interposer technology. The calibration methodology is explained. A good correlation between simulation and measurement has been found at different stages during the processing. Secondly, a model combining all the interposer features was used to simulate the bowing induced at wafer and thin die level. Finally, the model is used to provide some recommendations to mitigate the interposer bowing without any drastic change in the structure or impact onto its performances.

Introduction

Today, the choice of the packaging technology used to interconnect the chips has more and more impact onto the system performances. Emerging technologies such 3D silicon interposer are very promising candidates to increase the number of connections between the chips while shortening critical electrical paths through the device using TSV (Trough Silicon Via), leading to faster operation. Silicon interposers are being used for high power/performances applications to stack chips side-by-side. Consequently, thick Back End Of Line (BEOL) layers are preferred to decrease the interconnect resistance and optimize the power consumption. The combination of TSV (especially 100 μm deep) and thick BEOL can lead to high bowing at wafer and die level. Highly bowed wafers can generate handling issues during processing and impact the production. Interposer die bowing can impact yield and reliability during assembly and packaging processes. Bowing management is becoming even more critical for very large interposer dies.

In this paper we will present bowing simulation applied to imec's silicon interposer technology. The silicon interposer includes 10x100 μm TSV, four thick damascene BEOL layers, Cu bumps and redistribution layers (RDL) front side and back side. First, the analytical model used for the study will be briefly described. This model permits to calculate the bowing at wafer or thin die level based on the residual stress of each layer. This layer can be a uniform (blanket) layer, but it can also be a mixed layer consisting of two different materials with their own residual stress (patterned). In order to calibrate the model, the residual stress of each material used within imec interposer technology has been evaluated experimentally. The methodology will be explained. In a next step, bowing simulation and experimental data will be compared to review their good matching and validate simulation output. Finally, the calibrated model will be used to investigate the impact of the design and structure onto bowing of a full interposer at wafer and thin die level. Efficient solutions to mitigate the bowing without any drastic change of the interposer structure or impact onto its performances will be identified.

Analytical model description

The objective of the analytical model is to calculate the wafer warpage based on the residual stresses of each layer. This layer can be a uniform (blanket) layer, but it can also be a mixed layer consisting of two different materials with their own residual stresses (e.g. mixture of Cu and low-k or SiO_2 dielectric). Besides that, the model should also be able to calculate the impact of thinning the wafer, resulting in much higher warpage for the same stress levels in the different layers.

From the residual stress σ_{res}, the equivalent zero stress temperature (EZST) for the film material (E_{film}, v_{film}) on top of a full thickness wafer ($E_{substrate}$, $CTE_{substrate}$) is calculated as follows:

$$EZST = \frac{\sigma_{res}(1 - v_{film})}{(CTE_{film} - CTE_{substrate})E_{film}}$$

The evolution of the wafer warpage is calculated adding - step by step - the layer at the EZST, as shown in the figure 1.

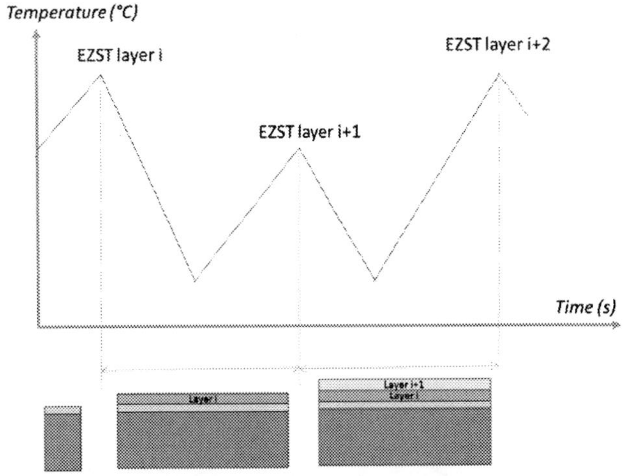

Figure 1. EZST evolution during processing.

The warpage is calculated for a multilayer structure with n layers of which layer 1 is the silicon wafer. Each layer is defined by its thickness h_i, the EZST T_i, CTE α_i and elastic modulus E_i.

The warpage of a specific build-up is calculated from following equations:

- Force equilibrium

$$\sum_{i=1}^{n} F_i = 0 \qquad \text{(1 equation)}$$

- Curvature equations:

$$\frac{1}{\rho} = \frac{M_i(1-\nu)}{E_i I_i} \quad \text{for i = 1 to n} \quad \text{(n equations)}$$

- Displacements in the neutral fiber of each layer

$$u_i = \alpha_i.(T_i - T_{ref}).L + \frac{F_i.L.(1-\nu)}{E_i h_i} \quad \text{for i = 1 to n}$$

(n equations)

- Displacements compatibility between neighboring layers

$$u_i - \frac{h_i}{2}\frac{L}{\rho} = u_{i+1} + \frac{h_{i+1}}{2}\frac{L}{\rho} \quad \text{for i = 1 to n-1 (n-1 equations)}$$

- Moment equilibrium (around neutral axis of layer 1)

$$\sum_{i=1}^{n} M_i = F_2\left(\frac{h_1+h_2}{2}\right) + F_3\left(\frac{h_1+2h_2+h_3}{2}\right) + ...$$
$$+ F_n\left(\frac{h_1+2h_2+...+2h_{n-1}+h_n}{2}\right) \quad \text{(1 equation)}$$

These 3n+1 equations can be solved to the following unknowns: F_i, M_i, u_i and ρ.

When a layer consists of two materials – e.g. Cu and dielectric (oxide of low-k) – and these materials have a different residual stress after processing, equivalent properties are derived. Suppose that a layer consists of α of material 1 (E_1, residual stress σ_1) and 1-α of material 2 (E_1, residual stress σ_1), the equivalent E-modulus E_{eq} and equivalent residual stress σ_{eq} are calculated as follows:

$$\sigma_{eq} = \alpha \frac{E_{eq}}{E_1'}\sigma_1 + (1-\alpha)\frac{E_{eq}}{E_2'}\sigma_2$$

$$E_1' = \frac{E_1 + E_{subs}}{2}$$

Where $E_2' = \frac{E_2 + E_{subs}}{2}$

$$E_{eq} = \left[\frac{\alpha}{E_1'} + \frac{1-\alpha}{E_2'}\right]^{-1}$$

With E_{subs} the elastic modulus of the substrate, i.e. the silicon wafer. These equations can be used to calculate equivalent properties for patterned layers, but also for wafers with Cu TSV's with specified depth.

Model calibration for interposer BEOL technology

The materials used for the interposer BEOL and corresponding thermo-mechanical properties are summarized in table 1. The calibration of the model consists in evaluating the residual stress of all the layers used to build the interposer. Residual stress in thin films can originate from the difference in thermal expansion between the layer and the substrate.

Material	Purpose	Thermal budget (°C)	Young's modulus (MPa)	CTE (ppm/°C)
Si	Substrate	NA	169000	2.6
SiO/SiC	PMD stack	430	70000	0.5
SiN	IMD/Passivation/ MIM insulator	370	210000	0.5
SiO	IMD/PMD/ Passivation	370	70000	0.5
SiCN/ SiCO	Passivation	370	100000	0.5
Al	Metal passivation	350	70000	21
TaN	MIM electrodes	370	350000	6.3
Cu	BEOL metal	180	120000	17

Table 1 : List and properties of materials used in interposer BEOL. The dielectric materials were deposited by Plasma-Enhanced Chemical Vapor Deposition (PECVD), Al and TaN by Physical Vapor Deposition (PVD) and Cu was electro-plated. Poisson's ratio of 0.3 was considered for all materials.

The origin can also be intrinsic depending on the deposition method and conditions. For that reason, we decided to evaluate the residual stress of each material at room temperature experimentally.

We first have considered in first approximation that for all the materials except Cu, the residual stress of the layer was stable within current interposer BEOL thermal budget (i.e. the residual stress of a layer was not modified when processing the next layer), meaning that there is no inelastic deformation (plasticity or creep) is occurring during the thermal processing. The methodology to evaluate the residual stress of all materials except Cu is first described: after measuring the bowing of the bare silicon 300 mm wafer, each layer has been deposited (one layer by wafer) following the same method and conditions than used in interposer BEOL. Then the wafer bowing was measured after deposition. Each material residual stress was derived at 20°C using the model, when the simulation was matching the measured post-pre bowing value (see table 2).

The residual stress values of the Cu was not considered as constant. Indeed, interposer BEOL is processed using damascene technology. The dielectric structures are filled with Cu using electroplating process. The plating step is followed by an anneal at 180°C. Finally, a Chemical Mechanical Polishing (CMP) step is used to remove the Cu in excess at the surface. Once the next damascene layer is processed, the Cu will see higher temperature (typically 370°C for Inter Metallic Dielectric (IMD) or passivation stack deposition), which will strongly impact its residual stress. Then, the methodology to evaluate its residual stress was different from previous part. After measuring the bowing of bare silicon 300mm wafers, a single damascene layer using interposer technology was built. The residual stress of all materials used to build the layer was known at that stage (except the Cu, that we wanted to evaluate). The wafer bowing has been measured again after the damascene layer CMP step.

978-1-4799-2408-0/14 $31.00 © 2014 IEEE

Deposition method & material	Residual stress @20°C (Mpa)
Si (bulk)	0
SiO/SiC	-25
SiN	-170
SiO	-170
SiCN/SiCO	-200
Al	150
TaN	-1200

Table 2 : Resulting material residual stress values @20°C.

Then, the layers used for the next IMD or passivation were deposited onto the single damascene layer. The bowing was measured once more afterward. The Cu stress values at 20°C, being the only unknown parameter, were then derived by simulating the different structures using the model. Each Cu residual stress value has been evaluated at the different stages of the process matching simulated and measured bowing values. The results are shown in table 3.

In order to validate the calibrated model for interposer BEOL, the bowing generated by interposer front side (without TSV) at full thickness wafer level (300 mm Ø and 775 μm thickness) have been simulated. A schematic cross-section of the interposer structure is shown on Figure 2. More information about the interposer structure and its process flow are available elsewhere [2].

Deposition method & material	Residual stress @20°C (MPa)
Cu after CMP	280
Cu after pass. deposition	580
Cu after IMD deposition	590

Table 3 : Cu residual stress values @20°C deduced by processing/measuring/simulating a Cu damascene layer.

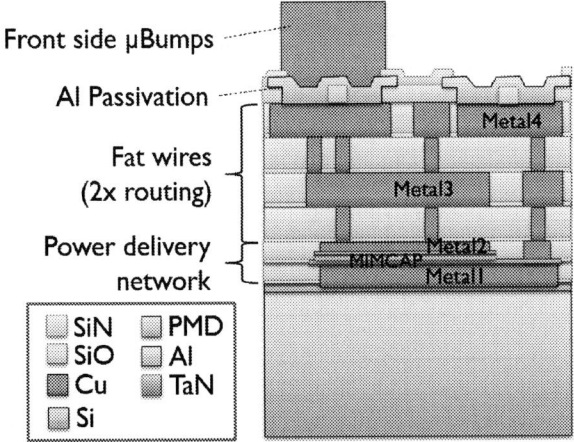

Figure 2. Schematics silicon interposer BEOL structure overview.

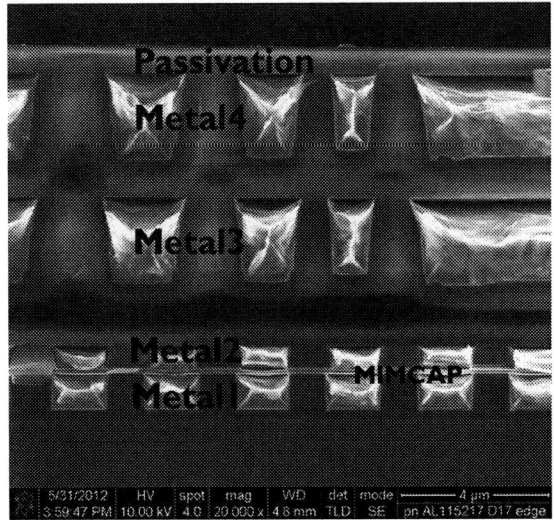

Figure 3. Cross-section of the interposer BEOL.

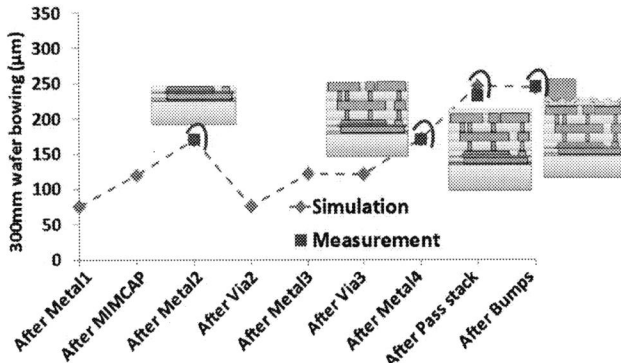

Figure 4. Comparison between simulated and measured 300mm wafer bowing at different stages of interposer BEOL processing.

Simulated and measured bowing are compared at different stages of the process. Layer thicknesses measured on a SEM cross-section (figure 3) were used as input for the simulation, for a better precision. A good agreement between simulation and measurement is observed (see figure 4), validating the calibrated model for interposer BEOL on full thickness wafer.

Special case of TSV and thin dies

The 10x100 μm TSV has been simulated at full wafer thickness level using a model with four layers (see figure 5). The first layer is used is uniform for the silicon substrate (675μm thickness). The layers 2 to 4 are mixed layers between Cu and silicon (100 μm), Cu and PMD stack (380 nm) and Cu and liner oxide (600 nm). As mentioned elsewhere [2], the liner is not removed during TSV CMP because the interposer is fully passive (no need to open any contacts).

In order to calibrate the TSV model, the residual stress of the plated Cu was required. The residual stress was deduced following the same approach than for the Cu used in BEOL layers, except that a 10x100μm TSV was processed instead of a single damascene layer. Wafers with 10x100 μm TSV were processed following the same process flow than presented in reference [2]. The Cu is annealed at 420°C during 20 min after plating, before CMP.

After the deposition of a thin capping layer at room temperature on top of the TSV, the full thickness wafers with 10x100 µm TSV were diced in 25 mm² coupon samples. The capping layer is used to avoid a free deformation in the vertical direction of the TSV Cu, to be closer from actual stack in which Metal1 is present on top of the TSV. In order to understand the evolution of TSV Cu residual stress during further BEOL processing, the bowing has been evaluated as a function of the temperature and can be seen in figure 6 (up to 370°C). The bowing at 20°C is increasing after further processing at 370°C despite TSV was annealed 20min at 420°C after plating and becomes stable after the first thermal cycle. Consequently, Cu residual stress value at 20°C was extracted after the first thermal cycle.

The TSV bowing model and the impact of BEOL integration were validated by comparing simulated and measured wafers, in which 10x100µm TSV were processed with Metal1/MIMCAP/Metal2 integrated on top of it (see figure 7). Wafers without (6.A) and with TSV (6.B) were processed and compared for reference. The results are summarized in table 4. The bowing values measured with and without TSV are very similar to the values predicted by the simulation. This result is validating the bowing prediction model for 10x100 µm TSV.

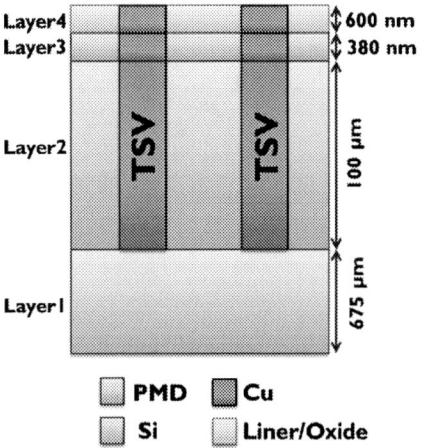

Figure 5. Schematic overview of the model and layers build-up used to simulate the bowing induced by 10x100µm TSV.

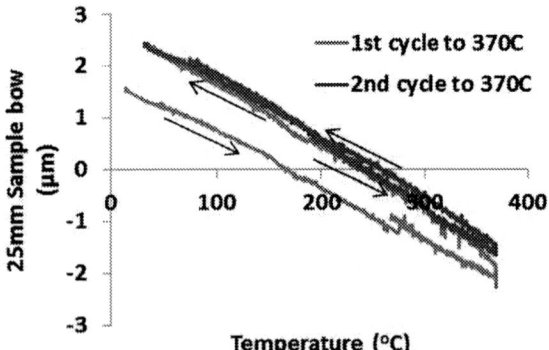

Figure 6. Bowing measurement on a 25mm coupon sample with 10x100µm TSV (full thickness silicon substrate) as a function of temperature. TSV was capped with a thin layer.

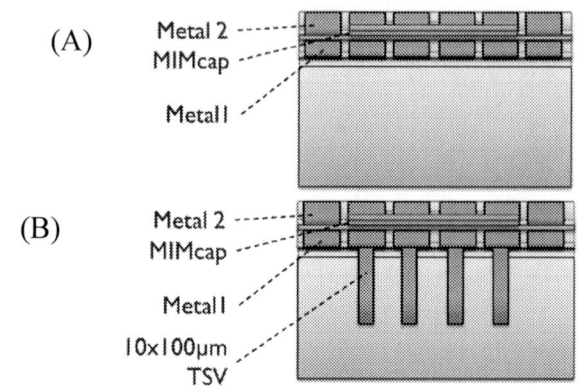

Figure 7. Schematics of the test structure without (A) and with TSV (B) with interposer BEOL technology (2 Metal layer, MIM capacitor and 10x100µm TSV).

Structure	Bowing from simulation (µm)	Bowing from measurement (µm)
Structure A (no TSV)	130	120
Structure B (with TSV)	263	260

Table 4 : Comparison between bowing obtained by simulation and by measurement on a test structure with and without TSV.

The last step for the full validation of the bowing model for interposer technology was to benchmark simulated and measured thin dies. To do so, the previous wafers with TSV/Metal1/MIMCAP/Metal2 were thinned using temporary bonding approach [3], summarized in figure 8. After bonding, the wafer is grinded down to a target of 107 µm. Then the silicon is recessed using wet etch down to 105 µm. SiN layer was deposited at low temperature (180°C) to mimic back side passivation. Then the temporary carrier is debonded after wafer taping, following by a dicing in pieces of 30x30 mm².

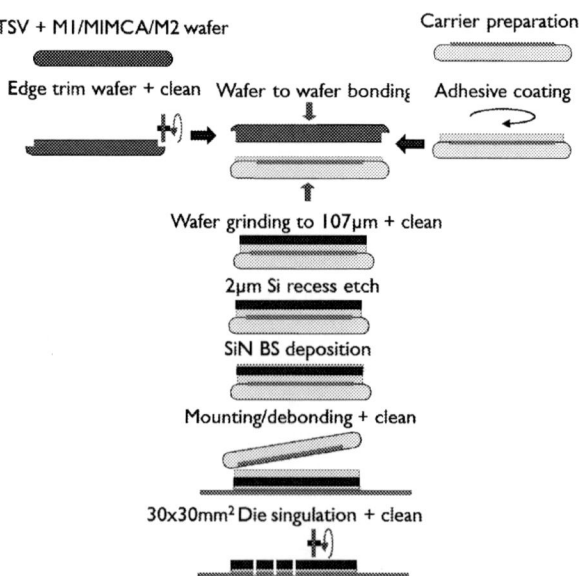

Figure 8. Thinning – debonding – dicing process flow used to prepare 30x30 mm² 105 µm thin dies with TSV/Metal1/MIMCAP/Metal2.

978-1-4799-2408-0/14 $31.00 © 2014 IEEE

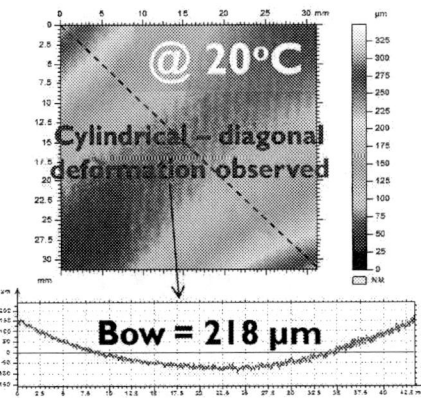

Figure 9. Bow mapping measured on 30x30 mm 105µm thin dies with TSV/Metal1/MIMCAP/Metal2. Maximum bowing of 218 µm measured in the diagonal.

The bowing of the thin die have been measured. The mapping is available on figure 9. A cylindrical – diagonal deformation of the die was observed, with a maximum bowing value of 218 µm measured in the diagonal.

In a case of thin die, we assumed for the simulation that the part of the TSV in the silicon was not contributing to bowing anymore, because the substrate Cu/Si composite structure is now symmetric over the thickness. Because of the low density of the TSV (0.8%), we assumed that the substrate thermo-mechanical properties were similar to pure silicon. The value obtained by simulation was 210 µm bow for a 30x30 mm^2 Metal1/MIMCAP/Metal2 die of 105µm thickness, including SiN on the back side. A very good correlation between the value predicted by the model and the value obtained by measurement is observed.

Full interposer wafer and die bowing simulation

The next step was to simulate the bowing induced by the full imec interposer structure which is presented figure 10. The interposer includes the front side BEOL (similar to figure 2), 10x100µm TSV (via middle) and 3 µm Cu redistribution layer (RDL) plus 50 µm Ø/thickness Cu pillar on the back side.

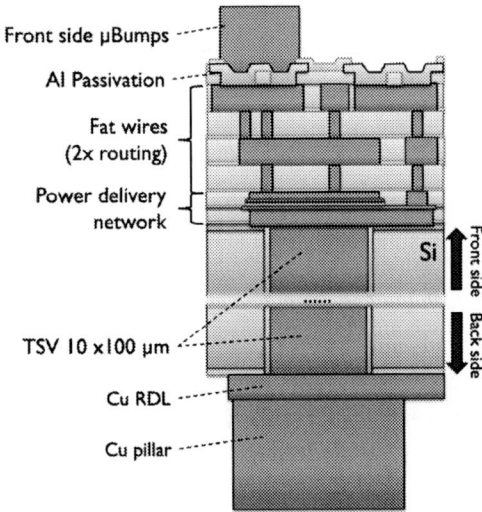

Figure 10. Schematics full silicon interposer structure overview (including TSV and back side).

The evolution of 300mm wafer bowing as a function of interposer front side processing with TSV is presented on figure 11. Bowing values as large as 400 µm would be obtained at the end of the front side processing. Such a value can be problematic in term of wafer handling in 300mm equipment, impacting interposer production. The main contributors are the TSV and Metal1/Metal2. Metal1/Metal2 are thick damascene planes (around 1µm) with high Cu density (75%) which are used as power and ground planes [2]. A tensile layer deposited at the back side of the wafer can be used to compensate the bowing in some extend and enable front side processing. This layer has to be removed before grinding, and consequently cannot be used for back side processing.

Before simulating the bowing induced by the full interposer stack at die level, the bowing induced by the back side modules (back side passivation, RDL and Cu pillars) were evaluated first on full thickness silicon wafer. The results are shown on figure 12. It can be seen that the compressive stress induced by the silicon nitride used as back side passivation and the tensile stress induced by the Cu RDL and pillar almost compensate. The bowing contribution of the back side module is then negligible compare with the bowing induced by the front side modules. Consequently, the back side modules were not directly taken into account to perform the simulation at thin die level, for simplification.

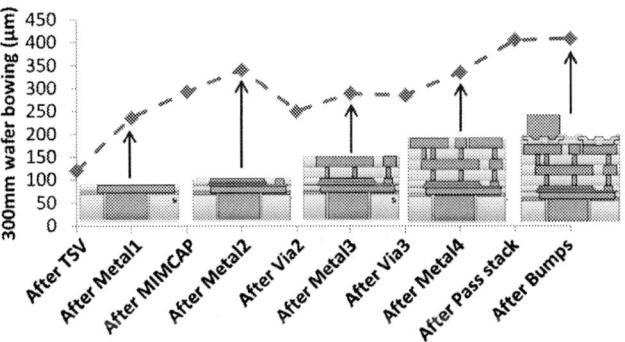

Figure 11. 300mm full thickness wafer bowing simulation at different stage of interposer front side processing (including TSV).

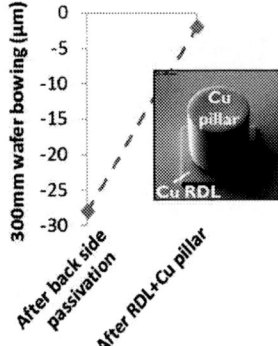

Figure 12. 300mm full thickness wafer bowing simulation at different stage of interposer back side processing.

978-1-4799-2408-0/14 $31.00 © 2014 IEEE   37

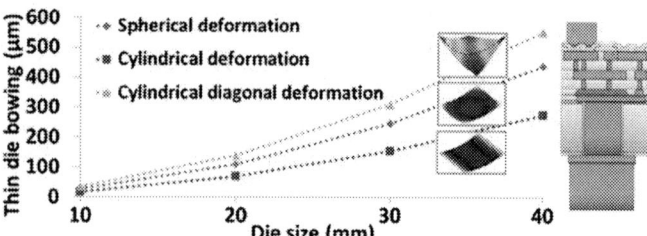

Figure 13. Bowing evaluation by simulation of interposer thin die (100μm silicon) as a function of die size for three different deformation modes.

The bowing evolution of thin (100μm silicon) dies based on simulation as a function of die size and deformation mode is shown on figure 13. The deformation mode of the die (cylindrical, spherical...) is link to the level of minimum energy. So far, we have no analytical solution derived which determines the preferred deformation. The bowing increases with the interposer die dimension.

The full interposer has been processed (front side and back side, see figure 14). The bowing was measured on the thin dies after wafer debonding and dicing. The interposer die is rectangular shape, with a length of 20mm in the X direction and a length of 10mm in the Y direction. The measurement results are shown in figure 15. It was deduced from the measurement that the deformation mode of the die was spherical. In the X and Y directions, a bowing of 130 μm and 30 μm were measured respectively. By comparing with model prediction (table 5), it can be seen that these values are slightly larger but very similar to the bowing values expected based on the model.

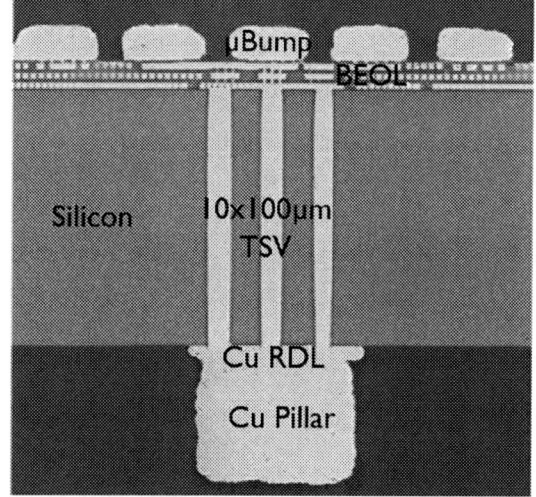

Figure 14. Cross-section of the interposer die after front side and back side processing.

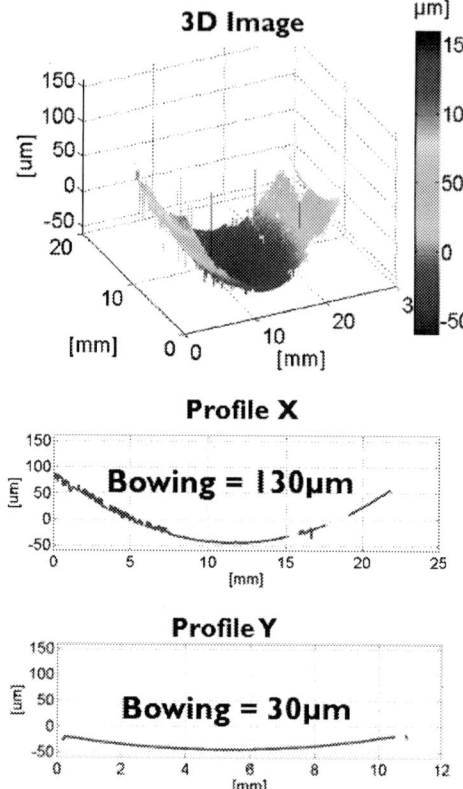

Figure 15. Evaluation of interposer thin die bowing (3D and X-Y profile). X profile correspond to 20mm length, Y profile to 10mm length.

Die dimension	Bowing value by Simulation (μm)	Bowing value by measurement (μm)
10 mm length	27	30
20mm length	109	130

Table 5 : Comparison of bowing values obtained by simulation and measurement of interposer thin die as a function of die size.

Bowing mitigation strategy definition

The equipment used to process interposer front side and back side (300mm imec pilot line) do have a limited capability to handle highly bowed wafers. As mentioned in previous part, stress layer can be deposited on wafer back side to compensate the bowing induced by interposer BEOL on wafer front side. However, the layer on the back side can be easily scratched during wafer handling, generating vacuum leakage when chucking in equipment using such a chuck. It can also generate particles. The layer on the back side can also crack if the stress becomes too important. Finally, this is not a solution usable for back side processing. On top of this, such high values of bowing can be problematic at die level for the top dies stacking onto the interposer in a die to die approach and/or for interposer packaging (especially using reflow process). To solve this problem, the model has been used to investigate structural modification of the interposer. The goal was to provide bowing mitigation strategy efficient at wafer and die level.

Three modifications were investigated:

- Replace standard Pre-Metal-Dielectric (PMD) layer by a thicker and more compressive insulator,
- The use of thinner Metal1 and Metal2,
- The use of a more compressive oxide in the BEOL.

Simulations were carried out replacing the standard PMD layer (300 nm/80 nm SiO/SiC layer) by a thick PECVD oxide (-170 MPa compressive stress according to table 2). In practice, this mainly consist in modifying the buildup in the part of the model dedicated to simulate the TSV, as shown in figure 16. The impact on bowing during front side processing on full thickness wafer can be observed on figure 17. A bow reduction of around 150 µm is observed (-37% bowing). At die level, a bowing value of around 45 µm (-59% bowing) is predicted by the model for a 20x20mm² assuming a spherical deformation mode.

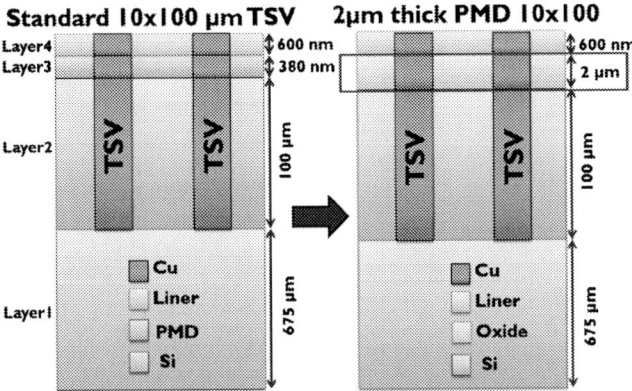

Figure 16. Schematic overview of the model and layers build-up used to simulate the bowing induced by 10x100µm TSV and its modification to compensate some tensile stress.

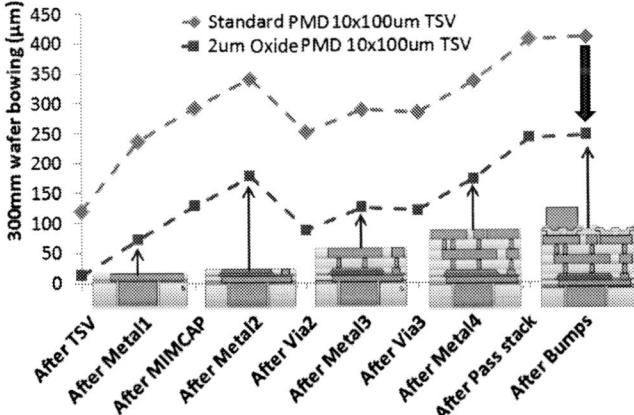

Figure 17. 300mm full thickness wafer bowing simulation at different stage of interposer front side processing using a standard PMD (blue curve) or replacing it by a 2µm compressive oxide (red curve).

The use of thinner Metal1 and Metal2 will increase the sheet resistance of the two layers and consequently may impact the electrical performances of the interposer. First, Metal2 is used as ground layer in a µstrip configuration for signal transmission using Metal3 or Metal4 as signal line. It has been shown by simulation that the impact on signal integrity was negligible for a ground layer thickness reduced from 1 to 0.1µm. An increase in Metal1 and Metal2 sheet resistance would increase the serial resistance of the MIM capacitor integrated in between the two planes. High serial resistance can impact the performances of the capacitor at high frequency. Figure 18 is reporting (red line) first order simulation of the capacitance at 1GHz as a function of the thickness of Metal1 and Metal2. It can be seen that the impact is negligible except for thickness below 0.2-0.3µm. On the same figure, the impact on bowing (at wafer level after full thickness processing) evaluated by simulation is also reported. We can see that reducing the thickness of the two planes can effectively reduce the bowing and that a thickness of around 0.4µm could be a good trade-off between bowing and performances.

The last modifications investigated was to replace the oxide material used in the BEOL by a new version with higher compressive stress. The effect has been simulated and is reported on figure 19, at wafer and thin die level (previous process of reference is and oxide of -170MPa of residual stress). It can be seen that a small modification of the stress of the oxide can be very efficient to decrease the bowing at wafer but also thin die level. It can also be seen that for an oxide of around -245 MPa of residual stress, a flat die could be expected at 20°C according to simulation (whatever the size). The wafer is not flat at this value because there is still the impact of the TSV (substrate structure not uniform before thinning).

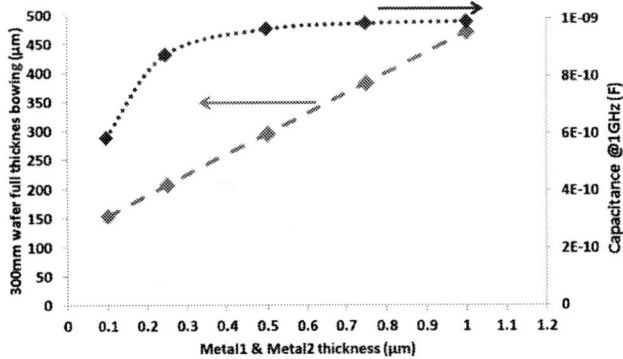

Figure 18. Impact of Metal1 and Metal2 thickness onto full thickness wafer bowing (after front side processing) (blue curve) and onto the capacitance at 1GHz of the integrated MIM capacitor (Based on a first order simulation of a 1nF decoupling capacitor).

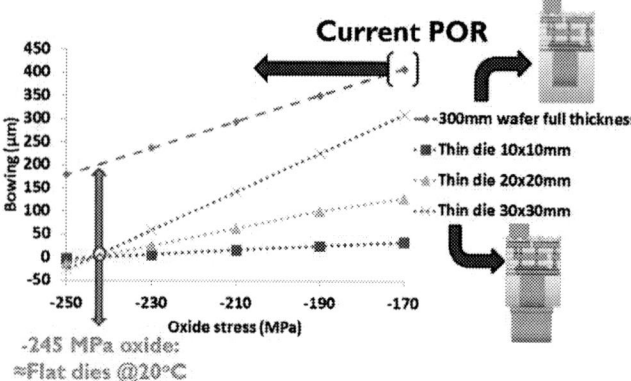

Figure 19. Simulation of the impact of higher compressive residual stress of the oxide used to build the interposer BEOL at wafer and die level.

Conclusions

A simple analytical model to simulate the bowing at wafer or (thin) die level has been described. This model permits to calculate the bowing based on the residual stress of each layer, which can be uniform (blanket), but also a mixed layer consisting of two different materials with different residual stress (patterned).

The model was then applied to imec interposer technology and the methodology used to calibrate the model has been described: the residual stress of each material used within the interposer has been evaluated experimentally. The good correlation between the bowing values from simulation and from measurement have been verified step by step for the BEOL, the TSV and simplified thin die.

The model was then apply to the full interposer structure in order (combining TSV, BEOL and thin die) to predict the evolution of the bowing during processing at wafer level and also after completion, at thin die level. A good correlation between the value expected and the one measured was observed at thin interposer level.

The proposed interposer structure generates large bowing values at wafer ($\approx$400μm bow after front side processing) or die level ($\approx$130μm bow for a 20x10mm^2 die). This can be problematic for the production of such interposer (wafer handling issue, impact on stacking and/or packaging yield...). The model has been used to identify and quantify parameters which can be modified within the interposer structure to efficiently reduce the bowing. The use of a more compressive and thicker PMD insulator layer, a reduction in Metal1/Metal2 thickness, the use of more compressive oxide within the BEOL, are promising and easy to implement solutions to reduce interposer bowing with a limited impact onto its performances.

Acknowledgments

This work was done in the frame of IMEC 3DIIAP program and has been strongly supported by IMEC partners, IMEC FPS units, 3D technology, Reliability and Modelling teams. Special thanks to the co-authors for their fundamental contribution.

References

1. B. Banijamali, "Advanced reliability study of TSV interposers and interconnects for the 28nm technology FPGA," in Proc. IEEE Electronic Components and Technology Conference *(ECTC)*, Lake Buena Vista, FL, May 31 - 3June , 2011, pp. 285–290.

2. M. Detalle, "Interposer technology for high band width interconnect applications," in Proc. IEEE Electronic Components and Technology Conference *(ECTC)*, Las Vegas, NV, May 28-31, 2013, pp. 323–328.

3. A. Jourdain, "Integration of the ZoneBOND™ temporary bonding material in backside processing for 3D applications," in Proc. IEEE Electronic System-Integration Technology Conference *(ESTC)*, Amsterdam, Netherlands, September 17–20, 2012, pp. 1–4.

High Performance IPDs (Integrated Passive Devices) and TGV (Through Glass Via) Interposer Technology using the Photosensitive Glass

Jong-Min Yook, Dongsu Kim, and Jun Chul Kim
Packaging Research Center, Korea Electronics Technology Institute,
68 Yatap-dong, Bundang-gu, Seongnam-si, Gyeonggi-do, 463-816, Korea
radio@keti.re.kr

Abstract

In this paper, a new TGV interposer technology is introduced in which the through via could be made by using a photolithography and chemical wet-etching. To make fine and accuracy via-holes, etching properties of the glass are studied in various UV-exposure times. It is possible to make TGVs from 60 μm to 20 μm with a 4:1 aspect ratio. Based on the TGV process, a high-aspect-ratio metal is made and high-Q spiral inductors can be realized by using the technology. The fabricated inductor has a very thick signal height more than 80 μm and its Q factor is more than 30 at 2 GHz. To demonstrate the process technology, a 0.9 GHz LPF, which is can be used for the RF front-end of the GSM band, is designed and fabricated using the inductor. The realized LPF has very low insertion loss of only 0.31 dB at the pass band, and this value is an improvement of more than 32% compared with the normal LPF.

Introduction

As increment of I/O numbers of ICs, silicon interposer technology becomes a key element for IC packages [1], [2]. However, conventional silicon interposers have some problems. The TSV process is rather complicated generally. It needs vacuum processes to make deep via-hole, and an insulation layer and diffusion barrier should be coated to minimize the signal loss and prevent Cu contamination of the body silicon. Nevertheless, this interposer technology is not sufficient for high frequency or high performance applications. To solve these problems, some papers introduced TGV interposer technologies [3]-[5]. Glass has a low dielectric constant and high signal isolation compared to the conventional silicon wafer, therefore it has some advantages to integrate RF devices. However, it is not easy to make fine via-hole in the glass wafer. Only some institutes or companies could make the fine via using the excimer UV laser which is very expansive.

Some researches show the photosensitive glass has some merits in micro-fabrications due to possibility of chemical wet-etching [6], [7]. There are three process procedures to make the glass via-hole by using the photosensitive glass. Fist, the glass should be exposed to UV lights (250-350nm range). Second, it needs an annealing process to re-crystallize the glass at approximately 580 °C and then, the glass wafer dipped in a diluted HF solution. According to reference [6], the selectivity between the bulk and crystalized glass is more than 20.

In this paper, the RF characteristic and wet-etching properties of the photosensitive glass are studied, and then a TGV process and advanced IPD process are developed using the glass.

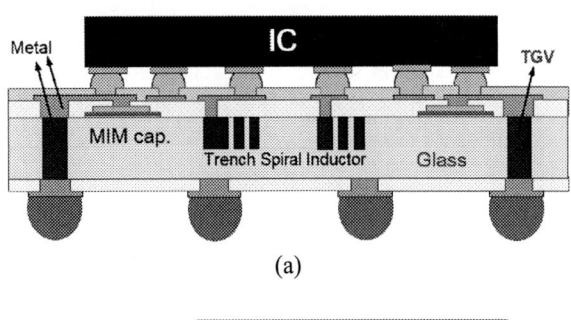

(a)

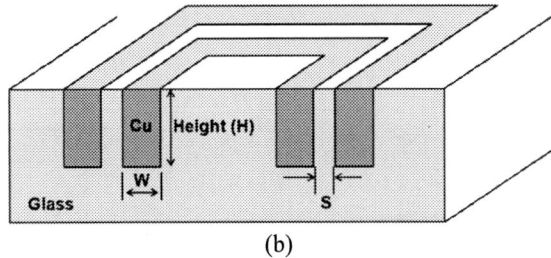

(b)

Fig. 1 (a) IC package structure using a glass interposer. (b) The proposed structure of the embedded spiral inductor.

Electrical Properties of the Photosensitive Glass

Figure 1(a) shows the design structure of the IC package using the proposed TGV glass interposer. There are fine through vias for interconnection between the front and back-side of the substrate, and thin film devices such as a NiCr resistor, MIM capacitor, and spiral inductor can be integrated in the interposer surface. As shown in Fig. 1(b), the proposed trench inductor can be embedded in the interposer for high power or high performance systems.

Previously, trench inductors using silicon substrate had been studied and their results shows that thick metal of the inductor could improve RF performances greatly [8]. However, all results were based on not a conventional lossy silicon but a HRS substrate which is very expansive. In the trench inductor structure, the resistivity of the substrate is very important because the spacing is narrow and the embedded signal area is very large. Hence, the electrical properties of the photosensitive glass (TS technology Inc., South Korea) was extracted and verified using a ring-resonator [9]. Generally, a photosensitive glass contains ZnO, Ce, and Ag atoms [6].

Figure 2 shows the fabricated ring-resonator. The thickness of the photosensitive glass is 1 mm and s-parameters were measured by on-wafer probing. The extract dielectric constant and loss tangent was 5.53 and 0.0036 at 2.1 GHz, respectively.

978-1-4799-2408-0/14 $31.00 © 2014 IEEE — 41 — 2014 Electronic Components & Technology Conference

Dielectric Constant (ε_r)	5.53 @ 2.1 GHz
Loss Tangent	0.0036 @ 2.1 GHz

Fig. 2 Extracted dielectric properties of the photosensitive glass using a ring-resonator method.

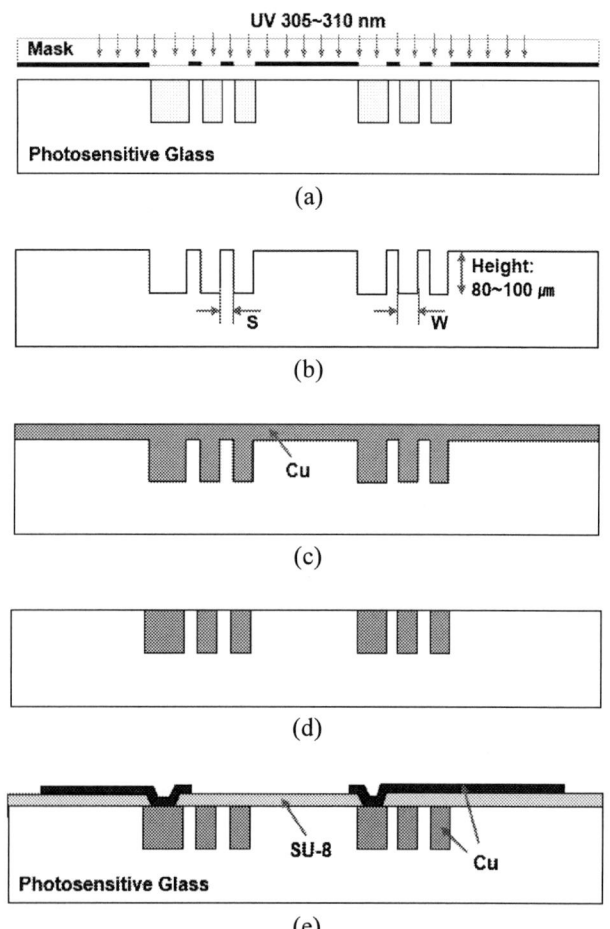

Fig. 3 Fabrication process of the TGV and trench spiral inductor. (a) UV lithography. (b) Glass wet-etching. (c) Via filling with copper using electroplating. (d) Planarization process. (e) Signal isolation and bridge line process.

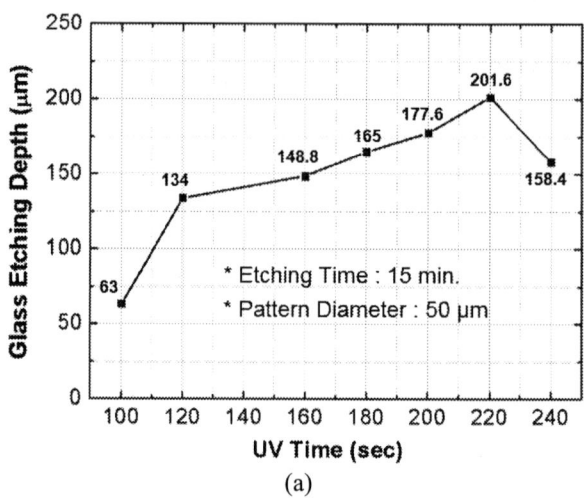

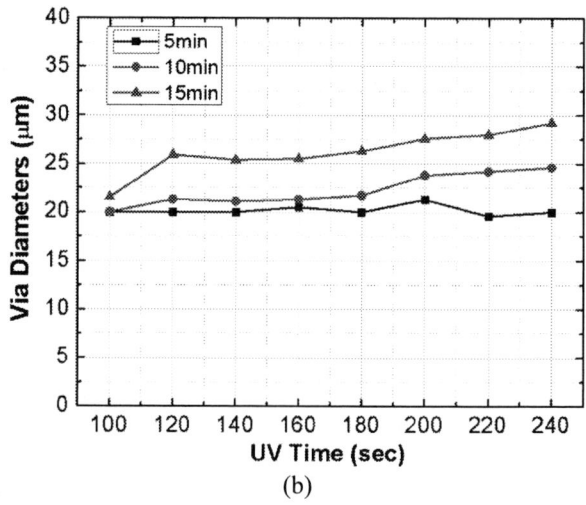

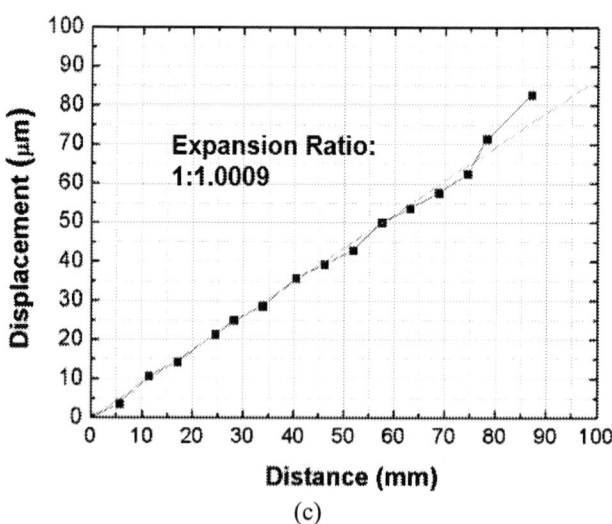

Fig. 4 Analysis of process conditions to make fine TGVs. (a) Etch depths for various UV-exposure times. (b) Via diameters for various UV-exposure times. (c) Pattern displacement after via formation.

978-1-4799-2408-0/14 $31.00 © 2014 IEEE

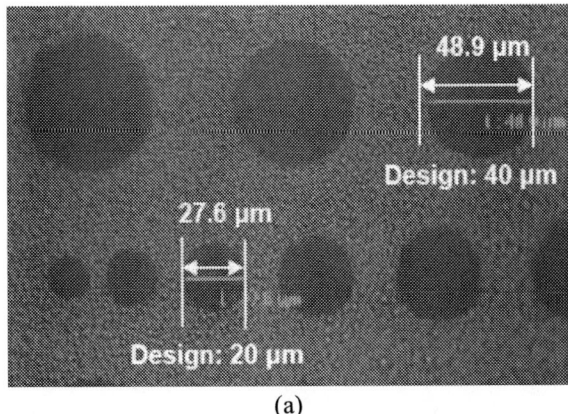

(a)

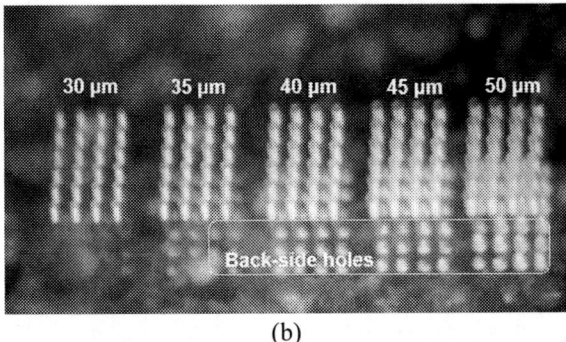

(b)

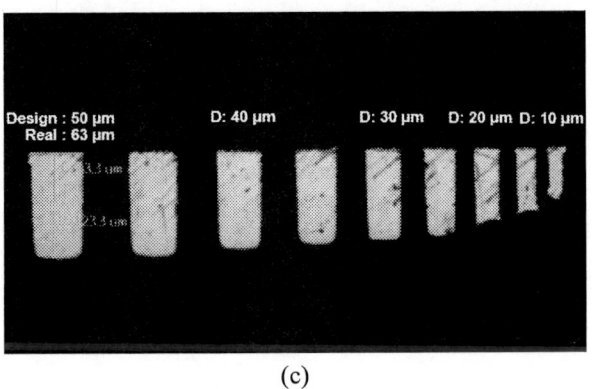

(c)

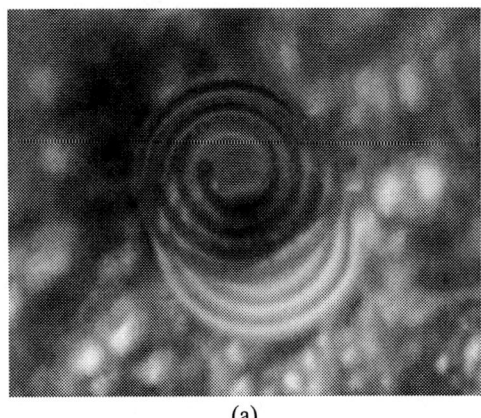

(a)

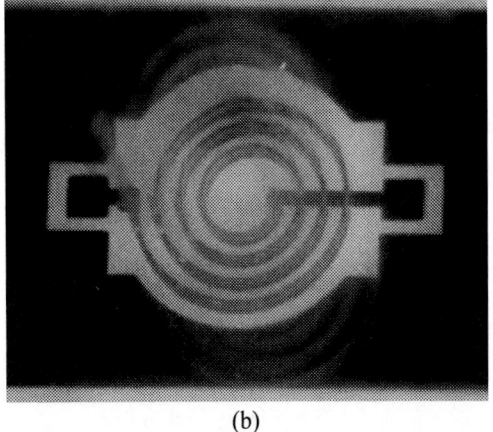

(b)

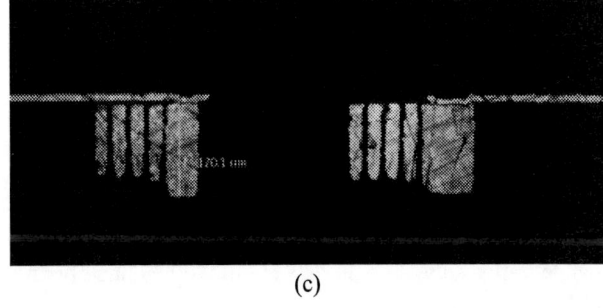

(c)

Fig. 5 Photographs of the fabricated TGV. (a) Glass surface after wet-etching. (b) Via-hole array. (d) Cross section of the TGV after the Cu filling and CMP process.

Fig. 6 Photographs of the fabricated trench-spiral inductor. (a) Front of the trench spiral after CMP process (W/S=10/20 μm). (b) Back of the fabricated inductor (W/S=10/50 μm). (c) Cross section of the fabricated inductor (W/S=10/10 μm).

TGV and Trench Spiral Inductor

As shown in Fig. 2, extracted electrical properties of the photosensitive glass are very good in RF applications. Therefore, it is possible to make through via or trench inductors without additional insulation layers.

Fig. 3 shows the fabrication process of the TGV and trench spiral inductor. A quartz mask was used to lithograph the via-hole and trench-spiral. In the exposed area, there is a release of electrons from the Ce^{2+} donor ion. These electrons then recombine with the Ag^+ ions to form Ag atoms in a annealing process at 580 °C (1 hr.) where the exposed parts are crystallized causing the atoms of silver to agglomerate into small fragments [6].

A 5% HF solution was used as an etchant. The etching process was done in an ultrasonic-bath at the atmosphere temperature (25 °C). Fig. 4(a) shows the etching rate of the photosensitive glass for various UV-exposure times. This graph shows that the etching rate could be improved by increasing the UV exposure time and there is a maximum point of the etching rate. Fig. 4(b) shows the via-hole diameter after the etching process for various UV-exposure times. It is very important to make the fine TGV or trench-spiral. As shown in this graph, a long exposure time caused an increment of the diameter in the same mask pattern.

978-1-4799-2408-0/14 $31.00 © 2014 IEEE

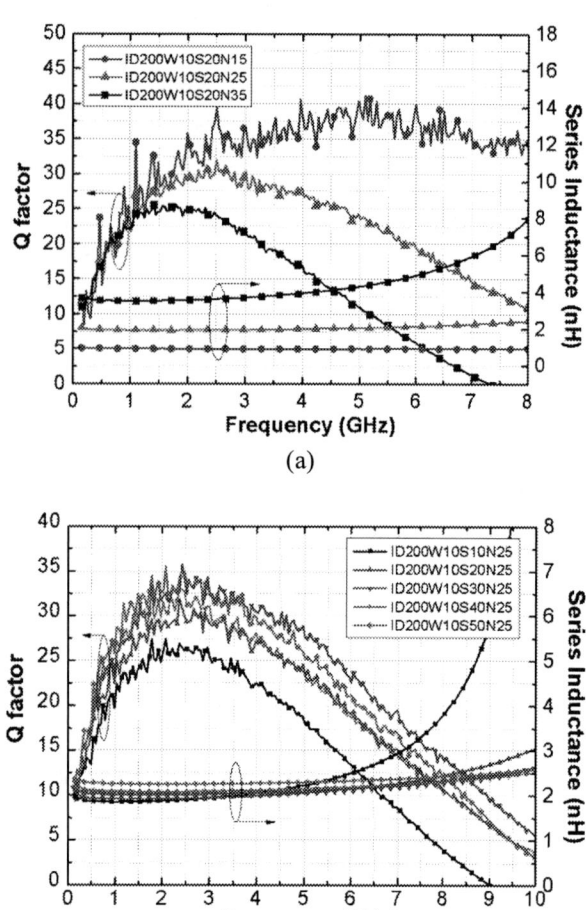

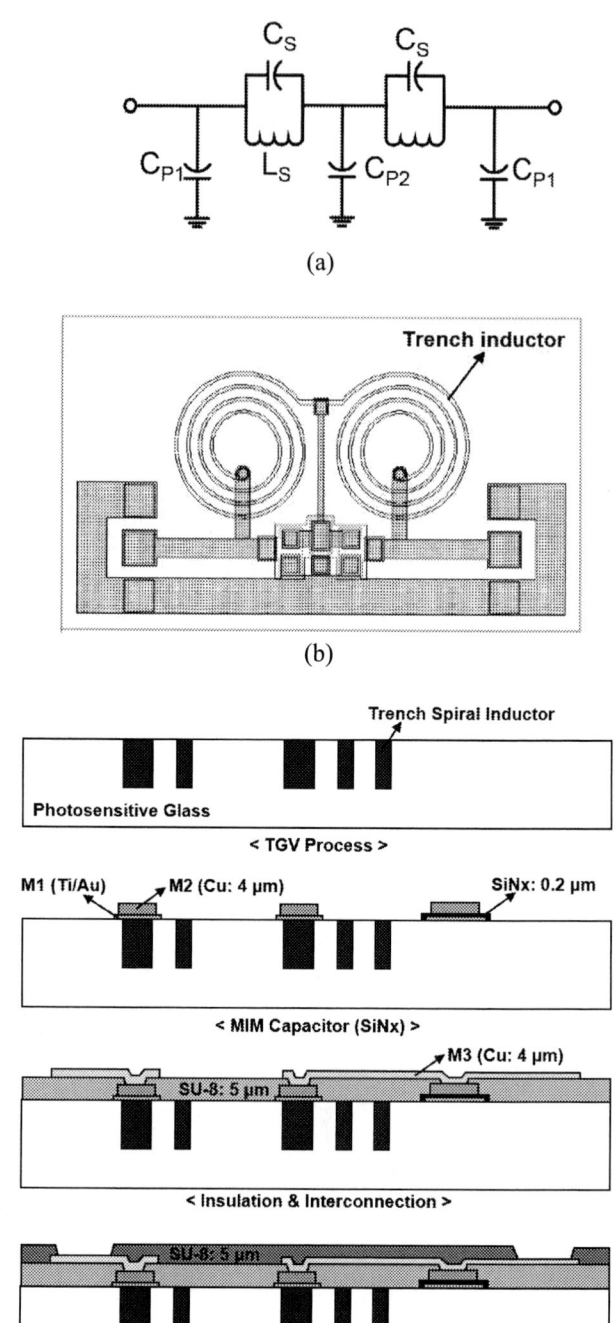

Fig. 7 Measurement results of the fabricated inductor. (a) Q factors and series inductances for various numbers of turns. (b) Q factors for various signal spacings.

Fig. 4(c) shows the lactation error of the via-holes. There was some location error after annealing process but it was predictable in a same mask pattern. The glass wafer was expanded after annealing process about 0.09 % linearly and it could be changed by a portion of the exposure area.

Fig. 5 shows the photograph of the glass via-hole after etching and Cu filling process. It was possible to make a 20 μm TGV with a 4:1 aspect ratio and the depth of the TGV was different for the diameter of the via-hole. Using the TGV process, a trench-spiral inductor can be made and its structure is shown in Fig. 1(b) and 3(e).

Fig. 6 shows the fabricated trench inductor. There are some image patterns in the back-side of the glass. It was possible to make a very fine spiral-pattern, a 20 μm pitch (W+S), with a high aspect ratio as shown in Fig. 6(c).

Fig. 7(a) shows the Q factor and series inductance of the fabricated trench-inductor for various numbers of turns. The measured Q-factor of the inductor is more than 30 at 2 GHz owing to the thick metal structure. Fig. 7(b) shows the Q factor for the increment of the signal spacing. This graph shows that the Q of the trench inductor is strongly dependent on the signal spacing.

Fig. 8 (a) Circuit design of the 0.9 GHz LPF. (b) The layout design of the proposed LPF (W/S=10/30 μm). (c) Fabrication processes.

Although the series inductance remained the almost same, the Q factor was greatly improved (more than 32%) when the spacing was changed from 10 to 40 μm. The reason of the Q improvement is the decrease in parasitic capacitance [10] and the proximity effect [11] between signal lines. As previously reported, the signal width does not affect the Q factor in this structure [8].

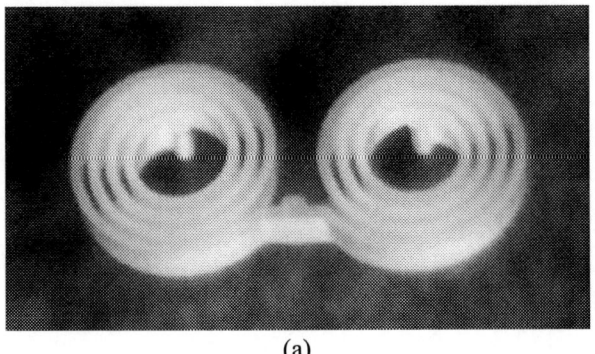

(a)

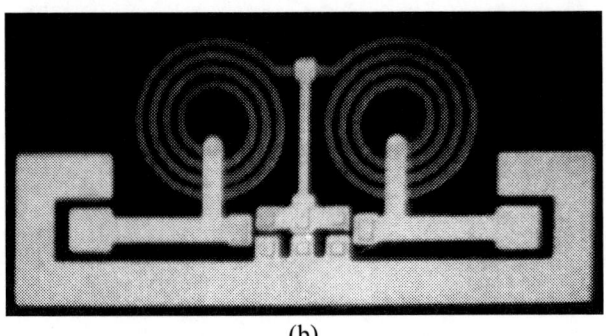

(b)

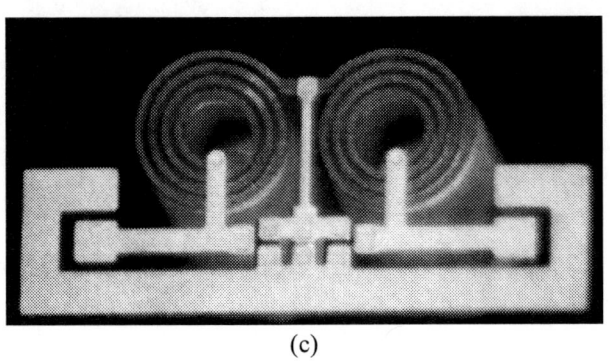

(c)

Fig. 9 Photographs of the fabricated trench-spiral inductor. (a) The trench-spiral shape after glass etching. (b) Normal LPF (W/S=30/10 μm). (c) The proposed trench-spiral LPF (W/S=10/30 μm).

Low-loss IPD LPF using the Trench Inductor

A low-loss LPF was designed and fabricated to demonstrate the trench-inductor process. Figure 8(a) and 6(b) show the circuit diagram [12] and layout design of the proposed IPD LPF. The LPF was designed for an RF front-end module of 0.9 GHz global system for mobile communication (GSM) band applications. To minimize the insertion loss, two inductors of the LPF were designed with the trench structure. The signal width and spacing of the designed inductor were 10 and 30 μm, respectively. To verify the effect of the trench inductor, a normal LPF was also designed with a standard IPD process where the LPF had almost the same area as the proposed LPF. The signal width and spacing of the normal LPF were 30 and 10 μm, respectively.

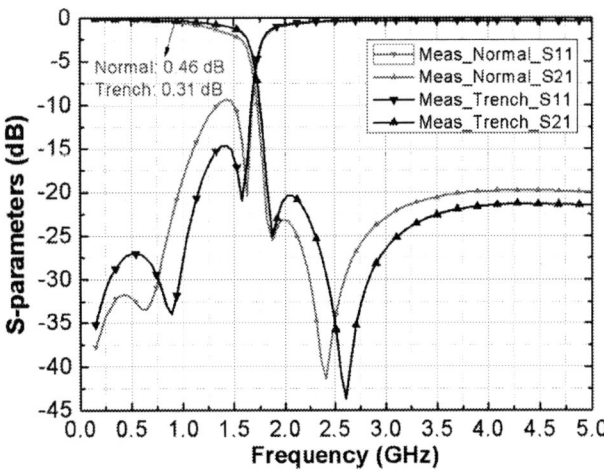

Fig. 10 Measured s-parameters of the fabricated LPF.

Fig. 8(c) shows the fabrication process of the LPF. Before standard IPD processes, trench inductors were realized on the glass (via-first process). A thin SiNx layer, which was deposited by PECVD, was used as the insulator of the MIM capacitor. M2 metal was used to make the top layer of the MIM capacitor as well as the spiral inductors of the normal LPF. The SU-8 epoxy photoresist was used to isolate signals from the inductor bridge line and also as a passivation layer.

Fig. 9 show the photographs of fabricated LPFs. As shown in Figs. 9(b) and 9(c), the two fabricated LPFs are almost the same, except for the inductor structure. The trench structure of the inductor is shown in Fig. 9(a). Fig. 10 shows the measurement results of normal and proposed LPFs. As shown in this graph, the proposed LPF has an improved RF performance with respect to the insertion loss. The proposed LPF has very low insertion loss of only 0.31 dB at the pass band, and this value is an improvement of more than 32% compared with that of the normal LPF.

Conclusions

A new TGV and trench-spiral inductor process were developed by using the photosensitive glass. In advance, the RF properties of the photosensitive glass were studied and etching properties of the glass were researched for various UV-exposure times. It was possible to make a fine via-hole of a 20 μm diameter with a high aspect ratio using the photosensitive glass. For the advanced IPD technology, a high-Q spiral inductor which is a trench-shaped signal was realized using the glass and it was also made a LPF by using the inductor. Owing to good signal isolation and thick metal structures, the realized spiral inductor and LPF had very good RF performances. The maximum Q factor of the fabricated trench inductor was approximately 33 at 2 GHz and the insertion loss of the LPF was 0.31 dB, respectively.

Acknowledgment

This work was supported by the Global R&D program of MOTIE/KIAT. [N0000686, Heterogeneous System IC Integration Process Technology Using IPD Si-substrate]

978-1-4799-2408-0/14 $31.00 © 2014 IEEE

References

1. K. Zoschke et al., "TSV based Silicon Interposer Technology for Wafer Level Fabrication of 3D SiP Modules," *In Proc. of 61th, Int. Conf. on ECTC*, 2011, pp. 836-843.

2. Seung Wook Yoon, et al., "Reliability of a Silicon Stacked Module for 3-D SiP Microsystem," IEEE Trans. on advanced packaging, Vol. 31, No. 1, February 2008.

3. Michael Töpper et al., "3-D Thin Film Interposer Based on TGV (Through Glass Vias): An Alternative to Si-Interposer," *In Proc. of 60th, Int. Conf. on ECTC*, 2010, pp. 66-73.

4. Vijay Sukumaran et al., "Design, Fabrication and Characterization of Low-Cost Glass Interposers with Fine-Pitch Through-Package-Vias," *In Proc. of 61th, Int. Conf. on ECTC*, 2011, pp. 583-588.

5. Chun-Hsien Chien et al., "Performance and Process Characteristic of Glass Interposer with Through-Glass-Via(TGV)," *In Proc. of 3D systems Integration Conference,* 2013, pp. 1-7.

6. C J Anthony *et al.*, "Microfabrication in Foturan™ Photosensitive Glass Using Focused Ion Beam," *In Proc. of the World Congress on Engineering*, 2007, pp. 1335-1339.

7. K. Feindt et al., "3D-structuring of Photosensitive Glasses," *In Proc. of Micro Electro Mechanical Sys.*, 1998, pp. 207-210.

8. Jong-Min Yook et al, "High Power and High Q Spiral Inductors using TSV Processes," *in Proc. of Int. Conf. on Solid State Devices and Materials,* 2013, pp. 126-127.

9. Gang Zou et al., Characterization of Liquid Crystal Polymer for High Frequency System-in-a-Package Applications," IEEE Trans. on Advanced Packaging, Vol. 25, No. 4, 2002.

10. C. Patrick Yue et al., "Physical Modeling of Spiral Inductors on Silicon," *IEEE Trans. on Electron Devices*, Vol. 47, No. 3, 2009.

11. William B. Kuhn et al., "Analysis of Current Crowding Effects in Multiturn Spiral Inductors," *IEEE Trans. on Microwave. Theory and Tech.*, Vol. 49, No. 1, 2001.

12. Kai Liu and Robert C Fry, "Small Form-Factor Integrated Passive Devices for SiP Applications," in IEEE MTT-S Int. Microwave Symp. Dig., 2007, pp. 2117 – 2120.

Challenges and Opportunities of Chip Package Interaction with Fine Pitch Cu Pillar for 28nm

Andy Bao, Lily Zhao, Yangyang Sun, Michael Han, Geoffrey Yeap,
Steve Bezuk, Pat Holmes, Cecille Alcira, Xuefeng Zhang, Kenny Lee

Qualcomm, Inc.
5775 Morehouse Drive, San Diego, CA 92111
Email: abao@qti.qualcomm.com

Abstract

As device dimension shrinks less than 65nm, the propagation delay, crosstalk noises, and power dissipation due to RC (Resistance Capacitance) coupling becomes significant. Cu and LK (Low-k dielectric) material have been introduced to reduce such delays and allow higher device speed and better performance. However, since dielectric material with low-k value usually possesses large amount of porosity, its mechanical properties are degraded significantly which leads to fragile silicon backend structure. This in turn brings in reliability issues like LK cracking due to CPI (Chip Package Interaction). The application of flip-chip packaging introduces significant amount of mechanical stress on BEOL (Back-End-Of-Line) at chip-attach processing step due to CTE mismatch, and makes CPI much more challenging and critical for silicon integration.

At advanced technology nodes, increasing performance demand of mobile processors coupled with SoC integration is one major driver of bump pitch reduction [1]. Higher I/O count can be achieved with finer bump pitch since die size very likely stays constant if not shrinking further. Cu pillar and ELK material have been introduced in 28nm to realize the pitch reduction and performance gain. Small UBM structure is required with fine pitch Cu pillar which introduces large amount of stress in BEOL layers. On the other hand, while k-value of ELK is reduced by ~20% compared to LK used in previous technologies, its hardness and mechanical modulus have been reduced by ~30%, resulting in major reduction of ELK material strength.

In this paper, we present our key learnings from 28nm CPI development with fine pitch Cu pillar. Empirical data based on CPI TV as well as mechanical stress simulations are discussed. UBM dimension which is a critical factor with Cu pillar from CPI perspective is searched at fine pitch, and our data shows CPI robustness limits pitch reduction with Cu pillar if using standard mass reflow process. ELK robustness is also tested at different process corners, including UBM size, bump height and Cu etching module. Some ELK marginality issues are discovered at certain process corner combinations. CPI margin at 28nm with fine pitch Cu pillar is then assessed by correlating mechanical stress simulation to thermal shock testing data. It is shown that min ~15% ELK margin in terms of max ELK stress is necessary to ensure no ELK delamination happening at process corners. Impact of IMC (Intermetallic Compound) and Ni barrier are also studied. It is found that growth of IMC is critical for ELK integrity with mass reflow process. Once IMC is fully grown between Cu pillar and substrate bonding pad, since its stiffness is 2~3X higher than Lead-free solder, mechanical stress on ELK layers increases dramatically. Additional work is carried out to minimize the growth of IMC. It is confirmed that addition of Ni barrier effectively suppresses IMC growth, and increases CPI margin at process corners by considerable amount. Detailed data is presented and final recommendations on fine pitch Cu pillar conclude the paper.

Introduction

CPI has drawn significant attention in semiconductor industry ever since LK material was introduced into BEOL due to its large amount of porosity and degraded mechanical properties. Wang et al presented their experimental measurement of interfacial fracture energy for LK interfaces, and multi-level sub modeling results using MVCC techniques [2]. Yeoh et al demonstrated Intel's HVM electroplated copper die bumps and ILD stress reduction with BEOL design optimization to mitigate CPI risk [3]. Farooq et al studied CPI with both Sn/Pb and Lead-free flip chip packages; also examined excessively reliability behavior with various underfill material [4]. Zhang et al summarized impacts on CPI margin of LK & solder material by 3D simulation, explored both interconnect and package structure optimization to improve reliability [5]. Gonzalez et al measured residual stress of 3D stacked packages with dedicated FET arrays as CPI sensors. They showed that FET current shifts can be used to measure the stress in the surface of the chip. A multi-scale simulation methodology was also demonstrated to capture detailed BEOL structure [6]. It is a general understanding that CPI solution including bump cell structure and assembly optimization shall be thoroughly studied during process and

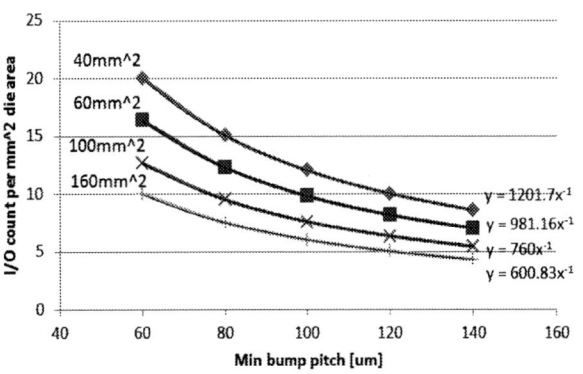

Fig. 1: I/O count per mm^2 die area increases as bump pitch shrinks. The effect is more prominent for smaller die. SoC die size is in the range of 40~60mm^2.

package development as its failure impacts product yield and performance. Furthermore, early CPI development is critical to define the bump cell and design rule manuals to meet aggressive tape-out schedule of mobile chipsets. The emerging trend of IC and package co-design also requires bump placement and dimension related design rules to be ready which is part of deliverables from CPI development.

SoC integration of mobile processors requires much dense IO in shrinking chip area as technology node advances. Performance requirement like DDR speed is also one of the main drivers for more IO by further reducing bump pitch. Fig. 1 shows I/O count per mm^2 chip area vs bump pitch. Number of I/O count roughly doubles when pitch going from 120um down to 60um. Power law fitting equations of various die size are also presented. The slope of each curve if plotting on a log scale becomes steeper as chip area shrinks, which is saying that reducing pitch is more efficient to increase I/O count at smaller die size. Today SoC chips favor die size at 40~60mm^2 range so bump pitch reduction is necessary. Bump cell dimensions would have to shrink together with this pitch reduction trend.

Bump Cell Optimization to Ensure Reliability at Process Corners

CPI induced mechanical stress in ELK layers increases significantly as bump cell UBM area gets smaller (Fig. 2). Modeling results of two BEOL metal stacks are plotted, one having eight layer metals while the other one nine layer metals. About 10% stress reduction is estimated due to adding one top metal layer. Given fixed bump pitch, UBM dimension shall be maximized as much as possible if searching for robust CPI margin. However, large UBM in fine pitch Cu pillar brings other assembly process concerns. As shown in Fig. 3, minimum distance between bump and a neighboring escape trace must be maintained; otherwise shorting risk becomes high which damages assembly yield. Assuming min bump pitch is 100um or 110um, modeling shows chip attach Cpk from trace shorting point of view decreases sharply from >1.33 to <0.8 when UBM diameter goes from 40um to 65um (Fig. 4). It is clear that CPI study shall not only solve BEOL delamination concern but also meet assembly process requirement. Assembly process optimization may be required in parallel to solve CPI challenges.

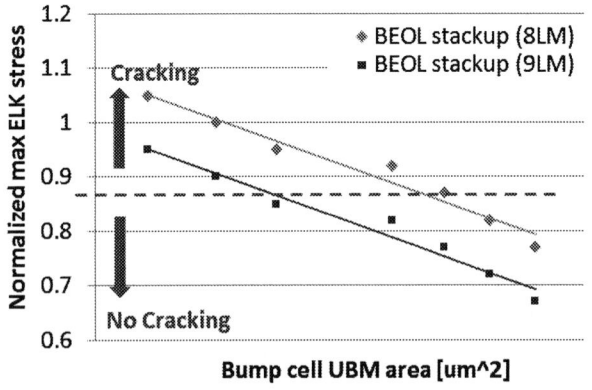

Fig. 2: Bump cell UBM area impacts ELK stress significantly as shown in simulation.

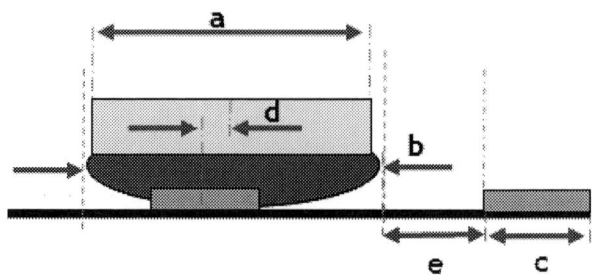

Fig. 3: Bump cell UBM dimension (a), solder diameter due to solder extrusion (b), neighboring trace width (c) and chip attach alignment offset (d) together determine the clearance (e) which is critical to reduce assembly shorting risk. A nominal clearance of 15~20um is often suggested.

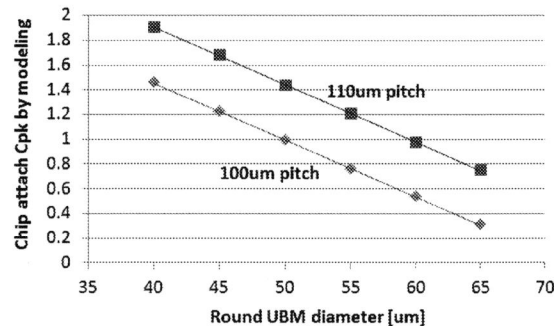

Fig. 4: While large UBM reduces ELK delamination risk, assembly chip attach Cpk from shorting point of view decreases sharply as UBM diameter increases. RSS based modeling takes into account critical dimensions shown in Fig. 3 and predicts Cpk. A bump cell that works at 110um pitch may not be a viable solution at 100um pitch. Assembly process optimization may be required in parallel to solve CPI challenges.

Conditions	Ni barrier	CPI hammer testing		Pacakge reliability testing		
		Multi-reflow w/o underfill	Quick temperature cycling w/o underfill	TC'B'	HTS	uHAST
Control	-	pass	pass	pass	pass	pass
	Yes	pass	pass	pass	pass	pass
Wafer level reflow corner	-	fail	fail	-	-	-
	Yes	pass	pass	pass	pass	pass
Bump height corner	-	fail	pass	-	-	-
	Yes	pass	pass	pass	pass	pass
Solder cap height corner	-	fail	fail	-	-	-
	Yes	pass	pass	-	-	-
Cu etching corner	-	fail	fail	-	-	-
	Yes	pass	pass	-	-	-

Table 1: Proposed bump cell structure fails CPI hammer testing at process corners without Ni barrier introduced between Cu pillar and solder cap. FA found IMC fully grown between Cu bump and substrate bonding trace. Stiffer IMC transfers larger stress in BEOL than solder material. Adding Ni barrier suppresses IMC growth efficiently and improves CPI margin at those process corners considerably.

Table 1 summarizes 28nm CPI TV data. The control leg shows robust behavior under CPI hammer and package

978-1-4799-2408-0/14 $31.00 © 2014 IEEE

reliability testing. However, legs at process corners fail hammer testing and do not meet qualification criteria. FA shows IMC is fully grown between Cu pillar and package bonding trace. Since IMC is much stiffer than solder material, it presents less plastic deformation during chip attach module and transmits larger amount of stress into BEOL which causes ELK delamination. Adding Ni barrier between Cu pillar and solder effectively suppresses IMC growth, and reduces its thickness by ~13% after chip attach (Fig. 5). As shown in Table 1, DOE legs at various process corners with Ni barrier pass CPI hammer and package reliability testing. A more robust CPI solution is achieved.

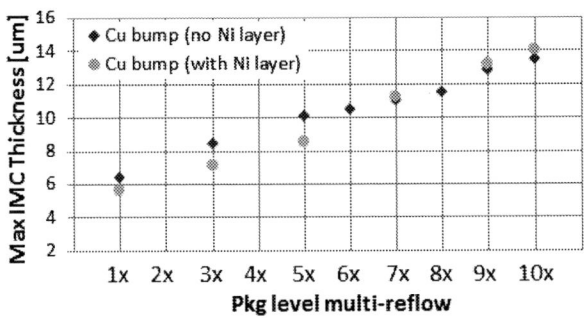

Fig. 5: Total IMC thickness between bump and substrate bonding trace reduces when adding Ni barrier between Cu pillar and solder cap, ~13% less at chip attach step.

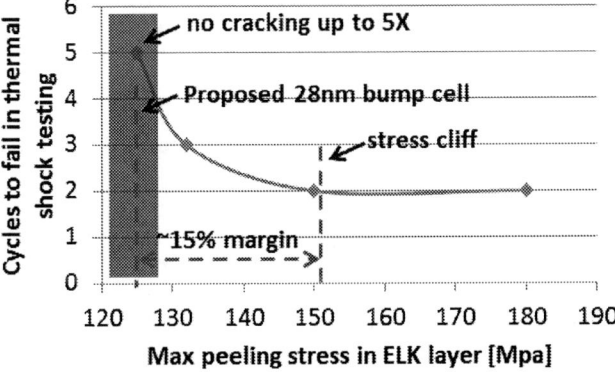

Fig. 6: Relative max ELK stress level from modeling is correlated to cycles to fail in liquid-to-liquid thermal shock testing. 28nm ELK material cliff is defined in terms of max ELK stress at 150MPa. Proposed 28nm bump cell has ~15% margin which seems sufficient to cover design and process corners.

Stress modeling needs to be validated before exercising modeling for bump cell structure optimization. In present work, liquid-to-liquid thermal shock testing is carried out for different UBM structures, and the results are used to calibrate stress model in a relative sense. Structures with high stress as calculated in the stress model show 'white bump' under CSAM earlier in thermal shock testing in terms of cycle to fail (Fig. 6). That provides confidence that the stress model can

predict reasonably well for 'A' to 'B' comparison. Based on thermal shock testing data and correlation of stress modeling results, 28nm ELK material cliff is defined in terms of max ELK stress. Product ELK margin can then be determined by stress simulation. Bump cell defined in previous section shows ~15% margin, which seems sufficient to cover process corners.

Conclusions and Acknowledgments

Large UBM area is required to prevent ELK delamination and is one of the most important knobs in terms of CPI margin. However, bum plating limitation, die preparation process, assembly yield and shorting risk etc. put a boundary on the maximum UBM dimension possibly allowed. As bump pitch further shrinks, it is more challenging to search for a robust CPI solution with smaller UBM area. Current assembly mass reflow process may need to be optimized. Thermal compression flip chip provides a promising alternate but with certain trade-off. The ELK delamination risk is lowered significantly due to the presence of pre-applied film or paste material during chip attach. Substrate manufacturing technology shall also be examined to reduce the CTE mismatch which is the root-cause of ELK cracking concern. In this work, key learnings from 28nm CPI development with fine pitch Cu pillar is presented and proposed bump cell has been implemented on products.

Authors would like to thank both internal and external collaborators for their great dedication and outstanding assistance on fine pitch Cu pillar CPI development for 28nm.

References

1. G. Yeap, "Smart Mobile SoCs Driving the Semiconductor Industry: Technology Trend, Challenges and Opportunities", *IEDM Tech. Dig.*, pp.13-16, 2013
2. G. Wang, C. Merrill, J. Zhao, S. Groothuis, and P. Ho, "Packaging effects on reliability of Cu/low-k interconnects", *IEEE Transactions on Device and Materials Reliability*, vole 3, Issue: 4), pp. 119-128, 2003
3. A. Yeoh et al., "Copper Die Bumps (First Level Interconnect) and Low-K Dielectrics in 65nm High Volume Manufacturing", *56th Electronic Components and Technology Conference*, San Diego, CA, May 2006
4. M. Farooq et al, "Chip Package Interaction for 65nm CMOS Technology with C4 Interconnections", *Proceedings of the IEEE 2007 International Interconnect Technology Conference*, pp. 196 – 198, 2007
5. X. Zhang, Y. Wang, J. Im, and P. Ho, "Chip–Package Interaction and Reliability Improvement by Structure Optimization for Ultralow-k Interconnects in Flip-Chip Packages", *IEEE Transactions on Device and Materials Reliability*, volume 12, issues 2, pp.462-469, 2012
6. M. Gonzalez, B. Vandevelde, A. Ivankovic, V. Cherman, B. Debecker, M. Lofrano, I. De Wolf, G. Beyer, B. Swinnen, Z. Tokei and E. Beyne, "Chip Package Interaction (CPI): Thermo Mechanical challenges in 3D Technologies", *Electronics Packaging Technology Conference (EPTC), IEEE 14th*, pp.547-551, 2012

Electromigration for Advanced Cu Interconnect and the Challenges with Reduced Pitch Bumps

Nokibul Islam, Gwang Kim, KyungOe Kim
STATS ChipPAC
Email: nokibul.islam@statschippac.com

Abstract

Cu Column bump has seen growing adoption in both high–end and low-cost mobile devices as well as in consumer, computing and networking devices. Higher input/output (I/O) density and very fine pitch requirements are driving very small feature sizes such as small bump on a narrow pad or bump on lead (BOL), while higher performance requirements are driving increased current densities, thus making electromigration (EM) performance a real and serious concern.

As the fine pitch and bump sizes decrease (in both mass reflow and thermo compression bonding processes) and the current density through the bump increases, EM reliability is becoming an alarming issue across the industry. High current density in Cu column bump combined with Joule heating may easily cause an early EM failure in field applications. Many researchers [1-3, 11-13] have published copper column EM data with a number of studies and EM variables, but no data has been published for robust BOL/bump design rules for high temperature and high current conditions which is indigent in high performance packages. This project has been initiated to resolve EM challenges in the industry by identifying the BOL/ bump design with regards to the temperature/current conditions so a robust design rule/process window can be established for next generation packages.

Introduction

As flip chip continues its rapid advancement in the demanding markets, the complexities of devices are increasing exponentially. Challenges that the packaging industry faces include the need to deliver cleaner power to devices provide enough input/output (I/O) to accommodate the volume of data in high speed devices and still satisfy all other requirements without compromising reliability and/or cost. Bundling this much functionality into a single piece of advanced silicon requires a large interconnect gap between the silicon and the system printed circuit board (PCB) to be bridged.

Adoption of a copper (Cu) column fcCuBE® (flip chip package with Cu column bump, BOL interconnection and enhanced process) in place of flip chip solder bumps has number of potential benefits including bump pitch reduction, possible design rule relaxation by using wider line and space for signal routing in a given design, removal of tight solder registration, and the removal of solder on pad (SOP) on the substrate, all of which result in a low cost flip chip package solution [4-10]. The reduction in pitch capability is simply driven by the fact that bump to bump spacing can be controlled better with finer pitch Cu column as compared to a standard solder bump due to the bump geometry, spherical solder shape vs. cylindrical for Cu, and differences that exist in the bump geometry post reflow process. At the reflow step, solder collapses and an increase in bump diameter is observed, whereas Cu column will not go through any such transformation and will not experience any dimensional changes. Furthermore, the collapse height, which defines the die to substrate gap (stand-off height), can be better controlled by Cu column as the column height can be modulated to provide the required stand-off height without any increase in bump diameter, whereas any increase in stand-off height for solder bumps is associated with corresponding increase in bump diameter. Such an increase in bump diameter is not desirable as it would reduce the bump to bump spacing, resulting in potential bump bridging and electrical shorts. Additionally, the reduced bump to bump spacing would create issues for Capillary Underfill (CUF) flow and lead to underfill voids. Alternatively, it would create more voiding problem for the Molded Underfill (MUF) process due to a larger filler size used by the MUF material.

A motivation for using fine pitch Cu column bump is to improve the EM performance of the device due to the higher current carrying capability of Cu. However, Cu column on a very narrow pad becomes an issue due to the higher device power and current density which can be further aggravated due to the Joule heating effect ultimately leading to early EM failure.

Semiconductor companies are very concerned with the issue since it is a major source of failures today. An imminent solution is needed in order to overcome the industry-wide problem. In this study, EM test vehicles were designed with fine pitch Cu column and BOL design pad. The bump structures are exactly the same as actual product. The current flow direction in and out (with current pushing through three bumps and out one bump and vice versa) was designed in such a way that both the die side and substrate side failure can be captured in an actual EM test. The first degree of parameters such as BOL pad width, current condition, temperature conditions, BOL vs BOC (bond on capture pad, SMD pad type) pad type, etc. were considered in the DOE. EM failure data was collected and analyzed with statistical tools. Very comprehensive BOL/bump design rules and an optimum assembly process window were established to design a robust next generation Cu column package.

Package Design

Both the actual product and EM test vehicle followed the same design rule. In this particular device design, 40nm silicon was used with peripheral array bumps with 150um pitch. Package body size was a 17 x 17 mm package with 425 solder balls of 0.35mm diameter and 0.8mm pitch in four layers Plated through Hole (PTH) substrate. The die size was approximately 5.2 x 5.7 mm. Substrate core material was chosen as Low CTE material to control the package warpage/coplanarity in addition to extreme low-K (ELK) die protection. The gap between the bump to nearest trace is the key for the fcCuBE design. Too narrow of a gap can cause assembly related issues such as solder bridging, shorting, etc.

Bumping process included PI re-passivation, Ti/Cu under bump metallization (UBM), and Cu column plating with a SnAg solder cap on top. As a result of this bumping structure, the peripheral bumps were located on the Al pads while the center mechanical bump array was located on the top passivation layer (no electrical contact). Figure1 shows the fcCuBE Cu bump and SnAg solder cap along with BOL trace detail for a given bump pitch design. Overall Cu column and solder cap height were optimized in order to create the optimum stand-off height required for successful CUF/MUF process in the assembly.

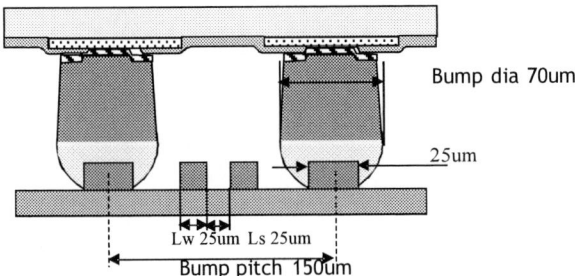

Figure 1: Bump pattern, and Cu pillar bump dimension detail w/ BOL pad

Assembly Process

The assembly process included several design iterations for bump height in order to optimize the CUF/MUF flow underneath the die. The original design with 42um column height with 35um solder cap encountered significant MUF voids. A modified design had 60um column height with 35um cap which enables a significant gap height increase. Higher column over solder cap ratio increased die level stress results in ELK crack (white bump) in actual product during the chip attach process. Significant ELK damage was experienced with taller column design even though it gives a better CUF/MUF process in the actual product. Figure 2 below shows white bump with taller column height with the actual product. Extensive simulation has been conducted to understand the safe limit of column/solder cap ratio. Finally, a design with 42um column and 35um solder cap with full open SR was introduced which maintained a smaller bump height to address the white bump issue. On the other hand, it increased the gap height significantly for void free MUF process.

Figure 2: white bump w/ taller bump (left picture), and no white bump w/ smaller bump (right picture)

The detailed assembly process including flip chip attach, under-filling, overmold, ball attach, and singulation were fully

optimized for assembly. The critical areas in the assembly process were identified as chip attach, molding, and ball mount processes. Additionally, an optimum amount of flux is needed in the chip attach process to make a good joint for very fine size/pitch Cu column. Die placement also plays a crucial role. If by any means the die are misaligned, solder bridging, non-wet, etc. might occurred in the chip attach process. Another important concern is white bump (bump delamination) for low K/ELK die. The white bump risk is much lower with the BOL pad versus BOC type pad. Having a smaller BOL pad helps to resolve the die level stress during the chip attach process by shifting the stress from die side to substrate side.

MUF process characterization focused on void free molding underneath the die. In this study void free MUF was one of the biggest challenges due to a finer diameter and a smaller gap height column. Moreover, the MUF filler size is much coarser than CUF, making the challenge even bigger. Today there are some finer filler MUF available in the market, but they have not yet achieved a preferred status for cost sensitive packages. Several iterations such as fine filler, taller bump height, and two step height solder resist, and an open solder resist under the die were used to fix the voiding issue. A comprehensive hammer test, MRT, and temperature cycles were conducted to authenticate void free design and process.

Electro-Migration Test

EM of bumps is a failure mechanism that leads to increased resistance, sometimes occurring with events such as formation of IMC, voids and cracks that can interrupt the bump joint and silicon, and/ or package metallization leading into the bump. The resistance increase can ultimately lead to a complete open in the device. The stress drivers for this type of failure mechanism are current density and elevated temperature. A motivation for using Cu column is to improve the EM performance of the device due to the higher current carrying capability of Cu. However, Cu column on a very narrow pad becomes an issue due to the higher device power and current density which can be further aggravated due to the Joule heating effect, ultimately leading to early EM failure

Bump level EM tests were performed both at in-house and a 3rd party vendor. EM test vehicle body size, die size, and bump structures were very similar to actual product. Current flow direction in and out (pushing through three bumps and out one bump and vice versa) was designed in such a way so both die side and substrate side failures can be captured in an actual EM test. Figure 3 shows the typical EM bump schematic for the three to one current flow condition. A dummy bump is attached between the functional bump to mimic the actual bump pattern in the product, moreover, dummy bumps help to normalize the "joule heating" effect during EM test.

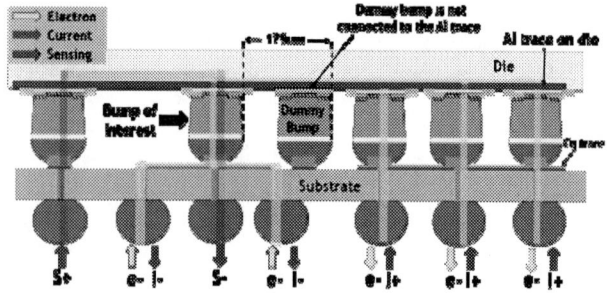

Figure 3: Typical EM bump schematic for 3 to 1 current flow condition

Actual electro-migration tests have been conducted both at in-house lab and a 3rd party vendor. Harsher condition legs were considered in the 3rd party vendor DOE than in house DOEs. The test matrix DOE is shown in Table 1 below. All samples have Cu column with 70um diameter and 52um gap height. Solder caps are 35um with 3 um thin Ni barriers between solder caps to Cu column. EM TV is very similar to actual product with 17X17mm body, 4 layer substrates with Cu OSP finish. The effect of the bump current flow in and out (3 in 1 out vs. 1 in 3 out) and BOL vs. BOC (bond on capture pad shown in Figure 4 below) pad were also considered in the DOE matrix. In order to create a Black's Model fit, a combination of at least five legs were used in the study. The devices under test (DUTs) were tested at constant current and temperature conditions. Actual device temperature will always be higher than oven temperature due to higher stress current and temperature (Joule heating effect). Therefore, Joule heating effect must be investigated and incorporated in the EM analysis. In this study, actual bump temperatures were derived using temperature coefficient of resistance (TCR) method. The average temperature increments due to Joule heating were added experimentally in each leg. Figure 5 shows a representative TCR result of the BOL structure at test with the condition of 150°C and 500 mA. The red box in Figure 5 shows the Joule heating on one DUT. Joule heating values of all the test conditions were less than 3°C

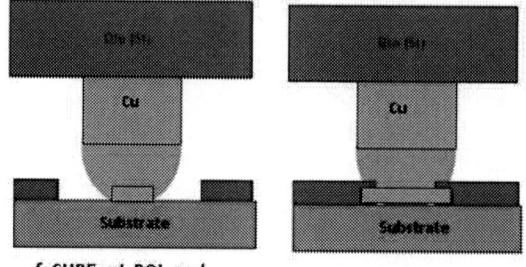

| fcCUBE w/ BOL pad | fcCUBE w/ BOC pad |

Figure 4: fcCuBE w/ BOL vs. BOC (SMD) Pad

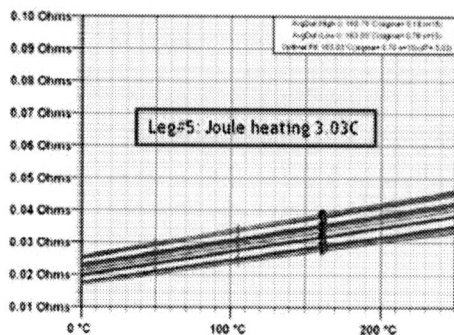

Figure 5: Typical TCR plot for BOL structure at 150°C, and 500 miliAmps stress current

The EM failure criterion was defined as the time at which a 15% increase in electrical resistance was observed. EM data was collected for over 7000 hours under accelerated conditions as indicated in Table 1. All samples have been tested over 7000 hours and no interconnection failures have been reported, however, we investigated interconnection morphologies induced by EM effects through cross-sectional analysis and analyzed the interfacial reaction characteristics between BOL and BOC structures for various stress conditions. The bump microstructures used in this test were observed with scanning electron microscopy (SEM) in the backscattered electron (BSE) mode, and the compositions of the resulting IMCs were measured by energy dispersive spectrometry (EDS).

EM Results

Over 7000 hours of EM test resistance shift data was plotted against stress time (hours). Figure 6 shows the resistance shift data for various temperature and current conditions. Based on resistance shift data, no failure was observed till 7000 hours which confirmed the robustness of the BOL bump structure in an fcCuBE package. The main objective was to collect adequate failure data to construct Black's equation which can be used as a tool for future package design optimization. However, no single failure was observed from any of the stress conditions.

Failure Analysis

Comprehensive failure analysis (FA) was conducted after 7000 hours of EM test to make sure there was no significant anomaly or crack in the interconnection area. A maximum 3% resistance shift was observed even with 650 miliAmps current at 160°C which confirms the robustness of fcCuBE bumps.

Test site	Leg#	Bump Composition	Bump net type	Temperature (C)	Stress current (mA)	Target testing hours	Comments
3rd party lab	1	Cu pillar w/ BOL pad	3 in 1 out	160	400	5000	DOE for Black's model fit
	2	Cu pillar w/ BOL pad	3 in 1 out	135	500		
	3	Cu pillar w/ BOL pad	3 in 1 out	150	500		
	4	Cu pillar w/ BOL pad	3 in 1 out	160	500		
	5	Cu pillar w/ BOL pad	3 in 1 out	160	650		
In house lab	6	Cu pillar w/ BOL pad	3 in 1 out	125	500	5000	DOE for Black's model fit
	7	Cu pillar w/ BOL pad	3 in 1 out	135	500		
	8	Cu pillar w/ BOL pad	3 in 1 out	150	300		
	9	Cu pillar w/ BOL pad	3 in 1 out	150	400		
	10	Cu pillar w/ BOL pad	3 in 1 out	150	500		
	11	Cu pillar w/ BOL pad	1 in 3 out	150	500		compare w/ leg#10
	12	Cu pillar w/ BOL pad	1 in 3 out	150	300		compare w/ leg#8
	13	Cu pillar w/ BOC pad	3 in 1 out	125	500		compare w/ leg#6

Table 1: EM Test DOE

978-1-4799-2408-0/14 $31.00 © 2014 IEEE

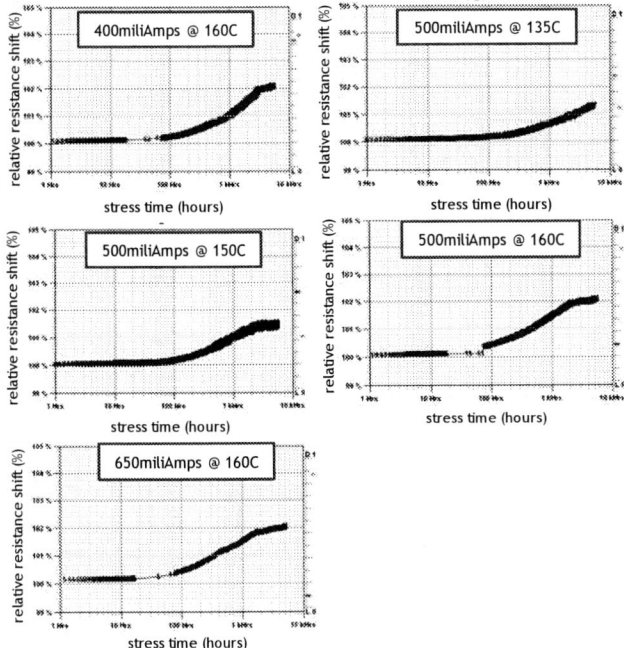

Figure 6: BOL resistance shift after 7000 hours for various test conditions

Two units from each leg were selected for destructive FA (failure analysis) and intermetallic morphology analysis. Figure 7 shows the location of the bump of interest and other bump structures in the EM package. No noticeable anomaly has been observed in the bumps after 7000 hours, as shown in Figure 7. Some minor voids due to solder diffusion were observed in the cross-section images. Detailed cross-section images for various legs were shown in Figure 8. Very thick IMC formed after 7000 hours of test. In the BOL structure very little solder was present in the bump compared to Cu Column. Almost the entire solder converted to IMC after 7000 hours of test. According to some literatures [3, 11] thicker IMC enhances EM performance. Some Kirkendal voids were also observed in the substrate pad to IMC interface (figure 8). Investigation found that Kirkendal void sizes have not been changed much over time. IMC thickness for each leg was also studied and analyzed per EM conditions. Over time SnAg solder was consumed by Cu. IMC thickness before and after EM were measured and plotted in Figure 9.

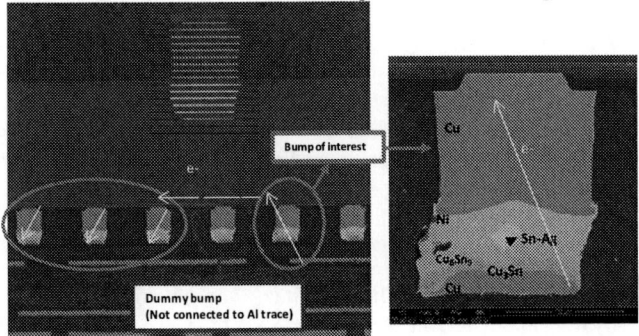

Figure 7: Detail X-section image of a unit (500 miliAmps @ 150^{0}C) after 7000 hours

The entire solder converted to IMC during EM testing. Very little IMC was observed in the Cu column interface due to presence of Ni barrier layer. No Cu diffusion took place in the column/solder interfaces due to Ni layer.

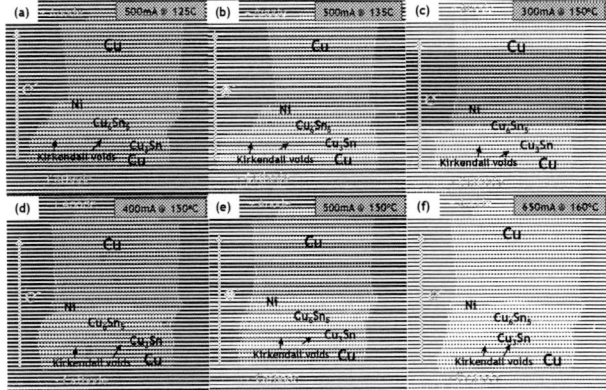

Figure 8: Cross-sectional images on each test condition after 7000 hrs: (a) 500 mA @ 125^{0}C, (b) 500 mA @ 135^{0}C, (c) 300 mA @ 150^{0}C, (d) 400 mA @ 150^{0}C, (e) 500 mA @ 150^{0}C, (f) 650 mA @ 160^{0}C

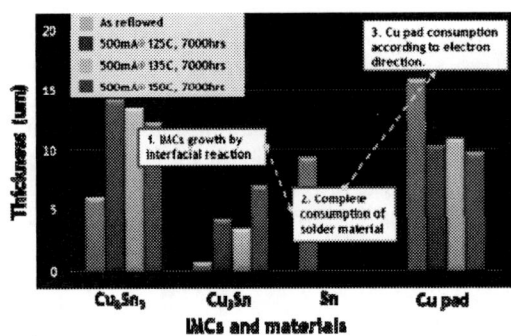

Figure 9: IMCs growth behaviors with three different temperatures (125^{0}C, 135^{0}C, and 150^{0}C) and 500 mA current conditions

The IMC growth mechanism in this study is illustrated in Figure 10. At reflow stage, Ni-Sn IMC was formed at the interface between the Ni layer and solder, and Cu-Sn IMCs were formed at the bonding interface between Cu pad and solder. During EM testing, the thickness of the Cu_6Sn_5 IMC increases until almost all Sn in the solder is consumed. Since the BOL structure has a limited amount of Sn and an infinite supply of Cu, Cu_3Sn IMC starts to grow thicker at the expense of Cu_6Sn_5 IMC. On the other hand, even though Ni_3Sn_4 IMC was formed at the interface between Ni/Sn after the reflow process, the Ni is a barrier layer to Cu and was not fully consumed during the EM test. This means that the Ni_3Sn_4 IMC barely grew; therefore Cu-Sn IMCs is the thicker IMC in interconnect.

BOL vs BOC Structure

EM occurs when the current density is sufficiently high enough to cause the drift of metal ions in the direction of the electron flow, and this is characterized by the ion flux density. Very high current can lead to a temperature gradient which is increasingly problematic and increasingly susceptibility to electro migration failures. Over design is one of the sources for higher current density. Current density effect has been included and analyzed in the study. A naturally bigger pad is better for EM performance due to larger area. Current density is smaller for a bigger pad (BOC). There are a number of studies [4-10] with BOL pad in flip chip packages, but they do not compare EM performance between pad types.

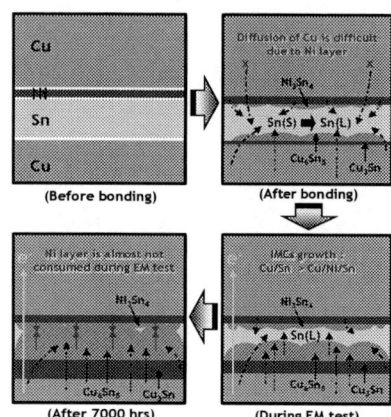

Figure 10: IMC formation mechanism and growth in the BOL structure combing Cu column and shallow solder bump

To compare EM characteristics between BOL and BOC Cu pad structures, an EM test on the BOC structure was also conducted under the same condition of 125°C and 500 mA (leg#13, Table#1). No EM failure was observed in either BOL or BOC pad (till 7000 hours). Similar resistance shift behaviour was observed in BOC structure as well. However, an extensive bump cross-section has been conducted to analyse the interconnection degradation of BOL and BOC structure. Figure 11 illustrates the side by side bump interconnection degradation comparison between BOL and BOC structures. In the BOC structure, SOP was used on the substrate to attach the flip chip die. Hence, the BOC structure much more solder as compared to the BOL structure. In the BOC structure, Cu_6Sn_5 dominates everywhere due to huge amount of solder compared to Cu. The solder phase almost converted into Cu_6Sn_5 IMC. Furthermore, in the BOC structure, the substrate side Cu pad is entirely consumed by solder. Typically, current crowding during EM test significantly occurs at the cathode edge area and also the relatively uneven consumption morphology of the cathode Cu pad will cause incremental current crowding which was the case of BOC structure.

It was found that sufficient Sn in the BOC structure will induce high Cu consumption because Cu atoms from the pad can easily migrate into the widespread solder area. While BOL results showed that a significant number of Cu atoms remain due to limited solder area. In other words, the incidence of interfacial void and crack at BOC is more likely to occur at a higher rate than with the BOL structure (shown in Figure 11). Finally, both BOL and BOC structures used in this test did not show any electrical failure which means better EM reliability with an fcCuBE bump even with small BOL pad width.

Finally, current study with some legs as small as 25um BOL pad width showed no EM failure or interconnection abnormality till 7000 hours. Comprehensive failure analysis

shows that even with 25um BOL width with as high as 600 miliAmps, stress current outperforms the BOC structure. One question remains unanswered; how small/narrow the BOL structure can be without sacrificing the EM reliability of the bump. A new study is being conducted with very narrow BOL pad (~15um) and high stress current (>500 miliAmps at 150°C) which would answer the above question. Too high of current density can lead to early EM failure due to excessive Joule heating, metal migration, and brittle failure of bumps at the IMC (entire pad consumed by solder and become brittle IMC). A design limit of BOL pad size and stress current is being investigated to overcome the issues for very fine pitch Cu column bumps in the future very fine pitch packages.

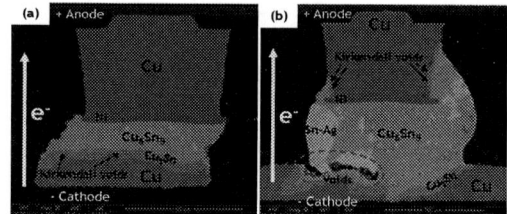

Figure11: Cross-sectional images of BOL (a) and BOC (b) under stress condition of 500 mA @ 125°C after 7000 hrs

Conclusion

EM tests were performed on fine pitch Cu column BOL interconnections and BOC pad structure for various temperature and current conditions. Over 7000 hours of EM test were conducted and no failures were observed. Very insignificant resistance shifts were observed irrespective of current stress and temperatures. However, partial crack, solder diffusion, IMC thickness formation, etc were observed in the interconnection during EM tests. Thicker IMC forms on the substrate side (bond side) than Ni barrier side (Cu column side).

EM performance was compared between BOL and BOC pads for a given current stress and temperature. None of the pad structures failed. However, more interconnection degradation, IMC conversion, solder voiding, and partial cracks were observed in the BOC pad than BOL pad. It was found that the sufficient Sn in the BOC structure will induce high Cu consumption because Cu atoms from the pad can easily migrate into the widespread solder area. BOL results showed that a significant number of Cu atoms remain due to limited solder area. Typically, current crowding during EM tests significantly occurs at the cathode edge area and also the relatively uneven consumption morphology of the cathode Cu pad will cause incremental current crowding. Finally, BOL EM data proved that fcCuBE bump is much more robust for fine pitch high performance packages. Future programs with much higher current density (smaller BOL pad) for very fine pitch Cu column package is currently being conducted.

Acknowledgments

The authors would like to thank Jae Myeong Kim, Eric Ouyang, and Dr. Raj Pendse of STATS ChipPAC for their continued guidance in the study. The authors want to express gratitude to the individuals of STATS ChipPAC RnD team in Korea, our partner companies that helped design the advanced packages; including actual EM tests.

978-1-4799-2408-0/14 $31.00 © 2014 IEEE

References

[1] Nokibul Islam et al, "Application of fcCuBE Technology to Enable Next Generation Consumer Device", *Electronic System Technology Conference, 2013. ESTC 2013,* Las Vegas, Neveda, May 20th-23rd, 2013

[2] Jae Myeong Kim et al, "Comparisons of Interfacial Reaction Characteristics on Flip Chip Package with Cu Column BOL Enhanced Process (fcCuBE) and Bond on Capture Pad (BOC) under Electrical Current Stressing", *15th International Conference on Electronic Materials and Packaging,* EMAP *2013. Seoul, Korea,* Oct 6-9th, 2013

[3] Yasumitsu Orii et al, "Electromigration Analysis of Peripheral Ultra Fine Pitch C2 Flip Chip Interconnection with Solder Capped Cu Pillar Bump", *Electronic Components and Technology Conference, 2011. ECTC 2011. 61st,* Lake Buena Vista, Florida, pp 340-245, May 31st – June 1st, 2011

[4] Hamid Eslampour et al, "Low Cost Cu Column fcPOP Technology", *Electronic Components and Technology Conference, 2012. ECTC 2012. 62nd,* San Diego, CA, pp. 871-876, May 29th-June 1st, 2012

[5] US Patent Nos. 7368817, 7700407, 7901983, 7973406, 8076232 and 8188598. Bump-on-lead Flip Chip Interconnection, Raj Pendse, Nov. 2004

[6] Joshi, M. et al, CuBoL (Cu Column on BoL) Technology: A Low Cost Flip Chip Solution Scalable to High IO Density, Fine Bump Pitch and Advanced Si Node. *Proc 61st Electronic Components and Technology Conf,* FL, May 2011.

[7] Eslampour, H. et al, Next Generation PoP Technology, Advanced Interconnect Technologies, *IMAPS Conference,* July 13, 2011.

[8] Pendse R, et al, "Low Cost Flip Chip (LCFC): An Innovative Approach for Breakthrough Reduction in Flip Chip Package Cost", *60th Electronic Components and Technology Conf,* Las Vegas, Ca, June 2010.

[9] Eric Oyuang, et al, "Improvement of ELK Reliability in Flip Chip Packages using Bond-on-Lead (BOL) Interconnect Structure", *IMAPS Conference,* October 2010

[10] Pendse R., et al, "Bond-on-Lead: A Novel Flip Chip Interconnection Technology for Fine Effective Pitch and High I/O Density", *Proc 56th Electronic Components and Technology Conf,* San Diego, Ca, May. 2006. pp. 16-23

[11] R. Labie, F. Dosseul, T. Webers, C. Winters, V. Cherman, E. Beyne, and B. Vandevelde, "Outperformance of Cu pillar Flip Chip Bumps in Electromigration Testing", *IEEE Electronic Components & Technology Conference,* Lake Buena Vista, FL, USA, 2011. p. 312

[12] Ahmer Syed et al, "Flip Chip Bump Electromigration Reliability: A Comparison of Cu Pillar, High Pb, SnAg, and SnPb Bump Structures", *IMAPS Device Packaging Conference 2010,* Scottsdale, AZ, pp. 166-171, March 9-11, 2010

[13] JH Yoo et al, "Analysis of Electromigration for Cu Pillar Bump in Flip Chip Package", *Electronics Packaging Technology Conference, 2010. EPTC 2010. 12th,* Singapore, pp. 129-133, December 8-10th, 2010.

Electromigration Performance of Cu pillar Bump for Flip Chip Packaging With Bump on Trace by Using Thermal Compression Bonding

Kuei Hsiao Kuo (Frank)[1*], Jason Lee[1], F.L. Chien[1], Rick Lee[1], Cindy Mao[1] and John Lau[2]

[1]Siliconware Precision Industries Co., Ltd. (SPIL), [2]ITRI

[1]No. 153, Sec. 3, Chung Shan rd., Tantzu, Taichung, Taiwan, R. O. C.

[1*]Tel: +886-4-25341525 ext. 7278. [1*]E-mail: frankkuo@spil.com.tw

[1]Business unit 3, Process Integration Engineering DEPT.1

Abstract

In this study, electromigration (EM) performance of 60μm pitch Cu pillar bumps assembled with bump on trace (BOT) process by using thermal compression bonding with non-conductive paste (TCNCP) is investigated. Emphasis is placed on the EM experimental measurement and analysis of the Cu pillar bump on trace. The test temperature ranges from 140-160°C and the current of 500-900mA are applied, which corresponding to the current density of 50-90kA/cm². The substrate finish is organic solderability preservative (OSP) on the Cu trace. The Cu pillar height is 45μm including 28μm Cu post + 2μm Ni + 15μm lead free SnAg solder cap. The bump size is 30 x 40μm (elongated).

The experiment result shows the joint solder is transferred to intermetallic compound (IMC) soon after EM test due to small solder joint height (7~8 um) after bonding. The EM performance of Sn-Cu IMC bumps outperform standard lead-free solder bump. The electromigration life with 4X tightened failure criteria of the Cu pillar bumped SnAg solder joints on trace is 3X better than that of the conventional lead-free SnAg or SnCu solder joints. The EM failures of Cu pillar bumped on trace are caused by void and crack due to solder creeping Cu post side wall. It shows the thermal driven failure is the dominant factor and the defects will be enhanced or reduced based on the electron flow direction. It's different to the conventional lead free solder-joint failure mode with serious cracks between the bulk solder and the $(Cu,Ni)_6Sn_5$ intermetallic compound.

Introduction

The increasing demanding for more device functionalities (high I/O density) and smaller die size drives the flip chip bump pitch decreasing from typical 150-180μm to 80-110μm. These are out of the limitation of traditional solder bump approach for flip chip interconnection because of the difficulty to insert underfill and high risk of solder bridging due to small solder joint stand-off height and bump space.

The flip-chip CSPs (Chip-Scale Packages) of Cu pillar bump on trace (BOT) structure can meet the fine pitch (<130μm) requirements due to higher stand-off height and lower bridge risk (small solder volume). Currently Cu pillar bump pitch is down to 60μm from typical 80-110μm. This drives the bump size down to 30-40μm with small solder cap height (15-20μm) and the requirements of tightened die bonding accuracy. Cu pillar on trace interconnected by using thermal compression bonding with non-conductive paste is selected due to tightened bonding accuracy, better warpage (non-wetting) control and ELK protection as compared to traditional mass reflow bonding process.

The smaller bump size designed in fine pitch devices will increase higher current density and thus increase the concerns of EM failure. For example, the current density is reached to 10^4 A/cm² when input 80mA current to a 30μm Cu pillar bump which is beyond the typical solder bump application with current density ~ 5×10^3 A/cm². In this study, the EM performance of 30x40μm Cu pillar bump on trace by thermal compression bonding is investigated. The samples are tested under different acceleration conditions with current density and temperature up to, respectively 10^5 A/cm² and 160°C. The EM performance is modeled by Black' equation based on test results. And then, the failure mode including the intermetallic compounds (IMC) formation and evolution, Cu and Ni diffusion, solder creeping and void/crack formation are discussed to understand the key failure mechanism and its effect to EM life with Cu pillar bump on trace structure.

Test Vehicle Description

A daisy-chain bump structure consisting of Kelvin structure was used in this test vehicle to monitor resistance change. Figure 1 shows a cross-section view of how the bump on the stress is fed. The current was forced up from center bump (P1) which was stressed with higher current density and flowed down from 5 adjacent bumps (G1~G5) which were stressed with lower current density. The test vehicle chosen for fine pitch Cu pillar bump on trace study was packaged with 19x19 mm² flip chip CSP package by using thermal compression bonding. The TV die used was 10.3 x10.4 mm² with 60μm bump pitch and lead free (SnAg) solder cap. The Cu pillar bump size is 30x40μm (elongated) and bump height 45μm including 28μm Cu post + 2μm Ni + 15μm lead free SnAg solder cap formed by electro-plating process. The bumping trace width on substrate is 20μm, OSP finish was applied to the exposed trace to prevent the oxidation of Cu trace. The configuration of EM test vehicle is listed in table 1.

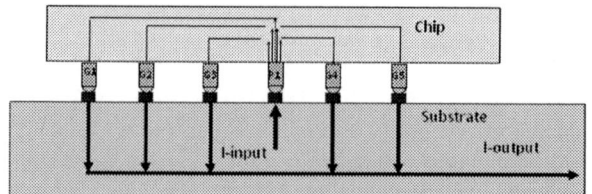

Figure 1. Schematic diagram of test structure

TV Feature	Description
Process	Plated Cu pillar bump
Solder Material	SnAg 1.8
UBM stackup	Sputtered Ti/Cu
UBM Diameter(um)	30 x 40 μm
Bump Height(um)	45μm (28μm Cu+2μmNi+15μm solder)
Surface Finish	Cu trace with OSP
Trace width	20μm

Table 1 Configuration of EM Test Vehicle

Experiment Plan

The EM test matrix was planned to 5 legs under different ambient temperatures and stress currents. The accelerated temperature condition includes 140°C, 150°C, 160°C and the current of 500mA, 700mA and 900mA were applied which corresponded to current density of 4.97×10^4 A/cm^2, 6.96×10^4 A/cm^2, and 8.95×10^4 A/cm^2, respectively, in relation to 30x40μm (elongated) UBM size. The loop resistance of DUT (device under test) was monitored in-situ by Quali-Tau MIRA system, and the failure criterion of the test is 5% resistance increase from initial values.

In addition to the monitor of resistance change, cross-section to time zero and failure analysis of failed parts by Scanning Electron Microscopy (SEM) and Energy Dispersive X-ray (EDX) are used.

Results and EM life prediction

Cu pillar bump on trace were tested at three different temperatures and input stress currents; detail leg conditions and results are showed in table 2. The junction temperatures caused by Joule heating are correlated from thermal coefficient of resistance (TCR) curve in each DUT before test. The experiment with 500mA (4.97×10^4 A/cm^2) at an ambient temperature of 160°C leads to a final EM test temperature of 167°C, while the experiment with 900mA (8.95×10^4 A/cm^2) results in a EM test temperature of 182°C started at the same ambient temperature. The bump resistance is real time monitored; the typical relative resistance degradation versus time is showed in Figure 2(a). The resistance keeps constant after start and gradually increases after a certain stress time. An interesting thing found is the resistance decreased to a certain value after 10~30% resistances increase from initial value; then it increase again. This is different to the conventional lead free solder bumped joint, shows in figure 2(b) which resistance increases continuously and fails with sudden surge in resistance. The possible mechanism of Cu pillar joint is IMC transformation and thermal migration which will be discussed in later failure analysis.

A lognormal distribution plot showed by cumulative distribution of failure (CDF) data of Cu pillar bump on trace are presented in Figures 3 and 4 for statistical analysis for EM life estimation. All sigma values showed in lognormal plot are kept tight within 0.94~0.98 which is statistically meaningful.

Leg	Current (mA)	Current density (KA/cm^2)	Oven Temp(C)	Bump temp from TCR(C)	Fail/Test ed	Duration (hrs)	t50(hrs) @5%
1	900	90	140	162	17/20	850	80.73
2	900	90	150	172	14/20	850	47.45
3	900	90	160	182	11/14	850	16.30
4	700	70	160	173	7/14	850	83.46
5	500	50	160	167	3/11	850	304.61

Table 2 Test matrix conditions and results

Figures 3 and 4 show the EM performance between different input currents (fixing temperature at 160°C) and oven temperatures (fixing current density at 90KA/cm^2). Fixing the same change percentage; the data indicate that the relative EM performance changes more significantly when input current changes as compared to temperature. The t50 lognormal life of 160°C-500mA (50KA/cm^2) is ~18X longer than that of 160°C-900mA (90KA/cm^2) caused by higher joule heating under high current density

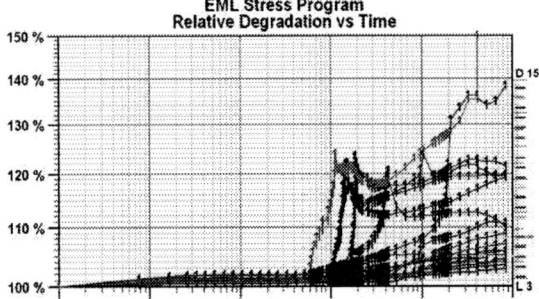

Figure 2(a). Resistance degration vs. time of Cu pillar bump on trace by TC bonding.

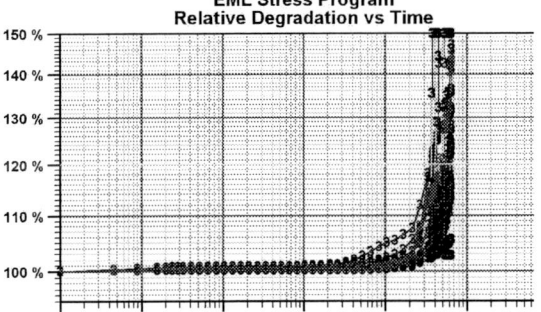

Figure 2(b). Resistance degration vs. time of typical lead-free solder bump (solder on pad).

978-1-4799-2408-0/14 $31.00 © 2014 IEEE

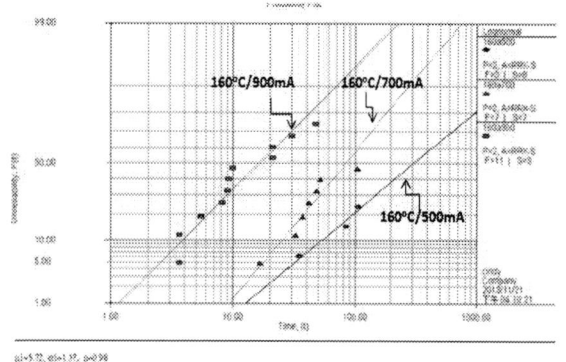

Figure 3. The EM performance between different input currents while fixing oven temperature to 160°C

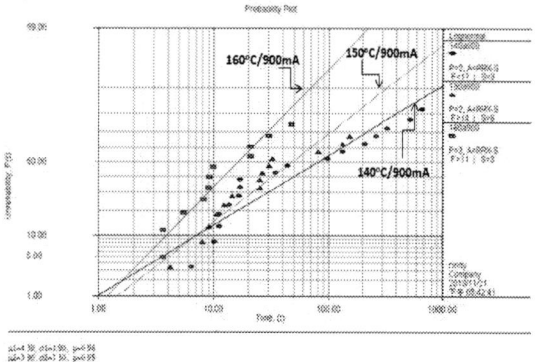

Figure 4. The EM between different oven temperatures while fixing input current to 900mA.

The EM life of Cu pillar bump on trace by thermal compression bonding is predicted by using Black's equation as below.

$$MTTF=AJ^{-n}exp(Ea/kT)$$

Where MTTF is the median time to failure, T is model parameter relating to temperature, J is model parameter relating to current density, A is the constant to be determined experimentally, n is the exponent associated with current density, Ea is the activation energy, and k is Boltzmann constant. The activation energy obtained based on this experiment is 1.36 eV which is slightly higher than reported in other researchers in the range of 1.11~1.20 eV for lead free solder bumps [1], [7]. This is due to all the solder is transferred to IMC in small solder volume joint and void formation caused by solder creeping (detail failure analysis will be given next). It is different to typical UBM depletion or crack propagated through IMC and bulk solder interface. The n value from this experiment is 2.93 which is aligned to typically published data 2.25~3.53 in micro Cu pillar bump with small solder volume [9].

The predict EM life of Cu pillar bump on trace by thermal compression bonding based on the calculated n, Ea can be modeled by Black' equation. To provide a practical assessment, maximum allowable input current design rules at various operation temperatures and commonly used UBM sizes required to meet 10 years under 0.1% cumulative failures based on the criteria of 5% resistance increase are presented in Figure 5. The maximum allowable current density (when the operation temperature is 100°C) shows in Figure 5 is 0.16 mA/um² which is >3X better than the typical LF solder bump (~0.048 mA/um²) in the same temperature [1], even though the failure criteria is 4X tightened from 20% to 5% resistance increase.

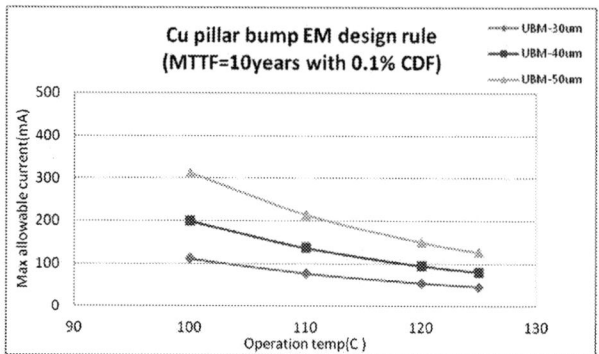

Figure 5. Maximum input current design rule for 10 years under 0.1% cumulative failures for Cu pillar bump on trace by thermal compression bonding.

Failure analysis and discussion

The typical Cu pillar solder joint after thermal compression bonding is showed in below Fig. 6. The Solder joint height between UBM Ni layer and Cu trace is 7~8 um. The inclusion defects related to thermal compression bonding are observed at the bonding interface. A large part of solder (Sn) was translated to Sn-Cu IMC from EDX results and slightly solder wetting Cu post side wall was observed due to flux is included in under fill material for oxide removal during bonding process. Fig. 8 shows no current stress solder joint post 850 hrs from 140°C/900mA EM test die, almost all the solder is converted to Sn-Cu IMC. Cu trace was slightly consumed during thermal aging but UBM Ni layer is almost intact. Obvious solder creeping to Cu post side wall and caused void and solder depletion in bulges at the edge of the bumps were found.

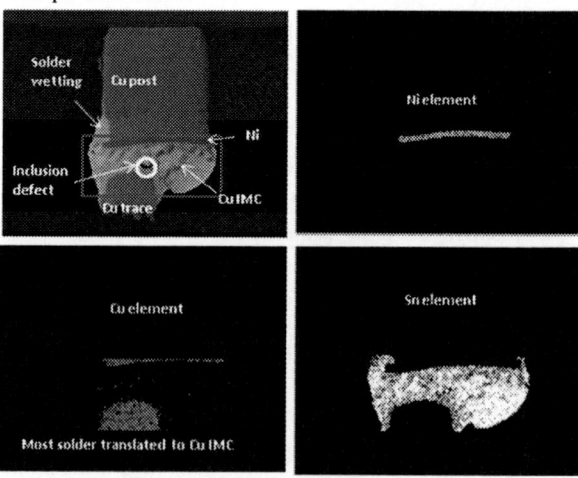

Figure 6. Cu pillar solder joint X-section and EDX mapping after ass'y (before test).

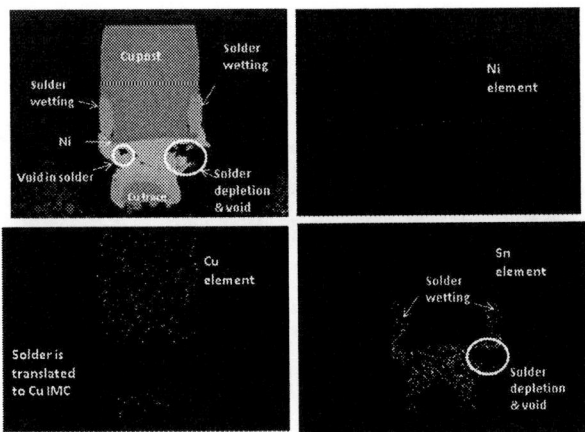

Figure 7. Cu pillar solder joint X-section post 850 hrs under 140°C temperature (no current input).

Figure 8 is the comparison of the solder joint with different UBM dimensions: 90um versus 30 um after finishing assembly process (before test). It can be seen that the IMC/solder ratio is much higher in Cu pillar solder joint due to small solder joint height, which means all the solder in the Cu pillar joint is transferred to IMC soon after EM test. With the diffusion between Cu, Ni and Sn, more $(Cu,Ni)_6Sn_5$ compounds will be formed until all the Sn was consumed which caused the resistance increase in the beginning. After solder consumption, the phase translation and form the Cu_3Sn between Cu and Cu_6Sn_5 would be dominant, so the resistance will be slightly lower or almost constant [3]. This explains the resistance decreased to a certain value or keep constant after 10~30% resistances increase from initial value shown in Figure 2(a). But as compared to typical lead-free bumps with 60~70 um joint height, the IMC/solder ratio is much lower, the $(Cu,Ni)_6Sn_5$ IMC will be continued growth due to infinity Sn source during EM test; finally it caused UBM/Cu pad over-consuming and cracks between bulk solder.

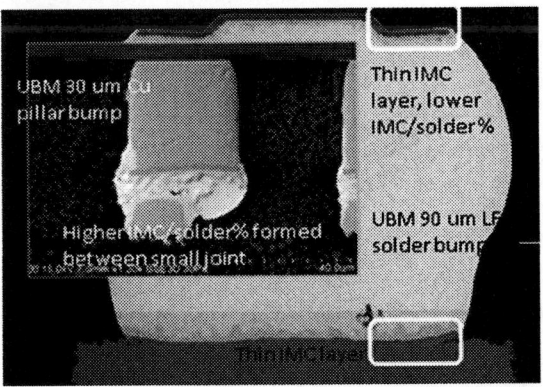

Figure 8. Solder joint with different UBM dimensions: 90 um versus 30um result in different stand-off heights and IMC/solder ratios.

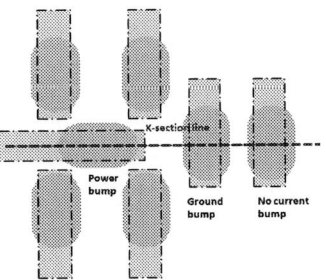

Figure 9. Illustration of Cu pillar solder joint X-section of EM test failed bump.

The illustration of Cu pillar solder joint X-section of EM test failed bump is showed in Figure 9. The X-sections were perpendicular to the long direction of pad for power bump and to the short direction of pad for ground bump and adjacent thermal reference bump. Figure 10 shows the failure mode of solder joint X-section after 900 mA current stress with temperatures, 140°C, 150°C and 160°C post 850hrs stress time. The amount of voiding, cracks (small cracks appeared between the particles which are entrapped in bonding surface) and solder creeping have significantly increased compared to the initial stage when temperature is increasing. Nevertheless, this degradation only slightly effect the monitor resistance; for example, there is only 6% resistance increase found in 160°C/900mA failed sample (Figure 10 (G), (H)) post 850hrs stress time. From the bump X-section comparison; it shows the thermal driven failure is the dominant factor and void, cracks will be enhanced or reduced based on electron flow direction. Higher joule heating temperature is expected in power bumps due to the current density of power bump is 5X higher than ground bump, but only minor void, cracks formed in it. This indicates that solder creeping and IMC formation on Cu post side wall will be reduced when electron flows from die to substrate side and increased in the opposite direction found in ground bump.

After EM test, all of the Sn solder was translated to $(Cu,Ni)_6Sn_5$ IMC and the growth of Cu_3Sn IMC layers were observed on substrate trace area. No obvious Kirkendall voids/plane formed between Cu and Cu_3Sn which are commonly found in Cu pillar bump on trace joint by mass reflow. The material dissolutions occurred on different sides of joints is depending on the electron flow direction. In the power bump; slightly Ni layer consumption was observed due to the electron wind from die side to Cu trace side reduced the dissolution rate of Cu atoms from Cu trace, but in the ground bump, the Ni layer is almost intact due to more Cu flux move into solder from Cu trace and reduce the Ni_6Sn_5 IMC formation .

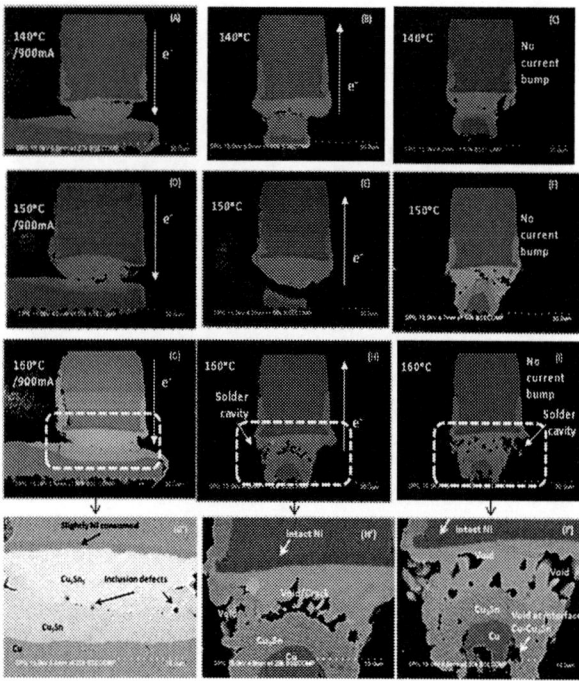

Figure 10. The Cu pillar solder joint X-section of EM test failed part under 140°C/150°C/160°C post 850 hrs (fixing 900 mA current input).

Conclusions

New features of Cu pillar bumps assembled with bump on trace (BOT) process by using thermal compression bonding have been observed, some of the important results are summarized in the followings.

1. Based on the experimental results for a fixed temperature (160°C) or current (900mA), it has been found and expected that, the lower the current or oven temperature the better the EM performance. The joule heating is significant increased when input current density is up to 90KA/cm^2 (~22°C higher than oven temperature) and it significantly reduces t50 lognormal life as compared to 50KA/cm^2 leg under the same oven temperature.

2. Based on the tighten failure criterion with 5% resistance increase from initial value, the activation energy (Ea) obtained from the experimental results is 1.36 eV and the exponent associated with current density (n) is 2.93.

3. The electromigration life with 4X tightened failure criteria of the Cu pillar bumped SnAg solder joints on trace has been found to be 3X better than that of the conventional lead-free SnAg or SnCu solder joints based on black' equation based from these experiments. The reference design rule of Cu pillar bump on trace by thermal compression bonding has also been provided.

4. The joint resistance of Cu pillar bumped on trace increases gradually after test started due to (Cu,Ni)$_6$Sn$_5$ IMC formation; but it started to decrease a little or

keep almost constant after 10~30% resistance increase from the beginning due to the growth of Cu$_3$Sn and the diffusion reached equilibrium state. However, the presence of the void/crack formed in solder joint becomes more pronounced with increasing operation temperature and current density, so the resistance increased again. The thermal driven failure is the dominant factor and void/cracks will be enhanced or reduced based on electron flow direction.

Acknowledgments

The authors would like to thank members of EC and RA team of SPIL for their help in sample preparation and experimental setup.

References

1. Kuei Hsiao Kuo, Jason Lee, Stan Chen, F.L. Chien, Rick Lee, and John Lau, "Electromigration Performance Of Printed SnCu0.7 Bump With Immersion Tin Surface Finish For Flip Chip Application", pp. 698-702 2012 Electronic Components and Technology Conference.

2. R Labie, P Limaye, KW Lee, CJ Berry, E Beyne, and I De Wolf "Reliability testing of Cu-Sn intermetallic micro-bump interconnections for 3D-device stacking", pp. 1-5, 2013 Electronic System-Integration Technology Conference.

3. Da-Quan Yu, Tai Chong Chai, Meei Ling Thew, Yue Ying Ong, Vempati Srinivasa Rao, Leong Ching Wai, John H. Lau, "Electromigration study of 50um pitch micro solder bumps using four-point Kelvin structure", pp. 930-935, 2009 Electronic components and technology conference.

4. Yasumitus Orii, Kazushige Toriyama, Sayuri Kohara, Hirokazu Noma, Keishi Okamoto, Daisuke Toyoshima, Keisuke Uenishi, "Effect of preformed IMC layer on electromigration of peripheral ultra fine pitch C2 flip chip interconnection with solder capped Cu pillar bump", pp.206-209, 2011 International Microsystems, Packaging, Assembly and Circuits Technology Conference.

5. Ahmer Syed, Karthikeyan Dhandapani, Robert Moody, Lou Nicholls, and Mike Kelly, "Cu pillar and μ-bump electromigration reliability and comparison with high Pb, SnPb, and SnAg bumps", pp.332-339, 2011 Electronic Components and Technology Conference.

6. Lou Nicholls, Robert Darveaux, Ahmer Syed, Shane Loo, Tong Yan Tee, Thomas A Wassick, Bill Batchelor, "Comparative electromigration performance of Pb free flip chip joints with vary board surface condition", pp.914-921, 2009 Electronic Components and Technology Conference.

7. Seung-Hyun Chae, Xuefeng Zhang, Huang-Lin Chao, Kuan-Hsun Lu, Paul S. Ho, "Electromigration lifetime statistics for Pb-Free solder joints with Cu and Ni UBM in plastic flip-chip packages", pp.650-656, 2006 Electronic Components and Technology Conference.

8. Riet Labie, Frank Dosseul, Tomas Webers, Christophe Winters, Vladimir Cherman, Eric Beyne, Bart Vandevelde, "Outperformance of Cu pillar flip chip bumps in electromigration testing" pp.312-316, 2011 Electronic Components and Technology Conference.

9. Ha-Young You, Yuchul Hwang, Jung-Woo Pyun, Young-Gyun Ryu, Hyoung-Sub Kim, "Chip package interaction in micro bump and TSV structure" pp.315-318, 2012 Electronic Components and Technology Conference.

10. S. Lee, Y.X. Guo, C.K. Ong, "Electromigration effect on Cu-pillar(Sn) bumps", pp.135-139, 2005 Electronic Components and Technology Conference.

Flip-chip Bonding Alignment Accuracy Enhancement using Self-aligned Interconnection Elements to Realize Low-temperature Construction of Ultrafine-pitch Copper Bump Interconnections

Bui Thanh Tung,[a,b] Naoya Watanabe,[a] Fumiki Kato,[a] Katsuya Kikuchi,[a] and Masahiro Aoyagi[a,b]

[a] Nanoelectronics Research Institute (NeRI)
[b] Institute for Photonics-Electronics Convergence System Technology (PECST)
National Institute of Advanced Industrial Science and Technology (AIST)
Tsukuba Central 2, 1-1-1 Umezono, Tsukuba, Ibaraki, 305-8568 Japan
tung.bui@aist.go.jp, Tel. +81-29-862-6510

Abstract

In this paper, the integration accuracy of conventional flip-chip bonding is effectively enhanced by means of automatically maintaining the alignment between the chip and substrate during the time that offsets may take place, i.e., bonding conditions applying period. The conventional bonding bump and pad elements have been appropriately modified to construct a concave-convex pair, i.e., self-aligned interconnection elements (SIEs). By this way, the post-bond offsets are determined by the SIEs, aiming at highly reproducible sub-micron range bonding precision. Moreover, because the post-bond offsets are guaranteed by the SIEs, ultrasonic assisted technique can be applied to make reliable bonds at acceptably low temperatures, while still maintaining the integration accuracy. Ultrafine-pitch (i.e. down to 10 µm) copper bump interconnections were realized using the proposed integration approach.

Introduction

The flip-chip bonding (FCB) with thermal compression method has been a promising technique for high density interconnection in 3D interconnection technologies which is emerging and playing important roles in furthering electronics miniaturization (Figure 1(a)). However, the final bonding accuracy of bonding technique depends not only on the alignment technology but also on the stacking process [1]. Post-bond accuracies of presently commercialized FCB machines have been limited to the range of 2 µm – 5 µm, because of unavoidable effects such as thermal-induced misalignment caused by the thermal expansion mismatching between the bonding chip and substrate, and by horizontal shifting of the press tool due to the shear force generated from the large down force during pressing. Therefore, for emerging applications, such in photonics-electronics convergence system tech-

nology, where the performance of a system is highly dependent on the coupling efficiency affected by assembly deviations [2–5], the integration accuracy of the FCB approach needs to be improved significantly.

The improvement of the integration accuracy has been investigated by several research groups [5–10]. Attempting to improve the alignment accuracy of the FCB method through increasing the resolution of the alignment stage motion and increasing the magnification of the IR image camera is not sufficient for high-precision bonding. Additionally, high-precision alignment systems have high costs, as well as very long process times. The key point to achieve high-precision bonding results is to maintain the alignment of the chip and the substrate during the bonding process [11–13]. To do so, the conventional bonding bump and pad elements have been appropriately modified to construct a concave-convex pair, i.e., self-aligned interconnection elements (SIEs). In this way, after being coarsely aligned, under the application of an appropriate aligning force, bonding bumps tend to slide into the hollow pads, and the misalignment, which may have happened in the alignment process, is self-corrected and maintained at that state during application of bonding forces. With this novel bonding approach, the post-bond offsets are determined by the SIEs, aiming at highly reproducible sub-micron range bonding precision.

In our previous works, we have succeeded in making reliable 10 µm-diameter Au bump interconnections with submicron range bonding accuracy [14–17]. Copper, due to its superior electrical conductivity, superior electromigration resistance, lower cost, as well as superior mechanical properties, such as yield stress and Young's modulus, is preferred over Au to form electrically and mechanically compliant interconnect elements [18–20]. The proposed bonding approach has been demonstrated and Cu/Au connections with submicron

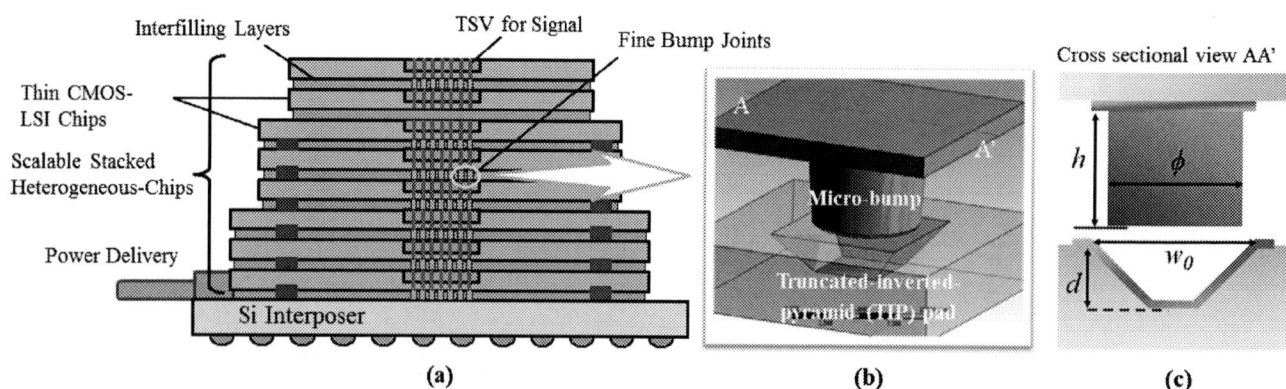

Figure 1. Schematic of the 3D integrated LSI system (a) using self-aligned interconnection elements (b, c).

978-1-4799-2408-0/14 $31.00 © 2014 IEEE

range bonding accuracy have been presented in our previous works [21–23]. In addition, the application of the proposed FCB approach to realize reliable fine-pitch metal-to-metal interconnects, which are suitable for high density interconnections and high accuracy chip stacking applications are also in progress. The initial results on developing of 15-μm-pitch interconnects have been reported in [22, 23]. In this paper, the development of fine-pitch copper bump interconnects, including the pitches down to 10 μm, is presented.

Self-aligned Interconnection Elements (SIEs) for Flip-Chip Bonding

Flip-chip bonding (FCB) or controlled collapse chip connectio (C4) is a promising technique for high density interconnection and heterogeneous integration (see Figure 2). According to a report on bumping interconnect technology roadmap (Yole Développement), fine-pitch interconnections for flip-chip technology are projected to decrease to 30 μm by 2016 [24]. In a conventional bonding process, a planar bonding pad is utilized (e.g., Figure 2 (a)). Accordingly, even with highly precise alignment approaches, the final results reveal variations due to the application of bonding conditions (e.g., Figure 2 (d, e)). To overcome this problem, appropriate modifications in interconnection elements have been conducted to have SIEs, a convex and concave pair (Figure 1 (b, c)). During the bonding process, under the application of a bonding force, bumps are aligned to get into the hollow pads, and the misalignment, which may have happened in the alignment

process, is self-corrected and maintained at that state. By this method, the alignment between a chip and a substrate is improved efficiently.

It is noted that both the substrate and the chip are held respectively by vacuum to the stage and bonding head, during bonding process. The substrate is much larger than the chip, and held by surrounding guides. Given this, we have ensured the substrate to be properly anchored to the stage and the chip has to slide with respect to the bonding head, when the misalignment correction phenomenon occurs.

Low-temperature Construction of Ultrafine-pitch Copper Bump Interconnections

Because of the post-bond offsets are guaranteed by the SIEs, ultrasonic assisted technique can be applied to make reliable bonds at acceptably low temperatures, while the integration accuracy still being guaranteed. The bonding process was demonstrated by the mounting process of a test chip to a substrate. In particular, fine-pitch electroplating Cu microbumps, and TMAH-etched truncated inverted pyramid (TIP) bonding pads, which are served as the SIEs, can be fabricated on (100) Si substrate. The fabrication process of electroplating bumps and TIP pads on test samples is briefly shown in Figure 3.

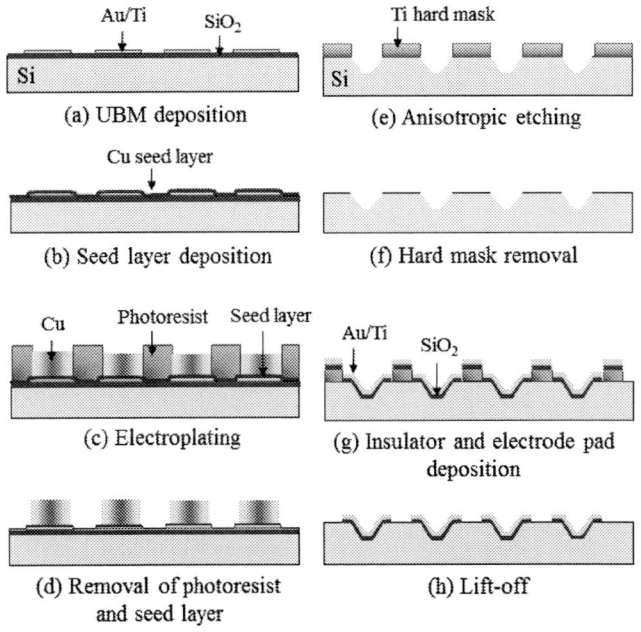

Figure 3. Fabrication process of MSCEs: fabrication process of electroplating bump (a-d), and metalized hollow pad (e-h).

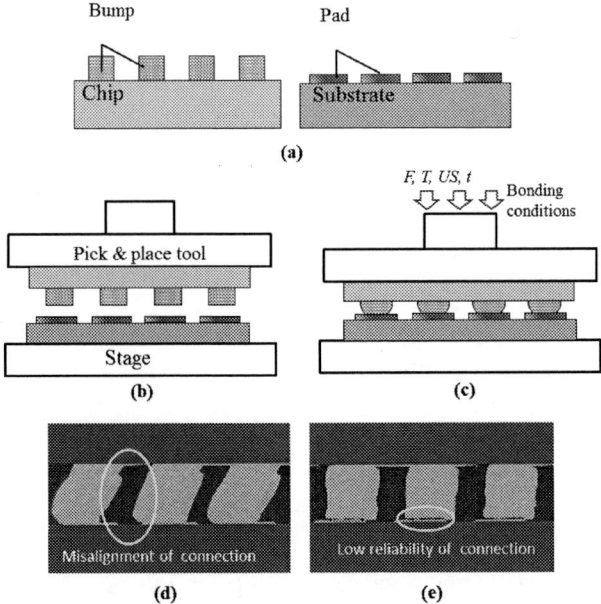

Figure 2. Conventional flip-chip bonding procedure. First, the bonding chip, held by a pick-and-place tool, is aligned with the substrate, which is fixed to the stage (b). Bumps on the chip and pads on the substrate are placed in contact with each other and bonding conditions (i.e., force, temperature, ultrasonic energy, for period of time) are applied through the bonding head to form microwelds (c). The common faults in flip-chip bonding, including misalignment after application of bonding conditions and low reliability of connection due to bump height deviation are showed in (d) and (e).

Bump diameter and cut-out width were decided so that bump can be inserted into the hollow pads even if the misalignment between them is in the range of micrometers. The conditions for electroplating were as follows: the plating bath was ETN (80 g/l CuSO$_4$–5H$_2$O, 200 g/l H$_2$SO$_4$, 50ppm Cl-), Uyemura Co., Ltd., the bath DC voltage was 50mV, the electric current density was 2 – 2.5 A/dm^2, the temperature of plating bath was 25 °C, and the plating time was 30 minutes. Metalized TMAH-etched inverted pyramid cavities were formed on the (100) Si substrate. Etching process was con-

978-1-4799-2408-0/14 $31.00 © 2014 IEEE 63

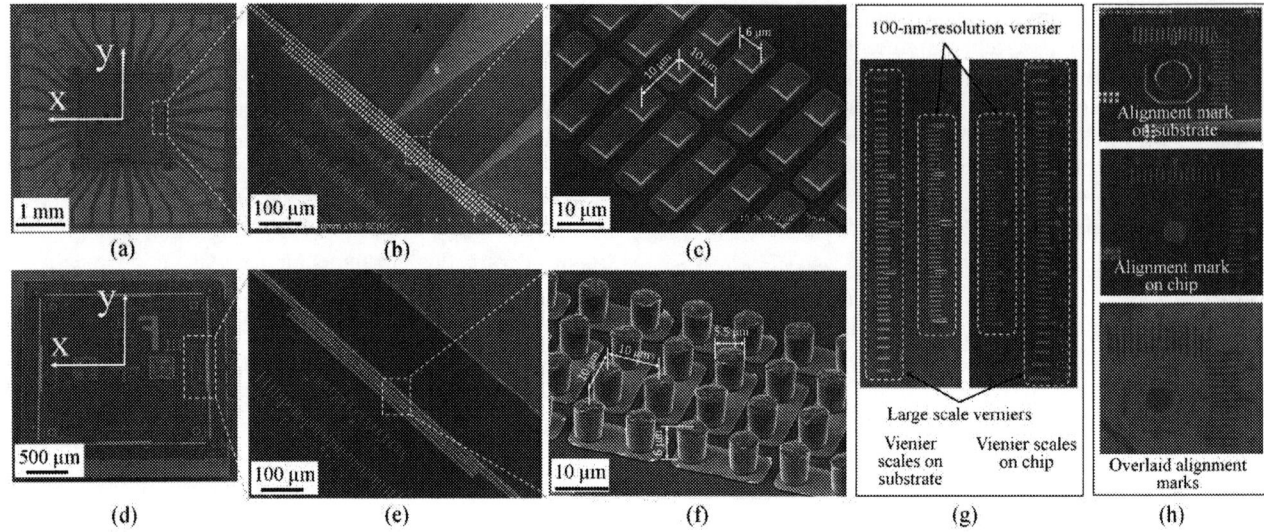

Figure 4. Fabrication results showing the TIP bonding pads on Si substrate ((a) – (c)), Cu bumps on test chip ((d) – (f)), vernier scale to evaluate the bonding accuracy (g) and alignment marks (h).

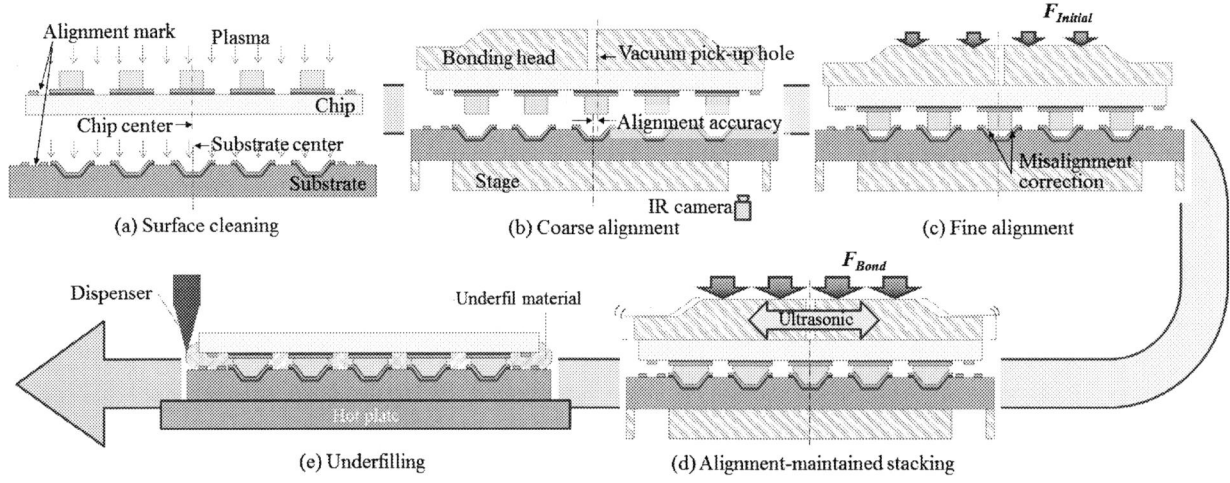

Figure 5. Schematic representation of the proposed modified US-FCB.

ducted using TMAH 22% at 70 °C for 8 min. Parameters of these samples are listed in Table 1. Bumps and pads were connected to form a daisy chain for the electrical testing of the interconnections. In order to evaluate the bonding accuracy, specially designed vernier scales were formed during the fabrication process, in the x- and y-axes. The fabricated chip specimen and substrate are shown in Figure 4. Figure 4 (a-c) presents the layout of TIP bonding pads on a substrate. Layout of Cu bumps on a test chip is presented in and Figure 4 (d-f). Vernier scales to evaluate the bonding accuracy and alignment marks for alignment process are presented in Figure 4 (g) and Figure 4 (h), respectively.

The mounting process was implemented using a flip-chip bonder CA-300 (Bondtech Co., Ltd.). The original thermal compression bonding head of the bonder was replaced with an ultrasonic horn (Adwelds Co., Ltd.). The modified TS-FCB

process is briefly described in Figure 5. The accuracy of the bonding approach is based on the self-alignment mechanism of SIEs while the bonding process can be obtained at low temperature is due to the assistance of ultrasonic energy. The optimized bonding parameters are listed in Table 2.

Table 1: Test chip features.

Characteristic	Flip chip
Die size	2.5 mm × 2.5mm
Pitch	10 μm
Bump diameter	5.5 μm
Bump height	6 μm
Bump count	2048 (=2^{11})

First, the chip and substrate surface are treated with plasma cleaning (Figure 5 (a)). Then, the chip is flipped and coarsely aligned with the substrate (Figure 5 (b)). This process is done using the normal aligning function of the bonding machine. A safety gap is remained between the lowest part of the top chip and the highest part of the substrate during this period. After that, the chip is moved downward to approach the bump to the hollow pad surfaces. An initial downward force is applied to the top chip. Self-alignment effect takes place to correct the misalignment if there is one (Figure 5 (c)). The correction of the misalignment is based on the mechanical guide coming from the sloped side wall of the hollow pad. Subsequently, bonding energies are applied to construct a strong connection (Figure 5 (d)) and the alignment-maintained strong bond is obtained. The bonded chip is finally filled with an underfilling material, a resin strengthening agent, and is heated to be hardened (Figure 5 (e)).

Table 2: Bonding parameters.

Parameter	Value
Loading (gf/bump)	4
Ultrasonic amplitude (μm)	2.2
Ultrasonic frequency (kHz)	48.5
Bonding time (s)	0.5
Bonding temperature (°C)	RT
Chip size ($l \times w \times t$ mm^3)	2.5 × 2.5 × 0.2
Substrate size ($l \times w \times t$ mm^3)	10 × 10 × 0.38

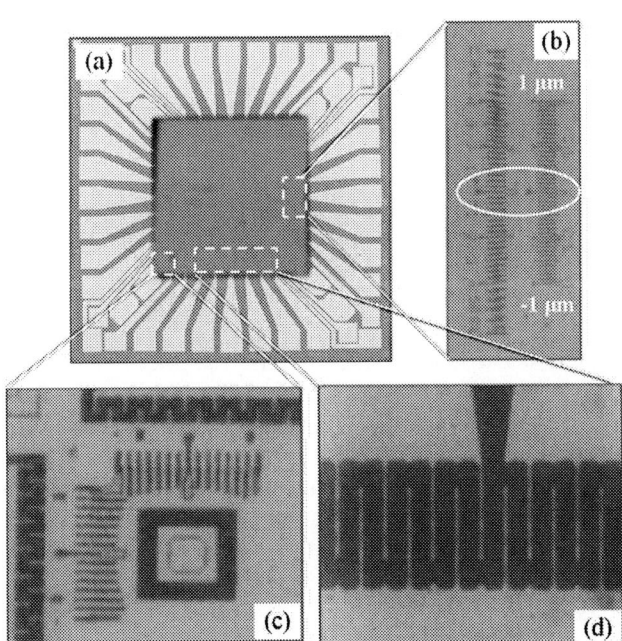

Figure 6. Bonded chip (a) and IR images illustrating post bond tolerances (b, c, d) showing the perfect matched of the 250-nm- and 100-nm-resolution vernier scales (b) and no bridging between the neighboring interconnections (c, d).

The stacked samples were inspected for bonding accuracy as well as properties of bonds. By using the designed vernier

scales, we could measure the offsets in the x and y directions with the resolutions of 100 in the range of ±1. Figure 6 shows a bonded chip on a substrate. By examining the overlaid vernier scales using an infrared (IR) camera, we confirmed that the chip was bonded with the substrate with the help of bonding bump-pad pairs, resulting in high-precision bonding, i.e., less than 500 nm in in-plane offsets. Although, the bonding process time was the same as that of a typical ultrasonic bonding process (i.e., without the use of the modified bonding bump-pad pair), the bonding accuracy was much improved.

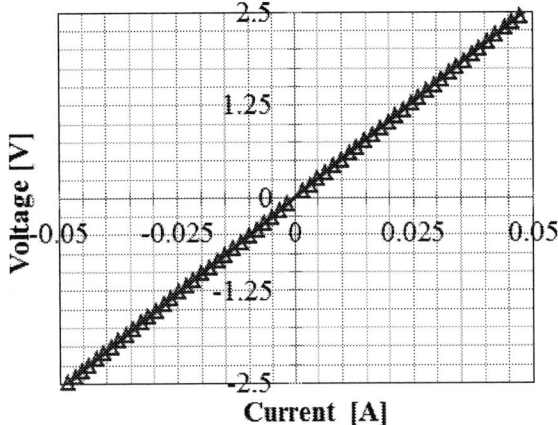

Figure 7. Current-voltage characteristics of connected daisy chain with 200 bump joins. The linearity depicts ohmic contacts of the interconnections.

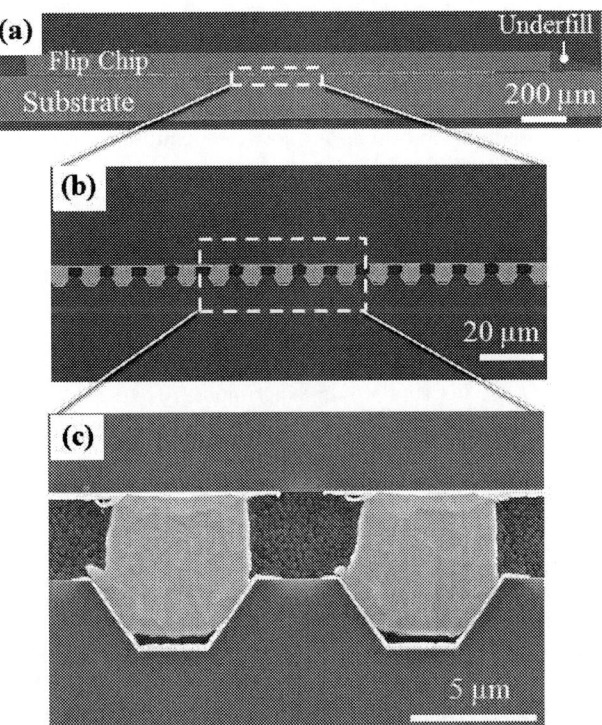

Figure 8. Cross sectional view of the as-stacked chip (a) and close-up view at interconnects (b, c).

In addition, the interconnection electrical resistances of the joints ware measured. The current-voltage characteristics of 200 connections daisy-chain are presented in Figure 7. The average resistance of the Cu bumps and hollow pad connection, including the wiring part, was calculated to be approximately 0.243 Ω. The linearity of the I–V characteristics confirmed the ohmic contact of the interconnections.

Owning to the presence of the modified bump-pad elements, the offsets between flip chip and substrate are corrected and maintained during stacking, stimulating for ultrafine pitch fine connection. Besides, critical alignment process is not required and conventional bonders can be utilized, which means time efficiently and economically. Moreover, since bumps collapse more easily at their edges when interfacing with TIP pads, bump height deviation induced problems, which would be the problem in conventional planar bonding pad (e.g., Figure 2 (e)), are avoidable. Furthermore, a large friction is considered to be generated when the cone bumps enter the pyramid hollow pad. This forms a clean interface by removing surface contaminants and, as a result, reliable connection scan be obtained event at lower temperature, i.e., room temperature.

Conclusions

A bonding process using a modified Au bonding pad and an electroplated Cu bump was introduced. The conventional bonding pads were modified to construct, in combination with bumps, self-aligned interconnection elements (SIEs), aiming at highly reproducible sub-micron range bonding precision for ultra-fine pitch interconnections. The experimental validation results reveal that the alignment accuracy of flip-chip bonding technique was significantly enhanced and ultrafine-pitch copper bump interconnects were constructed at room temperature.

The method proposed here is also scalable to fine pitch enabling future chip stacking applications and facilitating the development of heterogeneous integration systems, particularly those composed of non-heat-resistant materials as well as photonic-electronic convergent applications, where requirements for high precision, and low-temperature processing is compulsory.

Acknowledgments

This research is granted by the Japan Society for the Promotion of Science (JSPS) through the "Funding Program for World-Leading Innovative R&D on Science and Technology (FIRST Program)," initiated by the Council for Science and Technology Policy (CSTP). The authors would like to thank N. Igawa for his supports in experiment. A part of this work was conducted at the Nano-Processing Facility, supported by NPF, AIST.

References

1. S. H. Lee, K.-N. Chen, and J. J.-Q. Lu, "Wafer-to-Wafer Alignment for Three-Dimensional Integration: A Review," *J. Microelectromech. Syst.*, vol. 20, no. 4, pp. 885–898, Aug. 2011.
2. Y. Urino, T. Shimizu, M. Okano, N. Hatori, M. Ishizaka, T. Yamamoto, T. Baba, T. Akagawa, S. Akiyama, T. Usuki, D. Okamoto, M. Miura, M. Noguchi, J. Fujikata, D. Shimura, H. Okayama, T. Tsuchizawa, T. Watanabe, K. Yamada, S. Itabashi, E. Saito, T. Nakamura, and Y. Arakawa, "First demonstration of high density optical interconnects integrated with lasers, optical modulators, and photodetectors on single silicon substrate," *Opt. Express*, vol. 19, no. 26, pp. B159–B165, Dec. 2011.

3. Y. Arakawa, "Photonics-electronics convergence system technology (PECST) as one of the thirty FIRST projects in Japan," in *16th OptoeElectronics and Communications Conf. (OECC)*, 2011, p. 836.
4. K.-W. Lee, A. Noriki, K. Kiyoyama, T. Fukushima, T. Tanaka, and M. Koyanagi, "Three-Dimensional Hybrid Integration Technology of CMOS, MEMS, and Photonics Circuits for Optoelectronic Heterogeneous Integrated Systems," *IEEE Trans. Electron Devices*, vol. 58, no. 3, pp. 748–757, Mar. 2011.
5. K.-W. Lee, A. Noriki, K. Kiyoyama, S. Kanno, R. Kobayashi, W.-C. Jeong, J.-C. Bea, T. Fukushima, T. Tanaka, and M. Koyanagi, "3D heterogeneous opto-electronic integration technology for system-on-silicon (SOS)," in *IEDM Tech. Dig.*, 2009, pp. 531–534.
6. M. Koyanagi, "3D integration technology and reliability," in *Int. Reliability Physics Symp. (IRPS)*, 2011, pp. 328–334.
7. M. Esashi, "Wafer level packaging of MEMS," *J. Micromech. and Microeng.*, vol. 18, p. 073001, Jul. 2008.
8. C. G. Tsai, C. M. Hsieh, and J. A. Yeh, "Self-alignment of microchips using surface tension and solid edge," *Sens. Actuators, A*, vol. 139, no. 1–2, pp. 343–349, Sep. 2007.
9. S. Kawashima, M. Imada, K. Ishizaki, and S. Noda, "High-Precision Alignment and Bonding System for the Fabrication of 3-D Nanostructures," *J. Microelectromech. Syst.*, vol. 16, no. 5, pp. 1140–1144, Oct. 2007.
10. C. Wang and T. Suga, "Moire method for nanoprecision wafer-to-wafer alignment: Theory, simulation and application," in *Int. Conf. Electronic Packaging Technology and High Density Packaging*, 2009, pp. 219–224.
11. L. Jiang, G. Pandraud, P. J. French, S. M. Spearing, and M. Kraft, "A novel method for nanoprecision alignment in wafer bonding applications," *J. Micromech. Microeng.*, vol. 17, no. 7, pp. S61–S67, Jul. 2007.
12. S. H. Lee, F. Niklaus, J. J. McMahon, J. Yu, R. J. Kumar, H. Li, R. J. Gutmann, T. S. Cale, and J.-Q. Lu, "Fine Keyed Alignment and Bonding for Wafer-Level 3D ICs," *MRS Proc.*, vol. 914, pp. F10–05, 2006.
13. A. H. Slocum and A. C. Weber, "Precision passive mechanical alignment of wafers," *J. Microelectromech. Syst.*, vol. 12, no. 6, pp. 826 – 834, Dec. 2003.
14. B. T. Tung, L. Ma, M. Suzuki, F. Kato, S. Nemoto, N. Watanabe, and A. Masahiro, "Sub-micron-accuracy Gold to Gold Interconnection Flip-Chip Bonding Approach for Electronics-Optics Heterogeneous Integration," in *Int. Conf. Solid State Devices and Materials*, Kyoto, Japan, 2012, p. 1174.
15. B. T. Tung, M. Laina, M. Suzuki, F. Kato, S. Nemoto, and M. Aoyagi, "High-precision heterogeneous integration based on flip-chip bonding using misalignment self-correction elements," in *2012 International Conference on Optical MEMS and Nanophotonics (OMN)*, Banff, Canada, 2012, pp. 93 –94.
16. B. T. Tung, S. Motohiro, K. Fumiki, W. Naoya, N. Shunsuke, and Aoyagi Masahiro, "A Prospective Sub-micron Range Integration Approach for Photonics-Electronics Heterogeneous Convergence Applications,"

presented at the 2nd International Symposium on Photonics and Electronics Convergence (ISPEC2012), Tokyo, Japan, 2012, p. 53.

17. B. T. Tung, M. Suzuki, F. Kato, S. Nemoto, N. Watanabe, and M. Aoyagi, "Sub-Micron-Accuracy Gold-to-Gold Interconnection Flip-Chip Bonding Approach for Electronics–Optics Heterogeneous Integration," *Jpn. J. Appl. Phys.*, vol. 52, no. 4, p. 04CB08, 2013.

18. S. A. Khan, A. Choudhury, N. Kumbhat, M. R. Pulugurtha, V. Sundaram, G. Meyer-Berg, and R. Tummala, "Multichip Embedding Technology Using Fine-Pitch Cu-Cu Interconnections," *IEEE Transactions on Components, Packaging and Manufacturing Technology*, vol. 3, no. 2, pp. 197–204, 2013.

19. J. Fan, D. F. Lim, L. Peng, K. H. Li, and C. S. Tan, "Low Temperature Cu-to-Cu Bonding for Wafer-Level Hermetic Encapsulation of 3D Microsystems," *Electrochem. Solid-State Lett.*, vol. 14, no. 11, pp. H470–H474, Jan. 2011.

20. C. S. Tan and R. Reif, "Silicon Multilayer Stacking Based on Copper Wafer Bonding," *Electrochem. Solid-State Lett.*, vol. 8, no. 6, pp. G147–G149, Jun. 2005.

21. B. T. Tung, S. Motohiro, K. Fumiki, W. Naoya, N. Shunsuke, K. Katsuya, and A. Masahiro, "Modified thermosonic flip-chip bonding based on electroplated Cu microbumps and concave pads for high-precision low-temperature assembly applications," in *Electronic Components and Technology Conference (ECTC), 2013 IEEE 63rd*, 2013, pp. 425–430.

22. B. T. Tung, F. Kato, N. Watanabe, S. Nemoto, K. Kikuchi, and M. Aoyagi, "15-µm-pitch Cu/Au Interconnections Relied on Self-aligned Low-temperature Thermosonic Flip-chip Bonding Technique for A dvanced Chip Stacking Applications," in *Extended Abstracts of the 2013 International Conference on Solid State Devices and Materials*, Fukuoka, Japan, 2013, pp. 986–987.

23. B. T. Tung, F. Kato, N. Watanabe, S. Nemoto, K. Kikuchi, and M. Aoyagi, "15-µm-pitch Cu/Au Interconnections Relied on Self-aligned Low-temperature Thermosonic Flip-chip Bonding Technique for Advanced Chip Stacking Applications," *Jpn. J. Appl. Phys*, vol. 53, no. 4, 2014 (in press).

24. http://www.i-micronews.com/advanced-packaging/reports/, "I-Micronews - ADVANCED PACKAGING - Reports.

Development of Second-level Connection Method for Large-size CPU Package

Shunji Baba[*1], Masateru Koide[*1], Manabu Watanabe[*1], Kenji Fukuzono[*1], Tsuyoshi Yamamoto[*1],
Seiki Sakuyama[*2], Kozo Shimizu[*2], Keishiro Okamoto[*2], Daisuke Mizutani[*2]

[*1] Fujitsu Advanced Technologies Ltd.,
1-1, Kamikodanaka 4-chome, Nakahara-ku, Kawasaki-shi, Kanagawa 211-8588, Japan
[*2] Fujitsu Laboratories Ltd.,
10-1 Morinosato-Wakamiya, Atsugi-shi, Kanagawa 243-0197, Japan

Abstract

This paper reports on second-level interconnection development for a large-scale Ball Grid Array (BGA) package. Generally, control of warpage becomes a problem as BGA packages become larger. To solve this problem, the following two measures were executed. The first was adoption of a low-temperature solder, and the second was warpage control using a heat spreader as a fixture. We were able to decrease the reflow temperature to 200°C by applying the low-temperature solder, and the effect was a warp reduction of 200 μm. Moreover, the shape of the heat spreader was optimized through a thermal-stress simulation, obtaining a warp reduction of 100 μm. Verification with a test vehicle was executed, no short/opening was observed, and the results of a thermal cycle test and simulation confirmed there was no problem in reliability.

1. Background

For performance gains in computer systems, the CPU has more I/O signal pins. Moreover, for the shielding of high-speed transmission signal pins, the number of ground/power supply pins has increased. The increase in the number of CPU pins has led directly to an increase in package size. Consequently, warpage control during the reflow process has become the biggest problem in the second-level connection process.

Figure 1. Appearance of developed BGA package.

Figure 1 shows the appearance of the BGA package that we developed at this time. The specifications of the BGA package are summarized in Table 1. The number of BGA pads has reached 4,384, and the size of the package is 3,969 mm².

Table 1. Specifications of developed BGA package.

Item	Specification
LSI Size	728mm² (26mm/28mm)
Substrate Size Substrate Material	3,969 mm² (63mm/63mm), thickness=1.45 mm Organic (6-4-6 build-up)
Number of Pins	4,384 pins
BGA Pad Pitch	0.8 mm

Figure 2 shows the relationship between the amount of warping and the package substrate size at the temperature of the reflow process, as calculated by a thermal-stress simulation. The amount of warping, δ, (mm) is shown on the vertical axis, and the size of the package substrate is shown on the horizontal axis. The simulation model is a simple two-layer model consisting of the LSI and the substrate.

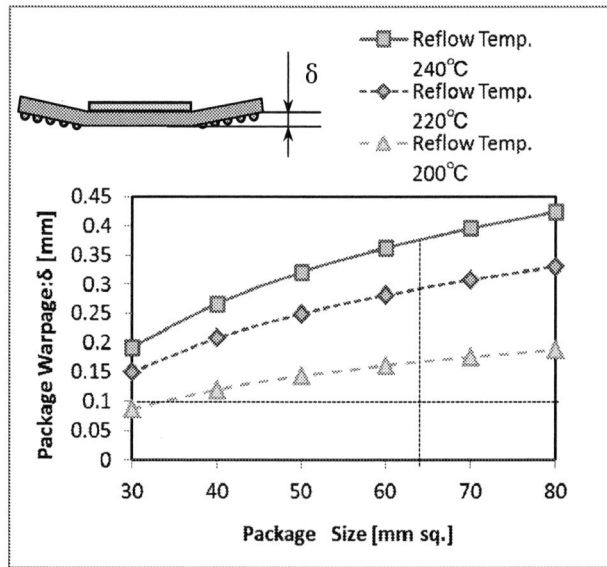

Figure 2. Relationship between package size and warpage.

In contrast, the necessary allowance for warping can be calculated geometrically in consideration of differences in the volume of the solder ball, the amount of pre-solder, etc. We estimate that the amount of warping should be 0.1 mm or less for a 0.8 mm pitch/3,969 mm² BGA package. Clearly from Figure 2, it is necessary to decrease the warp amount by about

2014 Electronic Components & Technology Conference

0.3 mm at the reflow peak temperature of 240°C, for the usual lead-free reflow process.

An effective approach for the warp reduction is to lower the peak temperature during the reflow process. Recently, solder application of a Sn-Bi alloy system was adopted for practical use, and the reliability of this alloy system has been actively researched in subsequent investigations [1-3]. Generally, 0.5-1.0% Ag is added to Sn-Bi alloy solder to improve ductility. The melting point of SnBiAg solder is about 138°C.

However, even when a low-temperature solder is applied and the reflow temperature decreases to 200°C, as understood from Figure 2, it is necessary to further decrease the warping by about 0.05 mm. The technique generally used for correcting warping is a jig in the reflow process. However, to make a reflow jig for this exclusive use is not preferable from the perspective of cost and productivity, so we have developed a process where the warping is controlled using a heat spreader as a warp suppression jig and the BGA joint and heat spreader joint are formed at the same time. We have already put this technology to practical use as a BGA package manufacturing process [4,5]. We applied the same technology to this BGA package development, which is described in detail in the following section.

2. Details about Warp Reduction Methods

2-1. Low-temperature solder material selection

Sn-57Bi-1.0Ag solder was selected as the first candidate in consideration of affordability. In recent years, a technique for improving mechanical characteristics has been to add a small amount of metal to Sn-Bi system solder. The solder material, especially that with Sb and Zn added, has an excellent impact resistance property [2,3]. Therefore, we selected Sn-57Bi-0.5Sb-0.5Zn solder as the second candidate. The physical property values of each solder are summarized in Table 2. Sn-57Bi-1.0Ag solder and Sn-57Bi-0.5Sb-0.5Zn solder has almost same melting point and coefficient of thermal expansion. Young's modulus of Sn-57Bi-0.5Sb-0.5Zn solder is about 4GPa lower than of Sn-57Bi-1.0Ag solder.

Table 2. Material properties of solders.

	Melting Point	Coefficient of Thermal Expansion	Young's Modulus	Yield Point
Sn-57Bi-1.0Ag	139°C	13.8 ppm	19 GPa	12.8 MPa
Sn-57Bi-0.5Sb-0.5Zn	140°C	13.8 ppm	15.2 GPa	12.9 MPa
Sn-Ag-Cu (Ref.)	217°C	20.8 ppm	20.2 GPa	12.4 MPa

2-2. Heat spreader design

Figure 3 shows the structure of the developed BGA package. The heat spreader is connected to the upper part of the LSI. The heat spreader extends to the BGA area, and the effect of warpage control is achieved with the extended part connected to the package substrate.

Clearly, the thickness of the heat spreader should be thick to reduce the warping. However, it should be as thin as possible to reduce thermal resistance. Therefore, the thickness of the heat spreader needs to be the minimum thickness effective for warpage control.

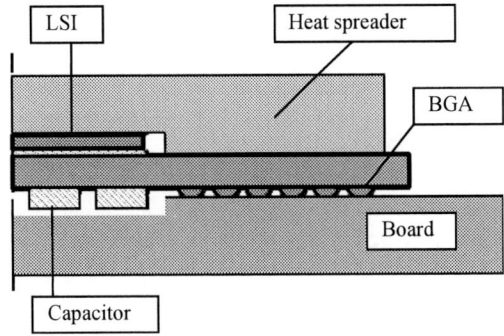

Figure 3. Structure of developed BGA package.

As mentioned above, at the size of this BGA package, it is necessary to suppress the warping to 100 μm or less in the opposite angle. The outer edge of the package curves up at high temperatures as shown in Figure 4. Consequently, there is a possibility of an open failure occurring at an outer edge. On the other hand, open failures at the inner edge must be considered in the cooling process. Therefore, it is necessary to consider both the warpage at high temperatures and warpage after cooling. Figure 5 shows the simulation results of the amount of warping in a thermal-stress simulation under both conditions. From the results, we assumed the thickness of the heat spreader should be 3 mm.

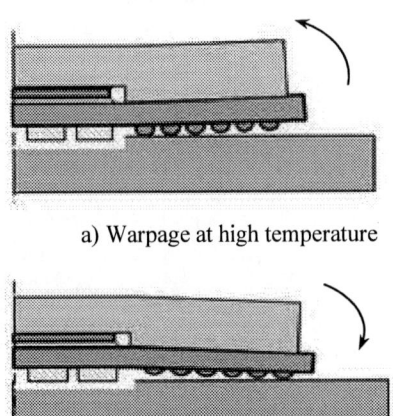

a) Warpage at high temperature

b) Warpage during cooling process

Figure 4. Schematic diagrams of warp direction.

978-1-4799-2408-0/14 $31.00 © 2014 IEEE

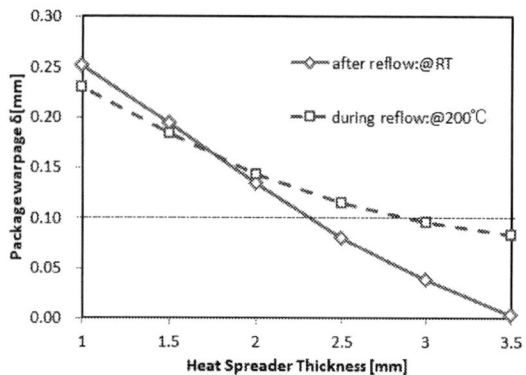

Figure 5. Simulation results of BGA package warpage.

3. Evaluation Tests and Results

3-1. Manufacturing tests

Table 3 shows the results of manufacturing tests. No short/open failure as an effect of the warpage control adopting the methods discussed in Section 2 was observed.

Table 3. Manufacturing test results.

Solder	Appearance after Ball Mount Reflow	Short Failure after BGA Reflow	Open Failure after BGA Reflow
Sn-Bi-Ag	Unacceptable*	0/20	0/20
Sn-Bi-Sb-Zn	Good	0/20	0/20

(Failure count/Sample count)

However, as shown in Figure 6, anomalies in the shape of solder bumps occurred after the solder ball mount reflow process. This deformation was observed only in the case of the SnBiAg solder ball mount reflow process.

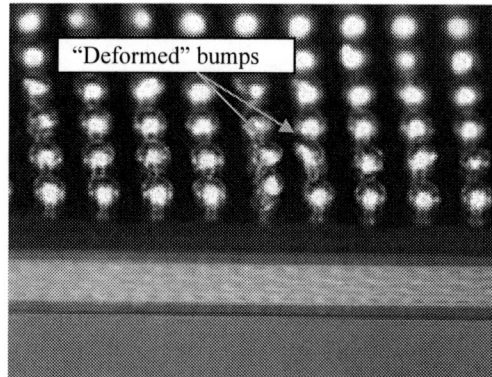

Figure 6. Appearance of SnBiAg solder bumps after ball mount reflow.

Figure 7 shows the SEM observation image of a "self-deformed" solder bump and the EPMA results of the surface of the bump. Segregation of Au and Pd was found on the surface. Moreover, crystalline growth was observed on the surface of the deformed part too. (Au and Pd originate in the pad surface plating of the substrate.)

We presumed that surface segregation of PdSn and an AuSn intermetallic compound acted as the starting point of the crystalline growth, a big crystalline structure formed on the surface, and then the deformation occurred. As is generally known, the segregation has a correlation with the cooling rate and the reflow peak temperature. Therefore, we performed an examination that changed the reflow peak temperature and the cooling rate, to investigate the incidence of deformation. Table 4 shows the results of the survey. No deformation was observed at the reflow peak temperature of 220°C or more.

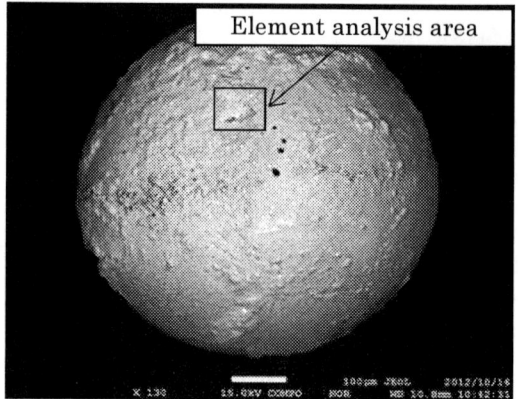

a) SEM image of "deformed" bump surface
b) EPMA results of bump surface

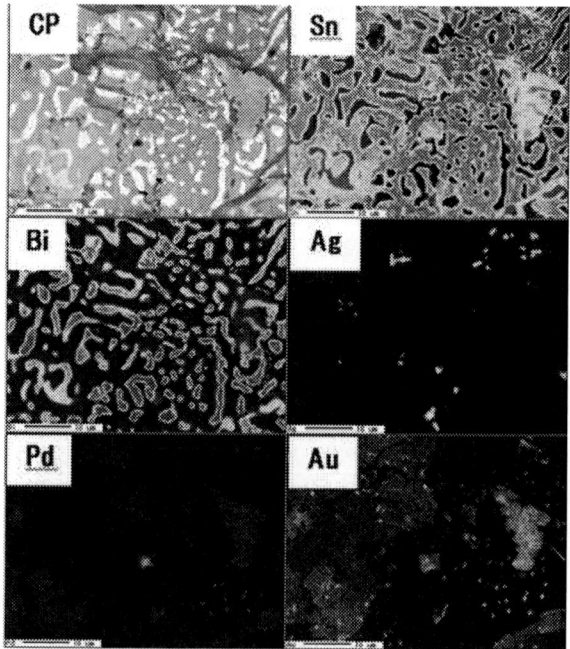

Figure 7. SEM image and EPMA results of bump surface.

Also, no deformation was observed under the conditions where the cooling rate was 0.65°C/sec.

The reflow temperature was considered to have the effect of preventing the Au segregation caused by the peak temperature exceeding the Au-Sn solid phase temperature (which is 217°C). However, a detailed mechanism of the effect of the cooling rate is not understood. If the Au segregation were the main cause of the deformation, a similar phenomenon would surely occur in SnBiSbZn solder (Au originates in substrate plating), but there was no deformation observed in the case of the SnBiSbZn solder under our test conditions. It is necessary to investigate the relationship between the cooling rate and the deformation and the effects of the trace element in detail.

Table 4. Ball mount reflow temperature and cooling rate vs. incidence of deformation.

Ball Mount Reflow Peak Temperature (°C)	Cooling Rate (°C/sec)		
	0.65	0.31	0.25
180	0%	**5.8%**	**24.2%**
210	0%	**1.7%**	**2.7%**
220	0%	0%	0%
230	0%	0%	0%

* Number of failed bumps/All BGA bumps × 100 (%)

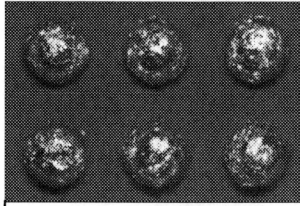 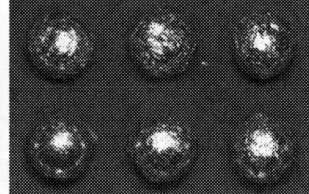

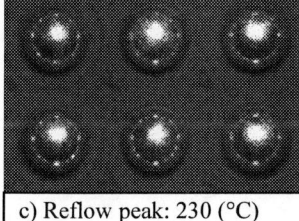

Figure 8. Appearance after ball mount reflow test.

3-2. Reliability tests

Finally, we were able to confirm that both the SnBiAg and SnBiSbZn solders were adequate in terms of manufacturability as discussed above, so BGA connection reliability verification tests were performed. The conditions of the reliability tests and the results are summarized in Table 5. The test board had an ordinary daisy-chain pattern, and DC resistance was continually monitored. The criterion of failure was a 10% change of DC resistance.

The preliminary solder was the same kind of solder for each solder ball (SnBiAg/SnBiAg and SnBiSbZn/SnBiSbZn).

Table 5. BGA connection reliability test conditions and results.

Item	SnBiAg	SnBiSbZn
Thermal cycle: -40°C to 100°C	No failure detected after 3,650 cycles	No failure detected after 1,400 cycles

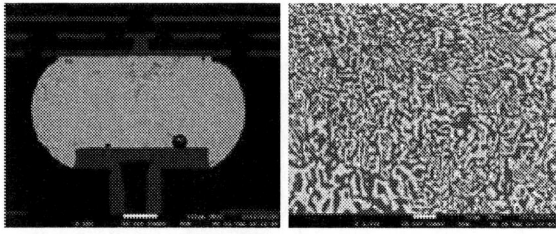

a) SnBiAg initially

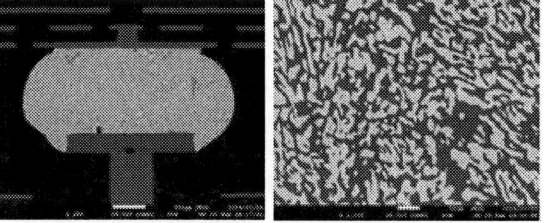

b) SnBiAg after 1,000 cycles

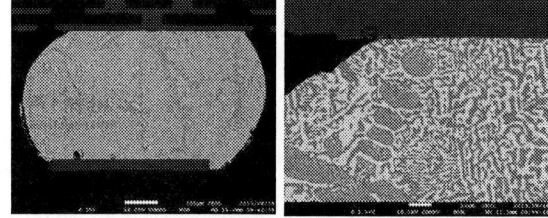

c) SnBiSbZn initially

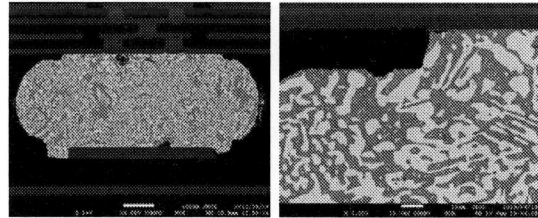

d) SnBiSbZn after 1,000 cycles
Figure 9. Cross sections of BGA joint.

Figure 9 shows the cross sections initially and after a thermal cycle test. Grain coarsening was observed in each solder, but there was no substantial problem.

4. Prediction of BGA Joint Fatigue Life

The fatigue life of the BGA joints was predicted based on the reliability examination results in Section 3-2. Destruction of a BGA joint is a cohesive failure and considered to obey the Coffin-Manson law.

$$\Delta\varepsilon = \alpha \cdot Nf^n \ \dots \ ①$$

$\Delta\varepsilon$: Inelastic strain range
Nf: Number of failure cycles
α: Fatigue ductile constant
n: Fatigue ductile exponent

Generally, the material constants α and n are determined through a thermal cycle test under different conditions. However, no failure occurred in the SnBiAg solder, as previously stated, in as many as 1,000 thermal cycles. Also, if a high temperature is set as a thermal cycle test condition, there is a possibility that the failure mechanism changes because of the change in the physical properties. For these reasons, it was difficult to adopt a general method. We thus decided to evaluate the fatigue life by using the following flow.

1) Acquire the material constant n in mechanical cyclic strain tests (see Figure 10).
2) Calculate $\varepsilon 1$, the strain during the temperature cycle test, and $\varepsilon 2$, the strain during the power cycle (actual operation), by thermal-stress simulation (see Figure 11).
3) Plot the intersection of the failure cycles of Nf1 and $\varepsilon 1$, and plot a line by using the n value acquired in the mechanical cyclic strain tests.
4) Plot the intersection of $\varepsilon 2$ and the line, and acquire an estimated value for the actual failure cycles of Nf2.
5) Check whether Nf2 is larger than the necessary fatigue life, Nfn.

The material constant value of each solder is shown in Figure 10.

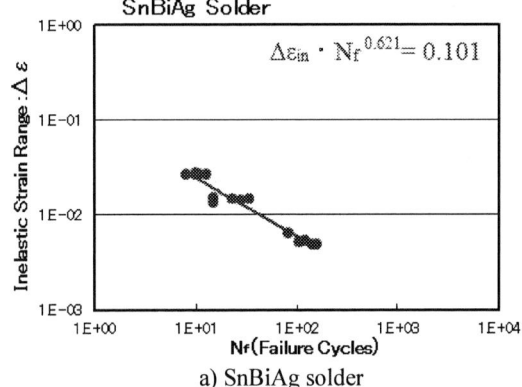

a) SnBiAg solder

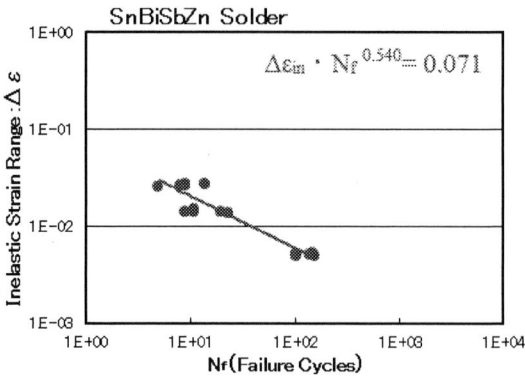

b) SnBiSbZn solder

Figure 10. Mechanical cyclic strain test results.

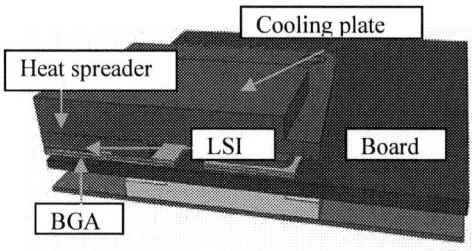

a) Simulation model

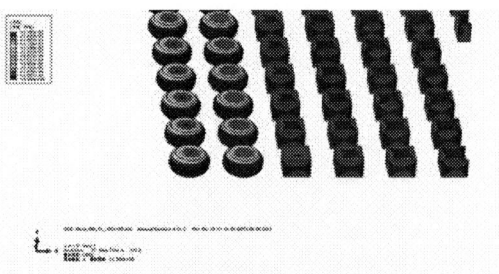

b) Strain during thermal cycle test

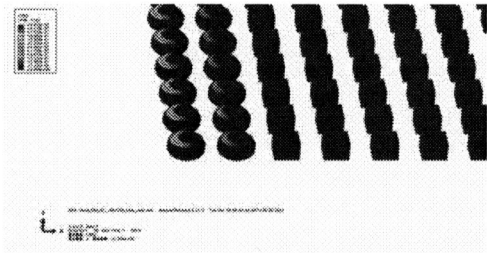

c) Strain during power cycle (actual operation)

Figure 11. Example of thermal-stress simulation.

Figure 11 shows an example of the thermal-stress simulation. The strain during the thermal cycle test and the strain during the power cycle are calculated for each kind of solder. This simulation considers the temperature dependence and the nonlinearity of each physical property value. We have already established a technique that uses the temperature distribution in actual operation as an initial condition of the stress simulation [6,7]. The same technique was applied to this simulation.

Figure 12 shows the results of plotting using this flow. As mentioned in Section 3-2, it is uncertain whether the failure cycles of Nf1 are accurate for both kinds of solder. Therefore, we used 3,650 cycles and 1,400 cycles as temporary values. However, they are understood to have a sufficient margin for the necessary failure cycles of Nfn. It is also understood that the fatigue life of each solder is long enough.

However, like in Figure 9, a tendency to collapse was observed in the SnBiSbZn solder after the temperature cycle test. The likelihood of this tendency affects the fatigue life. The verification of this problem is under consideration.

Figure 12. Fatigue life plot.

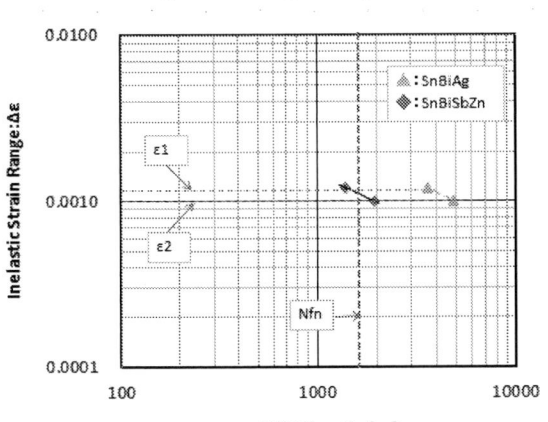

Conclusions

A second-level connection process for a large-size BGA package has been established with the application of a low-temperature solder and a warpage control method that uses a heat spreader as a warp suppression jig.

In the case of SnBiAg solder, the problem of ball deformation occurred after the ball mount reflow process, but has been solved by control of the process conditions. Both the SnBiAg solder and SnBiSbZn solder are understood to have sufficient reliability. However, two items remain to be investigated in Sn-Bi system solder application:

1) the detailed mechanism of SnBiAg solder deformation, especially the effect of trace materials, and

2) the effect of a tendency to collapse on the fatigue life of the SnBiSbZn solder.

References

1. M. Mostofizadeh, J. Pippola, and L. Frisk, "Reliability and Shear Strength of 42Sn-57Bi-1Ag (Wt.%) Lead-Free Solder Joints after Thermal Aging and Salt Spray Testing," 63rd Electronic Components & Technology Conference (2013), pp. 1010-1017

2. Keishiro Okamoto, Kenji Nomura, Shuichi Doi, Toshiya Akamatsu, Seiki Sakuyama, and Keisuke Uenishi, "Effects of Sb and Zn Addition on Impact Resistance Improvement of Sn-Bi Solder Joints," 46th International Symposium on Microelectronics (IMAPS 2013), pp. 104-108

3. Toshiya Akamatsu, Nobuhiro Imaizumi, Seiki Sakuyama, Tetsuhiro Nakanishi, Yumehiko Hattori, and Keisuke Uenishi, "Influence of Zn Addition to Eutectic Sn-Bi Solder on Joint Reliability with Cu Electrode," International Conference on Electronics Packaging (2011), pp. 566-571

4. Kenji Fukuzono, Masateru Koide, Shunji Baba, Manabu Watanabe, and Yuki Hoshino, "Development of High Efficiency Cooling Structure for Processor Package" (in japanese), 19th Symposium on Microjoining and Assembly Technology in Electronics, pp. 389-394

5. Joji Fujimori and Masateru Koide, "LSI Packaging Development for High-end CPU Built into Supercomputer," 61st Electronic Components & Technology Conference (2011), pp. 2028-2032

6. Kenji Fukuzono, Masateru Koide, and Manabu Watanabe, "Efficiency Improvement of Solder Joint Fatigue Life Evaluation for a High-end Pb-Free Glass Ceramic BGA Package"(in japanese), Proceedings of the 24th JIEP Annual Meeting (2010), pp. 196-197

7. Naoaki Nakamura, "A Study on Lifetime Prediction Method for Pb-free Glass Ceramic BGA Package"(in japanese), Proceedings of the 23rd JIEP Annual Meeting (2009), pp. 185-186

Development Fine Pitch Area Array Cu Pillar/Lead Free Solder Bumps for Large 28nm Die in Large Organic Flip Chip Packages

John Osenbach[1], Sue Emerich[1], S.Cate[3], D. Brady[2], Seung Min Hwang[4], J. Dang[5], and D. Crouthamel[1]

1LSI Corporation (USA), 1110 American Parkway NE Lehigh Valley Central, Allentown, PA 18109, USA
2 Amkor Technology (USA), 1900 S. Price Rd, Chandler, AZ 85286
3LSI Corporation (USA), 1320 Ridder Park Drive,San Jose, CA95131, USA
4 Amkor Technology (Korea), 100, Amkor-ro, Buk-gu, Gwangju 500-733
5 Kyocera America Inc., 472 Kato Terrace, Fremont, CA 94539

Abstract:

The viability of flip chip packages that incorporate small diameter (65um-80um) Cu pillar bumps on large die (up to 500mm^2) in large packages (up to 55mm x 55mm) is demonstrated. To do so, bump formation and microstructure must be controlled. In particular, avoidance of bump tearing defects, which were found on assembly with die $\geq$410mm^2, die > must be avoided. Fundamentally bump tearing results from uneven solidification because of temperature gradient in the assembly process during cool down, less than ideal solder volume, die pad – to – substrate pad offsets, increasing die size, and decreasing core thickness. These defects are further exaggerated by die designs that have regions of low and high bump density. Aided by finite element molding of the bump joints, a process was developed that eliminated tearing defects.

The reliability of Cu pillar bumps on large die/package flip chip devices was also demonstrated using temperature cycling. In this work, both mechanical test die as well as active 28nm/40nm devices were evaluated. Thin core 45mm x 45mm packages containing ~250mm^2 die using 65um diameter bumps passed 1500 cycles -55C to 125C with no signs of bump fatigue, die or package cracks. These devices had no bump tearing. Likewise, thick core 52.5mm x 52.5mm packages containing ~500mm^2 die using 80um diameter Cu pillars were subjected to temperature cycling. In this case bump tearing was present to some minor degree on all devices. Packages with mechanical test die passes 1500 cycles with no signsof bump fatigue, whereas packages with active die passed 700 cycles with no signs of bump fatigue, die or package cracks.

Introduction

The benefits of copper die-side bumps relative to solder bumps for flip chip packages include enhanced electrical and thermal performance; enhanced electro migration resistance; reduction in bump pitch; and the potential for lower cost substrate. All of these properties make Cu pillar an attractive technology for the growing flip chip market. Although Cu pillar bumping may have several advantages over its conventional solder bump counterparts, there are also challenges that must be overcome to enable implementation of this technology. The first and most studied and modeled challenge area is that related to the mechanical property difference between Cu-pillar bumps and more traditional solder bumps. In particular, Cu has a yield stress and modulus that are at least three times greater than that for Pb/Sn solder. Thus for a given design, Cu pillar bumps lead to higher interconnect driven stress than traditional eutectic Pb/Sn or lead free Ag/Sn solder bumps. Zhang, et.al.[1],quantified this effect with finite element modeling of a 42.5mm x 42.5mm, 0.8mm core organic package flip chip package containing a 14.5mm x14.5mm bumped die. Three different bump materials were used in these simulations, Cu pillar, eutectic Pb/Sn and lead free Sn/Ag solder. They found the interconnect stress for Cu pillar bumps was 10-15% higher than for lead free solder bumps and approximately 20-30% higher than that for eutectic Pb/Sn bumps. This higher interconnect stress combined with porous low-K and/or ultra-low k dielectrics which tend to have inferior mechanical properties (i.e. lower modulus, lower strength, lower fracture toughness, and worse adhesion) used in leading edge silicon nodes can potentially lead to failures due to die interconnect stress related damage[2]. This is further exaggerated by decreases in bump pitch and/or increases in die size.

The second potential problem with Cu-pillar bumps is related to the fact that Cu does not melt during the assembly process. As a result Cu pillar bumps are more susceptible to defective joints due to bump height variations. Thus, there is an increased probability that variations in the height of bumps and/or substrate pads as well as die and/or substrate warpage will negatively impact bump quality and assembly yield. As is the case for interconnect stress, decreases in bump pitch or increases in die size further amplifies this effect.

In this paper, assembly and reliability results for organic substrates flip chip product with fine pitch (130-150um) with half pitch diameter Cu pillar/lead free solder bumps, die up to 500mm^2, and packages up to 55mmx55mm is presented. This data shows that highly reliable devices with fine pitch Cu pillar/lead free bumps, for die up to 500mm^2 and packages up to 55mmx55mm can be manufactured if the bump structure, BOM and assembly processes are optimized.

Experimental Procedure:

A schematic of the device used in this work is shown in Figure 1. Three different test vehicles were used in this work, Table 1. The test vehicles were all made in 40 and 28nm silicon technology with 10 layers of metal. The die sizes ranged from ~235mm^2 to ~ 500mm^2 and the package sizes ranged from 45mmx45mm to 55mm x55mm. The test vehicles had transistors, interconnect and active circuitry. One of the test vehicles was specifically designed to look at design/process interactions. Both thin core, 0.4mm, and thick core, 0.8mm, organic substrates were used in this work. The substrate design and die design were the same for both substrate core thicknesses. All test vehicles were originally

978-1-4799-2408-0/14 $31.00 © 2014 IEEE

designed for lead free bumps at 180um pitch. Design modifications were made to the test vehicles to to accommodate Cu pillar bumps with diameters between 65um and 80um in diameter. These bump diameters are compatible with both 130 and 150um pitch bumps designs. Given that the bump density and thus the total cross sectional bump interconnect area on the test vehicles is some <40% of that present on a 130-150um bump pitch design, these test vehicles will experience a higher local stress than would be experienced on actual 130-150um pitch products. That is to say, once developed for these test vehicles, the process and BOM will likely produce devices at 130um pitch with even better reliability.

Top-Hat Lidded FlipChip package drawing

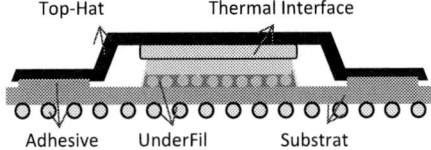

Figure 1: Schematic of flipchip package used in this work

	~ die size (mm²)	TV Pitch (um)	Bump Dia. (um)	Pkg Sixe (mm)	Core (um)
TV1	235	180	65	45x45	400
TV2	410	180	80, 65	55x55	800
TV3	500	180	80	52.5x52.5	800
TV4	500	180	80	52.5x52.5	400

Table 1: Test Vehicle (TV) matrix

The work was done in three phases, Table 2. Phase0 used die with the same bump configuration as the active die, however, they were did not contain any active circuitry. In this paper, these TVs are referred to as probe card die. The probe card die are composed of patterned aluminum metallization on top of SiO_2. The patterned aluminum metallization was then passivated with a standard plasma deposited silicon nitride-oxide layer. The passivation was patterned exposing the Al bond pad. Polyimide was then deposited on top of the silicon nitride/oxide passivation and patterned. Cu pillar bumps with a Sn/Ag solder tip were then plated using standard photolithography and plating technology. The probe card die were used to develop the process and BOM.

Phase 1 used die with active circuits. These die had the same aluminum/passivation/polyimide/Cu pillar-Sn/Ag materials and design that the probe card die had. The packaged live die were assembled with the process and BOM developed in Phase 0. The object of this phase was to build a small number of packages to verify and validate the applicability of the process and BOM to active die made in a 10L metal 40 or 28nm node technology. Here particular attention was paid to the stability and reliability of the lowK/ELK dielectric stack both as made and after exposure to JEDEC like thermo-mechanical testing. In this phase, process and/or BOM modifications could also be incorporated if assembly or reliability effects were observed. Any such modifications would then be incorporated in Phase 2 of the work.

Phase 2 live die packages are then built with the optimized process and BOM. Here a full complement of devices needed to show compliance with JEDEC environmental testing were run. This phase is used to confirm the process and BOM stability and reproducibility over a larger sample size and built at different times with different engineering oversight and operators. It is also used to show the process and BOM are compatible with the thermal mechanical tests required by JEDEC [3].

Phase	What is built	Expected outcome
0	probe card die	• Process and BOM development • Initial assessment of environmental stability (MRT/TC/uHAST)
1	live die	• Validation of process and BOM • Extended mechanical stability testing – insure ELK and BE stack have sufficient margin against process window • Validate JEDEC like technology robustness qual
2	live die	• Confirm reproducibility of process BOM • Validate and verify with larger sample size full JEDEC technology

Table 2: Phased process used for this work

The assemblies were made using standard industry practices and processes for die attach, capillary underfill, and lid attach. The underfill, thermal interface material, and lid/substrate attach materials used for all TVs in all phases of the work were the same. All of these materials are commercially available and widely used by the industry for lead free flip chip packages. The substrates used in these studies had a solder on pad (SOP) surface finish. The SOP was formed with Cu/Sn micro-balls.

After the devices from each phase were assembled, they were subjected to moisture level 4 exposure followed by 3x reflow at 245C. Subsequently, the lot was split into sub-lots. One sub-lot was subjected to extended temperature cycling and the other to unbiased HAST(uHAST). In all cases temperature cycling was done using JEDEC condition B (-55C to +125C) and the uHAST condition 130C and 85%RH. The devices were subjected to these stresses and periodically removed for re-testing. At each pull point a few devices were removed from the overall device population for physical analysis.

The physical analysis consisted of: i) cross sectioning followed by scanning electron microscopy (SEM); and ii) scanning acoustic microscopy, CSAM. The cross sections were used to determine the bump microstructure, including if crack, no wets, or other abnormalities were present. In addition for those phases of the study that used live die, functional testing was used to determine if there were any cracks or damage in the die stack, especially in those interconnect layers that contained porous dielectrics.; and ii)CSAM was used to evaluate the quality of the underfill/die and underfill/substrate interface as well as to insure no delamination in the dielectric stack.

Results and Discussion

Assembly and bump microstructure

Figure 2 shows typical cross sections showing the bump microstructure of the as assembled phase 0 -TV1, TV2 and TV3 respectively. Also shown in the figures are schematics of the die layout and the locations within the die were cross sections were taken. As shown in this study, cross sections were taken along all outer edges of the die as well as in the middle of the die, this was done to insure DNP and both local and global warpage effects were captured and fully quantified. The particular cross sections shown in the figures were those taken along the bottom edge of the die which represented the worst case or equivalent worst case bumps.

As shown, the Cu pillar bumps for TV1 are well formed with no defects. The solder wets the outer edge of both the left, L, and right, R, hand corner bumps with no such effect observed for the center, C, most edge bump. As can be seen the outer most corner bumps of all three TVs show this side wetting effect. This effect is related to the fact that the substrates were not designed to accommodate the mismatch in CTE between the die and the substrate. As a result the substrate pads are significantly misaligned, offset, with respect to the bump at solder reflow temperature, this drives the solder to exceed the wetting angle of the side of the bump, thus leading to side wetting. In contrast are the as made bump microstructures of TV2 and TV3. For both TV2 and TV3, the bumps on both outer corners, L and R, of the die have defects in the Sn/Ag solder potion of the joint. We refer to these defects as bump tear defects because they appear to the authors as an analog to hot metal tear defect that are found in the metal casting industry [4-6]. Such metal tear defects are not uncommon in the metal forming industry and because such defects can result in failure initiation sites during service, the phenomena has been studied for many decades. Previous work shows that hot tearing is a complex phenomenon, it l at the intersection of heat flow, fluid flow and multiphase mass flow. Fundamentally it is generally accepted that hot tearing occurs due to solidification shrinkage and thermal deformation developed during solidification. When the liquid metal solidifies there is a volume contraction. If at the same time the geometry of the system is constrained and prevents contraction, then a tensile force develops during solidification. When this tensile force is sufficiently high it can lead to separation of the liquid region that has not yet solidified, i.e. tearing. The affect is further complicated by alloys with large liquidus temperature ranges. This is because when in the liquidus range, the alloy contain both solids and liquids. Although Sn/Ag and Sn/Ag/Cu lead free solder alloys used in the IC industry do not have thermal dynamically driven excessively large liquidus temperature ranges, the fact that these alloys can exhibit a large degree of undercooling, leads to the postulation that such undercooling can also negatively affect the degree of solder tearing in Cu pillar joints. [7]

Although the data shown in Figure 2 indicates a clear dependence of tearing on die size, bump tearing must also be influenced by things other than die size as TV2, 410mm^2, is smaller than TV3, ~500mm^2, however TV2 has a much large bump tearing defects than TV3.One such effect appears to be related to the bump density and its distribution across the die, that is to say both global and local bump coverage percentages. In the case of TV2, the bump density, as measured by die surface area covered with bumps, varied from greater than 25% in many locations to less than 3% in other locations. The low surface coverage being in the lower left and right hand side of the TV where the largest solder tears were observed. In contrast, , the bump density for TV1 and TV3 tended to be in excess of 20-25% It is postulated that this bump density effect is fundamentally related to buoyancy and restoring forces that any individual bump joint experiences during solidification. Because the buoyancy and restoring forces in any region of the die just prior to solidification are proportional to the integrated solder surface area in that region($f\alpha\gamma*$ total solder surface area, where f is the force,γ is the solder surface tension), the forces on any individual bump will be dependent upon the density and geometry of bumps in that region. Thus when there are regions with low bump density and other regions with high bump density, the global forces in the low bump density area will be lower than those in the high bump density area. This combined with substrate-, and possibly die-, warpage just before solidification is expected to results in variation in the gap interconnect height. The variation in bump height is equivalent to variations in the tensile, elongated bumps, and compressive forces, compacted bumps, acting on the solidifying joints. Those in tension can be thought of as solder starved whereas those in compression can be thought of as solder enhanced. The solder starved joints will show the highest propensity for tearing as this adds to the lack of sufficient solder volume, where those with enhanced solder volume will tend not to have tears.

In addition to substrate to die pad CTE driven misalignment driving Cu pillar side wetting towards to outer edges of the die, it impacts solder tearing, this effect is shown in Figure 3. Here the area of the tear (solder gap) as measured in the cross section and the distance between the substrate pad center to that of the die (CuP-Pad Shift) was measured on a number of TV2 assemblies, for substrates with two different core materials with two different CTE'sCTE1 < CTE2. As shown there is a non-linear dependence of the tear area, missing solder area, on the die pad- to substrate pad offset. The dependence of tear area on pad offset is understandable in the context of the above discussion on solder volume. The offset essentially increases the solder volume needed to form the joint, the larger the offset the more the joint is solder starved, thus the more tearing propensity.

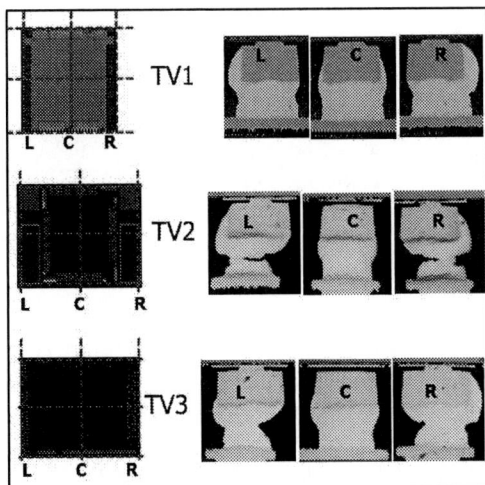

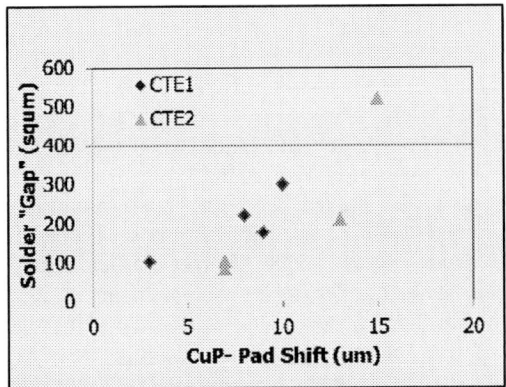

Figure 2: Schematic of TV including locations were cross sections were taken (dotted lines). Cross sections of the Cu pillar bumps on as assembled devices from phase 0

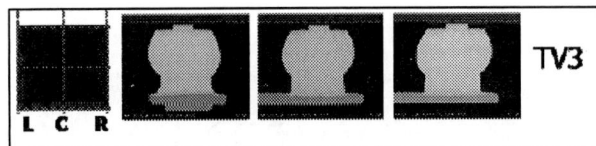

Figure 4:Cross sections of as assembled TV3 devices from phase 0 with 80um diameter lead free solder bumps

Reliability

Probe card die based TV1 and TV3 were subsequently subjected to moisture pre-conditioning, followed by 3X/245C reflow followed by 1500 cycles TCB and 96hrs uHAST. The results are given in Table 3. As shown there were no failures as measured by CSAM and cross sectional analysis up to and including 1500 cycles of TCB. Figure 5 shows typical cross sections of TV1 and TV3 after 1000 cycles of TCB. There does not appear to be any indication of cracks or other fatigue driven degradation in the Cu pillar bumps for either TV1 or TV3. This is somewhat surprising for TV3 given that there are some rather obvious solder tear defects in the corner Cu pillar bumps as shown in Figure 2 and 5. Apparently the Cu pillar joint with the underfill used in this work is sufficiently robust to prevent solder fatigue failures from occurring in joints with rather large solder tear defects. This result indicates that 80um diameter circular Cu pillar solder bump do not pose a reliability threat related to bump fatigue for flip chip packages made with die as large as 500mm^{2}in thick core $\leq$52.5mm x 52.5mm packages. The data also indicate bump fatigue degradation is not an issue for 45mm x 45mm thin core packages with die $\leq$ 235mm^2 made with 65um diameter Cu-pillar-Sn/Ag cap bumps.

Figure 3: Dependence of solder tear area on die to substrate pad offset

Because the solder tearing was excessive for TV2 and the bump distribution on this TV is not representative of actual products, work on this device was suspended pending improvements in the assembly process to eliminate or limit the degree of solder tearing.

To determine if the bump type, lead free Sn/Ag solder only versus Cu pillar-Sn/Ag cap, had an effect on bump tearing, TV3 devices were also made with 80um diameter Sn/Ag solder bumps and compared to the Cu pillar-Sn/Agcap bumps. To ensure that design, process and or BOM variations did not come into play, the lead free solder only bump devices were made with the same assembly process and BOM as that used for the Cu pillar-Sn/Ag cap bumps devices. Figure 4 shows a series of cross sections of the lead free solder only joints. As can be easily observed, bump tearing did not occur in the lead free solder only. Multiple cross sections on multiple assemblies were taken to verify this finding. In all cases, independent of where the cross section was taken or what assembly lot it was taken from, bump tearing was not observed. This result clearly indicates that the bump tearing observed in this work is related to the use of Cu pillar bumps. This effect will be briefly discussed in a later section of this manuscript.

	Type	MRT/ 245C	TCB 500	TCB 1000	TCB 1500	uHAST 96hr
TV1	PC	0/38	0/23	0/18	0/13	0/10
TV3	PC	0/38	0/23	0/18	0/13	0/10

Table 3: Summary of probe card reliability test results for probe card builds of TV1 and TV3.

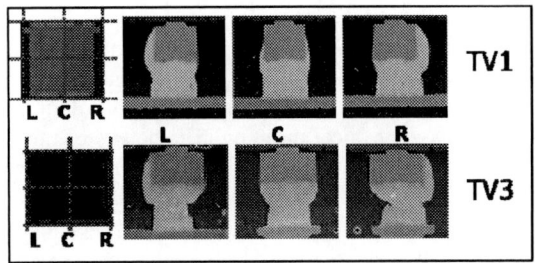

Figure 5: cross sections of TV1 and TV2 probe card die after 1000 cycles TCB.

Following this work, phase 1 assemblies of active die TV1 and TV3 were assembled using the same process and BOM developed in phase 0. The results of this phase are given in Table 4 and Figures6 and 7. As was the case for the probe card wafers there was no indication of bump fatigue or

978-1-4799-2408-0/14 $31.00 © 2014 IEEE   77

lowK/ELK damage after 1500 cycles and 700 cycles respectively for TV1 and TV3.

	Type	MRT/ 245C	TCB 500	TCB 700	TCB 1000	TCB 1500	uHAST 96hr
TV1	active	0/86	0/59	NA	0/59	0/59	0/22
TV3	active	0/86	0/59	0/59	pending		0/22

Table 4: Summary of active die reliability test results for probe card builds of TV1 and TV3.

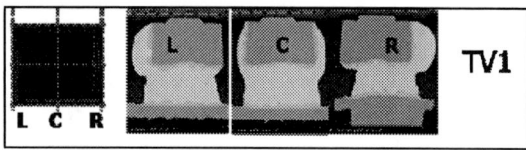

Figure 6: cross sections of TV1 TV3 active die after 1500 cycles TCB.

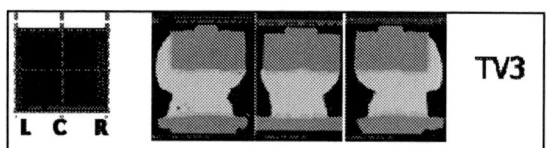

Figure 7: Cross sections of TV3 active die after 700 cycles TCB

Process Improvements to Eliminate Bump Tearing
Bump formation modeling

In parallel with the reliability work discussed above, assembly process work aimed at improving the Cu pillar bump joint microstructure was initiated. Solder joint modeling was done using a public domain finite element program developed by Ken Brakke [8]. The finite element model is set up as an interactive program for the study of fluid surfaces shaped by surface tension and other energies (e.g. gravity, external forces, etc.). It uses total system energy minimization to predict final shape of the solder joint, and the principle of virtual work to calculate the force displacement curves from a series of simulations Although very powerful, the finite element program requires the modeler to develop the overall geometry and boundary conditions by a trial and error method. This requires a significant amount of physical metallurgy and materials science knowledge along with programing and modeling skills to insure the solutions are tractable and have physical meaning. The details of the work are in a separate publication and are only summarized in this paper [9].

Figure 8 is a schematic that details the energies considered in the model, how they are related to the actual bump structure and a visualization of a typical output for a specific set of boundary conditions representative of Cu pillar-Sn/Ag cap bump joints used in this work. As shown, energies considered include solder surface tension, the wetted Cu to solder interfacial energy, the weight of the die-solder and cu pillar and the non-wetting boundary conditions at the solder mask opening.

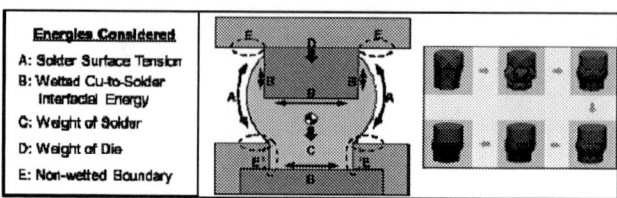

Figure 8: Energies considered, schematic of joint and representative output of a particular simulation run for the surface evolver finite element model

Figure 9 is a summary of a series of simulation aimed at developing a better fundamental understanding of the force displacement curves for both lead free solder bumps and Cu pillar-Sn/Ag cap bumps. Each line represents a single bump joint. In this plot simulation results are shown for both bump types, lead free and Cu pillar-Sn/Ag cap. Included are results for perfectly aligned die bump and substrate pads, zero offset, and 20um offset die and substrate pads. An offset of 20um was used as it represented the nominal worst case offset observed in this work. The calculation is done by pre-defining the gap height and die pad – to - substrate pad offsets for a given bump type and geometry, then calculating the resultant forces via the principle of virtual work. The range of height values were chosen to represent the worst case variations +/- 50%. A key indicating when the bump is in tension or compression is also added. When in tension the joint becomes solder starved, the higher the tension the more solder starved the join is. In contrast joints in compression effectively have excess solder. As per earlier discussion, tension tends toward solder tearing whereas compression does not. As shown, the dependence of the force on bump height for the lead free solder bumps is significantly different than that for the Cu pillar bumps. There are two fundamental differences between the force displacement curves of Cu pillar and lead free solder bumps joints: 1) lead free solder bump shave a force height curve that is well behaved, no inflection points within the joint height and pad-pad offsets simulated, whereas the Cu pillar bumps show multiple inflection points; 2) the 20um offset has almost no impact on the force-height curve of the lead free solder bumps, whereas the 20umoffset completely changes the functionality and shape of the force-height curve of the Cu pillar bumps. The difference in force-height-offset relationships for lead free solder and Cu pillar is fundamentally related to the fact that at specific heights and offsets solder will tend to wet the side/s of the copper pillar thus altering the force for a given joint height and offset, whereas no such effect is present in lead free solder.

A flip chip device will have many 10's to 10's???of thousands of solder joints. The height of each solder joint is defined by the distance between the die and substrate for that particular joint. When considering the totality of all joints in a flip chip device when the solder is in the liquid state, the joints are effectively bounded by two rigid surfaces, the die and the substrate. Force equilibrium requires that the sum of the forces acting over all of the joints must equal the force due to the weight of the die (which can be approximated as zero for the case where there are many solder joints). Further, with each joint having a distinct force-height response curve as determined by its degree of offset, the equilibrium requirement of zero net force implies some joints will be in tension while other will be in compression.

Due to heat transfer dynamics in the reflow chamber environment, the solder joints along the outer regions of the die will tend to cool faster and solidify before the joints toward the center of the die. Considering the case where substrate warpage does not play a role in the stress state of solder, for lead free solder bumps the cooling effect has relatively little effect on the stress state of the liquid joint (tension or compression), since the offset has relatively little impact on bump height This effect along with the larger solder volume in the lead free solder bumps are the fundamental reasons why no solder bump tearing is observed in lead free bumps shown in Figure 4. In contrast, for Cu pillar bumps, at the start of solidification, these outer joints will be in tension due to their higher degree of offset. Further, solder solidification is accompanied by volumetric shrinkage, which will tend to increase the tension in the outer joints before they fully solidify. Collapse of the outer joints (i.e. reduction in joint height in response to their tensile forces) is prevented by the opposing compressive force of the joints towards the center of the assembly which are still in the liquid state because of the temperature gradient in the assembly during cooling. In this scenario, the outer joints become starved of solder as they solidify and will tend to a higher propensity of tearing. One can see from figure 9that, for a given height, a copper pillar joint with pronounced offset will tend to be in tension, while a joint with minimal offset will tend to be in compression at that same height. This trend holds over a wide range of practical height and offset values.

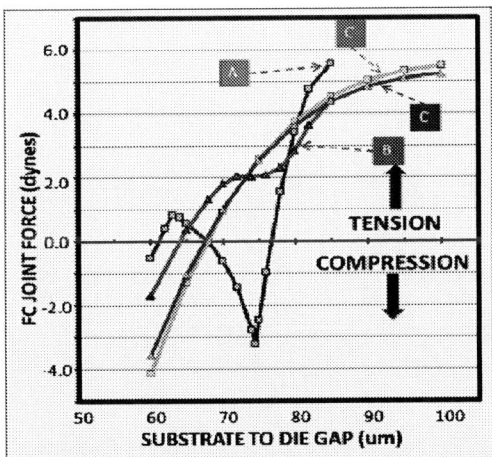

Figure 9: typical force vs. joint (die to substrate gap) height curve output from surface evolver finite element model for bumps with an 80um diameter: A) is a Cu pillar bump with die to substrate pads misaligned of zero (perfectly aligned) ; B) is a Cu pillar bump with a die pad to substrate pad misalignment of 20um; C) is a lead free Sn/Ag solder bump with die to substrate pads misaligned of zero (perfectly aligned); and D)is a lead free Sn/Ag solder bump with a die pad to substrate pad misalignment of 20um.

These affects are further exaggerated as the substrate planarity decrease. In general increased die and substrate size, increased layer count, and reduced core thickness will tend to

increase warpage. The decreased core thickness effect is shown in figure 10 which is a series of cross section micrographs of TV4 which is identical in all respects to TV3 with the exception it is in a 400um core substrate.

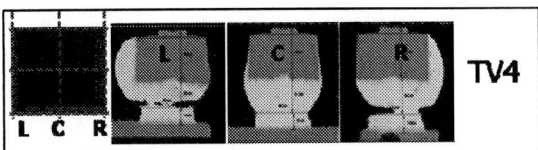

Figure 10: cross sections of TV4, TV3 in a 400um thick core.

Bump formation Improvements

With the understanding gained from the finite element modeling of the bumps, an improved process was developed to eliminate bump tearing. The results of this new process are shown in Figure 11. Comparison of the results in Figure 11 to Figures 2 and 10clearly show the vast improvement in the solder shape. The new process results in Cu pillar-Sn/Ag cap bumps with no solder tearing. There is ongoing work to prove reliability of devices made with the improved process; as the UF remains unchanged in the new process, it is expected the devices will reliability, similar to Figure 4 results

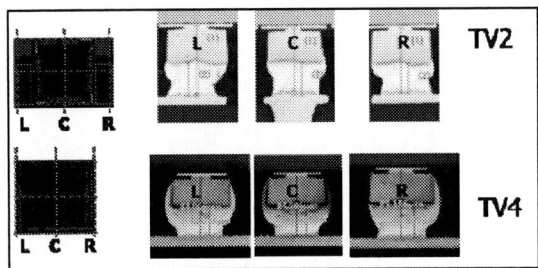

Figure 11: Cross sections of TV2, TV4 with new assembly process.

Conclusions

The viability of flip chip packages that incorporate small diameter (65um-80um) Cu pillar bumps on large die (up to 500mm²) in large packages (up to 55mm x 55mm) is demonstrated. To do so bump formation and microstructure must be controlled. In particular, avoidance of bump tearing defects, which were found on assembly with die ≥410mm², die > must be avoided. Fundamentally bump tearing results from uneven solidification because of temperature gradient in the assembly process during cool down, less than idea solder volume, die pad – to – substrate pad offsets, increasing die size, and decreasing core thickness. These defects are further exaggerated by die designs that have regions of low and high bump density. Aided by finite element molding of the bump joints, a process was developed that eliminated tearing defects.

The reliability of Cu pillar bumps on large die/package flip chip devices was also demonstrated using temperature cycling. In this work both mechanical test die as well as active 28nm/40nm devices were evaluated. These devices had no bump tearing. Thin core 45mm x 45mm packages containing ~235mm²die using 65um diameter bumps passed 1500 cycles -55C to 125C with no signs of bump fatigue, die or package cracks. Likewise, thick core 52.5mm x 52.5mm packages containing ~500mm^{2}die using 80um diameter Cu pillars were subjected to temperature cycling. In this case bump tearing was present to some degree on all devices. Packages with mechanical test die passes 1500 cycles with no signs of bump fatigue, whereas packages with active die passed 700 cycles with no signs of bump fatigue, die or package cracks

References

1. X.R. Zhang*, W.H. Zhu, RP. Liew, M. Gaurav, "Copper Pillar Bump Structure Optimization for Flip Chip Packaging with CuLow-K Stack" 11th. Int. Conf. on Thermal. Mechanical and Multiphysics Simulation and Experiments in Micro-Electronics and Micro-Systems, EuroSimE 2010

2. X. Zhang, S. H. Im, R. Huang, and P. S. Ho, Chapter 2 in Integrated Interconnect Technologies for 3D Nanoelectronic Systems (Editors: M. Bakir and J. Meindl), Artech House, Norwood, MA, 2008

3. JEDEC Standards JESD51-2 Integrated Circuits Thermal Test Method Environment Conditions Natural Convection (December 1995) and JESD51-9 Test Boards for Area Array Surface Mount Package Thermal Measurements(July 2000).

4. I. I. Novikov, "GoryachelomkostTsvetnykh MetalloviSplavov (Hot Shortness of Non-Ferrous Metals and Alloys)," pp. 299. (1966)

5. G. K Sigworth, "Hot Tearing of Metals," AFS Trans, vol. 104, pp. 1053-1062. (1996)

6. D.G. Eskin, K, Suyitno, and L. Katgerman, "Mechanical Properties in the Semi-Solid Sate and Hot Tearing of Aluminum Alloys," Progress in Materials Science, vol. 49, pp. 629-711. (2004)

7. Sung K. Kang, Moon Gi Cho, Paul Lauro, and Da-Yuan Shih, "Critical Factors Affecting the undercooling of Pb-free, Flip-Chip Solder Bumps and In-situ Observation of Solidification Process", Proceedings of 57[th]Electronic Components and Technology Conference, 2007.

8. Ken Brakke, The Surface Evolver, Version 2.70, August 15, 2013

9. D. Crouthamel and J. Osenbach, to be published

ELK Delaminate Improvement Methodology
on Cu Pillar interconnect BOP Structure

Nistec Chang, Albert Lan, Mark Liao, Eason Chen
Siliconware Precision Industries Co., Ltd. No. 153, Sec. 3, Chung-Shan Rd. Tantzu
Taichung 427, Taiwan, R.O.C.
Email: nistec@spil.com.tw

Abstract

During last couple years, the market of IC package have successful to implement Cu Pillar Flip Chip Technology as a mainstream of high density flip chip solution in each of portable markets(mobile phone, tablet & lots of portable entertainment solution). Moreover, concerning about high end product application which required 10*10mm above die area on larger flip chip ball grid array product, the substrate and bump dimension design will be an important factor to release ELK stress during huge CTE mis-match between silicon and organic substrate .

After reviewed the product market trend of CuBOP package (Cu pillar bump attach on bump on pad substrate structure), interconnection methodology have stayed on a perfect position to serve lots of high end zone on networking , graphic and specific ASIC devices by benefit of high density (compared with solder bump) bump layout design .Compared with solder bump structure, Cu pillar bump owned better electron migration performance . In order to well define optimun construction while utilizing Cu FC Package on BOP substrate structure, the paper plan to align the stress simulation index & failed event point of view to define bump dimension and solder mask resist opening (SRO) relationship for future advance, high end product and large die size implementation.

The analysis will utilize simulation methodology & popular reliability testing (Temperature Cycle Test, High Temperature, unbias HAST) result as a ranking for certain ratio definition between UBM dimension of bump design and SRO of substrate design. Based on stress simulation theoretical demonstration and actual practice result to find out that ELK wafer structure qualification index for real product application.

Introduction

Recently, more and more Hi-END networking and graphic products which required larger die size 15*15mm above square area interconnect on substructure pre-solder side in order to offer higher electrical and longer thermal cycling performance

This paper stated that interconnection key factors – UBM and SRO dimension will be dominated and affected quick temperature cycling test (-40°C to 60 °C) and whole reliability test items (moisture soaking level 4 , un-bias HAST 264hrs and -55 °C to 125 °C 1000 cycle and -150 °C high temperature storage test 1000 hrs) pass or not.

In this paper, assembly integration and it's impact on device characteristic will be discussed for coming couple pages. Before design bump mask to determinate Cu pillar bump dimension on assembly characterization test, author will rely on lower Low K stress index of stress simulation to choose proper Cu pillar bump dimension and start chip interconnect on BOP substrate experiment. Following page will utilize "CuBOP" abbreviation to state development activity.

Test Vehicle & Assembly Flow

To demonstrate assembly process and functionality are reliable or not , trigger Cu pillar test vehicle on conventional solder on pad substrate construction afterward .The bump and package basic information as Table 1 ; CuBOP packaging and assembly flow refer Fig1.The paper study procedure will be separated four portions to describe how to start the experimental matrix and how to accommodate interconnection in between Cu pillar bump and BOP substrate construction.

Table 1. Configuration of Test Vehicle Information

TV Features	Description
Si Node	ELK wafer
Chip Dimension	16*16*0.79 mm^3
Min. Bump Pitch	180um
Bump Height	65um
PKG Size	35*35 mm^2
Substrate C4 Design	BOP(w/ pre-solder)
Substrate Layer	3/2/3 Layer

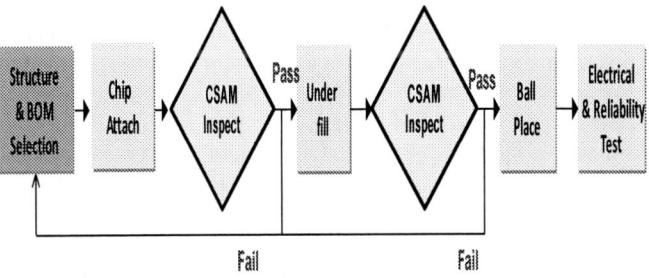

Figure 1. CuSOP package characterization test flow.
Two SAT CSAM quality inspection gate post die attach and underfill dispense process

Paper Study Test Procedure

Portion I. Cu Pillar Bump to BOP Substrate Interconnect Methodology

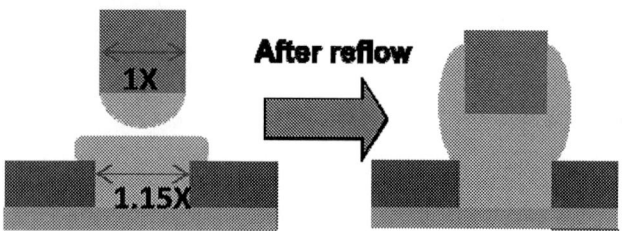

Figure 2. Substructure SRO > UBM diameter , solder easy to climb and creep Cu pillar surface

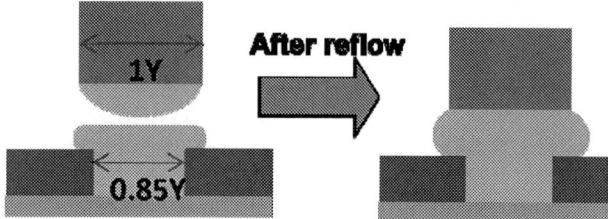

Figure 3. Substructure SRO < UBM diameter , solder difficult to climb and creep Cu pillar surface

There are two wetting schematic what we learned from several process validation. The fig2.showed solder is not only wetting on substrate SOP side but also along Cu pillar side wall to climb majority pillar height ; the fig3. showed solder only wetting on substrate SOP side . The major discrepancy came from solder wetting location to induce different stress level then cause different quality performance. Fig 4 appeared chip attach(without underfill protection) which found out that 2/10 samples white bump quality problem and moved forward package cross section to find out wafer ELK layer delamination serious quality issue . See the fig 4-5 CSAM image and cross section.

In order to find out best design window in between bump and substrate side, execute low stress simulation to research potential root cause and design optimum experiment matrix for further process validation. Portion II to III will describe how to minimize low K stress impact and solder creep situation. Then the last portion is a quick screen method which defined robust and reliable package structure and bill of material in advance. The test method cycle time is shorter than standard moisture and thermal cycling test. It's also a popular test methodology for new PKG and bill of material combination quality pre-screen procedure.

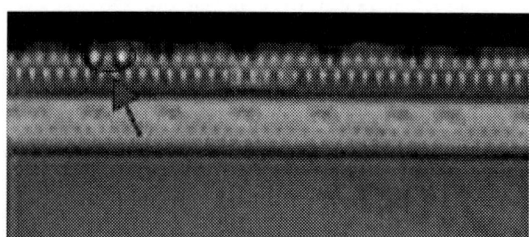

Figure 4. White bump image were detected by CSAM inpsection

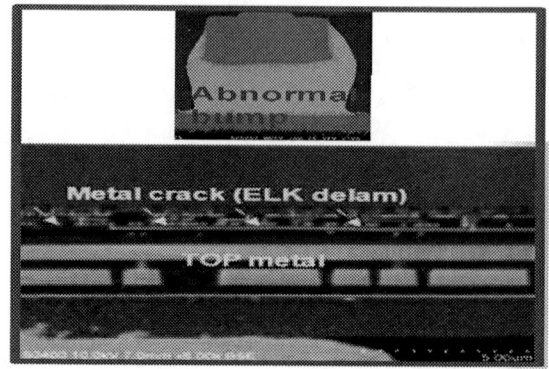

Figure 5. Serious solder creep on Cu pillar surface ELK delam was observed at same chip location by cross section inspection

Portion II. Cu pillar Flip Chip ELK Stress Simulation

Because of white bump occurred after chip attach process. It means huge stress generation during mass reflow process. According to ELK layer delamination phenomenon image found out that solder climbed whole pillar height. In order to fix this problem, approaching two directions to minimize stress impact of this stress simulation. One is solder creeping grade effect and the other is UBM dimension of Cu pillar bump structure. Firstly, to fixed UBM dimension to simulate solder creep effect after chip attach process. The low K peeling stress result showed 7~10% grow up after compared to no solder creeping situation. Refer to fig 6. local model from ANSYS analysis software and low K peeling stress simulation result on table2.

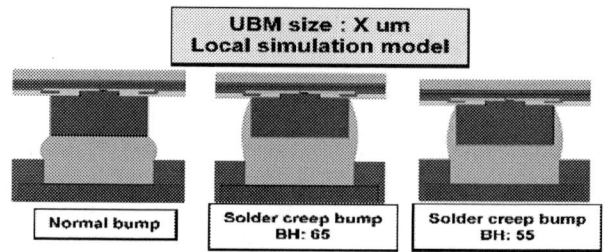

Figure 6. Peeling stress local model with three kinds of solder creeping to induce different bump height and Low K stress level

Table 2. Different low K peeling stress impact act on different solder creeping situation

UBM size (um)	X		
SRO (um)	1.15X		
Bump structure	Normal bump	Solder creep 100%	Solder creep 100%
Bump height (um)	65	65	55
Low-k peeling stress ratio	*Assume 100%*	*110%*	*107%*

On the another hand , base on lower low stress point of view , to increase UBM dimension to symbol "Y" which reduced stress value since mechanics formula P= F/A and larger bump area can disperse stress effect during chip attach thermal cycling stress test. Hence, simulate different solder creeping grade from 20% up to 100% to see any improvement space to solve white bump problem . According to table 3 simulation result , increased UBM dimension can reduce 13% low K peeling stress after compare to 100% solder creeping situation.

The stress simulation result is able to guide us to determine suitable UBM dimension for process characterization study then continuous collect inline process quality data and package reliability performance afterward.

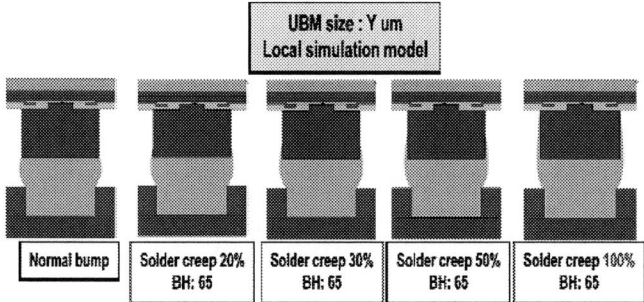

Figure 7. Peeling stress local model under "1Y" large UBM with five kinds of solder creeping o induce different bump height and Low K stress level

Table 3. Under "1Y" larger UBM dimension , different Low K peeling stress impact act on different solder creep on pillar situation

UBM size (um)	Y				
SRO (um)	0.85Y				
Bump structure	Normal bump	Solder creep 20%	Solder creep 30%	Solder creep 50%	Solder creep 100%
Bump height (um)	65	65	65	65	65
Low-k peeling stress ratio	94%	94%	94%	94%	97%

Portion III. Solder Creeping Process Characterization

Because of stress simulation indicated larger UBM implementation can decrease around 13% low K peeling stress during die attach mass reflow this duration time . At same time, to minimize solder creeping situation from 100% to 50% also can decrease 3% peeling stress. As regards solder creeping situation improvement method included followings key factors of this experiment :

a. Interconnect interface (bump to pre-solder) specific ratio optimization : To fix substrate SRO dimension and adjust UBM dimension to collect solder creeping performance. Need to watch CSAM image if it's relate to white bump defect or not

b. Cu pillar surface cleanliness : To dirty pillar surface to reduce solder creeping on Cu pillar surface chance.

c. Chip attach flux volume : To reduce flux volume to weaken chip attach clean mechanism then made melting solder difficult to climb on pillar surface.

Depended on above adjustable input factors to design assembly DOE matrix then proceed chip attach and collect solder creeping data for further JMP analysis. JMP analysis result is able to guide us to find out dominated key factor for solder creeping condition improvement. In other word, Cu pillar bump from die corner which gathered maximum stress. After that , checked data output from JMP software carefully. The samples moved forward cross section offline inspection to ensure no quality relate problem during CuBOP assembly packaging. There are two samples will be checked one straight line bump wetting situation by cross section inspection. The bump creeping height is measured by ruler scale after each assembly leg completed whole assembly packaging process (refer fig.8 creeping measure metrology).

Figure 8. Ruler scale to measure creeping height after cross section

Collected and exported all creeping value on JMP analysis software afterward. According to variability chart indicated two findings of this process validation (refer to fig 9.)

a. Solder creeping is easy to happen when design smaller UBM dimension no matter Cu pillar surface cleanliness and DA flux volume.

b. Solder creeping situation could be controlled within 30% once implemented "Y" larger UBM dimension. This is a significant major key factor what we learned from this experiment.

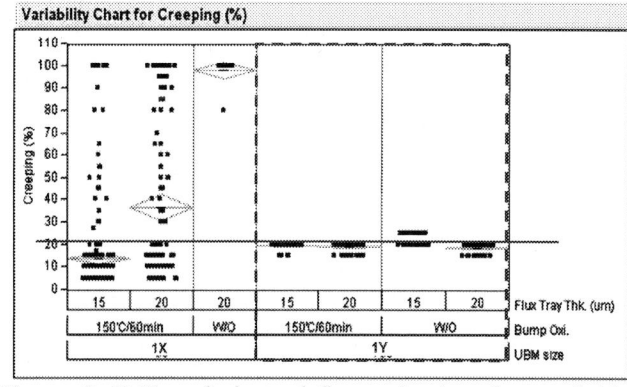

Figure 9. JMP analysis to define major key factor of this experiment

Portion IV. Quick Temperature Cycling for CuBOP PKG Structure Validation

In order to save time for PKG reliability test and prevent structure failure again during longer reliability whole test items. Hence, pre-screen the optimum UBM dimension and chip attach process recipe robustness are really important during CuBOP structure and process characterization phase. We performed quick temperature cycling test on this CuBOP structure experiment. The method is un-underfill semi- parts put into chamber with -40°C to 60 °C thermal cycling till to 30 cycles and also extend 50 cycles to prove PKG structure robustness or not. Then utilize CSAM inspection with high frequency (180MHz or 200MHz) transducer to detect white bump occurrence after every 5 times thermal cycling test is completed. Regarding this optimum PKG structure is not only passed QTC30 cycles but also passed QTC50 cycles. No white bump quality and PKG related side effect were observed. Also take 2 pcs samples after QTC30 cycles then dispense underfill for cross section inspection purpose. At the last row bump location we inspected 56 bump through SEM machine. According to SEM image judgment on corner bump location that didn't observe any metal crack situation. So it can prove this package structure is reliable for later standard reliability test .

Result and Discussion

The optimum bump and substrate structure packaging unit continuous moved forward package level reliability test to monitor electrical performance for following test conditions

   a. Moisture soaking test(30°C 60%RH) 96hrs : 90ea sample passed .

   b. un-bias test (130°C 85%RH,33.3 psi)264hrs : 45ea sample passed .

   c. Thermal cycling test (-55~125°C)1000cycles :45ea sample passed .

   d. High temperature storage (150°C)1000hrs : 45ea sample passed .

The test vehicle through electrical test afterward. The result showed no quality issue on those packaging units. It's also proof structure optimization is quite improvement on Cu pillar bump structure, especially on larger (>10*10 mm^2) Cu pillar die size and package size.

Conclusions

We presented Cu pillar bump interconnect BOP substrate development works. The process optimization results successfully demonstrated integration of specific ratio in between UBM and SRO ratio before tool up new Cu pillar mask and new substrate design. Finally, Cu pillar bump structure design (UBM dimension, structure and bump height)quite sensitive for assembly packaging point of view. Utilizing the optimum Cu pillar UBM dimension and substrate's SRO specific ratio to start development work can shrink packaging characterization cycle time   and accommodate high IO count and high performance, such as graphics , networking and FPGA product applications in the near future.

Acknowledgments

The authors would like to thank Daniel Yu and Ray Ho for packaging characterization study, foundry and substrate supplier partners for test chip and substrate manufacturing.

References

1. H. S. Jimmy Chew, *Chip and Manufacture Method Thereof*, U.S. Patent Application Publication (no#US 2008/0088013 A1), 2008.

2. Kim Hwee Tan,Han Shen..etc, *Die Pillar Structures and a Method of Their Formation* , U.S. Patent Application Publication(no#7,462,942 B2) , 2008.

Transient Dynamics Model and 3D-DIC Analysis
of New-Candidate For JEDEC JESD22-B111 Test Board

Pradeep Lall [(1)], Kalyan Dornala [(1)], Di Zhang [(1)],
Dongji Xie [(2)], Andy Zhang [(3)]
[(1)] Auburn University
NSF-CAVE3 Electronics Research Center
Department of Mechanical Engineering
Auburn, AL 36849
[(2)] Nvidia Corp., 2701 San Tomas Expressway, CA 95050, USA
[(3)] Texas Instruments Inc., 13020 TI Blvd, Dallas, TX 75243 USA
Tele: (334) 844-3424
E-mail: lall@auburn.edu

Abstract

The existing configuration of the JESD22-B111 test board does not impose identical strains during drop test on all the 15-components on the test board, requiring a large numbers of boards to develop meaningful life distributions. Two new candidate designs intended to serve as replacements for the JEDEC JESD22-B111 test board have been analyzed. Configuration-A includes four components located symmetrically on a square 3-inch x 3-inch printed circuit board. Configuration-B includes one-component located symmetrically on a square 3-inch x 3-inch printed circuit board. In this paper, explicit finite element models along with high speed imaging in conjunction with 3D DIC measurements have been used to capture transient strain histories at various board locations to quantify the symmetry of loading during a 1500g 0.5 ms shock pulse and 2900g, 0.3 ms shock pulse. The symmetry of the transient mode shapes and interconnect strains has also been quantified.

Introduction

Electronics in portable products may be subjected to shock and accidental drop during normal usage. In order to determine the survivability of electronic components under accidental drop, the industry has adopted the JEDEC Test Standard JESD22-B111. The test standard prescribes the application of a 1500g, 0.5ms shock pulse to the board assembly. The test board for the JESD22-B111 shock test has 15 components in a 3x5 component-array configuration on a 132 mm x 77 mm test board [JEDEC 2003]. The test method widely used to assess the board level drop reliability of components for handheld electronic products. Previous researchers have shown that the asymmetric array configuration of the components on the JEDEC test board imposes different stresses at different component locations [Gu 2006; Park 2007; Farris 2008; Xie 2010]. Owing to the different stress histories, a large number of test boards are required to develop meaningful life distributions, since components from different locations cannot be plotted on the same Weibull graph.

In this study, two new test board designs have been studied for symmetry of the shock pulse, strain and displacement distributions using a combination of high-speed imaging with 3D-DIC and simulation methods. Previously, Digital Image Correlation (DIC) has been widely used in a full field measurement by industry and researchers. [Zhou

2001, Yogel 2001, Hang 2005, Xu 2006] have studied the stresses in solder interconnects of BGA packages under thermal loading with DIC technique. Lall [2007, 2008, 2012] has measured deformation gradients and full field displacement in electronics subjected to drop and shock using digital image correlation. The use of 3D-DIC has the advantage of providing full-field strain and displacement data.

The new test board is 3-inch x 3-inch in size. Configuration-A includes four components located symmetrically on the test board. Configuration-B includes one component per test board located symmetrically at the center of the test board. Both configuration-A and configuration-B of the board assemblies have been studied under 1500g, 0.5ms shock-pulse. In addition, the one-component Board-B has been studied under 2900g, 0.3ms shock-pulse as well. Explicit finite element models have been used to study both the strain and displacement response of the test assemblies at the board center and the package corners. Modal analysis has been used to characterize the mode shapes of the test board. High speed imaging along with 3D digital image correlation has been used to study the full-field transient strain and displacement distributions at the package corners on both the one component board and the four component board assemblies. All the components have been monitored in-situ for resistive opens using high-speed data acquisition system. Previously, Xie [2013] studied the strain response of the two board configuration using stain gages at discrete locations. This study's acquisition of full-field displacement and strain data with high speed imaging to study transient mode shapes, use of explicit finite element models to capture the transient dynamic response of the board assemblies, and use of high speed data acquisition for in-situ damage monitoring of all the components is new. Results shown in the paper can be used to assess the potential of the new test board design to provide uniform strain distributions on the test board for all components under test.

Test Vehicle

Two test boards have been used to characterize the strain field levels and the transient dynamic behavior. Test Assembly-A is a 3" x 3" square boards with four-BGA packages. Test Assembly-B is a 3-inch x 3-inch square board with one-BGA package mounted in the center (Figure 1). The BGA is 13mm x 13mm in length and width and the PCB has 6 layers with a thickness of 1-mm. The BGAs in both test assemblies are mounted in a symmetric configuration with respect to the board center. There is 10-pin connector at each corner and one screw hole at each corner of PCB. The X-ray image of the package footprint is shown in Figure 2. All the packages are identical and have the configuration as shown in the Table 1

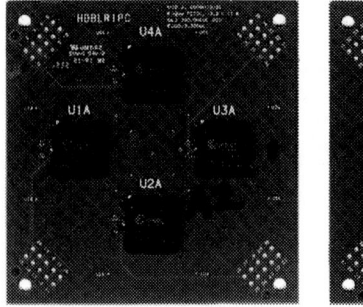 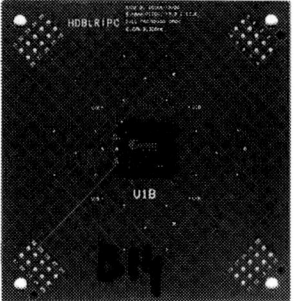

(a) Test Board-A   (b) Test-Board-B

Figure 1: (a) Test Board A with four BGA packages (b) Test Board B with single BGA package

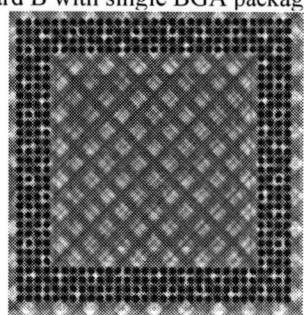

Figure 2: Package Foot Print

Table 1: Package Architecture for 13mm BGA

Ball I/O Count	432
Ball Pitch	0.4 mm
Chip Size	8mm x 8mm
Body Size	13mm x 13mm
Ball Matrix	31 x 31 Partial-Array
Package Type	CVBGA
Ball Diameter	0.25 mm
Substrate pad	SMD

Experimental Setup

The drop test was setup as per the layout shown in Figure 3. The test boards were subjected to controlled drop under 1500g, 0.5ms and 2900g, 0.3ms test conditions on a Lansmont Model 23 Shock Tower. The required sine wave pulse was obtained using pulse shapers on the shock tower.

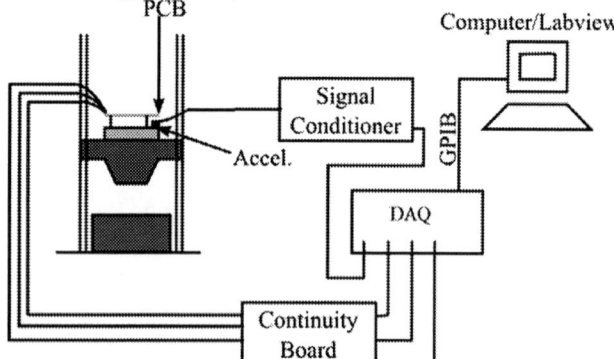

Figure 3: Layout of the Drop-Test Setup

An accelerometer of sensitivity 0.103 mV/g was used to measure the acceleration levels during the shock event. Drop heights for 1500g and 2900g levels were 14.2 inches and 20.6 inches respectively. The shock tower used an assisted gravity drop to attain the g-level required for the shock event. Package continuity was monitored during the transient dynamic shock event using a high speed data acquisition system at a sampling rate of 5 MHz. The data acquisition system was triggered to monitor for loss of continuity for every shock event. Each test package on the test assembly was in situ monitored for any high resistance events during the transient shock event to identify loss of continuity because of interconnect failure.

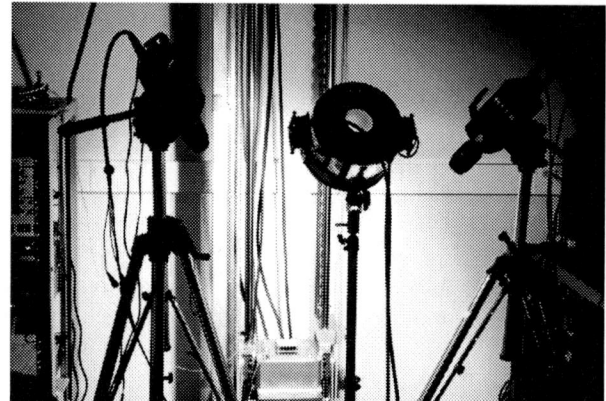

Figure 4: 3D-DIC Setup for Strain Measurement.

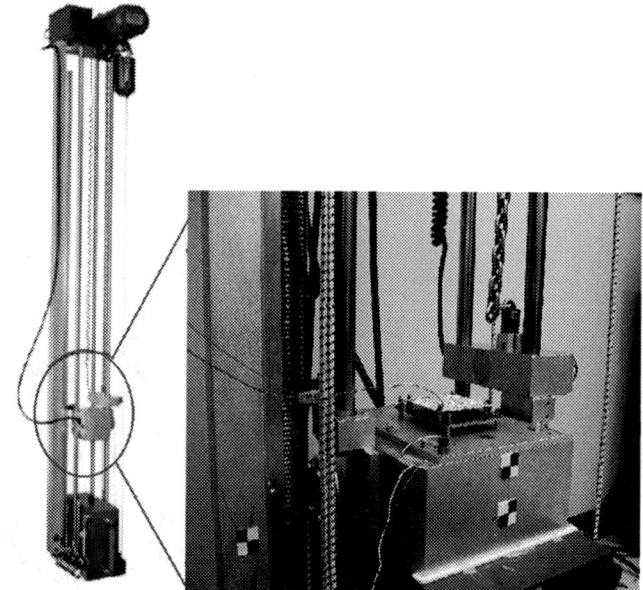

Figure 5: Close-up View of the 3D-DIC Setup for Strain Measurement.

Figure 4 shows the drop tower test setup and high speed camera setup. The test board was speckle coated on the top surface facing the high speed cameras. The speckles were used to measure the full-field strain and displacement of the test assembly during the shock event. The board assemblies were mounted to the shock-base with the components in the face down configuration as specified in the JEDEC JESD22-B111 test standard. Targets were mounted on the front face of the test assembly to measure the velocity prior to impact of

978-1-4799-2408-0/14 $31.00 © 2014 IEEE

the drop base. The velocity of the target points was measured from Image Analysis software. Figure 5 and Figure 6 shows the close-up views of the speckle coated test board mounted on the shock-tower. The drop event was simultaneously captured by two high speed cameras at 25,000 fps. The images extracted from the high speed video cameras were used to perform the 3D DIC analysis.

Figure 6: Speckle Coated Board Mounted on the Drop Base

3D Digital Image Correlation

Digital Image Correlation (DIC) has been used to measure the full field displacement and strain of the test assemblies during the transient shock event. The high speed video from the shock test has been used to track the motion of the speckles before, during and after the shock event. The two-camera configuration has been used to measure both the in-plane and out-of-plane deformation. Previously the feasibility of DIC for 3D measurement of displacements during the transient shock events has been demonstrated by [Lall 2007, 2008, 2012].

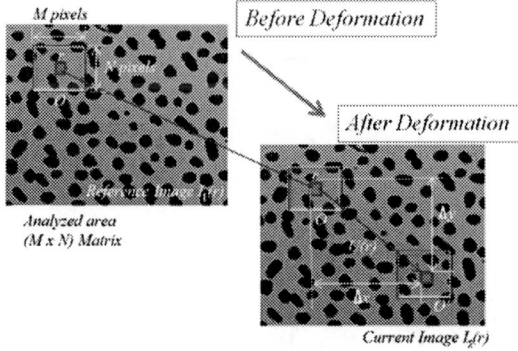

Figure 7: Digital Image Correlation Principle

The undeformed referred to as the original image and deformed images versus time have been captured using the high speed cameras (Figure 7). Since a single pixel is not a unique signature of a point hence a neighboring pixels are used. The collection of pixels is called a subset. Uniqueness of the pattern has been assured by using a non-repetitive random high-contrast speckle pattern. A subset of pixels around a reference pixel O in the reference image has been compared with the subset corresponding to pixels in the

deformed image using a predefined correlation function to describe the difference of the two digital sub images. Deformation of the subset during transient deformation has been accomplished through displacement mapping using a subset shape function. The subset has been stepped through the image to measure the displacement of the complete board assembly. An algorithm based on the mutual correlation coefficient or other statistical functions are used to correlate the change in a reference pixel in the original image and the corresponding reference pixel in the deformed image. Three such typical correlations include absolute difference, least square and cross correlation. There are three typical correlation functions which are used and are defined as follows:

Absolute Difference

$$C_A(r') = 1 - \frac{\iint_\Omega |I_2(r+r') - I_1(r)| \, dr}{\iint_\Omega I_1(r) \, dr} \tag{1}$$

Least Square:

$$C_L(r') = 1 - \frac{\iint_\Omega [I_2(r+r') - I_1(r)]^2 \, dr}{\iint_\Omega I_1^2(r) \, dr} \tag{2}$$

Cross-Correlation

$$C_C(r') = 1 - \frac{\iint_\Omega I_1(r) I_2(r+r') \, dr}{\left[\iint_\Omega I_1^2(r) \, dr \iint_\Omega I_2^2(r+r') \, dr\right]^{\frac{1}{2}}} \tag{3}$$

where I_1 and I_2 are the original and the deformed images, Ω (M x N) is the area of the sub image around reference pixel r, r` is the current pixel, CA(r`) is the current absolute correlation function, CL(r`) is the current least square correlation function, and CC(r`) is the current cross correlation function. The high speed cameras have been mounted on rigid tripod mounts to prevent movement of the cameras during the shock event relative to the shock tower or relative to each other. Two targets have been attached to the base to track the rigid body motion before impact (Figure 8). Software called Motion Plus has been used to calculate rigid body displacement of the whole drop event before impact. The initial velocities have been calculated through the first derivative of the displacement function. The velocities just before the impact for 1500g and 2900g drop levels were measured to be 2.417 m/s and 3.280 m/s respectively.

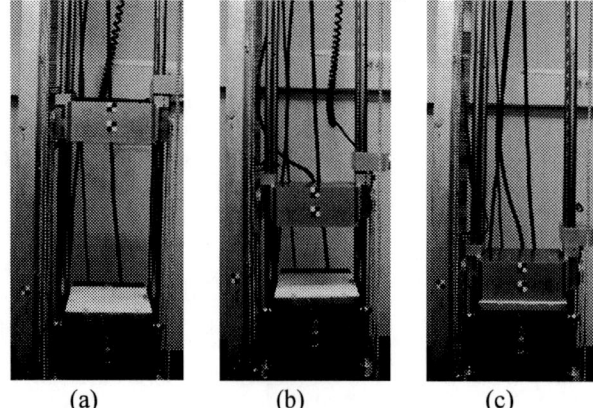

(a)   (b)   (c)

Figure 8: Drop base position at (a) Time, t=0 (b) during drop (c) at Impact

Explicit Finite Element Model

The transient dynamic response of the board assemblies during the shock event has been modeled using explicit finite element models. Reduced integration elements have been used for model both reduced integration shell element S4R for the printed circuit board, reduced integration 8-node solid element for the package, and the Timoshenko beam element for the solder interconnects. Figure 9 shows the finite element models for the four component test assembly and the one component test board assembly.

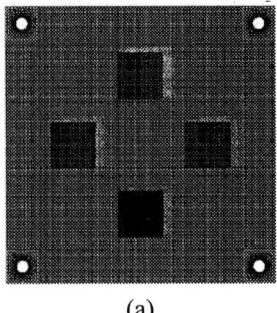 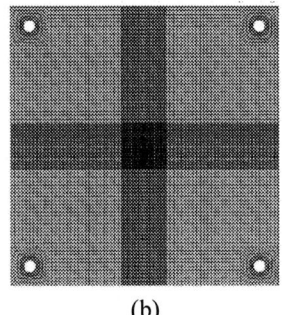

(a)   (b)

Figure 9: Finite Element Model of the (a) Four-Component Test Assembly-A (b) One-Component Test Assembly-B.

The governing differential equation of motion for the system consisting of board assembly mounted on the shock table system can be expressed as

$$[M]\{D\}_n + [c]\{\dot{D}\}_n + [K]\{D\}_n = \{R^{ext}\}_n \quad (4)$$

where $\{D\}$ is the d.o.f. vector, the "." on top represents time differentiation, and subscript "n" represents the time-step. $\{R^{ext}\}$ is the external force. The degree of freedom vector has been discretized in time using a taylor series. Taylor Series expansion of the $\{D\}$ vector at time step at n+1 and n-1

$$[D]_{n+1} = \{D\}_n + \Delta t \cdot \{\dot{D}\}_n + \frac{\Delta t^2}{2}\{\ddot{D}\}_n + \frac{\Delta t^3}{6}\{\dddot{D}\}_n \quad (5)$$

$$[D]_{n-1} = \{D\}_n - \Delta t \cdot \{\dot{D}\} + \frac{\Delta t^2}{2}\{\ddot{D}\}_n - \frac{\Delta t^3}{6}\{\dddot{D}\}_n \quad (6)$$

Equation (5) and (6) have been combined to calculate the velocity and acceleration of the body. Higher order terms have been omitted. So velocity and acceleration can be written as

$$[\dot{D}]_n = \frac{1}{2 \cdot t}(\{D\}_{n+1} - \{D\}_{n-1}) \quad (7)$$

$$[\ddot{D}]_n = \frac{1}{\Delta t^2}(\{D\}_{n+1} - 2\{D\}_n + \{D\}_{n-1}) \quad (8)$$

The equation of motion has been updated by substituting the expressions for velocity and acceleration from Equations (7) and (8) into Equation (4) to derive the Equation (9),

$$[\frac{1}{\Delta t^2}M + \frac{1}{2 \cdot t}C]\{D\}_{n+1} = \{R^{ext}\}_n + [K]\{D\}_n \quad (9)$$
$$+ \frac{1}{\Delta t^2}[M](2\{D\}_n - \{D\}_{n-1}) + \frac{1}{2 \cdot \Delta t}[C]\{D\}_{n-1}$$

For initial condition (n=0), $[D]_{-1}$ has been calculated by given starting conditions for the board assembly mounted to the shock table including initial displacement $\{D\}_0$, velocity $\{\dot{D}\}_0$ just before impact. $[\ddot{D}]_0$ has been computed by equation of motion

Figure 10: Isometric View of the Finite Element Model of the One-Component Board Assembly-B.

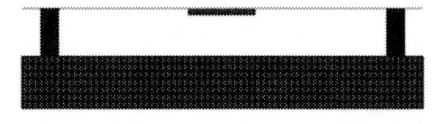

Figure 11: Front View of the Finite Element Model of the One-Component Board Assembly-B.

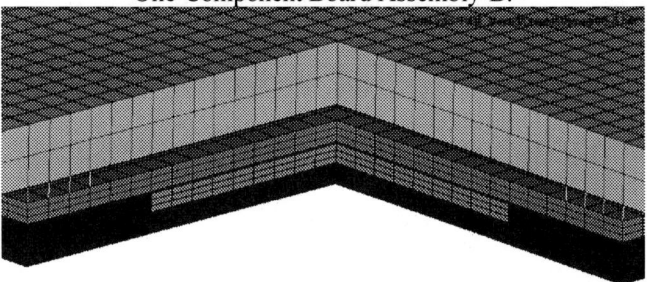

Figure 12: Cut-Section of the BGA Finite Element Model of the One-Component Board Assembly-B.

Figure 13: Isometric View of the Finite Element Model of the Four-Component Board Assembly-A.

$$[\ddot{D}]_0 = \{M\}^{-1}(\{R^{ext}\})_0 - [K]\{D\}_0 - [C]\{\dot{D}\}_0) \quad (7)$$

Thus, based on (6) $[D]_{-1}$ can be obtained by ignoring the higher order term Δt^3.

$$[D]_{-1} = \{D\}_0 - \Delta t \cdot \{\dot{D}\}_0 + \frac{\Delta t^2}{2}\{\ddot{D}\}_0 \qquad (8)$$

The degree of freedom vector at the next time step, $[D]_{n+1}$ requires the computation of the internal force vector $\{R^{int}\}_n$. The internal force vector $\{R^{int}\}_n$ has been obtained from the displacement vector at the prior time step $\{D\}_n$ and constitutive laws for the materials in the board assembly and shock table. Therefore, strains at $n \cdot \Delta t$ are known. The PCB has been modeled with reduced integration S4R shell elements. The components have been modeled with solid C3D8R elements and B31 Timoshenko beam elements. The concrete floor has been modeled with rigid R3D4 elements. A reference node has been placed behind the rigid wall for application of constraints. Node to surface contact has been used for impact between the shock table and the floor. An event length of 5ms after impact has been modeled. Time history has been monitored at a time period of 0.1ms at the corner solder joints of all BGAs in the test assembly.

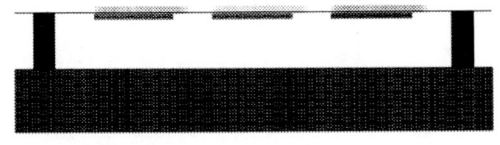

Figure 14: Front View of the Finite Element Model of the four-Component Board Assembly-A.

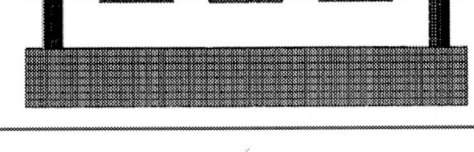

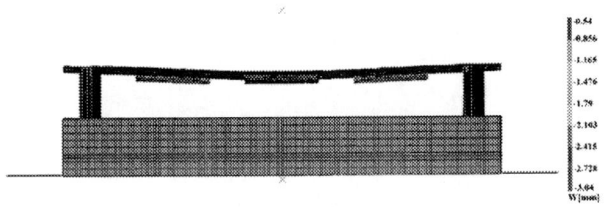

Figure 15: Transient dynamic model response before and after impact for four-component test assembly A.

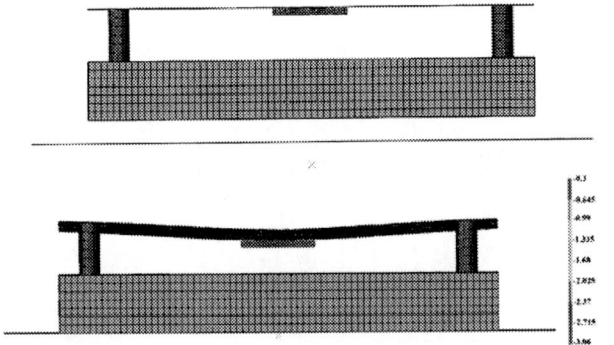

Figure 16: Transient dynamic model response before and after impact for one-component test board B.

Transient dynamic behavior of the populated board assemblies has been modeled under transient shock from two different initial velocities to get acceleration level of 1500Gs and 2900Gs have been simulated using ABAQUS Explicit. The experimentally measured velocity prior to impact using high speed imaging has been assigned as the initial conditions for the finite element model. The velocities before the impact were 2.417 m.s⁻¹ and 3.280 m.s⁻¹ for the 1500g and 2900g levels respectively.

Material Properties

The simulated weight of the model for all components and test board closely approximates the actual weight. Table 2 shows the material properties used as inputs for development of the smeared property models. Table 3 shows the simulated and actual weights for test boards A and B. Solder joints have been modeled using high strain rate properties from [Lall 2012] shown in Figure 17.

Table 2: Material properties for various package elements

	Elatic Modulus (GPa)	Possion's Ratio	Density (Kg/m^3)
Solder Ball (SAC105)	42	0.37	7400
Mold Compound	23.5	0.25	1650
Die	162	0.28	2329
Die Attach	2.76	0.35	7800
Epoxy Film	0.649	0.35	2100
BT Substrate	17.4	0.28	1800
Bare FR406	22.4	0.35	1850

Table 3: Comparison of actual and simulated masses

Board	Actual (gm)	Simulated Model (gm)
Test Board A	13.72	13.98
Test Board B	12.67	12.94

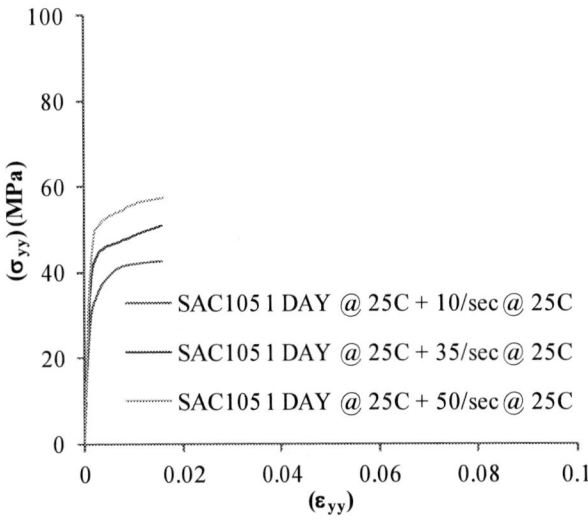

Figure 17: Stress vs. Strain for 1-day room temperature (25°C) aged SAC105 alloys at strain rates of 10 per sec, 35 per sec and 50 per sec [Lall 2012].

Modal Analysis

Natural frquencies and mode shapes for both the test board assemblies have been extracted from modal analysis of the board assemblies. The first six-modes correspond to three-translations and three- rotations and have been ignored. Figure 18 to Figure 25 show the deformation mode shapes 1, 2, 3, 4, 5 for both test assembly A and B. Mode shapes 2 and 3 are identical as the test board is symmetric. For the four component test board-A, the eigen frequencies for deformation mode shapes 1 thru 5 are 454.44 Hz, 848.64 Hz, 981.97 Hz and 1610.8 Hz. For the one-component test board-B, the eigen frequencies for mode shapes 1 thru 5 are 468.92 Hz, 854.65 Hz, 978.45Hz and 1596.2 Hz (Table 4). The transient mode shape is dominated by mode 1 and mode 4. While all the components on the four component board experience the same deformatoin in mode-1; it can be seen that the deformation history is different for mode shapes 2-5 for the four component locations.

Table 4: Natural frequencies for the Four-Component Test Board and the One Component Test Board

	Four-Component Test Board-A	One-Component Test Board-B
Natural Frequency-1	454.44 Hz	468.92 Hz
Natural Frequency-2	848.64 Hz	854.65 Hz
Natural Frequency-3	981.97 Hz	978.45 Hz
Natural Frequency-4	1610.8 Hz	1596.2 Hz

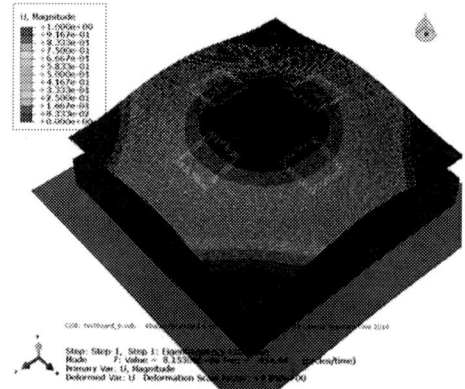

Figure 18: Mode-Shape-1 for Four Component Test Assembly-A

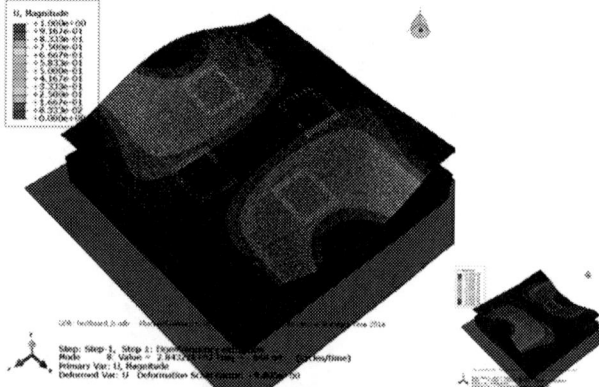

Figure 19: Mode-Shape-2 and 3 for Four Component Test Assembly-A

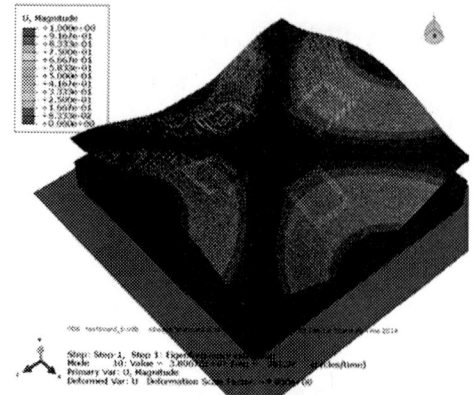

Figure 20: Mode-Shape-4 for Four Component Test Assembly-A

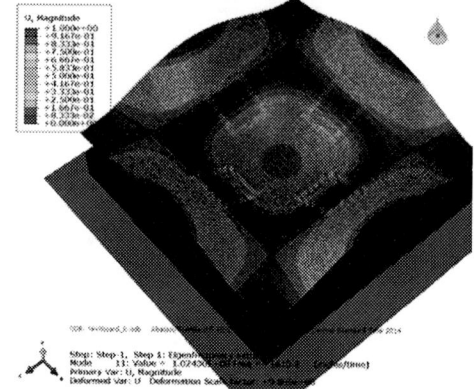

Figure 21: Mode-Shape-5 for Four Component Test Assembly-A

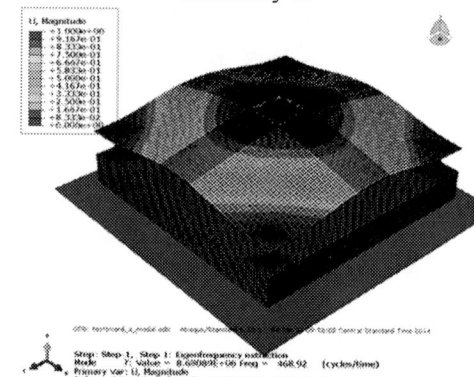

Figure 22: Mode-Shape-1 for One-Component Test Assembly-B.

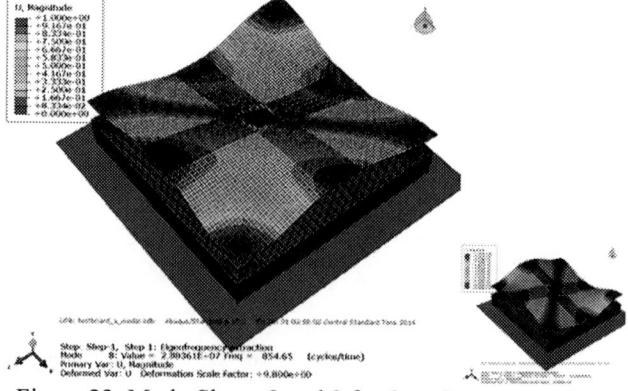

Figure 23: Mode-Shape-2 and 3 for One-Component Test Assembly-B.

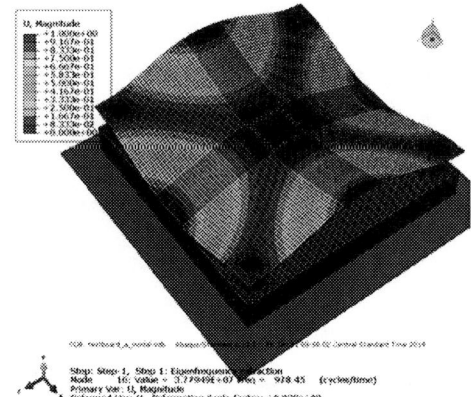

Figure 24: Mode-Shape-4 for One-Component Test Assembly-B.

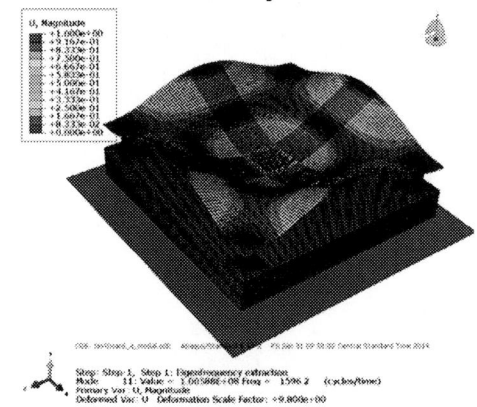

Figure 25: Mode-Shape-5 for One-Component Test Assembly-B.

Model Predictions

In this section, the model predictions from the explicit finite element transient dynamic model have been presented for both 1500g and 2900g shock pulses.

Printed Circuit Board Displacement Field

The peak board deflection and strains along with solder joint stresses have been extracted for both board assembly A and B subjected to the shock levels of 1500g and 2900g. The symmetry of the strain field on the test board at the various component locations and at the corner interconnects of each component have been evaluated.

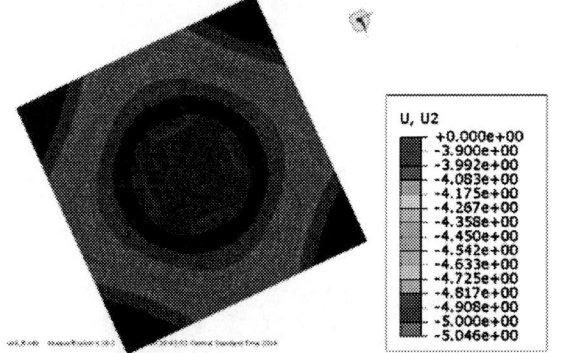

Figure 26: Model Predictions of Peak Deflection for Four-Component Board-A under Shock-Impact at 1500g.

The location of the peak stress in interconnects has also been identified. Figure 26 shows the deformation field under a

1500g transient shock is symmetric and the four components on Board Assembly-A experience the same displacement field on the printed circuit board after impact. The one-component test board-B experiences the symmetric transient deformation on the PCB for both 1500g and 2900g shock levels. Figure 27 shows the deformation field under 1500g for the one-component test board-B. Figure 28 shows the deformation field under 2900g for the one-component test board-B.

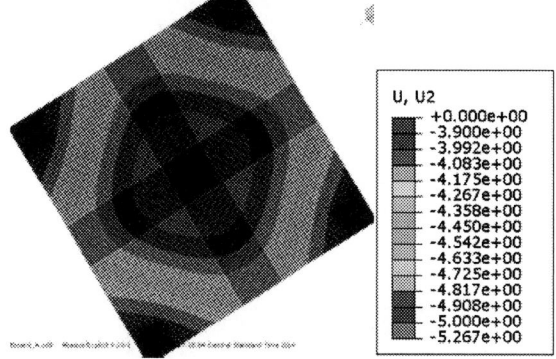

Figure 27: Model Predictions of Peak Deflection for One-Component Board-B under Shock-Impact at 1500g.

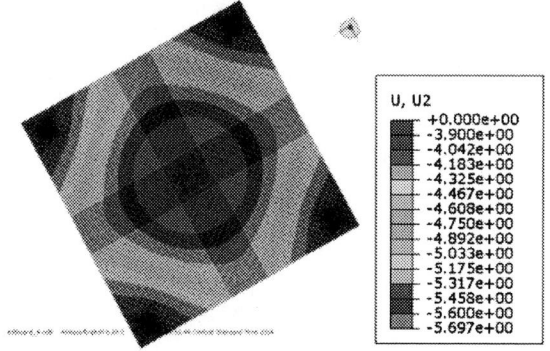

Figure 28: Model Predictions of Peak Deflection for One-Component Board-B under Shock-Impact at 2900g.

Solder Joint Transient Strain History

The solder joint stress have been extracted from the corner interconnects for each of the component corners on the four component test assembly-A, and for all corners of the one-component test board-B.

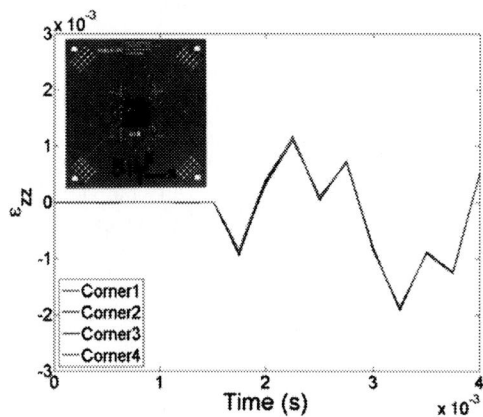

Figure 29: Interconnect Strain at location of the corner interconnects of one-component test board-B under 1500g shock-impact

978-1-4799-2408-0/14 $31.00 © 2014 IEEE   91

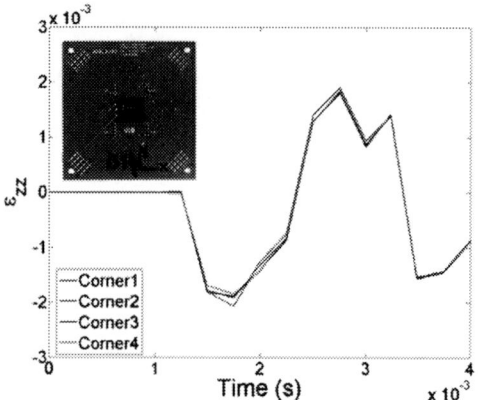

Figure 30: Interconnect Strain at location of the corner interconnects of one-component test board-B under 2900g shock-impact

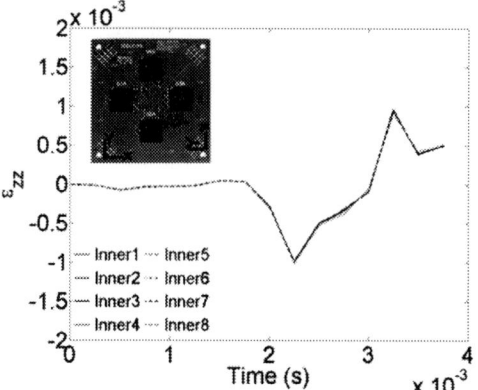

Figure 31: Interconnect Strain and in the INNER corner interconnects of four-component test board-A under 1500g shock-impact

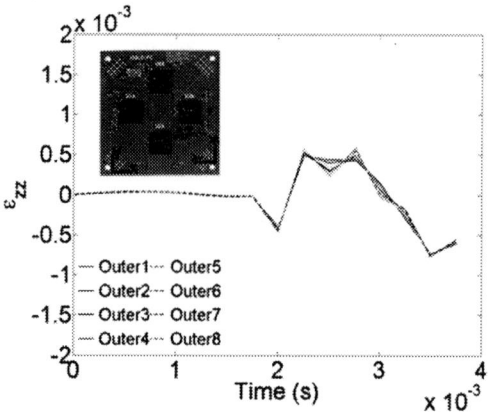

Figure 32: Interconnect Strain in the OUTER corner interconnects of four-component test board-A under 1500g shock-impact

Model predictions indicate that the strain experienced by the corner solder joints along the length of the beam (z-direction) in the corner interconnects are symmetric shown by the overlapping strain histories of ε_{zz} strain in Figure 29. The symmetry of the strain histories in the corner interconnects is shown to be valid even for the 2900g shock impact for the one-component test assembly-B (Figure 30).

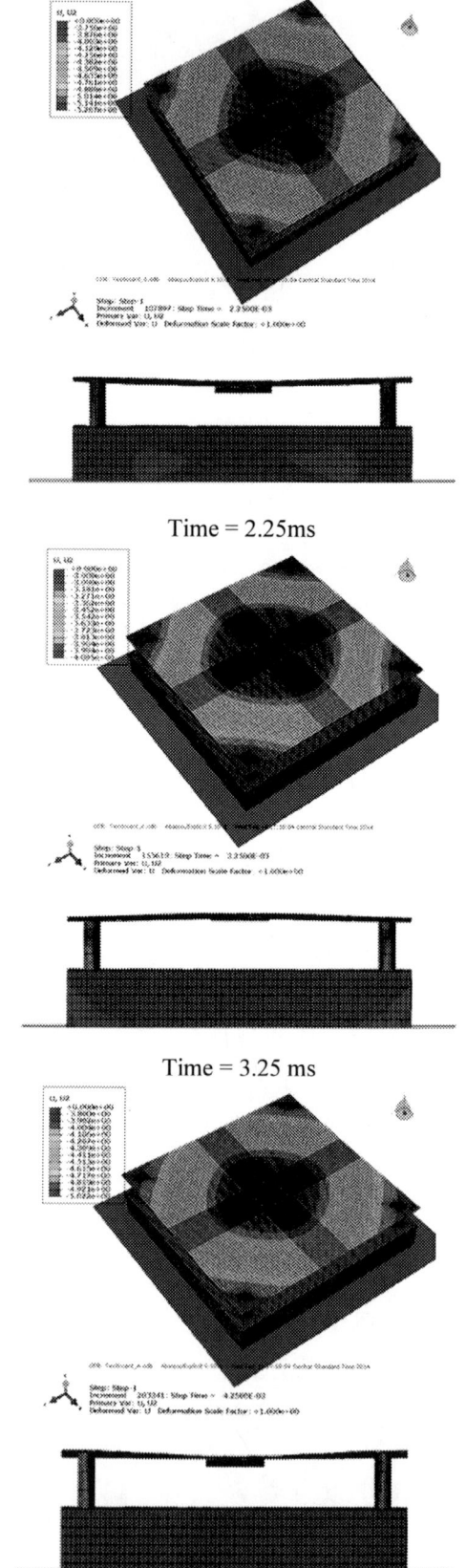

Figure 33: The out of plane deflection at (a) 2.25ms , (b) 3.25ms and (c) 4.25ms of the one component test board-B under 1500g shock.

978-1-4799-2408-0/14 $31.00 © 2014 IEEE   92

Further, for the four component test assembly-A, all interconnects at the inner corners show symmetry in strain histories (Figure 31). The outer corner interconnects also show symmetry of strain histories between the four components of the test assembly-A (Figure 32). The out of plane deflection contours of the one component test board-B under 1500g shock-impact at various times during the shock event is shown in Figure 33 showing symmetry of deformation field during the shock event.

Symmetry of Board Strain at the Corner Interconnect Locations

The strain in the PCB at the corner interconnects of both the test boards have been extracted. The strains of the one component testboard-B at the four corners along x and y for 1500g and 2900g drop impact is shown in Figure 34 and Figure 35 respectively. It is observed that all the four corners experience nearly the same strain levels in the PCB shown by overlapping plots from corner 1 thru corner 4. As the component location is at the center and symmetric from four corners of the board, the strains along x direction are nearly identical as that in y direction. Figure 34a and Figure 34b show nearly identical behavior and further this is true for Figure 35a and Figure 35b. In the case of four component testboard-A the x-strains in the PCB near all eight inner corners (Figure 36a) are nearly identical. A similar trend is seen in the y-strain as well indicated by the overlapping curves in Figure 36b.

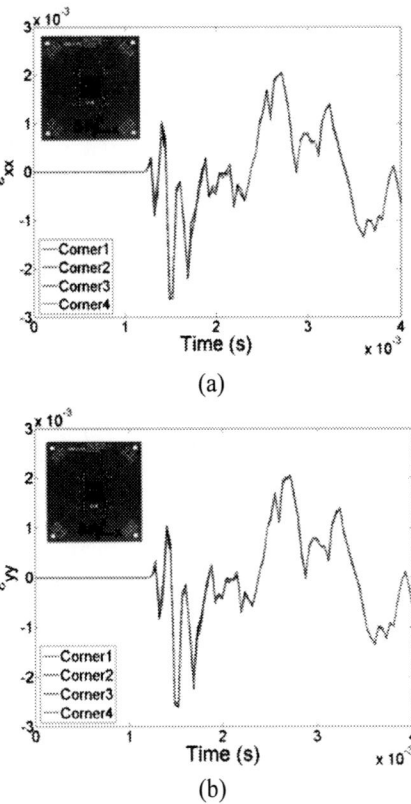

Figure 35: Strain in the PCB at location of the corner interconnects of one-component test board-B under 2900g shock-impact (a) ε_{xx} and (b) ε_{yy}

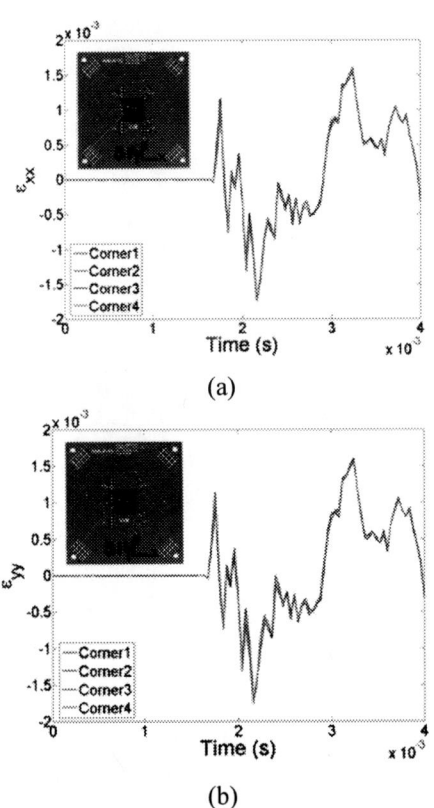

Figure 34: Strain in the PCB at location of the corner interconnects of one-component test board-B under 1500g shock-impact (a) ε_{xx} and (b) ε_{yy}

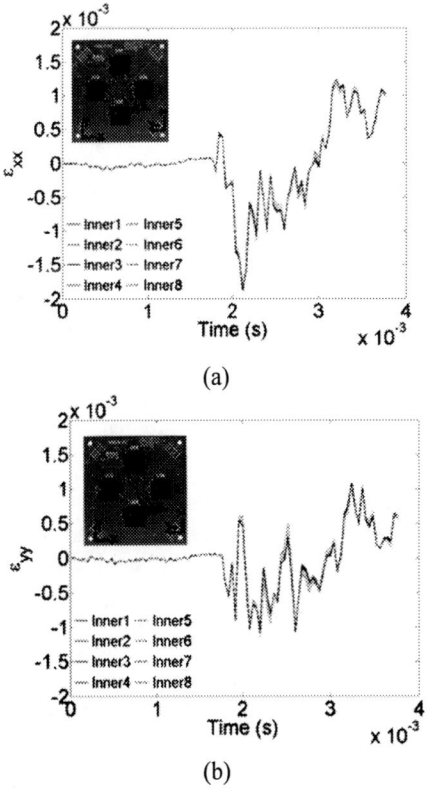

Figure 36: Strain in the PCB at location of the INNER corner interconnects of four-component test board-A under 1500g shock-impact (a) ε_{xx} and (b) ε_{yy}

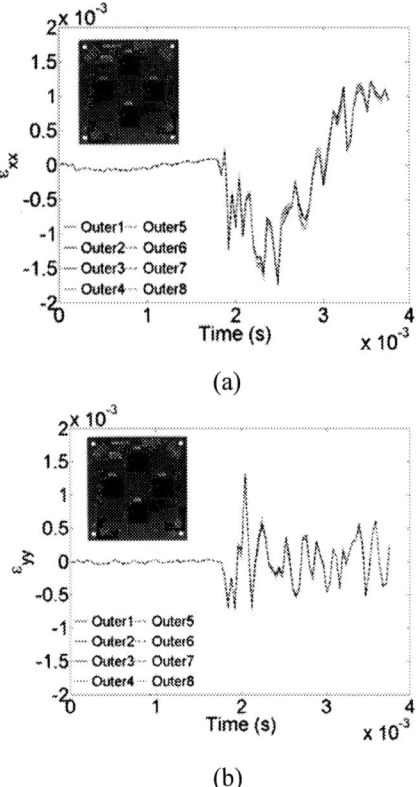

(a)

(b)

Figure 37: (a) ε_{xx} and (b) ε_{yy} Strain in the PCB at location of the OUTER corner interconnects of four-component test board-A under 1500g shock-impact

However, the x-strain and the y-strain are not equal, since in the four component configuration, the location of the corners are not symmetric w.r.t. edges of the board. The lack of equality in the x-strain and y-strain is seen near the outer corners (Figure 37) of the components as well. However, the outer-corners have nearly identical x-strain value between the four component locations – shown by the overlapping curves in Figure 37a. Further, the outer-corners have nearly identical y-strain value between the four component locations – shown by the overlapping curves in Figure 37b.

Experimental Data on Transient Deformation during Shock-Impact

In this section experimental measurements of PCB transient strains using 3D-DIC have been presented.

Printed Circuit Board Displacement Field

From 3D-DIC measurements, the peak board deflection and strains have been extracted for both board assembly A and B subjected to the shock levels of 1500g and 2900g. The symmetry of the strain field on the test board at the various component locations and at the corner interconnects of each component have been evaluated. The location of the peak strains in printed circuit board has also been identified. Figure 38 shows the deformation field under a 1500g transient shock is predominantly symmetric with small degree of asymmetry. The four components on Board Assembly-A experience the approximately the same displacement field on the printed circuit board after impact.

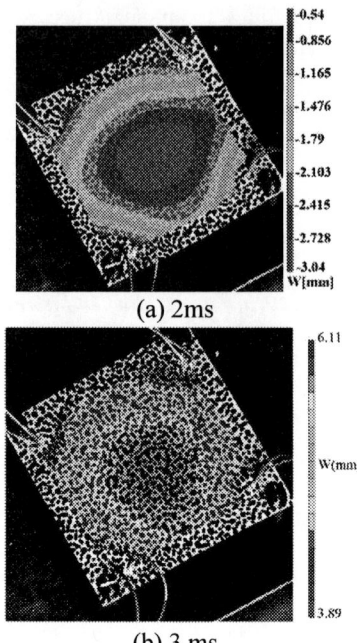

(a) 2ms

(b) 3 ms

Figure 38: Deflection for Four-Component Board-A under Shock-Impact at 1500g (a) 2 ms after impact (b) 3ms after impact.

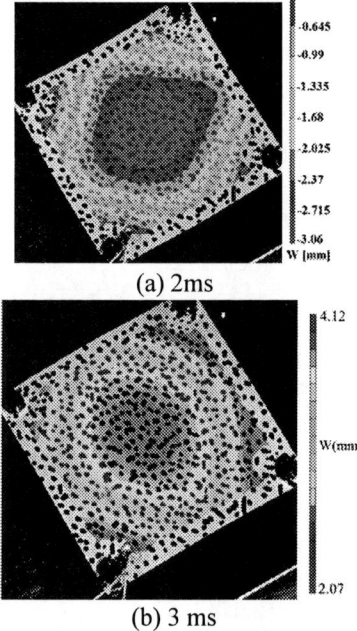

(a) 2ms

(b) 3 ms

Figure 39: Deflection for one-Component Board-B under Shock-Impact at 1500g (a) 2 ms after impact (b) 3ms after impact.

The slight degree of asymmetry in the displacement field indicated by the offset in the peak deformation at the center of the board assembly is attributed to board assembly's manufacturing variability and tolerances. Experimental data in Figure 38a, b, also indicates that the symmetry of the displacement field continues throughout the transient deformation of the board assembly. Figure 39 and Figure 40 show that the predominant symmetry of the board assembly's displacement field is valid for the 1500g and the 2900g shock impact condition for the one-component test assembly-B as well. The out of plane deflection of the one component test board-B under 1500g shock- impact at various events is shown in Figure 41. The images were captured using a high speed camera operating at 25,000 fps.

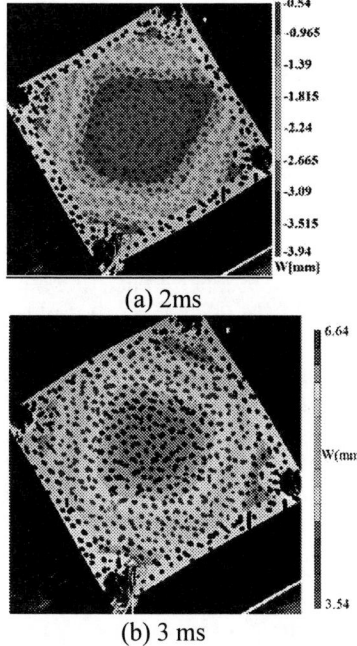

(a) 2ms

(b) 3 ms

Figure 40: Peak Deflection for single Component Board-B under Shock-Impact at 2900g (a) 2 ms after impact (b) 3ms after impact.

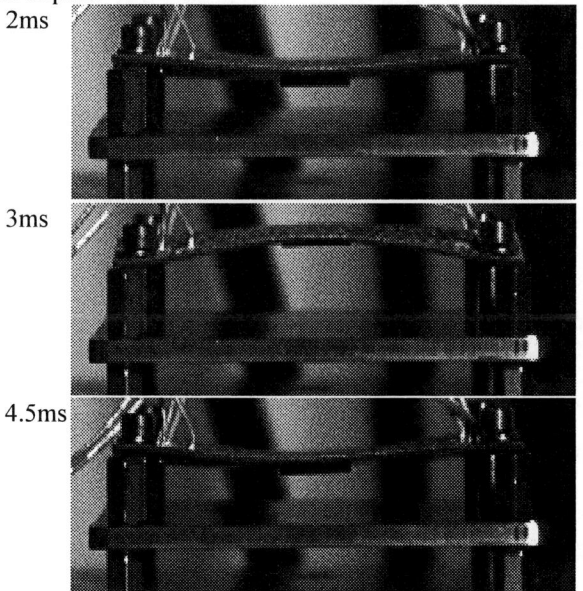

Figure 41: Deformation History for one-component test board-B under 1500g shock-impact at time 2ms, 3ms and 4.5ms

Symmetry of Board Strain at the Corner Interconnect Locations

Strains have been extracted in the PCB near the corner interconnects of both the test assemblies A and B to evaluate the peak value and its symmetry. Figure 42 shows the PCB x-strain and PCB y-strain time histories of the one component test assembly-B under 1500g drop impact.

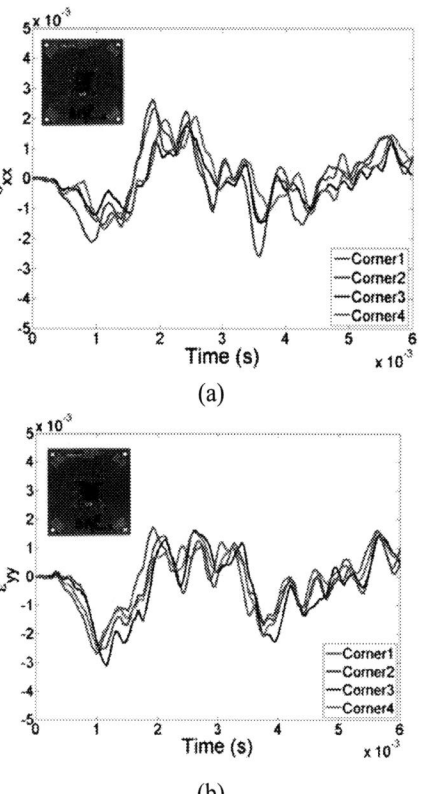

(a)

(b)

Figure 42: (a) ε_{xx} and (b) ε_{yy} Strain in the PCB at location of the corner interconnects of one-component test board-B under 1500g shock-impact

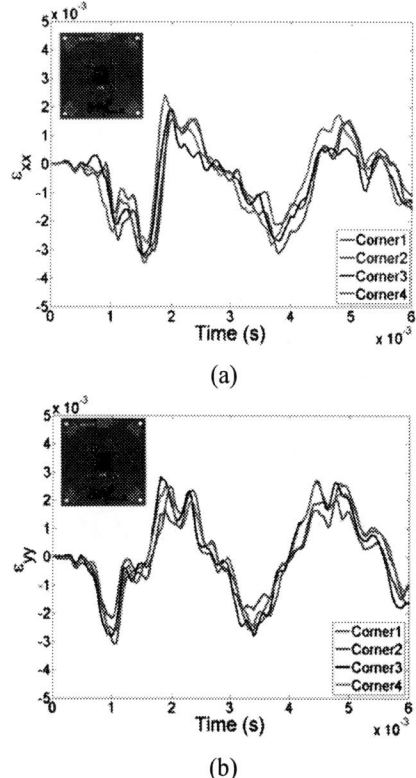

(a)

(b)

Figure 43: (a) ε_{xx} and (b) ε_{yy} Strain in the PCB at location of the corner interconnects of one-component test board-B under 2900g shock-impact

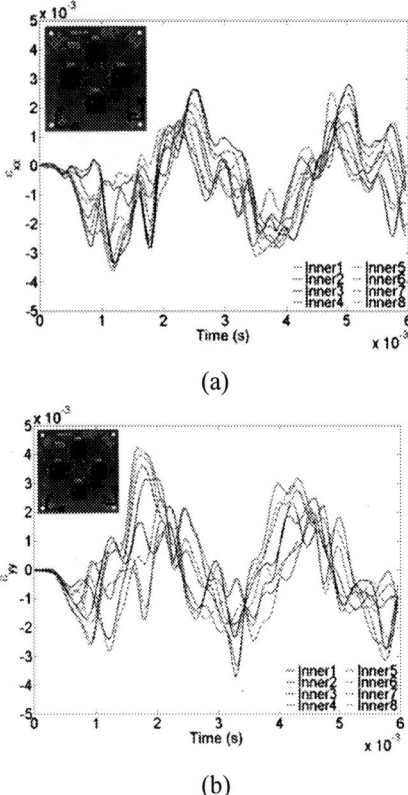

(a)

(b)

Figure 44: (a) ε_{xx} and (b) ε_{yy} Strain in the PCB at location of the INNER corner interconnects of four-component test board-A under 1500g shock-impact

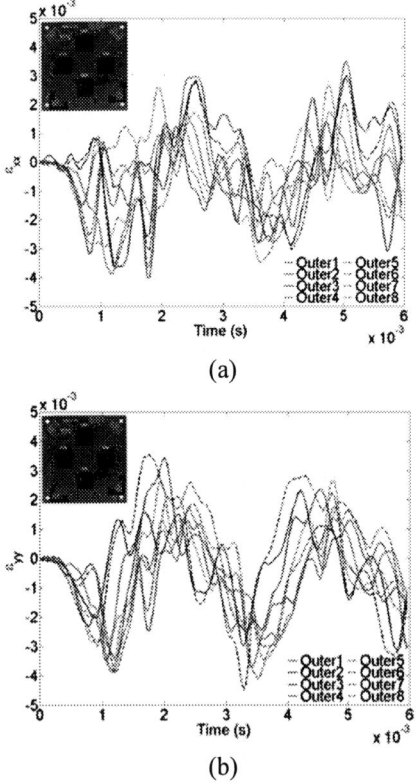

(a)

(b)

Figure 45: (a) ε_{xx} and (b) ε_{yy} Strain in the PCB at location of the OUTER corner interconnects of four-component test board-A under 1500g shock-impact

The strain distribution along x direction near the four corner interconnects are very similar as they overlap. Similarly along y direction the four corner regions experience almost same strain levels. The strains are symmetric along x and y and both the strain distributions behave in a similar fashion. The test assembly experiences approximately same peak value of strain in both the x and y direction. This observation of the symmetry of the strain and its amplitude can also be seen in the 2900g drop impact of one component test assembly-B in Figure 43. Figure 44 shows the PCB x-strain and PCB y-strain time histories near the inner corner interconnects under 1500g drop impact of test assembly-A. The strain distributions at all the inner corners are similar but not identical.This offset in the strain distributions of inner corners in both x and y directions can be attributed to board assembly's manufacturing variability and tolerances. Figure 45 shows the PCB x-strain and PCB y-strain time histories near the outer corner interconnects under 1500g drop impact of test assembly-A. The strain distributions at all the outer corners are similar but not identical.This offset in the strain distributions of outer corners in both x and y directions can be attributed to board assembly's manufacturing variability and tolerances. The peak amplitude of the outer x-strain and outer y-strain is slightly higher than that at the inner corners.

Validation of Model Predictions with Experimental Measurements

Model prediction of board strains have been validated with experimental measurements of board strains from high speed 3D digital image correlation. Figure 46 and Figure 47 show the transient strain histories at the package corners for the one component test board-A under 1500g and 2900g shock impact respectively. The peak strain is in the neighborhood of 2500$\mu\varepsilon$ at the package corners under a 1500g shock impact. Figure 47 shows the same behavior where the peak strain is in the neighborhood of 3000$\mu\varepsilon$ at the package corners under a 2900g shock impact. The mean experimental value of the strain from the four corners has been considered for validating with the model prediction. The strain histories from the experimental measurements and the model prediction show a good correlation.

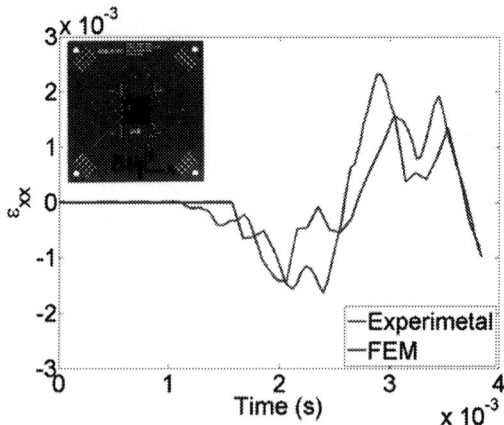

Figure 46: Correlation of model predictions of board strain versus experimental measurements of board strain from high speed 3D-DIC at CORNER-3 under 1500g shock impact for the one-component test assembly-B.

978-1-4799-2408-0/14 $31.00 © 2014 IEEE

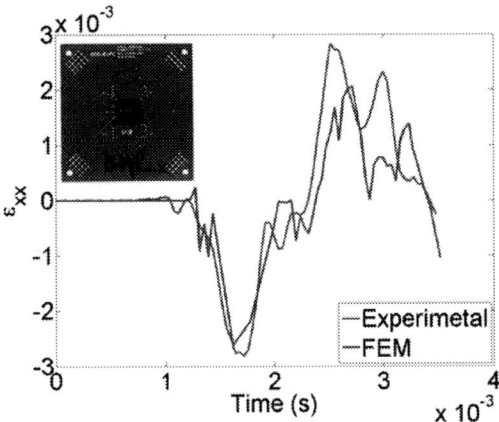

Figure 47: Correlation of model predictions of board strain versus experimental measurements of board strain from high speed 3D-DIC at CORNER-3 under 2900g shock impact for the one-component test assembly-B.

Failure Analysis

Board assemblies have been tested to failure to determine the drops to failure under both 1500g shock and 2900g shock test. No-failures have been observed in the one component test assemblies under 1500g shock impact till 19-drops. However, complete failure of the four component test board assembly has been observed at 19 drops (Figure 48). Figure 49 shows the close up image of the failure near U1A component location on the PCB and the Component. Thus, the four component test assembly configuration imposes higher stresses on the components than the one component test configuration under a 1500g, 0.5ms impact. Figure 50 shows still images from a high speed video of the failed package number U4A of test board A. Figure 51 shows the Weibull failure distribution plot of the test board A under 1500g, 0.5ms drop. The failure data is consistent with the model predictions and experimental measurements of board-strain at location of the corner interconnects, which is higher for the four-component test assembly-A compared to the one-component test assembly-B.

(a)

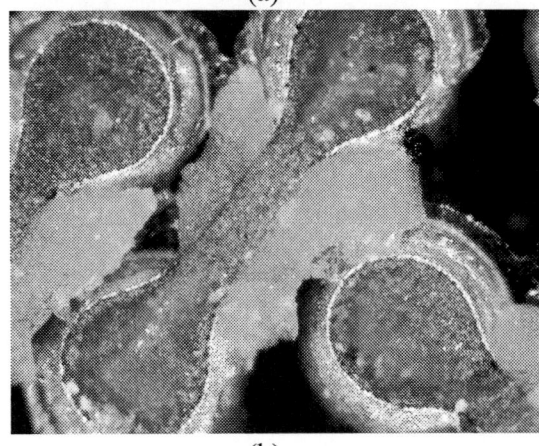

(b)

Figure 49: Close up image of the failure site at (a) U1A PCB footprint and (b) U1A BGA package.

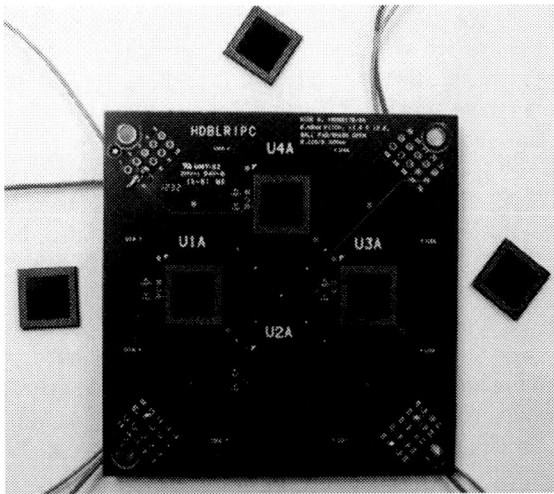

Figure 48: Failed components of the four component test assembly-A under 1500g drop. The drop numbers on the board assembly numbers indicate the number of drops to fail for the respective component location.

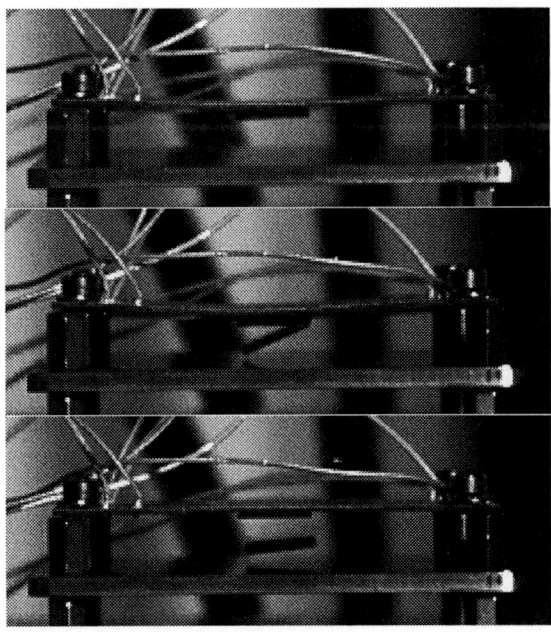

Figure 50: High speed video image capture of failed package number U4A of test board A under 1500g drop

978-1-4799-2408-0/14 $31.00 © 2014 IEEE 97

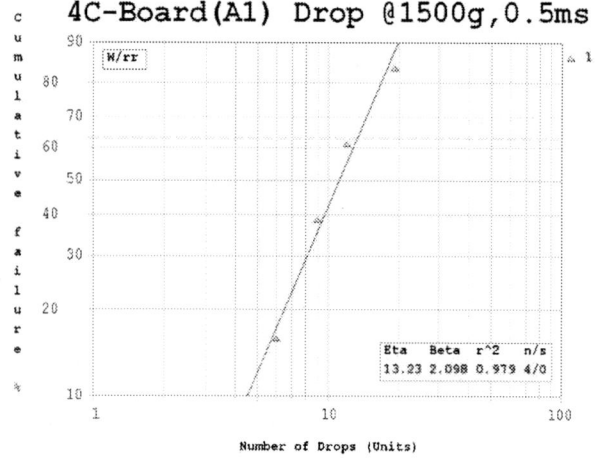

Figure 51: Weibull failure distribution plot of the test board A under 1500g, 0.5ms drop.

Conclusions

Explicit finite element models and high speed 3D-DIC imaging has been used to study two candidate board design intended to be replacements for the JEDEC JESD22-B111 test board. For the one-component test board-B, model predictions experimental measurements of strain and deformation indicate that the strain field is symmetric at the corners of the component. For the four-component test board the model predictions and experimental data yielded mixed results. Model predictions indicate that the four-component test board-A has symmetric strain field and symmetric displacement field. However modal analysis indicates that not all the mode shapes would apply identical board deformation to the four components on the test board. Further, the intended symmetry of strain and displacement for the four component test board is realized in a limited manner because of manufacturing variability. Experimental data from high speed 3D-DIC measurements indicates that the strain traces the four-component inner corners have limited similarity but are not identical. Furthermore, strain traces the four-component outer corners have limited similarity also and are not identical.

Acknowledgments

The research presented in this paper has been supported by NSF-FRS-1127913 and members of NSF-CAVE3 Electronics Research Center. The work was conducted as part of CAVE3 and Auburn University participation in the JEDEC B111 Revision Task Group. Parts used for testing were supplied by Amkor Technology and Huawei Technologies provided fabricated PCBs and board assemblies.

References

Farris, Andrew, Jianbiao Pan, Albert Liddicoat, Brian J. Toleno, Dan Maslyk, Dongkai Shangguan, Jasbir Bath, Dennis Willie, David A. Geiger, Drop Test Reliability of Lead-free Chip Scale Packages, Proceedings of the ECTC, pp. 1173-1180, 2008.

Gu, Yu, and Daniel Jin, Drop Test Simulation and DOE Analysis for Design Optimization of Microelectronics Packages, Proceedings of the ECTC, pp. 422-426, 2006.

JESD22-B111, Board Level Drop Test Method of Components for Handheld Electronic Products, Published by the JEDEC Solid State Technology Association, pp. 1-22, 2003

JESD-B110A, Subassembly Mechanical Shock, Published by the JEDEC Solid State Technology Association, pp. 1-16, 2004

Lall, P. Panchagade, D., Liu, Y., Johnson, W., Suhling, J., Smeared Property Models for Shock-Impact Reliability of Area-Array Packages, ASME Journal of Electronic Packaging, Volume 129, pp. 373-381, December 2007[d].

Lall, P., Choudhary, P., Gupte, S., Health Monitoring for Damage Initiation & Progression during Mechanical Shock in Electronic Assemblies, 56th ECTC, San Diego, California, pp.85-94, May 30-June 2, 2006[a].

Lall, P., Choudhary, P., Gupte, S., Suhling, J., Health Monitoring for Damage Initiation and Progression during Mechanical Shock in Electronic Assemblies, IEEE Transactions on Components and Packaging Technologies, Vol. 31, No. 1, pp. 173-183, March 2008[a].

Lall, P., Choudhary, P., Gupte, S., Suhling, J., Hofmeister, J., Statistical Pattern Recognition and Built-In Reliability Test for Feature Extraction and Health Monitoring of Electronics under Shock Loads, 57th Electronics Components and Technology Conference, Reno, Nevada, pp. 1161-1178, May 30-June 1, 2007[a].

Lall, P., Gupte, S., Choudhary, P., Suhling, J., Darveaux, R., Cohesive-Zone Explicit Sub-Modeling for Shock Life-Prediction in Electronics, 57th ECTC, pp. 515-527, Reno, NV, May 29- Jun 1, 2007[f]

Lall, P., Gupte, S., Choudhary, P., Suhling, J., Solder-Joint Reliability in Electronics Under Shock and Vibration using Explicit Finite Element Sub-modeling, IEEE Transactions on Electronic Packaging Manufacturing, Volume 30, No. 1, pp. 74-83, January 2007[b].

Lall, P., Gupte, S., Choudhary, P., Suhling, J., Solder-Joint Reliability in Electronics Under Shock and Vibration using Explicit Finite-Element Sub-modeling, 56th ECTC, pp. 428 – 435, 2006[b].

Lall, P., Hande, M., Bhat, C., Islam, N., Suhling, J., Lee, J., Feature Extraction and Damage-Precursors for Prognostication of Lead-Free Electronics, Microelectronics Reliability, Vol. 47, pp. 1907–1920, December 2007[e].

Lall, P., Iyengar, D., Shantaram, S., Pandher, R., Panchagade, D., Suhling, J., Design Envelopes and Optical Feature Extraction Techniques for Survivability of SnAg Leadfree Packaging Architectures under Shock and Vibration, 58th ECTC, Orlando, FL, pp. 1036-1047, May 27-30, 2008[d].

Lall, P., Iyengar, D., Shantaram, S., S., Gupta, P., Panchagade, D., Suhling, J., KEYNOTE PRESENTATION: Feature Extraction and Health Monitoring using Image Correlation for Survivability of Leadfree Packaging under Shock and Vibration, EuroSIME, Freiburg, Germany, pp. 594-608, April 16-18, 2008[c].

Lall, P., Panchagade, D., Choudhary, P., Gupte, S., Suhling, J., Failure-Envelope Approach to Modeling Shock and Vibration Survivability of Electronic and MEMS Packaging, IEEE Transactions on Components and

Packaging Technologies, Vol. 31, No. 1, pp. 104-113, March 2008[b].

Lall, P., Panchagade, D., Choudhary, P., Suhling, J., Gupte, S., Failure Envelope Approach to Modeling Shock and Vibration Survivability of Electronic and MEMS Packaging, 55th ECTC, pp. 480-490, 2005.

Lall, P., Panchagade, D., Iyengar, D., Shantaram, S., Suhling, J., Schrier, H., High Speed Digital Image Correlation for Transient-Shock Reliability of Electronics, 57th ECTC, Reno, Nevada, pp. 924-939, May 29 – June 1, 2007[c].

Lall, P., Panchagade, D., Liu, Y., Johnson, R. W., and Suhling, J. C., Smeared-Property Models for Shock-Impact Reliability of Area-Array Packages, Journal of Electronic Packaging, Vol. 129(4), pp. 373-381, 2007.

Lall, P., Panchagade, D., Liu, Y., Johnson, W., Suhling, J., Models for Reliability Prediction of Fine-Pitch BGAs and CSPs in Shock and Drop-Impact, 54th ECTC, pp. 1296-1303, 2004.

Lall, P., Shantaram, S., Suhling, J., Locker, D., Effect of High Strain-Rate on Mechanical Properties of SAC105 and SAC305 Leadfree Alloys, Proceedings of the 62nd ECTC, pp. 1312 - 1326, May 29-June 1, 2012.

Lall, P., Shantaran, S., Angral, A., Kulkarni, M., Explicit Submodeling and Digital Image Correlation Based Life-Prediction of Leadfree Electronics under Shock-Impact, 59th ECTC, San Diego, CA, pp. 542-555, May 25-29, 2009.

Park, Seungbae, Chirag Shah, Jae Kwak, Changsoo Jang and James Pitarresi, Transient Dynamic Simulation and Full-under Drop / Impact, Proceedings of the ECTC, pp.914-923, Reno, NV, 2007.

Xie, Dongji, David Geiger and Dongkai Shangguan, Billy Hu and Bill Werner, A New Drop Test Vehicle for a Uniform Shock Response, Proceedings of the ECTC, pp. 902-907, Las Vegas, NV, 2010

Xu, L., Pang, H., Combined Thermal and Electromigration Exposure Effect on SnAgCu BGA Solder Joint Reliability, Electronic Components and Technology Conference, pp. 1154-1159, May 2006.

Yogel, D., Grosser, V., Schubert, A., Michel, B., MicroDAC Strain Measurement for Electronics Packaging Structures,Optics and Lasers in Engineering, Vol. 36, pp. 195-211,2001.

Zhang, F., Li, M., Xiong, C., Fang, F., Yi, S., Thermal Deformation Analysis of BGA Package by Digital Image Correlation Technique, Microelectronics International, Vol. 22, No. 1, pp. 34-42, 2005.

Zhou, P., Goodson, K. E., Sub-pixel Displacement and Deformation Gradient Measurement Using Digital Image-Speckle Correlation (DISC), Optical Engineering, Vol. 40,No. 8, pp 1613-1620, August 2001.

Interconnect Reliability Prediction for Wafer Level Packages (WLP) for Temperature Cycle and Drop Load Conditions

Tong Cui, Ahmer Syed, Beth Keser, Rey Alvarado, Steven Xu, Mark Schwarz
Qualcomm Technologies, Inc.
5775 Morehouse Drive, San Diego, CA 92121
tongc@qti.qualcomm.com, 858-651-6246

Abstract

Interconnect reliability of wafer level packages (WLP) is one of the major concerns because of the direct connection of die to board without any substrate interposer. The dominant failure modes due to temperature cycling and drop include cracks in bulk solder, crack at pad to IMC interface, and RDL cracking at UBM interface. A number of factors affect this reliability; such as UBM/Pad size, bump density, bump depopulation, and die size and thickness. In addition, WLPs come in different flavors (UBM vs. No UBM,) which also have implications on interconnect reliability. With increasing die size, it is becoming critical to investigate this reliability very early in the product design cycle using simulations. Although simulations are very helpful in performing relative comparison to investigate design options, their utility can be further enhanced by developing life prediction models to determine if a certain design will meet customer reliability requirements for the end use application.

This paper discusses a finite element modeling based approach to establish such a life prediction model. Failure data on various test vehicles were collected using board level temperature cycle and drop test methods. This data covered a large range of WLP designs including various die sizes, bump pitches, bump densities, UBM/Pad sizes, fab nodes, and WLP structures. Each of these data points were then simulated using a detailed 3-D finite element modeling approach to compute strain energy density (SED) per cycle for temperature cycle conditions. Similarly, drop tests were simulated to determine drop damage parameters (Stress, Strain, Strain energy density) at various interfaces and RDL trace. The models incorporate every detail of package geometry including die size, solder bump dimensions from measurement, detailed structure below and above solder bump such as back end of line (BEOL), polymer layers, redistribution layers (RDL), UBM, and copper pad. Material properties were also measured of test boards to improve model accuracy and published creep constitutive equations were used for simulating non-linear behavior of solder joints due to temperature cycling and drop loading conditions.

The values of damage parameters determined from finite element modeling were then plotted against the mean life (50% failure rate) to establish the fatigue life prediction model. A power law curve fitting with a correlation co-efficient of greater than 95% resulted in fatigue life exponent of approximately -1 for solder fatigue, which is consistent with previous life prediction models. Since the intent was to use the model to predict first failures and 5% failure rate, Weibull analysis and test database was used to determine ratios of lower limits of 5% life at 90% confidence to mean life. The combined use of highly accurate finite element modeling and statistical analysis of test database provides greater confidence in predicting reliability and thus a useful tool for assessing and optimizing design very early in the design phase.

Introduction

WLP's are widespread in consumer mobile wireless devices due to their small size and low cost. WLP is used for power management, RFIC, bluetooth, WiFi, and other mobile applications. Image sensors, microcontrollers, memory, discretes, and integrated passive devices also use WLP packages. WLP eliminates the interconnects found in flip chip and wirebond packages as well as the leadframe or package substrate. The limitation of a WLP is its small real estate. All the power, ground, and signal pins must fit within the die area at the pitch of a BGA. Currently, 0.5mm and 0.4mm pitch wafer level packages are the most common in the industry [1-4].

Mobile applications are consuming wafer level packages at a rapid pace. The cumulative annual growth rate of WLP from 2012 to 2017 is predicted to be 11% growing from 21.8B units to 36.8B.[5]. However, WLPs are being challenged by the need for lower cost, finer pitch, larger die sizes, and more robust structures. To lower cost, the UBM can be removed—jeopardizing board level reliability. To add pin count to a fixed area, pitches are shrinking from 0.4mm pitch to 0.35mm pitch. To increase functionality, nodes are shrinking from 65 and 40nm to 28nm. To increase the reliability of larger die sizes, more balls per area are required resulting in less depopulation. All of these scenarios can be tested empirically [6, 7], but a predictive model is essential to save time and money when making product decisions.

Solder joint reliability has been extensively studied and reported in the literature. For temperature cycling life prediction, Lee et al [8] reviewed various solder fatigue approaches based on different deformation mechanisms, such as creep strain, plastic strain, and strain energy. For SnAgCu (SAC) solder, Syed proposed [9] both creep strain and energy based models and showed better correlation with that strain energy based model. Syed later updated these models [10, 11] for different representation of creep behavior of solder while eliminating most of the modeling assumptions. For wafer level packages, Bao et al [12] correlated the inelastic strain energy density with 4 layer wafer level package temperature cycling empirical data. Similarly, a number of approaches have been published for board and product level drop simulation [13-17].

Typical WLPs use a Cu RDL metal structure to connect the die pad to the solder ball (Figure 1) rather than a ball on pad or ball on repassivation structure where the solder ball sits directly on the die pad or a polymer passivation opening over the die pad, respectively. An additional RDL type WLP is shown in Figure 2. This structure does not have a UBM.

978-1-4799-2408-0/14 $31.00 © 2014 IEEE

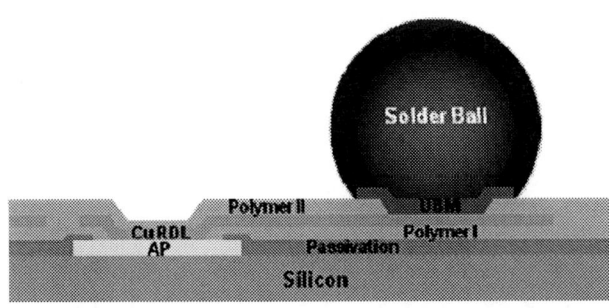

Figure 1: RDL type Wafer Level Package with UBM.

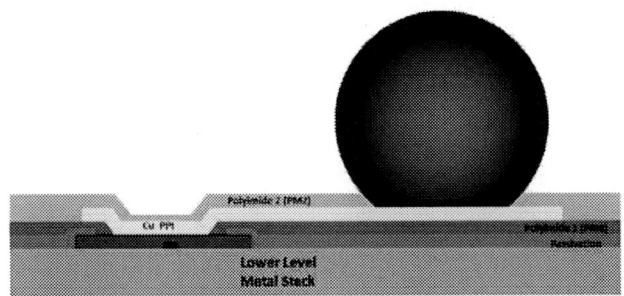

Figure 2: RDL type Wafer Level Package with no UBM

Because of obvious differences in package construction and interconnect structures, the life prediction models previously published for BGA type packages and interconnects may not be applicable to WLP. Similarly, this interconnect structure introduces new failure modes, especially under drop loading conditions which are brittle in nature and the models based on bulk solder cracking are not applicable.

This paper presents solder fatigue life prediction model for WLP under temperature cycling conditions while incorporating package and interconnect construction and material details into finite element models. Similarly, previously published drop simulation methodology [15] is modified by using component mode synthesis (CMS) modeling technique and IMC related failure mode for WLP.

Experiment Details

JEDEC JESD22-B111[18] criteria specifies the fabrication of the test boards used in this study. 8-layer (1-6-1) PCB boards use a standard core material with dimensions of 132mm x 77mm. 15 units were mounted in a 3 x 5 matrix on every board. Probe locations are added to the board for failure isolation. The PCB pads on the test board are copper with an organic solderability preservative (OSP) finish. All tests were run with non-soldermask defined (NSMD) pads.

A SAC305 solder paste is printed on the board using a 100um thick stencil. The components are soldered to the PCB in a reflow oven with a peak temperature of 245C. Visual and X-ray inspection is conducted after assembly to monitor placement, voiding, bridging quality, etc.

Temperature cycle (TC) on board experiments were run using JESD22-A104C criteria [19]. For each leg, 2 fully populated boards (30 samples) were tested up to 1000 cycles. Condition G, mode III was used with in-situ monitoring of the daisy chain devices. As seen in figure 3 & figure 4, failure

modes were similar for wafer level packages with or without an UBM. In both cases the failure occurred in the bulk solder near the WLP side of the solder joint.

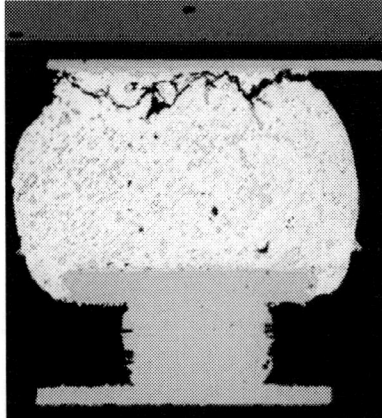

Figure 3: TCT Bulk Solder Failure on WLP without UBM

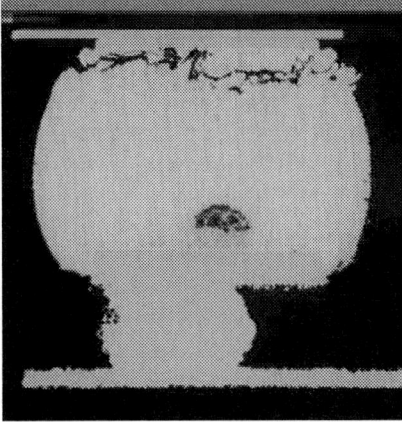

Figure 4: TCT Bulk Solder Failure on WLP with UBM

Figure 5: Drop Shock IMC Failures on WLP without UBM

JESD22-B111 method was used for drop shock (DS) testing[18]. For each leg, 4 fully populated boards (60 samples) were tested up to 200 drops. The desired shock wave was 1500g 0.5ms half sine event. Components were mounted facing down (Z- orientation) with continuity and table acceleration recorded for every drop. Figure 5 and figure 6 illustrate the drop shock failure modes for wafer level packages with or without an UBM. The primary failure mode for both structures was at the IMC on the package side (UBM-SOLDER or RDL-SOLDER interface). In some cases the WLP with UBM also revealed an additional failure mode of RDL cracking near the UBM, however, only a few test legs showed this failure mode and this failure mode is not considered here for drop life prediction.

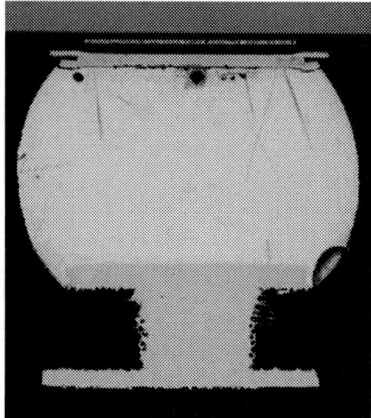

Figure 6: Drop Shock IMC Failures on WLP with UBM

The test data covered a large range of WLP designs including various die size, bump pitches, bump densities, UBM/pad size, fab nodes and WLP structures. Detailed measurements were conducted to obtain the real geometry of tested samples. Note that due to the geometry variation introduced during manufacturing process, a large number of samples were collected and the averaged dimensions were adopted. Generally these dimensions includes die size, solder bump dimensions after SMT, BEOL thickness, polymer layers thickness, redistribution layers thickness, UBM thickness and copper layer thickness. Board properties (modulus, CTE) were also measured on different PCBs to improve model accuracy.

Modeling and Life Prediction Methodology

Temperature Cycling: A previously established [10, 11] modeling and life prediction methodology for temperature cycle reliability is implemented here for wafer level packages (WLP). The modeling methodology consists of creating a 3-D finite element model of the whole package and a section of board to which package is attached to through solder joints. Since WLP sometime have ball depopulation schemes which are not symmetric with respect to X and Y axes, a full 3D model is created for each package-board assembly incorporating all solder joints. A global-local modeling approach is used to reduce the model size without affecting the response of critical joints. The global model has coarser mesh for package and board with non-critical joint solder joints modeled as cylinders with a coarse mesh for each joint. This assumption has been shown to not affect the response of critical solder joints [10]. The local models of critical solder joints are then inserted at multiple locations. Each of these local models use much more finer mesh and captured solder joint shapes as well as detailed structure below and above solder joints such as back end of line (BEOL), polymer layers, redistribution layers (RDL), UBM, and copper pad on the

board. The intermetallic compounds (IMCs) between solder and UBM (or RDL) and board pads were also incorporated in the model using a uniform IMC layer thickness at both interfaces.

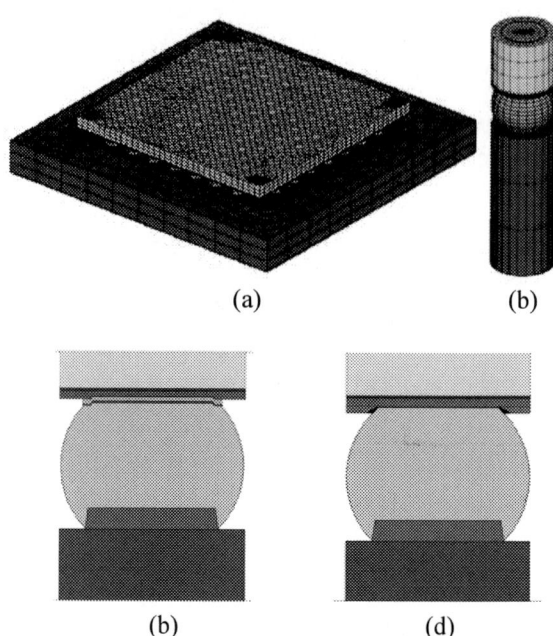

Figure 7: Global and local models for both 4Layer and 3Layer WLP options: (a) global model, (b) local critical joint model, (c) cross-sectional view of 4L (with UBM) joint model, and (d) cross-section view of 3L (no UBM) joint model.

Examples of global and local models for both 4Layer and 3Layer WLP options are shown in Figure 7. The models are created in such a way that there is no mesh discontinuity at the interface of global and local models, thus eliminating the need for constraint equations at these interfaces.

Since it is well known that creep is the primary deformation mechanism of solder joint during temperature cycling, all solder joints (global and local) are modeled using steady state creep constitutive equation. All other materials were assumed to behave in elastic manner. The material properties were also measured of representative test boards and incorporated in the models to improve model accuracy.

The finite element models for each package were analyzed by simulating two temperature cycles as per the conditions shown in the previous section. Using automatic time stepping technique, the models are solved at multiple steps throughout the temperature cycle and creep strain energy density (SED) for each element were saved in the output file. Finally, a volume averaged accumulated SED value was determined by post-processing a 25um thick layer of solder both at UBM/RDL and board pad interface. The details of this averaging are described elsewhere [9].

$$ N_{50} = C_1 \left(SED_{solder} \right)^{C_2} \qquad [1] $$

The life prediction model for temperature cycling is shown in Equation 1, which uses the accumulated for 25um thick layer of solder (SED_{solder}) from 2nd cycle and mean life (N_{50}) from the test data. The constant C_1 and exponent C_2 are determined by curve fitting the mean life for each test data point with the corresponding accumulated SED for the same package from finite element model. Since creep is used as the primary deformation mechanism here, it has been shown previously [9, 10] that the exponent C_2 should be very close to -1.

Drop Modeling and Life Prediction: Unlike temperature cycling, the drop reliability of a component is dependent on board size, board support scheme, number of components attached to the board, and the location of each of these components. While the first three factors determine the dynamic response of the board due to shock loading, the last factor determines how a component and its associated interconnects to the board respond to this dynamic behavior of the board. According to JEDEC board level drop test standard [17], a rectangular board (77x132mm) for four (4) screw supports and 15 components on the board at specific locations is used to evaluate the drop reliability of the components. The test standard also divides the component locations into different groups for reliability data analysis as the deformation behavior of the board is not uniform throughout the board. All of these factors are important factors and need to be incorporated into modeling methodology for board level drop simulation. This has been achieved in a drop simulation methodology published earlier [15]. In this methodology, the large mass method with rigid elements are used to translate shock input into dynamic response of the board and then detailed component and solder joint level models are used to quantify the response of solder joints to this dynamic behavior of the board.

The same methodology is used here for simulating drop reliability of WLP but with two major modifications: i) a more accurate method for incorporating the effect of components on the dynamic response of the board, and ii) the response of solder interconnect at IMC/UBM or IMC/RDL interface as that has been shown as the primary failure mode for WLPs.

In the previous methodology, the component effects were accounted for by increasing the stiffness and density of the board locally at the component locations [15]. Although this is still an acceptable method to capture the effect of components, the accuracy is further improved by incorporating Component Model Synthesis (CMS) option available in ANSYS™. Component mode synthesis is a powerful technique for dynamic analysis of a structure to capture local details in a much larger global model. This is especially useful when a local structure is repeated a number of times in a larger structure. Since board level drop board as per JESDB111 requires the same component located at 15 different locations on the board, component mode synthesis method provides an excellent option to incorporate into board level drop simulation. The implementation requires creating a component level model – same as the one shown in Figure 7 – and analyzing it using CMS. The CMS generation pass condenses a group of finite elements into a single Superelement, which includes a set of master degrees of freedom (DOFs) and truncated sets of normal mode in generalized coordinates. The master DOFs serve to define the interface between the superelements and other elements. Once a superelement is generated, it can be attached at

multiple locations in the global structure. The implementation of this technique for board level drop simulations with 15 components on the board is shown in Figure 8. This global board level model is then analyzed using modal superposition analysis technique to quantify the dynamic deformation of the board due to shock loading. The details of this methodology are provided elsewhere [15].

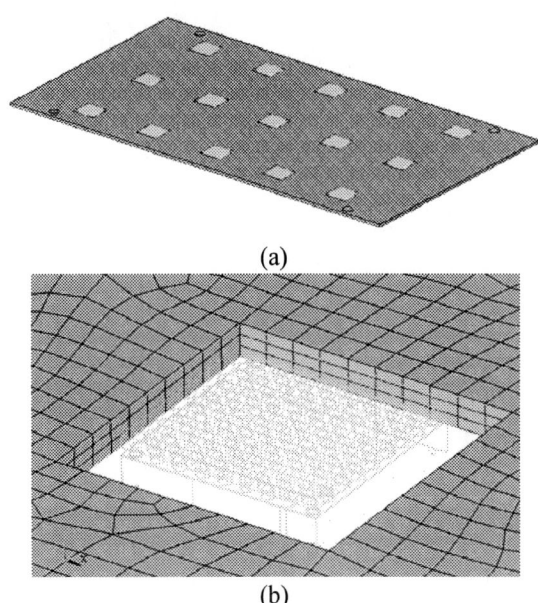

(a)

(b)

Figure 8: Drop model: (a) global board level model with components, (b) detailed view of one component level (superelement) model attached to interfacing board section.

The validity of this approach was tested using measured strain on the board. Figure 9 shows the 5 locations where strain was measured on the back side of the board with WLP mounted on the front side. The corresponding strain values in X direction (longer side of the board) are also shown. The maximum strain was measured under the component located next to screw supports (Gage #1 and #2). This is also the locations (all four corners next to screw supports) where failure occurs first for WLPs, thus indicting a direct relationship between the board strain and component failure locations. It is interesting to note that the bending behavior of the board at locations 3, 4, & 5 is out of phase compared to that at location 1 & 2 as can be observed by comparing the peak strain values.

The same behavior can be observed from Figure 10, showing the predicted strain from finite element simulations using this drop simulation technique. The strain values are highest next to the screw supports and change from tensile to compressive towards the center of the board.

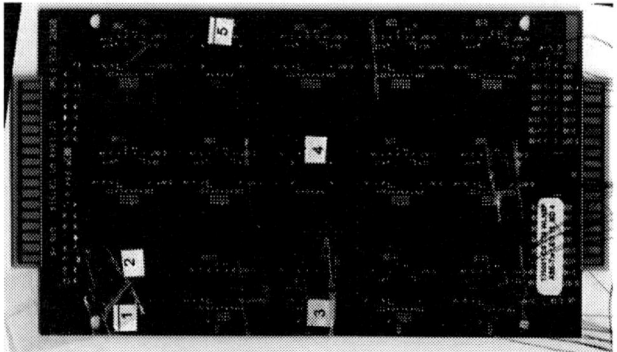

(a) Locations for Strain Measurement

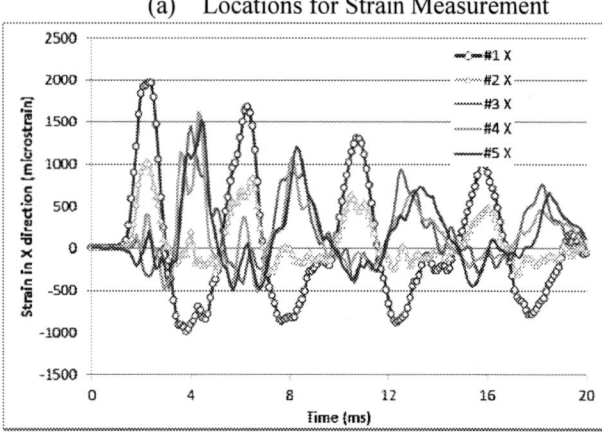

(b) Measured Strain Values in X-direction

Figure 9: Measured Strain at 5 locations on the board.

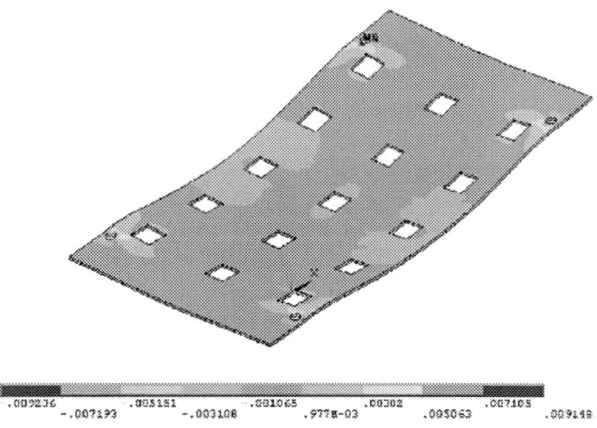

Figure 10: Strain distribution on the board from simulations

Once the dynamic response of the board is determined, it is then used as an input boundary condition to the component level model at different time steps using submodeling approach. As the failure in actual drop testing is found at UBM-IMC or RDL-IMC interface, the critical joint level model incorporates a layer of IMC at UBM or RDL interface, as shown in Figure 7, and post processing is done to determine elastic SED for IMC layer and interfacial stress at IMC-UBM or IMC-RDL interface. The model also incorporated elastic-plastic properties for Cu UBM and RDL

and creep properties of solder. The fatigue life for drop can then be determined using the following equation:

$$N_{50} = C_3 \left(SED_{IMC} \right)^{C_4} \qquad [2]$$

Data Correlation for Life Prediction

The primary purpose of an absolute life prediction method is to determine if a particular package will meet typical reliability requirements. In electronic industry, board level reliability requirements are typically based on either 1st failure or 5% life. However, the life prediction from simulation is typically based on mean life (50% failure rate, N50). This is primarily because simulation models are typically based on nominal dimensions, material properties, and assembly processes and the failure metric used in life prediction is representative of mean life behavior. In this section, the correlations are first provided for mean life and then a methodology is described to use this predicted mean life to estimate either the 1st failure or 5% life.

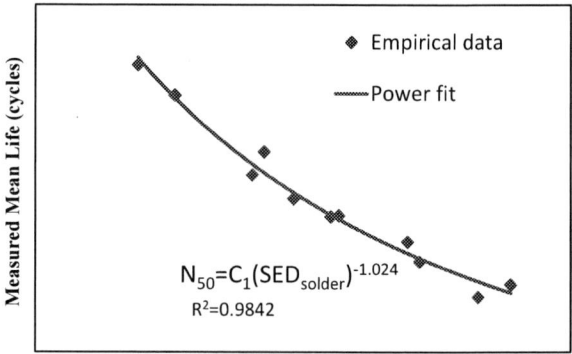

Figure 11: Correlation between predicted SED and measured N50 for Temperature Cycle test

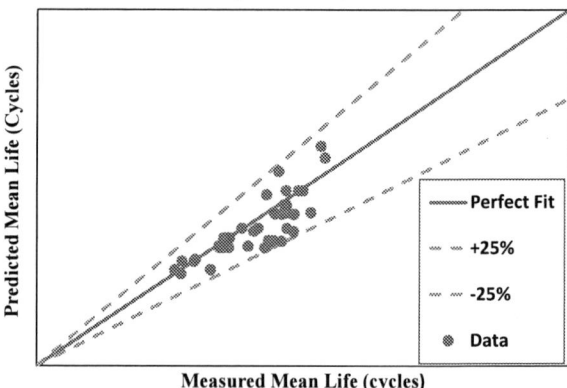

Figure 12: Model prediction and measured data comparison

Figure 11 shows the correlation for temperature cycling where the mean life (N50) from test data is plotted against the predicted inelastic strain energy density (SED_solder) of solder joint. The mean life from test data is determined by using the characteristic life (η) and slope (β) of Weibull cumulative failure distribution (CDF) for each test case. The curve fitting of Equation 1 shown in Figure 9 indicates a very good correlation between these two parameters with R^2 value of 0.98. As expected, the exponent C_2 of life prediction model is found to be very close to -1, the expected value for creep dominated failure mechanisms. Figure 12 shows the mean life comparison between model prediction and empirical data. It can be seen that most data falls within within ±25% error band.

Figure 13 shows the volume averaged Strain Energy Density in IMC layer (SED_{IMC}) during the first 9 milliseconds of drop simulations. The highest SED_{IMC} value was predicted to occur at about 4 millisecond at the corner joint closest to the screw support (Lower Left -LL corner of U1 location on the board). This is the same location that shows highest measured strain on the board (Figure 9). Further, the failure analysis also shows this joint to fail earlier than the other joints. This shows a good qualitative correlation between the predicted highest stressed joint and the observed highest strain and failure location. Note that the next highest value of SED_{IMC} occurs at the diagonally opposite corner joint (Upper Right – UR), while other two corner joints significantly lower values of SED_{IMC}.

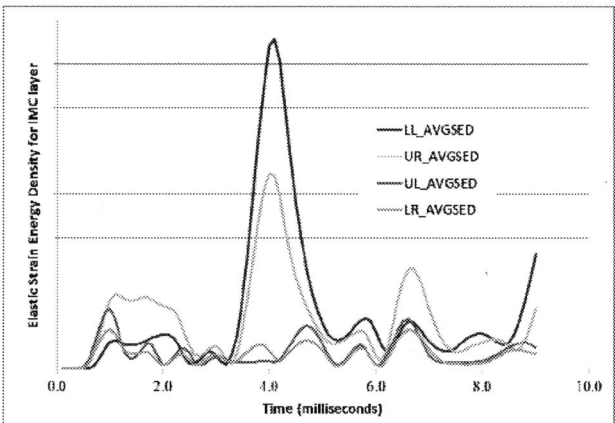

Figure 13: Time history of SED_{IMC} for different joint locations.

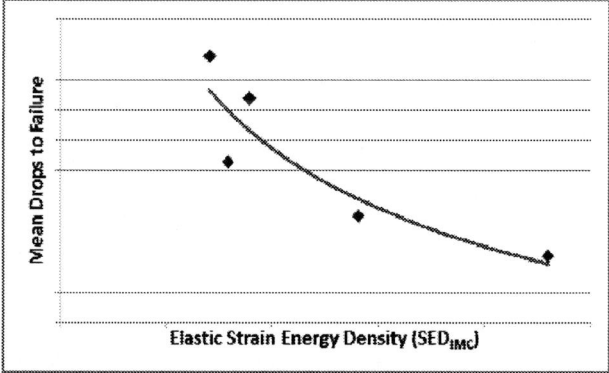

Figure 14: Correlation between predicted SED and measured Drop life

The curve fitting for drop life prediction is shown in Figure 14. Here the mean drops to failure are plotted against peak elastic strain energy density (SED_{IMC}) for the IMC layer at UBM or RDL interface. Although a good trend is predicted with higher SED_{IMC} resulting in lower drop life, there is a much larger scatter in the correlation. It should be noted that even though more than 30 legs/packages were tested in drop, the tests were usually stopped after 200 drops and not all legs

resulted failures, thus only a few legs were available for correlation.

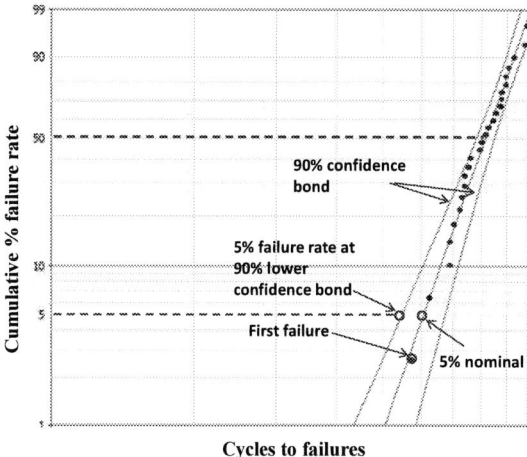

Figure 15: Statistical analysis of weibull distribution of fatigue failure data in temperature cycling

Figure 15 shows a typical Weibull plot of failure distribution from temperature cycle test. A Weibull plot is fitted as denoted by the blue curve. Most test data follow the curve but not exactly on it. A 90% confident interval defines the data variation range according to the empirical data distribution. It can be seen that the interval is very tight at 50% failure rate, and it becomes much wider at lower failure rates. It indicates that empirical data is more likely to have more scatter at lower failure rates, and thus mean life (failure rate of 50%) is a better parameter to correlate with fatigue failure metric.

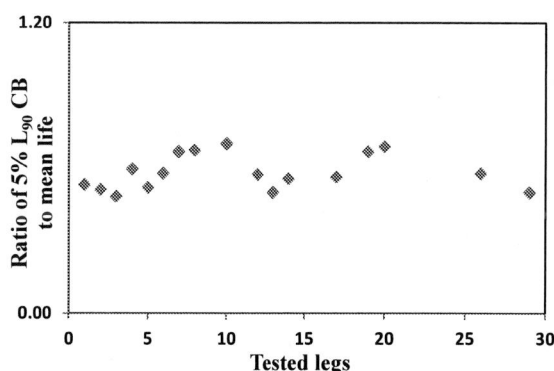

Figure 16: Ratio variation between 5%lower rang and mean life

To estimate the nominal 5% life from the predicted mean life, Weibull Cumulative Distribution Function (CDF) can be used which relates any failure rate with characteristic life η and slope β (characteristic life is related to mean life through β and Γ function). An average value of β can be used from the test database for this purpose. However, since typical reliability requirements are based on <u>5% life at 90% lower confidence bound (5%L_{90}CB)</u>, additional statistical analysis is needed to determine this 90% lower confidence bound life from the nominal 5% life. Alternatively, the 5%L_{90}CB life can be directly extracted out from Weibull analysis for each test leg using 90% confidence bounds. One can then use the ratio of this estimate to the mean life to determine 5%L_{90}CB life from the predicted mean life. This alternative approach is adopted here and this ratio is determined for each test leg. This is shown in Figure 16 for a number of test legs. As can be seen from the figure, this ratio is not a constant and varies from test to test within a tight band. For pure predictions without any test data for a particular package, this ratio is not known a priori but an average value from the database can be used without being overly conservative in 5% L_{90}CB life estimation.

The estimation for cycles to first failure (CFF) is more complex and is not as reliable as 5% life. CFF depends on the sample size and corresponds to different failure rate if the sample size is too small or too large. In addition, CFF typically represents the extremes of dimensional, material, and manufacturing tolerances which can be different each time samples from different lots are tested. In some cases, this first failure is an outlier compared to the rest of the distribution because of premature failure due to manufacturing defects, handling impact, etc. However, if CFF is to be used as a criteria, similar approach as 5% life can be used by calculating the ratio of CFF to the mean life from test database after removing all outliers from Weibull analysis.

Summary and Conclusions

This paper discussed the board level reliability prediction of WLP. Empirical data were collected to cover various die sizes, bump pitches, bump densities, UBM/pad size, fab nodes, and WLP structure. Failure analysis was performed to identify the failure modes for solder joint fatigue failure and drop shock failure. Detailed finite elements models are developed with specific geometry and structures of each package. Inelastic and elastic strain energy density are used as the failure metric for solder joint fatigue failure for temperature cycling and drop loading conditions, respectively. Statistical analysis was performed on the test data, and 90% confidence interval was added in a fitted Weibull distribution. The mean lives for different packages are then correlated with failure metrics to fit power law curves. An excellent fit was found between for temperature cycle and a fatigue model is established with similar parameters as previously developed models for BGAs [11]. For drop reliability predictions where failure mode was IMC failures at UBM/RDL interface, a good trend was found between elastic strain energy density (SED) and drops to failure. However, because of limited data points available for correlation, the data shows significantly more scatter and more data is needed to have a better model.

Acknowledgments

The authors would like to thank Qualcomm's Packaging Lab for support of the board design, procurement, and execution of the drop shock and temperature cycle tests and failure confirmation. Also, the authors would like to thank the Qualcomm Failure Analysis Lab for the cross-sections completed for this study.

References

1. Tong Yan Tee, Long Bin Tan, Rex Anderson, Hun Shen Ng, Jim Hee Low, Choong Peng Khoo, Robert Moody, Boyd Rogers "Advanced Analysis of WLCSP Copper Interconnect Reliability under Board Level Drop Test," *Proc. IEEE Electronics Packaging Technology Conference,* 2008, pp. 1086-1095.

2. Tong Yan Tee, Long Bin Tan, Rex Anderson, Hun Shen Ng, Jim Hee Low, Choong Peng Khoo, Robert Moody, Boyd Rogers "Advanced Analysis of WLCSP Copper Interconnect Reliability under Board Level Drop Test," *Proc. IEEE Electronics Packaging Technology Conference,* 2008, pp. 1086-1095.

3. Xuejan Fan and Qiang Han, "Design and Reliability in Wafer Level Packaging," *Proc. IEEE Electronics Packaging Technology Conference,* 2008, pp. 834-841.

4. L. Cergel, L. Wetz, B. Keser, J. White, "Chip Size Packages with Wafer Level Ball Attach and their Reliability," *Proc. of 4th International Conference on Advanced Semiconductor Devices and Microsystems,* 2002, pp. 27-30.

5. D. Yang. X. Ye, F. Xiao, D. Chen, L. Zhang, "Reliability of Fine Pitch Wafer Level Packages," *International Conference on Electronic Packaging Technology and High Density Packaging,* 2012, pp. 1097-1101.

6. TechSearch International, Inc., "2013 Flip Chip and WLP: Recent Developments and Market Forecasts", 2013, p. 9. Prismark Semiconductor Package Report Q1 2012, p16.

7. P. Yadav, S. Kalchuri, B. Keser, R. Zang, M. Schwarz, B. Stone, "Reliability Evaluation on Low k Wafer Level Packages," *Proc. IEEE Electronic Components and Technol. Conf. (ECTC),* 2011, pp. 71-77.

8. S. Xu, B. Keser, C. Hau-Riege, S. Bezuk, Y. W. Yau, "A Study of Wafer Level Package Board Level Reliability," *Proc. IEEE Electronic Components and Technol. Conf. (ECTC),* 2013, pp. 1204-1209.

9. Lee, W., Nguyen, L., Selvaduray, G., 'Solder joint fatigue models: review and applicability to chip scale packages', Microelectronic Reliability, Vol. 44(2000), pp. 231-244

10. Syed, A., "Accumulated creep strain and energy density based thermal fatigue life prediction models for SnAgCu solder joints," 54th ECTC, 2004, pp 737 – 746

11. Syed, A., "Updated Life Prediction Models for Solder Joints with Removal of Modeling Assumptions and Effect of Constitutive Equations," EuroSime 2006

12. Syed, A., "Updated Solder Fatigue Life Prediction Models For SnAgCu Solder Joints", SMTA Conference, 2006

13. Bao, Z., Burrell, J., Keser, B., Yadav, P., Kalchuri, S., & Zang, R., 'Exploration of the design space of wafer level packaging through numerical simulation', Proc 61st Electronic Components and Technology Conference, 2011, pp. 761-766

14. Tee, T.Y., Luan, J.E., Pek, E., Lim, C.T., and Zhong, Z.W., "Advanced Experimental and Simulation Techniques for Analysis of Dynamic Responses During Drop Impact," ECTC 2004, pp. 1089 –1094.

15. Zhu, L., "Modeling Technique for Reliability Assessment of Portable Electronic Product Subjected to Drop Impact Loads," 53rd ECTC Conference Proc., 2003, pp. 100-104.

16. Syed A., Kim Mo Seung, Lin Wei, Khim Young Jin, Song, Sook Eun, Shin, Hyeon Jae, Panczak Tony, "A Methodology for Droop Performance Prediction and Application for Design Optimization of Chip Scale

Packages," 2005 Electronic Components and Technology Conference

17. Harpreet S. Dhiman, Xuejun Fan, Tiao Zhou, "Modeling Techniques for Board Level Drop Test for a Wafer-Level Package," ECTC Proceedings, 2008

18. JESD22-B111, JEDEC Standard, "Board Level Drop Test Method of Components for Handheld Electronic Products", July 2003.

19. JESD22-A104C, JEDEC Standard, "Temperature Cycling", May 2005.

A Novel Drop Test Methodology for Highly Stressed Interconnects in Automotive Electronic Control Units

Shirangi M.H.*[1], Tsebo Simo G.[1], Wang Z.[2], Unnikrishnan R.[3], Heinrich T.[1]

[1]Robert Bosch GmbH, Automotive Electronics, Engineering Assembly and Interconnect Technology, Stuttgart, Germany
[2] Bosch Automotive Products (Suzhou) Co., Ltd, Suzhou, China
[3]RWTH Aachen University, Institut für Allgemeine Mechanik, Aachen, Germany

*Corresponding Author: Dr.-Ing. Hossein Shirangi
Robert Bosch GmbH, Department AE/EAI, Postfach 300240, 70442 Stuttgart, Germany
Hossein.Shirangi@de.bosch.com

Abstract

Drop events are infrequent on Electronic Control Units (ECUs) such as powertrain control modules, however they can cause significant damage at interconnects due to high stress levels, as these ECU designs utilize heavy components onto Printed Circuit Boards (PCB) that are in return built onto larger size of electronic products in the automotive electronics.

This paper introduces the efforts to develop a new drop test method for ECUs under high shock condition that reconciles expected PCB vibration observed in the myriad of different end-use applications that ECUs can experience. The test method attempts to replicate the PCB deflection and its inherent solder joint stress that have been shown by numerical and analytical studies. Additionally, the test method also considers the various types of loading conditions and component weight by characterizing changes in applied load (via shock pulse amplitude) at different heights and evaluating effect of gradual weight increase under constant shock test.

The paper outlines the benefits of this new test method as it utilizes symmetric loading for equivalent component response and stresses; it also allows for different setups that represent end-use conditions better. Moreover, it provides versatility for customized weight placement (center to edge) for increased PCB deflection, and greater control on the PCB vibration behavior minimizing concerns with test site miscorrelations. The latter section of this study illustrates how the use of proper daisy-chain test vehicles can enhance the statistical output as greater resolution can be achieved via the quadrant concept.

Results from numerous experiments and test characterization such as effect of pulse parameter, stress analysis via input-G, responses to drop height changes and weight shall be presented. Finally, the authors demonstrate how this test method can be universally applied for both automotive and consumer products; thereby effectively expanding the horizon of JEDEC handheld test (JESD22-B111).

1- Introduction

Electronic Control Units (*e.g.* for body electronics, brake control systems, and engine management) are widely used in automobiles for optimization of car performance as well as increasing the passenger comfort. The engine control unit collates all requirements for the engine, prioritizes and then implements them. Examples of requirements include accelerator pedal position and exhaust-system requirements in relation to the mixture composition. Figure 1 shows an ECU of a passenger car used for engine management [1].

(a)

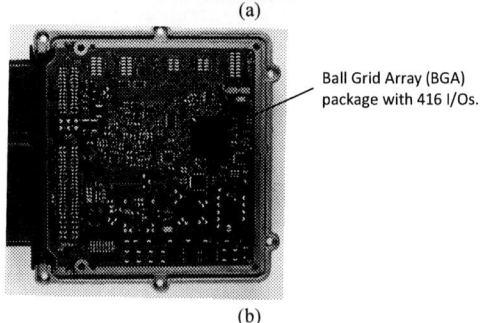

Ball Grid Array (BGA) package with 416 I/Os.

(b)

Figure 1. (a) Example of an Electronic Control Unit (ECU) used for engine management. BGA package is usually mounted on the opposite side [1], (b) Backside of the PCB and the mounting position of BGA.

ECUs such as powertrain control modules are rarely exposed to drop events which might occur due to human mistakes during their transport, handling and assembly. However, even in the rare case of an accidental drop, the stress level at the interconnections of an ECU can be much higher than that of handheld electronics, because of the larger size of electronic products in the automotive electronics together with heavier components mounted on Printed Circuit Board (PCB).

In a harsh drop scenario (e.g. free horizontal fall of an ECU on concrete ground from higher than one meter) some electronic devices such as Ball Grid Array (BGA) components may lose their structural integrity and fail even after a few drop events if not designed for such severe conditions. The situation becomes very critical when an electrically failed engine control unit is installed in the car resulting in zero kilometer failure. Consequently, drop qualification test methods for the release of BGA components for automotive electronics are needed and recently requested by some of automobile manufacturer.

This work aims at introducing a new drop test setup for electronic components at very high shock conditions. The goal is to provide a test methodology which mimics the behavior of PCB vibration of an engine control unit. Numerical and analytical investigations have approved that the deflection of PCB during a drop event has direct influence on the damage of the solder joints or the PCB laminate under solder balls. The variety of ECUs used in automotive is so diverse that the loading conditions may be completely different for different ECUs due to variation, distribution, and weight of heavy components such as capacitors and coils. Consequently, a new test setup was designed in such a way that the PCB loading conditions can be varied in two ways: 1) the first way is changing the applied load (shock pulse amplitude) using different drop heights. 2) The second way is to keep the applied shock constant while varying addition weights in the middle of PCB in order to increase the stresses at interconnects.

The advantages of the new testing method can be summarized as follows:

1- The test board is symmetrically loaded with respect to the PCB center, so that all four components mounted on the PCB have the same stress level. Moreover, the test PCB mounting situation in drop tester is more similar to that of ECU product by clamping the PCB between two aluminum jigs.

2- In order to increase the stress level of interconnects at a certain drop height, it is possible to increase the additional weight at the PCB center which increase the PCB deflection, hence increasing the interconnect stresses.

3- By using a suitable daisy-chain it is possible to get a lifetime for each corner of BGA. In this work, the "quadrant concept" was used; it means that from each BGA four lifetimes (from each test board 16 lifetimes) were extracted. This is a suitable method for providing statistically reasonable testing results using fewer test boards.

4- The new test method allows for the control of vibration behavior of the test specimen directly in the middle of the PCB. This is an important advantage of the new testing method in comparison to other available methods. By controlling the PCB vibration directly at the PCB it is possible to get comparable testing results from different labs.

5- The connector is located outside the frame and the cable which connects the PCB to event detector does not affect the vibration behavior of PCB.

2- Sensitivity Analysis of Base Excitation at Drop Table

Drop testers and free fall testers are widely used for performing drop tests on electronic products to investigate their reliability. However, the drop conditions that these equipments provide are often different from real drop test conditions of end products. Consequently, it is important to understand the real load these instruments produce in order to get a better correlation with field conditions. It is known that the acceleration level at drop table is an important indicator of the shock amplitude and frequency exerted to the test vehicle.

Figure 2 shows the schematic of a half-sine shock measured at drop table according to the JEDECB111 [2] standard. The accelerometer type used was 8704B5000 produced by Kistler Instrument Corporation and measures mechanical shock or vibration up to 5000 G's accurately (G: earth gravity). Acceleration measurements were performed at various drop heights using a chloroprene sponge rubber as strike surface with a fixed thickness of 4mm. Pulse duration was defined as the time interval between the instant when the acceleration first reached 10% of its specified peak level and the instant when the acceleration first returns to 10% of the specified peak level after having reached that peak level.

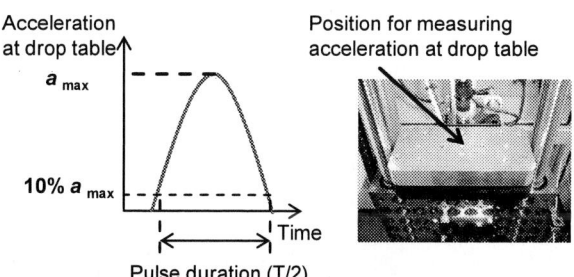

Figure 2. Shock diagram (left) and position for impulse measurement at drop table (right)

Figure 3 shows the measured peak value of acceleration and pulse duration at drop table according to the method shown above. It can be postulated that the peak value of impulse shock increases linearly with increasing the drop height, whereas the pulse duration decreases exponentially. The decrease in pulse duration (increase of excitation frequency) at elevated heights will results in excitation of more vibration modes at elevated drop heights. It should be noted that this non-linear behavior of measured excitation frequency is due to the hyper-elastic material behavior of the strike surface and depends on the temperature, shape, thickness and material of the strike surface.

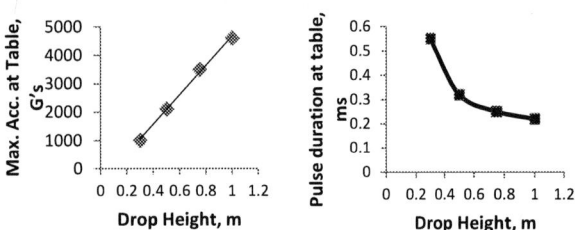

Figure 3. Effect of drop height on the peak acceleration and pulse duration at drop table

3- Drop Test Methodology Using the New Test Vehicle

As explained earlier the fixation situation of the PCB in an electronic control unit (ECU) together with large components mounted on PCB cause large deflection of the PCB during a drop event. The test vehicle which is developed for automotive applications should simulate the same behavior of PCB fixation in an ECU and also be flexible for various ECUs used in automotive electronics. These ECUs have different dimensions and, more importantly, have different mounted devices.

The mounting steps of the test vehicle are shown in Figure 4. Two aluminum jigs are used for the fixation of the PCB assembly. The lower frame is fixed to the drop table using strong bolts which practically enable a rigid connection between lower jig and drop table. Then the PCB assembly is mounted on the lower frame with BGA packages facing down. It should be noted that the lower jig has a cut-out of around 1mm deep and 4mm width which lets PCB rest in this gap easily. The upper jig is then mounted on top of the PCB and fixed using 8 bolts with a pre-defined fastening torque. It should be noted that the PCB has a thickness of 1.6 mm; it means that the outer boundaries of PCB are pressed between the two jigs and cannot move in out-of-plane direction.

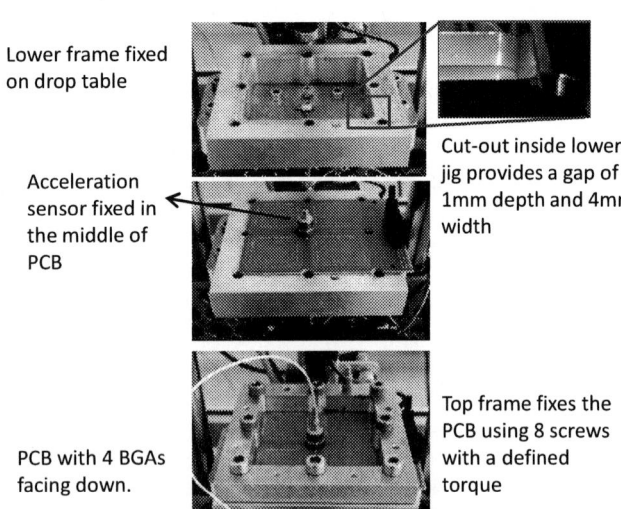

Lower frame fixed on drop table

Acceleration sensor fixed in the middle of PCB

PCB with 4 BGAs facing down.

Cut-out inside lower jig provides a gap of 1mm depth and 4mm width

Top frame fixes the PCB using 8 screws with a defined torque

Figure 4. Mounting steps of the new test vehicle

Acceleration Measurement in PCB

The deflection of PCB due to a drop event is the primary reason of failures in BGA interconnects [3,4]. Consequently it is essential to understand the parameters that influence the vibration behavior of the PCB. Since direct measurement of PCB warpage during a drop event requires expensive and time consuming facilities it might be adequate to use other characteristics which are easier to measure. One of the easiest ways to investigate the vibration behavior of PCB during a drop event is the direct measurement of acceleration at PCB. It should be noted that the addition of sensors alters the vibration behavior of the PCB by reducing the resonance frequency. This may not be a desired phenomenon for handheld electronics with usually low-weight components on the PCB. However, for automotive application, where usually heavy components such as coils and capacitors are mounted, the addition of more weight will simulate the real ECU drop more realistically.

Fig. 5 shows the acceleration measured at the PCB center over time for a bare PCB (4-layers copper, 1.6mm thickness) at different drop heights. It can be observed that the peak-value of acceleration increases with increasing drop height while the vibration frequency remains unchanged. This is an expectable fact, because the PCB vibrates at its own resonance frequency which is independent of external load provided by elevated drop height. Note that the acceleration sensor including screw nut weighs 10g.

An interesting result from the acceleration curves is that the acceleration at the PCB may be higher than that at the drop table depending on the additional weights and drop height as will be seen also in Figure 8. It is the kinetic energy of the drop tester which is transmitted to the PCB vibration. Depending on the energy provided to the PCB during a drop event and the weight and fixation of the PCB, the acceleration level may be higher or lower.

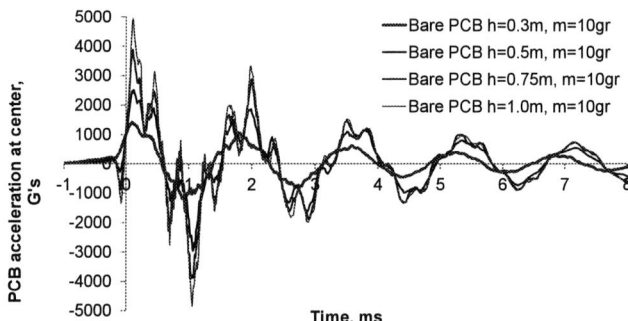

Figure 5. Effect of drop height on the acceleration of PCB (PCB response to external load)

It turned out that the measurement of acceleration at PCB is easy, reproducible, and consistent and does not require additional facilities, because most of commercial drop testers are equipped with accelerometers.

Strain Measurement on PCB

In order to measure the strains during and after a drop impact strain gages (general-purposed strain gages of type C2A-06-062WW-350 produced by Vishay precision group) were

978-1-4799-2408-0/14 $31.00 © 2014 IEEE   110

mounted on several positions of PCBs as shown for example in Figure 6.

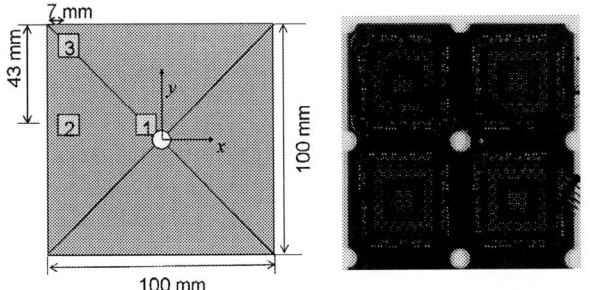

Figure 6. Dimensions of PCB and the positions of strain gages

Strain gages 1 and 3 were exactly mounted on the diagonal line of the PCB. It was found that generally strain values at point 3 are significantly smaller than those at point 1. Moreover, the strain values at point 1 in both x and y direction were almost identical which implies that the new test vehicle can provide symmetric bending with respect to the PCB center. Additionally it was found that the strain results at point 2 were different in x and y direction. While the strain values at point 2 in y direction were very small the strain values in x direction were much higher which is due to fixation situation of PCB.

Effect of Additional Weight on Acceleration and Strain at PCB

As stated earlier different ECUs used in automotive electronics may have different PCB dimensions. Moreover, the number and distribution of heavy components such as coils and capacitors may be very different for various ECU applications. This makes it necessary to provide a load carrying capability of each BGA type in order to preselect the package types to fulfill a specific reliability requirement.

Figure 7 shows how an additional weight is mounted in the middle of PCB in order to increase the PCB warpage, thereby increasing inconnect stresses during a drop event.

Figure 7. Increasing the PCB bending during a drop event by fixing additional weight at the center of PCB

Figure 8 shows the effect of additional weight on the acceleration-time curve of a bare PCB. In all these curves the drop height was fixed at 0.5m and all the drop conditions were kept constant. It can be observed that the additional weight results in a significant decrease in PCB acceleration due to an increase in the inertia of the PCB assembly. Moreover, the vibration frequency of the PCB decreases with increase in additional weight which is reasonable since the vibration frequency is inversely proportional to the square root of mass.

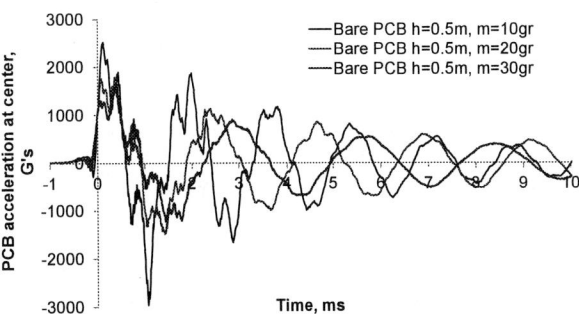

Figure 8. Additional weight results in reduction of peak acceleration and vibration frequency of PCB

Another interesting finding in the effect of additional weight is that the damping ratio depends strongly on the additional weight and decreases from 0.08 for m=10g to 0.05 for an additional weight of m=50g.

Figure 9 shows strain measurement results at point 1 which is located along the diagonal of PCB and is close to PCB center. The measurements were all done at drop height of 0.5m. It can be observed that the peak value of strain increased due to higher deflection of the PCB assembly as more weights were added to the PCB.

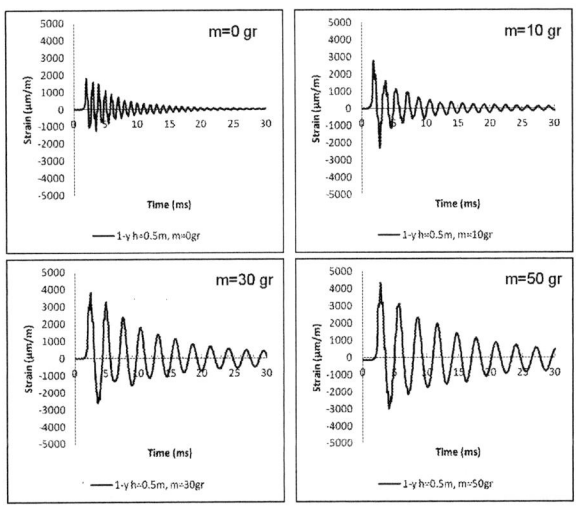

Figure 9. Effect of additional weight on in-plane strain (strain gage 1). Additional weight results in increase in peak strain and decrease in damping ratio.

4- Life Measurements

Drop tests on several BGA types were performed and the results show that in all cases the first ball which fails was located at one of four corners of the BGA. Since the PCB fixation inserts a high local strain gradient at the PCB close to the aluminum frames in most cases the first ball failed was located at the edge of frame as shown in Figure 10 highlighted with red squares.

Based on these observations a daisy-chain of board layout was designed in such a ways that each corner ball can be monitored during the drop event by using a quadrant daisy-chain concept which requires 16 channels for monitoring four BGAs at each corner. An electrical discontinuity of resistance greater than 100 ohms lasting for 1 microsecond or longer was selected as an interconnect failure.

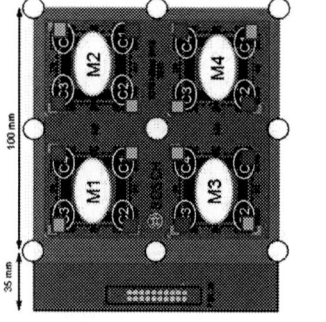 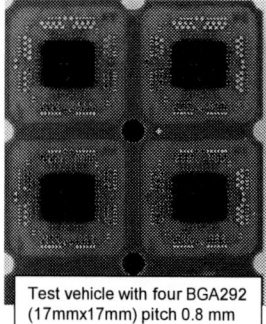

Test vehicle with four BGA292
(17mmx17mm) pitch 0.8 mm

Figure 10. Board design with 4 BGA292 packages. All BGA corners were monitored by an event detector.

Drop tests at various drop conditions showed that the first electrical discontinuity was at the edge balls (red zone in Fig 10 left) followed by the corner balls (blue zone) and center balls (yellow zone). This method illustrates how the use of a suitable daisy-chain test vehicle can enhance the statistical output as greater resolution can be achieved via the quadrant concept. From a single PCB assembly 16 life values were extracted. It should be noted that the number of data-points at the most critical location (8 edge balls) was twice the two other locations (4 corner and 4 center balls per PCB assembly) due to symmetry conditions with respect to PCB center.

Figure 11 shows how increasing the additional weight affected the drop lifetime of a BGA package. Two PCB assemblies were used for each test condition. In all cases the drop height was fixed at h=0.5m and the drop tests were repeated until all interconnects failed electrically. The initial value of PCB weight was measured (m_0) and the additional weight was increased by 10g increments. It can be observed that the edge balls always failed at first. The corner balls show a higher lifetime than center balls without additional weight. By increasing the additional weight the lifetime of all three zones decreased significantly. When the additional weight was increased to larger values of 30 grams, the lifetime of all zones showed similar smaller values.

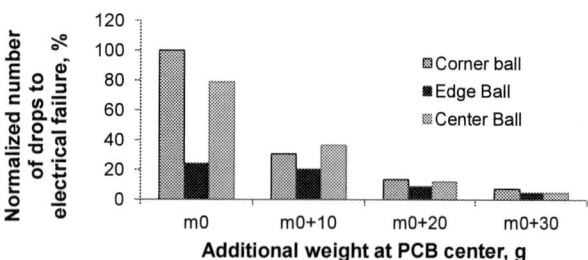

Figure 11. Number of drops to electrical failure as a function of additional weight

5- Stress Analysis using Finite Element Method

Finite Element Analysis (FEA) was employed in order to evaluate local stresses at the failure position and generate a lifetime model for the specific BGA package.

Lifetime model describes the relationship between a damage parameter and the number of cycles to failure. In order to identify the damage parameter which best describes the lifetime of BGA components under repeating drop tests, several parameter were studied. The maximum principal stress was found to be an ideal parameter which predicts the failure position and shows a linear relation with the number of drops to electrical failure in logarithmic scale. The stress evaluation at failure position (PCB laminate under Cu Pad) was assessed with special care as shown in Figure 12a. Fine mesh with node-to-node connectivity between solder ball, copper pad, and PCB laminate enabled accurate stress results at locations with high stress gradients. Especially at PCB pad where the failure happened (Figure 12b) the stress evaluation was challenging due to stress singularities at the interface. In order to avoid mesh dependency of stress results and minimize the stress singularity effects on the PCB/copper interface the method suggested in Ref. [4] was employed. In this method the stresses were evaluated at the second element layer below the copper pad, where the stress singularity effect is attenuated. The simulations were performed using the commercial finite element (FE) software ANSYS v15. An implicit time integration scheme was used and the strain and acceleration at PCB were calibrated against experimental results in order to validate the simulation approach as explained in detail in Ref [4].

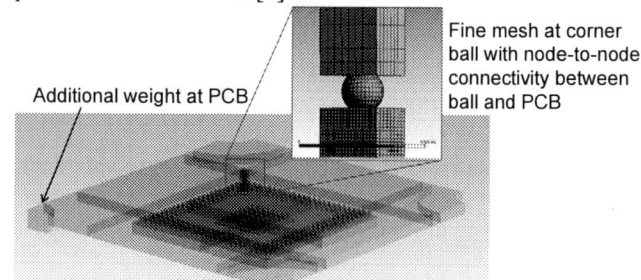

FEM Model (1/4th of PCB) with BGA facing down

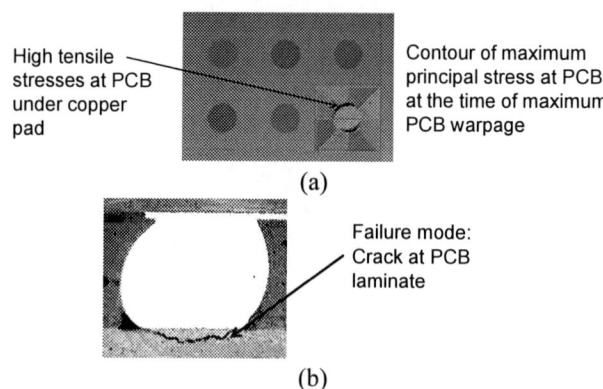

Figure 12. (a) FEA model of test vehicle with node-to-node connectivity at critical locations [3] (b) failure analysis shows pad crater is the dominating failure mode

Determination of Lifetime Model

In order to predict the reliability of BGA packages in an electronic control unit it is essential to determine a lifetime model which relates the stresses from field loading to lifetime results from board level reliability tests. The stress at each monitored interconnect (corner, center, and edge balls) of

978-1-4799-2408-0/14 $31.00 © 2014 IEEE   112

BGA package was evaluated using the finite element analysis. The stress results analyses were done at different drop heights with different additional weights.

Figure 13 shows the lifetime model of the BGA292 investigated in this study. The stress results were normalized to the peak value of stresses and the lifetime results were normalized to the highest lifetime measured at drop condition h=0.3m with no additional weight. Each point on Figure 12 represents the characteristic lifetime of at least 8 data points. The points on the left hand of the diagram represent drop conditions at elevated drop heights and/or large additional weights. On the contrary, points with high number of drops to electrical failure represent low stress regimes where the drop height was low and the additional weight was little.

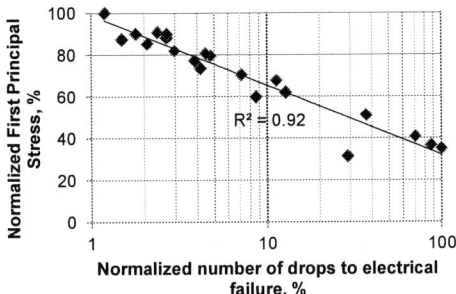

Figure 13. Lifetime model for BGA292 (first principal stress at PCB under copper pad vs. number of drops to electrical failure)

The lifetime model demonstrates how this test method can be universally applied for both automotive and consumer products; effectively expanding the horizon of JEDEC handheld test (JESD22-B111). By increasing the additional weights the stresses at interconnect are increased, making it possible to employ release tests that produce similar stress levels to those at product level tests. By removing the additional weight the stress situation gets more similar to stress levels of packages used in handheld electronics and mobile application.

6- Conclusions

A new drop test methodology was developed which allows for increasing the PCB deflection similar to a product level drop test at which the ECU is dropped from 1m height on a cement surface. The main results from the drop tests are listed below:

Peak value of excitation amplitude at drop table increases while the excitation frequency decreases.

The PCB response to increasing drop height can be summarized as follows:
- Peak value of acceleration increases due to higher kinetic energy provided to PCB assembly at drop event.
- Peak value of strain increases due to higher PCB warpage.

- Vibration frequency of PCB remains constant because it only depends on the PCB assembly and not external forces.
- Damage at interconnect increases.

The additional weight at PCB does not affect the excitation amplitude and frequency at drop table because the mass of drop table is much higher than the mass of test vehicle.

The PCB response to increasing additional weight in the middle of PCB can be summarized as follows:
- Peak value of acceleration decreases due to inertia effects.
- Peak value of strain increases due to higher PCB warpage.
- Vibration frequency of PCB decreases because it is inversely proportional to the square root of mass.
- Damage at interconnect increases.

An important factor that makes this test condition superior to its current JEDEC counterpart is that the method can be employed for both consumer and automotive electronics. By using a small weight in the middle of the PCB the deflection of the PCB is similar to that of a cell phone or tablet PC devices. By fixing more weight at the PCB the situation gets more similar to that in an ECU drop with heavy components. It is therefore possible to generate an S-N curve of interconnects over the whole spectrum of applications.

Acknowledgments

The great effort of our university student's colleagues is greatly appreciated. Authors would like to thank Guillaume Causse, Charlotte Alves, and Mayank Sharma for their contribution to this project during their thesis or internship.

Authors would like to thank their colleagues at JEDEC22B111 revision group for the fruitful discussions.

References

1. Robert Bosch GmbH, <u>Automotive Electrics Automotive Electronics</u>, John Wiley & Sons, 2007.
2. JESD22-B111, Board Level Drop Test Method of Components for Handheld Electronic Products.
3. Dongji Xie, Ife Hsu, Yingliang Zhou, et-al."A New and Effective Drop Test Evolution To Next-Gen Handheld Applications" *Prod. 2013 Electronic Components & Technology Conference (ECTC)*, 2013, USA
4. Tsebo Simo G., Shirangi M.H., Nowottnick M., Dudek R., Kaulfersch E., Rzepka S., Michel B., "Drop Impact Simulations for Lifetime Assessment of PCB/BGA Assemblies regarding Pad Cratering," *Proc. IEEE International Conference on Thermal, Mechanical and Multi-Physics Simulation and Experiments in Microelectronics and Microsystems (EuroSime)*, Ghent, Belgium, 2014 (in Press).

Early-State Crack Detection Method for Heel-Cracks in Wire Bond Interconnects

Michael Krüger, Stefan Trampert*, Andreas Middendorf, Stefan Schmitz*, Klaus-Dieter Lang
Technische Universität Berlin, Germany
*Fraunhofer Institute for Reliability and Microintegration (IZM), Berlin, Germany
michael.krueger@becap.tu-berlin.de, +49 30 46 403 706

Abstract

Reliability of electronic systems is finally limited due to thermo-mechanical fatigue of interconnections. Besides soldered interconnections wire bonding is one of the most commonly used interconnection technology in electronics. Due to thermo-mechanical loading wire bond technology suffers from cracking in the heel region and delamination in the interface. To increase lifetime and lower ecological impact of electronic systems, a condition monitoring concept is needed, which is able to determine the remaining lifetime of an interconnection. The scope of this paper lies in the development of a parameter measurement system for early-state crack detection in the heel region of wire bond interconnections. This parameter measurement system uses signal components generated by cyclic opening and closing of growing cracks. So it becomes possible to determine the remaining lifetime of the interconnection, which is directly connected to the lifetime of the whole system. Furthermore an analytical model is presented, which supports the experimental setup. Measured cracks are investigated by metallographic cross-sectioning of wire bonds and focused ion beam (FIB).

Introduction

Heavy wire bond connections are used in modern power electronic modules e.g. with insulated gate bipolar transistors (IGBT) to conduct the strong electrical currents. These modules are used in traction, power transmission and distribution as well as electro-mobility, often operating under harsh environmental and loading conditions. Nevertheless, lifetimes of about two decades are expected from the industry. The lifetime of power modules is mainly limited due to thermo-mechanical fatigue of electrical connections [1, 2]. Wire bonding of aluminum is still a common interconnection technology in electronics packaging. Due to external loads a wire bond is one weak point in a package. Vibrations or temperature changes can cause relative motion between different components connected via bond wire. There are two main failure mechanisms that lead to loss of electrical conductivity. First, there is the delamination between bond wire and pad (interface), which is commonly known as wire lift-off. The reason therefore lies in mismatch of the coefficient of thermal expansion (CTE) of the used materials. Heel cracking is the second failure mechanism, resulting from thermo-mechanical stresses in the wire due to heat transfer and Joule self-heating. Another reason is cyclic displacement between the two bond pads due to global CTE mismatch and vibration loads [3]. Since the heel of the wire is weakened additionally due to the bonding process, cracks are initiated in this region.

One common way of analyzing electrical conductive connections is to measure the resistance. However, the influence of cracks on the measured resistance only becomes significant if almost the whole connection cross-section is lost [4]. A parameter is needed that behaves differently.

Parameter shift can take place in two different ways (Figure 1). The first way produces sudden failures. By using this kind of parameter, a detection of early degradation states is not possible. The second way of parameter change is a failure due to degradation, where the observed parameter changes continuously until a failure criterion is reached. For detecting interconnect degradation parameters have to be identified that change continuously until failure.

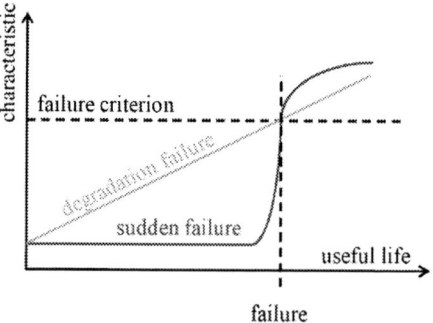

Figure 1. Variants of parameter shift until a failure criterion is reached.

Destructive investigations like shear or pull tests are used to characterize the quality of wire bond interconnections. However, there is no non-destructive, electrical measurement technique available to detect small cracks in bond wires [5].

Concepts of condition monitoring

In the following, different approaches to condition monitoring will be introduced. A distinguishing feature to classify these approaches is based on the relationship they set up between external load, system behavior, parameter shift and failure. The respective relationship for each approach is visualized in block diagrams in Figure 2. Failure is defined as the end of the system's functionality. This means that at least one parameter or one characteristic function does no longer meet the requirements. To evaluate the condition of the system with regard to an expected failure, these concepts may be applied [6]:

- Methods for measuring important system parameters (Parameter Test Method)
- Methods for estimating the condition by calculation (Life Cycle Unit)
- Methods for monitoring fuses or weak spots in the system (Condition Indicator)

978-1-4799-2408-0/14 $31.00 © 2014 IEEE

Parameter BIST	Condition Indicator	Life Cycle Unit

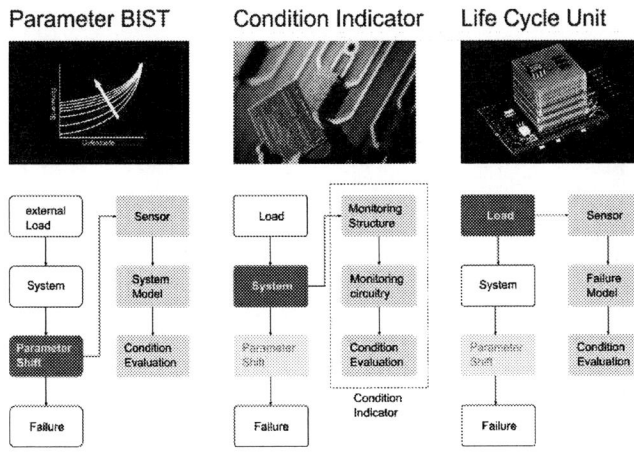

Figure 2. Three condition monitoring concepts consisting of parameter build-in self-test (BIST), condition indicator and life cycle unit.

The parameter test method evaluates the shift of constitutive parameters either of the whole system or of single components. For each of these parameters a threshold value is defined as failure criterion. Considering the rate of shift of important parameters, time of failure of the whole system can be predicted. In Built-In-Self-Tests (BIST), a specific version of the parameter test method, voltages and currents occurring in the system are measured and rated. The parameter test method does not work with fast changing parameters (sudden failure). In case of sudden failure, the time of failure cannot be estimated. Furthermore, in complex systems it may be difficult to identify significant parameters that can be used for condition monitoring. A major advantage of the parameter test method is its direct reference to the system's function. Even with complex relationships between external loads, failure mechanism and parameter shift, condition evaluation may be achieved. The challenge is to identify suitable parameters for life time prediction. Therefore, slight shifts of these parameters must be recognizable long before the system fails.

The concept of Life Cycle Units uses sensors to measure relevant environmental loads continuously. The system's life time is predicted applying previously defined failure models to the measured data. Application areas for Life Cycle Units are limited by their need for relatively complex infrastructure. Therefore, reasonable applications are condition monitoring of high-value systems and the monitoring of systems that already include most of the required hardware in their functional design. In such cases models for condition monitoring can be added for synergetic use of the existing hardware. Possible applications are control units in the aviation, train and automobile sectors.

The concept of Condition Indicators uses so called monitoring structures. The condition (working / failed) of these monitoring structures is periodically evaluated by a monitoring circuitry. A Condition Indicator consists of monitoring structures, monitoring circuitry and the algorithm to evaluate the system's condition. The monitoring structures can be seen as "fuses" or "early warning system". Knowledge about their condition can be used to predict the time to failure of the complete system. The monitoring structures are built in

a similar technology as the weakest component of the monitored system. For this purpose the system's most critical external loads, e.g. temperature cycles, high temperature, vibration or humidity must be identified. After identifying the weakest functional component, a suitable failure model for its technology is defined. The monitoring structures should be designed as simple as possible but close to the weakest component's technology. According to the scalable sensitivity parameters of the failure model, the predicted life time of these structures is defined shorter than the life time of the functional structure so that each monitoring structure has a dedicated life time.

Concentrating of wire bond interconnections only the concept of parameter measurement is usable for a detection of small cracks at the early degradation state. The other two concepts are only suitable for system monitoring, but not monitoring of individual components.

Analytical modeling of growing heel cracks

Investigations of cracked heel areas in wire bonds show always cracks at the top and bottom of the heel. They start from the surface and grow towards the center of the wire cross-section. Finite element analysis shows two areas in the heel region of plastic deformation due to bending during the bond process. One area is at the top of the wire, the other at the bottom [7].

One way of modeling cracks in interconnections analytically is to use rectangular shapes. Then, the depth of the interconnection plays a subordinate role. The main advantage is an easy understanding of the influence of the crack concerning electrical conductivity and parameter variations. The basis for modeling is shown in Figure 3. Here are h the height of the investigated interconnection, a the assumed height of a crack, w the width, d the depth of the interconnection and finally z the length of the crack consisting of the two parts x and y that are not necessarily equal. The arrow shows the direction of the mechanical cyclic load. With growing x and y the crack growth is modeled, which reduces the overall cross-section of the interconnection. The shape of the wire is here for the benefit of simple mathematical description set as rectangular.

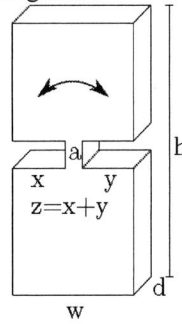

Figure 3. Model of a wire bond interconnection for analytical consideration.

When cyclic mechanical loads are applied the two cracks are closed an opened alternating, so that the cross-sectional area of the interconnection is always smaller than in the initial non-cracked state. Knowing that the electrical resistance is not sensitive enough for detecting the difference between these two states, the difference between two loading conditions is modeled as resistance difference because of higher sensitivity by removing the influence of the intact interconnect region. Referring to Figure 3 there are two main areas of the electrical resistance. The first is defined by the two intact areas with height h-a and the second is the defective area with height a. The sum of both parts gives the resulting electrical resistance.

$$R = R_{intact} + R_{defect} \tag{1}$$

R_{intact} can be seen as constant because it is not affected by the crack and therefore not changing. With resistivity ρ, the resistance can be calculated as follows.

$$R_{intact} = \frac{\rho}{d} \cdot \frac{h}{w} \tag{2}$$

For calculating R_{defect}, three load states have to be distinguished. If the partial crack with length x is closed ideally by bending the model virtually to the left, R_{defect_1} consists of height a and cross-section $d(w$-$y)$. If the model is straight as seen in the figure R_{defect_2} has height a and cross-section $d(w$-$z)$ and finally in the case of closed partial crack y, R_{defect_3} has the same height and $d(w$-$x)$ as cross-sectional area. In the following we calculate the difference between R_{defect_3} and R_{defect_1} and define it as ΔR. The overall crack depth is always z the sum of x and y.

$$\Delta R = R_{defect\_3} - R_{defect_1} = \frac{\rho \cdot a}{d} \cdot \left[\frac{1}{w-z+y} - \frac{1}{w-z+x}\right] \tag{3}$$

Assuming that there is a tiny crack at the beginning and x and y grow with different velocities as it is very likely in field use, ΔR is expected not only to rise, but to fall slightly until z is dominating the behavior. Additionally equation [3] is independent from the height h and has the resistivity, crack height and depth of the interconnection as factor.

The propagation of the crack z can be modeled with three parts. The first part gives the initial tiny crack that appears at an arbitrary point in the lifetime of the interconnection. Only one of the two variables (x or y) behaves like this. In the discussed case x has an initial, fast growing behavior. It is here assumed to happen at the beginning of the consideration. Part two consists of rising x and y. To confirm the model, the case is considered when both partial cracks have the same length with different crack propagation velocities. In the third part one of both partial cracks dominates the resistance change usually by causing a lift-off and loss of electrical and mechanical connection.

The following Figure 4 shows all these parts and the calculated ΔR as given from the model in equation 3. The values are normalized in order to discuss the dependent behavior of the parameters. In the figure the mentioned case is shown, where x and y have the same value at 80 % constriction.

The green dashed line expresses the partial crack length x and the pink dotted line the partial crack length y. The first is assumed to grow linearly with low increase in depth, after the initial cracking has taken place. Finally, x dominates the cracking behavior by higher rise. The second crack part y starts at about 20 % reduced cross-section to grow exponentially until 80 % of reduced cross-section and stops growing at this point. The sum of both parts is shown as z with the blue dash-dotted curve. Both, x and y are modeled in a simple way to show the behavior of the model and explain later on the experimentally observed signals.

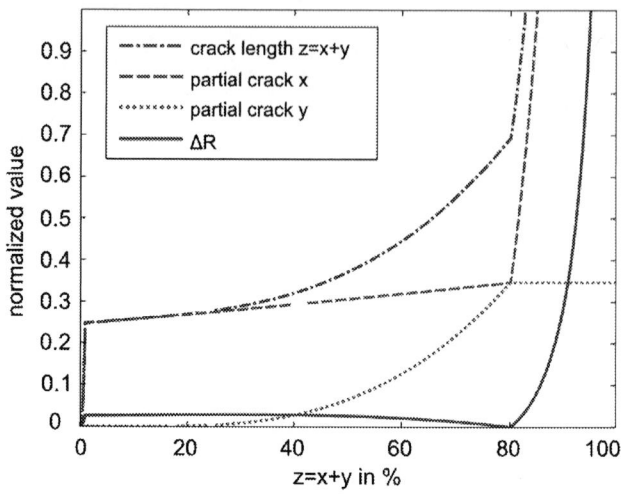

Figure 4. Development of crack parameters and resistance difference plotted over crack depth z.

Which one of both parts is growing faster than the other or dominating the cracking behavior does not matter because of the sums in equation 3. The resulting difference in resistance ΔR is shown in the red curve. It jumps to a detectable value at the beginning and is almost constant in the wide range until 80 %. Depending on the behavior of the two parts of z, the difference in resistance is even falling, as shown between about 40 and 80 %. As expected the value of ΔR becomes zero when x equals y. This behavior is not dependent on the individual way of crack propagation in x and y. Even if they are growing linearly, ΔR is decreasing. These considerations are necessary to explain the experimental observations and the cross-sections of the wires later on.

Experiment

Mechanical setup:

The experiment is set up on a piezo actuator that is used for bending the wire. This movement simulates a cyclic displacement between the two bond pads, which occurs in power devices. For the experiment, one wire is bonded between two printed circuit test boards (Figure 5). As wire material aluminum is used with a diameter of 300 μm. The wires are bonded with loop factor 15.

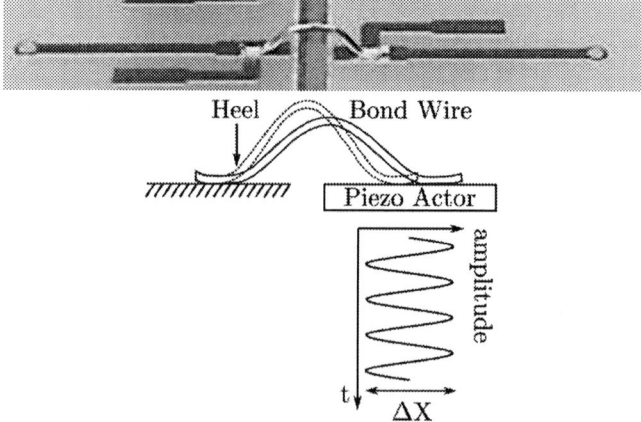

Figure 5. Mechanical setup for degrading wire bonds.

First tests revealed that this thickness lead to a maximum lifetime of about 4000 cycles with a cycle frequency f_1 of 1 Hz and cyclic displacement of 70 µm. This setup is suitable for keeping the test time short by not generating too fast degradation. The displacement of the actuator follows a sinusoidal function. As a consequence of controlling the displacement, it is force independent. Due to the displacement, the maximal stress level is localized in the two heel areas.

Electrical setup:

For measuring the ΔR as part of a modulated signal a lock-in voltmeter is used. A current of 0.5 A is applied on the bond wire as basis for the modulation. A 4-wire setup was structured with the lock-in voltmeter and the current source. By stimulating the wire with the piezo actuator at frequency f_1=1 Hz, the resistance R changes in the case of a present crack at the same frequency. The difference between the above discussed cases is modulated as signal to the frequency of the stimulation as ΔR. The lock-in voltmeter acts as extremely narrow band-pass filter. The stimulation frequency is provided as a reference for the center frequency of the filter. The measurement setup is shown in Figure 6.

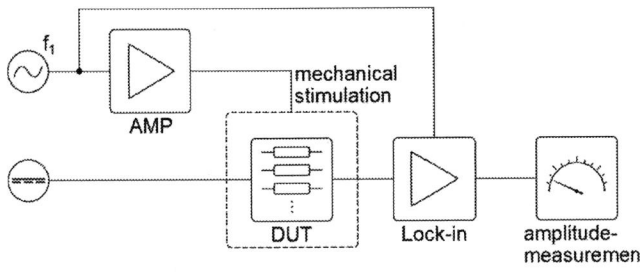

Figure 6. Measurement setup for detecting small cracks in wire bonds.

In the experiment, a digital-signal-processor-based dual-phase lock-in voltmeter is used. The phase difference between stimulation and reference signal is constant; nevertheless a single-channel lock-in amplifier could be used instead. The direct current through the bond wire causes a voltage drop. As a modulation base, this direct current offset is needed, but for the lock-in amplifier it is disadvantageous. A high-pass filter is set to a frequency of 0.2 Hz to suppress the non-alternating signal components. It avoids the following gain stage from being overdriven and allows a high pre-amplifier gain of 78 db. This setup results in a maximum input voltage of 2.5 µV at the stimulating frequency. The time constant of the lock-in voltmeter acts reciprocally to its bandwidth. So, 5 s time constant is chosen to realize an adequately smoothing to the output and reach high accuracy. Experiments are made without mechanical stimulation to widely exclude wrong results caused by slowly drifting equipment. The drift of the test equipment is lower than 2.5 nV in a period of 10 hours, which is less than 0.1 % of the full scale.

Results

Since the output signal of the lock-in voltmeter expresses the amplitude of the sinusoidal changing resistance difference,

it expresses the condition of the wire bond very well. In the measurement, signal amplitudes around 1 µV are expected. For measurement, four equally bonded wires are used. To investigate the crack depths and validate the presented model, the degradation is stopped at different stages beginning at 488 cycles over 868 and 1257 till 3293 cycles. The results are shown in Figure 7. The first thing that is noteworthy is a good reproducibility of different experiments. At the beginning, there is always a small region with an instant rise to about 0.45 µV. The section with falling signal can be observed at all four curves. A change in behavior takes place after 2500 cycles. Now the crack growth dominates the behavior of the signal, leading to sudden failure. The spikes in the blue dotted curve show sudden opening and closing directly before the sudden failure.

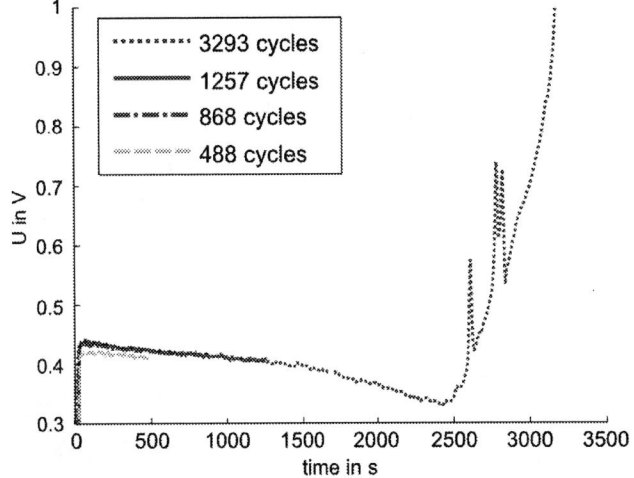

Figure 7. Measured cyclic resistance difference, expressing the condition of the investigated wire bond interconnection.

The falling curves can be explained by different crack-growth velocities of the two parts in z (Figure 3). One commonly used failure threshold is a signal rise of 20 % according to e.g. IPC 9701A. This threshold, beginning at the first maximum of the signals, limits the useful lifetime of the investigated bond wire to 2750 cycles. The falling signal curves can be used for an estimation of the remaining lifetime. Defining a threshold of 90 % of the initial maximum, gives a reproducible point in time where about 50 % of lifetime remains. Every other definition shifts this value in time. If the minimum at about 2450 cycles is reached and the signal begins to rise, only 10 % of lifetime remains until a failure occurs. This point can be used as last and most urgent indicator for replacing a monitored system.

To correlate the resulting crack depths with the measured signals, cross-sections of the heel regions are made. Due to not perfect symmetry of the wires, the side with more plastic deformation from the bond process fails first. As shown in Figure 7 the wire with more than 3000 load cycles has no longer electrical contact. So only three wires are cross-sectioned and investigated by optical analysis. Figure 8 shows the three wires. From left to right the wires with 488, 868 and 1257 cycles are shown. As expected the cracks in the heel region are growing.

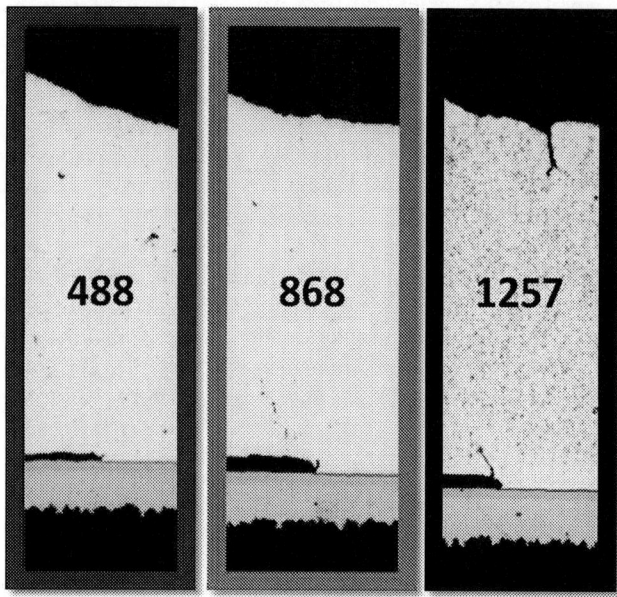

Figure 8. Cross-sectioned wire bonds for optical analysis.

Further FIB investigations show larger crack lengths as optically observed. Figure 9 shows that the crack depth is more than twice as deep as the optical observation has revealed.

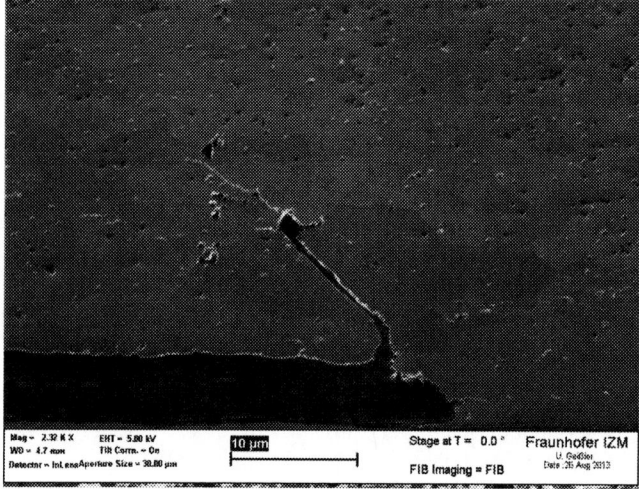

Figure 9. FIB investigation shows a larger crack depth, than observed optically.

Conclusions and Outlook

In this paper a parameter measurement based condition monitoring approach is developed that is able to detect cracks in the heel region of bond wires in a very early stage. The observed curves follow all the same trend and are exactly reproducible due to consequently developed experimental setup. This novel measurement technique shows at the beginning a large rise to a value around 0.45 µV. That can be explained by initially cracking of the surface near material. Another reason for the initially rising curves of ΔR could be an elongation of the whole bond wire during the first loading

cycles due to the applied shear force. In contrast to the commonly used direct current resistance measurement a wide area of falling signal amplitude can be observed. The reason lies in two cracks, growing into opposite directions starting at the top and bottom of the bond wire. In the case of different cracking velocities the observed signal trend can be described analytically. This behavior can be used to determine the remaining lifetime of a wire bond through prominent points in the graph. The first would be the local maximum at the beginning, the second the local minimum during the development of the curve and the third prominent point would be a change of e.g. 20 %, which indicates the failure according to common standards. Best suited for condition monitoring is the local minimum of the measured signal. The time span left to failure from this point on has always the same normed length to failure. If this behavior changes, an easy indicator for an upcoming failure is found.

The correlation between real crack depths, which can be observed by FIB analysis and measured signal amplitude builds the basis for further investigations.

Acknowledgments

The authors would like to thank Jan Höfer, Ute Geißler and Amrita Bohn for their support. The experimental work was done in the Electronics Condition Moinitoring Lab at the Fraunhofer IZM Berlin. This work is part of the federal founded BMBF (Federal Ministry of Education and Research) project RoBE and has been supported by the Technische Universität Berlin.

References

1. Cova P, Fantini F., "On the effect of power dycling stress on IGBT modules," *Microelectronic Reliability*, vol. 38, pp. 1347-1352, 1998
2. Mainka K., Thoben M., Schilling O., "Lifetime calculation for power modules, application and theory of models and counting methods," in *Proc. European Conference on Power Electronics (EPE)*, Birmingham, GB, Sept. 2011.
3. Cippa M., "Selected failure mechanism of modern power modules," *Microelectronics Reliability*, vol. 42, pp. 653-667, 2002
4. Krüger, M., Nissen, N.F., Reichl, H., "Intermodulation Distortion as Indicator for Interconnect Degradation," in *Proc. IEEE Electronic Components and Technol. Conf. (ECTC)*, Las Vegas, NV, May 2010.
5. Wilde, J., *Methoden zur Zuverlässigkeitsqualifizierung neuer Technologien in der Aufbau- und Verbindungstechnik*, 2006
6. Bochow-Ness, O., Eckert, T., Jaschke, J., Middendorf, A., Nissen, N.F., Reichl, H., "Condition Monitoring of Microsystems Supporting Sustainability," in *Proc. Electronics Goes Green Conf. (EGG2008+)*, Berlin, GER, 2008
7. Merkle, L., Sonner, M., Petzold, M., "Lifetime prediction of thick aluminium wire bonds for mechnical cyclic loads," *Microelectronics Reliability*, vol. 54, pp. 417-424, 2014

Accelerated Vibration Reliability Testing of Electronic Assemblies Using Sine Dwell with Resonance Tracking

Quang Su[1], James Pitarresi[1], Mohammad Gharaibeh[1], Aaron Stewart[1], Gaurang Joshi[1], and Martin Anselm[2]

[1]Department of Mechanical Engineering, Binghamton University

[2]AREA Consortium, Universal Instruments Corporation

Abstract

In this work a sinusoidal vibration test method with resonance tracking is employed for reliability testing of circuit assemblies. The system continuously monitors for changes in the resonant frequency of the circuit board and adjusts the excitation frequency to match the resonant frequency. The test setup includes an electrodynamic shaker with a real-time vibration control, resistance monitoring for identifying electrical failures of interconnects, and vibration logging for monitoring changes in the dynamic response of the assembly over time. Reliability tests were performed using both the resonance tracking sinusoidal test method and the traditional fixed-frequency method for assemblies, each consisting of a centrally mounted BGA device assembled with SAC305 solder. These tests show that the resonance tracking method gives more consistent failure times than the fixed frequency method. Failure analysis for the tested devices shows the primary failure mode is trace failure with evidence of solder fatigue. A finite element model, correlated with experimental modal analysis, is shown to accurately estimate the circuit board deflection estimated from the harmonic vibration data. This provides a means of estimating the stresses in the electronic interconnections while for accounting for the variability between test parts. These fine-tuned vibration measurement techniques and related finite element models provide the building blocks for high cycle solder fatigue plots (i.e., S-N curves).

Introduction

In order to establish vibration-driven reliability metrics and failure modes, testing of electronic assemblies is often performed using fixed-frequency sinusoidal excitation at the resonant frequency of the system identified at the start of the test. Throughout the test duration, the resonant frequency of the assembly can drift, resulting in significant changes in the vibration response of the system, resulting in wide variability in the test data. It is expected that the typical reliability scatter often observed in high-cycle fatigue testing of assemblies can be reduced by adjusting the input vibration to account for changes in the dynamic system.

Chen et al. [1] performed solder vibration life testing on circuit boards with solder ball grid arrays (BGAs) by driving the shaker system at the circuit board first resonant frequency, keeping the driving frequency and base vibration level constant for the duration of each test. Eckert et al. [2], in their life testing of circuit boards with flip chip components, also drive the shaker system with a constant excitation frequency, but use a laser vibrometer and a reference accelerometer to obtain control signals for the shaker controller to maintain the circuit board deflection at the center of the board during the tests. Che and Pang [3] performed life testing on flip chip on board (FCOB) assemblies, and Yu et al. [4] on circuit boards

with BGAs, using repeated sine sweeps near the first resonant frequencies of their test boards until solder joint failure. The sine sweep insures that the circuit board resonant frequency is still driven during the test even if the resonant frequency drifts, but does not account for changes in the overall vibration response level.

Standards are available for vibration testing. JESD22-B103B [5] describes using sine sweeps for vibration qualification of electronic components. MIL-STD-810G Method 514.6-Vibration [6] provides vibration testing guidelines for material development, reliability, and qualification purposes. These standards do not provide any vibration control strategies to address changes in resonant frequency that can occur during the duration of tests.

In this work a sinusoidal vibration test method with resonance tracking is employed that continuously monitors for changes in the resonant frequency of the system and adjusts the excitation frequency to match the resonant frequency. The test setup includes an electrodynamic shaker with a real-time vibration control, resistance monitoring for identifying electrical failures of interconnects, and vibration logging for monitoring changes in the dynamic response of the assembly over time. Reliability tests were performed using both fixed-frequency and resonance tracking sinusoidal test methods for an assembly consisting of a centrally mounted BGA device assembled with SAC305 solder.

In addition to testing, finite element (FE) analysis was also performed to help understand the identified failure modes. Researchers often use FE models to predict the stresses at solder joints and interconnections that occur during the vibration tests since it is not feasible to measure the solder joint strain due to their small size and their locations within FCOB assemblies and BGAs [1,2,3,4]. One difficultly in building an effective FE model of a board-level package is due to complex structure and range of geometric scale of the various parts. For example, the PCB contains copper layers, woven fabrics, plated-through holes, and so forth. In addition to the modeling demands of including such details, the mechanical behavior will typically be non-isotropic. Therefore, it is often acceptable to obtain equivalent orthotropic material properties and use them in the simulation. Similar complexities arise from modeling of the package and solder joints. Consequently, approximations and simplifications of both the geometry and material properties are often exploited in order to reduce the complexity of the resulting finite element models [1-4,7-10].

A robust methodology for improving finite element models of the assemblies, developed in ANSYS, with experimentally derived modal characteristics (mode shapes and natural frequencies) was used. This is a two-step process where in the first step focuses on the material properties and the second step targets the boundary conditions. By fine-tuning the models in this fashion, the effects of various boundary conditions (e.g., standoff screws used in drop testing) can be easily isolated from the general material properties so that critical performance parameters, such as stress and strain, may be computed.

Presented in this paper are our fine-tuned vibration measurement techniques and related finite element models to provide the building blocks for high cycle solder fatigue plots (i.e., S-N curves).

Test Methods

A schematic of the test setup is included in Figure 1. The vibration control loop consists of a vibration controller and analyzer (Spectral Dynamics Puma), power amplifier (Crown Techron 5530), electrodynamic shaker (Vibration Test Systems VTS100), accelerometer dual mode amplifiers (Kistler type 5010) and two accelerometers. The reference accelerometer (PCB Piezotronics 352C22) is mounted to the shaker head, and the response accelerometer (PCB Piezotronics 352C23) is attached to the circuit assembly at the board center using bee's wax. Also included in the test setup are a data acquisition system for vibration logging (National Instruments USB-6229 with LabView Signal Express), and an electronic event detector for monitoring circuit connectivity (Analysis Tech 256STD with WinDatalog v3.5.0).

decrease in natural frequency for the system. The vibration logging system measures both accelerometer signals and records the fundamental frequencies, amplitudes, and phase angles every second for the entire duration of the test. Time stamps are acquired with the vibration logs and event detection records, allowing the calculation of vibration cycles-to-failure.

Photographs of the physical test setup are included in Figure 2. Figure 2(A) shows the entire test, and Figure 2(B) is the shaker head with the circuit board mounted to standoffs. The torque setting for the mounting screws is 6in·lbs (96in·oz). The circuit board is mounted with the component side down and the response accelerometer is placed at the center of the board on the opposite side of the component. Figure 2(C) shows the component side of the assembly. The solder balls are daisy chained into two runs with resistance measurement points on the circuit board near the left and right edges of the component. Each daisy chain is split into two resistance monitoring channels with a common ground for a total of four solder ball quadrants, marked as Q1-Q4 in Figure 2(C). Strain gauge wire is used to connect the event detector to the circuit board. Light gauge wire and a reduced number of resistance monitoring channels minimize vibration mass loading.

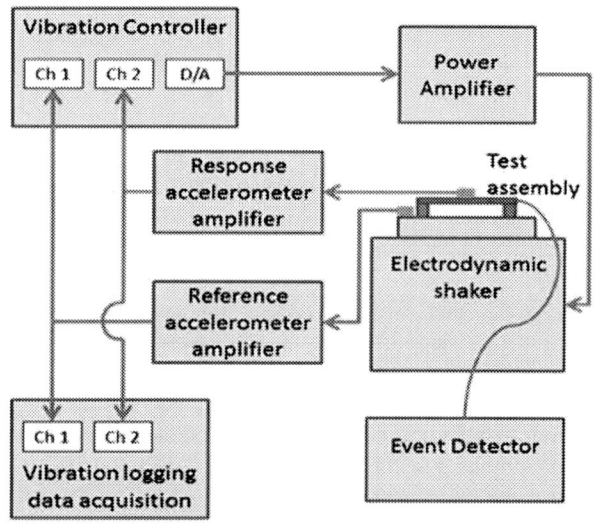

Figure 1. Test setup schematic with resonance tracking vibration control loop, vibration logging, and event detection.

The vibration frequency response of the assembly is measured with the vibration controller using a sine sweep at the start of the vibration reliability test. The resonant frequency is identified and a sine wave dwell test is initialized at that frequency. The phase relationship between the reference and response accelerometers is constantly monitored by the control system and the driving frequency is adjusted to maintain the phase angle at the start of the test. This keeps the driving frequency at the resonant frequency because at resonance the phase angle is roughly -90 degrees (depending on the level of damping), and a detected increase or decrease in phase angle corresponds to an increase or

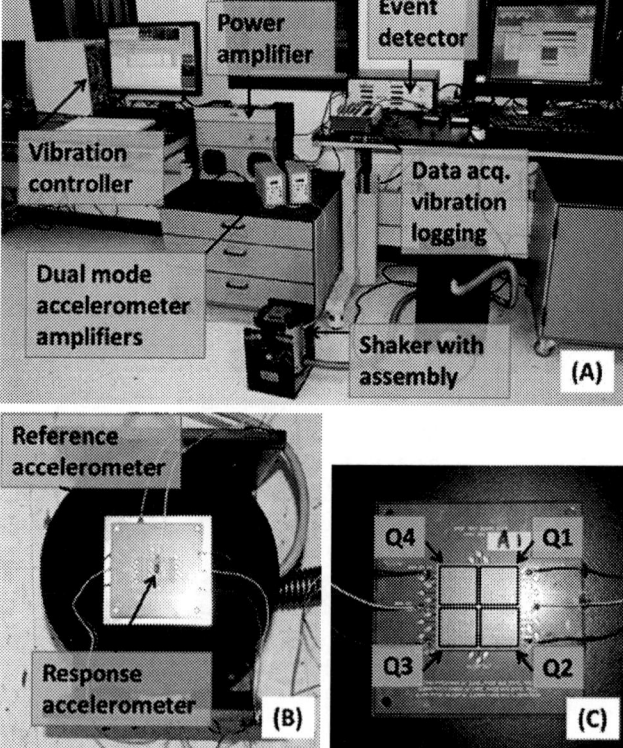

Figure 2. Photographs of test setup. (A) Physical test setup for schematic in Figure 1. (B) Circuit assembly mounted component side down to standoffs on shaker head with accelerometers mounted to the shaker head and circuit board center. (C) Component side of assembly with daisy chained solder balls. The daisy chains are wired for resistance monitoring of four solder ball runs. Light gauge wire is used to reduce vibration mass loading.

Circuit Assembly Details

The test boards include dummy BGA components, daisy chained to allow solder joint resistance monitoring with an event detector. The PCB's are 3 inch X 3 inch (7.62 cm X 7.62 cm) FR4, 40 mil thickness, with non-solder mask defined (NSMD) pads. The dummy component is 3 cm X 3 cm, 1.5 mm thick FR4 material with no die, and 1.27 mm pitch. I/O: 256 count 20 X 20, 4 row perimeter array.

A schematic of the PCB and component interconnections is shown in Figure 3. There are two isolated daisy chain runs at the left and right of the BGA, which are then split into quadrants with a common ground between each run. For example, the green box marks quadrant Q1, where ground is at "BGA Q1 & Q2 Out" and the resistance is measured at the upper right grace, "BGA Q1 In". Quadrant Q2 is part of the same daisy chain as Q1 and its resistance is measured at the lower right trace, "BGA Q2 In".

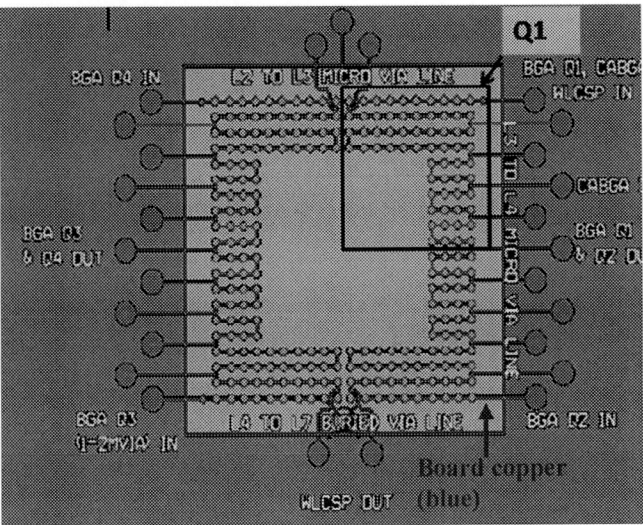

Figure 3. Schematic of component-board daisy chain (blue=board copper, red=component copper). The box at the upper right of the array shows one resistance monitored channel, Q1, which is part of the same daisy chain as Q2. Common ground for Q1 and Q2 is at "BGA Q1 & Q2 Out". All of the solder ball connections are monitored using a total of four resistance measurement channels.

Demonstration of Sine Dwell with Resonance Tracking

An 8 hour sine dwell test was performed on an assembly with event detection and vibration data logging. The input acceleration for the test was 3 g. The logged vibration data for this test is shown in Figure 4. The top plot is the circuit board to base transmissibility plotted versus time in hours. The middle plot is the corresponding phase angle between the reference and response accelerometers. The bottom plot is the identified frequency of the shaker and circuit board vibration. The data in Figure 6 show that the driving frequency is changed by the controller during the test to maintain the phase angle between -80 and -85 degrees and keep the transmissibility close to 90. The driving frequency at the start of the test was around 523 Hz and was changed to around 516 Hz near the end of the 8 hour test.

Note that the electrical failure times from the event detector are also marked on Figure 6. The first two failures occurred within the first hour of the test, during the time where there is the largest change in transmissibility level.

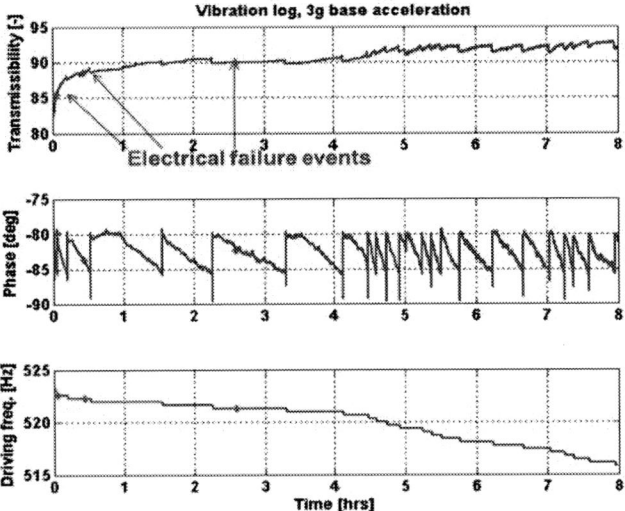

Figure 4. Logged vibration data for an 8 hour sine dwell with resonance tracking test at 3 g input acceleration. The assembly transmissibility, corresponding phase angle, and identified driving frequency are all plotted versus time in hours. During this test the controller incrementally decreased the driving frequency to maintain the phase angle between -80 and -85 degrees and keep the transmissibility around 90. The detected electrical failure times are marked on the plots.

Another sine sweep response measurement was performed on the assembly after the 8 hour test to obtain an accurate assessment of the change in the system. This is compared to the original response in Figure 5. The frequency corresponding to the peak of the transmissibility curve decreased in frequency from about 522 Hz to about 515 Hz, consistent with the change in driving frequency during the sine wave dwell test with resonance tracking. The transmissibility level at the peak also increased from about 80 to 90, which is consistent with the logged vibration data. This data shows that the resonance tracking system successfully kept the driving frequency at the resonant frequency of the system.

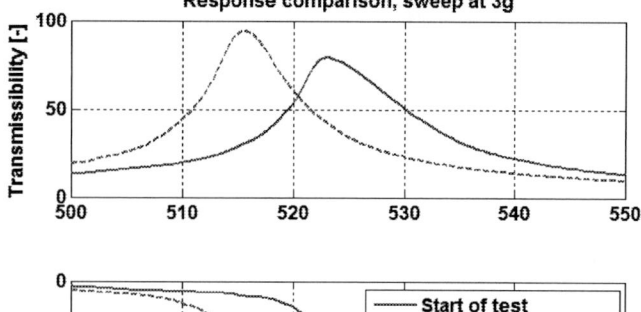

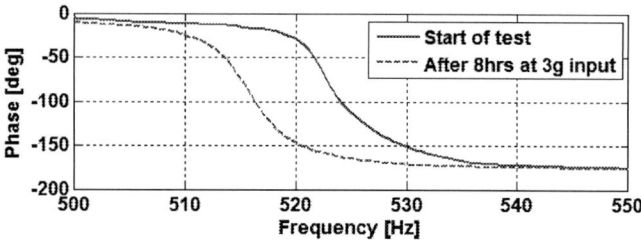

Figure 5. Comparison of assembly vibration response before and after an 8 hour 3g input sine dwell test with resonance tracking.

The circuit board deflection (circuit board displacement minus shaker displacement) can be estimated from the measured input and response vibration accelerations. From the data in Figure 4, the average center board deflection over the 8 hour test is 250 um with a standard deviation of 5.63 um. The deflection amplitude spectra were also calculated from the data in Figure 5 for the 3 g acceleration input, shown in Figure 6. At 522 Hz the board deflection amplitude is around 220 um at the start of the test. After the 8 hour test at 522 Hz the deflection is only around 100 um. This comparison clearly shows that keeping the driving frequency at 522 Hz for the entire duration of the test would result in larger variability with much lower deflection levels than what is obtained with resonance tracking, assuming the same change in the vibration response occurs in the fixed frequency test.

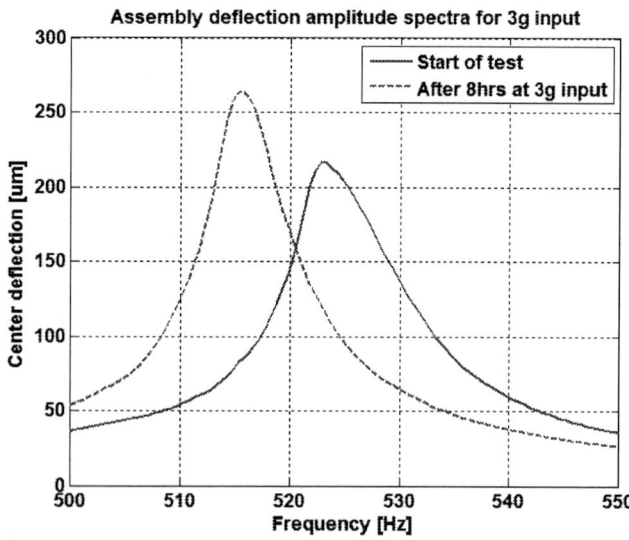

Figure 6. Amplitude spectra for the center of circuit assemble calculated from the data in Figure 7.

Vibration Reliability Test Results

Failure times for ten assemblies are reported in Table 1. Results using the resonance tracking sine dwell method are shown (B2RT-B7RT). Also included are results obtained using fixed frequency sine dwell (B8FF-B12FF), with the driving frequency fixed at the resonant frequency identified at the start of each test. The input base acceleration amplitude was equal to 2 g for all of the tests.

For the resonance tracking vibration tests, five of the six assemblies all had detected failures occurring after around six hours of testing. For the fixed frequency tests, there is no clear trend in failure time, with failure times ranging from around two hours, to no detected failures with up to 51 hours of testing. These test cases show that the resonance tracking test method gives more consistent failure times than the fixed frequency method.

Table 1. Summary of Assembly Vibration Tests. RT – Resonance Tracking; FF – Fixed Frequency.

Assembly	Failure times	Input acceleration amplitude [g]	Total Test duration
B2RT	Q2-7min Q4 - 1hrs,16min	2	3 hrs
B4RT	Q1 -5hr,46min,9s	2	6 hrs
B5RT	Q2 - 5hrs,50min	2	6 hrs
B6RT	Q2 - 5hrs,34min	2	6 hrs
B7RT	Q1 - 6hrs,25s	2	6hrs,50s
B8FF	Q2 - 14hrs,59min,20s	2	15 hrs
B9FF	Q2 - 3hrs,48s	2	3hrs,1min
B10FF	Q2 - 2hrs,9min,30s	2	3 hrs
B11FF	No failures	2	51 hrs
B12FF	No failures	2	45 hrs

Failure Analysis Results

Failure analysis of the assemblies indicates the primary failure mode created by these tests is trace failure at the circuit board side at the corner solder joints. Figure 7 represents the typical failure condition for the devices. Included are a global view of the interconnection and a close-up view at the failure location. A crack can be seen partially across the solder ball, and all the way through the lower circuit board pad.

Also, some solder joint fatigue cracks at the component and board sides were also detected. Typically board side solder fatigue was favored with longer crack lengths than those on the component side. These solder cracks were rarely seen in these tests and did not result in complete electrical failures.

978-1-4799-2408-0/14 $31.00 © 2014 IEEE

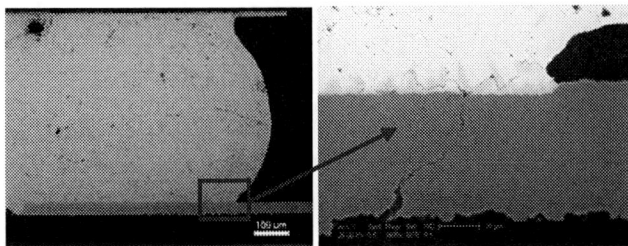

Figure 7. Typical failure mode for these tests is PCB trace failure at corner solder joint.

Finite Element Model

The element type used in the FE model is ANSYS element "Solid45". This element is defined by eight nodes; each node has 3 translational degrees of freedom. This element was attributed to PCB, solder joints, copper pads and component.

The modeling procedure was started by building a "unit solder" model as shown in Figure 8. This unit was copied as needed for the geometry of the assembly (20×20 Perimeter, 256 I/O package). The remaining portions of the component and PCB are then extruded up to the desired dimension. The total number of elements for this model is 147,468. Figure 9 shows the isometric view of the full model.

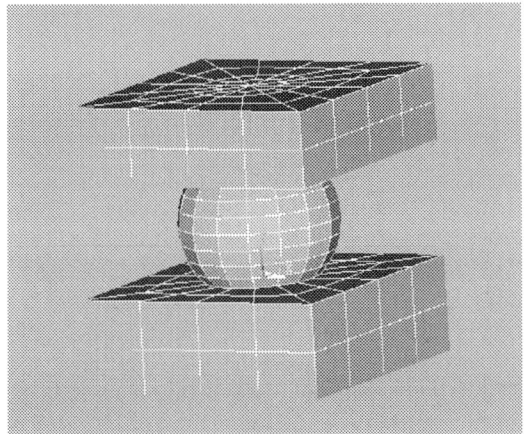

Figure 8. Typical solder joint unit.

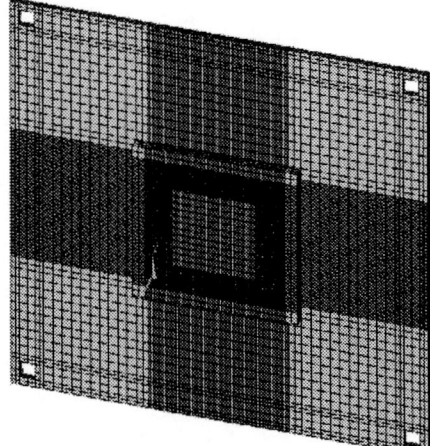

Figure 9. Full model mesh.

Correlation of FE Model with Experimental Modal Analysis

In order to build a FE model of the assembly that faithfully captures the essential dynamic characteristics, i.e., mode shapes and natural frequencies, it is necessary to employ an appropriate model "adjustment" technique so that any changes made to the model can be traced to a physical source of justification. The two primary reasons for making adjustments to the model include uncertainty with the material behavior and the boundary conditions.

The correlation methods used herein are based on systematically comparing the modal characteristics (i.e., the natural frequencies and mode shapes) of the first few lower modes of the assembly from both the FE model and dynamic measurements from physical samples. Specifically, the method is employed in two steps. The first step correlates the modal characteristics for a "free-edge" boundary condition (i.e., the test piece is modeled and measured with free boundary conditions on all edges). This effectively eliminates any uncertainty regarding the degree of fixity of the boundary conditions from the model so that any adjustments made are focused on material property changes in the FE model. Second, after the material properties are adjusted as best as possible, the test piece is mounted in the testing fixture (e.g., screw fasteners through holes drilled in the PCB). This second round of correlation is therefore focused on the effect of the boundary conditions on the modal characteristics. Again, adjustments are made to the "fixity" of nodes in the FE model to improve the mode shape and natural frequency correlation. Note: in this phase, the material properties are not adjusted – the focus of the second phase is on the boundary conditions only.

The modal impulse hammer testing method was used to measure the mode shapes and natural frequencies of the assembly. Figure 10 shows the tested boundary conditions. The "free-edge" boundary condition was approximated by hanging the assembly with fishing line. The assembly is also shown mounted to the fixture with standoffs for the "fixed" boundary condition. A miniature instrumented impulse force hammer (PCB Piezotronics 086E80) excited the system over 7x7 measurement grid, while the response accelerometer (PCB Piezotronics 352C23) remained fixed. The measured transfer functions were stored and processed using modal analysis software (Spectral Dynamics STAR 7).

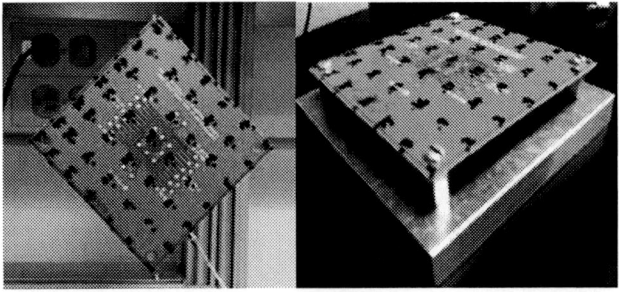

Figure 10. Tested boundary conditions. Left picture: "Free-edge" approximated by hanging from fishing line. Right picture: "Fixed" by mounting onto fixture with standoffs.

The correlation results are summarized in Table 2 for the "free-edge" boundary conditions, and the results for the "fixed" boundary conditions are summarized in Table 3. The natural frequencies from FE analysis and measurements correlate well with low percent error. The Modal Assurance Criterion (MAC) calculations for the mode shapes are also listed in Tables 2 and 3. MAC can be thought of as the dot product or linear regression coefficient between the FE and measured modal vectors. It takes the range between zero (no correlation) and one (perfect correlation). The FE model correlates well with the experimental modal analysis results for both "free-edge" and "fixed" boundary conditions.

Table 2. Natural frequencies comparison and MAC for assembly with "free-edge" boundary conditions.

Mode #	FEA Freq. (Hz)	Exp. Freq. (Hz)	%Error	MAC
1	525	504	4.1	0.98
2	760	771	-1.4	0.94
3	827	822	0.6	0.77
4	1048	1048	0.0	0.82

Table 3. Natural frequencies comparison and MAC for assembly with "fixed" boundary conditions.

Mode #	FEA Freq. (Hz)	Exp. Freq. (Hz)	%Error	MAC
1	541	540	0.2	1.00
2	961	965	-0.4	0.90
3	1320	1215	8.6	0.88
4	1736	1717	1.1	0.89

Additional fine tuning of the model is performed to account for part variability. The natural frequencies and damping ratios for all of the tested assemblies were extracted from the vibration transmissibility data. The different natural frequencies are accounted for by small modifications to the Young's Modulus of the PCB in the model. Table 4 shows comparisons of the measured and predicted center board deflection for the assembly subjected to 2 g base excitation at resonance. This shows that the model can accurately predict the experimentally determined deflection amplitudes.

Table 4. Comparison of assembly center-board deflection obtained experimentally and from finite element analysis.

Young's modulus (GPa)	Frequency (Hz)	Experimental deflection (µm)	FEA deflection (µm)	%Error
31	517	176	178	1.1
33	529	189	192	1.5
35	542	176	176	0.0
37	553	169	169	0.0
40	570	127	128	0.8
43	586	148	150	1.3

Stress Estimation by Finite Element Analysis

With a well-correlated and tuned FE model, it is now possible to simulate the dynamic response of the assembly to various types of loads. This provides a means for estimating the critical stresses within the electronic interconnections. For this paper, we focus our FE analysis on the PCB trace located at the corner of the solder ball array to match the identified failure mode from the reliability tests.

Figure 11 shows a corner PCB trace and solder ball pad from the FE model. The volume average stress at the trace failure location is calculated over the location marked on the figure.

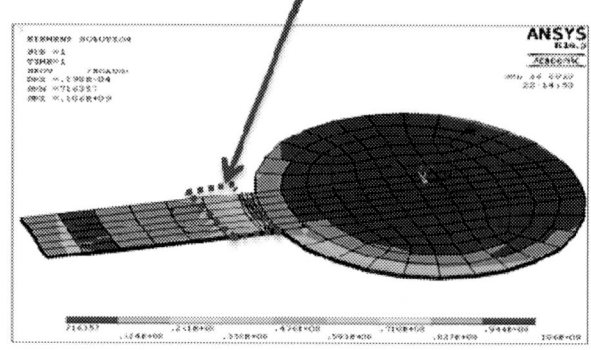

Figure 11. Corner PCB trace and solder ball pad from the FE model.

The model was used to obtain a relationship between the stress in the PCB trace and measured quantities. The result is the (Stress/Deflection) vs. Frequency curve shown in Figure 12. The frequency is the first resonant frequency of the assembly, which for our reliability tests is also the excitation frequency. Knowing this frequency and an estimate of the center board deflection, we can quickly estimate the stress amplitude at the PCB trace for any frequency within the range shown without needing to perform additional FE analysis.

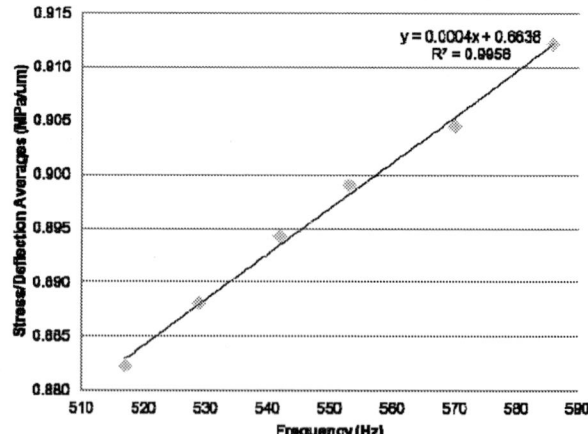

Figure 12. Stress divided by center board deflection vs. first resonant frequency from FE analysis.

978-1-4799-2408-0/14 $31.00 © 2014 IEEE

Conclusions

Changes in the resonance frequencies were observed over the duration of these tests, and the resonance tracking system was able to adjust the driving frequency accordingly. This resulted in earlier overall failure times for the assemblies tested with resonance tracking vibration control, compared to tests performed with the same excitation level with fixed driving frequency. Failure analysis was performed to verify the detected electrical failures. The primary failure mode for these assemblies tested is trace failure, with some observed solder fatigue cracking.

A FE model was assembled and tuned to experimental vibration and modal analysis data. This allows us to estimate the critical stresses within the electrical interconnections.

The improved resonance-tracking test method combined with high-fidelity FE models, combine to give high-accuracy insight into resonance failures for the assembly. In addition, data from this approach can be used to build S-N fatigue curves for various failure modes.

Acknowledgments

This work was funded by the Universal Instruments AREA Consortium. We thank M. Meilunas for test vehicle design and assembly.

References

1. Y. Chen, C.Wang, Y. Yang, "Combining vibration test with finite element analysis for the fatigue life estimation of PBGA components," Microelectronics Reliability 48 (2008) 638–644.

2. T. Eckert, W. Muller, N. Nissen, H. Reichl, "A solder joint fatigue life model for combined vibration and temperature environments," in: Electronic Components and Technology Conference, 2009. ECTC 2009. 59th, IEEE, pp. 522–528.

3. F. Che, J. Pang, "Vibration reliability test and finite element analysis for flip chip solder joints," Microelectronics Reliability 49 (2009) 754–760.

4. D. Yu, A. Al-Yafawi, T. Nguyen, S. Park, S. Chung, "High-cycle fatigue life prediction for Pb-free BGA under random vibration loading," Microelectronics Reliability 51 (2011) 649–656.

5. JEDEC Solid State Technology Association. 2006. Vibration, Variable Frequency, JESD22-B103B. Arlington, VA: JEDEC.

6. United States Department of Defense Test Method Standard. 2008. Environmental Engineering Considerations and Laboratory Tests, MIL-STD-810G, Method 514.6 Vibration.

7. J. Pitarresi, "Modeling of Circuit Cards Subject to Vibration," IEEE proceedings of the circuits and systems conference, pp. 2014-2107, 1990.

8. J. Pitarresi et al., "The smeared Properties Approach to FE Vibration Modeling of Printed Circuit Cards," ASME J Electron Pack, Vol. 113, pp.250-257, 1991.

9. J. Pitarresi, A. Akanda "Random Vibration Response of a Surface Mounted Lead/Solder Joint," ASME proceedings international electronics packaging conference, pp. 207-217, 1993.

10. B. Zhang, P. Liu, H. Ding, and W. Cao, "Modeling of Board-Level Package by Finite Element Analysis and Laser Interferometer Measurements," Microelectronics Reliability, Vol. 50, pp.1021-1027, 2010.

Crack Monitoring and Life Modeling Technique Towards High Thermal Cyclic and Mechanical Reliability of fcBGA Solder Joint

Dongji Xie*, Zhongming Wu#, Min Woo* and Tom McMullen*
*Nvidia Corp., 2701 San Tomas Expressway, CA 95050, USA
Nvidia Corp., Hi-tech Middle 2nd Road, Shenzhen Hi-Tech Industrial Park, Nanshan District, Shenzhen, China 518057
Contact E-mail: Dongjix@nvidia.com; Tel: +408 486 8630; Fax: +408 486 2919

Abstract

High reliability at the board level is challenging for a large flip chip ball-grid-array (fcBGA) where large die and stiff substrate are used. For those BGA solder joints, the difficulty is to achieve high reliability in both thermal cycling and mechanical dynamic tests. This paper presents experimental work on an fcBGA with a die size of 25x15mm, a body size of 40x40mm and over 1700 ball count. The reliability tests include thermal cycling from -40C to 85C up to 7,500 cycles, mechanical drop at 1500G and 9-point cyclic bending test run to failure. To develop a good reliability model, the integrity of both the solder joint and substrate copper traces are monitored using in-situ resistance measurement in combination with frequent cross-sectioning and dye-and-pry (DnP) to understand the progression of cracking in the solder joints. Using Finite Element Analysis (FEA), the solder joint failure mechanism was verified and a new failure mode in mechanical reliability testing has been confirmed.

Introduction

FcBGA is a main stream package format and widely used in the high performance microprocessor and graphic chips. A large fcBGA usually has a large die (15~25mm) so that the thermal stress between the die to the substrate and Fr-4 board is large during the thermal cycling. A substrate core with a lower thermal expansion value is used to ensure reliability at the first level solder bump interconnection. In some cases, the Coefficient of Thermal Expansion (CTE) value for a substrate core can be as low as 5~7ppm/C to match the silicon die [1]. On the other hand, to achieve high second level reliability a substrate with high CTE is desirable to reduce the thermal mismatch between the substrate and motherboard. Some studies has shown that a substrate with a CTE value of 10~12ppm/C may be ideal to achieve high reliability at bump and solder ball level [2~4]. In those cases, thermal cycling fatigue life may surpass 6000 cycles for air-to-air cycling (ATC) from 0 to 100C in certain packages [3].

Mechanical reliability during mechanical shock and bending is also important. Underfill or edge bond are the two methods most commonly used to achieve high mechanical reliability. Underfilling the whole BGA with epoxy provides full protection for the solder joints but is less desirable because of cost and re-work concerns. Edge bond to the component edge and corner can be used at lower cost but care must be taken to not touch the solder ball in order to allow re-work.

This paper presents experimental results including thermal cycling from -40C to 85C, mechanical drop and 9-point cyclic bending test run to failure. To develop a good

reliability model, the integrity of both solder joint and substrate copper trace are monitored using in-situ resistance measurement in combination with periodic cross sectioning and DnP to understand the progression of cracking in the solder joints. DnP is a very effective way to assess failure modes of solder joints at the board level where failure modes include solder ball cracks, intermetallic failures, pad cratering and pad lifting [5~7]. FEA was also employed and validated using the experimental data. Global and sub-modeling techniques were used to identify and predict the fatigue life of the weakest area. Studies have shown that both peel stress and strain energy density can used as an indicator for failures in the global model [8].

Mechanical reliability was checked using mechanical shock and cyclic bending tests. The failure mode found by experiment is actually a combined failure mode where the solder joint may fail at either the trace/pad or within the solder depending on the pad/trace widths and the stress levels of those structures. An interesting finding from authors' previous work [6] shows that, contrary to earlier understanding, increasing the trace width does not necessarily increase the number of drops to failure. The reason may be due to the failure shift identified from the FEA study. Further studies show that the trace design has to be optimized for critical solder joints to achieve a failure free solder joint in both thermal cycling and mechanical drop test [9].

Methodology

The package used in this study is a lid less fcBGA with a die size of 25x15mm, a body size of 40x40mm and over 1700 ball counts. Thermal cycling was performed in an air-to-air thermal chamber from -40 to 85C per JEDEC condition N with a ramp rate of approximately 10C/min. In-situ monitoring was used to check the continuity of daisy chains of solder joints up to failure or 7,500 cycles for all units. To verify the cracking of solder joints, all failures were pulled out for detailed failure analysis including DnP or cross-sectioning. Passing units were also pulled out periodically for DnP or cross-sectioning to assess the progress of solder joint cracking.

The mechanical shock test was performed per JEDEC 22B110 service condition B [5, 6] with acceleration of 1500G and pulse duration of 0.5ms. The mechanical shock test was performed face down, i.e., –Z axis only, representing the worst case for this test. The units were monitored in-situ and the tests were continued until failure. All failures were probed and analyzed using DnP.

The 9-point bending test was performed using a tensile machine with cyclic function. The bending was controlled to 500ue measured from 5mm away from the outer corner of the fcBGA. A typical strain curve is shown in Fig. 1. The tests were continued until failure. Failure analysis was performed after the tests to confirm the failure modes.

978-1-4799-2408-0/14 $31.00 © 2014 IEEE   126   2014 Electronic Components & Technology Conference

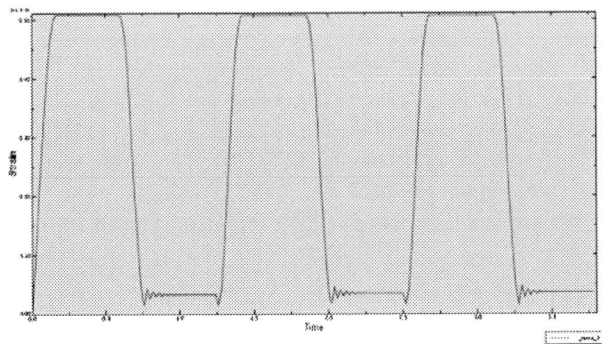

Fig. 1 PCB strain during bending test measured from 5mm away from the outer corner of the fcBGA. Strain unit in microstrain (ue).

In order to quantitatively assess the cracking, a mapping system was applied as shown Fig. 2. The failure modes were classified into six types dependent on the failure mode.

Six types of failure modes using color coding:
• Brown = pad crack at the BGA side
• Red = solder crack at BGA side
• Green =Solder crack in the middle
• Pink = solder cracked at PCB side
• Gray = Pad cratering on PCB side
• White/blank=no crack/failure (default)

The crack size was measured by area percent based on the pad area and quantified using a code (0~9) to represent 0~100% of the pad area. In this coding system, 0 represents no crack and 9 represents a full crack.

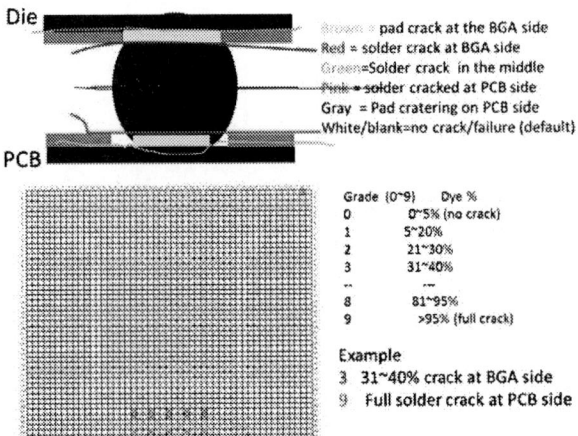

Fig. 2 Crack coding in Dye-and-Pry (DnP)

In parallel with experimental work, (FEA) was employed to further understand the failure mechanisms. Because the design is symmetric a quarter model was used as shown in Fig. 3 [8]. The center areas of the test vehicle are covered by the die shadow. The global model uses a relatively coarse model and is helpful to determine the critical balls which could be weakest and fail earlier than other balls. From previous experience, critical solder balls include the balls in the corners and inner corners. For that purpose, both maximum peeling stress and maximum creep

energy dissipation may be used to evaluate the failure risk. Once the critical solder balls are located, sub-modeling with fine meshed solder balls was performed at that location using the same thermal cycling profile; creep fatigue analysis was completed to predict the life. This sub-modeling is repeated in various locations for all critical solder balls to assess the fatigue life.

For mechanical bending, a sub-structure modeling method was used as shown in Fig. 4. A FEA model for mechanical shock is also illustrated. To better represent the solder joint, both 2D and 3D fillet was used for comparison.

For thermal cycling test, an elastic-plastic-creep study was employed to assess the creep-fatigue of the solder joints. The materials properties used in this FEA are listed in Table 1. For solder materials, an Anand creep model for SAC alloy is employed as described by Darveaux [9]. The creep parameters for lead-free alloy were further refined by Bhate et al [10] and are shown in Table 2. The life prediction using creep energy density, as described by Schubert et al. [11], is shown in Equation 1.

$$N_f = 345\Delta W^{-1.02} \tag{1}$$

where N_f is the characteristic life of the solder joint and ΔW is the cyclic creep strain energy density (averaged) along the crack path, calculated by FEA.

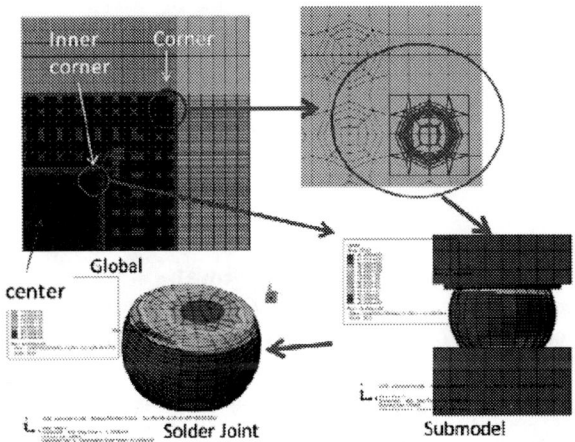

Fig. 3 Global-submodels of FEA for thermal cycling test.

For cyclic bending fatigue, the life prediction is characterized by the cyclic plastic strain and may follow Equation 2 [12].

$$N_f = 42.66\varepsilon_p^{-1.09} \tag{2}$$

where ε_p is the plastic strain range during a bending cycle.

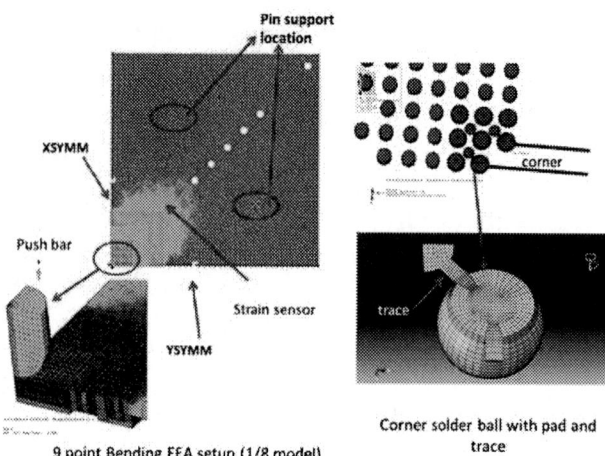

Fig. 4 FEA models for bending test (a) and mechanical shock (b).

Results and Discussion
Thermal Cycling Results and Crack Growth

Thermal cycling was performed from -40 to 85C for 7500 cycles. In-situ monitoring on corner balls and other critical balls did not detect any failure for those balls To validate if there were any cracks in the solder joints, a DnP test was performed and the results are shown in Fig. 5. Fig 5 illustrates the DnP image after 1000 thermal cycles.

DnP did reveal cracks in the solder joints under the die area as thermal cycling continued. (Due to the design of the test vehicle these solder joints were not covered by in-situ monitoring.) It was found that the crack began at around 2400 cycles at the center area under the die. This can be explained by the FEA results as shown in Fig. 6. FEA results indicate that the solder balls under the die area experience the highest peeling stress and strain energy dissipation (CENER) suggesting that these are the weakest solder balls and may fail first. Fig. 6 also shows that the distribution of peeling stress and maximum CENER do not exactly match. The maximum peeling stress occurred in the inner corner ball while the maximum CENER occurred in the third row from the inner corner. From the DnP results, the inner corner ball did not crack while the balls in the third row cracked. Cracks through >50% of the pad area were found in the balls that are entirely under the die

after about 5,000 cycles. This demonstrates that the maximum CENER may be more useful as an indicator of failure probability. This may be because peeling stress, which includes compression stress, may not contribute to cracking [13]. The cracks in the solder joints under the die shadow grew with the cycle count as shown in Fig. 7 and listed in Table 3. The crack growth of these solder joints may be expressed as an equation by

$$C = 10^{-16} * y^{4.372} \qquad (3)$$

Table 1 Material properties used in FEA.

Item	Mat.	Density (ton/mm3)	Modulus (MPa)	Poisson ratio	CTE, ppm/C
Solder ball	SnAg	7.8E-9	44400 (T=293), yield stress=45.1 MPa (273K)	0.36	23 (273K)
PCB	FR4	1.8E-9	E1=17689 E2=17689 E3=7715	Nu12=0.11 Nu13=0.39 Nu23=0.39	E1=18 E2=16 E3=60
Subs.	BT	1.8E-9	28000	0.27	E1=14 E2=14 E3=45
Die	Si	3.1E-9	150000	0.17	2.9
Under-fill		1.8E-9	14226 (T=273)	0.3	24 (273K)
PAD & trace	Cu	8.0E-9	128000, yield stress=75MPa	0.34	16.7
Stiffener & punch	steel	7.8E-9	200000	0.3	8.0
Edge bond	Glue	1.05E-9	490	0.35	45

Table 2 Parameters used in Anand Creep model (in mm-N-second system) [14].

Constant	Parameter	SAC387 (Bhate, 2007)
C1	S0	3.2992
C2	Q/k	9883
C3	A	15.773
C4	CSI	1.0673
C5	m	0.3686
C6	h0	1.08E+03
C7	S^	3.1505
C8	n	0.0352
C9	a	1.6832

978-1-4799-2408-0/14 $31.00 © 2014 IEEE

where C=crack area percentage and y is the cycle count. FEA life prediction is also shown in Table 3 in comparison to the test results. As shown in Table 3, the predicted fatigue life is about 4,610 and 5,610 cycles for the inner corner and outer corner respectively. This shows that the life prediction is relatively conservative.

TCT 1000cycles—No Glued unit DnP Results

No solder ball crack/pad cratering was found cross whole package.

Fig. 5 DnP image showing no crack found in solder joints at 1000cycles from thermal cycling.

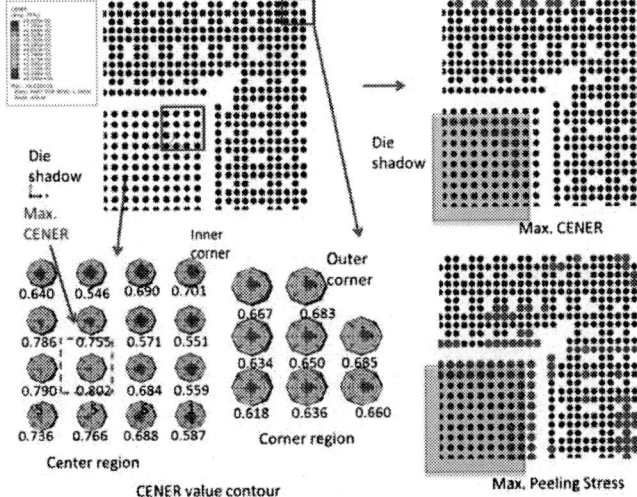

Fig. 6 FEA results showing the weakest balls are in the center area under the die (balls with high CENER or peeling stress are highlighted in red)

Table 3 Fatigue life of solder joints of fcBGA from thermal cycling tests

Location	1st failure	Fatigue Life (cycles)	
		Experimental	FEA
Corner	None	>7,500	5,610
Inner corner	None	>7,500	4,610
Center	2,400 (crack initiated)	5,147	4,776

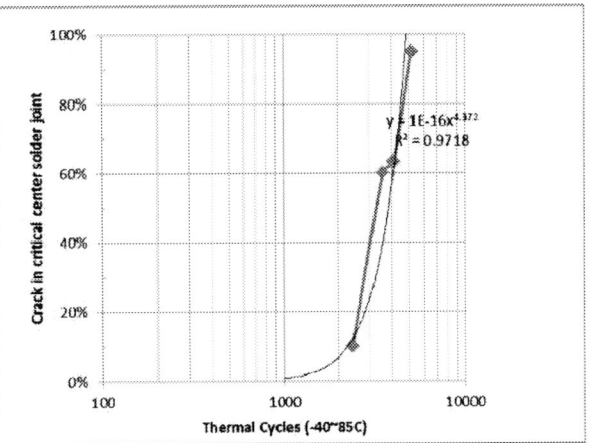

Fig. 7 Crack growth in solder joints during thermal cycling.

Cyclic Bending Test

Cyclic bending test was performed in a tensile machine where one push pin is in the center and 8 support pins are spread uniformly on a circle. The results for FcBGAs with and without glue are plotted in Fig. 8. Without edge glue the solder joints have a life of 767 cycles when the PCB strain is limited to 500ue. With edge glue solder joints have a fatigue life of 2,091 cycles. This suggests that the edge glue effectively enhances the solder joint reliability under cyclic bending. To understand the failure mechanism, units were analyzed by DnP as shown in Fig. 9. Interestingly, there are two typical failure modes: pad cratering and solder crack. Pad cratering was the main failure mode for FcBGA units without edge glue and solder failure was primarily found for fcBGA units with edge glue. However, a third failure, the partial solder crack, was found in a small solder ball area where the solder intersects the pad and trace as shown in Fig. 10. According to FEA results shown in Table 4, partial solder cracking occurred around 2,180 and 4,320 cycles respectively for fcBGA without and with edge glue. Those are much higher than the cycles for trace cracking (319 and 827 cycles) but much lower than that of through solder cracks. In summary, for fcBGA without edge glue, two major failure modes are (A) pad cratering and (B) combination of trace crack and partial solder crack. On the other hand, for fcBGA with edge glue, both types of failure modes exist but the primary mode is type B—combination of trace crack and partial solder crack. In the experimental work, the bending cycle to failure will depend on the probability of occurrence, i.e., which types of failure is dominant. The failure cycle count could be higher when Type B failure mode occurs. For fcBGA with edge glue the failure cycle count is lower where pad cratering occurred, as was the case for the first failure in this group. The reason for the difference is that the pad cratering is a brittle failure which involves cracking from traces, laminate and interfaces of fiber glass bundles. This type of failure is driven by stress and usually a short but unpredictable life is expected. Solder cracks are ductile, usually controlled by the cyclic fatigue, and could take longer to develop and grow. Comparing the FEA and experimental work, the trace cracking predicted by FEA is coincident with the 1st failure from the experimental work for both groups. The experimental mean life to failure, however, follows in-between the cycles to trace crack and partial solder failures.

Mechanical Shock Test

Mechanical shock test was performed using a shock table. The shock pulse is 1500G 0.5ms which is typically used for qualifying small BGAs in mobile applications [5, 6]. For large BGAs (>23mm), a typical acceleration for the drop shock is only 140G or below. The mechanical shock test result was plotted in Fig. 11. For test vehicles with no edge glue the first failure is at 1 drop with additional units failing on progressive drops. Units with edge glue survived 18 drops. The failure mode for mechanical shock was mainly pad cratering with some solder cracking in the case of no edge bond. The failure will shift to solder failure if edge glue is used. Those findings are in consistent with the cyclic bending results shown in Fig. 9. FEA results for mechanical shock are listed in Table 5. As shown in Table 5, a FEA model using 3D fillet will significantly enhance the solder failure life but the trace failure does not change too much as compared to that of 2D fillet. Adding edge glue increased drop cycles to failure 1~3 times for different components. In the experimental result as reference, the mean drop cycles to failure are 11 for no glue and 85 for glue. Those numbers fall between the predicted trace and partial solder failure, 2.7~57 drops (no glue) and 4.3~143 (glue) respectively. This suggests that edge glue is needed if the drop test requirement is 1500G.

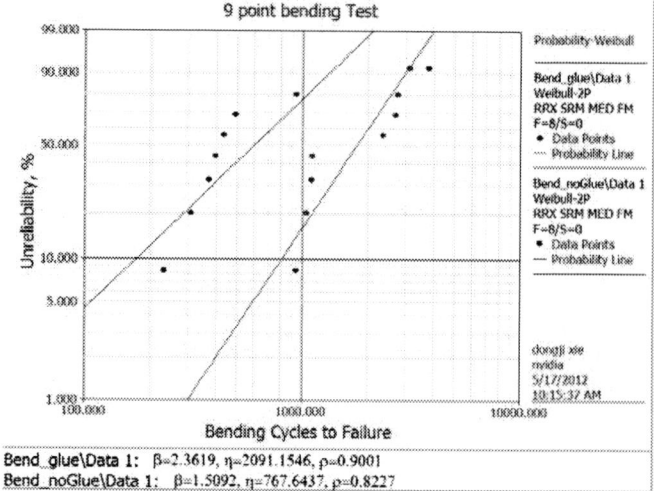

Bend_glue\Data 1: β=2.3619, η=2091.1546, ρ=0.9001
Bend_noGlue\Data 1: β=1.5092, η=767.6437, ρ=0.8227

Fig. 8 The Weibull plot of cyclic bend test to failure for fcBGA with and without glue.

Table 5 FEA results for mechanical shock test (1500G0.5ms at z-axis)

FEA model	2d-fillet, non glue	3d-fillet, no glue	3D fillet, with glue
LE sensor, ue	1543	1543	1604.72
Center Point, ue	664	618	618
Shock time considered	15ms	15ms	15ms
Copper failure cycles	3.6	2.7	4.3
Partial failure, cycles	21.4	57.4	143.5
Experiment, no glue	11.6		
Experiment, glue (calculated)			85

Table 4 Cyclic Bending Results (FEA vs. experimental)

Cyclic bending	No-glue	Glue
Push distance, FEA, mm	0.495	0.495
PCB strain, FEA, ue	450	493
Pad trace crack, , FEA cycles	319	827
Partial solder ball failure, FEA, cycles,	2,180	4,320
Through Ball (solder) failure, FEA cycles,	9,130	39,900
Nf-1st failure, Experimental, cycles	231 (trace failure)	935 (trace failure)
Nf-last failure, Experimental, cycles	3,801 (solder failure)	3,104 (solder failure)
Nf-63.2% failed, Experimental, cycles	767	2,091

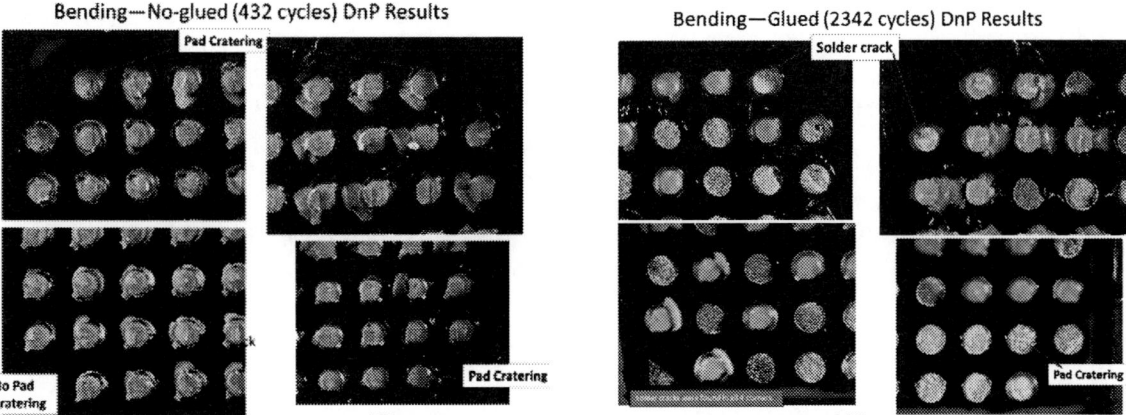

Fig. 9 DnP image showing pad lifted for no glue units (a) and solder crack in glue units (b). Note that each picture shows one corner of the fcBGA.

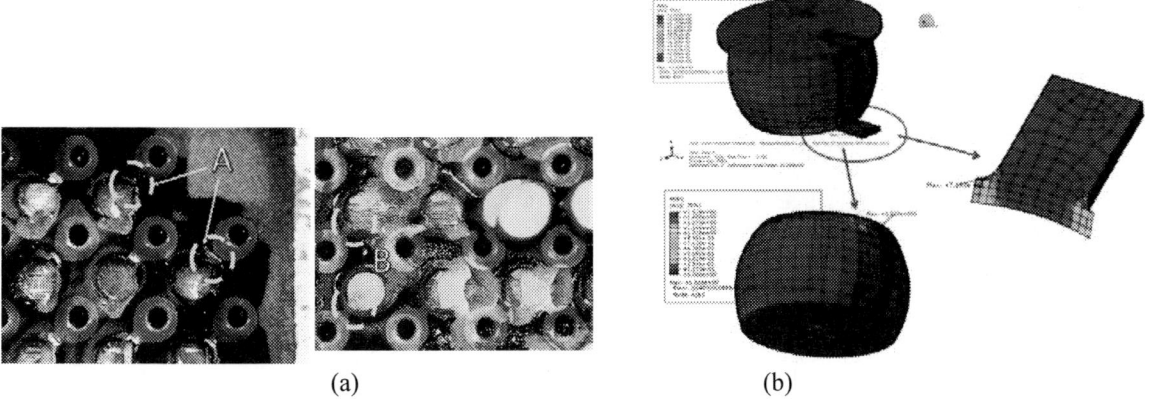

(a)   (b)

Fig. 10 DnP results from cyclic bending test. (a) Two types of failure modes: (Type A: pad cratering only and type B combination of trace and partial solder cracking. (b) FEA showing the plastic strain contours in the solder ball and trace.

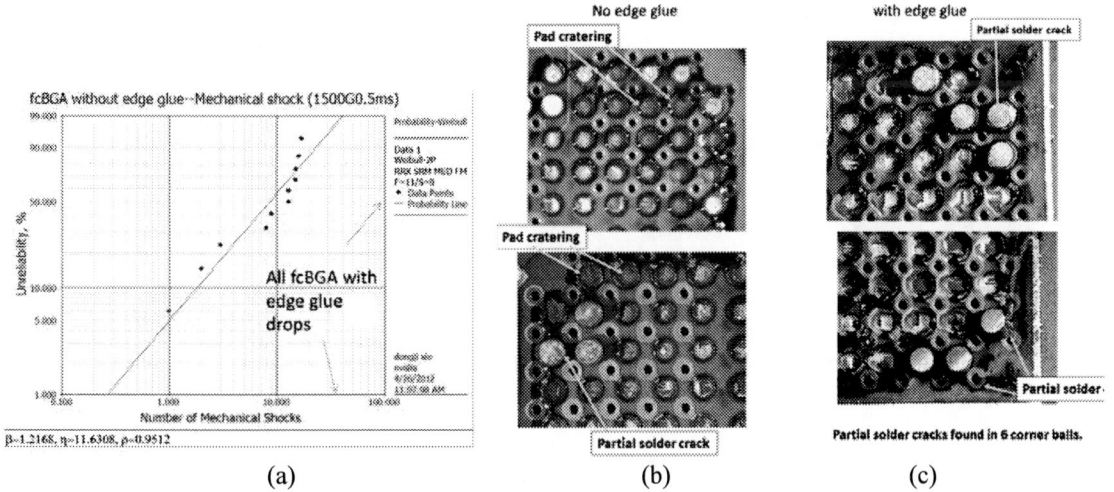

(a)   (b)   (c)

Fig. 11 Mechanical shock results (1500G 0.5ms). (a) Weibull plot showing no failures if edge glue is applied. (b) DnP image for no edge glue and (c) DnP image with edge glue.

Conclusions

Reliability of the solder joints of a large fcBGA test vehicle with a body size of 40x40mm2 has been studied. The results show that the board level reliability of this package has surpassed 7,500 cycles from -40 to 85C with cracks occurring only in the area under the die. Mechanical reliability experiments show that cyclic bending generates three types of failure modes. The dominant failure is a combination of trace crack and partial solder failure with the trace crack being the main reason for earlier failures. Use of edge glue effectively enhances mechanical reliability. This fcBGA test vehicle is able to pass very harsh mechanical shock tests at 1500G0.5ms when edge glue is used. FEA results show that strain energy density may be a more accurate predictor of failure during thermal cycling than peeling stress for this type of package. The life prediction of the solder joints from FEA is conservative for thermal cycling but in relative agreement with the trace failure in bending tests. For both cyclic bending and mechanical shock, the failure cycle counts from experimental follows the FEA prediction for trace failure and partial solder failures.

Acknowledgment

The author would like to thank Hung Nguyen in Nvidia for his assistant in preparing the test setup and failure analysis.

Reference

[1] Hitachi Chemical, http://www.hitachi-chem.co.jp/english/products/bm/b02/006.html.

[2] Dongji Xie, Vadim Gektin, David Geiger, "Reliability Study of High-end Pb-free CBGA Solder Joint under Various Thermal Cycling Test Conditions", Proc. 2009 ECTC, San Diego, USA.

[3] Andrew Mawer1, Thomas Koschmieder and David Mendez, Assembly and Reliability of High CTE Ceramic Land Grid Array, SMTAI, 2007.

[4] Rajiv Dunne, Kejun Zeng, Sergio Martinez, Venu Chauhan, Mike Estebar, Masood Murtuza and Steve Besuk, Board-level thermal cycling and mechanical Bend Reliability of HiCTE Flip-chip BGAs with Multiple BGA Pad Surface Finishes, SMTAI 2006, pp588~593.

[5] Dongji Xie, David Geiger and Dongkai Shangguan, Billy Hu and Bill Werner, A New Drop Test Vehicle for a Uniform Shock Response, ECTC2010, Las Vegas, NV. USA

[6] Dongji Xie, Ife Hsu, Yingliang Zhou, Andy Zhang, Min Woo, Ramgopal Uppalapati, Zhongming Wu and Tom McMullen, "A New and Effective Drop Test Evolution To Next-Gen Handheld Applications", ECTC2013, Las Vegas, NV. USA

[7] Dongji Xie, David Geiger and Dongkai Shangguan, et al, "Failure Mechanism and Mitigation of PCB Pad Cratering", Proc. 2010 ECTC, Las Vegas, NA, USA.

[8] Xuejun Fan, Min Pei, and Pardeep K. Bhatti, Effect of Finite Element Modeling Techniques on Solder Joint Fatigue Life Prediction of Flip-Chip BGA Packages, ECTC2010, Las Vegas, NV. USA

[9] R. Darveaux, "Effect of Simulation Methodology on Solder Joint Crack Growth Correlation," Proc 2000 ECTC, pp. 1048-1058.

[10] Dhruv Bhate, Dennis Chan, Ganesh Subbarayan, Chiu Tz-Cheng, Vikas Gupta and Darvin Edwards, Constitutive Behavior of Sn3.8Ag0.7Cu and Sn1.0Ag0.5Cu Alloys at Creep and Low Strain Rate Regimes, IEEE Transaction on CPMT, 2007.

[11] A. Schubert, R. Dudek, E. Auerswald, A. Gollhardt, B. Michel, H. Reichl, Fatigue Life Models for SnAgCu and SnPb Solder Joints Evaluated by Experiments and Simulation, Proc 2003 ECTC, p609.

[12] Ahmer Syed, Amkor Technology, 2001, TMS 2001.

[13] Mitul Modi, Carolyn McCormick, and Norman Armendariz, "New Insights in Critical Solder Joint Location", ECTC2005, , NV. USA

Fatigue Properties of Lead-free Solder Joints in Electronic Packaging Assembly Investigated by Isothermal Cyclic Shear Fatigue

Huili Xu[1,2], Tae-Kyu Lee[3], and Choong-Un Kim[1,4]

[1] Department of Materials Science and Engineering, University of Texas at Arlington
501 West First Street, Box 19031, Arlington, Texas 76019
[2]Present: Ball Attach and Interposer Attach Module, Intel Corporation,
5000 West Chandler Ave., Chandler, AZ
[3] Interconnect Technology Team (Manufacturing Technology Group), Cisco System Inc.
170 West Tasman Drive, SJC-D-2, San Jose, CA 95134
[4] Contact e-mail: choongun@uta.edu

Abstract

This paper reports the fatigue properties of Sn-Ag-Cu (SAC) and Pb-Sn solder alloys determined from isothermal shear fatigue testing and analysis of resulting data using modified Coffin-Manson fatigue model. In our study, a series of cyclic shear fatigue testing was conducted on the solder joints in PBGA assembly with variation in testing temperature, strain range, and strain rate (or cycle frequency). The number of cycles to the first joint failure in the assembly was determined from the resistance of each solder joint and taken as the fatigue life of the assembly. Analysis of the resulting data reveals that isothermal fatigue behavior of both SAC (Sn-Ag-Cu) and Pb-Sn alloys follow the classic Coffin-Manson fatigue model and provide constants indicative of fatigue properties of the alloy, namely fatigue ductility coefficient and ductility exponent. These fatigue constants are consistent with what is generally expected from a metallic fatigue system except for their dependence on temperature. This deviation, attributable to the involvement of the thermal strain that is added to the applied mechanical shear strain, suggests that consideration of the thermal strain effect needs to be included in the fatigue analysis of solder joints even if the major fatigue mode appears to be pure mechanical shear. The fatigue constants gained from our study provide insights helpful in understanding the mechanism behind variation in fatigue reliability with solder alloy compositions and solder microstructure.

Keywords: fatigue; lead-free solder; fatigue reliability; Coffin Manson fatigue model;

Introduction

Fatigue failure of solder joints in electronic assemblies has been the subject of extensive studies as it is a leading cause of reliability failure of modern electronic devices. Continuing miniaturization paralleled with expansion of mobile electronics subjects solder joints to more punishing mechanical loads, making the fatigue an even greater reliability threat. In addition, solder alloys used for modern electronic assembly are Pb-free, which are relatively new and less well understood materials. Therefore, there is a growing demand for better understanding the fatigue properties of Pb-free solder joints. [1-6] While there exist numerous insightful studies on the fatigue behaviors of the lead-free solder alloys, they offer limited assistance in correctly identifying fatigue parameters, essential for understanding fatigue properties of solder alloys, because the fatigue testing in those studies is done using accelerated thermal fatigue. Continuously varying properties of solder alloys with temperature and dynamically

evolving microstructure makes the fatigue behavior too complex to be analyzed in the frame of conventional fatigue model that assumes no abrupt change in microstructure or mechanical properties of solder alloys. A more advantageous testing method in terms of understanding fatigue properties of solder alloys is, therefore, isothermal fatigue. Isothermal fatigue induces fatigue failure by applying pure mechanical load at a fixed temperature, making the resulting data a more direct reflection of the solder joint properties and simpler to analyze and extract key fatigue parameters. An added benefit of isothermal fatigue is the fact that the result may bear relevancy to the reliability prediction of the emerging failure mechanisms such as the failure induced by vibration and shock. [7-10]

The apparent benefits of isothermal fatigue testing have instigated several studies on the isothermal mechanical fatigue behavior of solder alloys in recent years. Although these studies reveal a few key aspects of fatigue properties of solder joints, lack of data and testing consistency among studies make it difficult to extract all fatigue parameters necessary to establish a reasonably comprehensive fatigue model for solder alloy joints in an electronic assembly. [11-16] This makes some of key fatigue related properties difficult to explain. For instance, it is difficult to understand why SAC105 shows good shock resistance yet it is known to be prone to vibration induced fatigue failure. A simple approach to this question is to view the shock and vibration as two fundamentally different failure mechanisms governed by different aspects of solder properties. However, it may be possible to explain by extending the fatigue properties to an extreme strain range and rate (shock) and an extreme frequency (vibration) if a suitable model for the extension becomes available. The very first step toward such an extension should be the establishment of a comprehensive fatigue property model that includes the influence of all possible variables in fatigue conditions. For past several years, we have been conducting a series of isothermal shear fatigue tests with an aim of better understanding the fatigue behavior of solder alloys in the electronic assembly. [10] Our work reveals that the fatigue property follows the classic fatigue behavior of metallic systems although the model needs some adjustments to account for specific mechanical constraints imposed by the assembly on the joint. This paper presents results and understanding gained from our investigations, with emphasis on the validity of Coffin-Manson model, need for its modification to account for fatigue condition change due to involvement of CTE mismatch, and finally variation in the

2014 Electronic Components & Technology Conference

fatigue parameters, namely fatigue ductility coefficient and ductility exponent, with solder alloys and microstructure.

Experimental Method and Sample Configuration

Shear Tester system

As described in our previous paper [10], cyclic shear test allows evaluation of fatigue reliability with great consistency and accuracy. In our system, the displacement of the solder joints is achieved by the displacement of the chip in reference to the PCB board. In order to displace the chip with a controlled magnitude and speed, a shear head assembly is connected to a step-motor driven gear system. Figure 1 displays the shear-head assembly. As is shown, the shear head assembly is equipped with two grip blades that hold the chip. In our testing, cyclic shear fatigue testing was typically conducted by applying symmetrical displacement to the chip with a constant displacement rate without pause period.

The tests were carried out under various strain ranges, temperatures, frequencies, solder alloys, and thermal aging conditions. Thermal aging of the samples were conducted in the commercial oven at 100°C and 150°C for 500 hours, 1000 hours, and 1500 hours. The solder alloys investigated in our study were SAC305, SAC105 and 60Sn40Pb. Plastic strain ranged from +/-1% to +/-5%, frequency from 0.1Hz to 5Hz. The temperature for testing was set at 25°C, 50°C, 75°C and 100°C.

Figure 1 A picture showing a shear tester head assembly

Shear Testing Specimen

The samples used in our fatigue test were based on PBGA package assembly. As shown in Figure 2, there are 160 solder joints daisy chained into several segments that allow tracking of solder joints continuity through resistance measurement. The size of solder ball used in this assembly was ~600μm, and a cross-sectional image of the solder joint is shown in Figure 2 (c). These joints were made with an industry standard Si chip and PCB metallization structure that consists of Cu UBM/ENiG at chip side and Cu on PCB side. In order to detect failure, electrical resistance of these balls were tracked continuously with a measurement resolution higher than micro-Ohm under 100mA current.

Results and Discussion

Resistance change with the number of cycle to fail

Solder joints with different compositions and thermal aging conditions were tested under cyclic shear load with variation in strain range, temperature, and frequency. The

resistances of the solder joint in different rows were recorded to detect failure. Figure 3 shows two example cases of the resistance change recorded for 1st and 2nd row solder arrays as a function of cycle number. Note that the resistance rapidly increase to failure with cycle after the incubation period. This rapid increase in resistance with cycle is believed to reflect the rapid crack growth in the failing joint. This result may seem odd because crack growth is expected to proceed by Mode II (crack shear) which induces slow crack growth. However, as is analyzed in our previous paper, the small aspect ratio of the solder joint (height vs. width) makes the solder to rotate and changes the crack growth mode to the crack opening (Mode I). This is the reason why rapid crack growth occurs in our shear fatigue testing.

It is consistently observed that the first row solder joints (located at outmost array) always fail earlier than the others regardless of the solder composition and test conditions. This result seems strange because all joints in the assembly are under identical fatigue condition, and therefore, failure location is expected to be random. It is our belief that the residual thermal stress effect that is largest at the outmost solder array and leads to the observed results. Details of its mechanism will be discussed in future papers.

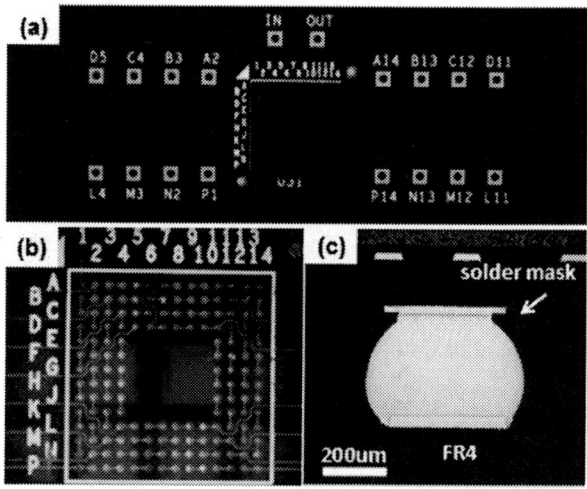

Figure 2. Picture showing (a) the PBGA assembly used for shear fatigue testing, (b) solder alloy configuration, and (c) cross-sectional view of the 600mm solder ball

Coffin-Manson Fatigue Model

For a systematic evaluation of fatigue properties of solder, the failure data collected from the cyclic shear fatigue is analyzed within the framework of the fatigue theories of metallic materials. It is found that the Coffin-Manson model works reasonably well for our data. [17-18] Its application to the fatigue analysis of solder alloys has been rarely attempted because the required number of tests to complete the model are extensive and maintaining the same failure mechanism with variation in test condition is difficult with other testing methods.

The Coffin-Manson model assumes that fatigue failure occurs when the total plastic strain reaches the fracture strain. At each cycle, metallic materials are subjected to a plastic deformation that induces work-hardening. As long as the fracture strain (or stress) remains greater than the applied strain (stress), the amount of elastic strain increases and the amount of plastic strain decreases during successive cycles. As work-hardening continues with each cycle, there is a competition between the decreasing fracture strain and the rate of increase of the elastic range and the applied strain range. Within the elastic range, failure does not occur. The Coffin-Manson fatigue model attempts to relate influence of such process to the fatigue fracture by taking a form that the total number of cycles to failure, N_f, is dependent on the plastic strain amplitude, $\Delta \varepsilon_f$, the fatigue ductility coefficient, ε'_f, and the fatigue ductility exponent, C:

$$\frac{\Delta \varepsilon_p}{2} = \varepsilon'_f (2N_f)^C \qquad (1)$$

$$\log \frac{\Delta \varepsilon_p}{2} = \log \varepsilon'_f + C \log(2N_f) \qquad (2)$$

$$|C| = 1/(1 + 5\alpha)$$

(3)

Equation 2 is the log form of Equation 1.

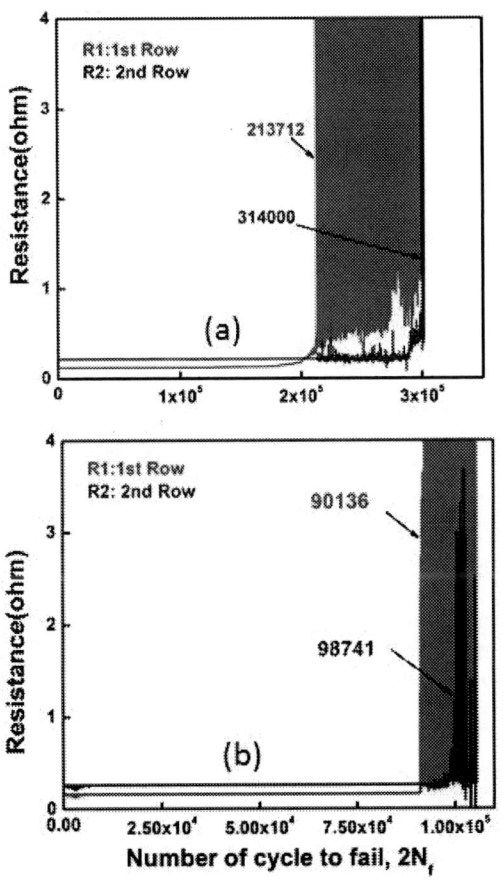

Figure 3. 1st and 2nd row solder resistance change with the number of cycle to fail under +/-2.4% strain (a) 150°C/500h thermal aged SAC105, 1Hz, 25°C, and (b) as received SnPb, 1Hz, 25°C.

In these equations, the fatigue ductility coefficient is approximately equal to the true fracture ductility because it is defined to be the strain that results in fracture of given material at the first ½ cycle. In other words, the ductility coefficient is conceptually similar to fracture strain under shear or tensile test. Also, Morrow suggests that the cyclic ductility exponent can be related to the strain-hardening rate as shown in Equation 3, where α is the cyclic strain hardening exponent. [19-20]

Effect of fatigue strain on fatigue lifetime

In order to understand the fatigue mechanism and the factors contributing to fatigue life, samples were tested under various strain conditions at different temperatures, thermal history, and frequencies for various solder alloys. Figure 4 shows the fatigue life of SAC105, SAC305, and Pb-Sn alloys as a function of displacement (plastic strain range) at the specified temperature and frequency. Note that the data fits extremely well to the Coffin-Mason model described in Equations (1)-(2). The data also indicates that the fatigue failure of solder alloys tested in our study is governed by a single mechanism. This consistency in fatigue and failure mechanism is found to be related to the fact that the failure location within the joint does not vary with the testing condition. The crack growth path is always fixed at the solder matrix near the UBM interface. Therefore, the fatigue constants, that is, the fatigue ductility coefficient and ductility exponent, in the Coffin-Manson model can be determined from the experimental data shown in Figure 4, and they can be safely connected to the metallurgical condition of the alloys.

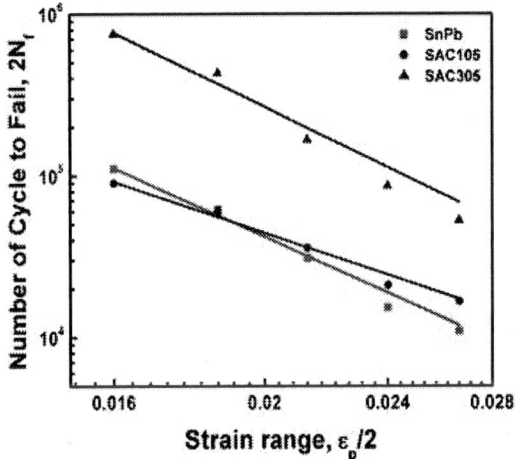

Fig.4. Applied strain range vs. number of cycles to failure for SnPb, SAC105 and SAC305 samples tested at 25°C, 1 Hz.

The results shown in Figure 4 indicate that SAC305 shows far better fatigue resistance than SAC105 and SnPb alloys, at least within the strain ranges used in our testing, which is in good agreement with findings made in other studies. [1-3, 5] This result is generally attributed to the higher yield strength of SAC305 (thus less plastic deformation) due to high density Cu-Sn IMC (intermetallic compound) phases in the solder matrix. A more accurate interpretation may be realized by inspecting the fatigue constants of each solder alloy. From the data shown in Figure 4, the ductility coefficient and ductility exponent can be determined; the x-axis intercept, that is, the strain at $N_f=1$, yields the ductility coefficient and the slope of N_f vs. strain is the inverse of the ductility coefficient. Table 1 lists the fatigue constants of the alloys determined in

978-1-4799-2408-0/14 $31.00 © 2014 IEEE

that manner. Note that SAC305 has the highest strain hardening exponent (α) and the lowest ductility coefficient. What this indicates is that SAC305 is essentially more brittle than SAC105, meaning that its fatigue resistance is not necessarily better at all strain ranges. In fact, the fatigue resistance can be worse than SAC105 at high strain levels. On the other hand, SAC305 provides better fatigue resistance at low strain ranges because the higher work-hardening rate reduces the degree of plastic deformation. This result suggests that high fatigue resistance of SAC305 may not be a result of IMC induced strengthening of solder matrix but may be a result of effective work hardening of solder due to the presence of IMCs.

Table 1. Coffin Mansion constants of SnPb, SAC105 and SAC305 measured at 25°C

Fatigue Parameters	SnPb	SAC105	SAC305
ε'_f	0.25	0.56	0.22
C	-0.23	-0.31	-0.18
α	0.67	0.45	0.91

Also, the results shown in Figure 4 and Table 1 may provide explain why SAC105 shows superior shock resistance yet inferior vibration resistance than SAC305. If shock and vibration induced fracture is considered as the fatigue at high strain extreme (shock) and at low strain/high frequency extreme (vibration), the results suggest that the vibration reliability of SAC105 is inferior while it resistance better against the shock induced fracture as SAC105 is the alloy with higher ductility coefficient and lower work-hardening exponent).

Effect of thermal aging on fatigue lifetime

Fatigue resistance is expected to be affected greatly by the microstructural details of the solder alloy. One such factor is the size and distribution of IMC phases in the solder matrix because they are likely to change the ductility and work-hardening rate of the alloys. This expectation can be tested by comparing fatigue resistance of the same assembly with variation in aging time. Aging is known to induce coarsening of existing IMC phases in the solder matrix as well as growth of interfacial IMCs, and therefore the variation in fatigue resistance with aging time is expected to reflect the influence from such change.

Figure 5 shows the fatigue life of SAC305 alloys tested at room temperature with variation in thermal aging time. Note that the fatigue life decreases with increasing aging time, which is consistent with observations made in many other studies. Figure 5 clearly shows the aging does not affect fatigue resistance of SAC305 alloys at low strain range. The aging effect becomes significant in the high strain condition. This means that a SAC305 based electronic assembly does not experience degradation of fatigue resistance with aging in low-strain fatigue conditions such as vibration. The results also suggest that the very reason that fatigue resistance of SAC alloys decreases with aging is due to the loss of ductility. Notice that the cyclic work hardening exponent (proportional to the slope) increases, while the fatigue ductility coefficient (the intercept to the strain range axis) decreases with aging.

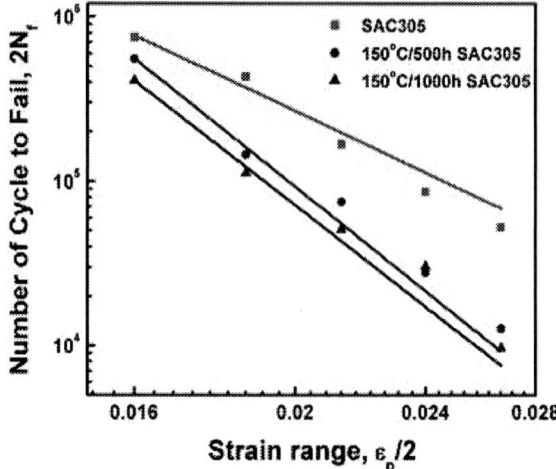

Figure 5. Fatigue lifetime versus strain range at 1Hz, 25°C.

The exact mechanism by which the loss of ductility occurs with aging is presently unknown. One possible source of the ductility loss may be the growth of interfacial IMCs as is suggested in various studies. [21-23] These studies suggest that the growth of IMC layer makes the solder joint more fatigue prone because the brittle IMC phase provides an easy path for crack growth. Also suggested is the possibility that the interface between solder and IMC is subjected to excessive level of stress concentration due to the presence of stress singularity. However, these suggestions cannot explain the result shown in Figure 5 because the fatigue crack induced in our study is found to proceed in an area distant from the interface. A more plausible explanation may be the IMC free zone seen in the solder matrix near the IMC layer. The growth of interfacial IMC drains nearby IMCs in the solder matrix, resulting in a creation of IMC free-zone. Applied fatigue strain and the resulting plastic deformation is likely concentrate on this narrow IMC free-zone as it is soft and is situated between the interfacial IMC and the IMC filled solder matrix. This condition may also affect the cyclic work hardening in a similar way.

Effect of temperature on fatigue lifetime

As discussed in previous sections, the isothermal shear fatigue test produces results that are in good agreement with general fatigue behavior of metallic materials and provides useful guidance in understanding fatigue properties of solder joints in an electronic assembly. Nonetheless, the isothermal shear fatigue testing is not without surprise, and it is found in the temperature dependence of the fatigue life. Since dislocation glide as well as the dynamic recovery becomes more active with increase in temperature, an increase in ductility and decrease in work-hardening rate is expected at higher temperatures. The results do not match exactly with this expectation.

Figure 6 shows fatigue life of SAC105 and PbSn alloys as a function of strain range measured at 4 different temperatures. The data shows that the fatigue life of both alloys decreases with increasing temperature at all strain ranges, which is not unexpected. What is unexpected is the decrease in the ductility coefficient (intercept of strain range axis) and increase in the work hardening exponent (slope) with temperature. This behavior is not limited to these two alloys but consistently seen in all alloys including the aged.

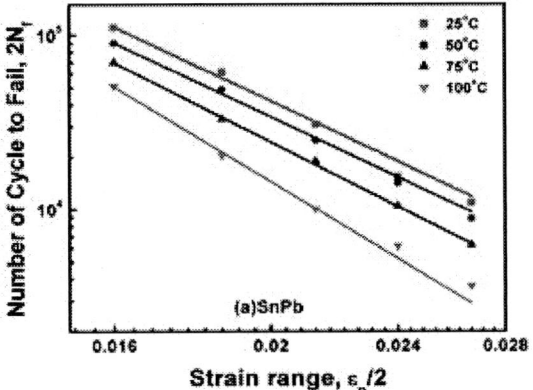

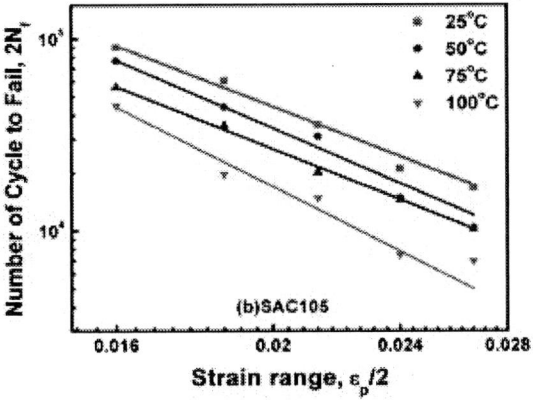

Figure 6. Fatigue lifetime versus strain range at 1Hz for different temperatures (a) SnPb and (b) SAC105.

The fact that unconventional change in fatigue constants with temperature is seen all alloys used in our investigation suggests that its origin is probably rooted to mechanical constraints unique to our testing method or test assembly. Although the exact source needs further investigation, we believe that variation in thermal strain at the different test temperatures is responsible for the unexpected temperature dependence of fatigue constants. The solder joint in the assembly experiences additional strain resulting from the mismatch in thermal expansion between the substrate and silicon chip. Since this thermal strain is constant at a given temperature and becomes larger at higher temperature (after residual strain is removed by annealing process), the actual fatigue strain that the joint is subjected to becomes more asymmetric at higher temperature. In one direction, the thermal strain adds to the applied shear strain, and in the other direction it subtracts. As fatigue life is dictated more by the largest strain within a cycle, the effect of thermal strain on the fatigue life is equivalent to conducting shear testing at higher strain range. It is then possible for fatigue life to appear shorter (low ductility) and also less sensitive to work

hardening. Therefore, more proper Coffin-Mansion model for the electronic assembly may be:

$$\frac{\varepsilon_t + \Delta \varepsilon_p}{2} = \varepsilon_f^{'} (2N_f)^C \qquad (4)$$

where ε_t represents the thermal strain.

When thermal strain possible in our PBGA assembly is calculated using Finite Element Analysis with known CTE (coefficient of thermal expansion) of materials in the assembly, it is estimated to be approximately 2um, 3.7um and 5.4um at 50, 75 and 100°C, respectively. Subsequent correction of data shown in Figure 6 using Equation 4 produces results of increasing ductility coefficient and decreasing work-hardening rate with increasing temperature. This result appears to support our interpretation and suggests that inclusion of thermal strain effect is necessary in the analysis of fatigue results even if the testing is conducted at an isothermal condition.

Conclusions

Our investigation attempts to find a way to generalize fatigue behavior of solder alloys in electronic assemblies through isotheral shear fatigue testing of a model PBGA structure and analyzing the results using the Coffin-Manson fatigue theory. As highlighted in this paper, our study produces results indicating that the fatigue property of solder alloys can be described most effectively using the frame of Coffin-Manson model even if the alloys are in the assembly structure. A more important finding is the fact that the fatigue constants gained from the model offer great assistance in understanding variation in fatigue reliability with alloy compositions and microstructural changes. For instance, the fatigue constants suggest that the fatigue resistance of SAC305 alloy is not necessarily superior to SAC105 or Pb-Sn alloys in all strain ranges. It is the low strain range where SAC305 shows superior fatigue resistance. This is against the general belief that SAC305 offers better fatigue reliability regardless of fatigue condition. What it suggests is that SAC305 would suffer from poor shock reliability compared to SAC105. In addition, enhancement of fatigue resistance of SAC305 may be achieved by dispersing IMCs more finely in the solder matrix to increase the work hardening rate while maintaining or increasing ductility of the solder.

One surprising but important finding is the involvement of thermal strain on the fatigue behavior even under isothermal condition. The thermal strain effect exists because the solder joint is in the assembly where CTE mismatch produces sizable strain. This effect is often ignored, but our results indicate that consideration on its effect is necessary in order to find fundamental fatigue properties of solder alloys. In reliability evaluation, its influence may be beneficial because it creates conditions closer to the end use. As long as its involvement is included in the data analysis, the fatigue constants gained from the isothermal fatigue can be valid and useful.

Acknowledgement

One of the authors, Huili Xu, is financially supported by Texas Instrument Fellowship for Woman PhD Student.

References

1. Shi.X.Q., et al., "low cycle fatigue analysis of temperature and frequency effects in eutectic solder alloy", International Journal of Fatigue, Vol. 22 (2000), pp. 217-228.

2. John. H.L., et al., "Low cycle fatigue study of lead free 99.3Sn-0.7Cu solder alloy", International Journal of Fatigue, Vol. 26 (2004), pp. 865-872.

3. Kanchanomai, C., et al., "Effect of temperature on isothermal low cycle fatigue properties of Sn-Ag eutectic solder", Materials Science and Engineering A, Vol. 381(2004), pp. 113-120.1.

4. Kim C.-U. , Bang W.-H. , Xu H. , Lee T.-K., "Characterization of solder joint reliability using cyclic mechanical fatigue testing ", JOM, Volume 65, Issue 10, October 2013, Pages 1362-1373.

5. H. Xie , Q. Tan , Y.C. Lee, "Electronic Packaging: Thermal, Mechanical, and Environmental Durability", Encyclopedia of Materials: Science and Technology (Second Edition), 2001, Pages 2715–2725.

6. Lau, J.H. et al., "Design for lead-free solder joint reliability of high-density packages", Solde. Surf. Mount Technol., Vol. 16 (2004), pp. 12-26.

7. Lau, J.H., "Solder joint reliability of flip chip and plastic ball grid array assemblies under thermal, mechanical, and vibrational conditions", *IEEE Trans. Comp., Pack. Manufact. Tech., Part B,* Vol. 19 (1996), pp. 728-735.

8. Leicht, L. et al., "Mechanical cycling fatigue of PBGA package interconnects", *Microelectronics Reliability,* Vol. 40 (2000), pp. 1129-1133.

9. Bang, W.H., et al., "Rate Dependence of Bending Fatigue Failure Characteristics of Lead-Free Solder Joint", *Electronic Components and Technology Conference,* 2009, pp.2070-2074

10. Xu, H.L., et al., "Fracture mechanics of Lead-free Solder Joints under Cyclic Shear Load", ECTC., 2010, pp. 484-489.13. Chaosuan, K., et al., "Low-cycle fatigue behavior and mechanisms of a lead-free solder 96.5Sn/3.5Ag", *Journal of Electronic materials,* Vol. 31 (2002), pp. 142-151.

11. Zhu, H.L. , Weng, L. , Can, Y. , Li, Y., "Fatigue life prediction of electronic chips based on singular edge field", Applied Mechanics and Materials, Volume 390, 2013, Pages 601-605.

12. Dasgupta, A., "Failure mechanism models for cyclic fatigue", IEEE Transactions on Reliability, 42(4)(1993), pp.548-555.

13. Yu, D., Yafawi, A. A., Park, S. and Chung, S., "Finite element based fatigue life prediction for electronic components under random vibration loading," in Proc. 60th Electron. Components Technol. Conf., 2010, pp.188–193.

14. Knecht, S. and Fox, L., "Constitutive relation and creep-fatigue life model for eutectic tin-lead solder", IEEE Transactions Components, Hybrids and manufacturing technology, 13(2)(1990), pp.424-433.

15. Syed, A., "Accumulated creep strain and energy density based thermal fatigue life prediction models for SnAgCu solder joints", 54th Electronic Components and technology Conference, 2004, pp.737-746.

16. Lee, W.W. et al., "Solder joint fatigue models: review and applicability to chip scale packages", *Microelectronics reliability,* Vol. 40 (2000), pp. 231-244.

17. Manson, S.S., "Fatigue: a complex subject — some simple approximations", *Exp Mech.,* Vol. 5 (1965), pp. 193-226.

18. Coffin. L.F., "A study of the effects of cyclic thermal stress on a ductile material", ASME Trans., Vol. 76 (1954), pp.931-950.

19. Morrow, J.D., "Internal Friction, Damping and Cyclic Plasticity", *ASTM-STP,* Vol. 378 (1965), pp. 45-52.

20. Solomon H.D., "Fatigue of 60/40 solder", *IEEE Trans CHMT,* Vol. 9 (1986), pp. 423-433.

21. Fix, A.R., et al., "Microstructural changes of lead free solder joints during long-term aging, thermal cycling and vibration fatigue", Soldering & Surface Mount Technology, 20(2008), pp. 13-21.

22. Peng, W., Monlevade, E., Marques, M. E., "Effect of thermal aging on the interfacial structure of SnAgCu solder joints on Cu", Microelectronics Reliability, 47(12)(2007), pp.2161–2168.

23. John H. L., Pang, T. H., Low, B. S., Xiong, L., "Thermal cycling aging effects on Sn–Ag–Cu solder joint microstructure, IMC and strength", Thin Solid Films, 462–463(2004), pp.370–375.

24. Kanda, Y., et al., "Influence of cyclic strain hardening exponent on fatigue ductility exponent for a Sn-Ag-Cu micro-solder joint", *J. Electron. Mater.,* Vol. 41(2012), pp. 580-587.

25. Kanchanomai, C. and Mutoh, Y., "Effect of temperature on isothermal low cycle fatigue properties of Sn-Ag eutectic solder", *Mater. Sci. Eng. A,* 381 (2004), pp. 113-120.

MEMS-Based Implantable Heart Monitoring System with Integrated Pacing Function

Fjodors Tjulkins*[1], Anh Tuan Thai Nguyen[1], Erik Andreassen[1,3], Nils Hoivik[1], Knut Aasmundtveit[1],
Lars Hoff[1], Ole Johannes Grymyr[2], Per Steinar Halvorsen[2], Kristin Imenes[1]

[1]HBV – Buskerud and Vestfold University College, IMST – Dept. of Micro and Nano Systems Technology,
Raveien 205, 3184 Borre, Norway
[2]Oslo University Hospital Intervention Centre
[3]SINTEF Materials and Chemistry, Oslo, Norway
E-mai: ft@hbv.no

Abstract

Miniaturized accelerometer-based implantable heart monitoring systems may increase safety during cardiac surgery. Such systems can provide sustained monitoring compared to imaging techniques. During ischemia, changes in acceleration occur instantly and before changes can be observed by electrocardiography. This may allow earlier detection of adverse events like graft occlusion. Recent designs have been made to meet requirements for postoperative use by easy and safe removal. However, the implantation procedure has been too complicated. The system presented in this paper is a reworked package in which a 3-axis accelerometer is placed inside a stainless steel capsule fabricated by additive manufacturing (3D printing). A needle attached to the package allows for easy implantation. In addition, the metallic capsule makes it possible to use the device as a pacing lead. The system demonstrates easy implantation. Furthermore, in conjunction with a second electrode, pacing and electrical potential sensing functionality is demonstrated.

Introduction

Cardiac ischemia is a condition where the heart muscle does not receive enough blood to function. It can occur due to graft occlusion after coronary artery bypass grafting, and it needs to be detected early in order to prevent myocardial infarction. Up to 4% of the patients have been reported to suffer graft occlusions immediately after the patient's chest was closed [1]. Hemodynamics and electrocardiography (ECG) has limited value for predicting myocardial infarction [2]. Accelerometer-based heart monitoring systems have been developed, and these have subsequently demonstrated their capability in early detection of regional ischemia [3] [4]. These systems performed well in peri-operative monitoring (monitoring during the surgery), but were not usable for post operative monitoring. This limitation was imposed by the size of the system – it was not possible to remove the device from the patient without additional surgery. Advances in the field of MEMS (MicroElectroMechanical Systems) accelerometers have lead to highly compact sensors becoming widely available. A system that utilized myocardial fixation and could be extracted from the patient without additional surgery, enabling post operative use, has been demonstrated [5]. The system performed well in preliminary trials, but clinicians' feedback identified certain factors that would prevent the system from wide clinical deployment. The chief drawback was the complicated implantation procedure. The second drawback was a consequence of the first: the channel made in the tissue was at times too large, so that the fixation would be compromised. In this paper a repackaged system is presented. This new system addresses the issue of complicated implantation and enables pacing and electrical potential sensing functionality when an additional electrode is used.

The approach taken to simplify the implantation and to make it more robust was to emulate the design and implantation of temporary myocardial pacing leads (heartwires) to a greater extent. The previous design merely emulated the placement whereas the current one also emulates the actual implantation procedure: a curved needle is attached to the capsule by a thread, and with this needle a channel is made in the cardiac tissue. Then the capsule is pulled by the thread into the channel. Compared to using a catheter, a needle gives less implant channel dilatation. This should improve the fixation of the capsule, and be less traumatic.

In addition to sensing the motion, it is also desirable to be able to perform pacing with the device. This combination of motion monitoring and pacing in one device will reduce the number of devices in the cardiac tissue compared to having separate devices for each function.

The metallic capsules described this paper were fabricated of 316L stainless steel, a well known biocompatible material. The capsule is a cylinder with a "leveled/flattened" top and bottom. The motivation for designing the capsule in this particular way was twofold. First, this was the smallest capsule that could house the sensor – the less tissue that is displaced, the less invasive is the system. Second, a clearly identifiable top and bottom are desired for easier sensor axis orientation.

The needle, for making the channel in the tissue, was taken from a suture. The sensor in this study has been described earlier [5], but the electrical connection was revised. The flat cable-substrate (i.e. the one-part cable and substrate for the sensor and electronics) was redesigned with two layers. This made it possible to narrow down the cable part to 0.7mm. To provide extra strength and electrical insulation, the part of the cable exiting the capsule was overmolded with a silicone compound. One of the 6 leads of the cable-substrate was connected to the inside wall of the capsule, allowing the package to act as a unipolar pacing lead. A schematic representation of the assembly is shown in Figure 1.

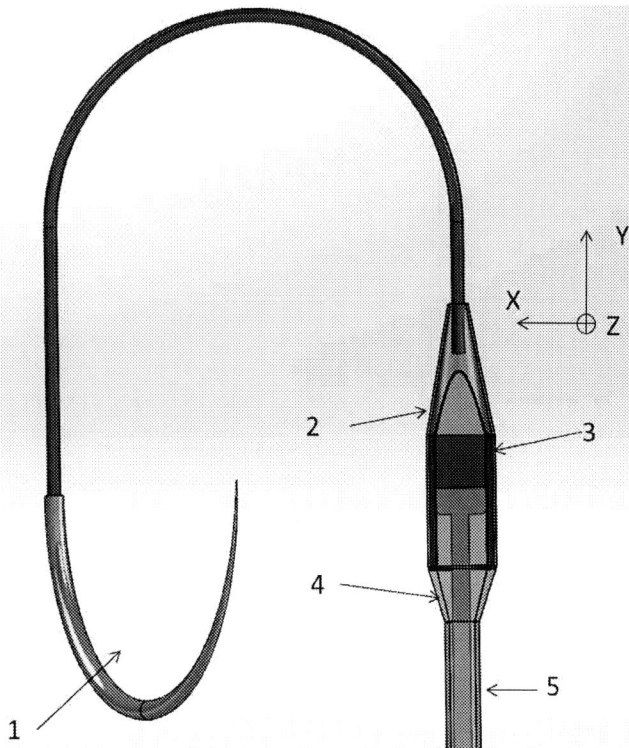

Figure 1 Schematic drawing of the assembled device. 1 – needle with a thread taken from a suture. 2 – stainless steel capsule containing the sensor. 3 – 3-axis accelerometer bonded to the flat cable-substrate. 4 – Joint between cable-substrate and capsule. 5 – Silicone-overmolded part of the cable-substrate.

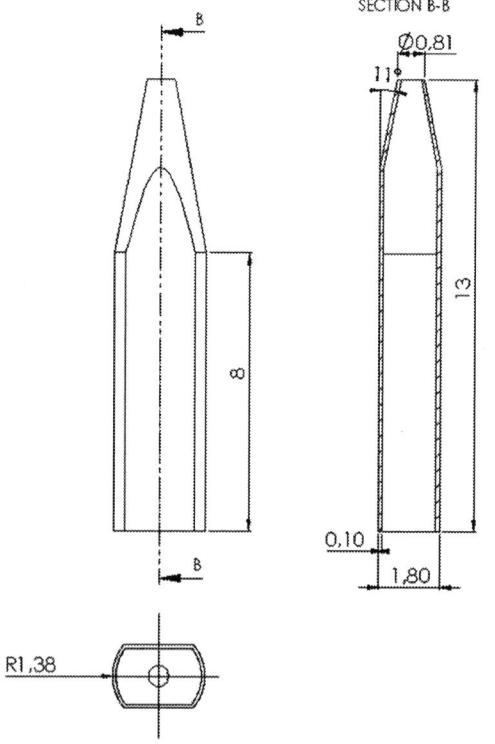

Figure 2 A drawing of the capsule. All dimensions in mm. After printing, 3mm (the printer base plate) were removed from the lower part of the capsule.

Methods and materials

The final device was assembled from several components, of which some needed additional post-processing. The individual components are described below:

Capsule design

The metal capsules were made by additive manufacturing (3D printing) using a Concept Laser M2 Cusing. The parts were printed on a baseplate and had to be cut from it after they were printed. Because of this, the capsule length for the printing process had to be increased by 3mm (to 13 mm). The capsule is shown in figure 2.

Capsule processing

The capsules as printed had a rather rough surface. This made it necessary to polish the capsules as a post processing step. The surface roughness was characterized using a DEKTAK 150 surface profiler with a 12.5 µm radius stylus.

Polishing was conducted by a two-step procedure. The first step was to polish the capsules on a Struers Knuth Rotor grinding station. The capsules were first polished with a P1500 grit SiC paper, then with a P2100 grit paper.

The abrasive removal of the grainy surface layer left "pits" in the capsule surface. These surface defects sensitize the metallic part to pitting corrosion [6]. The 316L formulation of stainless steel is vulnerable to pitting corrosion [7]. The risk of such corrosion was reduced by performing electrochemical polishing (ECP) [8-10]. In addition to making the surface smoother, electrochemical polishing also preferentially removes iron atoms compared to chromium, leaving a chromium-enriched and more corrosion resistant layer on the surface [9]. The ECP procedure was similar to the procedure presented in [11]: a 90 second long etching in a solution of H_3PO_4 (42 wt%), glycerol (47%) and H_2O with 10V current and 1-1.2A anodic current.

Sensor

The sensor and the decoupling capacitors employed for this device are the same as in [5]. The CMA3000A (Murata Electronics Oy, Japan) 3-axis accelerometer is no longer in production. However, at the time of writing, it is still the smallest commercial sensor ($2\times2\times0,95$ mm^3), alongside the BMA250 (Bosch, Germany). An even smaller 3-axis accelerometer has been announced (BMA355, Bosch, Germany), and subsequent generations of accelerometer based heart monitoring will take advantage of the trend towards miniaturization.

Electrical connection

To provide power to the sensor and read out the signals from it, an integrated cable-substrate was developed. The substrate layout follows that described in [5]. A contact pad was added on the bottom to provide a connection between cable and capsule that would be used for pacing, see figure 3. In addition, the cable-substrate in [5] was redesigned to be narrower in the region behind the substrate, making it more flexible.

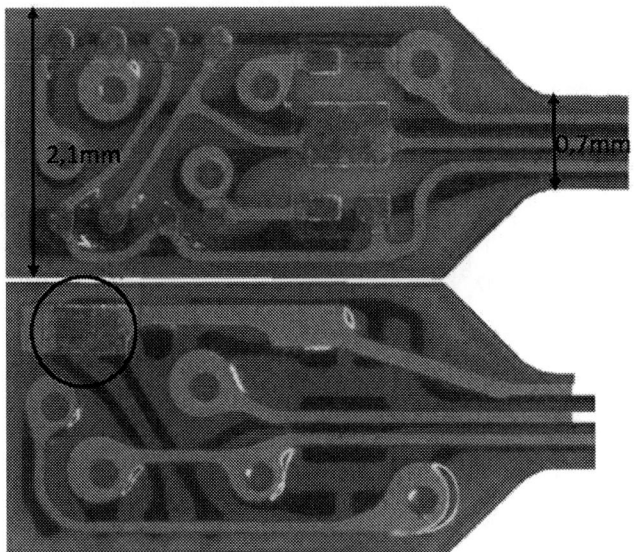

Figure 3 Top and bottom halves of the cable-substrate. A contact pad used to connect to the capsule circled in black.

To supply power and read-out signals from the sensor, 5 connectors are necessary. The 6th conductor is connected to the metallic capsule and used solely for pacing or electrical sensing when an external pulse generator is connected to it. The connector end of the cable-substrate is shown in figure 4.

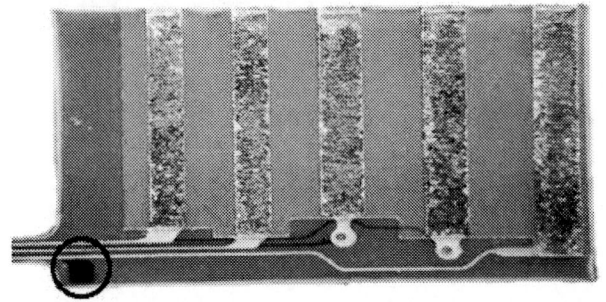

Figure 4 The connector part of the cable-substrate with 5 contact pads dedicated for power and I/O. The contact pad for pacing is encircled in black.

To strengthen the cable-substrate in the narrow region, 0.7mm wide part, it was overmolded with a silicone compound. To achieve this, a mold, consisting of a long channel (1.2mm in diameter), was fabricated. Clamps were used on both sides of the channel to center the cable substrate. A two component liquid silicone MED-4211 (NuSil, USA) was injected into the channel. The silicone was cured at room temperature for 48 hours, and after that the cable was removed from the mold and placed into an oven for post-curing for 2 hours at 80°C. The MED-4211 silicone meets the requirements of the ISO 10993-5 standard for cytotoxicity.

Needle and thread

The needle and thread for the assembly were taken from a non-absorbable, monofilament polypropylene suture; Surgipro II V-20 (Covidien, USA). The needle (V-20) is a half-circle needle ending in a taperpoint, with needle body length 26mm. The length of the thread was 100mm.

Potting compound

In this design, the sensor is secured inside the encapsulation by an adhesive that takes up the entire volume of the metallic capsule. The capsule is filled with the adhesive first and the sensor is placed inside. The displaced adhesive is removed.

A potting compound suitable for this device must be biocompatible; the adhesive bond must be sufficiently strong to keep the sensor inside the package during handling, implantation and extraction. A cyanoacrylate adhesive, Cyanolit 203TX (Panacol, Germany), was used. According to the manufacturer, this compound has met the requirements for USP class VI certification. The adhesive was cured according to the recommendations from the manufacturer, see table 1.

Table 1. Curing conditions for Cyanolite 203TX adhesive, per specification.

Appearance when cured	Curing	Post curing
Transparent	30 seconds at room temperature	24 hours at room temperature

Leakage current

An implant such as the system described in this paper would need to meet the requirements set by the International Electrotechnical Commission for leakage current limits. The IEC-60601-1 standard ascribes that under normal conditions the leakage current should not exceed 0.01 mA, and for a single-fault incidence the limit is 0.05 mA.

The sensor package was placed in a beaker with 0.9% saline solution and the beaker was put in a thermal chamber (Heraeus T6200, Thermo Scientific Germany) with temperature set to 37°C. The leakage current was measured with an electrometer (6430 Sub-femtoampere SourceMeter Instrument, Keithley Instruments Inc., Cleveland, OH, USA) providing a 3.3 V DC voltage.

Mechanical testing

It is important to know what forces would act on the device during implantation. To evaluate these forces a series of 15 implantations and pull-outs were conducted on a tissue phantom. A round capsule with an equally tapered tip was used in the same manner for reference. A soft tissue phantom (Blue Phantom, USA) was used to approximate the cardiac tissue. It is recognized that the phantom is different from actual cardiac tissue – no heart contractions, no fine blood vessels etc. The phantom is assumed to give higher forces than living tissue, partly because it is dry and therefore has higher friction. The forces were recorded by a force gauge (Model M4-2, Mark-10, USA). Capsule cross-sections and and the test method are illustrated in figure 5.

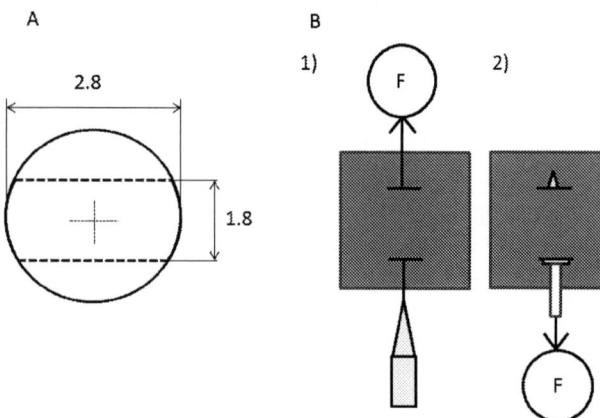

Figure 5 A: Profiles of two capsules used for comparison, B: pull-in and pull-out force measurement illustration.

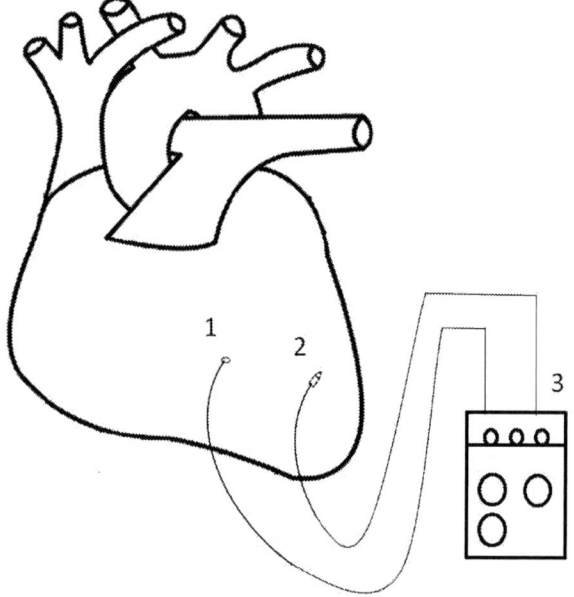

Figure 6 Schematic representation of the pacing set-up. 1 – heartwire, 2 – accelerometer based heart monitoring device, 3 – pacing pulse generator.

Sensor test in an animal model

The testing was approved by the Norwegian Animal Research Authority (NARA) (FOTS id4101) All researchers directly involved in the animal experiment were certified with FELASA category C (European Convention for the protection of Vertebrate Animals (ETS No. 123).

The pig, a 46.5 kg male, fasted overnight, with free access to water. Anesthesia was induced by an intramuscular injection of ketamine 20 mg/kg, azaperone 3 mg/kg and atropine 0.02 mg/kg, and was maintained by intravenous pentobarbital 2-3 mg/kg and morphine 0.5-1.0 mg/kg until tracheotomy was completed. General anesthesia was continued with inhaled isoflurane (1.0-1.5%) and morphine infusion (0.15-0.2 mg/kg). Mechanical ventilation was performed by a Leonplus anesthesia machine (Heinen + Löwenstein, Germany). Inspired oxygen fraction was 0.35 and tidal volume and ventilation rate were adjusted to keep arterial PCO_2 ~5.5 kPa.

A median sternotomy was performed and the pericardium opened. One 3-axis accelerometer based heart monitoring device was inserted apically on the left ventricle.

The chest and the pericardium were left open and the pig was placed in dorsal supine position.

The system described in the paper provides unipolar pacing – an additional pacing lead was used to act as a second electrode. A handheld pacing pulse generator, Medtronic 5375 (Medtronic, USA) was connected to a heartwire and the accelerometer based system and used to establish a pacing threshold and a sensing threshold. The pacing set-up is shown in Figure 6.

Results and discussion

Assembly and Mechanical testing

The use of a room temperature curing adhesive to secure the sensor inside the capsule simplified the final assembly process.

Capsules that underwent the polishing procedure appeared shiny and with smooth surface. The "pits" left after the abrasive polishing were present, but barely visible to the naked eye. Measured depth of remaining pits was 1-4μm, The roughness parameter R_a of the rest of the capsule was 0.5μm. A close-up of an assembled device is shown in Figure 7.

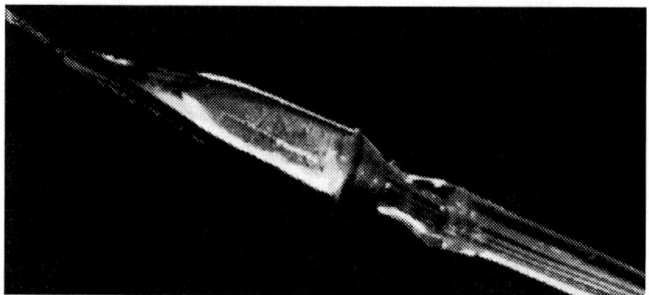

Figure 7. A close-up of an assembled sensor, showing the capsule, the joint between capsule and cable and the cable (lower right).

Figures 8 and 9 show the comparison between peak forces recorded for the device and the reference cylinder. The pulling force experiment demonstrated that of the two capsule designs tested, the smaller flattened requires less force to insert and pull-out.

The flattened capsule demonstrated larger dispersion in pull-in forces; however the average is well below that of a reference cylinder. The same is true for the pull-out forces.

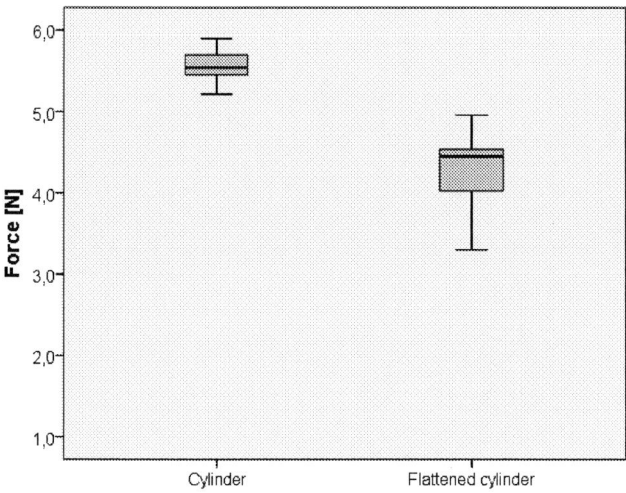

Figure 8 Measured force require to pull the capsule into the channel for the reference cylinder and the device with flattened capsule.

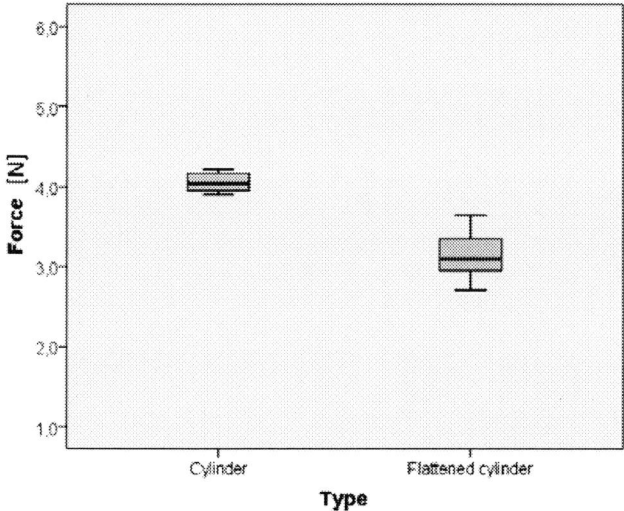

Figure 9 Measured force required to pull the capsule out of the channel for the reference cylinder and the device with flattened capsule.

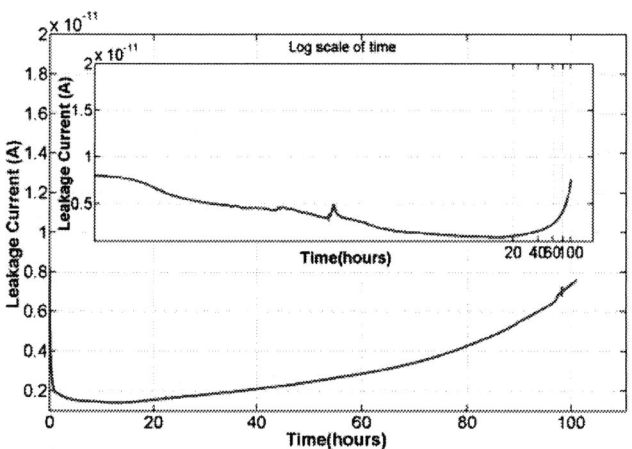

Figure 10 Leakage current measurements result over 100 hours in normal and logarithmic time scales.

Leakage current

After a 100 hours long leakage current measurement in 0.9% saline solution, the current remained orders of magnitude below the limit set by the IEC-60601 standard (0.01 mA).

Animal trial results

Acceleration signals recorded from a beating pig heart are shown in figure 11. These results are similar to what has been observed in previous studies using other sensors [3, 5]. The implantation was accomplished in one step and the sensor remained inside the channel for the duration of the experiment. The device was able to successfully pace the heart and sense the polarization of the left ventricle.

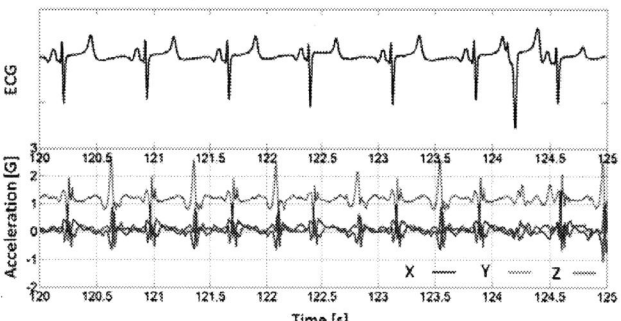

Figure 11 Acceleration recording. Natural heartbeat frequency.

Once the successful recording of acceleration signals was established, a temporary myocardial pacing lead was implanted next to the sensor. A pacing rhythm of 100 beats per minute was set. Successful pacing was achieved at a threshold of 10mA, and successful sensing at a threshold of 20mV. Note, that the sensing ability of the handheld pacing generator is limited to an indicator light lighting up with the frequency of the sensed signal. During the pacing part of the experiment, the indicator light frequency was observed to be consistent with the pacing rhythm settings and the rhythm recorded by a limb lead ECG and acceleration signals. Based on this, we conclude that the sensing and pacing using this sensor and electrode combination was successful.

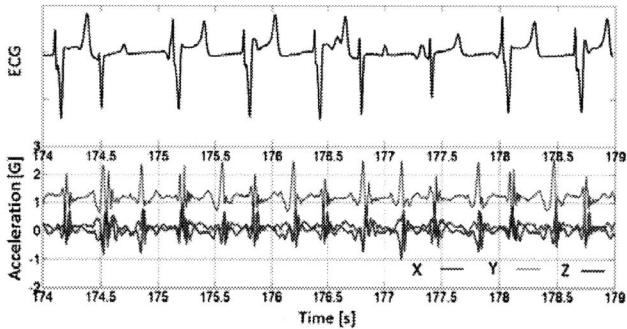

Figure 12 Acceleration recordings while pacing. Heart beat frequency 100 beats per minute.

Conclusions

This system, with an accelerometer inside a metallic capsule made by additive manufacturing, has demonstrated sensing via acceleration, and pacing when connected to a pulse generator. While the pacing was successful, more research is needed. The area of the capsule is much greater than that of a temporary pacing lead. This will lead to the current density being lower at the same supplied current. No retraction, bleeding or arrhythmias were observed upon sensor implantation. This indicates that the organ was not compromised by the implant.

The pull tests on the phantom demonstrated lower forces. As a first approximation, the smaller, flattened capsule can be seen as less damaging to the tissue.

The leakage current was roughly 80 pico ampere, far below the requirements set forth by the standard. A slight tendency for the leakage current to increase was observed.

Further research is necessary to develop and optimize the polishing steps after the additive manufacturing, and to conclude whether or not the material is sensitized to corrosion. In the present short-term animal experiments and leakage current test, where the device was subjected to corrosive environments, no corrosion was observed.

Acknowledgments

The Oslofjord Regional Research Fund in Norway is gratefully acknowledged for support through the regional institution project # 208933 "New Packaging Methods for Smart Implantable Microsensors". IMST lab engineers and employees of the Intervention Centre for their help in assembling and testing the device.

References

1. Hol, P.K., et al., *Intraoperative angiography leads to graft revision in coronary artery bypass surgery.* Annals of Thoracic Surgery, 2004. **78**(2): p. 502-505.
2. Jain, U., et al., *Electrocardiographic and hemodynamic changes and their association with myocardial infarction during coronary artery bypass surgery. A multicenter study. Multicenter Study of Perioperative Ischemia (McSPI) Research Group.* Anesthesiology, 1997. **86**(3): p. 576-91.
3. Elle, O.J., et al., *Early recognition of regional cardiac ischemia using a 3-axis accelerometer sensor.* Physiol Meas, 2005. **26**(4): p. 429-40.
4. Imenes, K., et al., *Assembly and packaging of a three-axis micro accelerometer used for detection of heart infarction.* Biomedical Microdevices, 2007. **9**(6): p. 951-957.
5. Tjulkins F., et al. *3-axis MEMS Accelerometer-based Implantable Heart Monitoring System with Novel Fixation Method.* in *Electronic Components & Technology Conference.* 2013. Las Vegas Cosmopolitan: IEEE.
6. Black, J., *Biological performance of materials : fundamentals of biocompatibility.* 2006, CRC Taylor & Francis: Boca Raton. p. pp. 60-61.
7. Sivakumar, M., U.K. Mudali, and S. Rajeswari, *Investigation of Failures in Stainless-Steel Orthopedic Implant Devices - Pit-Induced Fatigue Cracks.* Journal of Materials Science Letters, 1995. **14**(2): p. 148-151.
8. Weldon, L.M., et al., *The influence of passivation and electropolishing on the performance of medical grade stainless steels in static and fatigue loading.* Journal of Materials Science-Materials in Medicine, 2005. **16**(2): p. 107-117.
9. Hryniewicz, T., R. Rokicki, and K. Rokosz, *Corrosion characteristics of medical-grade AISI Type 316L stainless steel surface after electropolishing in a magnetic field.* Corrosion, 2008. **64**(8): p. 660-665.
10. Nazneen, F., et al., *Electropolishing of medical-grade stainless steel in preparation for surface nano-texturing.* Journal of Solid State Electrochemistry, 2012. **16**(4): p. 1389-1397.
11. Zhao, H., et al., *Electrochemical polishing of 316L stainless steel slotted tube coronary stents.* Journal of Materials Science-Materials in Medicine, 2002. **13**(10): p. 911-916.

Archipelago Platform for Skin-mounted Wearable and Stretchable Electronics

Yung-Yu Hsu, Cole Papakyrikos, Milan Raj, Mitul Dalal, Pinghung Wei, Xianyan Wang, Gil Huppert,
Briana Morey, and Roozbeh Ghaffari

MC10 Inc.

9 Camp St., 2nd Floor, Cambridge, MA 02140, USA

yhsu@mc10inc.com, (617) 234-4448

Abstract

In this investigation, the "archipelago" design is presented as a platform for skin-mounted wearable and stretchable electronics. The electronic components of the design were distributed between islands connected by stretchable serpentine structures. The analytical results show that at 20% overall elongation, the serpentines stretch 60% due to the rigidity of the islands. This 20% elongation is defined as the system stretchability. The 60% elongation on the serpentines is defined as the effective stretchability. At 60% effective stretch, the calculated equivalent plastic strain in a serpentine interconnect is 0.67%, which is well below the fracture limit of copper. Elongation experiments show that the archipelago structure has the system stretchability up to 76% for one-time-stretching, translating to 228% of the effective stretchability on the serpentines. Fatigue-tension experiments show that at 20% system stretch, the archipelago structure can withstand on average 71,950 cycles without electrical or mechanical degradation.

Introduction

Most electronic systems that power our digital world are rigid. Our skin and muscles are soft and curved. Electronic systems that can bend, twist, and stretch have many applications, particularly in skin-mounted wearable electronics. Examples range from wearable biomedical devices such as skin-mounted sweat monitors [1, 2] to wearable consumer electronics such as stretchable solar panels [3]. In order to create a highly deformable and reliable system, a design comprised of flexible islands linked by flexible and stretchable interconnects is necessary. Here stretchable is defined as the ability of a design to lengthen in one axis and flexible is defined as the ability to bend only. These flexible islands support the surface-mounted, packaged components and limit mechanical strain on them due to stretching. Additionally, the islands provide structural stability and make handling during the assembly process manageable. The design with multiple islands linked by stretchable interconnects, is termed the archipelago platform design. This platform maximizes the degree of freedom for which engineers can use to build functional electronic systems while maintaining the system's stretchability.

Previous stretchable electronic systems use one or two large islands for component population. This system design is limited to a small bending radius due of the size and rigidity of the islands. Consequently, this limits its applications. Researchers have broken down the large islands into small islands in a random distribution format in order to maximize stretchability and flexibility. However this format does not achieve uniform strain distribution throughout the islands and serpentines. As a result, it tends to fail at particular points in the system. These randomly distributed islands, also, create layout complexity and create a difficult environment to build an electronic system around.

In this paper, the archipelago design is presented as a platform for stretchable electronic system. Passive 0402 resistors are populated on the islands to monitor potential failures due to shear strain from the elastomer substrate and analytical analysis evaluates and determines the archipelago design guidelines. Numerical analyses give insight into the structure's mechanics and deformation mechanisms. Finally, electromechanical experiments were conducted to evaluate the structure's performance.

Design Concept of Archipelago Platform

To achieve a high level of comfort for wearability, the patch design must be as small and compliant as possible. Therefore the design of the archipelago platform required discrete islands for component population interconnected by stretchable serpentines. Electrical components were separated into groups according to the layout requirements and populated on the islands. The islands were arranged in a matrix format and interconnected by stretchable serpentines, as shown in figure 1a (with the elastomer encapsulation) and 1b (without the encapsulation for visualization). This matrix format allowed all the serpentine interconnects to have the same stress and strain distribution while stretching. Figure 1c and 1d show the top view of the serpentine design and the associated dimensions used in the archipelago platform. As mentioned in the previous section, there were 0402 passive resisters populated in the center of the islands to monitor potential component damages due to shear strain in elastomer substrate during experimentation. In this paper, three dimensional non-linear finite element analysis (FEA) was implemented to investigate the system's mechanics during elongation.

Figure 2 shows the three dimensional iso-view of the archipelago structure and the bottom right inset shows the detailed dimensions and material types. In the FEA, a 60% uniaxial elongation was applied to the system in correspondence with experimental conditions. ANSYS®, the commercial finite element code, was used to simulate the deformation of the structure and to calculate stress and strain distribution on the structure. The copper metal used in this study is assumed to be a plastically deformable solid, obeying the bi-linear kinematic hardening rule, with the elastic Young's modulus E_0=117 GPa, the yielding point σ_y=0.1723 GPa, and the Tangent Module E_t=1.0342 GPa. The elastomeric substrate is assumed to be an incompressible hyper-elastic Mooney-Rivlin solid [4] with coefficients of C10=0.026576 MPa and C01=-0.0020901 MPa.

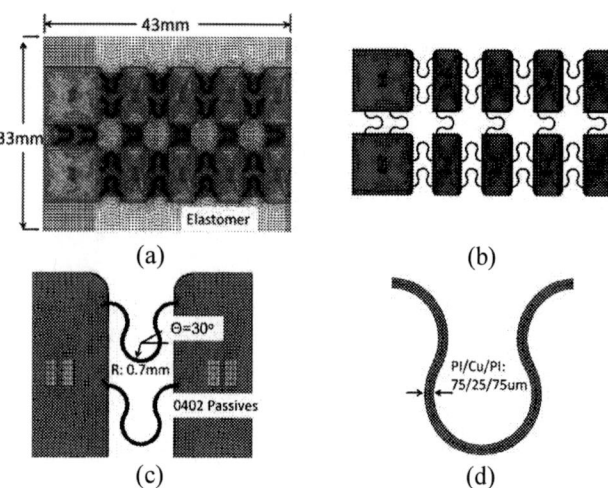

Figure 1. Archipelago structure (a) with encapsulation; (b) without encapsulation; (c) and (d) top view of the serpentine interconnect with dimensions.

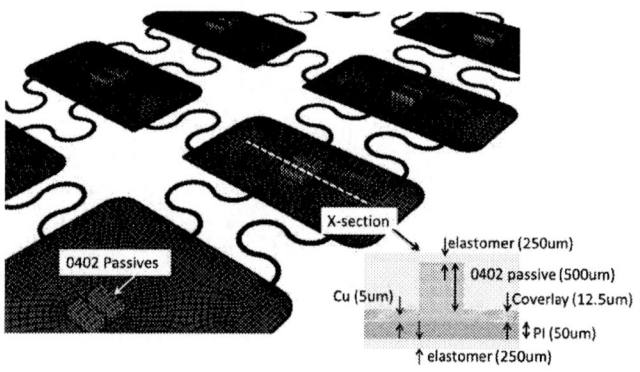

Figure 2. Iso-view of the archipelago structure. Bottom right inset shows the cross section dimensions.

Analytical Analysis on the Archipelago Design

When a stretchable electronic system was subjected to elongation, the serpentine stretches while the islands experience minimum elongation. To evaluate the stretchability, the original length of the archipelago (L_0) is defined as the gauge length and ΔL is defined as the amount of elongation, as shown in figure 3. System stretchability in this case is defined in equation (1). The percent elongation of the serpentines was greater than the system elongation due to the rigidity of the islands that prevented stretching. As a result, the serpentine stretchability is defined as the effective stretchability, and is shown in equation (2). By combining equation (2) and equation (1), the relation between the effective and the system stretchability is derived in equation (3). The effective stretchability is the amount of elongation that the serpentine has to accommodate. To prevent the failure, the effective stretchability of archipelago should be below the yielding point of the serpentine. If the gauge length (L_0) is a fixed number, effective stretchability can be

minimized by reducing the length of the island (L_i) and increasing the number (a) or the length (L_s) of the serpentine gap.

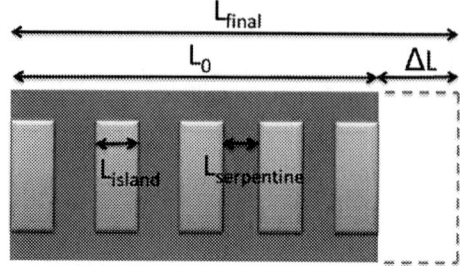

Figure 3. Archipelago measurements used in modeling the elongation of the design.

$$\%_{system} = \Delta L \big/ L_0 \tag{1}$$

$$\%_{effective} = \left(\Delta L / a\right) \big/ L_s \tag{2}$$

$$\%_{effective} = \%_{system} \times \left(L_0 \big/ a \times L_s \right) \tag{3}$$

$$\text{where } L_0 = a \times L_s + \left(a+1\right) \times L_i \tag{4}$$

a = number of serpentine gaps
L_i = length of island
L_s = length of serpentine gap

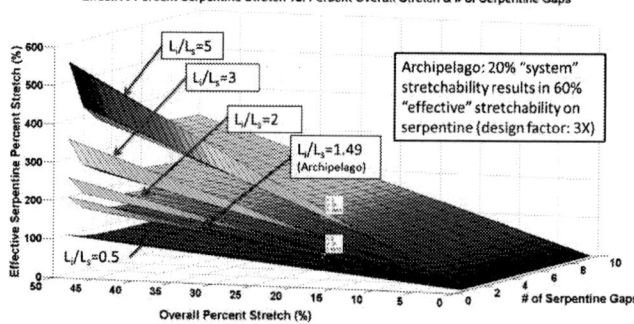

Figure 4. Three dimensional plot of equation three displaying effective serpentine stretch versus overall percent stretch and number of serpentine gaps.

Figure 4 shows the three dimensional plot of equation (3). The surfaces of the plot are a function of effective stretchability, system stretchability, number of serpentine gaps, and the ratio of length of an island and serpentine gap (L_i/L_s). As described in the previous section, the effective stretchability can be optimized by reducing the length of the island or increasing the length of the serpentine. For the archipelago design the ratio of the length of the island versus the serpentine gap (L_i/L_s) is 1.49. This value could be substituted into equation (3) and could be used to generate a design factor that describes the ratio between effective and system stretchability. For the archipelago the design factor is 3. The effective stretchability of the serpentine is three times

978-1-4799-2408-0/14 $31.00 © 2014 IEEE

higher than the system stretchability. For example, if the archipelago structure is stretched at a 20% elongation (system stretchability), the serpentine elongation is 60% (effective stretchability). From the design stand point, effective stretchability should be considered when designing the serpentine interconnects. The effective stretch cannot be above the yield point of the copper.

Numerical Modeling on the Deformation Behavior

Numerical modeling was used to validate the analytical model as described in the previous section, investigate the deformation behavior of the archipelago structure, and examine the potential failure mechanisms such as shear or plastic strain concentration in the components or serpentine. Figure 5 shows the top view of the FEM model. To validate the analytical model, the archipelago structure is stretched for 20% and the gap length (L_s) is show in figure 5b. The gap distance is measured from one end of the island to the opposite end of the adjacent island. In the FEM model, the effective stretchability of the gap distance is (a) 63%, (b) 64%, (c) 64%, and (d) 63%, which confirms and matches the analytical analysis from the previous section.

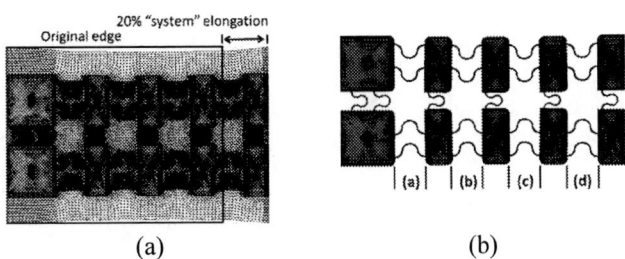

(a) (b)

Figure 5 (a) Top view of the FEM model; (b) gap distance for effective stretchability.

Figure 6 (a) shows the plastic strain distribution from the FEM model and figure 6 (b) shows the corresponding deformation mechanism from the experiment. In figure 6 (a), it is found that the plastic strain is localized on the serpentine at 60% effective stretchability when the archipelago is stretched at 20% system stretchability. There is no plastic strain on the islands, which confirms the assumption in the previous section that the islands do not contribute to the overall stretchability due to its rigidity. In addition, it is found that there is no plastic strain at the center area of the islands where the 0402 resistors reside on. This shows that the design, at 20% system elongation, limits the damage to the components. Furthermore, the plastic strain concentrates on the crests of the serpentine and at the transitions from the rigid island to the compliant serpentine. These two locations are identified as the potential failure locations in the experiment, which will be discussed in the later section.

Equivalent plastic strain is used as a measure of the serpentine stretchability and reliability. Figure 7 shows the equivalent plastic strain as a function of the effective stretchability on serpentine. It is shown that when serpentine is stretched for 8%, there is no plastic damage in the serpentine metal. When the serpentine is stretched up to 60% (effective stretchability), the plastic strain increases non-

linearly to 0.67%, which is below the yielding point of the metal. As a result, it is proved by the FEM that the archipelago design can be stretched for 20% system stretchability without severe mechanical degradation. The 20% elongation is used for performance evaluation for a skin-mounted wearable and stretchable electronic system, because the skin starts to tear after 20% strain according to the literature.

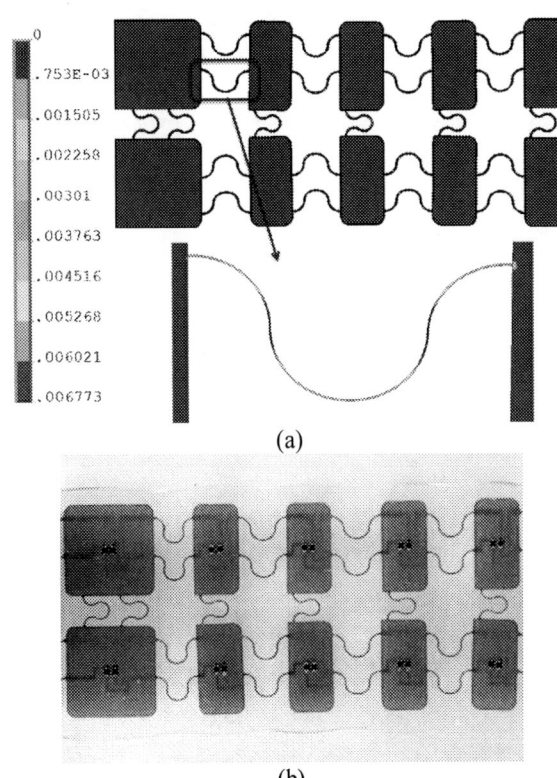

(a)

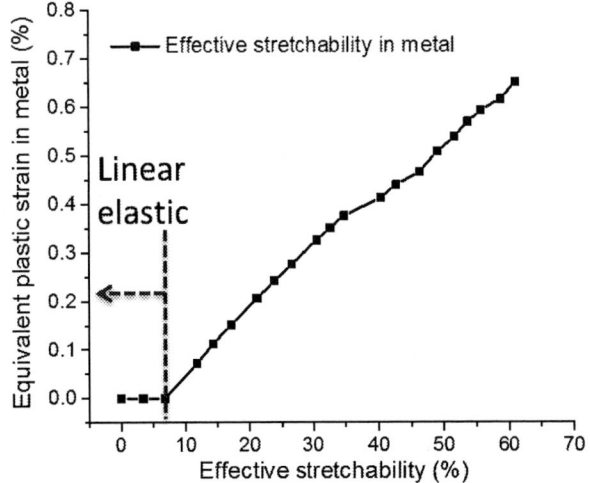

(b)

Figure 6 (a) FEM model as seen from the top view at 20% elongation. Enlarged is a serpentine showing the plastic strain concentrated at the crest and transitions between serpentine and island; (b) experiment corresponding to the FEM model.

Figure 7. Maximum plastic strain in metal as a function of elongation up to 50% elongation

In-Situ Electromechanical Measurement

To validate the FEM model and investigate the electromechanical behavior, the archipelago structure is fabricated using the flex process. Figure 8 shows the archipelago test structure, corresponding to the FEM model. There are four channels with leads extending from both ends of the archipelago for in-situ electrical measurement. Each channel consists of a serpentine connected in series with 0402 resisters. By applying resistors, the resistance could be used to monitor the electromechanical performance. If there is any mechanical or material degradation to either the serpentine or resisters, the resistance is expected to increase during elongation.

Five samples were stretched until failure to determine the maximum stretching capabilities of the archipelago structure. A 0.5% per second strain rate was applied to the samples during testing. Samples were clamped at the outermost edges of the exterior islands. This allows the inner most islands and the channels to elongate uniformly. Failure was studied by sampling resistance at a rate of 100 Hz as the sample was stretched.

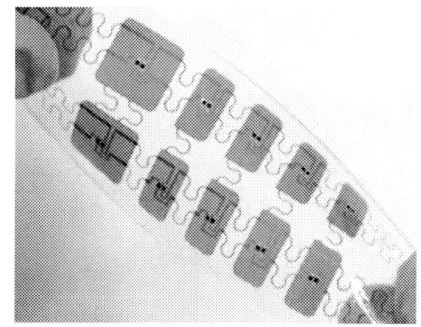

Figure 8. Archipelago test structure for FEM validation.

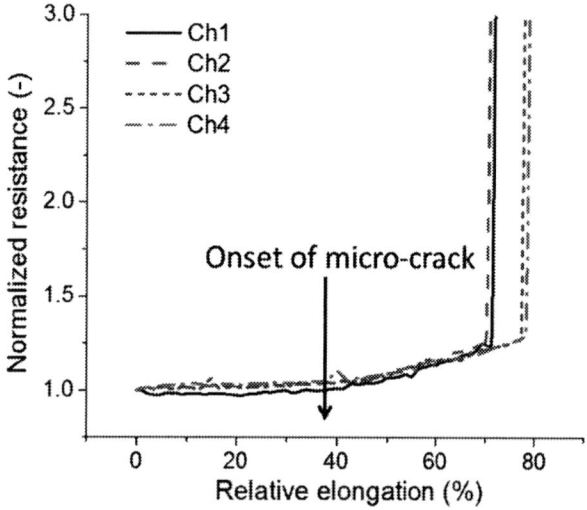

Figure 9. Normalized resistance as a function of elongation up to failure.

The resulting curve of one example is shown in figure 9. The Y-axis in figure 9 is the normalized resistance without units, which is determined by dividing subsequent resistance measurements by the initial resistance value. The system stretchability is defined as the relative elongation in the X-axis. It is found that the normalized resistance increases after 38% relative elongation as the archipelago structure is stretched. The pattern of the curves are similar to the previous studies [5], therefore, it is suspected that the major failure mode is on the serpentine, instead of the resisters. Since there are four channels per sample, there are four curves as shown in figure 9. Channel 1 and 2 are similar and channel 3 and 4 are similar. Both groups have identical trends except the elongation at the ultimate failure (infinite increase in resistance). This difference was due to experimental error that was produced when the sample slipped from the clamp during elongation. To determine the onset of micro-crack formation in the serpentine, the methodology [5] is implemented. It is found that the onset of the micro-cracks for both groups of serpentine is 38%. The onset of micro-crack formation could be translated into the system stretchability, which is about 38%.

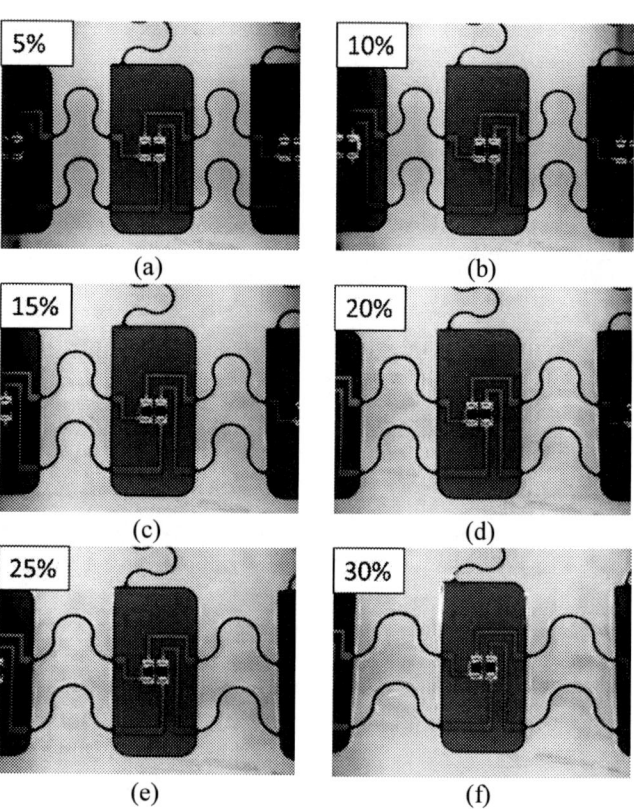

Figure 10. Progression of the stretching of archipelago structure from (a) 5%, (b) 10%, (c) 15%, (d) 20%, (e) 25%, and (f) 30%.

Figure 10 shows the progression of the stretching of archipelago structure, from 5% through 30% system stretchability. It is shown that the serpentine elongates whereas the islands remain the same shape. This deformation mechanism further confirms the analytical assumption and the numerical FEM model, as discussed in the previous sections. Additionally, delamination (Figure 10e and 10f) is detected around the perimeter of the islands and on the edge of the serpentine as seen in previous publications [6]. The stretching of the system has no obvious negative impact on the resisters residing on the center of the islands. This indicates that the components are well protected by the islands.

Figure 11 (a) shows the sample at relaxed state after stretching to failure for 78% system stretchability. The archipelago displays plastic deformation on the serpentine while the island and the populated resisters on the island remain intact. Figure 11 (b) shows the enlarged view on a plastically deformed serpentine. The serpentine has been completely separated from the elastomer substrate due to delamination, and plastically deformed due to twisting of the metal. Figure 11 (c) shows the detailed view on the fracture serpentine. The breaking point is at the transition from the rigid island to the compliant serpentine, which correlates and confirms the FEM prediction as shown in figure 6 (a).

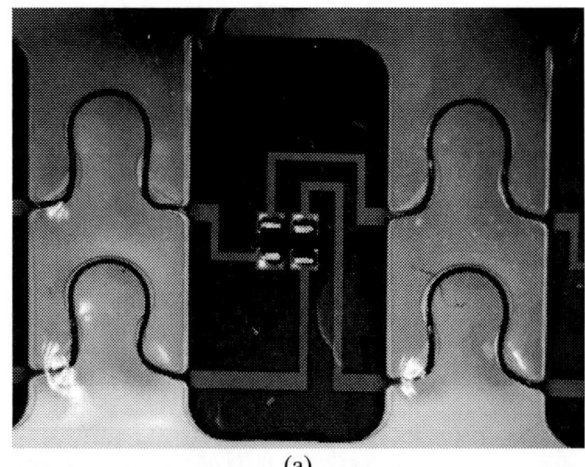

(a)

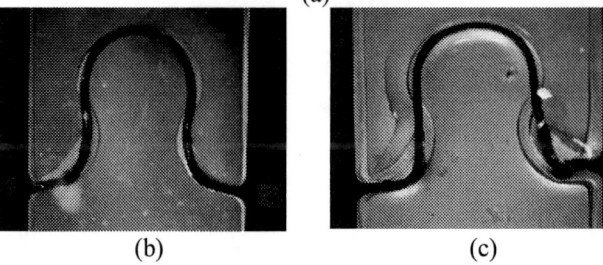

(b)   (c)

Figure 11. (a) Failed sample at relaxed state after stretching for 78%; (b) enlarged view of the plastically deformed serpentine; (c) fracture serpentine.

Fatigue life of the design is critical for the wearable applications. The electronics have to withstand stretching fatigue cycles on various parts of the body. If the stretchable electronic system is applied to a human thigh, the electronics have to withstand repeated stretching due to muscle movement. Therefore, the reliability of the archipelago structure was determined with fatigue life testing. Five different elongations (20%, 25%, 30%, 35%, 40%) were applied experimentally at the frequency of 2Hz. Figure 12 shows the linear fitting of elongation versus number of stretching cycles to failure (E-N curve). Every data point in this figure indicates the average number of five repeated

experiments. The linear fitting has a R^2 of 0.97. It is observed that the archipelago survives more than 71,000 cycles at 20% system stretchability. As the system stretchability increases to 40%, the number of stretching cycle to failure drops rapidly to 1,219 cycles. This phenomenon can be explained by the accumulated plastic strain in the metal. A higher elongation results in more accumulated plastic strain in the serpentine. As soon as the accumulated plastic strain in the metal reaches the intrinsic fracture strain, the cracks initiates in the metal and eventually ruptures.

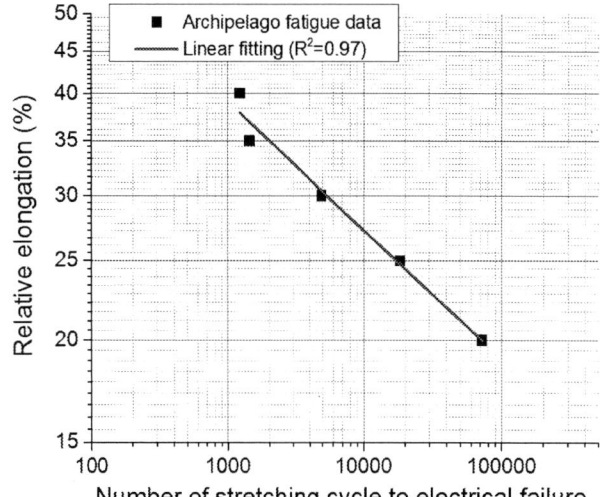

Figure 12. Fatigue life E-N curve of archipelago structure.

Development of the Skin-Mounted Wearable and Stretchable electronics

The design and testing data of the archipelago platform for wearable and stretchable electronics are reported in the previous sections. Because of the archipelago design, the components are protected from mechanical damage. Therefore, for applications, the archipelago has to incorporate components such as battery, micro controller, and memory onto the islands. Figure 13 (a) shows a completed wearable and stretchable system using archipelago platform. The system has a verity of sensing modalities, such as temperature and EMG, communicating with devices through Bluetooth Low Energy (BTLE) technology. The system is built on top of the stretchable and flexible archipelago platform (figure 13b) so that they may be affixed to the human body to measure physiological data.

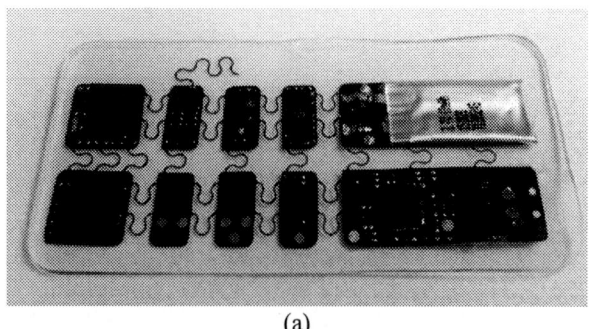

(a)

(b)

Figure 13. Skin-mounted wearable and stretchable electronic system at (a) relaxed state and (b) during stretching.

Conclusions

An archipelago platform for skin-mounted wearable and stretchable electronics is proposed in this paper. The analytical model provides the design guidelines and the design factor between system and effective stretchability. The system stretchability is defined as the overall elongation of the electronic system, whereas the effective stretchability is defined as the actual elongation on the serpentine. The effective stretchability on the serpentine is multiple times higher than the system stretchability. In this paper, the archipelago structure has a design factor of three between system and effective stretchability. Numerical models validate the analytical model and provide insight to the mechanics during stretching. There is limited strain on the islands that result protection of the electrical components placed on them. The serpentine is subjected to a majority of the deformation.

The archipelago platform shows its promising potential applications for skin-mounted wearable and stretchable electronics. The platform paves an avenue for design engineers to create stretchable electronic systems.

References

1. Hsu, Y.-Y., Hoffman J., Ghaffari R., Ives B., Wei P., Klinker L., Morey B., Elolampi B., Davis D., Rafferty C., and Dowling K. ,"Epidermal Electronics: Skin Sweat Patch," in *Proc. IEEE 7th IMPACT*, 2012, pp. 228-231.
2. Huang X., Yeo W. H., Liu Y., Rogers J. A.,"Epidermal Differential Impedance Sensor for Conformal Skin Hydration Monitoring," *Biointerphases*, Vol. 7: 52, 2012.
3. Lee, J., Lee J., Wu J., Shi M., Yoon J., Park S., Li M., Liu Z., Huang Y., and Rogers J. A., "Stretchable GaAs Photovoltaics with Designs That Enable High Areal Coverage," *Advanced Materials*, Vol. 23, pp. 986-991, 2011.
4. ANSYS User Manual V14.5, ANSYS, PA, 2012.
5. Yung-Yu Hsu, Kylie Lucas, Dan Davis, Rooz Ghaffari, Brian Elolampi, Conor Rafferty, Kevin Dowling, "A Novel Strain Relief Design for Multi-Layer Thin Film Stretchable Interconnects," *IEEE Transactions on Electron Devices*, Vol. 60, No. 7, pp. 2338-2345, 2013.
6. Yung-Yu Hsu, Mario Gonzalez, Frederick Bossuyt, Fabrice Axisa, Jan Vanfleteren, and Ingrid De Wolf, "Polyimide-enhanced Stretchable Interconnect: Factorial Design, Fabrication, and Characterization," *IEEE Transactions on Electron Devices*, Vol. 58, No. 8, pp. 2680-2688, 2011.

Inkjet Printing in Manufacturing of Stretchable Interconnects

Toni Liimatta, Eerik Halonen, Hannu Sillanpää, Juha Niittynen, and Matti Mäntysalo

Tampere University of Technology, Department of Electronics and Communications Engineering

Korkeakoulunkatu 3, B.O. Box 692, FI-33101 Tampere, Finland

matti.mantysalo@tut.fi

Abstract

Stretchable circuits have the potential to enable integrating electronics in everyday objects, but also skin-like, imperceptible electronic applications. However, manufacturing stretchable electronics requires developing novel manufacturing methods and using novel materials at least as substrate. Since the elastic materials for stretchable electronics are relatively soft, using traditional manufacturing methods becomes more problematic, whereas contactless material deposition by inkjet-printing is unaffected by such material properties. This study concentrates on feasibility analysis of using inkjet printing in manufacturing of stretchable electronics. First, printing related challenges are evaluated by manufacturing test structures with inkjet-printer using silver nanoparticle ink on elastic thermoplastic polyurethane substrate and sintering structures in convection oven. Adhesion between ink and substrate, but also sheet resistance, is evaluated. A minimum sheet resistance approx. of 26 mΩ/$\square$ was obtained, and peak strains of inkjet-printed conductors are found to be between 1.0 % and 1.5 %, but conductivity is observed to be almost fully reversible when strain is released.

Introduction

Flexible circuits have become a key enabling technology for dynamic and flex-to-fit applications in today's electronic devices. At the same time there has been growing interest for new manufacturing technologies which could enable cost-effective, fast and environmentally friendly production of flexible circuits. Printed electronics hold a potential to provide these benefits due to printing technology's additive nature and feasibility for mass production [1]. Inkjet-printing in particular is an emerging technology which enables fast adaptation to different circuit layouts and contactless material deposition [2].

Inkjet printing enables deposition of highly conductive tracks at relatively low temperatures with low material wastage and even on non-planar surfaces [3, 4]. Since melting of metal particles is size dependent [4], nano-scale metal particles can be sintered together at temperatures below softening point (T$_g$) of many common polymers. Niittynen et al. compares different sintering methods [6]. Solution based deposition of metal conductors using for example gold nanoparticles [7], silver nanoparticles [8] or copper nanoparticles [9] has been demonstrated on wide variety of flexible polymer substrates. When compared to conventional manufacturing methods of flexible circuits, inkjet printing involves fewer process steps and due to digital control of printing process, it is also more versatile than etching processes. Inkjet-printed conductors are typically thinner than conventional copper foils and have a somewhat porous structure, which have been shown to significantly improve performance in bending tests [10]. In addition, flexible circuit board manufactured with inkjet technology has been shown to be capable of replacing an existing flexible circuit board manufactured with conventional methods for example in a mobile phone [11, 12]. Therefore inkjet printing is a feasible alternative for manufacturing flexible electronics.

Making stretchable electronics is mainly branched in two directions, which both have their pros and cons. The first method is to use intrinsically stretchable materials, like conductive polymers or organic semiconductors, which have relatively poor electrical performance compared to conventional inorganic electronic materials. The second method is to use conventional electronic materials and make the system stretchable. This way good electrical performance is achieved, but the stretching is more challenging, since conventional semiconductors, like silicon, are hard and brittle. If rigid areas with silicon-based semiconductors are kept small enough and the interconnections between them are made stretchable, resulting system appears stretchable macroscopically. Both approaches involve using new materials, i.e. elastomers, as substrate. [13]

This paper focuses on strain properties of inkjet-printed nano-particle silver conductors. The performance of inkjet-printed conductors is studied by measuring the resistance of inkjet-printed lines with increasing strain levels until conductivity is considered to be lost. These results are then compared to test samples manufactured with evaporated silver.

Printability and fabrication process

Since thermoplastic polyurethane and silicone rubber are the most researched substrate candidates for stretchable electronics, elastomers chosen for experiments are thermoplastic polyurethane (TPU) Platilon U 4201 AU from Epurex Films GmbH & Co.KG and silicone rubber NM60 from New Metals and Chemicals Ltd. with thicknesses 50 μm and 500 μm respectively. Conductive ink selected for experiments is Harima NPS-JL from Harima Chemicals Group, which is silver nanoparticle ink with mean particle size of 7 nm. This ink can be sintered at temperature as low as 120 °C, which makes it suitable for a wide range of polymer substrates. Inkjet printers used in these experiments were Dimatix DMP-2831 with 16-jet printhead producing nominal 10 pl drop volume and iTi XY MDS 2.0 with 128-jet Spectra S-class printhead producing nominal 30 pl drop volume. The initial small scale testing was started with Dimatix DMP-2831 to evaluate printability of these two ink-substrate pairs and based on these results printing process was optimized for pilot scale inkjet printer iTi XY MDS 2.0. TPU and silicone rubber were cut into rectangular approximately 150 mm x 70 mm sized samples and attached to an aluminum carrier plate with Kapton® tape to avoid substrate movement during printing.

Experiments were started by printing test patterns with Dimatix DMP-2831 on both untreated substrate films. Test patterns included straight lines with different lines widths

2014 Electronic Components & Technology Conference

from single pixel to ten pixels, drop matrix and large continuous areas. The substrate temperature was 60 °C, nozzle plate temperature 43 °C, jetting frequency 7 kHz and 23 V ejecting amplitude. The best printing results were achieved with resolution 1270 dpi, but on TPU printing results were inconsistent and large areas suffered from dewetting due to low surface energy as can be seen in Figure 1.

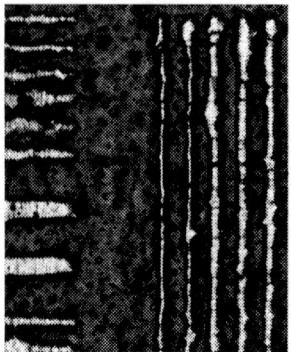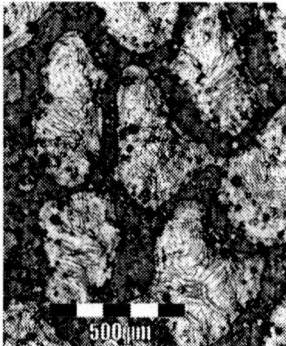

Figure 1. Fine lines on TPU (left) and dewetting of large areas (right).

Generally, low surface energy increases the contact angle of ink on substrate and consequently reduces drop size [14]. Although this enables printing finer line widths due to small drop size, in large continuous areas it causes problems, such as partial dewetting and pinholes. On silicone rubber printed patterns suffered from wrinkling and cracking as seen in Figure 2, but pattern geometries were very accurate. Both substrate materials were observed to warp during printing of large areas.

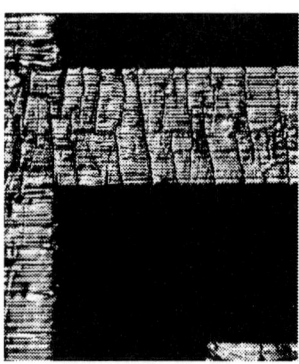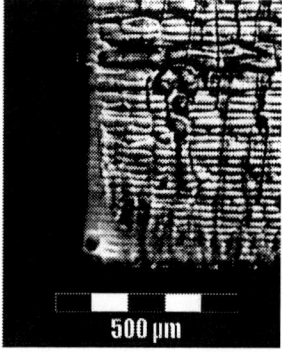

Figure 2. Wrinkled and cracked patterns on silicone rubber.

To enhance wetting next samples were treated by wiping substrate surface with cleanroom cloth soaked with n-tetradecane and dried in convection oven at 140 °C for 15 minutes. Tetradecane was chosen since it is the principal solvent of Harima NPS-JL ink. Significant swelling was observed in both materials with exposure to n-tetradecane, but swelling faded away during drying step. In order to avoid swelling induced warping during printing process, samples were pre-stretched approximately 10% when attached to carrier plate. Since swelling increases dimensions of the

substrate, sufficient tension can balance out this effect. In addition, this tension can help to balance out stresses caused by thermal expansion in subsequent sintering phase. However, it was observed that pre-stretching did not completely prevent substrate warping in printing process, but surface treatment with n-tetradecane enhanced wetting sufficiently so that continuous patterns could be printed with aforementioned printing parameters and resolution of 1270 dpi on TPU. On silicone rubber printed patterns were still wrinkled and cracked, which is probably due to higher swelling along axis perpendicular to surface. Some flooding and ink flow towards pattern centers was also observed on TPU with 1270 dpi.

Sheet resistance of printed structures was determined by printing Greek cross patterns, sintering samples in convection oven at 150 °C for 60 minutes and measuring sheet resistances with 4-point probe station and Keithley 2425 SourceMeter. Sheet resistance of patterns printed on TPU with DMP-2831 and resolution 1270 dpi was approximately 65.0 mΩ/□ with standard deviation of 14.7 mΩ/□. Patterns printed on silicone rubber were not conductive due to cracking, and adhesion between sintered ink and substrate was observed to be insufficient. Since sheet resistance variation on TPU was quite high, new samples were printed with resolution 1016 dpi and 2 layers, which gives nominally 27% higher ink volume per surface area and more even ink distribution due to lower ink flow on surface. Sheet resistance of these samples was on average 21.5 mΩ/□ with standard deviation of 0.2 mΩ/□. However, it was observed that with these parameters large printed areas were prone to crack formation even before sintering phase, which was presumed to be caused by thicker ink layer and higher swelling due to larger ink volume per surface area.

At this point printing process was scaled up for pilot scale inkjet printer iTi XY MDS 2.0 with Spectra S-class printhead. Printing was started with nozzle plate temperature 34 °C, substrate temperature 60 °C, printing speed 80 mm/s and jetting waveform amplitude 55 V. Resolution was 700 dpi, which is nominally close to ink volume per surface area as with DMP-2831 and resolution 1270 dpi, since droplets with Spectra S-class printhead are nominally three times larger in volume. Significant dewetting issues were seen between nozzle spacing due to different printing sequence of iTi XY MDS 2.0. Whereas with DMP-2831 printing was done with one nozzle, iTi XY MDS 2.0 uses all 128 nozzles and therefore printing sequence is different. Therefore 'stacked-coin' structure [15], in which every droplet dries separately, was seen as a potential solution. Print image was divided into four sub-images [14] and substrate temperature was raised to 70 °C in order to increase the ink evaporation rate. Samples printed with four sub-images and 700dpi resolution showed sheet resistance of approximately 51.0 mΩ/□ with standard deviation 14.6 mΩ/□, which is similarly diverse result as seen before with DMP-2831 and resolution 1270 dpi. Next samples with same print sequence, but higher resolution of 800 dpi, showed sheet resistance of approximately 28.2 mΩ/□ with standard deviation 1.9 mΩ/□. This sequential printing and enhanced solvent evaporation due to higher substrate temperature also solved cracking issues caused by swelling.

Finally an image masking algorithm, similar to Koskinen et al. [12] used, was introduced to further optimize the

978-1-4799-2408-0/14 $31.00 © 2014 IEEE   152

printing process and to enable printing 'stacked-coin' structure with only one image file. This masking algorithm produced ink coverage equivalent to approx. 830 dpi and samples printed with this algorithm showed sheet resistance values approx. of 25.8 mΩ/$\square$ with standard deviation of 0.8 mΩ/$\square$. Sub 100 μm lines and large continuous areas were printed successfully on both substrate materials with high yield, but on silicone rubber sintered patterns wrinkled and partially delaminated when pre-stretch was released, whereas on TPU this was not seen due to thermoforming of TPU during sintering phase. Sheet resistances are summarized in Figure 3.

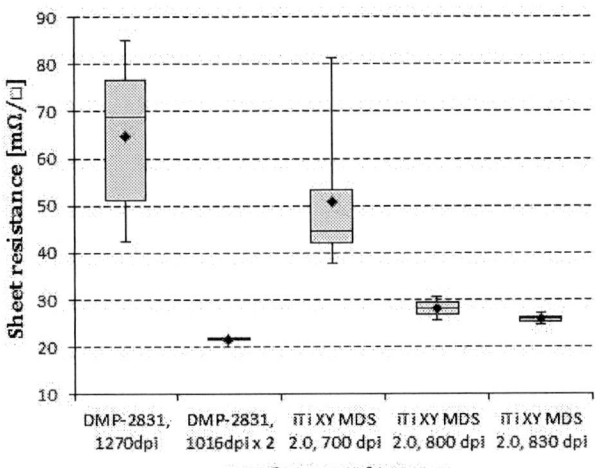

Figure 3. Measured sheet resistances of samples printed with different printers and different parameters. Grey boxes illustrate the standard deviation and horizontal lines across boxes are the median values. Vertical lines illustrate minimum and maximum values of measurements and the black dots are the averages.

Strain testing of inkjet-printed interconnects

Dog bone shaped patterns with 61 mm length and 1 mm line width were printed with iTi XY MDS 2.0 using aforementioned image masking algorithm on TPU with Harima NPS-JL silver nanoparticle ink. Samples were sintered in convection oven at 150 °C for 60 minutes and afterwards measuring wires (30 AWG) were attached to samples' ends with isotropic conductive adhesive Creative 124-08C, which was cured in convection oven at 80 °C for 90 minutes.

Fabricated samples were attached to custom made test bench shown in Figure 4. Attachment was done so that measurement wires were routed through the clamp fixtures and backside of contact pads was stiffened with Kapton® tape to focus strain only on printed lines. Linear movement resolution of test bench is 15 μm and resistance values are measured automatically after every step as an average of 100 repeated measurements. Strain rate was limited to maximum of 1.5 %/min to avoid shock effects caused by rapid changes in strain.

Inkjet-printed conductors were stretched until every printed line was observed to be broken. Measurement results

are shown in Figure 5. As can be seen, the achieved strains are fairly low. The highest measured peak strain was 1.50 %, but average peak strain of all samples was 1.00 % with standard deviation of 0.42 percentage point. It was observed that some of the conductors regained conductivity at some point after having lost conductivity at lower strain, i.e. conductivity was intermittent. In addition, strain-resistance curves showed notable hysteresis as strain was released after being stretched to 5 %, which means that samples regained conductivity at higher strain values as their measured peak strain at first stretch. For example the conductor, which lost conductivity at strain of 1.50 %, regained conductivity at strain of 3.18 %, but with resistance reaching up to 20 times the initial resistance. Once strain had been fully released, the average resistance of samples was 15.9 % higher than the initial value with standard deviation of 10.5 percentage points.

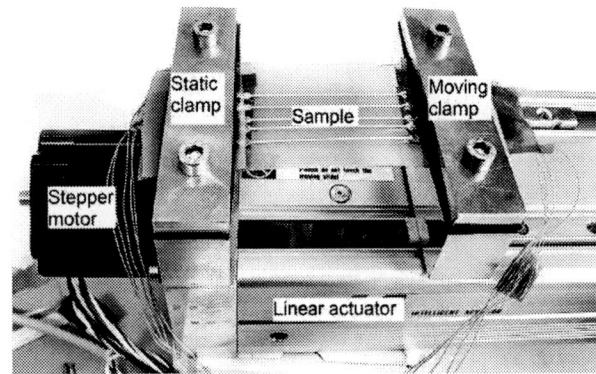

Figure 4. Custom made test bench for strain testing.

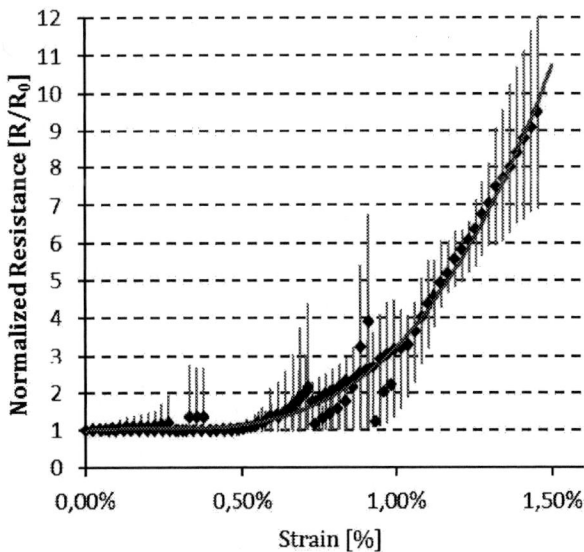

Figure 5. Normalized resistance of 1 mm wide inkjet-printed conductors measured under tensile strain. Black data points are averaged from all measured values and the vertical lines illustrate minimum and maximum values. Infinite values in measurement data were ignored to enable calculation of averages.

978-1-4799-2408-0/14 $31.00 © 2014 IEEE

Figure 6 shows crack formation in inkjet-printed conductors under approx. 5 % tensile strain. Crack formation is relatively scarce, when compared to microcrack formation in thin gold films [16, 17]. Formation of these cracks is assumed to cause the observed resistance hysteresis on the first strain cycle, since hysteresis was not as apparent on following strain cycles. It is presumed, that these cracks serve as strain relief and cause the lines to regain conductivity at higher strain levels than originally without cracking.

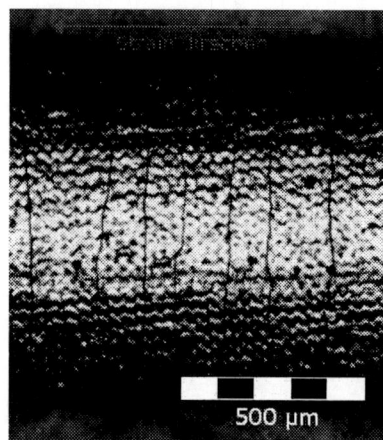

Figure 6. Microscope image of 1 mm inkjet-printed conductor under approx. 5 % tensile strain.

For comparison, samples with same shape, but evaporated 50 nm thick layer of silver on TPU, were also manufactured and tested. The resistance of evaporated samples is approx. 33 times higher than the resistance of printed samples. Measurement results of evaporated samples as a function of strain are shown in Figure 7. The actual peak strain of these conductors could not be determined, because conductivity was lost due to rupture of TPU film at contact line of TPU and rigid conductive adhesive, which was used to attach measurement wires. Although samples with thicker inkjet-printed conductors were stretched up to 5 %, same problem did not occur, which probably due to better mechanical support offered by inkjet-printed patterns. As can be seen in Figure 7, the evaporated samples could withstand significantly larger strain levels with smaller increase in resistance than inkjet-printed samples. Measurement results are similar as reported before by Robinson et al [16] with 50 nm thick evaporated gold films on PDMS and therefore it is probable that similar microcracking occurs also with thin silver layers.

Cyclic strain testing and time dependent recovery

Since conductivity of inkjet-printed samples was observed to be almost fully reversible after peak strain testing, it was decided to expose these samples to cyclic stretching. Measurements were performed so that samples were first stretched 5 % in test bench, strain was fully released and after 10 seconds samples' resistance was measured as an average of 100 sequential measurements. Measurement results are shown in Figure 8.

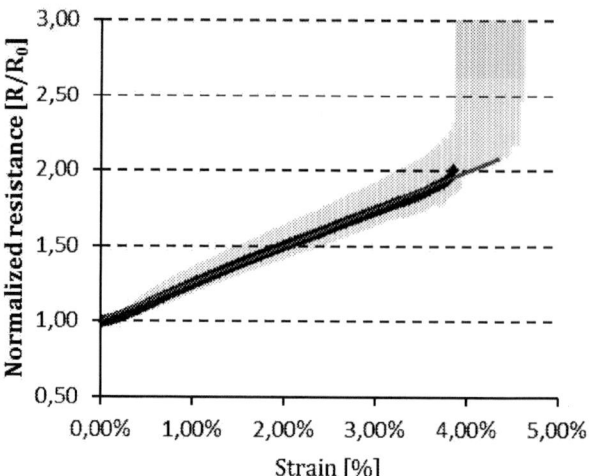

Figure 7. Normalized resistance of 1 mm wide and 50 nm thick evaporated silver conductors measured under tensile strain. Black data points are the calculated averages from all samples and grey area illustrates minimum and maximum values

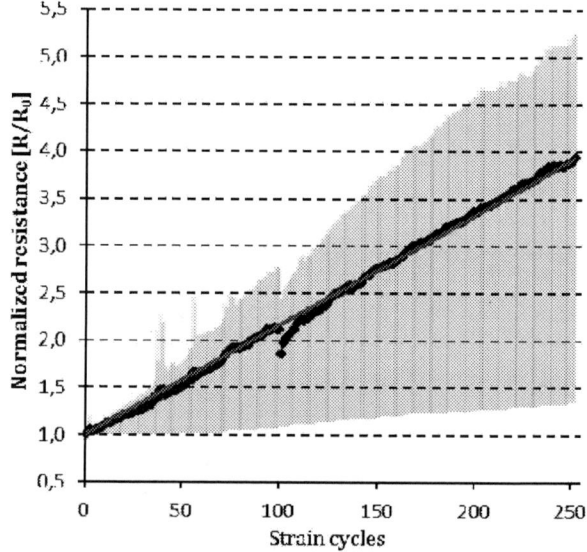

Figure 8. Normalized resistance of 1 mm wide inkjet-printed conductors in relaxed state after a number of strain cycles with 5 % peak strain. Black data points are the calculated averages from all samples and grey area illustrates minimum and maximum values

As can be seen in Figure 8, after 250 strain cycles the resistance of samples in relaxed state was as an average of 3.93 times their initial resistance with standard deviation of 1.41. However, after testing run had ended, it was observed that resistance of samples continued to decrease, i.e. recovery was time dependent. This recovery was recorded with measurement interval of 5 seconds from one set of samples, and results are shown in Figure 9.

As can be seen in Figure 9, resistance of samples follows logarithmic curve over time after the strain has been released. Resistance values were considered to be stable after one hour and the average final resistance was 1.30 times the initial resistance with standard deviation of 0.19. These values are considerably lower than initially measured in cyclic strain testing, which also means that results of cyclic strain testing are highly dependent on the timeframe in which resistance is measured. The time dependent recovery is probably caused by viscoelastic nature of thermoplastic polyurethane and further research will be needed to better analyze lifetime of inkjet-printed conductors in cyclic stretching.

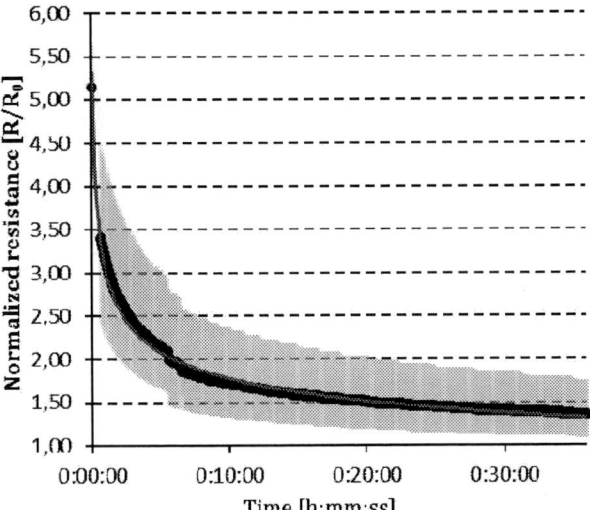

Figure 9. Normalized resistance of 1 mm wide samples along time after 250 strain cycles. Black data points are the calculated averages from all samples and grey area illustrates minimum and maximum values.

Conclusions

In this paper we have demonstrated a feasible method of manufacturing highly conductive circuits on elastic substrate by utilizing inkjet printing. The main challenges in printing on elastomer substrate were identified as hydrophobic nature of elastomer surface, CTE mismatch between metal and polymer, and swelling of substrate material caused by solvent absorption. These issues were solved by introducing a moderate pre-stretch to substrate material during manufacturing process and enhancing solvent evaporation by increasing substrate temperature and modifying printing sequence with image masking algorithm. A moderate pre-stretch keeps substrate material under tension during manufacturing process, which helps to keep it flat and also helps to balance out the change in print pattern dimensions due to thermal expansion or solvent absorption induced swelling of substrate material. Enhanced solvent evaporation and modified printing sequence cause every printed droplet to dry before adjacent droplets are deposited, which produces a 'stacked-coin' structure and minimizes the amount absorbed solvent per surface area thereby reducing substrate swelling.

Good printing results were achieved on thermoplastic polyurethane and silicone rubber, but adhesion on silicone rubber was observed insufficient. Also wrinkling occurred in printed patterns on silicone rubber when pre-stretch was released, whereas similar wrinkling was not seen on thermoplastic polyurethane, because it thermoforms during sintering phase. Image algorithm enabled printing of large areas without ink wetting problems as well as printing fine line widths in same process phase, but with slight decline in surface morphology of printed patterns due to 'stacked-coin' structure.

Measured peak strains for these inkjet-printed silver conductors on thermoplastic polyurethane were modest, averaging at 1.00 %, whereas significantly thinner evaporated silver conductors on same substrate used here for comparison showed similar strain-resistance response as evaporated gold films reported before by Robinson et al [16]. The measured peak strains of inkjet-printed conductors are comparable to results reported by Kim et al [18] with screen printed silver conductors, which gives a reason to presume that peak strain close to 10 % can be achieved with optimized planar patterning scheme. Although inkjet-printed conductors lost conductivity at relatively low strains, the conductivity was observed to be almost fully reversible and this recovery was observed to be time dependent, following a logarithmic curve, due to viscoelastic nature of substrate material.

Acknowledgments

This work is financially supported by Finnish Funding Agency for Technology and Innovation under Grant 40150/12. M. Mäntysalo is supported by Academy of Finland under Grant 251882

References

1. K. Hecker (ed.), OE-A Roadmap for Organic and Printed Electronics, 3rd ed., VDMA Verlag GmbH, 2009.
2. V. Pekkanen, M. Mäntysalo, and P. Mansikkamäki, "Design consider-ations for inkjet-printed electronic interconnections and packaging," in *Proc. 40th Int. Symp. Microelectron.*, San Jose, CA, USA, Nov. 2007, pp. 1076–1083.
3. M. Mäntysalo, et. al. "System Integration of Smart Packages using Printed Electronics" in *Proc. Electronic Compon. Technol. Conf.*, San Diego, CA, USA, May–Jun., 2012, pp. 997-1002.
4. Li Xie, et. at. "Heterogeneous integration of bio-sensing system-on-chip and printed electronics", *IEEE Journal on Emerging and Selected Topics in Circuits and Systems.*, Vol. 2, no. 4, 2012, pp. 672-682.
5. Buffat, P. and Borel, J. P. "Size effect on the melting temperature of gold particles." *Physical Review A*, Vol. , no. 6, 1976, pp. 2287-2298.
6. J. Niittynen, et al., "Alternative Sintering Methods Compared to Conventional Thermal Sintering for Inkjet Printed Silver Nanoparticle Ink", *Thin Solid Films*, Article in Press., 2014, 10.1016/j.tsf.2014.02.001
7. Huang, D., et. al., "Plastic-compatible low resistance printable gold nanoparticle conductors for flexible electronics", *Journal of the electrochemical society*, Vol. 150, no. 7, 2003, pp 412-417.
8. Kim, D., and Moon, J., "Highly conductive ink jet printed films of nanosilver particles for printable electronics.",

Electrochemical and Solid-State Letters, Vol. 8, no. 11, 2005, pp. 30-33.

9. Jang, S., et. al., "Sintering of inkjet printed copper nanoparticles for flexible electronics." *Scripta Materialia*, Vol. 62, no. 5, 2010, pp. 258-261.

10. Halonen, E., et al. "Dynamic bending test analysis of inkjet-printed conductors on flexible substrates. *in Proc. Electronic Compon. Technol. Conf.*, San Diego, CA, USA, May–Jun., 2012, pp. 80-85.

11. Koskinen, S., Pykari, L., and Mantysalo, M., "Inkjet printed flexible user interface module." *in Proc. Electronic Compon. Technol. Conf.*, San Diego, CA, USA, May–Jun., 2012, pp. 1009-1014.

12. Koskinen, S., Pykari, L., and Mantysalo, M., " Electrical Performance Characterization of an Inkjet-Printed Flexible Circuit in a Mobile Application", IEEE Trans. on Comp. Packaging and Manufacturing Technol. Vol. 3, no. 9, 2013, pp. 1604-1610.

13. Someya, T. "Stretchable Electronics. Weinheim", Wiley-VCH Verlag GmbH & Co. KGaA., 2013, 484 p.

14. Mäntysalo M. and Mansikkamäki P., "Inkjet-deposited interconnections for electronic packaging", *NIP 23 23rd International Conference on Digital Printing Technologies, and Digital Fabrication*, AK, USA, Sept. 16th-21st, 2007, pp. 813-817.

15. Pekkanen, V., *et al*, "Utilizing inkjet printing to fabricate electrical interconnections in a System-in-Package", *ELSEVIER Microelectronic Engineering*, Vol. 87, Issue 11, 2010, pp. 2382-2390.

16. Robinson, A. P., et. al. "Microstructured silicone substrate for printable and stretchable metallic films", *Langmuir*, Vol. 27, No. 8, 2011, pp. 4279-4284.

17. Lambricht, N., Pardoen, T., and Yunus, S. "Giant stretchability of thin gold films on rough elastomeric substrates", *Acta Materialia*, Vol. 61, 2013, pp. 540–547.

18. Kim, K. S., Jung, K. H., and Jung, S. B. "Design and fabrication of screen-printed silver circuits for stretchable electronics" *Microelectronic Engineering*, Article in Press. 2013, http://dx.doi.org/10.1016/j.mee.2013.07.003

Ultra Small Hearing Aid Electronic Packaging Enabled By Chip-In-Flex

John Dzarnoski, Susie Johansson
Starkey Hearing Technologies
Eden Prairie, Minnesota 55344
952-941-6401
john_dzarnoski@starkey.com
susie_johansson@starkey.com

Abstract

There has been enormous worldwide effort to increase the volumetric efficiency of electronic packaging. Much of this effort has been driven by the telecommunications industry that has succeeded in reducing cell phone size while simultaneously increasing functionality. The hearing aid business has always had the need to use extremely small electronic packaging because hearing aids pack electronics into the ear canal. In fact, in 1952 the first commercial device to make use of transistors was the hearing aid. In recent years hearing aid microelectronic packaging has been moving from ceramic hybrid-based packaging to flexible circuit based technologies. The introduction of wireless systems in hearing aids has sharply increased component count from less than 20 per device in 2005 to more than 70 in 2010. Due to the size and shape of a multitude of types of hearing aids, flexible circuits need to be folded and bent to fit inside hearing aid cases. All available space is essentially used. Additionally, more powerful processors and more memory are enabling sophisticated algorithms that are able to greatly improve sound quality. There continues to be a strong marketing desire to add more features to hearing products while at the same time making them smaller and less visible. These added features are further increasing component count and are the driving force behind the need to make smaller electronic assemblies.

This paper will examine the use of embedded die packaging (or chip-in-flex) to drive significant further size reduction in custom and standard hearing instruments over what can be achieved using chip-on-flex or ceramic hybrid based technologies. The performance improvement, size reduction, changes in supply chain, impact on wafer test, impact on device test and challenges of working with wafers instead of die will be discussed. This paper will also discuss the results of extensive reliability testing including accelerated aging, thermal shock, pad integrity, drop tests, moisture sensitivity, ESD testing, light sensitivity and hearing aid assembly solder simulation testing.

Introduction

By simply looking at the evolution of cell phones, it is obvious that consumers of today's electronic devices are constantly searching for the smallest, thinnest and most powerful electronics available. The telecommunications industry has been the leader for the majority of these advancements over the recent years, continuously trying to fulfill the need to reduce cell phone size while simultaneously increasing functionality. These demands are no longer limited to cell phones, but now apply to nearly all electronic devices.

Consequently, electronics companies worldwide now share the common need to develop methods to increase the volumetric efficiency of electronic packaging.

A generally unacknowledged driver is the hearing aid industry, which has always concentrated on the efficiency and size of microelectronics due to the extremely limited space of the ear canal into which the hearing aid electronics must fit. Hearing aids were leading the market for reduced-size electronics long before cell phones were even a concept. In fact, the first commercial application for the transistor, a groundbreaking new device at the time, was a hearing instrument produced by Sonotone in 1952. It was a hybrid two vacuum tube, one transistor hearing instrument measuring 3 inches by 2-3/4 inches by ½ inch thick. This was even before the first commercialized transistor AM radio was introduced in 1954. Since then the technology used in hearing aids has continued to advance and successful hearing aid manufacturers have been quick to adopt the latest in microelectronic packaging technology.

Today's hearing aids, such as Starkey's 3-Series product, are far removed from their vacuum tube ancestors. Modern hearing aids have significant computing power, running complex hearing algorithms that have enormous impacts on a patient's quality of life. The current industry trend is continued advancement by adding more memory, more signal processing capability and more wireless capability into the hearing aids to increase functionality and to improve performance. In order to continue to achieve this level of performance increase, the hearing business has had to again take the lead in electronics size reduction by making the most of emergent and leading edge technologies in order to develop and execute novel 3D packaging well ahead of other industries.

Background

The mainstream hearing aids of the late 1990's and early 2000's primarily used thick film ceramic hybrid technology to make exceedingly small hearing amplifiers that consisted of digital signal processors (DSP) tied together with memory (EEPROM's) and the required capacitors. Total part count was typically in the neighborhood of 12 components and the pad count was typically less than 16. Both custom and standard hearing aids were made in this fashion, with the custom aids having wires soldered to pads on hybrids and the standard hearing aids using hybrids that were terminated with solder pads in a ball grid array format (BGA). These BGA's were often SMT soldered to moderately complex flexible circuits [1]. Hybrids intended for standard hearing aids were typically more complex and had additional components SMT attached to the BGA surface and on occasion required cutouts on the flexible circuits to enable mounting.

Around 2006 ceramic hybrid technology was becoming extremely complex with the implementation of vertical interconnects (VIC) to connect multiple ceramic interconnect surfaces. Concurrently, the addition of wireless technology into hearing aids had begun. The thick film ceramic technology was no longer capable of supporting the rapidly increasing part count and required interconnect density necessary to attach 75 or more components without increasing the size to unacceptable dimensions. The dawn of chip-on-flex for hearing aids had begun [2].

Many of the first chip-on-flex devices used 4-layer flexible circuits having 100 um lines and spaces with 100 um vias and 200 um capture pads. These circuits enabled wireless hearing aids at various frequencies without thick film hybrids [3]. Since this time additional features have continued to be added and the limits of flexible circuit technology are consequently pushed to the degree that few companies can produce what is needed. Current products are now being designed with 50 um lines and spaces with 50 um vias and 150 um capture pads in 4 layers. Next generation chip sets will require even more aggressive (and expensive) routing density, down to 25 um lines and spaces with 50 um vias and 125 um capture pads with 5 or 6 layers.

The process of embedding components into circuit boards has been possible for a number of years. Significant performance improvements have been demonstrated by embedding capacitors adjacent to surface mounted die in very close proximity; however, cost has always been a drawback. Embedding processes have continued to improve, most recently with the development of embedding processes for active components in rigid and flexible circuits [4,5]. The capability to embed both active and passive components now provides the advantage of significant circuits size reduction.

For the hearing aid industry, the combination of moving complexity into an ultra-small embedded die module, referred to herein as chip-in-flex (CIF), and delaying migration to even more complex flex is an ideal solution. This paper will discuss the use of embedded die packaging (or chip-in-flex) to drive significant further size reduction in custom and standard hearing instruments over what can be achieved using chip-on-flex or ceramic hybrid based technologies. The size reduction, performance improvements, changes in supply chain, impact on wafer test, impact on device test, and challenges of working with wafers as an alternative to singulated die will be discussed. This paper will also discuss the results of extensive reliability testing including accelerated aging, thermal shock, pad integrity, drop tests, moisture sensitivity, ESD testing, light sensitivity and hearing aid assembly solder simulation testing.

Experimental Data

The embedded die module is fabricated with six copper layers with 40 um lines and spaces and 100 um vias. The die internal to the module is 3,633 um by 3,300 um by 85 um thick. The top surface of the flexible circuit has active and passive surface-mounted components attached. The entire top surface was transfer molded. This device is a stand-alone hearing amplifier that can be used directly in invisible-in-the-canal (IIC) or completely in the canal (CIC) hearing instruments without additional circuit components. The completed module measures 4.39 mm by 3.44 mm by 1.325 mm thick. The two figures below show a previous generation part using chip-on-flex technology with 4-layer copper flexible circuit and the new 6-layer chip-in-flex technology.

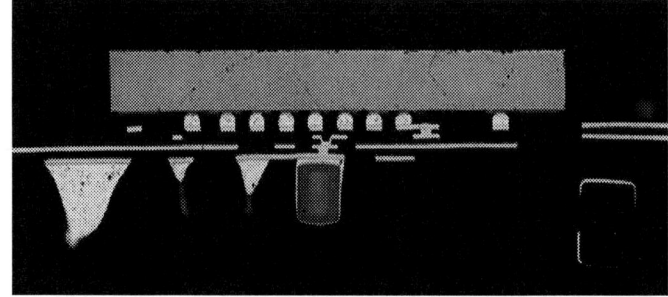

Figure 1. Cross-section of previous generation part using chip-on-flex technology.

Figure 2. Cross-section of new multi-chip module using 6-layer chip-in-flex technology.

Size reduction realized

To show the degree of size reduction possible using embedded die chip-in-flex technology a previous generation ceramic hybrid-based product, produced continuously in high volume since 2009, was redesigned using the embedded die technology. The CIF multi-chip module became a functional equivalent to the ceramic hybrid. The original device was 5.74 mm by 3.45 mm by 2.43 mm thick. The new device was 4.39 mm by 3.44 mm by 1.325 mm thick, resulting in a momentous 60% volumetric size reduction. Figure 3 shows a side-by-side comparison of the two devices in relation to a US one cent coin.

Due to the tremendous size reduction that results in increased fit rates and improved ease of manufacturing compared to the previous hybrid technology, this device was targeted for 2014 production in IIC hearing aids. Previously, the IIC has been using a hybrid that was originally designed for use in CIC hearing aids. The hybrid had required modification by chamfering the top mold material using a custom made miniature router to enable the hybrid to fit adequately inside the smaller IIC shell. Even with this

978-1-4799-2408-0/14 $31.00 © 2014 IEEE

Figure 3. Left, circa 2009 Hybrid (top) and chip-in-flex module (bottom); right, circa 2009 Hybrid (right) and chip-in-flex module (left).

modification the original IIC has a less-than-desirable fit rate in patient ears. Additionally, the required modifications resulted in greater difficulty during device assembly. Implementation of the chip-in-flex module resulted in substantial IIC size reductions and a greatly simplified assembly process. Figure 4 below shows a model comparing the old hybrid technology and the embedded die multi-chip module.

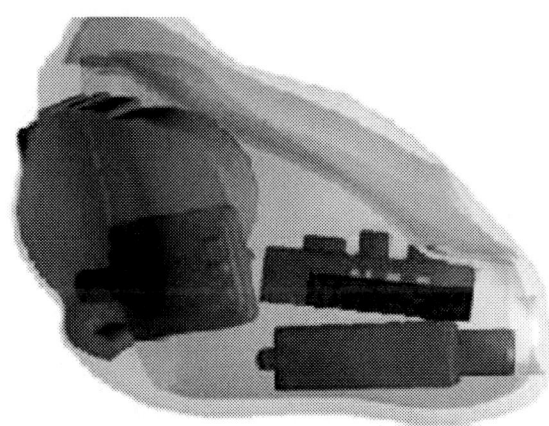

Figure 4. 3D rendering of IIC comparing hybrid (red) and embedded (green) technologies.

Performance improvement realized

A noise comparison was made to determine the impact of moving from the older silver/palladium thick film ceramic technology to the newer, higher quality copper/polyimide embedded technology. An Agilent DSA90804A Infiniium High Performance Oscilloscope was used to analyze noise. The embedded die module was connected to power, ground, microphone, receiver and communication bus. The noise on power rails and audio path was measured and simultaneously compared to the same measurements on a previous generation ceramic hybrid circuit connected in the same way. Noise was

measured from 200 Hz to 10,000 Hz, the frequency range relevant to hearing aids. The new embedded die technology consistently had a lower noise floor; for example, a 47% lower mean VDDC noise was observed. See Figure 5 for screen shot of a noise comparison.

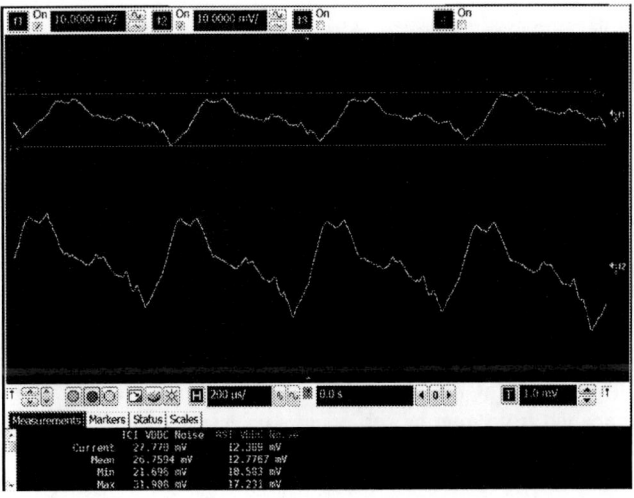

Figure 5. Example oscilloscope screen shot comparing noise of embedded die module to that of a ceramic hybrid.

Changes in supply chain

There were significant supply chain challenges in moving from chip-on-flex to chip-in-flex. Chip-on-flex devices were simple and easy to manufacture, utilizing off-the-shelf die and passives. The die are typically ordered in tape-and-reel format with BGA solder bumps and all other needed parts are ordered with standard SMT terminations. The chip-on-flex parts are then assembled with standard SMT equipment, followed by in-house testing. Infrastructure has been in place for a long time to support this assembly approach. In contrast, chip-in-flex is a wafer-based technology. It is necessary for the integrated circuit supplier or foundry to alter their process flows and testing methods to supply full thickness tested wafers without solder bumps.

An additional redistribution layer (RDL) needed for the embedding process was also necessary and had to be added to the wafers after receiving them from the foundry. The DSP chip has separate analog and digital sections that significantly impact routing of the RDL. Design of the RDL can negatively impact noise performance. This work had to be completed without the help of the chip supplier. Following application of the RDL the wafers needed to be thinned to 85 um and then diced. There are a limited number of back-end wafer processing companies that can do this processing. Module assembly then requires that the chips be picked from the dicing frame rather than from tape-and-reel. The die are very fragile when thinned to 85 um and picking from frame is the only practical handling method. Die handling was now associated with thousands of very small parts (typically around 10 mm^2 each) on frame compared to one at a time in tape and reel format. Only good die are picked based on the electronic wafer map that must also be supplied from the silicon test facility. Additionally, the process flow for 200 mm wafers differed from that of 300 mm wafers due to the limited number of places with 300 mm wafer handling capabilities. The cost to participate in this technology went from $5 to $15 for a die (typical die cost) to $20,000 or more for 200 mm wafers and to more than $45,000 for 300 mm wafers.

Impact on testing

Previous generation hybrids and surface mount assemblies were tested internally immediately following assembly, either in panel form (panels sizes ranging from about 6 cm square to about 13 cm square) or after dicing. The embedded die modules are also produced in panel form, but not internally and not with the same panel sizes previously used. They were now in long strips with many more parts per strip. A new generation vision system based tester had to be designed and built and then consigned to the supplier. It was also necessary that consigned testers load encrypted hearing aid-specific firmware. Internally, device firmware revision control is managed from internal servers in a private network. A means to maintain confidentiality, prevent pirating, and to not disrupt internal company control methodology needed to be designed, developed and deployed. After extensive collaborative development work, a tester was deployed that appeared to operate as if it were internal to our company, but was actually operated at a subcontractor by the subcontractor's operators. Tester support also needed to be arranged.

Embedded die module reliability

Extensive reliability testing was performed on this new technology. The evaluation included accelerated aging testing, thermal shock testing, pad integrity testing, drop tests, moisture sensitivity testing, ESD testing, light sensitivity testing and hearing aid assembly solder simulation testing; the results are discussed in the following sections. Additional tests that were performed included packaging and labeling characterization, device electrical testing, plated via reliability testing, visual inspection, dimensional analysis, internal assessment of layers, and construction analysis.

Accelerated aging

Accelerated aging testing of electronic components evaluates the field integrity and longevity of the electronics. The test simulates operation of the circuit in a climate of extreme heat and humidity over a period of several years, monitoring for dendrite growth and changes in bias current. Wires were soldered to a positive pad and a ground pad of the test circuits. These wires were soldered to the appropriate connections on a test board.

A sampling of 28 circuits was placed in an environmental chamber and a bias voltage of +1.25 V applied to the parts via a sealed cable port. The chamber parameters were set to 85 °C, 85% RH. A chart recorder was connected to record heat and humidity levels during the test. The current of each circuit was monitored every 7-10 seconds, regularly checking for anomalies, such as dendritic growth. All samples were run for 192 hours, then continued to 1000 hours. One circuit exhibited low drain following 196 hours. The remaining circuits which passed 196 hours remained in the tester for a total of 1000 hours of exposure. During this time one additional circuit failed, also due to low drain. All other circuits passed the electrical testing and had no visual evidence of external dendritic growth following 1000 hours of exposure. See Chart 1 for the current vs. minutes chart for the first 24 hours.

Both low drain circuits were subjected to failure analysis and were found to have an open in a layer-to-layer via connection that was very close to the module edge. This finding was traced back to a process change made during transition from development to manufacturing where the singulation operation changed facilities and equipment. This change caused a pad edge to module edge design rule violation that could be corrected by increasing the module final size by 40 um to accommodate the manufacturing equipment set.

Thermal shock

Thermal shock testing is used to demonstrate the thermal and mechanical robustness of typical assemblies. There are no definitive "pass/fail" criteria for this test, rather it is used for comparison testing of old versus new components and for benchmarking competitive products. De-ionized water was used for all testing. A sampling of 28 CIF modules were first exposed to an ice-cold water bath for a minimum of one minute, then immediately transferred into a boiling water bath for a minimum of one minute. This cycle was repeated 20 times. A visual inspection of the 28 modules, post-exposure, showed no evidence of delamination or damaged solder joints. There was also no variation in the electrical tests of the modules pre- versus post- exposure. See Figures 6 and 7.

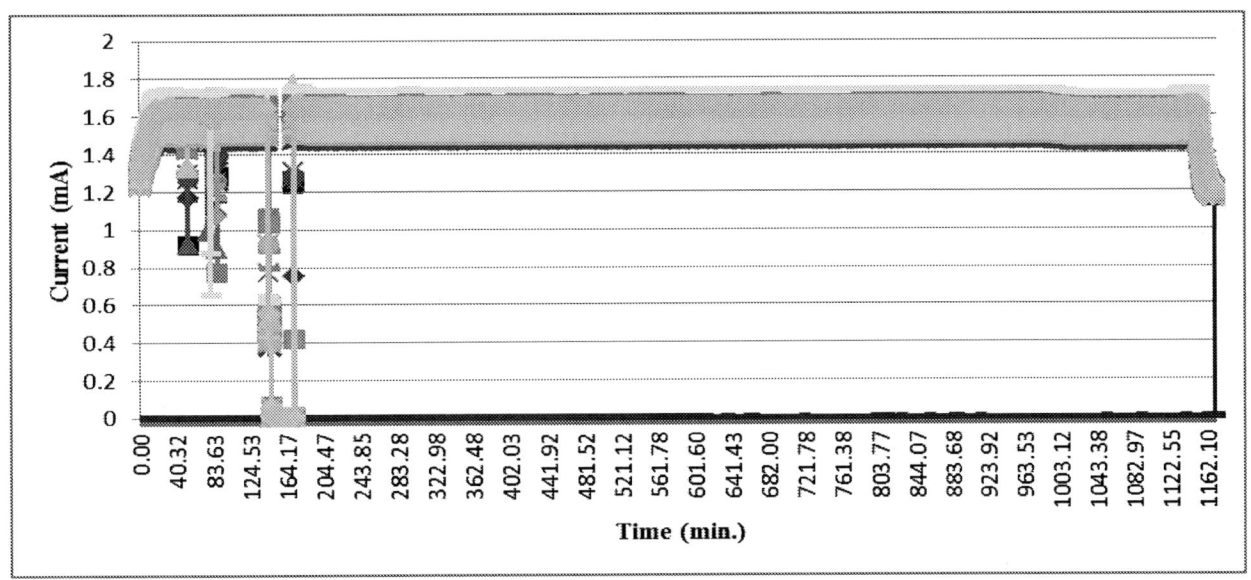

Chart 1. Example current vs. time chart for 24 hours.

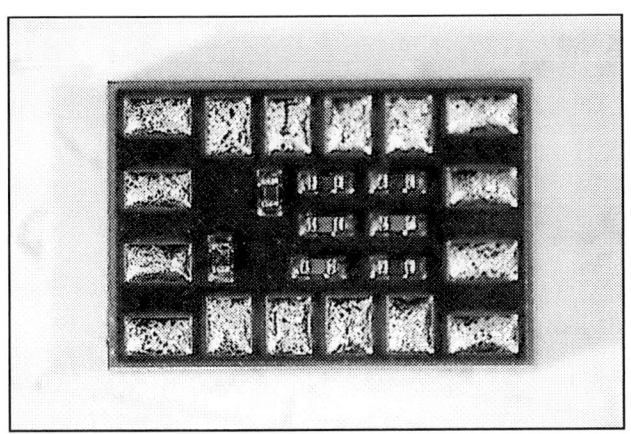

Figure 6. CIF module following thermal shock testing. No anomalies were found on the pad side after 20 cycles in a sampling of 28 modules.

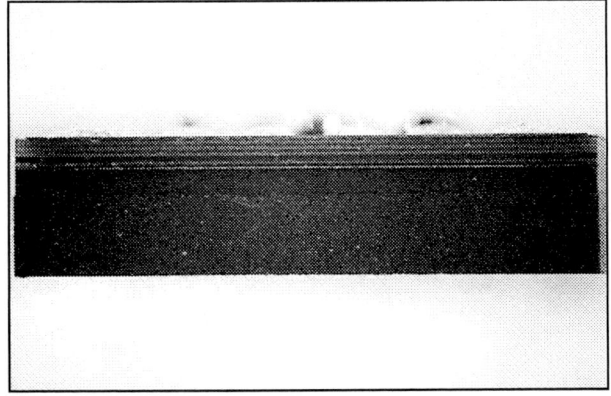

Figure 7. CIF module following thermal shock testing. No cracking was observed on either the short side or long side following 20 cycles in a sampling of 28 modules.

Pad integrity

Pad integrity testing simulates the mechanical and thermal stresses experienced by the solder pads during hearing aid production where wires are directly attached to the module. Mechanical stresses from wires and leads can cause pad lifting and repeated solder iron application can cause pad leaching. Wires were soldered to each solder pad and the wire looped and hooked onto a force gauge. The gage was set to a travel speed of 0.50 in./min. and the pad pulled to failure. The tensile force is recorded at failure. Three CIF modules were tested with 10 pads tested on each. Values of 0.5 lb. or greater were considered acceptable. All pads passed with the exception of one having a value of 0.37 lb. The cause for this failure was unclear and deemed as an anomaly. The testing will be repeated with a subsequent group of samples to check for similar low values in order to determine the frequency of the issue and ensure it was an isolated case. See Figure 8.

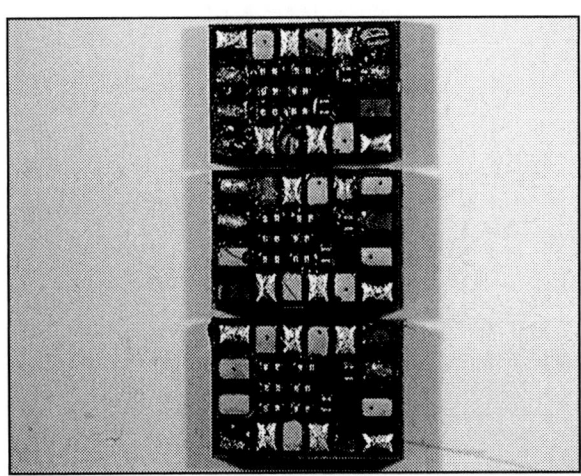

Figure 8. Three CIF modules following pad integrity testing.

To further evaluate the pad integrity, a soldering iron at 700°F was applied to selected pads to test for leach resistance. The soldering iron was applied for three seconds, removed for three seconds and the process repeated for 20 cycles. No leaching was observed.

Drop tests

The drop test is used to determine the effect of mechanical shock on assemblies. A passing module is determined as one that does not show any visual damage or electrical testing variation pre- versus post- drops. The CIF modules were dropped from a height of approximately 32 inches on to a vinyl tile floor. In each set of drops, six sets total (1/axis), the circuit was dropped three consecutive times. A total of 18 drops for each circuit were performed, followed by an electrical test and visual inspection. Twenty-two circuits were tested for a total of 396 drops. All circuits passed visual examination and electrical testing.

Moisture sensitivity

Moisture absorption has the potential to cause delamination or other damage when the assembly is exposed to solder heat. A sampling of 10 CIF modules was used for this testing. It was first necessary to remove any moisture absorbed prior to the start of the test. An initial weight for each module as received was recorded. The parts were weighed directly after opening the packaging provided by the supplier. There was no desiccant, indicator card or moisture barrier bag used with these parts. Following the initial weighing, the parts were placed in an oven at 120°C for 24 hours after which the parts were then re-weighed following the bake. Most of the moisture was gone after 1.5 hours. See Figure 9 for the desorption curve.

After completing the moisture removal procedure and weighing of samples, the CIF module samples were then placed in a 40°C/95% RH chamber. The parts were removed from the oven at specific intervals to characterize the moisture

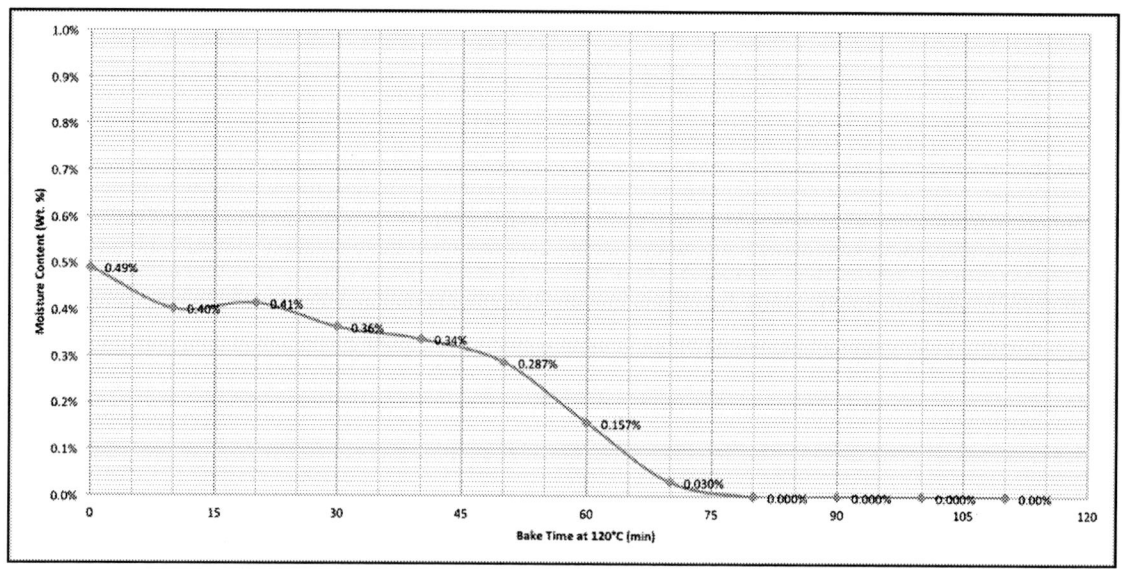

Figure 9. Moisture desorption rate of CIF modules as received.

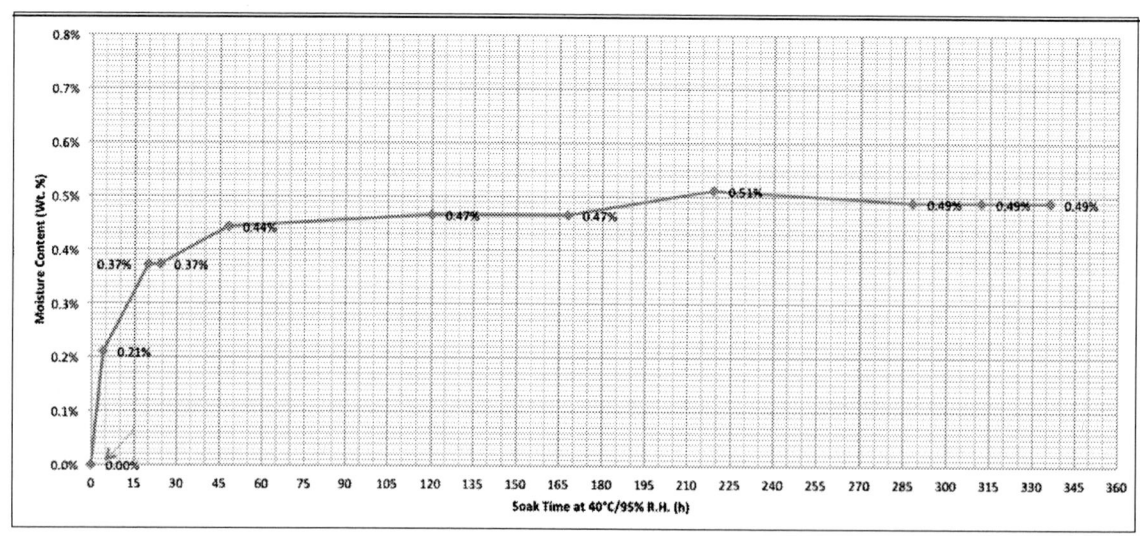

Figure 10. Moisture absorption rate of CIF modules.

absorption curve over time. Most of the moisture was absorbed within one hour, and a steady state was achieved within four hours. See Figure 10 for the absorption curve.

ESD testing

Electrostatic discharge (ESD) can cause device failure and this test is used to characterize the susceptibility of an electronic component to ESD damage. MIL-STD 883F, Method 3015.7 was used for all tests. A sampling of three modules was used and all passed without issue.

Light sensitivity

The light sensitivity test is designed to determine if the electric and/or acoustic functions of a device is adversely affected by exposure to sunlight. Three devices were wired so that they were configured for current drain and acoustic testing. A baseline measurement of the current drain and acoustic response was measured and recorded without a light source. The devices were then exposed for one minute to a light source of 12,000+/-500 foot candles as recorded by a light meter. During the exposure the current drain and acoustic response were again measured and recorded. No distortion was observed in the current drains or acoustic responses of the devices, indicating the CIF modules did not exhibit any light sensitivity. See Figure 11.

Hearing aid assembly/solder simulation testing

The purpose of the hearing aid assembly/solder simulation test is to simulate the stresses (thermal and/or chemical) that a device will be exposed to during the actual SMD and hearing aid assembly. The devices were baked for up to two hours at 110°C to remove any moisture. Flux was then applied to the user pads, and a soldering iron at 700°F touched each pad for two seconds, and then removed from the pad for two seconds,

repeating this cycle twice. In a sampling of 22 parts three CIF modules exhibited electrical testing errors following the soldering process. One error was the result of the soldering iron tip knocking off a capacitor during the testing process. The CIF module pads are smaller in size than the soldering iron tips currently used for testing and assembly; see Figure 12 for visual comparison. Consequently, new soldering iron tips are under evaluation and the passive components will be underfilled during subsequent builds to ensure a robust attachment. Additional assembly training will be necessary as well to guarantee uniform quality.

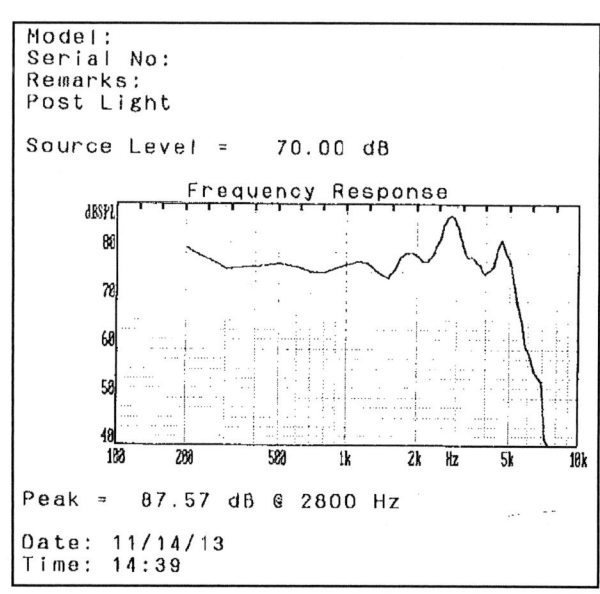

Figure 12. Pad size vs. solder tip. The CIF module pad is smaller in size than the soldering iron tip used in hearing aid simulation testing.

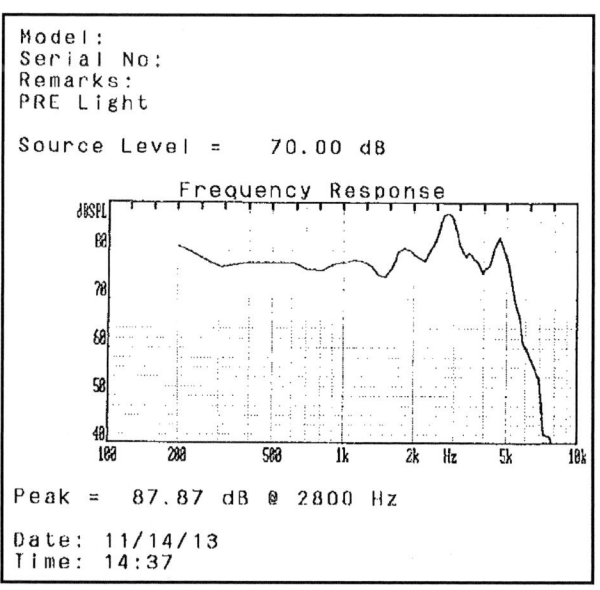

Figure 11. Acoustic response of device prior to light exposure (left) and during light exposure (right).

The remaining two CIF modules failed testing due to the errors exhibited during the final electrical test. Failure analysis of these samples involved, firstly, analysis by acoustic scanning, followed by de-encapsulation of the modules, and finally, cross-sectioning. The process revealed that there was die edge chipping of the surface mounted die within for both modules; see Figures 13 and 14. Tighter "pass/fail" criteria will be established for the incoming die, and the die placement process will be reviewed to check for any deficiencies that may lead to a less-than meticulous placement process.

Figure 13. Acoustic scan image of failed module with location of die chipping indicated.

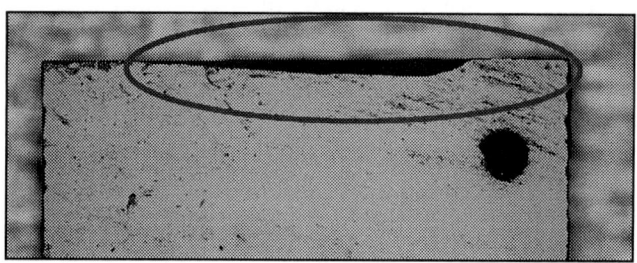

Figure 14. Surface mounted die with obvious chip-out following de-encapsulation.

Conclusions

We have presented and discussed the use of embedded die in flexible circuits for hearing aid applications. We have shown that significant size reduction is possible in combination with performance increase. We have also shown that the circuits are extremely reliable. In addition to what we have shown, these CIF modules passed all of Starkey's additional tests defined earlier as packaging and labeling characterization, device electrical testing, plated via reliability testing, visual inspection, dimensional analysis, internal assessment of layers, and construction analysis. We expect that this technology will be part of mainstream hearing aid production for years to come. Recent announcements have shown this technology is only in its beginning stages and many improvements are likely to come over the next few years [6].

References

1. J. Dzarnoski, D. Link., "Thick Film Ceramic Chip Packaging For Hearing Aids," in *Proc. of the 5th International Conference on Ceramic Interconnect & Ceramic Microsystems Technologies*, Denver, CO, April 21-23, 2009, pp. 63-70.

2. J. Dzarnoski, D. Link., High Density Hearing Aid Chip Packaging," in *Proc. of the 42nd International Symposium on Microelectronics*, San Jose, CA, November 3, 2009, pp. 18-25.

3. J. Dzarnoski, "Miniaturization of Microelectronics for Hearing Aids," in *Proc. of Life Science Alley – Advancing Medical Technology*, Minneapolis, MN, December 7, 2011.

4. M. Beesley, "Embedded Components on the Way to Industrialisation," in *Proc.of the SMTA Interational Conference*, Austin, TX, October 16-20.

5. K. Itoi, "Laminate Based Fan-out Embedded Die Packaging Using Polyimide Multilayer Wiring Boards", in *Proc. of the 8th International Wafer Level Packaging Conference,* Santa Clara, CA, October 3-6, 2011.

6. K. Itoi, T. Tessier, et al, "Polyimide PCB Embedded with Two Dies in Stacked Configuration," in *Proc. of the 10th International Wafer Level Packaging Conference*, San Jose, CA, November 5-6, 2013.

Fabrication of Silicon Based Microfluidics Device for Cell Sorting Application

Bivragh Majeed, Chengxun Liu, Lut Van Acker, Robert Daily, Tomokazu Miyazaki[1], Deniz Sabuncuoglu and Liesbet Lagae

Imec
Kapeldreef 75, Leuven 3001, Belgium
[1]JSR Micro NV, Technologielaan 8, B-3001 Leuven, Belgium

Bivragh.majeed@imec.be, 003216287735

Abstract

In this paper we report on a novel silicon fabrication process of a microfluidics device used for cell sorting application. The process uses wafer level fabrication technique that allows for a small form factor device in combination with a new photopatternable polymer with excellent properties for microfluidics on silicon. The ability to process on wafer level distinguish this from other processes, whose yields are limited to few test samples. The device fabrication includes: processing of micro-heaters, definition of polymer microfluidic channels and collective die-to-wafer bonding of glass substrate onto the polymer channels. The paper will focus on the development of a wafer scale process flow and initial characterization of the available devices.

Introduction

Fluorescence activated cell sorting (FACS) is an important diagnostic tool in microbiology to separate different type of cells at a very high throughput. In this technique, the charged cell droplets pass through a long channel, illuminated by laser and deflected by high electric field. Currently, there are numerous commercial products that are available for cell sorting, however they are quite bulky and expensive[1,2]. Microfluidic implementations of FACS instruments are therefore considered to be an important new direction of portable and low sample-volume applications. The microfabricated channels on silicon in lab-on-a-chip are used to analyze and sort the cells.

We are developing a technology for rapidly detecting and separating different type of cells. The device would have high through-put, low cost and ease of use. The device has three main components: a CMOS imager, microheaters and polymer microfluidics channels. The operating principle is based on separating cell with the help of a bubble jet flow. When a solution is loaded into the chip, it moves through the main polymer microfluidics channel. A light illuminates the cells and results in a fringe image. The fringe image is reconstructed into a 2D image to get the information about different type of cells. To sort the cell, microheater located downstream to the detector generates a rapid jet flow. The jet flow deflects the main flow and thus directs the cell in the right channel. The sorted cell than can further analyzed. In the future, these channels can be parallelized to from a very high throughput device.

The main distinguisher for our approach is the use of production friendly approach that is essential for the mass production of this technology. In standard reported application[3, 4] PDMS (polydimethylsiloxane, a silicon-based organic polymer) is widely used. PDMS is difficult material to be integrated with standard CMOS fabrication facility and it is not very easy to bond Si and glass with this material. Furthermore, PDMS is known to bare much higher autofluorescence compared with glass. Last but not least, in the case of cell sorting, a rigid material (i.e. high Young's modulus) such as glass is very important to achieve high system frequency response.

Here, we report on a new polymer material that overcome these issues. Photo-sensitive Adhesive material (PA), developed by JSR corporation Japan, is a new negative tone photo-patternable adhesive. It not only defines the microfluidics channels but also acts as bonding material. It has excellent resolution, has low curing temperature and has straight patterning profile essential for microfluidics channels. PA is also a biocompatible material. It can be used for direct thermal bonding for die to die or die to wafer bonding of Si onto glass.

In the paper we will report on a lab-on-chip prototype cell sorter that can process up to 2000 cells per second. One of the targeted applications is detection of circulating tumor cells in human blood. The paper will focus on the technology for the prototype and processing that will allow for wafer level processing. We will report on microheater, the automation, metrology and processing parameters for polymer adhesive material targeting a fabrication friendly process. The final device showing the cell sorting will be demonstrated.

Process flow

The processing starts with the formation of specifically designed microheaters on a 200mm silicon wafer. After growth of a 400nm thermal oxide (fig 1a), the heater stack consisting of Ti/AlCu/TiTiN layers with a 20/200/20/100 thickness(fig 1b) is deposited in AMAT_ENDURA tool. AlCu is the conducting layer while TiTiN acts as a barrier layer for the outside solution in the final device. The heater structures are patterned with lithography using 1200nm OIR resist. The metal stack is then etched (fig 1c) in LAM-Alliance A6 FEOL cluster with chloride based chemistry. The heaters are capped with an oxide (fig 1d) in AMAT-CVD chamber. Second lithography step defines the bond-pad and electrode opening (fig 1e). Oxide is then dry etched with a CF_4 and O_2 chemistry in a in LAM-Alliance A6 FEOL cluster. The resist is stripped and a third lithography is done to define only the bond pad opening (fig 1f). The top TiTiN layer from the bond-pad is then etched to allow for good wire bonding to the AlCu bond pad. In the next step, PA is coated on to the wafer. In the current device the minimum channels are 30µm wide and thickness of the polymer is also 30µm (fig 1g). The last step of the fabrication process is the bonding of glass die onto the si wafer (fig 1h). The glass is pre-punched allowing the polymer microfluidics channels connection to the outside testing equipment. The bonding is done in two step, firstly individual glass dies are bonded onto Si with a pick and place tool. Secondly, after populating the wafer collective bonding under force and temperature is performed to complete the processing.

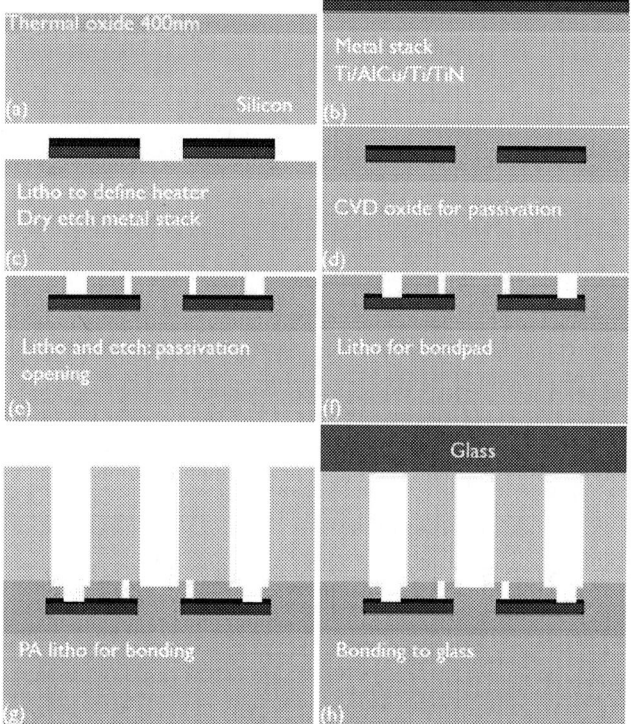

Figure 1. Schematic process flow for the formation of cell sorter device

Results

Figure 2 shows the results for the formation of the heater element. Figure 2a shows a top image of the full heating element, figure 2b shows the tip of the heater. The tip is very sharp that guarantees a very good bubble generation. Figure 2c shows the cross-section image of the metal stack after etch. The images shows a very clean etch and very selective metal etch landing on oxide.

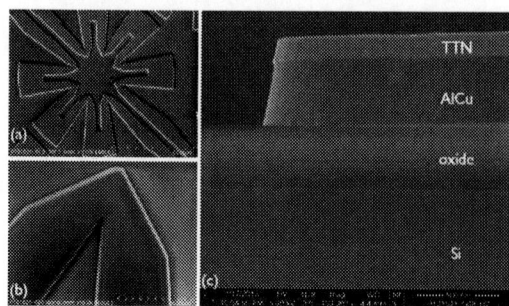

Figure 2. (a) Top view of heater element, (b) top view of heater tip, (c) Cross-section SEM image of etched metal stack

In the next step, the opening of bond pad and electrode was done followed by etching of the top TiTN layer. Figure 3a shows the XSEM image of the bond pad opening where top TTN is etched, landing on Al. The etch is very selective and no damage to the Al layer is observed. Figure 3b shows the electrode area where TTN is still present. These steps complete the processing of the microheater element for the cell sorter chip.

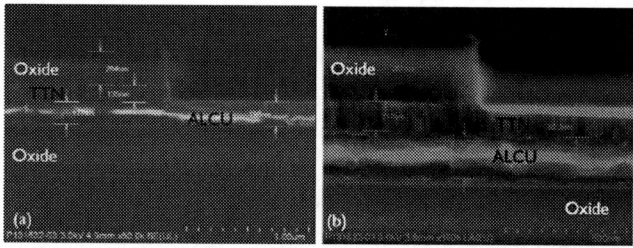

Figure 3. (a) SEM image of bond pad opening landing on ALCu. (b) SEM image of the electrode with TTN protection

PA processing

This section will focus on the processing of the PA material. The advantages of PA material are already discussed in the introduction, while here we describe the implementation of PA on wafer scale production. The parameters that were analyzed were automation and compatibility in our production line, check metrology set-up, check two different PA material: PA-S321 and PA-S321H capabilities for required application and check exposure and development process parameters. In the current device the minimum channels are 30µm wide and thickness of the polymer is also 30µm. In the first experiments, it was found that PA is compatible with backside rinse with RER500 and can be done on ACT12MTM coating tool in out production line. The thickness of the coated film is checked on the automatic Senduro (49 points) tool. Figure 4 shows the thickness of the PA material as a function of the spin speed. It shows that with PA_s321 with a single coat a maximum thickness of 15µm while with PA_s321H, a spin in a single coat a 25um layer is possible. PA_ s321 should be coated 2 layers at 1000rpm PA_s321H should be coated 2 layers at 2450rpm. Due to higher viscosity for PA_s321H, it is easier to use the thin version and all the subsequent development was done with PA_s321

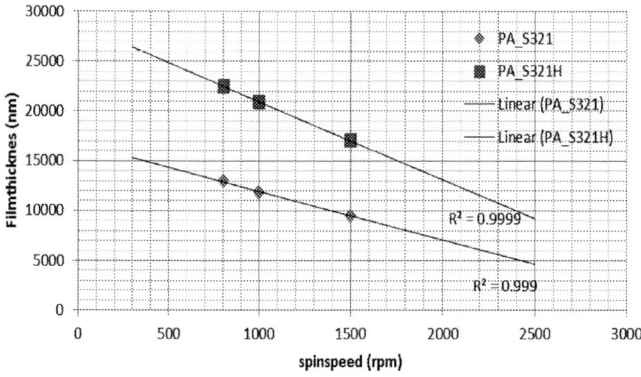

Figure 4: PA Film thickness versus spin speed for two different PA material.

After finalizing the PA thickness, the next step was optimization of exposure and development of PA. The exposure was done on an Ultratech AP300 tool with a test mask. The dose used was 1500-1700 with a step 20mJ. The development involved post exposure bake (PEB) of 110°C for 300sec and two times 60sec puddle with TMAH(2.38%), followed with 60sec rinse. The process finished with B-stage bake. From our experiment it was concluded that material is not very focus sensitive, exposure dose of between 1500 and 1700mJ i-line is sufficient and a dehydration bake has a positive effect on the performance by reducing wrinkles. With these conditions the first devices were fabricated. Figure 5 shows the results of formation of polymer microfluidics channels with PA material. Fig 5a shows very clean surface after PA processing and very straight profile, fig 5b shows the required thickness of 30μm of the PA material and fig 5c shows portion of cell sorter formed with PA material indicating different elements including flow channel, microheater and the exit channels.

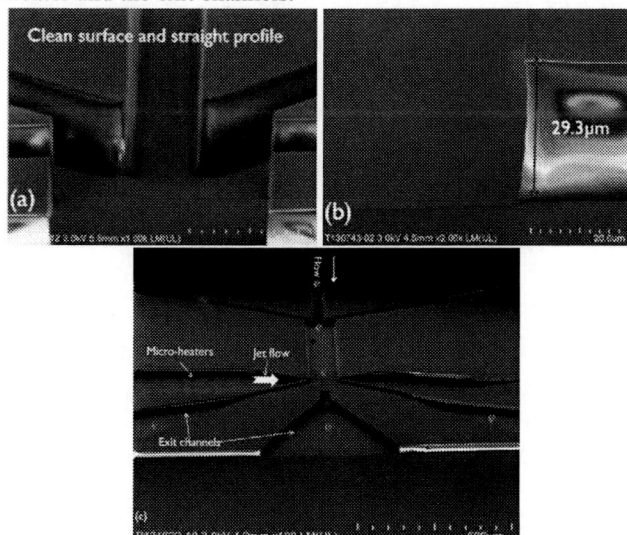

Figure 5. (a) Microfluidics channels formed in PA (b) PA material channel with the 30μm channel, (c) XSEM image of the full cell sorter chip showing different components.

The next step in the adaptation of PA material was to check the processing of PA for a full batch of 23 wafers. In the first experiments, there was a delay of 30 minutes between exposure and development. The results were quite interesting as given in fig 6. A delay of 30 minutes resulted in bridging of PA as well residues in the channels. It was estimated that at least 150min delay is needed between exposure and development for a full batch of wafers to be processed. A design of experiments was carried with exposure, PEB time and temperatures as the main parameters. The exposure was checked between 700-4000mJ, PEB time of 300-420sec and temperature from 80-120°C were tested. The results are summarized in figure 6 which showed that if the exposure dose is very low:

- trenches are closing
- the polymer start wrinkling

and if the exposure dose is too high:

- the crust problem gets worse

In another parameter, four puddles for development was tested but this results in the edges being attacked and after 30min delay the crust residues are not removed. The best solution is for lower PEB, higher energy resulting in long delay time without detrimental effects. The final conditions for processing were determined to be the exposure energy of 4000mJ and PEB at 80°C for 300 sec., which allowed for a delay of up to 3 hrs.

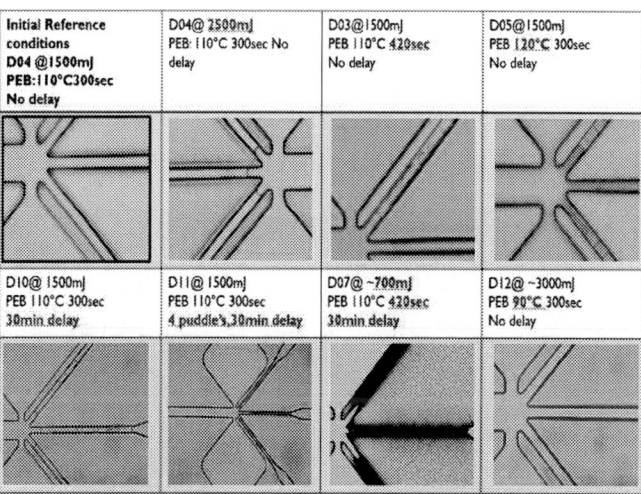

Figure 6. PA optimization DOE parameters: Higher energy, longer development time, longer PED time and higher PEB time

The last step of the fabrication process is the bonding of glass die onto the si wafer. The glass is pre-punched allowing the polymer microfluidics channels connection to the outside testing equipment. The bonding is done in two step, firstly individual glass dies are bonded onto Si with a pick and place tool. Secondly, after populating the wafer collective bonding under force and temperature is performed. The population step was done on BESI 2200 EVO FlipChip Bonder. During the bonding, the bottom wafer is mounted on a carrier and the glass wafer is remounted wafer into a 300mm tape on the ring. The bottom chuck is heated to a temperature 100°C, while top chip is not heated and roughly 500gF is applied for 2 sec. The step is repeated till the whole wafer is populated. The collective bonding is done on a EVG Bonder 200. The top chuck did not need any compliant material. Two different setting for temperature from 180°C to 230 °C for 2 hours with a force of 20N to 80N per die are applied with top and bottom chuck temperature setting are the same. The force and temperature are applied simultaneously to complete the process.

Figure 7 shows the XSEM image of the bonded device with PA. It shows that during bonding there is no deformation of PA after the bonding. There is no delamination between the PA, si and glass substrate even after cleaving the die. This indicate very good bonding and this will be tested with a microfluidics test discussed in the last section. Figure 8 shows the first fully processed and populated wafer with 20 cell sorter die.

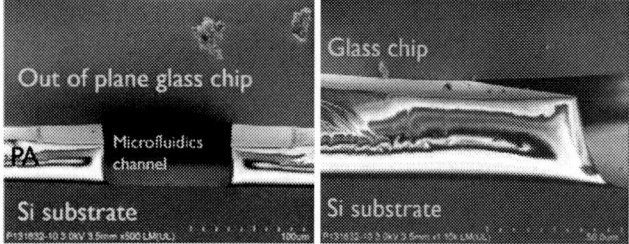

Figure 7. SEM image of a bonded die cross-sectioned

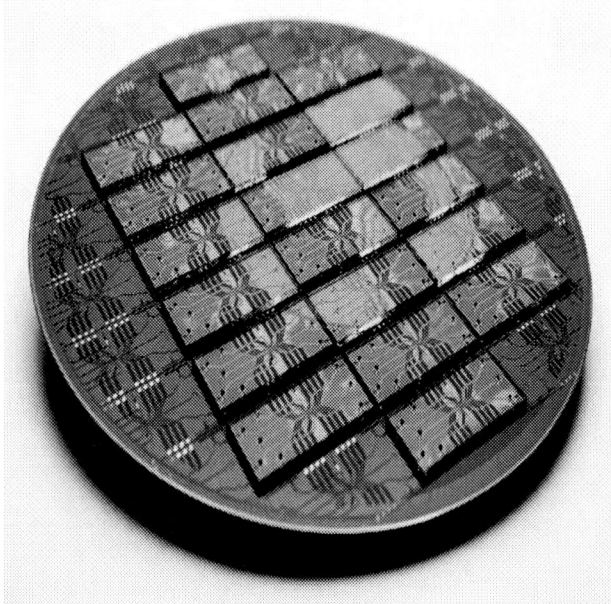

Figure 8. A fully processed wafer with 20 dies bonded onto a wafer

PA characterization and cell sorting

In the first test, the bond strength of PA material to glass and si substrate is tested with a leak test. A test chip is attached to external pumps and pressure is measured as function of flow rate. Preliminary result on a test sample are shown in figure 9. It shows that our devices can withstand a pressure in excess of 3 bar which is more than sufficient for most microfluidic applications such as cell sorting.

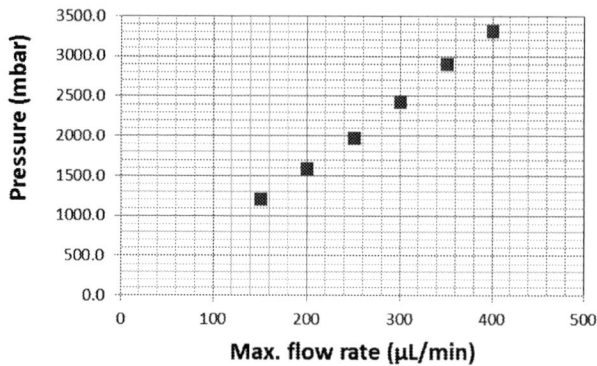

Figure 9. Pressure as a function of flow rate for one die bonded with PA

To test the frequency response of the jet flow, the full system consisting of cell sorting device was tested with electrical and fluidic connection to peripheral apparatus. The process involved detection of incoming cells, generation of heat pulse, formation of jet flow and resulting cell sorting. Figure 10a shows the initial state of the cell sorter device where cells flow in the center of the channel by hydrodynamic focusing. After detection of cell, bubbles are created in heater. The steam bubbles have a minimum life time of less than 10 µs. Figure 10b shows the jet flow after 15 µs heating pulse, picture taken by stroboscopic technique. Thus we demonstrate a successful fluorescent cell detection for human white blood cells.

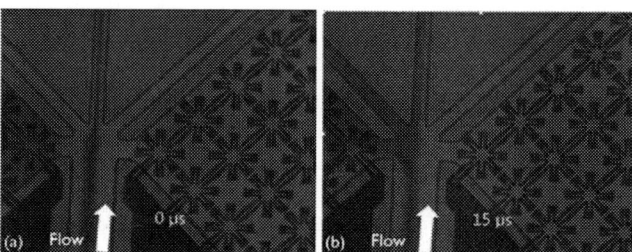

Figure 10. (a) initial state of the cell sorter (b) after bubbles are generated and flow is diverted.

Conclusions

We reported on an integrated microfluidic fluorescence cell sorter technology (µFACS) based on on-chip micro jet flow. The lab-on-chip prototype can process up to 2,000 cells per second. We showed the first fully processed wafers with 20 microfluidic devices. Process optimization of PA bonding material have been reported and initial results of working principle of our cell sorter was shown. Silicon technology is a cornerstone of eHealth and with wafer level PA integration optimization we have reached an important milestone towards the future.

Acknowledgments

The authors would like to thank JSR technical team for their useful discussion and for providing the PA test material.

We appreciate the help of Tania Dupont and Rita Verbeeck of imec with the help in cross-sectional SEM images.

References

1. A. Bhagat, H. Bow, H. Hou, S. Tan, J. Han, C. Lim, "Microfluidics for cell separation," *Med Biol Eng Comput,* vol. 48, pp.999–1014, 2010.

2. M. Eisenstien, "Devide and conquer," technology feature Cell sorting, Nature, Vol 441, pp. 1179-1185, Jun. 2006

3. H. Hoefemanna, S. Wadlea, N. Bakhtinaa, V. Kondrashova, N. Wanglerb, R. Zengerlea, "Sorting and lysis of single cells by BubbleJet technology" *Sensors and Actuators B: Chemical,* vol 168, pp. 442–445, June 2012.

4. P. Quinto-Su, H. Lai, H. Yoon, C. Sims, N. Allbritton and V. Venugopalan, "Examination of laser microbeam cell lysis in a PDMS microfluidic channel using time-resolved imaging", *Lab Chip*, Vol. 8, pp. 408–414, 2008.

A Novel 3D Neural Probe with Integrated Channel and Its Package

Xingming Fu[1], Yong Xu[2], Yuefa Li[2], Jinsheng Zhang[4,5], Xiaobing Luo[3], Sheng Liu[1*],

1 Cross-disciplinary Institute of Engineering Sciences, School of Power and Mechanical Engineering, Wuhan University,
Wuhan, Hubei, 430072, China
2 Department of Electrical and Computer Engineering, Wayne State University, Detroit, MI 48202 USA
3 Huazhong University of Sci & Tech, Wuhan, Hubei, 430074, China
4Department of Otolaryngology, Wayne State University, Detroit, MI
5Department of Communication Sciences & Disorders, Wayne State University, Detroit, MI
Victor_liu63@126.com

Abstract

Due to the extreme complexity of neuroscience and neurosurgery, there is naturally a need for probing systems with more functions which can probe in three dimensions. For instance, nerves can size differently in different locations and can present three dimensional neuro imaging with multiple recording and stimulating points, calling for the new developments in this emerging field. It is also highly desirable for drug delivery function and other functions to be integrated with needles and electrodes. This article presents MEMS based 3D neuro imaging probing that can precisely control electrodes size, shape, and density, and may be able to integrate micro-fluidics based drug delivery. It is also perceived that the micro-needles can also be integrated with fiber optics laser stimulation, chemical stimulation, thermal actuators to pump in thermal energy in various wave length range, such as those burning of folium artemisiae argyi, a traditional Chinese medicine, often used in acupuncture, electrical impulse, etc. This article will present the developments of neuro needles in terms of features of packaging and interconnect, with particular interest in 3D needle array packaging and interconnect. The demonstration of a prototype used in the actual animal testing will also be given.

Introduction

Neural probe is a useful tool for neurophysiology research and neural prostheses by detecting neural signals from the brain and recording extracellular neural signals or stimulating neurons. Many efforts have been done on metal wire and glass microprobe development [1-8]. However, due to the disadvantages of big size, poor reproducibility and difficult to realize multi-point recording, their applications are limited in neural research. Recently, due to the biocompatibility and the mature fabrication process of silicon, the MEMS based neural probe is attracting increasing attention.

MEMS micro fabrication technology can precisely control electrodes size, shape, and density, and may be able to integrate micro-fluidics based drug delivery [9-10]. Some exciting work such as integrating the needles with fiber optics laser stimulation, thermal actuators to pump in thermal energy in various wave length ranges, will all depend on the development of 3D needle array packaging and interconnect. MEMS technology, by using the photolithography, can provide the precision size control of electrodes, needles, etc. 3D array packaging can be realized by stacking, bonding, and

integration of TSV. In our group, we proposed to use folded technology to form 3D needle array [11]. In the following, we will review some key technology developments associated with features of packaging and interconnect. Finally, some results based on 2x3 array will be demonstrated and preliminary results will be presented.

The most three common neural probes are the Michigan Probe, Utah Array Probe and Micro-wire Array Probe. Although Micro-wire Probe, which typically consist of one to four insulated micro-wires with exposed tips, are simple, it is limited in their geometries and reproducibility, causing considerable insertion damage, and tending to splay out in tissue, making exact site placements uncertain[12]. Michigan produced a variety of microprobes in single-shaft, multi-shaft, and 3D stacked layouts based on MEMS technology [12-14]. Their packaging was also mainly based on silicon technology, where the interconnection was made with polysilicon which is a 4-5 μm thick silicon cable. However, poor reliability often occurs in this packaging technology: the cable is easy to break, provides low yield for longer lengths because of the high aspect ratio, and is not robust enough. The Utah probe arrays are built in the direction of the probes [15-16]. While its interconnection was realized by a bundle of insulated metal wires bonded on the back of the array, and the stiffness of the metal wire bundle makes these unfeasible for chronic implantation, especially for the high-density electrode arrays.

Extensive literature on MEMS technology based probes, such as those well-known Michigan probes and Utah probes, have been reviewed [11] and there is no need to present here. The current paper will focus on the packaging and interconnect designs to form various neuro needle systems. It is hoped that the procedure is general enough and high density needle array will be realized. Preliminary animal test using this novel 3D neural probe will be presented briefly in this paper. Spikes from two neurons are recorded from the primary auditory cortex of a rat and the data is then processed by a unit sort program.

Packaging and interconnection of 3D Microprobe

Two planar devices consisting of 3 silicon islands, two shank islands with three 2.8mm length probes in each island and one interfacing island with several bonding pads on it, are fabricated using MEMS technology and a flexible skin process. The silicon island with probes is shown as Fig.1.

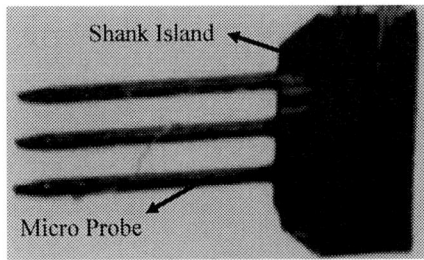

Figure 1. Device island and micro probe.

In fact, the shank island shown in Fig. 1 is a 2D device with only 3 probes. To realize the 3D packaging, a folding process which stacks the two shank islands by back-to-back method is used in this paper.

The two shank islands are aligned and then bonded together. Good alignment of micro probe is an important criterion on the quality of packaging. In this work, it is realized by manually aligned through an optical microscope. Useing of alignment holes or marks will get a better aligment results. The conventional glue method is used for bonding. Sheng Liu [17] have pointed that the properly soft adhesive material (small Young's modulus) is better choice for stress release caused by CTE (Coefficient of thermal expansion) mismatch when the devices undergo temperature changing.

To avoid the poor interconnection reliability shortage mentioned above to the packaging of Michigan probes, flexible cables is used for connecting silicon islands in this paper, as presented in Fig. 2. Flexible cables that connect these silicon islands could increase the robust of the connection. Parylene is widely used for implantable material due to its good biocompatibility. Therefore, these silicon islands are connected by a parylene layer, in which metal interconnects and microchannels are embedded. However, this kind of cables will experience stress concentration at the edge of the silicon islands and fracture the metal traces. Yong Xu [18] have proposed a micro cushion structure based on parylene micro-channels to minimize the stress concentration.

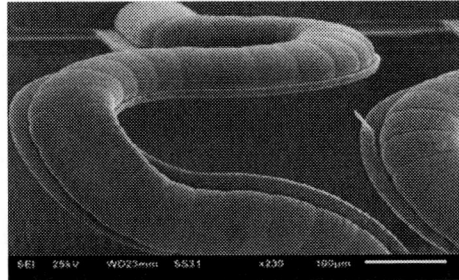

Figure 2. Detailed picture of a serpentine cable.

The assembled 3D neural probe with 2×3 array of electrodes is shown in Fig. 3. In this packaging, two islands are back-to-back stacked and then connected to the interfacing island by parylene layers in which metal wires are embedded. With the aim of preventing the top parylene layer from peeling off when it is used in implantation and also improving the electrical isolation of metal traces, parylene thin film is used to encapsulated on the sidewalls of the probe shanks.

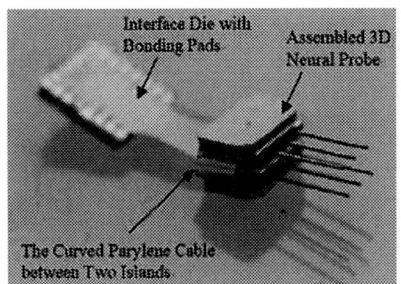

Figure 3. The shank island and micro probe [11].

Preliminary Animal Study Results

In order to evaluation the packaging and interconnection performance and the efficacy of the recording capability of 3D devices, the 2×3 array of electrodes was implanted in the auditory cortex of an adult normal rat. We first made the rat under general anesthesia state via isoflurane, and then generated a burr hole in its skull above the auditory cortex in which the device was then penetrated into it. The interface die of the device was bonded on a custom printed circuit board to a TDT (Tucker Davis Technologies) recording system which have been used in prior studies [19-20]. The picture of devices and test board are shown as Fig.4.

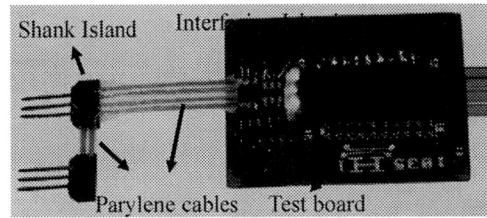

Figure 4. Devices and test board

A noise pulse system which could induce noise bursts was applied in the external auditory canal of a normal rat. The electrode array successfully detected the neural activity associated with the sensation of hearing in the rat. Two distinct neural signals were recorded from the electrode sites as shown in Fig.5 (a). Then the spikes from the two neurons were processed by a unit sorter program, the processed waveform could be seen in Fig. 5. It can be seen in Fig. 5 that the amplitude and spike rate between the two neurons are different. This study demonstrated the ability that the neural devices are able to acquire and isolate two distinct neural signals.

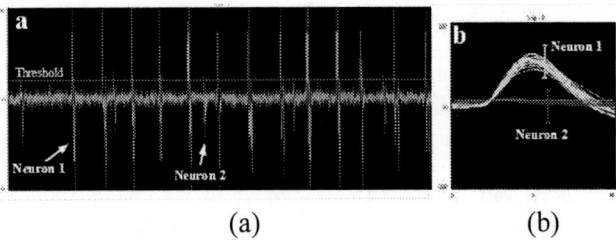

(a)   (b)

Figure 5. (a) Spikes from two neurons recorded from the primary auditory cortex of a rat. (b) Spikes from the two neurons were differentiated using a unit sorter program [11].

978-1-4799-2408-0/14 $31.00 © 2014 IEEE

Conclusions

In this paper, 2D devices consisting of neural probe islands and interfacing island are fabricated based on MEMS technology. Then the two neural probe islands were assembled into 3D probe array by folding process using the parylene cables to connect with each other. After folding process, the two islands were glued together by back to back type. The alignment of islands and adhesive selection are paid more attention for better reliability needs.

The evaluation board was setup and the prototype was used in an adult normal rate. Spikes from two neurons are recorded from the primary auditory cortex of a rat and the data is then processed by a unit sort program. The results demonstrated the ability of the neural devices to be able to acquire and isolate two distinct neural signals.

Acknowledgments

The support of Hightech Proram (863) with a contract number of 2012AA040501 and Basic Research (973) with a contract number of 2011CB309504 of Ministry of Science and Technology of PR China is highly appreciated.

References

1. R. C. Gesteland, B. Howland, J. Y. Lettvin, and W. H. Pitts, "Comments on Microelectrodes," Proceedings of the IRE, vol. 47, pp. 1856-1862, 1959.

2. K. Frank and M. C. Becker, "Microelectrodes for recording and stimulation," Physical Techniques in Biological Research, W. L. Nastuk, Ed. New York: Academic, vol. 5, 1964.

3. D. A. Robinson, "The electrical properties of metal microelectrodes," Proceedings of the IEEE, vol. 56, pp. 1065-1071, 1968.

4. O. F. Schanne, M. Lavallee, R. Laprade, and S. Gagne, "Electrical properties of glass microelectrodes," Proceedings of the IEEE, vol. 56, pp. 1072-1082, 1968.

5. Lin C W, Lee Y T, Chang C W, et al. "Novel glass microprobe arrays for neural recording". Biosensors and Bioelectronics, Vol. 25, 475-481, 2009.

6. Lee Y T, Lin C W, Lin C M, et al. "A pseudo 3D glass microprobe array: glass microprobe with embedded silicon for alignment and electrical interconnection during assembly," Journal of Micromechanics and Microengineering, Vol. 20: 025014, 2010

7. S. Musallam, M. J. Bak, P. R. Troyk, and R. A. Andersen, "A floating metal microelectrode array for chronic implantation," Journal of Neuroscience Methods, vol. 160, pp. 122-127, 2007.

8. T. K. Chowdhury, "Fabrication of extremely fine glass micropipette electrodes," Journal of Physics E: Scientific Instruments, vol. 2, pp. 1087-1090, 1969.

9. K. C. Cheung, K. Djupsund, Y. Dan, and L. P. Lee, "Implantable multichannel electrode array based on SOI technology," Microelectromechanical Systems, Journal of, vol. 12, pp. 179-184, 2003.

10. R. Rathnasingham, D. R. Kipke, S. C. Bledsoe, and J. D. McLaren, "Characterization of Implantable Microfabricated Fluid Delivery Devices," IEEE TRANSACTIONS ON BIOMEDICAL ENGINEERING, vol. 51, pp. 138-145, Jan 2004.

11. John J, Li Y, Zhang J, et al. "Microfabrication of 3D neural probes with combined electrical and chemical interfaces," Journal of Micromechanics and Microengineering, 21(10): 105011, 2011.

12. Wise K D, Anderson D J, Hetke J F, et al. "Wireless implantable microsystems: high-density electronic interfaces to the nervous system". Proceedings of the IEEE, 92(1), pp: 76-97, 2004.

13. K. D. Wise, J. B. Angell, and A. Starr, "An integrated circuit approach to extracellular microelectrodes," in the 8th Int. Conf. Engineering Medicine and Biology, pp. 14-15, 1969.

14. K. Najafi, "Solid-state microsensors for cortical nerve recordings," Engineering in Medicine and Biology Magazine, IEEE, vol. 13, p. 375, 1994.

15. P. K. Campbell, K. E. Jones, R. J. Huber, K. W. Horch, and R. A. Normann, "A silicon-based, three-dimensional neural interface: manufacturingprocesses for an intracortical electrode array," IEEE Transactions on Biomedical Engineering, vol. 38, pp. 758-768, 1991.

16. S. Suner, M. R. Fellows, C. Vargas-Irwin, G. K. Nakata, and J. P. Donoghue, "Reliability of signals from a chronically implanted, silicon-based electrode array in non-human primate primary motor cortex," IEEE Transactions on Neural Systems and Rehabilitation Engineering, vol. 13, 2005.

17. Liu S, Liu Y. "Modeling and Simulation for Microelectronic Packaging Assembly: Manufacturing, Reliability and Testing," John Wiley & Sons, 2011.

18. Kim E, Tu H, Lv C, et al. "A robust polymer microcable structure for flexible devices," Applied Physics Letters, 102(3): 033506. 2013

19. Hoa, M., et al., "Tonotopic responses in the inferior colliculus following electrical stimulation of the dorsal cochlear nucleus of guinea pigs," Otolaryngology - Head and Neck Surgery. 139(1): p. 152-155. , 2008

20. Olsson, R.H., III, M.N. Gulari, and K.D. Wise. "Silicon neural recording arrays with on-chip electronics for in-vivo data acquisition," in Microtechnologies in Medicine & Biology 2nd Annual International IEEE-EMB Special Topic Conference on. 2002.

CMOS Multiplexer for Portable Biosensing System with Integrated Microfluidic Interface

Tetiana Voitsekhivska, Eike Suthau and Klaus-Juergen Wolter
Technische Universität Dresden, Electronics Packaging Laboratory
D-01062 Dresden, Germany
E-Mail: voitsekhivska@avt.et.tu-dresden.de, Phone: +49 351 463 36107, Fax: +49 351 463 37035

Abstract

A portable biosensing system is presented in this work. The system consists of a measuring block featuring an integrated microfluidic interface on a secure digital (SD)-card form factor substrate. This SD substrate incorporates a biochip and a custom designed analog CMOS multiplexer. The integrated multiplexer is intended to connect to each sensor-FETs on the biochip individually, allowing a hot-plug connection to the measuring block. A low-cost PDMS glob top forms the package for protection of all wire-bonded integrated circuits and defines the microfluidic channel. Thus facilitating the delivery of the analyte to the biochip via a simple microfluidic interface without additional sealing.

Employing the proposed CMOS multiplexer allows simpler, more rapid and more accurate measurements comparing with the previous lab-scale setup. Achievable performance and required calibration steps are discussed towards the CMOS multiplexers ON-resistance, leakage currents, and dynamic characteristics. Overall size, cost and reliability are compared to a discrete solution.

The microfluidic part of the biosensing system incorporates a permanent magnet linear positioning drive and linear position sensor for effortless and dependable mechanical interfacing. The portable system is designed as a part of a PC-based setup and primed for stand-alone operation.

1. Introduction

The integration of traditional CMOS circuits into microfluidic systems employing novel biosensing packaging promises significant improvements of system performance, cost and reliability. Better understanding of the interaction between packaging technologies and constituent devices is required for further improvements. Commonplace CMOS technology opens up a new capability in the development of biosensing systems [1]. Simultaneous measurement of sensor elements calls for the development of the portable system for the screening of multi-segment nanostructures [2, 6].

In our previous work we explained a portable measurement system for the biosensing applications [3]. One constituent of the system is a CMOS integrated multiplexer, which allows to connect a multitude of the sensing elements to the measuring unit. In this work we present a static and dynamic characteristic of the CMOS integrated multiplexer.

The measurement of the biosensing elements requires stable and reliable signal acquisition. In this work the proposed CMOS analog multiplexer is compared with a discretely assembled analog multiplexer according to its electrical parameters.

Additionally, we present the microfluidic interface of the portable system, which simplifies the measurement process of the biochips, and when contacting to the analyte controls mounting. The sealing method of the test substrates offers a new and simplified approach for protecting all electrical contacts, while at the same time forming a simple microfluidic channel.

Using standard, low-cost components such as SD-card connectors is the basis for a reliable, user-friendly and commercially viable portable data acquisition system.

2. Materials und methods
2.1 Test of the CMOS integrated multiplexer

The integrated multiplexer consists of the 96 analog transfer gates and 32 shift registers with serial interface control pins NRES, DO, DI, CLK, LATCH. Each of 32 analog channels has 3 analog switches corresponding to the gate, source, and drain potentials of the measured SiNW FETs. The chip's bond pads have a size of $85 \times 85 \mu m^2$ [2].

For obtaining detailed electrical parameters of the CMOS multiplexer a measuring station connecting to a wafer probe station was developed and assembled (Figure 1).

Figure 1. The measuring station connected to wafer prober.

The measuring station consists of an Agilent/ HP 54833A Oscilloscope, a Keithley 2400 Source meter, a Keithley 6221 DC Current Source and custom designed Probe Card (Figure 2). The Probe Card allows to connect the different voltage and current sources to the analog multiplexer using a matrix of relays addressed by shift registers. All measurements were controlled via a LabVIEW user interface.

2014 Electronic Components & Technology Conference

Figure 2. Probe Card for the testing.

2.2. Biosensing chip measurements

The portable measurement system (Figure 3) was used for the measurement of the transfer characteristic of SiNW FETs – based biosensors. System functionality was first verified using different types of dummy-substrates with SMD resistors and MOS transistors. All measurements are performed using LabVIEW control programs.

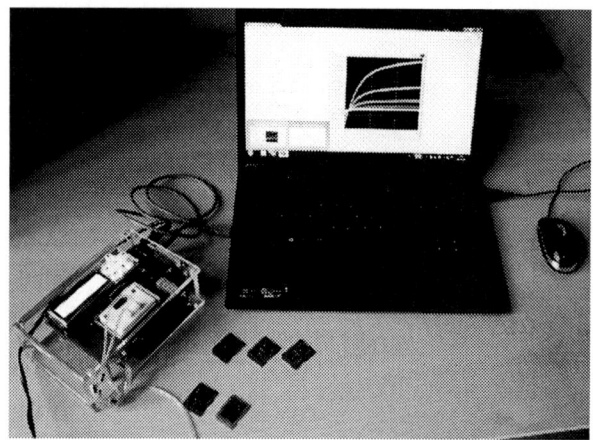

Figure 3. Portable biosensing system facilities.

The SiNW FET-based chip is provided by Max-Bergman Centre of Biomaterials Dresden. The sensor elements are bottom-up manufactured as parallel arrays of Schottky barrier SiNW FETs [4].

The reference transfer characteristic of SiNW FET-based biosensor is shown in Figure 4. The current around zero of gate voltage corresponds to the threshold region of the unipolar transistors on the biochip [5]. For gate voltages ranging from -2 to +2V the sensing elements exhibit their peak conductance, shown as a red box in Figure 4.

The biosensing platform is developed for label-free, real-time electrical signal measurement of chemicals and biosensors using silicon nanowires based Schottky FETs with APTES modified surface. For the test of the transfer characteristic of sensitive elements the biochips are assembled on the test substrate with the CMOS analog multiplexer (Figure 5).

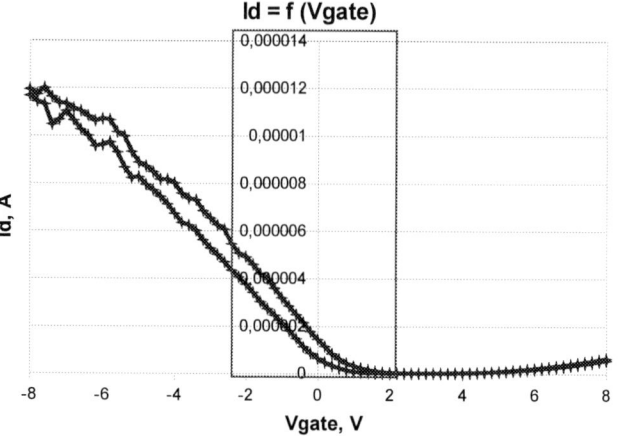

Figure 4. SiNW FETS – based biosensor dry state transfer characteristic, Vdrain = 2V.

The test board has a standard SD card size and contact area. The PCBs are manufactured by LeitOn GmbH according the following specification: FR4; 2 layers; 1.55 mm thickness; 35 µm copper thickness; 6 mil NiAu plating. The chips are connected to the board using a Universal automatic bonder F&K Delvotec 56xx, wedge-wedge bonding with 25 µm diameter aluminium wires. The transfer characteristic measurements of the FET transistors Id = f (Vg) is performed using LabVIEW. Since the peak transconductance measurement as most important part of the characteristic the gate voltage is in the range -2..+2 V.

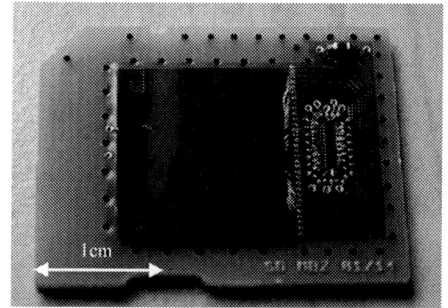

Figure 5. Test substrate with SiNW FETs-based biochip and CMOS integrated analog multiplexer.

2.3. Microfluidic interface

For the wet measurement of the biosensors a microfluidic interface was developed as part of the portable measurement system. The microfluidic interface consists of Permanent Magnet Linear Positioning Drives LSP2575M0506 by Nanotec with a torque up to 18 nN and a linear position sensor 9605 by BEI Sensors. The transport of the analyte to the biochip on the test card is performed through Polytetrafluoroethylene (PTTE)

978-1-4799-2408-0/14 $31.00 © 2014 IEEE

stamp with Mini hose fittings for fine pitch thread UNF 1/4"−28 by Reichelt Chemietechnik GmbH+Co. The 3D model of the device its microfluidic interface is shown in Figure 6. This interface allows to control the position of the stamp relative to the test substrate with the microfluidic channel. The capability of the sealing gasket is measured using a hydrostatic head. The sealing withstands up to 200 mbar of static pressure.

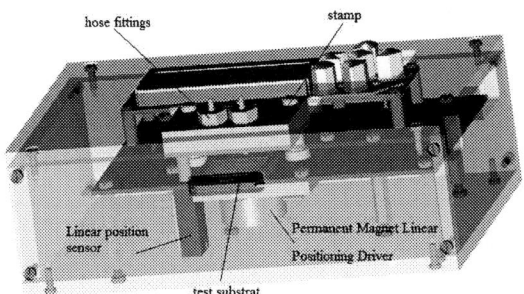

Figure 6. 3D model of the microfluidic interface.

For the forming of the microfluidic channel the test substrates are sealed with 184 Silicone Elastomer by Dow Corning. Polydimethylsiloxane (PDMS) is widely used for the protection of devices and building microfluidic channels due to its bio-compatibility, optical transparency, gas permeability and electrical insulation.

In order to protect the test board a sealing mold is designed, and manufactured (Figure 7). The sealing mold consists of two parts which the SD card test substrate is placed in between. The first part is made of PTTE and defines the position and form of the microfluidic channel on the biochip. The second aluminium form is used to press test board and form from PTTE together with four screw-bolts.

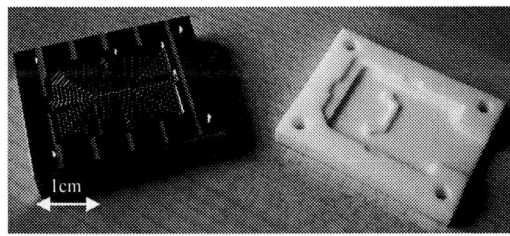

Figure.7. Sealing mold: left – aluminium part, right – part from PTTE.

The two compounds of silicone elastomer are mixed manually with a 10:1 mix ratio. Then this silicone is placed into a vacuum desiccator VWRI467-0088 for 10 min for de-airing. Subsequently, test board is placed between parts of the sealing form. Elastomer is dispensed into the sealing form using a Loctite® Analog Syringe Dispenser and cured at 125°C for 20

minutes. Finally, the transfer characteristics of the SiNW FETs on the biochip with silicone protection layer are measured using the portable measurement biosensing system.

3. Results and Discussion
3.1. Electrical parameters of the CMOS multiplexer

Table 1 gives an overview of the measured parameters for the CMOS analog multiplexer, test conditions, and test circuits. The measured data are derived from the test results of 32 CMOS analog multiplexers.

Table 1. Measuring parameters and test conditions.

Parameter, Unit	Parameter description Test conditions	Test circuit
R_{ON} single, Ω	Resistance of single channel in ON state, all other channels are OFF $V_{DD} = 5/3.3/2,5/1.8$ (V) $Isw = 100\mu/1\mu/10\mu$ (A) $V_{COM} = 0..V_{DD}$ (V)	
$I_{L,SGL,OFF}$, nA	Leakage current measured for single channel, with the corresponding channel in OFF state $V_{DD} = 5$ (V) $V_{COM} = 0..V_{DD}$ (V) $V_{SGL} = 0..V_{DD}$ (V)	
$I_{L,COM,OFF}$, nA	Leakage current measured at the common pin, with all channels in OFF state $V_{DD} = 5$ (V) $V_{COM} = 0..V_{DD}$ (V) $V_{SGL} = 0..V_{DD}$ (V)	
$I_{L,SGL,ON}$, nA	Leakage current measured with the single channel in ON state, while the common pin is not connected $V_{DD} = 5$ (V) $V_{SGL} = 0..V_{DD}$ (V)	
$I_{L,COM,ON}$, nA	Leakage current measured at the common pin with all corresponding channels in ON state, while all single pins are not connected $V_{DD} = 5$ (V) $V_{COM} = 0..V_{DD}$ (V) $V_{SGL} = 0..V_{DD}$ (V)	
t_{ON}, ns t_{OFF}, ns	Turn-on and turn-off time of the analog switch, measuring the propagation delay between digital input LATCH and analog output signal $R_L = 10$ MΩ $C_L = 1.5$ pF $V_A, V_B, V_C = $ VDD=5(V)	

978-1-4799-2408-0/14 $31.00 © 2014 IEEE   175

V_{COM} – Voltage at common pins A/B/C
V_{SGL} –Voltage at single channel pins $A_N/B_N/C_N$, $_{N}$ =0..31
V_{DD} – Supply voltage of the CMOS multiplexer
Isw – Switch current during resistance measurements

Figure 8 shows the ON resistance changing with different supply voltage V_{DD} and common channel voltage V_{COM}. The parameter is measured using four wire method at Isw =100µA. For standard supply conditions, V_{DD}=5V, R_{ON} averages approximately around 7 Ω.

The CMOS process yields good P-channel and N-channel MOSFETs. Connecting the PMOS and NMOS devices in parallel forms a basic bidirectional CMOS switch. The peak of the plot in Figure 8 corresponds to the RON mismatch between NMOS and PMOS transistor of the transfer gate.

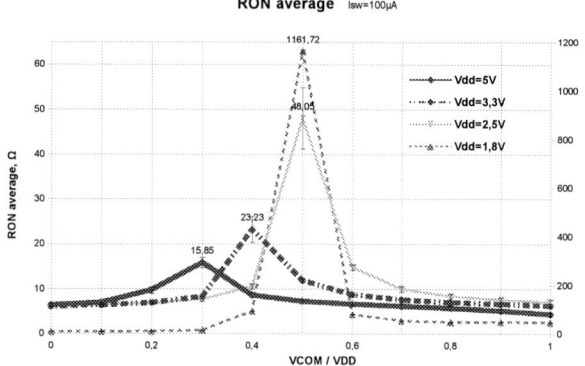

Figure 8. Average R_{ON} vs V_{COM} and V_{DD}.

In Figure 9 R_{ON} resistance for each channel at 5 V supply voltage and 2.5 V common voltage of the channel is shown. As it can be seen on the diagram, the resistance flatness is 1.9 Ω. The significant deviations of the measured R_{ON} are due to the imperfect contact of the measurement needles of the Probe Card.

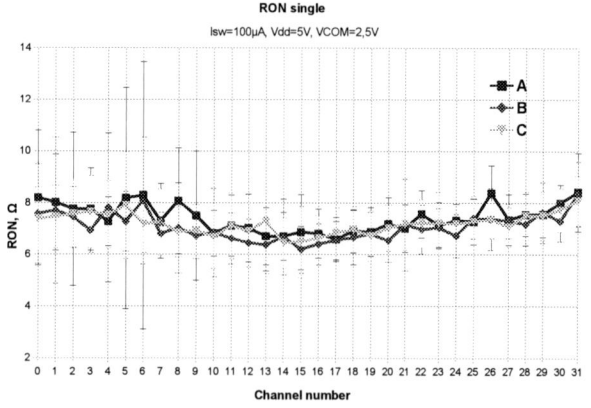

Figure 9. R_{ON} single vs channel number.

$I_{L,SGL,OFF}$ is measured for a single channel while applying different voltages at the switch. Figure 10 shows the leakage currents for channel A15 calculated from 32 chips. The maximal current leakage of the single channel has low switch voltage dependence and reaches values around 1.5 nA.

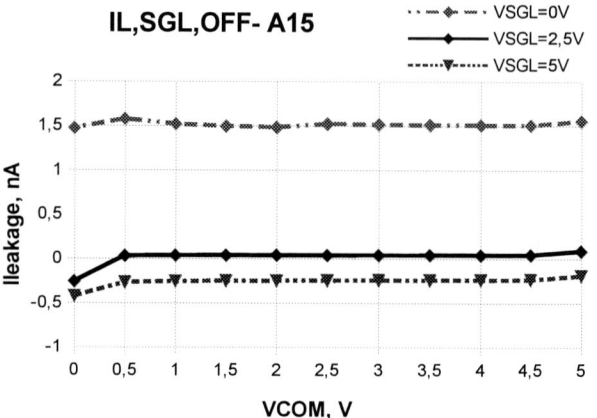

Figure 10. $I_{L,SGL,OFF}$ of the channel A15 vs V_{COM} and V_{SGL}.

Figure 11 demonstrates the leakage current at the common pin A in OFF state as a function of different single and common pin voltages. The maximum observed leakage current is around 45 nA. The plot shows weak single voltage dependence.

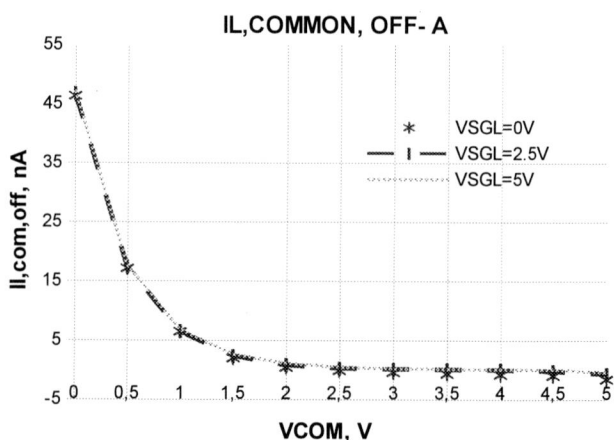

Figure 11. $I_{L,COM,OFF}$ of the common channel A vs V_{COM} and V_{SGL}.

To minimize leakage currents our portable system employs a virtual ground which all drain potentials of the FETs are connected to when they are not being measured. This diverts leakage currents to virtual ground and increases test accuracy.

978-1-4799-2408-0/14 $31.00 © 2014 IEEE   176

Common channel leakage currents is larger than single leakage current, and is caused by the bulk leakage current of the analog switch`s PMOS transistor.

Single switch ON state leakage current is largely due to the same mechanism and shown in Figure 12 for channels A00, B15 and C31.

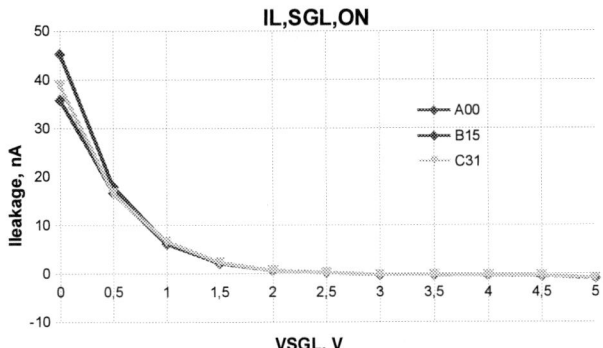

Figure 12. $I_{L,SGL,ON}$ of the single channels A00, B15 and C31 vs $V_{COM.}$

Common ON leakage currents are shown in Figure 13. The leakage currents depend on common voltage and their values vary from 10 nA to -7.56 nA. At 5V supply voltage turn-on and turn-off delays of < 20 ns and < 15 ns have been measured.

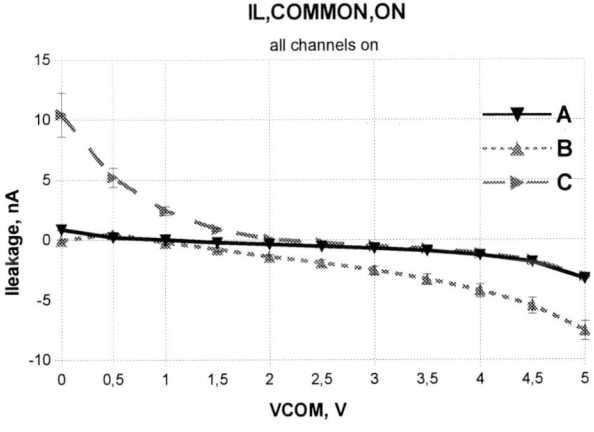

Figure 13. $I_{L,COM,ON}$ of the common channel A/B/C vs $V_{COM.}$

A discretely assembled analog multiplexer has been designed for comparison. The discrete solution has the same pin-out and logic as the integrated multiplexer. Table 2 summarizes the key parameters. For the discrete multiplexer a TS5A1066 analog transistor by Texas Instruments and 74HCT595 8/bit shift registers are chosen. The micro-edge vertical connector by TE

Connectivity AMP with 72 contacts with 1.27 mm pitch is chosen to connect the biochip to the multiplexer.

From Table 2 it can be seen that the CMOS multiplexer is the superior solution for the biosensing application due to its excellent electrical parameters, small size and low cost.

Table 2. Electrical parameters of the multiplexers at V_{DD}=5V.

Parameter	Discrete assembled analog multiplexer	CMOS integrated analog multiplexer
R_{ON} single typ, Ω	7.5	7
$I_{L,SGL,OFF}$, A	$0,1 \times 10^{-3}$	1.5×10^{-9}
$I_{L,COM,OFF}$, A	0.05×10^{-3}	45×10^{-9}
$I_{L,SGL,ON}$, A	0.1×10^{-3}	40×10^{-9}
$I_{L,COM,ON}$, A	0.05×10^{-3}	10×10^{-9}
t_{ON}, ns	9.5	<20
t_{OFF}, ns	2	<15
Number of connector contacts	72	9
Overall size of the test substrate, cm (Multiplexer+ biochip+connector)	15 x 20	3,2 x 2,4
Costs of multiplexer assembly, €	20	0.50

3.2. The biochips measurements

The measured transfer characteristic of the SiNW FETs-based biosensor using our portable system is shown in Figure 14 (Vdrain= 2V).

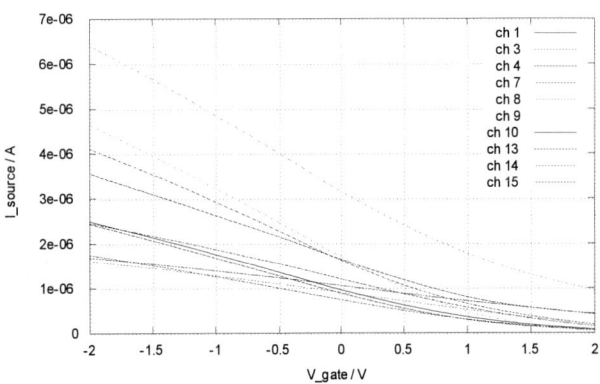

Figure 14. Transfer characteristic of SiNW FETs of the biochip in dry state.

On the plot it can be seen that the transfer characteristic of a sensor chip with 16 μm distance between source and drain contacts matches the reference data. After PDMS molding the transfer characteristic of the biochip was measured again. Figure 15 shows the SD card format test substrate after molding.

Figure 15. Test substrate after PDMS molding.

Figure 17 shows the transfer characteristic after the PDMS molding for a SiNW FETs-based biochip at Vdrain = 2V. It can be seen that the drain current of the sensing devices is increased for some channels while being decreased for others, retaining its principle dependence on gate voltage. Future work will follow to investigate the root causes.

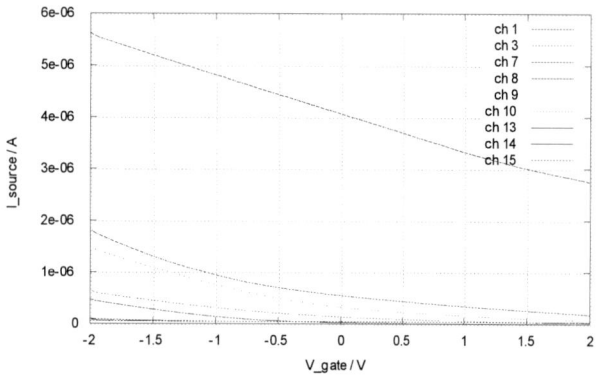

Figure 16. Transfer characteristic of SiNW FETs-based chip after PDMS molding.

Conclusions

The typical performance of the CMOS analog multiplexer is superior when compared to the discretely assembled counterpart. The evaluation of the test results of 32 integrated multiplexers shows a 7 Ω typical ON resistance, fast switching, and very low leakage currents.

The results show that the CMOS multiplexer is suitable for the highly sensitive, accurate and fast measurement of the biosensing structures. Due to the small overall size and very low static power the multiplexer could easily be applied in different applications.

The transfer characteristics of the SiNW FETs - based biosensor have been measured using the portable measurement system and CMOS multiplexer. The results matches the reference plots measured using standard lab-scale setup.

The transfer characteristic of the test substrates with PDMS molding are measured and exhibit a change of drain current and must be subject of future investigations.

The portable system using CMOS multiplexer has thus been shown to have promise for the measurements of biosensors, gas - and chemical sensors. Further development of the portable system will go into the design of a liquid gate electrode, and the miniaturization of the microfluidic channel.

Acknowledgments

This work was carried out within the framework of the research training group "Nano- and Biotechnologies for Packaging of Electronic Systems" funded by the German Research Foundation (DFG 1401/1).

References

1. X. Zhang, W.M. Wong, Y. Zhang, Y. Zhang, F. Gao, R. Nelson, J. Larue, "Design of a CMOS - based multichannel integrated biosensor chip for bioelectronic interface with neurons," *IEEE Engineering in Medicine and Biology Society.EMBS 2009 Conference, Minneapolis, Sep. 3 – 6, 2009, pp. 3814 – 3817.*

2. L. Novak, P. – Neuzil, J. Soon Bo Woon, Y. Wee "Pocket - size multiplexed electrical detection of bio - substances by ultra sensitive nanowire nanosensors," *Sensors, 2009 IEEE, Christchurch, Oct. 25 – 28, 2009, pp. 405 – 407.*

3. T. Voitsekhivska, E. Suthau, K.-J. Wolter, L. Baraban, F. Zörgiebel and G. Cuniberti, "Portable measurement system for silicon nanowire field - effect transistor-based biosensors," *36th International Spring Seminar on Electronics Technology (ISSE), Alba Iulia, May°8 – 12, 2013, pp. 349 – 354.*

4. S. Pregl, F. Zörgiebel, L. Baraban, G. Cuniberti, T. Mikolajick, W.M. Weber, "Channel length dependent sensor response of Schottky - barrier FET pH sensors," *Sensors 2013, IEEE, 3 – 6 Nov. 2013, pp. 1 – 4.*

5. S. Pregl, W.M. Weber, D. Nozaki, J. Kunstmann, L. Baraban, J. Opitz, T. Mikolajick and G. Cuniberti "Parallel arrays of Schottky barrier nanowire field effect transistors: Nanoscopic effects for macroscopic current output,", *Nano Research*, vol. 6, no. 6, pp. 381 – 388, June 2013.

6. M.H. Lee, S.W. Jung, W. Seong, "Silicon Nanowires for High - Sensitivity CRP Detection", *IEEE SENSORS 2010 Conference*, pp. 415 – 418.

Assembly of Mechanically Compliant Interfaces between Optical Fibers and Nanophotonic Chips

Tymon Barwicz[1*], Yoichi Taira[2], Hidetoshi Numata[2], Nicolas Boyer[3], Stephane Harel[3], Swetha Kamlapurkar[1],
Shotaro Takenobu[4], Simon Laflamme[3], Sebastian Engelmann[1], Yurii Vlasov[1], and Paul Fortier[3]

[1]IBM T.J. Watson Research Center, 1101 Kitchawan Rd., Yorktown Heights, NY 10598 USA
[2]IBM Research – Tokyo, 7-7 Shin-Kawasaki, Saiwai-ku, Kawasaki, Kanagawa, 212-0032 Japan
[3]IBM Bromont, 23 Boul. de l'Aeroport, Bromont, QC J2L 1A3 Canada
[4]Asahi Glass Co., AGC Electronics, Technol. Gen. Div. 1150 Hazawa-cho, Kanagawa-ku, Yokohama, 221-8755 Japan
*tymon@us.ibm.com

Abstract

Silicon nanophotonics may bring disruptive advances to datacom, telecom, and high performance computing. However, the deployment of this technology is hampered by the difficulty of cost efficient optical inputs and outputs. To address this challenge, we have recently proposed a low-cost, mechanically compliant polymer interface between standard single mode fibers and nanophotonic waveguides. Our concept promises better mechanical reliability and better optical performance than existing technology. To manage the cost of assembly, we show here that self-alignment features can be effectively used to bridge the gap between the accuracy required by single-mode optics (1-2 um) and the capability of high-throughput microelectronic assembly equipment (~10 um). We describe the complaint interface, the assembly strategy, and the design of our re-alignment features. We demonstrate experimentally that misalignments at assembly as large as +/-10 um are re-aligned by our self-alignment structures to +/-1 to 2 um. Our approach enables existing microelectronics equipment to be used for single-mode optics assembly.

Introduction

Significant progress has been reported in fabrication of silicon nanophotonics in microelectronic manufacturing environments [1,2]. However, the difficulty of packaging those photonic chips remains a large barrier to their deployment. Existing approaches to interfacing silicon nanophotonic waveguides to standard single-mode fibers use specialized equipment and show high cost and difficult

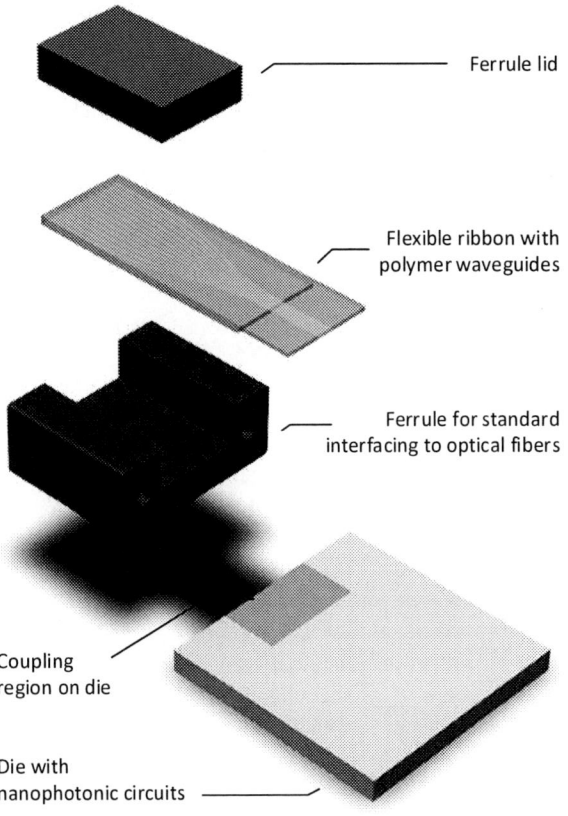

Figure 2. Exploded schematic of the compliant polymer interface. Polymer waveguides are defined on a flexible ribbon, assembled to a ferrule for compatibility with fiber connectors and capped with a lid. The compliant interface is then picked and placed on a nanophotonic die.

scalability to large port count. To achieve the cost-efficiency of microelectronic packaging, we believe the best path forward is to leverage the existing high-throughput microelectronic packaging tooling for photonic packaging. The key challenge in doing so is the tight assembly alignment required by single-mode optics, whose requirements are notably beyond microelectronic assembly tools capability. A self-alignment scheme to bridge this gap is demonstrated here.

Compliant polymer interface

We have recently proposed a low-cost, mechanically compliant polymer interface between standard single-mode fibers and nanophotonic waveguides [3]. It includes a standard fiber interface and integrated polymer waveguides.

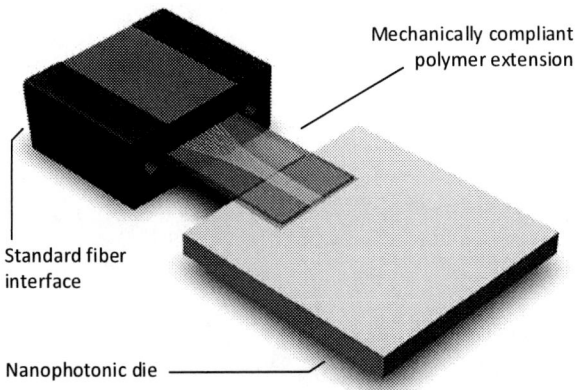

Figure 1. Schematic of a compliant polymer interface assembled to a nanophotonic die. The interface includes a standard fiber connection on one end and a compliant extension with polymer waveguides on the other end.

The optical power is butt coupled from fibers to polymer waveguides, routed through a mechanically compliant extension, and adiabatically coupled to silicon nanophotonic waveguides on a chip. As with all single-mode optics, the optical performance is identical for both forward and backward directions. This compliant interface shows much higher optical bandwidth than approaches based on vertical couplers [4]. In addition its compliant nature mechanically decouples the chip from the package promising better reliability than approaches based on rigid or direct fiber to silicon connection, which exacerbate chip to package interaction.

Various implementations of this concept are possible. The implementation discussed here is described in Figs. 1 and 2. Twelve polymer waveguides are lithographically defined on a flexible polymer ribbon. This ribbon is then assembled to a ferrule that enables detachable fiber connection using a standard mechanical transfer (MT) fiber interface. A 12x1 MT ferrule is meant to be used on the fiber cable side here. It includes 12 fibers on a 250 um pitch and metal pins for the ferrule to ferrule connection alignment. A lid is used on top of the polymer ribbon to improve reliability and symmetrize the mechanical characteristics of the interface. Finally, the compliant interface formed of the lid, ribbon, and ferrule is picked and placed on a nanophotonic die. The ribbon is permanently attached to the chip with UV epoxy. The polymer waveguides are spaced by 250 um on the fiber side to match the fiber distribution of an MT ferrule and by 50 um on the nanophotonic die interface to reduce the critical area of the ribbon to die assembly.

The alignment accuracy between the polymer ribbon and the ferrule as well as the alignment accuracy between the polymer ribbon and the nanophotonic die are critical to the optical performance of the interface. Ferrule to ribbon alignment impacts the accuracy of the fiber to polymer waveguide alignment. Ribbon to nanophotonic die alignment impacts the accuracy of the polymer waveguide to nanophotonic waveguide alignment. For the desired optical performance, an alignment accuracy of 1-2 um is desired in both cases. This is demonstrated here using self-alignment

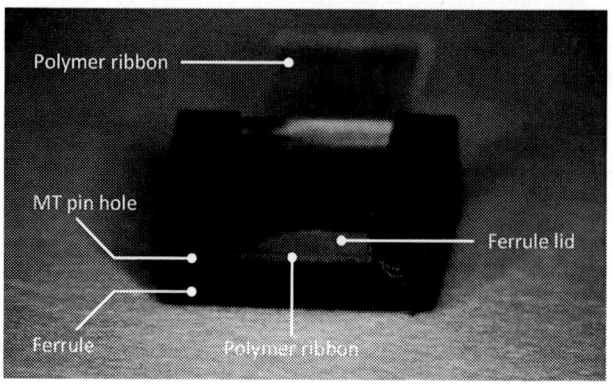

Figure 3. Compliant polymer interface viewed from the side of the fiber connector.

features to enable low-cost assembly in standard low-accuracy pick and place microelectronics tooling.

Polymer ribbon to ferrule self-aligned assembly

The polymer ribbon and ferrule are assembled to realize a compliant optical interface between the nanophotonic chip and an external parallel fiber cable. A picture of such interface is presented in Fig. 3.

The polymer waveguide ribbon is aligned and attached to a ferrule. The ferrule is injection molded and has a dimensional accuracy better than 1 um at critical positions. It is designed to provide a detachable MT interface to a fiber patch cable. Unlike common MT ferrules for glass fiber ribbons, this ferrule has a flat surface where the polymer ribbon top surface sits. The surface has two parallel ridges with a trapezoidal cross-section near the two large holes required for the alignment pins of the MT-ferrule interface. The polymer waveguide ribbon has two grooves with a rectangular cross-section to mate with the ridges on the ferrule's flat surface. Optical micrographs of the compliant interface viewed from of the fiber connector side are shown in Fig. 4. One can see the ferrule, the polymer ribbon, a laminated polymer backing to the ribbon and a lid added for robustness and for symmetrizing the mechanical properties of

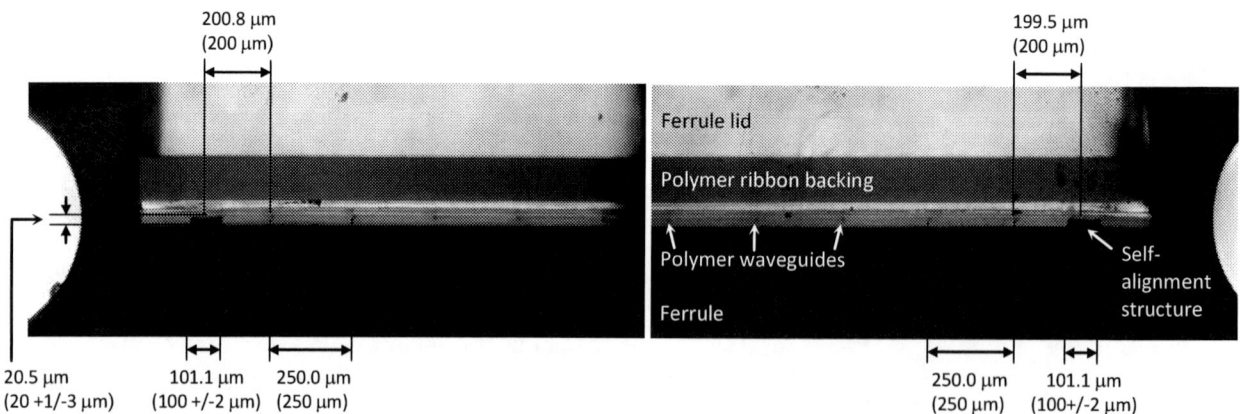

Figure 4. Optical micrographs of a compliant interface viewed from the fiber connector side. Self alignment structures are formed of a pair of trapezoidal ridges on the ferrule with mating grooves on the polymer ribbon. Measured dimensions are shown with expected dimensions in parenthesis.

978-1-4799-2408-0/14 $31.00 © 2014 IEEE

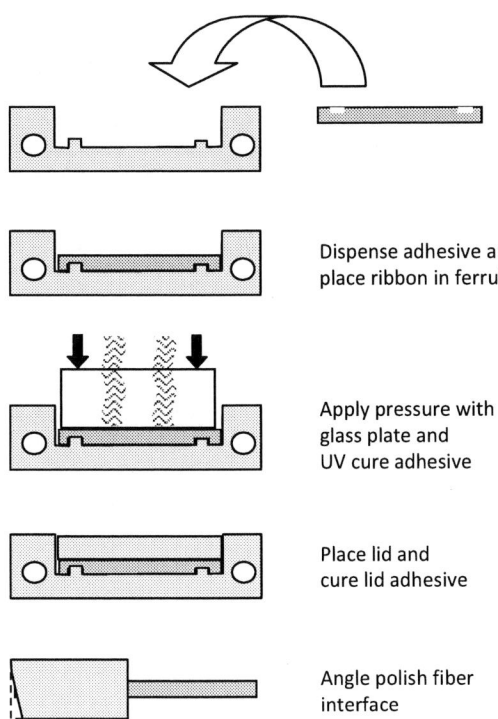

Dispense adhesive and place ribbon in ferrule

Apply pressure with glass plate and UV cure adhesive

Place lid and cure lid adhesive

Angle polish fiber interface

Figure 5. Assembly of the ferrule, ribbon, and lid to form the compliant interface

the fiber interface.

The trapezoidal ridges on the ferrule mate with the rectangular grooves on the polymer ribbon to self-align the parts at assembly. For the self-alignment mechanism to work effectively, the placement and dimensions of the self-alignment features on the ferrule and the polymer ribbon must match within designed-in tolerances.

The polymer ribbon can be accurately patterned using i-line projection lithography. If the polymer layer is thick enough, one can use this pattern for mechanical alignment. In multimode waveguides, one can use the polymer waveguide core pattern for alignment as the core layer thickness can exceed 30 μm. With single mode waveguides, however, one must use a separate alignment layer as the core layer is too thin to be used directly for mechanical re-alignment. Here, we pattern the waveguide top cladding layer, which is about 23 μm thick, to define the self-alignment grooves on the polymer ribbon. The overlay accuracy between the polymer waveguides and the polymer cladding grooves will fundamentally limit the alignment accuracy between fibers and polymer waveguides. Hence, projection lithography is desired over contact lithography for polymer ribbon fabrication as it provides significantly better alignment accuracy between the re-alignment and guiding layers (+/- 0.25 um versus +/- 1-2 um).

The polymer ribbon fabrication process is the following. After an under cladding layer is deposited and cured on a substrate, the core layer is deposited, lithographically exposed, and developed. The top cladding layer is then deposited, exposed with aligned projection lithography, and developed. The core pattern includes waveguide cores and

registration marks for visual alignment. The top cladding pattern includes grooves for self-alignment to the ferrule, ridges for self-alignment to the nanophotonic chip, and a large opening exposing the waveguide cores for adiabatic coupling to the nanophotonic chip as will be discussed below.

The desired ferrule accuracy can be accomplished by injection molding. Our ferrule design is U-shaped with a flat surface shifted down from the alignment pin center height to account for the offset between the polymer waveguide core center and the polymer cladding surface in contact with the ferrule. This ~ 21 um offset is required to match the vertical position of fiber cores in MT connectors where they are aligned with the center of the alignment pins.

A lid is placed on the polymer ribbon in the ferrule U-shape opening to improve robustness and symmetrize the mechanical properties. In the current experiment, the lid is a diced acrylate plate. However, it is preferred to be made of the same material as the ferrule for best thermo-mechanical stability, which is desired for reliability in real-world applications and will be implemented in our next iteration.

For assembly, the critical dimensions of our ferrule design are the center flat surface height and the width and center positions of the two ridges. For the polymer ribbon, the critical dimensions are the width and the center positions of the grooves as well as the cladding thickness and the waveguide core pitch. The dimensions of the lid are not critical.

The dimensional accuracy of both the ferrule and the ribbon were assessed. The ferrule design calls for trapezoidal alignment ridges that are 22.5 μm high, 100 μm wide at the bottom, and 80 μm wide at the top. The spacing between the centers of the ridges is 3.15 mm by design. The accuracy of the critical dimensions was measured on molded parts to be typically within 1 μm of design.

The polymer ribbon has shown slightly more dimensional variability than the molded ferrule as would be expected from a process including spin coating and development. In addition, its dimensions need to be pre-compensated at fabrication for a repeatable shrink of the polymer ribbon at anneal, which creates pitch variability that is not usually seen in lithographic processes. After suitable development, our current process shows 1-2 um control on dimensions that are critical to assembly. This control is sufficient for the current purpose. The flexibility of the polymer ribbon makes its dimensional control less critical than the ferrule's as it allows some level of compliance to the ferrule and nanophotonic chip.

The polymer ribbon and the ferrule are assembled as shown in the Fig. 5.

1. The ferrule is set on the assembly tool.
2. A small amount of UV adhesive is poured on the flat inner surface of the ferrule.
3. The waveguide ribbon is picked and roughly positioned on the inner surface of the ferrule so the two mating re-alignment pairs latch on.
4. The waveguide ribbon is pressed with a glass plate, then UV light is illuminated to cure the adhesive.

978-1-4799-2408-0/14 $31.00 © 2014 IEEE

5. The lid is placed on top of the ribbon, UV adhesive is poured into the space between the lid and U-shape ferrule, and the adhesive is cured by UV irradiation.

6. The end surface of the ferrule is angle polished at 8 degrees to match standard MT fiber interfaces.

Figure 4 shows the positional accuracy of the core positions relative to the ideal core position. In these micrographs, we see that positioning is better than +/- 1 µm along the horizontal and vertical direction of the inner flat surface of the ferrule. Considering achievable fabrication tolerances on the ferrule and the ribbon, we find that the waveguide positioning accuracy becomes structure limited below a positioning error of +/- 1.5 um, as it is in Fig. 4.

The reproducibility of the ferrule to ribbon assembly was investigated. In this assembly process, we found that appropriate pressure has to be applied onto the glass plate for good control of positioning. The adhesive material has to be sufficiently expelled from the ferrule to ribbon interface to ensure that the vertical position of the polymer cores is accurate within the thickness of the polymer ribbon cladding, which separates the waveguide layer from the ferrule surface.

Polymer ribbon to nanophotonic die self-aligned assembly

A top-down optical micrograph of a ribbon assembled to a nanophotonic die is presented on Fig. 6. Optical epoxy was dispensed on the nanophotonic die, the compliant interface was picked by the ribbon end, placed on the die, and the epoxy was UV cured while maintaining adequate pressure on the ribbon-to-die assembly. Twelve polymer waveguides adiabatically couple to twelve silicon nanophotonic waveguides near the center of the ribbon-to-die coupling region. The nanophotonic waveguides are too small and the polymer waveguides are too transparent to be seen in this view. The twelve lines visible in the center of the micrograph are the silicon fill exclusions around each nanophotonic waveguide. Fill structures are used to improve uniformity at chemical-mechanical polishing (CMP) but can scatter light and need to be avoided near the optical paths.

The self-alignment structures are the two wide lines on each side of the waveguide array of Fig. 6. They are formed of 200 um wide polymer ridges on the ribbon, and matching grooves with slanted sidewalls on the silicon die. A cross-section of a self-alignment feature after assembly is shown on Fig. 7. The silicon alignment groove is deeper than the polymer ridge is tall to provide tolerance for possible particle contamination on the groove floor at dicing. As described above, the polymer ribbon is made of a lithographically patterned polymer layer laminated to a polymer ribbon backing for robustness.

The rigidity of the polymer alignment ridge must be properly designed for best performance. On one hand, some give is desired as too rigid a structure will exacerbate the required fabrication tolerances on the self-alignment structures for successful nesting. On the other hand, too soft a structure will result in deformation of the alignment ridge at assembly instead of ribbon re-alignment.

A cross-sectional optical micrograph of the polymer waveguide to nanophotonic waveguide coupler is shown on Fig. 8. In the ribbon-to-die coupling region, the polymer

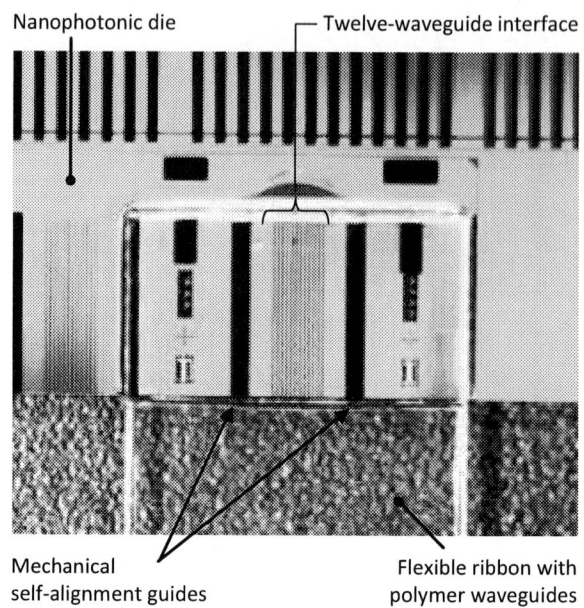

Figure 6. Top-down optical micrograph of a polymer ribbon to nanophotonic die connection. The lithographically defined features on the polymer ribbon are not visible here due to their transparency.

waveguide core is a ridge of slightly higher refractive index than the polymer cladding to which it is attached. The index contrast between the polymer core and polymer cladding is too small to be visible in this micrograph. The silicon nanophotonic waveguide has a cross-section of about 200 x 200 nm and appears as a microscope-resolution-limited bright spot near the middle of the micrograph. Optical power is adiabatically transferred from the polymer waveguide to the silicon waveguide by slowly increasing the width of the silicon waveguide using a non-linear width versus position profile. The optical design is described in [3]. This type of mode evolution transition is remarkably tolerant to fabrication

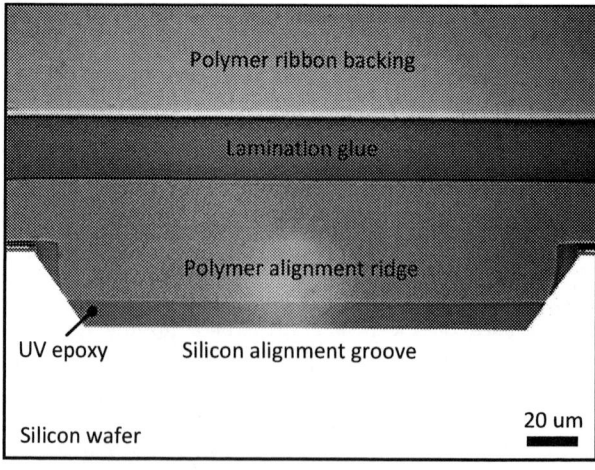

Figure 7. Cross-sectional optical micrograph of ribbon-to-silicon self-alignment guides after assembly.

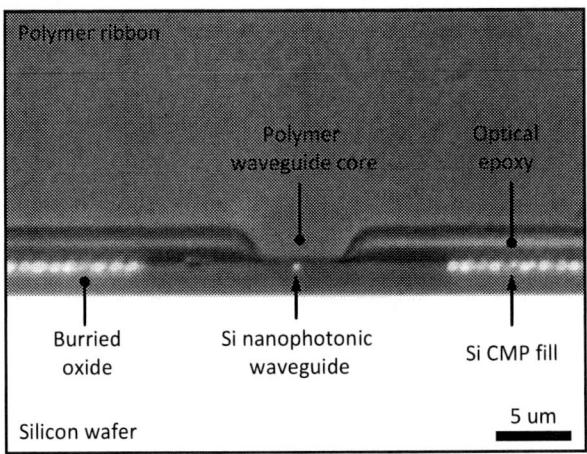

Figure 8. Cross-sectional optical micrograph of an adiabatic coupler between a polymer waveguide and a nanophotonic waveguide. The polymer ribbon to silicon die alignment must be within +/- 2 um for optimal optical performance.

and assembly imperfections. The longer the transition, the more tolerant it is. To enable +/- 2 um alignment and typical fabrication and assembly tolerances, the adiabatic transition should be 0.5 to 2 mm long. The optical bandwidth of the compliant interface is exceptionally wide with only a 0.2 dB penalty over a 200 nm wavelength bandwidth centered at 1.55 um [3]. By comparison, the typical bandwidth of a polarization independent vertical coupler results in a ~1dB penalty over a ~30 nm bandwidth [4].

To assess the performance of the self-alignment scheme, we have performed several assemblies with purposefully induced misalignment. The resulting polymer waveguide to nanophotonic waveguide alignment was measured at various

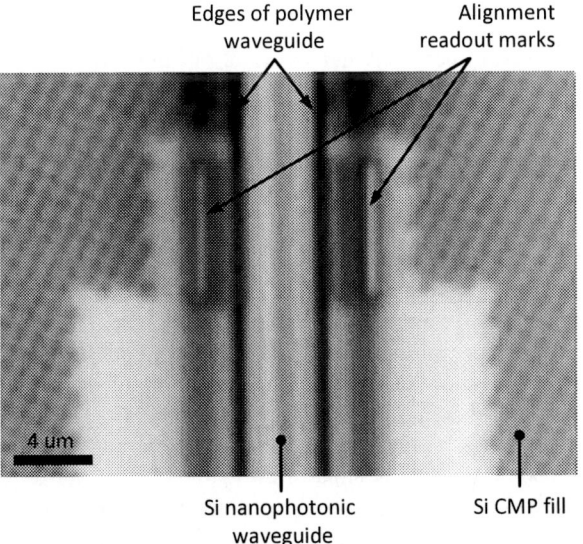

Figure 9. Top-down optical micrograph of a polymer to nanophotonic waveguide coupler region. Alignment accuracy was assessed from alignment readout marks and the edges of the CMP fill.

positions and analyzed. Two methods were used for measuring the misalignment. The core of the data was obtained from top down optical micrographs as shown in Fig. 9. A subset of the data was validated with cross-sectional micrographs as shown on Fig. 8. The nanophotonic waveguide is too small to be clearly visible in top-down micrographs. However, specially designed alignment readout marks and the edges of the CMP fill could be efficiently used to assess alignment.

The performance of the self alignment structures is presented on Fig. 10. Measurements were performed in all four corners of the waveguide array as circled in Fig. 10(a). The brut measurement data is shown in Fig. 10(b) for 12 assemblies. The die bonding tool used for this experiment has a claimed typical placement accuracy of +/- 0.5 um. We show that purposeful misalignments of up to +/-10 um, the typical accuracy of high throughput microelectronic assembly tools, are realigned to better than +/-2 um, the accuracy required for optimal optical performance.

A noticeable spread of data points from the various corners of a single assembly suggests non-negligible angular misalignment. The rotation could not be directly calculated as only one-dimensional misalignment could be accurately measured while two-dimensional data would have been required. The alignment longitudinal to the waveguide was not of interest as the optical coupler tolerances are much larger there than the typical +/-10 um alignment accuracy. Hence, only lateral misalignment was initially planned to be investigated. To assess the contribution of rotational misalignment, we have performed a multi-dimensional fit between the one-dimensional data and a geometric model using a BFGS optimization algorithm [5]. The resulting angular misalignment between the polymer waveguides and the nanophotonic waveguides is shown in Fig. 11(a). The lithographic alignment between the nanophotonic waveguide and the re-alignment groove in silicon as well as the lithographic alignment between the polymer waveguide and the polymer re-alignment ridge were performed with i-line steppers. Such tools have typical alignment accuracy of about +/-0.25 um over a ~20 mm field resulting in an angular alignment on the order of +/-0.001 to 0.002 degrees between the waveguides and their corresponding self-alignment features. The rotation shown in Fig. 11(a) is an order of magnitude higher and is hence attributed to the assembly process and not to the structure accuracy. As the assembly process is in cause, one could wonder if the angular misalignment is correlated to the purposefully induced misalignment. However, no such correlation is seen in Fig. 11(a). The residual angular misalignment contributes by only 0 to 0.8 um to the lateral misalignment. Hence, it may represent the limit of the angular re-alignment capability of the self-alignment structures based on the chosen rigidity of the polymer ridge. The lack of apparent correlation between the residual angular alignment and the purposeful lateral misalignment may indicate that the residual angular alignment is likely related to the initial angular misalignment and not to the initial lateral misalignment. The initial angular and lateral misalignments are not correlated in our study.

978-1-4799-2408-0/14 $31.00 © 2014 IEEE

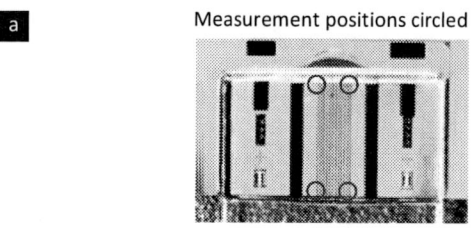

Measurement positions circled

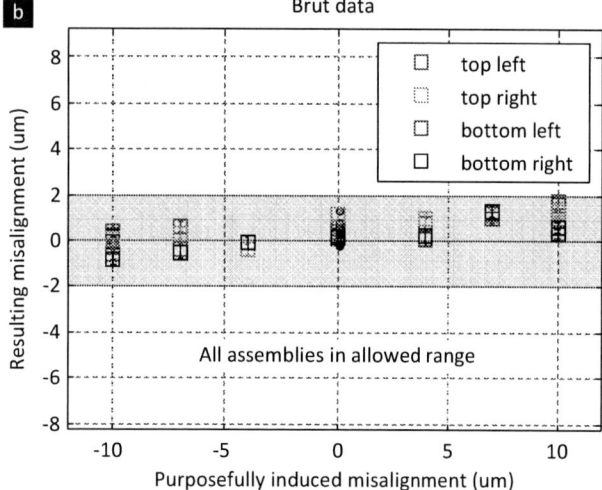

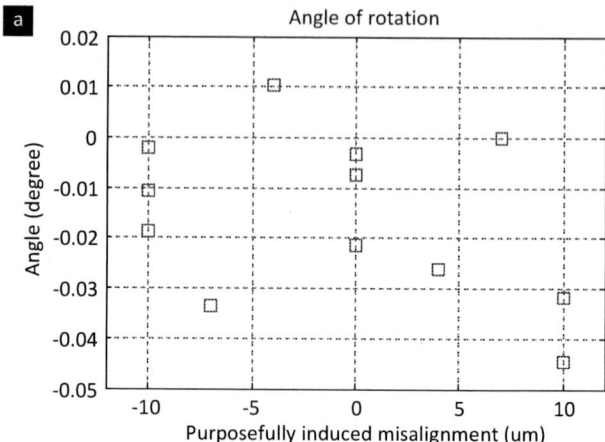

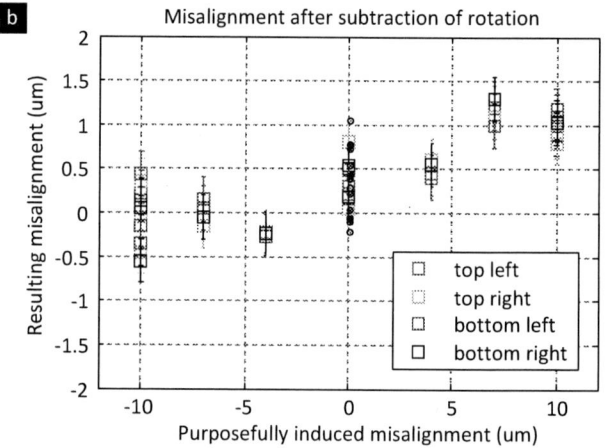

Figure 10. Performance of self-alignment between the polymer ribbon and the nanophotonic die. Final misalignment was measured at positions shown in (a) for purposeful initial misalignments up to +/- 10 um. The brut data is shown in (b) and demonstrates re-alignment to better than +/- 2 um.

To better understand the lateral re-alignment, we have subtracted the rotation of Fig. 11(a) from the data of Fig. 10(b) to obtain Fig. 11(b). We see that data from a single assembly (as best seen at -7, -4, +4, and +7 um purposeful misalignments) is tightened with respect to Fig. 10(b). However, a +/- 0.25 um spread remains and is consistent with our estimated measurement accuracy. This suggests that all the spread of misalignment data within a single assembly can be attributed to a combination of angular misalignment and measurement accuracy. Hence, non-linear in-plane deformations of the flexible ribbon are not noticeable in our data.

Two additional features transpire in Fig. 11(b). First, data at 0 um of purposefully induced misalignment is centered near 0.5 um and not near 0 um of residual misalignment. This offset can be partially attributed to the structural accuracy of the nanophotonic die and the polymer ribbon. As pointed out above, the self-alignment structures on both components were lithographically defined using i-line steppers. This results in a typical alignment accuracy of +/-0.25 um between the self-alignment structure and its waveguides. Hence, one should not expect better than +/- 0.5 um alignment from structural accuracy. Second, a correlation between residual misalignment and purposefully induced misalignment is apparent in both Figs. 10(b) and 11(b). This correlation is

Figure 11. Decomposition of brut misalignment into angular and lateral components. The residual angular misalignment is shown in (a) and the residual lateral misalignment is shown in (b).

attributed to increased deformation of the corners of the polymer alignment ridges at larger misalignments.

Conclusions

We have described the compliant interface, the assembly process, and the design of the self-alignment features. We have experimentally demonstrated that misalignment of up to +/- 10 um can be realigned to +/- 1-2 um. The limit of re-alignment set by the lithographic accuracy of the structures is of around +/- 0.5 um for the silicon to ribbon connection. On the other side of the compliant interface, the ferrule and ribbon fabrication accuracy limit the positioning accuracy of the polymer waveguides at the fiber interface to about +/- 1.5 um. We have analyzed the silicon to ribbon re-alignment data and decomposed the residual misalignment into angular and lateral components. This decomposition has shown that the initial lateral misalignment is not correlated to the residual angular misalignment but is correlated to the residual lateral misalignment. We have attributed this correlation to deformation at the corners of the polymer self-alignment ridges for large misalignments. Increasing the rigidity of the polymer ridge would further improve the residual alignment

978-1-4799-2408-0/14 $31.00 © 2014 IEEE

but would tighten the required fabrication control for adequate nesting of the re-alignment structures. The approach demonstrated here enables existing high-throughput microelectronics equipment to be used for single-mode optics assembly.

Acknowledgments

We would like to thank Masato Shiino for overseeing the ferrule injection molding, Guy Brouillette for operating the die bonding tool and Genevieve Beaulieu for acquiring the cross-sectional micrographs.

References

1. C. Gunn, "CMOS photonics for high-speed interconnects," *IEEE Micro*, vol. 26, no. 2, pp. 58-66, March-April 2006.

2. S. Assefa, S. Shank, W. Green, M. Khater, E. Kiewra, C. Reinholm, S. Kamlapurkar, A. Rylyakov, C. Schow, F. Horst, H. Pan, T. Topuria, P. Rice, D. M. Gill, J. Rosenberg, T. Barwicz, M. Yang, J. Proesel, J. Hofrichter, B. Offrein, X. Gu, W. Haensch, J. Ellis-Monaghan, and Y. Vlasov, "A 90nm CMOS Integrated Nano-Photonics Technology for 25Gbps WDM Optical Communications Applications," in *Proc. of IEEE International Electron Devices Meeting*, San Francisco, CA, Dec. 10-13. 2012, pp. 33.8.1 - 33.8.3.

3. T. Barwicz and Y. Taira "Low-cost interfacing of fibers to nanophotonic waveguides: design for fabrication and assembly tolerances," *submitted to Optics Express*, 2014.

4. A. Mekis, S. Gloeckner, G. Masini, A. Narasimha, T. Pinguet, S. Sahni, and P. De Dobbelaere "A Grating-Coupler-Enabled CMOS Photonics Platform," *IEEE J. Sel. Topics Quantum Electron.* vol. 17, no. 3, pp. 597-608, May-June 2011.

5. W. H. Press, S. A. Teukolsky, W.T. Vetterling, B. P. Flannery, *Numerical Recipes in C++: The Art of Scientific Computing*, Cambridge University Press, 2002.

Proposal of Integrated-optic Wavelength-selective Modulator Based on Coupling-efficiency Control of Distributed Bragg Reflector in Straight Waveguide

Shogo Ura[1], Tetsunosuke Miura[1], Satoshi Kawanami[1], Kenji Kintaka[2], Kosuke Asai[1], Kenzo Nishio[1], and Yasuhiro Awatsuji[1]

[1]Department of Electronics, Kyoto Institute of Technology
Matsugasaki, Sakyoku, Kyoto 606-8585, Japan
E-mail: ura@kit.ac.jp, phone: +81-75-724-7424
[2]National Institute of Advanced Industrial Science and Technology
1-8-31 Midorigaoka, Ikeda, Osaka 563-8577, Japan

Abstract

A new optical wavelength-selective waveguide modulator consisting of a simple distributed Bragg reflector (DBR) integrated in a two-mode straight waveguide with ITO thin film is proposed and theoretically discussed. The waveguide has double guiding core layers and support two TM guided modes. One is used as a transmission mode whereas the other is as a modulation mode. A wave of a specified wavelength among incident WDM waves of the transmission mode is contra-directionally and wavelength-selectively coupled by the DBR to the modulation mode. The coupling efficiency depends on a coupling length of the DBR. The coupling length varies effectively according to a propagation loss of the modulation mode. Absorption of the ITO layer is controlled to vary the propagation loss. As a result, the transmittance of the DBR is modulated wavelength-selectively. The operation principle is illustrated in detail and design examples are theoretically discussed.

Introduction

High-density broadband signal transmission by optical interconnection is one of the key technologies for constructing future ultrahigh-performance signal-processing units [1], [2]. A utilization of a wavelength-division multiplexing (WDM) technique in optical interconnections is very attractive in order to realize the bandwidth over terabits per second with high wiring density. Means for converting electrical signals to optical ones include direct modulation of a light source and modulation using an external modulator. In order to construct of WDM interconnection system, architectures of integration of a laser array of different wavelengths near LSI chips and direct modulation of each laser have been investigated [3]–[7], while use of an external WDM light source located separately from LSI chips and integration of optical waveguide modulators near LSI chips have bee also reported [8]–[12]. The latter configuration would be more tolerant to temperature change due to heat generation of LSI chips, and then has the advantage in stable operation of lasers on wavelength and power.

Two approaches have been investigated to modulate different wavelengths of WDM light independently as shown in Fig. 1. One is a combination of a demultiplexer (DeMUX), broadband optical waveguide modulators, and a multiplexer (MUX) [8], while the other is a cascaded integration of wavelength-selective waveguide modulators for different wavelengths in a single waveguide [9]–[16]. The cascaded integration of wavelength-selective modulators in a single waveguide would be much preferable for high-density wiring in WDM optical interconnection.

Microring modulators have been investigated extensively because of their small footprint and small power consumption as well as wavelength selectivity [9-14]. A microring modulator is a kind of wavelength-tunable filters, and its bandwidth is generally as narrow as 0.1 nm. Therefore, the resonant wavelength of each modulator needs to be dynamically adjusted to the wavelength of each channel of the WDM source. In addition, the total bandwidth of a microring based WDM optical interconnection is limited by the free spectral range (FSR) of the microring resonator, which is inversely proportional to the circumference of the ring.

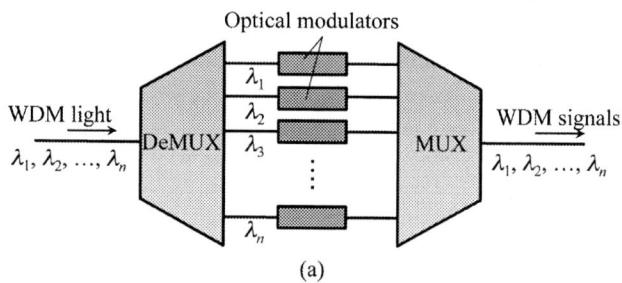

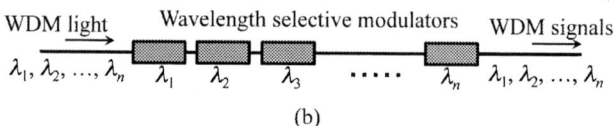

Figure 1. Configurations of optical modulation of an external WDM light: (a) combination of multi-/demultiplexers and modulators, and (b) cascaded integration of wavelength-selective modulators for different wavelengths.

Another type of optical waveguide modulator consisting of an integrated-optic in-line interferometer has been proposed and its operation principle was demonstrated by some of authors [15]. The proposed in-line interferometric device can modulate a light at specified wavelength with bandwidth of the order of nm, and allows light at all other wavelengths to pass through. Therefore, this device does not needs precise wavelength tuning and shows low crosstalk. The device utilizes four guided modes, namely TE_0 and TE_1 modes propagating forward and backward. Three kinds of distributed Bragg reflectors (DBRs) coupling two of the four modes are integrated to construct a kind of Michelson interferometer. The interference wavelength, that is the modulation wavelength, is sensitive to the difference of mode indices, and then stabilization of mode indices is important issue for practical applications.

Another new concept for integrated-optic in-line wavelength-selective modulation has been reported [16]. In this configuration, two guided modes are utilized and the coupling between them is controlled by an integrated-optic absorption modulator. This scheme is attractive because of its simple structure resulting in higher robustness in comparison to other above-mentioned methods. However, investigation of absorption-modulation materials has still remained and it is a key issue to apply this method practically.

On the other hand, there have been some reports on application of indium-tin-oxide (ITO) thin films to absorption modulation in plasmonic waveguides [17-21]. Capability of easy fabrication of the film by normal deposition process is very attractive besides of high-speed modulation and large refractive index change even though the modulation layer is very thin.

In this paper, we propose a new optical waveguide modulator based on the concept of Ref. [16] with combination of an ITO thin film modulator. ITO thin films deposited on SiO_2 substrate were characterized and modeled to predict its index modulation performance. A design example of the modulator is theoretically discussed.

Basic Configuration and Operation Principle

A basic concept of the proposed modulator is a wavelength filter with modulation function. A cross-sectional view of a typical configuration is illustrated in Fig. 2. A waveguide consisting of double guiding core layers is utilized to support two guided modes. One is used as a transmission mode whereas the other is a modulation mode. Absorption-modulation layer is formed where the field of the modulation mode is large and the filed of the transmission mode is small so that the attenuation of only the modulation mode is controlled by an extinction-coefficient variation of the absorption-modulation layer.

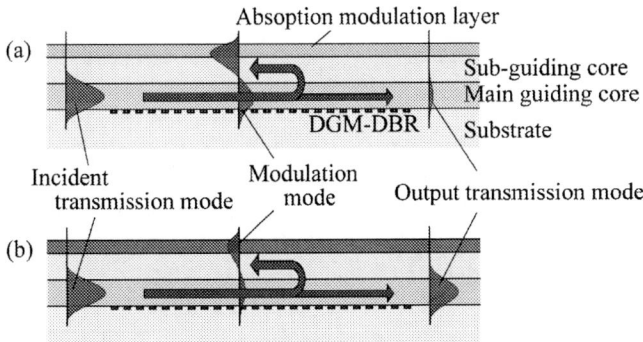

Figure 2. Cross-sectional views of a typical configuration of the proposed in-line wavelength modulator utilizing DGM-DBR with low (a) and large (b) losses of the modulation mode.

The transmission mode of a coupling wavelength is contra-directionally coupled by a different-guided-mode coupling DBR (DGM-DBR) to the modulation mode. The coupling wavelength is determined by a grating period of the DGM-DBR and two guided-mode indices. When the attenuation of the modulation mode is small, the DGM-DBR

works well and large coupling from transmission to modulation modes occurs resulting in low transmission, as shown in Fig. 2(a). On the other hand, the DGM-DBR does not work well when the attenuation of the modulation mode is large, and the resultant insufficient coupling allows high transmission, as shown in Fig. 2(b). In other words, the transmittance can be controlled by the absorption modulation. Incident waves of other wavelengths than the coupling wavelength can transmit without coupling. Thus the proposed structure serves as a wavelength filter with variable transmittance.

Coupling characteristics can be predicted by theoretical consideration based on coupled mode analysis. The coupled mode equations for the contra-directional coupling between different guided modes with propagation loss are summarized in Appendix.

Preliminary Experiments on Operation Principle

In order to confirm the operation principle of transmission variation due to the attenuation of the modulation guided mode, two test devices of different attenuations were fabricated and characterized. A radiation loss was utilized instead of the absorption loss for obtaining the different attenuations. A grating coupler (GC) was integrated to make the radiation. Two structures with and without GC were designed at coupling wavelength of 1540 nm and fabricated. A cross-sectional structure of the designed test device with GC is depicted in Fig. 3.

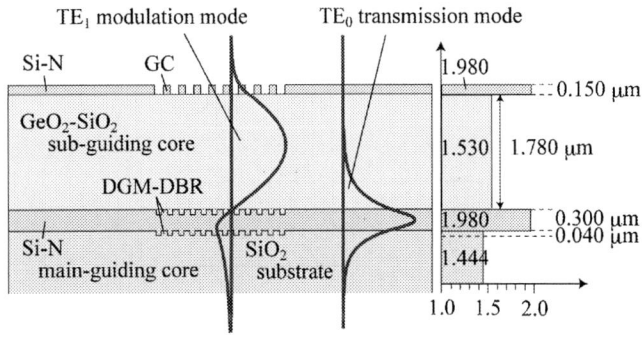

Figure 3. Cross-sectional view of a test device with GC.

A waveguide consisting of a Si-N main guiding core layer, a GeO_2-SiO_2 sub-guiding core and a Si-N grating layer is formed on a SiO_2 substrate. A grating corrugation for DGM-DBR is formed by etching of the substrate and transferred to the boundary between main and sub-guiding core layers. Another grating corrugation for GC is formed in the top Si-N grating layer. TE_0 and TE_1 modes were utilized as the transmission and modulation modes, respectively, and their mode indices were calculated to be 1.680 and 1.487, respectively. The grating period of the DGM-DBR was calculated to be 486.2 nm. The grating period of the GC was chosen to be 700.0 nm. The coupling coefficient of the DGM-DBR and the radiation-decay factor of GC were calculated to be 4.9 mm^{-1} and 10.6 mm^{-1}, respectively. In the sample without GC, the top layer consisted of a resist instead of the Si-N to have the same mode indices as the sample with GC. The thickness of the resist was 0.160 μm. When the coupling lengths of the DGM-DBR and GC were chosen to be 500 μm, the extinction ratio and the insertion loss of the DGM-DBR were predicted to be 8.0 dB and 7.2 dB, respectively.

The grating corrugation of the DGM-DBR was formed in a SiO$_2$ substrate by electron beam (EB) direct writing and reactive ion etching. A Si-N main guiding layer was deposited by sputtering. A GeO$_2$-SiO$_2$ sub-guiding layer was deposited by plasma-enhanced chemical vapor deposition. A Si-N layer was deposited and GC was formed by the EB lithography. In the sample without GC, the resist was spin-coated instead of Si-N deposition and GC integration. Input/output GCs for exciting or out-coupling the transmission mode were also integrated on the Si-N main-guiding core layer by the same fabrication process of the DGM-DBR.

A wave from a wavelength-tunable laser diode was coupled to the transmission mode in the waveguide by the input GC. The excited guided wave was partially coupled by the DGM-DBR to the modulation mode. The transmitted wave was coupled out by the output GC and collected by an objective lens (10 x, $NA = 0.25$) to an optical fiber connected to an optical spectrum analyzer. Measured transmission spectra are shown in Fig. 4. Transmittances of the devices with and without GC were −2.3 dB and −4.9 dB, respectively. It was confirmed that transmission of the DGM-DBR was varied by the attenuation of the modulation mode.

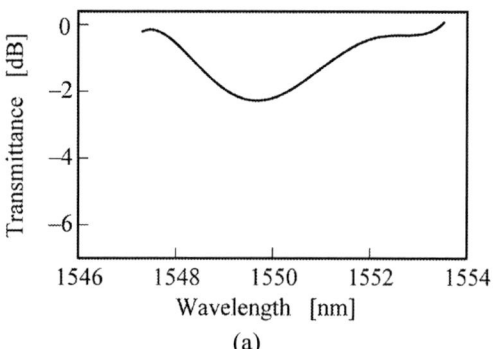

(a)

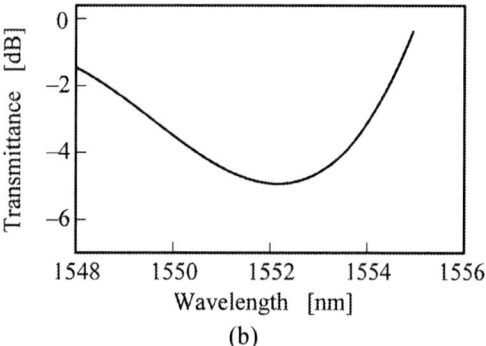

(b)

Figure 4. Measured transmission spectra of the test devices with GC (a) and without GC (b).

Characterization of ITO thin films

ITO thin film was considered as the absorption modulation layer. In order to design a wavelength selective modulator with it, it is necessary to study a variation of the absorption coefficient of the ITO film caused by carrier injection. This time we investigated the dependence of the complex refractive index $\hat{n} = n_r - j\kappa$ on the carrier density n_c. At first we measured the refractive index n_r and extinction coefficient κ of ITO thin films with different n_c. By assuming that n_c generated in fabrication gives the same effect onto $\hat{n} = n_r - j\kappa$ as n_c due to carrier injection does, we can predict a modulation performance. Two kinds of ITO thin films with resistivity of 1.04×10^{-6} and 2.03×10^{-5} Ωm were prepared on SiO$_2$ substrates and their complex refractive indices were measured by spectroscopic ellipsometry. According to Drude model, a dielectric constant can be written as

$$\varepsilon_r = \varepsilon_\infty - \frac{\omega_p^2}{\omega(\omega - j\gamma_c)}$$
$$= \varepsilon_\infty - \frac{\omega_p^2}{\omega^2 + \gamma_c^2} - j\frac{\gamma_c \omega_p^2}{\omega(\omega^2 + \gamma_c^2)} \equiv \varepsilon_1 - j\varepsilon_2 \quad (1)$$

where ε_∞ is the background constant, ω is the angular frequency, ω_p is the plasma frequency, and γ_c is the relaxation frequency. The plasma frequency ω_p is given by

$$\omega_p^2 = \frac{n_c e^2}{\varepsilon_0 m_c^*}, \quad (2)$$

where e is the elementary charge, ε_0 is the vacuum permittivity, and m_c^* is the carrier effective mass. In current study, we assume m_c^* is the same as the rest mass of electron. From the relation of $\varepsilon_r = \hat{n}^2$, n_r and κ are expressed by

$$n_r = \sqrt{\frac{\varepsilon_1 + \sqrt{\varepsilon_1^2 + \varepsilon_2^2}}{2}}, \quad \kappa = \sqrt{\frac{-\varepsilon_1 + \sqrt{\varepsilon_1^2 + \varepsilon_2^2}}{2}} \quad (3)$$

The parameters of ω_p, γ_c and ε_∞ were measured by the spectroscopic ellipsometry to be 3.1×10^{15} s^{-1}, 1.3×10^{14} s^{-1}, and 3.82, respectively, for an ITO film of lower resistivity, while those were 8.1×10^{14} s^{-1}, 4.3×10^{13} s^{-1}, and 4.19, respectively, for a higher-resistivity film. Experimentally obtained n_r and κ are shown Fig. 5. Large differences in n_r and κ can be seen between both films. The density n_c of the lower-resistivity ITO film was calculated from eq. (2) to be 3.0×10^{27} m^{-3}, while n_c of the higher-resistivity was calculated to be 2.1×10^{26} m^{-3}. Dependences of n_r and κ on n_c at a wavelength of 1.54 μm were theoretically estimated by introducing a simulation model and interpolating the experimentally obtained results and depicted in Fig. 6. In the model, ε_∞ and γ_c are assumed to be inversely proportional to n_c with appropriate backgrounds. This time we chose two carrier densities of 2.0×10^{27} and 1.3×10^{27} m^{-3} where the complex indices were calculated to be $0.30 - j\,0.68$ giving higher absorption and $1.06 - j\,0.11$ giving lower absorption, respectively. The change of carrier density is realized by making a capacitor with use of an ITO electrode and by injecting charge to the electrode.

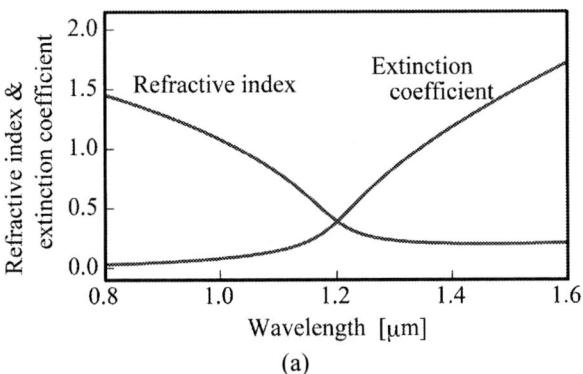

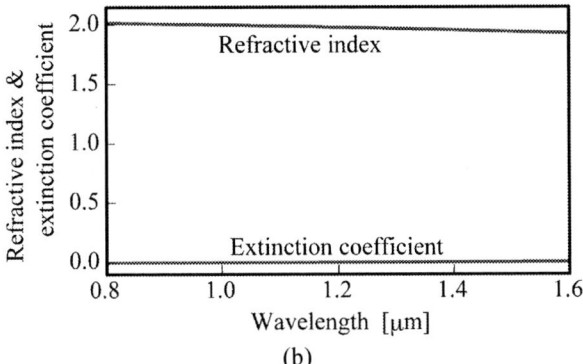

Figure 5. Measured refractive index and extinction coefficient of ITO films of low resistivity (a) and high resistivity (b).

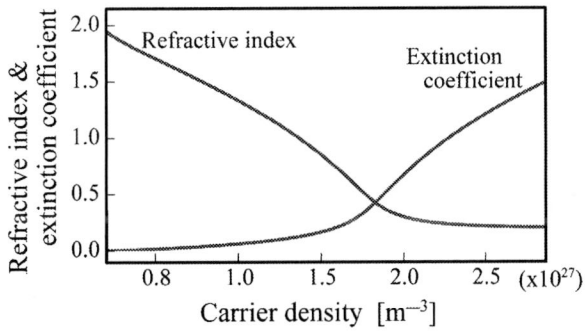

Figure 6. Expected dependence of refractive index and extinction coefficient upon carrier density at wavelength of 1.54 μm.

Design Example of the Proposed Modulator

Schematic cross-sectional structures of a design example of the proposed modulator are depicted in Fig. 7. A wavelength band centered at $\lambda = 1.54$ μm is considered. A waveguide consists of a Si-N main guiding core layer and a GeO_2-SiO_2 sub-guiding core layer on a SiO_2 substrate. An absorption modulator is formed on it and consists of an ITO layer, a SiO_2 insulating layer, and the top Al electrode. The ITO layer is used as the bottom electrode as well as the absorption modulation layer. When negative voltage is applied to the ITO electrode against the Al electrode, electron

carriers are induced and accumulated near the boundary between the ITO and the SiO_2 insulating layer. The thickness of the carrier accumulation layer is estimated to be about 5 nm [17]-[19]. We assumed that excess carriers are injected only into the 5-nm layer and the increase of n_c is uniform within the layer. Figure 7(a) shows the on-state case of higher absorption of $\hat{n} = 0.30 - j\,0.68$ obtained by carrier injection, and Fig. 7(b) shows the off-state case of lower absorption of $\hat{n} = 1.06 - j\,0.11$ without excess carriers. A two-dimensional model is considered here, and the waveguide supports two TM guided modes. TM_0 mode serves as the transmission mode while TM_1 mode takes a role of the modulation mode. Calculated distributions of the electric fields of the two modes are also schematically shown in Fig. 7. Complex mode indices of TM_0 and TM_1 were calculated to be $N_{0on} - j\,\alpha_{0on} = 1.84 - j\,9.4\times10^{-7}$ and $N_{1on} - j\,\alpha_{1on} = 1.54 - j\,3.3\times10^{-3}$, respectively, for Fig. 7(a), and $N_{0off} - j\,\alpha_{0off} = 1.84 - j\,4.3\times10^{-7}$ and $N_{1off} - j\,\alpha_{1off} = 1.54 - j\,1.3\times10^{-3}$, respectively, for Fig. 7(b).

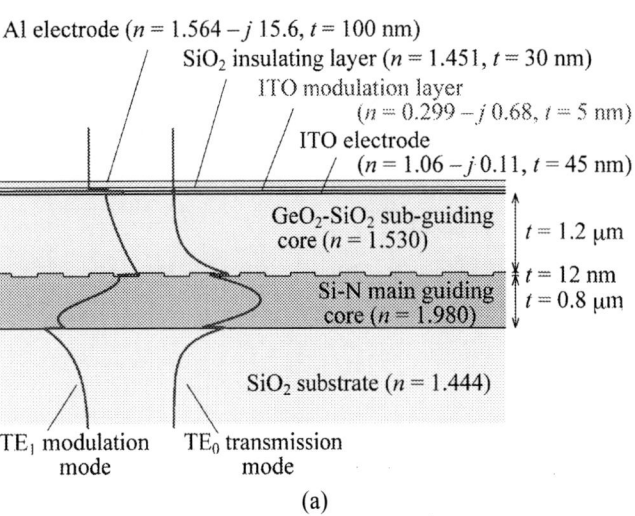

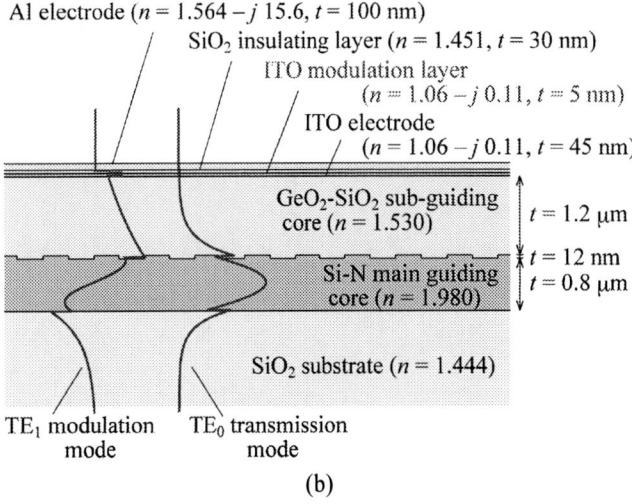

Figure 7. Schematic cross-sectional structures and calculated electric field distributions of TM guided modes for the on-state with higher absorption (a) and for the off-state with lower absorption (b).

A DGM-DBR is formed by corrugation of Si-N layer and embedded by GeO$_2$-SiO$_2$. The grating period Λ of the DGM-DBR was determined by

$$\Lambda = \frac{\lambda}{N_{0off} + N_{1off}} \cdot \qquad (4)$$

to be 455 nm.

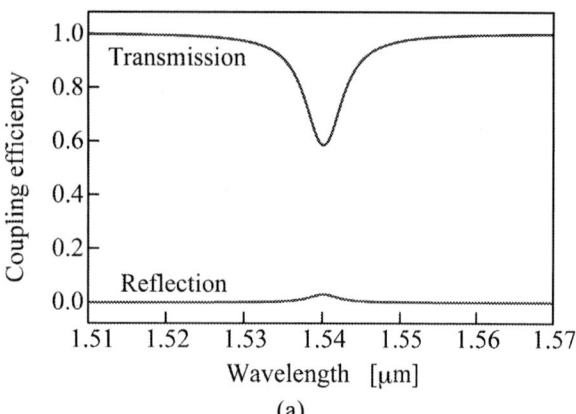

(a)

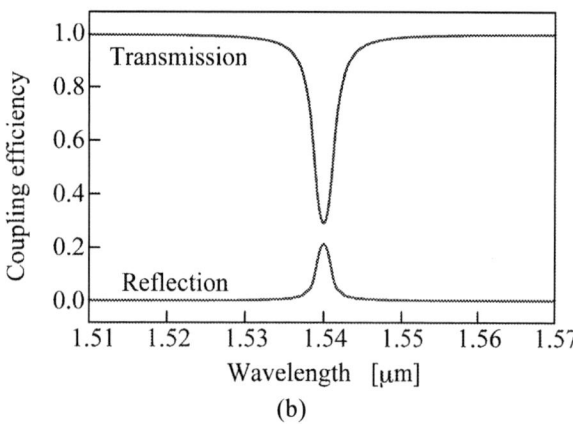

(b)

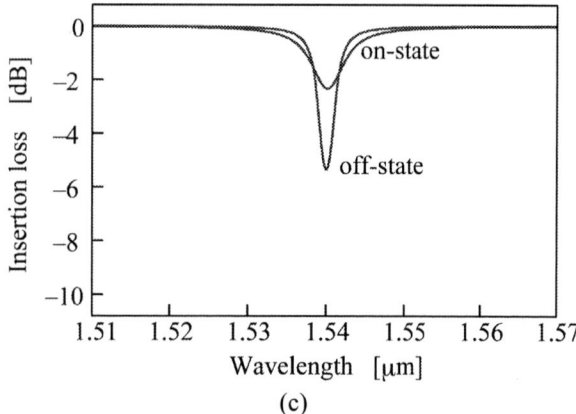

(c)

Figure 8. Calculated wavelength dependences of transmittance and reflectance of the DGM-DBR for the on-state (a) and off-state (b). Both transmittance curves are summarized in (c).

Transmittance and reflectance of the DGM-DBR were calculated by using Eqs. (A-17) and (A-18), respectively. Calculated wavelength dependences of the transmittance and reflectance of the DGM-DBR of 500-µm coupling length are summarized in Fig. 8(a) for the on-state and Fig. 8(b) for the off-state. Both transmittance of the DGM-DBR for the on-state and off-state are summarized in Fig. 8(c). The insertion loss and the extinction ratio at 1540 nm were predicted to be 2.3 dB and 3.0 dB, respectively. These values are just examples, and we can predict lower insertion loss and higher modulation depth by using a longer coupling length or different carrier densities.

Conclusions

A configuration of integrated-optic device was proposed and discussed for in-line wavelength-selective modulation. Two different guided modes in a double-core waveguide are utilized and a DGM-DBR coupling those modes is integrated. Transmission efficiency depends on the complex refractive index of a modulation layer on the waveguide. Utilization of an ITO film was considered as the modulation layer. Dependence of its complex refractive index on the carrier density was theoretically predicted based on spectroscopic-ellipsometry data of two kinds of ITO films with some assumptions. A device example was design at 1540 nm wavelength with a coupling length of 500 µm to give 2.3 dB insertion loss and 3.0 dB modulation depth even though the modulation layer was very thin of 5 nm. We can predict lower insertion loss and higher modulation depth by using a longer coupling length or different carrier densities. We may also expect better performance if we can enhance the effect of absorption by stacking ITO films and increasing effective thickness of the modulation layer. Anyway, the device must be optimized under consideration of required specifications for practical applications.

Appendix

The coupled-mode equations in the waveguide with absorption layer were used to design integrated-optic in-line wavelength-selective modulators. We summarize here the equations.

We consider a canonical wave guiding structure described by a distribution of the dielectric constant $\varepsilon_r = \varepsilon_r(x,y)$ and the relative permeability of $\mu_r = 1$. The electric field $\boldsymbol{E}^{(0)}$ and magnetic field $\boldsymbol{H}^{(0)}$ of a wave propagating in the canonical structure satisfy Maxwell's equations

$$\nabla \times \boldsymbol{E}^{(0)} = -j\omega\mu_0\boldsymbol{H}^{(0)}, \ \nabla \times \boldsymbol{H}^{(0)} = j\omega\varepsilon_0\varepsilon_r\boldsymbol{E}^{(0)}. \qquad (A-1)$$

We consider another structure including a conductivity σ and a modification or perturbation $\Delta\varepsilon_r = \Delta\varepsilon_r(x,y,z)$ in the dielectric constant. Arbitrary fields $\boldsymbol{E}$ and $\boldsymbol{H}$ in the structure satisfy

$$\nabla \times \boldsymbol{E} = -j\omega\mu_0\boldsymbol{H}, \ \nabla \times \boldsymbol{H} = \sigma\boldsymbol{E} + j\omega\varepsilon_0\big(\varepsilon_r + \Delta\varepsilon_r\big)\boldsymbol{E}. \qquad (A-2)$$

Using vector formulae from Eqs. (A-1) and (A-2) leads to

$$\nabla \cdot \left(E \times H^{(0)*} + E^{(0)*} \times H \right) = -\sigma E^{(0)*} E - j\omega\varepsilon_0 E^{(0)*} \Delta\varepsilon E .$$
(A-3)

Integration of this relation in a volume having an infinitely large area parallel to the xy-plane and infinitely small thickness along z-axis leads to the relation

$$\iint \frac{\partial}{\partial z} \left[E_t \times H_t^{(0)*} + E_t^{(0)*} \times H_t \right]_z dxdy$$
$$= -j\omega\varepsilon_0 \iint E^{(0)*} \left(\Delta\varepsilon - j\frac{\sigma}{\omega\varepsilon_0} \right) E dxdy$$
(A-4)

We take a mode labeled by m for $E^{(0)}$, $H^{(0)}$ and write it as

$$E_t^{(0)} = E_{tm}(x,y) \exp(-j\beta_m z),$$
$$H_t^{(0)} = H_{tm}(x,y) \exp(-j\beta_m z).$$
(A-5)

Modal expansion of E, H with use of Eq. (A-5) leads to

$$E_t = \sum_n c_n(z) E_{tn}(x,y) \exp(-j\beta_n z),$$
$$H_t = \sum_n c_n(z) H_{tn}(x,y) \exp(-j\beta_n z).$$
(A-6)

By substituting Eqs. (A-5), (A-6) to Eq. (A-4), we obtain

$$\sum_n \left\{ \frac{dc_n}{dz} - j(\beta_n - \beta_m) c_n \right\} \exp\left\{ -j(\beta_n - \beta_m) z \right\}$$
$$\times \iint \left[E_{tn} \times H_{tm}^* + E_{tm}^* \times H_{tn} \right]_z dxdy$$
$$= -j\omega\varepsilon_0 \sum_n c_n \exp\left\{ -j(\beta_n - \beta_m) z \right\}$$
$$\times \iint E_m^* \left(\Delta\varepsilon - j\frac{\sigma}{\omega\varepsilon_0} \right) E_n dxdy$$
(A-7)

Orthonormal relation can be written as

$$\frac{1}{4} \iint \left[E_{tn} \times H_{tm}^* + E_{tm}^* \times H_{tn} \right]_z dxdy = \pm\delta_{mn} ,$$
(A-8)

where δ_{mn} is Kronecker delta. The sign in the right-hand side is meaningful only for $m = n$, and a plus sign should be taken for $\beta_m = \beta_n > 0$ and a minus sign for $\beta_m = \beta_n < 0$. Then, Eq. (A-7) can be rewritten as

$$\pm \frac{dc_m}{dz} = -j \sum_n \kappa_{mn} c_n \exp\left\{ -j(\beta_n - \beta_m) z \right\} ,$$
(A-9)

$$\kappa_{mn} = \frac{\omega\varepsilon_0}{4} \iint E_m^* \left(\Delta\varepsilon - j\frac{\sigma}{\omega\varepsilon_0} \right) E_n dxdy .$$
(A-10)

The sign in the left-hand side in Eq. (A-9) should be plus for $\beta_m > 0$ and minus for $\beta_m < 0$.

We consider a contra-directional coupling, and rewrite c_1 and c_2 to $A(z)$ and $B(z)$ with $\beta_A > 0$ and $\beta_B < 0$. The coupled mode equations can be expressed by

$$\frac{dA(z)}{dz} = -j\kappa_{AA} A(z) - j\kappa_{AB} B(z) \exp(-2j\Delta z) ,$$
$$-\frac{dB(z)}{dz} = -j\kappa_{AB}^* A(z) \exp(+2j\Delta z) - j\kappa_{BB} B(z) .$$
(A-11)

where $2\Delta = \beta_B - \beta_A + K$ with use of the grating vector size K. The solutions of Eq. (A-11) are

$$A(z) = a_+ \exp\left\{ -j(\Delta + s')z \right\} \exp(\alpha z)$$
$$\quad + a_- \exp\left\{ -j(\Delta + s')z \right\} \exp(-\alpha z)$$
$$B(z) = b_+ \exp\left\{ +j(\Delta - s')z \right\} \exp(\alpha z)$$
$$\quad + b_- \exp\left\{ +j(\Delta - s')z \right\} \exp(-\alpha z)$$
$$s' = \frac{\kappa_{AA} - \kappa_{BB}}{2} ,$$
$$\alpha = \frac{\sqrt{4\left(|\kappa_{AB}|^2 - \Delta^2\right) - (\kappa_{AA} + \kappa_{BB})(\kappa_{AA} + \kappa_{BB} - 4\Delta)}}{2} .$$
(A-12)

The coefficients a and b are not independent from each other and satisfy the following relations

$$b_+ = \frac{\Delta - s + j\alpha}{\kappa_{AB}} a_+ , \quad b_- = \frac{\Delta - s - j\alpha}{\kappa_{AB}} a_- ,$$
$$s = \kappa_{AA} - s' = \frac{\kappa_{AA} + \kappa_{BB}}{2} .$$
(A-13)

Using initial conditions of

$$A(0) = a_+ + a_- = 1, \quad B(0) = b_+ \exp(\alpha L) + b_- \exp(-\alpha L) = 0$$
(A-14)

leads to the solutions

$$A(z) = \frac{\alpha \cosh\left\{ \alpha(L-z) \right\} - j(\Delta - s)\sinh\left\{ \alpha(L-z) \right\}}{\alpha \cosh(\alpha L) - j(\Delta - s)\sinh(\alpha L)} ,$$
$$\times \exp\left\{ -j\left(\Delta + \frac{\kappa_{AA} - \kappa_{BB}}{2} \right) z \right\}$$
$$B(z) = \frac{-j}{\kappa_{AB}} \frac{\left\{ (\Delta - s)^2 + \alpha^2 \right\} \sinh\left\{ \alpha(L-z) \right\}}{\alpha \cosh(\alpha L) - j(\Delta - s)\sinh(\alpha L)} .$$
$$\times \exp\left\{ j\left(\Delta - \frac{\kappa_{AA} - \kappa_{BB}}{2} \right) z \right\}$$
(A-15)

When we choose the canonical structure so that the self-coupling coefficients due to $\Delta\varepsilon$ are 0, we can summarize coupling coefficients from Eq. (A-10) to

$$\kappa_{AA} = \frac{\omega\varepsilon_0}{4} \iint E_A^* \left(-j\frac{\sigma}{\omega\varepsilon_0} \right) E_A \, dxdy \ ,$$

$$\kappa_{BB} = \frac{\omega\varepsilon_0}{4} \iint E_B^* \left(-j\frac{\sigma}{\omega\varepsilon_0} \right) E_B \, dxdy \ ,$$

$$\kappa_{AB} = \frac{\omega\varepsilon_0}{4} \iint E_A^* \left(\Delta\varepsilon - j\frac{\sigma}{\omega\varepsilon_0} \right) E_B \, dxdy \ , \qquad \text{(A-16)}$$

Then, the coupling efficiencies for transmission T and reflection R are expressed by

$$T = \left| \frac{A(L)}{A(0)} \right|^2 = \left| \frac{\alpha}{\alpha\cosh(\alpha L) - j(\Delta - s)\sinh(\alpha L)} \right.$$

$$\left. \times \exp\left\{ -j\left(\Delta + \frac{\kappa_{AA} - \kappa_{BB}}{2} \right)L \right\} \right|^2 ,$$

$$\text{(A-17)}$$

$$R = \left| \frac{B(0)}{A(0)} \right|^2 = \frac{1}{\kappa_{AB}} \left| \frac{\left\{ (\Delta - s)^2 + \alpha^2 \right\}\sinh(\alpha L)}{\alpha\cosh(\alpha L) - j(\Delta - s)\sinh(\alpha L)} \right|^2 .$$

$$\text{(A-18)}$$

Acknowledgments

The authors would like to thank Prof. H. Kikuta and Dr. S. Kameda in Osaka Prefecture University for their support on EB direct writing process of DGM-DBR.

References

1. N. Savage, "Linking with light," *IEEE Spectrum*, vol. 39, no. 8, pp. 32–36, Aug. 2002.
2. A. F. Benner, M. Ignatowski, J. A. Kash, D. M. Kuchta, and M. B. Ritter, "Exploitation of optical interconnects in future server architectures," *IBM J. Res. Dev.*, vol.49, no.4/5, pp.755–775, Jul./Sep. 2005.
3. S. Ura, "Selective guided mode coupling via bridging mode by integrated gratings for intraboard optical interconnects," Proc. SPIE, vol.4652, pp.86–96, San Jose, USA, Jan. 2002.
4. S. Ura, K. Shinoda, K. Kintaka, C. Ito, D. Nii, K. Nishio, and Y. Awatsuji, "Gigabits-per-second signal transmission from singlemode vertical-cavity surface-emitting laser via thin-film waveguide for wavelength-division-multiplexing optical interconnect board," *Jpn. J. Appl. Phys.*, vol.47, no.8, pp.6664–6666, Aug. 2008.
5. K. Kintaka, J. Nishii, K. Shinoda, and S. Ura, "WDM signal transmission in a thin-film waveguide for optical interconnection," *IEEE Photon. Technol. Lett.*, vol. 18, no. 21, pp. 2299-2301, Nov. 2006.
6. K. Kintaka, J. Nishii, S. Murata, and S. Ura, "Eightchannel WDM intraboard optical interconnect device by integration of add/drop multiplexers in thin-film waveguide," *J. Lightw. Technol.*, vol. 28, no. 9, pp. 1398-1403, May 2010.
7. S. Ura, K. Shimizu, Y. Kita, K. Kintaka, J. Inoue, and Y. Awatsuji, "Integrated-optic free-space-wave coupler for package-level on-board optical interconnects," IEEE J.

Sel. Top. Quantum Electron., vol.17, no.3, pp.590–596, May/June 2011.
8. A. Liu, L. Liao, Y. Chetrit, J. Basak, H. Nguyen, D. Rubin, and M. Paniccia, "Wavelength division multiplexing based photonic integrated circuits on silicon-on-insulator platform," *IEEE J. Sel. Top. Quantum Electron.*, vol.16, no.1, pp.23–32, Jan./Feb. 2010.
9. Q. Xu, B. Schmidt, J. Shakya, and M. Lipson, "Cascaded silicon micro-ring modulators for WDM optical interconnection," *Opt. Express*, vol.14, no.20, pp.9431–9436, Oct. 2006.
10. S. Manipatruni, L. Chen, and M. Lipson, "Ultra high bandwidth WDM using silicon microring modulators," *Opt. Express*, vol.18, no.16, pp.16858–16867, Aug. 2010.
11. B. G. Lee, B. A. Small, Q. Xu, M. Lipson, and K. Bergman, "Characterization of a 4 x 4 Gb/s parallel electronic bus to WDM optical link silicon photonic translator," *IEEE Photon. Technol. Lett.*, vol.19, no.7, pp.456–458, Apr. 2007.
12. N. Ophir, C. Mineo, D. Mountain, and K. Bergman, "Silicon photonic microring links for high-bandwidth-density, low-power chip I/O," *IEEE Micro*, vol.33, no.1, pp.54–67, Jan.-Feb. 2013.
13. P. Rabiei, W. H. Steier, C. Zhang, and L. R. Dalton, "Polymer microring filters and Modulators," *J. Lightwave Technol.*, vol.20, no.11, pp.1968–1975, Nov. 2002.
14. K. Debnath, L. O'Faolain, F. Y. Gardes, A. G. Steffan, G. T. Reed, and T. F. Krauss, "Cascaded modulator architecture for WDM applications," *Opt. Express*, vol.20, no.25, pp.27420–27428, Nov. 2012.
15. T. Miura, R. Mori, K. Kintaka, K. Nishio, Y. Awatsuji, and S. Ura, "Proposal of waveguide interferometer for in-line wavelength-selective modulator," in *Tech. Digest Conf. Lasers Electro-Opt. Pac. Rim (CLEO-PR)*, Kyoto, Japan, June 30-July 4, 2013, paper WL3-3.
16. S. Kawanami, R. Mori, T. Miura, K. Kintaka, K. Nishio, Y. Awatsuji, and S. Ura, "Different-guided-mode-coupling DBR for in-line wavelength-selective modulator," in *Tech. Digest Optoelectron. Commun. Conf.*, Busan, Korea, July 2-6, 2012, pp. 89-90.
17. E. Feigenbaum, K. Diest, and H. A. Atwater, "Unity-order index change in transparent conducting oxides at visible frequencies," *Nano Lett.*, vol. 10, pp. 2111-2116, May 2010.
18. V. J. Sorger, N. D. Lanzillotti-Kimura, R.-M. Ma, and X. Zhang, "Ultra-compact silicon nanophotonic modulator with broadband response," *Nanophotonics*, vol. 1, no. 1, pp. 17-22, May 2012.
19. A. Melikyan, N. Lindenmann, S. Walheim, P. M. Leufke, S. Ulrich, J. Ye, P. Vincze, H. Hahn, Th. Schimmel, C. Koos, W. Freude, and J. Leuthold, "Surface plasmon polariton absorption modulator," *Opt. Express*, vol. 19, no. 9, pp. 8855-8869, April 2011.
20. F. Michelotti, L. Dominici, E. Descrovi, N. Danz, and F. Menchini, "Thickness dependence of surface plasmon polariton dispersion in transparent conducting oxide films at 1.55 μm," *Opt. Lett.*, vol. 34, no. 6, pp. 839-841, March 2009.

21. V. E. Babicheva and A. V. Lavrinenko, "Plasmonic modulator optimized by patterning of active layer and tuning permittivity," *Opt. Commun.*, vol. 285, no. 24, pp. 5500-5507, Nov. 2012.

Porous Silicon Technology, a Breakthrough for Silicon Photonics: From Packaging to Monolithic Integration

M. Balucani[1,*], A. Klyshko[1], K. Kholostov[1], A. Benedetti[2], A. Belardini[2], C. Sibilia[2], M. Izzi[3], M. Tucci[3], H. Bandarenka[4], V. Bondarenko[4]

[1]Sapienza University of Rome-DIET, Via Eudossiana, 18 - 00184 Roma ITALY
[2]Sapienza University of Rome-SBAI, Via A. Scarpa, 16 - 00161 Roma ITALY
[3]ENEA Casaccia Research Centre Rome, Via Anguillarese 301, 00123, ITALY
[4]Belarusian State University of Informatics and Radioelectronics, P. Brovki str., 220013, Minsk, BELARUS
* balucani@diet.uniroma1.it

Abstract

Low cost concept based on the porous silicon technology is shown to be well suitable for integrating monolithically the photonic devices on a standard silicon wafers by using localized SOI structures fabricated by electrochemical anodization of silicon wafers followed by thermal oxidation of porous silicon. The new approach consists in realizing buried localized porous oxidized silicon by exploiting two different routes: n⁻ epi/n⁺/n⁻ structures on p-type wafers and ion-implantation on standard CMOS/BiCMOS wafers. The peculiarities of the developed approach, including anodization and thermal oxidation regimes to form oxidized porous silicon regions with the required properties are presented. The advantages of the proposed approach in realizing the fiber-to-chip and power-over-fiber coupling are discussed.

Introduction

Photons, unlike electrons, typically do not interact with each other and so they are almost immune to crosstalk. Furthermore, since light has a much greater bandwidth, or carrying capacity, photonic signals can carry far more bits per second than electronic signals, while dissipating much less power. Currently, the approach is to develop photonic devices as a fully integrated optoelectronics system and the forecast view looks towards a fully integrated optical system. The fascinating concept of 'Integrated Optics', with the meaning of optical devices and circuits realized in a patterned planar waveguide format, was first described by Stewart Miller in a historic special issue of the Bell System Technical Journal (BSTJ), as long ago as 1969 [1]. Miller envisaged an analogous development to Integrated Electronics with a continuously growing density of integration (Moore's law). However, it turned out that the growth of integration density in the optical domain was much slower than in electronics. There are several reasons that explain the different rates of development. An important one is that a single material – silicon - has an almost completely dominant role in Integrated Electronics, as the base material for more and more complex IC. The situation in the domain of Integrated Optics is quite different. There is a wide range of materials used for specific applications. As an example, integrated (Mach-Zehnder type) optical modulators in the ferroelectric material LiNbO₃ are among the key components that enable the worldwide optical internet. Interestingly, silicon – currently only in the form of silicon-on-insulator (SOI) – has emerged during the last few years as a strong contender for various applications of Integrated Optics. The first multi-Gb/s capable silicon photonics modulator was reported in 2004 [2]. Since then, silicon photonics has seen an explosion of interest, with low-loss waveguides [3], detectors [4], and grating couplers [5] all separately demonstrated. The challenge is the development of a platform that supports all of these key components simultaneously with attractive performance, and achieves drive voltages compatible with CMOS processes. Luxtera demonstrated a photonics platform with both integrated modulators and detectors working at 10 Gb/s in 2006 [6]. Since that Luxtera has announced 25 Gb/s capability [7] and recently reported a 4x28G link [8], though the technical details remain unpublished, and key device geometries and performance characteristics have been kept a trade secret. Kotura has achieved 40 Gb/s for various devices [9], but has not reported a common platform with both high-speed modulators and detectors on the same wafer, at the same time. Intel and IBM have reported individual high-performance devices [10, 11]. Meanwhile, ePIXfab has been providing fabrication services to the photonics community [12], and has published results on modulators and detectors for 40 Gb/s speeds separately [13,14], though not on results achieved simultaneously as a part of an integrated platform. Recently, was demonstrated a wafer-scale silicon photonics platform that achieves 15 GHz bandwidths, and is capable of 25 Gb/s data rates using a SOI wafer with a handle wafer of very high resistivity (i.e. 750 Ω·cm) [15].

Silicon photonics (SP) technology envisions a clear road map fitting the needs and the challenges of the future ICT (Information Communication Technology) systems and services. Four key applications will drive the evolution: intensive computing, broadband communication, mass storage and consumer multimedia, but still many key technological breakthroughs are needed to enable the intra-chip interconnections, as presented in [16].

Generation IV SP, forecasted for 2020 [16], promises a low cost solution by enabling close-proximity integration of photonics with electronics. The ultimate goal of silicon photonics is to monolithically integrate the photonic layer with optical functions onto silicon IC chips [17]. Several full integration schemes are considered, mainly based on back-end post-processing based on wafer-to-wafer molecular bonding [18]. The main objective of the generation IV SP, for 300 mm wafers and even more so for the forecast 450 mm wafer technology, will be to integrate all the photonic devices on a CMOS/BiCMOS wafer by using integration techniques, that do not depend on the specific node used to produce the silicon wafer - and by using only standard CMOS/BiCMOS processes.

Today, silicon photonics is mainly a hybrid technology. Integrated circuits (i.e. CMOS, SiGe, SiGe:C) are usually built with standard silicon wafers or, where SOI (Silicon On Insulator) wafers are used for CMOS, the top silicon (TopSi) thickness is lower than 100 nm and the buried oxide (BOX) is less than 200 nm. Meanwhile, silicon photonics has essentially been based on SOI wafers - that are different from the SOI wafers that are used to realize conventional CMOS electronics: TopSi 220 nm and BOX 3000 nm or 340 nm TopSi with 2000 nm BOX [19] or 220 nm/2000nm for, respectively, TopSi and BOX [20]. The silicon photonic waveguides (SPWs) in the SOI wafers described above are typically photonic wires that have significantly polarization dependent properties and, in order to produce polarization independent SPWs, the TopSi will typically be required to have a greater thickness: possibly as much as 1 µm [21]. High frequency operation for silicon photonics poses another issue for SOI wafers. The SOI handle wafer must have a very low doping level (i.e. 750 Ω·cm), thus increasing the cost of the wafer (i.e. x4). Such low doping is necessary in order to allow high frequency modulation (e.g. 25GHz) with low losses for the transmission line driving the SP modulator [22]. Attempts to integrate IC processors, memories and/or ASIC devices on silicon photonics SOI substrates will be quite challenging, if not impossible, due to the different device properties, due to thermal dissipation issues and due to the requires tight control of SP SOI cross-wafer uniformity [23].

To sum it up, currently, there is a lack of a technology that is capable of integrating monolithically silicon photonics and integrated circuits - and does not depend on the specific node used to produce the electronic devices. A suitable technology that could match the requirement for both ICs and SP would use localized SOI structures (with the eventual possibility of modulating the TopSi thickness). It would employ localized SOI structures where SP devices are required and bulk CMOS/BiCMOS and/or ultra-thin SOI wafers where ICs are to be realized.

The idea of realizing localized SOI structure by masked oxygen implantation has been tried, but without success (e.g. figure 1 from reference [24]). This concept, if it could be made to work, could offer a way utilizing the energy modulation of the implanted oxygen atoms, in order to obtain SOI wafers with different TopSi thicknesses.

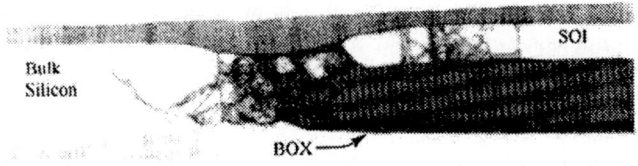

Figure 1. Masked oxygen implantation: unsuccessful SOI structure.

Current silicon photonics must also address another big issue: Fiber-to-Chip Coupling. Standard single-mode fibers (SMF) are very large in size by comparison with silicon

photonic strip or slot waveguides. The typical core diameter size of a standard SMF is approximately 8 µm. In consequence, coupling light from an optical fiber to a silicon photonic waveguide that has dimensions in the submicron range, becomes a challenging task. This is shown in figure 2 [25], where an optical fiber core and a silicon photonic SOI waveguide are drawn on the same scale.

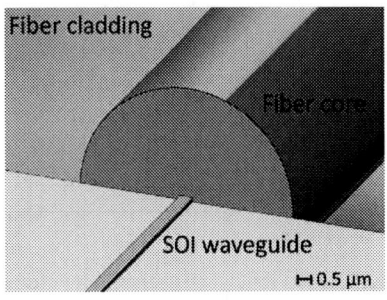

Figure 2. Fiber to SP waveguide coupling issue.

The large difference in dimensions between the fiber and the waveguide causes a considerable mismatch between their optical modes. Due to this mismatch, a coupling structure is needed to adapt a wide fiber to a narrow silicon waveguide and improve (i.e. reduce) the coupling losses. Typically, direct butt coupling between a SMF (mode field diameter MFD=10 µm) and a silicon photonic waveguide (MFD <1 µm) leads to more than 20dB insertion loss. The insertion loss between an optical fiber and a silicon photonic circuit is definitely a key issue, since it is directly linked with performance matters such as the link reach, the signaling rate, the receiver sensitivity, and so on. Moreover, in order to be compatible with functions for Fiber-to-the-x (FTTx), or wavelength division multiplexing (WDM) applications for instance, a good coupling structure is also required to be both broadband and polarization insensitive. Meanwhile, considering the packaging cost, the footprint of the coupling structure must be kept small and it must have sufficient alignment tolerances.

The conclusions of the European project HELIOS [26] give a clear overview of the fiber-to-chip coupling state-of-the-art. More technical details for each single type of coupling can be found in [25]. Meanwhile, NEC Corporation has presented a new vertical-coupling optical interface with a grating coupler for transmitting and receiving optical signals between single-mode optical fibers and photonic waveguides, with a coupling efficiency of about 50% at 850 nm. They found that tolerance misalignment of ± 1.5 µm and ± 2 µm in the x- and z-directions, respectively, leads for a coupling loss increase of 1 dB and they claim they could reach the same value for 1550nm [27]. From the packaging point of view the most simple and cost effective assembly technique is a V-groove pigtail. Following the state-of-the-art described in [26], the V-groove technique, due to alignment tolerances that are typically no better than ±5 µm, seems to be usable only in multimode applications or single-mode applications with a low optical loss budget.

978-1-4799-2408-0/14 $31.00 © 2014 IEEE   195

Technology

Localized Oxidized Porous Silicon

Our research group, in collaborations also with industrial partners, is following several different approaches to prove the manufacturability of a key enabling technology (KET) for the next generation of silicon photonics monolithically integrated with electronic circuits - independently of the IC technology node and solving the fiber-to-chip coupling issue.

The new approach proposed consists of realizing buried localized porous oxidized silicon (and also dense oxide: similar to thermal SiO2) by exploiting two different routes:

a) n^- epi/n^+/n^- structures on p-type wafers;
b) ion-implantation on standard CMOS/BiCMOS wafers

The second route, that is still under development (i.e. ion implantation simulation and testing), will provide a unique opportunity to realize, on the same wafer, CMOS/BiCMOS IC and silicon photonic structures that do not suffer from conflicting demands of silicon thickness and buried oxide (porous or dense) will be presented in a future article.

The first direction (i.e. route a), is based on localized oxidized porous silicon (OPS) technology. In figure 3 is presented the process concept.

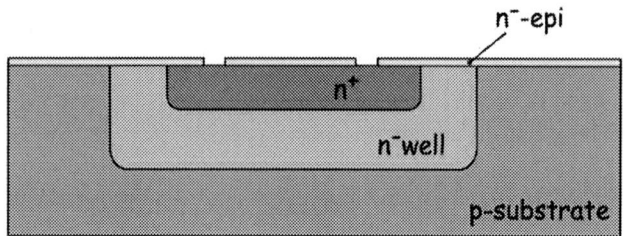

Figure 3. Starting structure for the formation of the localized OPS regions, converting n^+ region in PS and oxidizing it.

Starting from a standard p-type wafer, the localized SOI structure where the SP devices are to be realized is obtained by defining an n^--well in the p-type wafer and within the n^--well the future buried oxide zone is formed by defining a n^+-well. Once defined all the localized buried oxide zones needed for the SP devices, n^--epi layer is grown on the wafer.

Technological process of OPS formation consists of three main steps: (i) the anodic treatment of Si in HF-containing electrolyte to convert silicon into the porous state (porous silicon (PS)); (ii) the chemical cleaning of PS; (iii) the thermal oxidation of PS. The technological aspects of each step have been developed to succeed in OPS based technology.

Several principal problems have been solved on the way from fundamental research of PS to industrial technology of OPS. Development of any new process at a high-technology electronic company needs the equipment suited to operate under clean room conditions. So, the first major task has been to design suitable equipment for the formation of PS layer of uniform thickness and porosity. The new equipment is based on dynamic liquid drop/meniscus concept that allows to process the wafer locally with high uniformity touching the wafer only where is necessary to form the PS structures. Such

new equipment concept is presented in [28, 29]. It has also been necessary to develop the methods of chemical cleaning as well as defect-free oxidation of $1 - 10$ μm thick PS layer.

Solutions based on H_2O_2-NH_4OH-H_2O are been totally unsuited for PS chemical cleaning since they are harmful to PS. With evidence provided by secondary-ion mass spectroscopy (SIMS) and neutron activation analysis (NAA), the cleaning of PS has been best achieved by rinsing in de-ionized water for 20-25 min boiling in the solution based on H_2O_2-HCl-H_2O (1:1:3) for $10 - 15$ min, and rinsing in de-ionized water for $20 - 25$ min, in succession. It has been determined that the rising in de-ionized water should be performed immediately after PS formation. Once chemical cleaning has been performed, PS should be oxidized as quickly as possible.

Due to the large surface/volume ratio, PS has a very high rate of oxidation. This has allowed the oxidation of thick porous layer in a short time. However, when heated to temperature higher than 773K, PS undergoes irreversible transformation in the original microstructure (sintering) [30]

The sintering prevents the PS layer from the oxidation. But it can be easily avoided by pre-oxidation of PS thermally at low temperature (i.e. 673K) in dry oxygen. Temperatures of $1073 - 1173K$ have been used to oxidize PS. A densification step at 1323-1423K in wet oxygen followed by annealing in nitrogen has been found necessary to form an OPS equivalent to thermal SiO_2 (i.e. refractive index n=1.46). Optical characteristics (i.e. refractive index) of OPS are strongly dependent on PS porosity and oxidation regimes [31].

The method of fabricating the localized silicon on insulator structures, as shown in figure 4, is based on the preferential anodization of n^+-layer within the n^--epi/n^+/n^- structure. Such high selective anodization is possible because substantially higher voltage is required to transform lightly doped n^- silicon into PS in comparison with that for heavily doped n+ layers (i.e. doping higher than 10^{18} atm/cm^3). So, at low forming voltage anodization process stops as soon as the whole n^+ layer has been converted into PS.

Standard p-type Si (100) wafers were used as initial substrates. N^- well and n^+ layer zones were defined by ion implantation followed by RTA. Then, a 0.6 μm thick epitaxial was grown on the front of the wafer by Chemical Vapour Deposition (CVD) at 1248K from dichlorsilane ambient. An α-Si/Si_3N_4 film was deposited upon the epitaxial layer as a mask layer. Projection photolithography using reactive ion etching of both the mask and the epitaxial layer was used to define three-dimensional pattern of islands wherein the SP device should be formed. The anodic treatment of the structure in HF-containing electrolyte has selectively, by dynamic liquid drop/meniscus technique, converted the n^+ layer into PS underneath silicon islands to be isolated. Once oxidation of the PS layer has been performed, the Si islands isolated from the substrate with the OPS layer are ready for SP device fabrication.

In figure 4 is presented a SEM image where is shown the n^--epi/n^+/n^- structure after oxidation and densification.

978-1-4799-2408-0/14 $31.00 © 2014 IEEE

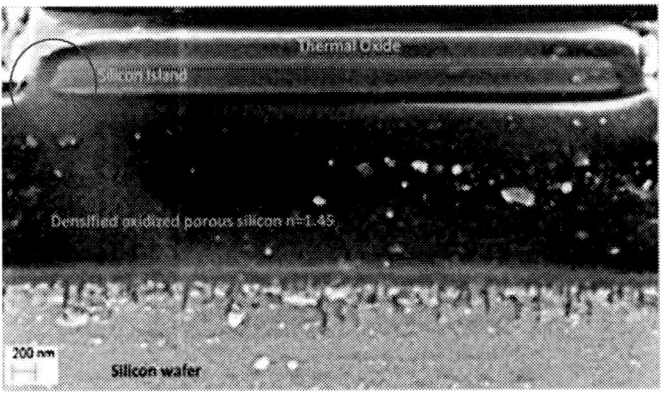

Figure 4. Cleaved cross-section of silicon on densified localized oxidized porous silicon.

In figure 5 a zoomed inset of figure 4 (red circle) showing the very low roughness of the interface between the silicon and the oxide. The buried oxide thickness is 1.2 μm.

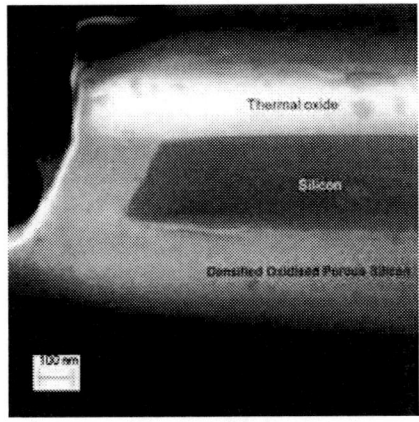

Figure 5. Zoomed part of figure 5 showing a detailed view of Si-OPS interface

It is well known that propagation losses depend on the interface roughness due to scattering losses, but if we consider that localized OPS waveguide based on a different technology and with higher roughness showed propagation losses of about 1 dB/cm at 1550 nm [32], a quite better result is expected with losses in the range of 0.1dB/cm as for standard SOI wafers. Simulation and design of SP devices and waveguides are under development and measurements are planned for the May 2014.

Thick oxidized PS is used to demonstrate that, within the same wafer, it is possible to realize thick OPS layers - enabling decoupling of the RF from the silicon substrate properties. A thick OPS structure obtained by converting p-type standard CMOS/BiCMOS wafer in OPS is under characterization by RF transmission line coupling losses on at RF. frequencies up to 60GHz - and preliminary measurements show that it is possible to reduce the coupling losses through the silicon substrate from the usual 20dB/mm level to less than 2dB/mm at 40GHz – by means of a thick OPS decoupling layer, without the requirement for expensive high-resistance wafers. Different thicknesses of OPS are under RF

characterization (i.e. 5, 10, 20, 80 micron). Figure 6 shows OPS layer fabricated with a thickness of about 80 μm.

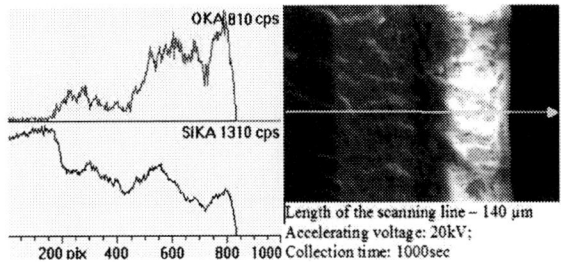

Figure 6. Thick OPS layer obtained by anodizing the standard p-type wafer. On the left the spectra and on the right the OPS cross section are presented.

Fiber-to-Chip Coupling and power over fiber

Thanks to the guiding properties of OPS in the visible and infrared spectra [33, 34], the ready availability of single-mode lasers at 850 nm with optical power up to 100mW, the possibility of realizing highly efficient silicon solar cells for the 850 nm single wavelength (i.e. > 32%) - and the fact that standard SMF (e.g. 8/125 μm) is able to withstand power levels up to 17 dBm without distortion for 1Gbps data rate, we are following a new approach for the fiber-to-chip coupling and opening a new concept sensors using power over fiber based all in silicon technology.

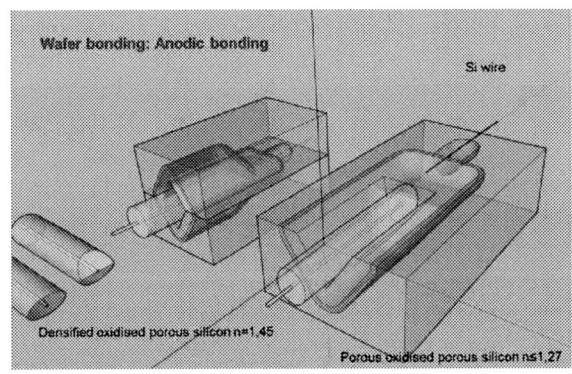

Figure 7. V-groove pigtail approach for SP wire coupling.

In figure 7, is presented the new concept of V-groove pigtail using OPS and in figure 8 is presented a SEM picture with superimposed the core and cladding of a SMF of the preliminary structures under realization.

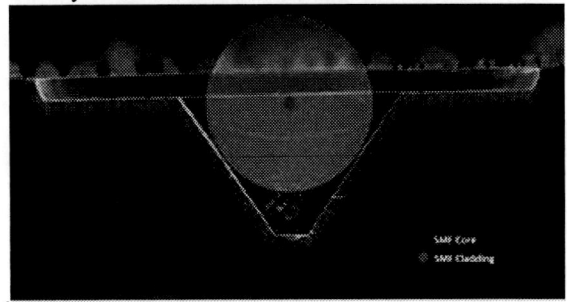

Figure 8. V-groove pigtail with OPS and superimposed SMF.

The approach to solve frequency splitter for the fiber over power (i.e. 850 nm) and optical signal (i.e. 1310 or 1550 nm) is presented in figure 9. There are also shown the principle concepts of the new approach followed to solve the fiber coupling and the power over fiber that has to be expanded trough a power lens toward a silicon solar cell..

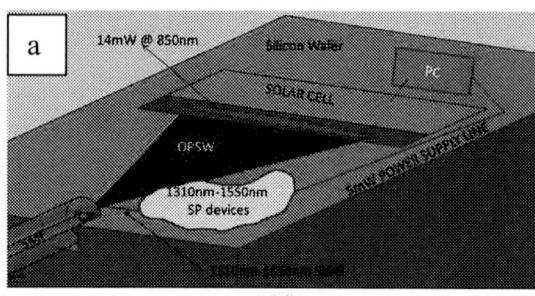

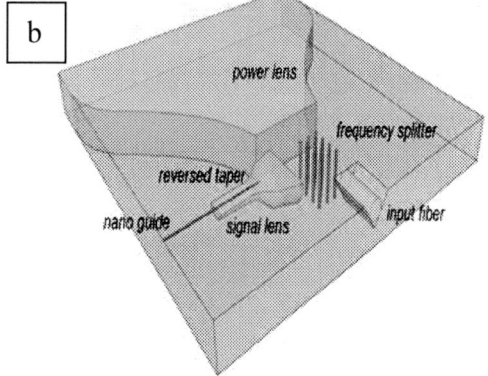

Figure 9. (a) Structure of signal, power over fiber coupling and solar cell; (b) schematic details of the coupling structures of the fiber over power concept.

To treat distinctly the two field, we aim to take advantage of the Photonic Crystals (PC) technology [35 – 38], which allows a separation of the power and signal fields and a redirection toward different sites.

In figure 10, is shown a preliminary structure showing macro-porous silicon, that will be used for the realization of PC frequency splitter, and the buried micro-porous silicon that will be used for the cladding layer for decoupling the silicon substrates..

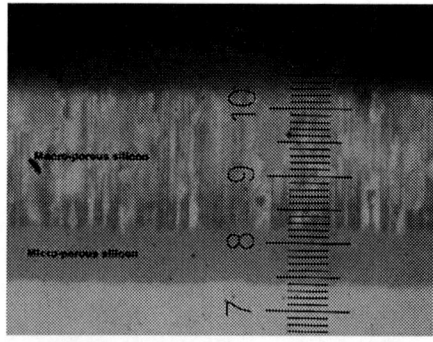

Figure 10. Macro-porous silicon on micro-porous silicon structure used to realize photonic crystals

In our approach the signal will be collected leaving the optic fiber by using the reversed taper grown upon a substrate. At the basis of the reversed taper technology there is the natural property of the energy to adapt to the surrounding environment and follow the most suitable path at its disposition. More precisely, any distribution of EM. field tries to fill the surrounding environment following the natural diffraction process of the radiation, thus an injected mode of an EM. wave progressively modifies its shape according to the EM. properties (such as the electric permittivity and magnetic permeability) of the environment. Abrupt changes of these properties give rise to unwanted scattering processes and reflections, with a consequent propagation losses and coupling losses when different devices need to be connected, for example when an optical fiber has to be connected to a SOI waveguide.

The reversed taper (RT) technology is the key element to reduce the fiber-to-chip coupling problem in silicon photonics. An inverted taper is basically known in literature as a SOI waveguide whose width decreases with the propagation direction from chip to fiber. While the width of the taper increases, the effective index of the mode slowly varies in order to match the SOI waveguide effective index, but the effective index of the mode at the tip has to be as close as possible to the effective index of the optical fiber mode.

Simulations

The evaluation of the coupling efficiency requires highly accurate numerical calculations. We have checked the RT's coupling ability by performing tridimensional (3D) numerical simulations by accurate simulations environment. For our estimations, we adopted Lumerical FDTD Solutions, which is a powerful instrument to simulate any electromagnetic process by using a fast finite-difference-time-domain (FDTD) algorithm. The FDTD method requires two distinct grids of points for the electric and magnetic fields, and is based on sequential updates of the same following Maxwell's equations for all the pre-set time domain; a certain attention must be paid while setting the spatial and time definition both for the incoming pulse and the time domain to allow the EM. field to reach the limits of the region under investigation. A precise selection and setting about the boundary conditions ensures an almost perfect deletion of any spurious reflection.

To predict the efficiency of our structure, in the numerical set-up we used perfect matching layers (PML) at the exterior bounds of our structure, which grant an efficient absorption of the outgoing EM. field to simulate an almost infinite environment outside the interested region. The main geometric blocks of the structure have been designed both by the use of basic elements provided by the FDTD Solution's CAD and by importing some particular blocks formerly designed (and exported) from Matworks Matlab. We used a mode-solver to set the emitted EM. field by a fiber glass system and injected this field toward the Silica\Silicon (SiO$_2$/Si) structure.

To allow the injection of the highest percentage of EM field, we adopted a distributed lens with a 3D profile for the index of refraction, which is able to make light come closer to the RT. This procedure will be described later.

978-1-4799-2408-0/14 $31.00 © 2014 IEEE   198

We evaluated the efficiency of the EM spatial conversion process by the use of several monitors, which allowed both to show directly the EM insertion inside the nano-WG for a qualitatively interpretation of the phenomenon, and to export the necessary data to perform calculations and quantitatively evaluate the insertion losses.

According to the results given, a sequential set of similar simulations where performed to gradually adjust the geometric parameters of the several blocks in order to improve the conversion performances and the several losses due to incorrect matching between the many component of the investigated structure, till reaching some optimal configurations grating an insertion loss around 0.5 dB.

In figure 11 is shown the square modulus profile for 1550 nm wavelength.

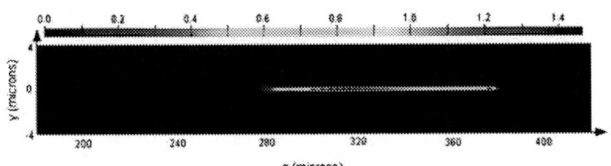

Figure 11. Square modulus profile of the electric field in the sagittal cross-section.

Due to the strong concentration of energy in the taper-guide system, the energy outside is close to 0. The lens effect beneath the reversed taper allows an insertion loss less than 0.6dB. The use of a thin silicon nano-waveguide (WG) tapered on its tip allowed to collect a huge amount of the injected EM field throughout the smooth transformation of its 2D section profile in the allowed mode by taking advantage of its sharp tip. The input field is progressively directed toward the interface between the silicon nano-taper and the glass substrate until it is completely contained in a tight region around it and then again inserted in the regular silicon waveguide, whose section is perfectly matched with the one possessed by the taper at the taper-WG connection point.

To evaluate the efficiency of the confinement, we evaluated the dependence of the power losses per unit length measured in dB/cm as a function of the thickness of the confining layer for 2 different cases of refractive index, 1.27 and 1.46.

Indeed, together with the reduction of the insertion losses, one of the main goal is to strongly reduce the unwanted power dispersion through the buried diffusive substrate. The presence of this latter is the consequence of the development of a confining layer for the beam leaving the optic fiber, and it represents the remains of the unchanged silicon layer beneath the transparent substrate.

We esteem that a depth of 800 nm is an excellent value to provide a strong confinement such that diffusion losses reduce to 0.1dB/cm for the TE0 mode, in case of 1.27 as the index value, and a slight increase of approximately 50 nm is required for the index value of 1.46. The TM0 mode confinement is more difficult because of the relative field distribution, since it is extended much more along the normal direction, and for this mode a thickness of 2250 nm is required

in order to reach the same losses per unit length as for the TE0 mode.

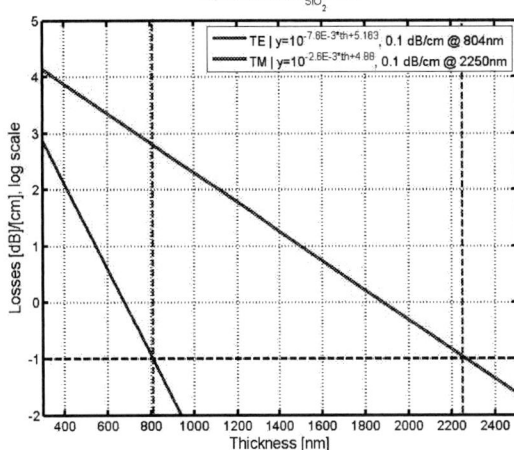

Figure 12. TE/TM losses in dB/cm function of the cladding thickness in the case of OPS with refractive index n=1.27.

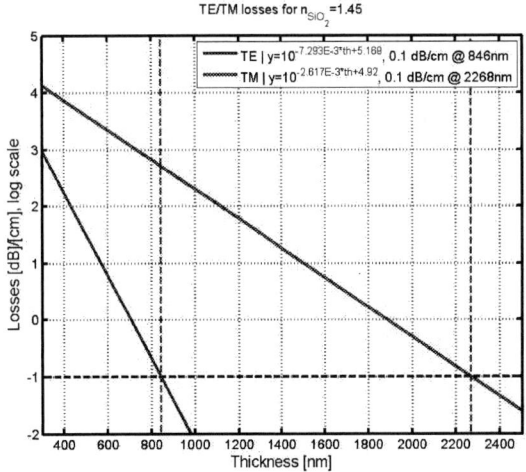

Figure 13: TE/TM losses in dB/cm function of the cladding thickness in the case of OPS with refractive index n=1.46.

For the power field to be converted into electric source, an optical device for the proper power collection is required. Since we need the maximal spread of EM. field to avoid excessive focusing toward the capturing cell, we need to tailor the input field into a different output profile by the use of another distributed lens with different index values inside.

We managed to determine the proper bidimensional (2D) distribution of refractive index values by adopting the ray-tracing method, which provides an accurate tailoring of the optical path by bending the front waves along the optical path in the modeled region [39, 40].

The technique basically requires the definition of many curvilinear optical rays, whose density is proportional to the emitted power across the input side. Along each ray, the refractive index is defined adopting a further discretization to subdivide the same ray into several points, where the refractive index is progressively defined through the use of a classic finite difference method and the locally smooth profile approximation.

We followed the eikonal equation for the each ray:

$$\frac{d}{dt}\, n\hat{t} = \vec{\nabla}n \rightarrow \frac{dn}{dt}\hat{t} + n\frac{d\hat{t}}{dt} = \vec{\nabla}n \qquad (1)$$

where t is the tangent curvilinear axis, n is the refractive index. We started from a known value, and progressively updating the values' series. The curvilinear axis is locally proportional to the differential displacement, and we require the total optical path along the single ray to be a fixed value for all the rays:

$$\int \frac{dS}{dt}\, dt = \int n\; t\; dt = \frac{\Delta\Phi}{k_0} = const \qquad (2)$$

with $k0$ being the scalar wave number. This part is granted by imposing a fixed differential value for the amount of phase growth tax while crossing the single rays.

We assume the relation between the normal vector and the osculating radius to be:

$$\vec{s} = \frac{d\hat{t}}{dt} = \left|\frac{d^2\vec{r}}{dt^2}\right| = \hat{s}\frac{\left|\dfrac{d\vec{r}}{dt}\right|^2}{R\; t} = \frac{\hat{s}}{R\; t} \qquad (3)$$

Ray after ray, the 2D profile is then defined by a final interpolation of the segmented distribution. Finally, final tests are performed to evaluate if the engineering of the device has been correctly completed.

The results of simulations are presented in figure 14 and figure 15 .On figure 15 the index values range from 1.27 to 1.46, divided into 16 different values. This profile is a fit for a Gaussian profile with an angular waist of 16 degrees.

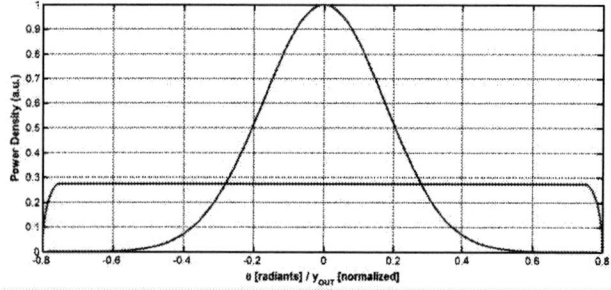

Figure 14. Profile. In blue, sharp gaussian profile of the input source field. In red, flat profile evaluated at the output panel of the power lens.

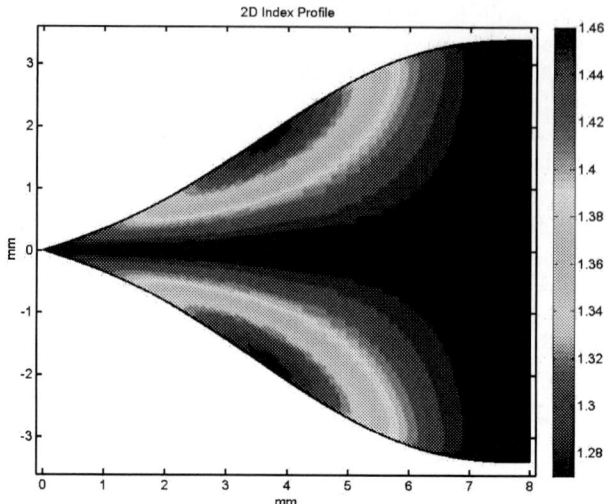

Figure 15. 2D index distribution to redirect the input beam and split its total power in order to make it acquire a flat output profile.

Realizing the structure with the simulated refractive indexes distribution will alow to obtain the flat profile of the power density on the output of the power lens which is necessary to obtain an optimal coupling with the concentrated silicon solar cell. The structure will then be optimized depending on the input and output power requirements not to exceed the power when the Auger recombination in silicon decreases the overall efficiency of the solar cell (figure 16)..

On figure 16 the results of simulation of silicon solar cell at 850 nm are presented. Following the simulation the optimal conditions are reached at around 100mW/mm2.

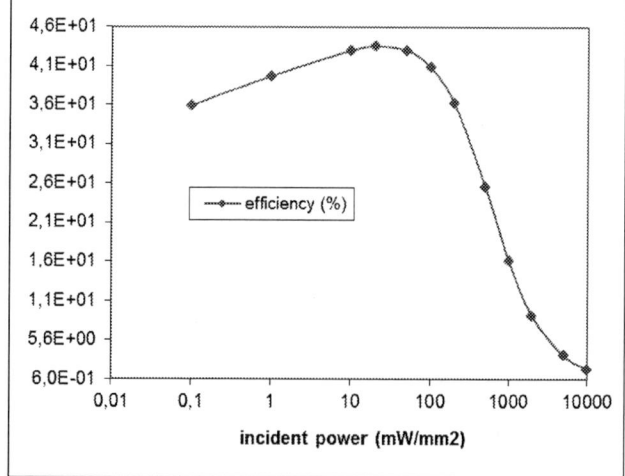

Figure 16. Silicon solar cell efficiency function of the incident power density at 850 nm.

Conclusions

A new approach for the monolithic integration of SP and IC was presented. The technology based on localized oxidized porous silicon, by using a new localized chemical treatment, is able to perform, depending on the structures that must be realized different porous silicon morphologies (i.e. macro and micro-PS). Isolated silicon islands where SP are to be realized were presented as also the new concept of fiber to chip coupling with the possibility to integrate also sensors with a new concept of power over fiber. The foundation for a new key enabling technology for SP signing a breakthrough for a real monolithic integration is posed. In the near future more details of such technology will be presented and as soon as the first SP devices will be completed, a more deep comparison with the actual SOI SP technology will be possible.

References

1. S. E. Miller, Integrated Optics: An Introduction, Bell Syst. Tech. J.,48 (7), 2059–2069 (1969)

2. A. Liu, et al. "A high-speed silicon optical mod. based on a metal-oxide-semiconductor capacitor," Nature 427, 615-618 (2004)

3. T. Baehr-Jones, M. Hochberg, C. Walker, and A. Scherer, "High-Q optical resonators in silicon-on-insulator-based slot waveguides," Applied Physics Letters 86, 081101 (2005)

4. C. T. DeRose et al. "Ultra compact 45 GHz CMOS compatible Germanium waveguide photodiode with low dark current," Optics Express 19, 24897-24904 (2011)

5. G. Roelkens, et al. "Bridging the Gap Between Nanophotonic Waveguide Circuits and Single Mode Optical Fibers Using Diffractive Grating Structures," Journal of Nanoscience and Nanotechnology 10, 1551-1562 (2010)

6. C. Gunn, "CMOS photonics for high-speed interconnects," IEEE Micro 26, 58-66 (2006)

7. A. Mekis, et al. "A Grating-Coupler-Enabled CMOS Photonics Platform," IEEE J. of S. T.in Quantum Electronics 17, 597-608(2011)

8. http://www.luxtera.com/

9. M. Asghari, and A. V. Krishnamoorthy, "Silicon photonics: Energy-efficient communication," Nature Photonics 5, 268-270(2011).

10. A. Liu, et al. "Wavelength Division Multiplexing Based Photonic Integrated Circuits on Silicon-on-Insulator Platform," IEEE Journal of Selected Topics in Quantum Electronics 16, 23-32 (2010)

11. M. B. Ritter, Y. Vlasov, J. A. Kash, and A. Benner, "Optical technologies for data communication in large parallel system," Journal of Instrumentation 6, C01012 (2011)

12. P. Dumon, W. Bogaerts, R. Baets, J.-M. Fedeli, and L. Fulbert, "Towards foundry approach for silicon photonics: silicon photonics platform ePIXfab," Electronics Letters 45, 581-U2 (2009)

13. D. J. Thomson, et al. "High contrast 40Gbit/s optical modulation in silicon," Optics Express 19, 11507-11516 (2011)

14. L. Vivien, et al. "42 GHz p.i.n Germanium photodetector integrated in a silicon-on-insulator waveguide," Optics Express 17, 6252-6257 (2009).

15. Tom Baehr-Jones, et al. "A 25 Gb/s Silicon Photonics Platform" arXiv:1203.0767

16. M.Zuffada: "The industrialization of the Silicon Photonics: technology road map and applications" Solid-State Device Research Conference (ESSDERC), 2012 Proceedings of the European, 17-21 Sept. 2012, pp. 7-13

17. Fedeli-J-M, "Si and InP integration in the HELIOS project", Optical Communication, 2009. ECOC '09. 35th European Conference on, 20-24 Sept. 2009, pp. 1 - 3

18. L.Fulbert & J.M. Fedeli: "Photonics-Electronics Integration on CMOS", Solid-State Device Research Conference (ESSDERC), 2011 Proceedings of the European, 12-16 Sept. 2011, pp. 13 – 18

19. http://www.amo.de/photonic_components.0.html

20. http://www.epixfab.eu/index.php?option=com_content&view=article&id=50&Itemid=60

21. G T Reed, B D Timotijevic, F Y Gardes, G Z Mashanovich, W R Headley, and N G Emerson, "Waveguides and devices in Silicon Photonics: Polarisation independence", Optoelectronic Integrated Circuits IX, edited by Louay A. Eldada, El-Hang Lee, Proc. of SPIE Vol. 6476, 647602, (2007) • 0277-786X/07

22. Baehr-Jones, T.; Pinguet, T.; Jing Li; Harris, N.C.; Streshinsky, M.; Li He; Novack, A.; Lim, A.E.; Liow, T.; Teo, S.H.-G.; Guo-Qiang Lo; Hochberg, M. "A high-speed silicon photonics platform" Photonics Conference (PHO), 2011 IEEE, 9-13 Oct. 2011 pp. 1-2

23. Rosseel, E., et al. " SOI thickness uniformity improvement using corrective etching for silicon nano-photonic device", 8th IEEE International Conference on Group IV Photonics (GFP), 2011, pp 71-73

24. H. Bernstein, N.J. Rohrer "SOI Circuits Design Concept" IBM Microelectronics, Kluwer Academic Publishers 2000

25. Jose Vicente Galan (2010) "Addressing Fiber-to-Chip Coupling Issues in Silicon Photonics" (Doctoral dissertation),riunet.upv.es/bitstream/handle/10251/9196/tesisUPV3450.pdf

26. http://www.helios-project.eu/Download/Public-deliverables document: D010 State of the art, last update

27. Hirohito Yamada et al., " Vertical-coupling optical interface for on-chip optical interconnection" OPTICS EXPRESS (2011) Vol. 19, No. 2 pp. 698-703

28. M. Balucani, K Kholostov, P Nenzi, R Crescenzi, D Ciarniello, D Bernardi, L Serenelli, M Izzi, M Tucci; "New Selective Processing Technique for Solar Cells", Energy Procedia (2013) 43, 54-65

29. M Balucani, D Ciarniello, P Nenzi, D Bernardi, R Crescenzi, K Kholostov, "New selective wet processing", Electronic Components and Technology Conference (ECTC), 2013 IEEE 63rd, 28-31 May 2013, pp. 247-254

30. V. Labunov, V. Bondarenko, L. Glinenko et al. Thin Solid Film (1986) 137, 123

31. E.V. Astrova, V.A. Tolmachev, "Effective refractive index and composition of oxidized porous silicon films" Materials Science and Engineering B69–70 (2000) 142–148

32. E. J. Teo, A. A. Bettiol, P. Yang, M. B. H. Breese, B. Q. Xiong, G. Z. Mashanovich, W. R. Headley, and G. T. Reed "Fabrication of low-loss silicon-on-oxidized-poroussilicon strip waveguide using focused proton-beam irradiation", OPTICS LETTERS (2009) 34, 5, pp. 659-661

33. M. Balucani, V. Bondarenko, A. Klusko, A Ferrari, " Recent progress in integrated waveguides based on oxidized porous silicon" Optical Materials (2005) 27, 5, pp. 776-780

34. Yakovtseva V., Dolgyi L., Vorozov N., Kazuchits N., Bondarenko V., Balucani M., Lamedica G., Franchina L., Ferrari A., "Oxidized porous silicon: From dielectric isolation to integrated optical waveguides", Journal of Porous Materials (2000) 7, 1, pp. 215-222

35. J. D. Joannopoulos, R. D. Meade, J. N. Winn, Photonic Crystals, Princeton Univ Press (1995).

36. J. D. Joannopoulos, P. R. Villeneuve, and S. Fan, "Photonic crystals: putting a new twist on light", Nature 386,143–149 (1997).

37. E. Moreno, D. Erni, and Ch. Hafner, "Modeling of discontinuities in photonic crystal waveguides with the multiple multipole method", Phys. Rev. E 66, 036618 (2002).

38. M.I. Hussein, "Reduced Bloch mode expansion for periodic media band structure calculations", Proceedings of the Royal Society A 465 (2109): 2825–2848 (2009).

39. A. Mandatori, A. Benedetti, C. Sibilia, M. Bertolotti, "Application of ray-path geometry modification to the design of a collimating structure for LEDs", Opt. Comm. Vol. 281, Issue 23, p. 5674-5682 (2008).

40. A. Mandatori, A. Benedetti, C. Sibilia, M. Bertolotti, "Application of ray-path geometry to the design of an optical cloaking structure", JOSA B, Vol. 25, Issue 10, pp. 1580-1584 (2008).

High Power Density LED Modules with Silver Sintering Die Attach on Aluminum Nitride Substrates

Marc Schneider[1], Benjamin Leyrer[1], Christian Herbold[2], Stefan Maikowske[3]
[1] Institute for Data Processing and Electronics (IPE)
[2] Light Technology Institute (LTI)
[3] Institute for Micro Process Engineering (IMVT)
Karlsruhe Institute of Technology (KIT), Kaiserstrasse 12, 76131 Karlsruhe, Germany
marc.schneider@kit.edu

Abstract

Current research studies deal with the investigation of the thermal and optical properties of four LED modules on different substrate materials. The LED modules consist of arrays of 98 blue emitting LEDs with an emission wavelength of 457 nm within an area of 2.11 cm^2. The modules are based on aluminum oxide or aluminum nitride substrates and the LED chips are attached by using a soldering or a pressureless silver sintering process. The modules are mounted on a high performance microstructured heat exchanger. By using the water driven cooler a maximum optical power density of 106.2 W/cm^2 at a forward current of 2100 mA and 837.5 W electrical input power is achieved. A saturation of the optical power density over the input current due to thermal degradation is not observed.

Introduction

Light engines consisting of a large number of densely packed high-power LED chips require sophisticated heat management for efficient and high performance operation. It is quite easy to achieve high optical power densities with small arrays of rows of LED chips just two chips wide due to the simple electrical connection on both sides of the rows and the possibility to spread the generated heat. For large area light sources with high optical power densities this design is unsuitable because of the inhomogeneous light emitting area. With regard to a small optical system with beam forming a small light emitting area is beneficial. These light sources could be used for flood lights in stadiums, TV studios, architectural lighting or automotive head lamps. But larger areas of tightly packed LED chips with high light densities present a technological challenge, because of the integration of a sufficient heat spreading technology for cooling, the required high fill factor of LEDs, and their electrical connections between the chips.

On that account, a layout with a very high possible chip density on a substrate and enough space to wire bond between the chips and compensate for fabrication tolerances was developed [1]. The evolution of the packaging technology includes two aspects, a new substrate material and a new technology for bonding the LED chips onto the substrate. All combinations of these LED modules with new and old substrate materials and different bonding technologies were fabricated and characterized.

Blue emitting LED chips were used in current investigations in contrast to former research studies [1, 2] where modules using UV-LED chips emitting in the near ultraviolet

range were investigated. The results are transferable to other emission wavelengths as the main challenge is the cooling of the LED chips. The presented module with the old substrate and old bonding technology is comparable to the UV-LED module in [2] with 465 W electrical input power and 97 W optical output power as it just differs in the used LED chips.

Substrate Technology

The aluminum oxide (Al$_2$O$_3$) ceramics substrate was identified as the main disadvantage to thermal performance of former LED modules due to the low heat conductivity. Therefore, modules are built on 0.63 mm thick aluminum nitride (AlN) substrates with a thermal conductivity of 180 W/(m·K) which is ten times higher than that of aluminum oxide with 18-24 W/(m·K). The substrates were lapped to provide a flat surface for thick film processing and for a good thermal connection to a heat sink. For fabrication of conducting structures on the AlN substrate the thick film silver palladium paste CL40-10141 from Heraeus was used. The paste was screen printed to the substrates and fired according to the datasheet. An insulation layer of IP9117 (Heraeus) was additionally printed and fired for protection of the silver from environmental influences. Additionally, the insulation layer

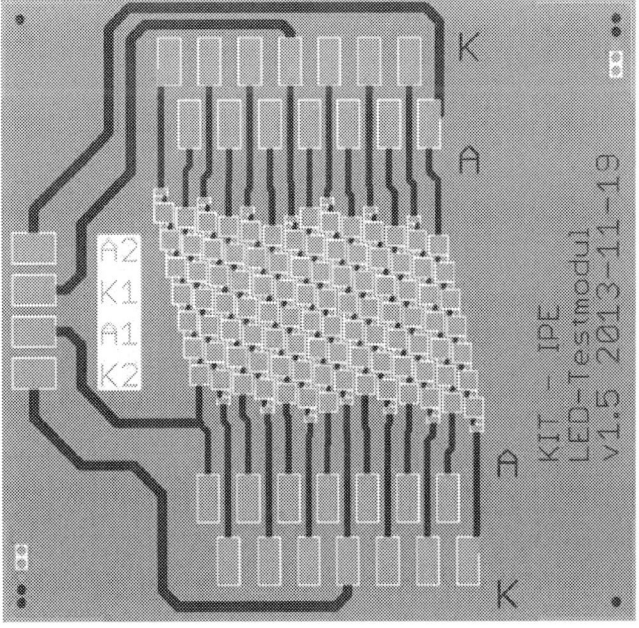

Figure 1: Layout of the LED modules with a size of 40×38 mm^2.

layout provides solder stops for soldered LED chips at respective modules.

The printed layout is presented in Fig. 1. Bond pads for 98 LED chips (each 1×1 mm^2) in a rhomboid pattern are located in the center with an area of only 2.11 cm^2. This pattern has already been used and discussed in [3]. The design enables a very tight packing of the LED chips resulting in a fill factor of 46% and the possibility to continue the pattern gapless and uninterrupted. Electrical connections are placed on the top, bottom, and left side. They provide individual access to series connected groups of seven or 49 LEDs. For experimental characterization, all 98 LED chips are connected in series by jumper wire bonds on the top and bottom contact pads and a jumper wire on the left contact pads.

In addition to the AlN substrates Al$_2$O$_3$ substrates were prepared for comparison. The silver platinum paste C4729H from Heraeus was used as thick film conductor for module substrates with LEDs attached by silver sintering. On the other hand, silver palladium paste C2060 from Heraeus was used for module substrates with soldered LED chips. Finally, the substrates got a protective layer of IP9117, too. Both layouts of the Al$_2$O$_3$ and AlN substrates are identical.

The used LED chips (Osram ODB40UX3) with an area of 1×1 mm^2 present a typical emission wavelength of 457 nm, which is located in the deep blue spectral region. The chips are equipped with a single bond pad at the front side and a full size back side contact.

Two different techniques were used for LED attachment. A conventional class 4, lead free, no-clean SAC305 solder paste (Kester NXG33) was used for soldering in a reflow oven after stencil printing the paste and placing the LED chips.

Beside this, silver sintering was used as more advanced LED attachment technique. The silver sintering paste LTS295 from Heraeus, containing micrometer sized silver particles and still in heavy development, was screen printed onto the substrates. The LED chips were placed into the wet paste and the paste was sintered in a box-type furnace at 230 °C for 75 min. No special preheating ramps, settling temperatures and cooling ramps were used. A comparison between this hot-in-hot-out method and special temperature profiles resulted in similar shear forces and failure mechanisms for the LED chips.

While fabricating the AlN based LED module with silver sintered chips a problem arose with this process. Due to some delays while placing the chips, the rough surface of the thick film silver, and the short drying time of the silver sintering paste, the silver sintering paste dried out. As a consequence the paste could not properly squeeze out of under the chips, did not distribute evenly, and the thickness of the silver sintering bond became too large. This also resulted in decreased heat transfer as well as in degraded electrical performance as described later.

After placement, the LEDs were connected in series by wire bonding using gold wire with a diameter of 32 µm, bonded by a wedge-wedge bonder. Each LED chip was connected by two wires to ensure a sufficient current handling capacity.

Figure 2: The four fabricated LED modules with 98 LED chips each. Left half: Al$_2$O$_3$ substrates, right half: AlN substrates, top half: soldered LED chips, bottom half: silver sintered LED chips.

Table 1: LED module types; numbering refers to Fig. 2.

	Al$_2$O$_3$	AlN
Soldered chips	#1	#3
Silver sintered chips	#2	#4

The whole bonding process was characterized by using test samples. Bonding strength of soldered and sintered LED chips was measured by shearing-off LED chips and measuring the required force.

Shear forces of around 66 N/mm^2 were measured on soldered chips, often resulting in a breakdown of the chip rather than a failure of the bond. The shear forces of the sintered chips were much lower and depend on the used thick film silver material system. On C4729H thick film silver, used for the Al$_2$O$_3$ substrate, an average shear force of 16.5 N/mm^2 was achieved while on CL40-10141, used for AlN substrates, the average shear force was only 10.0 N/mm^2. Low shear forces affect the thermal conductivity of the interface between thick film silver and silver sintering material. Therefore, further research activities will be made dealing with this important topic.

Fig. 2 shows the different LED modules, each with 98 LED chips. The modules built on Al$_2$O$_3$ substrates are on the left hand side (#1 and #2) while the modules with AlN substrates (#3 and #4) are presented on the right hand side. The modules #1 and #3 are built with soldered and the modules #2 and #3 are with silver sintered LED chips. Table 1 recaps the numbering of the LED module versions.

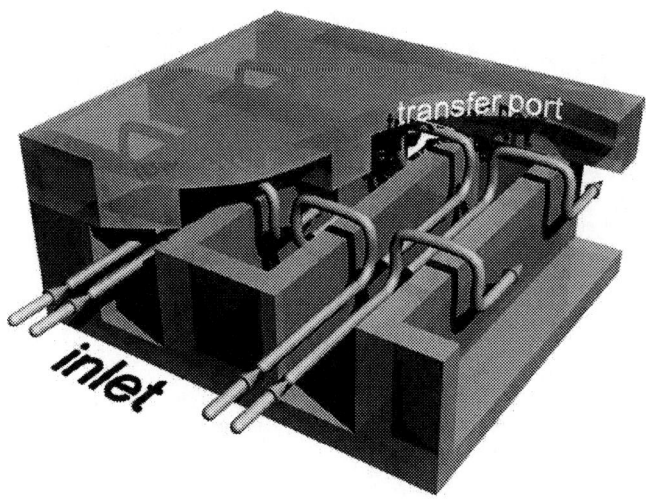

Figure 3: Schematic view of the used surface micro cooler: microstructure and principal water flow.

Thermal Characterization

Experimental characterization of the different modules was done by using a high performance microstructured heat exchanger which is single phase operated by liquid water. The used surface micro cooler was made from copper with internal micro structures to provide an excellent heat transfer to the cooling water [4]. Fig. 3 presents a schematic drawing of the cooler. The inlet and outlet ports interdigitate and are connected by a large number of narrow and short transfer ports. This design combines very high heat transfer capabilities of the microchannels and a low pressure drop, caused by many parallel and short microchannels [5, 6].

The cooler was operated with an internal system pressure of about 8.5 bar (at the device inlet) while the pressure drop is 3.3 bar at a liquid water mass flow of 360 kg/h with a temperature of 10.0-10.7 °C.

Fig. 4 shows a detailed view of the measurement setup with the LED module mounted on the surface micro cooler in the center. Consistent and uniform pressure was provided by a silicone sheet which is placed between module and mounting frame. A room temperature liquid metal alloy, based on gallium, as thermal interface material (TIM) provides a thermal path with very high heat transfer capabilities with a low temperature drop.

The LED modules were thermally characterized by measuring the pn-junction temperature with a thermal camera FLIR Thermacam PM575. Electrical power was provided by a Delta Elektronika SM 400-AR-8 power supply. Voltage and current was measured by a Fluke 289 and a Fluke 87 V digital multimeter respectively. The measurements were done by using a custom built thermal performance test rig with pressure, temperature and mass flow controls.

A spectrometer was used for measuring the spectral irradiance of the LED modules. The input optics of the spectrometer with its light guide is also presented in Fig. 4 but not aligned to the LED array. The distance between LED module and spectrometer input optics was measured for each module for correct optical calculations of the optical power density.

The thermal images of the modules at the highest possible currents for the individual modules are presented in Fig. 5. Module #1 presents an uneven temperature distribution with hotspots on both sides while the majority of LED chips have comparable temperatures. The hotspot on the left-hand side could be caused by an air bubble in the TIM layer and the resulting locally increased thermal resistance to the cooler. The temperature of the hotspot on the right-hand side is significantly higher. The related TIM layer seemed to be continuous and free of bubbles. Therefore, a void in the solder connection between LED chip and substrate could have caused the hotspot. This could be resolved without destruction by X-ray imaging the module. The temperature of the hotspot was 143 °C at a current of 1001 mA and a voltage of 347.0 V, while the other LED chips show a temperature of 102 °C at an overall electric input power of 347.3 W. Fig. 6 shows the measured temperature increase over the electrical input power for all four LED modules. Additionally, two curves of the UV-LED module presented in [2] are shown as gray lines with open triangles. These curves have a comparable trend to the curve of module #1, shown as blue curve with filled squares. The temperature difference of both gray curves underline the importance of the thermal interface between LED module and cooler, because the two measurements were conducted with the same module with differently assembled setups. As the gray curve for the UV-LED module with a cooling water mass flow of 360 kg/h (downward pointing trian-

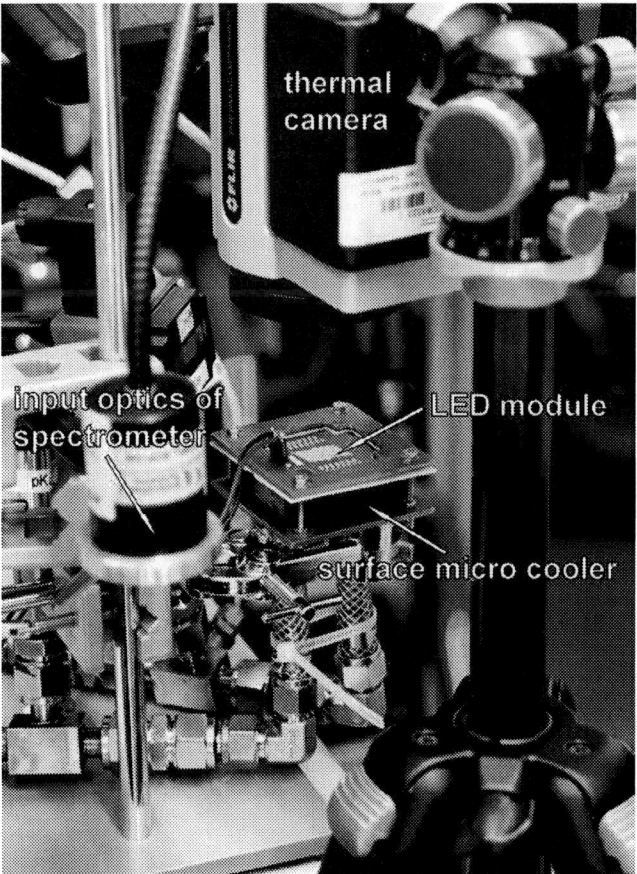

Figure 4: Close up view of the measurement setup.

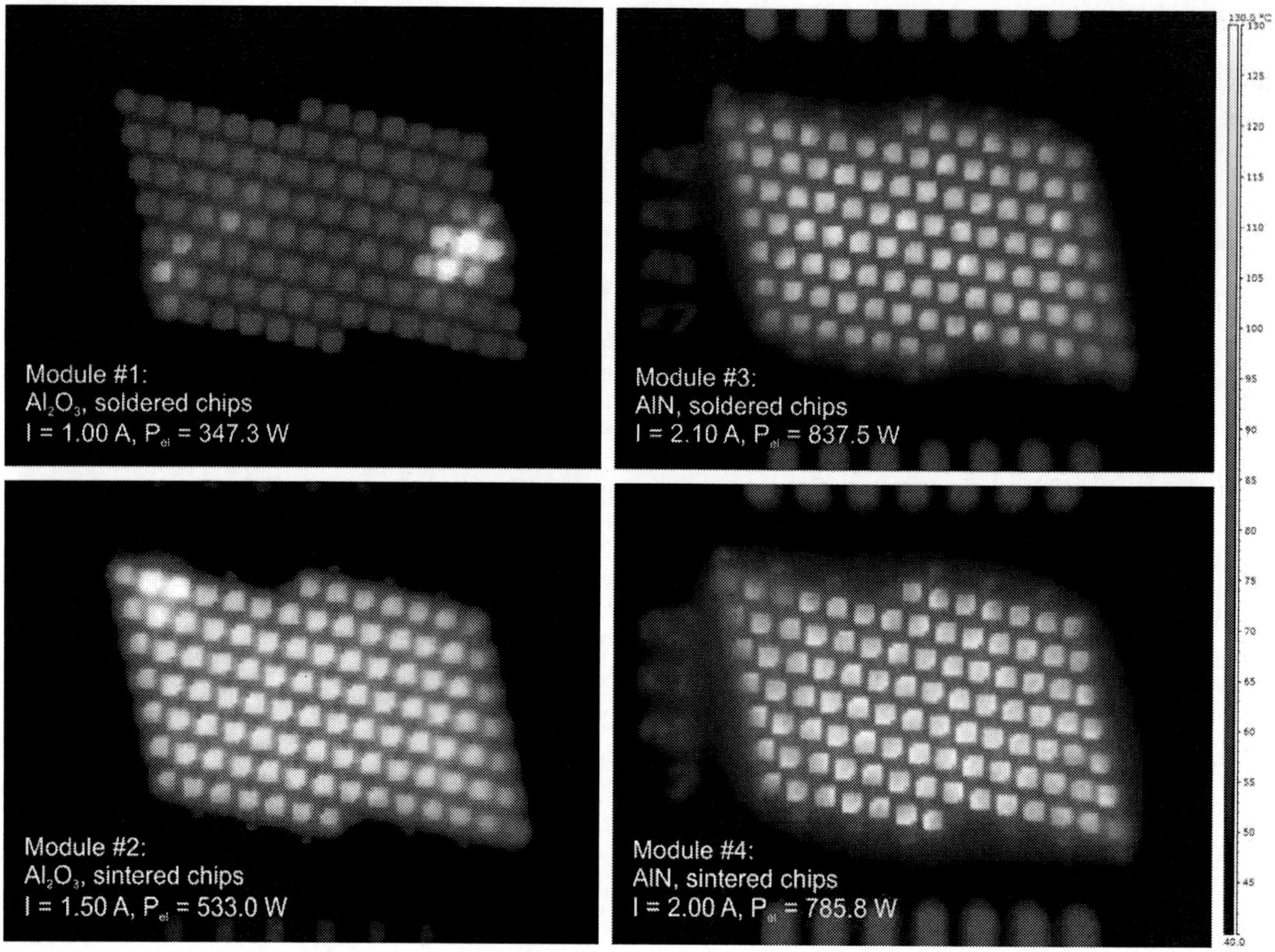

Figure 5 thermal images:

Module #1:
Al$_2$O$_3$, soldered chips
I = 1.00 A, P$_{el}$ = 347.3 W

Module #3:
AlN, soldered chips
I = 2.10 A, P$_{el}$ = 837.5 W

Module #2:
Al$_2$O$_3$, sintered chips
I = 1.50 A, P$_{el}$ = 533.0 W

Module #4:
AlN, sintered chips
I = 2.00 A, P$_{el}$ = 785.8 W

Figure 5: Thermal images of all new LED modules on surface micro cooler.

gles) is almost congruent with the curve for module #1, the curve for the same module with a mass flow of 316 kg/h (upward pointing triangles) shows a smaller temperature increase with increasing electrical input power which indicates better heat conduction. However, the similarity of both curves underlines the comparability of the thermal measurements of both kinds of modules with blue and ultraviolet emitting LED chips.

Module #2 with silver sintering die attachment presents a more homogeneous temperature distribution, but also a hotspot. The hotspot temperature is just 17 K higher than the hottest of the other chips, which is a much smaller temperature difference than for module #1, when the current was 1.60 A at an electrical input power of 569.8 W. At this power the highest junction temperature was 156.4 °C at the center of the hotspot and 139.6 °C at the hottest other chip. The optical output power did not saturate at this high temperature as presented later. For a better comparison Fig. 5 shows the thermal image at 1500 mA. The curve of module #2 in Fig. 6 (green line with filled circles) presents a smaller slope in comparison to module #1. Therefore, the silver sintering die attach leads to higher heat conduction from the chips to the substrate which results in lower junction temperatures at a given electrical input power.

The modules built on aluminum nitride (AlN) substrates present a nearly homogeneous temperature distribution without hotspots (Fig. 5: right-hand side). The most obvious difference between the thermal images of LED modules with

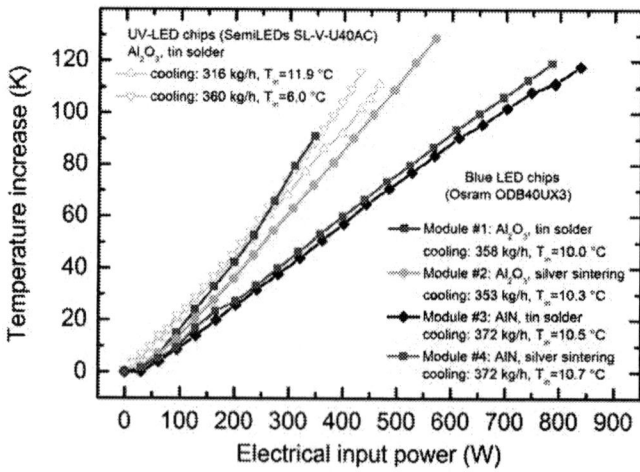

Figure 6: Measured LED junction temperature increase versus electrical input power for all new LED modules and the UV-LED module already presented in [2] for comparison.

Al$_2$O$_3$ substrate and AlN substrate is the halo around the LED chips, meaning the surroundings of the LED chip area have a higher temperature at AlN based modules. This behavior becomes clear when considering the different thermal conductivities. The TIM has a thermal conductivity of 32.6 W/(m·K) which is higher than the thermal conductivity of the Al$_2$O$_3$ ceramics (18 W/(m·K)) of the first two modules. Heat is transferred basically vertical through the substrate and does not spread out. In contrast to this, AlN ceramics has a much higher thermal conductivity of 180 W/(m·K) and therefore, the TIM becomes a thermal barrier. Hence, heat from the LED chips can spread out in the substrate. However, the narrow placed dies prevent efficient heat spreading of the whole array. Most of the heat is conducted through the substrate to the cooler. In Fig. 6 the curves for AlN based modules present smaller slopes compared to the curves of Al$_2$O$_3$ based modules. The slope for the AlN module with soldered chips is the smallest (module #3, black line with filled diamonds). This results in the highest possible electrical input power of all modules. At a current of 2101 mA a maximum junction temperature of 128.5 °C was measured. With a forward voltage of 398.6 V the electrical input power is 837.5 W and the power density is 396.9 W/cm^2.

Despite the considered better thermal conductivity of the silver sintering die attachment, the respective module performs slightly worse than the module with soldered chips. The curve for module #4 (red line with filled squares) is higher than the curve of module #3 in Fig. 6. This behavior is a result of the prolonged processing time, the rough surface of the thick film silver, and the short drying time of the silver sintering paste, as described above. The sintered LED module enables a current of 2002 mA and an electrical input power of 785.8 W with a maximum junction temperature of 129.6 °C. A faster processing of the silver sintering paste would result in a higher thermal performance and an increased power handling capacity.

Optical Characterization

The LED modules were optically characterized in the same setup. The spectral distribution of the emitted radiation for wavelengths between 350 nm and 550 nm was measured with a spectral resolution of 0.25 nm. An array spectrometer CAS 140CT from Instrument Systems with the fiber coupled input optics EOP 146 was used for the measurements. The sensor head with a diffuse input window and an aluminum aperture was placed 23 mm above the top surface of the LED module. Spectral irradiances were measured for currents between 100 mA and, depending on the LED module, up to 2100 mA. The measured values were converted to optical power densities through the solid angle projection. A nearly lambertian emission of an LED into the half-space and the module as one homogenously radiating area were assumed for the calculations.

Fig. 7 shows the calculated optical power densities of all modules. Additionally, the optical power density of the UV-LED module presented in [2] is shown as gray line with open triangles for comparison. The slope of the UV-LED module curve decreases with increasing current and saturates at

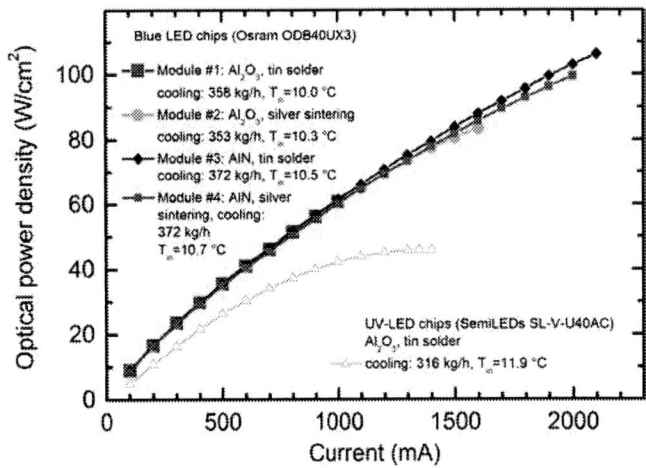

Figure 7: Calculated optical power density for all new LED modules and the UV-LED module already presented in [2] for comparison.

1350 mA at an optical power density of 45.9 W/cm^2. For higher currents, the optical power density decreases due to thermal degradation.

The curves of the blue emitting LED modules present a different behavior. Increasing current leads to a proportional increase of the optical power density with a slight negative bending. At low currents of up to 1000 mA, the curves are almost congruent and only small variations can be observed at higher currents. The thermal degradation could impair the optical efficiency but the observed effect is just minimal up to junction temperatures of about 125 °C. The similar behavior of the curves underlines that degradation by high current densities is more significant. Closer examination of the results shows that some thermal degradation occurs at high LED temperatures equivalent with high currents close to the end of the individual curve.

The highest optical power density was achieved with module #3 (AlN substrate, soldered chips). The highest possible operating current is converted to the highest optical power density of 106.2 W/cm^2 at 2102 mA resulting in a radiated optical power of 224.1 W with an electrical input power of 837.5 W and an efficiency of 26.8 %.

The maximum optical power density of module #4 with AlN substrate and sintered chips is just a little bit lower with 99.3 W/cm^2 at 2002 mA. The overall efficiency with 26.7 % is also slightly lower, due to the higher junction temperature.

These investigations indicate that the LED modules could be driven up to much higher junction temperatures to achieve even higher power densities. A thermal degradation up to declining optical emission was not observed. Further experiments should clarify the maximum output power limit. According to the application, other temperature limits such as temperature stability of wavelength conversion dyes, which may limit the usable temperature range have to be taken into account.

The optical power density of the blue LED chip modules is between 35 % and 40 % higher than of the UV-LED module. This behavior is caused by the LED material itself and the difficulty to control it for lower emission wavelengths. Hence the efficiencies of UV-LEDs are currently lower compared to blue LEDs. Additionally, the used blue LED chips are current state of the art chips, while the UV-LED chips of the comparison module represent the development status of 2009. Currently, better binnings with higher efficiencies are available. Beside this, blue LED chips have higher commercial importance due to their use in white emitting LEDs for common lighting. Therefore, much more research and development dealing with blue LED chips has been performed.

Conclusions

By using a ceramic substrate with high thermal conductivity, it was possible to build a high power LED module with an optical power density of more than 106 W/cm^2. An even higher optical power density can be expected by further increasing the LED current and by using silver sintering as die attachment method. The corresponding module presented in this paper suffers from a problem in thermal conductivity caused during the fabrication process of the module. Another module should present even better thermal and optical performances.

Acknowledgments

We would like to thank the German Federal Ministry of Education and Research for funding parts of this work within the joint research project "ProPower". Further thanks go to Heraeus Precious Metals GmbH & Co. KG and Heraeus Materials Technology GmbH & Co. KG for supporting us with thick film and silver sintering pastes and to Osram GmbH for providing the LED chips.

References

1. M. Schneider, B. Leyrer, C. Herbold, K. Trampert, J. J. Brandner, "Thermal Improvements for High Power UV LED Clusters," *Proc 61st Electronic Components and Technology Conf*, Orlando, FL, May 2011, pp. 1636-1641.

2. M. Schneider, B. Leyrer, C. Herbold, S. Maikowske, "Very High Power Density LED Modules on Aluminum Substrates with Embedded Water Cooling," Proc. 63rd Electronic Components and Technology Conf, Las Vegas, NV, May 2013, pp. 529-534.

3. M. Schneider, C. Herbold, K. Messerschmidt, K. Trampert, J.J. Brandner, "High Power UV-LED-Clusters on Ceramic Substrates," *Proc. 60th Electronic Components and Technology Conf*, Las Vegas, NV, June 2010, pp. 679-685.

4. M. Schneider, B. Leyrer, C. Herbold, S. Maikowske, J. J. Brandner, "Index Matched Fluidic Packaging of High Power UV LED Clusters on Aluminum Substrates for Improved Optical Output Power," *Proc. 62nd Electronic Components and Technology Conf*, San Diego, CA, May 2012, pp. 187-193.

5. F. Brighenti, N. Kamaruzaman, J.J. Brandner, "Investigation of self-similar heat sinks for liquid cooled electronics," Applied Thermal Engineering, vol. 59, pp. 725-732, Sep. 2013.

6. J.J. Brandner, N. Kamaruzaman, S. Maikowske, "A Microstructure Device for Single Phase Surface Cooling," *Proc. ASME/JSME 2011 8th Thermal Engineering Joint Conference AJTEC2011*, Honolulu, HI, USA, March 2011, p. AJTEC2011-44038.

Effect of Optical Design on the Thermal Management for the Smart TV LED Backlight Systems

Kivanc Karsli[a] and Mehmet Arik[b]
[a] Vestel Elektronik Sanayi ve Ticaret A.S.
Manisa, Turkey
kivanc.karsli@vestel.com.tr
[b] Department of Mechanical Engineering Ozyegin University
Istanbul, Turkey
mehmet.arik@ozyegin.edu.tr

Abstract

Due to recent advances in electronics, lighting and communication technologies, SMART Televisions (TV) have become more affordable, and are rapidly replacing old-fashioned LCD (liquid crystal display) TV technologies. These TVs are nearly ten times thinner; they include much brighter displays, true-color qualities, at least two times faster refresh rates and smart communication features. While providing such advanced features, lighting has become a critical part of technology advancement. While the backlight unit consumes over 50% of total energy in TVs, light emitting diodes (LEDs) have rapidly replaced the conventional CCFL (Cold Cathode Fluorescent Lamp) based LCD backlight units due to their low energy consumption.

The present study identifies the impact of optical solutions with high-power LEDs located at the corner of the backlight unit on the optical efficiency for an advanced TV system. Various optical designs are created and impacting metrics for novel LED backlight units are determined. A computational and experimental study has been performed to identify the optical-thermal interactions in a tight-space LED packaging for a slim TV. Optical modeling has been performed via Light Tools optical simulation software, while thermal modeling was performed via Icepak CFD software. Smart optical packaging resulted in an effective light distribution of minimum 75% brightness uniformity, with no MURA effect problem on the panel The thermal challenge was found to be immense; so a smart thermal packaging was vital. An experimental validation of computational models was also performed. While the target is obtaining a uniform light distribution generated by HB (high brightness) LEDs and reduction of cost by having fewer LEDs, optical structure is found to be very critical in terms of both optical efficiency and pattern design of the light guide plate (LGP), which is an essential part of slim LED TVs. The study is concluded with a detailed discussion of the impact on luminance uniformity and cost for advanced SMART TVs.

Introduction

LED light sources offer long lifetime, low energy consumption, do not contain harmful substances, have an exceptional optical performance, well controlled color quality and availability, as well as the possibility of slim design [1]. As a result of their fast efficacy amelioration, designers tend to reduce the number of LEDs and LED light bars for easier assembly and lower component cost. However, due to high-power LEDs' high heat dissipation and the difficulty of obtaining optical uniformity with fewer numbers of LEDs, it

is very challenging to decrease in the number of LEDs while meeting optical, thermal and reliability performance metrics. In addition to these difficulties, reliability and optical performance depend heavily on the LED operating temperature due to their solid state nature. To be able to reduce the heat generation of the LEDs by decreasing the power consumption of the LEDs, an efficient optical design is one of the most critical design factors for LCD backlight units.

LED TV backlight units (BLU) are separated into two types as edge-lit LED and direct-lit LED. Edge types have the advantage of slimmer TV design compared to direct types. However, edge types have optical and thermal disadvantages because of the tight placement of LED packages. Due to the location of the LEDs, light distribution of the BLU depends heavily on LGP - LED package system efficiency. While distributing the light from the edges to the entire panel, light efficiency decreases because of long distance light travels in the LGP. In addition to optical difficulties, due to small volume placement, thermal management of the LEDs becomes more challenging than direct type BLUs [2]. If the thermal distribution is not sufficient optical films and LGP will be warped which causes picture quality problems [3].

When the market and development trends of LCD mobile devices and LCD TVs are examined, LCD TV backlight systems are following the LCD mobile backlight systems. According to this phenomenon, it is expected that the LED placement of LCD TV's will be moved to the corners of the system [4]. The main motivation of this change is decreasing the number of LEDs per system to be able to have an easier assembly process and decrease the material cost of the system. However, due to large size of the TVs and difficulties of edge type BLU, optical and thermal challenges of corner-lighted LCD backlight units are much more difficult. From the optical design point of view, the efficiency of the LGP will be lower than current edge type structures and achieving the desired luminance uniformity will not be easy. Different LED packages and optical structures are computed to analyze their optical effects on the system.

To dissipate the high heat from high power LEDs conventional metal-based (such as Aluminum) heat spreaders are not sufficient for this application. Advanced thermal materials with high thermal conductivity and low thermal resistance have to be investigated. Graphite based or CVD diamond like materials may offer advantages compared to conventional heat spreader alternatives.

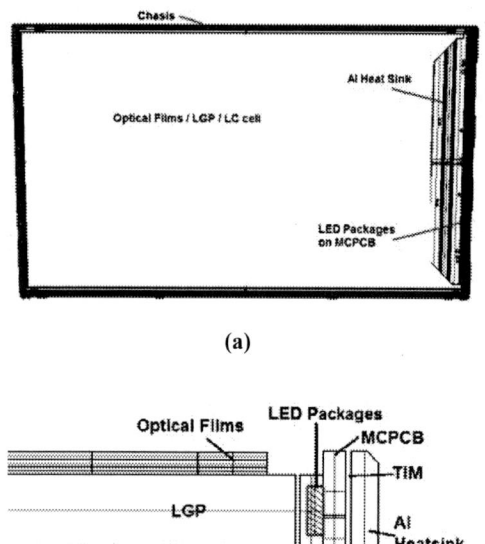

(a)

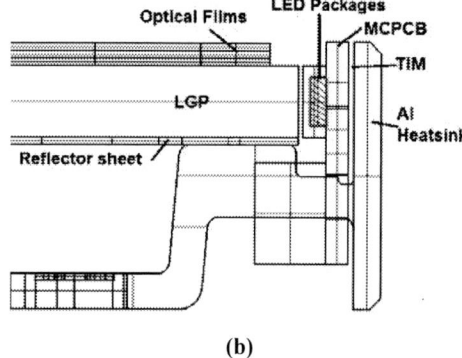

(b)

Figure 1. Edge type backlight unit structure. (a) Front view of the LCD panel. (b) Placement of the components (side view).

Optical Design Architecture

Before starting the optical design of a standard 32" corner-lit BLU, the target specs of the LCD TV are defined. The most important optical specifications (specs) of an LCD TV and the design targets are shown in Table 1. Contrast ratio, response time and viewing angle specs are related with Liquid Crystal (LC) cell which is not within the scope of this study. Others can be controlled by BLU specifications. The most critical targets of optical design are center luminance and brightness uniformity of the TV. According to the market trends, center luminance of the LCD TV is decreasing. Entry-level (low-cost) model LED TVs' center luminance are starting from 250 nits after 2012 [4]. The expected luminance uniformity is higher than 75% without any MURA problem.

	Target	Unit
Contrast Ratio	LC cell spec	-
Response Time	LC cell spec	ms
Center Luminance of White	250	nit
White Brightness Uniformity	75	%
Color Chromaticity	9300	K
Color Gamut	68	% NTSC
Viewing Angle	LC cell spec	

Table 1. Target optical specs of the TV.

Before starting the LED characterization, various optical film structures and different LC cells are investigated to choose the most optimal structure. According to the experimental results, if the luminance gain of the optical sheets after LGP was around 600% and the LC cell transmittance is 7.2%, then a total of 812 lumen output from the LEDs is sufficient to meet the design targets. To create an 812 lumen output, various LED packages were investigated. Specifications of the LED packages and their design parameters are presented in Table 2. For choosing the most suitable LED package, the most important factors are viewing angle of the LEDs, thermal resistance of junction-to-solder point and total power consumption. Viewing angle of the LED package affects the light distribution which is a very crucial for LGP pattern design. Thermal resistance of the package and the total power consumption of the LEDs affect the LED junction temperature and thermal management of the system. Therefore, using two commercial LED packages (Cree XM-L2) will meet the system requirements. However, the packages which enable using only two LEDs in the system have geometric disadvantages. The reason is having a target to keep the edge type TV depth and current edge type BLUs include 3 mm-thick LGPs inside. When LEDs with dimensions larger than 3 mm are placed in the corner of the BLU, it is not possible to direct the light into the LGP without any secondary optical component.

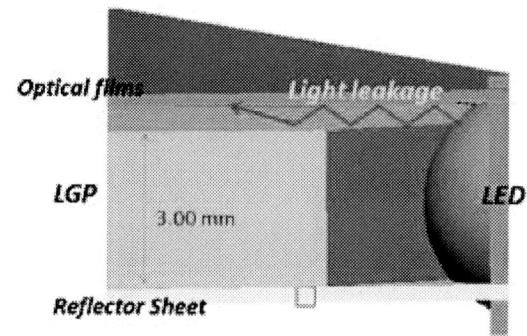

Figure 2. Corner type backlight unit structure with big sized HB LED package (side view).

Fig. 2 displays the corner placement of large LEDs. When the LEDs are placed directly at the edge of the backlight, light leaks from top and bottom sides of the LGP, which cannot be controlled by the LGP pattern; this creates corner light leakages on the panel. To solve this problem, a light guide tube (LGT) is defined to distribute the light along the edge of the LGP and to direct the light into the LGP. LGT material is defined as PMMA, same as the LGP. It helps the light propagate through the tube with high efficiency. To direct the light into the LGP, necessary parts of the LGT are covered with reflective material. At the LED side of the tube, a cylindrical reflector is designed to keep the light directional and inside the tube. In Fig.3 an optimal LGT and LED placement is presented. According to the optical (Light Tools) modeling results central luminance and uniformity target 75% is achieved with this design structure. Luminance distribution of the panel can be seen in Fig.4.

Characteristics	Unit	XML	XP-G2	XT-E	XM-L2	XBD	Luxeon-Z-ES
Thermal resistance, junction to solder point	°C/W	2.5	4	5	2.5	6.5	3
Viewing angle (FWHM)	degrees	125	115	115	125	115	120
Temperature coefficient of voltage	mV/°C	-2.1	-1.8	-2.5	-1.6	-2.5	-1.6
DC forward current (Max.)	mA	3000	1500	1500	3000	1000	1050
LED junction temperature (Max.)	°C	150	150	150	150	150	135
Forward voltage	V	3.03	3.02	3.32	2.9	3.05	2.85
Forward current	mA	1200	1230	1370	970	650	700
Number of LED	-	2	2	2	2	4	4
Luminous flux	lm	812	812	812	812	812	812
Power consumption	W	7.27	7.42	9.1	5.63	7.93	7.98

Table 2. LED package design parameters.

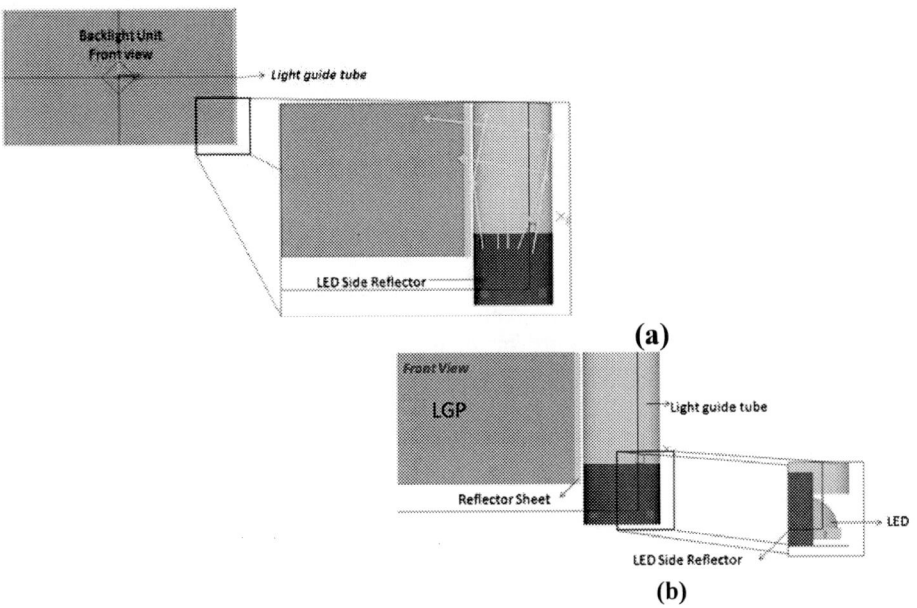

(a)

(b)

Figure 3. Light Guide Tube (LGT) solution for corner-lit backlight. (a) Placement of LGT in the backlight. (b) LED placement in the LGT

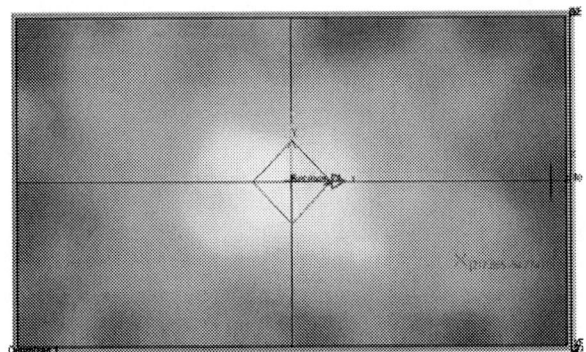

Figure 4. Luminance distribution of corner lit BLU with LGT

simulations are conducted for 4 LED package solutions using Cree XBD and Philips Luxeon Z-ES packages. The aforementioned placement configuration of the LEDs is displayed in Fig.5, where only the lower right corner of the four LED design is shown. According to the simulation results with the same level of LED lumen output, panel luminance is almost 20% higher than LGT based (with secondary optics) design. The luminance distribution of the system is presented in Fig.6.

To compare with- and without secondary optics (LGT) solution, smaller LED packages should also be investigated. However, when the package size gets smaller, the lumen output of the package also decreases. Therefore, the number of LED packages should be increased from 2 to 4. Light Tools

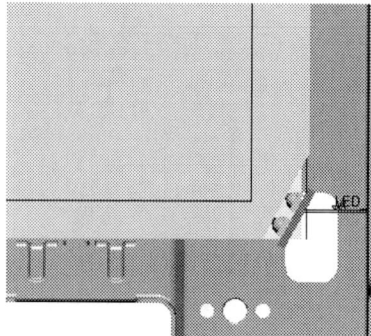

Figure 5. Lower right corner LED placement of the four-LED design.

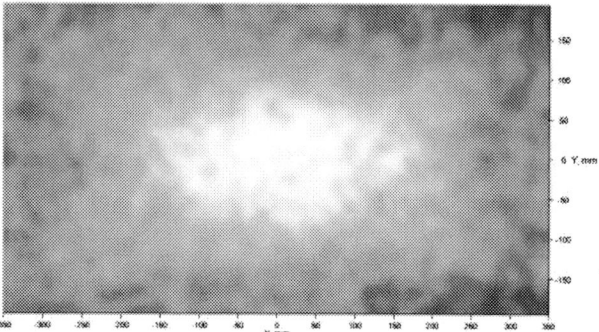

Figure 6. Luminance distribution of the 4-LED design.

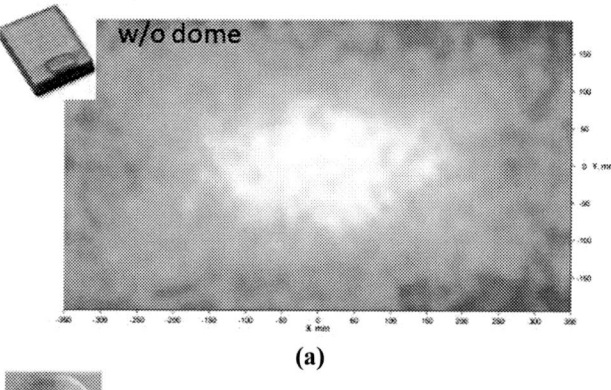

(a)

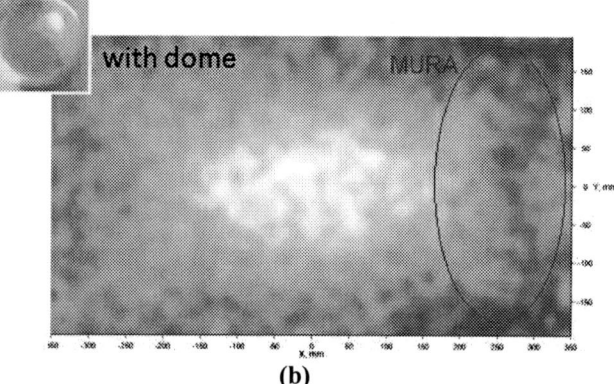

(b)

Figure 7. Luminance distribution results. (a) LED flat top shaped. (b) LED dome top shaped.

While conducting simulations for these two LED packages, i.e., Cree XBD and Philips Luxeon Z-ES, another critical design point is the top shape of the LED package. Cree XBD package has a dome-shaped top surface while Philips

Luxeon Z-ES has a flat top. This difference changes the distance between the LED and LGP which affects the optical efficiency of the system. If the LED has a dome-shaped top, LED-LGP distance increases by 0.2 mm. According to the Light Tools modeling results, flat-top LED packages lead to a 7% higher central luminance on the panel. Fig.7 shows the TV luminance distribution results of two LEDs with same LGP pattern. According to these results central luminance of TV is not the only parameter which is affected by top shape of the LED. Picture quality is also affected due to MURA (luminance discontinuity) problem at the right hand side of display.

	Dome shaped	Flat shape
Central luminance	93%	100%
Luminance Unif.	75%	78%

Table 3. Optical simulation result comparison between dome shaped and flat shaped LED packages.

According to simulation results, LEDs placed to the corner without light guide tube (LGT) have 20% higher luminance than LEDs with light guide tube. The loss happens in the Light guide tube and at the enterence of the LGP. This luminance difference will be more severe when the real product is produced because of mechanical imperfections. In addition to optical efficiency advantage, without LGT system has fewer number of components which means lower cost and easier production process. In corner lit systems, the top shape of the LED package affects the optical performance of the system. According to simulation results, because the flat top shaped LEDs can be placed closer to LGP, their optical efficiency is 7% higher than dome structure. Due to the dome shape, it creates some LCD panel optical performance changes too. The luminance uniformity becomes 3% lower than flat shaped LEDs and the LED side of the LCD has some luminance discontinuity (MURA) problem when the same pattern applied on the LGP.

Thermal Management of Compact Optical Packaging

LEDs are highly temperature sensitive. The lumen output, light quality and lifetime strongly depends on LED junction temperature. While corner-LED structure provided some inherit optical benefits, it has to be coupled with a smart thermal management as well.

In thermal management point of view, the critical issue of LED backlight system is the junction temperature of the LEDs. If the LED junction temperature is high, LGP and optical films can be damaged and LED life time cannot meet the TV life time. Current edge type backlight units use Aluminum heat sink to dissipate the heat. Due to their low volume placement and close internal structure for dust sealing, there isn't any space for natural convection. Therefore, dissipation is primarily driven by conduction. Current 32" LED TV's LED packages dissipate around 0.35 W heat, but in corner LED system it is 1.5 W. Therefore, conventional cooling system will not be enough to achieve 85°C junction temperature. Thermal pyrolitic graphite (TPG) is a good candidate for cooling system with its high thermal conductivity in two dimensions.

978-1-4799-2408-0/14 $31.00 © 2014 IEEE   212

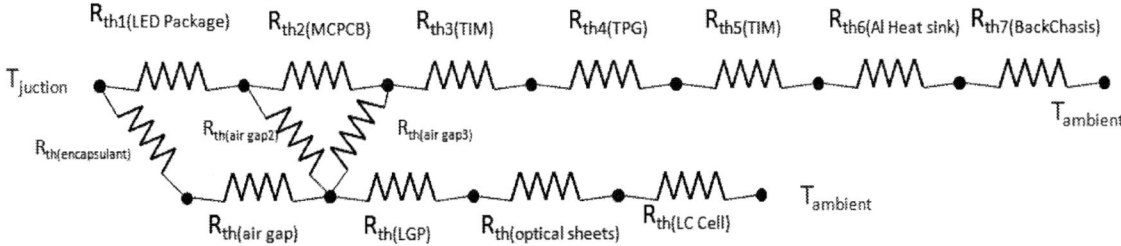

Figure 8. Thermal network diagram of corner lit backlight unit with TPG heat sink

In the first step, thermal network diagram of the idealized system is created for corner lit backlight unit with TPG heat sink which can be seen in Fig.8. As it can be seen from the thermal resistance network and from mechanical drawings in Fig.9, LEDs are mounted on a MCPCB. MCPCB is attached to TPG by using a thermal interface material (TIM) and TPG is then attached to an aluminum heat sink. The purpose of the TPG is conducting and spreading the heat to the Aluminum heat sink with a larger surface area.

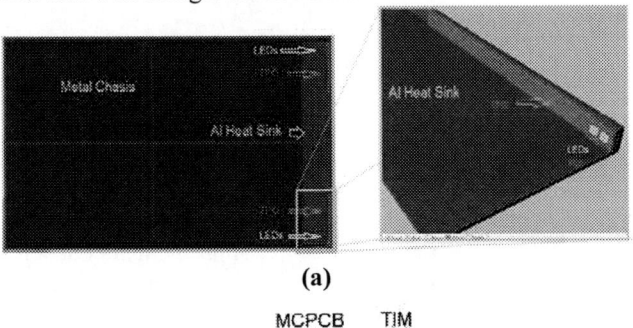

(a)

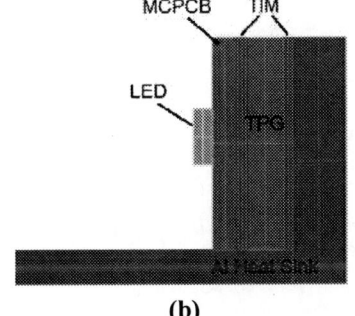

(b)

Figure 9. Mechanical structure of corner lit backlight. (a) The placement of Al heat sink and the LED packages (b) cross sectional view of LED and TPG on Al heat sink.

	k (W/mK)	Thickness (mm)	Area (mm²)	Rth (K/W)
Copper	400	0.035	34	2.57E-03
Dielectric layer	10	0.085	51	1.66E-01
Alluminum	177	0.6300	51	6.97E-02
MCPCB	61.5	0.75	51	2.39E-01

Table 4. MCPCB and its inner structure specifications.

In this structure, there are some critical points to be considered. For instance, the dielectric layer (D/L) of MCPCB is one of them. Because the MCPCB placed between TPG and heat source, it should have a low thermal resistance. Therefore thermal conductivity of D/L is taken as 10W/m-K. When the

copper and aluminum parts are included, total PCB thermal conductivity is calculated as 61.5 W/m-K. Thermal conductivity and thickness values can be seen in Table.5.

Thermal pyrolitic graphite (TPG) is an appropriate heat sink as well as heat spreader due to its high thermal conductivity. Its thermal conductivity is more than 3 times that of copper and its weight 20% lower than aluminum. Different thickness and length of TPG is computed in this study. Table 6 presents the material properties and thicknesses of various design components.

Components	k (W/mK)	Thickness (mm)
Back Chasis	45	0.60
Al Heat Sink	177	1.60
LED	58.8	0.59
MCPCB	61.5	0.75
TIM	1.2	0.25
TPG (x/y/z)	1350/1350/10	1.10

Table 5. Corner lit backlight unit's critical heat path specs.

A computational study has been performed with ICEPAK CFD software. Figure 10 shows the computational domain. Two LED chips are attached on the PCB. No thermal interface resistances are accounted for and perfect bonding is assumed in the models.

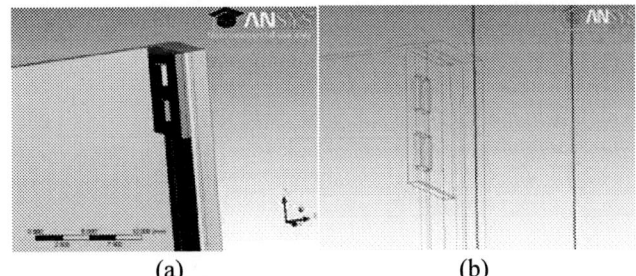

(a)   **(b)**

Figure 10. Computational domain (a) Design Modeler (b) ICEPAK

A mesh sensitivity study has been performed. Finally, number of elements was selected to be 1356472 and number of nodes was 1464668. LED parts of the system were defined as assembly parts and fine meshing is applied at the chips.

978-1-4799-2408-0/14 $31.00 © 2014 IEEE   213

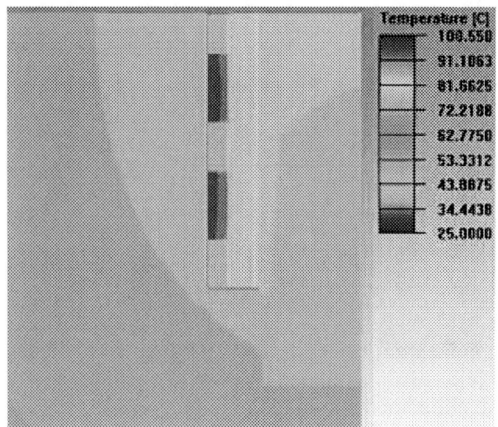

Figure 11. Temperature distribution of the LED on PCB – Al spreader

Figure 11 shows the temperature distribution when Al heat sink is used. It is found that maximum junction temperature is 100.5 °C leading to a junction to ambient resistance of 50.3 K/W.

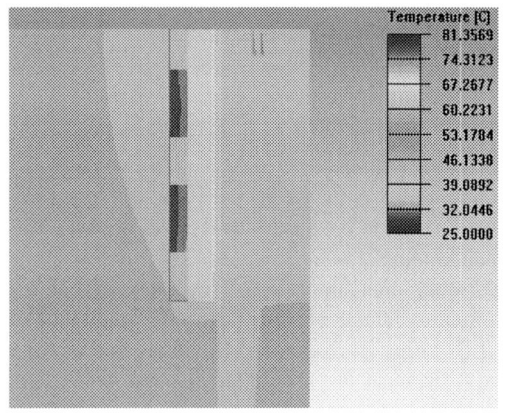

Figure 12. Temperature distribution of the LED on PCB – TPG spreader

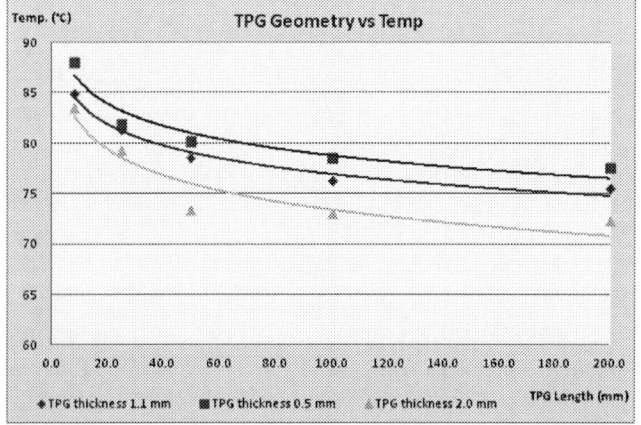

Figure 13. Junction temperature for various thickness and length of TPG

Figure 12 shows the temperature distribution when TPG heat sink is used. It is found that maximum junction temperature is 81.3°C leading to a junction to ambient

resistance of 37.8 K/W. There is a clear benefit of implementing TPG as a heat spreader for SMART TV applications.

Figure 13 presents the junction temperature for various thickness and lengths of TPG as a heat spreader for corner-LED application. According to the results, after 100 mm of length the effect of TPG is almost saturated. Therefore it is not efficient to use TPG longer than 100 mm due to high cost. When the thickness is increased, junction temperature is decreases as a result of reduction in conduction thermal resistance. However increasing the thickness from 0.5 mm to 2 mm which increases the material amount 4 times, it changes the temperature about 5 °C.

Conclusions

While advanced TV systems provides unique advantages such as picture quality, low energy consumptions and compactness, they also pose challenges such as optical and thermal management. Novel technologies and packaging approaches have to be developed. Optical and thermal parameters of a 32" corner-lit edge LED back light unit were defined and evaluated. Computational results showed that package mechanical dimensions, top shape of the package and placement of high brightness LEDs play a critical role in optical design domain. In addition to LED package properties, secondary optics' negative effect on luminance efficiency is revealed. In thermal management, it was seen that using only Al heat sink is not enough to dissipate the heat to achieve for the safe LED junction temperature. High thermal conductive material TPG has to be used to achieve reliable temperature levels. Thickness and length variations of TPG affect the thermal performance of the system. According to the thermal needs of the system it should be defined accordingly.

Acknowledgments

Authors would like to thank to Turkish Ministry of Science, Industry and Technology SANTEZ program for providing the financial support for this program under the contract number 1427.STZ.2012-1. Authors also would like to acknowledge for providing partial financial support for Vestel Elektronik Sanayi ve Ticaret A.S..

References

1. M. Arik, A. Setlur, S. Weaver and J Shiang, *Energy Efficient Solid State Lighting Systems*, Encyclopedia of Thermal Packaging, World Scientific Publishing, Ed. A. Bar-Cohen, 2012

2. Sung Ki Kim, "Analysis on Thermal Management Schemes of LED Backlight Units for Liquid Crystal Displays", *IEEE Transactions on Components, Packaging and Manifacturing Technology*, Vol.2, No.11, pp. 1838–1846, November 2012.

3. Chung-Yi Chu, Min-Chun Pan and Johnson Ho, "Thermal Analysis and Experimental Validation on TFT-LCD Panels for Image Quality Concerns", in *IEEE Electronics Packaging Technology Conference (ECTC)*, Dec 2006, pp. 543-548,

4. *Q3'12 Quarterly LED & CCFL Backlight Cost Report, October 2012, Display Search AN NPD Group Company.*

Wafer Level LED Packaging with Optimal Light Output and Thermal Dissipation for High-brightness Lighting

Liang Wang, Gabe Guevara, Hala Shaba, Roseann Alatorre, Rey Co, Ron Zhang
Invensas Corporation
3025 Orchard Parkway, San Jose, CA 95134
lwang@invensas.com , (408)324-5150

Abstract

Market size of high-brightness LED lighting has been rapidly growing upon the continual improvement of quantum efficiency and light extraction. However some key breakthroughs must be made before this technology can be fully adopted into the broad market, such as efficient thermal dissipation and low manufacturing cost. A major portion of cost of an LED module falls in the packaging processes after the emissive device stack has been fabricated. Also given the thin thickness of device stack, the packaging structure holds the bottleneck for thermal dissipation. We address these two key challenges with a novel wafer-level packaging structure integrated into the device stack, which enables maximal thermal dissipation rate from active device stack to substrate while allowing high aperture ratio and optimal light output. Our approach applies full wafer-level batch process from epitaxial growth all the way down to packaging for light extraction and wavelength conversion, in order to achieve high throughput and high yield at low cost. Initial prototypes of GaN based blue LED with big chip size have been fabricated without selective electrodes for minimized contact resistance, exhibiting high brightness at relatively low drive voltage (3.5V). As one key step in wafer level packaging, the wafer bonding process was characterized with Moiré patterning to understand the temperature-dependent warpage profile, with simulation performed in guidance to final solution for compensating the warpage profile along the bonding process and afterwards. Different approaches were applied in learning the most effective bonding technique for this packaging structure. Further development is ongoing to improve the overall power efficiency and color quality, including optimal materials for ohmic contacts at both electrodes, large-area light extraction structure, and integrated phosphor material. This wafer-level packaging technology is scalable to large wafer size for achieving superior thermal and optical performance at high-throughput and low cost.

1. Introduction

High-quality, high-brightness and long-lifetime lighting sources offer comfortable visibility and safety both indoors and outdoors. On the other hand, the energy for general lighting within residential and commercial buildings, over parking lot and public area, as well as on streets and roads accounts for up to 22% of the total electricity consumption in United States [1]. A significant portion of the electric energy enabling general illumination was converted to heat. Additional energy must be consumed to cool down the heat generated by these inefficient lighting sources. Therefore there is a critical need nationwide to sufficiently improve the overall power efficiency for general lighting purpose.

Correspondingly there undergoes a transition on general lighting market from traditional inefficient incandescent light bulbs to more energy-efficient solid-state lighting devices, such as light emitting diodes (LEDs) and organic light emitting diodes (OLEDs). Upon the incoming regulatory mandates and incentives around the globe, the market adoption of solid state lighting devices is expectedly on an accelerating curve.

Compared to other lighting technologies, GaN-based LEDs are advantageous in terms of efficient energy consumption, long lifetime and high brightness. The invention of the first GaN-based blue LED by Nakamura et. al. [2] paved the road for high-efficiency high-brightness LEDs which soon led to the development of the white LEDs with wavelength conversion approaches such as yellow phosphor coating and color mixing (RGB). This has enabled the successful implementation of white LEDs in a variety of applications such as backlighting for display, automotive headlights and retrofit light bulbs. The total global shipments of LEDs for general lighting are projected to grow from 440 million units in 2012 to 2.7 billion units by 2016 [3-4]. High-brightness LED lighting has gained high attention in the industry and its market share for general lighting has been rapidly expanding, mainly owing to the continued technological advances on improving internal quantum efficiency, light extraction and wavelength conversion [5-8]. In spite of promising developments, there remain some key breakthroughs to be made before LED lighting technology can be fully adopted into the broad market, with emphasis on efficient thermal dissipation and low manufacturing cost in order to improve both total luminous flux and cost-per-lumen ratio. Lower cost per lumen is necessary for LEDs to be competitive to traditional light sources like fluorescent lights in market pricing. In parallel, high brightness requires LEDs to be operated at high current density which result in higher power loss into heat due to droop behavior, and thereby highly efficient mechanism of thermal dissipation is needed to timely conduct heat away from the high-power LED chip.

Packaging of these high-power LED chips plays a key role for both cost reduction and thermal dissipation. About 50% of the total LED production cost falls in packaging [3-4] where a lot of room for cost reduction remains. Since the thickness of the LED device stack is only a few microns, the packaging structure and its interface to the LED device stack is the bottleneck for thermal dissipation. Si based wafer-level packaging is advantageous to address these two critical challenges. Silicon-based wafer level packaging utilizes mature processes in semiconductor industry with ease of chip-packaging integration at wafer level, enabling the manufacturing of a large amount of packages on one silicon wafer in parallel and thereby considerably lowering the production cost. The thermal conductivity of silicon packaging is 149 W/(mK) which enables efficient heat removal from LED chip. Additionally owing to silicon's low coefficient of thermal expansion of $2.6 \times 10^{-6}/°C$, there occurs low strain incorporation within the LED packaging structure upon operational temperature cycling which secures high reliability. Further improvement in thermal dissipation and cost reduction can be achieved using Through-Silicon-Vias (TSV) to increase the maximum amount of chips per wafer, as suggested by John Lau et. al. [9]

2. LED device structure design for wafer level packaging

Owing to the correlation between multi-fold properties (electrical, optical, thermal and mechanical) of an LED, one change for improving one performance (e.g., thermal), probably will affect all the other performances (e.g., electrical, optical and mechanical). Therefore there is an increasing need for comprehensive and systematic design where all the physical properties of every key component are considered coherently, built into one model preferably. This way, in consideration of any change in the system for improving one performance, one can investigate the consequences on all the other properties and thus have an integrated view of its effects on the whole system.

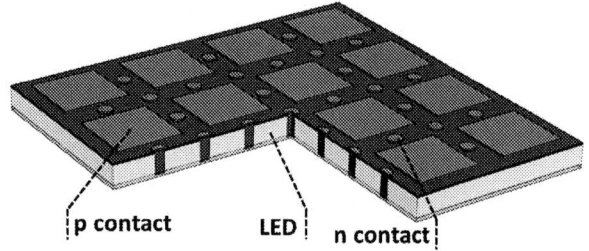

Figure 1. Wafer-level packaging structure for both thermal and optical performance

We address multiple technical challenges with a novel wafer-level packaging structure integrated into the device stack, which enables maximal thermal dissipation rate from active device stack to substrate while also allowing high aperture ratio and optimal light output. Fig. 1 depicts this design where the area of p contacts is maximized and the n contacts are multiple embedded vias surrounding each LED device. The n contacts can be extended along the periphery of each LED device to form walls wrapping around the LED device, to fully improve both optical and thermal performance by harvesting side emission of light and maximizing the area of thermal path. This design applies full wafer-level batch process from epitaxial growth all the way down to packaging for light extraction and wavelength conversion, in order to achieve high throughput and high yield at low cost.

For proof-of-concept of this design, LED wafers were fabricated based on this design with n contacts in both forms of via and wall. Si interposer wafers were fabricated for interconnecting and then bonded to LEDs. Si interposer dies were electrically connected to PCBs through wire bond, followed by underfill and lens encapsulation.

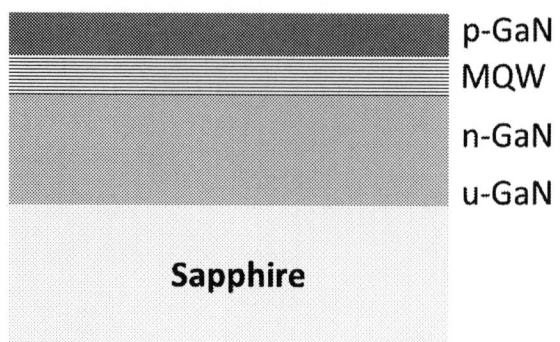

Figure 2. Cross-section diagram of the LED device stack

3. Process flow of LED wafer and interposer wafer

The LED devices of different sizes up to 2mm were fabricated on 2" sapphire substrate starting with MOCVD process to grow epitaxial layers of u-GaN/n-GaN/MQW/p-GaN as the device stack, as show in Fig. 2. The wafers were then processes with photolithography and inductively coupled plasma (ICP) reactive-ion etching (RIE), in order to isolate LED devices from each other by complete removal of epitaxial layer on defined street lines.

To enable electrical contact to n-GaN layer, vias or walls were patterned around the periphery of the defined LED area, with photolithography and ICP RIE again. Other than necessary isolation between n and p contacts, most of the p-GaN area was defined as p contact, for maximizing thermal and optical performance. For the simplicity of the proof of concept, same metal stack Cr/Au was deposited with e-beam evaporation over the defined p and n contacts, which will slightly increase the contact resistance, raise the drive voltage for visible light emission and reduce the LED's power efficiency at high current density. A thin layer (500nm) of SiO_2 was then deposited with plasma-enhanced chemical vapor deposition (PECVD) and patterned with photolithography and RIE to expose the p and n contacts. The LED fabrication was accomplished by patterning wafer bonding pads over p contact area and filling the n contact via or wall with metal materials, which serves three purposes including parallel extraction of heat from LED, optical reflection for light output, and wafer bonding pads to the Si interposer.

Si interposer was built for interconnecting LED device so that both n and p electrodes of LED device can be fabricated on one side of the epitaxial stack, leaving the other side fully utilized for light output with aperture ratio > 90% and enabling full-area backside electrical contacts for maximum heat removal. Si wafer with thermally grown oxide serves as the substrate of interposer. Al redistribution layer (RDL) was patterned with sputtering deposition, photolithography and wet etching. Thin TiN layer was applied on both sides of Al as adhesion and barrier layers. Another layer of SiO_2 was then deposited with PECVD and patterned with etching as the passivation to RDL. Then photolithography with thick resist was applied for subsequent electroplating of metal/solder bump over the RDL openings.

978-1-4799-2408-0/14 $31.00 © 2014 IEEE   216

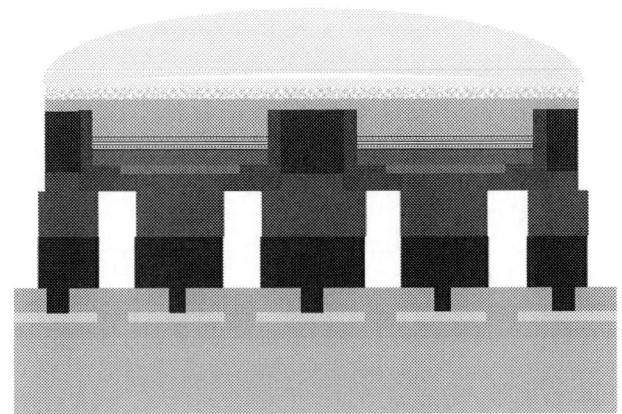

Figure 3. The LED packaging diagram after bonding LED to Si interposer, with lens encapsulation.

4. LED singulation-first packaging

LED wafers were thinned down to ~200um on sapphire side prior to singulation. No phosphor coating was applied in this study since this prototype focuses on interconnecting technologies for LED packaging. After surface cleaning and flux dipping, LED dies were then aligned and bonded to Si interposer with a reflow profile appropriate to the metal stacks of bonding pads on both LED and Si interposer, followed by underfilling process to provide environmental protection and absorb and distribute the mechanical and thermal stress induced in subsequent processes and device operation. The singulated flip-chip dies were then assembled onto PCBs through die attach. The pads on Si interposer were electrically connected to PCB through wire bond. Lens molding was accomplished as the final assembly step, serving as the encapsulation to keep the LED package away from environmental contamination and stress, as well as the secondary optics to increase the light output efficacy of the LED and direct the light emission uniformly into the hemispherical zone.

Figure 4. Example of blue light emission from one prototype of LED package

For simplicity of proof-of-concept, lens molding was implemented with materials based on commercially available non-yellowing epoxy encapsulant, applied at 140-180°C to seal the LED chip, underfill and wire bonds onto the PCB into single housing.

5. Characterization of assembled LED prototype

The resultant prototypes of LED package emit blue light visible to human eyes starting at ~3.5V, as displayed by the optical image in Figure 4. Six LED devices of different contact types and device sizes up to 2mm were fabricated on each LED chip. Current-voltage characteristics of all the 6 devices were measured using Agilent B1500A Semiconductor Device Analyzer. Figure 5 shows one example of the current-density-voltage characteristics.

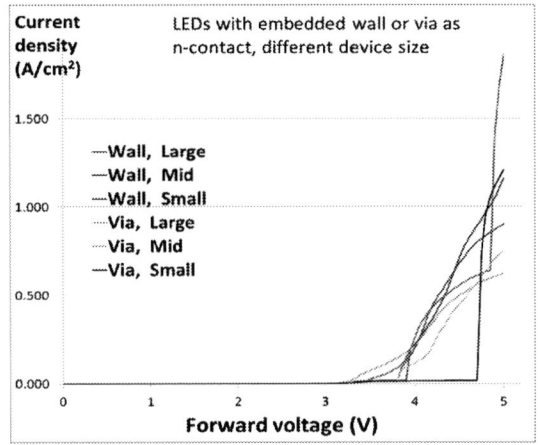

Figure 5. Current density-voltage characteristics of one prototype of packaged LED.

As one can observe in Fig. 5, embedded vias versus walls as the electrical contacts to n-GaN cause poorer electrical performance, i.e., lower current density at given forward voltage. This is due to relatively smaller contact area of embedded vias compared to walls. At relatively high voltage (>4.5V), devices tend to give rise to an abrupt increase in current density, probably attributed to surpassing the Schottky barrier at the metal-GaN interface. The non-ideal metal stack of Cr/Au interfacing both n-GaN and p-GaN as electrodes is responsible for the Schottky barrier and the consequential poor electrical contact. Full characterization of these LED prototypes with integrated sphere is ongoing at independent test source, and the results will be communicated elsewhere.

6. Wafer bonding approach

In addition to this LED singulation-first method, another approach in pursuit of wafer level packaging was taken where the LED wafer without singulation was bonded to Si interposer wafer so that in subsequent processes the sapphire substrate of LED wafer can be separated by laser liftoff technique [10] and then the LED-mounted Si interposer wafer can be singulated and assembled. The LED wafers and Si interposer wafers used for this approach were fabricated in the same batch as the LED singulation-first approach. After surface cleaning and flux dipping, LED wafer was aligned to Si interposer for wafer bonding with a reflow profile

978-1-4799-2408-0/14 $31.00 © 2014 IEEE 217

appropriate to the metal stacks of bonding pads on both LED and Si interposer.

One key challenge for wafer bonding is the flatness of the pair of wafers to be bonded. Owing to the ever increasingly finer pitch of the interconnection for higher performance and smaller form factor in consumer electronics, solder bump diameter and height are set to decrease accordingly. This roadmap of advanced packaging for integrated circuits might not directly impact solid-state lighting, however, key requirements on device performance and manufacturability still impose some critical challenges on packaging high-brightness LEDs at wafer level. In order to achieve high aperture ratio for maximal area of light emission, in the flip-chip device structure as shown in Fig. 1 and Fig. 3, the area of the via or wall as electrical contacts to n-GaN should be relatively small, which translates into small diameter and low height for solder bumps if no costly redistribution layer is added to allow for space adjustment. Even with bigger and higher solder bumps at coarse pitch, wafer bonding could be difficult if the wafer size is 4" and larger as the roadmap of the industry shows for high-brightness LEDs [3,4]. In order to enable large scale wafer bonding, the total warpage across the wafer for both LED wafer and sub-mount wafer must be tightly controlled so that the wafer-level mismatch of warpage between the pair of wafers is smaller than the height of solder bumps at solder reflow temperature and along the cooling down path to room temperature.

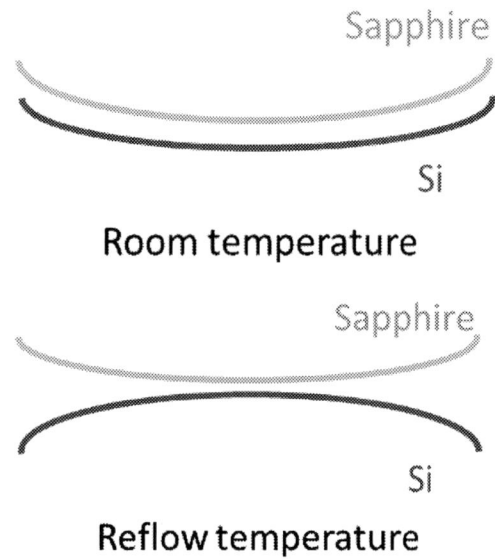

Figure 6. Warpage profile of a pair of LED wafer (Sapphire) and sub-mount wafer (Si) at room temperature and reflow temperature, based on Moiré measurement.

Prior to wafer bonding, shadow Moiré interferometry measurement was taken on both LED wafer (Sapphire substrate) and sub-mount wafer (Si substrate) in order to investigate the warpage profile of each wafer to be bonded from room temperature heating up to reflow temperature and cooling down to room temperature. The result is shown by the schematics in Figure 6. The warpage profile of sapphire-based wafer remains almost same along the entire course of heating up and cooling down, whereas the warpage profile of Si-based wafer significantly changes with temperature, even the bending direction switches at the temperature around the middle of heating up or cooling down path, as the schematics in Figure 6 depicts. The warpage profile of a fabricated wafer and its relation to temperature is attributed to the CTE mismatch of the materials comprising the wafer and the process history of the wafer. LED wafers started with a flat sapphire substrate which was then heated to >900°C for epitaxial growth of a few microns' GaN stack with metalorganic chemical vapour deposition (MOCVD). The resultant GaN/sapphire wafer was stress free at the high temperature for MOCVD growth, whereas mismatched contraction occurred when the wafer was cooled down to room temperature. Because the coefficient of thermal expansion (CTE) of sapphire is higher than that of GaN [11], sapphire tends to shrink more than GaN upon cooling down to room temperature and thus the LED wafer forms a warpage profile of bending towards sapphire side. Furthermore, since the reflow temperature is much lower than the temperature for MOCVD growth, this warpage profile was locked in the LED wafer and remained unchanged during the reflow thermal cycle. Therefore it is quite unmanageable to correct the warpage of LED wafer after MOCVD growth is finished. In order to achieve wafer-level flatness, a layer of material with melting point higher than MOCVD temperature and CTE lower than that of sapphire can be deposited at the calculated thickness as a compensation layer onto the backside (opposite to GaN side) of sapphire substrate before MOCVD growth of epitaxial GaN so that the materials on two sides of sapphire (compensation layer and GaN stack) are balanced in thermal expansion to flatten potential warpage. This approach was not taken in this work and will be implemented in our continual work to develop wafer level LED packaging technologies.

The warpage of sub-mount wafer results from the CTE mismatch between Si substrate and the metal routing traces fabricated over Si substrate. The metal traces were fabricated at temperatures up to 100°C through physical vapor deposition (PVD) and electroplating processes. Upon cooling down from these processing temperatures, the metal traces experienced more contraction than the underlying Si substrate due to higher CTE and thus led to a warpage profile at room temperature bending away from Si substrate, as shown in Figure 6. When the sub-mount wafer was heated to the reflow temperature which is higher than the processing temperatures for metal traces, metal traces were subject to more thermal expansion than the underlying Si substrate due to higher CTE which caused the sub-mount wafer to bend towards Si substrate side, as shown in Figure 6. This warpage profile of sub-mount wafer in combination with that of Led wafer makes wafer bonding difficult, especially the gap between the pair of wafers is too large around wafer edge. Therefore the warpage profile of sub-mount wafer must be corrected so that at reflow temperature through cooling down path to room temperature, the gap between the pair of wafers are within the tolerance given by the solder bump height. The warpage profile on sub-mount wafer was corrected by two implementations. A layer of Al was coated on the backside of Si substrate at the thickness calculated using the thermal simulation to compensate the thermal expansion of metal traces patterned on the front side of Si substrate. Additionally after careful investigation through fabrication processes, the metal ring coated along the periphery of the wafer was identified as another source of imbalance of thermal expansion, which was then removed mechanically. As a result of this correction, the warpage profile of sub-mount wafer was greatly reduced over the entire range from room temperature to reflow temperature, as shown by Figure 7.

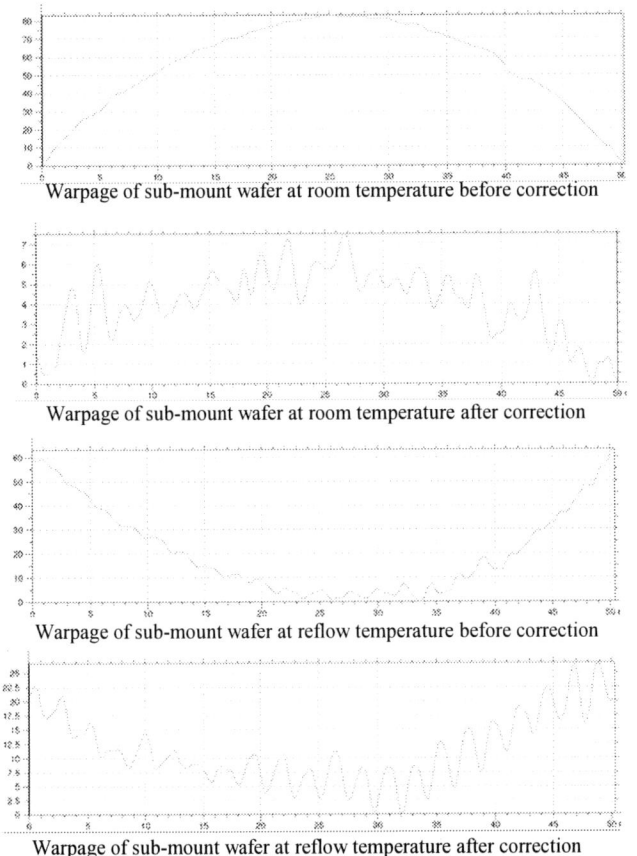

Warpage of sub-mount wafer at room temperature before correction

Warpage of sub-mount wafer at room temperature after correction

Warpage of sub-mount wafer at reflow temperature before correction

Warpage of sub-mount wafer at reflow temperature after correction

Figure 7. The effect of warpage correction on a sub-mount wafer, Moiré measurement data at room temperature and reflow temperature, viewed from backside of Si wafer.

After warpage correction, the gap between the LED wafer and the sub-mount wafer appeared manageable for wafer bonding. A series of processes including surface cleaning, flux dipping, wafer alignment, reflow and cool down were implemented and the LED wafer and sub-mount wafer were successfully bonded via solder bump reflow. X-ray was used to investigate the quality of wafer bonding across the wafer. The solder bumps were successfully reflow-bonded to the under bump metallization (UBM) patterned on LED wafer, across the wafer area from center to edge. Fig. 8 shows a group of these joints viewed at an angle from sub-mount side.

Figure 8. X-ray tomography of LED wafer bonded onto sub-mount wafer, viewed at an angle from sub-mount side.

7. Laser Liftoff

Process development of laser liftoff was performed on the bonded LED-submount wafer pair with a 248 nm excimer laser radiating into the sapphire side, using the laser liftoff technique known to the industry [10]. Sapphire is transparent to the laser at this wavelength so that the radiation energy is mostly absorbed by the epitaxial GaN layer next to sapphire. In fact the majority of photon-matter interaction occurs within 100nm of epitaxial GaN interfacing sapphire. The UV photon energy is absorbed by GaN which is then decomposed. Fundamentally speaking a type of plasma is created to separate the epitaxial GaN and sapphire. The sapphire side of the wafer pair was exposed to single shot laser with a square beam of approximately 250µm x 250µm area. Process parameters and laser flux level (energy density) was varied in an attempt to determine the conditions for successful lift off. For a typical laser liftoff process, the affectation of the epitaxial GaN gradually changes as the laser flux increases above a certain threshold. The gradual alternation of the exposed areas is indicative of laser lift off successfully applied to the underlying devices.

In our preliminary development work, the single shot tests demonstrated that as the laser flux increases there occurs an abrupt change from no affectation to an explosive debonding of the epitaxial GaN layer. The explosive debonding is due to the fact that the epitaxial GaN layer is not bonded to a planar surface which would otherwise provide sufficient back support, instead the epitaxial GaN layer contacts a topology of separated solder bumps with air gap in between, as shown in Fig. 3. This topology does not provide adequate backing support and thereby there is no resistance on the backside of the epitaxial GaN layer. Therefore when high pressure builds up during the laser exposure plasma process, forces generated at the sapphire/GaN interface pushes the epitaxial GaN layer forward, causing rupture of the film. This fragmentation of the epitaxial layer was observed on the test wafer pair during the initial attempt of laser liftoff. In order to perform successful laser liftoff on the LED wafer pair, continual development is ongoing to fill the air gap between solder bumps through wafer-level underfill process after wafer bonding and before laser liftoff.

978-1-4799-2408-0/14 $31.00 © 2014 IEEE

8. Conclusions

Wafer-level LED packaging structure was designed with metal contact in forms of via and wall embedded in LED device stack for maximal thermal dissipation and optical output. Initial prototypes of GaN based blue LED with big chip size exhibited high brightness at relatively low drive voltage. Si interposer was applied for multi-chip packaging at wafer level, with various benefits including enhanced thermal conduction, ease of chip-packaging integration at wafer level, high-throughput packaging by parallel process, reduced production cost, and scalability to larger wafer size. To improve the overall power efficiency and color quality, further developments continue to be implemented, including selective materials for ohmic contacts at p and n electrodes, current-spreading layer, large-area light extraction structure, and integrated phosphor material.

In order to address the key challenges for large scale wafer level packaging of LED, two important measures must be taken: 1) the total warpage across the wafer for both LED wafer and sub-mount wafer must be tightly controlled so that the wafer-level mismatch of warpage between the pair of wafers is smaller than the height of solder bumps at solder reflow temperature and along the cooling down path to room temperature; 2) Any air gap between solder bumps must be filled with supporting material (before or after wafer bonding) in achieving planarization as the back support to the thin GaN layer during laser liftoff separation of sapphire from GaN.

Acknowledgments

The authors thank Ilyas Mohammed, Bel Haba, Laura Mirkarimi, Rajesh Katkar and Tami Harchysen for their insightful advices, valuable suggestions and helpful assistance.

References

1. *A map for Solid-State Lighting*, Office of Science, United States Department of Energy.
2. Nakamura, S., Mukai, T. and Senoh, M., "High-power GaN p-n junction blue-light-emitting diodes", *Jpn J. Appl. Phys*, Vol. 30, p. L1998.
3. *LED Packaging, Market & Technology report*, Yole Developpement, January 2013
4. *Lighting the way: Perspectives on the global lighting market*, McKinsey&Company, 2nd Ed., 2012
5. Umit Ozgur, Huiyong Liu, Xing Li, Xianfeng Ni, Hadis Morkoc, GaN-Based Light-Emitting Diodes: Efficiency at High Injection Levels, *Proceedings of the IEEE*, Vol. 98, No. 7, 1180 (2010)
6. Martin F. Schubert et. al., Polarization-matched GaInN/AlGaInN multi-quantum-well light-emitting diodes with reduced efficiency droop, *Appl. Phys. Lett.* 93, 041102 (2008)₁
7. Lee Y.J., Kuo H.C., Wang S.C., Hsu T.C., "Increasing the extraction efficiency of AlGaInP LEDs via n-side surface roughening," *IEEE Photonics Technology Letters*, Vol 17, Issue 11 (2005), pp. 2289-2291.
8. Steigerwald. D.A., Bhat. J.C., Collins. D., Fletcher. R.M., Holcomb. M.O., Ludowise. M.J., Martin. P.S., Rudaz. S.L., "Illumination with solid state lighting technology," *IEEE Journal of Selected Topics in Quantum Electronics*, Vol 8, Issue 2, (2002), pp. 310-320.
9. John Lau, Ricky Lee, Matthew Yuen and Philip Chan, 3D LED and IC wafer level packaging, *Microelectronics International*, 27/2 (2010) 98–105
10. Chen-Fu Chu, et. al., Study of GaN light-emitting diodes fabricated by laser lift-off technique, *J. Appl. Phys.*, Vol. 95, No. 8, 3916(2004)
11. M. Leszczynski, et. al., Thermal expansion of gallium nitride, *J. Appl. Phys.*, 76 (8), 4909 (1994)

High Power Laser Packaging Challenges and Standardization

Eric Zhou, Jeffrey Morris
LDX Optronics, Inc.
1729 Triangle Park Dr.
Maryville, TN 37801, USA
ericzhou01@yahoo.com

Hanguo Wang
Department of Physics, UCLA
hanguo@ucla.edu

Abstract

The initial study of the electromagnetic acoustic resonance effect in high power semiconductor laser packages is reported here, along with the first detailed investigation on the thermal impact of the pins on some standard high power laser packages.

This paper reviews the challenges of packaging high power lasers and fiber optic modules, including electrical, thermal, mechanical, and different material requirements. Computer modeling and experimental studies have been done on some of the packages currently available on the market. The studies have found that high power lasers operated at quasi-continuous wave (QCW) input current above a few amps can emit acoustic noises generated by electromagnetic forces, which may excite the resonance frequencies of the laser package. We call this phenomenon electromagnetic acoustic resonance (EMAR). An input QCW frequency was swept from 100 Hz to 20,000 Hz, and audible noises were detected from several standard high power laser packages, such as HHL (high heat load) and TO-3. Additional findings are reported in this paper including the fact that standard high power laser packages, such as Butterfly fiber optic package and HHL package, can generate heat energy when high current is passed through their pins. This unwanted heat energy can in turn feed back to the laser, which would impact the laser performance. The temperature on the pin of a Butterfly package, for example, can rise to 80°C when a 3 amp current passes through it. This paper provides both experimental and theoretical studies on these issues. Based on the study results, some recommendations on changes are proposed at the end of the paper in order to improve the performance of laser packaging. These recommendations are also intend to promote standardization of the packages of high power laser in the future, which could reduce the cost of high power laser packaging.

Introduction

High power semiconductor laser products have been on the market since the early 90's. The advantages of higher powered lasers with smaller sized robust packages would enable these products to fulfill many new applications in industrial, medical, military, and other fields. These applications are increasingly demanding higher performance and reliability of the diode lasers, which raises more challenges in the packaging of the lasers: high thermal conductivity to remove the heat from a highly concentrated heat source, such as the laser diode or array, to control the laser operating temperature; thermal and mechanical structural

stability; and the ability to handle high electrical current density. There are, however, not many standard packages available on the market today that meet these demands, since specialized applications limit the popularity of these packages. In comparison, the packages found in telecom or datacom applications, such as TO can laser packages and Butterfly fiber optical packages, are much more cost effective. Standardized package designs and specifications, coupled with high volume demands, brought the cost of manufacturing these packages down significantly. Today, when those in the high power laser industry attempt to tap into the supply source of communication industries to reduce the cost of their own packaging, they find that it is very difficult to directly adopt these packages due to special requirements for high power laser packaging, such as different electrical, thermal, mechanical, optical, and material requirements. Standardization of laser packages will soon become an important issue across the high power laser industry.

So far research on high power laser packaging has been focused on how to dissipate the highly concentrated heat away from the chip or array [1~3]. In low power laser applications, the majority of the heat is generated from the laser chip and other internal components. However, in high power lasers, little research has been published concerning the heat generated from high current (a few amps to a few hundred amps), passing through the pins, or electrical terminals, of the package. Many "standard" packages on the market were not designed to handle such high currents. For example, some packages use steel pins to carry electrical interconnection between laser and power supply, and these steel pins have a much higher resistance compared to copper pins. This generates tremendous heat when a high current passes through them. The unwanted heat will be looped back into the laser chip through the electrical path. In this paper, we will discuss the thermal impact of the pins on high power laser packaging.

Many of the high power laser diodes or arrays operate under quasi-continuous wave (QCW) condition [1~2]. However, in our recent research, we have found that high power laser devices, operating at QCW condition, can emit acoustic noise generated by electromagnetic forces, which may excite the resonance frequencies of the laser packages. We call this phenomenon electromagnetic acoustic resonance (EMAR). It is well known that vibrations, combined with elevated temperature, can cause solder joints to undergo early failure in electronic devices [4]. So far, little research has been published on this subject for photonic devices. In this paper, we will discuss how QCW current can cause electromagnet acoustic resonance in a variety of high power laser packages, such as HHL and TO-3 packages. Based on our research, we provide some recommendations on how to reduce the acoustic resonance effect.

Electromagnetic Acoustic Resonance Effect

Electromagnetic acoustic resonance (EMAR) phenomena occur in a variety of settings. The fundamental physics is clear: any electrical current will create a magnetic field surrounding the current path. This magnetic field will interact with any other magnetic materials or magnetic fields generated by other electrical current paths nearby. The interaction between these magnetic fields will, therefore, form electromagnetic forces on all components involved. When the initial driving current changes its direction (AC current) or amplitude (QCW current), the interaction between the electromagnetic fields will change the forces that are applied to the magnetic materials or to the conductors that carry the electrical current. These forces will generate vibrations, leading to acoustic noise. When the vibration frequency is coupled with the resonant modes of the package, the acoustic noise will peak. This phenomenon has been found in many electrical devices ranging from electrical transformers, where it is known as mains hum, to high-pressure-sodium lamps [5~7]. Here, we first report our study on the EMAR phenomena in the case of high power laser devices driven by quasi-continuous wave (QCW) current. We have found that EMAR can cause a variety of high power laser devices to emit acoustic noises. We have conducted some experimental studies, as well as performed theoretical modeling, of the acoustic resonance effect when high power lasers are operated at QCW condition. Here, we provide study results on two types of HHL package, as well as a TO-3 package, but we have found that this phenomenon is generic and could occur in a variety of packages.

Experiment Setup

Measurements of acoustic effects were conducted on semiconductor laser packages driven by the quasi continuous wave (QCW) current, see figure 1. A Function Generator (Leader LFG-1300S) was used to modulate a Laser Diode Controller (ILX Lightwave LDC-3900) which fed the QCW current to the test sample. A Two Channel Oscilloscope (Tektronix TDS 320) was used to monitor the signal modulation. The test sample was secured to a base that acts as a temperature controlled heat sink and which prevents lateral

movement of the test sample. A microphone was placed flush against a surface of the test sample. In the case of high heat load (HHL) package (see Figure 2), the lid of the package was removed, and the microphone was placed on the top surface of the wall holding the pins of the package. In the case of TO-3s (see Figure 3), the cap of the package was removed, and the microphone was placed against the inside surface of the base of the package. An acoustic analyzer provided a real-time acoustic frequency distribution, allowing measurement of acoustic noise peak frequency and loudness.

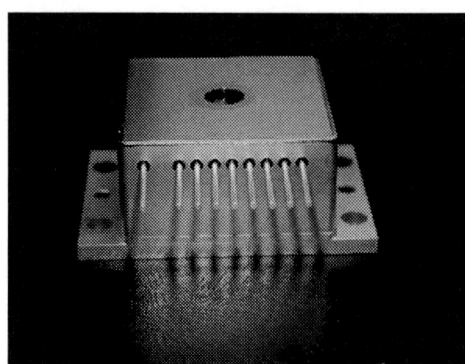

Figure 2. High power laser diode in high heat load (HHL) package (Pin #2 is absent).

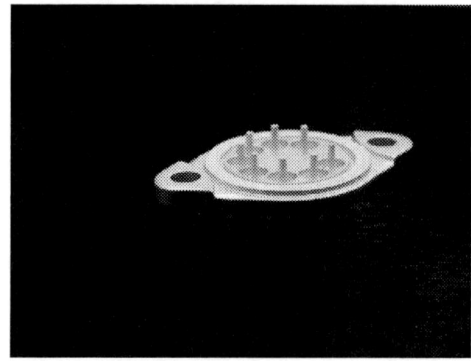

Figure 3. TO-3 header (Window can cover removed)

Measurements and Analysis

We first tested an HHL Laser device, λ = 808 nm, emitter size = 650 μm, 7W, driven by a 1.5 A QCW current. The modulation signal used was a square wave, biased so that the current driving the test sample was always positive. This signal was swept from 100 Hz to 20,000 Hz. We found that the laser device emitted audible noises, and that the noise peaked at the following input (QCW) frequencies in Table1.

Table 1. HHL-A

Input (QCW) Freq. (kHz) to Cause Noise Peaks	
2.0	6.1
2.5	6.3
4.5	8.0
5.3	11.0

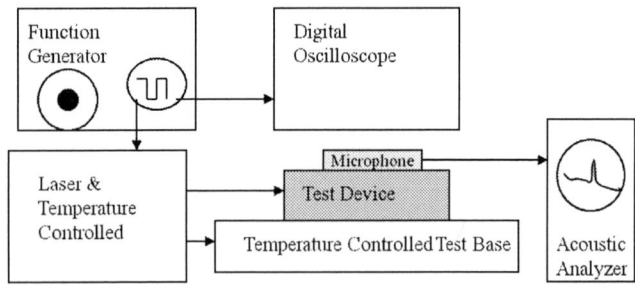

Figure 1. Schematic of experimental setup for high power laser package electromagnetic acoustic noise tests.

To further determine the source of the noise, we tested an empty HHL package (see Figure 2), which we will call "HHL-A," which does not include the laser, TEC, or any other components. We directly connected the current carrying pins, Pin #4 the anode (+) and Pin #7 the cathode (-), on the inside of the package with a copper hardwire (AWG 22, length = 15 mm). A QCW current of 5.6 A was supplied to HHL-A, with frequency swept from 100 to 20,000 Hz. We discovered that an empty HHL package does in fact emit acoustic noise. We also discovered that the frequency of the acoustic noise generated is not always the same as the frequency of the input (QCW) current. Data collected on the input (QCW) frequency, acoustic frequency (peak frequency), and loudness of the acoustic peaks are summarized in Figures 4, 5, and 6. Figure 4 shows the acoustic peak frequency versus the corresponding input (QCW) frequency. Figure 5 shows the acoustic peak loudness vs. the acoustic peak frequency.

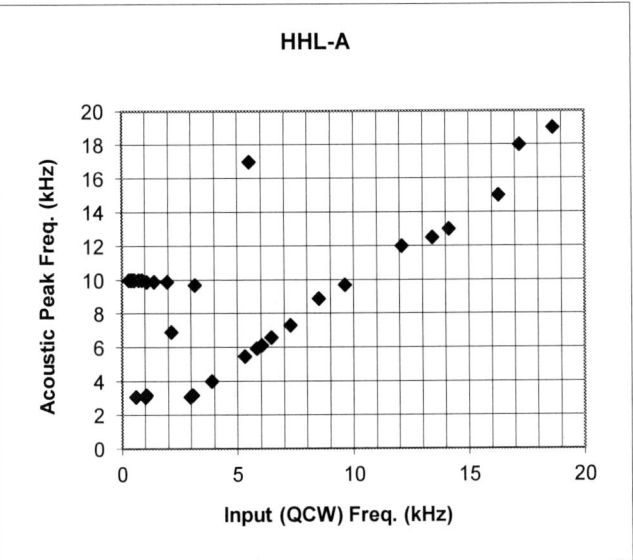

Figure 4. HHL-A, Acoustic Peak (Resonance) Frequency (kHz) vs. Input (QCW) Frequency (kHz).

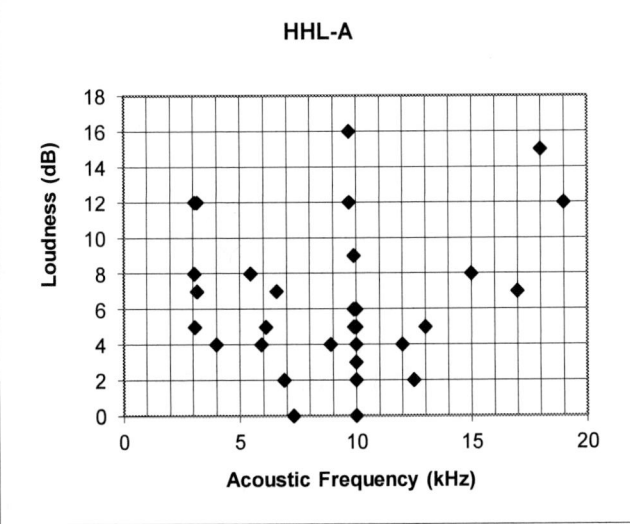

Figure 5. HHL-A, Acoustic Peak Loudness (dB) vs. its Frequency (kHz) obtained at different QCW frequencies.

From this, we can see that certain frequencies exist, for example at 3 and 10 kHz, where the sound will peak at the same acoustic frequency given different input QCW frequencies. This indicates that those acoustic frequencies are resonant frequencies of the package.

Further analysis shows that there are some harmonic relationships between the acoustic peak frequency and the correspondent input QCW frequency, see Figure 6 below.

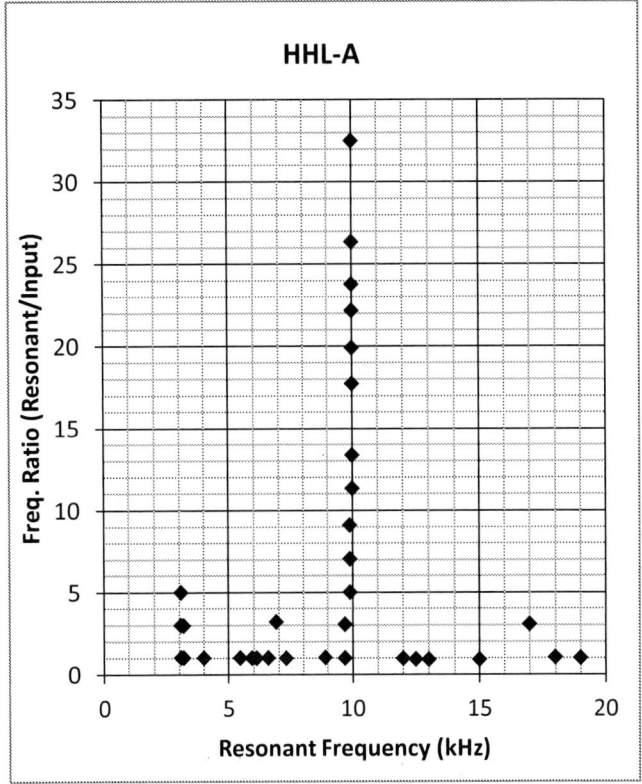

Figure 6. HHL-A, Ratio of Acoustic Peak (Resonance) Frequency / Input (QCW) Frequency distribution over different Acoustic Peak (Resonant) Frequencies (kHz).

The results show clear evidence that the acoustic noise has a resonant effect on the HHL package. The ratios of the acoustic peak frequency versus the corresponding input (QCW) frequencies are almost always integer numbers of minimum 1. This indicates that the input (QCW) frequency is driving the acoustic noise at harmonic modes, and this further supports the fact that certain acoustic peaks, such as at 3 kHz and 10 kHz, are resonant modes of the package.

We were also interested in seeing if this phenomenon occurred in other packages. Another package that we tested was the TO-3. Knowing that noise can be generated in an empty HHL package, we tested an empty TO-3 package. Inside the TO-3, Pin #4 and 5 were directly wired to form a loop out, to see if it would produce sound. Input QCW electrical current was a square wave of 5.6 A. Again, the input QCW frequency was swept from 100 Hz to 20,000 Hz.

We found that the TO-3 package does produce audible noise as well. Figure 7 shows the acoustic peak frequencies versus the corresponding input (QCW) frequencies. Although

978-1-4799-2408-0/14 $31.00 © 2014 IEEE

the peaks generated from the TO-3 are different than from the HHL package, they show a similar behavior in that they have resonant modes.

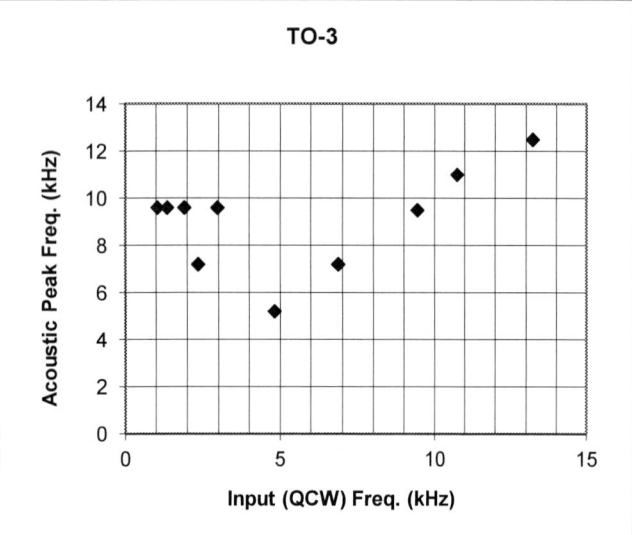

Figure 7. TO-3, Acoustic Peak (Resonance) Frequency (kHz) vs. Input (QCW) Frequency (kHz).

From this we concluded that the production of acoustic noise / vibration, driven by QCW current, is a common phenomenon in high power semiconductor laser packages.

Finite Element Analysis - Acoustic

In order to clearly understand the acoustic noise and the peaks detected in our experiments, finite element models were created to simulate the resonant modes of the HHL package. We used a fine mesh and constrained the bottom of the package to replicate experimental conditions. The simulation revealed that the resonant modes of the HHL package are very complicated. There are many contributors to the resonant modes, and therefore to the acoustic peak frequencies; some of the contributions come from the package housing, some come from the inside wire, and some come from the pins. This is why we detected so many acoustic peaks in our experiments. To simplify the modeling, we removed the pins from the package HHL-A, so this would allow us to simulate the contribution to the resonance modes from the case only. The following table, Table 2, lists those resonant modes obtained from the simulation.

Table 2. Resonant Modes of HHL-A Package Simulation

Resonant modes (kHz)	Contributor	Resonant modes (kHz)	Contributor
3.106	Pins	10.493	Case
3.138	Pins	10.567	Case
4.194	Pins	10.851	Case
7.315	Case, Wire, Pins	11.292	Case, Wire
7.878	Case	15.251	Case
7.892	Case	18.348	Case
9.708	Case	19.372	Case, Wire, Pins

The following figures (Figure 8, 9, and 10) show some of the typical simulation results where they graphically illustrate the displacement of the package parts during excitation of different resonant modes of the package.

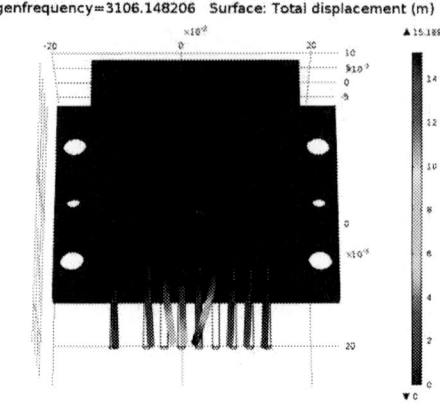

Figure 8. HHL-A simulated displacement of one resonant frequency 3.106 kHz contributed by the pins.

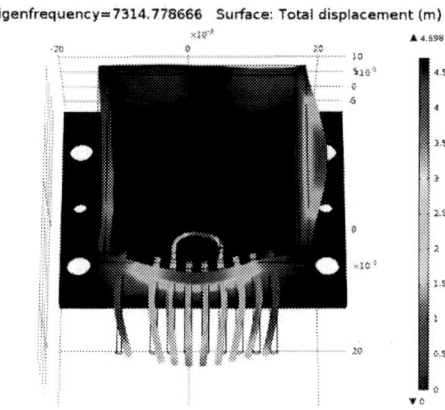

Figure 9. HHL-A simulated displacement of one resonant frequency 7.314 kHz contributed by the HHL-A housing, pins and the wire.

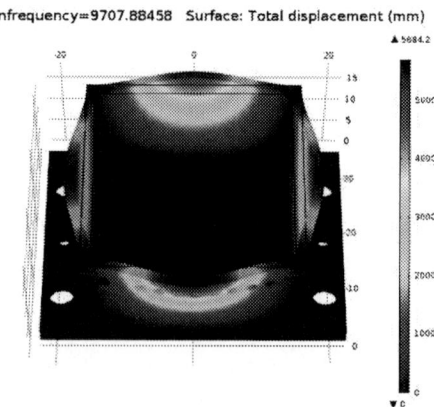

Figure 10. HHL-A simulated displacement of one resonant frequency 9.707 kHz contributed by the HHL-A housing.

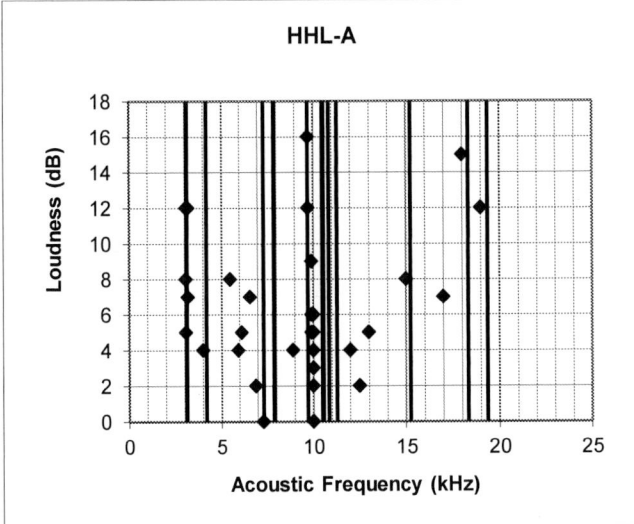

Figure 11. HHL-A, illustration of simulated "Resonant Modes" (black lines) overlaid onto the plot (Figure 5) of Acoustic Peak Loudness (dB) vs. its Frequency (kHz) obtained at different input QCW frequencies.

In order to better understand the correlation between the theoretical simulated resonant modes and the acoustic noise peaks obtained in our experimental study, we graphed these simulated resonant modes as lines onto Figure 5, the graph of peak acoustic noises versus its acoustic peak frequency detected under different input QCW frequencies for the empty HHL-A, see Figure 11. Almost all of the resonant modes predicted correlate to experimentally detected acoustic noise peaks, within testing error. There are a few detected acoustic peaks, such as peaks near 6 kHz, 12.5 kHz and 17 kHz, that were not shown in the simulation results. This indicates the limitation of the simulation tool and shows the complication of the resonant mode structure of the laser package. Otherwise, the results show a very close correlation between simulation resonance modes and experimental data. The theoretical simulation results provide a clear picture on the overall resonance effect.

Finite Element Analysis – Electromagnetic Field Distribution

From the above experiments and the analysis of resonant frequencies, we know that the acoustic peaks correspond to the resonance frequencies of the packages of high power lasers. The acoustic force or the source of the noise comes from the electromagnetic energy that results from the interaction between the package materials and the electromagnetic field created by the input QCW electrical current, which can be described by the Lorentz force equation:

$$f = \rho E + J \times B \qquad (1)$$

where **f** is the force density, ρ is the charge density, **J** is the current density, **E** is the electric field, and **B** is the magnetic field. Both **E** and **B** interact with each other in the media governed by the Maxwell equations:

$$\nabla \times E = -\frac{\partial B}{\partial t} \qquad (2)$$

$$\nabla \times B = \mu J + \mu \varepsilon \frac{\partial E}{\partial t} \qquad (3)$$

where ρ is the charge density, as a function of time and position, ε is the permittivity of the material, μ is the permeability of the material, and **J** is the current density vector, as a function of time and position.

Therefore, analyzing the electromagnetic field distribution in the package is crucial for understanding the root cause of the issue. The HHL-A package is made of several materials: the bottom is CuW, the wall and the pins are Kovar, and the pin feed-through is sealed by glass. We simulated the electromagnetic field generated from a 5A QCW current through the appropriate pins, Pin #4 (anode) and Pin #7 (cathode). Note that Pin #2 is absent. Figure 12 shows the magnetic flux density (B field) around the package.

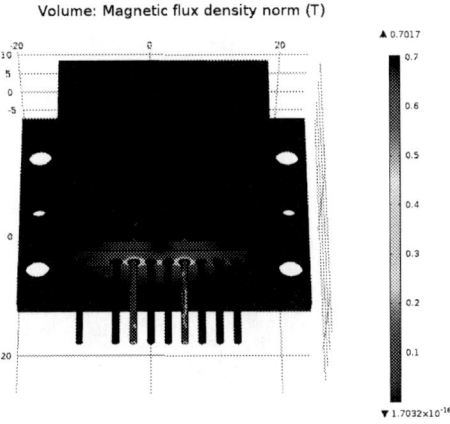

Figure 12. HHL-A magnetic flux density (B field) with input QCW 5A on Pin #4 (+) and #7(-) (Pin #2 absent).

The electromagnetic field is more concentrated on Pin #4 and Pin #7 since they carry the QCW current. The magnetic field on the package wall near pins # 4 & 7 is very strong too. The changing of the electromagnetic field, and therefore the changing of the force applied on the pins and the package wall, generates the vibration, the source of the acoustic noise. A higher magnetic field will generate a stronger force to create greater acoustic noise. The maximum value of the simulated magnetic field strength of the sample HHL-A, as seen in Figure 12, is around 0.7 teslas (T).

It is well known in electronic packaging technology that vibration plus elevated temperature will greatly reduce the solder joint lifetime, and therefore reduce the long term reliability of devices [4]. The acoustic noise could be a potential threat to the long-term reliability of high power lasers. Our study indicates that the key to reducing the acoustic noise is the reduction of the induced electromagnetic field around the laser package.

We propose that using a package with different materials and geometry will reduce the noise generated. Specifically, we will look at a fiber optical HHL package made by a different manufacturer, see Figure 13, with copper pins and an

978-1-4799-2408-0/14 $31.00 © 2014 IEEE

alumina sealing "bar" for the pin feed-through. The bottom and wall materials are similar as HHL-A, CuW and Kovar respectively. We will call this sample "HHL-B."

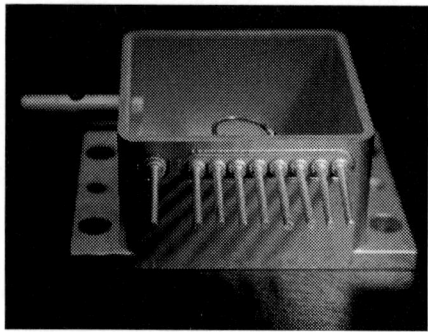

Figure 13. HHL-B sample, fiber optical HHL package.

Figure 14 shows the results of a simulation of magnetic flux density (B field) with the same driving parameters, QCW 5A, as the one run for HHL-A. In the model, the fiber tube was ignored on the HHL-B package for the simplicity of modeling.

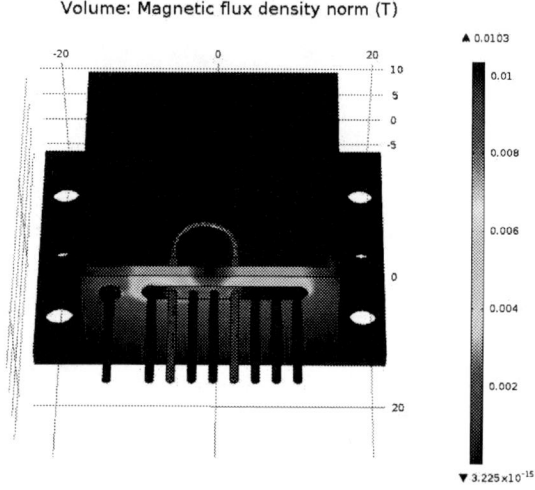

Figure 14. HHL-B magnetic flux density (B field), the input QCW 5A on Pin #4 (+) and #7(-) (Pin #2 absent).

The magnetic field strength on the HHL-B package is significantly less than that of the HHL-A (maximum values of 0.01 T and 0.7 T, respectively). This means that the electromagnetic force on the package is smaller for the HHL-B package, and thus should produce a quieter noise. From our simulation results, it can be clearly seen that the copper pins and ceramic sealing bar play very important roles in reducing the electromagnetic field strength, and therefore reducing the force generating the acoustic noise.

This is experimentally verified by comparing the peaks loudness produced by HHL-B with that of HHL-A. Figure 15 shows the experimental results of the HHL-B, under the same test conditions as for the one earlier for HHL-A.

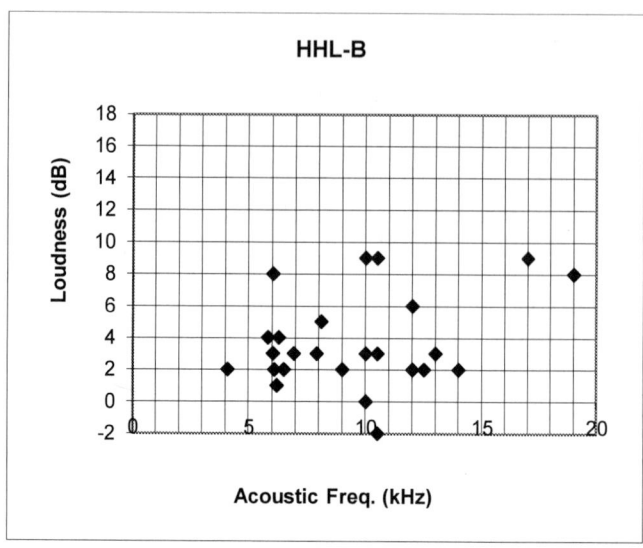

Figure 15. HHL-B Acoustic Peak Loudness (dB) vs. its Frequency (kHz) obtained at different input QCW frequencies. Same testing conditions as that of HHL-A in Figure 5.

When compared with HHL-A, Figure 5, both the range and the loudness of the acoustic noise peaks for HHL-B are smaller. The maximum noise peak level reduced from 16 dB in HHL-A to 9 dB in HHL-B. This validates our proposal that the HHL-B package will produce less noise.

Thus, we recommend that for QCW operated high power laser devices, manufacturers should avoid using ferromagnetic materials in the package, especially for the pins and materials that surround the electrical current path. For example, in HHL type packages, the HHL-B type package would be better than HHL-A type to reduce the acoustic noises in the applications involving QCW operation.

Thermal Effects of Laser Package Pins

The thermal behaviors of multiple semiconductor laser packages were investigated. In each case, a steady current, CW, beginning with 1A and increasing in increments, was passed through the package, and the temperature at various points on the electrical I/O pins of the package were measured.

A laser diode controller (ILX Lightwave LDC-3900) controlled the input current to the test sample, and a digital thermometer (Fluke 2160A) with a micro thermal-couple probes the temperature of the test sample. The base of the package was secured to a thermoelectric cooler (TEC) controlled by a temperature controller (ILX Lightwave LDC-3722) held at 25°C.

The first sample tested was a 14-pin Butterfly package, see Figure 16. This package is commonly used in the communication industry as a type of fiber optical module. The package is made of several materials: the bottom is CuW, the sidewall is Kovar, the Pins are Kovar with a cross section of 0.51mm × 0.25mm and 13mm long, and the pin feedthrough is ceramic. The terminal pins were connected via a clip, with the current feeding into pin #2, which is connected to the case. The case of the package was grounded. Table 3 is the test

978-1-4799-2408-0/14 $31.00 © 2014 IEEE   226

results from experiments, which show the temperature distribution on the pins of the butterfly package as the result of running CW input current at 1A, 2A, and 3A, respectively. From the test results, we can see that pin #2 gets very hot (80°C) at 3A input current. This is not only because of the small cross section of the pin, but also because the pin material, Kovar, has a higher electrical resistivity, 294 micro Ohm-cm, and lower thermal conductivity, 0.17W/cm°C, than copper, which has resistivity 1.68 micro Ohm-cm, and thermal conductivity 2.87W/cm°C.

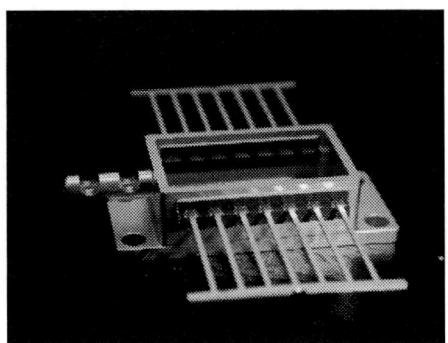

Figure 16. 14-pin Butterfly fiber optic package.

Table 3. Temperature Distribution on the Pins of Butterfly Fiber Optical Package.

Input I(A)	Probe Location on Pins	Pin #1	Pin #2 (+)	Pin #3	Pin #4	Top Edge
1	Pin End	26	26	26	25	25
1	Pin Mid	27	31	27	26	25
1	Pin Tip	27	30	27	26	25
2	Pin End	26	27	26	26	26
2	Pin Mid	30	49	29	28	26
2	Pin Tip	28	45	30	27	26
3	Pin End	27	28	27	27	27
3	Pin Mid	32	80	37	29	27
3	Pin Tip	30	68	35	30	27

According to Joule's law, the dissipated power of heat at a conductor can be calculated from the following equation:

$$P = I^2R = I^2\rho\frac{\ell}{A} \qquad (4)$$

Where I is the current, ρ is the electrical resistivity of the material, ℓ is the length of the conductor, and A is the cross section area of the conductor. Taking into account material properties, cooling systems, and ambient temperatures, the following models were created showing heat distribution of a stationary package fed with 1A (Figure 17 a), 2A, and 3A (Figure 17 b) CW currents.

Note that there is a clear temperature gradient on pin #2 of the package. For a 3A input, the temperature at the midpoint of pin #2 is above 80°C, while the end of the pin near the package is near 30°C. These results agree with experiment test results. But in the simulation, the tips of the pins are significantly hotter (95.2 °C) than the experiment test result

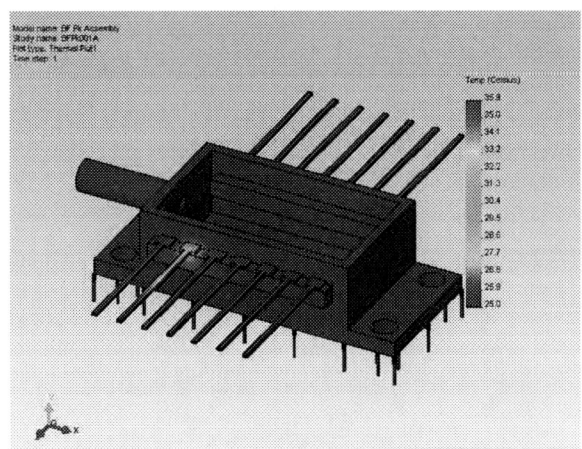

Figure 17 a. 1A CW current input to14-pin Butterfly fiber optical package. The color scale is from 25.0°C (blue) to 35.9°C (red).

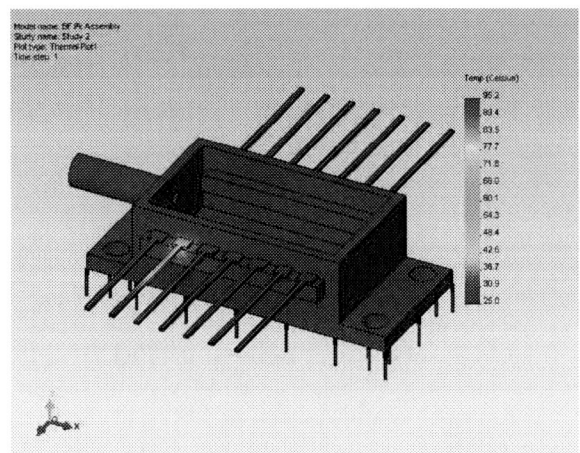

Figure 17 b. 3A CW current input to14-pin Butterfly fiber optical package. The color scale is from 25.0°C (blue) to 95.2°C (red).

(68°C), see table 3. This is because the heat at the tip of the pin #2 was dissipated away from the package through the connection of the clip wire.

The second sample tested was a HHL package with Kovar pins like the one tested for acoustics earlier in the paper (HHL-A). The terminal pins were connected via clip wires, with pin #4 the anode, and pin #7 the cathode. Table 4 shows the results of tests run at 1A, 3A, and 5A

Table 4. Temperature Distribution on the Pins of HHL-A Sample Package

Input I(A)	Probe Location	Pin #3	Pin #4 (+)	Pin #5	Pin #6	Pin #7 (-)	Pin #8	Wire Mid
1	Pin Mid	26	27	26	26	26	26	26
3	Pin Mid	28	34	29	27	31	27	29
3	Pin End	27	30	27	28	30	27	29
5	Pin Mid	30	53	33	33	45	29	37
5	Pin End	30	39	30	30	38	29	37
7	Pin Mid	36	79	40	40	69	33	44

The following models were created showing heat distribution of a stationary package fed with 1A, 3A (Figure 18 a), 5A, and 7A (Figure 18 b) CW currents.

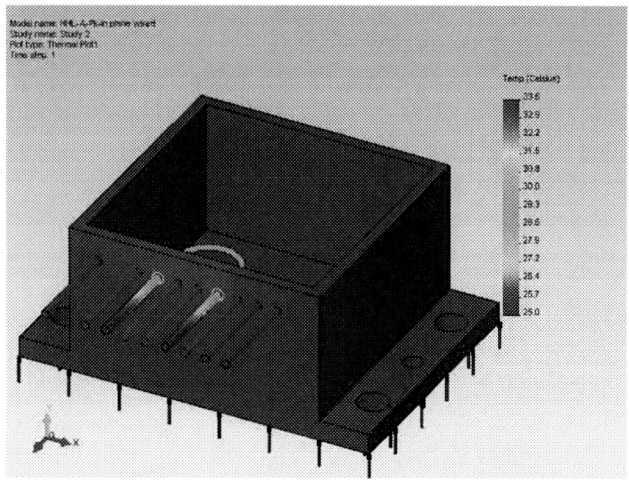

Figure 18 a. 3A CW current input to HHL package, HHL-A sample. The color scale is from 25.0°C (blue) to 33.6°C (red).

Note that, like the butterfly package, there is a clear temperature gradient in the pins of the HHL-A package. The middle of pins #4 and #7 are significantly hotter than the ends of the pins close to the body. For 3 A, at pin 4 (anode), the temperature is 33°C at the middle versus 30°C at the end. This is also easily seen in the temperature distribution simulations. The temperature of the wire inside of the package is 29°C. When the input current was increased to 7A, pin #4 and #7 generated so much heat that they could bring the temperature of the wire inside the package to over 45°C, which would affect the laser significantly.

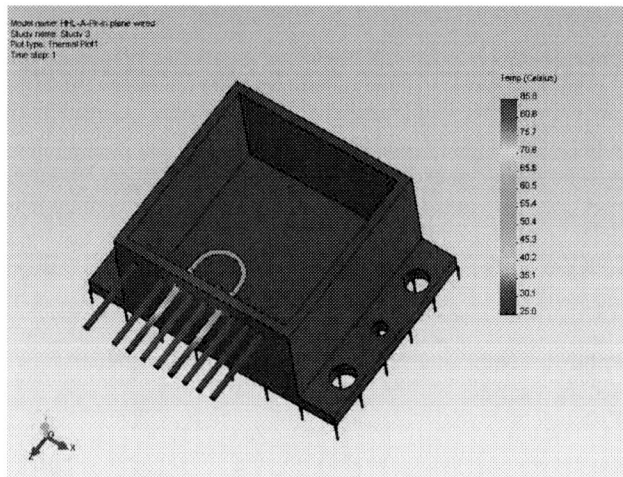

Figure 18 b. 7A CW current input to HHL package, HHL-A sample. The color scale is from 25.0°C (blue) to 85.8°C (red).

The third sample tested was an HHL fiber optic package that has copper pins instead, like the one tested for acoustics earlier in the paper (HHL-B). These pins are similar in size as those in HHL-A. The terminal pins were connected via a clip,

with pin 4 the anode, and pin 7 the cathode. Table 5 shows the results of tests run at 1A, 3A, 5 A, and 7A.

Table 5. Temperature Distribution on the Pins of HHL-B Sample Package

Input I(A)	Pin #3	Pin #4 (+)	Pin #5	Pin #6	Pin #7 (-)	Pin #8	Probe on Pin	Wire Mid
1	25	25	25	25	25	25	Pin Mid	25
3	25	26	26	26	27	26	Pin Mid	26
3	26	26	26	26	26	26	Pin End	26
5	27	29	28	28	30	28	Pin Mid	28
5	27	28	27	27	28	27	Pin End	28
7	28	34	30	30	34	29	Pin Mid	30
7	29	32	30	30	31	29	Pin End	30

The following models were created showing heat distribution of a stationary package fed with 1A, 3A (Figure 19 a), 5A, and 7A (Figure 19 b) CW currents.

Note that, unlike the butterfly package and HHL-A package, the temperature gradient on pin #4 and Pin #7 of HHL-B is much smaller and the overall temperature is much lower too. The temperature is nearly the same throughout the entire pin #4 and #7. This temperature distribution is also seen in the experiment results, listed in Table 5. This is because the copper pins in HHL-B have less electrical resistance (1.68 micro Ohm-cm) and higher thermal conductivity (2.87W/cm°C) than the Kovar pins in HHL-A. Based on the study, it is clear that Kovar pins are not suitable for high power laser packages. Although Kovar pins were originally adopted by the communication industry due to early manufacturing technology availability and less impact to the performance of communication devices, for high power laser packages, however, Kovar pins should be replaced by pins with lower resistance materials such as copper or copper alloys.

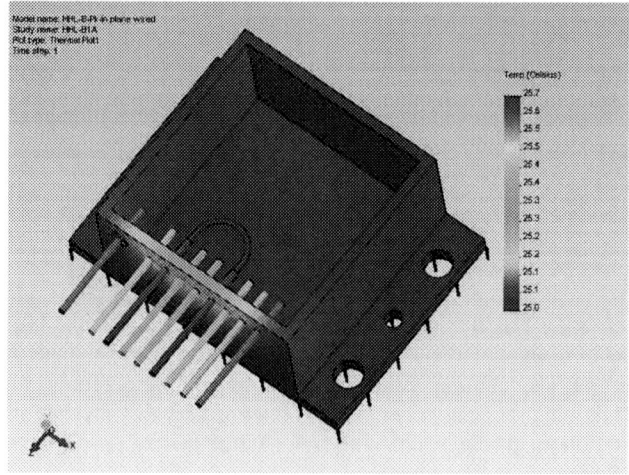

Figure 19 a. 3A CW current input to HHL package, HHL-B sample. The color scale is from 25.0°C (blue) to 25.7°C (red).

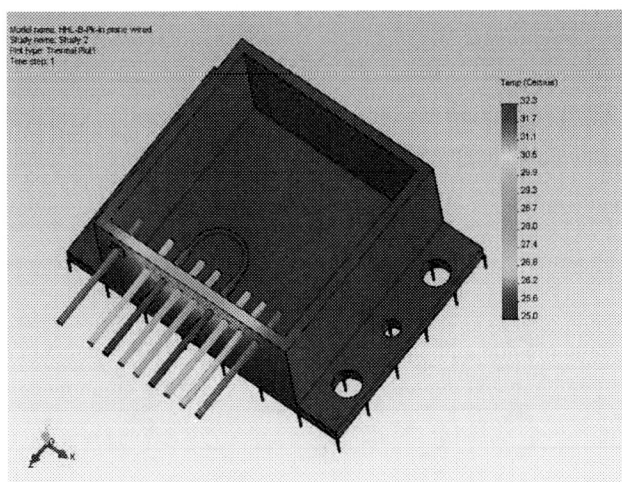

Figure 19 b. 7A CW current input to HHL package, HHL-B sample. The color scale is from 25.0°C (blue) to 32.3°C (red).

Conclusions

We have investigated the acoustic noise emitted from various high power semiconductor laser modules operated at QCW conditions. We found that electromagnetic induced acoustic resonance EMAR in the laser packages is the cause of the noises, and this effect is a general issue in various high power laser packages operating at QCW conditions. Our study has clearly indicated that the ferromagnetic materials (Kovar) used in the pins and walls of these packages have enhanced the EMAR effect; therefore, we recommend replacing them with non-ferromagnetic materials such as copper or ceramic materials to reducing the potential acoustic noise that may be generated by the laser devices when operated in QCW condition. In addition, we found that the steel pins used on some standard packages, such as Butterfly fiber optical module and HHL (like HHL-A sample) package, will generate higher heat energy due to higher resistance. The unwanted heat will feed back inside of the package, which will increase the temperature of the lasers, especially at a higher input CW current situation. Therefore, we recommend that in high power laser package, manufacturers should avoid any poor conductor, such as Kovar, being used in the pins or electrical terminals that carry high current.

Acknowledgments

The principle author, Eric Zhou, is currently a senior student of Santa Susana High School in Ventura County, California. He appreciates LDX Optronics Inc. and the Department of Physics at UCLA for providing him with student internships and for the resources for the experimental and theoretical research in this paper.

References

1. Farsad, E., Abbasi, S.P., Zabihi, M.S., and Sabbaghzadeh, J., "Thermal Analysis of 300W-QCW Diode Laser with Copper Foam Heatsink," *2011 Symposium on Photonics and Optoelectronics (SOPO)*, Wuhan, 16-18 May 2011, pp. 1-4.
2. Jingwei Wang, Zhenbang Yuan, Yanxin Zhang, Entao Zhang, Di Wu, and Xingsheng Liu, "250W QCW conduction cooled high power semiconductor laser," *International Conference on Electronic Packaging Technology & High Density Packaging, 2009, ICEPT-HDP '09*. Beijing, 10-13 Aug. 2009, pp. 451-455.
3. Jong Hwa Choi, and Moo Whan Shin, "Thermal analysis of the open type of laser diode," *17th International Workshop on Thermal Investigations of ICs and Systems (THERMINIC), 2011*, Paris, 27-29 Sept. 2011, pp. 1-5.
4. Zhao, Y., Basaran, C., Cartwright, A., and Dishongh, T. "Thermomechanical behavior of BGA solder joints under vibrations: an experimental observation," *The Seventh Intersociety Conference on Thermal and Thermomechanical Phenomena in Electronic Systems, 2000. ITHERM 2000.* Vol.2, Las Vegas, NV, 23-26 May 2000, pp. 349 – 355.
5. Walter Kaiser, Alexander Fernandez Correa, and Ricardo Paulino Marques, "Sound emissions from pulse operated high-pressure-sodium lamps," *IEEE Transactions on Industry Applications*, Vol.43, No.5, pp. 1199 - 1206, 2007.
6. Jorge García-García, Jesús Cardesín, Javier Ribas, Antonio J. Calleja, Emilio L. Corominas, Manuel Rico-Secades, and Jose Marcos Alonso, "New Control Strategy in a Square-Wave Inverter for Low Wattage Metal Halide Lamp Supply to Avoid Acoustic Resonances," *IEEE Transactions On Power Electronics*, VOL. 21, NO. 1, pp. 243-253, JANUARY 2006.
7. Kung, P., Lutang Wang, and Comanici, M.I., "A Comprehensive Condition Monitoring Solution for the Transformer," *XXth International Conference on Electrical Machines (ICEM), 2012*, Marseille, 2012, pp. 1520-1525.

Novel Highly Moisture Resistant Optical Adhesives and their High Power Resistivity

Seiko Mitachi [†] and Kazushi Kimura [‡]

[†] Tokyo University of Technology   1404-1 Katakura, Hachioji, Tokyo 192-0982 Japan

[‡] The Yokohama Rubber Co. Ltd.   2-1, Oiwake, Hiratshuka, Kanagawa 254-8601, Japan

E-mail: [†] mitachi@stf.teu.ac.jp, [‡] k.kimura@hpt.yrc.co.jp

Abstract

We have developed highly moisture resistant optical adhesives for optical communication. These cyanogen-free, silane modified adhesives are less allergenic and more environmentally friendly than conventional ones. These adhesives consist of organic and inorganic hybrid structure and their median times to failure by deterioration are 11-132times longer than those of Epotek-353ND. In the temperature-humidity cycle test, heat cycle test, and pressure cooker test these adhesives showed better reliability and higher power resistivity than 353ND.

1. Introduction

Highly moisture-resistant optical adhesives need to be developed for optical devices to create optical devices for outside-environment use [1]. We have studied ways to ensure the reliability of optical devices [2-4] since 2001. In 2005, we reported reliability test results for physical contact between optical connectors assembled with different material ferrules [2]. In 2006, we reported the development of odorless and optical adhesives that was highly durable against water and further, we have reported reliability test results for optical devices assembled with this adhesive [3]. Since 2008, we have reported silane modified, highly moisture-resistant optical adhesives [4-6] as alternatives to a conventional epoxy resin adhesive, Epotek-353ND, which is widely used for optical device assemblies. We have already reported on long-term reliability test results for optical connectors, attenuators, and polarization maintaining fiber (PMF) pigtail devices assembled with the novel optical adhesives and have shown that they have better polarization characteristics due to their lower strain property [7].

We report various reliability tests and high power test results for optical connectors and attenuators assembled with the novel optical adhesives, and they are shown to have more stable optical properties, and a much longer life-time, and better high-power resistivity.

2. Experiments

2.1 Preparation of new adhesives

Our new adhesives are cyanogen free, and their main ingredients are silane modified epoxy resins synthesized by a chemical reaction like-formula (1). By changing the ratio of silane modification and/or changing the type of epoxy resins, ingredients such as those in formula (2) and formula (3) can also be used. Our adhesives also contain other ingredients such as silane coupling agents, epoxy resins, and imidazole

curing agents for viscosity control. We lined-up several versions of these adhesives, Type Y (Y-41 ~ Y-43).

Table 1. Formulations of cured blends

		Y-14	Y-41	Y-42	Y-43
Epoxy resin	BPA type epoxy resin 1)	50			
	BPF type epoxy resin 2)		17	20	23
	Triglycidylized epoxy resin 3)		50	40	30
	Total (Epoxy resin)	*50*	*67*	*60*	*53*
Silane coupling agent	Secondary amino silane 4)	50	18	22	26
	Epoxy functional silane 5)		15	18	21
	Total (Silane)	*50*	*33*	*40*	*47*
A-part: Base		100	100	100	100
Water/Organotin catalyst		1/10			
B-part: Curing agent	Imidazole		10	10	10

The main ingredient of Y-14 is the silane modified epoxy resin with a glycidyl and a silanol group and synthesized by using a secondary amino silane addition reaction to BPA epoxy resin, 1) and 4) as shown in Table 1. This can be cured by using a water and organotin silanol condensation catalyst with no glycidyl reaction. Versions Y-41 to Y-43 consist of silane modified epoxy resins with a glycidyl group and a silanol group, synthesized by secondary amino silane addition reaction to BPF epoxy resin 2) and 4) in Table 1, a multifunctional epoxy resin, 3), and an epoxy functional silane, shown by 5). These can be cured by imidazole at 120°C for one hour with no silanol condensation catalyst.

2.2 Preparation of bonded samples

In the bonded sample preparation process, the resin part is mixed with the hardener part and then well stirred, the adhesives are sandwiched between floating glass plates, and 25 sets are prepared. Adhesives were around 20 μm thick.

Commercial float glass plates ($5.0^t \times 25 \times 40$ mm) were used as adherents and were immersed in acetone and cleaned in an ultrasonic cleaning machine for 10 minutes. The side that did not touch a tin molten bath was used as the bonding plane. The bonded area was around 1 cm^2 [4 x 25] mm.

For both 80°C-85% RH and 90°C-85% RH long-term reliability tests, 25 sets were prepared. To provide a uniform thickness of the glue, 200 g weights were put onto the adhesive surface and cured at 100°C-120°C for 1 hr. After the solidification of adhesion, test specimen thickness and bonded area were measured.

The newly developed optical adhesives used in the experiment are shown in Table 2. These are all silane-modified two-component epoxy adhesives. We also used Epotek-353ND (353ND) as a reference.

Table 2. Newly developed optical adhesives used for experiments

Optical adhesives		Mix ratio of weight for base resin and hardener	Curing conditions	
Adhesives customized for optical devices	Y-14	100 : 10	100℃ ~ 120℃	1 hr
	Y-41			
	Y-42			
	Y-43			

To complete the temperature-humidity cycle test, heat cycle test, and pressure cooker test, the Y-45R newly developed optical adhesive for polarization maintaining optical devices was additionally tested.

2.3 Preparation of optical devices for high power test

The SC connectors patch codes were assembled in Honda Tsushin Kogyo Co., Ltd., in conditions in which three sets of the patch code with one side assembled with silane modified adhesives Type-Y43 and the other side with 353ND and another three sets both sides assembled with Type-Y43. Additionally, ten conventional SC connectors patch codes assembled both side SC connectors with 353ND as references.

Each three of the four (5dB, 10dB, 15dB and 20dB) optical attenuators of metal doped type which were assembled with type-Y41, Type-Y43 and 353ND were also prepared in Honda Tsushin Kogyo Co., Ltd.,

2.4 Accelerated deterioration test

High temperature and high humidity tests at 80 °C - 85 % and 90 °C - 85 % were done for hundreds of hours up to several tens of thousands of hours, and the shear strength at compression tests was measured by using a benchtop materials testing machine, STA-1150. Deterioration was defined as a level of strength less than 20 kgf/cm^2.

The samples used were a commercially available adhesive, Epotek 353ND (Type-E), and four types of silane modified series (Type-Y14 and customized productsY-41, 42, 43).

Reliability parameters were calculated on the assumption that a log-normal distribution of time to failure can be applied to the deterioration of the adhesives. Time related to failure distribution is expressed as formula (4),

$$f(x) = \frac{1}{t\sigma\sqrt{2\pi}} exp\left[-\frac{1}{2}\left\{\frac{ln(t/t_m)}{\sigma}\right\}^2\right] \text{--------(4)}$$

where $t_m = t_{50}$ is the median time to failure (MTF), taken for 50% of samples to fail, and σ is the standard deviation of ln(t).

Median time to failure of every sample degradation is referred to as MTF. For variance in the deterioration, time to failure (TTF) of the logarithmic value is plotted on the Y-axis and the variance value (1/ρ) is plotted on the X-axis, and by the least squares approximation, y = ax + b was sought. The median is calculated by e^b.

2.5. Temperature-humidity cycle test

The treatment of starting temperature 23 deg C - humidity 93%, 2hrs at -10 deg C, 2.5 hrs at 65 deg C – humidity 93%, 2hrs at -10 deg C, and 2hrs at 23 deg C – humidity 93% was cycled 20 times in a total of 240 hrs.

2.6 Heat cycle test

The treatment of the starting temperature of 22 deg C, 2hrs at -40 deg C, 2 and hrs at 75 deg C, and going back to 22 deg C, was cycled 21 times.

2.7 Pressure cooker test

The pressure cooker test was done with TR-24S of ALP for the bonded samples when temperature was at 121 deg C, humidity at 100% RH, vapor pressure was 2 atm, and processing time was100 hrs.

2.8 High power test procedure

A high-power test was done by launching one watt LD light of 1.55 μm for 40 min to concatenated SC-SC codes assembled with Type-Y43 and 353ND and to the optical attenuators of metal doped type assembled with type-Y41, Type-Y43 and 353ND.

Before and after the test, optical attenuation was measured to confirm insertion loss was always less than 0.5 dB and the attenuation change was less than 0.2 dB. Reflection loss was measured to confirm that it was always more than 40 dB. The endface parameters of the connectors and the optical attenuator's male side before and after the high power test were measured with a ZX-1 zoom interferometer to check the change of fiber height (permanent fiber withdrawal). Further temperature changes of devices were measured by using a thermocouple.

3. Results
3.1 Accelerated deterioration test results
3.1.1 One example of test results for optical adhesive Type Y-41

In 2010, we reported MTF on Y-41 from the extrapolation from 60% degradation [6]. We present more exact MTFs from over 96% deterioration results. Time to failure of adhesive Y-41 and variance value under the test of high-temperature high-humidity accelerated aging is shown in Table 3. After 11,214 hr at 80°C-85% RH, 23 out of 24 samples have been deteriorated, and the test is now currently continuing. Also, after 25,181 hr at 90°C-85% RH, 24 out of 24 samples were all deteriorated, and the test was finished.

On the basis of the results in Table 3, Figure 1 shows variance of time to failure of deterioration of Y-41. The failure time distribution can be well approximated linearly by a logarithmic normal distribution formula (4). From the intercept of the straight line on the vertical axis, MTFs for Y-41 were 4,133 h at 80°C-85% RH and 9,855 h at 90°C-85% RH. These MTFs are 11-56 times longer than previously reported MTFs, which averaged 376 h at 80°C-85% RH and 177 h at 90°C-85% RH for Epotek-353ND [4, 5, 6].

Table 3. Median time to failure in aging by high-temperature and high-humidity conditions.

Sample \ Condition	80°C-85%	Average	90°C-85%	Average
Type-E(353ND-①)	569 hrs		226 hrs	
Type-E(353ND-②)	235 hrs	376 hrs	136 hrs	177 hrs
Type-E(353ND-③)	323 hrs		170 hrs	
Type-A(UV-cure-①)	1115 hrs		784 hrs	
Type-A(UV-cure-②)	1027 hrs	1043 hrs	688 hrs	743 hrs
Type-A(UV-cure-③)	986 hrs		756 hrs	
Type-Y-14-①	1281 hrs		1991 hrs	
Type-Y-14-②	3165hrs	3063 hrs	6156 hrs (Under testing [24/25])	3730 hrs
Type-Y-14-③	4743 hrs (Under testing [24/25])		3044 hrs	
Type-Y(Y-41)	4133 hrs (Under testing [23/24])		9855hrs	
Type-Y(Y-42)	7601 hrs (Under testing [23/25])		17454 hrs	
Type-Y(Y-43)	7729hrs (Under testing [20/25])		23332 hrs (Under testing [21/25])	
Type-Y(Y-44)	4788hrs (Under testing [24/25])		10805 hrs (Under testing [22/24])	

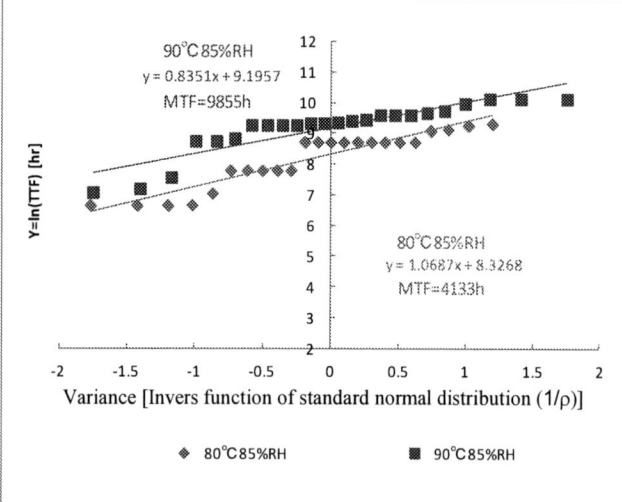

Figure 1. Lognormal distribution of adhesion failure time of optical adhesive Y-41

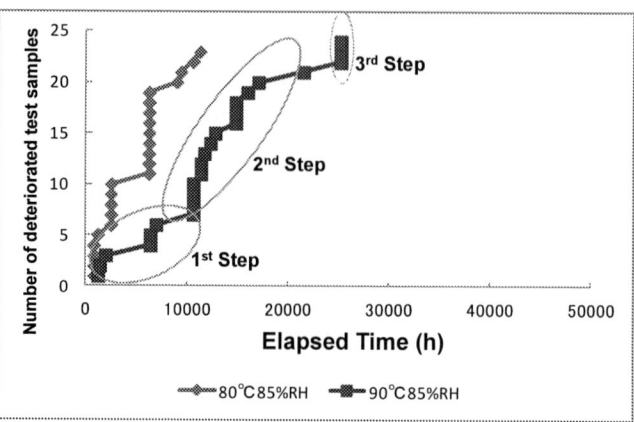

Figure 2. Time dependent cumulative distribution of deterioration

The time-dependent cumulative distribution of deterioration on optical adhesive Y-41 is shown in Figure 2. Deterioration speed at 90°C-85% RH is slower than at 80°C-85% RH. Here, Y-41 displays three-step deteriorations, in particular at 90°C-85% RH. That is, the first step is a mixture of two steps: initial deterioration and dominant hardening reaction. The second step is a balance step between deterioration and hardening. The final step is the deterioration dominant step. Much more detailed discussions concerning this matter will be seen in the Journal of Materials Engineering and Performance by Springer which is now in print [8].

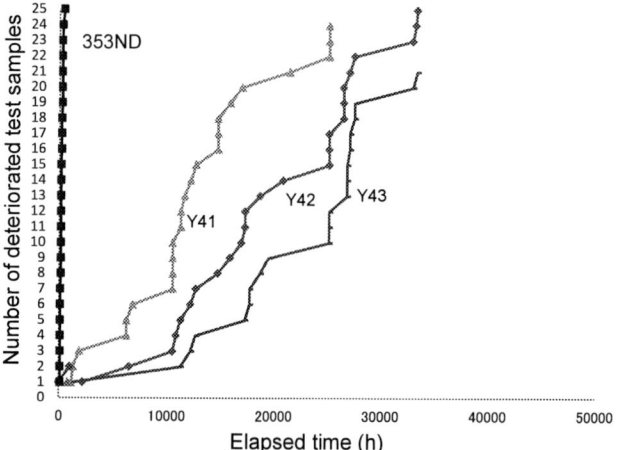

Figure 3. Time dependent cumulative distribution of deterioration in various optical adhesives at 90°C-85% RH

Time-dependent cumulative distribution of deterioration in various optical adhesives is shown in Figure 3. The cumulative distribution of deterioration is remarkably different between 353ND and novel optical adhesives. The deterioration speed is much slower in the case of Y-41, Y-42, and Y-43. In the initial deterioration and hardening dominant reaction step, Y-43 and Y-42 is little bit slower than Y-41. The second and third steps seem to be similar in Y-41, Y-42, and Y-43, with Y-41 taking the least time, Y-42 being in the middle, and Y-43 taking the longest.

The MTFs of the developed optical adhesives and conventional 353ND are compared in Table 4. All developed adhesives have MTFs over ten times longer than that of

353ND. The MTF of Y-43 is 21 times longer at 80°C-85% RH and 132 times longer at 90°C-85% RH.

Table 4. Ratio of MTF for novel optical adhesives to 353ND

Adhesive	80°C-85% R		90°C-85% RH	
Y-14	13	times	35	times
Y-41	11	times	56	times
Y-42	20	times	99	times
Y-43	21	times	132	times
353ND	1	time	1	time

3.2 Temperature-humidity cycle test results

The initial compression shear strength distribution of each adhesive is shown in Figure 4. The number by the markers in the graph indicate overlapping of the samples. And the arrow indicates that strength is over 100 kgf/cm². For example in Y-41, five samples were all over 100 kgf/cm², and 353ND was over 100 kgf/cm² in two samples.

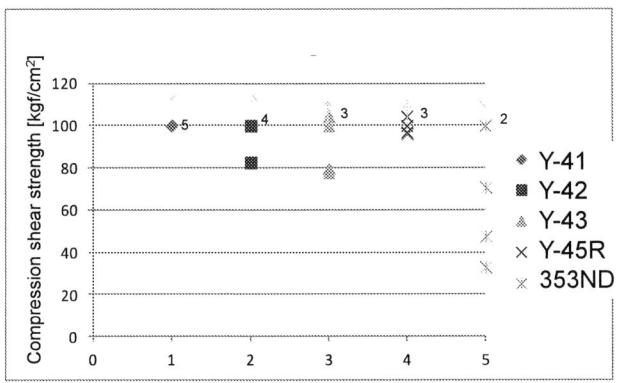

Figure 4. Initial strength of adhesives after curing (n=5)

The strength change after the temperature-humidity cycle test is shown in Figure 5. In the case of Y-45R, four samples maintained over 100 kg/ cm², and there was little variation in strength. In 353ND, almost all samples were destroyed by interfacial peeling and strengths were decreased. In the cases Y-41, Y-42 and Y-43, strengths were little bit decreased but not so much as 353ND and interfacial peeling was not found. All were destroyed by cohesive failure.

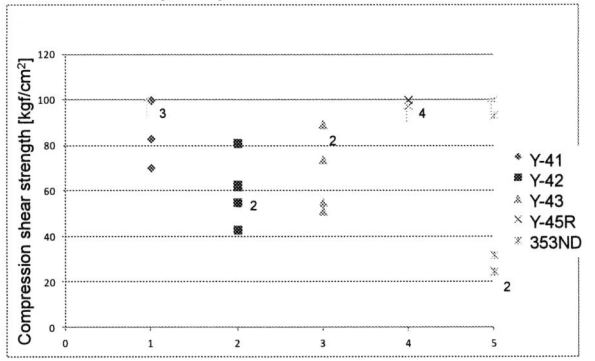

Figure 5. Strength change of adhesives after temperature-humidity cycle test (n=5)

3.2 Heat cycle test results

The strength change after the heat cycle test is shown in Figure 6. In Y-41, Y-42, Y-43, and Y-45R, a glass substrate break was dominant and there was no interface peeling. On the other hand in 353ND, there was three instances of interface peeling.

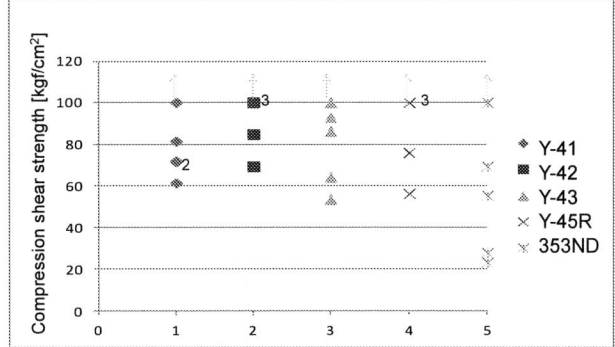

Figure 6. Strength change of adhesives after heat cycle test

3.4 Pressure cooker test results

Strength change of adhesives after the pressure cooker test is shown in Figure 7. After the test all samples of 353ND were peeled off even before the compression shear test. In Y-41, all samples were over 100 kgf/cm², and Y-42, Y-43 and Y-45R maintained good strength. The newly developed Y-type adhesives have better resistivity against heat and humidity than 353ND.

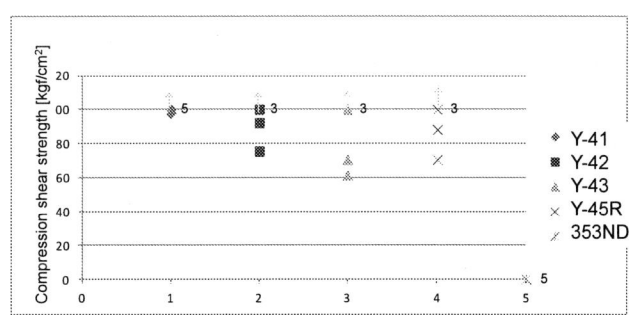

Figure 7. Strength change of adhesives after pressure cooker test (n=5)

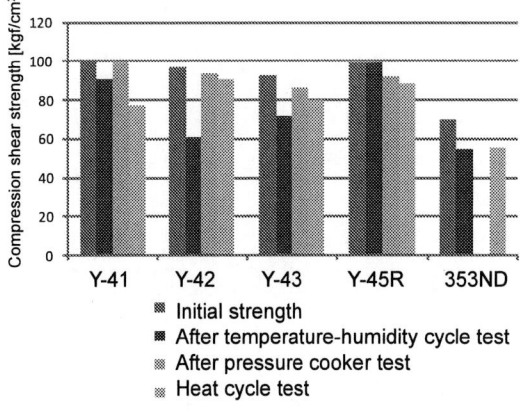

Figure 8. Average compression shear strength change after various tests (n=5)

The change in average compression shear strength after various tests is shown in Figure 8. The newly developed Y-41, Y-42, Y-43, and Y-45R remained stronger than 353ND. In particular, Y-41 and Y-45R show good durability.

A photograph of a pig-tail optical connector ferrule after the pressure cooker test is shown in Figure 9. For the ferrule assembled with Y-42, there was no problem with the integrity of the ferrule. On the other hand, for a ferrule assembled with 353ND, detachment of bonded part was found. This phenomenon indicates that after detachment of optical fibers and ferrule water molecules will penetrate into the naked fiber part and water molecules will attack the crack of the optical fiber surface and it will cause a break in optical fibers. This breakage presents a very dangerous situation for an optical fiber network.

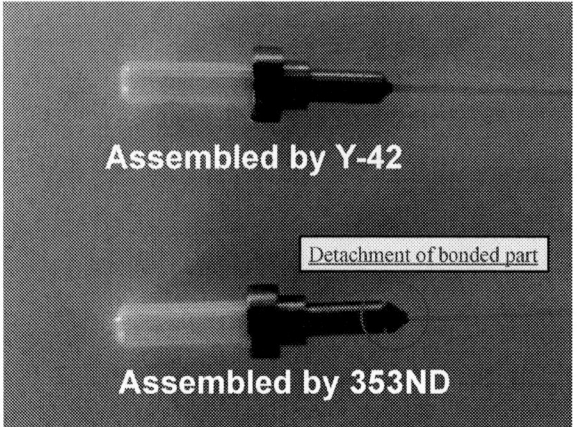

Figure 9. Optical connector ferrule after pressure cooker test

3.5 High power test results

One watt high-power, 1.55-mm LD light launching tests were done for 40 min for each of the three SC-type 5-dB, 10-dB, 15-dB, and 20-dB fixed attenuators assembled with 353ND, Y-43 and Y-41 by connecting both sides with SC connectors. The attenuation change after one watt high power test is listed in Table 5. The average attenuation change is in 0.12 dB to 0.15 dB and there is no difference among them.

Table 5. Attenuation change after high power test for optical attenuators

Attenuation (dB)		353ND			Y43			Y41		
		Before	After	Difference	Before	After	Difference	Before	After	Difference
5dB	No.1	4.74	4.77	0.03	5.04	4.96	0.08	4.87	4.81	0.06
	No.2	4.8	4.77	0.03	4.93	4.93	0	4.94	4.98	0.04
	No.3	4.94	4.93	0.01	4.86	4.84	0.02			
10dB	No.1	10.38	10.27	0.11	19.62	19.73	0.11	19.43	19.46	0.03
	No.2	10.36	10.26	0.1	10.59	10.54	0.05	10.39	10.31	0.08
	No.3	10.4	10.31	0.09	10.8	10.69	0.11			
15dB	No.1	14.83	14.77	0.06	15.1	15.01	0.09	14.76	14.71	0.05
	No.2	15.09	15.09	0	15.22	15.15	0.07	14.72	14.55	0.17
	No.3	14.91	14.85	0.06	14.97	14.77	0.2			
20dB	No.1	20.42	20.07	0.35	20.51	20.11	0.4	20.08	19.88	0.2
	No.2	20.27	20.07	0.2	20.72	20.47	0.25	21.07	20.66	0.41
	No.3	20.69	20.34	0.35	20.93	20.48	0.45			
			av.	0.12		av.	0.15		av.	0.13

One watt high-power, 1.55-mm LD light launching tests were done for 40 min to SC connectors themselves concatenation assembled with 353ND and Y-43 (n=3). The average insertion loss change was 0.0027 dB for 353ND and 0.050 dB for Y-43 and both were always less than 0.2 dB.

There is no difference between them and there was no problem. The average reflection loss change was 0.7dB for 353ND and 1.07 dB for Y-43 and both were always over 50 dB and there was no problem.

The average temperature changes after a high-power test for optical attenuators are listed in Table 6. The temperature increased with 19.8 deg C in 353ND, 22.1 deg C in Y-43, and 20.8 deg C in Y-41. All temperatures were less than 60 deg C and there is no problem.

The average temperature increases after high power test for SC connectors' concatenation were only 0.57 deg C in 353ND and 1.33 deg C in Y-43. The change is very small and thus presents no problem.

Table 6. Temperature change after high power test for optical attenuators

Temperature (deg C)		353ND			Y43			Y41		
		Before	After	Difference	Before	After	Difference	Before	After	Difference
5dB	No.1			0	28	48.5	20.5	30.1	48.8	18.7
	No.2	24.1	47.5	23.4	28.6	47.7	19.1	30	49.7	19.7
	No.3	27.7	43.7	16	28	49.8	21.8			
10dB	No.1	33.6	51.5	17.9	28	55.5	27.5	33.7	51.1	17.4
	No.2	31.8	47.9	16.1	28.7	48.7	20	32.4	49.5	17.1
	No.3	30.5	46.4	15.9	28.3	51.8	23.5			
15dB	No.1	27.9	47.3	19.4	28.7	51.9	23.2	32.2	54.2	22
	No.2	30.1	49.5	19.4	27.6	48.3	20.7	33.2	56.4	23.2
	No.3	28.5	47.1	18.6	27.2	48.9	21.7			
20dB	No.1	28	52	24	28	56	28	32.8	54.8	22
	No.2	28.7	44.7	16	28.7	48.3	19.6	29.5	54	24.5
	No.3	26.4	55.2	28.8	31.9	51.2	19.3			
			av.	19.6		av.	22.1		av.	20.6

The change in fiber height after a high-power test for optical attenuators is shown in Table 7. The average fiber height change was 28.9 nm for 353ND and 14.4 nm for Y-41. There was a two-fold difference between them. This seems to be a significant difference and in particular, over a 50-nm change that occurred in four samples in 353ND, but not in Y41.

Table 7. Fiber height change after high power test for optical attenuators

Fiber height		353ND (nm)			Y41 (nm)		
		Before	After	Difference	Before	After	Difference
5dB	No.1	-5.9	-23	17.1	-4.4	5.6	10
	No.2	-8.5	-10	1.5	-1.1	3.8	4.9
	No.3	-8.2	-12.6	4.4			
10dB	No.1	-9	-34.6	25.6	-18.7	-15.6	3.1
	No.2	-9	-31.1	22.1	-5.5	-12.1	6.6
	No.3	-12.3	-32.8	20.5			
15dB	No.1	-9	-65.4	56.4	1.5	-21.1	22.6
	No.2	-13.1	-82.6	69.5	0.9	-25	25.9
	No.3	-9.5	-70.6	61.1			
20dB	No.1	-11.3	-72.6	61.3	-28.8	-68.6	39.8
	No.2	-6.1	-2.4	3.7	-3.7	-5.8	2.1
	No.3	-9.2	-5.8	3.4			
			av.	28.9		av.	14.4

Figure 10 shows fiber height change after a high power test for fixed metal dope type optical attenuators. In 353ND, the difference fiber height reached over 60 nm at 15-dB and 20-dB, while in Y-43, the difference is always less than 40 nm. The average difference is 28.9 nm in 353ND and 14.4 nm in Y-41. These results mean that Y-41 had a better high-power resistivity than 353ND and performed similarly in the cases of damp heat and water immersion tests [1], [2].

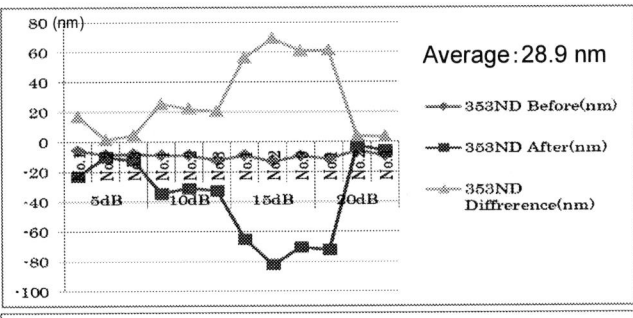

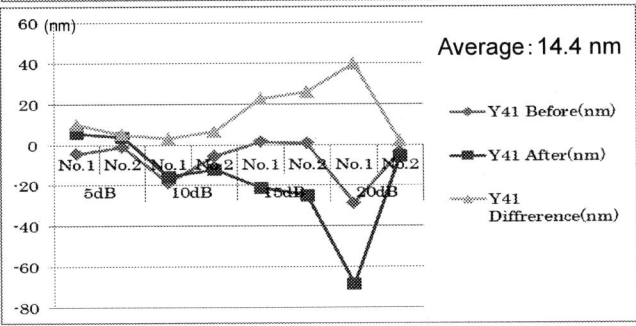

Figure 10. Fiber height change after high power test for fixed metal dope type optical attenuators 5-, 10-, 15-, and 20-dB

Differences in fiber height after high temperature-high humidity test, after water immersion test, and high power test are summarizes in Table 8. This table indicates that in type-Y adhesives permanent fiber withdrawal is always less than half that of in 353ND.

Table 8. Differences in fiber height after high temperature-high humidity test, after water immersion test [7], and high power test

	After high-temp & High-humid test (85°C-85%, 960 h)	After water immersion test (45°C, 168 h, 6 cycles)	After high-power LD light launching test (1.55 μm, 1 W, 40 min)
Type-E	**101-103 nm** (353ND) [SC connectors concatenations]	**31-72 nm** (353ND) [SC connectors concatenations]	**Avg. 28.9 nm** (353ND) [5, 10, 15, 20-dB Attenuators]
Type-Y	**41-88 nm** (Y-41, Y-42, Y-43) [SC connectors concatenations]	**12-24 nm** (Y-41, Y-42, Y-43) [SC connectors concatenations]	**Avg. 14.4 nm** (Y-41) [5, 10, 15, 20 dB Attenuators]

4. Conclusions

In conclusion, our novel silane modified highly moisture-resistant optical adhesives have much better reliability in the accelerated deterioration test, temperature-humidity cycle test, heat cycle test, and pressure cooker test. Furthermore, they have higher-power resistivity for optical devices than that of currently used epoxy adhesives.

5. Acknowledgments

This research was partially supported by the Japan Science and Technology Agency, Grant-in-Aid for the Adaptable & Seamless Technology Transfer Program (A-STEP) through Target-driven R&D) High-risk challenge, AS2414075M, used in 2012-2014. We also thank to Mr. Norio Murata for his technical discussion and advice.

6. References

1. S. Mitachi, Adhesion in optical devices, Journal of The Adhesion Society of Japan, Vol. **37**, no. 11, pp. 450-458, 2001, in Japanese

2. S. Mitachi, M. Shioda, S. Nakajima and H. Takeuchi, Reliability Test Results for Physical Contact between Connectors Assembled with Different Material Ferrules, in Proc. IEEE LEOS 2005, Sydney, October 23–27, 2005, WH4, p511-512.

3. S. Mitachi, T. Nakamura, M. Shimizu, C. Umeyama, and Y. Kawata, Reliability Test Results for Optical Devices Assembled with a Non-Smell & Highly Moisture Durable New Optical Adhesives, Proc IEEE LEOS 2006, Montreal, Octtober 29-November 2, 2006, TuB2, pp. 170-171.

4. S. MITACHI, Development of Highly Moisture Durable Optical Adhesives for Optical Devices, Proc. The 58th Electronic Components and Technology Conference, Lake Buena Vista, Florida, USA, May 27-30, 2008, pp. 207-212.

5. S. Mitachi, A. Harada, I. Ko, Y. Hiramoto, and A. Kikkawa, Reliability Improvement of Optical Devices by Using Newly Developed Highly Moisture Durable Optical Adhesives, in Proc. International Conference on Electronics Packaging 2009, Kyoto, April 14-19, 2009, pp. 108-113.

6. S. Mitachi, A. Harada, Y. Asano A. Kikkawa, and K. Kimura, Development and Reliability Evaluation of Highly Moisture Durable New Optical Adhesives", Proceedings of the 2010 IEICE Society Conf. Sakai, September, 14-17, 2010, C-3-62,, p. 183, in Japanese.

7. S. Mitachi and K. Kimura, Reliability and Polarization Characteristics of Optical Devices by using Silane Modifned Highly Moisture-Resistant Optical Adhesives, ICEP 2013 Proceedings, Osaka, April 10-12, 2013, TD3-2, pp. 340-344.

8. S. Mitachi, A. Ito, and K. Kimura, ""Development of Highly Moisture Resistant Silane Modified and Cyanogen-Free Type Optical Adhesives and Their Reliability", Journal of Materials Engineering and Performance, to be published in 2014.

Engineered Thermal Interface Material

Lyndon Larson[1], Yin Tang[1], Loren Durfee[1], Cassandra Hale[1], David Plante[1], Sushumna Iruvanti[2],
Rebecca Wagner[2], Taryn Davis[2], Hai Longworth[2], Annique Lavoie[3], Richard Langois

Dow Corning Corporation. 2200 W. Salzburg Road, Midland, MI 48686-0994, USA
IBM Microelectronics Division, 2070 Route 52, Hopewell Junction, NY, 12533, USA
IBM Canada, 23 Boulevard de l'Aéroport, Bromont, QC, J2L 1A3, Canada
E-mail: lyndon.larson@dowcorning.com

Abstract

The power dissipation and device junction temperature control in high end processors, stacked and hybrid packages, test and burn-in systems, LED devices, etc. present challenges in cooling. Many types of consumer devices and sensors are proliferating. All these applications require an ongoing improvement in thermal management. A key aspect of electronic package cooling is the thermal interface material used between the heat generating component and the heat spreader or heat sink. High performance thermal interface materials enable Tj reduction, device performance improvement and/or lower power operation. Organic laminate packages are especially vulnerable to package failures driven by CTE mis-match driven stresses and strains. Choice of TIM is therefore critical in addressing not only the thermal challenges, but also the mechanical weaknesses of a laminate package. Often a polymeric TIM with adequate compliance to address the mechanical issues and yet having a high thermal performance is desired. The properties of the TIM, such as the modulus, elongation, adhesion to both surfaces and thermal impedance, have to be carefully selected for optimum performance in a package. In this paper, we report the development of an industry leading, high performance thermal interface material. The project involved engineering the matrix polymer properties to systematically vary the composite modulus and die shear strength and meet the desired TIM property objectives. Methodical material property characterizations were carried out for feedback and formulation improvement. A few formulations were developed with TIM1 impedance in the range of 0.04-0.07 cm²C/W. The thermal performance was measured on thermal test vehicles. Material and process parameters were investigated to minimize voiding. Material characterization and thermal performance results are presented in this paper.

Introduction

As the electronics industry needs more and more advanced semiconductors to meet the demanding requirements of current and future end-uses, there is considerable engineering effort behind-the-scenes to ensure these advanced semiconductors can maintain performance for their given environments and expected lifetimes. In particular, a major consideration is the overall robustness of the semiconductor package. The semiconductor package must provide the following functions: protection, power distribution, signal distribution, design and test, and thermal management [1]. This paper focuses primarily on package-level thermal management. A thermal interface material (TIM) used between the die and the heat spreader enables effective heat removal from the die to ensure its performance and reliability. The thermal interface material and the interfaces are subjected to mechanical stresses and strains in the package, degrading their integrity and resulting in loss of performance. Therefore, the chosen thermal interface material must not only meet the thermal objective, but also be able to withstand the mechanical stresses and strains to maintain its performance through the life of the package without significant degradation.

In order to engineer an advanced thermal interface material, one must keep in mind the big picture and understand trade-offs that will occur during the evaluation process. Much of this study focuses on analyzing such tradeoffs and some of the characterization challenges involved. The type of package selected for this evaluation was a single die flip chip BGA thermal test vehicle (TTV).

This engineered thermal interface material formulated by Dow Corning is a TIM1 application -that is to say connecting directly to the silicon and a heat spreader [2]. An example of this is depicted in Figure 1. However, this TIM could be useful in other such applications requiring similar properties.

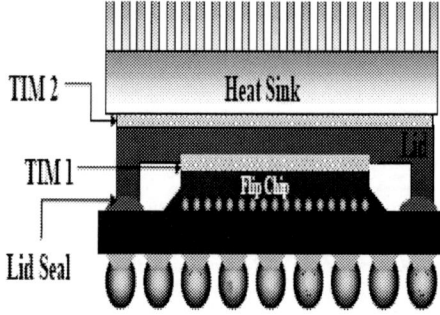

Figure 1. Schematic illustration for lidded flip chip

Common thermal interface materials include pastes, gels, and metals [3]. The type of material and its inherent properties have a significant impact on the performance and reliability of the package. Very rigid materials and non-curable materials might not survive thermal cycling due to various failure mechanisms reported previously [4]. In some cases, very soft materials are preferred, either with or without self-priming adhesion.

Silicone-based materials are commonly used for this type of application since the physical properties such as good wetting, low modulus, and ability to heavily load with thermally conductive fillers can be suitable for providing the needed reliability and thermal performance [5].

The engineered thermal interface material must balance all of these formulation permutations to maintain the proper level for key properties without excessively sacrificing a countering property. For example, increasing modulus tends to decrease elongation, or higher filler loading tends to increase hardness.

Note that this package uses a reinforcing adhesive around the perimeter of the laminate substrate to help hold the lid in place and to maintain package integrity during the assembly process and subsequent stress testing [2]. This perimeter adhesive is also commonly a silicone-based material. Figure 2 shows a simplified assembly process for a typical flip chip manufacturing line [6]. The subsequent process steps and the materials of construction of the package also play a critical role in the overall package robustness.

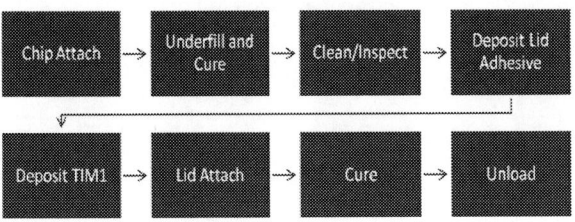

Figure 2. Flip chip assembly process steps

Characterization Challenges

The first step of this project was to identify which physical properties were the most relevant for this application. The following properties were selected: modulus, elongation, hardness, and adhesion, in addition to bulk thermal conductivity.

The second step was to experiment with methods that may be useful to characterize the set of properties identified. For example, modulus can be measured in different ways and selecting the most appropriate methodology can depend on what aspect of the material performance is of interest. In this case, both storage modulus and flexural modulus were studied. As mentioned previously for this set of formulations, there is a trade-off between certain properties, for example elongation has an inverse relationship with modulus, while adhesion tends to be proportional to modulus. This is depicted in Figure 3. This interdependence can cause some limitations when trying to systematically control one variable without changing the others.

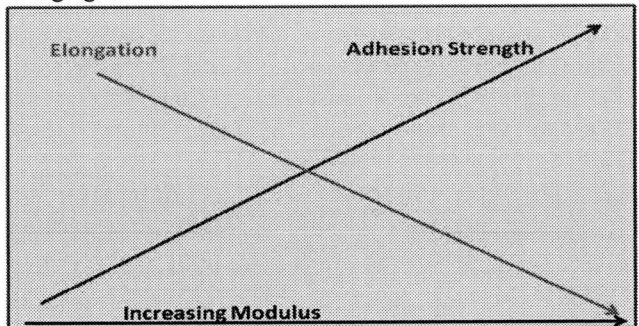

Figure 3. Relationship of elongation and adhesion strength with modulus

One of the challenges was the measurement of elongation to break. For most of the early stages of the project, elongation testing was done following a standard test such as ASTM D412-C. The standard elongation measurement techniques require a dog-bone sample (on the order of 2mm thick) to be pulled in tension. Figure 4 shows the relationship between modulus and elongation by this elongation measurement method. However, for TIM1 applications, it is not unusual to have thickness dimensions on the order of 0.030 to 0.070 mm. Therefore it can be misleading to extrapolate such bulk property-level testing from standard tests to situations where much thinner films are used. An alternate method of elongation testing was developed to address this thin film elongation, which will be discussed later.

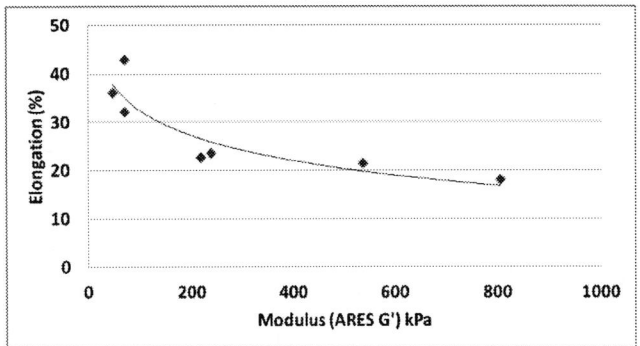

Figure 4. Elongation versus storage modulus

Additionally, many of the prototype formulations had very low modulus values and thus, were very fragile -which posed problems during sample preparation, and when clamping and pulling at high speeds in tension. Therefore, some special techniques had to be developed such as molding individual dog-bones instead of stamping them from a slab, creative de-molding methods, utilizing alternate means of clamping, and changing the test speed to slower than the standard technique. Figure 5 shows an alternate clamping method which involved using pieces of the same material to help sandwich the dog-bone and proved beneficial to improve data quality.

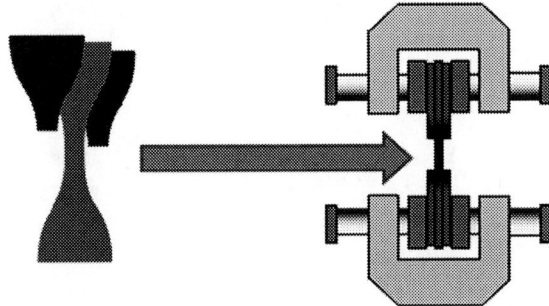

Figure 5. Improved clamping technique.

Adhesion testing was the second major challenge for this project. Typically for a silicone material, the preferred failure mode is to have cohesive failure so that the adhesion of the material to each interface is stronger than the cohesive strength of the material [7]. This can be validated by inspecting both substrates that were bonded together after a destructive test. A cohesive failure mode means both substrates exhibit residual material on them. Conversely, when there are areas with no material left on the substrate, then that is indicative of adhesive failure. Adhesive failure mode can range from partial failure to either substrate or complete adhesive failure to one of the substrates. Examples of these failure modes are shown in Figures 6 and 7.

Figure 6. Cohesive failure   **Figure 7.** Adhesive failure

One difficulty encountered when assessing failure mode is selecting the type of stress to apply in order to drive the test specimens to failure. Some of the most common methods involve applying shear forces; the lap shear test or die shear method are examples of this. The die shear specimen typically consists of a square piece of bare silicon (10x10mm) bonded using the silicone TIM to a representative substrate, in this case a metal lid commonly used in flip chip applications. The silicon/silicone/lid sandwich is cured and tested. Depending on the apparatus, the silicon die is then sheared off using a pushing mechanism, either horizontally or vertically and the stress at break is recorded. The silicon and lid pieces can then be inspected to assess the failure mode. As mentioned previously, the die shear adhesion strength trends along with modulus (shown in Figure 8).

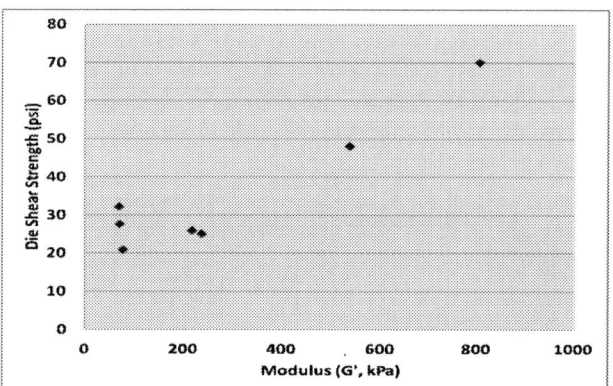

Figure 8. Die shear adhesion versus modulus

Further modifications were subsequently made to enable improved consistency for the die shear adhesion test results. These items included optimizing cleaning techniques, applying a consistent force when compressing the specimens, curing under controlled conditions, trimming the fillets before testing, and using uniform testing parameters. One key improvement was utilizing silicon pieces that are more representative of the actual size of the test vehicles. For example, instead of using small 10x10 mm silicon squares mentioned previously, the improved die shear specimens had a

larger surface area which required modified tooling to test accurately. Die shear adhesion strength was successfully demonstrated using these much larger silicon pieces. Some examples of this set-up are shown in Figure 9 below.

Figure 9. Large die shear specimens, before and after die shear testing

The die shear test proved to be valuable for assessing long-term stress effects on the thermal interface. An example relationship for high temperature bake stress conditioning of four different formulations can be seen in Figure 10.

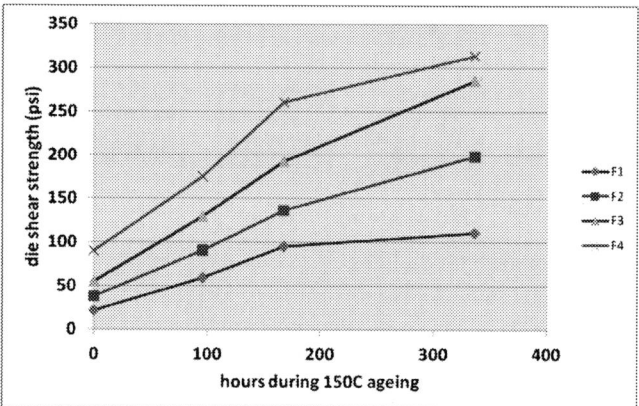

Figure 10. Die shear test to characterize effects from high temperature exposure

Additionally, after consulting with experts, it was determined that although shear forces play a major role for the types of stresses this type of package experiences, other forces may be relevant. Therefore, some experiments were performed using a stud pull testing technique to introduce tensile force destructive testing in order to observe any differences in break stress and/or failure mode. Figure 11 shows some of the results from this testing.

In general, the die shear method consistently results in lower values for this type of material, but typically is less noisy and more efficient and is therefore commonly followed Note however that there is a difference in failure mode between the methods for batch B. This is an example of a false positive with die-shear testing only for TIM adhesion evaluation. Complementary data from stud pull may provide data better suited to the application.

	Die Shear		Stud Pull	
	Max load (psi)	Failure mode	Max load (psi)	Failure mode
Batch A	13	CF	20	CF
Batch B	27	AF	49	CF

Figure 11. Die shear versus stud pull results

Another interesting aspect of the stud pull method is that one can measure the stress strain curve and calculate the strain to break. Since the sample is a similar geometry as a real application with a more realistic bond line thickness (nominally 50 microns used here) than the typical elongation method mentioned previously (2mm) it is thought to be perhaps a more appropriate method of characterizing elongation. Comparing the two techniques using one formulation, the standard method indicates a value of about 50%, whereas the thin film technique is closer to the 190% range. Figure 12 shows examples of these results. More work is planned to further investigate this technique.

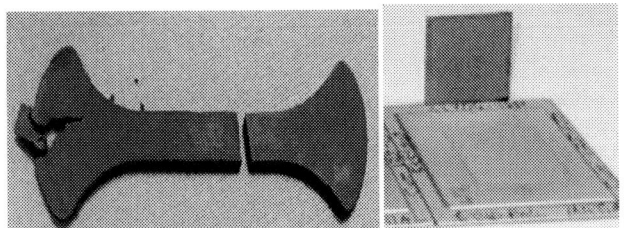

Figure 12. Examples of standard (left) and modified (right) elongation test specimens

In order to be useful to the end-customer, the material must also be easy-to-use. It was designed to be a one-part, cold storage material for use on industry-standard dispense equipment. The trade-off between loading with high amounts of filler in order to maximize thermal performance has to be countered against the ability to maintain easy dispensability.

Numerous trials were performed to ensure the various formulations of interest would not exceed this range. Additionally, the material was designed to be void-free. Voiding can originate from many sources such as poor dispense set-up, incorrect dispense patterns, improper cure profile, etc. When the dispense process is fully optimized, the cure profile can contribute to voiding and perhaps even delamination. Voiding and/or delamination can be assessed using a non-destructive technique called Scanning Acoustic Microscopy (SAM) [2].

Voiding and delamination issues can be difficult to trouble-shoot since they can manifest long after the original root cause of the problem. This issue was encountered when trying to optimize the cure profile for one such formulation. The SAM images were taken after the test vehicles were assembled and they showed similar desirable images between the two different cure profiles. However, after reflow, the SAM images show totally different results, with the standard (non-optimized) cure process exhibiting some unattractive SAM images. Figure 13 shows an example of this.

However, a potential source of voiding that end-users will usually assume, may come from the formulation itself. Therefore this material must be engineered to minimize as many sources of voiding as possible. In order to achieve zero or near-zero voiding, the formulation must be highly purified and one technique to assess this is Thermogravimetric Analysis (TGA).

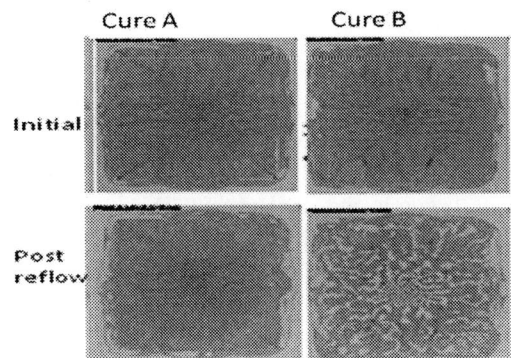

Figure 13. Cure A (left) vs Cure B (right)

At the earliest stages of this project, it was recognized that high purity raw materials will be needed, and extensive analytical work to verify this by TGA was conducted. Figure 14 shows a set of overlaid isothermal TGA graphs of one high purity formulation prototype. Conditions include both air and in nitrogen atmospheres and material in the cured and un-cured states (for > 3000 minutes).

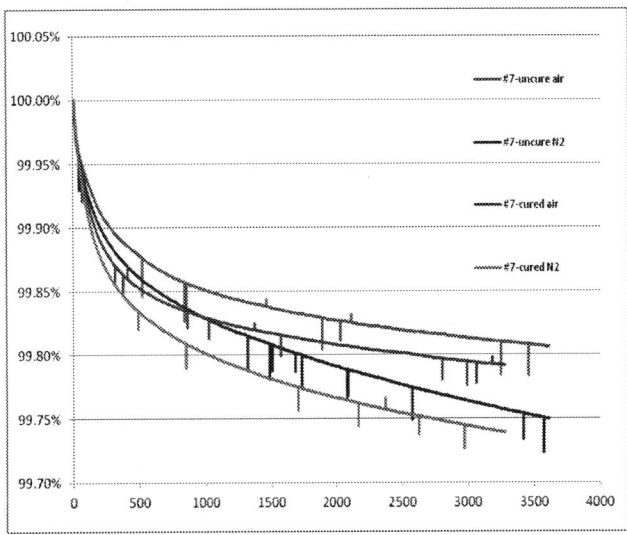

Figure 14. Thermogravimetric Analysis

Thermal Testing and TTV Validation

One of the controversial topics in the field of thermal interface material evaluation is how much to rely on bulk thermal characterization of the material itself. Many end-users rely solely on the published bulk thermal conductivity data from suppliers to pre-screen materials for their thermal applications. While this is generally a good protocol, there are potentially many pitfalls in following this approach since bulk thermal conductivity may not be representative of real-life applications. Additionally, there are many different techniques to characterize thermal conductivity.

For this project, versions of ASTM D5470 were used. This method is a steady state technique that involves placing a solid sample of fixed dimension between two temperature-controlled plates. One plate is heated while the other is cooled and temperatures of the plates are monitored until they stabilize. The steady state temperatures, the thickness of the sample and the heat input to the hot plate are used to calculate the thermal conductivity (see Figure 15). Results based on this characterization method for a typical formulation are shown in Figure 16.

Dow Corning also uses a laser flash method, but the majority of this project focused on the 1-D conduction method described above.

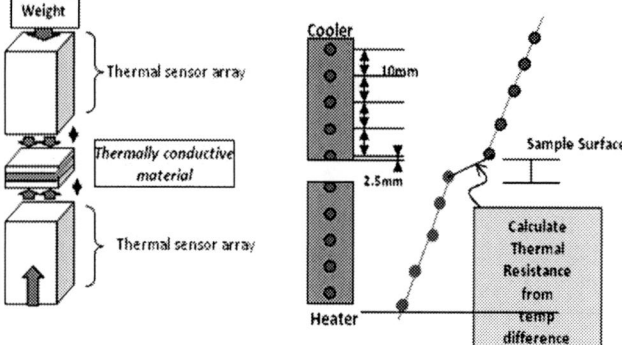

Figure 15. Thermal test set-up

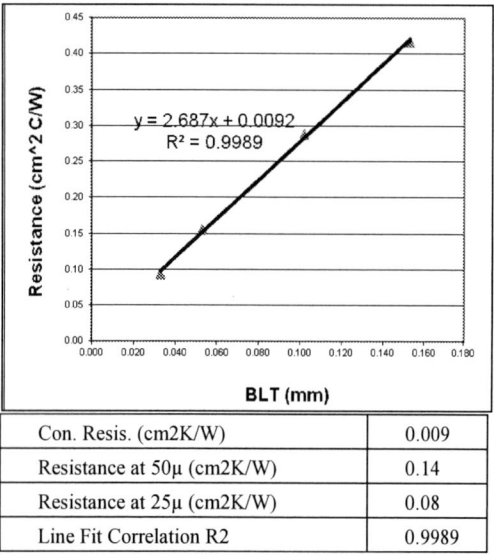

Con. Resis. (cm2K/W)	0.009
Resistance at 50µ (cm2K/W)	0.14
Resistance at 25µ (cm2K/W)	0.08
Line Fit Correlation R2	0.9989

Figure 16. Thermal testing results

However, as emphasized earlier, these thermal measurements offer only a guidance as to whether the material might perform well enough in the application -if it can survive the assembly process. To know how it actually performs initially and after being exposed to various stress conditions, the best practice is to have thermal test vehicles. Once several rounds of formulation range-finding experiments were carried out to target the desired properties, several formulations were selected for TTV testing.

These TTVs were assembled using realistic manufacturing conditions and subjected to industry-standard, JEDEC stress conditions (shown in Figure 17).

Preconditioning Simulated Reflow	JEDEC level 1-3
GroupB testing	-40-60C T/S Shock and Vibration
Thermal Cycling	-55-125C for 1000 cycles
Temperature Humidity	85C/85%RH for 1000 hours
Temperature Aging	150C for 1000+ hours

Figure 17. Stress conditions conducted for TIM studies

There were extensive TTV builds completed to date, and some of the results from seven different formulations are shown in Figure 18. Note that these results are at time-zero, or before subjected to stress conditions mentioned above. Selected formulations were then identified for additional stress conditions to assess changes in performance.

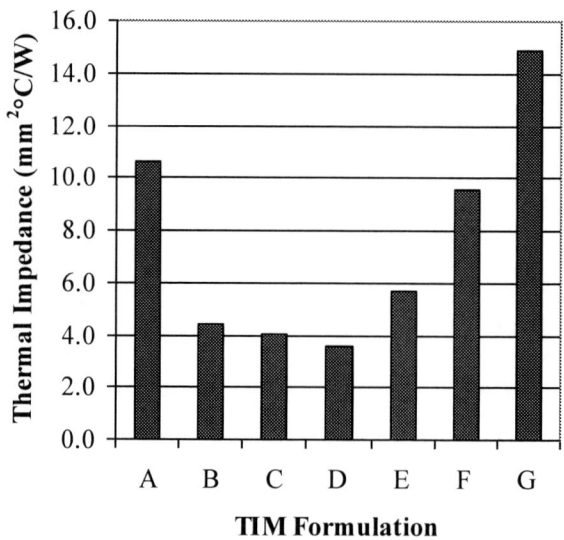

Figure 18. Initial (prior to stress) thermal performance of seven different TIM formulations in TTVs

Several of these formulations show reasonably good thermal performance. Two formulations ((material prototypes B and D) were selected for additional stress testing. Both of these formulations exhibited thermal stability when carried through for the thermal cycling stress conditions as well as for high temperature bake stress condition. The results of which are shown in Figures 19 and 20, respectively.

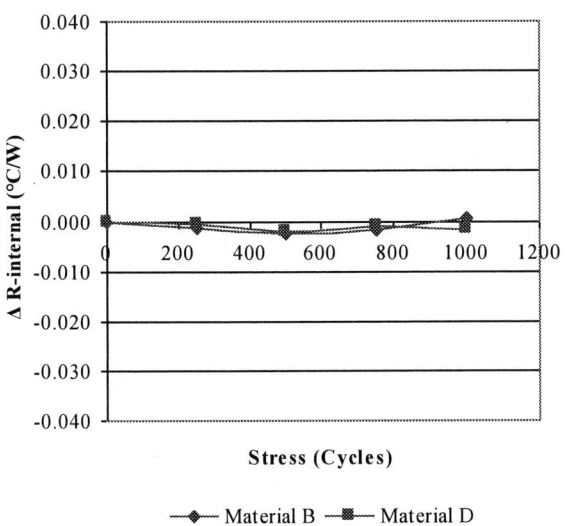

Figure 19. Thermal stability of selected TIM formulations in TTVs for thermal cycling stress condition.

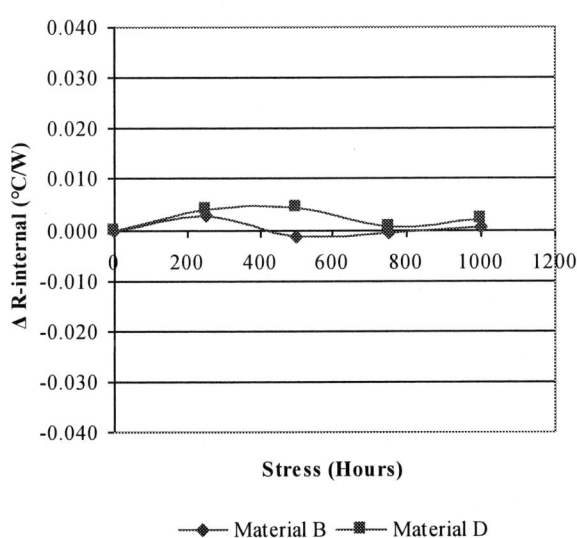

Figures 20. Thermal stability of selected TIM formulations in TTVs for high temperature bake stress condition

Conclusions

The project team has identified a set of physical properties that represents important performance indicators and subsequently has developed techniques using existing and modified instrumentation and procedures to characterize these properties. Systematic sets of prototype formulations targeting the range of material properties desired were prepared and a select set of these samples were assembled into thermal test vehicles and tested. Many of the property

assumptions were validated and went into the learning cycle for additional iterations of studies. This rigorous approach to understanding the property and performance was instrumental in order to improve upon the target properties for an engineered thermal interface material.

This engineered TIM exhibits balanced adhesion, elongation, modulus, and thermal impedance performance in the range of 0.04 to 0.07 cm^{2}C/W with a tailored rheology for easy dispensing and optimized for other favorable processing characteristics.

Acknowledgments

The authors would especially like to thank additional people who contributed to the project: Dorab Bhagwager, Adriana Zambova, Marilyn Jensen, Afrooz Zarisfi from Dow Corning and Tuhin Sinha and Kamal Sikka from IBM. The authors are also thankful for all the contributions and the great lab support at IBM and Dow Corning throughout this effort.

References

1. C. A. Harper, *Electronic Packaging and Interconnection Handbook*, McGraw-Hill, New York, 2000, pp. 1-1.
2. M. Stern et al, "Evaluation and Inspection of Lid Attach for Advanced Thermal Packaging Materials", IPACK2007-33629, ASME, July, 2007.
3. Perfecto, E., *Flip Chip Fabrication and Interconnection*, ECTC 2011 PDC #14, p.3.
4. Harper, C. A., *Electronic Packaging and Interconnection Handbook*, McGraw-Hill, New York, 2000, pp. 2-37 to 2-41.
5. R.S. Prasher and J.C. Matayabas, "Thermal Contact Resistance of Cured Gel Polymeric Thermal Interface Material, *IEEE Trans. CPMT*, vol. 77, no. 4, pp. 702–709, Dec. 2004
6. Perfecto, E., *Flip Chip Fabrication and Interconnection, ECTC 2011 PDC 14*, p.85.
7. J. Tonge, et al. "Adhesion and Cure Mechanism Studies for Advanced Lidded Flip Chip Applications," IMAPS Proceedings, 2013, pp. 101-105.

Degradation Mechanisms in Electronic Mold Compounds Subjected To High Temperature in Neighborhood of 200°C

Pradeep Lall[1], Shantanu Deshpande[1], Yihua Luo[1],
Mike Bozack[1], Luu Nguyen[2], Masood Murtuza[3]

[1]Auburn University
NSF-CAVE3, Electronic Research Center
Department of Mechanical Engineering
Auburn, AL. 36849
[2]Texas Instruments, Santa Clara, CA, 95052
[3]Texas Instruments, Stafford, TX, 77477
Tele: (334)-844-3424, Email: lall@auburn.edu

Abstract

Plastic encapsulated microelectronics (PEMs) has found wide spread applications in automotive environments for varied roles. Transition to hybrid electric vehicles and fully electric vehicles has increased the trend towards greater integration of electronics in automotive under hood environments. Electronics in such applications may be mounted directly on engine and on transmission. Electronics under hood may be subjected to temperatures in neighborhood of 200°C. Commercially available PEMs are able to operate in the neighborhood of 175°C. However, sustained operation at temperatures of 200°C or higher is beyond the state of art. Materials and processing techniques needed for sustained high temperature operation for 10 years and 100,000 miles of vehicle operation are yet unknown. There is need for studies for understanding the failure mechanisms of PEMs at sustained high temperature. In this paper, new approach is discussed to study physical and chemical stability of molding compound when it is subjected to very high temperature for prolonged duration. Four mold compound candidates were selected for test purpose. They were subjected to thermal aging at 200°C and 250°C, for 5000 hours. For degradation study, bulk mold compound specimens as well as 20 pin SOIC devices, encapsulated with MC candidates were used. Test vehicle was bonded with gold wires, and Pd coated Al pad. For bulk mold compound samples, weight loss test, DMA, FTIR, XPS tests were performed at fixed time intervals. To study integrity of SOIC devices, resistance spectroscopy, x-ray inspection and current leakage tests were selected. Another set was subjected to 120 hours of aging at 130°C/100%RH condition to check leakage current. Performance of MC candidates at high temperature was evaluated using all these tests. Sensitivity of each test towards detecting degradation of EMC's is also discussed and most effective tests are suggested.

Introduction

Plastic encapsulated microelectronics (PEMs) has found wide spread applications in automotive environments because of their cost, availability and functionally advantages [McCluskey 2000]. These have driven electronic manufacturers to use them for varied roles including collision avoidance systems, antilock braking systems, lane departure warning systems, and cruise control. Transition to hybrid electric vehicles and fully electric vehicles has increased the trend towards greater integration of electronics in automotive under hood environments. Electronics in such applications may be mounted directly on engine and on transmission. Electronics under hood may be subjected to temperatures in neighborhood of 200°C. Military and aerospace applications require reliable operation over an extended period of time at high temperatures. Electronic mold compounds (EMC) in addition to providing protection for the silicon chip, are required to meet U-94, plastic flammability standards. In order to meet these conditions flame retardants are added into molding compounds, which at high temperatures can work as catalyst and accelerate process of failure of bond wires [Harman 1997; Noolu 2004; Uno 2000]. Presence of oxygen in MC's causes thermo-oxidation decomposition and accelerates decomposition rate [Rappoport 1995]. Commercially available PEMs are able to operate in the neighborhood of 175°C. Thermal degradation causes major changes in physical and chemical properties of molding compound, which can severely affect reliability of electronic devices [Gojun Hu 2006; Teverovsky A 2005]. Performance of molding compound and its effect on ball shear strength was evaluated at 150°C and 200°C for 1000 hours by [Chen 1995]. Thermal degradation of MC causes breakdown of long polymer chains and releases free radicals in the process which may corrode the internal package architecture or act as a catalyst in the process [de Vreugd 2009; Ng 2007]. During the degradation, oxidation layer formed in the molding compound has a significant effect on the mechanical properties of EMC, accompanied with the build-up of thermal stresses in packages [Dandong Ge 2011; Daoguo Yang 2011]. Interface between EMC and lead frame, when subjected to high temperature bias, due to presence of high stresses has been found to form micro-cracks. These micro-cracks degrade package reliability by eventually propagating into major cracks in the molding compound causing catastrophic failures [Ma 2007; Chen 2011]. However, package architectures capable of reliable sustained operation at temperatures of 200°C or higher is beyond the state of art. Materials and processing techniques needed to design for sustained high temperature operation for 10 years and 100,000 miles of vehicle operation are yet unknown. There is need for studies for understanding the failure mechanisms of PEMs at sustained high temperature. The development of high temperature PEMs is further complicated by the fact that the automotive industry is in the middle of a transition to copper-aluminum wirebond system. In this paper, new approach is discussed to study physical and chemical stability of molding compound when it is subjected to very high temperature for prolonged duration. Four mold compound candidates were selected for test purpose. Different tests were performed on bulk mold compound samples and SOIC devices encapsulated with MC candidates. Physical and chemical changes were tracked down with different techniques. From the results of tests, best performing molding compound is reported.

Test Vehicle

Two types of test vehicles were used in this study. The first test sample geometry was a bulk mold compound sample, and second test sample geometry was 20 Pin SOIC devices molded with mold compound candidate. In all, four mold compounds were evaluated, labeled as A, B, C, D. The first sample geometry was fabricated in two versions including round and square. The round sample shown in **Figure 1** was 50 mm in diameter, and 3mm in thickness. The round samples were used for weight loss analysis. Square samples were 25.4 mm in length, 25.4 mm in width, and 0.6 mm in thickness as shown in **Figure 1**. Mold compound-A is a high temperature epoxy molding compound capable of low warpage, and excellent reflow crack resistance. Mold compound-B is a higher temperature version of the mold compound–A. Mold compound-C is a multi-aromatic epoxy with multi-aromatic hardener and low-alpha filler with a filler content of 85%. Mold compound-D is a silicon encapsulation material capable of high electrical resistivity, high thermal and mechanical stability.

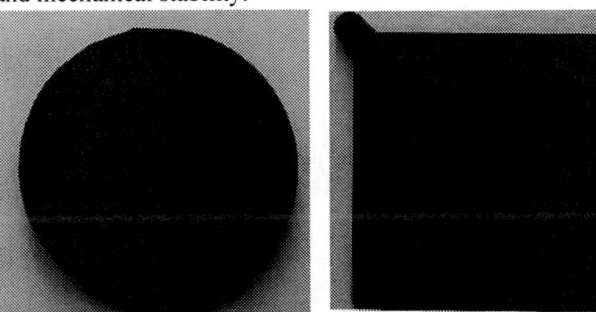

Figure 1 . Mold Compound Samples.

Figure 2. 20 pin SOIC device.

To study effect of molding compound degradation, 20 pin SOIC devices, were molded with mold compounds A, C and D. Devices were wire-bonded with gold wirebond on a Nickel-Palladium coated Aluminum pad. The test devices were partially daisy chained. Only 8 out of 20 legs were wire bonded and connected to each other for resistance monitoring, while others legs were used to monitor leakage current if any. Optical microscopic images are shown in **Figure 2**. X-ray image is shown in **Figure 3**. 3D reconstructed model is shown in **Figure 4**.

Figure 3. X-ray Image of SOIC Device.

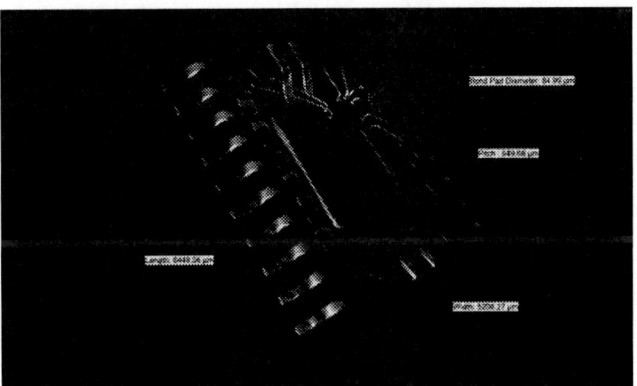

Figure 4. µCT Reconstruction of SOIC Device.

Approach

The polymeric encapsulants intended for high temperature operation must possess adequate mechanical strength, adhesion, chemical resistance, electrical resistance, high moisture and temperature resistance for prolonged periods of time. To study effect of material parameters on high temperature storage reliability of packaged devices, molded packages (20 pin SOIC) were manufactured with the EMC candidates. In order to isolate the effect of mold compound bulk molded samples were studied as fabricated as well. Both were subjected to thermal aging for prolonged duration and properties, including weight loss, change in modulus, changes in chemical composition of bulk MC were measured. Wirebond continuity, changes in microstructure, MC cracking and delamination were tracked in 20 pin SOIC devices over the test duration.

Test Matrix

One set of test vehicles was subjected to thermal aging at 200°C, and another at 250°C. Weight loss analysis of bulk mold compound was performed with weighing scale of resolution 0.1mg at a reading interval of one week. Samples were taken out at the interval of 1000 hours for FTIR (Nicolet IR 100 model) and XPS (Kratos XSAM 800 model) test from thermal chamber. Both tests were performed to track changes in bond structure of molding compound. DMA test was performed to measure change in Tg and modulus. The 20 pin SOIC devices were cross-sectioned at the interval of 1000 hours, to check wirebond integrity. Changes in microstructure of molding compound were monitored at the same time. In order to study cracking, and delamination, visual inspection, X-ray analysis were used. Resistance spectroscopy technique, a highly sensitive method was used to measure resistance of devices when they were subjected to thermal aging at the interval of 500 hours. After aging for 5000 hours, the packages were subjected to 130°C/100%RH condition for 120 hours and tested for leakage current.

Bulk Mold Compound Test-Results

(1). Weight Loss Test

Weight loss test was performed on bulk molding compound samples. Round samples were used for the weight loss test. Prolonged exposure to very high temperature results in degradation of the binding material in molding compound producing a gradual loss of weight. Figure 5 and Figure 6 show plots of percentage weight loss and aging duration, for mold compound candidates for thermal aging at 200°C and 250°C. Test on mold compound candidate "A" was stopped after 888 hours because of excessive weight reduction. For mold compounds B, C, and D, changing slopes of graphs shows that more than one mechanism is causing weight loss. For all molding compounds for both 200°C and 250°C aging conditions, initial rate of weight loss is higher as compared to rate of weight loss at later stages. After approximately 1000 hours of aging, weight loss trend follows Fickian diffusion model. It is hypothesized that the initial rapid weight reduction might be because of combination of diffusion and loss of volatile matter present in the molding compound. Exposure to 200°C, for 5000 hours of aging showed that candidate "C" had minimum weight loss of 0.95%, while B and D had lost 2.08% and 1.35% of original weight. Exposure to 250°C for 5000 hours of aging at produced higher weight loss with mold compounds B, C and D exhibiting weight loss of 14.1%, 7.47%, 1.75% respectively. Rate of weight loss of epoxy mold compound A was highest, followed by its high temperature variant mold compound B for both the test conditions. Initial weight loss rate of non-aromatic epoxy mold compound-C was lower than the silicone mold compound-D, but after 750 hours of aging, this rate decreased rapidly for D, which exhibited the lowest weight reduction. Thus, the silicone molding compound "D" was much more stable than other candidates for HTSL.

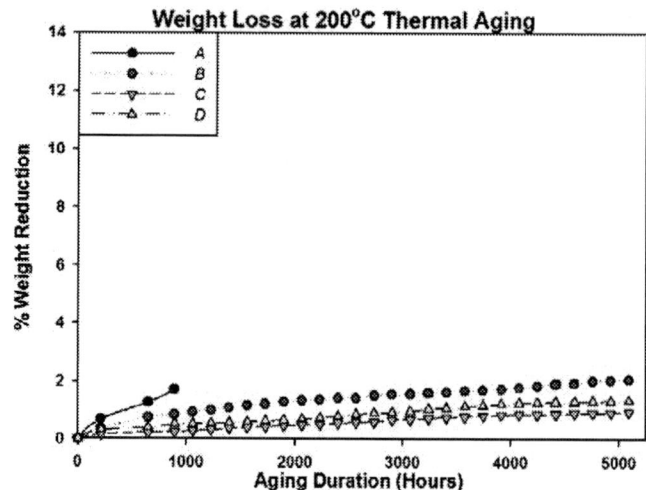

Figure 5 – Weight loss of Epoxy Mold Compounds A, B, C, and D at 200°C

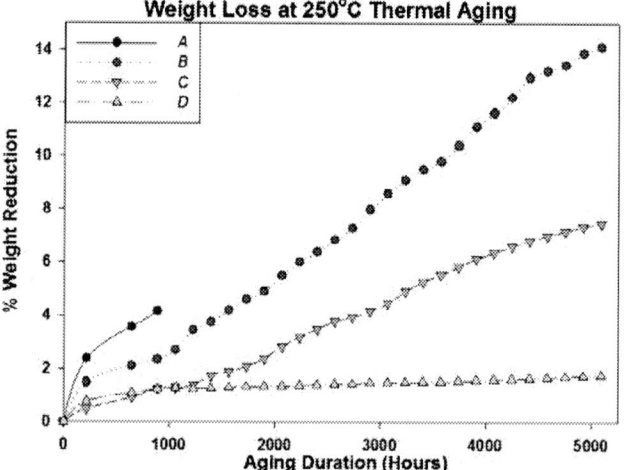

Figure 6 – Weight loss of Epoxy Mold Compounds A, B, C, and D at 250°C

(2). FTIR Test

FTIR technique has been used by previous researchers to distinguish mold compound types and find structural changes during curing process [Ng 2006]. The FTIR for the EMC has been run by preparing the mold compound in KBr. EMC samples were powdered and then mixed with KBr powder to form pallets, on which actual test was performed. Then the spectrum of KBr has been run to provide the background spectrum and the KBr peaks subsequently removed from the composite spectrum. This technique was also used to find moisture content in molding compounds [Chen 2011]. The peaks of the FTIR spectrum have been identified in Fourier Transform Infrared Spectroscopy (FTIR) technique was used to track changes in bonding structure of molding compound up to 5000 hours of aging (Figure 8, Figure 9, Figure 10).

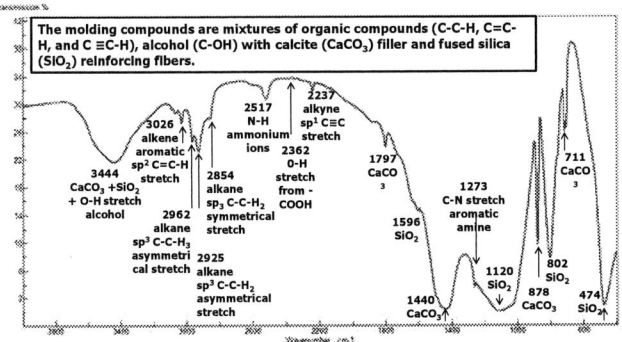

Figure 7 – Peak identification in the FTIR spectrums of EMC candidates.

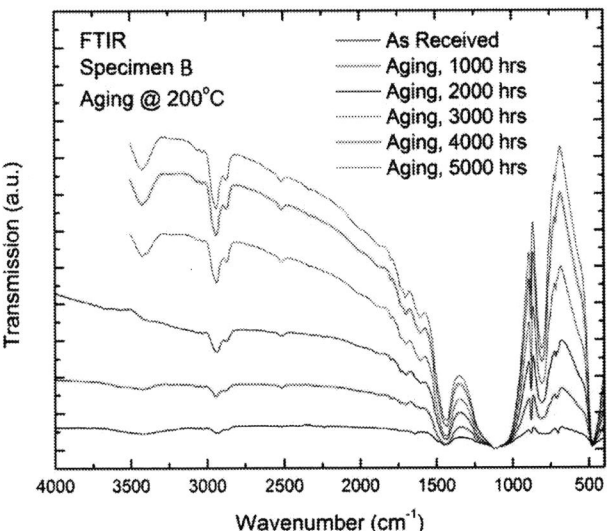

Figure 8 – FTIR spectrums of EMC Candidate-B subjected to thermal aging at 200°C.

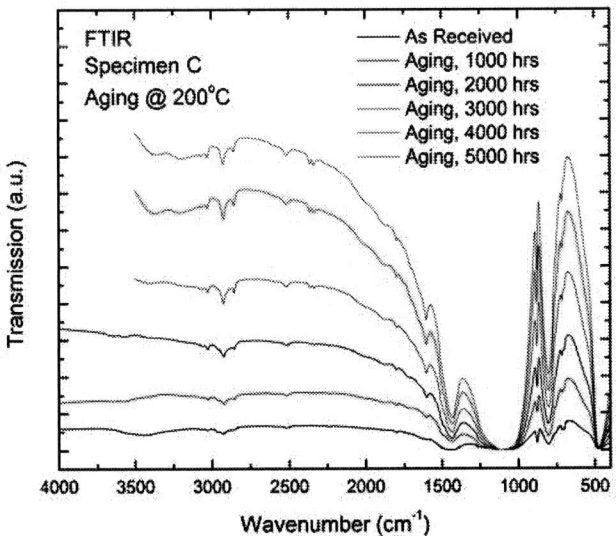

Figure 9 – FTIR spectrums of EMC Candidate-C subjected to thermal aging at 200°C.

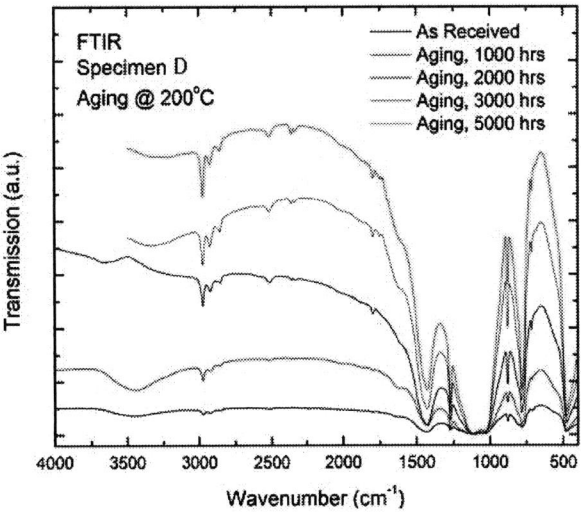

Figure 10 – FTIR spectrums of EMC Candidate-D subjected to thermal aging at 200°C.

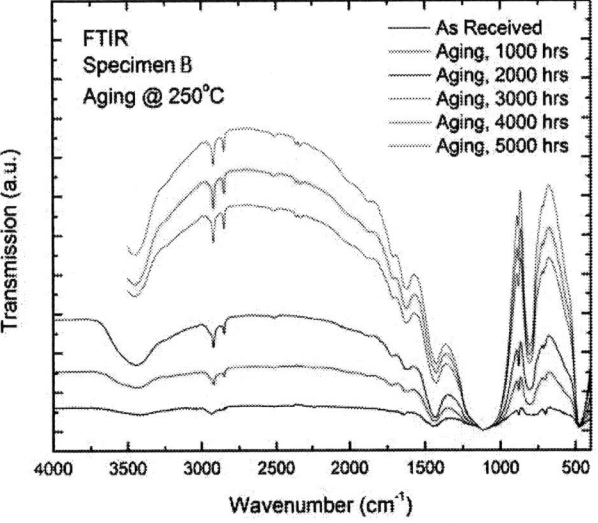

Figure 11 – FTIR spectrums of EMC Candidate-B subjected to thermal aging at 250°C.

The FTIR uses an interferometer with a beam splitter which takes an incoming infrared beam and divides it into two optical beams. One beam reflects from a stationary mirror and the second reflects from a mirror mounted to a mechanism that is capable of moving a few millimeters away from the beam splitter. The two beams are recombined in the beam splitter. The path of one beam is fixed and that of the second beam is constantly changing. The signal called the interferogram from the interferometer is result of interference of the two beams. The interferogram has been decoded using a fourier transformation of the input signal. Every bond has its own natural frequency. When the external excitation frequency (IR light) matches with natural frequency, resonance occurs, that causes rapid vibration of bond. There is specific energy bond for different modes of vibration for different bonds. During the resonance, this specific energy is absorbed by material. This phenomenon has been used to find changes in bonding structure on the molding compound.

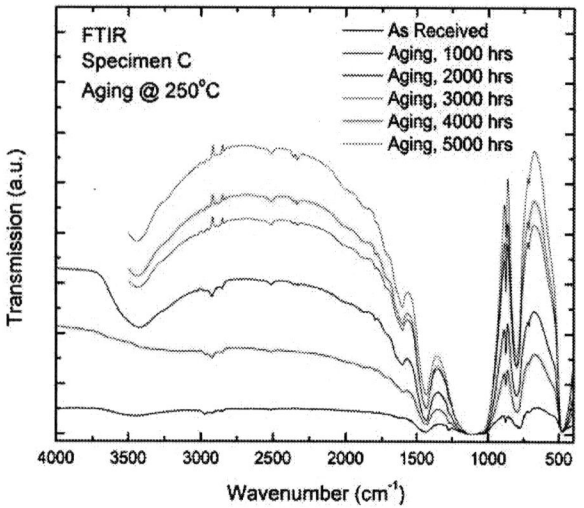

Figure 12 – FTIR spectrums of EMC Candidate-C subjected to thermal aging at 250°C.

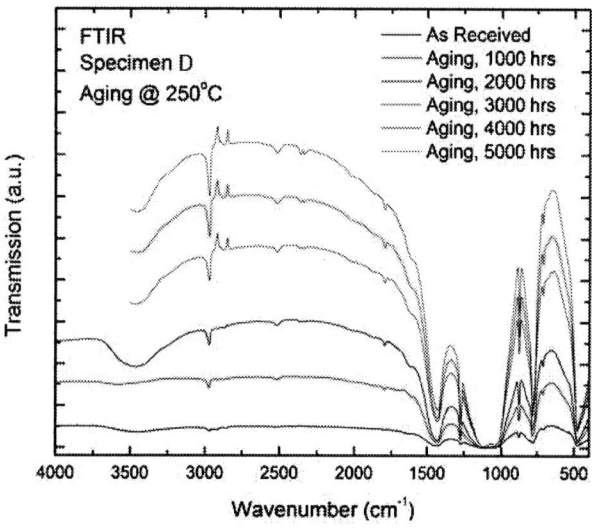

Figure 13 – FTIR spectrums of EMC Candidate-D subjected to thermal aging at 250°C.

Figure 8 to Figure 10 shows FTIR spectrum of MC candidates B, C, and D, subjected to thermal aging at 200°C respectively. Figure 11 to Figure 13 shows FTIR spectrum of MC candidates B, C, and D, subjected to thermal aging at 250°C. For both cases, 200°C and 250°C thermal aging, no major changes were observed in the $CaCO_3$ and the SiO_2 peaks. Concentration of alkane group was found to be increasing, while other, like carbonyl, calcite, and fused silica did not show any considerable change. Experiment was repeated multiple times to ensure accuracy of FTIR plots. Weight gain tests show that the degradation of the mold compound has started before any additional changes in the bonding structure are visible in the FTIR spectrum.

(3). DMA Test

Dynamic mechanical analyzer was used to track down change in glass transition temperature and modulus of molding compound, as a function of aging time. Researchers in past have used DMA technique to evaluate thermo-mechanical properties of molding compound [Hu 2006].

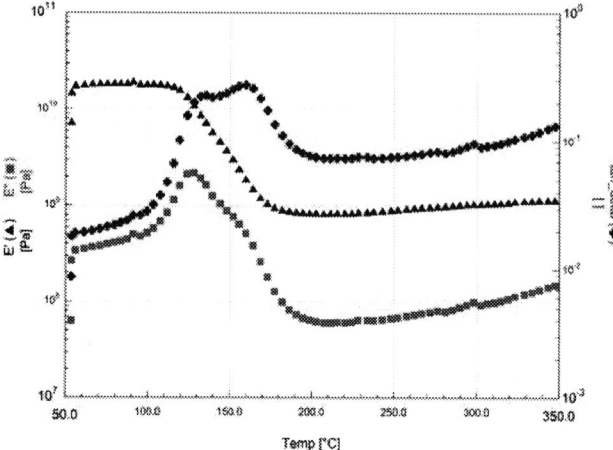

Figure 14 – DMA results for mold compound "C" for pristine samples.

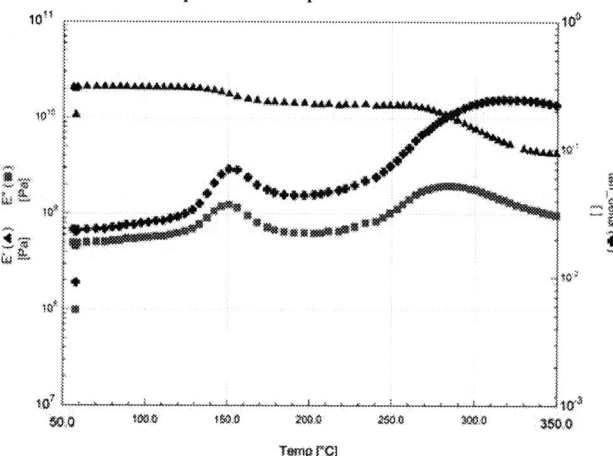

Figure 15 – DMA results for mold compound "C" thermal aging at 200°C for 1000 hours.

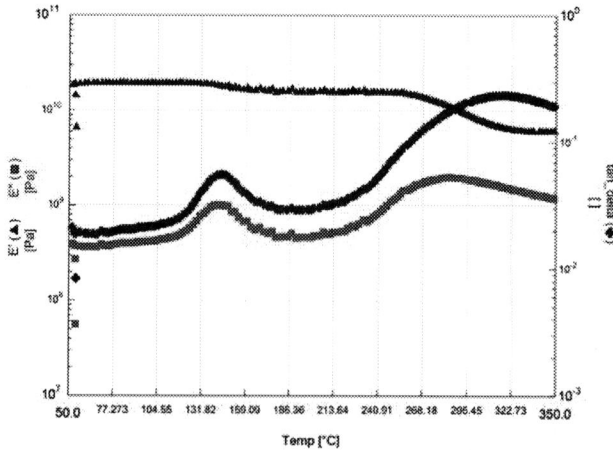

Figure 16 – DMA results for mold compound "C" thermal aging at 200°C for 2000 hours.

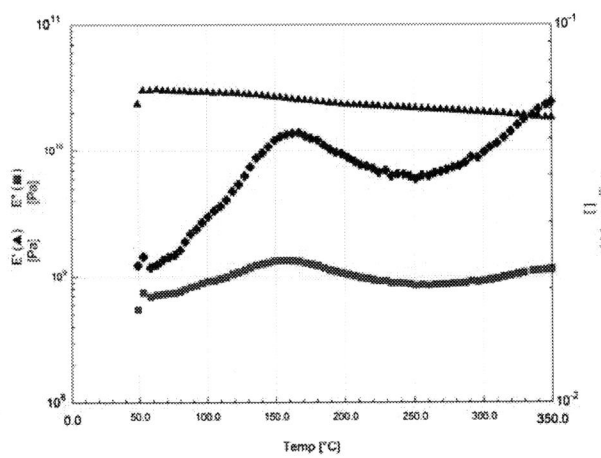

Figure 17 – DMA results for mold compound "C" thermal aging at 250°C for 1000 hours.

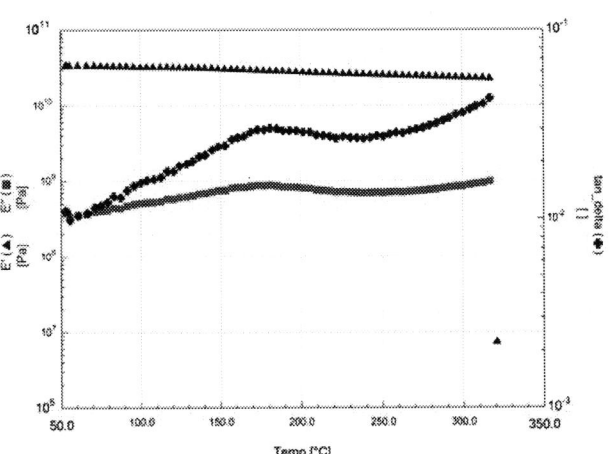

Figure 18 – DMA results for mold compound "C" thermal aging at 250°C for 2000 hours.

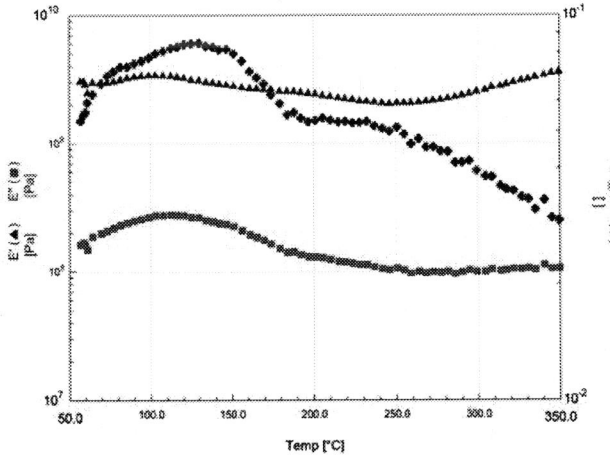

Figure 19 – DMA results for mold compound "D" for pristine samples.

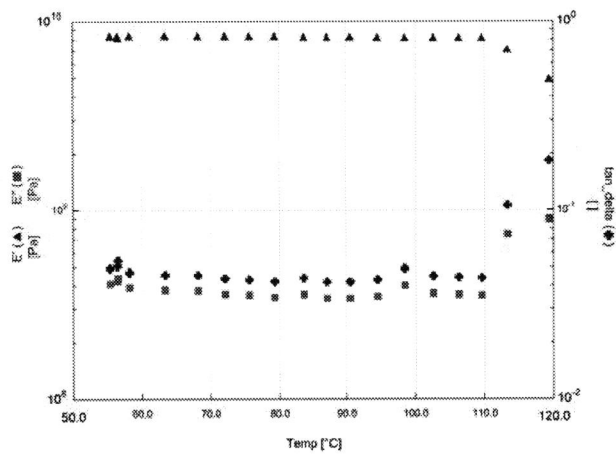

Figure 20 – DMA results for mold compound "D" thermal aging at 200°C for 1000 hours.

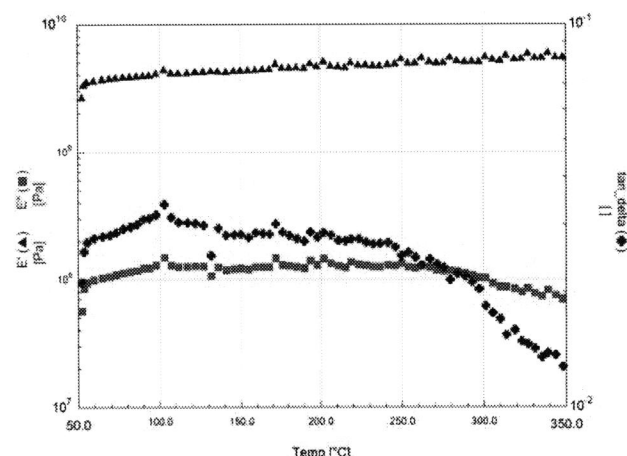

Figure 21 – DMA results for mold compound "D" thermal aging at 200°C for 2000 hours.

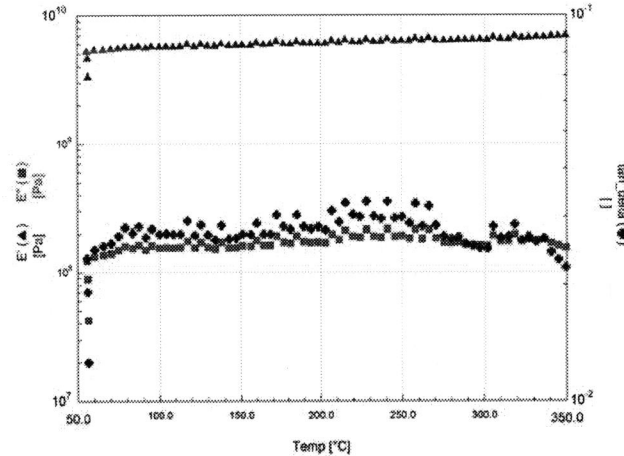

Figure 22 – DMA results for mold compound "D" thermal aging at 250°C for 1000 hours.

978-1-4799-2408-0/14 $31.00 © 2014 IEEE

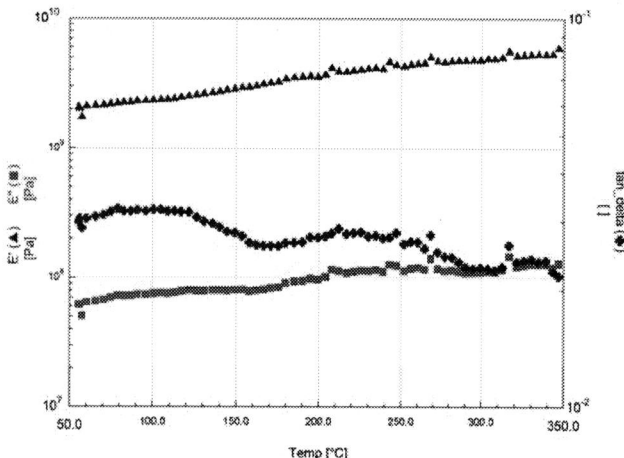

Figure 23 – DMA results for mold compound "D" thermal aging at 250°C for 2000 hours.

TA instruments RSA-III was used for the purpose of DMA measurements. Test was performed on mold compound samples C and D because they were performing better than A and B. Three point bending test, at constant frequency of 1Hz was used for all runs. Figure 14 to Figure 18 shows test result for mold compound C, subjected to 200°C and 250°C thermal aging. Figure 19 to Figure 23 shows test result for D, subjected to same conditions. In most cases, for aged samples, initial modulus found to be more than that of time-zero samples. This increase was higher for 250°C thermal aging than 200°C. For mold compound-C, two glass transition temperatures were detected, at 130.5°C and 159.4°C corresponding to the two peaks in the tan-delta plot for 200°C in pristine samples. The glass transition temperature of mold compound-C shifted to 150°C and 320°C after 1000 hours of aging at 200°C. The glass transition peak drifted further to 147°C and 322.73°C after 2000 hours of aging at 200°C. In the case of 250C thermal aging, the glass transition temperature shifted to 154.6°C after 1000 hours and then to 170°C after 2000 hours. The second T_g was not observed till 350°C for mold compound C. The drop in storage modulus above the glass transition temperature was observed for mold compound C samples aged at temperature of 250°C. For mold compound-D the initial T_g was detected at 132.6°C in pristine samples. In the case of aged samples of mold compound D, the T_g was not detected. The silicone based mold compound did not exhibit a drop in modulus above the glass transition temperature; rather the modulus increased with temperature. Test was stopped after 2000 hours of aging. It is hypothesized that at higher temperatures, the binding material detereorates while filler material (silicon oxide particles) remain stable. Once the glass transition temperature has been exceeded, the mold compound property changes because of changes in property of binding material. Breakdown of binding material causes increase in percentage silicon content, and promotes crosslinking of polymers - causing change in glass transition temperature, and increase in modulus, along with reduction of drop in modulus above the glass transition temperature. Mold compound-D exhibited no drop in modulus above the glass transition temprature, and the least weight reduction. The continuous increase in modulus

can be because of further crosslinking of polymers, and strong bonding between filler and binder material.

(4). XPS Test

X-ray photoelectron spectroscopy was used to track changes in bonding structure of surface of molding compound. Typically XPS is used to analyze surface composition, with depth of 10 to 100 Å. Mono-energetic X-ray beam is irradiated on sample which is kept in ultra-high vacuum. X-rays penetrate the surface, and dislodge one of the core electrons. The dislodged electron leaves the surface of object with certain kinetic energy.

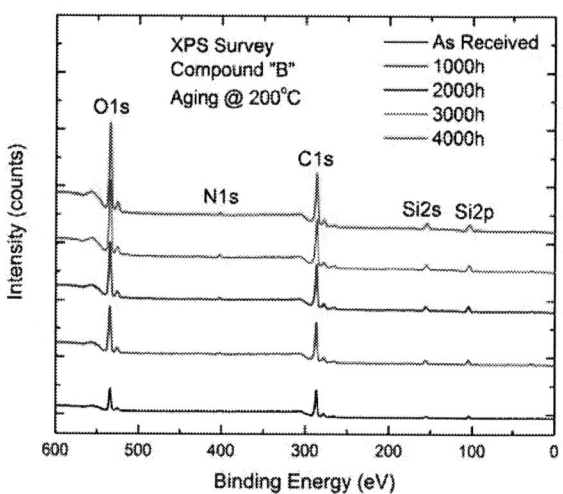

Figure 24– XPS results for mold compound-B subjected to 200°C

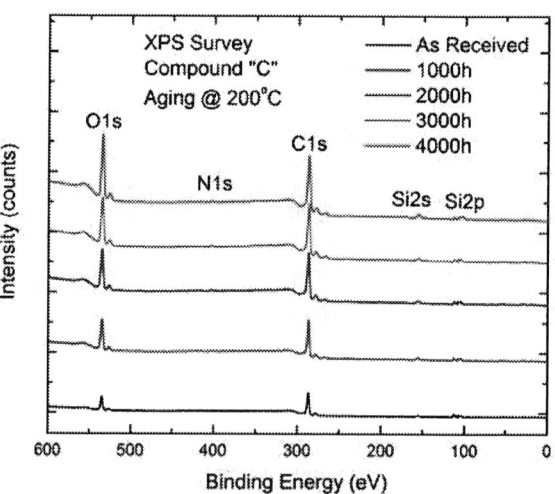

Figure 25– XPS results for mold compound-C subjected to 200°C

The kinetic energy of electron is measured by detector, and plot of number of counts (number of electrons with specific energy) is plotted against kinetic energy. Every elements core electron has its associated characteristic energy which is documented in reference manuals.

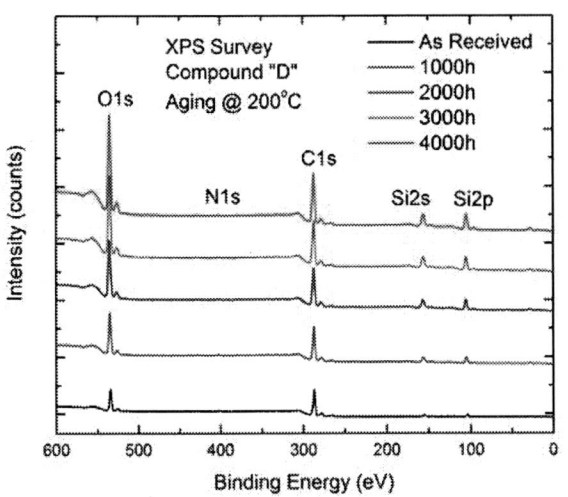

Figure 26– XPS results for mold compound-D subjected to 200°C

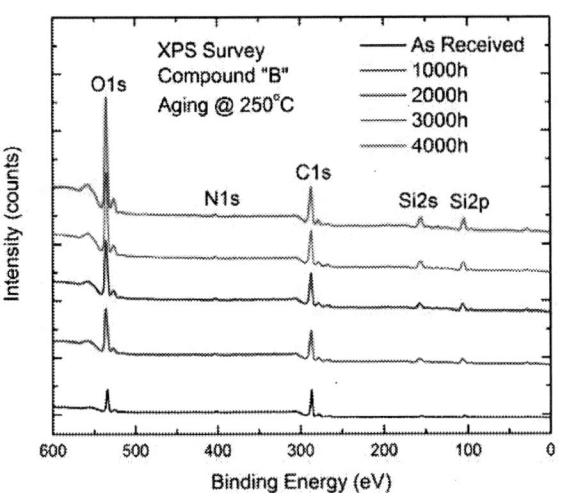

Figure 27– XPS results for mold compound-B subjected to 250°C

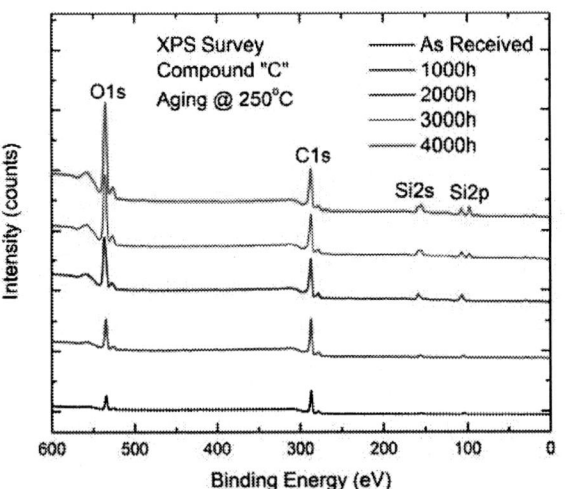

Figure 28– XPS results for mold compound-C subjected to 250°C

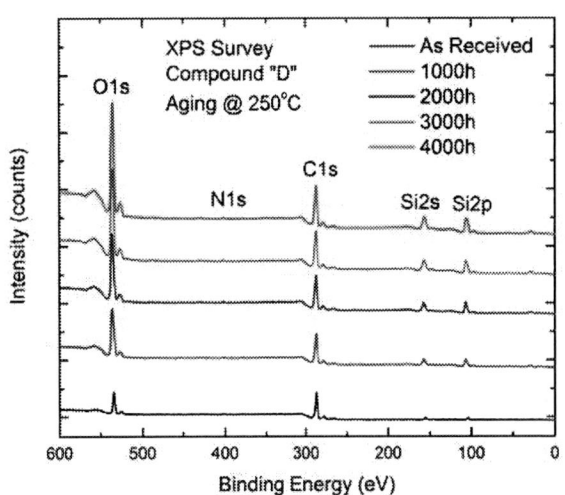

Figure 29– XPS results for mold compound-D subjected to 250°C

Table 1 – Composition Change in EMC candidate B Under Thermal Aging

Aging Time (Hrs)	200°C (Atomic %)			250°C (Atomic %)		
	% Oxygen	% Carbon	% Silicon	% Oxygen	% Carbon	% Silicon
Time 0	17	75	8	17	75	8
1000	30	57	13	42	38	20
2000	31	56	13	42	36	22
3000	36	52	12	53	22	25
4000	44	39	17	60	14	26

Table 2 - Composition Change in MC candidate C Under Thermal Aging

Aging Time (Hrs)	200°C (Atomic %)			250°C (Atomic %)		
	% Oxygen	% Carbon	% Silicon	% Oxygen	% Carbon	% Silicon
Time 0	16	80	4	16	80	4
1000	18	77	5	24	69	7
2000	25	67	7	49	29	22
3000	22	72	6	53	21	26
4000	38	46	16	55	19	26

Depending on the energy peaks, composition and bonding structure of the surface of the specimen can be determined. [Van Der Heide 2011] XPS test was conducted on MC samples at the interval of 1000 hours of thermal aging. Figure 24 to Figure 26 show XPS results from MC candidates B, C and D respectively subjected to 200°C. Figure 27 to Figure 29 show XPS results from MC candidates B, C and D respectively subjected to and 250°C isothermal aging respectively.

Table 3 - Composition Change in MC candidate D Under Thermal Aging

Aging Time (Hrs)	200°C (Atomic %)			250°C (Atomic %)		
	% Oxygen	% Carbon	% Silicon	% Oxygen	% Carbon	% Silicon
Time 0	31	45	24	31	45	24
1000	32	43	25	37	39	24
2000	39	37	24	45	28	27
3000	41	38	21	51	21	28
4000	40	38	23	50	25	25

Table 1, Table 2 and Table 3 shows change in composition of mold compound sample because of thermal aging. General trend for all molding compounds for both temperature conditions is loss of carbon, and gain in Oxygen and Silicon. Loss of carbon can be attributed to loss of binding material; resulting in weight loss of molding compound. Increase in oxygen is driven by oxidation of byproducts of polymer chain. Byproducts formed after degradation of polymer chain could be chemically reactive. The byproducts might react with atmospheric oxygen to produce oxides. Additional reason for the relative increase in oxygen and silicon is the lowering of carbon fraction.

Table 4 – Tabulated result of loss of carbon and gain in oxygen after 4000 hours of thermal aging

Mold Compound Specimen	% Loss of Carbon Content		% Gain in Oxygen Content	
	200°C	250°C	200°C	250°C
B	48	81	158	252
C	42	76	137	244
D	14	44	29	61

Table 4 shows percentage loss of carbon and gain in oxygen in molding compound samples after 4000 hours of aging when compared to time zero samples. Maximum loss of carbon was observed in case of MC candidate B, which was 48% and 81% for 200°C and 250°C aging conditions. MC candidate D showed minimum loss of carbon, while candidate C showed intermediate loss. Percentage gain in oxygen content was highest in MC candidate B, followed by C and least in D. Result of XPS test are in good agreement with other tests. After cross-sectioning the molding compound, D showed least amount of oxidation compared to others (shown in following sections), and in weight loss test (described in earlier section) D (least loss of carbon) performed better than C and B.

20-Pin SOIC Test-Results

(1). Resistance Spectroscopy Test

The resistance spectroscopy (RS) measurement is highly accurate method to measure resistance. It consists of a modified Wheatstone bridge and a probe station. RS is designed to detect and measure very small AC signals all the way down to a few nanovolts. It uses phase-sensitive detection technique so as to single out the component of the signal at a specific reference frequency and phase. The resultant signal is passed through the modified Wheatstone bridge in which the package has been hooked. The RS method is described in detail in [Lall 2011, 2012].

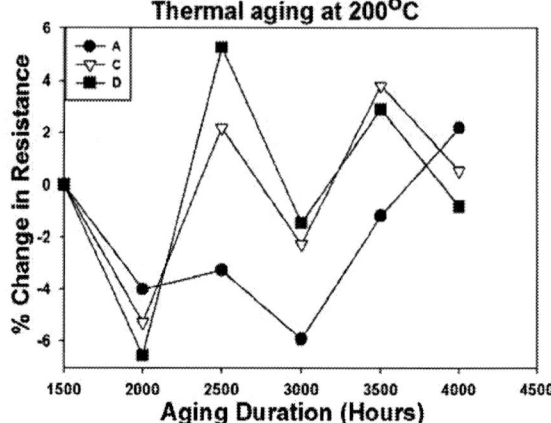

Figure 30 – Resistance Change in packaged 20-pin devices at 200°C thermal aging

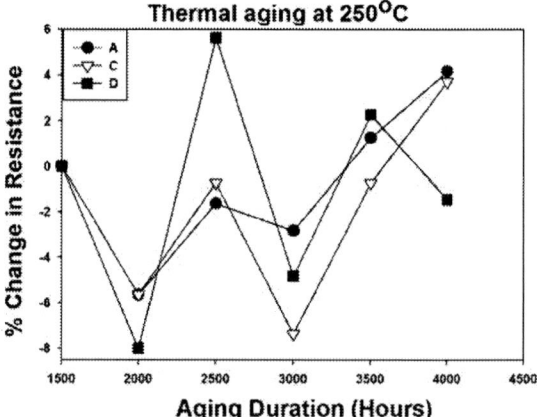

Figure 31 – Resistance Change in packaged 20-pin devices at 250°C thermal aging

Figure 30 and Figure 31show the percentage change in resistance of wire bond during thermal aging at 200°C and 250°C. Package would have been considered as electrically failed if an increase in resistance is more than 20% of its original resistance. No measurable difference in resistance was found till 1500 hours of aging in both cases. None of the devices electrically failed after 4000 hours of aging. Thus, the data indicates that in spite of degradation in the mold compound during operation at high temperature, the wire bond interconnects continue to function.

(2). Current Leakage Test

Thermally aged specimens were subjected to elevated temperature-humidity bias of 130°C-100%RH. Leakage current was not physically measured; instead resistance of open legs was monitored. Test was conducted for 120 hours. Samples used for this test were aged at 200°C and 250°C for 5000 hours. These samples were tested for leakage current before subjecting them to humidity environment, and no leakage was found. After performing temp-humidity test resistance drop was observed for devices molded with candidate C. No such drop was observed in case of MC candidate D. For both molding compounds, there was no increase in resistance of wire bonded pairs.

(3). Failure Analysis

During the degradation of the mold compound, chemical changes may occur in the EMC releasing functional groups. The released functional groups might attack metals inside package and causing failure. In order to study the effect of mold compound degradation on wirebond pairs, parts were withdrawn from thermal chamber at interval of 1000 hours. The parts were cross-sectioned and polished till the cross-section plane near center of the wirebond.

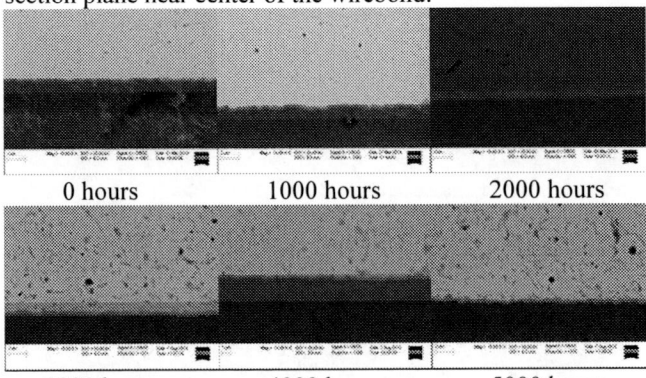

0 hours	1000 hours	2000 hours

3000 hours	4000 hours	5000 hours

Figure 32 – Au-Al Wirebond interface in Package with Molding Compound "C", 200°C

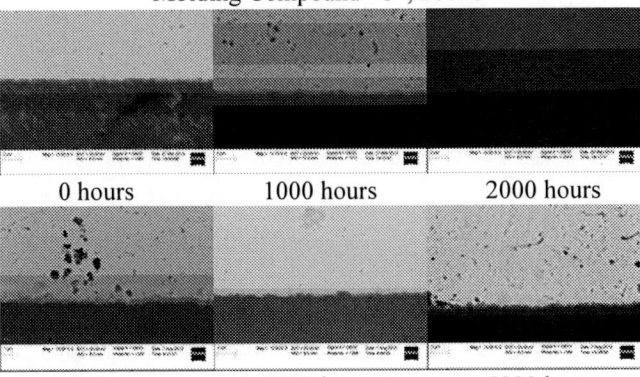

0 hours	1000 hours	2000 hours

3000 hours	4000 hours	5000 hours

Figure 33 - Au-Al Wirebond interface in Package with Molding Compound "C", 250°C

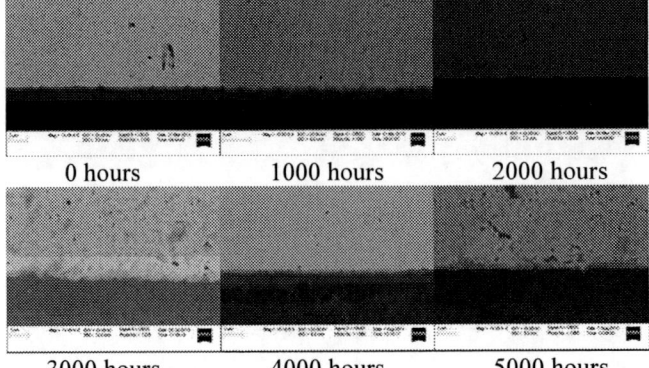

0 hours	1000 hours	2000 hours

3000 hours	4000 hours	5000 hours

Figure 34 - Au-Al Wirebond interface in Package with Molding Compound "D", 200°C

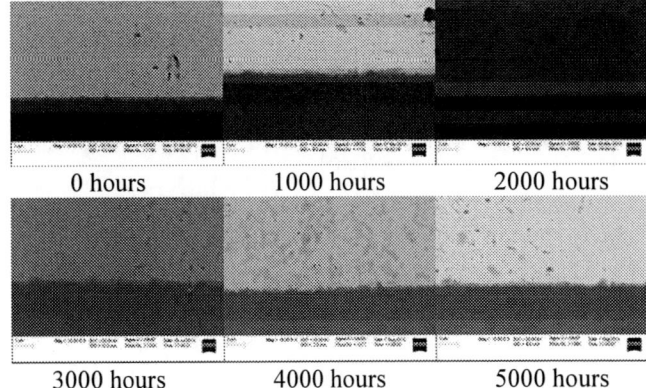

0 hours	1000 hours	2000 hours

3000 hours	4000 hours	5000 hours

Figure 35 - Au-Al Wirebond interface in Package with Molding Compound "D", 250°C

Using scanning electron microscopy, wirebond interface was analysed. Gold wirebond bonded on Nickel-Palladium coated Aluminum pad forms a solid diffusion layer at the interface. EDS was used to find composition of this layer. Figure 32, Figure 33, Figure 34, and Figure 35 show the test results. No major changes in diffusion layer thickness, as well as composition of layer were detected. For all molding compounds, and all aging cases, microstructure of wirebond found to be unchanged. Change in microstructure of molding compound was also monitored, using SEM. Oxidised layer (white-colored layer) in molding compound can be seen in thermally aged samples. Figure 36 shows EDS spot analysis of degraded layer. In this layer the filler material shows a loss of binding material. High oxygen content of the outer EMC layer indicates that the EMC has been oxidised. The decrease in the Carbon content indicates a loss of weight and binding material. Due to the loss of the binding material, the relative silica content of the oxidized layer was found to be higher for the degraded part.

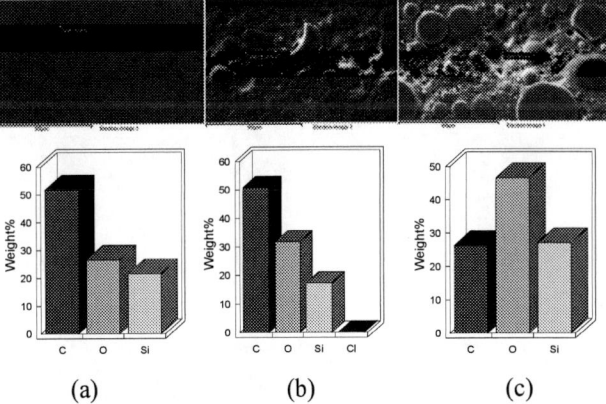

Figure 36 EDS Analysis of EMC Candidate-A – (a) Time 0 (b) Intermediate Oxidation (c) complete Oxidation

In the case of mold compound-A, rapid growth of oxidation layer was observed as shown is Figure 37 and Figure 38. Most of the area in case of 200°C aging, and complete area in case of 250°C was oxidized after 2000 hours of aging. Mold compound-A exhibited the highest rate of weight loss, and mold compound oxidation. EMC-A became very brittle after 2000 hours, that its handling was major issue. Mold compound cracked easily, and broke during normal handling.

978-1-4799-2408-0/14 $31.00 © 2014 IEEE 251

Due to these issues further testing on A was terminated and it was not considered as potential candidate. EDX analysis was performed on white oxidation layer in the mold compound. Results showed higher oxygen content and lower carbon content in degraded layer as compared to time 0 sample. Loss of carbon is because of evaporation of products of decomposition of long polymer chains. Higher oxygen content shows that some of the products are oxidised. Results are shown in Figure 36.

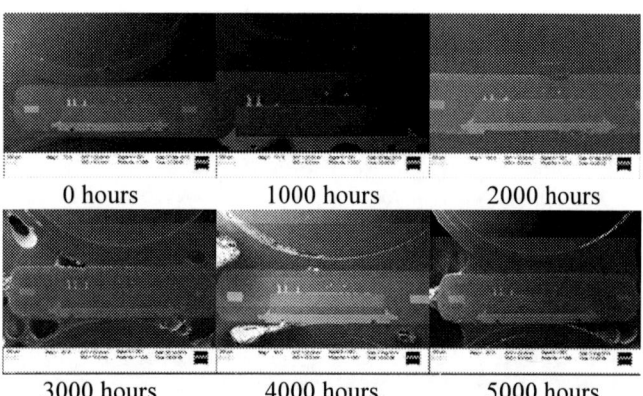

| 0 Hours | 1000 Hours | 2000 Hours |

Figure 37 Molding Compound "A", 200°C

| 0 Hours | 1000 Hours | 2000 Hours |

Figure 38 Molding Compound "A", 250°C

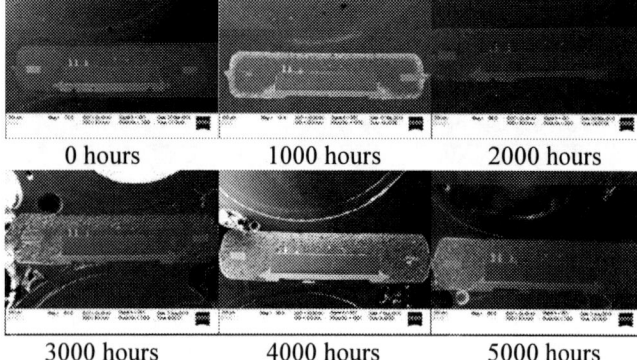

| 0 hours | 1000 hours | 2000 hours |

| 3000 hours | 4000 hours | 5000 hours |

Figure 39 Molding Compound "C", 200°C

| 0 hours | 1000 hours | 2000 hours |

| 3000 hours | 4000 hours | 5000 hours |

Figure 40 Molding Compound "C", 250°C

For mold compound-C, oxidation layer was observed after 2000 hours of aging at 200°C, and 1000 hours of aging at 250°C, shown in Figure 39 and Figure 40. Oxidation layer is at along edges of package, and from there it propagates towards center of the package. For 200°C case, a large cross-

section of the mold compound was oxidised after 5000 hours, and at 250°C, complete cross-section area was oxidised. For EMC candidate D, results were very different, and are shown in Figure 41 and Figure 42. There was no oxidation area observed at edges of the package in this case; but there were oxidation spots spread throughout the cross-section in all cases. Density of the oxidation spots increased for the aged samples, but even after aging for 5000 hours, however most of the area did not exhibit oxidation.

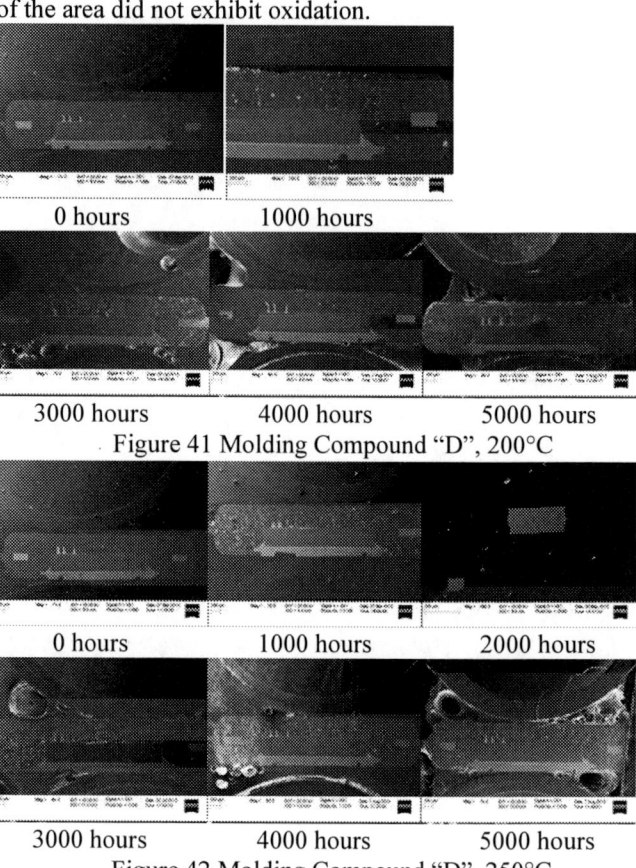

| 0 hours | 1000 hours |

| 3000 hours | 4000 hours | 5000 hours |

Figure 41 Molding Compound "D", 200°C

| 0 hours | 1000 hours | 2000 hours |

| 3000 hours | 4000 hours | 5000 hours |

Figure 42 Molding Compound "D", 250°C

(4). X-Ray Test

X-ray technique was used to find delamination, cracking in packages non-destructively. For candidate C, cracks were found after 3000 hours of aging at 200°C and 2000 hours of aging at 250°C.

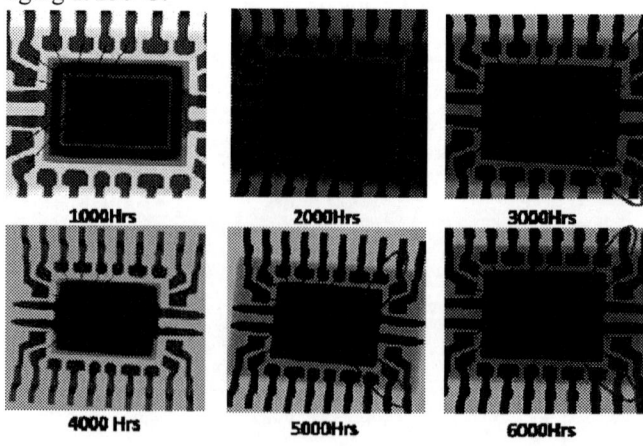

Figure 43 – X-ray images of MC "C", 200°C

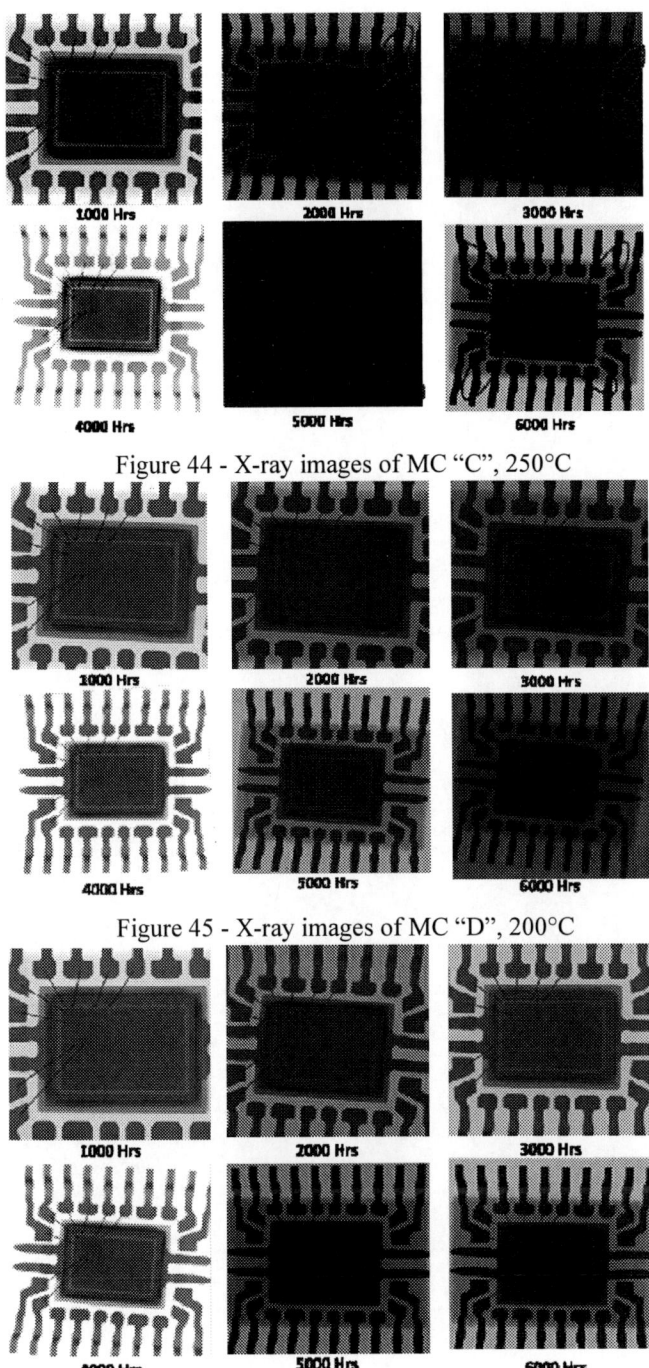

Figure 44 - X-ray images of MC "C", 250°C

Figure 45 - X-ray images of MC "D", 200°C

Figure 46 - X-ray images of MC "D", 250°C

Location of crack was common; all the cracks were at the corner of package, propagating diagonally. For EMC-D, no crack was found after aging for 6000 hours in both conditions. Optical microscopic images taken were in good agreement with X-ray images.

Summary and Conclusions

Multiple tests were performed on bulk mold compound samples and 20 pin SOIC devices molded with four different EMC candidates including two high temperature epoxy molding compounds, non-aromatic mold compound, and silicone encapsulant. The EMCs code-named A, B, C and D

were analyzed under a high temperature thermal aging at 200°C and 250°C for up to 5000 hours using a variety of analytical test techniques. Analystical techniques used to assess the high temperature performance of the EMC candidates included weight loss analysis, Fourier Transform Infrared Spectroscopy (FTIR), Dynamic Mechanical Analyzer (DMA), X-ray photoelectron spectroscopy (XPS), and Resistance Spectroscopy. Results indicate that the silicone based encapsulants (EMC-D) exhibited the best high temperature performance out of all the candidate molding compounds tested in this study. The high temperature epoxy mold compounds (EMC-A) exhibited the poorest performance out of all the molding compunds tested in this study. Out of all tests, some tests such as weight loss analysis and DMA test for bulk molding compound, and cross-sectioning, X-ray analysis, resistance measurement were found to be more effective compared to other tests. Weight loss of molding compound can be co-related to loss of binding material. Loss of material can be visually observed in cross-sections of 20 pin devices.

Acknowledgments

The research results presented in this paper are based on projects supported by industrial members of the NSF-CAVE3 Electronics Research Center at Auburn University.

References

Chen A.S. and Randy H.Y. Lo. "High Temperature Applications for IC Encapsulated Packaging", Proceedings of 45th Electronic Components and Technology Conference, pp. 889-893, 1995.

Chen Y, Ping Li, "The "Popcorn Effect" of Plastic Encapsulated Microelectronic Devices and the Typical Cases Study", Proceedings of International Conference on Quality, Reliability, Risk, Maintenance and Safety Engineering, pp 482-485, 2011.

de Vreugd, J., A. Sánchez Monforte, Jansen K.M.B., Ernst L.J. C. Bohm, A. Kessler, H. Preu, "Effect of Postcure and Thermal Aging on Molding Compound Properties", Proceedings of 11th Electronics Packaging Technology Conference, pp 342-347, 2009.

Ge, Dandong , Chai Chee Meng, lan K.L.L, Walter M, "A Comparison Study of Thermal Aging Effect on Mold Compound and Its Impact on Leadframe Packages Stress", International Symposium on Advanced Packaging Materials (APM), pp 403-409, 2011.

Harman G, *Wire Bonding in Microelectronics: Materials, Processes, Reliability, and Yield.*" 2nd edition, Published by McGraw-Hill Professional, 1997.

Hu G, Andrew A. O. Tay, Yongwei Zhang, Wenhui Zhu, Characterization of Viscoelastic Behaviour of a Molding Compound with Application to Delamination Analysis in IC Packages, Proceedings of 8th Electronics Packaging Technology Conference, pp 53-59, 2006.

Lall P, Lowe R, Goebel K, Particle filter models and phase sensitive detection for prognostication and health monitoring of lead-free electronics under shock and vibration, Proceedings of 61st IEEE Electronic Components and Technology Conference (ECTC), pp 1097-1109, 2011.

Lall P, Lowe R, Goebel K, Prognostication Based on Resistance-Spectroscopy and Phase-Sensitive Detection for Electronics Subjected to Shock-Impact, Journal of Electronic Packaging, Volume 134, Issue 2, June 2012

Lu Xiuzhen, Li Xu, Huaxiang Lai, Xinyu Du, Johan Liu, Zhaonian Cheng "Studies on Microstructure of Epoxy Molding Compound (EMC)-Leadframe Interface after Environmental Aging", International Conference on Electronic Packaging Technology & High Density Packaging (ICEPT-HDP), pp 1051-1053, 2009.

Ma Lilli, Shengxiang Bao, Dechun Lv, zhibo Du. "Application of C-mode Scanning Acoustic Microscopy in Packaging", Proceedings of 8th International conference on Electronics Packaging Technology, pp 1-6, 2007.

McCluskey Patrick, Kofi Mensah, Casy O'Connor, Fabian Lilie. "Reliability of Commercial Plastic Encapsulated Microelectronics at Temperatures from 125°C to 300°C", Proceedings of the IEEE Aerospace conference , pp 445-450, 2000.

Noolu N, N. Murdeshwar, K. Ely, J. Lippold, and W. Baeslack, "Phase Transformations in Thermally Exposed Au-Al Ball Bonds", Journal of Electronic Materials, Volume 33, pp. 340-352, 2004.

Rappaport N. and A. Efros, Monte carlo simulation of polymer high-temperature degradation and fracture, in High temperature properties and applications of polymeric materials, Edited by H.M. M.Tant, M.Rogeers, ACS Washington, DC, pp. 155-165, 1995.

Teverovsky Alexander, Effect of Environments on Degradation of Molding Compound and Wire Bonds in PEMs, NASA Electronic Parts and Packaging Program (NEPP), October 2005

Uno T and K. Tatsumi, Thermal reliability of gold-aluminum bonds encapsulated in bi-phenyl epoxy resin, Journal of Microelectronics reliability, Volume 40, pp. 145-153, 2000.

Yang, Daoguo, Zaifu Cui, The Effect of Thermal Aging on the Mechanical Properties of Molding Compounds and the Reliability of Electronic Packages, Proceedings of International Conference on Electronic Packaging Technology & High Density Packaging, 2011, pp 1030-1033, 2011.

Yan-Shan Ng, Wan-Keong Hooi, FTIR Method to Distinguish Mold Compound Types and Identify the Presence of Post-Mold-Cure (PMC) Treatment on Electronic Packaging, Proceedings of 9th Electronics Packaging Technology Conference, pp 652-656, 2007.

Time, Temperature, and Mechanical Fatigue Dependence on Underfill Adhesion

Joseph Cremaldi[1], Michael Gaynes[2], Peter Brofman[2], Noshir Pesika[1], Eric Lewandowski[2*]

[1] Engineering Tulane University New Orleans, LA 70118
[2] IBM T. J. Watson Research Center, Yorktown Heights, NY 10598
*E-mail: eplewand@us.ibm.com, Phone: 914-945-1159

Abstract

The observations of the present study show that underfill properties can vary greatly after the initial cure and that these changes can have a significant effect on adhesion. Temperature studies illustrate that underfill adhesion is a strong function of temperature near its glass transition temperature and reveal the importance of adhesion tests at the temperature extremes of the operating conditions and/or reliability thermal cycling tests. Subcritical crack growth results demonstrate that subcritical strain energy release rates do not necessarily scale with critical strain energy release rates. These results are used to create a new adhesion screening methodology that more closely mimics the time, temperature, and mechanical fatigue conditions that an underfilled 1st level package will typically experience.

Introduction

The solder interconnects between a silicon die and an organic laminate are enabled by introducing underfill into the solder interconnect region. Without underfill, the solder joints will rapidly fatigue during temperature on/off cycles due to concentrated thermo-mechanical stresses on the solder interconnects. Underfill prevents the concentration of stress onto just the solder interconnects by more evenly distributing the thermo-mechanical stresses throughout the whole solder interconnect region.

The first organic packages, in the early 1990s, had chips that were on the order of seven mm on a side, had a robust BEOL (back end of line) chip structure, used a ductile lead based solder, and had a large interconnect pitch/gap. During this period, the constraints on the underfill were modest and the major emphasis on underfill mechanical properties was simply placed on tailoring the underfill coefficient of thermal expansion (CTE) to match the predominantly high lead solder joint, around 28 ppm/C, and a high underfill modulus, greater than 10 GPa, to relieve stress from the solder.[1] However, over the last 20 years chips have increased in size, lead based solders have been phased out, BEOLs have become more fragile, and the interconnect pitch/gaps have decreased. These new, challenging constraints have led to a much more complex set of ideal underfill material properties.[2] Contemporary underfill properties of interest include CTE, modulus, glass transition temperature (Tg), thermal conductivity, bulk fracture toughness, and interfacial fracture toughness. It is difficult to make a case that any one underfill property is more important than another; however, in many instances package failure has been linked to underfill failure that originated due to poor underfill adhesion.[2,3]

Underfill interface fractures are typically driven by thermo-mechanical stresses; however, other environmental or mechanical stresses can also drive interface fractures. Generally, the underfill delamination initiates at the points of highest mechanical stress. In a 1st level package, this region is typically the corner of the die. Even a small delamination near the chip corner can concentrate the mechanical stress, leading to further delamination or the propagation of cracks into the surrounding regions.[3] Figure 1 shows common failure modes associated with an underfill delamination. Delamination can lead to solder fatigue cracks, inter-layer dielectric (ILD) cracks, substrate cracks, or bulk underfill cracks. Due to the range of fractures associated with underfill delamination, maintaining good underfill adhesion, especially at the areas of highest mechanical stress, is extremely important. Consequently, a great deal of effort has been placed on characterizing underfill adhesion through computational modelling and experimentation.

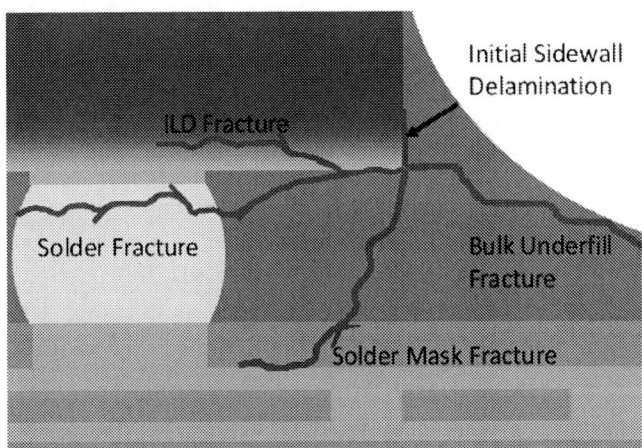

Figure 1: Possible failure modes stemming from an underfill delamination

Currently, computational modeling of a 1st level package is effective at predicting the effects of linear elastic materials on the 1st level package. Computational parametric studies of linear elastic materials have provided valuable trends, which have led to coarse adjustments in package dimensions and materials selection.[3,4] Unfortunately, the modeling of time, temperature, and strain dependent materials, such as a nonlinear viscoelastic underfill, is still in its infancy at the 1st level package scale. Over the last decade, modelers have begun to tackle pieces of this problem and have produced some very informative results.[5,6] However, to accurately predict something as intricate as an underfill delamination one needs time, temperature, and strain dependent material data, as well as accurate material models, for all the 1st level package materials. Another issue with modeling complex materials, such as an underfill, is that it takes significantly more computational resources than a basic elastic model. At this point in time, such comprehensive modeling of underfill adhesion is just not feasible at the development and manufacturing level.

To date, the most accurate method of determining whether an underfill will delaminate is still the module level thermal cycle; however, it is not practical to screen every underfill with a module level test. Module level thermal cycling requires expensive module hardware, and the test cycle from build to post-test failure analysis is a minimum of six months (See Table 1). To minimize module level testing, it is common that underfills are screened with simpler and less expensive characterization tests so as to narrow the possible underfill candidates that enter module level thermal cycling. Usually, screening is based on the values of a relative benchmark material(s). These are normally underfills that have successfully passed thermal cycling on similar hardware. The use of a relative ranking is commonly employed because of the lack of detailed computational modeling required to produce absolute adhesion values.

Possible Processing Treatments			
Treatment	Temperature (C)	Cycles	Time (hrs)
Underfill Cure	125-165	—	1-3
Process Capping	125-165	—	0.2-2
Process Die Marking	125-165	—	0.5-1.5
Humidity Bake Outs	125	—	8-24
BGA Attach Reflow	225-255	--	—
Possible Thermal Cycling Treatments			
Treatment	Temperature (C)	Cycles	Time (hrs)
Thermal/Shock (T/S)	-40/60	5	5
Thermal/Age (T/A)	125	—	24
Thermal/Humidity(T/H)	30 @ 60% RH	--	192
JEDEC Reflow	225-255	3	3
Thermal Cycle Stressing 1	-55/125	500-1500	250-750

Table 1: Possible processing, pre-conditioning, and thermal cycling treatments a 1st level package might experience. Not every module will experience all of these treatments. Some of the process treatments may be repeated multiple times, such as the humidity bake out.

Common underfill adhesion tests used to screen 1st level electronic packaging adhesives include the cantilever beam (CLB), dual cantilever beam, four point bend, blister, button or die shear, wedge, and lap shear tests. The adherends in these tests are often made of or coated with surfaces common to a 1st level package such as silicon, polyimide, silicon nitride, solder mask, etc. These adherends may also be subjected to processes common to a 1st level package assembly. The tests are convenient because the data analysis is relatively simple, the testing apparatus is basic, and the sample preparation is efficient. These adhesion screens provide a time zero adhesion strength and a fracture mode, and can be used as a coarse delamination predictor as well as a process optimization guide. However, these tests by themselves can easily lead to a false characterization when they do not factor in the effects of aging, temperature, and subcritical fracture.

More advanced underfill adhesion techniques exist such as the Sandwich Brazil Nut (SBN) test and other mixed mode geometries. These techniques can provide fracture mechanic parameters for a range of mode mixity, but the testing usually requires significant design, more complex crack monitoring, more detailed computational analysis, and time, temperature, and strain dependent data/models for the other 1st level package relevant materials. The already complex apparatus, design, and analysis can get much more complicated when one factors in mechanical fatigue and temperature. While the material parameters these advanced tests provide are necessary for the ideal all-encompassing 1st level package simulation, which may one day lead to an all virtual qualification, the added value gained by these more advanced techniques is not necessary for a screening process based on a relative rankings. For the majority of underfill screening, a study of a single fracture mode as a function of age, temperature, and subcritical stress should be enough to successfully rank capillary underfills.

Underfill cure kinetics and degradation are a function of temperature and time. Underfills can continue to cure or degrade after the manufacturer's suggested cure schedule. If the additional cure or degradation is not accounted for during the material property evaluation, there can be significant deviations between the measured and the actual property values during thermal cycling or during operating conditions.[7] Depending on how the properties change after the additional curing (for better or worse), one may rank the underfill lower or higher than its realistic value. For example, a typical module experiences many temperature deviations after the initial underfill cure, such as moisture bake outs, lidding, die marking, and BGA attach. Additional thermal excursions include any pre-conditioning procedure and the high temperature deviations during thermal cycling (See Table 1). These temperature deviations can lead to further cure/degradation that can affect many properties such as CTE, modulus, and Tg, as well as directly or indirectly affect interfacial and bulk fracture properties.

Underfill mechanical properties are a strong function of temperature. Usually the CTE, loss, and storage modulus are measured as a function of temperature during the initial underfill screening process, but adhesion and bulk fracture properties are not. Not accounting for temperature effects during the screening procedure can lead to a gross over estimate in the underfill's adhesion properties. The mechanical properties' temperature dependencies are even more pronounced near the Tg of the underfill. Near the Tg, underfills and polymers in general can experience large reductions in their interface[8,9] and bulk fracture properties, as well as the frequently mentioned increase in CTE and drop in modulus.

The majority of underfill delaminations are not critical in nature. Delaminations typically occur during cyclic fatigue at some sub-critical strain energy release rate, and unfortunately, the underfill's subcritical fracture properties do not have to scale with the underfill's critical fracture properties. A lot of characterization on adhesive joints in fatigue has been performed in the literature[10,11], but very little of this work has been applied to underfills.[12,13] Additional screening based on subcritical fracture properties should mitigate the risk of underfill delamination during thermal cycling. In addition, although subcritical fracture measurements require more time than a typical critical fracture measurement, the time invested is minimal compared to the time wasted in a failed module level thermal cycling test.

In this paper, the subcritical fracture data is presented in the form of crack growth rate vs the strain energy release rate. Particular focus is paid to the crack growth rate, because the lack of accurate modeling and in situ stress measurements makes predicting a characteristic stress, strain, or fracture parameter range difficult. However, based on typical thermal cycling delamination failures, one can guess an approximate crack growth rate region. For a module level thermal cycle test, the number of cycles is near 1000 and the characteristic side wall length of a non-thinned die is 10^{-4} to 10^{-3} m. This equates to a characteristic crack growth rate near 10^{-6} to 10^{-7} m/cycle. The maximum strain energy release rate was chosen as the fracture parameter because it is the fracture parameter most often used in fatigue tests of bonded polymeric joints.

The goal of this paper is to present a new adhesion screening method that more closely mimics the time, temperature, and mechanical fatigue conditions that an underfill will typically experience during subsequent process and module level thermal cycling. First, thermo-mechanical properties of the underfill are examined as a function of different aging conditions, paying close attention to shifts in the glass transition temperature. Next, critical and sub-critical strain energy release rates are measured as a function of temperature with aged and un-aged samples. Lastly, the results from the study are used to create a more accurate adhesion screening methodology.

Experimental

Two commercial underfills, denoted as A and B in this paper, were chosen as candidates for this study due to an inconsistency found between their material properties and their performance during thermal cycling. Initial underfill adhesion screening, using the CLB adhesion test, showed that underfill B had a much higher critical strain energy release rate (G_C) than underfill A, while all other customary pertinent material properties for the two underfills were similar (Table 2). Conversely, in extended thermal cycling of large die on organic laminates, underfill B would fail and underfill A would pass. A failure analysis of the underfill B modules concluded that underfill delamination from the sidewall of the die was the root cause. Examination of the chip side wall showed little to no resin attached in the delamination region. This large discrepancy between the initial adhesion results and the thermal cycling performance makes these two underfill candidates an interesting and very informative case study.

Isothermal aging experiments were performed on the two underfill samples to examine the effect of aging. All samples, adhesion samples, and samples for bulk material characterization were initially cured with the manufacturer's suggested cure in a nitrogen (N_2) environment. All subsequent aging dwells were also performed in an oven with an atmosphere of N_2. After aging, the samples were either tested immediately or stored in N_2 at 25C until testing. No samples were stored for more than two weeks. Bulk material analysis of the underfill's thermo-mechanical properties was

performed to examine the effect of aging and temperature on material properties. Thermo-mechanical properties were measured with a TA Instruments Q200 differential scanning calorimeter (DSC), a TA Instruments Q400 thermal mechanical analysis (TMA), and a TA Instruments Q800 dynamic mechanical analysis (DMA). The Tg of the underfill was determined by two separate means: the peak in the loss modulus (DMA) and the inflection point of the heat flow vs temperature curve (DSC).

Property	Underfill A	Underfill B
Glass Tranisition Temperature via DSC	103 °C	102 °C
Coefficient of Thermal Expansion < Tg	29 ppm/ °C	27 ppm/ °C
Coefficient of Thermal Expansion > Tg	95 ppm/ °C	100 ppm/ °C
Storage Modulus @ 1Hz and 25 °C	11 Gpa	11 Gpa
G_c on SiO2 @ 25 °C	230 J/m²	495 J/m²
1500 cycles of JEDEC Thermal Cycling B	Passed	Did Not Pass

Table 2: Summary of time zero material property values for underfills A and B. Material properties based on manufactures suggested cure only.

Monotonic loading adhesion tests were performed to examine how the critical strain energy release rate varied with age and temperature. The adhesion tests used the CLB geometry, with a SiO_2 passivation on the deflected adherend. The CLB sample preparation, geometry, and analysis all followed the protocol detailed by Paquet et al.[2] Monotonic loaded samples were tested on an Instron 5566 frame with an Instron 3119-506 environmental chamber and a 1kN static load cell. A controlled pre-crack with the Instron was performed on all CLB samples prior to adhedsion characterization. For experiments at temperatures other than room temperature, the environmental chamber, sample fixture, and frame were allowed to equilibrate for two hours.

Cyclic loading fatigue tests were performed to examine the effect of age, temperature, and subcritical strain energy release rate on underfill adhesion. The same sample preparation and geometry used in the monotonic loading experiments were also used in the cyclic fatigue experiments. The samples were tested on an Instron E3300 frame with an Instron CP100557 evironmental chamber and 250N dynamic load cell. The tests were performed in constant displacement mode, with a displacement ratio of 0.1. By selecting the constant displacement to correspond to a strain energy release rate near G_C of a CLB sample with a known compliance (already determined by the monotonic loading experiments and the Instron controlled precrack), one can easily obtain the whole fatigue curve. Obtaining a full fatigue curve via a constant force displacement is considerably more difficult because one would need a force that equated to a strain enrgy release rate just above the the threshold value. Acquiring the threshold strain energy release rate via a quick monotonic loading experiment is not possible. Initially samples were tested at multiple frequencies, 1Hz, 3Hz, and 6Hz. All frequencies gave similar results for both underfills, but 3Hz was selected because it gave the best trade off for high signal to noise at low forces and reduced testing time.

The benefits of using the CLB sample geometry described in Paquet et al. are the straightfoward sample preparation and the economical sample size; however, the small size makes a direct measurement of crack length very difficult. For this reason crack length was not directly measured, but instead analytically backed out via the sensitve optical encoder displacement measurement and dynamic load cell force measurement. This indirect measuremnt was also made possible because of the highly controlled geometry and material properties of the CLB adherends. The results using this method of crack measurement were very repeatable even at the low crack growth rate range. A testing method that does not require the extra complexity of a direct crack measurement techinque is also more conducive to development and manufacturing screening tests.

Fatigue experiments were performed at 25°C and 125°C. To separate the effects of aging and subcritical fracture properties, all tests performed at 125°C were performed with only aged samples. The time scale of one fatigue test is on the order of days which is comparable to the aging time scale, so in-situ aging of initally un-aged samples would convolute the sub-critical fracture screening.

It was difficult to perform fatigue experiments at temperatures below room temperature, due to the long time scales associate with fatigue testing, the condensation at temperatures below room temperature, and the small sub-critical forces. Condensation on the sample, fixture, or extension leads to false force readings, making accurate measurements impossible. All attempts at preventing significant condensation during the fatigue experiment were unsuccessful, and therefore, no accurate data at -55°C was recorded. Future work will focus on accurately making these sub-ambient measurements.

As discussed in the introduction, the crack velocities associated with delaminations of underfill from the sidewall are in the range 10^{-6} to 10^{-7} m/cycle. For this reason the fatigue experiments were often terminated in 10^{-8} to 10^{-9} m/cycle range. Consequently threshold strain energy release rates were not always clearly defined in the fatigue curves. In some fatigue experiments extrapolating G_C may also be difficult, because of error in the predicted constant displacement amplitude. The origin of this error is variance in the initial sample pre-crack.

Results and Discussion

The goal of underfill screening is to select candidates with the best chance of passing thermal cycling by measuring material properties; therefore, underfill properties should be measured under conditions similar to those seen prior to and during the thermal cycling process. One set of conditions that is very important, but often overlooked in underfill studies, is the underfill aging history. Before any thermo-mechanical measurements, underfills should receive a pre-conditioning treatment that mimics the typical temperature-time history leading up to module level thermal cycling. The objective of this aging study is to select an accelerated temperature-time profile that is representative of the aging an underfill would experience prior to JEDEC thermal cycling.

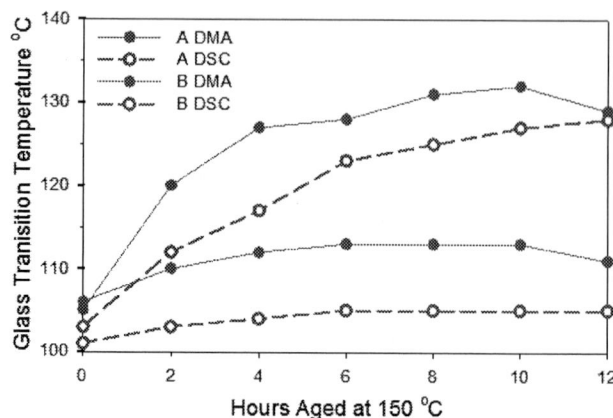

Figure 2: Tg as a function of time at 150C. Tg was measured via the DMA (peak of loss modulus) and DSC (inflection point).

Based on typical post underfill cure processing and JEDEC pre-conditioning temperatures (Table 1), two isothermal aging temperatures, 125°C and 150°C, were initially selected for the aging study. To characterize the effect of aging, changes in Tg as a function of time at an elevated temperature were measured with the DMA and DSC. Figure 3 demonstrates how the Tg for underfills A and B varies with aging time at 150°C. With just the manufacturer's recommended cure (time equals zero on figure 3), the glass transition temperatures of both underfills are similar: about 102°C via the DSC and 106°C via the DMA. Additional aging at 150°C generates increases in the glass transition temperatures for both underfills, but the aging effect is greatest for underfill A. The Tg for underfill A increases with time aged at 150°C until about 10hrs, where the Tg levels off at 126°C, measured via the DSC. The Tg of underfill B also increases with time aged at 150°C, but levels off at only 4 hrs with a Tg value of 105°C, measured via the DSC.

Similar isothermal aging measurements were also performed at 125°C. A comparison between the 150°C data and 125°C data revealed that the effects of aging at 150°C for 10 hrs were similar to those of aging at 125°C for 24 hrs. Based on the two isothermal aging studies, the plateau of Tg as a function of age, and the desire for accelerated aging, it was concluded that aging at 150°C for 10 hrs would be a good representative accelerated aging period for all future experiments in this paper. In addition to isothermal aging at 150°C, some samples were subjected to a simulated 3X reflow in order to mimic the effect of a typical BGA JEDEC pre-condition test. The effects of a 3X reflow were negligible on all the material properties including adhesion. For the remainder of this paper an "aged" sample will refer to a sample isothermally aged at 150°C for 10hrs, unless otherwise noted.

Figures 4a and 4b show the effects of aging on the storage modulus and the coefficient of thermal expansion, respectively. Just as shifts in the Tg as a function of aging were larger for underfill A than underfill B, so too are shifts in the storage modulus and the coefficient of thermal expansion as a function of aging. These large shifts in thermo-mechanical properties can have large effects on the 1st level package stresses and are essential in understanding package reliability because they are the drivers behind underfill failures, such as delaminations. Unfortunately, it is difficult to predict how underfill stress/strains will change based solely on changes in modulus and/or CTE since the package is such a complicated 3D model, the mechanical properties of many of the materials are time, temperature and strain dependent, and additional aging can shift the zero stress/strain points of the materials. To enable a prediction of changes in the stress/strain state of the underfill as a function of aging would require advanced FEM in conjunction with complex time, temperature, and strain dependent material measurements. Although discussion of aging effects on package stresses is outside of the scope of this paper, we wanted to draw attention to the strong possibility that aging can also significantly alter the stress/strain drivers behind underfill delamination as well as the adhesion strength.

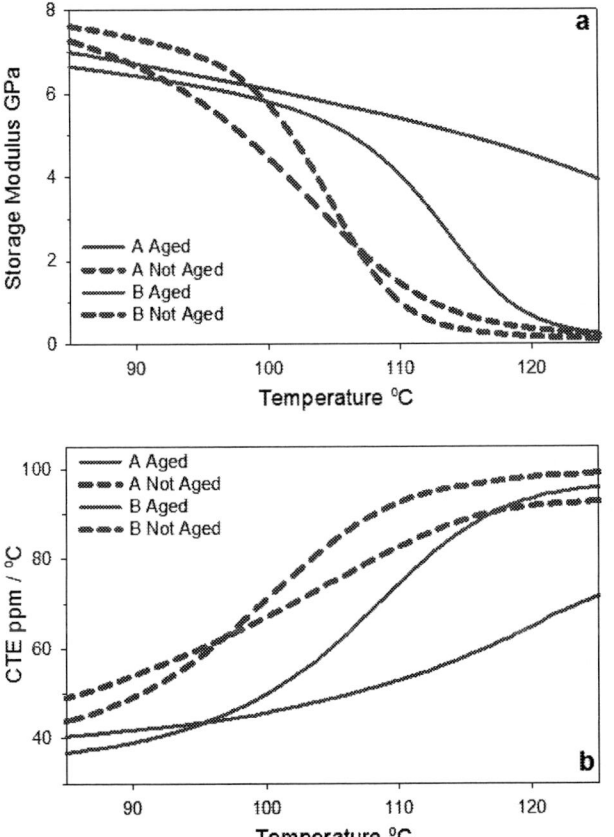

Figure 3a shows the storage modulus as a function of temperature for un-aged and aged underfill samples. Figure 4b shows the coefficient of thermal expansion as a function of temperature for un-aged and aged underfill.

The critical strain energy release rate was measured as a function of temperature and age. Figure 5 summarizes the results of these monotonic loading tests. Three principal results can be extracted from these experiments: the effect of temperature, the effect of aging, and the interaction of temperature and age.

The effect of temperature on adhesion is clearly present with both underfills. There is a large drop in G_c for both

underfills that were un-aged at 125°C when compared to the un-aged samples at 25°C. For these samples, 125°C is above the Tg of the un-aged underfills (figure 3), and the large decrease in critical strain energy release rate agrees well with the literature.[8,9] These results highlight the importance of operating below the glass transition of the underfill. The decrease in adhesion strength may not lead to a failure if the package stress/strains are small, such as in a small die, but in the case of 1st level packages with large stress/strains, one should generally operate below the Tg of the underfill. Lower temperatures seem to have a minimal effect on underfill adhesion. Decreases in G_c at -55 °C compared to 25°C are small for un-aged samples, and in the case of aged underfills are non-existent.

The effect of aging on the critical strain energy release rate is greatest for underfill A, and this also corresponds well to the earlier Tg studies. The critical strain energy release rate for underfill A at 25°C decreases from 230 J/m^2 to 170 J/m^2 when aged, an approximate 25% decrease. The importance of this result cannot be overstated; not accounting for aging can lead to inaccurate screening results.

The interaction of aging and temperature on the critical strain energy release rate is most clearly seen at 125°C. The critical strain energy release rate increases significantly at 125°C for underfill A when aged, approximately 100%. Again this change in adhesion corresponds well with the change in Tg of underfill A. There is also a smaller relative increase in G_C of underfill B at 125 °C for the aged sample compared to the un-aged sample, and the effect of -55°C is negligible.

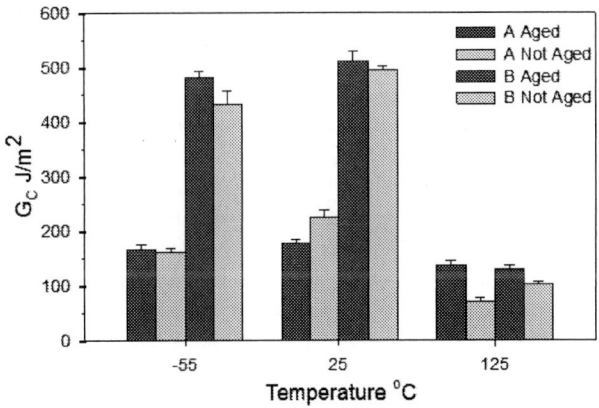

Figure 4: Critical strain energy release rates at multiple temperatures for underfills A and B. Error bars represent one standard deviation.

Figure 6 shows a plot of the critical strain energy release rate for aged samples as a function of temperature. The finer granularity of the strain energy release rate as a function of temperature around the glass transition of underfill B clearly shows that adhesion decreases sharply near the Tg. The values at 130°C highlight another consideration during adhesion screening. Testing at the temperature limits of JEDEC thermal cycling may not be representative of actual thermal cycling. Typically, module level thermal cycles will overshoot the JEDEC limit to ensure that the temperature limits are actually met as well as to decrease cycle time. Since the goal of screening is to select an underfill with the best chance of passing thermal cycling, it may be best to test at the extreme temperatures of the actual thermal cycle, not the minimum JEDEC specifications. In the case of underfill A and B, the temperature seen during thermal cycling was above 130°C.

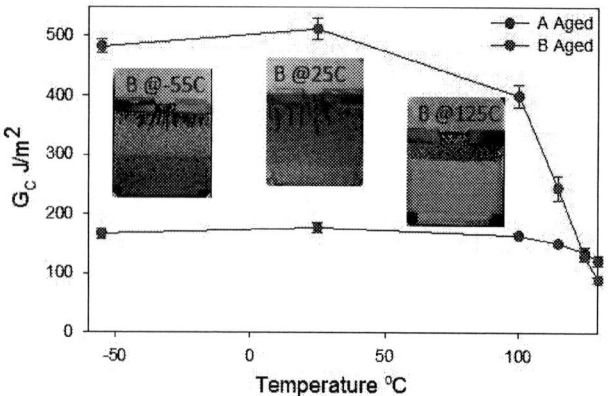

Figure 5 shows G_C as a function of temperature for aged underfills. Error bars represent one standard deviation. The insets show the fracture modes of aged underfill B at -55°C, 25°C, and 125°C. The more cohesive fail near the polyimide tape is due to the controlled pre-crack at room temperature, and the direction of propagation is in the downwards direction. The region below the pre-crack region is the G_C testing zone.

Along with the critical strain energy release rate, changes in the fracture mode can also illustrate adhesion problems. The fracture mode for aged underfill A is near interface cohesive over the entire temperature range. The fracture mode for underfill B is typical cohesive or near interface cohesive for all samples between -55°C to 115°C; however, at temperatures $\geq$125°C the failure mode for underfill B is significantly more interfacial with little to no resin on the adherend (Inset figure 6). This failure mode at temperatures $\geq$125°C more closely resembles that seen during the FA on the failed underfill B thermal cycling modules.

G_C and fracture mode as a function temperature and age highlight the complex dependencies of underfill adhesion and demonstrate the need to account for these dependencies when attempting to predict thermal cycling performance. Not accounting for higher temperatures greatly overestimates the adhesion performance of underfill B and greatly underestimates the adhesion performance of underfill A.

Fatigue testing was performed on the underfills to examine the effect of subcritial strain energy release rates on crack propagation. Figure 7 shows a comparison of the two un-aged underfills at 25°C. The critical strain energy release rates for underfill A and B can be approximated from these curves at 200 J/m^2 and 500 J/m^2, respectively, and correspond well with monotonic loading tests. The threshold strain energy release rates for underfill A and B are not as clear, but are near 30 J/m^2 and 60 J/m^2, respectively. The Paris law exponentes for underfills A and B are 3.6 ± 0.2 and 4.2 ± 0.2, respectively. The relative positon of the curves clearly illustrates the fact that underfill B performs much better in fatigue than underfill A for un-aged samples at 25°C. The entire fatigue curve of underfill B lies significantly to the right

of underfill A, meaning that for any given strain energy release rate, the crack growth rate is significantly smaller. These results agree well with the un-aged G_C measurements at 25°C. The inset in figure 7 shows the effect fatigue has on the the fracture mode. At slower crack rates (lower strain energy relese rates) the crack is driven toward the SiO_2 adherend surface. All fatigue samples, irresepective of underfill, age, or temperature, showed this facture mode behavior at slower crack rates. It is interesting to note that driving the crack to the interface has been used as a means to probe surface treatments,[11] and a similar test could possibly be used as a means to probe surface contimantion during processing, such as flux cleaning.

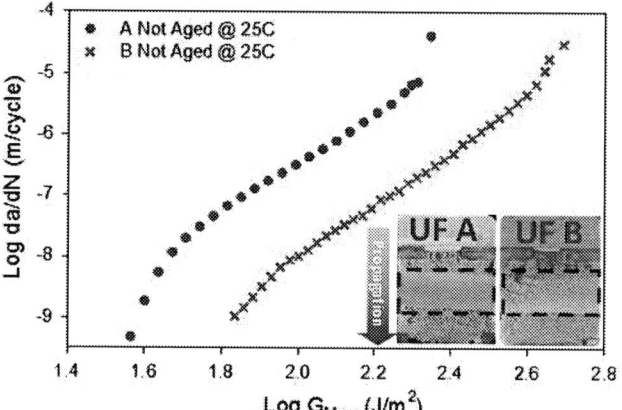

Figure 6 shows the fatigue curves for not aged A and B underfills at 25°C. Inset shows the fracture mode of underfills A and B. The fatigue test region of each sample is outlined in a dotted black rectangle and the green arrow to the left shows the direction of crack propagation.

Figures 8a and 8b show the effect of aging on the underfill. Aging appears to have very little effect on the underfill adhesion of underfill B, agreeing well with the previous G_C and T_g results. The effect of aging is much more pronounced for underfill A at higher G_{Max}, but dimishes at lower G_{Max}.

The effect of temperature on both underfills can be seen in figure 9a and 9b. For underfill A, the fast fracture is minimally affected by temperature; however, the lower threshold region is significantly affected. Changes in G_C as a function of temeprature seem to underestimate the temperature effects on subcritical strain energy release rates. Underfill B shows a very large drop in adhesion performance in fatigue at higher temperatures, similar to the G_C measurements. Also, the typical sigmoidal shape is now distorted especially in the range 10^{-5} to 10^{-7} m/cycle. When comparing the fatigue behavior of underfills A and B at 125 °C, inset figure 10, one sees that the fatigue behavior is very similar throughout the whole crack rate range.

Figure 10 is a direct comparison of the two aged underfills at 130°C under fatigue. In the 10^{-5} to 10^{-7} m/cycle range, underfill A lies significantly to the right of underfill, meaning that underfill A should have better performance at elevated temperatures over this range. The trend of decreasing adhesion performance for underfill B with elevated

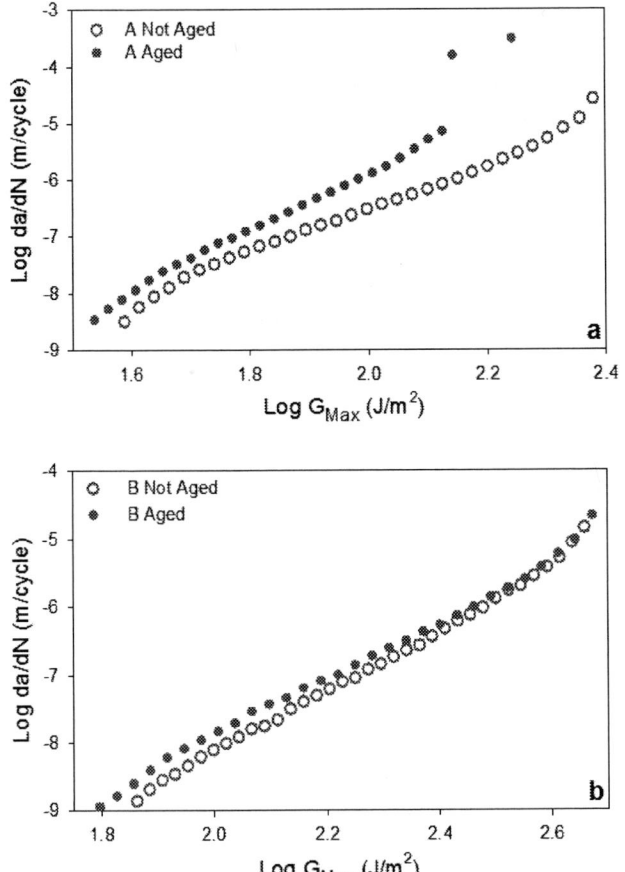

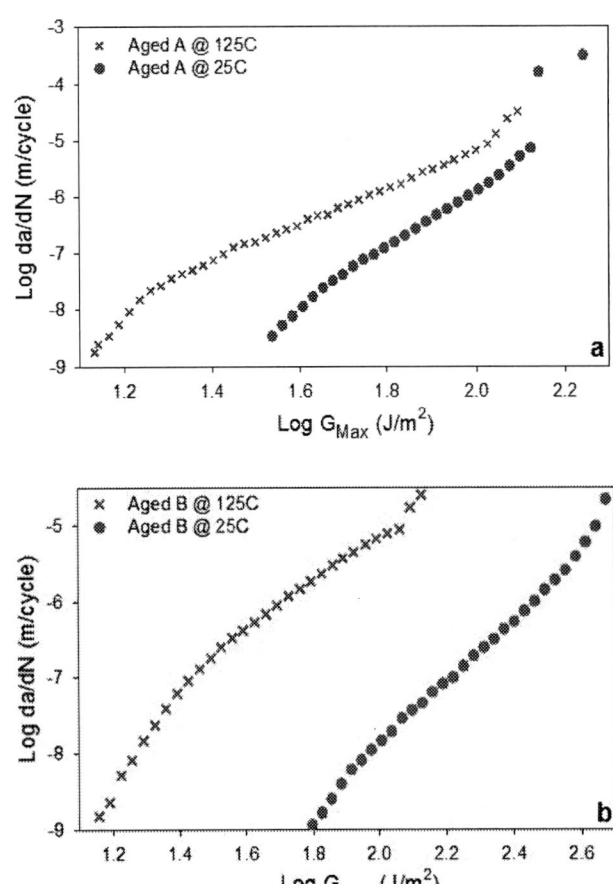

Figure 7 shows the change in fatigue behavior as a function of age for underfills A (8a) and B (8b).

Figure 8 shows the change in fatigue behavior as a function of temperature for aged underfills A (9a) and B (9b).

temperature is similar to that seen in the G_C experiments and strongly suggests that operating at temperatures above its Tg caused the delaminations during module thermal cycling.

The results of the aging, temperature, and fatigue study do not unquestionably say that underfill A should have performed better than underfill B in module level thermal cycling performance. Such a prediction is difficult for two reasons. One, the stress/strain drivers are very complicated and cannot be estimated solely based on the measured material properties, as mentioned earlier in the text. Two, the thermal cycling test results for underfill B were not dismal; it only failed in extended thermal cycling, meaning that the uncertainty in such a prediction would have to be very small.

What can be said from this case study is that the large difference in Gc for the un-aged underfills at 25C clearly overestimates the relative performance during module level thermal cycling, and that more representative testing, such as the one outlined in this paper, provides a more accurate picture of adhesion during thermal cycling. The G_C measurements at 130°C (Figure 6), the fracture mode at higher temperatures (Figure 6), and the mechanical fatigue performance at 130°C (Figure 10) are much more in line with the actual thermal cycling performance than the more typical room temperature G_C measurements for un-aged underfills.

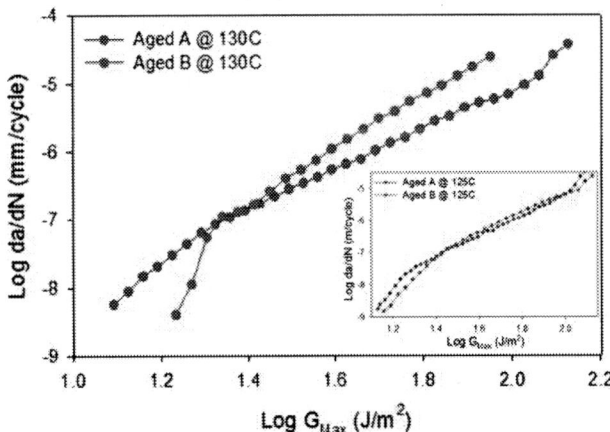

Figure 9 is a comparison of aged underfill A and B at 130 °C in fatigue. Inset is a comparison of aged underfill A and B at 125 °C in fatigue.

These experiments also point to the root cause of the delaminations seen during thermal cycling of underfill B and demonstrate the importance of not exceeding the glass transition temperature of the underfill.

Conclusions

Depending on their chemistry, underfills can age significantly after the manufacturer's suggested cure. This can have large effects in terms of their thermo-material properties. Underfill adhesion is a strong function of temperature, especially when approaching the Tg of the underfill. Above the Tg, underfill adhesion can reduce significantly; therefore, adhesion testing should account for these possible transitions. Also, because underfills fail in fatigue, it is important to rank underfills on the fatigue properties which do not necessarily scale with their critical fracture properties. In this case study, the initial screening results grossly overestimated underfill B's adhesion performance under thermal cycling relative to underfill A's performance; however, when a more representative adhesion screening methodology is applied the screening results more closely resemble the performance of the module level thermal cycling tests.

This new adhesion test methodology is summarized below:

1. First examine the effect of expected temperature history: how do the material properties change during a representative build process? Pay specific attention to changes in Tg and change in the shape of the loss modulus. If the properties change significantly, perform a representative thermal age on all samples before measuring the underfill's thermo-mechanical properties.

2. Examine the polymer transitions, i.e. Tg, of the underfill, and if these occur within or near the thermal cycling extremes, then testing at the temperature extremes is necessary. Also, the real limits of the thermal cycle might exceed those set in the JEDEC spec: the temperatures might be even more extreme in order to ensure meeting the JEDEC spec.

3. Perform G_c measurements. If underfill transitions occur within the thermal cycling extremes, then monotonically test at those temperature extremes.

4. Perform fatigue tests. Use the monotonic G_C testing to set the upper bounds for constant displacement fatigue testing. Again, if the polymer transitions occur near the thermal cycling temperature range, test under fatigue at those extremes

Acknowledgments

Thank you to the Yorktown model shop for fabricating many of the fixtures to make these experiments possible. Also thank you to Marie-Claude Paquet, Catherine Dufort, Claude Blais, and Helene Lavoie at IBM Bromont, as well as Thomas Lombardi and David Questad from IBM East Fishkill, for their many productive discussions.

References

[1]   K. Puttlitz and P. Totta, Area Array Interconnection Handbook, Boston, MA: Kluwer Academic, 2001, p. 453.

[2]   M. Paquet, M. Gaynes, E. Duchesne, D. Questad, L. Bélanger and J. Sylvestre, "Underfill Selection Strategy for Pb-Free, Low-K and Fine Pitch Organic Flip Chip Applications," in 56th ECTC, San Diego, CA, 2006.

[3]   M. Paquet, J. Sylvestre, E. Gross and N. Boyer, "Underfill Delamination to Chip Sidewall in Advanced Flip Chip Packages," in 59th ECTC, San Diego, CA, 2009.

[4]   C. Zhai, Sidaharth, R. Blish and R. Master, "Investigation and Minimization of Underfill Delamination in Flip Chip Packages," IEEE Transactions of Device and Materials reliability, vol. 4, pp. 86-91, 2004.

[5]   N. Chhanda, J. Suhling and P. Lall, "Implementation of a Viscoelastic Model for the Temperature Dependent Material Behavior of Underfill Encapsulates," in 13th IEEE ITHERM Conference, Orlando, FL, 2012.

[6]   L. Niu, D. Yang and G. Zhang, "Characterization Modelling and Parameter Sensitivity Study on Electronic Packaging Polymers," in International Conference on Electronic Packaging Technology and High Density Packaging, Beijing, China, 2009.

[7]   C. Lin, J. C. Suhling and P. Lall, "Isothermal Aging Induced Evolution of the Material Behavior of Underfill Encapsulants," in 59th ECTC, San Diego, CA, 2009.

[8]   L. Shijian and C. Wong, "Influence of Temperature and Humidity on Adhesion of Underfills for Flip Chip Packaging," IEEE Transactions on component and packaging technologies, vol. 28, no. 1, pp. 88-94, 2005.

[9]   X. Shi, Y. L. Zhang and W. Zhou, "Determination of Fracture Toughness of Underfill/Chip Interface with Digital Image Speckle Correlation Technique," IEEE Transactions on components and packaging, vol. 30, no. 1, pp. 101-109, 2007.

[10]   I. Ashcroft and S. Shaw, "Mode I Fracture of Epoxy Bonded Bomposite Joints 2. Fatigue Loading," International Journal of Adhesion and Adhesives, vol. 22, no. 2, pp. 151-167, 2002.

[11]   S. Azari, M. Papini, J. Schroeder and J. Spelt, "Fatigue Threshold Behavior of Adhesive Joints," International Journal of Adhesion and Adhesives, vol. 30, no. 3, p. 145–159, 2010.

[12]   K. Nagarajan and R. Dauskardt, "Adhesion and reliability of Underfill Substrate interfaces in Flip Chip BGA Packages: Metrology and Characterization," in Proc. Pan Pacific Symposium Conference, 2002.

[13]   B. McAdams and R. Pearson, "Initiation and Propagation of Delaminations at the Underfill/Passivation Interface Relevant to Fhip-Chip Assemblies," IEEE Transactions on device and materials reliability, vol. 4, pp. 169-175, 2004.

Study on Isotropic Electrically Conductive Adhesive for Medical Device Applications

Shawn Shi, Scott Sleeper, and Chunho Kim
Medtronic, Inc.
2343 W. Medtronic Way, Tempe, AZ 85281, USA
shawn.shi@medtronic.com, Phone: 480-929-5614

Abstract

In implantable medical device applications, quality and reliability are critical aspects of the product. There are some unique mechanical and electrical challenges to the silver filled isotropic electrically conductive adhesive (ECA) used in these devices. The mechanical challenge is the result of the very complicated system design which involves a wide variety of integrated circuits directly bonded and over-molded onto the double-sided printed wiring board (PCB). Only a very small percentage of ECAs are capable of meeting the mechanical strength requirements for this application. These mechanically acceptable ECAs must also be thoroughly studied to ensure that a strict resistance stability requirement can be met. It is imperative to fully understand the fundamentals of the ECA material chemistry and performance relationship before it can be used in medical applications.

A series of studied results will be presented in this paper. The system design challenge on ECA formulation is presented. The numerical and experimental study results on PCB design and its impact on bond pad-adhesive delamination are reported. The electrical performance of epoxy-based and acrylate-based conductive adhesives was comparatively studied. The device level critical electrical performance of conductive adhesives was found not to be correlated to bulk resistivity of cured ECA, but significantly correlated to the interfacial resistance between the cured adhesive and bond pad. Unfolding the atomic level phenomenon occurring at this interface during adhesive curing and the device burn-in process is one of the main focuses of this paper. The fundamental atomic and molecular level mechanism of the assembly process and the material chemistry impact on resistance stability will be discussed. The mechanism includes how the conductive path was formed on the adhesive and gold pad interface during adhesive cure and how the interface evolves during the burn-in process. Also discussed are ingredients in the adhesive formulation that may impact the resistance instability, and the solution space for stable resistance.

Part 1: Factors on ECA delamination

The microelectronic modules used in high power implantable medical devices are usually two sided to save space. On one side, all the electronic components are die attached and wire bonded onto a printed circuit board (PCB), and then all the components are transfer molded using epoxy molding compound. This process is called printed wiring board (PWB) assembly process in this paper. The PWB assembly process sequentially includes the following process steps: Die attach (DA) adhesive dispense, die pick and place, adhesive cure, wire bonding (WB), transfer molding, and post transfer mold cure (PMC). On the other side, many electronic components are soldered onto the same PCB board using standard surface mount technology (SMT). This process is called DCA assembly process and includes the following steps: solder paste print, component pick and place, solder reflow, underfill (UF) dispense, underfill cure, and flux residue cleaning. A finished microelectronic module is shown in Figure 1.

Molded side   SMT side

Figure 1: The appearance of a finished microelectronic module.

Since both sides of the PCB have design features such as copper traces and solder resist openings, they interact with each other and increase the PCB design complexity. One of the major issues from the interaction is die attach adhesive and wire bond pad delamination.

Molding stresses and delamination mechanisms

In the transfer molding process during mold clamping and epoxy mold compound (EMC) injection, components are subjected to a variety of mechanical stresses. These stresses can affect both the silicon dice and the die attach adhesive used to bond the dice to the PCB. If the distribution of stresses applied during the transfer molding process exceeds the distribution of the die attach adhesive strength (or the attachment interfaces), die attach delamination can occur, either in whole or in part. This phenomenon has been observed on different PCB designs and can be successfully mitigated by PCB design and molding process changes as described in more detail below. To date this phenomenon has only been observed on some dice bonded with conductive die attach adhesive. The primary metrology used to detect die attach delamination is acoustic imaging, augmented with cross-sectioning to positively identify the interface.

Several die attach delamination initiators have been identified and observed during development. These include layout\topology interactions, stress inducing process excursions, and surface quality issues on incoming material.

Typical layout/topology issues include surface features on the opposing side of the PCB, such as solder mask openings and surface metal traces. During transfer molding, the non-molded side of the PCB is supported by the flat surface of the mold tool. Solder mask openings for components are not supported as they are slightly recessed compared to most of the PCB surface. By contrast, metal surface traces appear as high points compared to the surrounding PCB due to lack of planarization of the solder mask coating. As the PCB is clamped in the mold, the high points deflect towards the mold side, which in turn deflects the die and creates a local stress point. As the molding progresses, the deflected die is molded in place. Subsequent processes eventually allow the PCB to stress relax, returning towards the initial condition, causing stress at the die attach layer and possible delamination. A similar phenomenon occurs if there is foreign material (FM) in the mold chase. FM acts as a high point on the side opposite of the mold, and produces a similar effect as the surface metal trace. Delamination due to solder mask openings is created by a different mechanism, where the high pressure packing phase of the mold process applies pressure to both dice and the PCB surface. Localized deflection occurs where the solder mask is unsupported by the mold tool and if a die is located over or near the deflecting PCB, shear and tensile stresses develop that can lead to delamination. Design rules can mitigate the delamination risk by reducing solder mask opening sizes or proximity to die on the opposing side. Surface metal traces can also be moved so they do not overlap die locations.

FEA model of die delamination

A non-linear FEA model was created using MSC Software Corporation's Patran® and Marc® to evaluate die delamination at a specific PCB location noted for delamination during the molding process. Figure 2 shows both sides of the FEA model. A mold pressure of 7.85 MPa was applied to the die and PCB surfaces with a uniform mold temperature of 175 °C. The nominal die shear yield strength at the molding temperature is 12 MPa. The FEA model was limited to the die, die attach adhesive, die attach pad, underlying surface mount pads and surrounding PCB and solder mask material. The loading conditions represent the transfer mold material packing phase only (no pre-stress from prior processes), where the die side of the PCB is subjected to high pressure due to epoxy mold compound (EMC) injection at high temperature. Cross section results combined with acoustic imaging indicate that this process step is responsible for the delamination due to evidence of mold compound filling the delaminated gap post mold and no delamination prior to molding.

All polymer materials used in the model are temperature dependent. In the case of the die attach adhesive, limited material property data is available so the elastic perfectly plastic constitutive model had to be extrapolated from die shear data and the supplier's data sheet.

The model consists of ~42,000 8 node hex elements with a congruent mesh between the PWB, die attach pad, and die attach adhesive. The die, solder mask and surface mount pads use glued contact. The solder mask surface is constrained in the Z direction to simulate the mold chase surface and the X and Y boundaries of the model are constrained with symmetry surfaces.

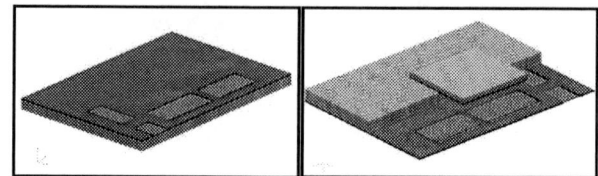

Figure 2: FEM for die overlapping solder mask openings, some PCB elements removed for clarity

FEA model results

Figure 3 compares the equivalent Von Mises stress in the die attach adhesive to an optical photograph of the delaminated region of an actual die from the same PCB location and an acoustic image of the location. The red color in the fringe plot indicates stress equal or greater than yield strength. The conservative assumption is applied that yield strength is equated to delamination. A 50X deformation fringe plot is shown in Figure 4 (deformation is greatly exaggerated to indicate deformation direction) and it indicates that the die is caused to bend as the PCB deforms over the unsupported solder mask openings on the opposite side of the board. As transfer mold pressure is applied the PCB "sinks" into the opening and the die follows.

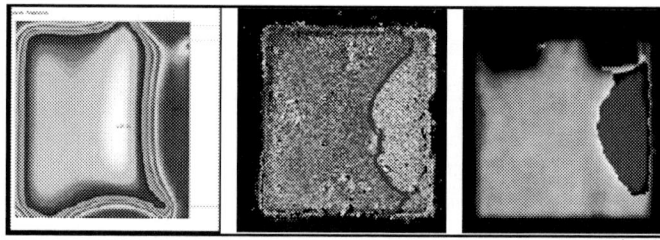

Figure 3: FEA die attach adhesive stress, optical and acoustic images of delamination

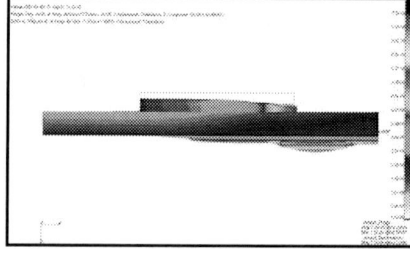

Figure 4: Deformed die and PCB shown at 50X

The FEA model provides insight into the delamination mechanism including why surface features on the opposing side could lead to die attach stress and delamination. It provides an estimate of delamination area via correlation with the acoustic imaging and destructive test results. Finally, the model can be used to test potential design changes intended to reduce or eliminate delamination.

Experimental Study

A series of experiments were performed to study the effect of transfer mold process parameters on DA and WB pad delamination. The key transfer mold parameters studied and their ranges are listed in Table 1.

Table 1: Transfer Mold Parameters Studied

Parameter	Low	High
Transfer Pressure (kN)	4.0	8.0
Mold Time (sec)	120	300
Mold Temperature (°C)	160	180
Clamping Force (kN)	150	295

The first DOE was performed to determine the significant transfer mold parameters among the four shown above. Only transfer pressure and temperature are significant and the other two parameters are not significant in the studied ranges. DA and WB pad delamination count were measured after soaking the molded samples in the MSL2a standard moisture condition (60°C/60%RH for 120 hours) then transferring the samples through reflow/UF cure thermal profile. Note that PMC was skipped to make the effect of transfer mold conspicuous. Figure 5 shows that more DA and WB pad delamination were induced by higher mold pressure and temperature. The effect of mold pressure is well explained by the FEA modeling described above. It seems that the effect of temperature is related to CTE mismatch. Higher mold temperature is likely to cause the higher CTE mismatch experienced through the transfer mold, reflow and UF cure processes.

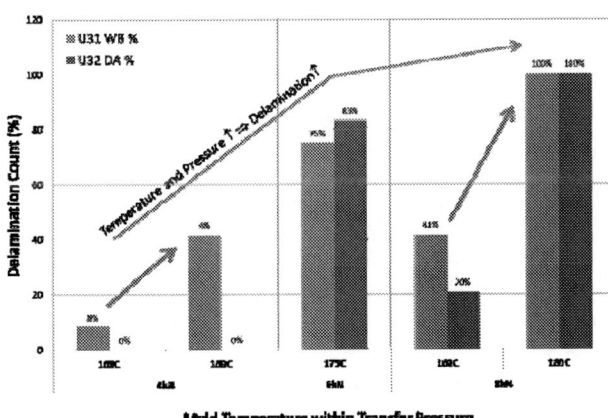

Figure 5: Delamination vs. Mold Pressure and Temperature

Another DOE was performed to study the effect of moisture interacting with the transfer mold conditions on DA and WB pad delamination. Note PMC was skipped for the same purpose described in the previous DOE except the control leg. The molded samples were soaked at four different soaking times, 0 hour, 20 hours, 40 hours and 120 hours at 60°C/60%RH humidity. Then the samples went through reflow/UF cure furnaces with thermal exposure only and finally were inspected for delamination. Delamination was counted for each of U31 WB and U32 DA. The percentage of delamination samples for each of U31 WB and U32 DA is shown in Figure 6. It should be noted that (1) both DA and WB pad delamination increase with moisture soaking duration, i.e. moisture content and (2) the lower transfer mold temperature/pressure results in the less delamination. There was zero delamination observed with 4kN mold pressure and 160°C mold temperature for this application.

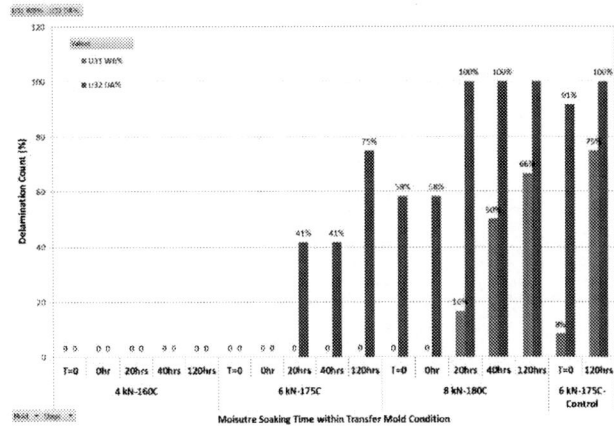

Figure 6: Moisture Effect of Delamination

Although optimization of geometric design and process parameters can reduce delamination, selection of right die attach ECA is equally important to minimize the delamination issue. Five commercially available die attach adhesives were selected for adhesive delamination evaluation. Two types of test vehicles (TV1 and TV2) were built using each of the five adhesives. TV1 was built using design-optimized PCB I under nominal settings of mold pressure (6KN) and clamping force (343KN) with 175°C molding temperature. TV2 was built using design-optimized PCB II under extreme settings of mold pressure (10KN) and clamping force (390KN) with 175°C molding temperature. The purpose of using the extreme molding condition is to observe how sensitive each adhesive is in responding to different levels of stress at the adhesive bonding interface. All other PWB assembly parameters and incoming dice used in building all the TVs were the same. Acoustic imaging was used to detect adhesive delamination after the following process steps: transfer mold (TM), post mold cure (PMC), and underfill cure (UC). All the images were obtained using Transmission Scanning Acoustic Microscope (TSAM).

The TSAM images from the test vehicles made under nominal setting of mold pressure and clamping force are shown in Figure 7 below. Under this process condition, no adhesive delamination was observed across all five adhesives post transfer mold (TM). However, after post mold cure (PMC), severe delamination was observed on adhesive C while no delamination was observed for the other four adhesives. Adhesive C was removed from further evaluation.

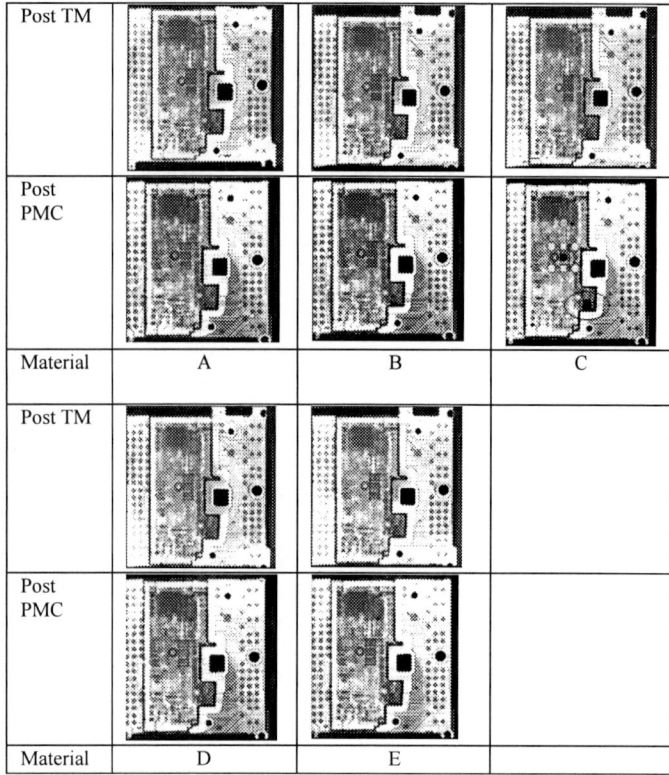

Post TM			
Post PMC			
Material	A	B	C
Post TM			
Post PMC			
Material	D	E	

Figure 7: Delamination assessment from TV1 of five adhesives after TM and PMC.

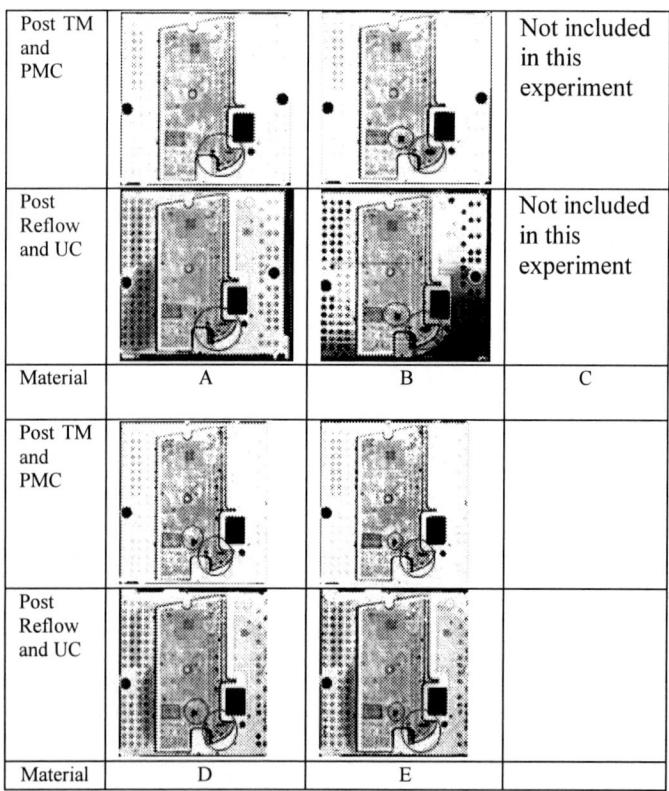

Post TM and PMC			Not included in this experiment
Post Reflow and UC			Not included in this experiment
Material	A	B	C
Post TM and PMC			
Post Reflow and UC			
Material	D	E	

Figure 8: Delamination assessment from TV2 of four adhesives after PMC and UC.

The TSAM images from the test vehicles made under extreme settings of mold pressure and clamping force were shown in Figure 8 above. Under this process condition, all adhesives showed some level of adhesive delamination. Among them, adhesive A has least delamination, followed by adhesive D, E, and F. The result revealed that adhesive A has the best performance under increased stress. For all the four adhesives, the delamination formed through PMC did not grow in the later SMT process steps (i.e. solder reflow and underfill cure (UC)).

Die warpage change with DA cure condition is shown in Figure 9. Lower cure temperature and time leads to lower warpage

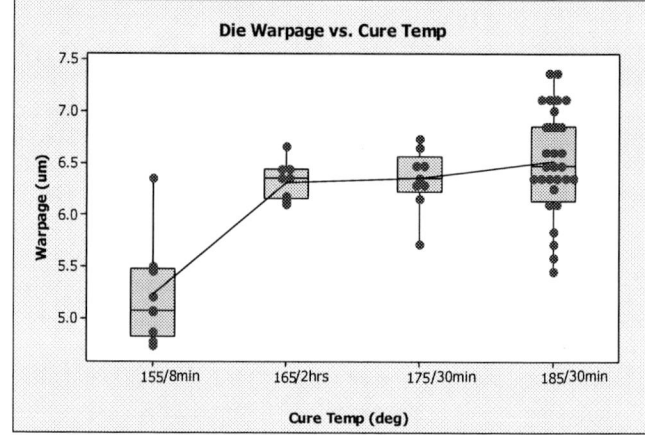

Figure 9: Die Warpage vs. Cure Condition

At the end of these studies, adhesive A performed best in terms of delamination. This adhesive was selected for the next phase electrical performance assessment.

Part 2: Electrical Performance Study

The conductive adhesive used in the medical device functions as the electrical interconnect to achieve device output. In Figure 10, the backside of a field-effect transistor (FET) is the drain of the FET which is attached to the die attach metal pad using an ECA. The source and gate of the FET are wire-bonded onto other metal pads on the same PCB (for simplicity, the wire bonding from gate to pad is shown, but the wire bonding from source to pad is not shown). To make the entire device work properly, the total resistance from the die attach metal pad to the source needs to be in the milliohms range when the gate is on, and the resistance change before and after burn-in needs to be less than half a milliohm .

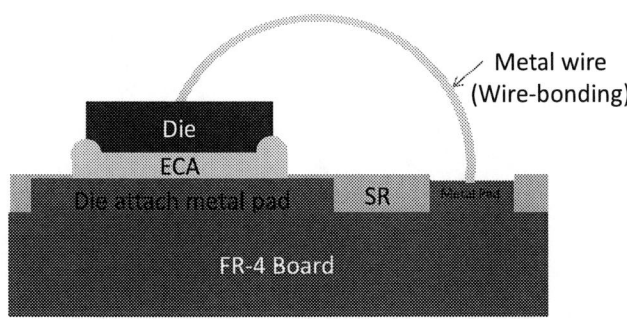

Figure 10: A typical assembly of an FET on PCB using ECA and wire bonding.

During development, it was observed that some devices had an abnormal resistance change during burn-in and current surge as shown in Figure 11 (A). The red triangle represents the measured resistance after each burn-in event, and the blue diamond represented the measured resistance after each current surge event. There were two trends. One is that the burn-in process increased the resistance and current surge reduced the resistance. The other is that the magnitude of burn-in incurred resistance increase continuously reduced with the increase of the burn-in cycle numbers. For the passing unit, the extended burn-in process actually resulted in the decrease of the resistance as shown in Figure 11 (B).

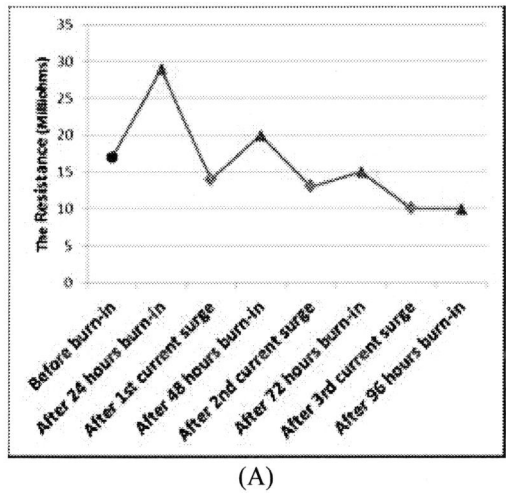

(A)

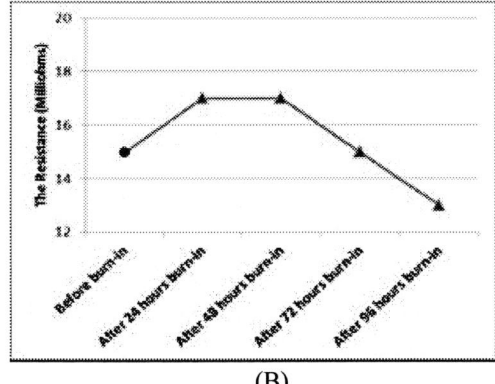

(B)

Figure 11: Resistance change during burn-in and current surge.

To understand the fundamental mechanism behind this phenomenon, the first step is to identify where the problem source is. From the die attach metal pad to the FET source, there are seven possible locations. They are the die attach metal pad itself, the interface between die attach metal pad to ECA, ECA itself, ECA to the die interface, die itself, die to wire-bonding, and the wire itself. Intensive failure analysis (FA) was conducted and the results pointed to one location which is the die attach metal pad to ECA interface.

An extensive literature search was then conducted and no existing study was found to clearly explain the observed phenomenon. An in-depth study was required to elucidate the root cause and address the yield and potential quality issue.

The first step was to scrutinize the details of the die attach metal pad to ECA interface. Scanning electron microscope (SEM) was used to investigate the interface integrity after burn-in. No micro delamination was observed at any interface between copper to nickel, nickel to palladium, and gold to ECA. No other abnormalities were found. The typical SEM pictures of the interface for both passing units and units that did not meet the specification (failing units) are shown in Figure 12.

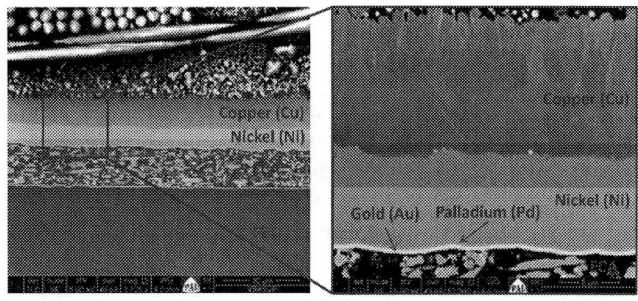

Figure 12: SEM picture of the die attach metal pad to ECA interface.

The next step is to zoom in on the palladium to gold, and gold to ECA interface. Both the passing unit and the failing unit were comparatively studied. The SEM photos of the interface morphology are shown in Figure 13. The SEM photos were taken under 150,000X magnification. Figure 13 (A) and (B) were the typical observed interface morphology of passing units, and (C) and (D) were the typical observed interface morphology from failing units. There was no difference of the Ni to Pd interface and Pd to Au interface between passing and failing units, and these metallic interfaces showed no abnormality. At the gold to ECA interface, two types of gold surfaces were observed from both passing and failing units. One type of gold surface was simply clean as shown in photo (A) and (C). The other type of gold surface was coated by an unknown layer as shown in photo (B) and (D). There were two differences between passing units and failing units. One difference was that the unknown layer was much more frequently observed in the passing units than in the failing units. The other was that the unknown layer from the passing units was often smooth and uniform as shown in photo (B), but the unknown layer from the failing units often had small void-like black dots in it as shown in photo (D).

978-1-4799-2408-0/14 $31.00 © 2014 IEEE   267

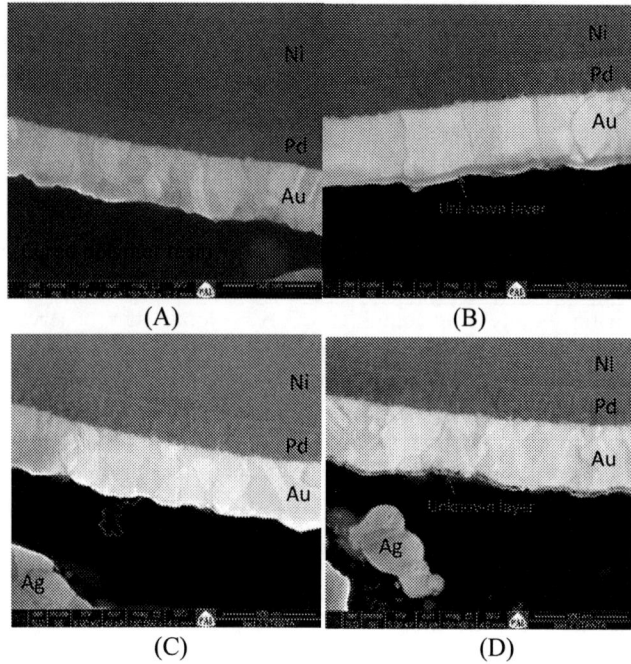

(A) (B)

(C) (D)

Figure 13: High magnification SEM photos of the gold to die attach adhesive interface from passing and failing units.

Scanning transmission electron microscope (STEM) was used to analyze the chemical nature of the unknown layer. The elemental information of the unknown layer is shown in the Figure 14. Figure 14 (A) and (B) are the STEM photos and EDS and EELS diagrams from two locations of one passing unit. Figure 14 (C) is the STEM photo and EDS and EELS diagrams from one location of one failing unit. All the EDS and EELS data indicated that the unknown layer mainly consists of gold and silver elements. It is a layer of gold silver alloy with an atomic concentration gradient along the line profile. High gold concentration is close to the gold surface and higher silver concentration is close to the ECA surface. An important finding was that the gold-silver alloy layer in Figure 14 (B) and (C) contains sulfur. However, the sulfur was found at the location very close to the gold surface for the failing unit while the sulfur was found at the location away from the gold surface for the passing unit.

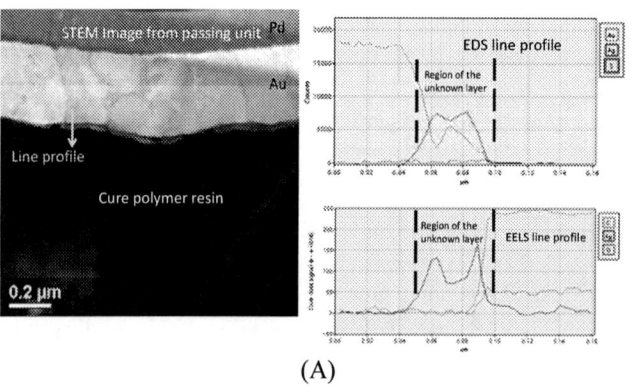

(A)

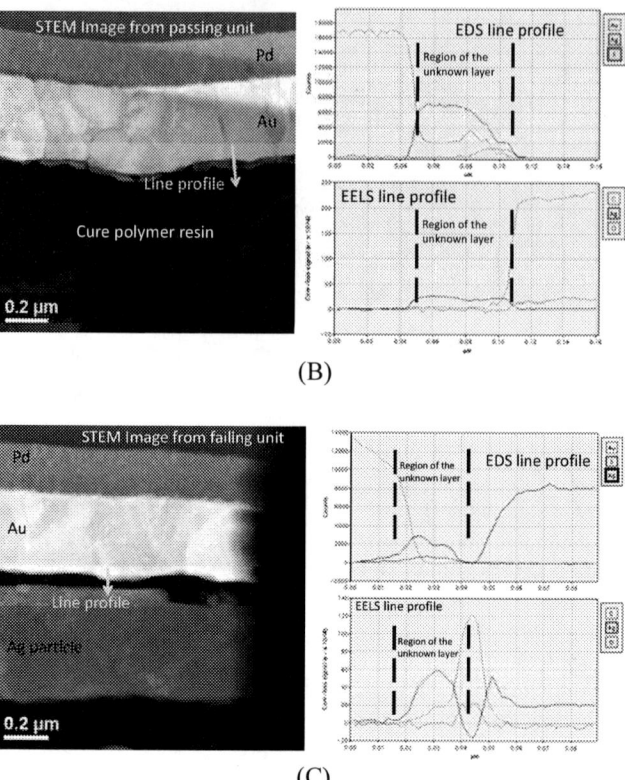

(B)

(C)

Figure 14: STEM image and EDS and EELS diagrams on two different locations of a passing unit (A) and (B), and one location of a failing unit (C).

The next step was to study the formation of gold silver alloy and the source of sulfur.

The SEM images at 80,000X and 150000X magnification shown in Figure 15 (A) and (B) provide a clue on how the gold silver alloy layer was formed. This finding fundamentally reveals a mechanism of the establishment of an interfacial conductive path between gold pad and silver flake in the ECA. The electrically conductive path between the silver flake in the ECA and gold finished metal pads was formed through silver-gold interfacial diffusion, but not through mechanical contact. As shown in Figure 15 (B), the silver-gold interfacial diffusion at least involves surface diffusion and grain-boundary diffusion. The diffusion mainly happened during the die attach adhesive cure and subsequent elevated temperature process steps such as transfer mold (TM), post mold cure (PMC), solder reflow, underfill cure (UC), and burn-in processes. This type of diffusion was accumulative and enhanced by increasing the temperature. The diffusivity of different types of silver to gold diffusion at different temperatures (surface diffusion, grain-boundary diffusion, and lattice diffusion) was well studied by Hwang et al[1].Based on the diffusivity reported by Hwang et al, up to 100nm thick diffusion layer could be formed during the entire assembly process, which was consistent with the thickness of silver-gold alloy layer observed in this study.

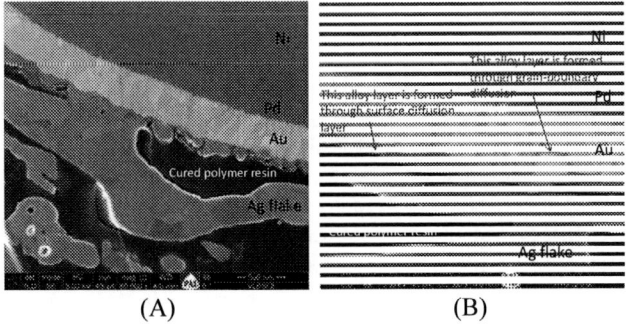

(A)   (B)

Figure 15: High magnification SEM images showing the silver-gold alloy layer formation through surface and grain-boundary diffusion.

The STEM image and EDS and EELS diagrams shown in Figure 16 pointed to the source of the sulfur. The sulfur came from the silver flake inside the cured electrically conductive adhesive (ECA). The ECA supplier reported that no sulfur was detected on their incoming silver flake.

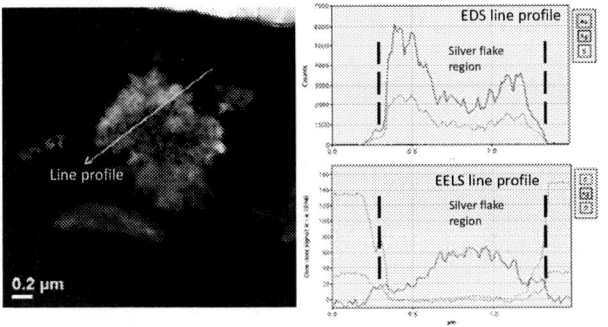

Figure 16: STEM image and EDS and EELS diagrams from a silver flake.

An extensive search of the sulfur source was conducted both at Medtronic and the ECA supplier. The search ended with focus on one compound called tetrasulfide silane. This substance was used in the ECA to improve the adhesion between the ECA and gold surface. One typical tetrasulfide silane molecular formula is shown in Figure 17. Based on the chemical name, each molecule was supposed to contain four sulfur atoms. However, the sulfide silane synthesis process usually produces sulfide silane with a varied number of sulfur atoms in one molecule. Each sulfide molecule contains two to nine sulfur atoms with every molecule having four sulfur atoms on average.

Sulfide silane is widely used in vulcanization of rubbers. How it functions as an adhesion coupling agent is depicted in Figure 17. The polysulfide linkage in the sulfide silane molecule breaks and releases out sulfur atom when it is heated up to an elevated temperature. The higher the number of x is, the lower the temperature is needed to break the polysulfide linkage. The remaining alkoxysilylalkylsulfide attaches to the gold surface through coordination interaction between sulfur atom and gold and silver surface and functions as adhesion coupling agent.[2] The released atomic level sulfur is very reactive, and it can quickly react with silver atom and form

silver sulfide. The resistivity of silver sulfide is 0.1 to 10 ohm-m[3]. It is about eight orders higher than the resistivity of pure silver which is 1.8×10^{-8} ohm-m[4].

Incoming Tetrasulfide silane

$$(CH_3CH_2O)_3Si\text{-}(CH_2)_3\text{-}S_x\text{-}(CH_2)_3\text{-}Si(OCH_2CH_3)_3 \ (x=2\text{-}9)$$

$$\downarrow \Delta \ (heat)$$

$$(CH_3CH_2O)_3Si\text{-}(CH_2)_3\text{-}S_{x\text{-}y}\text{-}(CH_2)_3\text{-}Si(OCH_2CH_3)_3 \ (y=1\text{-}7)$$

$$+$$

S (atomic level activity)

$$\downarrow + 2Ag$$

Ag$_2$S

Figure 17: The reactions of sulfide silane during adhesive cure.

The data from Figure 16 also indicated that the sulfide silane coupling agent went through phase separation during adhesive cure as a significant amount of sulfur element was found at the silver particle, but not in the cured resin region. The mechanism behind the phase separation is shown in Figure 18. At the beginning of polymerization, all the molecules in the ECA formulation were uniformly distributed. With the progress of polymerization, the molecular chain propagated and formed a secondary structure (i.e. conformal structure) and started to exclude the sulfide silane molecule. The continuous growth and propagation of the polymer chain finally formed the cross-linking structure and many sulfide silane molecules were excluded to non-polymerized domains such as the gold surface and silver particle surface. Since the sulfide molecule exclusion process is impacted by both thermodynamic and kinetic factors, small amount of sulfide silane molecules could be trapped in the cross-linked polymer network, especially when the curing process is fast.

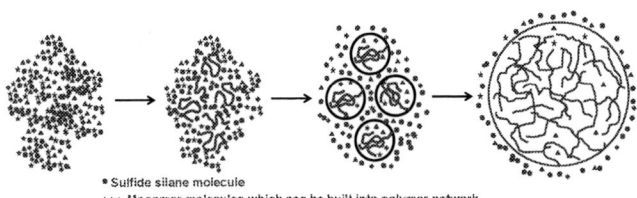

* Sulfide silane molecule
*** Monomer molecules which can be built into polymer network

Figure 18: Schematic drawing of phase separation during polymerization of ECA formulation.

All this data supports the following proposed mechanism behind the resistance observation shown in Figure 11. Based on the ingredients provided by the ECA supplier, the sulfide silane does not have a functional group to react with the monomer and become part of the cross-linked network. During the ECA cure (i.e. polymerization), the sulfide silane molecule is gradually squeezed out from the polymer matrix and moves to the Au pad and silver particles. At the same time, the polysulfide linkage is broken and sulfur atoms are

978-1-4799-2408-0/14 $31.00 © 2014 IEEE   269

released from the sulfide silane molecule. The sulfur atom then reacts with the silver and forms silver sulfide which is not electrically conductive. During the ECA cure, reflow, underfill cure, and burn-in processes, there are two competing events happening at same time. One event is that the surface and grain-boundary silver atom diffuses along gold surface. This event develops the electrical conductivity from ECA to gold pad. The other event is that the sulfur atom reacts with silver and forms silver sulfide right at the gold-silver interface layer. The formed silver sulfide increases the resistance of the layer. Since the silver sulfide formation rate is faster than silver diffusion rate when sufficient amount of sulfur atom is available at that surface, the interface layer resistance increases during the first several burn-in cycles. But the amount of available sulfur compound at the gold-silver layer is limited. Once it is depleted, the silver atom diffusion takes control. This explains the fact that the extended burn-in process reduces the layer resistance as shown in Figure 11 (B).

As a final step, the study results were communicated to the ECA supplier and the supplier was asked to strictly control the added concentration of sulfide silane, the quality of the incoming sulfide silane, and the curing kinetics of the formulated adhesive. With the cooperation of the ECA supplier, stable resistance before and after burn-in was achieved.

Conclusions

Finite Element Analysis was found to be a useful tool to investigate delamination mechanisms. It was able to demonstrate how surface features on the opposing side of the PCB could lead to die attach stress and delamination. FEA models are predictive and can estimate delamination area when validated by acoustic imaging and destructive test results. Finally, the model can be used to test potential design changes intended to reduce or eliminate delamination. The experimental study showed that more DA and WB pad delamination were induced by higher mold pressure and temperature. The effect of mold pressure agreed well with the FEA modeling prediction. This trend was shown more evident with increased moisture contents. Therefore it is recommended to use lower mold temperature and pressure to reduce delamination rate. The electrical conductivity path was created through surface and grain-boundary diffusion of silver to gold during ECA cure and subsequent process steps with elevated temperature. The silver diffusion onto gold is an accumulative process. A sulfide silane coupling agent is often used in ECA formulation to increase the adhesion between ECA and gold surfaces. The sulfide silane coupling agent can release sulfur atoms during ECA curing and the sulfur atoms react with silver and cause instable ECA/gold interface resistance issues. To achieve the stable electrical performance of cured ECA, the added concentration of sulfide silane, the quality of the incoming sulfide silane, and the curing kinetics of the formulated adhesive need to be strictly controlled.

Acknowledgments

Special thanks to Molly McGuire for lab and acoustic imaging data collection, Mark Ricotta for his guidance in delamination DOE builds, Karen Dye for all the SEM and TEM analysis, Brian Roy for electrical failure mode analysis, Jeff Yates for resistance data collection during burn-in and current surge, and all the other engineers and technicians for their direct and indirect contribution to this paper. Without their hard work and expertise, this study could not be completed.

References

1. J.C.M. Hwang, J.D.Pan, and R.W. Balluffi, "Measurement of grain-boundary diffusion at low temperature by the surface-accumulation method. II. Results for gold-silver system," *Journal of Applied Physics*, vol. 50(3), pp. 1349-1359, March 1979.

2. Abraham Ulman, "Formation and Structure of Self-Assembled Monolayers," *Chemical Reviews*, vol. 96, No.4, pp. 1533-1554, 1996.

3. M.H.Hebb, "Electrical Conductivity of Silver Sulfide," *The Journal of Chemical Physics*, vol. 20, issue 1, pp. 185-190, January 1952.

4. D. Giancoli, *Physics: principles with applications*, 4th edition, Prentice Hall, London, 1995.

Effect of Aligned Nanofiber in Nanofiber Solder Anisotropic Conductive Films (ACFs) on the Solder ball Movement for Flex-on-Flex (FOF) Assembly

[1]Tae-Wan Kim, [1]Sang-Hoon Lee and [1]Kyung-Wook Paik

[1]Nano Packaging and Interconnect Lab. (NPIL)
Department of Materials Science and Engineering
Korea Advanced Institute of Science and Technology (KAIST)
291 Daehak-ro, Yuseong-gu, Daejeon, 305-701, Korea

phone: +82-42-350-3375, fax:+82-42-869-8624
e-mail: tkim31@kaist.ac.kr

Abstract

Anisotropic conductive films (ACFs) have been widely used as an interconnection adhesive materials for decades due to its high resolution, light weight, thin structure, and low power consumption. However, the high demands for miniaturization caused interconnection issues such as short circuits when it comes to fine pitch applications. By introducing nanofiber into the ACF system, nanofiber not only suppresses the solder ball movement during the bonding processes but also forms an insulating layer around the solder ball preventing short circuit between neighboring electrodes to occur. Even though the benefits of nanofiber were proven, the effect of the nanofiber orientation on the solder ball movement suppression ability has not been clarified.

In this study, nanofiber/solder ACFs using nanofibers that are aligned in various directions were investigated with respect to solder ball capture rate, electrical/mechanical properties, and reliability. Nanofiber/solder ACFs consists of adhesive resin, solder ball, and nanofiber material. The solder ball incorporated nanofibers were first obtained through electro-spinning technique. In order to align the nanofibers, drum type receiver was used where the receiver was capable of rotating up to 2500 rpm during the electro-spinning process. This allows the nanofiber to deposit in an aligned fashion. After the solder ball incorporated aligned nanofiber layer was obtained, the solder incorporated nanofiber layer was then laminated between two non-conductive films (NCFs) in order to produce the finalized aligned nanofiber/solder ACFs. Here Sn58Bi (58%Sn and 42%Bi) solder ball was used whichhave a low melting temperature of 138 °C. For the bonding method, vertical ultrasonic bonding method was used to make fresh solder joint on metal electrodes. Solder ball normally contains an oxide layer which prevents stable joint to form. Vertical ultrasonic vibration allows us to break the oxide layer during the bonding process forming a stable metallurgical solder joint. The solder movement analysis, electrical/mechanical properties, and reliability measurements were done to confirm the aligned

nanofibereffect.

As a result, aligning the solder incorporated nanofibers using the electrospinning technique was successfully performed. Since the role of the nanofiber is to suppress the solder movement during the resin flow, it was shown that the solder ball capture rate increases as the aligned nanofibers are parallel to the resin flow direction. Also, by introducing nanofiber and Sn58Bi solder ball, excellent electrical properties as well as reliability was observed.

1. Introduction

Anisotropic conductive films (ACFs) are adhesive interconnection material, which consist of conductive particles and adhesive polymer resin. This adhesive films are used for various applications such as interconnection material for flat panel displays (flex to PCB, chip-on-chip, and chip-on-glass) and flip chip semiconductor packaging applications due to its fine pitch capability and simple process [1][2][3]. Due to the fast development of electronic devices, the demands for further miniaturization and multi-functionalization are unavoidable. However, when it comes to fine pitch packages, the main limitations is their unstable electrical properties. During the bonding process, the resin tends to flow out of the system allowing the conductive particles to migrate. Because of this conductive particles migration, the conductive particles have the tendency to agglomerate creating an electrical bridge between neighboring electrodes resulting short circuit. Regarding to the short circuit issue, nanofiber incorporated ACFs were introduced in the past in our group [4][5][6].

In this study, we introduce solder incorporated aligned nanofiber into the ACF system in order to maximize the nanofiber effect on the solder ball suppression while eliminating any electrical short circuit issue. This is done by using electrospinning technique along with drum type receiver where the receiver is capable of rotating. [7].

Moreover, we will introduce Sn58Bi solder ball which has a low melting temperature of 138°C and ultrasonic bonding to achieve low temperature, fast resin curing bonding and stable joint formation.

2. Experimental

2.1. Material preparation & equipment

2.1.1. Test vehicle preparation

Polyimide base 100 μm pitch flexible printed circuits (FPCs) with a dimension of 30 mm X 24 mm X 0.05 mm were designed as shown in figure 1. The main bonding area of the substrate was 2 mm X 10 mm. The electrodes of the FPCs were mainly copper electrodes with a height of 12 μm and the electrodes were finished treated with 100 nm organic solderability preservative (OSP) material in order to prevent any additional oxidation. Figure 2 shows the cross-sectional SEM image of the OSP finished Cu electrode. For electrical property measurement, the bottom FPC contains 6 regions to measure the contact resistance, and 4 regions to measure the insulation resistance.

Top substrate   *Bottom substrate*

Figure 1.Thephotographic image of polyimide base 100 μm pitch flexible printed circuits (FPCs)

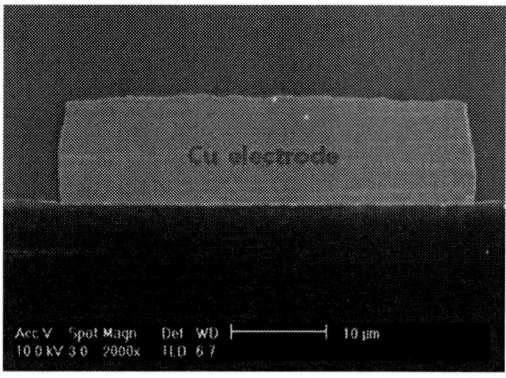

Figure 2.The SEM image of the copper electrode of 100 μm pitch FPCs

2.1.2. Aligned nanofiber/Sn58Bi solder ACF formation

Sn58Bi solder ball containing nanofibers were obtained by using the electrospinning equipment shown in Figure 3. When high electric field is applied to the syringe needle, this generates charge to the polymer solution. When it reaches its critical point, a stream of polymer solution erupts from the tip and migrates to the ground receiver in a Taylor cone shape resulting nanofibers. The polymer used in this experiment was polyvinylidenefluoride (PVDF). In order to produce Sn58Bi solder ball containing PVDF

nanofibers, 15 wt. % of PVDF, 42 wt. % dimethylacetamide (DMAC), and 28 wt. % acetone along with 15 wt. % Sn58Bi solder balls were mixed thoroughly while heating at 50°C . The Sn58Bi solder ball size used in this experiment ranged from 5 μm to 15 μm. After the polymer solution is well mixed, the polymer solution was then electrospun through electrospinning equipment. PVDF nanofiber was electrospun at 10 KV while working distance, pumping rate, and needle diameter were fixed as 10 cm, 50 μL/min, and 250 μm. In order to align the nanofibers, drum type receiver was used where the receiver was capable of rotating up to 2500 rpm during the electro-spinning process.Three different types of solder incorporated nanofibers were electrospun for comparison: random, parallel to the electrode, perpendicular to the electrode. Once the solder incorporated nanofibers were fabricated, the fiber layer was laminated between two non-conductive films (NCFs) with a thickness of 10 μm using the roll and vacuum laminator resulting the finalized PVDF nanofiber/Sn58Bi solder ACFs with a thickness of around 25 μm.

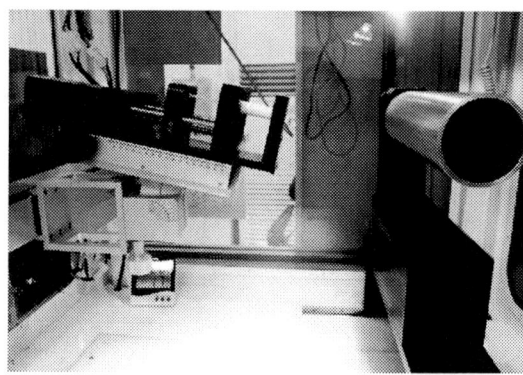

Figure 3.The photographic image of Electrospinning apparatus

2.1.3. Vertical ultrasonic bonding apparatus& bonding condition

Solder ball normally contains native oxide around the solder ball surface, which prevents stable solder joint to form. Applying vertical ultrasonic vibration allows us to break this oxide layer during the bonding process forming stable metallurgical solder joint. The vertical ultrasonic bonding apparatus shown in figure 4 includes 3 main regions: bonding region, preparation region, and aligning region. The longitudinal vibration frequency of the bonder is 40 KHz and the output power is 400 W. The vibration is mainly applied on the vibration horn and the U/S vibration amplitude ranges from 4 to 13 μm. In order to prevent vibration horn damage and maintain the vibration uniformity, 200 μm thick rubber interposer was used between the horn and the flex substrate. The schematic of the vertical ultrasonic bonding process is shown in figure 5.

In this experiment the bonding condition was separated into two parts: pre-bonding condition and main-bonding condition. The bonding conditions were decided based on the resin curing temperature and Sn58Bi solder ball melting temperature. The pre-bonding was performed bellow the resin curing temperature ($\leq$ 80 °C) allowing the solder ballto be positioned between the top and bottom electrodes before the solder melting occurs. The optimized pre-bonding condition was set to 6 MPa 3.8 μm amplitude (78 °C) 10 second bonding. The mainbonding was performed so that the Sn58Bi solder ball melting occurs while the resin is fully cured. The main-bonding condition was set to 3 MPa 11.2 μm amplitude (200 °C) 5 second bonding.

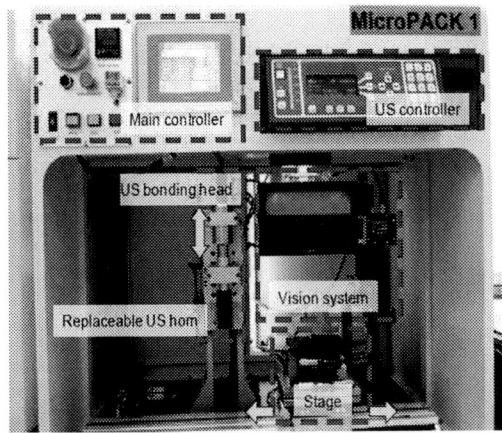

Figure 4.The photographic image of vertical ultrasonic bonding apparatus

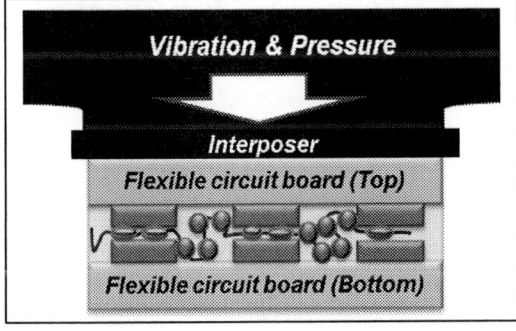

Figure 5.The schematic image of vertical ultrasonic bonding process

2.2. Characterization of nanofiberorientation effects on ACF joint properties

2.2.1. Tensile strength analysis

The tensile test of three different nanofibers (random, parallel to the electrode, perpendicular to the electrode) were performed using the micro tensile test apparatus shown in figure 6(a) along with the D638 condition of ASTM standard. All three nanofibers were cut into type V specimen dimension of D638, which has a gage length of 7.62 mm, width of 3.18 mm, and overall length of 63.5 mm shown in figure 6(b). Figure 7 shows the SEM image of each nanofiber after it was cut into type V specimen dimension.

Since the role of the nanofiber is significant during the

pre-bonding process, the tensile test was performed at pre-bonding temperature (~80°C)using the furnace with a cross head speed of 10mm/min.

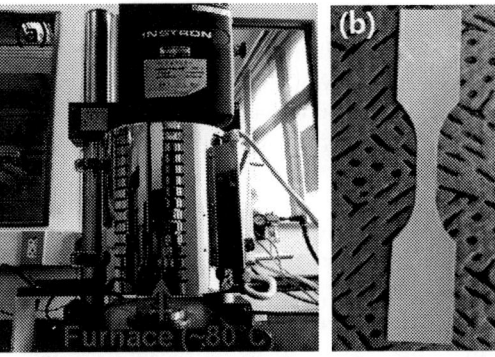

Figure 6.The photographic image of (a) micro tensile test apparatus and (b) nanofiber specimen

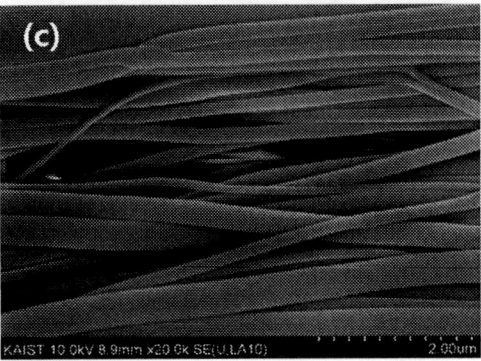

Figure 7.The SEM images of three different types of

nanofibers : (a) random, (b) parallel to the electrode, and (c) perpendicular to the electrode after preparing micro tensile test specimen

2.2.2. Solder ball movement analysis

During the pre-bonding process, solder balls tend to migrate due to polymer resin flow. The solder ball movements were observed before and after pre-bonding through the optical microscope. Since, it is difficult to observe the solder balls directly through the metal electrodes using the optical microscope, 26 μm thick transparent PI film was bonded on top of the 100 μm pitch FPC in order to observe the particle movement through the microscope. The number of solder ball was counted before and after the pre-bonding process to confirm the nanofiber orientation effect on the solder movement. The number of captured solder ball was then converted intocapture rate for comparison.

2.3. Electrical property analysis & Reliability evaluations on vertical U/S bonded ACF joints

2.3.1. Contact resistance analysis

After performing the vertical ultrasonic bonding, the contact resistances were measured to characterize the electrical properties of the three differently oriented nanofiber/solder ACFs (random, parallel to the electrode, perpendicular to the electrode). This was done by using the 4 point probe analyzer. Current and voltage were applied through the electrical path shown in figure 8(b), where contact resistance value comes out as an output.

2.3.2. Insulation resistance analysis

As mentioned before, the solder balls will migrate during the bonding process due to the resin flow causing agglomeration of solder balls which eventually results short circuit to occur. The insulation property of the three differently oriented nanofiber/solder ACFs (random, parallel to the electrode, perpendicular to the electrode) were measure by using themultimeter. The electrical path is shown in Figure 8(a).

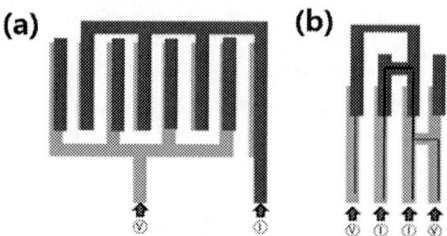

Figure 8.The schematic image of electrical path of the flex substrate (a) insulation resistance and (b) contact resistance

2.3.3. Pressure Cooker Test

The pressure cooker test (PCT) was performed in order to evaluate the reliability of the FOF assembly using the three differently oriented nanofiber/solder ACFs (random, parallel to the electrode, perpendicular to the electrode). The test was performed at 2 atm, 121 °C, 100 % RH for 48 hours. The contact resistances were measure every 24 hours to confirm any electrical failure (open circuit).

3. Results and discussion

3.1. Aligned nanofiber/Sn58Bi solder ACF formation

Sn58Bi solder incorporated nanofiber layer was first produced using the electrospinning technique. As mentioned previously in the experimental section, each polymer solution concentrations were optimized and then electrospun at an optimized spinning condition to produce Sn58Bi solder incorporated nanofiber layer. Also in order to fabricate aligned nanofibers, drum receiver was rotated at a fast speed up to 2500 rpm. It is shown in figure 9 and figure 10 that as the rotation speed increases, the nanofiber angle distribution becomes smaller and smaller. At 2500 rpm, over 90 % of the nanofiber showed an average nanofiber angle of 5.8 degrees considering the y axis zero degree. Also, when controlling the needle to receiver distance while fixing the drum speed at 2500 rpm, 10 cm needle to receiver distance showed an average nanofiber angle of 6.8 degrees which is shown in figure 11 and figure 12. When the distance is larger than 10 cm, majority of the nanofibers were lost due to the backwind produced by the drum rotation. After producing the PVDF nanofiber, the nanofiber layer was laminated between two 10 μm thick NCF films using roll and vacuum laminator to produce the finalized nanofiber/solder ACFs. These finalized nanofiber Sn58Bi solder ACFs were used for U/S bonding.

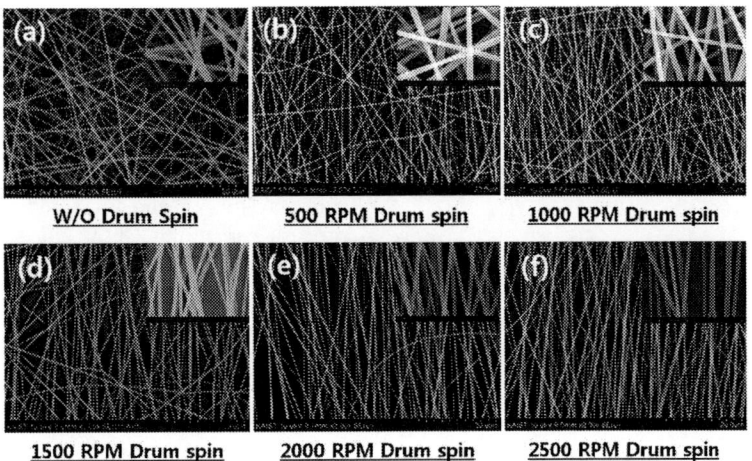

Figure 9.The SEM images of nanofibers depending on the drum receiver speed (a) without drum spin (b) 500 rpm (c) 1000 rpm (d) 1500 rpm (e) 2000 rpm (f) 2500 rpm

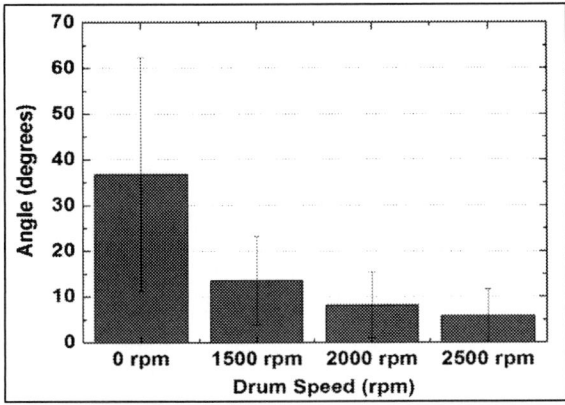

Figure 10.The nanofiber angle distribution depending on the drum receiver speed

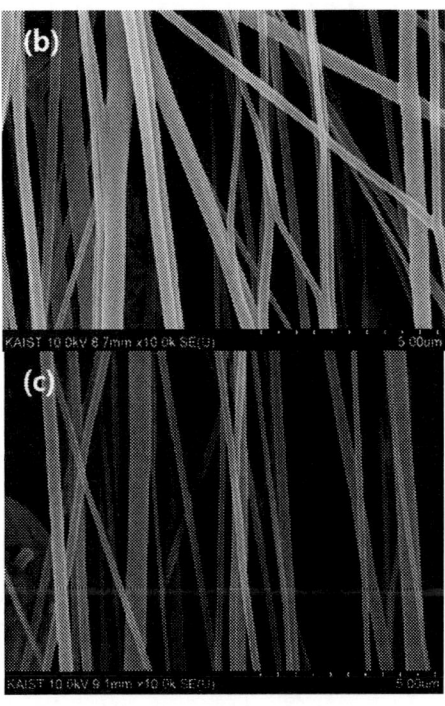

Figure 11.The SEM images of aligned nanofibers depending on the needle to receiver distance: (a) 5 cm, (b) 7 cm, and (c) 10 cm (set rotation speed: 2500 rpm)

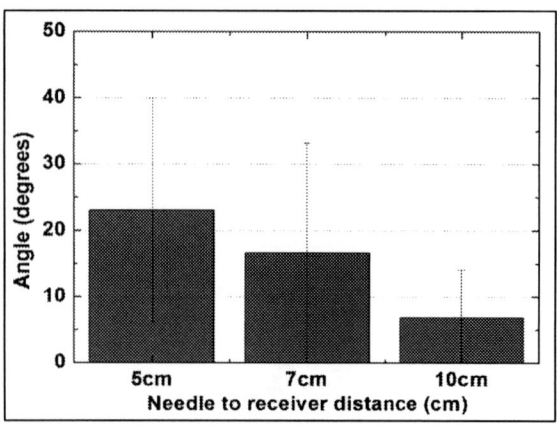

Figure 12.The nanofiber angle distribution depending on the needle to receiver distance

3.2. Characterization of nanofiberorientation effects on ACF joint properties

3.2.1.Tensile strength analysis

The tensile strength analysis results are shown in figure 13. Nanofibers that were aligned parallel to the pull axis showed the highest ultimate tensile strength of 27MPa. However, when the nanofibers were aligned perpendicular to the pull axis, the ultimate tensile strength value was only 0.24 MPa. For the randomly oriented nanofiber, the ultimate tensile strength was 15 MPa which was almost the average value.

When looking at the strain point of view, nanofibers that were aligned perpendicular to the pull axis showed the highest value of 6.3 while nanofibers that were aligned parallel to the pull axis showed the lowest value of 2.1. The reason why nanofibers aligned perpendicular to the pull axis can have such a high strain is because the nanofiber structure is formed by a single connected nanofiber rather than a bundle of nanofibers.

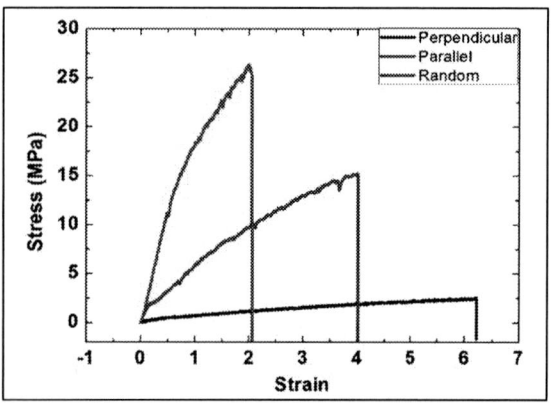

Figure 13.Tensile strength test results of three different types of nanofibers : random, parallel to the electrode, and perpendicular to the electrode

3.2.2. Solder ball movement analysis

The results of solder ball movement analysis are shown in figure 14. For nanofiber/solder ACF using random nanofibers,

the number of solder balls located on the electrode before pre-bonding were around 45 for 10000 μm^2 area. However after the pre-bonding process, the number of particles reduced down to 35 resulting 75.9 % capture rate. Nanofiber/solder ACFs using parallel and perpendicular nanofibers had 27 and 28 solder balls before the pre-bonding process and 22 and 14 solder balls after the pre-bonding process, resulting 82.9 % and 50.4 % capture rate respectively. Since the major resin flow occurs parallel to the electrode direction, by using nanofibers that are parallel to the electrode, the solder ball capture rate was increased 7.0 % compared nanofiber/solder ACF using random nanofibers. On the other hand, when using nanofibers that are perpendicular to the electrode, the capture rate drops even lower than the nanofiber/solder ACF using random nanofibers.

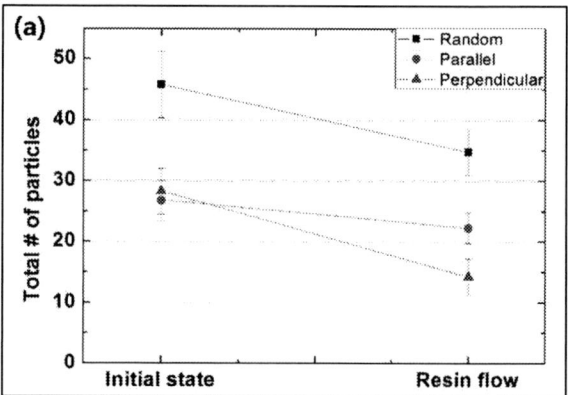

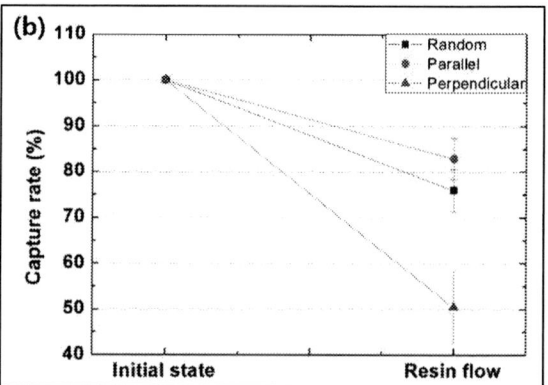

Figure 14.Solder ball movement analysis results of three different types of nanofiber/solder ACFs : random, parallel to the electrode, and perpendicular to the electrode.(a) number of particles before and after resin flow (b) capture rate

3.3. Electrical property analysis &Reliability evaluations on vertical U/S bonded ACF joints

3.3.1. Contact resistance analysis

The average contact resistances of nanofiber/solder ACFs using random, parallel to the electrode, and perpendicular nanofiberswere15.1mΩ, 15.3mΩ, and 15.3mΩ respectively as shown in figure 15. All three nanofiber/solder ACFs showed lower contact resistances compared to the conventional ACF, but almost no difference between the three nanofiber/solder ACFs. For conventional ACF, the conductive particle forms a physical contact with the electrodes preserving it shape, while for all three nanofiber/solder ACFs, the Sn58Bi solder ball spreads out forming a metallurgical joint with the electrodes allowing larger contact area. This is why the three nanofiber/solder ACFs showed lower contact resistances compared to the conventional ACF contact resistance.

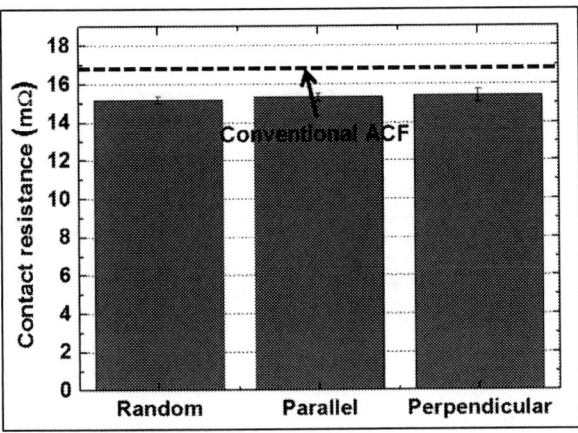

Figure 15. Contact resistance results of three different types of nanofiber/solder ACFs : random, parallel to the electrode, and perpendicular to the electrode

3.3.2. Insulation property

In addition to the movement suppression capability, nanofiber forms an insulation coating around the solder ball which acts as a protection layer. The insulation property results for three different types of nanofiber/solder ACFs are shown in figure 16. For conventional ACF, conductive particles tend to agglomerate after the bonding process due to the poor conductive particlesuppression ability. This agglomeration forms bridge between the electrodes resulting short circuit. One the other hand, no short circuit was observed for the all three nanofiber/solder ACFs showing excellent insulation property regardless of the nanofiber orientation. This eventually proves the excellent insulation property of nanofiber/solder ACFs.

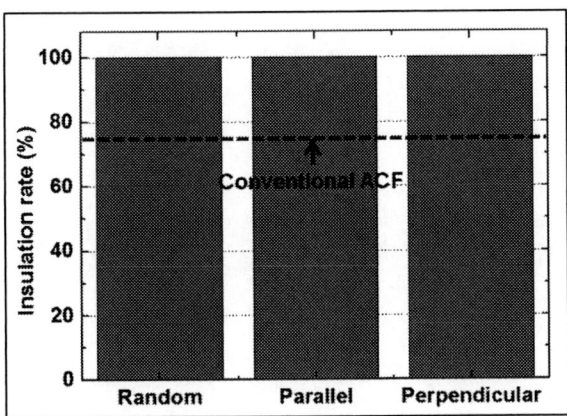

Figure 16. Insulation resistance results of three different types of nanofiber/solder ACFs : random, parallel to the electrode, and perpendicular to the electrode

3.3.3. Reliability evaluations on U/S bonded ACF joints

After the bonding process, joint stability can be tested for reliability issues. The pressure cooker test (PCT) results of the three different types of nanofiber/solder ACFs are shown in figure 17. For all three types of nanofiber/solder ACFs, there was a minor increase (~20 mΩ) in the contact resistance but no open circuits were observed even after 48 hoursshowing stable solder joint formation.

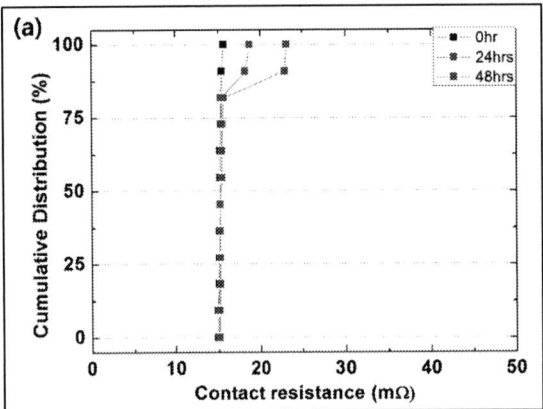

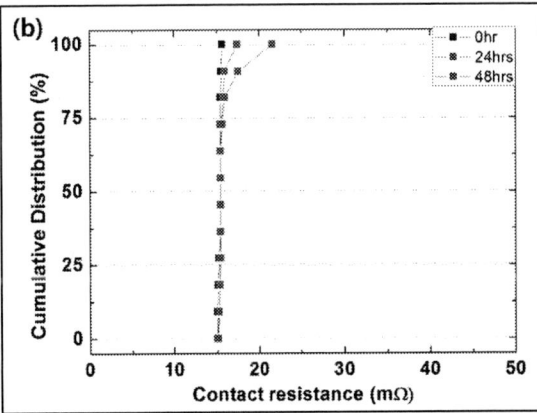

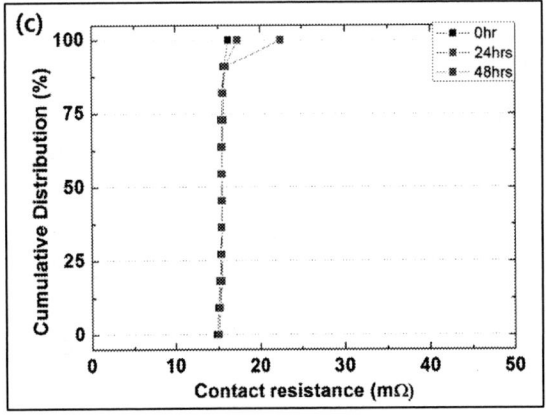

Figure 17. PCT results of three different types of nanofiber/solder ACFs : (a) random, (b) parallel to the electrode, and (c) perpendicular to the electrode

4. Conclusion

Fine pitch FOF assembly was successfully demonstrated using aligned nanofiber Sn58Bi solder ACF and ultrasonic bonding. Due to the fine pitch issue such as short circuit, the concept of nanofiber was introduced. When aligning the nanofibers, rotating the drum receiver at 2500 rpm while maintaining the needle to receiver distance to 10 cm showed the lowest nanofiber angle distribution. Also, micro tensile test results showed that nanofiber parallel to the pull axis had the highest tensile strength value of 27 MPa. However, nanofiber perpendicular to the pull axis had the lowest tensile strength value of 0.24 MPa . When looking at the movement analysis of nanofiber/solder ACF using nanofibers parallel to the electrode showed a 7 % conductive particle capture rate improvement compared to the nanofiber/solder ACF using random nanofibers with perfect insulation property. On the other hand, nanofiber/solder ACF using nanofibers perpendicular to the electrode showed a 25 % decrease in the conductive particle capture rate compared to the nanofiber/solder ACF using randomly oriented nanofibers which prove that the resin mainly flows parallel to the electrode. Lastly, all three nanofiber/solder ACFs showed stable contact resistance as well as reliability due to the stable metallurgical solder joint formation.

Overall, aligned nanofiber Sn58Bi solder ACF using nanofibers parallel to the electrode showed excellent solder ball suppression ability as well as electrical and mechanical performances even at 200 °C bonding temperature by incorporating aligned nanofiber and low temperature Sn58Bi solder balls.

5. References

[1] K. W. Lee, "Ultrasonic Anisotropic Conductive Films (ACFs) Bonding of Flexible Substrates on Organic Rigid Boards at Room Temperature", *in Proc. 57th Electron.Compon. Technol. Conf.,* Reno, NV, 2007, 480-486.

[2] Yim, M.J., Paik, K.W., "Recent advances on anisotropic conductive adhesives (ACAs) for flat panel displays and semiconductor packaging applications," *IEEE Transactionson Advanced Packaging*, Vol. 26, No. 5 (2006), pp. 304-313.

[3] Kristiansen, H., Liu, J., "Overview of conductive adhesive interconnection technologies for LCD's,"*IEEE Transaction on Components & Packaging Technologies*, Vol. 21, No. 2 (1998), pp. 208-214.

[4] K. L. Suk, "Nanofiber anisotropic conductive adhesives (ACAs) for ultrafine pitch chip-on-film (COF) packaging," *in Proc. 61st Electron.Compon. Technol. Conf.,* Lake Beuna Vista, FL, 2011, pp. 656–660.

[5] Lee, S.H., Suk, K.L., Paik, K. W., "Study on FinePitch

Flex-on-Flex Assembly Using Nanofiber/Solder Anisotropic Conductive Film and Ultrasonic Bonding Method", *IEEE Transactionson Advanced Packaging*, Vol. 2, No. 12 (2012), pp. 2108-2114.

[6] Kim, T.W., Suk, K.L., Lee, S.H., Paik, K. W., "Low Temperature Flex-on-Flex Assembly Using Polyvinylidene Fluoride Nanofiber Incorporated Sn58Bi Solder Anisotropic Conductive Films and Ultrasonic Bonding," *Journal of Nanomaterials*, Vol. 2013, pp. 1-10.

[7] Meng, Z.X., Wang, Y.S., Ma, C., Zheng, W., Li, L., Zheng, Y.F., "Electrospinning of PLGA/gelatin randomly-oriented and aligned nanofibers as potential scaffold in tissue engineering," *Materials Science and Engineering: C*, Vol. 30, Issue 8, pp. 1204-1210.

Adhesive Enabling Technology for Directly Plating Metal on Molding resin

Kwonil Kim, Kenichiroh Mukai, Brian Eastep, Lee Gaherty, Anirudh Kashyap,
Lutz Brandt
Atotech USA Inc.
369 Inverness Parkway, #350, Englewood, CO, 80112, USA
+1(303)217-5376,

Abstract

This paper aims to introduce a new wet chemical process for adhesion enhancement of plated metal to epoxy molding compounds (EMC). The approach is based on an innovative combination of mechanical anchoring and chemical adhesion.

The new approach broadens possible application ranges and replaces more limited processes such as sandblasting and sputtering, which have cost and technical drawbacks.

Introduction

Plating on epoxy molding compounds or molding resins is a relatively new field which could open new design and business opportunities. One such opportunity is the increased use of electromagnetic interference (EMI) shielding for closely spaced active components. Currently, this is accomplished mostly by metallic shells or "caps." However, this technique also increases the spatial requirements and reduces flexibility of component layout on the PCB.

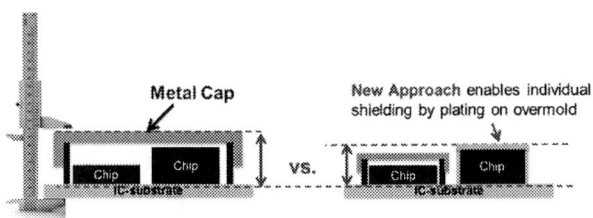

Figure 1: Concept of plated shields.

An alternative and more space saving approach is a plated EMI shield using a wet-chemically deposited Cu or Ni seed layer. While in some instances, sandblasting and sputtering a metallic seed layer in preparation of plating is feasible and can give adequate adhesion – it has the technical drawbacks of poor sidewall coverage and low metal layer thickness. As a result, scale- up for mass production is relatively difficult and costly. Classical electroless and electrolytic metallization are much more desirable but have been so far limited by insufficient adhesion.

In this paper we will present a new "Adhesion Enhancement" approach, where components encased by molding resin are directly coated with an electroless copper or nickel layer by using an adhesion promoter. By this method spatial requirements are minimized. Also, this process fits into the existing infrastructure of the PCB industry.

A second possible application is the "direct" circuitization of molding resins. This application aims at replacing standard build-up resins while at the same time embedding active

components. This approach has the merit of being comparatively cost effective due to the cheaper materials.

There are several challenges in treating molding resins with wet chemistry. Firstly, molding resins have high filler content (70-90% wt.) with a wide size distribution ranging from a few nm to tens of μm. The roughness created due to the large filler sizes used does not allow for the "direct" circuitization of molding resins. Secondly, molding resins, unlike build-up resins have not been optimized for adhesion to plated metal and even contain waxy release agents which counteract adhesion. For this reason, classical desmear followed by electroless seeding can get a maximum adhesion of up to 2N/cm on typical molding resins. This adhesion appears to be insufficient to prevent delamination in the subsequent processes.

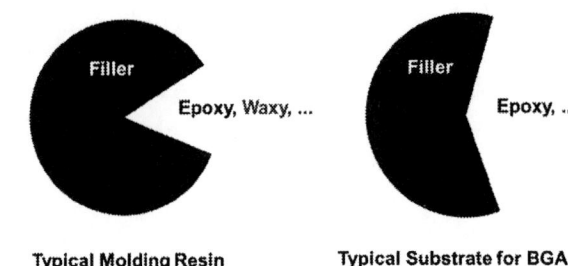

Figure 2: Different contents between molding resin and substrate for BGA, CSP.

In contrast, with the new approach presented in this paper, we are able to obtain up to 5N/cm of peel strength. The plated layer stands up well to the thermal reliability tests such as reflow shock (260°C) and HAST treatment; there is no delamination or significant loss of adhesion – key requirements for this application.

The technology is based on a new "Adhesion Enhancement" step providing a thin organic layer, which interacts with the molding resin and primes it for plating.

Results obtained so far suggest that these surface treatments are applicable to various kinds of molding resins.

EXPERIMENTS AND RESULTS

EMC substrates from various suppliers were investigated. The substrates were first cleaned in deionized water or a standard cleaning bath to remove loose debris and particles from the surface. The "Adhesion Enhancement" chemistry was then sprayed as a thin and uniform layer onto the substrate. During a brief bake, the organic components then selectively interact with the substrate surface.

A permanganate based oxidative treatment finally removes both excess chemistry and treated molding resin surface. Adhesion then results from a combination of mechanical and chemical bonding.

A general process flow is given below:

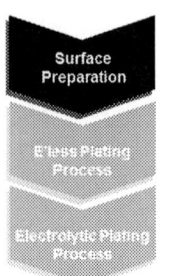

Adhesion Enhancement

Electroless Metal Plating

Electrolytic Metal Plating (option)

Figure 3: Adhesion and Metallization Process Overview.

Typical process conditions for the new Adhesion Enhancement approach are given below.

Process	Time (mins)	TEMlp (°C)
Surface Preparation Spray	3 - 4	RT
Baking	2 - 15	90-120
Oxidization Process	5 - 35	78 - 82
Reduction Process	1 - 2	25 - 35

Figure 4: Process Details "Adhesion Enhancement."

Process	Time (mins)	TEMlp (°C)
Adhesion Booster Treatment (Spray)	3 - 4	RT
Baking	8-12	80 - 120

Figure 5: Process Details Optional "Adhesion Booster."

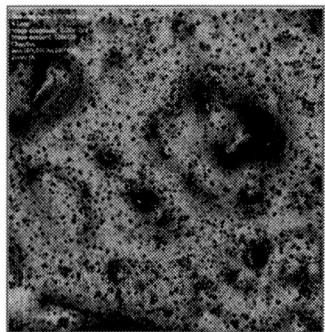 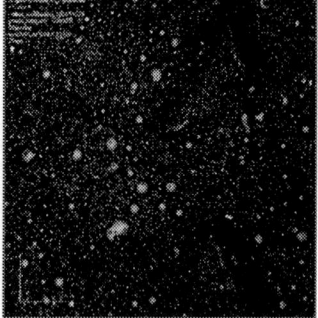

Figure 6: LEXT 4000* image "as- received" (left) and after the Reduction Process (right).

*2D 120x120μm high resolution confocal laser microscope image

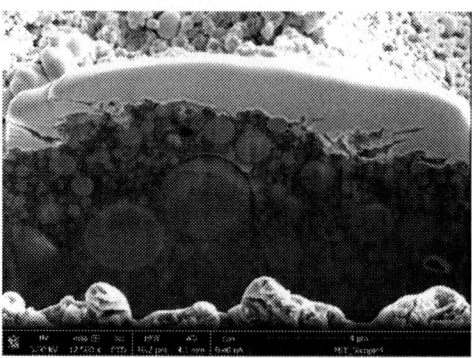

Figure 7: SEM image of FIB X-section after Reduction.

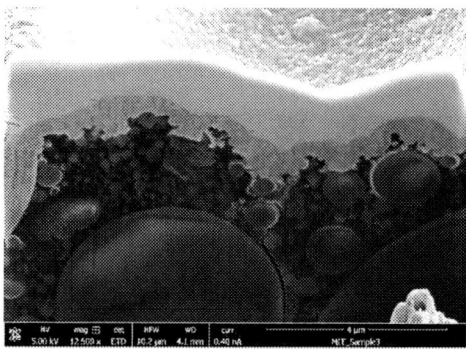

Figure 8: SEM image of FIB X-section after E'less Cu.

Examples shown in Figures 6-8 are meant to showcase the high filler content causing a relatively high average surface roughness (SRa) in the order of several microns.

The surface is then seeded with Pd activator for electroless copper or nickel plating. The thickness of those layers is typically in the order of several microns for EMI shielding applications. The exact sequence and thickness depends on the specific shielding requirement.

After electroless metallization the samples are rinsed and dried. This is followed by annealing at 150°C for 30 minutes.

If thicker metallization is required, additional electrolytic plating can be applied again followed by annealing.

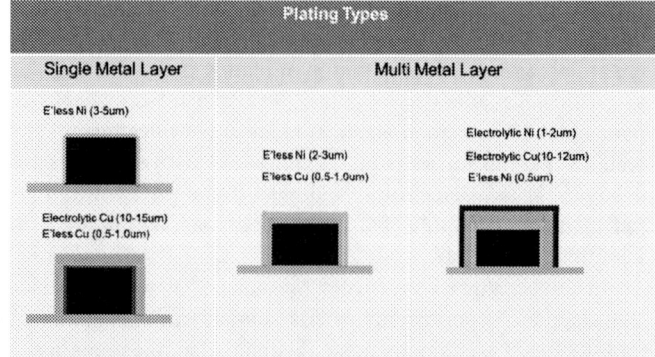

Figure 9: Various types of metal layer build-up on molding compounds.

For direct circuit formation on the molding resins surface, smaller filler sizes (<<10 μm) are preferred as the filler size directly impacts surface roughness and therefore also fine line resolution capability.

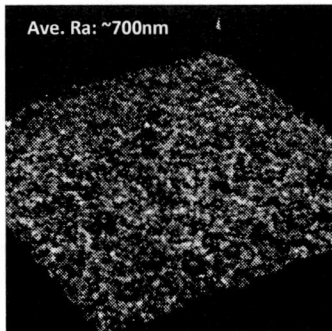

Figure 10: LEXT 4000 image.
*after Reduction for 10μm filler EMC

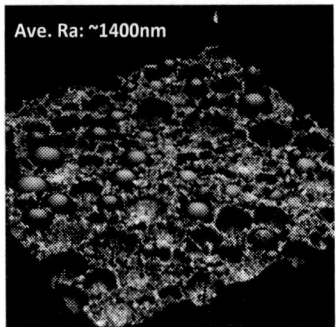

Figure 11: LEXT 4000 image.
*after Reduction for 20μm filler EMC

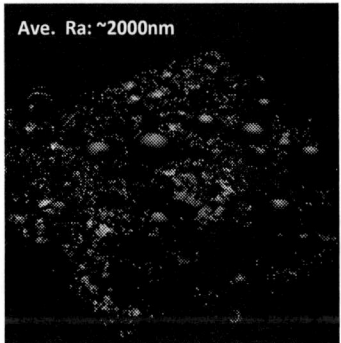

Figure 12: LEXT 4000 image.
*after Reduction for 45μm filler EMC

The above 3D images demonstrate how average roughness is dictated by filler size. Smaller filler sizes allow for a smoother surface with higher fine line (L/S) capability.

ADHESION & RELIABILITY TESTING

Adhesion quality for thin metal layers in the order of several N/cm was determined by Cross Hatching & Tape Testing (IPC-TM-650 Test Method 2.4.1).

Approximately 1.6x1.6 cm² substrates were nickel plated, cross-hatched, and checked for adhesion quality using a ~2N/cm adhesive tape (3M Scotch 600).

Typical failure modes are shown in the table below. Only the process based on the new "Adhesion Enhancement" passed the cross-hatching tape test.

The New Approach	Oxidation	Results	
-	-		Failed(Blister)
-	Yes		Failed
Yes	Yes		Passed

Figure 13: Tape test results comparison between processing w/ and w/o the new approach for metal combination with electroless copper(0.5μm) and electroless nickel(1.0μm).

As shown below, the use of the "Adhesion Enhancement" allows for the deposition of various sequences of metal layers with passing tape test results. This good adhesion opens a wide range of on chip circuitization and EMI shielding applications.

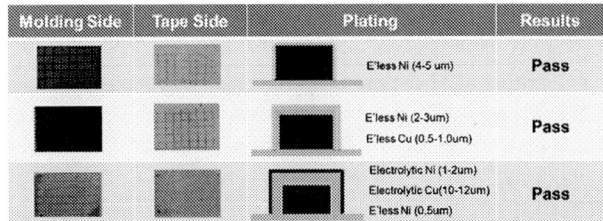

Figure 14: Cross-hatch tape test result on various metal sequences.

For EMI shielding applications, the full manufacturing process has to be considered because the incoming strips/sub-diced strips (w/ chip embedded) generally have solder resist, gold metal finishing or solder balls on their backside. This side has to be protected from the plating solutions. In manufacturing, this protection is a key requirement and can be achieved for instance by the application of a peelable ink or gaskets.

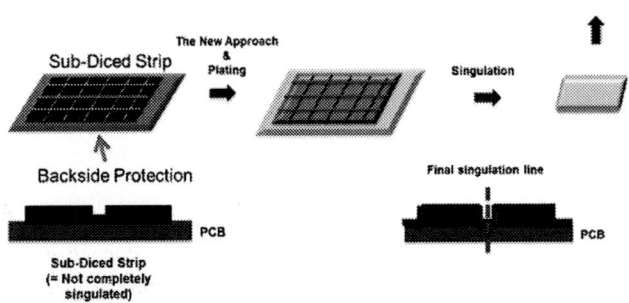

Figure 15: Process flow for manufacturing of EMI shielding.

Furthermore, for direct copper trace formation on the molding resin surface, an optional "Adhesion Booster" has been developed to even further enhance adhesion to the EMC.

The "Adhesion Booster" is a true chemical adhesion promoter which can be applied in combination with the described "Adhesion Enhancement" approach.

Performance of the "Adhesion Booster" is strongly dependent mple is given below.

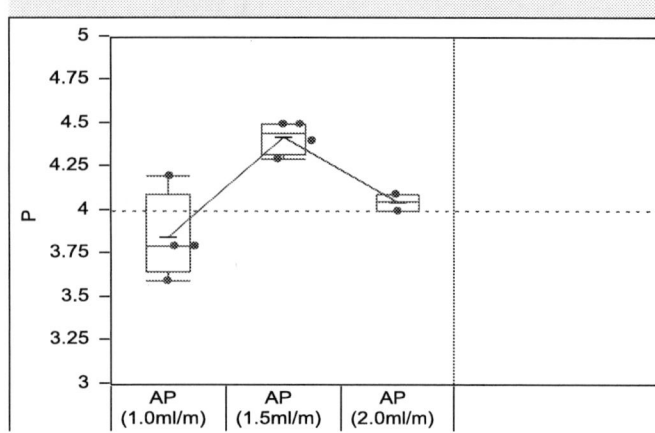

Figure 14: Initial peel strength vs. relative spray rates on EMC type A.

For this application, a total copper thickness in excess of 25 microns was applied. Adhesion of such metal layers is typically evaluated by a 90° Peel Strength (IPC-TM-650) evaluation whereby a 1cm wide strip of the plated layer is peeled from the substrate while recording the force.

This peel strength measurement tool is widely used in the PCB industry to check the adhesion strength between substrate and plated copper layer. Adhesion values above 3-4N/cm are considered to indicate adequate adhesion, that is the ability to survive thermal stress applied during component attachment or "reflow" without delamination or significant loss of adhesion.

Figure 15: Peel Strength Measurement Tool.

On several molding resins intended for circuitization, peel strength values of 4-5N/cm could be consistently achieved and maintained after (solder) reflow.

Material	(N/cm)			
	Mean	Max	Min	Pst Rfl
Supplier A - Type 1	4.3	4.6	3.9	4.8
Supplier A- Type 2	4.6	5.0	4.0	4.9
Supplier B - Type 1	4.0	4.2	3.7	3.8
Supplier B - Type 2	3.4	3.6	3.2	3.5
Supplier B - Type 3	3.7	3.8	3.6	4.1

Figure 16: Peel strength evaluation of different resin types. Reliability of adhesion was tested by "solder-reflow" heating to 260°C /30secs (repeated 5 times) as well as highly accelerated stress humidity testing (HAST: 96h, 130°C, 85% RH). No delamination or significant peel strength degradation occurred.

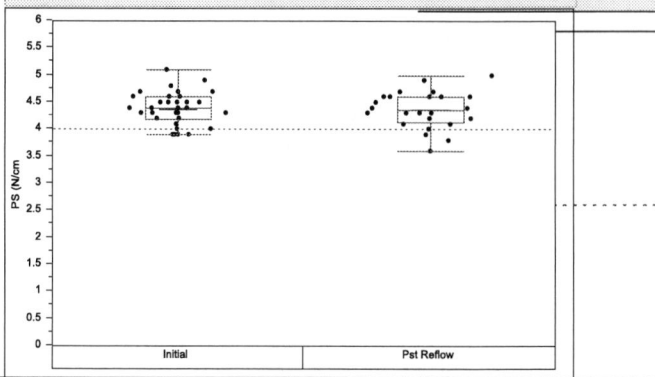

Figure 15: Peel Strength - initial and post reflow.

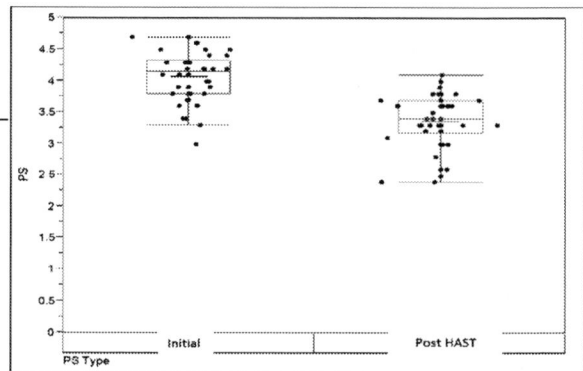

Figure 16: Peel Strength - initial and post HAST.

In comparison, classical desmear surface treatment of the same material could only achieve 1.5-2.5 N/cm of adhesion, while without any surface treatment there is no adhesion. A further advantage of the new "adhesion enhancement" approach is also the lower roughness compared to the classical desmear process (Figure 17).

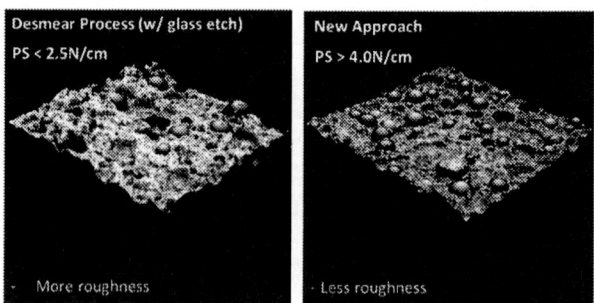

Figure 17: Direct comparison for peel strength & roughness between classical Desmear and "Adhesion Enhancement."

To determine the failure modes of the "Adhesion Enhancement" and "Adhesion Booster" processes, the peeled off copper and resin layers were investigated. Typically, cohesive failure is observed, which is the preferred mode over interfacial. For cohesive failure, resin and filler and filler fragments are observable on the peeled copper foil.

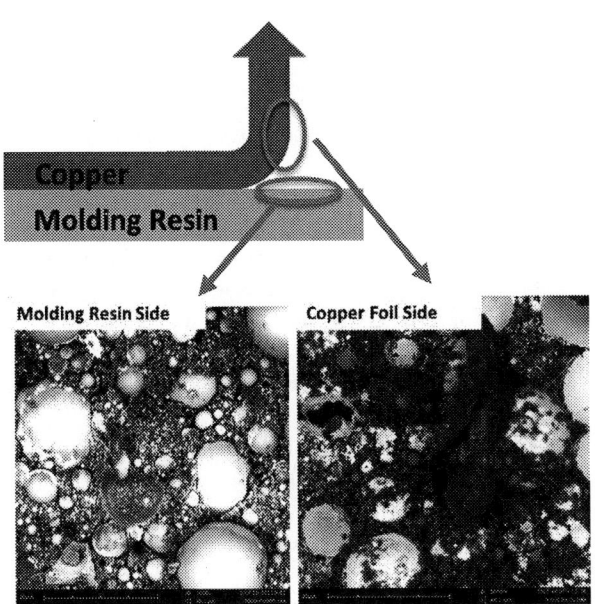

Figure 18: Peeled strip image – resin and copper side.

Figure 18 shows resin filler on the peeled copper foil consistent with cohesive failure. For reference, an image of the copper foil (Figure 19) is shown after interfacial failure. There is only little resin and filler residue present. Interfacial failure typically indicates low adhesive strength.

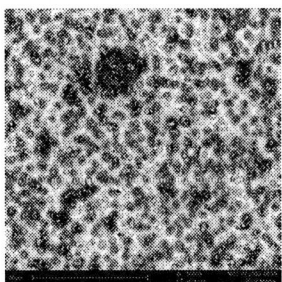

Figure 19: Copper foil after peeling with interfacial failure and low adhesion (<1 N/cm).

One major concern with respect to the adhesion of the plated metal to the EMC resin is the internal stress of the metal layer. High internal stress reduces the peel strength and can even cause local delamination or blisters. To release the internal stress of the metallic layer, typically an annealing step of 100 - 200°C is applied.

The optimal annealing process temperature depends on the targeted performance, the resin Tg, or the acceptable level of oxidation. An example is given in Figure 20 exploring annealing temperatures and times vs. adhesion.

In this example, annealing at 160°C provides the tightest peel strength distribution. Above 180°C, gradual degradation of adhesion as indicated first by a widening distribution and eventually by a significant drop of adhesion becomes apparent.

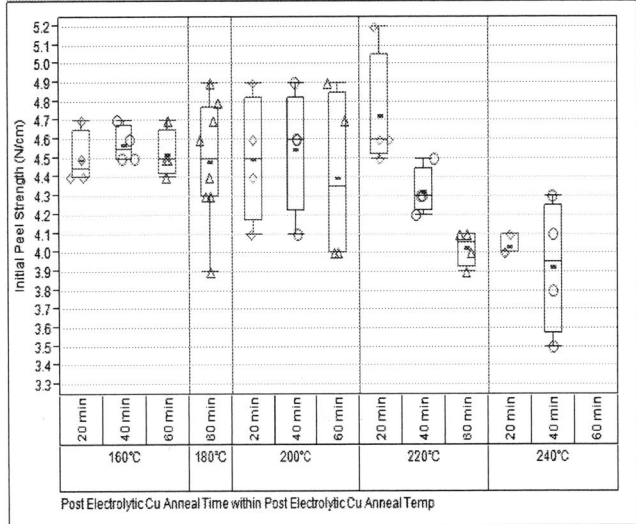

Figure 20: EMC annealing study – peel strength vs. temperature and time.

Conclusions

It has been shown that with the new "adhesion enhancement" process, the wet-chemical processing and the seeding of plated metal on molding resins becomes feasible. This allows the facile application of EMI shielding to closely space active components as well as the "direct" creation of conductive traces on EMC embedded chips.

References

1. Ishida Masaaki, Yamada Keiju, Yamazaki Takashi, *Electromagnetic Shielding Technologies for Semiconductor Packages,* Toshiba review Vol.67 No.2 (2012) http://www.toshiba.co.jp/tech/review/2012/02/67_02pdf/a03.pdf [hiperlink]

2. Scott Moris, Eric Schonthal, *Shielding RF Components at the Package Level*, AN RFMD® WHITE PAPER(2010) P10 [White Paper]

Modeling, Design, and Demonstration of Low-temperature Cu Interconnections to Ultra-thin Glass Interposers at 20 μm Pitch

Tao Wang, Vanessa Smet, Makoto Kobayashi[+], Venky Sundaram, P Markondeya Raj*, and Rao Tummala, Fellow, IEEE
3D Systems Packaging Research Center, Georgia Institute of Technology, Atlanta, USA
+ Namics Corporation, Niigata, Japan
* Email: raj@ece.gatech.edu

Abstract

This paper reports the first design and demonstration of a manufacturable 20 μm pitch Cu interconnection technology to ultra-thin glass interposers. Bonding is accomplished at temperatures below 200 °C without the need for solders. Manufacturability challenges such as substrate warpage, bump noncoplanarity and assembly throughput with low bonding times are addressed with this technology. The modeling and experimental results indicate that the ultra-fine pitch Cu interconnection offsets more than 3 μm non-coplanarity. Bonding interfaces were characterized to show that metallurgical bonding microstructure is formed even with a bonding time of 5 seconds, with superior electrical properties. A mechanism for low-temperature metallurgical bonding is proposed based on the characterization results.

Introduction

Off-chip interconnections are predicted to reach 20 micron pitch over the next few years to meet the miniaturization and performance demands by 2.5D and 3D stacked packages. Traditional flip-chip and copper pillar-solder cap interconnections, in spite of their many advantages due to liquid solder formation at the reflow temperatures, face challenges in meeting these pitch and power requirements. Therefore, solid-state bonding-enabled Cu interconnection technology is becoming a very important and necessary off-chip interconnections technology. The key benefits with this approach are: (1) Solid-state bonding enables smaller bumps for fine-pitch interconnections without the risk of bridging; (2) Cu has high electrical and thermal conductivities bringing high-speed signal transmission and high power-density handling; (3) Reliability challenges are minimized because no unstable interfaces associated with intermetallics and solders are present; (4) Cu plating is a standard BEOL process with well-established infrastructure.

In the last decade, Cu-Cu direct bonding has been one of the most extensively researched approaches to enable ultra-fine pitch and high performance chip-to-chip and chip-to-interposer interconnections. However, the need for careful Cu surface preparations and removal of residual oxides, combined with high-temperature and long annealing times for interdiffusion and recrystallization created several manufacturing challenges. Further, these bonding approaches are not tolerant to substrate warpage and bump non-coplanarities that are invariably present during IC assembly onto interposers and packages [1~3]. Ultra-thin glass interposers present yet another challenge, due to their brittle nature.

This paper comprehensively addresses these challenges resulting in a novel low-temperature Cu interconnection process for high thermo-mechanical reliability and high power-handling applications. The assembly technology also accounts for substrate warpage and bump non-coplanarity, and yet achieves ultra-short Cu interconnections at ultra-fine pitch, at low temperatures, without solders. The low-temperature assembly process is performed below the Tg of low CTE laminates, minimizing warpage issues during assembly. Table 1 summarizes the key objectives of Georgia Tech program with this innovative copper interconnection technology.

Table1. Objectives of the low-temperature copper interconnection technology at Georgia Tech

Metric	Parameters
Pitch Height	20-30 μm 10 μm
Properties	Resistance: < 10 mΩ per interconnection 1000 cycles of TCT (-55 °C ~ 125 °C), >192 hours of U-HAST (85% RH, 130 °C)
Reliability	Electromigration: > 1000 hours @ 10^6 A/cm^2, 130 °C
Manufac. process	Bonding temperature: < 200 °C Bonding time: <5 seconds Non-coplanarity offset: > 3 μm Bonding environment: in air

These objectives are accomplished with a patented process [5] consisting of ICs having 5~10 micron height Cu bumps, interposers or substrates with bonding pads having traditional ENIG (electroless nickel immersion gold) surface finish on both Cu bumps and Cu pads, pre-applying a unique B-stageable non-conductive underfill (BNUF) on the interposer or substrate, and thermocompression bonding to form the Cu interconnections. Thin Ni-Au layers were formed on the bonding surfaces as shown in Figure 1.

The key benefits with this approach are: (1)Low-temperature bonding without the need for solders, (2) Ultrahigh current-handling because of stable metallurgical interfaces; (3) High thermo-mechanical reliability because of pre-applied polymer adhesive; (4) High throughput by forming metallic bonding in less than 5 seconds. This paper describes the finite element modeling, fabrication and assembly of dies on glass interposers, as well as the characterization of the interfaces.

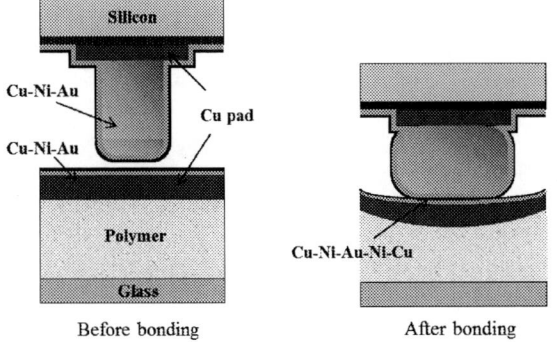

Figure 1. Schematics of Cu bump to Cu pad bonding with Ni-Au surface finishes on both sides

Finite Element Modeling

The first step in studying 20 μm pitch Cu interconnections is to model the deformation of the Cu bump during thermocompression bonding. The diameter of Cu bump was chosen as 8 μm, slightly less than 10 μm, to accommodate the bump widening effects during thermocompression bonding, as shown in Figure 2. The Cu bump height in the model was 12 μm, slightly larger than 10 μm, to accommodate the bump collapse. The thicknesses of Cu pads on the silicon or interposer or substrate were designed to be 2 μm and 3 μm, respectively. In addition, a 10 μm-thick polymer was assumed on either sides of the glass, before the formation of copper pads.

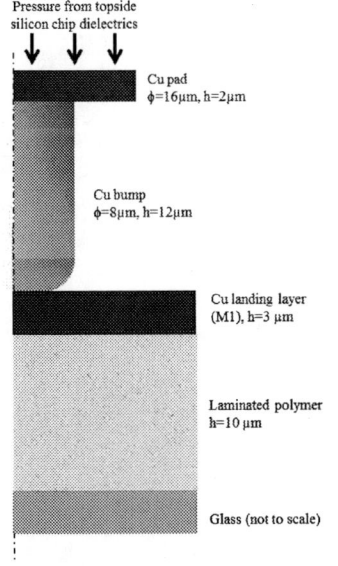

Figure 2. Finite-element model for the thermocompression bonding of Cu bump to Cu landing pad

Copper is considered an elasto-plastic material, described by its Young's modulus of 110 GPa, Poisson ratio of 0.34, and ultimate stress of 210 MPa. The true stress-true strain curve of the oxygen-free copper at 175 °C and a strain rate of 10^{-4} was used. [6] The frictional coefficient between the Cu surfaces (bump and pad) was set to be 0.75 [7]. The viscosity of the laminated polymer is neglected here, and it is considered a purely elastic material, described by its Young's modulus of 0.3 GPa. The Young's modulus of glass is 77 GPa and the Poisson ratio equals to 0.22. The modeling results are shown in Figure 3 and discussed next.

Bump and pad deformation: Stresses concentrate at the edge of the contacting area between the bump and pad. Plastic deformation of Cu initially starts at the stress-concentration area and then spreads towards the bump center. This leads to metal plastic-flow along the radial direction as well as longitudinal direction, emulating the mechanism of bump collapse while forming Cu interconnections. Collapse is defined as the vertical displacement of on-chip copper pads. The three states captured from the bonding modeling results are shown in Figure 3, in which collapse = 2.16 μm in state (a), collapse = 4.14 μm in state (b), and collapse = 6.39 μm in state (c).

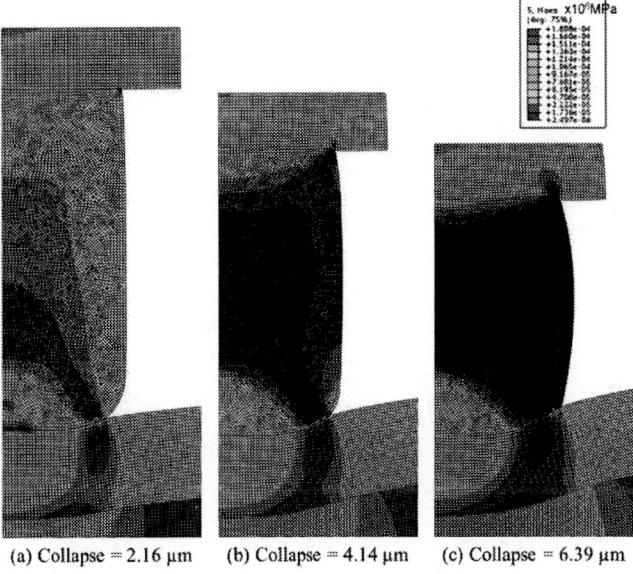

Figure 3. Bump and pad deformation and stress contour at various steps during compression bonding

Even though the bump diameter would increase due to the lateral deformation of the metal, it does not exceed so as to result in bridging between the two adjacent bumps. In Figure 4, the bump diameter was plotted as a function of the top-pad collapse, indicating that the bump diameter would increase by less than 2 μm when it collapses by 6.3 μm. Based on the theory of time-independent plasticity, the bump and pad deformation, as well as the resulting collapse can be accomplished in much less than one second. Therefore, these two factors provide the capability of ultra-fast die bonding with ultra-fine pitch off-chip interconnections.

Stress distribution on the on-chip Cu pads: Minimal stresses are desired on the on-chip dielectrics in order to avoid dielectric failures. Figure 4 plots the maximum von Mises stress on the dielectrics with the collapse in the top pad. It can be seen that the stress reaches an upper limit of 205 MPa after the large plastic strain. After achieving the bonding state shown in Figure 3(c), the maximum stress imposed on silicon dielectrics is 130 MPa, and occurs at the outer edge of the top pad-surface.

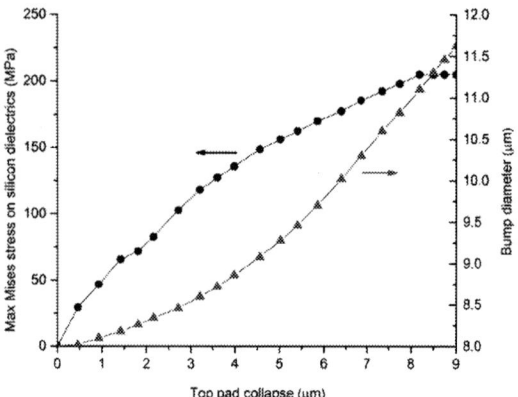

Figure 4. The evolution of bump diameter and maximum von Mises stress on silicon chip dielectrics with the top pad collapse

Fabrication

Silicon wafers were plated with a routing layer and Cu microbump layer, followed by ENIG surface finish. Dry-film photoresist lithography was used for semi-additive Cu electrolytic plating in both layers. Commercial electrolyte formulations from Atotech were used for the Cu plating and ENIG finish. The thickness of Ni varies from 200 to 500 nm, while that of Au is close to 100 nm. Regarding Cu bumps non-coplanarity, the average surface roughness (Sa) was 0.3 μm, while the maximum peak-to-valley height (Sz) was ~2 μm. Bump height variations within a single die was less than 3 μm.

The fabrication procedures for glass interposers consist of polymer lamination on a thin glass substrate, electroless Cu plating, photoresist patterning, electrolytic Cu plating, and ENIG surface finish. Glass substrates with 100 μm thickness were used, which brings handling issues due to its native fragility. The laminated polymer was 10 μm thick ZIS-100 from Zeon Corporation.

The Cu interconnection without solders does not lead to any solder-bridging or intermetallics formation. These attributes enable scaling to ultrafine pitch and high reliability. Both Au and Cu have FCC structures, and are the two most widely used metals in wire bonding, which requires excellent malleability and ductility properties. Au is chemically noble and can be deposited easily with either dry or wet process. Extra-high stress is not required to break down native oxides to form direct Au-Au bonding [8, 9]. Therefore, Au, in combination with Cu interconnections, is the optimum metal for direct bonding, compared to other alternatives including Al, Ni, Ti, and Pd [10-15].

A diffusion barrier layer must be deposited between Cu and Au to prevent the interdiffusion of Cu and Au atoms. Electroless-plated Ni layer was applied to act as the adhesion and solid-diffusion barrier layer between Cu and Au. The thickness of Ni layer can be adjusted to 200~500 nm, much less than 5.5 μm for standard C4 interconnections, due to the elimination of intermetallics issues. Electroless or electrolytic plating Au can also be applied prior to immersion Au if a thicker Au layer is required. The Ni and Au surface finish processing were applied on both Cu microbumps on silicon die and the outer surface of Cu layer on glass interposers.

Assembly on Glass Interposer

Assembly was performed with thermocompression bonding. All the thermocompression bonding experiments were performed with silicon dies on 100 μm thick glass interposers at a bonding temperature of 200°C. Prior to bonding, the BNUF or NCF was dispensed on the glass interposer and baked at low temperatures. The BNUF adhesive had a material formulation to prevent Cu oxidation on microbump and pad surfaces [16].

Figure 5 shows the images of glass interposers before and after die assembly. The alignment between the Cu microbumps and landing pads were checked with X-ray microscopy. The standard bonding profile for the Cu interconnection comprises of 365 MPa bump pressure for 60 seconds at 200°C. Other bonding conditions were also studied by adjusting the bump pressure from 90 MPa to 365 MPa, and the bonding time from 3 seconds to 10 min.

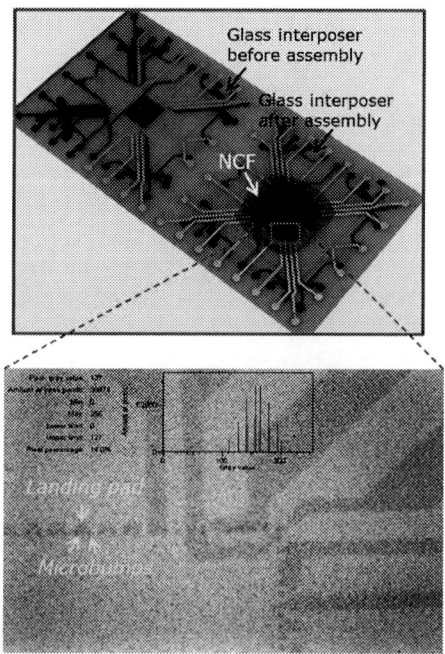

Figure 5. Top: Glass interposers before and after die assembly; Bottom: X-ray microscopy of the die-to-glass interposer assembly

As described in the modeling section, lateral plastic deformation of Cu occurs with the interconnection collapse, resulting in an increase of the contact area between the microbump and the pad. This implies that the effective pressure applied on the contact area decreases until it reaches the strength of the strain-hardened metal. The deformation stops at this point.

To observe the Cu microbump deformation after thermocompression bonding, the die was detached from the glass interposer after the assembly step. Figure 6(a) shows the SEM image of Cu microbumps prior to bonding, while Figure 6(b) shows the SEM image of Cu microbumps after bonding. The bump height decreased from 12.5 μm to 9.0 μm due to the bump deformation. Usually, thermocompression bonding process designed for fine-pitch and short interconnection height faces major challenges associated with non-coplanarity arising from bump height variations, interposer warpage and bond head tilt. Considering the deformation of Cu pad, the described Cu interconnection technology can address all the non-coplanarity challenges by offsetting more than 3 μm non-coplanarites but not introducing any lateral bridging problems. This attribute promises the described Cu interconnection technology for manufacturable applications as die-to-interposer and die-to-package interconnections in the ultra-fine pitch era.

attributed to low-temperature metallurgical bonding in technologies that are similar, such as friction-stir welding, solid-state welding and thermosonic bonding [17-18]. When high pressures are applied, the metal layers plastically deform under pressure at temperatures below 200 °C, due to their native malleability and ductility. In addition, plastic deformation takes place due to high stress concentration that occurs at the rough surface regions along the bonding interfaces, resulting in local plastic strain variation which helps to bring the top Au surface and the bottom Au surface into intimate contact, although no planarization process was performed on the bonding surfaces prior to bonding. Therefore, the seams and void sizes were much smaller than the root mean square values of plated bumps and pads.

Interdiffusion or self-diffusion of gold is expected to occur during thermocompression bonding after the two Au surfaces were pressed to attain intimate contact. Self-diffusion speed of Au atoms is accelerated at elevated temperatures, resulting in a robust and continuous Au direct bonding layer, which shows high reliability performance. This has been confirmed in a separate thermal aging study by Kumbhat et al. [4]. Moreover, the outward diffusion of Au atoms into Cu was blocked by the Ni barrier layer, stopping the formation of Kirkendall voids arising from Au-Cu interdiffusion [19].

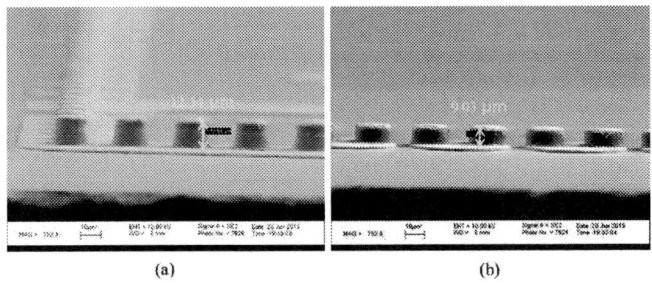

Figure 6. Cu microbump array: (a) before thermocompression bonding, (b) after thermocompression bonding

Bonding Interface Characterization and Mechanisms

The cross-sectional SEM image of Cu interconnection after ion milling is shown in Figure 7. The top side shows the Cu bump on silicon die, and the bottom side shows the Cu landing pad on the polymer dielctric on glass interposer. Both the Cu bump and the Cu pad were plastically deformed under thermocompression. The 10μm- thick polymer lying under the Cu pad has sub-GPa modulus and therefore undergoes continuous viscoelastic deformation at the bonding temperature, leading to significant pad-deflection as well as bump-collapse. During thermocompression bonding, local metallic contacts are formed at the bonding interface. The surrounding underfill polymer further enhances the bonding strength of the joint and thermomechanical reliability performance.

Metallic bonding happens due to solid state diffusion and by plastic deformation at the bonding interfaces. While the relative contributions of each factor requires more extensive modeling and TEM characterization, both factors are

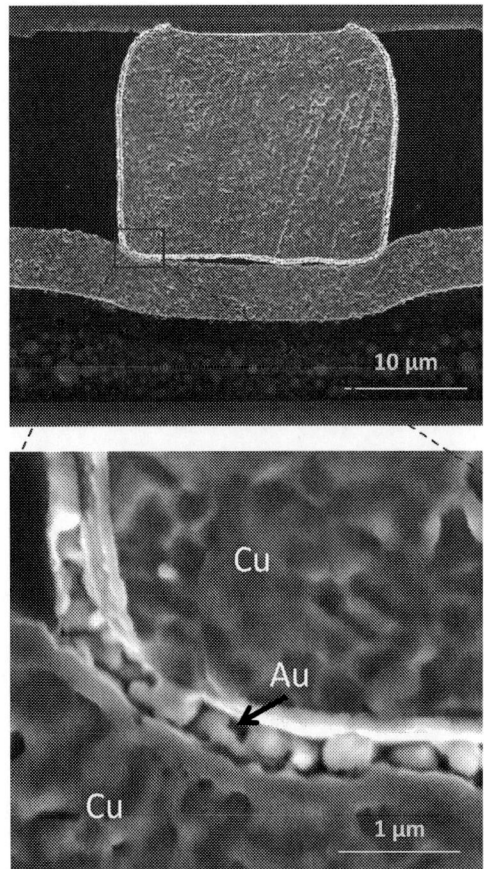

Figure 7. Cross-sectional SEM images of the Cu interconnection and the bonding interface

Experimental correlation between the bonding times and temperature for Au direct bonding has been illustrated by Tong [9]. Their work concluded that the activation energy E_A for low-temperature (room temperature to 250°C) Au-Au bonding is estimated to be ~0.41 eV. There are 3 major kinds of self-diffusion mechanisms in pure gold: lattice diffusion, grain boundary diffusion, and surface diffusion. Activation energies for these 3 kinds of diffusion are 1.71 eV [20], 0.88 eV [21], and 0.40 eV [22], respectively. Following Arrhenius relation, excellent correlation was seen between the experimental low-temperature Au-Au bonding results [9] and surface self-diffusion of pure Au. Therefore, it can be surmised that interdiffusion is driven by surface and grain boundary migration processes that are limited by the surface self-diffusion of the metal atoms.

For direct bonding between two polished atomically-flat metal surfaces, diffusion-induced grain boundary migration, along with the reduction of interfacial energy is the main reason that causes the bonding interfaces to diminish [23]. However, with two rougher metal surfaces, after the initial plastic deformation stage, surface diffusion bonding is the slowest step that dominates the bonding mechanism of bridging the bonding interfaces until they disappear. This implies that the activation energies of direct bonding with Au interfaces are determined by the Au surface self-diffusion.

The combined effects of local plastic deformation at the interface, and gold interdiffusion results in good metallurgical bonding along the bonding interface. In conjunction with the polymer adhesive that bonds the chip and the substrate, this results in high thermomechanical and electromigration reliability as previously shown by Kumbhat et al. [4]. The bonding conditions such as load, temperature and cycle time determine the extent of metallurgical bonding and overall reliability. To achieve a manufacturable process with microscale roughness and non-uniform copper surfaces, bonding pressure is the most critical parameter to accommodate such non-planarities. The bonding pressure can be further lowered with better bump process control, softer and more reactive interfaces created with specific bump geometry and interface material designs.

Conclusions

A breakthrough in low–temperature Cu interconnection technology, that also overcomes the warpage and non-coplanarity issues, to achieve solder-free bonding for highly reliable, ultra-fine pitch off-chip interconnections is presented. Ni-Au layers were applied to the copper bonding interfaces prior to the thermo-compression bonding. The bonding pressure at deformation temperature brings the two Au surfaces into intimate contact, followed by solid interdiffusion to achieve the metallurgical bonding. The thin continuous Au bonding layer and the nickel diffusion-barrier layer on top of thick Cu, lead to high current-handling performance. The novel Cu interconnection technology at 30 μm pitch enables 2.5D integration on glass interposers with unprecedented I/O density that is extensible to 20 μm pitch and below, high speed and low-loss signal transmission, and high power-density.

Acknowledgments

This research is supported by the Interconnections and Assembly (I&A), and Low-Cost Glass Interposer and Packaging (LGIP) focused programs at Georgia Tech PRC. The authors would like to thank the full members and supply chain partners for their funding and intellectual support. The authors also thank Akira Mieno, Robin Taylor and the team from Atotech for their surface finish expertise, and Taiji Sakai from Fujitsu Laboratories Ltd for valuable technical discussions.

References

1. B. Swinnen, W. Ruythooren, and P. De Moor et al., "3D integration by Cu-Cu thermo-compression bonding of extremely thinned bulk-Si die containing 10 μm pitch through-Si vias," in *Proc. Electron Device Meeting. International (IEDM)*, Dec 11-13, 2006, pp. 1-4.

2. A. Fan, A. Rahman, and R. Reif, "Copper wafer bonding," *Electrochemical and Solid-State Letters*, 2(10), pp. 534-536, 1999.

3. A. Sitaram, "Scaling 2.5D/3D: the next R&D challenge," in *2nd Annual IEEE Global Interposer Technology Workshop*, Atlanta, GA, Nov 14-16, 2012.

4. N. Kumbhat, A. Choudhury, and G. Mehrotra et al., "Highly reliable and manufacturable ultrafine pitch Cu-Cu interconnections for chip-last embedding with chip-first benefits," *IEEE Trans. CPMT*, Vol. 2, No. 9, pp. 1434-1441, Sept. 2012.

5. "Interconnect Assemblies and Methods of Making and Using Same", US Patent 2012/0104603 A1.

6. R. Sandstrom, J. Hallgren, and G. Burman, "Stress strain flow curves for Cu-OFP", Svensk karnbranslehantering (SKB), 2009.

7. Rigney D. A., and J. P. Hirth, "Plastic deformation and sliding friction of metals," *Wear*, Vol. 53, Issue. 2, pp. 345-370, April. 1979.

8. G. G. Zhang, X. F. Ang, Z. Chen, and C. C. Wong, "Critical tempereatures in thermocompression gold stud bonding", *Journal of Applied Physics*, Vol. 102, No. 6, pp. 063519, 063519-7, Sept. 2007.

9. Q.-Y. Tong, "Room temperature metal direct bonding", *Applied Physics Letters*, Vol. 89, No. 18, pp. 182101, 182101-3, Oct. 2006.

10. M. Nimura, J. Mizuno, and A. Shigetou et al., "Study on hybrid Au-underfill resin bonding method with lock-and-key structure for 3-D integration", *IEEE Trans. CPMT*, Vol. 3, No. 4, pp. 558-565, Apr. 2013.

11. S. Mark Spearing, Christine H. Tsau, and Martin A. Schmidt, "Gold thermocompression wafer bonding", in *Proc. Advanced Materials for Micro- and Nano-Systems (AMMNS)*, 2004.

12. H. Oppermann, and L. Dietrich, "Nanoprous gold bumps for low temperature bonding", *Microelectronics Reliability*, Vol. 52, No. 2, pp. 356-360, 2012.

13. Lauren E. S. Rohwer, and D. Chu, "Thin gold to gold bonding for flip chip applications", in *Proc. Electronic Components and Technology Conference (ECTC)*, May 31-June 3, 2011, pp. 907-910.

14. Yu-San Chien, Yan-Pin Huang, and Ruoh-Ning Tzeng et al., "Low temperature (<180°C) wafer-level and chip-level In-to-Cu and Cu-to-Cu bonding for 3D integration", in *Proc. Electronic Components and Technology Conference (ECTC)*, May 28-31, 2013, pp. 1146-1152.

15. F. Marion, B. Goubault de Brugiere, and A. Bedoin et al., "Aluminum to Aluminum bonding at room tempreature", in *Proc. Electronic Components and Technology Conference (ECTC)*, May 28-31, 2013, pp. 146-153.

16. V. Sundaram, Q. Chen, and T. Wang et al., "Low cost, high performance, and high reliability 2.5D silicon interposer", in *Proc. Electronic Components and Technology Conference (ECTC)*, May 28-31, 2013, pp. 342-347.

17. A. Oosterkamp, L. Djapic Oosterkamp, and A. Nordeide, "Kissing bond phenomena of solid-state welds in Al alloys", *Supplement to theWelding Journal*, August 2004, pp. 225-231.

18. H.Y. Chen, J. Cao, J. K. Liu, X.G. Song, J.C. Feng, "Contributions of atomic diffusion and plastic deformation to the diffusion bonding of metallic glass to crystalline aluminum alloy", *Computational Materials Science* 71 (2013) pp. 179–183.

19. D. F. Lim, J. Wei, K. C. Leong, and C. S. Tan, "Surface passivation of Cu for low temperature 3D wafer bonding", *ECS Solid State Letters*, Vol. 1, Issue. 1, pp. 11-14, 2012.

20. S. M. Makin, A. H. Rowe, and A. D. LeClaire, "Self-diffusion in gold", *Proceedings of the Physical Society. Section B*, Vol. 70, No. 6, 1957.

21. D. Gupta, "Grain-boundary self-diffusion in Au by Ar sputtering technique", *Journal of Applied Physcis*, Vol. 44, No. 10, pp. 4455-4458, Oct 1973.

22. H. Gobel, and P. von Blanckenhagen, "A study of surface diffusion on gold with an atomic force microscope", *Surface Science*, Vol. 331-333, Part B, pp. 885-890, July 1995.

23. M. Martinez, M. Legros, and T. Signamarcheix et al., "Mechanisms of copper direct bonding observed by in-situ and quantitative transmission electron microscopy", *Thin Solid Films*, Vol. 530, pp. 96-99, March 2013.

Low-Cost TSH (Through-Silicon Hole) Interposers for 3D IC Integration

John H. Lau, Ching-Kuan Lee, Chau-Jie Zhan, Sheng-Tsai Wu, Yu-Lin Chao, Ming-Ji Dai, Ra-Min Tain, Heng-Chieh Chien,
Chun-Hsien Chien, Ren-Shin Cheng, Yu-Wei Huang, Yuan-Chang Lee, Zhi-Cheng Hsiao, Wen-Li Tsai, Pai-Cheng Chang,
Huan-Chun Fu, Yu-Mei Cheng, Li-Ling Liao, Wei-Chung Lo, and Ming-Jer Kao
Electronic and Optoelectronic Research Lab, Industrial Technology Research Institute
Rm.168, Bldg.14, No.195, Sec. 4, Chung Hsing Road Chutung, Hsinchu 310, Taiwan
886-3591-3390; johnlau@itri.org.tw

ABSTRACT

In this investigation, a SiP (system-in-package) which consists of a very low-cost interposer with through-silicon holes (TSHs) and with chips on its top- and bottom-side (a real 3D IC integration) is studied. Emphasis is placed on the fabrication of a test vehicle to demonstrate the feasibility of this SiP technology. The design, materials, and process of the top-chip, bottom-chip, TSH interposer, and final assembly will be presented. Shock and thermal cycling tests will be preformed to demonstrate the integrity of the SiP structure.

INTRODUCTION

One of the applications of 3D IC integration is passive interposer or called 2.5D IC integration [1-3]. In general, it consists of a piece of device-less silicon with TSVs (through-silicon vias), RDLs (re-distribution layers), and/or IPDs (integrated passive devices) supporting one or more high-performance, high-density, and fine-pitch chips without TSVs [4-12]. This is schematically shown in the top drawing of Figure 1. It can be seen that the TSV/RDL interposer is supporting Chip-1 and Chip-2 side-by-side on its top surface. Another design is shown in the bottom drawing of Figure 1. It can be seen that the interposer is supporting these two chips on its top- and bottom-side. In this case, the size of the interposer can be smaller (or more chips can be placed on the same size of interposer), and the electrical performance can be better because the chip-to-chip interconnects are face-to-face instead of side-to-side [13-22]. Also, it is truly a 3D IC integration with a passive interposer, which will be the focus of this study.

Fig. 1 3D IC integration with a passive interposer.

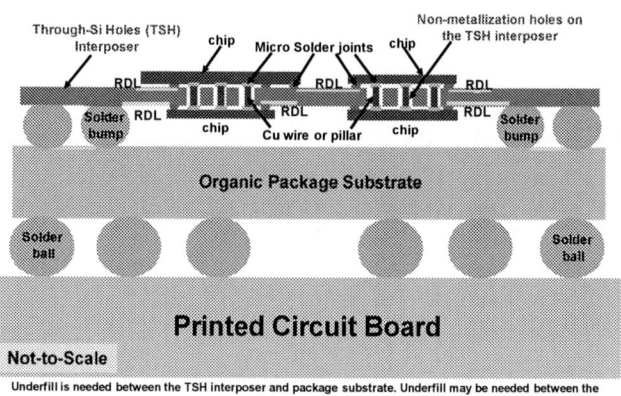

Underfill is needed between the TSH interposer and package substrate. Underfill may be needed between the TSH interposer and chips.

Fig. 2 A SiP which consists of a TSH interposer supporting chips with Cu pillars on its top-side and chips with solder bumps on its bottom-side.

TSV is the heart and most important key enabling technology of 3D IC integration. Usually, there are six key steps in making a TSV, namely: (a) via formation by either deep reactive ion etch (DRIE) or laser drilling, (b) dielectric layer by plasma-enhance chemical vapor deposition (PECVD), (c) barrier and seed layers by physical vapor deposition, (d) via Cu-filling by electroplating, (e) chemical-mechanical polishing (CMP) to remove the overburden Cu, and (f) TSV Cu reveal by backgrinding, Si dry-etching, low-temperature passivation, and CMP. Thus, how to make low-cost TSVs is one of the important research topics for 3D IC integration. In this study, a class of very low-cost interposer with through-silicon holes (TSHs) and with chips on its both sides (a real 3D IC integration) is developed.

Figure 2 schematically shows a SiP with a TSH interposer supporting a few chips on its top- and bottom-side. The key feature of TSH interposers is there **is not** metallization in the holes and the dielectric layer, barrier and seed layers, via filling, CMP for removing overburden copper, and Cu revealing are not necessary. Comparing to the TSV interposers, TSH interposers only need to make holes (by either laser or DRIE) on a piece of silicon wafer. Just like the TSV interposers, RDLs are needed by the TSH interposers.

The TSH interposers can be used to support the chips on its top side as well as bottom side. The holes can let the signals of the chips on the bottom-side transmit to the chip on the top-side (or vice versa) through the Cu pillars and solders. The chips on the same side can communicate to each other with the RDLs of the TSH interposer. Physically, the top chips and bottom chips are connected through Cu pillars and micro solder joints. Also, the peripherals of all the chips are

978-1-4799-2408-0/14 $31.00 © 2014 IEEE   290   2014 Electronic Components & Technology Conference

soldered to the TSH interposer for structural integrity to resist shock and thermal conditions. In addition, the peripherals of the bottom-side of the TSH interposer have ordinary solder bumps which are attached to a package substrate. It has been shown in [23] that the electrical performance of the TSH interposer is better than that of the TSV interposer.

In this study, a very simple test vehicle is fabricated to demonstrate the feasibility of this SiP with TSH interposer technology. The design, materials, and process of the top-chip with Cu pillars, bottom-chip with solder bumps, and TSH interposer will be presented. The final assembly of the SiP test vehicle which consists of the chips, interposer, package substrate, and PCB will also be provided. Shock and thermal cycling tests will be preformed to demonstrate the integrity of the SiP structure.

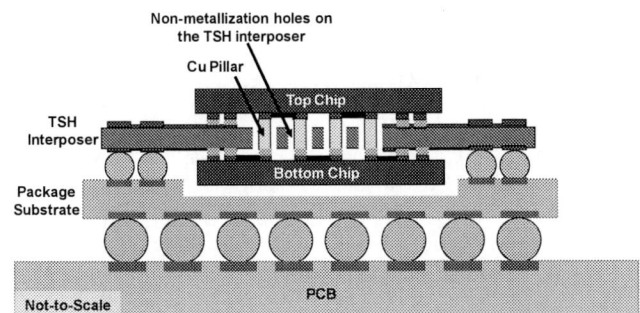

Fig. 3 SiP test vehicle which consists of a top-chip with Cu pillars and a bottom-chip with solder bumps on a TSH interposer.

TEST VEHICLE

The test vehicle is shown in Figure 3. It can be seen that it consists a TSH interposer, which is supporting a top-chip with Cu pillars and a bottom-chip with UBM and solder. The interposer module is connected to a package substrate and then attached to a PCB.

The dimensions of the top-chip are 5mm x 5mm x 725μm. The chip has (16 x16 = 256) Cu pillars at its central portion and two-row (176) of Cu UBM/pads at its peripherals. The diameter of the Cu pillars is 50μm. They are 100μm-tall

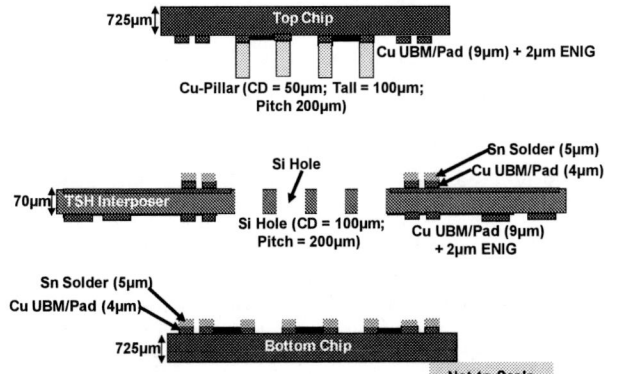

Fig. 4 Geometry, dimension, and interconnects of the top-chip, bottom-chip, and TSH interposer.

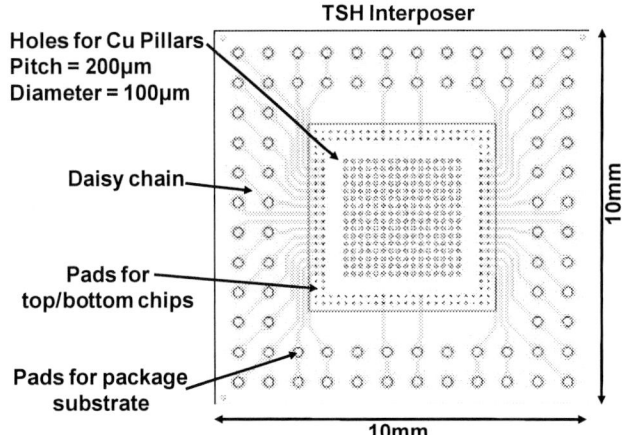

Fig. 5 Layout of the TSH interposer.

and on 200μm-pitch as shown in Figure 4 and Table 1. The thickness of the peripheral Cu UBM/pads is 9μm and with electroless (2μm) Ni and immersion (0.05μm) Au (ENIG).

The dimensions of the bottom-chip are also 5mm x 5mm x 725μm. The chip has 432 Cu UBM/pads (4μm) and coated with Sn solders (5μm). The central 256 are for the interconnections of the Cu pillars of the top-chip.

The dimensions of the TSH interposer are 10mm x 10mm x70μm (Figures 4 and 5). It has 256 holes at its central portion to let the Cu pillars to pass through. The diameter of the holes is 100μm and the pitch of the holes is 200μm. There are two-row (176) of peripheral Cu UBM/pads (4μm) coated with Sn solders (5μm) on the top-side of the TSH interposer for the interconnections of the top-chip. On the other hand, there are two-row (176) of peripheral Cu UBM/pads (9μm) with ENIG (2μm) on the bottom-side of the TSH interposer for the interconnections of the bottom-chip.

Table 1 Key elements of the SiP and their dimensions.

Top-Chip with Cu-Pillar	
Dimension	5mm x 5mm x 725μm
Cu-Pillar	CD = 50μm; Tall = 100μm; Pitch = 200μm
Cu UBM/Pad	9μm + 2μm (ENIG)
Bottom-Chip without Cu-Pillar	
Dimension	5mm x 5mm x 725μm
Cu UBM/Pad	4μm
Sn Solder	5μm
TSH interposer	
Dimension	10mm x 10mm x 70μm
TSH	CD = 100μm; Pitch = 200μm; Depth = 70μm
Top-Side: Cu UBM/Pad and Sn Solder	4μm (Cu) and 5μm (Sn)
Bottom-Side: Cu UBM/Pad	9μm + 2μm (ENIG)
Package Substrate (Cavity)	15mm x 15mm x 1.6mm (6mm x 6mm x.9mm)
PCB	132mm x 77mm x 1.6mm

The dimensions of the organic package substrate are 15mm x 15mm x1.6mm. There is a cavity (6mm x 6mm x 0.5mm) on the top-side of the substrate for the bottom-chip. The dimensions of the printed circuit board (PCB) are 132mm x 77mm x 1.6mm, which is a standard size of the JEDEC (JESD22-B111) specification [24].

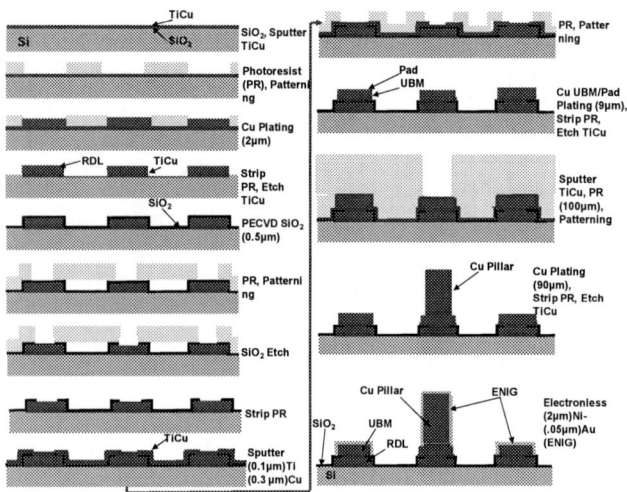

Fig. 6 Process flow in fabricating the top-chip with RDL, Cu UBM/pad, and Cu pillars.

TOP-CHIP WITH UBM/PAD and Cu-PILLAR

The process flow of fabricating the top-chip with Cu pillars is shown in Figure 6. Since this is not a device chip but a piece of silicon, thus the daisy chain (RDL) will be fabricated first. A SiO_2 is deposited on a Si wafer with PECVD provided by Novellus (now Lam Research) at 200°C. Then, (0.1μm) Ti and (0.3μm) Cu are sputtered by the MRC machine. It is followed by spin coating a photoresist and patterning with a mask using a photolithography technique (align and expose). Electroplate the Cu (2μm) RDL. Then, strip off the photoresist and etch off the TiCu. A SiO_2 (0.5μm) is deposited on top of the whole wafer with PECVD. Again, photoresist and patterning. Then, dry etch the SiO_2 by RIE (reactive ion etch). It is followed by stripping off the photoresist and now it is ready for wafer bumping.

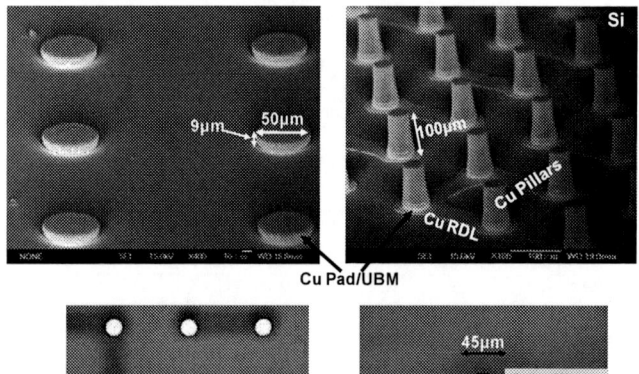

Fig. 7 SEM images of the Cu UBM/pads and Cu pillars (diameter = 50μm at the bottom and = 45μm at the top).

First, sputter the seed layer: (0.1μm) Ti and (0.3μm) Cu. Then, photoresist (9.5μm from JSR) and patterning of all 432 pads. It is followed by electroplating the Cu UBM/pad (9μm). Strip off the photoresist and etch off the TiCu seed layer.

Then, a positive type of photoresist (100μm) provided by HD Microsystems is spin coated on the wafer and it is followed by patterning only the central 256 pads, and Cu plating (90μm) with SEMITOOL (now Applied Materials) at room temperature. Then, the top surface of the wafer is flattened by DISCO's fly cut. It is followed by stripping off the photoresist and etching off the TiCu. Finally, electroless Ni (2μm) and immersion Au (0.05μm). Figure 7 shows the SEM (scanning electron microscopy) images of the Cu pillars, Cu UBM/pads, and Cu RDLs (daisy chains) on the top-chip. The smaller diameter (45μm) on top of the Cu pillars is because of the smaller opening on top of the dry film photoresist.

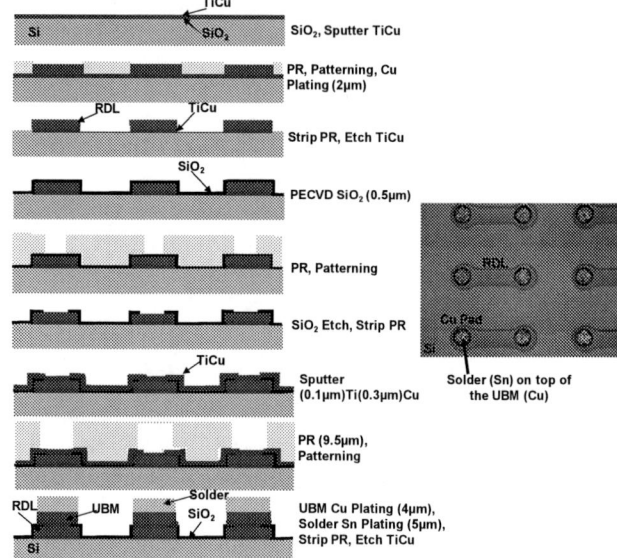

Fig. 8 Process flow in fabricating the bottom-chip with RDL, and Cu UBM/pad + Sn solder

BOTTOM-CHIP WITH UBM/PAD/SOLDER

The process flow in fabricating the bottom-chip with UBM/pad and solder is shown in Figure 8. It can be seen that the process in making the Cu RDL is the same as those of the top-chip. For most of the wafer bumping processes, they are the same except the photoresist thickness and solder. After photoresist (9.5μm) and patterning all (432) pads, it is followed by electroplating the Cu UBM/pad (4μm) and electroplating the Sn solder (5μm). Strip off the photoresist and etch off the TiCu. A photo of the RDL, Cu UBM/pad, and Sn solder cap on the bottom-chip is shown in Figure 8.

TSH INTERPOSER

The process flow in fabricating the TSH interposer with Cu UBM/pad + Sn solder on its top-side and Cu UBM/pad on its bottom-side is shown in Figure 9. It can be seen that, the RDL and Cu UBM/pad on the bottom-side of the interposer will be fabricated first and the processes are basically the same as those for the top-chip. Except after the strip off of the photoresist and etch off of the seed layer (TiCu) of the 9μm-thick UBM/pad, it is followed by ENIG (2μmNi-0.05μmAu). Then, the bottom-side of the interposer wafer with UBM/pad is temporary bonded with an adhesive to a 750μm-thick silicon supporting wafer (carrier). It is followed by thinning the top-side of the interposer wafer down to 70μm. Then,

978-1-4799-2408-0/14 $31.00 © 2014 IEEE

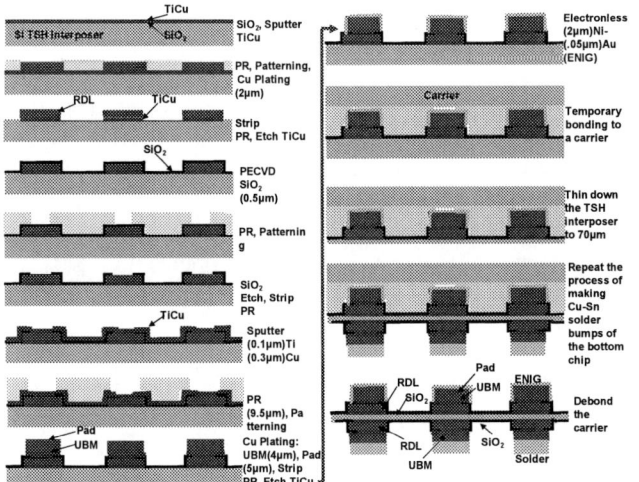

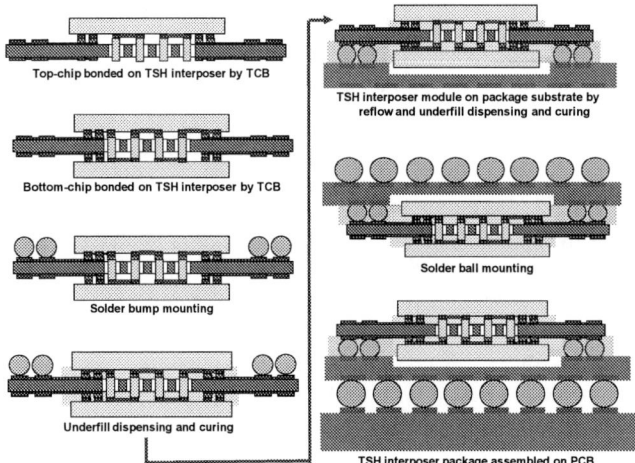

Fig. 9 Process flow in fabricating the TSH interposer with Cu UBM/pad + Sn solder on its top-side and Cu UBM/pad on its bottom-side.

Fig. 11 Process flow in the final assembly of SiP.

repeat all the process steps in fabricating the UBM/pad + Sn solder of the bottom-chip mentioned earlier. Finally, debond the carrier wafer from the interposer wafer at room temperature. At this stage, the interposer wafer with 176 peripheral UBM/pads + Sn solders on its top-side and 176 peripheral UBM/pads at its bottom-side as shown in Figures 4 and 5 is achieved. The 256 holes are fabricated with UV laser drilling by the Siemens MicroBeam 3205. The power is 3400mW. Figure 10 shows the photo images of the 70μm-thick TSH interposer wafer with RDLs, pads for the package substrate, peripheral pads for the chips, and 100μm-diameter holes on 200μm-pitch.

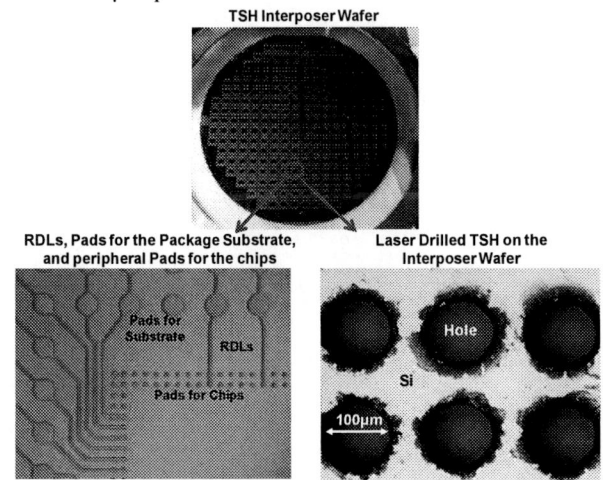

Fig. 10 TSH interposer wafer (T) with laser drilled holes (R) and RDL and pads for chip (L).

FINAL ASSEMBLE

The process flow of the final assembly of the SiP with TSH interposer test vehicle is shown in Figure 11. First, the top-chip with the 176 peripheral UBM/pads is thermocompression (TC) bonded (with the SuSS FC-150 bonder) to the peripheral UBM/pads + Sn solders on the top-side of the THS interposer. (The Cu pillars passed through the holes of the THS interposer.) The TC bonding conditions are

shown in Figure 12. It can be seen that: (a) the maximum bonding force is 1600g, (b) the maximum temperature of the chuck is 150°C; (c) the maximum temperature of the head is 250°C; and (d) the cycle time is 120 seconds.

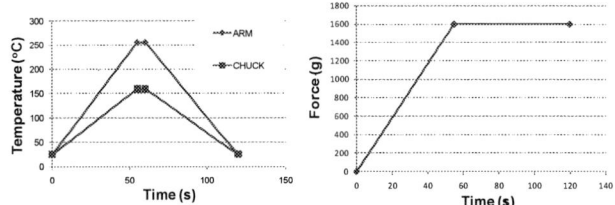

Fig. 12 TC conditions in bonding the top-chip to the top-side of TSH interposer: temperature (L) and force (R).

Then, all the 432 UBM/pads + Sn solders on the bottom-chip are TC bonded to the tip of the 256 central Cu pillars and 176 peripheral UBM/pads on the bottom-side of the TSH interposer. The bonding conditions are basically the same as those of the top-chip except the bonding force is reduced to 800g. It is followed by solder (Sn3wt%Ag0.5wt%Cu) bump (350μm-diameter) mounting on the bottom-side of the TSH interposer. Then, a capillary-type underfill with 50% filler contents (average filler size = 0.3μm and maximum filler size = 1μm) is dispensed along two adjacent sides of the top-chip. After the underfill (a) fills the gap between the top-chip and the TSH interposer; (b) flows through the holes of the TSH interposer; and (c) fills the gap between the bottom-chip and the TSH interposer, then it is cured at 150°C for 30 minutes.

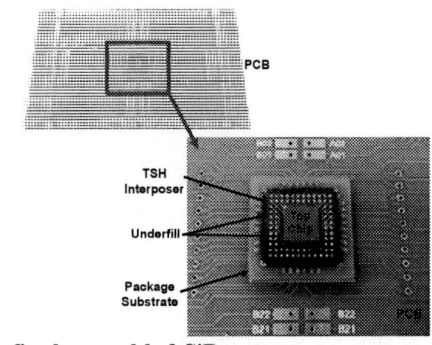

Fig. 13 The final assembled SiP.

The whole TSH interposer module is solder reflowed on the package substrate by a standard lead-free temperature profile with a maximum at 240°C. In order to enhance the solder joint reliability, the underfill is applied between the TSH interposer and the organic package substrate. It is followed by solder (Sn3wt%Ag0.5wt%Cu) ball (450μm-diameter) mounting on the bottom-side of the package substrate. Finally, the whole SiP package is solder reflowed on the PCB with the same lead-free reflow temperature profile just mentioned. The final assembled SiP test vehicle is shown in Figure 13. It can be seen that the PCB is supporting the package substrate, which is supporting the TSH interposer, which is supporting the top-chip. The bottom-chip is blocked by the TSH interposer and cannot be seen.

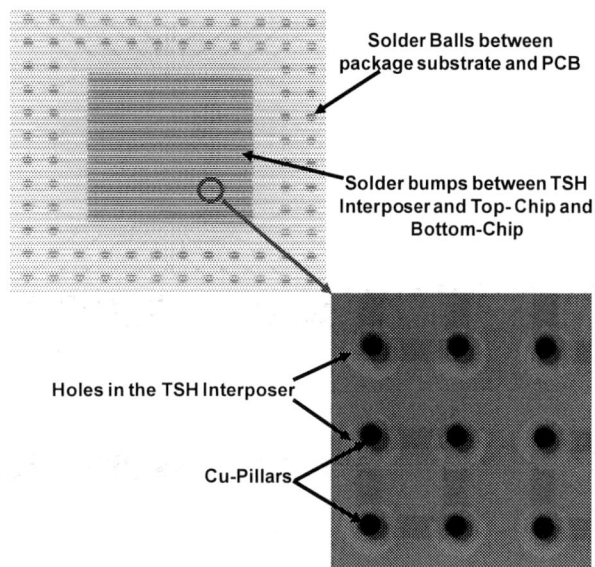

Fig. 14 X-ray images showing the location of Cu pillars and TSHs of the SiP which consists of the top-chip, TSH interposer, bottom-chip, package substrate, and PCB.

Figure 14 shows the x-ray images of the final assembled SiP. It can be seen that: (a) the Cu pillars are not touching the side-wall of the TSH, and (b) the Cu pillars are almost at the center of the TSH.

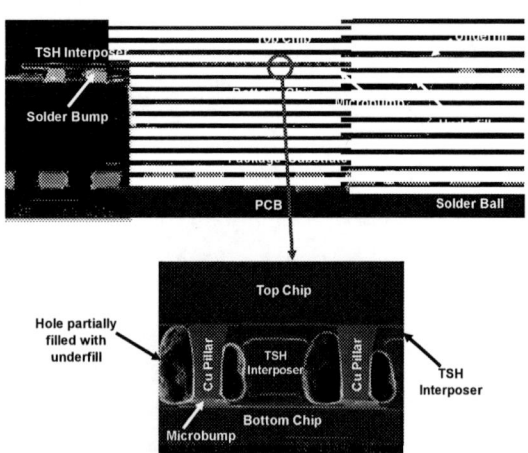

Fig. 15 SEM image showing a cross-section of the SiP which consists of the top-chip, TSH interposer, bottom-chip, package substrate, and PCB.

Figure 15 shows the SEM image of a cross-section of the SiP, which includes all the key elements such as the top-chip, TSH interposer, bottom-chip, package substrate, PCB, micro bumps, solder bumps, solder ball, TSH, and Cu pillars. It can be seen through the x-ray and SEM images that the key elements of SiP structure are properly fabricated.

RELIABILITY ASSESSMENTS

In this study, the reliability assessments to verify the structural and thermal integrity of the assembled SiPs are drop test and thermal cycling test.

Shock (Drop) Test and Result

The drop test board and set-up are based on JESD-B111 [24]. The SiPs are facing upward during the test. Four standoffs on the fixture provide the support and the spacing for deflections of the PCB during impacts as shown in Figure 16. The drop height is 460mm, which leads to the acceleration = 1500g as shown in Figure 17. After 10 drops, there is not any failure, i.e., no resistance change of the daisy chains.

Fig. 16 Drop test of the SiPs set-up.

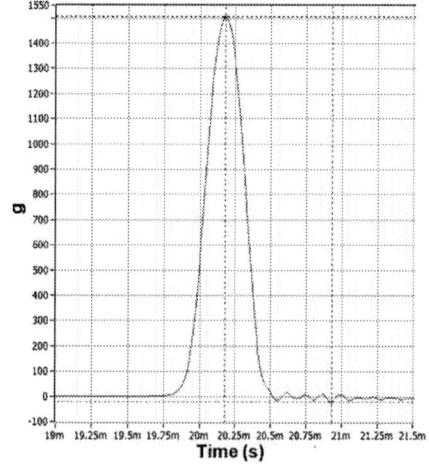

Fig. 17 Drop test profile according to JESD22-B111.

Thermal Cycling Test

The thermal cycling test conditions are: -55°C ↔ 125°C, and one hour cycle (15 minutes ramp up and ramp down and 15 minutes at dwells). At present, there is not any failure and the test is ongoing. Because there is not any thermal expansion mismatch between the chips and the TSH interposer, the micro bump solder joints should last for a very long time. With underfill between the chips and TSH interposer, the micro solder joints should last even longer. Because of the underfill between the TSH interposer and the organic package substrate, the solder bump joints should also be last for a long time.

SUMMARY AND RECOMMENDATIONS

A SiP which consists of a TSH interposer with chips on its top- and bottom-side has been developed. Drop and thermal cycling tests have been performed to demonstrate the integrity of the SiP structure. Some important results and recommendations are summarized as following.

➤ The RDL (daisy-chain) top-chip with 256 100μm-Cu-pillar+2μm-ENIG at its central portion and 176 9μm-UBM/pad+2μm-ENIG around the peripherals has been properly fabricated.

➤ The RDL (daisy-chain) bottom-chip with 432 4μm-UBM/pad+5μm-Sn-solder has been properly fabricated.

➤ The RDL (daisy-chain) TSH interposer with 256 central holes (drilled by laser), 176 4μm-UBM/pad+5μm-Sn-solder around the peripherals on its to-side, and 176 9μm-UBM/pad+2μm-ENIG around the peripherals on its bottom-side has been properly fabricated.

➤ The final assembly of the top-chip, bottom-chip, TSH interposer, package substrate, and PCB has also been properly fabricated. These are evidenced by the SEM and x-ray images.

➤ The structural integrity of the SiP has been demonstrated by drop test. Based on the JEDEC specification, after 10 drops and there is not any failure.

➤ The thermal cycling test conditions are: -55°C ↔ 125°C, and one hour cycle (15 minutes ramp up and ramp down and 15 minutes at dwells). At present, there is not any failure and the test is ongoing.

➤ The chips (especially the bottom-chip) should be thinned down to less than 200μm. In that case, the cavity in the package substrate is not needed.

ACKNOWLEDGMENTS

The authors would like to thank the financial support of MOEA, Taiwan. The strong support of the 3D IC Integration program by Dr. C. T. Liu, VP and Director of Electronics & Optoelectronics Research Lab is greatly appreciated. The authors would also like to thank the Ad-STAC members, Hitachi Chemical and DISCO for their support on fabricating the test vehicle.

REFERNCES

[1] Lau, J. H., *Through-Silicon Via for 3D Integration*, McGraw-Hill Book Company, New York, NY, 2013.

[2] Lau, J. H., *Reliability of ROHS-Compliant 2D and 3D IC Integration*, McGraw-Hill Book Company, New York, NY, 2011.

[3] Lau, J. H., C. K. Lee, C. S. Premachandran, and A. Yu, *Advanced MEMS Packaging*, McGraw-Hill Book Company, New York, NY, 2010.

[4] Banijamali, B., S. Ramalingam, K. Nagarajan, and R. Chaware, "Advanced Reliability Study of TSV Interposers and Interconnects for the 28nm Technology FPGA", *Proceedings of IEEE/ECTC*, Orlando, Florida, June 2011, pp. 285-290.

[5] Chaware, R., K. Nagarajan, and S. Ramalingam, "Assembly and Reliability Challenges in 3D Integration of 28nm FPGA Die on a Large High Density 65nm Passive Interposer", *Proceedings of IEEE/ECTC*, May 2012, San Diego, CA, pp. 279-283.

[6] Banijamali, B., S. Ramalingam, H. Liu and M. Kim, "Outstanding and Innovative Reliability Study of 3D TSV Interposer and Fine Pitch Solder Micro-bumps", *Proceedings of IEEE/ECTC*, May 2012, pp. 309-314.

[7] Xie, J., H. Shi, Y. Li, Z. Li, A. Rahman, K. Chandrasekar, D. Ratakonda, M. Deo, K. Chanda, V. Hool, M. Lee, N. Vodrahalli, D. Ibbotson, and T. Verma, "Enabling the 2.5D Integration", *Proceedings of IMAPS International Symposium on Microelectronics*, September 2012, San Diego, CA, pp. 254-267.

[8] Li, Z., H. Shi, J. Xie, and A. Rahman, "Development of an Optimized Power Delivery System for 3D IC Integration with TSV Silicon Interposer", *Proceedings of IEEE/ECTC*, San Diego, CA, May 2012, pp. 678-682.

[9] Banijamali, B., C. Chiu, C. Hsieh, T. Lin, C. Hu, S. Hou, S. Ramalingam, S. Jeng, L. Madden, and D. Yu, "Reliability Evaluation of a CoWoS-enabled 3D IC Package", *Proceedings of IEEE/ECTC*, May 2013, pp. 35-40.

[10] Lin, Larry, Tung-Chin Yeh, Jyun-Lin Wu, Gary Lu, Tsung-Fu Tsai, Larry Chen, and An-Tai Xu, "Reliability Characterization of Chip-on-Wafer-on-Substrate (CoWoS) 3D IC Integration Technology", *Proceedings of IEEE/ECTC*, Las Vegas, NV, May 2013, pp. 366-371.

[11] Chuang, Yi-Lin, Chung-Sheng Yuan, Ji-Jan Chen, Ching-Fang Chen, Ching-Shun Yang, Wei-Pin Changchien, Charles C.C. Liu, and Frank Lee, "Unified Methodology for Heterogeneous Integration with CoWoS Technology", *Proceedings of IEEE/ECTC*, San Diego, CA, May 2013, pp. 852-859.

[12] Kwon, W., M. Kim, J. Chang, S. Ramalingam, L. Madden, G. Tsai, S. Tseng, J. Lai, T. Lu, and S. Chin, "Enabling a Manufacturable 3D Technologies and Ecosystem using 28nm FPGA with Stack Silicon Interconnect Technology", *IMAPS Proceedings of International Symposium on Microelectronics*, October 2013, pp. 217-222.

[13] Li, L., P. Su, J. Xue, M. Brillhart, J. H. Lau, P. Tzeng, C. Lee, C. Zhan, M. Dai, H. Chien, and S. Wu, "Addressing Bandwidth Challenges in Next Generation High Performance Network Systems with 3D IC Integration," *IEEE/ECTC Proceedings*, San Diego, CA, May 2012, pp. 1040-1046.

[14] Chieh, H. J. H. Lau, Y. Chao, M. Dai, and R. Tain, "Thermal Evaluation and Analyses of 3D IC Integration SiP with TSVs for Network System Applications," *IEEE/ECTC Proceedings*, San Diego, CA, May 2012, pp. 1866-1873.

[15] Wu, S., J. H. Lau, and H. Chien, "Nonlinear Thermal

Stress and Creep Strain Analyses of a 3D IC Integration SiP with TSVs for Network System Application," *IEEE ICEP Proceedings*, Japan, April 2012, pp. 681-688.

[16] Wu, S. T., J. H. Lau, H. Chien, Y. Chao, R. Tain, L. Li, P. Su, J. Xue, and M. Brillhart, "Thermal Stress and Creep Strain Analyses of a 3D IC Integration SiP with Passive Interposer for Network System Application", *Proceedings of the 45th IMAPS International Symposium on Microelectronics,* September 2012, pp. 1038-1045.

[17] Chien, H. C., J. H. Lau, T. Chao, M. Dai, and R. Tain, "Thermal Management of Moore's Law Chips on Both sides of an Interposer for 3D IC integration SiP," *IEEE ICEP Proceedings*, Japan, April 2012, pp. 38-44.

[18] Lau, J. H., P. Tzeng, C. Zhan, C. Lee, M. Dai, J. Chen, Y. Hsin, S. Chen, C. Wu, L. Li, P. Su, J. Xue, and M. Brillhart, " Large Size Silicon Interposer and 3D IC Integration for System-in-Packaging (SiP)", *Proceedings of the 45th IMAPS International Symposium on Microelectronics,* September 2012, pp. 1209-1214.

[19] Ji, M., M. Li, J. Cline, D. Secker, K. Cai, J. H. Lau, P. Tzeng, C. Zhan, C. Lee, "3D Si Interposer Design and Electrical Performance Study", *Proceedings of DesignCon*, January 28-31, 2013, Santa Clara, CA, pp. 1-WA2: 1-23.

[20] Wu, S. T., H. Chien, J. H. Lau, M. Li, J. Cline, and M. Ji, "Thermal and Mechanical Design and Analysis of 3D IC Interposer with Double-Sided Active Chips", *IEEE/ECTC Proceedings*, Las Vegas, NA, May 2013, pp. 1471-1479.

[21] P. J. Tzeng, J. H. Lau, C. Zhan, Y. Hsin, P. Chang, Y. Chang, J. Chen, S. Chen, C. Wu, C. Lee, H. Chang, C. Chien, C. Lin, T. Ku, M. Kao, M. Li, J. Cline, K. Saito, M. Ji, "Process Integration of 3D Si Interposer with Double-Sided Active Chip Attachments", *Proceedings of IEEE/ECTC*, Las Vegas, NV, 2013, pp. 86-93.

[22] Lau, J. H., P. Tzeng, C. Lee, C. Zhan, M. Li, J. Cline, K. Saito, Y. Hsin, P. Chang, Y. Chang, J. Chen, S. Chen, C. Wu, H. Chang, C. Chien, C. Lin, T. Ku, R. Lo, and M. Kao, "Redistribution Layers (RDLs) for 2.5D/3D IC Integration", *IMAPS Transactions, Journal of Microelectronic Packaging*, Vol. 11, Issue 1, 2014 (in press).

[23] Wu, S., J. H. Lau, H. Chien, J. Hung, M. Dai, Y. Chao, R. Tain, W. Lo, and M. Kao, "Ultra Low-Cost Through-Silicon Holes (TSHs) Interposers for 3D IC Integration SiPs", *IEEE/ECTC Proceedings*, San Diego, CA, May 2012, pp. 1618-1624.

[24] JESD22-B111, *Board Level Drop Test Method of Components for Handheld Electronic Products*, JEDEC Standard, July 2003

Cu Pattern Density Impacts on 2.5D TSI Warpage Using Experimental and FEM Analysis

C.T. Yeh, C.Y. Wu, C.F. Lin, K.M. Chen, M.J. Lin, Y.C. Lin, C.L. Kuo,
United Microelectronics Corporation,
Inc.,No. 3, Li-Hsin Rd. II, Hsinchu Science Park, Taiwan 30078, R.O.C.,
E-mail: hamy_yeh@umc.com

Abstract

Through Silicon Interposer (TSI) needs to fulfill multi-die stacking in one packaging which can bring high integration density, short interconnection length and small size for next generation devices. Die stacking is a key process in the TSI manufacturing flow, and within that process, die warpage is of central concern. This is because the large warpage of the Si-interposer induces poor joining of μ-bump interconnection, lithograph missing and die breakage during the assembly process. According to our experience, the Cu metal density of redistribution layer (RDL) significantly effects on TSI warpage.[2] The work proposes to design four different Cu metal densities, namely code A (10~25%), B (25~40%), C (40~55%) and D (55~70%) and combine the different metal density to investigate the relationship between Cu metal pattern density and TSI warpage under four different TSI size. Each unit of Cu metal density area is 12 x 12 mm^2. The TSI size is ranging from unit 1 x 1 (12 x 12 mm^2), 1 x 2 (12 x 24 mm^2), 2 x 2 (24 x 24 mm^2) and 2 x 3 (24 x 36 mm^2).

In this work, the TSI warpage modeling methodology has been developed and carried out to find the correlation between pattern density and warpage behavior using the finite element method (FEM).

A simplified pattern-inclusion model has been constructed to obtain its warpage response under a thermal loading. Furthermore, the equivalent model has been established to obtain the effective mechanical properties of the equivalent layer (i.e., the composite layer of Cu traces and IMD) through adapting the warpage results to correctly match the simplified pattern-inclusion model. Therefore, the TSI warpage can be calculated during the reflow process after acquiring the effective mechanical properties of different Cu pattern densities. The simulation warpage variation of 1 x 1 (12 x 12 mm^2) TSI die size for code A, B, C and D are 8 μm, 19 μm, 32 μm, and 47 μm, respectively. The simulation results indicate that large TSI with high Cu metal density possesses high warpage.

After fabricating the different Cu metal density wafer, the wafer was grinded to 100 μm for dicing, then using thermal shadow moiré to measure the TSI warpage after one time reflow process. The experimental TSI warpage variation of 1 x 1 (12 x 12 mm^2) TSI die size with code A, B, C and D are 6~10 μm, 21~26 μm, 31~36 μm, and 43~70 μm, respectively. The experimental results imply that large TSI with high Cu metal density possesses high warpage. After measuring the warpage of those dies during the thermal curve, we compared the measurement data with our simulation result and explored 3 subjects: different pattern density, die size, and symmetry. We found the TSI warpage between simulation and measurement result are consistent. The finite element simulation results in the work correlate well with the experimental test results. Therefore, we are highly confident that this finite element model can help generate design guidelines for wafer patterns and predict the die warpage variation during the assembly process.

Introduction

To control the warpage of TSI dies after wafer thinning is critical.[1] Warpage tends to be most extreme after wafer thinning, as shown in Figure 1. A high degree of warpage will impact the process yield for 2.5D TSI MEoL & the BEoL process. Therefore, the Cu metal density of the redistribution layer (RDL) significantly effects TSI warpage.[2] To accurately controll the warpage of the TSI die during our research, we used the finite element method (FEM) to ascertain the correlation between Cu pattern density and warpage behavior; then we used the experimental results to verify the simulation model.

For the purpose of our test, we designed a test vehicle with different Cu metal densities in a single wafer. One wafer can be split into 4 different die sizes, and after we did this, the wafer was ground to 100 μm for dicing. To get our die warpage measurements, we used thermal shadow moiré to measure the warpage of the TSI die and to monitor the warpage variation during reflow.

By comparing the measurements and our simulation data, we have been able to establish a model for predicting the relation between pattern density and variation in the amount of warpage. This has allowed us to propose an adequate guideline for TSI product design to eliminate the potential negative effects of warpage.

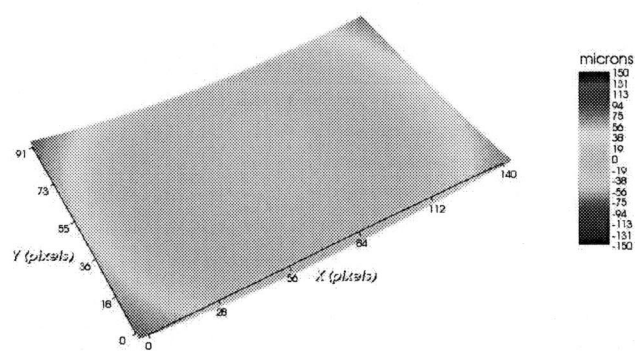

Figure 1. Die warpage.

Experiment

The Test Pattern

To verify the correlation between pattern density and warpage, we designed four different Cu metal densities, namely code A (10~25%), B (25~40%), C (40~55%) and D (55~70%). In addition, we also prepared four TSI sizes; each unit of Cu metal density area is 12 x 12 mm^2, and there are four sample sizes; 1 x 1 (12 x 12 mm^2), 1 x 2 (12 x 24 mm^2), 2 x 2 (24 x 24 mm^2) and 2 x 3 (24 x 36 mm^2).

In this work, we chose 14 test patterns to measure the warpage. The components are showed to Figure 2. In order to simplify the variation of the wafer process and Cu pattern density impact, the wafers only have one layer of Cu routing without passivation. This helps us to understand how Cu pattern density influences the warpage behavior.

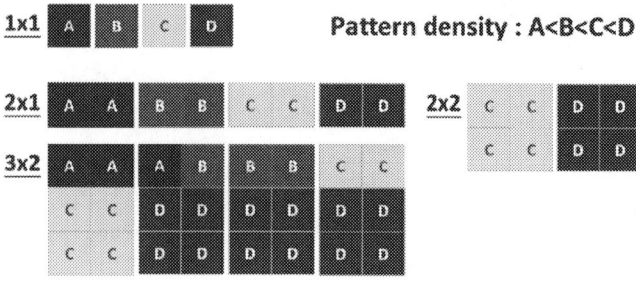

Figure 2. Test pattern components

Process Flow and Warpage Measurement

In this work, our main purpose is to explore the relationship between the Cu pattern density of TSI wafers and the resulting variation in warpage. Therefore, the sample preparation should be similar to that of actual products. For the 2.5D Si interposer, the required thickness of the product wafer is approximately 100µm, and back of the wafer is coated with organic/inorganic film for isolation. This film on the reverse side of the wafer will impact die warpage. To prevent the noise of backside film is needed.

As a result, the die thickness in our samples is ground down to 100µm, and then sawed to achieve the required die size (12 x 12 mm^2, 24 x 12 mm^2, 24 x 24 mm^2, 36 x 24 mm^2). Dies larger than 36 x 24 mm^2 were not used because there is no adequate tooling for the pick & place process. A pick & place tool must be used for die pick-up because various problems can result from manually handling thin dies; a large die may crack or chip due to manual manipulation. In order to simplify the experiment and eliminate the impact of other factors, thee reverse side of the wafer receives no passivation after it is ground. Steal dicing was used for wafer sawing. This reduces the chance of dicing failure due to a mechanical saw.

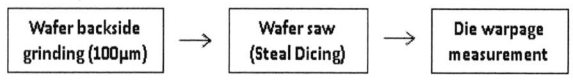

Figure 3. Sample preparation flow

The warpage measurement of the samples was performed many times because we found that the value of warpage measurement only becomes stable after the reflow is done many times. The remaining stress of dies after wafer backside grinding is suspected. Many different mechanisms are used to clean and reduce stress to the opposite side of the wafer during grinding. Therefore, the impact on the back of the wafer during grinding is also an important factor to consider in die warpage measurement.

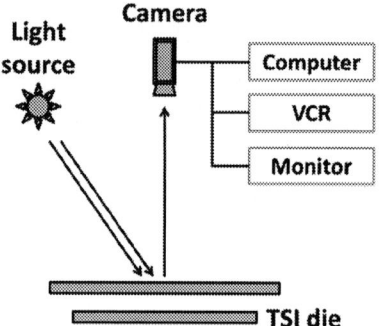

Figure 4. Shadow moiré measurement mechanism

Die warpage measurements were taken using thermal shadow moiré. The mechanism is shown in Figure 4. As *JEDEC* defines it, the surface deviation of the die was measured through high temperature reflow. In order to get sufficiently high revolutions for the warpage measurement, grafting of 100~300 lines per inch was selected.

The warpage measurement of samples has conducted for many times. It is found that the value of warpage measurement is stable after reflow for many times. The remaining stress of dies after wafer backside grinding is suspected. For wafer backside grinding process, it has many different mechanisms for backside clean and stress relief. Therefore, to consider the impact of wafer backside grinding is also important for die warpage measurement.

TSI Warpage Modeling Methodology

In this research, finite element analysis (FEA) modeling was carried out to simulate the TSI warpage during the reflow profile. However, TSI warpage modeling is known to be difficult and challenging with Cu traces and Inter-layer dielectrics (IMD) due to its complicated layout and large differences in the scale between layout thickness and TSI die size. To overcome these issues, a methodology to simulate the TSI warpage was developed. Figure 5 shows the simulation flow for TSI warpage modeling.

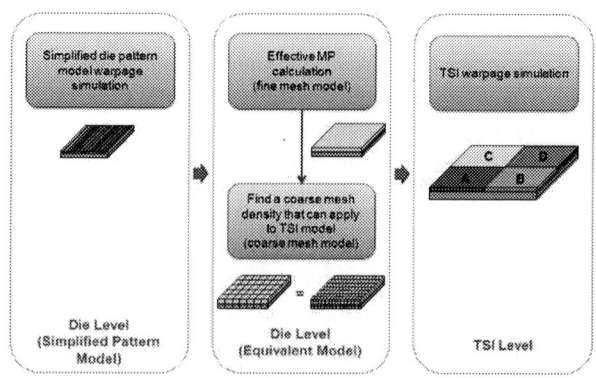

Figure 5. Simulation flow

978-1-4799-2408-0/14 $31.00 © 2014 IEEE   298

First, a simplified pattern-inclusion model was constructed to obtain its warpage response under thermal loading, as shown in Figure 6. The pattern-inclusion model is defined as a die size of 3 mm × 3 mm and a thickness of 100 μm that is covered with 1 μm-thick Cu traces and 1 μm-thick oxide. The Cu traces are assumed to be 1 μm line spaces (L/S) parallel with each other. The pitch of the Cu traces is defined by the metal density. In this paper, there are four metal densities (A, B, C, D) used for calculating the effective mechanical properties. Due to symmetrical geometry, only a quarter three-dimensional FE model was constructed for the area, as seen in Figure 6.

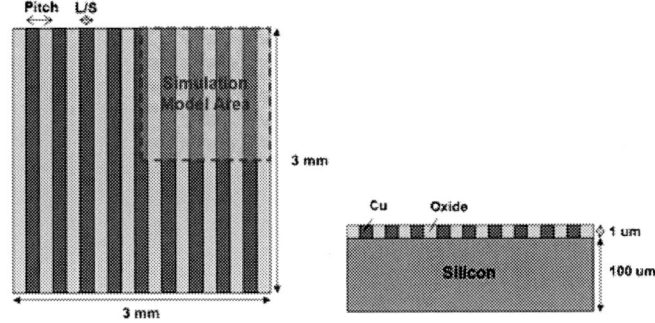

Figure 6. Pattern-inclusion model

The FE model and its boundary conditions are shown in Figure 7, and the material properties are presented in Table 1. There are around 2.7, 2.7, 4.1 and 5.4 million FE elements for metal densities A, B, C and D respectively. The pattern-inclusion models were simulated under a thermal loading and the warpage (z-directional deformation response) can be calculated. Figure 8 shows the warpage contours of different metal densities. The max. warpage is located at the die corner and their values are 0.3 μm, 0.7 μm, 1.22 μm and 1.79 μm respectively. These results reveal that the max. warpage increases as the metal's density is increased.

The next step in our study is to establish equivalent model to obtain the effective material properties of the equivalent layer (i.e., the composite layer of Cu traces and IMD). The equivalent model, as shown in Figure 9, was constructed for replacing the pattern-inclusion mode. All dimensions of the equivalent model are the same as the pattern-inclusion model. The only difference between the two is that the equivalent layer is used to replace the composite layer of Cu and oxide. The equivalent model was meshed with both fine and coarse mesh, as shown in Figure 5. The fine mesh model has the same mesh density as the pattern-inclusion model. From our design, the die size of TSI is from 12 × 12 mm² to 24 × 36 mm². It's difficult to simulate the TSI warpage with Cu traces and IMD because of the large scale difference between the layer thickness and TSI dimension. In addition, adopting the fine mesh density in the TSI model requires a lot of time and memory space to calculate the solution. Hence, the coarse mesh model was constructed and used to verify that its density can be applied to TSI model.

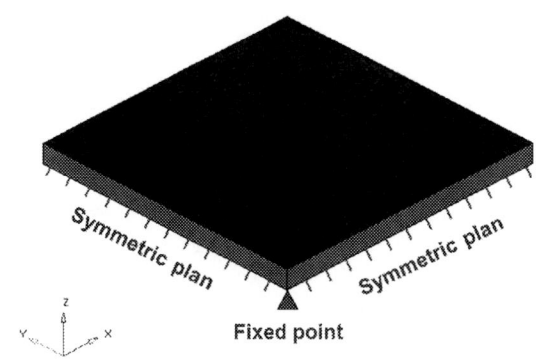

Figure 7. Finite element model and boundary condition

The effective mechanical properties of the equivalent layer were calculated by considering the same warpage response for both the pattern-inclusion model and the equivalent model. The maximum warpage had been determined from the previous results of the pattern-inclusion model, after which the effective mechanical properties of the equivalent layer could be calculated by adapting the volume fraction equation of mechanical properties (Young's modulus, CTE and Poisson' ratio) and fitting them to correctly match the warpage response of the simplified pattern-inclusion model. Table 2 shows the fitting results of mechanical properties for different metal densities. The comparison of the warpage results for the pattern-inclusion model, fine mesh equivalent model and coarse mesh equivalent model is displayed in Figure 11. The same mesh and boundary conditions were used for both the pattern-included model and fine mesh equivalent model, and both have the same warpage results. In addition, the same warpage results are also achieved from the fine mesh equivalent and coarse mesh equivalent models. This indicates that the coarse mesh equivalent model can replace both the pattern-inclusion model and the fine mesh equivalent model. Therefore, the coarse mesh density can be adapted to the TSI model.

Figure 8. Warpage contour of different metal densities

Table 1. Material properties used in the pattern-inclusion model

Material	E (GPa)	CTE (ppm/°C)	Poisson's ratio
Silicon	131	2.8	0.278
Cu	121	16.3	0.34
Oxide	72	0.5	0.16

Table 2. Fitting results for different metal density

Material	E (GPa)	CTE (ppm/°C)	Poisson's ratio
Equivalent Layer for metal density A	85.8	5.0	0.211
Equivalent Layer for metal density B	93.1	7.3	0.238
Equivalent Layer for metal density C	100.8	9.8	0.266
Equivalent Layer for metal density D	107.8	12.0	0.291

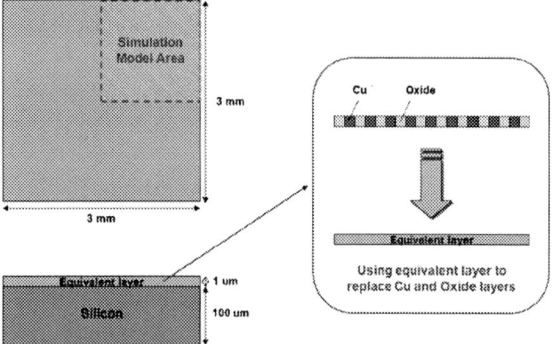

Figure 9. Equivalent model

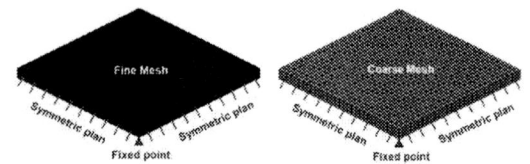

Figure 10. Fine mesh and coarse mesh equivalent model

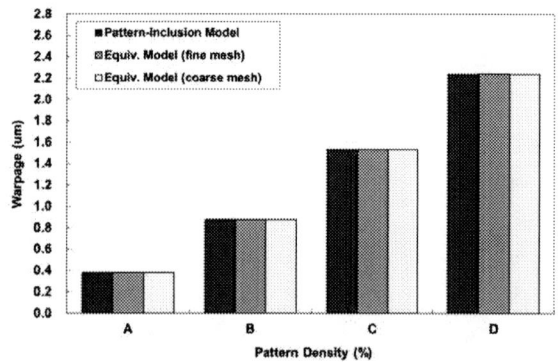

Figure 11. Comparison of warpage results for pattern-inclusion model, fine mesh equivalent model and coarse mesh equivalent model

Results and Discussion

Average pattern density vs Die Warpage

Figure 12 shows the results of the die warpage for different pattern densities and different areas. The warpage value of the incoming wafer is about 65 μm (concave); this is reasonable because the CTE of Cu is ~16.3 ppm/°C, and Si is 2.8 ppm/°C. The higher average pattern density has a larger variation in the amount of its die warpage. For the 1x1 die size, the warpage value increases ~100% when the Cu pattern density is increased by 10~20%. Our test vehicle has only one layer Cu, with neither front-side or back-side PASV. This highlights the close co-relation between Cu pattern density and warpage variation.

It is also important to note that area is a factor of die warpage variation. When the average pattern density is the same but the area is different, the smaller die will have less warpage. It shows that if the TSI die has the same average Cu pattern density, the warpage value of the die will increase 100% when the die size increases 100% (see figure 12). The sets of A to AA and B to BB (average Cu pattern density is same) clearly manifest these results. If the test vehicle wafer is very pure, with only one layer of metal and no TSV or front side PASV, the warpage variation will be simplified. However, for the sets of C to CC and D to DD, the Cu pattern density is higher. As the die size increases, the warpage value increases to be less than 100%. The root cause should be from measurement revolution and deviation.

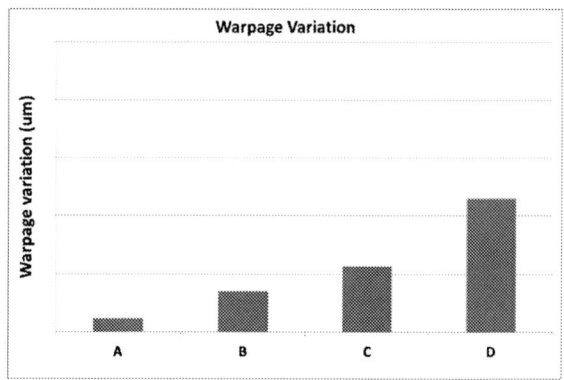

(a) Die warpage variation vs pattern density

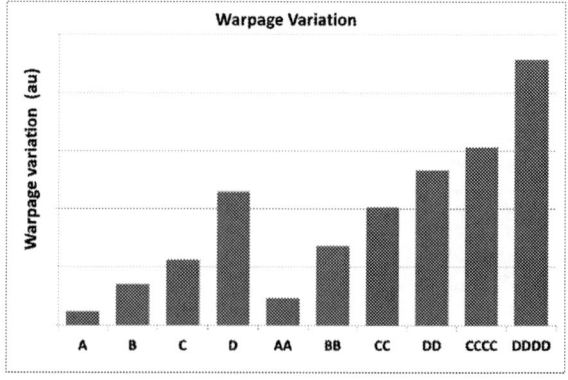

(b) Die warpage variation vs area

Figure 12. The correlation of die warpage variation between pattern density and area

To explore the influence of pattern symmetry, in this work, we don't have more uniform Cu pattern density area because of design limitation in the test vehicle. The major limitation is due to a stitching issue for Cu pattern drawing and the die size arrangement. Figure 13 shows that the warpage variation still follows the pattern density, and the symmetry influence is not obvious. For larger die sizes (36x24 mm^2), the warpage variation is increased by having a higher Cu pattern density area. It can be concluded from a 3x2 die of A+B+4D, 2B+4D and 2C+4D. The area of higher Cu pattern density will have more thermal budget because the Cu metal adsorbs more heat than Si during the reflow process. This causes more warpage variation in the area. Regarding 2x2 die (24x24 mm^2), resource restraints prevented the gathering of this data; allocation of more research resources will be considered after of the new test vehicle is designed to get a more uniform Cu pattern density.

Pattern density vs Through thermal warpage variation

We measured the die wafer during the re-flow process. Regarding thermal warpage variation, the lower the pattern density is better. Figure 14 shows this phenomenon and indicates that if we would like to control the warpage during the thermal process, we should reduce the pattern density.

Simulation model calibration

TSI warpage modeling methodology has been developed and carried out to find the correlation between pattern density and warpage behavior. The warpage comparisons of both the experiment and the simulation are shown in Figure 15 for 1×1 TSI. From the experiment results, the warpage variations are 7 μm (code A), 21 μm (code B), 34 μm (code C) and 69 μm (code D). From the simulation results, the warpage variations are 6 μm (code A), 15 μm (code B), 26 μm (code C) and 38 μm (code D). The simulation shows a trend similar to that found in the experiment: higher pattern density has larger warpage variation. This is to be expected because warpage is primarily induced through a mismatch of CTE and Cu and the silicon substrate. A higher metal density must induce more warpage variation.

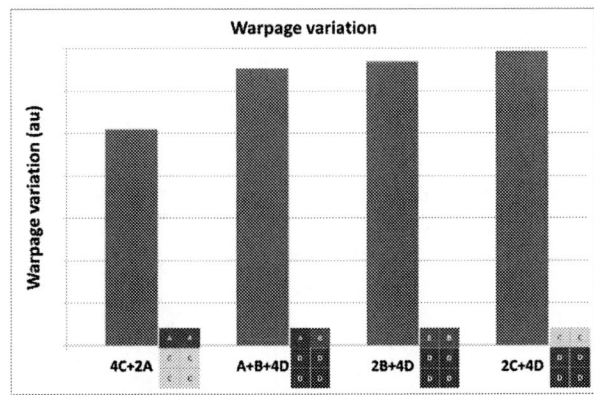

Figure 13. The correlation of die warpage variation by different 3x2 pattern density components.

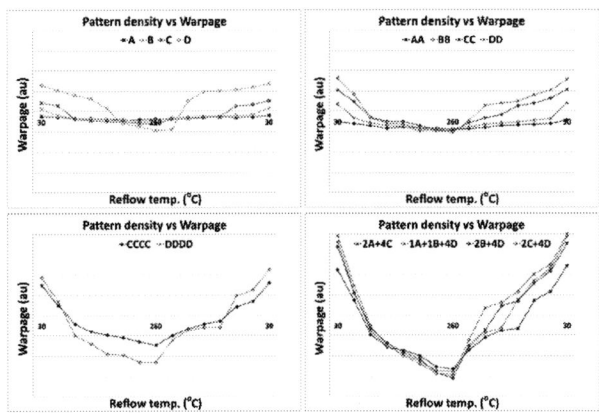

Figure 14. Through thermal die warpage variation

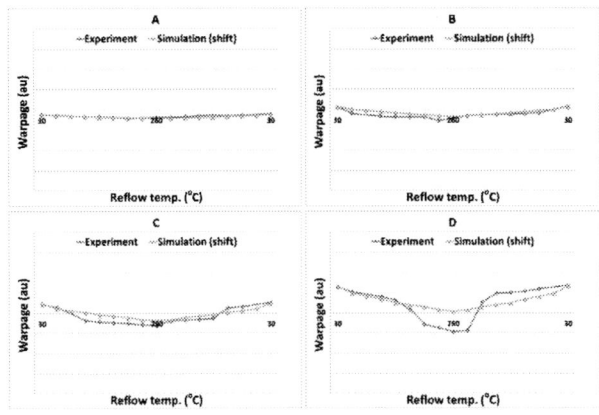

Figure 15. Warpage comparison of experiment and simulation for 1×1 TSI

Figure 16 displays the R-squared comparisons of the experiment and simulation for 1×1 TSI. The simulation results are consistent with the experiment of 1×1 TSI with the pattern density of codes A, B, and C, but not for code D. This may be because of the resolution of our Shadow moiré measurement machine. Code A, B and C present smoother curves during the reflow process, where code D has a sharper curve at 200~217°C.

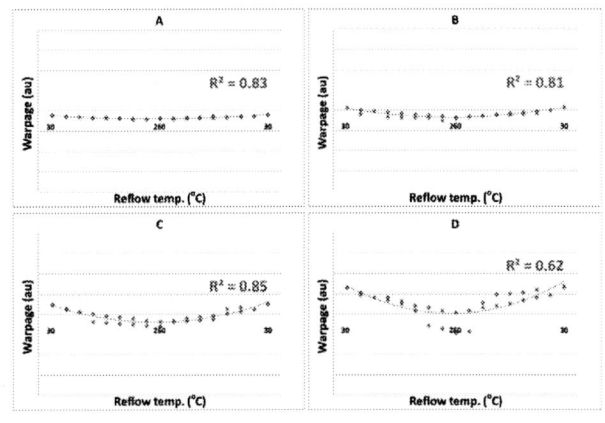

Figure 16. R-squared comparison experiment and simulation for 1×1 TSI

Similarly, the warpage comparisons of experiment and simulation are shown in Figures 17, 19 and 21 for 2×1, 2×2 and 3×2 TSI respectively. For 2×1 TSI, the warpage variations are 14 μm (code AA), 40 μm (code BB), 61 μm (code CC) and 80 μm (code DD) from the experiment and 16 μm (code AA), 37 μm (code BB), 64 μm (code CC) and 93 μm (code DD) from the simulation. For 2×2 TSI, the warpage variations are 92 μm (code CCCC) and 137 μm (code DDDD) from the experiment and 103 μm (code CCCC) and 150 μm (code DDDD) from the simulation.

For 3×2 TSI, the warpage variations are 153 μm (code 2A+4C), 196 μm (code A+B+4D), 199 μm (code 2B+4D) and 208 μm (code 2C+4D) from the experiment and 138 μm (code 2A+4C), 200 μm (code A+B+4D), 207 μm (code 2B+4D) and 225 μm (code 2C+4D) from the simulation. These also show that higher metal density induces larger warpage variation. In addition, the R-squared comparisons of experiment and simulation for 2×1, 2×2, 3×2 TSI are presented in Figure 18, 20, 22. The simulation results in good accord with the experiments, indicating that the simulation methodology is feasible and correlates well with the experimental test results. Therefore, we are highly confident that this finite element model can generate improved results in the future.

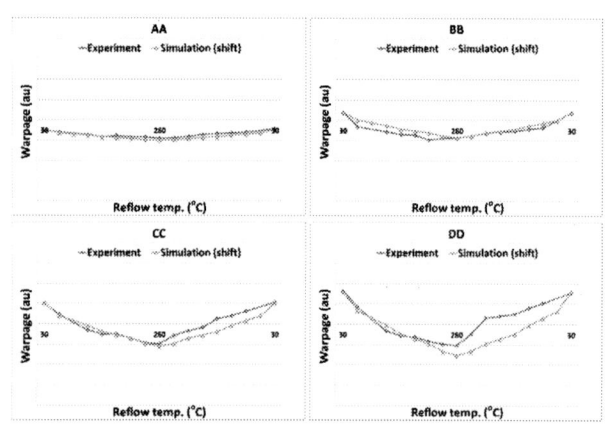

Figure 17. Warpage comparison of experiment and simulation for 2×1 TSI

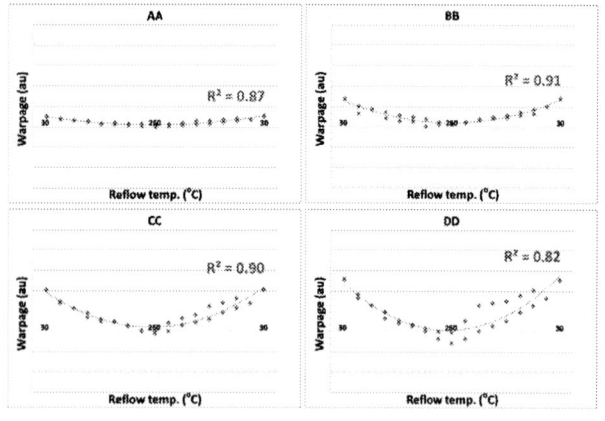

Figure 18. R-squared comparison experiment and simulation for 2×1 TSI

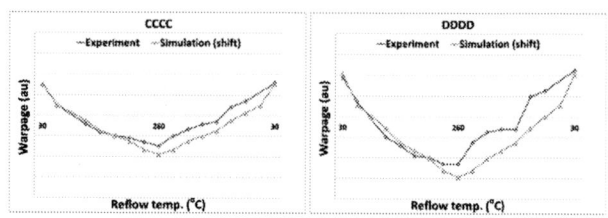

Figure 19. Warpage comparison of experiment and simulation for 2×2 TSI

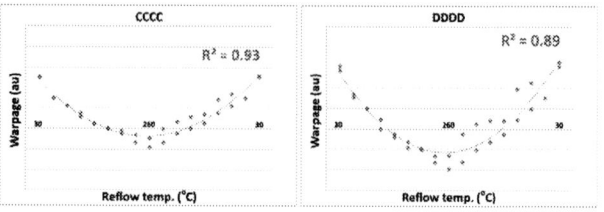

Figure 20. R-squared comparison experiment and simulation for 2×2 TSI

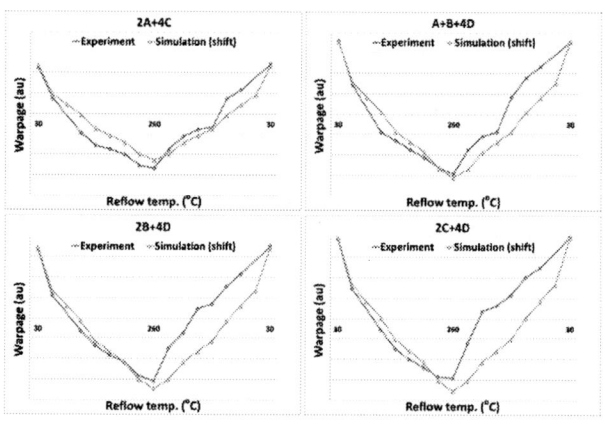

Figure 21. Warpage comparison of experiment and simulation for 3×2 TSI

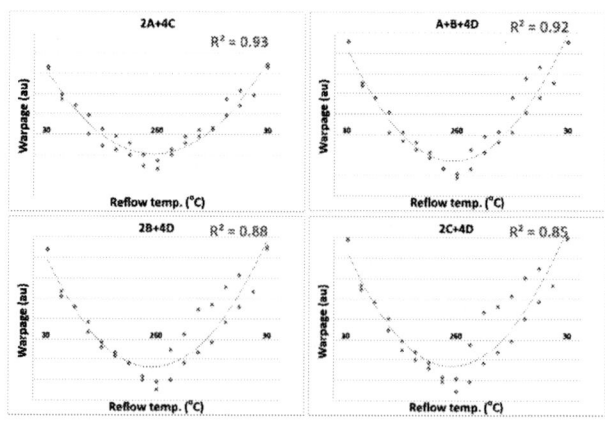

Figure 22. R-squared comparison experiment and simulation for 3×2 TSI

Conclusion

In summary, we have proposed a new FEM model showing the relation between pattern density and warpage behavior, and planned our test vehicle with different pattern densities and areas to verify our modelled results.

From our experiments we found that the TSI warpage between the simulation and measurement data are less than 20~30%. The finite element simulation results in the work correlate well with the experimental test results. Therefore, we are highly confident that this finite element model can help generate design guidelines for wafer patterns and predict die warpage variation during the assembly process. In the future it will be important to increase the multi-layer metal Cu and PASV on the wafer to check warpage variation. The issue of the establishment of complicated simulation models will be a key topic for future research.

Acknowledgments

The warpage measurement has been carried out from MTI (Macrotech Technology Inc.).

The authors wish to thank all the staff of MTI who supported this work.

References

1. Seung Wook YOON, Dae Wook YANG*, Jae Hoon KOO, "3D TSV Processes and its Assembly/Packaging Technology", page 1-5, 3D System Integration, 2009. 3DIC 2009. IEEE International Conference on.

2. X. F. Pang, T. T. Chua, H. Y. Li, E.B. Liao, W. S. Lee, F. X. Che, "Characterization and Management of Wafer Stress for Various Pattern Densities in 3D Integration Technology", page1866 – 1869, Electronic Components and Technology Conference (ECTC), 2010.

3. Hebb J.P. and Jensen K.F., "The effect of patterns on thermal stress during rapid thermal processing of silicon wafers," IEEE Transactions on Semiconductor Manufacturing, 11(1998), pp. 99-107.

4. Nathan R.D., Liu J.J. and Tom J., "Experimental investigation of bare silicon wafer warp," IEEE Microelectronics and Electron Device Workshop, 2004, pp. 120-123.

5. Kim Y., Kang S-K. and Kim S.E., "Study of thinned Si wafer warpage in 3D stacked wafers," Microelectron Reliab (2010), doi:10.1016/j.microrel.2010.05.006.

6. Thakur R.P.S., Chhabra N. and Ditali A., "Effects of wafer bow and warpage on the integrity of thin gate oxide," Appl Phys Lett 1994;64(25):3428-30.

7. F. X. Che, H. Y. Li, Xiaowu Zhang, S. Gao, and K. H. Teo, "Wafer level warpage modeling methodology and characterization of TSV wafers," 61st Electronic Components and Technology Conference (ECTC), 2011, pp. 1196-1203.

A Resilient 3-D Stacked Multicore Processor Fabricated Using Die-level 3-D Integration and Backside TSV Technologies

[1]K-W. Lee, [1]H. Hashimoto, [1]M. Onishi, [1]Y. Sato, [1]M. Murugesan, [1]J-C Bea, [1]T. Fukushima, [2]T. Tanaka, [1]M. Koyanagi

[1]New Industry Creation Hatchery Center (NICHe), Tohoku University, Sendai, Japan
[2]Department of Biomedical Engineering, Tohoku University, Sendai, Japan
kriss@bmi.niche.tohoku.ac.jp, 81-22-795-4119

Abstract

A highly dependable 3-D stacked multicore processor with TSV self-test and self-repair functions for highly area-efficient TSV repair has been proposed. The prototype 3-D stacked multicore processor with two layer structure is implemented using die-level 3-D integration and backside Cu TSV technologies. The basic functions of tier boundary scan and self-repair circuits via TSVs between each layer in the 3-D stacked multicore processor are successfully evaluated. X-ray computed tomography (X-ray CT) scanning technology is proposed as a non-destructive failure analysis method to characterize high-density TSVs integration, and bump joining qualities in the 3-D stacked multicore processor.

Introduction

Recently, high functionality and availability of LSI becomes increasingly important. However, conventional 2D LSI has serious technical limitation of slowdown device-scaling rate, consequently, increasing chip size and global wire length, and reducing production yield. To solve these problems, a three-dimensional (3-D) integration technology is indispensable [1-7]. Many LSI chips with smaller size are vertically stacked and electrically connected by using high-density and short-length vertical through silicon vias (TSVs). In a 3-D stacked LSI, smaller chip size improves the production yield and TSV interconnects dramatically reduce the global interconnection length and increase the wiring density. The 3-D LSI technology also allows LSI to adopt redundant or spare modules in order to improve its availability and dependability. Therefore, it will be possible to simultaneously satisfy both high functionality and high availability for realizing 3-D VLSI system with self-restoration function as shown in Fig. 1. To realize such 3-D VLSI system comprising many memories and processor layers, one of the most important matters is to enhance the stacking yield by increasing the connectivity of vertical connection between the stacked layers. To enhance the connectivity, it is indispensable to apply TSV repair circuit or redundant TSVs. Various TSV repair technologies have been proposed [8-9]. However, it has large area penalty issue due to many redundant TSVs. These technologies need huge numbers of redundant TSVs, consequently the area for redundant TSVs are much larger than that for logic gates. Therefore, it is important to reduce the number of redundant TSVs for minimizing the

TSV area penalty with keeping its high repairability. In order to meet those requirements, we proposed a highly dependable 3-D stacked multicore processor with TSV self-test and self-repair circuits for highly area-efficient TSV repair to improve the reliability and the yield of 3-D VLSI system [10]. In this study, to evaluate the usability of highly efficient TSV repair technology, the prototype 3-D stacked multicore processor of two tiers structure is implemented using die-level 3-D integration and backside TSV technologies.

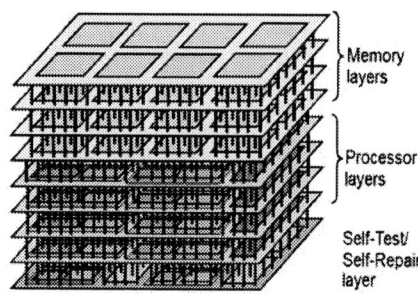

Fig. 1. 3-D VLSI system with self-restoration function

A Resilient 3-D Stacked Multicore Processor

Fig. 2 shows the block diagram of highly dependable 3-D stacked multicore processor with self-test and self-repair functions. To achieve high-performance and highly dependable 3-D multicore processor, several single-core processor chips are vertically stacked and electrically connected by vertical bus using TSVs.

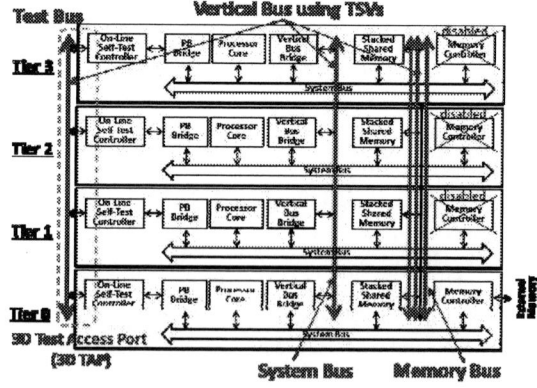

Fig. 2. The block diagram of highly dependable 3-D stacked multicore processor with self-test and self-repair functions

2014 Electronic Components & Technology Conference

In the 3-D stacked multicore processor, processor cores are classified into three types – system coordinator core layer, processing core layer, and spare core layer. The system coordinator core controls the whole system – task scheduling, online hardware self-test scheduling, and self-repair of the damaged core, etc. Processing core execute the tasks allocated by the system coordinator core. Each spare core is in a cold-standby state. When the damaged core is found in online self-test sequence, the failed core is swapped with one of the spare cores by logical ID reallocation and the system continues to execute tasks in the system as shown in Fig. 3 (failover).

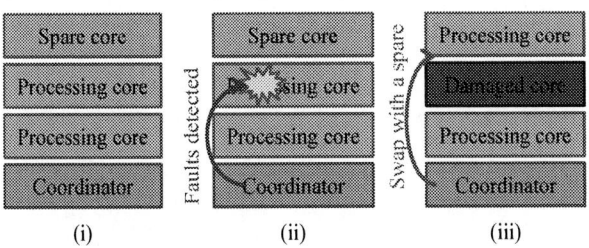

Fig. 3. Failover in the resilient 3-D stacked muluticore processor system. (i) shows the initial state of the system with stacked four tiers. If any fault is detected in self-test sequence (ii), the system coordinator core swaps the damaged core with a spare core to keep operations in the system (iii)

To achieve built-in self-test and self-repair circuit for TSVs in the 3-D stacked multicore processor, it is important to simplify the test and the repair method including repair algorithm and circuit, because complex one causes increasing variation of signal delay and the circuit size. The built-in self-test and self-repair circuit for TSVs consists of the test pattern and expectation generator, inter-tier boundary scan test circuit, comparator between test result and expectation, repair circuit to assign signals uniquely to TSVs.

Implementation of 3-D Stacked Multicore Processor

(A) Design of Processor Core Chip

To implement the 3-D stacked multicore processor for highly efficient TSV repair, we designed and fabricated 2-D processor core chip with self-test and self-repair functions using 90-nm CMOS technology as shown in Fig. 4. The self-reparable TSVs are implemented for the vertical system bus and the vertical bus for the stacked shared memory in the processor. Each bonding pad has quadruple TSV. Selectors, 16 signals of 20 TSVs, are adopted for the vertical system bus and the stacked shared memory in order to implement the self-repair function. Therefore, the processor chip has totally 1,920 TSVs – 960 TSVs for bonding pads, 540 TSVs for the vertical system bus, and 420 TSVs for the stacked shared memory [11].

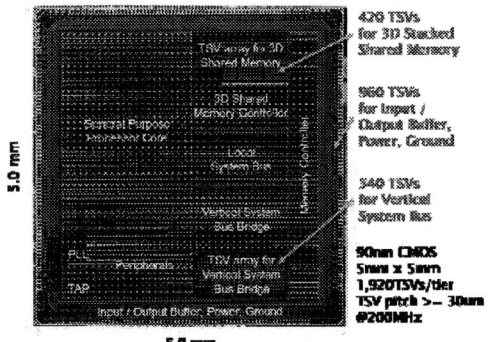

Fig. 4. Photograph of fabricated 2-D processor core chip with self-test and self-repair functions

(B) Fabrication of 3-D Stacked Multicore Processor using Die-level 3-D Integration Technology

The 3-D stacked multicore processor was fabricated using the die-level 3-D integration and backside TSV technologies as shown in Fig. 5 [12-14]. At first, the processor core chip is thinned down to 200um thickness by mechanical grinding following CMP treatment. Cu/Sn bumps are formed on the chip surface in die-level. The processor chip with Cu/Sn bumps is glue-bonded temporally to a supporting wafer with the alignment mark by the high-accuracy bonder and cured in the vacuum chamber. The alignment accuracy within 2-µm is required to avoid the impact of misalignment on the subsequent backside lithography process for via formation. Then, the processor chip is thinned down to around 50-µm thickness by mechanical grinding following chemical mechanical polishing (CMP) treatment. After cleaning on the thinned surface by DI water, plasma-TEOS dielectric layer of 1-µm thickness is deposited on the backside surface as a hard mask. After the backside TSV patterning with high alignment accuracy within 2-µm, via holes of 10-µm diameter are etched from the backside of Si substrate until to exposure metal 1 layer in the processor chip using BOSCH process with SF_6 and C_4F_8 gases. P-TEOS liner of 1-µm thickness is deposited into via holes and the bottom dielectric liner in via hole is contact-etched by dry etching using the dielectric hard mask. Barrier and seed layers of Ta/Cu are carefully deposited into via holes by sputtering. Then, Cu electroplating is used to completely fill via holes. Cu/Sn bumps are formed on by electroplating as electrodes and Cu RDL (re-distribution line) is formed by wet etching patterning. All processes to form TSVs and Cu/Sn bumps are performed in die-level.

The first-layer processor chip is face-up thermo-mechanically bonded to Si interposer with high alignment accuracy within 2-µm and electrically connected via Cu TSVs and micro bump-joining. After open/short and simple functional test to confirm the joining quality between the processor chip and Si interposer, the stacked sample is dipped into a stripper and the supporting wafer is easily de-bonded from the chip surface by removing the glue layer using a stripper. After cleaning the chip surface by a cleaner, it is ensured that the second layer chip is bonded to the bottom layer chip. After the evaluation of the joining quality, the supporting wafer is de-bonded again by removing the glue layer. By repeating these processes for the other processor chip, a known-good 3-D stacked multicore processor chip is fabricated.

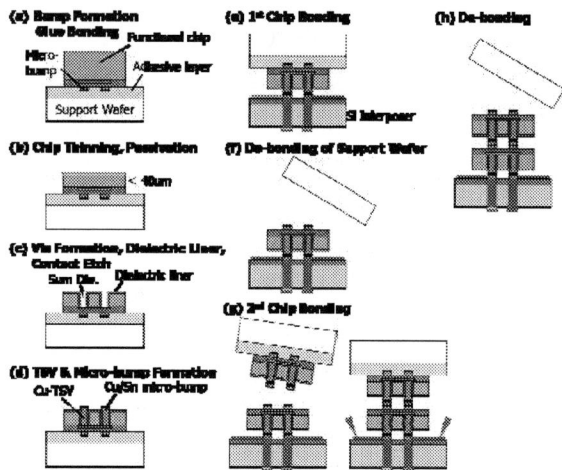

Fig. 5. Photographs of the fabricated Cu/Sn bumps and Cu RDL on the backside surface in die-level

Backside Cu TSV technology is developed for realizing low-cost die-level 3-D integration [9-10]. Via-last backside Cu TSV approach has high flexibility to commercial device chip/wafer. It also has better reliability from Cu TSV-induced stress and Cu contamination issues, when compared with via-middle Cu TSV approach. However, backside Cu TSV has some technological challenges such as alignment accuracy to metal 1 layer, notch-free Si via etching, and good electrical contact between backside Cu TSV and metal 1 layer [14]. To achieve compact-sized 3-D stacked system, the fine-sized backside TSV is designed, in which the Cu TSV of 10-μm diameter is connected to metal 1 layer of 18-μm width in BEOL passed through n+ well area of 14-μm width, as shown in Fig. 6(a).

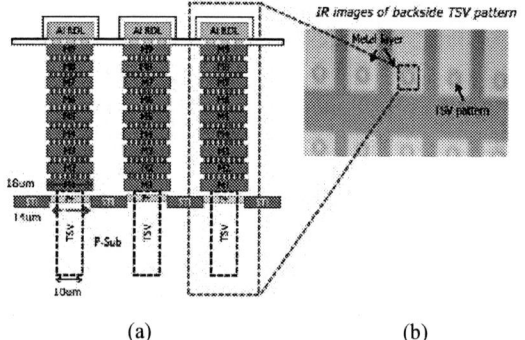

(a)   (b)

Fig.6. Design rule of backside TSV (a) and IR images after the backside TSV patterning to M1 layer (b)

The TSV patterning on the backside surface of the thinned

chip is performed using the alignment mark on the supporting wafer by a conventional mask aligner. Fig. 6(b) shows the infrared (IR) images after the backside TSV patterning, which indicates that TSV patterns on the backside are well located within metal 1 layer in the front side with the average alignment accuracy of within 2-μm.

Following Si via etching from the backside of Si substrate, a dielectric layer underneath the metal 1 layer in BEOL emerge when Si via etching is completed. However, the dielectric layer is not etched by Si etching gases, and as a result, significant sidewall etching (notch) of Si occurs at the surface of the dielectric layer during Si over-etching because of accumulated positive ions. Time-modulation bias technology that periodically turns RF (radio frequency) bias on and off to control plasma ion energy is applied to minimize notch phenomenon during Si etching.

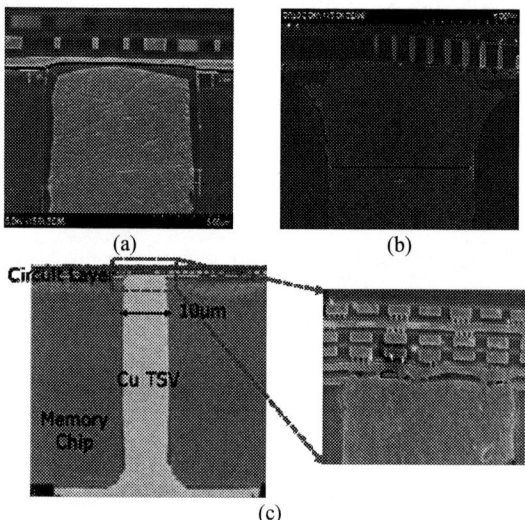

Fig.7. SEM cross-sectional images of before the optimization; liner under-etch (a), notch (b), and after the optimization (c) of backside Cu TSV formation

Fig. 7 shows the scanning electron microscopy (SEM) cross-sectional images before the optimization; liner under-etch at the bottom region in TSVs (a), notch phenomenon (b), and after the optimization (c) of backside Cu TSV formation. After the optimization, the notch is less than 100-nm at the surface of dielectric layer even after 50% Si over-etching. This value is not an serious issue to form TSV. The fine-sized backside Cu TSV of 10-μm diameter was successfully fabricated. Good electrical contact between Cu TSV and metal 1 layer is a key parameter for the 3-D IC fabrication, because TSV is an important factor to determine 3-D IC performance. As the dielectric liner at the bottom of the hole is contact-etched by dry etching, the bottom surface of metal 1 layer is exposed to atmospheric environment and easily gets oxidized. Before the formation of barrier/seed layers, thin metal oxide layer is removed by Ar back-scatter sputtering to confirm good electrical contact. The electrical characteristics between Cu TSV and metal 1 layer are evaluated using TSV daisy chain patterns in the chip, as shown in Fig. 8.

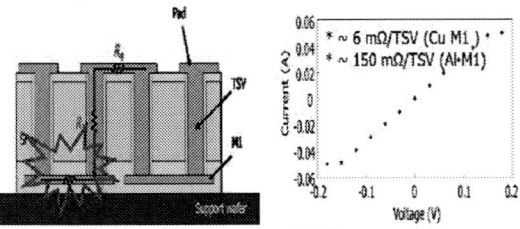

(a) Conceptual structure of TSV chain (b) Contact resistance to metal layer

Fig.8. Contact resistances between backside Cu TSV and metal 1 layer

The contact resistances between Cu TSV and metal 1 layer are approximately 6 mΩ (M1 Cu) and 150 mΩ (M1 Al), respectively. We assumed that Al layer is more easily oxidized, when compared with Cu layer. Thick Al oxide layer is difficult to remove than Cu oxide layer, and therefore, it raises concern of inducing relatively high contact resistance. TSV formation process needs to be optimized in the future to reduce the contact resistance to Al layer, although it has adequately low value for TSV application. The electrical characteristics of vertical interconnection have been evaluated using TSV and Cu/Sn bump daisy chain pattern in the chip. The resistances of a pair of Cu-TSV, Cu/Sn micro-bump are approximately 70 mΩ (M1 Cu) and 240 mΩ (M1 Al), respectively. Although the resistance to Al metal 1 layer is relatively high, the resistance values of the both cases are adequately low for high-performance 3-D ICs.

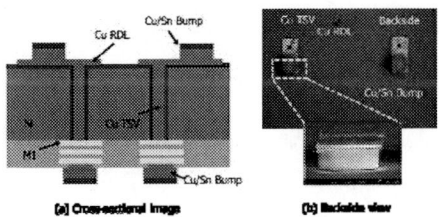

Fig. 9. Cross-sectional image (a) and the photographs of Cu/Sn bumps and Cu RDL on the backside surface

To flexibly connect different functional chips and avoid the stress effect induced by Cu TSV, Cu/Sn bumps are formed depart from Cu TSV by using redistribution metal lines (RDL). Fig. 9 shows the cross-sectional image (a) and the photographs of Cu/Sn bumps and Cu RDL (b), which are formed on the back surface of each chip in die-level.

Fig. 10 shows the conceptual image of the prototype 3-D stacked multicore processor with two layer structure on Si interposer. Fig. 11 shows the top view (a) and SEM cross-sectional image (b) of the fabricated 3-D stacked multicore processor, where two layers of processor core chip with

50-μm thickness (including 10-μm BEOL layer and 40-μm Si substrate) are bonded on Si interposer. The backside Cu-TSVs are good contacted to metal 1 in each layer without notching phenomenon.

Fig.10. The conceptual image of the prototype 3-D stacked multicore processor with two layer structure on Si interposer

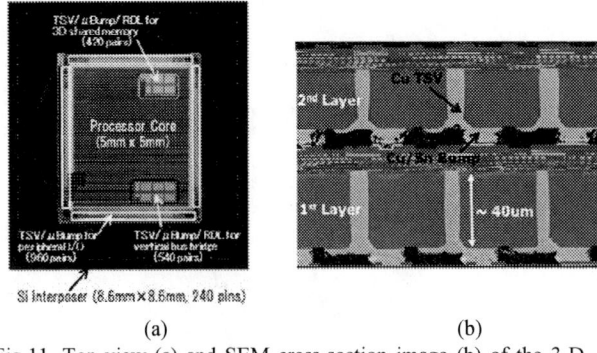

(a)   (b)

Fig.11. Top view (a) and SEM cross-section image (b) of the 3-D stacked multicore processor

(C) Characterization of 3-D Stacked Multicore Processor

We evaluated the basic functions of tier boundary scan and self-repair circuits for TSVs through TSVs between two processor layers in the 3-D stacked multicore processor as shown in Fig. 12. The prototype 3-D stacked multicore processor is successfully implemented by die-level 3-D integration and backside TSV technologies.

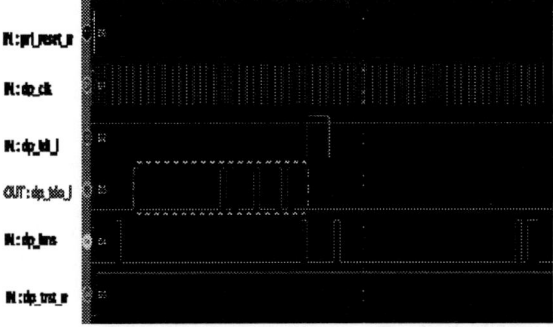

Fig.12. Measured output waveform from the 3-D stacked multicore processor

Fig. 13 shows X-ray computed tomography (CT) scanning images before the optimization (a) and after the optimization (b) of the fabrication process for the 3D stacked multicore processor with two-layer structure. Many numbers of Cu TSVs, RDLs, metal bumps, and BEOL interconnects are clearly seen from the 3-D stacked multicore processor. We can analyze the failures such as voids in Cu TSVs and bump joining qualities. X-ray CT scanning technology is useful method as a non-destructive 3-D failure analyzing to characterize high-density TSVs integration and metal-bump joining in 3-D stacked LSIs.

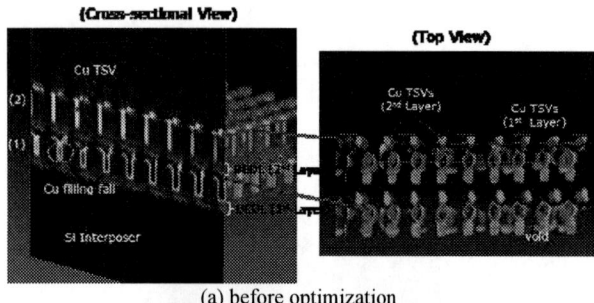

(a) before optimization

(b) after optimization

Fig. 13. X-ray CT-scan images before optimization (a) and after optimization (b) of the fabricated 3D stacked multicore processor with two-layer structure

Conclusions

The 3-D stacked multicore processor with two layer structure is fabricated using die-level 3-D integration and backside Cu TSV technologies. The basic functions of tier boundary scan and self-repair circuits via TSVs between each layer in the 3-D stacked multicore processor are successfully evaluated. X-ray CT scanning technology is proposed as a non-destructive failure analysis method to characterize high-density TSVs integration, and bump joining qualities in the 3-D stacked multicore processor.

Acknowledgments

This research was supported by the Dependable VLSI Project of Core Research for Evolutional Science and Technology (CREST) of Japan Science and Technology Corporation (JST).

References

[1] M. Koyanagi, "Roadblocks in Achieving Three-Dimensional LSI," Proc. 8th Symposium on Future Electron Devices, pp.50 – 60. 1989.

[2] T. Kunio, K. Oyama, Y. Hayashi, and M. Morimoto, "Three dimensional ICs, having four stacked active device layers,"

IEEE International Electron Devices Meeting (IEDM), pp.837–840, 1989.

[3] M. Koyanagi, H. Kurino, K-W. Lee, K. Sakuma, N. Miyakawa, H. Itani, "Future system-on-silicon LSI chips," IEEE MICRO, 18 (4), pp.17 – 22. 1998.

[4] S.J. Souri, K. Banerjee, A. Mehrotra, and K.C. Saraswat, "Multiple Si layer ICs: Motivation, performance analysis, and design implications," in Proc. 37th ACM Design Automation Conf., pp.873 – 880, 2000.

[5] P. Ramm, D. Bonfert, H. Gieser, J. Haufe, F. Iberl, A. Klumpp, A. Kux, R. Wieland, "Interchip via technology for vertical system integration," Proc. IEEE Int. Interconnect Technology Conf. (IITC), pp.160 – 162, 2001.

[6] Mekis A, Abdalla S, Analui B, Gloeckner S, Guckenberger A, Koumans K, Kucharski D, Liang Y, Masini G, Mirsaidi S, Narasimha A, Pinguet T, Sadagopan V, Welch B, White J, and Witzens J, "Monolithic Integration of Photonic and Electronic Circuits in a CMOS Process", Proc. SPIE, pp. 68970L, 2008.

[7] Tze-Chiang Chen, "Device Technology Innovation for Exascale Computing", Proc. VLSI Symposium, pp 8– 11, 2009.

[8] U.S. Kang, H.J. Chung, S.M. Heo, S.H. Ahn, H. Lee, S.H. Cha, J.S. Ahn, D.M. Kwon, J.H. Kim, J.W. Lee, H.S. Joo, W.S. Kim, H.K. Kim, E.M. Lee, S.R. Kim, K.H. Ma, D.H. Jang, N.S. Kim, M.S. Choi, S.J. Oh, J.B. Lee, T.K. Jung, J.H. Yoo, and C.G. Kim, "8Gb 3D DDR3 DRAM Using Through-Silicon-Via Technology," proceedings of IEEE Solid-State Circuits Conference 2009, pp. 130 – 131.

[9] L. Jiang, Q. Xu, and B. Eklow, "On Effective Through-Silicon Via Repair for 3-D-Stacked ICs," IEEE Transactions on Computer-Aided Design of Integrated Circuits and Systems, 2013, pp. 559 – 571.

[10] M. Koyanagi, "Three-Dimensional VLSI System with Self-Restoration Function" 2nd JST International Symposium on Dependable VLSI Systems (DVLSI 2013), Dec. 6-7, 2013, Tokyo, Japan.

[11] H. Hashimoto et al., "Highly Efficient TSV Repair Technology for Resilient 3-D Stacked Multicore Processor System", IEEE 3D System Integration Conference, pp.978-1-4673-6484-3, 2013

[12] Y. Ohara, K.W. Lee, T. Fukushima, T. Tanaka, M. Koyanagi, "Development of Versatile Backside Via Technology for 3D System on Chip", Int. Conf. On. Solid State Devices and Materials (SSDM), Tokyo, 237-238, 2010.

[13] Yuki Ohara, Kang-Wook Lee, Jicheol Bea, Takafumi Fukushima, Tetsu Tanaka, and Mitsumasa Koyanagi, " Novel Detachable Bonding Process with Wettability Control of Boning Surface for Versatile Chip-Level 3D Integration", IEEE International 3D System Integration Conference, Osaka, pp. 3.2.1-4, 2012.

[14] K-W Lee, Y. Ohara, K. Kiyoyama, S. Konno, Y. Sato, S. Watanabe, A. Yabata, T. Kamada, J-C Bea, H. Hashimoto, T. Fukushima, T. Tanaka, and M. Koyanagi, "Characterization of Chip-level Hetero-Integration Technology for High-Speed, Highly Parallel 3D Stacked Image Processing System", IEEE International Electron Devices Meeting (IEDM), pp.785-788, 2012

3D Stacking Induced Mechanical Stress Effects

V. Cherman, G. Van der Plas , J. De Vos, A. Ivankovic*, M. Lofrano, V. Simons, M. Gonzalez, K. Vanstreels, T. Wang, R. Daily, W. Guo, G. Beyer, A. La Manna, I. De Wolf*, E. Beyne

IMEC

Kapeldreef 75, 3001, Leuven, Belgium

*Also at Dept. MTM, Fac. Engineering, KULeuven, Belgium

vladimir.cherman@imec.be

Abstract

In this work the effects of 3D stacking technology on the performance of devices are systematically studied. For this study a special chip consisting of a number of stress sensors and vertical interconnect loops was designed and manufactured in 65nm technology. Local variations of stress with a magnitude of up to 300 MPa are detected at different locations along the chip and are being characterized using finite element modeling and micro-Raman spectroscopy measurements.

Introduction

3D integrated circuits (3D-IC) are gaining popularity due to the unique properties and improved performance they can potentially offer to the new generations of microelectronic components. Those are lower power consumption and better signal integrity and bandwidth due to shorter interconnects, smaller footprint, possibility of heterogeneous integration of different technologies in one 3D-IC and many others [1]. Three key components of the high density 3D chip stacking are the through silicon vias (TSVs) inside the chip(s), the inter-chip interconnects established by solder joints or metal pillars and the glue or underfill (UF) material which improves the mechanical stability of the assembly [2,3]. As in any complex system, introduction of additional mechanical and electrical joints into the integrated circuit may have an impact on its performance and reliability [4,5]. One can distinguish two main mechanisms which control this impact. First is the performance and the reliability of the electrical interconnects and second is the effects of additional mechanical stress induced in different parts of 3D-IC on FEOL and BEOL structures. The performance and the reliability of TSVs have been intensively addressed and a lot of progress has been made in understanding and improving the TSV technology [2] and its effects on devices [5,6,7,8] and interconnects. On the other hand the influence of stacking has not received adequate attention so far though some particular cases were studied and / or modeled in the past [9,10]. To fill-in this gap and to provide the experimental validation to the complex finite element models, the effects of the 3D stacking are systematically studied in this work.

Test Vehicle

To study the effects of the 3D stacking a special chip was designed and manufactured in 65nm technology. The chip contains different test structures in FEOL and BEOL which are used as electrically measurable sensors. These sensors are grouped in 240µm x 240µm square cells. The cells are regularly repeated in 32 rows and 32 columns within the area of the 8mm x 8mm test chip. The sensors available on the test chip are differently oriented n- and p-FETs and diodes which are used as stress and temperature gauges, respectively. Metal-insulator-

metal (MIM) parallel-plate capacitors formed in the BEOL are used as additional stress gauges, which are sensitive to vertical stresses. The test chips are realized with two different options. In one case the die is thinned down to 200µm and it carries 5µm Cu, 3.5µm Sn with 15µm in diameter electroplated micro-bumps. This die is marked as "Tier 1" in Fig.1. In the second option, the via-middle technology was used to incorporate 50µm deep 5µm in diameter TSVs. This 50µm thick die (Tier 2 in Fig.1) contains a 3µm thick copper redistribution layer (RDL) and 7µm Cu micro-bumps, 25µm in diameter, processed on the back side. It also contains 5µm thick 50µm in diameter copper pillars plated on the front side passivation layer. Thermo-compression bonding in combination with applied in-situ no-flow underfill was used to individually assemble the pairs of single dies together in the back-to-face 3D stacks. The micro-bumps between the dies and the copper pillars on top of the stack are arranged in a regular array pattern (Fig.2a) with a pitch of 40µm and 240µm, respectively.

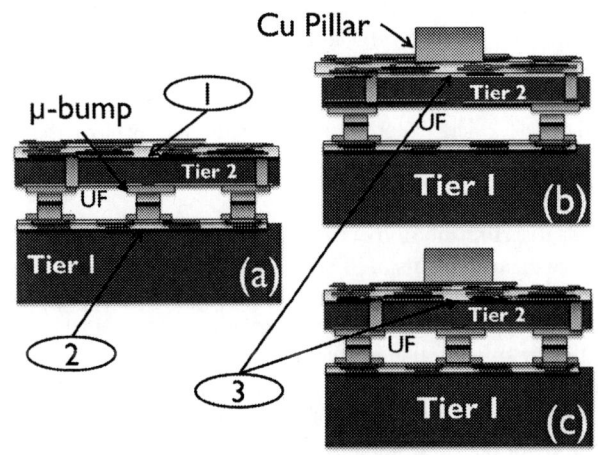

Figure 1. Cross sections at three different locations inside the 3D-stack. (a) 40µm pitch array of µbumps, (b) Cu-pillar on top of the thin die, (c) combination of the micro-bumps and the Cu-pillar. "UF" is the underfill. "1" - "3" indicate the positions of the FEOL stress sensor arrays.

The FE stress sensors used in this work are n-FETs with the channel dimensions 4µm x 4µm. The FETs are grouped in the 7x8 arrays (Fig.2) and they are used to measure local stress or its variations near the micro-bumps or the Cu pillars or a combination of both on two tiers (as in Fig.1a-c). Similar arrays of transistors have already been used in the past to measure the local stress distributions around the TSVs [5,8] and the micro-bumps [10,11]. Additional 8 sensors are located outside the

978-1-4799-2408-0/14 $31.00 © 2014 IEEE

area of the array and are placed under eight peripheral micro-bumps (as in Fig.2a).

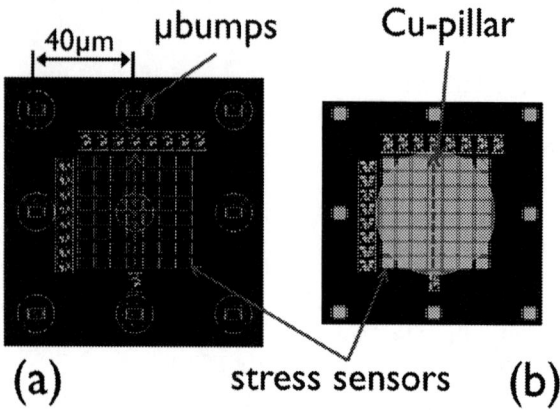

Figure 2. Arrangement of the FEOL stress sensors (a) around the micro-bumps and (b) the copper pillar. The green squares indicate the positions of the n-FET stress sensors. The pitch of the micro-bump array is 40μm. The red arrows indicate the positions of the sensors located along the symmetry line of the micro-bump / Cu pillar.

Experimental Results and Validations

The saturation currents of the sensors (I_{ON}) are measured at different locations in the 3D stack. The distribution of these currents within each FET array (ΔI [%]) are calculated as $\Delta I=(I_{ON}-I_{ON,ref})/I_{ON,ref} \cdot 100\%$. The minimal currents measured within every array are used as the reference ($I_{ON,ref}$). As a result of the performed experiments, three local effects of the 3D stacking have been identified and they are discussed in the following sections.

Stress distribution above the micro-bump on Tier 2.

The measurements of the stress sensors array located on the Tier 2 above the micro-bump (as in Fig.1a, position 1 and Fig.2a) indicate that the stress variation at that location is negligibly small or that it is homogenous. This is seen from the surface plot of ΔI_{ON} currents measured within the array (Fig.3a) or from the 1-D plot (Fig.3b) indicating maximum of 1% variation of the ΔI_{ON} along the red line.

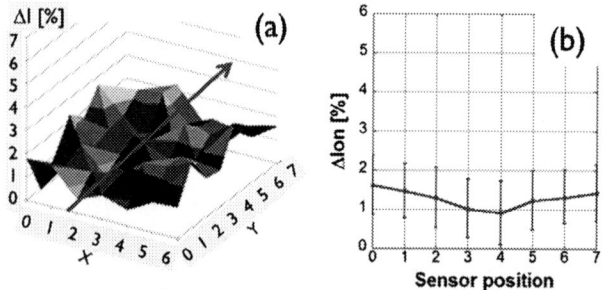

Figure 3. Surface plot of the relative I_{ON} currents of the stress sensors within the array located on Tier 2 above the micro-bump (a) and the variation of the currents along the red arrow (b).

The error bars indicated in figures 3b-5b represent standard deviations calculated based on the measurements of 120 3D stacks.

On the first sight, such a behavior is in conflict with the previous experimental observations [10, 12] and also with the results of the finite elements modeling (FEM) [13] which indicate that during the cooling phase of the thermal-compression bonding (TCB) the shrinking UF pulls the thin Si die while the micro-bumps constrain this force, causing wavelike deformation of the top die [11] and thus inducing in-plane and out-of-plane mechanical stress in it. This mechanism is possible due to the differences in the thermal expansion coefficients between the UF and the micro-bump materials. The main difference between the previously published results and this work is that in the past, the stress was measured and simulated around a single stand-alone micro-bump. In contrast, in this work the micro-bumps are arranged in the regular array pattern with a pitch of 40μm (as in Fig.2a). This pitch size is comparable with the thickness of the top Si die, which is 50μm. The combination (or the ratio) of the thickness of the Si die(s) and the micro-bumps pitch size largely determines the warpage of Si. By increasing this ratio the stress induced in Si due to bending can be minimized.

To support this hypothesis an FE model of the experimental 3D stack (as in Fig.1a) has been developed. The model includes two bare Si dies, the UF and the micro-bumps (Fig.4a).

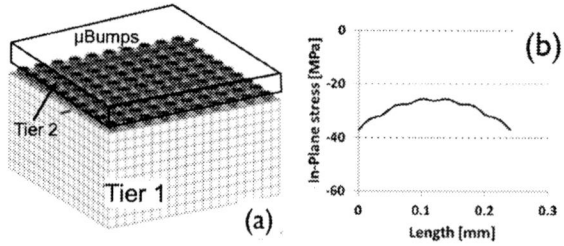

Figure 4. FE model of the fragment of the 3D stack with the regular array of the micro-bumps (a) and the simulated distribution of the in-plane stress along the solid blue line (b).

In this model the assumption has been made that the curing temperature of the underfill (in our case 270°C) is the stress-free temperature, i.e., at this temperature all the components of the 3D stack (the underfill, Si, the micro-bumps) are at zero stress. The model simulates the stress induced in the stack after it has been cooled down to the room temperature.

The distribution of the x-component of the in-plane stress along the solid blue line (Fig.4a) is shown in Fig.4b. As it can be seen, the average in-plane stress on the top surface of the top thin die is below 40MPa and no local variations of stress at the positions of the micro-bumps can be observed. The simulated vertical stress was negligible and hence is not reported in this paper.

The simulated results are in the good agreement with the experimental observations and confirm the hypothesis that the in-plane stress induced by bending of the dies in the 3D stacks can be suppressed by increasing the die-thickness to bump-pitch ratio.

978-1-4799-2408-0/14 $31.00 © 2014 IEEE

Stress distribution below the micro-bump on Tier 1.

The bottom Tier 1 die has a thickness of 200μm. It is very unlikely that the micro-bumps on top of this die (Fig.1a, location 2) can cause its local bending and as the result have any impact on the stress distribution in the thick Si. However, the analysis of the electrical measurements of the array of sensors below the micro-bumps on Tier 1 indicates that very high local stress is induced in the area below the micro-bump.

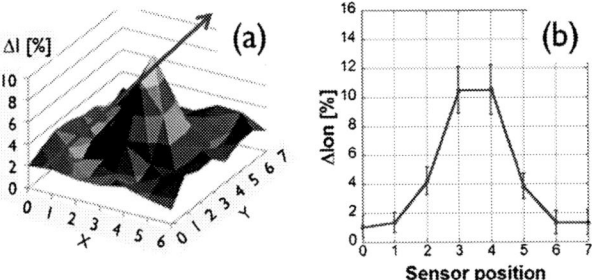

Figure 5. 2D distribution of the relative I_{ON} currents of the stress sensors within the array located on Tier 1 below the micro-bump (a) and the variation of the I_{ON} currents along the red line (b).

The sensors located below the micro-bump show 10% increased I_{ON} in comparison with the sensors located outside the micro-bump area as shown in Fig.5. It is worth noting that the eight standalone sensors located below the peripheral micro-bumps (as in Fig.2a) also show a similar trend. To understand this phenomenon the FE model introduced in the previous sub-section has been used.

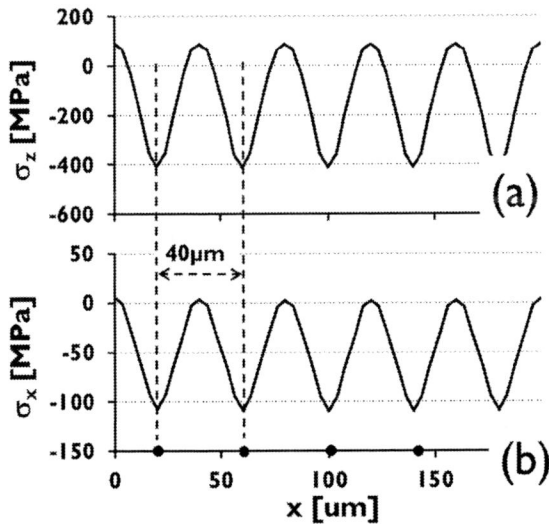

Figure 6. Vertical, σ_z, (a) and In-plane, σ_x, (b) - components of stress modelled for the top surface of the Tier 1 die along the line connecting the centers of the micro-bumps (dashed blue line in Fig.4a.). Black dots indicate the locations of the micro-bumps.

The in-plane and vertical components of stress (Fig.6) are simulated at the locations along the line connecting the centers

of the micro-bump landing pads on the bottom Tier 1 die (dashed blue line on Fig.4a). As it can be seen, there is a high vertical compressive stress (>400MPa) at the locations below the centers of the micro-bumps (Fig.6a). In the areas between the micro-bumps the vertical stress goes to zero and changes sign. It reaches 100MPa tensile stress at the symmetry points between two bumps. The in-plane stress follows a similar trend as the vertical stress. The x-component of the in-plane stress is shown in Fig.6b. It goes to 100MPa compressive stress at the positions below the micro-bumps and it reduces to zero in the areas between the micro-bumps. One may assume that the high vertical stress is due to the mismatch in the thermo-expansion coefficients between the UF and the micro-bump materials. During the thermo-compression bonding the 3D stack is heated up to 270°C. Then it cools down and the underfill cures and solidifies. As the result of this process the UF material shrinks more than the micro-bump. This, in-turn, induces the vertical stress acting from the micro-bump towards silicon. To support this hypothesis the local arrays of stress sensors located below the micro-bumps were measured at different temperatures varying in the range between +25°C and +75°C. If we assume that the stress is indeed built-in during cooling of the 3D stack then the magnitude of this stress should be temperature dependent. A linear decrease of the stress with temperature is demonstrated in Fig.7 where the relative change of I_{ON} for the sensors located below the micro-bumps (sensors with coordinates 3,4 and 4,4 in Fig.5a and also eight sensors below the peripheral micro-bumps from Fig.2a) is plotted as the function of temperature.

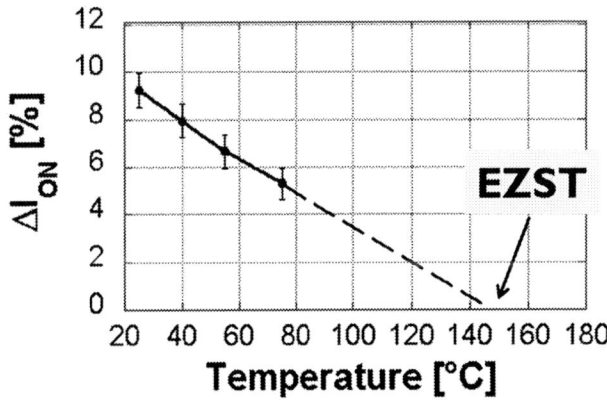

Figure 7. Median ΔI_{ON} of 10 sensors located below micro-bumps as function of temperature. The magnitude of the error bars indicate the calculated standard deviations of the currents at every temperature. The dashed line is the linear extrapolation of the ΔI_{ON} vs. T curve. EZST is the extrapolated equivalent zero stress temperature of the 3D stack.

The linear dependency can be extrapolated to the temperature where the stress becomes zero (dashed line in Fig.7). This temperature is called the equivalent zero stress temperature (EZST) of the 3D stack and in our case it is 140±7°C. This extrapolated temperature is very close to the glass transition temperature (Tg) of the underfill material (Tg=156°C) used in the 3D stacks.

The variations of the saturation currents of the n-FET sensors within the array can be translated into the mechanical stress. For this transformation, the sensitivity of the I_{ON} to the vertical stress is to be measured. These measurements were performed using a nanoindenter, where a spherical tip with the radius of 250µm (Fig.8a) was used to push on the copper pillar, 50µm in diameter, located above the array of sensors (as in Fig.2b). For this calibration experiment a single thick die and not a 3D stack was used. The die was mounted and wire-bonded inside the DIL-40 package to enable the electrical measurements of the sensors in-situ during application of the vertical stress. As it can be seen from Fig.8b the I_{ON} current of one of the sensors located below the copper pillar is increasing linearly with increasing vertical compressive stress. The stress reported on the horizontal scale in Fig.8b is calculated as the ratio of the force measured by the nanoindenter and the area of one copper pillar. The slope of the ΔI_{ON}-σ_{33} curve is 300ppm/MPa. Considering that the applied stress is compressive and it is commonly treated as negative stress (in contrast to the tensile stress usually considered as positive), the gauge factor of the n-FETs for the vertical stress is - 300ppm/MPa.

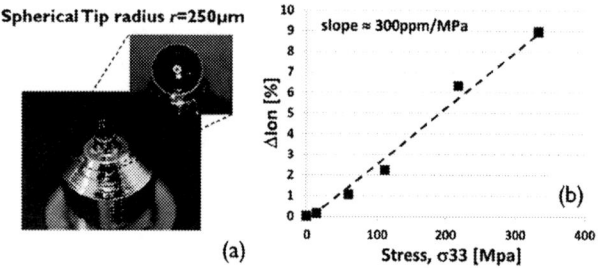

Figure 8. Microphotograph of the spherical tip of the nanoindenter (a) and the ΔI_{ON} as the function of the vertical stress (b).

The calculated gauge factor of the sensors in combination with the electrical measurements of the array of sensors from Fig. 5 can be used to calculate the vertical stress induced by the micro-bump on the top surface of the Tier 1 die. This transformation is only possible under the assumption that the vertical stress is dominating over the other components of stress in the considered case. This assumption is not entirely true considering the significant in-plane stress simulated by the FEM (Fig.6b), however, giving the fact that the physical nature of the compressive in-plane stress in Si under the bump is not fully understood, a rough estimation of vertical stress under the bump is still possible.

The measured vertical stress on the top surface of the Tier 1 die (along the red line in Fig.5a) is shown in Fig.9. The maximal stress is detected below the center of the micro-bump and it is 300MPa more compressive in comparison with the stress outside of the micro-bump area. The absolute sign and the magnitude of the stress is difficult to calculate as the particular dies used for the 3D stacks were not measured before stacking.

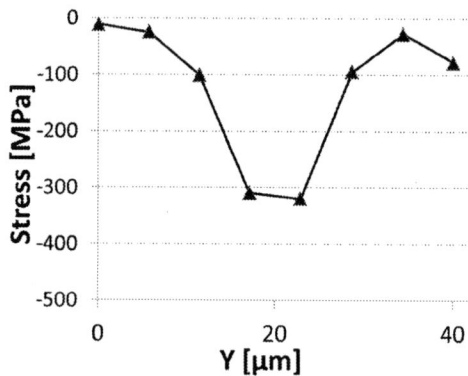

Figure 9. Vertical stress on the top surface of Tier 1 along the line below the center of the micro-bump (red line in Fig.5a). The gauge factor of the n-FETs for the vertical stress - 300ppm/MPa is used to calculate the stress.

The stress below the micro-bump on Tier 1 was also measured using Raman spectroscopy [14]. For this experiment, the 3D stack was embedded in epoxy and cross sectioned. The plane of the cross section coincides with the location of the micro-bump as is shown in Fig.10a. The channels of eight transistors separated by shallow trench isolation (STI) are also clearly seen in this image. The shift of the Raman frequency is proportional to the stress induced in Si and is plotted in Fig.10b as a function of the position along the red line. It has a wavy shape and the peaks and the dips of eight waves correspond to the positions of eight FET channels and the STI zones, respectively. This is due to stress from the STI.

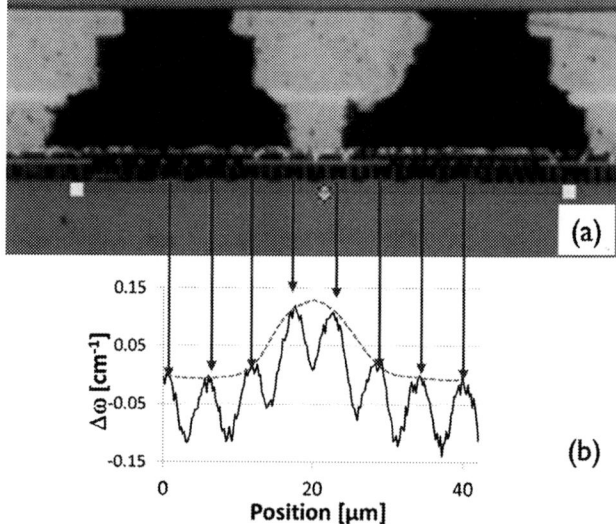

Figure 10. Cross section of the 3D stack (a) and the shift of the Raman frequency along the red line (b). Green spot indicates the position and the diameter of the laser beam used in the Raman spectrometer. The dotted line in (b) connects the Raman frequency shift measured at the positions of the stress sensors.

The measured Raman frequency shift has also one global peak (dotted line in Fig.10b). This global peak coincides with the center of the micro-bump and indicates that the stress below the bump is more compressive than the stress outside of the bump area.

The shape of the Raman frequency shift is compared with the measurements of the sensors and also with the results of the FE simulation in Fig.11. For a visual comparison the vertical scale of the Raman response was adjusted. It has to be pointed out that the FEM technique simulates the absolute stress and the different components of stress can be individually modelled. On the other hand, Raman spectroscopy and the electrical sensors always require the reference (stress free) measurements and calibrations to be able to calculate absolute stress levels. In absence of stress-free measurements, the vertical position of the experimental curve and the one of the Raman shift are arbitrary, as is the case here. Nevertheless, from Fig.11 one can see that the two experimental results (electrical and Raman spectroscopy) and also the results of the FE analysis indicate high compressive stress below the micro-bump. From the FE model one can also see that this stress goes to zero and becomes slightly tensile outside of the micro-bump area.

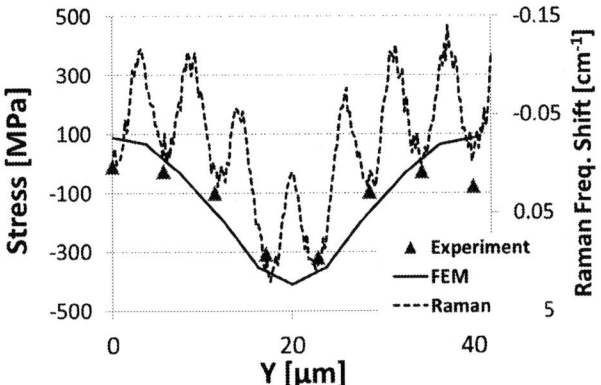

Figure 11. Mechanical stress on the top surface of Tier 1 along the line below the center of the micro-bump (red line in Fig.5a). Black triangles are the results of the electrical measurements of the stress sensors, solid line is the result of the FEM and the dashed line is the frequency shift of the Raman signal.

Stress distribution below the Cu pillar on Tier 2.

A third location at which the local distribution of stress is measured and analyzed is the area below the copper pillar on the top Tier 2 die (location 3 in Fig.1b,c). One should expect no significant variation of stress at that area as the copper pillar is free standing on the top surface of the 3D stack and there are no external vertical or lateral forces acting on the pillar. The variations of ΔI_{ON} current of the stress sensors located along the red line from Fig.2b under the Cu pillar without and with the supporting micro-bump between the dies are shown in Fig.12. As in all previous experiments, the measured absolute I_{ON} currents of the sensors within one array are normalized by the minimal current from the same array. Thus, only the relative variations of stress within one array can be analyzed by this method.

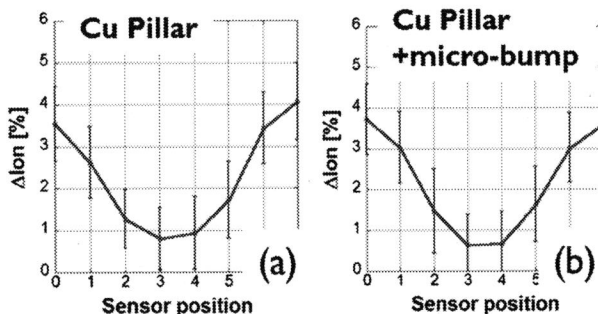

Figure 12. Variation of ΔI_{ON} current of the sensors located along the red line from Fig.2b (a) under the Cu pillar and (b) under the pillar in combination with the micro-bump between the two dies.

As one can see from Fig.12, in contrast to the expectation, the response of the sensors varies within 3-4% and indicates that the stress below the center of the Cu pillar is different from the stress at the periphery of the pillar. It can be pointed out that the measurements performed at the same locations on top of the Tier 2 die before stacking do not show such variations of the currents. The possible explanations for this phenomenon can be found by analyzing the details of the process of the 3D assembly. The critical step in this process is the thermo-compression bonding which is carried out at high temperature and high vertical force. This force is applied to the copper pillars which are located at the top surface of the Tier 2 die.

Firstly, under such stress the thin (50μm thick) top die can locally bend down and freeze in this position upon curing of the underfill. This would explain the measured variation of the I_{ON} currents below the pillar for the case shown in Fig.1b. However, the sensors located between the pillar and the micro-bump (as in Fig.1c) show a similar trend. In the latter case the bending of the top die is not likely to happen due to the supporting action of the central micro-bump.

Secondly, the copper pillars may deform plastically at high force and high temperature. After the bonding force is released and the assembly is cooled down to the normal temperature, the in-plane stress can locally be induced between copper and the Si die. To explore this hypothesis the single full-thickness die with the copper pillars was placed in the nanoindenter. The vertical force was applied to the pillar and (unlike the calibration measurements reported in the previous paragraph) the entire array of sensors was measured. First, the distribution of the I_{ON} currents was measured at zero stress and it is shown as the 3-D surface plot in Fig.13a. Then the stress was swept between 0 and 600MPa with the step of 150MPa. The currents measured at zero stress are used as the reference to calculate the relative changes of the currents at different stress levels. The insets in Fig.13 show the measured variations of the relative currents of the sensors located at 8 positions along the red line. It can be seen that the currents of the sensors below the copper pillar increase linearly with the increasing vertical stress. The change of the currents reaches 20% at the stress of 600MPa (Fig.13e). The local distribution of the currents within the array also changes. The stress at the periphery of the pillar is increasing faster than that at the center of the pillar.

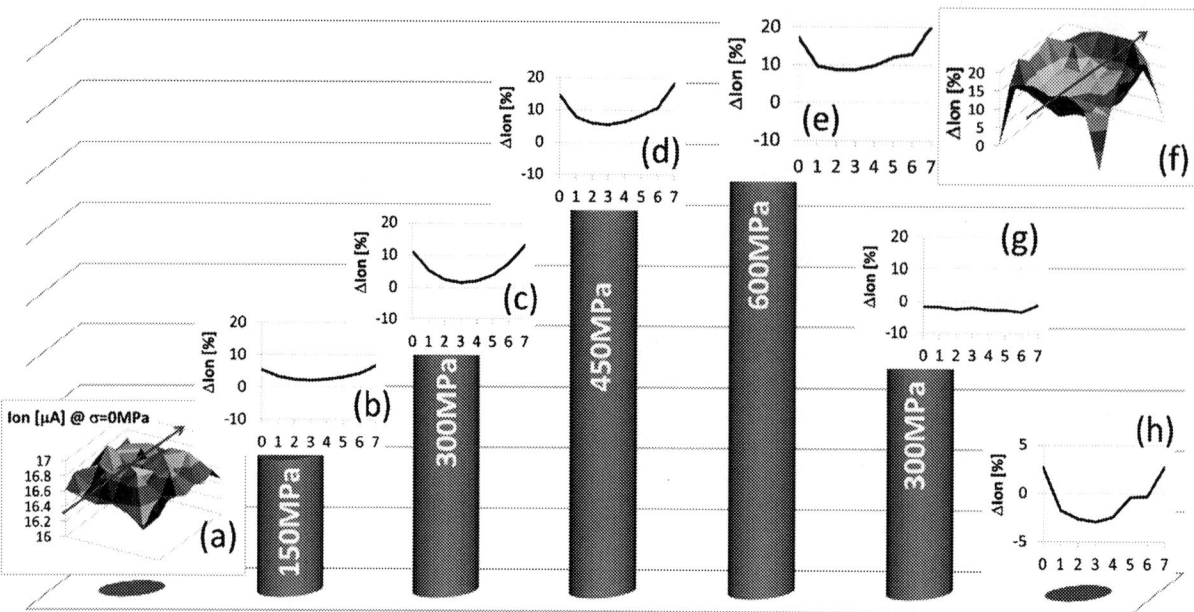

Figure 13. Variation of ΔI_{ON} current in the array of sensors below the copper pillar as the function of the vertical stress applied to the pillar. Red arrow on the 3-D surface plot indicates the positions of the sensors measured and reported for the different stress levels.

The difference of the currents within the array measured at the stress of 600MPa is 10% indicating large gradients of stress under the pillar. From Fig.2b it can also be seen that four sensors at the corners of the array are not covered by the pillar and these sensors are not affected by the vertical stress as is shown in Fig.13f. This indicates that the effect of the vertical stress applied to the pillar is very local and does not propagate outside the area of the pillar.

The ramp down of the stress was done in two steps. First, the sensors were measured at the stress of 300MPa, where the distribution of the currents became very uniform and in average 1% lower than their initial values measured in the beginning of the experiment (Fig.13g). In the final step, the sensors were measured at zero stress (Fig.13h). As it can be seen, at the absence of the vertical load, the average change of the current is reduced to zero, while the variation of the currents within the array (from -2% at the center of the pillar to 2% at the periphery) indicates that the stress below the pillar became non-uniform. This non-uniformity can be explained by the plastic deformation of copper under the high vertical stress.

The plastic deformation of copper was also observed directly during the nanoindentation test. The results of the nanoindantation measurements with a conical tip with flat end, in combination with the finite element modelling were used to determine the mechanical properties of the copper pillar such as Young's modulus, E, hardening, E_T, and the yield stress (Fig.14). The last one is found to be around 300MPa at room temperature. The indentation mark which is clearly seen on the optical image of the copper pillar taken after the nanoindentation test (inset in Fig.14) additionally confirms the hypothesis that the copper pillar deforms plastically at the high levels of vertical stress.

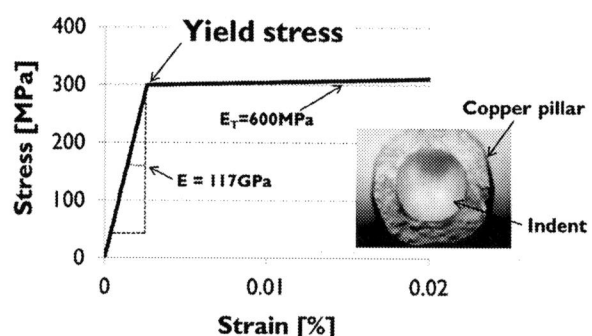

Figure 14. Graphical representation of the mechanical properties of the copper pillar and the image of the plastically deformed pillar after the indentation. E is the Young's modulus and E_T is the hardening.

The remaining profile of the stress (its shape and the amplitude as in Fig.13h) measured by the sensors after the nanoindentation test is similar to that measured in the 3D stack (Fig.12). This supports indirectly the hypothesis that the stress induced in the Tier 2 die under the copper pillar is the result of the vertical stress applied to the pillar during the thermo-compression bonding. If this stress is higher than the yield stress of copper at the bonding temperature of 270°C the latter will start deforming plastically. This state can be considered as stress free. After removing the external force the in-plane stress is built in the die and in the copper pillar.

Conclusions.

Custom 3D-stacking enabled test chips were used to investigate the effects of the assembly process on the local re-distribution of the mechanical stresses in the 3D-ICs. The planar transistors located on both levels of the 3D stack are used as the stress sensors. As a result of the performed experiments,

978-1-4799-2408-0/14 $31.00 © 2014 IEEE

three local effects of the 3D stacking have been identified: (1) no expected stress above the micro-bump is detected, (2) sensors below the micro-bumps are greatly affected by the 3D stacking, (3) significant local stress below the copper pillar is observed. All three effects have been supported by additional experiments such as the mechanical calibration of the sensors, nanoindentation, Raman spectroscopy, and by the finite elements modeling. It was shown that the sensors below the micro-bumps are affected due to the high vertical compressive stress build-in during the thermo-compression bonding while the fine pitch of the micro-bumps which is comparable with the thickness of the thin die provides good mechanical support to the latter thus limiting its vertical deformations (bending) and minimizing in-plane stresses. It is also shown that the copper pillars on top of the 3D stack deform plastically during the thermo-compression bonding. This plastic deformation induces local residual stresses and affects the performance of the FE devices.

Acknowledgments

The authors would like to thank core partners of the imec Industrial Affiliation Program (IIAP) on 3D Integration Technology.

References

1. Beyne, E. The rise of the 3rd dimension for system intergration. Interconnect Technol. Conf. 2006, 1–5 (2006).
2. Plas, G. Van Der & Limaye, P. Design issues and considerations for low-cost 3-D TSV IC technology. Solid-State Circuits, 46, 293–307 (2011).
3. Lin, J., Chiou, W. & Yang, K. High density 3D integration using CMOS foundry technologies for 28 nm node and beyond. (IEDM), 2010 IEEE 22–25 (2010).
4. Lin, L. et al. Reliability characterization of Chip-on-Wafer-on-Substrate (CoWoS) 3D IC integration technology. 2013 IEEE 63rd Electron. Components Technol. Conf. 366–371 (2013).
5. Cherman, V. O. et al. Impact of through silicon vias on front-end-of-line performance after thermal cycling and thermal storage. IEEE Int. Reliab. Phys. Symp. 2B.3.1–2B.3.5 (2012).
6. Yu, C., Chang, C. & Wang, H. TSV process optimization for reduced device impact on 28nm CMOS. … 2011 Symp. 45, 2010–2011 (2011).
7. Cho, S. et al. Impact of TSV proximity on 45nm CMOS devices in wafer level. 2011 IEEE Int. Interconnect Technol. Conf. 1–3 (2011).
8. Mercha, A. et al. Comprehensive analysis of the impact of single and arrays of through silicon vias induced stress on high-k / metal gate CMOS performance. 2010 Int. Electron Devices Meet. 2.2.1–2.2.4 (2010).
9. Lemke, B. Experimental determination of stress distributions under electroless nickel bumps and correlation to numerical models. Sensors Journal, 11, 2711–2717 (2011).
10. Ivankovic, A. et al. Analysis of microbump induced stress effects in 3D stacked IC technologies. 2011 IEEE Int. 3D Syst. Integr. Conf. (3DIC), 2011 IEEE Int. 1–5 (2012).
11. Ivankovic, A. et al. FET arrays as CPI sensors for 3D stacking and packaging characterization. 2012 IEEE Int. Reliab. Phys. Symp. 2E.3.1–2E.3.9 (2012).
12. Lemke, B. Experimental determination of stress distributions under electroless nickel bumps and correlation to numerical models. Sensors Journal, 11, 2711–2717 (2011).
13. Gonzalez, M. et al. Chip package interaction (CPI): Thermo mechanical challenges in 3D technologies. 2012 IEEE 14th Electron. Packag. Technol. Conf. 547–551 (2012).
14. De Wolf. I. Micro-Raman spectroscopy to study local mechanical stress in silicon integrated circuits. Semicond. Sci. Technol. 11, 139–154 (1996).

Six-Die Stacking: Three-Dimensional Interconnects Using Au and Pillar Bumps

Fei-Jain Wu, Lung-Hua Ho, Chih-Ming Kuo, Chia-Jung Tu, Chih-Hsien Ni, Shih-Chieh Chang, Chuan-Yu Wu, Kung-An Lin, Wei-Hsin Wu, and Yung Shen Wu

Chipbond Technology Corporation

No.3, Li Hsin 5 Rd., Science Park, Hsinchu 30032, Taiwan, ROC

Jasonh@chipbond.com.tw

Abstract

Fine-pitch three-dimensional (3D) interconnects have become a favored choice for next-generation packaging technology, helping to meet industry demands for small form factors, large bandwidth, and low power requirements in electronic devices. Critical changes are being made to meet fine-pitch and narrow-gap joint specifications, such as changing the solder to a micro pillar bump, using thermal compression instead of mass reflow to connect chips, and replacing capillary underfill (UF) material with wafer-level UF material. Consequently, a chip can be connected to another chip or a Si interposer, subsequently transforming the entire joint to an intermetallic compound (IMC).

This paper proposes an alternative fine-pitch 3D chip-stacking technology that involves using a bump material, bonding method, and UF material that are similar to those of the current chip-on-film inner-lead bonding technology typically used in the liquid crystal display driver IC industry. In applying the Au-rich AuSn eutectic characteristic, one side of the joint is using Au, and the other side is bumped using a Cu pillar.

A thermal compression bonding process is performed in an ambient atmosphere without the presence of flux. Subsequently, the UF material is applied to fill the gap. A simulated 30-μm pitch, 6-die stacked structure was examined in this study. Critical parameters affecting the joint structure and reliability were evaluated. A robust interconnect was fabricated using optimal bump structures, bonding parameters, and capillary UF material without a filler. The proposed joint structure exhibits an increase in resistance of less than 20% after performing a critical high-temperature storage test. The thermal stability of the joint is comparable with that of a normal C4 joint and improves upon the Cu_3Sn/Cu_6Sn_5 IMC or AuSn joints that are fabricated using a Au stud bump and solder. Furthermore, the thermal stability of the joint increases when the number of bonding increases, making it particularly effective for die-stacking operations.

1. Introduction

Recently, three-dimensional (3D) IC packaging technologies have made major progress in three architectures: the TSMC chip-on-wafer-on-substrate (CoWoS), the HMC Consortium hybrid memory cube (HMC), and the AMD/Hynix high-bandwidth memory (HBM) [1]-[6]. In these architectures, a micro joint is formed by bonding the Cu pillar bump with a solder cap on one chip to other structures (e.g., a Cu pillar bump, Cu pad, or NiAu pad) on another chip. Reliable joints are attained by converting these joints, at various degrees of free Sn content, into complete

Cu_3Sn/Cu_6Sn_5 IMC joints [7]. Traditional reflow is employed when the bond pitch and gap are large. Thermal compression bonding is favorable when the bond pitch decreases and flux cleaning becomes difficult. Subsequently, capillary UF material can be applied to fill the joint gap when the filler particle in the UF is sufficiently small. Alternatively, wafer-level UF (WLUF) can be applied before the bonding process. WLUF acts simultaneously as a flux and UF, thereby eliminating the need for flux cleaning.

The proposed approach was developed using Au and pillar bumps in fine-pitch 3D chip-stacking [8]. At temperatures above the melting point of solder, the rapid reaction between Au and solder expels the oxide from the joint surface. This eliminates the need for flux, making it an effective approach for fine-pitch interconnect fabrication. Because of the compliant nature of Au bump and the minimal coefficient of thermal expansion (CTE) mismatch between two chips, a low viscosity UF material without filler can be used. This technology is similar to chip-on-film (COF) bonding technology, where the Au bump is bonded to the Sn-coated Cu trace on a polyimide tape. Characteristics of the Au-rich AuSn joint include the higher melting alloy and thermal stability of the joint when excess Au further reacts with the AuSn. Thus, the sequential stacking of chips is possible, where a single joint is bonded multiple times. Subsequently, a UF or molding process can be performed after stacking and electrically testing all of the chips. Another benefit of the proposed approach is that bump material and bonding processes are readily available. The UF material and process remain identical to those used in current bond-then-fill technologies, without the possible side effects that can occur when WLUF is adopted. Bump pitch, diameter, and joint gap remain similar to those attained using current COF technologies, without the potential scaling effects that occur in transition from a C4 bump to a micro bump. All of these aspects frame the technology as a production-friendly process.

This paper presents an update on previous reliability and structural studies that have used this approach.

2. Experiments

Target structure, stacking method, test vehicle, and simulated structures

Fig. 1 depicts the targeted chip-to-chip (C2C) stacked structure employed in this study. The C2C interconnects are formed sequentially from the bottom to top until all of the chips are stacked. Subsequently, the UF is applied. The metal pads (not shown in the figure) at the bottom of the substrate or Si interposer are assumed to maintain their solderability to the next-level interconnect, such as C4 solder bumps after the stacking. To simulate AuSn joints subjected to multiple bonding processes in the stacked structure, identical bonding conditions were applied multiple times to the single-bonded joints. All of the joints in the targeted stacked structure were assumed to undergo thermal, mechanical, and metallurgical changes between one-time and five-time bonded joints. For example, the second joint from the bottom of the C2C stacked structure undergoes four times as many thermal, mechanical, and metallurgical changes than that of a one-time bonded joint at the top of the C2C structure.

Test vehicles were fabricated as described in [8]. A total of four daisy chain arrays, each comprising 6×134 bumps, were tested on each chip.

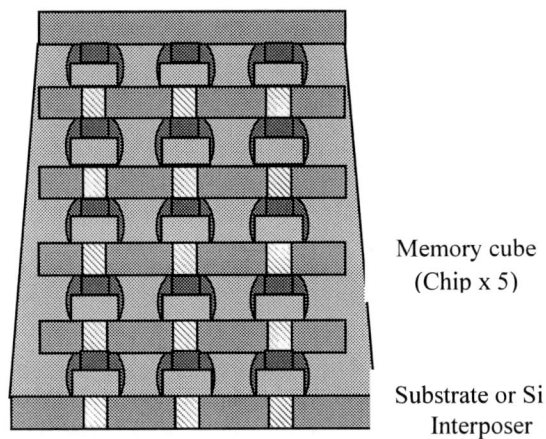

Memory cube
(Chip x 5)

Substrate or Si
Interposer

Fig. 1. Schematic drawing of the target structure.

Micro bump fabrication, bonding, and UF process

Au bump, e-plated, and e-less Sn or SnAg-plated Cu pillar bumps were fabricated, as described in [8], in which an e-plated Cu pillar bump was capped with e-plated solder, and a Cu pillar plated with e-less had a thin layer of e-less-plated solder covering both the top and sidewall of the Cu pillar. The bump sizes ranged from 12 to 20 μm for all three bump types.

Bonding was performed using a thermo-compression flip chip bonder (placement accuracy, ±3 μm). The tool head supplied the major heat to the Au-bump side, whereas the bonding stage provided the minor heat to the pillar-bumped side during the bonding.

Using a semi-automatic dispensing machine (position accuracy, ±10 μm), UF was applied to the C2C-bonded structure with a substrate at the bottom and a chip at the top.

Shear, confocal scanning acoustic microscopy, electromigration, and reliability tests

Shear tests were conducted using a Dage 4000 shear tester (shear height, 400 μm; shear speed, 100 μm/s; test load, 500 g). The test samples were C2C-bonded structures with a substrate at the bottom and a chip at the top. Confocal scanning acoustic microscopy (CSAM) was performed using a Sonascan Gen 6 (at 230 MHz) and a Sonix ECHO VS model UHF7 machine (at 200 MHz).

An electromigration (EM) test was conducted using a four-point probe on a C2C stacked structure consisting of a planar Kelvin structure. The samples were stressed under a constant electrical current in an oven at 110 °C. The current and current density were calculated based on the smallest cross-section area of the joint after the bonding. Resistance was monitored in-situ by an Agilent 34980A measurement unit. Chip temperature was monitored by a thermocouple. A pair of bumps was connected together in the planar Kelvin structure, and current was allowed to flow from the Au-to-pillar bump at one of the joints, and from the pillar-to-Au bump at the other joint. A 10% increase in resistance was designated as the failure criteria.

Temperature cycling (TC), high-temperature storage (HTS), and unbiased highly accelerated stress testing (μHAST) were conducted using a JEDEC Level 3 precondition (5 TC cycles from -40 °C to 60 °C, baked for 24 h at 125 °C, soaked at 30 °C/60%RH for 192 h, and reflowed 3 times at a peak temperature of 260 °C). C2C-bonded daisy chain dice, all with a substrate at the bottom, were measured to determine the resistances. A sample screening was performed, after bonding and the preconditioning test, before being subjected to the next step of the tests. A 20% increase in resistance was designated as the rejection criteria.

3. Results and Discussion

Bonding parameters, bump parameters, and joints

Previous report has revealed that two types of joint structure could be obtained by combining appropriate bumps (e.g., bump size, height of the Au and pillar bumps, and solder cap thickness) and bonding (e.g., bonding temperature, bonding time, tool temperature, and stage temperature) parameters [8]. The joint structure depicted in Fig 2b has a bottom gap (i.e., the distance from bottom of the bump to the passivation surface) that equals to zero. It is beneficial to reduce the bonding pressure in the second to the fifth bondings during the stacking process, because the bonding force is distributed over a larger area. The drawbacks of this type of joint structure, however, are the formation of sidewall voids, larger bump deformation, and decreased space between joints. While both e-plated and e-less pillars can be fabricated to either type of joint structure, joints thus formed develop their unique characteristics at the degree of bump deformation, joint shape, void formation at sidewall of pillar, and consistency of the UF process. For example, the joint structure displayed in Fig. 2b tends to have more UF voids than does the joint structure displayed in Fig. 2a. In comparisons, the joint structure in Fig. 2a offers more flexible selections of bump and bonding parameters.

Combining a high solder cap thickness and low Cu pillar height in the e-plated pillar was necessary to obtain the type of joint shown in Fig. 2b. A combination of thinner solder and a higher Cu pillar led to the type of joint depicted in Fig. 2a. The formation of different joint structures correlates with the extent of the AuSn reaction, the volume of space between the bottom of Au and the passivation surface that is to be filled by

AuSn alloy from the AuSn reaction, and the amount of Au and Cu pillar bump deformations during the bonding process.

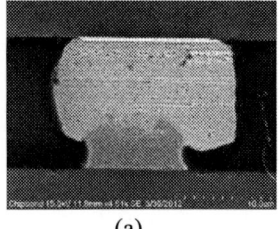
(a)

(b)

Fig. 2. Types of C2C joint.

Reliability testing results and thermal stability of the joint

Table 1 is a summary of the reliability test conditions and results for the joint structures with a bottom gap equaling zero. The cost and process were considered based on a 6-μm (height) Au bump. The performance of two UF materials was compared; one without filler (UF A), and the other with filler (UF B). The results indicate a consistent and successful μHAST and TC test, regardless of the number of bondings, pillar types, and UF materials (Table 1a and 1b). Inconsistencies were observed, however, following the HTS tests (Table 1c). Apparently, only the one-time and five-time bonded samples from the combination of the e-plated pillar and UF A can pass the HTS tests. Both are required to ensure that the proposed stacked structure works.

The continuous increase in resistance during the HTS was attributed to excess amounts of Sn or SnAg alloy, a phenomenon reported by IBM when a Au stud bump and solder were used to form a AuSn joint [9]. This result is also consistent with the observations of a previous study on the effect of the thickness of the SnAg cap on joint deterioration after HTS [8].

# of bonding	Pillar	UF	Reject/Total
1-time	E-plated	A	0/12
1-time	E-plated	B	0/12
5-time	E-plated	A	0/11
5-time	E-plated	B	0/12
1-time	E-less	A	0/15
1-time	E-less	B	0/15
5-time	E-less	A	0/14
5-time	E-less	B	0/13

(a). 1000-cycle TC test results.

# of bonding	Pillar	UF	Reject/Total
1-time	E-plated	A	0/12
1-time	E-plated	B	0/12
5-time	E-plated	A	0/12
5-time	E-plated	B	0/10
1-time	E-less	A	0/12
1-time	E-less	B	0/13
5-time	E-less	A	0/12
5-time	E-less	B	0/12

(b). 96-h uHAST test results.

# of bonding	Pillar	UF	Reject/Total
1-tme	E-plated	A	0/12
1-time	E-plated	B	3/12
5-time	E-plated	A	0/15
5-time	E-plated	B	3/15
1-time	E-less	A	6/15
1-time	E-less	B	6/15
5-time	E-less	A	0/15
5-time	E-less	B	0/14

(c). 1000-h HTS test results.

Table 1. Reliability test result summary for C2C structure where Au bump is in immediate contact with the passivation surface after the bonding; UF A: No filler, UF B: With filler.

The relationship between solder thickness and the increase in resistance after 1000 h of HTS was evaluated for the one-time and five-time bonded C2C structures. Fig. 3 demonstrates the effectiveness of reducing the solder thickness on reducing the increase in resistance after HTS. The figure also indicates the unique characteristics of a smaller increase in resistance for the five-time bonded samples compared with that of the one-time bonded samples. The difference between increases in resistance for these two types of sample is larger when the solder is thicker, and is smaller when the solder thickness is below 0.5 μm. Such a small difference in increased resistance is highly desirable for chip stacking operations, and is in accordance with the assumption that the joint becomes more thermally stable because of further AuSn reaction accompanying the excess Au that is still present in the joint. This difference also indicates that the joint does not have to be protected by UF before the next chip is stacked on top of it, adding substantial flexibility to the chip assembly process.

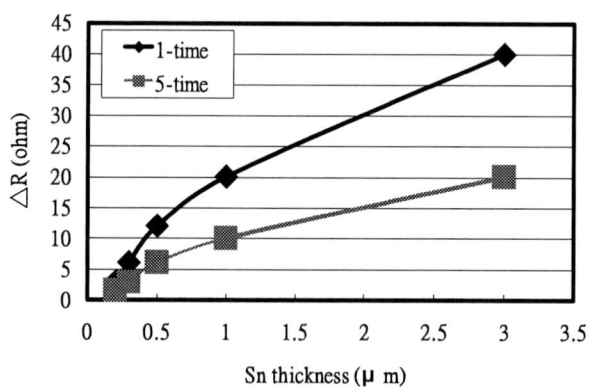

Fig. 3. Resistance increase after 1000-h HTS vs. Sn thickness; e-plated pillar (solder ≥ 1 μm) or e-less pillar (solder <1 μm).

A significant increase in resistance was observed after preconditioning and at the early stages of the HTS, before returning to zero for the one-time bonded samples (Fig. 4). Resistance remained unchanged for the five-time bonded samples throughout the precondition and the duration of the HTS test. This phenomenon further demonstrated that the five-time bonded sample has higher thermal stability than the one-time bonded sample. It also implies that the two-time bonded samples are more thermally stable than the one-time bonded samples. Therefore, two-time and six-time bonded samples were used to replace the one-time and five-time bonded samples, respectively, in the reliability tests, for the purpose of achieving a more robust thermal stability window in the HTS test. Based on using optimal material and bonding parameters, the reliability test results (Table 2) indicate that the simulated C2C structures meet the standards of the basic stress tests. UF without filler is sufficient for passing the reliability tests, which is generally not the case in a Cu/Sn IMC system. Eleven-time bonded samples, which simulate a 10-die stacked structure, also pass the reliability test (Table 2), further verifying the high thermal stability of the AuSn system.

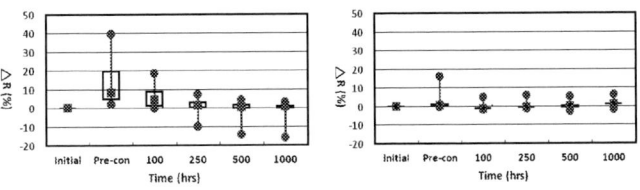

(a). One-time bonded   (b). Five-time bonded

Fig. 4. Increase in change of resistance, comparing one-time to five-time bonded samples using e-less Sn pillar bump.

# of bonding	Test	UF	Reject/Total
2-time	HTS	A	0/14
2-tme	TC	A	0/12
2-time	uHAST	A	0/12
6-time	HTS	A	0/15
6-time	TC	A	0/12
6-time	uHAST	A	0/12
11-time	HTS	A	0/12
11-time	TC	A	0/13
11-time	uHAST	A	0/12

Table 2. Reliability test results for two-time, six-time, and 11-time bonded C2C structures.

The long-term HTS test was extended to 3500 h, and the results were compared with those of IBM's long-term stability test for the AuSn system by using a Au stud and solder. The IBM test underwent increases in resistance from the beginning of the test, a 40% increase after 1000 h of testing, and a 60% increase in resistance after 2000 h of testing. The current Au-rich AuSn system has a near-zero increase in resistance after 1000 h, less than 5% after 2000 h, and less than 10% after 3500 h of HTS for both two-time and six-time bonded samples (Fig. 5).

The thermal stability was reflected in the shear mode and shear force test results. Fracture surfaces greater than 60% were observed primarily at the IMC near the pillar-Au interface, as indicated by the mixed Cu, Sn, and Au signals in the energy dispersive X-ray spectrometry (EDX). Fig. 6 demonstrates that fracture interfaces also occurred inside the Au bump, pillar bump, passivation, and at the Al pad-

passivation interface. These fracture modes were considered normal, and were attributed primarily to the brittle nature of IMC and the weak adhesion between the Al and the passivation layer. A considerable increase in shear force was observed when comparing two-time to six-time bonded samples, presumably resulting from the continuous increase in AuSn or Cu/Au/Sn IMC in the joint during the multiple bonding process (Fig. 7). A slight decrease in shear force was observed after the 3x reflow, which may have resulted from the release of free Sn from the Cu/Au/Sn IMC formed during the reflow process.

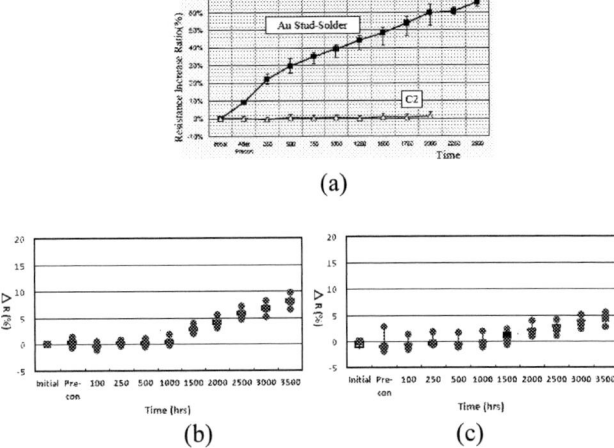

(a)

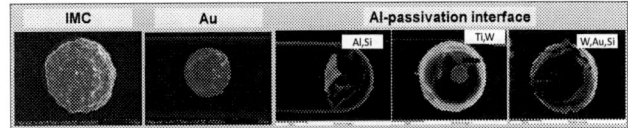

(b)   (c)

Fig. 5. Long-term HTS test results: a. IBM Au stud + solder; b. two-time bonded samples using e-less SnAg pillar; c. six-time bonded samples using e-less SnAg pillar.

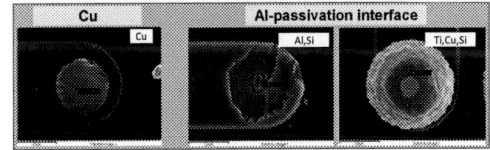

(a). Substrate side with Au-bump.

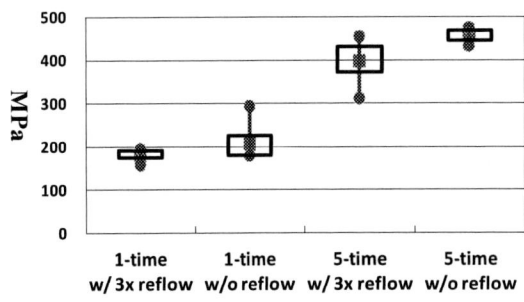

(b). Chip side with pillar.

Fig. 6. Shear modes of the C2C joint.

Fig. 7. Effect of multiple bonding and 3x reflow on shear force; Au bump being larger than the pillar bump.

The thermal stability of the joint was examined under various bonding conditions to accommodate wafers that are tolerant to different thermal history. For example, bonding at 400 °C caused a passivation crack and lifted bump (Fig. 8) for one wafer source but not for another. Thus, low bonding temperatures are necessary. The bonding and reliability test results demonstrate that system characteristics can be maintained when the bonding temperature is reduced to 360 °C. No passivation crack or lifted bump was observed afterward.

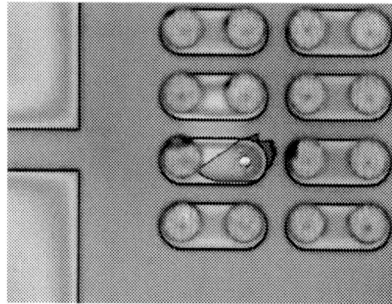

Fig. 8. Passivation crack observed during bonding at 400 °C.

In contrast with the HTS test results, the µHAST and long-term TC (up to 3500 cycles) test results (Fig. 9) demonstrate that stable resistance is attainable, even when UF A was used. These phenomena were observed for all of the tested samples for all bump and bonding parameter combinations. Thus, these two tests were not the reliability constraints of the system.

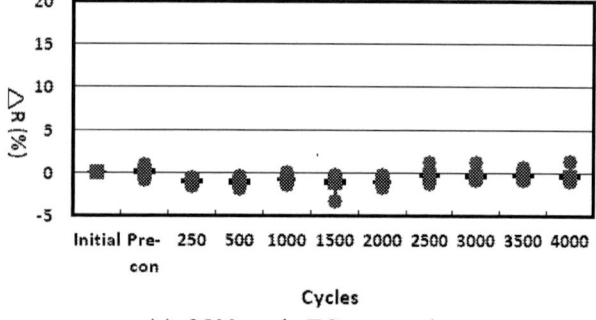

(a). 3500-cycle TC test results.

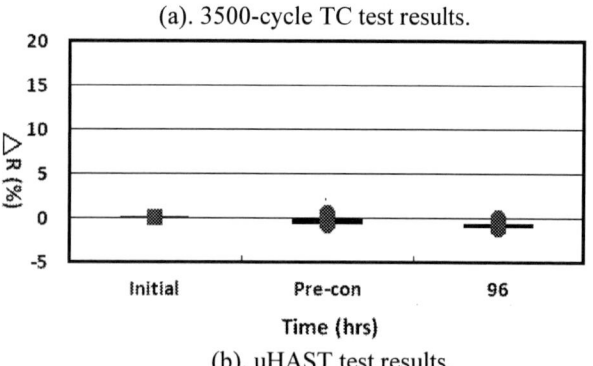

(b). uHAST test results.

Fig. 9. Long-term TC and uHAST test results.

Bonding window evaluation

For reducing costs, the samples with a Au bump that was smaller than the pillar bump were tested for RA, and the results were positive. Fig. 10a shows a cross-sectional-view of the joint fabricated using Au bump that is larger than the pillar bump, and Fig. 10b shows a cross-sectional-view of the joint fabricated using Au bump that is smaller than the pillar bump. The following implications can be made based on comparison.

First, using a Au bump that is smaller than the pillar bump facilitates the extension of the technology into a finer pitch because the Au is less deformed outside the diameter of the pillar. Second, the UF gap can be increased in a controlled manner and the UF process is simpler because it can be increased by simply increasing the pillar height without increasing the height of the Au bump or affecting the joint integrity. However, this is not the case when the Au bump is larger than the pillar bump.

For example, increasing the gap height by increasing the height of the Au bump would increase the cost of materials. Moreover, increasing the gap height by increasing the pillar height would increase the risk of the bump sliding while bonding (Fig. 11). This bump sliding behavior was not observed when the Au bump was smaller than the pillar bump, even when the pillar height was increased. In another cost reduction measure, both the Au bump size and bump height were reduced simultaneously. However, the reliability test results demonstrate that the bonded structures were marginal, which was attributed to the Au bump being too short to accommodate the overall coplanarity requirements during the bonding process, and that a smaller die than the current test vehicle might achieve a greater success rate when using a lower and smaller Au bump.

 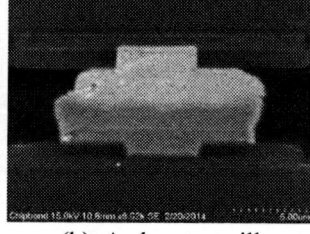

(a). Au bump > Pillar   (b). Au bump < pillar

Fig. 10. Joint shape after bonding.

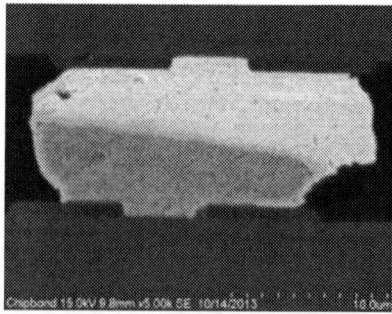

Fig. 11. Bond sliding.

978-1-4799-2408-0/14 $31.00 © 2014 IEEE

Based on the open/short of the daisy chains, the bonding yield was high even when all of the bumping processes were identical to the current production line. For example, no chemical mechanical polishing (CMP) was applied after the bump was fabricated. The current COF bonding and UF processes also remained unchanged. Fig. 12a shows the minimal and maximal values extending from of the boxes after the precondition test and during HTS test, for the samples that were fabricated using the dice from the wafer edge. This phenomenon was not observed for samples that were fabricated using the same wafer, but not from its wafer edge (Fig. 12b).

Furthermore, the box shapes, consisting of the medians and quartiles of the box plots for both samples, remained in the same range of values throughout the test. This indicates that no fundamental change occurs in the joint behavior, regardless of whether the dice were from the wafer edge. The samples in Fig. 12a were therefore attributed to the use of a wafer edge die with poor control of the bump height, which is typical of the bumping processes. Additional bumping process adjustments are necessary to improve the bonding yield. Before the improvement is made, samples can be screened based on resistance data after joint fabrication, or after a preconditioning test.

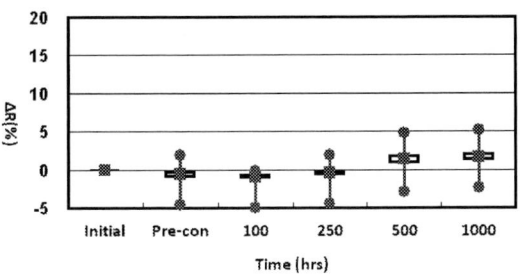

(a). Samples fabricated using die from wafer edge.

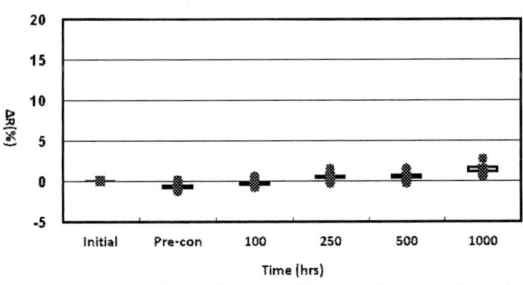

(b). Samples fabricated using die not from wafer edge.

Fig. 12. HTS test results comparison.

Stacking consideration

A critical factor of sequential stacking using the AuSn alloy (Fig. 1) is that a chip with both Au and pillar bumps on both sides of the die can be retrieved from the Au-bumped side (with the pillar-bumped side facing downward), and subsequently bonded to the stack (with the Au-bumped side facing upward). Repeating this bonding process multiple times allows the stack to be built layer by layer. Although the test results demonstrate that picking up die from the Au-bumped side is not a concern, bonding the chip onto the stack exerts pressure onto the un-bonded Au bump, which is similar to a coining process. To verify this assertion, the Au bump was compressed once and five times under typical bonding conditions. For a top-to-bottom stacking processes (Fig. 1), one-time coining occurs before the Au bump is bonded with the pillar. The scanning electron microscopy (SEM) images in Fig. 13 demonstrate that the coining causes a 3.5% reduction in bump height and a 4.6% enlargement in bump diameter. Slightly greater height reduction and diameter enlargement were observed after performing the coining process five times. However, these numbers are small in comparison with the possible measurement errors caused by the roughness of the Au bump surface before coining. These effects could also be compensated for by controlling specific parameters, such as Au bump hardness and the ratio of Au bump diameter to pillar bump diameter, and by using a smaller but taller Au bump. Finally, using a Au bump that is larger than the pillar bump could minimize the coining effect because the bonding pressure is based on the diameter of the pillar bump rather than that of the Au bump.

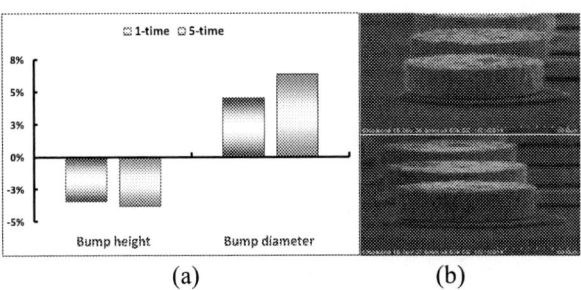

(a)   (b)

Fig. 13. a. Multiple coining effect on the height and diameter of the Au bump; the Au bump is larger than the pillar. b. Au bump before (top) and after (bottom) coining.

The consistency of the UF gap is a critical parameter for controlling the success of the UF process, because it determines the amount of resin that is used to fill the gap. Thus, constant layer-to-layer UF gaps in the stacked structure are desirable. SEM was employed after each bonding to measure and compare the UF gaps at the corners of the C2C structure. The results in Fig. 14 indicate that the largest gap reduction (approximately 1.6 μm) occurred during the second bonding. The mean gap reduction after the second bonding was 0.3 μm per bonding. Thus, the total volume of the UF material to be used in the stack can be accurately estimated based on each gap, even when the UF gaps at each layer are inconsistent.

The space between joints was measured to indicate the degree of bump deformation and the feasibility of extending the technology into a finer pitch. Fig. 14 demonstrates that the mean space becomes 0.25 μm smaller after each bonding. This space reduction is anticipated to be less when the Au bump is smaller than the pillar bump. Therefore, extending technology into a finer pitch is highly feasible.

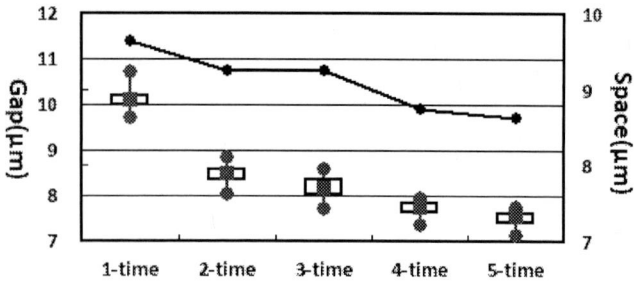

Fig. 14. Multiple bonding effects on UF gap and space between joints.

The coplanarity of the C2C structure is another consideration in multiple bonding processes. Coplanarity is the combined effect of individual coplanarity mainly based on the chip thickness, Au bump height, pillar bump height, tool head, and bonding stage. Strict within-die coplanarity is necessary in a Cu/Sn IMC joint system because of the rigid nature of the bumps. In the AuSn system, this requirement was less strict because of the soft nature of the Au. Fig. 15 displays the coplanarity results of the C2C-bonded structure after multiple bonding processes. The coplanarity remained relatively constant (mean: 0.35 μm; maximum: 0.8 μm) throughout the stacking process, and are comparable to those typically observed in COF processes.

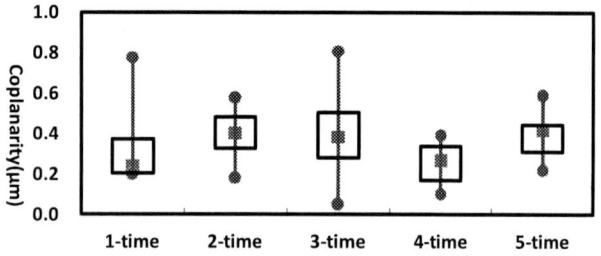

Fig. 15. Effect of multiple bonding on C2C coplanarity.

EM test

The simulated C2C structure was subjected to EM tests at 110 °C at with current densities of 5.2×10^3 and 6.5×10^4 A/cm²; the latter being higher than that of $1\sim3 \times 10^4$ A/cm² requested by the customer [7]. Fig. 16 demonstrates that the resistance decreases very slightly after 1000 h, indicating the electromigrational robustness of the joint and potential metallurgical changes. The resistance stability of the joint throughout the EM test differed from that of the Cu/Sn IMC joint system, in which resistance initially increased abruptly and continued to increase to a level higher than the typical 20% pass/fail criteria used for a C4 joint [7].

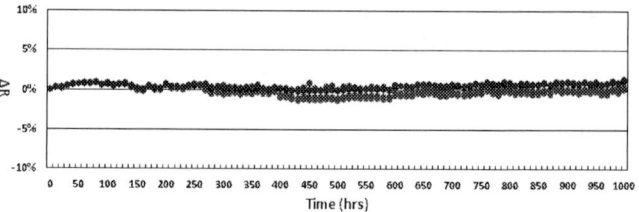

(a). Two-time bonded samples, 6.5×10^4 A/cm²

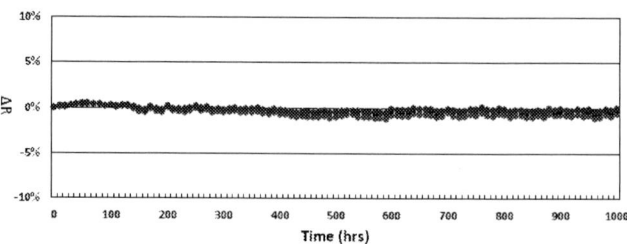

(b). Six-time bonded samples, 6.5×10^4 A/cm²

Fig. 16. EM test results; both with Au bump larger than the pillar bump.

Application

The high thermal stability exhibited by the AuSn makes it a suitable candidate for C2C chip-stacking technologies, such as those used in HMC architecture in which a stacked memory cube is bonded to a larger logic die at the bottom (final C2C structure in Fig. 18). By replacing the interposer in Fig. 1 with a logic die, HMC could be fabricated by using the bottom-to-top stacking sequences. For example, a memory chip is bonded to the logic die first, and all the other memory chips are stacked sequentially.

Alternatively, the memory cube can be stacked from bottom-to-top by using a bottom chip with a Au bump on both sides. The memory chips in the middle are fabricated with Au and pillar bumps on both sides of the chip. Finally, the top chip for the stack requires a pillar bump only. The complete memory cube can then be flipped and bonded to a logic chip with a pillar bump on one side and a solderable pad at the other side. Subsequently, the stacked hybrid memory cube can be molded or underfilled. Fig. 18 depicts the bonding sequence. The overall thermal changes that the chip undergoes are typically low, which is a critical factor when comparing the bonding sequence with the chip-to-wafer (C2W) process. Similar C2C bonding sequences can be employed to develop a AuSn system to fabricate AMD/Hynix's HBM.

To obtain superior throughput, C2W fabrication processes, such as those employed in CoWoS architecture or those that are combined with C2C process to fabricate HBM or HMC, are preferred over C2C bonding processes. Because wafers undergo extensive thermal changes at temperatures greater than 300 °C, C2W bonding is not recommended.

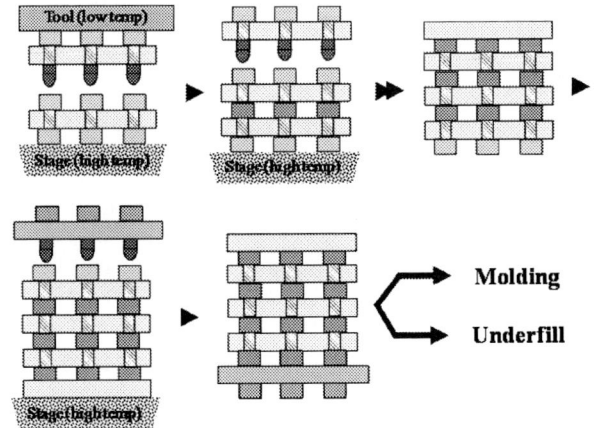

Fig. 18. HMC fabricated using the AuSn system.

Conclusion

In this study, 30-μm pitch micro pillars and Au bumps were examined for fabricating interconnects for 3D-IC stacking applications. Thermal compression bonding and the stable attributes of AuSn alloy facilitated the evaluation of a fluxless and UF process, without filler, based on the proposed joint structure. These distinct characteristics facilitate the fabrication of the proposed stacking structure, and demonstrate the potential for extending this technology into finer-pitch and more complex structures.

References

1. Kirk Saban, "Xilinx Stacked Silicon Interconnect Technology Delivers Breakthrough FPGA Capacity, Bandwidth, and Power Efficiency", White Papers: Virtex-7 FPGAs, WP380(v.1.1), October 21, 2011, http://www.Xilinx.com.

2. Larry Lin, Tung-Chin Yeh, Jyun-Lin Wu, Gary Lu, Tsung-Fu Tsai, Lary Chen, An-Tai Xu, "Reliability Characterization of Chip-on-Wafer-on-Substrate (CoWoS) 3D IC Integration Technology," Electronic Component and Technology Conf., 2013, pp. 366~371.

3. Gopal Raghavan, "Five Emerging DRAM Interfaces You Should Know for Your Next Design,", Cadence Design Systems; http:// www.cadence.com.

4. "Hybrid Memory Cube Specification 1.0," Hybrid Memory Cube Consortium, http://hybridmemorycube.org/files/SiteDownloads/HMC_S pecification%201_0.pdf.

5. "High Bandwidth Memory,", JESD235, October 2013, http://www.jedec.org/standards-documents/docs/jesd235

6. "AMD working with Hynix for Development of High-Bandwidth 3D Stacked Memory", http://wccftech.com/amd-working-hynix-development-highbandwidth-3d-stacked-memory/.

7. Hsiao-Yun Chen, Da-Yuan Shih, Cheng-Chang Wei, Chih-Hang Tung, Yi-Li Hsiao, Douglas Cheng-Hua Yu, Yu-Chun Liang and Chih Chen, "Generic rules to Achieve Bump Electromigration Immortality for 3D IC Integration," Electronic Component and Technology Conf., 2013, pp. 49~57.

8. Fei-Jain Wu, Lung-Hua Ho, Chih-Ming Kuo, Chia-Jung Tu, Chin-Tang Hsieh, Chih-Hsien Ni, Shih-Chieh Chang, Chuan-Yu Wu, Hui-Yu Huang, Kung-An Lin, and You-Ming Hsu, "Three Dimensional Interconnect using Au and Pillar Bumps," Electronic Component and Technology Conf., 2013, pp. 1933~1939.

9. 25. Yasumitsu Orii, Kazushige Toriyama, Hirokazu Noma, Yukifumi Oyama, Hidetoshi Nishiwaki, Mitsuya Ishida, Toshihiko Nishio, Nancy C. LaBianca, Claudius Feger, "Ultrafine-pitch C2 Flip Chip Interconnections with Solder-capped Cu Pillar Bumps," Electronic Components and Technology Conf., 2009, pp. 948~953.

TSV-less 3D Stacking of MEMS and CMOS via Low Temperature Al-Au Direct Bonding with Simultaneous Formation of Hermetic Seal

S. L. Chua[1,^], A. Razzaq[1,^], K. H. Wee[2], K. H. Li[1], H. Yu[1], and C. S. Tan[1, *]

[1] School of Electrical and Electronic Engineering, Nanyang Technological University, 50 Nanyang Avenue, Singapore
[2] DSO National Laboratories, 20 Science Park Drive, Singapore
*Phone: +65-6790-5636; Email: tancs@ntu.edu.sg; ^ Equal contribution

Abstract

3D integration has been widely recognized as the next generation of technology for integrated microsystems with small form factor, high bandwidth, low power consumption, and possibility of heterogeneous More-than-Moore integration. Heterogeneous integration of MEMS and CMOS is critical in future development of multi-sensor data fusion in a low-cost chip size system. MEMS/CMOS integration was primarily done using monolithic and hybrid/package approaches until recently. In this work, 3D CMOS-on-MEMS stacking without TSV using direct (i.e. solder-less) metal bonding is demonstrated. This MEMS/CMOS integration leads to a simultaneous formation of electrical, mechanical, and hermetic bonds, eliminates chip-to-chip wire-bonding, and hence presents competitive advantages over hybrid or monolithic solutions. We present the fabrication flow and verify the performance of the stacked MEMS/CMOS microsystem in this paper.

The micro-electro-mechanical system (MEMS) accelerometer is fabricated using bulk micromachining technology with silicon-on-insulator (SOI) substrate. The CMOS readout circuit for the capacitive MEMS accelerometer, implemented in $0.35\mu m$ (2P4M) process, consists of three essential circuit blocks: (i) low noise, band-pass gain stage, (ii) synchronous demodulator and (iii) off-chip, low-pass filter. The MEMS chip is metallized with a single layer of Au for electrical contact and sealing ring prior to bulk etching and release. Au is chosen to withstand harsh process conditions. The CMOS chip contains four Al metal layers. The top most layer is patterned for electrical contact and sealing ring that match those of the MEMS chip. In order to ensure proper operation, the delicate micro-structures (MEMS) should be protected from the ambient. In this approach, a hermetic seal ring is formed simultaneously during face-to-face stacking of CMOS on MEMS and hence eliminating the need for post-processing hermetic encapsulation. Face-to-face stacking also eliminates the need for chip-to-chip wire bonding. Effectively, the CMOS layer acts as an "active cap". In addition, I/Os to the MEMS chip are routed through the CMOS metal layers to simplify the MEMS process. Since the I/O count is low, TSV is not used and electrical feed-through is achieved by peripheral pads. As no solder is applied, the top passivation layer of the CMOS chip is partially recessed to expose the CMOS metal layer for ease of direct bonding with the MEMS metal layer. The metal surfaces are carefully treated and bonded. The bonded samples are packaged inside 44-pin J-leaded ceramic package for testing.

The functionality of the readout circuit is verified first, using off-chip MEMS, followed by verification of the bonded CMOS/MEMS chip. In both cases, the MEMS is excited by anti-phase sinusoid carriers within the frequency range 50kHz-1MHz. The variation in the peak-to-peak amplitude of the gain stage output is observed as the MEMS is flipped between -1g/+1g orientations. In addition, the phase of the output signal agrees with the flip direction.

Testing for hermeticity of the Al-Au thermo-compression bond is done in accordance with the MIL-STD-883E standard. Passive sealed cavities with volume of about 1.4×10^{-3} cm^3 are fabricated and bonded at 300 °C under a bonding pressure of around 8.4MPa for 10 mins. The cavity chips are created by deep reactive ion etching (DRIE) on silicon followed by the deposition of 50 nm of Cr adhesion layer and 150 nm of Au with electron beam evaporation method. The blanket capping chips are deposited with 100 nm of SiO$_2$ by plasma-enhanced chemical vapor deposition (PECVD) and 150 nm of Al by sputtering. Bubble test is used for gross leak and helium test is used for fine leak inspection. For shear strength measurement, test samples are prepared using the same processes for the samples in hermetic test, without the creation of cavities. The bonding parameters are kept the same. Results from the hermeticty and shear tests are presented and discussed.

Introduction and Motivations

Advancements in packaging co-design, low-cost materials and reliable interconnect technologies are critical in enabling the innovative packaging solutions required to help drive the electronic industry forward. Packaging semiconductor devices is becoming a challenge for the industry. As process technologies become smaller as a result of rigorous Moore's Law scaling, the performance and power limitations in interconnects and packaging became prominent. The traditional interconnect and packaging solutions starts to slow down signal transmission and consumers substantial amount of power. 3D packaging has been identified as a technology platform to ensure continuous performance enhancement ("More Moore") and functional diversification ("More than Moore") in future integrated systems.

3D integration for heterogeneous More-than-Moore integration of MEMS and CMOS provides competitive advantages over hybrid or monolithic solutions. In this work, 3D CMOS-on-MEMS stacking without TSV using direct (i.e. solder-less) metal bonding is demonstrated. This MEMS/CMOS integration leads to a simultaneous formation of electrical, mechanical, and hermetic bonds, eliminates chip-to-chip wire-bonding. The CMOS die also serves as a capping medium to isolate the MEMS from the ambient removing the need of hermetic packaging, thus reducing the thickness of the overall package. Both the MEMS and CMOS dies are fabricated with standard industry processes. SOI MEMS will be used in this work and will be tested after the integration with the CMOS die to ensure that the thermo-compression

978-1-4799-2408-0/14 $31.00 © 2014 IEEE

bonding does not damage the device. We present the fabrication flow and verify the performance of the stacked MEMS/CMOS microsystem in this paper.

Low-temperature (300 degrees Celsius) Cu-Cu thermo-compression bonding has been demonstrated in previous study to obtain hermetic packaging of 3D microsystems in accordance with the MIL-STD-883E standard[1]. 3D integration of MEMS and CMOS via low-temperature (300 degrees Celsius) Cu-Cu thermo-compression bonding has also been proposed[2]. However it is difficult to produce SOI MEMS dies with Cu pads as the insulator release process will also strip away common Cu barrier materials.

In this paper, Al-Au thermo-compression bonding will be used to simultaneously form of hermetic seal and connections as shown in Figure 1. Au is chosen as it is a common pad material for MEMS dies and Al-Au bonding is commonly found in wire-bonding. In this paper, Al-Au thermo-compression bonding will be evaluated according to MIL-STD 883 for shear strength [3] and leak test [4]. SOI MEMS are created with the sealing ring for direct bonding with the CMOS sealing ring. The integrated 3D system is evaluated for its functionality and reliability.

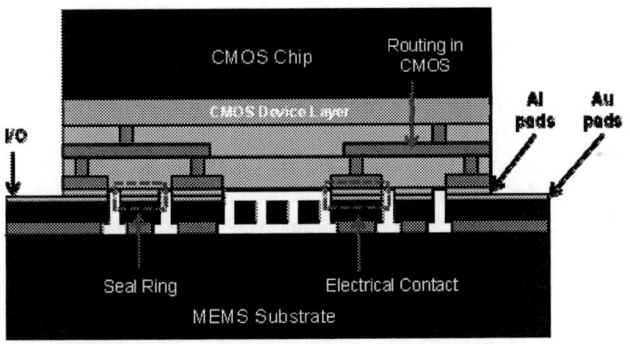

Fig. 1 3D stacking of MEMS and CMOS

Al-Au Thermo-compression Shear Strength Experiment

Au-Al die level thermo compression bonding are produced and tested for die shear strength. Silicon wafers (600μm thick) are used in the thermo-compression bonding experiment. The wafer is deposited with 50 nm of Cr followed by 150 nm of Au. The Cr is deposited to ameliorate the adhesion between Au and Si. Dicing of the wafer is done to produce 5 mm × 5 mm size dies. A second blank wafer is PECVD deposited with 500 nm of SiO₂ and sputtered with a layer of Al of 150 nm thickness. This wafer is diced into 10 mm × 10 mm dies.

The 5mm× 5 mm die with Au and 10mm× 10 mm die with Al are placed with the metal layers facing each other. They are bonded with a die bonder with an applied force of 210 N at different bonding temperature for 10 minutes. Subsequently, the dies are annealed at the same temperature as the respective bonding temperature. The bonding temperatures and annealing temperature used are 350, 300, 290 and 280 °C, respectively. The dies are shear strength tested and the results are shown in Figure 2. The points in Figure 2 are the mean values (sample size of 5) of the shear strength at the corresponding bonding temperature while the standard deviations are depicted in the bars. The requirement of shear strength indicated in MILSTD-883E standard is 5 kg of force.

From Figure 2, it is shown that with a bonding temperature at 290 °C and above, the produced bonding strength is acceptable. When the samples are bonded with 280 °C bonding temperature, the average shear strength of the dies is below 5 kgf. Figure 3 shows the C-SAM image of the samples bonded at 290 °C. The C-SAM shows that good bonded areas are presented in only half of the dies. This non-uniformity of bonding is due to the non-uniform application of force from the die bonder.

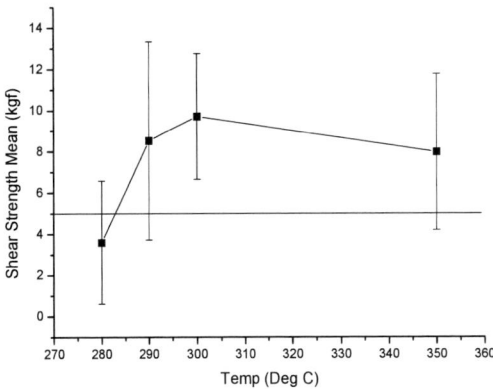

Fig. 2 Shear strength test results with varying bonding temperature

Al-Au Thermo-compression Hermeticity Experiment

Au-Al die level thermo-compression bonding are produced and tested for hermeticity. Silicon wafers are used in the thermo-compression bonding experiment. Deep reactive-ion etching (DRIE) is used to form the cavities and the seal rings to a depth of 120 μm using photo-resist as an etching mask (Figure 4). All cavities are designed and etched to a volume of 1.4×10^{-3} cm³. The surrounding air channel is formed to separate the sealed cavities from the dummy area and to provide a path for helium gas flow during bombing and leak test. After DRIE, the patterned wafer is deposited with 50 nm of Cr and 150 nm of Au. Dicing of the wafer is done to produce 10 mm × 10 mm sized dies with the cavity in the center.

Fig. 3 Shear strength test results with varying bonding temperature C-SAM image of samples bonded at 290°C

A blank wafer is PECVD deposited with 500 nm of SiO₂ and sputtered to form Al layer of 150 nm. This wafer is diced into 15 mm × 15mm dies. The 10 × 10 mm die with Au and 15 × 15 mm die with Al are placed with the metal layer facing each other. The samples are bonded using a bonding force of 2100N and bonding temperature of 300 degree Celsius. The samples are held under 2100N and bonding temperature for 10 minutes and annealed at 300°C for 1 hour in wafer bonder. The schematic of the formation of sample dies for hermetic testing is shown in Figure 5.

978-1-4799-2408-0/14 $31.00 © 2014 IEEE

After the dies are bonded, they are tested for hermeticity according to MILSTD-883E standard. Helium leak test have been performed on the bonded dies. The bonded samples are placed in a chamber filled with helium gas at a pressure of 75 Psia (~0.52 MPa) for an exposure time over 2 hr (helium bombing). Then the samples are tested for helium leak using a mass spectrometer within 1 hr. The leak rates for the samples are less than the rejection limit (5×10^{-8}atm-cc/s) stated in the MILSTD-883E standardwith a mean leak rate of 2.7×10^{-8}atm-cc/s.

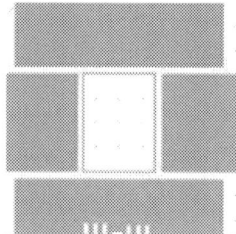

Fig. 4 Seal cavity mask pattern for DRIE for hermetic test dies

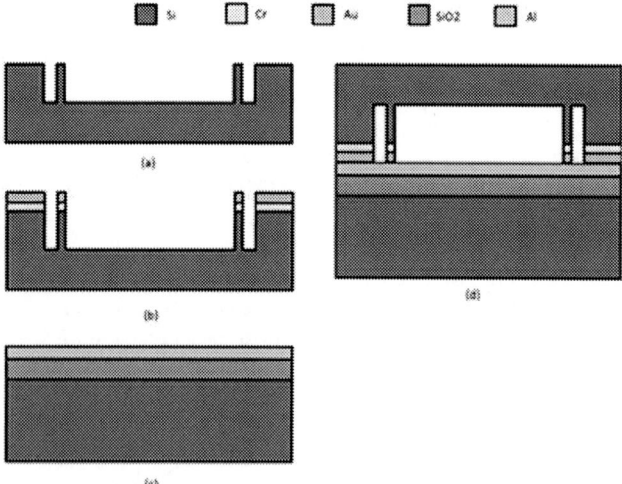

Fig. 5 Schematics of the formation of sealed cavity for helium leak rate detection: (a) forming of cavities, seal rings and air channel using DRIE etching; (b) sequential deposition of Cr layer and Au bonding layer; (c) deposition of SiO2 isolation and Al bonding layer; (d) Au-Al thermo-compression bonding of the cavity wafer to the capping wafer.

Subsequently, perfluorocarbon gross leak test is performed by submerging the dies in fluorocarbon FC-72 (C_6F_{14}) in a pressurized chamber of 75 Psia for 3 hours. The dies are taken out and allowed to dry for 2 minutes. The dried dies are place in a beaker of fluorocarbon FC-43 ($C_{12}F_{27}N$) at 125 degrees Celsius. If a stream of small bubbles (shown in Figure 6) or a single big bubble is observed, the sample would have failed the gross leak test. The samples have to pass both the helium fine leak test and the gross leak test. Only the samples that passed the gross leak test are considered in the mean fine leak rate calculated in the previous paragraph. Figure 7showsthe C-SAM image of a few samples and the sealing rings are visible and continuous.

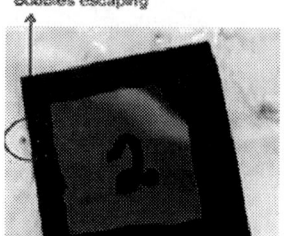

Fig. 6 Example of a stream of bubbles observed in gross leak test

Fig. 7 C-SAM image of hermetic test dies

TEM and EDX Observation of Al-Au Interface

A different sample with dies specification similar to shear strength tested dies is bonded using a bonding force of 1200N and bonding temperature of 300 degree Celsius. The samples are held under 1200N and bonding temperature for 10 minutes and annealed at 300 degree Celsius for 1 hour in wafer bonder. This sample is inspected with TEM and EDX. Figure 8 shows the location of observation marked by the dotted line.

Figure 9 shows the TEM image of the Al-Au interface of the sample at the observation location. There is no distinct interface observable. EDX line scan is performed in the location shown in Figure 10a and the result of the line scan is shown in Figure 10b. It is observed that Al and Au are evenly distributed throughout the original Au and Al layer. This shows that all the original Al and Au layers have mixed and formed intermetallic.

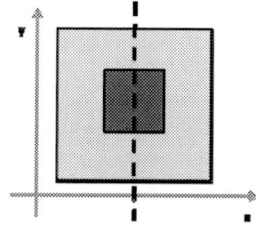

Fig. 8 Location of TEM and EDX observation on the sample

CMOS Readout Circuit

The circuit for MEMS readout comprises of a low noise, band-pass gain stage, a fully differential synchronous demodulator and an off-chip, low-pass filter. The block diagram for the readout circuit is shown in Figure 11.

978-1-4799-2408-0/14 $31.00 © 2014 IEEE

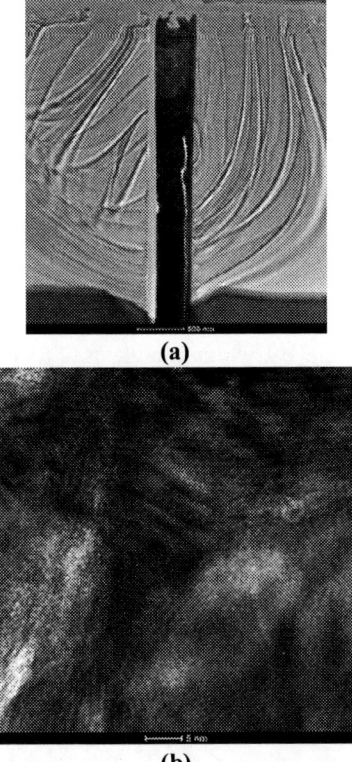

(a)

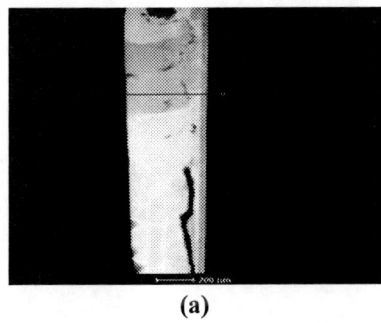

(b)

Fig. 9 TEM image of a sample with increasing magnifications from (a) to (b)

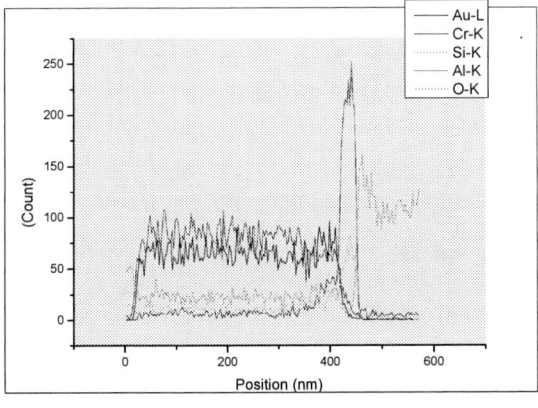

(a)

(b)

Fig. 10 EDX line scan location (a) and line scan results (b)

The low noise, band-pass gain stage is realized by using two identical single-ended output amplifiers, as shown in Figure 12. These single-ended output amplifiers are based on folded-cascode topology, which offers the advantages of a much improved common-mode input range and an increased output swing. This allows more flexibility in selecting the gain factor and which can be conveniently set externally via tunable feedback capacitance C_f. The flexibility in selecting the gain factor enables the same readout circuit design to work well with a wider range of accelerometer designs, in terms of output capacitance and sensitivity. It is also desirable to have very large feedback resistance R_f since the lower corner of the band-pass gain stage is determined by $1/2\pi R_f C_f$. Using a very large feedback resistor however means an inefficient utilization of on-chip space. This can be taken care of by using pseudo-resistors in place of real resistors which can provide very high feedback resistance, of the order of MΩ, while consuming minimal space. The unity gain bandwidth of the folded-cascode single-ended output amplifiers decides the upper corner frequency of the band-pass gain stage. The lower and the upper corner frequency limits of the gain stage thus determine the working carrier frequency range, which are 10 kHz and 18 MHz respectively for the reported readout design. The first amplifier of the gain stage boosts up the MEMS sensor's input while the second amplifier works as an inverting stage to generate a second differential input for the synchronous demodulator.

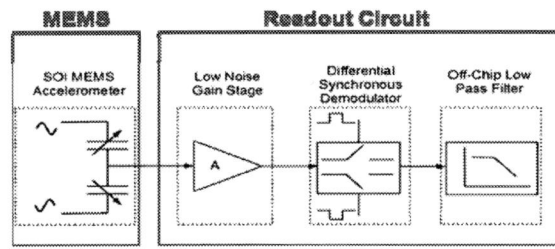

Fig. 11 System block diagram for the readout circuit. The gain factor 'A' is determined by Cf. Implementing the low-pass filter off-chip saves on-chip space

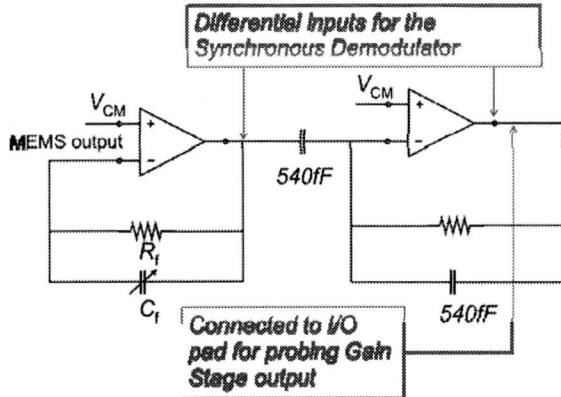

Fig. 12 Schematic diagram of the low noise, band-pass gain stage which consists of two single-ended output amplifiers based on folded-cascode architecture. Pseudo-resistors are used in the feedback path of the amplifiers. Tunable feedback capacitance allows variable gain.

A fully differential synchronous demodulator is the second building block of the readout circuit. Demodulation is done through envelope detection, by implementing a four-switch

978-1-4799-2408-0/14 $31.00 © 2014 IEEE

full-wave rectifier. The demodulator helps in easing the filtering requirements and reduces the second order harmonic distortion at the same time. The fully differential synchronous demodulator makes it possible to move the low-pass filter off-chip and which greatly helps in saving on-chip space. Moreover, as a result of ease in filtering requirements, there is no longer any need of using an active filter and this minimizes the overall circuit's power consumption. The implemented filter is a passive, second order low-pass filter with a cut off frequency of 200 Hz.The die diagram is shown in Figure 13.

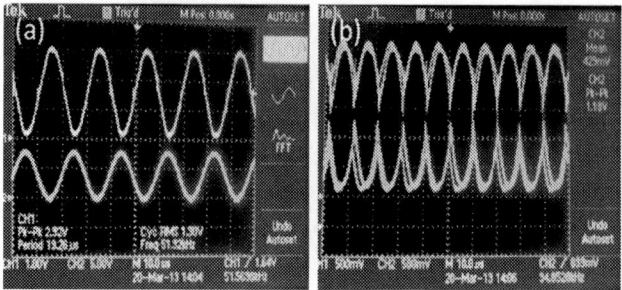

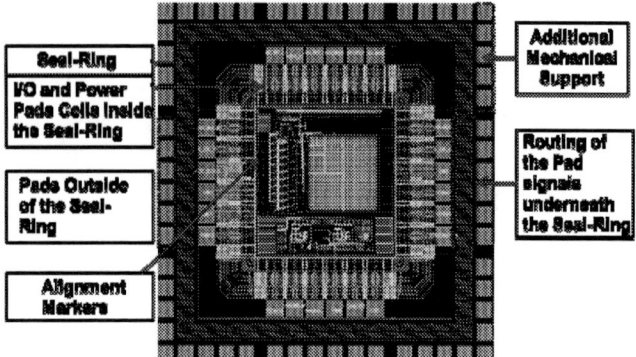

Fig. 13 Die Micrograph of the readout circuit fabricated through MPW (0.35µm, 2P4M process). Key components are highlighted and specific ones are seal ring, mechanical support and alignment marks to assist the MEMS-CMOS bonding.

Standalone Validation of the Readout Circuit – Results and Discussion

The working of the readout circuit was verified with a commercial MEMS accelerometer chip. The accelerometer is driven by differential sinusoid excitation carriers and its corresponding output, based on change in differential capacitance, drives the CMOS readout.

The results shown in Figure14 below are obtained when the MEMS chip is excited by anti-phase but unequal amplitude excitation carriers, 5Vpp and 7.5Vpp, at 50 kHz carrier frequency and 0g acceleration.

Using sinusoid carriers gives a sinusoid gain stage output. The peak-to-peak amplitude of this output is dependent on:
- The amplitude of the excitation carriers. Large amplitude carriers result in a large amplitude gain stage output.
- The frequency of the excitation carriers. Attenuation at carrier frequencies near the corners or outside the flat-band region of the band-pass gain stage significantly diminishes the output.
- The gain factor of the gain stage. Using smaller feedback capacitance C_f gives a larger gain factor.
- The tilt angle of the accelerometer axis with respect to the horizontal. Larger tilt results in larger change in capacitance and hence an enhanced input to the gain stage.

Differential sinusoid inputs to the synchronous demodulator from the gain stage along with an externally supplied CLOCK input generates fully-rectified sinusoids at its outputs.

Fig. 14 (a) The band-pass gain stage output (yellow, upper trace) during the standalone testing of the CMOS readout, when one of the carrier amplitude is 7.5Vpp (blue, lower trace) at 50 kHz frequency and 0g acceleration. (b) Fully-rectified sinusoids at the synchronous demodulator outputs, verifying that the readout is working as intended.

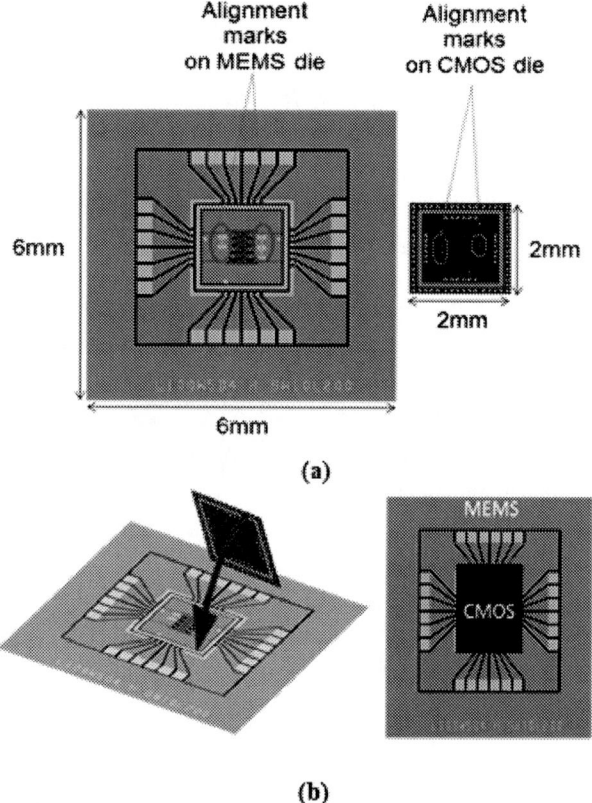

Fig. 15 (a) The alignment marks on MEMS and CMOS die are used for orienting the dies before stacking. (b) The CMOS die is vertically stacked on top of MEMS die

Vertically Stacked CMOS-on-MEMS Chip - Packaging

A standard CMOS chip has a passivation layer that is typically over 2µm thick, which can pose problems with thick metal deposition on the MEMS side. The CMOS dies are post-processed, in that the passivation layer is etched in depth controlled manner so that the metal on the MEMS can be easily reached.

The stacking process is illustrated in Figure15. Alignment marks on the dies aid in orienting the dies properly before stacking. This TSV-less approach of stacking CMOS on top of MEMS results in a simultaneous formation of the hermetic

978-1-4799-2408-0/14 $31.00 © 2014 IEEE   328

seal-ring and thereby eliminates any need for post-stacking hermetic encapsulation.

The bonded MEMS/CMOS dies are then packed inside 44 pin J-leaded surface mount ceramic chip packages for verifying the functionality of the bonded chips as shown in Figure 16.

Vertically Stacked CMOS-on-MEMS Chip – Functionality Testing

The functionality of the bonded MEMS/CMOS microsystem (Figure 17) is verified by conducting the -1g/+1g flip test. The results presented in Figure 18 are obtained using fully-differential sinusoid carriers of 1Vpp amplitude and 50 kHz frequency. The peak-to-peak amplitude of the band-pass gain stage output shows variation that is proportional to the sine of the tilt angle between the accelerometer axis and the horizontal. Using a single-axis accelerometer restricts the total rotation to 180° (tilt angles between -90° to +90°) that corresponds to an acceleration range of -1g to +1g.

The minimum amplitude case in Figure18(a) is when the chip is at 0g orientation. The peak-to-peak amplitude of the gain stage grows out-of-phase and in-phase with respect to carrier in –g and +g flip directions respectively. The mean of the positive rectified sinusoid at the demodulator output is plotted against g at various carrier frequencies in Figure 19. The maximum peak-to-peak amplitudes observed in the two flip directions are roughly equal thereby implying an approximately symmetrical behavior of the accelerometer.

Fig. 16 The CMOS chip is bonded face-to-face on the MEMS chip. The bonded chip is then wire bonded to the package for electrical testing. The vertically stacked CMOS and MEMS chip has a thickness of 1155µm.

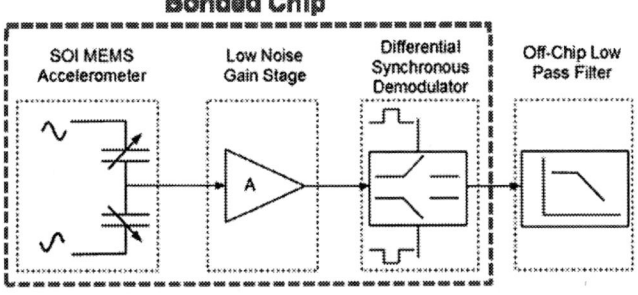

Fig. 17 System block diagram for the bonded MEMS/CMOS microsystem. The MEMS and CMOS have been seamlessly integrated to form a single bonded chip. There is no wire-bonding from chip to chip. Chip to chip electrical connections are done using direct metal bonding without solder.

Noise performance of the bonded MEMS/CMOS microsystem was assessed using spectrum analyzer. The analyzer output at 50 kHz carrier frequency and +1g acceleration is shown in Figure 20, when carrier amplitude is 1Vpp. Likewise, the FFT spectrum at other carrier frequencies and flip orientations was used for computing the signal-to-noise ratio (SNR) and output voltage noise. Reduction in SNR at increasing carrier frequencies is due to an increase in the noise floor level and a simultaneous gradual reduction in the gain stage output at higher frequencies (refer to Figures 21 and 22). Other specifications are listed in Table I.

Table I Summary of the specifications of the vertically bonded MEMS/CMOS chip.

Parameters	Measured Values
Supply Voltage	3.3 V
Power Consumption	1.491 mW
Input Referred Noise of the readout circuit	32.663 nV/$\sqrt{Hz}$
Total Harmonic Distortion	0.380% at 50 kHz carrier frequency and +1g acceleration
Minimum Detection Signal	± 0.139g
Resonant Frequency	136 kHz
Technology	SOI bulk-micromachining for MEMS and AMS 0.35µm (2P4M) for the CMOS readout

Vertically Stacked CMOS-on-MEMS Chip – Reliability Testing

The bonded MEMS/CMOS chip was subjected to 500g mechanical shock test for testing the reliability of the metal-metal contact at the MEMS-CMOS interface. The setup for this test is shown in Figure 23. Under this setup, the bonded chip experiences 10 repetitive vertical vibrations, where each vibration lasts for a short period of 1.03 ms and exerts a maximum vertical acceleration of 503.55g on the chip(Figure 24). This test is in accordance with the industrial standard JESD22-B104C.

The working of the bonded chip was again checked by conducting the -1g/+1g flip test and the chip was still functional after going through the shock test. It can thus be concluded that the low temperature Al-Au contact at the MEMS-CMOS interface remained intact and successfully survived the shock test.

978-1-4799-2408-0/14 $31.00 © 2014 IEEE

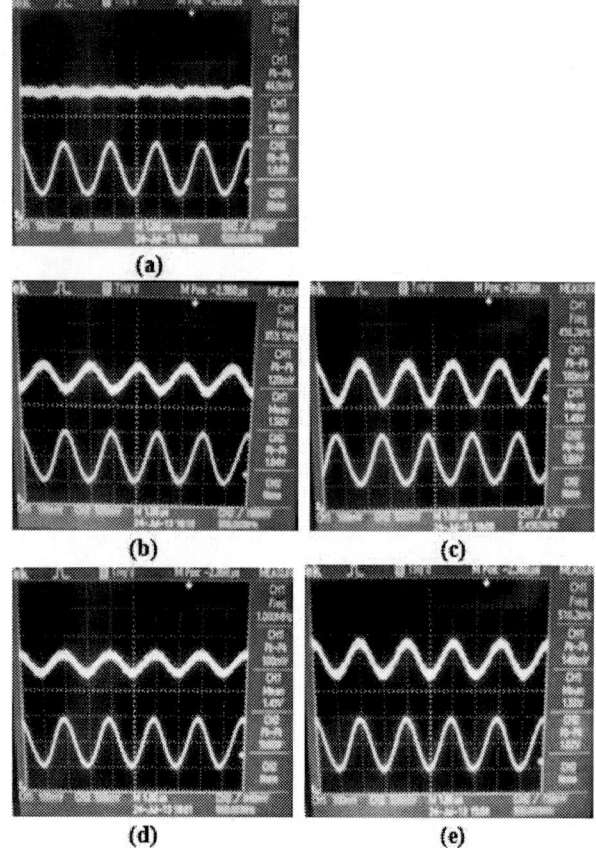

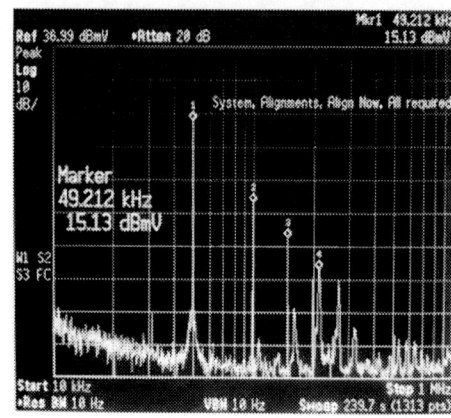

Fig. 20 FFT spectrum showing fundamental peak at the carrier frequency and higher order harmonics at integer multiples of carrier frequency.

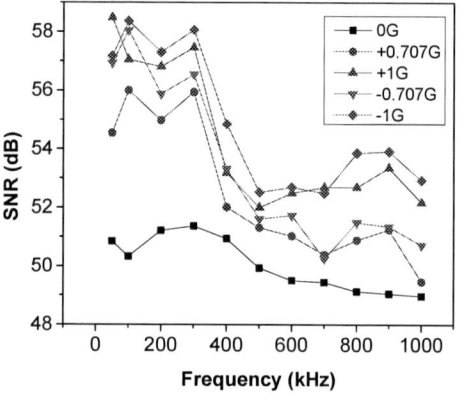

Fig. 18 The gain stage output (yellow, upper trace) when the excitation carrier amplitude is 1Vpp (blue, lower trace) at 50kHz frequency and the bonded chip is tilted at: (a) 0g, (b) -0.707g, (c) -1g, (d) +0.707g and (e) +1g orientation. Acceleration at a tilt angle θ is given by g sin(θ).

Fig. 21 SNR as a function of carrier frequency at various g orientations, when carrier amplitude is 1Vpp.

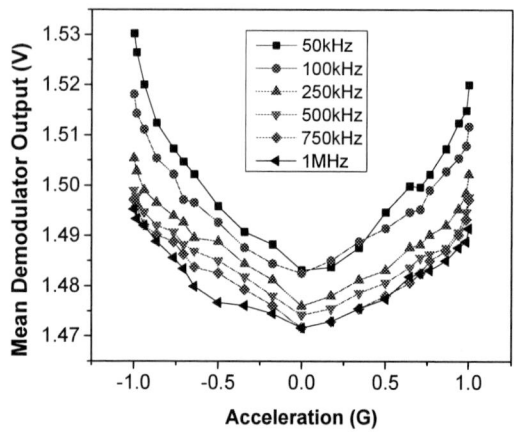

Fig. 19 Variation in the mean demodulator output with g at various carrier frequencies for the bonded chip.

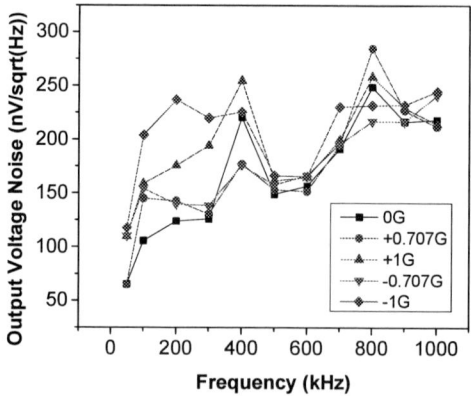

Fig. 22 Output voltage noise as a function of carrier frequency at various g orientations, when carrier amplitude is 1Vpp.

978-1-4799-2408-0/14 $31.00 © 2014 IEEE   330

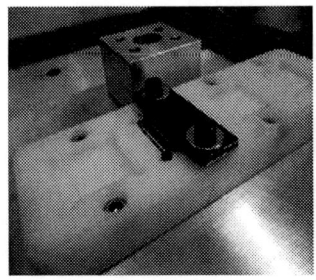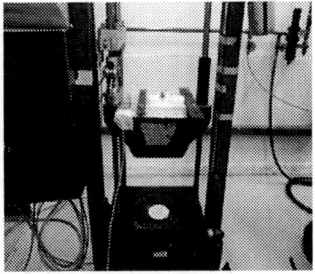

Fig. 23 Setup for the 500g shock test. The chip is subjected to a maximum vertical acceleration of 503.55g for a short duration of 1.03 ms for a total number of 10 times.

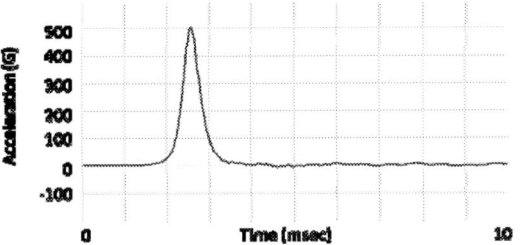

Fig. 24 General control profile during the 500g and 1 ms mechanical shock test.

Conclusion

Al-Au thermo-compression bonding have been tested and verified with reasonable results. Helium leak test demonstrates that the samples achieve an acceptable helium leak rate of average 2.7×10^{-8} atm.cc/sec, which is lower than the reject limit of $5 \times 10_{-8}$ atm.cm3/sec defined by the MILSTD-883E standard. The hermetic test samples are also tested for gross leak and have passed the perfluorocarbon gross leak test. A CMOS readout circuit has been designed in a 0.35μm CMOS process employing chopper stabilization. It has a sensitivity of 5.1mV/g. Working sample of the integrated 3D CMOS-on-MEMS bonded chip using Al-Au thermo-compression has been shock tested and demonstrated that the Al-Au bonding is reliable mechanically. Since Al-Au thermo-compression bonding can provide electrical, mechanical and hermetic bonds in one step, this TSV-less 3D integration method will simplify the fabrication process and improves yield.

References

[1] J Fan, et al. (2012). Wafer-level hermetic packaging of 3D microsystems with low-temperature Cu-to-Cu thermo-compression bonding and its reliability. *J. Micromech. Microeng., 22*(10). doi:10.1088/0960-1317/22/10/105004

[2] R. Nadipalli, et al. (2012). 3D Integration of MEMS and CMOS via Cu-Cu Bonding with Simultaneous Formation of Electrical, Mechanical and Hermetic Bonds. *3D Systems Integration Conference* (pp. 1-5). IEEE International.

[3] MIL-STD-883E. (1996). *METHOD 2019.5.*

[4] MIL-STD-883E. (1996). *METHOD 1014.9.*

[5] G. K. Fedder, et al. (2008). Technologies for Cofabricating MEMS and Electronics. *Proceedings of the IEEE*, Vol. 96, No. 2

[6] J. H. Smith, et al. (1995). Embedded Micromechanical Devices for the Monolithic Integration of MEMS with CMOS. *International Electron Devices Meeting*. IEEE.

Chip Package Interaction: An Experiment Study on White Bump Mitigation Using Flat Laminates

Yi Pan, Jeffrey A. Zitz, David L. Questad, Kamal K. Sikka
IBM Microelectronics Division
2070 Route 52, Hopewell Junction, NY, 12533, U.S.A
E-mail: ypan@us.ibm.com, phone: (845) 894-5938

Abstract

Chip-Package Interaction (CPI) related failure risk has increased in organic laminate-based electronic packages fueled in part by certain industry-segments' requirements for larger silicon die size and signal performance. CPI-related stresses increase directly with radial distance from the center of the die, known as distance from the neutral-point (DNP). The resulting risk of local delamination (White Bumps, WBs) in the silicon back-end-of-line (BEOL), fracture of C4 interconnections and failure of underfill and chip ultimately impact electronic package reliability. Named from the notable visible white halo or circle around a C4 in a CSAM (C-Mode Scanning Acoustic Microscopy) image, WBs are the prime indicator of a CPI-related failure event in an organic laminate-based electronic package.

Laminate and BEOL design, specifically wiring proximal to the C4 interconnection, are known to influence WB reliability. In this paper we specifically study the effect of laminate and BEOL design on the generation of white bumps. Four different chip designs, some of which are more susceptible to WBs than others are mated with both product-design and "flat" laminates, or laminates consisting of only a core and single copper layer on each side. Our ultimate goal is to highlight the silicon & laminate designs' ability to modulate WB occurrences, to create design ground-rules thereby relaxing the cost and facilitating the development of package assembly solutions.

A comprehensive experiment was carried out to achieve these goals. As companion test vehicles to product laminates, four different flat laminate (0-2-0) designs were fabricated with several different core materials to both segregate the effect of laminate wiring design and modulate the overall laminate coefficient of thermal expansion (CTE). Four separate chip designs were used to study the effect of the back end of line design.

Thermal stress was induced in the experiment using a high-cooling rate or "hammer" chip-join (HCJ) reflow profile which drives a higher C4 shearing stress than with the nominal profile. The occurrence of WBs was then monitored while the test vehicles were repeatedly subjected to the hammer profile in a controlled hammer thermal cycling (HTC) process. Diagnostic measurements including 3D Surface Metrology and CSAM were performed after each HTC cycle to monitor both laminate warpage and WB occurrence respectively. White Bump evolution was statistically analyzed and correlated to each die design, laminate cross-sectional design and CTE. Results demonstrated the importance of laminate design in that the flat laminates consistently lead to fewer WBs than product laminates. Lower-CTE laminates also result in fewer WBs than product laminates.

Introduction

Chip-package interaction thermo-mechanical stresses (CPI Stress) exist in electronic packages due to the coefficient of thermal expansion mismatch among the different materials used to fabricate the package and critical temperatures (curing) at which these materials are processed. Such stresses can reach threshold levels and lead to failure of the electronic package during manufacturing process and field application. Accelerated-life stress testing such Accelerated Thermal Cycling (ATC) and Deep Thermal Cycling (DTC) is typically utilized to assess the risk of stress-related failure modes.

A recent summary of CPI failure modes including delamination of BEOL dielectric layers, passivation cracks, metal shift, inter-metallic shorts, damage under the bump can be found in [1] by van Driel. White Bumps are the prime indicator of localized cracks or delamination in the dielectric layers in an organic laminate-based electronic package. This failure mode is sensitive to the structure at the BEOL, bump dimension, dielectric material properties (adhesion and fracture toughness), die fabricating process, CTE mismatch between the die and the substrate, and cooling rate during chip joint [2-7].

The use of Ultra-Low K (ULK) dielectric material in BEOL for advanced circuitry design posts a significant challenge to chip packaging reliability since the ULK film material is weak in both adhesion and fracture toughness versus conventional dielectric materials. Much work has been done to mitigate CPI stress and resulting WBs by improving adhesion and fracture toughness of dielectric films [8], improving the via passivation design [5], developing a low CTE organic substrate [9], improving the chip joining process (slow cooling rate post-reflow) [6], developing differential heating/cooling chip joining method [7]. CPI Stresses in the BEOL can be further reduced by using an epoxy compound as an underfill (encapsulation) material for C4s in a flip chip package which helps redistribute and thus mitigate the stress in the C4s.

The location of WB occurrence has also been observed to vary from expected "high-stress" locations within the BEOL on product chip designs as compared to simpler test chips. Copper wiring density proximal to the C4 interconnection within either the laminate (organic chip carrier) or the chip BEOL has been observed to influence WB occurrence intensity and location. In this paper we specifically study the effect of laminate and BEOL design on the generation of white bumps. By using a high–cooling rate "hammer thermal cycling" (HTC), a repetitive thermal stress at an enhanced level was induced to more clearly expose susceptibility to WBs which were then easily and quickly observed in a CSAM image.

2014 Electronic Components & Technology Conference

Experimental Description

To carry out the experiment, a total of 16 types of testing vehicles (TVs) were built using four different chip (die) designs, Chips A, B, C, and D, as tabulated in **Table 1**. There were two different types of BEOL in these chip designs. The Type II BEOL design had been proved to be more resistant to delamination than the Type I BEOL design from testing within IBM. Each chip design was mated with both product-design chip-carrier organic laminates (4-2-4, or 3-2-3 laminate construction, denoted as BU1) and three other "flat" (0-2-0, denoted as BU2) laminates consisting of only a core and single copper layer on each side. The core type (material and thickness) for the product laminate was denoted as CO1. As a control, one of the three flat laminate also used CO1 as detailed in **Table 2**.

The overall CTE of the laminates utilized are summarized in **Figure 1**. The product-design laminate CTE (3-2-3 or 4-2-4 design) is the highest at 18-19ppm/C. The composite CTE of the 0-2-0 laminate built with the same CO1 core is reduced from 18-19 ppm/C to 16-17 ppm/C. To establish stress cliffs, the CTE of the flat laminates were further reduced to 14-15ppm/C by using two additional lower-CTE core materials, denoted as CO2 and CO3, as showed in **Figure 1**. The quantity of samples populated in each test vehicle cell is 8.

TV	TV-A	TV-B	TV-C	TV-D
Chip	A	B	C	D
Size	14.5x14.5	15.3x15.3	17.5x17.5	19.0x19.0
BEOL	Type I	Type I	Type I	Type II
I/Os	4960	4959	5932	7137

Table 1: Testing vehicles: chips

TV	TV-A	TV-B	TV-C	TV-D
Size	47.5mm	37.5mm	40.0mm	42.5mm
Product laminate	4-2-4 / BU1-CO1	3-2-3 / BU1-CO1	4-2-4 / BU1-CO1	4-2-4 / BU1-CO1
Flat laminate 1	0-2-0 / BU2-CO1	0-2-0 / BU2-CO1	0-2-0 / BU2-CO1	0-2-0 / BU2-CO1
Flat laminate 2	0-2-0 / BU2-CO2	0-2-0 / BU2-CO2	0-2-0 / BU2-CO2	0-2-0 / BU2-CO2
Flat laminate 3	0-2-0 / BU2-CO3	0-2-0 / BU2-CO3	0-2-0 / BU2-CO3	0-2-0 / BU2-CO3

Table 2: Testing vehicles: laminates

In the chip join process, a HCJ reflow profile, which drives a higher C4 shearing stress than the nominal profile, was implemented. Similarly, the C4 interconnects were not underfilled as the epoxy material will redistribute and mitigate the stress of C4s. A baseline CSAM image of each individual test vehicle was recorded immediately after HCJ to record the initial WB status prior to thermal cycling tests.

After the post-HCJ CSAM measurements, the test vehicles were then subjected to HTC (-55°C - 125°C) utilizing a high-rate cooling rate for a total of 5 cycles. Diagnostic measurements including 3D Surface Metrology and CSAM were performed after each HTC cycle to monitor both laminate warpage and WB occurrence respectively.

Results

The upper left quarter of a typical CSAM image showing WBs is illustrated in **Figure 2**. Multiple WBs were observed at the die corners and along the die edge regions. Its cross section view from failure analysis confirmed that the observations are due to delamination that occurred in the BEOL dielectric layers.

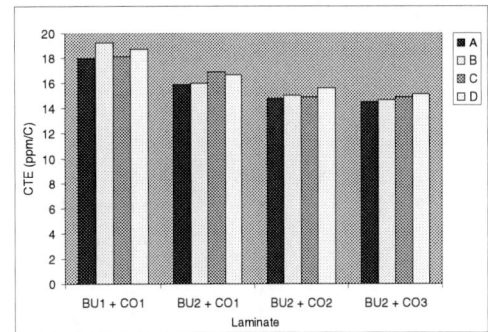

Figure 1: CTE of all laminates. Laminate As are for Chip A, Bs for Chip B, Cs for Chip C, and Ds for Chip D.

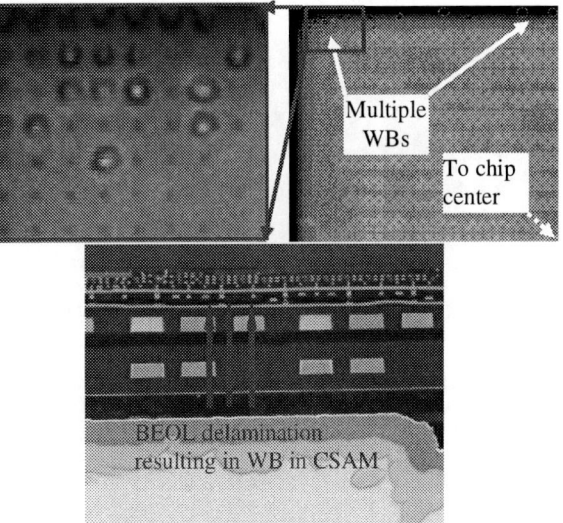

Figure 2: A CSAM image showing multiple WBs at the upper left corner and mid-edge of the die post HTC-5 (top) and its cross-section view (bottom).

The number of parts of each test vehicle detected with WBs is summarized in charts as illustrated in **Figure 3**. The product laminate designs utilized for this experiment are known to be susceptible to WB occurrence, thus at Post HCJ, several test vehicles built with either product-design or flat laminate were detected with existing WBs. At this point however, there was no explicit trend showing whether WB occurrence is influenced by either type of laminate. As HTC cycles accumulate, the number of part with WB increases in all TVs form factors. At the completion of the experiment (post 5 HTCs), the number of parts with WBs in product laminate TVs surpassed that of all flat laminate TVs with the exception of TV-A where only 4 parts built with product laminate were detected with WBs.

978-1-4799-2408-0/14 $31.00 © 2014 IEEE   333

For a more direct comparison of WB occurrence, the total number of WBs was normalized with respect to the TV's number of I/Os (number of possible failure sites) as illustrated in **Figure 4**. It is noted that the number of WBs detected in the two smaller-chip TVs (TVs A & B) was much smaller than that in the two larger chip TVs (TVs C & D). The charts shown in **Figure 4** are plotted in four different scales in order to show the wide range of data.

For each TV, test results confirm that the laminate wiring layer design influences the frequency and location of WBs. Elimination of the wiring layers by utilizing TVs with flat laminates reduces the WB frequency, except for TV-A. Furthermore, by comparing flat laminates utilizing progressively lower CTE cores, data clearly shows that reducing the core CTE can reduce CPI stresses and thus result in a lower rate of WB generation.

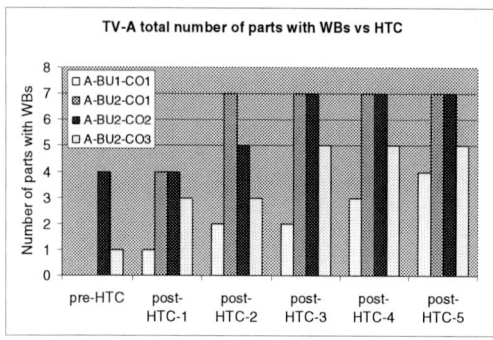

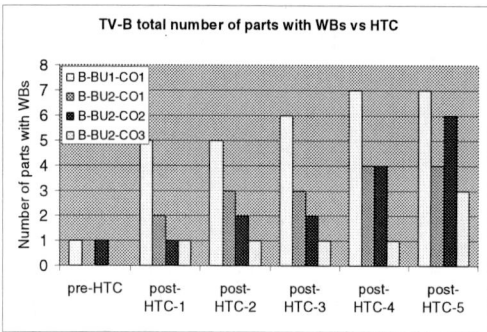

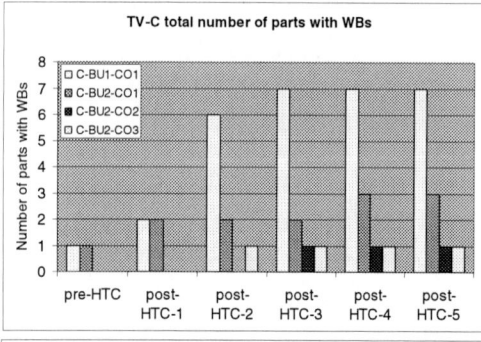

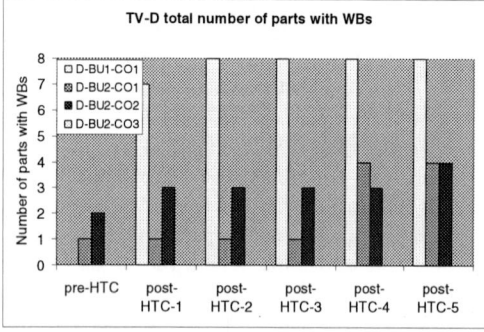

Figure 3: Number of parts detected with WBs

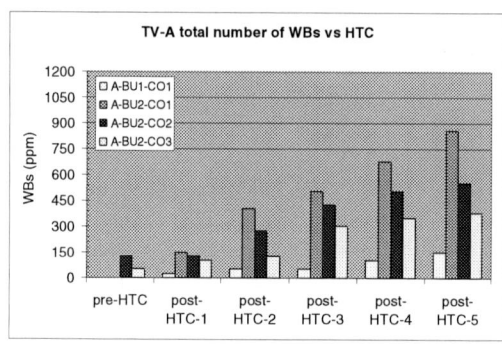

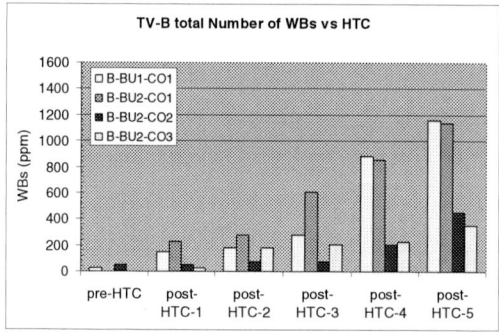

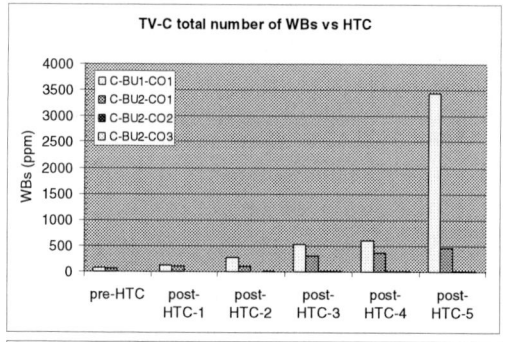

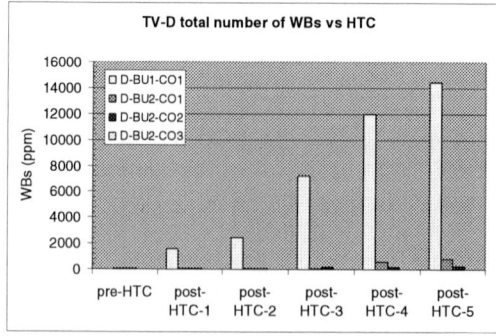

Figure 4: Normalized number of WBs (part per million C4s)

978-1-4799-2408-0/14 $31.00 © 2014 IEEE   334

The frequency and location of white bumps occurring on all 8 parts of each TV for Chip D at the end of stress (post 5 HTCs) were compiled from the CSAM images and are denoted by a "+" in **Figure 5**. Each small circle represents a C4 location. Multiple WB occurrences at the same location were colored coded by frequency of occurrence. Two trends are visually obvious; a) WBs occur in more locations and more frequently at a particular location using TVs made with product laminates (**Figure 5a**) as compared to flat laminate TVs (**Figs. 5b – 5d**) and b) WBs do not occur in the central region of the chip.

During a change in temperature, shear stress in the BEOL and interconnect layers increases radially from the chip center to corner due to CTE mismatch between the chip & laminate and is commonly known as the DNP effect. However, we observe multiple WB occurrences near the center of the chip edge (**Figs. 5a-5b**), an area where the shear stress, although elevated, would be equivalent to areas near the chip corner where no WBs are observed.

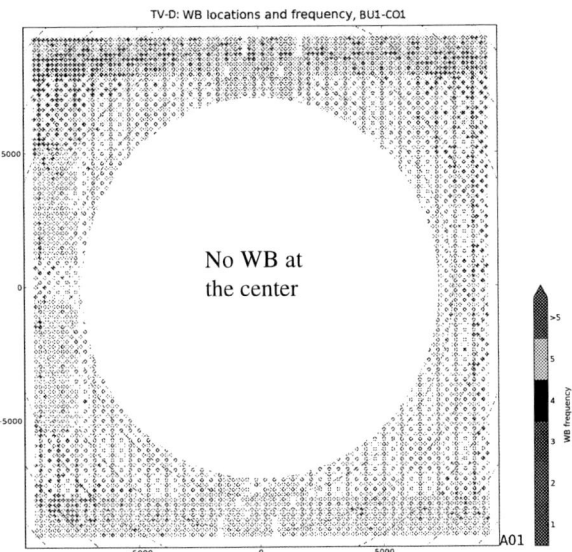

Figure 5a: WB location and frequency of TV-D with **BU1** and **CO1** post HTC-5.

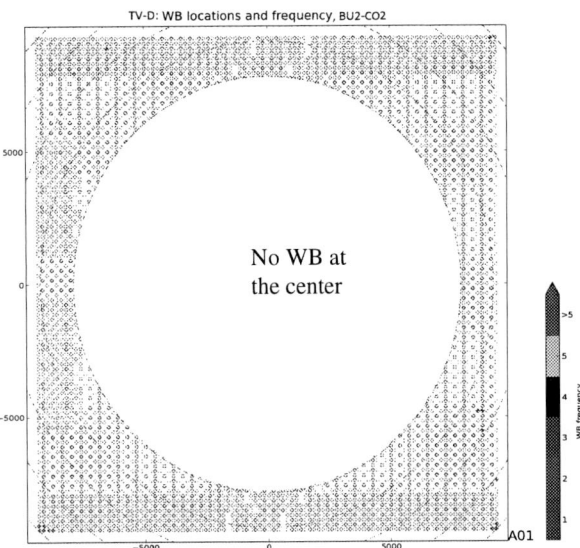

Figure 5c: WB location and frequency of TV-D with **BU2** and **CO2** post HTC-5.

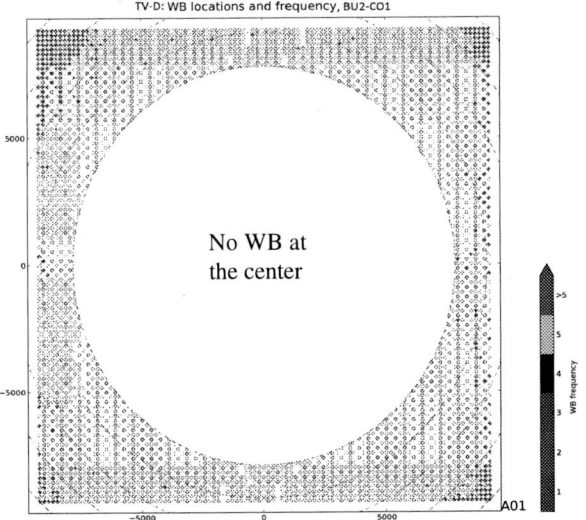

Figure 5b: WB location and frequency of TV-D with **BU2** and **CO1** post HTC-5.

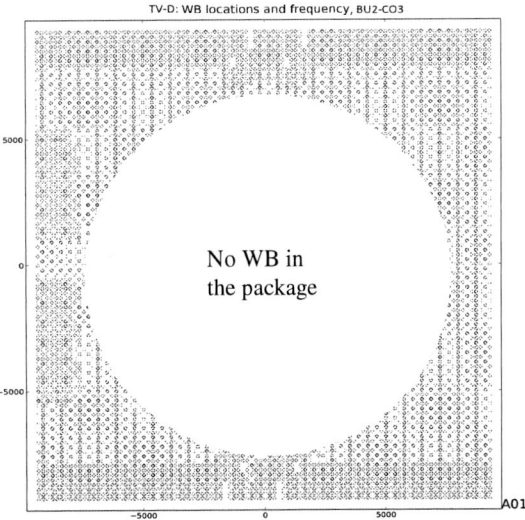

Figure 5d: No WB: TV-D with **BU2** and **CO3** post HTC-5.

A finite element simulation is utilized to help explain this observation. An explicit quarter-symmetric model of the chip, laminate and all individual c4s of TV-D was generated. The combined stress state within the C4s near the mid-side edge of the chip was assessed as the package was cooled from the solder solidification temperature to the lowest temperature of the HTC. Results indicate that the out-of-plane stress (S33) is also very high at the peripheral region of the chip as shown in **Figure 6**. The locally-elevated out-of-plane stress (S33) in concert with the existing DNP-related shear stress is responsible for the additional delamination risk of the BEOL and resulting WBs in this region.

The normalized maximum out of plane stresses in the large-die TVs (TV-C, TV-D) of product and flat laminates are compared in **Figure 7**. It is illustrated that the maximum out of plane stress is reduced in the flat laminates where the wiring build up layers in the laminate are removed. It is also shown that the maximum out-of-plane stress decreases as the CTE of the core material is reduced.

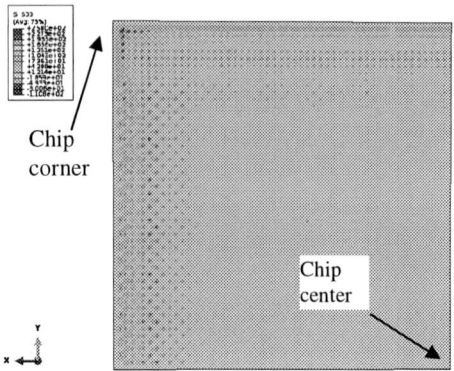

Figure 6: Finite element simulation results show high tensile stress (S33) occurs near the mid-side edge of the chip as well as at the chip corner.

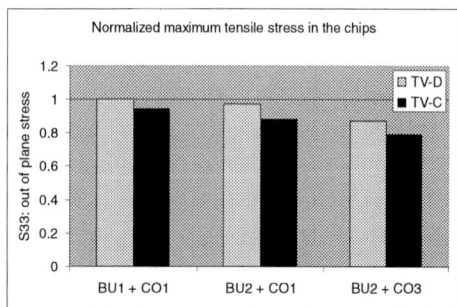

Figure 7: Tensile normal stress in the chips for TV-C and TV-D (normalized with respect to the product TV-D).

Discussions

Our quarter-symmetric finite element model predictions of the high stress regions correlated well with the experimental results of WB frequency and location. Temperature variation within the package during cooling down post chip joint was neglected. Chiliro and Lee [10] showed that different heat capacitance between a large and small die may affect the temperature distribution during cooling after chip join reflow. This will also change the stress distribution among the C4 bumps and can also help explain why there are multiple WBs observed at region near the chip edges, since the peripheral C4 bumps solidify sooner during the cooldown than the interior

C4 bumps and thus accumulate stress over an extended change in temperature.

Unexpectedly, the results did not indicate that the Type II BEOL structure (utilized in Chip D) is more resistant to delamination (WBs) than the Type I BEOL structure utilized in Chip C. Any beneficial effect of the Type II BEOL structure may have been overwhelmed by the higher stress levels (larger chip size) and additional failure locations (C4s) of the TV-D as compared to TV-C. The presence of these competing variables likely nullified insight to the BEOL material or structural effects. Thus, for further studies concentrating on BEOL design robustness to WBs, it is recommended that experimental variables be limited to BEOL variables only while chip, C4 count and laminate structure remain invariant.

Conclusions

White Bump evolution was statistically analyzed and correlated to each die design, laminate cross-sectional design and CTE. Results demonstrated the importance of laminate design in that the flat laminates consistently lead to fewer WBs than product laminates. Lower-CTE laminates also result in fewer WBs than higher-CTE laminates. The experimental results correlated well to quarterly-symmetric finite element simulation results.

Acknowledgments

The authors gratefully acknowledge the assistance provided by E. Kastberg, J. Sylvestre, H. Zhang, J. Rosa, D. Ramos, H. Liu, J. Jacobi, T. Childress, C. Reynolds, B. Chan, G. Osborne.

References

1. W.D. van Driel, "Reliability consequences of the chip-package interactions" in *Proc. of IEEE Electronics Packaging Technol. Conf. (EPTC)*, 2009, pp. 406–411.

2. M. Farooq, et al, "Chip Package Interaction for 65nm CMOS Technology with C4 Interconnections", in *Proc. IEEE Intl Interconnect Technol. Conf. (IITC)*, 2007, pp. 196–198

3. S. Sankaran, et al., "A 45 nm CMOS node Cu/Low-k/ Ultra Low-k PECVD SiCOH (k=2.4) BEOL Technology", in *Proc. Electron Devices Meeting (IEDM)*, 2006 , pp. 1–4.

4. S. W. Lee, et al., "A study on the chip-package-interaction for advanced devices with ultra low-k dielectric", in *Proc. IEEE Electronic Components and Technol. Conf. (ECTC)*, 2012, pp. 1613–1617.

5. E. Misra, et al., "Novel design and integration enhancements in the final polymeric passivation for improved mechanical performance and C4 electromigration in lead-free C4 products", in *Proc. IEEE Electronic Components and Technol. Conf. (ECTC)*, 2012, pp. 571–576.

6. H. Nakajima, "High-speed, low-power device trend and low-k layer delamination", in *Proc. Electronic Materials and Packaging (EMAP)*, 2012, pp. 1–4.

7. L. Sakuma, "Differential heating/cooling chip joining method to prevent chip package interaction issue in large

978-1-4799-2408-0/14 $31.00 © 2014 IEEE

die with ultra low-k technology", in *Proc. IEEE Electronic Components and Technol. Conf. (ECTC)*, 2012, pp. 430–435.

8. T. Usami et al., "Mechanical properties of SiCOH film related to CPI and High Load Indentation test of real Cu/Low-k structure with bumps", in *Proc. IEEE Interconnect Technol. Conf. (IITC)*, 2012, pp. 1–3.

9. T. Yamada, et al. "Development of a Low CTE chip scale package", in *Proc. IEEE 63rd Electronic Components and Technology Conference (ECTC)*, 2013, pp. 944–948.

10. C.J. Uchibori and M. Lee, "Impact of chip package interaction on Cu/Ultra low-k interconnect delamination in Flip Chip Package with large die", in *Proc. IEEE Interconnect Technol. Conf. (IITC)*, 2009, pp. 21–219.

Design and Package Technology Development of Face-to-Face Die Stacking as a Low Cost Alternative for 3D IC Integration

Zhe Li, Yuan Li, John Xie
Altera Corporation

Abstract

F2F stacking provides an alternative 3D packaging solution for multi-chip integration without use of TSV. High density interconnection can be achieved with direct Face-to-Face (F2F) stacking to enable high bandwidth die to die interface. Simplified stacking process and lower development cost make F2F stacking an attractive solution for cost sensitive applications. A comparative study of performance was performed on F2F stacked Field Programmable Gate Array (FPGA) die in a flip chip organic package. The paper first presents thermal analysis to address power density increase, hot spot and temperature variations in the F2F package. Next the paper focuses on electrical performance validation including both IO and power delivery analysis. With appropriate chip design and optimization, we demonstrate that F2F stacking induced thermal and electrical impacts can be controlled to meet speed and performance specs equivalent to 2D system. The manufacturing design rules have been optimized to meet yield requirements as well as ensuring product reliability.

Introduction

In recent years, 3D IC integration has attracted a great interest in semiconductor industry as a "More Than Moor" approach to break the silicon scaling trend [1]. Increasingly, traditional silicon process scaling alone can no long meet system performance, throughput and power requirements. 3D stacking enables more functional integration than single chip solution. Using through silicon via (TSV) to build 3D stacks of chips makes it possible to eliminate long run of interconnect lines in a 2D system. Power consumption is also reduced because of short interconnect path between active devices and lower IO drive strength to drive the short interconnects [2]. 2.5D and 3D packaging are being actively investigated. In order to achieve manufacturability readiness and cost effectiveness, the entire 3D fabrication sequence requires seamless integration from foundry to OSAT back-end process.

Though TSV technologies and 3D integration hold much promise, the industry also realized that the new technology will take time to get established in supply chain and drive volume production. While TSV and 3D technology development continues to improve, a whole variety of interim interconnect/packing solutions are being developed to address the bandwidth and scaling density demands. Direct Face-to-Face stacking through high density micro joints provides an alternative 3D packaging solution for multi-chip integration with low cost [3]. F2F stacking offers silicon grade interconnect density IO driver and reduced wiring parasitics for a high performance chip-to-chip interface. F2F packaging also has a time-to-market advantages compared to 3D IC which typically requires a longer lead time to implement design and longer cycle time due to the complexity of 3D manufacturing process.

F2F stacking provides additional benefits for compact form factors. There are some existing packaging solutions, such as side-by-side multi-chip package (MCP), package-on-package (POP) and package-in-package (PIP) platforms that feature footprint miniaturization and package height reduction. These packaging technologies allow for two or multi-chip to be bonded together. However, they do not offer the same degree of connectivity, scaling density, and bandwidth that F2F stacking can provide to meet the requirements of next generation products.

Figure 1. Face-to-Face stacked die on a BGA substrate as a low cost 3D IC solution without using TSV

This paper provides an overview of process flow and manufacturing activities for F2F stacking integration in FPGA applications. F2F stacking is a packaging technology designed to have two or more die assembled together. As shown in Figure 1, typically for two-die stacking, the mother die (top die) is larger than the daughter die (bottom die). Key feature of this technology is chip-on-chip bonding through fine pitch copper pillar micro-bumps. F2F stacking process flow is relatively simple. The critical steps are bumping and chip-on-chip assembly by Thermal Compression bonding with Non-Conductive Paste (TCNCP). TCNCP yield is sensitive to CoC alignment which depends on the planarity and topology of copper pillar micro-bumps [4]. Face-to-Face stacking is at early phase of technology development. The entire flow must be optimized to deliver overall yield and quality performance for manufacturing readiness and volume production

This paper is targeted at F2F package thermal and electrical design optimization. Package structural change from monolithic to stacked die presents unique challenges in the F2F package design. Thermal performance of F2F stacked die is discussed and compared to monolithic die in a flip-chip package. We also present a co-design approach in which RDL is employed together with die pinout and substrate routing to address the interconnect density and high speed channel bandwidth performance needs. High speed memory channel SI analysis is demonstrated to meet the channel loss and jitter requirements. Lastly, we discuss the PDN design and simulation methodology to accommodate F2F stacking impacts.

F2F Assembly Process Flow

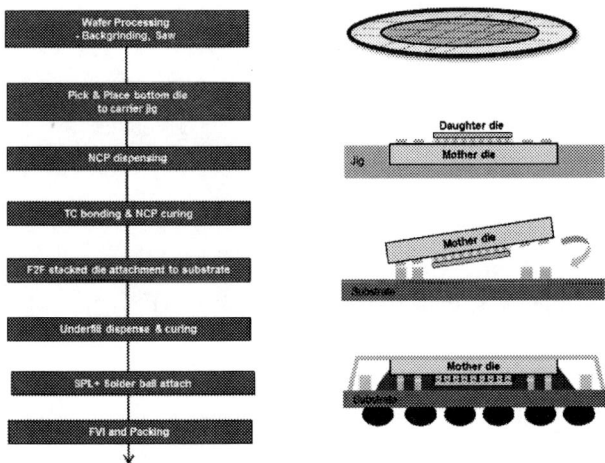

Figure 2. Process flow for Face-to-Face die stacking and substrate assembly

Face-to-Face assembly process is relatively simple. Figure 2 presents the process flow for F2F chip-on-chip bonding and then chip-on-substrate (CoCoS) assembly. After bumping is done on to the mother die wafer and daughter die wafer, back grinding and dicing are performed. The daughter die is then joined to the mother die through fine pitch copper pillar micro bumps. Thermal compression bonding with non-conductive paste (TCNCP) is employed for CoC assembly. As compared to the standard reflow process, TCNCP is chosen for CoC fine pitch assembly because of its accurate alignment control along with thermal compression to assist copper pillar joint formation. TC bonding consumes process time, but also provides process flexibility needed for F2F assembly to balance material stress for high yield and reliability. After completion with CoC assembly, the stacked dice are then flip-chip attached to the substrate through a mass reflow process. Capillary underfill is needed at this stage to ensure no air gaps form between die and substrate. The assembly is then followed by lid attach, ball attach, laser marking and test.

Package Structure

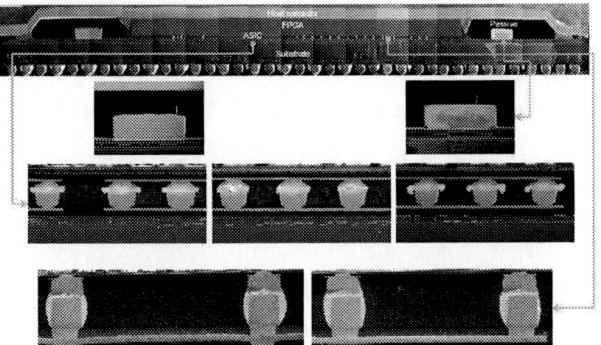

Figure 3. Cross section of test chip with FPGA die and daughter die face-to-face stacked with copper pillar joints

The F2F stacking process, interconnection material and package structures have been optimized for yield, reliability, electrical and thermal performance. Figure 3 shows the cross section and close-up view of the F2F assembled package and copper pillar joint structures. Two bump designs with different sizes and geometries are featured on the mother die. Fine pitch micro bumps are located in the CoC die pad region while the peripheral region is populated with large bumps in 185um pitch. An adequate amount of standoff clearance is needed so that die thickness variation and substrate warpage can be accommodated in CoS assembly. In F2F stacking, the daughter is thinned down to 75um. The CoC collapsed gap typically range from 25 to 35um depending on the bump designs. Total standoff height between die and substrate is on the order of 100 to 150um.

Cu Pillar Technology

Fine pitch copper pillar joint is the enabling technology for CoC assembly. For current face-to-face configuration, a taller copper pillar bump is created on the daughter die and a shorter height bump is formed on the larger mother die. The tall copper pillar bump is made up of 25um non-reflowable copper post with 15um SnAg solder cap. Table 1 shows more bumping options on both mother die and daughter die to make the fine pitch copper pillar joint. Pull-in and Pull-out designs are referring to the passivation opening relative to UBM size. If it is pull-out design, passivation opening is smaller than the UBM. Pull-in design has passivation opening larger than the UBM size. Both designs have been implemented and test out in DOE studies to reduce joint failures due to interconnect material stress. DOE result shows that pull-in design provides more bump contact area to silicon metal pads and thereby improves bump shear strength. Pull-out design has better protection on RDL layer and provides more design capability to reduce bump pitch to 40um and below.

Table 1. Fine pitch bumping technologies

	Daughter Die		Mother Die	
Die Thickness	75um		780um	
Bump Height	40um		3~5um	
Bump Composition	Cu Pillar/ LF Solder	Cu Pillar/ LF Solder	Ni/Au	Ni/Solder
Bump/PBO	Pull in	Pull Out	Pull In	Pull Out
Bump Structure				

Product Prototype Test Vehicle

To address the design and manufacturing challenges, we have designed a prototype F2F TV for FPGA integration with a high performance processor or memory. The F2F TV has a 0.18um 4mmx4mm CPLD chip mounted on a 40nm 17mmx22mm FPGA device through copper pillar joints in 50um bump pitch. The CoC stacked die is assembled on a 40mm organic substrate with 185um pitch copper bumps. The F2F TV contains an extensive set of structures for modeling and characterization of stacked die configuration and its impact to electrical and thermal performance. Characterization structures comprise of resistance, capacitance and S-parameter test structures for RDL, micro-bumps and substrate interconnects. F2F TV also contains varying critical area structures for yield and reliability characterization.

Thermal Analysis

In this work, the focus is on the comparison between F2F and monolithic packages. A F2F package and a monolithic single-die flip chip package are thermally analyzed using CFD (computationally fluid dynamics) models. In order to have fair assessment, both packages have the same package size, similar die size and the same total power consumption (26W). Both packages are modeled under the following conditions:

- JEDEC 2s2p PCB
- Ambient temperature 50 °C
- Al heat sink, same size as the package, plated fin, 20 fins, fin height 10 mm, thickness 1 mm, base thickness 1.5 mm
- Air flow rate 2 m/s

Table 2. T_j comparisons

	Tj of FPGA
F2F	101.1
Monolithic	98.8

The junction temperatures are listed in Table 2. The F2F package shows thermal performance similar to the monolithic package. With 26-W power, the difference in the junction temperature between the two packages is less than 3 °C. The higher Tj of F2F can be attributed to the smaller die size compared to the monolithic package. The temperature contours are shown in Figure 4.

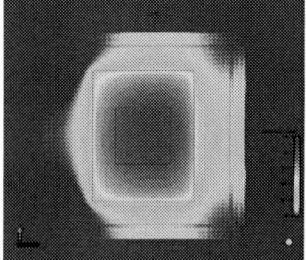 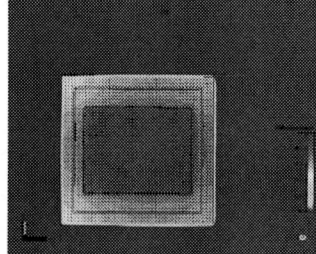

Figure 4. Die temperature contour on the monolithic (left) and F2F (right)

For the F2F package, under the conditions used, about 16W exits through the top and about 9.6 W exits through the bottom, or 60% out through the top and 37% the bottom

Thermal modeling has shown that F2F has comparable thermal performance to monolithic packages.

IO Electrical Performance

IO channel in F2F package has added 3 components, fine pitch copper pillar joints, redistribution layer (RDL) and tall copper joints between mother die and substrate. On substrates and below, IO channel routing is same as the conventional package. Figure 5 shows the RDL routing that connects IO channel from fine pitch bumps to large copper bumps. Due to stacked die spacing requirement, there are about 2~3mm routing length on RDL to fan out and get access to the substrate bumps.

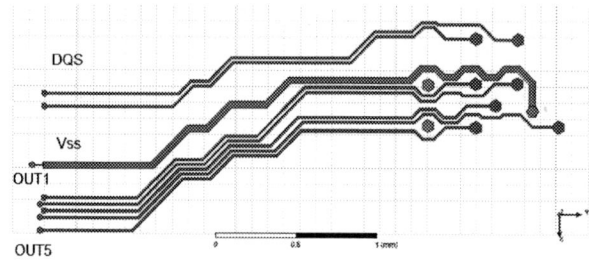

Figure 5. RDL routing for DDR channel in F2F stacked die package

DDR electrical validation was performed on data bus Read and Write operations for F2F stacked die package in comparison with conventional DDR channel design. The goal of this study was to validate the healthiness of DDR interface and to understand the performance difference between F2F stacked die and conventional packages. Package 3D and 2D models (Ansoft Q2D, Ansof HFSS) were carried out to simulate the DDR channel performance and then link simulation is used to run design of experiments (DOE) analysis to identify the weakest link in the design. The worst case bit pattern is generated to simulate the eye width and eye height. DOE analysis is a good way to assess the voltage and timing margins on DDR interface.

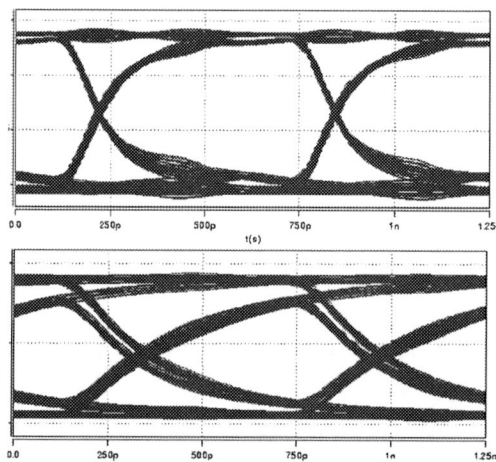

Figure 6. RDL impact to DDR channel eye opening (Top: DDR in flip-chip package, bottom: DDR in F2F package)

Link simulation of one DQ group data lines and DQS differential pair is carried out to capture the ISI, crosstalk and SSN impact from RDL routing and die-to-substrate interconnects. A snapshot of results of this study is shown in Figure 6 on DDR channel ISI due to RDL insertion loss. It is shown between F2F and conventional package using optimized routing and bump pattern, DDR performance degradation is about 8% in terms of voltage and timing margin. DDR performance degradation is mainly caused by RDL resistance. Another observation that can be made is that channel crosstalk in the F2F stacked die package are on par with conventional package. Channel crosstalk can be controlled by adding ground shielding to avoid potential coupling between active data lines

Power Integrity

Power integrity is one of the important issues in high speed system design. Power and current levels are expected to increase with corresponding to logic and I/O increment in F2F configuration. Some of the conventional wisdoms on power delivery network (PDN) analysis can still be applied to identify worst case power supply noise scenario.

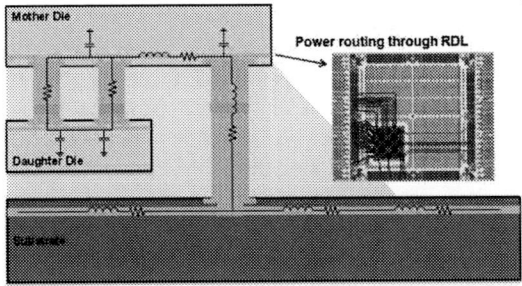

Figure 7. Power delivery network in F2F package

F2F package pose a new challenge on PDN design for its stacked die configurations and tightly coupled PDN system. The proposed face-to-face die stacking with two levels of copper pillar based joints and RDL interconnects is presented in Figure 7. The RDL has been discussed in the paper as a physical routing layer to link IO from the daughter die to C4 bumps and substrate. RDL also serves as the routing layer for power delivery. With limited routing space, metal stripping on RDL have to be balanced to meet both IO and power performance specifications. Power bus are connected from all four sides of the daughter die to minimize the potential DC resistance on power and ground.

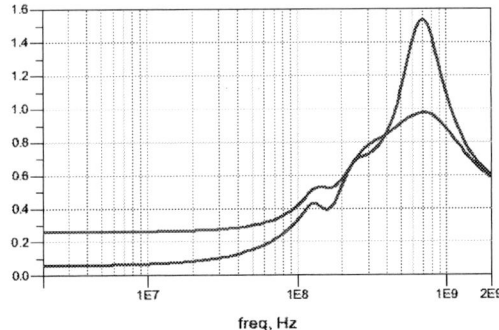

Figure 8. PDN impedance comparison between conventional package(red) and F2F package (blue)

In this study, interconnect parasitic impacts to PDN performance has been investigated in different packaging configurations. For power distribution, the preferred design is to reduce the interconnect resistance so that IR drop on power delivery network is small and DC voltage level at transistor power nodes is maintained. But it is not always feasible in practical designs, especially in F2F stacked die packages. RDL routing increases the DC resistance. Tall copper joint also adds resistance to the PDN. DC resistance needs to be managed in F2F package. Resistance impact to power stability is two folded. At high frequencies, resonance peak is reverse proportional to the PDN resistance. Usually the resonance peak occurs at a frequency determined by the on-die capacitance (ODC) and the inductance found in the package

and PCB system. High resonance peaks give rise to large dynamic voltage droop in response to simultaneous switching activities. Previous work has demonstrated the issues caused by PDN impedance peaks. Methods have been developed to measure the PDN response to logic switching activities and correlate these with simulation results [5].

The parasitics in F2F stacked die package are extracted for comparison with conventional flip chip package and PDN impedance simulations are shown in Figure 8. The red line shows the PDN impedance in a conventional flip chip package design. The blue line shows the first pass power delivery performance of F2F stacking package as indicated in the test vehicle design. The simulation results confirm with our theoretical analysis. F2F stacked die package has a high DC resistance and low resonance peak than conventional packages. When considering the PDN design, balance the PDN performance from DC to high frequencies is the key for complete power delivery. In this study, RDL metallization, assignment of power and ground balls at several locations have been optimized to meet system power integrity requirements.

Conclusions

F2F stacking as a compelling and scalable package integration technology is demonstrated to address system bandwidth, low cost and small form factor needs. Such a new technology includes innovations in thin wafer handling, copper pillar micro-joint, die-to-die stacking as well as and die-to-package interconnections. Die-to-die interconnect is joined at 50µm pitch through TCNCP assembly. Key variables such as bonding force, bump co-planarity and warpage control have been optimized for F2F stacked die assembly. Test vehicles were designed and fabricated for interconnect reliabilities, thermal and electrical performance validations. Thermal and electrical modeling is carried out to verify F2F stacking impacts and make product design trade-offs.

Acknowledgments

The authors would like to thank Jemmy Sutanto, DongHe Kang, Michael Oh, Kwang-Seok Oh and their advanced product development team from Amkor for their support.

References

1. Knickerbocker, J.U., et al "3-D silicon intergration and silicon packaging technology using silicon through-vias," *IEEE Journal of Solid-State Circuits,* Vol. 41, No.8 (2006), page 1718- 1725

2. Hong Shi, Zhe Li, Yuan Li, Jenny Jiang, John Xie, "Performance Challenge and Opportunity of TSV Silicon Interposer in 3D System," *DesignCon* (2012)

3. Jemmy Sutanto, etc., "Development of Chip-on-Chip with Face to Face Technology as a Low Cost Alternative for 3D Packaging," *IEEE ECTC* (2013)

4. Zhe Li, SC Tan, YH Yew, PT Teh, MJ Lee, John Xie, "Flip-chip packages with Periphery Cu Pillar Bumps as Wire Bond Replacement- Design, Modeling & Characterization," *IMAPS* (2013)

5. Zhe Li, John Xie, Arif Rahman, "Development of an Optimized Power Delivery System for 3D IC Integration with TSV Silicon Interposer," *IEEE ECTC* (2012)

From C4 to Micro-Bump: Adapting Lead Free Solder Electroplating Processes to Next-Gen Advanced Packaging Applications

Julia Woertink, Yi Qin, Jonathan Prange, Pedro Lopez-Montesinos,
Inho Lee, Yil-Hak Lee, Masaaki Imanari, Jianwei Dong, Jeffrey Calvert
Dow Electronic Materials
455 Forest St., Marlborough, MA 01752
woertink@dow.com

Abstract

SnAg solder is the industry standard for lead-free wafer bumping. SnAg electroplating chemistry for C4 bumping must be capable of achieving tight performance standards over a wide range of applications: mushroom or in-via electroplating, a diverse range of die designs and pattern densities, and over a wide range of plating rates, including high speeds to enable higher throughput. Beyond C4 applications, SnAg electroplating chemistries must also deliver strong performance on emerging micro-bumping and Cu pillar capping applications, leading to new materials challenges and new demands on both the solder and copper electroplating chemistries.

Electrodeposited SnAg solder and Cu pillar performance is controlled by the specially formulated additives used during the electroplating process, paired with optimized high purity, low-alpha emitting inorganic metal salts and acids. Selection of robust, versatile SnAg and Cu plating chemistries capable of operation at high plating rate is critical to achieving optimized high volume, high throughput, and cost effective manufacturing. Improper additive selection can lead to a number of defects, which can manifest after plating, after reflow, or during reliability testing. These defects include as-plated defected or rough surface morphology, post-reflow macro-voiding, micro-voiding, bridging, and Ag_3Sn plate formation, and poor height uniformity control across the die and wafer. These defects can lead to critical failures that lower yield and require wafer rework or wafer scrapping.

To reduce the frequency of solder bumping, micro-bumping and Cu pillar capping defectivity, solder and Cu electroplating processes must be selected to enable wide process versatility and stability. In this paper, a wide range of bumping defects is introduced, root causes are discussed, and defect mitigation strategies are presented. Further, next-generation SnAg electroplating chemistries are presented that are designed to minimize solder defectivity and provide high performance, high yield, wide process windows, and high throughput options over a wide range of applications.

Introduction

Solder joints are a key component in electronic interconnection. Historically, tin-lead (SnPb) solder was used due to its desirable electrical, mechanical, and reliability properties [1,2]. Recent environmental concerns have necessitated the development of lead-free solders, however, and tin-silver (SnAg) solder is now the most widely used lead-free solder for advanced packaging applications [3]. SnAg is typically deposited by electroplating, due the lower cost and greater reliability and SnAg solder bumps deposited by electroplating have now been in mainstream manufacturing for over five years in traditional C4 bumping applications.

Industry trends towards higher performance semiconductor devices with shrinking form factor have led to significant materials challenges for advanced packaging metallization processes [4]. A wide range of advanced packaging applications, including flip chip, WLP, and 3D-TSV have emerged and traditional C4 bumps are being replaced by micro-bumps and Cu pillars capped with SnAg solder to meet new performance demands [5]. These applications present new challenges to the solder plating process to compensate for new geometries, metallurgy, and new process integration schemes. Increasing demands on performance have been coupled to requirements for higher plating rates to enable higher process throughput, and stricter requirements on yield, requiring low defectivity and high process capability. Suddenly a wide range of defects have emerged as SnAg processes are applied to these new applications. These effects threaten the yield of Cu pillar and TSV technology.

Experimental

Test samples were prepared by electroplating on 300mm patterned test wafers with either horizontal or vertical wafer plating tools. Test wafers for micro-bump testing were patterned with 20μm or 30μm diameter vias, while test wafers for SnAg bumping tests were patterned with 75μm diameter vias. For SnAg C4 plating, electroplated Ni UBM was deposited first using NIKAL™ BP Ni plating chemistry, followed by SnAg plating either with SOLDERON™ BP TS6000 SnAg plating chemistry, or with alternative additive systems. For micro-bump plating, Cu was deposited with either INTERVIA™ 8540 Cu plating chemistry, or with alternative additive systems, followed either by direct SnAg plating or sequential Ni and then SnAg plating using the chemistries described above.

After electroplating, the photoresist was stripped and the samples were reflowed using either a flux-based belt furnace reflow process or a flux-free reflow process. WID and WIW were calculated from full die height mapping results using optical profilometry and samples were visualized by SEM or optical microscopy. FIB-SEM cross-section was used for micro-void analysis, and X-ray inspection was used for macro-void analysis.

Results and Discussion

1. Height uniformity control

Control of height uniformity across an electroplated die and across a 300mm patterned wafer is critical to achieving successful interconnection. Height non-uniformity is generally reported as WID ("Within Die"), which is a measure of the range of heights across a die divided by twice the average height of the bumps in the die. WIW ("within wafer") is a measure of the range of heights across a wafer, divided by twice the average bump height across the wafer. Having tight control of WID and WIW enables facile alignment and bonding, minimizing I/O faults.

During the process of electroplating patterned wafers, SnAg is deposited in vias through patterned photoresist. During this process, current distribution varies significantly both across single die and across the full 300mm wafer. As potential is applied to the wafer during electroplating, the generated electric field is distributed non-uniformly across the die, significantly affected by the varying regions of conductive (exposed metal seed layer at base of the patterned vias) and nonconductive (un-patterned photoresist regions) areas. Larger variations in current cause more significant height variation (larger WID and WIW), as the local plating rate of each SnAg bump is directly related to its local current density during plating [1,6].

Several factors affect the magnitude of this effect. Current density has a significant effect by increasing the magnitude of the variation, and therefore higher plating rates, which also enable higher throughput for plating, present more significant challenge for WID/WIW control. Wafer and die design also have significant effects on the magnitude of current uniformity, as this uniformity is highly sensitive to variations in density. As such, fine pitch die and varying pitch die, present the largest challenge in WID/WIW control, with bumps at the edge of dense die or in more isolated regions generally depositing at higher relative rates. Industry trends towards high performance devices require higher I/O densities and demands to improve process throughput requires faster electroplating rates. The fundamental challenge of WID and WIW height control has increased significantly, and this has also been coupled to tighter specifications on WID/WIW, particularly for micro-bumping, to increase yield. Poor control of height distribution can lead to a number of defects, including bump bridging (Figure 1) and incomplete overlap during bonding.

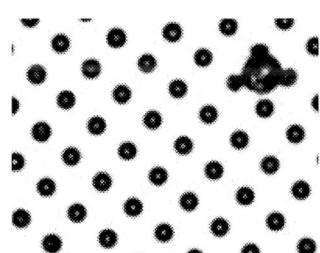

Figure 1. Top down optical image of post-reflow SnAg bump bridging due to within die height variation

In this work, control of WID and WIW height uniformity for challenging die at high plating rates was observed by the selection of a solder electroplating additive formulated to counteract the effect of the current non-uniformity. The WID performance using a suppressing additive (S1) with weak electrochemical polarization strength was compared to the WID performance using a suppressive additive with strong electrochemical polarization strength (S2). The significantly larger polarization effects of S2 in electrochemical testing correlated to a significant improvement in WID compared to the performance when using S1. S2 was found to suppress

SnAg plating by shifting the plating potential and dampening current fluctuations, while suppressing plating over a wide range of current densities. By applying the additive S2 during electroplating, SnAg bumps with tight WID uniformity control were observed and a significant improvement in process capability was achieved in 300mm wafer level plating (Figure 2).

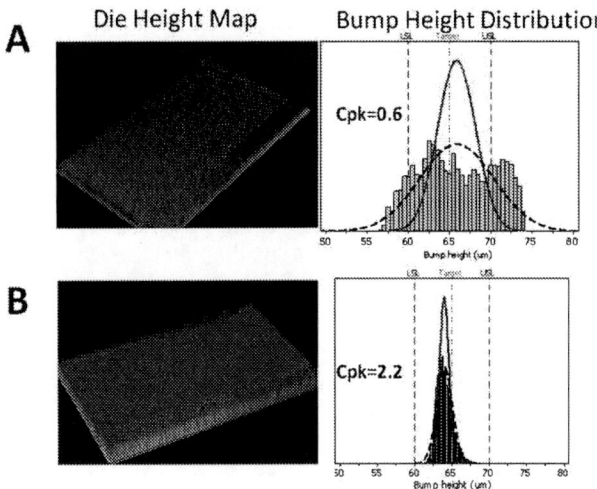

Figure 2. Die height maps (left) and bump height distribution and capability summaries (right) for SnAg bumps plated with the S1 (A) or the S2 (B) additive.

Another important characteristic of solder electroplating additives is their ability to control the solution wetting properties of the patterned photoresist vias during plating. In this study, two wetting additives were evaluated: additive W1 provided relatively weak solution wetting properties, while additive W2 enabled facile photoresist wetting and solution transport. For wafers plated with additive W1, small bumps were observed, indicative of slower nucleation and plating in select vias due to insufficient wetting properties of the SnAg chemistry (Figure 3A). In contrast, wafers plated with additive W2 showed defect-free performance, with no small bumps detected (Figure 3B). Control of small-bump defectivity is an important factor in increasing yield, as they can significantly affect the range of height and lead to high non-uniformity and low yield. Wetting additive design must also be paired with robust process control, both with plating tool settings and process parameters, to promote solution wetting and uniform, concurrent electroplating initiation for all vias across a wafer.

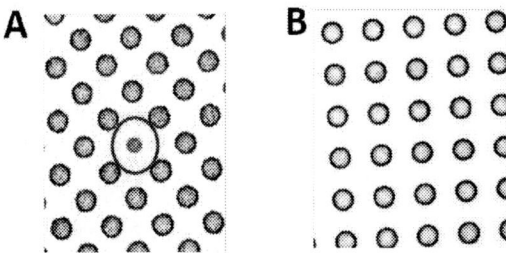

Figure 3: Top-down optical microscope image of as-plated SnAg bumps, electroplated with wetting additive W1 (left), showing the small-bump defect or W2 (right), showing defect-free performance.

2. Surface morphology control

In electrolytic SnAg baths, grain structure is controlled by additives designed to suppress plating locally, at the growing grain boundaries of the tin, which adjust microstructure during plating. Without the addition of grain refining additives, the grain structure of electroplated SnAg is rough and dendritic, and highly prone to defects during reflow. Strong grain refinement is particularly important for SnAg micro-bumps and Cu Pillar capping applications and conventional grain refining additives used in SnAg electroplating are often not capable of delivering smooth morphology for these challenging applications, where even low roughness values represent a significant proportion of the total surface area and height of a micro-bump. For these applications, roughness also significantly affects WID uniformity, as variations in plating height caused by inconsistent morphology have a large impact relative to the low target plating heights.

In this study, two grain refining SnAg electroplating additives were compared for performance in SnAg C4 bumping and micro-bumping: R1, an additive with low surface activity and weak surface suppression strength, is compared to R2, a strongly surface-suppressive grain refiner. With the R1 grain refiner, rough morphology is observed for C4 plating, though with consistent roughness observed throughout the bump. Extending this formulation to the micro-bumping application, however, a significant increase in roughness relative to solder volume was observed, and inconsistencies in overall surface structure are also observed (Figure 4A). In contrast, with the R2 grain refining additive, smooth as-plated morphology is observed for C4 bumping, and upon extension to micro-bumping applications, consistently smooth morphology is also observed (Figure 4B). By selecting for smooth as-plated morphology, versatility across a wide range of applications was achieved, with increased process control.

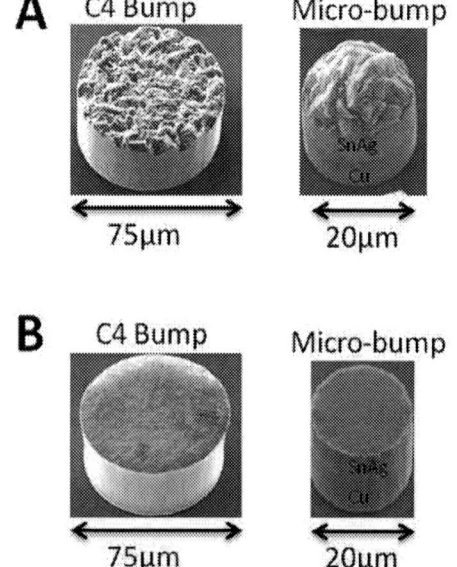

Figure 4. As-plated SEM of SnAg C4 bump (left) and micro-bump (right) electroplated with grain refining additive R1 (A) or R2 (B).

After the deposition process, SnAg bumps, micro-bumps, and SnAg capped Cu Pillars are generally reflowed to generate smooth hemispherical bumps and caps for bonding. A wide range of defects can manifest during the reflow process if the as-plated surface morphology of the SnAg is rough or otherwise defected. Rough or defected surface morphology allows for the entrapment of organic and particle contaminants from the plating bath, from the photoresist removal process, and from the reflow process. These impurities can have an adverse effect on the deposit quality and can result in voiding in the solder during reflow. Rough or defected surface morphology also presents a challenge to the reflow process, and can lead to molten solder sputtering around the reflowed bumps, observed as satellite solder spheres, caused by inconsistent melting and control of molten solder surface tension (Figure 5)

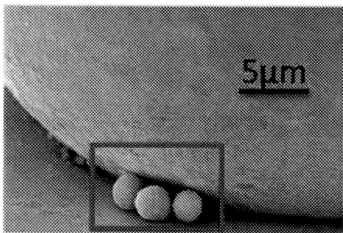

Figure 5. Satellite solder spheres observed at the base of a SnAg bump post reflow.

3. Ag% composition control

Control of alloy composition is another important factor in SnAg electroplating. Generally Ag% targets within the 1-2.5% range are selected for C4 bumping and micro-bumping, depending on application. Once a desired target Ag% is selected, maintaining control of the Ag% over the wafer, within the die, and within the bump is vital, as Ag% in a SnAg alloy has a significant effect on both melting point and mechanical properties. Inconsistencies in Ag% within a wafer or within a die can result in regions of high local Ag% composition, which can result in a significant defect during reflow: Ag_3Sn plate formation. Ag_3Sn intermetallic compounds (IMCs) are generally formed during the reflow process, and under optimized process conditions, these IMCs are distributed uniformly throughout the solder, providing mechanical strength to the deposit. At high levels of Ag% composition within a SnAg bump, Ag_3Sn IMCs can form in large plates during the reflow process [7]. This IMC formulation is irreversible and the plates increase in size and can protrude from the solder bump over multiple reflow and over the thermal lifetime of the bump. Ag_3Sn can be visualized both optical microscope, when they extend from the reflowed solder bumps, but can also be visualized by X-Ray inspection, as being present both within the solder bumps and also at the surface (Figure 6).

978-1-4799-2408-0/14 $31.00 © 2014 IEEE   344

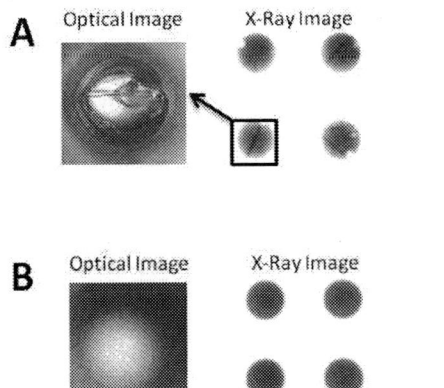

Figure 6. Ag$_3$Sn Plates visualized post reflow by optical microscope (A, left) and by X-ray inspection (A, right), compared to smooth, defect-free post reflow morphology by optical microscope (B, left) and X-ray inspection (B, right).

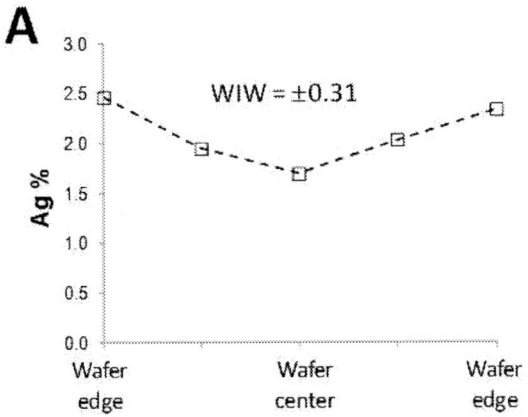

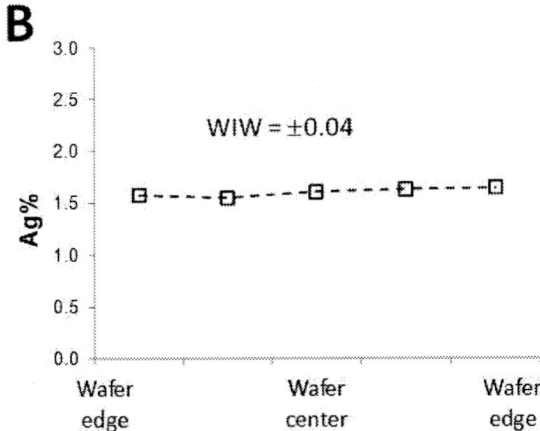

Figure 7. WIW Ag% distribution summary for 300mm wafers plated with SnAg using additive C1 (A) or C2 (B).

Ag$_3$Sn plates can lead to bump bridging, as they extend out from the reflowed bump towards the next neighbor, growing to relieve deposit stresses. As they grow, these significant stresses forming in the bump lead to microstructural defects in the solder, which result in macro-voids that can be visualized at the nucleation points of the plates (Figure 6A, right). In addition to bridging, which leads

to electrical faults, Ag$_3$Sn plates can also significantly hinder bonding and lower yield. The risk of bridging and the yield cost of Ag$_3$Sn plates is most significant in fine pitch and micro-bump arrays, where high I/O density is critical to achieving performance targets, and where next-neighbor distance is minimal. As such, mitigating the risk of Ag$_3$Sn plates, by fine control of Ag% distribution has increased importance in emerging high performance and micro-bumping applications.

The average Ag% over a die or over a wafer can generally be controlled by plating tool parameters, plating rate, and the concentration of Ag$^+$ in a SnAg plating bath. Ag% uniformity, however, defined as the variation of Ag% over a bump, die, or wafer, is strongly affected by the Ag$^+$ complexing electroplating additive used. In this study, two electroplating complexing additives were compared for Ag% distribution control across a 300mm wafer. By changing only the complexing additive, a significant change in Ag% distribution is observed. The wafer plated in a bath containing C1 showed a Ag% variation within the wafer (WIW Ag%) of ±0.31% (Figure 7A). In contrast, the wafer plated with the C2 complexer showed a very tight WIW Ag% distribution, of ±0.04% (Figure 7B).

4. Solder macro and micro-voiding control

SnAg solder voids are a significant concern for device reliability and yield and affect many aspects of performance. Solder voids affect the mechanical integrity of the joint, with macro-voids in the bulk solder leading to brittleness of the solder and micro-voids formed at metallic interfaces leading to cracking and crack propagation within IMC layers [8]. Solder voids can also cause poor WID and WIW height uniformity post-reflow, as reflow voids can significantly and non-uniformly increase the diameter of the reflowed bumps. Finally, solder voids negatively impact yield for bonding, preventing the needed overlap volume of metallic layers to form a reliable metallurgical bond.

There are a wide range of mechanisms that lead to solder voiding, many linked to other defectivity discussed in this paper, such as Ag$_3$Sn formation and rough or defected as-plated surface morphology. Deposit quality, linked to microstructure and deposit purity, also significantly affect solder voiding. In this paper, macro-voids are characterized by their visibility by X-Ray inspection, and generally have a diameter >2µm. In contrast, micro-voids are visible most readily be FIB-SEM and are generally observed at the interfaces between metal layers, particularly in and directly in contact with growing IMC layers.

In next-gen SnAg applications, such as micro-bumping, control of macro and micro voiding is particularly important. For macro-voiding, the low solder volume in these applications reduces the ability to compensate for voided regions, as even a relatively small 2µm diameter macro-void in the bulk solder represents a significant fraction of the total solder volume. Micro-voids also present particular concern, as micro-bumping generally involves small diameter Cu Pillars capped with SnAg with or without thin barrier layers, such as Ni. The formation of IMCs is much faster at a Cu/SnAg interface than for a Ni/SnAg interface and the risk of micro-voiding increases as a consequence. IMCs become a

978-1-4799-2408-0/14 $31.00 © 2014 IEEE   345

significant fraction of a micro-bump volume post reflow, and the integrity of this IMC layer is directly connected to the reliability of the bump and to yield.

Macro-voids depend heavily on the SnAg deposit quality and performance. Micro-voids also significantly depend on the quality of the under-bump metallurgy. A common mechanism of micro-void formation relates to the diffusion of one metal into another, leading to vacancies that manifest as voids. For example, Kirkendall voids can form between Cu/SnAg layers in a Cu Pillar capped with a SnAg cap, due to Cu diffusion into the solder and formation of IMC layers [9]. The frequency of these voids and the magnitude of the voiding phenomenon are highly dependent on the quality of the Cu Pillar deposit, and the compatibility of the Cu Pillar metallurgy with the SnAg cap. High levels of impurities incorporated into the Cu Pillar can significantly impact the diffusion of the Cu into the solder, which in turn impacts voiding. This is particularly critical for micro-bumping applications, where IMC plays a particularly significant role, as it makes up a significant proportion of the micro-bump volume and impacts overall mechanical properties.

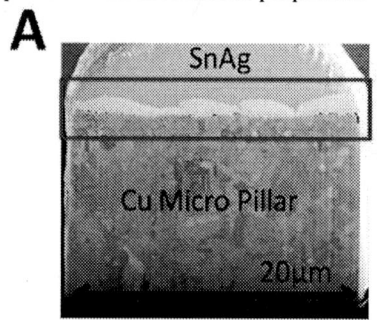

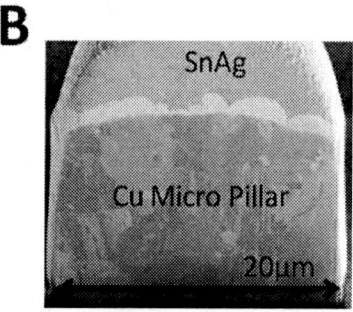

Figure 8. Micro-bump FIB-SEM after reflow. Cu micro pillar electroplated with additive package A1 (A) shows significant micro-voiding at the Cu/Sn interface and IMC layer, while Cu micro pillars electroplated with additive package A2 (B) are micro-void free in the micro-bumping application with no Ni barrier post reflow.

The impurity level and deposit microstructure in an electroplated Cu Pillar deposit is largely controlled by the additives used during the electroplating process. In this study, two Cu Pillar additive packages were compared for solder compatibility in a micro-bumping application. After sequential Cu Pillar and SnAg plating and a single reflow cycle, FIB-SEM was performed to visualize the Cu/Sn interface. Cu micro-pillars electroplated with additive package A1 showed significant micro-voiding in the Cu/Sn IMC layer

after a single reflow cycle. In contrast, Cu micro-pillars electroplated with additive package A2 showed micro-void free performance after reflow in the micro-bumping application (Figure 8). The impact of UBM quality on micro-voiding and interfacial compatibility is significant, and selection of compatible Cu and SnAg electroplating chemistries is critical to achieving reliable micro-bumps.

Achieving a void free Cu Pillar bump or micro-bump after reflow is critical, but maintaining void-free performance and mechanical integrity over repeated thermal stresses is also needed, to ensure device reliability. Cu Pillars electroplated with optimized plating additives, as described above, were thermally stressed both by thermal cycling testing and by multiple (10X) reflow cycling to evaluate Cu/SnAg compatibility and interfacial integrity and micro-voiding over significant thermal stresses. Micro-void free performance was maintained after 1000 thermal cycles and also after 10X reflow, despite the significant increase in IMC thickness during the testing [10].

In addition to UBM quality, micro and macro-voiding can also be observed due to poor solder deposit quality. In this study, SnAg bumps were electroplated on the same UBM, comparing two different solder electroplating additive packages, formulation 1 (F1) and formulation 2 (F2). Upon reflow, macro-voids were observed by FIB-SEM and by X-Ray inspection for SnAg bumps plated with additive formulation 1, but not for additive formulation 2 (Figure 9). Deposits formed with additive formulation 1 were also observed to have significantly more total incorporated impurity level than those formed with additive formulation 2. As was previously observed for Cu Pillar quality, impurities incorporated into the electroplated metal deposit and as-plated deposit quality play critical roles in void formation upon reflow.

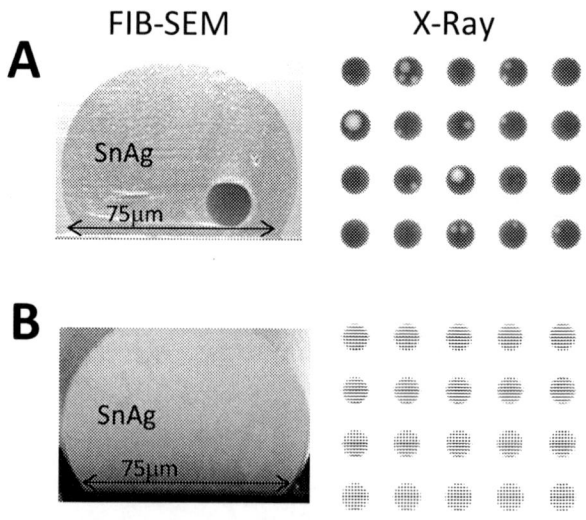

Figure 9. SnAg bump FIB-SEM (left) and X-Ray inspection (right) after reflow comparing performance of additive formulation F1 (A), where macro voids are observed, to additive formulation F2 (B), with void-free performance.

978-1-4799-2408-0/14 $31.00 © 2014 IEEE

Conclusions

Trends towards higher performance semiconductor devices with shrinking form factor have led to the emergence of a wide range of advanced packaging applications, including flip chip, WLP, and 3D-TSV, which are coupled to the replacement of traditional C4 bumps with micro-bumps. These applications present new challenges to the SnAg electroplating process and necessitate the design of additives capable of minimizing defectivity, maximizing performance, and increasing throughput.

A suite of additives is necessary to meet this challenge, each designed with individual performance, including Ag^+ complexation, electrochemical suppression, wetting properties, and grain refinement. These additives must also work in concert to prevent void formation and enable compatibility with under bump metallurgy. When combined into a single formulation, as for the SOLDERON™ BP TS6000 SnAg electroplating product, robust performance over a wide range of applications, include C4 bumping in both in-via and mushroom plating applications, Cu Pillar capping and micro-bumping can be achieved over a wide range of plating rates (Figure 10).

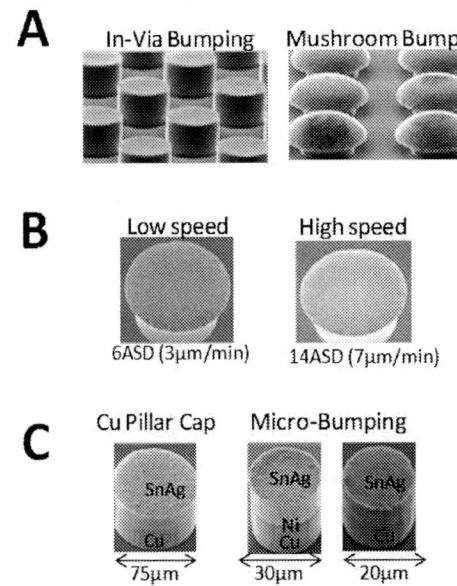

Figure 10. SEM comparing as-plated performance of SOLDERON BP TS6000 SnAg deposits for C4 applications (A), at low and high plating rates (B), and for Cu Pillar capping and micro-bumping applications, with and without Ni barrier layer (C).

Acknowledgments

The authors like to acknowledge Jui-Ching Lin for FIB-SEM data collection and analysis in support of this work.

™ Trademark of The Dow Chemical Company ("Dow") or an affiliated company of Dow.

References

1. M. Schlesinger, M. Paunovic, *Modern Electroplating*, John Wiley & Sons, Inc., Hoboken, NJ, 2010, pp. 139-204.
2. R. Asgari, "Semiconductor backend flip chip processing, inspection requirements and challenges" *IEEE/CPMT/SEMI International Electronics Manufacturing Technology Symposium*, San Jose, CA, July 17-18, 2002, pp. 18-22.
3. J. Dong, Y. Luo, D. Praseuth, S. Christian, J. Calvert, "Lead-free Solder Bump Electroplating Chemistry for Wafer-level Packaging," *14th Annual Green Chemistry & Engineering Conference*, Washington D.C., 2010.
4. J. Calvert, "Enabling Materials Technology for 3D-TSV Packaging," *SEMI Technology Symposium, SEMICON Korea*, Seoul, Korea, February 12-13, 2014.
5. P. Garrou, "The Future of Packaging: A Look from 50,000 Feet," *Advanced Packaging Magazine*, May 25, 2013.
6. J. Prange, J. Woertink, Y. Qin, P. Lopez-Montesinos, I. Lee, Y-H Lee, M. Imanari, J. Dong, and J. Calvert, "Next-Generation Lead-Free Solder Plating Products for High Speed Bumping, Capping and Micro-Capping Applications", *IMAPS Conference Proceedings*, Orlando, FL, September 30-October 3, 2013.
7. D. Frear, J. Jang, J. Lin, C. Zhang, "Pb-Free Solders for Flip-Chip Interconnects," *Journal of Materials*, vol. 53. no. 6, pp. 28-32, 2001.
8. H. Lee, M. Chen, H. Jao, T. Liao, "Influence of interfacial Intermetallic compound on fracture behavior of solder joints," *Materials Science and Engineering A*, vol. 358, 134-141, 2003.
9. K. N. Subramanian, *Lead-free Solders: Materials Reliability for Electronics*, John Wiley and Sons Ltd, West Sussex, United Kingdom, 2013.
10. Y. Qin, J. Calvert, J-C Lin, J. Dong, M. Imanari, I. Lee, P. Lopez Montesinos, J. Prange, E. Reddington, W. Tachikawa, and J. Woertink, *"Reliable Interconnection with Electroplated Cu Pillars and SnAg Solder Caps"*, IWLPC Proceedings, San Jose, CA, November 6, 2013.

Development of New 2.5D Package with Novel Integrated Organic Interposer Substrate with Ultra-fine Wiring and High Density Bumps

Kiyoshi Oi*, Satoshi Otake*, Noriyoshi Shimizu*, Shoji Watanabe*, Yuji Kunimoto*, Takashi Kurihara*, Toshinori Koyama*, Masato Tanaka*, Lavanya Aryasomayajula** and Zafer Kutlu***

*SHINKO ELECTRIC INDUSTRIES CO., LTD. RESEARCH and DEVELOPMENT DIV.
36, Kita Owaribe Nagano-shi, 381-0014, Japan
Email: kiyoshi_ooi@shinko.co.jp
** GLOBALFOUNDRIES, Wilschdorfer Landstrasse 101, 01109 Dresden, Germany
*** GLOBALFOUNDRIES, 2600 Great America Way Santa Clara, CA 95054 USA

Abstract

2.5D packaging technology utilizing silicon interposers is being developed and used for high-performance applications as the demand for miniaturization and higher density continues to increase. Silicon interposers enable very high density interconnects using standard semiconductor fabrication process technology, but are challenged as size increases. An alternative solution, high density laminate interposer, can offer advantages in cost, form factor, and infrastructure over other interposer options available today. However, it is difficult to manufacture high density wiring and flip chip bump pads on laminate interposer because conventional build-up technology limits fine line and via diameter. Our solution is the combination of an integrated organic interposer substrate with high density interconnects and thermo-compression flip chip bonding. Our solution eliminates the backside integration process for silicon interposer and assembly of interposers onto substrates. Conventional assembly processes can be utilized for assembling dies onto the integrated organic interposer substrate. Feasibility of this 2.5D package has been demonstrated by assembling dual-die with 40um pitch copper pillar bumps onto this novel integrated organic interposer substrate with 2μm line and space.

Introduction

Recent advancement in mobile applications such as smart phones and tablets are progressing rapidly. The demand on miniaturization of form factor and high performance are becoming much more important factors. Die partitioning enables miniaturization of small form factor and high performance without the yield issues of other methods. The focus of this paper is die partitioning technology conducted by fine pitch flip chip interconnect and die to die connection through fine lines.

The materials under consideration for such an interposer are silicon, glass, and organic dielectrics. Silicon is a leading candidate for 2.5D packaging [1]-[3], but is challenged as size increases. Glass offers the promise of performance at lower cost than silicon, but there is no manufacturing infrastructure. Organic materials used in substrates have a prevalent supply chain, but cannot meet 2.5D line and space requirements that are finer than conventional laminate.

By processing a fine-metal layer interposer directly on top of a conventional organic substrate, we have designed an integrated organic substrate and organic interposer. Our development progress has confirmed some basic properties [4], though we have not yet fully evaluated the die assembly.

Key Points:
- High accuracy alignment of die to interposer.
- CTE mismatch and warpage control during soldering.
- Elimination of underfill void.

We have also measured thermal warpage of the organic interposer with die. We will report on the feasibility of our design for 2.5D packaging focusing on the assembly process.

The Concept of Organic Interposer " i-THOP"
(Integrated Thin film High density Organic Package)

Figure 1 shows a schematic cross-sectional image of i-THOP with die. Dual-die were assembled side by side and fine copper traces were used for die to die connection. A schematic image of die to die connections through fine line and fine pitched bump is shown in Figure 2. The pattern consisted of a thin film layer which is built on a conventional laminate substrate.

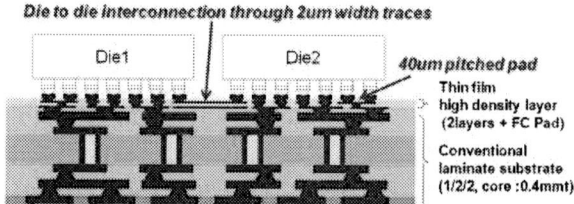

Figure 1. Schematic cross sectional image of i-THOP

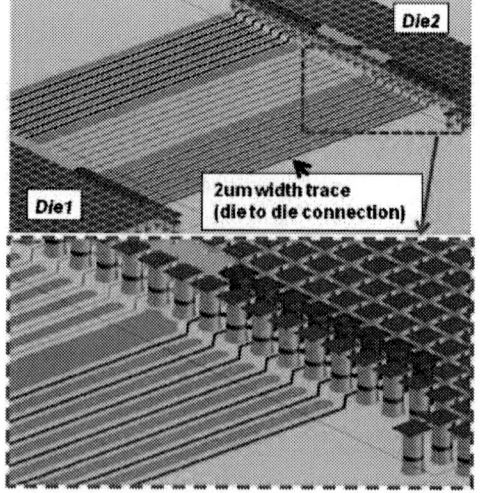

Figure 2. Die to die connection through 2μm width copper trace.

Design specifications of i-THOP used in this study are shown in Table 1. Two dies were placed side by side with a 1mm gap. Die to die connections were achieved through 2μm width (minimum) line. The insulation resin between fine layers was formed with 3μm thickness, and vias were opened with 10μm diameter. 40μm pitch bonding pads for flip chip were formed with 25μm diameter by electrolytic copper plating. Cu-OSP (Organic Solderability Preservation) and thin ENEPIG (Electro-less Ni/Pd/Au plating) were selected as two surface finishes in this study.

Table 1. Design specifications of i-THOP

Package size	21 x 15 x t0.66 mm
Structure	3 (thin film high density layer) +1/2/2(conventional substrate)
Minimum line & space	2μm / 2μm
Die layout	Dual-die / package
FC pad counts / die	18,272
FC pad pitch	40μm Pad pitch
FC pad diameter	25μm
Surface finish	Cu-OSP or thin ENEPIG (E'less Ni/Pd/Au)

Fabrication of Test Vehicle
Fabrication of i-THOP

Figure 3 shows schematic images of fabrication process of i-THOP. The fabrication process was separated into four parts, which are described as follows:

1) Preparation for conventional laminate substrate:
   Core substrate and laminated layers were fabricated using established build up technologies.

2) Preparation for thin film layer:
   In order to prepare the thin film layer, insulation resin of the conventional substrate was smoothed by CMP (Chemical Mechanical Polishing) process. Surface roughness of insulation resin was approximately Ra 20nm after CMP treatment.

3) Thin film layer build up:
   Initially, insulation resin for thin film layer was formed with spin coating process. Micro vias were opened by photolithography. Ti/Cu was sputtered to form the seed layer. Then photo resist was formed and patterned using stepper. The copper trace was formed by electrolytic plating. Finally, the seed layer was removed.

4) Surface finish:
   Cu-OSP or thin ENEPIG were deposited on copper pads.

Figure 4 shows the outward appearance of i-THOP, which utilizes 40μm pitch bumps and 2μm width copper traces. The SEM (Scanning Electron Microscopy) images shown in Figure 5 highlight the thin film layers that were added on a conventional build up substrate. Vias with 10μm diameter, bonding pad with 25μm diameter, and fine copper traces with 2/2μm line/space were constructed.

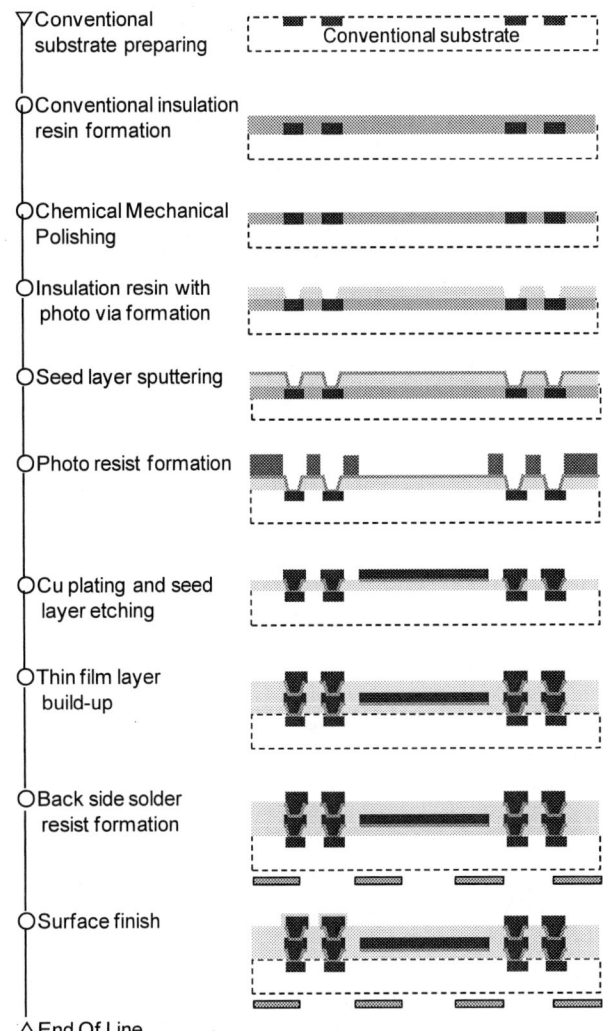

Figure 3. Schematic images of i-THOP process.

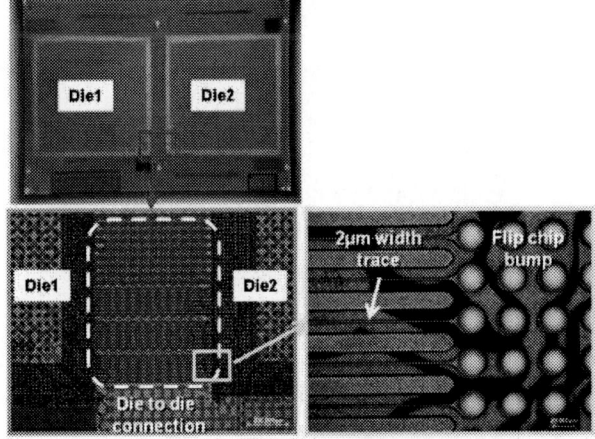

Figure 4. Die patterns on i-THOP

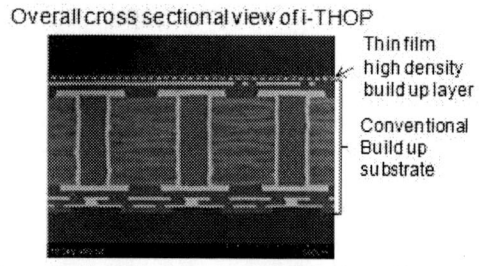

Overall cross sectional view of i-THOP

Thin film high density build up layer

Conventional Build up substrate

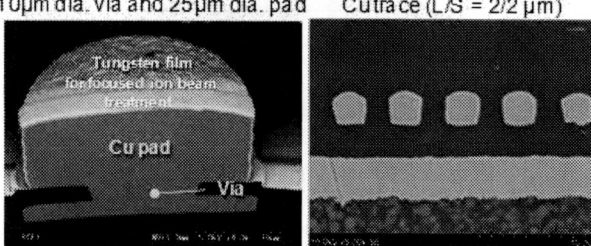

10μm dia. via and 25μm dia. pad   Cut trace (L/S = 2/2 μm)

Tungsten film ion focused ion beam treatment

Cu pad

Via

Figure 5. SEM image of i-THOP cross section

Figure 6A is an SEM image of the 40μm flip chip bonding pads. In Figure 6B, both Cu-OSP and ENEPIG surface finishes can be seen to be well-deposited on the copper pads. The XPS (X-ray Photoelectron Spectroscopy) measurements in Figure 7 show the plating thickness of the nickel, palladium, and gold layers for the ENEPIG surface finish.

Fabrication of Test Die

The structure of the test die is shown in Figure 8. The bumps are copper pillars with a Ni barrier and SnAg solder caps. The die has a daisy chain pattern formed in an aluminum routing layer.

A) 40μm pitch flip chip bonding pads

B) Enlarged image

Cu-OSP   Thin ENEPIG

Ra:0.30μm   Ra:0.31μm

Figure 6. SEM images of flip chip bonding pad.

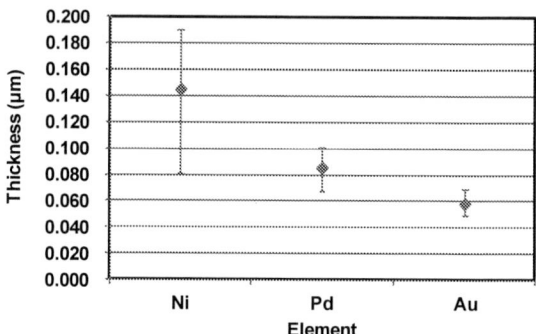

Figure 7. Thickness of thin ENEPIG plating.

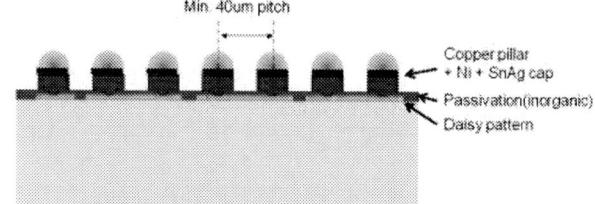

Min. 40um pitch

Copper pillar + Ni + SnAg cap

Passivation(inorganic)

Daisy pattern

Figure 8. Schematic image of the test die.

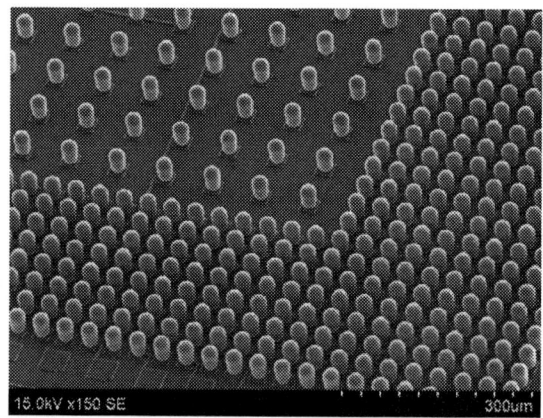

Figure 9. SEM image of test die

Flip Chip Processing

Thermo-compression bonding was used for the flip chip attachment, which included two different process flow options, shown in Figure 10. One was pre-applied underfill process using NCP (Non-Conductive Paste), and the other was a post-applied CUF (Capillary Underfill) process. The NCP contained a flux agent that removed oxidation from the metals during thermo-compression bonding.

1) Pre-applied underfill process:

Figure 11 shows cross sectional images of solder joints formed with different bonding forces. When bonding force was low at 0.55gf/bump, the solder at the top of copper pillar did not make sufficient connection to the pad. When the bonding force was high at 1.09gf/bump, undulation of the copper pad and copper trace under a pad caused via cracks. Taking into account the results, we considered that it was difficult to have enough process margin for the connection. In addition, an optimization of the NCP material would be required.

978-1-4799-2408-0/14 $31.00 © 2014 IEEE   350

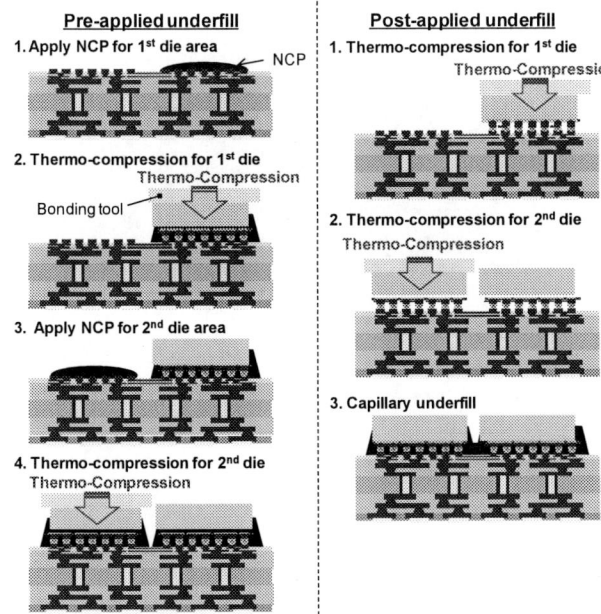

Figure 10. Evaluation process

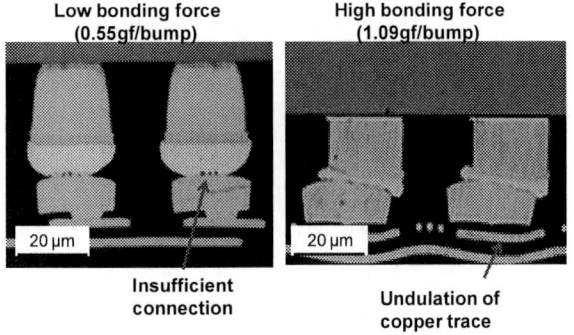

Figure 11. Cross sectional images of solder joint using the post-applied underfill process.

2) Post-applied underfill process:

In case of post-underfill processing for Cu-OSP surface finish, the oxide film of the copper pad must be removed during soldering. Initially, flux was applied on the copper pads. Then the substrate was heated on a vacuum stage. Almost all of the flux evaporated at this time. After that, dual-die were bonded one by one. Finally, capillary underfill was applied. In the case of thin ENEPIG surface finish, flux was not needed for bonding because of its good wettability.

Figure 12 shows cross sectional images of solder joints using the post-applied underfill process. A good solder joint was confirmed for both surface finishes. As a result, the best method of flip chip bonding for this structure was determined to be thermo-compression bonding with a post-applied underfill process.

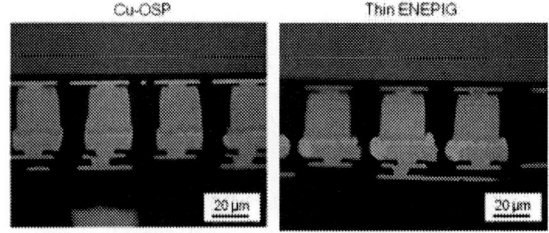

Figure 12. Cross sectional images of solder joint using post applied underfill process.

3) Process optimization for flip chip bonding:

Solder thickness between copper pillar and copper trace was found to be one of the most important factors to form a good solder joint. If the solder thickness was out of control, insufficient solder wetting or solder bridging could occur. Figure 13 shows the relationship between the post-bond solder thickness and bonding force. The solder thickness was found to decrease with increasing bonding force. When bonding force was higher than 0.4gf/bump, the solder thickness saturated around 3.0μm. There was no difference in the solder thickness between the two surface finishes tested.

Figure 14 shows the results of a solder bridging evaluation with different bonding forces. For bonding force higher than 0.56gf/bump, solder bridging was observed in X-ray inspection and cross sections. When bonding force was lower than 0.28gf/bump, no solder bridging was observed. Good solder joint was achieved with bonding force at 0.28gf/bump. Figure 15 shows the solder joints at various points around the perimeter of the two test die, which were bonded with optimized conditions. Good solder joints were confirmed at all areas of the substrate.

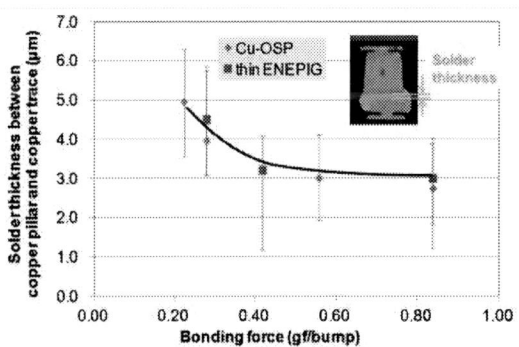

Figure 13. The Solder thickness distribution as a function of bonding force.

Figure 14. X-ray inspection and cross sectional images with different bonding forces.

Figure 16 shows the results of underfill void inspection by SAT (Scanning Acoustic Transmission) and parallel lapping. There were no underfill voids found for either surface finish.

Finally, Figure 17 shows an outward appearance of the completed flip chip assembly produced with the i-THOP concept.

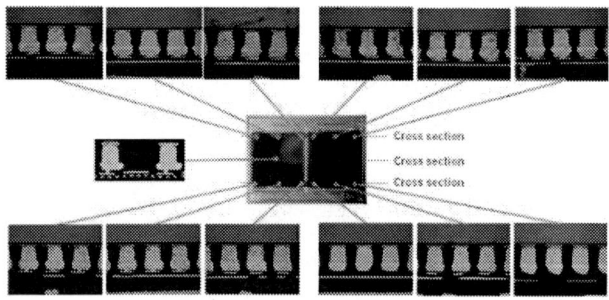

Figure 15. Solder joints across the test die area.

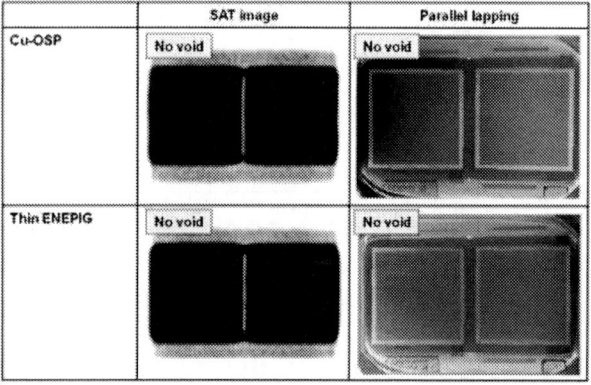

Figure 16. Inspection results of underfill void inspection by SAT and parallel lapping.

Figure 17. Attached dice on i-THOP

Thermal Warpage

Figure 18 indicates the results of thermal warpage characterization of assembled i-THOP measured by Shadow Moiré. Temperature was raised from room temperature to 260°C, and then cooled back down to room temperature. Warpage was measured in the 19.5 x 13.5mm area from the BGA side. At room temperature, warpage was observed to be concave in the areas corresponding to the two dies, and convex in the area between them. Figure 19 shows the co-planarity of i-THOP at various temperatures. Warpage at room temperature was found to be roughly 30μm. When the temperature was increased to 200°C, the concave warpage shape changed to become almost flat. The package shape became slightly convex at 260°C with warpage of roughly 16μm. When temperature decreased to room temperature, the package returned to its initial shape. The assembled i-THOP showed small thermal warpage throughout a reflow temperature range that is not likely to impact board level assembly.

Conclusions

In this study, flip chip assembly of i-THOP was optimized, and the characteristics of assembled organic interposer i-THOP were evaluated.

An i-THOP organic interposer was fabricated using thin film technologies on a standard core material, that contained 2μm minimum width lines for the die to die connections and 40μm pitch bump pads.

A comparison was made between thermo-compression flip chip bonding processes using pre-applied underfill and post-applied underfill. It was found that the post-applied underfill process was better suited for i-THOP since it enabled a low force bonding process with a higher process window.

Thermal warpage of an i-THOP assembly was characterized. Maximum warpage was found to be approximately 30μm at room temperature, and becoming nearly flat throughout a typical reflow temperature range with a maximum of 260°C temperature. Based on these results, it is not likely that board-level assembly will be impacted.

Through this study, the feasibility of 2.5D package using i-THOP as an alternative solution for silicon interposer was confirmed. It is believed that this technology will become a high impact 2.5D packaging solution for the industry.

Acknowledgments

We would like to thank Amkor Technology for Cu pillar fabrication and wafer dicing for the test die wafers.

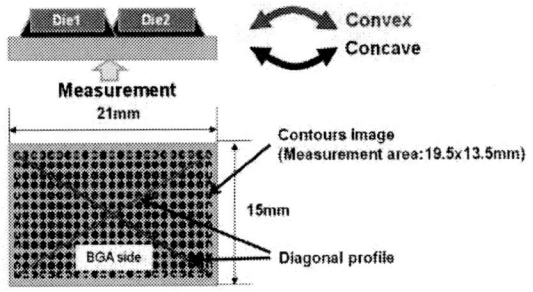

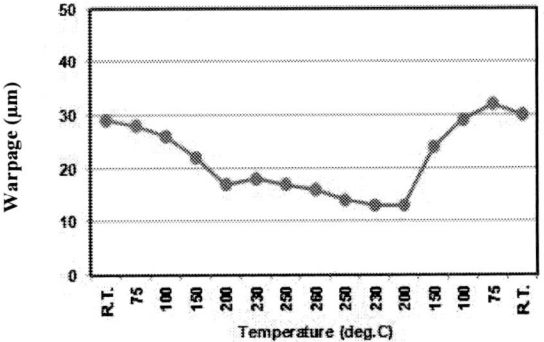

Figure 18. Color contours image and diagonal profile of thermal warpage measured by Shadow Moiré.

Figure 19. Warpage of i-THOP with dies.

References

1. Kei Murayama, Mitsuhiro Aizawa, Koji Hara, Masahiro Sunohara, Ken Miyairi, Kenichi Mori, Jean Charbonnier, Myriam Assous, Jean-Philippe Bally, Gilles Simon and Mitsutoshi Higashi, " Warpage Control of Silicon Interposer for 2.5D Package Application", in *Proc. IEEE Electronic Components and Technol. Conf. (ECTC)*, Las vegas, Nevada, May 28–31, 2013, pp. 879–884.

2. Bahareh Banijamali, Raghunandan Chaware, Suresh Ramalingam and Myongseob Kim, "Quality and Reliability of 3D TSV Interposer and Fine pitch Solder Micro-bump for 28nnm Technology", in *Proc. International Symposium on Microelectronics (IMAPS)*, Long Beach, California, Oct 9–13, 2011, pp. 189–192.

3. Masahiro Sunohara, Hedeaki Sakaguchi, Akihito Takano, Rie Arai, Kei Murayama and Mitsutoshi Higashi, "Studies on Electrical Performance and Thermal Stress of a Silicon Interposer with TSVs", in *Proc. IEEE Electronic Components and Technol. Conf. (ECTC)*, Las vegas, Nevada, Jun 1–4, 2010, pp.1088–1093.

4. Noriyoshi Shimizu, Wataru Kaneda, Hiromu Arisaka, Noyuki Koizumi, Satoshi Sunohara, Akio Rokugawa and Toshinori Koyama, "Development of Organic Multi Chip Package for High Performance Application", in *Proc. International Symposium on Microelectronics (IMAPS)*, Orland, Florida, Sep30–Oct3, 2013, pp. 414–419.

Package Embedded Decoupling Capacitor Impact on Core Power Delivery Network for ARM SoC Application

GaWon Kim, Max (Sunghwan) Min, Melinda (Ling) Yang, Anil Gundurao, Eileen You, Harpreet Gill,
Seungyong Cha*, Younghoon Kim*, Se-ho You*, Seungbae Lee** and Woonghwan Ryu***

Samsung Semiconductor Inc. System LSI SoC Bay Area R&D (SBR)
85 West Tasman Dr., San Jose, CA 95134, USA
*Samsung Electronics Corporation (SEC), Package Development Team, Semiconductor R&D Center
**Samsung Electronics Corporation (SEC), System LSI Division, Quality Team, EMI/EMS Lab
***Samsung Electronics Corporation (SEC), System LSI Division, Design Technology Team
gawon.kim@ssi.samsung.com

Abstract

In this paper, a BGA package having a ARM SoC chip is introduced, which has component-type embedded decoupling capacitors (decaps) for good power integrity performance of core power. To evaluate and confirm the impact of embedded decap on core PDN (power distribution network), two different packages were manufactured with and without the embedded decaps. The self impedances of system-level core PDN were simulated in frequency-domain and On-chip DvD (Dynamic Voltage Drop) simulations were performed in time-domain in order to verify the system-level impact of package embedded decap. There was clear improvement of system-level core PDN performance in middle frequency range when package embedded decaps were employed. In conclusion, the overall system-level core PDN for ARM SoC could meet the target impedance in frequency-domain as well as the target On-chip DvD level by having package embedded decaps.

I. Introduction

As ARM SoCs grow in performance and capability, core power is the most important issue and it has inherently big challenges for lower power consumption and high operating frequency. To ensure high frequency operating frequency of core logics, the lowest simulataneous switching noise (SSN) of core power delivery network (PDN) is essential for ARM SoC application.

Embedded decoupling capacitor (decap) in package substrate is an advanced approach for dramatically shorter interconnection with better power integrity of core PDN compared to the surface-mounted-typ (SMT) decoupling capacitors on top or bottom package layers that have much longer interconnection path, which makes it ineffective as shown in Fig. 1. There is the 3rd option of the On-PKG decap, land-side decap which is mounted on bottom side of package substrate. However, it is obvious the embedded decap is much more effective than the land-side decap.

Embedded decap technology itself has been studied for several years. There are two types of embedded decap technology in package or printed circuit board (PCB) substrate. One of them is a thin-film planar-type embedded decap ([1]-[3]) and the other one is a passive component-type embedded decap ([5]-[6]). Previous researches in [1]-[3] mainly focused on the thin-film embedded decap technology which is made by thin dielectric film with high dielectric

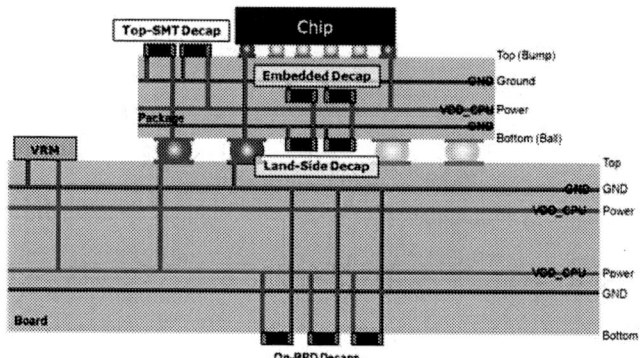

Fig. 1. On-PKG decoupling capacitor (decap) options: Top-SMT decap, Embedded decap, and Land-side decap

constant between power and ground. However, the thin-film embedded decap needs additional layers, special dielectric material, and brand new assembly process. On the other hand, the passive component-type embedded decap can be implemented by general capacitor component and normal substrate with extra hole and SMT process which means cost effective method for better electrical performance. This is the reason why the package or PCB designers are more interested on the component-type embedded decap in package or PCB substrates. The embedded decap in this paper means component-type.

In this paper, a BGA package design having ARM SoC chip will be introduced, which has component-type embedded decaps for good power integrity performance of core power. The core PDN in package substrate was designed first then it was optimized by sweeping the number of the embedded decaps to maximize the electrical performance and cost effectiveness. Then, the package substrates were manufactured with and without the embedded decaps to evaluate and confirm the embedded decap impact on core PDN.

II. Analysis of Core PDN (Power Delivery Network) with/without package embedded decap in frequency-domain

In actual SoC product development, package should be prepared before the silicon delivery. It means that package should be designed before SoC tape out or almost at the same time considering SI/PI simulation analysis of package performance, package substrate manufacturing. So, the

978-1-4799-2408-0/14 $31.00 © 2014 IEEE   354   2014 Electronic Components & Technology Conference

package PI engineers can get only preliminary estimation results of power consumption and On-chip decap amount in package design and SI/PI simulation stage. Also, usually exact board layout and On-board decap information are not available in package design stage. These are the difficulty of the best package design and package SI/PI optimization for the real SoC product development.

In the new design stage of a package for ARM SoC application, the embedded decap was adopted for core PDN because of the high power consumption for stable power integrity. The core PDN in package substrate was designed first then it was optimized by sweeping the number and density of the embedded decaps to maximize the electrical performance and cost effectiveness.

However, there are a few requirements to use package embedded decap in package substrate. In order to assemble component-type embedded decap in package substrate, a big aperture is needed. The aperture for package embedded decap should be located directly underneath of chip, which is the best position for shortest interconnection path and the most effective in power integrity view point. Also, to place that aperture, package design should consider the placement of the aperture, which means any vias and signal/power routing cannot be placed that aperture area. The last thing is the core thickness limitation of package substrate stack-up. For mass production yield and lower cost of package substrate, it is preferable to assemble package embedded decap in one core dielectric layer which is the most thick layer. In this case, the core dielectric thickness is dependent on the total thickness of package embedded decap after soldering.

Fig. 2 shows the self impedance graphs of core PDN in bare package and with several embedded decap options. The used embedded decaps decreased the self impedance of core package PDN in middle frequency range and they cannot affect the self impedance in high frequency range, over 1GHz. The final chosen embedded decap option is the circle-symbol red solid graph having 6ea 1uF package embedded decaps.

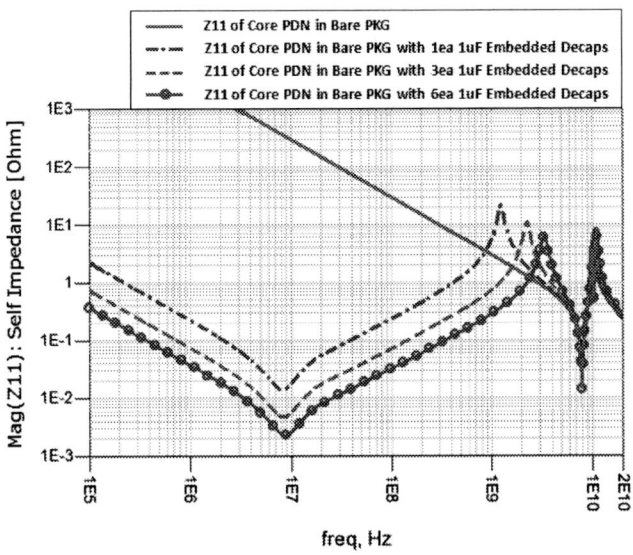

Fig. 2. Simulated self impedance graphs of core PDN in bare package and with several package embedded decap options

The amount of the package embedded decap is quite high (6uF) and the interconnection between the package embedded decaps and ARM SoC is extremely short, which means the package embedded decaps are integrated very effectively.

Embedded decap in the package substrate looks very effective for core PDN in the package. However, in the system point of view, we need to check the effectiveness of the package embedded decaps in the entire system-level core PDN including On-chip PDN, package PDN, and board PDN together.

Based on the preliminary power consumption of core power and the voltage noise spec, the target impedance can be calculated from simple equation (1) [4].

$$Z_{TARGET} = \frac{VDD \times ripple\ (\%)}{I_{average}}\ (\Omega) \qquad \text{eq.(1)}$$

However, the constant target impedance is too severe a design guideline, and there is a risk that this will cause overdesigned system-level core PDN, which is not cost effective. The exact frequency dependent target impedance is the best way but it is very difficult to get this information in the pre-layout stage which means before SoC tape out. Therefore, three-level target impedances ($Z_{TARGET1}$=10mΩ, $Z_{TARGET2}$=30mΩ, and $Z_{TARGET3}$=60mΩ) were setup for this core PDN of ARM SoC.

After SoC design stage, CPM (Chip Power Model) can be provided from the On-chip PI team. CPM model is generated based on the core operating mode and a specific mode was chosen as it is believed to be representative and close to worst case power scenario. Using this On-chip CPM model which is included On-chip PDN and On-chip decap information with power noise current profile, the system-level core PDN was simulated in frequency-domain. For better understanding, the separated self impedances (Z11 parameter) of core CPM (On-chip PDN) and bare package (meaning any On-PKG decaps) are shown in Fig. 3(a). The solid dark blue graph is the self impedance of core CPM itself and the dashed purple curve shows the self impedance of core bare package PDN. The dotted pink line represents the self impedance of merged core CPM and the bare package PDN which is same as pink solid graph in Fig. 3(b). As plotted in Fig. 3(a), the impedance resonant peak occurs at 95MHz due to the On-chip capacitance and the package inductance. Fig. 3(b) and 3(c) describe the self impedance of the system-level core PDN at bump side with CPM (On-chip PDN), package PDN with/without package embedded decap, and with/without board PDN. In both Fig. 3(b) and 3(c), the circle-symbol graphs are the case with package embedded decaps.

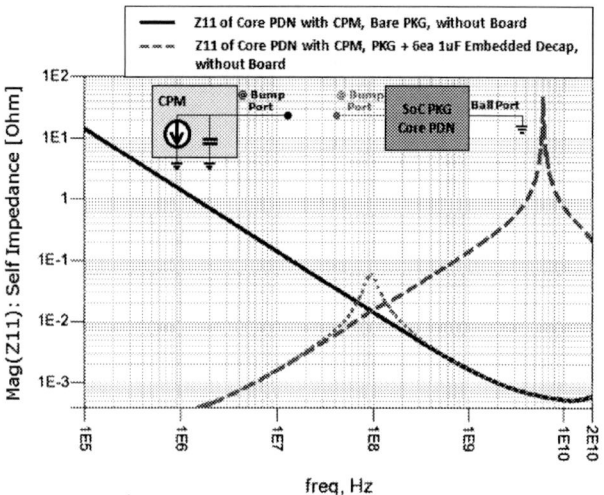

(a) Self impedance (Z11 parameter) of core CPM (On-chip PDN) and bare package, separately

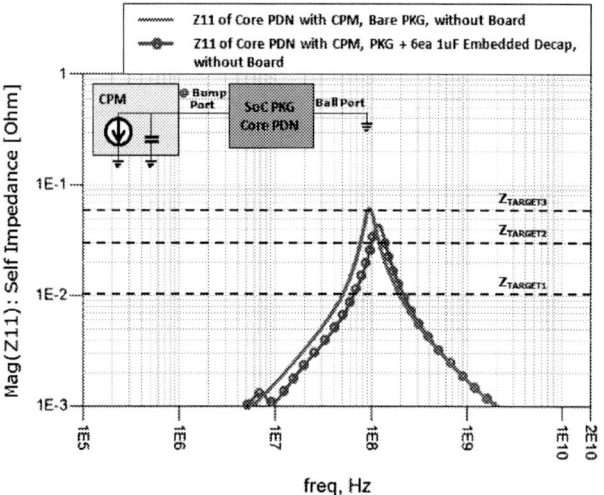

(b) Simulated self impedances of system-level core PDN at bump point with CPM (On-chip PDN), PKG with/without package embedded decap, and without Board

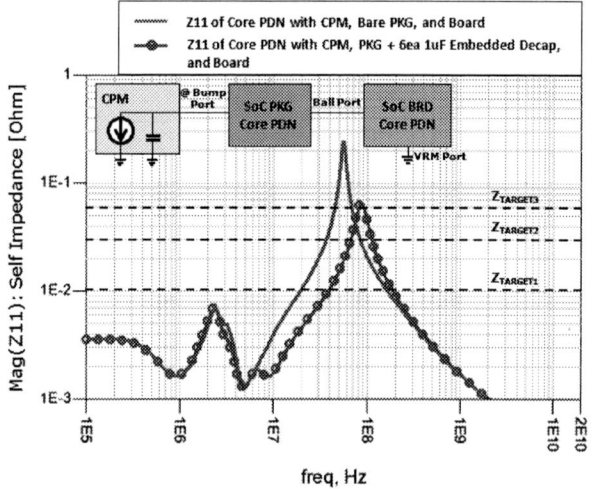

(c) Simulated self impedances of system-level core PDN at bump point with CPM (On-chip PDN), PKG with/without package embedded decap, and with Board PDN

Fig. 3. Simulated self impedances of system-level core PDN at bump side with CPM (On-chip PDN), package PDN with/without package embedded decap, and with/without Board PDN

Some PI engineers especially having On-chip background misunderstand often that board PDN doesn't affect on the time-domain CPM simulation result or On-chip DvD (Dynamic Voltage Drop) simulation result. So, they think the board PDN doesn't need to be included in those simulations. However, based on the analysis of this ARM SoC project, it is not always true. Depending which mode is adopted for CPM generation or On-chip DvD simulation, board core PDN could be one of important factors with package core PDN. To show those effects, the self impedance of CPM model, package PDN with/without package embedded decap, and without board PDN was plotted in Fig. 3(b).

As shown in Fig. 3, there is clear improvement of self impedance in system-level core PDN due to package embedded decap in middle frequency range, around 10MHz to 100MHz for both cases with/without board PDN. However, the impedance values and the frequency of resonant peaks are quite different in Fig. 3(b) and 3(c). As shown in Fig. 3(b),

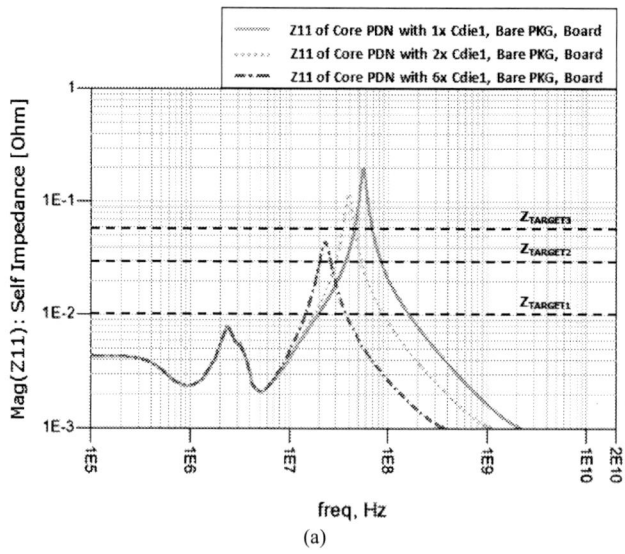

(a)

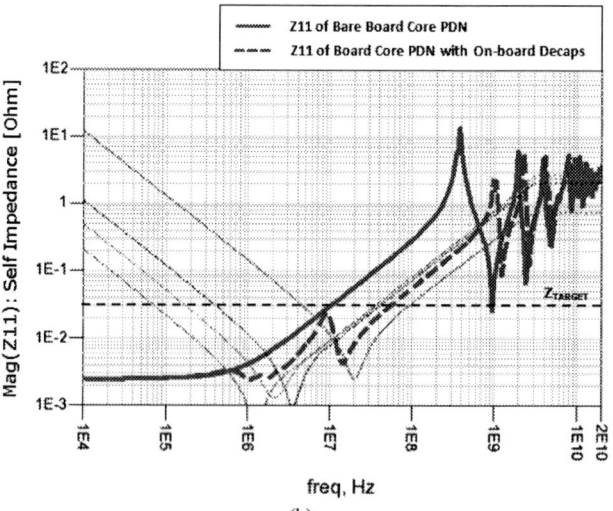

(b)

Fig. 4. (a) Simulated self impedances with On-chip decap variation at bump side in frequency-domain for On-chip decap assessment in SoC pre-layout stage (b) Simulated self impedances with On-board decap variation in frequency-domain for On-board decap assessment in SoC pre-layout stage

978-1-4799-2408-0/14 $31.00 © 2014 IEEE

the impedance resonant peak of the pink solid graph is at 95MHz with 61mΩ impedance while the impedance peak of red circle-symbol curve with 6ea 1uF package embedded decaps is at 117MHz with 44mΩ impedance. If board core PDN is not included as Fig. 3(b), the self impedance appears to almost to meet almost the highest target impedance ($Z_{TARGET3}$=60mΩ) at resonant peak (61mΩ) even without package embedded decaps. And, the impact of package embedded decap is not big, just 17mΩ decrease in Fig. 3(b). However, including board core PDN, there is no chance to meet the highest target impedance ($Z_{TARGET3}$=60mΩ) in the self impedance at resonant peak without package embedded decaps as depicted in Fig. 3(c). Including board core PDN, the impedance resonant peak of the orange solid graph is at 55MHz with 244mΩ impedance while the impedance peak of blue circle-symbol curve with 6ea 1uF package embedded decaps is at 85MHz with 66mΩ impedance. And, the package embedded decaps decrease more impedance value (178mΩ) than the case without board core PDN (17mΩ). Also, the impedance resonant frequencies with/without board core PDN is quite off (95MHz vs. 55MHz: 117MHz vs. 85MHz), which means the board core PDN should be included in order to capture a complete system response correctly.

In SoC pre-layout stage, these PI analysis results can provide the guideline for required On-chip decap amount and design guideline for board PDN and On-board decap amount and density. For example, On-chip decap assessment results and On-board decap assessment in frequency-domain are shown in Fig. 4(a) and Fig. 4(b), respectively. The dotted curves are the self impedances of the On-board decap candidates in Fig. 4(b).

III. Analysis of Core PDN (Power Delivery Network) with/without package embedded decap in time-domain

To analyze system-level core PDN in time-domain, On-chip Dynamic Voltage Drop (DvD) simulations were performed [7]. As mentioned in previous session, one of core operating modes was selected for this ARM SoC project to evaluate the system-level core PDN voltage noise. Since the effective On-chip decap amount changes depending on which mode the core operates in, the entire core PDN system characteristics is also dependent on this core operating mode. Also, as shown in Fig. 5, the current load frequency content of the chosen core operating mode is very important factor for On-chip DvD simulation results. Therefore, On-chip DvD simulation results in this paper are not absolutely true but they are conditionally true under the assumption of one core operating mode.

For correlation of the frequency-domain analysis for On-chip decap variation in Fig. 4(a), the On-chip DvD simulation results in time-domain are plotted in Fig. 6. As increase of On-chip decap amount, the peak-to-peak value of voltage noise decreases as well. Especially, the sharp drop in Fig. 6 clearly states the monotonic decrease due to on-chip decap amount increase. In Table1, the comparison of On-chip DvD simulation results is summarized as On-chip decap variation.

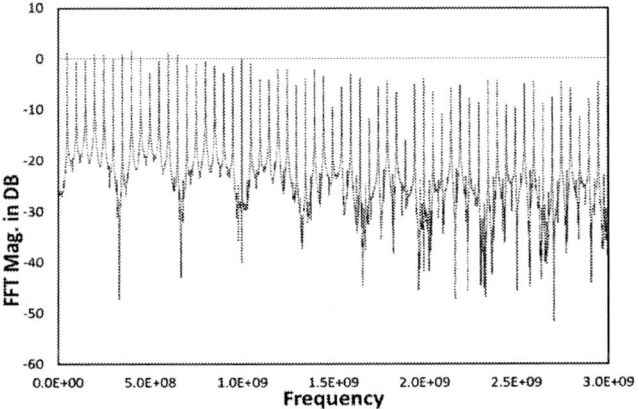

Fig. 5. Current Load Frequency Content of core operating mode for On-chip DvD simulation

Table1. Comparison of On-chip DvD simulation results as On-chip decap variation

On-chip decap Variation	Peak-to-Peak Voltage Droop
1x C_{die1}	100%
2x C_{die1}	86.60%
6x C_{die1}	84.80%

To verify the analysis and optimization of core PDN performance in frequency-domain, On-chip Dynamic Voltage Drop (DvD) simulations were performed with/without package embedded decap. Fig. 7(a) shows the comparison of On-chip DvD results with/without package embedded decap for without board core PDN case. The dotted graph shows the case without package embedded decap and the dash-dot curve represents the case with package embedded decap. As previously mentioned in Fig. 3(a), without board core PDN, the package embedded decap effect is very small and it is hard to tell which case is better. However, as described in Fig. 7(b), On-chip DvD voltage noise of core PDN with package embedded decap including board PDN improves significantly, and these results are consistent with the frequency-domain analysis in Fig. 3. Dashed graph shows the case without

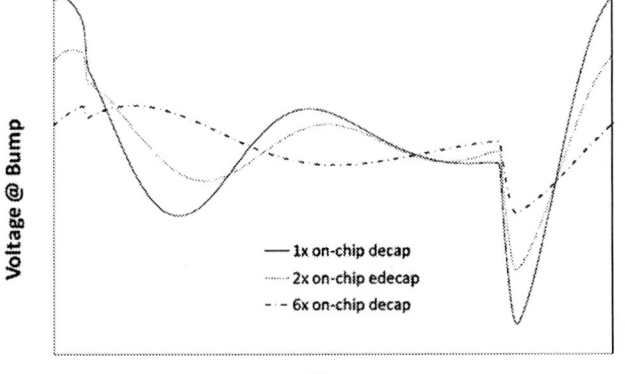

Fig. 6. Simulated On-chip DvD voltage noise at bump point with On-chip decap variation in time-domain for On-chip decap assessment in SoC pre-layout stage

978-1-4799-2408-0/14 $31.00 © 2014 IEEE

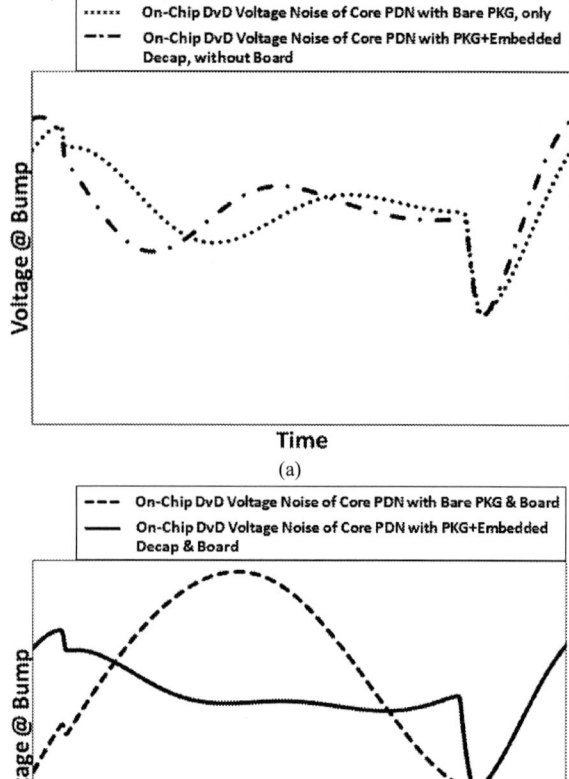

(a)

(b)

Fig. 7. On-chip Dynamic Voltage Droop (DvD) simulation results of core power at bump point (a) Voltage noise of core PDN with/without package embedded decap, without board (b) Voltage noise of core PDN including board PDN with/without package embedded decap

package embedded decap and the solid curve is with package embedded.

IV. Measurement and verification of package embedded decap performance in frequency-domain

To evaluate and verify the package embedded decap impact of core PDN in ARM SoC application, two different package substrates with and without 6ea 1uF package embedded decaps were manufactured and the self impedances of those bare package and SoC-assembled package with/without 6ea 1uF package embedded decap were measured in frequency-domain. The measured self impedances are compared with the simulation results from 2.5D commercial EM tool. Fig. 8 shows the measurement setup for Z-parameter in frequency-domain using VNA (Vector Network Analyzer, Agilent E5071C model) with probe station. The measured frequency was swept from 100 kHz to 8 GHz due to fine accuracy in middle frequency range to ensure the package embedded decap impact.

In Fig. 9(a), the measured and simulated self-impedances (Z11 parameters) of core power are described for bare package substrate with 6ea 1uF package embedded decaps, which means SoC is not assembled. The solid dark blue graph

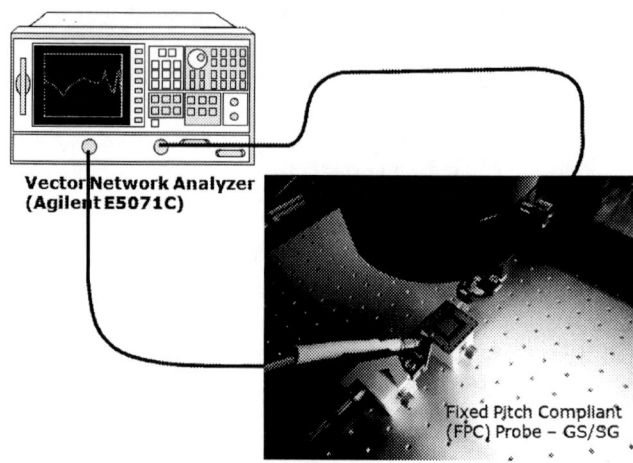

Fig. 8. Z-parameter Measurement Setup in frequency-domain using VNA (Vector Network Analyzer: Agilent E5071C) with Probe Station

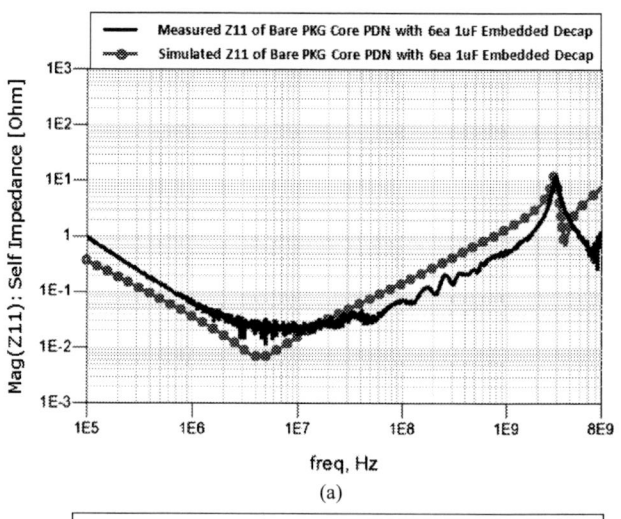

(a)

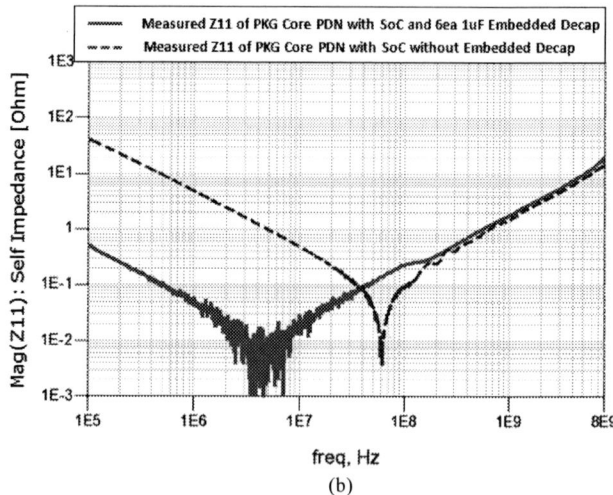

(b)

Fig. 9. (a) Simulated and measured self impedances (Z11 parameters) of bare package core PDN with package embedded decap (b) Measured self impedances (Z11 parameters) of SoC assembled package with/without package embedded decap

is the measured self impedance and the dark orange circle-symbol curve represents the simulated self impedance. The measured self-impedance of core PDN in bare package with 6ea 1uF package embedded decaps shows overall good

978-1-4799-2408-0/14 $31.00 © 2014 IEEE   358

agreement with the simulated self-impedances. Especially, the resonant peaks of both impedance graphs have good correlation at 2.91GHz and 2.9GHz and the peak impedance levels are also very similar as 11.7mΩ and 13.6mΩ. However, the package embedded decap characteristics (the capacitance value, ESL (equivalent series inductance), and ESR (equivalent series resistance) values) in the simulation and measurement show some difference meaning the package embedded decap model provided from decap vendor was not a validated model from the measurement. Especially, the calculated capacitance amount from the measured impedance of package PDN with 6ea 1uF package embedded decaps shows only 3.25uF. This value is an approximately 25% decrease of the expected capacitance value (4.27uF) from the simulation using the package embedded decap model. One lesson from this correlation and verification is we essentially need the measurement-verified package embedded decap model. It will be included for next project as important input deliverable for package and system-level PI optimization.

For more verification of the package embedded decap performance from the measurement, two measured self impedance curves are plotted in Fig. 9(b). The solid blue graph is the measured self impedance of SoC-assembled package including 6ea 1uF package embedded decaps while the dashed green curve represents the measured self impedance of SoC-assembled package without any package embedded decap. The performance of package embedded decaps is clearly shown to reduce big amount of the self impedance in middle frequency range as depicted in Fig. 9(b). Therefore, the much better PI performance can be expected due to the package embedded decaps based on these frequency-domain measured results.

V. Conclusions

In this paper, one BGA package design of ARM SoC was introduced, which has component-type embedded decaps for good power integrity performance of core power. The core PDN in package substrate was optimized by sweeping the number of the embedded decaps to maximize the electrical performance and cost effectiveness. To verify the impact of package embedded decap on frequency-domain system-level core PDN, the self impedances of system-level core PDN were simulated in frequency-domain using the On-chip CPM (Chip Power Model), package core PDN with/without package embedded decap, and board core PDN. There is clear improvement of self impedance of system-level core PDN due to package embedded decap in middle frequency range for both cases with/without board PDN.

Furthermore, in order to confirm the impact of the package embedded decap on time-domain system-level core PDN performance, the On-chip DvD (Dynamic Voltage Drop) simulations were performed including package core PDN with embedded decap models, with/without board PDN. The analyzed DvD simulation results showed the component-type embedded decap in package substrate improves the core PDN performance a lot, especially in case including board core PDN.

Then, the package substrate was manufactured with/ without the embedded decaps to evaluate and confirm the embedded decap impact on core PDN. The fabricated package substrates were measured for the self-impedance (Z-parameter: Z11) in frequency-domain to verify the impact of package embedded decap on core power performance. The simulated self impedance results of the package substrates with/without the embedded decaps in frequency-domain were compared with the measured results, and the comparison result shows overall consistent alignment with the measured self-impedance to confirm the improvement due to the package embedded decap from power integrity point of view.

In conclusion, the overall system-level core PDN of ARM SoC could meet the target impedance in frequency-domain, as well as the target On-chip DvD level, by using certain amount of the package embedded decaps, since the amount of the package embedded decap is optimized the overall system-level core PDN including On-chip CPM and the interconnection between the package embedded decaps and ARM SoC is extremely short, which means the package embedded decaps are integrated very effectively.

References

1. M. Xu and T. H. Hubing, "Estimating the power bus impedance of printed circuit boards with embedded capacitance", in IEEE Transaction of Advanced Packaging, vol. 25, no. 3, pp. 424-432, Aug. 2002.

2. Hyungsoo Kim, et. al., "Suppression of GHz range power/ground inductive impedance and simultaneous switching noise using embedded film capacitors in multilayer packages and PCBs", in IEEE Microwave Wireless Component Letter, vol. 14, no. 2, pp. 71-73, Feb. 2004.

3. P. Muthana, K. Srinivasan, A. E. Engin, M. Swaminathan, R. Tummala, V. Sundaram, B. Wiedenman, D. I. Amey, K. H. Dietz, and S. Banerji, "Improvements in noise suppression for I/O circuits using embedded planar capacitors", in IEEE Transaction of Advanced Packaging, vol. 31, no. 2, pp.234-245, May 2008.

4. Swaminathan, M. and Eugin, A. E., Power Integrity Modeling and Design for Semiconductors and Systems, Prentice hall, 2007.

5. Kouichi Tanaka, et al., "Warpage and Electrical Performance of Embedded Device Package, MCeP", in Proceedings of 2011 IEEE Electronic Component and Technology Conference, pp. 1377-1383.

6. Yongki Min, et al., "Embedded Capacitors in the Next Generation Processor", in Proceedings of 2013 IEEE Electronic Component and Technology Conference (ECTC), pp. 1225-1229.

7. Melinda (Ling) Yang, et al., "SoC Power Integrity from Early Estimation to Design Sign Off", in Proceedings of 2014 DesignCon.

8. Dinh T. Tran, GaWon Kim, et al., "Analysis and verification of board power delivery network impact on DDR3L memory interface in ARM SoC application", in Proceedings of IEEE 22nd Electrical Performance of Electronic Packaging and Systems (EPEPS) 2013, pp. 215-218.

Embed Glass Interposer to Substrate for High Density Interconnection

Dyi-Chung Hu*, Yin-Po Hung*, Yu-Hua Chen*, Ra-Min Tain*, Wei-Chung Lo**
*Unimicron Technology Corp., ** Industrial Technology Research Institute
No.290, Chung-Lun Village, Hsin-Feng, Hsinchu, (304) Taiwan
e-mail: dchu@unimicron.com

Abstract

Current organic substrates are limited to lines/space 10/10 μm and via size around 50 μm. However, the semiconductor with advance node needs fine line/space of 5/5 or 3/3 and even 2/2 μm in the future. Interposer provides a high density interconnection with fine line and small via that cannot be matched by current laminate substrate technology. We have proposed a new structure that embedded interposer to organic substrate (Flip chip - Embedded Interposer Carrier, FC-EIC®). This structure has virtues of know good substrate and compatible with current packaging and assembly infrastructure. We have demonstrated the feasibility of integrating silicon interposer to a laminate substrate.

In this paper, we evaluated the feasibility of integration of a glass interposer into an organic substrate. The selection of glass, temporary bonding materials and built-up dielectric materials were established. The compatibility of lamination and built-up process with glass interposer was demonstrated. By using dielectric materials with small CTE and high modules together with an innovative structure, significant warpage reduction of the EIC-glass was achieved. For a 10 cm x 10 cm laminated carrier with a proper dielectric material set, the panel warpage of the EIC-glass structure can be reduced from 20 mm to less than 3 mm. The EIC-glass structure has been tested under 500 TCT cycles and no glass damage and delamination between glass and the built-up dielectric film was observed.

Introduction

Semiconductor technology is rapidly moving from 28 nm to 20 nm and even to 16 nm nodes. However the organic substrate, which is widely used as a substrate for silicon chips, is difficult to match the line shrinking speed of the semiconductor. Current organic substrate can support circuit pattern of line/space of 10/10 μm, but the industry is looking for line/space of 5/5 μm and even in some case 2/2 μm in the future. The material and tools to support this fine line in organic substrate is under intense development in the industry.

But for semiconductor industry, the infrastructure of fine lines and small vias are already there. 2.5D and 3D packaging integration using semiconductor technology and silicon are under intense development. Silicon interposer can meet fine line/space requirement easily by means of wafer level processes. However, the cost of processing is very high. This prevents the wide use of silicon interposer. We have proposed a new FC-EIC® structure which embedded the interposer to an organic substrate in ECTC 2013 [1]. The EIC structure has lower profile and possible better electrical performance by eliminating the solder joints between the interposer and the organic substrate. The integration of silicon interposer and

organic substrate was demonstrated [1]. However, one of the processing challenges is to reduce the warpage of the EIC-silicon structure during process. The CTE of silicon is around 3 ppm. However, current build-up dielectric materials are difficult to have CTE below 10 ppm without glass fiber. Hence an interposer material with CTE closer to 10 ppm is desirable to reduce the panel warpage during process. To further reduce the cost of the interposer, large panel process is needed as proven in the flat panel display industry. In this paper, we selected glass as interposer material to be integrated into the EIC structure. The selection of glass has several advantages compared to silicon, such as can be silicon matched CTE, high electrical resistivity and the availability in thin and large panel formats [2]. Moreover, the CTE of glass material can also be selected to satisfy different package structure requirement [3]. The thermal, mechanical, and reliability of glass had also been demonstrated [4]. Advanced applications such as Cu Microwire Arrays also have been addressed [5].

In this study, a photo-sensitive glass interposer with CTE compatible to the current advanced dielectric material was selected to be integrated into the EIC-glass structure. By the combination of the advantages of glass interposer and the FC-EIC® structure, this combination could be one of the low cost solutions to address industrial needs for high density interconnection substrates.

FC-EIC Package Structure

Figure 1 shows the basic FC-EIC® structure. Only top side of the interposer is exposed for the chip interconnection and the rest of the interposer is encapsulated with dielectric material. Multi-layer dielectric films are subsequently built up with micro vias and redistributed circuits, forming an interposer embedded IC substrate.

Figure 1. Basic structure of FC-EIC®.

One can view the EIC substrate as a hybrid substrate with fine circuit patterns on top to support fine pitch interconnect of chip and between chips. The coarser circuit on the bottom side of EIC is for the interconnections of EIC substrate to the system PCB. This structure can support one chip or multichip interconnection as shown in Figure 2(a) and 2(b).

2014 Electronic Components & Technology Conference

Figure 2. Concept of applying FC-EIC for (a) single die flip chip and (b) system in package (SiP).

The primary target of this study is to embed the glass interposer so as to form a glass interposer embedded carrier. In order to achieve this goal, various technologies from different industries such as wafer level process, PCB process, and semiconductor package technology are used.

The selections of carrier bonding adhesive and built-up dielectric materials are the key issue in this research. In this paper, important issues of integrating glass and laminate substrate during processing EIC-glass structure will be addressed separately by designing short loop experiments. Figure 3 summarized the important items to be addressed in this paper. 1. The compatibility of laser process to glass interposer. Laser is used in the formation of via that connects the bottom I/Os of the glass interposer to the laminated substrate. The thermal effect from laser may cause issues on glass, adhesive, and interconnections. 2. To reduce the warpage during the EIC process. Warpage causes processing issues especially in the photolithography process modules. The lost of co-planarity will cause uneven and out of focus exposure of circuit patterns and cause failures in circuit formation. The warpage after EIC structure released from the carrier will even cause chip joining issues during EIC package assembly process.

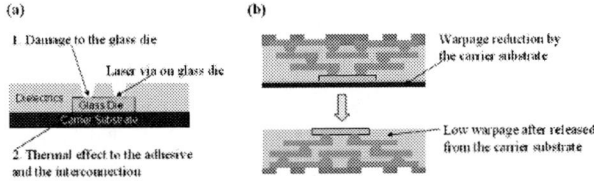

Figure 3. Items to be addressed in the process of EIC. (a) Compatibility of laser process. (b) Warpage reduction, in and after EIC process.

Compatibility of Laser Process with Glass Interposer

The process integration of EIC in this study used dummy glass dies of 100 μm thickness which were prepared by sputtering a thin layer of Ti/Cu (0.1/0.3 μm) on both sides of the photosensitive glass wafer. The general process flow is shown in Figure 4. The EIC built-up process started with a supporting carrier that was coated with temporary adhesive. The EIC panel can be released after the multi-layer built-up lamination process. (The evaluation of the adhesive materials on the temporary carrier will be addressed in the following sections.) Subsequently the glass dies were attached on the supporting carrier using flip chip bonder with high bonding accuracy (< ±10μm) [1]. For the test vehicles used in this study, three layers of RDL were design for fan-out the circuits on the interposer. The function of the first built-up material needs to encapsulate the dies but also form RDL pattern with

minimum line/space of 20/20 μm for the fan-out of the package. Laser via drilling was used for forming the blind vias with top/bottom via diameters of 60/45 μm. After via formation by laser, conventional semi-additive processes including de-smear, electro-less seed layer, photolithography, and electroplating were used to form the RDL circuit patterns. The second and third layers' built-up processes used the same approach but materials of higher modulus and lower coefficient of thermal expansion (CTE) were used. Finally, solder mask printing and opening followed by surface finish using ENIG were performed. The built-up structure can then be released from the supporting carrier. The release process depends on the material used as the temporary adhesive.

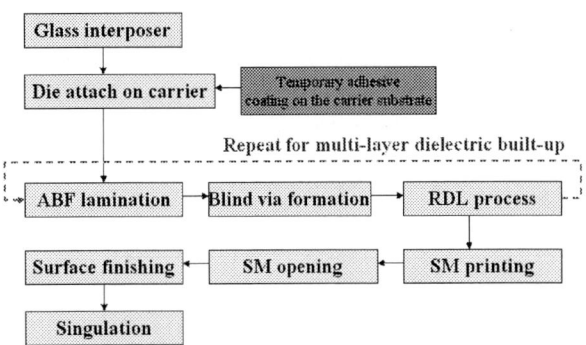

Figure 4. Process flow of EIC-glass.

By following the developed process sequences described above and applying suitable materials, the test vehicle that contains one single glass die embedded in EIC-glass structure was successfully been built. Glass dies with thickness around 100 μm were embedded. Laser drilled micro vias, which connect thin films on the glass and circuit on the laminate substrate, were performed. After a series of processing steps, no significant issues described by Figure 3(a) were found. This study indicates that the compatibility of PCB process, which include laser and lamination process for forming the EIC-glass structure, is feasible.

To successfully build-up the EIC-glass structure, there are two important process improvement directions need to be addressed. They are 1. The evaluation of suitable temporary adhesives for substrate carrier. 2. The selection of suitable dielectric materials together with an optimal structure of the laminated dielectrics.

Reduction of the Panel Warpage during EIC-glass Process
(a) The Selection of Temporary Adhesives

The requirement of the temporary adhesive is to fix the interposer during the lamination process and later can be removed after finishing the overall EIC-glass process. In this study, two different types of material were evaluated as the release layer, or said temporary adhesive layer. One type is the temporary bonding material used in the thin wafer handling system, and the other is a thermal release film, which is generally used in fan out wafer level package. For the first type of material, it can leave a residue free surface after releasing from the supporting carrier. However, the embedded die could be easily tilted during die attach and the subsequent lamination process. This could cause problems in the subsequent EIC-glass built-up processes. As to the thermal release type temporary bonding material, the release temperature is around 200°C. This type of adhesive film has less die tilt. However, it has residues remaining on the dielectric film and needs to be cleaned after the EIC-glass panel release process. Table 1 summarized the characteristics of these two types of materials.

Table 1. Comparison table of the two temporary bonding materials

	Temporary Bonding material for wafer handling system (type 1)	Thermal release film (type 2)
Material type	Liquid paste or Film	Film
Die attach temp.	120°C ~ 220°C	RT
Transparency	Good	Poor
Release temp.	RT	RT
De-bonder	Yes	No
Residual	No	Yes, clean solvent required
Remarks	1. Stable process 2. Die tilt 3. High bonding Temp. 4. Need de-bonding equipment 5. Large panel production issues	1. Simple process 2. No additional equipment required

Figure 5 shows the topography analysis on samples using type 1 and type 2 temporary adhesive materials. In Figure 5(a1), the image shows an uneven surface of using type 1 material as the temporary adhesive. The topography shown in Figure 5(a2) represents the dielectric material was deformed and the die was recessed into the dielectric material. Figure 5(a3) shows a schematic illustration of the structure after panel release. The height of the bulge relative to the die surface is roughly about 20 μm while the distance *b* represents the bulge to the dielectric surface is about 10 μm. As for type 2 material, Figure 5(b1) shows a relatively even surface but with residues after EIC-glass panel de-bonding. The topography figure is listed in Figure 5(b2), while the die surface is higher than that of the dielectric layer. The distance of the die to the dielectric material was measured to be less than 20 μm, as designated by *c* in Figure 5(b3). Type1 material is typically a low modulus material. In this evaluation, the elastic modulus of type 1 material is around 1.7 Gpa with Tg around 90°C. It may possibly be deformed in the subsequent built-up lamination process. Die tilt and uneven dielectric surface would be resulted from the 100°C lamination temperature and lamination pressure. As for type 2 material, the Tg of the film is relatively higher (>120°C). During lamination process, the type 2 film would not be severely deformed under the same conditions.

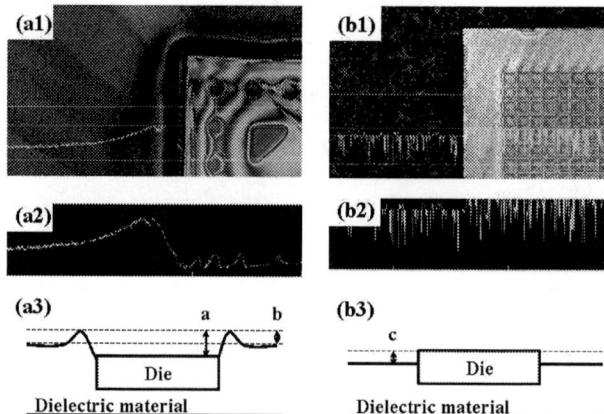

Figure 5. EIC-glass structure using the temporary bonding adhesives from thin wafer handling material (type1). (a1) surface image, (a2) surface topography and (a3) structure illustration. And EIC-glass structure using thermal release film (type 2). (b1) surface image, (b2) surface topography, (b3) structure illustration.

In summary, a high temperature stable, high modulus, with good adhesion strength temporary bonding materials is desirable in the EIC-glass process. It is also preferable that the temporary bonding materials leave a residual free surface on the EIC structure as shown in Figure 5(a).

(b) Selection of Dielectric Material and Structure

It is apparent that FC-EIC® is an asymmetrical structure and the final product might be warped due to the stress generated during the lamination of dielectrics in the built-up process. This innovative built-up dielectric structure consists of different types of dielectric material such that the warpage of the EIC-glass structure was well under controlled. In this study, several different combinations of dielectric materials were evaluated for the EIC-glass structure. The built-up dielectric layers were divided into three main categories. The first layer is the molding layer, which is used to embed the glass interposer. The second layer is designated for accommodating the fine line patterns. The third layer is the reinforce layer for enhancing the rigidity and reducing the warpage of the EIC-glass substrate. For these three different purposes of the dielectric layers, three corresponding dielectric materials were selected and designated as dielectric *a*, *b*, and *c*. They were used for the molding layer, fine line layer and reinforce layer, respectively.

Figure 6 shows the result of sample by using the above combination of dielectric layers comparing to the one using the dielectric material of the same type. For a 10 cm x 10 cm multi-layer built-up panel after panel release, the maximum warpage was reduced from 20 mm to less than 3 mm. Low CTE and high modulus are the primary considerations for the selection of the dielectric materials. The CTE of the material used in this study is in the range from 10 to 20 ppm. And the maximum elastic modulus is nearly 20 Gpa.

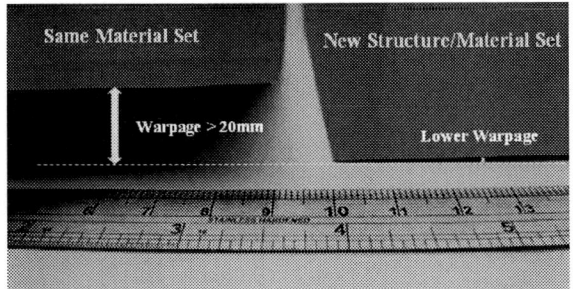

Figure 6. Comparison of the warpage of the two different built-up structures.

By applying the combination of the dielectric materials, final product with well-controlled warpage can be accomplished. Figure 7 shows the cross sectional view of the glass die embedded EIC-glass structure. The multi-layer structure comprising of materials *a*, *b* and *c* provides a feasible solution for the reduction of EIC-glass structure warpage. The glass dies were well integrated into the dielectric material without separation of individual dielectric layers. Subsequent pre-condition and thermal cycle tests were performed to verify the stability of the structure.

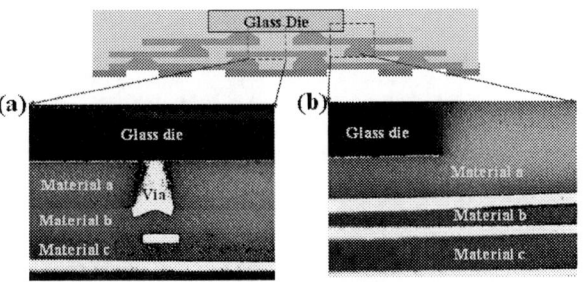

Figure 7. Cross section of glass die embedded into EIC-glass structure in (a) die center and (b) die edge.

Mechanical Compatibility of Glass Interposer in EIC-glass structure

The mechanical structure stability of the FC-EIC structure was evaluated using glass dies embedded in built-up dielectric materials. In previous study [6], feasibility and compatibility of the dielectric materials after lamination had been introduced. In this study, glass dies were introduced into the newly developed materials to form EIC structure for evaluating the mechanical stability. Typically, glass is a fragile material in which the yield and reliability are concerns. In this study, the selection of the photo-sensitive glass (CTE around 9 ppm) and the dielectric materials (CTE from 10 to 20 ppm) are important. The EIC-glass structure can encapsulate the glass die to form a protection structure. The test samples after pre-condition (MSL-3a) showed no significant crack in the glass die. After 500 thermal cycle tests (TCT), no defect was found in the EIC-glass structure as well as the embedded glass (Figure 8). The TCT test was still on-going when this paper was submitted. Results of more test cycles (>500 cycles) will be updated at the presentation of this paper.

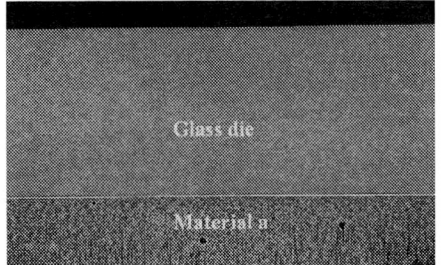

Figure 8. Cross section view of EIC-Glass sample after 500 cycles thermal cycle test (TCT).

Conclusions

In this study, various process and materials sets were identified for a successful integration of glass interposer into the EIC structure. Glass with CTE around 9 ppm was selected. A film type thermal release temporary adhesive was evaluated and selected to the EIC-glass built-up structure. By combining dielectric films with different properties and structure, the overall warpage of panel was reduced. This achievement suggested the foundation of the feasibility in large panel process. The compatibility of laser process with the EIC-glass process was also evaluated.

The mechanical stability of the EIC structure was tested. After 500 TCT cycles, no damage in the glass nor delamination of the dielectric layers were observed.

Based on the above individual process and materials evaluation, it is very promising to realize the EIC-glass process and structure. The preparation of a glass interposer test vehicle based on photosensitive glass is underway. We plan to integrate it to the EIC structure based on the process developed in this paper.

Acknowledgments

The authors would like to thank Unimicron's NBD team and ITRI's 3DIC team for their efforts in this study, also the support from Ministry of Economic Affairs and Unimicron's management team.

References

1. D. C. Hu, et al, "An Innovative Structure that Combines Interposer and Carrier", in *Proc. IEEE Electronic Components and Technol. Conf. (ECTC)*, Las Vegas, NV, May 28–31, 2013, pp. 1332–1335.
2. V. Sukumaran, et al, "Design, Fabrication and Characterization of Low-Cost Glass Interposers with Fine-Pitch Through-Package-Vias", in *Proc. IEEE Electronic Components and Technol. Conf. (ECTC)*, Lake Buena Vista, FL, May 31-June 3, 2011, pp. 583–588.
3. J. Keech, et al, "Fabrication of 3D-IC Interposers", in *Proc. IEEE Electronic Components and Technol. Conf. (ECTC)*, Las Vegas, NV, May 28–31, 2013, pp. 1829–1833.
4. K. Demir, et al, "Thermolmechanical and Electrochemical Reliability of Fine-Pitch Through-Package-Copper Vias (TPV) in Thin Glass Interposers and Packages", in *Proc.*

978-1-4799-2408-0/14 $31.00 © 2014 IEEE

IEEE Electronic Components and Technol. Conf. (ECTC), Las Vegas, NV, May 28–31, 2013, pp. 353–359.

5. X. Qin, et al, "Large Silicon, Glass and Low CTE Organic Interposers to Printed Wiring Board SMT Interconnections using Copper Microwire Arrays" in *Proc. IEEE Electronic Components and Technol. Conf. (ECTC)*, Las Vegas, NV, May 28–31, 2013, pp. 867–871.

6. Y. P. Hung, et al, "Process Feasibility of a Novel Dielectric Material in a Chip Embedded, Coreless and Asymmetrical Built-up Structure", in *International Microsystems, Packaging, Assembly, and Circuit Technology Conf. (IMPACT)*, Taipei, TW, Oct. 22-25, 2013, pp.508-511

First Demonstration of a Surface Mountable, Ultra-Thin Glass BGA Package for Smart Mobile Logic Devices

Venky Sundaram*, Yoichiro Sato^, Toshitake Seki[+], Yutaka Takagi[+], Vanessa Smet, Makoto Kobayashi[#], and Rao Tummala

3D Systems Packaging Research Center, Georgia Institute of Technology, Atlanta, GA 30332, USA
^Asahi Glass Company, Yokohama, Japan
[+]NGK Spark Plug Co. Ltd., Aichi Prefecture, Japan
[#]Namics Corporation, Niigata, Japan
*vs24@gatech.edu

Abstract

This paper presents the first demonstration of an ultra-thin glass BGA package that is assembled on to mother board with standard SMT technology. Such a package has many new advances that include ultra-thin glass, high speed through via hole formation and copper metallization, double-side RDL wiring with advanced 3 micron ground rules, and Cu-SnAg microbump assembly of a 10mm silicon test die. Glass, as a package, overcomes the shortcomings of organic packages in bump pitch, CTE mismatch to Si and warpage and silicon interposers in electrical performance and cost. Glass packages are being developed to manufacture both as wafers for improved performance over Si and as panels to improve bump pitch over organic packages. Glass, therefore, is not just a high performance and low volume technology, like silicon interposers, but a pervasive package technology with lower cost, higher performance and thinner than silicon and organic packages. Glass has compelling benefits in thickness and I/O pitch reduction and reliability for one of the highest volume applications, namely, the packaging of high I/O logic devices for smart mobile systems. This paper represents a paradigm shift in ultra-thin packages using large glass panels for future smart mobile and high performance devices, and the first demonstration of 100um thin glass packages with 50-80um chip-level I/O pitch and 18mm x 18mm body size surface mount assembly at 400um pitch.

Introduction

Glass packaging started at Georgia Tech and many other R&D groups as a lower cost and higher performance alternative to silicon interposers, due to the low loss and large panel availability of ultra-thin glass [1]. The Georgia Tech team began to demonstrate such an interposer with its industry partners addressing major barriers that include the handling of large, ultra-thin glass panels, forming large number of ultra-small through vias at small pitch, metallization of these small copper vias without defects and with good adhesion, forming 2-5 micron RDL wiring layers with bump pitch at 20-50 microns, assembly of chips to these brittle substrates, and improving its thermal conductivity. It became clear that glass can be a pervasive platform technology that is both a high performance and very low cost technology. As the glass packaging technology matures and is applied to high volume applications, it should be at the same or lower cost than organic packaging. Glass simplifies the material manufacturing compared to laminates, since FR-4 or low CTE laminates require four different materials and process technologies such as glass fiber, glass and ceramic fillers,

epoxy, flame retardant and all these put together to form the 5th material that is called core or prepreg. In the case of glass, it is one material and one process and is ultra-thin to start with, driven by touch displays. Glass also reduces the processing cost per package since it can be scaled to much larger manufacturing panels or roll to roll than FR-4, driven by its superior dimensional stability and higher modulus. Glass is a high temperature material, resistant to moisture and is available in many compositions with many CTEs, from below Si CTE to above GaAs CTE.

Glass is seen, therefore, not just as a high performance interposer technology, but as a pervasive low cost platform packaging technology that is suitable for packaging all devices that are packaged currently that include single chip logic and memory packaging, MEMS & Sensors Packaging, RF, Power & Analog Packaging. In addition, it is applicable and is being developed for packaging of 2.5D multichip packaging with ultra-high number of interconnections between chips in side-by-side configuration. Ultimately, it is being developed as a superior alternative to 3D ICs, but without TSVs, in what is called 3D interposers and packages. Other applications that are being explored include 3D IPDs for passives on both sides of ultra-thin glass, and 3DIPAC for ultra-thin passives and actives on both sides with shortest distance between actives and passives. The specific applications for 3D IPACs include cell phone cameras, RF and power modules, 3D optoelectronics replacing electronic TSVs with Photonics.

This paper focuses on one of the highest volume applications, namely, the packaging of high I/O logic devices for smart mobile systems. A schematic of the glass BGA package that can be SMT attached to FR-4 motherboard, is shown in Figure 1.

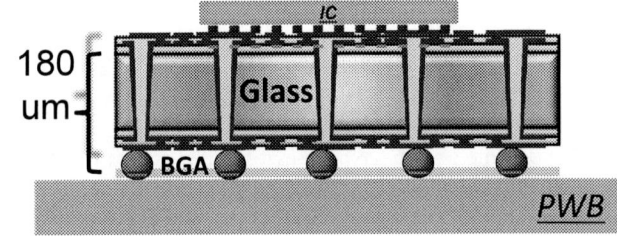

Figure 1. Glass BGA Package for Smart Mobile Application Processors

Previously, the authors have reported individual building block technologies such as through via hole formation, metallization and reliability, electrical characterization of glass substrates and initial board-level assembly and reliability [1, 2]. This paper, for the first time, integrates all the glass package building blocks into the first thin glass BGA package demonstration.

The second section describes some of the building block technology advances in glass packaging. The third section presents the first glass substrate process or record (POR) with a full process flow, materials and tools. The fourth section presents the first demonstration of a mobile glass BGA package including test vehicle design, glass substrate fabrication, chip and board level assembly. The fifth section analyzes the warpage and thermal behavior of glass BGA packages and the final section summarizes the research.

Glass Package Building Block Technologies

Initial advances in through package vias (TPV) and re-distribution layers (RDL) on glass have been previously reported [2]. This section describes the latest advances in TPV, RDL and glass assembly.

Through Package Vias (TPV) in Ultra-Thin Glass: The single biggest barrier to the use of glass as an interposer and package is its brittleness, resulting in cracking of glass during via hole formation of small diameter TPVs at small pitch. Addressing this barrier requires overcoming a number of fundamental challenges including handling of thin glass, defect-free via hole formation, and low-stress conductive via metallization at high speed. The glass interposer starts with ultra-thin glass (30-180μm) enabled by innovations in fusion draw and float glass formation technologies in defect-free panels and wafers. Thinner glass, not only enables the formation of ultra-small TPVs having TSV-like dimensions (<10μm via diameter) with low aspect ratios, it also significantly increases the throughput of via hole formation. Thin glass also reduces interconnect length, hence shorter signal path, leading to smaller latency.

Glass, however, is inherently brittle; and handling of ultra-thin glass during processing is, therefore, a key challenge. A double-sided, panel-based process was developed using polymer lamination of glass with low modulus and low electrical loss (tan δ) polymers. Polymer lamination, not only facilitates handling of thin glass, but also reduces laser impact on the glass surface during laser via formation and avoids direct metallization on glass surfaces. This novel handling method was previously demonstrated at 180um glass thickness, but has recently been applied successfully to fabricate double side ultra-thin glass panel substrates as thin as 30um thin drawn glass, without the use of bonded carriers, as shown in Figure 2.

Several methods have been investigated and four different methods have been demonstrated with via formation speed of greater than 1000 vias per second. Two of these methods are laser based, and the other two have been provided by Asahi Glass and Corning Glass respectively [3, 4]. The key to reliable TPV hole formation is to minimize heat and stress accumulation, and prevent micro crack formation. Metallization of TPVs was carried out using low cost wet electroless plating of thin copper, followed by electroplating of thick copper.

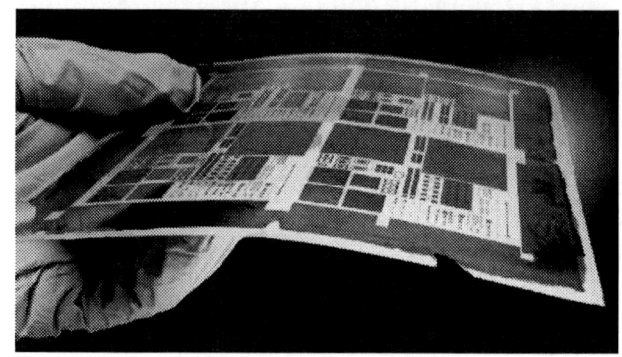

Figure 2. Demonstration of Handling of Ultra-Thin 30um Glass Through TPV and RDL Fabrication without Carriers

A low cost, package substrate-compatible, semi-additive-plating (SAP) approach was developed to metallize the TPVs and double-sided wiring simultaneously. Wet chemistry for metallization (as opposed to vacuum-based sputtering), enables large panel processing. Both conformal and via fill plating processes have been applied to thin glass TPVs. Figure 3 shows cross-section images of metallized TPVs formed by excimer laser ablation and the impact of thickness reduction on reducing the via diameter and pitch.

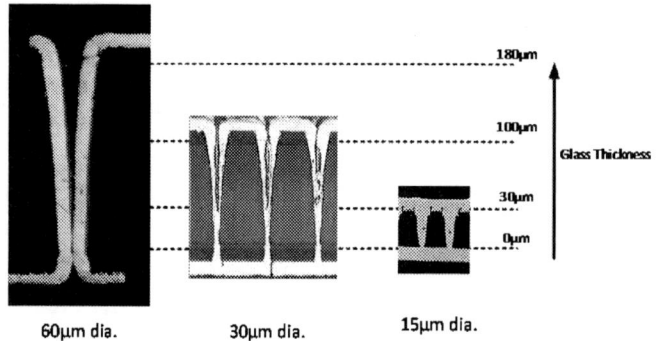

Figure 3. TPV in Thin and Ultra-Thin Glass by Excimer Laser Ablation and Semi-Additive Copper Plating Processes

3-5um Redistribution Layers (RDL) on Glass:

Fine pitch re-distribution layers were fabricated on both sides of the thin glass with TPV simultaneously using low cost panel level processes, without the use of any chemical-mechanical polishing (CMP) or vacuum deposition methods, resulting lower cost than wafer back end of line processes used on silicon interposers. The goal was to shrink the design rules for lines and vias in each layer, in order to minimize the total number of metal layers. For the mobile glass package, a 2 + 0 + 2 layer structure was targeted with ultra-fine lines on both metal layers on top and bottom of the glass. A schematic of the target design rules to support 50um I/O bump pitch is shown in Figure 4.

978-1-4799-2408-0/14 $31.00 © 2014 IEEE   366

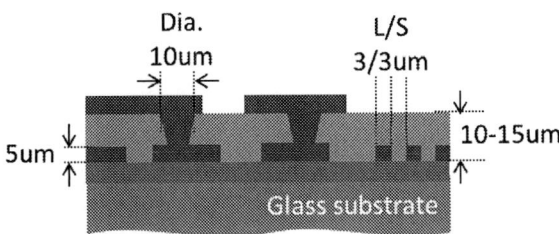

Figure 4. Schematic Cross-section of GEN 1 Target Dimensions for RDL Layer on Glass Interposer

Glass has an ultra-flat and smooth surface similar to silicon and better than organic laminate cores, and hence can extend the semi-additive plating process limits for line width reduction faced by organic substrates. Silicon interposers can achieve sub-micron line widths, but have line resistivity and cost challenges driven by single sided thin film processes and back end of line damascene processes on wafers. An ultra-smooth polymer dry film dielectric, ZEONIF™ ZS-100 from Zeon Corporation, Japan, with 10um and 15um thickness, was laminated on both sides of the thin glass to achieve a flat and smooth surface for electroless seed layer plating and lithography. A low cost and large panel scalable wet metallization process combined with large field size lithography and high resolution dry film photoresists was used to fabricate a test vehicle with line widths and spaces from 1um to 10um. A projection UV lithography tool, UX44101 from Ushio Japan, was used for imaging on a high resolution dry film resist from Hitachi Chemical. Major challenges in the semi-additive process, including seed layer adhesion to polymer dielectric, photoresist adhesion to smooth seed layer, and seed layer etching, have been overcome. A top view SEM micrograph and a cross-section of 5um lines and spaces used to route five rows of 50um I/O pitch bumps is shown in Figure 5.

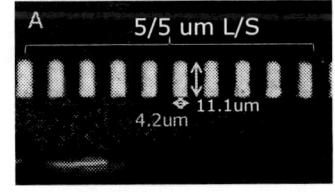

Figure 5. Top View SEM and Cross-section of 5um lines and spaces with 11um copper thickness fabricated on ZS-100 Polymer RDL Layer on Glass

Generation 1 Glass Substrate Process of Record (POR)

A Georgia Tech glass consortium involving a global team of industry end users and supply chain partners was launched in 2010 to address the technical challenges to the realization of ultra-thin glass interposers and packages. The team has demonstrated the first generation glass substrate with a process of record (POR) overcoming all the major barriers, namely, fine pitch and high speed through package vias

(TPV), thin glass handling, Cu TPV reliability and double side re-distribution layers at 50μm I/O pitch. The Gen 1 glass substrate, was demonstrated using 100μm and 150μm thin glass made by advanced float and fusion draw processes from Asahi Glass, Corning, and Schott Glass companies. The fine pitch through vias were formed using a high speed laser process, demonstrating high throughput in excess of 1000-2000 vias per second. Multi-layer and double side RDL with low cost materials, tools and processes using ultra-thin dry film polymers with lithographic ground rules down to 3-5um, discussed in the previous section were integrated on thin glass with fine pitch TPVs. A complete process flow, bill of materials, tool sets for manufacturing, and design rules have been compiled, ready for transfer to industry partners for prototyping and volume manufacturing. Extensive reliability tests have been conducted and the through via daisy chains at 120μm pitch, metallized with plated copper, have passed more than 1500 accelerated thermal cycle (-55°C to 125°C) tests with MSL-1 and MSL-3 pre-conditioning. A cross-section and top view of a four metal layer double sided glass substrate is shown in Figure 6.

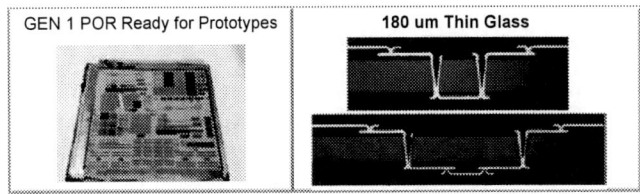

Figure 6. First Process of Record (POR) for Glass Substrate Fabrication Demonstrated by Georgia Tech and Its Partners

The demonstration of the first POR for glass substrates represents a major milestone in the glass technology development and has been subsequently used to optimize assembly processes for thermo-compression bonding at chip-level and BGA reflow at board-level. The POR was also used to design, fabricate and assemble the first mobile glass BGA package demonstrator targeted at application processors as explained in the following section.

Glass BGA Package Demonstrator for Mobile Application Processors

The smart mobile package test vehicle is the first demonstration of a fully integrated glass package with assembly of a silicon test die on an ultra-thin glass interposer with vias and double sided routing layers, with a fully populated BGA ready for SMT attach to the motherboard. This test vehicle consists of a mobile application processor package emulator with ~5,500 I/Os on a 10mm x 10mm test die, a 100μm thick glass substrate with 4 metal layers, and an SMT-compatible 18.4mm x 18.4mm BGA package at 400μm pitch.

Test Vehicle Design: The 10mm x 10mm test die has four peripheral rows of bumps at 80/40 μm pitch in a staggered configuration, and a central area array at 150 μm pitch. A 4-

metal-layer interposer structure with 120 µm pitch TPVs and BMVs, and 4-7um L/S was designed to escape route such a high I/O density down to the BGA for SMT interconnection to printed wiring board (PWB). Figure 7 illustrates an example of escape routing on the top metal layer used in this test vehicle design. The final design was panelized on a 150mm x 150mm quarter panel containing 24 individual BGA package units.

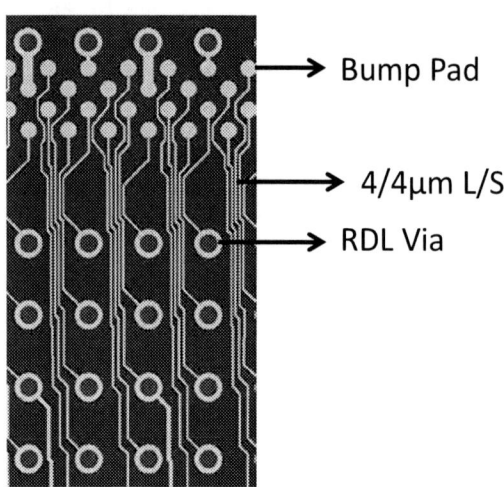

Figure 7. Schematic of Escape Routing with 4 µm Lines and Spaces connecting Microbump Landing Pads to RDL Vias on the Top Layer of Glass Substrate

Process Flow: A complete process flow starting with thin glass and covering the substrate fabrication, chip assembly, BGA balling, dicing and SMT placement on PWB is shown in Figure 8.

1. **Glass surface treatment**

2. **Double side polymer lamination**
   ZS-100 20µm

3. **TPV formation**
   Excimer or CO$_2$ laser

4. **TPV & 1st RDL metallization**
   Semi-Additive Process

5. **Double side polymer lamination**
   Cu treatment, 20µm ZS-100

6. **RDL micro via formation**
   UV laser

7. **RDL Via and 2nd RDL metallization**
   Semi-Additive Process

8. **Passivation & surface finish**

9. **Chip-level Interconnection**
   Panel-level process

10. **BGA balling & singulation**
    Ball drop & dicing

11. **Board-level SMT assembly**
    Assembly on PWB board
    Batch reflow

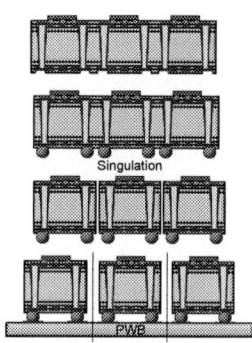

Figure 8. Complete Process Flow Starting with Bare Glass and Ending in SMT Attach of Glass BGA Package to PWB

Chip-Level Assembly: Die-to-panel assembly was achieved by low-pressure thermocompression bonding with a pre-applied polymer (BNUF, NCF or NCP) using a Finetech Matrix semi-automatic flip-chip bonder with a placement accuracy of 3um. A 6" x 6" panel can be accommodated on the stage of the flip-chip bonder to achieve panel-level assembly. The assembly sequence was as follows:

1- Dispensing pre-applied polymer on the substrate bonding areas, followed by B-staging of BNUF (skipped for NCF or NCP)

2- Panel placement on the stage of the flip-chip bonder, kept in place with vacuum, and heating the plate to 70-90degC to lower the viscosity of the polymer material, so as to not introduce any misalignment when placing the die

3- Pick and place and thermocompression bonding of each die individually on the panel. The heat transfer was unidirectional from the chip-side. After placement with a higher applied pressure (necessary to create contact through the viscous polymer), the pressure was reduced to 1.5MPa and temperature ramped up to 260C (lead-free reflow temp), maintained for 3 seconds to allow IMC formation with the ultra-short bumps. Ramping down the temperature before moving on to the next die.

4- After bonding of all dies, a batch-type post-cure in air was used to fully cure the polymer material for 1 hour at 165C for NCP, 3 hours at 165C for BNUF.

This bonding scheme enables bonding on a panel with low warpage, is scalable to thermo-compression bonders used in production, and allows the assembly of a single die without affecting the adjacent dies. Figure 9 shows a cross-section micrograph of the Cu-SnAg interconnects from chip to glass. One of the key benefits of glass for mobile processor packaging is the high modulus of glass core (~80 GPa) compared to that of low CTE organic core (~20-35 GPa), which reduces warpage during chip assembly. The high glass transition of temperature of glass at around 550C results in no change in the glass modulus during the entire assembly temperature ramp, hold and cool down cycle.

978-1-4799-2408-0/14 $31.00 © 2014 IEEE   368

Figure 9. Cross-section Micrograph of Fine Pitch Cu-SnAg Interconnections on Thin Glass Substrate by Thermo-Compression Bonding

The warpage behavior of thin glass was measured using a Akrometrix PS-200 Shadow Moire tool, and compared to the normalized warpage of thin core organic substrates for similar die size and package size available in recent published literature. The high and stable modulus of glass combined with low CTE of around 3.8 ppm/C resulted in very low warpage for the glass substrate as shown in Figure 10, which will enable scalability to thinner packages.

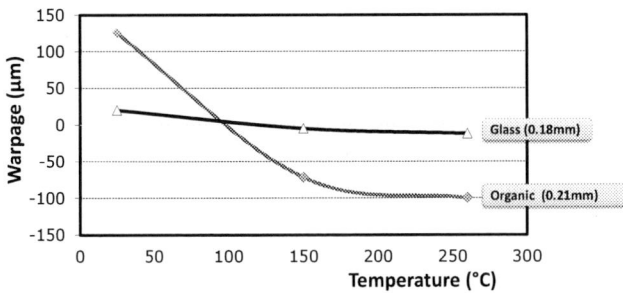

Figure 10. Reduced Warpage of High Modulus Glass Substrate Compared to Organics after Lead-Free Solder TCB Bonding and Underfill Cure

SMT Attach to PWB: Silicon interposers for high performance applications are typically assembled to an organic BGA package which is then mounted to the PWB. This results in much higher cost in the bill of materials (BOM) and added thickness as well additional sites for interconnection failures. For cost sensitive mobile applications, it is preferred to have only one level of packaging from chip level to board level. The glass BGA package was designed to be directly SMT attached to standard FR-4 PWBs using reworkable lead-free solder ball interconnections without the use of underfill at board level. Glass offers the ability to tailor the CTE from 3ppm/C to 10ppm/C with several CTE values available in between. This benefit of selecting glass CTE in between that of silicon and FR-4 enables board-level reliability of glass BGA packages extendable to larger body sizes [5]. Figure 11 shows the top view of a fully integrated glass BGA package for the mobile application processor emulator after SMT attach to board. A customized mechanical dicing process was developed in partnership with Disco, Japan to singulate the glass packages prior to SMT pick and place and reflow. The glass substrate has extremely low warpage even after three reflows, first at die bonding, second at BGA ball reflow on the back side of the glass substrate, and third at board level SMT attach.

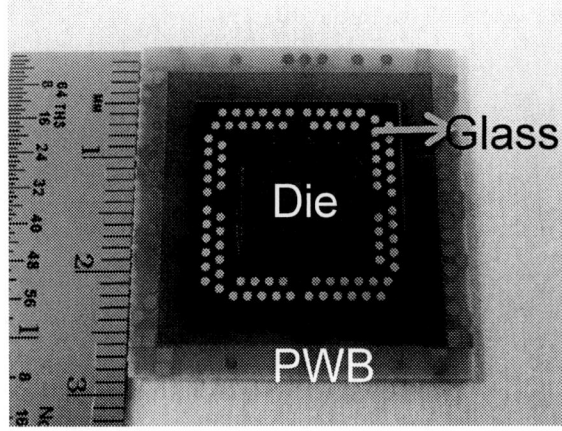

Figure 11. First Demonstration of Glass BGA Package for Mobile Application Processors with 18mm x 18mm body size and 400um BGA pitch using 100um Thin Glass

Conclusions

This paper presented the first demonstration of an ultra-thin glass BGA package for mobile application processors that is assembled on to mother board with standard SMT technology. Such a package integrated for the first time, a number of building block technologies in ultra-thin glass, high speed through via hole formation and copper metallization, double-side RDL wiring with advanced 3 micron ground rules, and Cu-SnAg microbump assembly of a 10mm silicon test die. Glass, as a package, overcomes the shortcomings of organic packages in bump pitch, CTE mismatch to Si and warpage and silicon interposers in electrical performance and cost. Glass, therefore, is not just a high performance and low volume technology, like silicon interposers, but a pervasive package technology with lower cost, higher performance and thinner than silicon and organic packages.

Acknowledgments

The authors would like to acknowledge the numerous contributions of the glass packaging team from GT PRC and its industry partners to this work. This work was supported by funding from the Low Cost Glass Interposers & Packages (LGIP) global industry consortium at Georgia Tech PRC.

References

1. Sukumaran, V.; Bandyopadhyay, T.; Sundaram, V.; Tummala, R., "Low-Cost Thin Glass Interposers as a Superior Alternative to Silicon and Organic Interposers for Packaging of 3-D ICs," *Components, Packaging and Manufacturing Technology, IEEE Transactions on* , vol.2, no.9, pp.1426,1433, Sept. 2012.

2. Sukumaran, V.; Bandyopadhyay, T.; Chen, Q.; Kumbhat, N.; Fuhan Liu; Pucha, R.; Sato, Y.; Watanabe, M.; Kitaoka, Kenji; Ono, M.; Suzuki, Y.; Karoui, C.; Nopper, C.; Swaminathan, M.; Sundaram, V.; Tummala, R., "Design, fabrication and characterization of low-cost glass interposers with fine-pitch through-package-vias,"

Electronic Components and Technology Conference (ECTC), 2011 IEEE 61st , vol., no., pp.583,588, May 31 2011-June 3 2011.

3. Takahashi, S.; Horiuchi, K.; Tatsukoshi, K.; Ono, M.; Imajo, N.; Mobely, T., "Development of Through Glass Via (TGV) formation technology using electrical discharging for 2.5/3D integrated packaging," *Electronic Components and Technology Conference (ECTC), 2013 IEEE 63rd* , vol., no., pp.348,352, 28-31 May 2013.

4. Shorey, A.; Pollard, S.; Streltsov, A.; Piech, G.; Wagner, R., "Development of substrates for through glass vias (TGV) for 3DS-IC integration," *Electronic Components and Technology Conference (ECTC), 2012 IEEE 62nd* , vol., no., pp.289,291, May 29 2012-June 1 2012.

5. Xian Qin; Kumbhat, N.; Sundaram, V.; Tummala, R., "Highly-reliable silicon and glass interposers-to-printed wiring board SMT interconnections: Modeling, design, fabrication and reliability," *Electronic Components and Technology Conference (ECTC), 2012 IEEE 62nd* , vol., no., pp.1738,1745, May 29 2012-June 1 2012.

Towards a Quantitative Mechanistic Understanding of the Thermal Cycling of SnAgCu Solder Joints

D. Schmitz[1], S. Shirazi[1], L. Wentlent[2], S. Hamasha[1], L. Yin[3], A. Qasaimeh[4], and P. Borgesen[1]

[1]Department of Systems Science & Industrial Engineering, Binghamton University, Binghamton, NY 13902
[2]Materials Science & Engineering, Binghamton University, Binghamton, NY 13902
[3]GE Global Research, Niskayuna, NY 12309
[4]Tennessee Tech University, Cookeville, TN 38505

Abstract

Microelectronics manufacturers continue to subject a wide range of products or representative test vehicles to accelerated thermal cycling tests. Most such testing is focused on comparisons, whether among alternatives or to an established requirement. However, more often than commonly recognized such comparisons may not reflect the relative performances in service. In fact, most current models have been shown to fail to account for important systematic trends as well as being inconsistent with our current understanding of the failure rate controlling damage mechanism.

An alternative mechanistically justified model for the thermal fatigue life of SnAgCu solder joints has been proposed. Damage and failure occurs by recrystallization of the large Sn grains across the high strain region of the joint, followed by crack growth along the resulting network of high angle grain boundaries. The recrystallization was shown to be the damage rate controlling mechanism, except for extremely high strain assemblies and/or harsh cycling conditions, i.e. if we can predict the recrystallization we can predict the number of cycles to failure. So far the model accounts for a variety of important trends and offers guidance as to the interpretation and generalization of accelerated test results. General extrapolations towards service conditions will, however, require the specific functional dependence of the rate of recrystallization on the stress and the precipitate distributions.

Another potential difficulty is that constitutive relations are extracted from single sided creep experiments while the dislocation cell structures built up under cyclic loading are certain to be different. Furthermore, the repeated 'annealing' during the high temperature dwell affects the hardening. Even if these effects could be ignored the ongoing evolution of the constitutive relations would still effectively prevent the extraction of the above-mentioned functional dependence from comparisons between thermal cycling results and FEM.

A special experiment is ongoing in which stresses and strains on the solder joints can be controlled and measured directly, allowing the testing of individual assumptions underlying the proposed model. Preliminary results are presented and compared to results of thermal cycling across different temperature ranges.

Introduction

Accelerated testing would rarely be meaningful without the assumption of at least a qualitative correlation with performance of the product under realistic service conditions. Such a correlation is however far from always obvious. At the extreme, cases have been identified where a mechanistic understanding would suggest that the alternative surviving the longest in testing might not do so in actual use.

Figure 1 shows two comparisons between four different SnAgCu alloys. Model BGA assemblies with 20 mil diameter solder balls were cycled between 0°C and 100°C with 10 minute heating and cooling ramps. The solid diamonds reflect the characteristic life, N_{63}, for each alloy in cycling with 10 minute dwells at both temperature extremes. These are very common accelerated test parameters in the microelectronics industry, and taken by themselves the results would suggest that alloy IV (a Zn doped alloy) is the best. However, the open circles show what happens when the dwell times were extended to 30 minutes each. Suddenly, alloy IV is the worst performer. The question would seem to be which test best represents realistic service conditions, but the fact is that either would almost certainly require major extrapolations.

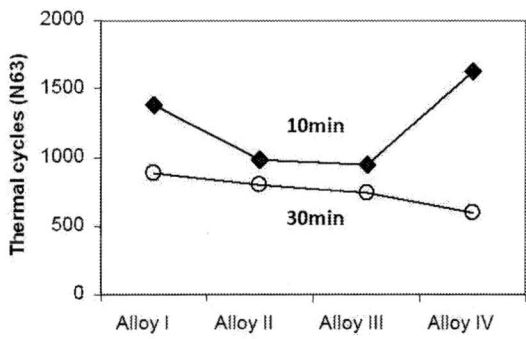

Figure 1: Characteristic number of thermal cycles to failure in 0/100°C cycling with two different dwell times at the temperature extremes [21].

Figure 2 shows two comparisons between four different designs of model CSP assemblies with SAC305 solder joints. The designs differed in terms of pitch, solder volume and pad size. Cycling between -40°C and 125°C suggested that designs C and D (the largest joints and pitches) would be the most reliable, but the strain ranges were clearly high. Cycling instead between 0°C and 80°C design A (the smallest joints and pitch) is clearly superior. It may be tempting to assume that the latter is closer to most realistic service conditions, but if we do not understand the reason for the different performance, we cannot be sure that the trend continues.

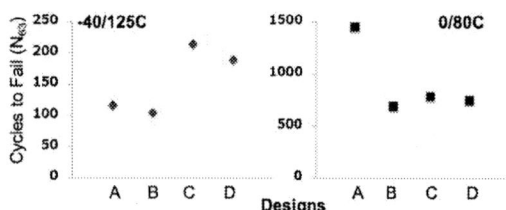

Figure 2: Characteristic number of cycles to failure for four different model CSP designs in -40/125°C cycling (left) and 0/80°C cycling (right). In both cases with 10°C/min ramp rates and 15 minute dwells [1].

The above examples illustrate at the very least the need for a mechanistic understanding of the effects of cycling parameters on damage evolution and failure. Only this will allow us to predict when accelerated testing may lead to the wrong conclusion.

As for a *quantitative* model, which would allow us to extrapolate accelerated test results to life in service, attempts at this are commonly based on a 2-stage approach:

1) the calculation of stresses, strains and parameters that can be derived from those versus temperature and time
2) the prediction of damage evolution based on this.

In the case of SnAgCu solder joints modeling is complicated by the ongoing changes in the constitutive relations during storage [2] and cycling [3]. Dutta et al. have developed microstructurally adaptive constitutive models to account for this [4], but accurate Finite Element Modeling (FEM) based on these would be computationally prohibitive without significant simplifying approximations.

A different question is to which extent these or any other constitutive relations allow for simulations of the most relevant parts of each thermal cycle. We shall address this briefly below.

As for a damage function, Borgesen et al. relied on an unusually comprehensive set of thermal cycling test results to argue that any of the current models in the literature could be strongly misleading [5]. They furthermore noted [1] that almost all the models were inconsistent with our mechanistic picture (below). Certainly, none of them seem able to explain the size dependent acceleration factors for SAC305 joints evident in Figure 2.

The present paper briefly outlines the model proposed by Borgesen et al. and addresses some of the underlying assumptions through both thermal cycling and a unique experiment simulating thermal cycling by alternating between isothermal loading and annealing.

Experimental

Samples for isothermal cycling were prepared by reflow soldering 30 mil (0.75mm) diameter spheres onto 22 mil (0.55mm) diameter Cu pads on typical BGA component substrates.

Soldering was accomplished by first printing a tacky flux through a stencil onto the substrate pads and then placing individual spheres through apertures in a separate 'bumping' stencil. Reflow was done in a nitrogen ambient with less than

50 ppm O₂ using a Vitronics-Soltec 10-zone full convection oven and a profile with a 245°C peak and 45-60 seconds above liquidus.

A row of 18 joints was first annealed for 48 hours at 100°C in order to coarsen and stabilize the secondary precipitates, then loaded simultaneously in an Instron tensile tester. The joints were loaded to the designated creep load at a rate of approximately 0.3MPa/s after which the load was maintained for 5 minutes. The loading was then reversed to bring the joints back to the original position over a period of about 5 minutes. This was followed by a one hour anneal at 100°C, after which the loading sequence was repeated. Figure 3 shows the loading-unloading part of a typical cycle.

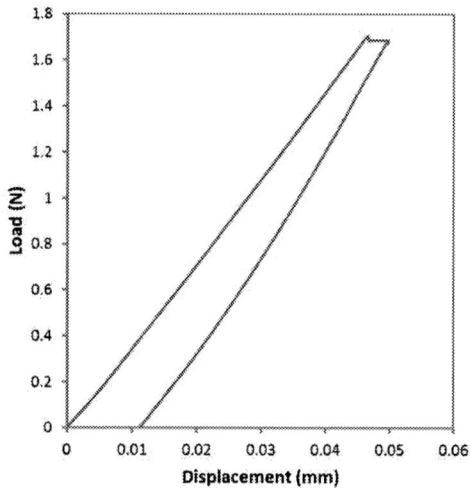

Figure 3: Load per joint vs. displacement during the first part of a typical isothermal cycle.

All thermal cycling experiments employed fully balanced model CSP components each consisting of a 20 mil thick silicon 'die' sandwiched between two 16 mil thick FR4 substrates using thin layers of a rigid flip chip underfill to fully encapsulate the die. In all cases the silicon was sized so that all solder pads were located under it. 16 mil (0.4mm) diameter SAC305 balls were soldered to solder mask defined Cu pads on the component. The corresponding 62 mil thick four layer board had OSP coated non-solder mask defined Cu pads.

One set of thermal cycling experiments employed components with 8 x 8 full arrays of 0.38mm diameter pads with five different pitches. The resulting assemblies were cycled in either 0/100°C or -20/100°C with 10°C/minute ramps and 10 minute dwells at either extreme.

Another set of thermal cycling experiments employed components with 12 mil (0.3mm) diameter pad openings arranged in a conventional partially depleted array of 256 pads on a 0.8mm pitch. The resulting assemblies were cycled in either 0/100°C or 0/80°C with 10°C/minute ramps and 15 minute dwells.

After cycling, solder joints were cross sectioned and the microstructure first characterized by cross polarizer microscopy. Selected samples were further analyzed by Electron Backscatter Diffraction (EBSD).

Understanding & Model

Realistic SnAgCu solder joints are invariably the results of single solidification events during cool-down from reflow, leading to either one or three distinct Sn grain orientations in each. In the latter case these orientations are a result of cyclic twinning [6]. Small solder volumes and pad sizes may allow for such strong undercooling that the twinned grains form an interlaced structure, but typical BGA and CSP scale joints end up with large Sn grains such as those shown in Figure 4.

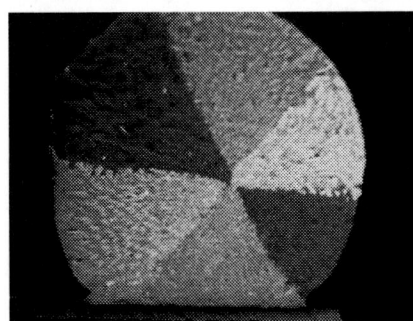

Figure 4: Cross polarizer image of typical SAC305 solder joint [7].

It has long been recognized [8 - 12] that the failure of such joints in thermal cycling tends to involve recrystallization within the Sn grains (Fig. 5). Qasaimeh et al. [13] showed crack growth to remain relatively slow until a continuous network of grain boundaries had been established across the high strain region of the joint, and Yin et al. [14] found that the completion of this network tended to take about 1/3 of the total life, independently of cycling parameters. This led to the suggestion that the life could be predicted based on a prediction of the rate of recrystallization.

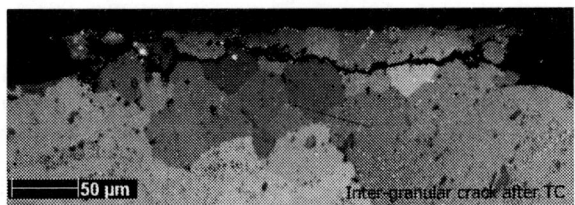

Figure 5: SAC305 solder with crack through recrystallized region near component pad after thermal cycling.

Korhonen et al. [15] showed that reproducing thermal cycling stresses and dwell times in isothermal cycling at any temperature between -25°C and 125°C was not sufficient to cause significant recrystallization. Borgesen et al. [16, 1] argued that the level of recrystallization shown in Figure 5 could only be achieved through the repeated alternations between the establishment of a dislocation cell structure at low temperature and the coalescence and rotation of these cells at a higher temperature characteristic of thermal cycling. Based on this, Borgesen et al. [1] have proposed a model according to which the rate of damage in thermal cycling varies with the density of dislocations and the high temperature dwell time. As far as the dislocation structure built up during the low temperature dwell is concerned, the subsequent recrystallization cannot be considered dynamic

and the density of importance is the one established by the end of the dwell. The authors argue that this must scale with the creep rate at that point. To the extent that the continuous addition of dislocations to this structure during the reversed loading at high temperature contributes to the recrystallization, this part can be considered dynamic. The authors argue that *this* contribution may be calculated based on the creep or work during the high temperature dwell *only*. Neither alternative is compatible with a scaling of life with the total work done in each cycle.

Based on such considerations and a number of simplifications Borgesen et al. proposed a first *very* simple approximation for the life in thermal cycling:

$$N_F = \psi / \{1 + \xi * t_{dwell}\} \qquad (1)$$

where ψ accounts for the dislocation density and the term in the parenthesis describes the coalescence and rotation during the high temperature dwell time, t_{dwell} [1]. While this approximation is practical the authors caution that it must be used with considerable care. Notably, ignoring initial precipitate distributions and subsequent coarsening the expression cannot account for effects of solder volume and pad size (Fig. 1), although these are predicted by the general model [1].

Results and Discussion

The results in Fig. 1 offer a particularly obvious example of the need for a quantitative mechanistic understanding of the behavior of SnAgCu solder joints in thermal cycling. The general picture outlined above allows us to rationalize a variety of systematic trends otherwise confounding our interpretation of accelerated test results [14, 1]. However, two important issues remain unresolved as far as the model is concerned.

The model relates the number of cycles to failure to the dislocation density established during the low and/or high temperature dwell, and thus to the near-steady state creep rate. The question remains as to what constitutive relations might be appropriate for the calculation of this creep rate. Constitutive relations are usually extracted from single sided creep experiments, while the dislocation cell structures built up under reversed cyclic loading are certain to be different. Furthermore, the repeated 'annealing' during the high temperature dwells affect the hardening, but as evidenced by the eventual recrystallization they do not completely eliminate the dislocation cell structures and the hardening associated with them. It is therefore not clear how best to account for contributions of primary creep in each cycle. We address this for use in the Borgesen model below.

The other issue is to which extent the recrystallization is in fact controlled by an ongoing production of dislocations during the high temperature dwell, i.e. whether the recrystallization process is dynamic. The relative importance of the high and the low temperature dwells is addressed in two different ways.

Isothermal Cycling Experiments

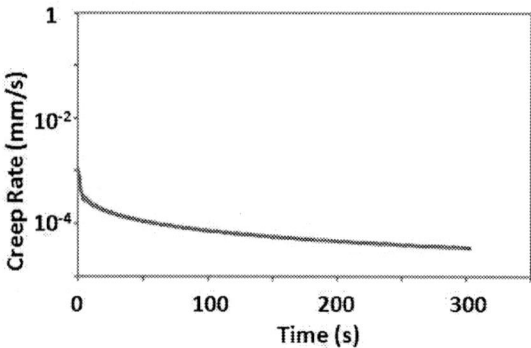

Figure 6: Creep rate vs. time under a constant load of 8MPa per joint in a typical cycle.

Our isothermal cycling experiments allow for the direct measurement of creep vs. time under conditions resembling those encountered in thermal cycling. Figure 6 shows an example for 0.75mm diameter SAC305 joints on Cu pads after a number of preceding cycles. The samples were loaded to 8MPa per joint at a rate of approximately 0.3MPa/s right after cool-down from annealing, and the load then kept constant for 5 minutes. Starting right after the 8MPa/joint is reached the room temperature creep rate is seen to drop rapidly over the first few seconds, but to level off substantially within less than 5 minutes.

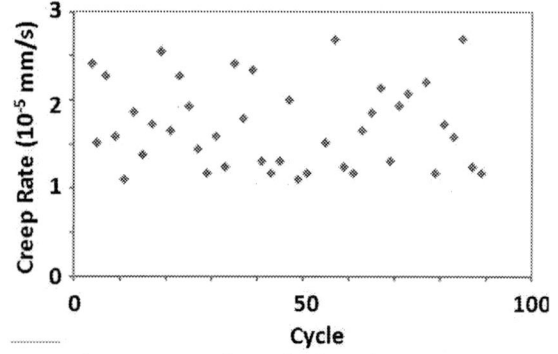

Figure 7: Creep rate at the end of 5 minutes under a constant load of 8MPa/joint vs. number of cycles.

Figure 7 shows the creep rate at the end of the 5 minute dwell as a function of number of cycles. For reasons associated with the current experimental procedure this measurement shows considerable scatter, but there is no long term trend. The same is true for the total displacement during each dwell (Fig. 8). We conclude that the initial 48 hour anneal tended to stabilize the precipitate distributions after which it only took a few cycles for the behavior to become largely repeatable. This is good news for FEM calculations as it may not be necessary to model every cycle after an initial anneal.

Calculations of, for example, the total work per thermal cycle would still require accurate accounting for major contributions from primary creep during both ramps and the early stages of each dwell. The Borgesen model does, however, rely only on calculations of the creep rate near the end of either the low temperature or the high temperature dwell, as this is expected to represent the relevant dislocation density. For dwell times longer than a few minutes these rates can be reasonably well approximated based on steady-state creep. Dutta et al. have developed microstructurally adaptive constitutive relations to account for the ongoing coarsening of the secondary precipitates during thermal cycling and aging [4]. We propose that these will be well suited for our purposes. Nevertheless, the extraction of functional dependencies in the model from comparisons between thermal cycling results and FEM are still expected to be very sensitive to the approximations still needed in the latter.

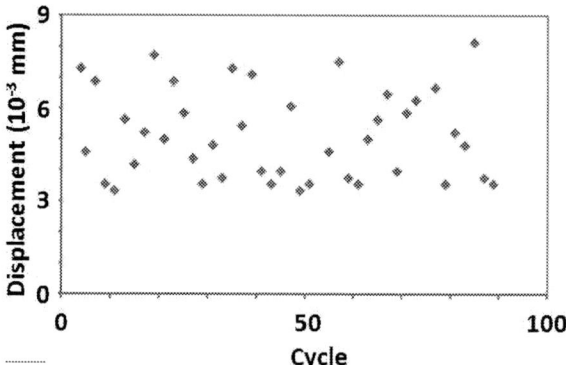

Figure 8: Total displacement at the end of 5 minutes under a constant load of 8MPa/joint vs. number of cycles.

Isothermal cycling is therefore ongoing to address dependencies on strain rates and temperatures, and if relevant work, directly. These experiments are extremely labor intensive and time consuming, but so far they appear to offer preliminary insight into the question of whether the recrystallization is in fact predominantly dynamic.

The experiments allow for the isolation of effects of individual parameters. Thus, an experiment was first conducted in which 0.75mm diameter SAC305 joints were loaded to 12MPa per joint for 5 minutes, and then loading was reversed and a displacement rate used that returned the sample to the original position within 5 minutes. This was then followed by a one hour anneal at 100°C before the loading sequence was repeated.

Figure 9: Cross polarizer images of near-pad region in two joints after 50 cycles to 12 MPa/joint.

Figure 9 shows cross polarizer images of two joints after 50 such cycles. Recrystallization is evident but still much less than what is commonly observed in thermal cycling (Fig. 5). This is, however, believed to be a result of competing damage mechanisms. Isothermal cycling of SAC305 usually leads to failure by transgranular crack growth before much recrystallization [15, 17, 18]. Similarly, in thermal cycling transgranular crack growth competes with the formation of a

978-1-4799-2408-0/14 $31.00 © 2014 IEEE   374

continuous network of grain boundaries across the high strain region in the joint. Joints may therefore also fail by transgranular crack growth in thermal cycling if the cyclic strain range is high enough.

Indeed, Figure 10 shows a cross polarizer image of a typical SAC305 joint in our highest strain CSP assemblies after failing in -40/125C cycling. The joint failed after only 149 cycles. The image shows a very limited amount of recrystallization compared to the level shown in Figure 5. Clearly, the sample failed before an obvious network of high angle grain boundaries had evolved across the joint.

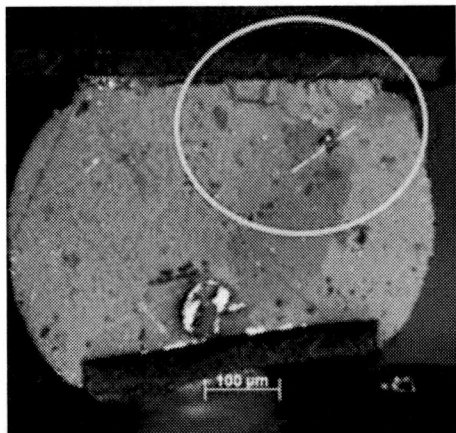

Figure 10: Cross polarizer image of 20 mil diameter SAC305 solder joint with crack after 149 cycles of -40/125°C.

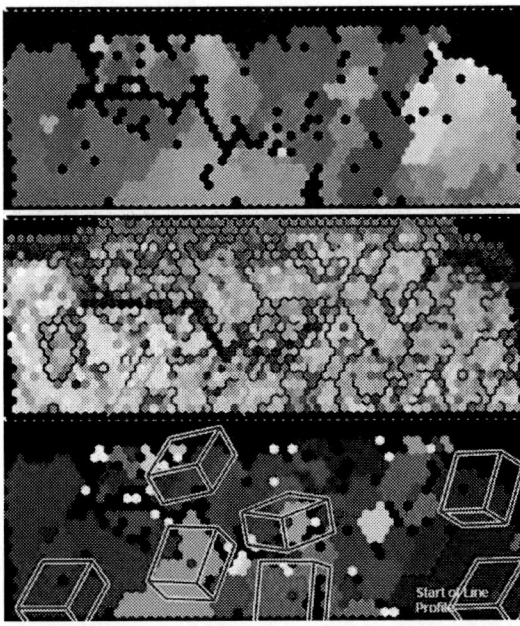

Figure 11: EBSD of the region circled in Figure 10. Top – color map showing different Sn orientations; middle – lines indicating boundary misorientations (red 2-5°, green 5-10°, blue >10°).

This does not mean that recrystallization was not involved. Figure 11 shows an EBSD image of the region circled in Figure 10. A large density of boundaries with misorientations in excess of 10° is observed in the area of the crack. Other

joints surviving a similar number of cycles showed more or less recrystallization. Work is in progress to assess whether this variability is consistent with expected effects of Sn grain orientations.

According to the current version of the Borgesen model [1] reproducing this in our isothermal cycling experiment will require a similar number of cycles and a similar dislocation density in each cycle. If controlled by transgranular crack growth the failure of a SnAgCu joint in isothermal cycling occurs upon the accumulation of a given amount of total work [19, 20]. This is only an approximation, the total work to failure increasing systematically with loading rate, and recent work suggests that a limited variation with cycling amplitude is associated with an increase in the amount of work going to heat due to friction between the opposing surfaces of large cracks toward the end of life [20]. Otherwise, the work to failure did not vary much with temperature, dwell times, or repeated variations in amplitude. Overall, as long as they are controlled by transgranular crack growth experimental life times can thus be extrapolated quite well to lower stresses based on the assumption of a constant work to failure.

In the experiment above cycling to 12MPa per joint led to an average work of 1.4×10^{-4}J per cycle and a total life of 50 cycles. Cycling to 6MPa per joint led to an average work of 1.4×10^{-5}J per cycle, suggesting a life of about 500 cycles, while cycling to 8MPa per joint led to an average of 2.7×10^{5}J per cycle, and thus an estimated life of 260 cycles.

These estimates are upper limits based on the assumption of failure by transgranular crack growth at the lower stresses as well. Recrystallization may cause faster failure than that. Cycling is ongoing at both loads to address that. The predicted life at 8MPa is about twice that of the joints in Figure 10 and 11. This may suggest that the low temperature strain rate is lower, but there may also be time for more recrystallization before failure.

Figure 12: Cross polarizer image of SAC305 solder joint after 100 cycles of 'simulated thermal cycling' (alternating isothermal cycling and annealing).

So far, cycling to 100 cycles with 8MPa per joint led to a level of recrystallization near the pad that is very limited (Figure 12), but seemingly of the same general magnitude as in the thermal cycling sample in Figure 10 even though the joints are expected to last longer. This would seem to support a picture in which the dislocation density established during the low temperature part of the thermal cycle controls the degree of coalescence and rotation during the high temperature dwell that eventually leads to recrystallization.

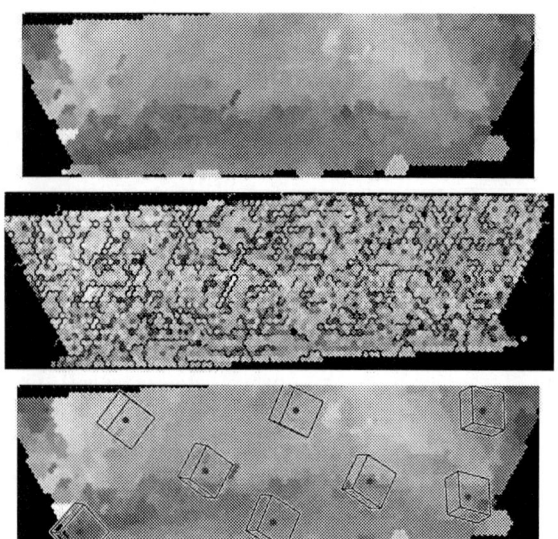

Figure 13: EBSD of the near-pad region in Figure 12. Top – color map showing different Sn orientations; middle – lines indicating boundary misorientations (red 2-5°, green 5-10°, blue >10°).

Figure 13 provides for a more accurate comparison to the EBSD image in Figure 11. At the top different Sn grain orientations are indicated by different colors. The center image shows small, medium and large boundary misorientations, and the image at the bottom shows a graphic representation of some of the grain orientations. Most misorientations are still small, meaning that this is still early in the dynamic recrystallization process, but the overall picture leaves little doubt that this represents the early stages of the kind of recrystallization seen in Figure 5.

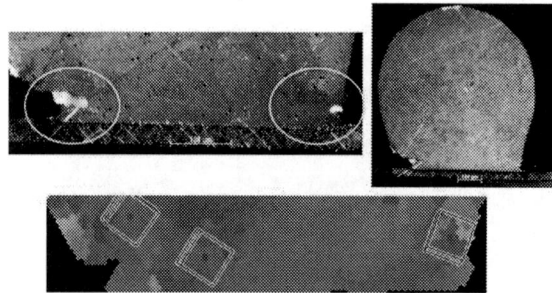

Figure 14: Cross polarizer and EBSD images of SAC305 solder joint after 100 cycles of 'simulated thermal cycling'.

Figure 14 finally shows another joint also subjected to 100 cycles. Unlike the one above this joint shows no indication of the beginning of 'global' recrystallization across the high strain region of the joint. Instead, it shows the onset of two cracks at either side of the joint.

The crack on the right side is very small, less than 10µm in length, and there is no significant misorientation near the tip (Figure 15). This would be consistent with the assumption

of a competition between transgranular crack growth and recrystallization, thermal cycling also often leading to a limited amount of the former before completion of a network of high angle grain boundaries by the latter [13, 14].

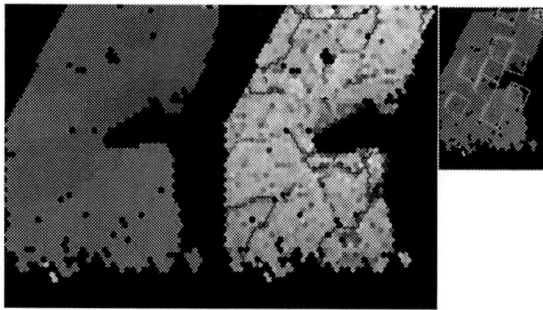

Figure 15: EBSD of the small crack at the right side of the joint in Figure 14.

The crack on the left is considerably longer, almost 50µm, and here there are a significant number of high angle boundaries near the tip (Figure 16). In fact, angles approach 60° right at the tip. This might support the assumption that the apparent transgranular crack growth mechanism also involves recrystallization [1]. Since it is not controlled by the formation of an entire network of high angle boundaries modeling of this, and thus prediction of life in isothermal cycling, will however still need to be different.

As reflected in the lower parts of Figures 13 and 14 the Sn grain orientations are quite similar. Work is ongoing to assess whether the differences in recrystallization are associated with the small difference in orientation or with the local distributions of secondary precipitates.

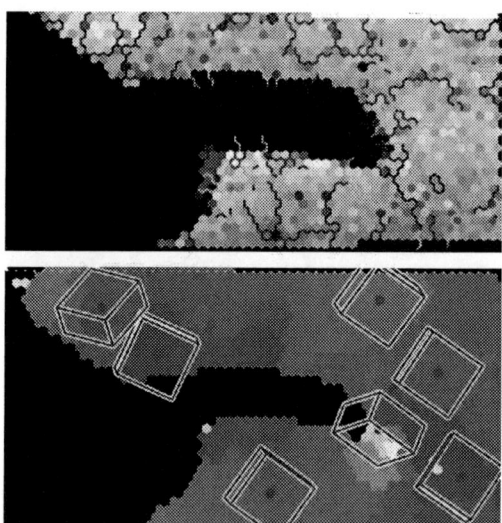

Figure 16: EBSD of the larger crack at the left side of the joint in Figure 14.

Thermal Cycling Experiments

The above was complemented by conventional thermal cycling of model assemblies. Figure 17 shows first a Weibull plot reflecting the effect of the high temperature dwell on the life of model CSP assemblies with 0.4mm diameter SAC305 solder joints. Lowering the dwell temperature from 100°C to 80°C while keeping the low temperature dwell (0°C), dwell times and ramp rates the same led to a slight broadening of the failure distribution and an increase in characteristic life, N_{63}, by a factor of 2.2.

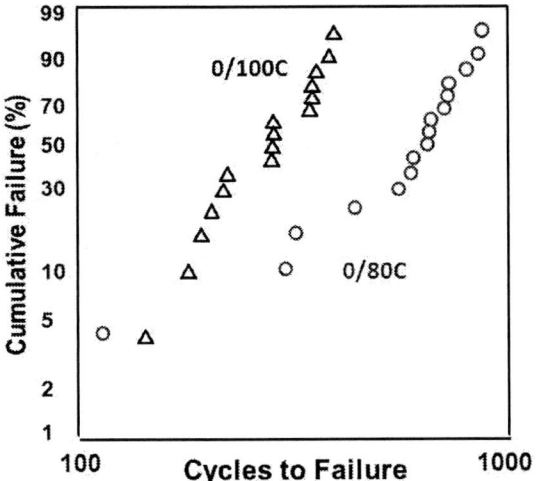

Figure 17: Cumulative failure distributions for model CSP assemblies in 0/100°C and 0/80°C thermal cycling with 15 minute dwells at both temperature extremes.

The effect of the low temperature dwell was much weaker. Figure 18 shows the life of 0.4mm diameter SAC305 solder joints in -20/100°C cycling plotted against the corresponding life in 0/100°C cycling for each of five different model CSP designs. The combined uncertainties of two independently measured N_{63} values leads to significant scatter, but there is no systematic trend with respect to cyclic strain range. On average, lowering the minimum temperature to raise the temperature range (ΔT) by 20% lead to a 10% reduction in life.

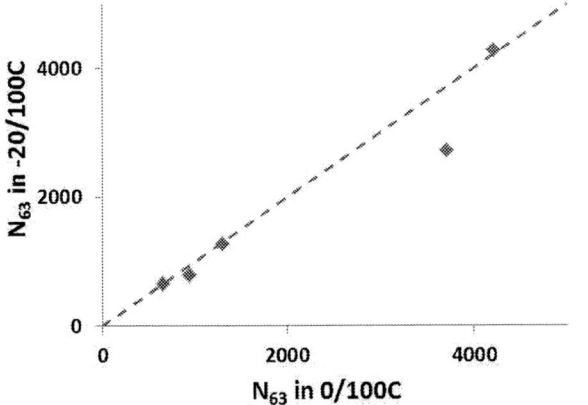

Figure 18: Characteristic life, N_{63}, in -20/100°C vs. life in 0/100°C for model CSP assemblies with 10 minute dwells at both temperature extremes. Broken line represents the life in the two cycles being equal.

The stronger sensitivity to the high temperature dwell is not a result of more effective coalescence and rotation of the dislocation cells there. The dwell time parameter ξ in Equation 1 was found to vary slowly, or not at all, with the dwell temperature. This would suggest that the dislocation density determining the rate of recrystallization during a given cycle, and thus the parameter ψ, depends significantly on the maximum temperature. While the dislocation cell structure established at low temperature is important, the contribution of dislocations created during the high temperature dwell may be dynamic, i.e. according to Borgesen et al. [1] this may actually depend on the work there. The stresses at the high temperature are of course sensitive also to the difference between this and the low temperature extreme, especially at the beginning of the dwell, affecting the creep rates as well. Nevertheless, it seems worth noting that according to the constitutive relations proposed by Dutta et al. [4] the steady state creep rate *at the same stress* is a factor of 2.2 higher at 100°C than at 80°C, in excellent agreement with the results in Figure 17.

Conclusion

Except in very high strain assemblies and/or harsh cycling conditions in which life is less than a couple of hundred cycles, it has been shown that the life of a SnAgCu solder joint in thermal cycling is controlled by the evolution of a continuous network of high angle grain boundaries across the high strain region of the joint.

Interpretations of thermal cycling experiments are hampered by the inability to vary one important parameter at a time, and by the lack of direct knowledge of the stresses and/or strain rates involved. Not surprisingly, current constitutive relations do not allow for calculation of the latter except perhaps after the first few minutes in each dwell.

Solder joints were loaded alternately with thermal annealing while measuring loads and creep rates directly. This allowed for simulation of thermal cycling, except that no load was applied at the high temperature. Preliminary results support a picture in which continuous recrystallization is a result of the build-up of dislocation cell structures during the low temperature dwell and the coalescence and rotation of these during the high temperature dwell. This would mean that the recrystallization cannot be dynamic.

Accelerated thermal cycling results do, however, show life in thermal cycling to be much more sensitive to the maximum than the minimum temperature. This would seem to suggest that contributions from dynamic recrystallization during the high temperature dwell are still important. Our mechanical loading and annealing experiments are being extended to include a reversed load at high temperature in order to assess this directly.

Acknowledgments

This research was supported by the National Science Foundation under Grant No. DMR 1206474. The assistance of Michael Meilunas, Universal Instruments, with sample design and preparation is gratefully acknowledged.

References

1. Borgesen, P., Wentlent, L., D. Schmitz, S. Shirazi, Yin, L., Qasaimeh, A., Hamasha, S., Meilunas, M., and Anselm, M., "On the Interpretation of Thermal Cycling Test Results for SnAgCu Solder Joints", in preparation

2. Zhang, Y., Cai, Z., Suhling, J. C., Lall, P., and Bozack, M. J., "The Effects of Aging Temperature on SAC Solder Joint Material Behavior and Reliability", Proc. ECTC 2008, 99-112

3. Dutta, I., Pan, D., Marks, R. A. and Jadhav, S. G., "Effect of thermo-mechanically induced microstructural coarsening on the evolution of creep response of SnAg-based microelectronic solders", Mater Sci. Eng. A 410-411, 2005, pp. 48-52.

4. Kumar, P., Huang, Z., Chavali, S., Chan, D., Dutta, I., Subbarayan, G., and Gupta, V., "A Microstructurally Adaptive Model for Primary and Secondary Creep of Sn-Ag-based Solders", IEEE Trans. Comp., Packag. & Manuf. Technol. 2 (2012) pp. 256-265

5. Borgesen, P., Meilunas, M., Jiang, J., Qasaimeh, A., Hamasha, S., and Anselm, M., "On the Applicability of Current Thermal Cycling Models to SnAgCu Solder Joints", in preparation

6. Lehman, L., Xing, Y., Bieler, T., and Cotts, E., "Cyclic twin nucleation in tin-based solder alloys", Acta Mater. 58, 3546 (2010).

7. Qasaimeh, A., Jaradat, Y., Wentlent, L., Yang, L., Yin, L., Arfaei, B., and Borgesen, P., "Recrystallization Behavior of Lead Free and Lead Containing Solder in Cycling", Proc. 61st ECTC (2011) pp. 1775-1781

8. Henderson, D. W., Woods, J. J., Gosselin, T. A., Bartelo, J., King, D. E., Korhonen, T. M., Korhonen, M. A., Lehman, L. P., Cotts, E. J., Kang, S. K., Lauro, P., Shih, D.-Y., Goldsmith, C., and Puttlitz, K. J., "The microstructure of Sn in near-eutectic Sn-Ag-Cu alloy solder joints and its role in thermomechanical fatigue", J. Mat. Res. 19 (2004) pp. 1608

9. Terashima, S., Kobayashi, T., and Tanaka, M., "Effect of crystallographic anisotropy of β-tin grains on thermal fatigue properties of Sn–1Ag–0.5Cu and Sn–3Ag–0.5Cu lead free solder interconnects", Sci. Technol. Weld. Join. 13 (2008) pp. 732

10. Telang, U., Bieler, T. R., Zamiri, A., and Pourboghrat, F., "Incremental recrystallization/grain growth driven by elastic strain energy release in a thermomechanically fatigued lead-free solder joint", Acta Mater. 55 (2007) pp. 2265

11. Sundelin, J., Nurmi, S., and Lepistö, T., "Recrystallization behaviour of SnAgCu solder joints", Mater. Sci. Eng. A-Struct. 474 (2008) pp. 201

12. Mattila, T., and Kivilahti, J., "The Role of Recrystallization in the Failure of SnAgCu Solder Interconnections Under Thermomechanical Loading", IEEE Trans. Compon. Pack T. 33 (2010) pp. 629

13. Qasaimeh, A., Lu, S., and Borgesen, P., "Crack Evolution and Rapid Life Assessment for Lead Free Solder Joints", Proc. 61st ECTC, 2011, pp. 1283-90

14. Yin, L., Wentlent, L., Yang, L., Arfaei, B., Qasaimeh, A., and Borgesen, P., "Recrystallization and Precipitate Coarsening in Pb-free Solder Joints during Thermo-mechanical Fatigue", J. Electronic Materials 41 (2012) pp. 241-252

15. Korhonen, T. K., Lehman, L., Korhonen, M. A., and Henderson, D. W., "Isothermal fatigue behaviour of the near-eutectic Sn–Ag–Cu alloy between -25 and 125C", J. Electron. Mater. 36 (2007) pp. 173–178.

16. P. Borgesen, L. Yang, A. Qasaimeh, L. Yin, and M. Anselm, "A Mechanistically Justified Model for Life of SnAgCu Solder Joints in Thermal Cycling", Proc. SMTA Pan Pacific Conf. 2013

17. Mayyas, A., Yin, L., and Borgesen, P., "Recrystallization of Lead Free Solder Joints – Confounding the Interpretation of Accelerated Thermal Cycling Results", Proc. ASME Int. 2009, IMECE2009-12749

18. Yang, L., Yin, L., Roggeman, B., and Borgesen, **P.**, "Effects of Microstructure Evolution on Damage Accumulation in Lead Free Solder Joints", *Proc. 60th ECTC* (2010) pp. 1518-1523

19. Qasaimeh, A., 'Study of the Damage Evolution Function for SnAgCu in Cycling', Ph. D. dissertation, Binghamton University, May 2012

20. Hamasha, S., Qasaimeh, A., Jaradat, Y., and Borgesen, P., "Correlation Between Solder Joint Fatigue Life and Accumulated Work in Isothermal Cycling", in preparation

21. P. Borgesen, E. Al-Momani, and M. Meilunas, "Effect of Dwell Time on the Life of Lead Free BGA Joints in Thermal Cycling", Proc. SMTA Int. 2009

Exploration of Aging Induced Evolution of Solder Joints
Using Nanoindentation and Microdiffraction

Mohammad Hasnine, Jeffrey C. Suhling,
Barton C. Prorok, Michael J. Bozack, Pradeep Lall
Center for Advanced Vehicle and Extreme Environment Electronics (CAVE[3])
Auburn University
Auburn, AL 36849
Phone: +1-334-844-3332
FAX: +1-334-844-3124
E-mail:jsuhling@eng.auburn.edu

Abstract

Due to aging phenomena, the microstructure, mechanical response, and failure behavior of lead free solder joints in electronic assemblies are constantly evolving when exposed to isothermal and/or thermal cycling environments. In our ongoing studies, we are exploring aging phenomena by nano-mechanical testing of SAC lead free solder joints extracted from PBGA assemblies. Using nanoindentation techniques, the stress-strain and creep behavior of the SAC solder materials are being explored at the joint scale for various aging conditions. Mechanical properties characterized as a function of aging include the elastic modulus, hardness, and yield stress. Using a constant force at max indentation, the creep response of the aged and non-aged solder joint materials is also being measured as a function of the applied stress level. With these approaches, aging effects in actual solder joints are being quantified and correlated to the magnitudes of those observed in testing of miniature bulk specimens.

In our initial work (ECTC 2013), we explored aging effects in single grain SAC305 solder joints. In the current investigation, we have extended our previous work on nanoindentation of joints to examine a full test matrix of SAC solder alloys. The effects of silver content on SAC solder aging has been evaluated by testing joints from SACN05 (SAC105, SAC205, SAC305, and SAC405) test boards assembled with the same reflow profile. In all cases, the tested joints were extracted from 14 x 14 mm PBGA assemblies (0.8 mm ball pitch, 0.46 mm ball diameter) that are part of the iNEMI Characterization of Pb-Free Alloy Alternatives Project (16 different solder joint alloys available). After extraction, the joints were subjected to various aging conditions (0 to 12 months of aging at T = 125 C), and then tested via nanoindentation techniques to evaluate the stress-strain and creep behavior of the four aged SAC solder alloy materials at the joint scale.

The observed aging effects in the SACN05 solder joints have been quantified and correlated with the magnitudes observed in tensile testing of miniature bulk specimens performed in prior studies. The results show that the aging induced degradations of the mechanical properties (modulus, hardness) in the SAC joints were of similar order (30-40%) as those seen previously in the testing of larger "bulk" uniaxial solder specimens. The creep rates of the various tested SACN05 joints were found to increase by 8-50X due to aging. These degradations, while significant, were much less than those observed in larger bulk solder uniaxial tensile specimens

with several hundred grains, where the increases ranged from 200X to 10000X for the various SACN05 alloys. Additional testing has been performed on very small tensile specimens with approximately 10 grains, and the aging-induced creep rate degradations found in these specimens were on the same order of magnitude as those observed in the single grain joints. Thus, the lack of the grain boundary sliding creep mechanism in the single grain joints is an important factor in avoiding the extremely large creep rate degradations (up to 10,000X) occurring in larger bulk SAC samples. All of the aging effects observed in the SACN05 joints were found to be exacerbated as the silver content in the alloy was reduced. In addition, the test results for all of the alloys show that the elastic, plastic, and creep properties of the solder joints and their sensitivities to aging are highly dependent on the crystal orientation.

The observed mechanical behavior changes in joints are due to evolution in the microstructure and residual strains/stresses in the solder material, and measurements of these evolutions are critical to developing a fundamental understanding of solder joint aging phenomena. As another part of this work, we have performed an initial study of these effects in the same SAC305 solder joints that were tested using nanoindentation. The enhanced x-ray microdiffraction technique at the Advanced Light Source (Synchrotron) at the Lawrence Berkeley National Laboratory was employed to characterize several joints after various aging exposures (0, 1, and 7 days of aging at T = 125 C). For each joint, microdiffraction was used to examine grain growth, grain rotation, sub-grain formation, and residual strain and stress evolution as a function of the aging exposure. The entire joints were scanned using a 10 micron step size, and the results were correlated with changes in the mechanical response of the joint specimens measured by nanoindentation.

Introduction

Solder joint fatigue is one of the predominant failure mechanisms in lead free electronic assemblies exposed to thermal cycling. Thus, accurate mechanical properties and constitutive equations for solder materials are needed for use in mechanical design, reliability assessment, and process optimization. Ma, et al. [1] has reviewed the literature on the mechanical behavior of lead free solders. The mechanical properties of a lead free solder are strongly influenced by its microstructure, which is controlled by its thermal history including solidification rate and thermal aging after solidification. Due to aging phenomena, the microstructure, mechanical response, and failure behavior of lead free solder joints in electronic assemblies are constantly evolving when exposed to isothermal aging and/or thermal cycling environments. Such aging effects are greatly exacerbated at higher temperatures typical of thermal cycling qualification tests. However, significant changes occur even with aging at room temperature.

The effects of aging on the constitutive and failure behavior of lead free solder has been extensively investigated by the authors [2-14]. In initial work, room temperature (25 C) and elevated temperature (50, 75, 100, and 125 C) aging were shown to severely degrade the mechanical properties and creep behavior of SAC alloys [2-5]. The measured stress-strain data demonstrated large reductions in stiffness, yield stress, ultimate strength, and strain to failure (up to 50%) during the first 6 months after reflow solidification. After approximately 10-20 days of aging, the lead free solder joint material properties were observed to degrade at a slow but constant rate. In addition, even more dramatic evolution was observed in the creep response of aged solders, where up to 10,000X increases in the secondary creep rates were observed for aging up to 6 months. The aged solder materials were also found to enter the tertiary creep range (imminent failure) at much lower strain levels than virgin solders (non-aged, tested immediately after reflow solidification). Cai and coworkers [6] have demonstrated that such aging effects can be reduced through the use of doped SAC solder materials (i.e., SAC-X).

Mustafa, et al. have studied the aging-induced changes occurring in the cyclic stress-strain behavior of lead free SAC solders for both tension/compression [7] and shear [8] loadings. The effects of several parameters were examined including aging time, temperature, strain/stress limits, and solder alloy composition. In addition, the evolution of the solder hysteresis loops with aging were characterized and empirically modeled. Similar to solder stress-strain and creep behavior, there was a strong influence of aging on the hysteresis loop size (and thus the rate of damage accumulation) in the solder specimens.

Zhang, et al. [9-10] have shown that prior aging causes large reductions in the reliability of BGA test assemblies subjected thermal cycling accelerated life testing. Motalab, et al. [11-12] have included aging effects in the Anand constitutive model and energy density based failure criterion for SAC solders, and then used these theories with finite element analyses to predict the thermal cycling life of aged BGA assemblies. Good correlations were achieved with the measured lifetimes from references [9-10]. Finally, Lall and coworkers [13] have investigated changes in the high strain rate behavior of SAC solders subjected to aging.

The prior work on aging discussed above has been based on uniaxial testing of miniature bulk solder tensile specimens. These samples were solidified in glass tubes under a controlled temperature profile in an effort to accurately match the microstructure of actual solder joints. Complementary studies by other research groups have verified aging induced degradations of SAC mechanical properties. In those investigations, mechanical testing was performed on a variety of sample geometries including lap shear specimens,

Iosipescu shear specimens, and custom solder ball array shear specimens.

There have been limited prior mechanical loading studies on aging effects in actual solder joints extracted from area array assemblies (e.g. PBGA or flip chip) [15-18]. This is due to the extremely small size of the individual joints, and the difficulty in gripping them and applying controlled loadings (tension, compression, or shear). Pang, et al. [15] have measured microstructure changes, intermetallic layer growth, and shear strength degradation in custom SAC single ball joint lap shear specimens subjected to elevated temperature aging. Darveaux [16] performed an extensive experimental study on the stress-strain and creep behavior of solder using specially constructed double lap shear specimens with a 10 x 10 area array solder balls. He found that aging for 1 day at 125 C caused significant effects on the stress-strain and creep behavior. For example, aged specimens were found to creep much faster than non-aged specimens by a factor of up to 20 times for both SAC305 and SAC405 solder alloys. Wiese, et al. [17] also studied the effects of aging on solder joint creep using custom assemblies with 4 flip chip solder balls, and found highly accelerated creep rates after aging at 125 C. Finally, Dutta and coworkers [18] used an impression creep technique with a cylindrical punch to study creep in PBGA solder balls that had been subjected to thermal-mechanical cycling.

Nanoindentation techniques have been widely used to probe the mechanical properties and deformation behavior of extremely small material samples [19]. Over the past decade, it has been applied by several investigators to characterize lead free solder joints and intermetallic compounds (IMC) in lead free solder joints [20-34]. In early studies by Rhee, et al. [20-21], Chromik, et al. [22], and Deng and coworkers [23-24], the elastic modulus E and hardness H of various regions in Sn-Ag and SAC lead free solder joints were explored by nanoindentation. In particular, the properties for the Sn-rich phase (β-tin) and the eutectic phase (containing β-tin and a mix of Sn-Ag and Sn-Cu intermetallics) were explored. Attempts were also made to indent individual Sn-Ag and Sn-Cu intermetallic particles [22-24], and then to compare the properties of the IMCs to those for the solder joint phases. Rhee, et al. [20] also measured changes in the mechanical properties after the joints were subjected to thermal-mechanical cycling.

Gao, et al. [25-26] have used nanoindentation to characterize the effects of loading rate on the modulus and hardness of Sn-Ag lead free solders. They also performed creep experiments for two different microstructures (bulk cast and reflowed), and examined the effects of elevated temperature aging on the hardness. Indentation experiments with a heated stage to control the solder sample temperature have been performed by Sun, et al. [27], Liu, et al. [28], Gao, et al. [29], and Han and coworkers [30]. The alloys tested in these studies included SAC387, 80Au-20Sn, SAC305, and SAC357, respectively. In all of these investigations, the temperature dependencies of the mechanical properties (E, H) of the solder matrix or individual solder phases were characterized. In addition, the sensitivity of the creep response of lead free solder to temperature has also been examined [28-30]. Han, et al. [31] have studied the

978-1-4799-2408-0/14 $31.00 © 2014 IEEE   380

indentation size effect on the creep behavior of SAC357 lead free solder.

The effect of thermal aging on the mechanical properties of intermetallic compounds at SAC solder joint interfaces have been explored using nanoindentation by Xu and Pang [32-33] and Song, et al. [34]. Significant drops in both the modulus and hardness were recorded for aged samples relative to non-aged samples. Xu and Pang [32] also characterized the mechanical properties of the various phases and IMCs in a SAC387 solder joint. Venkatadri, et al. [35] have studied the effects of aging on lead free solder joints using a micro-hardness test to perform single idents on joints.

As discussed above, there is an extensive literature that documents the large changes in the microstructure and mechanical behavior that occur in bulk lead free solder specimens during isothermal aging. There have also been some limited investigations on the effects of aging on mechanical properties and creep behavior of solder joint arrays. In addition, nanoindentation has been utilized to study aging induced changes in the mechanical properties of intermetallic compound layers in solder joints. However, there has been little work on aging effects on mechanical properties and creep behavior in individual solder joints. Such knowledge is crucial for the optimizing the design, manufacturing, and reliability of microelectronic packages. Characterization of individual joints is quite challenging because of their extremely small size, and the difficulty in gripping them and applying controlled loadings.

In our ongoing studies, we are exploring aging phenomena by nano-mechanical testing of SAC lead free solder joints extracted from PBGA assemblies. Using nanoindentation techniques, the stress-strain and creep behavior of the SAC solder materials are being explored at the joint scale for various aging conditions. In our initial work (ECTC 2013), we explored aging effects in single grain SAC305 solder joints [14]. In the current paper, we have extended our previous work on nanoindentation of joints to examine a full test matrix of SAC solder alloys. The effects of silver content on SAC solder aging has been evaluated by testing joints from SACN05 (SAC105, SAC205, SAC305, and SAC405) test boards assembled with the same reflow profile. In all cases, the tested joints were extracted from 14 x 14 mm PBGA assemblies (0.8 mm ball pitch, 0.46 mm ball diameter) that are part of the iNEMI Characterization of Pb-Free Alloy Alternatives Project (16 different solder joint alloys available). After extraction, the joints were subjected to various aging conditions (0 to 12 months of aging at T = 125 C), and then tested via nanoindentation techniques to evaluate the stress-strain and creep behavior of the four aged SAC solder alloy materials at the joint scale.

The observed aging effects in the SACN05 solder joints have been quantified and correlated with the magnitudes observed in stress-strain and creep testing of miniature bulk specimens performed in prior studies, as well as with new test results for tests performed on very small tensile specimens with approximately 10 grains. In addition, the influence of the silver content of the SACN05 solder joints on the observed aging degradations has been explored in detail.

The measured mechanical behavior changes in joints are due to evolution in the microstructure and residual strains/stresses in the solder material, and measurements of these evolutions are critical to developing a fundamental understanding of solder joint aging phenomena. As another part of this work, we have performed an initial study of these effects in the same SAC305 solder joints that were tested using nanoindentation. The enhanced x-ray microdiffraction technique at the Advanced Light Source (Synchrotron) at the Lawrence Berkeley National Laboratory was employed to characterize several joints after various aging exposures (0, 1, and 7 days of aging at T = 125 C). For each joint, microdiffraction was used to examine grain growth, grain rotation, sub-grain formation, and residual strain and stress evolution as a function of the aging exposure. The entire joints were scanned using a 10 micron step size, and the results were correlated with changes in the mechanical response of the joint specimens measured by nanoindentation.

Experimental Procedure

Solder Joint Samples

The lead free solder joints in this study were extracted from PBGA assemblies (Amkor CABGA, 14 x 14 mm, 192 balls, 0.8 mm ball pitch, 0.46 mm ball diameter). The test boards were assembled as part of the iNEMI Characterization of Pb-Free Alloy Alternatives Project [36], and a variety of samples with 16 different solder ball alloys are being studied. In this work, the BGA ball alloys considered were SAC105 (98.5Sn-1.0Ag-0.5Cu), SAC205 (97.5Sn-2.0Ag-0.5Cu), SAC305 (96.5Sn-3.0Ag-0.5Cu), and SAC405 (95.5Sn-4.0Ag-0.5Cu). These 4 alloys have been collectively referred to as SACN05 in this paper, with N representing the integer percentage value (1, 2, 3, 4) of the silver content of the ball alloy of the unassembled BGA component. The test boards had ENIG surface finish, and SAC305 solder paste was used in the surface mount assembly process for all of the ball alloys. Thus, with the exception of the SAC305 components, the final joint compositions of the assembled BGA components were not exactly the same as the ball alloys in the unassembled parts.

The assembled PBGA components were cut out from the test boards and then cross-sectioned. These samples were mounted in an epoxy molding compound suitable for SEM microscopy, and then polished to a level appropriate for nanoindentation. Details of the sample preparation process include mechanical grinding with several SiC papers (#320 to #400, #600, #800 and #1200), and then final polishing with 1 µm diamond paste followed by 0.05 µm colloidal silica suspensions. This resulted in mirror finish samples suitable for nanoindentation tests and SEM microscopy.

Nanoindentation System and Test Procedures

The nanoindentation tests were conducted using an instrumented MTS Nanoindenter XP system with a Berkovich tip indenter. The load versus indentation displacement normal to the cross-section surface was measured during each indentation experiment, and the elastic modulus could then be extracted using the approach proposed by Oliver and Pharr [36-37] to process the measured slope of the load-displacement curve in the unloading phase. In addition, the

Continuous Stiffness Measurement (CSM) technique [38-39] was also used in all experiments to extract elastic modulus and hardness as a function of the distance from the surface (indentation depth). The yield stress of the solders was estimated from the measured hardness using the Tabor relationship [14, 40-41].

A typical cross-sectioned lead free solder ball sample after nanoindentation testing is shown in Figure 1, and a close-up view of an example permanent indentation mark is shown in Figure 2. For each set of experimental test conditions, an array of several indents spaced 30 μm apart were made (e.g. 2 x 3 array as shown in Figure 1), and the individual indent test values were averaged to obtain statistically relevant results and consistency of inspection. All tests were performed on single grain (Sn crystal) solder balls, so that there were no orientation effects caused by an indentation array covering two or more grain boundaries of grains with significantly different crystal orientation (different material properties).

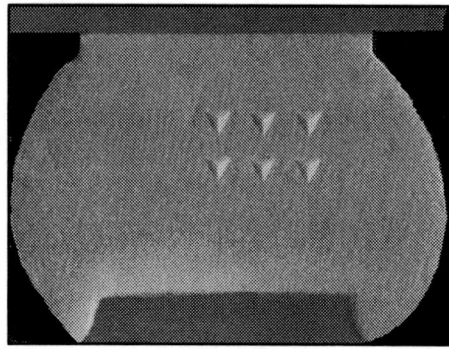

Figure 1 - Solder Ball after Nanoindentation Testing

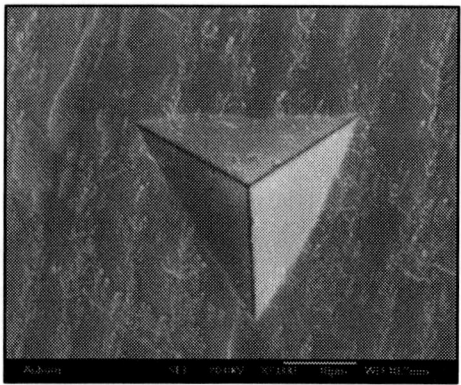

Figure 2 - Permanent Indentation after Testing

A maximum load of 30 mN was selected for the tests so that the indentation marks were large enough to cover all the phases of solder material (see Figure 2). Thus, the nanoindentation tests characterized the global mechanical properties of the solder joints, rather than the localized properties in the β-Sn phases (Sn-rich dendrites) or in the eutectic phases between dendrites that contain Sn-Ag and Sn-Cu intermetallics. Indentation experiments were conducted at constant indentation displacement rate of 10 nm/sec, corresponding to an effective strain rate of 0.05 sec^{-1}.

Calibration of the load and hardness measurements was performed on fused silica. A holding period of 60 sec was employed after 90% unloading to allow for thermal drift correction. In our work, the thermal drifts were kept smaller than 0.05 nm/sec, and the thermal drift effects were excluded from the resulting displacement data.

Indentation creep tests were performed holding the load constant at its maximum value of 30 mN for 500-1000 sec. After this dwell, the sample was unloaded at the same 10 nm/sec displacement rate used during loading. The method for calculating the creep strain rate proposed by Mayo and Nix [42-43] has been adopted in this study, as well as log hyperbolic tangent creep model proposed by Chhanda, et al. [44]. In addition, the approach presented by the authors in reference [14] has been utilized to predict (extrapolate) tensile creep strain rates for low stress levels using nanoindentation creep data measured at very high compressive stress levels.

The measurement and processing of nanoindentation load vs. displacement data for lead free solders are discussed in detail in our earlier paper [14]. Example measurement results are included for a SAC305 solder joint, as well as the calculations of the elastic modulus, hardness, yield stress, and creep strain rate vs. stress curves.

Aging Studies on SAC Solder Joints

Aging phenomena in lead free solder joints have been explored by nanoindentation testing of several SAC samples exposed to various aging conditions prior to testing. The SAC alloys considered were from BGA components fabricated using four different SACN05 ball alloys (SAC105, SAC205, SAC305, and SAC405). Using the nanoindentation methods detailed above, the stress-strain and creep behavior of the SAC solder joints have been explored for various aging conditions. Mechanical properties characterized as a function of aging include the elastic modulus and hardness, as well as the creep strain rate vs. applied stress response. With these approaches, aging effects in solder joints were quantified and correlated to the magnitudes of those observed in stress-strain and creep testing of miniature bulk specimens with hundreds of grains performed in prior studies [2-6]. In addition, new creep experiments with aging have been performed on very small tensile specimens with approximately 10 grains, and these data have also been correlated with the nanoindentation results.

The SACN05 lead free solder joints in this study are composed of 95+% Sn. Thus, the properties of Sn dominate the behavior of these alloys. Sn has a Body-Center Tetragonal (BCT) crystal structure, and each grain exhibits highly anisotropic characteristics in its mechanical, thermal, and diffusion properties. With lattice constants of a = b = 5.83Å, much larger than that of the c axis, c = 3.18 Å, the elastic modulus and hardness of Sn are highly dependent on grain orientation. For example, the modulus of elasticity of tin in the (001) direction (67.6 GPa) is three times higher than that in the (100) direction (23.6 GPa) [45]. Single BGA solder joints have been observed to typically have 1-6 grains after reflow [46-48]. The Sn grain/crystal orientations and the intermetallic phases present in a solder joint will play important roles in determining its mechanical behavior and the reliability. Due to the highly anisotropic nature of Sn crystals, a variety of mechanical properties can be found in a single joint that contains multiple grains.

To avoid having the nanoindentation results vary across a solder joint when making several indents under the same aging conditions, we have limited our testing to single grain solder joints. Even with this restriction, it is important to understand the precise grain orientation and the crystallographic direction normal to the polished surface in each tested joint so that results from multiple joints can be compared. Polarized light microscopy and Electron Back Scattered Diffraction (EBSD) techniques have been utilized for this purpose.

Experimental Test Matrix

In initial testing, the aging induced changes in mechanical behavior of two unique single crystals SAC 305 solder joints were examined. SEM photos of the two joints are shown in Figure 3. For each joint, 7 unique sets of aging conditions are being explored: no aging; and 10, 30, 90, 180, 270, and 360 days of aging at T = 125 C. For each aging condition, a 2 x 3 array of indentations was performed on each joint, and Figure 4 illustrates the conceptual map of the 7 indentation regions for the various aging conditions across each solder joint.

Solder Joint #1   Solder Joint #2

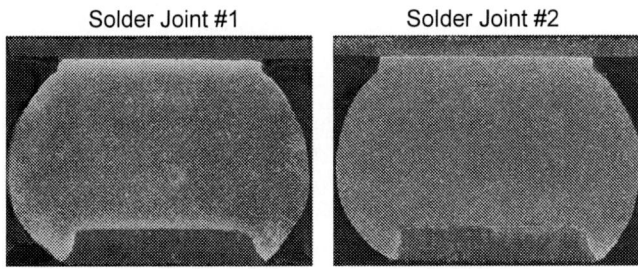

Figure 3 - SAC305 Solder Joint Samples for Aging Studies

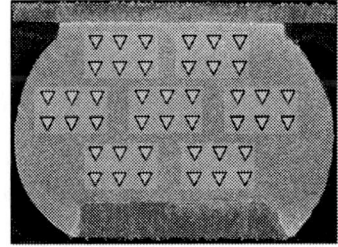

Figure 4 - Indentation Regions for Various Aging Conditions

In subsequent work, aging effects were evaluated using nanoindentation testing of two SAC105 joints, three SAC205 joints, and three SAC405 joints. In all of these cases, single crystal joints were again tested, and a 2 x 3 array of indentations was performed for each aging conditions. The aging conditions completed to date include: no aging; and 10, 30, 90 days of aging at T = 125 C. In our ongoing work, we are completing the experiments on these 8 joints for 180, 270, and 360 days of prior aging at 125 C.

All of the BGA test boards were stored in a freezer at T = -10 C prior to cross-sectioning to minimize any aging effects after board assembly. Aging of the epoxy mounted samples

was then performed in a box oven, and light polishing was performed to remove any oxides after each oven exposure. For each aging condition and solder joint, the individual and average results for the 6 indentations have been reported for the elastic modulus, hardness, and creep strain rate vs. stress curves.

Solder Joint Crystal Orientations

Using polarized light optical microscopy on an Olympus BX60 metallographic microscope system, it was verified that each of the 10 tested SACN05 solder joints were comprised of a single grain. The unique colors and contrasts obtained for the samples for each alloy under polarized light also indicated that the crystal orientations were different. To more rigorously understand the joint orientations, Electron Backscatter Diffraction (EBSD) was performed [46-48].

A scanning map with a step size 5 μm in each direction was obtained for each joint using an EBSD system (Oxford HKL Channel 5) attached to a JEOL JSM-7000F SEM. For each scanning point, the Electron Backscatter Diffraction Pattern (EBSP) or so-called Kikuchi bands were indexed, and an available phase database for β-Sn was used to find the three Euler angles relative to the crystallographic directions of the tetragonal crystal. The results showed that the crystal orientation was constant across each joint.

For the two SAC305 joints, the measured Inverse Pole Figure (IPF) plots are shown in Figure 5 and the Euler angles are tabulated in Figure 6. Based on Euler angle calculations, the Miller indices of the directions perpendicular to the cross-sectioning planes of each joint were (1 5 -6) and (1 5 10), for solder joints #1 and #2, respectively. Using the elastic modulus map for a β-Sn crystal shown in Figure 7, the orientation of solder joint #2 is seen to be slightly closer to the highest stiffness (001) direction. Thus, it was expected to have slightly higher modulus and hardness values, and slower creep rates for all aging conditions, even before our testing (see below) revealed that this was actually the case.

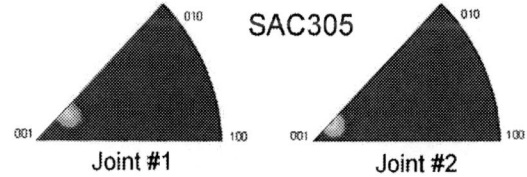

Figure 5 - Inverse Pole Figures for SAC305 Solder Joints

Joint	φ_1 [deg]	ϕ [deg]	φ_2 [deg]
#1	86.35	155.37	10.41
#2	124.28	16.18	11.16

Figure 6 - Crystal Orientation Euler Angles for SAC305 Solder Joints from EBSD

Analogous Inverse Pole Figure (IPF) plots for the tested SAC105, SAC205, and SAC405 solder joints are shown in Figures 8, 9, and 10, respectively. For each alloy, the joint orientations were unique for each joint in the set. Using the modulus map in Figure 7, it is expected that joint #1 will have the largest modulus and hardness for the SAC105 joints. Likewise, joint #1 is expected to have the largest modulus and hardness for the SAC205 joints, and joint #2 is expected to have the largest modulus and hardness for the SAC405 joints.

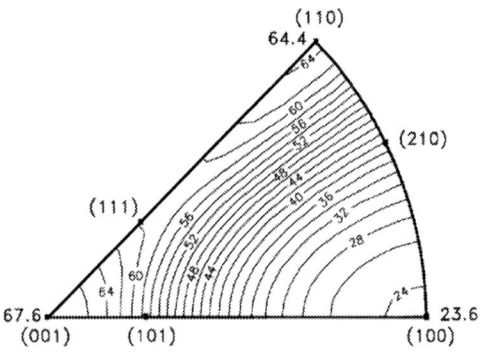

Figure 7 - Elastic Modulus of β-Sn Crystal as a Function of Orientation [45]

SAC105

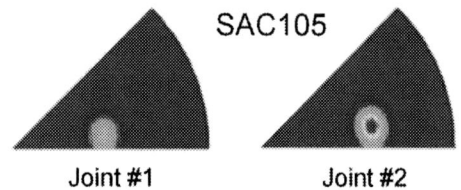

Figure 8 - Inverse Pole Figures for SAC105 Solder Joints

SAC205

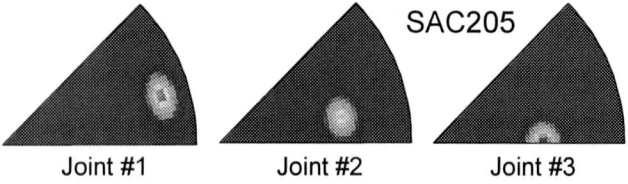

Figure 9 - Inverse Pole Figures for SAC205 Solder Joints

SAC405

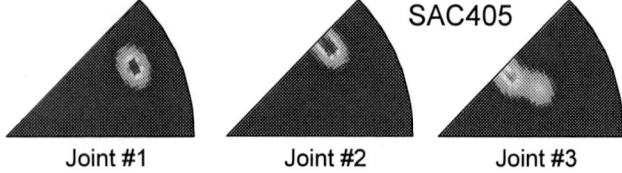

Figure 10 - Inverse Pole Figures for SAC405 Solder Joints

Effect of Aging on Elastic Modulus and Hardness

Using the nanoindentation procedures discussed above, the CSM method was used to measure elastic modulus and hardness values for the 6 indents performed for each aging condition on each solder joint. For example, the measured average values for the two SAC305 joints are tabulated in Figures 11 and 12. As expected from the crystal orientation results presented above in Figure 5, the mechanical properties of joint #2 were slightly larger than those for joint #1 for all aging times. It is observed that for 360 days of aging at 125

C, that the elastic modulus of solder balls #1 and #2 degraded by 28.2% and 31.4%, respectively. The corresponding degradations of the hardness over the same period were 41.6% and 43.2%. These percentage reductions with 360 days of aging at 125 C are very similar to those observed for the elastic modulus (34.4%) and yield stress (44.1%) of SAC305 measured by tensile testing of larger multi-grain uniaxial samples in our prior aging investigations [4-6]. The uniaxial samples from prior investigations have been examined using polarized light microscopy, and are known to contain hundreds of grains with a variety of orientations.

Aging Time (Days)	Elastic Modulus (GPa)	
	Joint #1	Joint #2
0	50.25 (0.54)	55.59 (1.98)
10	45.65 (1.45)	46.52 (2.32)
30	40.95 (0.45)	41.84 (1.33)
90	37.63 (1.79)	39.40 (1.48)
180	36.59 (1.89)	38.91 (0.91)
270	36.21 (1.62)	38.54 (1.41)
360	36.08 (0.48)	38.13 (2.06)

Figure 11 - Elastic Modulus Values from Nanoindentation (SAC305, Aging at 125 C) (Average and Standard Deviation)

Aging Time (Days)	Hardness (GPa)	
	Joint #1	Joint #2
0	0.310 (0.021)	0.340 (0.008)
10	0.257 (0.015)	0.264 (0.019)
30	0.210 (0.009)	0.220 (0.018)
90	0.198 (0.013)	0.202 (0.010)
180	0.191 (0.012)	0.198 (0.020)
270	0.186 (0.010)	0.195 (0.007)
360	0.181 (0.010)	0.193 (0.006)

Figure 12 - Hardness Values from Nanoindentation (SAC305, Aging at 125 C) (Average and Standard Deviation)

Plots of the mechanical property evolution with aging time for the two SAC305 joints are presented in Figures 13 and 14. The evolutions of the mechanical properties with aging were well fit using linear and exponential decay empirical models:

$$E = C_0 + C_1 t + C_2 e^{-C_3 t} \qquad (1)$$

$$H = C_0 + C_1 t + C_2 e^{-C_3 t} \qquad (2)$$

where C_0, C_1, C_2, C_3 are regression constants. The majority of the property degradations in the SAC305 solder joints occurred within the first 20-30 days of aging in an exponential manner. After this point, the variations are small and can be fit with a linear decay. These observations are in agreement with our aging results for testing of larger solder tensile specimens [2-6]. In addition to the nanoindentation results from the current paper, we have added our prior tensile testing results (blue curve) from multi-grain SAC305 samples to the elastic modulus data in Figure 13. It is observed that the modulus variations are qualitatively similar in shape. The percentage reductions are ~30-35% after 360 days of aging at 125 C for both the solder joint nanoindentation and miniature bulk tensile test results as discussed above. The magnitudes of the modulus at the joint-scale are significantly larger than those of the tensile samples for all aging times. This is not unexpected due to the single grain nature of the solder joints tested. As discussed above, the 2 tested SAC305 joints both had their out-of-plane orientations near the (001) direction of β-Sn (see Figures 5 and 7). This orientation has a modulus of 67.6 GPa, which is at the upper end of the range (23.6 < E < 67.6) of values possible when considering all crystal orientations. We have established that the miniature bulk tensile specimens (80-100 mm in length) are typically composed to hundreds of grains with a variety of orientations. Thus, the modulus of the tensile specimens will be significantly smaller and highly averaged, on the order the average/mid-point of the modulus spectrum for β-Sn.

Similar results were obtained for aging induced degradation of the three other solder alloys. For example, Figures 15 and 16 illustrate the modulus and hardness degradations of the two SAC105 joints with aging. Similarly, the modulus and hardness plots for the three SAC205 joints are contained in Figures 17 and 18, while the plots for the three SAC405 solder joints are in Figures 19 and 20. In all cases, the empirical models in Eqs. (1) and (2) fit the data well. In addition, the single crystal joint degradations for each alloy were on the same order as those observed in tensile testing of miniature bulk solder samples of that alloy.

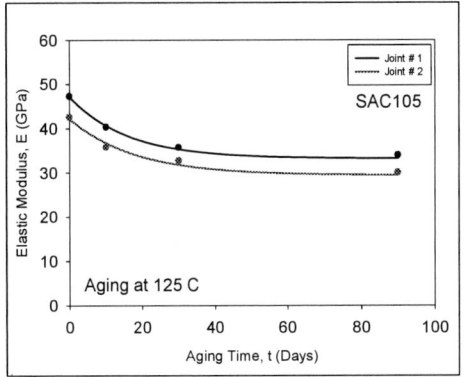

Figure 15 - Evolution of Elastic Modulus with Aging Time (SAC105, Aging at 125 C)

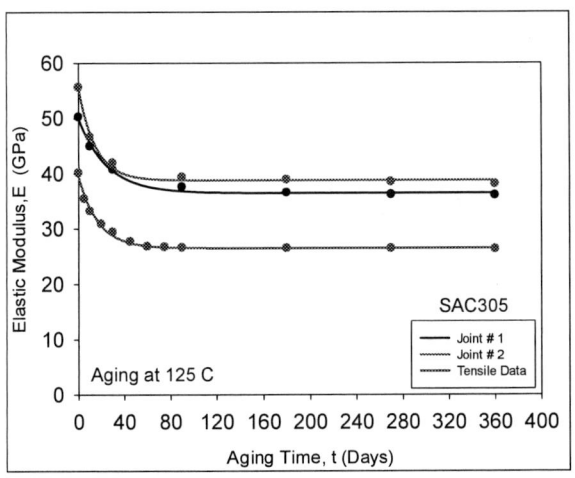

Figure 13 - Evolution of Elastic Modulus with Aging Time (SAC305, Aging at 125 C) Tensile Data from Reference [6]

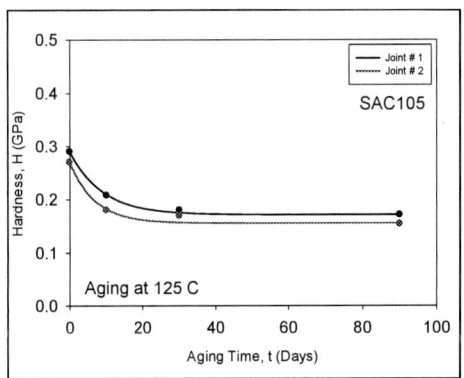

Figure 16 - Evolution of Hardness with Aging Time (SAC105, Aging at 125 C)

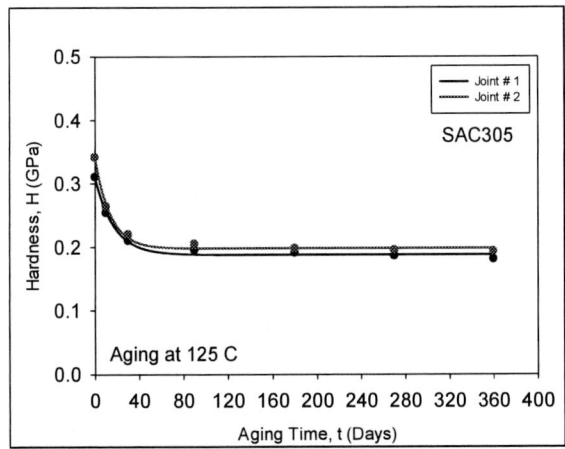

Figure 14 - Evolution of Hardness with Aging Time (SAC305, Aging at 125 C)

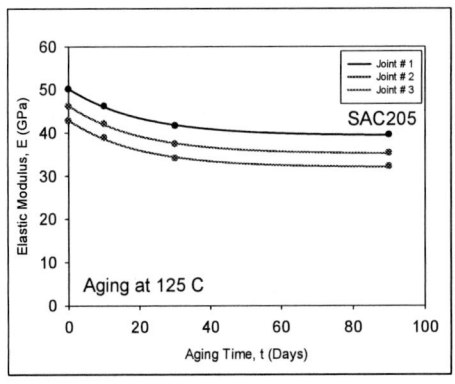

Figure 17 - Evolution of Elastic Modulus with Aging Time (SAC205, Aging at 125 C)

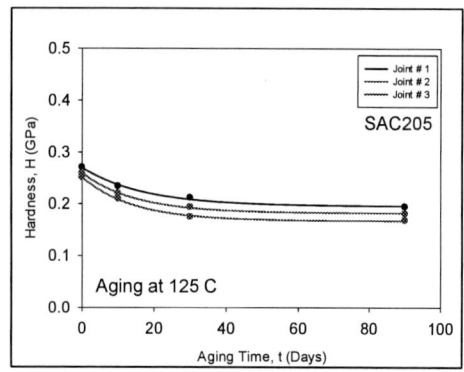

Figure 18 - Evolution of Hardness with Aging Time
(SAC205, Aging at 125 C)

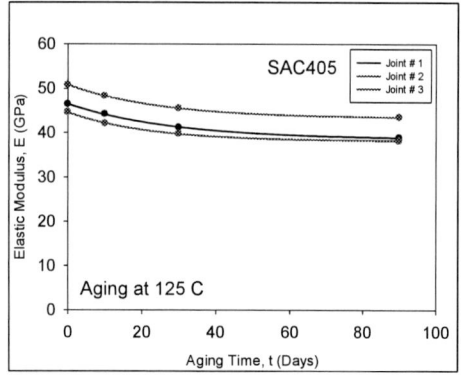

Figure 19 - Evolution of Elastic Modulus with Aging Time
(SAC405, Aging at 125 C)

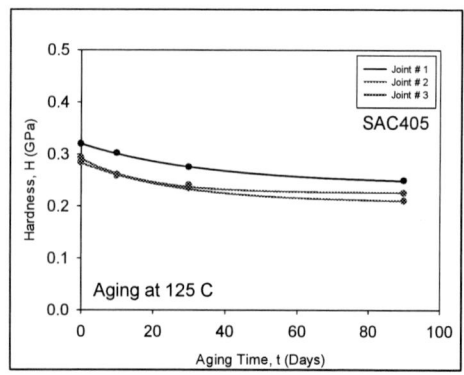

Figure 20 - Evolution of Hardness with Aging Time
(SAC405, Aging at 125 C)

Effect of Aging on Creep Response

Using the nanoindentation creep procedures discussed above, the evolution of the creep response with aging at 125 C was characterized for each of the ten SACN05 solder joints. For all of the tests, a constant peak load of 30 mN and a dwell time of 900 sec were used during the hold periods. Using the recorded creep displacement vs. time data, the creep strain rate vs. applied stress curves were generated as outlined in our prior work [14].

Figure 21 shows the aging dependence of the maximum creep displacements in the two SAC305 joints that occurred after 900 sec of constant load. It is observed that higher creep

displacements occurred in joint #1 relative to joint #2 for each of the aging times. This result agrees with the earlier observations that joint #1 will have poorer mechanical properties and creep response than joint #2 due to its single grain orientation normal to the polished surface being further away from the (001) direction of β-tin (see Figure 5).

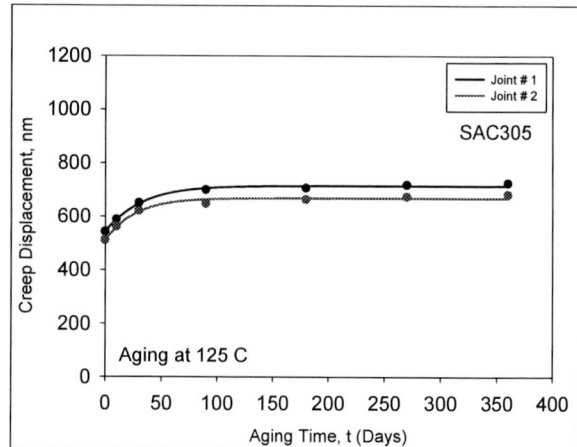

Figure 21 - Maximum Creep Displacement vs. Aging Time
(SAC305, Aging at 125 C)

Figures 22-23 contain the strain rate vs. applied stress responses for SAC305 joints #1 and #2, respectively. Each curve in these plots represents a fit of the exponential creep model to the nanoindentation creep data as outlined in reference [14]. Since the nanoindentation compressive stress levels during the creep experiments were typically above 50 MPa, these results are extrapolations for the stress levels below this level. The red curves in these graphs are for the no aging case, and represent the best (lowest) creep strain rates occurring in the joints. As aging progressed, the creep strain rates became much larger in each joint for every stress level, pushing the creep response curves significantly higher (blue, green, black, pink, light blue, and orange curves for prior aging at 125 C for 10, 30, 90, 180, 270, and 360 days). The largest changes occurred in the first 30 days of aging, and the changes between 30 and 360 days were significantly reduced.

The red, blue, green, black, pink, and light blue data points in each plot are creep strain rates measured by separate tensile testing of extremely small uniaxial specimens that were aged for 0, 10, 30, 90, 180, and 270 days, respectively. These specimens had a gage length of 10 mm, and polarized light microscopy revealed that they were composed of 10 or less grains. The agreement of the creep curves from nanoindentation with the small scale specimen tensile creep data is very good at the lower stress levels, suggesting that proposed extrapolation procedure has the potential to yield accurate predictions.

Numerical values of the creep strain rates have been extracted from Figures 22-23 for SAC305 joints with no-aging and with 10, 30, 90, 180, 270, and 360 days aging at 125 C, and for an applied stress level of $\sigma = 15$ MPa. The results are tabulated in Figure 24 along with the observed strain rates for $\sigma = 15$ MPa found through tensile creep testing of the small 10 mm uniaxial samples (data points in Figures

978-1-4799-2408-0/14 $31.00 © 2014 IEEE

22-23). In addition, the creep rates we measured in prior studies [5-6] by tensile testing of longer (80-100 mm) miniature bulk uniaxial samples are also included.

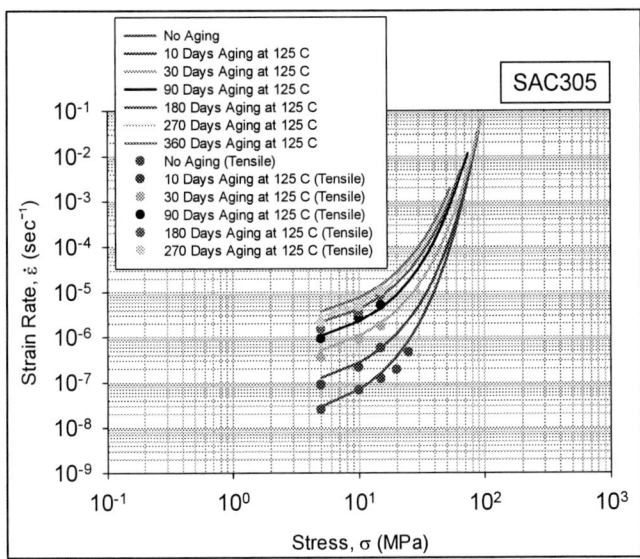

Figure 22 - Creep Strain Rate vs. Stress for Joint #1
(SAC305, Aging at 125 C)

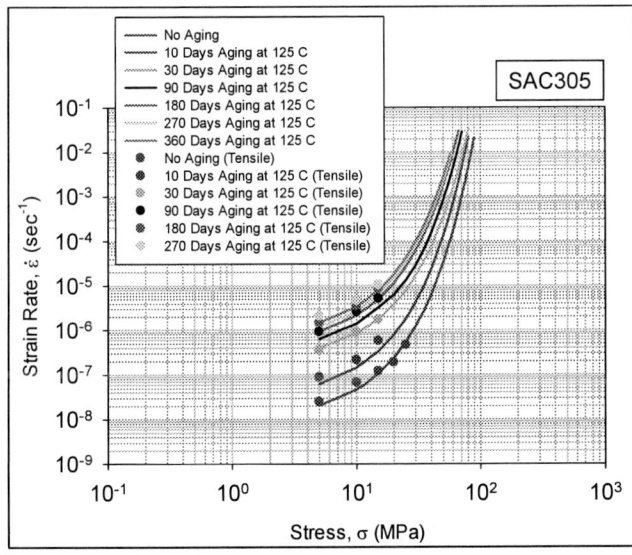

Figure 23 - Creep Strain Rate vs. Stress for Joint #2
(SAC305, Aging at 125 C)

The increases in the creep stain rates observed for the two joint samples were similar and on the same order as those measured for the small 10 mm tensile specimens. For example, after 90 days of aging the two joints had increases of 32.4X and 30.8X, while the creep rate of the small tensile specimens increased by 42.9X. However, dramatically different results were obtained for the longer miniature bulk tensile samples, where an increase in the creep rate of 226.1X was observed. Thus, there were significantly more aging induced degradations in the larger bulk solder tensile samples relative to the joint samples and small tensile specimens. As

discussed below, this can be explained from the single grain/crystal nature of the joint samples and the small number of grains (<10) present in the small tensile specimens that were tested. A direct comparison of the creep rate degradations for the SAC305 solder joints (average) and the SAC305 small tensile specimens is shown in Figure 25. Good agreement was observed for all aging times.

Aging Time (Days)	Creep Strain Rate (sec^{-1}) for σ = 15 MPa			
	Joint #1	Joint #2	Tensile (Small)	Tensile (Large) Refs. [5-6]
0	1.82×10^{-7}	1.05×10^{-7}	1.19×10^{-7}	0.36×10^{-7}
10	5.26×10^{-7}	3.38×10^{-7}	2.11×10^{-7}	2.57×10^{-7}
Increase	2.89X	3.21X	1.77X	7.14X
30	2.07×10^{-6}	1.79×10^{-6}	1.72×10^{-6}	5.90×10^{-6}
Increase	11.4X	17.0X	14.4X	164.9X
90	5.89×10^{-6}	3.23×10^{-6}	5.11×10^{-6}	8.14×10^{-6}
Increase	32.4X	30.8X	42.94X	226.1X
180	8.84×10^{-6}	4.98×10^{-6}	8.54×10^{-6}	12.9×10^{-6}
Increase	48.57X	47.44X	71.76X	358.3X
270	1.25×10^{-5}	6.34×10^{-6}	1.03×10^{-5}	21.0×10^{-6}
Increase	66.68X	60.38X	87.33X	583.3X
360	1.45×10^{-5}	7.90×10^{-6}	TBD	33.0×10^{-6}
Increase	77.54X	75.23X	TBD	916.7X

Figure 24 - Creep Rate Changes with Aging (SAC305)

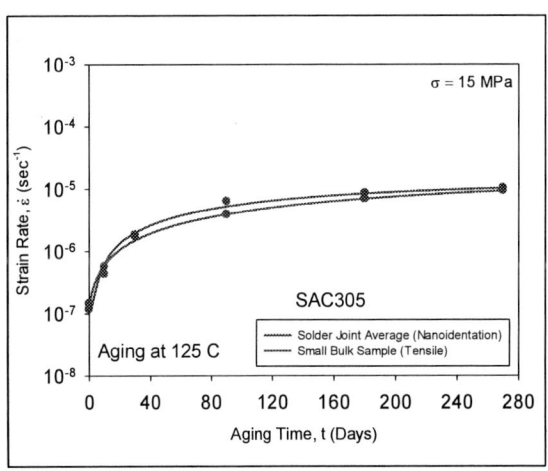

Figure 25 - Creep Strain Rate Degradations with Aging
(SAC305, Aging at T = 125 C)
(NI and SAC305 Small Tensile Specimens)

Typical sources of creep in metals are matrix creep and grain boundary creep. Since the single crystal SAC305 joints in question had no grain boundaries, they primarily experienced matrix creep. In particular, aging causes growth of the intermetallic particles (Ag_3Sn and Cu_6Sn_5) in the β-Sn matrix of a single grain SAC solder joint. As the intermetallic particles are coarsened, they cannot resist the movement of dislocations and thus there is a loss of strength. In addition, coarsened particles are considerably softer, and as a result, the

978-1-4799-2408-0/14 $31.00 © 2014 IEEE

dislocations can pass the particles more easily leading to increased creep deformations. The miniature bulk tensile samples have hundreds of grains, and experience significant grain boundary creep mechanisms such as grain boundary sliding. These effects can be significantly curtailed in lead free solders by migration of intermetallic particles to the grain boundaries providing pinning. While, the presence of only a single grain in a solder joint has been shown to significantly reduce the degradations in the creep response with aging, the magnitude of aging effects in multi-grain lead free solder joints remains to be quantified.

Similar results were obtained for the aging induced degradation of the creep response of the three other SACN05 solder alloys. For example, Figures 26 and 27 illustrate the creep rate vs. stress response of the two SAC105 joints for various aging times. These results are tabulated in Figure 28 for $\sigma = 15$ MPa, along with the observed strain rates measured in prior studies [5-6] by tensile testing of longer (80-100 mm) miniature bulk uniaxial samples. Similarly, the creep rate vs. stress response plots for the three SAC205 joints are contained in Figures 29, 30, and 31; and the tabulated values for $\sigma = 15$ MPa are in Figure 32. Finally, the creep rate vs. stress response plots for the three SAC405 joints are contained in Figures 33, 34, and 35; and the tabulated values for $\sigma = 15$ MPa are in Figure 36.

Aging Time (Days)	Creep Strain Rate (sec^{-1}) for σ = 15 MPa		
	Joint #1	Joint #2	Tensile (Large) Refs. [5-6]
0	4.69×10^{-7}	3.66×10^{-7}	1.0×10^{-7}
10	1.93×10^{-6}	2.48×10^{-6}	2.1×10^{-4}
Increase	4.12X	6.78X	2100X
30	4.68×10^{-6}	7.05×10^{-6}	3.2×10^{-4}
Increase	9.98X	19.26X	3200X
90	1.01×10^{-5}	1.82×10^{-5}	7.5×10^{-4}
Increase	21.53X	49.73X	7500X

Figure 28 - Creep Rate Changes with Aging (SAC105)

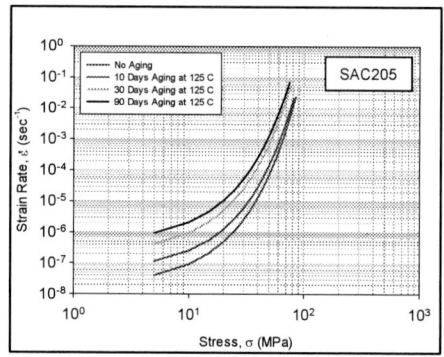

Figure 29 - Creep Strain Rate vs. Stress (SAC205, Joint # 1)

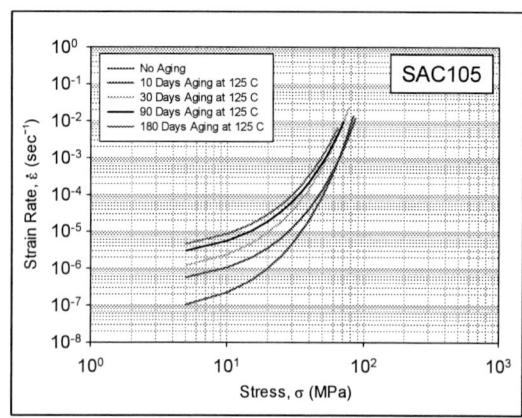

Figure 26 - Creep Strain Rate vs. Stress (SAC105, Joint # 1)

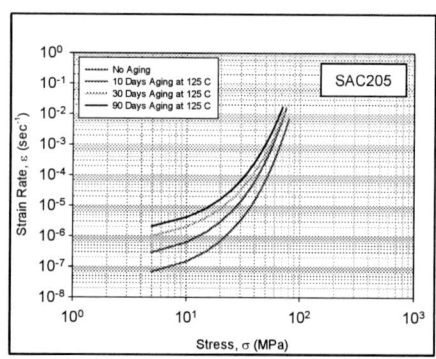

Figure 30 - Creep Strain Rate vs. Stress (SAC205, Joint # 2)

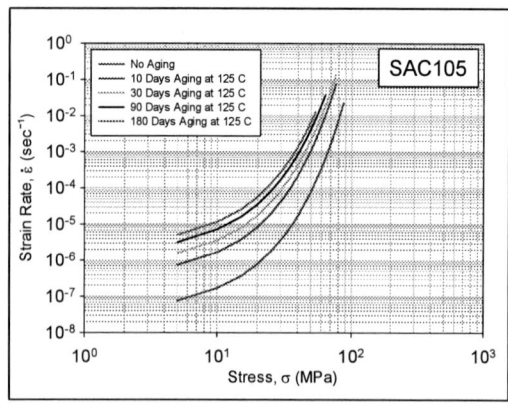

Figure 27 - Creep Strain Rate vs. Stress (SAC105, Joint # 2)

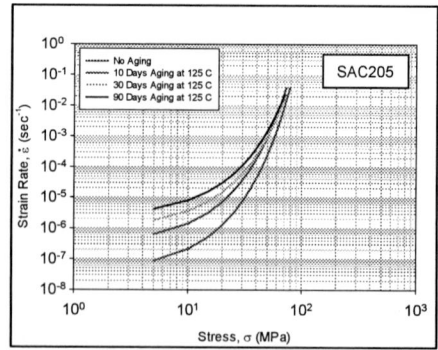

Figure 31 - Creep Strain Rate vs. Stress (SAC205, Joint # 3)

Aging Time (Days)	Creep Strain Rate (sec⁻¹) for σ = 15 MPa			
	Joint #1	Joint #2	Joint #3	Tensile (Large) Refs. [5-6]
0	2.10×10^{-7}	3.18×10^{-7}	5.10×10^{-7}	3.80×10^{-8}
10	5.56×10^{-7}	1.36×10^{-6}	3.16×10^{-6}	2.50×10^{-5}
Increase	2.65X	4.28X	6.19X	657.89X
30	2.16×10^{-6}	4.00×10^{-6}	7.29×10^{-6}	7.10×10^{-5}
Increase	10.28X	12.58X	14.29X	1868.42X
90	4.69×10^{-6}	8.29×10^{-6}	1.55×10^{-5}	8.50×10^{-5}
Increase	22.33X	26.06X	30.39X	2236.84X

Figure 32 - Creep Rate Changes with Aging (SAC205)

Aging Time (Days)	Creep Strain Rate (sec⁻¹) for σ = 15 MPa			
	Joint #1	Joint #2	Joint #3	Tensile (Large) Refs. [5-6]
0	1.06×10^{-7}	3.56×10^{-7}	5.11×10^{-7}	3.5×10^{-8}
10	2.23×10^{-7}	7.53×10^{-7}	2.59×10^{-6}	2.1×10^{-6}
Increase	2.10X	2.11X	5.06X	60.0X
30	5.05×10^{-7}	2.52×10^{-6}	5.63×10^{-6}	5.10×10^{-6}
Increase	4.76X	7.08X	11.01X	145.71X
90	8.55×10^{-7}	4.53×10^{-6}	8.38×10^{-6}	6.10×10^{-6}
Increase	8.06X	12.72X	16.39X	174.28X

Figure 36 - Creep Rate Changes with Aging (SAC405)

As with the results for the SAC305 samples, the creep rates for the other SACN05 alloy joints all increase by significant amounts (from 8-50X) with 90 days of aging, but the increases are far less than those found using tensile testing of the large miniature bulk uniaxial samples with hundreds of grains (175-7500X). The largest creep rate increase (50X) was found in SAC105 joint #2, while the smallest creep increase (8X) was found in SAC405 joint #1.

<u>Effect of Aging on Solder Alloy Compositions</u>

It is well understood from prior investigations that lowering the silver content of SAC alloy tends to decrease the stiffness and strength, and increase the creep strain rate. Our aging studies using large tensile specimens with hundreds of grains also have suggested the alloys with lower silver content are more susceptible to aging induced degradations in their mechanical properties and increases in their creep strain rates [5-6]. The same conclusions can be made qualitatively using the aging data for single crystal SACN05 solder joints presented above. However, it is difficult to make precise quantitative comparisons of joints for the four alloys because the properties are also strongly dependent on the crystal orientation. As seen from Figures 4, 8, 9, and 10, there does not exist a set of four tested joints, one from each alloy, that have similar crystal orientations. Thus, it is difficult to separate the relative performance of the joints from different alloys.

In the discussion below, we have compared the relative changes due to aging for four joints, one from each SACN05 alloy. Aging data for up to 90 days at T = 125 C have been analyzed. The particular joints considered were SAC105 joint #2, SAC205 joint #2, SAC305 joint #3 (new), and SAC405 joint #1. Figure 37 illustrates a plot of the percentage reduction in the elastic modulus occurring in the four joints as a function of the aging time:

$$\text{Reduction in Modulus } (\%) = \left| \frac{E_{Aged} - E_{No\,Aging}}{E_{No\,Aging}} \right| \times 100 \qquad (3)$$

Likewise, Figure 38 illustrates a plot of the percentage reduction in the hardness occurring in the four joints as a function of the aging time:

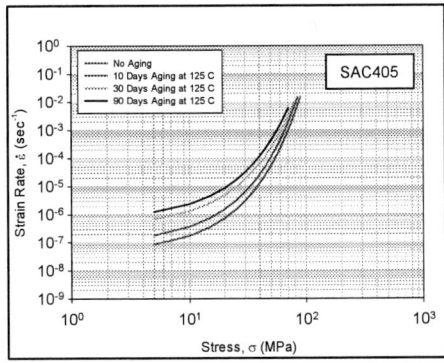

Figure 33 - Creep Strain Rate vs. Stress (SAC405, Joint # 1)

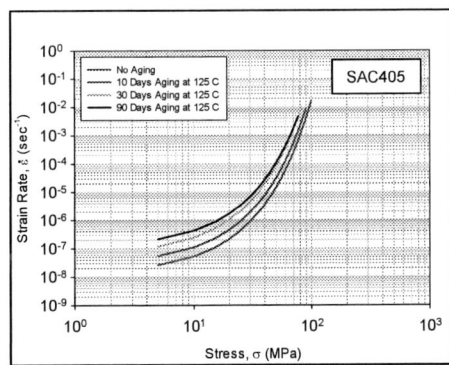

Figure 34 - Creep Strain Rate vs. Stress (SAC405, Joint # 2)

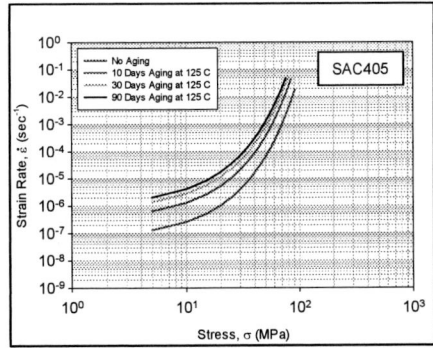

Figure 35 - Creep Strain Rate vs. Stress (SAC405, Joint # 3)

$$\text{Reduction in Hardness } (\%) = \left| \frac{H_{Aged} - H_{No\ Aging}}{H_{No\ Aging}} \right| \times 100 \quad (4)$$

The joint data presented in these graphs suggest that lowering the silver content tends to increase the normalized reduction in the mechanical properties (elastic modulus and hardness). The highest percentage changes (degradations) were observed for the SAC105 joint for all aging times, while the lowest changes (degradations) were seen in the SAC405 joint for all aging times.

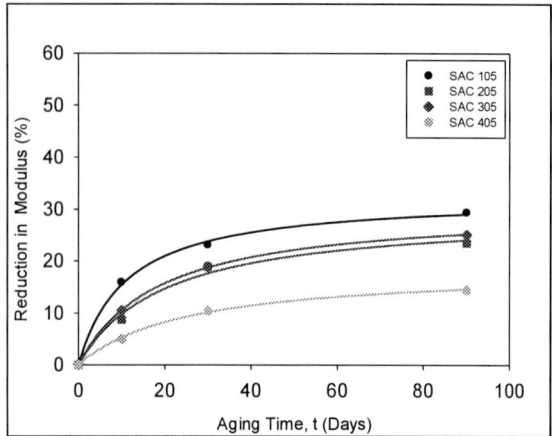

Figure 37 - Effect of Silver Content on SACN05 Joint Elastic Modulus Change (Aging at 125 C)

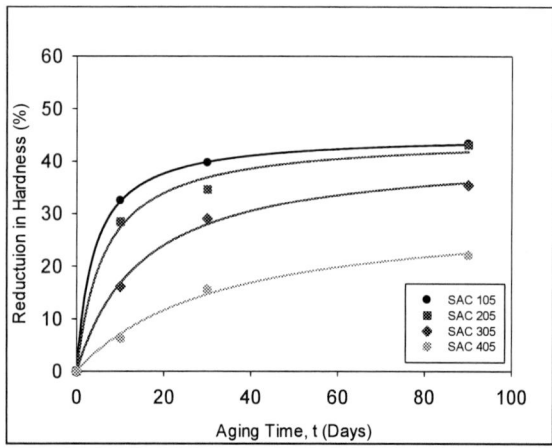

Figure 38 - Effect of Silver Content on SACN05 Joint Hardness Change (Aging at 125 C)

A similar comparison was made in Figure 39 for the degradation of the creep response of the four joints with aging. In this case the increase in the creep strain rate was plotted as a function of aging time:

$$\text{Creep Rate Increase} = \frac{\dot{\varepsilon}_{Aged}}{\dot{\varepsilon}_{No\ Aging}} \quad (5)$$

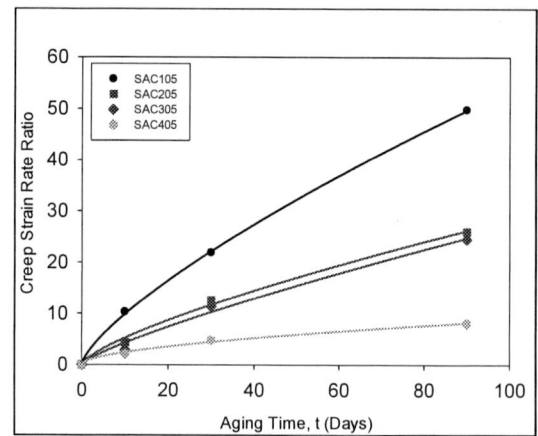

Figure 39 - Effect of Silver Content on SACN05 Joint Creep Strain Rate Change (Aging at 125 C)

As observed previously, the largest creep rate increase (50X) was found in SAC105 (joint #2), while the smallest creep increase (8X) was found in SAC405 (joint #1).

The presence of more Ag content in the SACN05 alloy gives additional precipitates in the form of Ag_3Sn IMC particles, which are responsible for hardening of Sn matrix. In addition, the lower silver content alloys (e.g. SAC105) showed larger coarsening of Ag_3Sn particles relative to the higher silver content alloys (e.g. SAC405).

Initial X-Ray Microdiffraction Study

Introduction

Synchrotron x-ray microdiffraction is a new powerful experimental technique available for studying mechanical properties of materials at small length scales, in the range of grains and defect interactions [49]. The method is capable of measuring the variation in internal (residual) stress, strain, grain orientation, grain rotation, and plastic deformations between grains and within individual grains. This can provide important information on changes occurring within a solder joint at small scales. The technique is based on very small submicron high brilliance x-ray beam, which is produced using focusing optics such as Kirkpatrick-Baez (KB) mirrors at synchrotron facilities. Data are recorded as raster scans of an array of Laue diffraction patterns by a large area charge couple device x-ray detector, and then subsequently analyzed to give grain orientation, stress/strain distributions, and dislocation densities.

Wu, et al. [50] and Chen, et al. [51] have used x-ray microdiffraction to study the microstructure and plastic deformation in microelectronic interconnect materials under electromigration. They have found the grain growth involves grain boundary migration and rotation of neighboring high resistance grains due to electromigration. In addition, the grain orientation and stress distributions were calculated. They also studied the bending in large grains by looking the shifting of the Laue pattern which provides grain rotation information. Microdiffraction provided the insight into the mechanism of plastic deformation by grain rotation [51].

Kunz, et al. [52] have studied the evidence of residual strain in deformed natural quartz by microdiffraction. In addition, they calculated the individual strain components of the strain tension along the sample axis. Sarobol, et al. [53] have used microdiffraction to study the effect of grain misorientation and elastic anisotropy on tin whisker and hillock formation. They reported that local elastic strain energy density, grain misorientation, and elastic strain gradients influence the formation process. In another study [54], subgrain formation and recrystallization were identified in thin solder films due to thermal cycling.

Microdiffraction Experiment

In this work, and initial study on three SAC305 solder joints from the BGA assemblies has been performed using microdiffraction. The x-ray microdiffraction measurements were performed on Beamline 12.3.2 at the Advanced Light Source (ALS) in Lawrence Berkeley Laboratory. The energy range of the x-ray beam varied from 5 keV to 24 keV. The x-ray beam was focused by Kirkpatrick-Baez (KB) mirrors, and the beam size was approximately 1 μm x 1 μm. The solder joint samples were placed on a high precision stage and scanned through the x-ray beam at its focus spot. Each sample was raster scanned with 10 μm step size and a 1 sec exposure time. After initial scanning, the each sample was isothermally aged at 125 C for 1 day and then scanned again. Finally, the samples were aged another 6 days at 125 C (for a total aging time of 7 days), and scanned a final time. At each scanning step, the Laue diffraction patterns were collected in reflection geometry, using a MAR133 x-ray CCD detector mounted at 90° to the incident beam. The Laue patterns were analyzed and post processed using the in-house XMAS software.

Example Microdiffraction Results

Figure 40 shows the evolution of the grain orientation map of one of the SAC305 solders with aging. From the results, it is observed that the grains did not grow significantly after 7 days of aging. However, it appears that some recrystallization occurred at the top of the joint near the BGA component copper pad. Due to recrystallization, stresses might relax at the surface grains, creating a stress gradient between new grains and deformed grains, and leading the process of crack initiation and growth in the solder joint. This is expected because recrystallization depends on dislocation movement on the slip system operating during deformation. By tracking a particular example grain, it was found that after 7 days of aging, the grain rotated from 0.2644 deg to 0.4602 deg, which is believed to due to shear stress gradient. Grain rotations from the high temperature aging are related to the rearrangement of dislocations that leads to subgrain formation and local plastic deformations.

Using the microdiffraction data, it was also possible to calculate the deviatoric stresses and strains in individual grains, and their evolution during aging. For example, Figure 41 shows a typical result for one of the deviatoric shear stress components. In all cases, the isothermal aging caused changes to occur in the various stress and strain components, which can lead to stress gradients and local changes in mechanical behavior.

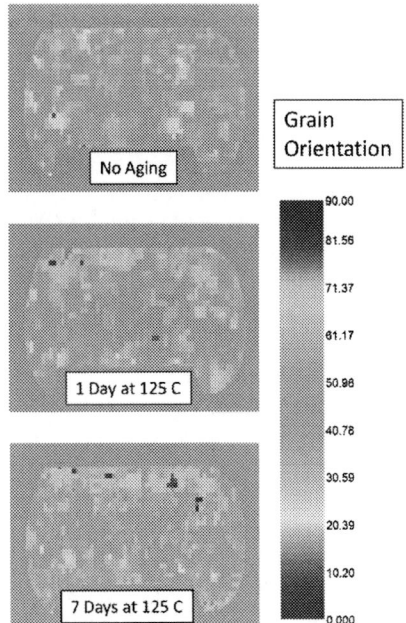

Figure 40 - Measured Grain Orientation Distribution from Microdiffraction (SAC305)

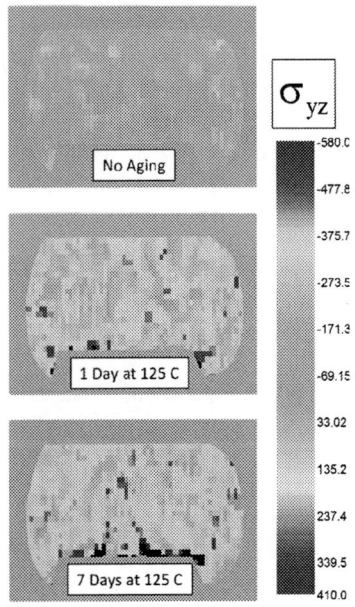

Figure 41 - Measured Deviatoric Stress Distribution from Microdiffraction (SAC305)

Summary and Conclusions

In this investigation, we have extended our previous work on nanoindentation of joints to examine a full test matrix of SAC solder alloys. The effects of silver content on SAC solder aging has been evaluated by testing joints from SACN05 (SAC105, SAC205, SAC305, and SAC405) test boards assembled with the same reflow profile. In all cases, the tested joints were extracted from 14 x 14 mm PBGA assemblies (0.8 mm ball pitch, 0.46 mm ball diameter) that are part of the iNEMI Characterization of Pb-Free Alloy

Alternatives Project (16 different solder joint alloys available). After extraction, the joints were subjected to various aging conditions (0 to 12 months of aging at T = 125 C), and then tested via nanoindentation techniques to evaluate the stress-strain and creep behavior of the four aged SAC solder alloy materials at the joint scale.

The observed aging effects in the SACN05 solder joints have been quantified and correlated with the magnitudes observed in tensile testing of miniature bulk specimens performed in prior studies. The results show that the aging induced degradations of the mechanical properties (modulus, hardness) in the SAC joints were of similar order (30-40%) as those seen previously in the testing of larger "bulk" uniaxial solder specimens. The creep rates of the various tested SACN05 joints were found to increase by 8-50X due to aging. These degradations, while significant, were much less than those observed in larger bulk solder uniaxial tensile specimens with several hundred grains, where the increases ranged from 200X to 10000X for the various SACN05 alloys. Additional testing has been performed on very small tensile specimens with approximately 10 grains, and the aging-induced creep rate degradations found in these specimens were on the same order of magnitude as those observed in the single grain joints. Thus, the lack of the grain boundary sliding creep mechanism in the single grain joints is an important factor in avoiding the extremely large creep rate degradations (up to 10,000X) occurring in larger bulk SAC samples. All of the aging effects observed in the SACN05 joints were found to be exacerbated as the silver content in the alloy was reduced. In addition, the test results for all of the alloys show that the elastic, plastic, and creep properties of the solder joints and their sensitivities to aging are highly dependent on the crystal orientation.

The observed mechanical behavior changes in joints are due to evolution in the microstructure and residual strains/stresses in the solder material, and measurements of these evolutions are critical to developing a fundamental understanding of solder joint aging phenomena. As another part of this work, we have performed an initial study of these effects in the same SAC305 solder joints that were tested using nanoindentation. The enhanced x-ray microdiffraction technique at the Advanced Light Source (Synchrotron) at the Lawrence Berkeley National Laboratory was employed to characterize three joints after various aging exposures (0, 1, and 7 days of aging at T = 125 C). For each joint, microdiffraction was used to examine grain growth, grain rotation, sub-grain formation, and residual strain and stress evolution as a function of the aging exposure. The entire joints were scanned using a 10 micron step size. Additional analysis and testing is underway to further our understanding of localized changes occurring in joints during aging.

Acknowledgments

This work was supported by the NSF Center for Advanced Vehicle Electronics and Extreme Environment (CAVE[3]). We thank the members of iNEMI for providing the joint samples tested in this work. We also gratefully acknowledge and thank Dr. Martin Kunz and Dr. Nobumichi Tamura for assistance with the microdiffraction measurements and data extraction, and for valuable discussions. We acknowledge a grant from the DOE for access to beamline 12.3.2 of the Advanced Light Source (ALS), which is supported by the Office of Basic Energy Science, Material Science Division, of the U.S. Department of Energy under Contact DE-AC02-05Ch11231 at Lawrence Berkeley Laboratory.

References

1. Ma, H., and Suhling, J. C., "A Review of Mechanical Properties of Lead-Free Solders for Electronic Packaging," *Journal of Materials Science*, Vol. 44, pp. 1141-1158, 2009.

2. Ma, H., Suhling, J. C., Lall P., Bozack, M. J., "Reliability of the Aging Lead-free Solder Joint," *Proceeding of the 56th Electronic Components and Technology Conference*, San Diego, California, pp. 849-864, 2006

3. Ma, H., Suhling, J. C., Zhang, Y., Lall, P., and Bozack, M. J., "The Influence of Elevated Temperature Aging on Reliability of Lead Free Solder Joints," *Proceedings of the 57th IEEE Electronic Components and Technology Conference*, pp. 653-668, Reno, NV, May 29-June 1, 2007.

4. Zhang, Y., Cai, Z., Suhling, J. C., Lall, P., and Bozack, M. J., "The Effects of Aging Temperature on SAC Solder Joint Material Behavior and Reliability," *Proceedings of the 58th IEEE Electronic Components and Technology Conference*, pp. 99-112, Orlando, FL, May 27-30, 2008.

5. Zhang, Y., Cai, Z., Suhling, J. C., Lall, P., and Bozack, M. J., "The Effects of SAC Alloy Composition on Aging Resistance and Reliability," *Proceedings of the 59th IEEE Electronic Components and Technology Conference*, pp. 370-389, San Diego, CA, May 27-29, 2009.

6. Cai, Z., Zhang, Y., Suhling, J. C., Lall, P., Johnson, R. W., Bozack, M. J., "Reduction of Lead Free Solder Aging Effects using Doped SAC Alloys," *Proceedings of the 60th Electronic Components and Technology Conference*, pp. 1493-1511, 2010.

7. Mustafa M., Cai Z., Suhling J. C., Lall P., "The Effects of Aging on the Cyclic Stress-Strain Behavior and Hysteresis Loop Evolution of Lead Free Solders," *Proceedings of the 61st Electronic Components and Technology Conference*, pp. 927-939, 2011.

8. Mustafa, M., Cai, Z., Roberts, J. R., Suhling, J. C., Lall, P., "Evolution of the Tension/Compression and Shear Cyclic Stress-Strain Behavior of Lead-Free Solder Subjected to Isothermal Aging," *Proceedings of ITherm 2012*, pp. 765-780, San Diego, CA, May 30 - June 1, 2012.

9. Zhang, J., Hai, Z., Thirugnanasambandam, S., Evans, J. L., Bozack, M. J., Sesek, R., Zhang, Y., Suhling, J. C., "Correlation of Aging Effects on Creep Rate and Reliability in Lead Free Solder Joints," *SMTA Journal*, Volume 25(3), pp. 19-28, 2012.

10. Zhang, J., Hai, Z., Thirugnanasambandam, S., Evans, J. L., Bozack, M. J., Zhang, Y., Suhling, J. C., "Thermal Aging Effects on Thermal Cycling Reliability of Lead-Free Fine Pitch Packages," *IEEE Transactions on Components and Packaging Technologies*, Vol. 3(8), pp. 1348-1357, 2013.

11. Motalab, M., Cai, Z., Suhling, J. C., Zhang, J., Evans, J. L., Bozack, M. J., Lall, P., "Improved Predictions of Lead Free Solder Joint Reliability that Include Aging Effects," *Proceedings of the 62nd IEEE Electronic Components and Technology Conference*, pp. 513-531, San Diego, CA, May 30 - June 1, 2012.

12. Motalab, M., Cai, Z., Suhling, J. C., Zhang, J., Evans, J. L., Bozack, M. J., Lall, P., "Correlation of Reliability Models Including Aging Effects with Thermal Cycling Reliability Data," *Proceedings of the 63rd IEEE Electronic Components and Technology Conference*, pp. 986-1004, Las Vegas, NV, May 28-31, 2013.

13. Lall, P., Shantaram, S., Suhling, J., Locker, D., "Effect of Aging on the High Strain Rate Mechanical Properties of SAC105 and SAC305 Leadfree Alloys," *Proceedings of the 63rd IEEE Electronic Components and Technology Conference*, pp. 1277-1293, Las Vegas, NV, May 28-31, 2013.

14. Hasnine, M., Mustafa, M., Suhling, J. C., Prorok, B. C., Bozack, M. J., Lall, P., "Characterization of Aging Effects in Lead Free Solder Joints Using Nanoindentation," *Proceedings of the 63rd IEEE Electronic Components and Technology Conference*, pp. 166-178, Las Vegas, NV, May 28-31, 2013.

15. Pang, J. H. L., Low, T. H., Xiong, B. S., Xu, L., and Neo, C.C., "Thermal Cycling Aging Effects on Sn–Ag–Cu Solder Joint Microstructure, IMC and Strength," *Thin Solid Films*, Vol. 462-463, pp. 370-375, 2004.

16. Darveaux, R., "Shear Deformation of Lead Free Solder Joints," *Proceedings of the 55th IEEE Electronic Components and Technology Conference*, pp. 882-893, 2005.

17. Wiese, S., and Wolter, K. J., "Creep of Thermally Aged SnAgCu Solder Joints," *Microelectronics Reliability*, Vol. 47, pp. 223-232, 2007.

18. Dutta, I., Pan, D., Marks, R. A., Jadhav, S. G., "Effect of Thermo-Mechanically Induced Microstructural Coarsening on the Evolution of Creep Response of SnAg-based Microelectronic Solders," *Materials Science and Engineering* A, Vol. 410-411, pp. 48-52, 2005.

19. Fischer-Cripps, A. C., *Nanoindentation*, Third Edition, Springer, 2011.

20. Rhee, H., Lucas, J. P., Subramanian, K. N., "Micromechanical Characterization of Thermo-Mechanically Fatigued Lead-Free Solder Joints," *Journal of Materials Science: Materials in Electronics*, Vol. 13, pp. 477-484, 2002.

21. Lucas, J. P., Rhee, H., Guo, F., Subramanian, K. N., "Mechanical Properties of Intermetallic Compounds Associated with Pb-Free Solder Joints Using Nanoindentation," *Journal of Electronic Materials*, Vol. 32(12), pp 1375-1383, 2003.

22. Chromik, R. R., Vinci, R. P., Allen, S. L., Notis, M. R., "Measuring the Mechanical Properties of Lead Free Solder and Sn-Based Intermetallics by Nanoindentation," *Journal of Metals*, pp. 66-69, 2003.

23. Deng, X., Chawla, N., Chawla, K. K., Koopman, M., "Deformation Behavior of (Cu,Ag)-Sn Intemetallics by Nanoindentation," *Acta Materialia*, Vol. 52, pp. 4291-4303, 2004.

24. Deng, X., Chawla, N., Chawla, K. K., Koopman, M., "Young Modulus of (Cu,Ag)-Sn Intemetallics by Nanoindentation," *Materials Science and Enginering A*, Vol. 364, pp. 240-243, 2004.

25. Gao, F., Taekmoto, T., "Mechanical Properties Evolution of Sn-3.5Ag Based Lead Free Solders by Nanoindentation," *Materials Letters*, Vol. 60, pp. 2315-2318, 2006.

26. Gao, F., Nishikawa, H., Takemoto, T., "Nanoscale Mechanical Response of Sn-Ag Based Lead Free Solders," *Proceedings of the IEEE Electronic Components and Technology Conference*, pp. 206-210, 2007.

27. Sun, Y., Liang, J., Xu, J. H., Wang, G., Li, X., "Nanoindentation for Measuring Individual Phase Mechanical Properties of Lead Free Solder Alloy," *Journal of Materials Science: Materials in Electronics*,Vol. 19, pp. 514-521, 2008.

28. Liu, Y. C., Teo, J. W. R., Tung, S. K., Lam, K. H., "High Temperature Creep and Hardness of Eutectic 80Au/20Sn Solder," *Journal of Alloys and Compounds*, Vol. 448, pp. 340-343, 2008.

29. Gao, F., Nishikawa, H., Takemoto, T., Qu, J., "Mechanical Properties Versus Temperature Relation of Individual Phase in Sn-3.0Ag-0.5Cu Lead Free Solder Alloy," *Microelectronics Reliability*, Vol.49, pp. 296-302, 2009.

30. Han, Y. D., Jing, H. Y., Nai, S. M. L., Xu, L. Y., Tan, C. M., Wei, J., "Temperature Dependence of Creep and Hardness of Sn-Ag-Cu Lead Free Solders," *Journal of Electronic Materials*, Vol. 39(2), pp.223-229, 2010.

31. Han, Y. D., Jing, H. Y., Nai, S. M. L., Xu, L. Y., Tan, C. M., Wei, J., "Indentation Size Effect on the Creep Behavior of SnAgCu Solder," *International Journal of Modern Physics B*, Vol. 24(1-2), pp. 267-275, 2010.

32. Xu, L., Pang, J. H. L., "Nanoindentation on SnAgCu Solder Joints and Analysis," *Journal of Electronic Materials*, Vol. 35(12), pp. 2107-2115, 2006.

33. Xu, L., Pang, J. H. L.,"Nanoindentation Characterization of Ni-Cu-Sn IMC Layer Subjected to Isothermal Aging," *Thin Solid Films*, Vol. 504, pp. 362-366, 2006.

34. Song, J. M., Huang, B. R., Liu, C. Y., Lai, Y. S., Chiu, Y. T., Huang, T. W., "Nanomechanical Response of Intermetallic Phase at the Solder Joint Interface-Crystal Orientation and Metallurgical Effects," *Materials Science and Engineering A*, Vol.534, pp. 53-59, 2012.

35. Venkatadri, V., Yin, L., Xing, Y., Cotts, E., Srihari, K., Borgesen, P., "Accelerating The Effects of Aging on Reliability of Lead Free Solder Joints in a Quantitative Fashion," *Proceedings of the IEEE Electronic Components and Technology Conference*, pp.398-405, 2009.

36. Henshall, G., "iNEMI Lead-Free Alloy Alternatives Project Report: Thermal Fatigue Experiments and Alloy Test Requirements," *Proceedings of the SMTAI*, pp. 317-324, 2009.

37. Pharr, G. M., Oliver, W. C., Brotzen, F. R., "On the Generality of the Relationship Among Contact Stiffness, Contact Area, and Elastic Modulus During Indentation,"

Journal of Materials Research, Vol. 7(3), pp. 613-617, 1992.

38. Oliver, W. C., Pharr, G. M.,"An Improved Technique for Determining Hardness and Elastic Modulus Using Load and Displacement Sensing Indentation Experiments," *Journal of Materials Research*, Vol. 7(6), pp. 1564-1583, 1992.

39. Hay, J., Agee, P., Herbert, E. "Continuous Stiffness Measurement During Instrumented Indentation Testing," *Experimental Techniques*, Vo. 34(3), pp. 86-94, 2010.

40. Tabor, D., *Hardness of Metals*, Oxford University Press, 1951.

41. Zhang, P., Li, S. X., Zhang, Z. F., "General Relationship between Strength and Hardness," *Materials Science and Engineering A*, Vol. 529, pp. 62-73, 2011.

42. Mayo, M. J., Nix, W. D., "A Micro-Indentation Study of Superplasticity in Pb,Sn and Sn-38 wt%Pb," *Acta Materialia*, Vol. 36(8), pp. 2183-2192, 1988.

43. Mayo, M. J., Nix, W. D., "Mechanical Properties of Nanophase TiO2 as Determined by Nanoindentation," *Journal of Materials Research*, Vol. 5(5), pp. 1073-1082, 1990.

44. Chhanda, N., Suhling, J. C., Lall, P., "Experimental Characterization and Viscoplastic Modeling of the Temperature Dependent Material Behavior of Underfill Encapsulants," *Proceedings of InterPACK 2011*, ASME, Paper No. IPACK2011-52209, pp. 1-13, 2011.

45. Lee, B. Z., Lee, D. N., "Spontaneous Growth Mechanism of Tin Whiskers," *Acta Materialia*, Vol. 46(10), 1998.

46. Bieler, T. R., Jiang, H., Lehman, L. P., Kirkpatrick, T., Cotts, E. J., Nandagopal, B., "Influence of Sn Grain Size and Orientation in the Thermomechanical Response and Reliability of Pb-Free Solder Joints," *IEEE Transactions on Components and Packaging Technologies*, Vol. 31(2), pp. 370-381, 2008.

47. Telang, T. U., Bieler, T. R., "The Orientation Imaging Microscopy of Lead Free Sn-Ag Solder Joints," *Journal of Metals*, pp. 44-49, 2005.

48. Yang, S., Tian, Y., Wang, C., "Investigation on Sn grain Number and Crystal Orientation in the Sn-Ag-Cu/Cu Solder Joints of Different Sizes," *Journal of Material Science: Materials in Electronics*,Vol. 21, pp. 1174-1180, 2010.

49. Kunz, M., Tamura, N., Chen, K., Macdowell, A. A., and Ustundag, E., "A Dedicated Superband X-Ray Microdiffraction Beamline for Materials, Geo-, and Environmental Sciences at the Advanced Light Source," *Review of Scientific Instruments*, Vol . 80, 2009.

50. Wu, A. T., Tamura, N., Lloyd, J. R., Kao, C. R., and Tu, K. N., "Synchrotron X-ray Microdiffraction Analysis on Microstructure Evolution in Sn under Electromigration," *Material Research Society Symposium Proceedings*, Vol. 863, 2005.

51. Chen, K., Tamura, N., Valek, B. C., and Tu, K. N., "Plastic Deformation in Al(Cu) Interconnects Stressed by Electromigration and Studied by Synchrotron Polychromatic X-ray Microdiffraction," *Journal of Applied Physics*, Vol. 104, 2008.

52. Kunz, M., Chen, K., Tamura, N., and Wenk, H. R., "Evidence of Residual Elastic Strain in Deformed Natural Quartz," *American Mineralogis*, Vol. 94, pp. 1059-1062, 2009.

53. Sarobol, P., and Chen, W. H., "Effects of Local Grain Misorientation and β-Tin Elastic Anisotropy on Whisker and Hillock Formation," *Journal of Materials Research*, Vol. 28(5), 2013.

54. Sarobol, P., Koppes, J. P., Chen, W. H., Su, P., Blendell, J. E. and Handwerker, C. A., "Recrystallization as a Nucleation Mechanism for Whiskers and Hillocks on Thermally Cycled Sn-alloy Solder Films," *Materials Letters*, Vol. 99, pp. 76-80, 2013.

Accessing Adhesive Induced Risk for BGAs in Temperature Cycling

Guruprasad Arakere, Milena Vujosevic, Min Pei
Intel Corporation
1900, Prairie City Road, Folsom, CA 95630
guruprasad.arakere@intel.com

Abstract

BGA components in mobile systems are often found with adhesives applied at the package corners. The primary objective of these adhesives is to improve shock margins to prevent failures of solder joints in the case of sudden drop of the device. . Different adhesives with a wide range of thermo-mechanical material properties have been in use with limited understanding and quantification of their impact on the temperature cycling reliability of the BGA. The objective of this study is to comprehend the impact of temperature cycle on adhesives applied in the corner of BGA components and to define a range of acceptable properties of adhesives for appropriate material selection. The study utilizes a computational mechanics based Finite Element Analysis (FEA) and a Response Surface methodology to perform a detailed numerical Design of Experiments (DOE). Results were validated with test data. The results of the study indicate that for many adhesives, accelerated temperature cycling tests can lead to wrong conclusions about adhesive performance in the field. Moreover, the impact of corner glue on SJ reliability in temperature cycling is strongly dependent on assembly parameters (board thickness, substrate thickness, pitch, etc.) which has not been typically accounted for in the past. In addition, impact of any individual glue thermo-mechanical material property (Glass transition temperature, Young's modulus, Coefficient of thermal expansion) cannot be considered independently of other glue properties. All these interdependencies across geometric and material properties, as well as the temperature range the overall assembly is exposed to have been accounted for in defining adhesive property limits that would not compromise BGA performance in temperature cycling.

1. Introduction

Proliferation of mobile devices such as Ultrabook, smartphones and tablets creates unique set of concerns for shock reliability of BGA components. These concerns are a result of the use conditions these devices are exposed to as well as design features characterized by thin boards and smaller solder joints. Often times the enhancement of mechanical shock/drop performance is achieved through application of different types of polymer based adhesives. These adhesives are usually applied in the BGA corners in the form of corner glue dot or corner edge bond as shown in Figure 1. The adhesives prevent differential bending between the package substrate and board, thereby mitigating the risk to the corner solder joints under dynamic loads and improving the structural performance of packages [1-2]. The adhesive material shock mitigation capability depends on the material modulus/stiffness and its adhesion strength of the substrate/board material interface [3].

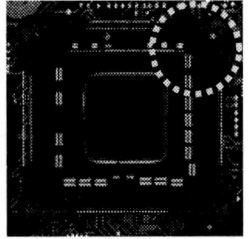

Figure 1. Typical corner glue adhesives in the form of dot and edge bond.

Although adhesive materials are used mostly to enhance mechanical shock/drop performance of packages, the solder joint reliability (SJR) under temperature cycling (TC), needs to be understood in the presence of adhesives. The adhesives commonly used in industry come in different ranges of thermomechanical properties, i.e. glass transition temperature (T_g), coefficient of thermal expansion (CTE) and Young's modulus (E). Adhesive material chemistry, filler materials, material curing process (e.g. thermal and ultra-violet) etc. significantly influence their thermomechanical properties. The thermomechanical properties of adhesives can significantly influence the response of adhesives under TC conditions [4-6]. The thermal expansion (CTE) and Young's modulus (E) of a typical adhesive material as a function temperature is shown in Figure 2. It is evident that the thermal expansion and mechanical behavior are temperature dependent. At temperatures below the material T_g, the CTE (defined as CTE_1) is low and modulus (defined as E_1) is high, while, at temperatures above T_g, CTE is high (defined as CTE_2) and modulus (defined as E_2) is low.

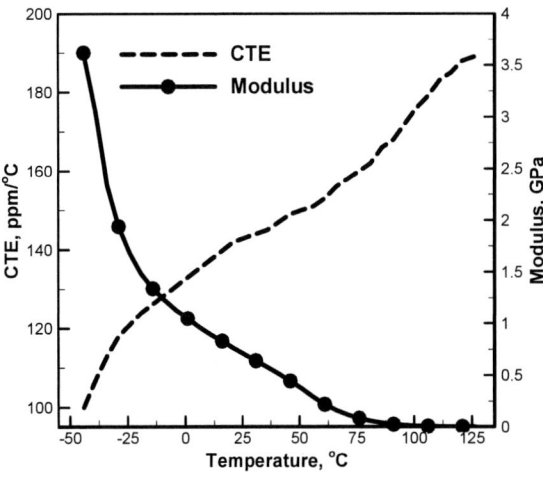

Figure 2. Thermal expansion and mechanical property behavior of a typical adhesive material. (with T_g of 70°C)

In the absence of adhesive materials, it has been well established that in FCBGA components (without thermal enabling load) under TC conditions, the SJs at highest risk are in the die shadow region [7-8] and package capability in TC is controlled by the these solder joints. Addition of adhesive in package corner region can change the location of critical solder joints from the die shadow to package corner region. The explanation for this is due to the fact that adhesive materials generally have higher CTEs compared to solder material (with CTE of ~22ppm/^{0}C) and when the package is subjected to a uniform temperature cycle (TC), the corner glue material undergoes cyclic thermal expansion (at high temperature) and contraction (at low temperature). This cyclic expansion and contraction of the adhesive results in local bending of the substrate in the package corner thereby causing the package corner solder joints (neighboring the adhesive) to experience cyclic tensile/compressive stresses (due to axial force, F_a) and shear stresses (due to shear force, F_s) as shown in Figure 3, [9]. The corner solder joints experiencing these cyclic stresses can ultimately result in their fatigue failure. The extent of expansion/contraction behavior will depend on the CTE of the adhesive and the temperature range. It has been shown in [9] that out-of-plane deformation (due to axial force, F_a) of the joints is mostly due to the CTE property of the adhesive material, while, high modulus adhesive materials tend to reduce the shear deformation (due to shear force, F_s) in the joints in the case of low CTE adhesives. This suggests that, in the presence of certain adhesive materials, the corner joints could be at highest risk under TC and therefore fail first before the die shadow joints. As a result, if during TC certification tests the capability of the component is defined without adhesives, it's capability could reduce if adhesives are applied during the board assembly process.

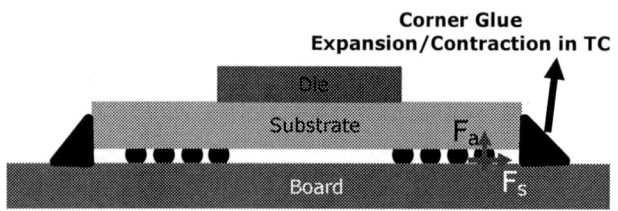

Figure 3. A schematic showing the effect of corner glue on BGA package joints during TC conditions.

This study identifies the adhesive material properties that would not compromise performance of BGA component under TC loading condition. In this paper the focus is on BGA components without thermal enabling load and without mold.

The basic assumption was that the impact of adhesives depends on the adhesive material properties, TC conditions and package assembly parameters. Understanding the interaction between these parameters and their effect on package SJR is essential to enable selection of adhesive materials which would perform well under TC conditions. This is accomplished by utilizing mechanics based numerical models to conduct extensive design of experiments (DOE)

analysis. The DOE results are further used to construct a response surface model (RSM) based tool via a multi-linear regression analysis.

3. Model

A brief description of the computational mechanics based finite element model and design of experiments approach used is provided in this section.

Finite Element Model

A numerical finite element (FE) model is utilized to compute the BGA package SJ stresses and damage under TC conditions. A typical FE model of a BGA package assembly with corner glue adhesive in the package corner is shown in Figure 4(a). The developed model is a quasi-static implicit analysis model using commercial FE software. The quasi-static nature of the problem arises due to the presence of time-dependent creep material constitutive model for the solder material [10]. All package/board materials are modeled using a linear-elastic and homogeneous material constitutive model with temperature dependent mechanical properties.

A global/local modeling approach is used to solve the TC problem. Within this approach, a coarsely meshed quarter symmetric FCBGA package global model as shown in Figure 4(a) is initially solved. Details of the package assembly are provided in the following section. The results from the global model are used to solve finely meshed individual solder joint (SJ) local models as shown in Figure 4(b). Displacement results from the global model are used to drive the substrate and board regions within the local model. Three complete TCs are modeled in order to obtain converged model response. All regions in the global (including the adhesive) and local models are modeled using second order brick elements. The corner glue adhesive is represented as an edge-bond (L-shaped) in the package corner region. The glue is assumed to not be in contact with the solder joints. The corner glue length is 5mm. The height of the glue depends on the substrate thickness and is assumed to be up to 75% of the substrate thickness. The ratio of height to width is assumed to 1.0. The adhesive is assumed to be rigidly tied to the board/substrate surfaces in all six degrees of freedom. It is modeled as a linear-elastic isotropic and homogeneous material with thermal properties. The thermal expansion and modulus below (CTE_1 and E_1, respectively) and above (CTE_2 and E_2, respectively) the material T_g are assumed to be constant for the adhesive material.

The SJ damage under TC conditions is accessed using the Inelastic Strain Energy Density (ISED) as a metric[11]. It is assumed that SJ damage/failure under TC is dominated by solder material creep. The ISED is a scalar quantity that represents the amount of energy per unit volume dissipated in the solder material due to creep deformation. It is computed using the deviatoric (which cause shape change) stress components within the solder material. The ISED is computed in two regions (each 25µm thick solder layers) neighboring the solder-substrate and solder-board interfaces as shown in Figure 4(b). It should be noted that within each of these solder regions, two element layers of solder material are used to compute ISED.

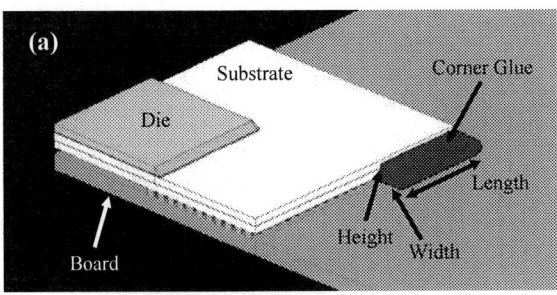

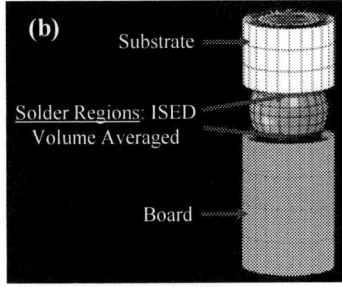

Figure 4. (a) A typical FCBGA package global TC model with corner glue; (b) Local TC model of a single solder joint.

Design of Experiments

One of the key objectives of the work is to comprehend and evaluate the acceptable adhesive material property limits from package SJR perspective. Towards this end, a numerical modeling based approach is used to conduct extensive DOE [12] to understand and quantify the effects of: (a) package assembly parameters (e.g. pitch, board thickness etc.) and (b) TC conditions (i.e. accelerated test-conditions and use-conditions) on the acceptable adhesive material property limits. A summary of all the DOE factors/parameters and their levels are shown in Table 1. In the DOE, the solder ball-size and solder resist opening (SRO)/pad sizes are scaled with varying pitch. The ranges for adhesive material properties are selected based on known material properties for materials commonly used in industry. Two TC conditions are considered, one is the accelerated test-condition with temperature range from -40°C to 100°C and other is the use-condition from 25°C to 75°C. It should be noted that both conditions have uniform temperatures without any temperature gradient within the package assembly. For each numerical experiment, the numerical model outputs, namely the maximum ISED in the die shadow and package corner regions are used as responses within the DOE model.

Further, the results from the DOE are used to perform a multi-linear regression analysis to construct a response surface model (RSM). In the RSM, the maximum adhesive induced SJ ISED is represented as an analytical function of package assembly parameters, adhesive material properties and TC conditions. The RSM enables assessment of package SJR risk due to the presence of adhesive in a time-efficient manner without the need for further modeling within the DOE ranges.

Table 1. Summary of DOE study.

Package Assembly Parameters				
Substrate Thk. (mm)		Board Thk. (mil)		Pitch (mm)
0.3, 0.8		32, 62		0.4, 0.65
Corner Glue Material Properties				
T_g (°C)	CTE_1 (ppm/°C)	CTE_2 (ppm/°C)	E_1 (GPa)	E_2 (GPa)
50, 75, 100	40, 70	100, 200	1, 6	0.005, 0.4
Temperature Conditions				
Accelerated Test-condition		Use-condition		
Range: -40°C to 100°C		Range: 25°C to 75°C		

An FCBGA package without thermal enabling load is considered within this study. The package design parameters are defined in Table 2. The package design parameters are as follows: (a) package size: 27×27 (mm); (b) die thickness: 0.5mm; (c) die size: 13.5×13.5 (mm); and (d) SJ type: metal defined pads on board side. A rectangular BGA pattern with a central cavity (an area in the die shadow without BGAs) under the die shadow region is considered as shown in Figure 5. The innermost SJs around the cavity are aligned with the die edges. Also, in the package corner region a single row of BGA is depopulated.

Table 2. Summary of FCBGA package considered.

Package Size (mm)	Die Size (mm)	Die Thk. (mm)
27×27	13.5×13.5	0.5

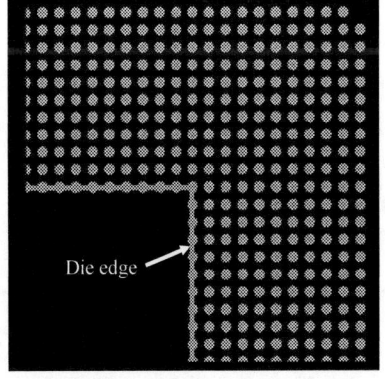

Figure 5. Package BGA map with BGA cavity aligned with die edge.

4. Results and Discussion

This section explains and quantifies the effects of package assembly parameters, adhesive material properties and their interaction on TC performance of BGAs with corner glue. It also discusses the impact of TC conditions assembly is exposed to with special focus on the accelerated vs. use condition temperature ranges. It uses developed understanding to define acceptable material property limits for corner glue which will not reduce package SJR in TC.

Corner Glue Length Effect

The effect of corner glue adhesive length on the maximum damage/ISED in the package corner SJs in an FCBGA package is evaluated. The board thickness, pitch and substrate thickness are 32mil (0.8128mm), 0.65mm and 0.8mm, respectively. Five different corner glue lengths from 1mm to 5mm in increments of 1mm are considered. The effect of adhesive length on the maximum damage/ISED in the package corner SJs is displayed in Figure 6. Results indicate that an increase in adhesive length significantly increases the maximum package corner SJ ISED. The expansion of a larger volume of adhesive material (in the case of longer adhesives which have higher geometric stiffness) results in larger forces in the package corner SJs causing higher TC risk. It is seen that the maximum SJ damage tends to saturate when the adhesive length is 5mm. Therefore, a constant adhesive length of 5mm is used in the entire study. It should be noted that for all the cases the maximum package corner SJ ISED was larger than that in the die shadow joints.

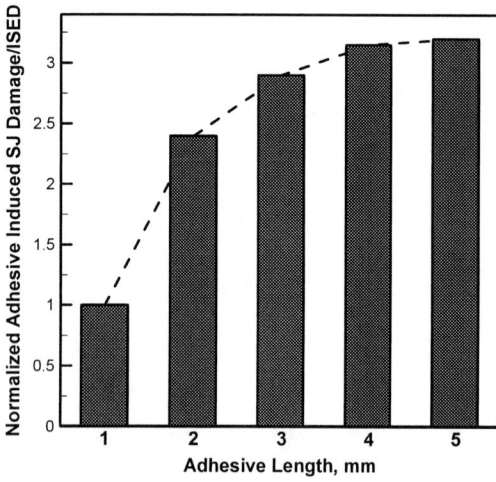

Figure 6. Effect of corner glue length on the maximum damage/ISED in the package corner SJs.

Impact of Corner Glue on Die Shadow SJ Risk

The impact of package assembly and adhesive material parameters on the maximum damage/ISED in the die shadow SJs is presented in this section. A Pareto chart showing the effect of above parameters on the maximum damage in the die shadow SJs is shown in Figure 7. The key results are: (a) package assembly parameters have the largest effect (in decreasing order of board thickness, pitch and substrate thickness) on the maximum damage in the die shadow SJs. Significant interaction between parameters is also evident; and (b) Impact of adhesive material properties on the die shadow SJ damage is negligible. The results show that increasing board thickness and decreasing pitch tends to increase die shadow SJR, while the impact of adhesive material properties is negligible. This implies that the die shadow SJR in TC is dominated by stresses induced due to CTE mismatch between the board-substrate-die materials.

Further, since adhesives are farther away from the die shadow joints in the package corners, their impact on die shadow joints is negligible

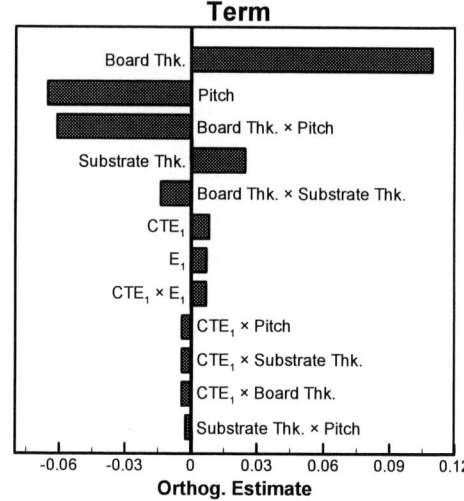

Figure 7. Effect of parameters on die shadow SJR risk under TC conditions.

Impact of Corner Glue on Package Corner SJ Risk

The impact of package assembly and adhesive material properties on the maximum damage/ISED in the package corner SJs is discussed below. A Pareto chart showing the effect of above parameters on the maximum damage in package corner SJs is shown in Figure 8. The key results are as follows: (a) Package assembly parameters: Pitch has the largest effect followed by the substrate thickness on the maximum damage in package corner SJs. The effect of board thickness is negligible and therefore not shown; and (b) Adhesive material properties: Material CTEs and T_g have a significant effect on the package corner SJ damage/ISED. However, the adhesive material modulus has a relatively smaller effect.

The effect of pitch can be attributed to the higher load carrying capacity in the case of larger SJs (associated with larger SRO/pad-sizes and ball sizes in large pitch) which reduce the package corner SJ stresses/damage. The effect of substrate thickness can be attributed to thick (stiffer) substrates being able to transfer high loads to the corner SJs during adhesive material expansion/contraction in TC conditions. Regarding the impact of adhesive material properties, results indicate that decreasing material CTEs and increasing T_g reduce the maximum damage in the package corner SJs. The relative impact of material modulus is relatively low. Also, the high CTE (CTE_2) above the adhesive material T_g can cause significant acceleration of SJ damage in the package corner joints. This implies that the impact of adhesive materials on package corner SJR also depends on the temperature conditions which the adhesives experience above the material T_g. Therefore, using high T_g and low CTE adhesive materials can significantly improve package corner SJR.

In summary, adhesive material properties can significantly impact the SJR of package corner joints under TC conditions.

978-1-4799-2408-0/14 $31.00 © 2014 IEEE

Therefore, selection of adhesive materials with suitable material properties which will not reduce (in FCBGA packages without thermal enabling load) package SJR is critical. From the above results it is evident that in addition to the adhesive material properties, package assembly parameters also have significant impact on package corner SJR. Also, it is clearly seen that the impact of corner glue adhesive is local to the package corner joints and does not impact the die shadow joints.

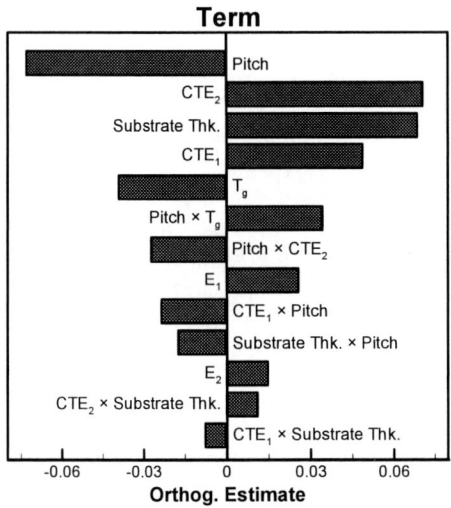

Figure 8. Effect of parameters on BGA package corner risk under TC conditions.

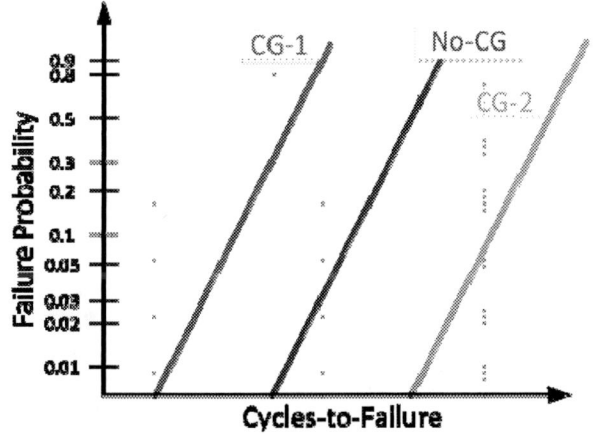

Figure 9. Test data comparing the package corner SJR in FCBGA packages with two corner glues and baseline case.

The effect of adhesive material properties (T_g and CTE) on package corner SJR presented above is in good agreement with experimental results. The experimental results comparing the package corner SJR for an FCBGA package with two different corner glues (CG-1 and CG-2) with different material properties and a baseline without (No-CG) corner glue case is shown in Figure 9. The CG-1 material is a low T_g (50^0C) and high CTE (CTE_1=150, CTE_2=208) adhesive, while, CG-2 is a high T_g (105^0C) and low CTE (CTE_1=33, CTE_2=65) adhesive. The packages are tested under accelerated test conditions with temperature range from -40^0C to 100^0C. In agreement with the modeling results, it is

seen that the package corner SJR for CG-2 is significantly higher compared to CG-1 and baseline case. This clearly shows that adhesive materials with higher T_g and lower CTEs can improve package corner SJR.

Acceptable Adhesive Material Property Limits

Without corner glue adhesives the considered FCBGAs (that do not have thermal enabling load applied) the first SJ failure is in the die shadow SJs and the maximum SJ damage/ISED is in these joints. However, from the above discussion it is evident that with adhesives (depending on adhesive material properties), the first SJ failure location could shift to the package corner joints and the maximum SJ damage/ISED could be in these joints. Therefore, presence of adhesive will influence the TC capability of the package corner joints and thereby affect package SJR.

In the FCBGA packages considered, the adhesive induced package SJR risk is accessed using ratio of the maximum package corner SJ damage/ISED with adhesive to maximum package SJ damage/ISED (which is in the die shadow region) without adhesive. The results of the modeling DOE are combined in a simple Adhesive Decision Making Tool that based on geometric parameters of the assembly and material properties of adhesives computes a decision metric that defines acceptability of the adhesive. In this paper, this ratio will be referred to as the *acceptance metric* or simply as metric. This metric is a dimensionless scalar quantity and a value less than 1.0 implies that the adhesive material does not reduce package SJR, while a value greater than 1.0 means that the adhesive reduces package SJR. A value greater than 1.0 also implies that first SJ failure will be in the package corners joints due to the adhesive induced stresses. Ideally an adhesive good for package SJR in TC should have a metric value less than 1.0. The effect of package assembly parameters and temperature cycling conditions on the acceptable adhesive material property limits (defined using the *acceptance metric*) is discussed below.

Effect of Package Assembly Parameters on Acceptable Adhesive Material Property Limits

The developed RSM based tool is used to evaluate the adhesive induced package corner SJ damage as a function of package assembly parameters, adhesive material properties and TC conditions. The effect of board thickness on the acceptable adhesive material property limits is evaluated. An FCBGA package on two board thicknesses, 32mil and 62mil, and subjected to accelerated (-40^0C to 100^0C) TC conditions is considered. In Figure 10(a) and (b), the metric is plotted for corner glue adhesives whose material properties (E_1 vs. CTE_1) are within a certain range when the board thickness are 32mil and 62mil, respectively. It should be noted that Figure 10(a) and (b) are for a constant adhesive material T_g of 100^0C. Also, the following parameters are constant: substrate thickness=0.8mm, pitch=0.65mm, CTE_2=150ppm/^{0}C and E_2 =5MPa.

As mentioned earlier, the acceptable adhesive materials in TC are those whose metric is less than 1.0. From Figure 10 it is seen that the acceptable adhesive materials which do not reduce package SJR in TC are those below the 1.0 line, since, the die shadow joints are at higher risk than the package corner joints and therefore fail first. However, adhesive materials outside the acceptability window (above the 1.0 line) reduce package SJR though higher risk in the package corner SJs compared to the die shadow joints. From Figure 10(a) and (b), it is evident that the acceptable adhesive material property limits in the case of a 62mil thick board is much wider compared to that for a 32mil board. In the case of a 32mil board, the die shadow joints are at lower TC risk compared to a 62mil board due to the board thickness effect. Therefore, in the case of 32mil board, the package corner joints could become higher risk joints (compared to die shadow joints) even with lower CTE adhesives and fail first. This implies that the metric value will be higher with 32mil board compared to 62mil board, resulting in narrower adhesive material property limits and therefore smaller set of acceptable adhesive materials. The results clearly show that the acceptable adhesive material property limits under TC conditions will strongly depend on package assembly parameters. For purpose of brevity, only the effect of board thickness is provided here. However, substrate thickness and pitch also influence the acceptable adhesive material limits as indicated earlier.

Effect of Temperature Cycling Conditions on Acceptable Adhesive Material Property Limits

As mentioned earlier, the thermal (CTE) and mechanical (E) properties of adhesive materials vary significantly below and above the material T_g. Therefore, the temperature range adhesives are exposed to during accelerated test-conditions and use-conditions can influence the adhesive material response and therefore package SJR risk. For example, adhesive materials with T_g inside the accelerated test-condition temperatures can cause significant acceleration of SJ failures due to higher adhesive CTEs. The effect of two uniform TC conditions (i.e. accelerated test condition and use-condition) on the acceptable adhesive material property limits is evaluated. An FCBGA package with 32mil thick board, 0.8mm substrate thickness and 0.65mm pitch is considered. Figure 11(a) and (b) show the acceptable adhesive material property (E_1 vs. CTE_1) limits when the package is subjected to both test and use-conditions, respectively. It should be noted that the adhesive material T_g is constant at 50^0C and inside the temperature range for both TC conditions. Also, the figures are for a constant $CTE_2=150ppm/^0C$ and $E_2=5MPa$.

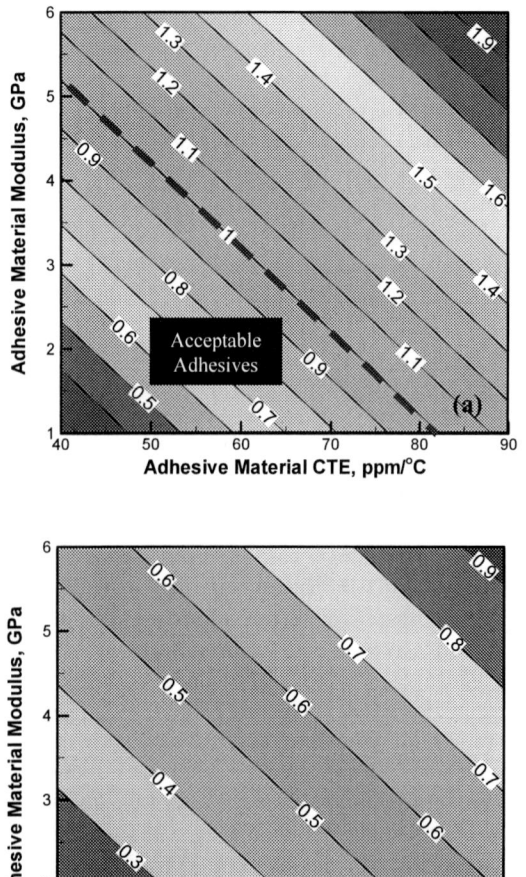

Figure 10. Acceptable adhesive material property limits in the case of: (a) 32mil thick board; (b) 62mil thick board.

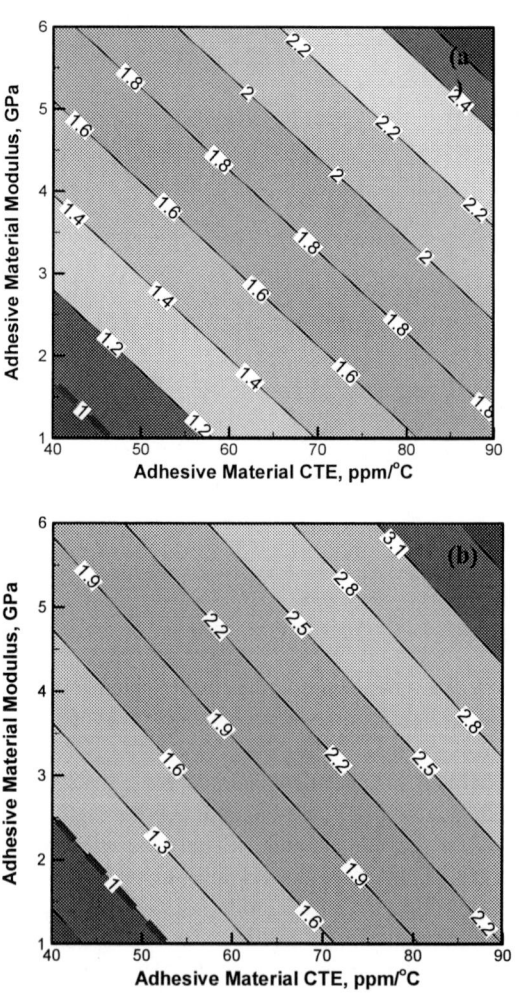

Figure 11. Acceptable adhesive material property limits for: (a) accelerated test-condition (-40^0C to 100^0C); (b) use-condition (25^0C to 75^0C) for material T_g of 50^0C.

The results indicate that the acceptable adhesive material limit is wider under use-condition TC than under accelerated test-condition. The lower peak temperature in use-conditions causes less adhesive induced SJ stresses/damage in the package corner joints resulting in a wider acceptability window for adhesive materials. This implies that certain adhesive materials which are unacceptable (adhesives will reduce package SJR) under accelerated test-conditions and would fail the test could be acceptable under use-conditions. This sensitivity of adhesive materials to temperature ranges/conditions is important since temperatures encountered in accelerated test-conditions can artificially (not seen in use-conditions) increase the adhesive induced SJ damage and failure. This shows that the test conditions during package TC tests (with adhesives) can impact adhesive material selection. Test conditions in which the adhesive material T_g is inside the test temperature range can cause significant acceleration of the adhesive induced SJ damage resulting in misleading test results.

The TC conditions considered in this study are uniform (no thermal gradient) temperature conditions in which the package corner joints and adhesive will reach the specified uniform temperatures. However, under actual processor power cycling conditions, a significant thermal gradient is present in the die region, while, the temperatures in the package corner region is much lower since the package corner joints are away from the die. It was established through a modeling study that under power cycling conditions the maximum die shadow ISED is ~1.4 times that under uniform test-condition due to the presence of a thermal gradient. It should be noted that the factor of 1.4 is not constant and will depend on package assembly parameters. For the same given package assembly parameters and adhesive material parameters given above, the acceptable adhesive material property limits under power cycling conditions is displayed in Figure 12. Results show that the acceptable adhesive material property limits are much wider compared to uniform temperature conditions due to the lower adhesive induced SJ damage in the package corner regions caused by lower temperatures.

The above results are for a given adhesive material T_g of 50^0C. The acceptable adhesive material property limits for the above mentioned package with a constant material T_g of 75^0C is shown in Figure 13(a) and (b) for both test and use-conditions, respectively. The material T_g is outside the use-condition temperature range but inside the test-condition range. In agreement with prior results, the acceptable adhesive material property limits are wider under use-condition TC than under test-condition. The lower SJ stresses under use-conditions coupled with non-linearly in solder material response result in wider material limits. The acceptable adhesive material property limits when the adhesive material T_g is constant at 100^0C is shown in Figure 14(a) and (b) for both test and use-conditions, respectively. The material T_g is outside both the use-condition and test-condition temperature ranges. Similar to prior results, the acceptable adhesive material property limits for adhesives are wider under use-condition TC than under test-conditions. This indicates that for adhesives with any T_g the above difference in acceptability limits between use-condition and test-condition is present.

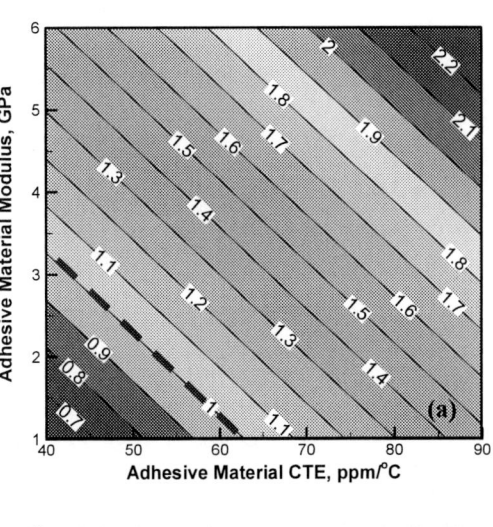

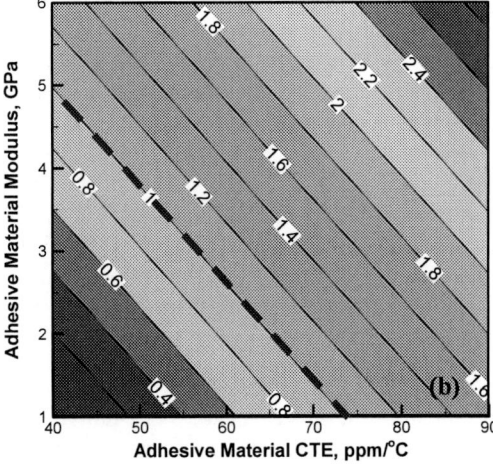

Figure 13. Acceptable adhesive material property limits for: (a) accelerated test-condition (-40^0C to 100^0C); (b) use-condition (25^0C to 75^0C) with material T_g of 75^0C.

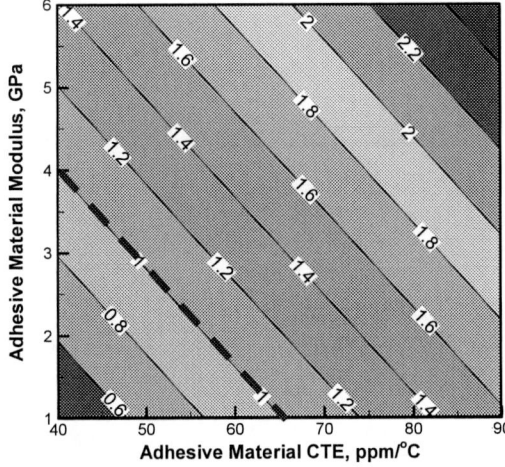

Figure 12. Acceptable adhesive material property limits under power cycling conditions for material T_g of 50^0C.

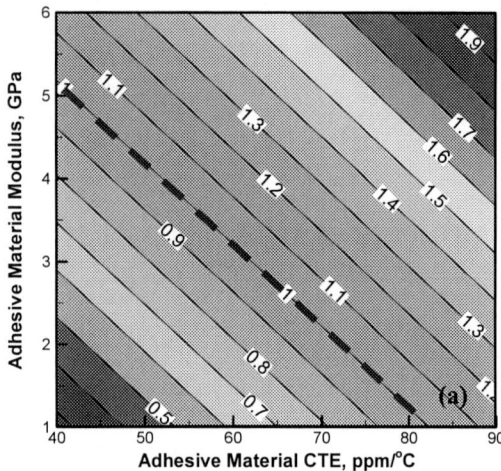

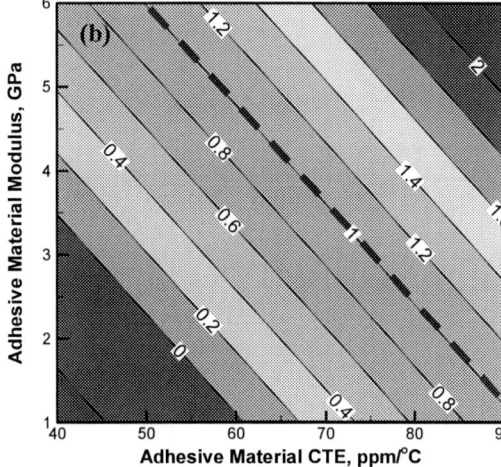

Figure 14. Acceptable adhesive material property limits for: (a) accelerated test-condition (-40°C to 100°C); (b) use-condition (25°C to 75°C) with material T_g of 100°C.

The above results show that temperature range during TC can significantly influence the acceptable adhesive materials from package SJR consideration. The acceptable adhesive material limits under use-conditions are much wider than that under accelerated test-condition which implies that adhesives which fail under accelerated test-condition could perform well under use-condition. Using high T_g and low CTE corner glue adhesive materials with T_g outside the TC range can significantly protect the package corner joints under TC conditions. Such adhesives could ensure that the first joints to fail in the case of FCBGA packages (without load) are those in the die shadow region and therefore will not reduce package SJR under TC conditions.

5. Implications

Understanding the impact of TC conditions and package assembly parameters on acceptable adhesive materials from a package SJR perspective is one of the key objectives of this effort. A response surface model/tool has been developed to predict the maximum adhesive induced SJ damage as a function of package design parameters, TC conditions and

adhesive material properties. As TC testing of packages with adhesives to access package SJR is generally time consuming and expensive, the developed tool can be used to predict adhesive induced package SJR risk in a few minutes.

An example of ranking a given set of commonly used corner glue adhesive materials based on the acceptability metric under accelerated TC conditions is given in Table 3. Based on the metric, the adhesive *CG-1* performs best and does not negatively impact package SJR. It is seen that adhesives with high T_g and low CTE are good for package SJR.

Table 3. An example of ranking adhesive materials based on package SJR performance under accelerated test-conditions.

Name	T_g ($^{\circ}$C)	CTE_1 (ppm/$^{\circ}$C)	CTE_2 (ppm/$^{\circ}$C)	E_1 (GPa)	E_2 (GPa)	Metric
CG-1	100	34	70	7.5	1	0.9
CG-2	100	68	100	2.5	2	1.05
CG-3	50	70	140	5	0.1	2.21
CG-4	40	100	170	3	0.1	3.0
CG-5	70	150	250	1	0.1	4.5

The results from the study show that the temperature ranges during TC significantly influence the adhesive induced package SJR risk. The adhesive material property limits in use-conditions are wider compared to that under accelerated test-conditions. Under actual power cycling conditions when the temperatures in the package corners are lower, the acceptable material property limits are much wider. This implies that certain adhesive materials which are unacceptable (adhesives will reduce package SJR) under accelerated test-conditions and fail the test may perform well under use-conditions (power cycling) conditions. An example showing the adhesive induced SJ risk (for the same corner glue materials as in Table 3) under use-conditions is shown in Table 4.

Table 4. An example of ranking of adhesive materials based on package SJR performance under use-conditions.

Name	T_g ($^{\circ}$C)	CTE_1 (ppm/$^{\circ}$C)	CTE_2 (ppm/$^{\circ}$C)	E_1 (GPa)	E_2 (GPa)	Metric
CG-1	100	34	70	7.5	1	0.5
CG-2	100	68	100	2.5	2	0.6
CG-3	50	70	140	5	0.1	0.9
CG-4	40	100	170	3	0.1	1.45
CG-5	70	150	250	1	0.1	1.6

Comparing the results from Table 3 and 4, it is seen that the adhesive materials which reduce package TC performance under accelerated test-conditions however perform well and do not reduce package SJR in use-conditions.

The study demonstrates that the package assembly parameters also impact the acceptable adhesive materials under TC. The acceptable adhesive material property limits for thin boards are narrower compared to thick boards which implies that impact of adhesives on thin boards is larger. Therefore, selecting thin boards for accessing adhesives is important.

6. Summary and Conclusions

When adhesives are used to improve shock performance of BGAs, it is very important to evaluate the impact they will have on TC performance. Some adhesives can compromise performance in TC and even change the location of critical solder joints. To determine the acceptable range of adhesive material properties that would not reduce TC margins a modeling based design of experiment evaluation was conducted in this work. The results indicate that there is an interaction between adhesive material properties and geometric parameters of the assembly as well as among the key thermomechanical properties of the adhesive. The modeling results are confirmed by experimental findings implying that modeling can be an effective tool for these types of risk assessments. The results of the modeling DOE are combined in a simple Adhesive Decision Making Tool and based on geometric parameters of the assembly and material properties of adhesives, it computes a decision metric that defines acceptability of the adhesive.

A very important part of this study is the evaluation of adhesives not only for accelerated temperature range but also for field conditions. The key conclusion is that accelerated tests with adhesives can be misleading because some adhesives that fail in accelerated tests will perform well in the field.

Understanding the established impact of corner glue material properties on package SJR is very beneficial for: a) evaluating TC risks due to a given adhesives and/or b) developing new adhesives with beneficial material properties. Such adhesive materials with optimized material properties may be required for potential use with packaging technologies scaling to smaller and thinner form factors with tighter pitches, where any loss of performance margin can be detrimental.

It is important to note that results presented in this paper are limited to FCBGA components without mold and without thermal enabling load. Similar studies were done for these two configurations and will be published in future publications.

Acknowledgments

The authors would like to thank Muffadal Mukadam (Intel) for sharing his experimental data and Ru Han (Intel) for useful discussions.

References

1. K. Meyyappan, A. Mcallister, M. Kochanowski, "Effects of Glue on the Bend Performance of Flip Chip Packages", *IEEE Trans. on Components and Packaging Technologies*, vol. 31, no. 3, Sep. 2008.

2. W. K. Loh, "Solder joint reliability prediction of flipchip packages under shock loading environment," *in Proc. InterPACK '05 Conf.*, San Francisco, CA, Jul. 17–25, 2005.

3. L. Fan, C. K. Tison, C. P. Wong, "Study on Underfill/Solder Adhesion in Flip-Chip Encapsulation," *IEEE Trans. On Advanced Packaging*, vol. 25, no. 4, 2002.

4. B. V. Chheda, S. M. Ramkumar, R. Ghaffarian, "Thermal Shock and Drop test Performance of Lead-free Assemblies with No-underfill, Corner-underfill and Full-underfill," *IEEE ECTC Procee. 60th*, pp. 935-942, 2010.

5. J-Y. Lee, T-K. Hwang, J-Y Kim, M. Dreiza, "Study on the Board Level Reliability Test of Package on Package (PoP) with 2nd Level Underfill," *IEEE ECTC*, 2007.

6. P. Borgesen, D. Blass, M. Meilunas, "Effects of Corner/Edge Bonding and Underfill Properties on the Thermal Cycling Performance of Lead Free Ball Grid Array Assemblies," *ASME, Journl. Of Elec. Packaging*, vol. 134, 2012.

7. P. Bhatti, M. Pei and X. Fan, "Reliability Analysis of SnPb and SnAgCu Solder Joints in FC-BGA Packages with Thermal Enabling Preload," *IEEE ECTC*, 2006.

8. L. Garner, C. Zhang, K. S. Beh, K. Helms, Y. L. Tan, "Effect of Compressive Loads on the Solder Joint reliability of Flip Chip BGA Packages" *IEEE ECTC*, pp. 692 – 698, 2004.

9. M. Pei, R. Han, Y. Ge, S. Goyal, V. Rajarathinam, M. Mukadam, "The effect of corner glue on BGA package temperature cycling performance: A modeling study," *IEEE ECTC*, pp. 1494 – 1499, 2013.

10. S. Wiese, M. Roellig, K.-J. Wolt, "Creep of Eutectic SnAgCu in Thermally Treated Solder Joints," *Proc. IEEE Electronic Components and Technol. Conf. (ECTC)*, vol. 2, pp. 1272 – 1281, 2005.

11. A. Syed, "Accumulated Creep Strain Energy and Energy Density Based Thermal Fatigue Life Prediction Models for SnAgCu Solder Joints", *Proc. IEEE Electronic Components and Technology Conference*, pp. 737-746, 2004.

12. D. C. Montgomery. G. C. Runger, N. F. Hubele, "Engineering Statistics", *John Wiley & Sons, Inc.*, 1998.

Characteristics of Ceramic BGA using Polymer Core Solder Balls

Hiroya Ishida, Kiyoto Matsushita
Sekisui Chemical Co., Ltd.
Koka, Shiga, Japan
ishida030@sekisui.com, +81-748-62-7110

Abstract

The thermal reliability of lead-free solder joints between ceramic substrates and FR4 printed circuit boards (PCBs) is a major concern when designing ceramic ball grid arrays (CBGAs), particularly for thicker or larger substrates, since their CTE is much smaller than that of PCB. In order to achieve high reliability for critical applications, polymer core solder balls (PCSBs) are a promising solution. PCSBs enable high reliability without the need for other materials such as underfill resin or modification of package design to ensure reliability. In this study, the characteristics of PCSBs were evaluated using a variety of techniques in order to assess their adaptability to CBGA applications. The solder joint condition and the presence of intermetallic compounds (IMCs) were investigated, and the board-level reliability and electrical performance were compared to those of conventional SAC305 solid solder balls. Thermal cycling tests from −40 to +125 °C were carried out using an alumina CBGA with a 25 × 25 mm^2 package size and a 1.0 mm ball pitch. As a result of a larger solder bump height and a finer IMC microstructure, PCSBs exhibited a board-level reliability more than three times higher than that for SAC305 solder balls. PCSBs also exhibited stable electrical performance, with a similar initial resistance to that for SAC305 solder balls. Electromigration tests were carried out at a temperature of 150 °C and a current of 1.8 A, and lifetimes of more than 700 hours were found for both types of solder balls. The PCSB failure mode was fatigue cracking in the solder at the interface with IMCs on the ceramic side, which is similar to the case for SAC305. The use of PCSBs in large-sized HITCE CBGAs was also investigated to assess their potential for increasing the demand for larger package applications.

Introduction

Ceramic ball grid arrays (CBGAs) are often chosen for high-end products because their multi-layer ceramic substrate has better heat resistance and higher electric performance than organic substrate. They have been used for various applications such as aerospace technologies, communication modules, radio frequency modules, field-programmable gate arrays, CMOS image sensors (CISs), high-end computer servers, and industry equipment. Both their usage and functionality are increasing. For instance, CISs are being in various demands for smartphones, industry equipment, inspection apparatuses, medical and automotive devices, and other applications that require highly sensitive sensors and highly reliable packages. For example, CISs are increasingly used for automotive safety applications, often being mounted on the car body outside. Therefore, a very high level of reliability is required for automotive CISs. Another example of increasing functionalities for CBGAs is server. As a large-scale data is exchanging with various devices, our information-driven society depends on big data being handled at high speeds. Servers play a key role in supporting cloud-computing infrastructure. To meet market requirements, advanced servers having more functionality are required, and their package sizes are becoming larger, from 40 × 40 mm^2 to 60 × 60 mm^2 or 70 × 70 mm^2. Even 100 × 100 mm^2 package sizes are expected in the near future. Some evaluations of CBGAs for servers have been reported [1, 2]. However, there is still limited second-level interconnect reliability for large-size CBGAs. In fact, in all of the above examples, there are potential reliability problems in CBGA applications.

Board-level temperature cycling reliability is one major concern in CBGAs, since the use environment for CBGAs is very strict. Because of the different coefficients of thermal expansion (CTE) between the ceramic substrate and the printed circuit board (PCB), more thermal stress is generated at the solder joint in CBGAs. Although underfill or corner bonding can be used to improve reliability, device manufacturers are reluctant to use underfill because it makes it more difficult to rework packages. It is also desirable to omit the underfilling process in order to reduce manufacturing costs and shorten lead time.

Solder selection is a key factor in the selection of package materials and the design of the structure, after shifting solder alloy from lead solder to lead-free solder for consumer applications. Polymer core solder balls (PCSBs) are promising solder materials because their polymer cores keep standoff height and absorb generated stress over the solder joint interface, which may help ensure sufficiently high reliability without using underfill. Several studies have been investigated [3-9]; however, the performance of PCSBs for CBGAs has not yet been proven sufficiently, and their capabilities are still not well understood.

Thus, in this study, CBGAs using PCSBs were prepared, and their reliability and electric performance were assessed to determine their suitability for CBGA applications.

Configuration of PCSBs

The PCSB used in this study is shown in Figure 1. The final ball diameter is approximately 670 μm, which consists of a polymer core and several metal plating layers. The polymer core is spherically foamed to 550 μm by co-polymerization of divinylbenzene. The CTE of the core is 40 ppm at 20°C. The first nickel layer of the core is a primer for plating the next layer. The first copper layer, which is 20 μm, contributes to higher conductivity. A second layer of thin nickel is then added to prevent serious copper diffusion from the first copper layer and to secure the pure solder for PKG assembly. The thin 5% silver solder layer is formed by plating, but it

2014 Electronic Components & Technology Conference

smoothly changes to tin-silver-copper SAC alloy by diffusing a thin second copper layer under the solder layer. Mottled nickel doping on the solder surface is used to realize a finer intermetallic compound (IMC), $(Cu,Ni)_6Sn_5$, at the electrode interface. A fine IMC helps anchor the solder connection, which is also expected to improve reliability. In addition, nickel-doped PCSBs exhibited improved reliability [10, 11].

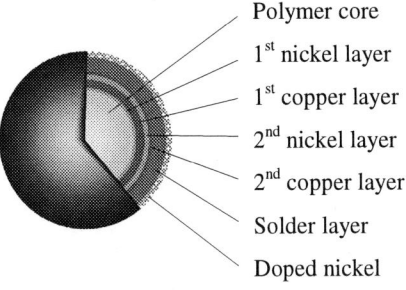

Polymer core
1st nickel layer
1st copper layer
2nd nickel layer
2nd copper layer
Solder layer
Doped nickel

Figure 1. Schematic diagrams of PCSBs

Figure 2 shows top views of the solder bumps after balling on PKG. The surfaces of the PCSB bumps are smoother since their solder shrink is less due to a smaller amount of soldering. Therefore, oxidization of the solder surface is suppressed, and solderability in PKG assembly is ensured. In addition, the lower oxidization is expected to reduce the risk of generating voids inside the solder that decrease the reliability, because solder oxide reacts with the flux and generates water vapors during the reflow process, and they may become voids if they remain inside solder.

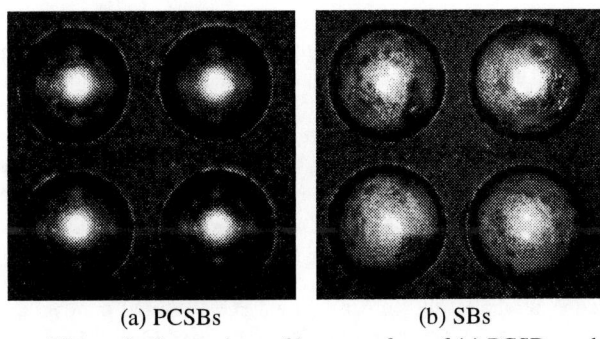

(a) PCSBs   (b) SBs

Figure 2. Comparison of bump surface of (a) PCSBs and (b) SBs SAC305 after balling

The melting temperature of solder is approximately 224°C, which is the same as that of solder balls (SBs) SAC305, as shown in Figure 3. We viewed the condition sidewise under the reflow in PKG assembly with just the flux. PCSBs kept the standoff height without depressing by PKG weight, unlike SBs.

(a) PCSBs

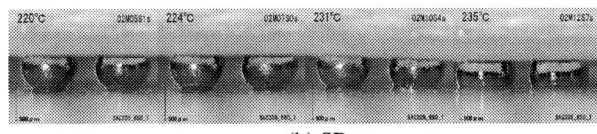

(b) SBs

Figure 3. Comparison of solder melting temperatures of (a) PCSBs and (b) SBs SAC305

Experimental Procedures

Test Element Group (TEG) Design

Two types of daisy chain devices as CBGAs were used for PCSB characterization: a 25×25 mm^2 PKG body size made of alumina with a 1.0 mm ball pitch, and a 61×61 mm^2 PKG body size made by Kyocera high thermal coefficient of expansion (HITCE) with a 1.0 mm ball pitch. Table 1 lists the design dimensions and appearances of these daisy chains. Solder bumps were formed in a full-grid array to evaluate the temperature cycling (TC) reliability. Ball mounting was conducted by using a SAC305 solder paste, which was printed with a 200-μm-thick stencil. Reflow was performed in a nitrogen atmosphere with a peak reflow temperature of 250°C. The PCB is made of FR-4, and its dimension for the PKGs is 216 mm × 116 mm × 2 mm. The PCB pad design is non-solder mask defined (NSMD). The SB pad and solder mask opening have dimensions of 550 μm and 650 μm, respectively. An organic solderability preservative (OSP) was used as the PCB surface finish. Table 2 lists key elements of the PCB design. PKG assembly was conducted by using a SAC305 solder paste as well, which was printed with a 200-μm-thick stencil. Reflow was performed in air using the same temperature profile as for ball mounting. Figure 4 shows a cross-sectional SEM image after balling and PKG assembly. Figure 5 shows the interfacial IMC microstructures of the solder joint at the PKG side and the PCB side after PKG assembly, respectively. EDX analysis revealed that the IMC at the PCB side was $(Cu,Ni)_6Sn_5$. This implies that the doped nickel was incorporated in the solder and reached the copper electrode during solder convection in reflow. The finer IMC is expected to better anchor the electrode and improve reliability.

Table 1. CBGA TEG parameters

Parameter	Description1	Description2
PKG Size	25×25 mm^2	61×61 mm^2
PKG Thickness	1.0 mm	2.0 mm
Material	Alumina	HITCE™
Material CTE (room temp)	7.1 ppm/ °C	11.7 ppm/ °C
Bump Pitch	1.0 mm	1.0 mm
Pad Opening	600 μm	600 μm
Pad Surface Finish	Ni/Au	Ni/Au
Solder Ball Count	420 bumps	3588 bumps
Appearance		

Table 2. PCB parameters

Parameter	Description
Board Size	$216 \times 116 \text{ mm}^2$
Thickness	2.0 mm
Material	FR-4
Material CTE (room temp)	14 to 17 ppm/°C
Land Size	550 μm
Land Type	NSMD
Solder Resist Opening Size	650 μm
Solder Resist Thickness	20 μm
Surface Finish	Copper-OSP

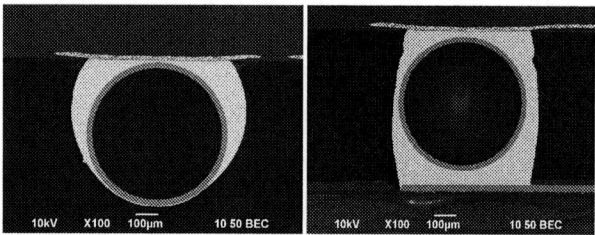

Figure 4. Cross-sectional SEM images after (left) balling and (right) PKG assembly

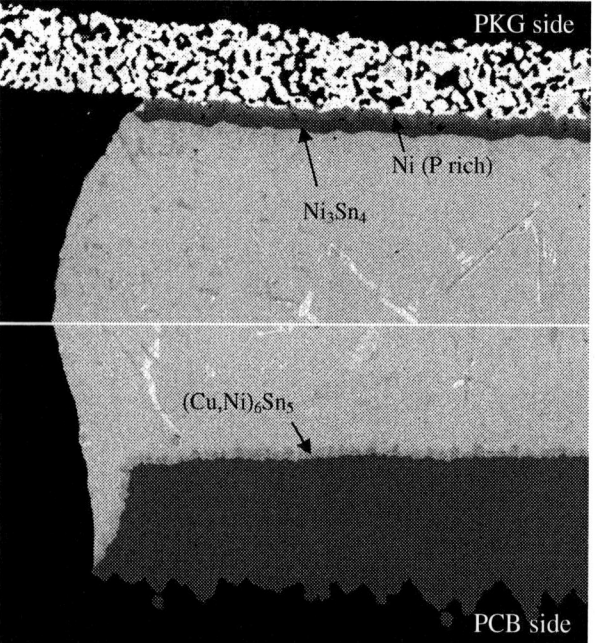

Figure 5. SEM images of IMC morphology

Board-Level Reliability (BLR) Test

Board-level TC tests (TCTs) for $25 \times 25 \text{ mm}^2$ CBGAs were conducted in accordance with JEDEC specifications JESD22-A104D (condition G). The temperature was cycled from −40 to 125°C with a 10 min dwell on each end. The transition time on each side was approximately 2 min. Board-level TCTs for $61 \times 61 \text{ mm}^2$ CBGAs were conducted in accordance with JEDEC specifications JESD22-A104D (condition J). The temperature was cycled from 0 to 100°C with a 30 min dwell on each end. The transition time on each side was approximately 5 min. The resistance of each side was monitored in real time. A failure was considered to have occurred when the resistance increased by 10% from the initial resistance. Distributions were evaluated in terms of the shape parameter m and the characteristic life η (the minimum duration for 63.2% accumulative failure). Optical microscopy was used to analyze the failure mode.

In addition, a mechanical bend test was conducted. The specimen board was placed on two supports 120 mm apart (L), and the actuator applied a force with a deformation of 4 mm by 1 Hz at the center of the two supports (L/2).

Electromigration Test

The same TEG of $25 \times 25 \text{ mm}^2$ used for the BRL test was used for the EM test. On the TEG, the balls are connected to a Ni/Au pad, which is 600 μm wide and 4–7 μm Ni/~0.5 μm Au thick. On the PCB, the balls are connected to a 550-μm-diameter Cu pad of 20 μm thickness. Solder joints were stressed with 1.8 A DC current at 150°C. The current density was $1.59 \times 10^2 \text{ A/cm}^2$, calculated on the basis of the pad opening diameter. Failure was considered to be a 10% rise in resistance. After current stressing, we observed the cross section of the solder joints by scanning electron microscopy.

Results and Discussion

Comparison of Reliabilities of PCSBs on Different CBGAs

We evaluated PCSB reliability with different CBGAs, as shown in Table 1. In addition to the PCSBs, a solid SB SAC305 of 650 μm was evaluated as a reference. The results reveal that PCSBs have a much higher reliability than SBs with the TEG of $25 \times 25 \text{ mm}^2$ using the alumina substrate, as shown in Figure 6. The TC reliability of PCSB is more than 300% higher than that of the conventional SB SAC305. The crack generated at the location farthest from the center during TCT laterally propagates from the weakest interface because of the lateral stress caused by CTE mismatch between the ceramic substrate and the PCB. Therefore, the stress is concentrated at the corner of the TEG. We used the dye and pry method to see the failure locations, as shown in Figure 7. The failure occurred in the weakest solder connection at the interface. Figure 8 shows cross-sectional optical microscopy images of the failure mode. These images reveal that the crack occurred along the IMC interface near the pad on the PKG side and proceeded into the solder, while there were no other cracks around the equatorial of the PCSB bump, as in previous reports [6]. We propose three reasons why PCSB shows higher reliability than SB and the generation of cracks only at the IMC interface near the pad on the PKG: 1) higher standoff height, 2) higher Young's modulus of the polymer core, and 3) higher CTE of the polymer core.

Stress mainly generated at the age of the solder connection because of the lateral stress caused by the CTE mismatch between the ceramic substrate and the PCB. In the PCSB bump, the lateral stress is smaller than the larger vertical stress because of the higher standoff height. Therefore, the generated

978-1-4799-2408-0/14 $31.00 © 2014 IEEE

stress at the solder connection in PCSBs is smaller than that generated in SBs.

In addition, since the Young's modulus of the polymer core (4.7 GPa) is smaller than that of the solder (53.9 Gpa), the polymer core disperses the stress more and also weakens the stress at the edge of bump.

Moreover, since the CTE of the polymer core (40 ppm/°C) is higher than that of the alumina ceramic substrate (7.1 ppm/°C) and PCB (approximately 16 ppm/°C), the core expands more. This implies that more thermal stress concentrates at the center of the PCSB bump and weakens the stress at the edge of the bump. However, cracks often generate around the equatorial area of PCSB bumps, and the copper layer breaks as well since the solder at the equatorial of the bump is thin [6]. In this study, there was no crack around the equatorial area of the PCSB bumps, as shown Figure 8. This implies that the nickel barrier strengthens around the copper layer and keeps the flexibilities of the copper layer and solder by suppressing the copper diffusion into the solder.

In the TEG of 61 × 61 mm² using the HITCE substrate, PCSBs show around 200% higher TC reliability, as shown in Figure 9, but the reliability improvement by PCSBs is not as good as in the TEG of 25 × 25 mm². The reason for this appears to be the CTE difference between the HITCE substrate and the alumina substrate. In the TEG of 61 × 61 mm², the CTE difference between the HITCE substrate (11.7 ppm/°C) and PCB (16 ppm/°C) is smaller than the difference in the TEG of 25 × 25 mm², where the alumina ceramic sustrate is 7.1 ppm/°C and PCB is 16 ppm/°C. Thus, the above CTE effect of the polymer core (40 ppm/°C) for the HITCE substrates (11.7 ppm/°C) is not as large as that of the alumina (7.1 ppm/°C). The failure locations were around the corners of the TEGs, as shown in Figure 10. Figure 11 shows cross-sectional optical microscopy images of the failure mode. These images reveal that the crack also occurred along the IMC interface near the pad on the PKG side, and proceeded into the solder without other cracks. Unlike SAC, the PCSBs kept the cylindrical bump fillet without collapsing although the TEG size is much larger and heavier.

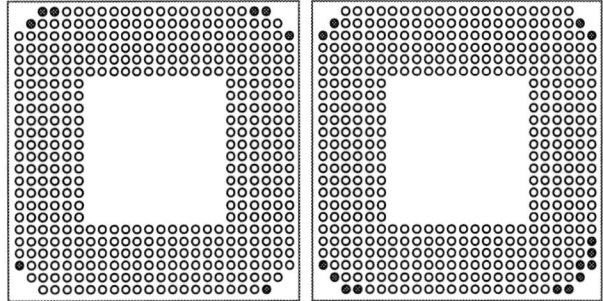

Figure 7. Dye and pry TCT failure with TEG of 25 × 25 mm²

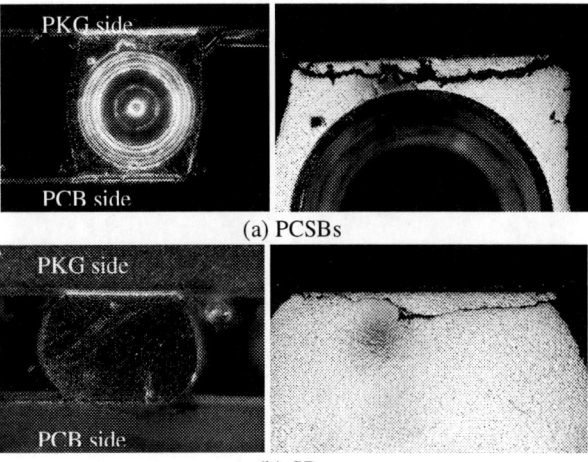

(a) PCSBs

(b) SBs

Figure 8. Cross-sectional optical microscopy images of TCT failure mode with TEG of 25 × 25 mm²

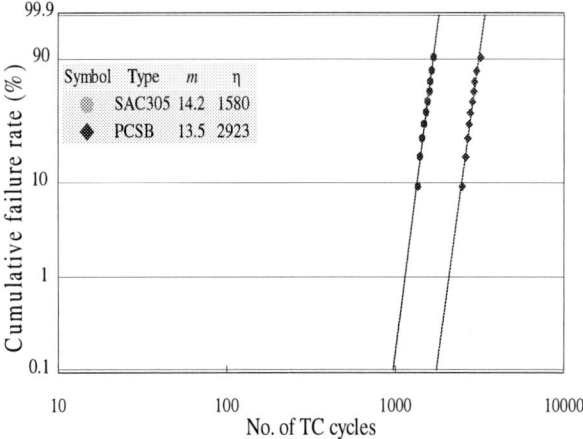

Figure 9. Weibull plot of TCT with TEG of 61 × 61 mm²

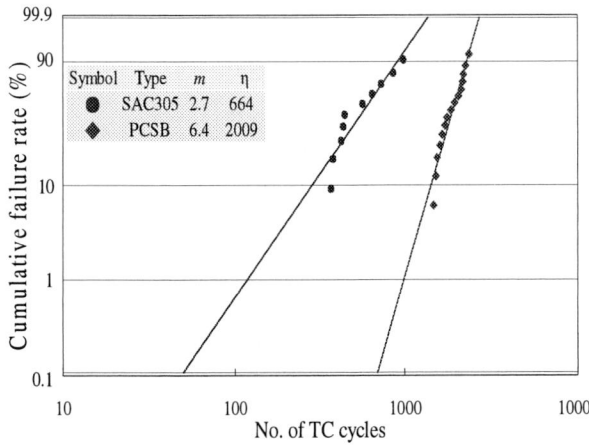

Figure 6. Weibull plot of TCT with TEG of 25 × 25 mm²

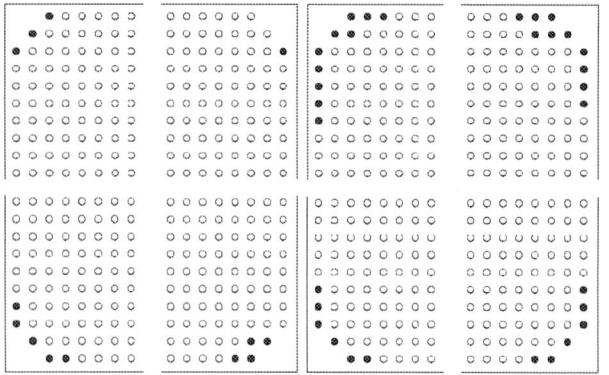

Figure 10. Dye and pry TCT failure with TEG of 61 × 61 mm²

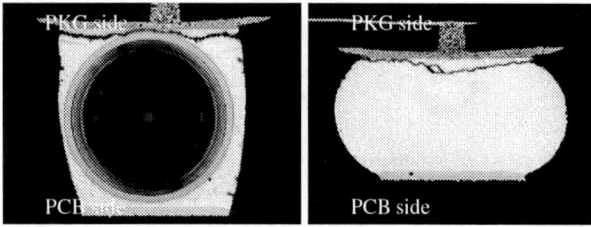

Figure 11. Cross-sectional optical microscopy images of TCT failure mode with TEG of 61 × 61 mm²

We also performed bending tests, which revealed that PCSB has approximately 200% higher reliability than SAC305, as shown in Figure 12. This seems to be due to a higher dumper effect of the PCSB bump against vertical stress, since the PCSB bump has higher standoff and is more elastic because of the smaller Young's modulus of its core.

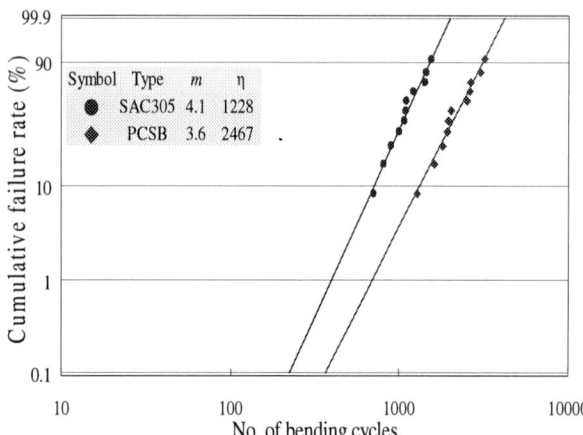

Figure 12. Weibull plot of bending test with TEG of 25 × 25 mm²

Comparison of Electric Performance of PCSBs

Figure 13 shows the resistance traces of five PKG samples during electromigration testing. The average initial resistance in 420 PCSB bumps including circuit is 1.286 ohm, which is almost the same as the average of 1.281 ohm in SBs SAC305. The resistance in PCSBs gradually increases and then sharply rises between 780 and 880 h, which is also the same as the EM performance of SAC305. These performances are thought to be due to the 20 µm copper layer of PCSB that contributes to improving the bump conductivity. Figure 14 and 15 shows cross-sectional optical microscopy images of the failure mode. These images reveal that PCSB has the same failure mode as SAC305. The voids are formed by electromigration and occur at the areas of high current density or high resistance. Therefore, the voids in PCSBs are formed at the IMCs of the pad, and they are also formed the around the copper layer at the equatorial area, as reported [12]. We see that voids are formed at the edge of the solder joints along the IMCs at the pads of the PKG on the cathode side and the PCB anode side. At the PKG cathode side, Sn atoms seem to migrate because of the electron flow into the solder from Ni_3Sn_4 IMCs, while at the PCB cathode side, Cu atoms seem to migrate into the solder from Cu–Sn IMCs while consuming the copper of the pad and simultaneously increasing the Cu–Sn IMCs. Thus, the copper pad is thinner on the PCB cathode side than on the PCB anode size. On the other hand, there is no significant IMC growth or voids on the copper layer of the PCSB bump, unlike the previous report [12]. It seems that the conductivity of PCSB is maintained by the pure copper that is protected by the nickel barrier layer, and the IMC didn't affect the EM performance.

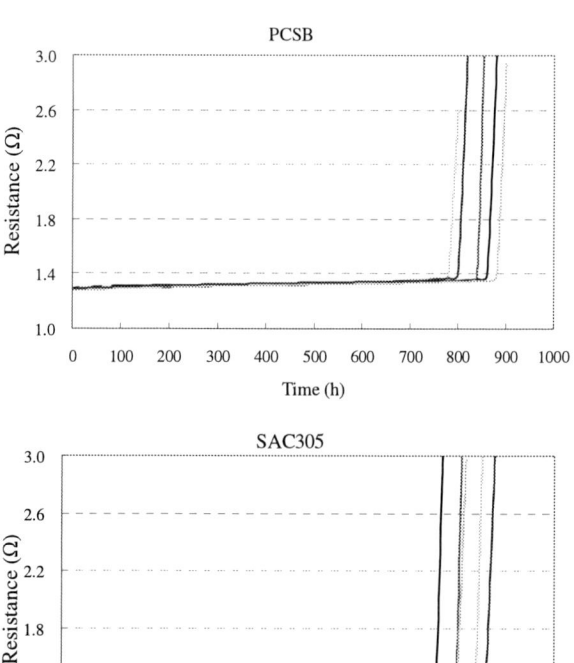

Figure 13. Comparison of EM test results for PCSB and SB SAC305

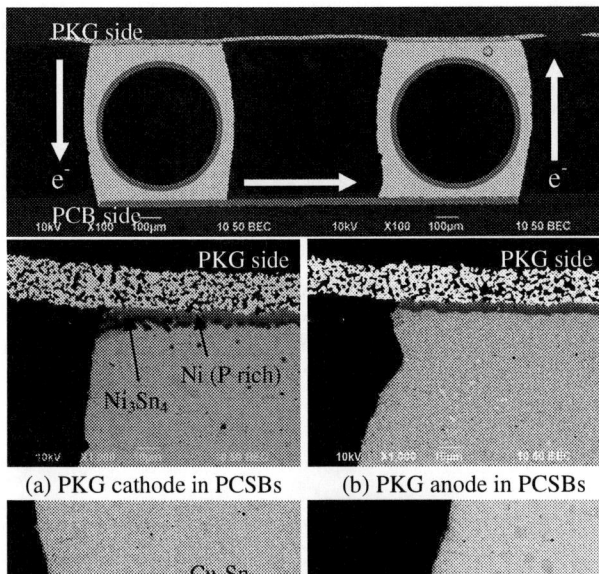

Figure 14. Cross-sectional optical microscopy images of EM failure mode in PCSBs

(a) PKG cathode in PCSBs   (b) PKG anode in PCSBs

(c) PCB anode in PCSBs   (d) PCB cathode in PCSBs

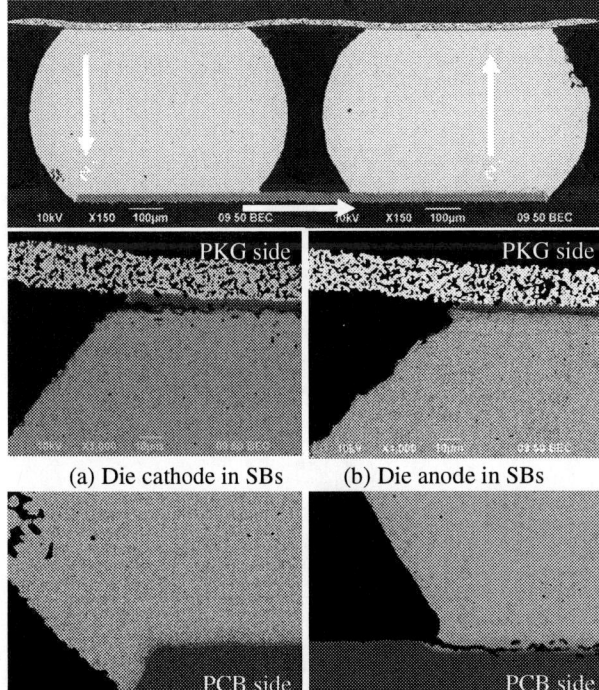

(a) Die cathode in SBs   (b) Die anode in SBs

(c) PCB anode in SBs   (d) PCB cathode in SBs

Figure 15. Cross-sectional optical microscopy images of EM failure mode in SBs SAC305

Conclusions

PCSBs showed over 200%–300% higher TC reliability with CBGAs than SAC305, allowing over 2,000 temperature cycles to be performed without using underfill, and kept the stand-off height without depressing the bump, even for a large CBGA. In addition, PCSBs achieved a 200% higher bending reliability than SAC305. These findings suggest that it is feasible to handle CBGAs with PCSB under mechanical stress environments such as vibration in transfer. Moreover, there was no significant difference in EM performance of PCSBs and SBs SAC305. In future work, the EM performance of PCSBs with a thicker copper layer will be considered because the thicker copper may improve performance. Moreover, the high frequency characteristic will be also studied.

This study's results suggest that PCSBs can enable CBGA application in severe environments without significant capital increase and complicated processes to ensue reliability.

References

1. X. (sam) Dai, N. Pan, A. Castro, J. Culler, M. Hussain, R. Lewis, and T. Michalka, "High I/O Glass Ceramic Package Pb-free BGA Interconnect Reliability," in *Proc. IEEE Electronic Components and Technol. Conf. (ECTC)*, Lake Buena Vista, FL, May 31–June 3, 2005, pp. 23–29.

2. D. Xie, V. Gektin, and D. Geiger, "Reliability Study of High-end Pb-free CBGA Solder Joint under Various Thermal Cycling Test Conditions," in *Proc. IEEE Electronic Components and Technol. Conf. (ECTC)*, San Diego, CA, May 26–29, 2009, pp. 109–116.

3. J. Miettinen, J. Tanskaneu, and E. O. Ristolainen, "Stacked 3-D MCP with Plastic Ball Vertical Interconnections," in *Proc. IEEE Electronic Components and Technol. Conf. (ECTC)*, May 27–30, 2003, pp. 1101–1105.

4. M. Sumikawa, R. Murayama, M. Ogawa, and H. Matsubara, "Reliability of a Wafer Level Packaging Method with Plastic-core Solder Bumps: -Utilizing Sn-Ag solder at 0.3 mm diameter-," *IMAPS*, vol. 5288, pp. 693–698, Nov. 2003.

5. J. Galloway, S. Ahmer, K. WonJoon, K. JinYoung, C. Jeff, K. YunHyeon, K. SeungMo, K. TaeSeong, L. GiSong, and R. SangHyun, "Mechanical, Thermal and Electrical Analysis of a Compliant Interconnect," *IEEE*, vol. 28, no. 2, pp. 618–623, Jun. 2005.

6. S. Movva and G. Aguirre, "High Reliability Second Level Interconnects Using Polymer Core BGAs," in *Proc. IEEE Electronic Components and Technol. Conf. (ECTC)*, June 1–4, 2004, pp. 1443–1448.

7. T. Kangasvieri, O. Nousiainen, J. Putaala, R. Rautioaho, and J. Vähäkangas, "Reliability and RF performance of BGA solder joints with plastic-core solder balls in LTCC/PWB assemblies," *Microelectron. Reliab.*, vol. 46, no. 8, pp. 1335–1347, August 2006.

8. Y. P. Wang, L. Y. Hung, D. S. Jiang, C. C. Chang, Y. P. Wang, and C. S. Hsiao, "High Drop Performance Interconnection: Polymer Cored Solder Ball," in *Proc. IEEE Electronic Components and Technol. Conf. (ECTC)*, Lake Buena Vista, FL, May 27–30, 2008, pp. 1208–1211.

9. O. Gaillard, "Board Level Reliability Testing of High-Rel Microprocessors Asociated to Rohs Packaging Solutions," *IMAPS*, 2008, pp. 737–744.

10. R. D. Sun, N. Okinaga, K. Matsushita, and M. Okuda, "Study on Improving the Drop Impact Reliability of Plastic Core Solder Balls," in *Proc. ICEP*, 2009, pp. 869–874.

11. H. Ishida and K. Matsushita, "Marked Reliability Increase of Plastic-Cored Solder Ball for Large Size Wafer-Level CSP," in *Proc. IWLPC*, 2012.

12. R.L.J.M. Ubachs, "Electromigration in WLcSP solder bumps," *Miroelectronics Reliability*, 2010, vol. 50, pp. 1678–1683.

Lifetime Prediction of Cu-Al wire Bonded Contacts for Different Mould Compounds

Rene Rongen, G.M. O'Halloran, Amar Mavinkurve, Leon Goumans, Mark-Luke Farrugia
NXP Semiconductors
Gerstweg 2, 6534AE, Nijmegen, The Netherlands
rene.rongen@nxp.com, Telephone: +31 24 353 4060

Abstract

Large scale conversion of gold to copper wiring in microelectronics can only become successful when all the failure mechanisms that can be encountered during reliability testing, or during product application life are understood.

One of these mechanisms is corrosion of the contact between the copper (Cu) ball and the aluminum (Al) bond-pad, consisting of various intermetallic compounds (IMCs), which are more sensitive to corrosion compared to gold (Au) Al IMC.

This study elaborates on three corrosion mechanisms present in the Cu-Al system: interfacial Cu-Al IMC corrosion, bulk Cu-Al IMC corrosion and Al bond pad corrosion. For the first mechanism, which is dominant, an empirical corrosion model is introduced.

To gather data for this model, a recently developed method for analyzing the IMC contact area to study the dynamics of the dominant mechanism is used. Data was collected from various devices, which were exposed to accelerated aging conditions. The focus of this paper is on unbiased conditions using a wide temperature and humidity range.

In total four epoxy molding compound types have been investigated and are compared to each other, using the empirical model proposed in this paper.

Finally, it is shown that the model allows the prediction of the life time of Cu-Al ball contacts for different application conditions and also allows the selection of an appropriate mould compound type.

Introduction

It is well accepted that copper ball bonds show better reliability at elevated temperatures compared to gold bonds [1-3] but inferior performance under the influence of moisture [4-7].

Despite the huge amount of literature published in recent years on copper wire bond studies there is a lack of agreement and understanding on the specific corrosion mechanism in the Cu-Al ball bond contact. Studies reported in literature focus on improving reliability by selection of materials used; wire type [6, 8-12] and epoxy molding compound [6, 13-15]. In most of these studies the corrosion of the Cu-rich phases in the intermetallic compound stack, for example Cu_9Al_4, play a key-role in the degradation of the Cu contact [6-7,16-17]. Only one study [18] mentions the existence of a thin and extremely Al-poor layer, just above the "regular" IMC stack. In the work presented here, we show that the degradation of this very layer is the dominant mechanism that determines the life time of Cu-Al contacts.

To the best of our knowledge, none of the published studies mention an acceleration or aging model, which can be used for the prediction of the life time of the Cu-Al ball bond contact. In general the well known Peck model for Al bond pad corrosion, in the presence of chlorine ions, is still referred to and/or used for lifetime prediction calculations [19, 20]. However, for the dominant mechanism presented in this paper, the Peck power law model does not apply, and the use of the exponential humidity model is proposed [21, 22]. The exponential model has been successfully applied to real case studies using four different epoxy molding compounds. The IMC contact analysis was performed with a recently developed technique which allows 2D planar view of the contact area [23]. This technique is preferred compared to the traditional cross-sectioning through the ball bond/bond pad method which gives information on the dominant degradation mechanism as explained below, in one dimension only.

Failure Mechanism

Three different corrosion mechanisms in and around the IMC contact are observed. Halogen ions act as a catalyst in the chemical reaction and are assumed to trigger these mechanisms sequentially. We propose that this is related to an increasing concentration of halogen-ions around the Cu-Al IMC contact. This is in agreement to results reported by Fraunhofer Institute [24] on Au-Al IMC, where higher halogen concentrations lead to a higher degree of corrosion.

The first mechanism is corrosion along the Cu-ball to IMC interface as shown in Figure 1. This interfacial IMC corrosion is initiated by small amounts of halogen ions above a first threshold value and is accompanied with a typical crack or small gap due to delamination of the contact along the corrosion front. The corrosion front is propagating along the Cu-ball just above the IMC phases underneath.

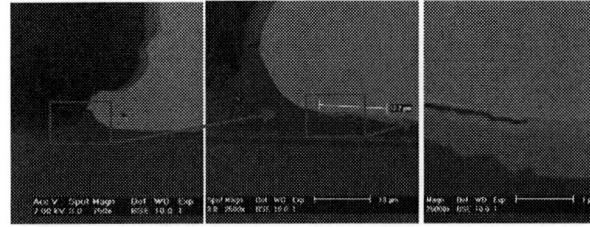

Figure 1. First corrosion mechanism in aged Cu-Al ball bond: interfacial IMC corrosion

The observation of this very specific crack, needs a more detailed explanation. While the IMC contact is formed during wire bonding, small amounts of Al are dissolved in the bulk Cu-ball just adjacent to the contact interface between the ball and the IMC phase richest in Cu. This small amount of Al (in the order of a few at%) increases the corrosion sensitivity of the freshly formed contact [18,25]. To prove the existence of this tiny layer, a sample was prepared where a Cu-ball is formed on a very thick Al layer (> 500 μm) without any molding compound encapsulation. This contact was aged extensively by exposing it to 300 °C for 100 hours to develop various IMC layers of different compositions. A cross section is shown in Figure 2, in which well developed IMC phases are clearly visible.

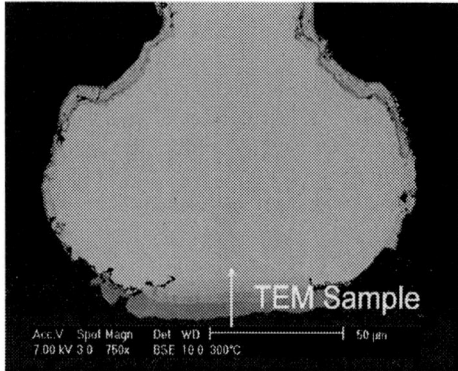

Figure 2. Extensively aged Cu-Al contact in free air environment (no molding compound); the arrow indicates the direction of the TEM analysis shown in Figure 3)

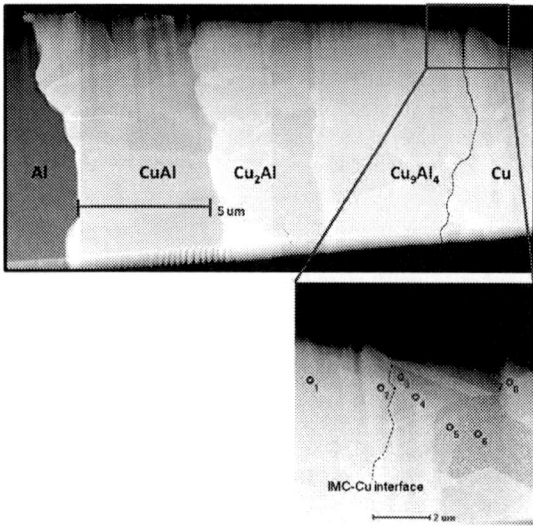

Figure 3. TEM micrograph at location shown in Figure 2

Table 1 – Cu-Al composition of the locations in Figure 3, and location 9 even further into the ball (~15 μm)

	At% @ location No.											
	CuAl	Cu₂Al	Cu₉Al₄	O_1	O_2	O_3	O_4	O_5	O_6	O_7	O_8	O_9
Cu	46	62	73	78	78	94	94	98	99	98	99	100
Al	54	38	27	22	22	6.1	6.2	2.3	0.9	2.0	1.4	0

A TEM sample was prepared to determine the composition of the different IMC phases (see Figure 3). The observed IMC phases were analyzed with respect to their composition. The results are shown in Table 1. The region just above the IMC-Cu interface (to approximately position O_5 in

Figure 3) was found to consist of only 6 at% Al. This layer is very thin, less than 2 μm, compared to the IMC phases identified in Figure 3 with typical thickness of a few μm, after this severe storage test. According to Cu-Al phase diagrams available this corrosion sensitive layer is most likely the solid solution phase. The higher Al content of the CuAl, Cu₂Al and Cu₉Al₄ IMC phases result in a higher corrosion resistance [2].

As soon as the halogen concentration at the corrosion front exceeds a second threshold due to thermal or hydrolytic degradation of the compound, the second mechanism will start. This is bulk IMC corrosion, typically starting with the copper rich phase as observed in various corrosion studies [5, 6, 16]. In cross sections this second mechanism appears as black pockets of oxidized IMC at the Cu-ball to IMC interface (see Figure 4). This type of corrosion is also observed in severely aged Au-Al IMC contacts [4, 5].

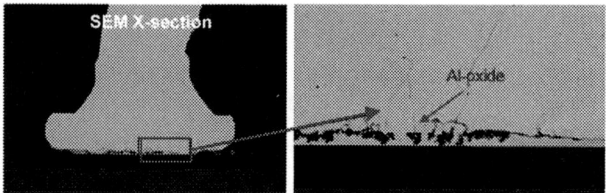

Figure 4. Second corrosion mechanism in aged Cu-Al ball bond: bulk IMC corrosion

The third mechanism is classical bond pad corrosion as shown in Figure 5, initiated only when the halogen concentration exceeds a third threshold value.

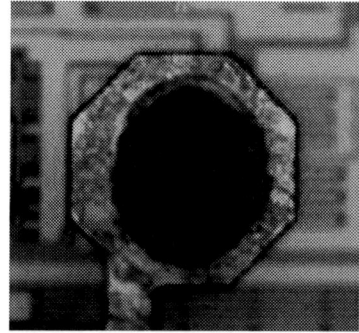

Figure 5. Third corrosion mechanism close to Cu-Al ball bond: "classical" bond pad corrosion

Each of the three mechanisms is accompanied with a reaction of the Al present in the layers attacked by corrosion, in which the halogen ions play the role of catalyst. Even in the case of bulk Cu-Al IMC corrosion, aluminum oxide is formed and copper is released. The two reaction schemes are:

$$4Al + 3O_2 + catalyst \rightarrow 2Al_2O_3 + catalyst \quad (1a)$$
$$4Cu_xAl_y + 3yO_2 + catalyst \rightarrow 2yAl_2O_3 + 4xCu + catalyst \quad (1b)$$

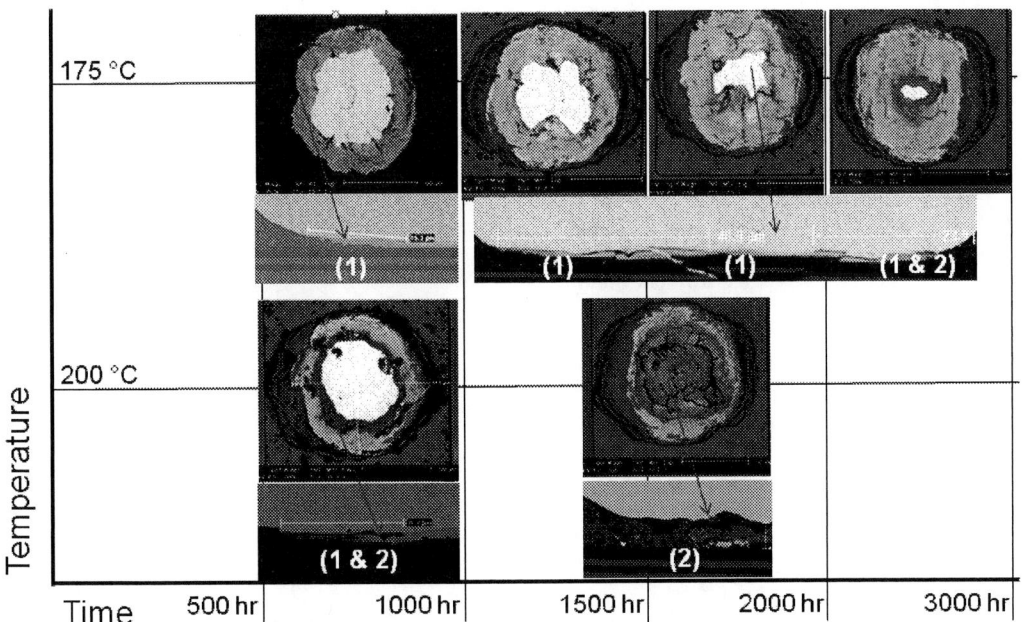

Figure 6. Corrosion progression for aged Cu-Al ball bonds in time for different temperatures using planar analysis and cross-sections; mechanism in brackets: (1) Cu-Al IMC interface corrosion, (2) Cu-Al IMC bulk corrosion

Once the corrosion reaction initiates, the halogen ions reform, which leads to continuation of the reaction until all aluminium in the intermetallics are eventually consumed [26].

In Table 2 the intermetallic phases involved in the corrosion mechanism and prerequisites are summarized.

Table 2 – Overview of the three corrosion mechanism observed in and close to the Cu-Al IMC contact area

Corrosion	Composition	Prerequisite
Cu-Al IMC Interface	$Cu_{>9}Al_{<1}$	O_2 and lowest catalyst threshold concentration
Cu-Al IMC bulk	Cu_9Al_4, Cu_2Al, $CuAl$, $CuAl_2$	O_2 and median catalyst threshold concentration
Al Bond pad	Al	O_2 and highest catalyst threshold concentration

Experimental

Samples

Four different epoxy mould compounds were used in the study, in combination with different package types; HLQFP, HVQFN, DIP 16 and HSOP. In all cases bare Cu wire was used. An overview of the mould compounds and packages is shown in Table 3.

Table 3 – Overview of used Molding Compounds

Type	Characteristic description	Package Type
A	Brominated OCN-PN based epoxy from Supplier 1	DIP16, HSOP
B	Halogen free Multiaromatic epoxy from Supplier 1	HSOP
C	Halogen free Multiaromatic/ Biphenyl epoxy from Supplier 2	DIP16
D	Halogen free Multiaromatic from Supplier 2	DIP16, HVQFN

Test Conditions

Unbiased humidity test (uHAST) and high temperature storage test (HTSL) according to JEDEC test methods [27, 28] were performed. For surface mount device (SMD) package types, the samples were first exposed to pre-conditioning test. Pre-conditioning is used to model conditions during soldering of the device to the PCB board. This test is performed according to JEDEC specification [29].

The reliability test conditions are shown in Table 4. At each read point analysis is performed to measure the degradation of the copper-aluminium contact.

Table 4 – Reliability Test Conditions

Test Name	Condition (Temp/RH)	Read points (hours)
HTSL	150 °C	500, 1000, 2000, 3000, 4000, 5000
	175 °C	100, 500, 1000, 1500, 2000, 3000
	200 °C	20, 50, 100, 500, 1000, 2000, 2500
	225 °C	1, 2, 4, 8, 20, 50, 100, 200, 500, 1000
(u)HAST	130 °C /75%	96, 192, 288, 384
	130 °C /85%	96, 192, 288, 384
	130 °C /95%	96, 192, 288, 384, 576
	110 °C /85%	96, 192, 288, 384, 576

Analysis Method

The samples were removed from the reliability chamber at the selected read points. Samples tested under moist conditions were dried at 120 °C before analysis. Initially the samples were cross-sectioned to show the degradation within the IMC contact, as shown in Figure 1. Besides being very labor intensive, this method was not very satisfactory as only one position in the cross-section could be examined and it was not always clear if the ball was cross-sectioned in the exact center of the contact.

An improved method to examine degradation was developed [23]. Using this method the Cu ball is etched off the bondpad and the amount of intermetallic remaining under the ball can be easily examined by means of a back-scatter

SEM detector. A back-scatter detector will allow sufficient contrast between the IMC that is oxidized and the non-oxidized IMC. Figure 6 shows a comparison between a cross-sectioned and planar analysis of the same contact. The grey donut shaped area is the corroded IMC, around a bright region in the center which is still intact. The measurement uncertainty, originating from spread and measurement accuracy, is about ± 2 µm.

From Figure 6 it can be clearly seen that different failure modes are present. A distinction can be made between interfacial corrosion (mechanism 1) which appears as a crack between the ball and intermetallic, and intermetallic corrosion (mechanism 2), which appears as a thick oxide in the cross section and in the planar view. In this case the entire intermetallic layer turns into oxide.

Results

The planar analysis of an aged IMC contact can be used to investigate the corrosion dynamics. The starting point is the identification of different areas as depicted in Figure 7.

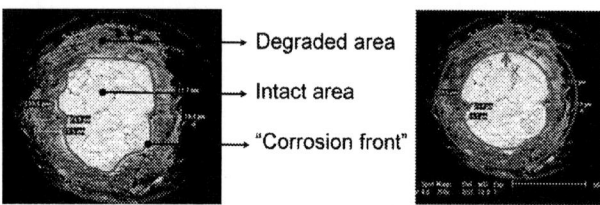

Figure 7. The different areas and corrosion front in an aged Cu-Al ball bond, definition of crack length "x"

The corroded, donut shaped, area is actually the progression of the corrosion front over a distance length "x" from the periphery of the initial IMC contact inwards, which is the length of the crack as observed in cross sections as shown in Figure 1. The procedure used to define the circular peripheries is described in previous work [23]

The dynamic behavior of this crack is studied over time, as a function of environmental driving factors such as temperature and humidity. It is expected to be influenced by material properties, and is therefore expected to be different for the four types of compounds that are used. Because the corrosion front is progressing along with the diffusion of halogen ions (catalyst in equations 1), the crack length is expected to grow with the square root over time, according to:

$$x \propto \sqrt{D \cdot t} \qquad (2)$$

where "D" is the Diffusion Coefficient and "t" is time.

Temperature

For the four types of compounds, the dynamic behavior over time is studied for at least three temperatures, by applying the planar analysis method on devices taken from HTSL experiments at various read points (Table 4). By plotting the obtained crack lengths versus the square root over time, the graphs as shown in Figure 8 are obtained.

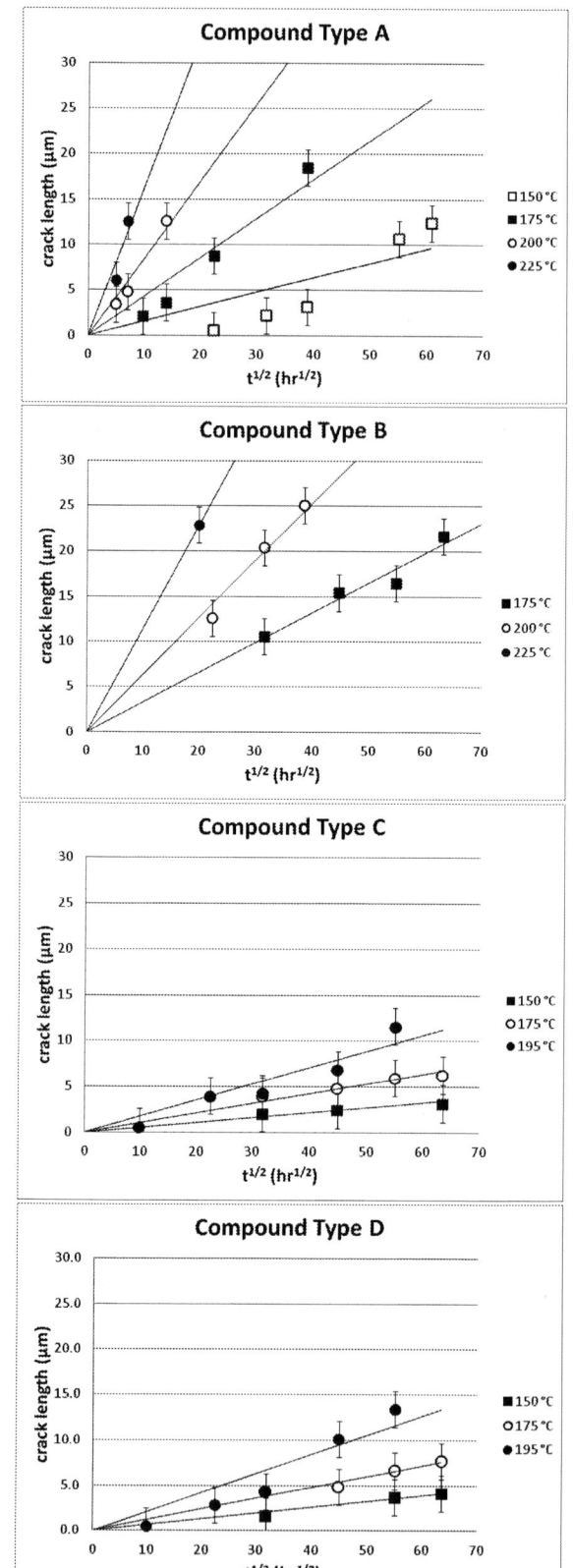

Figure 8. Square root over time plots of the crack length "x" for the four different compound types (Table 3) at different temperatures (measurement uncertainty is ± 2 µm).

Compounds A and B reveal a significantly faster degradation than compounds C and D. Comparing the degradation at 175 °C after 3600 hours, the crack for compound A and B propagated a distance of about 20 to 25 μm, whereas for compound C and D this is only 5 to to 10 μm.

According to eq. (2), the slope "s" of any line derived from linear regression in Figure 8, is proportional to the square root over the Diffusion coefficient

$$s \propto \sqrt{D} \qquad (3)$$

In addition, it is to be expected, that the temperature dependence of the Diffusion coefficient follows Arrhenius law

$$D \propto \exp(-Ea/k_B T) \qquad (4)$$

where "E_a" is the Activation Energy and "k_B" is the Boltzmann constant (8.617×10^{-5} eV/K).

For each of the four compound types, the Arrhenius plot is shown in Figure 9. The activation energies, derived from linear regression, are comparable for all types and range from 0.9 to 1.1 eV. Also according to this plot, the compounds can be split into 2 groups: type A and B vs. type C and D.

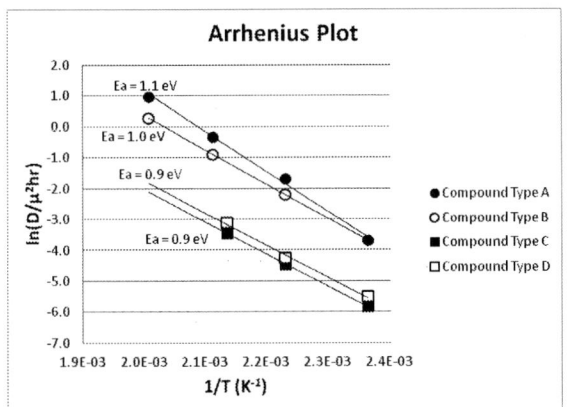

Figure 9. Arrhenius plot of the diffusion coefficients derived from the data in Figure 8.

Humidity

Next, an extensive study was made for (unbiased) exposure to relative humidity at elevated temperatures. Compound types B, C and D do not reveal significant degradation under the humidity conditions used in this study (see Table 4). For compound type A however, this is not the case and the crack length versus the square root over time is depicted in Figure 10 for the various relative humidity values at the two temperatures measured.

At 130 °C the degradation is much faster than at 110 °C. Comparing the actual degradation under 95 % RH after 625 hours, the crack length at 110 °C is with 4 μm about 3 times shorter than at 130 °C.

Similar to the dependency on temperature discussed in the previous paragraph, an assumption can be made by selecting the most likely relationship between the diffusion coefficient and the relative humidity, which is an exponential dependency:

$$D \propto \exp(a \cdot RH) \qquad (5)$$

where "a" is the Humidity coefficient.

A logarithmic plot of "ln(D/μm²/hr)" versus "RH %" allows the determination of this humidity coefficient. The result is shown in Figure 11.

Note, that the data for 0 % RH are based on the Arrhenius relationship in Figure 8 for compound A, by calculating the value for the diffusion coefficient at 110 °C and 130 °C. This is related to the selection of the dependency in eq. 5. The humidity acceleration according to exponential dependency [21, 22] allows a direct link with the Arrhenius relationship for the HTSL data, which is not possible with the standard "Power-law" Peck Model [19], because of a singularity for 0% RH. In conclusion, for mould compound type A, we determine a humidity constant with a value of 0.035/%RH. For the other mould compound types B, C and D, which show no degradation due to moisture, the humidity coefficient is equal to 0/%RH.

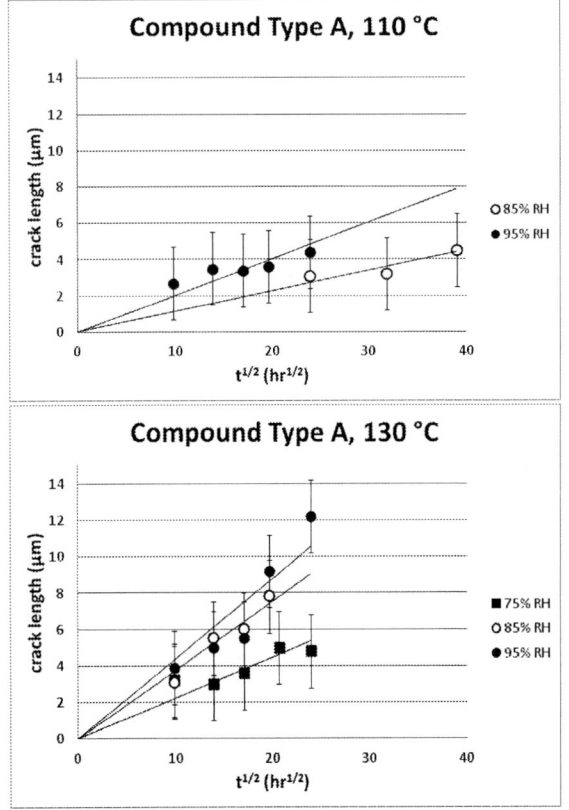

Figure 10. Square root over time plots of the crack length "x" for compound type A for various relative humidity conditions at 110 °C and 130 °C (measurement uncertainty is ± 2 μm).

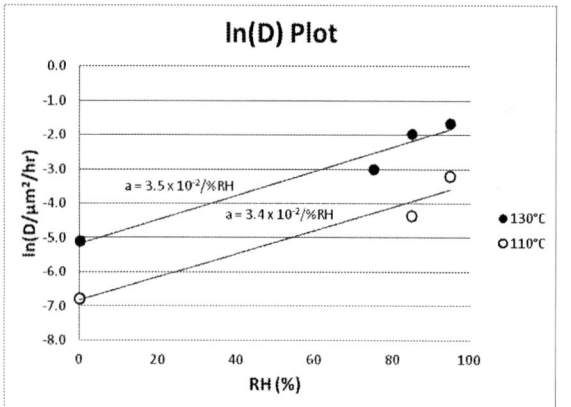

Figure 11. Logarithmic plot of $\ln(D/\mu m^2/hr)$ vs. RH for 110 °C and 130 °C.

Degradation model

In the previous section, by studying the crack growth over time we found two empirical relationships between the dynamical behavior of the corrosion front, one with the environmental temperature (eq. 4) and the other with the relative humidity (eq. 5). By combining these empirical equations, the following model for the crack growth over time is proposed, based on starting equation (eq. 1)

$$x = \sqrt{A \cdot \exp(a \cdot RH) \cdot \exp(-Ea/k_B T) \cdot t} \qquad (6)$$

where:
A = Pre-exponential constant ($\mu m^2/hr$)
a = Humidity coefficient (/%RH)
Ea = Activation Energy (eV)
k_B = Boltzmann constant (8.617×10^{-5} eV/K)
T = Temperature (K)

As a next step, we make a direct fit to this model using the method of least squares. This is done by calculating the square of the difference "δ" between each measured crack length "x_m" and the predicted value "x_p" using (eq. 6), i.e.,

$$\delta^2 = (x_m - x_p)^2 \qquad (7)$$

then taking the square root of the sum of all squared errors, i.e.,

$$\Delta = \sqrt{\sum_i \delta_i^2} = \sqrt{\sum_i (x_{m,i} - x_{p,i})^2} \qquad (8)$$

which is finally minimized. This is repeated for every compound type, taking all the data in Figures 8 and 10, using "A" and "E_a" (and "a" for compound type A) as fit parameters. Finally in Figure 12, the model fit data is compared to the measured data. Table 5 gives an overview of the fit parameters which allows comparison of the compound types.

Applicability of the model

Figure 12 shows measured and calculated values for the crack length with the linear regression result for a straight line

through the origin. It is concluded that there is a fair agreement between measured and calculated crack lengths; taking into account that the optimum relationship is a straight line with slope 1 through the origin. Deviations are the largest for small crack lengths for which the measurement itself shows a relatively large uncertainty.

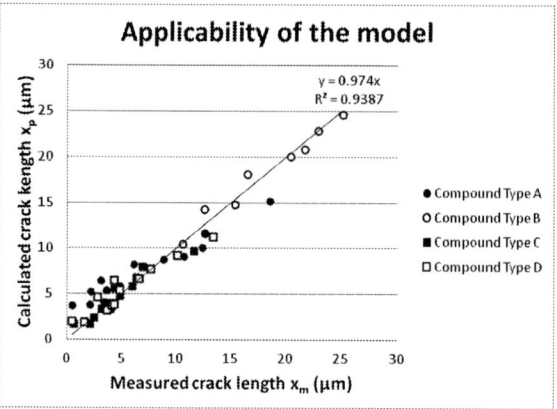

Figure 12. Measured versus calculated crack length using the proposed model (eq. 6).

Compound type comparison

In Table 5, the compound type specific parameters "A", "a" and "E_a" are shown. This overview allows to draw the following three main conclusions:

1) The activation energy of the four compound types are very similar and in the range from 0.9 to 1.1 eV
2) The main differentiator for the compound types is the Pre-exponential constant. This parameter predominantly defines the corrosion rate and varies over a large order of magnitude from 10^8 to 10^{11} $\mu m^2/hr$
3) Compound type A is the worst in terms of Cu-Al IMC interface corrosion. Compound type C and D are the best and quite comparable.

Table 5 – Compound type parameter comparison

Type	Pre-exponential constant "A" ($\mu m^2/hr$)	Humidity coefficient "a" (/%RH)	Activation Energy "Ea" (eV)
A	6.3×10^{11}	3.5×10^{-2}	1.1
B	6.1×10^{9}	0.0	1.0
C	2.0×10^{8}	0.0	0.9
D	6.3×10^{8}	0.0	0.9

Discussion

The main parameter distinguishing the compound types from one another is the Pre-exponential constant "A". The dimension of this constant is $\mu m^2/hr$, the same as that for the diffusion constant and therefore expected to be characteristic for, among others, mobility and concentration of oxidizing agents like halogen ions in a compound. Note that the mobility of halides under given conditions will depend on physical factors like the free volume of the compound (which is related to the glass transition temperature), and chemical factors like the pH and presence of ion trappers. This is the subject of ongoing work which makes an attempt to link key

978-1-4799-2408-0/14 $31.00 © 2014 IEEE

mould compound characteristics with the empirical model in order to give it a better physical basis.

The benefit of this empirical model and the derived compound specific parameters will be illustrated by the following example. Assuming a 0 hr IMC contact area of 2500 μm^2, meaning a ball diameter contact of 63μm, and a criterion that a minimum contact with a diameter of only 5 μm is considered to be a fail, the expected life time can be calculated. This is shown in Table 6 for two cases: (1) a typical temperature of the contact area of 90 °C (e.g. in a consumer application), and (2) a temperature of 125 °C (e.g. in an automotive under-the-hood application.

Table 6 – Expected life time of a 63 μm IMC contact

Type	Consumer: 90 °C (year)	Automotive under-the-hood: 130 °C (year)
A	322	9
B	173	8
C	1240	69
D	1153	58

From this overview it can be concluded that all four compound types are suitable for consumer applications, but only types C and D show enough robustness margin for high temperature operation such as automotive under-the-hood application.

Conclusion and outlook

In this paper we have presented the dominant degradation mechanism in a Cu-Al IMC contact, which is corrosion along a thin and Al-poor ($Cu_{>9}Al_{<1}$) layer, which is already present directly after wire bonding. The dynamics of corrosion of this layer under different temperature and humidity conditions has been structurally studied and analyzed. As a result a degradation model is derived, which allows the prediction of the lifetime of the Cu-Al IMC contact for different application conditions. The lifetime strongly depends on the halogen ion concentration and mobility in the epoxy molding compound. This and the corrosion under biased humidity condition, which adds a driving force to the halogen ion transport via the imposed electrical field, is under investigation.

Acknowledgements

Thanks to all the people who contributed to this work; in particular Louis Chen, Albert Yang and Peter Koch for careful preparation of many of the samples used, Eric Meeuwsen for performing all the cross-sections and Leon van Nimwegen for the TEM analysis.

The research leading to these results has received funding from the ENIAC Joint Undertaking and from Senter-Novem in the Netherlands under Grant Agreement number 120009 and from NXP Q&R Technology Development Project Funding.

References

1. C.L. Gan et al, "Superior performance and reliability of copper wire ball bonding in laminate substrate based ball grid array", *Microelectronics Int.* vol 30/3, pp. 169-175, 2013

2. Y. Zeng, et al., "Thermodynamic study on the corrosion mechanism of copper wire bonding", *Microelectronics Rel.,* vol 53, pp. 985-1001, (2013),

3. S.H. Kim, et al., "The Interface Behavior of the Cu-Al Bond System in High Humidity Conditions", in *Proc. IEEE 2010 12th Electronic Package Technology Conference* (EPTC), Singapore, Dec 8-10 2010, pp. 545-549

4. C.D. Breach et al, "Failure of gold and copper ball bonds due to intermetallic oxidation and corrosion", *in Proc. 18th Int. Symp. Phys. & Fail. Analysis of Integrated Circuits* (IPFA), July 4-7 2011, Incheon, Korea

5. C.L. Gan, et al., "Extended reliability of gold and copper ball bonds in microelectronic packaging", *Gold Bull* vol 46, pp. 103-115, (2013)

6. H. Abe, et al., "Cu wire and Pd-Cu wire package reliability and molding compounds"*, in Proc. IEEE Electronic Components and Technology Conference (ECTC)*, San Diego, May 29-June 1 2012, 2012, pp. 1117-1123

7. M. Tsuriya, et al., "Copper wire bonded package characterization and reliability for QFN package from iNEMI collaborative project", in *Proc. IEEE 15th Electronic Package Technology Conference (EPTC)*, Dec. 11-13 2013, Singapore, pp. 249-254

8. D. Stephan, et al., "Reliability of palladium coated copper wire", in *Proc. IEEE Electronic Components and Technology Conference (ECTC)*, May 31- June 3 2011 Orlando, FL, pp. 1508-1515

9. M.C. Han et al., "The role of Cu-Al IMC coverage and aluminium splash in Pd-copper wire HAST performance", in *Proc. IEEE 15th Electronic Package Technology Conference (EPTC)*, Dec. 11-13 2013, Singapore, pp. 394-398

10. C.L. Gan et al, "Comparative reliability studies and analysis of Au, Pd-coated Cu and Pd-doped Cu wire in microelectronics packaging", *PLOS One*, vol 8, pp. 1-8, Nov. 2013

11. I. Singh et al., "Pd-coated Cu wire bonding technology: chip design, process optimization, production qualification and reliability test for high reliability semiconductor devices", in *Proc. IEEE Electronic Components and Technology Conference (ECTC)*, San Diego, May 29-June 1 2012, pp. 1089-1095

12. C.W. Tan et al, "Corrosion study at Cu-Al interface in microelectronics packaging", *Appl. Surf. Sci.*, vol 191, pp. 67-73, 2002

13. P. Su et al, "An evaluation of the effects of molding compound properties on reliability of Cu wire components", in *Proc. IEEE Electronic Components and Technology Conference (ECTC)*, May 31- June 3 2011 Orlando, FL, pp. 363-369

14. Y.H. Tian, et al. "Reliability and failure analysis of fine copper wire bonds ecapsulated with commercial epoxy molding compound", *Microelectronics Rel.* vol. 51, pp. 157-165, 2011

15. H. Seki et al, "Study of EMC for Cu bonding wire application", in *Proc. IEEE CPMT Symp.*, Tokyo, Japan, Aug 24-26, 2010

16. Boettcher, T. et al, "On the Intermetallic Corrosion of Cu-Al wire bonds" in *Proc 12th Electronics Packaging Technology Conf (EPTC),* Singapore, Dec. 8-10 2010. pp. 585-590.

17. C.L. Gan et al., "Technical barriers and development of Cu wirebonding in nanoelectronics device packaging", *J. Nanomaterials*, pp. 1-7, 2012

18. G.H.M. Gubbels, et al. "Characterization of intermetallic compounds in Cu-Al ball bonds: thermo-mechanical properties, interface delamination and corrosion", in *Proc. IEEE 4th Electronic System Integration Technology Conf. (ESTC),* Amsterdam, Sep 17-20, 2012

19. D.S. Peck, *"A Comprehensive Model for Humidity Testing Correlation"*, in *Proc. IEEE International Reliability Physics Symposium (IRPS)*, 1986, pp. 44-50.

20. F.C. Classe, S. Gaddamraja, "Long term isothermal reliability of copper wire bonded to thin 6.5μm aluminum", in *Proc. IEEE International Reliability Physics Symposium (IRPS)*, April 10-14, Monterey, CA, 2011, pp. CP1.1 - CP 1.5

21. C. Dunn et al., "Recent Observations on VLSI Bond Pad Corrosion Kinetics", *J. of the Electrochemical Society*, vol. 135, Issue 3, pp. 661-665,1988

22. J. McPherson et al., "VLSI Corrosion Models: A Comparison of Acceleration Factors", in *Proc. of 3rd Internatonal Symposium on Corrosion and Reliability of Electronic Materials and Devices*, Electrochemical Society, Miami, FL, Oct. 9-14 1994, p. 270.

23. G.M. O'Halloran, et al., "Planar Analysis of Copper-Aluminium Intermetallics", in Proc. International Symposium for testing and Failure Analysis (ISTFA), San Jose, CA, Nov 3-7 2013, pp.297-300

24. R. Klengel, et al., "Novel investigation of influencing factors for corrosive interface degradation in wire bond contacts", in Proc. European Microelectronics and Packaging Conference (EMPC), Grenoble, Sept 9-12 2013

25. G. Plascencia, T.A. Utigard, "High temperature oxidation mechanism of dilute copper aluminium alloys", *Corrosion Science,* vol 47, pp. 1149-1163, 2005

26. P. Chauhan et al., "Copper wire bonding concerns and best practices", J. Electronic Materials, vol 42, pp. 2415-2434, 2013

27. JESD22-A118, Accelerated Moisture Resistance - Unbiased Hast, March 2011

28. JESD22-A103, High Temperature Storage Life, Dec 2010

29. JESD22-A113, Preconditioning of Plastic Surface Mount Devices prior to Reliability Testing, Oct 2008

The Corrosion Performance of Cu Alloy Wire Bond on Al Pad in Molding Compounds of Various Chlorine Contents under Biased-HAST

Ying-Ta Chiu[a],*, Tzu-Hsing Chiang[a], Yin-Fa Chen[a], Ping-Feng Yang[a], Louie Huang[a], Kwang-Lung Lin[b]

[a]ASE Group

No. 26, Chin 3rd Road, Nantze Export Processing Zone,

Kaohsiung 81170, Taiwan (R.O.C.)

[b]Department of Materials Science and Engineering, National Cheng Kung University

No.1, University Road, Tainan 70101, Taiwan (R.O.C.)

*E-mail: Chandler_Chiu@aseglobal.com

Abstract

The present article investigated the performance and interfacial behavior between Cu wire bond and Al pad under molding compounds of different chlorine contents. The epoxy molding compounds (EMCs) were categorized as ultra-high chlorine, high chlorine and low chlorine, respectively, with 24, 7.6, and 4.3 ppm chlorine contents. The ball bonds were stressed under $130^{\circ}C/85\%RH$ with biased voltage of 10V. The interfacial evolution between Cu wire bond and Al pad was investigated in EMC of three chlorine contents after the biased-HAST test. The Cu bonding wires used in the plastic ball grid array (PBGA) package include bare Cu wire (4N Cu) and Pd coated Cu alloy wire (98Cu2Pd). The as bonded wire bond exhibits an average Cu-Al IMC thickness of ~0.12 μm in both types of Cu wire. Two Cu-Al IMC layers, Cu_9Al_4 and $CuAl_2$, analyzed by EDX were formed after 100h of biased-HAST test. The joint failed in 192h and 1296h, respectively, under ultra-high and high chlorine content EMC. The joint lasts longer than 2000h with low chlorine content EMC. The corrosion of IMC formed between Cu wire and Al pad, occurs in the ultra-high and high chlorine molding compound. The results of EDX analysis indicate that the chlorine ion diffuses from molding compound to IMC through the crack formed between IMC and Al pad. Al_2O_3 was formed within the IMC layer. It is believed the existence of Al_2O_3 accelerates the penetration of the chlorine ion and thus the corrosion.

Introduction

Gold wire bonding is the most reliable technology for producing electrical interconnections between functional chip and external substrate. Due to high gold price, Cu wire has become one of the most cost effective wire bond candidates. Hence, the wire bonding system Cu/Al has been replacing Au/Al system in a number of semiconductor packaging. However, the Cu wire bonding is suffering from some reliability issues, such as die pad damage, corrosion, etc. [1,2] One of the major corrosives in industrial products is the chlorine content of epoxy molding compound (EMC). [3,4] Many literature have been reported that the Cu wire bonding suffered chlorine ion attack on Cu-Al IMC after thermal annealing, pressure cooker test (PCT), highly accelerated stress test (HAST) or biased-HAST. [5,6,7] In Cu-Al binary system, five intermetallic phases could form during $300^{\circ}C$. However, only Cu_9Al_4, CuAl and $CuAl_2$ have been identified at the interface between Cu wire bonding and Al pad metallization. [8-10] The dissociated chlorine ion could diffuse from molding compound to attack on Cu-Al IMC, such as Cu_9Al_4 and $CuAl_2$.

Lin et al. reported Pd diffuses from the interior of Pd/Cu ball toward the bond interface. [11] The Pd-rich layer acted as a diffusion barrier between the Cu and Al to reduce the IMC growth. [12] As a result, it is of great interest to understand the corrosion of the Cu-Al IMC of bare Cu and Pd/Cu wire (Pd coated Cu) under biased-HAST. However, the corrosion was not observed in the low chlorine molding compound of PBGA package.

Experiment setup

Bare Cu wire (4N Cu) and Pd coated Cu wire (98Cu2Pd, EX1) were bonded in the plastic ball grid array packages (PBGA) by using a Kulicke and Soffa (K&S) IConn automatic ball bonder. The diameter of wire is 0.8 mil and the aluminum metallization pad (Al, Cu0.3%) on the silicon chip had 1μm thickness. Bonding process recipes are bond temperature: $165^{\circ}C$, bond time: 15 ms (bare Cu) and 18 ms (Pd coated Cu), bond power: 68 mA (bare Cu) and 85 mA (Pd coated Cu), bond force: 7 g (bare Cu) and 8 g (Pd/Cu). After bonding process, the bonded samples were encapsulated with three kinds of epoxy molding compound for biased-HAST. The ball bonds were stressed under $130^{\circ}C/85\%RH$ with biased voltage of 10V. The two chlorine contents of molding compound, ultra- high chlorine (24 ppm) and high chlorine (7.6 ppm), are commercially available green resins. The third molding low chlorine content (4.3 ppm) is made of the green resin by the electrical kinetic (EK) method. EK experiments were conducted in a cell of 15 cm (W) x 30 cm (L) x 15 cm (H), consisting of three compartments: cathode reservoir with 5 cm in length, anode reservoir with 5 cm in length, and soil specimen chamber with 20 cm in length. The processing fluid was initially placed into both anode and cathode reservoirs and replenished in the anode reservoir every half day. After EK test under potential of 24 V for 5 day duration were conducted to evaluate the removal mechanisms of chlorine ion (Cl^-) in molding compound and the Cl^- concentration from 7.6 down to 4.3 mg/L. The investigation of Cu-Al IMC was using a filed-emission scanning electron microscope (FESEM) with energy dispersive x-ray spectroscopy (EDX). The observation of interface of Cu/Al was utilized the dual-beam focus ion beam (DB-FIB) by precise etching.

2014 Electronic Components & Technology Conference

Results

The Table 1 lists the summary result of biased-HAST data with three molding compounds. Figure 1 shows the cross-sectional backscatter electron images of bare Cu and Pd/Cu wire after as-received. The bare Cu ball width is 37.5um and height is 22.5um as shown in Figure 1(a). Figure 1 (c) is the enlarged area of 1 (a). The IMC morphology shows grayer than Al layer and discontinuous at the interface between Cu ball and Al pad. The IMC is $CuAl_2$ analyzed by EDX. In Cu-Al binary system, the $CuAl_2$ (θ phase) is the first formation IMC after thermal sonic bonding. [8] The visible maximum thickness of $CuAl_2$ is close to 0.2 um. This is the very thin IMC layer but a few amount of the aluminum layer leaves on the SiO_2 layer after bonding. The Pd/Cu ball has a 36.7um in width and a 20um in height, as shown in Figure 1(b). Figure 1 (d) is the enlarged area of 1(b). The $CuAl_2$ shows a continuous layer between Cu and SiO_2 layer and has a thickness of ~0.18 um. A few amount of aluminum layer only piled up on the right hand side.

Table 1 Summary of biased HAST data with three molding compounds.

Molding Compound	Wire	Biased-HAST							
		96h	192h	288h	384h	960h	1056h	1296h	1344h
Ultra-high Cl (24 ppm)	Bare Cu	3/5	5/5	--	--	--	--	--	--
	Pd_Cu	0/5	2/5	5/5	--	--	--	--	--
High Cl (7.6 ppm)	Bare Cu	0/5	0/5	0/5	0/5	2/5	5/5	--	--
	Pd_Cu	0/5	0/5	0/5	0/5	1/5	3/5	5/5	--
Low Cl (4.3 ppm)	Bare Cu	0/5	0/5	0/5	0/5	0/5	0/5	0/5	0/5
	Pd_Cu	0/5	0/5	0/5	0/5	0/5	0/5	0/5	0/5

A bare Cu wire ball with ultra-high Cl molding compound failed after 192h of biased-HAST, as shown in Figure 2. No obvious damage is found at the interface between Cu wire and Al pad. The thickness of $CuAl_2$ is still very thin up to 192h of biased-HAST. The cross section of the failed bare Cu wire ball was analyzed further by FIB. Figure 3(a) is the SEM image of the failed bare Cu wire after FIB etching. Figure 3(b) is the magnified image of the area of the interface in Figure 3(a). The crack formed at the interface between Cu wire and Al pad. The $CuAl_2$ became disintegrative and dispersive in the Al layer. An EDX analysis at point 1 reveals in Figure 4. Based on the EDX spectrum, the Cl peak is small but the other (Cu, O and Al) peaks are high. Presumably the $CuAl_2$ should be corroded during biased-HAST testing. [8,13] The corrosion behavior will discuss later.

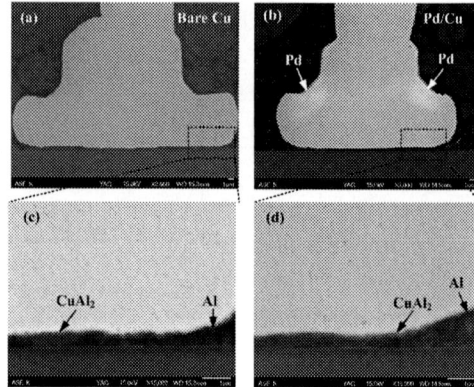

Figure 1 The cross section images of as-bonded (a)、(c) bare Cu and (b)、(d)Pd/Cu wire.

Figure 5 shows a Pd/Cu wire ball failed with ultra-high Cl molding compound after FIB etching. A crack formed between Cu wire ball and $CuAl_2$ layer. Al layer is absent due to complete consumption. The EDX analysis at point 2 shows in Figure 6. The Ti signal is from Ti layer but the oxygen and Si are from SiO_2 layer. The peak of element Cl is very small and the peaks of Cu and Al are high.

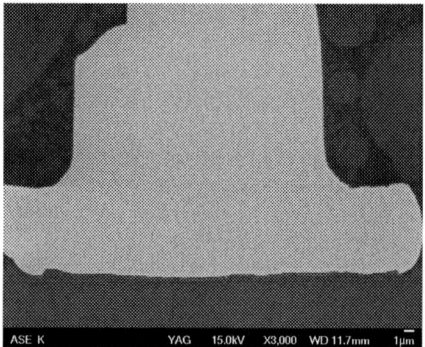

Figure 2 The failed bare Cu wire ball with ultra-high Cl compound after 192h of biased-HAST.

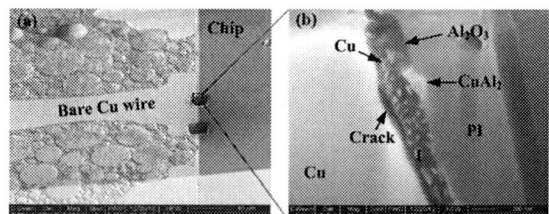

Figure 3 (a) The SEM image of bare Cu wire with ultra-high Cl compound after FIB etching (b)the enlarge etching area of (a).

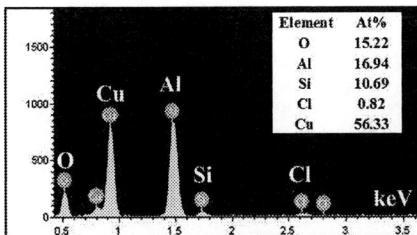

Element	At%
O	15.22
Al	16.94
Si	10.69
Cl	0.82
Cu	56.33

Figure 4 EDX spectrum taken from point 1 in Figure 3(b)

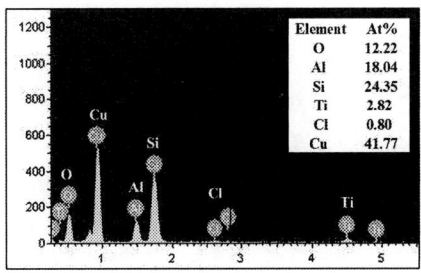

Element	At%
O	12.22
Al	18.04
Si	24.35
Ti	2.82
Cl	0.80
Cu	41.77

Figure 6 EDX spectrum taken from region 2 in Figure 5

Figure 7 represents the cross section images of failed bare Cu wire and Pd/Cu wire with high Cl$^-$ molding compound after 1056h of biased-HAST. A very severe damage happen at the interface between bare Cu wire and SiO_2 layer as shown in Figure 7(a). Some big segments of Cu distribute at the two edges of bare Cu ball. The black area in the middle of the interface almost is Al_2O_3 layer. Figure 7(c) shows the magnified image of the selected area in Figure 7(a). Some of the Cu segments formed in the upper left area adjacent to Cu ball but the big Cu segment was in the lower left area. The EDX result indicates the Cu segments should be CuO. A lot of Cu particle distributed in the Al_2O_3 layer. This is because $CuAl_2$ was corroded by Chlorine from molding compound resulting in Cu reduction. [13] However, no obvious damage was found at the interface between Cu and Al after 1056h of biased-HAST as shown in Figure 7(b). Figure 7(d) is the enlarged image of the selected area of Figure 7(b). The bright layer is Pd distribution. [11] The crack was found at the interface between Pd distribution and Al_2O_3 layer. The $CuAl_2$ attached to the SiO_2 layer and the Al_2O_3 layer was above the $CuAl_2$. Interestingly, the Al metallization pad remains a little next to $CuAl_2$ and Al_2O_3 layer was above the Al metallization pad. This indicates that $CuAl_2$ was not corroded completely and Al pad was not consumed entirely.

In order to understand the microstructure inside the Cu/Al interface, the failed bare Cu and Pd/Cu wire ball were etched by FIB further. Figure 8(a) displays the SEM image of failed bare Cu wire ball with high Cl$^-$ compound after FIB process. The dotted rectangle area of Figure 8(a) is magnified as shown in Figure 8(b). A large number of voids appear in the Cu_9Al_4 layer, with the majority of them being to closer to the Cu/Cu_9Al_4 interface. Some tiny Cu particles dispersed in the Al_2O_3 layer. The dendrite of CuO also was included in the Al_2O_3 layer.

Figure 9(a) shows the SEM image of the failed Pd/Cu wire with high Cl$^-$ compound after FIB etching. The dotted rectangle area of Figure 9(a) is magnified as shown in Figure 9(b). A crack formed at the interface of Cu/Al_2O_3 and the Pd distribution layer attached the Al_2O_3 layer. A large number of Cu particles dispersed in the Al_2O_3 layer. The EDX analysis result of point 3 is presented in Figure 10.

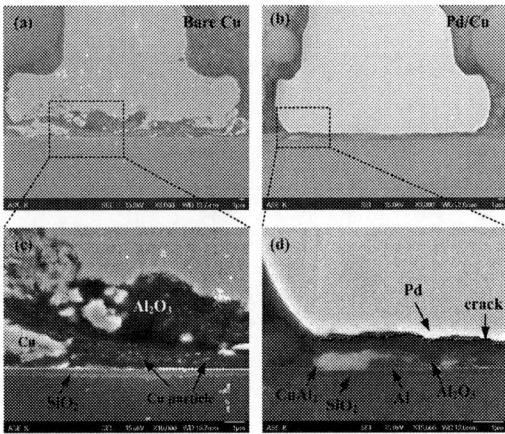

Figure 7 SEM images of (a) 、 (c)failed bare Cu wire and (b) 、 (d)failed Pd/Cu wire with high Cl$^-$ molding compound after 1056h of biased-HAST

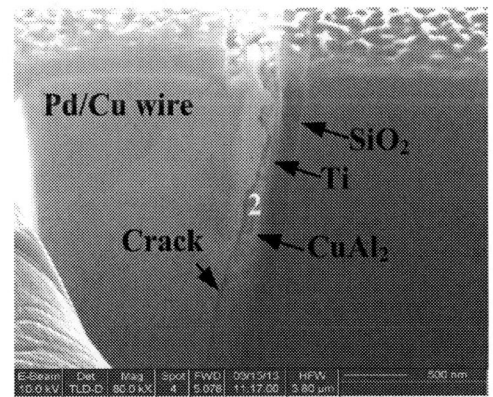

Figure 5 The microstructure image of Pd/Cu wire with ultra-high Cl$^-$ compound after FIB etching.

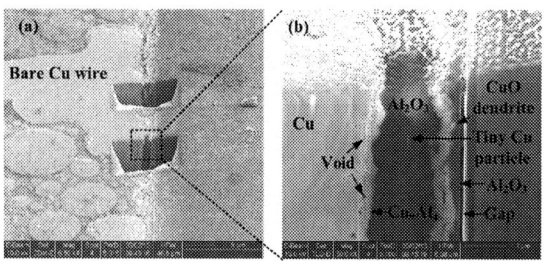

Figure 8 (a) The SEM image of bare Cu wire with high Cl⁻ compound after FIB etching (b)the enlarge image of the selected area of (a).

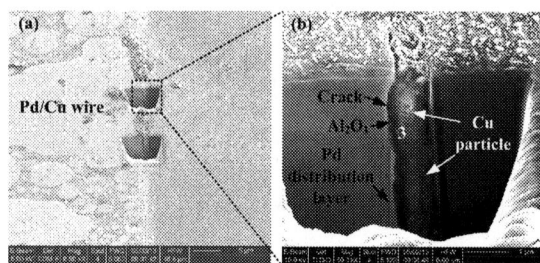

Figure 9 (a) The SEM image of Pd_Cu wire with high Cl⁻ compound after FIB etching (b)the enlarge image of the selected area of (a).

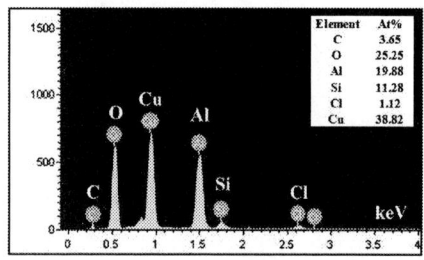

Figure 10 EDX spectrum taken from point 3 in Figure 9

Although the bare Cu and Pd/Cu wire are not failed with low Cl⁻ compound after 1344h of biased-HAST, the microstructures need to clarify. Figure 11 displays the SEM microstructure of bare Cu and Pd/Cu wire with low Cl⁻ molding compound for up to 1344h of biased-HAST. No harsh damage was found in the bare Cu wire and Pd/Cu wire even at the interface of ball/pad as shown in Figure 11(a) and 11(b). A small amount of Cu_9Al_4 is observed at the interface between Cu and Al as shown in Figure 11(c). The thickness of Cu_9Al_4 reaches ~0.4 µm and the Al pad is incompletely consumed. This indicates the IMC growth rate is very slow due to the low temperature (130°C). A micro crack formed at the interface of Cu_9Al_4/Cu. However, no crack was found at the interface of Cu_9Al_4/Cu in Figure 11(d). The thin Cu_9Al_4 layer attached to the SiO_2 layer. Compared with the Cu_9Al_4 of bare Cu, the morphology is totally different. In addition, Pd-rich layer can retard Al and Cu atoms to interdiffuse. [12] This indicates the growth rate of Cu_9Al_4 in the Pd/Cu wire ball could be slower than in the bare Cu wire. To further realize the IMC morphology inside the interface of bare Cu/Al pad, Figure 12 shows the SEM image of bare Cu wire with Cl⁻

molding compound after FIB process. The crack formation was along the interface between Cu_9Al_4 and bare Cu. A gap exists between Al pad and Ti layer. The thickness of Cu_9Al_4 reaches ~0.4 µm and the remnant of Al pad is also very thin. The Al layer is almost depleted. The $CuAl_2$ transferred to Cu_9Al_4 needs more Cu and Al atoms during aging. The Cu atoms are sufficient from Cu wire but Al atoms are inadequate due to only 1 um thick Al pad. Therefore, Cu_9Al_4 cannot grow thicker because of inadequacy.

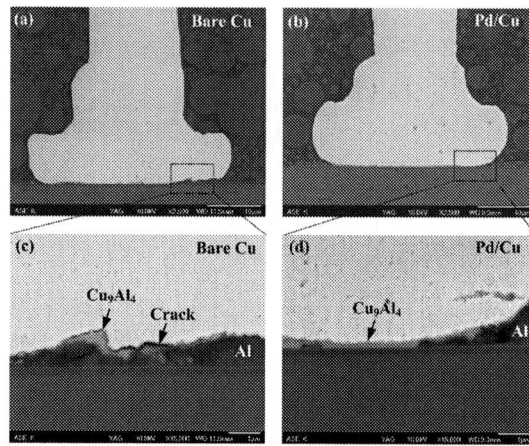

Figure 11 SEM images of (a)、(c) bare Cu wire and (b)、(d)Pd/Cu wire with low Cl⁻ molding compound after 1344h of biased-HAST

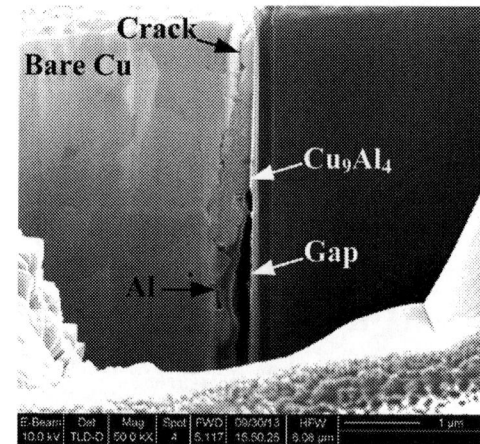

Figure 12 The SEM image of bare Cu wire with low Cl⁻ compound after FIB etching.

Interestingly, the SEM image of Pd_Cu wire after FIB etching is quite different from bare Cu wire (Figure 12), as shown in Figure 13. The Cu_9Al_4 layer of Pd/Cu wire is thinner than bare Cu wire. No gap or crack was found at the interface of Pd/Cu/Ti layer. Figure 14 is the EDX spectrum of point 4 in Figure 13. The EDX analysis of point 4 confirms Pd element dispersed in the Pd/Cu wire adjacent to Cu_9Al_4 layer.

978-1-4799-2408-0/14 $31.00 © 2014 IEEE

Figure 13 The SEM image of Pd/Cu wire with low Cl-compound after FIB etching.

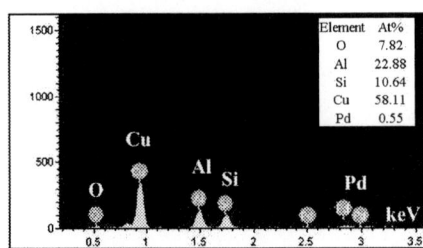

Figure 14 EDX spectrum taken from region 4 in Figure 13

Discussion

According to the results shown above, the failure rate can increase with increasing chlorine content in the molding compound during biased-HAST. Two events should be noticed, IMC corrosion and Cu oxidation. At first, the initial IMC, $CuAl_2$, forms at the interface between Cu ball and Al pad after bonding. During biased-HAST, chlorine ion will be dissociated by the humidity environment and move to the anode. In the literature, the CuAl IMC would be corroded chlorine ion driven by electrical potential. Thus, it is of great interest to understand the failure mechanism caused by corrosion.

The corrosion of $CuAl_2$ is suggested to be a chemical reaction. The following sequence of reaction could be considered a possible mechanism.

$$CuAl_2+6Cl^- \rightarrow 2AlCl_3+Cu+6e^- \qquad (1)$$
$$AlCl_3+3OH^- \rightarrow Al_2O_3+3HCl+3e^- \qquad (2)$$

Due to the chlorine ion fast attack, $CuAl_2$ is decomposed before phase transferring to Cu_9Al_4. The intermediate, $AlCl_3$, is considered unstable phase. This is because $AlCl_3$ is absent at the interface of Cu/Al. Therefore, $AlCl_3$ reacts with hydroxyl from moisture and becomes Al oxide. The final products are precipitated Cu particle and Al_2O_3. Besides, the depletion of $CuAl_2$ will accompany crack formation. This is because the weak bonding between oxide (Al_2O_3) and metal

(Cu). However, it is reported the Pd distribution could retard Chlorine diffusion. [12] This may conclude Pd_Cu wire has a better humidity reliability test result than bare Cu.

The other failure mechanism is Cu oxidation. Although Cu wire is encapsulated in the package, Cu oxidation still could occur for a long time after IMC corrosion. In this study, the Cu oxidation is not observed in both failed bare Cu and Pd/Cu wire with ultra-high molding compound. This indicates corrosion rate is faster than oxidation rate. However, The diffusion of oxygen may accompany moisture absorption by molding compound. [14] After IMC corrosion, the void or crack forms at interface. Oxygen can diffuse through void or crack and react with Cu. Hence, CuO forms at the interface between Cu and Al_2O_3 layer, as shown in Figure 7(a) and 7(c).

On the other hand, the observation of void formation in Cu_9Al_4 indicates the phase transformation of Cu-Al IMC. Figure 15 shows a schematic illustration of void formation in Cu_9Al_4. After bonding, the initial IMC is $CuAl_2$. Cu_9Al_4 appears when diffusing Cu and Al atoms inter diffuse through $CuAl_2$ layer. Many literatures pointed out that Cu_9Al_4 grew faster than CuAl due to kinetics is dominant. [8] Therefore, Cu and Al atoms diffuse more in Cu_9Al_4. On the assumption, the low melting point metal diffusing specimen (Al, 660°C) has a faster diffusivity in high melting point metal (Cu, 1083°C). The diffusing Al atoms diffuse faster than Cu resulting in the void formation in Cu_9Al_4 layer. Besides, the formation of Cu_9Al_4 results in shrinking ~5% the volume. [8] This indicates the void formation may accompany phase transformation of $CuAl_2$ into Cu_9Al_4. However, no void was found in the Pd/Cu wire as shown in Figure 9(b). This is because Pd distribution layer is a diffusion barrier layer for Cu and Al atoms. Cu_9Al_4 could not form even until $CuAl_2$ was corroded by chlorine.

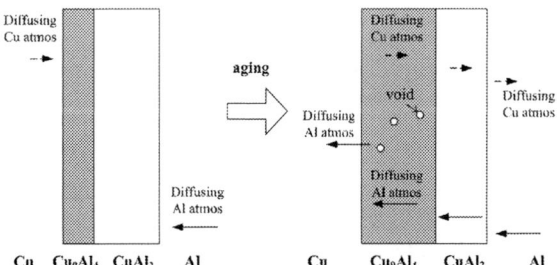

Figure 15 A schematic illustration of void formation in Cu_9Al_4

Conclusion

Two types of Cu wire (bare Cu & Pd/Cu) using three kinds of molding compound were investigated on the corrosion performance of Cu alloy wire under biased-HAST. For Cu wires, Pd/Cu wire has a better performance than bare Cu wire. Pd element can retard chlorine ion to corrode IMC. Meanwhile, Pd distribution layer formation could delay oxidation occurrence. For molding compound, the low chlorine content has the best resistance against biased-HAST among ultra-high and high chlorine compound. The lifetime of ultra-high chlorine molding compound against biased-HAST

978-1-4799-2408-0/14 $31.00 © 2014 IEEE   423

is 4.6 times than low chlorine molding compound. Higher chlorine content molding compound will result in more dissociated chlorine ions. More chlorine ions will speed up the occurrence corrosion.

References

1. H. Clauberg, P. Backus, and B. Chylak, "Nickel-palladium bond pads for copper wire bonding," Microelectron. Reliab. 51, pp.75-80, 2011.

2. Y.H. Tian, C.J. Hang, C.Q. Wang, G.Q. Ouyang, D.S. Yang, and J.P. Zhao, "Reliability and failure analysis of fine copper wire bonds encapsulated with commercial epoxy molding compound," Microelectron. Reliab. 51, pp. 157–165, 2011.

3. H. Seki, P. Chen, H. Nakatake, S.i. Zenbutsu, and S. Itoh, "Study of EMC for Cu bonding wire application," IEEE CPMT Symposium Japan, 2010, pp. 1–3.

4. H. Abe, D.C. Kang, T. Yamamoto, T. Yagihashi, Y. Endo, H. Saito, T. Horie, H. Tamate, Y. Ejiri, N. Watanabe, and T. Iwasaki, "Cu Wire and Pd-Cu Wire Package Reliability and Molding Compounds", IEEE 62nd Electronic Components and Technology Conference (ECTC), 2012, pp. 1117–1123.

5. T. Uno, "Bond reliability under humid environment for coated copper wire and bare copper wire," Microelectron. Reliab. 51, pp. 148–156, 2011.

6. D. Stephan, F.W. Wulff, and E. Milke, " Reliability of palladium coated copper wire," 12th Electronics Packaging Technology Conference (EPTC), 2010, pp. 343–348.

7. C. F. Yu, C. M. Chan, L. C. Chan, K. C. Hsieh, "Cu wire bond microstructure analysis and failure mechanism," Microelectron. Reliab. 5, pp. 119-124, 2011.

8. H. Xu, C. Liu, V. V. Silberschmidt, S. S. Pramana, T. J. White, Z. Chen, V. L. Acoff, "Behavior of aluminum oxide, intermetallics and voids in Cu–Al wire bonds," Acta Mater. 59, pp. 5661-5673, 2011.

9. P. Ratchev, S. Stoukatch, and B. Swinnen, "Mechanical reliability of Au and Cu wire bonds to Al, Ni/Au and Ni/Pd/Au capped Cu bond pads," Microelectron. Reliab. 46, pp. 1315–1325, 2006

10. 10. K. Hyoung-Joon, L. Joo Yeon, P. Kyung-Wook, K. Kwang-Won, J. Won, C. Sihyun, L. Jin, M. Jung-Tak, and P. Yong-Jin, "Effects of Cu/Al intermetallic compound (IMC) on copper wire and aluminum pad bondability," IEEE Trans. Compon. Packag. Technol. 26, pp. 367–374, 2003.

11. Y. W. Lin, R. Y. Wang, W. B. Ke, I.S. Wang, Y. T. Chiu, K. C. Lu, K. L. Lin, and Y. S. Lai, "The Pd distribution and Cu flow pattern of the Pd-plated Cu wire bond and their effect on the nanoindentation," Mater. Sci. Eng. A 543, pp. 152–157, 2012.

12. Y. W. Lin, R. Y. Wang, W. B. Ke, I.S. Wang, Y. T. Chiu, K. C. Lu, K. L. Lin, and Y. S. Lai, "The influence of Pd on the interfacial reactions between the Pd-plated Cu ball bond and Al pad," Surface and Coatings Technology, Vol. 231, no. 25, pp. 599–603, Sept. 2013.

13. J. Osenbach, B. Wang, S. Emerich, J. DeLucca, and D. Meng, "Corrosion of the Cu/Al Interface in Cu-Wire-Bonded Integrated Circuits," IEEE 63rd Electronic Components and Technology Conference (ECTC), 2013, pp. 1574–1586.

14. P. Su, H. Seki, C. Ping, S. Itoh, L. Huang, N. Liao, B. Liu, C. Chen, W. Tai, and A. Tseng, "Effects of Reliability Testing Methods on Microstructure and Strength At the Cu Wire-Al Pad Interface," IEEE 63rd Electronic Components and Technology Conference (ECTC), 2013, pp. 179–185.

The Effect of Nickel Microalloying on Thermal Fatigue Reliability and Microstructure of SAC105 and SAC205 Solders

Richard Coyle
Alcatel-Lucent, Murray Hill, NJ, USA
richrad.coyle@alcatel-lucent.com
Richard Parker
iNEMI, Tipton, IN, USA
rddlparker@gmail.com
Babak Arfaei
Universal Instruments, Conklin, NY, USA
Francis Mutuku
Binghamton University, Binghamton, NY
Keith Sweatman and Keith Howell
Nihon Superior Co., Ltd., Osaka, Japan
Stuart Longgood
Delphi, Kokomo, IN, USA
Elizabeth Benedetto
Hewlett-Packard Co., Houston, TX, USA

Abstract

This study explores the effect of a nickel (Ni) microalloy addition on the thermal fatigue performance and microstructure of two low Ag content, Pb-free solder alloys, Sn-1.0Ag-0.5Cu (SAC105) and Sn-2.0Ag-0.5Cu (SAC205). The alloy performance was evaluated using two different area array component test vehicles, an 84-pin chip scale package (CSP) and a 192-pin fine pitch ball grid array (BGA). The baseline alloy microstructures were characterized using polarized light microscopy and scanning electron microscopy with backscattered electron imaging for phase identification. Thermal fatigue performance was assessed with accelerated thermal cycling (ATC) using four temperature cycling profiles with distinct temperature ranges (ΔT) and temperature extremes. Additionally, each temperature profile used a standard 10 minute dwell time or an extended 60 minute dwell time. A microalloy addition of 0.05% Ni was found to alter the base microstructures of the SAC105 and SAC205 alloys. Generally, the Ni addition improved the thermal fatigue life but the improvement was not consistent in both alloys, both components, and across all thermal cycling profiles. The most consistent response was with the 84CTBGA component, which showed improved reliability with the Ni addition in all of the thermal cycles.

Key words: Pb-free solder, thermal fatigue, Ni microalloying, solder microstructure.

Introduction

Significant innovations in Pb-free solder alloy formulations are being driven by volume manufacturing and field experiences. This has resulted in an increase in the number of Pb-free solder alloy choices beyond the common near-eutectic Sn-Ag-Cu (SAC) alloys first established as replacements for Sn-Pb. Much of the new Pb-free alloy development is in response to shortcomings of near-eutectic SAC such as poor mechanical performance, particularly resistance to drop shock or impact loading. However, for an alloy to replace near-eutectic SAC it also should provide acceptable thermal fatigue resistance to satisfy requirements for long term solder attachment reliability [1].

The drop shock reliability of solder joints is influenced by the combination of bulk solder and interfacial intermetallic layer properties. One of the prominent drop shock solder joint failure modes is characterized by cracking at the intermetallic solder bond interface. This cracking is accompanied by minimal deformation in the bulk solder. The absence of bulk solder deformation during high strain rate drop shock loading is due to the strain rate sensitivity of the solder. Eutectic SnPb solder is less sensitive to strain rate and more likely to deform in the bulk. SAC solders are more sensitive to strain rate with the yield strength increasing more rapidly at higher strain rate, resulting in minimal plastic deformation in the bulk. This prevents the drop shock energy from attenuating in the bulk solder and allows more dynamic stress to be transferred to the solder bond interface, which results in interfacial cracking [2].

It is recognized that the microstructural characteristics of most alloys are responsible for their mechanical performance [3-6] and SAC solders are no exception. The most straightforward modification for improving drop shock reliability of SAC solder alloys is to reduce the Ag content well below near-eutectic levels [1, 7]. Lowering the Ag content in a SAC solder alters the microstructure by reducing the volume fraction of Ag_3Sn intermetallic (IMC) precipitates that serve as the primary strengthening agent in SAC solders. As the Ag_3Sn volume fraction decreases, the modulus and yield strength decrease, the ductility increases, and less stress is transferred to the interfacial layer under dynamic loading. Thermal fatigue resistance in contrast, is promoted by a higher Ag content and a larger volume fraction of Ag_3Sn precipitates [8-13]. Thus a reliability dilemma exists where lowering the volume fraction of Ag_3Sn precipitates improves drop shock performance but compromises the thermal fatigue resistance.

Another method for improving SAC drop shock reliability is to add a fourth alloying element to the solder. Some common fourth element additions include Al, Bi, Co, Mg, Ni, Sb, and Zn. It is assumed that these alloying elements improve the ductility and toughness of the interfacial intermetallic layer by modifying its composition and morphology [14-18]. Quaternary alloy formulations also are known to promote intermetallic precipitation in the bulk solder. Because solder thermal fatigue likewise is structure sensitive, modification of the bulk solder microstructure has the potential to improve thermal fatigue resistance. Some of the alloying elements used in SAC solders are effective when added in amounts so small that they would be considered trace impurities if found in a traditional SnPb eutectic solder. This process is called microalloying, a term borrowed from ferrous metallurgy that refers to the process of adding alloying elements other than the major constituents to produce a positive effect on the behavior and performance of an alloy [19].

Nickel (Ni) is perhaps the most common microalloying element used in commercial SAC solder formulations. The beneficial effects of microalloying with nickel, which were identified first in the tin-copper (Sn-Cu) eutectic system, extend also to SAC alloy joints [20]. Ni microalloying is believed to increase the resistance to interfacial fracture during high strain rate, drop shock loading by improving the mechanical properties of the interfacial IMC [15, 17, 18]. Ni also has been shown to improve the mechanical properties of the bulk SAC solder including its ductility, yield strength, and tensile strength [21-29]. The relationship between Ni microalloying and mechanical properties of Sn-based Pb-free solders has been attributed to microstructural refinement. The addition of Ni can also decrease the amount of Sn undercooling which alters solidification but the reported extent of these effects varies widely throughout the literature. There is disagreement regarding the optimum Ni content and the exact effect on solidification and the specifics of microstructural refinement. Further, the data developed using traditional mechanical test methods may not be wholly applicable to microelectronic applications. The major concern is that large bulk solder samples required for accurate mechanical property measurements have significantly different microstructures than those found in the small solder joints typical of microelectronic assemblies [30, 31]. Such experimental limitations must be considered when reviewing the basic findings from the literature.

Hammad studied SAC0507 with 0.05 and 0.1 wt. % additions of Ni. He reported a relatively small decrease in undercooling of 6 °C in SAC0507 with 0.05 wt. % Ni but did not find a similar effect on undercooling with the higher Ni content. The SAC0507 with 0.05 wt. % Ni had the smallest β-Sn cell or dendrite size, the highest yield and tensile strengths, and best creep performance [21, 22]. The alloy with lower Ni content also contained a more lamellar Ag_3Sn precipitate morphology, which has been associated with better thermal fatigue performance [32].

El-Daly et al. studied Ni microalloying in both SAC105 and SAC205 and showed a reduction in undercooling of 8-10 °C with the addition of 0.05 wt. % Ni [23, 24]. El-Daly observed a factor of two improvement in room temperature creep strength with Ni compared to the base alloys. In contrast to the work of Hammad, Ni caused a slight reduction in the β-Sn dendrite size in SAC105 and no increase in the β-Sn dendrite size in SAC205. El-Daly suggests that Ni improves

the creep strength by promoting the formation of $(Cu,Ni)_6Sn_5$ intermetallic particles that serve as heterogeneous nucleation sites for β-Sn and subsequently strengthen the Sn matrix by precipitation hardening. This hypothesis may have merit based on the work of Seo et al. on Ni additions in a Sn-1.2Ag alloy [25]. In the absence of Cu, Seo found that the β-Sn dendrite size increased with increasing Ni content and the maximum reduction in undercooling was measured at 0.15 wt. % Ni, which is a much different result than that of Hammad and El-Daly. These findings suggest that the role of Ni in the solidification is much different in the presence of Cu and the precipitation of β-Sn could be influenced by the formation of $(Cu,Ni)_6Sn_5$ intermetallic particles. While the heterogeneous nucleation of β-Sn by $(Cu,Ni)_6Sn_5$ has not been established conclusively, it should be noted that the Cu and Ag IMC phases can nucleate with minimum undercooling ahead of the β-Sn, even in nominally eutectic liquid SnAgCu interconnects [26]. However, even if the heterogeneous nucleation hypothesis applies to the room temperature creep behavior reported by El-Daly, it may not apply to thermal creep fatigue behavior in accelerated thermal cycling (ATC), where performance is controlled by the Ag content and Ag_3Sn intermetallic volume fraction.

Che et al. in a comparison of the tensile properties of SAC105 with 0.02 and 0.05 wt. % additions of Ni, found that the 0.05 wt. % Ni alloy exhibited better ductility as characterized by larger elongation, lower elastic modulus and lower yield stress [27]. They suggested the better ductility was because Ni facilitates nucleation of the Cu_6Sn_5 or $(Cu,Ni)_6Sn_5$ intermetallic phases but they provided no microstructural analysis to support this supposition.

Two studies of Ni additions to SAC305 yielded significantly different results. Cheng and coworkers concluded that Ni microalloy additions of 0.05 and 0.10 wt. % had almost no effect on the undercooling, reduced the β-Sn dendrite size only slightly, did not affect the tensile strength, and lowered the ductility [28]. Their findings were largely independent of the Ni content. In contrast, the results from the study by Kim et al. [29] with 0.10 wt.% Ni in SAC305 showed a reduction in undercooling of almost 15 °C, a reduction in the β-Sn dendrite size, increased the tensile strength and yield strength and improved elongation (ductility). Kim suggests that Ni increases IMC formation, which suppresses undercooling, refines microstructure, and improves ductility without degrading strength. Cheng and Kim report very different results for properties and microstructure of the same alloy, which illustrates the challenges of comparing test data from the literature. A plausible explanation for the ostensibly different results from these two similar studies is that differences in sample dimensions and fabrication create differences in solidification and microstructure that were manifested as the observed differences in property measurements [30, 31].

There are some conflicts in the literature between Ni additions, properties, and microstructure, but existing data suggest that Ni microalloying generally improves the tensile, shear, and creep properties of various Sn-based, Pb-free alloys. Unfortunately, very limited thermal fatigue data have been developed using SAC-Ni microalloys and commercial

978-1-4799-2408-0/14 $31.00 © 2014 IEEE

components. Terashima et al. performed thermal cycling experiments and found that a 0.05 wt. % Ni addition resulted in a measurable improvement in fatigue resistance of a Sn1.2Ag0.5Cu alloy [33]. In addition, the fatigue performance of the 1.2 Ag alloy with Ni was only slightly lower than SAC305 in the same test. The authors attribute the improved fatigue life with Ni microalloying to refinement of both the β-Sn dendrite size and the $(Cu,Ni)_6Sn_5$ particle size, which slows microstructural coarsening. Although exact strengthening mechanisms can be debated, the results from this study are a positive indication that Ni microalloying can improve the thermal fatigue reliability of a SAC solder.

A major electronics industry consortium has been conducting an extensive investigation using accelerated thermal cycling (ATC) and two commercial area array test vehicles to evaluate thermal fatigue performance of multiple Sn-based, Pb-free solder alloys [34]. This paper presents the findings from a subset of the temperature cycling test matrix targeting the effect of a Ni microalloying on the thermal fatigue performance of Sn-1.0Ag-0.5Cu (SAC105) and Sn-2.0Ag-0.5Cu (SAC205) Pb-free solder alloys. Thermal fatigue was assessed using four temperature cycling profiles with distinct temperature ranges (ΔT) and temperature extremes. Additional test cells were provided by using a standard 10 minute dwell time and an extended 60 minute dwell for each temperature cycling profile. The baseline microstructures of the alloys were characterized using polarized light microscopy and scanning electron microscopy with backscattered electron imaging for phase identification.

Experimental

Test Vehicles and Alloys

The printed circuit board (PCB) test vehicle and the two area array components (192CABGA and 84CTBGA) are shown in Appendix A and their attributes are listed in Table 1. Both components are considered chip scale packages with the 192CABGA having a larger die to package ratio. The PCB laminate material is LG451HR.

Table 1: Area array and printed circuit board (PCB) test vehicle attributes.

BGA Package		
Designation	192CABGA	84CTBGA
Die Size	12x12 mm	5x5 mm
Package Size	14x14 mm	7x7 mm
Ball Array	16x16	12x12
Ball Pitch	0.8 mm	0.5 mm
Ball Diameter	0.46 mm	0.3 mm
Pad Finish	Electrolytic Ni/Au	Electrolytic Ni/Au
PCB		
Thickness	2.36 mm (93mils)	
Surface Finish	High temp OSP	
No. Cu Layers	6	
Pad Diameter	0.356 mm	0.254 mm
Solder Mask Dia.	0.483 mm	0.381 mm

The two base alloys used in this study were SAC105 and SAC205. For comparison, the test included the base alloys modified with a microalloy addition of 0.05 wt. % Ni (500 ppm). The test vehicles were assembled using Type 3 no-clean SAC305 solder paste. The higher Ag content SAC305 solder paste used for surface mount assembly resulted in an increase in Ag concentration in both components. The effective Ag

content after board level attachment is shown in Table 2. The exact analyzed alloy compositions and calculations of effective Ag content have been documented previously [34].

Table 2: Nominal solder alloy compositions and effective Ag content after surface mount assembly with SAC305 solder paste.

Alloy	Nominal Composition (wt. %)				Effective Composition* (wt. %)	
	Sn	Ag	Cu	Ni	Ag (BGA)	Ag (CSP)
SAC205	97.5	2.0	0.5	0.00	2.10	2.20
SAC205+Ni	97.5	2.0	0.5	0.05	2.10	2.20
SAC105	98.5	1.0	0.5	0.00	1.20	1.50
SAC105+Ni	98.5	1.0	0.5	0.05	1.20	1.50

*estimated % Ag after board attach w/SAC305 paste

Accelerated Temperature Cycling

The components and the circuit boards were daisy chained to allow electrical continuity testing after surface mount assembly and in situ, continuous monitoring during thermal cycling. Accelerated temperature cycling was done in accordance with the IPC-9701A industry test guideline [35]. The ATC test matrix consisted of four different temperature cycling profiles, each with standard 10 minute dwell times and extended 60 minute dwell times as shown in Table 3.

Table 3: The temperature cycling profiles used in the experimental test matrix.

Cycle Number	Minimum Temp.	Maximum Temp. (°C)	Temp. Range (°C)	Dwell Time (min.)
1	0	100	100	10
2	0	100	100	60
3	-40	100	140	10
4	-40	100	140	60
5	25	125	100	10
6	25	125	100	60
7	-15	125	140	10
8	-15	125	140	60

The solder joint resistance was monitored with either an event detector or a data logger set at a resistance threshold of 1000 ohms, also described elsewhere [34]. The failure data are reported as characteristic life η (the number of cycles to achieve 63.2% failure) and slope β from a two-parameter Weibull analysis. Complete details of the experimental plan, test vehicles, test protocols, chamber profiling, and surface mount assembly are provided in reference [34].

Microstructural and Failure Analyses

A baseline microstructural characterization was performed prior to temperature cycling on representative board level assemblies from each of the component and alloy test cells. This process documented differences in the microstructures resulting from differences in alloy composition. Microstructural characterization and failure analysis was done using optical metallography (destructive cross-sectional analysis), polarized light microscopy (PLM), and scanning electron microscopy (SEM). Using methods developed previously, the SEM operating in the backscattered electron imaging (BEI) mode was used to differentiate phases in the SAC microstructures. Backscattered electron imaging (BEI) has been shown to be particularly useful in previous studies for differentiating phases in the SAC microstructures [8, 10, 32].

978-1-4799-2408-0/14 $31.00 © 2014 IEEE

Results and Discussion

Test Results – Expected Trends and Data Integrity

The thermal cycling results are summarized in Tables 4 and 5, the bar charts in Figures 1 and 2, and the Weibull plots in Appendix B. The tables include the characteristic lifetimes (N63) and Weibull slopes (β) to enable comparisons between the base alloys (SAC105 and SAC205) and those with the 500 ppm Ni microalloy additions. Data are presented for four different temperature cycles, each with 10 or 60 minute dwell times.

The IPC 9701 guideline for attachment reliability evaluations mandates a sample size of 32 to minimize the impact of experimental deviations or outliers on the Weibull statistics [35]. This study used a sample size of 16 per cell due to resource limitations imposed by the large number of alloys and cycles required in the experimental plan [34]. There are variations in Weibull slope (β) across the data sets and these variations should be taken into consideration when making characteristic lifetime comparisons between data sets.

Table 4: Summary of ATC failure statistics for the 192CABGA and 84CTBGA with SAC105 and SAC105+Ni.

Temperature Cycle	Dwell Time (min)	192CABGA SAC105		192CABGA SAC105 + Ni	
		Characteristic Life (cycles)	Slope β	Characteristic Life (cycles)	Slope β
0/100°C	10	4910	5.4	4707	6.6
-40/100°C	10	3054	7.1	2956	5.1
25/125°C	10	3144	3.5	3413	4.1
-15/125 °C	10	2666	10.0	3447	9.0
0/100°C	60	2558	6.5	2818	4.5
-40/100°C	60	2301	9.6	2804	8.1
25/125°C	60	2479	6.5	2145	4.1
-15/125 °C	60	1473	7.3	1705	9.1

Temperature Cycle	Dwell Time (min)	84CTBGA SAC105		84CTBGA SAC105 + Ni	
		Characteristic Life (cycles)	Slope β	Characteristic Life (cycles)	Slope β
0/100°C	10	6826	7.9	7683	7.1
-40/100°C	10	4280	9.8	5451	6.2
25/125°C	10	5771	7.7	7346	6.7
-15/125 °C	10	4099	8.8	6165	8.5
0/100°C	60	4137	4.5	5348	5.3
-40/100°C	60	3369	10.3	4435	8.6
25/125°C	60	3778	6.9	5072	7.0
-15/125 °C	60	2448	6.8	3302	8.1

Table 5: Summary of ATC failure statistics for the 192CABGA and 84CTBGA with SAC205 and SAC205+Ni.

Temperature Cycle	Dwell Time (min)	192CABGA SAC205		192CABGA SAC205 + Ni	
		Characteristic Life (cycles)	Slope β	Characteristic Life (cycles)	Slope β
0/100°C	10	5312	10.7	4839	4.1
-40/100°C	10	3119	8.3	3140	5.2
25/125°C	10	3893	6.9	3590	8.6
-15/125 °C	10	2557	10.3	3189	7.9
0/100°C	60	3518	7.4	3241	4.9
-40/100°C	60	3574	3.8	2619	10.8
25/125°C	60	2742	7.2	2383	9.5
-15/125 °C	60	1736	6.4	1663	6.8

Temperature Cycle	Dwell Time (min)	84CTBGA SAC205		84CTBGA SAC205 + Ni	
		Characteristic Life (cycles)	Slope β	Characteristic Life (cycles)	Slope β
0/100°C	10	9062	7.1	9095	8.9
-40/100°C	10	5241	10.3	6067	6.3
25/125°C	10	6525	5.2	7477	8.4
-15/125 °C	10	4256	7.6	6340	6.9
0/100°C	60	5206	3.3	5881	9.9
-40/100°C	60	4304	4.2	4602	15.6
25/125°C	60	4584	8.2	5380	8.7
-15/125 °C	60	3100	7.8	3526	7.9

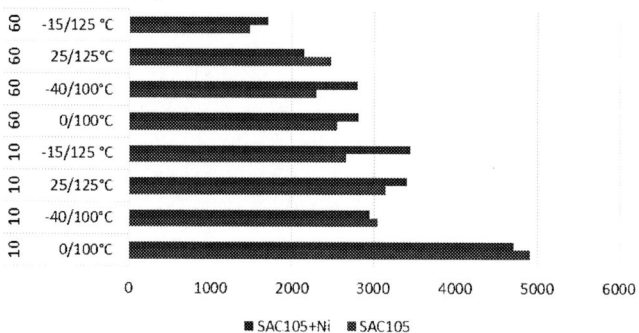

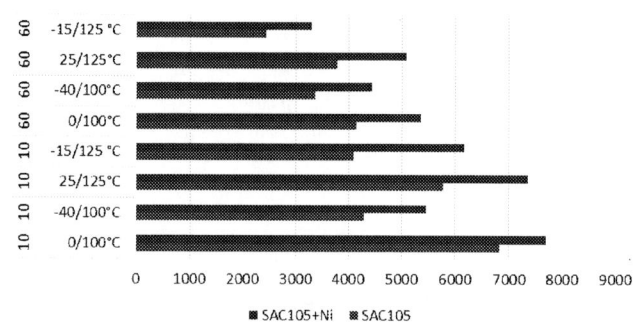

Figure 1: Bar charts illustrating the reliability effect of a 500 ppm Ni microalloy addition to SAC105 for 192CABGA and 84CTBGA components.

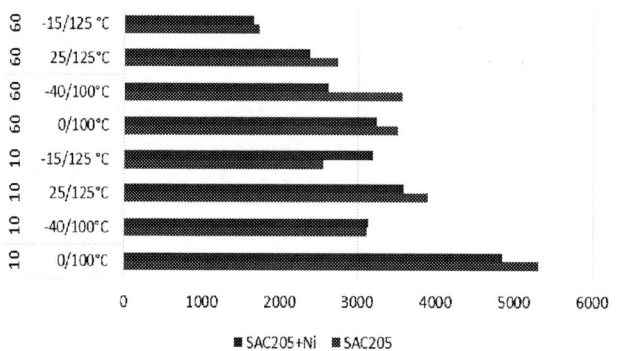

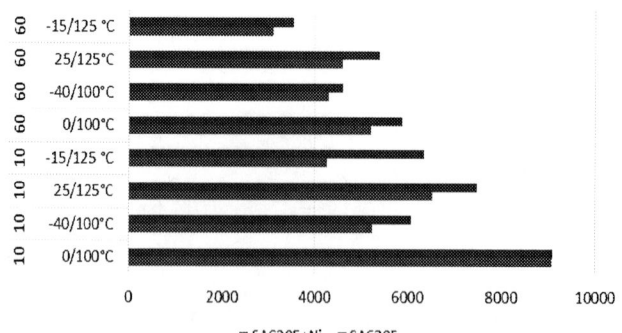

Figure 2: Bar charts illustrating the reliability effect of a 500 ppm Ni microalloy addition to SAC205 for 192CABGA and 84CTBGA components.

There are four anticipated trends that can be used to gauge the overall quality and integrity of the data sets. First, the results from previous studies have shown that the smaller 84CTBGA outperforms the larger 192CABGA by a wide margin [8, 36-38]. The 192CABGA package construction and high die to package ratio generates higher shear strains in the solder joints resulting in earlier failures. The better performance of the 84CTBGA with all alloys and cycling profiles is reconfirmed by the data in Tables 4 and 5. Second, the thermal fatigue reliability of SAC alloys increases with higher Ag content [8-13]. This is validated by the consistently better performance of SAC205. Third, the best reliability performance is expected with the least aggressive, 0/100 °C cycle and poorest is expected with the most aggressive, -15/125 °C cycle. This anticipated result is observed universally for all combinations of alloy and temperature cycle. Finally, the Weibull statistics confirm that the increase in dwell time from 10 to 60 minutes decreases the thermal cycling reliability of the Pb-free solders [36]. Collectively, these trends confirm the expected results for component type, Ag content, temperature cycle, and dwell time and validate the general integrity of the temperature cycling data sets.

Test Results – Effect of Ni Microalloying

The general trend throughout the reliability data (measured by characteristic lifetime) indicates that the 500ppm Ni microalloy addition improves the thermal fatigue life. However, a review of the data in Figures 1 and 2 shows that Ni microalloying did not improve the fatigue life consistently in both alloys, both components, and across all thermal cycling profiles from the test matrix. Most of the irregularity in the data is associated with the results for the 192CABGA component. When Ni is added to solder used with the 192CABGA component, the result is a moderate increase in reliability in most SAC105 test cells. However with the same Ni addition in SAC205, some test cells are relatively unaffected while others exhibit a slight decrease in reliability. The response to Ni was far more consistent with the 84CTBGA component, which showed improved reliability with the Ni addition in all of the thermal cycles. Overall, the incremental reliability improvement with the Ni addition was greater in SAC105 than it was in SAC205. With few exceptions, the improvement was not affected significantly by the extending the dwell time to 60 minutes.

Microstructural Characterization

Before discussing the effects of Ni microalloying on microstructure, it is helpful to review some relevant microstructural features of SAC solders. Figures 3a and 3b show characteristic, low magnification metallographic cross sections of BGA interconnects made with a SAC solder. The boundaries between the contrasting sections of the polarized light micrograph in Figure 3b are high angle Sn grain boundaries with orientation differences exceeding 15° [26]. Large scale BGA solder joints generally display one to three of these large Sn grains. Often this is called the "beach ball" morphology. Depending on the solidification conditions, there may be deviations from the beach ball morphology leading to formation of a fine grain interlaced twinned morphology. The polarized light micrograph in Figure 3d contains regions with the large grain, beach ball morphology as well as those containing the small grain, interlaced twinned morphology. The fine grain interlaced twinned morphology is of high interest because it has been associated with better thermal fatigue performance [39-41].

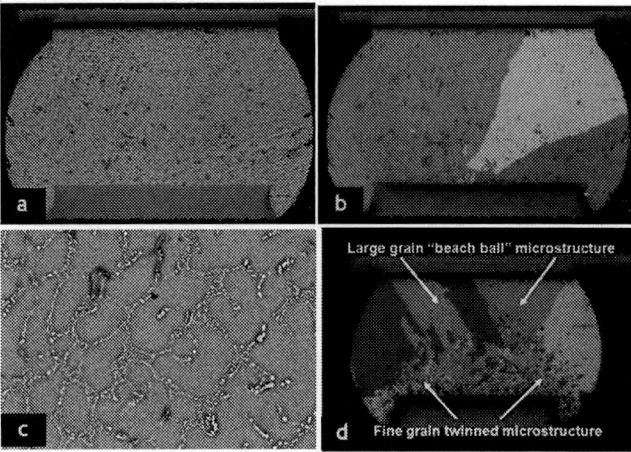

Figure 3: Micrographs illustrating various microstructural features in Pb-free SAC solders.

Within each of the large Sn grains, there are multiple smaller solidification cells or dendrites. In a backscattered image such as Figure 3c, the Sn appears as the gray (background) phase, the Ag_3Sn is the white phase, and the Cu_6Sn_5 is the less dense dark phase. These Sn dendrites tend to solidify with the same crystallographic orientation but in contrast to the large grains, have low angle orientation differences [42]. In most cases there are binary eutectic ($Sn+Ag_3Sn$) regions surrounding the Sn dendrites. These microstructural features can be resolved at higher magnification as shown in Figure 3c. Some Cu_6Sn_5 intermetallic particles may be present at the low angle boundaries or within the Sn dendrites but ternary eutectic decomposition ($L \rightarrow Sn + Ag_3Sn + Cu_6Sn_5$) is less common at these boundaries even in near-eutectic SAC alloys. All three phases can solidify in different morphologies depending on cooling rate, solder volume, surface finish metallization, and nominal base solder composition. The **Introduction** reviewed the complex relationships between microstructure and mechanical properties and the attempts to refine microstructure to improve properties.

Figures 4 through 6 show a series of backscattered electron images of the time zero baseline solder joint microstructures for the eight test vehicle combinations including 192CABGA and 84 CTBGA packages and four Pb-free solder alloys (SAC105, SAC105+Ni, SAC205, and SAC205+Ni). These images show microstructures adjacent to the soldered interface at the component side of the solder joint, in the location where the first fatigue crack generally initiates and propagates to failure. Most of the details of the dendritic microstructure cannot be resolved in the lower magnification micrographs shown in Figure 4 (original magnification 500X), but the Cu-based intermetallic particles (dark phase) can be resolved and appear to be more prominent in both Ni microalloys. Ni microalloying may be increasing nucleation of the $(Cu,Ni)_6Sn_5$ phase as suggested by Seo and El-Daly [24, 25] or it could be promoting growth of those particles by altering undercooling and solidification. Intermetallic spalling was observed in the SAC205 microstructure of the 84CTBGA shown in Figures 4 and 6. Snugovsky et al. have studied this reaction and have indicated that the term spalling is a convenient but incorrect characterization [43]. Spalling actually results from solidification conditions in the bulk solder rather than a reaction causing spalling (separation) from the metallization interface. In previous thermal cycling tests using the same alloy and component, spalling did not have a measureable effect on thermal fatigue life [8].

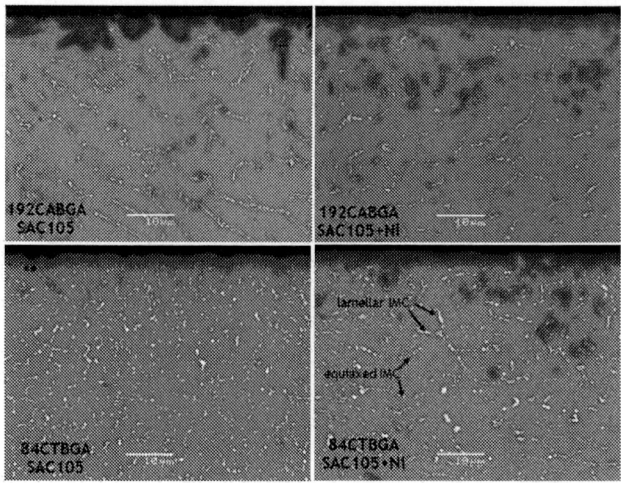

Figure 5: Higher magnification backscattered images showing baseline BGA solder joint microstructures for the 192CABGA and the 84 CTBGA components with SAC105 and SAC105+Ni alloys.

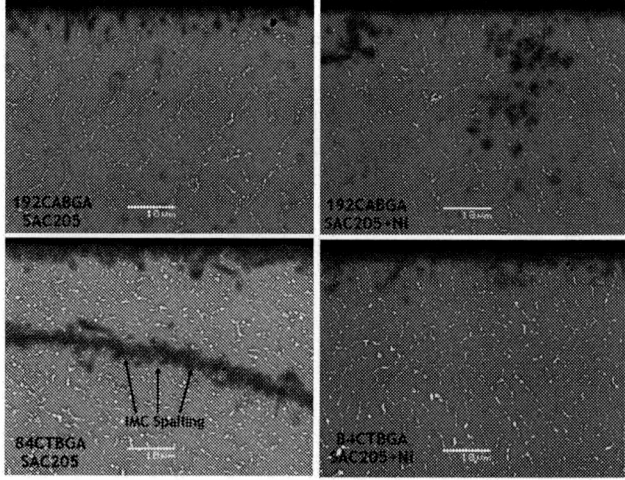

Figure 6: Higher magnification backscattered images showing baseline BGA solder joint microstructures for the 192CABGA and the 84 CTBGA components with SAC205 and SAC205+Ni alloys.

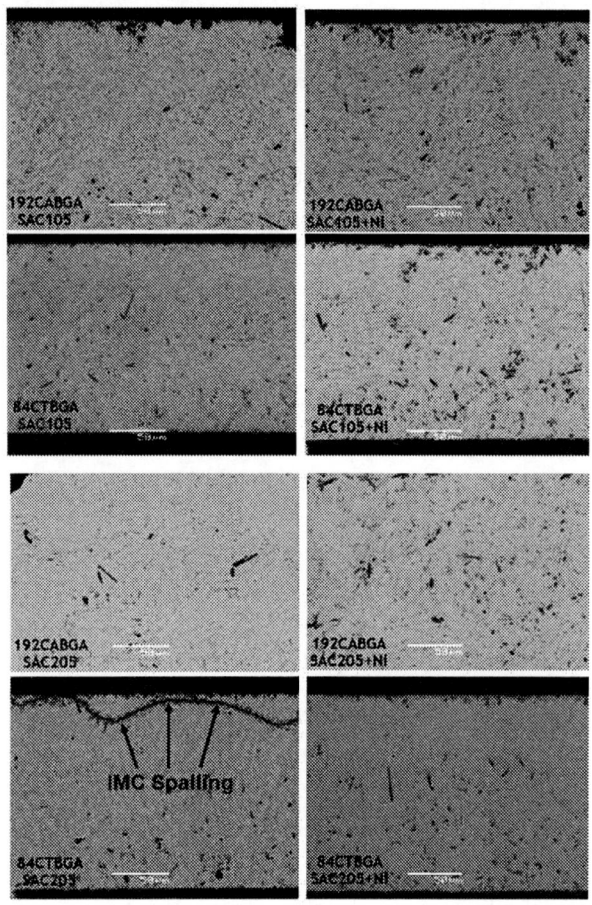

Figure 4: Backscattered electron micrographs showing moderately higher Cu-based IMC particle density (dark particles) in solders microalloyed with Ni. Also note IMC spalling in the 84CTBGA with SAC205 solder.

Figures 5 and 6 show higher magnification backscattered images (original magnification 2000X) of the baseline BGA solder joint microstructures for the 192CABGA and the 84 CTBGA components. When Ni is added to the SAC105 alloy (Figure 5), the solder solidifies with a larger Sn dendrite size with both BGA components. In the case of the SAC105+Ni alloy, there is some Ag_3Sn precipitation in lamellar rather than equiaxed morphology. The average Ag_3Sn particle size also appears to be greater with the Ni addition but extensive quantitative metallography would be required to confirm that observation.

The images in Figure 6 show that a 500 ppm Ni addition to SAC205 results in a slightly larger Sn dendrite size, particularly in the 84CTBGA component. The effects on Ag_3Sn precipitate morphology and particle size in SAC105 are less obvious in SAC205, perhaps due to the larger number of particles with the higher Ag content alloy.

Many of the studies cited in the **Introduction** discuss the improvements in mechanical properties associated with Ni microalloying. Those improvements typically are attributed to microstructural refinement that includes reduced Sn dendrite size and increased precipitation of the Ni-modified intermetallic phase, $(Cu,Ni)_6Sn_5$. The current results appear to contradict those findings since Ni microalloying is found to increase rather than reduce the Sn dendrite size. Fine microstructural features generally improve mechanical properties although Kang has attributed a moderate increase in thermal fatigue life to larger Sn dendrites [44]. It should be noted that the Kang study used a large ceramic BGA with large solder spheres and those results may not be applicable to the current study. The Cu-based IMC particles appear more prominently when Ni is added (Figure 4). However, it does not seem that the small number of particles and the random particle distribution would be conducive to significant strengthening. Unless the Cu-based IMC particles are aligned along the Sn dendrite boundaries, they would not be expected to promote thermal fatigue resistance effectively.

Ni microalloying affected the as-reflowed microstructures of the SAC105 and SAC205 alloys in both components used in the current study. The most prominent effect with Ni was an increase in Sn dendrite size which typically, has been associated with a loss in strength. The measured increases in thermal fatigue reliability of the Ni microalloys (Figures 1 and 2) cannot be explained easily on the basis of the increase in Sn dendrite size, results from the literature, or from known metallurgical structure property relationships. The addition of Ni also seemed to introduce some lamellar precipitation of Ag_3Sn. It has been hypothesized that lamellar Ag_3Sn promotes better fatigue performance by being more resistant to particle coarsening but there is not a large body of work to support this suggestion [32]. In summary, microstructural changes were observed with Ni microalloying, but there is no clear correlation between the observed microstructural features of the Ni-modified alloys and their improved thermal fatigue performance.

The variation in Sn grain morphology revealed with polarized light microscopy (Figures 3b and 3d) is another microstructural feature that is believed to affect thermal fatigue reliability. Solder joints were examined in the outer rows of the components and four types of Sn grain morphologies were identified as presented in Figure 7. Quantitative measurements were made on the individual solder joint microstructures using the following arbitrary criteria to distinguish interlaced morphologies. An interlaced morphology was identified when the interlaced region exceeded more than 50% of the total joint cross sectional area. A partially interlaced morphology was identified when the interlaced region was between 10% and 50% of the total joint area. The other two morphologies were identified as single grain and multiple grain.

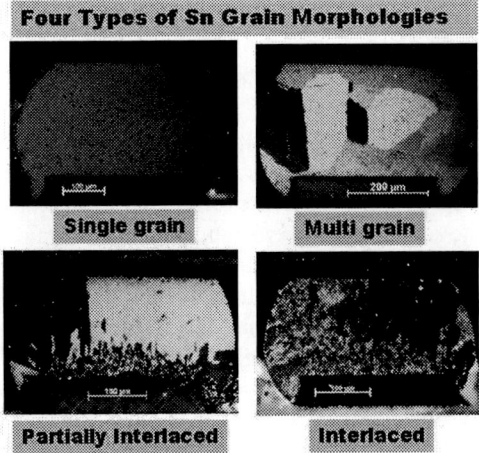

Figure 7: Examples of the four types of Sn grain morphologies identified in the BGA solder joints.

The results of the Sn grain measurements using polarized light microscopy are summarized in the bar charts in Figure 8 for SAC105 and SAC105+Ni and in Figure 9 for SAC205 and SAC205+Ni. Several series of polarized light micrographs are provided in Appendix C. For the 192CABGA, the addition of Ni to SAC105 increases the number of multiple grain joints but produces no interlaced joints. For the same component with Ni added to SAC205, there are no single grain joints, some multiple grain joints, and a significant number of interlaced joints. With the smaller 84CTBGA component, the addition of Ni to SAC105 has no distinguishable effect on the type or distribution of grain morphologies. However, the effect of the Ni microalloy addition is dramatic in this component with the SAC205, where Ni results in more than half of the joints being interlaced and the remainder being partially interlaced and multiple grain.

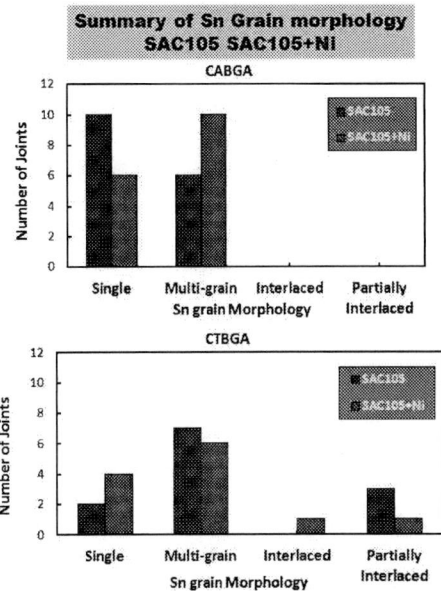

Figure 8: Bar charts summarizing the Sn grain morphology measurements for the SAC105 and SAC105+Ni alloys.

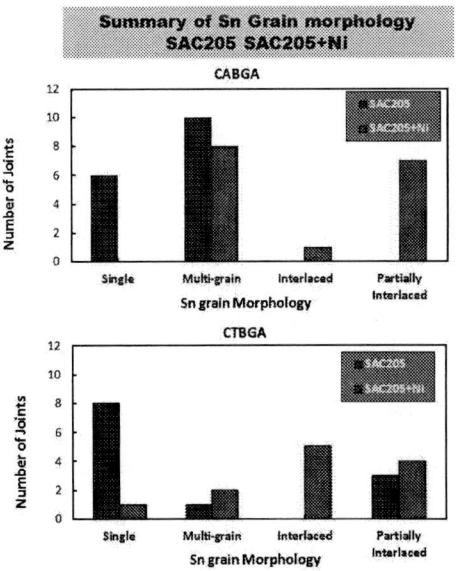

Figure 9: Bar charts summarizing the Sn grain morphology measurements for the SAC205 and SAC205+Ni alloys.

It is evident that the addition of Ni changes the Sn grain morphology of both solders when used with either component. Interlaced twinning appears to be promoted more with the higher Ag content alloy and smaller solder volumes, which is consistent with work published previously by the current authors [36]. There is a greater tendency for the interlaced and partially interlaced morphologies in the 84CTBGA with either solder. Since the most consistent improvement in reliability was observed with the 84CTBGA component (see Figures 1 and 2), there would appear be at least a nominal connection between the addition of Ni, the appearance of the interlaced morphologies, and the increase in thermal fatigue reliability. However, there is an inconsistency in this argument because the large incremental change in morphology in SAC205+Ni did not result in a comparable increase in reliability compared to SAC205 without Ni. On the other hand, SAC105+Ni had a minimal change in morphology but the most significant increase in reliability over the base SAC105 alloy.

To summarize, Ni microalloying produces distinct changes in the solder microstructures and the Sn grain morphologies of SAC105 and SAC205 solders with each BGA component. The microalloy addition of Ni also improves the thermal fatigue reliability of the two solders. However, some inconsistencies in the test data make it difficult to establish direct correlations between the Sn dendrite microstructures and Sn grain morphologies produced by microalloying and the ensuing improvement in thermal fatigue performance. Although it is plausible that the interlaced twinning introduced by Ni microalloying plays a role in improving reliability, it cannot be the sole factor or account for all the experimental observations. The Ag content and the resultant Ag_3Sn IMC morphologies are acknowledged as dominant factors in thermal fatigue reliability of the family of SAC alloys. The analyses of the dendritic microstructures (Figures 5 and 6) indicate some lamellar Ag_3Sn precipitation and a possible increase in average Ag_3Sn particle size with the Ni addition. These microstructural features would be expected to resist fatigue damage and perhaps work in concert with the interlaced morphology to account for the improvement in thermal fatigue performance.

Test Results – Failure Analysis

Metallographic failure analysis was performed to confirm the thermal fatigue failure mode. Only minimal additional post-failure microstructural analysis was performed due to intrinsic limitations imposed by test vehicle design and original experimental plan [34, 36]. Because the test board was populated with both components (see Appendix A) and the characteristic lifetime of the 84CTBGA is substantially longer (e.g., test data in Tables 4 and 5), the 192CABGA components were exposed to a large number of cycles beyond the final failure in that set. Consequently, significantly more fatigue and microstructural damage accumulated throughout the solder ball array of the 192CABGA packages.

The optical photomicrographs in Figures 10 and 11 show examples of thermal fatigue cracking in the 192CABGA and 84CTBGA components selected from two of the thermal cycles. The fatigue cracking is located predominantly at the package side of the solder joints but full fractures have been observed occasionally at the board side. The failed solder joints are characterized by a variety of crack paths and fracture features such as crack branching, recrystallization, and cavitation at boundary triple points. These are common fracture characteristics for SAC alloy thermal fatigue failures and are consistent with those reported first by Dunford [45] and confirmed in previous studies that employed the same test vehicle [8, 37]. In both components, the greatest amount of damage is observed at the die shadow and it is assumed that the initial failures occur most often at the package corners or at the die shadow. The absolute first failure typically cannot be identified unambiguously by the metallographic analysis due to the multiple complete fractures and the overall extent of the thermal cycling damage in the microstructure.

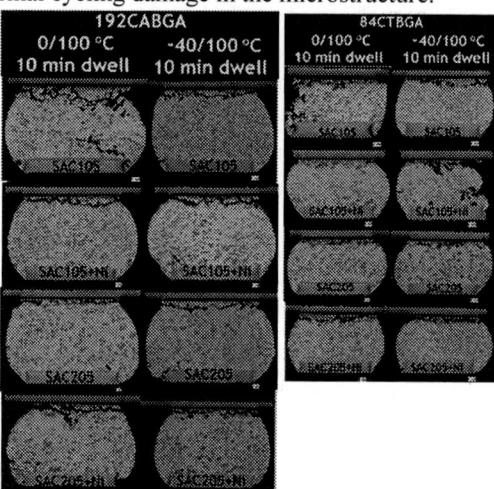

Figure 10: A series of optical photomicrographs showing examples of thermal fatigue cracking in the 192CABGA and 84CTBGA components with the four Pb-free solder alloys. Results are shown for two different thermal cycling profiles.

Intermetallic coarsening (primarily Ag_3Sn) and microstructural evolution in SAC alloys during thermal cycling have been discussed in numerous publications [8-11]. The backscattered electron images in Figure 11 illustrate the accelerated intermetallic particle coarsening in the strain-localized region of the solder joint where fatigue crack propagation occurs. The combination of strain and temperature in this region promotes recrystallization and intergranular fatigue crack propagation.

The electron micrographs shown in Figures 12 and 13 compare the microstructures of baseline (time zero) BGA samples to those that have evolved during thermal cycling tests. The micrographs were captured in the strain localized regions surrounding a fatigue crack or near a propagating crack tip. These images illustrate the dramatic extent of particle coarsening that occurs during cycling. The resultant microstructures in this region contain very few large, coarsened Ag_3Sn particles at the conclusion of cycling. The relatively small critical strain-localized area and the paucity of Ag_3Sn IMC particles present a serious if not impossible challenge for quantifying the microstructural evolution. In some cases, the intermetallic particle coarsening immediately surrounding the crack is strikingly similar regardless of the component type or alloy. Therefore, it is not possible to determine the effect of Ni on intermetallic particle coarsening, recrystallization, or crack propagation based on the current investigated samples.

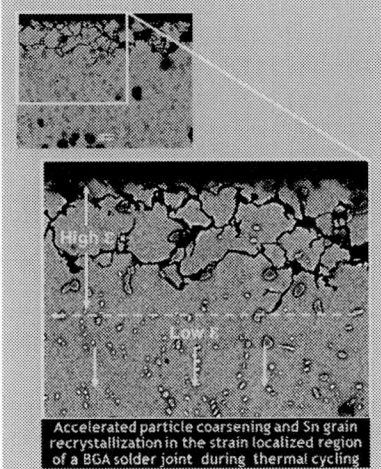

Figure 11: Accelerated intermetallic particle coarsening, recrystallization, and fatigue cracking in the strain-localized region of a SAC solder joint.

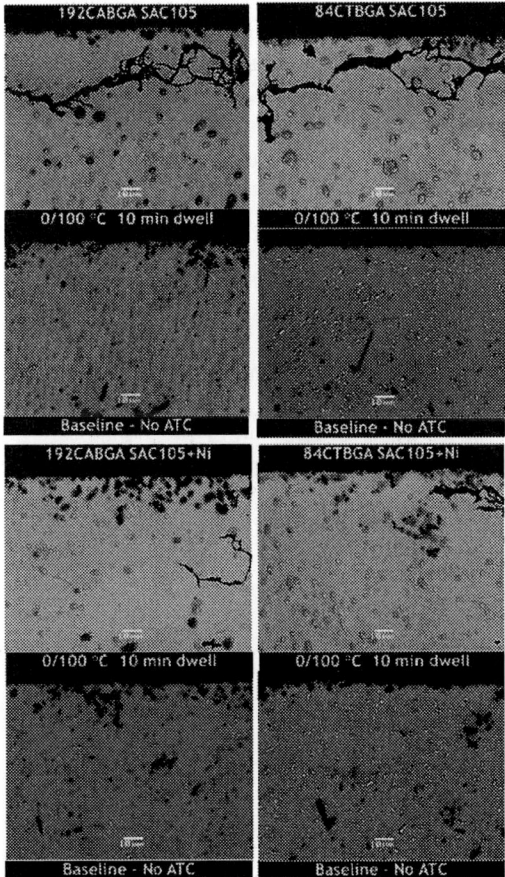

Figure 12: Backscattered electron micrographs comparing SAC105 and SAC105+Ni microstructures of baseline (time zero) BGA samples to those that have evolved during thermal cycling tests. The ATC profile was 0/100 °C with a 10 minute dwell time.

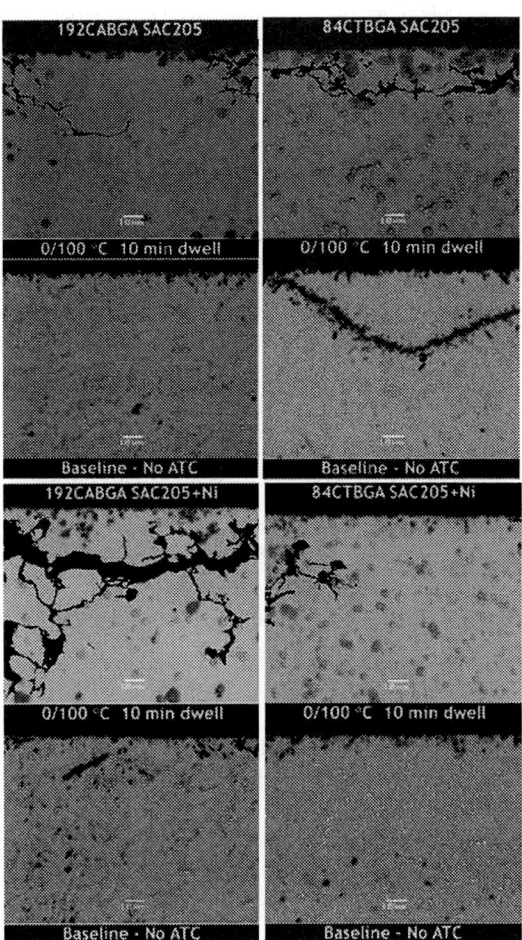

Figure 13: Backscattered electron micrographs comparing SAC205 and SAC205+Ni microstructures of baseline BGA samples to those that have evolved during thermal cycling tests. The ATC profile was 0/100 °C with a 10 minute dwell.

Suggestions for Additional Work

In order to understand the full impact of Ni on thermal cycling performance, it is important to determine the effect of Ni microalloying on all of the microstructural constituents including the Ag_3Sn IMC precipitates, the Cu-based IMC precipitates, and the Sn grain morphology. Quantitative metallography should be performed on the baseline samples used in this study to assess the effect of Ni microalloying on the number of Ag_3Sn precipitates and the average Ag_3Sn precipitate size. It may be difficult to establish a clear correlation due to the inherent differences in effective Ag content shown in Table 2 due to surface mount assembly with the higher Ag content SAC305 solder paste and the difference in solder ball volume between the two components. Therefore, it would be preferable to repeat this experiment using matching solder paste and solder ball compositions to remove any ambiguity introduced by the unequal Ag content in the two components. The quantitative metallographic analysis also should include measurements of the Cu-based intermetallic precipitates to determine if Ni microalloying promotes additional precipitation or growth of the Cu intermetallic phase. The location of those precipitates is important since it has been suggested that they can serve as heterogeneous nucleation sites for β-Sn and subsequently strengthen the Sn matrix by precipitation hardening. The chemical composition of the Cu intermetallic phases should be determined for SAC alloys with and without Ni additions. If more thermal cycling experiments are planned, it would be prudent to increase the sample size so that samples could be removed at intervals prior to complete cracking for analysis of microstructural evolution. The Sn grain analysis using polarized light has limitations and a more quantitative analysis might be accomplished with orientation imaging microscopy (OIM) and electron backscattered diffraction (EBSD).

Although there is some agreement that the optimum Ni level needed to improve the interfacial intermetallic integrity is 500 ppm, comparable data do not exist for optimizing thermal fatigue reliability. The results from many experimental studies in the literature [14-18, 21-29] suggest that the thermal fatigue performance of the SAC alloys should be evaluated with both 500 and 1000 ppm levels of Ni. For completeness, the work should also incorporate quantitative metallographic analysis to explore the effect of Ni content on solidification and microstructure.

In addition to the microstructural and thermal cycling studies suggested above, there are some other materials experiments that could clarify the relationships between microalloying, solidification, microstructure, and reliability. During the initial solder ball attachment to the component substrate and subsequent surface mount reflow assembly, small, unknown amounts of Ni and Au are dissolved from the package surface finish into the solder balls [46]. The magnitude of these unintended and unquantified microalloy additions are not necessarily identical for both components due to the different package attributes. Because these additions are likely to have some contribution to the final microstructure, it would be useful to know the exact Ni content of the solder balls following assembly. Likewise, it would be useful to quantify the increase in Cu content in the solder balls due to dissolution from the PCB pad during reflow soldering. It is an experimental challenge to make accurate quantitative compositional measurements of trace elements on such small samples. The analytical method described by de Sousa et al. [47] requires destructive extraction of the solder balls from the assembly and would appear to require considerable skill and patience. Perhaps there is an alternate, non-destructive spectroscopic technique that could be used to make the measurements.

Numerous published studies show that Ni microalloying can decrease the amount of Sn undercooling, thereby altering the solidification and microstructure. The magnitude of the change in undercooling and the extent of these effects vary widely and at least some of this variation likely is due to the difference in size (mass) of the samples tested [48]. Rather than attempting to interpolate findings from the literature, it would be better to measure the undercooling of the current alloys using the exact solder spheres used in the study.

Conclusions

The microalloy addition of 500 ppm (0.05 wt. %) Ni to SAC105 and SAC205 generally improved the thermal fatigue life of two different BGA components evaluated using multiple accelerated temperature cycling profiles. However, the improvement in fatigue life was not consistent in both alloys, both components, and across all thermal cycling profiles. There was considerable irregularity in the data for the 192CABGA component, where the Ni addition produced a moderate increase in reliability in most SAC105 test cells but no consistent, positive effect on the reliability in the SAC205 test cells. The response to Ni was far more consistent with the 84CTBGA component, which showed improved reliability with the Ni addition in all of the thermal cycles. Overall, the incremental reliability improvement with the Ni addition was greater in SAC105 than it was in SAC205. With few exceptions, the improvement was not affected significantly by extending the dwell time to 60 minutes.

Ni microalloying produced several changes in the as-reflowed microstructures of the SAC105 and SAC205 alloys in both components used in the current study. The Ni addition resulted in a noticeable increase in Sn dendrite size in both solders used with either component, but this finding is not consistent with the observed improvement in thermal fatigue performance, which more often would be associated with a decrease in dendrite size or microstructural refinement. The addition of Ni also introduced some lamellar precipitation of Ag_3Sn which may contribute to better fatigue performance by increasing resistance to particle coarsening. With Ni microalloying, the Cu-based intermetallic particles are more prominent in both alloys but there is no evidence in the literature that Cu-based IMC particles promote improved thermal fatigue performance directly.

There is clear evidence in this study that Ni microalloying increases the frequency of the multiple grain and interlaced twinned morphologies in SAC105 and SAC205 solders when they are used with either component. Interlaced twinning occurs more frequently with the higher Ag content alloy and there is a greater tendency for the interlaced and partially interlaced morphologies in the 84CTBGA with either solder. The most consistent improvement in reliability was observed with the 84CTBGA component, and this suggests a nominal correlation between the addition of Ni, the appearance of the interlaced morphologies, and the increase in thermal fatigue reliability. While it appears that the interlaced twinning introduced by Ni microalloying plays a role in improving reliability, it cannot be the only factor nor can it account for all the experimental observations. The reliability improvement more likely is due to the combined effects of Ni on solidification temperature and thus different aspects of microstructure such as Sn grain morphology, Sn dendritic size, and the resultant Ag_3Sn IMC precipitate size and morphology. Additional work is needed to quantify the effect of Ni microalloying on the Ag_3Sn precipitate size and morphology and on its role in the precipitation of Cu-based intermetallic particles.

References

1. G. Henshall, R. Healy, R. S. Pander, K. Sweatman, K. Howell, R. Coyle, T. Sack, P. Snugovsky, S. Tisdale, and F. Hua, "iNEMI Pb-free Alloy Alternatives Project Report: State of the Industry, *SMT Journal*, vol. 21, no. 4, pp. pp. 11-23, 2008.

2. Dhafer Abdulameer Shnawah, Suhana Bintimohd Said, Mohd Faizul Mohd Sabri, Irfan Anjum Badruddin, and FA Xing Che, "High-Reliability Low-Ag-Content Sn-Ag-Cu Solder Joints for Electronics Applications," Journal of Electronic Materials, vol. 41, No. 9, pp. 2632-2658, 2012.

3. R.F. Smallman and A. H. W. Ngan, Modern Physical Metallurgy, 8th ed., Butterworth- Heinemann, Oxford, UK, 2014.

4. P. Haasen and B. L. Mordike, Physical Metallurgy 3rd ed., Cambridge University Press, Great Britain, 1996.

5. R. M. Brick, R. B. Gordon, and A. Phillips, Structure and Properties of Alloys, McGraw-Hill, New York, NY, 1965.

6. Bruce Chalmers, Physical Metallurgy, John Wiley & Sons, New York, NY, 1959.

7. J. Bartelo, S. Cain, D. Caletka, K. Darbha, T. Gosselin, D. Henderson, D. King, K. Knadle, A. Sarkhel, G. Thiel, C. Woychik, D. Shih, S. Kang, K. Puttlitz and J. Woods, "Thermomechanical Fatigue Behavior of Selected Pb-Free Solders, IPC APEX 2001, LF2-2, January 14-18, 2001.

8. Richard Parker, Richard Coyle, Gregory Henshall, Joe Smetana, and Elizabeth Benedetto, "iNEMI Pb-Free Alloy Characterization Project Report: Part II – Thermal Fatigue Results for Two Common Temperature Cycles," *Proceedings of SMTAI 2012*, pp. 348-358, Orlando, FL, October 2012.

9. Richard Coyle, John Osenbach, Maurice Collins, Heather McCormick, Peter Read, Debra Fleming, Richard Popowich, Jeff Punch, Michael Reid, and Steven Kummerl, "Phenomenological Study of the Effect of Microstructural Evolution on the Thermal Fatigue Resistance of Pb-Free Solder Joints," *IEEE Trans. CPMT*, vol. 1, no. 10, pp. 1583-1593, October 2011.

10. Richard Coyle, Peter Read, Heather McCormick, Richard Popowich, and Debra Fleming, "The Influence of Alloy Composition and Temperature Cycling Dwell Time on the Reliability of a Quad Flat No Lead (QFN) Package," *Journal of SMT*, Vol. 25, Issue 1, pp. 28-34, January-March 2011.

11. Richard Coyle, Heather McCormick, John Osenbach, Peter Read, Richard Popowich, Debra Fleming, and John Manock, "Pb-free Alloy Silver Content and Thermal Fatigue Reliability of a Large Plastic Ball Grid Array (PBGA) Package," *Journal of SMT*, Vol. 24, Issue 1, pp. 27-33, January-March 2011.

12. G. Henshall. J. Bath, S. Sethuraman, D. Geiger, A. Syed, M.J. Lee, K. Newman, L. Hu, D. Hyun Kim, Weidong Xie, W. Eagar, and J. Waldvogel, "Comparison of Thermal Fatigue Performance of SAC105 (Sn-1.0Ag-0.5Cu), Sn-3.5Ag, and SAC305 (Sn-3.0Ag-0.5Cu) BGA Components with SAC305 Solder Paste," *Proceedings APEX*, S05-03, 2009.

13. S. Terashima, Y. Kariya, Hosoi, and M. Tanaka, "Effect of Silver Content on Thermal Fatigue Life of Sn-xAg-0.5Cu Flip-Chip Interconnects," *J. Electron. Mater.* vol. 32, no. 12, 2003.

14. F.X. Che, W.H. Zhu, Edith S.W. Poh, X.W. Zhang, X.R. Zhang, "The study of mechanical properties of Sn–Ag–Cu lead-free solders with different Ag contents and Ni doping under different strain rates and temperatures," *Journal of Alloys and Compounds*, vol. 507, pp. 215–224, 2010.

15. Weiping Liu, Ning-Cheng Lee, Adriana Porras, Min Ding, Anthony Gallagher, Austin Huang, Scott Chen, and Jeffrey ChangBing Lee, "Achieving High Reliability Low Cost Lead-Free SAC Solder Joints Via Mn or Ce Doping" *Proceedings of 59th ECTC 2009*, pp. 994-1007, San Diego, CA, 2009.

16. T. Laurila, V. Vuorinen, J.K. Kivilahti, "Interfacial reactions between lead-free solders and common base materials," *Materials Science and Engineering*, R 49, 1-60, 2005.

17. K.S. Kim , S.H. Huh, and K. Suganuma, "Effects of fourth alloying additive on microstructures and tensile properties of Sn–Ag–Cu alloy and joints with Cu," *Microelectronics Reliability* , vol. 43, pp. 259–267, 2003.

18. K Sweatman, S. Suenaga and T. Nishimura, "Strength of Lead-free BGA Spheres in High Speed Loading" Proceedings Pan Pac, 2008.

19. H. M. Cobb, *Dictionary of Metals*, pp. 145, ASM International, 2012.

20. K. Nogita, J. Read, T. Nishimura, K. Sweatman, S. Suenaga and A. K. Dahle, "Microstructure control in Sn-0.7mass%Cu alloys", Materials Transactions, Vol. 46, No. 11 (2005) 2419-2425.

21. A.E. Hammad, "Evolution of microstructure, thermal and creep properties of Ni-doped Sn–0.5Ag–0.7Cu low-Ag solder alloys for electronic applications," *Materials and Design*, vol. 52, pp. 663–670, 2013.

22. A.E. Hammad, "Investigation of microstructure and mechanical properties of novel Sn–0.5Ag–0.7Cu solders containing small amount of Ni," *Materials and Design*, vol. 50, pp. 108–116, 2013.

23. A.A. El-Daly and A.M. El-Taher, "Evolution of thermal property and creep resistance of Ni and Zn-doped Sn–2.0Ag–0.5Cu lead-free solders," *Materials and Design*, vol. 51, pp. 789–796, 2013.

24. A.A. El-Daly, A.E. Hammad, A. Fawzy, and D. A. Nasrallh, "Microstructure, mechanical properties, and deformation behavior of Sn–1.0Ag–0.5Cu solder after Ni and Sb additions," *Materials and Design*, vol. 43, pp. 40–49, 2013.

25. Sun-Kyoung Seo, Moon Gi Cho, Sung K. Kang, Jaewon Chang, and Hyuck Mo Lee, "Minor Alloying Effects of Ni or Zn on Microstructure and Microhardness of Pb-free Solders," *Proceedings of Electronic Components and Technology Conference*, pp. 84-89, 2011.

26. Toni T. Mattila and Jorma K. Kivilahti , "The Failure Mechanism of Recrystallization – Assisted Cracking of Solder Interconnections," Chapter 8 in **Recrystallization**, Krzysztof Sztwiertnia, ed., pp. 179-206, InTech, March 3, 2012.

27. F.X. Che, W.H. Zhu, Edith S.W. Poh, X.W. Zhang, X.R. Zhang, "The study of mechanical properties of Sn–Ag–Cu lead-free solders with different Ag contents and Ni doping under different strain rates and temperatures," *Journal of Alloys and Compounds*, vol. 507, pp. 215–224, 2010.

28. Fangjie Cheng, Hiroshi Nishikawa, and Tadashi Takemoto, "Microstructural and mechanical properties of Sn–Ag–Cu lead-free solders with minor addition of Ni and/or Co," *Journal of Material Science*, vol. 43, pp. 3643–3648, 2008.

29. K.S. Kim , S.H. Huh, and K. Suganuma, "Effects of fourth alloying additive on microstructures and tensile properties of Sn–Ag–Cu alloy and joints with Cu," *Microelectronics Reliability* , vol. 43, pp. 259–267, 2003 .

30. Hongtao Ma, "Charaterization of Lead-Free Solders for Electronic IC Packaging," PhD Thesis, pp. 25-34, Auburn University, Auburn, Alabama, May 10, 2007.

31. Martin Wickham, Jaspal Nottay, and Christopher Hunt, "A Review of Mechanical Test Method Standards for Lead-Free Solders," NPL Report MATC(A)69, National Physics Laboratory, Teddington, Middlesex, UK, TW110LW, 2001.

32. Richard Coyle, Michael Reid, Claire Ryan, Richard Popowich, Peter Read, Debra Fleming, Maurice Collins, Jeff Punch, and Indraneel Chatterji, "The Influence of the Pb free Solder Alloy Composition and Processing Parameters on Thermal Fatigue Performance of a Ceramic Chip Resistor," *Proceedings of Electronic Components Technology Conference*, pp. 423-430, San Diego, CA, 2009.

33. Shinichi Terashima, Yoshiharu Kariya, and Masamoto Tanaka, "Improvement on Thermal Fatigue Properties of Sn-1.2Ag-0.5Cu Flip Chip Interconnects by Nickel Addition," *Materials Transactions*, vol. 45, no. 3 pp. 673-680, 2004.

34. Gregory Henshall, Jian Miremadi, Richard Parker, Richard Coyle, Joe Smetana, Jennifer Nguyen, Weiping Liu, Keith Sweatman, Keith Howell, Ranjit S. Pandher, Derek Daily, Mark Currie, Tae-Kyu Lee, Julie Silk, Bill Jones, Stephen Tisdale, Fay Hua, Michael Osterman, Bill Barthel, Thilo Sack, Polina Snugovsky, Ahmer Syed, Aileen Allen, Joelle Arnold, Donald Moore, Graver Chang, and Elizabeth Benedetto, "iNEMI Pb-Free Alloy Characterization Project Report: Part I – Program Goals, Experimental Structure, Alloy Characterization, and Test Protocols for Accelerated Temperature Cycling," *Proceedings of SMTAI 2012*, pp. 335-347, Orlando, FL, 2012.

35. IPC-9701A, "Performance Test Methods and Qualification Requirements for Surface Mount Solder Attachments," IPC, Bannockburn, IL, 2006.

36. Richard Coyle, Richard Parker, Michael Osterman, Stuart Longgood, Keith Sweatman, Elizabeth Benedetto, Aileen Allen, Elviz George, Joseph Smetana, Keith Howell, and Joelle Arnold, "iNEMI Pb-Free Alloy Characterization Project Report: Part V – The Effect of Dwell Time on Thermal Fatigue Reliability," *Proceedings of SMTAI 2013*, pp. 470-489, Ft. Worth, TX, October 2013.

37. Richard Coyle, Richard Parker, Gregory Henshall, Michael Osterman, Joe Smetana, Elizabeth Benedetto, Donald Moore, Graver Chang, Joelle Arnold, and Tae-Kyu Lee, "iNEMI Pb-Free Alloy Characterization Project Report: Part IV - Effect of Isothermal Preconditioning on Thermal Fatigue Life," *Proceedings of SMTAI 2012*, pp. 376-389, Orlando, FL, October 2012.

38. Joe Smetana, Richard Coyle, Thilo Sack, Ahmer Syed, David Love, Danny Tu, Steve Kummerl, "Pb-free Solder Joint Reliability in a Mildly Accelerated Test Condition," *Proceedings of APEX 2011*, S28-01, Las Vegas, NV, March 2011.

39. B. Arfaei, L. Wentlent, S. Joshi, A. Alazzam, T. Tashtoush, M. Halaweh, S. Chivukula, L. Yin, M. Meilunas, E. Cotts, and P. Borgesen, "Improving the Thermomechanical

Behavior of Lead Free Solder Joints by Controlling the Microstructure," *Proceedings of ITHERM*, 392-398, 2012.

40. B. Arfaei, L. Wentlent, S. Joshi, M. Anselm, and P. Borgesen, "Controlling the Superior Reliability of Lead Free Assemblies with Short Standoff Height through Design and Materials Selection," *Proceedings of IMECE,* 2012

41. B. Arfaei, N. Kim and E.J. Cotts, "Dependence of Sn Grain Morphology of Sn-Ag-Cu Solder on Solidification Temperature," *J. Electronic Materials,* vol. 41, no. 2, 362-374, February 2012.

42. T. Bieler, H. Jiang, L. Lehman, T. Kirkpatrick, and E. Cotts, "Influence of Sn Grain Size and Orientation on the Thermomechanical Response and Reliability of Pb-free Solder Joints," *Proceedings of Electronic Components and Technology Conference*, pp. 1462- 1467, San Diego, CA, 2006.

43. L. Snugovsky, P. Snugovsky, D.D. Perovic, and J. W. Rutter, "Spalling of SAC Pb free solders when used with Ni substrates," Materials Science and Technology, vol. 25, no. 10, pp. 1296-1300, 2009.

44. S.K. Kang, Paul Lauro, Da-Yuan Shih, Donald W. Henderson, Timothy Gosselin, Jay Bartelo, Steve R. Cain, Charles Goldsmith, Karl J. Puttlitz, and Tae-Kyung Hwang, "Evaluation of Thermal Fatigue Life and Failure Mechanisms of Sn-Ag-Cu Solder Joints with Reduced Ag Contents," *Proceedings of Electronic Components and Technology Conference 2004*, pp. 661-667, Las Vegas, NV, June 1-4, 2004.

45. S. Dunford, S. Canumalla, and P. Viswanadham, "Intermetallic Morphology and Damage Evolution Under Thermomechanical Fatigue of Lead (Pb)-Free Solder Interconnections," *Proceedings of Electronic Components Technology Conference,* pp. 726-736, Las Vegas, NV, June 1-4, 2004.

46. Richard Coyle, Richard Parker, Babak Arfaei, Keith Sweatman, Keith Howell, Stuart Longgood, and Elizabeth Benedetto, "iNEMI Pb-Free Alloy Characterization Project Report: Part VI – The Effect of Component Surface Finish and Solder Paste Composition on Thermal Fatigue of SN100C Solder Balls," *Proceedings of SMTAI 2013*, pp. 490-414, Ft. Worth, TX, October 2013.

47. Isabel de Sousa, Donald W. Henderson, Luc Patry, and Robert Martel, "Implementation of Increased Cu Levels (1%) in SAC Alloys for PBGA Applications," *Proceedings of SMTAI 2008*, pp. 435-443, Orlando, FL, October 17-21, 2008.

48. Sung K. Kang, Moon Gi Cho, Paul Lauro, and Da-Yuan Shih, "Critical Factors Affecting the Undercooling of Pb-free, Flip-Chip Solder Bumps and In-situ Observation of Solidification Process," *Proceedings of Electronic Components Technology Conference,* pp. 1597-1603, Reno, NV, May 29-June 1, 2007.

Appendix A
PCB Test Vehicle Populated with BGA Components

Appendix B
Weibull Plots

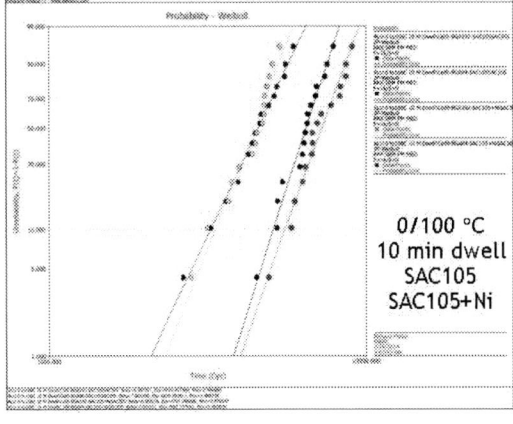

0 / 100 °C
10 min dwell
SAC105
SAC105+Ni

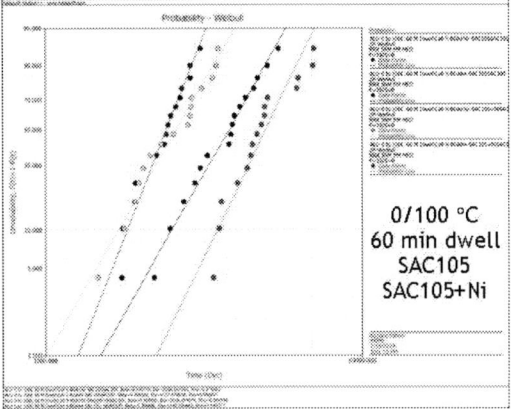

0 / 100 °C
60 min dwell
SAC105
SAC105+Ni

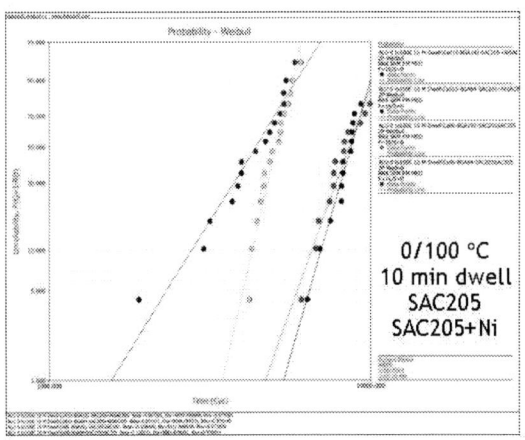

0 / 100 °C
10 min dwell
SAC205
SAC205+Ni

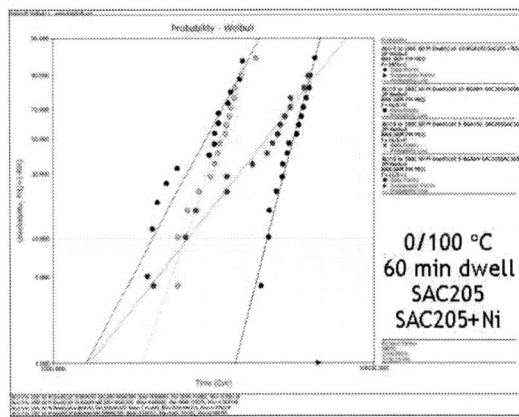

0 / 100 °C
60 min dwell
SAC205
SAC205+Ni

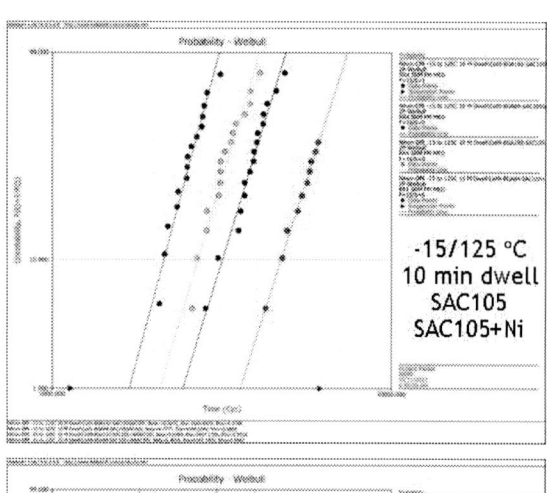

-15/125 °C
10 min dwell
SAC105
SAC105+Ni

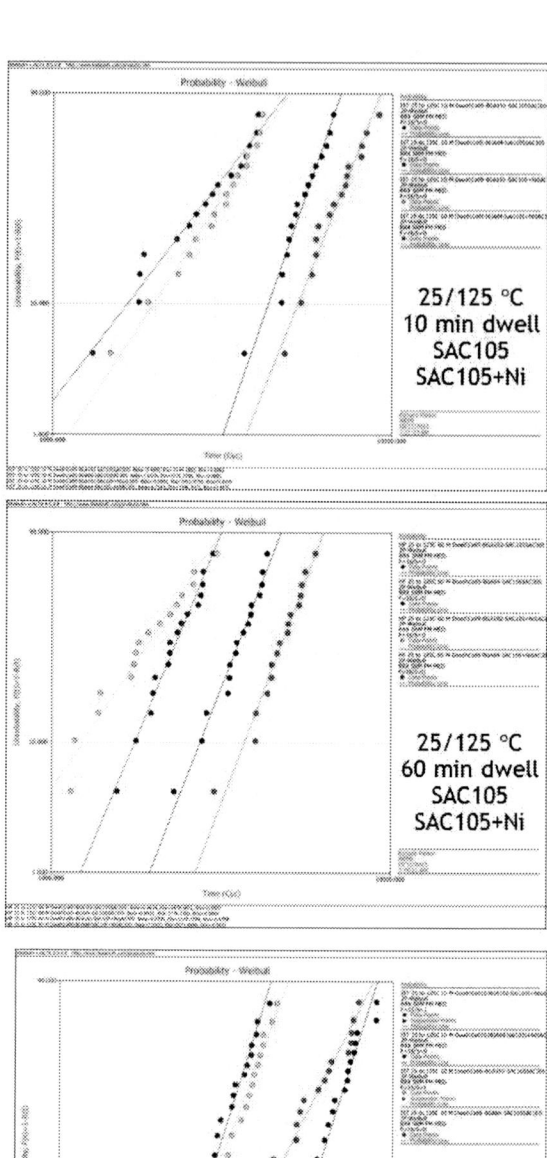

25/125 °C
10 min dwell
SAC105
SAC105+Ni

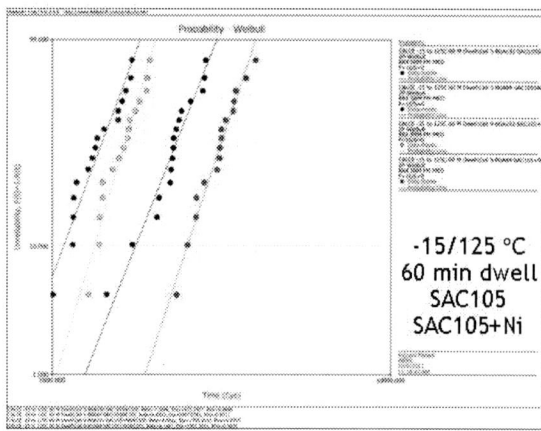

-15/125 °C
60 min dwell
SAC105
SAC105+Ni

25/125 °C
60 min dwell
SAC105
SAC105+Ni

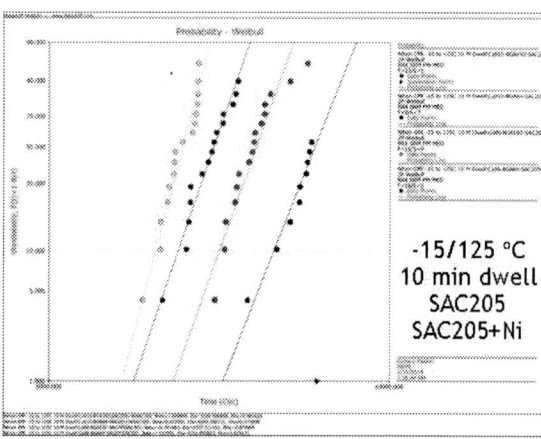

-15/125 °C
10 min dwell
SAC205
SAC205+Ni

25/125 °C
10 min dwell
SAC205
SAC205+Ni

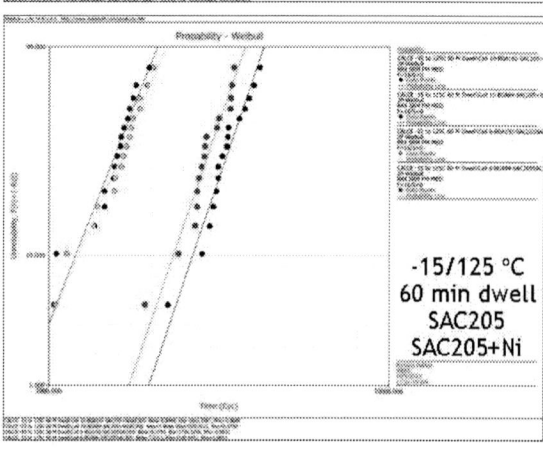

-15/125 °C
60 min dwell
SAC205
SAC205+Ni

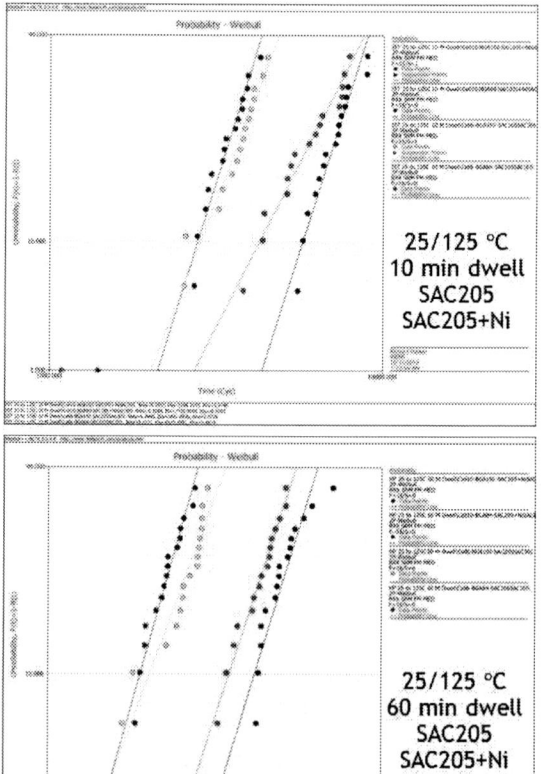

25/125 °C
60 min dwell
SAC205
SAC205+Ni

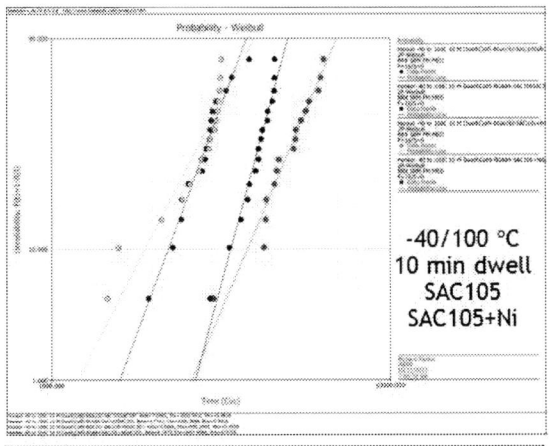

-40/100 °C
10 min dwell
SAC105
SAC105+Ni

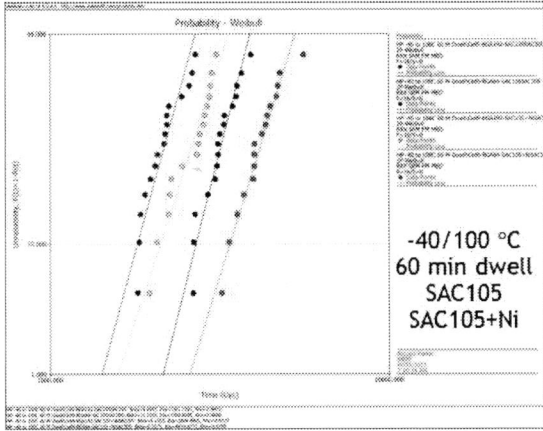

-40/100 °C
60 min dwell
SAC105
SAC105+Ni

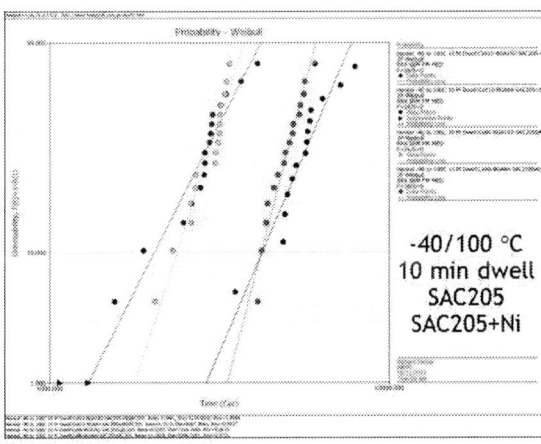

-40/100 °C
10 min dwell
SAC205
SAC205+Ni

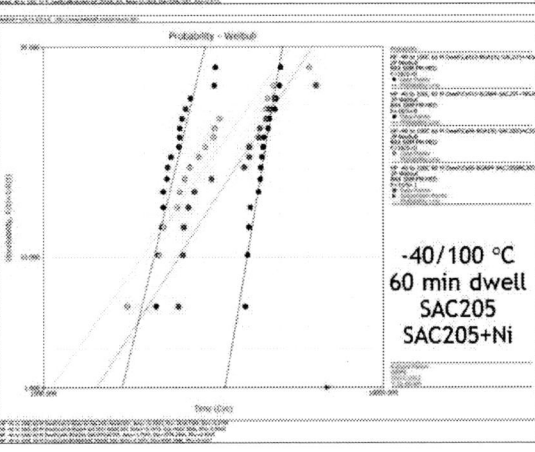

-40/100 °C
60 min dwell
SAC205
SAC205+Ni

Appendix C
Polarized light Images of Sn Grain Morphology

192CABGA SAC105 - as reflowed, no ATC

- More than 60% of the joints are single grain
- About 40% are multi grain
- No partially interlaced or interlaced joints observed

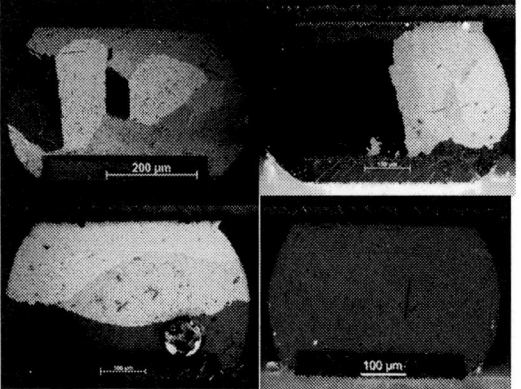

192CABGA SAC105+Ni - as reflowed, no ATC

- More than 60% of the joints are multi grain
- About 40% are single grain

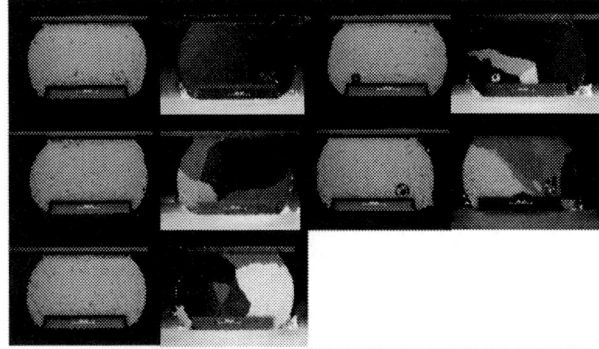

84CTBGA SAC105 - as reflowed, no ATC

- 58% of the joints are multi grain
- About 25% are partially interlaced, the rest are single grains
- Partial interlacing was observed on component and board sides

84CTBGA SAC105+Ni - as reflowed, no ATC

- More than 50% of the joints are multi grain
- The rest of the joints are either single grain, interlaced or partially interlaced

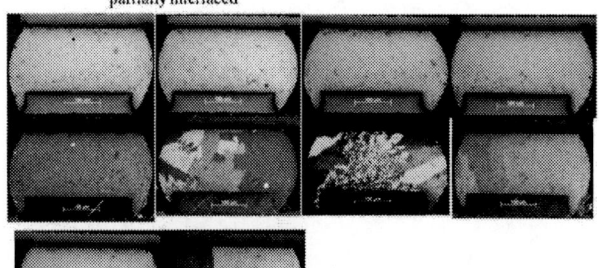

978-1-4799-2408-0/14 $31.00 © 2014 IEEE

192CABGASAC205 - as reflowed, no ATC

- All joints are either single or multi grain
- No interlaced or partially interlaced joints observed

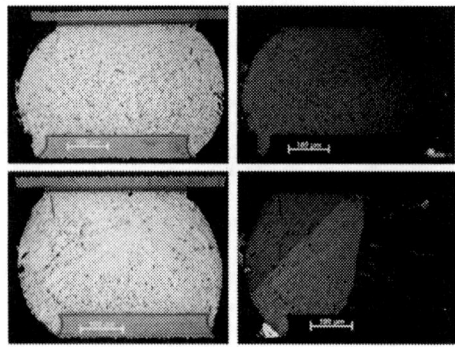

192CABGASAC205+Ni - as reflowed, no ATC

- No single grains are observed
- All joints observed are either multi grain, partially interlaced or interlaced
- More than 50% of the observed joints are partially interlaced

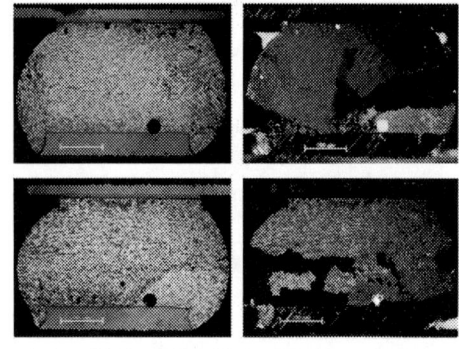

84CTBGASAC205 - as reflowed, no ATC

- More than 66% of the joints are single grain
- A few joints are either interlaced or partially interlaced

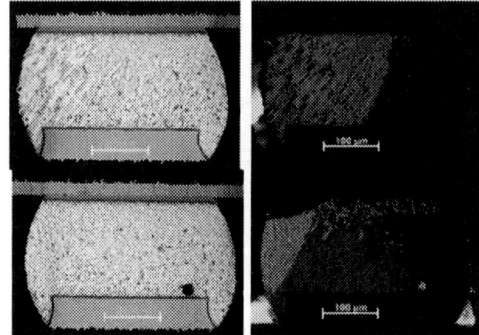

84CTBGASAC205+Ni as reflowed, no ATC

- More than 80% of the joints are either partially interlaced or interlaced and the rest are multigrain
- Less than 10% are single grain

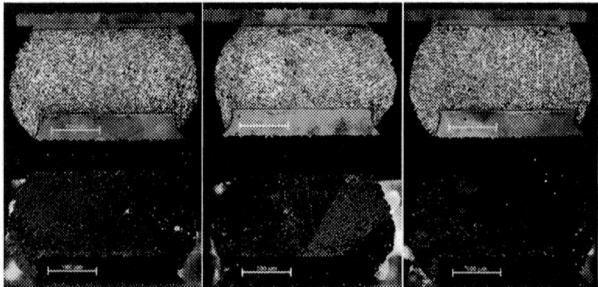

Flexible Non-volatile Cu/Cu$_x$O/Ag ReRAM Memory Devices Fabricated Using Ink-jet Printing Technology

Simin Zou and Michael C. Hamilton

Electrical and Computer Engineering, Auburn University, AL 36849

E-mail: mchamilton@auburn.edu

Phone: 334-844-1879

Abstract

Flexible electronic devices are an emerging kind of electronics in a technological field that is attracting an increasing amount of attention. Flexible resistive random access memory (ReRAM) devices have great potential to replace conventional nonvolatile RAM. Here, we have investigated ink-jet printed Cu/Cu$_x$O/Ag ReRAM devices fabricated on flexible Kapton substrate with a 20μm×20μm cell size. The electroformed memory cells exhibit stable bipolar resistive switching (BRS) behavior under low-range direct current sweep in the temperature range from 255K to 355K. The Cu/Cu$_x$O/Ag ReRAM devices switch to high resistance state (HRS) and low resistance state (LRS) at opposite polarities of the applied voltage bias. Furthermore, the ReRAM device has excellent switching endurance and data retention performance and has ability to operate well even after over 1000 flexes. The good ductility of ink-jet printed silver and electroplate copper electrodes and simple cross-point structure of the memory cell result in excellent flexibility and mechanical robustness, indicating great potential for future flexible nonvolatile RAM applications.

Introduction

With a rapid development of information technology, high-speed high-data-density low-power inexpensive ReRAM devices have attracted wide attention due to their functional and powerful nonvolatile memory applications, neuromorphic and biological circuit applications, signal processing and programmable logic applications [1]. Resistive switching (RS) behavior was observed in a large variety of material systems, including organic materials [2], ferromagnetic materials [3], perovoskite oxides [4] and binary metal oxides such as HfO$_x$ [5]-[6], TiO$_x$ [7]-[9], NiO [10], TaO$_x$ [11], ZnO [12] and Al$_2$O$_3$ [13]. The ReRAM devices reported have one or several excellent properties, such as high HRS/LRS ratio [9]-[12], high speed switch [14], long retention [15], high switch endurance [11], low-voltage operation [14], low power operation [9], small cell area [16], and multibit operation [17]. Nowadays, the demand for flexible ReRAM [9], [18]-[19] has increased due to their advantages of foldability, human-friendly contact, portability and wearability. On the fabrication level, flexible electronic systems have merits such as low-temperature fabrication and processing, as well as simple and inexpensive manufacturing process.

A ReRAM device consists of a two-terminal metal-oxide-metal (MOM) sandwich cross structure. There are two kinds of resistive switching behaviors have been reported in ReRAM, which are unipolar resistive switching (URS) and bipolar resistive switching (BRS). The switching direction of URS depends on applied voltage amplitude which means it shows SET and RESET at the same polarity [8].

Alternatively, BRS depends on the polarity of applied voltage, and SET and RESET occur at opposite polarities.

In this study, we present characterization of the BRS in Cu/Cu$_x$O/Ag (1<x<2) memory devices for flexible ReRAM applications. This memory cell is based on copper and silver, which are relatively abundant in natural world. The fabricated ReRAM devices with the characteristics: high resistive switching repeatability of more than 500 times, low-voltage operation (less than 1 V), reliable switching endurance of over 100 cycles, data retention for nearly 2 weeks, low-temperature fabrication, HRS/LRS ratio of greater than 20 and operated well after being flexed over 1000 times. Moreover, we use ink-jet printing technology during the fabrication process which has extensive potential for roll-to-roll processing in the industrial production. In addition, based on the electrical experimental results, the physical mechanism of BRS characteristics of the device is also discussed.

Experimental Details

A photograph of fabricated Cu/Cu$_x$O/Ag memory cells is given in Fig. 1. The inset shows one single memory cell test circuit.

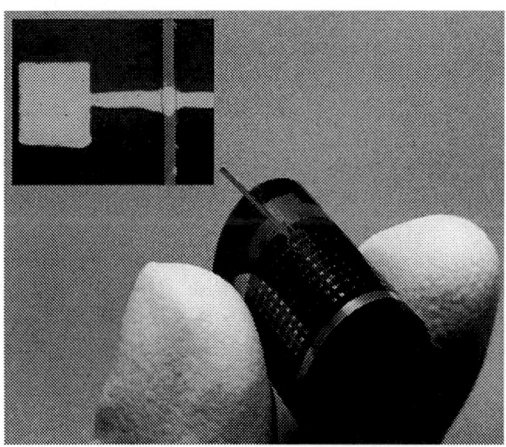

Figure 1: Cu/Cu$_x$O/Ag ReRAM devices fabricated on flexible Kapton substrate. The inset shows a single memory cell.

The BRS ReRAM device was fabricated as a two-terminal crossbar MOM capacitor structure, where the Cu$_x$O was used as a thin oxide layer of the capacitor. A Ag nano-particle seedlayer was ink-jet printed on Dupont Kapton polyimide substrate. After thermally curing the printed sample at 200 °C for 2 h, the sample was electroplated in copper plating solution to form a Cu layer with a thickness of approximately 1 μm. To prepare the thin oxide film, the Cu coated sample was placed on the hot plate to anneal the structure at a

978-1-4799-2408-0/14 $31.00 © 2014 IEEE   441   2014 Electronic Components & Technology Conference

temperature of 200 °C for 3 h in air. To electrically contact Cu_xO active area, another terminal Ag top electrode (TE) was formed by printing a second Ag nano-particle layer. Next, a 1% HCl solution was used to remove the oxide in areas away from the Ag TEs to open contact areas to the Cu bottom electrode (BE). More fabrication details could be found in our previous work [20]

The memory cell has a cell size of 20 μm×20 μm, which is the smallest deposit feature limit of the cartridge. In the cell-size dependent test, different cell-size memory devices have also been fabricated in the same process. The ink-jet printer we used is FUJIFILM DMP (Dimatix Material Printer) 2831 with a piezoelectric print head cartridge. Fig. 2 shows the photograph of the ink-jet printer.

Figure 2: FUJIFILM Dimatix Material Printer.

Result and Discussion

There were previous publications of the copper oxide resistive switching characteristics [21]. However, we fabricated $Cu/Cu_xO/Ag$ ReRAM with an ultrathin oxide layer, which was estimated to be approximately 23 Å. Fig. 3 shows the X-ray photoelectron spectroscopy (XPS) of the original Cu_xO film and the same sample after sputter etching for 2 min and 4 min. In addition, the XPS surface elemental composition versus sputter time is shown in Table 1. The high resolution scans over the Cu2p peak in the black line indicates a native oxide state, which is a CuO over Cu and this also proved by the roughly 50/50 surface element composition, indicating the initial surface is CuO on top of Cu metal. During sputter etching (etch rate = 25 Å/min), there is a transition from CuO to Cu_2O/Cu, shown by the Cu2p shape in the red line and the blue line. There is probably a very thin layer of Cu_2O at the interface. The approximate 75/25 atomic ratio of Cu and O probably indicates that the analysis volume at this depth samples both Cu and Cu_2O. Since sputtering induces damage, artifacts, and can reduce the oxide, a more reliable method to measure the thickness of the CuO overlayer is to use an algorithm used in surface studies based on the attenuation and inelastic mean free path of electrons as their proceed through the oxide overlayer [22] [23]. The calculated thickness of Cu_xO is 23 Å. Fig. 4 shows the SEM image of a cross-sectional view of $Cu/Cu_xO/Ag$ MOM memory device fabricated on Kapton substrate.

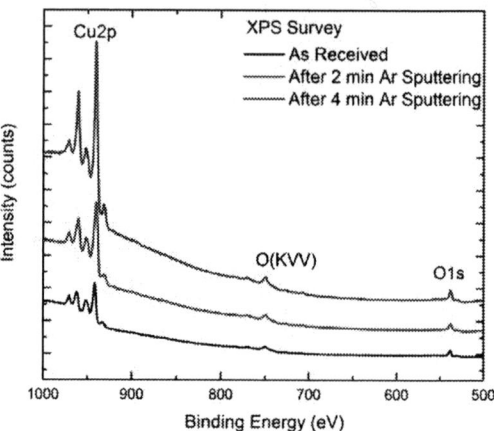

Figure 3: XPS spectra of Cu_xO film before and after 2 min/4min sputter etching.

Table 1: XPS surface elemental composition (at%) versus sputter time.

Time	Cu	O
0	47.9	52.1
2 min	63.7	36.3
4 min	69.6	30.4

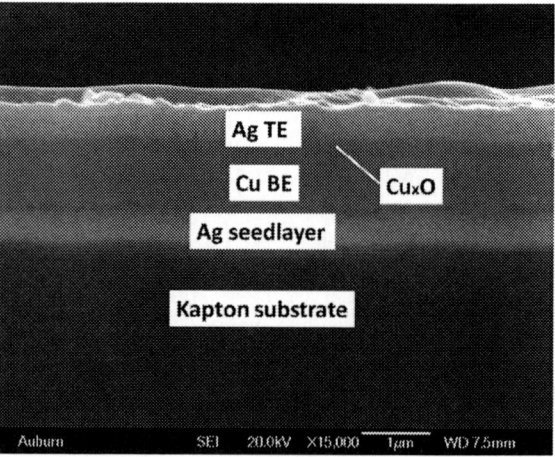

Figure 4: SEM image of the cross section of the $Cu/Cu_xO/Ag$ ReRAM device.

A typical current-voltage (I-V) characteristic on a semi-log scale of BRS in $Cu/Cu_xO/Ag$ MOM memory cell is shown in Fig. 5. An initial electroforming process was performed. This pretreatment process is believed to activate resistive switching in the switching material layer and lead to create the conductive objects (e.g. filaments) [24]. The electroforming process is labeled as (1) in the Fig. 5. By applying a voltage sweep from 0 V to +2 V as the arrow showed, a large increase (corresponding to a dramatic reduction in resistance) occurs in the memory cell's current at approximately +1.8 V (electroforming voltage), indicating that strong 'electrical stress' may lead to the formation of conducting filaments in the insulting matrix. The arrows in the figure indicate the voltage sweeping direction (-1 V→ 0 V→ +1 V→ 0 V→ -1 V), and the I-V hysteresis curve is obtained. During the voltage scanned from -1 V to -0.7 V (SET voltage), a sharp current decrease is observed before step (2), which is called the SET process, resulting in a significant change of resistance in the ReRAM memory cell from low resistance state (LRS) to high resistance state (HRS). Next, when the sweeping voltage increases to +0.7 V (RESET voltage), a rapid increase of current is observed at the end of step (3), which is called RESET process. During RESET process the resistance of the ReRAM device is switched from HRS to LRS. The magnitude of both the SET voltage and RESET voltage are less than 1 V, which shows the low-voltage operation feature of the $Cu/Cu_xO/Ag$ memory devices.

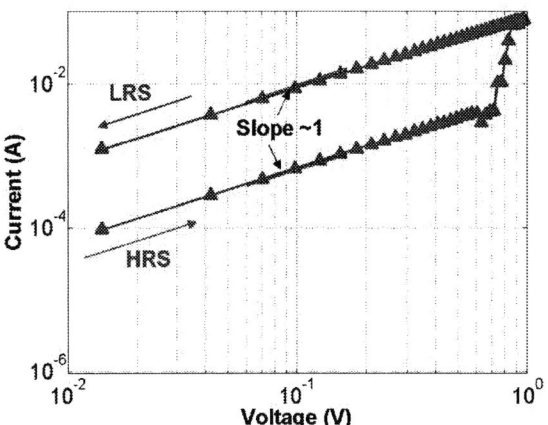

Figure 6: Logarithmic plot of I-V characteristic of positive voltage region.

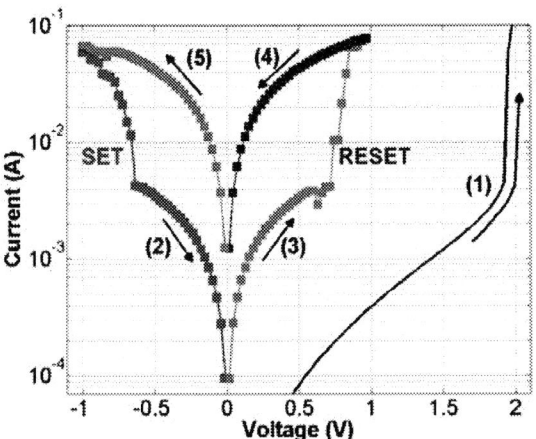

Figure 5: Electrical forming process and I-V characteristics of the $Cu/Cu_xO/Ag$ ReRAM device cell measured by double DC voltage sweep from -1V to +1 V and back to -1V.

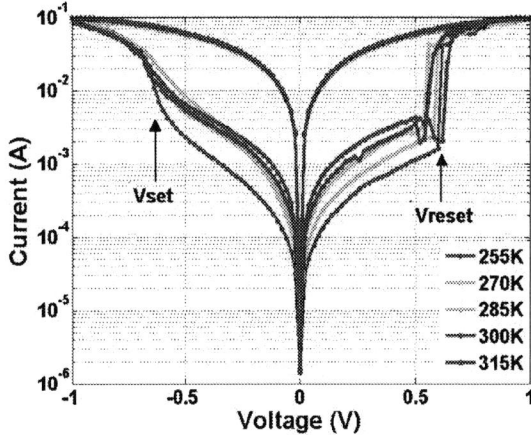

Figure 7: Semi-log scale of I-V curves at temperature range from 255K to 315K with 15K steps.

To explore the electrical-conduction mechanism of $Cu/Cu_xO/Ag$ memory devices, the double logarithmic plot of I-V curves for the positive region was replotted, and is shown in Fig. 6. As indicated in Fig. 6, in the positive voltage bias region, starting from 0V, the I-V behavior of $Cu/Cu_xO/Ag$ memory cell in HRS is very close to linear and the linear dependence for applied voltage is smaller than +0.60 V. In this low-voltage region, the slope of I-V curve is very close to 1, in agreement with Ohm's law. Next, a dramatic increase of current is observed at approximately +0.7 V (V_{RESET}) due to the formation of conductive filaments through the device [24],[25], which does not follow space charge limited current (SCLC) theory [26]. Finally, the Ohmic conduction behavior was observed again in the decreasing voltage bias scan in the LRS. In addition, temperature dependence of BRS characteristics after electroforming is investigated, which is shown in Fig. 7. There is no strong temperature dependence on the switching voltages V_{SET} and V_{RESET}, which are approximately -0.7 V and +0.7 V respectively.

To further clarify the resistive switching mechanism of $Cu/Cu_xO/Ag$ memory cells, cell-size dependent test is performed. Fig. 8 shows the cell-size dependence of the HRS and LRS of $Cu/Cu_xO/Ag$ memory cells. Different cell-size $Cu/Cu_xO/Ag$ memory devices were fabricated in the same condition. The cell sizes were from $20{\times}20~\mu m^2$ to $800{\times}800$ μm^2 All devices show reproducible BRS behavior during DC voltage scan. The best linear fitting lines are also shown in the figure. With the increase of cell size, the ratio of HRS/LRS deceases roughly from 30 to 10. The resistance of HRS decreases from approximately 264 Ω to 46 Ω, whereas the resistance of LRS only shows slight dependence of cell sizes, decreasing from 8.32 Ω to 4.46 Ω. The result reveals that the resistive switching behavior is a local phenomenon in the $Cu/Cu_xO/Ag$ memory cell, dominated by the local conductive filament path, further indicating the physical mechanism of the $Cu/Cu_xO/Ag$ MOM cross-point structure is attributed to the formation and rupture of the conductive filament path.

978-1-4799-2408-0/14 $31.00 © 2014 IEEE

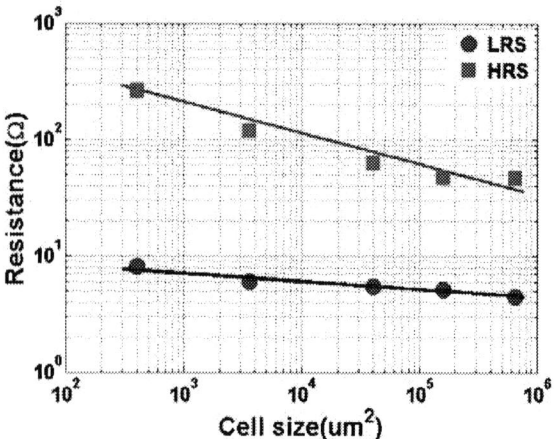

Figure 8: Resistance of HRS and LRS versus cell size of Cu/Cu$_x$O/Ag ReRAM devices.

Fig. 9 shows the data endurance at 295 K. For the switching cycle measurements, -0.8 V and +0.8 V were applied to the device as the SET voltages and RESET voltages, respectively, switching the device to HRS and LRS. The resistances of HRS and LRS were recorded at low voltage range from -0.01 V to +0.01 V. During 100 switching cycles, resistance value of HRS is approximately 120 Ω and LRS is around 6 Ω with slight fluctuations. The resistance value of HRS and LRS remained stable with a ratio of 20. Fig. 10 shows the retention characteristics of the device at room temperature. After switching the device to one resistive state (HRS or LRS), no electrical power was needed to maintain the resistance in this state. As the Fig. 11 shown, the resistance at both HRS and LRS were remained stable for nearly 2 weeks, which further proved the nonvolatile nature of the Cu/Cu$_x$O/Ag ReRAM device.

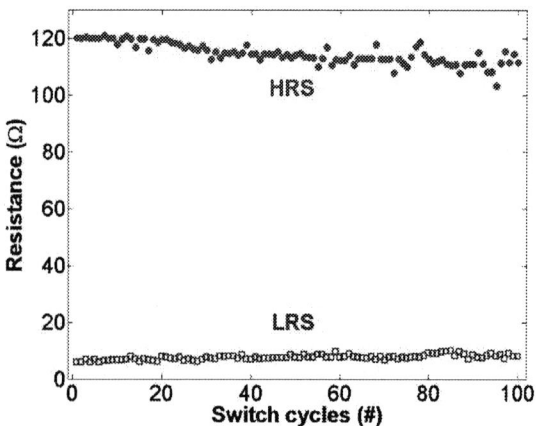

Figure 9: Switching endurance test of the Cu/Cu$_x$O/Ag ReRAM device during 100 cycles. The resistances of HRS and LRS were read from -0.01 V to +0.01 V.

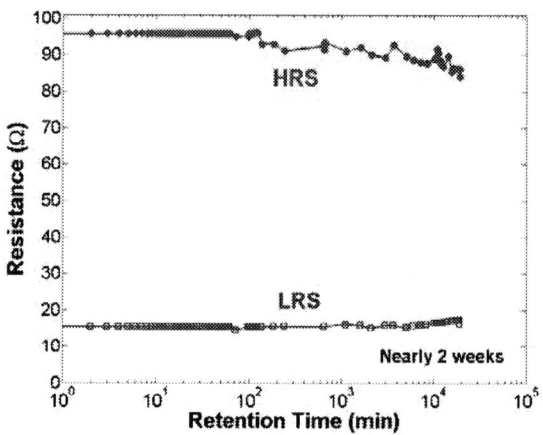

Figure 10: Data retention performance of the Cu/Cu$_x$O/Ag ReRAM device. Both the resistance at HRS and LRS are remained stable nearly 2 weeks.

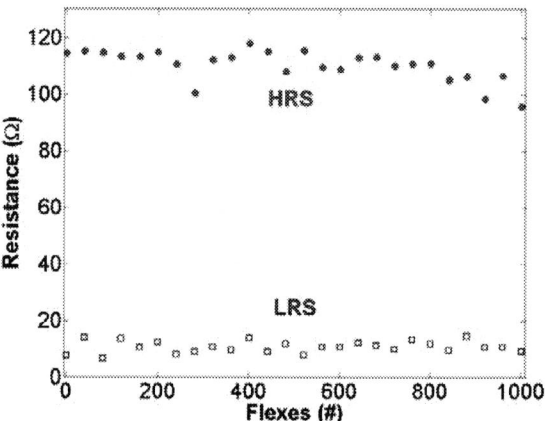

Figure 11: Resistance of HRS and LRS of the Cu/Cu$_x$O/Ag ReRAM memory cell during 1~1000 flexes.

Good electrical characteristic under bending condition is crucial for practical flexible electronic devices. This required property of our Cu/Cu$_x$O/Ag memory cells was studied in detail. The statistical data of mechanical robustness test for the Cu/Cu$_x$O/Ag ReRAM memory cell is shown in Fig. 11. The device was physically flexed for over 1000 cycles. For one flex cycle, the 25-mm-long device was bent into a convex shape and then to a concave shape with a radius of 8mm (convex shape shown in Fig. 1). During 1000 flex cycles, the ratio of HRS and LRS was approximately 20 and remained relatively stable. Moreover, we investigated the HRS/LRS ratios as a function of the bending radius of our Cu/Cu$_x$O/Ag memory cells. The device was bent to form a convex shape with different radius of 55 mm, 35 mm, 26 mm, 17.5 mm, 11.5 mm and 5.5 mm and the device were measured while being flexed, which shows in the Fig. 12. The devices exhibit stable I-V hysteretic behavior under bending condition. With the increase of curvature, the resistances at HRS and LRS remained stable with a ratio of greater than 7, which is shown in Fig. 13 (a). In particular, we investigated the distribution of

SET voltages and RESET voltages according to the bending radius, which is shown in Fig. 13 (b). The V_{SET} was at a range from -0.58 V to -0.73 V and the V_{RESET} was from +0.64 V to +0.79 V. There was no overlap between SET voltages and RESET voltages observed during the bending process, indicating the device can be controlled precisely and operated well at low voltage, even under significant flexing conditions.

Conclusions

In conclusions, the characteristics of the $Cu/Cu_xO/Ag$ BRS RRAM device have been studied and presented. These results are in line with the suggested resistive switch mechanism of $Cu/Cu_xO/Ag$ ReRAM devices is attributed to the local conductive filament paths. The observed trend of increased ratio of HRS to LRS as the cell area decreases can be regarded as a benefit of device scaling. Moreover, the memory devices were fabricated based on Cu and Ag, which are abundant and easy to process. The fabricated devices exhibited reproducible and reliable BRS characteristics, excellent switching endurance and mechanical robustness properties, as demonstrated by the flex-testing results presented in this paper. The simple, flexible inexpensive low-temperature-fabricated low-voltage operation ink-jet printed $Cu/Cu_xO/Ag$ memory cell is expected to provide opportunities for flexible memory electronics and neuromorphic electronic devices.

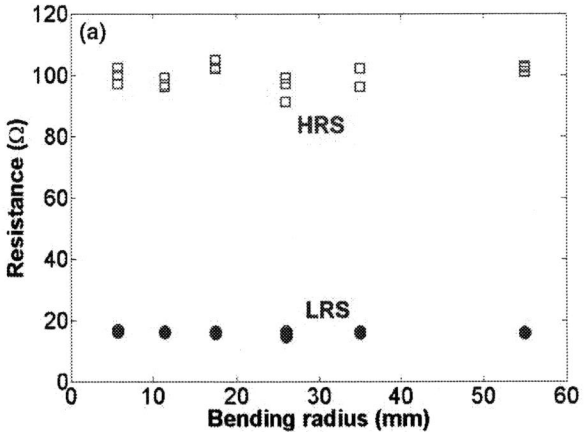

Figure 12: Photograph of the device bent at radius = 11.5mm.

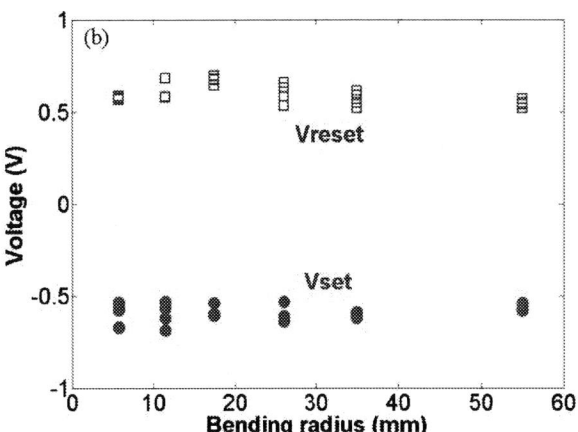

Figure 13(a): Resistance of HRS and LRS of the $Cu/Cu_xO/Ag$ ReRAM memory cell as a function of the bending radius. (b) V_{SET} and V_{RESET} of the $Cu/Cu_xO/Ag$ ReRAM memory cell as a function of the bending radius.

Acknowledgments

The authors would like to thank Dr. Michael Bozack for his help with the X-ray photoelectron spectroscopy (XPS) analysis for the Cu_xO thin film.

References

1. Jo, S. H., Chang, T., Ebong, I., Bhadviya, B. B., Mazumder, P., and Lu, W., "Nanoscale memristor device as synapse in neuromorphic systems," *Nano Lett.*, vol.10, no.4, pp. 1297-1301, March 2010.

2. Verbakel, F., Meskers, S. C., Janssen, R. A., Gomes, H. L., Colle, M., Buchel, M., and de Leeuw, D. M., "Reproducible resistive switching in nonvolatile organic memories," *Appl. Phy. Lett.*, vol. 91, no. 19, pp. 192103-192103, November 2007.

3. Liu, S. Q., Wu, N. J., and Ignatiev, A., "Electric-pulse-induced reversible resistance change effect in magnetoresistive films," *Appl. Phy. Lett.,* vol. 76, no. 19, pp. 2749-2751, May 2000.

4. Beck, A., Bednorz, J. G., Gerber, C., Rossel, C., and Widmer, D., "Reproducible switching effect in thin oxide films for memory applications," *Appl. Phy. Lett.,* vol. 77, no. 1, pp. 139-141, July 2000.

5. Chen, Y. S., et al., "Highly scalable hafnium oxide memory with improvements of resistive distribution and read disturb immunity," *IEEE International Electron Devices Meeting (IEDM),* Baltimore Maryland, December 2009, pp. 1-4.

6. Lee, H. Y., P. S. Chen, T. Y. Wu, Y. S. Chen, C. C. Wang, P. J. Tzeng, C. H. Lin, F. Chen, C. H. Lien, and M-J. Tsai. "Low power and high speed bipolar switching with a thin reactive Ti buffer layer in robust HfO_2 based RRAM," *IEEE International Electron Devices Meeting (IEDM),* San Francisco, CA, pp. 15-17 December 2008, pp. 1-4.

7. Rohde, C., et al., "Identification of a determining parameter for resistive switching of TiO_2 thin films," *Appl. Phy. Lett.,* vol. 86, no. 26, pp. 262907-262907-3, June 2005.

8. Jeong, D. S., Schroeder, H., and Waser, R., "Coexistence of Bipolar and Unipolar Resistive Switching Behaviors in a Pt/TiO$_2$/Pt Stack," *Electrochemical and solid-state Lett.,* vol. 10, no. 8, G51-G53, 2007.

9. Gergel-Hackett, N., Hamadani, B., Dunlap, B., Suehle, J., Richter, C., Hacker, C., and Gundlach, D., "A flexible solution-processed memristor," *IEEE Electron Device Lett.,* vol. 30, no. 7, pp. 706-708, July 2009.

10. Huang, Y. C., Chen, P. Y., Chin, T. S., Liu, R. S., Huang, C. Y., and Lai, C. H., "Improvement of resistive switching in NiO-based nanowires by inserting Pt layers," *Appl. Phy. Lett.,* vol. 101, no. 15, pp. 153106-153106, October 2012.

11. Yang J J, Zhang M, Strachan J P, Miao F, Pickett M D, Kelley R D, Medeiros-Ribeiro G and Williams R S., "High switching endurance in TaO$_x$ memristive devices," *Appl. Phys. Lett.,* vol. 97, pp. 232102, December 2010.

12. Zhang, J., Yang, H., Zhang, Q. L., Dong, S., and Luo, J. K., "Bipolar resistive switching characteristics of low temperature grown ZnO thin films by plasma-enhanced atomic layer deposition," *Appl. Phys. Lett.,* vol. 102, no. 1, pp. 012113-012113, January 2013.

13. Wu, Y., Lee, B., and PHILIP WONG, H. S., "Al$_2$O$_3$-based RRAM using atomic layer deposition (ALD) with 1-μA reset current," *IEEE Electron Device Lett.,* vol. 31, no. 12, pp.1449-1451, 2010.

14. Lee, D., et al., "Excellent uniformity and reproducible resistance switching characteristics of doped binary metal oxides for non-volatile resistance memory applications," *IEEE International Electron Devices Meeting (IEDM),* San Francisco, CA, December 2006, pp. 1-4.

15. Baek, I. G., et al., "Highly scalable nonvolatile resistive memory using simple binary oxide driven by asymmetric unipolar voltage pulses," *IEDM Technical Digest. IEEE International,* December 2004, pp. 587-590.

16. Chien, W. C., et al. "Unipolar Switching Behaviors of RTO WO$_X$ RRAM," *IEEE Electron Device Lett.,* vol. 31, no. 2, pp. 126-128, February 2010.

17. Wang, M., W. J. Luo, Y. L. Wang, L. M. Yang, W. Zhu, P. Zhou, J. H. Yang et al., "A novel Cu$_x$Si$_y$O resistive memory in logic technology with excellent data retention and resistance distribution for embedded applications", *VLSI Technology (VLSIT), Symp., Honolulu,* June 2010, pp. 89-90.

18. Ji, Y., Cho, B., Song, S., Kim, T. W., Choe, M., Kahng, Y. H., and Lee, T., "Stable switching characteristics of organic nonvolatile memory on a bent flexible substrate," *Adv. Mater.,* vol. 22, no. 28, pp.3071-3075, July 2010.

19. Kim, S., Yarimaga, O., Choi, S. J., and Choi, Y. K., "Highly durable and flexible memory based on resistance switching," *Solid-State Electron.,* vol. 54, no. 4, pp. 392-396, April 2010.

20. Zou, S., Xu, P., and Hamilton, M. C., "Resistive switching characteristics in printed Cu/CuO/(AgO)/Ag memristors," *Electron. Lett.,* vol. 49, no. 13, pp. 829-830, June 2013.

21. Han, J. W., and Meyyappan, M., "Copper oxide resistive switching memory for e-textile," *AIP Adv.* 1, no. 3, pp. 032162-032162, September 2011.

22. Biesinger, M. C., Lau, L. W., Gerson, A. R., and Smart, R. S. C. , "Resolving surface chemical states in XPS analysis of first row transition metals, oxides and hydroxides: Sc, Ti, V, Cu and Zn," *Appl. Surf. Sci.,* vol. 257, no. 3, pp. 887-898, November 2010.

23. Carlson, T. A., et al., "Electron Spectrosc," *Proc. Int. Conf.,* vol. 207, 1972.

24. Jeong, D. S.,et al., "Emerging memories: resistive switching mechanisms and current status," *Rep. Prog. Phys.,* vol. 75, no. 7, pp. 076502, 2012.

25. Wang, S. Y., Huang, C. W., Lee, D. Y., Tseng, T. Y., and Chang, T. C., "Multilevel resistive switching in Ti/Cu$_x$O/Pt memory devices," *J. Appl. Phy.,* vol. 108, no. 11, pp. 114110-114110, December 2010.

26. Dong, R., et al. "Reproducible hysteresis and resistive switching in metal-Cu$_x$O-metal heterostructures," *Appl. Phy. Lett.,* vol. 90, pp. 042107, 2007.

Ultra-High Refractive Index LED Encapsulant

Chia-Chi Tuan[1], Ziyin Lin[1], Yan Liu[1], Kyoung-Sik Moon[1], Sehoon Yoo[2], Myong-Gi Jang[3], and Ching-Ping Wong[1,4*]

[1]School of Materials Science and Engineering, Georgia Institute of Technology,
771 Ferst Drive, Atlanta, GA 30332
[2]Micro-Joining Center, Korea Institute of Industrial Technology, Korea
[3]EI Lighting Co. Ltd, Korea
[4]Department of Electronic Engineering, Chinese University of Hong Kong, Hong Kong
[*]Corresponding author: 404-894-8391, cp.wong@mse.gatech.edu

Abstract

In this paper, we report a method for preparing a silicone-based nanocomposite material for high brightness light emitting diode (LED) packaging. High refractive index (RI) encapsulant is desired in order to reduce the RI contrast between the LED chip and air, thus increasing the light extraction efficiency (LEE). In our method, TiO_2 nanoparticles were added to the silicone resin to prepare the nanocomposite, and the RI was improved from 1.56 of neat silicone to 1.73 at the GaN-emitting wavelength of 460 nm. The nanocomposite also exhibits high relative transparency in the visible light wavelength of 300-800 nm. At 460 nm, the material containing 10 wt% filler loading shows above 90% relative transparency. Accelerated reliability test was conducted on the high-performance nanocomposites, and less than 2% degradation was measured in both RI and relative transparency.

Introduction

LEDs offer many advantages over conventional incandescent and fluorescent light sources, including longer lifetime, higher luminous efficiency, smaller volume and better reliability. However, low LEE is a major limitation in the current LED packaging technology. Due to the large difference in RI between LED and air, much of the emitted light is internally reflected, causing loss of brightness and efficiency [1]. Moreover, the light trapped in the device may convert into heat, which further lowers the conversion efficiency, and can also degrade device reliability.

LED encapsulants are often used in the packages to increase the LEE by reducing the RI contrast. However, epoxy and silicone, which are materials commonly used for the encapsulant, have RIs ranging from 1.45 to 1.55 in the wavelength range of emitted light. Since GaN-based LEDs that emit the blue light have RI of 2.47 at the emission wavelength (460 nm), the RI difference between the LED chip and the neat polymer encapsulant still limits the LEE.

It has been reported that, by increasing the RI of the encapsulant material, the LEE can be improved significantly. For the GaN LED coated with a single layer of encapsulant with uniform RI, LEE increases with increasing RI of the encapsulant, until it reaches saturation when encapsulant RI approaches 2.0 [2]. Defining extracting efficiency ratio as the LEE of light coupled into encapsulant material compared to that of light coupled directly to air, the ratio is nearly doubled when the RI of the encapsulant is increased from 1.5 to 2, as shown in Figure 1. Even more light can be extracted from the package when an additional layer of lower RI encapsulant is coated between the high-RI layer and air, since the secondary layer further reduces Fresnel reflection at the encapsulant-air interface [3].

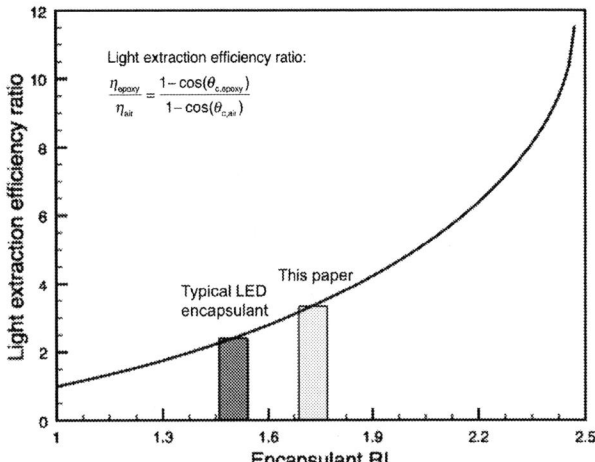

Figure 1. Effect of encapsulant RI on extraction efficiency ratio for GaN LED (RI = 2.47).

In composite materials, the dispersion of fillers in the matrix has always been a serious issue, as non-uniform dispersion results in unsatisfactory material properties in the thermal, electrical, and mechanical aspects. For optical applications, composites should also have controlled filler-induced light scattering. Rayleigh scattering can be minimized by decreasing the dimension of the filler particles to an order of magnitude below the wavelength of light.

Epoxy- and silicone-based composite materials prepared using nanoparticles have been reported as encapsulant materials. Silica particles have been used as fillers in epoxy-based LED encapsulants, and above 90% transmittance values were achieved. But SiO_2 has a known RI between 1.4 and 1.55, which are comparable to our silicone resin, and little improvements in the composite RI can be attained [4]. ZnO nanoparticles synthesized via a hydrolysis route has been incorporated into silicone resin by Liu *et al.*, and the 8 wt% loading nanocomposite increased the RI by 0.02, to 1.52 [5]. ZnO is a direct band-gap semiconductor with a wide band gap of 3.37 eV, and its nanocomposites are expected to be highly transparent due to minimal light absorption in the visible range [6]. However, the intrinsic RI of ZnO is only 2.1 at 460 nm, and is not sufficiently high to produce an encapsulant material for high light extraction purposes. Although ZnS has a RI that is slightly higher, large filler loading levels are still required for its composites to show significant RI improvements [7].

PbS exhibits an ultrahigh RI above 4, and Weibel *et al.* have shown that 90 wt% PbS-polymer composites displayed RI as high as 3, which would reach the LEE saturation range for GaN LED [8]. But PbS has a direct band gap at 0.41 eV, and absorbs in all of the visible light range [9]. At the reported loading of 90 wt%, it is reasonable to conclude that the composite is completely opaque and cannot be used towards LED encapsulation.

TiO_2 crystals have an ideal combination of properties for optical applications. All three of its crystalline phases has large band gaps larger than 3 eV (Table 1), suggesting minimal light absorption in the visible, and two of its three crystalline phases exhibit high RI up to 2.90 [10], [11]. Sol-gel methods have been used in the preparation of TiO_2 nanoparticles, which were mixed with epoxy to prepare composite materials. 30 wt% loading composite achieved RI of about 1.7 at 633 nm, but and RI of 1.8 would require about 60 wt% loading [12]. Such filler loading level may lead to difficulties in material processing and may alter the mechanical properties of the material. Tao *et al.* also reported preparation of TiO_2-epoxy composites. The polymer chains were grafted onto anatase TiO_2 by phosphate ligand engineering and alkyneazide click chemistry. The nanocomposites showed high transparency, and the RI was improved from 1.5 to 1.8 in a 30 wt% loading material [13].

Table 1. Band gap values of TiO_2 crystalline phases [10], [11].

TiO_2 phase	Anatase	Brookite	Rutile
Band gap (eV)	3.20	2.39	3.03

We have previously developed a method to synthesize 10-15 nm TiO_2 nanoparticles via a simple, low-cost hydrothermal synthesis, and to use the nanoparticles in a silicone-based nanocomposite for LED encapsulation [14], [15]. Although the 5 wt% filler loading sample prepared by the method achieved RI of 1.63 at 460 nm, it fell short at optical transparency in higher-loading samples.

In this paper, we report a method to prepare high RI TiO_2-silicone encapsulant material. By *in-situ* surfactant modification in the TiO_2 synthesis, we produced nanoparticles highly dispersible in the silicone matrix. The rutile and brookite phases of TiO_2 have RI's of 2.90 and 2.79, respectively. By adding such nanoparticles into the silicone resin, we achieved significant enhancement in the RI of the material. Our encapsulant material also shows high optical transparency in the wavelength range of interest.

Materials and Methods

TiO_2 synthesis and nanocomposite preparation

TiO_2 nanoparticles were synthesized via a method modified from a previously published work [16]. The method has been used to prepare high RI composite materials, but no transparency data were reported [17]. A reaction mixture was prepared using isopropanol, organic surfactant, and formic acid. Titanium alkoxide was used as the nanoparticle precursor and was added to the mixture. An isopropanol-water

solution was added to the mixture at 60ºC, and the reaction was allowed to complete overnight. The formation of TiO_2 by the hydrolysis reaction can be observed readily as the solution color changed from clear to white. The nanoparticles were cleaned with isopropanol and collected by centrifugation.

The synthesized TiO_2 nanoparticles were added to a silicone resin at desired filler loading level. The composite material was made homogeneous by sonication and stirring, followed by a solvent removal process. The nanocomposite was cured at 150ºC for two hours before material characterization.

Characterization

The dimension of the nanoparticles was determined using transmission electron microscopy (TEM) images, which were acquired using a JEOL 100CX TEM. X-ray diffraction (XRD) analysis was performed using Cu Kα radiation (45 kV and 40 mA) on a Philips X-pert alpha-diffactometer. A Woollam M2000 ellipsometer was used to determine the RI of the nanocomposite. Optical transparency of coated samples in the wavelength range of 300-800 nm was measured by a Shimadzu Co. UV-2450 UV-vis spectrophotometer.

Results and Discussion

We used TEM to characterize the morphology of the as-synthesized TiO_2 nanoparticles. As seen in Figure 2, the nanoparticles are well-dispersed in an isopropanol solution. The nanoparticles are about 10 nm in diameter, which is much smaller than the wavelength of the LED light, thus minimizing light scattering. The XRD analysis indicated that the nanoparticles are of rutile and brookite TiO_2 phases. Both of the crystalline phases exhibit much higher RI's than that of the neat silicone, and both contribute to the significant RI improvement in the encapsulant material that will be disused.

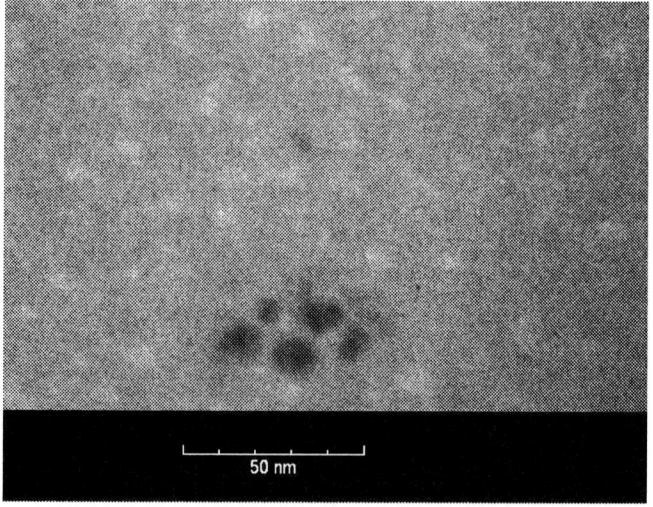

Figure 2. TEM image of the as-synthesized TiO_2.

Using the *in-situ* surfactant modification method, combined with organic surfactant that is highly compatible with the silicone resin, we prepared highly transparent nanocomposite material. Figure 3 shows the optical transmittance of the encapsulant material in the wavelength rage of 300-800 nm. Table 2 summarizes the transmittance at of our material with different filler loading levels at 460 nm.

At 10 wt% TiO_2 loading, the nanocomposite still exhibited 95% relative transmittance. Because the material is highly transparent, it is suitable as a LED encapsulation material.

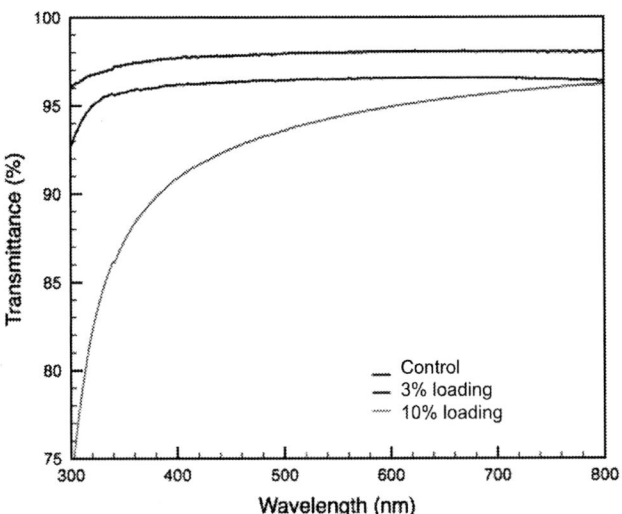

Figure 3. Optical transmittance of TiO_2-silicone nanocomposites at different loading levels.

Table 2. Relative transmittance of TiO_2-silicone nanocomposites at different loading levels.

Sample	Control	3% TiO₂	10% TiO₂
Relative transmittance at 460 nm	100%	99%	95%

In addition to the high optical transparency, the nanocomposite should also exhibit high RI in order to improve the LEE of the LED device. We determined the RI of our material by ellipsometry, where a linearly polarized light source is directed at the nanocomposite film coated on a silicon substrate, and the changes in the wave amplitude and phase of the reflected beam are analyzed. At the wavelength of 460 nm, neat silicone has RI of 1.56. With as little as 3 wt% of filler, our nanocomposite improved the RI to 1.67, which represents a 7% increase. The encapsulant material with 10 wt% filler loading further improved the RI to 1.73, which is a > 11% increase compared to the silicone resin (Figure 4).

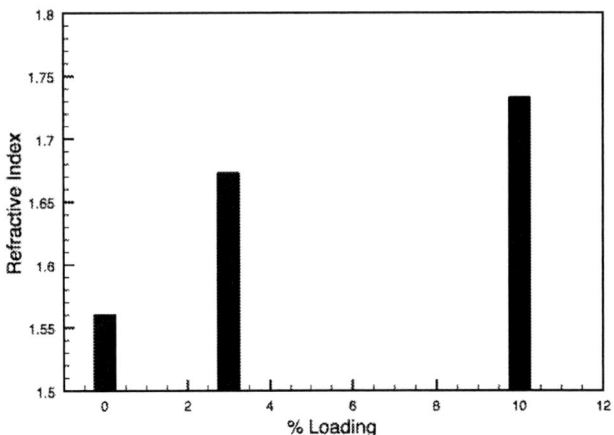

Figure 4. RIs of TiO_2-silicone nanocomposites at different loadings.

A high quality encapsulant material should maintain its performance throughout the lifetime of the LED device. We conducted accelerated reliability test by the JEITA ED-4701/200 203 standards on the nanocomposites. The samples underwent 300 hours of reliability cycles at 90% relative humidity. The cycles consisted of temperature soaking from 25°C to 65°C, to -10°C, and back to 25°C, totaling 24 hours per cycle (Figure 5). The encapsulant material was again characterized using UV-vis spectrophotometry and ellipsometry after the reliability conditioning cycles. The optical transparency of the material was hardly affected after the reliability test. Less than 2% change was measured at 460 nm (Figure 6). On the RI part, the nanocomposite material showed a < 1.5% decrease after the temperature-humidity test.

Figure 1 COMPOSITION OF ONE CYCLE

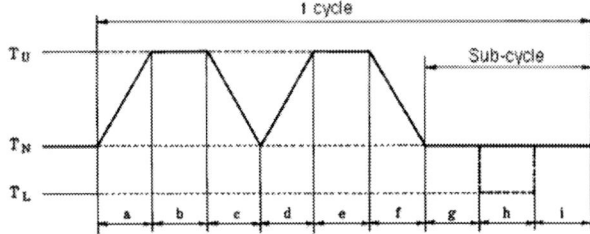

T_U, T_N, and T_L indicate the various temperatures.

Table 1 TEMPERATURE AND HUMIDITY CYCLE CONDITIONS

Step \ Conditions	Time (h)	Temperature (°C)	Relative humidity (%)
a	2.5	-	90 ~ 96
b	3.0	65±2	90 ~ 96
c	2.5	-	80 ~ 96
d	2.5	-	90 ~ 96
e	3.0	65±2	90 ~ 96
f	2.5	-	80 ~ 96
g	1 ~ 4	25±2	90 ~ 96
h	3.0	-10^{+3}_{-5}	Arbitrary
i	1 ~ 4	25±2	90 ~ 96
1 cycle	24	-	-

Figure 5. Reliability test cycle conditions.

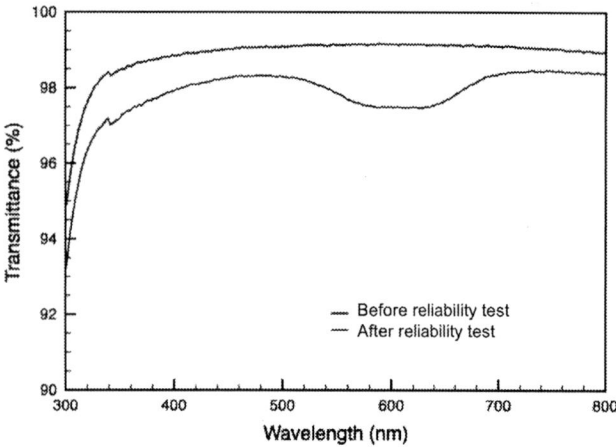

Figure 6. Optical transmittance of TiO$_2$-silicone nanocomposites before and after reliability test.

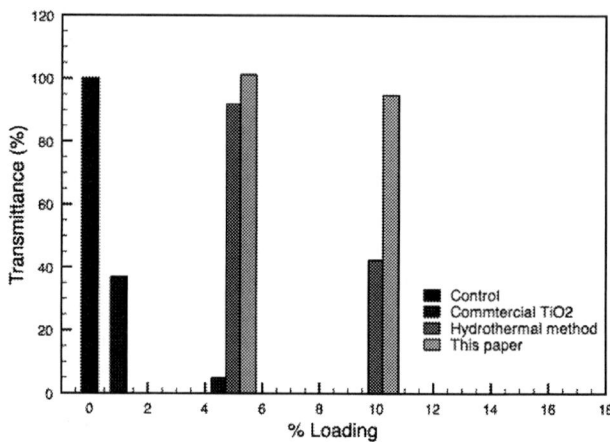

Figure 8. A comparison of the transparencies of TiO$_2$-silicone nanocomposites prepared from different methods.

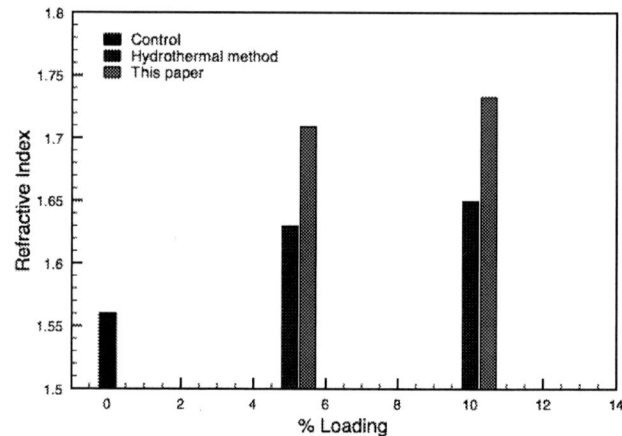

Figure 9. RI of encapsulant materials prepared using different methods.

As mentioned previously, we have developed a hydrothermal TiO$_2$ synthesis method to produce 10-15nm nanoparticles. The 5 wt% loading TiO$_2$-silicone nanocomposites prepared using such nanoparticles were reported to exhibit an RI of 1.63, and relative optical transparency of 92% (5 μm thickness) at 460 nm [14], [15]. However, at higher loading of 10 wt%, the nanocomposite material failed to maintain its transparency feature, showing only 42% relative transmittance at the corresponding wavelength. The RI enhancement was also unsatisfactory, providing only a 0.03 improvement compared to its lower-loading counterpart. Nanocomposite material prepared using commercial TiO$_2$ nanoparticles of 40 nm was even worse. Due to the large nanoparticle dimension, heavy scattering resulted in nearly 0% transparency at 5 wt% loading (Figure 7). Using the method reported in this paper, we have improved both transparency and RI of the encapsulant material significantly, as shown in Figures 8 and 9.

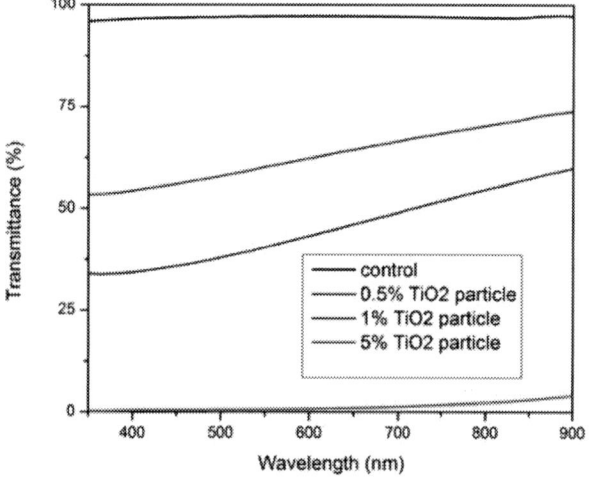

Figure 7. The transmittance of nanocomposites prepared from commercial TiO$_2$ with different loadings [14].

Conclusions

We prepared a TiO$_2$-silicone nanocomposite material for the encapsulation of LED chips. The new TiO$_2$ synthesis method significantly improves nanoparticle dispersion in the silicone resin. The nanocomposite prepared by our method can achieve high refractive index with 10 wt% filler loading, while maintaining high relative transparency above 90%. The nanocomposite also shows high moisture and temperature resistance under accelerated reliability test. This high RI nanocomposite material can be used as LED encapsulant to enhance the LEE of the device.

Acknowledgments

We would like to acknowledge financial support from the Korea Institute for Advancement of Technology (KIAT).

References

[1] M. R. Krames, O. B. Shchekin, R. Mueller-Mach, G. O. Mueller, L. Zhou, G. Harbers, and M. G. Craford, "Status and Future of High-Power Light-Emitting

Diodes for Solid-State Lighting," *J. Disp. Technol.*, vol. 3, no. 2, pp. 160–175, Jun. 2007.

[2] M. Ma, F. W. Mont, X. Yan, J. Cho, E. F. Schubert, G. B. Kim, and C. Sone, "Effects of the refractive index of the encapsulant on the light-extraction efficiency of light-emitting diodes.," *Opt. Express*, vol. 19 Suppl 5, no. September, pp. A1135–40, Sep. 2011.

[3] F. W. Mont, J. K. Kim, M. F. Schubert, E. F. Schubert, and R. W. Siegel, "High-refractive-index TiO2-nanoparticle-loaded encapsulants for light-emitting diodes," *J. Appl. Phys.*, vol. 103, no. 8, p. 083120, 2008.

[4] T. Li, J. Zhang, H. Wang, Z. Hu, and Y. Yu, "High-performance light-emitting diodes encapsulated with silica-filled epoxy materials.," *ACS Appl. Mater. Interfaces*, vol. 5, no. 18, pp. 8968–81, Sep. 2013.

[5] Y. Liu, Z. Lin, X. Zhao, S. Yoo, K. Moon, and C. P. Wong, "ZnO quantum dots-filled encapsulant for LED packaging," in *2012 IEEE 62nd Electronic Components and Technology Conference*, 2012, pp. 2140–2144.

[6] Y. Yang, Y. Li, S. Fu, and H. Xiao, "Transparent and Light-Emitting Epoxy Nanocomposites Containing ZnO Quantum Dots as Encapsulating Materials for Solid State Lighting," *J. Phys. Chem. C*, vol. 112, no. 28, pp. 10553–10558, Jul. 2008.

[7] C. Lü, Z. Cui, Z. Li, B. Yang, and J. Shen, "High refractive index thin films of ZnS/polythiourethane nanocomposites," *J. Mater. Chem.*, vol. 13, no. 3, pp. 526–530, Feb. 2003.

[8] M. Weibel, W. Caseri, U. W. Suter, H. Kiess, and E. Wehrli, "Preparation of Polymer Nanocomposites with 'Ultrahigh' Refractive Index," *Polym. Adv. Technol.*, vol. 2, pp. 75–80, 1991.

[9] W. Scanlon, "Intrinsic Optical Absorption and the Radiative Recombination Lifetime in PbS," *Phys. Rev.*, vol. 109, no. 1, pp. 47–50, Jan. 1958.

[10] L. Miao, P. Jin, K. Kaneko, a Terai, N. Nabatova-Gabain, and S. Tanemura, "Preparation and characterization of polycrystalline anatase and rutile TiO2 thin films by rf magnetron sputtering," *Appl. Surf. Sci.*, vol. 212–213, pp. 255–263, May 2003.

[11] A. Di Paola, G. Cufalo, M. Addamo, M. Bellardita, R. Campostrini, M. Ischia, R. Ceccato, and L. Palmisano, "Photocatalytic activity of nanocrystalline TiO2 (brookite, rutile and brookite-based) powders prepared by thermohydrolysis of TiCl4 in aqueous chloride solutions," *Colloids Surfaces A Physicochem. Eng. Asp.*, vol. 317, no. 1–3, pp. 366–376, Mar. 2008.

[12] J. L. H. Chau, C.-T. Tung, Y.-M. Lin, and A.-K. Li, "Preparation and optical properties of titania/epoxy nanocomposite coatings," *Mater. Lett.*, vol. 62, no. 19, pp. 3416–3418, Jul. 2008.

[13] P. Tao, Y. Li, A. Rungta, A. Viswanath, J. Gao, B. C. Benicewicz, R. W. Siegel, and L. S. Schadler, "TiO2 nanocomposites with high refractive index and transparency," *J. Mater. Chem.*, vol. 21, no. 46, p. 18623, 2011.

[14] Y. Liu, Z. Lin, X. Zhao, K. Moon, S. Yoo, J. Choi, and C. P. Wong, "High refractive index and transparency nanocomposites as encapsulant for high brightness LED packaging," *2013 IEEE 63rd Electron. Components Technol. Conf.*, pp. 553–556, May 2013.

[15] Y. Liu, Z. Lin, Z. Zhao, C.-C. Tuan, K. Moon, S. Yoo, J. Choi, and C. P. Wong, "High Refractive Index and Transparent Nanocomposites as Encapsulant for High Brightness LED Packaging Yan," *Trans. Components, Packag. Manuf. Technol.*, 2014.

[16] T. C. Monson, M. a. Rodriguez, J. L. Leger, T. E. Stevens, and D. L. Huber, "A simple low-cost synthesis of brookite TiO2 nanoparticles," *J. Mater. Res.*, vol. 28, no. 03, pp. 348–353, Nov. 2012.

[17] T. C. Monson and D. L. Huber, "High Refractive Index TiO2 Nanoparticle/Silicone Composites."

A Novel Methodology for Wafer-Specific Feed-Forward Management of Backside Silicon Removal by Wafer Grinding for Optimized Through Silicon Via Reveal

Tyson Alvanos[2], John Garant[1], Yu Iijima[2], Richard Indyk[1], Christopher Rosenthal[4]
Osamu Sato[3], Naoki Sugase[3], Hideo Takizawa[3], Frank Wei, Ph.D.[2]
[1]IBM Corp., 2070 Route 52, Hopewell Junction, New York
[2]Disco Hi Tec America 3270 Scott Blvd, Santa Clara CA
[3]Lasertec Corp., 2-10-1 Shin-Yokohama, Kohoku-ku, Yokohama Japan
[4]Lasertec USA Inc., 2025 Gateway Place Ste. 430, San Jose CA

Abstract

As 3DIC with through silicon vias (TSV) approaches high volume manufacturing readiness the importance of precision backgrinding has become increasingly more evident. Active management of the backgrinding process has multiple benefits in that it reduces the risk of wafer backside contamination due to premature contact with vias, it enables optimization of the post-thinning residual silicon thickness and the final via reveal process. It can compensate for poor via fabrication depth uniformity and less than ideal temporary bonding total thickness variation (TTV). In this paper we will demonstrate the utility of two tools that when used together can systematically produce thin TSV containing wafers to their optimal thickness while protecting the wafers from particulate contamination. The first instrument in this process scheme is a metrology tool that utilizes IR reflectance to measure the silicon thickness remaining between the bottom of the TSV and backside surface of the wafer. These measurements are then transferred to a special grinding tool that can interpret the data and make changes to the grinding depth within the wafer so as to leave thickness of silicon above the TSVs as uniform as possible. Having removed the bulk of the Si through mechanical grinding and Chem/Mech polishing, the next step in the via-middle process is to remove the last few microns of Si overburden to expose the vias. By leaving a shallower Si layer above the TSV during the thinning process compared to current process, the final via reveal process time can be reduced. Also the need for rework in this process because of the wafer-to-wafer variability in the remaining silicon thickness above the TSV can be eliminated. In addition to measuring the pre-grinding Si thickness, the IR reflectance measurement tool can be used to verify the remaining silicon thickness post grind, establishing thinning process feedback and via reveal process feed-forward data.

Measurement Using IR Reflectance

This tool employs two optical technologies: an IR interferometer used to measure how deep objects are embedded in the silicon wafer and a sensor to measure the total thickness of the bonded wafer stack.

The IR interferometer optics make it possible to quickly and efficiently take measurements of critical components such as device wafer thickness, via depth (height), adhesive thickness, and remaining silicon thickness (RST) above the TSV. The tool's design makes it possible for both pre-grind (full thickness wafers) and post-grind wafers (ultra-thin wafers) to be analyzed without operator assistance or major alterations of the tool.

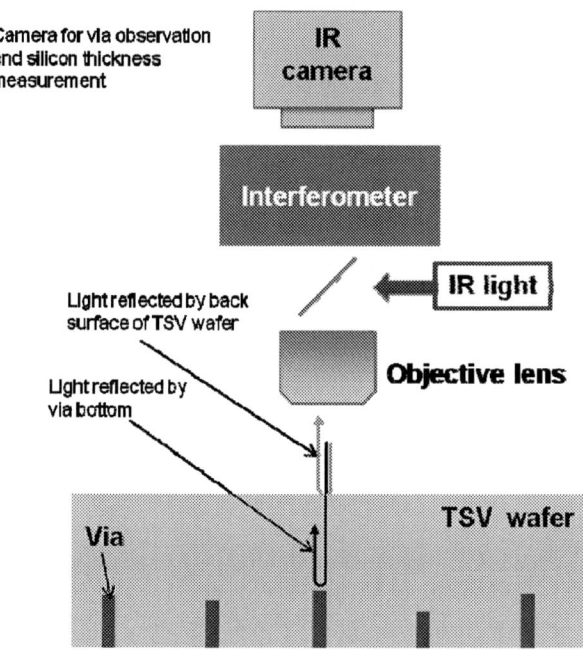

Figure 1. IR interferometer optics

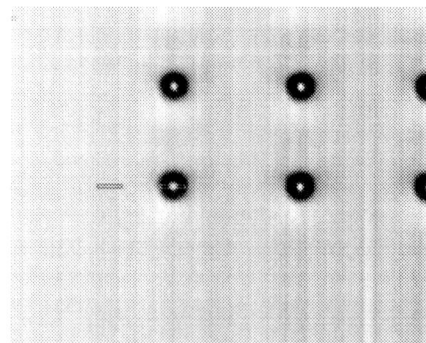

Figure 2. Target positioning image from IR Camera

Upon loading of a wafer onto the tool's stage, the tool utilizes optical pattern recognition to locate the pre-selected TSV to be measured. The location and number of TSV's to be measured can be defined by the user in a recipe. Once the machine has positioned itself over one of the pre-programmed TSV (Figure2) to be measured, four measurements are recorded by the tool: the Remaining Silicon Thickness ("RST" in Figure 3), the Via Depth ("VD" in Figure 3), the Device Wafer Thickness ("WT" in Figure 3) and the total thickness of the bonded wafer stack ("BWT" in Figure 3). Another value called "Base Height" ("BH" in Figure 3) is calculated by subtracting the RST from the BWT, and logged in a file with the other measurements for transfer to the grind tool.

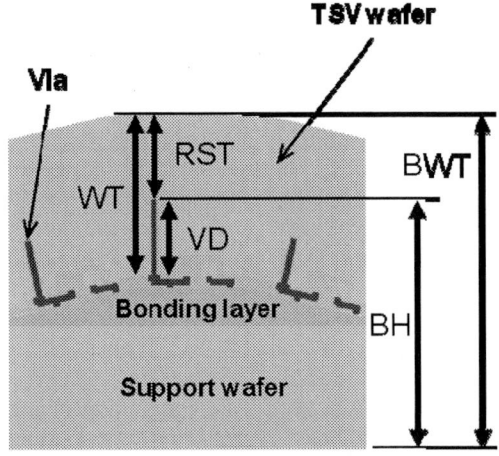

Figure 3. Measurement item for transfer to grinder

The measured values at each position programmed in the recipe can be plotted on a wafer map (Figure 4) for easy interpretation. The Base Height and Via Depth distribution information is also recorded in a file format that can be sent to the grinder to assist with the wafer thinning process.

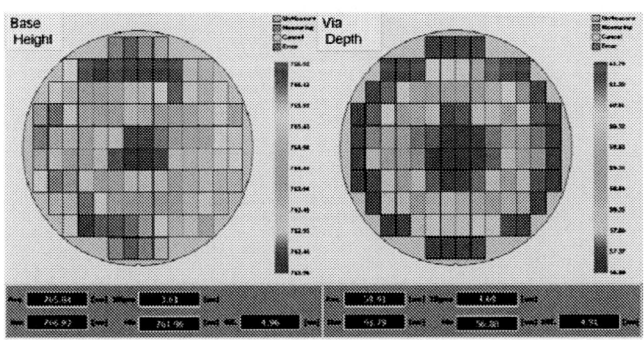

Figure 4. Base Height and Via Depth distribution map

After the wafers have been ground to the desired thickness they can be returned to the metrology tool to collect additional information of value for both quality control and downstream processing. Typically, this includes post-thinning RST measurements that can be fed forward for via reveal process optimizations as well as confirmation of any improvements or changes in the TTV of the wafer stack.

The accuracy of the Base Height calculated by measurements taken by the tool is well in the sub-micron range and for this reason more than sufficient for the intended use in this process.

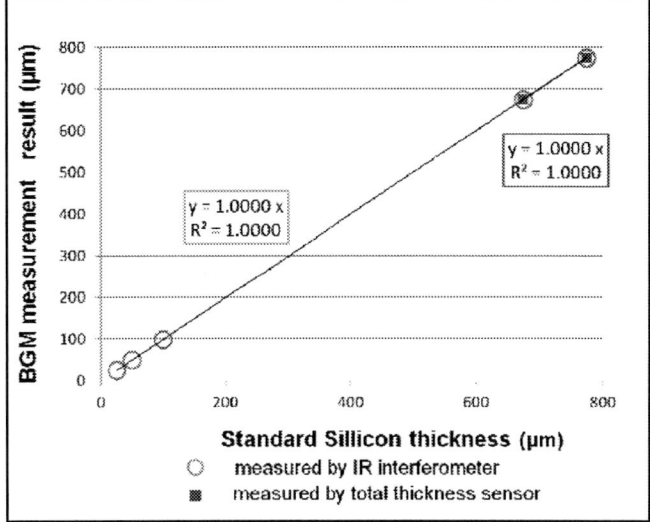

Figure 5. Standard thickness gauge measurement result

Figure 5 and Table 1 show the accuracy and repeatability of the interferometer and thickness gauge employed by the tool. The IR interferometer and total thickness gauges measured silicon substrates of known thickness as certified by an accredited third party and were compared.

Table 1. Measurement repeatability

Standard Si thickness (±0.05) (μm)	Measurement result			
	IR Interferometer optics measurement result		Total thickness gauge measurement result	
	Average(μm)	3σ	Average(μm)	3σ
25.28	25.25	0.02	-	-
50.39	50.36	0.10	-	-
99.46	99.35	0.06	-	-
674.57	674.63	0.05	674.42	0.02
774.96	774.94	0.02	774.86	0.01

TSV Wafer Thinning Process Module

Background

The via-middle process scheme for TSV 3DIC fabrication holds significant interests from the semiconductor industry [1]. In this scheme, blind vias are fabricated on the device-side of TSV wafers after completing the front-end processes. Then, the buried side of the vias is exposed by removing the backside Si to complete the TSV IC structures. Since the global-level TSV depths are only approximately 50um or less [2], a temporarily bonded support wafer is needed for handling the thinned wafer during subsequent processing. See Figure 3. Fixed-abrasive based, in-feed wafer grinding is the process of choice to remove the bulk amount of Si above the vias due to the fast throughput and low consumables usage.

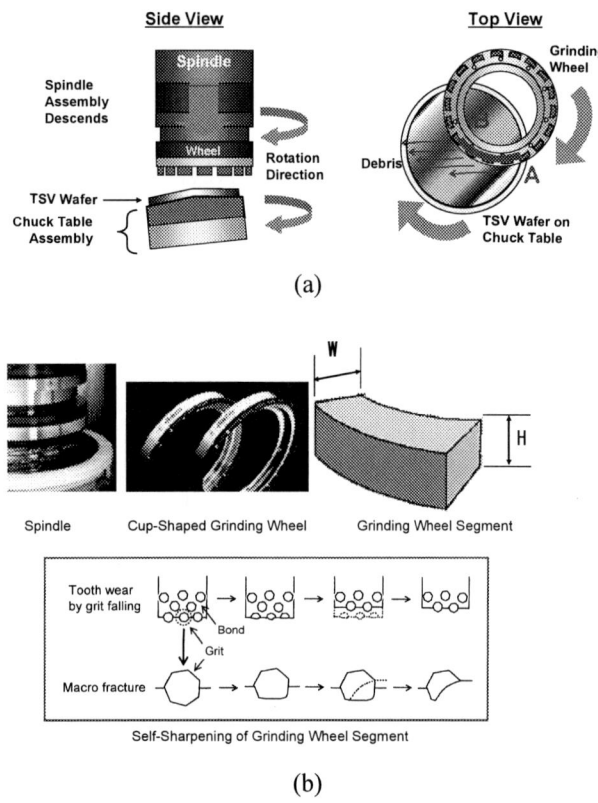

(a)

(b)

Figure 6. (a) Illustration of in-feed wafer grinding process. (b) Fixed abrasive stable grinding mechanism.

During in-feed grinding process, the chuck table of the grinder holds the wafer affixed by vacuum. A cup-shaped grinding wheel attaches to a high precision, high power spindle driven by an electrical motor. The top surface of the chuck table follows a slight conical shape that also translates onto the wafer surface, while the chuck table's rotational central axis maintains a slight tilt angle with respect to that of the spindle central axis – vertical to the machine's floor. See Figure 6(a). In this fashion, as the spindle assembly descends with both the spindle and the chuck table are rotating at high speeds, the grinder creates a process locus, arc-AB shown in Figure 6(a) where the grinding wheel make contact with the

back surface of the wafer. The slope of the conical chuck table allows cooling water and debris to escape.

The essence of the grinding process is similar to that of filing, removal of materials from the wafer surface in fine shavings up to several microns per second. The grinding wheel segments comprise of extremely hard abrasive grits, such as diamond, boron carbide, or boron nitride, as well as bonding materials to hold the segments' shape; hence, the term 'fixed abrasive' is used. The segments mimic the function of the teeth on a file with partially exposed abrasive grits while processing. In order to prevent dulling of the segments, by design, the bonding material erodes over time by the debris generated such that the dulled grits fall off the segment to expose fresh grits underneath – self-sharpening behavior. A stable process thus continues while the fixed abrasive segments are slowly consumed. See Figure 6(b).

Auto-TTV Grinding of TSV Bonded Wafers

When the tilt angle between the axes of the chuck table and the spindle assembly exactly compensates that of the slope of the chuck table surface, the grinder produces a perfectly flat wafer: The ground surface is parallel to the wafer's bottom surface; however, as the tilt angle varies, the wafer may take on a spectrum of linear superposition of radially symmetric 'V-' and 'W-like' shapes. Figure 7 shows examples of wafer surface profiles attainable on the grinder. For bonded wafers ground during TSV 3DIC fabrication, controllably reproducing such shapes can be advantageous in compensating for variations and/or non-uniformities induced by other processes such as via formation and support wafer bonding.

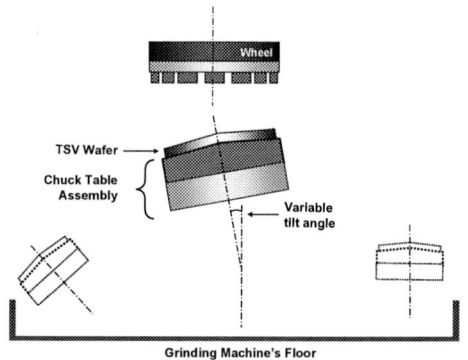

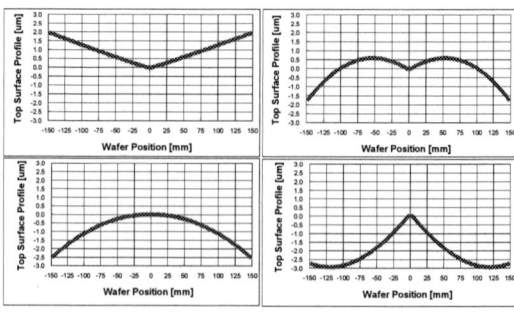

Figure 7. Examples of attainable wafer profiles by tilting the axes of the chuck table with respect to the grinding spindle.

The grind tool can controllably tilt and maintain a range of angles between the axes of spindle and chuck tables in an automated fashion. The machine also accepts feed-forward metrology results from upstream processes regarding the depths of vias and other constituents' thickness distributions across a bonded wafer. Data can be imported via SECS/GEM communications across the fab. Subsequently, the grinder calculates the optimal, wafer-specific tilting angle during process first and foremost, to maintain a minimum RST above the deepest vias to prevent blindly grinding into the via metallization, thus contaminating the tool and wafer, and secondly, to produce minimal RST variations across wafer as much as possible, thus compensating for any non-uniformities induced by the wafer bonding process. This process function is named *(re-)Auto-TTV Adjustment.* A controlled reduction in the variations of RST means potentially reduced RST targets in process flow design and/or improved uniformity in the downstream via reveal process step.

During auto-TTV enabled wafer grinding, the thinning process starts with the coarse grind wheel removing the bulk of the Si overburden above the TSV's. At a recipe-programmed wafer thickness, the tool automatically shifts the wafer to the fine grind process station. The fine grind station tilts its axes to match the wafer shape corrections calculated based on any feed-forward metrology data or user defined data. At this point, an IR non-contact, interferometric thickness gauge is engaged to track *in-situ* the thickness of the top TSV wafer only. At an intermediate thickness, the non-contact IR gauge can also swing across the wafer radius to confirm the desired top TSV wafer shape. The grinder can then further apply any corrections if needed, before reaching the final target wafer thickness. Ideally, a wafer-specific grind surface shape, conformal to the TSV via tip profiles across each wafer is desired. However, due to fundamental limitations of the grinder construction, non-radial symmetric profiles cannot be compensated. Such non-uniformities may limit the thinning targets' minimum RST above the deepest via depths; however, as via depths uniformities improve, reductions in the RST thinning target are possible.

Figure 8 illustrates the feasibility of the auto-TTV grinding process with feed-forward metrologies: Figure 8(a) contrasts the post-grinding RST results on TSV test wafers of the identical design and processes that have been purchased from a commercially available source. Wafers A and B were ground without any feed-forward metrology. The finishing thickness of the TSV wafer was a fixed target. Due to wafer-to-wafer variations in TSV fabrication depths, the resultant RST varied significantly from one to the other. Wafer A showed only 3um of RST remaining. On the other hand, wafers C and D had feed-forward metrologies post-bonding. Wafer-specific recipes were used. Wafer-to-wafer RST results were consistent in this case. Figure 8(b) shows an example of reduction in within-wafer RST variations. On this wafer, post-bonding metrology data was imported into the grinder, and appropriate adjustments were applied. The grinder was able to compensate for within-wafer variations in Base Height resulting in a reduction in RST variation near target thickness.

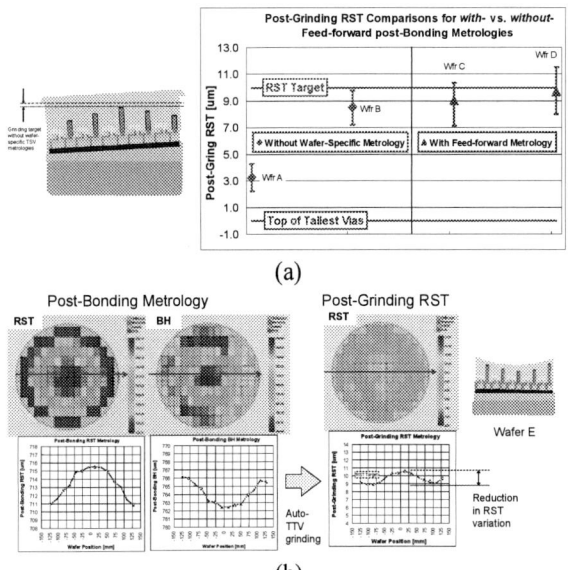

Figure 8. Examples of TSV wafer grinding using auto-TTV function with feed-forward metrology data.
(a) Feed-forward data reduced wafer-to-wafer variations in post-grind RST compared to case when no metrology present.
(b) Feed-forward metrology with auto-TTV grinding is able to reduce within-wafer RST variations post-grind.

For TSV wafers, after grinding, the grinding process-induced damage layer can be removed by CMP, among other means, leading to increased die strengths. Lastly, the back surface of the TSV wafer requires stringent cleaning step, in order to achieve a pristine state ready for the clean room environment. The new generation grinder incorporates all of the aforementioned processes, resulting in a self-contained, complete TSV wafer thinning process module.

Functional TSV Wafers Process Improvement

TSV Wafer Thinning Process Flow

IBM has been developing processes for the fabrication of 3DIC components over the past several years. One of the key developments in the wafer thinning process sector was the implementation of a change in the wafer-thinning tool and the wafer-thinning process methodology. Specifically, feed-forward TSV metrology, wafer-specific grinding corrections, and post-thinning high-level wafer cleaning features have been added. These steps have resulted in improvement in the process variations in post-thinning TSV RST and post-reveal via height uniformity, which are critical to the 3DIC manufacturing process technology development. A comparison of the general wafer thinning process flows, before and after this tool and process change is shown in Figure 9.

978-1-4799-2408-0/14 $31.00 © 2014 IEEE

WAFER THINNING PROCESS FLOW COMPARISON

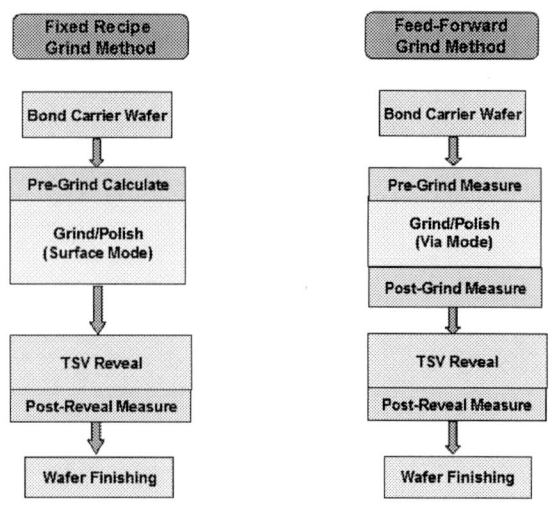

Figure 9. 3DIC Wafer Thinning Process Flow

The grind/polish wafer thinning process was originally performed in what is known as the 'surface mode' on the previous generation grind tool. In this process mode, post-bonding metrology processes were not available; rather, TSV depth data from the wafer fab was used in the manual grind calculations. The amount of Si to be removed was programmed into the process recipe as a fixed amount for any given wafer or, group of wafers if the amount of silicon to be removed was similar within the group. The grinder used the top surface of the bonded wafer as a process starting reference. In 'surface mode' grinding, wafer grinding required calculations of the Si removal amount based on the starting Si thickness of the wafer, the in-line average TSV depth data, and the target RST. The amount of material to be removed from the backside of the wafer is calculated as follows:

Removal Amount = Incoming Si Thickness – TSV Height – RST Target

Wafers within a job lot are then binned according to their respective calculated removal amounts and are processed on the grind tool according to those removal amounts on a sub-job basis.

Prior to processing, the grinder had been set up such that the ground surface was parallel to the wafer bottom surface within the manufacturer's specifications. As wafers were thinned under such a 'fixed' recipe mode, compensation for variations and/or non-uniformities within a wafer could not be made. These variations include:

1. Thickness variations of the carrier
2. Via depths variations
3. Bonding adhesive material thickness variations

Furthermore, as mentioned in the previous section, pristinely clean wafer surfaces post-thinning are required for the TSV wafers to continue in downstream processing. It was desired to improve upon the cleanliness of the wafers exiting the grind tool, which performs an inherently dirty process,

thus enabling high-yielding downstream wafer finishing processes.

In order to improve on the issues mentioned above, the current IR metrology/wafer grind tool set was introduced to the Wafer Thinning process module and has since been used exclusively. After bonding, each wafer is measured for the RST, via depth, device wafer thickness and total wafer stack thickness. The distance from the base of the bonded wafer to the top of the TSV is automatically calculated from the other measurements. Seventeen metrology measurement locations per wafer are used as shown in Figure 10.

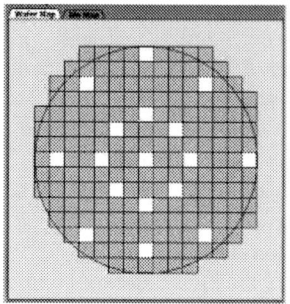

Figure 10. Post-Bonding Metrology TSV Measurement Locations

The metrology data is then formatted, stored on a server, and shared with the grinder. In operation, the wafer ID's are read by the grinder and are used to point to the appropriate measurement data set. Using this data, wafer-specific grinding recipe corrections are created by the grinder. These corrections consist of axes' tilting angle calculations such that the final RST above the wafer locations shown in Figure 10 are as uniform as possible.

Process Improvement Characterizations

After the Wafer Thinning module, the wafers are characterized again using the IR metrology tool to monitor the post-thinning RST – a feedback point for the wafer thinning step and a feed-forward point for the via reveal process step. The post-thinning RST run chart after the implementation of the feed-forward process change is shown in Figure 11. Well-controlled RST over time is observed. Such a consistent result over time implies that the post-thinning RST target can safely be reduced in the future without high concerns of grinding into the deepest vias and contaminating the TSV wafer and grind tool with TSV metal. Reduction in the wafer's RST would naturally lead to improved efficiency in the downstream via reveal process.

978-1-4799-2408-0/14 $31.00 © 2014 IEEE

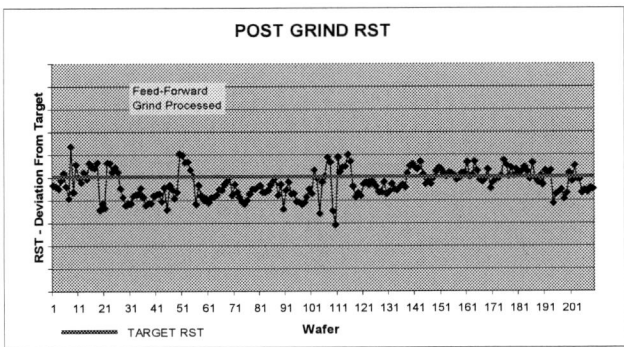

Figure 11. Post-thinning RST. All data was collected in the feed-forward grind process timeframe showing the consistency of the process methodology.

Accompanying the improved control on the process variations in post-thinning RST, the downstream post-via reveal height uniformity also improved significantly compared to that seen before the toolset and process change. Figure 12 shows the distribution, per process lot, of the via protrusion height after the via reveal step in the Wafer Finishing process module. After the tool and process methodology improvement, the average range of via protrusion distribution decreased by approximately 60%. This is a vital process improvement and has lead to positive impacts on downstream processes which are critical to overall 3DIC process yields and device reliability.

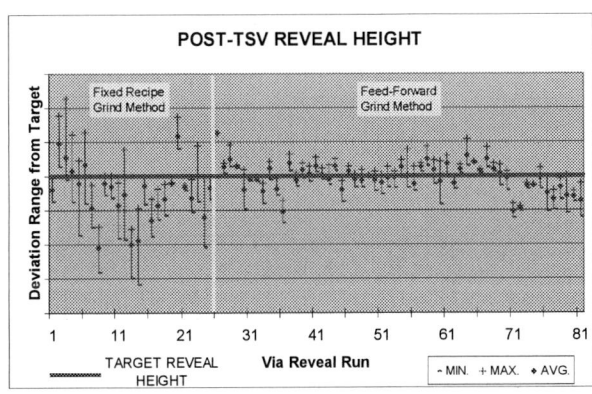

Figure 12. Post-via reveal TSV height range by lot showing the improvement in post-TSV reveal height uniformity.

Monitoring of Other Salient Processes

Prior to the process change, post-thinning RST measurements were not taken since an appropriate metrology tool was not available. In order to rule out the possibility of the downstream process improvements seen with the new toolset/process being caused by any other factors, salient process monitors were checked and no indications of other process shifts were observed.

First, Figure 13 shows the within-wafer range of incoming TSV via depths before and after the process change in the Wafer Thinning module. Consistent degrees of variation in via depth uniformity across a wafer are observed.

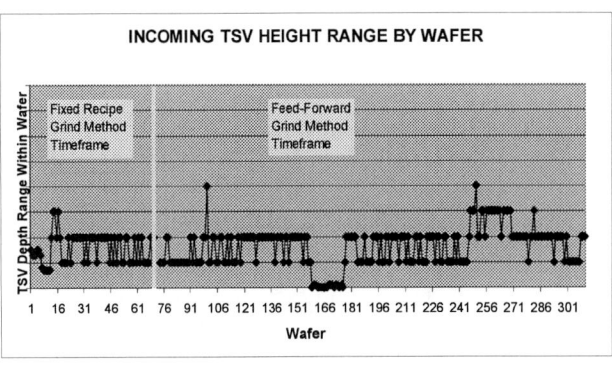

Figure 13. Range of incoming TSV via depth range per wafer. No changes in the data are seen in the timeframe straddling the implementation of feed forward grinding.

Second, Figure 14 shows the historical monitor of the via reveal process. Stable and consistent performance of this process is also confirmed. It can be concluded then, that the process improvement seen in via protrusion variations must be enabled by the implementation of the feed-forward, auto-TTV control in the TSV thinning process.

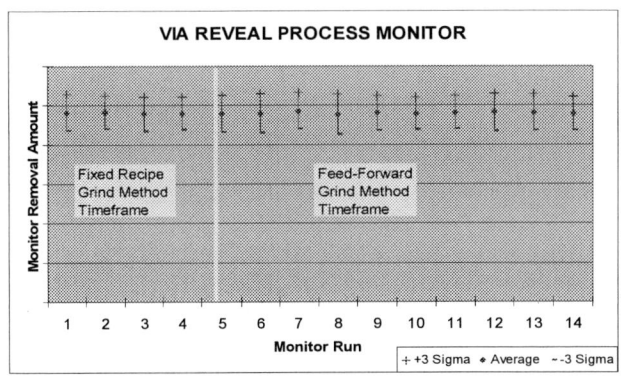

Figure 14. Via reveal process monitor. Consistent performance is seen before and after the implementation of the feed-forward wafer grind toolset and process.

Improvement in Wafer Surface Cleanliness

After both the coarse and the fine grinding steps, the grinder transfers the wafer to the polishing station in the tool to remove grinding surface damage and prepare it for final cleaning. Compared to the surface condition resulting from the thinning process performed with the previous generation grind tool, a clear improvement in surface cleanliness is accomplished in the new grind tool. This improvement has been enabled by the incorporation of post-polish wafer buffing, wafer edge clamping, sponge scrub cleaning, ozonated water rinsing, and water spraying within the tool. A comparison of representative wafer particle counts performed on wafers processed on both the previous and the new toolsets is shown in Figure 15. A significant improvement in the number of particles detected in all size bins is observed. This improvement also correlates to a significant decrease in the occurrence of yield losses associated with foreign material in downstream processes which were occurring prior to the implementation of the new wafer grind tool.

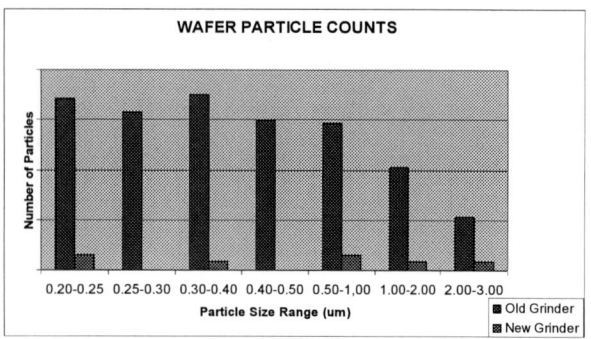

Figure 15. Comparison of post-thinning wafer surface cleanliness. Clear improvement in all size bins using the new grind tool is shown.

Conclusions

The development of feed-forward post-bonding TSV thickness metrology and wafer-specific, auto-TTV grinding adjustment through the use of the IR reflectance metrology tool and the new wafer grind tool enabled a significant reduction in post-thinning RST variations. Within-wafer thickness variations of the constituents in a pair of bonded wafers as well as via depths distribution in a TSV wafer can be captured in the metrology step. The measurement data are fed forward to the wafer-thinning module. Based on the wafer-specific data, the grinder generates compensation during grinding to minimize the variations in post-thinning RST from wafer to wafer. During process development at IBM, such a TSV wafer thinning methodology resulted in significant uniformity improvement in via protrusion amounts after the via reveal process step. Additional benefits resulting from improved post-thinning wafer cleanliness were realized through incorporation of processing features on the new grind tool. All of these tool and process enhancements have proved critical for 3DIC TSV device fabrication consistency.

Acknowledgments

The authors would like to thank Operation V Department at DISCO Corp. Japan for the support and internal discussions, specifically Susumu Hayakawa, Yoshihiro Tsutsumi, Yuki Inoue, and Shinya Watanabe for their contributions, also TOK Co., LTD for support of temporary bonding and de-bonding of sample TSV wafers.

References

1. M.G. Farooq *et al.*, "3D copper TSV integratino, testing and reliability," in *Proc. IEEE International Electron Devices Meeting. (IDEM)*, Washington D.C., Dec 5-7, 2011, pp. 7.1.1 – 7.1.4

2. International Technology Roadmap for Semiconductors, 2011 Edition, Interconnect Chapter
http://www.itrs.net/Links/2011itrs/2011Chapters/2011Inter
connect.pdf

978-1-4799-2408-0/14 $31.00 © 2014 IEEE

Thermal Characterization of Power Devices Using Graphene-based Film

Pengtu Zhang[1,2], Nan Wang[1], Carl Zandén[1], Lilei Ye[3], Yifeng Fu[3] and Johan Liu[1,4]

[1]SMIT Center, Department of Microtechnology and Nanoscience (MC₂),Chalmers University of Technology, Kemivägen 9, SE 412 96, Gothenburg, Sweden

[2]East China University of Science and Technology, Meilong Road 130, Shanghai 200237, China

[3]SHT Smart High Tech AB, Fysikgränd 3, SE 412 96 Gothenburg, Sweden

[4]Key Laboratory of New Displays and System Integration, SMIT Center and School of Mechatronics and Mechanical Engineering, No 149, Yan Chang Road, Shanghai 200072, China

Email: jliu@chalmers.se

Abstract

Due to its atomic structure with sp^2 hybrid orbitals and unique electronic properties, graphene has an extraordinarily high thermal conductivity which has been reported to be up to 5000 W/mK. As a consequence, the use of graphene-based materials for thermal management has been subject to substantial attention during recent years in both academia and industry. In this paper, the development of a new type of graphene-based thin film for heat dissipation in power devices is presented. The surface of the developed graphene based film is primarily composed of functionalized graphene oxide, that can be bonded chemically to the device surface and thus minimize the interface thermal resistance caused by surface roughness. A very high in-plane thermal conductivity with a maximum value of 1600 W/mK was detected by laser flash machine regarding to the graphene-based films. To investigate the structure of the graphene-based films, scanning electron microscopy (SEM) and raman spectroscopy were carried out. Finally, LED demonstrators were built to illustrate the thermal performance of graphene-based film and the functional layers. IR camera recorded a 5°C lower temperature of a LED demonstrator with SHT G1000 as the binding layer instead of a commercial thermal conductive adhesive,

1. Introduction

Integration and power density of microelectronic systems have been continuously increasing for decades. Efficient systematic thermal management solutions are an immediate requirement to dissipate the large amount of heat generated by the integrated chips and devices. Aiming for highly thermal conductive, cost-effective and eco-friendly processes, intensive efforts have been put into the research of novel thermal management materials in both academia and industry during recent years.

Graphite has been developed in industry as heat dissipation material for many years, because of its relatively high thermal conductivity and low price. However, the typical poor contact between the surfaces of substrates and graphite film causes a high thermal interface resistance, which limits its performance and thus also the use in high power density applications. Recently, nano-scaled carbon materials, in particular graphene[1], have attracted much attention due to their promise of extraordinary performances in respect to both electrical and thermal properties.

Composed of one or few atomic layers of sp^2-bound carbon, phonon transfer inside graphene becomes extremely efficient, which lead to an outstanding in-plane thermal conductivity of about 5000 W/mK[2]. Consequently, different approaches of utilizing graphene for heat dissipation have been investigated and reported in literature. One example is the finding that direct transfer of CVD-grown graphene sheets onto targeted substrate surfaces could efficiently remove heat from hotspots and thus help decrease the working temperature of active devices[3]. Another example is about graphene/polymer composite materials that has been developed and utilized for thermal management[4]. However, these applications have their own limitations, such as complex and expensive process or limited thermal conductivity.

In this paper, we developed a new type of graphene-based thin film aiming for heat dissipation in power devices. The thin film is composed of horizontally aligned graphene sheets which are fabricated by a chemical exfoliation process. To optimize the contact to the device surfaces, a graphene dip-coating and self-assembling process was developed. Additionally, in order to minimize the thermal contact resistance, the surface of the film which contacts with device surfaces was functionalized with small molecules. The small molecule, acting as the functional agent, could both react with the epoxy- and hydroxyl groups, located at the plane and the edge of graphene sheets, and also form covalent bonds with silicon dioxide or metal oxide surfaces. It results in the formation of molecular bridges between the grapheme-based surface and the device substrates. A sketch of the chemical bonding is shown in Fig. 1.

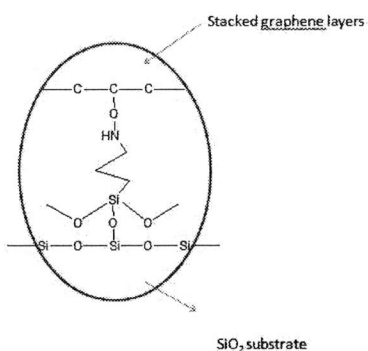

Figure 1. Sketch of chemical bonds between graphene-based surface and silicon dioxide substrate surface.

2. Experiment

2.1 Material preparation

2.1.1 Preparation of the graphene-based films

In typical experiments, graphite oxide dispersion was prepared following Hummer's method and exfoliated by ultrasonication to make graphene oxide. The dispersion was then centrifuged and washed by deionized water for several times to remove the residual acid. Finally, the dispersion was dried up at 98°C to remove all the water and followed by dissolving again into deionized water with calculated amount to form 1 mg/mL of graphene oxide (GO) dispersion .

The aqueous graphene dispersion was prepared by reducing graphene oxide with L-Ascorbic Acid (LAA)[5]. In a typical experiment, 40ml of the prepared GO dispersion was mixed with 1 g of LAA and stirred for 1 h at 98°C. The dispersion was then centrifuged and washed three times by deionized water to remove the residual LAA. Polyvinyl alcohol (PVA) was added into the dispersion to obtain a uniform suspension.

The graphene-based films were then prepared by filtration of the aqueous graphene dispersion with polycarbonate filter paper. After filtration, the free standing graphene-based film was obtained by dissolving the filter paper in acetone. To avoid uneven surface and cracks, only small amounts of the dispersion was filtrated each time. By controlling the amount of the dispersion that has been filtrated, films of different thicknesses could be prepared.

2.1.2 Preparation of the functionalized graphene oxide

Aiming to diminish the interface thermal resistance between different substrates by chemical bonding, a silane functionalized GO suspension was prepared[6,7]. Here, 20 mg of GO powder and 5 mg of Dicyclohexylcarbodiimide (DCC) were added into 30 ml of the functional agent and followed by 2 h of ultrasonication. The suspension was then heated up to 100°C and stirred for 3 h to perform the functionalization.

2.1.3 Preparation of the sample for thermal resistance measurements

To evaluate the effects of the functionalized graphene-based material on decreasing the interface thermal resistance, a sandwich testing model was assembled as shown in Fig. 2. Copper plates with a size of 8×8×1mm were used as substrates. A layer of silicon dioxide (70nm) and aluminum oxide (60nm) were coated on the two substrates respectively in order to simulate the device interface. After sputtering with silicon dioxide and aluminum oxide, the substrates were first modified with the functionalized graphene oxide and then dip-coated with the aqueous graphene-based dispersion. A small pressure was applied to stack the two coated substrates together. The sample was then heated up to 120°C and kept for 4h to form the bonding between the multilayered graphene and the corresponding substrate. For comparison, copper plates were also treated and stacked by following the same procedure but without the modifying process. By controlling the amount of the dispersion, the bonding layers between two substrates had a thickness of 30 ± 5 μm.

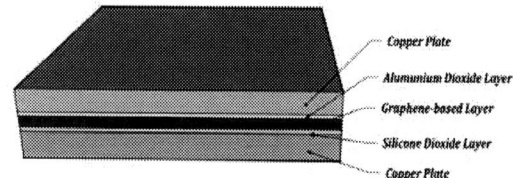

Figure 2. Model of the sandwich structure.

2.1.4 Preparation of demonstrators

A demonstration of the thermal performance in application was carried out by using the graphene based films and a reference material to assemble LEDs on a silicon dioxide substrate. As a reference, one of the LEDs was attached to the silicon dioxide substrate by commercial thermal conductive adhesive (TCA in Fig. 3a). The other one was assembled by the graphene-based film as described above (SHT G1000 in Fig. 3b). During the assembly, graphene-based film was stacked between the two substrates pre-coated with the functionalized graphene oxide and then dried together under pressure at 120°C for 6 h.

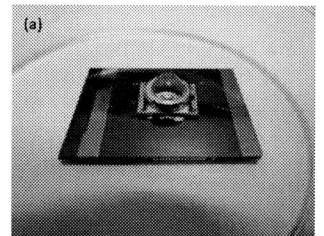

Figure 3. Layout of the demonstrators of TCA (a) and SHT G1000 (b).

2.2 Characterization

Thermal properties of self-developed graphene-based material were measured by Laser-flash machine with a model of LFA 447/2. A Raman spectrometer (Horiba Jobin Yvon Xpolora) was employed to identify the quality of the graphene-based film. Scanning election spectroscopy (JEOL JSM-6301F Scanning Electron Spectro-scope) was applied to check the inner-layered structure of the graphene-based film. FTIR measurement (Perkin Elmer Ultra Two) was carried out to identify chemical groups on the functionalized graphene oxide. An infrared camera with a type of Therm CAM PM595 NTSC was utilized to record the temperature change of different demonstrators.

3. Results and discussion

Table 1 shows the in-plane thermal conductivity of the graphene-based films. The thermal conductivity of 60 μm film was found to be around 1000 W/mK, while the thermal conductivity of natural graphite only reached 140-500 W/mK[11]. However, the in-plane thermal conductivity of the film was measured to reach 1600 W/mK, when thickness was decreased to 20μm. It can be seen that as the thickness of the films increases, the in-plane thermal conductivity reduces. As an explanation, with the increase of graphene layers, the number of conduction channels also increases, but the q phase available for Umklapp scattering increases even more, thus leading to a lower thermal conductivity[10].

Table 1. In-plane thermal conductivity of graphene-based films of different thicknesses.

Thickness of the film	In-plane thermal conductivity
20μm	1600W/m·K
40μm	1200W/m·K
60μm	800W/m·K

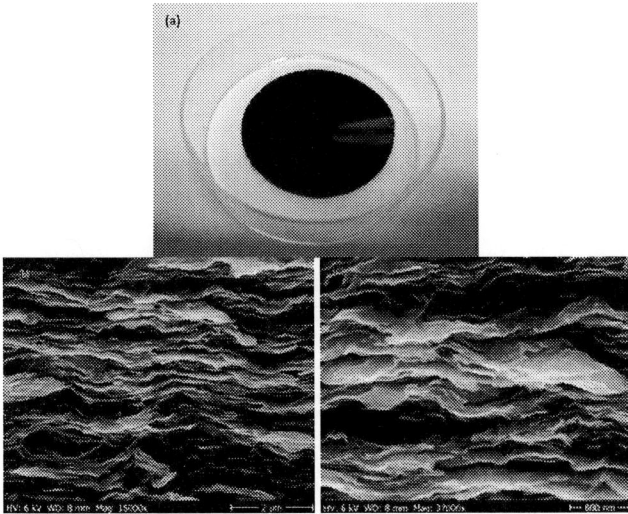

Figure 4. Layout (a) and SEM image (b,c) of the graphene-based film.

The SEM image of the graphene-based film in Fig. 4 indicates that the graphene layers are well aligned along horizontal orientation when being filtrated. This structure could explain the comparatively high in-plane thermal conductivity of the graphene-based films.

Raman shift of the graphene-based film is shown in Fig. 5. The D band at 1340cm⁻¹, G band at 1570cm⁻¹ and the 2D band at 2630cm⁻¹ was observed. The D/G peak ratio, which is used to detect the defects and disorders in sp^2 network, is nearly 1. It means that some defects still remained on the graphene-based film after chemical reduction. The double peak of the G band indicates that the oxygen containing groups (epoxy, carboxyl, hydroxyl) also formed some interlayer connections in the graphene-based film. From the 2D/G ratio, it could be concluded that the graphene structure in this film is multilayered[9].

FTIR analysis of silane modified graphene oxide was performed to confirm the functionalization. Fig. 6a shows the FTIR analysis results of pure graphene oxide film. Peaks at 3204 cm⁻¹ and 1028 cm⁻¹ represent -OH stretching, the peak at 1720 cm⁻¹ represents C=O vibration stretching and the peak at 1610 cm⁻¹ represents C=C stretching. In Fig. 6b, after functionalization, the vibration stretching of -C-H- in -CH₂- of the functionalizing material was detected at peak 2856 cm⁻¹ and 2924 cm⁻¹. At the same time, peaks around 1012 cm⁻¹ and 684 cm⁻¹ indicate the presence of -Si-O-Si- and -Si-C-, which provide more evidences for a successful functionalization[6,8].

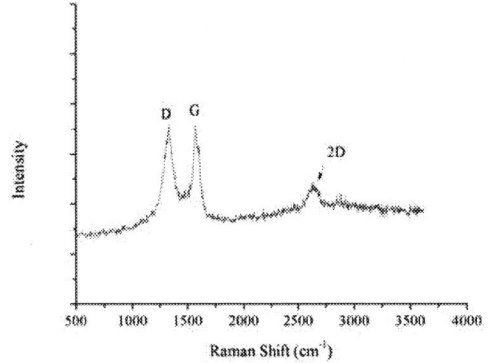

Figure 5. Raman shift of the graphene-based film.

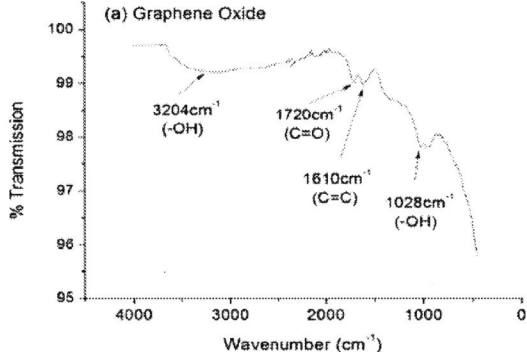

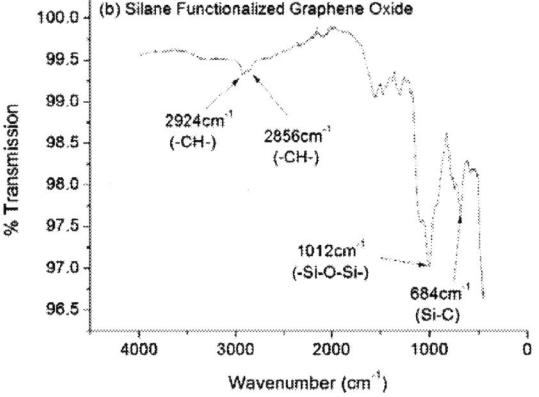

Figure 6. FTIR spectra of graphene oxide (a) and silane functionalized graphene oxide (b).

The combined thermal contact resistance was measured by Laser-flash and the results are shown in Fig. 7. The thermal resistance of two pre-sputtered copper plates with dry contact reached as high as 756 K mm²/W. After the copper plates were bonded with a 30 ± 5 μm of graphene-based layer without functionalization, the combined thermal contact resistance was reduced to 482 K mm²/W. However, with a same thickness of functionalized graphene-based bonding layer, the combined thermal contact resistance was further reduced to 98.7 K mm²/W. It is clear that the contact resistance between the different surfaces was reduced significantly by the silane functionalized graphene oxide. As an explanation, the interfacial heat transfer can be enhanced by the strong chemical bonding between the graphene-based

978-1-4799-2408-0/14 $31.00 © 2014 IEEE

layer and the silicon dioxide or aluminum oxide substrate surfaces.

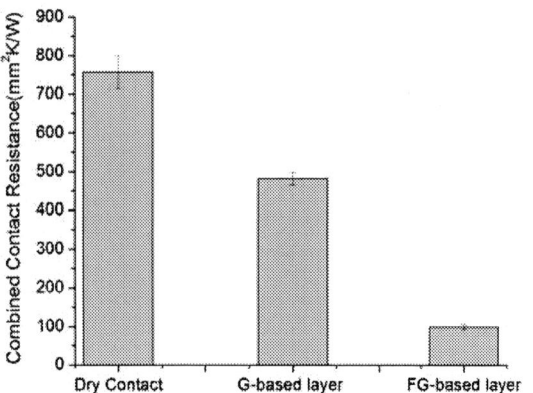

Figure 7. Combined thermal contact resistance of different bonding layers.

The result of the IR measurement is shown in Fig. 8. The temperature change of the LED surface was recorded by IR camera until the temperature is stable under the power density of approximately 300 W/cm². The peak temperature of the LED demonstrator bonded with commercial thermal conductive adhesive (TCA) reached 58°C after lighting for 3 minutes. However, the other demonstrator bonded with the self-developed graphene-based film (SHT G1000) and the functionalized layer showed a 5°C lower temperature compared to previous one, which has a peak temperature of 53°C. Additionally, the temperature increase of the SHT G1000 bonded demonstrator was observed to be slower than the one bonded with the adhesive. A lower peak surface temperature and a slower temperature increase indicate a more efficient heat transfer through SHT G1000 bonding, and illustrate a possibility for alleviating thermal management issues.

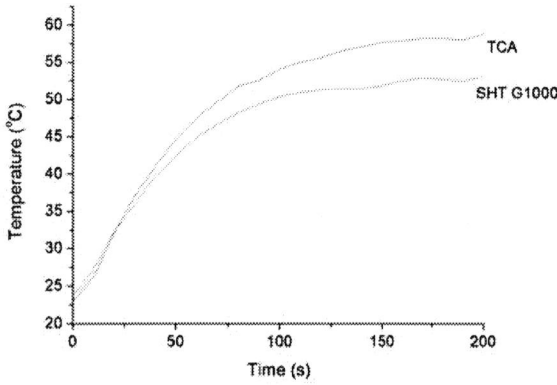

Figure 8. Temperature transient of TCA and SHT G1000 bonded demonstrators.

4. Conclusions

Increasing power density of electronic devices and thermal management of the systems has received substantial attention. How to effectively transfer the heat and make the systems cooler has become a hot topic in research. In this paper, we introduced a productive and cost-effective way to produce graphene-based films for thermal management. The developed graphene-based film (SHT G1000) showed very high in-plane thermal conductivity, and was integrated to a system for improved thermal management. A functionalized graphene oxide layer was also utilized to decrease interface thermal resistance between different substrates. IR camera measurements showed a 5°C lower temperature of a LED based demonstrator when using SHT G1000 instead of a thermally conductive adhesive, which indicates the promising potential of this graphene-based film for thermal management applications.

Acknowledgments

This work was supported by EU programs "Nanotherm", "Eranet Nano-TIM", and the SSF program "Scalable Nanomaterials and Solution Processable Thermoelectric Generators", contract no. EM11-0002. This work was also carried out as a part of the Sustainable Production Initiative and the Production Area of Advance at Chalmers. In addition to this, the work was also supported by the Shanghai Science and Technology Program (12JC1403900) and NSFC (51272153).

References

1. A. K. Geim, "Graphene: Status and Prospects," *Science*, vol. 324, pp. 1530–1534, June. 2009.

2. A. A. Balandin et al, "Superior Thermal Conductivity of Single-Layer Graphene,"*Nano Lett,* vol. 8, no. 3, pp. 902-907, Feb.2008.

3. Z. L. Gao, Y. Zhang, Y. F. Fu, M. F. Matthew, J. Liu, "Thermal Chemical Vapor Deposition Grown Graphene Heat Spreader for Thermal Management of Hot Spots,"*Carbon*, vol. 61, pp. 342-348, Sep.2013.

4. Khan M. F. Shahil , Alexander A. Balandin, "Graphene-Multilayer Graphene Nanocomposites as Highly Efficient Thermal Interface Materials,"*Nano Lett*, vol.12, pp. 861-867, Jan.2012.

5. J. Zhang, H. Yang, G. Shen, P. Cheng, J. Zhang and S. Guo, "Reduction of Graphene Oxide via L-ascorbic acid,"*Chem Comm*, vol.46, pp. 1112-1114. Dec.2010.

6. H. Yang, F. Li, C. Shan, D. Han, Q. Zhang, L. Niu, and A. Ivaska, "Covalent Functionalization of Chemically Converted Graphene Sheets via Silane and its Reinfrocement, "*Journal of Material Chemistry*, vol.19, pp. 4632-4638, May.2009.

7. R. M. Pasternack, S. R. Amy, Y. J. Chabal, "Attachment of 3-(Aminopropyl)triethoxysilane on Silicon Oxide Surfaces: Dependence on Solution Temperature,"*Langmuir*, vol.24, pp. 12963-12971, Oct.2008.

8. Y. Lin, J. Jin, M. Song, "Preparation and Characterisation of Covalent Polymer Functionalized Graphene Oxide,"*Journal of Material Chemistry*, vol.21, pp. 3455-3461, Oct.2010.

9. N. W. Pu, C. A. Wang, Y. M. Liu, Y. Sung, D. S. Wang, M. D. Ger, "Dispersion of Graphene in Aqueous Solutions with Different Types of Surfactants and the Production of Graphene Films by Spray or Drop Coating,"J*ournal of*

Taiwan Institute of Chemical Engineers, vol. 43, pp. 140-146, June.2011.

10. S. Ghosh, W. Bao, D. L. Nika, S. Subrina, E. P. Pokatilov, C. N. Lau, A. A. Balandin, "Dimentional crossover of thermal transport in few-layer graphene," *Nature Materials*, vol.9, pp. 555-558, July.2010.

11. ASEM InterPACK '05, Proceeding of IPACK2005, "Thermal Performance of Natural Graphite Heat Spreaders, http://graftechaet.com/CMSPages /GetFile.aspx?guid=94ab371b-d460-433a-8817-593bcc3eedfc.

High Performance Phase Change Thermal Interface Materials
Based on Porous Graphitic Carbon Spheres-Paraffin Wax Composite

Zhihua Cao[a,b], Kai Zhang[c], Guoping Zhang[a], Matthew M.F. Yuen[c], Ping Gu[b], Xianzhu Fu*[a], Rong Sun *[a], and C.P. Wong[d]

[a]Shenzhen Institutes of Advanced Technology, Chinese Academy of Sciences, Shenzhen 518055, PR China, [b]Department of Modern Mechanics, University of Science and Technology of China, Hefei 230027, PR China, [c]Department of Mechanical Engineering, Hong Kong University of Science and Technology, Hong Kong, P R China, [d]Department of electronic Engineering, The Chinese University of Hong Kong, Hong Kong, PR China
Email: xz.fu@siat.ac.cn, rong.sun@siat.ac.cn Tel: 86-755-86392151 Fax: 86-755-86392299

Abstract

Amorphous carbon sphere precursor with low thermal conductivity was first prepared by hydrothermal reaction of glucose. After treatment with KOH and $Fe(NO_3)_3$ at high temperature, highly porous and graphitic carbon spheres was obtained. The absorption of paraffin wax increases as the porosity increases, and the thermal conductivity of carbon increases with the graphitic degree. The graphitic degree of the carbon sphere was enhanced more than 90%. Paraffin wax could impregnate into the porous graphitic carbon spheres even more than 5 times mass. The porous graphitic carbon spheres-paraffin wax composite exhibited high thermal conductivity and latent heat as well as good shape stability, which is a promising thermal interface material (TIM).

Keywords: graphitic carbon spheres, thermal interface materials, phase change material, paraffin wax, porous

Introduction

As the power and packaging density of electronic devices greatly increases, thermal interface materials (TIMs) become more and more important for their thermal management in recent years.[1-5] Phase change TIMs with large latent heat absorption could be more effectively used to delay or modify the temperature rise of the electronics comparing to conventional TIMs. Paraffin waxes are usually used as phase change materials (PCMs) due to their advantages such as high latent heat, chemically stable, and commercially available at low cost, etc [6-10]. However, paraffin wax has low thermal conductivity and might overflow when it melts.

Herein, we developed a porous graphitic carbon microspheres-paraffin wax composite to enhance the thermal conductivity and prevent the overflow of paraffin wax as high performance phase change TIMs. The structure and properties of the porous graphitic spheres were also investigated.

Experiment

Preparation of carbon microspheres

A certain quality of glucose and $Fe(NO_3)_3.9H_2O$ were dissolved into aqueous solution respectively and mixed into 60 ml solution. After ultrasonic dispersing, the mixed solution was put into PTFE lining and hydrothermal reaction was happened when the hydrothermal synthesis reactor was heated in drying oven. The dark brown colloid was washed with anhydrous ethanol and deionized water after hydrothermal reaction. Then, carbon microspheres were obtained after drying and grinding.

Preparation of porous and graphitic carbon microspheres

As-prepared carbon microspheres were mixed with KOH solution and dried in drying oven. Then the drying powders were put into ceramic crucible. The drying powders were calcined in the tube furnace under nitrogen atmosphere; the heating rate was 5 °C/min. The porous carbon microspheres were got after washing the calcined product with HCl and deionized water. The preparation of porous graphitic carbon spheres was similar to porous carbon microspheres except for original material were porous carbon microspheres and $Fe(NO_3)_3.9H_2O$.

Preparation of Phase Change Materials (PCMs)

As-prepared porous carbon microspheres, or graphitic carbon microspheres and paraffin were mixed with petroleum ether and heated. Then, phase change thermal interface materials were obtained based on porous graphitic carbon spheres-paraffin wax composite.

Characterization and measurement

The morphology of samples was characterized by field emission scanning electron microscopy (FE-SEM, FEI Nova Nano SEM 450). The X-Ray diffraction (XRD, Rigaku D/Max 2500, Japan) with Cu-Kα radiation was taken to measure the crystallographic structure of the products. Adsorption–desorption measurements were conducted on a Micromeritics ASAP 2020 BET apparatus with liquid nitrogen at 77K. The thermal conductivity was investigated by TIM thermal resistance & conductivity measurement apparatus (LW-9389). DSC was used to measure the phase change temperature and the latent heat of the phase change materials.

Results and Discussions

As shown in Figure 1, carbon microspheres as precursor were obtained from hydrothermal reaction with glucose. There were some different between the products with and without $Fe(NO_3)_3.9H_2O$ addition in the preparation process. When the glucose solution without $Fe(NO_3)_3.9H_2O$, the carbon microspheres were smaller (300 nm) and linked together. As the glucose solution with $Fe(NO_3)_3.9H_2O$, the carbon microspheres were bigger (about 5000 nm), good degree of sphericity and apart from each other. It might be attributed to $Fe(NO_3)_3.9H_2O$ playing a catalytic role in the carbon microspheres formation reaction process. The bigger carbon microspheres were selected in the following process and the optical solution composed of 55 ml 0.5 mol/L glucose and 5 ml 0.3 mol/L $Fe(NO_3)_3$, the hydrothermal temperature is 180 °C and the soaking time is 480 min.

978-1-4799-2408-0/14 $31.00 © 2014 IEEE

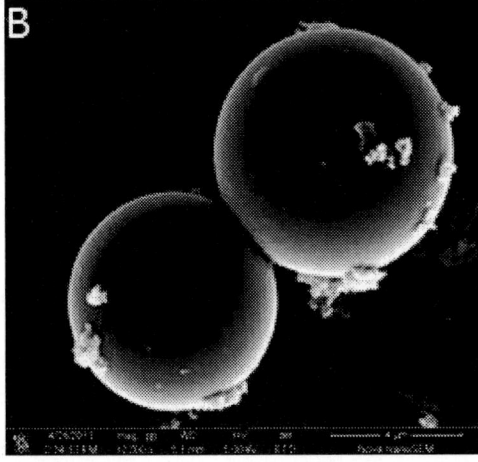

Figure 1. SEM of carbon microspheres: liquor of hydrothermal reaction without (A) and with (B) Fe(NO$_3$)$_3$.9H$_2$O.

Figure 2 illustrated the morphology of porous carbon microspheres prepared through calcinations with different mass ratio of KOH and carbon microspheres precursor. The temperature of calcination was 700 °C, the heating rate was 5 °C /min and the calcination time was 180 min. When the mass ratio of KOH and carbon microspheres precursor increased to 1.5:1, the BET specific surface area of carbon microspheres increased to 1381 m^2/g, which was about 83 times larger than the carbon microspheres precursor of 16.7 m^2/g (Figure 3). And the prepared samples remained good spherical and dispersity. However, the carbon microspheres were destroyed and divided into small parts when the mass ratio of KOH and carbon microspheres exceeded to 3. As shown in reaction process (1), the mixture of as-prepared carbon microspheres and KOH would react at high temperature [11-13]. The carbon was corroded by the KOH to form pores in the microspheres then the surface area became larger at high temperature. But the reaction would be excessive then the carbon microspheres were destroyed if the quantity of KOH were too much. To obtain large surface area and with good sphericity for phase change materials support, we chose the porous carbon microspheres prepared through calcinations with KOH: carbon microspheres precursor = 1.5:1 in the following experimental. The XRD pattern of the prepared porous carbon microshperes was shown in Figure 4. It was

obviously that the porous carbon microspheres were amorphous and very low graphitic.

The main reaction process is:

$$4KOH + C \rightarrow K_2CO_3 + K_2O + 2H_2 \qquad (1)$$

And the specific processes are:

$$2KOH \rightarrow K_2O + H_2O \qquad (2)$$

$$C + H_2O \rightarrow H_2 + CO \qquad (3)$$

$$C + K_2O \rightarrow 2K + CO \qquad (4)$$

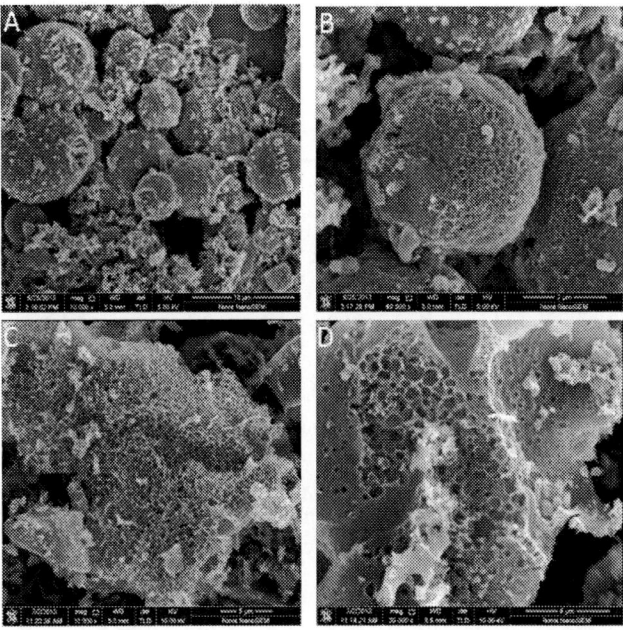

Figure 2. SEM of porous carbon microspheres prepared through calcinations with KOH: carbon microspheres precursor = 1.5:1 (A, B); KOH: carbon microspheres precursor = 3:1 (C, D).

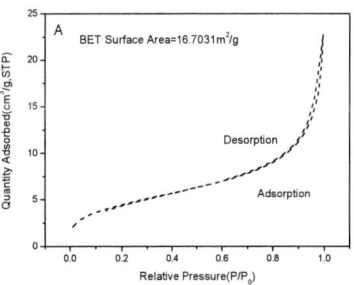

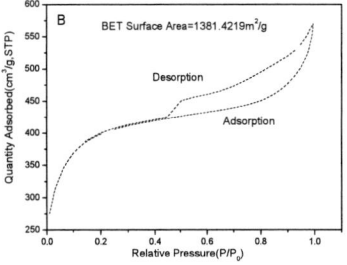

Figure 3. Nitrogen adsorption–desorption isotherms of carbon microspheres before (A) and after (B) KOH treatment at high temperature.

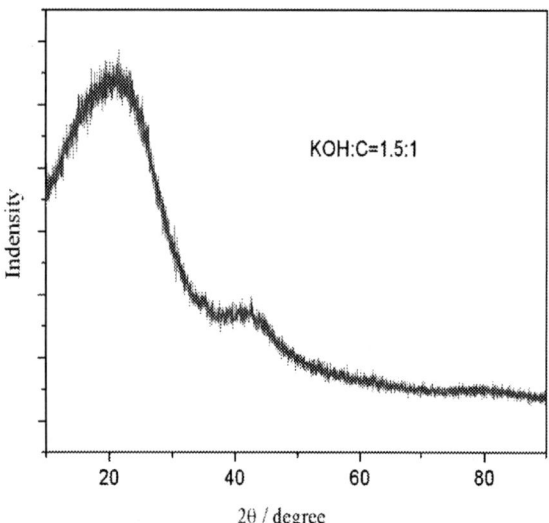

Figure 4. XRD curve of porous carbon microspheres prepared through calcinations with KOH: carbon microspheres precursor = 1.5:1.

The thermal conductivity of carbon is directly related with the graphitic degree, the amorphous carbon usually exhibits very poor thermal conductivity. To obtain highly graphitic porous carbon microspheres, the porous carbon microspheres absorbed $Fe(NO_3)_3$ solution and then heated at high temperature under N_2 atmosphere. Figure 5 displayed XRD patterns of the porous carbon microspheres-$Fe(NO_3)_3$ mixture after heating at 700 ^{o}C, and HCl solution treatment to remove the formed metal oxides. The graphitic degree of carbon could be calculated from the XRD pattern according to the formula (5). The degree of graphitization g:

$$g=(0.3440-d_{002})/(0.3440-0.3354) \qquad (5)$$

(002) surface layer spacing $d_{002}=\lambda/(2\sin\theta)$,

λ is X-ray wavelength ,

θ is (002) plane diffraction angle.

The degree of graphitization gradually improved with the ratio of $Fe(NO_3)_3.9H_2O$ and porous carbon microspheres increasing. The graphitic degree was about 80% as the mass ratio was 1.5:1, and reached to 90% when the mass ratio of $Fe(NO_3)_3.9H_2O$ and porous carbon microspheres increased to 3:1. The graphitic degree also could be reached to 90% for the $Fe(NO_3)_3.9H_2O$: carbon ratio of 1.5:1 when the heating temperature was 1400°C (Figure 6). $Fe(NO_3)_3.9H_2O$ was used us an effective catalyst for carbon graphitization here, thus highly graphitic carbon could be formed at relatively low heating temperature [14-16]. It could be observed in Figure 7 that the morphology of the porous carbon microspheres still remained the sphericity after heating with $Fe(NO_3)_3$ at high temperature. The spherical morphology is an advantage when the porous graphitic carbon as substrate of the phase changes materials for the thermal interface materials.

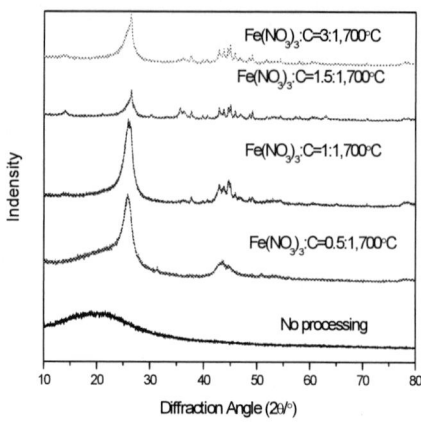

Figure 5. XRD patterns of porous graphitic carbon microspheres prepared through calcinations at 700 ^{o}C with different ratios of $Fe(NO_3)_3.9H_2O$.

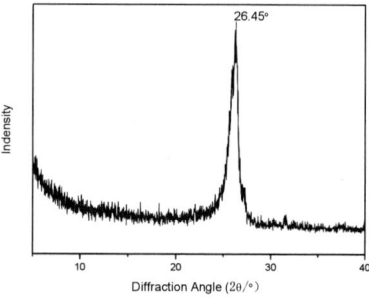

Figure 6. XRD curve of porous graphitic carbon microspheres prepared through calcinations at 1400 ^{o}C with $Fe(NO_3)_3.9H_2O$: porous carbon microspheres = 1.5:1.

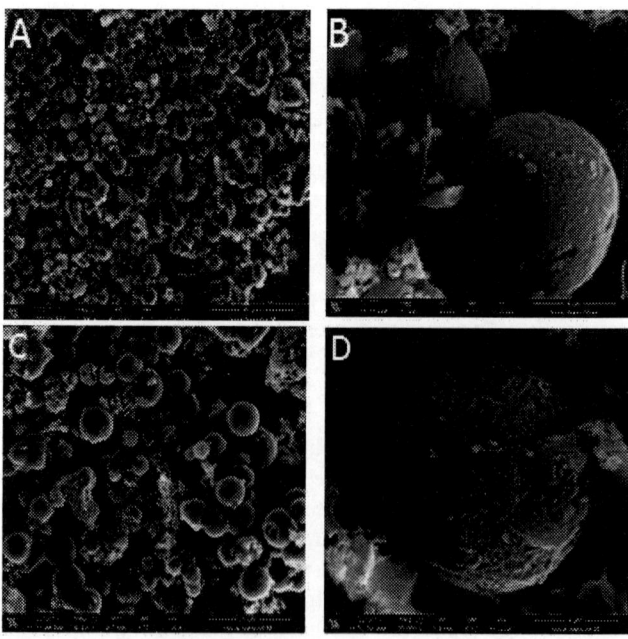

Figure 7. SEM of porous and graphitic carbon microspheres：$Fe(NO_3)_3.9H_2O$: porous carbon microspheres (mt) = 1:1; $Fe(NO_3)_3.9H_2O$：porous carbon microspheres (mt) = 1.5:1.

Phase change materials of paraffin wax have two problems for utilization as thermal interface materials for low thermal conductivity and easily overflow. Porous and graphitic carbon microspheres solve both problems simultaneously. The absorption mass of paraffin wax increased as the area surface of porous graphitic carbon spheres increased and the thermal conductivity of carbon increased with the graphitic degree. As shown in Figure 8, porous and graphitic carbon microsphere which absorbed 3-fold or 5-fold (mt) paraffin wax remained well dispersed and even when the PCMS were heated to 90 °C (higher than the melting point of paraffin wax), no liquid paraffin wax overflowed. Figure 9 showed that the characteristic peak of paraffin wax was 21.36 appeared in XRD of PCMs, so paraffin wax had been absorbed in the porous and graphitic carbon microspheres well and no liquid paraffin overflowed when the temperature was higher than the melting point of paraffin wax.

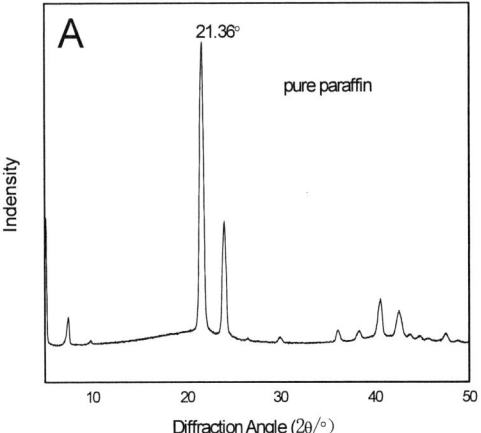

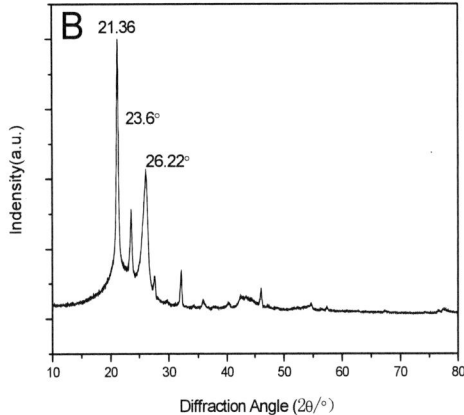

Figure 9. XRD of paraffin wax (A) and phase change thermal interface materials based on porous graphitic carbon spheres-paraffin wax composite (B, paraffin wax: porous and graphitic carbon microspheres (mt) = 5:1)

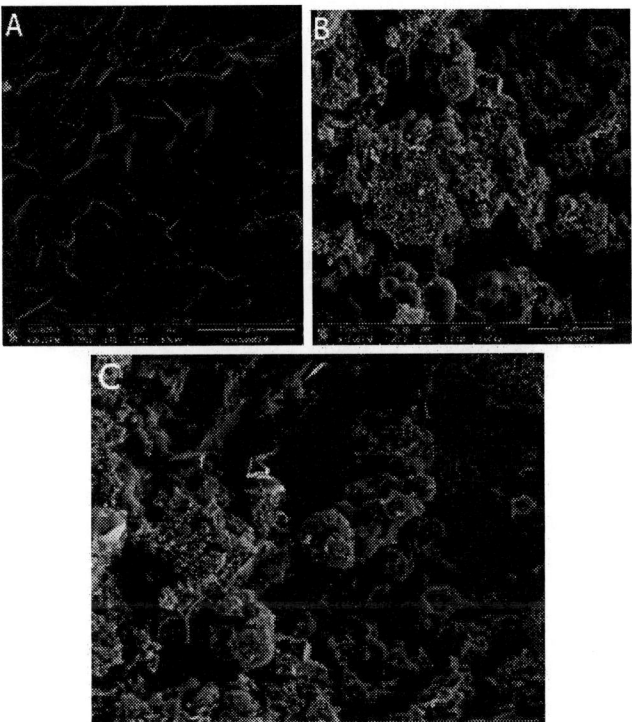

Figure 8. SEM of Phase Change Materials (PCMs): A. paraffin wax; B. paraffin: porous and graphitic carbon microspheres (mt) = 3:1; C. paraffin wax: porous and graphitic carbon microspheres (mt) = 5:1.

Figure 10 showed the DSC curves of pure paraffin wax and PCMs. The melting temperature of paraffin wax was 50.44 °C and the enthalpy of solid-liquid phase change (H_{pw}) was 126J/g. the melting temperature of PCMs was 51.95 °C and the enthalpy of solid-liquid phase change (H_{PCMs}) was 102J/g. The melting temperature became higher but the peak temperature became lower due to addition of porous graphitic carbon microspheres. By the contribution of porous graphitic carbon microspheres, the thermal conductivity of PCMs was improved and the time of paraffin's melting was short, so the peak temperature was lower. According to Formula (6), phase change enthalpy of PCMs changes with the size of phase transition enthalpy and organic matter content.

$$H_{PCMs} = H_{pw} \times mt \qquad (6)$$

where H_{PCMs} is the enthalpy of solid-liquid phase change of PCMs; H_{pw} is the enthalpy of solid-liquid phase change of paraffin wax; mt is mass fraction of paraffin in PCMs.

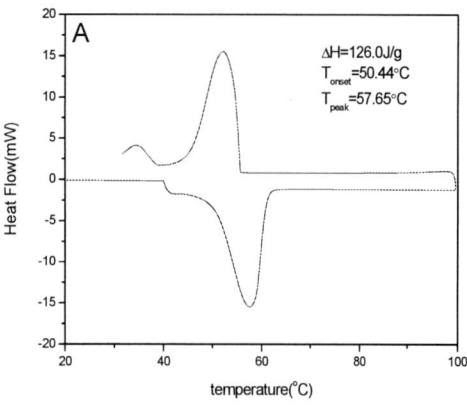

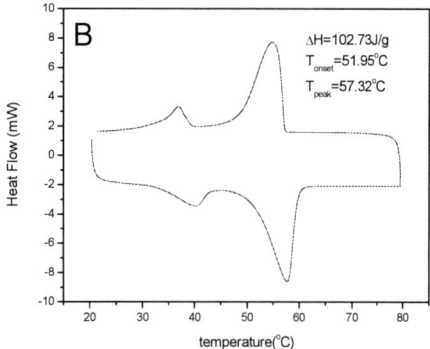

Figure 10. DSC of paraffin wax (A) and phase change thermal interface materials based on porous graphitic carbon spheres-paraffin wax composite (B, paraffin wax: porous and graphitic carbon microspheres (mt) = 4:1)

Figure 11 showed the thermal conductivities of pure paraffin wax and porous graphitic carbon microsphere-paraffin wax composite (80% paraffin wax). The thermal conductivity of pure paraffin was only 0.429W/m °C. And the thermal conductivity of PCMs containing 700 °C graphitized carbon microspheres and paraffin wax (80 wt%) was 1.312 W/m. °C which had been increased for about 300%. The over thermal performance of phase change energy storage material has been greatly improved by adding carbon microspheres and increasing the graphitic temperature.

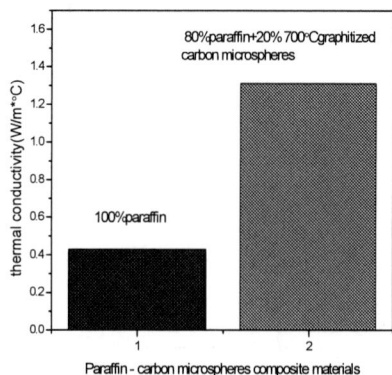

Figure 11. Thermal conductivity of phase change energy storage composite materials.

Conclusion

Highly porous and graphitic carbon microspheres were prepared as high thermal conductive substrate for phase change materials. Carbon microspheres precursor was synthesized by glucose hydrothermal process. After treatment with KOH and $Fe(NO_3)_3$ at 700°C, the carbon microspheres exhibited more than 1400 m^2/g BET surface area and 90% graphitic degree. The porous graphitic carbon microspheres had excellent adsorption ability of paraffin wax, even more than 4 times and no liquid overflowed when the temperature was higher than the melting point of paraffin wax. The high degree graphitization of carbon microspheres enhanced the thermal conductivity of paraffin wax phase change materials more than 200% and demonstrated high enthalpy of solid-liquid phase change. High performance porous graphitic carbon microspheres-paraffin wax composite was a promising thermal interface material.

Acknowledgments

This work was financially supported by the National Natural Science Foundation of China (No. 21203236), Guangdong and Shenzhen Innovative Research Team Program (No.2011D052, KYPT20121228160843692), and Shenzhen Electronic Packaging Materials Engineering Laboratory.

Reference

1. Wu,X.F. et al, "The importance of anharmonicity in thermal transport across solid-solid interfaces", Journal of Applied Physics, Vol. 115, (2014), pp.014901-014901-8
2. Vanga-Bouanga, C. et al, "Modification of polyethylene properties by encapsulation of inorganic material", Dielectrics and Electrical Insulation, Vol. 21, (2014), pp.1-4
3. Yao, Y. et al, "High-quality vertically aligned carbon nanotubes for applications as thermal interface materials", Components, packaging and manufacturing technology, Vol. 4, (2014), pp.232-239
4. Nowak,T. et al, "Approach for reliability of thermal interface materials in battery cell sensors", Thermal investigations of ICs and systems, (2013), pp.348-351
5. Ravi P., C. P. Wang. et al, materials for advanced packaging, Georgia Institute of Technology, 2009,pp.437-458
6. Sttella P. J. et al, "Heat transfer characteristics in latent heat storage system using paraffin wax", Journal of mechanical science and technology, Vol. 26, (2013),pp.959-965
7. Wang S. H. et al, "Rapid prototype mold for wax patterns with the help of phase change materials", Int J. Adv. Manuf. Thechnol.,Vol.62,(2012),pp.35-41
8. Garg H. P. et al, "Latent heat or phase change thermal energy storage", solar thermal energy storage , Centre of energy studies , Indian institute of technology , New Delhi, India, pp.154-291
9. Cemil A. et al, "Preparation, thermal properties and thermal reliability of form-stable paraffin/polypropylene composite for thermal energy storage", J Polym. Environ. ,Vol.17,(2009),pp.254-258
10. Jisoo J. et al. "Application of PCM thermal energy

storage system to reduce building energy consumption", J Therm. Anal. Calorim. Vol.111,(2013),pp.279-288

11. Qu D. Y. et al, "Studies of activated carbons used in double-layer capacitors", J Power Source, Vol.74,(1998),pp.99-107

12. Ishii C. et al, "Anomalous magnetism of activated carbon having ultra high surface area", Tanso. Vol.193,(2000),pp.218-222

13. Eheburger P. et al, "Active surface area of microporous carbons", Carbon, Vol.30,(1992),pp.1105-1109

14. Dhakate S. R. et al, "Catalytic effect of iron oxide on carbon/carbon composites during graphitization", Carbon, Vol.35,(1997),pp.1753-1756

15. Oya A. et al, "Electron microscopic study on the turbostratic carbon formed in phenolic resin carbon by catalytic action of finely dispersed nickel",Carbon,Vol.17,(1979),pp.71-76

16. Oya A. et al, "Influence of particle size of metal on catalytic graphitization on non-graphitizing carbons", Carbon, Vol.19,(1981),pp.391-400

High Sensitivity In-Plane Strain Measurement Using a Laser Scanning Technique

Hanshuang Liang[1], Teng Ma[2], Cheng Lv[2], Hoa Nguyen[1], George Chen[1], Hao Wu[1], Rui Tang[1], Hanqing Jiang[2], Hongbin Yu[1]

[1]School of Electrical, Computer and Energy Engineering, Arizona State University, Tempe, AZ 85287, USA
[2]School for Engineering of Matter, Transport and Energy, Arizona State University, Tempe, AZ 85287, USA
yuhb@asu.edu

Abstract

Unevenly distributed thermal strain, among materials with different coefficients of thermal expansion (CTE) within electronic packages under operation, affects the working performance for the whole unit. The points that experience the highest strain are most likely to be the failure locations. Knowledge of the strain distribution across the package is in great need in order to analyze the failure mechanism and consequently to improve the design of the packaging. The well-known strain mapping techniques presently in use include Moiré and digital image correlation (DIC) techniques. The former is featured as a full-field and highly sensitive strain sensing technique but is inadequate for spatial resolution when the strain variation of the interested area is very small. The latter is capable of achieving high strain sensitivity and high spatial resolution, but compromises the field of view. This work is to develop a strain sensing technique with high strain sensitivity and high spatial resolution while simultaneously achieving a large field of view. High strain sensitivity is validated by comparing the measured CTE values with reference values from homogenous materials. The system currently in use is capable of making measurements at the 10-micro strain scale which was proven by accurately measuring the CTE of Si. The spatial resolution has been studied by performing a one-dimensional scan across the pre-defined patterns with specific feature sizes. The designed SU-8/Si pattern is to model electronic packages with a simplified structure. Further studies will be continued on electronic packages.

Strain Sensing Techniques

Following the accelerated miniaturization trend, microelectronic packages have been pushed forward to be smaller and smaller in order to meet the demand of the market. The inevitable consequences include increased heat dissipation among layers of different packaging materials, that results in various degrees of deformations in the layers, which is due to a CTE mismatch. The deformation brings in non-uniform distributions of strain across the whole package, where the spots experiencing the highest strain are most likely to fail. There is a strong need in the electronics packaging community to investigate thermal induced strain spreading among layers at a fine scale. Therefore it is imperative to develop a strain sensing technique to map the in-plane strain distribution with high strain sensitivity, high spatial resolution, and a large field of view. The often used strain sensing techniques capable of strain mapping include micro Moiré [1-5], optical digital image correlation (DIC) [6-10] and scanning electron microscope (SEM) DIC [11-14], etc. Micro Moiré has been proven to be a highly sensitive, full-field in-plane sensing technique. However, the illuminated area for generating a Moiré pattern needs to be large enough

to detect small strain; consequently, it lacks the ability to resolve strain with small spatial variations. DIC techniques can achieve a high spatial resolution with high in-plane displacement resolution. However, the field of view is compromised since a large optical magnification is required, and becomes a limiting factor when detailed strain mapping in a large area is needed. The proposed strain sensing technology in this work is to fill a technology gap that requires: high strain sensitivity and high spatial resolution while maintaining a large field-of-view.

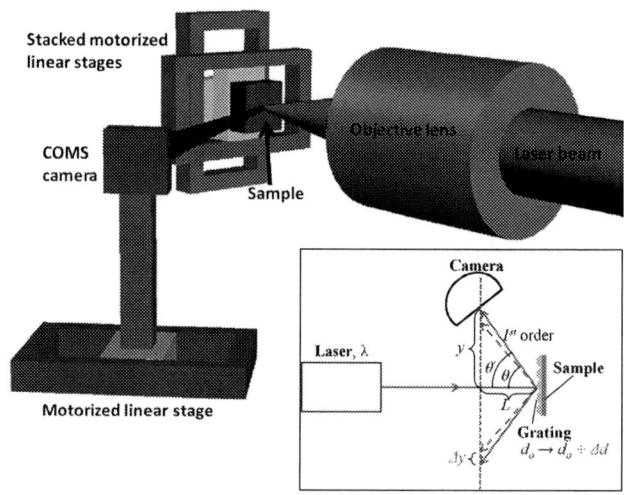

Figure 1. Schematic of the optical setup and the working principle (inset).

The schematic in Fig. 1 shows the working principle for the strain sensing technique used in this work. A collimated laser beam is focused on the grating which is fabricated onto the target substrate. The strain information is carried by the grating, in terms of grating wavelength since it varies according to the volume change of the underlying substrates. The camera is mounted in order to capture the 1st order diffraction light profile. By heating or stretching the grating, the diffraction angle changes due to the grating period variation, denoted as Δd, which is a result of the strain. This leads to a displacement in the peak position of the diffracted light, which is the measured parameter Δy. Two equations are used to derive the relationship between Δy and Δd. They are grating equation,

$$d_o \sin\theta = m\lambda,$$

where $m = 1$ for the 1st order diffraction, and the geometric relationship between the incident laser spot and the detector position,

$$\tan\theta = y/L.$$

Note, the laser wavelength λ and grating period d are in the order of micrometers, while y and L are on the order of tens of centimeter. Combining these two equations gives the relationship between Δy and Δd [15], which has a linear dependency with a magnification factor of as large as 10^6, when the proper initial values are chosen for λ, d_o and L. Therefore the optical setup is designed to amplify the nanoscale change of the grating period by almost six orders of magnitude into a diffraction peak shifting on the order of several millimeters on the detector side. This significantly amplifies the small displacement on the grating fabricated on the sample so that the desired sensitivity and accuracy is achieved.

Experimental studies have been conducted to demonstrate the strain sensitivity and spatial resolution of this laser scanning technique. Following that, this work also evaluates the feasibility of applying the strain sensing technique to composite samples, in terms of spatial resolution and thermal strain sensitivity. The tested sample is comprised of a composite of SU-8 and Si.

Demonstration of Strain Sensing Using Laser Scanning Technique: Strain Sensitivity and Spatial Resolution

The spatial resolution is studied using a specially designed grating pattern on a polished silicon substrate made with electron beam lithography (EBL) in this work. Alternatively, for a large area, sub-micron periodic grating structure, soft material contact lithography can be used. [16, 17] The grating wavelength is spatially varied on the substrate to model the strain distribution, where the strain is defined as $\Delta d/d_o$. By scanning the laser beam across the whole grating pattern, one can obtain the grating wavelength d versus the laser spot's position on the sample. By comparing the extracted grating wavelength distribution from the experiment with the original design, one can study the spatial resolution and accuracy of the strain sensing. The strain sensitivity is validated by extracting the coefficient of thermal expansion from silicon.

The EBL defined pattern, shown in Fig. 2 is designed to mimic the strain distribution across composite structures, such as SU-8/Si strips, with an exaggerated strain variation (around 3% difference in grating wavelength). The pattern is assembled with multiple domains, covered with a grating of constant wavelength or a gradient of wavelengths. The domains covered with a gradient of gratings are defined to be around 20 μm wide, while domains covered with a constant grating are defined to be 100 μm wide. From the left side of the pattern, as shown in Fig. 2, the grating wavelength d starts at 825 nm. After the 1st 100 μm wide domain, it ramps down gradually to 800 nm over a 20 μm wide area, and remains at the 800 nm wavelength for the following 100 μm domain. It then ramps back up to 825 nm and repeats the previous patterns. The whole grating pattern has a width of 460 μm and a length of 100 μm. The optical image of the grating shows the high quality of the EBL defined grating pattern. Since the designed grating wavelength variation is so small, one can hardly observe the grating wavelength change through a microscope even at 1000x magnification. The gradient strain distribution is expected at the interface of the two materials with mismatched CTEs, which is why a gradient grating is incorporated in the pattern. Although the strain gradient can be large and sharp at the interface of two materials in real cases, it is designed to be a much smoother gradient in the EBL pattern for simplified modeling.

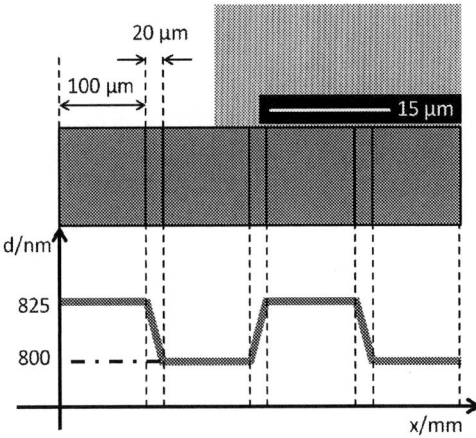

Figure 2. Design of the EBL pattern (schematic), and an optical image of the grating (top right corner with scale bar).

The one-dimensional scan was performed with a sample stage step size of 2 μm and a scanning distance of 520 μm, which was sufficient to scan across the whole pattern. The laser spot size on the grating surface is ~50 μm in diameter, which is smaller than the width of domains covered by the 100 μm constant grating pattern. The results are plotted in the contour plot as shown in Fig. 3 with a linear scale and in Fig. 4 with a log scale of the diffracted light intensity. For each sample position, the camera captured the profile of the 1st order diffraction profile, which is the combination of Gaussian peaks from the illuminated domains. It is plotted as a column for the corresponding sample position in the contour plot, where the diffracted light intensity can be seen in the varying signal intensity. Four plots of the diffraction light profile are displayed in Fig. 3, with enlarged images of the individual scan profiles at four positions. Scanning the sample gives a series of columns containing the diffraction light profile. The contour plot is formed by stitching these columns together. The y-axis of the contour plot displays the grating wavelength, which is calculated from the peak position of the diffracted beam from the camera side. From the contour plot, one observes that the diffraction peaks from 800 nm and 825 nm grating patterns are well separated at the camera side. The diffraction light intensities from all the constant grating domains are at a relatively stable level. The size of the domains with constant grating patterns are observed to be around 100 μm wide, however, the gradient domains between the constant domains turned out to be larger than the designed width of 20 μm. This is because the gradient domain areas are partially overlapping with the constant domains. In addition, an asymmetry is observed for the upwards and downwards transient domains, in terms of the light intensity and the width of the diffraction light profile. The next step is to extract the grating wavelength from the strongest peak at each sample position.

978-1-4799-2408-0/14 $31.00 © 2014 IEEE   471

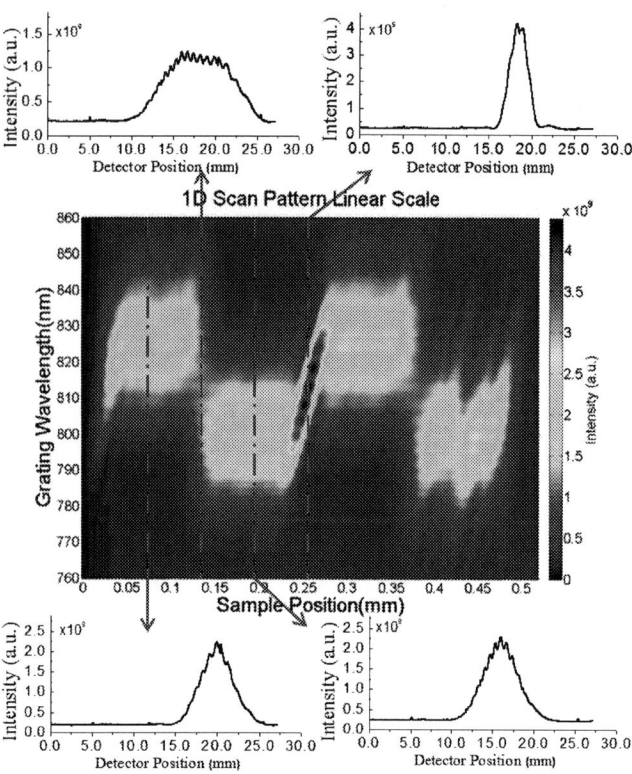

Figure 3. Demonstration of the strain mapping on an EBL defined gratings on Si substrate. Four plots of diffraction light profiles at four individual sample positions are displayed around the contour plot.

As displayed in Fig. 4, the extracted grating wavelength is plotted versus sample position as the black curve. It is superimposed onto the 1D contour plot which has a light intensity displayed in a log scale. The extracted grating wavelength from the 100 µm wide domains are 800 nm and 825 nm periodically, which is in agreement with the designed values. However, the flat region shows the constant grating wavelength is around 80 µm wide, narrower than the actual domain size. At the interfaces of the different domains, the grating wavelength shifts exhibit certain tendencies. The transition from 825 nm to 800 nm has a steeper slope compared to the transition from 800 nm to 825 nm. This is dependent on the laser spot size and the initial distance between the camera and the sample. This can be compensated by adding corrections to the scanning results after further investigation of this effect.

The fourth constant domain from the left side of the pattern displays imperfections in the grating pattern. The grating wavelength variation from the imperfect spots is only a few nanometers, which is barely noticeable with a high magnification optical microscope. On the other hand, it is highly pronounced from the results produced by the laser scanning technique. As marked on Fig. 4, the smallest resolvable feature size verified by the scanning results is 10 µm, which indicates its spatial resolution of 10 µm when applying a much larger laser spot size of 50 µm in diameter.

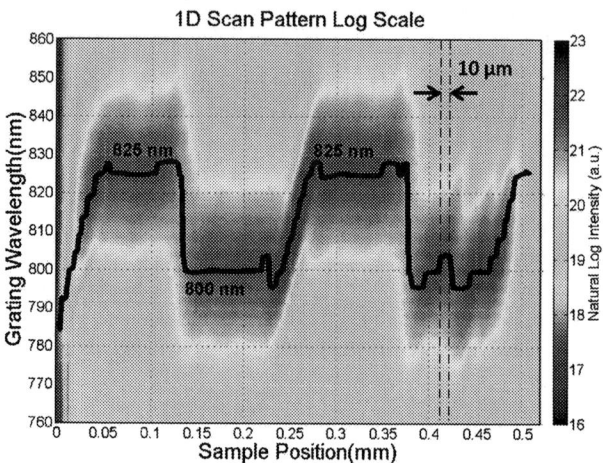

Figure 4. Contour plot of 1D scan results across the EBL defined pattern using a log scale, superimposed with the extracted grating wavelength from the contour plot.

Previous works have demonstrated the strain sensitivity with a PDMS grating bonded on a target substrates by applying this laser scanning technique [15]. In order to verify the strain sensitivity for a zero-thickness grating, (where the grating is directly fabricated onto the substrate without applying an additional adhesive layer,) on the target substrate, elevated temperature measurements were taken on a Si substrate that contained EBL defined grating patterns. The Si sample was mounted on a copper block which could be heated by inserting heating cartridges. The temperature was ramped from 25°C to ~50°C. At each temperature, a scan was performed at the camera side to record the profile of the diffraction signal. The peak position extracted from the diffraction light profile was utilized to calculate the grating period, $d = d_o + \Delta d$. Consequently, the strain was calculated as $\Delta d/d_o$ and is displayed on the y-axis in Fig. 5. Then the CTE of Si was extracted from the slope to be 2.74 ppm/°C, which matches the reference value of 2.6 ppm/°C. The small strain measured from the silicon sample is on the order of 10 micro-strain scale, and thus demonstrates the high strain sensitivity of this technique.

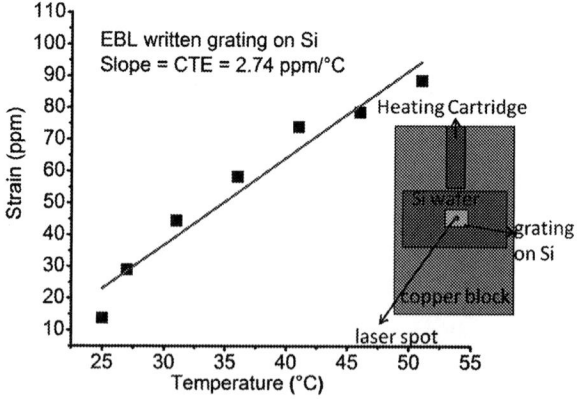

Figure 5. CTE extraction from the thermal measurement on a Si substrate covered with an EBL written grating. The schematic shows the thermal measurement setup.

Fabrication and Finite Element Analysis (FEA) of A Planarized Junction of SU-8/Si

In order to obtain the strain information at the junction of two dissimilar materials with different CTEs upon thermal loading, we fabricated a globally planarized junction composed of SU-8/Si. The fabrication of the SU-8/Si starts from a silicon on insulator (SOI) wafer. The top silicon layer is 10-20 μm thick and is patterned into silicon strips using a standard lithography process. The width of the silicon strips and the spacing are in the range of tens of microns to several hundreds of microns. Then an SU-8 layer is spin-coated on top to fill in the trenches completely. After hard-baking the SU-8 layer, deep reactive ion etching (DRIE) is used to etch the silicon substrate from the backside until the SiO_2 etch stop layer, which is then removed at the subsequent step using hydrofluoric acid (HF). Fig. 7 shows the scanning electron microscopy (SEM) image (a) of a SU-8/Si junction and the optical image (b) of a grating on the junction fabricated with EBL. The optical image doesn't focus well for the up-right-hand corner area since the junction surface is not perfectly flat and exhibits slight amounts of warping induced by the fabrication process.

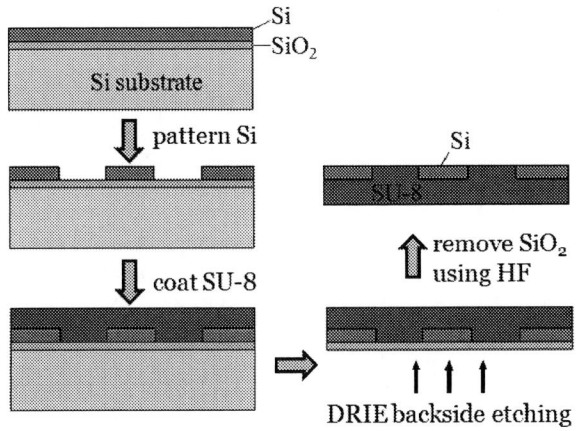

Figure 6. Fabrication flow of SU-8/Si junction

We should point out that this silicon surface is coming from the unpolished side of the device layer and is relatively rough compared to the polished silicon surface. As a result the rough surface can abate the reflection of the laser light. Also, the sidewalls of the silicon strips are relatively deep (>10 μm) and therefore are neither vertical nor perfectly smooth due to isotropic dry etching. After filling with the SU-8 and baking it is possible that cracks or delamination at the interface of the sidewalls are present due to the large CTE mismatch between SU-8 and Si.

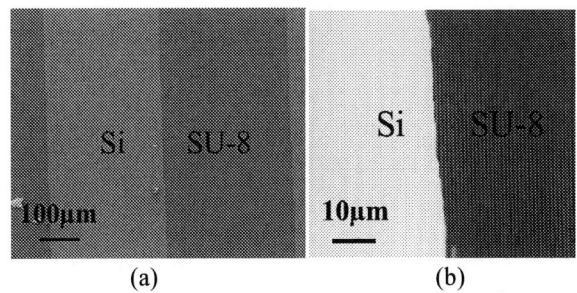

Figure 7. SEM image of SU-8/Si junction (a) and optical image of the gratings written on SU-8/Si junction using EBL (b).

The commercial finite element package ABAQUS is used to simulate the thermal deformation of the junction structure of SU-8/Si when subjected to temperature changes. Fig. 8(a) shows the model, including three 300-μm-width by 20-μm-thick silicon strips embedded in a 200-μm-thick by 2000-μm-long SU-8 substrate with 300 μm spacing between the strips. The thermal strain analysis is conducted by introducing a uniform temperature change ΔT in the whole domain. The silicon and SU-8 are modeled by a 4-node bilinear plane strain element (CPE4) for two different cases. In case one, the ideal bonding, the SU-8/Si interface is treated as shared nodes which indicates the perfect bonding between the silicon strips and the SU-8 substrate on both the bottom and the two sides. In the latter case, the weak bonding case, the silicon strips are connected to the SU-8 substrate only through the bottom using TIE constraint and there is no bonding with the SU-8 on the two sides of each silicon strip. The weak bonding case is used to simulate a scenario when the bonding is less than ideal and there is delamination on the two sides. The following material parameters are used in the analysis: $E_{SU-8} = 2$ GPa, $v_{SU-8} = 0.3$, $\alpha_{SU-8} = 52 \times 10^{-6}/°C$, $E_{Si} = 130$ GPa, $v_{Si} = 0.3$, $\alpha_{Si} = 2.6 \times 10^{-6}/°C$, $\Delta T = 45°C$, where E, v and α are Young's modulus, Poisson's ratio and the CTE, respectively. We assume that the surface is ideally flat and there is no warpage existing.

Strain contours in the horizontal direction on the top surface of the structure are shown in Fig. 8(b) for the ideal bonding case and (c) for the weak bonding case. Fig. 8(d) shows the strain as a function of horizontal distance on the top surface of the junction structure for the two cases. The strain on the SU-8 area is much higher than that of the silicon strips and when subject to a temperature change of 45 °C, the junction experiences a sudden strain change due to the CTE mismatch. The strain on the silicon surface fluctuate slightly and the two ranges are at the same level for both cases: $1.3 \times 10^{-4} \sim 2.9 \times 10^{-4}$ for the ideal bonding case and $8 \times 10^{-6} \sim 2.3 \times 10^{-4}$ for the weak bonding case. But the strain ranges differ greatly on the surface of SU-8, ranging from 3.7×10^{-3} to 6.4×10^{-3} while exhibiting very sharp peaks for the ideal bonding case. However, in the weak bonding case, the strain ranges from 3.8×10^{-3} to 4.1×10^{-3} while exhibiting blunt peaks This is reasonable since the surface of the SU-8 pattern has more constraints on the two sides from the silicon strips while under thermo-mechanical loading, and therefore has a steeper strain gradient cross the SU-8 surface along with a sharp strain jump on the edges for the ideal bonding case.

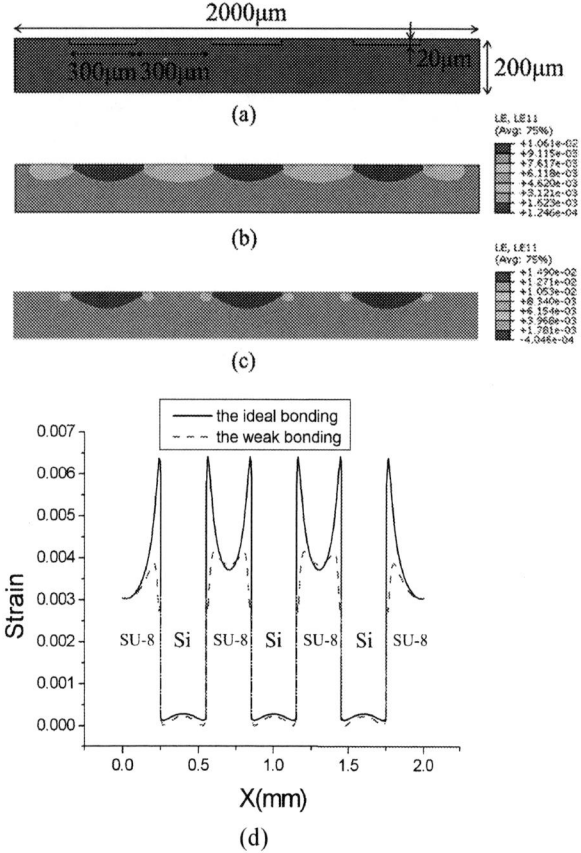

Figure 8. (a) Schematic of the SU-8/Si junction structure. (b) Strain contours in the horizontal direction on the surface for the ideal bonding case and (c) for the weak bonding case. (d) Strain as a function of the horizontal distance on the top surface of structure. Here the temperature change ΔT is 45°C.

Strain Mapping on SU-8/Si Composite Sample

Electronic packages are typically integrated with various different materials with a mismatched CTE, which leads to a complicated distribution of strain across the whole package. To evaluate the applicability of this laser scanning technique for strain mapping, we tested a composite sample as described in the preceding section to model a simplified case of strain distribution. The results were compared with thermal strain distributions calculated by FEA for a similar sample structure. The comparison between the measurement and simulation results evaluates the feasibility of applying this laser scanning technique towards advanced applications of electronic packages.

We first fabricated a uniform grating structure with an 800 nm wavelength across consecutive SU-8/Si strip samples using EBL and metal lift-off techniques. Ideally, for a uniform material, the grating wavelength will not change across the sample. After heating the composite sample made of CTE mismatched materials, such as SU-8/Si strips as we fabricated and discussed above, the sample will experience unevenly distributed thermal strain depending on the thermal expansion from the underlying substrates and the constraints from the surrounding materials. By measuring the degree of expansion at the surface, one can map the strain distribution on the composite sample. The grating pattern fabricated onto the composite sample contracts and expands to follow the expansion of the underlying substrate, and thus, records the strain information. By spatially scanning the sample with a small laser spot size, the grating wavelength variation is captured by the camera as a shift in the diffraction peak position, which can be translated back into strain information through data processing. Thus the SU-8/Si composite samples that were fabricated with a grating were tested to validate the capability of this strain sensing technique.

Figure 9(a) shows the optical image of one grating pattern fabricated onto the SU-8/Si composite sample. The whole grating pattern covers an area of 1 mm by 0.5 mm. As observed from the optical image, the grating lines are not perfectly aligned with the SU-8/Si strips, which affects the captured diffraction light signal when the laser is shining on the interface of the two different strips. Observing the high magnification optical image of the grating area confirms that the grating quality of the Si strips is not as good as the grating on the SU-8 strips. This is likely caused by the EBL lithography process that involves developing and lift-off in solutions that can distort the 300 μm thick SU-8 sample. The other direct observation from the optical image is that the whole SU-8/Si composite sample is not flat after fabrication with the EBL defined grating pattern, which can be the result of a non-flat surface from the epoxy that was used to attach the thin sample to the sample stage. In addition, the sample exhibits warping due to the thermal heating and immersion in acetone during the fabrication process of making the gold grating patterns. Therefore even at room temperature, the grating wavelength does not appear to be at a constant 800 nm across the entire pattern, even though the original intention was to design a uniform grating pattern. This can be seen in the contour plots for measurement results in Fig. 9. However, this non-uniform strain observed from the sample illustrates the power of this laser scanning technique which can delineate strain variation at a very high spatial resolution.

The laser beam was scanned across the whole grating region by increments of 5 μm, as demonstrated in Fig. 9(a) with the scanning area and scanning direction at room temperature (23°C). The testing results are depicted on the contour plot in Fig. 9(b). The diffraction light intensity from the SU-8 substrate is indeed stronger than the signal from the Si substrate, which is likely due to the wavering quality of grating patterns on those regions: the SU-8 surface is smoother than the unpolished side of the Si surface from the original SOI wafer. At room temperature, the extracted grating wavelength already appears to show small amounts of variances versus the laser spot position. The information recorded from the room temperature scan is a combination of factors that alter the diffraction angle from the grating pattern. Some of the factors can be explained, while others cannot be fully interpreted using the current system setup.

978-1-4799-2408-0/14 $31.00 © 2014 IEEE 474

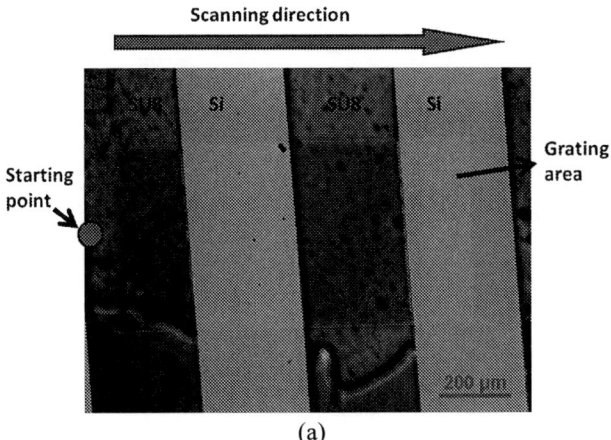

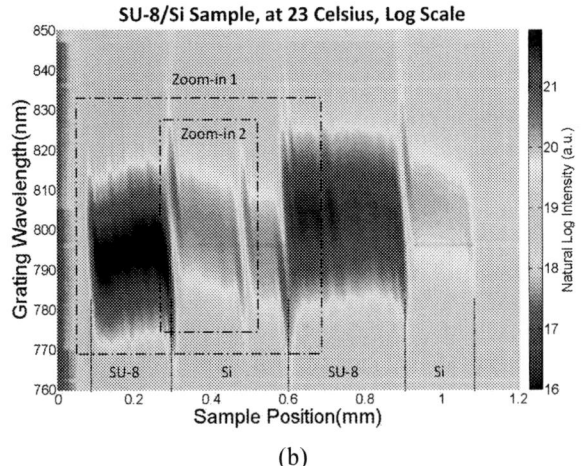

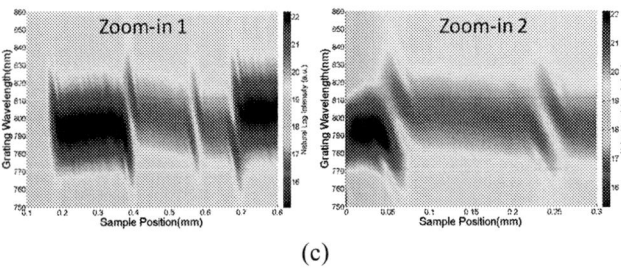

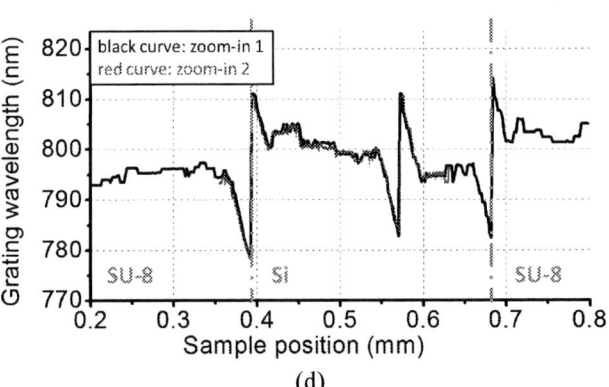

Figure 9. (a) Optical image of the grating area on the SU-8/Si substrate, marked with the scanning area and direction. (b) Contour plot of the 1D scan across the SU-8/Si composite

structure, using a linear scale. (c) Contour plots with smaller sample scanning step size, 2 μm (left) and 1μm (right), for the highlighted region in (b). (d) Superimposed plots of the extracted grating wavelength versus sample position from contour plots in (c).

In order to verify the repeatability of the recorded grating wavelength variation, two additional 1D scans with smaller step sizes were conducted at the same temperature, which also shows more detailed information. Both scans displayed in Fig. 9(c), cover the Si region where there is a discontinuity in the grating pattern which comes from a noticeable imperfection in the grating pattern. By superimposing the extracted grating wavelength versus sample position plots from the two zoom-in scans, one observes that the three small steps within the silicon region repeats itself in those two independent measurements. The height of the steps is also about 2~4 nm, which confirms that it is a signal coming from the sample instead of random noise. Note that the previous CTE extraction of silicon sample validates that even 10 micro-strain is detectable from this laser scanning technique. The dimension of the steps is between 20-30 μm. This indicates that the resolvable feature size from the laser scanning technique is at least 20 μm, thus demonstrating the spatial resolution. One strategy to improve the spatial resolution is to further reduce the laser spot size. Hence, with a sufficient number of sample steps and fine step sizes along with a proper laser spot size, one should be able to compose a comprehensive and accurate strain map.

The same pattern has been scanned at an elevated temperature of 68°C to examine the thermal strain and its spatial variation. The results are plotted in Fig. 10(a) as a contour plot. The diffraction signal shifts up, which indicates that the grating periodicity increases as a result of thermal expansion. The whole pattern exhibits a lateral shift as well, which is caused by the thermal drift in the sample. Quantitative comparison is performed by superimposing the plots of the extracted grating wavelength versus sample position under two different temperatures, which can be seen in Fig. 10(b). The SU-8 region expands more under thermal loading than the Si regions, which is consistent with the simulation predictions. Something interesting is that the width of the heated up pattern is wider than the width of the pattern before heating. Therefore, compensations have been applied for the calculation of the grating wavelength difference, but are not displayed in the grating wavelength versus sample position plot in Fig. 10(b). The strain distributions within the highlighted regions are plotted. From left to right on the SU-8 strip, the strain varies between 6×10^{-3} and 3×10^{-3}. On the Si strip, the strain variation is between 6×10^{-4} and 1×10^{-3}. The experimental results from SU-8 region are slightly smaller than the FEA predictions while the experimental results from Si region are larger than the finite element simulations. However, comparing the experimental results between the different regions matches the simulation work.

978-1-4799-2408-0/14 $31.00 © 2014 IEEE

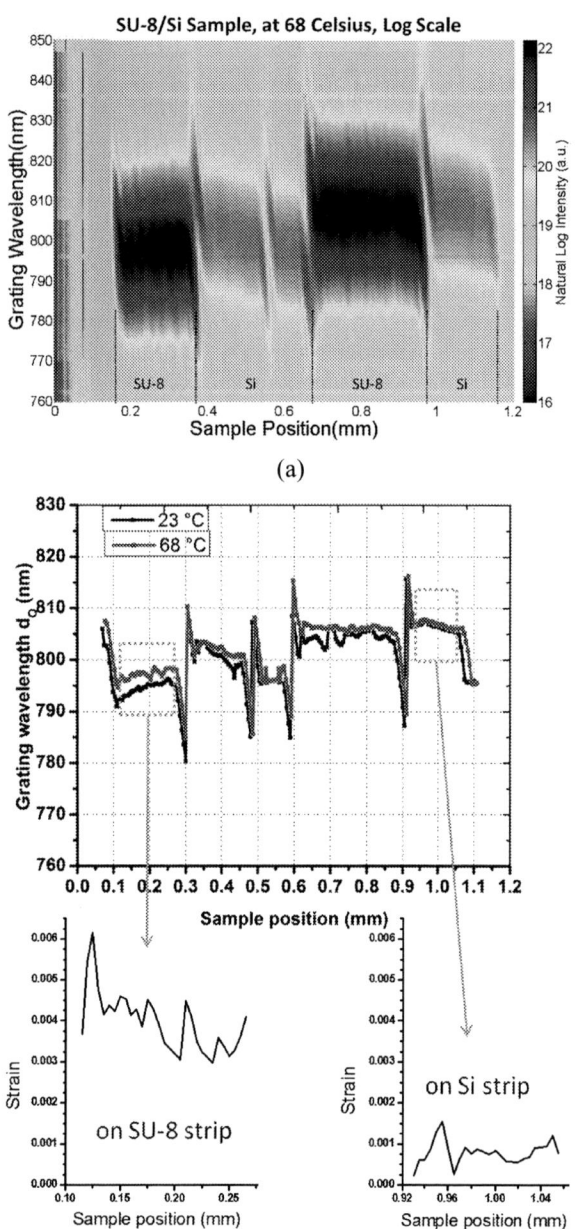

Figure 10. Contour plot of SU-8/Si composite sample at 68°C is presented in (a). The corresponding extracted grating wavelength is plotted as red curve in (b), while the extracted grating wavelength at 23°C (Fig. 9(b)) is plotted as the black curve. Strain is calculated for SU-8 and Si regions, based on the difference between the two temperatures.

Strain distribution analysis isn't done on the middle region. This is because the strain information in these regions is dominated by other effects, such as warping and imperfections on the grating patterns. Although these factors may impact the regions where analysis is done, they are minor effects compared to the effects discussed in the FEA work.

Conclusions

The laser scanning technique is demonstrated to have both high strain sensitivity and high spatial resolution on pre-

defined samples. Note that this technique is capable of detecting localized strain, unlike Moiré techniques (which relies on a sufficient field of view to form Moiré pattern).The CTE measurement for silicon proves that the detectable strain to be as small as 0.0001. Even though the entire elevated temperature range is only 30°C, it indicates the capability for monitoring small CTE materials such as Si. This has many applications in different packaging processes, such as reflow, which typically occurs at over 200°C. Scanning the EBL defined pattern validates the resolvable feature size to be as small as 10 μm. This technique in principle, is able to scan an unlimited field of view, which is determined by the traveling distance of the translation stage. The investigation of strain distribution on the composite sample under thermal loading validates the feasibility of applying this technique towards electronic packages. Future work includes plans for optimizing the system for advancing the strain mapping capability as well as to test this technique on electronic packages.

Acknowledgments

Hanshuang Liang and Teng Ma contributed equally to this work.

We acknowledge the support from Intel Corporation through an ASU Consortium, Connection One, and stimulating discussions with Ravi Mahajan, Benny Poon, and Min Tao. TM acknowledges the financial support from the China Scholarship Council. GC and HN would like to thank the Fulton Undergraduate Research Initiative (FURI) program at Arizona State University for providing partial funding. HJ acknowledges the support from NSF under Grant No. CMMI-0700440. HY acknowledges the support from NSF under Grant No. ECCS-0926017. HY and HJ acknowledge the support from NSF under Grant No. IIP-1343474.

References

1. J. D. Wood, *et al.*, "Measurement of microstrains across loaded resin-dentin interfaces using microscopic moire interferometry," *Dental Materials,* vol. 24, pp. 859-866, Jul 2008.
2. B. Chen and C. Basaran, "Automatic full strain field Moire interferometry measurement with nano-scale resolution," *Experimental Mechanics,* vol. 48, pp. 665-673, Oct 2008.
3. B. Han and Y. Guo, "Thermal deformation analysis of various electronic packaging products by Moire and microscopic Moire interferometry," *Journal of Electronic Packaging,* vol. 117, pp. 185-191, Sep 1995.
4. C. M. Liu and L. W. Chen, "Digital atomic force microscope Moire method," *Ultramicroscopy,* vol. 101, pp. 173-181, Nov 2004.
5. J. McKelvie, "Moire strain analysis: an introduction, review and critique, including related techniques and future potential," *Journal of Strain Analysis for Engineering Design,* vol. 33, pp. 137-151, Mar 1998.
6. Y. H. Zhou, *et al.*, "Large deformation measurement using digital image correlation: a fully automated approach," *Applied Optics,* vol. 51, pp. 7674-7683, Nov 2012.
7. C. Cofaru, *et al.*, "Pixel-level robust digital image correlation," *Optics Express,* vol. 21, pp. 29979-29999, Dec 2013.

8. D. Spera, *et al.*, "Application of Stereo-Digital Image Correlation to Full-Field 3-D Deformation Measurement of Intervertebral Disc," *Strain*, vol. 47, pp. E572-E587, Jun 2011.

9. M. A. Sutton, *et al.*, "The effect of out-of-plane motion on 2D and 3D digital image correlation measurements," *Optics and Lasers in Engineering*, vol. 46, pp. 746-757, Oct 2008.

10. H. Lu and P. D. Cary, "Deformation measurements by digital image correlation: Implementation of a second-order displacement gradient," *Experimental Mechanics*, vol. 40, pp. 393-400, Dec 2000.

11. J. Kang, *et al.*, "Microscopic strain mapping using scanning electron microscopy topography image correlation at large strain," *Journal of Strain Analysis for Engineering Design*, vol. 40, pp. 559-570, Aug 2005.

12. B. Winiarski, *et al.*, "Surface Decoration for Improving the Accuracy of Displacement Measurements by Digital Image Correlation in SEM," *Experimental Mechanics*, vol. 52, pp. 793-804, Sep 2012.

13. N. Li, *et al.*, "Full-field thermal deformation measurements in a scanning electron microscope by 2D digital image correlation," *Experimental Mechanics*, vol. 48, pp. 635-646, Oct 2008.

14. H. Jin, *et al.*, "Micro-scale deformation measurement using the digital image correlation technique and scanning electron microscope imaging," *Journal of Strain Analysis for Engineering Design*, vol. 43, pp. 719-728, Nov 2008.

15. T. Ma, *et al.*, "Micro-strain sensing using wrinkled stiff thin films on soft substrates as tunable optical grating," *Optics Express*, vol. 21, pp. 11994-12001, May 2013.

16. K. Chen, *et al.*, "Facile large-area photolithography of periodic sub-micron structures using a self-formed polymer mask," *Applied Physics Letters*, vol. 100, Jun 2012.

17. C. J. Yu, *et al.*, "Tunable optical gratings based on buckled nanoscale thin films on transparent elastomeric substrates," *Applied Physics Letters*, vol. 96, Jan 2010.

978-1-4799-2408-0/14 $31.00 © 2014 IEEE

Biophysicochemical Evaluation of Passivation Layers for the Packaging of Silicon Microsystems in Medical Devices

Jorge Mario Herrera Morales [1], Jean-Charles Souriau [1], David Ratel [1], François Berger [1], Gilles Simon [1]

[1]CEA Leti, MINATEC Campus; 17 rue des Martyrs; 38054 Grenoble - France

Abstract

This paper is dedicated to the research and development of hermetic and biocompatible packaging of microelectronic devices for medical applications. Materials such as Al_2O_3, HfO_2, TiO_2, ZnO, SiN, SiO_2, SiOC, SiC, a-CH, and BN which had been reported as biocompatible and compatible with fabrication in standard clean rooms were studied. As there were many candidate materials, a method for selecting the best passivation layers was implemented. Our methodology consisted of screening materials by accelerated aging tests in Phosphate-Buffer Saline (PBS, a saline solution that resembles the human blood serum), gas permeability tests and in-vitro cytotoxicity tests. To quantify the stability of each material in PBS, the corrosion rate (defined as the amount of corrosion loss in thickness over time) of each packaging layer was monitored over time by Variable Angle Spectroscopic Ellipsometry (VASE). Helium gas permeability measurements of selected packaging layers are also provided. Finally, candidate layers were confirmed to be non-cytotoxic according to the norm for in-vitro evaluation of medical devices ISO10993-5.

Introduction

Facing an aging population, global demand for implantable devices and in-situ medical diagnostics is expected to greatly increase in the coming years. Nonetheless, medical implantable devices face two challenges: miniaturization and biocompatibility. Miniaturization could be addressed by 3D silicon technology that allows for a drastic reduction in the volume of electronics needed for portable applications like mobile phones, tablets, nomadic handsets, etc. This high integration technology could be adapted to medical implants and related electronic devices (Fig.1).

However, research and development of biocompatible and hermetic encapsulants for the packaging of 3D silicon devices are more recent. The study and development of biocompatible coatings must be addressed early on for the biomedical use of 3D microelectronic devices to become a reality [1]. This will enable the spreading of 3D silicon technology to existing medical applications, and could participate in novel applications such as advanced neurology, functional implants, smart prosthesis, etc.

In the medical electronics area, the ISO 10993 battery of tests represents the minimum biocompatibility requirements that must be met for the pre-clinical and clinical evaluation of medical devices [1]. Moreover, the Food and Drug Administration (FDA and other medical regulatory bodies around the world demand as well that whenever novel Si devices are used inside a medical device, they must be coated with a biocompatible encapsulant that protects the patient in case the outer plastic or metallic housing containing the device fails or is infiltrated by external biofluids. Therefore, this biocompatible encapsulant must not allow the leaking of harmful substances into or out of the Si device in case there is direct contact with chemically aggressive environments such as blood. To summarize, the biopackaging layer of 3D silicon devices must be biocompatible, biostable and act as good bi-directional barrier to the diffusion of toxic or corrosive substances from or into the device. In addition, it should be compatible with the fabrication process flow of silicon devices in standard clean rooms in order for this technology to be economically interesting compared to traditional approaches for the packaging of medical devices such as glass encapsulation or hermetic welded titanium canisters.

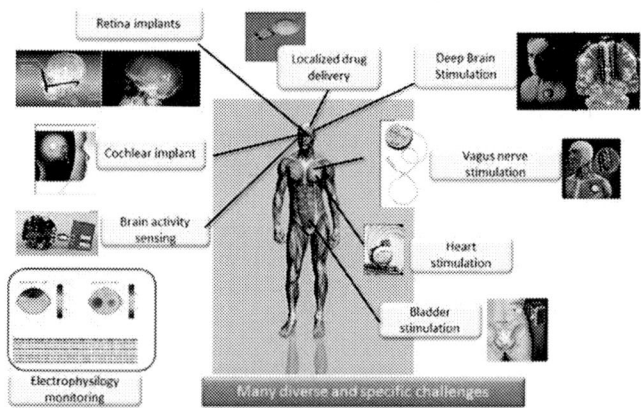

Figure 1. Diverse and specific challenges for the use of 3D silicon technology in smart medical devices.

Strategy and screening methodology

Our strategy for screening encapsulants consisted of choosing packaging materials which had been reported as biocompatible in the literature and could be readily deposited by chemical vapor deposition in our clean room facilities. Our candidate materials were Al_2O_3, HfO_2, TiO_2, ZnO, SiN, SiO_2, SiOC, SiC, a-CH, and BN [1-5].

As packaging layers from these materials having high conformality and density (which correlates with hermeticity) were needed, there were two deposition methods readily available for us: Plasma-enhanced Chemical Vapor Deposition (PECVD) and Atomic Layer Chemical Vapor Deposition (ALD). ALD produces thin films of high density and excellent conformality at relatively low temperatures (80-250°C), but only one atomic layer is deposited per deposition cycle. One deposition cycle lasts typically from 5 to 60 sec, depending on the temperature of deposition and whether plasma is used or not. For this reason, it is impractical to grow layers larger than 50 nm by ALD. On the other hand, PECVD can rapidly produce films with high density and good conformality larger than 50 nm, but it does so at middle range temperatures (300-400°C). The choice of deposition method will depend finally on the thermal budget of the intended application (e.g. encapsulation of CMOS, OLEDs, OPV, MEMS, etc.) and the intended level of hermeticity.

On the other hand, as there were many candidate packaging layers, a screening method for selecting the most biocompatible, biologically stable and hermetic layer was necessary. Our screening methodology consisted of evaluating selected packaging layers by:

(1) Accelerated aging tests in a saline solution that simulates the chemical environment found inside the human blood serum, such as Phosphate-Buffer Saline (PBS).

(2) Gas permeability tests to assess the hermeticity of the packaging layers to external agents like moisture or oxygen.

(3) Cytotoxicity and morphology tests to confirm the biocompatibility of the selected layer.

The passivation layers shown in Table 1 were deposited on Si and Kapton coupons by thermal Atomic Layer Deposition (ALD) in a Fiji F200 machine from Cambridge Nanotech Inc. or Plasma-enhanced Chemical Vapor Deposition (PECVD) in a Centura 5200 from Applied Materials Inc. Deposition details can be found elsewhere [1-5].

Table 1. Packaging layers deposited on Si[a] and Kapton[b] substrates.

Material	Thickness /nm	Fabrication
Al_2O_3	30	ALD 250°C
HfO_2	14	ALD 250°C
TiO_2	14	ALD 250°C
ZnO	30	ALD 250°C
SiN	100	PECVD 400°C
SiO_2	100	PEVCD 400°C
SiOC	100	PECVD 400°C
SiC	100	PECVD 350°C
a-CH [c]	100	PECVD 400°C
BN	50	PECVD 400°C

[a] Si monitor p-type (001), 1-50 Ω•cm, 725 μm thick, size 2.0x1.5 cm.

[b] Kapton 200HN supplied by DuPont, thickness 50.8 μm, diameter 65 mm.

[c] Amorphous Hydrogenated Carbon films (50% C, 50% H), a.k.a DLC films.

These samples were then kept inside a nitrogen chamber until the different screening tests were performed.

Aging test

Si coupons of 2.0x1.5 cm were diced from 200 mm polished Si(001) p-type wafers in our clean room facilities. These coupons were then cleaned by rinsing with 18.2 MΩ•cm de-ionized (DI) water, followed by acetone, isopropyl alcohol and again DI water. A nitrogen blow gun was used to dry the coupons and then they were dehydrated on a hot plate at 125°C for 15 min. Packaging layers were then deposited on these coupons.

For the immersion part, 3-4 coupons of each packaging layer were placed on the bottom of a borosilicate petri dish and PBS solution pored over them until a fixed level of ~80% the total volume. The PBS solution was supplied by Euromedex Inc. as a 0.2 μm-filtered 10X solution containing KH_2PO_4 10.6 mM, Na_2HPO_4•$2H_2O$ 30.0 mM, NaCl 1.54 M and dissolved to 1X with bi-distilled water of 18.2 MΩ•cm. Heating was performed on a hot plate whose temperature was set to 54-60°C as measured by a portable electronic thermometer with an accuracy of ±0.1°C. This temperature range is reported here as 57°C.

The PBS solution was changed completely every week and losses due to evaporation of solution during the week were compensated by additions of DI water until reaching the original level of liquid.

Coupons were removed from the PBS solution and cleaned with DI water, a soft tissue and a nitrogen blow gun before all thickness measurements. Those materials showing problems of corrosion or a significant change in thickness over time at 57°C were further evaluated by aging tests at 37°C. The goal was to determine if such corrosion problems would occur at 37°C as well.

Aging tests by immersion in PBS at 37°C were performed on a hot plate. Here as well, it was difficult to precisely regulate the temperature of the hot plate. The temperature of the PBS solution was measured in the range of 34-40°C. Such temperature range is reported here as 37°C.

To quantify the stability of each packaging layer in PBS, the film thickness variation of each coupon was monitored over time by Variable Angle Spectroscopic Ellipsometry (VASE). All thickness measurements were performed using a computer-controlled variable angle of incidence spectroscopic ellipsometer of the rotating analyzer type by J.A. Woollam Co., Inc. The measurement area of 6x4 mm² was always taken at the center of the coupon with the help of a paper support with alignment marks (Fig. 2).

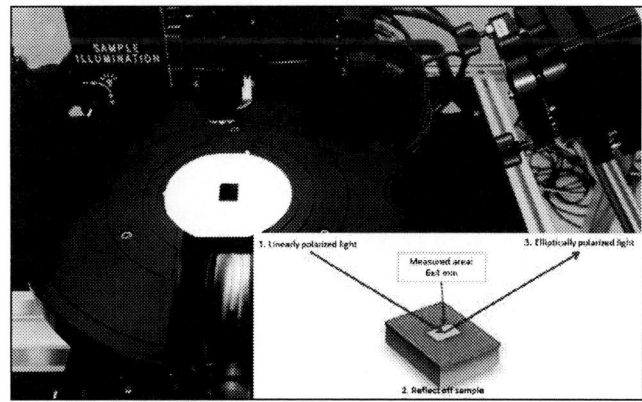

Figure 2. A typical measurement of film thickness by variable angle spectroscopic ellipsometry (VASE).

A typical measurement was made in the wavelength range of 400–1700 nm with 5 nm intervals at an incident angle of 65°. For the analysis the VASE software WVASE32 (version 3.774) was used. In most cases, the Mean Squared Error (MSE) of the thickness estimate by data fitting models was not larger than 5. The error estimates of film thickness vary between 0.1 and 1.0 nm. For example, the average thickness of 15 coupons coated with 14.0 nm ALD TiO_2 (according to the manufacturer of the Fiji F200™ machine) was 13.9±0.1 nm and all samples had a MSE smaller than 2. However, the main difficulty to monitor the change of film thickness over time was to measure exactly the same location on the coupon. It was noticed that shifting the measured area by 0.2 cm resulted in film thickness differences of 0.5-2.0 nm. Hence, a measurement error of ±2.0 nm can be expected when measuring the variation of film thickness over time.

The thickness variation of different packaging layers after 4-8 weeks of immersion in PBS at 57°C is plotted in Figure 3.

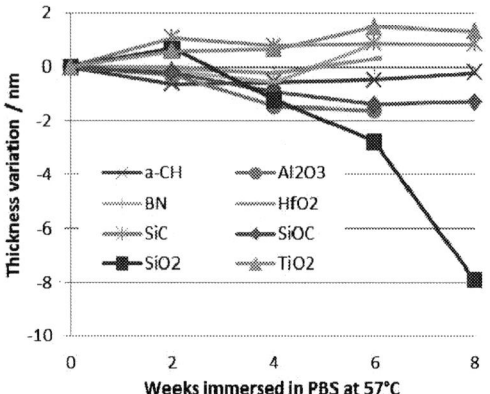

Figure 3. Thickness variation of different packaging layers immersed in 57°C PBS. SiN is off the chart (2 weeks,-97 nm).

The change of thickness was computed as the current minus the original (day 0) thickness for each coupon. Values plotted in Figure 3 are the average of three or four coupons measured per material. SiN coupons lost on average of 95-99 nm after two weeks of immersion in PBS at 57°C and are not plotted in the graph.

From Figure 3, it can be seen that there are packaging layers which pass our criteria of stability of a film thickness variation smaller than 2 nm after 6-8 weeks in PBS at 57°C. They are Al_2O_3, HfO_2, TiO_2, SiC, SiOC, BN and a-CH. In Figure 4, unspoiled coupons of TiO_2 and SiC after 8 weeks in 57°C PBS are shown. The small increase of thickness of some packaging layers in Figure 3 is probably due to precipitation of salts on top of the coupons or due to the accuracy of our measurements being ±2 nm.

Figure 4. TiO_2 (Top) and SiC (Bottom) coupons after 8 weeks immersed in PBS at 57°C. No apparent changes are visible.

Those packaging layers whose thickness loss was larger than two nanometers after 6-8 weeks of immersion were considered unstable. That is the case for SiN and SiO_2 packaging layers. They were therefore tested in PBS at 37°C to determine if a similar behavior would be observed. The results of these tests are plotted in Figure 5.

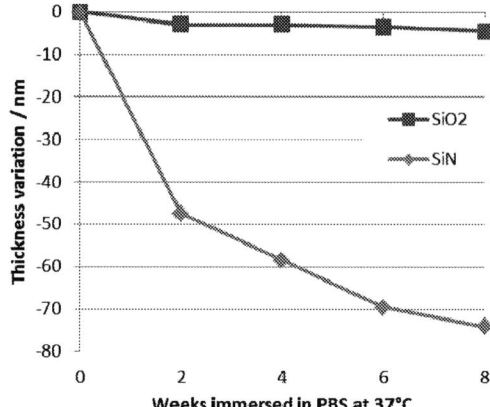

Figure 5. Thickness variation of SiN and SiO_2 packaging layers immersed in PBS at 37°C.

It can be concluded from Figure 5 that SiN layers are not stable in PBS at 37°C. Moreover, SiN coupons show signs of irregular corrosion after a few days of immersion in PBS (Fig.6). On the other hand, SiO_2 layers deteriorated slowly and more uniformly in PBS (Fig.7).

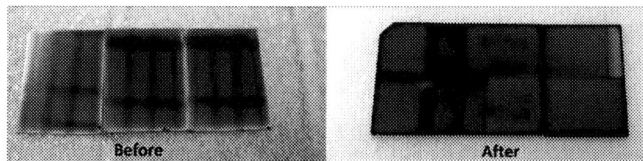

Figure 6. SiN coupons before and after 2 weeks of immersion in PBS at 37°C.

Figure 7. SiO_2 coupons after 8 weeks of immersion in PBS at 37°C. Few apparent changes are visible.

Although corrosion is typically associated with metals, it can be defined in general as the deterioration of materials by chemical interaction with its environment. The amount of corrosion loss in thickness over time is thus called the corrosion rate. Corrosion rates over a time period of one year were calculated for materials with a significant loss of thickness in the aging tests.

A corrosion rate of 20-100 nm per year in PBS at 37°C was found for SiO_2 packaging layers deposited by PECVD with SiH_4 as precursor at 400°C. In 57°C PBS, a corrosion rate of 40-140 nm/yr was found for deposited SiO_2 layers.

It can be seen from Figure 5 that the corrosion rate of SiN does not follow a straight line, but it trends toward a limit value. Such deviation from linearity is probably due to the precipitation of salts on top of the coupons which passivated the SiN layer from further chemical reactions. Calculated corrosion rates were actually very different from experiment to experiment and it was noticed that the frequency of renewal of the PBS solution is an important factor in its chemical aggressiveness. If the PBS solution is changed completely every week and losses due to evaporation of solution during the week are compensated by additions of DI water, a corrosion rate of 0.5-1.2 µm per year in 37°C PBS is expected for a SiN packaging layer fabricated by PECVD at 400°C. In 57°C PBS, a corrosion rate of 1.5-2.5 µm per year was calculated for fabricated SiN layers.

The previous corrosion rate values should be interpreted as average corrosion rates under the experimental conditions described here. Biological mediums such as blood or DMEM (Dulbecco's Modified Eagle Medium, a standard cell culture medium) are typically much more chemically aggressive [10]. Ideally, the packaging layer must not react at all with PBS and its thickness should remain stable as time passes.

As for ZnO coupons, they were rapidly dissolved or delaminated by PBS and there was little remaining after a couple of days (Fig.8). The most probable reason for this is the lack of a complete removal of the silicon native oxide present on the coupons before the deposition of ZnO ALD layers. In fact, ZnO thin films deposited on Si(001) substrates are known for being stable biosensors [5]. However, all packaging layers of Table 1 were deposited on a thin layer of native silicon oxide measuring between 1.5 and 2.5 nm as measured by VASE. Microscope observations of ZnO coupons suggest problems of adhesion between the ZnO film and such native silicon oxide (Fig.8). Further analysis by FTIR and XPS measurements are underway as of this writing.

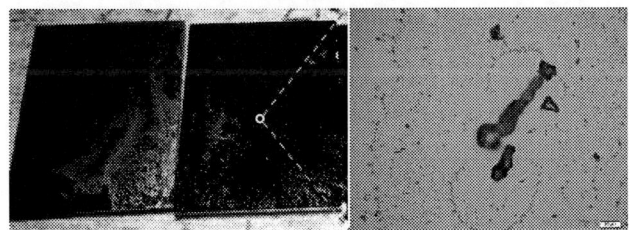

Figure 8. Left: ZnO coupons after 2 days of immersion in PBS at 37°C. Right: µphoto (50x zoom) of coupons at Left.

Gas permeability test

In order to compare different packaging layers using a gas permeability test, a permeable polymer is needed as a substrate. Typically, PET is used for this task but as our deposition processes occur at temperatures as high as 400°C, Kapton 200HN™ (thickness 50.8µm) was chosen as substrate. Kapton substrates were used as provided by the supplier (DuPont™) and cut inside a clean room with clean scissors to the diameter required by our gas permeameter (65 mm, Fig.9).

Figure 9. Al$_2$O$_3$ (Left) and SiOC (Right) films deposited on Kapton. Coverage problems are visible on the SiOC sample.

The hermeticity of thin-film protective layers is usually characterized by the permeation rates of oxygen and water vapor across the barrier layer. However, as WVTR measurements are time-consuming and far too complicated for the purpose of screening packaging layers, alternative methods using Helium gas and UHV mass spectrometers have been proposed [7-8]. The rationale behind this method is that packaging layers which are a bad barrier against helium gas are typically bad barriers against oxygen and water vapor as well. Therefore, helium transmission rates (HeTR) are coherent with WVTR for choosing best packaging materials.

Helium permeation measurements were performed with a patented homemade permeameter which consists of a cell containing the sample in which the upstream side is filled with the gas of which the permeability is to be measured and the downstream side is an ultra-high vacuum chamber encasing a quadripolar analyzer and an ion trap detector MKS E-Vision+ mass spectrometer. This cell is connected to a gas bench equipped with different permeating gas reserves and a primary vacuum unit, allowing control of the upstream side atmosphere (Fig.10) [8-9].

Results of the helium gas permeability measurements performed at INES (French National Solar Energy Institute) are shown in Table 2. Three samples per packaging layer were measured in the permeameter and the maximum BIF (Barrier Improvement Factor, defined as the ratio by which the barrier gas permeability is improved compared to the barrier reference material) values are shown in Table 2

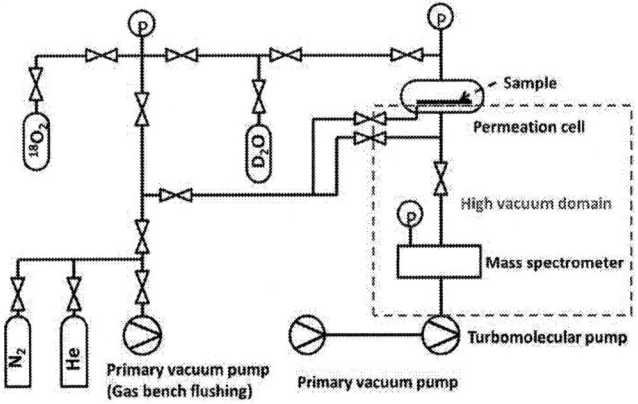

Figure 10. Scheme of permeameter developed by S. Cros *et al.* [9].

Table 2. Helium gas permeability of different packaging layers deposited on Kapton 200HN substrates.

Gas Barrier	Temp. (°C)	Helium BIF	HeTR (g m^{-2} d^{-1} mbar^{-1})
Kapton200HN (Reference)	40	1.0	67
Kapton200HN+14nmTiO$_2$	40	24	2.8
Kapton200HN+30nmAl$_2$O$_3$	40	32	2.1
Kapton200HN (Reference)	25	1.0	38
Kapton200HN+100nmSiN	25	24	1.6
Kapton200HN+100nmSiO$_2$	25	31	1.2

A helium BIF equal to or larger than 30 means that such a barrier typically has a WVTR equal to or smaller than 10^{-3} g/m^2/day and is therefore a good candidate for the packaging of Organic Solar Cells or Implantable Devices [7-8].

Although thin-film protective layers listed in Table 1 were deposited on Kapton™ 200HN polymer substrates, only layers listed in Table 2 had a BIF larger than one. A helium BIF close or equal to 1 means that the packaging layer has big defects, pinholes or cracks due to an imperfect coating process or cracking of the rigid inorganic layer on the flexible substrate [8]. Indeed, Kapton 200HN is highly flexible (Fig.8) but materials like SiC and a-CH are known for their hardness and brittleness. As for other layers such as SiOC, irregular coverage of the Kapton substrates was visible after deposition. In conclusion, it was not possible to measure the helium gas permeability of all packaging layers listed in Table 1 and further experiments with a different substrate are required.

Cytotoxicity test

The materials listed in Table 1 are well known for their biocompatibility. Nonetheless, biochemical contamination or handling with toxic tools can turn a previously biocompatible material into one incompatible for medical implantation. For that reason, all packaging materials must be tested in-vitro to assess their effects on cell viability and cellular morphology. As the packaging materials studied here are intended for subcutaneous implantation, it was decided to test the materials with a standard fibroblast cell culture (L929). A fibroblast is a type of cell that synthesizes the extracellular matrix and collagen, and plays a critical role in wound healing.

Si coupons with a full sheet of packaging layer were sterilized by immersion in ethanol at 70°C and irradiation with UV light (250 nm). Then, in-vitro cytotoxicity tests were performed by the extraction method (Fig.11). The packaging layer was placed in contact with a typical cell culture medium supplemented with serum for 24h at 37°C in an incubator. Each condition produces 3 mL of extract (resulting in a ratio of 3 cm^2 per mL of extract). Afterward, 100 µL of extract are placed in contact with L929 cells for 24h at 37°C. The cell viability was evaluated with the help of an indicator of metabolic activity. In parallel, the cellular morphology produced by contact with the different materials was evaluated by observation under a microscope with fluorescent markers that target the nucleus and cellular cytoplasm.

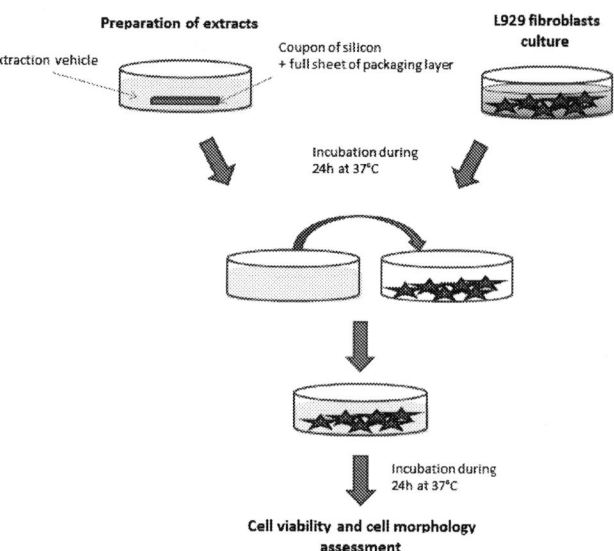

Figure 11. Scheme of the cytotoxicity test.

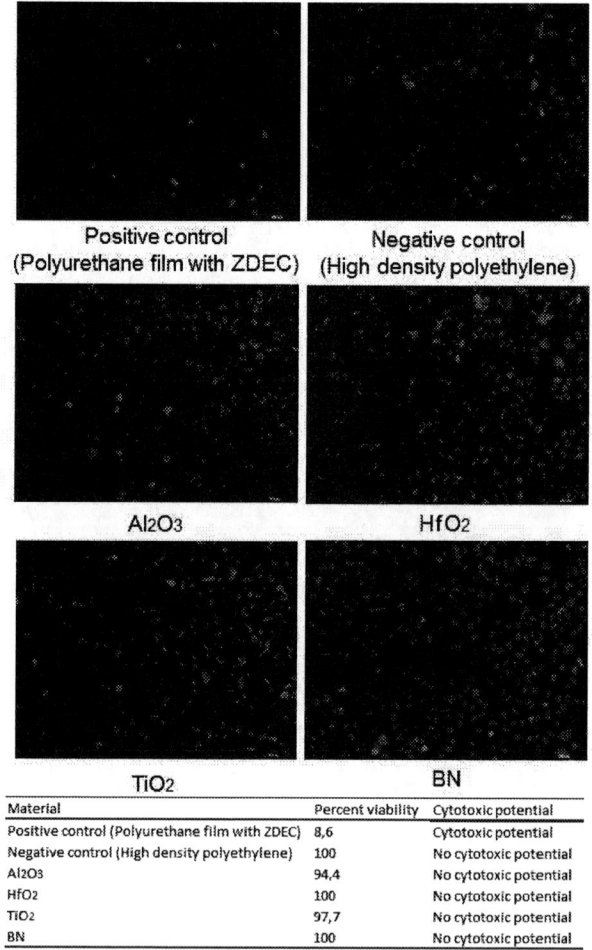

Material	Percent viability	Cytotoxic potential
Positive control (Polyurethane film with ZDEC)	8,6	Cytotoxic potential
Negative control (High density polyethylene)	100	No cytotoxic potential
Al2O3	94,4	No cytotoxic potential
HfO2	100	No cytotoxic potential
TiO2	97,7	No cytotoxic potential
BN	100	No cytotoxic potential

Figure 12. Results of the cytotoxicity test for selected packaging layers deposited on Si coupons.

Every material was evaluated twice by the cytotoxicity test with three points of measure per test (9 samples) and then submitted to the test of morphology (3 samples). In total, 12 coupons or more need to be measured per material.

As of this writing, four materials (Al_2O_3, TiO_2, HfO_2 and BN) were found to be non-cytotoxic to L929 cells (Fig.12). Cell morphology tests in parallel showed that these four packaging layers do not induce adverse biocompatibility effects. We did not observe any alteration of the cell morphology for these four materials. The other six packaging layers will be tested in-vitro for cytotoxicity and cell morphology soon.

Conclusions

Test methods

An aging test method to quantify the degradation of packaging layers in a solution that simulates the human blood has been proposed. This aging test method is well adapted for the screening of biocompatible packaging layers. Future work to measure the biodegradation of packaging layers in biological mediums such as FBS (Fetal Bovine Serum) are interesting to provide a good comparison of corrosion rates in-vitro and in-vivo.

The cytotoxicity and morphology tests proposed here are coherent with reported studies on the biocompatibility of candidate packaging layers and are well adapted for screening biocompatible materials.

The gas permeability test method as proposed here is not properly adapted for assessing the hermeticity of all inorganic packaging layers, because Kapton substrates are too flexible compared to rigid materials such as SiC. There were also problems of substrate coverage by our PECVD deposition processes.

Packaging materials

It was found that widely used packaging layers in microelectronics such as SiN and SiO_2 are corroded by PBS at 37°C. Corrosion rates in 37°C PBS of 20-100 nm/yr and 0.5-1.2 µm/yr are expected for SiO_2 and SiN packaging layers deposited by PECVD at 400°C, respectively. SiO_2 has good properties as gas barrier and it does not have a very high corrosion rate. Therefore, for certain short-term applications that are not in direct contact with chemically aggressive biofluids, the use of PECVD-deposited SiO_2 might be practical.

ZnO packaging layers have stability problems after a couple of days of PBS immersion. Adhesion problems between ALD-deposited ZnO layers and the native oxide surface of Si wafers are suspected.

Packaging layers of Al_2O_3, a-CH, BN, HfO_2, SiC, SiOC, and TiO_2 have good stability as proved by a film thickness variation smaller than 2 nm after 6-8 weeks of immersion in PBS at 57°C. Moreover, four of them were confirmed to be non-cytotoxic according to in-vitro tests outlined by the norm ISO10993-5. They are consequently good candidates as biocompatible packaging layers of medical electronic devices that are in contact with chemically aggressive environments.

Acknowledgments

This work has been performed with the help of the "Plateforme technologique amont" de Grenoble, with the financial support of the "Nanosciences aux limites de la Nanoélectronique" Foundation and the CNRS Renatech network. We would like to thank as well the different research institutes of CEA (Clinatec, INES) that collaborated with us on this project.

References

1. G. Kotzar, "Evaluation of MEMS materials of construction for implantable medical devices," *Biomaterials*, vol. 23, pp. 2737–2750, July 2002.
2. S. Lousinian, "Optical and surface characterization of amorphous boron nitride thin films for use as blood compatible coatings," *Solid State Sciences*, vol. 11, pp. 1801–1805, May 2009.
3. R. Matero, "Atomic layer deposited thin films for corrosion protection," *Journal de Physique IV France*, vol. 9, no. PR8, pp. 493–499, Sept. 1999.
4. R, Zhuo, "Silicon oxycarbide glasses for blood-contact applications," *Acta Biomaterialia*, vol. 1, issue 5, pp 583–589, Sept. 2005.
5. S. K. Arya, "Recent advances in ZnO nanostructures and thin films for biosensor applications: Review," *Analytica Chimica Acta*, vol. 737, pp. 1–21, August 2012.
6. J. K. Chapin and K. A. Moxon, *Neural Prostheses for Restoration of Sensory and Motor Function*, CRC Press LLC, Florida, 2001, pp. 75–100.
7. A. Hogg, "Ultra-thin layer packaging for implantable electronic devices," *J. Micromech. Microeng.*, vol. 23, no. 7, 075001, May 2013.
8. A. Morlier, "Gas barrier properties of solution processed composite multilayer structures for organic solar cells encapsulation," *Sol. Enegy Mater. Sol. Cells*, vol. 115, pp. 93-99, August 2013.
9. M. Firon, S. Cros, P. Trouslard, Method and Device for Measurement of Permeation, US Patent US2007186622, 2007.
10. M. Op de Beeck, "Design and characterization of a biocompatible packaging concept for implantable electronic devices", in *Proc. Of 44th Intern. Symp. on Microelectronics*, Long Beach, USA, Oct. 9-13, 2011, pp. 152-160.

A New Era in Manufacturable, Low-Temperature and Ultra-Fine Pitch Cu Interconnections and Assembly Without Solders

Vanessa Smet*, Makoto Kobayashi†, Tao Wang, Pulugurtha Markondeya Raj, and Rao Tummala
3D Systems Packaging Research Center, Georgia Institute of Technology, Atlanta, USA
†Namics Corporation, Niigata, Japan
* Email: vanessa.smet@prc.gatech.edu

Abstract

This paper presents the first demonstration of a high-throughput die-to-panel assembly technology to form Cu interconnections without solder at temperatures below 200°C. This interconnection technology, previously established with individual single-chip packages on both organic and glass substrates, at pitches down to 30μm, is brought up to a significant manufacturable level by two major innovations: 1) ultra-fast thermocompression bonding (TCB) process with pre-applied polymer, in air, and without any prior surface activation; 2) die-to-panel assembly process with heating from die side exclusively for reduced substrate warpage. The initial proof of concept reported in this paper consists of assembly of 15 silicon dies with Cu bumps at 100 μm pitch, on a 3" x 5" organic substrate, by sequential TCB at 210°C for 3 seconds, and 190°C for 10 seconds. X-ray analysis, C-SAM imaging, cross-section observation with optical microscopy and SEM, and electrical yield characterization indicate the formation of strong metallurgical interconnections. This pioneering technology addresses many manufacturability challenges presently hindering the technology-transfer of direct Cu-Cu bonding, the "holy grail" in the semiconductor industry, by offering a potentially low-cost, high-throughput solution, compatible with industry-standard assembly lines. Scalable to ultra-fine pitches onto low-CTE glass, silicon or organic packages, it has the potential to become a major enabler for the next two or more decades.

Introduction

The need for higher bandwidth between two or more ICs at low power for next generation mobile and high-performance systems is expected to drive off-chip interconnections pitches down to 10 μm and below by 2020. Reduction in pitch ultimately leads to scaling down the micro-bump standoff height and diameter, and, in case of solder-based technologies, to reduction in solder volume. Standard copper pillars with solder cap, currently dominating the market at pitches as low as 40 μm, face many fundamental challenges to meet the miniaturization and performance targets of emerging applications, including bridging, increased interfacial stress between intermetallic layers and residual solder due to the reduced solder volume, and poor current-handling capability. Therefore, what is required is a new class of ultra-short solder-free interconnection technologies that relies on solid-state bonding to achieve ultra-fine interconnection pitch without the risk of bridging and eliminating unstable interfaces associated with intermetallic compounds. Copper has high electrical and thermal conductivities enabling high-speed signal transmission and high power-handling capability, and is also compatible with back-end-of-line processes, benefiting from well-established

plating infrastructures. All-copper interconnections are consequently highly sought and extensively researched as the "holy grail" in the semiconductor packaging industry. However, the current approaches to direct copper-to-copper bonding often involve strenuous processing steps for surface activation and oxide removal, such as chemical-mechanical polishing, or complicated bonding processes at temperatures exceeding standard lead-free reflow, with long annealing times for interdiffusion or recrystallization, or requiring vacuum or a specific atmosphere [1-8]. The complexity of these solutions leads to high manufacturing cost, thus hindering technology transfer to high-volume assembly lines. Further, the proposed assembly technologies have low tolerance to substrate warpage, and bump and pad non-coplanarities, which are invariably present in wafers and substrates fabrication.

The low-temperature copper interconnection technology recently patented by the Georgia Tech 3D Systems Packaging Research Center [9] comprehensively addresses these challenges with the following key innovations: 1) preventing copper oxidation, typically present in Cu-Cu bonding, by standard ENIG (electroless Ni – immersion Au) or ENEPIG (electroless Ni – electroless Pd – immersion Au) surface finish, applied on micro-bumps and pads; 2) accommodation of micro-bumps and pads non-coplanarities by collapse of the interconnections under pressure during thermocompression bonding; 3) warpage reduction with assembly at temperatures as low as 180°C, below the glass transition temperature Tg of low-CTE (coefficient of thermal expansion) laminates; 4) reduction in bumping cost by eliminating solder plating and using ultra-short all copper bumps, about 10 μm in height. The superior thermomechanical reliability and electromigration resistance at 10^6 A/cm² of these advanced copper interconnections were demonstrated on low-CTE organic and glass interposers at pitches as low as 30 μm [10-12].

The current phase of this research aims at scaling this technology down to 20 μm pitch and below [13], and also bringing it to a manufacturable level with a high-throughput assembly strategy. This forms the key focus of this paper. The established process for low-temperature copper bonding consists of thermocompression bonding with a unique pre-applied polymer (e.g. B-stageable no-flow underfill – BNUF, nonconductive adhesive, film or paste – NCA, NCF or NCP), as capillary underfill is not applicable for such fine gap. Compared to a traditional flip-chip bonding sequence that comprises of die-to-substrate pick-and-place on a strip, followed by batch reflow, each die is generally assembled individually onto the substrate in thermocompression bonding, therefore penalizing assembly throughput and increasing processing cost. Reducing the time spent under the thermocompression tool is subsequently critical for manufacturability. This paper demonstrates an enhanced assembly technology that enables the formation of a strong metallurgical bond between micro-bumps and pads by pressure-induced plastic deformation of non-oxidized metal surfaces in a mere few seconds. The bonding time is found to be dependent on the polymerization of the epoxy-based pre-applied material, as is explained in the first section. To further increase the throughput, a panel-based process has been developed, relying on heating only from the die side, to limit warpage effects. Daisy chain test dies with copper interconnections ~20 μm in height, at 100 μm peripheral pitch were fabricated for this first proof-of-concept demonstration, and assembled on 3" x 5" organic substrates. The initial process conditions – 3 seconds at 210°C and 10 seconds at 190°C – were validated by several analytical methods, including electrical DC measurements, to evaluate the quality of the joints and interconnection yield. The preliminary results are discussed to assess the enhanced manufacturability of the proposed approach, applicable in standard assembly lines.

Test Vehicle Fabrication

A test-vehicle that incorporates daisy chain structures was designed to evaluate the yield and reliability of the copper interconnections after panel-level thermocompression bonding. Although this technology targets wide I/O, large-die and ultra-fine pitch applications, coarser design rules were chosen for this first process demonstration. The silicon die is 5 mm x 5 mm and 600 μm in thickness. It features 760 copper micro-bumps that are 30 μm in diameter and ~20μm in height, with 3 peripheral rows at 100 μm pitch, and a central area array at 250 μm pitch. The design includes 4 two-point probe structures for corner daisy chains, 8 half-edge daisy chains, while the central area array is broken into 4 individual daisy chains as illustrated in Figure 1.

The dies were fabricated from a 600 μm thick wafer using a standard semi-additive process. A 2 μm thick SiO2 layer was first deposited by Plasma-Therm PECVD, followed by sputtering of a 30 nm Ti – 500 nm Cu seed layer. A 2.5 μm thick Cu dogbone redistribution layer is then laid out on the wafer by electrolytic plating, followed by bumping photolithography and electroplating of the ~20 μm height Cu micro-bumps. After photoresist strip and seed layer etching, ENIG surface finish is applied on the Cu bumps to prevent oxidation. This surface finish consists of ~5 μm of electroless Ni and ~100 nm of immersion Au. Commercial electrolyte formulations from Atotech were used for the Cu plating and ENIG finish.

Although the application focus of this interconnection technology is primarily with low-CTE glass and silicon interposers, the semi-additive fabrication process is time-consuming, involving complex process steps such as dielectric film lamination, electroless Cu plating to form a seed layer, lithography followed by Cu electrolytic plating to build the redistribution layers, photoresist stripping, seed layer etching, and finally surface finish. Organic Cu-cladded FR4 substrates, in 6" x 6" size, patterned with the substractive method were thus used for this feasibility study. The 50 μm thick Cu film coating on both sides of the 1 mm thick organic core was first etched down to 10 μm thickness, before proceeding with

lithography and etch-back processes to build the Cu pattern without the need for a seed layer. After photoresist strip, ENIG surface finish is plated with 5.5–7 μm Ni and ~100 nm immersion Au. The individual single-die interposer unit represented in Figure 1 is about 20 mm x 20 mm in size.

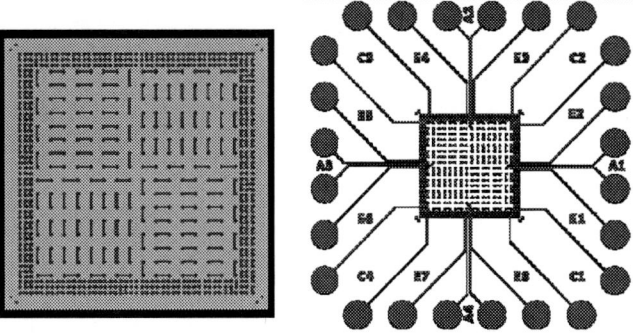

Figure 1. Test vehicle layout showing daisy chains for yield and reliability evaluation: 5 mm x 5mm die (left) and 20 mm x 20 mm substrate with probing pads (right).

Interface Characterization and Bonding Mechanism

Process-of-record (PoR) for low-temperature Cu interconnections without solders: The assembly process was previously demonstrated on glass and organic interposers with excellent reliability performance in high-temperature storage, unbiased highly-accelerated stress test and thermal cycling test obtained at 30 μm pitch [10-12]. A unique polymer material with a filler-free composition such as pre-applied underfill or non-conductive film is first dispensed on the substrate bonding area with good control of the viscosity and deposited volume. B-stageable no-flow underfill (BNUF from Namics Corporation) was used for this set of experiments. After dispensing, this epoxy-based material requires B-staging at 70°C for 1h in air. The Si die is then assembled onto the polymer-coated substrate by thermocompression bonding at 200°C, 365 MPa with a 60s dwell time at peak temperature, consistent with the established PoR conditions. The BNUF material reaches a low-viscosity point at temperatures ~100°C, allowing the excess material to be squeezed out from the bonding interface and flow, enabling contact between Cu micro-bumps and pads, then starts hardening at higher temperature. By the end of the bonding process, the BNUF material is fully cured and the assembled structure does not necessitate any post-treatment.

Interface analysis and proposed bonding mechanism: During thermocompression bonding, the pressure applied on the die transfers directly to the Cu micro-bumps, resulting in high stress concentration per unit area at the rough bump-to-pad interfaces. When the bonding stress exceeds the yield strength of copper, plastic deformation of the bumps and pads ensues, leading to partial collapse of the bumps. The pictures in Figure 2 show a scanning acoustic microscopy (SEM) analysis of micro-bumps before and after thermocompression with 365 MPa applied pressure. A vertical collapse of the bumps can be clearly observed, with a decrease of the bump height by ~3μm, resulting from deflection of the bumps and landing pads. Further, no significant lateral expansion of the bumps can be noted, with an increase in bump diameter by less than 1 μm. This bonding technology thus addresses major challenges faced by ultra-short interconnections due to non-coplanarities arising from bump and pad height variations and substrate warpage by offsetting ~3 μm non-coplanarities with no risk of lateral bridging.

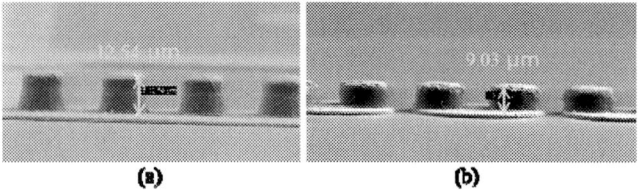

Figure 2. SEM pictures of Cu micro-bumps (a) as plated, (b) after thermocompression bonding in PoR conditions [13].

The SEM picture in Figure 3 shows the cross-section of a Si die (top) on a polymer-laminated glass interposer (bottom) after ion-milling. The pre-applied underfill fills the gap between die and interposer without any apparent voids. Both Cu bumps and landing pads are plastically deformed. The lack of a distinct bonding interface indicates the formation of a metallurgical bond between bumps and pads over a large portion of their surfaces. Pressure-induced local plastic strain brings the thin immersion Au layers from the ENIG surface finish into intimate contact, enabling solid-state interdiffusion of the Au atoms. The combination of 4 interdependent factors – oxide-free Cu surfaces; pressure to reduce diffusion distances and accelerate interdiffusion by enhancing dislocation and lattice defect densities at the interface; temperature to increase the diffusion rate; and time for interface recrystallization – leads to the formation of the observed metallic bond. Nickel acts as a barrier layer to prevent diffusion of Au into Cu, eliminating the risk of forming Kirkendall voids. The pre-applied underfill enhances the die shear strength by introducing compressive stress in the structure due to polymer shrinkage, and also prohibits oxidation and degradation of the Cu interconnections due to its built-in fluxing action. In addition, it may fill in any gaps, as a transient liquid, that exist between the two surfaces. The bonding mechanism is analyzed and detailed in [13].

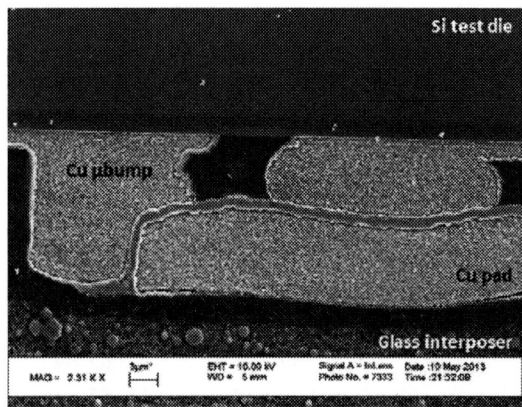

Figure 3. SEM picture of the ion-milled cross-section of a Si test die, assembled on an ultra-thin glass interposer with thermocompression bonding using a pre-applied underfill.

Considering the temperature-dependence of yield strength of metals, maximum plastic deformation of bumps and pads is permanently and instantaneously achieved upon reaching the bonding peak temperature, initiating metallic bonding. A few

seconds is enough to create a metallurgical bond, as confirmed by the SEM pictures in Figure 4 of two ion-milled cross-sections of Cu interconnections formed at 200°C – 365 MPa for 60s and 3s, respectively.

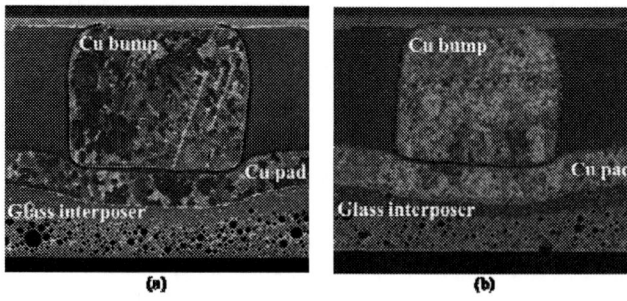

Figure 4. SEM pictures of ion-milled cross-sections of Cu interconnections formed by thermocompression bonding at 200°C – 365 MPa applied for (a) 60s (PoR conditions) and (b) 3s.

Temperature-dependence of polymer curing: Curing kinetics of the pre-applied underfill is another factor that constrains the bonding time at peak temperature. While it is not necessary to achieve full cure of the material during assembly, the bonding pressure can only be released after polymerization initiation. The graph in Figure 5 illustrates the temperature-dependence of the curing behavior of epoxy, given as an example, as most BNUF, NCF or NCP are epoxy-based materials. The curing rate increases with temperature and a shorter time is required to kick off the hardening phase at a higher temperature. Temperature impacts the curing kinetics with an exponential Arrhenius-type relation rather than a simple linear dependence. For instance, while BNUF polymerization starts after only 3s at 210°C, 10s are required at 190°C. A decrease in temperature by 10% thus leads to a 3x increase of the time under the thermocompression tool, dramatically impacting the assembly throughput. The polymer material properties should subsequently be optimized to enable fast initiation of the hardening phase at temperatures below 200°C.

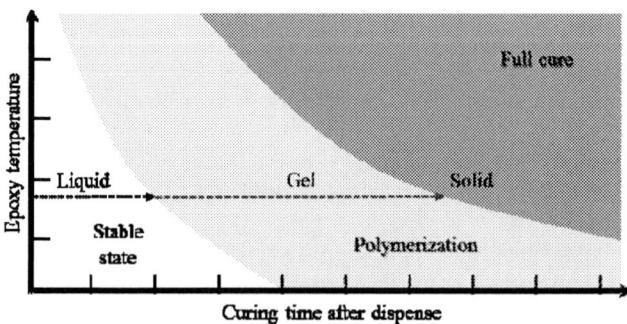

Figure 5. Illustration of the temperature-dependence of the curing behavior in an epoxy-based system.

Once the polymer reaches a gel-like firmness, the bonding pressure can be released. A batch-type post-cure might be necessary to reach the polymer solid state. The BNUF used here requires further baking for 3h at 165°C in air. This additional curing phase also enhances joint recrystallization, improving the quality of the metallurgical bond.

High-throughput Die-to-Panel Assembly Demonstration

Apart from large 2.5D high-performance interposers where the dies are mounted on singulated units for cost and yield reasons, assembly in industry-standard low- to high-volume lines is usually performed on a substrate strip to facilitate BGA balling for board-level interconnections, for both traditional flip-chip bonding or low-pressure thermo-compression Cu pillar with solder cap technology. In the latter case, two strategies can be considered: (1) tacking the die by pre-cure of a non-conductive paste followed by gang bonding where pressure is applied on multiple dies at once to reflow the solder and fully cure the NCP; (2) bonding of each die individually using a thermocompression tool, potentially followed by a batch-type process to cure the polymer. Although the first scenario could potentially have higher throughput since the bonding time is now divided by the number of dies on a strip, it is more difficult to implement as it requires pressure uniformity on a strip-scale, which can only be achieved with high-accuracy equipment with micron-scale parallelism between the press plates. The second approach involves a specific assembly sequence where the heat transfer is dominated by the thermocompression flip-chip tool. The plate onto which the substrate is placed is maintained at a constant temperature below laminate T_g, to limit substrate warpage and to prevent any degradation of the joints of the mounted components during subsequent assemblies. Unidirectional heat transfer may jeopardize uniform flowability and polymerization of the pre-applied polymer (NCP, BNUF) during bonding as this class of materials is essentially thermally insulating. Thin dies with ultra-short interconnections resulting in a smaller gap and thinner polymer layer are thus desirable for strip-level thermocompression assembly.

Figure 6. Illustration of panel-level assembly capability: 6" x 6" glass panel accommodating 25 dies, sitting on the flip-chip bonder stage after assembly.

A panel-level process was developed using a Finetech Matrix semi-automatic flip-chip bonder with a placement accuracy of 3 μm and controllable temperature and force profiles. The stage can accommodate a 6" x 6" substrate size as illustrated in Figure 6, and has thus a very low heat capacity with a temperature ramp rate limited to 2°C/s. A customized 6 mm x 6 mm vacuum-locked spring gimbal tool head, enabling pre-leveling of the die, was used to prevent any die tilt during assembly. BNUF was first dispensed on a 3" x 5" organic substrate that can accommodate 15 dies, and B-staged. The substrate was then placed onto the stage and maintained at 70°C during the entire assembly sequence to slightly decrease BNUF viscosity to prevent misalignment of the die during placement. Each die is eventually picked, placed and bonded at 210°C applied for 3s. The temperature ramp rate being limited to 6°C/s for the tool head, the assembly cycle for a single chip lasts ~90s, but the 365 MPa bonding pressure is released before the cool-down phase. After assembly of all dies, the panel is subjected to an oven cure at 165°C for 3h in air to ensure that the BNUF material would reach its final solid stage.

Interconnection Yield Characterization

The assembly process with heating only from the top was first validated at substrate-unit level. The resistances of the unit daisy chains were measured prior to oven cure, to assess the interconnection yield after thermocompression bonding. All daisy chains were yielded with resistance values within the expected range (~1–4 Ω). The samples were then mounted in epoxy molds and polished to prepare them for cross-section analysis. The optical images displayed in Figure 7 confirm adequate joint quality and underfilling.

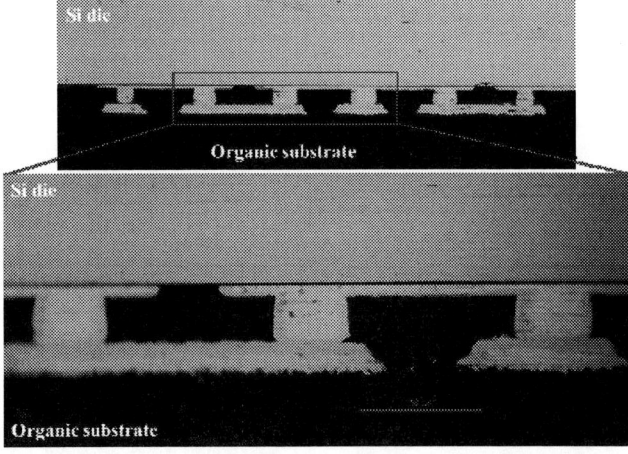

Figure 7. Optical image of cross-section of assembled Si die on organic substrate with Cu interconnections formed by thermocompression bonding with heating from die side only at 210°C, 365 MPa with a 3s dwell time at peak temperature.

The first panel-level demonstration was conducted on the 3" x 5" organic substrate of Figure 8 (a) onto which 15 Si dies were assembled by thermocompression bonding at 210°C, maintained for 3s and an applied force of 200 N, using the process described in the previous section. Additional assembly trials were performed at a lower temperature – 190°C for 10s – on a smaller substrate with 4 dies introduced in Figure 8 (b). The assembled structures were screened by X-ray analysis and C-SAM acoustic imaging to evaluate the alignment accuracy and quality of underfilling, respectively. Close-up pictures of a typical sample are displayed in Figure 9 for both bonding conditions, confirming that the placement accuracy is within the 3 μm tolerance of the flip-chip bonder.

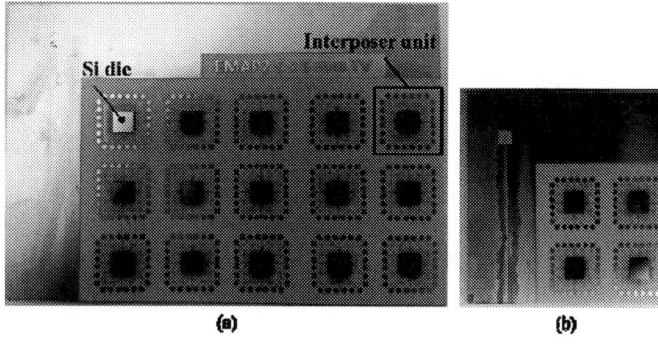

Figure 8. Optical picture of organic substrate with multiple 5mm x 5mm Si dies assembled by panel-level thermo-compression bonding with 365 MPa applied pressure and at (a) 210°C for 3s, and (b) 190°C for 10s.

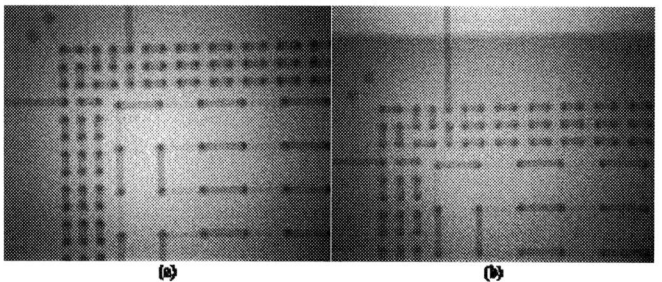

Figure 9. Close-up X-ray imaging of a typical die assembled at panel-level on organic substrate with (a) 210°C for 3s, and (b) 190°C for 10s, showing accurate alignment.

No voiding was observed by C-SAM analysis of the assembled panels, as shown in Figure 10. Good quality underfilling was thus obtained despite the unidirectional heat transfer, due to the unique flowability and curing properties of the BNUF material developed by Namics Corporation.

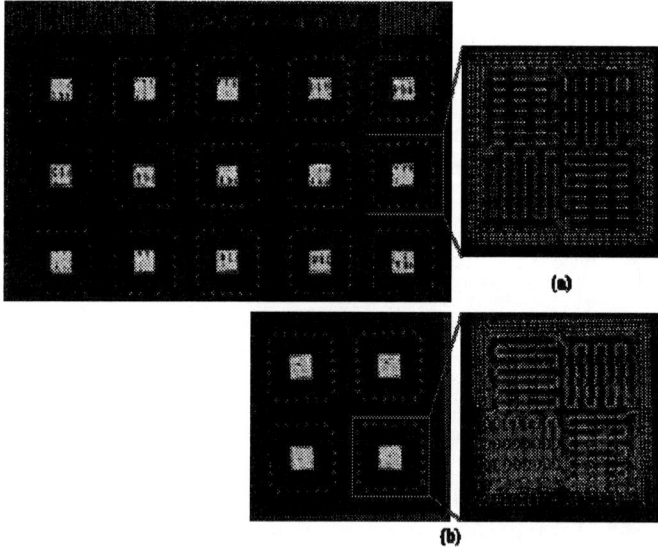

Figure 10. C-SAM imaging of assembled structures with (a) 210°C for 3s, and (b) 190°C for 10s showing void-free underfill layer at panel and unit levels.

The interconnection yield was assessed by DC measurements of the daisy chains resistance after

thermocompression bonding and post-cure. The number of functional chains, over the 16 designed, is reported for each individual unit in Tables 1 and 2, respectively corresponding to each bonding conditions. Apart from one unit with only 50% functional structures, which was found to be due to poor alignment, the yield exceeded 75% for all 18 assembled units, validating the process conditions.

Table 1. Electrical yield evaluation of the 15 as-bonded dies assembled on a 3" x 5" organic substrate at 210°C for 3s.

Number of functional daisy chains / 16				
14	14	16	14	15
14	13	13	14	12
16	12	16	14	8

Table 2. Electrical yield evaluation of the 4 as bonded dies assembled on a 2" x 2" organic substrate at 190°C for 10s.

Number of functional daisy chains / 16	
14	16
14	15

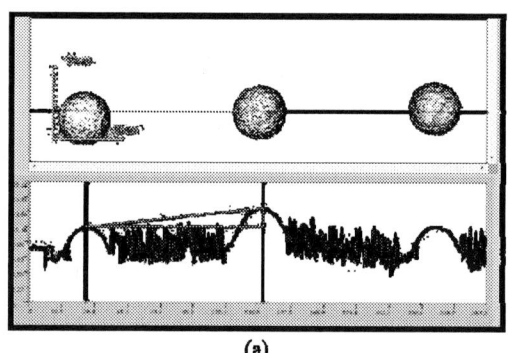

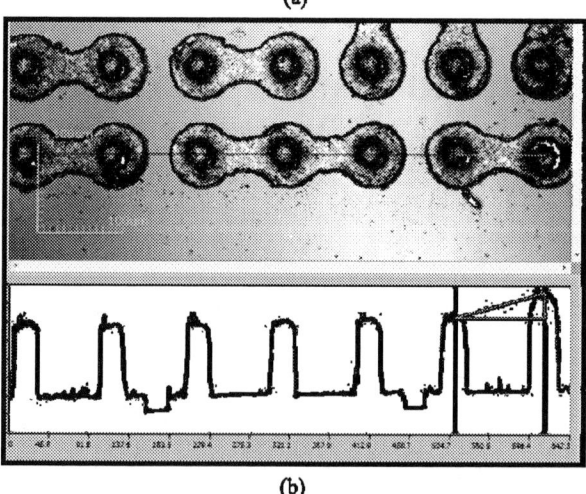

Figure 11. 3D microscopy images of a typical Si test die showing bump height variations of (a) ~3 μm and (b) ~9 μm across 3 and 7 Cu interconnections, respectively.

Possible causes for a non-perfect yield include die tilt, bump non-coplanarities and substrate warpage. As no specific orientation was noticed in the failed daisy chains, and no significant die tilt was observed in the cross-sections, flatness and parallelism of the bonding tool head were ruled out as the causes for yield loss. The bump co-planarity was analyzed with 3D microscopy on a Si die randomly picked from the wafer. The results are summarized in Figure 11, with large discrepancies in the measured bump height across a die, up to 9 μm. These height variations exceed the non-coplanarities that can be offset by pressure-induced plastic deformation, resulting in non-land of the bumps adjacent to a larger one. This directly translates into an open-circuit and a nonfunctional daisy chain reading. Optimization of the Cu electrolytic bath conditions and plating densities is required for enhanced bump height uniformity and optimum interconnection yield.

Conclusions

This paper demonstrated, for the first time, an innovative low-temperature, ultra-fast, die-to-panel Cu interconnection and assembly process, compatible with standard thermocompression bonding tools. The bonding mechanism is metallurgical in nature as confirmed by interface characterization. The main limiting factor in reducing the bonding time further was found to be the hardening of the pre-applied polymer, which is highly temperature-dependent. Two sets of temperature and time for which the BNUF exhibited initial polymerization – 210°C for 3s and 190°C for 10s – were selected for initial trials. A panel-level assembly sequence was developed and demonstrated with heat-transfer from the die side exclusively. Preliminary trials were conducted on 2" x 2" and 3" x 5" organic substrates accommodating 4 and 15 Si dies with Cu interconnections at 100 μm pitch, respectively. Good alignment and underfilling quality were achieved. The interconnection yield was found to exceed 75%, excessive non-coplanarities of bumps accounting for the failed daisy chains. These preliminary results are a promising proof-of-concept demonstration of the Georgia Tech PRC patented technology. Further optimization of the bumping plating process is required to achieve better interconnection yield. Further reduction of the bonding time, combined with decreasing the thermocompression pressure, will bring this advanced interconnection process to a more manufacturable level, offering a compelling solution for the future generations of ultra-fine pitch high-performance systems, starting a new era without solders.

Acknowledgments

This research is supported by the Interconnections and Assembly (I&A) program at Georgia Tech PRC. The authors would like to thank the full members and supply chain partners for their funding and intellectual support, in particular Atotech for their plating chemistries, and Namics Corporation who specifically developed the underfill material for the targeted application. The authors are very grateful to Anne Matting and Anna Stumpf, interns from TU Dresden, for their support with fabrication and cross-sectioning. Finally, special thanks to Xian Qin for her help with 3D microscopy, and Saumya Gandhi and Florian Nebe for their polishing skills.

References

1. A. Shigetou, T. Itoh, K. Sawada, and T. Suga, "Bumpless interconnect of 6-μm pitch Cu electrodes at room temperature," in *Electronic Components and Technology Conference, 2008. ECTC 2008. 58th*, 2008, pp. 1405-1409.

2. R. Reif, A. Fan, K. N. Chen, and S. Das, "Fabrication technologies for three-dimensional integrated circuits," in *Quality Electronic Design, 2002. Proceedings. International Symposium on*, 2002, pp. 33-37.

3. C. S. Tan, D. F. Lim, S. G. Singh, S. Goulet, and M. Bergkvist, "Cu–Cu diffusion bonding enhancement at low temperature by surface passivation using self-assembled monolayer of alkane-thiol," *Applied Physics Letters,* vol. 95, pp. 192108-192108-3, 2009.

4. C. M. Whelan, M. Kinsella, L. Carbonell, H. Meng Ho, and K. Maex, "Corrosion inhibition by self-assembled monolayers for enhanced wire bonding on Cu surfaces," *Microelectronic Engineering,* vol. 70, pp. 551-557, 2003.

5. W. Yang, H. Shintani, M. Akaike, and T. Suga, "Low temperature Cu-Cu direct bonding using formic acid vapor pretreatment," in *Electronic Components and Technology Conference (ECTC), 2011 IEEE 61st*, 2011, pp. 2079-2083.

6. W. Blair and Ziptronix, "A Path to 3D Integration using Silicon Vias and DBI™ (Direct Bond Interconnect)," in *EMC-3D SE Asia Technical Symposium*, 2007.

7. A. Sitaram and SEMATECH, "Scaling 2.5D/3D: the next R&D challenge," in *2nd Annual IEEE Global Interposer Technology Workshop*, Georgia Institute of Technology, Atlanta, GA, USA, 2012.

8. A. Tay, M. Iyer, R. Tummala, V. Kripesh, E. Wong, M. Swaminathan*, et al.*, "Next generation of 100-mu m-pitch wafer-level packaging and assembly for systems-on-package," *Ieee Transactions on Advanced Packaging,* vol. 27, pp. 413-425, MAY 2004 2004.

9. "Interconnect Assemblies and Methods of Making and Using Same ", US Patent 2012/0104603 A1.

10. A. Choudhury, N. Kumbhat, P. M. Raj, R. Zhang, V. Sundaram, R. Dunne*, et al.*, "Low temperature, low profile, ultra-fine pitch copper-to-copper chip-last embedded-active interconnection technology," in *Electronic Components and Technology Conference (ECTC), 2010 Proceedings 60th*, 2010, pp. 350-356.

11. S. A. Khan, N. Kumbhat, A. Goyal, K. Okoshi, P. Raj, G. Meyer-Berg*, et al.*, "High current-carrying and highly-reliable 30μm diameter Cu-Cu area-array interconnections without solder," in *Electronic Components and Technology Conference (ECTC), 2012 IEEE 62nd*, 2012, pp. 577-582.

12. N. Kumbhat, A. Choudhury, M. Raine, G. Mehrotra, P. M. Raj, R. Zhang*, et al.*, "Highly-reliable, 30μm pitch copper interconnects using nano-ACF/NCF," in *Electronic Components and Technology Conference, 2009. ECTC 2009. 59th*, 2009, pp. 1479-1485.

13. T. Wang, "Modeling, Design, and Demonstration of Low-temperature Cu Interconnections to Ultra-thin Glass Interposers at 20 μm Pitch," presented at the Electronic Components and Technology Conference, 2014, Orlando, USA, 2014.

Enabling Fine Pitch Cu & Ag alloy Wire Bond Assessment for 28nm Ultra Low-k Structure

John D. Beleran*, Ninoy Milanes II, Gaurav Mehta, Dr. Nathapong Suthiwongsunthorn
United Test and Assembly Center Ltd (UTAC), 5 Serangoon North Ave 5
Singapore 554916

Ranjan Rajoo, Chan Kai Chong
Global Foundries Singapore Pte. Ltd, 60 Woodlands Industrial Park D Street 2
Singapore 738406

E-mail: john_beleran@utacgroup.com*, ranjan.rajoo@globalfoundries.com

Abstract

The use of copper wire in IC packaging has been growing steadily driven by cost effectiveness. However, there are concerns and issues that prevent or delay copper wire bonding technology qualification. Copper wire inherent hardness properties induces higher stress on bond pad and underlying pad structure, which results to inevitable pad crack or damage if pad structure is not robustly design or thin aluminum pad thickness of $\leq 0.8um$ for copper wire bonding. Silver alloy wire on the other hand showed great potential in IC packaging around ~20% softer compared to harder copper wire. Copper and silver wire are also cost competitive than gold wire. Both wires may potentially take over gold wire bond especially in handheld devices in the coming years.

Compared to gold, silver alloy wire is superior in terms of electrical and thermal conductivity and their mechanical properties are quite similar. Silver and copper wire has similar thermal conductivity and both wires still require inert gas environment like pure nitrogen (N_2) and forming gas (95% N_2, 5%H_2) during free-air-ball (FAB) formation.

An axisymmetric transient nonlinear dynamic model for fine pitch copper wire bonding was developed with FEM simulation software ABAQUS/Explicit in order to assist in understanding the copper and silver wire bond process focusing on stress or strain investigation at aluminum and low k layers. FEM was focused on Cu wire due to its inherent hardness properties compared to softer silver alloy wire. Simulation result showed that majority of ball bond and Al pad deformation is completed in the very beginning when bond force is applied. Similar to the highest stress on low-k layer is also in the very beginning of bond force application and near the bond pad center area.

This study aimed to share the learning, challenges and success story for Cu and silver alloy wire bond characterization using different fine wire diameter sizes for 28nm ultra low-k wafer technology on a FBGA 9x9mm package. Fine wire diameter sizes of 18um, 15um and 13um for both palladium coated Cu (PdCu) and silver (Ag) alloy wire types where characterized and compared based on the output responses results gathered during DOE assessment.

Over all wire bond DOE results showed good interaction result with no pad crack/damage for optimum wire bond recipe and validated by FIB cut showing no abnormality on low-k layer. Stress test reliability performances for both wire types were assessed and showed no abnormality.

Sliver alloy wire sample wet acid decapsulation process has been a concern once failure analysis is required to ascertain bonding wire connection integrity due to wire corrosion, cracking, thinning and wire easily breaks. With the adoption of laser decapsulation process similar to laser type used in laser grooving of wafer and without using any acid have shown a very good result as depicted in figure 12. A more consistent intermetallic assessment for Ag/Al IMC at ball bond side and easier to quantify similar to conventional Au/Al IMC is recommended as compared to bond pad side having lower Al/Ag IMC with wider variance.

Finally, with the combined simulation analysis data, experimental DOE's and initial reliability data result, FBGA 9x9mm package was assembled using a more challenging harder wire type of palladium coated Cu with 18um (0.7mil) wire size. Package reliability qualification per industry JEDEC standard for laminate packages passed with no failure as summarized in table 11.

Introduction

The semiconductor industry is being driven to evaluate lower cost alternatives as a cost reduction method. Implementing Cu and silver wire bonding in HVM for fine pitch package applications presents a series of technical challenges especially for 28nm ultra low-k technology that must be overcome. Due to mechanical integration and inherent weak interfacial adhesion of cu/low-k materials, there are increased challenges for assembly manufacturer especially for wire bonding process. The mechanical stress from wire bonding can inevitably cause bond pad crack or damage during wire bonding or at reliability test stage. Careful and proper wire bond DOE's and controls are very crucial in reducing stress on the cu/low-k bond pad structures that are robustly design for Cu and Ag alloy wire bonding. There are many types of interconnect wires commercially available in the market today such as Ag alloy, coated Cu wire and bare Cu wire that will soon potentially replaced Au wire bond in the coming years with Cu and Ag alloy wire bonding becoming more popular.

Palladium coated Cu has inherent hardness properties than bond pad aluminum, thus this may increase the stress level on the bond pad like Al splashing as well as the underneath pad structures of the bond pad during wire bonding process. Weak pad structure and thin Al pad thickness of $\leq 0.8um$ posed greater challenge for Cu wire bonding due to inevitable pad crack or damage concern. Hence, silver alloy wire position itself as an alternative option for Cu wire bonding due to its softer properties in between Au and Cu wire. Silver alloy wire have a lot of advantages compared to Cu; (a) wider 2nd bond workability window, (b) does not require high end wire bond machine, (c) UPH improved by 5~10%, (d) 30% reduction for N_2/forming gas consumption, (e) wire bonding

978-1-4799-2408-0/14 $31.00 © 2014 IEEE

is straightforward making conversion from Au to silver alloy wire process easier, (f) enables easier reverse bonding (SSB/BSOB) or stitch over-bump process.

This study was carried out on three different 28nm ultra low k devices with Al bond pad thickness of 1.55um and varying bond pad opening and pad pitches as shown in the table 1. Three different ultra fine wire sizes were selected for this study as shown in the table 1. KnS Iconn wire bond machine platform equipped with Cu kit and forming gas (95%N_2, 5%H2) with flow rate of 0.5/0.7 liter/minute was used for this fine pitch study and with machine capability at 35um bond pad pitch.

Device Information	Device A	Device B	Device C
Wire Diameter	0.7mil (18um)	0.6mil (15um)	0.5mil (13um)
Package Type	FBGA 9x9mm	FBGA 9x9mm	FBGA 5x5mm
Wafer Technology	28nm	28nm	28nm
Bond Pad Opening	48um	40um	38um
Bond Pad Pitch	53um	45um	45um
Bond Pad Material	Al	Al	Al
Al Pad Metal Thickness	1.55um	1.55um	1.55um
Wire Tpes	Pd plating Cu wire	Pd plating Cu wire	Pd plating Cu wire
	Ag alloy wire	Ag alloy wire	-

Table 1 – package and devices information

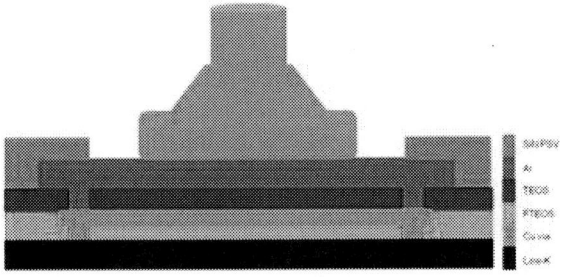

Figure 1 – Ball bond and pad schematic structures illustration

From our previous study, we have assessed through FEM analysis that figure 1 bonding pad structure was a good choice for low k design [1] having the least stress at low-k layer amongst other pad structures compared for Cu wire bonding process. Thus, this design structure was selected to further validate its performances for fine pitch Cu and Ag alloy wire bond.

Finite Element Model and Simulations

An axisymmetric transient nonlinear dynamic model for fine pitch copper wire bonding was developed with FEM simulation software ABAQUS/Explicit in order to assist in understanding the copper wire bond process focusing on stress or strain investigation at aluminum and low k layers. FEM was focused on Cu wire due to its inherent hardness properties compared to softer silver alloy wire. Free-air-ball (FAB) hardness of palladium coated Cu and silver alloy wire were 42% and 16% respectively harder than that of Au, thus increases more stress especially on low k pad structure.

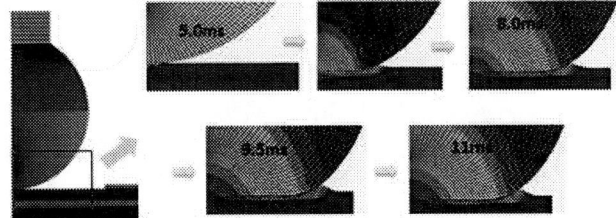

Figure 2 - Dynamic response of the wire bond simulation

Dynamic response of the wire bond simulation at 6.5ms bond time and bond force of 60gram is applied to the ball through capillary, and a significant deformation is seen with Al starts to squeezed out until 11ms, the deformation is no longer significant [1] as shown in figure 2. This dynamic illustration shows the wire bonding deformation is mostly completed in the very beginning. The application of ultrasonic power helps to scrub the Al surface for better adhesion.

The displacement of the lowest and highest points (node A and node B) on the Al pad is plotted in figure 3 with deformation mostly completed around first 10ms bonding time.

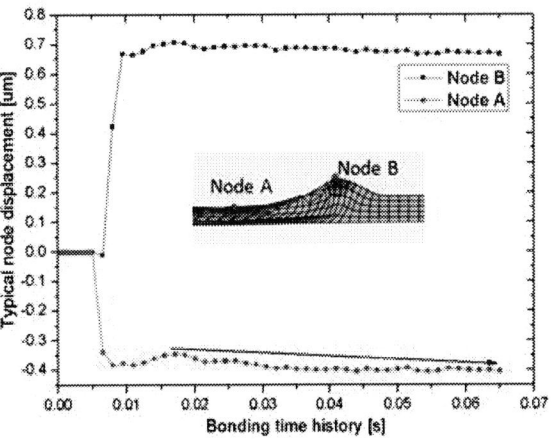

Figure 3- Nodal displacement of Al pad.

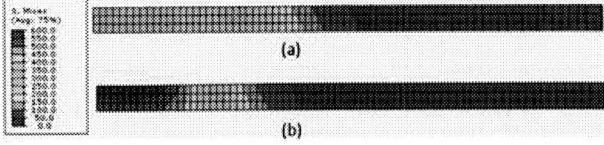

Figure 4 - Von Mises stress on low k layer, (a) wire bond last step, Max=361.3MPa, (b) When under highest stress in all steps, Max = 598.5MPa.

Figure 4 showed low-k layer Von Mises stress analysis, (a) the stress near end of bonding time while (b) is the stress when the bond force is applied. This figure means that the largest stress of the low-k layer was not at the end of wire bond process but, just happens when bonding force was applied as plotted in figure 5.

978-1-4799-2408-0/14 $31.00 © 2014 IEEE

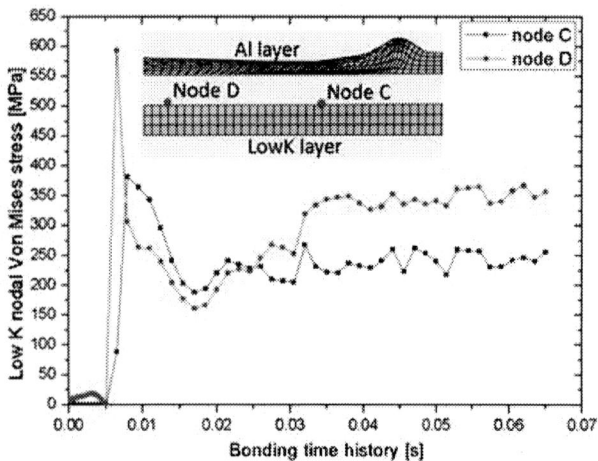

Figure 5 - Von Mises stress on low-k layer at 2 different locations.

Two stress points at the low-k layer was plotted as shown in figure 5 to further understand the stress distribution. Near center area is node D while node C is the point where has it has the lowest displacement. The high stress is near the pad center in the beginning while there is only small amount of ball bond deformation similar to contact area and once bonding force is applied it induced high stress. When the force comes to node C, the contact area is already quite large. Therefore, node D stress is larger than node C.

Experimental Simulations

The test chips samples were all mounted into FBGA 9x9mm laminate package as shown in Figure 6. The die bond pad opening and pitch size are shown in table 1 for the three different wire sizes.

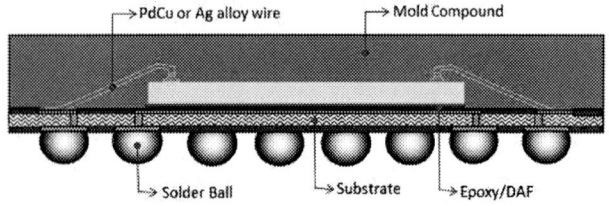

Figure 6 – FBGA 9x9mm device

All equipment and process set-up used during the evaluation process are listed in Table 2. In order to evaluate the Cu and Ag alloy wire bonding process quality the following parameters are investigated: non-destructive analysis involved ball bond diameter and height measurement. And destructive analysis include ball bond shear, ball pull, pad cratering test, intermetallic coverage (IMC), ball bond cross-section in order to measure Al remnant thickness, and FIB cut analysis to check for any internal micro-cracks damage.

Wire Bonder	KnS Iconn equip with cu kit
Bond Placement Accuracy	+/- 2um
Gas Type	5%H$_2$, 95% N$_2$ (forming gas)
Ball Shear/Pull	Dage 4000 series
Ball Height/Optical Inspection	Hisomet Microscope
Cross-section	Struer

Table 2 – Equipment and process set-up

Device Name	Wire Diameter	1st Bond DOE
A	18um (0.7mil) Palladium coated Cu wire	DOE 1
	18um (0.7mil) Ag alloy wire	DOE 2
B	16um (0.6mil) Palladium coated Cu wire	DOE 3
	16um (0.6mil) Ag alloy wire	Validate DOE 3
C	13um (0.5mil) Palladium coated Cu wire	DOE 4

Table 3 – Wire bond DOE characterization matrix

Wire bond DOE's were separately performed for device A B and C as shown in table 3. Device A comes with two separate 1st bond DOE's for palladium coated copper wire (DOE1) and silver alloy wire (DOE2) in order to assess wire bond output responses and compare the derived optimum 1st bond parameter settings. Device B established optimum 1st bond process parameter settings (DOE 3) using palladium coated Cu at 15um (0.6mil) wire size will be used to validate for silver (Ag) alloy wire type. Lastly, for device C, though having almost same test vehicle as device B will have a separate 1st bond DOE (DOE 4) set up in order to achieve the required specification as stated in table 4 using 13um (0.5mil) wire diameter size.

All DOE results were analyzed using SAS-JMP software to predict the optimum range of parameters. This predicted optimum range of parameters will then be validated and fine tune if necessary in order to achieve the desired optimum set of parameters and, overall results must passed the corresponding requirements as stated in table 4.

Different output responses were collected and measured during wire bond DOE experiment like ball pull and ball shear strength, ball bond diameter and height, % intermetallic (IMC), pad damage after crater test, ball bond cross-section for aluminum remnants thickness and flatness. All these responses were fed into the DOE model in order to analyze and established the optimum process parameters. Shown in table 4 are the wire bond output responses, requirements and sample size for palladium coated Cu and Ag alloy wire bond process.

Output Responses	Sample Size	Device A	Device B	Device C
Wire Diameter	-	0.7mil (18um)	0.6mil (15um)	0.5mil (13um)
Ball Size	20 balls/2 units	34 ~ 42um	30 ~ 36um	28 ~ 32um
Ball Size + Al Splash	20 balls/2 units	34 ~ 44um	28 ~ 38um	28 ~ 34um
Ball Height/Thickness	20 balls/2 units	7 ~ 13um	7 ~ 13um	6 ~ 12um
Ball Shear Strength	25 balls/2 units	7.0 gf (min)	7.0 gf (min)	7.0 gf (min)
Ball Pull Strength	25 wires/2 units	2.5 gr (min)	2 gr (min)	2 gr (min)
Ball bond Lift	25 wires/2 units	0	0	0
Pad Crater Test	2 units/all pads	no crack	no crack	no crack
IMC Coverage	1 unit	80% min	80% min	80% min
Ball bond Cross-section (Al pad thickness remnant)	1 ball bond	Ball bond flatness / min 10% Al remnant	Ball bond flatness / min 10% Al remnant	Ball bond flatness / min 10% Al remnant
Focus Ion Beam (FIB)	1 ball bond	no metal damage	no metal damage	no metal damage

Table 4 – Wire bond output responses and requirements

Test Results		Specs	PdCu Wire	Ag alloy Wire	Remark
			Nom Parameter	Nom Parameter	
Ball Size + Al Splash (um)	Min	34~44um	40	39.4	pass
	Ave		41.5	39.8	pass
	Max		42	40.4	pass
	Stdev		0.7	0.3	pass
Ball Height (um)	Min	7~13um	9	9.1	pass
	Ave		9.7	10.1	pass
	Max		10	10.9	pass
	Stdev		0.5	0.6	pass
Ball Pull (gf)	Min	4 gr (min)	6.0	5.3	pass
	Ave		6.9	6.0	pass
	Max		7.6	6.7	pass
	Stdev		0.4	0.3	pass
	Cpk		4.2	3.1	pass
Ball Shear (gr)	Min	7 gr (min)	12.3	14.2	pass
	Ave		15.9	15.5	pass
	Max		21.4	16.9	pass
	Stdev		1.5	0.7	pass
	Cpk		2.7	3.7	pass
% NSOP		0	0	0	pass
% Ball lift			0	0	pass
% Pad Peel			0	0	pass
Pad Crater Test		no crack	0	0	pass
IMC (%)		80% (min)	90.1%	87.2%	pass
Al thickness Remnants after bond		10% (min)	35%	38%	pass

Table 5 – 0.7mil PdCu and Ag alloy wires summary of results

Wire Bond Characterization

a) Device A and 18um (0.7mil) wire size: Palladium coated Cu (PdCu) and silver (Ag) alloy wires 1st bond DOE used a full factorial 3 factors and 2 level design with 13 legs. Prior to 1st bond DOE, second bond parameter settings and free-air-ball (FAB) were baseline and established before embarking for 1st bond assessment.

Summary of 1st bond results comparison is shown in table 5 with both wire type passing all output responses and requirements stated in table 4. Failure analysis results in figure 7 showed no pad crack/damage after ball bond etch, >80% IMC achieved, Al remnant <50% of the original Al thickness and focus ion beam (FIB) cut validation showed no abnormality.

Shown table 6 is the three key factors for 1st bond DOE and its optimum value settings. Obviously for softer Ag alloy wire requires lesser 1st bond parameter in comparison with harder PdCu wire however, both set of parameters passed all the bonding requirements as summarized in table 5.

Intermetallic (IMC) Coverage Assessment

Typical chemical solutions were applied to etch away Cu ball bond and Al pad prior to intermetallic (IMC) coverage assessment. Prior to IMC inspection, samples were baked at 175°C for 4 hours in order to promote IMC formation that is visible and easier to quantify after chemical etched completed. The Cu/Al IMC can be identified through optical inspection on the bond pads [3] where darker brown color areas represent the IMC formation. However, silver alloy wire samples Ag/Al IMC were done on both bond pads and ball bond side similar to conventional Au/Al IMC assessment for comparison.

1st Bond Parameters	PdCu wire	Ag alloy wire	% Decrease for Ag wire
Constant Voltage (CV), [mils/s]	0.3	0.25	17%
USG power [mAmp]	50	43	14%
Bond Force [grams]	13	10	23%

Table 6 – Optimum 1st bond parameter for 0.7mil PdCu and Ag alloy wire types

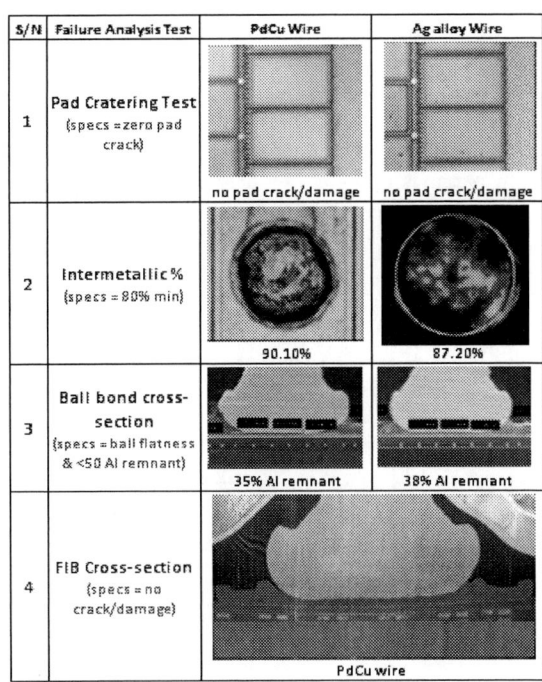

Figure 7 – 18um (0.7mil) wire size failure analysis results

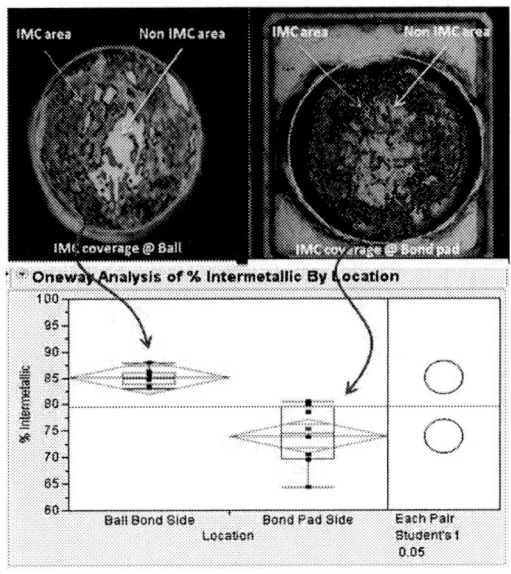

Figure 8 – Ag/Al IMC coverage comparison at bond pad vs. ball bond side

Figure 8, showed the Ag/Al IMC comparison taken from bond pad and ball bond side. The % IMC along the ball bond side are more consistent and higher in value compared to bond pad side % IMC using commercially available software to quantify % IMC. Hence, for Ag alloy wire the Ag/Al IMC inspection should be done at ball bond side for more consistent data and easier to quantify similar to conventional Au/Al IMC quantification. Further validation for this Ag/Al IMC on ball bond side exhibited good results during the initial reliability assessment plotted in figure 12 & 13.

b) Device B and 15um (0.6mil) wire size:

Only palladium coated copper (PdCu) wire was used for the DOE utilizing full factorial 2 factors, 3 level design with 9 legs. Two significant parameter factors considered were ultrasonic (USG) and bond force parameters. Once set of optimum parameters were validated and established, this set of 1st bond parameter will be use to validate for silver (Ag) alloy wire as per planned in table 3. Shown in table 7 is the 1st bond DOE summary for PdCu wire type.

Test Results		Leg 1	Leg 2	Leg 3	Leg 4	Leg 5	Leg 6	Leg 7	Leg 8	Leg 9
Ball Height (um)	Min	7	7	7	7	7	7	7	7	6
	Ave	8	7.5	7.8	8.2	7.8	7.5	7.4	7.5	6.8
	Max	9	8	9	9	9	8	8	8	8
	Stdev	0.67	0.53	0.79	0.79	0.63	0.53	0.52	0.53	0.63
Ball Size + Al Splash (um)	Min	32	30	33	33	33	32	31	34	32
	Ave	33.2	33	33.7	33.7	33.6	33.6	33.2	34.6	34.6
	Max	34	34	34	34	34	35	36	36	36
	Stdev	0.92	1.49	0.48	0.48	0.52	0.97	1.75	0.7	1.35
Ball Pull (gf)	Min	3.2	3.3	3.7	4.1	4.1	4.3	4.2	4.5	4.5
	Ave	4.7	4.6	4.8	5.2	5.2	5.3	5.3	5.3	5.3
	Max	6.4	7.6	8.0	6.3	6.2	6.0	6.4	6.2	6.3
	Stdev	0.9	1.0	0.9	0.6	0.6	0.5	0.6	0.5	0.5
	Cpk	1.0	0.9	1.1	1.9	1.8	2.2	1.8	2.3	2.4
Ball Shear (gr)	Min	9.3	9.8	9.3	13.0	13.4	13.1	12.4	13.4	15.2
	Ave	14.8	14.1	14.1	15.7	15.8	15.9	16.1	17.4	19.1
	Max	16.9	18.3	18.5	18.2	18.2	19.6	20.4	24.2	24.6
	Stdev	1.7	1.9	2.7	1.2	1.2	1.5	1.9	2.7	2.9
	Cpk	2.1	1.8	1.3	3.2	3.4	2.6	2.1	1.7	1.8
% NSOP		0	8	20	0	0	0	0	0	0
% Ball lift		53%	73%	37%	0	0	0	0	0	0
% Pad Peel		0	0	0	0	0	0	0	0	0

Table 7 – 0.6mil PdCu wire 1st bond DOE summary and result

Ball bond lift failures was observed for leg 1, 2 and 3 during wire pull test, whilst latter two legs also encountered non-stick on pad (NSOP) failures. All these failures were product of low parameter combination settings. Set out in table 9 is the 1st bond validation summary result using optimum nominal parameter settings from PdCu wire type. All wire bonding requirements stipulated on table 4 for 16um (0.6mil) were all achieved.

Some differences in output responses were observed as shown in the table 9 after Ag alloy wire validation. Obvious from the results that only ball bond diameter size increased (10%) for Ag alloy on same set of 1st bond parameter due to softer Ag wire properties. Observed ball bond height, pull and shear strength with 12%, 20% and 10% decreased respectively for softer Ag alloy wire as summarized in table 8.

1st Bond Responses	PdCu Wire	Ag alloy Wire	% Change
Ball Bond Diameter Size	lower	higher	10% ↑
Ball Bond Height/Thickness	higher	lower	12% ↓
Ball Bond Pull Strength	higher	lower	20% ↓
Ball Bond Shear Strength	higher	lower	10% ↓

Table 8 – 0.6mil wire diameter size responses summary comparison for PdCu and Ag alloy wire

Test Results		Specs	PdCu Wire Nom Parameter	Ag alloy Wire Nom Parameter	Remark
Ball Size + Al Splash (um)	Min	30~38um	32	35	pass
	Ave		32.5	35.2	pass
	Max		33	36	pass
	Stdev		0.5	0.4	pass
Ball Height (um)	Min	7~13um	11	9	pass
	Ave		11.6	10.3	pass
	Max		12	11	pass
	Stdev		0.5	0.7	pass
Ball Pull (gf)	Min	2 gr {min}	4.1	3.3	pass
	Ave		4.9	4.0	pass
	Max		6.1	5.1	pass
	Stdev		0.6	0.5	pass
	Cpk		1.7	1.4	pass
Ball Shear (gr)	Min	7 gr {min}	9.1	8.1	pass
	Ave		10.4	9.4	pass
	Max		11.6	10.9	pass
	Stdev		0.6	0.8	pass
	Cpk		3.5	2.4	pass
% NSOP		0	0	0	pass
% Ball lift			0	0	pass
% Pad Peel			0	0	pass
Pad Crater Test		no crack	0	0	pass
IMC (%)		80% {min}	87.3%	86.3%	pass
Al thickness Remnant after bond		10% {min}	39%	47%	pass

Table 9 – 0.6mil PdCu and Ag alloy wire validation summary result using nominal parameter setting

Figure 9 – 15um (0.6mil) wire size failure analysis representative images for both wire types

Similar to 18um wire size, failure analysis results showed no pad crack/damage after ball bond etch, >80% IMC achieved, Al remnant <50% of the original Al thickness and focus ion beam (FIB) cut showed no abnormality.

Test Results		Specs	PdCu Wire	Remark
			Nom Parameter	
Ball Size + Al Splash (um)	Min	28~34um	32	pass
	Ave		33.5	pass
	Max		34	pass
	Stdev		0.6	pass
Ball Height (um)	Min	7~13um	7	pass
	Ave		7.9	pass
	Max		9	pass
	Stdev		0.5	pass
Ball Pull (gf)	Min	2 gr (min)	2.8	pass
	Ave		3.4	pass
	Max		4.1	pass
	Stdev		0.3	pass
	Cpk		1.5	pass
Ball Shear (gr)	Min	7 gr (min)	10.3	pass
	Ave		12.5	pass
	Max		14.9	pass
	Stdev		1.2	pass
	Cpk		1.5	pass
% NSOP		0	0	pass
% Ball lift			0	pass
% Pad Peel			0	pass
Pad Crater Test		no crack	0	pass
IMC (%)		80% (min)	93.3%	pass
Al thickness Remnants after bond		10% (min)	35%	pass

Table 10 – 0.5mil PdCu wire 1st bond summary result using nominal parameter settings

S/N	Failure Analysis Test	PdCu Wire
1	Pad Cratering Test (specs = zero pad crack)	no pad crack/damage
2	Intermetallic % (specs = 80% min)	93.30%
3	Ball bond cross-section (specs = ball flatness & <50 Al remnants)	35% Al remnant
4		FIB Cross-section (spec = no pad crack/damage)

Figure 10 – 15um (0.6mil) wire size failure analysis representative images for both wire types

c) Device C and 13um or 0.5mil wire size:

Only palladium coated copper (PdCu) wire DOE using full factorial 2 factors, 3 levels design with 9 legs similar to 0.6mil wire size DOE was conducted for device C. Same

range of wire bond parameter from device B was derived partly due to same test vehicle sample used for the assessment. All wire bonding output responses and requirements for 0.5mil wire size stated in table 4 passed with summary of results shown in table 10. Alternatively, failure analysis of sample units for pad crack/damage, IMC, Al remnant and FIB cut as shown in figure 10 validated achieved results in table 10.

Decapsulation Challenges for Ag alloy wire

Unlike Cu wire bond devices, decapsulation for molded or encapsulated sample using laser ablation first on areas of interest before applying wet chemical acid etch using standard sulfuric/nitric acid (1:1 ratio) in order to fully expose the die and wire loop profile prior to performing failure analysis has been a major technical barrier for Ag alloy wire molded sample. Various studies has been done for wet chemical acid etch using different nitric acid concentrations as well as mixed with sulfuric acid did not provide a good result. Silver alloy wire is easily consumed if not corroded by nitric/sulfuric or mix acid during chemical etching. Ag alloy SEM photo showed cracking and wire thinning causing the wire to easily break if incorrect decapsulation recipe is used i.e. using nitric acid at soaking temperature >100°C or when using mix acid. Another concern for this Ag alloy wire was the wire center hollow area if temperature applied during wet chemical etch is >100°C as shown in figure 11. No further hollow part was observed if soaking temperature was set <100°C during wet chemical acid etch.

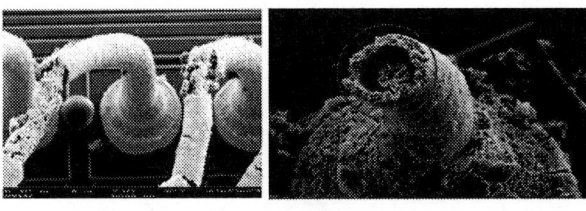

Figure 11 – Ag alloy wire decapsulation results with soaking temp >100°C

Figure 12 – Ag alloy wire laser decapsulation result

Even microwave induced plasma decapsulation process [2] that enables clean exposure of Cu/Al ball bonds while preserving the bond wire surface features was not able to successfully decapped the samples causing Ag wire corrosion, oxidation and wire break during microwave induced plasma decapsulation process.

Finally, using laser decapsulation process similar to laser type used in laser grooving of wafer and, without using any acid had shown a very good result for Ag alloy wire decapsulation without damaging the Ag alloy wire surface as shown in figure 12. Ball bond, loop profile and second bond are exposed fully and ready for any failure analysis requirement.

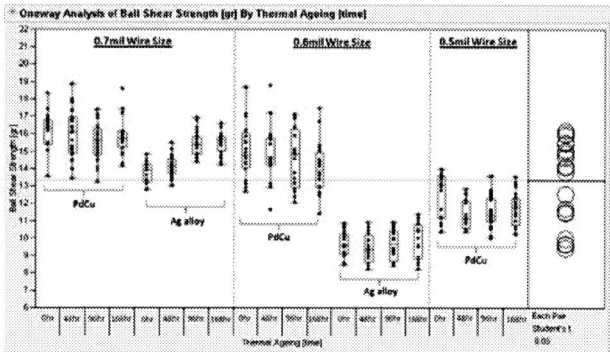

Figure 13 – PdCu & Ag alloy wire ball shear strength vs. thermal ageing results

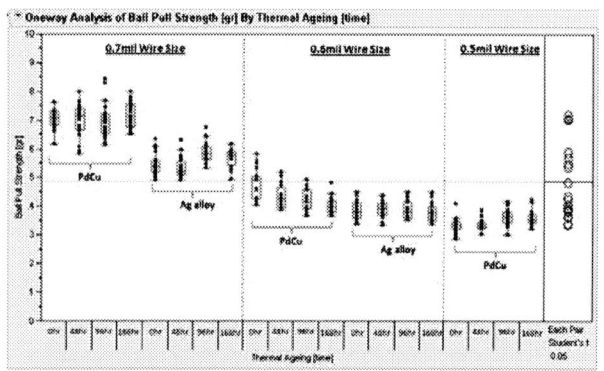

Figure 14 – PdCu and Ag alloy wire ball pull strength vs. thermal ageing results

Reliability Assessment and Results

Initial reliability assessments for PdCu wire bonded samples were subjected to high temperature storage test (HTS) at 175°C to 168 hours of continued thermal ageing and, with no mold encapsulation. Destructive analysis of ball pull and shear test were carried out at each read out points in order to assess 1st bond integrity and process robustness. Shown in figure 13 and 14 were the graph results for ball shear and pull strength for the three different wire sizes. No ball lift failures were observed during pull testing across the thermal ageing timeframe for all wire sizes. Ball bond shear strength for Ag alloy versus PdCu wire at 18um was reduced by 7% while for 15um wire size was greatly reduced to 30% partly due the softer properties of Ag alloy. This was further supported by the 20% and 10% ball pull strength reduction of

Ag alloy vs. PdCu wire at 18um and 15um wire sizes respectively.

TEST	Conditions	SPEC	Read Point	Sample Size	Results	Remarks
Precondition (MSL3)	L3 condition and reflow @ 260±0/5°C peak lead free	JESD22-A113B J-STD-020B	192 hrs	225	0 / 225	passed
Temp Cycle Test	-55°C to 125°C cond. B (precon parts)	JESD22_A104_A	500x	75	0 / 75	passed
			1000x	75	0 / 75	passed
Thermal Shock Test	-55°C to 125°C cond. C (precon parts)	JESD22_A106_A	300x	75	0 / 75	passed
			500x	75		passed
Unbiased HAST	130°C, 85% RH (precon parts)	JESD22-A118-B	96 hrs	75	0 / 75	passed
High Temp Storage Test (HTS)	Ta = 150°C (no precon)	JESD22-A103-A	500 hrs	75	0 / 75	passed
			1000 hrs	75	0 / 75	passed

Table 11 – Device A, FBGA 9x9mm using 18um PdCu qualification reliability summary result

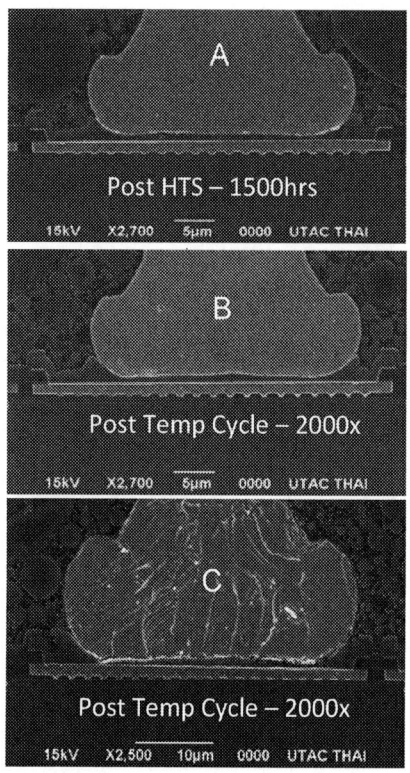

Figure 15 – Ball bond cross-section results after stress test reliability. (A) PdCu wire after HTS 1500hrs, (B) PdCu wire after Temp Cycle 2000x, (C) Ag alloy wire after Temp Cycle 2000x

Package Reliability Qualification

Device A using palladium coated Cu wire with 18um (0.7mil) wire size was assembled for full package qualification applying those previously established optimum wire bond process parameters. Other assembly robust processes learning from 40/45nm low k has been validated and applied to the 28nm ultra low-k especially for laser grooving followed by mechanical saw dicing process and mold or encapsulation process using green mold compound with controlled Cl (<20ppm) ionic content and, pH close to neutral condition. These were two critical parameters for mold compound that we notably implemented for Cu and Ag wire. Full package reliability test result is shown in table 11 with all read out points passed electrical test and no package delamination based from industry JEDEC standard requirements. Ball bonds cross-sectioned for extended reliability at high temperature storage up to 1500 hours and temperature cycles up to 2000 cycles for both PdCu and Ag alloy wire did not show any abnormality for the under pad structures as shown in figure 15.

Conclusions

Fine pitch wire bonding on 28nm ultra low-k chips using palladium coated Cu (PdCu) and silver (Ag) alloy wires have demonstrated good and reliable results. First bond DOE's were performed on three devices with different bond pad configurations and so far all wire bonding requirements were all satisfied and achieved. Validation though focus ion beam (FIB) cut analysis for ball bonded samples did not indicate any abnormalities caused by wire bonding process. Harder PdCu wire still managed to overcome and overwhelmingly exhibited good result especially for pad crack/damage on 28nm low-k structure coupled with thicker Al pad metal thickness of 1.55um acting as additional cushion during bonding process. A more consistent intermetallic assessment for Ag/Al IMC at ball bond side and easier to quantify similar to conventional Au/Al IMC compared to bond pad side Al/Ag IMC quantification. Technical challenges for Ag alloy wire molded sample decapsulation concerns has been resolved using laser decapsulation process with same laser type used for wafer laser grooving and, had shown very good result for decapsulation without damaging the Ag alloy wire surface as shown in figure 12. Ball bond, loop profile and second bond are exposed fully and ready for any failure analysis requirement. Further studies is underway to further remove those excess mold compound under the wire that are not fully or totally removed by laser decapsulation.

Finally, with PdCu (0.7mil) wire assembled on FBGA 9x9mm utilizing all robust and established assembly processes, managed to pass the package reliability qualification per JEDEC standard and even extended stress reliability test did not show any abnormality after ball bond cross-section.

Acknowledgments

The authors would like to thank all colleagues in UTAC, Corporate R&D and management staff for their support; GLOBAL FOUNDRIES Singapore for the collaborative development activities leading to the completion of this paper.

References

1.0 YB. Yang, Kumar N, John B, Hyman R.,"A dynamic study of pad structure impact on bond pad/low-K layer stress in copper wire bond", 2012 IEEE 14th Electronics Packaging Technology Conference

2.0 J.Tang, A.R.G.W. Knobben, E.G.J. Reinders, C.Th.A. Revenberg, J.B.J. Schelen, and C.I.M. Beenakker, "Microwave Induced Plasma Decapsulation of Thermally Stressed Multi-tier Copper Wire Bonded IC Packages", ICEPT 2013

3.0 Tracy Jia Lin Yap, Yin Kheng Au, Poh Leng Eu, "Importance of Cu/Al Intermetallic Coverage in Copper Wire Bonding with Sensitive Pad Structure" EMAP 2012, 14th International Conference

Assembly of Multiple Chips on Flexible Substrate Using Anisotropic Conductive Film for Medical Imaging Applications

Hoang-Vu Nguyen[1*], Trym Eggen[2], Bjørnar Sten-Nilsen[2], Kristin Imenes[1] and Knut E. Aasmundtveit[1]

[1] IMST – Department of Micro and Nano Systems Technology, HBV – Buskerud and Vestfold University College, Raveien 215, 3184 Borre, Norway

[2] GE Vingmed Ultrasound AS, Strandpromenaden 45, 3191 Horten, Norway

[*] Email address: hoang.v.nguyen@hbv.no, Phone: +47-33037924

Abstract

Flip-chip interconnection technology based on anisotropic conductive film (ACF) has been selected to assemble multiple chips (i.e. ASICs – Application-Specific Integrated Circuits) on a flexible substrate in an electro-acoustic module. The chips are relatively large with a high number of I/O pins distributed over its area (on the order of 100 mm²). In this work, the processes of bonding one single chip, two chips simultaneously and four chips simultaneously to a substrate using ACF were characterized and compared. The results show similar effect of bond force on electrical resistance of ACF interconnects for bonded samples with one single chip, two chips and four chips. High bond yield values (more than 99 %) are obtained for all groups of samples bonded with a normalized bond force varying from 50 N/chip to 500 N/chip. In general, higher bond force, hence higher bond pressure, provides better ACF interconnects in terms of lower electrical resistance and higher bond yield. This work has demonstrated the feasibility of bonding multiple chips, each of which has a relatively large area with a high number of I/O pins, to a flexible substrate using ACF.

Introduction

Medical imaging is an important tool to perform medical diagnosis. Ultrasound imaging technique has become one of the most utilized forms of diagnostic imaging available today. Ultrasound equipment is easy to use, mobile and relatively cheap compared to other techniques such as MRI (Magnetic Resonance Imaging) and x-ray CT (Computed Tomography). In order to obtain high-resolution 3D ultrasound images, the transducer inside an ultrasound probe must have a 2D array with an extremely high number of acoustic elements, typically several thousand. As each acoustic element needs to have a dedicated electrical connection to driving electronics, ASICs (Application-Specific Integrated Circuits) with a high number of I/O pins are thus required for ultrasound imaging applications. Short-distance electrical connections between the acoustic elements and the ASIC's I/O pins are demanded in order to improve temporal resolution, and hence real time imaging. In addition, an interface to the console requires a system substrate for external routing. An ultrasound Electro-Acoustic Module (EAM) therefore typically includes an array of acoustic elements closely integrated with an ASIC, which is connected to a system substrate [1, 2].

A potential approach to build an EAM is laminating acoustic elements to one side of a system substrate and an ASIC to the other side, as shown in Figure 1. The system substrate can be a flexible printed circuit with vias connecting corresponding bond pads on both sides of the substrate. The assembly for such an EAM includes bonding the acoustic transducer as well as the ASIC to both sides of the substrate. The assembly of ultrasound transducers and ASICs with thousands of electrical connections is challenging due to a demand of high bond yield and reliable interconnects under a tough assembly process.

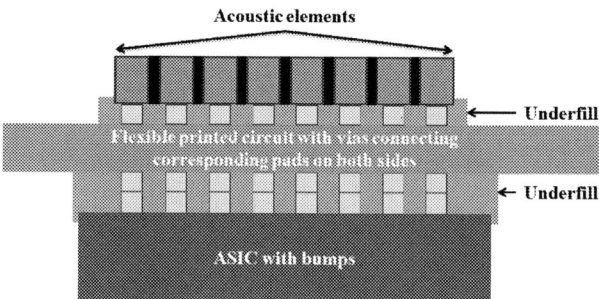

Figure 1: An Electro-Acoustic Module (EAM) in an ultrasound probe

The present work deals with the assembly of ASIC on flexible printed circuit, called flex substrate. In practice, manufacturing and bonding ASICs with a high number of connections have a high risk in terms of cost, yield, and process flexibility. Therefore, bonding individual smaller ASICs, instead of one very-large ASIC, to a substrate is a promising approach for the assembly of real EAMs. The main requirements of the bonding process of ASICs to flex substrates include:

- high bond yield;
- reliable interconnects throughout the entire life cycle of the EAMs;
- high-throughput bonding process that does not occupy manufacturing equipment (such as bonders) for an extensive amount of time.

Flip-chip interconnection technology based on anisotropic conductive adhesives (ACAs) has emerged as a potential solution for bonding ASIC to flex substrate in ultrasound applications. The main reason is because the technology can provide electrical connection, mechanical strength and sealing/underfilling in one quick process step [3, 4]. In addition, the ACA technology has potential to provide high bond yield [5, 6] and reliable interconnects for chip-on-flex applications [7-9].

2014 Electronic Components & Technology Conference

978-1-4799-2408-0/14 $31.00 © 2014 IEEE

ACA materials consist of a non-conductive adhesive matrix filled with conductive particles. The particles can either be solid metal spheres or metal-coated polymer spheres. In either case, the spheres must be monodisperse. The concentration of conductive particles is normally kept far below the percolation threshold (the minimum particle fraction where satisfactory conductivity in all directions first occurs) in order to prevent particles from contacting each other [3]. For ACA interconnects, the mechanical strength is provided by the adhesive matrix, while the electrical conduction is established when conductive particles are trapped between the interconnection bumps/pads on chip and substrate. In consequence, the electrical conduction of ACA interconnects is restricted to the Z-direction (perpendicular to the plane of chip/substrate), while the electrical insulation in the X- and Y-directions is maintained. ACAs can be either in paste form or in film-form, the latter being called an anisotropic conductive film (ACF). ACF is the most popular used form of ACAs, due to the ability to precisely control the volume of material, the density and the distribution of the particles within the sample, as well as the ease of handling.

In this work, the process of bonding multiple ASICs to a flex substrate using ACF is characterized. Test samples include dummy silicon (Si) chips and flex substrates that mimic the ASICs and the substrates in a real EAM. The processes of bonding one single chip, two chips simultaneously and four chips simultaneously to a substrate are compared. Bonded samples are characterized by means of electrical resistance of ACF interconnects and bond yield. The deformation of conductive particles trapped in ACF interconnects is also visually inspected by using cross-sectioned samples.

Experimental

Sample description

Dummy chips were manufactured, aiming at mimicking the geometry and the distribution of interconnection bumps on an ASIC. The chips are made of Si and are populated with gold (Au) bumps over an area on the order of 100 mm^2. The Au bumps were electroplated to an aspect ratio (height-to-diameter) of approximately 1:6. The bumps are distributed in the form of m-by-n matrix with bump pitch in the range of 100 – 400 μm.

Flex substrates are made of polyimide with bond pads populated identically on both sides, as illustrated in Figure 2. The corresponding pads on each side are connected by vias. The bond pads are composed of copper-nickel layers with a gold flash on top. The dimensions and the distribution of the pads are corresponding to those of the ASIC's bumps. Two groups of flex substrates were used in this work. One group, denoted F1, has bond area suitable for bonding one single chip and two chips. The other group, denoted F2, has larger bond area for bonding four chips. Note that the only difference between these two groups of substrates is the bond area.

Test samples include one, two or four dummy chips bonded to a flex substrate, being denoted as single-chip, dual-chip and quad-chip samples, respectively. The chips and the substrates are designed and configured to form daisy chains involving all interconnects in the samples. For each chip, there are 15 daisy chains with hundreds of interconnects in each chain.

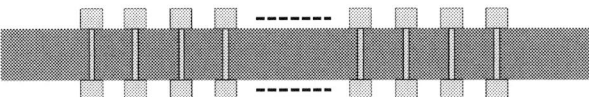

Figure 2: A sketch of a flex substrate with identical bond pads on both sides. The corresponding pads on each side are connected by vias.

Adhesive

The ACF used in this work is a single layer ACF from H&S HighTech Corporation. This ACF contains Ø5 μm Au-coated monodisperse polymer spheres and Ø0.8 μm SiO$_2$ particles. The ACF's thickness is selected in accordance with the dimensions and the distribution of the chip's bumps and the substrate's pads so that ACF interconnects can be obtained.

Flip-chip bonding process

The ACF bonding process of chips to substrates was performed using a flip-chip bonder FINEPLACER pico from Finetech GmbH & Co. KG. The bonder can apply a bond force up to 700 N. Three custom-made bond tools were used for "pick and place" and bonding of one single chip, two chips simultaneously and four chips simultaneously to a substrate. The tools have a ball joint structure, which allows a limited range of smooth movement in all directions, to ensure the co-planarity between chips and substrates during bonding. A transfer station was specially designed to support the picking up of two chips simultaneously and four chips simultaneously with the correct alignment. First, individual chips are picked, aligned and placed one by one on the station. The chips are held on the station by vacuum. Two or four chips are then picked up simultaneously using the corresponding bond tools when the vacuum on the station is disconnected. Bond line temperature corresponding to varying bond force values was characterized using flat thermal couples inserted between chip's bumps and substrate's pads during the bonding process without ACF applied.

The process of bonding dummy chip(s) to a flex substrate using ACF includes two main steps: pre-bonding and final bonding. Main parameters in both process steps include bond pressure, bond temperature and bond time.

In the pre-bonding, the ACF is first cut to an appropriate size and subsequently applied to bond area on the flex substrate. The ACF is then pre-bonded on the substrate using a light pressure and a low temperature recommended by the ACF manufacturer. The ACF is not cured during the pre-bonding since the temperature used is well below its (curing) onset temperature. After the pre-bonding, the protective film of the ACF is removed.

In the final bonding, the chip's bumps and the substrate's pads are first aligned while the flipped chip(s) are kept on a bond tool. The flipped chip(s) are then pressed onto the flex substrate and heat is applied to the chip(s) and the substrate.

978-1-4799-2408-0/14 $31.00 © 2014 IEEE

The target bond force is achieved before the bond line temperature reaches the onset temperature at which the ACF starts curing. After the main bond time, the bonded sample is cooled down, while still under pressure, to a temperature below the glass transition temperature of the full cured ACF. In the characterization of the ACF final bonding, the bond force (i.e. bond pressure) was the main varying parameter whereas the bond temperature and the bond time were selected based on the recommendations from the ACF manufacturer. Figure 3 shows examples of a dual-chip and a quad-chip sample after bonding.

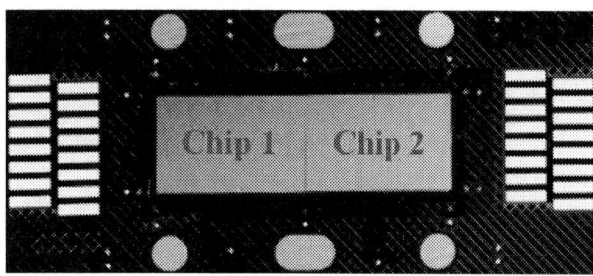

(a) A dual-chip sample with two chips bonded to a F1 flex substrate

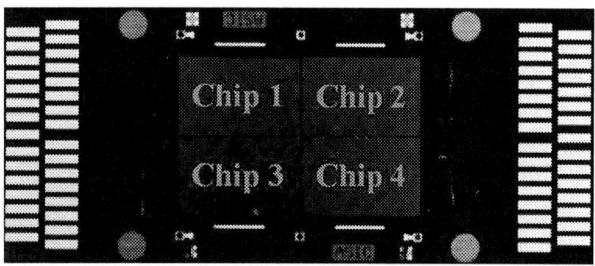

(b) A quad-chip sample with four chips bonded to a F2 flex substrate

Figure 3: Example of bonded samples after the ACF final bonding

Characterization method

Bonded samples were characterized by measuring the electrical resistance of daisy chains and by the bond yield. The resistance of the daisy chains was measured by a Keithley 2100 multimeter using the two-point probe method. The resistance of probe wires was found to be negligible compared to the measured resistance of each daisy chain. For each chip in a bonded sample, there are a total of 15 daisy chains being measured. The bond yield for each group (single-, dual- and quad-chip) of samples was determined by dividing the number of interconnects with sufficient conductance by the total interconnects of all samples in that group. Open interconnects in bonded samples were determined by probing the pads on the other side, with no chips bonded, of flex substrates. As the highest resolution of probing is two interconnects, both interconnects were claimed to be open if an open pair was found. A pair of interconnects was claimed as open when its measured resistance is higher than 5 Ω.

The deformation of conductive particles trapped in ACF interconnects was visually inspected by cross-section microscopy.

Results

The characterization of ACF bonding process for single-chip samples was performed with bond force varying from 50 N to 500 N. For dual-chip and quad-chip samples, the bond force was varied up to the maximum force that the flip-chip bonder can provide (i.e. 700 N).

Figure 4 shows effect of bond force on electrical resistance of ACF interconnects for single-chip, dual-chip and quad-chip samples. The resistance of one interconnect is determined by dividing the measured resistance of one daisy chain by the number of interconnects in that chain. This value, however, consists of resistance of track on the chip and track on the flex substrate, in addition to the resistance of the individual ACF interconnect. The results show similar effect of bond force on resistance of ACF interconnects for single-chip, dual-chip and quad-chip samples. The interconnect resistance decreases gradually with increasing bond force. The higher the bond force is, the lower and less scattered the interconnect resistance is. In addition, the interconnect resistance of single-chip, dual-chip and quad-chip samples bonded at a corresponding (normalized) bond force is comparable.

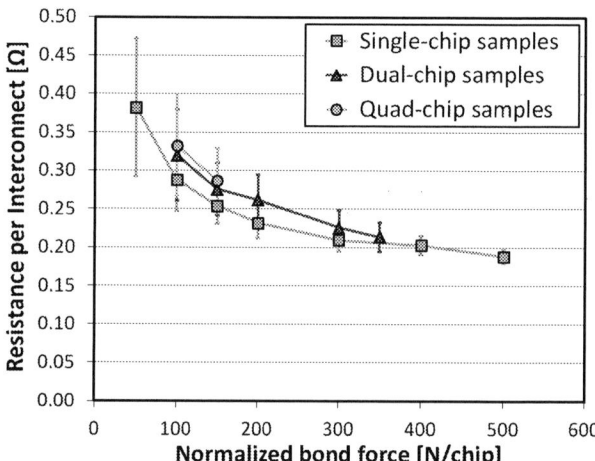

Figure 4: Effect of bond force on electrical resistance of ACF interconnects for single-chip, dual-chip and quad-chip samples. For dual-chip and quad-chip samples, the bond force presented is normalized to the number of chips in the samples. Each resistance value is the average of several thousand interconnects. Note that the resistance values shown in the figure also include the resistance of tracks on chips and flex substrates. The track resistance contributing to the result of each interconnect is estimated at about 0.16 Ω, based on practical resistance measurements for tracks on chips and substrates.

Bond yields corresponding to varying bond force values for single-chip, dual-chip and quad-chip samples are shown in Table 1. In general, high bond yield values (more than 99 %) are obtained for all groups of bonded samples. A 100 % bond yield is achieved with a normalized bond force value of 350 N/chip and above. However, the bond yield still retains extremely high values (more than 99.8 %) with a normalized bond force above 100 N/chip.

Table 1: Bond yield of single-chip, dual-chip and quad-chip samples bonded with varying bond force values. Each bond yield value is obtained based on a total number of several thousand interconnects.

Normalized bond force [N/chip]	Bond yield		
	Single-chip samples	Dual-chip samples	Quad-chip samples
50 N	99.27 %	–	–
100 N	99.85 %	99.73 %	99.46 %
150 N	99.93 %	99.85 %	99.86%
200 N	100.00 %	99.94 %	–
300 N	100.00 %	99.97 %	–
350 N	–	100.00 %	–
400 N	100.00 %	–	–
500 N	100.00 %	–	–

(–) : not available

Figure 5 shows a typical cross-sectional image of a sample bonded with a normalized bond force of 150 N/chip. In general, the bond line thickness of the interconnect varies at different areas due to the geometry of the bumps on chip and the pads on flex substrate. The bond line thickness achieves a lowest value of about 2 µm, corresponding to a deformation degree up to 60 % of the conductive particles.

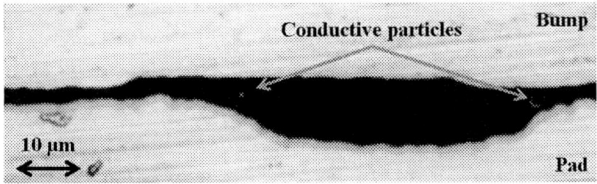

Figure 5: Optical micrograph of an ACF interconnect in a cross-sectioned sample bonded with a normalized bond force of 150 N/chip

Discussion

In this work, bond temperature and bond time used in the final bonding were selected based on the recommendations from the ACF manufacturer. The DSC (Differential Scanning Calorimetry) analysis provided by the manufacturer for the ACF's curing kinetics shows a high curing degree over 90 % when the ACF is cured with our selected bonding conditions. With such a high degree of curing, the electrical and the mechanical performance of interconnects based on conductive adhesives can achieve a stable level and will not be significantly improved by further increasing curing degree of the adhesives [3, 10]. Hence, bond force was selected as the main varying parameter in the characterization of the ACF final bonding.

The results from Figure 4 show comparable interconnect resistance of single-chip, dual-chip and quad-chip samples

bonded at the same normalized bond force. This indicates a linear relationship between the bond force needed for bonding and the number of chips in a sample in order to obtain ACF interconnects with similar electrical resistance. As all dummy chips are identical in this work, bond pressure calculated by dividing bond force by the total area of chips can be used as a universal parameter for bonding samples with varying number of chips, i.e. varying bond area of hundreds of mm².

As can be seen from Figure 4 and Table 1, higher bond force, hence bond pressure, provides better ACF interconnects in terms of electrical resistance and bond yield. Even though high bond pressure seems to be a good choice, it leads to a demand of a bonder with capability of providing very high bond force if chips with large area are to be bonded. The demand of such a special bonder will add more cost to and restrict the flexibility of the manufacturing process [5, 11]. The reliability of ACF interconnects is another factor that should be taken in to account. It is because using too high or too low bond pressure can significantly impact the reliability of ACF interconnects [12, 13]. Therefore, the trade-off between the bond pressure used and the bond quality in terms of electrical resistance, bond yield and reliability must be considered.

The cross-sectional image in Figure 5 shows a non-uniform bond line thickness within an ACF interconnect due to the geometry of the bump on chip and the pad on substrate. The conductive particles at the peripheral area of the chip's bump are the most deformed while the particles at the concave area of the bump are less deformed. No particles are deformed where both the concave areas of the chip's bump and the substrate's pad meet due to the separation distance being larger than the sphere diameter. Hence, the peripheral area of the bumps on chips is the main bond area of each ACF interconnect in our samples. The bond line thickness corresponding to this main bond area of ACF interconnects should thus be used to determine the deformation degree of conductive particles.

The root cause of open interconnects in samples bonded at relatively low normalized bond force is under investigation. In general, there are two main reasons for open ACF interconnects; 1) lack of conductive particles trapped between bumps on chips and pads on substrates; and 2) lack of sufficient contact between conductive particles and bumps on chips, as well as between conductive particles and pads on substrates. Factors contributing to these main reasons include:

- Local co-planarity misalignment between bumps on chips and pads on flex substrates. This happens when local deflection exists on flex substrates.
- Imperfectness of bumps and/or pads, such as non-uniformity or missing bumps/pads
- Non-uniform size of conductive particles in ACF
- Insulating layers that may exist between particles and bumps/pads
- Non-uniform distribution of conductive particles in ACF
- Non-uniform thickness of ACF

In this work, the number of open interconnects in bonded samples decreases, i.e. bond yield increases, with increasing bond force (Table 1). Hence, factors that are independent of the bond parameters would not be the main sources for more open interconnects in samples bonded with lower bond force. Those factors include missing bumps/pads on chips/substrates and ones related to the properties of the ACF, such as non-uniform size and distribution of particles in ACF, non-uniform ACF thickness. On the other hand, the local co-planarity misalignment between bumps and pads due to the local deflection on flex substrates as well as the non-uniformity of bumps/pads are relevant for the phenomenon observed. At low bond force values, these factors can cause a bond line thicker than the diameter of conductive particles, hence no particles are trapped or sufficiently deformed in some interconnects. This issue can be reduced, and further eliminated, when a higher bond force is applied. It is because a high bond force can compensate a certain degree of imperfectness between bumps and pads of bonded samples [14]. Further analyses such as cross-sectioning and visual inspection as well as x-ray inspection of failed interconnects are to be carried out to determine the true failure mechanism of open interconnects in samples bonded with a low bond force.

Conclusions

The assembly of multiple chips on a flexible substrate using anisotropic conductive film (ACF) was successfully developed. The chips mimic a real driving electronic used in an electro-acoustic module, and they have a high number of pins distributed over an area on the order of 100 mm^2. The processes of bonding one single chip, two chips simultaneously and four chips simultaneously to a substrate were characterized by means of interconnect resistance and bond yield. Bond force (i.e. bond pressure) was the main varying parameter in the bonding process while bond temperature and bond time were kept fixed. The characterization results of samples including one, two or four chips bonded to a flex substrate (being denoted as single-chip, dual-chip and quad-chip samples, respectively) were compared.

The effect of bond force on the electrical resistance of ACF interconnects is similar for single-chip, dual-chip and quad-chip samples. The interconnect resistance decreases gradually with increasing bond force. The higher the bond force is, the lower and less scattered the interconnect resistance is. In addition, single-chip, dual-chip and quad-chip samples bonded at a corresponding normalized bond force exhibit a comparable interconnect resistance.

High bond yield values (more than 99 %) are obtained for all groups (single-, dual- and quad-chip) of samples bonded with a normalized bond force varying from 50 N/chip to 500 N/chip. A 100 % bond yield is achieved for samples bonded with a normalized bond force of 350 N/chip and above. However, the bond yield still retains an extremely high value (~ 99.8 %) when a normalized bond force above 100 N/chip is used.

The results from this work are an important input for choosing a proper manufacturing process in which the trade-off between the bond force used and the bond quality in terms of electrical resistance, bond yield and reliability must be considered.

Acknowledgments

The present work was funded by the Oslo Fjord Regional Research Fund (Oslofjordfondet, Norway) through the project *Electro Acoustic Module (EAM) Technology for Advanced Ultrasound Applications* (project number 217689).

The authors would like to thank Warren Lee at GE Global Research Center for contributing ideas and offering discussion inputs.

References

[1] I. O. Wygant, X. Zhuang, D. T. Yeh, S. Vaithilingam, A. Nikoozadeh, O. Oralkan, *et al.*, "An Endoscopic Imaging System Based on a Two-Dimensional CMUT Array: Real-Time Imaging Results," in *IEEE Ultrasonics Symposium*, Rotterdam, The Netherlands, 2005, pp. 792-795.

[2] K. E. Aasmundtveit, T. T. Luu, T. Eggen, C. E. Baumgartner, N. Hoivik, K. Wang, *et al.*, "Thermosonic Bonding for Ultrasound Transducers: Low-temperature Metallurgical Bonding," in *The 4th Electronics System Integration Technology Conferences*, Amsterdam, The Netherlands, 2012, pp. 1-6.

[3] J. E. Morris and J. Liu, "Electrically Conductive Adhesives: A Research Status Review," in *Micro- and Opto-Electronic Materials and Structures: Physics, Mechanics, Design, Reliability, Packaging.* vol. 2, E. Suhir, Y. C. Lee, and C. P. Wong, Eds., ed: Springer Science+Business Media, Inc., 2007, pp. 527-570.

[4] R. S. Pai and K. M. Walsh, "The Viability of Anisotropic Conductive Film as A Flip Chip Interconnect Technology for MEMS Devices," *Journal of Micromechanics and Microengineering,* vol. 15, pp. 1131–1139, 2005.

[5] A. Larsson, F. Oldervoll, T. A. T. Seip, H.-V. Nguyen, H. Kristiansen, and Ø. Sløgedal, "Anisotropic Conductive Film for Flip-Chip Interconnection of A High I/O Silicon Based Finger Print Sensor," in *The 22nd Micromechanics and Micro systems Europe Workshop*, Tønsberg, Norway, 2011, pp. 186-189.

[6] M. Bigas, E. Cabruja, and M. Lozano, "Bonding Techniques for Hybrid Active Pixel Sensors (HAPS)," *Nuclear Instruments and Methods in Physics Research A,* vol. 574, pp. 392-400, 2007.

[7] P. Palm, J. Maattanen, Y. De Maquille, A. Picault, J. Vanfleteren, and B. Vandecasteele, "Comparison of different flex materials in high density flip chip on flex applications," *Microelectronics Reliability,* vol. 43, pp. 445-451, 2003.

[8] M. J. Yim, J.-S. Hwang, J. G. Kim, J. Y. Ahn, H. J. Kim, W. Kwon, *et al.*, "Highly Reliable Flip-Chip-on-Flex Package Using Multilayered Anisotropic

Conductive Film," *Journal of Electronic Materials,* vol. 33, pp. 76-82, 2004.

[9] M. A. Uddin, M. Y. Ali, and H. P. Chan, "Achieving Optimum Adhesion of Conductive Adhesive Bonded Flip-chip on Flex Packages," *Reviews on Advanced Materials Science,* vol. 21, pp. 165-172, 2009.

[10] M. Inoue and K. Suganuma, "Influential Factors in Determining The Adhesive Strength of ACF Joints," *Journal of Materials Science: Materials in Electronics,* vol. 20, pp. 1247–1254, 2009.

[11] S. Kardos, M. Somora, S. Slosarcik, J. Urbancik, and I. Vehec, "Anisotropic Conductive Joints Utilization for Joining of Flexible Cables at Rigid Substrates," *Journal of Electrical Engineering,* vol. 60, pp. 48-52, 2009.

[12] J.-W. Kim and S.-B. Jung, "Effect of Bonding Force on The Reliability of The Flip Chip Packages Employing Anisotropic Conductive Film with Reflow Process," *Materials Science and Engineering A,* vol. 452-453, pp. 267-272, 2007.

[13] L. Frisk, K. Saarinen, and A. Cumini, "Reliability of ACF Interconnections on FR-4 Substrates," *IEEE Transactions on Components and Packaging Technologies,* vol. 33, pp. 138-147, 2010.

[14] A. Seppälä, S. Pienimaa, and E. Ristolainen, "Flip Chip Joining on FR-4 Substrate Using ACFs," *The International Journal of Microcircuits and Electronic Packaging,* vol. 24, pp. 148-159, 2001.

High Frequency High Current Point of Load Modules with Integrated Planar Inductors

Wenli Zhang, Yipeng Su, David Gilham, Mingkai Mu, Qiang Li, and Fred C. Lee
Center for Power Electronics Systems
Bradley Department of Electrical and Computer Engineering
Virginia Polytechnic Institute and State University
Blacksburg, VA 24061
Phone: 540-231-8209, E-mail: wzhang11@vt.edu

Abstract

Planar inductors made by mixed laminates of commercially available low-fire Ni-Cu-Zn ferrite tapes and metal-flake composites developed by NEC-TOKIN have been used for the high density integration of point-of-load (POL) modules. The incremental permeability and core loss density were characterized on the toroidal samples under high dc bias. Measurement results demonstrate that both materials are suitable for applications in high frequency high current POL converters. In order to realize a high power density POL module, a multilayer ferrite inductor laminated with alternating layers of ESL 40010 and ESL 40012 in a 1:1 ratio has been fabricated and integrated with the active layer using modified low temperature co-fired ceramics (LTCC) processing. Meanwhile, the standard printed circuit board (PCB) processing has been adopted for the POL integration with a PCB embedded inductor using metal-flake composite materials. These developed 3-dimensional (3-D) integration approaches can be used to reduce the footprint and increase the power density for the high frequency high current POL converter. It is demonstrated that the power efficiency of POL modules with integrated planar inductors can achieve above 87% at operating frequency of 2 MHz and output current of 15 A. Additionally, no obvious efficiency degradation was observed on the integrated POL modules after certain numbers of thermal cycling from -40 ^{o}C to +150 ^{o}C.

Introduction

High power density point-of-load (POL) converters are highly desired to fulfill the requirements of miniaturization for the thriving portable electronic devices. Increasing switching frequency (>1 MHz) to reduce the size and weight of passives and integrating passives like capacitors and inductors are two common routes to increase the power density for POL converters. The emerging next-generation semiconductor power devices such as gallium nitride (GaN) transistors offer the potential benefits in various power conversion applications, which enables the high switching frequency up to several megahertz. In today's POL modules, magnetic components consume the largest footprint and become the most limiting factor for high density integration. With the increased demand to reduce the size of power inductors, magnetic core materials with low energy loss and high permeability under high dc bias as well as high operating frequency are exceedingly needed. Moreover, such magnetic materials have to be compatible with the conventional or newly developed integration techniques for power electronics applications [1], [2].

A series study has been performed on the realization of the high density integration of high frequency high current POL modules at the Center for Power Electronics Systems (CPES).

Low-profile planar inductor structures with a lateral flux pattern were studied for possible use in integrated POL converters [3]. The two-phase inversed coupled inductor design was also investigated for the sake of further core volume and loss reduction [4], [5]. The designed planar inductor was subsequently used to fabricate integrated POL modules. In the structure of integration, the magnetic inductor was designed as a substrate carrying the rest of the converter circuit to realize a truly 3-dimensional (3-D) integration which can save footprint of the inductor and fully utilize the available space [6], [7].

In this work, mixed low-fire ferrite laminates and metal-flake composite materials were firstly chosen for the fabrication of planar inductors and further high density integration of POL modules based on the comparison of core loss density and permeability with other candidate magnetic materials. Then, modified low temperature co-fired ceramic (LTCC) processing and hybrid integration techniques were used for the fabrication of the POL module with multilayer ferrite inductor substrate. Next, the POL module with printed circuit board (PCB) embedded metal-flake inductor was prepared using PCB technology. Finally, electrical performance and thermal reliability were tested on both high power density integrated POL prototype modules.

Characterization of Magnetic Materials

The high dc current is constantly applied to bias the inductors used in designed high frequency POL converters. It is, thus, important to consider the influence of the dc bias on the magnetic properties of core materials exposed in high excitation levels. The incremental permeability and core loss density under dc bias up to 4000 A/m were evaluated using a newly developed experimental setup by applying an additional winding and dc current to the test samples [8]. 4 high-frequency magnetic materials (mixed low-fire Ni-Cu-Zn ferrite laminates developed by CPES [9]; ESL 40010 low-fire Ni-Cu-Zn ferrite tapes purchased from ElectroScience; metal-flake composite magnetic materials obtained from NEC-TOKIN; 4F1 Ni-Zn ferrite cores purchased from Ferroxcube) were selected for investigation of the impact of superimposed dc bias on their magnetic properties.

The incremental permeability with applied dc bias of these 4 selected materials was measured at the frequency f of 1.5 MHz and compared with each other, as shown in Fig. 1. The incremental permeability continuously decreases as the superimposed dc bias H_{dc} increases up to 4000 A/m for all testing materials. It is demonstrated that the NEC-TOKIN's metal-flake composite materials have the highest permeability among all examined samples in the whole H_{dc} range, which is highly desired for the high current POL application. The mixed low-fire ferrite laminates exhibit lower incremental permeability than the NEC-TOKIN's materials but higher than the commercial ESL 40010 and 4F1 ferrites.

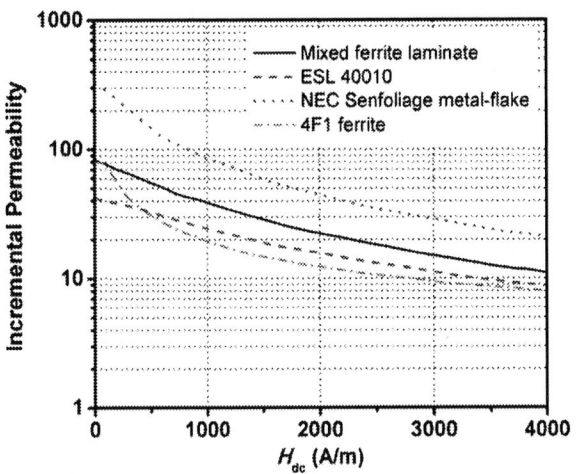

Figure 1. Variation of incremental permeability measured at 1.5 MHz under dc bias for selected magnetic materials.

The changes of core loss density mearsured at frequency f of 1.5 MHz and flux density B_m of 10 mT with superimposed dc bias for the chosen samples are illustrated in Fig. 2. The core loss density P_v for the commerical ESL 40010 and 4F1 ferrites gradually increases with the dc bias H_{dc} increasing, the latter of which shows a relatively lower P_v till the H_{dc} increases up to 2400 A/m. The P_v measured on the mixed ferrite laminates and NEC-TOKIN's metal-flake composite materials initially decreases at 400 A/m and then increases, however, these two materials have the lower P_v at H_{dc} above 800 A/m when compared with the other two materials. The NEC-TOKIN's materials exhibit the lowest P_v when the superimposed H_{dc} is higher than 2400 A/m.

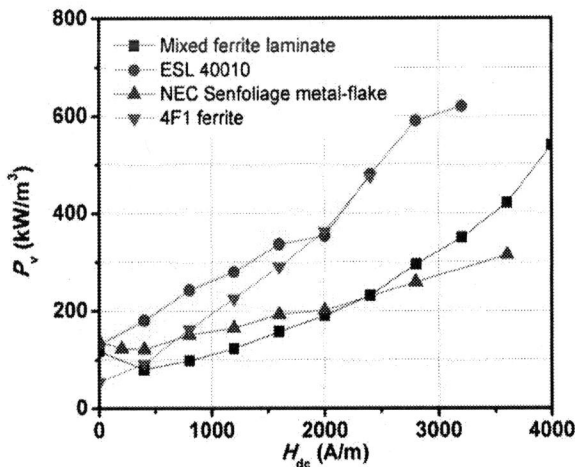

Figure 2. Comparsion of core loss density under dc bias for selected samples measured at f = 1.5 MHz and B_m = 10mT.

Considering the higher incremental permeability and lower core loss density especially at a high dc bias level, the

mixed ferrite laminates developed by CPES and metal-flake composite materials invented by NEC-TOKIN were used for the design and fabrication of planar inductors as well as high density integration of POL modules.

Fabrication of Integrated POL Modules with LTCC-Based Low-Fire Ferrite Laminates

As a well-established electronic packaging technology for three decades, LTCC materials and processing provide high density electrical interconnects and hermetic packaging for integrated circuits. From the beginning of this century, LTCC processing techniques have been demonstrated to be an effective approach for the fabrication of low-profile power inductors or transformers using low temperature sinterable ferrite tapes [10]-[12]. The LTCC-based low-fire ferrite tapes are normally produced in tape format by tape-casting a slurry on polymeric carrier substrates at varying thickness (50-350 μm). The winding (vias, lines, and pads) made by silver or gold conductive pastes can be co-fired with magnetic ferrites at a temperature around 900 °C. The developed LTCC-based inductors and transformers are reliable, compatible with surface mount technology, and do not require hand winding.

A 3-D integration concept that the whole converter is assembled in a vertical way has been realized in this study. A low-profile passive layer is designed as a substrate for the whole converter and the active layer is mounted above. This vertical configuration allows a saving in footprint and a full space utilization, which greatly increases the converter's power density. The key technologies in this integration solution can be summarized as low-profile magnetic component design and fabrication, high-frequency active layer design, and the integration of both passive and active layers. As mentioned previously, the mixed low-fire ferrite laminate has shown desirable magnetic performance under high dc bias and is also applicable for integrating with semiconductor devices. In the following sections, the fabrication of a low-fire multilayer ferrite inductor substrate and its further integration with the active layer will be introduced.

A. Preparation of Low-Fire Multilayer Ferrite Inductors

The modified LTCC processing technique was used in this study for the preparation of planar low-profile ferrite inductors. A process diagram of inductor fabrication is illustrated in Fig. 3. The ferrite tapes were firstly cut into 50 mm x 50 mm squares. A certain number of layers of the cut squares were then stacked alternatively one above another using two commercial low-fire tapes, ESL 40010 and ESL 40012 (ElectroScience, King of Prussia, PA). Next, the stacked layers were placed between heated platens (70 °C) and pressed at 10 MPa for 15 min using a uniaxial hydraulic press (2112; Carver, Wabash, IN). The designed inductor geometry and conduction vias were cut out of the mixed ferrite laminates using a CO_2-laser system (Resonetics, Nashua, NH) after lamination process. The additional via filling of silver paste (7740; DuPont Microcircuit Materials, Research Triangle Park, NC) was conducted. Finally, ferrite laminates and compatible silver pastes were co-fired at a peak temperature of 885 °C for 3.5 h on a flat porous zirconia setter in a box furnace (Vulcan 3-550; DENTSPLY International,

York, PA) in air atmosphere. The organic binder was burnout at 450 °C for 1 h. Both temperature ramp-up and cool-down processes were controlled at the same rate of 0.5 °C/min. During the sintering process an additional pressure of 0.5 kPa was applied on top of the mixed ferrite laminates. After sintering, 0.3 mm thick copper electrodes and two connection pins were soldered on the surface of sintered inductors to generate the conductive winding for testing.

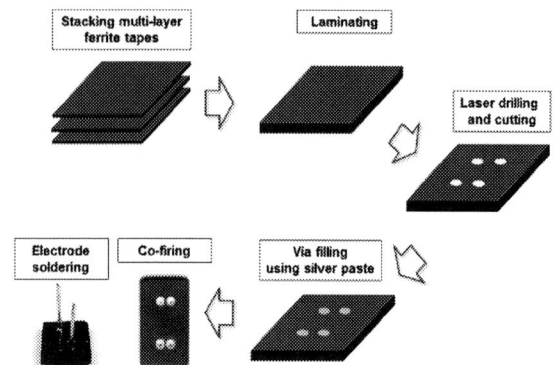

Figure 3. Process flow of planar inductor fabrication using low-fire ferrite tapes and LTCC technology. The sintering shrinkages of 20% in x-y direction and 15% in z direction for the mixed ferrite laminates have been taken into consideration in the design process. The prolonged sintering profile adjusted from manufacturers' recommendation was determined in order to reduce stresses developed in the mixed ferrite laminates due to the different sintering behaviors of each type of layers. The silver paste was selected based on its higher conductivity compared with other similar products.

B. Integration of Planar Inductors with Active Layers

The developed planar inductor substrate has to be connected with an active layer to realize the converter function. Direct bonded copper (DBC) and printed circuit board (PCB) are utilized for the fabrication of active layers. The DBC substrate provides better thermal performance while the PCB substrate provides a cost-effective solution. The PCB version is discussed in this section as an example to demonstrate the integration process.

The LTCC process for inductor fabrication introduced in previous section was slightly modified in order to facilitate further integration of inductor substrate with active layer. Ferrite laminates with laser drilled vias were first sintered alone in air. The patterned conductor vias were then filled with silver paste but two of those vias at the diagonal corners were left blank. Filled silver paste was finally post-fired on sintered ferrite laminates at 850 °C for 30 min. The copper electrodes were soldered on the one side of inductor substrate before bonding to PCB active layer.

PCB based active layers were manufactured based on the design and all chips as well as passives were soldered on the one side of active layer. Two copper pins were then soldered on the other side of the active layer. The pin soldering sites were positioned to accurately correspond to the empty holes left on the multilayer ferrite inductor. A solder mask was used to protect the substrate and prevent excessive solder paste from spreading during soldering.

The selected thermal interface material (SUPREME 10ANHT; MasterBond, Hackensack, NJ) in a paste form was applied evenly on to the active layer surface without components after pin soldering and solder mask removing. Inductor substrate was then attached with gentle pressure to the bonding layer. The inductor surface soldered with copper electrodes was in contact with that thermal interface material (TIM). Subsequently, the TIM was cured at 120 °C for 90 min. A bonding layer, electrically non-conductive but thermally conductive, was firmly formed between inductor substrate and PCB active layer after curing. Finally, another set of copper electrodes was soldered on the exposed surface of inductor substrate. An integrated POL with low-fire ferrite inductor substrate and PCB active layer has been fabricated in this manner and used for further electrical testing and thermal reliability evaluation. A schematic drawing and a photo image of the assembled POL module are illustrated in Fig. 4.

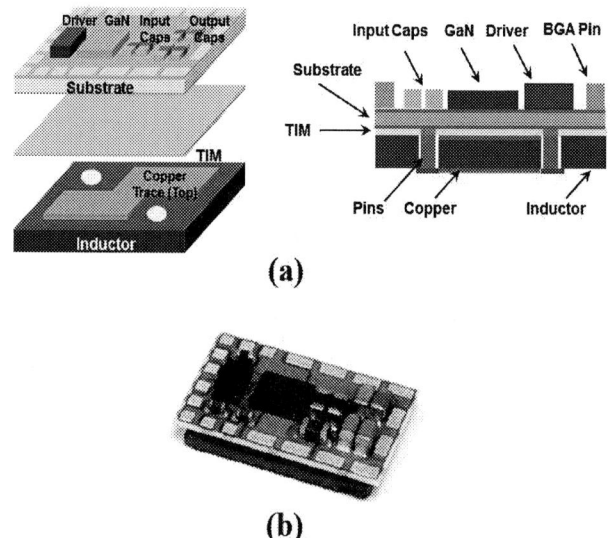

Figure 4. (a) Structural drawing of integrated POL module. This structure includes three basic parts: the top active layer with a GaN transistor, a driver, connection pins, and input/output capacitors, the bottom ferrite inductor layer with copper electrodes, and the adhesive TIM layer in the middle; (b) Photo of assembled POL module with LTCC-based low-fire ferrite inductor substrate.

Fig. 5 presents a microscopic image taken on the cut cross-section surface of an assembled module. No pores, cracks, or other major defects were observed on the cured TIM bonding layer, meanwhile, no delamination was inspected at the interfaces between TIM and copper on PCB or multilayer ferrite inductor. A fully assembled POL module is realized with all functional components integrated into one part. However, such a high-temperature ceramic process involved integration method is not cost effective and is difficult to achieve in automatic production. The other assembly approach studied integrates a planar inductor using a traditional PCB manufacturing technique.

978-1-4799-2408-0/14 $31.00 © 2014 IEEE

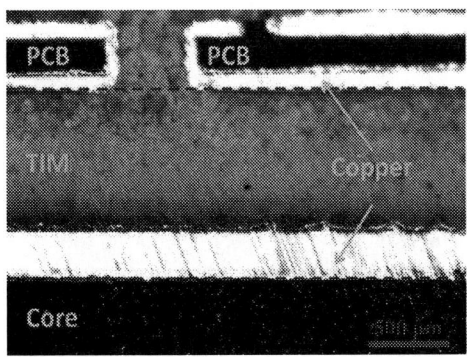

Figure 5. Optical microscope image of the cross-sectional view of the assembled POL module with LTCC-based ferrite inductor substrate. The sample was mounted in epoxy and then cut using a diamond blade saw (VC-50; LECO, St. Joseph, MI) after epoxy curing. The exposed cross-sectional surface was ground and polished for observation. Other samples shown in this work for optical microscope or scanning electron microscope (SEM) observation were prepared using the same method.

Processing of POL Modules with PCB Embedded Metal-Flake Inductors

Embedding passive or active components into the substrates (e.g. PCB and DBC) has gained increasing interest from both academia and industry. The growing attention toward embedded structure is not only because of the demand for higher power density but also the need for better electrical performance and reliability. Technologies for embedding passive components (e.g. inductors) into build-up layers of polymeric substrates have been investigated for more than a decade [13], [14]. Since most of the medium and high power converters use PCB as the substrate, current research on embedding technology focuses on utilizing standard PCB materials and processes.

It is reported in this work of fabrication of a prototype POL module with PCB embedded metal-flake composite inductor. As introduced in the magnetic characterization section, the metal-flake composite materials developed by NEC-TOKIN exhibit higher permeability and similar core loss density compared to some sintered ferrites because of the alignment improvement and the volume ratio increment of magnetic flakes. Moreover, such composite materials are capable of being either molded as magnetic cores or built into thick-film format by tape casting. The thickness of the laminates is able to build up to several millimeters. The laminated plate can be further processed by mechanical machining and no high temperature treatment is required. It is thus that the NEC-TOKIN's metal-flake composite materials are ideal to realize the PCB integration of the POL module with high output current, enabling a cost-effective and automatic production processing.

A. Preparation of PCB Embedded Inductor Substrate

The commercial PCB materials used in this work, FR406 epoxy laminates and prepregs, were obtained from Isola. The process flow of fabricating a multilayer PCB structure with embedded metal-flake composite magnetic core is illustrated in Fig. 6. Firstly, individual layers were prepared for building the multilayer PCB structure. The top and bottom layers were made of two C-stage fully cured copper cladded laminates. The surface winding on both layers were prepared by etching off extra copper using a bench-top etching machine. The middle layer was another C-stage laminate with an opening for inductor embedding. Two B-stage semi-cured prepregs were prepared as adhesive layers located in between every laminate layers. All layers were cut in the same dimensions. Alignment holes on the corners were laser drilled at every building layers and a rectangular-shape cavity fit for the geometry of embedded core was also formed in the middle layer using the same CO_2 laser machine. Then, the made layers were stacked in sequence with the assistance of alignment pins. The planar core made by NEC-TOKIN's metal-flake composite materials was embedded inside the multilayer PCB. Subsequently, the stack was laminated using a hydraulic press and the recommended heat and pressure profiles from manufacturer. The peak lamination temperature and pressure applied were 190 °C and 2 MPa, respectively. The semi-cured epoxy in prepregs was melted to adhere to different layers under certain heat and became fully cured after the lamination process. After lamination, the peripheral materials of the multilayer laminate were cut off and four holes going through both PCB layers and the magnetic core were drilled using a PCB milling machine (QC5000; T-Tech, Norcross, GA). Finally, the inductor winding was realized through pin soldering. A 2-turn planar inductor sandwiched into a two-layer PCB with a copper layer on the top and bottom is demonstrated in Fig. 6. The low-profile inductor substrate with lateral flux pattern can utilize the superiority of the metal-flake composite material, since the magnetic metal-flake is also laterally aligned.

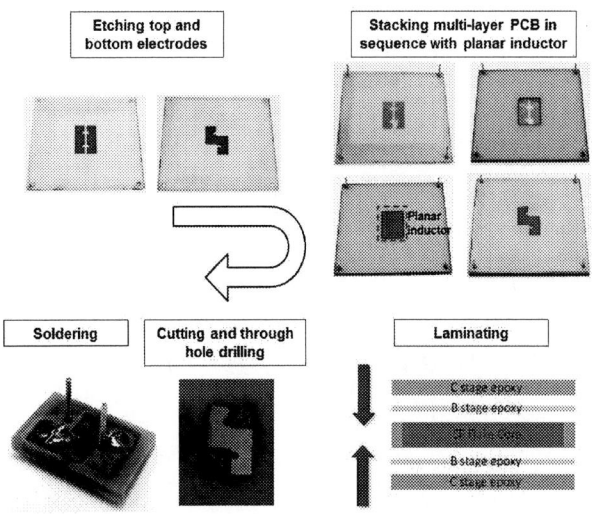

Figure 6. Fabrication process of PCB embedded planar inductor. The lamination parameters (pressure, temperature, and heating/cooling rates) need to be carefully controlled to avoid any damages to the embedding structure. A symmetric layout is recommended for a uniform pressure distribution in order to prevent the severe warpage of the multilayer PCB substrate.

The PCB embedded planar inductor was vertically sectioned to investigate the PCB lamination results. Fig. 7 shows the SEM image of a cross-sectional view of the multilayer PCB structure with the embedded core. A cohesive laminate was demonstrated by observing no obvious cracking on both PCB and core materials and delamination at their interfaces. As shown in Fig. 7, few small pores were found on the PCB layers next to the vertical side of the core material, which was possibly due to the volatiles that were generated during the lamination process and trapped afterwards. The application of the vacuum lamination technique would solve this problem.

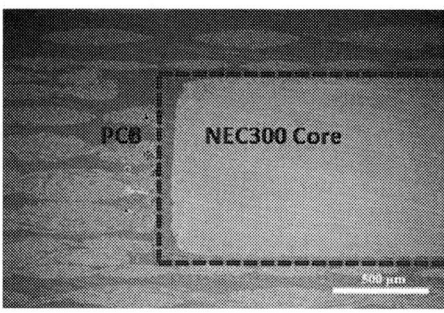

Figure 7. SEM micrographs of laminated PCB substrate with embedded metal-flake composite inductor.

B. Integration of PCB Embedded Inductor Layer with Active Layers

The fabrication of the active layer was performed using the same PCB materials and designed active circuitry. Similar to LTCC-based integration techniques, two connection pins were soldered on the backside of the prepared active PCB layer. One more lamination process was conducted to integrate a PCB embedded core substrate with the top active layer. An additional one prepreg layer was inserted in between the two parts. After lamination, the two connection pins through the PCB embedded core were truncated to be flush with the bottom and soldered to surface electrodes to generate the complete winding. All other components were finally surface mounted on top of the active layer using one-step reflow soldering.

The layout of this 4-layer PCB based module with embedded core is illustrated in Fig. 8(a). The laminated multilayer structure is expanded by intentionally enlarging the distance between each layer. All components except the inductor are placed on the top layer. The switching node point (VSW) and the output point (V_{out}) at the top layer are connected to the pads on the second layer and further to the lower surface winding of the inductor at the bottom layer through two conductive pins. The upper surface winding of the inductor at the third layer is connected to the lower surface winding at the bottom layer by two blind vias. The planar inductor is embedded between the third layer and the bottom layer. Part of the second layer is functional as a shielding layer. The eddy current induced in the shielding layer creates the opposite flux to cancel the flux caused by the parasitic inductance of the high frequency switching loop, which reduces the loop inductance and switching loss. A completely fabricated high frequency POL module with PCB embedded planar inductor is displayed in Fig. 8(b). The integrated module was used for further electrical and thermal characterizations.

Performance of High Power Density Integrated POL Modules

In order to maintain a desirable efficiency, the switching frequency of the POL is continuously decreased from multi-megahertz to several hundreds of kilohertz, with the increasing current level. This is mainly because of the limitation of available power semiconductor devices and suitable magnetic materials. It is difficult to achieve both high efficiency and high frequency operation of the POL converter with a high output current load.

In this study efficiencies were tested for a 12 V_{in} to 1.8 V_{out} conversion at different frequencies and load levels for both integrated POL modules. As illustrated in Fig. 9(a), the power efficiency measured at full load (15 A) for the POL module with low-fire ferrite inductor substrate can achieve 89% at 2 MHz and 87% at 5 MHz. At light load condition (5 A), the efficiency measured at 2 MHz and 5 MHz increases to 93% and 90%, respectively. It can be seen from Fig. 9 (b) that the POL module with the PCB embedded core can achieve above 84% efficiency in the load current range from 5 A to 20 A and the peak efficiency is as high as 89% with an output current of 10 A tested at 1.5 MHz.

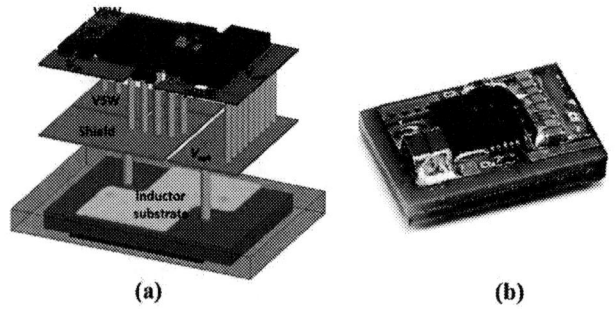

(a)   **(b)**

Figure 8. (a) Diagram of an expanded view of the multilayer PCB structure with embedded planar inductor; (b) Photo of a prototype of the integrated POL module with PCB embedded inductor.

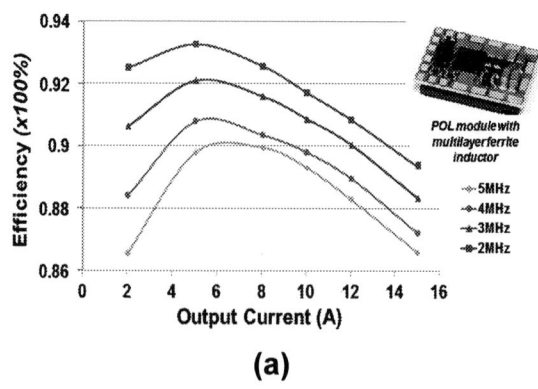

(a)

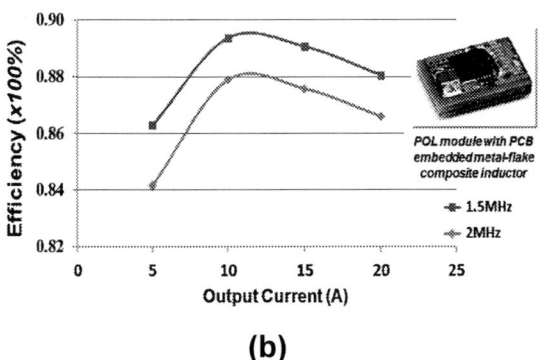

(b)

Figure 9. Efficiency data of integrated 12 V_{in}, 1.8 V_{out} single-phase POL modules with (a) ferrite inductor substrate and (b) PCB embedded metal-flake composite inductor measured at different frequencies and output current levels.

For both two modules tested at 2 MHz operating frequency and 15 A output current, the efficiencies are over 87% that is competitive to the same rating industry products.

Thermal Reliability Test

Thermal reliability of the fabricated hybrid power modules is a major concern in industry applications where significant stresses may occur at interfaces between different materials due to the temperature swing. Such thermally induced stresses can result in defects, such as cracking and delamination, which eventually fail the module. Therefore, it is important to examine the thermal withstand capability of interconnects and the whole modules.

A. Thermal Cycling Experimental Setup

Thermal cycling testing has been used to assess the reliability of high-density integrated POL modules developed by both routes. The temperature profile was determined according to the JEDEC standard JESD22-A104-D [15]. The temperature extremes were set at -40 °C and +150 °C. The cycle time was 45 min and the dwelling time at both temperature extremes was 10 min. A customized test setup was applied in order to satisfy the standard requirements, as shown in Fig. 10.

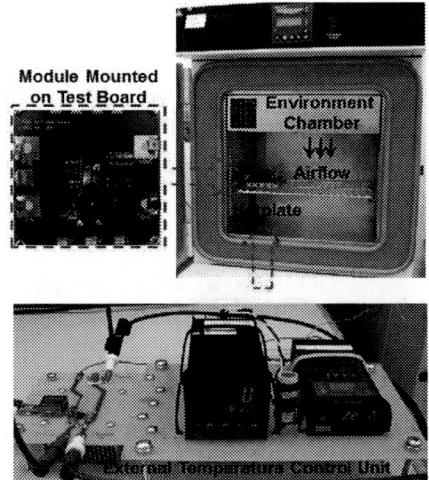

Figure 10. Experimental setup of thermal cycling test. The testing module was placed on top of a hotplate which was used to heat up the sample inside the environmental chamber (Tenny TUJR; Thermal Product Solutions, White Deer, PA). The temperature of the chamber was constantly controlled at -40 °C for a sufficient cooling rate. During the experiment, the temperature ramp-up rate monitored on the testing module was carefully adjusted using an external control unit so as to create the required thermal profile.

B. Testing Results

After 150 thermal cycles of the POL module integrated with LTCC-based low-fire ferrite inductor, no cracking, delamination, and other defects were found based on the visual inspection using an optical microscope, as illustrated in Fig. 11(a). Although there existed a large coefficient of thermal expansion (CTE) mismatch among different materials, the TIM bonding layer still securely attached to both the PCB active layer and inductor substrate layer without any signs of separation. In addition, the copper traces were not delaminated from the PCB and ferrite inductor substrates. It is also demonstrated in Fig. 11(b) that the efficiencies measured before and after the thermal cycling test on this module are approximately the same. No performance degradation was observed by virtue of the reliabilities of all components and the developed integration technique.

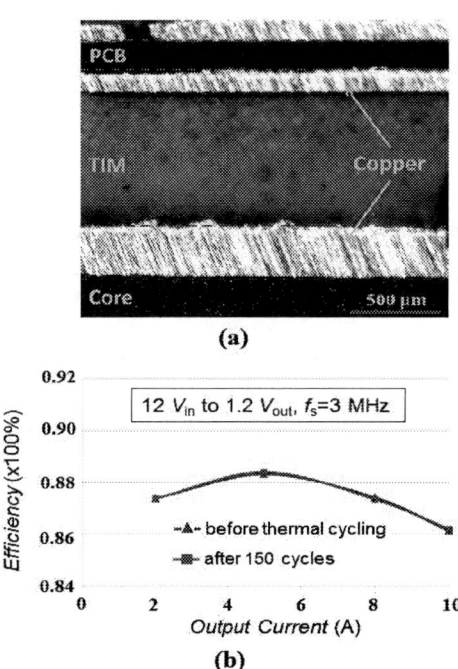

Figure 11. (a) Optical microscopic observation of ferrite inductor integrated POL module after thermal reliability test; (b) Comparison of efficiency tested before and after thermal cycling. All efficiency data were measured at room atmosphere.

As the SEM image of the cross-sectional view is shown in Fig. 12(a), no delamination was observed at all interfaces in the module after 600 thermal cycles. Only a slight bending was found on the bottom PCB layer. Additionally, few small pores and cracks were examined in the metal-flake composite material embedded inside multilayer PCB. However, the efficiency of the module has a negligible change after thermal cycling, as shown in Fig. 12(b). The thermal reliability of this integrated module with multilayer PCB embedded magnetic core has been validated.

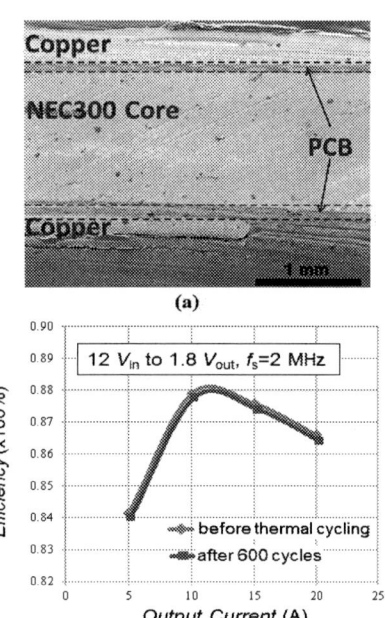

Figure 12. (a) SEM image of POL module with PCB embedded core after 600 thermal cycles; (b) Impact of thermal cycling test on the power module efficiency. All efficiency data were obtained at room atmosphere.

Conclusions

The bulky conventional discrete magnetic component becomes the bottleneck for the high frequency high current power conversion in a high density format. In this study, the planar inductors that worked as passive substrates were prepared and integrated into the POL modules in order to improve the power density. Two approaches were used: (1) hybrid integration of the LTCC-based low-fire ferrite inductor substrate with a PCB active layer and (2) embedding planar magnetic core in a PCB multilayer structure using associated technology. The efficiency of the integrated POL modules, as demonstrated by prototypes, can achieve more than 87% at high frequency (2 MHz) and high output current level (15 A). The determined power densities of a 15 A, 5 MHz POL module integrated with ferrite inductor substrate and a 20 A, 2 MHz module assembled using a PCB embedded inductor are 1000 W/in^3 and 800 W/in^3, respectively. Both assembled POL modules survive after hundreds of thermal cycles, validating the reliability of all integrated components and the compatibility of magnetic materials with traditional PCB materials. Additionally, fabrication of PCB embedded inductor POL modules using standard PCB technology, in contrast to making modules with multilayer ferrite inductors, may reduce the manufacturing cost due to the possible mass-production and relatively low processing temperature. These developed assembly technologies overcome the challenges in the integration of high power density POL modules enabling high frequency high current applications, which satisfy the continuous demand for miniaturization of POL modules.

Acknowledgments

This work was supported in part by the U.S. Department of Energy, ARPA-E project "Power Supplies on a Chip (PSOC)" under Grant DE-AR00000106. The authors would like to thank Dr. Dongbin Hou for his assistance in the magnetic property measurements. The authors would also like to thank Dr. Zheng Chen for his help on the thermal experimental set-up. A special thanks to Isola for supporting PCB materials and CPES industry member NEC-TOKIN that provided metal-flake composite materials used in this research work.

References

1. C. O. Mathuna, P. Byrne, G. Duffy, W. Chen, M. Ludwig, T. O'Donnell, P. McCloskey, and M. Duffy, "Packaging and integration technologies for future high-frequency power supplies," *IEEE Trans. Ind. Electron.*, vol. 51, no. 6, pp. 1305–1312, December, 2004.
2. F. C. Lee and Q. Li, "High-frequency integrated point-of-load converters: overview," *IEEE Trans. Power Electron.*, vol. 28, no. 9, pp. 4127–4136, September, 2013.
3. Q. Li and F. C. Lee, "High inductance density low-profile inductor structure for integrated point-of-load converter," in *Proc. IEEE Appl. Power Electron. Conf. Expo.*, 2009, pp. 1011–1017.
4. Q. Li, Y. Dong, F. C. Lee, and D. J. Gilham, "High-density low-profile coupled inductor design for integrated point-of-load converters," *IEEE Trans. Power Electron.*, vol. 28, no. 9, pp. 547–554, September, 2013.
5. Y. Su, Q. Li, and F. C. Lee, "Design and evaluation of a high-frequency LTCC inductor substrate for a three-dimensional integrated DC/DC converter," *IEEE Trans. Power Electron.*, vol. 28, no. 9, pp. 4354–4364, September, 2013.
6. A. Ball, M. Lim, D. J. Gilham, and F. C. Lee, "System design of a 3D integrated non-isolated point of load converter," in *Proc. IEEE Appl. Power Electron. Conf. Expo.*, 2008, pp. 181–186.
7. S. Ji, D. Reusch, and F. C. Lee, "High-frequency high power density 3-D integrated gallium-nitride-based point of load module design," *IEEE Trans. Power Electron.*, vol. 28, no. 9, pp. 4216–4226, September, 2013.
8. M Mu, Y. Su, Q. Li, and F. C. Lee, "Magnetic characterization of low temperature co-fired ceramic (LTCC) ferrite materials for high frequency power converters," in *Proc. Energy Conv. Cong. Expo.*, 2011, pp. 2133–2138.
9. W. Zhang, M. Mu, D. Hou, Y. Su, Q. Li, and F. C. Lee, "Characterization of low temperature sintered ferrite laminates for high frequency point-of-load (POL) converters," *IEEE Trans. Magn.*, vol. 49, no. 11, pp. 5454–5463, November, 2013.
10. G. Slama, "Low-temp co-fired magnetic tape yields high benefits," *Power Electron. Technol.*, vol. 29, no. 1, pp. 30–34, January, 2003.

11. R. L. Wahlers, C. Y. D. Huang, M. R. Heinz, A. H. Feingold, J. Bielawski, and G. Slama, "Low profile LTCC transformers," in *Proc. SPIE International Society for Optical Engineering*, 2002, pp. 76–80.

12. R. Hahn, S. Krumbholz, and H. Reichl, "Low profile power inductors based on ferromagnetic LTCC technology," in *Proc. Electron. Comp. Technol. Conf.*, 2006, pp. 517–523.

13. J. Y. Park and M. G. Allen, "Low temperature fabrication and characterization of integrated packaging–compatible, ferrite-core magnetic devices," in *Proc. IEEE Appl. Power Electron. Conf. Expo.*, 1997, pp. 361–367.

14. Y. Zhang and S. Sanders, "In-board magnetics processes," in *Proc. IEEE Power Electron. Special. Conf.*, 1999, pp. 562–567.

15. Temperature Cycling, JEDEC Standard No. 22-A104D, 2009.

Integrated Microprobe Array and CMOS MEMS by TSV Technology for Bio-Signal Recording Application

Lei-Chun Chou[1], Shih-Wei Lee[1], Po-Tsang Huang[1], Chih-Wei Chang[2], Shang-Lin Wu[1],
Jin-Chern Chiou[1,3*], Ching-Te Chuang[1], Wei Hwang[1,4], Chung-Hsi Wu[4],
Kuo-Hua Chen[4], Chi-Tsung Chiu[4], Ho-Ming Tong[4], Kuan-Neng Chen[1**]

[1]National Chiao Tung University, Hsinchu, Taiwan; [2]University of California, Los Angeles, USA; [3]China Medical University, Taichung, Taiwan; [4]Advanced Semiconductor Engineering Group, Kaohsiung, Taiwan

Tel: *+886-3-571-2121#31881; ** +886-3-571-2121#31558;

Email: * chiou@mail.nctu.edu.tw; ** knchen@mail.nctu.edu.tw

Abstract

Bio-signal probes that provide stable observation with high-quality signals are crucial for understanding how the brain works and how the neural signal transmits. Because bio-signals are weak and noisy, the length of the string connecting the sensor and Complementary Metal–Oxide–Semiconductor (CMOS) circuit significantly impacts bio-signal quality. The collected weak signals from the sensor must pass through a series of interconnections and interfaces that introduce noise and lead to bulky packaged systems. This work uses through-silicon via (TSV) technology to connect the μ-probe array bio-sensor and CMOS circuit located on opposite sides of a chip for brain neural sensing applications. With the elimination of wire bonding and the reduction of the soldering, bio-signal quality can be significantly improved.

Introduction

In the past decade, advances in micromachined/assembled micro probe arrays with electrical stimulation/recording ability have played an essential role the exploration of central neural systems. Simultaneous observation of a large number of cell activities is required to understand the nervous system [1]. Microelectrode arrays give a method for accessing numerous neurons simultaneously with high spatial resolution [2]. Extracellular action potentials are recorded by surgically implanting neural probes into neurons of interest, which result from neural activities [3]. Probes that could insert a large number of recording sites into neural tissues with minimal tissue damage are therefore needed. Additionally, the design of probe arrays should be optimized for experiment such purpose that an electrode diameter of a few micrometers could support single-unit recording [4].

Since the 1950s, microelectrodes combined with electronic recording and signal processing began to allow for meaningful studies of the central nervous system at the cellular level [5]. A great deal was learned gradually about how single neurons work. Serially moving sharpened wire electrodes in tissue also acquired considerable amounts of information about nervous system function at the circuit level, especially in sensory areas. However, arrays of electrodes, and perhaps large arrays, were clearly needed to fully understand signal processing in complex neural networks. Early experiments glued individual electrodes together or used cutoff wire bundles to record simultaneously from many points with some success, but were limited by

their geometries and reproducibility [6,7]. Furthermore, they caused considerable insertion damage, and typically spread out in tissue, making exact placement difficult. Since they were fabricated easily with technology, microwire electrode arrays are still used extensively for both acute and chronic extracellular recording [8 -10]. However, the length of the connected string between the sensor and CMOS circuit has significant impact on the quality of the inherently, weak and noisy bio-signal. The collected weak signals from the sensor need to pass through a string of interconnections and interfaces that introduce noises and lead to bulky packaged systems.

Highly integrated and miniaturized neural sensing microsystems that provide stable observation, a small form factor and biocompatible properties are crucial for brain function investigation and neural prostheses realization by acquiring accurate signals from an untethered subject in his/her natural habitat [11,12]. Such biomedical devices usually comprise sensors and Complementary Metal–Oxide–Semiconductor (CMOS) circuits for biopotential acquisition, signal conditioning, processing, and transmission. Many approaches for solving these problems have been developed, including stacked multichip [13,14], a microsystem with separated neural sensors [15], and monolithic packaged microsystem. Regardless of which scheme is used, signals collected from a sensor are weak and must pass through a string of interconnections, including wire bonds, flip-chip bonds and welded or soldered bonds to processing circuits. The excessive number of interfaces and connections introduce noise and result in bulky packaged systems.

Integrated Bio-signal Recorder by TSV Technology

Fig. 1 shows the overall system architecture. The neural signal comes from the brain tissue is modeled as a current source, collected by the neural probe array on the back-side of the chip. The raw signals pass through an electrode-tissue interface, which is modeled by a Randles Cell [16]. The signal is then transferred by TSV array through the chip to the CMOS front- end circuits on the front- side of the chip for signal conditioning and processing. On the back-side of the chip, 480 neural probes are divided into a 4x4 array. A total of 16 channels are designed in a chip for local area mapping. For each channel, 42 TSVs are lumped together for low impedance connection. It is possible to use one single TSV for one recording channel. However, in this paper, the main focus is to demonstrate the feasibility of double-side

2014 Electronic Components & Technology Conference

single-chip integration concept, and therefore a group of TSVs are lumped together to mitigate the variance caused by laboratory fabrication process and to improve yield. 16 parallel analog front-end circuits are designed for 16 channel signal inputs. In the measurement setup, another common reference electrode on scalp is required for referencing. Besides, the proposed structure still allows stacking of other CMOS chips fabricated with different technologies onto the circuit side using 3D IC techniques.

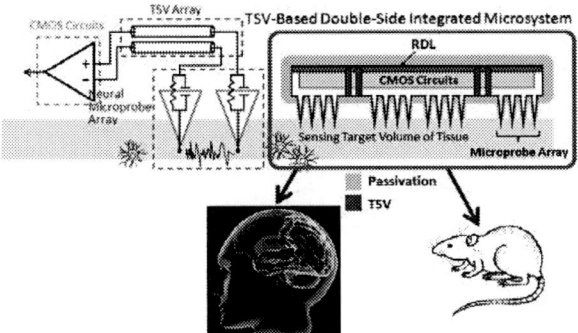

Fig.1. Structure of the integrated microsystem.

Fig. 2 illustrates the detailed physical structure of CMOS circuits, TSV and neural probe array. The CMOS pads on the front-side of the chip are connected to TSV by two layers of redistribution layers (RDLs). A passivation layer is used to protect the RDL, TSV and CMOS circuitry. On the back-side of the chip, parylene-platinum/titanium-parylene structure is designed to form the connection between TSVs and neural probes. The metal layer on the tip is exposed by etching process to serve as a sensing material of the probe.

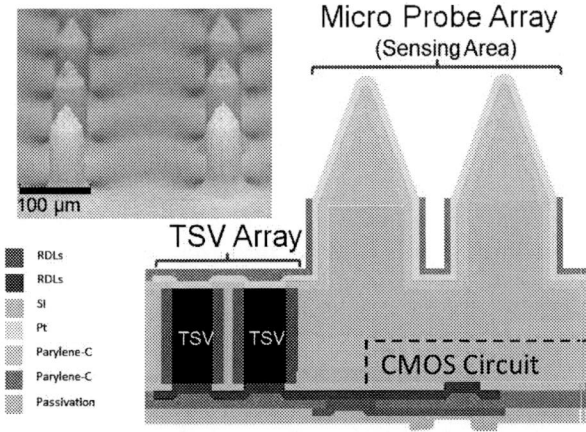

Fig.2. Cross-section view of the structure.

Fabrication

Figure 3 shows the detailed process flow. In the front process, CMOS circuits are fabricated using United Microelectronics Corporation (UMC) 0.18-μm process technology on an 8-inch Si wafer. Then a front-side, via-last, and fully-filled Cu plating process is executed to fabricate Cu

TSVs that are 200μm deep (height) and 25-30 μm in diameter. Next, the RDL is fabricated for connections between TSV arrays and circuit input pads. An Inductive Coupled Plasma (ICP) etching process is then applied on the back side of the wafer to form the microprobe array.

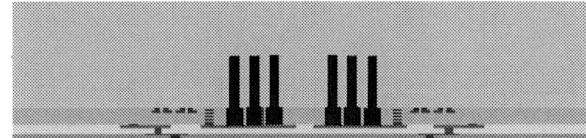

(a) Finish the CMOS circuit & TSV array

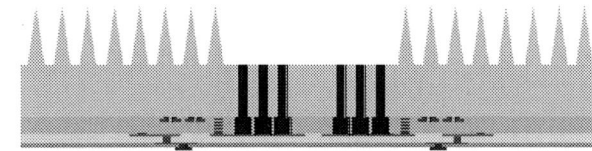

(b) Probing by ICP

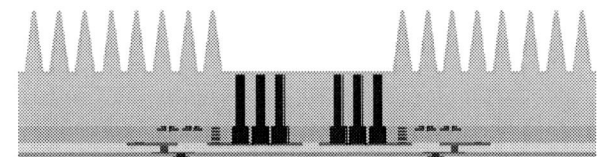

(c) Deposited parylene-C

(d) Open TSV area by O₂ plasma etching

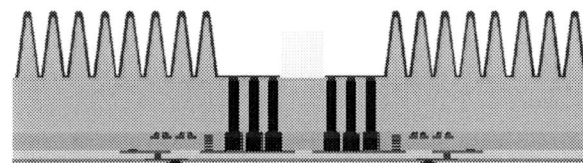

(e) Define channel

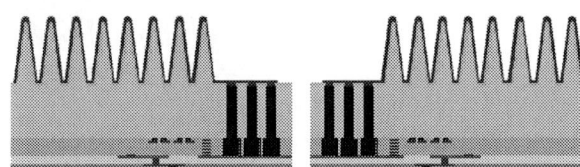

(f) Dicing & package

Fig.3. Detailed process flow including post processing.

At the start of the ICP etching process, isotropic etching is used to etch the probe tip area into a hill (Fig 4(a)). Then, anisotropic etching is used to etch the probe height to 150 μm (Fig 4(b)). Finally, isotropic etching is applied again to etch the tip of the probe. To insert the probe into the *in vivo* brain, tip diameter must be <5μm (Fig. 4(c)).

978-1-4799-2408-0/14 $31.00 © 2014 IEEE   513

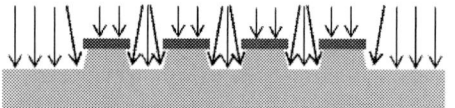

(a) Isotropic etching first create the hill structure.

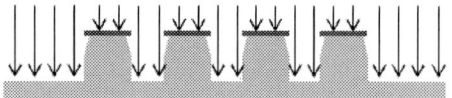

(b) Anisotropic etch to 150 um height.

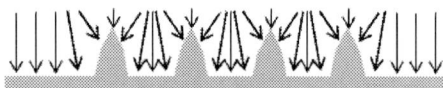

(c) Isotropic etching to create the probe tip

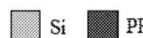

Fig.4. ICP etching process flow

After the ICP step of the post process flow, a 5-μm parylene-C is deposited on the structure to isolate different channels. Parylene-C is biocompatible, and is commonly used *in vivo* body. The area of TSV must be open to transfer the signal from the probe side to the circuit side using O_2 plasma. However, with the same etching rate for photoresist (PR) and parylene-C, the area, especially that of parylene-C, near the probe tip is over etched. A hard mask solves this problem (Fig. 5). The hard mask is implemented using a standard 4-inch glass mask. The TSV open area is drilled by laser to let the O_2 plasma pass the hard mask. The tip area of the probe is protected by the hard mask, such that only the TSV open area can be etched by O_2 plasma.

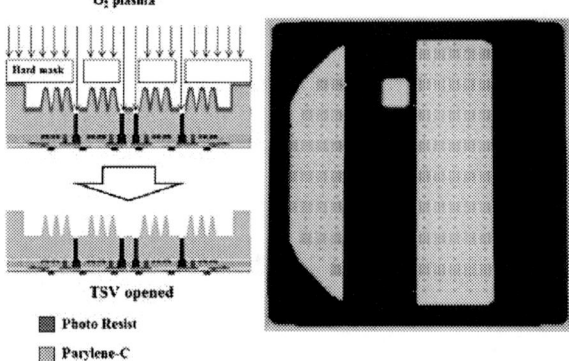

Fig.5. Hard mask design for O_2 plasma etching.

After the TSV is opened by O_2 plasma, a 3000-Å platinum is sputtering and lifted off to define different channels. In this step, platinum (Pt) is used instead of gold because the body may erode the gold, breaking the biosensor. In lift-off processing, the ultrasonic cleaner is used to decrease lift-off processing time.

Figure 6 is a cross-sectional view of the device before post-processing. The CMOS circuit and TSV has already implemented by the United Microelectronics Corporation (UMC) and Advanced Semiconductor Engineering (ASE) Group. Figure 7 is a cross-sectional view of the device with a probe array and TSV.

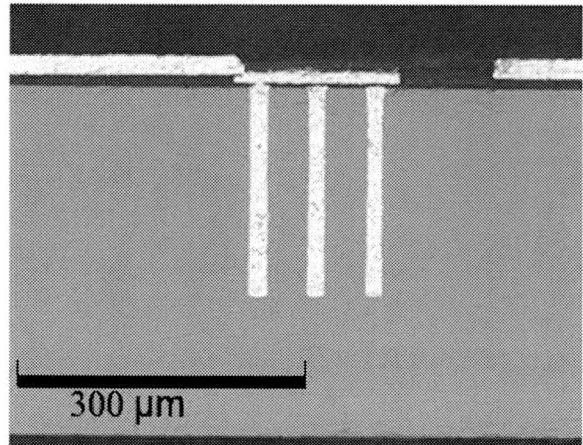

Fig.6. Cross-section view of the device before the post processing.

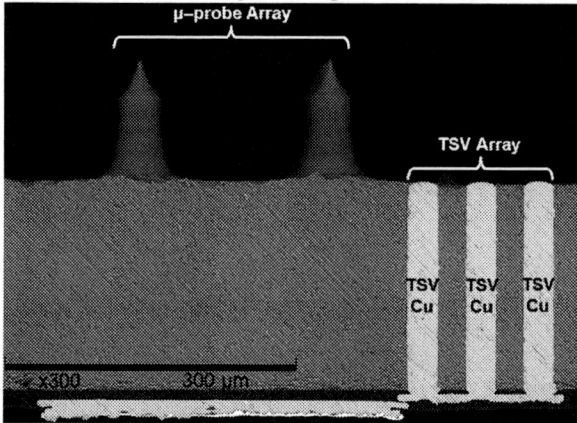

Fig.7. Cross-section view of the device with TSV array and probe array.

Figure 8 shows the integrated microsystem, including probe arrays, a Printed Circuit Board (PCB), and connector. This work creates 4-channel and 16-channel designs. Total area is 6 mm × 6 mm.

Fig.8. Photograph of the integrated probe microsystem.

978-1-4799-2408-0/14 $31.00 © 2014 IEEE

Electrical Characteristics

Daisy chain and comb structure are fabricated to investigate the electrical characteristics of 25-μm and 50-μm-diameter TSV. Figure 9 lists resistance measurements of 30 Cu TSV arrays. The measured daisy chain resistance keeps stable under current stressing. To ensure that the insulating capability of the sidewall TSV is adequate, the comb structure is designed to measure capacitance and current leakage.

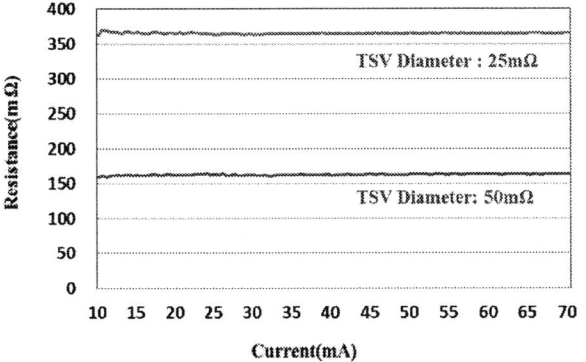

Fig.9. Two-point resistance measurement of an array with 30 Cu TSVs.

Capacitance is measured from -10 to 10V (Fig. 10). Average measured capacitances for 25-μm-diameter Cu TSVs are 0.74 pF and 0.88 pF, respectively. Current leakage from the TSV structure is low and in the nA scale (Fig. 11). These experiment results validate the electrical properties of the TSV sidewall and filling.

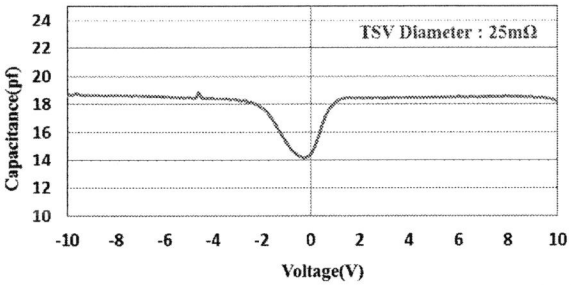

Fig.10. C-V sweeping of 100 TSVs using comb structure.

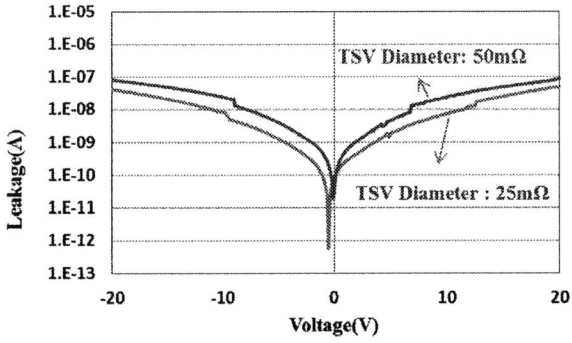

Fig.11. Leakage current measured between +/- 10V on 100 Cu TSVs.

X-ray microscopy images show no visible voids inside Cu TSVs of the bio-chip, indicating TSVs are filled fully with Cu (Fig. 12). These experiment results show that TSV sidewall insulation, electrical performance, and fabrication have adequate quality.

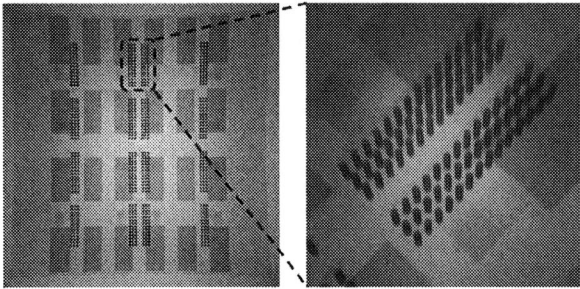

Fig.12. The bio-chip and TSV array under X-ray microscopy

Figure 13 is the impedance measurement result, including micro probe array and PCB/connector, in 0.9% saline which emulates the in-vivo environment. Fig. 13(a) shows 4-channel results and Fig. 13(b) shows 16-channel results, respectively. The impedance for 4 channels is 441.1265Ω with standard deviation 56.5515Ω at 1KHz; for 16 channels is 1136.76Ω with standard deviation 691.89Ω at 1KHz.

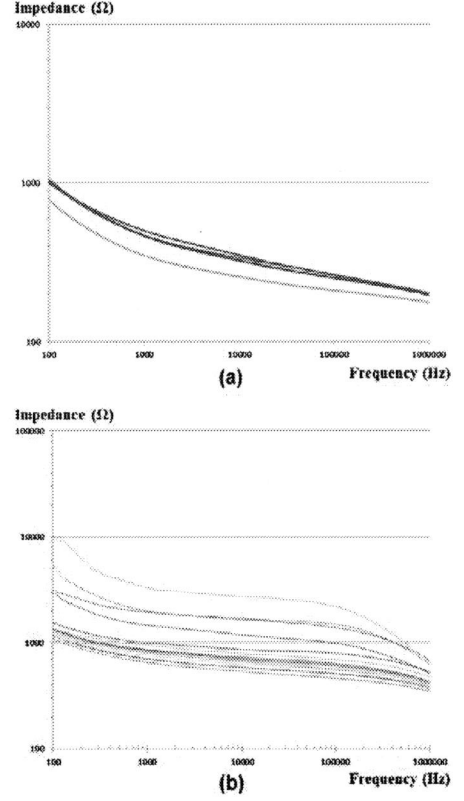

Fig.13. Impedance measurement results of (a) 4 channels and (b) 16 channels.

Besides, in order to ensure function of all channels of device operate well. Figure 14(a) & 14(b) are signal recording results of the micro probe array from a 1 V sine wave (peak to peak) source signal. Notably, sine wave here is used to ensure that all channels of the bio-signal package operate well only.

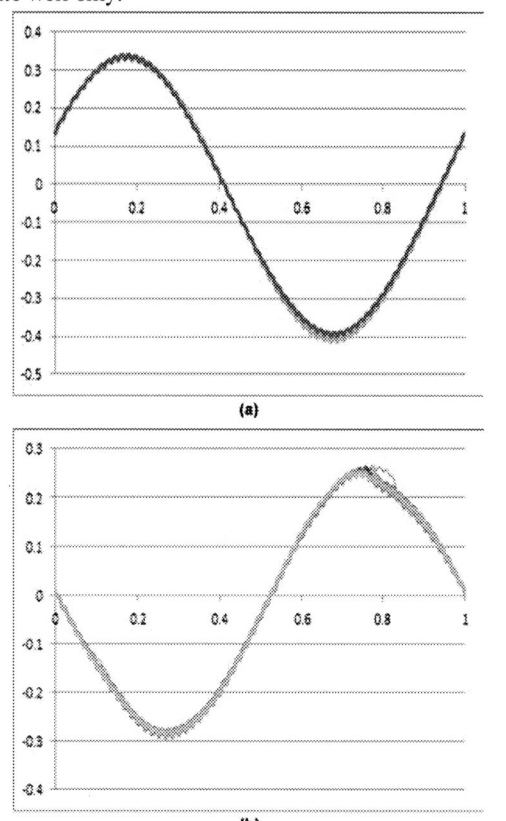

Fig. 14. Characteristics of (a) the 4-channel and (b) 16-channel μ-probe array.

Conclusions

In this paper, a special TSV-based double-side bio-signal recording device is designed and tested. By TSV technology, the signal travels the shortest path to the CMOS device. By eliminating conventional wire bonding, bio-signal quality can be improved significantly. The chip has only a size of 5 mm × 5 mm with 3 × 8 TSV arrays for each channel. Therefore, the rat survival rate increased due to the small size of the device. There are 30 × 16 microprobes in 16 channels die and 140 × 4 microprobes in 4 channels die. Since multiple channels can acquire different neural cell signals, this design benefits the neural-signal analysis. For this recording scheme, all the post processes have been developed and the micro system is ready for bio-signal investigation.

Acknowledgments

This work was supported in part by the National Science Council, Taiwan, R.O.C. under Contract No. 102-2221-E-

009-160, No. 102-2220-E-009-014, No. 102-2220-E-009-002, and "Aim for the Top University Plan" of the National Chiao Tung University and Ministry of Education, Taiwan, R.O.C.. This work was also particularly supported by R&D Piloting Cooperation Projects between Industries and Academia at Science Parks under Contract Number: 100A20 and the UST-UCSD International Center of Excellence in Advanced Bioengineering sponsored by the Taiwan National Science Council I-RiCE Program under Grant Number: NSC-101-2911-I-009-101. The authors would like to thank National Chip Implementation Center (CIC) for chip fabrication.

References

1. M. Chcurel, "Windows on the brain," *Nature*, Vol.412, 2001, pp.266–268.
2. Y. Yao, M. N. Gulari, J. A. Wiler and K. D. Wise, "A microassembled low-profile three-dimensional microelectrode array for neural prosthesis applications," *Journal of Microelectromechanical Systems*, Vol.16, 2007, pp.977-988.
3. K. Frank, M. C. Becker, "Electrodes for extracellular recording and stimulation," *Physical Techniques in Biological Research*, 1964, pp.22-87.
4. J. D. Green, "A simple microelectrode for recording from the central nervous system," *Nature*, Vol.182, 1958, 962.
5. R. C. Gesteland, B. Howland, J. Y. Lettvin, and W. H. Pitts, "Comments on microelectrodes," *Proceddings of The IRE*, Vol.47, 1959, pp.1856-1862.
6. C. A. Terzuolo and T. Araki, "An analysis of intra-versus extra-cellular potential changes associated with activity of single spinal motoneurons," *Annals of the New York Academy of Sciences*, Vol.94, 1961, pp.547-558.
7. M. Verseano and K. Negishi, "Neuronal activity in cortical and thalamic networks," *Journal of General Physiology*, Vol.43, 1960, pp.177-195.
8. M. A. Nicolelis, A. A. Ghazanfar, B. M. Faggin, S. Votaw, and L. M. Oliveira, "Reconstructing the engram: Simultaneous, multisite, many single neuron recordings," *Neuron*, Vol.18, 1997, pp.529-537.
9. J. C.Williams, R. L. Rennaker, and D. R. Kipke, "Long-term neural recording characteristics of wire microelectrode arrays implanted in cerebral cortex," *Brain Research Protocols*, Vol.4, 1999, pp.303-313.
10. I. Porada, I. Bondar,W. B. Spatz, and J. Kruger, "Rabbit and monkey visual cortex: More than a year of recording with up to 64 microelectrodes," *Journal of Neuroscience Methods*, Vol.95, 2000, pp.13-28.
11. K. C. Smith, et al., "Through the Looking Glass: Trend Tracking for ISSCC 2012," *IEEE Solid-State Circuits Magazine*, Vol.4, No.1, 2012, pp.4-20.
12. K. Arimoto, et al., "What's next in Robots? Sensing, Processing, Networking Toward Human Brain and Body," *ISSCC Dig. Tech. Papers*, 2012, pp.514.

13. B. Gosselin, et al., "A Mixed-Signal Multichip Neural Recording Interface With Bandwidth Reduction," *IEEE Transactions on Biomedical Circuits and Systems*, Vol.3, No.3, 2009, pp.129-141.

14. B. K. Thurgood, et al., "A Wireless Integrated Circuit for 100-Channel Charge-Balanced Neural Stimulation," *IEEE Transactions on Biomedical Circuits and Systems*, Vol.3, No.6, 2009, pp. 405-414.

15. A. M. Sodagar, et al., "A Wireless Implantable Microsystem for Multichannel Neural Recording," *IEEE Transactions on Microwave Theory and Techniques*, Vol.57, No.10, 2009, pp.2565-2573.

16. W. Franks, I. Schenker, P. Schmutz, and A. Hierlemann, "Impedance characterization and modeling of electrodes for biomedical applications," *IEEE Transactions on Biomedical Engineering*, Vol.52, 2005, pp.1295-1302.

Material Characterization of a Novel Lead-Free Solder Material - SACQ

Tak-Sang Yeung, Henry Sze, Keith Tan, Javed Sandhu, Chong-Wei Neo, Edward Law
Broadcom Corporation
5300 California Ave, Irvine, California 92617
dtsyeung@broadcom.com

Abstract

The wafer level ball grid array (WLBGA), a silicon die-size package, offers the small form factor and high-performance packaging solution. Good board-level reliability under drop impact is achieved on account if its light weight. There is, however, a limitation on the reliability of the solder joint at board level, subject to thermal cyclic loading. This places a limit on the silicon die-size window in which WLBGA packaging can be applied. Unlike other plastic BGAs, a WLBGA is a silicon chip directly mounted onboard without a plastic interposer. This causes a larger coefficient of thermal expansion (CTE) mismatch between silicon and organic printed circuit board (PCB), which leads to higher solder joint stress. This scenario gets even worse with increasing die size where corner solder joints are further away from neutral points. This study considers methods for extending the board- level reliability window in the face of increasing die-size requirements. One way to improve the reliability of the solder joint is by applying underfill epoxy material to protect the joints. Another possibility is by adopting lower CTE board material for the PCB. These methods are, however, not cost-effective. A more cost-effective solution is introduced by developing a new solder alloy with better creep and mechanical properties to enhance the solder joint integrity. A novel solder alloy material was successfully developed and proven to enhance the solder joint reliability, when subject to thermal loading, without degrading the board-level drop-test reliability. This study presents the characterization of the new solder alloy. Its visco-plastic constitutive behavior was tested by arranging tensile tests under constant strain-rate loading at various temperatures. The constitutive model is constructed by fitting the experimental data collected through a 9-parameter Anand model. The validity of the material model was verified by comparing numerical analyses with the experimental results.

Keywords:

WLBGA, thermal cycling, reliability

Introduction

Wafer-level ball grid array packages provide good package and board-level reliability performance. However, the high complexity of integrated circuits and dramatic increase in functionality of new technologies is driving more stringent requirements like larger package size, finer pitch (and hence small ball size together with shorter solder-joint height) pushing the reliability margin of the WLBGA package to its limit, especially board-level reliability (BLR) under thermal cyclic (TC) loading. Improvement of TC BLR can be achieved through optimization of the package design as well as selection of proper package materials. Package design parameters involve die thickness, under-bump metallurgy

(UBM) dimensions, including thickness and size, passivation opening, pad dimension, and redistribution layer (RDL) structure. Material variables are composed of factors like UBM structure type, passivation material and solder-alloy materials. The current study only addresses the improvement and development work through the application of a novel solder alloy.

Materials exhibit strong time-dependent plastic and time-dependent creep behavior when their homologous temperatures, defined as ratio of field temperature (in K) and melting point (in K), is greater than 0.5. Solder interconnections form mechanically and electrically when integrated-circuit (IC) packages are mounted on PCBs. These solder joints are subject to deformation and damage, dominated by creep under the field-application environment like shock and fluctuating temperature conditions since melting points of solder-alloy materials are typical in the range of 473K. Packages that use solder material with improved creep resistance therefore deliver better reliability of the solder joint under adverse loading. The Anand model, being a unified constitutive material model used in many other studies, is adopted here to describe both the plastic and creep responses of the solder joints [1].

Constitutive Material Model

The Anand model was first introduced to model metal deformation under high temperature by Anand. Such model may be applicable to solder alloys as their homologous temperatures are greater than 0.5 due to their relatively low melting point. There are two major characteristics in the model: First, it requires no explicit yield condition and no loading/unloading criterion. The plastic strain is assumed to occur at all nonzero stress states. Second, a single internal state variable is used to represent the isotropic resistance to plastic flow. The flow equation is given by:

$$\dot{\varepsilon}_p = A \left[\sinh\left(\frac{\xi \sigma}{s} \right) \right]^{1/m} e^{\left(\frac{-Q}{RT} \right)}; \sigma < s \qquad (1)$$

where the uniaxial inelastic strain rate and equivalent stress are denoted by $\dot{\varepsilon}_p$ and σ. A, ξ and m are material constants and T is the absolute temperature. Q and R denote the activation energy and the universal gas constant. The state variable, as denoted by s, describes the deformation resistance and is expressed in the evolution equation as below:

$$\dot{s} = \left\{ h_0 \left| 1 - \frac{s}{s*} \right|^a sign\left(1 - \frac{s}{s*} \right) \right\} \dot{\varepsilon}_p; a > 1 \qquad (2)$$

where $s* = \hat{s} \left[\frac{\dot{\varepsilon}_p}{A} e^{\left(\frac{Q}{RT} \right)} \right]^n \qquad (3)$

h_0, a, n, $\hat{s}$ are the material constants. Saturation value of s is denoted by s^*. s is also directly proportional to the stress with another material constant c (<1) as below

$$\sigma = cs \qquad (4)$$

where $c = \dfrac{1}{\xi}\sinh^{-1}\left[\left(\dfrac{\dot{\varepsilon}_p}{A}e^{\left(\frac{Q}{RT}\right)}\right)^m\right]$ \qquad (5)

Therefore the Anand model is composed of nine material parameters of A, ξ, m, Q, h_0, a, n, $\hat{s}$ and s_0 (initial value of s).

Material Characterization

A novel solder alloy is discussed. The solder was melted and cast in a dog-bone shape as shown in Fig. 1. The test sample has a cross-section of 5.3 mm x 5 mm with a reduced section length of 25.4 mm. All samples were stored at −3°C before testing to ensure solder microstructure stability.

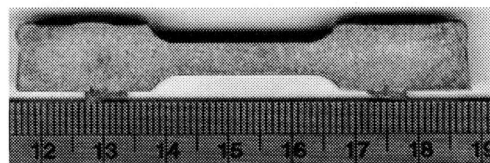

Figure 1. Dog-Bone Specimen

Figure 2. Test Arrangement

The test specimens were tested under an Instron 5565 tester equipped with environment chamber as shown in Fig. 2. All experiments were conducted and measured by using low-force load-cell type of 500N or 5kN. The axial strain was measured by a dynamic extensometer with a gauge length of 12.5mm. Constant strain-rate experiments on the solder specimens were carried out with different constant strain rate of 1×10^{-4} s^{-1}, 1×10^{-5} s^{-1} and 1×10^{-6} s^{-1} while the chamber was maintained at constant test temperature of 25°C, 75°C, 125°C, and 150°C. All samples were loaded in the test chamber 100 minutes before conducting the test to ensure thermal equilibrium. Results of the constant strain-rate monotonic tests are shown in Fig. 3. The saturation stress increases as strain rate increases. On the other hand, the stress decreases with increasing temperature.

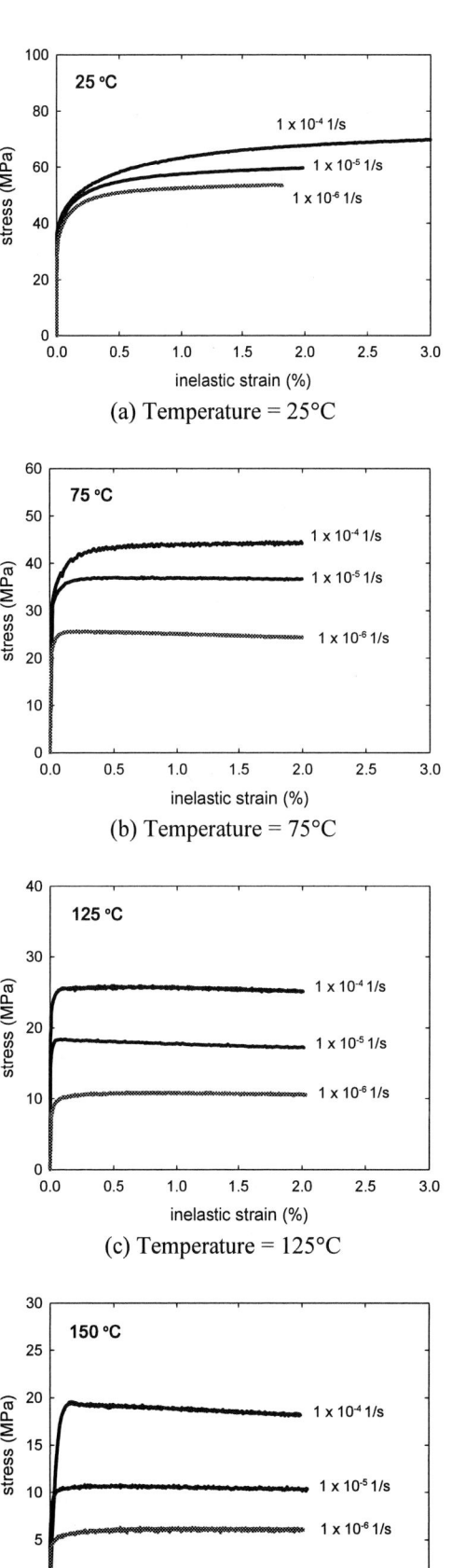

(a) Temperature = 25°C

(b) Temperature = 75°C

(c) Temperature = 125°C

(d) Temperature = 150°C

Figure 3. Constant Strain Rate Behavior of New Solder

Anand Material Model Fitting and Verification

Saturation establishes when steady-state plastic flow occurs. Equation (1) can be rewritten as:

$$\dot{\varepsilon}_p = \dot{\varepsilon} = A\left[\sinh\left(\frac{\xi\sigma^*}{s^*}\right)\right]^{1/m} e^{\left(\frac{-Q}{RT}\right)} \qquad (6)$$

where $\dot{\varepsilon}$ is the steady state or applied strain rate, σ^* and s^* are the saturated values of stress and state variables. The saturated stress can therefore be rewritten from Equations (3) & (6) as

$$s^* = \frac{\hat{s}}{\xi}\left[\frac{\dot{\varepsilon}}{A}e^{\left(\frac{Q}{RT}\right)}\right]^n \sinh^{-1}\left\{\left[\frac{\dot{\varepsilon}}{A}e^{\left(\frac{Q}{RT}\right)}\right]^m\right\} \qquad (7)$$

The model parameters are determined using two steps: determination of five parameters A, $\hat{s}/\xi$, Q/R, m and n first, and then followed by the remaining four parameters h_0, a, s_0, ξ. The first group of five parameters is estimated from Equation (7) by fitting the data pairs of saturated stress and applied strain rate against temperature through a nonlinear least-square fit approach. By substituting Equation (4) into (2) and performing integration on both sides, the stress of the model can be represented as:

$$\sigma = \sigma^* - \left\{(\sigma - cs_0)^{(1-a)} + (a-1)\left[ch_0(\sigma^*)^{-a}\dot{\varepsilon}_p\right]\right\}^{1/(1-a)} \qquad (8)$$

Determination of the second group of parameters, h_0, a, s_0, ξ can therefore be done by fitting Equation (8) with the stress-strain experimental data. Finally, $\hat{s}$ is calculated from $\hat{s}/\xi$ and ξ. The nine parameters for this new solder material, determined through experimental data fitting, are listed in Table 1.

Constant strain-rate response of the material is shown in Figure 4. Good correlation between Anand model fitting data and experimental data is achieved, especially under low strain-rate conditions. Figure 5 shows the steady-state creep behavior of the material. Again, it indicates that there is a good agreement between the Anand model prediction and the experimental data.

Table 1 Anand Model Parameters

A	s⁻¹	2.45e8
Q/R	°K	13509
m		0.36
n		0.0056
h_0	MPa	686
s_0	MPa	1.55
ξ		0.224
$\hat{s}$	MPa	2.12
a		1.25

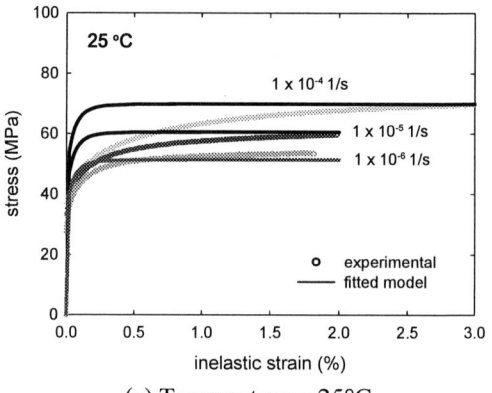

(a) Temperature = 25°C

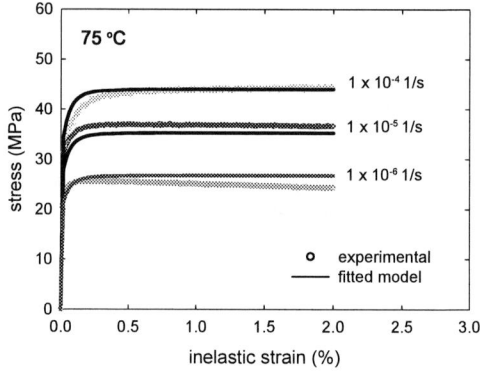

(b) Temperature = 75°C

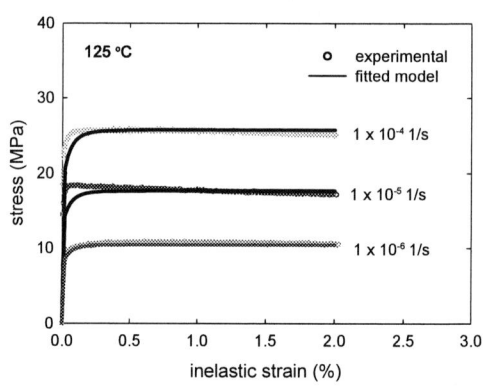

(c) Temperature = 125°C

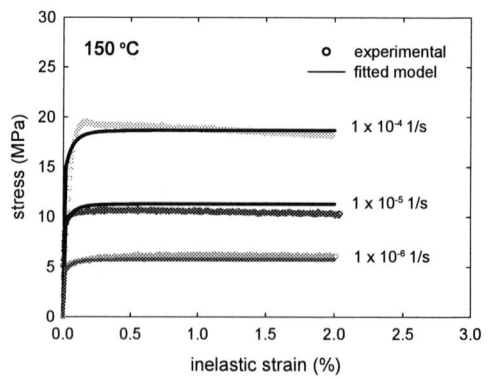

(d) Temperature = 150°C

Figure 4. Constant Strain Rate Response

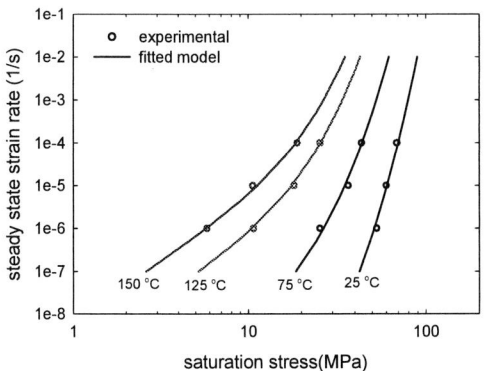

Figure 5. Steady-State Response

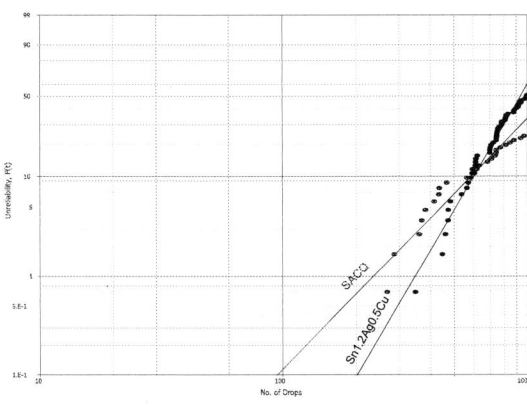

(a) Drop Test Performance

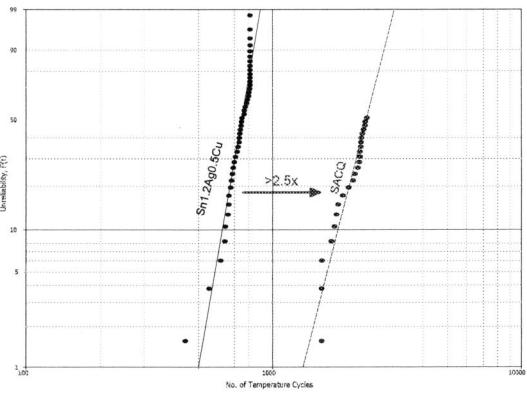

(b) Temperature Cycling Test Performance
Figure 6. Weibull Plot of a WLBGA

Board Level Reliability and Prediction

Figures 6 (a) and (b) show reliability test results for board-level drop and temperature cycling on a Broadcom WLBGA device. The package size is 5mmx5.3mm with 133 balls. It is a regular ball matrix but with 11 balls removed. Its ball pitch is 0.4mm. The results compare two different alloys: Sn1.2Ag0.5Cu and the new solder, SACQ. It is clearly shown that the drop test performance is comparable among these two solder alloys while significant improvement is found in the device with the new solder.

The reliability of the solder joint under temperature-cycling load is further investigated through finite element

simulation. For simplicity, a WLBGA model of package size 5mmx5mm is used but with a full ball matrix of 12x12. A quarter model is formed with consideration of symmetry. The whole board assembly is subjected to temperature cycles from -40°C to 125°C. A stress-free temperature of 125°C is assumed in the analysis. Only solder material is assumed to be viscoelastic in the Anand material model; other materials are assumed to be linear elastic. In addition, only the corner joint, which is expected to be the worst solder joint, is modeled with finer mesh in the study. The model is shown in Figure 7.

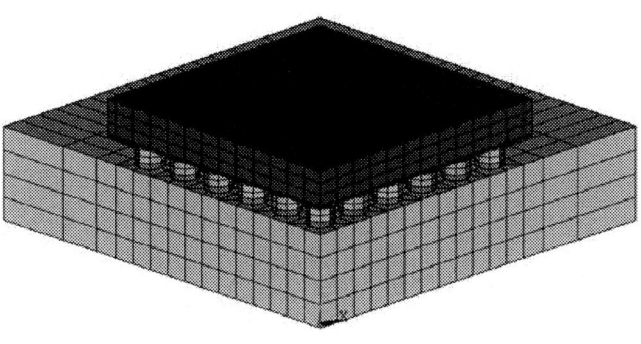

(a) Quarter Model

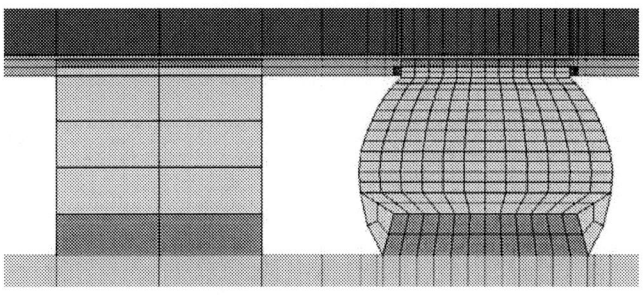

(b) Cross-Section of the Corner and Nearby Joints
Figure 7. Finite-Element Model of a WLBGA

Plastic work density accumulated per temperature cycle at the interface between the solder joint and copper pad is used to evaluate the reliability of the solder joint. A layer of 25 μm solder is used in the energy calculation. Since the Anand nine parameters for solder Sn1.2Ag0.5Cu are not available, properties of Sn1.0Ag0.5Cu are used instead as listed in [2]. The plastic works of the corner solder joints are shown in Figure 8. Higher plastic energy in Sn1.0Ag0.5Cu joints implies that the solder joint fails earlier in Sn1.0Ag0.5Cu than in SACQ. In addition, as discussed in [3], once the accumulated energy density per cycle ΔW is extracted at the numerical simulation, the crack initiation N_0 and cracking growth rate da/dN can be estimated through Equations (9) and (10), respectively. The empirical constants k_1, k_2, k_3, and k_4 are obtained through curve fitting of failure data and the typical values of k_2 and k_4 are near to -1 and +1, respectively. The life prediction can therefore be considered to be inversely proportional to ΔW. Table 2 summarizes the values of ΔW, extracted at the end simulation for the two solder alloys. The energy density per cycle accumulated at Sn1.0Ag0.5 is

978-1-4799-2408-0/14 $31.00 © 2014 IEEE

approximately 2.5x more than SACQ and the result correlates to the actual temperature-cycle test data as shown in Figure 5(b).

$$N_0 = k_1 (\Delta W)^{k_2} \qquad (9)$$

$$\frac{da}{dN} = k_3 (\Delta W)^{k_4} \qquad (10)$$

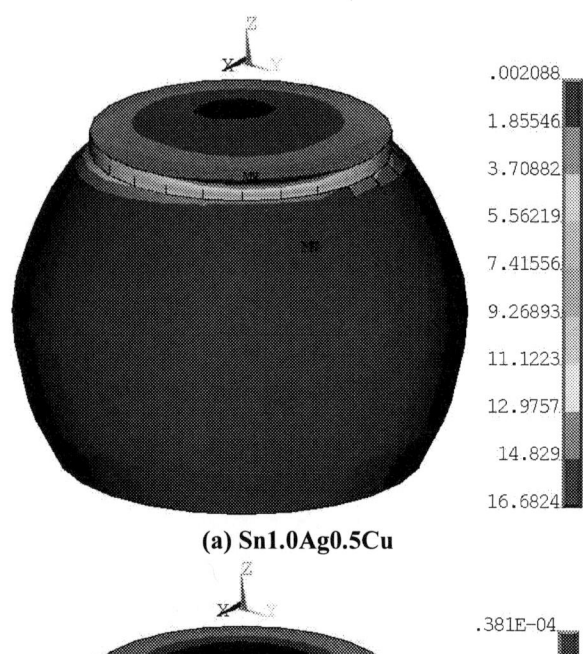

(a) Sn1.0Ag0.5Cu

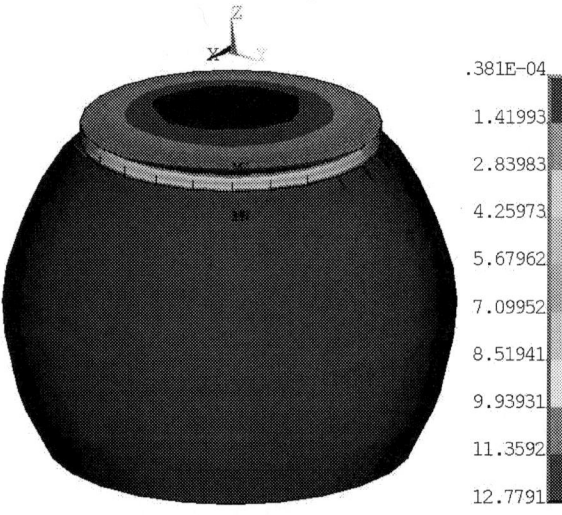

(b) SACQ

Figure 8. Plastic Work of the Corner Joints

Table 2. Plastic Work Density per Cycle

Solder Alloy	ΔW (MPa)
Sn1.0Ag0.5Cu	0.613
SACQ	0.248

Conclusions

A new solder alloy has been developed to improve board-level reliability. The Anand material model is used to fit the test data from constant strain-rate experiments. Nine parameters are presented here and the material model is used in finite-element simulation in a WLBGA mounted onboard under temperature cyclic loading. Numerical prediction correlates well with the actual test failure data.

Acknowledgments

The authors would like to acknowledge the helpful support and discussion from Accurus.

References

1. L. Anand, "Constitutive equations for the rate-dependent deformation of metals at elevated temperature", *J. Eng. Mater. Technol., vol. 104, no. 1*, pp. 12-17, 1982
2. G. Subbarayan, et al., "Constitutive and failure behavior of SnAgCu solder alloys", InterPACK 2011, SnAgCu Solder Tutorial
3. R. Dareaux, "Effect of simulation methodology on solder joint crack growth correlation", *Proc. IEEE Electronic Components and Technol. Conf. (ECTC)*, 2000, Las Vegas, NV, pp. 1048–1058.

Lithography Challenges for 2.5D Interposer Manufacturing

Klaus Ruhmer

Philippe Cochet, Roger McCleary, Rich Rogoff, Rajiv Roy

Rudolph Technologies, Inc.
16 Jonspin Rd.
Wilmington, MA 01887
+1 978-253-6200
Klaus.Ruhmer@Rudolphtech.com

I. ABSTRACT

In recent years, 2.5D packaging has quickly turned from a buzzword into an Advanced Packaging reality. Not dissimilar to the Multi-Chip-Modules (MCMs) of the past, 2.5D packages utilizing high density interposers with favorable electrical characteristics can be a cost efficient and high performance alternative to significantly more complex 3D or SOC integration schemes.

Dictated by the trend towards ever thinner, smaller and higher integrated and more capable devices, high density interposer technology is required to enable 2.5D packages. Patterning of these kinds of substrates, regardless if manufactured on silicon, glass or other suitable materials requires relatively advanced lithography systems. Simple contact or proximity exposure is no longer up to the task.

This paper specifically lists the various patterning / lithography challenges which are being encountered when manufacturing high density 2.5D interposers. Typical back-end lithography requirements regarding minimum resolution, overlay, maximum sidewall angle capability in relatively thick resist and depth of focus are established and discussed. In addition, the application-specific lithography challenges such as a large exposure field size, IR backside alignment capability for TSV (Through-Silicon-Via) or TSG (Through-Glass-Via) definitions and warped wafer or substrate handling are being reviewed and characterized.

As with all back-end processes, interposer manufacturing must be extremely cost efficient and high yielding. A middle ground between costly front-end processes and more robust, faster and lower cost back-end processes has to be found. This paper also discusses potential cost reduction via economy of scale. A lithography cost analysis for glass interposer manufacturing on large panels is being offered.

Key words:

Advanced Packaging, 2.5D, 3D, TSV, TSG, Stepper, Panel, Glass Interposer, Si Interposer, thick resist, IR alignment, overlay, sidewall angle, backside alignment, Through Silicon Via, Through Glass Via, RDL, Redistribution Layer, MCM, Multichip Module

II. INTRODUCTION

Semiconductor packages must be thin, small and cost-effective. These requirements for the semiconductor back-end have become a "given" in our industry. As front-end advances in terms of scaling and device performance are slowing down, the demands for the back-end extend beyond size and cost. Packaging today also plays a critical role when it comes to device and system performance. A new term has even been coined: System Scaling Technology – the combination of front-end, middle-end and back-end to advance microelectronic systems. Many different advanced packaging approaches are being pursued, one of which being the concept of 2.5D packaging.

The term "2.5D packaging" has not always been used consistently throughout literature. The definition used for the purpose of this paper can be summarized as following [1]: A 2.5D package utilizes an interposer between multiple silicon dice and a SiP (System in Package) substrate, where this interposer has through vias connecting the metallization layers on its front and back surfaces.

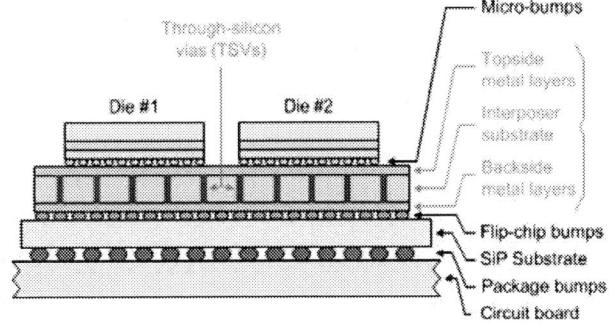

Fig. 1: A 2.5D IC/SiP using an interposer and through-vias

Interposers are further differentiated based on their level of sophistication. The industry commonly refers to passive (metallic), active and optical interposers, the latter being in early R&D stages at the time of this publication. Interposer substrates are often made out of silicon as this material offers some inherent advantages. The thermal coefficient of expansion (TCE) matches that of the dice itself, hence any thermally induced stress can be minimized. Additionally, all processing and manufacturing infrastructure for Si is already established in semiconductor facilities. Currently though, silicon interposers are very expensive and not viable for high-volume mobile devices from a cost perspective. Si interposers today are built on mature front-end lines (e.g. 65nm node). That is overkill for the relatively large dimensional line/space and via requirements of interposers. Unfortunately, there is not an easy way to reduce cost out of Si interposers. Most new 300mm tools are being developed at the advanced nodes. Front-end lines used for Si interposers are unbalanced with respect to capability and cost. Hence the search for alternative integration technologies such as those based on fan-out wafer level packaging or alternative interposer materials. Glass, for

978-1-4799-2408-0/14 $31.00 © 2014 IEEE   523   2014 Electronic Components & Technology Conference

example, offers similar benefits with regards to its ability to match TCE while also exhibiting significantly improved electrical characteristics; for high-end, high-frequency devices, this can be critical [2]. On the other hand, glass manufacturing infrastructure, although established in the flat panel display industry, is not as readily accessible to semiconductor manufacturers. Besides silicon and glass, the industry is also exploring other interposer material options such as high-density laminate substrates or even ceramic substrates.

2.5D integration is being considered for several different applications. Besides networking, 2.5D is expected to be used for high-end graphics as well as multi-core / memory integration for mobile devices [4].

Fig. 2: 2.5D application examples: High-end graphics and multi-core processors

The manufacturing challenges from a patterning / lithography perspective are fundamentally similar regardless of the interposer material chosen. A detailed review of these challenges is the topic of this paper.

LITHOGRAPHY CHALLENGES

Common interposer manufacturing processes are comparable to typical wafer level advanced packaging processes such as wafer bumping, copper pillars [5] (Fig. 3), Cu redistribution layers for WLCSP or Fan-Out. Metal seed layers have to be patterned and etched. Plating molds need to be defined either in thick photoresist or in photo definable dielectric materials.

Additionally, the creation of vias either in silicon (Through-Silicon-Vias or TSV) or in glass (Through-Glass-Vias or TGV) requires front-to-back registration capability. Last but not least, 2.5D interposers require processing of RDL (redistribution layers) on both sides. Double side processing tends to impact the physical behavior of wafers or substrates. It will lead to (in some cases severe) warpage.

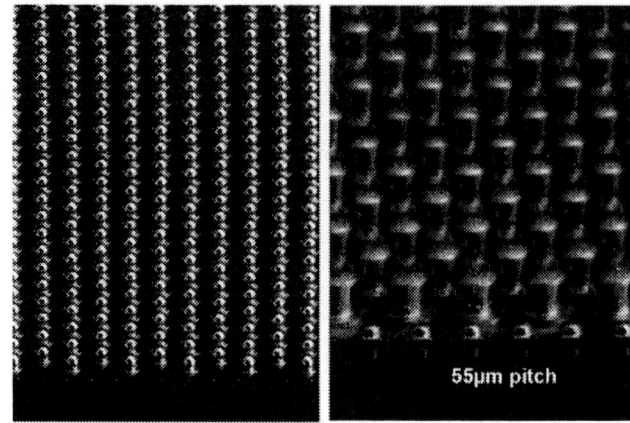

Fig. 3: Copper pillars with solder caps

When analyzing the various lithography steps, the following key challenges can be identified:

- Minimum resolution
- Overlay accuracy (from layer to layer)
- Sidewall angle and CD control
- Depth of focus (DOF)
- Exposure field size
- Infrared alignment capability (through Si)
- Warped wafer or substrate handling

A. Minimum resolution

2.5D interposers are, in essence, high-density substrates which interconnect two or more individual dice and potentially carry integrated passive devices within them. Although the resolution requirements are in the single digit micron range (Fig. 4), sub-micron resolution is not needed for these devices. The optical characteristics of suitable lithography systems should offer an N/A of 0.1 to 0.15 in order to meet the line/space (L/S) resolution requirement while also preserving a reasonable depth of focus.

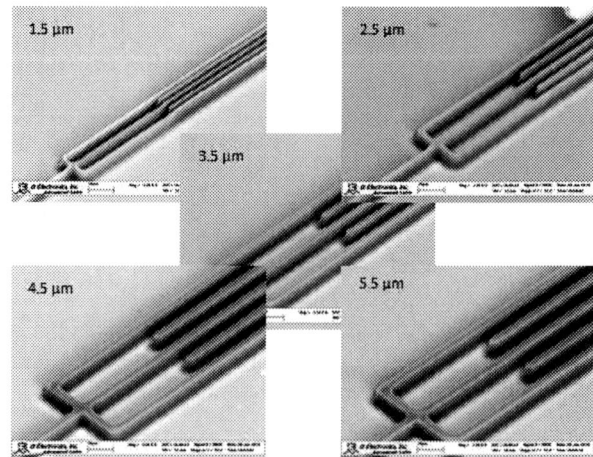

Fig. 4: Imaging resolution <2μm

B. Overlay accuracy

The ability to accurately align layer to layer is critical for any semiconductor patterning process. This is also the case for interposers. Depending on the complexity of the interposer, five or more lithography layers per substrate side can be required. Overlay errors compound and can ultimately cause yield loss. As a rule of thumb, the overlay accuracy should be about one third of the resolution limit of the optical system. If the resolution limit is 1.5μm, a 0.5μm overlay specification is reasonable. Full-field exposure systems as well as 1x steppers are particularly limited when it comes to overlay accuracy. This has to do with the inability to compensate for magnification errors. A single sided telecentric lens system with a reduction characteristic (e.g. 2x) combined with a freely adjustable reticle position is a suitable solution to the overlay challenge.

C. Sidewall angle and CD control

When processing thick photoresist or photo definable dielectrics, well controlled sidewall angles are a critical requirement and in some cases a challenge. Although sidewall angles are primarily a function of the photosensitive material and its processing (pre-bake, post-bake, developing, etc.), the exposure system plays an important role. Accurate focus control across the wafer or substrate is required to achieve tight CD control and consistent sidewall behavior.

Fig. 5: AZ4620 resist, 10μm via opening

D. Depth of focus (DOF)

One of the most prominent reasons why typical sub-micron front-end lithography systems such as DUV steppers and scanners are not suited for interposer manufacturing is their limited depth of focus, a tradeoff of the relatively high numerical aperture of such systems. Without a DOF in the 10μm or greater range, patterning of very thick layers is challenging at best.

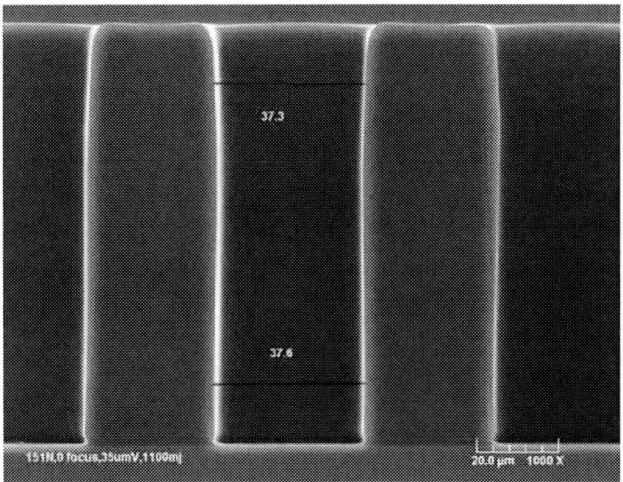

Fig. 6: THB 151N resist, 100μm thick, 35μm feature size

E. Exposure field size (FoV)

The exposure field size (sometimes referred to as field of view or FoV) is yet another important parameter for interposer manufacturing. One could think of the interposer as the actual die. The size of the exposure field should at least cover one die (interposer) in order to avoid stitching. Stitching, a common lithography strategy in the flat panel display and x-ray sensor industries, is inherently slow and difficult and should therefore be avoided. Although there are no standard interposer sizes, they tend to get larger and larger as they combine and interconnect more chips as well as passive components.

F. Infrared alignment capability

One of the main characteristics of any interposer are its through-via interconnects from the top side to the bottom side. There are different approaches to form such vias. A dry etch batch process (Deep Reactive Ion Etch – DRIE) is a fairly common and efficient way. Via locations are lithographically defined and through holes are then formed in bulk. Naturally, the lithography step to define via locations must be referenced to the other side of the wafer or substrate. So called front-to-back alignment is needed.

In the case of glass interposers, alignment to the backside of the substrate is relatively easy as a microscope can obtain an image of an alignment feature through the glass substrate. In the case of silicon, it is not possible using visible light. An infrared microscope option can be a good solution. IR penetrates silicon and can be used to pick up an image from the wafer backside.

G. Warpage:

IDMs and foundries alike have started to build Si interposers on mature front-end lines (e.g. 65nm node). Given the fact that minimum feature sizes on interposers are still an order of magnitude larger than the capability of such lines, this appeared to be the lowest cost option available. Besides some of the lithography challenges mentioned above, the issue of wafer or substrate warpage due to double-side processing is not easily dealt with on these established and mature fab processing lines. The unique equipment features to handle warped wafers are not typically available or installed on any processing equipment within these type of lines. This can lead to yield loss, breakage and extended downtime and therefore negatively impact cost.

Fig. 7: Warped interposer wafer [8]

Specifically for the lithography step, the issue of limited Depth of Focus adds to the problem. Equipment solutions with warped handling features such as switchable and compliant vacuum gaskets on chucks and handlers are needed for lithography and other processing steps.

COST CONSIDERATIONS

As with all back-end processes, manufacturing cost is a critical concern [3]. Front-end processes are designed to work on wafers, which are necessarily round and difficult to make larger. With 2.5D interposer-based integration schemes, there is an opportunity to move away from round wafers and onto larger, square or rectangular substrates – economy of scale is a proven way to reduce cost. In fact, whole industries already exist, such as flat panel displays and solar panels, which use similar manufacturing processes on large rectangular substrates. Consistent with the topic of this paper, the following cost considerations specifically refer to the lithography portion of the manufacturing process. Other, non-litho cost aspects of 2.5D interposer manufacturing are not being discussed here [6]. Some of these non-litho cost aspects are yield, via formation (drill & fill, handling), availability of commercial interposer suppliers, cost effective probe & test strategy, alternative interposer materials, amongst others [7].

In front-end photolithography processes, where square die first met round wafers, there is an inherent inefficiency near the wafer's edge, where squeezing as many die as possible onto the wafer inevitably results in part of the exposure field falling uselessly in the exclusion zone or off the wafer entirely. With an appropriately-sized rectangular substrate the rectangular pattern from the mask could fit perfectly, ultimately increasing the average number of interposers per exposure and thereby, the throughput of the exposure process. Likewise, using a larger substrate also increases throughput by reducing the nonproductive time spent exchanging substrates. Moreover, the same considerations that have historically driven increases in wafer size should also apply to non-round, non-wafer substrates, potentially providing substantial gains from using large panels throughout the manufacturing process. The flat panel display industry has increased its panel sizes over the years from 400mm x 500mm (Gen 2) to 2400mm x 2800mm (Gen 9) and beyond.

In an effort to understand the potential economic benefits, we constructed a model to compare the throughput of a 650mm X 550mm panel-based lithography process with a 300mm wafer-based lithography process. The model considered 8mm square die or interposers with 100μm streets and a 5mm wafer edge exclusion. We looked at two different mask (reticle) configurations containing 48 interposers (8 X 6) and 49 interposers (7 X 7) that could be exposed using the 84mm diameter field of the 2x reduction JetStep™ Panel Lithography System (Rudolph Technologies). Both mask configurations resulted in 947 die per wafer. Since the square exposure field (7 X 7) required only 23 exposures, 6 less than the 29 exposures required by the rectangular field (8 X 6), all subsequent comparisons use the square field. The panel required 120 exposures resulting in 5,214 die / interposers. It should be noted that either exposure field configuration was implemented on a standard 6-inch square, .25-inch thick reticle.

The most obvious advantage in the panel process accrues from the more than 5X greater number of substrate exchanges required by the wafer process, resulting primarily from the larger size of the panel substrate. Less obvious, but also important, are two different "square peg in a round hole" effects. The first is the decrease in the number of exposures required that results from the better fit of the rectangular field and the rectangular panel. The second is the increase in surface utilization that results from the better fit between the rectangular die and the rectangular substrate: 947 8mm die cover 86% of the surface of a 300mm wafer, whereas 5214 8mm die fill 94% of the surface of a 650mm X 550mm panel. A potential disadvantage of the panel process is the requirement for more alignment because of the increased substrate size.

For a more precise comparison, we calculated throughput in die per hour, assuming each wafer exchange took 15 seconds (including WEP) and each panel exchange took 13 seconds. Fig. 8 compares the results for 8mm X 8mm die, including evaluation at two different doses (600 mj and 800 mj) and four different numbers of alignment points (4, 9, 16, and 18 points). In all cases, the panel process demonstrated approximately 2X (die per hour) throughput advantage over the wafer process. Predictably, increasing the dosage or the number of alignment points reduced the throughput for both wafer and panel processes. The decrease in throughput associated with increase in number of alignment points had an impact, for example, going from 9 to 16 points at the 600 mj dose reduced throughput by 8.8%.

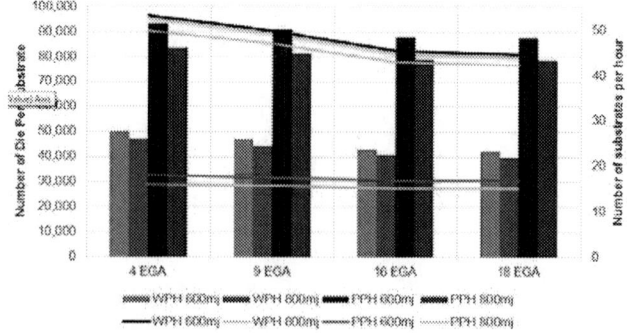

Fig. 8: Throughput comparison – wafer vs. panel

Next, we estimated the cost-of-ownership for wafer and panel lithography processes, comparing three different wafer exposure systems (1X stepper, 2X JetStep wafer stepper and 2X JetStep panel stepper). Fig. 9 shows the parameters used for the comparison. Fig. 10 shows the relative cost per 100 die calculated for 8mm X 8mm die using 9 alignment points at 600 mj and 800 mj doses. At the 600 mj dose, cost per die decreased by approximately 18% for the JetStep wafer stepper and nearly 40% for the JetStep panel stepper, when compared to the 1X stepper. Similarly, at the 800 mj dose, cost per die decreased approximately 13% and 35%. We saw estimated cost savings of similar magnitude for smaller (3mm X 3mm) and larger (16mm X 16mm) die.

	1X Stepper	JetStep Wafer	JetStep Panel
Exposure wavelength	ghi	ghi	ghi
Field size in mm²	68x26	66x52 or 59.4x59.4	66x52 or 59.4x59.4
Substrate size	Wafer 300mm	Wafer 300mm	Panel 650x550mm²
Dose use in mj	600 & 1000	600 & 1000	600 & 1000
Reticle size	6"	6"	6"
Lamp power	2 lamps at 1.2KW	3.5KW	3.5KW
Production parameters	90% uptime and 24/7	90% uptime and 24/7	90% uptime and 24/7
Number of exposures per substrates	44	23	120
Die sizes in mm²	3x3, 8x8, 16x16	3x3, 8x8, 16x16	3x3, 8x8, 16x16

Fig. 9: Cost of ownership comparison parameters

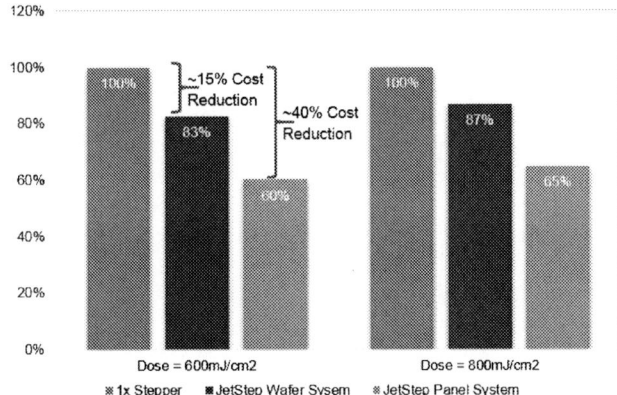

Fig. 10: Relative cost comparison for litho process

CONCLUSION

2.5D packaging is a novel implementation of an already well-established concept which used to be referred to as Multi-Chip-Modules (MCMs). A high-density interposer is used to efficiently interconnect two or more chips. From a lithography perspective, the manufacturing of these interposers poses some unique challenges such as resolution, overlay, sidewall, depth of focus, exposure field size, infrared alignment and substrate warpage. Addressing these challenges calls for lithography equipment technology which is specifically geared towards this application.

From a cost perspective, moving from round wafers to rectangular panels ("panel-ization") saves corner space, delivering a roughly 10% improvement in surface utilization. The larger size of the substrate and the improved fit between the mask and substrate reduce the transfer overhead by a factor of five. The potential reduction in throughput resulting from an increase in the number of alignment points is more

than offset by the improvements in throughput. Compared to a 1X stepper on wafers, panel-based processes can reduce lithography cost per die by as much as 40%.

Clearly, there are many aspects of "panel-ization" that must be addressed before these processes gain broad acceptance. It is worth noting that panel lithography is not new. It is widely used in related industries, such as the manufacturing of flat panel displays and photovoltaic solar panels. The potential economic benefits of panel-based lithography are significant. The model discussed here evaluates relatively modest sized panels. Larger panels may offer even greater benefits. Clearly, the transition to panel-based processes for advanced packaging applications bears serious consideration.

Acknowledgment

The author wishes to thank Roger McClearly, Steve Gardner, James Webb, Rajiv Roy and several other Rudolph personnel around the world that have contributed to generating the data.

References

[1] Clive Maxfield, EE Times Article, 4/8/2012: "2D vs. 2.5D vs. 3D ICs 101"

[2] Aric Shorey, Alan Huffman, John Keech, Matt Lueck, Scott Pollard, Philippe Cochet, Klaus Ruhmer (Corning, RTI, Rudolph): Advancements in Fabrication of Glass Interposers, ECTC 2014, Orlando, FL.

[3] Rich Rogoff, Rudolph Technologies: A Square Peg in a Round Hole - The Economics of Panel-based Lithography for Advanced Packaging Applications, Rudolph internal publication, February 2014

[4] Ron Huemoeller, AMKOR Technology, Architectural Success of 2.1D & 2.5D, SEMI 3D Summit, Grenoble, France, January 2014

[5] M. Juergen Wolf, Fraunhofer IZM ASSID, 3D Technology as a Holistic Approach, SEMI 3D Summit, Grenoble, France, January 2014

[6] Eric Beyne, IMEC, Cost analysis 2.5D and 3D system integration, SEMI 3D Summit, Grenoble, France, January 2014

[7] Michael Thiele, GlobalFoundries, Enabling 2.5D Technology for Commercialization, SEMI 3D Summit, Grenoble, France, January 2014

[8] N. Sillon, CAE LETI, From 3D Toolbox to 3D Integration: Examples of Successful 3D Applicative Demonstrators, SEMI 3D Summit, Grenoble, France, January 2013

Package Embedded Inductors for Integrated Voltage Regulators

William J. Lambert, Michael J. Hill, Kaladhar Radhakrishnan, Leigh Wojewoda, Anne E. Augustine
Intel Corporation, Assembly & Test Technology Development
5000 W. Chandler Blvd.
Chandler, AZ 85226

Abstract

Intel® 4th generation Core™ microprocessors are powered by high frequency integrated switching voltage regulators. The inductors required to implement these regulators were constructed using the routing layers of conventional organic flip chip packaging. This paper provides an overview of the simulation and measurement of these embedded inductors including representative results from production packages.

Introduction

Increased emphasis on fine grain power management has led to a proliferation of independent power supplies for the different logic blocks in a microprocessor die. Traditionally these supplies have been implemented as multiple discrete voltage regulators on the motherboard. However, adding more voltage regulators to the motherboard adds cost, increases area, and requires extra package pins. To avoid this problem, Intel's® 4th generation Core™ microprocessors [1] feature Fully Integrated Voltage Regulators (FIVR) [2], which are buck regulators integrated into the microprocessor die and the package. A two-stage power conversion scheme is used in these microprocessors as shown in Figure 1. A single first stage step down VR, which is on the motherboard, converts from the power supply or battery voltage (12-20V) to approximately 1.8V, which is distributed across the microprocessor die. Each major component on the die is then powered by its own independent FIVR. Depending on the product segment, there are anywhere from 10 to 30 voltage domains on the die, each of which is powered by a FIVR having up to sixteen phases.

Figure 2 shows the simplified schematic of a single phase synchronous buck regulator as implemented in a FIVR. Multiple phases, consisting of duplicate inductors and MOSFETs in parallel, are usually implemented but are not shown for simplicity. The power MOSFETs and feedback and PWM control circuitry are implemented on microprocessor die. The decoupling capacitance is also implemented on the die using MIM capacitors from the 22nm [3] process, and is sometimes supplemented by discrete package capacitors. The inductors are implemented in the bottom routing layers of the organic flip chip package. A high level overview of the system is available in [2]. This paper provides an overview of the implementation and characterization of the inductors that were used, which are mentioned only briefly in the previous reference.

The high switching frequency (140 MHz) of a FIVR is necessary to make the passive components small enough to integrate on the die and package. While the inductors could theoretically have been implemented as discrete components, the total count (up to 60 on consumer products) made the option cost prohibitive. The inductance targets were low enough that the inductors could be implemented using only package routing, but the performance and reliability of this approach needed to be established. This paper describes the simulations and measurements that were used to study the performance of the package embedded inductors and to determine the designs used in high volume products. Simulation results matched passive measurements very closely, and demonstrated that the product inductor designs are not only capable of meeting specified performance targets, but also have few drawbacks compared to discrete components.

Implementation

The acceptable range of inductance for each FIVR domain on the die was bounded by performance requirements such as

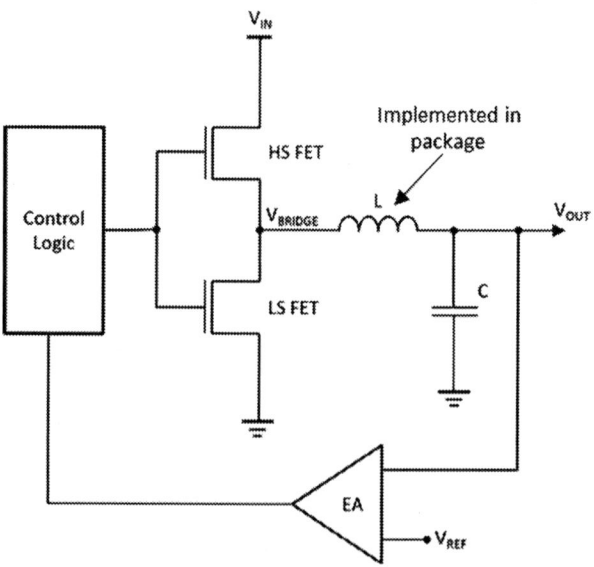

Figure 1: Power conversion scheme used in Intel's® 4th generation Core™ microprocessors.

Figure 2: Circuit Representation of a Buck Regulator

2014 Electronic Components & Technology Conference

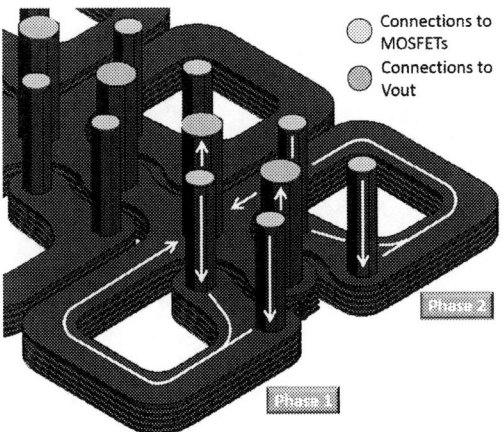

Figure 4: Isometric view of the embedded inductors showing PTHs through the package core and bottom layer routing.

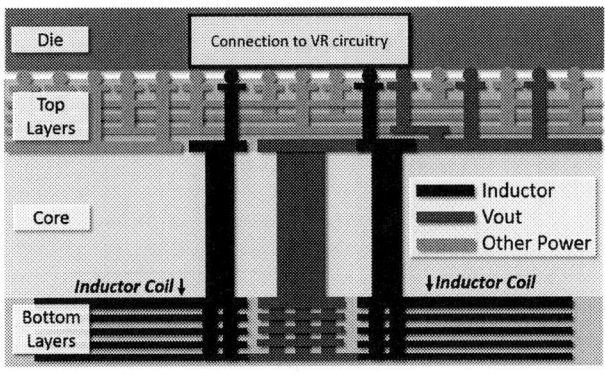

Figure 5: Cross-sectional view of the inductor embedded in the package.

voltage ripple, transient response, and current handling capability. Domain specific inductance limits varied but all fell under 20nH. Several topologies were considered in the development of the inductors but in most cases a straightforward solenoid provided the best combination of performance, reliability, and volume utilization. A representative inductor is shown Figure 4. The power MOSFETs on the die are connected to the solenoidal turns through buildup layer microvias (not shown) and finally a plated through hole (PTH). The majority of the inductance is realized by the turns on the bottom layers of the package; the low-loss core material above and air below the turns ensure a high quality non-magnetic inductor is realized. After the turns are completed, the inductor is connected to a shorting bar which is shared among all other inductors used for the same domain. This shorting bar is then connected to the output power plane routing on the upper layers of the package via PTHs, which connect to the corresponding voltage domain on the die. This arrangement is shown by the cross-sectional view in Figure 5. A disadvantage of this approach is that nearly the entire routing area on the bottom package layers under the die is consumed, as shown in the picture of the bottom side of an LGA desktop package in Figure 3. However, placing the inductors immediately below the FIVR circuitry on the die and the voltage domain they are supplying ensures that the primary current flow path is vertical through the package. This minimizes the losses due to lateral current flow on the package through a reduced number of routing layers.

The solenoidal implementation offered straightforward control over inductor parameters. Increasing the area of the void for the inductor turns increased inductance approximately linearly. Connecting the turns in parallel ensured a very low DC resistance, whereas connecting some of the turns in series produced a much larger inductance (roughly proportional to the square of the number of turns).

Simulation

Accurate simulation models were particularly important for design because any changes to the inductors would have required the re-design and re-manufacturing of the package. The frequency range of interest for the simulations spanned from DC to several times the switching frequency (f_{sw}) (which

is necessary to accurately capture voltage ripple and power loss at harmonics of the switching frequency). Early analysis showed that resistance at f_{sw} and above was very sensitive to the details of the inductor routing, including small features in the upper layers, which were not easily accounted for by analytical models. Therefore full 3D extractions of the package routing were used for modeling.

Both full-wave and quasistatic 3D simulators were evaluated. With correct settings both types were capable of predicting key parameters across the frequency range of interest and both types resulted in similar run times. However, full wave solvers provided options for configuring excitations, boundaries, and solver settings that were not easily duplicated in the quasistatic solvers that were evaluated. Therefore a full-wave solver was used for analysis.

To match the configuration of passive measurements (discussed in the next section) lumped port excitations were defined at the same points on the package surface that were contacted by the RF probes. This required all of the package layers from the surface to the bottom to be included in the extraction. The X-Y extents of the model generally included all of the inductor routing and output power delivery routing, though in some cases portions of the power delivery routing

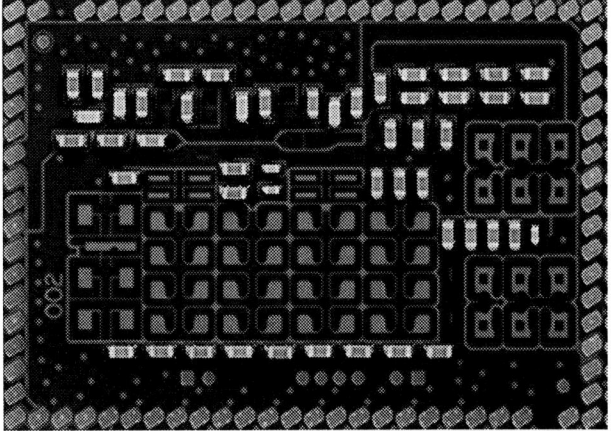

Figure 3: The bottom of a desktop LGA package showing the embedded inductors.

978-1-4799-2408-0/14 $31.00 © 2014 IEEE

could be truncated to reduce run times. Figure 6 shows an isometric view from the bottom of the package of the 3D geometry for a typical simulation model.

Early results showed good correlation to measurements for the inductance, but sometimes substantial errors in resistance, even when model extents were much larger than the inductors being simulated. Two main sources of error were identified. The first source of error was full-wave simulator settings. The basis order used by the solver was observed to have strong impact on the simulated resistance, because the thin package metal layers typically were meshed with only a single layer of tetrahedra, across which the fields varied significantly. When a low order basis was used (for example, "0th" order in a commercially available tool, which solves six unknowns per tetrahedron) the resulting resistance matches well at low frequencies (the dashed red curve in Figure 7), but the resistance is significantly under-predicted at high frequencies. Using 1st order basis functions (the solid blue curve in Figure 7) produced results that closely matched measurements. Surface impedance boundaries used for conductors also introduced error because the skin depth of copper near f_{sw} is between 5µm and 10µm, which is similar to the total thickness of the metal. As a result, the current density distribution in the solenoidal turns falls in the transition region between the DC and skin depth limited cases. To accurately capture losses in this region the fields must be solved inside the copper. When a surface impedance boundary was used instead, the effective thickness was overestimated, yielding low resistance predictions at lower frequencies (as illustrated by the dotted green curve in Figure 7, where a surface impedance boundary was used with 1st order basis functions).

The second effect that had to be considered was the losses due to eddy currents induced by the embedded inductors on other nearby nets. In Figure 6 the actual inductor routing is colored red and metal on other nets (including inductors on other nets) is colored blue. In early simulations some or all of the metal on other nets was removed in order to reduce run times, particularly on larger models and in the upper layers where its impact on inductor characteristics was assumed to be negligible. Later simulations showed that the losses due to induction on the surrounding metal were not negligible, and that it had to remain the simulation in order to produce a correct result.

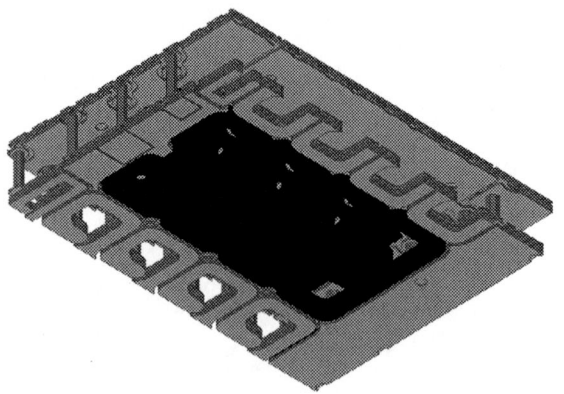

Figure 6: Geometry from a typical 3D simulation.

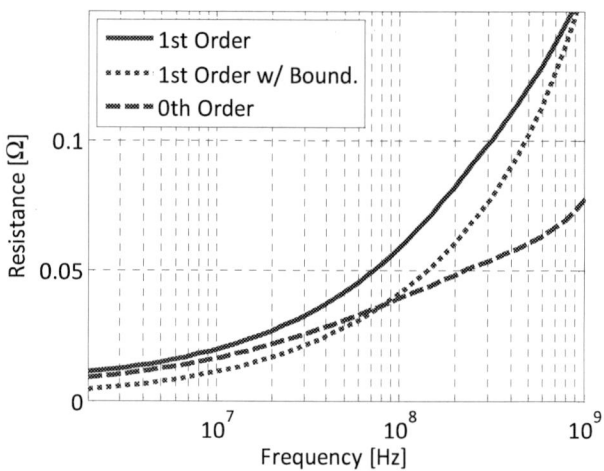

Figure 7: Simulated resistance versus frequency curves for different simulator settings.

Using a higher basis order, solving the fields in the metal, and leaving extra routing in in the model all increased simulation time, but allowed the methodology to produce consistently accurate results without repeated runs or fine tuning of simulator parameters.

Package Integrated Inductor Measurement

When considering techniques for parameter extraction it is important to consider the impedance ranges and frequencies of the interest of the devices under test (DUT) in order to determine the proper measurement equipment and technique. Based on nominal design targets the absolute value of the impedance of the embedded inductors at and below f_{sw} was expected to be less than 3Ω, while the real part of the impedance could be under 50mΩ. L-C-R Meters are commonly used to characterize discrete inductors; however, with the addition of the required probe hardware for embedded inductor measurements, and given the frequencies of interest, it was found that L-C-R meters did not provide the level of accuracy or reproducibility required.

Vector Network Analyzers (VNA) are also commonly used for high frequency device characterization [4], but their accuracy is highly dependent on the configuration of the measurement. Figure 8 shows three VNA measurement configurations that could have been chosen for the embedded inductor measurement: (a) two port series, (b) one port shunt, and (c) two port shunt. Of the three configurations, the two-port series or one-port shunt are superficially the most attractive options, as they provide the easiest configuration for RF probing and match the configuration of the embedded inductors in the LC circuit. However, due to the 50Ω characteristic impedance of the VNA, the two-port shunt configuration provides higher measurement sensitivity.

To examine the range of DUT impedances appropriate for each topology one can plot the magnitude of the key S-parameter for each configuration vs. the magnitude of the DUT impedance as shown in Figure 9. From the figure it can be seen that the one port measurement topology is only sensitive to

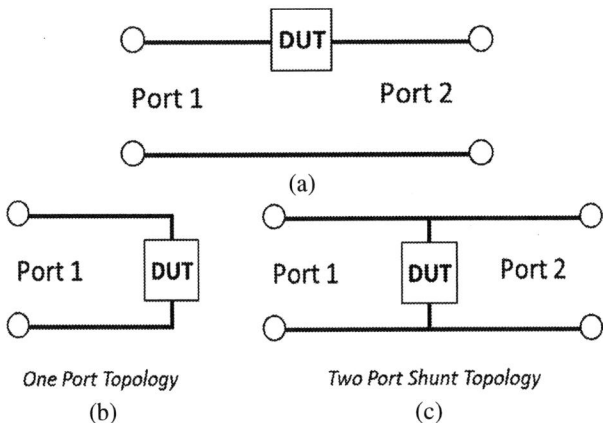

Figure 8: (a) Two-port series (b) one-port shunt and (c) two-port shunt measurement configurations.

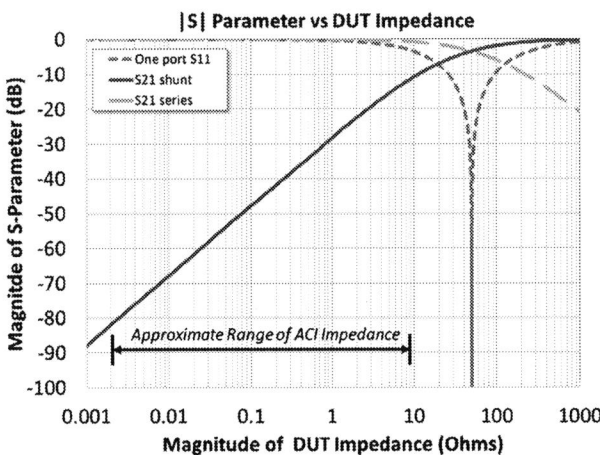

Figure 9: Key S-parameter magnitude as a function of DUT impedance.

impedances in the ~10Ω to ~200 Ω range, well above the expected impedance of the embedded inductors in the frequency range of interest. Likewise, impedances below approximately 10Ω result in values of $|S_{21}|$ very close to 0dB when the 2 port series configuration is used. Conversely, the 2 port shunt configuration produces $|S_{21}|$ values that nicely span the VNA's dynamic range when the DUT impedances are in the sub-10Ω range, while maintaining $|S_{21}|$ well above the instrument's noise floor. This is the same impedance range where the embedded inductors fall over the frequency range of interest. Thus, the 2-port shunt configuration was chosen for characterization of the embedded inductors.

Probing the Embedded Inductors

Each inductor in the package substrate is connected to the CPU die through an array of die bumps. The layout of the bumps and inductors was conducive to the use of commercially available GS/SG 150um RF probes. Although it is typically desirable to land RF probes on a well-designed gold plated pad, this was not an option as the inclusion of such a pad would prevent the connection of the inductors to the CPU die. Thus it was necessary to land the RF probes directly on the die bumps (Figure 10). The impact of this non-ideal landing was studied and compared to landings on test vehicles containing more ideal landing pads. It was found that with proper technique these landings could be made with acceptable reproducibility by multiple operators, and that the numerical results agree well with the more controlled test cases with proper landing pads.

In addition to individual inductor measurements, the ability to land probes on die bumps enabled the examination of more complicated inductor to inductor coupling mechanisms. The inductor blocks typically consist of a number of inductors sharing an output rail (see Figure 6). Because of this the inductor to inductor coupling is of interest, as is the bank to bank coupling (Figure 11). Both of these quantities can be measured using the VNA by proper selection of the probe locations (Figure 12).

Once the measured S-parameters are collected from the VNA they can be converted to an impedance matrix using (1), and if desired, an inductance matrix by using (2), where *[I]* is the identity matrix, *[S]* is the measured S-parameter matrix, *[Z]* is the impedance matrix, *[L]* is the inductance matrix, and ω is

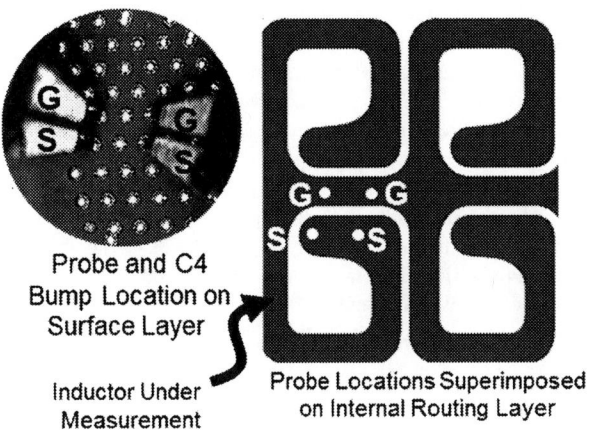

Figure 10: Landing RF probes on die bumps for inductor measurements

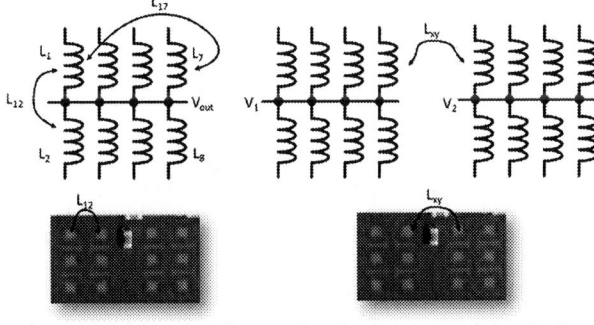

Figure 11: VNA Measurements of inductor to inductor coupling can be performed.

the angular frequency of the measurement. Either of these matrices can then be used to directly compare to simulations of the inductors, or it can be used as input to a larger simulation of the FIVR circuitry to estimate other FIVR performance metrics.

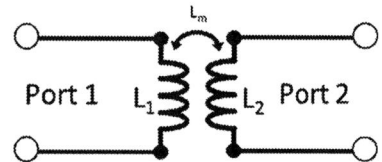

Figure 12: VNA Configuration for inductor to inductor coupling measurements.

$$[Z] = ([I] - [S])^{-1}([I] + [S]) \qquad (1)$$

$$[L] = \frac{1}{\omega}\text{Im}\{[Z]\} \qquad (2)$$

DC Performance

The DC resistance of the embedded inductors is an important element of the voltage regulator efficiency. While the low end of the VNA's frequency range (under 100kHz) effectively provides a DC measurement, the sensitivity of even the two-port shunt configuration is degraded enough to limit the accuracy of the results, particularly for inductors with DC resistances under 20mΩ. For improved accuracy 4-wire DC resistance measurements are a better choice for characterizing the DC resistance. Four needle probes were used along with a commercially available micro-ohm meter to characterize the DC resistance of each inductor using a similar probing configuration to the AC measurements. Because the 4-wire measurement utilizes two probes for sinking and sourcing the test current, and two different probes for measuring the induced voltage, this well-known technique avoids the inclusion of contact resistance into the measurement results.

Automation & Reproducibility

Each CPU package utilizing FIVRs contains many individual inductors. For example, a single Intel Xeon server package may contain more than 150 inductors. To understand high volume manufacturing effects, including within package and cross lot variations, it is not practical to collect large volumes of inductor measurements by hand. For this reason the inductor measurements were automated using a motorized wafer probing station. This probing station allows RF or DC probes to be landed on different substrate locations automatically. A custom software tool was developed to import the measurement probe locations, land the probes, trigger the VNA or DC meter and download the data. In addition to simplifying the collection of data this also allows for better reproducibility by improving the uniformity of the probe landings.

Within Intel the study of the repeatability and reproducibility of a metrology is referred to as a metrology capability analysis (MCA). In this context repeatability quantifies the variability in the measurement data collected on an unchanging DUT over multiple measurement cycles in one measurement session, while reproducibility refers to the variability in the measurement of an unchanging DUT when all regularly occurring environmental and equipment variables are considered. For example, a reproducibility study includes variations caused by the introduction of different operators, variations caused by performing the measurement on different days, and variations due to the influence of the operator's ability to calibrate the VNA, as well as site variations that occur in normal operation such as variations in temperature and humidity. In an MCA the repeatability and reproducibility are characterized statistically so that any variations seen in the measurement data can be compared to the variation in the metrology to understand if the observed variation is a real effect or if it could be due to expected variations in the metrology. An example of the quality of the reproducibility of the embedded inductor metrology is shown in Figure 13. This figure shows the results of the measurement of the same inductor by three different operators over three different days. The standard deviation across the nine measurements is less than 4pH, which indicates the metrology is capable of resolving very small difference in inductance between different structures.

Results

Figure 14 shows the modeled versus measured inductance for two different representative embedded inductor designs on a microprocessor product. The error between simulation using nominal dimensions and measurement is less than 2% near f_{sw} and remains under 5% across the entire simulated frequency range.

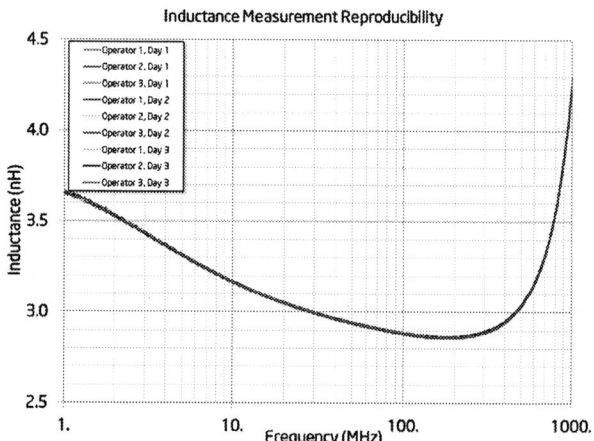

Figure 13: Example showing the reproducibility of the embedded inductor inductance measurement metrology.

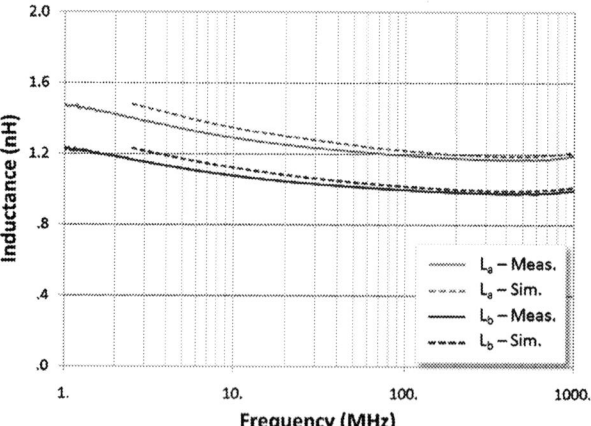

Figure 14: Measurement versus simulation for two embedded inductor designs.

978-1-4799-2408-0/14 $31.00 © 2014 IEEE

A third example is shown in Figure 15 where the inductance mismatch near f_{sw} is larger at 5%. The resistance for the same inductor is shown in Figure 16 where the mismatch is larger at 11%. Following the measurement the unit was cross-sectioned, and the simulation re-run with identical settings and material properties, but with dimensions taken from the cross section. This reduced the model to measurement error at 100MHz to under 2% for inductance and under 8% for resistance.

In Table 1 the inductor from Figure 15 is compared to a low-cost discrete 0402 inductor from a major supplier. While the footprint of the 0402 component itself is smaller than that of the embedded inductor, the total footprint on the package required for 0402 inductor pads given competitive assembly rules is similar in size to the embedded inductors. It is also worth noting that the thickness of the discrete inductors would prevent them from being mounted under BGA packages, which represent a significant portion of the mobile market. The data shows that the performance of the embedded inductor is very similar to the discrete option.

Table 1: Comparison of an embedded inductor and 0402 inductor at 100MHz

	LAC [nH]	RAC [mΩ]	RDC [mΩ]
0402 L	2.35	90	40
Embedded L	2.51	98	36

Table 2 and Table 3 compare the measured and simulated inductance matrices at 108MHz for a coupled inductor structure with four inductors. The largest error in the self-inductance terms is under 6%. The dominant coupling terms also match extremely well (both the 400pH of intended negative mutual inductance between coupled pairs and the 400pH of unwanted positive mutual inductance between two of the inductors). Other much smaller mutual terms are also present in the matrix and match very well to simulations, with 20pH or less error.

Larger errors of up to 18% on the diagonal are present in the resistance matrices seen in Table 3. The simulation used nominal dimensions, so the error would likely have been reduced if a cross-section was performed. The small mutual terms also match well, with less than 2mΩ errors.

Table 2: 4x4 Simulated and Measured Resistance Matrices for a Coupled Inductor

Sim. [L] [nH] - 108 MHz				Meas. [L] [nH] - 108 MHz			
1.34	-0.41	0.02	0.04	1.39	-0.42	0.02	0.03
-0.41	1.32	0.40	0.04	-0.42	1.40	0.40	0.04
0.02	0.40	1.33	-0.39	0.02	0.40	1.38	-0.40
0.04	0.04	-0.39	1.39	0.03	0.04	-0.40	1.47

Table 3: 4x4 Simulated and Measured Resistance Matrices for a Coupled Inductor

Sim. [R] [mΩ] - 108 MHz				Meas. [R] [mΩ] - 108 MHz			
73.7	-9.2	1.4	2.1	62.4	-8.1	0.7	1.7
-9.2	71.8	7.5	0.5	-8.5	63.0	6.0	1.2
1.4	7.5	74.0	-8.6	0.4	6.3	62.5	-7.1
2.1	0.5	-8.6	72.1	1.5	1.2	-7.5	64.7

Every unique inductor across every distinct Intel® 4th generation Core™ product was measured on multiple packages and achieved similar correlation to pre-fabrication simulations. Figure 17 provides an example of these results for resistance for several different structures spread across an LGA package.

High Volume Manufacturing Study

To quantify the high volume manufacturing (HVM) variability of package embedded inductors, packages from multiple lots manufactured by two different package vendors were measured. A total of 2800 inductance and AC resistance measurements were performed to characterize the variability. In addition to quantifying HVM variability of package embedded inductors, another motivation behind this study was to quantify the likelihood of catastrophic defects in inductor fabrication. It was shown through simulations that DC resistance of the inductor can be used as a metric to monitor the quality of inductor manufacturing. To quantify DC resistance

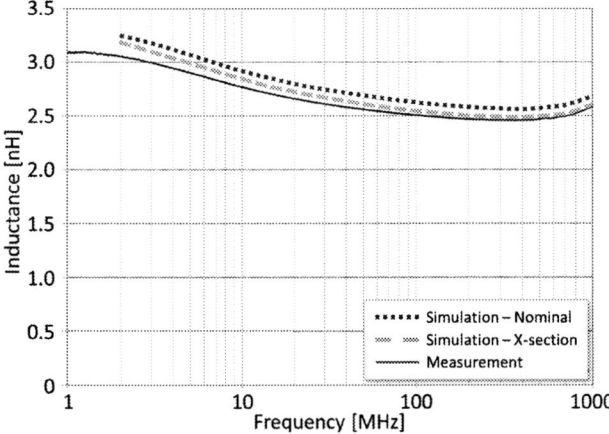

Figure 15: Comparison between the measured and simulated inductance of an embedded inductor

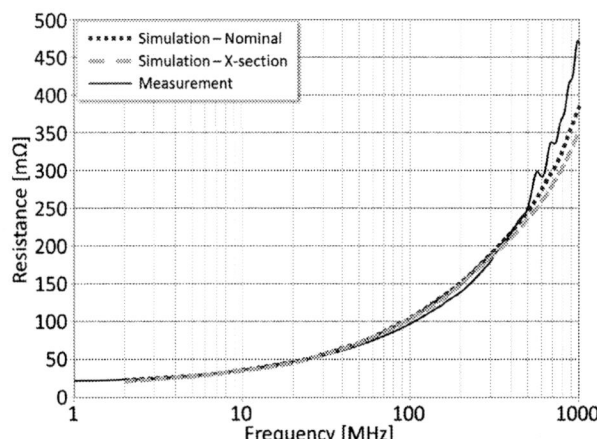

Figure 16: Comparison between the measured and simulated resistance of an embedded inductor

978-1-4799-2408-0/14 $31.00 © 2014 IEEE

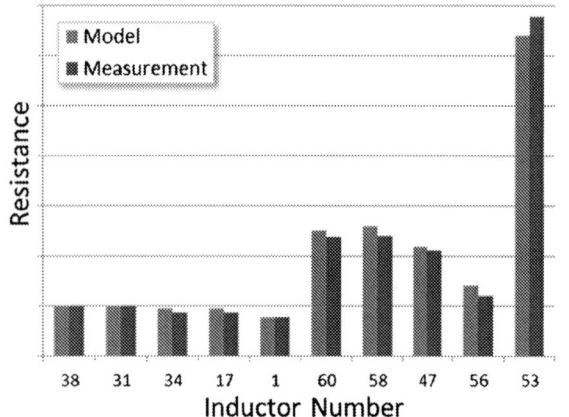

Figure 17: Normalized modeled versus measured resistance at 100MHz for different inductors across a package.

variability, measurements were performed on 131 packages, for a total of 13,984 DC resistance measurements.

Table 4 summarizes the variations in the values of inductance and AC resistance measurements. The variation is quantified in terms of three standard deviations divided by the mean value of the measurement ($3\sigma/\mu$). It can be seen from Table 4 that AC resistance is more sensitive to manufacturing variations than inductance. No significant difference was observed between the vendors. The overall variation was found to be 2.3% for inductance and 5.4% for AC resistance. For AC resistance and inductance of the package embedded inductors measured in this study, manufacturing variations were found to be low and it was concluded that the measured variation values are not expected to hinder system performance.

Table 4: Summary of Averaged $3\sigma/\mu$ Values of Inductance and AC Resistance Measurements (Rac is at 100MHz)

Vendor	Vendor to Vendor Variation $3\sigma/\mu$		Overall Variation $3\sigma/\mu$	
	L	Rac	L	Rac
A	1.5%	4.1%	2.3%	5.4%
B	2.3%	3.3%		

Table 5 summarizes $3\sigma/\mu$ values of DC resistance measurements. The overall variation for DC resistance was found to be 15.6%. There was a discernable difference in the DC resistance values between the two suppliers, which was attributed to differences in the metal dimensions. By comparing the range of values in Table 4 and Table 5, it can be said that DC resistance values of the inductors are more sensitive to manufacturing variations as compared to AC resistance and inductance values, as expected from simulations.

Table 5: Summary of $3\sigma/\mu$ Values of DC Resistance Measurements

Vendor	Vendor to Vendor Variation $3\sigma/\mu$	Overall Variation $3\sigma/\mu$
A	6.8%	15.6%
B	10.3%	

Conclusions

Package embedded inductors were an important part of enabling integrated voltage regulation on Intel® 4th generation Core™ microprocessors. Integrating the inductors into the package carries some risks because of the difficulty in making changes late in the design, but this paper has demonstrated that well designed simulation and measurement procedures can accurately predict and measure the inductors, which removes much of this risk. Furthermore, the results achieved show that the inductors that were manufactured had very good performance, even compared to discrete options.

Aside from reducing cost, the embedded inductors have the additional benefit of allowing the designer to choose the precise value, footprint, and location of the inductor. Affordable discrete passives would have to have been selected from a limited set of values and form factors, and correspond to chip shooter compatible placement rules. A further advantage comes from the extremely tight routing tolerances required on the package for controlled impedance high speed IO signaling. These tolerances produce embedded inductors with minimal variation between packages even between separate manufacturers, as shown in the previous section.

Because the inductors are non-magnetic the inductance that can be achieved for a given footprint will be limited, which may eventually require new technologies to be examined as silicon dimensions continue to decrease.

Acknowledgments

The authors would like to thank Sean Christ and Nevin Altunyurt for their extensive assistance in this work. The design of the embedded inductors was heavily influenced by Ted Burton. Many others at Intel were involved in the design and routing mentioned in this paper.

References

[1] N. Kurd, et al., "Haswell: A Family of IA 22nm Processors," in *ISSCC 2014*, San Francisco, 2014.

[2] T. Burton, G. Schrom, F. Paillet, J. Douglas, W. J. Lambert, K. Radhakrishnan and M. J. Hill, "FIVR – Fully Integrated Voltage Regulators on 4th Generation Intel® Core™ SoCs," in *Advanced Power Electronics Conference*, Fort Worth, TX, 2014.

[3] C. Auth, et. al, "A 22nm High Performance and Low-Power CMOS Technology Featuring Fully-Depleted Tri-Gate Transistors, Self-Aligned Contacts, and High Density MIM Capacitors," in *2012 Symposium on VLSI Technology*, Honolulu, HI, 2012.

[4] L. Wojewoda, D. Athreya and M. J. Hill, "Use Condition Characterization of Package Components," in *Proceedings of the ASME 2011 Pacific Rim Technical Conference & Exposition on Packaging and Integration of Electronic and Photonic Systems*, Portland, OR, 2011.

978-1-4799-2408-0/14 $31.00 © 2014 IEEE

Power Supply Filter for PLL Circuit in Digital Systems

Nam Pham, Faraydon Pakbaz, Zhenrong Jin, Lloyd Walls
IBM Corporation, Austin Texas
npham@us.ibm.com, (512) 286-8011

Abstract

This paper presents an effective design approach for the power supply filter of a phase lock loop (PLL) based clock generator in a multi-core ASIC. The noise sensitivity of different types, filter design, system design issues, and measurement techniques for verification and understanding of jitter behavior on power supply noise are discussed

Introduction

Phase-lock-loops (PLL) are widely used in ASIC chips for network equipment, computers, or electronic devices for generating high-precision and low jitter reference clocks, maintaining synchronization, or recovering signal from noisy communication channels. A board filter is usually required to filter out board noise due to interference between on-board switching supplies or noisy digital ASICs. Usually the filter and its component values are specified in data sheets or application notes for a PLL, and it is usually specified for the worst case to cover all different PLL configurations. The specification thus can be a barrier for system filter optimization for a certain PLL configuration. In addition, parasitic effects, variations in component selection, placement, and connectivity have to be considered in practical board designs. All of the above can lead to an awkward board design to just meet the filter specification. To optimize the filter in the system, one needs to look at the filtering as a whole from understanding noise sensitivity of a PLL to understanding the impact of the package design to filter performance, and then to optimize the board design.

This paper systematically describes the filtering requirements of a PLL from chip design, package design and finally to board design. The hardware verification approach is also discussed. The authors hope to bridge the gap between chip and board design, and provide insight into filter optimization of the PLL to help mitigate noise in a system.

Sensitivity of the PLL to Supply Noise

The PLL circuit employs a voltage control oscillator to generate a clock which is sensitive to fluctuation of the power supply voltage. A typical analog PLL clock generator circuit is shown in Figure 1 [1]. Since the VCO is the most sensitive component of the PLL to supply noise, other components such as the output buffer, are not discussed in detail in this paper.

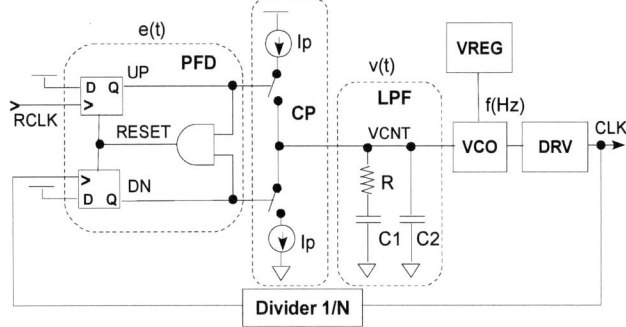

- Phase-Frequency Detector (PFD)   • Clock Driver (DRV)
- Charge-Pump (CP)   • Feedback Divider (FBDIV)
- Low-Pass Filter (LPF)   • Voltage Regulator (VREG)
- Voltage-Controlled Oscillator (VCO)

Figure 1: PLL Circuit

A PLL can be modeled as a complex domain (s-domain) with linear system assumption for the steady state stage of operation [2]. Figure 2 shows the PLL phase model.

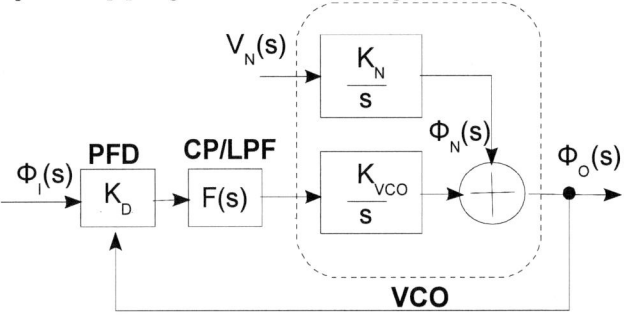

Figure 2: PLL Phase Model

Assuming the power supply noise, V_N, is injected into the PLL VCO, and the divider is kept to 1, the PLL closed loop transfer function from $V_N(s)$ to $\varphi_O(s)$ is given by [2]:

$$\frac{\phi_O(s)}{V_N(s)} = \frac{s \times K_N}{(s + \omega_Z) \times (s + \omega_{3dB})} \qquad \text{Eq. 1}$$

Here ω_{3dB} is the PLL 3dB bandwidth, ω_z is the PLL zero frequency, and $\omega_z \ll \omega_{3dB}$. This equation demonstrates that in the PLL clock generator, the power supply noise is rejected by 20dB/dec, when the power supply interference (PSI) frequency is greater than the PLL 3dB bandwidth. For frequencies between ω_z and ω_{3dB}, the output clock phase varies with the PSI amplitude as:

$$\frac{\phi_O(s)}{V_N(s)} \cong \frac{K_N}{2\pi \times f_{3dB}} \qquad \text{Eq. 2}$$

As an example, Figure 3 shows the power supply rejection PSNR characteristics of a PLL for two different setting of PLL's 3dB bandwidth

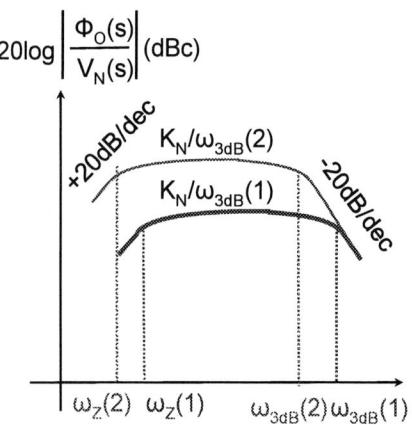

Figure 3: PLL Noise Rejection

Measuring Jitter from Power Supply Spectrum

The frequency domain phase error can be converted to jitter [3]. When a single-tone sinusoidal signal, f_M, representing PSI, is applied to the power supply of a PLL, a power spectrum analyzer can be used to display the graph shown in Figure 4A. The f_O is the clock output of the PLL and the f_M is the fundamental sideband tone of the PSI. If x represents the level difference between them, then peak-to-peak jitter DJ can be computed as

$$Dj = \frac{2 \times 10^{-\left|\frac{x}{20}\right|}}{\pi \times f_O(Hz)} 10^{12} \qquad \text{Eq.3}$$

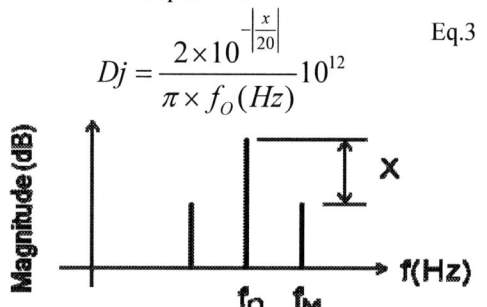

Figure 4A: Power Spectrum Analyzer Measurement

The jitter can also be measured by the phase noise analyzer. The PSI manifests itself as a phase spur relative to the carrier, and is computed as

$$\Delta\phi = \frac{2 \times 10^{\frac{y}{20}}}{\pi \times f_O(Hz)} 10^{12} \qquad \text{Eq. 4}$$

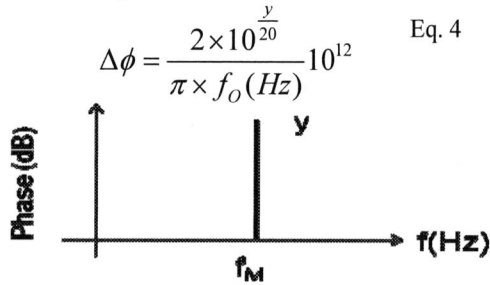

Figure 4B: Phase Noise Analyzer Measurement

PSNR Characteristic of Analog and Digital PLL

A typical ASIC consists of both Analog (APLL) and Digital PLL (DPLL) for different purposes. The DPLL is an APLL with all or part of the noise-sensitive analog circuits replaced by digital circuitry. Replacing analog circuitry with digital circuitry increases the PLL design portability and testability, which scales naturally with nm technology, reduces chip area required by the capacitor filter, and improves behavioral simulation. However, it is often insufficient for phase tracking applications. Therefore, the APLL is often used in the clock generation of the high-speed buses while the DPLL is used for the core logic of the chip. An example of analog VCO is shown in Figure 5 which shows an APLL operation based on the resonance principle of an LC tank circuit where f=1/2πsqr(LC). When the reactance of the tank circuit is equal to an externally induced stimulus, the oscillation is sustained. Figure 5 shows an implementation of an LC tank VCO. The desired frequency can be obtained and sustained by digitally controlling the charging of the varactor (variable capacitor).

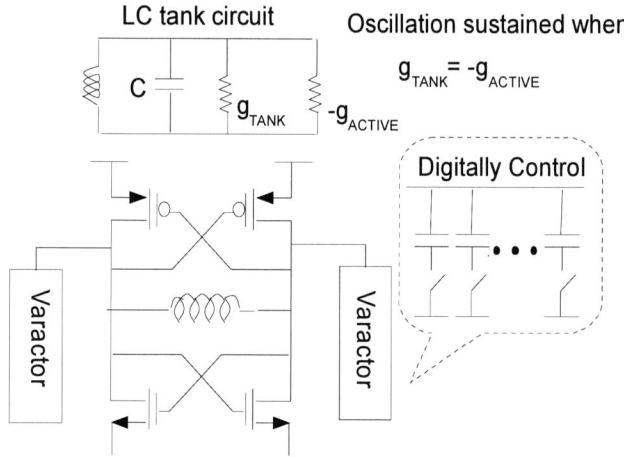

Figure 5: LC Tank VCO Implementation

The VCO can be converted to a digital circuit by replacing the LC tank circuit above with a series of inverters connected as shown in Figure 6. Each digital control signal can selectively activate/deactivate a row of the ring oscillator. By selectively activating/deactivating the oscillator elements in the ring oscillator, the control logic changes the fill factor of the ring oscillator. The fill factor is the ratio of active oscillator elements to the total number of oscillator elements in the array. The frequency of the of the ring oscillator array increases in proportion to the number of active oscillator elements [4].

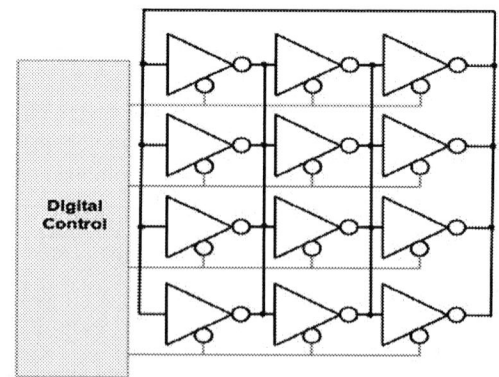

Figure 6: Digital VCO Implementation

Table 1 below shows a comparison of the two types of VCOs [5].

Ring Oscilator VCO	LC Tank VCO
Advantage	
1) Highly integrated in VLSI 2) Low power 3) Small die area occupancy 4) Wide tuning range	Excellent phase noise and jitter performance at high frequency
Disadvantage	
As frequency increase, performance degrades because of phase noise or jitter increases	1) Inductor and varactor have large die area occupancy 2) High power consumption 3) Small tuning range

Table 1: Digital and Analog VCO Comparison

On-Chip Voltage Regulator

The performance requirements due to jitter for PLL's are more stringent for core and high speed IO applications. As the silicon process advances and L_{eff} becomes smaller, it becomes much more difficult to achieve high quality analog properties in the PLL, such as power supply noise rejection. Therefore, the demands placed on on-chip linear voltage regulators (VR) become more substantial. Today's on-chip linear voltage regulators are typically required to have a very low supply rejection ratio of less than 35dB [6]. In addition, the VR's secondary function is to set a precise DC voltage level for the PLL power supply through a reference voltage which is independent of process variations, normal changes in power supply voltages, and chip temperature. Figure 7 shows a circuit schematic of a typical linear regulator. The reference voltage Vbg should be supplied by a reference voltage source such as the band-gap reference with Miller cap circuit [7], [8].

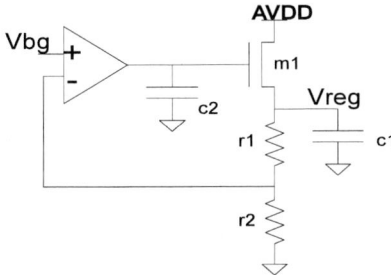

Figure 7: Voltage Regulator Circuit

A typical power supply noise transfer function of the VR is shown in Figure 8 as the VR(PSNR) curve. In this example, the supply noise is significantly attenuated at low frequency, but eventually passes through to the regulated output at high frequency ($\omega \gg \omega_{VR}$), where ω_{VR} is the regulator PSNR bandwidth. In practice, this bandwidth is typically in the order of a few MHz. As a result, a circuit connected to the regulator VR output is well protected from supply noise below that bandwidth. On the other hand, at high frequencies, the VCO is immune to noise due to averaging effect at nominal voltage levels as shown as PLL(PSNR) curve. The combination of the PLL close-loop supply noise rejection and the regulator noise rejection gives the PLL a great deal of protection over a wide frequency range. However, designing the VR so that the bandwidth is well beyond the PLL bandwidth requires

more power and chip area. Therefore, inevitably there exists a small gap in the PSNR transfer function which allows noise to pass through. Figure 8 shows the combined PLL and VR PSNR curve with the sensitivity frequency ω_L located between ω_{VR} and ω_{3dB}. Apparently, the goal of a system filter is to provide protection and plug this gap.

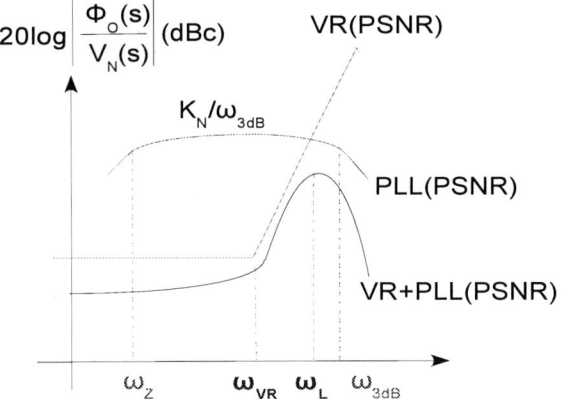

Figure 8: VR and PLL Noise Rejection

System Design Issue

A generic low pass filter circuit such as the one shows in Figure 9, is typically used in a system power supply design. The filter is very effective for suppressing noise generated from the system board, but not as effective in rejecting noise generated from the ASIC itself. The plotted simulation data shows the response at the PLL AVDD pin when a AC coupled signal generator is used to inject a signal at the input Vin or to the output AVDD to mimic the noise coming from a system or a chip, respectively. The techniques used in the lab to apply noise will be discussed in the next section. If inductor L1 becomes zero, the filter characteristics are greatly affected. This scenario presents a trade-off between chip and system noise rejection ability.

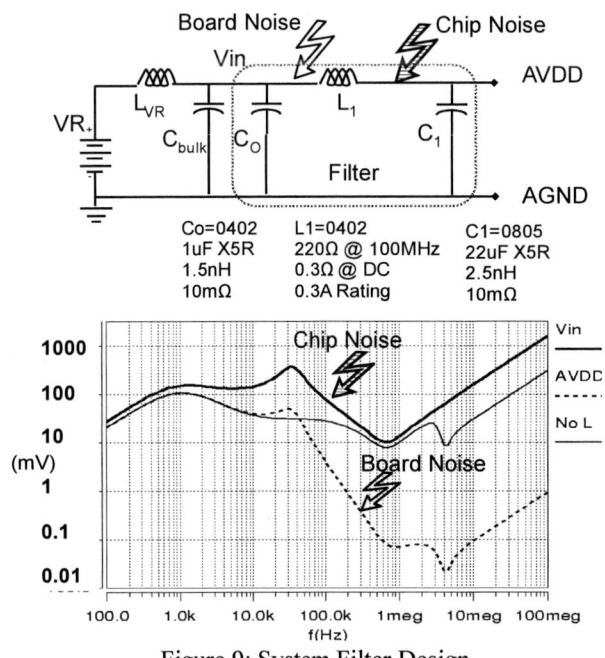

Figure 9: System Filter Design

The noise response of the filter should be designed to avoid any peaking at the power supply switching frequency or any resonance point that can cause oscillation and system instability. These power supply noise suppressing techniques are extensively discussed in other literature.

In a multi-core design, the ASIC requires multiple clock domains, and hence many PLLs. If each PLL requires a filter, the design becomes unmanageable. PLLs of the same type should be grouped into using a common filtered power domain. However, this approach has its own problems since the power domain which ties the PLL circuits together is now part of the filter structure. The LC network of the package, the coupling from other PLL, and impedance of the VR as seen from the AVDD pin can cause oscillations [6]. Careful design and simulation must be done to ensure that the resonance of the structure does not coincide with the 3dB bandwidth frequency of the PLL.

The use of an AVDD and AGND differential pin pair are used to supply voltage to an ASIC PLL. The filter circuit for the AVDD and AGND supply resides on the system card or board. As pressure to reduce the number of package pins increase and the demand for simpler power supply wiring increase, the more attractive a single-ended PLL power supply routing scheme becomes.

Figure 10 show how the differential and single-ended routing scheme differed in the experiment. Substrates DD1.0 and DD2.0 are identical except DD2.0 has a short wire placed between chip AVSS and VSS to simulate the single PLL supply while DD1.0 uses AVDD and AGND to supply voltage to the PLL circuit. The current return path in Figure 10A, is confined between the wire pair while, in Figure 10B, the current is returned through the VSS due to its lower inductance path than at AGND.

Substrate Ground Noise Measurement

Figure 11 shows the oscilloscope traces for the two cases. The scope was set to AC coupled mode. A 4.0 mV of noise was observed between the AVDD and ground of DD1.0. The probe was a single-end coax cable with ground shield soldered to the ground of the substrate. The measurement effectively grounded the negative terminal of the capacitor C1 to ground. The same measurement was applied to case B and the scope recorded 8.8 mV. To quantify the interference noise from the environment which could have contaminated the measurement, 2 mV noise was observed when both tips of the probe were soldered to ground. This 2mV represents the error of the measurement. To further validate the measurement, additional probing from AGND to system ground shows very similar values of 9.3 versus 8.8 mV which are within the 2 mV error of the measurement, which validates the effect of large capacitor, C1. With 22uF, C1 practically fixes the AC voltage difference between AVDD and AGND to be the same value. Therefore, the real difference of 4.8 mV from the two cases is the difference between 8.8 to 4.0 mV which is the voltage drop across VSS to ground pins of the substrate, V(Lgnd). This substrate ground noise is the sum of the power supply noise at the input of the filter when routing the filter as single-ended. With the large number of substrate ground pins in this design, the noise generate through VSS is considerably small. However, for a substrate that may have far fewer

ground pins, the noise could be substantial and the filter may have to be routed differentially.

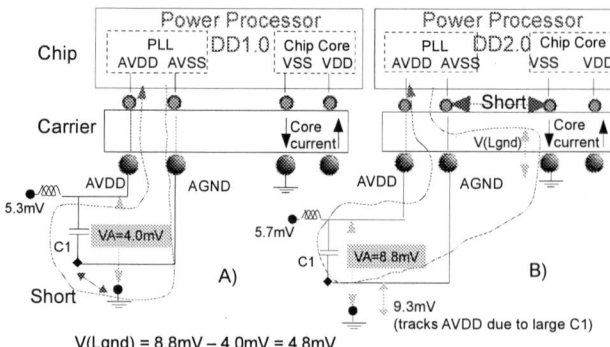

$$V(Lgnd) = 8.8mV - 4.0mV = 4.8mV$$

Figure 10: Current Return Path for Differential and Single-ended Filter

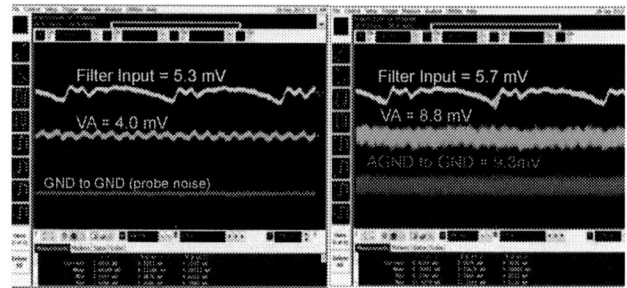

Figure 11: Ground Noise of Single-ended Filter

Noise Injection Technique

Artificial noise can be injected into the power supply to characterize the PLL PSNR. A probe point is required which must be placed inside the ASIC to bring out the output of the PLL into the system board. Probing too far down the path of the clock tree, passing too many MUXes, and gates, and at the system level can yield erroneous results due to unquantifiable interference. The noise can be generated from a signal generator as a single tone sinusoidal signal superimposed on a DC level equivalent to the PLL's AVDD voltage supply while the normal power to the PLL is disconnected from the system. While this approach is a straight forward one, it poses a problem during system power on. The system board usually enables several power supplies in a sequential order and is strictly followed to satisfy the requirement of the system power up. Turning the AVDD power supply on in the correct order in the sequence can be difficult in a lab environment.

Another approach might be to inject noise by AC coupling it into the AVDD pin. Figure 12 shows the set up of such an approach. Using a DC blocking capacitor of 2uF, the signal generator such as the HP 3324A is connected to the AVDD to provide voltage noise at varying frequencies. A high bandwidth scope such as the Infinium 9000 was used to measure the jitter at the PLL output and a second scope, a DSO 9254A, was used to monitor the voltage and frequency of the injected noise.

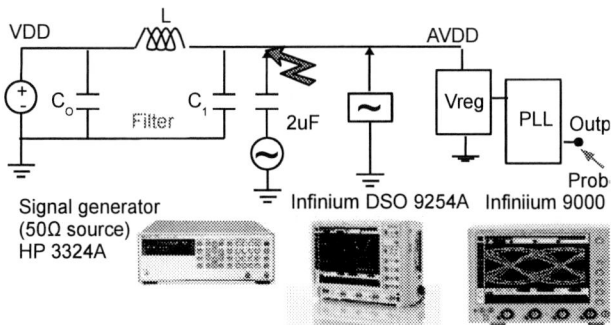

Figure 12: Noise Injection Set-up

The capacitor C1 must be removed in order to inject a large noise amplitude into the AVDD pin. For an example, Figure 13 shows cases with and without C1 capacitor. Without capacitor C1, hundreds of mV of noise can be injected, from KHz to 10 MHz.

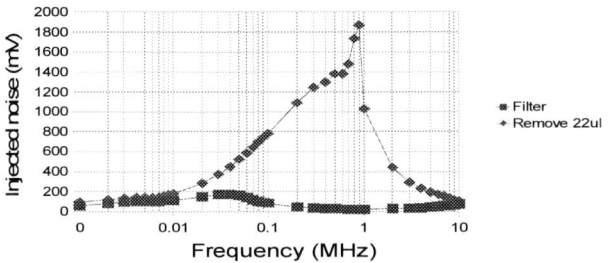

Figure 13: Possible Injected Noise Level

To inject a constant amplitude noise to the AVDD pin, the signal generator should be adjusted at each frequency to keep the AVDD noise level constant over the range of frequencies of interest. The second scope DSO 9254A is used for this purpose which monitors the amplitude and frequency at AVDD pin.

Each PLL clock generator provides a clock signal to different parts of the chip which require a unique clocking behavior as discussed earlier. Phase jitter is the deviation of the VCO output edges from ideal placement in time. This jitter affects I/O bus applications because it reduces the timing margin of the transmit and received signal within a bit time. The APLL works best for this case due to its phase jitter. Period jitter is the deviation from one cycle to the next. This may be important for core CPU operation due to its logical operation in clock cycle timing. The DPLL, therefore, is a suitable choice for this job. The time interval error (TIE) jitter which records the jitter from a long string of clocks and compares it against the idea one, was chosen in this lab measurement because it was a worst case scenario due to its cumulative effect and was easy to observe on the oscilloscope as a result of large injected noise level.

The following lab measurement results revealed the characteristic of the typical APLL and DPLL. Figure 14a shows the TIE jitter values as a function of frequency for different noise amplitude levels. AVDD is injected with 32, 47, and 75 mV over the frequency range of 0.1 to 10MHz. At 5 MHz, the jitter reached its maximum. As the noise

amplitude increased, as predicted by equation 2, jitter increased linearly as shown by Figure 14b. At lower frequencies, from 10 to 100KHz, the PLL showed a less sensitivity, even as large noise amplitudes were injected as shown in Figure 14c. Figure 14d compares the APLL and DPLL characteristic which have different frequency sensitivities. As expected, the APLL was typically less sensitive to noise than the DPLL.

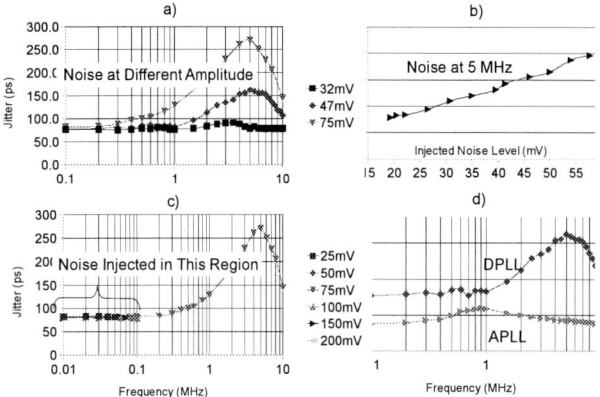

Figure 14: PLL Noise Rejection Curve

Measurement and Simulation Correlation

Figure 15 shows the simulation and measurement result of the DPLL. The effectiveness of the VR is apparently limited as the cut-off frequency occurs earlier than the PLL PSNR range resulting in a gap at 5 Mhz where the PLL is vulnerable to noise.

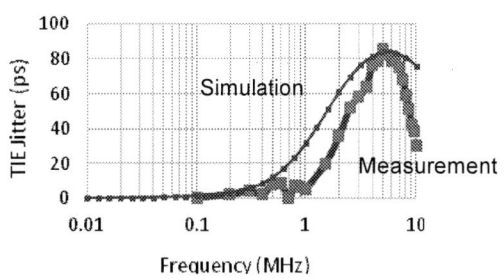

Figure 15: Measurement and Simulation Correlation

In this scenario, a system filter should be designed to cover this short coming. Figure 16 shows a lab measurement of the new filter design intended to lower the noise at this sensitive range. A constant AC coupled noise as described earlier was applied to the AVDD pin while a probe was also placed at the pin to monitor the voltage variation. Two ceramic capacitors, Co and C2, 0402 1uF are used to place a resonance near 5 MHz, and C1 is still kept a 22uF to control the resonances at lower frequencies. The new filter achieves a flat response up to 10MHz as compared to the old one as shown. Notice that this filter does not need the ferrite bead inductor because, as shown in Figure 9 earlier, the function of inductor L was to block noise coming from the system voltage regulator which is in the low frequency range. For system with noise coupled from high frequency activity, the filter is

978-1-4799-2408-0/14 $31.00 © 2014 IEEE   539

not as effective. In the multi-core ASIC, chip noise is primary source of interference that this filter becomes highly effective. The NoL curve is a compromise between the two noise sources, and it is optimized for the sensitive frequency of the PLL. In this design, as more 0402 1uF capacitors are added, the 5MHz frequency resonance is further suppressed to protect the PLL from both system and chip noise. The resonance at 5 MHz for the new filter was not a sharp dip as in the simulation curve of Figure 9 because the probe interference noise was masked the scope from detecting very low voltages in this measurement.

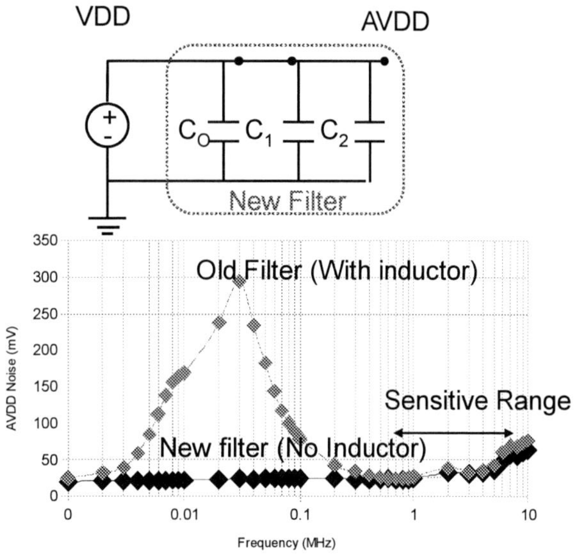

Figure 16: New Filter Design

Conclusions

PLL is known to be sensitive to power supply noise. The transfer function of noise injection to jitter can be computed base on theory or circuit simulation. A spectrum analyzer or real time scope can be used to measure PLL jitter as a result of power supply noise. When an artificial noise is injected, the real characteristic of the PLL in hardware is revealed which gives insight to a better method of designing the system filer. The approach was discussed and the technique was presented to support this methodology.

Acknowledgments

Jay Carman for lab measurement, James Jordan and Sungjun Chun for technical discussions

References

1. Dennis Fischette, website: www.delroy.com
2. John Abcarius, Sharon Wang. "Assess Power-Supply Noise Rejection in Low-Jitter PLL Clock Generators – Application note 4457, Maximintegrated Jun 03, 2009, www.maximintegrated.com
3. Faraydon Pakbaz "Determining-Simulating-Tx-Rx-TF-Jitter.pdf" IBM ASIC Application Note
4. Amr Hafez & Chih-Kong Yang "Design and Optimization of Multipath Ring Oscillators", IEEE Trans. On Circuits and Systems – Regular paper, Vol. 58, No. 10, Oct. 2011
5. Takahito Miyazaki et "A Performance Comparison of PLLs for Clock Generation Using Ring Oscillator VCO and LC Oscillator in a Digital CMOS Process" Proceedings of the 2004 Asia and South Pacific Design Automation Conference
6. J. S. Shor " Voltage Regulator Circuits for Low-Jitter PLL's with High PSSR (>40dB) in a Purely Digital Process 65nm Process", Proceedings of IEEE COMCAS 2008, May 13-14, 2008, Tel Aviv, Israel
7. Tzung-Je Lee and Chua-ChinWang "A Phase-Locked Loop with 30% Jitter Reduction Using Separate Regulators". Hindawi Publishing Corporation VLSI Design, Volume 2008, Article ID 512946, 8 pages, doi:10.1155/2008/512946
8. Hoh Heo & Hang Jeong "An Accurate Analysis of a VCO Voltage Regulator" 2008 International SoC Design Conference, page I-84

Coaxial Through-Package-Vias (TPVs) for Enhancing Power Integrity in 3D Double-side Glass Interposers

Gokul Kumar, P. Markondeya Raj, Jounghyun Cho*, Saumya Gandhi, Parthasarathi Chakraborti, Venky Sundaram, Joungho Kim* and Rao Tummala

3D Systems Packaging Research Center, Georgia Institute of Technology

* Terahertz Laboratory, KAIST, Daejeon, South Korea

Email: gokul.kumar@gatech.edu

ABSTRACT

Double-sided 3D glass interposers and packages, with through package vias (TPV) at the same pitch as TSVs in Si, have been proposed to achieve high bandwidth between logic and memory with benefits in cost, process complexity, testability and thermal over 3D IC stacks with TSV. However, such a 3D interposer introduces power distribution network (PDN) challenges due to increased power delivery path length and plane resonances. This paper investigates the use of coaxial through-package-vias (TPVs) with high dielectric constant liners as an effective method to deliver clean power within a 3D glass package, and provides design and fabrication guidelines to achieve the PDN target impedance. The Coaxial TPV structure is simulated using electromagnetic (EM) solvers and a simplified equivalent circuit model to study via impedance and parasitics. Test vehicles with anodized tantalum oxide capacitors were fabricated in ultra-thin, 100μm thick glass interposers to demonstrate process feasibility, with a capacitance density of 5 nF/mm^2. Self-impedance (Z11) of a 3D glass interposer containing the coaxial TPVs was analyzed with variations in (a) Via location, (b) Number of coaxial vias, and (c) Via capacitance and stack-up, to provide optimal PDN design guidelines. Based on the above parameters, the added decoupling vias achieved more than 30% impedance suppression over multiple resonance frequencies between 0.5-6 GHz, providing an effective and flexible PDN design method for double-side 3D glass interposers.

I. INTRODUCTION

Logic and memory stacking with through silicon vias (TSV) to form 3D ICs, with much higher interconnect density than the current package-on-package (PoP) stacking, has been proposed and is being developed to meet ultra-high bandwidth (25-100GB/s) demands of smart mobile systems with lower power consumption and miniaturization. However, the adoption of these 3D ICs has been delayed by many challenges that include thermal issues, testability and high cost. A simpler approach to achieve this high bandwidth using ultra-thin 3D glass packages has been proposed and fabricated by Georgia Tech [1] [2], which results in elimination of complex and costly TSVs in the logic die, as shown in Figure 1. Such a 3D glass approach uses an ultra-thin, 30-50μm glass interposer with stacked memory on one side and logic IC on the other side, interconnected by through package vias (TPVs) at the same pitch as TSVs in Si, and SMT mounted onto the printed wiring board (PWB) through solder ball interconnections [3]. While this technology offers cost, testability and thermal advantages over 3D IC stacking, it introduces a new challenge in power delivery, directly attributable to the long PDN path through lateral power-ground (P/G) planes. In addition, multi-mode plane resonances in glass interposers, that are common to high-

resistivity substrates including organic packages, require careful PDN design.

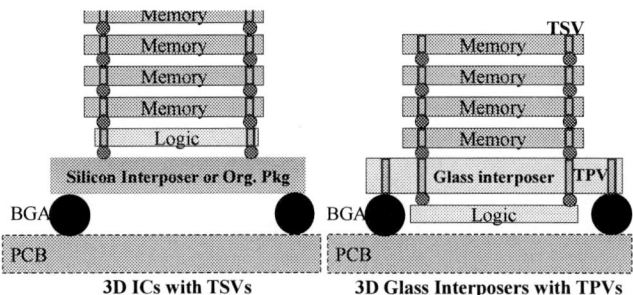

Figure 1 : Approaches for high bandwidth 3D-integration

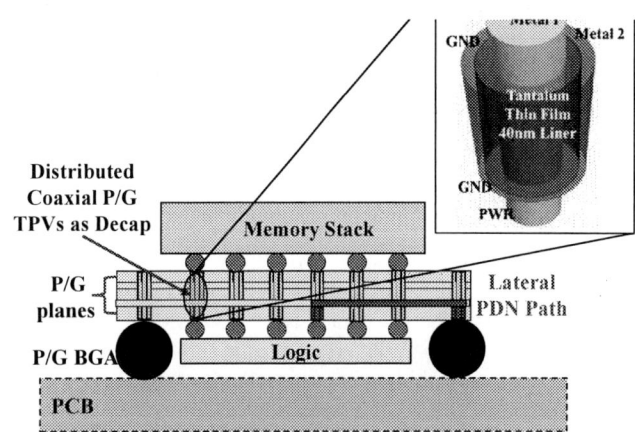

Figure 2 : 3D Glass interposer PDN with coaxial Vias as distributed decoupling capacitors

This paper, for the first time, integrates high-dielectric constant materials into TPVs in ultra-thin glass interposers to improve the PDN performance of double sided 3D interposers, and the techniques presented here can also be applied to single-chip 2D packages and multi-chip 2.5D interposers. Figure 2 illustrates the major power delivery challenges in 3D glass interposers and the proposed structure with coaxial TPVs to address some of these challenges. The coaxial TPVs with high dielectric constant thin films between the power and ground via conductors form ultra-miniaturized decoupling capacitors along the package power path. This distributed capacitance, placed very close to the active die, acts as charge reservoirs and presents an improved power delivery solution without ESL limitations or additional space requirements.

The authors previously reported the PDN characteristics of 3D double-side interposers with reduced power/ground ball

978-1-4799-2408-0/14 $31.00 © 2014 IEEE   541   2014 Electronic Components & Technology Conference

grid arrays (BGAs) due to the placement of logic die at the bottom of the substrate [4]. The effects of noise coupling between signal nets and power/ground planes at resonant frequencies , including the impact of signal return path discontinuity were studied [5]. Recent literature compared the effects of simultaneous switching noise and signal discontinuity in silicon and glass interposers using M-FDM methods [6, 7]. Embedded and die-integrated decoupling schemes have been proposed with glass and silicon interposers to improve their power delivery profile [8-10]. Recent studies employed three-dimensional P/G coaxial vias (TSVs) in 3DIC stacks for added coupling capacitance [11]. However, effect of via placement and physical geometry in the power and signal integrity of double-sided flip-chip interposers has not been investigated. This paper goes beyond published literature to present a detailed study of the PDN impedance profile in 3D glass interposers including the effects of via interconnection capacitances. In addition, it explores and demonstrates the use of coaxial through-package-vias as an effective design technique to improve power delivery networks.

Table 1: 3D interposer system dimensions

Coaxial TPV Diameter	30-60 μm
Tantalum Liner Thickness	40nm
Ultra-thin Core Thickness	30-100μm
Interposer/Package size	17mm x17mm
PWB size	25mm x25mm
BGA pitch	400 μm

Following this introduction, Section II describes the modeling and design of coaxial vias. Parasitics for different technologies are examined and extracted using full-wave electromagnetic (EM) solvers, forming building blocks for the subsequent analysis. An equivalent circuit model is proposed to identify the key parameters contributing to the capacitance, and the results are tabulated. Section III deals with a discussion of various dielectric options, selection of the front-up materials, and an initial process demonstration for the TPV with capacitors in 3D glass interposers. Section IV studies the benefit of the proposed coaxial vias to address PDN challenge by modeling the PDN response with and without decoupling vias from the previous sections. Based on this investigation, design guidelines for an optimal P/G TPV physical design and dimensions are provided. Section V summarizes the results from this study.

II. ELECTRICAL DESIGN OF COAXIAL THROUGH PACKAGE VIAS

This section examines via parasitics for different coaxial TPV process methods and stack-up technologies, and presents a simplified circuit model to analyze coaxial via interconnects. The parasitic properties of the coaxial vias that are most critical for PDN design, including the inductance, capacitance and total via impedance are analyzed first. Both analytical equations and an electromagnetic solver (Q3D) is used for this study for parametric analysis. Since the overall resistance of thick coaxial vias are extremely small, they are not focused in this section. Integration of high-k thinfilm materials such as

tantalum oxide and barium titanate between the inner conductor and outer shell of coaxial TPVs in an ultra-thin glass substrate is studied provide up to 50x higher via decoupling capacitance. Figure 3 shows the electrical block diagram of the 3D interposer PDN including coaxial TPVs.

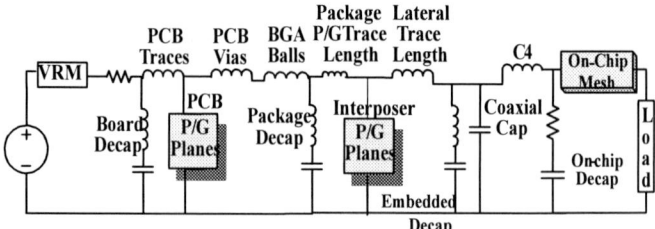

Figure 3: Schematic Diagram of 3D interposer PDN

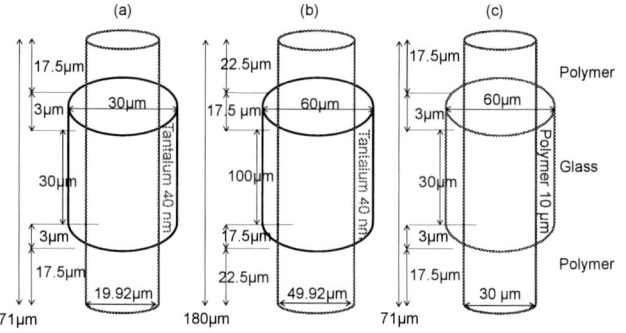

Figure 4: Co-axial Via Dimensions considered in this Study - 40nm tantalum oxide liner with (a) 30μm Glass thickness (b) 100μm thickness, 15μm polymer liner with (c) 30μm glass thickness

Table 2: Summary of coaxial via capacitance

Liner material	ε	Glass Thickness (μm)	Theoretical Value (pF)	Q3D (pF)
Tantalum Oxide	25	30	12.4	12.7
		100	117	120
Polymer	3.01	30	0.0117	0.018

Table 3: Summary of coaxial via loop inductance

Glass Thickness (μm)	Outer Radius (μm)	Inner Radius (μm)	Theoretical Value (pH)	Q3D (pH)
30	10	9.96	0.028	0.022
100	25	24.96	0.042	0.032
30 (polymer)	25	15	3.63	4.7

Considering current technology and future miniaturization, three coaxial TPV physical configurations are chosen as shown in Figure 4. Two variations of total TPV height (71 μm and 180 μm) in ultra-thin glass substrates (including dielectric and glass sections) is considered with both moderate-K(Tantalum oxide) and polymer liner configurations. The target metallization thickness was defined to be 5 μm.

978-1-4799-2408-0/14 $31.00 © 2014 IEEE   542

From literature, simplified analytical expressions for capacitance and loop inductance per unit length of a coaxial via can be given as

$$C_L = \frac{2\pi\varepsilon_0\varepsilon_r}{ln\left(\frac{D}{d}\right)} \qquad F/m$$

$$L_L = \frac{\mu}{2\pi} \ln(D/d) \qquad H/m$$

D is the outer via radius

d is the radius of the inner via

$\varepsilon_0, \varepsilon_r$ is the permittivity of free space and TPV liner

μ is the permeability

Table 2 presents a summary of various capacitance values that are extracted for coaxial vias. As expected, it can be seen that the maximum value of capacitance is achieved with the 100μm glass via having the tantalum liner due to the highest capacitance density and surface area. Table 3 details the loop inductance of coaxial TPVs. It can be seen that the loop inductance of the p/g via is extremely small due to the thin coaxial liner, when compared with the loop inductance of a polymer based coaxial TPV. These results indicate useful PDN applications with the proposed P/G TPV configurations. Based on the extracted values, a simplified equivalent circuit model for glass TPVs was constructed as shown in Figure 5. It can be seen that the dominant capacitance is from the thin high-K liner material that is separated by the power and ground conductors, and has a theoretical value in several tens of pico farads.

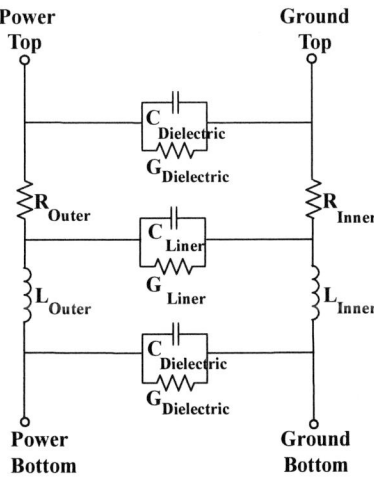

Figure 5: Circuit model for Coaxial P/G Glass TPVs

The impedance profile between the power and ground vias of a single coaxial TPV from different configurations was simulated using a full wave EM solver (HFSS) as shown in Figure 6. The self-impedance is measured between via pads of the top two metal layers as shown. It was seen that lowest impedance was observed with 100μm coaxial TPV due to its large surface area and length of the capacitor, followed by the ultra-thin 30 μm TPV. The impedance of polymer based coaxial vias were several orders higher than the other configurations, with smaller capacitive values due to larger separation between the metal plates. The effective capacitance of coaxial vias can be increased by connecting a number of the coaxial P/G vias in parallel, similar to the configuration present in interposer applications. The higher capacitance vias exhibit

decoupling at lower frequencies, enabling selective tuning to meet target impedance.

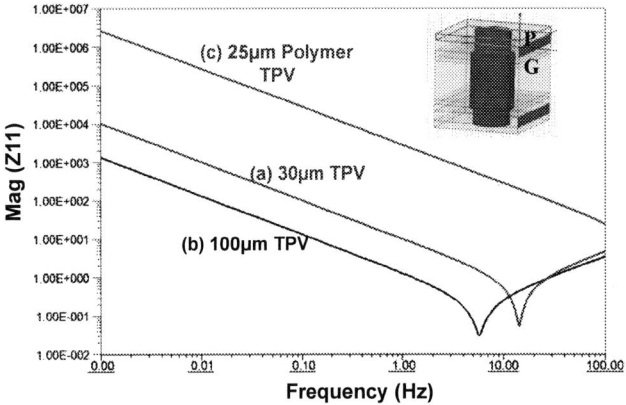

Figure 6: Self-Impedance profile of single P/G coaxial via

III. POWER-GROUND TPV: MATERIALS AND PROCESSES

3D interposers need to support low impedance power-ground coaxial TPVs for efficient power supply. Based on the simulations shown in the previous section, the power-ground needs a capacitance in the order of 50-100 pF per via. This section deals with a discussion on various dielectric options, selection of the front-up materials, and an initial process demonstration for the TPV capacitors in 3D glass interposers.

Table 4: Dielectric options for TPV capacitors in 3D interposers, and their merits and demerits

Dielectrics	Film properties	Merits	Demerits
Oxides Oxynitrides	$\varepsilon = 5\text{–}7$ t = 20 nm 2.5 nF/mm²	High BDV allows 30 nm films Standard semiconduct or tools	Expensive semiconductor tools, Lower capacitance densities;
Moderate K paraelectrics -Tantalum oxide	$\varepsilon = 25$ t = 40 nm 5 nF/mm²	Simpler processing compared to high K films	Moderate capacitance densities
High K super paraelectrics - BST	$\varepsilon = 300$ t = 150 nm 15-20 nF/mm²	Dielectric constant low but stable with lower loss	Conformal dielectric is a challenge with standard processes
High K ferroelectrics - Barium titanate	$\varepsilon = 3000$ t = 1 μm 30 nF/mm²	High density	Higher temperature processing, Lower reliability and leakage, Conformal dielectric is a challenge with standard processes

Dielectric options

Low k films: In silicon trench and TSV capacitors, oxides, nitrides and oxynitrides of silicon with permittivies of 5-7 are widely used. The Breakdown Voltages (BDVs) of these films range from 0.8-1 V/nm which represent the highest BDV of all materials. Even a 20 nm film can stand voltage of 15 V.

978-1-4799-2408-0/14 $31.00 © 2014 IEEE   543

Furthermore, these materials present minimal reliability concerns because of their intrinsic structure and properties. This leads to a planar capacitance density of about 2-3 nF/mm². Moreover, these technologies are dependent on very standard front-end semiconductor tools such as Plasma Enhanced Chemical Vapor Deposition (PECVD), Liquid Phase Chemical Vapor Deposition (LPCVD). The deposition temperatures are less than 300 C, and no issues of adhesion and thermomechanical compatibility arise because of their strong covalent bonding to the electrode and good CTE match with glass. However, higher capacitance densities with low-cost panel processes are sought for 3D glass interposers.

Moderate k films: Oxides of tantalum, hafnium, zirconia, titania etc. show permittivities of 15-80. Some of these have breakdown strengths of 500-800 V/micron. They can be deposited thinner down to 30-50 nm showing capacitance densities approaching 5-10 nF/mm². Moderate k films represent 2-3 X improvement in capacitance density compared to traditional oxides and nitrides. The most common technique for depositing these thin films in trenches is Atomic Layer Deposition (ALD), which also has cost and through-put limitations.

High k films: Ferroelectrics show permittivities of 1000-5000 in 1-3 micron thickness, with a capacitance density of about 30 nF/mm². Films with these properties require extremely high temperature processes (~1000 ℃). The permittivity reduces dramatically for thinner films, and therefore the capacitance density does not improve much. With glass TPV-compatible processes (<650 ℃), using lead-free dielectrics, films in thicknesses of 100-200 nm and permittivities of 200-300 are feasible. The capacitance density is less than 15-20 nF/mm² with these systems. However, limitations of high-k film technologies arise from several reasons:

- Non-conformal deposition processes such as sputtering. Standard processes result in non-uniform dielectric thickness as shown in Figure 7a and 7b.
- Higher temperature for crystallization
- Lower BDV requiring thicker films
- Thickness-dependent permittivity which results in diminishing returns of capacitance density with thickness reduction
- Intrinsic reliability issues from lattice defects and interfaces

The capacitance densities achieved with various TPV dielectrics are compiled in Table 4. Based on this summary, anodized dielectrics provide the best combination of capacitance density, process-compatibility with glass, and scalability to large panels. Hence, anodized tantalum oxide TPV capacitors are chosen as the front-up option to be demonstrated.

TPV Capacitor Process:

Silicon trenches and vias are formed from dry-etching processes using inductive-coupled plasma with fluorine gas, which then diffuses to the low-pressure etching chamber, where silicon is directionally etched. The most common etching is done with Bosch process. In the Bosch process, sidewall passivation with a polymer is utilized to prevent lateral etching and give directionality to the etching process.

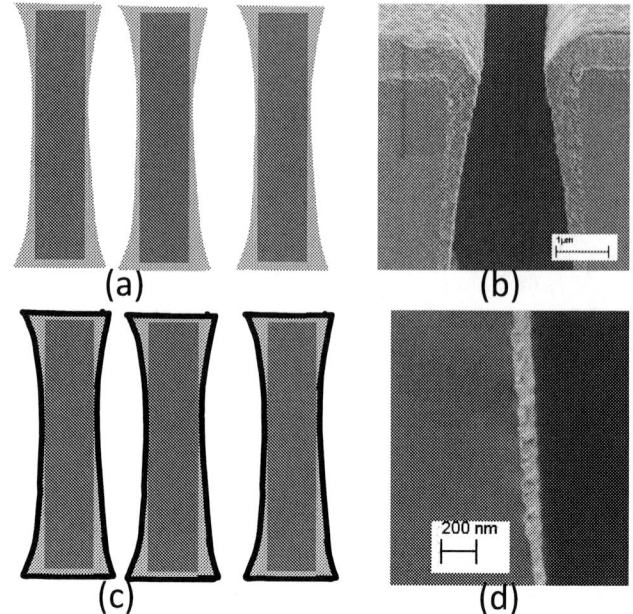

Figure 7: Sputtered dielectrics with non-uniform thickness (a), cross-section of sputtered barium titanate (b). Sputtered metal with conformal and uniform anodized films (c) and cross-section of anodized tantalum oxide with uniform dielectric thickness (d).

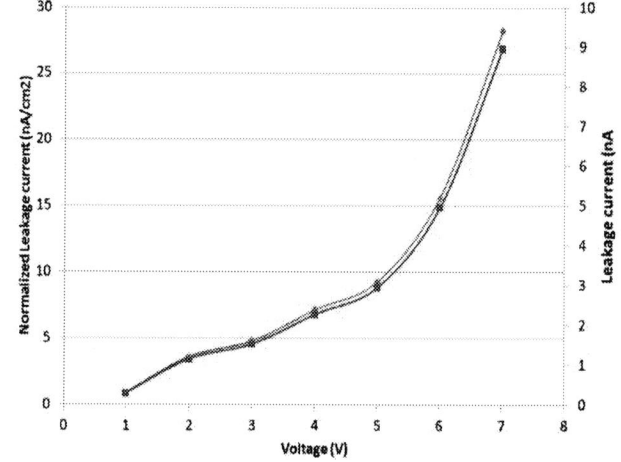

Figure 8: Leakage current with anodized tantalum dielectrics.

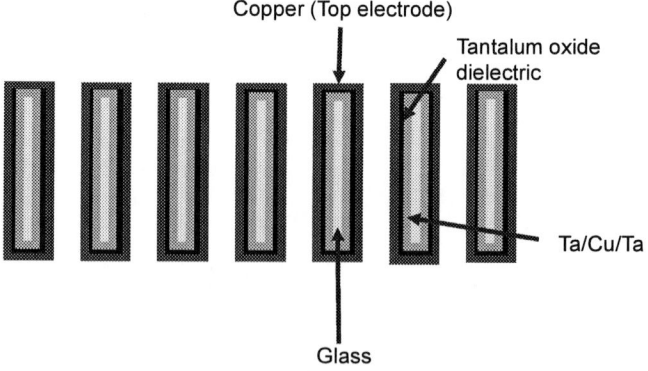

Figure 9: Schematic of fabricated coaxial via in glass interposers

978-1-4799-2408-0/14 $31.00 © 2014 IEEE   544

For the glass TPVs, GT-PRC and its partners have demonstrated excimer laser-based via formation process with much lower cost, followed by electroless or sputtered copper metallization. The process can achieve aspect ratios of 5-10.

To form tantalum oxide dielectrics in TPVs, tantalum is sputtered with a DC-magnetron. A pressure of 6 mtorr and power of 350 W is used in the current work, to yield a sputtering rate of 1-2 A/sec. The coverage with tantalum can be further improved with an inductively-coupled power supply. The tantalum film is subsequently anodized in citric acid to form conformal tantalum oxide films, as shown in Figure 7c and 7d. The resulting film has a dielectric constant of 21-28. For anodization at 8 V in potentiostatic mode, this yields a conformal dielectric of about 30-35 nm, with a capacitance density of 5-6 nF/mm^2. The capacitance densities, though not as high as that for ferroelectric films, are stable at GHz frequencies due to paraelectric behavior and can effectively address broadband decoupling needs. The films show a low leakage current of 3-5 nA/cm^2 at the typical operating voltage of 2-3 V as shown in Figure 8. A cross-section of the TPV structure is shown in Figure 9. The top electrode is sputtered and subtractive-patterned to yield the TPV capacitor test-structures as shown by the process test-vehicle in Figure 10.

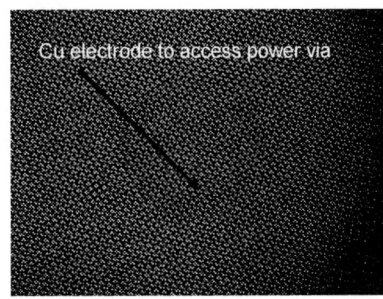

Figure 10: Top View of the fabricated sample

IV. PDN IMPEDANCE ANALYSIS FOR 3D GLASS INTEROSERS

This section describes the benefits of the proposed coaxial vias for PDN applications, by modeling the PDN response with and without decoupling vias from previous sections.

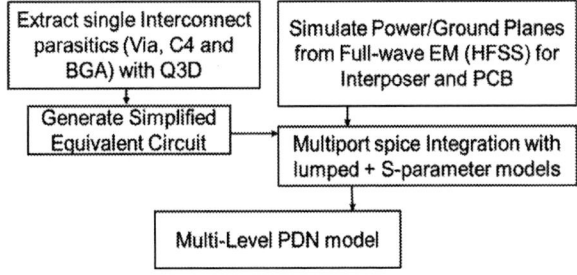

Figure 11: Flow Chart for the segmented PDN approach

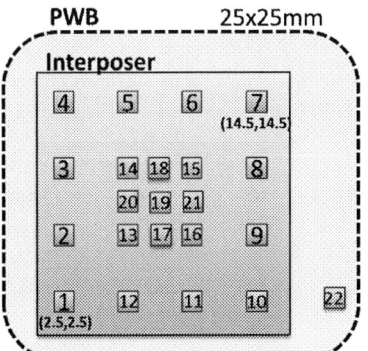

Figure 12: PDN Port Locations

System PDN Impedance Modeling

The simulated system configuration consists of a four-metal layer glass interposer with dimensions as shown in Table 1. Since modeling the entire PDN including vias and PWB interconnects is challenging due to simulation time, a combination of segments from full-wave simulations and spice integration is used to perform PDN system analysis. This method, described in detail in [4] and figure 11, was used to compute the total PDN impedance profile in frequency domain, through integration of distributed ports across the package and PWB planes. The PDN impedance of individual package and PWB planes is modeled using a 3D EM solver (HFSS). The multi-port P/G plane impedance models are then integrated with BGA and via lumped parasitics at appropriate ports depending on the configuration. Figure 12 presents numbered port locations for BGA and coaxial via group net placement. In this PDN study, the interposer is considered to be always connected to the PWB through BGAs and Vias, in order to model the impact of TPVs and BGA variations. The assumptions for BGA placement in this study are the same as [4] and table 6, and the vias are placed from ports 1-20 on both the interposer and PWB, with the impedance profile studied at port 19 of the interposer.

Table 5: Dielectric properties of substrate materials

Material	Dielectric Constant, ε@ 1MHz	Loss tangent, tan δ
Borosilicate Glass	5.3	0.004
Polymer Laminate	3.01	0.005
Tantalum oxide	25	0.001

Table 6: BGA Port Configuration

Configuration	No. of P/G BGAs	PKG to PWB port connections	L(pH) /port
Full BGA array	700	1-16	7.69
3D Interposer	300	1-12	4.7

PDN impact due to 3D interposers

Previously [4], it was shown that the loop inductance for this double-side 3D interposer-package was trace-dominated and the reduction of interposer P/G BGAs due to the bottom die placement had minimal impact on the system PDN self-impedance at low frequencies. Figure 13 shows the PDN profile comparison between glass interposers with fully populated BGA arrays and centrally depopulated BGA's with reduced second level P/G interconnections to represent the 3D interposer configuration. The 3D interposer PDN generates high frequency (8-10 GHz) impedance peaks from to parallel resonance modes. This phenomenon is proposed to occur due to multi-mode coupling between the additional lateral plane inductance on the interposer package and the package capacitance.

PDN impact due to Decoupling Coaxial Vias

A comparison of the PDN profile before and after the addition of decoupling vias is presented in Figure 14. For the purpose of this study, and based on the analysis from section II, decoupling vias with a capacitance of 30pF/via are considered. Based on the system configuration, 2000 of these coaxial vias are assumed to be present in parallel, as a part of the interposer PDN. Both the full-array interposer and 3D interposer PDN have shifted series resonances occurring at 100 MHz due to the series coupling between the decoupling capacitance and the plane capacitance. The addition of decoupling vias completely suppresses the 3D interposer resonances around 10 GHz and shifts the resonance to lower frequencies, presenting a clear benefit in power delivery design.

PDN impact due to Decoupling Coaxial Vias Configuration

Figure 15 presents the comparison of the PDN profile based on the variation of decoupling vias distribution on the 3D interposer. For the purpose of this study, we consider three types of arrangements based on the coaxial via distribution; (a) uniform distribution across the entire P/G plane, (b) Staggered distribution at the center and periphery, and (c) coaxial vias on the periphery only. All the three cases have the same amount of decoupling vias (2000), but are only different in their distribution as shown in table 7. It can be seen that the uniform distribution presents the lowest impedance profile, followed by staggered and peripheral configurations. This is due to the minimization of loop inductances due to localized capacitance action.

Table 7: Coaxial Via Port Configuration

Via Distribution	PKG to PWB port connections
Uniform	1-20
Staggered	1-16
Peripheral	1-12

PDN impact due to Decoupling Coaxial via Capacitance

Figure 16 presents the comparison of the PDN profile based on the variation of decoupling vias capacitance on the 3D interposer due to different process and stack-up technologies. For the purpose of this study, three types of arrangements are considered, - (a) 30pF/via, representing a high capacitance

density P/G via in glass (b) 3pF/via, representing current generation P/G capacitance from tantalum deposition on glass, and (c) 25fF/via, representing a polymer based coaxial vias. It can be seen that each capacitance technology will provide decoupling capacitances at specific frequencies based on their resonance frequencies, and the highest PDN benefit can be obtained by having the largest capacitance near the chip probing location.

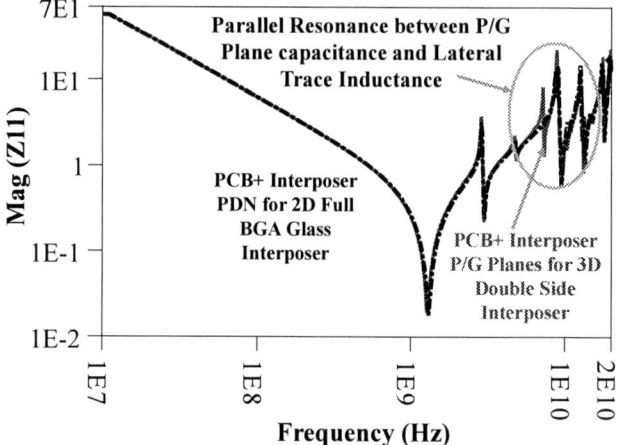

Figure 13: PDN impact due to 3D interposers

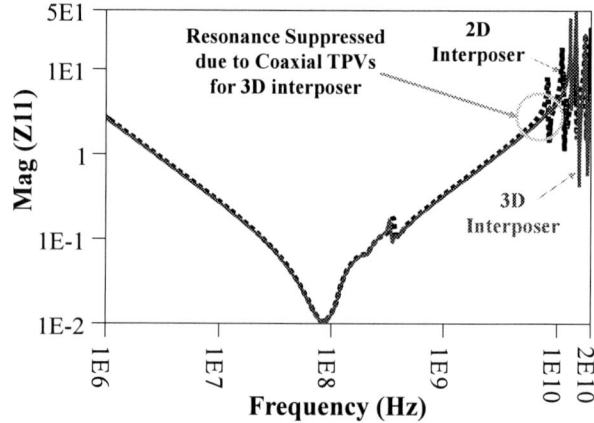

Figure 14: Resonance suppression due to coaxial vias

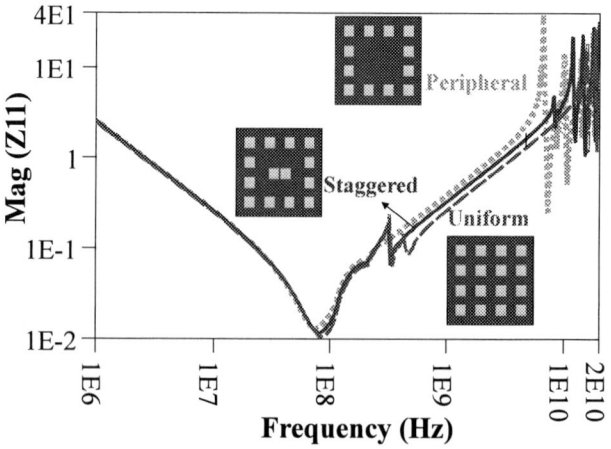

Figure 15: PDN impact due to via placement variations

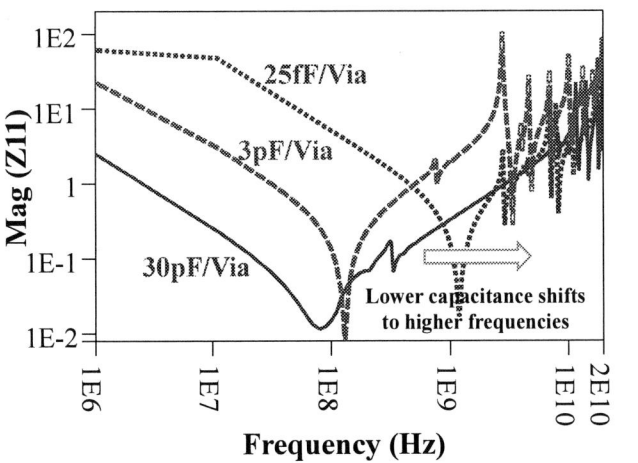

Figure 16: PDN impact due to decoupling via capacitance variation

V. SUMMARY AND CONCLUSION

This paper examines the modeling, analysis, design, and fabrication of coaxial TPVs to improve power integrity of 3D glass interposers. The addition of co-axial TPVs generated a decoupling network between the P/G planes, thus reducing the magnitude of PDN resonances. Based on the proposed structure, a uniform grid distribution of coaxial vias was found to have the maximum benefit to suppress 3-D system self-impedance. Cross-sections from the process test vehicles based on ultra-thin 100μm glass samples containing coaxial TPVs were fabricated. Overall, the proposed coaxial method enables effective PDN design in 3D interposers, offering a simpler and scalable approach to improve power integrity.

Acknowledgments

The authors would like to acknowledge the LGIP consortium industry members for their support and Srikrishna Sitaraman and Vijay Sukumaran for useful discussions.

References

[1]   G. Kumar, T. Bandyopadhyay, V. Sukumaran, V. Sundaram, L. Sung-Kyu, and R. Tummala, "Ultra-high I/O density glass/silicon interposers for high bandwidth smart mobile applications," in *Electronic Components and Technology Conference (ECTC), 2011 IEEE 61st*, 2011, pp. 217-223.

[2]   V. Sukumaran, G. Kumar, K. Ramachandran, Y. Suzuki, K. Demir, Y. Sato*, et al.*, "Design, Fabrication, and Characterization of Ultrathin 3-D Glass Interposers With Through-Package-Vias at Same Pitch as TSVs in Silicon," *Components, Packaging and Manufacturing Technology, IEEE Transactions on,* vol. PP, pp. 1-1, 2014.

[3]   V. Sundaram et al, ""Ultra-thin interposer assembly with through vias", May 3 2012.

[4]   G. Kumar, S. Sitaraman, C. Jonghyun, K. Sung Jin, V. Sundaram, K. Joungho*, et al.*, "Power delivery network analysis of 3D double-side glass interposers for high bandwidth applications," in *Electronic Components and Technology Conference (ECTC), 2013 IEEE 63rd*, 2013, pp. 1100-1108.

[5]   C. Jonghyun, K. Youngwoo, K. Joungho, V. Sundaram, and R. Tummala, "Analysis of glass interposer PDN and proposal of PDN resonance suppression methods," in *3D Systems Integration Conference (3DIC), 2013 IEEE International*, 2013, pp. 1-5.

[6]   V. Sridharan, M. Swaminathan, and T. Bandyopadhyay, "Enhancing Signal and Power Integrity Using Double Sided Silicon Interposer," *Microwave and Wireless Components Letters, IEEE,* vol. 21, pp. 598-600, 2011.

[7]   X. Biancun and M. Swaminathan, "Modeling and analysis of SSN in silicon and glass interposers for 3D systems," in *Electrical Performance of Electronic Packaging and Systems (EPEPS), 2012 IEEE 21st Conference on*, 2012, pp. 268-271.

[8]   W. Yushu, X. Shu, M. R. Pulugurtha, H. Sharma, B. Williams, and R. Tummala, "All-Solution Thin-film Capacitors and Their Deposition in Trench and Through-Via Structures," *Components, Packaging and Manufacturing Technology, IEEE Transactions on,* vol. 3, pp. 688-695, 2013.

[9]   S. Gandhi, X. Shu, P. M. Raj, V. Sundaram, M. Swaminathan, and R. Tummala, "A low-cost approach to high-k thinfilm decoupling capacitors on silicon and glass interposers," in *Electronic Components and Technology Conference (ECTC), 2012 IEEE 62nd*, 2012, pp. 1356-1360.

[10] S. Gandhi, P. M. Raj, V. Sundaram, H. Sharma, M. Swaminathan, and R. Tummala, "A new approach to power integrity with thinfilm capacitors in 3D IPAC functional module," in *Electronic Components and Technology Conference (ECTC), 2013 IEEE 63rd*, 2013, pp. 1197-1203.

[11] N. H. Khan, S. M. Alam, and S. Hassoun, "Power Delivery Design for 3-D ICs Using Different Through-Silicon Via (TSV) Technologies," *Very Large Scale Integration (VLSI) Systems, IEEE Transactions on,* vol. 19, pp. 647-658, 2011.

Modeling of Switching Noise and Coupling in Multiple Chips of 3D TSV-Based Systems

Huanyu He[1], Xiaoxiong Gu[2], Jian-Qiang Lu[1]

[1]Rensselaer Polytechnic Institute, Troy, New York 12180, USA

[2]IBM T.J. Watson Research Center, Yorktown Heights, NY 10598, USA

heh2@rpi.edu, 1-(518)-951-4812

Abstract

This paper reports on modeling of simultaneous switching noise (SSN) in power distribution network (PDN) for 3D systems, where multiple IC chips are stacked and connected by through-silicon vias (TSVs). The noises generated by current switching during the transition from idle state to active state are analyzed with both on-chip and off-chip PDNs. SSN is decomposed into different frequency components and their characteristics are discussed. The switching noises in active and silent chips are extracted to evaluate the noise interference in a 3D chip stack. Decoupling capacitors and stagger intervals are used to suppress the noise. Modeling the PDN to minimize the switching noise effects based on our hybrid approach is demonstrated to be effective to analyze the 3D PDN and understand the design tradeoffs in 3D architectures.

I. Introduction

Design and optimization of power delivery in 3D integration is increasingly important for design of various 3D integrated systems. A key element for 3D integration is the through silicon via (TSV), which provides massive interconnections between stacked chips, and shortens the signal and power delivery paths, hence further enhances data bandwidth and system performance [1,2].

To realize the tremendous benefits in 3D integrated systems, it is critical to understand power integrity issues due to the accumulated current consumed by multiple ICs, such as large di/dt noises caused by the high slew rate and parasitic inductance, small voltage margins caused by the continuingly scaling supply voltage, and *IR* drops caused by the component resistance [3,4]. Several critical issues of the power delivery in 3D chip stacks have been addressed [5-7].

This paper discusses the characteristics of simultaneous switching noise (SSN) in 3D systems at the presence of off-chip PDN (e.g., package and board). The voltage noise is generated by the IC's transition from idle state to active state in addition to the transistors' switching. A number of factors are considered including decoupling capacitors, current switching patterns, and power consumption configurations in 3D chip stacks, especially the arrangement for high-power IC and low-power IC and their interference. A hybrid approach of combining full-wave modeling and circuit simulation is adopted to model both on-chip and off-chip PDNs, supporting the simulation for lower frequency and longer duration.

II. 3D IC Architecture and Models

Power delivery has a hierarchical structure with different levels of PDNs that differ in size and impedance response and spread out at various locations in the system [8].

The baseline 3D TSV-based architecture used in this work is shown in Fig. 1 with emphasis on the 3D chip stack, where four ICs/chips are vertically stacked and interconnected with

TSVs and micro-bumps. Flip-chip bonding is used to connect the 3D chip stack and a silicon interposer. The chip stack and interposer are packaged and mounted on a PCB board. Each chip could be a microprocessor, memory, or other types of chips for different 3D architectures, and hence with different power consumption features.

In general, the on-chip power grid is distributed over multiple metal layers [9]: the upper two metal layers are the thickest ones with lowest resistance, which are modeled as orthogonal copper power/ground (P/G) lines as shown in Fig. 2. In fact, the power grid must extend down to metal1 or metal2 in order to maintain supply voltage for the on-chip devices. For the sake of simplicity, the effects of short metal lines and vias in lower metal layers are omitted. The on-chip grid mesh is 512-µm-by-512-µm in dimension, which contains 130×130 orthogonal P/G grid lines. P/G grid lines are connected through local Cu vias. Power grid meshes in adjacent chip layers are connected with P/G TSVs and micro-bumps. The TSV has a diameter of 5 µm with 50 µm in length. Each TSV is surrounded by a 0.1-µm-thick silicon dioxide isolation layer.

A hybrid modeling approach is used, which combines full-wave electromagnetic (EM) modeling and SPICE simulation in two steps [10]. In the partitioning step, the basic components of the on-chip PDN such as TSV, local via, P/G grid line, and micro-bump are extracted and modeled with full-wave EM solver. The EM solver obtains the impedance, admittance, and scattering matrices among the ports for each component. In the assembly step, the complete circuit network of the on-chip PDN is assembled with the equivalent circuit model of each component according to the specific 3D power delivery architecture. The off-chip PDN model is built based on a commercial micro-processor test board [11], which directly connects to all the solder balls beneath the silicon interposer.

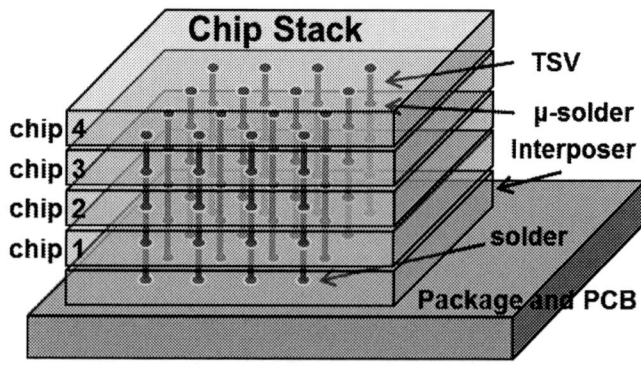

Figure 1. The baseline 3D stacked IC structure. Four identical Si ($\rho = 10\ \Omega\cdot$cm) chips are vertically stacked on a silicon interposer with Cu TSVs ($\rho = 1.82\ \mu\Omega\cdot$cm) and micro-bumps ($\rho = 14.3\ \mu\Omega\cdot$cm).

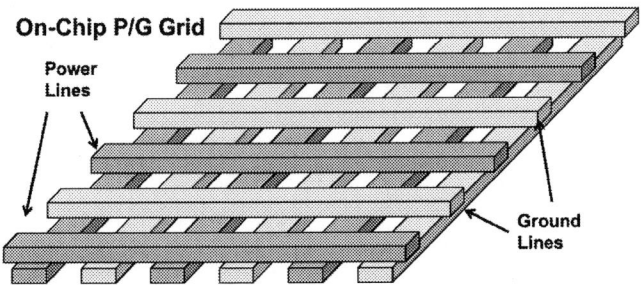

Figure 2. Illustration of on-chip power grid (partial). Power grid consists of Cu lines (width = 2 μm, pitch = 4 μm, thickness = 1 μm) and local Cu vias (1 μm × 1 μm × 1 μm).

In the PDN circuit model, the on-chip P/G grid wires and local vias are represented by series of inductance and resistance. On-chip devices are simplified in terms of unit cells, each of which consists of a decoupling capacitor (and its parasitic inductance and resistance) and a current load attached to each node of the power grid mesh as shown in Fig. 3. Power grid meshes in different layers are connected through P/G TSVs. All the solder balls beneath the interposer TSVs connect with the same node on the off-chip PDN, which consists of the lumped element models of package, PCB, and VRM, etc. The scheme of a full-PDN circuit model is shown in Fig. 4 with detailed off-chip PDN.

Distributed-element models are used for the on-chip PDN with each component modeled by full-wave EM solver. In such way, the circuit model can accurately capture the components' impact on the resonant behaviors among different chip layers or within a single chip layer.

The simplified lumped-element model of the off-chip PDN does not impair the accuracy in modeling of the impedance response at high frequencies. The currents at high frequencies are confined within the on-chip PDN so only the on-chip PDN matters. This can be attributed to the large

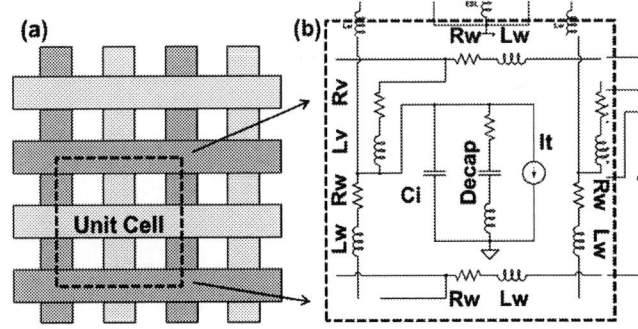

Figure 3. Unit cell model includes the on-chip P/G grid wires, P/G vias, decoupling capacitor (Decap) and current load (It).

inductance of the off-chip PDN that in fact isolates the on-chip PDN at high frequencies due to skin effect and parasitic inductance. At lower frequencies, the wavelength is much larger than the dimension of the 3D chip stack, which suggests that the whole chip stack shares the same phase and behaves as a few lumped elements. As shown in section IV and V, a simplified lumped-element model of the off-chip PDN is good enough for evaluating the impact of the on-chip PDN components.

III. Modeling of Switching Activity and Noise

The simultaneous switching noise strongly depends on the current switching pattern and the PDN impedance. The current drawn from PDN contains different frequency components due to transistor switching and idle-to-active transition. The switching current waveform depends on the IC function, workload, application, performance, etc. Without loss of generality, in our model each chip could be the aggressor chip or the victim chip, which only differ in the current/power consumption.

We assume a triangular current waveform for each current load synchronized by a constant clock signal (3 GHz). Within

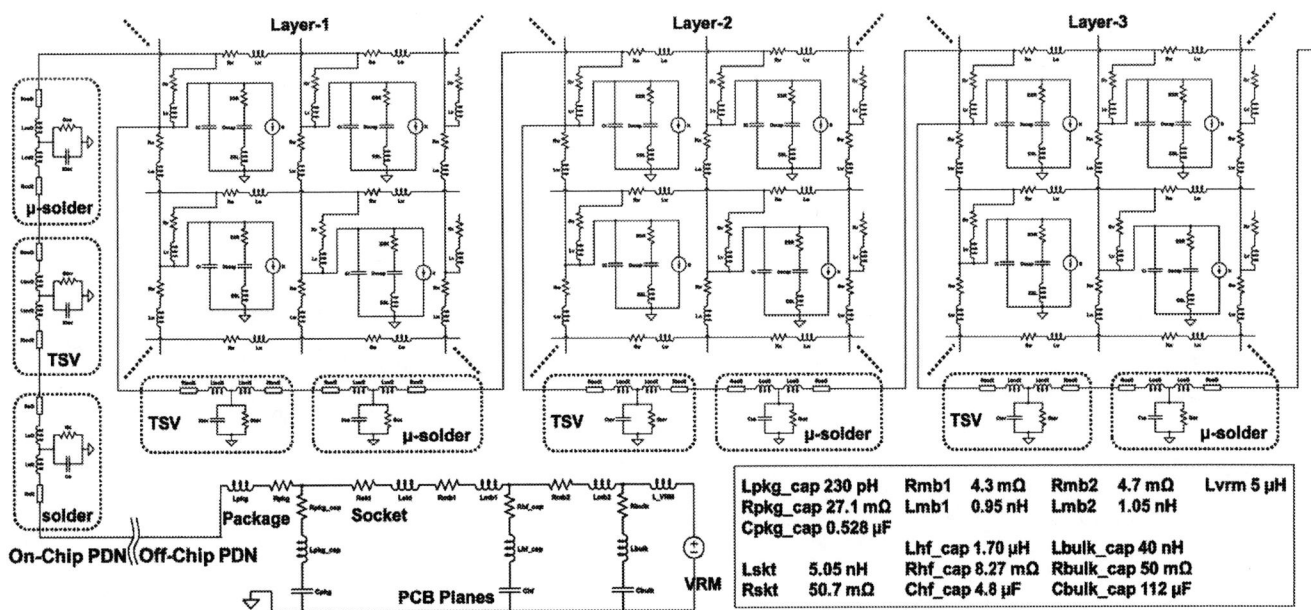

Figure. 4. Scheme of full-PDN equivalent circuit model. The on-chip PDN uses a distributed element model for multiple chip layers, whereas the off-chip PDN uses a lumped element model. The value for each lumped element is shown as well.

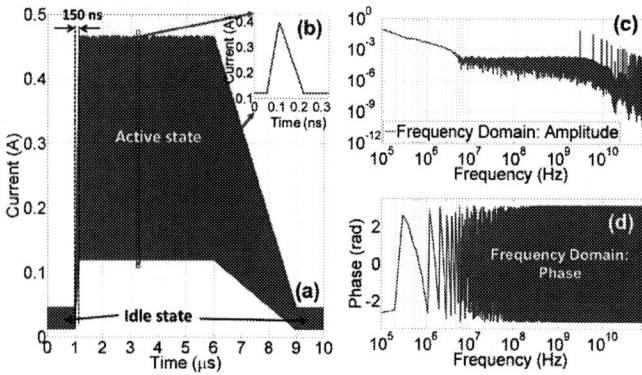

Figure 5. Switching current when IC transiting between idle state and active state in the time-domain (a) - (b), and in the frequency domain (c) - (d).

a clock cycle, the switching current waveform has a rising time of 55 ps and a falling time of 111 ps. For the active chip layer, the maximum current density is 1.5 A/mm^2, which results in a maximum total current of 0.4 A, The minimum total current is 0.12 A. The current waveform is shown in Fig. 5(b).

Microprocessors often adopt power-saving techniques such as clock gating, power gating and dynamic supply voltage, which can significantly reduce power consumption [10]. Using these techniques, the microprocessor dithers between different states of performance and power consumption to save more power. Consequently, the IC transition from idle state to active state also causes transient noise along with the transistor switching. The IC state transition is modeled as a linear increasing from 10% to 100% within 150 ns starting at $t = 1\mu$s, as shown in Fig. 5(a). The simulated duration is 10 μs. The time-domain waveform of the switching current is transformed into frequency domain with amplitude and phase shown in Fig. 5(c) and (d), respectively. The time-domain waveform has 10^7 data points with a time interval of 1 picosecond. The frequency resolution is 0.1 MHz with a bandwidth of 500 GHz. The discrete amplitude peaks are correlated with the clock frequency and its harmonics.

Since the PDN circuit model does not reach down to the logic level, PDN impedance can be used to estimate the voltage noise generated by the switching current [8]. Assuming the active chip layer consumes power uniformly, all the current loads (on that chip) are identical and perfectly synchronized. PDN impedance is simulated from the PDN voltage with unit current consumption. It reveals the frequency response of the voltage versus current.

Consider a scenario that the top chip layer consumes 1-A current uniformly (other chips do not draw current). This means that each load port on the top chip layer connects to an identical current load while the other ports are open. The voltage can be probed at various ports, which gives the PDN average impedance, as shown in Fig. 6. The location of the probing port barely makes a difference, which indicates that the voltage noises within one chip layer are roughly the same if the current consumption is uniform in that chip layer.

Fig. 7 shows the switching voltage noise derived from PDN impedance (Fig. 6) and the current switching pattern

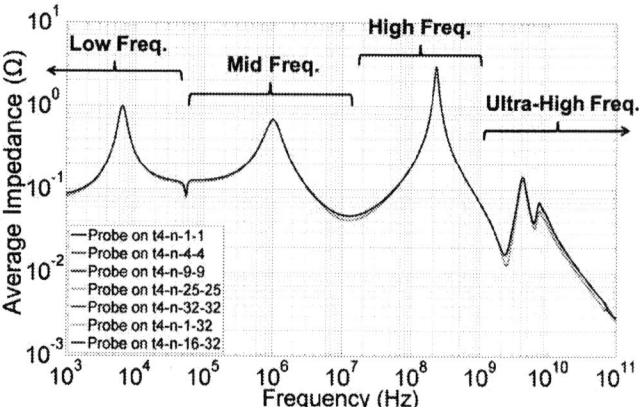

Figure 6. Simulated PDN average impedance. Assume the top chip to have a uniform current distribution.

(Fig. 5). The largest voltage noise occurs right after the moment when the IC transits from idle state to active state. The voltage noise includes three distinguishable components among mid-, high- and ultra-high frequency regions, associated with different resonant loops (i.e., PCB, package, and on-chip PDN). Their waveforms are extracted as shown in Fig. 8(a)-(c) with spectrum shown in (d). The idle-to-active transition evokes rapidly rising of all these frequency components, which generally make up the largest voltage noise.

The ultra-high frequency noise (above 1 GHz, Fig. 8(a)) is determined by the PDN impedance at the clock frequency and its harmonics. The resonance between the inductive P/G TSVs and capacitive ICs takes place within this frequency range, resulting in impedance peaks, which can possibly affect SSNs in the active IC and other idle ICs. The high-frequency noise (about 100 MHz, Fig. 8(b)) is attributed to the impedance peak associated with package/socket/board resonance. The mid-frequency noise (about 1 MHz, Fig. 8(c)) coincides with the impedance peaks due to the LC resonant loop formed by package PDN (part of off-chip PDN) and active IC.

IV. Distribution of Switching Noise in 3D Chip Stack

The PDN is both linear and reciprocal system at any frequency given the passive PDN component model. The

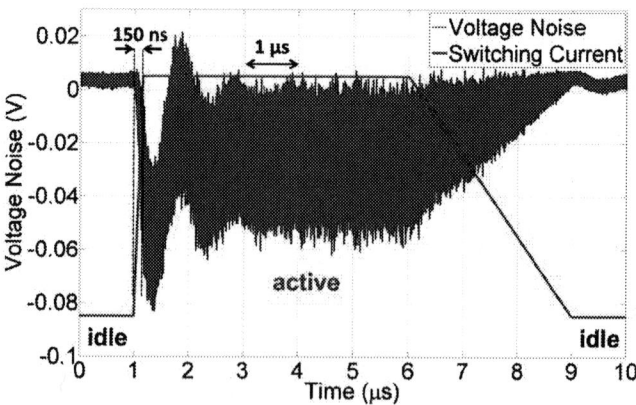

Figure 7. Simulated PDN switching noise. The red curve shows the switching current profile associated with the IC transition from idle state to active state.

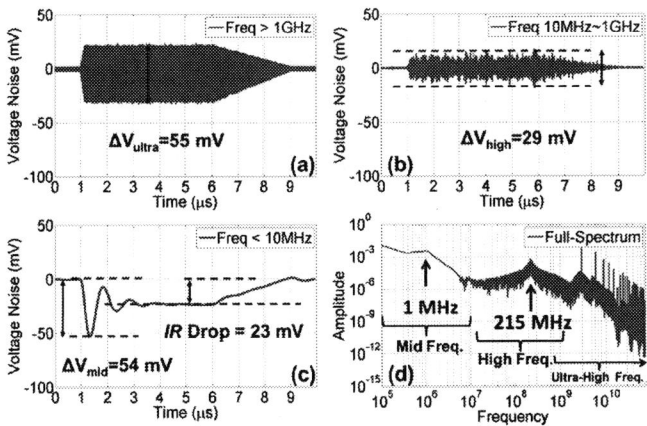

Figure 8. Decomposition of switching voltage noise in different frequency regions (a) - (c) and the spectrum (d). (a) - (c) emphaze ultra-high frequency noise (3, 6, 9… GHz, $\Delta V_{PP} \sim 55$ mV), high-frequency noise (215 MHz, $\Delta V_{PP} \sim 29$ mV), and mid-frequency noise (1 MHz, $\Delta V_{PP} \sim 54$ mV), respectively. The *IR* drop is 23 mV.

correlation of the switching noise and current of the 3D chip stack can be described using an impedance matrix $\{Z_{ps}(\omega)\}$. In the frequency domain, Z_{ps} gives the voltage noise on the probing chip layer (*p*-th layer) when only the *s*-th chip layer consumes 1-A current uniformly. The overall voltage on the probing chip is $V_p(\omega)$:

$$V_p(\omega) = \sum_{s=1}^{N} Z_{ps}(\omega) \cdot I_s(\omega),$$

where $N = 4$ is the total number of the chip layers, $I_s(\omega)$ is the current consumed by *s*-th layer. Z_{ps} is equal to Z_{sp} due to the reciprocal network.

The impedance matrix element Z_{ps} is plotted in Fig. 9 for the 4-layer chip stack. The impedance matrix elements below 1 GHz are very close, which means all chip layers experience similar low-, mid- and high-frequency noises in spite of which chip layer is active. However, the ultra-high frequency noises are quite different in each chip layer depending on the corresponding impedance Z_{ps} at the clock frequency and its harmonics.

The complexity of the ultra-high frequency noise comes from the LC resonance loops formed by the TSVs and

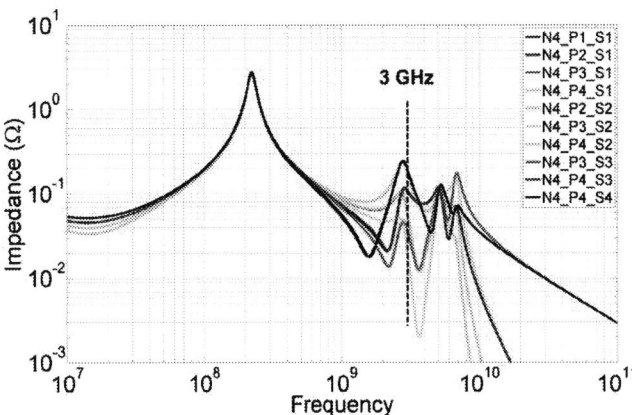

Figure 9. Simulated PDN impedance matrix.

multiple chip layers. Different resonance modes exist in the chip stack and create disturbance of the impedance matrix in the ultra-high frequency region as shown in Fig. 9. Here, it is not necessary that the nearest chip layer endures the largest coupling noise. For instance, the voltage waveforms of all four chips are shown in Fig. 10. Although only the 4-th chip layer is switching, layer 1 experiences larger switching noises than layer 2 and layer 3. It is also remarkable that the switching noises in the PDNs of silent chip layers are in the same order of magnitude with that in the active chip layer. This is different from the switching noise in the signal network, where the voltage in the victim is generally smaller than that in the aggressor.

The similarity in the switching noise pattern indicates the entire chip stack shares the same mid-frequency noise, which is extracted along with the high- and ultra-high frequency components as shown in Fig. 11. Interestingly, although the high-frequency noise is induced by the package resonance, it is also independent of the active layer location. This is because the resistance and inductance of the vertical interconnects (i.e., TSV, solder, and micro-bump) are much smaller than those of the package. In this frequency region (10 MHz – 1 GHz), only the overall capacitance of the chip stack matters, as well as the inductance of the package. With 3-GHz clock signal, the processor chip generates relatively larger switching noises when it is on the layer 1 or layer 4. However, when the active chip is on layer 2 or layer 3, the switching

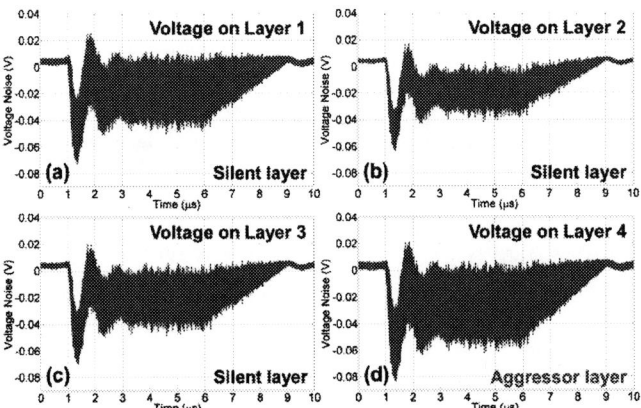

Figure 10. The waveform of switching noise in each chip layer when only the top (the 4th chip layer) consumes current.

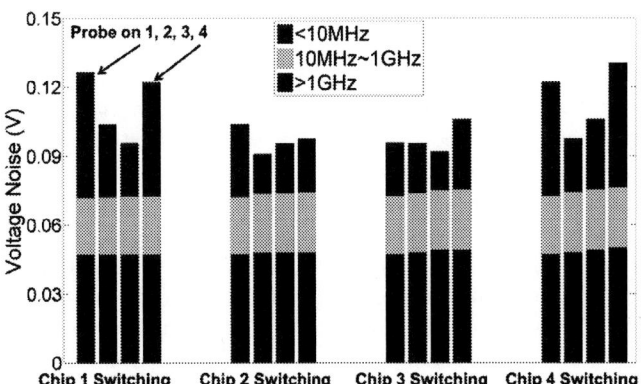

Figure 11. Different frequency components of switching noises in each chip layer with one of the chip layers transiting from idle state to active state with a 3-GHz clock rate.

978-1-4799-2408-0/14 $31.00 © 2014 IEEE

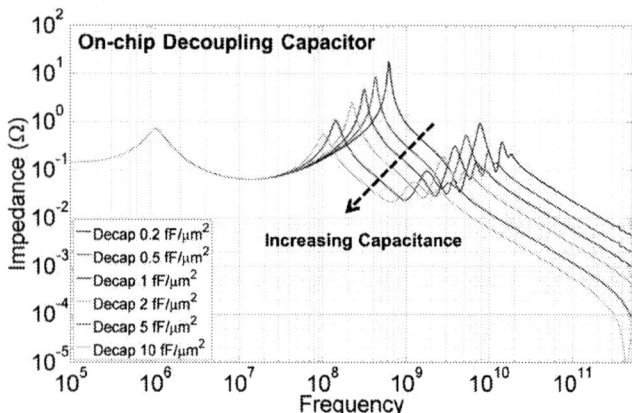

Figure 12. Simulated PDN average impedance with varying on-chip decoupling capacitor (decap) density.

noise is even smaller in the active chip than in the silent chip.

V. On-chip Decoupling Capacitor Impact

On-chip decoupling capacitors (decaps) are the capacitors that are closest to the devices. They play a critical role in shunting the high frequency current and reducing the AC noise. The decap density is usually determined by technology and limited by the available on-chip area. The use of deep trench capacitor may be able to further increase the capacitance density, but the on-chip decoupling capacitor is still expensive in general.

The decap density has a significant impact on PDN impedance in the high- and ultra-high frequency regions. More specifically, the decap density affects the resonance between the capacitive chip stack and the inductive package (in the high frequency region), and the resonance between the capacitive chip layers and inductive TSVs (in the ultra-high frequency region).

In addition to driving these resonance-induced impedance features towards lower frequency and reducing the overall on-chip impedance, increasing the decap density can also smooth these impedance peaks and valleys, as shown in Fig. 12. The frequency components of the switching noise are compared for different decap density in Fig. 13(a). Consequently, the high-frequency and ultra-high frequency switching noises can be considerably reduced by adding more capacitors, as compared in Fig. 13(b) and (c). However, the impedance at mid-frequency is barely affected by on-chip decoupling capacitors. Therefore, the voltage damping at $t = 1$ μs is still significant as shown in Fig. 13(c). Improving PDN performance in the mid-frequency region relies on the off-chip PDN optimization, which can be further addressed with package design and technologies (beyond the scope of this paper).

The high-frequency noise is only sensitive to the overall on-chip decoupling capacitance, which can take the advantage of 3D integration. The space for decoupling capacitors may be tighter for some certain ICs (e.g., processor/logic) than others (e.g., memory). We can increase the decoupling capacitors in non-space-tight ICs to reduce the high frequency noise. Three scenarios are compared in Fig. 14, which has 2-fF/μm^2 decap

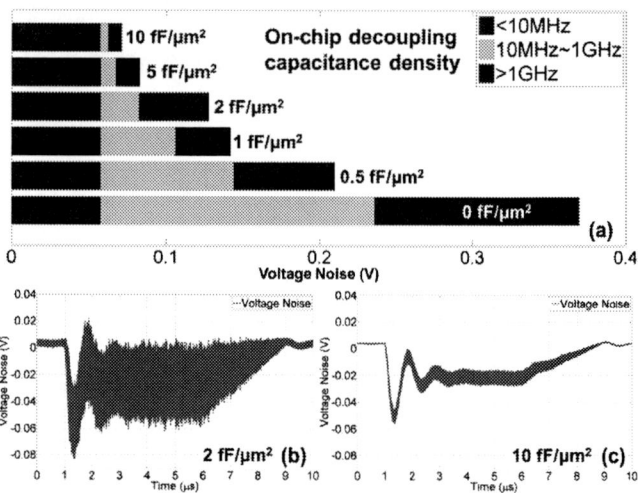

Figure 13. Frequency components in switching noise with various decap density (a), and the waveform of the switching noise in PDN with a decap density of 2fF/μm^2 (b) and 10 fF/μm^2 (c), respectively.

for all four chip layers, 6-fF/μm^2 for layer 1-3 with 2-fF/μm^2 for layer 4, and 5-fF/μm^2 for all chips respectively. The results verify that both the high- and ultra-high frequency noises can be reduced by adding on-chip decap to a single chip layer, but the decap distribution needs more careful design to achieve optimal reduction for the ultra-high frequency noise.

VI. Switching Interval

The modification of on-chip PDN has limited impact on the mid-frequency noise. The largest voltage noise occurs right after the IC transiting from idle state to active state, when the mid-frequency oscillation is aroused and rapidly damped after a few cycles. The mid-frequency noise is very sensitive to the transition time. Compared to the noise waveform from $t = 6$ μs to $t = 9$ μs (Fig. 13(c)), when the active chip gradually transits to idle sate, the mid-frequency noise is much smaller than that during $t = 1$ μs to $t = 3$ μs.

An alternative method can be used when there are multiple chips are transiting from idle state to active state. Because the entire 3D chip stack shares mid-frequency noise, a stagger interval can be introduced into the transition of multiple chips to partially cancel out the mid-frequency noise. The noise

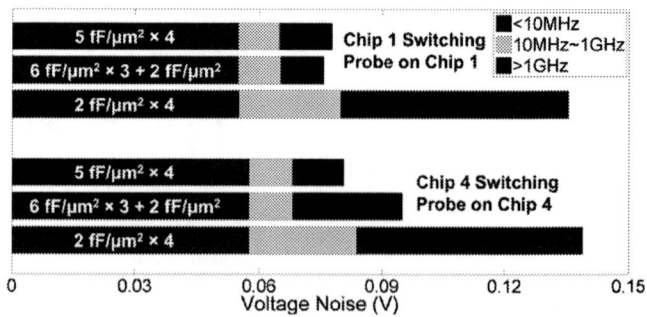

Figure 14. Simulated frequency components of PDN switching noise with different on-chip decap distributions, where "6 fF/μm^2 × 3 + 2 fF/μm^2" represents 6-fF/μm^2 decap for chip layer 1-3, and 2 fF/μm^2 decap for layer 4.

978-1-4799-2408-0/14 $31.00 © 2014 IEEE   552

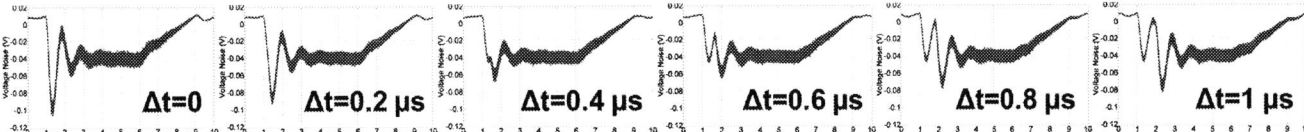

Figure. 15. Waveform of switching noise in chip layer 4 with different stagger intervals for layer 3 and layer 4 transiting from idle state to active state. Chip 3 transits at $t = 1$ μs and chip 4 transits at $t = 1$ μs $+ \Delta t$.

cancel-out is simulated with different delay time as shown in Fig. 15, where chip 3 and chip 4 are assumed to transit from idle state to active state. The optimal delay time is close to 0.4 μs, which is roughly a half of the mid-frequency oscillation period. A high decap density (10 nF/μm^2) is used here to emphasize the mid-frequency.

Conclusions

The quantitative modeling and analysis provide important design implications for TSV-based 3D power delivery. The switching voltage noise includes three distinguishable components among mid-, high- and ultra-high frequency regions, associated with different resonant loops, which have to be addressed.

The ultra-high frequency noise is determined by the PDN impedance at clock frequency and its harmonics. Within this frequency range, the resonance between the inductive P/G TSVs and capacitive ICs takes place, resulting in SSN in PDNs of the active IC and other idle ICs.

The high- and mid-frequency noises coincide with the impedance peaks due to the LC resonant loop formed by package PDN and active IC, as well as the one associated with package/socket/board resonance. The entire IC stack shares similar noise in phase and amplitude. Consequently, the IC stack suffers from similar mid- and high-frequency noise in PDN in spite of being active or idle.

Increasing the on-chip decoupling capacitance can reduce anti-resonance peaks and alleviate SSN, especially to reduce high- and ultra-high frequency noises. Introducing stagger intervals for multiple chip transition can help reduce the mid-frequency noise.

The modeling results can be used as design guidelines for reducing switching noise in TSV-based 3D power delivery and optimizing the 3D PDN performance.

Acknowledgments

This work was supported by SRC GRC Interconnect and Packaging Sciences Program (Contract #: 2012-KJ-2253).

References

1. J.-Q. Lu, "3-D Hyperintegration and Packaging Technologies for Micro-Nano Systems," *Proceedings of the IEEE*, vol. 97, no. 1, pp.18-30 Jan. 2009.

2. J. U. Knickerbocker, P. S. Andry, B. Dang, R. R. Horton, C. S. Patel, R. J. Polastre, K. Sakuma, E. S. Sprogis, C. K. Tsang, B. C. Webb, and S. L. Wright, "3D Silicon Integration," in *Electronic Components and Technology Conference (ECTC)*, Lake Buena Vista, FL, 2008, pp. 538-543.

3. H. He and J.-Q. Lu, "Compact Models of Voltage Drops in Power Delivery Network for TSV-Based Three-Dimensional Integration," *Electron Device Letters*, IEEE, vol. 34, no. 3, pp.438-440 2013.

4. Q. K. Zhu, Power Distribution Network Design for VLSI: Wiley, 2004.

5. H. He, Z. Xu, X. Gu, and J.-Q. Lu, "TSV Density Impact on 3D Power Delivery with High Aspect Ratio TSVs," in *Advanced Semiconductor Manufacturing Conference (ASMC)*, 2013 24th Annual SEMI, Saratoga Springs, NY, 2013.

6. K. Kim, C. Hwang, K. Koo, J. Cho, H. Kim, J. Kim, J. Lee, H.-D. Lee, K.-W. Park, and J. S. Pak, "Modeling and Analysis of a Power Distribution Network in TSV-Based 3-D Memory IC Including P/G TSVs, On-Chip Decoupling Capacitors, and Silicon Substrate Effects," *Components, Packaging and Manufacturing Technology, IEEE Transactions on*, vol. 2, no. 12, 2057 - 2070 2012.

7. H. He, Z. Xu, X. Gu, and J.-Q. Lu, "Power Delivery Modeling for 3D Systems with Non-Uniform TSV Distribution," in *Electronic Components and Technology Conference (ECTC)*, 2013 IEEE 63rd Las Vegas, NV, 2013.

8. M. Swaminathan and A. E. Engin, *Power Integrity Modeling and Design for Semiconductors and Systems*: Pearson Education, Inc., 2007.

9. N. H. E. Weste and D. M. Harris, *CMOS VLSI Design A Circuits and Systems Perspective*, Fourth Edition ed., 2011.

10. Z. Xu, Q. Wu, H. He, and J.-Q. Lu, "Electromagnetic-Simulation Program with Integrated Circuit Emphasis Modeling, Analysis, and Design of 3-D Power Delivery," *Components, Packaging and Manufacturing Technology, IEEE Transactions on*, vol. 3, no. 4, 2013.

11. "Intel Pentium 4 Processor in the 423 pin package / Intel 850 Chipset Platform Design Guide," 2002.

Characterization of On-die Power Supply Noise in FCBGA (Flip-Chip Ball Grid Array) Packages

Hyunho Baek and William R. Eisenstadt
University of Florida
1064 Center Dr. NEB 505, Gainesville, FL 32611
hhbaek@ufl.edu

Abstract

In this work, the authors demonstrate Core-type and IO-type on-die noise characterization to enhance CPU PDN (Power Delivery Network) performance in microprocessor. The two representative power supply noise characteristics are simulated using excitation models and an impedance modeled of the power delivery network. Then, the design metrics for the PDN are analyzed based on excitation. Behavior of PDN current activity is also estimated in order to be anticipated the power supply noise efficiently.

I. Introduction

Power fluctuation affects performance and creates functionality degradation in the chip, package and board as shown in Figure 1. This figure depicts a typical supply noise waveform measured on the die due to rapid, large changes in supply and I/O current and, current through resistivity, IR events. Note that this includes four different droops from each resonance frequency in the impedance of the power delivery network as described in [1].

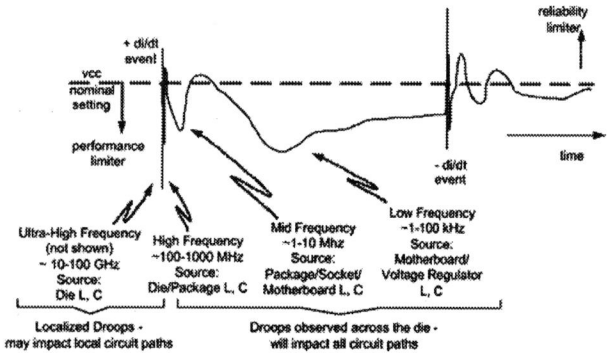

Figure 1. Microprocessor Supply Voltage fluctuations caused by interaction of parasitics with changes in current demand [1].

Therefore, all levels of a system such as microprocessor IC which is lowest-level, package, printed circuit board and VRM have to be analyzed [2] since the device operation is directly affected by the on-chip power supply noise which is caused by the transients of the power delivery network. Power Distribution Network (PDN) designers usually verify the performance of the PDN design as shown in the equation (1)-(3). PDN designers can get the Z_{PDN} from the package and board model. And, circuit designers can provide the current profile which is an excitation ($I_{EXCITATION}$) of the circuit activity to PDN designers. The current profile ($I_{EXCITATION}$) on-die can be separated into two representative circuit activities; one is core-type circuit activity and the other is I/O-

type circuit activity. Then, Vnoise can be derived from the impedance profile (Z_{PDN}) and excitation model ($I_{EXCITATION}$).

$$V_{NOISE} = Z_{PDN}(f) * I_{EXCITATION}(f) \qquad (1)$$

$$V_{NOISE} \leq V_{TARGET} \qquad (2)$$

$$Performance = function (V_{NOISE}, etc.) \qquad (3)$$

When V_{NOISE} is revealed based on the equation (1), the PDN designers will be verified whether the V_{NOISE} is met their specification, V_{TARGET} or not in (2). At this point, V_{TARGET} has to be determined carefully since non-optimal V_{TARGET} can result in over- or under-designed PDNs. The impact of the over-designed PDNs increases cost. On the other hand, the impact of the under-design PDNs degrades performance. Therefore, PDN cost can be reduced if other factors of system performance compensate for V_{NOISE}. Next generation product decisions can be better informed by quantifying VNOISE sensitivity and evaluating package performance.

In this work, the authors demonstrate that Core-type and IO-type on-die noise characterization enhances CPU PDN performance in FCBGA packages. Two representative power supply noise elements are simulated using the excitation models and the impedance of the modeled power delivery network. Then, the design metrics for the PDN designers are analyzed based on the excitation since the PDN risk on all CPU circuits and logic can be addressed quickly. The behavior of PDN current activity is also characterized in order to anticipate the power supply noise efficiently.

II. Noise Sources

In Figure 2, the Intel Sandy bridge microprocessor [3] is composed of Core processers, I/O, L3 cache, Graphic processor and memory controller. Each block of the microprocessor has circuits and the current profile will be determined by those circuit activities. Among the circuit activity of the blocks, Core-type circuit activity and I/O-type circuit activity are the main issues that create the power supply noise that affects to both circuit performance and PDN performance.

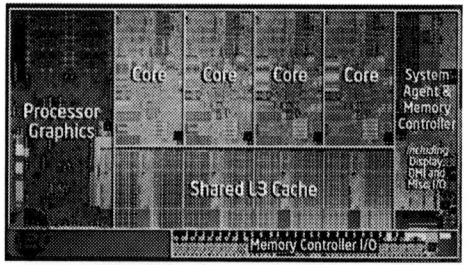

Figure 2. Intel Sandy bridge microprocessor [3]

978-1-4799-2408-0/14 $31.00 © 2014 IEEE   554   2014 Electronic Components & Technology Conference

The elements of CPU PDN noise sources are categorized based on the core-type circuits and I/O-type circuits in Table 1. Example of the circuit block, physical characteristic and PDN stress are presented in Table 1.

Table 1. Elements of CPU PDN Noise Sources

Type	Example	Physical Characteristic	Factors that affects PDN Stress
Core	-CPU -Graphics Processor	- Core block takes large area on the die - Core block can be operated by low voltage from process	- Clock gating starts or stops the clock tree and logic - Power gating powers up or down large C_{die}. - Data patterns
IO	-Parallel -Serial	- I/O block is defined by industry standard - I/O block area is placed along chip periphery - Output drivers capable of the load capacitance	- Data patterns - Periodic burst at PDN resonant frequency - Power is controlled by power management

Note that the noise sources can be affect the PDN stress depending the level of each activity, let's take a look at the each type of the excitation model both core-type and I/O-type in Table 2 and 3.

Table 2. Noise Source on the Core Block

Activity	Description
Clock Gating [4]	Clock and logic signal power are saved by temporarily disabling the clock signal on registers whose outputs do not affect circuit outputs Actual implementation may have multiple clock domains, more than one clock each on different phases.
Activity Change	This is a change in activity, either increasing activity or decreasing activity. This activity change might be associated with voltage and frequency scaling.
Power Gating [5]	Power gating is a technique used in integrated circuit design to reduce power consumption, by shutting off the current to blocks of the circuit that are not in use. Possible power gating modes are Power On-state, Off-state, Up and Down.
Cyclic Activity	This is a repetitive behavior and probably has low likelihood of occurring in actual system use.

Table 3. Noise Source on the I/O Block

Activity	Description
Data patterns	Sequential I/O is typical of large file reads and writes by "0" or "1", and typically involves operating on one block immediately after its neighbor. This can stress the I/O PDN with I/O PDN resonance.
Activity Change	Depending on the tasks, I/O circuit activity can change such as "wake-up" sequence when the tasks from a near-sleep state to high activity.
Periodic Burst Pattern [6]	Defined as a burst of circuit switching following with the same number of idle states. When a periodic burst pattern repeats at the PDN resonant frequency, it will stimulate and reinforce the PDN resonance remarkably.

Hence, the excitation model, Icc(t) is needed to be defined for the noise sources so that the power supply noise can be generated by each type of the excitations and the impedance of the power delivery network.

III. Power Delivery Network in Microprocessor

The power delivery network (PDN) can be simplified into a lumped RLC network including a decoupling cap to be made lower than targeted impedance based on the die, package and board in a microprocessor system. Ideally, the PDN provides sufficient voltage and current for the IC bias the instant that the transistors switch. However, simultaneous switching of several devices, Ldi/dt, chip-package resonance, and inadequate decoupling capacitance can create the power supply noise. The inductance elements can increases the impedance of the power deliver network.

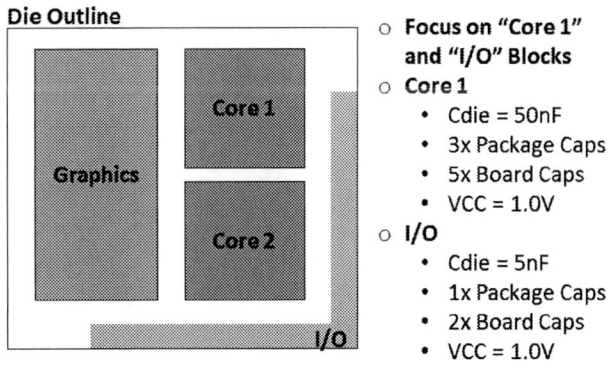

Figure 3. Modeled Die Floor Plan

In order to simulate the power supply noise through the combination of the excitation models and the impedance of the power delivery network, the author has been created a simple die model as shown in Figure 3 and a SPICE model which corresponds to the die floor plan to be representing a power delivery network as shown in Figure 4.

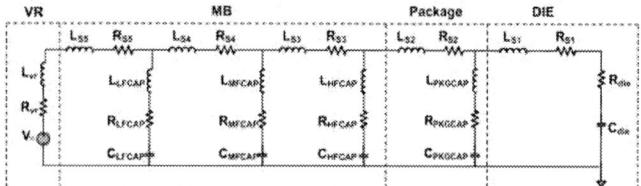

Figure 4. SPICE model of the die floor plan PDN

In this simulation, the spice model of the PDN is defined with the parameters as shown in Table 4. For convenience, the motherboard parameters are excluded since the power fluctuation is dominant by 1st and 2nd resonant frequencies in the network.

Table 4. Elements of Power Delivery Network

List	Core	IO	List	Core & IO
C_{pkgcap}	3×100nF	1×100nF	R_{S1}	0.1mΩ
R_{pkgcap}	3×35mΩ	1×35mΩ	L_{S2}	100pH
L_{pkgcap}	3×250pH	1×250pH	R_{S2}	0.1mΩ
R_{hfcap}	5×20mΩ	2×20mΩ	R_{VR}	3mΩ
L_{hfcap}	5×200pH	2×200pH	L_{VR}	150pH
C_{hfcap}	5×600nF	2×600nF	R_{die}	400uΩ
C_{die}	50nF	5nF		

Using the SPICE model of the power delivery network, the core-type and I/O-type impedance profiles can be derived as shown in Figure 5 and Figure 6 so that they can generate power supply noise with the excitation models. The resonant frequency of the impedance profiles are 101MHz, 105MHz for Core-type, IO-type respectively. The ranges of the resonance frequency are roughly matched with 100 MHz range in the last few process technologies since chip packages have not changed significantly. [7]

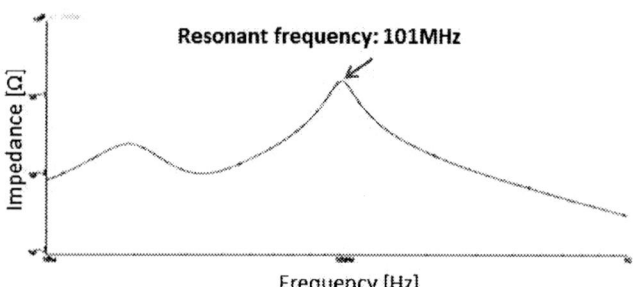

Figure 5. Impedance profile of the Core-type PDN

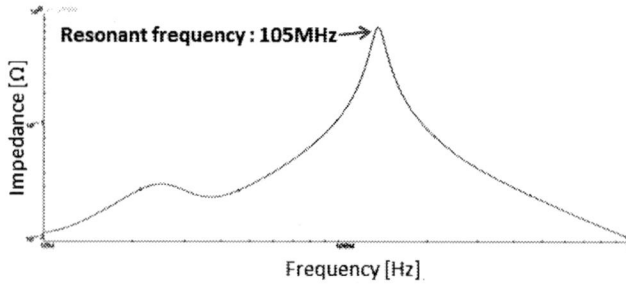

Figure 6. Impedance profile of the IO-type PDN

IV. Simulated Power Supply Noise of the Excitation Model

The power supply voltage can be generated using the modeled impedances and the activities correlating to Table 2 and 3. For reference, the defined current profile of the activities in Figure 7 and Figure 8 for Core and IO type is shown their representative behavior, not scaled by a specific process or a technology.

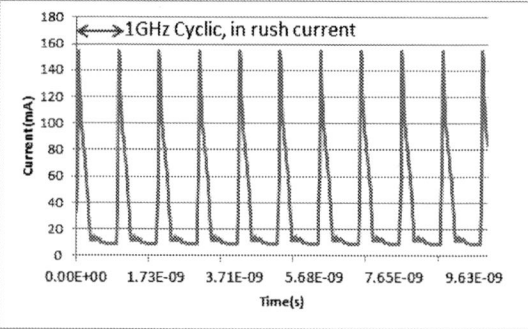

A) Power up – Power Gating
(In rush current: ex. 1GHz Cyclic), Icc(t)

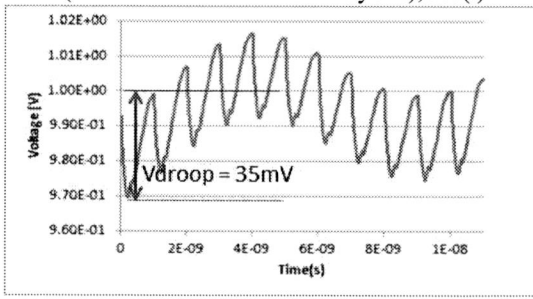

B) Power up - Noise waveform, Vcc(t)

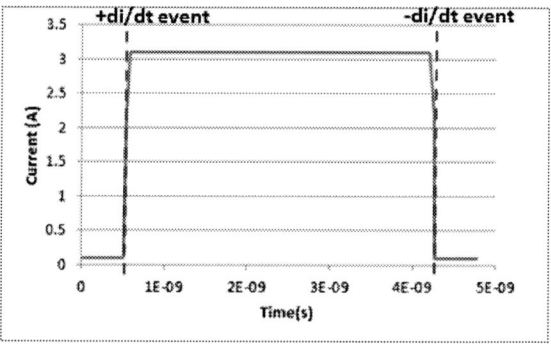

C) Data Pattern, Icc(t)

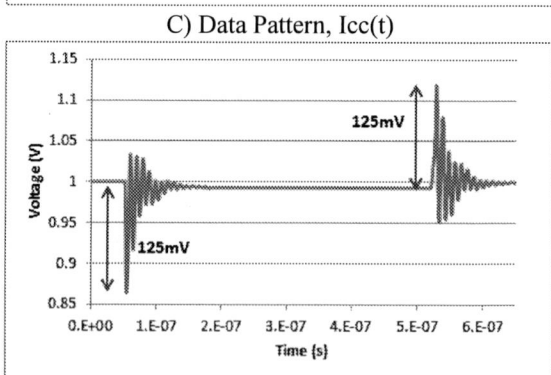

D) Data Pattern – Noise waveform, Vcc(t)

Figure 7. Power supply noise based on the Core type Icc(t) and Vcc(t).

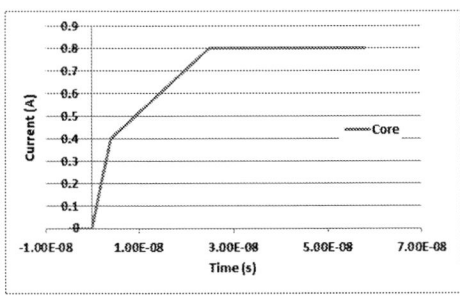

E) IccMax, Icc(t)

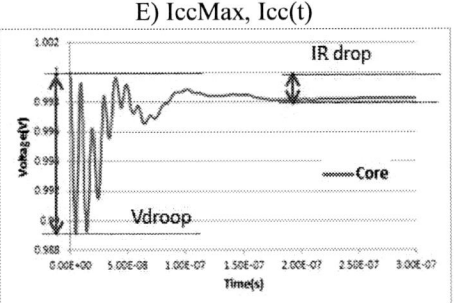

F) IccMax - Noise waveform, Vcc(t)

Figure 7. Continued

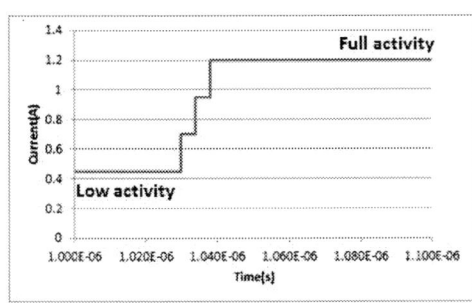

A) Power up - Staggering, Icc(t)

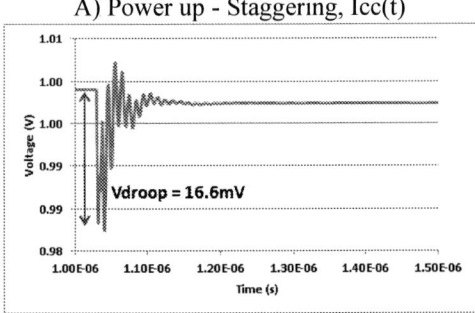

B) Power up – Noise waveform, Vcc(t)

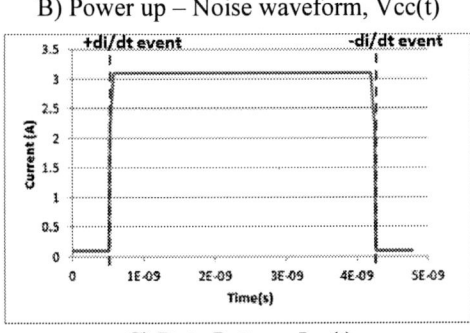

C) Data Pattern, Icc(t)

Figure 8. Power supply noise based on the I/O type Icc(t) and Vcc(t).

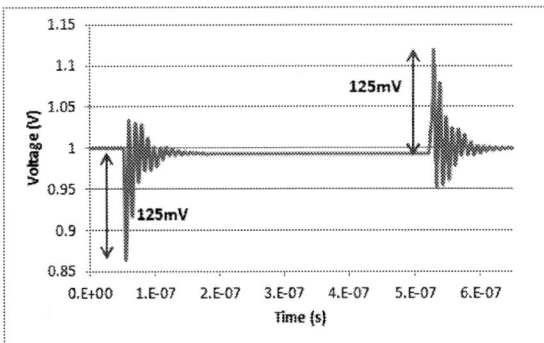

D) Data Pattern – Noise waveform, Vcc(t)

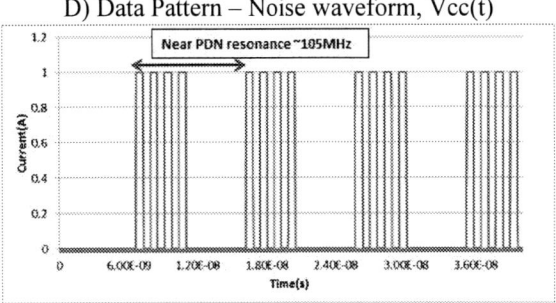

E) Protocol - Periodic burst pattern stimulus, Icc(t)

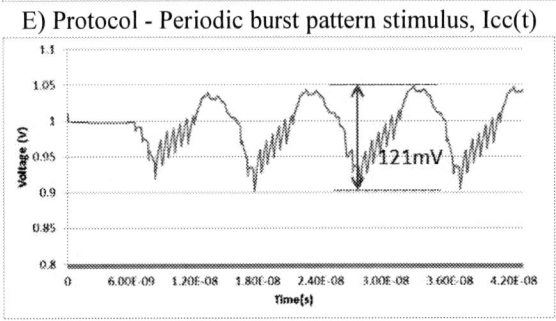

F) Protocol – Noise waveform, Vcc(t)

Figure 8. Continued

Based on the simulation results of the power supply noise, not only the range of the noise but also the design metrics can be determined as shown in Table 5. The frequency of each noise is determined by how fast the voltage droop recovers to the nominal voltage. And, PDN designers can examines the design metrics of the PDN in order to address the PDN stress.

Table 5. Design Metrics of Each Excitation

List	Core-Type		IO-Type		Metric
	Freq (Hz)	Voltage (V)	Freq (Hz)	Voltage (V)	
Power Up	1G	35m	117M	16.6m	Vdroop
IccMax	106M	14m			Vmin (DC)
Data Pattern	102M	125m	102M	125m	Vdroop
Periodic burst Pattern			105M	121m	Vdroop

In addition to, the behavior of the PDN Icc(t) can be summarized as shown in Figure 9. Both voltage droop and

978-1-4799-2408-0/14 $31.00 © 2014 IEEE   557

Table 6. Summary of PDN Current Icc(t)

Icc(t)	Current Step	Periodic Current Impulse	Complex Current Impulse
Description	- Single or Multiple current steps cause power supply noise on the other circuit blocks that remain powered up for shared voltage rails. - Can be caused by power and clock gates - Power Up - Power Down - Circuit blocks size can vary - PMIC (Power Management Integrated Circuit)	- More realistic behavior and better model of clock gating and activity factor - Data Bursts - Periodic Burst pattern - Overlaps high-frequency noise on PDN resonance behavior	- Will be used to test the validation design system - Varying magnitudes - Varying frequency content
Core Type	- Multiple steps can be separated by significant time	- Current swing can be very large depending on circuit block size - GHz clock implies large di/dt	- Magnitude can vary depending on time, x, y
I/O Type	- Multiple steps can be separated by significant time	- Current swing per I/O is small so total current depends on number of bits	- Protocol determined by encoding scheme can reduce the noise.

voltage overshoot are occurred simultaneously, but the voltage droop is a more critical issue since it affects the performance and functionality degradation of the chip than the voltage overshoot. The voltage overshoot is normally considered for the electrical overstress (EOS) of the system and represent the power supply exceeding its voltage.

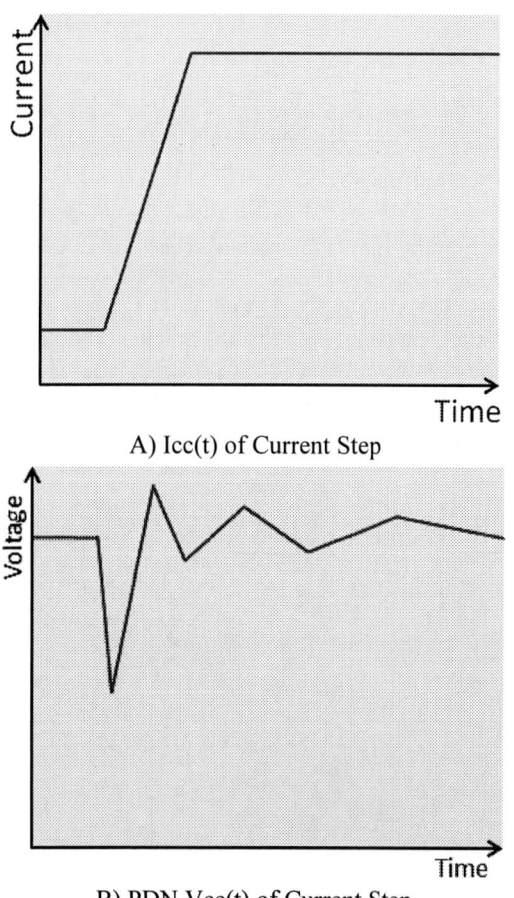

A) Icc(t) of Current Step

B) PDN Vcc(t) of Current Step

C) Icc(t) of Periodic Current Impulse

High Frequency Noise Inducing on PDN resonance frequency

D) PDN Vcc(t) of Periodic Current Impulse

Figure 9. Behavior of PDN current Icc(t) and Vcc(t)

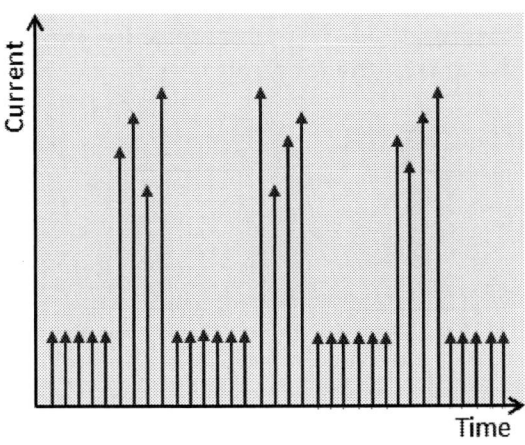

E) Icc(t) of Complex Impulse Train

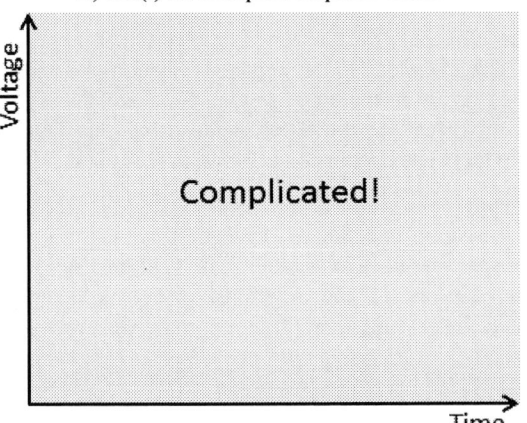

F) PDN Vcc(t) of Complex Impulse Train

Figure 9. Continued

Therefore, two representative noise sources; Core-type circuit activity and I/O-type circuit activity can be defined in terms of the behavior of the PDN Icc(t) and the noise waveforms as shown in Table 6 and the PDN voltage Vcc(t) can be anticipated with only the behavior of the excitation model. This work contributed to the effective design of PDNs.

V. Conclusions

In conclusion, Core-type and I/O-type noise are demonstrated and characterized using both excitations and noise waveforms. PDN current Icc(t) can be characterized with the Core-type and I/O-type noise in terms of current steps, periodic current "impulse" and complex "impulse" trains. The differences of the noise characteristic are demonstrated between Core-type and the I/O-type circuit as well as design metrics of each current profile are analyzed. A CPU PDN performance can be enhanced by using noise characterization for Core and I/O-type circuit activity on die.

References

1. Kurd, N.; Mosalikanti, P.; Neidengard, M.; Douglas, J.; Kumar, R., "Next Generation Intel Core™ Micro-Architecture (Nehalem) Clocking," *Solid-State Circuits, IEEE Journal of*, vol.44, no.4, pp.1121,1129, April 2009

2. Dan Oh et al., (2010) "In-situ characterization of 3D package systems with on-chip measurements," *Electronic Components and Technology Conference (ECTC)*, 2010 Proceedings 60th , vol., no., pp.1485,1492, 1-4 June 2010

3. Intel Sandy Bridge Review (2011) from http://www.bit-tech.net/hardware/cpus/2011/01/03/intel-sandy-bridge-review/

4. G. Yeap, Practical Low Power Digital VLSI Design. Boston, MA: Kluwer Academic Publishers, 1998.

5. http://en.wikipedia.org/wiki/Power_gating

6. Smith, L.; Shishuang Sun; Boyle, P.; Krsnik, B., "System power distribution network theory and performance with various noise current stimuli including impacts on chip level timing," Custom Integrated Circuits Conference, 2009. CICC '09. IEEE , vol., no., pp.621,628, 13-16 Sept. 2009

7. Ankur Agrawal (2010) "Design of High Speed I/O Interfaces for High Performance Microprocessors" Ph.D. dissertation, Dept. Elect. Eng., Harvard Univ., Cambridge, MA, 2010.

An Enhanced Power Integrity Analysis Flow Based on the Interdependence between Simultaneous Switching Output Noise and Static IR Drop

Minghui Han, Amir Amirkhany, and Wei Xiong
Samsung Display
217 Devcon Drive, San Jose, CA 95112
Contact Email: minghui.han@samsung.com

Abstract

This paper presents an in-depth study on how the magnitude of simultaneous switching output (SSO) noise is affected by on-chip supply grid resistances. A key observation of the study is that under certain circumstances, increasing the resistance of certain parts of a supply grid can be very effective in reducing SSO noise, and the gain from SSO noise reduction can significantly outweigh the resulted increase in static IR drop. Based on this observation, an enhanced power integrity analysis flow is proposed for high speed interface design. Unlike conventional practices, our proposed flow considers SSO noise and static IR drop as two closely interrelated issues, and addresses them in a co-design manner throughout the design process.

I. Introduction

For high speed I/O interface designs, having sufficient voltage margin has become a very challenging task. Moreover, the situation will likely exacerbate as the technology continues to advance and the operating voltages keep shrinking. In general, the loss of voltage margin in a chip comes from two sources: the DC loss due to static IR drop, and the AC loss due to transient fluctuation of current flowing through the supply grid. For AC loss, simultaneous switching output (SSO) noise, which occurs when all the I/O circuits toggle at the same time, is usually the dominant contributor [1, 2]. In common practices, the control of static IR drop is done by making the supply grid as least resistive as possible. On the other hand, the control of SSO noise focuses on maximizing the placement of on-chip decoupling capacitance as well as minimizing the inductance of supply rails inside the package [3-6]. To our best knowledge, the importance of supply grid resistances in determining the magnitude of SSO noise has not been fully appreciated. As a result, in typical power integrity analysis flows, static IR drop and SSO noise are often treated separately.

In this paper, we present an in-depth study on how supply grid resistances can affect the magnitude of SSO noise. The study starts with simple analytical models of power delivery network (PDN), in order to establish appropriate conceptual framework for the effects of supply grid resistances. The simple analytical models are then replaced by a comprehensive layout-based PDN modeling flow, which can capture the effects of any arbitrary change on the layout of a supply grid. With this flow, rigorous investigation is conducted on how changing the resistances of a supply grid would affect SSO noise. The investigation concludes that an overly aggressive static IR drop target can be detrimental to SSO noise control and hence the overall voltage margin. The proper goal in design practices should be to achieve optimal trade-off between SSO noise and static IR drop. To accomplish this, the recommended power integrity flow is an iterative process. First, some initial guess on appropriate IR drop and SSO noise target is made based on a pre-layout PDN model. Second, along the way to the completion of chip layout, the two targets are continuously revised using a reliable layout-based PDN model.

The remaining of this paper is organized as follows. Section II focuses on theoretical analysis with several simple RLC circuit-based PDN models. Section III presents a rigorous layout-based PDN model. Section IV describes the result of virtual layout experiments using this layout-based PDN model. The conclusion is given in section V.

II. Theory

This section is dedicated to theoretical analysis of simple RLC circuit-based PDN models. The goal here is to elucidate various aspects of the relationship between supply grid resistances and voltage margin loss on a given supply.

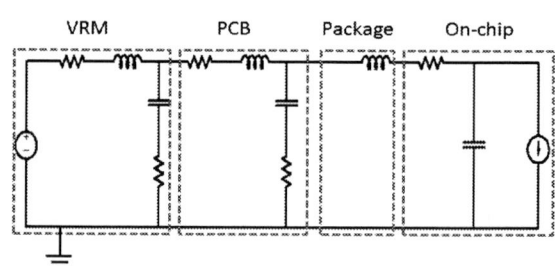

Figure 1. A pre-layout PDN model that uses lumped RLC circuits to model each section of the PDN system.

Despite their inaccuracy, RLC circuit-based PDN models are widely used in supply voltage simulations [7-9], especially at the pre-layout stages of a design. One popular example is given in Fig. 1, where a complete PDN system is modeled by several LRC circuits. In particular, the on-chip PDN section, i.e. the metal interconnects and the decoupling capacitors, is represented by a lumped RC circuit, whereas I/O toggling activity is described by an ideal current source. Assuming that the off-chip PDN sections are properly designed, the model in Fig.1 can be reduced to a single LRC circuit for frequencies above several MHz, as shown in Fig.2 (a). Using this simple model, deriving the PDN impedance seen by the toggling I/O cells is straightforward

$$Z = \frac{R + j\omega L}{1 - \omega^2 LC + j\omega RC} \tag{1}$$

According to (1), the PDN impedance Z tends to resonate at certain frequency due to the presence of L and C. In particular, when R is small, the peak magnitude $|Z|_{max}$ is roughly L/RC, which suggests that increasing R has similar effectiveness in SSO noise reduction as increasing C or reducing L, given that R is small enough. In other words, R acts as some damping mechanism for the PDN resonance. Furthermore, if R becomes too large, the PDN resonance will disappear entirely.

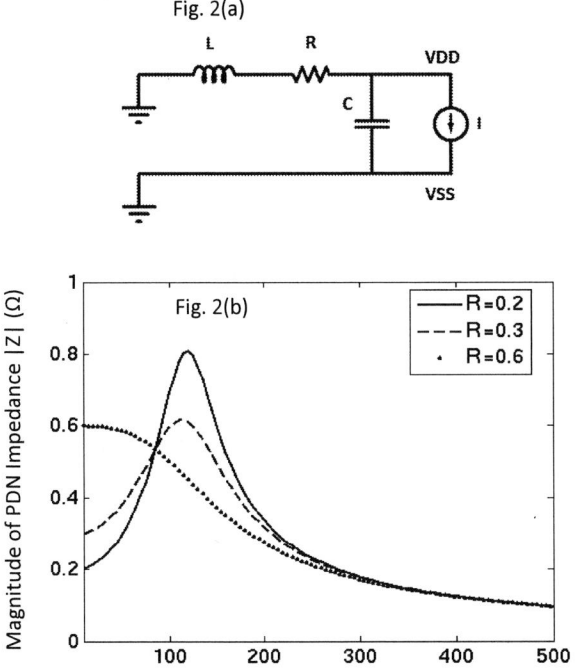

Fig. 2(a)

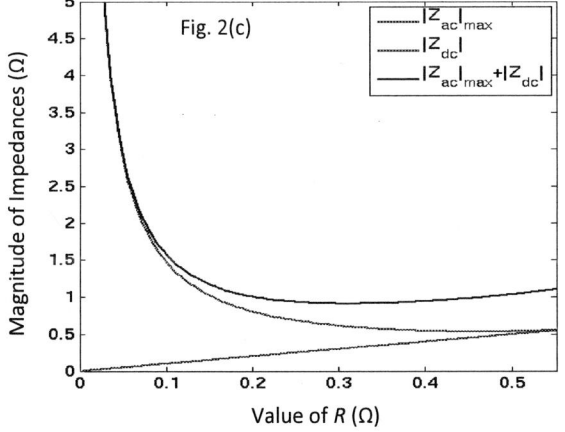

Figure 2. (a) A further simplified PDN model compared to that in Fig. 1, where R represents the effective on-chip supply grid resistance, C represents the effective on-chip decoupling capacitance, and L represents the inductance of package PDN. I/O toggling activities are described by an ideal current source. (b) Magnitude of PDN impedance $|Z|$ versus frequency with three different values of R. (c) Peak magnitude of PDN impedance versus R.

Several numerical examples are provided in Fig. 2(b). Here it is assumed that $L = 0.5$nH, $C = 3.5$nF. As can be seen, when $R = 0.2\Omega$, Z resonates at around 120MHz, and the peak magnitude $|Z_{ac}|_{max}$ is 0.82Ω. Increasing R to 0.3Ω reduces $|Z_{ac}|_{max}$ to 0.62Ω, as well as slightly lowers the resonance frequency. Increasing R to 0.6Ω, however, completely

destroys the resonance. The peak magnitude of Z is now happening at DC, equal to 0.6 Ω.

A more complete picture of the impact of R is presented in Fig. 2(c). The three curves in the figure describe respectively how R affects $|Z_{ac}|_{max}$, $|Z_{dc}|$, and the sum of the two. It should be noted that the value of $|Z_{ac}|_{max}$ becomes increasingly sensitive to the change of R when R gets smaller, while in contrast the dependence of $|Z_{dc}|$ on R is always linear. As a result, $|Z_{ac}|_{max} + |Z|_{dc}$ reaches its minimum value when R is in the range of 0.2Ω and 0.3Ω. From voltage margin perspective, this means that one should balance between DC voltage margin and AC voltage margin, and there exists an optimal value of R with respect to the net supply voltage margin.

Yet the model in Fig.2 (a) implicitly assumes that R is between C and L. Therefore the observation above only applies to the supply grid paths located between decoupling capacitors and power/ground pads. In reality, the supply grid paths between decoupling capacitors and I/O cells can also be fairly resistive. To study the impact of this type of supply grid resistance, we introduce a modified PDN model in Fig. 3(a), where an extra resistor R_c is added between current source I and capacitor C. In consequence, the PDN impedance seen by toggling I/O cells now is

$$Z = R_c + \frac{R + j\omega L}{1 - \omega^2 LC + j\omega RC} \qquad (2)$$

It is clear from (2) that R_c affects Z very differently compared to R. Unlike the anti-resonance effect possessed by R, increasing R_c will always lead to a higher $|Z_{ac}|_{max}$. This is consistent with intuition that decoupling capacitors should be placed as close as possible to active circuits in order to be effective. Specific numerical examples about R_c are given in Fig. 3(b) and Fig. 3(c). Here it is again assumed that $L = 0.5$nH, $C = 3.5$nF. R is fixed at 0.2Ω; R_c can choose different values. Fig. 3(b) shows that when R_c changes from 0.0Ω to 0.1Ω, $|Z|$ increases at all frequencies, which in turn leads to Higher SSO noise. Fig. 3(c) summarizes how $|Z_{ac}|_{max}$, $|Z|_{dc}$, and $|Z_{ac}|_{max} + |Z|_{dc}$ varies when R_c sweeps from 0Ω to 0.55Ω. Compared to Fig. 2(c), here they all monotonously increase as R_c gets bigger. Hence to obtain larger voltage margin, R_c should be minimized as much as possible.

Another important type of supply grid resistance is associated with the supply grid paths between different I/O cells. Its impact on supply voltage margin can be analyzed with the model shown in Fig. 4(a), which extends the single LRC model in Fig. 2(a) by introducing another LRC circuit and connecting the two with a resistor R_{12}. The two current sources in the model represent two adjacent I/O cells. Both can generate noise on VDD supply that will affect not only themselves but also their neighbor. Different from the previous two models, here a 2×2 matrix is needed for PDN impedance characterization. Specifically, Z_{11} and Z_{22} measure the amount of VDD noise seen at each I/O cell due to the current generated from its own toggling activity; Z_{12} and Z_{21} measure the amount of VDD noise seen at each I/O cell due to the current generated from toggling activity of the other I/O cell. The detailed expressions are given in (3).

978-1-4799-2408-0/14 $31.00 © 2014 IEEE

$$Z_{11} = \left.\frac{V_1}{I_1}\right|_{I_2=0} = \frac{Z_{10}(R_{12}+Z_{20})}{R_{12}+Z_{10}+Z_{20}}$$

$$Z_{22} = \left.\frac{V_2}{I_2}\right|_{I_1=0} = \frac{Z_{20}(R_{12}+Z_{10})}{R_{12}+Z_{10}+Z_{20}}$$

$$Z_{12} = \left.\frac{V_1}{I_2}\right|_{I_1=0} = \frac{Z_{10}Z_{20}}{R_{12}+Z_{10}+Z_{20}}$$

$$Z_{21} = \left.\frac{V_2}{I_1}\right|_{I_2=0} = \frac{Z_{10}Z_{20}}{R_{12}+Z_{10}+Z_{20}}$$

where

$$Z_{10} = \frac{R_1+j\omega L_1}{1-\omega^2 L_1 C_1+j\omega R_1 C_1}$$

$$Z_{20} = \frac{R_2+j\omega L_2}{1-\omega^2 L_2 C_2+j\omega R_2 C_2} \qquad (3)$$

Figure 3. (a) A modified PDN model for analyzing the impact of supply grid resistance between active circuits and decoupling capacitors, which is represented by R_c. (b) Magnitude of PDN impedance $|Z|$ versus frequency with two different values of R_c. (c) Peak magnitude of PDN impedance versus R_c.

Fig. 4(a)

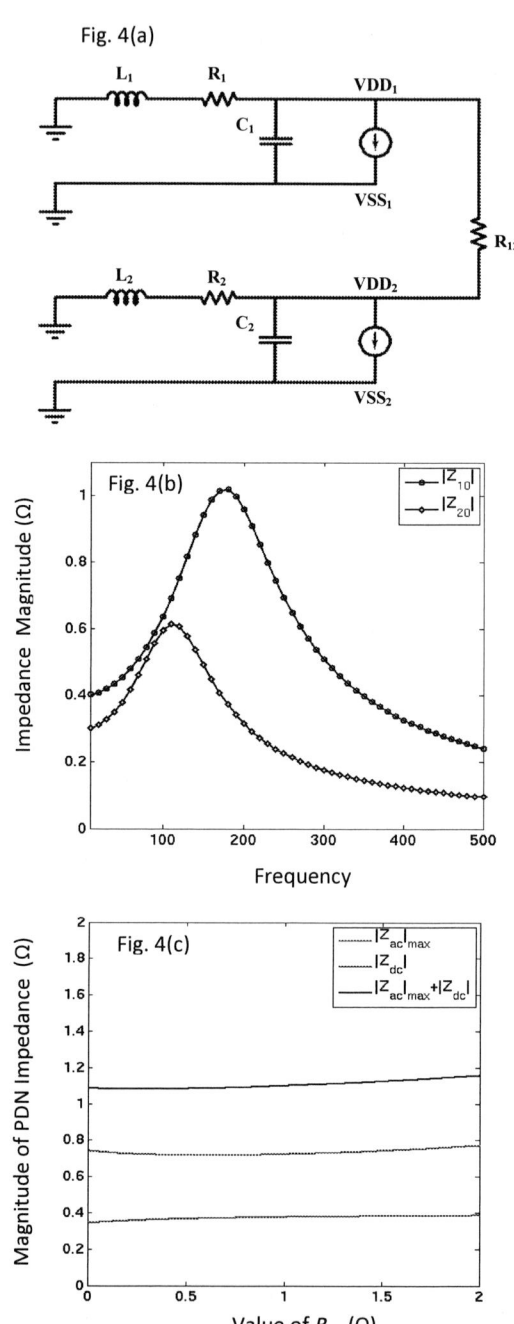

Fig. 3(a)

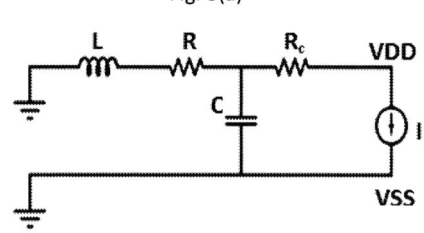

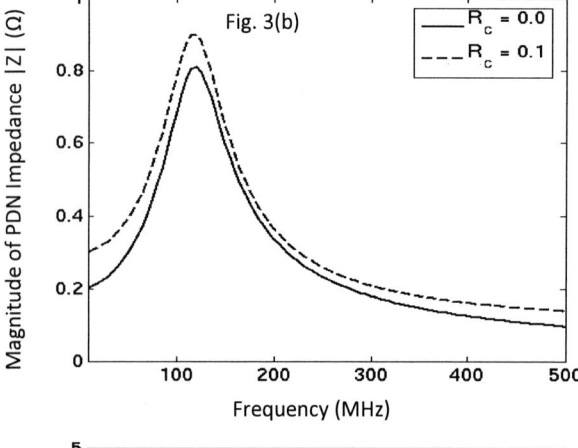

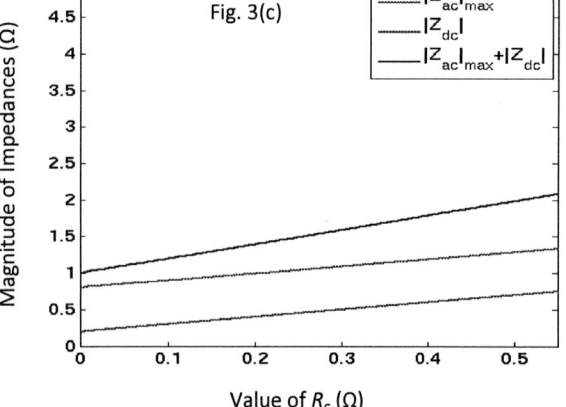

Figure 4. (a) A modified PDN model for analyzing the impact of supply grid resistance between different I/O cells, which is represented by R_{12}. (b) An example of very different Z_{10} and Z_{20}. (c) Impact of R_{12} on the peak magnitude of PDN impedance, where Z_{10} and Z_{20} are the same as depicted in (b).

978-1-4799-2408-0/14 $31.00 © 2014 IEEE

As (3) indicates, the voltage drop at an I/O cell comes from the activity of both itself and its neighbor. Without loss of generality, let us focus on the I/O cell represented by I_1. Increasing R_{12} will cause $|Z_{12}|$ to decrease, but on the other hand will result in a larger $|Z_{11}|$. The two effects offset one another to certain extent. Consequently, the dependence of $|Z_{11}+Z_{12}|$ on R_{12} tends to be quite modest. An extreme case is when $Z_{10} = Z_{20}$. According to (3), we then have $Z_{11} + Z_{12} = Z_{10}$. It means that R_{12} does not affect the supply voltage of I_1 at all. On the other hand, even when Z_{10} significantly differs from Z_{20}, the impact of R_{12} on the supply voltage of I_1 can still be small, as demonstrated in Fig. 4(b, c).

To summarize, in this section we present a theoretical analysis on how supply grid resistances can affect the voltage margin. The analysis covers three major types of supply grid resistances. Strictly speaking, the models used in the analysis are oversimplifications of actual PDN. Nonetheless they help yield important insights: first, there is indeed a strong interdependence between SSO noise and static IR drop; second, this interdependence is largely caused by the resistance of supply grid paths between on-chip decoupling capacitors and power/ground pads.

III. Layout-based PDN Model

To further validate the relationship between SSO noise and static IR drop, we develop a comprehensive layout-based PDN modeling flow. In our flow each PDN section, including the on-chip, the package, and the board, is characterized by a multi-port equivalent circuit or s-parameter model. The generation of those multi-port models is through extracting actual layout data with industry-leading EDA tools. For instance, Apache Design's Totem and Chip Power Module (CPM) [10] is our choice for extracting the equivalent circuit model from the layout of on-chip PDN.

The general methodology of our flow can be explained by examining how we model a × 32 memory interface. Fig. 5(a) gives a top view of the interface's layout implemented in a test chip. Largely following the wirebond pad assignment, the I/O circuits are distributed across entire interface rather than concentrating in one location. For example, the I/O circuits of each byte occupy a region of about 900μm wide. To ensure sufficient spatial granularity in the generated multi-port model, we define ports at the bit level. This is illustrated in Fig.5(b), where for each DQ and DQS bit in Byte 0, a port is defined for the local VDDIO traces on metal layer M1 (highlighted in red), and the local VSS traces (highlighted in white) are treated similarly. Totally there are 159 ports in the extracted on-chip PDN model, as Fig. 6 shows. Among them, 39 ports are defined at the locations of wirebond pads (11 for VDDIO, 9 for VDDR, 1 for VDDA, and 18 for VSS); 120 ports are defined at the locations of I/O bits (96 for DQ bits, and 24 for DQS bits). In addition, we replace the ideal current sources used in analytical PDN models with transistor-level driver circuits, which are connected to a post-layout channel model. In short, power integrity analysis and signal integrity analysis are seamlessly integrated in our flow.

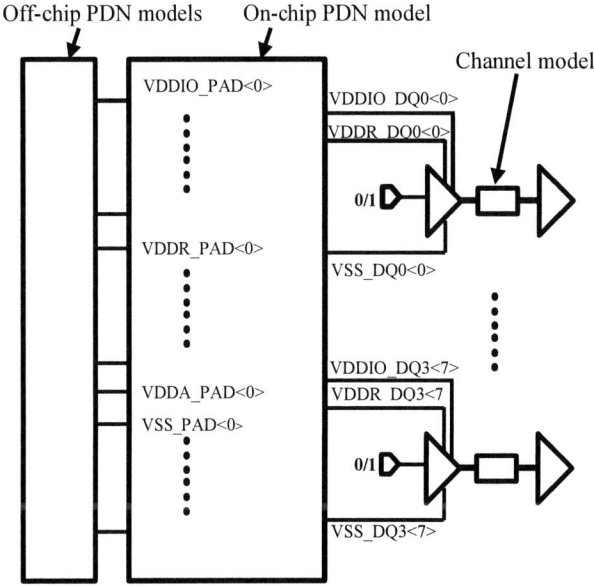

Figure 6. Schematics of our layout-based PI/SI co-simulation flow for the × 32 memory interface described in Fig.5.

Now let us verify the described flow with lab measurements. In our test chip, we have the option to set DQS bits independently of DQ bits, which offers a convenient access to probing on-chip supply noises. For example, we can set a chosen DQS_P to be always high, which then stays connected to the local VDDIO net via an approximately 50Ω resistance. Meanwhile, its complementary DQS_N is always low and stays connected to the local VSS net. Their voltage waveforms can be measured at the receiving end of the memory channel through SMA cables. A good correlation between the measured and simulated DQS voltages is demonstrated in Table 1 at three different data toggling frequencies: 100MHz, 200MHz, and 400MHz. Furthermore,

Fig. 5(a)

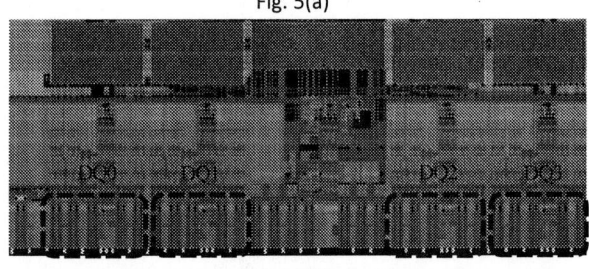

Fig. 5(b)

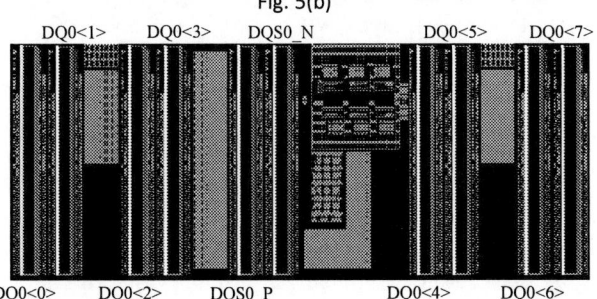

Figure 5. Layout of a × 32 memory interface implemented in a test chip using wirebond package. (a) Top view of the entire memory interface. (b) Zoomed-in view of the I/O circuits in byte DQ0.

978-1-4799-2408-0/14 $31.00 © 2014 IEEE

the shapes of simulated waveforms closely resemble those of the measured waveforms, as Figure 7 illustrates.

		DQS3_P (mv)	DQS3_N (mv)	Diff. DQS3 (mv)
100MHz	measured	67.7	62.7	50.9
	simulated	73.1	62.9	52.4
200MHz	measured	67.3	68.3	44.5
	simulated	66.8	70.2	49.8
400MHz	measured	56.4	56.7	26.1
	simulated	43.7	56.1	28.1

Table 1. Measured and simulated peak-to-peak voltages on DQS3_P/N and their difference at the receiving end of the memory channel. All DQ bits are toggling simultaneously at 100MHz, 200MHz, and 400MHz respectively.

Fig. 7(a)

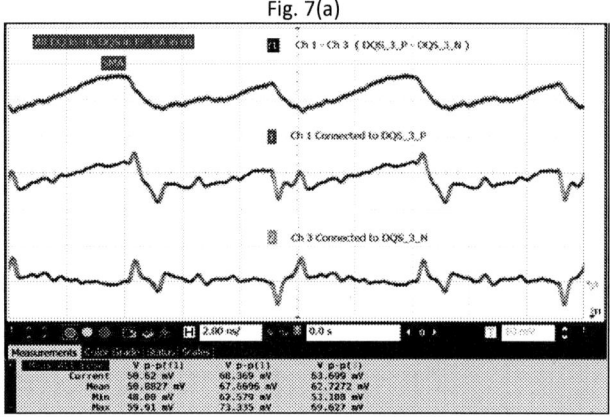

Fig. 7(b)

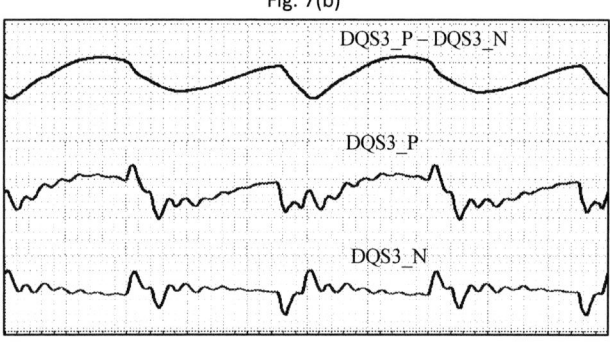

Figure 7. (a) Measured and (b) simulated voltage waveforms of DQS3_P/N at the receiving end of the memory channel. All the DQs are toggling at 100MHz data pattern.

IV. Analysis and Discussion

With the comprehensive layout-based PDN modeling flow described in section III, rigorous study of the impact of supply grid resistances becomes possible. That is, given a particular chip layout, one can modify the supply grid at any location and predict with confidence how static IR drop and SSO noise would change accordingly.

For example, in the memory interface illustrated in Fig. 5, there are six metal layers in total. The top metal layers M5

and M6 provide the global interconnects among I/O circuits, decoupling capacitors, and supply pads; the lower metal layers M1 and M2 provide the local connection between individual I/O circuits and the decoupling capacitors nearby. According to the analysis in section II, the impact of supply grid resistances strongly depend on their relative positions inside the chip. The resistance of those global interconnects on M5 and M6 is expected to affect the voltage margin very differently from the resistance of those local connections on M1 and M2.

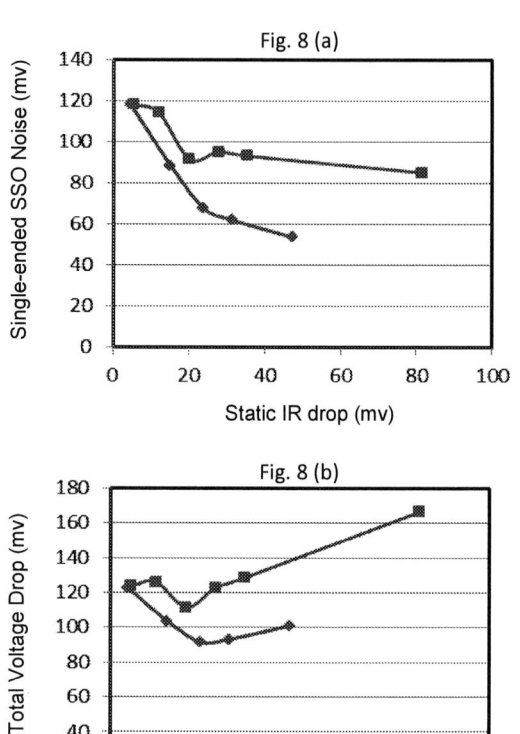

Figure 8. Result of virtual layout experiments on the memory interface described in Fig. 5. Blue curves represent the case of changing sheet resistance on top metal layers M5 and M6. Red curves represent the case of changing via resistance between lower metal layers M1 and M2. (a) Static IR drop vs. peak SSO noise (single-ended) (b) Static IR drop vs. worst total voltage drop.

To see if it is indeed the case, we perform two sets of virtual experiments on the memory interface's layout. In the first set of experiments, the only change introduced to the layout is the sheet resistance of M5 and M6; in the second set of experiments, it is instead the resistance of vias between M1 and M2. The result is summarized in Fig.8. First, it confirms that the effort of minimizing static IR drop can cause SSO noise to rise sharply. Second, it shows that the impact of supply grid resistances is truly location dependent. From voltage margin perspective, the supply grid optimization should be focused more on the global metal interconnects than on the local ones.

978-1-4799-2408-0/14 $31.00 © 2014 IEEE   564

V. Conclusion

It is demonstrated in this paper that supply grid resistances not only determine static IR drop but also strongly affect SSO noise. Therefore the conventional practice of treating static IR drop and SSO noise separately often results in sub-optimal performance of overall power integrity. A better approach is to address them in a co-design manner. Instead of trying to achieve lowest possible static IR drop or SSO noise, the goal should be to find the optimal tradeoff between the two. To accomplish this, one can start with a simple LRC circuit-based PDN model, such as those described in section II, to make initial guess of appropriate targets for static IR drop and SSO noise. These initial targets are then continuingly refined along the way of chip layout completion, with the help of a validated layout-based PDN modeling flow.

VI. Acknowledgment

The authors want to thank Dr. Gary Yip for his work on supply noise measurements.

References

1. R. Senthinathan, J. L. Prince, "Simultaneous switching ground noise calculation for packaged CMOS devices," *IEEE Journal of Solid-State Circuits,* vol. 26, no. 11, pp. 1724–1728, Nov. 1991.

2. K. T. Tang, E. G. Friedman, "Simultaneous switching noise in on-chip CMOS power distribution networks," IEEE Trans. VLSI Systems, vol. 10, no. 4, pp. 487–493, Aug. 2002.

3. H. Kim, B. K. Sun, J. Kim, "Suppression of GHz range power/ground inductive impedance and simultaneous switching noise using embedded film capacitors in multilayer packages and PCBs," *IEEE Microwave and Wireless Components Letters,* vol. 14, no. 2, pp. 71–73, Feb. 2004.

4. R. Schmitt, C. Yuan, "Power distribution design considerations and methodology for multi-gigabit I/Os ," in *Proc. 2005 International Symposium on Electromagnetic Compatibility,* Chicago, IL, Aug. 2005, pp. 660-665.

5. T. Sudo, "Behavior of switching noise and electromagnetic radiation in relation to package properties and on-chip decoupling capacitance," in *Proc. 17th International Zurich Symposium on Electromagnetic Compatibility,* Zurich, Feb. 27-Mar. 3, 2006, pp. 568-573.

6. E. Song, J. S. Pak, J. Kim, "TSV-based decoupling capacitor schemes in 3D-IC," in *Proc. IEEE Electronic Components and Technology Conference (ECTC),* San Diego, CA, May 29-June 1, 2012, pp. 1340-1344.

7. H. R. Cha, O. K. Kwon, "An analytical model of simultaneous switching noise in CMOS systems," *IEEE Trans. Advanced Packaing,* vol. 23, no. 1, pp. 62–68, Feb. 2000.

8. S. Chun, M. Swaminathan, L. D. Smith, J. Srinivasan, J. Zhang, M. K. Lyer, "Modeling of simultaneous switching noise in high speed systems," *IEEE Trans. Advanced Packaging,* vol. 24, no. 2, pp. 132–142, May. 2001.

9. H. Lan, M. Han, R. Schmitt, "Modeling and measurement of supply noise induced jitter in a 12.8Gbps single-ended memory interface," in *Proc. IEEE Electrical Performance of Electronic Packaging and Systems Conf. (EPEPS),* Tempe, AZ, Oct. 21-24, 2012, pp. 43-46.

10. Totem v13.2.1p1, *Apache Design, Inc. A subsidiary of ANSYS, Inc.*, San Jose, CA.

Improving the Target Impedance Method for PCB Decoupling of Core Power

Guang Chen and Dan Oh

Altera Corporation
101 Innovation Dr. San Jose, CA 95134
guchen@altera.com

Abstract

Decoupling core power for modern processors or SOCs is a challenging task due to large power consumption. The decoupling network designed by a commonly used target impedance approach is known to be very pessimistic and very difficult to implement. In this paper, a step surge current is identified as a major source of core power noise. By considering the ramp time of the surge current, we propose a modified target impedance method that significantly reduces the pessimism built into the original target impedance method. As an example, a real FPGA decoupling case is used to demonstrate the effectiveness of the new proposal.

I. INTRODUCTION

Target impedance method [1], introduced more than a decade ago, is still one of the most popular decoupling techniques used in these days. The target impedance approach defines a maximum impedance value based on allowable voltage tolerance and maximum current by applying Ohm's law in frequency domain as follows:

$$Z_{target} = \frac{V_{max\_allowed\_ripple}}{I_{dynamic\_current}} . \quad (1)$$

PCB decoupling process starts with comparing the impedance of power distribution network against the target impedance over the frequency of interest. Decoupling capacitors are added as needed to correct any violation.

The target impedance approach has been successfully used in designing power distribution network (PDN) for various applications including core power, analog interface supply, various digital blocks. However, as modern processors or SOCs require large power consumption, meeting the target impedance becomes extremely difficult. Recently, many authors identified that the target impedance may result in pessimistic design. In fact, in our latest large FPGA devices, the target impedance can no longer be met at the maximum power consumptions. Due to the introduction of a new process node and power saving considerations, the new FPGA devices run with reduced power supply voltage significantly decreasing allowed noise margin. However, the number of transistors built into FPGA core is drastically increased resulting in higher core power consumption and noise level.

The computation intensive applications, such as OpenCL, DSP, and GPGPU computing cases, can create a large current transition in a short period which generates large AC noise. Due to the large current change, the target impedance is very small and a large number of PCB decoupling capacitors are required. To illustrate this decoupling challenge, two different

PDN decoupling designs are compared in Table I. A total of 19 decoupling capacitors are required when the target impedance is 15mΩ. The number of capacitors required increases to 117 for the same PDN when the target impedance is reduced to 3mΩ due to the lower supply voltage and large current requirement. Designing cost effective PCB PDN is important for system designer as excessive capacitors raise BOM cost and increase difficulties of component placement.

Table I. PCB decoupling examples for two FPGA devices

	Power Requirement	Target Impedance (mΩ)	Decoupling Capacitors
Case A	1.5V, 5A, 5% ripple	15	19 caps (5x100uF, 2x2.2uF, 1x1uF, 1x0.47uF, 2x0.22uF, 5x0.1uF, 3x47nF)
Case B	0.9V, 15A, 5% ripple	3	117 caps (45x470uF, 35x330uF, 10x2.2uF, 27x1uF)

To overcome this decoupling challenge or remove some of pessimism built in the Target impedance method, experimental approaches have been applied in practice. In [2], the relaxed impedance border line is defined based on measurements using real applications. In [3], the current spectrum is first simulated for the worst case switching activity, and then, the maximum passing PDN impedance curve is measured by varying decoupling capacitors and power balls. Using the current spectrum and the maximum PDN curve, a new target ripple spec in the frequency domain is derived. In [4], time domain voltage noise is used to a new target impedance boundary with a maximum allowable inductance curve in addition to a conventional dc flat impedance line. The maximum allowable inductance is calculated based on the worst case rising time of IC switching current. However, actual switching current waveforms of real systems are very complicated since the current spectrum spreads over a wide frequency range. The maximum allowable inductance curve, therefore, cannot be applied easily in practice.

In this paper, we focus our study on core power delivery. The conventional target impedance works well for small IP blocks and interface power but it tends to be pessimistic or even impossible to decouple for core power applications where large current is required. In Section II, we discuss key critical aspects of core power noise and identify that the major core power delivery issue is decoupling of large surge current.

Then, we propose the modified target impedance curve to account for the characteristics of the surge current in Section III. Finally, a case study is performed to demonstrate a real world application of our modified target impedance concept.

II. CORE POWER DISTRIBUTION NETWORK DESIGNS

Before we propose our new modified target impedance concept, a few key aspects of core noise activity are considered in this section. Figure 1 shows a conceptual plot of the core current activity pattern.

During the low activity period, core uses very little current and induced supply noise is not a concern anymore. The most of current is due to leakage and clock tree which does not create supply noise which impacts the core performance. During the high activity period, large switching current is consumed and significant power noise is expected. Fortunately, as a core block contains many circuitries, the switching activities are randomized resulting in clock like patterns as shown in [5]. Our measurement and simulation for different applications also produce the similar behavior as shown in Figures 2 and 3. Noise profile contains a few high-frequency tones and low and medium frequency components are very small.

The high-frequency noise is well controlled by on-chip decoupling capacitor (ODC). Typical processor or SOCs place sufficient caps (including symbiotic caps) and high frequency noise should not be significant. Moreover, the high-frequency noise is less detrimental than the medium or low frequency noise in terms of timing impact perspective [6].

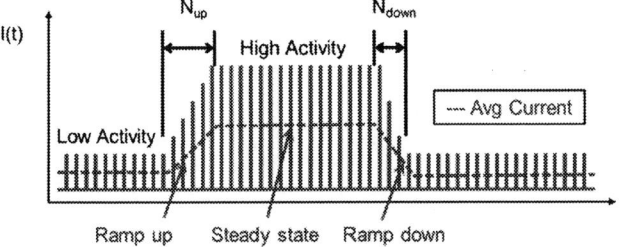

Figure 1. Conceptual representation of low and high core activity pattern

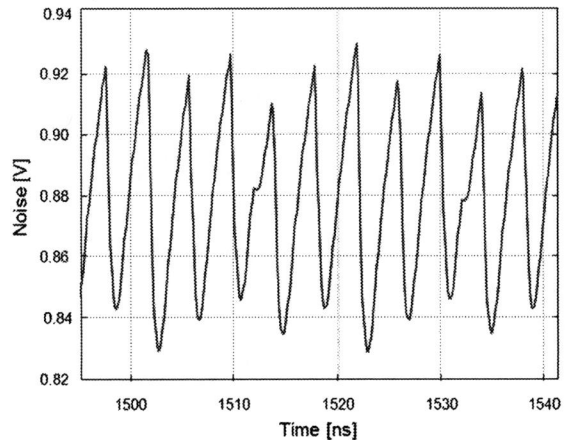

Figure 2. Simulated core noise waveform for OpenCL application

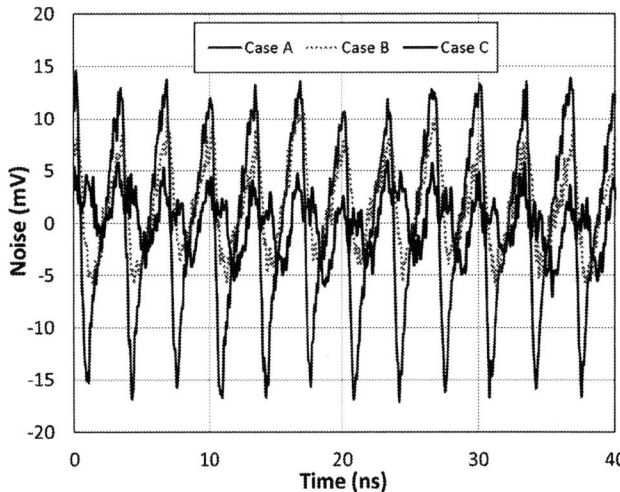

Figure 3. Measure noise waveforms for various applications

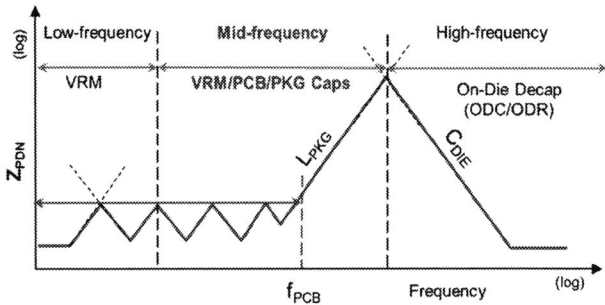

Figure 4. Typical PDN profile and major decoupling contributors for different frequency regions

Figure 4 illustrates the major contributors to the PDN impedance profile over different frequency regions. At the low frequency region, PDN voltage fluctuation is mainly regulated by voltage regulator module (VRM) and the low impedance region in PDN impedance reflects VRM's effective range in noise regulation. At mid-frequency, the impedance profile is determined by the interaction among decoupling capacitors, PCB and package design. PDN impedance profile has multiple peaks formed by PCB and package parasitic inductance and PCB decoupling capacitors in this frequency range. And these peaks are regulated by PCB decoupling capacitors. The PCB decoupling has no impact on PDN impedance profile at high frequency where the impedance profile is determined by resistance and capacitance of the on-chip PDN network. ODC is highly effective for decoupling this high frequency noise.

The transition frequency between low frequency and mid-frequency is defined as the frequency where bulk capacitors become dominant in determining PDN impedance. The location of the transition frequency depends on the type of VRM used, the load, and noise requirements. This frequency can only be identified through PDN modeling. With design improvement, modern VRM designs are capable of responding to faster slew rate and this helps reducing the bulk decoupling capacitors required. The generic VRM model, however, is conservative. One experiment shows that the measured ripple magnitude is still within spec after removing

50% of bulk capacitors in the original scheme for a PCB decoupling design. More accurate VRM model is needed in PCB decoupling design to take advantage of the design improvement.

The transition frequency between mid-frequency and high frequency is defined as the highest impedance peak formed by vertical inductance (PCB via inductance and package inductance) and on-chip capacitance (on-package decoupling capacitors and on-die capacitance). The transition frequency can be calculated based on the vertical inductance and on-chip capacitance. For some core power rails that have large on-chip capacitance or on-package decoupling capacitors (OPDs), the frequency of this boundary can be as low as a few MHz.

The impedance boundary is set by the vertical impedance (PCB via and package resistance and inductance). Adding PCB decoupling capacitors does not reduce vertical resistance and inductance. Therefore, decoupling impedance target set below this boundary cannot be achieved. The highest effective decoupling frequency for PCB PDN design, shown as f_{PCB} in Figure 4, is defined by the frequency of the crossing point of the target impedance and vertical inductance. As results, it is important that the system designers do not overdesign their decoupling design to cover the frequency range which is outside the effective PCB decoupling region. Device manufactures should provide sufficient information to avoid the potential overdesign.

Now, going back to the core activity pattern in Figure 1, the major core noise component which needs a good decoupling scheme is the transition period from the low activity to high activity. In the following section, we derive a modified target impedance curve to decouple surge current during the transition.

III. MODIFIED TARGET IMPEDANCE FOR MODELING SURGE CURRENT

Modeling of surge current highly depends on the transition time. In a conventional target impedance approach, the PDN impedance is compared against the target impedance over the frequency of interest. This assumes that the worst case current draw could show up at any point within this range which is very pessimistic in practice. This assumption is reasonable for interface power rails such as ones for parallel interface, memory IO, and various serial links because the interface link can be activated or deactivated in random fashion. Hence, there is a good chance of hitting the worst case current draw at any frequency point. However, it is extremely unlikely for core powers to draw a large current at any frequency points. The actual current spectrum can reduce this pessimism as shown in [3] but its application to processors or SOCs is limited due to general application usage.

As described in the previous section, core activity can be divided into two different states, the steady state and the transition state. During the high activity state, all transistors switch randomly and the number of switching transistors does not have large variations and this results in an steady averaged current draw at different clock frequencies as shown in Figures 3 and 4. As these noise components are well

decoupled by on-die decoupling capacitors, only the surge current needs to be decoupled by PCB and package.

The noise due to the surge current is a strong function of ramp time. If the ramp time is sufficiently large, a very minor power noise is generated; whereas, the ramp time is small, a large PDN ringing is generated. This is demonstrated in Figure 5. In the early 2000s, Intel's used the fast transition analysis (called the first-dip analysis) and predicted much worse performance degradation than actual system performance [7]. Intel later identified the gap is due to the lack of modeling jitter tracking between data and clock paths [8]. We believe the gap may come from inaccuracy in modeling the transition time.

The transition current consists of a wide range of frequency content and excites noise ranging from the first droop to PDN resonances depending on the transition time. The actual simulated noise waveforms for a typical first-dip analysis using different rising times are shown in Figure 6. As demonstrated in Figure 5, the step response of the first-dip analysis shows a strong dependency on the rising time of the transition current. In graphics applications, the reported number of the worst case current ramp cycles was 70 clock cycles where as for the recent general-purpose computing on graphics processing units (GPGPU), the worst case current ramp worsened to only 15 clock cycles [9]. In FPGA applications, we are expecting a large number of cycles.

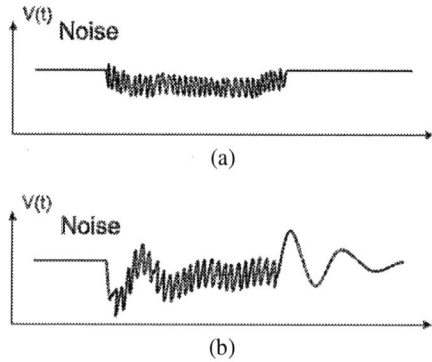

Figure 5. Noise waveforms of the surge current in Figure 1 for (a) a large transition time and (b) small transition time

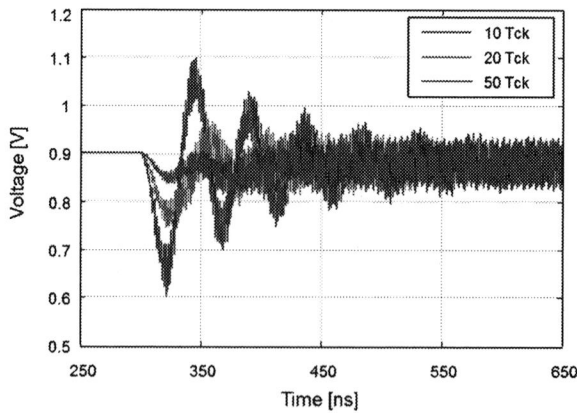

Figure 6. Simulated noise waveforms for different rising times of the transition core current

To model the impact of the transition time in conventional target impedance concept, we first plot the step response in the frequency domain using the following frequency domain expression:

$$i(t) = \frac{I_o}{T_r}(r(t) - r(t - T_r)) \Leftrightarrow I(s) = \frac{I_o}{T_r} \cdot \frac{1}{s^2}(1 - e^{-T_r s}) . \quad (2)$$

where $r(t)$ is the ramp function, I_o is the current magnitude, and T_r is the risetime. The frequency domain plot for different transition times are shown in Figure 7. The transition time is changed from 0.1ns to 1us. As shown in the figure, the current step with zero transition time is a straight line with a -20dB slope. Increasing the transition time introduces dips at the harmonics of the step transition. Furthermore, it adds an additional -20dB attenuation to the envelope of the current content starting from the knee frequency, f_{knee}, defined by

$$f_{knee} = \frac{0.35}{T_r} . \quad (3)$$

For example, the roll-off starts at approximately 350KHz if the transition time of the signal is 1µs.

As a conventional target impedance is valid for the step response with a zero transition time. Additional -20dB slope due to the finite rise time allows additional room to relax the target impedance curve. Figure 8 illustrates this concept. First, the solid black curve shows the fast step response with conventional flat target impedance and the resulting frequency domain voltage noise. The blue dotted curve shows the step response with finite rise time and the relaxed target impedance. The net voltage spectra for both cases are same.

Now, a new target impedance curve can be defined to capture the transition time to avoid overdesigning PDN beyond the knee frequency. The new proposed target impedance consists of two regions: the flat line region and slope region. The flat line region is the same as that of the original target impedance. In the slope region, the target impedance increases in a 20dB/dec ratio which is the same as the impedance increase by inductor. Hence, the relaxed slope can be defined by the equivalent inductance which is equal to the risetime:

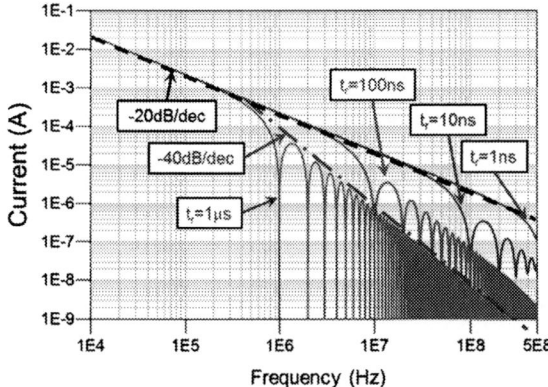

Figure 7. Frequency domain current distribution of a step current with different transition times

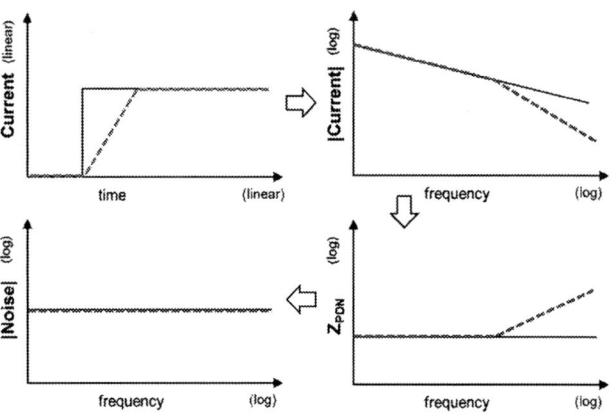

Figure 8. Illustration of the different current profiles and relaxed target impedance curves meeting the same voltage noise target

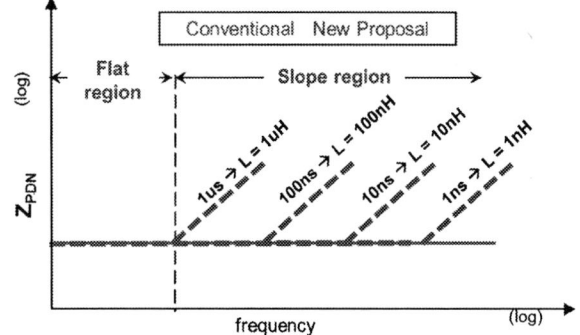

Figure 9. Modified target impedance curve with rise time models

$$L_{eq} = T_r . \quad (4)$$

The new target impedance for a few different transition times is shown in Figure 9. With the transition time impact being modeled in the target impedance, the proposed impedance curve eliminates some of pessimism in the conventional target impedance approach. Based on our new target impedance curve, the system designer can perform trade-off analysis between the current amplitude, risetime, and derive desired PDN decoupling solutions. In [4], the similar expression and target impedance proposal are derived by assuming the current profile is a triangle and calculating the voltage noise purely in time domain.

IV. PCB DECOUPLING OPTIMIZATION EXAMPLE

In this section, PCB decoupling of FPGA device is considered to illustrate how to optimize the PDN and system parameters to meet noise requirement in conjunction with our new target impedance proposal. In our test case, FPGA core draws 4A current in 16ns. Our target for voltage ripple is 10mV. The target impedance therefore is 2.5mΩ. Figure 10 shows the original PDN impedance profile. As shown in the original impedance curve there are two spots that violate the target impedance specification. The first violation below 1MHz is fixed by adding additional 10x47uF PCB decoupling capacitors. The resulting impedance is shown as the solid blue line in Figure 8. The second violation above 2MHz is hard to fix as it is caused by the vertical inductance and cannot be

978-1-4799-2408-0/14 $31.00 © 2014 IEEE   569

reduced via PCB decoupling. There is no solution to meet the noise requirement based on the conventional target impedance approach.

Figure 11 shows alternative solutions by considering the transition times. Three different target impedance profiles are depicted in the figure. All of them are defined to meet the ripple requirement of 10mV. The original PDN impedance meets the target impedance requirement if the transition frequency locates at 300kHz. The calculated rise time corresponding to the transition frequency is around 1.1μs. Hence, the original PDN design can support 4A of current as long as the transition time of surge current longer than 1.1μs. The time-domain voltage waveforms for the original waveform using the ramp time of 16ns and 1.1μs are shown in Figure 12.

For the improved PDN design, the transition frequency is raised to 2MHz and the shortest transition time allowed is reduced to 170ns. On the other hand, if the transition time cannot be changed, the maximum current change has to be reduced to 1A relaxing the target impedance to be 10mΩ.

The time-domain waveforms for all cases are shown in Figure 12. The original PDN resulted in 31mV ripple failing the spec significantly. Both the original and improved PDN designs met the spec if the proper transition times are used. For the reduced current case showed only 7mV noise due to slight over design of decoupling in mid-frequency range. In general, both current and transition time should be optimized simultaneously to achieve the best balanced performance.

V. CONCLUSIONS

In this paper, we first studied the core activity and associated current draw pattern and concluded that PCB decoupling needs to focus on the surge current during a transition state. A new target impedance curve is proposed based on the surge current distribution in frequency domain. The proposed target impedance concept accurately captures the finite risetime of surge current to the noise and significantly reduces pessimism of PCB decoupling design based on the conventional target impedance method.

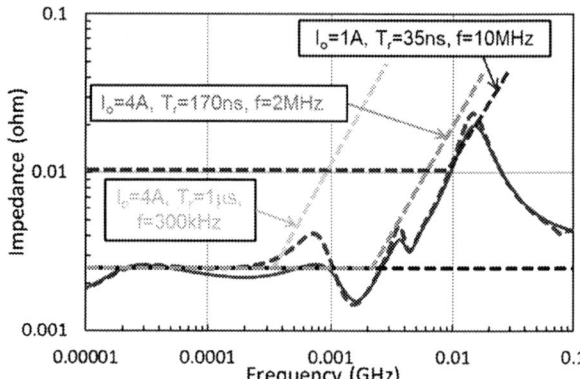

Figure 11. PDN profile before and after decoupling improvement

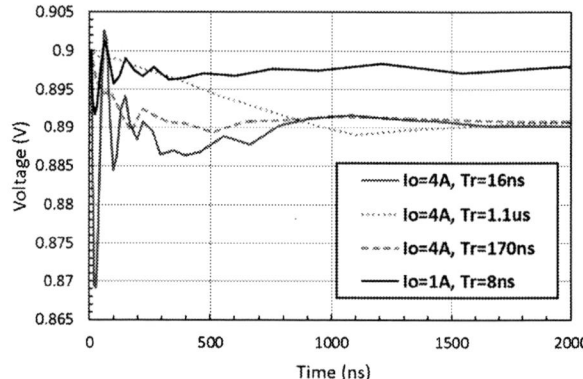

Figure 12. Simulated transient votlage noise waveform for different settings

As shown in the case study, the proposed modified target impedance allows PCB designers to explore design space and optimize decoupling design based on the needs for current and transition time. Modern processors often use power gating or clock gating to save power consumption. Sudden changes in power or clock network create large supply noise as we have demonstrated. Our new target impedance can be effectively used to determine the shortest ramp up time allowed for these gating operations. Many global chip-level operations, which simultaneously operate over a large number of blocks including power and clock gating, are not performance critical and do not need to be decoupled for the worst case. Such decoupling schemes would be very expensive or even impossible to be implemented. An architecture or circuit level solution is better a way to address these issues.

Finally, it should be noted that the proposed approach is valid for core power decoupling where large current is drawn by many circuitries. For other small IP or IO interface power rails, the conventional target impedance may be still the best approach. When the worst case current profile is known, the existing target impedance can be relaxed based on the current spectrum content.

Acknowledgments

Authors would like to thank Mr. Cuong Nguyen and Dr. Yujeong Shim for simulation data and helpful discussion and Mr. Wern Shin Choo for measurement data.

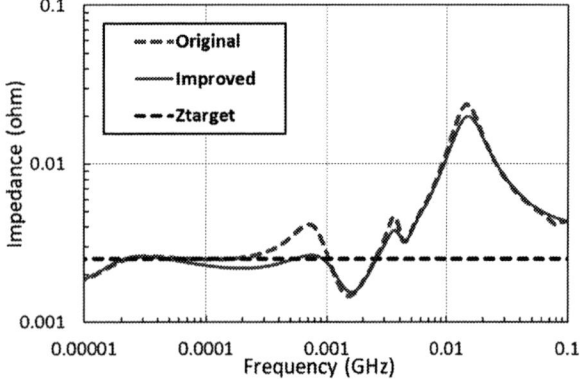

Figure 10. PDN profile before and after decoupling improvement

References

1. L. D. Smith, R. E. Anderson, D. W. Forehand, T. J. Pelc, and T. Roy, "Power distribution system design methodology and capacitor selection for modern CMOS technology," *IEEE Trans. on Advanced Packaging*, pp. 284-291, Aug. 1999.

2. M. S. Tanaka, M. Toyama, H. Nakashima, J. Yamada, M. Haida, and I. Osshima, "Chip oriented target impedance for digital power distribution network design," in *Proceedings of IEEE Electrical Performance of Electronic Packaging and Systems Conference*, Oct. 2012, pp. 220-223.

3. J. Lee, J. K. Paek, K. Park, J. Lim, J. Yun, and J. Lee, "Decoupling capacitor optimization by spectral current-based power delivery network impedance formulation for SSD/eMMC/uSD card," presented at the *IEC DesignCon*, 2014.

4. J. Kim, S. Wu, H. Wang, Y. Takita, H. Takeuchi, K. Araki, G. Feng, and J. Fan, "Improved target impedance and IC transient current measurement for power distribution network design," *IEEE International Symposium on Electromagnetic Compatibility*, pp. 445-450, 2010.

5. S. Lin, M. Nagata, K. Shimazaki, K. Satoh, M. Sumita, H. Tsujikawa, and A. T. Yang, "Full-chip vectorless dynamic power integrity analysis and verification against 100uV/100ps-resolution measurement," in *Proceedings of IEEE Custom Integrated Circuits Conference*, Oct. 2004, pp. 509-512.

6. D. Oh, A. Razmadze, and K. Chandrasekar, "Power integrity analysis for core logic blocks," in *Proceedings of IEEE Electrical Performance of Electronic Packaging and Systems Conference*, Oct. 20123, pp. 79-82.

7. T. Rahal-Arabi, G. Taylor, M. Ma, and C. Webb, "Design and validation of the Pentium® III and Pentium® 4 processors power delivery," in *Symposium on VLSI Circuits Digest of Technical Papers*, June 2002, pp. 220-223.

8. K. L. Wong, T. Rahal-Arabi, M. Ma, and G. Taylor, "Enhancing microprocessor immunity to power supply noise with clock-data compensation," *IEEE Journal of Solid-State Circuits*, pp. 749-758, Apr. 2006.

9. W. Mao, Y. Zhou, A. Naik, and S. Sudhakaran, "Study of BGA package cap for high-performance computing GPU," *IEEE International Symposium on Electromagnetic Compatibility*, pp. 858-862, 2013.

Process Development to Enable 3D IC Multi-Tier Die Bond for 20μM Pitch and Beyond

Y.H. Hu[1], C.S. Liu[2], M.T, Chen[2], M.D. Cheng[2], H.J. Kuo[2], M.J. Lii[2], A. La Manna[3], K. J. Rebibis[3], T. Wang[3],
S.V. Huylenbroeck[3], R. Daily[3], G. Capuz[3], D. Velenis[3], G. Beyer[3], E. Beyne[3], Doug C.H. Yu[2]

[1]TSMC assignee at IMEC
[2]TSMC, 8, Li-Hsin Rd. 6, Hsinchu Science Park, 300-77 Hsinchu, Taiwan
[3]IMEC, Kapeldreef 75, B-3001 Leuven, Belgium.
[1]Phone: 32-16-288112, Fax: 32-16-281576; E-mail: yhhuc@tsmc.com

Abstract

We demonstrate for the first time 3D multi-tier (N=4) 50μm thin die bonding for 3D IC technology using low bonding temperature and pressure for Cu TSVs bonded on Cu bumps with a cost effective structure. Die-to-die (D2D) thermal compression bonding (TCB) process with scrubbing is carefully studied in order to improve the bump height TTV and surface roughness. The bonding temperature and pressure can also be reduced significantly to below 220C and 100MPa. The standalone thin die warpage initially 15μm is reduced to 5.4μm by applying the optimized TCB process.

The electrical characterizations show good daisy chain connections between each stacked chip and the resistances are very close to the theoretical values. The cross section SEM proofs good TSV alignment to Cu bump, and TSV nails deform and land nicely onto the Cu bump.

Finally, we propose to move forward to die-to-wafer approach and migrate to 10μm bump pitch for advanced package application.

Introduction

Through-silicon via (TSV) is a key technology to enable 3D IC development and to extend Moore's Law [1-3]. 3D IC using Cu TSV direct bond is cost effective approach as it allows omitting backside re-distribution layer (RDL) and backside bumping process [4-6].

In this study, we present 3D multi-tier (N=4) thin die bonding and stacking with good interconnections between TSV and Cu bump by using TCB process. Normally, TCB process is carried out under high bonding temperature, pressure, and longer bonding time. We study TCB scrubbing to overcome these issues and successfully implement scrubbing into the process. The influences of TCB scrubbing to bump height TTV, surface roughness, and bump shear force are discussed herein. Further, we revealed the warpage behaviors on standalone thin die and stacked chips. Lastly, electrical characterization and SEM inspection on TSV+ Cu bump will be shown.

3D Structure and Process Flow

The stacking structure is carried out via 50μm thin top die bonded on 200μm bottom die by D2D sequential bonding process. The bottom die has Al pad in the peripheral area for wire bonding package (Figure 1). The front side of top wafer has Cu bump with 12.5μm diameter and 5μm bump height, and backside has TSV with 5μm in diameter and 50μm in depth. The minimum bump pitch is only 20μm. The top wafer is handled by temporary bonding process to grind to 50μm thickness, and the TSV Cu nail is then revealed by wet recess etching process (Figure 2). The front side of bottom wafer has Cu bump on die center with 12.5μm

diameter and 5μm bump height, and wire bonding Al pad on die edge (Figure 3).

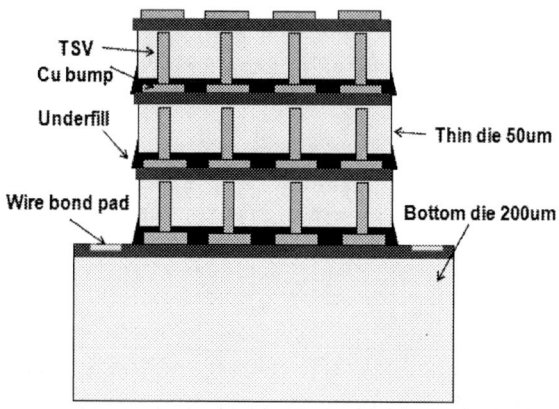

Figure 1: 3D multi-tier (N=4) stacking structure. The stacking process is carried out via back-to-face and die to die (D2D) sequential bonding.

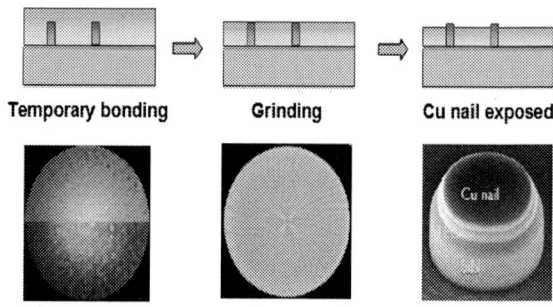

Figure 2: Top wafer PTCO process flow. The 20μm pitch TSV wafer is handled by temporary bonding and thinned down to 50μm by backside grinding. The 2μm Cu TSV nails received thin Sn layer as metal finish.

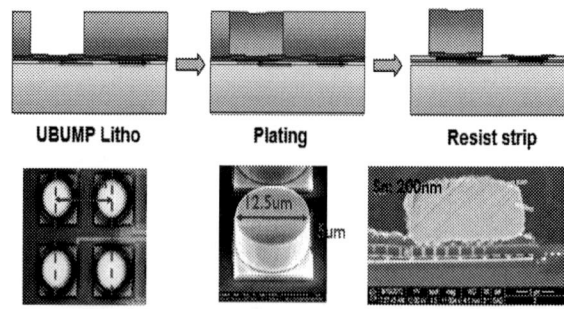

Figure 3: Bottom wafer PTCP process flow. The 20μm pitch Cu bump has 12.5μm diameter and 5μm bump height with 200μm thickness. The Cu bumps are covered by a thin Sn layer as a metal finish.

978-1-4799-2408-0/14 $31.00 © 2014 IEEE   2014 Electronic Components & Technology Conference

Electroless Sn is plated as metal finish on front side of Cu bumps and backside of TSV Cu nails for both top and bottom die prior to multi-tier thin die bonding by TCB. The bonding is performed via D2D and back to face sequential bonding approach (Figure 4).

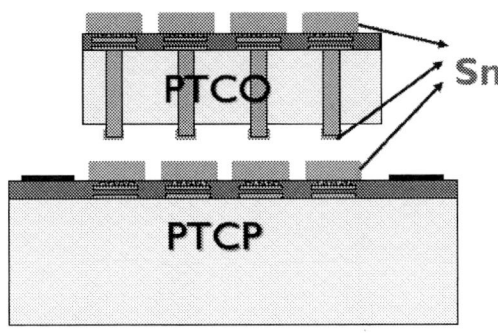

Figure 4: D2D bonding scheme. Both top and bottom die processed using 65nm CMOS technology with 2 metal layers. Electroless Sn plating are implemented on Cu bumps and backside Cu TSV nails.

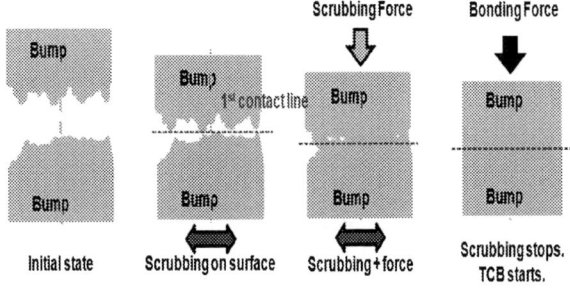

Figure 5: Principle of TCB scrubbing. Scrubbing is applied prior to TCB process, and enables the first contact on the bump surface. With the movement on the X-Y plane, the bump surface is smoothed prior to TCB process.

Figure 6: Surface roughness comparison before and after TCB scrubbing. Bump height TTV improved from 0.74μm to 0.52μm after scrubbing.

TCB Scrubbing

Conventionally, TCB process is carried out under high bonding temperature and pressure, inducing Cu bump oxidation and thin dies crack issues. We introduced scrubbing into TCB process to achieve low temperature <200C and pressure bonding <100MPa for multi-tier thin die process (Figure 5). Scrubbing with low scrubbing frequency is

used to create good bump height TTV and flatten the Cu bump surface (Figure 6). The key parameters, scrubbing time and force, are investigated. From Figure 7, we can conclude that longer time and smaller force results in better scrubbing performance (Figure 7).

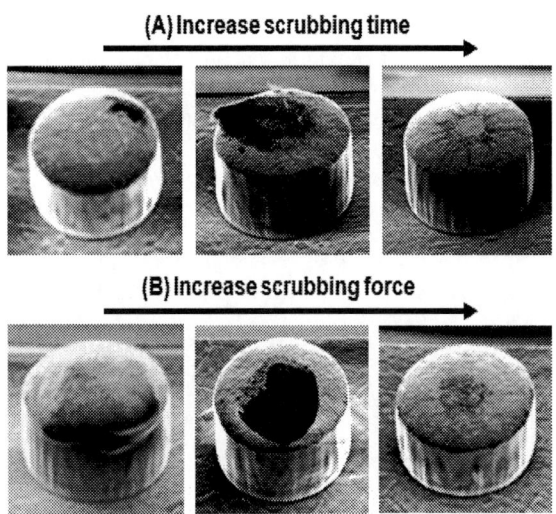

Figure 7: Top view of bump surface on TCB (A) scrubbing time and (B) scrubbing force. Longer scrubbing time and lower scrubbing force are preferred to have better scrubbing performance.

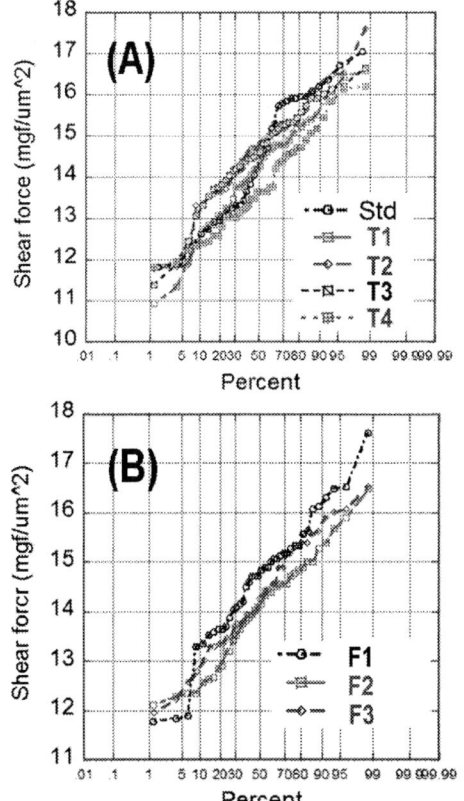

Figure 8: Bump shear force test on TCB parameters (A) scrubbing time and (B) scrubbing force. No significant differences with scrubbing time from T1 to T4, and scrubbing force from F1 to F4.

As scrubbing induces in-plane mechanical force on bump surface, it is necessary to check the bump shear force results to investigate the influence on scrubbing time and force. No significant differences are found on bump shear force (Figure 8).

Multi-tier Die Bonding

The warpage behaviors on a standalone thin die and stacked chips are first time revealed from N=1 to N=4 (Figure 9). The top thin die initially has 15μm high warpage with a smiling shape, and after TCB the warp reduces to 5.4μm on N=2 stack. The warpage remains 5μm when more top thin dies are bonded to N=4.

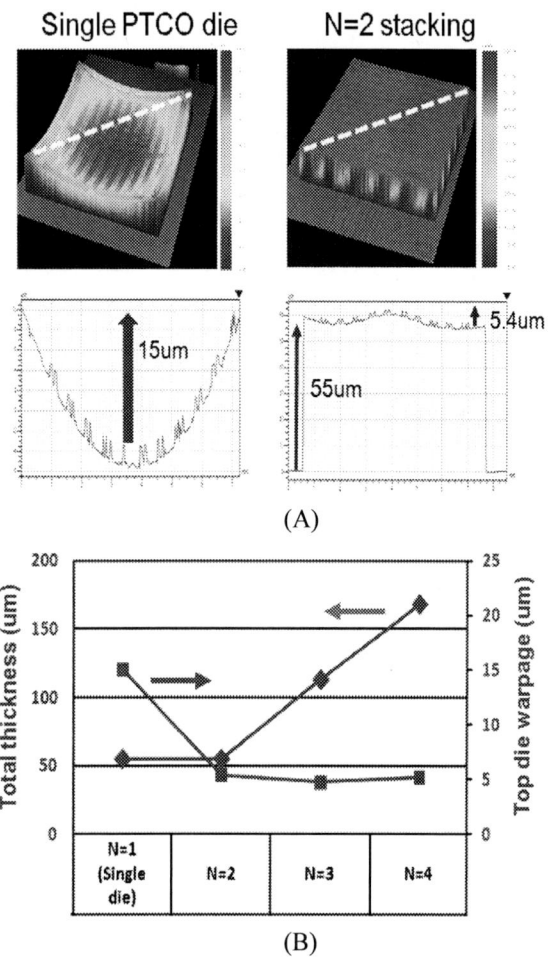

(A)

(B)

Figure 9: Warpage evaluation. (A) Single PTCO top die and N=2 stacked die. (B) Multi-tier stacked die from N=2 to N=4.

Electrical Characterization

The electrical characterization on 3D stacks is performed by measuring "TSV + bump" on Kelvin and daisy chain structures (Figure 10-12). The Kelvin structure resistances from C0 to C2 level have similar low value ~170mohm close to the theoretical value, indicating good interface bonding quality and good connections between TSV and Cu bump on each stacked die (Figure 10). The daisy chain resistances on C0-C2 level also fit well to the expected value (Figure 11-12), which means good interconnecting on TSV + bump.

Cross-section SEM shows that Cu TSVs aligned very well to Cu bumps on each stacked chip, and the bottom of TSV nails deform and connect nicely to the top of Cu bumps (Figure 13).

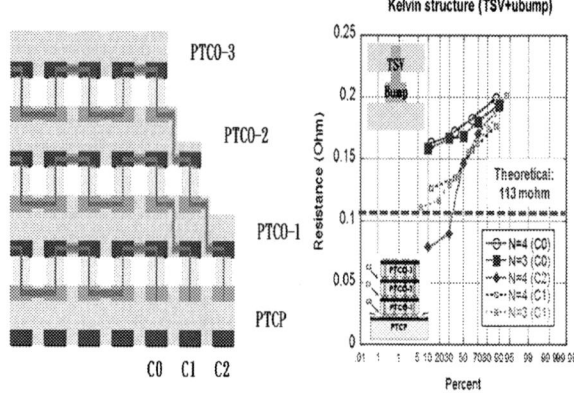

Figure 10: Electrical connection measurements on Kelvin structure. C0-C1-C2 level on three PTCO die back-to-face assembled on a PTCP substrate. The resistances are working and close to the theoretical value 113mohm.

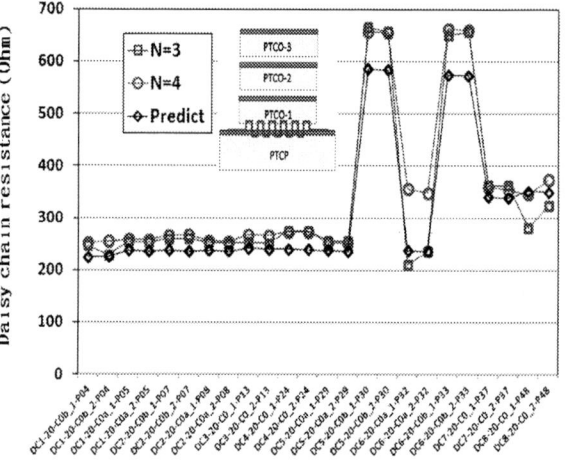

Figure 11: Electrical connection measurements on daisy chain on C0 level between bottom PTCP and top PTCO-1. There are 24 daisy chain interconnections and all the resistances fit well to the predict value on N=3 and N=4 samples.

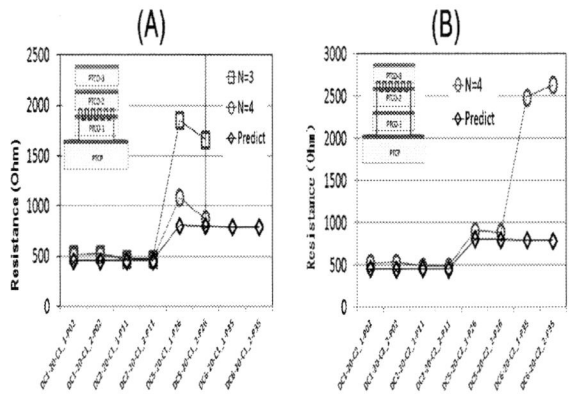

Figure 12: Electrical connection measurements on daisy chain on (A) C1 level between PTCO-1 and PTCO-2 and (B) C2 level between PTCO-2 and PTCO-3.

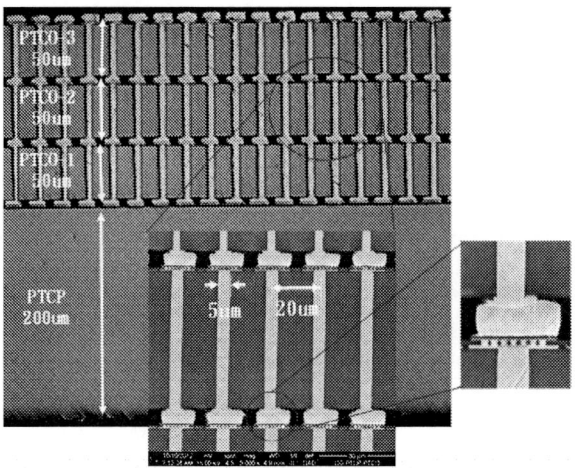

Figure 13: SEM cross-section on N=4 stacked chips. 20um pitch TSV nails deformed and connected nicely to the bottom of Cu bumps.

Conclusions

We developed a 3D multi-tier thin die stacking process for Cu TSVs directly bonded on Cu bumps. Scrubbing with low scrubbing frequency is implemented successfully to achieve good interconnection between TSV and Cu bump on each stacked die. Electrical resistances and SEM showed that the process is reliable and can be adapted for next generation fine pitch package application.

Outlook

Further work will focus on die to wafer (D2W) approach and migrate to 10um bump pitch bonding process.

References

1. Katti, G. et. al., "3D stacked ICs using Cu TSVs and die to wafer hybrid collective bonding," IEDM (2009) pp. 357-360.
2. Agarwal R, Zhang W, Limaye P, Labie R, Dimcic B, Phommahaxay A, Soussan P. Cu/Sn Microbumps Interconnect for 3D TSV Chip Stacking. ECTC 2010, p415.
3. Hu YH, Liu CS, Lii MJ, Rebibis KJ, Jourdain A, La Manna A, Beyne E, Yu CH. Cu-Cu hybrid bonding as option for 3D IC stacking, IITC 2012, p1-3.
4. Hu YH, Liu CS, Lii MJ, La Manna A, Rebibis KJ, Zhao M, Beyne E, Yu CH. 3D stacking using Cu-Cu direct bonding for 40um pitch and beyond, ESTC 2012, p1-5.
5. Hu YH, Liu CS, Lii MJ, Rebibis KJ, Jourdain A, La Manna A, Beyer G, Beyne E, Yu CH. 3D stacking using Cu-Cu direct bonding, 3D Systems Integration Conference (3DIC), 2011 IEEE International, p1-4.
6. Daily R, Wang T, Capuz G, Miller A. Micro scrubbing: an Alternative Method for 3D Thermo compression Bonding Cu Cu Bumps and High Bump Density devices with Low Force, Time and Temperature. ECTC 2013, p1-5.

Factors in the Selection of Temporary Wafer Handlers for 3D/2.5D Integration

Bing Dang, Bucknell Webb, Cornelia Tsang, Paul Andry, and John Knickerbocker
IBM T. J. Watson Research Center
1101 Kitchawan Road, Yorktown Heights, NY 10598
dangbing@us.ibm.com

Abstract

This paper reviews the recent progress in temporary wafer handling technologies for 3D/2.5D wafer integration. Several critical factors in the selection of a temporary wafer handler technology for 3D/2.5D integration will be discussed and some recommendations are made.

Introduction

3D/2.5D integration with through-silicon-via (TSV) usually involves wafer thinning below 100um thickness, which often requires a temporary wafer handler solution. In the recent years, many temporary bonding technologies have been developed to support TSV wafer processing [1-4].

Glass and silicon have been used for temporary wafer handlers and each has their advantages and limitations as a temporary wafer handler material. There are many critical requirements for mechanical, thermal, and electrical properties of the semiconductor wafers during TSV wafer manufacturing. For instance, a temporary wafer handler must help support the mechanical integrity of an ultra-thin TSV wafer. Also, the temporary handler needs to address the well-known wafer warpage issue of highly-stressed TSV wafers because standard electrostatic chucking (ESC) in semiconductor equipment can only tolerate a certain degree of warpage [5]. The use of glass or silicon handler may have different implications on the chucking technology because of compatibility issue.

Meanwhile, high-quality thin film deposition may require temperatures beyond 350 °C. Good thermal stability is a key to the selection of a temporary bonding adhesive, while chemical resistance is also necessary to support wet processing. In addition, wafer-level test and burn-in usually generate significant amount of heat, especially for high-power microprocessor products [6]. As a result, a temporary wafer handler solution needs to meet the heat dissipation requirement based on the power level.

Finally, as a temporary handler technology, a fast debonding method is critical to separate a TSV wafer from its handler wafer after all the processing is completed. The reported debonding mechanisms include "thermal-sliding" [1], "chemical-assisted release [2], "laser-assisted" release [3], and room-temperature "mechanical peeling" release [4]. Each of them has advantages and limitations. A debonding technology should be selected based on temperature and throughput requirements.

All the above factors need to be carefully considered in order to select a proper technology for certain products/applications involving thin TSV wafers. In this work, the key factors in selecting of a temporary wafer handler technology is reviewed and discussed. Silicon and glass will be compared as a temporary wafer handler material

from the perspective of inspection, chucking, warpage, heat dissipation, as well as debonding options.

Wafer warpage consideration for 3D/2.5D integration using a temporary handler

Wafer warpage is commonly observed even in conventional 2D semiconductor manufacturing [7]. It is exacerbated when excessive stress is induced by through-silicon-vias (TSVs), dielectric films as well as metal wiring on 3D/2.5D wafers [8]. Meanwhile, TSV integration typically involves thinning of silicon wafers that leads to significant reduction in their mechanical strength and bending stiffness. Therefore, temporary wafer handlers have been widely used to support the mechanical integrity of thinned device wafers throughout 3D wafer processing. Using a finite-element-analysis (FEA) simulation, wafer warpage can be approximately estimated based on the basic physical properties of the main materials in a thin wafer bonded with a temporary handler, as listed in Table 1.

Table 1. Physical properties of various materials for the estimate of warpage in a bonded wafer pair

	Si	Glass	SiO$_2$	Polyimide
Young's modulus (GPa)	165	84	70	3.2
CTE (ppm/°C)	3.1	3.2	0.55	55

In this study, a 32nm CMOS wafer with 13 levels of ILD/metal build-up and integrated TSV is used as an example to illustrate the change of wafer warpage. Figure 1 shows the wafer warpage as a function of thickness for various conditions including an non-supported CMOS wafer, an unsupported thinned wafer, a thinned TSV wafer supported with a glass handler, and a thinned TSV wafer supported with a silicon handler.

As indicated in Figure 1, as a wafer is thinned, the wafer warpage increases dramatically if it is not supported by any handler since mechanical strength become weaker. The wafer warpage can be more than 15 mm when a TSV wafer is thinned down to 50 μm post wafer fabrication. In fact, it is well known that the mechanical strength of a silicon wafer is proportional to the square of its thickness. Figure 2 illustrates the basic trend of the overall wafer warpage as a function of its thickness and TSV thickness in various conditions. For a non-thinned TSV wafer, plot A (red) indicates that the deeper the TSV is, the larger the warpage is observed for a non-thinned TSV wafer because a higher mechanical stress is introduced by the via-filling metal. Once such a TSV wafer is bonded with a handler, the overall wafer warpage becomes very complex because of the mechanical coupling between different materials such as silicon, glass, dielectric film, metal thin film, adhesive, and so on. Plot B and C (blue dotted line) indicate that there is an obvious difference between glass handler case and silicon handler case. In the glass handler case, the overall warpage is much higher than the original non-thinned TSV wafer. This may be explained by the fact that the modulus of the glass is only about half of that for silicon. The overall stiffness of a bonded wafer stack is actually lower than the original non-thinned TSV wafer. As the silicon portion of the bonded pair is thinned, the glass handler gradually dominates the overall mechanical strength. On the other hand, the silicon handler case shows reduced overall warpage because a silicon handler essentially adds additional stiffness to the original TSV wafer. As the TSV portion in such a wafer pair is thinned, the overall warpage is actually slightly reduced.

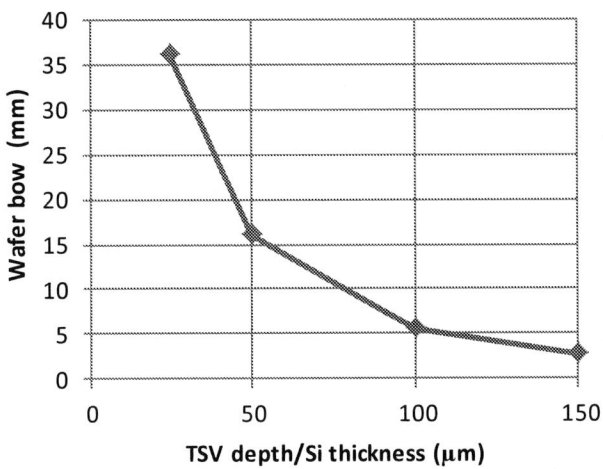

Figure 1. An example of simulated warpage of a CMOS wafer with 13 levels wiring and TSV as a function of its thickness without handler support.

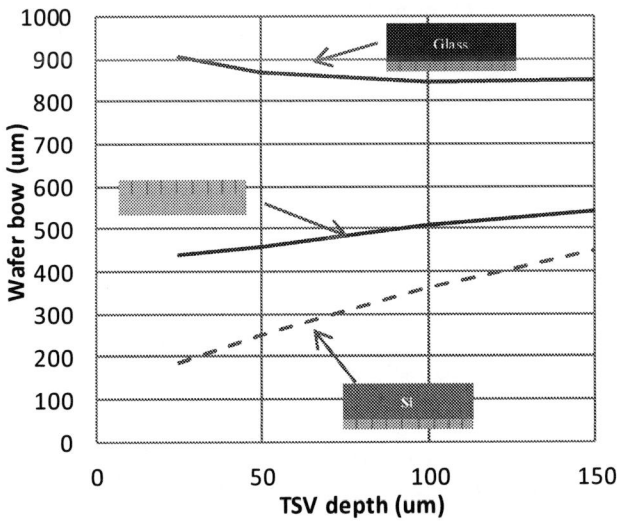

Figure 2. Simulated wafer warpage as a function of the TSV wafer thickness(or TSV depth) for a non-thinned TSV wafer (red), thinned TSV wafer bonded with glass handler (purple), thinned TSV wafer bonded with a silicon handler (blue dotted line).

Nevertheless, the above simulation is the very much simplified without consideration of process dependent stress just for the purpose of illustration. In practice, there are many methods to intentionally add or reduce stress in the bonded wafer pairs intentionally in order to balance the stress condition on the two sides of a wafer [8].

During semiconductor wafer processing, wafers need to be held stable and flat in semiconductor equipment. Electrostatic chucking (ESC) technology has been widely used for conventional semiconductor Si wafers [5,9] in vacuum chambers. Conventional ESC chucking technology is based on the electrostatic forces between the chuck plate and the bottom of a wafer since a capacitor is formed. Conductive or semiconductor wafers can be electrostatically clamped by a simple parallel plate capacitor configuration. In this case, the clamping force can be simply calculated as:

$$F = C \cdot \frac{U^2}{2d} = \varepsilon \frac{AU^2}{2d^2} \qquad (1)$$

As described previously, significant wafer warpage is usually present in 3D/2.5D TSV wafers that may impact the distance between the electrodes (d). Therefore, higher voltage may be required in order to generate sufficient attractive force to flatten a TSV wafer (or a bonded wafer pair). Another type of ESC chuck is based on two metal electrodes embedded in a dielectric bed, also called interdigitated or "bipolar" type chuck. Its advantage is that it does not require electrical connection to a wafer or substrate to be chucked. Therefore, an insulator substrate such as a glass wafer can be clamped by a "dipole" attraction mechanism. However, such an ESC chuck usually yields lower clamping force because of the reduced capacitor area [9]. Its clamping force can be calculated as follows:

$$F_{bp} = \varepsilon \cdot \frac{AU^2}{8 \cdot d^2} \qquad (2)$$

In summary, if a glass handler is used to support a device wafer, the bipolar configuration type ESC is required and a higher voltage may be needed to provide sufficient clamping force.

Wafer inspection consideration

Another equipment issue for a 3D/2.5D TSV wafer bonded on a handler is its inspection. Defects or voids may be present when a device wafer is bonded to a handler wafer using an adhesive. Therefore, proper tooling should be used for the inspection. The commonly available inspection options include optical microscope, infrared (IR) microscopes, and scanning acoustic microscopes (SAM). A glass handler can be easily inspected using optical microscopy because it is optically transparent (see Figure 3).

A glass handler is also compatible with IR microscopy or scanning acoustic microscopy (SAM). However, a Si handler will require special Infrared microscopy because it is not optically transparent. Figure 4 shows an example of IR absorption characteristics of typical Si wafers. With wavelength longer than 1.12um, a Si substrate becomes semi-transparent to IR light. Therefore, IR microscopy may be used for its inspection. However, IR microscopy will be hampered by the presence of any metallization.

978-1-4799-2408-0/14 $31.00 © 2014 IEEE

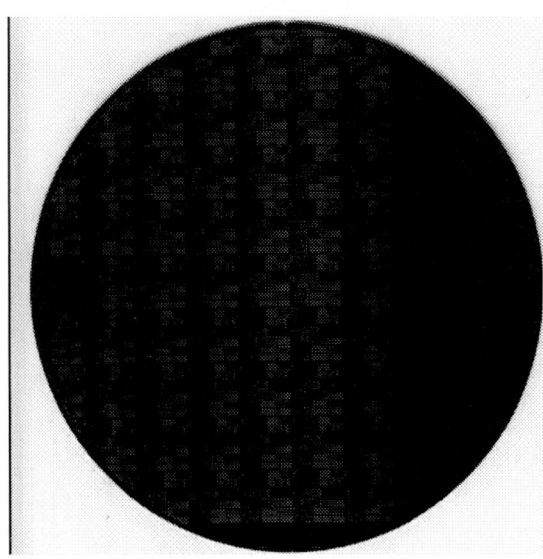

Figure 3 An example of optical inspection on the bonding adhesive through a glass handler

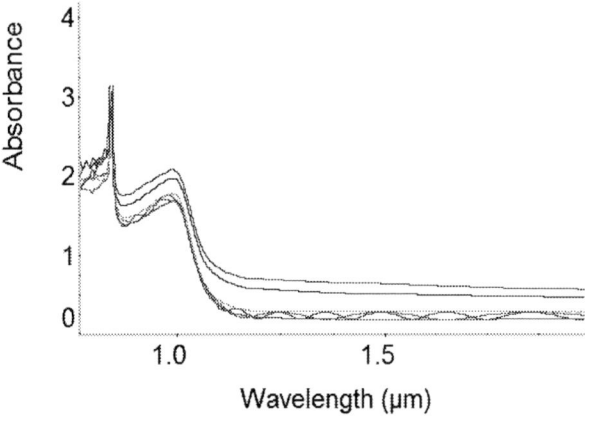

Figure 4. An example of optical absorption characteristics of various Si wafers.

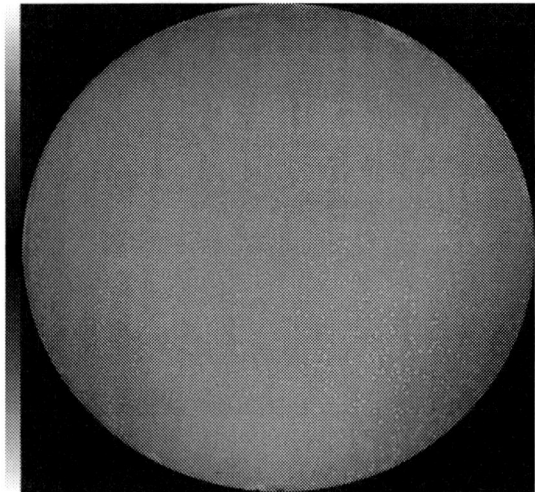

Figure 5. An example of CSAM inspection on the bonding adhesive through the silicon handler

In contrast, SAM seems to be an attractive option for a 3D wafer bonded on Si handler because it is usually compatible with metallization layers. However, the inspection through-put of a SAM tool depends on the spot resolution of the transducer. Figure 5 shows an example of SAM inspection result on a bonded wafer pair to illustrate some common bonding defects.

Thermal conductivity consideration for a temporary wafer handler during wafer level test

In addition to the mechanical integrity, thermal resistance is another important factor for a 3D wafer bonded on a handler wafer because heat dissipation may be required during processing as well as wafer-level test and burn-in [6]. In particular, good heat dissipation is critical for wafer-level test of high power microprocessor products. Water cooled heat sink (or namely cold plate) is commonly used during wafer-level test to assist the heat dissipation. The overall thermal resistance can be significantly changed when a silicon device wafer is bonded to another handler wafer because of the added thermal resistance from the adhesive and the handler.

Table 2. Thermal properties of various materials for a TSV wafer during wafer-level test and burn-in.

	Si	Glass	Adhesive A	TIM	Heat sink
Thermal resistivity (°C-mm/W)	7	833	2702	1786	1
Typical thickness (mm)	0.05	730	0.01~0.09	0.01	1

Table 2 lists thermal resistivity and thickness of various materials involved in typical temporary wafer bonding as well as wafer-level test. In general, glass handers are not a good thermal conductor because thermal resistivity of a typical glass is more than 100 higher than that of silicon. It is also noticeable that a typical bonding adhesive is also poor thermal conductor, which has more than 2 times higher thermal resistivity than that of glass.

With a simplified model, thermal resistance of primary components of a TSV wafer bonded with a glass handler or a silicon handler can be calculated then. Figure 6 shows the approximate thermal resistance contribution of primary components in a TSV wafer bonded with a glass handler or a silicon handler during wafer-level test. Because of its high thermal resistivity, the handler is the dominant component when a glass handler is used to support a thinned TSV wafer. In case of silicon handler, the bonding adhesive becomes the dominant component because of the low thermal resistivity of silicon. Depending on the TSV integration process requirement, the required adhesive thickness may vary significantly from one product to another. For example, if the final metallization (Ni/Au etc.) on a TSV wafer is relatively thin (<10μm), it does not require a very thick bonding adhesive to bond to a handler wafer. However, if conventional C4 solder bumps are already present on a TSV wafer prior to wafer thinning, a thick bonding adhesive is necessary (up to 70 – 90um) in order to embed the C4 solder bumps between the handler and the device wafer. Since the thermal resistivity of a typical polymeric adhesive is very high, the overall thermal resistance of the bonded wafer pair increases accordingly.

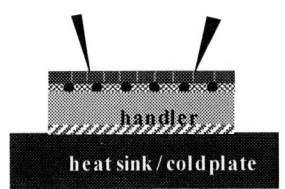

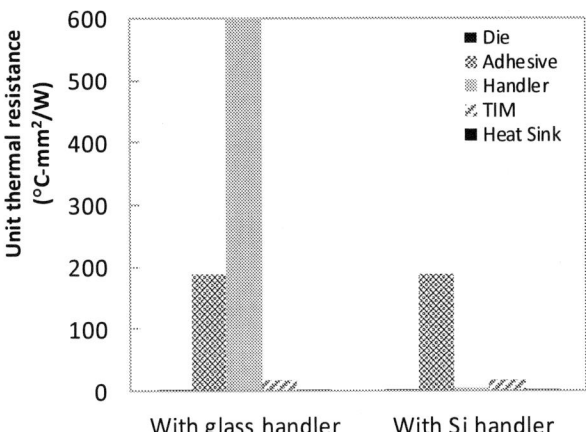

Figure 6. The primary thermal resistance components for bonded 3D wafer on glass handler or Si handler (assuming adhesive is 0.07mm thick) during wafer-level test and burn-in.

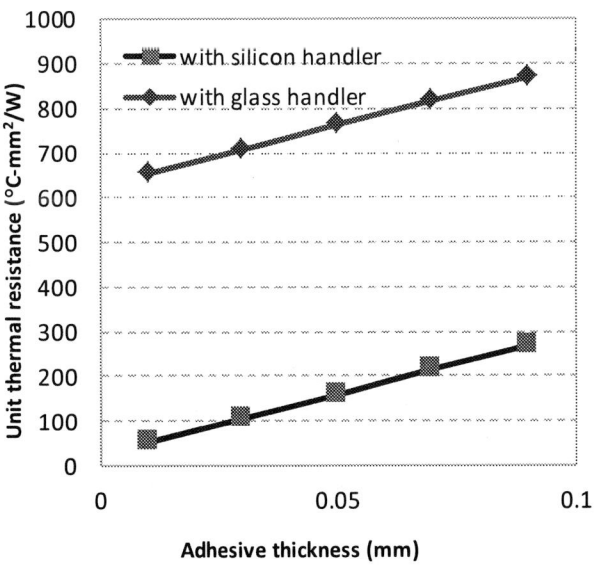

Figure 7. Calculated unit thermal resistance as a function of bonding adhesive thickness for a thinned TSV wafer bonded with a glass handler or a Si handler

Figure 7 illustrates that the calculated overall unit thermal resistance of a device wafer bonded on either a glass handler or a silicon handler increases linearly with the adhesive

thickness. In general, the overall thermal resistance of a TSV wafer bonded with a glass handler is nearly 10 times higher than that of a TSV wafer bonded with a silicon handler. For high power applications, the heat dissipation during wafer level test and burn-in might be a challenge. The silicon handler can offer lower thermal resistance, therefore, is a good candidate for high-power applications. However, when a silicon handler is used for the purpose of good thermal conductivity, the adhesive thickness should be carefully controlled because a thick adhesive will dominate the overall thermal resistance.

Recent progress in release technologies for temporary wafer handlers

In the past few years, a lot of progress has been made in temporary wafer handling technologies across the semiconductor industry. The "thermal sliding" technology is self-limiting in terms of thermal stability [1]. Therefore, this paper will focus on the other three main handler release options that are widely studied and reported [2-4]. As summarized in Table 3, TOK and Suss Micro-tec reported the use of chemical-assisted release [2, 10] with acrylic and polymide-based adhesives respectively. In this approach, a solvent dissolvable adhesive and a perforated handler are required to release a device wafer from the handler wafer, as shown in Figure 8. A handler may be a glass or silicon wafer as long as it is perforated to allow the penetration of dissolving chemicals. The drawback of such an approach is the relatively high cost and lower mechanical strength of the special perforated handlers.

Table 3. Summary on recent temporary wafer handler release technology options

	Chemical-assisted release	Laser-assisted release	Mechanical release
Handler type	Perforated Si or Glass	Glass	Si or glass
Operation Temperature	Low to medium temp.	Room temp.	Room temp.
Equipment specialty	Rinse chamber	Laser	"Peeling chuck"
Compatibility with dicing tape	Protection needed	Compatible	Compatible
Through-put	Low to medium	High	High
References	[2] [10]	[3] [10] [11]	[4,12]

Meanwhile, IBM, Suss Micro-Tech, and 3M [3,10,11] reported laser-assisted release technologies. In this approach, UV or near-IR laser has been used to radiate the adhesive or release layer to break the bonding interface or degrade the adhesion. Because the handler must be optically transparent for the selected laser wavelength, all these studies have focused on glass handler so far.

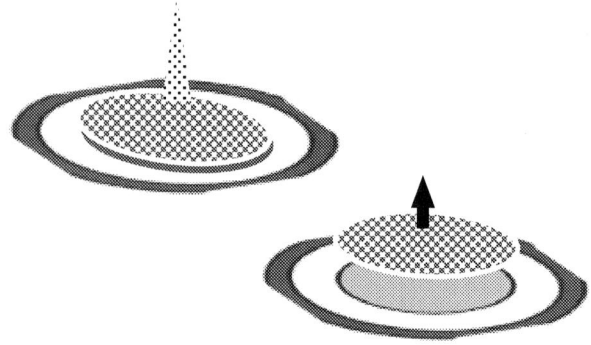

Figure 8. Schematic illustration of the "chemical assisted" release technology for temporary handler debonding with the use of perforated handler

Figure 9. Schematic illustration of the "laser assisted" release technology for temporary handler debonding.

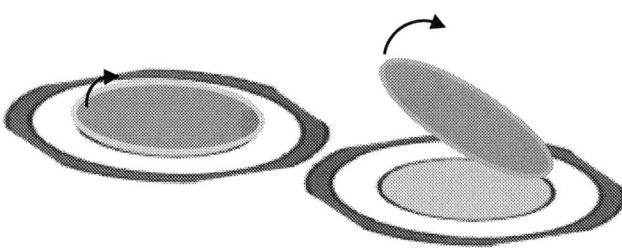

Figure 10. Schematic illustration of the "mechanical peeling" release technology for temporary handler debonding.

The third approach, mechanical release, has received a lot of attention recently [4,12]. The principle of this approach relies on a controlled low adhesion between a release layer and a handler. It requires special equipment with a "peeling" chuck to separate the handler wafer from a device wafer. During the release, the thinned TSV wafer side of a bonded wafer stack is attached onto a dicing tape frame and secured by vacuum chuck. The handler side is held by another vacuum chuck. A slight separation is first initiated at one edge and a debonding wave then propagates across the wafer area to release the entire wafer from the handler. An advantage of this approach is that it can accommodate both glass and silicon handlers. However, this technology requires special debonding chucking systems and significant mechanical impact may be imposed on the surface of circuitry because of the "peeling" force. The potential mechanical impact on low-k or ultra low-k interlayer dielectrics needs to be studied in order to qualify such a technology.

Conclusions

The selection of a temporary handler technology must consider the factors described below:

The thermal stability and the chemical stability of the temporary bonding adhesive must be compatible with all the wafer processing steps;

wafer warpage needs to be mitigated by a proper handler to meet the chucking requirement in semiconductor tools;

wafer inspection approach needs to be compatible with the selected handler to detect voiding or defects in the temporary bonding adhesive;

if wafer-level test is needed when a device wafer is attached on a temporary handler, the overall thermal conductivity of the bonded wafer pair needs to be controlled to meet the heat dissipation requirement;

finally, release technology must be reliable and fast for any temporary handler technology.

Acknowledgments

We would like to thank Thuy L. Tran-Quinn and Kurt Smith at the IBM S & T G for their discussion and technical support. We would also like to thank Ghavam Shahidi and Tze-Chiang Chen for their management support.

References

1. S. Pargfrieder, et al., "Temporary Bonding and DeBonding Enabling TSV Formation and 3D Integration for Ultra-thin Wafers," in Proc. 10th *Electronics Packaging Technology Conference (EPTC)*, pp.1301-1305, 2008.
2. K. Tamura, et al., "Novel adhesive development for CMOS-compatible thin wafer handling," in *Proc. 60th IEEE Electronic Components and Technology Conference (ECTC)*, pp.1239-1244, 2010.
3. B. Dang, et al., "CMOS compatible thin wafer processing using temporary mechanical wafer, adhesive and laser release of thin chips/wafers for 3D integration," in *Proc. 60th IEEE Electronic Components and Technology Conference (ECTC)*, pp.1393-1398, 2010.
4. A. Phommahaxay, et al, "Ultrathin wafer handling in 3D Stacked IC manufacturing combining a novel ZoneBOND™ temporary bonding process with room temperature peel debonding," in *Proc. IEEE International 3D Systems Intergation Conf. (3DIC)*, pp. 1 – 4, 2012.
5. K. Asano, et al., "Fundamental study of an electrostatic chuck for silicon wafer handling," *IEEE Transactions on Industry Applications*, vol.38, no.3, pp.840-845, 2002.
6. D. Gardell, "Temperature control during test and burn-in," in *Proc. 8th Intersociety Conference on Thermal and Thermomechanical Phenomena in Electronic Systems, (ITHERM)*, pp.635-643, 2002.
7. F.X Chen, et al., "Wafer level warpage modeling methodology and characterization of TSV wafers," in *Proc. 61st IEEE Electronic Component and Technology Conference (ECTC)*,pp.1196-1203, 2011.

978-1-4799-2408-0/14 $31.00 © 2014 IEEE

8. X. Pang, et al., "Characterization and management of wafer stress for various pattern densities in 3D integration technology," in *Proc. 60th IEEE Electronic Component and Technology Conference (ECTC)*, pp.1866-1869, 2010.

9. R. Wieland, et al., "Thin substrate handling by electrostatic force," in *Proc. 2nd European Conference & Exhibition on Integration Issues of Miniaturized Systems - MOMS, MOEMS, ICS and Electronic Components (SSI)*, pp.1-4, 2008.

10. K. Zoschke, et al., "Polyimide based temporary wafer bonding technology for high temperature compliant TSV backside processing and thin device handling," in *Proc. 62nd IEEE Electronic Component and Technology Conference (ECTC)*, pp.1054-1061, 2012.

11. K. Saito, et al., "Advances of 3M™ wafer support system," in *Proc. IEEE International 3D Systems Integration Conference (3DIC)*, pp.1-4, 2012.

12. A. Jourdain, et al., "Integration and Manufacturing Aspects of Moving from WaferBOND HT-10.10 to ZoneBOND Material in Temporary Wafer Bonding and Debonding for 3D Applications" in *Proc. 63rd IEEE Electronic Component and Technology Conference (ECTC)*, pp.113-117, 2013.

Optimization and Challenges on TSV MEOL Integration

DoHyeong Kim, DongHun Lee, YoungChul Seo, JungSoo Park, SeungChul Han, BoRa Jang, JooHyun Kim,
YoungSuk Chung, SeongMin Seo, and ChoonHeung Lee
Amkor Technology Korea, Inc.
151, Dongil-ro, SeongDong-gu, Seoul, Korea
dohyeong.kim@amkor.co.kr

Abstract

In the development of emerging 3D IC packaging using TSV, wafer back side treatment is one of key processes. The TSV back side process called MEOL (Middle End Of Line) is a newly introduced process which is performed after front side treatment/bumping, or before chip stacking assembly. The MEOL process adapts some conventional fab processes such dry etch, PECVD, and CMP but there are some differences in the process conditions such as non-mask pattern etch, low temperature deposition of dielectric film, Cu and dielectric film polish and etc [1]. Moreover traditional processes of semi-conductor packaging such as edge trim, wafer back grinding and dicing are also adapted with some modified conditions. For thin wafer handling, wafer temporary bonding and debonding processes are introduced as well. TSV MEOL process consists of the above processes and other supporting ones. These various processes should be aligned or integrated to achieve high yield thru the best optimization. So, it is very important to understand relationship among the processes how they interact each other. This paper is to describe how to optimize every single process with its process monitoring activities in terms of TSV MEOL integration and also some challenges to overcome for volume production of TSV packaging.

Introduction: Why process integration is important?

As many studies have been done over the last few years[2],[3],[4],[5], TSV MEOL process includes edge trim, wafer bonding, wafer back grinding, 1st via reveal with Si etch, passivation deposition with PECVD, 2nd via reveal with CMP, backside bumping and carrier debonding (Fig.1).

After bonding carrier with TSV wafer, this bonded pair should be processed through MEOL flow, which consists of wafer fab, and packaging, and some new processes are additionally required to be added. But although typical processes of this industry are applied, new approach and understanding of process is required.

Due to the combination of various processes mentioned above, there are lots of integration concerns in MEOL process. For example, Si TTV and process temperature could be critical items because of using temporary bonding adhesive onto bonded pair. This TTV should be minimized for perfect via revealing with Si etch process considering via depth variation which is formed at wafer fab. If via depth variation is wide, via revealing processes could be difficult to make it.

When process temperature would increase too high, the adhesive could be deformed and make serious defect in case of the adhesive material is thermoplastic.

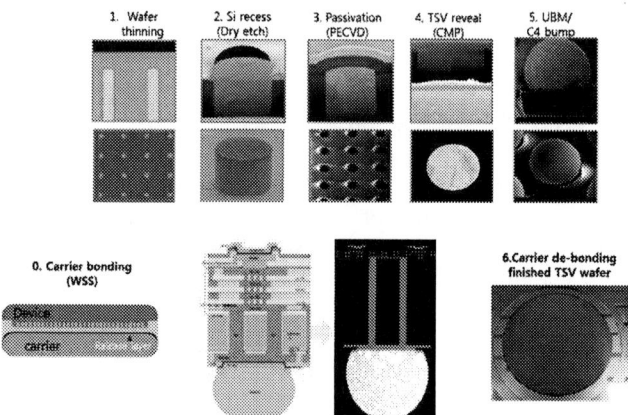

Figure 1. TSV MEOL process flow

The handling scale of MEOL process is also middle position between wafer fab and packaging, because it is almost submicron level which is bigger than fab but smaller than typical packaging. Thus specific metrology is required to measure Si thickness, passivation thickness and specific feature. Moreover, wafer backside process (MEOL), is obviously cost adder due to revealing via on backside of wafer. As of now, MEOL process cost could not be acceptable due to some expensive processes. In terms of integration for volume production, its process cost needs to be optimized to provide more cost competitiveness. Another key thing is to achieve high yield. To improve process yield like conventional manufacturing processes, in-depth understanding of each process and integration is required.

TSV MEOL process and approach of integration

A. MEOL Process scheme

Among process scheme, soft reveal which is not reveal Cu nail during wafer back grinding process is widely used. Flat reveal could be considered as an option if via depth variation is large or Si TTV on MEOL process is big. However, it has some concern of Cu contamination. So, in case it would not be issue on device function, the flat reveal can be one of MEOL process scheme. Fig.2 shows process flow of soft reveal and flat reveal.

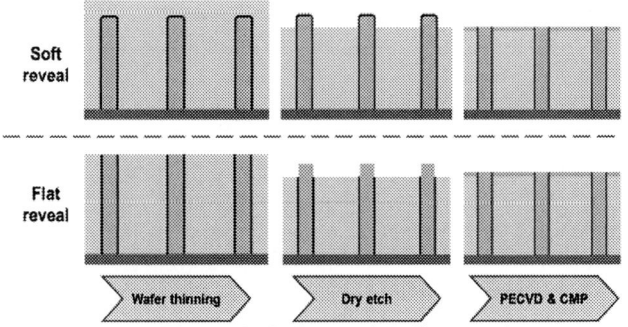

Figure 2. Soft reveal & flat reveal

TSV MEOL process can be categorized as Fig.3. The fab process such as dry etch, PECVD and CMP have been used for a long time on wafer fabrication. To apply them to MEOL process, some process condition changes like low temperature process are required, because temporary bonding adhesive used on the new process could be deformed by high process temperature and then it can induce critical defects. Based on currently used adhesive material, the recommended process temperature is less than 200 Celsius degree. It is required to develop low temperature PECVD condition for TSV MEOL process.

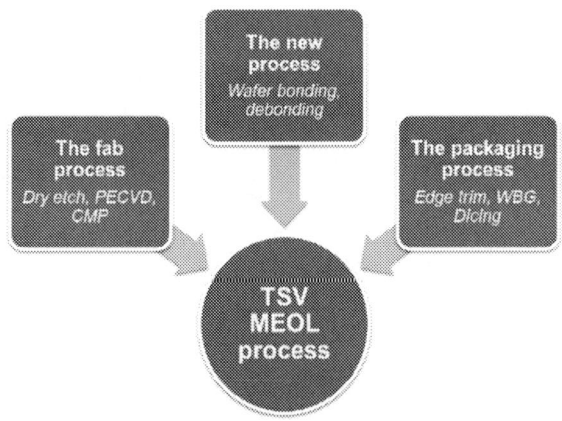

Figure 3. Process categories on MEOL

B. Wafer bonding

In order to process TSV wafer on MEOL, bonded pair should be used with carrier either Si wafer or glass wafer. As this bonded pair is processed, typical process equipment of the fab process and the packaging process needs to be modified to handle bonded pair. And process surface is also required to keep clean not to generate any defects on dry etch, PECVD, and CMP which are very sensitive with particles because packaging process could generate some particles affecting the fab process. This is one of important items to be considered during integration.

And, the new process, wafer bonding and debonding have some challenges, which require simple debonding process, really innovative temporary bonding adhesive and process compatibility with other process on MEOL flow. Recently various adhesives and process equipment is released to deal with difficulties. But, there is no champion as of now.

Therefore it is required to optimize process with current system and material introduced in the market.

To do this, there are some requirement such as lower adhesive TTV, less stress attack onto thin TSV wafer during debonding carrier to prevent wafer breakage and possible defects. So carrier pretreatment and complicated process steps on wafer bonding and debonding are used recently. So process equipment becomes expensive with complicated process steps such as coating, high pressure bonding, vacuum chamber, baking, curing, or UV irradiation. A lot of temporary bonding adhesives are also introduced, which are thermoplastic adhesive and thermoset adhesive. The former one is simply used on thermo-compression bonding at higher temperature than soften temperature of adhesive and then easy to be removed by solvent. This adhesive system becomes weak at high temperature because voiding issue into adhesive layer could occur. In case of thermoset material, it is very stable on higher temperature because of curing to form cross link of adhesive. But debonding could be complicated [6]. This will be explained at debonding section.

C. Wafer back grinding, Si thinning of device wafer

Low adhesive TTV after bonding is main factor for low Si TTV after wafer thinning. Fig.4 shows Si TTV improvement wafer bonding technology is changed.

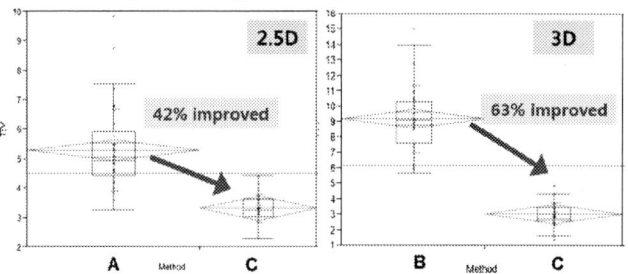

Figure 4. Si TTV on wafer thinning depending on wafer bonding technology

Soft reveal was designed for no Cu nail exposure during wafer thinning. So it is very important to be thin Si thickness as possible as close to Cu tip in the point of cost reduction. TSV wafer with good via depth uniformity helps to keep minimized RST, (remained Si thickness), to via. Based on experience, around 3 to 4um Si TTV after wafer thinning is enough to be close without any Cu expose. In terms of process tool, dedicated wafer back grinding machine is recommended because it can handle bonded pair not to touch backside, control Si TTV using embedded sensor to check Si thickness, and install wet polishing module as well as cleaning module to remove any particle and residue using specific chemicals during wafer thinning. After Si thinning, an integrated cleaning module is really helpful to clean out particles right after process. In general, if particle or some residue might be existed on polished Si surface, it might not be easy to remove it even for requested processes. So cleaning of TSV wafer backside is also very critical integration approach to prevent quality concerns at following processes. Si polishing using CMP used in wafer fab is not required actually because that tool is expensive and has higher COO.

978-1-4799-2408-0/14 $31.00 © 2014 IEEE 583

D. Dry etch, Si recess and via reveal

On backside of thinned TSV wafer with proper cleaning, via revealing is required through Si recess etch. Dry etch is very good application using SF6 without revealing Cu nail, because of high selectivity between Si and Si oxide of via linear. When optimized dry etch process with SF6 is applied, its selectivity showed about 180:1 between Si and Si oxide which is good enough to keep via liner without Cu nail exposure considering general via liner thickness. In this dry etch, etch rate of 5 to 6um/min can be applied using SF6 for soft reveal, which give good process throughput [7].

Si TTV is also key item to check process performance of dry etch. TTV of Si recess means actual die thickness of wafer so continuous TTV control from wafer bonding to dry etch is really important. Fig.5 shows this result of integration among processes. As shown in fig.5, the difference between TTV of adhesive on wafer bonding and Si TTV, about 4um, means morphology on front side of TSV wafer. And by controlling etch profile, Si TTV can be adjusted. And profile of Si thickness can be compensated. Using this approach, 1 to 2um of Si TTV can be adjusted and then final TTV after dry etch showed about 3 to 4um. This trend can be acceptable to be used in advanced package. The dry etch is mostly considered to reveal via. If Si thickness is enough to be thinner than minimum via depth, it could be thought all of via can be revealed perfectly. Via revealed height is also measured to check actual via depth as well as Si thickness after dry etch and this result is quite aligned well with via depth information from TSV wafer fab.

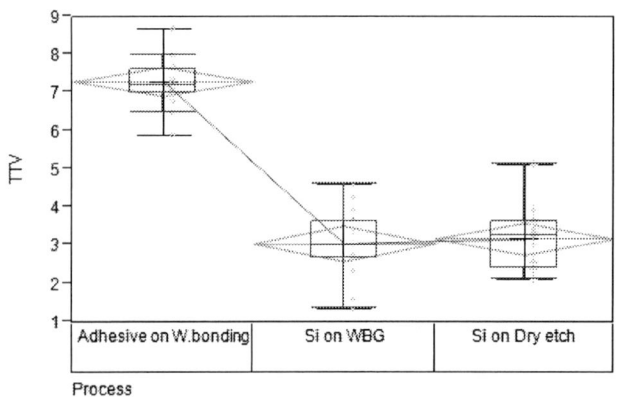

Figure 5. Comparison of TTV for process

When flat reveal is required to be processed, CF4 gas or NF3 gas could be used. In this case, etch rate is quit slower than that of SF6. CF4 showed 0.7um/min and NF3 about 3um/min. However, this application is still worried about Cu contamination onto Si surface. Thus this flat reveal should be used to the case without concern of Cu contamination.

Although dry etch looks clean, inorganic particles might be contaminated on etched Si surface which is activated. Some particle could become root cause of yield loss at following process, passivation deposition and CMP. So post dry etch cleaning may be required. In order to remove inorganic particles, proper chemical is needed. For better process in upcoming days, integrated equipment with dry etch and wet cleaning for this application will be required for HVM. And this kind of tool will be capable of minimizing particles and improving yield.

E. PECVD, Passivation deposition

Backside of TSV wafer needs to form passivation layer. PECVD is the suitable process to deposit inorganic passivation, such as SiN or SiO2. But there is another challenge. As described, bonded pair is formed with thermoplastic adhesive or thermoset adhesive. Basically PECVD is very high temperature process, almost 400 Celsius on wafer fab in general. However bonded pair cannot be exposed such a high temperature because of damage of thermoplastic adhesive. In this case, alternative PECVD is required at low temperature, less than 200Celsisus where thermoplastic adhesive can endure without any critical deformation. But some material could generate voids into adhesive layer so this kind of thermal behavior should be confirmed when deciding temporary bonding adhesive. When using thermoset adhesive, it is stable at higher temp because of cross linking. But if wafer bonding process might not be optimized using thermoset adhesive, adhesive void could be issue because of expansion of remained solvent inside adhesive. Therefore, in order to perform stable process without issue, the selection of temporary bonding adhesive is the key at low temperature PECVD process [8]. As explained above, passivation deposition is performed at low temperature then deposition rate could be slower than that of conventional high temperature deposition, which is one of concerns for throughput.

Generally, SiN is used as backside passivation. This film has good passivation performance such as good metal diffusion barrier, electrical properties and WER (wet etch rate) and durable mechanical property. Sometimes SiO2 is also used with SiN. Depending of property of SiN film after deposition with low temperature PECVD, deposition thickness could be tuned. Thinner layer is slightly helpful to increase process throughput. For example, SiN and TEOS based SiO2 are deposited continuously, SiO2 is removed by CMP, which is one of capping layer to distinguish SiN from stacked passivation structure, which would be helpful to apply EPD (end point detection) on CMP. And passivation stress is one of important factor on PECVD in terms of integration. Deposited film on die backside influences die warpage behavior during assembly process flow.

During flip chip interconnection, as the variation of temperature, TSV die warpage behavior would be varied. Generally, interconnection would not be easy for the TSV die with high warpage. So, depending on backside film stress, die warpage shows various behavior (Fig.6). Even though tuning film stress, this value means wafer level one based on wafer deformation after film deposition. Performing dicing of TSV wafer, all of TSV dice would not show similar behavior because local film stress is different depending on the position of wafer. Thus, when tuning film stress of passivation, stress distribution on wafer is also critical integration item since uniform back side passivation is required through optimization of PECVD [9].

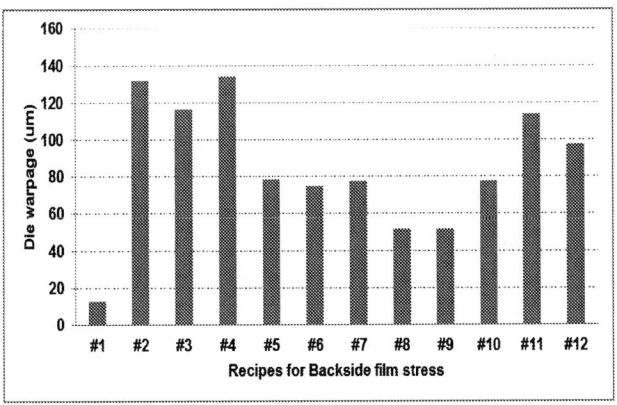

Figure 6. Behavior of die warpage with backside passivation stress

F. CMP, Planarization of backside passivation and Cu exposure

After deposition of passivation with SiN and Si oxide, CMP exposes Cu nail by polishing passivation surface and revealing Cu tip. This CMP process is close to Cu CMP because of removing Si oxide and Cu simultaneously. So slurry needs to have proper removal rate and selectivity to control Si oxide and Cu. Especially, depending on Cu density with via patterns, erosion or Cu dishing could occur. If selectivity might not be improper, serious erosion or dishing might happen. In order to control this process properly, the slurry optimized components such as chemicals and oxidizer is required to meet process specification such as WIWNU and WIDNU. In this case, it's considered 10% range is proper.

In some case, revealed via could be too high due to wide via depth variation. If CMP is applied to this case, tall via could be broken. It is considered this failure could be generated by bending via during polish. Evenly revealed via can distribute down force and polished almost at the same time, but for unevenly revealed via, especially, down force of polish head could press taller via by rotating head and platen. They could be bent and broken at the weakest point. In this case, aspect ratio of revealed via with deposited film on PECVD become less than 1:1 which had been found through some experiment.

During CMP, many kinds of defects could be found. Some of them just come from CMP itself. But most of others come from previous processes Defect control on CMP is the key challenge to success TSV MEOL integration. Details will be discussed on other section.

G. Backside RDL and bumping

As mentioned above, CMP is to expose Cu nail where electro rode is formed such as RDL and bump. Most of advanced packaging adopts flip chip technology so that bumping on wafer backside is also critical. Depending on chip stacking technology or packaging design, C4 bump or Cu pillar bump can be used [5]. Recently fine pitch bump interconnection is required so that Cu pillar bump is being applied. Bumping process consists of vacuum process and thermal process, where its process compatibility with bonded pair is also critical to prevent deformation of adhesive. Low temperature process need to be considered at less than 250

Celsius with convection heating condition. This is also integration approach to develop backside bumping for TSV product.

H. Carrier debonding

After backside bumping for interconnection, carrier should be finally detached through the debonding process. Depending on temporary bonding adhesive and debonding method, its own flow is decided. Thermal slide off just detach carrier with high shear force. But it is big concern to handle thin wafer without mounting on dicing tape, even though process flow is simple. Due to thin wafer handling after debonding until wafer mount, wafer breakage is the most critical concerns and shear force and high temperature to soften temperature of thermoplastic adhesive could induce bump deformation so very careful process optimization is required.

To avoid wafer broken and bump damage, many different methods were introduced in the market. Most of debonding method is focused on preventing wafer damage and defect after mounting dicing tape. When mounting TSV wafer, it will be helpful to handle thin wafer. Debonding should be secured without any critical defect, so considerations for debonding aredescribed as below.

The first, it is required mounting tape has good chemical resistance because chemical remove adhesive without any residue on the front side of wafer. And it is also considered debonding tool should protect tape area during cleaning.

The second, it is required to use low stress debonding method. Easier detaching carrier needs pretreatment on carrier side but, it could influence process cost and material cost regarding carrier recycle. Laser debonding is one of easy way. However, mechanical debonding needs more development with proper temporary bonding adhesive

The third, when removing adhesive after detaching carrier, bump on wafer front side could be damaged, if process load could be big. In case of using thermoplastic adhesive, solvent can be clean it on wafer without mechanical force. However, thermoset adhesive cannot be melt with any solvent so that it could be removed with mechanical way. During mechanical method such as peeling off can be also act on some stress to bumps. If it would not be optimized, bump damage or wafer crack could be induced. Thus, proper adhesive selection is important to meet designed MEOL process flow and its integration because there is tradeoff between two adhesive systems.

I. Wafer dicing, Stealth dicing for thin TSV wafer

After debonding process, TSV wafer should be diced to each die for chip attach process. Conventional wafer has only circuit and bumps at front side. However, TSV wafer has structures at front and back side so that adhesive layer of dicing tape should be laminated and cover all of structure on backside or front side in case of performing wafer flipping, if required. If that doesn't cover them, die chipping or cracking could occur during typical dicing process such as blade dicing or laser grooving and blade dicing. If adhesive layer would be thick to prevent these concerns, then die pick up on chip attach process could be hard after dicing. So easier die pick up is required even though adhesive layer of tape would be thick.

Or thinner adhesive layer of dicing tape could be used with alternative dicing method.

Stealth dicing has big merit to apply to this TSV wafer application because of almost no chipping process. And SD can be applied with typical dicing tape which is already confirmed on HVM. Laser irradiation can be performed through dicing tape with transparency even though tape doesn't fully covered structure on wafer side laminated tape.

Finally, SD can dice TSV wafer without any yield loss by process optimization as well as integration with debonding process not to occur serious defect such as die crack or wafer crack. Through this integration, almost perfect process can be achieved and then die pick up performance is also good (Fig.7). If debonding or dicing process could be unstable, it might be useless that prior other processes would be optimized well. The yield of wafer process is very critical at final stage, where everything might be lost when bad situation may happen. It is why integration with debonding and dicing is important.

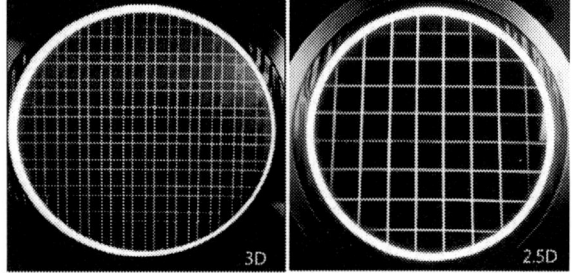

Figure 7. Result of Stealth dicing for TSV wafer

Metrology and process monitoring on MEOL

Through MEOL process, various processes are applied and many metrology technologies are also adopted. Basically, most of metrology of MEOL process was already used in the semiconductor industry, except handling of bonded pair [10]. So following MEOL process, metrology and key monitoring item are discussed.

Before wafer bonding, edge trim is processed to remove bevel area of wafer because sharped edge could not be formed after wafer thinning. In this process, trimmed width and depth using microscope can be measured simply and check any chipping is existed or not.

When wafer bonding is performed, interested things are adhesive TTV and void into adhesive layer. When using glass, it is easy to check void through glass side. But using Si carrier, adhesive voids need to be monitored with SAM. So glass carrier makes monitoring step simple. After bonding, adhesive thickness and its TTV is interesting item. Thus IR interferometer can measure its thickness through glass carrier or Si carrier. One of characteristics of IR is the transparency into Si as well as glass so IR can measure temporary bonding adhesive layer selectively. In wafer bonding, adhesive thickness and its TTV are main items for process monitor because distribution of adhesive thickness needs to be thicker than that of bumps formed at front side of TSV wafer. And its TTS is also affect to Si TTV on wafer thinning and Si recess process. IR interferometer can be also applied to monitor Si thickness and TTV on wafer back grinding and dry etch. This metrology can confirm that via doesn't revealed after wafer back grinding. And Si thickness and TTV of dry etch can also confirm via reveal is perfect across wafer. This data means whether via revealing process is acceptable or not so that this is very important monitoring items in terms that MEOL process integration.

Revealed via height also give important information. Using white light interferometer or laser confocal microscope, revealed height can be measured. This data means actual revealed via is acceptable after dry etch, as explained on previous section. Using data of revealed via and Si thickness, actual via depth can be confirmed during MEOL process.

PECVD process deposits inorganic passivation such as SiN and Si oxide film. General process monitoring items are film thickness, uniformity and RI using ellipsometer. In terms of integration on MEOL process, film stress is also critical measurement item, but this is not regular process monitoring item, because die level warpage by film stress is more interesting. So characterization of film stress is also required to optimize MEOL integration.

CMP is to polish passivation and revealed via together to form flat surface. So ellipsometer can measure polished film thickness, and white light interferometer or laser confocal microscope can see amount of via dishing by slurry selectivity. If this dishing is too deep, it is said that process control or slurry might have critical problem. And if polished film uniformity would show unexpected data, slurry selectivity and process condition for via density of TSV wafer might not be optimized. Thus, main monitoring items on CMP are film thickness, WIWNU, WIDNU and via dishing amount. Fig.8 shows one of example for backside passivation polish.

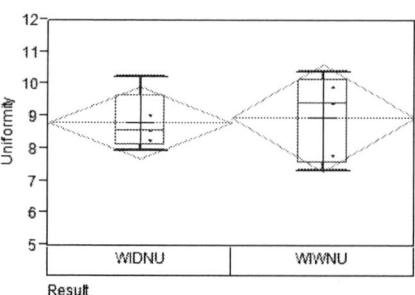

Figure 8. Non uniformity of backside passivation after CMP

After CMP, optical visual inspection is applied to find any defect. AOI is adequate tool for this purpose. Based on result of AOI, process yield could be defined. After debonding and dicing, AOI is also good tool. It is required to define KGD to apply assembly process. Next section will explain some consideration and challenges on defect and yield on MEOL process. For example, via crack could be found as Fig.9. This kind of defect can be only found at MEOL process. Based on this result, via formation process could be optimized to eliminate some defect.

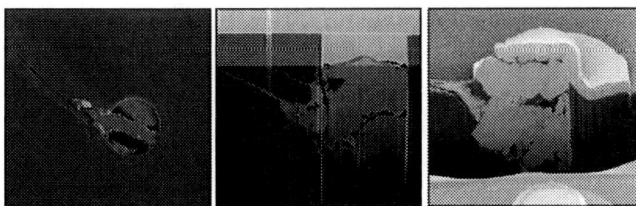

Figure 9. Defect of via crack

Defect and yield control through TSV MEOL integration

So far, it was explained how integration of MEOL process could be optimized based on effect and cause through all of process flow. During process, yield should be increased for HVM. As previously explained, integration approach of TSV MEOL could generate some defects which is very similar with typical one or something new on packaging area or really new one. Processes of MEOL are to coat adhesive, to remove Si, to deposit thin film and to remove thin film as well as metallic material, Cu. Going through the processes byproduct or particle could be produced, which is one of defect on process. If some defect is found on some process, root cause of that defect could come from prior processes. For example, if TTV of temporary bonding adhesive is too big, unexpected via reveal could happen during wafer thinning. Because this process is only designed for removing Si, automatic optical inspection, AOI, can't be implemented at wafer thinning. But if Cu nail might be exposed on wafer thinning, oxidation of Cu nail could occur during dry etching due to reaction with Sulfide ion. Then AOI at post dry etch could detect this defect. And, if AOI of post dry etch would not be applied, this defect could be found as irregular shape of via at post CMP inspection, actually next process, that is undesirable in terms of quality control. One more example is that weak thermoplastic adhesive is deformed on at high process temperature. This deformation is not easy to be seen due to embedded between carrier and TSV wafer. Even though adhesive would be deformed on some process, it would not be easily found. This deformation is finally found after CMP as blisters, or passivation defect or Si exposure (Fig.10).

Very small void of wafer bonding also causes blister defect after CMP. This relation means that integration is one of key issue, which can be also challenges for this integration and each process optimization.

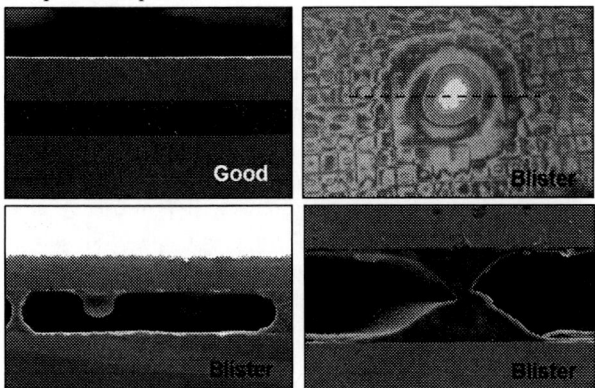

Figure 10. Defect by blister after CMP

And process scale of MEOL is much smaller than that of typical packaging process. Before introducing TSV application on packaging area, small size of interconnection was about 20um or so with Cu pillar bumps. Recently, exposed diameter of TSV is 10um or 5um. Over exposed Cu nail, RDL and bumps for interconnection are formed, which means particle control become tighter than that of assembly. A few micron or submicron level particles should be controlled during TSV MEOL process. This is also new challenge but need to be solved by using technology from wafer fab. Thus, enhanced cleaning process can be applied to MEOL process to keep more cleanness. Even though packaging area would not be good at high level wet cleaning with specific chemical, wafer fab have applied lots of cleaning method to catch particles, even extremely small nano scale particle. This is very good reference for packaging to handle chemical wet cleaning on TSV MEOL process. Fig.11 shows some example of particles /FM to detect as defect.

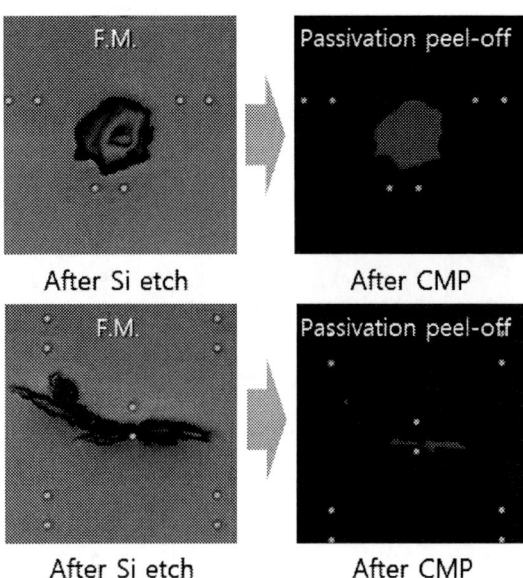

Figure 11. Defect after Si etch and after CMP

Process	Lot A	Lot B	Lot C
FS bump	98.40%	99.50%	99.50%
MEOL-1	86.30%	98.50%	99.50%
BS bump	95.90%	99.00%	99.50%
MEOL-2	99.00%	99.50%	99.50%
Overall Yield	80.6%	96.5%	98.0%

Figure 12. Example of yield calculation

And as described at introduction of this paper, TSV wafer processing consists of FS bumping, MEOL and Backside RDL and bumping, which means yield management could be also one of critical challenges to prepare HVM, especially TSV interposer for 2.5D TSV packaging application. The chip size of TSV interposer is very large so that 50 to 70 interposers are fabricated on 300mm wafers depending on its size. When only one defect would kill one die, yield loss could become serious. Fig.12 showed the example of yield trend with TSV interposer. MEOL1 means from edge trim to CMP and MEOL2 is debonding and wafer dicing. So this kind of sensitivity to yield become another challenge, however aggressive cleaning as well as process optimization and integration can get acceptable yield for HVM Readiness.

Process cost with MEOL integration

TSV MEOL process is performed on wafer backside, which means it is cost adder actually. Backside process of TSV wafer is inevitable because of revealing via and forming electrode for interconnection on packaging. Even though TSV device would show impressive performance with additional process cost, cheaper product cost could have competitiveness in the market. In order to achieve cheaper one, low cost approach is also preferable to the market requirement, generally.

Fig.13 shows normalized process cost and which process occupied large portion on MEOL process cost. Among them, wafer bonding and debonding looks critical however these process is newly emerged and has some limitation to minimize its cost. Carrier recycling is only one of key for reduction. Therefore, more basic understanding for process cost is required to minimize it. In order to improve throughput on each single process on MEOL, via depth variation is the start point. If via depth variation is too big, via revealing process, dry etch, could take long time. When minimizing via depth variation, etch amount can be reduced and finally its process cost could be minimized. And CMP is also one of critical process in terms of process cost because of many consumables. So, its material cost need to be reduced to minimize MEOL process cost, especially slurry for backside planarization of TSV wafer.

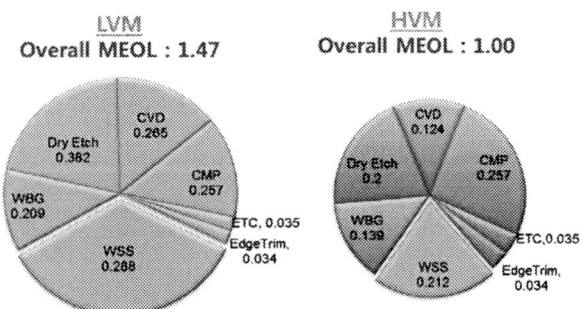

Figure 13. Normalized COO of MEOL

First of all, it is really required to achieve high throughput on TSV MEOL, this is the first requirement to reduce MEOL process cost by optimizing every single process based on understanding of MEOL integration.

Using TSV wafer with good via depth uniformity, good adhesive TTV on wafer bonding can generate good Si TTV on wafer back grinding process. Then remained Si can be closer via tip on backside of TSV wafer. In point of wafer back grinding, a few micrometer of Si thickness reduction could not affect to its throughput. But the throughput of dry etch can be much improved by grinding Si close to via tip as possible as we can, because etch amount is decreased. And amount of

CMP can be also decreased, which is also affirmative in terms of process cost.

Based on experience, dry etch is one of critical process to optimize process in terms of throughput. To get reasonable process cost on dry etch, faster etch rate is really required but faster ER might be worse Si TTV, thus optimized H/W and process are required. On the passivation deposition, its thickness is quit thicker than inorganic film using in wafer fab. So chamber cleaning is also one of critical item to improve throughput. Once thicker film is deposited, chamber cleaning time and its sequence also need to be optimized well. If not, about 30% of throughput loss can occur. When reviewing PECVD and its equipment, its configuration and capacity to handle thicker film and high speed chamber cleaning need to be checked. And deposition rate at low temperature is generally low. So deposition rate needs to be increased as much as we can. But In order to manage film stress tuning, stress uniformity control and increasing deposition rate at the same time, it is real challenge. So it is recommended critical item on PECVD is also optimized at first and then tuned step by step.

Various development activities are required to minimize cost of TSV MEOL process. If process cost could become proper cost to absorb into the end product cost, it becomes desirable as a standard MEOL process cost. However, it still looks high, which is the most challenge on TSV MEOL process as of now.

Conclusion

It is described what TSV MEOL process is and how process can be optimized and the goal can be achieved such as higher yield and proper process cost in terms of MEOL integration. To achieve the goal, lots of challenges still exist which should be overcome for HVM production. In order to overcome, it is required not only each single process is more optimized but also integration of MEOL is optimized. Finally, this approach can be helpful to launch TSV product successfully in the market. Many of TSV players already overcame many challenges but it looks still remained as below,

1) Wafer bonding and debonding technology to be simplified with very reasonable process cost and less sophisticated process and equipment.

2) Required to develop temporary bonding adhesive to endure more than 250C ideally which have MEOL process window wide and be easy to be removed and less stress to thin wafer regardless of adhesive system.

3) Yield management for TSV interposer for 2.5D TSV product due to huge die size. It is not easy to get overall yield to expect on HVM mode typically on wafer processing. Just losing one interposer, yield drop could be more than 1.5% in case of 70 TSV interposers on 300mm wafer.

4) Acceptable COO (cost of ownership) on TSV MEOL process. MEOL process cost is inevitably adder. But need to define which range could be really reasonable considering price of end product using TSV device, which could be another big challenge

through all of semiconductor companies such as fabless, wafer fab and OSAT.

5) In order to overcome challenges on TSV MEOL integration, innovative platform could be required.

As a result, it can be said lots of development for TSV product to enter the market is obviously meaningful for advanced packaging area, because the industry of semiconductor packaging starts to use the fab technology beyond traditional packaging technology. This means additional issues in terms of technology integration could be emerged such as challenges on TSV MEOL process, which is one of the technology convergence of semiconductor business. Therefore, it may mean new era that some boundary start to fade between packaging and wafer fab. If it might be real, that will be big challenges on the convergence which is just emerged. Among them, TSV packaging will be one of the beginning at new technology era.

Acknowledgments

Authors would like to thank engineers, technicians and operators who are developing TSV wafer process in Amkor Technology Korea.

References

1. B. Wu, A. Kumar, S. Ramaswami, "3D IC stacking technology," Mc Graw Hill Book Publication, 2011.
2. Bo Kai Huang, "Integration Challenges of TSV Backside Via Reveal Process", IEEE ECTC 2013, pp.915-917.
3. Seung Wook YOON, "2.5D/3D TSV Development and Assembly/Packaging Technology", IEEE 13th Electronics Packaging Technology Conference 2011, pp.336-340.
4. T. Mourier, "3D Integration challenges today : from technological toolbox to industrial prototypes", IEEE ECTC 2013.
5. Dong Wook Kim, "Development of 3D Through Silicon Stack (TSS) Assembly for Wide IO Memory to Logic Devices Integration", , IEEE ECTC 2013, pp.77-80.
6. Seung-Woo Lee, "Temporary Bonding & Debonding Adhesives for 3D Multichip Packaging", Polymer Science and Technology Vol. 24, No. 3, pp277-284.
7. K. Buchanan et al, "Etch, dielectrics and metal barrierseed for low temperature through-silicon via processing,"in Proc. IEEE International Conference 3D Systems
8. Kath Crook, "Dielectric Stack Engineering for Via-Reveal Passivation", IEEE ECTC 2013, pp.576-580.
9. Jean Charbonnier," Bow Management with Temperature for Thin Chips Integration", EMPC 2013, September 9 - 12, Grenoble; France.
10. Sandip Halder, "Metrology and Inspection Challenges for Manufacturing 3D stacked IC's", IEEE ASMC2013, pp.75-79.

TSV Integration on 20nm Logic Si: 3D Assembly and Reliability Results

Rahul Agarwal[1], Dave Hiner[2], Sukeshwar Kannan[1], KiWook Lee[2], DoHyeong Kim[2], JongSik Paek[2], SungGeun Kang[2], Yong Song[2], Sebastian Dej[1], Dan Smith[1], Sara Thangaraju[1], Jens Paul[1]
[1]GLOBALFOUNDRIES Inc. 2600 Great America Way, Santa Clara, CA 95054 USA
[2]Amkor Technology Inc., 1900 South Price Road, Chandler, AZ 85286
Rahul.agarwal@globalfoundries.com

Abstract

Each new technology node brings new design and technology challenges making it harder to maintain Moore's law in a cost effective way. Maintaining cost effectiveness is becoming a major challenge for IDMs, fabless companies and foundries. 3D/2.5D technologies offer some unique advantages over traditional scaling such as higher power efficiency, higher bandwidth and heterogeneous integration which can arguably lower design complexity and manufacturing cost. While advantages of 3D ICs are well known, adoption of this technology has been shifting out due to several technological challenges and manufacturing supply chain concerns. In this paper, 3D packages are realized by stacking mechanical Wide IO memory onto a 20nm low power mobile logic die with through silicon vias (TSVs). This architecture is very promising for mobile application as it can provide lower power consumption, higher bandwidth and faster communication between memory and logic with a smaller form factor. Various technical challenges that were addressed while building a 3D package along with its process and reliability results, both wafer level and package level, are discussed in this paper.

Introduction

Rapid growth of mobile market and increased complexity in the latest mobile devices are driving rapid innovations in DRAM developments. LPDDR3, LPDDR4, Hybrid Memory Cube (HMC), High Bandwidth Memory (HBM) and Wide I/O [1, 2] are some of the emerging memory technologies which are being developed to address the "memory wall" [3] issue. Each of these memory architecture are being developed to meet the needs of different market segments. This paper focuses on stacking a Wide I/O memory onto a logic device with TSV for 3D packaging. Latest Wide I/O2 standard specifies 4 channels with a bandwidth of 6.4Gbps with 64 I/Os per channel. This can result in a 25.6Gbps data rate with a provision of up to 128 I/O per channel and 51.2Gbps data rate. A Wide I/O configuration uses a significantly larger I/O pin count at a lower frequency and 3D stacking reduces interconnect length and capacitance, thus eliminating the need for on die termination. This results in the lowest I/O power for higher bandwidth. As discussed by Dong Wook et al. [4], other packaging techniques such as PoP, Multi die stack wire bonding etc. cannot fully utilize the large number of I/Os enabled by Wide I/O standard. TSV and 3D stacking is required to fully utilize the advantages provided by Wide I/O memory.

300mm logic wafers, for low power mobile applications, were fabricated with TSVs on 20nm node silicon (Si) in GLOBALFOUNDRIES Malta, NY. Bumping, backside integration and assembly was done in Amkor Technology®'s leading factory in Korea. Several test structures such as bump crack sensor, daisy chain and leakage structures were used to determine interconnect integrity, assembly yield and reliability yields.

A schematic of the configuration is presented in Figure 1. Fine pitch copper pillar bumps with SnAg caps were used to connect logic die to the substrate and Cu-Ni-SnAg microbumps were used to connect memory die to the backside of thinned logic die. In this paper, a brief summary of test vehicle and test structure is presented followed by key process considerations and details. Finally results from 3D stacking of memory on logic are presented. Key learnings are summarized and package level reliability data along with impact of thinning on key device performances are discussed.

Figure 1. Illustration of wide I/O memory on logic die with TSV.

Test vehicle and test structures

Table 1 shows key characteristics of the test vehicle used for this work. Via-middle approach was used to etch 6μm wide, 55μm deep TSVs in logic wafers. ~11x11mm^2 logic die with Wide I/O1 standard was used for this test vehicle which specifies a TSV pitch of 40μm in one direction and 50μm in another. Mechanical memory die of ~6x3.5mm^2 were fabricated with daisy chain structures in the wide I/O area and corners.

Table 1. Test vehicle specifications.

Feature	Specification
TSV Integration	Via-middle process
TSV dimensions	6um diameter, 55um deep
TSV Pitch	40um/50um
Logic Die size	11x11mm^2
Logic to substrate	Cu Pillar (Cu/SnAg)
Logic to substrate interconnection	TCNCP
Memory to Logic	Microbump (Cu/Ni/SnAg)
Logic backside finish	Ni/Au
Memory to Logic interconnection	TCNCP
Substrate	Laminate fcCSP
Package size	14x14mm^2, 135 IOs

978-1-4799-2408-0/14 $31.00 © 2014 IEEE

Figure 2 shows different daisy chain designs which were created in the Wide IO areas. Some daisy chains were closed through the backside metallization which was formed after backside via reveal (Figure 2 (i)) and some were closed through top mechanical memory die (Figure 2 (ii)). In addition, logic die also have fine pitch periphery bumps creating other test structures like bump crack sensors, leakage structures, etc. These periphery chains are closed through the substrate. During package level reliability tests all these daisy chain and leakage structures were monitored.

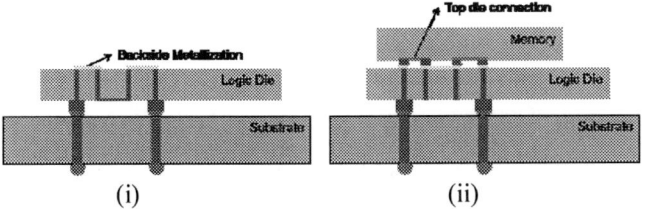

(i)   (ii)

Figure 2. Schematic drawing of daisy chain structures connected at various levels of interconnects.

Fabrication and Assembly

Fabrication and assembly is discussed in four major segments. First is via-middle integration of TSVs in a 20nm logic wafer, second is the bumping operations for both logic and memory wafers, third is backside integration to reveal the TSVs from the backside of logic wafer, and fourth is the assembly process where thinned logic die and mechanical memory dies were stacked onto a substrate to form a 3D package.

1. TSV Integration

A via-middle scheme was adopted for this work where TSVs were fabricated post gate and contact modules. A Bosch process was used to etch TSVs to the desired depth of 55μm. TSV depth variation of less than +/-0.6μm 3σ was maintained across the wafer on dense and isolated arrays. **Error! Reference source not found.** Figure 3 shows typical TSV depth gathered across a 300mm Si wafer using non-contact technology. During depth measurement an offset of 1μm is seen between this non-contact measurement and physical inspection. TSV profile was monitored post TSV etch using a non-destructive interferometry that harnesses the reflected white light and provides both bottom critical dimension (CD) and depth of the TSVs. TSV profile measurements are pivotal for both TSV integration as well as backside integration when the TSVs are revealed. This is because the TSVs are of a high aspect ratio and any undesirable change in profile may propagate down to TSV plating and chemical mechanical polishing (CMP) either causing voids in the TSVs or over-polish due to non-uniform stack on field. In this work, a highly conformal oxide liner with >80% step coverage with a PE-CVD cap to seal the liner from moisture absorption is used. Void free copper (Cu) filled TSVs were achieved with PVD Ta as the barrier metal and Cu is used for the seed layer. A target of more than 10nm of barrier metal on the bottom sidewall of the TSVs was maintained.

One of the challenges is to not erode the contacts and pre-metal dielectric layer (PMD) and to maintain a good dielectric uniformity from center to edge of the wafer during the TSV CMP step. In the integration scheme used by GLOBALFOUNDRIES, a cap layer is deposited to protect the contacts and provide a CMP stop layer above contacts and pre metal dielectric (PMD) during CMP of the TSV and isolation liner. Cu on field is removed by a CMP step along with the barrier/seed layers and liner, stopping at the capping layer. Figure 4 shows contacts which are 2μm away from TSVs and are protected via thin capping layer during the final CMP step. Both uniformity and erosion of contact cap layer and PMD at the dense and ISO TSV array and SRAM cells are monitored by a fixed ellipsometry model; if erosion or under-polish was observed at TSV CMP, this could result in opens or shorts of vias in metal 1 chains fabricated in back end of line (BEOL). After successful CMP, wafers are processed through BEOL stack and are finished with terminal metal before bumping.

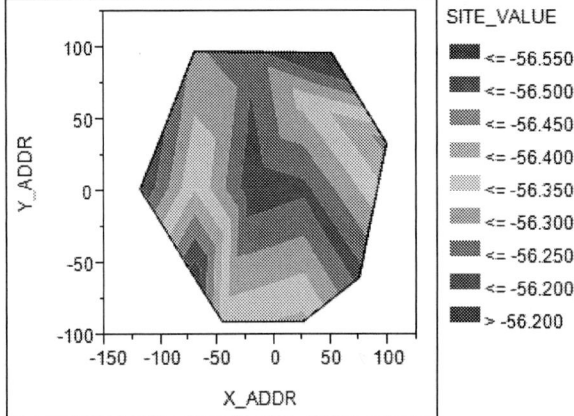

Figure 3: Typical contour map showing TSV depth variation across 300mm wafer, with 1μm offset coming from measurement tool.

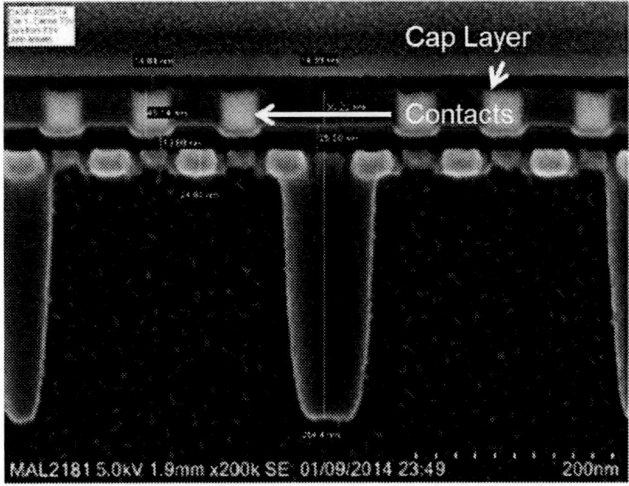

00275_Die1_DenseTSV_2umFromTSV_200k

Figure 4. Cross sectional view of the wafer after final TSV CMP is shown. The thin cap layer on top of the contact module prevents its erosion during CMP.

2. Bumping

3D TSV packaging relies heavily on bumping technology and capability derived from advanced flip chip CSP products.

978-1-4799-2408-0/14 $31.00 © 2014 IEEE

Having high density design rules and high reliability, copper pillar bumping and assembly are the standard for 3D TSV packaging.

Wafer bumping for the TSV bearing bottom die and the top die follow similar processing. The bottom die interface to substrate matches high volume manufacturing (HVM) rules with a periphery 40/80µm staggered pitch. The bump is comprised of plated Cu and a SnAg alloy cap. The top die interface has been aligned to the industry standard Wide I/O 40x50µm pattern and matches memory industry standards with a Cu/Ni/SnAg bump structure.

Wafer backside bumping is processed similar to traditional copper pillar bumping. To allow for robust assembly, solder based backside finishes were avoided. For this test vehicle a Ni/Au backside pad finish was selected.

3. Backside Integration

The backside integration scheme for via-middle is the key component linking wafer fabrication to package assembly. Backside integration takes into account key wafer characteristics including wafer stress, TSV depth, and TSV diameter and connects them to a middle end of line (MEOL) process providing high yielding end wafers suitable for package assembly. The key elements to this integration are contained within the primary set of backside process cells including: wafer supporting system (WSS) bonding, wafer thinning, silicon etch, backside passivation (PE-CVD), chemical mechanical polish (CMP), backside metallization, and WSS de-bond. For this test vehicle, an inorganic soft reveal process (ISRO) integration scheme was selected. Inorganic soft reveal is a process set defined when the Cu within the TSV liner is revealed after CMP, without exposing the Si to Cu.

The main criteria of backside integration quality is the measurement of total thickness variation (TTV) across the 300mm wafer. Figure 5 shows the average TTV for the processing of the test vehicle from WSS bonding to silicon etch.

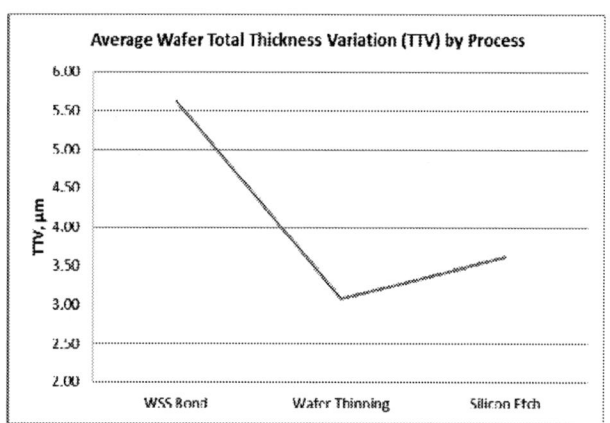

Figure 5: Average total thickness variation (TTV) by process.

Tight control of wafer TTV through these three key process steps (WSS bonding, wafer thinning and Si etch) ensures a high yielding process at the CMP. Figure 6 shows a cross sectional SEM image of the TSV, passivation, and final metallization for the test vehicle.

The final aspect of backside integration is the dicing of the TSV bearing wafer. The nature of the thinned TSV wafer and its properties require a close tie to backside integration. Dicing is completed after de-bonding of the logic wafer from the carrier system. Common methods for dicing include mechanical dicing, laser groove plus mechanical dicing, plasma etching, and stealth dicing. For advanced silicon nodes, such as this test vehicle, stealth dicing provides a proven process for control of dicing dimensions and prevention of damage to the fragile wafer. Figure 7 shows the stealth dicing result from this test vehicle (TV) wafer.

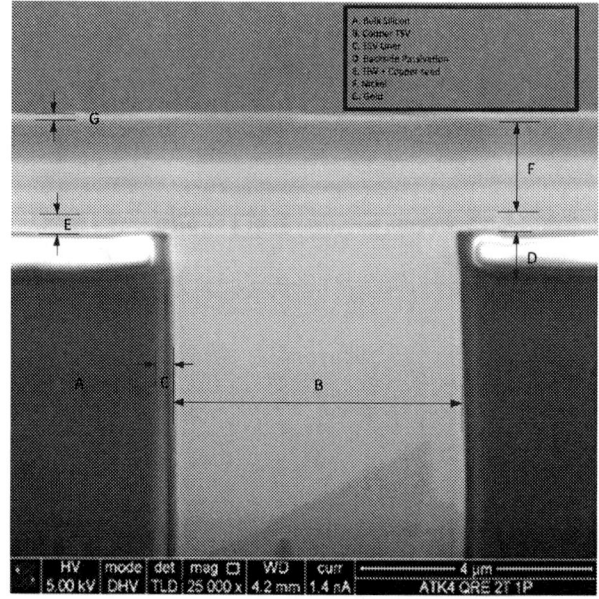

Figure 6: SEM image of final TSV with backside metallization.

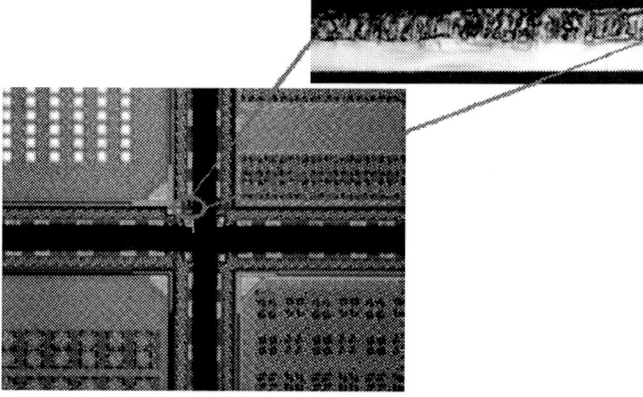

Figure 7: Optical and X-SEM image after stealth dicing.

4. Assembly and Packaging

Several TSV based assembly methodologies and process flows are practiced in the industry today. For this test vehicle a chip to substrate flow was selected. Thermo-compression bonded copper pillar interconnects with non-conductive paste (NCP) allowed for tight control of assembly yields and package characteristics. In this process the thinned TSV bearing logic is bonded to the substrate, followed by the

978-1-4799-2408-0/14 $31.00 © 2014 IEEE

bonding of the top die to the Wide I/O pattern on the first die backside (Figure 8).

Successful management of wafer properties in the backside integration process, assembly process, and substrate all play important roles in the management of package warpage. Package warpage was controlled to less than 75μm throughout the reflow profile. Figure 9 shows the test vehicle shadow moiré data from 2 different substrate suppliers.

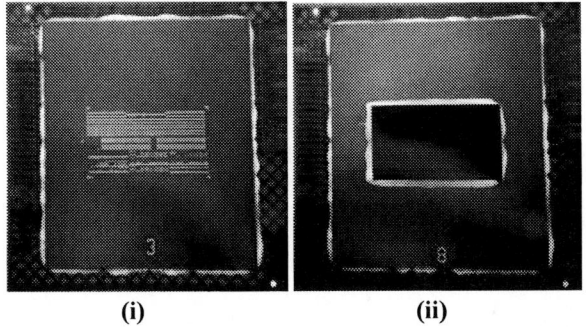

(i)   **(ii)**

Figure 8: Bonded die images from test vehicle. (i) After thin die attach to substrate, (ii) After top die attach to the backside of TSV die.

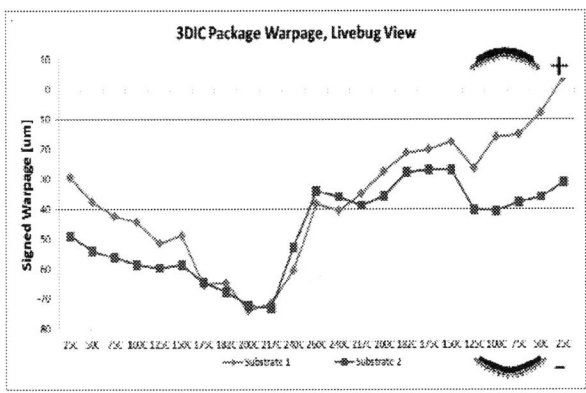

Figure 9: Shadow Moiré data from TV.

The 3D assembly process has been successfully employed across multiple TSV development programs and draws key areas of learning from high volume manufacturing including: copper pillar bumping, stealth dicing, thermo-compression bonding, non-conductive paste underfill, thin die handling, molding, and test. As a result, assembly yields for the test vehicle exceeded 98%.

Reliability Result

To characterize impact of the wafer thinning process on TSV liner integrity, dielectric breakdown tests were performed. Three different I-V curves are shown in Figure 10, comparing breakdown voltages of TSV liner on a full thickness wafer with that on thinned wafer (50μm) after completing backside integration and measuring breakdown voltage of backside dielectric layer. The breakdown reference current is ~1pA and all three I-V curves show the same breakdown voltage. This ensures that the wafer thinning process has no impact on TSV liner integrity, and good integrity of backside low temperature dielectric is maintained, resulting in a highly reliable packaging process flow.

Validating front-end of line (FEOL) device performance and ensuring minimal impact on these highly delicate and sophisticated devices because of backside integration, is an essential task before products can be taped out for 3D packaging. To understand this, various types of FEOL devices were characterized before thinning through key parameter measurements such as Vt_{lin} and Id_{lin}. These parameters were measured again after the wafer thinning process to study the impact of wafer thinning on device performance. Post thinning testing was performed with thin wafers on tape frame as discussed by Adam et al. [5].

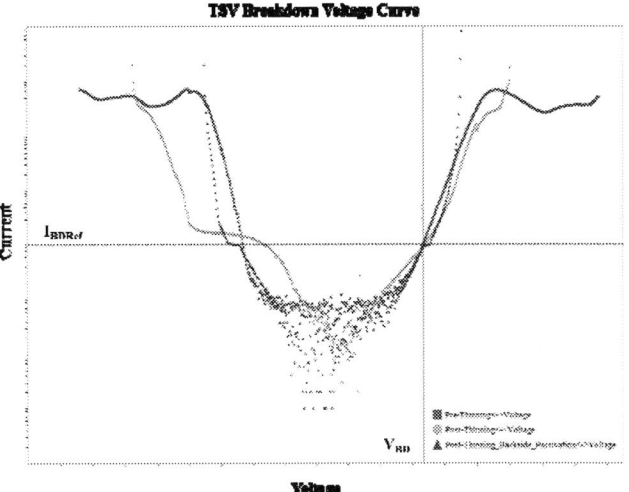

Figure 10: I-V curves comparing dielectric breakdown voltages for TSV liner on full thickness, thinned wafer and backside passivation.

The delta/difference in Vt_{lin} and Id_{lin} between pre-thinning and post-thinning (50μm) wafer is shown in Figure 11. In Figure 11(i) the Vt_{lin} and Vt_{sat} of long-channel nFETs decreases by ~1.7mV and ~1.2mV after thinning, while that of the short-channel nFETs decreases by ~0.3mV and ~3.1mV respectively. The Vt_{lin} and Vt_{sat} of long-channel pFETs decreases by ~0.5mV and ~0.4mV after thinning, while that of the short-channel pFETs decreases by ~1.6mV and ~1.65mV respectively. The Vt_{lin} and Vt_{sat} of ring oscillators using nFETs decreases by ~1.3mV and ~1.2mV after thinning, while that using pFETs decreases by ~0.1mV and ~1.05mV respectively. These shifts in Vtlin and Vtsat after the wafer thinning process are <1.5%, thereby resulting in negligible impact on the device performance.

The shifts in Id_{lin} and Id_{sat} measurements are shown in Figure 11(ii). The Id_{lin} and Id_{sat} of long-channel nFETs decreases by ~48μA and ~11μA after thinning, while that of the short-channel nFETs decreases by ~36μA and ~13μA respectively. The Id_{lin} and Id_{sat} of long-channel pFETs decreases by ~50μA and ~2μA after thinning, while that of the short-channel pFETs decreases by ~14μA and ~8μA respectively. The Id_{lin} and Id_{sat} of ring oscillators using nFETs decreases by ~31μA and ~13μA after thinning, while that using pFETs increases by ~11μA and ~8μA respectively. These shifts in Id_{lin} and Id_{sat} after the wafer thinning process are <0.25%, resulting in negligible impact on device performance.

This confirms that the wafer thinning and backside integration process has negligible impact on device performance. The shift in all the key parameters is below 1.5% which is well within the device tolerance limits.

After performing all wafer level characterization, as discussed above, assembly was performed to do package level testing. JEDEC standard environmental tests were performed on 3D packages by subjecting them to the following stress steps:

1. High Temperature Storage (HTS)
2. Pre-conditioning
   a. Unbiased Highly Accelerated Stress Test (uHAST)
   b. Thermal Cycling (TC)

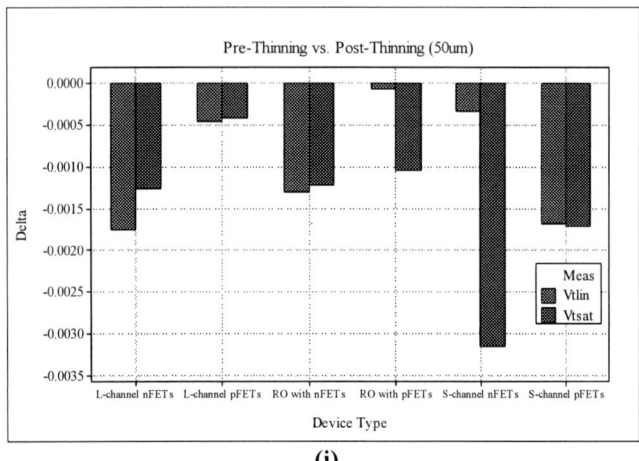

(i)

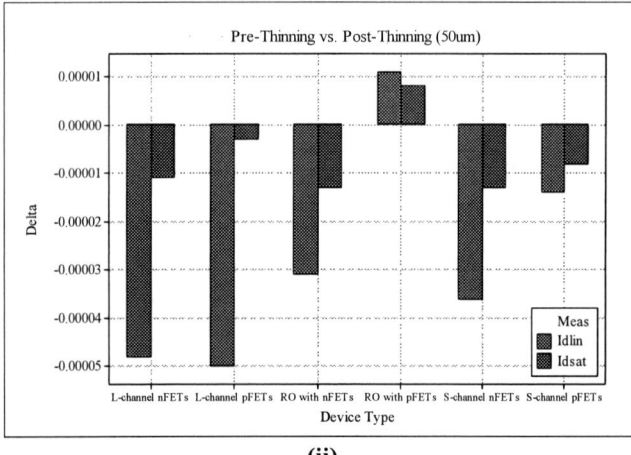

(ii)

Figure 11: FEOL device performance pre-thinning vs. post-thinning (i) Delta – Vt_{lin} and Vt_{sat} and (ii) Delta – Id_{lin} and Id_{sat}.

HTS was performed by storing the samples at 150°C up to 168 hours. Pre-conditioning was performed by initially baking the samples for 24 hours at 125°C followed by soaking at 30°C with 60% relative humidity for 192 hours, followed by three reflow cycles at 260°C. After pre-conditioning, samples were divided into two sets with one set subjected to uHAST and another set subjected to TC. uHAST was performed by baking the samples at 130°C and 85% humidity for 192 hours. TC was performed between -55°C

and 125°C for up to 1000 cycles with intermediate readout done after 500 cycles. The stress steps and conditions along with their corresponding electrical readouts are shown in Table 2.

Table 2: Environmental stress test conditions.

Stress Test	Conditions	Electrical Readout
High Temperature Storage (HTS)	150^{0}C	Readout @ 168 hrs.
Pre-conditioning (MSL3)	24 hrs. @ 125^{0}C 192 hrs. soak @ 30^{0}C/60% RH 260^{0}C reflow x3	Pre-conditioning readout
Unbiased Highly Accelerated Stress Test (uHAST)	130^{0}C/85% RH	Readout @ 192 hrs.
Thermal Cycling (TC)	-55^{0}C to 125^{0}C	Readout @ 500 cycles Readout @ 1000 cycles

The incoming assembly yield of the 3D packages was 98% and sample of packages have successfully completed all the environmental stress tests as described above in Table 2.

Summary and Discussion

In this paper we have presented 3D packaging results using 6x55µm Cu filled TSVs etched on 20nm LPM device wafers. High yielding and stable Cu pillar bumping, via reveal (MEOL process) and assembly processes were developed and demonstrated resulting is successful assembly of a mechanical memory die on top of thin TSV wafers. Major technical consideration used and issues addressed during fabrication processes are discussed. Reliability data both at wafer level, after thinning are presented. No impact is seen on liner breakdown voltage or on device performance after thinning. Finally package level reliability results are shown on this test vehicle where all the reliability targets are met for this test vehicle.

Acknowledgments

The authors would like to thank the 3D teams in GLOBALFOUNDRIES and Amkor Technology® for their continuous support for this work along with reliability groups and failure analysis group in both companies.

References

1. www.Cadence.com
2. http://www.jedec.org/sites/default/files/Barry%20Wagner_Mobile%20Forum_May_2013-Final-04232013.pdf
3. www.hybridmemorycube.org
4. D. W. Kim, "Development of 3D Through Silicon Stack (TSS) Assembly for Wide IO Memory to Logic Devices Integration," in *Proc. IEEE Electronic Components and Technol. Conf. (ECTC)*, Las Vegas, NV, May 28–31, 2013, pp. 77-80.
5. A. Beece, "Impact of Wafer Thinning on High-K Metal Gate 20nm Devices," in *Proc. IEEE Electronic Components and*

978-1-4799-2408-0/14 $31.00 © 2014 IEEE

Technol. Conf. (ECTC), Las Vegas, NV, May 28–31, 2013, pp. 1892-1897.

978-1-4799-2408-0/14 $31.00 © 2014 IEEE

TSV MEOL (Mid End of Line) and Packaging Technology of Mobile 3D-IC Stacking

Duk Ju Na, Kyaw Oo Aung, Won Kyung Choi, *Tsuyoshi Kida, *Toshihiko Ochiai, *Tomoaki Hashimoto, *Michitaka Kimura, *Keiichirou Kata, Seung Wook Yoon and Andy Chang Bum Yong

STATS ChipPAC Ltd. 5 Yishun Street 23, Singapore 768442

*Renesas Electronics Co., 1753, Shimonumabe, Nakahara-Ku, Kawasaki, Kanagawa 211-8668, Japan

seungwook.yoon@statschippac.com

Abstract

Increasing demand for advanced electronic products with a smaller form factor, superior functionality and performance with a lower overall cost has driven semiconductor industry to develop more innovative and emerging advanced packaging technologies. Memory bandwidth has become a bottleneck to mobile processor performance and lower power consumption for high performance computing needs. To reduce obstacles, a revolution in device architecture and package technologies is require. 3D TSV (Through Silicon Via) stacking is to be one of the technologies that can meet those requirements.

This paper mainly describes the 3D TSV MEOL process and packaging technology, especially TSV wafer process and thin die bonding process with Cu column bump on substrates. In order to prove the quality of this 3D package, some stress tests were conducted to evaluate the reliability on the package and board level.

The innovative TSV MEOL process and flip chip assembly with micro Cu column contributes to high density and reliable 3D/TSV integrations has been developed and demonstrated. The target package had two tier structures which consisted of a 28 nm logic device and Wide I/O memory. The logic device was fabricated by via-middle process and accompanied with over 1200 TSVs, and 40 μm / 50 μm bump pitch layout. Advanced 300mm backside via reveal (BVR) process was developed with thin wafer handling and temporary bonding/debonding process. Thermocompression bonding method with Cu pillar was applied to both connections between the memory die and the logic die and between the logic die and an organic substrate so that the high reliability could be achieved. As reliability test items, temperature cycling test, high temperature storage test, humidity test, unbiased highly accelerated stress test and pressure cooker test were performed and passed JEDEC standard reliability tests as well as board level reliability test. After functional test with stacked 3D TSV with logic and Wide IO memory, 12.8 GB/s transmission and drastic I/O power saving compared to LPDDR3 were observed.

INTRODUCTION

Moore's law is approaching physical limitations of CMOS scaling, and three dimensional (3D) integrations have been proposed as solutions [1,2]. The wide band transmission between the logic and the memory is becoming indispensable for not only mobile products, but also other products related to a network area such as servers and data centers. These days 3D integration with Through Silicon Via (TSV) is considered as the key solution, which brings benefits leading to low power consumptions and downsizing of products[3-5].

Advanced flip chip bonding process is one of the key technologies to realize 3D TSV integrations[6, 7], and to enable heterogeneous integrations which combines different functional devices (logic, memory, RF, MEMS etc.). Especially, to develop effective underfill methods for 3D is unavoidable to relieve mechanical stresses so that the reliabilities of interconnections can be enhanced [8-11].

This paper mainly describes 3D TSV packaging technology of mobile 3D-IC stacking, especially MEOL process, package assembly and it reliability. In order to prove the quality of this 3D package, component and board level reliability tests were conducted to evaluate the reliability on the package level. Finally, 28 nm logic device and Wide I/O DRAM were assembled into 3D structure for functionality demonstration with this new technology.

3D Advanced Wafer Level (WLP) Packaging Evolution

Increasing demand for more advanced, smaller and lighter mobile products with superior functionality and lower overall cost has driven the development of innovative and sophisticated packaging technologies, as shown in Figure 1. One of the exciting electronic market trends is the growing availability of mobile devices, such as smartphones, tablets and ultrabooks, which fully realize the dream of computing and communication convergence, with adequate bandwidth and speed to provide a rich user experience. It is particularly important that the next generation of WLP meets the increasing demand for higher bandwidth, improved thermal dissipation and enhanced reliability in cost-effective, scalable solutions to satisfy the growing mobile market.

The need for higher levels of integration, improved electrical performance / reduction of timing delays and shorter vertical interconnects is driving a shift from 2D to 2.5D and 3D package designs. 3D integration is proceeding on three fronts--moving from package level (die and package stacking) to wafer level (especially Fan-out WLP), and, more recently, to the silicon (Si) level with TSV and interposers. Today's new lightweight mobile computers are innovative devices providing true convergence with powerful computing functions, high speed communications and visual, sensing, and imaging technologies. This convergence is pushing traditional packaging well beyond its typical limits in the areas of form factor, reliability and performance.

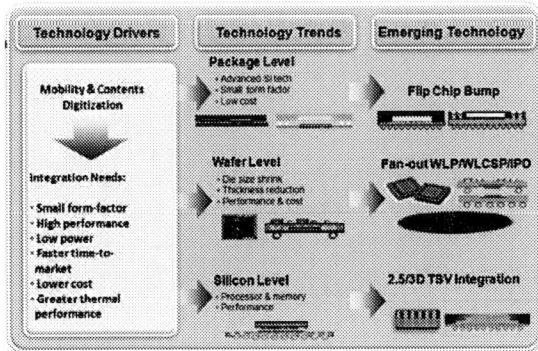

Figure 1: Emerging technology trends for the mobile marketplace.

3D TEST VEHICLE (TV) DESIGN AND DETAILS

The target 3D vehicle consisted of Wide I/O DRAM with micro-bumps, the logic device with high density TSVs and fine pitch Cu pillars and an organic substrate, as shown in Fig. 2. NCP was adopted as an underfill material to protect flip chip joints between the logic die and the substrate.

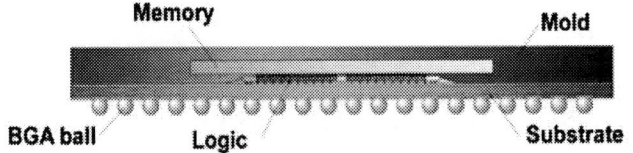

Figure 2. Schematics TSV 3D IC stacking.

Table I. Specifications of logic and memory TV

		Logic TV	Memory TV
Die size [mm]		6 x 6	9 x 9
Die thickness [um]		50	260
TSV	Diameter [um]	10	-
	Depth [um]	50	-
	Pitch [um]	40;50(x;y)	-
	Process	Via-middle	-
Front-side	Pad layout	Center, peripheral	Center
	Pad count	900	1200
	Pad pitch [urn]	100 um staggered	40;50(x;y)
Back-side	Pad layout	Center	-
	Pad count	1200	-
	Pad pitch [urn]	40;50(x;y)	-
Substrate	Thickness [mm]	0.22	
	Metal layers	4 metal layers (1/2/1 buildup substrate)	
Package	Size [mm]	14 x 14	
	Height [mm]	< 1.0	
	Structure	700 I/O BGA	

Table I summarizes major specifications of the logic TV, the memory TV and the substrate. They were designed to clarify behaviours of 3D structure in the following reliability tests. The logic TV was fabricated by so called via-middle process and accompanied with 1200 TSVs, a size of 6 x 6 mm², a thickness of 50µm and 40 µm/50 µm UBM pitch in x-y directions. Cu columns on the front-side were built along

peripheral four rows (100 µm pitch for each row) and on the center area of the die. Solder caps of Cu pillars were formed by electroplating with SnAg. The interface between the logic and the memory was designed by following JEDEC standards. Wide I/O DRAM had 260 µm-thickness, 9 x 9 mm2 size and micro-bumps in the center area of the die. The target of total package thickness including BGA balls was less than 1.0 mm.

3D TSV MEOL PROCESS

TSV MEOL process flow occurs between the wafer fabrication and back-end assembly process. MEOL processes support the advanced manufacturing requirements of 2.5D and 3D TSV as well as wafer level packaging, flip chip and embedded die technology.

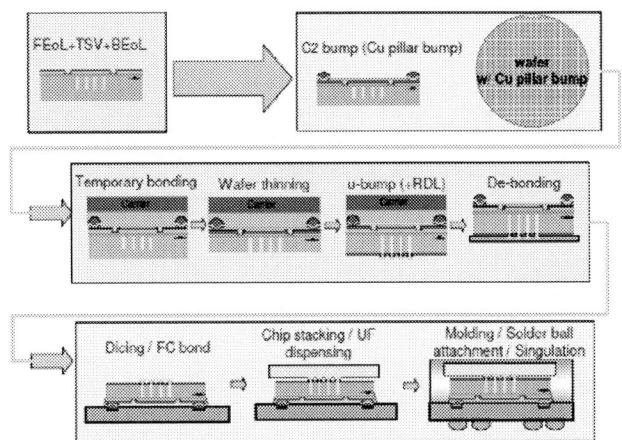

Figure 3. MEOL and package assembly steps in overall 3D TSV process flow.

Flip chip and wafer level packaging are important drivers of mid-end processing in addition to the anticipated growth in 3D solutions utilizing TSV technology, particularly with the integration of memory and logic devices at advanced technology nodes. The initial markets that are expected to embrace 3D TSV technology are mobile applications, memory stacking and high performance processors for the computing segment.

Cu COLUMN MICROBUMP PROCESS

Micro bumping technology where bump pitches are less than 50 micrometers using solder is explored extensively in industry for realization of miniaturized 3D IC integration. For Wide IO microbump in JEDEC 42.6 standards, it has 50/40um bump pitch in x/y direction, respectively. Cu column with solder cap micro-bumps have been studied with the objective to develop reliable fine pitch solder micro joints at low cost. Microbump fabrication is based on photolithography and electroplating processes, which is compatible with conventional IC fabrication. The fabrication process of wafer level process starts with bare Si wafer using Ultra Violet (UV) - light lithography of spin on dielectric material. Secondly, Redistribution line (RDL) layer plating to re-route the Al/Cu bond pads to microbump locations. Thirdly, passivation of RDL layer using spin on dielectric coating and UV

978-1-4799-2408-0/14 $31.00 © 2014 IEEE

lithography to open the RDL metal pads at the bump pads. Fourthly, deposition of Ti or TiW/Cu seed layer and patterning of thick photoresist film using lithography to copper pillar plating and then Ni/solder plating. SEM micrographs of microbump are shown in Fig.4 for 20μm diameter and 40μm height microbumps for 80/40μm staggered bump pitch.

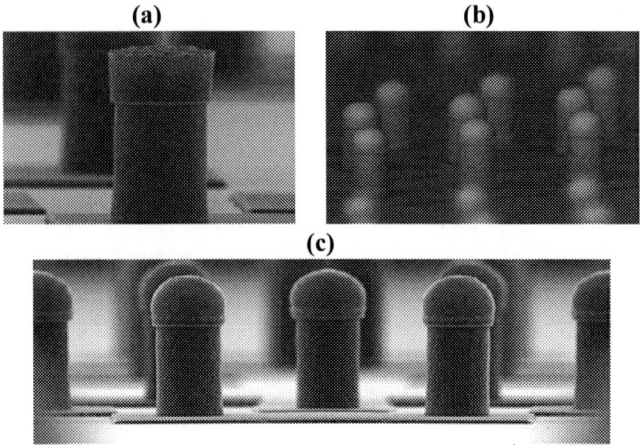

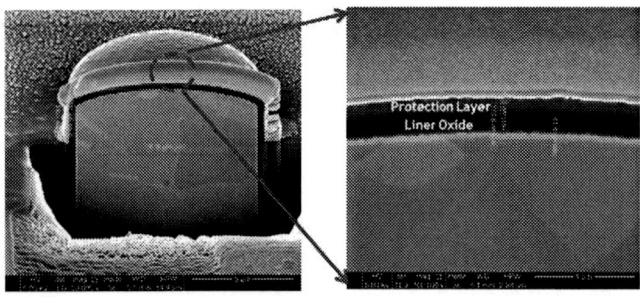

Figure 4. Micrographs of micro bump (a) after micro-bump fabrication and (b & C) after reflow

BACKSIDE VIA REVEAL (BVR) PROCESS

As shown in 3D TSV process flow of Fig.3, TSV is revealed to backside for 3D vertical interconnection after front-end TSV formation. With temporary bonding/debonding system, TSV wafer from fab is to be thin-down and Si etched to expose Cu via with fab process. TSV Cu was fully protected through high selective Si etch process to assure no Cu contamination. (Fig.5) There was also TOF SIMS (Time-of-Flight Secondary Ion Mass Spectroscopy) analysis for Cu contamination on Si wafer during CMP (Chemical Mechanical Polishing) process and verified non-detectable Cu content after chemical composition analysis along whole 300mm TSV wafer. Fig. 6 shows SEM cross-section view and photo of solid Cu filled TSV of 10um diameter and 50um depth after BVR process. Fig. 7 shows TTV (total thickness variation) of 3D TSV wafer after BVR process. It showed only a few microns of TTV values of 300mm test vehicles after process optimization with several DOEs.

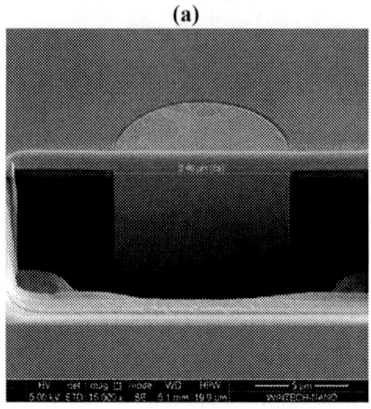

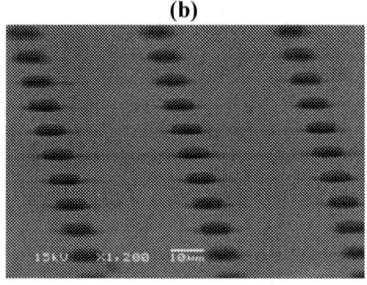

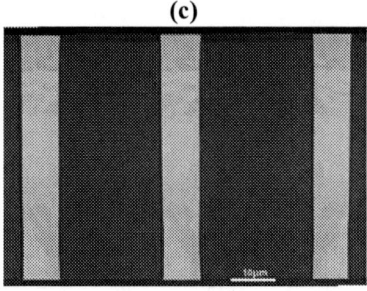

Figure 6. SEM micrographs of backside TSV via revealed after CPM and Si etching process in 12" 3D TSV wafer. (a) FIB titled view of TSV via revealed, (b) revealed Cu tips and (c) cross-sectional view of 10um diameter and 50um depth TSV

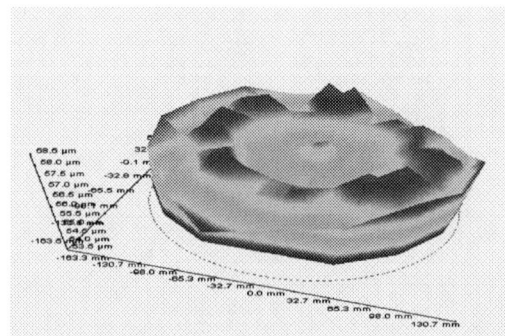

Figure 7. Total thickness variation (TTV) of revealed TSV wafer with temporary bonding.

3D TSV ASSEMBLY AND PACKAGING

Compared to conventional flipchip process, TSV assembly process is more complex due to TSV wafer as well as

Figure 5. FIB SEM micrographs of protective Cu TSV BVR process.

microbump. As shown in Fig. 8, there are additional materials, like as additional encapsulation in between bump and flipchip die or bump and TSV die. There are quite critical challenges for assembly view point both in materials and assembly process.

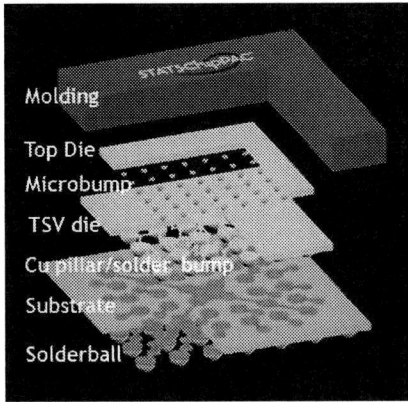

Figure 8. Schematics of TSV 3D IC packaging.

(a)

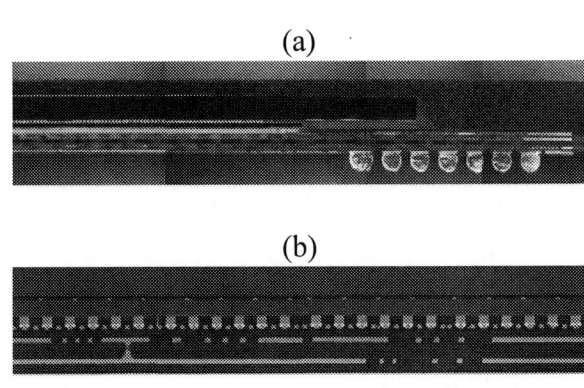

(b)

(c)

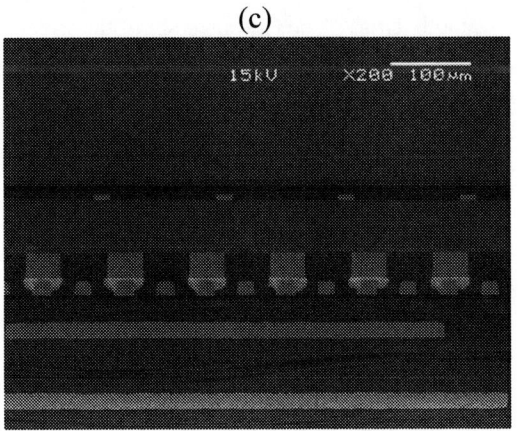

Figure 9. Micrographs of (a) cross-section of 3D IC stacking and (b,c) 40/80 μm pitch of chip-to-substrate bonding.

In advanced 3D TSV stacking technologies, one of the important steps is to develop and assembly fine pitch and high density solder microbumps. Solder microbumps for flip-chip

interconnections allow high wiring density in the Si-carrier, as compared to organic or ceramic substrates, and enable high-performance signal and power connections.

Flip chip assembly was carried out to establish bonding process and investigate the reliability with Cu pillar microbump. After microbump test vehicle fabrication with bump, the flip chip attachment was carried out. Several DOE (design of experiments)s were carried out to find optimized flip chip attach process conditions as functions of time, temperature and pressure. Assessments by checking fractural surface and mechanical shear strength were conducted to evaluate DOEs of bonding parameters.

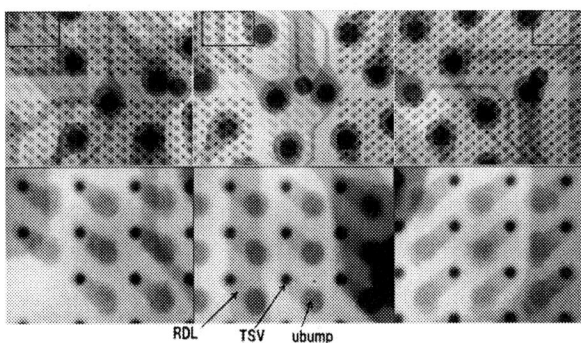

Figure 10. X-ray micrographs of 40/50μm pitch of chip-to-chip bonding of 3D TSV IC stacking

Fig. 9 shows the micrographs of cross-section of the chips joined for 40μm pitch microbumps. A misalignment of about <2μm was observed between the Si chip and 3D TSV chip after assembly. This misalignment was a result of accuracy limitation of the bonder equipment. After assembly, X-ray image was observed and found successful 3D TSV flipchip bonding without voids in between chip-to-chip, chip-to-substrate, respectively as shown Fig.10.

ELECTRICAL PERFORMANCE VALIDATION[1]

Largest benefit of 3D integration is the capability to realize System in Packages (SiP) which consist of different wafer technologies. To evaluate actual performances of the unique assembly process developed in this study, the 3D package which combined 28 nm logic and Wide I/O DRAM was assembled. This 3D package had the extremely small logic device as a bottom die and the large memory device as a top die. The logic device was fabricated with 28 nm technology, and accompanied with 1200 TSVs, a size of 2 x 6mm^2 and a thickness of 50 μm. The memory device was a 4 Gbit 512 DQs monolithic Wide I/O DRAM, and accompanied with a size of 9 x 9 mm^2 and a thickness of 260 μm. The developed assembly process could fill the space between the bottom die and the top die. After the assembly, electrical performances of the samples were evaluated with LSI testers. Not only the 1200 TSVs connectivity, but also DRAM bit quality, at speed test, current consumption were investigated carefully. In conclusion, it was proved that the test vehicle achieved 12.8 GB/s transmission and 89 % reduction of I/O power compared to LPDDR3.

978-1-4799-2408-0/14 $31.00 © 2014 IEEE

COMPONENT AND BOARD LEVEL RELIABILITY

The reliability of the target 3D package was evaluated by temperature cycling (TC), high temperature storage (HTS), humidity test (HT), unbiased highly accelerated stress test (uHAST) and pressure cooker test (PCT). Table II summarizes the stress conditions and results. Before the reliability tests, all samples were treated with the moisture sensitivity level 3 (MSL3) (30°C /60% RH, 168h) and three times reflow (a peak reflow temperature: 260°C). In this test vehicle, daisy chains were laid out to evaluate the connections among the logic die, the memory die and 1200 TSVs. The resistance of each daisy chain was measured periodically in all test items. And board level reliability tests were also carried out as shown below in Table II.

In these tests, the criterion of resistance increase was decided to be within 20 % of each initial value. The resistance changes in all reliability tests were less than ± 3 %. In conclusion, all the samples passed component and board level reliability.

Table II. Test items and results of package and board level reliability tests.

Test Item (Test condition)	End point	Results
1.Component Level Test		
Temperature Cycling (TC C -55/125degC)	1500 cycles	pass
High Temperature Storage / HTS (150degC)	1000h	pass
Humidity Test /HT (85degC/85%RH)	1000h	pass
Unbiased Highly Accelerated Stress Test /uHAST (130degC/85%RH)	500 h	pass
Pressure Cooker Test/PCT (121degC/100%RH)	300h	pass
2.Board Level Test		
Temperature Cycle Test JEITA	500 cycles	pass
Mechanical Drop Test JEDEC	30 drops	pass
Cyclic Bending Test JEITA	1500 cycles	pass
Board Bending Test JEITA	-	pass
Allowable Bending Test JEITA	-	pass
Impact Bending Test JEITA	-	pass

CONCLUSION

1. The innovative assembly of 3D TSV MEOL and package assembly technology has been developed for reliable 3D/TSV integrations.

2. Cu pillar bump, temporary bonding debonding and BVR process were optimized and qualified with 300mm daisychain and 28nm functional wafers.

3. Package assembly stacking process was established for chip-to-chip, chip-to-substrate interconnection with thermocompression bonding with 40/50um ultra-fine pitch microbump.

4. All of the samples passed 1500-cycle TC, 1000h HTS, 1000h HT, 500h uHAST and 300h PCT and board level reliability including TCoB and bending tests.

5. 28 nm logic device and Wide I/O DRAM were assembled into 3D structure with this technology.

6. 3D test vehicle showed 12.8GB/s transmission and 89% reduction of I/O power consumption compared to LPDDR3. As a result, the robust process for 3D integration was established.

ACKNOWLEDGMENTS

The authors would like to thank our colleagues of 3D TSV team for process development and characterization, Mr. Jang Tae Huan for bump and package design in STATS ChipPAC.

REFERENCES

1. Kentaro Mori, Yoshihiro Ono, et al. "High Density and Reliable Packaging Technology with Non Conductive Film for 3D/TSV", IEEE 3D Conference, US, Nov 2013.

2. RS Patti, "Three dimensional integrated circuits and the future of system on chip designs," Proceedings of the IEEE, vol. 94, No. 6, June 2006, pp. 1214-1224.

3. D. Milojevic, et al., "DRAM-on-logic Stack - Calibrated thermal and mechanical models integrated into PathFinding flow," in Proc. Custom Integrated Circuits Conference, San Jose, CA, September 2011, pp. 1-4.

4. Dong Wook Kim, et al., "Development of 3D Through Silicon Stack (TSS) Assembly for Wide IO Memory to Logic Devices Integration," in Proc. 63th Electronic Components and Technology Conference, Las Vegas, NV, May 2013, pp. 77-80.

5. J. Roullard, et al., "Evaluation of 3D interconnect routing and stacking strategy to optimize high speed signal transmission for memory on logic," in Proc. 62th Electronic Components and Technology Conference, San Diego, CA, May 2012, pp. 8-13.

6. M. Gerber, et al., "Next generation fine pitch Cu Pillar technology -Enabling next generation silicon nodes," in Proc. 61th Electronic Components and Technology Conference, Lake Buena Vista, FL, May 2011, pp. 612-618.

7. Y. Orii, et al., "Electromigration analysis of peripheral ultra fine pitch C2 flip chip interconnection with solder capped Cu pillar bump," in Proc. 61th Electronic Components and Technology Conference, Lake Buena Vista, FL, May 2011, pp. 340-345.

8. S. Kawamoto, et al., "The Optimization of the Composition of Non-Conductive Film and the Lamination to Wafer," in Proc. 63th Electronic Components and Technology Conference, Las Vegas, NV, May 2013, pp. 778-784.

9. T. Nonaka, et al., "Low temperature touch down and suppressing filler trapping bonding process with a wafer level pre-applied underfilling film adhesive," in Proc. 62th Electronic Components and Technology Conference, San Diego, CA, May 2012, pp. 444-449.

10. K. Honda, et al., "NCF for pre-applied process in higher density electronic package including 3D-package," in Proc. 62th Electronic Components and Technology Conference, San Diego, CA, May 2012, pp. 385-392.

11. Chien-Feng Chan, et al., "Development of thermal compression bonding with Non Conductive Paste for 3DIC fine pitch copper pillar bump interconnections," in Proc. 13th Electronics Packaging Technology Conference, Singapore, December 2011, pp. 329-332.

Thermally Enhanced 3 Dimensional Integrated Circuit (TE3DIC) Packaging

S. Snyder, J. Thompson, A. King, E. Walters, P. Tyler and M.R. Weatherspoon
Harris Corporation, Government Communications Systems Division (GCSD)
Melbourne, FL USA
Steven.Snyder@harris.com

Abstract

Commercial packaging roadmaps clearly depict the imminence of chip stacking utilizing through silicon via (TSV) technology (commonly referred to as 3-dimensional integrated circuits (3DIC)) as a means to improve system performance by reducing routing lengths, latency, and drive power while increasing functionality per unit volume. Roadmaps and packaging research focus areas also depict complex 3DIC and packaging architecture concepts to include heterogeneous materials, components and features [1-2]. This added diversity often exacerbates physical proximity effects such as thermal and electromagnetic (EM) coupling. To reduce unwanted thermal and EM coupling, we propose interleaving 25µm thick, flexible, high thermal conductivity (1600 W/m·K, in-plane) pyrolytic graphite sheets (PGS) [3] into 3DIC stacks. The PGS provides passive parallel thermal paths from each die to the package heat spreader with potential significant reduction in overall package thermal resistance (Θ_{JC}). This also provides design flexibility to thermally decouple sensitive components within the package from intermittent power sources by thermal routing. Interleaving PGS is also expected to impart EM shielding benefits between die and serve to reduce overall package emissions. This paper focuses on potential junction-to-case thermal resistance improvements of a Thermally Enhanced 3-Dimensional Integrated Circuit (TE3DIC) packaging solution. By utilizing a path-finding set of thermal models of a TE3DIC BGA package, design parameters are adjusted to assess various cases, promote design intuition, and ultimately lead to a final test vehicle design. This test vehicle includes three TSV die with both uniform and localized (fireball) Ni80Cr20 heater traces and include resistive platinum temperature sensors. These die are copper pillar bonded with PGS interleaved and flip chip soldered to a 37mm BGA carrier. The perimeter of the PGS is bonded to a terraced copper heat spreader. Detailed modeling estimates a 44% reduction in package thermal resistance with PGS interleaves in a three die stack compared with a similar direct bonded die stack. Reductions are greater for 3DIC designs with copper pillar bonding compared to covalently bonded die stacks.

Introduction

As TSV technology experiences wider adoption and 3DIC packages evolve, the fundamental challenges of physical proximity effects in electromagnetic (EM) and thermal coupling become greater. Considering the latter case, thermal mitigation was cited as a contributing factor in a commercial FPGA for not stacking logic die directly. A planar layout of multiple die using silicon interposers was favored instead. [4]. For extreme thermally challenged 3DICs, cooling using actively pumped and passive fluidic integration in 3DICs are under development [5] considering cases of high average power and hot spot cooling. Fluidic thermal management hardware accounts for a large fraction of the system volume, weight, and cost and undermines efforts to transfer emerging components to small form-factor applications [6]. Therefore, significant improvements using pure conduction techniques for 3DICs are still desired for comparatively easier design, fabrication, and maintainability. To this end, thermal TSVs (TTSV) have also been studied[7], however, alternative pure conduction techniques that route heat parallel to the stack could reduce the junction-to-case thermal resistances without consuming the additional die resources TTSVs require.

Industry roadmaps and "off-roadmap applications" [1] also highlight increasingly complex heterogeneous 3DIC architectures including mixtures of silicon (CMOS) devices, compound semiconductor devices, MEMS devices, integrated passives, etc. as well as interposers made from silicon or glass, with each combination thereof in pursuit of a particular application. In these cases power dissipation is not necessarily the only consideration. Localized temperature control within discrete components or sections may be required as well as compensation for potential thermal impediments such as thick polymer films, glass interposers, or even air gaps as often required by some low loss RF structures. The need for localized temperature control can arise from the close proximity of thermally sensitive devices like MEMS devices and sensors with respect to regions that dissipate transient power like power amplifiers or processor cores used intermittently. Thermally isolating sensitive structures using low thermal conductivity construction materials in conjunction with strategic placement of high thermal conductivity materials serves to route the heat to achieve localized thermal isolation where needed while maintaining high heat dissipation conductive paths.

Pyrolytic graphite sheets (PGS) are one type of commercially available [3] high thermal conductivity material that is thin (10µm - 100µm thick variants), flexible, and easily shaped. These features make it a good candidate for improved 3DIC power dissipation and thermal routing. The material is also electrically conductive with potential benefits in EM isolation of close proximity components provided adequate measures can be taken to prevent shorting within such dense electrical 3DIC routing. In contrast, the material can be challenging to work with in the areas of handling, bonding, and chemical compatibility. Prior work by the authors using PGS suggested that the potential benefits of PGS make these potential assembly issues a secondary consideration at least until the modeling revealed the potential magnitude of thermal enhancement.

In this paper we will discuss our primary interest in establishing the potential for PGS to reduce the package thermal resistance of 3DIC packages. Special thermal routing cases will not be addressed. Although the test vehicle has

978-1-4799-2408-0/14 $31.00 © 2014 IEEE

been designed to include RF radiating elements on each die, the EM shielding investigation will not be covered in this paper. The remainder of the paper describes the thermal modeling approach, discusses some of the design parameters, describes the final Thermally Enhanced 3-Dimensional Integrated Circuit (TE3DIC) test vehicle design, and predicts the improvement in junction-to-case thermal resistance as compared to various baseline cases without PGS interleaves.

Thermal Model Development

To vet the concept of PGS interleaves in a notional 3DIC package, an initial model was developed based on the 3DIC BGA package depicted in Figure 1. This simplified model suggested that nearly half of the heat flow could be directed through the PGS straps (see Figure 2) to the copper lid resulting in a junction-to-case thermal resistance improvements of 23%, 36%, and 42% for the die closest to the heat sink, the middle die, and the die attached to the carrier respectively versus the same configuration without

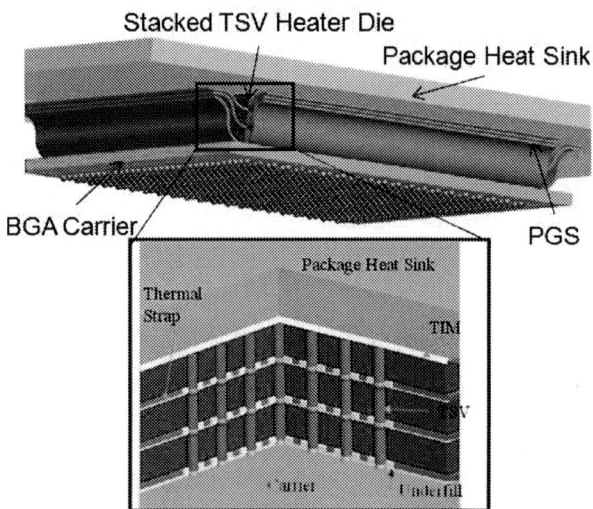

Figure 1. Initial concept of TE3DIC Package with PGS interleaves between die and cut-away details showing TSVs.

PGS interleaved between each die. Note that the PGS straps in the thermal model were spaced apart to provide more uniform spreading than the original concept in Figure 1.

Although this preliminary model lacked rigor it serves to illustrate the concept and suggest that this approach deserves a more detailed study. The details of this thermal model will not be discussed in favor of the following discussion of a

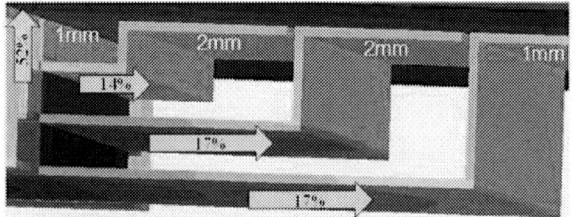

Figure 2. Heat flow through the die stack as well as in the PGS straps that are interleaved between die.

more rigorous model.

Since a primary goal of the project is to fabricate a TE3DIC test vehicle for thermal experimentation and ultimate reconciliation to the thermal model, the first step in creating the model was to create a baseline that supported practical considerations for materials, geometrical configuration, feasibility of fabrication, and test interfaces. Several what-if scenarios where modeled to explore the sensitivity and to steer the final design of the test vehicle.

A three TSV die stack was selected for modeling. Copper pillar bonding was selected as a compact interconnect to allow minimal PGS removal when interleaved and is seen as a trend in 3DIC construction because it simultaneously forms both mechanical and electrical connections [8-9]. Because NiCr resistors require electrical insulation above the silicon, uncertainty existed as to the thermal resistance within the die compared to a CMOS device (where heat is produced in the junctions within the silicon). This led to further uncertainty as to which side of the die to place the PGS and whether the simulated active surface should face up or down in the stack. A single die thermal model was created to help resolve this ambiguity.

Single Die Heat Flow Evaluation

Our purpose was to demonstrate the utility of interleaving PGS between vertically interconnected die to couple high heat flux components with high-thermal conductivity parallel paths directly to the heat sink. To accomplish this efficiently the PGS must be in contact with the hottest side of the die. In traditional CMOS processing, the heat is generated at the transistor silicon interface. The back end of line (BEOL) wiring consisting of several passivation layers provides a thermal insulator on top of the transistors. Accordingly the majority of the heat flux is dissipated through the back side of the die. In this example it is conceived that the PGS would be bonded to the back side of the die in a 3DIC CMOS chip stack.

The first step in developing a system level thermal model for the 3D Thermal Chip Stack Test Vehicle was to analyze the heat flow through a single heater die. Several plausible dielectric layer configurations were analyzed to reveal which side of the heater die possessed the highest heat flux. Figure 3 provides a cross-sectional illustration of the heater die that was modeled with the respective dielectric/silicon layers identified by letters that are defined in Table 1. Consideration was given to modeling a producible design consistent with in-house capabilities and for reconciliation with the thermal model. Accordingly, this study was performed on both 250μm and 400μm thick heater die bearing TSVs as illustrated in Figure 1. The analysis was performed on 33 different layer configurations (17 for 250μm and 16 for 400μm die thicknesses). Table 2 provides the layer combinations that were analyzed as well as the percentage of heat flow that was observed at the top surface of the die (the balance of the heat flowing through the back side of the die). Polyimide and SiO_2 were chosen as the dielectric materials since they are common in IC manufacturing. The heat flow through the die was modeled using FloTHERM®, computational fluid dynamics (CFD) software used extensively in the electronics industry for both macroscopic

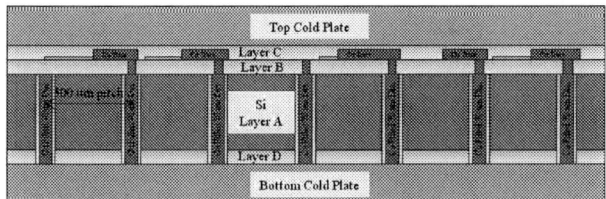

Figure 3. Die cross-section revealing the dielectric layer stackup of the layers to be modeled on a Si die bearing TSVs.

(system level) and microscopic (die level) analyses. Cold plates were fixed at 0°C to the top and bottom sides of the die. Adiabatic conditions were applied to the Cu TSV interfaces to prevent a thermal short circuit that would have skewed the results. A thermal dissipation of 1W was applied to the 350μm wide serpentine Ni80Cr20 whole die heater.

Several conclusions can be drawn from the data presented in Table 2 including: (1) in 29 of the 33 cases analyzed >50% of the heat flow was directed to the top surface of the die, (2) the thicker the die the more heat flows to the top side of the

Layer	Material	Thickness	Thermal Conductivity (W/m-K)	Thermal Resistance (K/W)
A1	Si	250 um	149	1.67E-06
A2	Si	400 um	149	2.68E-06
B1	SiO2	1 um	1.3	7.69E-07
B2	SiO2	0.1 um	1.3	7.69E-08
B3	SiO2	0 um	N/A	N/A
B4	Polyimide	5 um	0.52	9.61E-06
C1	SiO2	1 um	1.3	7.69E-07
C2	SiO2	2 um	1.3	1.53E-06
C3	Polyimide	5 um	0.52	9.61E-06
D1	SiO2	1 um	1.3	7.69E-07
D2	Polyimide	5 um	0.52	9.61E-06

Table 1. Die cross-section revealing the dielectric layer stackup of the layers to be modeled on a Si die bearing TSVs.

die, (3) the choice and thickness of active-side dielectric materials strongly influences the direction of the heat flow, (4) applying dielectric materials to the back side of a die can redirect the majority of the heat flow to the top surface. The later observation could be useful in some cases for thermal isolation and guiding the thermal path in 3DIC stacks.

In the majority of the cases analyzed the heating element was sandwiched between two dielectric layers. However, there were a total of 13 cases (5, 6, 12, 13, 17, 20, 23, 26, 29, 30, 31, 32, and 33) where the heating element was in direct contact with the high thermal conductivity Si (more representative of transistors in CMOS devices). In only 3 of these cases (i.e., 6, 29, and 31) was the observed heat flow greater at the bottom of the die. In two of these cases (i.e., 29, and 31) there was 5μm of Polyimide on top of the heating element and it stands to reason that the majority of the heat flow would be directed to the bottom of the die given that 5μm of polyimide has an 83% and 72% higher thermal resistance than 250μm and 400 μm thick Si, respectively. However, for case 6 there is only 2μm of SiO₂ on top of the heating element which is on the 250μm thick Si die which has an 8.3% higher thermal resistance than that of the 2μm thick SiO₂ layer but the majority of the heat (i.e., 57.1%) still flows out the bottom of the die. In contrast, if 5μm of Polyimide is

Table 2. Heat flow through a single heater die according to the die layer configurations outlined in Table 1.

Case	Die Layer Configuration	Heat Flow (%) Die Top	Heat Flow (%) Die Bottom
1	A1, B1, C1	64.8%	35.2%
2	A1, B1, C2	53.3%	46.7%
3	A1, B2, C1	55.9%	44.1%
4	A1, B2, C2	44.1%	55.9%
5	A1, B3, C1	54.6%	45.4%
6	A1, B3, C2	42.9%	57.1%
7	A1, B4, C3	50.9%	49.1%
8	A2, B1, C1	71.8%	28.2%
9	A2, B1, C2	61.2%	38.8%
10	A2, B2, C1	66.2%	33.8%
11	A2, B2, C2	55.0%	45.0%
12	A2, B3, C1	65.4%	34.6%
13	A2, B3, C2	54.2%	45.8%
14	A2, B4, C3	52.8%	47.2%
15	A1, B1, C1, D1	70.9%	29.1%
16	A1, B2, C1, D1	65.0%	35.0%
17	A1, B3, C1, D1	64.1%	35.9%
18	A1, B1, C2, D1	60.2%	39.8%
19	A1, B2, C2, D1	53.7%	46.3%
20	A1, B3, C2, D1	52.8%	47.2%
21	A2, B1, C1, D1	75.8%	24.2%
22	A2, B2, C1, D1	71.6%	28.4%
23	A2, B3, C1, D1	71.2%	28.8%
24	A2, B1, C2, D1	66.1%	33.9%
25	A2, B2, C2, D1	61.4%	38.6%
26	A2, B3, C2, D1	60.8%	39.2%
27	A1, B4, C3, D2	65.8%	34.2%
28	A2, B4, C3, D2	66.8%	33.2%
29	A1, B3, C3	12.4%	87.6%
30	A1, B3, C3, D2	50.3%	49.7%
31	A2, B3, C3	18.3%	81.7%
32	A2, B3, C3, D2	52.2%	47.8%
33	A1, B3, C2, D2	84.0%	16.0%
	400um thick Si Die		
	250um thick Si Die		

applied to the back side of the die (i.e., comparison of case 33 to case 6) then 84% of the heat flows out the top side of the die. Thus, applying 5μm of Polyimide to the backside of a typical CMOS die could act as a thermal control lever, and would increase the heat flow through the top side of the die by 41%.

In almost all of the cases the delta heat flow out the top side of the die is greater (range is 5%-11%) for the thicker die (ex. compare cases 1-6 to cases 8-13 and 15-20 to 21-26) because 400μm thick Si has a 37% higher thermal resistance than 250μm thick Si. However, comparing case 7 to case 14 the delta heat flow to the top surface is only 1.9%. This is because the heating element is sandwiched between 5μm of Polyimide on either side and the extra thickness of the Si has a minimal effect on the heat flow direction.

We chose to utilize the layer stack up defined in case 24 to build the system level thermal model for the 3D Thermal Chip Stack Test Vehicle because it most closely resembles the layer stack up for the TSV wafers to be fabricated. We also elected to utilize 400μm thick Si to ease the fabrication of double sided processing without the use of a process carrier wafer.

The TSV construction encases the Si within a 1μm thick SiO_2 passivation layer on all surfaces (including the via side walls). The NiCr and Pt resistive elements will be deposited over 1μm SiO_2 and passivated with 2μm of SiO_2. In this configuration, 66.1% of the heat flows out the top surface of the die. The die will be assembled in a flip chip fashion (i.e., the active side of the die bearing the NiCr and Pt layers will be facing the carrier substrate) and the PGS will be bonded to the active the active side of the die.

3D Thermal Chip Stack Test Vehicle Model

Four detailed thermal models were constructed to compare the package thermal resistance (Θ_{JC}) for a vertical stack of 3 heater die. These models compared the TE3DIC packaging approach and three control configurations. The control configurations included a covalently bonded die stack, gap filling with a thermal interface material (TIM), and slight air gaps between die due to Cu micro-bump bonding.

The CFD software was used to analyze each configuration. To accurately capture the thermal performance of each model all of the pertinent details of the die and the package were imported and modeled including the TSVs, Cu pillar intra die interconnects, circuit metallization, SiO_2 layers, all of the thermal interface materials, PGS, Cu heat spreader, and BGA carrier substrate. The analysis parameters were Steady State Conduction as this is the primary mode of heat transfer within a typical BGA package. The number of grid cells to accurately render and solve the models was in excess of 15 million.

The TE3DIC package is illustrated in Figure 4. The Si heater die are 400μm thick and encased within a uniform 1μm thick SiO_2 layer and another 2μm of SiO_2 on top of the resistive elements. The die are bonded in a flip chip configuration with PGS between each of the 3 die in contact with the hottest side of the die (i.e., the side bearing the resistive elements). For typical CMOS devices the PGS can just as easily be bonded to the back side of the die. The PGS is bonded to the die using a 5μm thick film material (1.5W/m-K) and to the Cu heat spreader with a 50μm thick TIM (3.5W/m-K). A low thermal conductivity material (0.5W/m-K) is used to bond the top die (die 3) to the Cu heat spreader (385 W/m-K) as well as the underfill between die 1 and the 1mm thick carrier substrate. For the carrier an in-plane thermal conductivity of 19 W/m-K and through-plane thermal conductivity of 0.3 W/m-K were used. Note that for this model, the PGS was derated slightly based on authors' estimates and prior history with lifetime testing of other pyrolytic graphite forms. For PGS, 1500W/m-K was used in this model.

Three separate PGS designs were also analyzed. In the first design, 200μm diameter holes are laser milled in the 25μm thick PGS on a 0.5mm pitch to allow the 70μm diameter 75μm tall Cu pillars to pass through the PGS and bond to the next die. The large holes help prevent shorting with the Cu pillars while still allowing 87.5% of the PGS to be in contact with the surface area of the die. However, this design would not be useful for high density interconnect

applications (ex. 10μm diameter TSVs on a 20μm pitch) where it would be impractical to impossible to mill such holes and have a sufficient amount of PGS in contact with the die for thermal spreading.

The second PGS design investigated the effect of creating a window frame in the 25μm thick PGS by milling out a square central area where the high density intra-die I/O interconnect could pass through having a 1mm area of the PGS in contact all around the perimeter of the die. In this case, 36% of the die surface area would be in contact with the PGS. The third PGS design is a slight variant of the first and second designs in that the PGS cross-sectional area for conduction is doubled by placing two pieces of 25μm thick PGS between the die in both the "Swiss cheese" and "window frame" approaches.

Figure 5 provides an illustrative example for the three control samples. In each of these controls the PGS is removed from the stack and either filled with a TIM, an air gap, or the die are directly bonded together. The controls all utilize the same heater die design as the TE3DIC package.

The first control represents a chip stack in which the die are covalently bonded at the oxide interface creating a solid cube of 3 heater die with zero air gaps. Ziptronix has demonstrated such a process (Direct Bond Interconnect DBI®) for covalently bonding two planarized surfaces at low temperatures [9]. The interfaces at the heat spreader and carrier substrate possess the same materials as the TE3DIC package.

In the second control a TIM with thermal conductivity of 0.5W/m-K replaces the PGS sheets. The 75μm gap is maintained to allow the TIM material to flow between the die. We chose such a TIM so as to be electrically isolating and prevent electrical shorting between the Cu pillars.

The third control represents a chip stack in which die are bonded together using thermocompression bonding of copper micro-bumps leaving behind a very thin 3μm air gap.

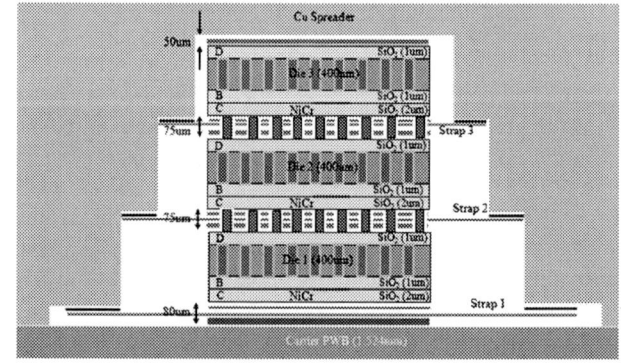

PGS (1500W/m-K – In-Plane; 15W/m-K – Thru-Plane) - 25um
TIM (1.5W/m-K) - 5um
TIM (1.5W/m-K) - 45um
Underfill (0.5W/m-K) - 50um
Insulating Material (0.5W/m-K) - 50um
TIM (3.5W/m-K) - 50um

Figure 4. TE3DIC package utilizing a terraced Cu heat spreader and laser milled holes in the PGS to allow intra-die interconnects to be made with Cu pillars. (Note: an alternative approach is to laser mill a square in the center portion of the PGS and bond the PGS to a 1mm ring around the die perimeter).

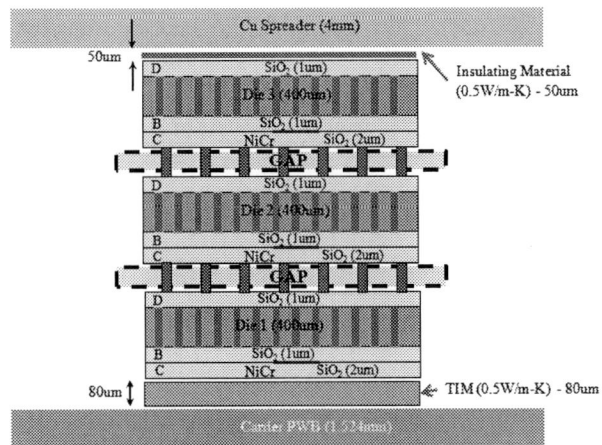

Control	Gap Height	Gap Material
1	0 μm	NA
2	75 μm	TIM (0.5 W/m-K)
3	3 μm	Air

Figure 5. Comparative control models representing a covalently bonded die stack, die bonded and underfilled with a TIM, and die bonded with a 3μm air gap between die.

All cases only consider the uniform whole die heater as the thermal source in which each die in the stack dissipates 1W for a total package power dissipation of 3W. This dissipation corresponds to a total heat flux of 5.58W/cm^2 per die.

The Cu heat spreader is considered to be in intimate contact with an isothermal cold plate set to 0°C for all simulations to simplify the computation of the Junction-to-Case thermal resistance. It is common practice when characterizing thermal resistance to utilize 1W (unity) as a known power dissipation and 0°C as a sink temperature. The corresponding junction temperature, T_J, is the thermal resistance based on the following relationship:

$$T_J = T_C + P_D \cdot \theta_{JC} \rightarrow \theta_{JC} = \frac{T_J - T_C}{P_D} = \frac{T_J - 0°C}{1W} = T_J$$

Equation 1. Die level package thermal resistance

where T_C is the case temperature, P_D is the power dissipated, and θ_{JC} is the junction to case thermal resistance. This rationale is used to compute the thermal resistance for each individual die in the 3 die stack. In addition, an overall package thermal resistance is also calculated as this would normally be the reported worst case impedance specified by IC manufacturers. The calculation is identical to that of Equation 1 but instead of the power dissipation being 1W the dissipation is the total package dissipation (in this case 3W). The junction temperature is the hottest die in the 3 die stack extracted from the thermal model. This hot spot always occurs on Die 1 which is the furthest from the Cu heat spreader and the cold plate. The junction-to-case thermal resistance is then calculated according to Equation 2.

$$\theta_{JC} = \frac{T_J - T_C}{P_D} = \frac{T_{J,Predicted} - 0°C}{3W}$$

Equation 2. 3DIC package junction-case thermal resistance

Results

Table 3 provides the comparative results for T_J and θ_{JC} for the four different TE3DIC PGS configurations as well as the three separate control samples. The reported T_J value for each of the 3DIC configurations was from Die 1 as mentioned earlier. Equation 2 was used to calculate the θ_{JC} values. The control samples in which there was a TIM material or an air gap between the die (i.e., Control 2 and Control 3) possessed the highest T_J and θ_{JC} values whereas the T_J and θ_{JC} values for all of the TE3DIC PGS configurations were lower than that of the covalently bonded chip stack. The lowest T_J and θ_{JC} values are observed for PGS configurations with the highest surface area in contact with the heater die.

Table 3. Comparative results of T_J and Θ_{JC} for the four different TE3DIC PGS versus the three control configurations.

		T_J (°C)	Θ_{JC} (°C)
TE3DIC PGS Configurations	25μm Thick PGS (Whole Die)	2.3	0.77
	50μm Thick PGS (Whole Die)	1.76	0.59
	25μm Thick PGS (Die Perimeter)	4.08	1.36
	50μm Thick PGS (Die Perimeter)	3.44	1.15
Control 1	Covalent Bond	4.14	1.38
Control 2	75μm Die-to-Die Gap + TIM	7.25	2.42
Control 3	3μm Die-to-Die Air Gap	6.49	2.16

Table 4 provides the percent reduction in θ_{JC} between the TE3DIC PGS configurations and the Control samples. It is observed that interleaving 50μm thick PGS between die produces the greatest reductions in θ_{JC} (ex. 57% reduction over covalently bonded die and 76% reduction over die bearing a TIM between bonded die). This is due to a greater cross-sectional area of conduction for the PGS. Reducing the thickness of the PGS between the die by half still provides a significant impact in reducing θ_{JC} and could be beneficial for high density 3DIC's. It is also observed that interleaving the 50μm thick PGS in the window frame approach where the PGS is in contact with a 1mm area around the perimeter of the die will provide a 17% improvement in reducing θ_{JC} over the

Table 4. Predicted percent reduction in Θ_{JC} among the TE3DIC PGS configurations and the Control samples.

	TE3DIC vs. Control Θ_{JC} Improvement	Control 1 (Covalent)	Control 2 TIM (75μm Gap)	Control 3 (3μm Gap)
TE3DIC PGS Configurations	25μm Thick PGS (Whole Die)	44%	68%	65%
	50μm Thick PGS (Whole Die)	57%	76%	73%
	25μm Thick PGS (Die Perimeter)	1%	44%	37%
	50μm Thick PGS (Die Perimeter)	17%	53%	47%

Table 5. Calculated junction temperature (T_J, °C) for each die and corresponding package configuration at 1W.

	Power (W)	25 μm Thick PGS (Whole Die)	50 μm Thick PGS (Whole Die)	25 μm Thick PGS (Die Perimeter)	50 μm Thick PGS (Die Perimeter)	Control 1 (Covalent Bond)	Control 2 (75 μm gap + TIM)	Control 3 (3 μm Air Gap)
Die 3	1	1.78	1.36	1.72	1.5	3.93	3.75	3.96
Die 2	1	2.14	1.63	3.46	2.93	4.05	6.17	5.66
Die 1	1	2.3	1.76	4.08	3.44	4.14	7.25	6.49

covalently bonded die stack. This implementation may be more producible in certain applications where high density interconnect fields are present.

Keeping in mind that the package θ_{JC} is based on the total 3DIC power over the worst case T_J in the stack, it is helpful to look at the T_J in all die for comparison. Referring to Table 5, even though the worst case T_J dictates the overall package thermal resistance, the PGS straps do have a significant impact on other die in the stack. For example, when the 25μm PGS / Die perimeter case is examined against Control 1, the T_J of Die 3 in the stack sees a 56% improvement, even though Die 1 only sees a 1% improvement.

The data presented in Table 6 provides the percent improvement in the θ_{JC} between the TE3DIC PGS configurations. There is a 23% improvement in θ_{JC} by doubling its thickness and a 49% improvement by covering the surface area of the die with the PGS. Decreasing the thickness of the PGS in contact with the majority of the die surface area to 25μm still has a 33% improvement over 50μm PGS that is bonded around the perimeter of the die.

The TE3DIC thermal model was also used to determine the heat flow path sensitivity of the TIM thermal resistance used between the heat spreader and die stack. This was evaluated because it is known that spreading resistances can vary greatly as thermal resistances that are closer to the sink vary. This phenomenon was seen as shown in Table 7. With a highly conductive material, such as a silver filled TIM, a 76/23 copper spreader/PGS heat flow ratio was predicted. An electrically insulating TIM with reduced thermal conductivity divides the heat flow 42/57 copper heat spreader/PGS.

Table 6. Calculated improvement in θ_{JC} between the four separate TE3DIC PGS configurations.

TE3DIC PGS Configurations	θ_{JC} Improvement
50μm Whole Die v 25μm Whole Die	23%
50μm Whole Die v 25μm Perimeter	56%
50μm Whole Die v 50μm Perimeter	49%
50μm Perimeter v 25μm Perimeter	15%
25μm Whole Die v 50μm Perimeter	33%
25μm Whole Die v 25μm Perimeter	43%

Table 7. Heat flow through the die stack as a function of TIM conductivity at the copper spreader interface.

	Heat Flow vs. Die Stack-Cu Spreader TIM	
Path	BL w/ 10W/m·K TIM	BL w/ 0.5W/m·K TIM
Die Stack - Cu Spreader	75.0%	40.6%
PGS Strap #3	5.7%	24.4%
PGS Strap #2	9.6%	19.4%
PGS Strap #1	7.1%	12.6%
Die Stack - Carrier	0.8%	1.1%
Ambient Loss	1.8%	1.9%
Total	100.0%	100.0%

3D Thermal Chip Stack Test Vehicle

A thermal chip stack test vehicle has been designed after seeking significant guidance from the thermal model that will be used to demonstrate the TE3DIC packaging approach. The test vehicle consists of a BGA package constructed with a 37mm x 37mm polyimide carrier substrate (1mm solder ball pitch), four thermal chips, structurally enhanced graphite sheets, and a terraced copper heat spreader as illustrated in Figure 6. The 3DIC is mounted to the carrier in flip chip fashion and the PGS interleaves thermally strap each die to a terrace on all four sides of the copper heat spreader. The terraced copper heat spreader is 3cm x 3cm x 0.28cm and is shown in Figure 7. The terrace is designed to be flush with the height of the die at each layer in the die stack and provides a bond interface to attach the PGS using a TIM as discussed earlier. The carrier substrate fans out the interconnect I/O from a 0.5mm pitch at the chip stack interface to a 1mm pitch at the test board interface and provides mechanical support for the chip stack.

Although, other package styles could serve equally well to study package thermal resistance, the thermally enhanced BGA package is a traditional high heat flux package that provides a convenient interface to cold plates and allows the PGS a sufficient span to spread heat as a parallel path to the heat sink. The package was designed to accommodate either a three or four die stack with one or two 25μm thick PGS sheets interleaved between each die. Control cases are based on the same BGA package with removal of the PGS and variations to the thermal interfaces between the stacked die.

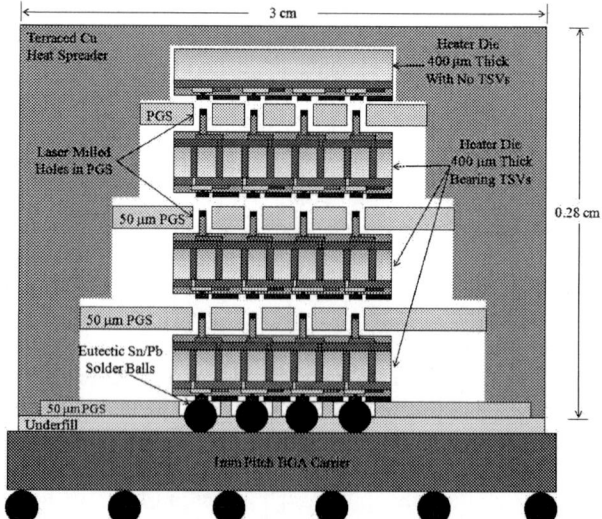

Figure 6. 3D thermal Chip Stack Test Vehicle.

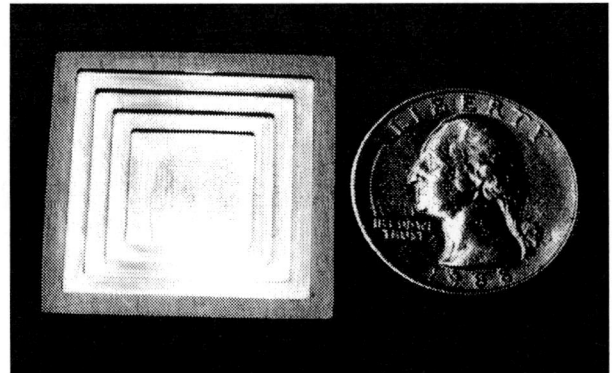

Figure 7. Terraced Cu heat spreader before polishing.

Heater Die Design

Heater die bearing TSVs to create the 3D Thermal Chip Stack Test Vehicle were used because they provide precise thermal measurements at the die level through the use of Pt resistive temperature detectors (RTD). The heater die were designed and fabricated on 0.4mm thick 150mm diameter high resistivity (1,000 – 10,000 Ω-cm) silicon wafers bearing 70μm diameter Cu filled TSVs on a 0.5mm generic grid array. Each die is 11mm x 11mm and possess a total of 484 TSVs, each having a unique layout. Figure 8 provides a generalized illustration of the heater die design. Each die bears metalized thin film resistive serpentine structures including one large Ni80Cr20 uniform die heater, two small Ni80Cr20 fireball heaters, and five Pt RTDs spaced over the surface of the die. Each four-terminal resistive structure is electrically connected to the TSVs with copper thin film interconnect traces for precision measurement through the BGA package. The resistors and sensors were designed to route away from the TSVs to prevent overlap and stress in the stack-up. Copper pillars (70μm diameter and 75μm tall) capped with tin (5-7μm thick) were electroplated on one side of the chip to provide the die-to-die interconnect for the 3DIC stack. A cross-sectional illustration of the die layer stack-up is provided in

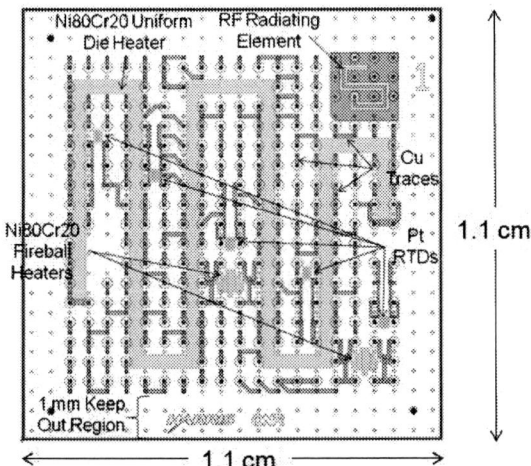

Figure 8. Design instance of a heater die fabricated on a TSV wafer bearing a single uniform whole die heater, two fireball heaters, five RTDs, and an RF radiating element.

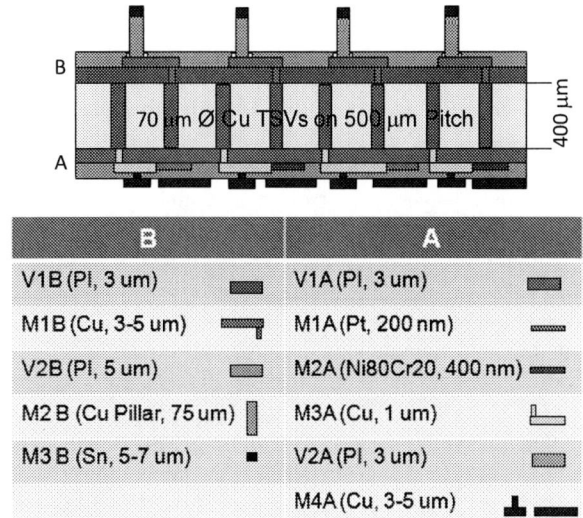

B		A	
V1B (PI, 3 um)		V1A (PI, 3 um)	
M1B (Cu, 3-5 um)		M1A (Pt, 200 nm)	
V2B (PI, 5 um)		M2A (Ni80Cr20, 400 nm)	
M2B (Cu Pillar, 75 um)		M3A (Cu, 1 um)	
M3B (Sn, 5-7 um)		V2A (PI, 3 um)	
		M4A (Cu, 3-5 um)	

Figure 9. Heater die cross-sectional layer stack-up.

Figure 9. Side A consists of six mask layers to create the resistive thermal and sensor elements while Side B consists of five mask layers to create the bonding interface structure.

To minimize mask cost and fabrication time, the mask set was designed with all four thermal die designs to create the 3D thermal chip stack test vehicle. A total of 90 die were laid out over the wafer. Figure 10 shows a picture of Side A and Side B of the completed wafers. The fabrication and assembly details of the 3D Thermal Chip Stack Test Vehicle along with the measured thermal performance will be reported in a follow up paper.

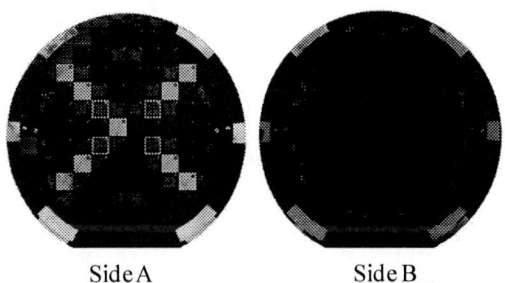

Side A   Side B

Figure 10. Side A and Side B of the heater die fabricated on 400mm thick 150mm diameter Si wafers bearing Cu filled TSVs on a 0.5mm pitch.

Conclusions

The effects on junction temperatures in a thermally enhanced 3DIC package were studied. Pyrolytic graphite sheets were interleaved in a three- die stack to provide parallel thermal paths to the package heat spreader. Thermal models were created to help define a TE3DIC test vehicle that will be used for future model validation and to demonstrate fabrication techniques.

The model compared the thermal effects of comingling the PGS within the copper pillar interconnect field vs. limiting the attachment of the PGS to the periphery of the die and avoiding the interconnect field while maintaining a 1mm overlap. Higher PGS coverage improved Θ_{JC} by 43% and

49% for one sheet of 25μm thick PGS and two sheets of 25μm thick PGS respectively. Increasing the PGS sheet quantity between die from one 25μm sheet to two 25μm sheets reduced the Θ_{JC} from 15% and 23% for perimeter PGS attachment and whole die PGS attachment, respectively.

The TE3DIC was also compared to control cases without the PGS interleaves. Control cases with gaps between die whether filled with a TIM or with an air gap revealed dramatic improvements in Θ_{JC} by using the PGS ranging from 37% to 76% depending on the specific case. The overall package improvement for the TE3DIC vs. the covalently bonded control sample was insignificant; however, there was a 56% improvement in T_J for the die closest to the heat spreader in the TE3DIC. This result suggests that in cases of die with dissimilar power dissipation in the same stack, a TE3DIC package design could benefit higher power devices placed closer to the heat spreader, compared to a covalently bonded stack.

The significance of the reductions in package thermal resistance can be viewed several ways. A 37% improvement in Θ_{JC} means that 37% more power can be dissipated w/o increasing junction temperatures. Alternatively, if no additional power is added to an enhanced package, it will operate 37% cooler. So for a 3DIC without thermal enhancement normally operating at 100°C, its lifetime can now be extended more than 3X following a general axiom that the lifetime of semiconductors doubles for every 10°C drop in junction temperature [10]. Another advantage is to reduce the volume of the heat rejection solution to the ambient environment. It could be expected that the volume of a finned heat sink could be reduced by about 37% in volume.

It is important to note that in the 3DIC stacked geometry presented here that package thermal resistance will depend on the power dissipated in each die. The additional parallel spreading resistances from the PGS in this design create this condition. So for applications where the proportion of power changes among the die during use, the package thermal resistance also changes. It is also worth noting that adding multiple parallel spreading resistances makes the effect of the spreading sensitive to thermal resistances the closer the resistances are to the heat sink. Thermal models are appropriate for these cases since intuition may fail the designer even when using such high conductivity materials. In this model, heat flow analysis showed the dramatic effects when the heat spreader-to-3DIC TIM was varied from 0.5 W/mK to 9.96 W/mK. In the former case, 42% of the heat was transported across the 3DIC to the heat spreader while 57% travelled through the PGS. In the later case, 76% travelled through the 3DIC-heat spreader interface and 23% through the PGS.

In addition to the reduced packaged thermal resistance in copper pillar bonded 3DIC stacks, the PGS appears to be an excellent candidate material where thermal routing is required or compensation is needed for heterogeneous packages that may introduce poor thermal conductors (such as polymer films, glass interposers, low temperature co-fired ceramics, and air gaps) in their design.

The TE3DIC packaging approach proposed above may have additional challenges in scaling in both feature sizes and production. The TSVs and copper pillars are relatively large compared to many commercial 3DICs. Decreasing these features for use with PGS in this manner would likely result in greater than 1:1 aspect ratio milling of PGS and placement accuracy of less than a few micrometers of the PGS to the die. The approach shown is also suitable only for chip-to-chip bonding techniques. Thermal modeling of the design proposed, however, does predict significant reduction in package thermal resistance for copper pillar bonded 3DICs.

Acknowledgments

The authors gratefully acknowledge the funding support of the Office of Naval Research (ONR).

References

1. S. C. Johnson, "Some 'off-roadmap' apps show natural affinity for 3D packaging," *3D Packaging*, pp. 6 – 10, Sept. 2011.
2. Georgia Tech 3D Systems Packaging Research Center http://www.prc.gatech.edu/research/technology/systemsint egration.shtml#pubs
3. Panasonic, "PGS Graphite Sheets," AYA0000CE2 datasheet, Dec. 2013.
4. A Glimpse Behind the Curtain at Xilinx, and other 2.5D Solutions, http://www.3dincites.com/2010/08/a-glimpse-behind-the-curtain-at-xilinx-and-other-2-5d-solutions/
5. A. Bar-Cohen, J. J. Maurer, and J. G. Felbinger, "DARPA's Intra/Interchip Enhanced Cooling (ICECool) Program," CS MANTECH Conference, New Orleans, LA
6. A. Bar-Cohen, "Thermal Packaging From Problem Solver to Performance Multiplier," *Electronics Cooling*, pp. 8-11, Dec. 2013.
7. H. Yu, Y. Shi, L. He, and T. Karnik, "Thermal Via Allocation for 3-D ICs Considering Temporally and Spatially Variant Thermal Power," *IEEE Trans. On Very Largescale Integration (VLSI) Systems*, vol. 16, No. 12, pp.1609-1619, Dec 2008.
8. P. Garrou, "3D Integration: Alternative to Continued Scaling," *in Proc. IEEE Electronic Components and Technol. Conf. (ECTC) Professional Development Course No. 4*, San Diego, CA, May 29 – June 1, 2012.
9. P. Garrou, C. Bower, and P. Ramm, *Handbook of 3D Integration Technology and Applications of 3D Integrated Circuits*, Wiley-VCH, Strauss GmbH, 2008, vol 2, pp. 487 – 502.
10. R. Viswanath, V. Wakharkar, A. Watwe, and V. Lebonheur, "Thermal performance challenges from silicion to systems," *Intel Technol. Journal*, Q3, pp. 1-16, 2000.

Filler Trap and Solder Extrusion in 3D IC Thermo-compression Bonded Microbumps

Yingxia Liu[1], Menglu Li[1], Dong Wook Kim[2], Sam Gu[2], Dilworth Y. Parkinson[3], Justin Blair[3], and K. N. Tu[1]*,

1. Department of MSE, UCLA. Los Angeles, CA 90095
2. Qualcomm, San Jose, CA 92121
3. Advance Light Source, Lawrence Berkeley National Laboratory, Berkeley, CA 94720
*Email: kntu@ucla.edu

Abstract

As the Moore's law is drawing to an end, there is a growing consensus that 3D stacking of Si integrated circuit chips is necessary to continue the current technological trend. In our work, a test 3D IC structure is successfully achieved and the effect of filler trap on microbump of solder joints will be discussed. In our test samples, the two Si chips are connected through thermo-compression bonding of 20 μm diameter microbumps. After bonding the two chips together, some underfill materials are found to stay on the interface of joining in the microbumps; the residual underfill in the microbump is defined as "filler trap". Together with the formation of filler trap, a ring-type solder extrusion also forms on the circumference of the microbump. Although the filler trap is an electrical insulator, it has been found that the solder extrusion is a good electricity conductor. But the filler trap in solder can cause a detour of electric current path in solder. The detoured current potentially increases current crowding in the joint which might be accelerating electromigration. To investigate this potential issue, specially processed test vehicle which intentionally has higher filler traps and solder protrusion was fabricated for easier detection. The synchrotron radiation x-ray tomography results show that the solder extrusion in microbumps have a random shape, which may cause a random distribution of early failure in electromigration.

1. Introduction

Mobile hand-held consumer electronic products have a rapidly growing market today, witnessed by the popularity of ipad and iphone. Most people make their first contact to internet, not by a PC, but rather by a smart phone. The phone provides various functions for instant communication and recorded entertainment. It seems that the functions we can have in the future hand-held devices are only limited by our imagination as well as by the physical size of the hand-held device. Today, 2-dimensional very-large-scale-integration of transistor circuits on a Si chip has reached the limit of Moore's law and the industry has been looking for ways to extent the limit. The way that seems most promising to continue current technological performance trend is to use 3D IC technology in which several semiconductor chips are stacked together, and it saves space in a device having a tight requirement of space [1], [2]. The stacking requires vertical interconnects. Microelectronic companies are developing Through-Si-Vias (TSV) and microbump (Cu/solder) technologies for the vertical interconnect.

In 3D IC, microbump joints can be formed by thermo-compression boding. Thermo-compression bonding is the procedure to bond two metals together by heating and compression. It was applied to wafer bonding through metals like Cu-Cu, Au-Au layer [3-5]. The condition is about 350 to 400 °C, 200 kPa and 30 to 60 min. Under this condition, the Cu layer or Au layer is in solid state. The interdiffusion of atoms between the two metal layers is not easy. But the diffusion is critical for them to join together. To ensure a good joint, it requires a longer bonding time, very smooth wafer surface condition and so on. The conditions are impractical. So the industry has turned to thermal compression bonding of solder joints. At the temperature of about 250 °C, solder microbumps can be joined together in a few seconds to minutes. This gives solder joint a promising future in the trend of minimization.

The cross-sectional structure of thermo-compression bonding of a solder microbump of is shown schematically in Fig.1.

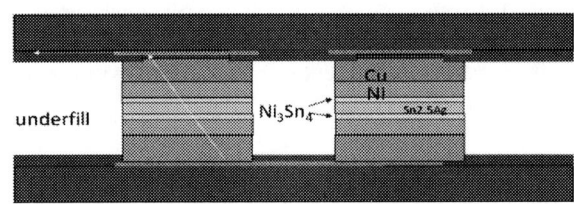

Figure 1. Schematic diagram of thermal compression bonding

Traditional underfill processes are based on local dispensing after solder bump reflow, i.e., capillary dispensing. In 3D IC, such process presents some limitations due to the need for a long dispensing distance with small gap. This is because typically the arrays of microbump are located far from the edge of the stacking chips [6]. Also, the microbumps and the space between microbumps are too small to allow a successful dispense underfill after bonding. Thus, underfill is dispensed before bonding in thermo-compression bonding. However, trapped underfill fillers after pre-applied underfill chip attach process could negatively impact on elecromigration (EM) resistance of the solder joint. We will discuss more details in the later part of this article.

EM is the mass transportation of a flux of atoms driven by the momentum exchange between a high density of moving charge carriers and diffusing atoms. The drifting electrons collide with the diffusing atoms, causing them to exchange positions with neighboring vacancies in the direction of electron flow. This results in the accumulation of atoms on the anode end and vacancies on the cathode end of the interconnection. Such behavior can lead to hillock and void

978-1-4799-2408-0/14 $31.00 © 2014 IEEE   609   2014 Electronic Components & Technology Conference

formation, which can become short and open failures, respectively [7].

However, in addition to the effect of electron wind force on EM due to high current density, the flip-chip solder joint configuration has a built-in effect of current crowding on accelerated EM, as shown in Fig.2, which is a schematic diagram of their cross section. The structure composed of the top Al line, a pair of solder bump and a thick Cu line connecting it to other bumps or to the outside circuit. The cross section of the Al line on the chip side is at least two orders of magnitude smaller than that of the solder bump. But the current passing through is the same in the Al line and the solder, so the current density is much bigger in Al line. Thus, there is a very large current-density change at the contact between the bump and the line, resulting in current crowding at the entrance or the departure area.

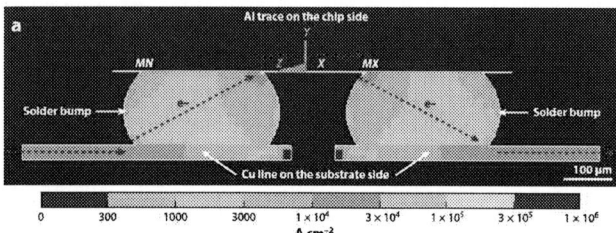

Figure 2. Cross-sectional view showing the current-crowding effect in solder bumps during current stressing. Peak current density occurs at the current entrances of the Al trace into the solder bump [7].

Current crowding will make the current density in the solder bump near the entrance point approximately one order of magnitude higher than the average current density in the middle of the bump. This high current density will enhance EM, thus has a bad effect on EM.

2. Filler trap found in 3D IC

In our work, the 3D IC structure is composed of two Si chips (without devices) and three layers of solder joints. An SEM image of a cross-sectional view of the test sample is shown in Fig. 3. The two triangular objects on the top and the bottom are part of the fixture used to hold the test sample. From the bottom, we see the largest size solder balls and they are the BGA on the bottom side of the substrate. On the upper sider of the substrate is the middle layer and middle size of C-4 flip chip solder joints of 100 μm in diameter. Above the flip chip solder joints there are two Si chips, between the chips are the array of smallest solder joints of 20 μm in diameter, so called the solder microbumps.

Fig. 4 shows a higher magnification of the cross-sectional view of a daisy chain of array of microbumps. We found most of the microbumps don't have perfect bonding. There are dark particles in the middle layer. After checking by FIB (focus ion beam), we can see clearly that the dark part in solder is the same with underfill material, as indicated by red arrow in Fig. 5.

Main reason of filler trap during thermo-compression bonding process is premature cure of the pre-applied underfill resin system which is mainly caused by not fully optimized process condition and/or bump geometry. Once resin cure started, solder wetting is no longer able to flow out filler and underfill from the joint due to the increased viscosity of pre-applied underfill.

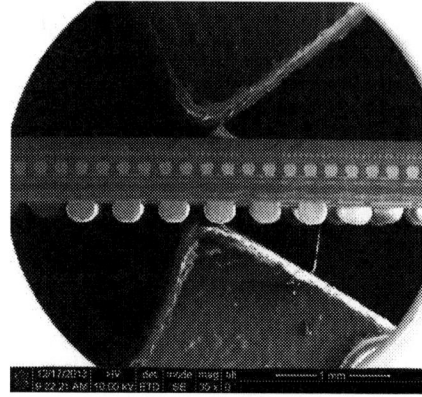

Figure 3. SEM image of the cross-sectional view of the 3D structure.

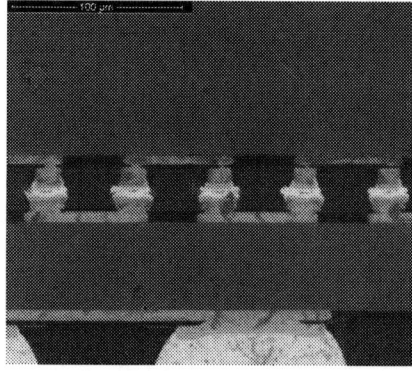

Figure 4. SEM image of five microbumps.

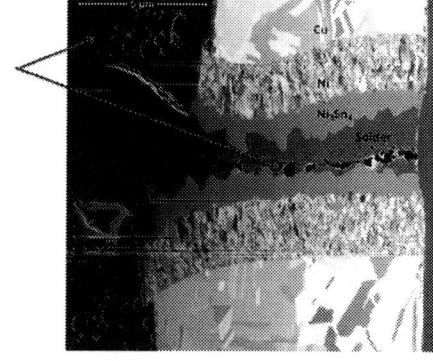

Figure 5. FIB image of the one microbump.

3. Solder extrusion in microbumps

As we can see in Fig. 5, there is a big solder extrusion at the side of the microbump. Its formation maybe because in thermo-compression bonding, the pressure will press the liquid solder out as extrusion. It is worth mentioning that, although the filler trap in solder is insulator, the solder extrusion provides a big enough area for current to pass through.

However, there are some potential issues in solder extrusion. The images we got by synchrotron radiation x-ray tomography showed the random shape of solder extrusion, as shown in Fig. 6 and Fig. 7. The experiments were conducted on the microtomography Beamline 8.3.2 at the Advanced Light Source of Lawrence Berkeley National Laboratory. The voxel size used was 0.9 μm. A total of 4097 images were taken in a single scan. Viewing and analysis of the images were performed using the commercial 3-D visualization software Avizo 6.0 [8].

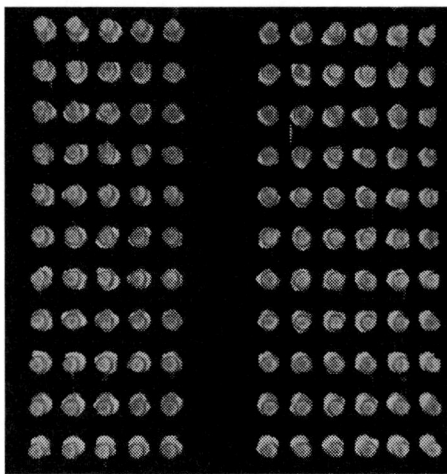

Figure 6. Synchrotron radiation x-ray tomography top view image of microbumps.

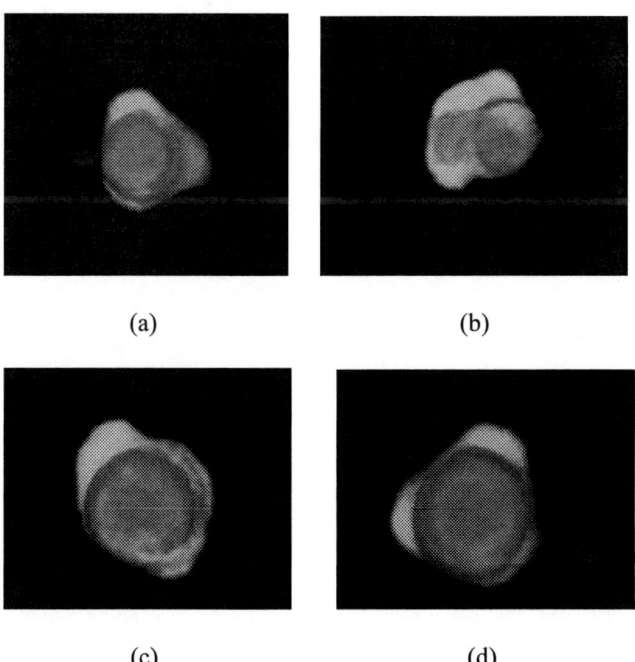

Figure 7. Synchrotron radiation x-ray tomography top view image of individual microbumps.

From Fig. 6 and Fig. 7, we can see a random distribution

of solder extrusion in microbumps. In microelectronic technology, it is important to have precision fabrication of every building block of device component to be the same; otherwise the difference in physical parameter can cause a difference in performance. In EM especially, as we already have filler trap in the microbumps, current passing through the microbumps will take a detour to avoid filler trap and this may cause current crowding. The different shape of solder extrusion will make the current density distribution different from different microbumps, and also the current crowding point will be different. The failure distribution or mean-time-to-failure may be wide. The experiment of EM will be carried out later to verify it.

4. Current crowding in microbumps with filler trap

Because the existence of filler trap and solder extrusion, current will take a detour to pass through the microbumps. We can expect to see current crowding occurs in the solder (Sn) layer surrounding the epoxy filler.

In flip chip solder joints, the void formation mechanism is dominated by current crowding at the cathode contact, and the typical failure mode is the propagation of pancake-type void across the cathode contact interface. In the microbumps, we expect that void formation also occurs at the current crowding area. Then voids may form at the extruded Sn layer, which will have a new failure mode. For example, if the voids start to form at the Sn layer and propagate around the epoxy filler, we may see a void ring at the middle of the solder. Since the cross-sectional Sn area around the filler trap is much smaller than the area of Sn without filler trap and also the filler trap is insulator, it may be much easier to cause an open failure around filler trap when there is a ring of void forms around the filler trap.

5. Conclusions

Filler trap and solder extrusion are observed in solder microbumps of 20 μm in diameter formed by thermal compression bonding to produce test samples of stacking Si chips in 3D IC packaging technology. Cross-sectional images of the microbumps have been studied by using SEM, FIB, and synchrotron radiation tomography. These results indicate that the filler trap may cause current crowding around it in electromigration and in turn, it could lead to a wide distribution of electromigration performance.

Acknowledgments

The authors would like to acknowledge the financial support of Qualcomm for this research and Advanced Semiconductor Engineering in Taiwan, ROC, for the supply of sample. We also would like to acknowledge the help of the Advanced Light Source. The Advanced Light Source is supported by the Director, Office of Science, Office of Basic Energy Sciences, of the U.S. Department of Energy under Contract No. DE-AC02-05CH1123. Also, we thank Noah Bodzin at Nano Lab, UCLA, and Sergey Prikhodko, Molecular & Nano Archaeology (MNA) Laboratory at UCLA, Dept. of Materials Science and Engineering for the convenience of using FIB and SEM.

References

978-1-4799-2408-0/14 $31.00 © 2014 IEEE   611

1. J. U. Knickerbocker, "Three-dimensional silicon integration," *IBM Journal of Research and Development*, vol. 52, no. 6, pp. 553-569, Nov. 2008.

2. K. N. Tu, "Reliability challenges in 3D IC packaging technology," *Microelectronics Reliability*, vol. 51, no. 3, pp. 517-523, March. 2011.

3. Eun-Jung Jang, "Annealing temperature effect on the Cu-Cu bonding energy for 3D-IC integration," *Metals and Materials International*, vol. 17, no. 1, pp. 105-109, Feb. 2011

4. C. S. Tan, "Cu–Cu diffusion bonding enhancement at low temperature by surface passivation using self-assembled monolayer of alkane-thiol," *Applied Physics Letters*, vol. 95, no. 19, pp. 192108-1-192108-3, Nov. 2009.

5. Reif R, "3-D interconnects using Cu wafer bonding: technology and applications," in *Advanced metallization conference (AMC)*. 2002.

6. A.L. Mamma, "Use of wafer applied underfill for 3D stacking," *44th Int. Symposium on Microelectronics*, Long beach, CA, 2011, pp.8-16.

7. Chen Chih, "Electromigration and thermomigration in Pb-free flip-chip solder joints," *Annual Review of Materials Research*, vol. 40, pp. 531-555, Aug. 2010.

8. Tian Tian, "Quantitative X-ray microtomography study of 3-D void growth induced by electromigration in eutectic SnPb flip-chip solder joints," *Scripta Materialia*, vol. 65, no. 7, pp. 646-649, Oct. 2011.

Correlation between Cu Microstructure and TSV Cu Pumping

Joke De Messemaeker[1], Olalla Varela Pedreira[1], Harold Philipsen[1], Eric Beyne[1], Ingrid De Wolf[1], Tom Van der Donck[2] and Kristof Croes[1]

[1]imec, Kapeldreef 75, 3001 Leuven, Belgium
[2]KULeuven - Department of Metallurgy and Materials Engineering, Kasteelpark Arenberg 44, 3001 Leuven, Belgium
joke.demessemaeker@imec.be, +3216281743

Abstract

Cu pumping is the irreversible extrusion of Cu from Cu-filled through-silicon vias (TSVs) exposed to high temperatures during back-end of line (BEOL) processing. The distribution of Cu pumping values over the TSVs of a single wafer has a large intrinsic spread. As potential BEOL reliability issues due to Cu pumping will first occur at the highest pumping TSVs, they can be mitigated if the fundamental cause for this large intrinsic spread is known and under control. This paper describes a clear correlation between Cu pumping and TSV Cu microstructure based on the grain size at the top of 5x50 μm TSV, disregarding twin boundaries. For the mitigation of TSV Cu pumping the ideal microstructure was shown to consist of a single grain spanning the whole TSV cross section, bringing down the highest measured Cu pumping value from 248 nm to 73 nm. This effect was attributed to the absence of rapid diffusion paths and grain boundary sliding ability.

Introduction

3D IC stacking allows decreasing the total interconnect length and so the power consumption of a chip, while increasing the on-chip density and enabling heterogeneous integration schemes. TSVs are introduced for the interconnection of multiple dies in a stack. Most often these are filled with copper. One possible integration scheme is the via-middle approach, where the TSVs are processed after front-end-of-line CMOS and contact processing, and before BEOL interconnect processing [1,2].

In the via-middle approach, Cu-filled TSVs are exposed to high temperatures during BEOL processing. Due to the large difference in coefficient of thermal expansion with the surrounding Si, high compressive stresses arise in the Cu TSV. These stresses are partly relaxed by irreversible extrusion of the Cu, a phenomenon known as 'Cu pumping', which may damage the BEOL layers on top of the TSV. In order to reduce the amount of Cu pumping during BEOL processing, commonly a post-plating anneal is applied prior to Cu CMP [3].

The distribution of Cu pumping values over the TSVs of a single wafer has a large intrinsic spread [4,5]. For 5x50 μm TSVs, the 99.9th percentile of the distribution was found to be a factor 8 higher than its median, independent of the applied post-plating anneal conditions [6]. As potential BEOL reliability issues due to Cu pumping will first occur at the highest pumping TSVs, they can be mitigated only if the fundamental cause for this large intrinsic spread is known and controlled.

In this paper a correlation between Cu pumping and the TSV Cu microstructure will be presented for 5x50 μm TSVs filled using 2 different plating chemistries A and B, by comparing top-down EBSD scans made at the top of the TSV with the amount of Cu pumping measured for individual TSVs.

Samples and Experimental Approach

For this investigation 2 wafers were produced with a TSV diameter of 5 μm and a TSV depth of 50 μm, where the TSVs were filled using 2 different commercial Cu plating chemistries A and B. The corresponding samples will be referred to as "ChemA" and "ChemB", respectively. The 300 mm wafer processing consisted of the following steps:

1. Blanket dummy FEOL deposition (stack of SiO_2 and Si_3N_4 films, total thickness ~ 650 nm)
2. TSV etch
3. Liner, barrier and Cu seed deposition
4. Cu plating for TSV fill
5. Post-plating anneal
6. CMP to remove the Cu overburden and the barrier/liner on field (stopped on the dummy FEOL)
7. Cap deposition (~ 180 nm SiO_2 + ~ 25 nm Ta)

Steps 5-7 are part of the Cu pumping measurement approach, shown in Fig. 1 [4].

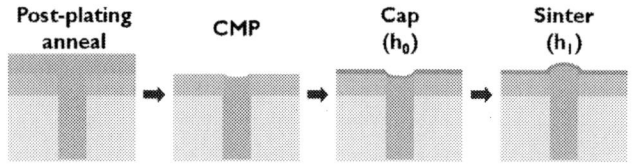

Figure 1. Cu pumping measurement approach used in this paper (adapted from [4]).

The plating process conditions were adapted to assure an appropriate TSV fill for each plating chemistry.

The post-plating anneal conditions, which are different for both wafers, are listed in Table 1.

Table 1. Sample description.

Sample	Plating chemistry	Post-plating anneal
ChemA	A	20 min at 180 °C
ChemB	B	none

The SiO_2 cap on top of the TSVs aims to mimic the presence of BEOL layers. This cap causes a mechanical back stress on the Cu in the TSV, creating a more even, dome-shaped protrusion instead of the rough appearance observed for uncapped TSVs. It also prevents Cu oxidation during the experiment. The thin Ta cap is required to assure good reflectivity of the top surface in the optical profilometer [4].

The capping processes impose a thermal budget of in total 3 min at 350 °C on the wafers.

After wafer processing, 3x3 cm² samples were cut from the wafers at a position between wafer center and edge (identical position for each wafer), and the heights of selected TSVs were measured before (h_0) and after (h_1) an anneal of 20 min at 420 °C in a nitrogen atmosphere (referred to as the "sinter", Fig. 1). Note that this approach characterizes the *residual* Cu pumping during the fixed sinter. The sinter aims to mimic the thermal budget of the final anneal step after BEOL processing at imec, which is also 20 min at 420 °C.

The TSV heights were measured using white light interferometry, which was demonstrated to be a valid technique for collecting large amounts of Cu pumping data [4]. Selected TSV arrays were scanned before and after the sinter, and for each TSV the average and maximum height of its top surface with respect to the field, resp. h^{avg} and h^{max}, were extracted (Fig. 2). The maximum residual Cu pumping values Δh^{max} were then calculated as

$$\Delta h^{max} = h_1^{max} - h_0^{avg} \qquad (1)$$

This maximum amount of Cu pumping is directly related to the potential damage to the BEOL layers.

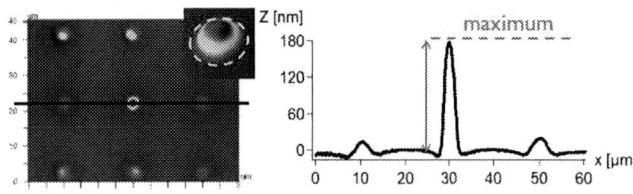

Figure 2. Example of a WLIR scan result for 5x50 µm, 20 µm pitch TSVs after the sinter. The concept of maximum Cu pumping height is illustrated on the line profile (right) extracted from the height map (left). The 3D image of a single TSV (inset) shows the typical shape after Cu pumping at the TSV top.

For the ChemA sample the Cu pumping was measured in TSV arrays with pitches of 10, 20 and 35 µm, while for the ChemB sample only 10 µm pitch arrays were included (TSV pitch is defined as the distance from the center of one TSV to the center of the nearest-neighbor TSV). On both samples these measurements were repeated on 8 different dies, yielding a total of 344 (ChemA) and 960 (ChemB) measured TSVs.

After the Cu pumping measurements, the Ta and oxide cap were removed in order to obtain high quality EBSD patterns from the TSV Cu. This was done by mechanical polishing using a solution with 0.04 µm oxide particles (OPS). By tilting the sample slightly off the position parallel to the polishing platen, lines emerge on the sample surface where an interface of the film stack emerges at the surface. With the help of these lines, the depth of polishing for the target TSV array could be tuned to within 0-0.3 µm below the initial TSV Cu surface. Finally, in order to minimize the charging effect in the SEM, a thin carbon coating was applied.

TSV cross sections were made with a focused ion beam (FIB). EBSD measurements were made at 20 kV.

Cu Pumping Measurement Results

The Cu pumping distributions for the 2 samples are shown on the lognormal probability plot in Fig. 3. For sample ChemA it ranges from 10 to 215 nm, and for sample ChemB from 2 to 248 nm. The distributions for the 2 samples are approximately lognormal, and fairly similar.

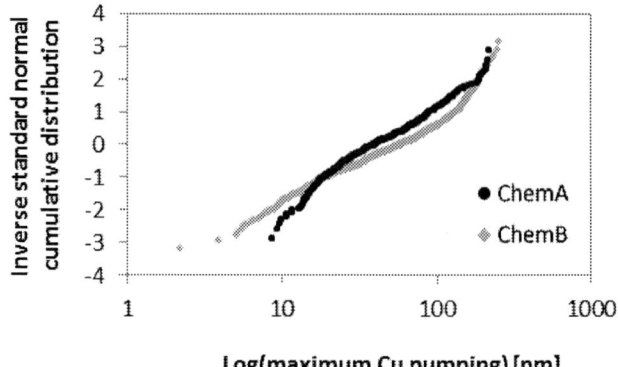

Figure 3. Lognormal probability plot of the maximum Cu pumping values for samples ChemA and ChemB.

In order to detect potential correlations between the Cu pumping value of a TSV and its microstructure and texture, for each Cu pumping distribution 5 bins are defined containing an equal amount of measured TSVs, as listed in Table 2.

Table 2. Calculation of TSV bins based on the minimum (min), maximum (max) and n.20 percentiles (pn.20) of the Cu pumping distributions in Fig. 3, and results for ChemA and ChemB samples. Numbers are in nm. The color coding in the first column is employed in Figs. 7-11.

Bin	Calculation		ChemA		ChemB	
	Min	Max	Min	Max	Min	Max
1	min	p20	10	22	2	22
2	p20 + 1	p40	23	31	23	42
3	p40 + 1	p60	32	48	43	73
4	p60 + 1	p80	49	80	74	116
5	p80 + 1	Max	81	215	117	248

As-Processed Microstructure and Texture

An orientation map of a TSV cross section for sample ChemA is shown in Fig. 4a. The apparent deformation of the TSV bottom, which in reality is flat, is an artifact from the FIB milling (curtaining effect). The Cu microstructure consists of large, annealed grains containing a large volume of twins. This indicates that the microstructure is recrystallized and may have undergone some grain growth, which could be expected based on the capping thermal budget of 3 min at 350 °C to which all TSVs are subjected.

Fig. 4b shows the corresponding texture, presented as the inverse pole figure for the TSV axis. This macroscopic reference was chosen because the TSV has a cylindrical symmetry, as during plating the current supply through the Cu seed has no preferred radial or circumferential direction. The TSV axis on the other hand is a potential preferred direction, because it is the growth direction during the bottom-up plating process. The texture in Fig. 4b shows 2 fibers parallel to the TSV axis, the stronger one at <001> and the weaker one near <210>.

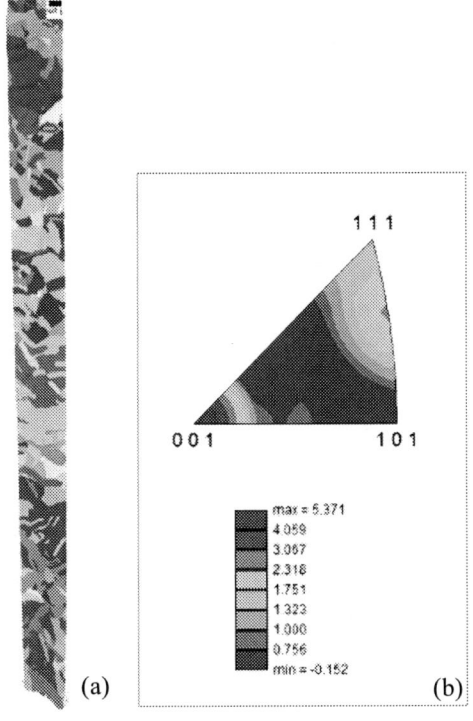

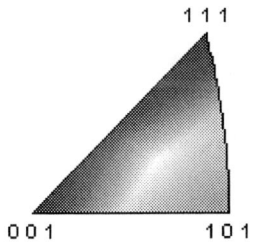

(a)   (b)

Figure 4. (a) Orientation map and (b) inverse pole figure for the TSV axis of a TSV from sample ChemA in as-processed condition (before sinter). The color scale used in (a) is shown in Fig. 5.

Figure 5. Inverse pole figure indicating the color scale applied in Figs. 4, 7 and 8 for the crystal direction parallel to the TSV axis.

Correlation between Cu Pumping and Cu Microstructure

For establishing a correlation between Cu pumping and the Cu microstructure, for both investigated samples top-down EBSD scans were made of 10 μm pitch TSVs of 1 of the 8 dies included in the Cu pumping measurements. This die was selected such that the Cu pumping values of its TSVs cover the 5 bins defined in Table 2. For the ChemA sample a random 5x5 array of TSVs on the die was selected. The height map for this array is shown in Fig. 6. For the ChemB sample, 50 TSVs from

a set of 120 were selected so that bin 1 and bin 5 have slightly higher statistics than the other bins, in order to more easily detect contrasts in microstructure between the highest and lowest pumping TSVs.

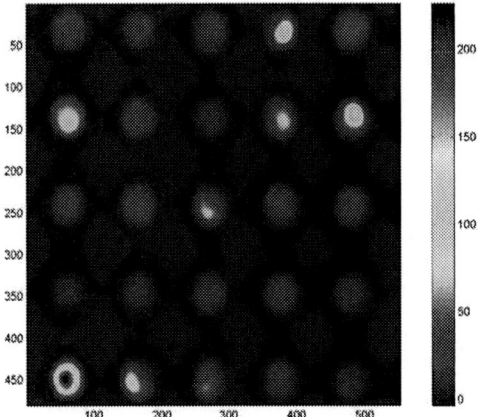

Figure 6. Height map of the 5x5 TSV array of ChemA selected for EBSD scans, obtained by optical profilometry after the sinter. Numbers along the sides of the map indicate distances from the corner, in nm. The color scale at the right indicates the height, in nm. At the bottom left a very high pumping TSV can be seen ($\Delta h^{max} = 215$ nm).

The orientation maps of the selected TSVs are shown in Fig. 7 and 8 for samples ChemA and ChemB, respectively. Charging of the TSV Cu, which is isolated from the surrounding Si by the SiO_2 liner, led to a deformation of the circular cross section of the TSV. In spite of this deformation, the indexed crystal orientation is unaffected and the grain structure and grain boundary type can correctly be assessed. In order to detect potential relations between the microstructure and the corresponding amount of Cu pumping, the background for each orientation map is colored according to the bins defined in Table 2.

In both samples the microstructure at the top of the TSV contains a high fraction of twin boundaries. From Figs. 7 and 8, no correlation between the grain structure and the corresponding Cu pumping value can be derived. Also no correlation was found between the Cu pumping values and the crystallographic orientations of the Cu grains at the top of the TSV. For ChemA this is exemplified by the inverse pole figures in Fig. 9, showing the combined orientations for all TSVs of bins 1, 2, 4 and 5. Clearly no correlation between orientation and Cu pumping emerges.

Twin boundaries have a significantly lower energy and mobility than random high angle grain boundaries. Therefore and in view of the large fraction of twin boundaries in Figs. 7 and 8, also a grain reconstruction ignoring twin boundaries was carried out. This results in the grain maps of Figs. 10 and 11, where each reconstructed grain of a TSV has a unique color. Now a correlation with Cu pumping does emerge, as there seems to be a clear preference for TSVs consisting of one single grain at the TSV top to display low Cu pumping values.

978-1-4799-2408-0/14 $31.00 © 2014 IEEE   615

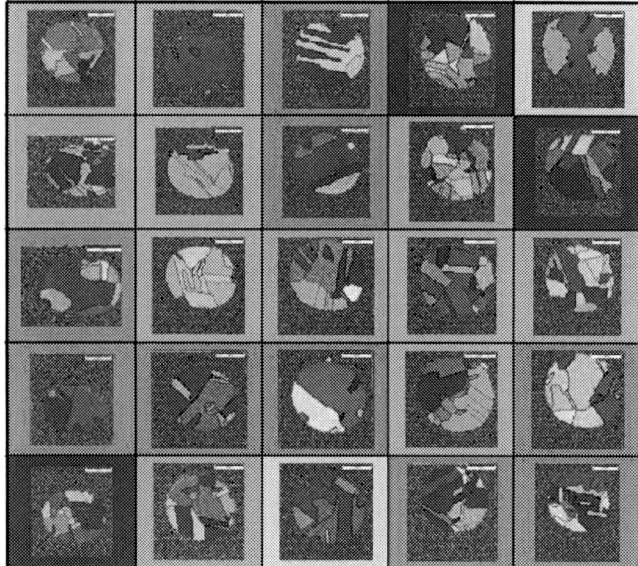

Figure 7. Orientation maps for the 5x5 TSV array of ChemA, corresponding to the height map of Fig. 6. Data points are colored according to the scale in Fig. 5. High angle grain boundaries are indicated with a red line, twin boundaries with a black line and low angle grain boundaries with a white line. The background color for each TSV corresponds to the bin color from Table 2.

Nevertheless some TSVs consisting of a single grain at the TSV top also fall in bin 3. In order to condense this correlation into a single figure, the microstructures of Figs. 10 and 11 were classified according to the number of grains across the TSV cross section. The occurrence of smaller grains at the TSV edge can be expected to contribute less to Cu pumping than larger grains situated closer to the TSV center. Therefore, for TSVs containing such smaller grains at the edge an additional range of microstructure classes was created, indicated by the number of large grains followed by a '+' sign. An overview of all microstructure classes so generated is shown in Fig. 12. Grouping these classes as indicated in Fig. 12, the histograms of Figs. 13 and 14 result. Here the number of TSVs in each bin was rescaled to the fraction of TSVs in that bin for the whole Cu pumping distribution, i.e. 20 % (cf. the bin definitions in Table 2), in order to reflect the shape of that distribution.

Figs. 13 and 14 show that TSVs containing a single grain at the TSV top (class 1) are exclusively associated with lower Cu pumping values (bins 1, 2 and 3: < 49 nm for chemA and < 73 nm for chemB). TSVs containing more than 3 grains across the TSV top (classes 3+/4/4+/5/5+) exclusively generate higher Cu pumping values (bins 4 and 5). While this correlation is very clear and independent of the plating chemistry used, the grain structure at the TSV top is not the only factor at play. Indeed, the intermediate microstructure classes 1+/2/2+/3 produce Cu pumping values which are spread out over the whole distribution. Therefore the microstructure at the TSV top is not the only predicting factor for Cu pumping. Nevertheless, the fact that the observed correlation holds for 2 different samples using different plating chemistries and post-plating anneals, suggests that the relation is fundamental.

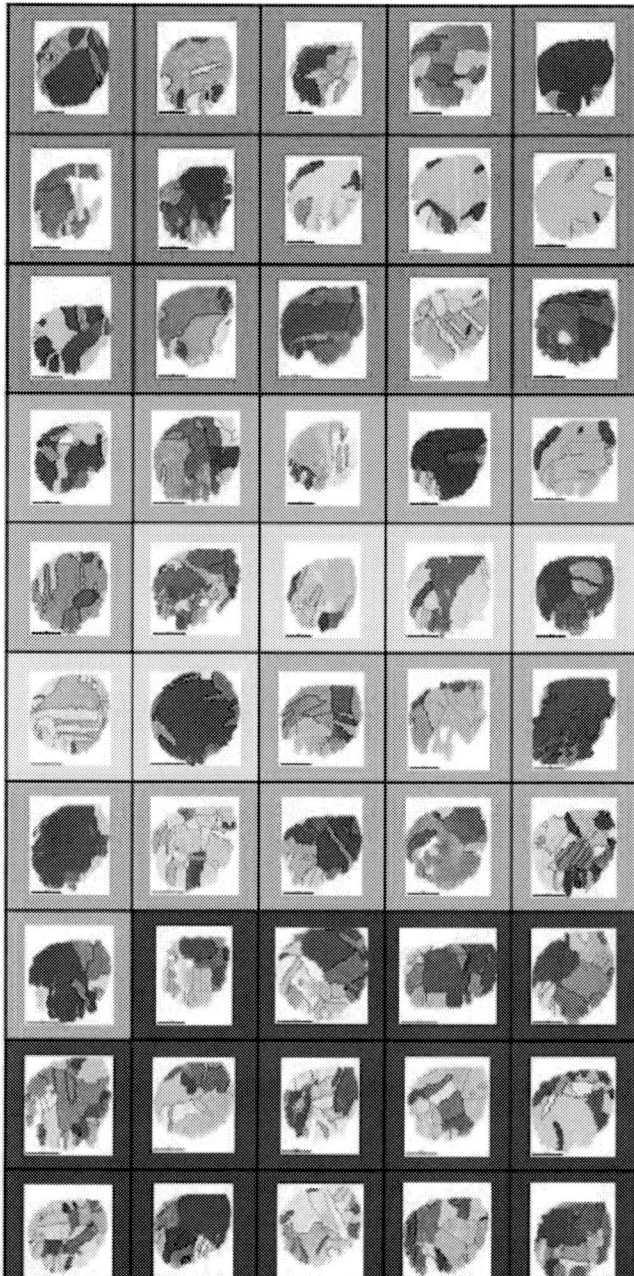

Figure 8. Orientation maps for the selected TSVs of ChemB. Data points are colored according to the scale in Fig. 5. High angle grain boundaries are indicated with a red line, twin boundaries with a black line and low angle grain boundaries with a white line. The background color for each TSV corresponds to the bin color from Table 2.

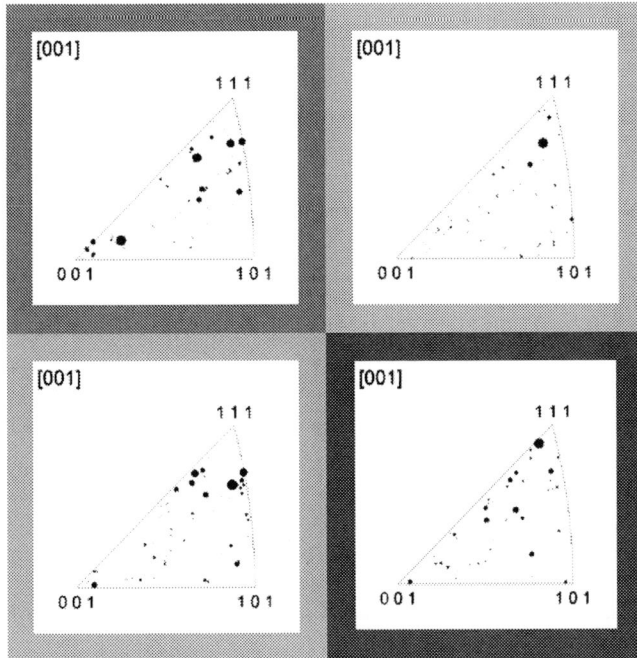

Figure 9. Inverse pole figures (IPFs) for the TSV axis, for the 5x5 array of selected TSVs for ChemA. The size of the dots is related to the corresponding grain size. The background color for each IPF corresponds to the bin color from Table 2.

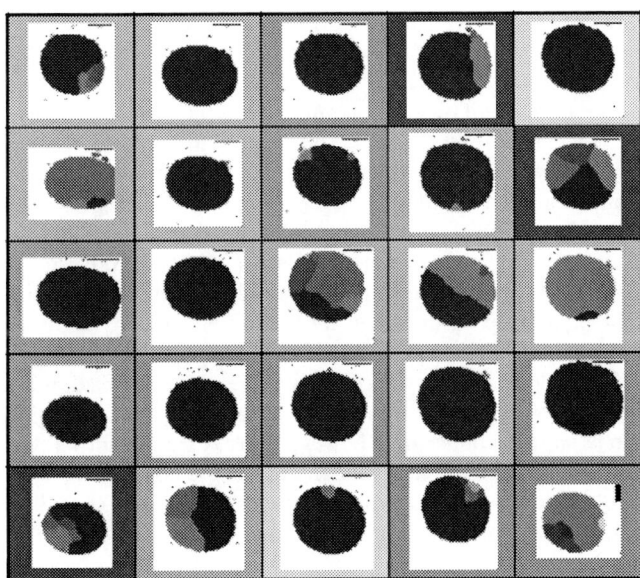

Figure 10. Grain maps for the 5x5 array of selected TSVs for ChemA, excluding twin boundaries. Per map each grain has a unique color, unrelated to its crystallographic orientation. The background color for each TSV corresponds to the bin color from Table 2.

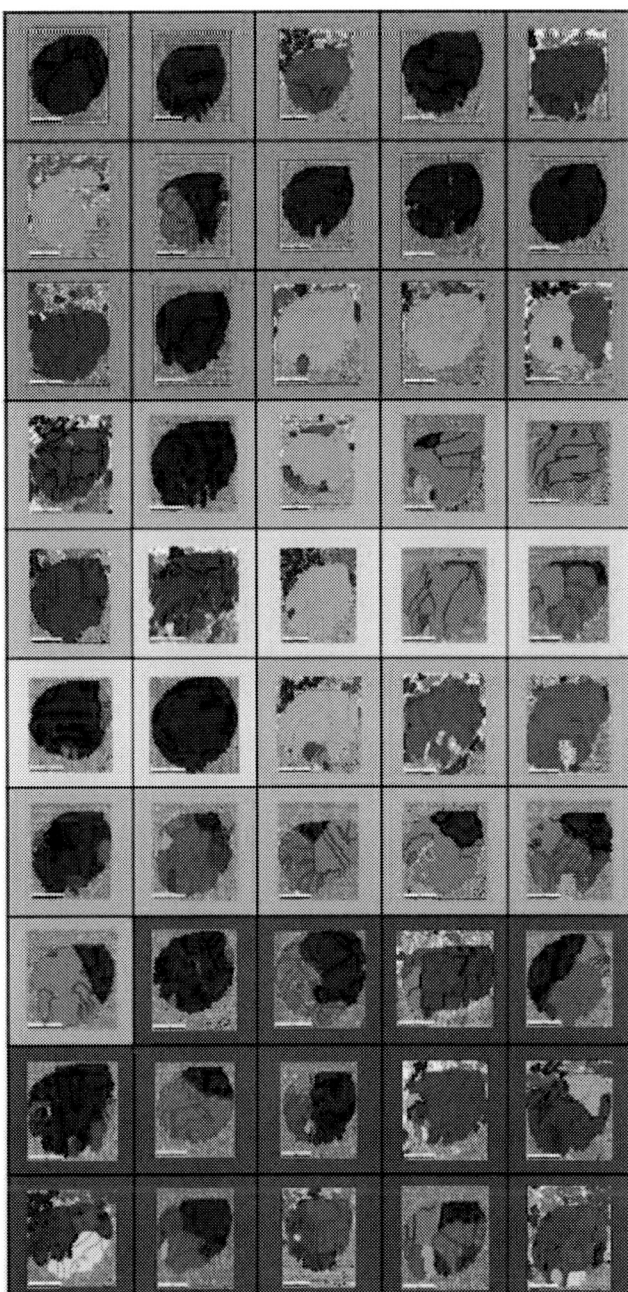

Figure 11. Grain maps for the selected TSVs for ChemB, excluding twin boundaries. Per map each grain has a unique color, unrelated to its crystallographic orientation. High angle grain boundaries are indicated with a red line, twin boundaries with a black line and low angle grain boundaries with a white line. The background color for each TSV corresponds to the bin color from Table 2.

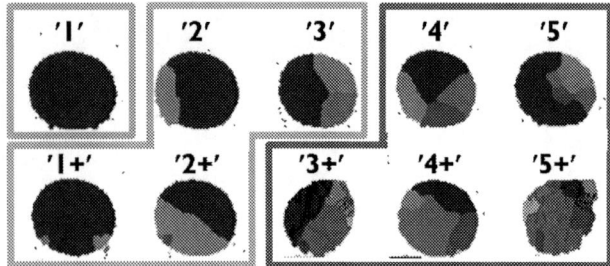

Figure 12. Microstructure classes defined by the number of large grains across the TSV cross section. The additional '+' sign indicates the occurrence of smaller grains at the TSV edge.

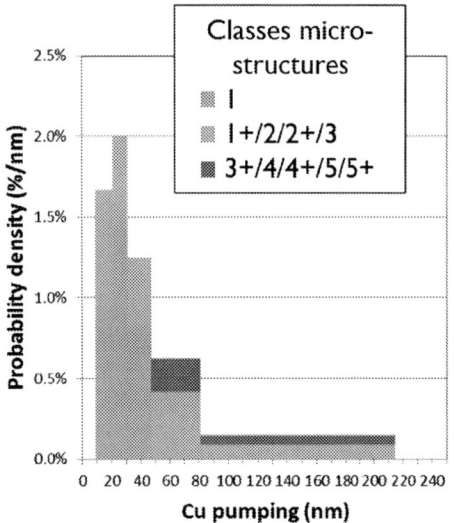

Figure 13. Probability density plot for ChemA, for 3 groups of the microstructure classes defined in Fig. 12.

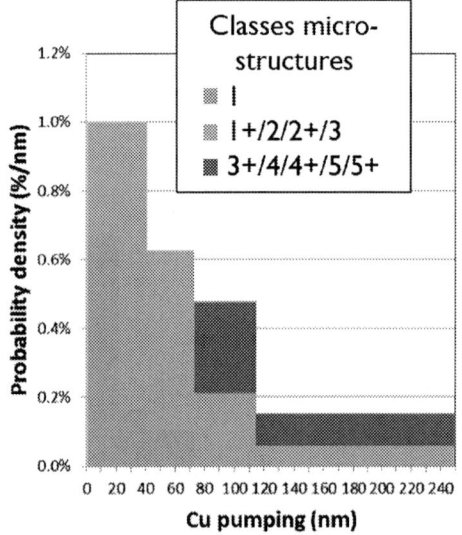

Figure 14. Probability density plot for ChemA, for 3 groups of the microstructure classes defined in Fig. 12.

Discussion

Cu pumping is the irreversible extrusion of Cu measured after cooling down to room temperature, corresponding to the plastic deformation occurring to relax the high compressive stress in the Cu at the sinter temperature. The following mechanisms can contribute to this deformation [7]:

1. Dislocation glide based plastic deformation
   a. instantaneous (yield)
   b. time-dependent (power-law creep, supported by dislocation climb)
2. Diffusion based time-dependent plastic deformation
   a. lattice diffusion (Nabarro-Herring creep)
   b. grain boundary diffusion (Coble creep)

Evidence for the involvement of time-dependent processes (1b and 2) was found in the dependency of residual Cu pumping during a fixed sinter, on the duration of a post-plating anneal at 420 °C [6]. No direct evidence was found for the occurrence of yield (no. 1a), but this cannot be excluded either.

Measurements and simulations show that most of the plastic deformation during Cu pumping occurs at the TSV top [8,9]. This in part explains why a correlation was found based on the microstructure at the TSV top only, even though this is only a very small part of the overall TSV microstructure.

The observed correlation between grain structure type defined in Fig. 12 and Cu pumping, means that the presence of random high angle grain boundaries at the top of the TSV enables and promotes the plastic relaxation processes at the sinter temperature, while the presence of twin boundaries does not contribute. This can be explained based on the possibly occurring mechanisms for plastic deformation listed above.

In the case of dislocation glide (either instantaneous, time-dependent or both), stress and strain differences between crystals with different orientations must be accommodated by grain boundary sliding. The grain boundary sliding rate for coherent twin boundaries is an order of magnitude lower than for random high angle grain boundaries, so that the presence of the latter will enable more plastic relaxation [10]. In the case of diffusion creep, at the sinter temperature of 420 °C the grain boundary diffusion rate is significantly higher than the lattice diffusion rate, so that significantly more plastic relaxation can be expected to occur in the presence of grain boundaries [7]. Coherent twin boundaries however have a significantly lower diffusion coefficient than random high angle boundaries, again explaining why the presence of twins would not contribute significantly to the plastic relaxation [10].

Conclusions

A clear correlation between Cu pumping and TSV Cu microstructure was found for 5x50 μm TSVs filled using 2 different plating chemistries A and B, based on the grain size at the top of the TSV when twin boundaries are disregarded. TSVs containing a single grain at the TSV top are exclusively associated with lower Cu pumping values, while TSVs containing more than 3 grains across the TSV top exclusively generate higher Cu pumping values. This correlation was independent of the plating chemistry used.

These results can be explained by considering that diffusion and grain boundary sliding are both much slower along twin boundaries than along random high angle grain boundaries. Therefore the presence of random high angle grain boundaries increases both the diffusion creep rate and the dislocation creep rate, leading to higher Cu pumping.

For the mitigation of TSV Cu pumping the ideal microstructure was shown to consist of a single grain spanning the whole TSV cross section, bringing down the highest measured Cu pumping value from 248 nm to 73 nm.

Acknowledgments

This work was done in the framework of the IMEC 3DIIAP program and would not have been possible without the combined efforts of the 3D team at imec. The authors also gratefully acknowledge the use of the EBSD equipment at MTM, KULeuven.

References

1. A. Mercha et al., "Comprehensive analysis of the impact of single and arrays of through silicon vias induced stress on high-k/metal gate CMOS performance", *2010 International Electron Devices Meeting (IEEE, 2010)*, pp. 2.2.1-2.2.4.

2. J. Van Olmen et al., "3D stacked IC demonstration using a through silicon via first approach", *2008 International Electron Devices Meeting* (IEEE, 2008), pp. 1-4.

3. C. Okoro et al., "Elimination of the axial deformation problem of Cu-TSV in 3D integration", in *Stress-induced phenomena in metallization: 11th international workshop*, Dresden, Germany, April 12-14, 2010, 214-220D.

4. I. De Wolf et al., "Cu pumping in TSVs: Effect of pre-CMP thermal budget", *Microelectronics Reliability*, vol. 51, no. 9-11, pp. 1856-1859, 2011.

5. A. Heryanto et al., "Effect of copper TSV annealing on via protrusion for TSV wafer fabrication", *Journal of Electronic Materials*, vol. 41, no. 9, pp. 2533-2542, 2012.

6. J. De Messemaeker et al., "Impact of post-plating anneal and through-silicon via dimensions on Cu pumping", *2013 Electronic Components & Technology Conference* (IEEE, 2013), pp. 586-591.

7. H.J. Frost and M.F. Ashby, "Deformation-mechanism maps: the plasticity and creep of metals and ceramics", Oxford: Pergamon, ISBN 0-08-029338-7 (1982).

8. T. Jiang et al., "Plasticity mechanism for copper extrusion in through-silicon vias for threedimensional interconnects", *Applied Physics Letters*, vol. 103, pp. 211906-1-5, 2013.

9. C. Okoro, "Thermo-mechanical characterization of copper through-silicon-via interconnect for 3D chip stacking", PhD thesis, KULeuven (2010).

10. R.W. Baluffi, "Grain boundary diffusion mechanisms in metals", *Metallurgical transactions B*, vol. 13B, pp. 527-553, 1982

TSV Reliability Model under Various Stress Tests

Ben-Je Lwo, Frank, M.-S. Lin and Kuo-Hsin Huang
Department of Mechatronic, Energy and Aerospace Engineering,
Chung-Cheng Institute of Technology, National Defense University, Taiwan.
lwob@ndu.edu.tw, 886-3-3900119

Abstract

Through Silicon Via (TSV) is the key technology for the 2.5D and 3D packaging, but reliability evaluations on the new technology products are limited because of the complexity of the environmental issues. To this end, reliability experiments on self-design TSV samples under combinations of several different environmental variables were proposed and performed in this study. The Weibull distribution model was next employed for reliability analyses and the parameters for each of the experimental results were extracted. After analyzing the Weibull parameters, factors that accelerate TSV unreliability are compared with discussions. This paper finally presents failure analyses through OM/SEM observations

Key Words: TSV (Through Silicon Via), Reliability, Stress Test.

Introduction

Due to the advantages such as higher density, faster signal transmission, lower power consumption, broader functionality, and the abilities on heterogeneous connections, 2.5D and 3D packaging integration with TSV (Through Silicon Via) technology have drawn the attention of the packaging industry in the past decade, and the TSV applications were fast expanding at the same time. As the maturity of the TSV manufacturing process and the diverseness of possible TSV application, reliability evaluations on the newly developed technology products were rapidly introduced in the literature in recent years [1-3]. However, because of the complexity from many different environmental effects, the published TSV reliability data are still limited so that further related studies are needed.

Based on our earlier works on TSV property analyses [4-5] and the first reliability study on a TSV structure [6], an advanced study on TSV reliability under several different environmental variables was proposed and performed in this paper. To this end, a typical TSV structure for CIS (CMOS Imaging Sensor) applications was first designed and made. To eliminate the parasitic resistance from the interconnecting wires and the contact resistance from the probing, four-point probe (4PP) measurements were employed for more accurate resistance measurements.

After sample preparations, stress tests with different variables were performed. That is, the TSV samples were tested under several different environmental test conditions which were combinations of current bias, thermal aging, temperature cycling, and thermal-humidity cycling according to the JEDEC/JEP standards, and the failure criterion is set as

10% of relative resistance change. For better statistic meaning on the experimental data, more than 50 useful samples were prepared for each of the stress test. The Weibull distribution model was next employed for data analyses and the parameters of the model were obtained from fitting through the Weibull++ software. The extracted parameters from each of the experiments were then compared so that factors on TSV unreliability due to different environmental variable are proposed with an empirical model. Finally, OM and SEM observations were carried out for failure analyses.

Test Sample Preparations and the Measurements

The 8-inch test wafers were made through a low-temperature, via-last process with commercialized 1P3M CMOS technology, and the process was finished with ball mounting. Chips for current bias testing were next cut from the wafers and mounted on a self-design printed circuit board (PCB) afterward to construct current paths through the TSV, as shown in Figure 1. Note that the connecting wires in Figure 1 can be used for both the current bias providing paths and the measurement paths. For samples without current bias, direct probing contacts on test chip balls were used during the measurements.

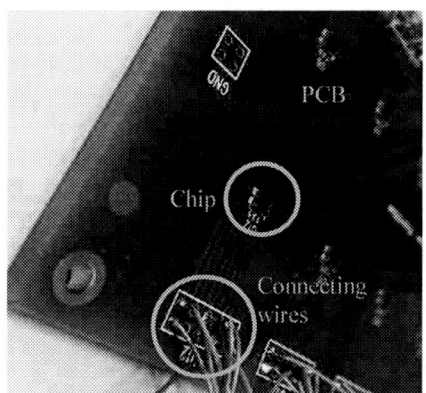

Figure 1. The Biasing Testing Chips on a PCB

The TSV test patterns in this study include: (1) The via-chain structures (VC) with 15 vias to illustrate the electrical continuity of the test samples, and (2) The Kelvin structure (KV) to measure the resistance between via top and via bottom of a single TSV. Figure 2 and Figure 3 respectively shows the via-chain structure and the Kelvin structure, and tests on both of the patterns were performed with 4PP measurements. 10% of relative resistance change is set as the

2014 Electronic Components & Technology Conference

978-1-4799-2408-0/14 $31.00 © 2014 IEEE

failure criterion in this work because circuits are generally designed to work with a maximum 10% on RC delay [7].

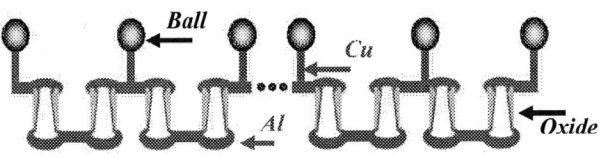

Figure 2: The Via Chain Structure

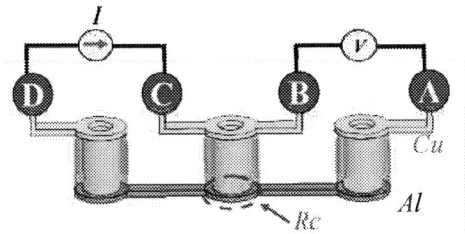

Figure 3: The Kelvin Structure with the 4PB Measurement

The Flow Chart and the Weibull Reliability Model

Figure 4 is the flow chart for the reliability experiments. After verification the basic sample properties, the bias samples were mounted on the PCB as described in the previous section. For non-bias samples, pre-conditionings were performed from room temperature to 240 ^{0}C in 6 min. with 50 second on peak temperature with free air cooling afterward to simulate the PCB mounting process so that both of the bias and non-bias samples are compatible. The initial resistance measurements provided the bases for the following reliability tests, and three stress tests were carried out for both bias and non-bias tests. That is, the reliability tests in this paper include:

(1) The Thermal Aging Tests for samples under 125^{0}C of storages temperature according to JESD22-A108C standard [8].

(2) The Temperature Cycling Tests (TCT) based on JESD22-A104D standard (condition J, soak mode 4) [9]. That is, test samples were put into a temperature-controlled oven with 120 minutes of temperature cycling. The temperature ranges were set between 0 ^{0}C and 100 ^{0}C, and the lower and upper soaking time were 15 minutes, respectively.

(3) The Temperature Humidity Cycling Tests (THCT) modified from JESD22-A100C standard [10]. In the tests, the temperature cycle was 8 hours with 30 ^{0}C ~ 85 ^{0}C temperature range, as Figure 5 shows, and the ramp-up and ramp-down periods were 2 hours, respectively. In addition, 95% was set to be the relative humidity (RH) during the cycling.

For biasing tests, we first observed cross sections on typical TSV samples through OM to calculate the required current magnitudes so that 2×10^4 A/cm^2 current density was provided to each sample [11]. Note that the current density we

picked is the upper limit from the JEP 154 standard. After the experiments, reliability analyses were performed with the Weibull distribution function:

$$F(t) = 1 - \exp\left[-\left(\frac{t}{\eta}\right)^{\beta}\right]$$

(1)

Where F(*t*), η and β are respectively the cumulative distribution (or the cumulative failure) function, the scale parameter which is the time for 63.2% of cumulative sample fail, and the shape parameter to predict the relationship between the lifetime and the failure rate . The Weibull++ software (version 6) was used for parameter extraction through fitting from the experimental data, and the strength of the fittings was checked through the correlation coefficients (ρ) calculated from the fitting.

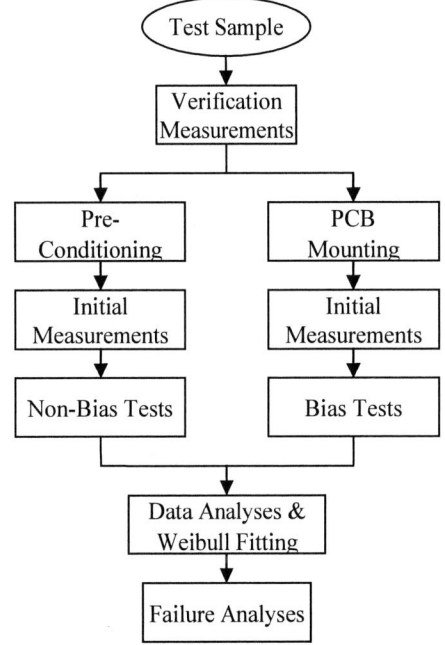

Figure 4: The Flow Chart

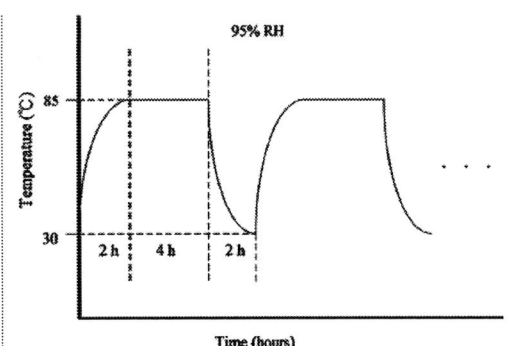

Figure 5: The THCT Process

Results and Discussions

The Preconditioning Results

Table 1 compares the average resistance measured before and after preconditioning for the non-bias samples, and 3.5% and 3.6% of resistance changes were found for the KV samples and the VC samples, respectively. Oxidation is believed to be the reason for the resistance rises after preconditioning.

Table 1: Average Resistance Before and After Preconditioning (unit: Ω)

	Kelvin	Via Chain
before	0.114	6.425
after	0.118	6.658

The Resistance Monitoring

Figure 6 and Figure 7 respectively shows typical resistance monitoring results for no-bias thermal aging samples and biased THCT (Temperature Humidity Cycling Test) samples. Similar plots were observed for all tests under their own environmental conditions (not shown). From the figures, it is found that sample resistances increased during the testing, and fluttering are observed on many samples when approaching the failure points. In addition, remarkable resistance increases are observed in many test samples after failure. Note that the biased samples have much short failure time than the non-bias ones in the plots.

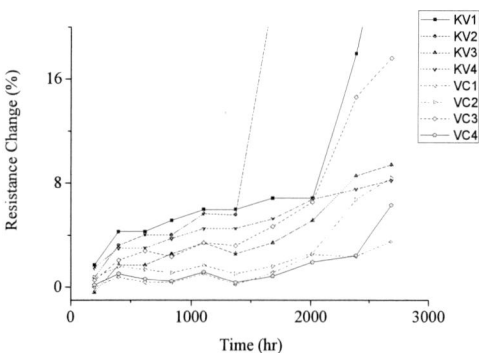

Figure 6: Typical Resistance Monitoring for Aging Samples without Bias (KV: Kelvin samples; VC: via chain sample)

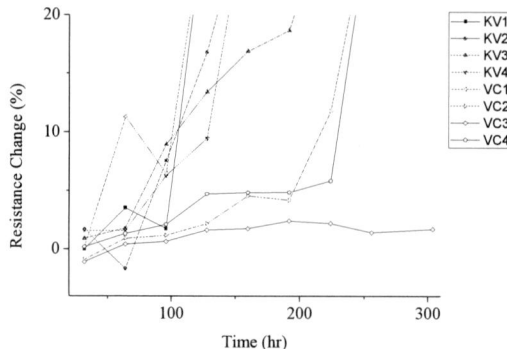

Figure 7: Typical Resistance Monitoring for THCT Samples with Bias (KV: Kelvin samples; VC: via chain sample)

The Cumulative Failure Results

Figure 8 plots the time versus cumulative failure results for THTC tests, including samples with (in red dots) and without (in blue dots) current bias. Similar data for all other tests were obtained (not shown). It is seen from the plots that current bias rapidly accelerates failure rate so that it is an extremely important factor for sample unreliability. In addition, via chain samples have smaller failure rates than the Kelvin samples because the individual TSV unreliability was smoothed by the other TSVs and the connecting metal wires in a chain structure.

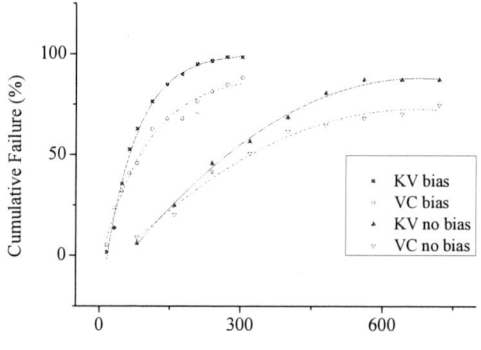

Figure 8: Cumulative failure curves for THTC samples. (KV: Kelvin samples; VC: via chain sample)

Figure 9 and Figure 10 respectively shows the reliability test results for non-bias and bias samples. From the figures, it is found that temperature humidity cycles are the worst environmental effect for the three reliability tests, and thermal cycles is slightly worse than thermal aging. Furthermore, better reliability performances are also noted on the via chain (VC) samples after comparing with the Kelvin (KV) samples because of the averaging behaviors.

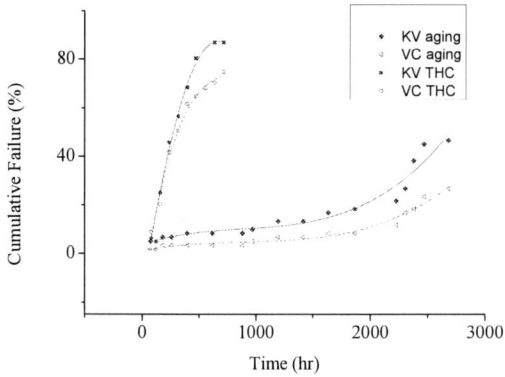

Figure 9: Cumulative Failure Curves for
the Non-bias Samples.

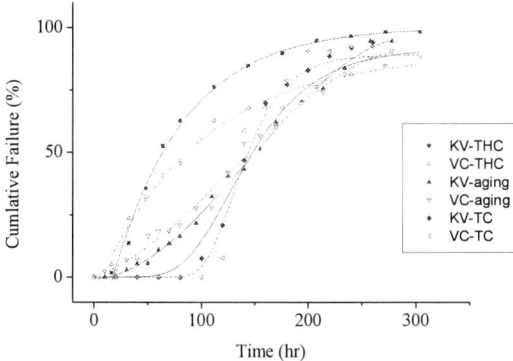

Figure 10: Cumulative Failure Curves for
the Biasing Samples.

The Weibull Analyses

Table 2 and Table 3 lists the Weibull parameters extracted through the Weibull++ software, respectively for the non-biasing and the biasing samples. There are many interesting comparisons from the data:

(1) Most of the VC samples have smaller β but larger η than the KV samples under the same testing environment. That is, reliabilities of the VC samples are better than the KV sample because of the average effects described in the previous subsection.

(2) $\beta > 1$ in most of the conditions indicates that the failure rates increased with time. That is, the "wear out" phenomena were implied. In addition, β for non-biased, thermal aging samples are approximately 1, which indicates constant failure rates through the lifetime.

(3) Similar β but about one third η values are found for the biased THTC sample by comparing with the non-biasing THTC samples. Similarly, the extremely large η for the non-bias aging samples indicate that current bias is a strong effect that rapidly reduce sample reliability.

(4) η for THCT are about 60% ~ 70% than the values for TCT and Aging. That is, THCT has a stronger unreliability

factor than the other two. These results consist with the reliability studies on wire bonding BGA packaging recently [12].

(5) Although biased TCT and biased aging have close η values which imply similar 63.2% failure time, the much larger β on TCT samples shows faster sample degrading at late lifetime.

Table 2: Weibull Parameters for the Non-biasing Samples

Test	Type	β	η	ρ
THCT	KV	1.60	379	0.986
	VC	1.24	484	0.982
Aging	KV	1.03	5878	0.919
	VC	1.15	10840	0.910

Table 3: Weibull Parameters for the Bias Samples

Test	Type	β	η	ρ
THCT	KV	1.69	104.2	0.968
	VC	1.10	138.0	0.973
TCT	KV	4.49	178.8	0.959
	VC	5.24	189.4	0.935
Aging	KV	2.66	170.0	0.970
	VC	1.73	175.1	0.997

According to the above statements, it is concluded that current bias is the strongest factor for unreliability, and thermal-humidity cycling is the second. Furthermore, unreliability due to thermal cycle effect is compactable with the thermal aging effect in the earlier lifetime, but the thermal cycle effect becomes stronger in the late lifetime. That is, time for 82% cumulative failure (the middle point between η rate and 100%) for aging is about 1.1 times longer than TCT. As a result, we postulate a simple empirical unreliability model as the form:

$$AU = F_{bias} \times F_{humid} \times F_{cycling} \qquad (2)$$

where AU represnets Accelerated Unreliability, and the factors in the above equation are listed in Table 4. Note that the factors are 1 if not effected by the speciaifc environmental conditions in the list.

Table 4: Factors for Accelerated Unreiability Model

Symbol	Source	Value
F_{bias}	Current Bias	3.5
F_{humid}	Humidity	1.5
$F_{cycling}$	Thermal Cycle	1.1

The failure analyses

After OM/SEM inspections, oxidation, delamination and flanking have been observed in the previous works for non-biasing samples [6], and similar results were also found for the biasing samples (not shown). In addition, voids as shown in Figure 11 due to EM (electromigration) were found at the current corner of the biasing samples. This phenomenon explains the reason why current bias fast reduces the sample reliability.

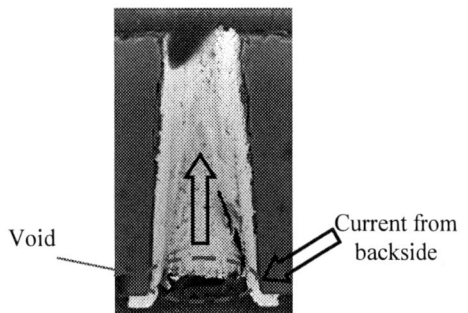

Figure 11: SEM Cross Section for a Typical Biasing Sample with Current Flow Directions

Conclusions

In this paper, reliability experiments on self-design TSV samples under different environmental variables were performed, and the reliability analyses based on the Weibull model were reported. After analyzing the extracted parameters, it is concluded that current bias is the strongest factor for unreliability, and thermal-humidity cycling is the second. Consequently, factors on TSV unreliability due to different environmental variable are finally proposed with an empirical model.

Acknowledgments

The authors would like to thank National Science Council of the government of the Republic of China for financial supports under grant NSC 101-2221-E-606 -005. Test sample preparations by UMC are greatly appreciated.

References

1. C. Cassidy, *et al*, "Through Silicon Via Reliability," *IEEE Trans. on Device and Material Reliability*, Vol. 12, No. 2, pp. 285-295, 2012,

2. T. Frank, *et al*, "Reliability of TSV Interconnects: Electromigration, Thermal Cycling, and Impact on Above Metal Level Dielectric," *Microelectronics Reliability,* Vol. 53, No. 1, pp. 17-29, 2013.

3. P. Ramm, *et al*, "Through Silicon Via Technology – Processes and Reliability for Wafer-Level 3D System Integration," *Proc. of the 58th ECTC*, pp. 841-846, 2008.

4. H. Chung, et *al*, "The Advanced Pattern Designs with Electrical Test Methodologies on Through Silicon Via for CMOS Image Sensor," *Proc. of the 60th ECTC*, pp. 297-302, 2010.

5. H. Chung, *et al*, "A Complete Resistance Extraction Methodology and Circuit Modelsfor Typical TSV Structures," *Int. J. of Electronics*, Vol. 100, No.9, pp. 1256-1269, 2013.

6. B.-J. Lwo, and C,-Y. Ni, "Reliability Analyses on a TSV Structure for CMOS Image Sensor," *Proc. of the 62th ECTC*, pp. 76-79, 2012.

7. T. Frank, *et al*, "Reliability Approach of High Density Through Silicon Via (TSV)," *Proc. of the 12th EPTC*, pp. 321-324, 2012.

8. Temperature, Bias, and operating Life, J*ESD 22-A108C standard*, 2005.

9. Temperature Cycling*, JESD22-A104D standard*, 2009.

10. Cycled Temperature-Humidity-Bias Life Test, *JESD22-A100C standard*, 2007.

11. Guideline for Characterizing Solder Bump Electromigration under Constant Current and Temperature Stress, *JEP 154 standard*, 2008.

12. M. Tsuriya, *et al.*, "The Reliability Performance of Copper Wire Bonding BGA Package by Way of HAST Methodology," *Proc. of the 2013 IMPACT-IAAC*, pp. 335-340, 2013.

Development of Process and Design Criteria for Stress Management in Through Silicon Vias

O. Hölck[1], M. Nuss[1], A. Grams[1], T. Prewitz[1], P. John[1], C. Fiedler[1], M. Böttcher[1], H. Walter[1],
M. J. Wolf[1], O. Wittler[1], K.-D. Lang[2]
[1]Fraunhofer Institute for Reliability and Microintegration (IZM),
Gustav Meyer Allee 25, 13355 Berlin, Germany
ole.hoelck@izm.fraunhofer.de
[2] Technical University Berlin, Berlin, Germany

Abstract

In this paper we report experimental results of through silicon vias (TSVs) at an early processing step which are used in a context of reliability assessment. Parameter studies and evaluation of stresses using finite element analysis contribute to an optimization of processing parameters. Our results comprise nanoindentations to characterize copper protrusion, elasticity and hardness from the top as well as on cross-sections into the depth of the TSVs, EBSD analysis of cross-sections to characterize grain size and orientation and seed-layer influence. FIB-images were used to identify failure modes and finite element studies show reasonable agreement to detected stresses and reliability hot spots. Detailed Raman maps are presented as an outlook for further investagtion of stress fields at TSV cross sections. Design criteria to improve the production process are discussed based on the obtained results.

Introduction

3D system integration has become an important field of semiconductor research and there is increasing demand from the semiconductor industry [1, 2]. The through silicon via (TSV) technique is considered the most promising approach for 3D stacking technology, used for both interposer integration and for direct contacts within active wafers. Vertical connections between wafer top-side and wafer back-side provide the opportunity for bringing different devices into one package. However, many reliability relevant issues are in need of further investigations. One of the challenges is the reliability of the barrier layer and the integrity of the oxide liner isolation, which impede copper migration into the silicon chip and prevent current leakage and interference with functional parts. Residual stresses induced by thermal mismatch between copper and silicon endanger the barrier and oxide layer and again may compromise functional parts in the silicon.

Especially for the global validation of material properties within TSVs and the interdependency with their surrounding areas the Via First Technology is the most flexible design, and is thus investigated here. After planarization the TSV formation step has been finished. Until this point, the process to create TSVs includes annealing steps to recrystallize copper within the TSVs at an early stage, in order to prevent later damage to the structure. Then, assuming all rules for thermal budget definition have been observed, structural changes due to further process steps are avoided and it becomes possible to simulate integration schemes for processing with different requirements of the TSVs.

TSV fabrication

A Fraunhofer IZM ASSID test vehicle (ASSID-TC1) with a die size of 7.5 x 7.5 mm^2 is used on 300mm wafers. The chip contains various electrical test structures and TSV arrays with varying diameter and density.

The following summary represents an overview of the applied process steps for TSV formation:

1) Lithography and dry etch

In order to transfer the TSV pattern to the wafer a positive photo resist is coated, exposed and developed. During the following step the applied structure is etched into the silicon substrate using an inductive coupled plasma (ICP) source reactor and executing a modified Bosch process developed for deep silicon etching.

2) TSV isolation

The electrical isolation of the TSVs is realized by a silicon oxide film deposited with a chemical vapour deposition process. The delivered conformity of the SiO_2-film inside the very small but deep TSV-structures is less than 50%. Due to known physical properties of the gas phase reaction the most critical and thereby the thinnest area is on the bottom sidewall inside the TSVs. The process is run at 350°C.

3) Barrier- and seed-film deposition

To ensure a reliable adhesion of copper-seed layer and for diffusion barrier purpose a titanium (Ti) layer is deposited on top of the isolation. This is done in a high-vacuum self-ionized plasma (SIP) chamber by physical vapor deposition (PVD). Moderate ionization rates allow good sidewall and bottom coverage for the process features.

4) TSV copper metallization

The copper TSV metallization is carried out by electrochemical deposition in an electrolyte with three additives present in the bath. These additives, a suppressor, an accelerator and a leveler, are necessary to enable a bottom-up fill mechanism which allows acceleration of copper plating at the via bottom and suppression of plating near the top.

5) Annealing and CMP

Because of recrystallization effect of the electrochemical deposited copper a change of the volume inside the TSVs occurs. In order to prevent this microstructural change at later steps in the process line, it is necessary to anneal the copper. The process temperature and time depends on the hottest step within the complete further manufacturing process. The temperature treatment step is performed on a batch furnace tool in forming gas atmosphere with optimized temperature ramp up and down steps and a steady treatment temperature for given time. In the following chemical mechanical polishing treatment the layers on the wafer top side are removed and the surface is planarized. The used slurry is a combination of chemical components for the etching part and mechanical components for the abrasive material removing part. The ratio of both components depends on the material that needs to be removed.

Figure 1 shows detailed FIB-images of TSV cross-sections with different aspect ratios after the annealing step (a): 430°C and b): 450°C for 2 hours).

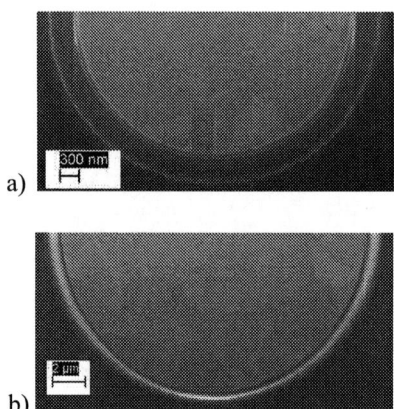

a)

b)

Figure 1: Detailed FIB-images of TSV-sections with aspect ratios of a) 5/50 and b) 20/100 [μm radius / μm depth]. Both TSVs were annealed for 2 hours at T = 430°C and T=450°C, respectively. Note the different scale.

Experimental characterization

Following the fabrication process as described above, batches of TSVs featuring three different geometrical parameter pairs were prepared and exposed to several annealing times and temperatures. The focus of most experimental work lies on TSVs of 20μm diameter and 100 μm depth and on annealing at 250°C for 30min and 450°C for 2 hours.

Nanoindentation

Nanoindentation was used to characterize the elastic modulus of the electroplated copper filling within the TSV at different annealing stages and at different depths. To this end, one sample annealed for 30 minutes at 250°C and one at 450°C for 2 hours were prepared in cross-sections. Nanoindents were set at the center of the TSVs along the length at positions indicated in Figure 2. For annealing at 250°C for 30 minutes, modulus and hardness were measured as E_{250}=162±20 MPa and H_{250}=2.3±0.4 and for 2 hours at 450°C E_{450}=172±10 MPa and H_{450}=2.0±0.1, leading to the conclusion that no significant differences in depth or annealing time could be found for the samples. Nano-indentations were performed using a Berkovich-tip with a force of 4 mN and depths ranging from 300-400 nm.

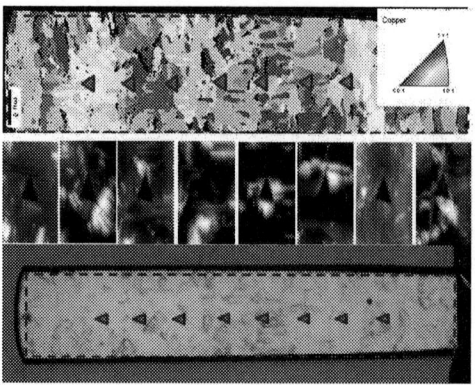

Figure 2: Nanoindentation along the length of a TSV annealed at 250°C for 30 minutes. Grain orientation was determined by EBSD (top). Scan images were performed after indentation (middle). Triangles in bottom picture indicate roughly the position of the indents.

Scans of the nanoindenter tip operated in an AFM-like mode were used to characterize the copper protrusion. The complete area of several TSV samples in three different annealing stages was measured. Representative scans are depicted in Figure 3 along with an average protrusion height of the copper. It can be seen, that Cu-protrusion increases with increasing annealing temperature in accordance to [3].

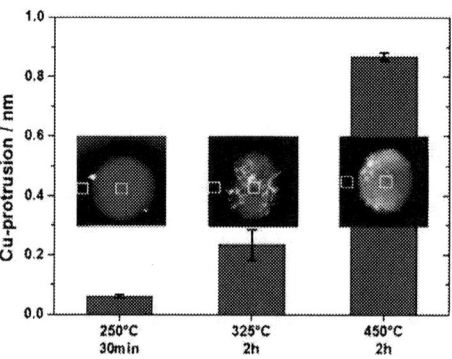

Figure 3: Copper protrusion from nanoindentation-scans after annealing at different temperatures. Squares indicate the area of averaging and reference baseline. Error bars indicate standard deviation.

FIB imaging

Pronounced copper protrusion at ambient temperature indicates plastic deformation due to CTE-mismatch in the beginning of the annealing step and subsequent re-cristallization of the copper while annealing. As stated in the former section, this step is performed early in the process in order to prevent damage at later processing steps when elevated temperatures are needed. Recrystallization leads to a stress-free state at the annealing temperatures. However, upon cooling, the CTE-mismatch between copper and silicon again leads to stresses within the TSV and its interface with the silicon chip, giving rise to failure. Using Focused Ion Beam (FIB) imaging on cross-sections of the TSVs, three different failure modes could be identified, as depicted in Figure 4: The red arrows indicate a cohesive crack in the SiO_2 top layer, presumably in crack opening mode (mode I); another crack propagates along the Si/SiO_2-interface and subsequently grows into the SiO_2 top-layer from below. The location of crack initiation could not be determined but irregularities stemming from the etching process seem likely. A third arrow indicates the position where copper filling and SiO_2- and Ti-barrier meet at the top of the TSV. All three failure modes have been observed only at high annealing temperature and time (2h@450°C).

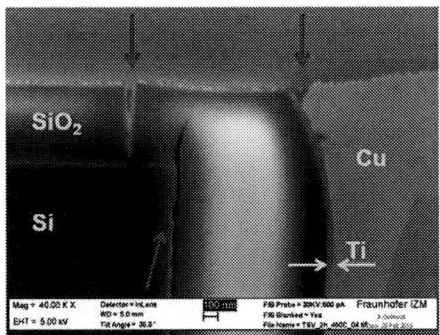

Figure 4: FIB-image of TSV cross-section showing three failure modes indicated by red arrows: Cohesive crack in SiO_2-layer, crack at Si/SiO_2-interface (growing into SiO_2-layer) and interface crack initiation at SiO2/Ti/Cu junction.

FIB-imaging was also employed to check the quality of the barrier layers along the length of the samples. To this end, a cross-section was milled at three different depths and detailed images were recorded as shown in Figure 5. At the position near the top of the TSV smooth layers of SiO_2 and Ti are observed (cp. also Figure 1). The SiO_2 layer becomes somewhat more irregular at lower positions (not shown) but both layers remain intact.

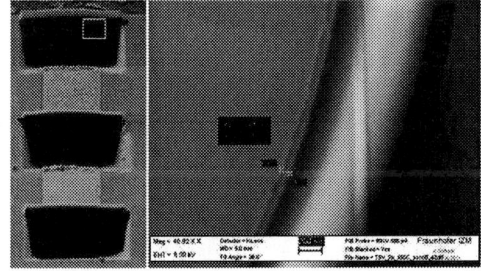

Figure 5: FIB-image detailing the barrier layer of a TSV annealed for 2 hours at 450°C.

EBSD analysis

The electron backscatter diffraction (EBSD) technique is used to analyze the influence of the annealing step on the microstructure of the electroplated copper filling the TSV. The technique is based on the detection of backscattered electrons from the first 50nm of the sample surface [4]. Backscattered electrons form so called Kikuchi patterns which are detected and analyzed. The patterns can be related to the local crystal orientation. If all crystal orientations within a scanned area are determined, local orientation maps can be generated. Inverse pole figure maps show the crystal orientations with respect to the sample coordinate system. The three coordinate axes of the sample system consist of the normal direction (ND) of the cross-sectioning plane,

transversal direction (TD, width of TSV) and rolling direction (RD, length of TSV). The sample coordinate system for the conducted EBSD measurements can be seen in Figure 6.

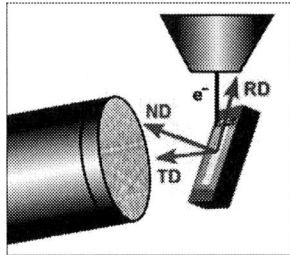

Figure 6: EBSD setup indicating the sample coordinate system.

The generated local orientation maps contain informations about grain size, grain orientation and grain boundaries. EBSD analysis was performed on three samples to investigate the influence of seed-layer-texture on the recrystallization during annealing. EBSD-analysis on two cross-sectioned TSV samples and different annealing times and temperature are shown in Figure 7. Grain boundaries are determined through a point-to point misorientation of 2° or larger. Colors in Figure 7 represent different grains only. The finer grain structure in the low-temperature annealing is obvious.

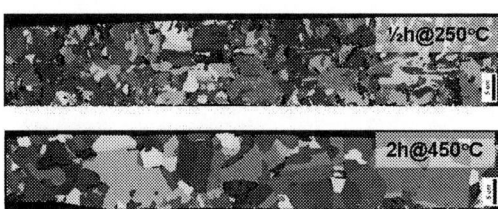

Figure 7: EBSD representation of grainsize. Top:TSV annealed for 30 minutes at 250°C. Bottom: TSV annealed at 450°C for 2 hours. See text.

In Figure 8 a grain orientation representation is shown (colors represent orientations as indicated in the inlets): on the left, the copper seedlayer prepared on a bare waver produces a fine structure with grain diameters smaller than 1 μm and a texture with a (111)-grain orientation along the plane normal (growth-direction of seedlayer). In the middle of Figure 8, a detail of the TSV annealed at 250°C for 30 minutes (shown in Figure 7) was analysed. Here, the orientation of grains relative to the in-plane-direction TD was analysed, corresponding to the growth-direction inside the vias. The red rectangle marks the area of the Cu-seedlayer. It can be seen, that no distinct texture can be observed in the former seedlayer region, indicating that by the process of recrystallization the preferred orientation of the seedlayer has mostly vanished. On the right side of Figure 8, a TSV annealed at 450°C for 2 hours was analysed correspondingly. Clearly the texture has completely vanished. Grain growth has progressed considerably and consumed smaller grains, leaving a coarse grained structure. It should be noted that black areas in all images correspond to areas of small confidence indices

978-1-4799-2408-0/14 $31.00 © 2014 IEEE 627

(not voids), where grain orientation could not satisfactorily be determined.

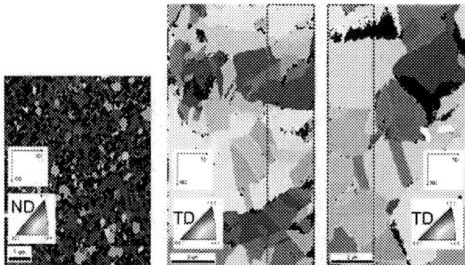

Figure 8: EBSD representation of grainorientation relative to growth-direction. Left: copper-seedlayer (normal to the wafer-plane, ND). Middle: TSV annealed at 250°C for 30 minutes in cross-section. Right: TSV annealed at 450°C for 2 hours in cross-section. Red-dotted rectangles mark the area of former seedlayers. See text.

In Figure 9 a grain size distribution is shown for TSVs featuring different aspect ratios of (20/100, 10/100, 5/50, μm/μm respectively) after annealing at 450°C and 430°C for two hours. The recystallized structure of the large TSV at the bottom and the smallest one at the top can be identified by the formation of large grains, while for the 10/100 TSV the result is ambiguous (but confirmed by EBSD-maps on two other TSVs of the same batch). Further analyses e.g. the investigation of the occurrence of twin boundaries (see Figure 10) in comparison are currently conducted to further interpret this result.

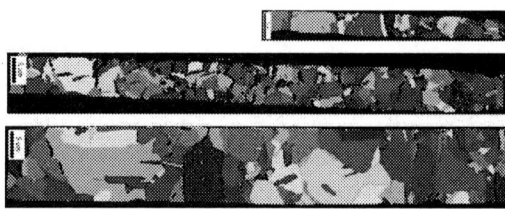

Figure 9: EBSD representation of grains for TSV with 3 different aspect ratios. Clearly a coarse grain size can be observed for the 20/100 (bottom) and the 5/50 (top) TSV, while the 10/100 shows ambiguous results. See text.

Figure 10: EBSD representation of twin grain boundaries for the TSV of aspect-ratio 10/100 in Figure 9. See text.

Simulation

Experimental investigations provide valuable information on microstructure, material properties and failure modes as was shown above. To help interpret the results as well as to optimize parameters and ascertain their limits, two finite element models were set up. The first one is a radially symmetric 2D-model as sketched in Figure 11 comprising the copper via filling, the Ti- and SiO₂-barrier layers and the silicon wafer.

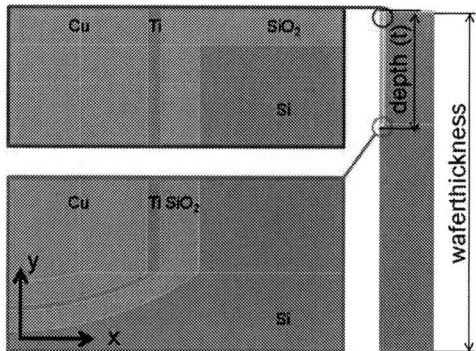

Figure 11: Sketch of the radially symmetric model of the TSV showing details of the top and bottom regions where different materials meet.

Geometrical parameters are compiled in Table 1 and material parameters (valid for all models) in Table 2. Boundary conditions were set as frictionless support for nodes at the symmetry axis and coupling in x-direction for nodes at the right edge of the model (at x = ½ pitch). Thermal load was a temperature drop of 425 K according to the cooling from annealing temperature at 450°C, where the stress free state was assumed. Results of the radial stresses (along x-axis) are shown in Figure 12. Stresses upon cooling occur due to thermal mismatch between the involved materials.

Table 1: Geometric parameters of radially symmetric model.

Parameter	Value [μm]
Wafer thickness	775
Via depth (t)	100
Via diameter (d)	20
Model radius =½ pitch	50
Cu-seedlayer thickness	1.75
Ti-layer thickness	0.4
SiO₂-layer thickness	0.7

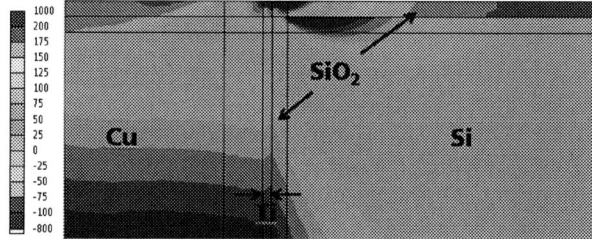

Figure 12: Radial stresses (MPa) in radial model upon cooling from 450°C (stress free state) to 25°C showing good agreement to the identified failure modes (cp. Figure 4).

978-1-4799-2408-0/14 $31.00 © 2014 IEEE   628

Table 2: Material parameters for finite element models.

Material	deformation model	CTE [ppm/K]	E-modulus [GPa]	shear-modulus [GPa]	Poisson-ratio	yield-stress [MPa]	tangent-modulus [MPa]
Si [8,9]	elastic	2,1 (250K) 3,61 (500K)	X:170 (200K) Y: 131 (200K) Z: 170 (200K)	XY: 79,83 (200K) YZ: 79,83 (200K) XZ: 51,13 (200K)	XY: 0,28 YZ: 0,36 XZ: 0,064	-	-
SiO_2 [8,10]	elastic	0,55	73	-	0,17	-	-
Cu [8,11]	elastic -plastic	16,11 (250K)	131 (250K)	-	0,34	59 (240K)	697
Ti [8,12,13]	elastic -plastic	8,1 (250K)	107 (240K) 104 (360K)	-	0,34	400 (243K) 142 (523K)	867

The calculated stress peaks agree well with the positions of the cracks in the FIB image in Figure 4. The peak of the radial stress on the top of the SiO_2 layer near the edge of the TSV is indeed loaded in loaded perpendicular to the crack face, similar to the interface crack initiation at SiO_2/Ti/Cu junction. Stress results from the interface of the SiO_2/Si layer from further below the surface, show that the stress is increasing with depth, suggesting crack initiation to start at irregularities further below the surface.

These investigations help to identify critical stresses for the development of cracks in the surface close to the TSV. Therefore, stresses were calculated by means of FE-simulations for the cooling down from different annealing temperatures. Since experimentally no cracks were observed at annealing temperatures up to 325°C but frequently at 450°C as shown in Figure 4, stresses occurring between the two temperatures can be assumed critical. For the model setup shown in Figure 11 these were determined to be between 400 MPa and 550 MPa for the crack in the SiO_2 and 250 MPa and 330 MPa for the crack between the copper and the titan layer

A parameter study using the 2D-model determined the impact of different geometry parameters. Via depth, wafer thickness and pitch size were found to have little influence on the resulting stresses in the area of crack formation within the investigated parameter field. On the other hand, via diameter and the thickness in particular of the SiO_2 layers have a significant (more than 10% increase) influence on the maximum stresses along evaluation paths covering the positions where cracks have been detected experimentally. A parameter combination can thus be found optimizing via fabrication with respect to the observed failure modes.

Outlook

In order to validate the results obtained from the 2D axisymmetric assumption of the above simulation model, further experimental and simulation efforts are undertaken:

To experimentally determine stresses in silicon, Raman scattering is frequently utilized [5-7]. First results are shown in Figure 13 for a top-sided Raman scan in backscattering geometry. The Raman-peak-shift across a TSV annealed at 450°C for 2 hours (see inlet) and annealed at 250°C for 0.5 hours (circles). The shift is notably larger for the higher annealing time indicating an increase in stresses with higher annealing time and temperature.

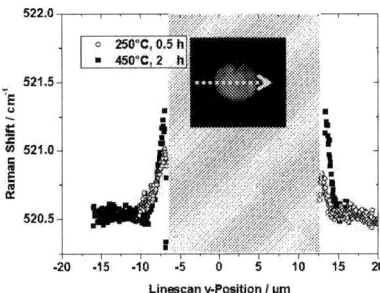

Figure 13: Linescan of the Raman-shift along a path crossing a TSV annealed at 450°C for 2 hours (see inlet) and annealed at 250°C for 0.5 hours (circles).

Conducting a series of line scans in the area around a TSV in cross-section (aspect-ratio 5/50 and annealed at 430°C for 2 hours), a map of the Raman-shift was recorded as shown in Figure 14. As was confirmed in repeated measurements and on other samples, a distinct change in the Raman-shift toward the bottom of the TSV can be made out, enveloping the TSV from a central position to the bottom.

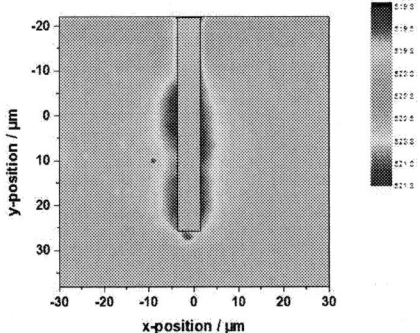

Figure 14: Map of the Raman-shift of a cross-sectioned TSV of aspect ratio 5/50 and annealed at T = 430°C for 2 hours.

The task of relating the shifts of the Raman-peak to stresses is tedious: While the result of the measurement is a scalar value, the stress tensor consists of several entries which do not linearly add up to the resulting shift. On the contrary, stresses of opposite sign may well lead to vanishing Raman-peak-shifts even though regions of high stresses exist. The result of the radial stresses of a 2D-model of a 20/100 TSV is shown in Figure 15. Similar to the field of the Raman-peak-shift, the stresses increase towards the center of the TSV, increasing in width, and enveloping the TSV.

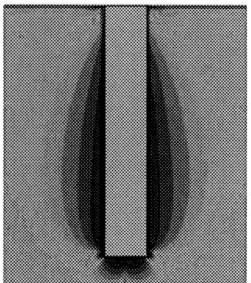

Figure 15: Radial stressintensity as obtained in a 20/100 TSV after temperature drop of 425 K.

It is unlikely that a simple radially-symmetric model can represent the individual stress components leading to the Raman-peak-shift sufficiently well. Therefore, a 3D model will be set up to calculate the Raman-shift from the stress-components. If such methods succeed, it will be possible to verify simulation models of TSVs by comparing calculated and measured Raman-shifts similar as in [5] for even better understanding and thus improved reliability assessment.

Conclusion

Processing TSVs for 3D-integration demands careful consideration and knowledge of material behaviour and failure, in order to optimize reliability and ensure the functionality of the system. With this work, we want to demonstrate tools to investigate, understand and optimize material behaviour of encapsulated copper TSVs after thermal treatment. This enables the definition of thermal budgets within specified (customer) process flows. Using the presented experimental analysis methods, possible failure modes were identified and correlated to structural changes within the TSV during annealing. Simulation models predict the hot spots where failure is likely to occur in good agreement to experimental results, and allow optimization of parameters with respect to an improved reliability of the TSV.

Acknowledgments

The authors thank the following colleagues for contributions to the results: M. Broll, A. Gollhardt, H. Kukuk-Schmidt and S. Huber.

References

[1] Lang, Klaus-Dieter; Wolf, M. Juergen; Thomasius, R.; Reichl, Herbert : Assembly and Integration Technologies for Miniaturized Sensor Systems Proceeding, ISBN 978-3- 9810993-2- 13. Sensor Kongress 2007, Nürnberg, 22.-24.5.2007 , 2007

[2] John H. Lau, (2011) "Overview and outlook of through-silicon via (TSV) and 3D integrations", Microelectronics International, Vol. 28 Iss: 2, pp.8 – 22

[3] P. Saettler, M. Boettcher, und K.-J. Wolter, „Characterization of the annealing behavior for copper-filled TSVs", in Electronic Components and Technology Conference (ECTC), 2012 IEEE 62nd, 2012, S. 619–624.

[4] Electron Backscatter diffraction in material science, A.J. Schwartz, M. Kumar, B.L. Adams, D. Field, 2nd ed. 2009

[5] P. Saettler, M. Boettcher, und K. J. Wolter, „µ-Raman spectroscopy and FE-analysis of thermo-mechanical stresses in TSV periphery", in 2013 14th International Conference on Thermal, Mechanical and Multi-Physics Simulation and Experiments in Microelectronics and Microsystems (EuroSimE), 2013, S. 1–7.

[6] P. Saettler, M. Boettcher, und K.-J. Wolter, „Characterization of the annealing behavior for copper-filled TSVs", in Electronic Components and Technology Conference (ECTC), 2012 IEEE 62nd, 2012, S. 619–624.

[7] I. De Wolf, V. Simons, V. Cherman, R. Labie, B. Vandevelde, und E. Beyne, „In-depth Raman spectroscopy analysis of various parameters affecting the mechanical stress near the surface and bulk of Cu-TSVs", gehalten auf der Electronic Components and Technology Conference (ECTC), 2012 IEEE 62nd, 2012, S. 331 –337.

[8] Cindas LLC online data base: https://cindasdata.com

[9] Matthew A. Hopcroft, William D Nix and Thomas W Kenny, "What is the Young's Modulus of Silicon?", Journal of Microelectromechanical Systems, vol. 19, no. 2, pp. 229 – 238, April 2010

[10] „Accuratus," [Online]: http://accuratus.com

[11] „Copper Development Association Inc.," [Online]. Available:http://www.copper.org

[12] „IFE Institute for Energy Technology," [Online]. <http://www.ife.no/en/ife/departments/materials_and_corrosion_tech/files/facts-and-figures-for-commonly-used-titanium-alloys/view>

[13] ASM International, Atlas of Stress-Strain Curves, Second Edition, ASM International, 2002.

High-speed Wet Etching of Through Silicon Vias (TSVs) in Micro- and Nanoscale

Liyi Li[1], Ching-Ping Wong[1,2*]

1. School of Materials Science and Engineering, Georgia Institute of Technology, Atlanta, GA, USA
2. Department of Electronic Engineering, The Chinese University of Hong Kong, Shatin, HongKong

*Corresponding author: cpwong@cuhk.edu.hk

Abstract

In this paper, a novel wet etching method, named metal-assisted chemical etching (MaCE), is applied for through silicon vias (TSVs) fabrications. The influence of key experimental parameters in MaCE, including catalyst, etchant and dimension of TSVs are discussed. Especially, the type, geometry and morphology of catalyst are varied and the results are compared. A high etching rate over 10 μm/min and a high aspect ratio over 100 are observed in sub-micron scale TSV etching. The presented data demonstrates that MaCE is a promising candidate for TSVs etching with high speed, high aspect ratio capability, submicron capability and low cost.

1. Introduction

Through silicon vias (TSVs) play a pivotal role in the current trend of three-dimensional miniaturization of electronic device. TSVs provide vertical interconnects between stacked components through chips or interposers, which significant shorten the length of interconnection compared to traditional two-dimensional interconnects [1]. Among the process flow for TSVs manufacturing, the overall geometry of TSVs through the 3D space is defined by the deep silicon etching step. Currently, TSVs are majorly etched by deep reactive ion etching (DRIE) technology. The high cost of DRIE has hindered the large-scale application of TSVs. Recently, we reported a novel wet etching method, named metal-assisted chemical etching (MaCE), that managed to etch vertical pores with width down to 30 nm and aspect ratio over 100 [2]. In MaCE, silicon substrates seeded with metal catalyst were etched in liquid etchant solution following the equation:

$$6HF + 2H_2O_2 + Si \xrightarrow{\text{Metal}} H_2SiF_6 + 4H_2O \qquad (1)$$

In MaCE, only the volume of silicon that was in contact with the metal catalyst is selectively etched. By using randomly deposited Au nanoparticles, anisotropic pores with diameters comparable to the catalysts were formed across the whole substrate. However, besides nanopores that align perpendicular to the substrate surface, some randomly winding pores were also observed, which is undesirable for TSVs application.

In this paper, in order to improve the uniformity of overall etching profile in MaCE, we study the effect of catalyst morphology and type in MaCE of randomly deposited catalyst nanoparticles. Then we used the uniform catalyst spots defined by electron beam lithography (EBL) to fabricate TSVs array with defined diameter and spacing. The results are summarized and compared. Based on the results, a guideline for choosing the correct set of experimental parameters is provided.

2. Discussion

The etching profile of a certain MaCE experiment is comprehensively determined by multiple experimental parameters, such as catalyst, etchant, temperature, substrate and etc. For each of these parameters, several variables exist [3-5]. A certain etching profile is the result of a whole set of parameters that are involved during the etching. We name a certain set of parameters as an input set. In order to understand the individual influence of every aspect of the parameters in an input set, we did a systematic investigation as shown in details in **Table 1**. Especially, we focus on two key parameters: catalyst and etchant. For the catalyst parameter, we varied the type, thickness and morphology of catalyst. After etching, we view the sample under scanning electron microscope (SEM). The top-view SEM images are taken by directly viewing the top surface of the as-etched silicon, while the cross-sectional images are taken from the mechanically-cleaved silicon substrate. The maximum depth with which metal catalysts penetrate into silicon is measured from cross-sectional image (named as "depth" in **Table 1**); the nominal etching rates are calculated by dividing the depth by etching time (named as "rate" in **Table 1**). For convenience of discussion, each test coupon is given a specific ID number in the first column of Table 1, and illustrated by a set of SEM images in the following part of discussion.

Table 1 MaCE test ID number and experimental parameters

ID	Catalyst	$\rho(\%)$	Time (min)	Depth (μm)	Rate (μm /min)
1	1mMAg10s 150°C	90	2	13.1	6.55
2	1mAAg10s 190°C	90	2	13.2	6.6
3	5mMAg30s 190°C	90	2	0.4	0.2
4	1mMAu10s 150°C	90	2	5	2.5
5	1mMAu10s 190°C	90	2	1.3	0.65
6	5mMAg30s 190°C	90	2	NA*	NA
7	Au100nm	90	3	NA	NA
8	Au50nm	90	15	162	10.8
9	Au50nm	40	10	27	2.7
10	Au50nm 190°C	90	5	NA	NA
11	Au10nm	40	10	9.6	0.96
12	Au100nm	90	10	5	0.5
13	Au100nm	90	20	8.3	0.415
14	Au100nm	90	40	19.6	0.49

*NA: non-applicable due to shallow etching.

It has long been known that metal particles can be deposited on silicon substrate by galvanic replacement reaction [6, 7] with the assistance of hydrofluoric acid (HF):

$$nHF + M^{n+} + Si \longrightarrow SiF_6^{2-} + M + H^+ \qquad (2)$$

The process is also known as electroless metal deposition (EMD). In experiment, silicon substrate is immersed in the plating solution containing metal ion and HF for a certain time. At the initial stage, metal ions (M^{n+}) are reduced on the silicon surface and form nuclei. As the plating proceeds, more M^{n+} are reduced on the nuclei, which gradually grow into metal nanoparticles (NPs). If noble metal ion are used, the noble metal NPs thuse deposited on silicon are well suitable as catalyst for MaCE process thereafter [8]. The particle size can be controlled by the concentration of metal and HF. Also, it is reported that the morphology of metal NPs can be adjusted by thermal treatment [9]. Since the dimension of the metal particles falls into nanoscale, effective morphology adjustment can be achieved under 200 °C, which is compatible with CMOS device. Based on this argument, we fabricate silver (Ag) and gold (Au) NPs for TSVs etching experiments in nanoscale by MaCE.

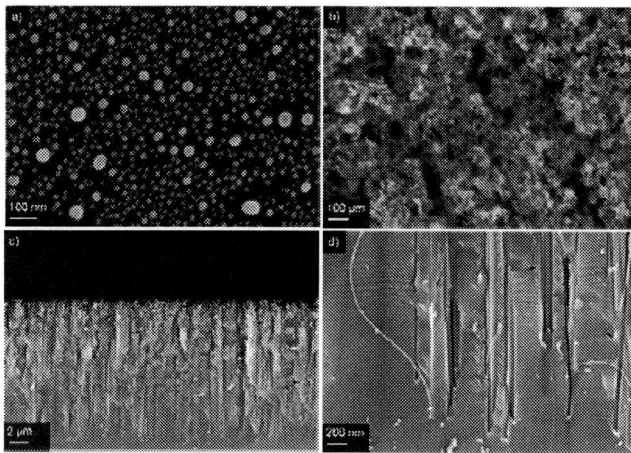

Figure 1 (a) Top view SEM image of silicon substrate plated by 1mM Ag plating solution for 10 sec and annealed at 150 °C. Top view (b) and cross-sectional (c) SEM images of Ag NPs-loaded silicon substrate shown in (a) after etching in ρ(90%) etchant for 2 min (Test ID#1). (d) shows the magnified SEM image of the bottom part in (c).

Figure 1 (a) shows the silicon surface plated by Ag NPs in 1mM Ag^+ solution for 10 sec and annealed at 150 °C (ID#1). Compared to the as-deposited Ag NPs [2], most NPs evolve into spherical shape and the irregular sharp corners disappear. After etching in solution composed of HF and H_2O_2, the top surface of silicon substrate is roughened and vertical nanopores can be found (Figure 1 (b) and (c)). Here we define the term $\rho = [HF]/([HF]+[H2O2])$, where [X] refers to the concentration of chemical X in mol/L. An etchant solution with ρ=90% (named ρ(90%) etchant) is used in this paper if not specified otherwise. In the magnified image (Figure 1 (d)), the diameter of nanopores is found to be 50-250 nm, significant larger than the diameter of original Ag NPs shown in Figure 1 (a). The fact that the diameter of nanopores etched by Ag NPs is larger than that of original Ag NPs may be caused by two factors: (1) extra etching on sidewalls of the pores; (2) several NPs form aggregates and etch as an entity. Another finding is that some randomly winding pores are also observed in Figure 1 (c), which is similar to the results in [2].

Since the contour of Ag NPs becomes more uniform and the irregular corners are removed/smoothened, the random etching found in [2] are less likely to be caused by the shape irregularity of catalysts themselves.

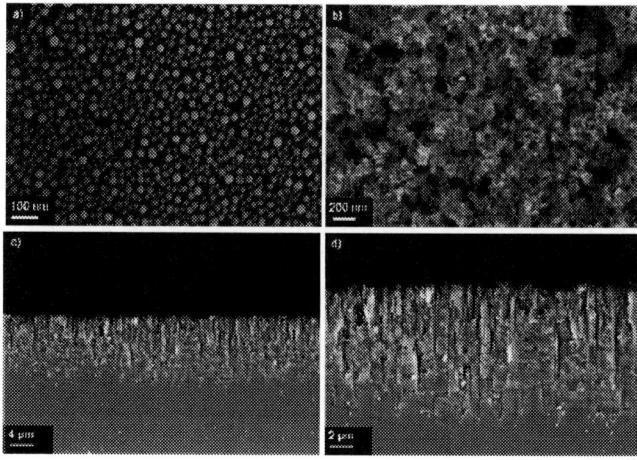

Figure 2 (a) Top view SEM image of silicon substrate plated by 1mM Ag plating solution for 10 sec and annealed at 190 °C. Top view (b) and cross-sectional (c) SEM images of Ag NPs-loaded silicon substrate shown in (a) after etching in ρ(90%) etchant for 2 min (Test #2). (d) shows the magnified SEM image of the bottom part in (c).

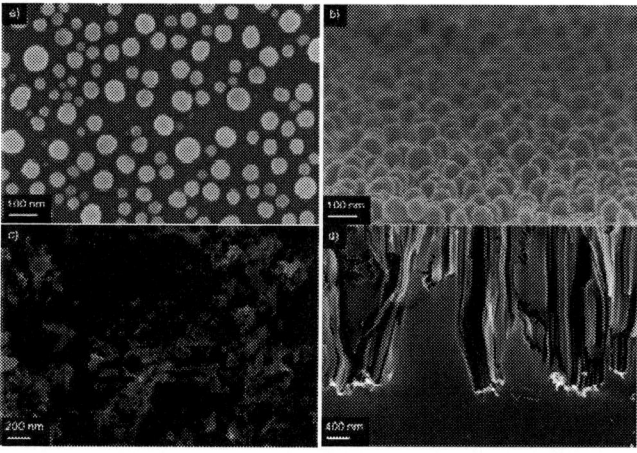

Figure 3 Top view (a) and bird-eye view (b) SEM image of silicon substrate plated by 5mM Ag plating solution for 30 sec and annealed at 190 °C. Top view (c) and cross-sectional (d) SEM images of Ag NPs-loaded silicon substrate shown in (a) after etching in ρ(90%) etchant for 2 min (Test #3). (d) shows the magnified SEM image of the bottom part in (c).

When the annealing temperature is increased to 190 °C, the size distribution of as-deposited Ag NPs becomes more uniform (Figure 2 (a), Test #2). However, the etching profile does not change essentially (Figure 2 (b) and (c)); the nominal rate is 6.6 μm/min, both randomly winding pores and vertical pores can be observed. Further, we increase the concentration of Ag^+ in the silver plating solution (Test #3). The diameter of Ag NPs becomes significantly larger and spherical contour of the NPs after annealing can be clearly observed (Figure 3 (a) and (b)). After etching, large aggregates of Ag NPs are found at the bottom of the etched pores. Most of these aggregates move vertically into the silicon, while some of them alter their etching direction during etching. The nominal etching rate is only 0.2 μm/min. Two statements can be proposed from Figure 2 and Figure 3: (1) randomly winding pores are found in as-deposited irregular NPs as well as spherical-shaped annealed NPs, thus shape irregularity is not a dominated cause for the issue of random etching in MaCE; (2) larger NPs aggregates etch slower than that of smaller NPs.

Besides Ag, we also studied the performance of Au as catalyst. Similar to Ag NPs, Au NPs are deposited on silicon using EMD. Figure 4 (a) and (b) shows the morphology of Au NPs plated in 1mM AuCl$_4^-$ plating solution and annealed at 150 $^{\circ}$C (Test #4) and 190 $^{\circ}$C (Test #5), respectively. The size of Au NPs in both images is significantly smaller than that of AgNPs. After etching, a considerable amount of winding pores are found near the top surface (Figure 4 (e) and (f)). The nominal rates are 2.5 μm/min for Test #4 and 0.7 μm/min for Test #5. The decrease of etching rate may be due to the enlarged aggregates size at the higher annealing temperature. Test #4 and #5 also illustrate the morphology of catalyst morphology plays an important role in determining the etching rate as well as the shape of etching profile, since Au has been proposed to have a higher catalytic activity for H$_2$O$_2$ decomposition and thus a higher etching rate as MaCE catalyst.

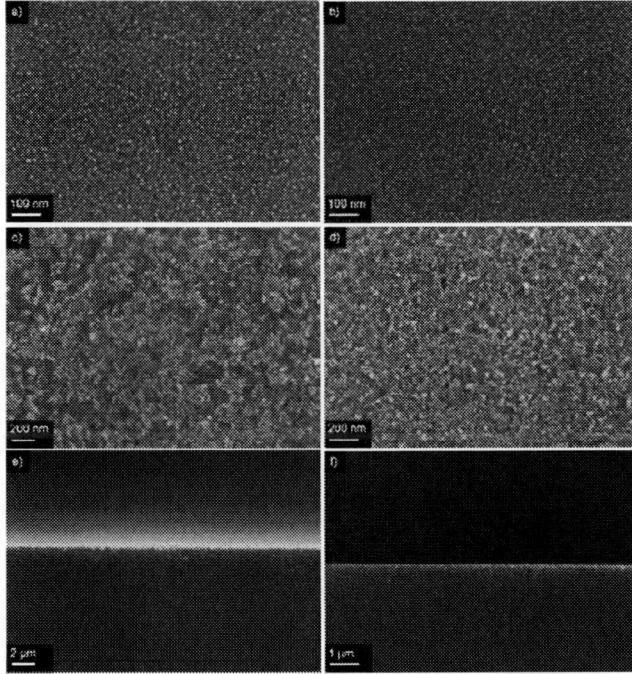

Figure 4 (a) Top view SEM image of silicon substrate plated by 1mM Au plating solution for 10 sec and annealed at 150 $^{\circ}$C. Top view (c) and cross-sectional (e) SEM images of Ag NPs-loaded silicon substrate shown in (a) after etching in ρ(90%) etchant for 2 min (Test ID#4). Top view (b) SEM image of silicon substrate plated by 1mM Au plating solution for 10 sec and annealed at 190 $^{\circ}$C. Top view (d) and cross-sectional (f) SEM images of Ag NPs-loaded silicon substrate shown in (b) after etching in ρ(90%) etchant for 2 min (Test #5).

In application, TSVs are etched after lithographic pattern to achieve precisely defined lateral position and geometry.

Here we use electron beam lithography (EBL) to study the etching behavior of patterned catalyst. First we use the EMD method to fabricate an array of Ag NPs with 100 nm diameter. Unlike the ones deposited on bare silicon surface mentioned above, the as-deposited Ag NPs with patterned EBL resist shows a uniform lateral diameter of ca. 120 nm with a widely-dispersed thickness of 150 -200 nm (Figure 5 (a) and (b)). The result indicates that although the lateral geometry of the Ag NPs during EMD process is limited by EBL resist, the vertical growth is unlimited and non-uniform across the silicon substrate. After annealing, the Ag NPs no longer maintain the pillar shape and slightly collapse into non-uniform spherical shape (Figure 5 (c), Test #6). After etching, the Ag NPs still stay on the top surface of the substrate and some shallow rough dents are formed around the Ag NPs. Interestingly, the Ag NPs that are originally separated by 500 nm spacing come into aggregation after etching (Figure 5 (d)). The aggregation of catalyst NPs in various sizes and types mentioned above

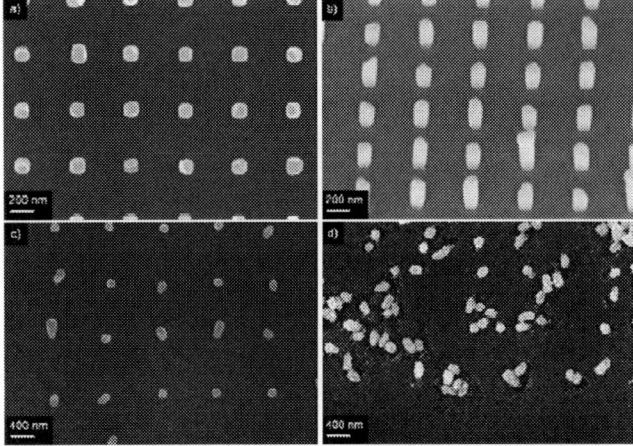

Figure 5 (a) Top view and (b) bird-eye view SEM images of silicon substrate plated by 5mM Ag plating solution for 30 sec patterned by EBL. (c) Top view SEM image of the catalyst shown in (b) annealed at 190 $^{\circ}$C. Top view (d) SEM images of Ag NPs-loaded silicon substrate shown in (c) after etching in ρ(90%) etchant for 2 min (Test #6).

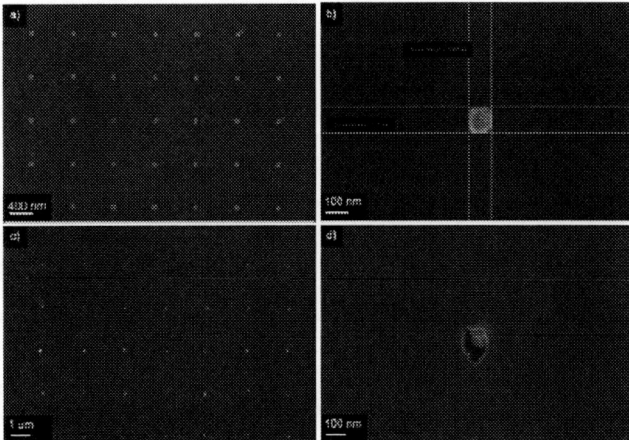

Figure 6 (a) Top view SEM image of silicon substrate patterned by EBL and deposited by 100 nm-thick Au. (c) Top view SEM image of the Au-loaded silicon substrate shown in (a) after etching in ρ(90%) etchant for 3 min (Test #7). (b) and (d) are magnified images of (a) and (c) on sites of Au dots.

978-1-4799-2408-0/14 $31.00 © 2014 IEEE

strongly indicates an attractive force between metal particles is involved during the etching.

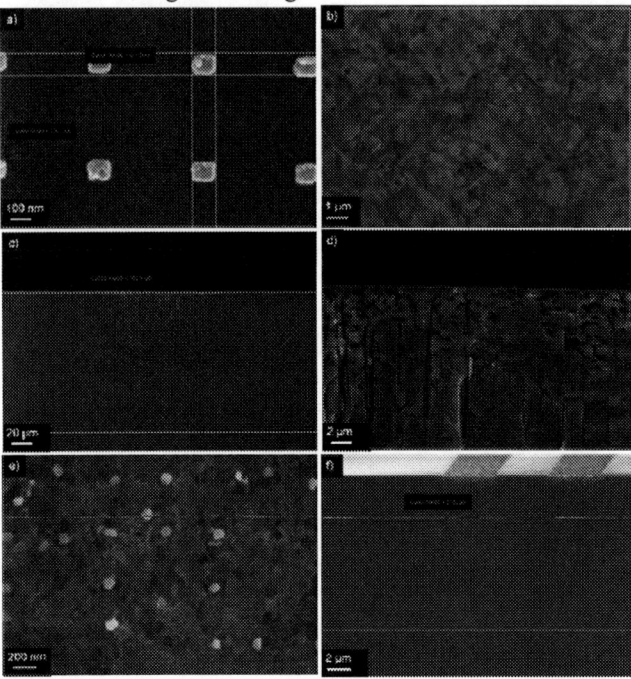

Figure 7 (a) Top view SEM image of silicon substrate patterned by EBL and deposited by 50 nm-thick Au. (b) Top view and (c) cross-sectional SEM image of the Au-loaded silicon substrate shown in (a) after etching in $\rho(90\%)$ etchant for 15 min (Test #8). (d) Top view and (e) cross-sectional SEM image of the Au-loaded silicon substrate shown in (a) after etching in $\rho(40\%)$ etchant for 10 min (Test #9). (f) shows the magnified image in the top area of (e).

We try to increase the etching rate of patterned catalyst by reducing their thickness, which may make the mass transfer of reactant easier. In this regard, we deposit 100 nm-thick Au on pattern silicon substrate by electron beam evaporation (EBE), which is known to produce a thin film with controllable and uniform thickness. Figure 6 (a) and (b) shows an array of 100 nm-thick Au dots. After etching, the Au dots etch into the substrate with well defined sidewalls (Figure 6 (c) and (d), Test #7). When the thickness of the catalyst is further decreased to 50 nm, the etching rate dramatically increases. Figure 7 (a) shows the morphology of 50 nm-thick Au dots. After etching, no Au catalyst can be observed from top (Figure 7 (b)); the maximum penetration depth is 162 μm after 15 min's etching and the diameter of the etching profile is 100-400 nm (Figure 7 (c),Test #8). The nominal etching rate is calculated to be 10.8 μm/min; the aspect ratio of vias is over 100. This super high aspect ratio of etching profiles in sub-micron scale together with the high etching rate clearly demonstrates the advantage of MaCE over traditional high aspect ratio etching method [10]. In addition, we demonstrate the high ρ value of etchant is necessary for the high etching rate. Figure 7 (e) and (f) show the etching result of the same substrate in $\rho(40\%)$ etchant (Test #9). After etching, a considerable amount of catalysts remain near the top surface of substrate, and the etching rate drops to 2.7 μm/min. Another interesting finding is that if the 50 nm-thick catalyst is annealed at 190 oC, the sharp corners of Au dots

disappeared (Figure 8 (a), Test #10), and only a randomly winding etching along the lateral direction is observed in $\rho(90\%)$ etchant (Figure 8 (b)). This result implies that although the overall shape of catalyst become more uniform, some channels inside catalyst may be annihilated during annealing and the vertical mass transfer of reactants through catalyst may be inhibited.

Based the results obtained from the etching of 100 nm-diameter catalysts, we try to fabricate TSVs with larger lateral geometry. Figure 9 shows the etching results of Au dots in $\rho(40\%)$ etchant (Test #11). The Au dots array possesses two set of diameters, i.e. 100 nm and 400 nm, and are loaded on the same wafer. From the top view, many etched shallow trenches are found on the top surface of the area with 100 nm-diameter Au dots. However, the number density of these surface etching features significantly decreased in the etching profile of 400 nm-diameter Au dots (Figure 9 (b)). Also, vertical etching can be clearly observed at the initial stage by 400 nm-diameter Au dots.

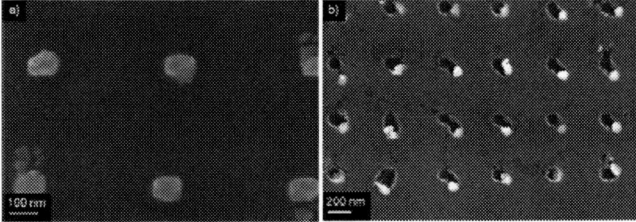

Figure 8 (a) Top view SEM image of silicon substrate patterned by EBL and deposited by 50 nm-thick Au. The sample is annealed at 190 °C. (b) Top view the Au-loaded silicon substrate shown in (a) after etching in $\rho(40\%)$ etchant for 5 min (Test #10).

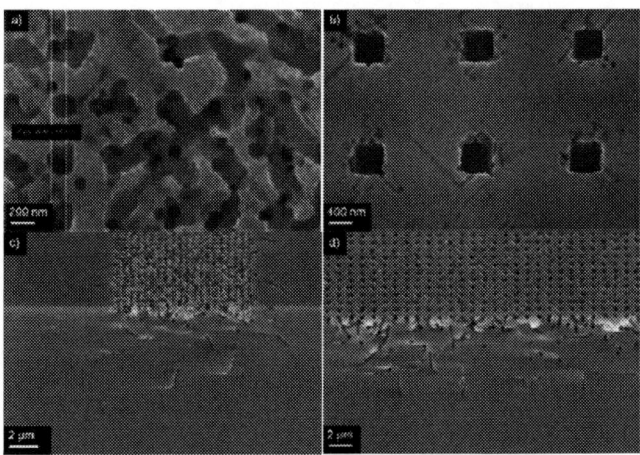

Figure 9 (a) Top view and (b) cross-sectional SEM image of silicon substrate patterned by EBL with 100 nm diameter, deposited by 10 nm-thick Au and etched in $\rho(40\%)$ etchant for 10 min. (c) Top view and (d) cross-sectional SEM image of silicon substrate patterned by EBL with 400 nm diameter, deposited by 10 nm-thick Au and etched in $\rho(40\%)$ etchant for 10 min (Test #11).

If the lateral size of the catalyst is further increased to 3 μm by photolithography patterning (PL), randomly winding etching profiles disappear and nearly all the catalysts etch into the silicon substrate (Figure 10, Test #12). Some vertical

978-1-4799-2408-0/14 $31.00 © 2014 IEEE

TSVs can be observed, while some others adopted tilted profiles. It should be noted that since Au catalyst lie on the bottom surface of TSVs after etching, it facilitates the bottom copper filling of the TSVs through electroplating (Figure 10 (d)), which is favorable for the post etching process in TSVs manufacturing. For even larger catalysts with 10 μm (Figure 11, Test #13) and 15 μm (Figure 12, Test#14) diameter, the vertical etching profiles become more dominant versus the tilted ones. The etching rates of Test #12, 13 and 14 are all around 0.5 μm/min. In comparison, in Test #7 where identical etchant is used, the etchant rate is significantly lower. This comparison indicates that for the 100 nm-thick Au catalyst, etching behavior is significantly different between microscale and nanoscale.

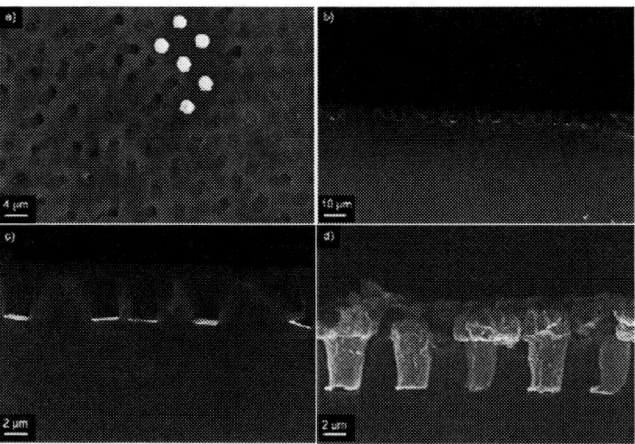

Figure 10 (a) Top view and (b) cross-sectional SEM image of silicon substrate patterned by PL with 3 μm diameter, deposited by 100 nm-thick Au and etched in ρ(90%) etchant for 10 min. (c) shows the magnified image in the middle area of (b). (d) shows the cross-sectional images of TSVs in (b) filled with copper (Test #12).

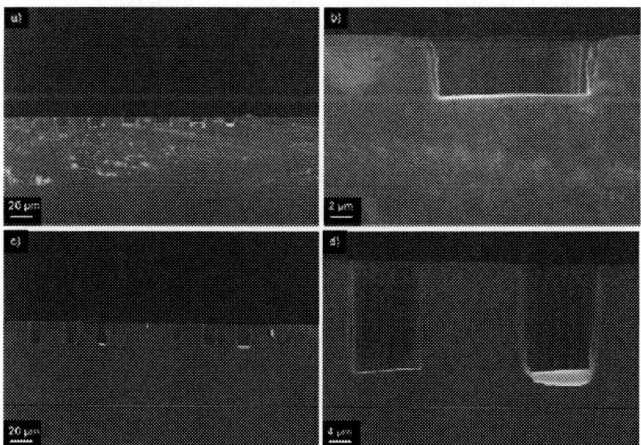

Figure 11 (a) Top view and (b) cross-sectional SEM image of silicon substrate patterned by PL with 10 μm diameter, deposited by 100 nm-thick Au and etched in ρ(90%) etchant for 10 min (Test #13). (c) Top view and (d) cross-sectional SEM image of silicon substrate patterned by PL with 15 μm diameter, deposited by 100 nm-thick Au and etched in ρ(90%) etchant for 10 min (Test #14).

Conclusions

In conclusion, we have investigated the MaCE for TSV etching. The comprehensive influence of etchant composition, catalyst type, morphology and lateral geometry on the etching profile is presented. Vertical vias from few-micron scale to nanoscale are etched by MaCE. Especially, a high etching rate over 10 μm/min and a high aspect ratio over 100 are observed in sub-micron scale TSV etching. The presented data demonstrates that MaCE is a promising candidate for TSVs etching with high speed, high aspect ratio capability, submicron capability and low cost.

Acknowledgments

The authors thank IEN at Georgia Tech for cleanroom facility training and usage. The authors thank CNC at Georgia Tech for SEM facility. The authors thank funding support from National Science Foundation (NSF CMMI #1130876).

References

1. Y. Morikawa, T. Murayama, Y. N. T. Sakuishi, A. Suzuki, and K. Suu, "Total cost effective scallop free Si etching for 2.5D & 3D TSV fabrication technologies in 300mm wafer," in *Electronic Components and Technology Conference (ECTC), 2013 IEEE 63rd*, 2013, pp. 605-607.

2. L. Li and C. P. Wong, "High aspect ratio sub-100 nm silicon vias (SVs) by metal-assisted chemical etching (MaCE) and copper filling," in *Electronic Components and Technology Conference (ECTC), 2013 IEEE 63rd*, 2013, pp. 2326-2331.

3. Z. Huang, N. Geyer, P. Werner, J. de Boor, and U. Gösele, "Metal-Assisted Chemical Etching of Silicon: A Review," *Adv. Mater.*, vol. 23, pp. 285-308, 2011.

4. L. Li, Y. Yao, Z. Lin, Y. Liu, and C. P. Wong, "Low-cost micrometer-scale silicon vias (SVs) fabrication by metal-assisted chemical etching (MaCE) and carbon nanotubes (CNTs) filling," in *Electronic Components and Technology Conference (ECTC), 2013 IEEE 63rd*, 2013, pp. 581-585.

5. L. Li, Y. Liu, X. Zhao, Z. Lin, and C.-P. Wong, "Uniform Vertical Trench Etching on Silicon with High Aspect Ratio by Metal-Assisted Chemical Etching Using Nanoporous Catalysts," *ACS Appl. Mater. Interfaces,* vol. 6, pp. 575-584, 2014/01/08 2014.

6. L. A. Nagahara, T. Ohmori, K. Hashimoto, and A. Fujishima, "Effects of HF solution in the electroless deposition process on silicon surfaces," *Journal of Vacuum Science & Technology A,* vol. 11, pp. 763-767, 1993.

7. P. Gorostiza, J. Servat, J. R. Morante, and F. Sanz, "First stages of platinum electroless deposition on silicon (100) from hydrogen fluoride solutions studied by AFM," *Thin Solid Films,* vol. 275, pp. 12-17, 1996.

8. K. Q. Peng, J. J. Hu, Y. J. Yan, Y. Wu, H. Fang, Y. Xu, S. T. Lee, and J. Zhu, "Fabrication of Single-Crystalline Silicon Nanowires by Scratching a Silicon Surface with Catalytic Metal Particles," *Adv. Funct. Mater.,* vol. 16, pp. 387-394, 2006.

9. Y. K. Mishra, S. Mohapatra, D. Kabiraj, A. Tripathi, J. C. Pivin, and D. K. Avasthi, "Growth of Au nanostructures by annealing electron beam evaporated thin films," *Journal of Optics A: Pure and Applied Optics,* vol. 9, p. S410, 2007.

10. Y.-C. Hsin, C.-C. Chen, J. H. Lau, P.-J. Tzeng, S.-H. Shen, Y.-F. Hsu, S.-C. Chen, C.-Y. Wn, J.-C. Chen, T.-K. Ku, and M.-J. Kao, "Effects of etch rate on scallop of through-silicon vias (TSVs) in 200mm and 300mm wafers," in *Electronic Components and Technology Conference (ECTC), 2011 IEEE 61st*, 2011, pp. 1130-1135.

Replacing the PECVD-SiO$_2$ in the Through-Silicon Via of High-Density 3D LSIs with Highly Scalable Low Cost Organic Liner: Merits and Demerits

Murugesan Mariappan, Takafumi Fukushima, JiChel Beatrix, Hiroyuki Hashimoto,
Yutaka Sato, Kangwook Lee, Tetsu Tanaka and Mitsumasa Koyanagi

NICHe, Tohoku University, 6-6-10, Aza-Aoba, Aramaki, Aoba-ku, Sendai, 980-8579, Japan
Phone: +81-22-795-4119; Fax: +81-22-795-4313; E-mail: murugesh@bmi.niche.tohoku.ac.jp

Abstract

A novel approach to suppress the conventional Cu-TSV induced thermo-mechanical stress in 3D-LSI chip is proposed, fabricated and tested. In this approach, a thermal-chemical-vapor-deposition grown organic poly-imide based polymer is conformally deposited along the side wall of the TSV. As-grown polymer was tested for its physical properties and mechanical properties, and was also evaluated for their role in minimizing the thermo-mechanical stress in vicinal and via-space Si. It was found that replacing the conventional SiO$_2$ dielectric liner (sandwiched between the via-metal and Si) with organic polymer greatly helps in suppressing the thermo-mechanical stress, and thus the keep-out zone.

Introduction

Though the highly sophisticated photo-lithography technology enables one to keep in pace with the scaling of Moore's law, but it is acheived at the expense of heavy cost. Not only that, but also the aggressive scaling of devices leads to several reliability concerns such as short-channel effect, a tremendous increase in off-state current, I_{off} or a large leakage current. Hence, in recent days, the three-dimensional heterogeneous integration of various functional chips has picked enormous momentum as an alternative route to the conventional 2D-VLSI/ULSI approach. The 3D-IC/Si integration can be achieved by vertically stacking several functional chips and connecting them with the through-silicon-vias (TSVs) and microbumps as shown in fig. 1[1]. This vertical stacking of various functional chips thus leads to a decrease in power consumption and an increase in speed due to the short interconnect length. In line with this, Silicon (Si) wafers with thickness less than 20 μm are attracting more and more interest especially in three-dimensional (3D) chip stacking in order not only to reduce the TSV length, but also to ease the TSV fabrication and lateral interconnection formation process [2-4]. Even in this 3D-LSI approach there exist several reliability problems 3, 5-7], one major problem is how to obtain a conformal and also a uniformly-thick good-quality dielectric layer along the TSV sidewall. By and large, the in-organic dielectric liner in the TSV is grown by PE-CVD method. Even though one can obtain conformal liner along the TSV sidewall, the thickness of the liner varies continuously from the top to the bottom. Also, in the case of high-frequency application the electrical loss is huge with the in-organic dielectric liner, SiO2. Therefore it is wise to replace the in-organic dielectric liner by the organic polymer, where the electrical loss is largely suppressed and it has been already shown by several authors [8, 9]. However they are mainly based on spin-on dielectric coating technique, where it is difficult to conformal coating of the organic material not only in the high-aspect ratio TSVs, but also in the TSV with smaller diameter.

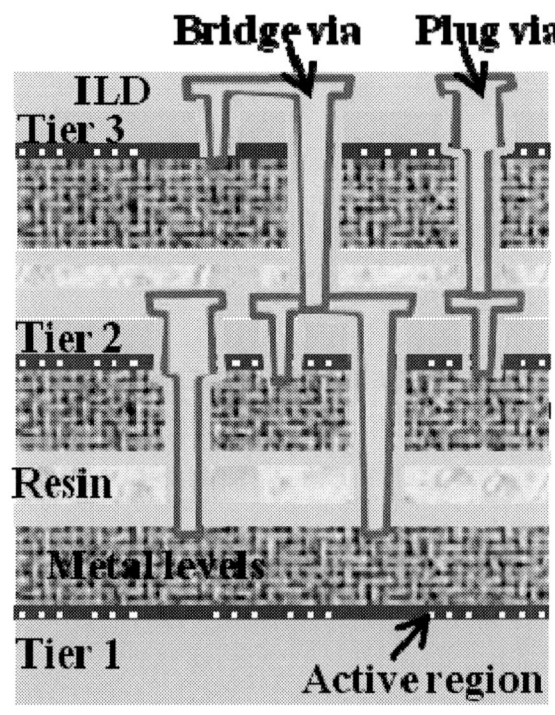

Figure 1: **Typical schematic view of three - dimensional large scale integrated (3D-LSI) system.**

In this study, we will be addressing the merits and demerits of the polyimide based organic liner in the TSV-side wall grown by low-temperature thermal chemical vapor deposition (TCVD), where the TSV aspect ratio is around 10 and the minimum TSV diameter is around 5 μm.

Experimental details

The co-polymerization of the 4,4'-oxydianilline and the pyromellitic-anhydride directly along the TSV sidewall was carried out at 195°C. From the FESEM images it was evident that conformal formation of 2 μm-thick PI liner inside the deep trench. As formed polymer thickness in 8"/12" wafers was monitored by reaction time and vacuum level. The liner coverage along the side-wall was examined for various TSV sizes ranging from 5 μm to 30 μm. We were able to achieve

978-1-4799-2408-0/14 $31.00 © 2014 IEEE

75 - 80% and 100% of PI coverage for via diameters <10 μm and >15μm, respectively. After the liner formation, both the barrier (Ta/Ti metal), Cu seed layers were deposited using ionized sputtering, followed by electro-plating. As obtained organic liner was examined for its physical properties such as glass-transition temperature, Tg, co-efficient of thermal expansion, CTE, mechanical strength such as tensile strength and Young's modulus, and finally dielectric constant ε and strength, and are listed in table 1.

The two dimensional thermo mechanical stress distribution in the vicinal-Si and via space region exerted by Cu-TSV without polyimide liner and with stress absorbing polyimide liner and the stress line profile was obtained by μ-Raman spectroscopy (μRS) using 488 nm laser as an excitation source. μRS is a versatile method to analyze the residual remnant strain/stress in the Si not only qualitatively, but also quantitatively. μR spectra were recorded in backscattering mode on a confocal microscope equipped with a CCD detector using Ar⁺ 488 nm laser. The details regarding stress evaluation by μRS are described elsewhere [10].

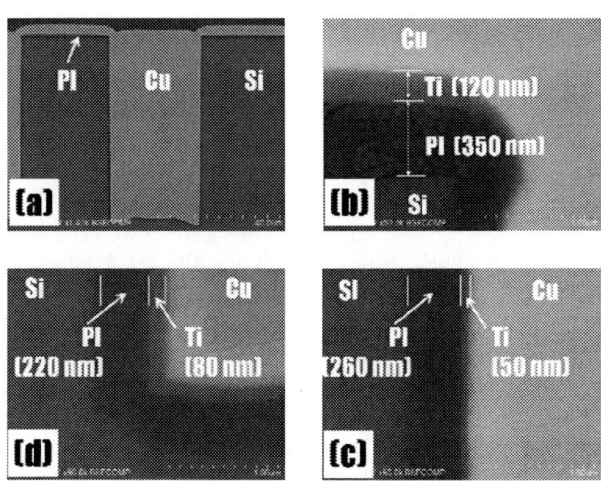

Figure 2: **Cross-sectional SEM images revealing the coverage ratios for PI liner, Ti barrier and Cu seed and electroplated layers.**

Table 1

Properties of PI®

Tg (°C)	> 360
CTE (ppm/K)	20
Tensile strength (GPa)	0.24
Young's modulus (GPa)	3
Dielectric constant ε	3.0
Dielectric strength (kV/mm)	276

Results and Discussion

Shown in fig. 2 are the scanning electron microscopic

images obtained on the cross-sectional sample containing Cu-TSVs, where in the liner material is TCVD-grown PI. Regardless of the aspect ratio (various a/r values used in this work ranges from 3 to 10), for the field thickness value of 350 nm, the coverage values of PI liner in via top-corner (fig. 2b), via middle (fig. 2c) and bottom-corner (fig. 2d) are found to be respectively, 260 nm, 260 nm, and 220 nm. Further optimization of TCVD process led to the coverage of PI at via bottom-corner as 75 to 80% and 100%, respectively for the via-diameter of <10 μm and >15μm.

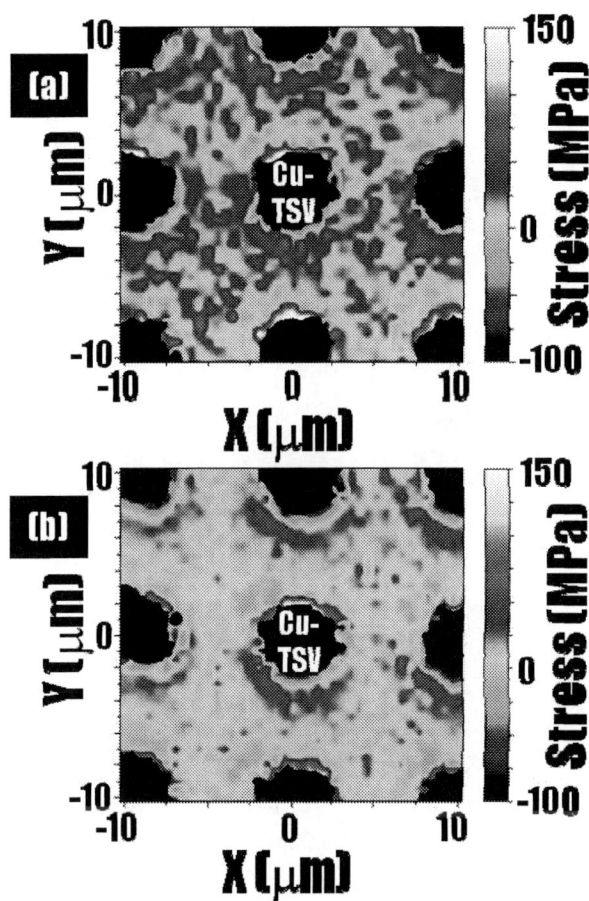

Figure 3: **2D-thermo-mechancial stress distribution surface profile obtained for the LSI die containing 5 μm-diameter Cu-TSVs with 500 nm-thick PI liner: as-filled state (a), and after annealing at 200 C for 30 min. (b)**

Shown in fig. 3(a) and 3(b) are the 2D-stress surface profile data for the LSI die with 5 μm-diameter Cu-TSVs containing 500 nm-thick PI polymer as dielectric liner obtained for the as-filled state and annealed at 200 C for 30 min. respectively. Shown in fig. 4 are the similar stress profile data obtained for the 5 μm-diameter TSV. For comparison, we have included the 2D-stress surface profile data taken for the conventional Cu-TSV in fig. 5. The ultra thin Si dies/wafers are highly susceptible to the induced thermo-mechanical stress caused by various 3D-integration processes. Such stress related issues become an important reliability concern in the 3D-LSI, since such stress in the package induces failures of electronic devices, e.g. die crack, wire break, etc. The main cause for the induced stress in the 3D-LSI Si wafers/dies is the co-efficient of thermal expansion mismatch between different materials involved in a package. In general, the thermo-mechanical stress in the 3D-LSIs can be classified into two types, (i) an interconnect induced stress (by the metals of the TSVs and microbumps) and (ii) the organic adhesives induced stress (injected in between the different layers of the stacked chips). By and large, the cure temperature of various organic adhesives used in the chip stacking process is well over the 150 °C. When the whole die cools down, the organic adhesives (epoxy material) shrink more than the Si die/wafer due to the large difference in CTE.

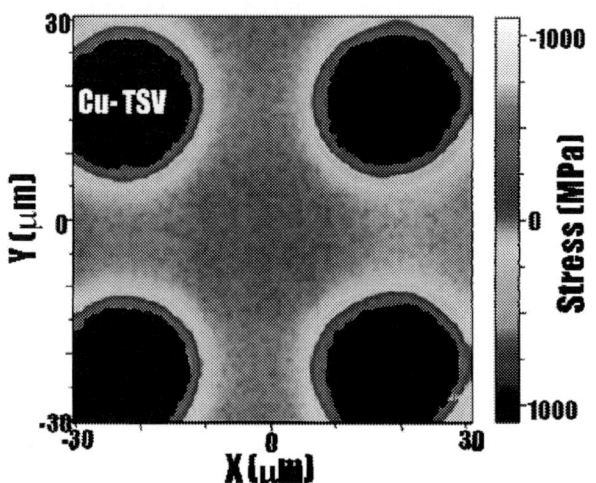

Figure 5: **2D-thermo-mechancial stress distribution surface profile for the LSI die containing conventional 20 μm-diameter Cu-TSVs with SiO$_2$ liner obtained after annealing at 200 C for 30 min.**

On the other hand in the conventional TSVs, the in-organic dielectric SiO$_2$ layer is sandwiched between the TSV metals and the bulk Si. The thermal expansion co-efficient (CTE) of Si (~3 x 10^{-6}/°C) significantly differs from the Cu CTE (~17 x 10^{-6}/°C). This large CTE mismatch induces strain in Si, which aggravates the reliability concern in 3D-LSIs after many cycles heating experiment. Therefore, it is worth to replace the liner that absorbs the thermal stress caused by the TSV metals. In figures 3 and 4, irrespective of the via diameter (φ=5 μm in fig. 3 and φ=30 μm in fig. 4), the Cu-TSV with TCVD-grown organic polyimide polymer as dielectric liner along the TSV side-wall induces about 50 MPa of compressive stress in the via-space region after annealing at 200 C for 30 min. Whereas in the as-filled state, a large tensile stress in the vicinity of TSV and relatively smaller tensile stress homogeneously distributed in the via-space region. However, the amount of compressive stress induced by the 20 μm-width conventional Cu-TSVs with SiO$_2$ liner is around 600 MPa and 100 MPa respectively in the vicinity of TSV and in the TSV-space region, as shown in fig. 5. Further, the compressive stress in the vicinity of TSV is followed by the tensile stress before turning into compressive stress in the TSV-space region. This large amount of induced thermo-mechanical stress by the Cu is largely absorbed by the soft organic layer in fig. 3 and 4. Therefore we were able to confirm stress reduction on the surface of LSI dies by replacing the in- organic SiO$_2$ line with the organic PI-based polymer.

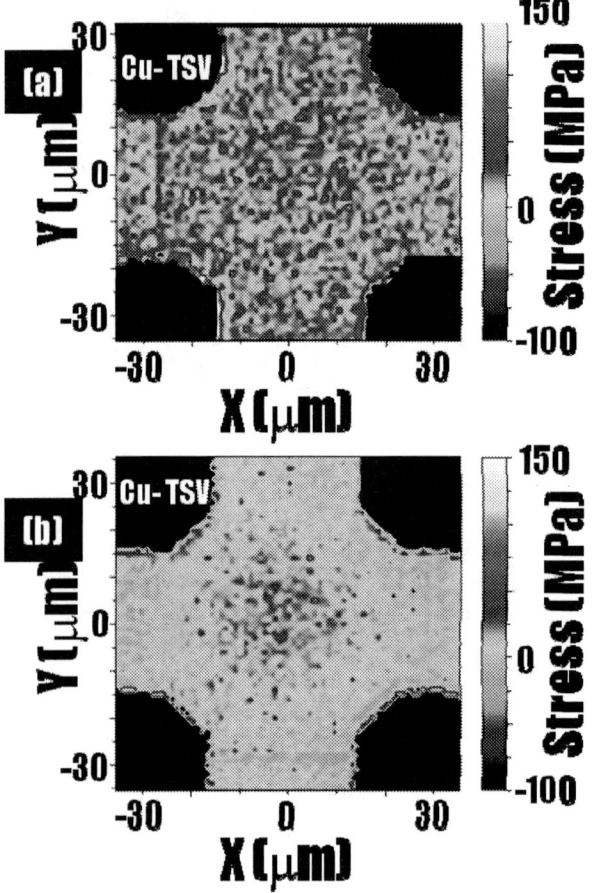

Figure 4: **2D-thermo-mechancial stress distribution surface profile obtained for the LSI die containing 30 μm-diameter Cu-TSVs with 500 nm-thick PI liner: as-filled state (a), and after annealing at 200 C for 30 min. (b)**

Line profile μ-Raman data (fig. 6) revealed that for TSV without polyimide liner, the stress values were around -350 MPa and -280 MPa for the as-filled state and the annealed one, respectively. Whereas in the case of TSV with stress-absorber, the compressive TM stress induced in the TSV space Si has been reduced to half (-210 MPa) and to one-third (-115 MPa) respectively for as-filled state and after ann. at 250°C for 20 min. It is that the much reduced thermo-mechanical stress in the Si induced by Cu-TSVs is owing to the low-modulus organic liner along the TSV sidewall which absorbs the CTE mismatch stress between the silicon of 3 ppm/deg and Cu of 17ppm/deg.

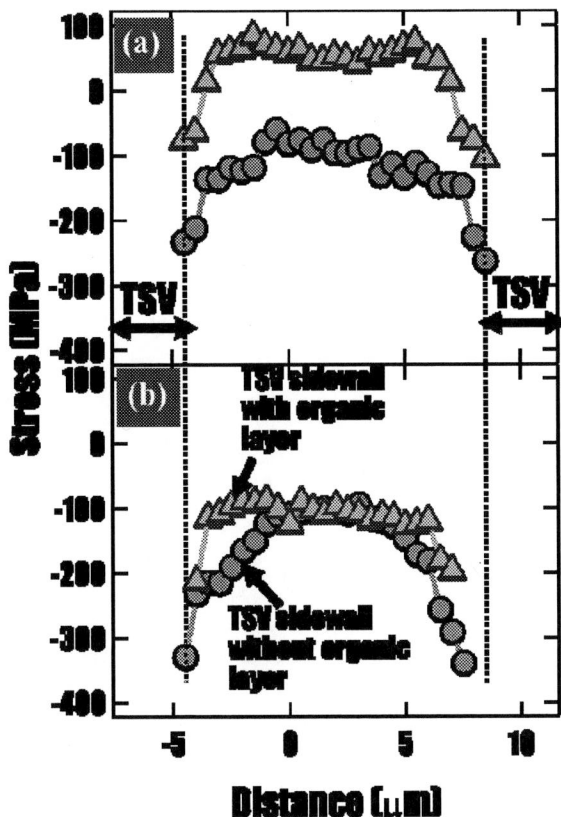

Figure 6: **Line stress profile obtained for the LSI die containing Cu-TSVs with (filled triangle) and without (filled circle) 500 nm-thick PI liner: after annealing at 200 C for 30 min. (a), and as-filled state (b).**

The heat resistance stability and the thermal reliability of polyimide liner in the TSV is evaluated respectively by the TGA, TDS and thermal cycle test for the interposer containing TSVs with 10 μm diameter and 20 μm pitch. The Si interposer underwent 1000 thermal cycles from -55°C to 125°C. From the TGA analysis it was found that TCVD grown polyimide is stable up to 400 °C. The mechanical characteristics such as Young modulus and hardness of the polyimide were studied by using nano-indentation technique for different film thicknesses. With increase in film thickness, we did observe a small decrease in the modulus which may be attributed to the incorporation of nitrogen as observed in the XPS results.

In order to confirm that the TCVD-formed polyimide is devoid of any contaminants, we have recorded the angle-resolved x-ray photoelectron spectroscopy analysis of C1s, O1s, F1s and N1s level for polyimide liner using Mg Ka X-rays was carried out for Irrespective of polyimide film thickness, the C1s core level spectrum revealed the presence

of carbonyl and imide ring structure. However, the intensity of N1s increases with an increase in the film thickness. As a reliability concern, in the event of the out-diffusion of this furious nitrogen in the long run that leads to a porous liner.

In conclusion, we successfully deposited a conformal polyimide liner along the 3um-30um-diameter TSV sidewall by TCVD and characterized for their dielectric and mechanical properties. Tremendous reduction of thermo-mechanical stress in the active Si caused by Cu-TSVs by replacing the conventional SiO2 liner with organic liner is proposed, tested and proved.

References:

[1] M. Koyanagi, T. Fukushima and T. Tanaka,"High-Density Through Silicon Vias for 3-D LSIs", *Proc. IEEE, 97*, 49 (2009).

[2] B. Swinnen; W. Ruythooren, P. De Moor; L. Bogaerts, L. Carbonell, K. De Munck, B. Eyckens, S. Stoukatch, D.S. Tezcan, Z. Tokei, J. Vaes, J. Van Aelst, E. Beyne, "3D integration by Cu-Cu thermo-compression bonding of extremely thinned bulk-Si die containing 10 μm pitch through-Si vias", *IEEE International Electron Devices Meeting Technical Digest,* pp.1-4 (2006).

[3] M. Murugesan, J.C. Bea, H. Kino, Y. Ohara, T. Kojima, A. Noriki, K.W. Lee, K. Kiyoyama, T. Fukushima, H. Nohira, T. Hattori, E. Ikenaga, T. Tanaka, and M. Koyanagi, "Impact of Remnant stress/strain and metal contamination in 3D-LSIs with through –Si vias fabricated by wafer thinning and bonding", *IEEE International Electron Devices Meeting*, pp. 361-364 (2009).

[4] M. Murugesan, J.C. Bea, T. Fukushima, T. Konno, K. Kiyoyama, W.C. Jeong, H. Kino, A. Noriki, K.-W. Lee, T. Tanaka, and M. Koyanagi, "Cu Lateral Interconnects Formed Between 100-μm-Thick Self-Assembled Chips on the Flexible Substrates", *Proc. 59th Electronic Components and Technology Conference.* pp.1496 (2009).

[5] M. Murugesan, H. Kino, H. Nohira, J.C. Bea, T. Fukushima, T. Tanaka, and M. Koyanagi, "Wafer thinning, bonding, and interconnects induced local strain/stress in 3D-LSIs with fine-pitch high-density microbumps and through-Si vias", *IEEE International Electron Devices Meeting*, pp. 30-33 (2010).

[6] M. Murugesan, H. Kino, H. Hashiguchi, C. Miyazaki, H. Shimamoto, H. Kobayashi, T. Fukushima, T. Tanaka, and M. Koyanagi, "High-Density 3D-LSI technology using W/Cu hybrid TSVs", *IEEE International Electron Devices Meeting*, pp. 30, 2010.pp. 139 2011.

[7] M. Murugesan, H. Kobayashi, H. Shimamoto, F. Yamada, T. Fukushima, J.C. Bea, K. W. Lee, T. Tanaka, and M. Koyanagi, "Minimizing the Local Deformation Induced around Cu-TSVs and CuSn/InAu-Microbumps in High-Density 3D-LSIs", *IEEE International Electron Devices Meeting*, pp. 657-660 (2012).

[8] D.S. Tezcan, F. Duval, H. Philipsen, O. Luhn, P. Soussan, B. Swinnen, "Scalable Through Silicon Via with polymer

978-1-4799-2408-0/14 $31.00 © 2014 IEEE

deep trench isolation for 3D wafer level packaging", *Proc. 59th Electronic Components and Technology Conference.* pp.1159 (2009).

[9] V. Sundaram Q. Chen, T. Wang, H. Lu, Y. Suzuki, V. Smet, M. Kobayashi, R. Pulugurtha, R. Tummala," Low cost, high performance, and high reliability 2.5D silicon interposer", *Proc. 63rd Electronic Components and Technology Conference, pp.* 342 (2013).

[10] M. Murugesan, H. Nohira, H. Kobayashi, T. Fukushima, T. Tanaka and M. Koyanagi, "Locally Induced Stress in Stacked Ultrathin Si wafers _ XPS and μ-Raman study", *Proc. 62nd Electronic Components and Technology Conference, pp.* 625-629 (2012).

Investigation of a TSV-RDL In-line Fault-Diagnosis System and Test Methodology for Wafer-level Commercial Production

Runiu Fang[1], Min Miao[2*], Xin Sun[1], Yunhui Zhu[1], Guanjiang Wang[3], Yichao Xu[3], Minggang Sun[2], Yufeng Jin[1]

1. National Key Laboratory of Science and Technology on Micro/Nano Fabrication, Peking University, Beijing, 100871, China.
2. Information Microsystem Institute, Beijing Information Science & Technology University, Beijing, 100101, China.
3. Peking University Shenzhen Graduate School, Shenzhen, 518055, China.
*Email: miaomin@ime.pku.edu.cn, miaomin@bistu.edu.cn , Phone: 86-10-62752536

Abstract

In the first three quarters of 2013, semiconductor industry witnessed a great multiplication of 12-inch TSV wafers mounting to 1 million plus scale. Despite this increasing popularity, TSV technology suffers from high cost due to yield loss caused by process defects. Poor insulation and connectivity are the major problems for TSV and RDL(Re-Distribution Line) structures. Without a cost-effective test system and methodology, the faulty TSVs may be stacked onto good ones and therefore bring forth an increasing chip cost. In this paper, the leakage current and pathway resistance are characterized for TSV and RDL structures, and a two-step in-line test methodology was proposed to pinpoint these defects and screen out KGDs (Known Good Dies) on the wafer. Also, a wafer-level fault-diagnosis system based on the proposed technology was built, including a probe station, an analyzer and a controller software, and two test instances were carried out on the test system. The test results demonstrated the capability of the test methodology and system, and proved the potential feasibility of the methodology for volume pre-bond test.

I. Introduction

Along with the rapid development of TSV technology, 3D IC and SiP integration are emerging as attractive alternatives to TSV-free schemes, namely die stacking with wire-bonding as vertical interconnects and package on package stacking. For high-end electronics, TSV technology is exploited to enhance performance and energy efficiency, as well as to reduce form factor and weight. For other applications, TSV technology is also capable of giving product additional advantages. For instance, by employing TSV scheme, the package size of CIS (CMOS Image Sensor) has been remarkably reduced, therefore valuable area is saved for other electronics in hand-held terminals[1, 2]. Through TSV interposers (2.5D IC integration), several smaller chips of limited capability are interconnected together to achieve a higher performance, and the inclusion of interposer will help to improve the overall thermal-mechanical stability for the package system[3].

Due to the complicated process flow of TSV technology (including RDL and micro-bump process), there are many kinds of potential defect mechanism related to TSVs. During the fabrication of TSVs, electroplating might induce voids, and increase possibilities of electrical opens and rise in resistance, while the poor deposition of oxide layer might precipitate shorts between TSVs. Also, flaws in removal of seed layer will cause wide-range shorts between TSVs and RDLs[4]. In order to avoid the case that defect-free dies or KGDs (Know Good Dies) will be bonded to the faulty ones,

the pre-bond test of the TSVs on wafer is necessary. Therefore, what the paper concerns is that how to pinpoint related defects during the manufacturing process, and screen out KGDs or at least dies with good TSV and RDLs for bonding, thereby reducing the yield loss.

Several solutions dedicated to pre-bond test have been explored. As the most widely used technology in device test, probe technology is able to directly test TSVs of large size and pitch(>20μm). The efficiency can be improved by utilizing probe cards. For small TSVs, however, probe tips need to be pressed on additional probe pads instead of TSV tips, because the current technology is insufficient to probe on vias of 5μm diameter and 10μm pitch(or smaller)[5]. The second solution is the on-chip DfT(Design for Testability) architecture. TSVs are accessed and evaluated through on-chip test circuits which are fabricated along with functional devices. The test circuits inevitably occupy relatively large die area, however the test efficiency is quite remarkable with a good DfT architecture [6-8]. The third possible solution is the employment of X-ray microscopy. Several tentative test results have shown that X-ray microscopy can provide a good profile imaging quality of copper pillars(TSVs), so plate-induced voids can be diagnosed instantly with the technique[9, 10]. However, the limited image resolution failed to pinpoint defects embedded in the oxide layer.

Obviously, probe technology combined with electric instrumentation has sustainable competitiveness with TSVs of large size and pitch compared with other pre-bond test solutions. And we should note that even with the trend of miniaturization, middle to large sized TSVs still enjoy popularity in many TSV-inside volume products. The rest of this paper is organized as follows. Section II investigates the TSV defects and proposes a two-step in-line test methodology dedicated to pinpoint the defects based on probe technology. Section III presents a TSV-RDL fault-diagnosis system intended for commercial production, to implement the proposed methodology and two test instances are carried out on the system. Finally, section IV concludes the paper.

II. The TSV defects and test methodology

A. Defects in TSV and RDL formation process and test objectives

The most representative manufacturing process for TSV w/o RDL is illustrated in figure 1. There are five key steps before wafer thinning, namely (1)via etching,(2)SiO_2 deposition,(3)barrier and seed layer deposition,(4)Cu electroplating and (5)CMP (chemical and mechanical polishing) to remove copper overburden. In the above steps, quality control of SiO_2 deposition and Cu electroplating still remains a challenge, especially for high aspect ratio TSVs.

978-1-4799-2408-0/14 $31.00 © 2014 IEEE

2014 Electronic Components & Technology Conference

The pinhole effect and unconformal deposition on sidewalls may lead to poor insulation. The pinch-off during electroplating and crack between a TSV and RDL combination may lead an increased resistance, and high electromigration risk or even an open. In spite of various defect mechanisms, it is noticed that the electrical effects of the defects all falls into two categories: poor insulation and poor connectivity. The poor insulation results in high leakage current, and the poor connectivity results in high pathway resistance. They are both fatal to TSV applications.

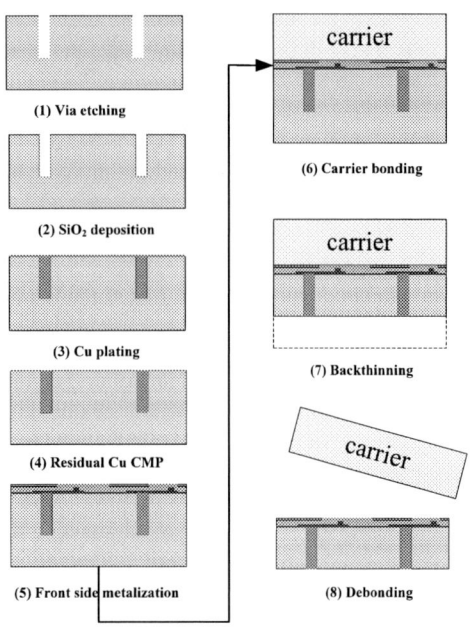

Figure 1 General process flow of TSV w/o RDL

B. Characterization of leakage current

In order to evaluate the quality of insulation, voltage sweeping test is usually applied to characterize leakage current between neighboring TSVs[11]. For TSV pairs with qualified oxide layers (with the thickness of 1μm), the equivalent resistance between TSVs should remain stable under a relatively low voltage(<10V) based on our measurement. But the magnitude of the resistance varied with oxide process and conformality of deposition. When voltage keeps increasing, the electric field in the thinner part of oxide layer ascends, combined with the acceleration effect of defects, i.e. particles and pin-holes, the breakdown occurs at a certain voltage[12]; at the same time, the l-V curve rises up almost vertically.

Figure 2 shows the test configuration of the leakage current test. The probe needles are probed in a coplanar way on adjacent TSV pads. Based on the measured I-V curves of neighboring TSVs, we concluded the leakage current test results (with extremely high voltage, i.e. 100V in our case, and the current limit is set to 100μA to avoid wafer overheating) into two types. The typical breakdown I-V curve is illustrated in figure 3(a), in which the abrupt rise of the curve signifies the breakdown. The breakdown is irreversible. If we re-test the TSV pairs that have achieved breakdown, we will not get a linear I-V curve, but curves illustrated in figure

3(b). A possible explanation for this is the breakdown of oxide makes TSV copper partially contact with the silicon, which generates the Schottky-like junction, and the I-V curve under this condition exhibits property of Schottky I-V curve(Current rises exponentially with the applied voltage). In order to differentiate two types of I-V curves illustrated in figure 3, the logarithmic current coordinate is applied in figure 4 for the same 5 TSVs. For TSVs that have just reached the breakdown, the steep rise of the slope is observed at the breakdown voltage, while for those have reached breakdown in the test before, the corresponded slopes of I-V curves are gradual and stable, as shown in figure 4(a) and (b) respectively.

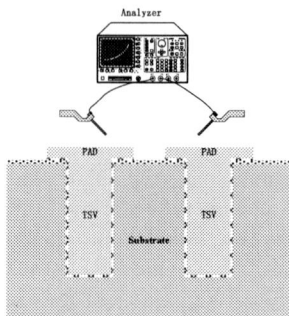

Figure 2 Test configuration of leakage current test

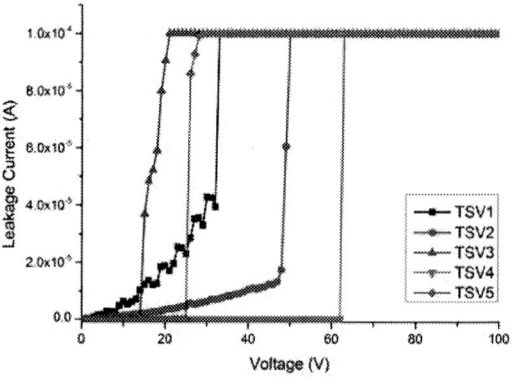

(a) Breakdown I-V curves of five TSV pairs

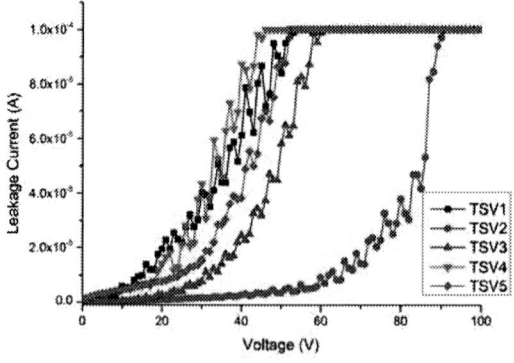

(b) Schottky-like I-V curves of five TSV pairs

Figure 3 two types of leakage current curves

978-1-4799-2408-0/14 $31.00 © 2014 IEEE   642

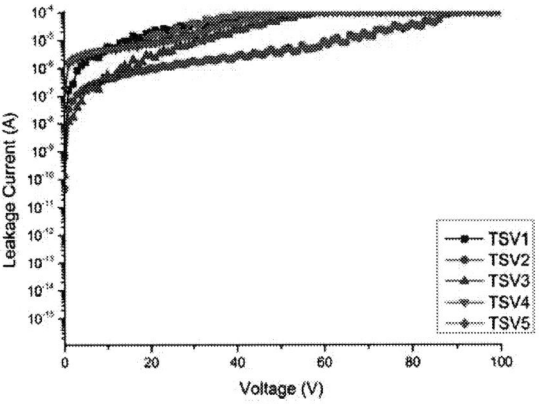

(a) Breakdown I-V curves in logarithmic current

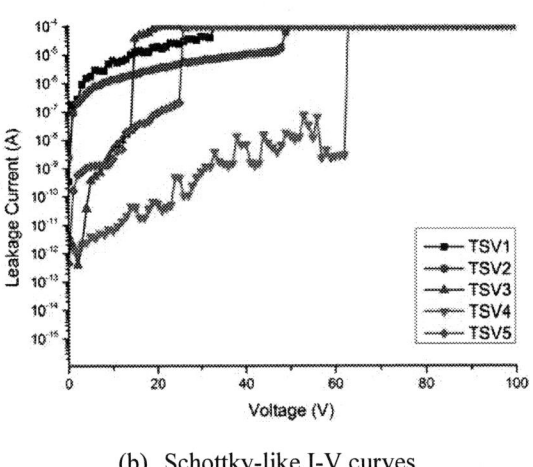

(b) Schottky-like I-V curves

Figure 4 two types of leakage current curves in logarithm

In some uncommon cases, two steep rises are observed in one I-V curve as shown in figure 5, which signifies two separate breakdowns in the TSV pair. This situation is explained as follows. The TSV pair includes two intact TSVs with possibly different breakdown voltages. With the increasing voltage, the breakdown of the weaker one of the two TSV is first achieved, then the other breakdown. However, the voltage gap between the two breakdowns is usually small. There are two reasons for this. The first one is that the breakdown voltages should be the same between adjacent TSVs because of limited nonuniformity of manufacturing process. The second one is that, after the first breakdown, the bias voltage is almost entirely applied on second oxide layer, and the abrupt rise of voltage will soon breaks down the second oxide layer.

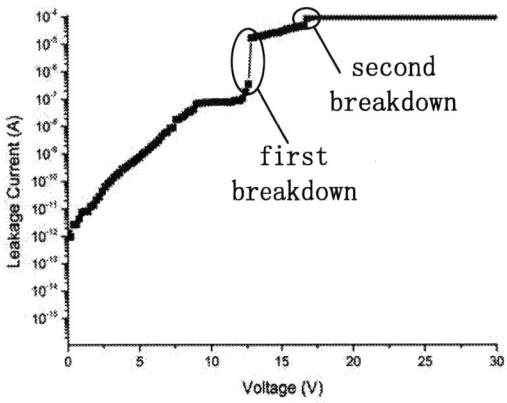

Figure 5 the double breakdown I-V curve in a TSV pair

Another uncommon and abnormal type of I-V curve is illustrated in figure 6. It clearly suggested that there are alternative short path between TSVs. This typically implies the selective removal of seed layer is ineffective.

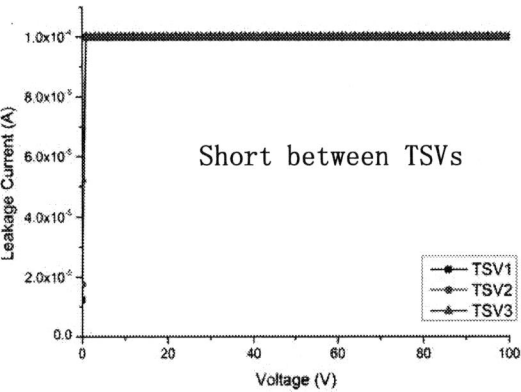

Figure 6 I-V curve between shorted TSVs

C. Characterization of pathway resistance

For a given manufacturing process, the resistance of TSVs and RDLs is determined by their structural parameters, namely diameter and height of TSVs, and length, width and thickness of RDLs. Kelvin method is usually applied to characterize resistance of the single TSV or RDL structure. A series of RDLs of different structural parameters were measured with Kelvin method and the results are shown in table 1. The resistance of a conductor is proportional to its length, and inversely proportional to its cross-sectional area. Therefore, if we set a TSV or RDL as a benchmark whose resistance as well as structural parameters are known, then any resistance of TSV or RDL of different structural parameters but fabricated with the same manufacturing process, should be determined. While in table 1, K01 was set as the benchmark for RDLs. Based on this method, theoretical values of other RDLs are calculated. The results demonstrated that the ratio difference between theoretical and measured value is less than ±10%.

978-1-4799-2408-0/14 $31.00 © 2014 IEEE

Table 1 The RDL resistance of different structural parameters

	Length(μm)	Width(μm)	Thickness(μm)	Measured value(Ω)	Theoretical value(Ω)	Measured/Therot cial
K01	500	10	3	0.458	Benchmark	100%
K02	500	20	3	0.222	0.229	96.9%
K03	500	30	3	0.144	0152	94.7%
K11	500	10	2.8	0.480	0.490	97.9%
K12	500	20	2.8	0.246	0.245	100.4%
K13	500	30	2.8	0.169	0.154	109.7%

D. The two-step in-line test methodology

In the general manufacturing process of TSV wafer, blind vias are fabricated in the first place, then followed by temporary bonding and backthinning process, after which the via backside is exposed, as shown in figure 1. The wafer with certain thickness can provide critical mechanical strength during the process. However, most height of TSVs are no more than 100μm (varied with diameters and other factors), while the thickness of a bare wafer is over 200μm. Therefore, TSVs are actually formed as blind vias before becoming through vias. Based on this conception, we proposed a 2-step in-line test methodology to identify defects during the process as quick as possible, and to screen out KGDs for the next process. Figure 7 illustrates the test methodology. After electroplating and residual CMP process, the TSV wafer is mounted to probe station and measured with leakage currents. The equivalent resistance of oxide layers in TSV should range from 1M to 100G[13]. A 10V stimulus is applied to the TSV with the substrate grounded. If the leakage current is less than 10μA (equivalent to 1M ohm in resistance), the TSV under test will be marked as qualified. After the automatic test on the whole wafer with such ATE-like technology, all TSVs are characterized with their leakage current test with all dies of poor insulation pinpointed. The number of defective dies is then calculated, to determine whether it is still cost-effective for the TSV wafer to proceed to the next process step. If the percentage of defective dies among total dies is relatively high, the TSV wafer should be abandoned; meanwhile, if the percentage is below the given threshold (which should be based on aspects of process quality and cost control), the wafer should proceed.For via-middle and via-last process, the subsequent process is BEOL metallization or top RDL fabrication, which wires TSVs to devices. Then temporary bonding and backthinning are carried out. As discussed in section II-C, the resistance of the TSV-RDL combination is determined by structural parameters. Thus, for every TSV-RDL combination, the theoretical value is calculated based on the benchmark resistance, and compared with the measured one. If the difference between the two is within ±20%, the pathway should be considered defect-free. While voids in TSV or crack between the TSV and RDL are expected with a relatively high measured resistance, and ineffectiveness in seed layer removal comes along with a relatively low measured resistance. However, for many TSV applications, TSVs are not looped together in one stack. Also, double-sided probing is expensive and problematic. So, even with the backside of TSV exposed, it is still hard to test the connectivity of a given TSV. Therefore, for the TSV wafer without RDLs, the conductive carrier wafer should be used for temporary bonding. With the help of the conductive wafer, all TSVs are testable on the backside with their resistance. After the pathway resistance test, all TSVs w/o RDLs with poor connectivity are pinpointed. Until now, the two-step test has screened out all defect-free dies, and the defective dies should be abandoned after dicing.

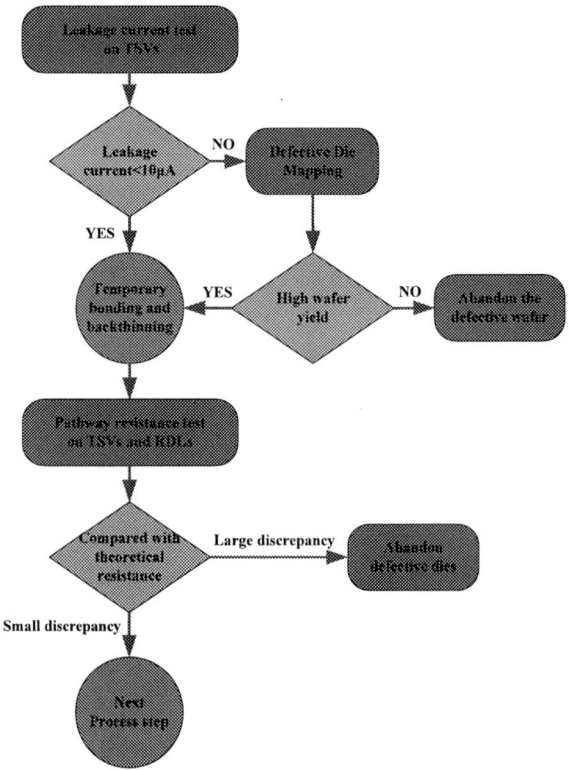

Figure 7 The flow diagram of the two-step in-line test methodology

III. The test system and test instances

In order to implement the proposed test methodology, we built up the test system including an automatic probe station and a semiconductor parameter analyzer. Besides, a controller software is developed to control the probe station and analyzer based on GPIB protocol. Through retrieving test data from the analyzer, the controller determines the existence of defects and maps all defective dies. Figure 8 shows the picture of the test environment and software controller interface.

In order to demonstrate the capability of the test methodology and test system, two test instances were

978-1-4799-2408-0/14 $31.00 © 2014 IEEE

performed. The first test instance is the TSV embedded passive interposer served as the integrator of 4 identical SRAM chips. Figure 9 shows the layout of the interposer. TSVs are located outside the ground and Vdd rings, and connected to top RDLs which are represented in red and blue lines in the layout. The related parameters as listed in table 2.

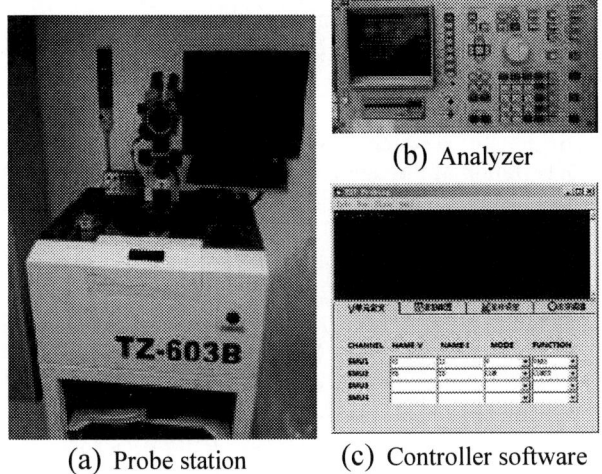

(b) Analyzer

(a) Probe station

(c) Controller software interface

Figure 8 Test system and software

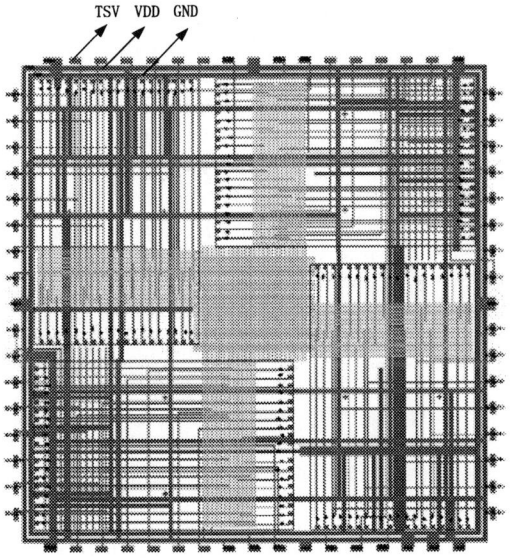

Figure 9 layout of the TSV interposer

Table 2 information about the TSV interposer wafer

Wafer size(mm)	150mm
Die size(mm)	17mm*17mm
Dies per Wafer	44
TSVs per Die	136
TSV diameter(μm)	100
TSV pitch(μm)	800

After the plating and copper CMP process, the interposer wafer was mounted to the test setup. The leakage current test was carried out automatically with all TSVs on more than half of total dies. The count of TSVs whose leakage current surpassed 10μA in a die is marked in figure 10. Most dies have defective TSVs based on the test results. Thus, the feasibility for proceeding process of this wafer should be reconsidered. In fact, after a more detailed leakage current measurement, it was found that part of the seed layer had not been effectively removed, which caused the short between TSVs in some dies.

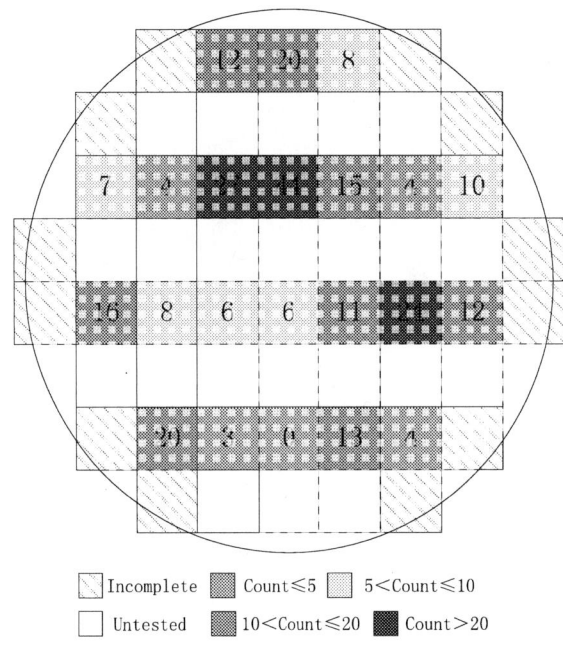

Figure 10 the mapping results of leakage current test

The second test instance was an 8 inch TSV wafer fabricated by a medium sized semiconductor design and manufacturing company. The potential application for the TSV wafer is CMOS image sensor. The wafer was thinned before via etching, thus we can measure the leakage current as well as pathway resistance all in one test. Before the resistance measurement, the benchmark resistance should be set up. Figure 12 shows the Kelvin structure on the wafer for TSV resistance characterization, and the test results are shown in table 3. The TSV has a relatively large resistance as it has an annular copper outside and insulated void inside. Because of some accidents during the manufacturing process, the wafer was not expected to own good quality. Still, the test was carried out, and the results revealed that most TSVs (over 80%) are not qualified, with either high leakage current or high pathway resistance.

Through the two test instances, the test system and methodology are demonstrated to be capable of pinpointing defective TSV dies, and proved of potential feasibility for volume pre-bond test.

978-1-4799-2408-0/14 $31.00 © 2014 IEEE

Figure 11 The 8 inch TSV wafer

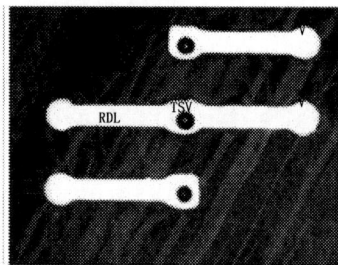

Figure 12 Kelvin structure for TSV measurement

Table 3 the measured TSV characterized with Kelvin method

Kelvin TSV	Resistance(Ω)	Pearson's r
K1	7.707	0.999982362
K2	7.730	0.999982525
K3	8.249	0.999635418
K4	7.882	0.999988966
K5	8.037	0.999898181
AVERAGE	7.921	-

IV. Conclusions

In the manufacturing process of TSV wafers, pre-bond test is usually necessary to avoid yield loss caused by faulty dies stacking on to KGDs. Based on the characterization of leakage current and pathway resistance of TSV and RDL structures, this paper proposed a two-step in-line test methodology based on probe technology, which was dedicated to pinpoint TSV-related defects during the manufacturing process as quick as possible, and to screen out all KGDs before die stacking.

In order to implement the methodology, a wafer-level fault-diagnosis system was built up, including a probe station, an analyzer and a developed controller software. Two test instances were performed, with the capability of the test methodology and system demonstrated. The results proved the potential feasibility of the methodology for volume pre-bond test.

Acknowledgments

The work presented is co-funded by National Science and Technology Major Project of China (Project No. 2009ZX02038), National Natural Science Foundation of China (Project. No. 61176102 and 60976083), Funding Project for Academic Human Resources Development in Institutions of Higher Learning Under the Jurisdiction of Beijing Municipality (Project. No. PHR201108257, "Research on technical basics for three-dimensional system-in-package") .

References

1. Coudrain P, Henry D, Berthelot A, et al. 3D Integration of CMOS image sensor with coprocessor using TSV last and micro-bumps technologies[C]//Electronic Components and Technology Conference (ECTC), 2013 IEEE 63rd. IEEE, 2013: 674-682.
2. Sekiguchi M, Numata H, Sato N, et al. Novel low cost integration of through chip interconnection and application to CMOS image sensor[C]//Electronic Components and Technology Conference, 2006. Proceedings. 56th. IEEE, 2006: 8 pp.
3. Zhang, X., T. Chai, J. H. Lau, C. Selvanayagam, K. Biswas, S. Liu, D. Pinjala, G. Tang, Y. Ong, S. Vempati, E. Wai, H. Li, B. Liao, N. Ranganathan, V. Kripesh, J. Sun, J. Doricko, and C. Vath, "Development of Through Silicon Via (TSV) Interposer Technology for Large Die (21x21mm) Fine-pitch Cu/low-k FCBGA Package", IEEE/ECTC, May, 2009, pp. 305-312.
4. Marinissen E J, Zorian Y. Testing 3D chips containing through-silicon vias[C]//Test Conference, 2009. ITC 2009. International. IEEE, 2009: 1-11.
5. Noia B, Chakrabarty K. Pre-bond probing of TSVs in 3D stacked ICs[C]//Test Conference (ITC), 2011 IEEE International. IEEE, 2011: 1-10.
6. Lou Y, Yan Z, Zhang F, et al. Comparing through-silicon-via (TSV) void/pinhole defect self-test methods[J]. Journal of Electronic Testing, 2012, 28(1): 27-38.
7. Cho M, Liu C, Kim D H, et al. Design method and test structure to characterize and repair TSV defect induced signal degradation in 3D system[C]//Proceedings of the International Conference on Computer-Aided Design. IEEE Press, 2010: 694-697.
8. Chen P Y, Wu C W, Kwai D M. On-chip TSV testing for 3D IC before bonding using sense amplification[C]//2009 Asian Test Symposium. 2009: 450-455.
9. Sekhar V N, Neo S, Yu L H, et al. Non-destructive testing of a high dense small dimension through silicon via (TSV) array structures by using 3D X-ray computed tomography method (CT scan)[C]//Electronics Packaging Technology Conference (EPTC), 2010 12th. IEEE, 2010: 462-466.
10. Sueoka K, Yamada F, Horibe A, et al. TSV Diagnostics by X-ray Microscopy[C]//Electronics Packaging Technology Conference (EPTC), 2011 IEEE 13th. IEEE, 2011: 695-698.
11. MIAO M, XU Y, WANG G, et al. An In-line Test Method for the TSV Insulation Integrity[J]. Journal of Test and Measurement Technology, 2012, 6: 002.
12. Lee J C, Ih-Chin C, Chenming H. Modeling and characterization of gate oxide reliability[J]. Electron Devices, IEEE Transactions on, 1988, 35(12): 2268-2278.
13. Tsai M, Klooz A, Leonard A, et al. Through silicon via (TSV) defect/pinhole self test circuit for 3D-IC[C]//3D System Integration, 2009. 3DIC 2009. IEEE International Conference on. IEEE, 2009: 1-8.

Bonding Technologies for Chip Level and Wafer Level 3D integration

Katsuyuki Sakuma[1,2], Spyridon Skordas[3], Jeffrey Zitz[1], Eric Perfecto[1], William Guthrie[4], Luc Guerin[5], Richard Langlois[5], Hsichang Liu[1], Koushik Ramachandran[1], Wei Lin[3], Kevin Winstel[3], Sayuri Kohara[6], Kuniaki Sueoka[6], Matthew Angyal[1], Troy Graves-Abe[1], Daniel Berger[1], John Knickerbocker[2], and Subramanian Iyer[1]

[1]IBM Semiconductor Research & Development Center, 2070 Route 52, Hopewell Junction, NY, 12533, U.S.A.
[2]IBM T. J. Watson Research Center, 1101 Kitchawan Road, Yorktown Heights, NY, 10598, U.S.A.
[3]IBM Semiconductor Research & Development Center, 257 Fuller Road, Albany, NY 12203, U.S.A.
[4]IBM Microelectronics Division, 1000 River St, Essex Junction, VT 05452, U.S.A.
[5]IBM Canada, 23 Blvd de L'Aeroport, Bromont QC J2L 1A3, Canada
[6]IBM Research Tokyo, Shinkawasaki, Kanagawa, 2120032, Japan
Email: ksakuma@us.ibm.com Tel: 1-845-894-5605

Abstract

This paper provides a comparison of bonding process technologies for chip and wafer level 3D integration (3Di). We discuss bonding methods and comparison of the reflow furnace, thermo-compression, Cavity ALignment Method (CALM) for chip level bonding, and oxide bonding for 300 mm wafer level 3Di. For chip 3Di, challenges related to maintaining thin die and laminate co-planarity were overcome. Stacking of large thin Si die with 22 nm CMOS devices was achieved. The size of the die was more than 600 mm^2. Also, 300 mm 3Di wafer stacking with 45 nm CMOS devices was demonstrated. Wafers thinned to 10 μm with Cu through-silicon-via (TSV) interconnections were formed after bonding to another device wafer. In either chip or wafer level 3Di, testing results show no loss of integrity due to the bonding technologies.

1. Introduction

Three-dimensional integration (3Di) offers benefits such as increased packaging density, heterogeneous integration, lower power, and high bandwidth due to shorter interconnect length [1,2,3]. Some key technologies needed to enable 3Di include the formation of through-silicon-vias (TSV), wafer bonding and wafer thinning. 3D integration can include a Si interposer, which is used to redirect circuitry between a chip carrier and one or more top chips [4,5]. High-precision alignment, positioning and bonding are required for the fabrication of various 3Di devices, especially for fine-pitch 3D. Interconnect density and bonding technology requirements are different between chip level and wafer level 3Di. The schematic cross-sections for chip level 3D and wafer level 3D are shown in Figure 1. Features of 3Di technologies for both chip level and wafer level are shown in Table 1.

Chip level 3Di has major challenges such as stacking, throughput, and warpage control of large thin Si die on an organic laminate. In chip level 3Di, the dice have to be aligned and placed on a carrier. The assembly cost depends on the assembly speed, required chip alignment precision, and other factors. Generally, one by one sequential die assembly is slower and more expensive than batch assembly process, however die pick and place of multiple die can aide productivity. The authors previously reported CALM, in which cavities were used as stacking templates for chips on a wafer [6]. CALM has good potential for effective,

inexpensive, and high throughput 3D fabrication. This technology greatly improves stacking throughput, while maintaining good bond alignment. New results are reported in this approach with a template of square cavities, in which chips were aligned and joined in a batch bonding step. Stacking experiments of 70 μm thick TSV chips were performed and in-plane positions of the solder joints were measured relative to die edges before and after stacking to evaluate alignment accuracy. The results show precise chip alignment accuracy with 2.0 μm (standard deviation of misalignment) for 3-layer stacking by CALM.

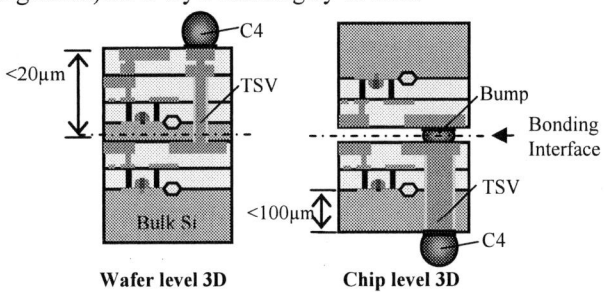

Figure 1. Schematic cross-section of chip and wafer level 3D.

Warpage control at bonding is a challenge in chip level 3Di [5,7,8]. A large interposer is required to achieve high speed signal transmission for large top die and/or multiple die applications. Such interposers present significant handling issues at bonding when joining a warped interposer on top of a warped laminate using known techniques. Successful assembly can be obtained with new bonding technology using existing tools, such as furnace reflow and thermo-compression (TC) bonder. Furnace reflow is a traditional method for 2D assembly which offers a high throughput batch process. With this technique, it is possible to join a thin, small interposer with a common pitch to a laminate and then to join another die with a finer pitch to the interposer. Although throughput is slower compared to furnace reflow, TC bonding can achieve fine-pitch bonding of warped components as it features placement accuracy of about ±2 μm. Also, C4 stresses and chip warpage can be reduced by differential heating/cooling chip join process using TC bonding [9]. In this paper, the bonding challenges for large die packaging are discussed. Finally, 3D chip level integration technology has been demonstrated with large Si die in 22 nm CMOS technology.

Wafer-scale bonding promises significant benefits for achieving 3Di, such as micro-scale interconnects (IC) with lower IC delay, lower power consumption, and very high data bandwidth due to the inherent higher IC density by wafer-scale processing [10]. Also, highest throughput is possible as chips are bonded in parallel and singulated later. Therefore significant cost reduction and high volume production in 3D application space can be realized. The main challenge in wafer scale 3Di is with respect to stack yields, as known good dies (KGDs) are not selectable therefore the resulting final yield is dependent on individual wafer yields and the bonding process yields. Therefore, wafer-scale 3Di may not be suitable for stacking chips of dissimilar size and with different IC pitches or for larger chip sizes which may produce very low stack yields with this approach.

Wafer-scale 3Di involves bonding layer preparation and permanent bonding of wafers requiring precise wafer level alignment. Wafer bonding is followed by thinning and formation of TSV Cu ICs to establish interconnection. With minimal additions to equipment fleets, existing fab infrastructure can be used for such integration if a suitable bonding approach is selected. Traditional solder based integration is not suitable for this technique, as the wafers are subjected to high temperature processes for backside processing and TSV formation after wafer stacking. The final chip stack can be processed as a single chip similar to 2D integration.

In this paper we report on successful wafer level 3Di by stacking of 300 mm wafers with 45 nm CMOS devices in a face-to-back scheme. This prototype emulates a processor-cache structure [11]. Overall the devices were not affected by the bonding process and the TSV interconnection formation. Key device properties and the inter-layer communication speed are reported to this effect.

Table 1. Features of 3D integration technologies.

	Chip level integration	Wafer level integration
Alignment required	Chip size	Wafer scale
Yields	High (KGD)	Depends on wafer yields Chip repair @ test is benefit
Bonding technology	Furnace reflow / TC / CALM	Oxide bonding
Throughput on stacking	High (CALM) Middle (Furnace reflow) Low (TC)	High
TSV & bump size	Large	Small
TSV formation	Before bonding	After bonding
Si thickness	< 100 µm	< 20 µm
Suitable for	Both common and different size (Furnace reflow, TC), Common size (CALM)	Common size

2. Comparison of bonding technologies

In this section we present several bonding technologies such as metal-metal bonding, SiO$_2$ bonding, and Hybrid bonding. Each approach comes with advantages and limitations. A summary comparison of bonding technologies is shown in Table 2.

2.1 Metal – metal bonding

This involves the formation of metal micro-bumps, pillars or studs on the bonding surface of each chip/wafer and bonding that establishes direct electrical connection. Various approaches for forming the metallic protrusions can be used.

In one example, there is C4 (controlled collapse chip connection) formation that utilizes the compositions of Sn based solder metallurgy [12]. This process has been used in manufacturing for decades [13]. It has important advantages, such as direct electrical connection, low soldering temperature, a self-aligning effect at bonding, excellent wetting behavior, good corrosion resistance, etc. It typically requires underfill to ensure mechanical support, especially for interconnections with larger coefficient of thermal expansion mismatch. However it suffers from limited scalability for wafer stacking due to the thermal instability of the bonds versus needed downstream processing. It is suitable only for chip level 3Di with large bump pitch ICs.

Another metal-metal bonding method involves the use of intermetallic compound (IMC) bonding [7,14]. In this example, a higher melting point metal (i.e., Cu) is combined with a lower melting point metal (i.e., In, Sn). The metals are plated or injected using IMS (Injected Molten Solder) in layers to form the bond structures. The lower melting point metal is placed between the higher melting point metal pillars. Once such structures are formed, a thermo-compression step is applied to promote diffusion of the low melting point metal into the higher melting point metal, with which it forms IMC states. The IMC melting point is higher than the standard solder reflow temperature. This is a desirable feature, because high thermal stability is needed to allow repetition of bonding processing and wafer processing for 3D integration process.

Another example is Cu-Cu TC bonding. This approach has the advantage of direct electrical connection, high thermal conductance and low electrical resistivity. However, bonding alignment needs to be very accurate, especially for small critical dimension (CD) features. Also, the thermal and compression stress can be a concern for process yield [15,16].

2.2 SiO$_2$ bonding

In SiO$_2$ bonding, wafers feature surface oxide films that are suitable for bonding. The bonding surfaces are activated by plasma exposure, then the hydrophilic surface is typically cleaned/terminated by use of a water rinse. The wafers are aligned, positioned and bonded after minimal initial contact is made between the wafers [17]. Oxide bonding adds flexibility in overcoming topography. Bonding alignment is accurate and stable as initial bonding, which locks the alignment, is at room temperature, followed by a thermal annealing to strengthen the bond. Besides throughput advantages, this technique avoids high thermo-mechanical stresses as experienced with TC bonding. Furthermore, as it does not establish direct electrical connections, it does not present a reliability challenge as much as metal-metal or hybrid bonding, where misalignment at bonding or insufficient underfill can result in Cu poisoning, migration, or diffusion. Lastly, oxide bonding is the most amenable to scaling the CD and pitch of 3D ICs via TSV scaling, on the order of 1 µm

978-1-4799-2408-0/14 $31.00 © 2014 IEEE 648

[11,17]. Key wafer to wafer integration challenges for good quality bonding, interconnection, high yield for future manufacturing include control of oxide films composition, planarity control without particles or contamination, precision alignment and productivity enhancements for cost reduction.

2.3 Hybrid bonding

This bonding requires special surface preparation such as chemical mechanical polish (CMP), plasma and/or wet cleans for surface cleaning and surface termination/activation. Typically, low surface topography and atomically smooth roughness is required. This approach features the advantage of direct electrical connection and high bond strength. However, the scaling of interconnections CD and pitch can be more challenging compared to SiO_2 bonding. Additionally, the thermal and compressive stresses can be a concern for process yield [18,19].

Table 2. Comparison of bonding technologies.

Bonding type	Metal bonding			SiO2 bonding	Hybrid bonding (metal and adhesive)
	C4 bonding	IMC bonding	Cu-Cu bonding		
Bonding temperature	~260°C	~260°C	~400°C	R.T.	~300°C
Heat tolerance	<260°C	<450°C *	<1084°C	<1400°C	<400°C *
Connectivity	Mechanical and Electrical	Mechanical and Electrical	Mechanical and Electrical	Mechanical	Mechanical and Electrical
Interconnection pitch	Low	Middle	Middle	High	Middle
Chip level applicability	High	High	Medium *	NA	Medium *
Wafer level applicability	Low	Medium *	Medium *	High	Medium *
Concerns	Low heat tolerance, large I/O pitch	IMC thickness control, reliability	Thermal & compression stress, flatness, cleaning	Flatness, voids/particles	High bonding pressure, processing integrity, flatness

* Different depends on the process, target application

3. Chip level 3D integration

3.1 Addressing throughput

One of the challenges with sequential stacking and bonding of chips in 3Di is low process throughput. In this study, we used the CALM process [6] to improve throughput for chip level integration. We fabricated chip stacks to evaluate alignment accuracy and thermal reliability characteristics of the CALM process. Figure 2 shows the fabrication flow. First, we cut test chips to proper size for the template cavity. We used a commercial wafer dicing machine with a diamond blade. During the dicing process, we continuously adjusted both the blade's cutting edge and the cut width for uniform chip dimensions. This step is important, as it defines stacking alignment accuracy. Next we aligned the cavity template at a predefined position on the wiring silicon substrate using an optical microscope with an X-Y-θ stage. The cavity template with a thickness of 0.5 mm was made from a metal plate with a small Coefficient of Thermal Expansion (CTE). The template has square cavities, in which the chips are aligned. After aligning on the Si substrate, the template was brought into contact with the substrate, and was fixed with a removable adhesive. Then, the precisely diced chips were placed into the cavities. Finally, the chips were pressed into the cavity. Once all chips were placed in proper cavities, they were joined in a batch bond step. To compensate for differences in stack heights in each cavity, we used Si spacers on top of the chips. The stacked chips were heated above and below through the spacers and the stage heater. Typical temperature and applied force were 250°C and 10 N, respectively. After removing the template from the substrate, the cavity stacking process was complete and the stacks were tested.

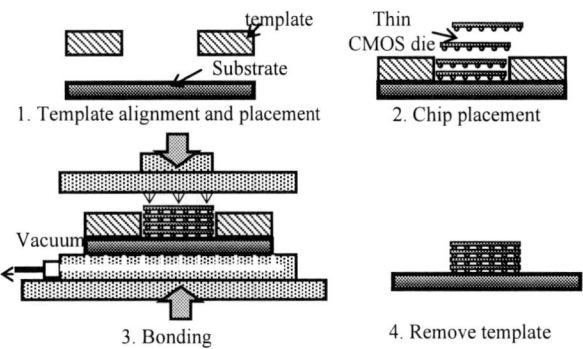

Figure 2. Cavity Alignment Method process flow.

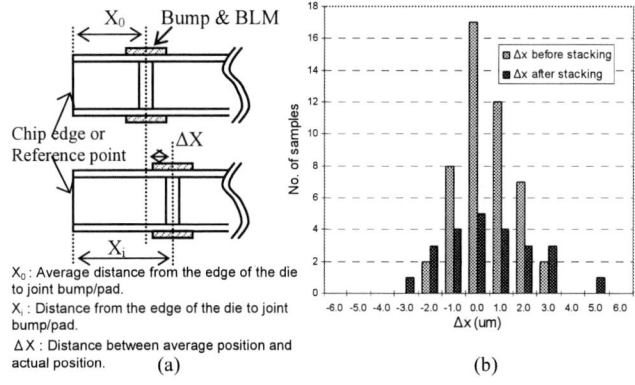

X_0 : Average distance from the edge of the die to joint bump/pad.

X_i : Distance from the edge of the die to joint bump/pad.

ΔX : Distance between average position and actual position.

Figure 3. Alignment of die cavity technology: (a) Schematic drawings of the dies showing the deviation Δx, (b) Histogram of the deviation from the average position of the bump before and after 3-layer stacking.

CALM Alignment accuracy

The in-plane positions of the solder joints relative to the die edges were measured before and after stacking to evaluate alignment accuracy, by using an optical microscope with digital position readout for these measurements. Measurement accuracy is estimated as ±1 μm. Prior to stacking, the cavity and chip sizes as well as the relative distance (x_i) from the die edges to joint bump/pads were measured. Both die edges were measured for each x and y direction and the average distance (x_0) for each die group to be stacked was obtained. For each die side, the deviation Δx of the distance x_i from x_0 was calculated (Figure 3(a)). Before stacking, standard deviation of Δx is 1.1 μm with a narrow distribution (Figure 3(b)). After the measurement, three dies were joined in a batch bonding step by CALM. For measuring the relative distance

978-1-4799-2408-0/14 $31.00 © 2014 IEEE   649

(x_i) of the joint bump/pads, chip stack samples were cut and the cross-sections were polished. The relative distance (x_i) for each stacked each layer was measured based on a reference position of the bottom layer. In each layer, the deviation Δx of the distance (x_i) from average distance (x_0) was obtained. Because the standard deviations of the Δx in each layer were <1µm, we expect that the rotational deviations were negligible. A histogram of the deviation Δx before and after stacking is shown in Figure 3. The standard deviation of Δx values after stacking is 2.0 µm. The CALM bonding has resulted consistent alignment accuracy of a few microns, even after 3-layer die stack. Higher alignment accuracy could be obtained if the dicing accuracy is improved.

3.2. Addressing warpage

3D assembly can be complicated by warped components. Warpage of thin TSV die is not ideal for conventional furnace reflow. Laminate substrates experience additional warping during heating because each laminate layer can have different CTE. Also, thin die can experience warping during 3D assembly; however, it can be more severe than the laminate substrate warpage. Figure 4 characterizes the warpage behavior of thin CMOS die by Shadow Moire method using shape inversion plots. Significant warpage difference between the peak and room temperature is evident. Thin die warpage during furnace reflow may result in non-wetting of C4 and/or bridging between C4 joints if the target is a large die that is highly warped. The influence of thin die warpage on 3Di becomes more significant as die size increases and component thickness decreases. Therefore, new assembly technology is needed to overcome such warpage challenges (Figure 5).

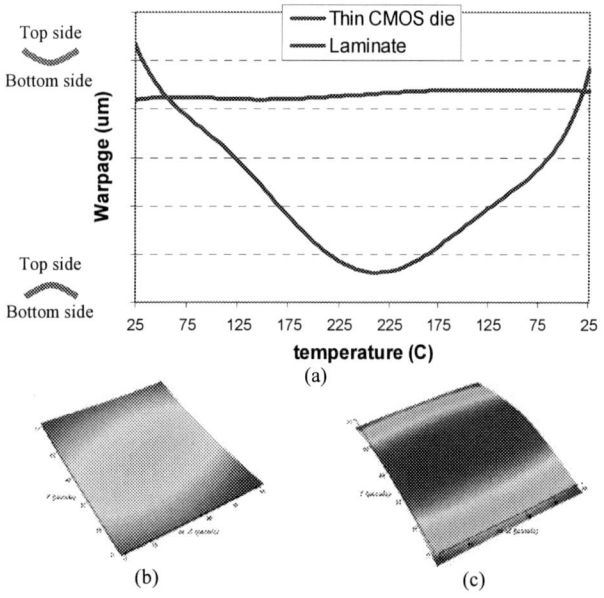

Figure 4. Warpage behavior change of thin CMOS die with size larger than 600 mm² (a), typical fringe pattern of shadow moiré warpage at 25ºC (b) and 250ºC (c).

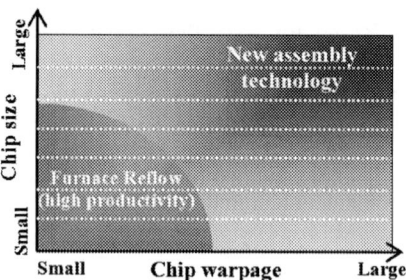

Figure 5. Suitable assembly method according to chip size and chip warpage.

There are two different assembly approaches for chip level 3Di (Figure 6): the "top-down" and the "bottom-up" approach. In the top-down case, a top CMOS die is joined to a bottom CMOS thin die by C4 or micro-bump and then the bottom die is bonded to a substrate by C4 [7]. In the bottom-up case, a bottom CMOS thin die is bonded to a substrate first and then the top CMOS die is joined. In this work, the assembly approach and the bonding technology scheme were optimized to overcome warpage during assembly, resulting in improved co-planarity compared to traditional assembly processes.

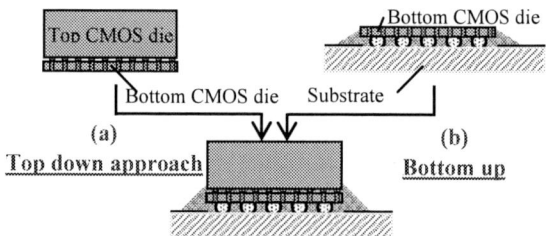

Figure 6. Process flow of chip level 3D integration.

3.3. Integration process

A 22 nm LSI chip with low-k ICs was used for chip level 3Di. The schematic representation of the chip level 3Di process is shown in Figure 7. Cu-TSV is introduced in the Back-End-of-Line (BEOL). The TSVs are formed by deep silicon reactive ion etching (RIE) that forms high aspect ratio vias. Conformal SiO_2 is deposited for TSV sidewall insulation with high-step-coverage then barrier and seed layers are deposited by sputtering. The vias are completed with bottom-up Cu electroplating to fill the TSVs. Additional BEOL levels are fabricated after TSV processing and subsequently planarized by CMP. Figure 8 shows a cross-section SEM of an integrated TSV and BEOL structure. Void-free TSVs were formed and no delamination of the BEOL dielectric stack was observed. Thick Cu wiring levels provide stable voltage for high performance applications but can also result in additional wafer/chip bow. Stress in the inter-layer dielectric (ILD) was tuned to minimize wafer and chip bow. After forming the Top-Surface-Metallurgy (TSM), the wafer is attached to a glass handler for grinding, polishing, and etching. The wafers are thinned to reveal the backside via metal, followed by oxide/nitride isolation layer deposition. A patterned electroplating process is used to define the wafer finish wiring, terminal metallurgy, the solder deposition. Two types of bumps were used for the interconnection: low solder volume with Cu pillars for the top die to the bottom die, and high solder volume for the bottom die to the laminate. In both cases the lead-free SnAg interconnection consisted of Ni-solder-Cu joins. The top die consisting of Cu and SnAg was joined to Ni/Au pads on the bottom die side, while the bottom die side had the traditional SnAg bump with terminal Ni joined to pre-solder on the laminate side. Both sides were joined sequentially using a new bonding methodology to overcome warpage issues. A thin bottom die is joined to an organic laminate and then a top 22 nm CMOS die is joined to the thinned 22 nm CMOS die in a face-to-face configuration. It is noteworthy that in the process practiced here underfill is applied under the bottom die before joining the top die. Figure 9 shows a cross-section SEM image of the assembled C4 joints. The gap between bottom die and laminate is stable and no joint failure was observed. Electromigration results will be covered in a separate paper [20], but these indicate that the selected Ni-solder-Cu structure meets the current product requirements.

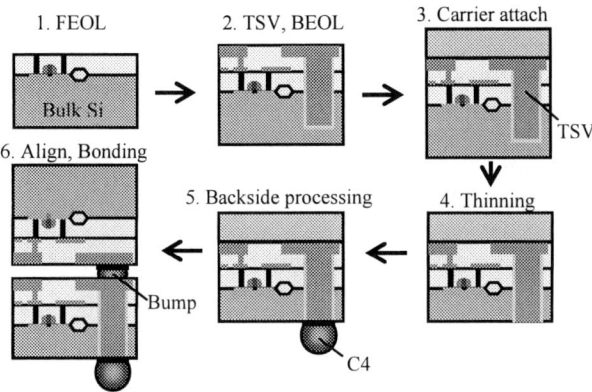

Figure 7. Schematic representation of face-to-face chip level 3D integration process.

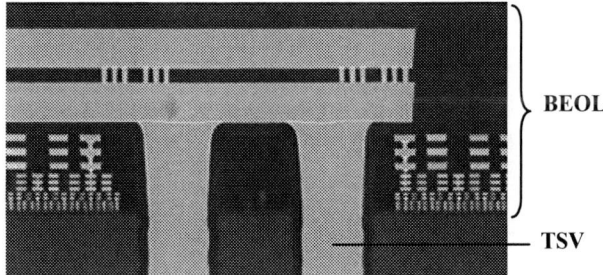

Figure 8. Cross-section SEM showing integrated TSV and BEOL structure (22 nm).

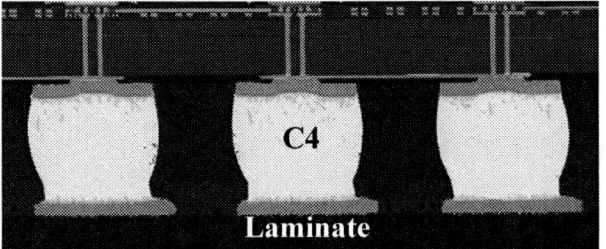

Figure 9. Cross-section SEM image of C4s assembled using chip level 3D integration process.

3.4. Hardware results

Dice were assembled on a laminate with 7 layers of build-up circuitry on each side of the core. Figure 10 shows cross-section SEM image of a 3D integration with a 22 nm CMOS die stack on a laminate. The stacked dice appear to be in good alignment. Table 3 shows the stress conditions that were run to evaluate the 3D modules and test results. The stresses were carried out to the durations specified by the JEDEC standard [21]. All stresses were carried out on land grid array (LGA) modules, except THB for which the modules were socketed. The sample size and number of lots that were sampled are also shown in Table 3. The tests during readouts included comprehensive continuity and leakage testing of various macros that were designed to evaluate the integrity of the upper-bump-TSV-lower-bump connection, the laminate wiring to those structures, and the integrity of the BEOL wiring levels in both chips in a 3D format.

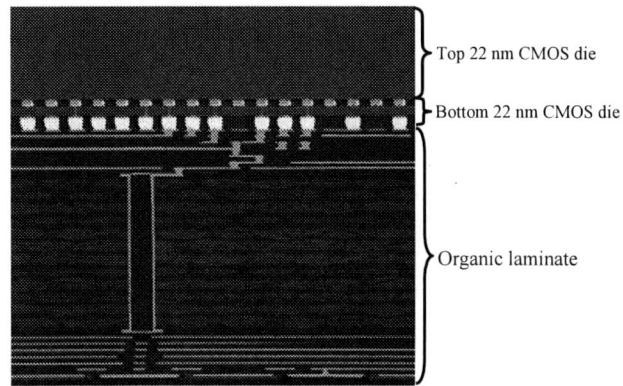

Figure 10. Cross section SEM image of face-to-face die stack on organic laminate fabricated with IBM's 22 nm process.

Table 3. Reliability stress conditions for chip level 3D modules and test results.

Cell	Stress	Condition	JEDEC Spec	Requirements	Qty (Lots)	Result
A	TC-G	-40/125°C	A104	850 cy / 0 Fail	13 (1)	Pass
B	THB	85°C /85% RH/3.6V	A101	1000 hr / 0 Fail	30 (2)	Pass
C	HTS	150°C	A103	1000 hr / 0 Fail	13 (2)	Pass

The thermal reliability of Thermal Interface Material Level-1 (TIM1) and the die-to-die interfaces is also evaluated by using the same test vehicle. The test vehicle was constructed with vertically-aligned heaters and 25 thermal sensors (Resistance Thermal Device, RTD) embedded in the BEOL layers of each die. A thermocouple was attached to the center outside surface of the lid as a package reference temperature sensor. The traditional interface from the top of the chip (stack) to the package lid through the TIM1, often labeled Theta J-C or Rint, is monitored using the 25 top-chip sensors referenced to the external lid-attached thermocouple. Unique to the stacked-chip structure is a new thermal interface in the interconnection level between the die as visible in Figure 10. The thermal resistance parameter associated with this interface is labeled as Resistance die-to-die (Rd-d) and is calculated using the temperature difference measured between each of the vertically-aligned temperature sensors embedded in each BEOL stack. As shown in Figure 11, both Rint and Rd-d were extremely stable throughout the duration of stressing with a maximum thermal degradation of less than 4°C per 1000W of power dissipation observed at the End-of-Stress readout.

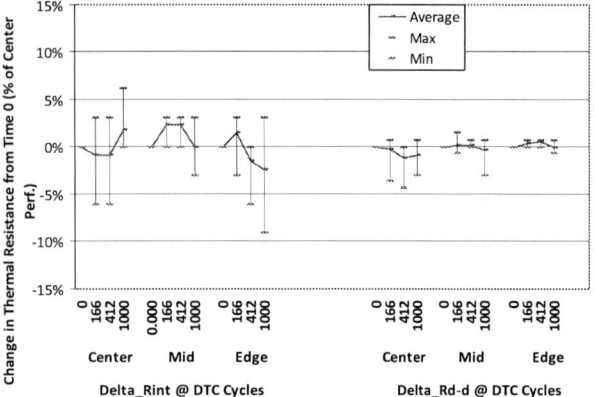

Figure 11. Thermal Reliability of TIM1 (Rint) and die-to-die (Rd-d) interfaces through 1000cy of DTC (-55/+125°C) Accelerated Stressing.

4. Wafer level 3D integration

A wafer-scale 3Di strategy involving oxide bonding was also pursued in this work. The wafers were stacked and thinned, and the interconnection between them was achieved using TSVs. In wafer-level 3Di, the starting yield must be high as the various wafers are fabricated separately. The wafers can be stacked using oxide bonding, which is compatible with BEOL integration and the final wafer stack is approximately the same thickness as normal wafers, which eliminates the need for new equipment for downstream processing.

4.1. Integration process: Wafer thinning and TSV integration

An optimized two-layer oxide bonding film was used for the wafer-scale 3Di in this work. Film depositions were accomplished by plasma-enhanced chemical vapor deposition (PECVD). The process for the first layer was optimized to overcome wafer topography, as bonding surface flatness and absence of topography is crucial for defect-free bonding. This bonding layer underwent special treatments to enhance bonding strength and ensure void-free bonding. A special polishing step was used to achieve sufficient bonding surface flatness, followed by a cleaning step. Subsequently, the second bonding layer was deposited. An optimized combination of thermal treatment, CMP and surface cleaning steps were used. Oxide bonding surfaces characterized using atomic force microscopy (AFM) are typically shown to be atomically smooth with root mean square (RMS) roughness values of 0.2 nm-0.4 nm.

Figure 12 shows a schematic representation of the wafer level 3D integration process. A bare Si wafer was bonded to the device wafer for use as handle using the low-temperature

oxide bonding method. The device wafer was then thinned on the backside to a thickness of 10-12 μm by using a combination of mechanical grinding, RIE, and wet chemical thinning, all these processes employing conventional silicon fabrication equipment and processes. An oxide bonding layer was then prepared on the backside of the device wafer and the stack was bonded to another paired full-thickness device wafer, with the handle wafer now at the top of the stack. The handle wafer was then removed by a combination of mechanical grinding, RIE, and wet thinning.

After removal of the handler, deep RIE was used to form the TSVs from the front side of the thinned wafer, a special process that was designed to handle the complex interlayer dielectric stacks, bond interface, and the thinned Si. The TSV insulator was deposited using a CVD process. Another RIE step was used to open the TSV insulator at the bottom of the TSVs. Then TSV metallization followed, which involved Cu liner and seed deposition with physical vapor deposition (PVD) and bottom-up Cu plating, so that TSVs with aspect ratio up to 5:1 were fabricated.

Three additional Cu levels were fabricated on the thinned wafer stack after TSV formation to complete this test vehicle. Low resistivity Cu TSVs integrated with Cu wiring are critical for containing the IR voltage drop for high-performance applications. All processing and tools are compatible with advanced metal gate to at least the 14 nm technology node. This special face-to-back stacking process is repeatable for multiple wafer stacking and integration.

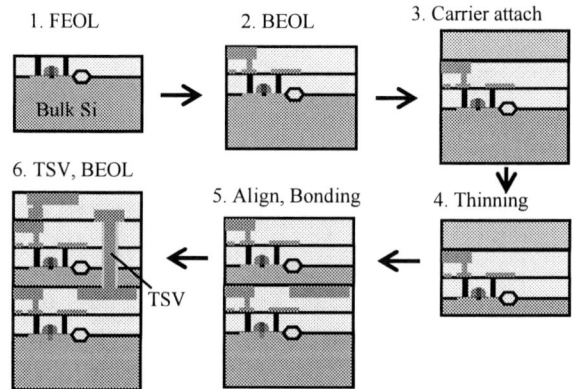

Figure 12. Schematic representation of wafer level 3D integration process.

4.2. Hardware results

Figure 13 shows cross-sectional SEM image of the finished bonded wafers. The TSV chain structures, with each link featuring two TSVs, exhibit resistance of 65±10mΩ/link including local wire resistance. This is suitable for high performance applications demanding current densities >1A/mm^2 with total TSV area <1%. The TSV capacitance was <40 fF. The leakage and capacitance of the TSV arrays to the substrate are linear versus the number of TSVs in the array, close to the limits of the tester. The extrapolated leakage is 1.25pA/TSV at 2V. For a 400 mm^2 size chip featuring up to 50K TSVs, the total TSV contribution to leakage would be 20μA. Key device characteristics, such as FET Ion/Ioff show no significant effect from the wafer

978-1-4799-2408-0/14 $31.00 © 2014 IEEE   652

stacking process (Figure 14). Also FETs in the proximity of TSVs show no effect due to TSV presence when compared to devices without TSVs nearby (Figure 15). This is significant as it points to the capability for 3D communication between two stacked chips that is equivalent to a 2D chip design employing 100 to 200 μm long wires for macro-to-macro signaling. This high-performance stacked eDRAM prototype integrated about 11,000 TSVs. The thinned and thick wafers emulate processor and cache memory, respectively. Memory functionality and intra-layer communication were tested by built-in-self-test (BIST) engines on the thick wafer chips that could access eDRAM on both wafers. Demonstration of 16Mb eDRAM functionality with intra-layer communication as high as 1.48GHz at 1.3V was achieved with eDRAM retention times greater than 200 μs. It is noteworthy that the maximum test frequency was limited as the test was performed at wafer level by using cantilever probes, resulting in higher voltage drop versus socket based module test.

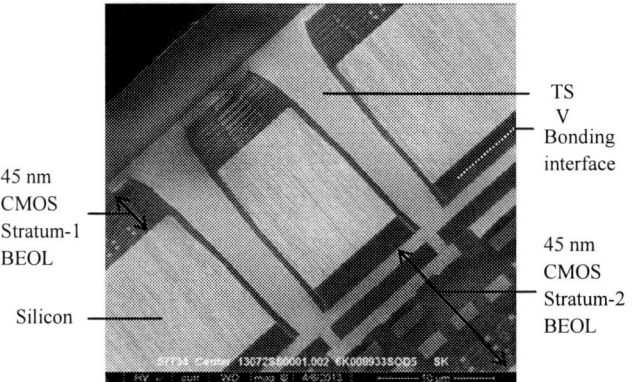

Figure 13. SEM cross section of donor and acceptor device strata after wafer level bonding and removal of handle wafer.

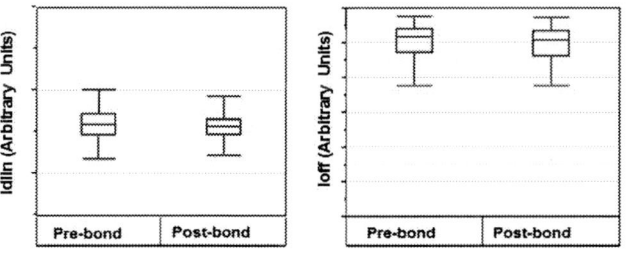

Figure 14. FET Ion/Ioff before and after bonding.

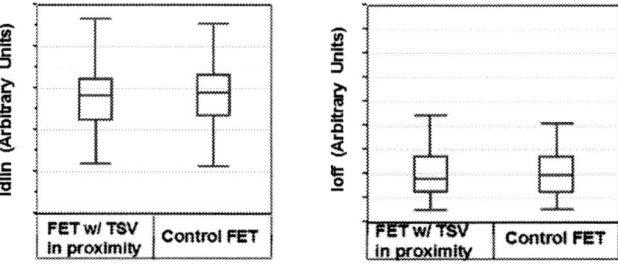

Figure 15. FET Ion/Ioff between FET in the proximity of the TSVs and control FET.

5. Summary and conclusion

In summary, this paper reports on advancements and comparison of several 3Di bonding technologies. The demonstration of CALM process indicated that dice were stacked within a few micron alignment accuracy. We have demonstrated successful stacking of a large top on a thin Si bottom die with Cu TSVs integrated in 22 nm CMOS technology. The size of the dice was more than 600 mm². The results of the eDRAM prototype with 45 nm CMOS devices fabricated through 300 mm oxide wafer bonding, wafer thinning, and TSV formation strongly suggest that this approach is appropriate for high performance applications such as stacking of high performance memory.

Acknowledgements

The authors would like to sincerely thank Thomas Wassick, Van Thanh Truong, Glenn Pomerantz, Thomas Lombardi, Hongqing Zhang, Marcus Interrante, Gregg Monjeau, Bill Sablinski, Pooja Batra, Nicolas Boyer, and Terri Childress-Carter of IBM Microelectronics Division. We also would like to acknowledge management support and encouragement from Brian Sundlof, Jean Trewhella, and T.C. Chen of IBM during this research.

References

1. S.S. Iyer, et al., "Process-DesignConsiderations for Three Dimensional memory Integration", *Proc. Symp. VLSI tech.*, June 2009, pp 60-63.
2. M. Koyanagi, H. Kurino, K.W. Lee, K. Sakuma et al, "Future System-on-Silicon LSI chips," *IEEE MICRO*, Vol 18, no. 4, pp.17-22, Jul/Aug. 1998.
3. K. Sakuma et al., "3D chip-stacking technology with through-silicon vias and low volume lead-free interconnections," *IBM J. Res. & Dev.,* vol. 52. No. 6, pp. 611-622 (2008).
4. J. U. Knickerbocker, et al., "Three-dimensional silicon integration," *IBM J. Res. & Dev.*, vol. 52. No.6, pp. 553-569 (2008).
5. W.C. Chiou, K.F. Yang, J.L. Yeh, et al., "An ultra-thin interposer utilizing 3D TSV technology," *Proc. Symp. VLSI Tech.*, pp107-108 (2012).
6. K. Sakuma, P.S. Andry, C.K. Tsang, K. Sueoka, et al., "Characterization of Stacked Die using Die-to-Wafer Integration for High Yield and Throughput," *Proc. 58th ECTC*, Lake Buena Vista, Fl, 2008. pp. 18–23.
7. K. Sakuma, K. Sueoka, S. Kohara, et al., "IMC bonding for 3D Interconnection," *Proc. 60th ECTC*, 2010, pp.864-871.
8. K. Murayama, et al., "Warpage Control of Silicon Interposer for 2.5D Package Application," *Proc. 63rd ECTC*, 2013, pp.879-884.
9. K. Sakuma, K. Smith, K. Tunga, et al., "Differential heating/cooling chip joining method to prevent chip package interaction issue in large die with ultra low-k technology," *Proc. 62nd ECTC*, pp.430-435, 2012.
10. N. Maeda, Y.S. Kim, Y. Hikosaka, H. Kitada, et al., "Development of Sub 10- μm Ultra-Thinning Technology using Device Wafers for 3D Manufacturing

of Terabit Memory," *Proc. Symp. VLSI Tech.*, pp105-106 (2010).

11. P. Batra, D. LaTulipe, S. Skordas, et al., "Three-Dimensional Wafer Stacking using Cu TSV integrated with 45nm high performance SOI-CMOS embedded DRAM technology," *S3S*, 2013.

12. N.G. Koopman, T.C. Reiley, P.A. Totta et al., "Chip-to-Package Interconnections," in *Microelectronics Pafkaging Handbook*, Van Nostrand Reinhold, p.361, (1989).

13. K. DeHaven, J. Dietz, "Controlled collapse chip connection (C4)-an enabling technology," *Proc. 44th ECTC*, 1994, pp.1-6.

14. N. Hoivik, and K. Aasmundtveit, "Wafer Level Solid-Liquid Interdiffusion Bonding", *Handbook of Wafer Bonding*, Wiley-VCH Verlag & Co, 2012, p.181.

15. K-N. Chen, et al., "Structure, Design and Process Control for Cu Bonded Interconnects in 3D Integrated Circuits," *Proc. IEDM*, 2006, pp.1-4.

16. R. Reif, A. Fan, K.N. Chen, and S. Das, "Fabrication technologies for three-dimensional integrated circuits," *Proc. of International Symposium on Quality Electronic Design*, 2002, pp. 33-37.

17. S. Skordas, D. LaTulipe, K. Winstel, et al., "Wafer-Scale oxide Fusion Bonding and Wafer Thinning Development for 3D Systems Integration," *LTB-3D*, p203, 2012.

18. L. Di Cioccio, "Cu/SiO2 Hybrid Bodning", *Handbook of Wafer Bonding*, Wiley-VCH Verlag & Co, Weinheim, 2012, p.237.

19. M. Nimura, J. Mizuno, K. Sakuma, and S. Shoji, "Solder/Adhesive Bonding Using Simple Planarization Technique for 3D Integration," *Proc. 61st ECTC*, 2011, pp.1147-1152.

20. M. Lu, C. Goldsmith, et al., "Effect of Ag and Cu Content in Sn Based Pb-Free Solder on Electromigration," to be published in *Proc. 64th ECTC*, 2014.

21. JEDEC Solid State Technology Association, Electronic Industry Association; see http://www.jedec.org/home.

Dependence of Solder Joint Reliability on Solder Volume, Composition and Printed Circuit Board Surface Finish

Babak Arfaei[1], Francis Mutuku[2], Keith Sweatman[3], Ning-Cheng Lee[4], Eric Cotts[2], Richard Coyle[5]

[1]Universal Instruments Corporation, Conklin, NY
[2]Physics Department and Materials Science Program, Binghamton University, Binghamton, NY
[3]Nihon-Superior, Osaka, Japan
[4]Indium Corporation, Clinton, NY
[5]Alcatel-Lucent, Murray Hill, NJ
babak.arfaei@uic.com

Abstract

Thermal fatigue and room temperature isothermal mechanical performance of various Pb-free and SnPb solder joints were examined. Various solder alloys doped with Ni (SN100C) and Mn (SAC105-Mn and SACM) were evaluated and compared to SAC105, SAC 205 SAC 305 and SAC 405 and eutectic SnPb alloys. Solder spheres ranging from 10 to 20 mils were reflowed on various printed circuit board (PCB) surface finishes such as copper organic solderability preservative (Cu-OSP) and electroless nickel immersion gold (ENIG). The mechanical behavior of these solder joints was evaluated in low speed and high speed shear tests, and also in shear fatigue test. The effect of isothermal aging was examined. Custom made ball grid array (BGA) packages with the same alloys and sizes were tested in accelerated thermal cycling (ATC) test. The results from room temperature mechanical tests were correlated to package level ATC results obtained for the thermal cycling profile of -40/125°C. The effect of solder volume and composition on the solidification temperature of each solder joint was carefully measured by differential scanning calorimeter (DSC). Precipitate sizes and distributions were analyzed using backscattered scanning electron microscopy (SEM). Sn grain morphology was characterized by polarized light microscopy (PLM) and electron backscatter diffraction (EBSD).

Investigation of the lifetimes of various solder joints in room temperature fatigue and accelerated thermal cycling tests showed distinct dependences of lifetime on solder composition. Distinct increases in lifetimes with increases in Ag content were observed. Results suggested the recrystallization and failure mechanism in Pb-free solder joints are strongly affected by Ag_3Sn precipitates. Combination of Cu/ENIG surface finishes generally resulted in some improvement in thermal fatigue performances. Results also showed that solder volume can greatly affect the microstructure and performance of SAC solder joints in mechanical and ATC tests. Larger samples generally solidified at higher temperatures and revealed different Sn grain morphologies than smaller samples, which generally undercooled more. Addition of dopants generally reduced the undercooling, resulting in different solder joint microstructures. The effect of variation in solder composition and volume and PCB surface finish on solder joint microstructure and lifetime was carefully evaluated.

Introduction

Pb-free solder alloys with low Ag content and various microalloy additions or concentrations of dopants are being developed as alternatives to standard Sn3Ag0.5Cu (SAC305) alloys for some applications. Elements such as Ni, Mn and Ti are used as microalloying elements in an attempt to achieve better performance in both drop and accelerated thermal cycling (ATC) tests. However, the effects of these elements on the microstructural features of solders and thus failure mechanisms are not well established. As the Pb-free transition has progressed, various limitations of high Ag alloys have been identified and this has resulted in a proliferation of new alloys to address emerging risks [1-3]. While the high cost of Ag is one important factor for developing low Ag alloys, there are other factors that prompted a move to lower Ag contents [2]. The high flow stress and elastic modulus of the high Ag solders creates assemblies with solder joints vulnerable to failure in drop impact, either by crack propagation through the solder joint itself or through the printed circuit board laminate (pad cratering). Lower Ag SAC alloys and those microalloyed with additional elements have shown better performance in mechanical tests such as drop. Liu et al. demonstrated better performance of SAC alloys doped with Mn, Ce or Ti in drop testing [4]. Lower Ag compositions were chosen initially by single digit adjustments to the SAC305, which led to alloys such as SAC205 and SAC105. A low Ag alloy based on the Sn-Cu-Ni system with the further addition of Ge (SN100C) now is one of the alloys most widely used for Pb-free wave soldering [1].

The full impact of newly developed alloys on reliability of printed circuit assemblies has yet to be determined, particularly with respect to thermal fatigue, which is the well-known wear-out mechanism of solder joints. Thermal fatigue combined with solder creep is considered a major source of failure of surface mount (SMT) components [2]. This is particularly a concern for high reliability applications, because low Ag alloys may not have sufficient fatigue resistance. The published data on low Ag alloys and those alloyed with a fourth element are limited. One of the most extensive works is being conducted by the Pb-free alloy characterization program sponsored by the International Electronics Manufacturing Initiative (iNEMI) [1-3]. The goal of such work is to develop reliability data for 12 different Pb-free alloys and to assess the influence of various temperature cycling profiles on reliability. Results of this work suggest that higher Ag content alloys, those with a higher Ag_3Sn particle density, generally perform better in thermal cycling test. The iNEMI work also demonstrated that increasing the dwell time results in a decrease in fatigue reliability. Those observations support the hypothesis that the fatigue resistance in SAC alloys is controlled by the Ag content and Ag_3Sn precipitate density. However, in that study the compositions of original solder spheres were changed due to assembly using SAC 305 solder paste, making it difficult to investigate the exact effect of alloys and their microstructure on thermal fatigue performance [1-3].

The microstructure of near eutectic SnAgCu (SAC) profoundly affects the properties of a Pb-free solder joint. The morphology of Ag_3Sn and Cu_6Sn_5 precipitates, their number and spacing, are important in determining the reliability of SnAgCu solder joints. Thermal history and sample geometry could affect the precipitate microstructure and mechanical properties such as creep. In Pb-free SAC solder alloys, Sn is the major component of the alloys and Sn generally displays large anisotropies in its thermomechanical properties [5-8]. The number and morphology of Sn grains in a SnAgCu solder joint also affect reliability. During the solidification of SnAgCu solder joints, the Sn often undercools considerably but only nucleates from the melt at a single point within each joint [6]. Several reports showed that in typical SAC solder joints the as-reflowed microstructure will consist of large β-Sn grains, typically with one or three grains [6, 7]. Often these grains appear to be arranged radially around a circle with clearly defined boundaries, defined as a "Kara's beachball" configuration (Fig. 1), due to cyclic twinning [7].

The beach ball Sn grain morphology is not the only as-reflowed microstructure observed in SAC solder joints. Often (particularly in smaller joints), a different structure, interlaced Sn grain morphology is observed (Fig. 2) [9, 10]. Unlike the beach ball structure, many smaller grain boundaries are present in interlaced structures. This interlaced Sn grain morphology also displays only three Sn grain orientations. The interlaced twinning structure was observed in a variety of SAC joints, with small amount of solder volume including land grid arrays (LGAs) and flip chips [10-12]. Previous work has shown that there is a correlation between solidification temperature and the Sn grain morphology in these near eutectic SnAgCu samples. Generally samples solidified at higher temperatures show beach ball structure, while smaller samples that undercool more show interlaced Sn grain morphology [6, 9-12]. Figure 1 shows examples of two different Sn grain morphologies. Smaller (100 μm diameter) flip chip sample, LF2 alloy, solidified around $170^{\circ}C$ and shows interlaced Sn grain morphology while the larger (500 μm diameter) SAC 205 sample solidified around $200^{\circ}C$ and shows three large Sn grains. Polarized light microscopy (PLM) and electron backscatter diffraction (EBSD) are two common techniques to investigate the Sn grain morphology of SAC solder joints (Fig. 1) [5-13].

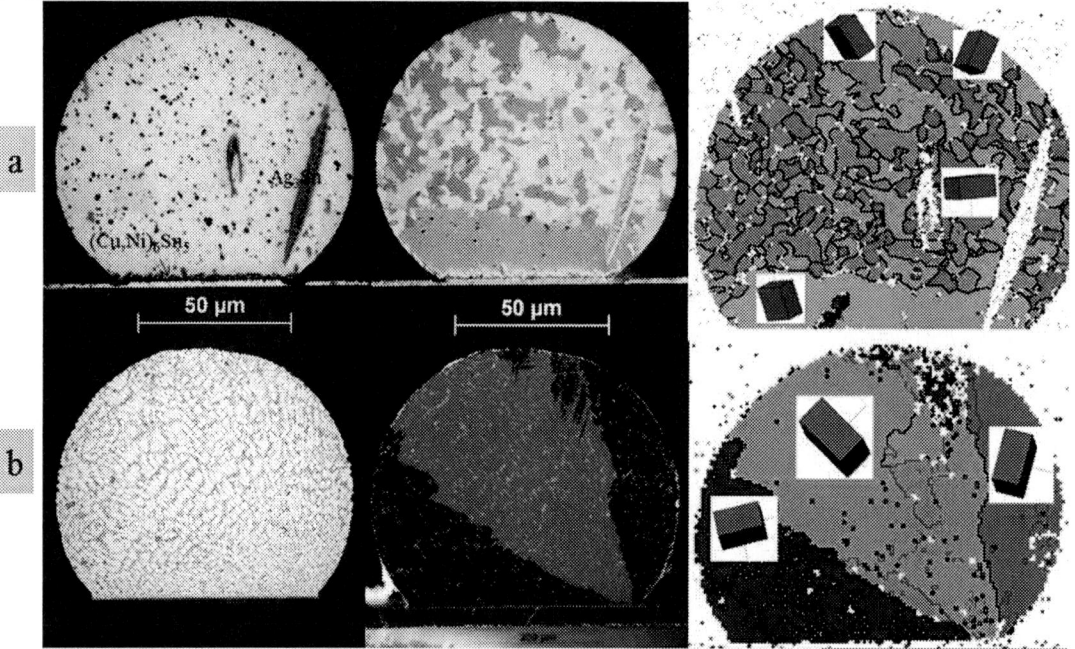

Figure 1. Bright Field Micrograph, Optical micrograph with crossed polarizers and EBSD map of (a) 100 μm diameter flip chip showing interlaced twinned microstructure; (b) 500 μm diameter SAC 205 sample showing three large Sn grains representative of beach-ball Sn grain morphology. (From Ref. 6)

There are several reports that highlight the effect of Sn grain morphology on mechanical and thermal cycling behavior of SAC solder joints [11, 13-14]. Results from some of the current authors have shown that single Sn grain solder joints of particular orientations failed unexpectedly early, as compared to multi-grain SnAgCu solder joints during isothermal fatigue test [5, 14]. Several published studies have shown that the Sn grain orientation directly affects the electromigration (EM) reliability of Pb-free SnAgCu solder

interconnects [15]. Differences in Sn grain morphology (i.e., beachball versus interlaced) have been shown to influence the thermal cycling performance of SAC solder joints. Results showed that SAC LGA assemblies with the interlaced structure exhibited a significantly longer lifetime in accelerated thermal cycling test as compared to the lifetime of BGA assemblies with the beach-ball Sn grain morphology [10-11, 16-17]. Figure 2 shows the results of -40/125°C thermal cycling test with 10 minute dwells. The LGAs outperformed the BGAs in the same test. This was not expected, as the standoff height is considerably smaller (by a factor of three) compared to BGA samples. Investigation of the corresponding Sn grain morphologies (Figure 2) showed that the LGAs displayed an interlaced Sn grain structure, while the BGAs comprised either a single Sn grain, or beach ball structures having three large Sn grains. At the lower pitch, 1.4 mm, the LGA samples surprisingly outperformed the BGA samples of the same pitch size. At the large pitch, 2.2 mm, (higher stress) the BGAs lasted longer than LGAs. LGAs, however, still showed better performance than what was expected. Of course, the Sn grain morphology is only one aspect of microstructure that can affect the reliability; other parameters such as precipitate morphology, size and spacing should be considered when evaluating reliability performance of solder joint.

The recrystallization of Sn during thermal cycling has been directly related to crack propagation in SAC solder joints [17-19]. Yin et al. have proposed a damage accumulation model that correlates the evolution of the solder microstructure and thermal cycling fatigue [18]. A fatigue crack then starts to propagate along the network of grain boundaries through the recrystallized area until failure (Figure 3). The recrystallization was mainly found close to high strain regions on the component side and seemed to vary systematically with the density of secondary precipitates in a solder joint [11, 18]. Mattila et al. also studied the failure mechanism of solder interconnections under various thermal cycling conditions [19-20]. A continuous network of high angle boundaries was observed to provide a path for fatigue cracks to propagate [11, 19].

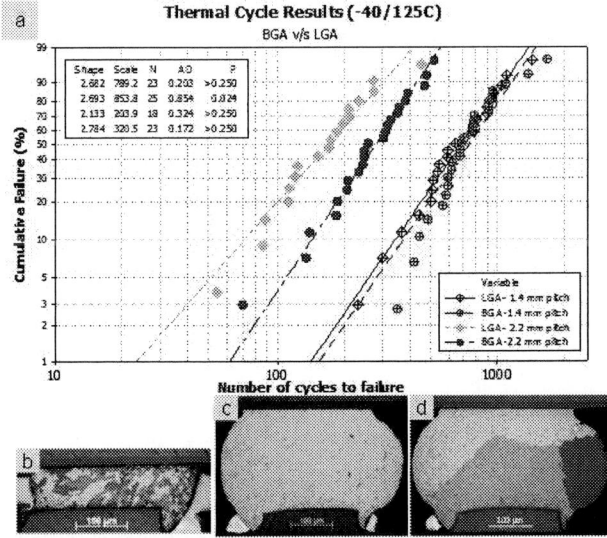

Figure 2. (a) Weibull plot showing the cycles to failure of 16 mil BGAs and 10 mil LGAs. Results for two different joint pitches, 1.4 mm and 2.2 mm, in -40/125°C thermal cycling test are plotted. The representative micrographs with crossed polarizers of the samples are shown. (b) 10 mil LGA samples show presence of interlaced twinned morphology. 16 mil BGA samples show either (c) single grain or (d) beach ball structure (From Ref. 11).

Generally, during a thermal cycling test, strain-enhanced precipitate coarsening was observed close to crack area (Figure 4a and 4b). Our previous results and those published by Matilla et al. show coarsening of Ag_3Sn precipitates during thermal cycling [11, 17-19]. It was clear that the microstructure in the strain concentration regions was completely different than as-reflowed microstructure further from crack area (Figure 4 (b)).

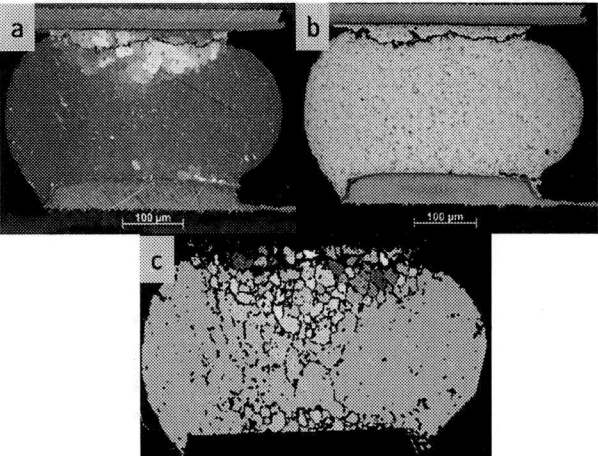

Figure 3. Optical micrographs of a 16 mil solder joint after thermal cycling, -40/125°C: (a) with cross-polarizer, (b) bright field, (c) EBSD map of the same joint. The recrystallized region near component side is clearly visible. Formation of new grains and intergranular fatigue cracking can be seen in the EBSD map.

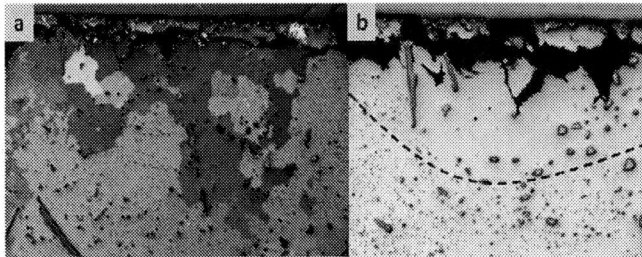

Figure 4. Optical micrographs of a solder joint after thermal cycling, 0/100°C: (a) cross-polarized (b) bright field, of the same joint. In the recrystallized region above the dashed line, larger and fewer precipitates are evident. The intergranular fatigue crack can be seen on the component side (From Ref. 17).

The sequence of failure was found to be different in smaller samples (that show interlaced Sn grain morphology) compared to that observed in larger samples. Results of EBSD studies of LGAs sample cycled to 25% of its projected life suggested that many new grains are formed, indicating that that the whole joint has been recrystallized [11, 17]. Examination of a failed LGA sample (Figure 5) confirmed that many new high angle grain boundaries were formed during cycling and that a crack propagated along high angle grain boundaries.

Previous investigations showed that the initial microstructure can also be affected by other parameters such as pad finishes, solder volume and solder composition [1, 6, 8]. Variations in composition of Pb-free solder, by addition of some small dopants (e.g., Ni or Al), have also been shown to have distinct effects on the solidification temperature of SAC alloys [9-10, 21]. Changes in composition and solidification temperature significantly affect the microstructure of solder joints. These changes in microstructure affect the isothermal mechanical and thermal fatigue behavior of a given solder joint [5, 11, 14]. The mechanical properties of the Pb-free solders are known to be strongly affected by their microstructural changes. Published data show the dependence of mechanical properties of various solder joint configurations on aging [22]. Coarsening of Ag_3Sn precipitates results in weaker solder and shorter lifetimes in room temperature shear fatigue [22]. Suhling et al. [23] observed that the creep rate increased strongly as the aging temperature and time increased. The sensitivity to aging was found to be increased as the silver content was reduced. These observations were attributed to coarsening of smaller (micron scale) precipitates. The small precipitates effectively act as barriers to the movement of dislocations and make the solder harder, thus enhancing the creep resistance. The studies on the effect of pre-aging on ATC reliability of solder joints are limited. In general, some level of degradation was observed during thermal cycling when the samples were isothermally pre-aged [24-25]. However, the results are not consistent for different packages/solder alloys. Recent results showed that pre-aging for 10 days at 125°C did not alter the failure modes, although this isothermal aging resulted in Sn grain growth in interlaced structures [11, 17].

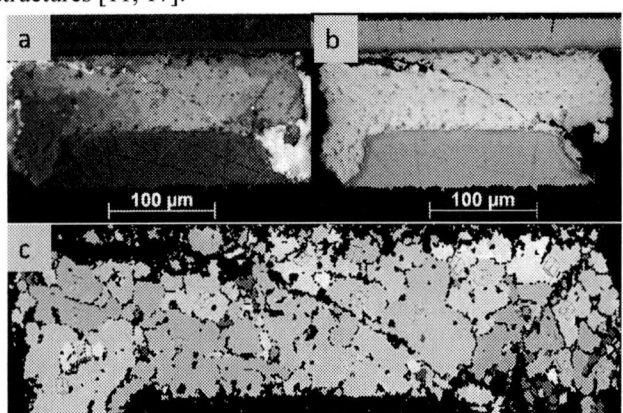

Figure 5. Optical micrographs of a 10 mil solder joint after thermal cycling, 0/100°C: (a) with cross-polarizer (b) bright field, (c). The whole joint is recrystallized. Formation of new grains and intergranular fatigue crack can be seen in the EBSD map (From Ref. 11).

There are many published works on mechanical behavior of Pb-free solders joints [23, 26] but most of those experiments were performed on solder specimens much larger than typical microelectronic solder joints. Therefore, the results may not be representative of commercial solder joints. The common solder joint level mechanical tests for the interconnection reliability in the microelectronic packaging industry are low and high speed shear test as well as cold and hot bump pull tests. While those tests are relatively simple and fast, there are little indications that those results can correlate with board level reliability tests such as drop or thermal cycling test. At the same time, assembly level testing under the conditions of most concerns such as drop, and ATC

are expensive and not always as easy to interpret as simpler joint level tests.

Variations in microstructure from solder to solder because of different degrees of undercooling, differences in various surface finishes, and differences in composition make it extremely complicated to investigate the solder microstructure and evaluate the effect of solder composition and addition of small alloy additions on thermomechanical properties of solder joints. In this work, we seek to establish the effect of microalloying elements on microstructure and reliability of various solder compositions and solder volumes, reflowed on different surface finishes, and compare the results to near eutectic SAC and eutectic SnPb alloys. The effects of solder volume and surface pad finish on isothermal mechanical and thermal cycling behavior of solder joints are discussed.

Experimental
Isothermal Mechanical Test

Solder joints of the different Pb-free alloys were assembled by reflowing preformed solder spheres of 20 mil (508 microns) diameters on 16 mil (406 microns) solder mask defined (SMD) pads. Flux was deposited on the pads by stencil printing process prior to placing the solder spheres. The test vehicles were reflowed in a forced convection oven containing a nitrogen atmosphere. The peak temperature recorded was 245°C for 60s. A total of 16 solder joints were tested for each test condition. Table 1 shows the various alloys investigated in this study.

Low speed (1mm/s) shear testing of solder joints was done with a DAGE4000 and high speed shear testing (1m/s) was done with a DAGE4000HS. The test is designed to transform the vertical compressive forces from the DAGE to a horizontal shear force which is applied parallel to the surface of the printed circuit board (PCB). The tests were conducted in accordance with the JEDEC JESD22-B117A standard [27]. A shearing pin is used to apply the shear force onto the solder joint. The position of the shearing pin is critical in getting an accurate estimation of the shear strength of the solder joint. The shear height was calculated for each solder ball according to the JEDEC standard. The shear height was taken as 10% of the solder ball height calculated and was in close agreement with JEDEC standard, which specifies less than 25% of the solder height [27]. The ball height for 20 mil spheres was approximately 540 microns and the shear height was 36 microns. Two pad metallurgies were evaluated: Cu-OSP and Electroless Nickel Immersion Gold (ENIG). The shear strength was determined for as-assembled solder joints and after isothermal aging at 125°C for 500 hours (approximately 3 weeks). The test was carried out at two different test speeds, low speed at 1mm/s and high speed at 1m/s. The two different speeds were chosen to examine the effect of strain rate on the failure modes of various SnPb and Pb-free alloys and because the strain rates in drop and thermal cycling applications are significantly different.

In order to perform shear fatigue tests, the solder joints were mounted on a testing block and loaded in shear at a height of 36 microns above the copper pads surface and a load rate of 350μm/s, up to a maximum load of 350g. The loading direction was reversed at the end of each half cycle.

978-1-4799-2408-0/14 $31.00 © 2014 IEEE

This controlled shear cycling was continued until the solder ball completely failed. Shear fatigue test was performed on 20 mil spheres reflowed on Cu-OSP surface finish. In all isothermal tests, the failure surfaces were inspected through the instrument's microscope. Higher magnification images were taken using FEI Environmental Scanning Electron Microscope (SEM). Figure 6 shows the setup used for low and high speed shear testing and shear fatigue testing.

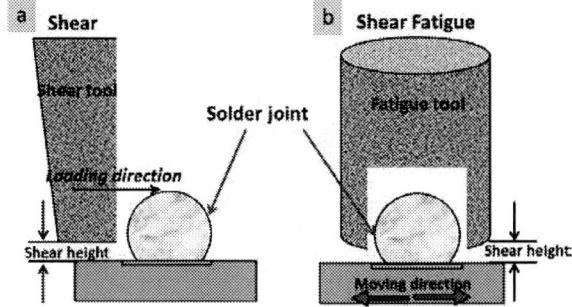

Figure 6. Schematic of (a) shear and (b) shear fatigue tests.

Accelerated Thermal Cycling

Thermal cycling experiments were conducted with components designed to minimize warpage and ensure lateral uniform properties. These packages were constructed with a 0.5mm thick silicon die sandwiched between two 0.4mm thick FR4 substrates with thin layers of a commercial flip chip underfill. The component pads were all solder mask defined. Test vehicles for this part of the experiment were manufactured from 10 and 16 mil diameter spheres. These were first fluxed and reflowed onto Cu-OSP coated pads on model BGA components with a 20x20 array of pads and then fluxed again and reflowed onto matching pads on printed circuit board.

Two different surface finishes, Cu Organic Solderability Preservative (OSP) and electroless nickel gold surface (ENIG), were used to evaluate the effect of surface finish. The board was a glass fiber reinforced FR4 substrate with eight signal layers and a total thickness of 2.52mm with provision for assembling two different pitch sized components (0.8mm and 1.0mm). The pads on the board were non-solder mask defined. In all cases, the silicon die is sized so that all solder bumps are located completely below the die shadow region. Two different surface finishes, Cu-OSP and ENIG, were evaluated. The daisy chain pattern used for each assembly was identical, although the pad pitch was varied. Four different alloys, SnPb, Sn0.5Ag1Cu-0.3Mn (SACM), SN100C and SAC 305 of 10 mil balls, were tested. Similar assemblies using 20 mil spheres were made with the above alloys except the Mn doped alloy (SACM). A flux printing process was used in both ball attach and assembly in order to minimize solder joint voiding and maintain a tightly controlled solder joint volume. This also resulted in no change in initial solder composition. Using a no clean tacky flux, test vehicles were reflow soldered in a forced convection oven containing a nitrogen atmosphere. The peak temperature and time above liquidus recorded were 242°C and 61s respectively. The test specimens were inspected after assembly by using an ohm-meter to check electrical continuity and with x-ray imaging to examine solder joint quality. X-ray

imaging revealed that an acceptable solder joint formed during the assembly process. No instances of solder bridges or solder balling were observed. Voiding was minimal, and was not expected to be a factor in the experiment.

The conventional accelerated thermal cycle condition, -40/125°C, included 15 minute dwell times at the temperature extremes and 6.5°C per minute transition rates between the extremes. These specimens were monitored using event detection with Anatech® testers set to record resistance exceeding 500 ohms and lasting 200 nanoseconds or longer; failures were defined by IPC-9701 criteria [28]. Generally, after failure was detected, samples were removed from the chamber for microstructural characterization. The summary of all the alloys investigated in different isothermal mechanical tests and accelerated thermal cycling tests are listed in Table 1. It is important to note that two different Mn doped alloys were tested; SAC105-Mn and SACM.

Table 1. Different alloys investigated in this study. In shear and shear fatigue test 20 mil spheres were tested. In ATC test 16 and 20 mil spheres were tested.

Alloy	ATC	Shear (low & high) speed	Shear Fatigue
SnPb	X	X	X
SN100C (Sn-0.7Cu-0.05Ni+Ge)	X	X	X
SACM (SAC051+0.03Mn)	X		X
SAC 105		X	X
SAC105Mn (SAC105+0.05Mn)		X	
SAC 205			X
SAC 305	X	X	X
SAC 405			X

The solidification temperatures of the selected samples were measured using Differential Scanning Calorimetry (DSC). The samples were heated at a rate of 60°C/min to a peak temperature of 250°C, annealed for 2 min, and cooled at 60°C/min to room temperature. Three reflow runs in the DSC were performed for each sample. Undercooling was calculated by subtracting solidification temperature from melting temperature.

In order to study the microstructure of the solder joints, the samples were mounted, cross sectioned, and polished using diamond and colloidal silica solutions. Various solder samples in the form of freestanding, on one substrate and assembled with two surface finishes were investigated. Selected samples after reflow, isothermal aging and after mechanical isothermal test and after failure in accelerated thermal cycling test were evaluated. Bright field and cross polarized optical microscopy was used to examine the Sn grain morphology. Sn exhibits birefringent properties, resulting in different colors for different crystal orientations under cross polarized imaging. A high resolution Zeiss 55 VP Scanning Electron Microscope (SEM) equipped with an Energy Dispersive Spectrometer (EDS) was used for imaging and chemical analysis. A commercial software package, OIM from EDAX-TSL, was used for crystallographic texture measurements.

Results

Room Temperature Shear Test

The comparison of low speed shear strength for five different alloys reflowed on two different surface finishes is shown in Figure 7(a). The results for samples reflowed on Cu-OSP show that the SAC305 alloy has the greatest shear strength while the eutectic SnPb alloy has the lowest strength. For reflow on the ENIG substrate, SAC 305 still has the highest strength while SAC105-Mn exhibits the lowest strength. The SnPb eutectic shows more sensitivity to surface finish with the strength of the alloy on the ENIG substrate showing an increase compared to samples reflowed on the Cu substrate. These results indicate that the as-reflowed alloys do not show large variations in shear strength irrespective of the surface finish. This suggests that variations in composition of alloys and thus differences in their microstructure are not sensitive to the low speed shear test. Thus, one can conclude that shear strength test is not a suitable technique to differentiate performance of the various solder alloys.

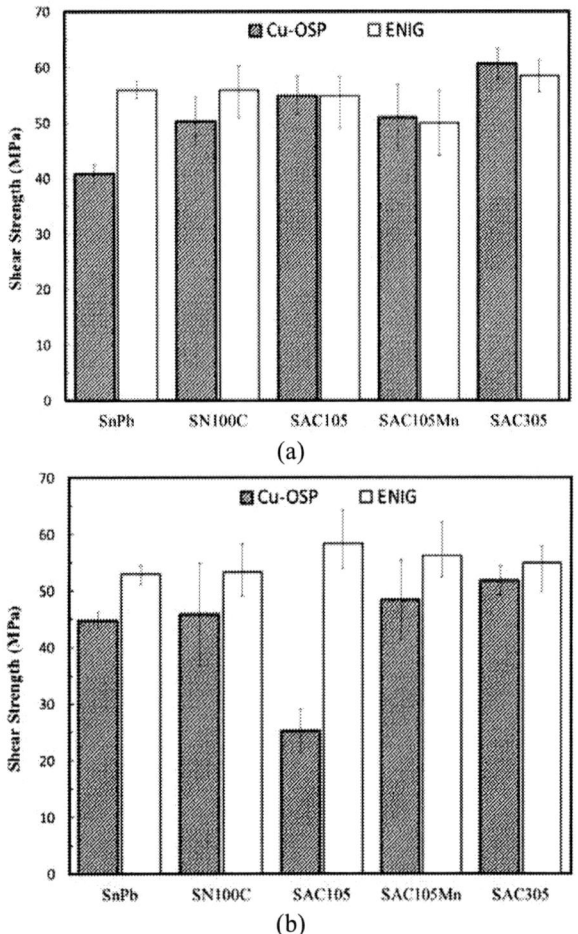

Figure 7. Low speed (1mm/s) shear strength of five different alloys reflowed on Cu-OSP and ENIP surface finishes, (a) as reflowed condition, (b) pre-aged at 125°C for 500h.

In order to evaluate the effect of isothermal pre-aging, the alloys were aged at 125°C for 500h prior to shear testing. The results for the pre-aged samples are shown in the bar chart in Figure 7(b). The results show a substantial reduction in the strength of SAC 105 reflowed on a Cu substrate after pre-

aging. The SAC305 alloy reflowed on the Cu substrate showed some reduction in strength and larger variation results, which might be the result of precipitates coarsening. For all other cases, change in shear strength after isothermal aging is relatively insignificant. These results further suggest that shear strength test is not very sensitive to the microstructure of solder alloy.

The failure surfaces of isothermally tested shear samples were analyzed. Failures were categorized one of four different modes as illustrated in Figure 8: 1) Ductile: when solder ball fractured at or above the surface of the solder mask within the bulk solder material; 2) Pad cratering: when solder pad lifts with solder ball and lifted pad might include ruptured base material; 3) Brittle or Interfacial break: when the break was at intermetallic compound (IMC);. 4) Mixed-mode failure was characterized as quasi-brittle or quasi-ductile. If the intermetallic compound covered less than 50% of failure surface, then the failure mode was considered as quasi ductile. When less than 10% of any other mode appeared on a surface, it was ignored. In low speed testing, irrespective of the surface finish and isothermal aging conditions, 100 percent bulk solder failures were recorded. However, different failure modes were observed after shear testing at high speed. Different failure modes are shown in Figure 8.

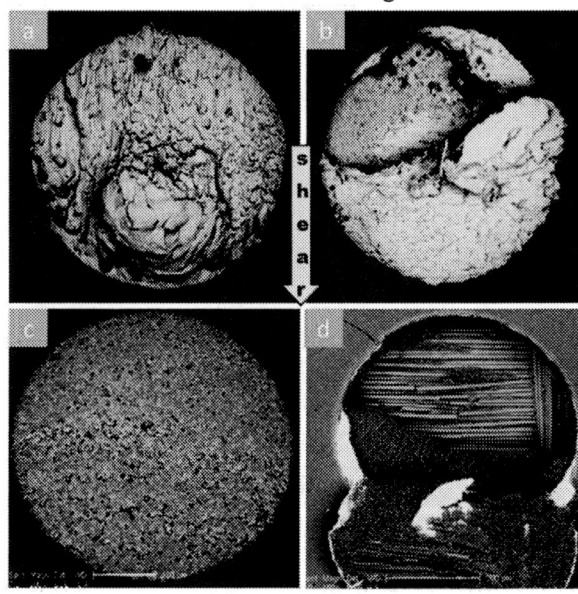

Figure 8. Different failure modes after isothermal testing: (a) Bulk Solder, (b) Quasi Ductile, (c) Intermetallic Compound (d) Pad Cratering. Shear direction is shown as well.

High speed shear testing (1m/s) was performed on alloys reflowed on Cu-OSP surface finish. Various failure modes for as reflowed and pre-aged samples were observed and summarized in Figure 9. The SAC 305 and SN100C alloys showed some percentage of complete IMC failure in the as-reflowed condition. After aging, there was more evidence of solder and quasi ductile fracture observed for SAC 105, SAC 305 and SAC 105-Mn. However, SN100C showed more brittle fracture after aging. The failure in the SnPb eutectic alloy always occurred within the bulk solder. Changes in the morphology and composition of intermetallic compounds in SAC alloys reflowed on Cu after aging could account for the change in the failure mode. The SAC-Mn alloy showed some evidence of IMC failure after it was pre-aged. In summary, the high speed shear test results indicated that various failure modes are observed with each of the different alloys. It is likely that changes in microstructure during isothermal aging are responsible for altering the failure mode of various solder alloys. Thus, it appears that high speed shear testing is more sensitive to microstructure compared to low speed shear testing.

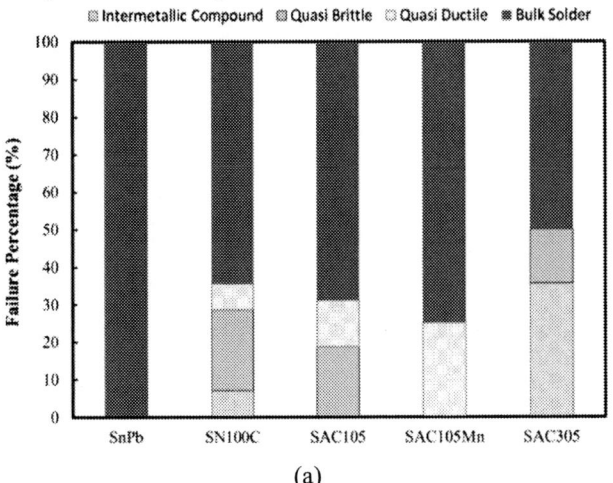

(a)

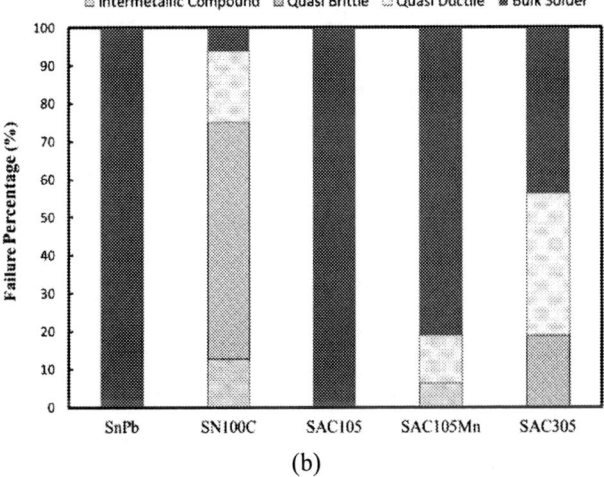

(b)

Figure 9. Different failure modes after high speed isothermal testing on five different alloys reflowed on Cu-OSP surface finish: (a) as-reflowed, (b) pre-aged at 125°C for 500h.

Room Temperature Shear Fatigue Test

The sensitivity of the room temperature fatigue lifetime to solder composition was examined. The cycles to failure were measured for 20 mil diameter spheres of each solder composition reflowed on Cu-OSP. These data are presented in the form of two parameter Weibull plots of cumulative percentage of failures versus cycles to failure. Fits to these data sets provide estimates of the characteristic lifetime or scale η, shape parameter or slope β, for each data set.

The results in Figure 10 show a clear dependence of room temperature fatigue life on solder composition of various SAC alloys. The characteristic lifetime of the SAC 405/Cu solder joints is more than six times that of the

SAC105/Cu in this test. The characteristic lifetime of the SAC205/Cu solder joints is fifty percent longer than that of SAC105/Cu solder joints, although the shape parameter was lower. Such results for room temperature fatigue tests are consistent with results from the literature for similar SAC compositions in accelerated thermal cycling (ATC) testing, where higher Ag alloys showed better thermal fatigue performance [1-3].

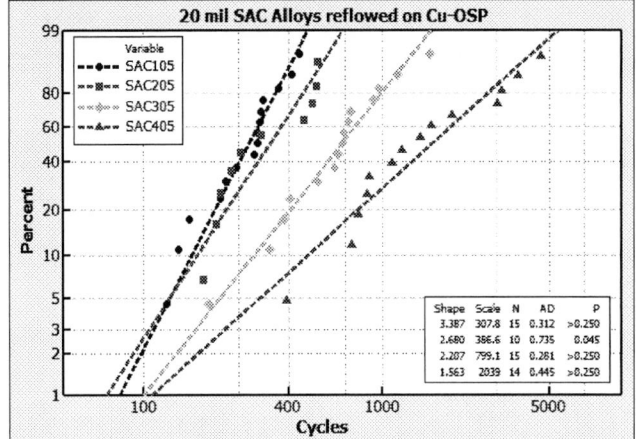

Figure 10. Weibull plots showing the room temperature shear fatigue results of 20 mil SAC alloys reflowed on Cu-OSP surface finish. As the amount of Ag in the alloy was increased, fatigue lifetimes increased as well.

Additional results of room temperature shear fatigue test on five different alloys including the SnPb are shown in Weibull plot of Fig. 11. As expected, the SN100C alloy, which contains no Ag, has the lowest lifetime and SAC 305 alloy has the longest lifetimes. The SAC 305 lifetime is nearly 12 times that of the SN100C alloy. The SAC 105 alloy with 0.5wt% more Ag than SACM alloy lasted 100% longer than SACM alloy. These observations reconfirm that a higher amount of Ag improves the fatigue lifetime of SAC alloys. It is interesting to note that the SnPb alloy outperforms all the low Ag, Pb-free alloys. However, SAC 305 still lasted 1.5 times longer than SnPb alloy. It is evident that room temperature shear fatigue test is sensitive to the composition of solders.

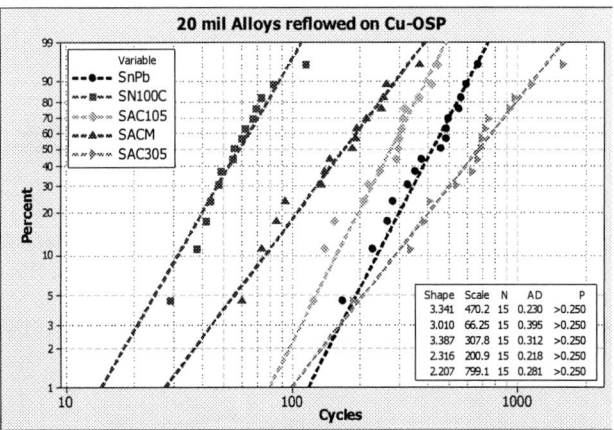

Figure 11. Weibull plots showing the room temperature shear fatigue results of 20 mil alloys reflowed on Cu-OSP surface finish. Shear fatigue results were sensitive to solder composition.

Accelerated Thermal Cycling Test

Thermal fatigue reliability of four different alloys of SnPb, SN100C, SAC305 and SACM was evaluated using a thermal cycling profile of -40/125°C with 15 min dwell time. Figure 12 shows a Weibull plot of the results of cycling for components constructed with 20 mil spheres and assembled on substrates using Cu or ENIG surface finishes (the component surface finish was Cu-OSP). SnPb and SAC 305 alloys built on Cu/ENIG surface finishes had somewhat better reliability when compared to those assembled with Cu/Cu surface finishes. However, SN100C performed better when it was assembled on Cu/Cu surface finishes. Higher values for the shape factors were obtained for SnPb and SN100C assembled with Cu/ENIG surface finishes. For both combinations of surface finishes, i.e. Cu/Cu and Cu/ENIG, SAC 305 shows higher lifetimes when compared to SnPb and SN100C alloys. SN100C could perform better than SnPb if Cu/ENIG surface finishes were used.

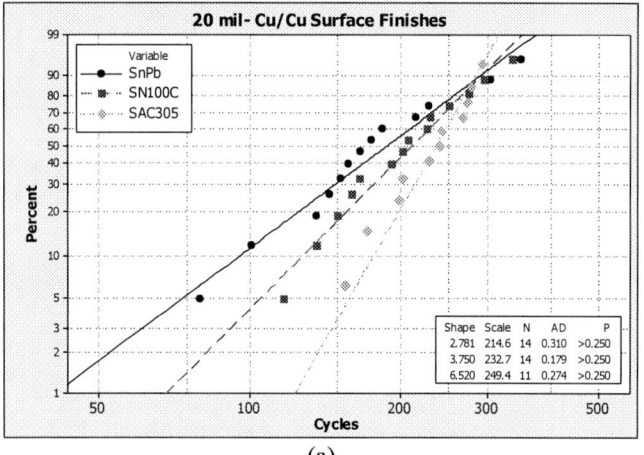

(a)

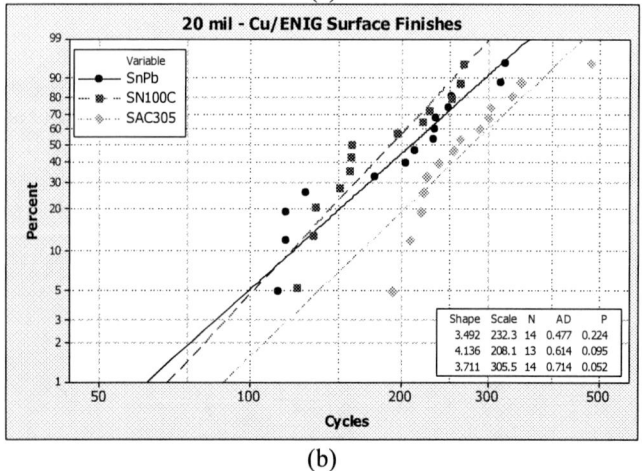

(b)

Figure 12. Weibull plots showing the accelerated thermal cycling results of components constructed with 20 mil solder spheres of SnPb, SN100C and SAC 305 and assembled on (a) Cu/Cu, (b) Cu/ENIG surface finishes. Thermal cycling profile was -40/125 °C.

Weibull plots for components constructed with 10 mil spheres are presented in Figure 13. In addition to SnPb, SN100C and SAC305, SACM were included in the experiment. The same trend seen in the larger components was observed here; the higher Ag alloy, SAC 305, outperformed all other alloys irrespective of surface finishes. The results confirm that higher Ag alloys show better reliability in accelerated thermal cycling test as reported by previous investigators [1-3]. Similarly to larger components (cf. Figure 12), the combination of Cu/ENIG surface finishes resulted in better reliability. However, the performance of SN100C was the same on Cu/Cu and Cu/ENIG finishes. As observed in larger components, SAC 305 showed some reduction in reliability when assembled on Cu/Cu surface finishes. Results indicate that SnPb alloy lasts longer or equal to SN100C and SACM on both surface finishes. Direct comparisons to SACM were limited because SACM has a lower shape factor on both surface finishes.

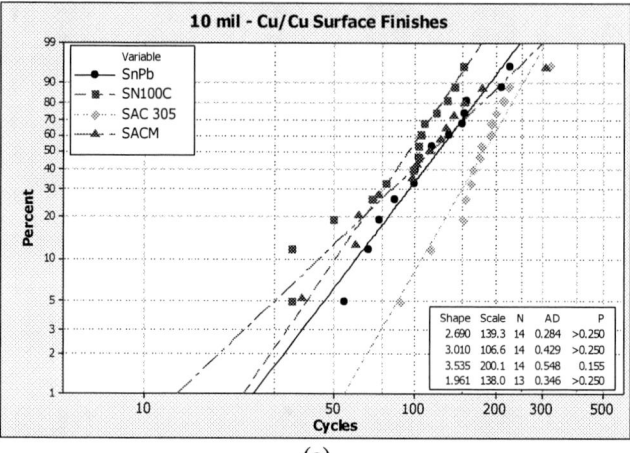

(a)

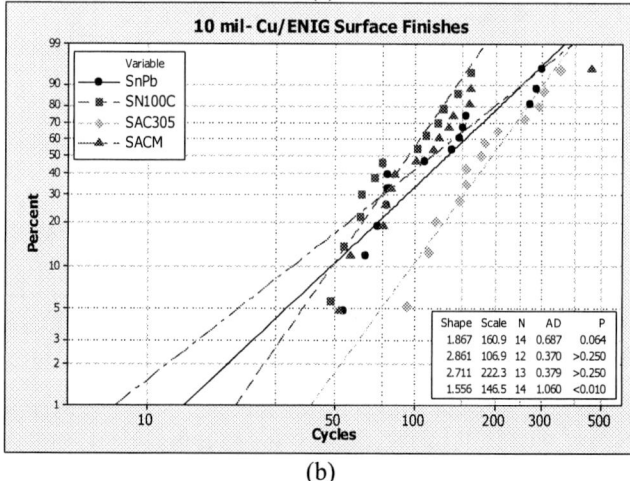

(b)

Figure 13. Weibull plots showing the accelerated thermal cycling results of homemade components constructed with 10 mil solder spheres of SnPb, SN100C and SAC 305 and assembled on (a) Cu/Cu, (b) Cu/ENIG surface finishes. Thermal cycling profile was -40/125°C.

Microstructural Analysis

Figure 14 shows DSC measurements of free standing spheres of various alloys obtained from commercial solder manufacturers. The degree of undercooling was measured for 10 and 20 mil solder spheres. The undercooling for SnPb is not different for 20 and 10 mil diameter spheres. However, in the case of SAC 305 and SN100C, smaller 10 mil spheres showed a higher degree of undercooling as expected. For the SAC 305 alloy, the 10 mil spheres undercooled much more than the 20 mil spheres (Fig. 14); the 10 mil solder balls undercooled approximately 75°C while the 20 mil spheres undercooled 9°C. Caution should be taken while comparing the undercooling values of different alloys and sizes to other published data. Our previous results have shown that there is a variation from solder to solder when samples are obtained from different manufactures or batches [12]. Variation in the trace impurity level from sphere to sphere might be the reason for different undercooling values obtained in different experiments performed on same nominal composition of solder alloys.

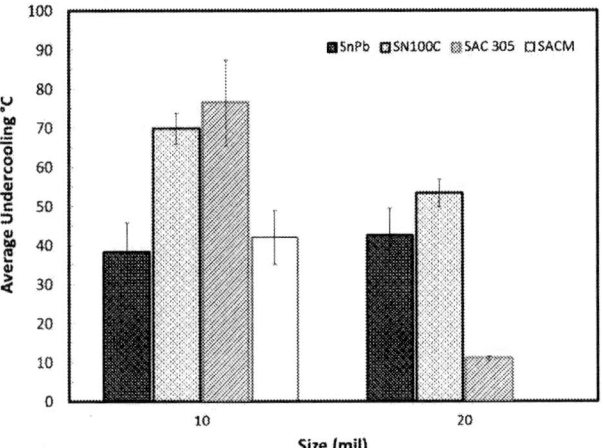

Figure 14. The variation of the undercooling (°C) for various solder alloys. Free standing samples of 10 and 20 mil were tested. Results showed the average of three runs on five different samples. The reflow peak temperature was 250°C and heating and cooling rates were 1°C/s.

It is well known that the microstructures of Pb-free alloys are sensitive to variations in composition and solidification temperature. Representative bright field micrographs of 10 and 20 mil free standing solder spheres are shown in Fig. 15. There is an obvious difference in microstructure of SnPb, SAC 305 and SN100C alloys. Smaller 10 mil samples that undercooled more (cf. Fig. 14) have smaller precipitates compared to larger 20 mil samples that solidified at higher temperatures. In the case of SN100C alloy, large Sn plates were observed, which is observed commonly in near eutectic Sn-Cu compositions.

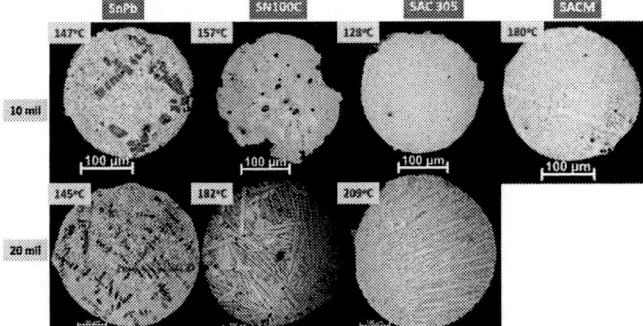

Figure 15. Optical images image of free standing 10 and 20 mil SnPb, SN100C, SAC 305. SACM was available only in 10 mil spheres. Solidification temperature of each sphere is indicated on the image.

For the same solder composition solidified at a lower temperature, smaller precipitates are expected. For example, 10 mil SAC 305 solidified at 128°C shows smaller Ag_3Sn precipitates compared to larger 20 mil SAC 305 that solidified at 209°C. The Sn dendrite sizes were also larger for the larger samples that solidified at higher temperatures. However, there is a variation in microstructure in different regions of the samples. This is expected because at a further distance from the nucleation point, Sn dendrites are expected to be larger as reported previously [5]. The amount of Mn in the SACM alloy is about 300 ppm. Several studies mention that this amount of Mn is not detectable with common EDS analysis. With a higher percentage of Mn, the presence of $MnSn_2$ phase is reported [30]. In this case, there was no significant effect of Mn on microstructure. More detailed analysis is needed to quantify the size and morphology of precipitates in order to determine the effect of Mn on microstructure. This work is in progress to address those aspects to better clarify the effect of small addition of Ni and Mn on microstructure of solder joints and their effect on reliability in thermal cycling tests.

It was also shown that solidification temperature affects the Sn grain morphology of solders [6]. The optical micrographs using polarized light of 10 and 20 mil free standing spheres of SAC 305 and SN100C are shown in Figure 16. The 10 mil sample that solidified at 128°C shows interlaced twinning structure while the larger 20 mil sample that solidified at 209°C shows beach ball structure commonly observed in BGA samples. Although the SN100C alloy shows a change in the Sn grain morphology as a function of undercooling, the Sn grain morphology looks different than SAC 305 alloy. Previous results by Lehman et al. confirmed that Sn-Cu alloys also show twining structure, with only one nucleation event similar to SAC alloys [7]. It has been proposed that Ag or Ni changes the nucleus structure and promotes the twinning structure [9, 21].

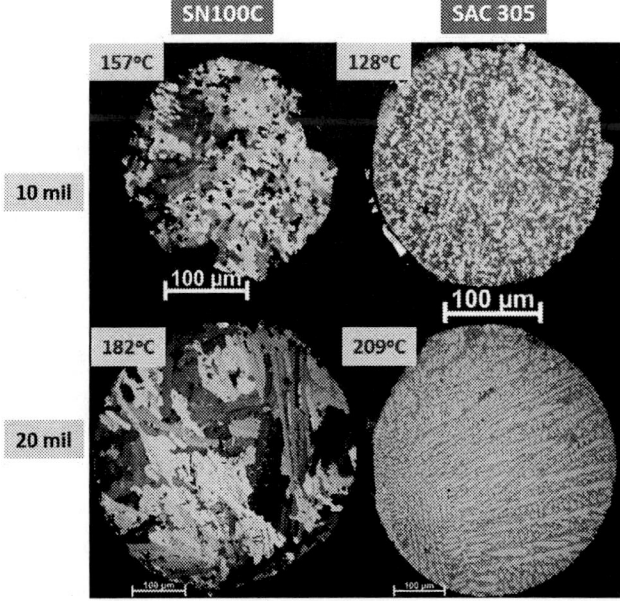

Figure 16. Micrographs with crossed polarizer of free standing 10 and 20 mil SAC305 and SN100C solder spheres. Solidification temperature of each solder ball is measured. Smaller, 10 mil spheres undercooled more and showed lots of grain boundaries, while larger, 20 mil showed large grains.

The optical micrographs of two 20mil SAC 305 samples that had completely failed after room temperature shear fatigue test are shown in Fig. 17. The room-temperature fatigue study revealed that the crack initiated in the Sn-Ag-Cu solder near the IMC and propagated through the bulk solder across the joint. The cycles to failure of the two joints were distinctively different; one of the joints lasted more than 20 times longer than the other joint. Although previous study revealed a relationship between the room temperature fatigue life of near eutectic Sn-Ag-Cu solder and the orientation of single Sn grains [5, 14], examination of Sn grain orientation of samples in Fig. 17 showed that this correlation does not extend to those two joints. In fact, the samples had essentially the same orientation, even though they failed at very different cycles. It is apparent that other effects, such as variations in precipitate size and number, may also be important, even for these single-Sn-grain samples. Quantitative metallographic analysis on x-section and z-section samples is needed to further clarify the effect of precipitates on lifetimes of solder joints in room temperature fatigue test.

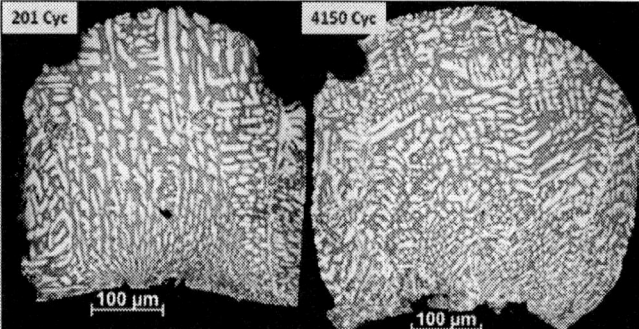

Figure 17. Bright field optical micrographs of the failed 20 mil SAC305/Cu joints after room temperature shear fatigue test. The cycles to failure were distinctively different.

Figure 18. Optical micrographs with cross polarizer image of components constructed with 10 mil spheres assembled on Cu/ENIG surface finishes. Cracks were observed near the component side. Recrystallization and intergranular crack propagation was observed for Pb-free alloys.

It is well known that the microstructure of Pb-free solder joints evolves during thermal cycling test [11, 7]. The optical micrographs of components assembled on Cu/ENIG

surface finishes of SN100C and SACM alloys after failure in ATC testing are shown in Fig. 18

Both alloys showed beach ball Sn grain morphology that originated from the original solidification event. Full cracks can be seen to subtend the joint, near the component. Formation of new grains close to crack was observed. The common failure mechanism was recrystallization and intergranular crack propagation along those newly formed gains, consistent with the description of the failure mechanism proposed by Yin [18]. While the failure mechanism for SN100C and SACM alloys seems to be similar to SAC 305 alloy, more work is needed to evaluate the effect of small additions of Ni and Mn on recrystallization behavior of those alloys.

Conclusions

The results of low speed shear strength testing of different solder alloys after reflow and after isothermal pre-aging on Cu and ENIG substrates showed bulk solder failures but minimal differences in shear strength between different alloys. This finding suggests that low speed shear test data is not sensitive to solder microstructure. It was found out that the failure mode observed in high speed shear testing depended upon solder composition. Variations in these failure modes were observed after the solder joints were subjected to isothermal pre-aging. The only exception was SnPb alloy, which did not show any variations; failures were found exclusively in the bulk solder.

The results of room temperature fatigue testing of various solder/Cu joints showed distinct dependences of fatigue lifetime on solder composition. Distinct increases in lifetimes with increases in Ag content were observed in SAC/Cu solder joints. SAC 305 alloy showed higher room temperature fatigue lifetimes when compared to SnPb, SACM and SN100C alloys.

The results of ATC tests showed similar trends as observed in the shear fatigue test. SAC 305 outperformed SnPb, SN100C and SACM irrespective of surface finish combination. Results suggest the recrystallization and failure mechanism in Pb-free solder joints are strongly affected by Ag_3Sn precipitates. The combination of Cu/ENIG surface finishes generally resulted in somewhat better thermal fatigue performances. Microstructural analysis of solder spheres confirmed that smaller spheres exhibit more undercooling. Because the solder volume affects the solidification temperature of the joint, the precipitate size and density as well as the Sn grain morphology of solder joints will be affected. All these aspects of microstructure are known to affect the thermal fatigue performance of solder joints.

Acknowledgments

Babak Arfaei is greatly thankful to Shantanu Joshi, Binghamton University, for performing mechanical tests on selected alloys, and Gaurang Joshi, Binghamton University, and Michael Meilunas, Universal Instrument, for their help performing the ATC testing.

References

1. Sweatman K., Howell K., Coyle R., Parker R., Henshall G., Smetana J., Benedetto E., Liu W., Pandher R., Daily D.,

Currie M., Nguyen J., Lee T-K., Osterman M., Miremadi J., Allen A., Arnold J., Moore D., and Chang G., "iNEMI Pb-Free Alloy Characterization Project Report: Part Iii - Thermal Fatigue Results For Low-Ag Alloys", *Proc. SMTA International* , pp. 359-375, Oct. 2012

2. Coyle R., Parker R., Longgood S., Sweatman K. Howell K., Arfaei B., "iNEMI Pb-Free Alloy Characterization Project Report: Part VI-The Effect of Component Surface Finishes and Solder Paste Composition on Thermal Fatigue of Sn100C Solder Balls," *Proc. SMTA International*, Oct. 2013,.

3. Coyle R., Parker R., Henshall G., Osterman M., Smetana J., Benedetto E., Moore D., Chang G., Arnold J., Lee T. K., "iNEMI Pb-Free Alloy Characterization Project Report: Part IV-Effect of Isothermal Preconditioning on Thermal Fatigue Life," *Proc. SMTA International* 2012.

4. Liu W., Lee N-C., Porras A., Ding M., Gallagher A., Huang A., Chen S., Lee J C-B, "Achieving High Reliability Low Cost Lead-Free SAC Solder Joints Via Mn or Ce Doping", *Proc. 59th Electronic Components and Technology Conference* (ECTC), pp. 994-1007

5. Arfaei B. and Cotts E., "Correlations Between The Microstructure and Fatigue Life of Near-Eutectic Sn-Ag-Cu Pb-Free Solders", *Journal of Electronic Materials*, Vol. 38, No. 12, pp. 2617-2627, 2009.

6. Arfaei B., Kim N., Cotts E.J., "Dependence of Sn Grain Morphology of Sn-Ag-Cu Solder on Solidification Temperature", *Journal of Electronic Materials*, vol. 41, no. 2, pp. 362-374, 2012.

7. Lehman L.P., Xing Y., Bieler T.R., and Cotts E.J., "Cyclic Twin Nucleation in Tin-Based Solder Alloys", *Acta Materialia*, Vol. 58, No. 10, pp. 3546-3556, 2010.

8. Mutuku F., Arfaei B., Anselm M., Cotts E., " Effect of Variation in the Reflow Profiles in Room Temperature Fatigue Tests" *Proc. SMTA International,* 2013

9. Parks, G., Arfaei, B., Benedict, M., Cotts, E., Minhua Lu, Perfecto, E. ," The dependence of the Sn grain structure of Pb-free solder joints on composition and geometry", *Proc. 62nd Electronic Components and Technology Conference (ECTC)*, pp. 703-709, May. 2012.

10. Arfaei B., Wentlent L., Joshi S., Anselm M., and Borgesen P., "Improving the Thermomechanical Behavior of Lead Free Solder Joints by Controlling the Microstructure". *Proc. 13th IEEE Intersociety Conference ITherm*, 2012.

11. Arfaei B., Anselm M., Joshi S., M-Shirazi S., Borgesen P., Cotts E., Wilcox J., Coyle R. "Effect of Sn Grain Orientation on Failure Mechanism and Reliability of Lead-Free Solder Joints in Thermal Cycling Tests" , *Proc. SMTA International*, pp. 539-550, Oct. 2013

12. Arfaei B., Wentlent L., Joshi S., Anselm M., and Borgesen P., "Controlling the Superior Reliability of Lead Free Assemblies with Short Standoff Height Through Design and Materials Selection", *Proc. of IMECE,* Nov 2012,

13. Bieler, T.R., Hairong Jiang, Lehman, L.P., Kirkpatrick, T., Cotts, E.J. and Nandagopal, B "Influence of Sn Grain Size and Orientation on the Thermomechanical Response and Reliability of Pb-free Solder Joints", *IEEE Transactions on Components, Packaging, and Manufaturing Technology*, Vol.31, No. 2, pp. 370-381, 2008.

14. Arfaei B., Xing Y., Woods J., Wolcott J., Tumne P., Borgesen P. , Cotts E., "The Effect of Sn Grain Number and Orientation on the Shear Fatigue Life Of SnAgCu Solder Joints", *Proc. 58th Electronic Components and Technology Conference* (ECTC), pp. 459-465, May. 2008.

15. Lu M., Shih D.Y., Lauro P., Goldsmith C., Henderson D.W., "Effect of Sn grain orientation on electromigration degradation mechanism in high Sn-based Pb-free Solders," *Applied Physics Letters,* 92, 211909 (2008).

16. Joshi S., Arfaei B., Singh A., Gharaibeh M., Obaidat M., Alazzam A., Meilunas M., Yin L., Anselm M., and Borgesen P., "LGAs Vs. BGAs – Lower Profile and Better Reliability" *Proc. SMTA International*, 2012.

17. Arfaei B., M-Shirazi S., Joshi S., Anselm M.,Borgesen P., Cotts E., Wilcox J., Coyle R. "Reliability and Failure Mechanism of Solder Joints in Thermal Cycling Tests", *Proc. 63rd Electronic Components and Technology Conference* (ECTC), pp. 125-132, Jun. 2013.

18. Yin L., Wentlent L. , Yang L., Arfaei B., Qasaimeh A. and Borgesen P., "Recrystallization and Precipitate Coarsening in Pb-Free Solder Joints During Thermomechanical Fatigue", *Journal of Electronic Materials*, Vol. 41, No. 2, pp. 241-252, Feb. 2012.

19. Mattila T.T. , Kivilahti J. K., The Failure Mechanism of Recrystallization-Assisted Cracking of Solder Interconnections, pp. 179-207, 2012.

20. Mattila T.T., Kivilahti J.K.,"The Role Of Recrystallization In The Failure Of SnAgCu Solder Interconnections Under Thermomechanical Loading", Transactions on components and packaging technologies, Vol. 33, PP. 629-635, 2010.

21. Arfaei, B., Benedict, M. ; Cotts, E.J. "Nucleation rates of Sn in undercooled Sn-Ag-Cu flip-chip solder joints," Journal of Applied Physics , vol.114, no.17, pp.,173506-10, 2013"

22. Arfaei B., Tashtoush T., Kim N., Wentlent L., Cotts E., Borgesen P., "Dependence of SnAgCu Solder Joint Properties on Solder Microstructure", *Proc. 61st Electronic Components and Technology Conference* (ECTC), pp. 125-132, Jun. 2011.

23. Zhang, Y., Cai, Z., Suhling, J. C., Lall, P., and Bozack, M. J., "The Effects of Aging Temperature on SAC Solder Joint Material Behavior and Reliability", *Proc. 58th Electronic Components and Technology Conference* (ECTC), May. 2008, pp. 99-112.

24. Smetana J., Coyle R., Sack T., Syed A., Love D., Tu D., Kummerl S., "Pb-free Solder Joint Reliability in a Mildly Accelerated Test Condition," *Proceedings of IPC/APEX*, Las Vegas, NV, April 2011

25. Smetana J., Coyle R., Read P., Popowich R., Fleming D., and Sack T., "Variations In Thermal Cycling Response Of Pb-Free Solder Due To Isothermal Preconditioning", *Proc. SMTA International*, pp. 641-654, Oct. 2011

26. Che F.X., Zhu W.H., Poh S.W., Zhang X.W., Zhang X.R., "The study of mechanical properties of Sn–Ag–Cu lead-free solders with different Ag contents and Ni doping under different strain rates and temperatures," *Journal of Alloys and Compounds*, vol. 507, pp. 215–224, 2010.

27. JEDEC Standard, "Solder Ball Shear", JESD22-B117A.

28. IPC-9701A "Performance Test Methods and Qualification Requirements for Surface Mount Solder Attachments".

29. Song J. M.; Lin L.W., Lee N.C., Lai Y.S., Chiu Y. T., "Metallurgical Perspective on Alloying Modification of Sn-Ag-Cu Solders," *Electronics Packaging Technology Conference, 2008. EPTC 2008.*, pp.1358,1363, 2008.

The Effects of Aging on the Fatigue Life of Lead Free Solders

Muhannad Mustafa, Jordan C. Roberts, Jeffrey C. Suhling, Pradeep Lall
Center for Advanced Vehicle and Extreme Environment Electronics (CAVE[3])
Auburn University
Auburn, AL 36849
Phone: +1-334-844-3332
FAX: +1-334-844-3124
E-Mail: jsuhling@eng.auburn.edu

Abstract

Solder joints in electronic assemblies are typically subjected to thermal cycling, either in actual application or in accelerated life testing used for qualification. Mismatches in the thermal expansion coefficients of the assembly materials leads to the solder joints being subjected to cyclic (positive/negative) mechanical strains and stresses. This cyclic loading leads to thermomechanical fatigue damage that involves damage accumulation, crack initiation, crack propagation, and failure. While the effects of isothermal aging on solder constitutive behavior (stress-strain and creep) have been examined in some detail, there have been few prior studies on the effects of aging on solder failure and fatigue behavior. Aging leads to both grain and phase coarsening, and can cause recrystallization at Sn grain boundaries. Such changes are closely tied to the damage that occurs during cyclic mechanical loading.

In this investigation, we have examined the effects of aging on the cyclic stress-strain and fatigue behaviors of lead free solders. Solder test specimens (SAC105 and SAC305) have been prepared and subjected to cyclic stress/strain loading at different aging conditions. Both uniaxial specimens subjected to cyclic tension/compression and Iosipescu lap shear samples subjected to cyclic positive/negative shear have been studied. A four-parameter hyperbolic tangent empirical model has been used to fit the entire cyclic stress-strain curve and the hysteresis loop size (area) was calculated using definite integration for a given strain limit. This area represents the energy dissipated per cycle, which is correlated to the damage accumulation in the joint. Samples were subjected to cyclic loading over a particular strain range until fatigue failure occurred, and then various popular empirical failure criteria such as the Coffin-Manson model and the Morrow model have been used to estimate the fatigue life. Fatigue failure was defined to occur when there was a 50% peak load drop during mechanical cycling.

Prior to testing, the specimens were aged (preconditioned) at 125 C for various aging times, and then the samples were subjected to cyclic loading at room temperature (25 C). It has been observed that prior aging dramatically decreases the mechanical fatigue life. It was also found that degradations in the fatigue/failure behavior of the lead free solders with aging are highly accelerated for lower silver content alloys (e.g., SAC105). Comparisons have been made between the fatigue lives under both uniaxial tension/compression and shear loadings, and good agreement was found. A microstructural adaptive fatigue model including aging effects has been

proposed, and shown to accurately predict the measured fatigue data for all aging conditions.

Introduction

Thermally cycling induced solder joint fatigue is a common failure mode in electronic packaging. When subjected to temperature changes, stresses in electronic assemblies are typically developed due to the mismatches in the coefficients of thermal expansion (CTE) of the soldered components and the PCB. Cyclic temperature changes, either due to external environment or power switching, can therefore lead to substantial alternating stresses and strains within the solder joints. During cyclic loading, micro cracks form within the solder material followed by macro cracks which leads to damage and ultimately to fatigue failure. Aging of solder materials degrades their mechanical and creep properties, and is expected to exacerbate the damage accumulation during cycling loading. Energy dissipation occurs during cyclic loading due to yielding and occurrence of viscoplastic deformations. The strain energy density dissipated per cycle can be calculated from the area of stress-strain hysteresis loops developed during cyclic loading.

Many prior researchers have studied the cyclic stress-strain behavior of Sn-Pb solder materials. For example, models to predict fatigue life of Sn-Pb solder alloys under thermal cycling or thermomechanical fatigue loading have been developed in several investigations [1-8]. Measurement of deformations during thermal cycling and stress-strain hysteresis in Sn-Pb solder joints has also been extensively examined. For example, Hall [9-10] measured deformations and hysteresis experimentally for the case of a leadless ceramic chip carrier soldered to an FR4 printed wiring board and subjected to thermal cycling. Pao [11-13] used a two-beam geometry to successfully measure the thermo-mechanical hysteresis Sn-Pb solder and other Sn-based alloys. Haacke, et al. [14] described a test assembly and reported both hysteresis and thermomechanical fatigue data for Sn-Pb solder.

The literature on the mechanical behavior of lead free solders has been reviewed by Ma and Suhling [15]. There are also several prior papers on the cyclic stress-strain behavior of lead free solder materials. For example, Raeder and co-workers [16] reported thermomechanical deformation behavior of Sn-Bi eutectic solder under fatigue loading in terms of plastic strain energy density. Dusek, et al. [17] studied the stress vs. strain behavior and evolution of the hysteresis loops of Sn3.5Ag0.5Cu solder during isothermal fatigue at several different temperatures (24, 30, 60, and 125

C). They reported that with slow cycling, where creep and stress relaxation play dominant roles, the number of cycles to failure increases with higher temperatures. Herkommer, et al. [18] have developed a damage model that is capable of predicting material behavior under both mechanical shear cycling and thermal cycling loading conditions. Zhang and Dasgupta [19-20] have discussed the mechanical and thermal cycling durability of selected lead-free solders (Sn3.9Ag0.6Cu, Sn3.5Ag, and Sn0.7Cu), and developed a mechanical fatigue damage criterion. They demonstrated good correlation between their damage criterion and cycles to failure under different strain rate and temperature conditions.

Whitelaw, et al. [21] have determined the parameters of the Bonder-Partom model from uniaxial cyclic stress-strain tests for both lead-free solders and Sn-Pb solder. They have also verified the model under isothermal thermo-mechanical cycling conditions. Korhonen and coworkers [22] conducted uniaxial cyclic tests under various working temperatures for near eutectic Sn-Ag-Cu alloy to understand the isothermal fatigue behavior. Nucleation and growth mechanisms of fatigue damage were also observed using optical microscopy, scanning electron microscopy (SEM), and the Electron Back-Scattering Diffraction (EBSD) method. Kanchanomai, et al. [23] performed uniaxial strain controlled cyclic test to investigate fatigue failure and mode II crack growth behavior of Sn-Ag eutectic solder alloy for different strain ranges, load drop parameters, and times to failure.

Fatigue failure behavior and cyclic creep deformations in lead free solders were also observed by Shang, et al. [24], where the fatigue lives of different lead-free solders were related to cyclic stress amplitude, number of cycles, stress ratio, loading frequency, temperature, alloy composition, and microstructure. Both uniaxial cyclic tests and shear cyclic tests were carried out by Andersson, et al. [25] to establish a comparison of isothermal mechanical fatigue properties of lead free solder joints and bulk solders. Finally, uniaxial cyclic tests were performed by Pang, et al. [26] for SAC387 (95.5Sn-3.8Ag-0.7Cu) and Sn-Cu (99.3Sn-0.7Cu) solder alloys. In their study, the Coffin-Manson model and the Morrow model were used to describe the low cycle fatigue behavior of those materials.

Several studies have been performed on degradation of solder material properties when the alloys are exposed to isothermal aging. For example, aging has been demonstrated to reduce BGA ball shear strength [27], solder elastic modulus [28], drop reliability [29], fracture toughness [30], and shear strength [31].

The effects aging on the constitutive and failure behavior of lead free solder has been extensively investigated by the authors [32-44]. In initial work, room temperature (25 C) and elevated temperature (50, 75, 100, and 125 C) aging were shown to severely degrade the mechanical properties and creep behavior of SAC alloys [32-35]. Cai and coworkers [36] have demonstrated that aging effects can be reduced through the use of doped SAC solder materials (i.e., SAC-X). In addition, Zhang, et al. [37-38] have shown that prior aging causes large reductions in the reliability of BGA test assemblies subjected thermal cycling accelerated life testing. Motalab, et al. [39-40] have included aging effects in the

Anand constitutive model and energy density based failure criterion for SAC solders, and then used these theories with finite element analyses to predict the thermal cycling life of aged BGA assemblies. Good correlations were achieved with the measured lifetimes from references [37-38]. Finally, Hasnine, et al. [41] has used nanoindentation to study the evolution of mechanical properties and creep rates in BGA solder joints subjected to aging, while Lall and coworkers [42] have investigated changes in the high strain rate behavior of SAC solders subjected to aging.

Mustafa, et al. have studied the aging-induced changes occurring in the cyclic stress-strain behavior of lead free SAC solders for both tension/compression [43] and shear [44] loadings. The effects of several parameters were examined including aging time, temperature, strain/stress limits, and solder alloy composition. In addition, the evolution of the solder hysteresis loops with aging were characterized and empirically modeled. Similar to solder stress-strain and creep behavior, there was a strong influence of aging on the hysteresis loop size (and thus the rate of damage accumulation) in the solder specimens.

While the effects of aging on solder constitutive behavior (stress-strain and creep) have been recently examined in some detail, there have been no prior studies on the effects of aging on solder failure/fatigue behavior. In the present investigation, we have examined the effects of aging on the cyclic stress-strain and fatigue behaviors of lead free solders. Solder test specimens (SAC105 and SAC305) have been prepared and subjected to cyclic stress/strain loading at different aging conditions. Both uniaxial specimens subjected to cyclic tension/compression and Iosipescu lap shear samples subjected to cyclic positive/negative shear have been studied. Samples were subjected to cyclic loading over a particular strain range until fatigue failure occurred, and then various popular empirical failure criteria such as the Coffin-Manson model and the Morrow model have been used to estimate the fatigue life. Prior to testing, the specimens were aged (preconditioned) at 125 C for various aging times, and then the samples were subjected to cyclic loading at room temperature (25 C). Comparisons have been made between the fatigue lives under both uniaxial tension/compression and shear loadings, and a microstructural adaptive fatigue model including aging effects has been proposed and evaluated.

Experimental Procedure

Uniaxial Test Sample Preparation

Solder uniaxial samples are often fabricated by machining of bulk solder material, or by melting of solder paste in a mold. In the current study, we have used a novel procedure where solder uniaxial test specimens are formed in high precision rectangular cross-section glass tubes using a vacuum suction process [32-36, 39-44]. The tubes are first cooled by water quenching. They are then sent through a SMT reflow at a later time to re-melt the solder in the tubes and subject them to any desired temperature profile (i.e. same as actual solder joints).

The solder is initially melted in a quartz crucible using a pair of circular heating elements (see Figure 1). A

978-1-4799-2408-0/14 $31.00 © 2014 IEEE

thermocouple attached on the crucible and a temperature control module is used to direct the melting process. One end of the glass tube is inserted into the molten solder, and suction is applied to the other end via a rubber tube connected to the house vacuum system. The suction forces are controlled through a regulator on the vacuum line so that only a desired amount of solder is drawn into the tube. The specimens are then cooled to room temperature using a user-selected cooling profile. For the reflow oven controlled cooling, the tubes are first cooled by water quenching, and they are then sent through a reflow oven (9 zone Heller 1800EXL) to re-melt the solder in the tube and subject it to the desired temperature profile as shown in Figure 2. Typical glass tube assemblies filled with solder and a final extracted specimen are shown in Figure 3. For some cooling rates and solder alloys, the final solidified solder samples can be easily pulled from the tubes due to the differential expansions that occur when cooling the low CTE glass tube and higher CTE solder alloy.

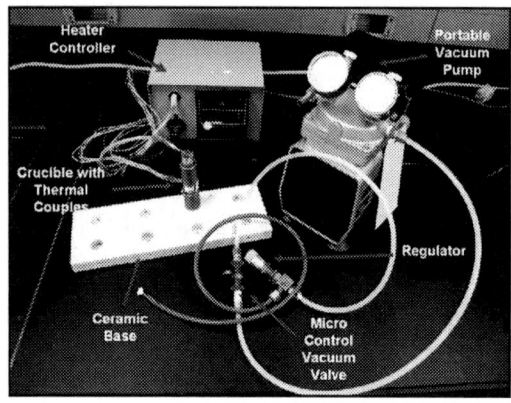

Figure 1 - Specimen Preparation Hardware

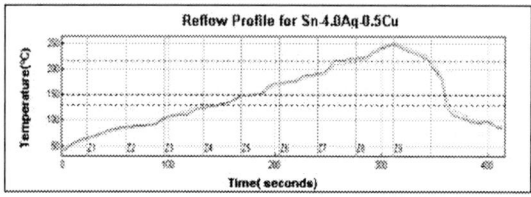

Figure 2 - Solder Reflow Temperature Profiles for SAC (105/205/305/405) Solder Specimens

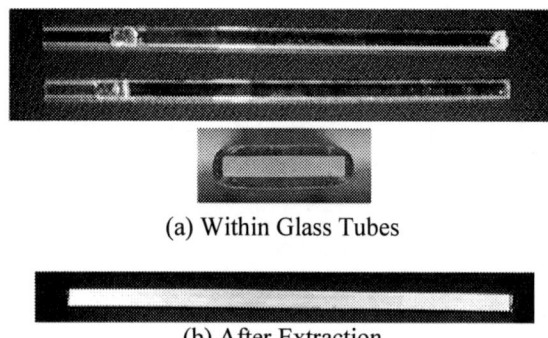

(a) Within Glass Tubes

(b) After Extraction

Figure 3 - Solder Uniaxial Test Specimens

In the current work, we formed uniaxial samples with nominal dimensions of 80 x 3 x 0.5 mm. A thickness of 0.5 mm was chosen because it matches the height of typical BGA solder balls. The specimens were stored in the aging oven immediately after the reflow process to eliminate possible room temperature aging effects.

Mechanical Testing System

A MT-200 tension/torsion thermo-mechanical test system from Wisdom Technology, Inc., as shown in Figure 4, has been used to test the samples in this study. The system provides an axial displacement resolution of 0.1 micron and a rotation resolution of 0.001°. Testing can be performed in tension, shear, torsion, bending, and in combinations of these loadings, on small specimens such as thin films, solder joints, gold wire, fibers, etc. Cyclic (fatigue) testing can also be performed at frequencies up to 5 Hz. In addition, a universal 6-axis load cell was utilized to simultaneously monitor three forces and three moments/torques during sample mounting and testing. Environmental chambers added to the system allow samples to be tested over a temperature range of -185 to +300 °C.

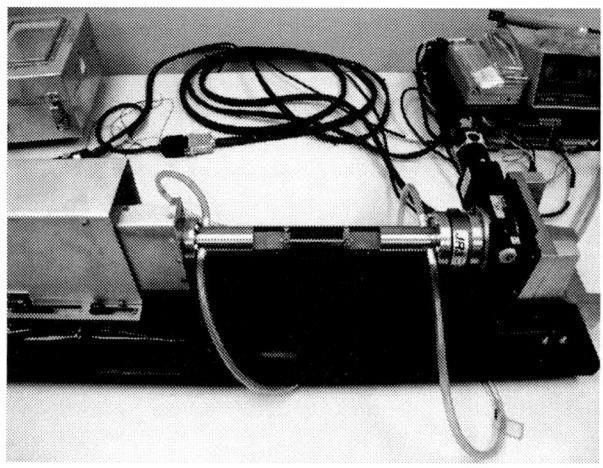

Figure 4 - MT-200 Testing System with Solder Sample

During uniaxial testing, forces and displacements were measured. The axial stress and axial strain were calculated from the applied force and measured cross-head displacement using

$$\sigma = \frac{F}{A} \qquad \varepsilon = \frac{\Delta L}{L} = \frac{\delta}{L} \qquad (1)$$

where σ is the uniaxial stress, ε is the uniaxial strain, F is the measured uniaxial force, A is the original cross-sectional area, δ is the measured crosshead displacement, and L is the specimen gage length (initial length between the grips). The gage length chosen in this study for cyclic stress-strain and fatigue testing was 10 mm. A short specimen length was necessary so that buckling could be prevented during the compressive part of the cyclic testing. The buckling force was estimated prior to testing using Euler's Formula:

$$F_{cr} = \frac{\pi^2 EI}{L_{eff}^2} \quad (2)$$

where E is initial elastic modulus of solder material, and the effective length $L_{eff} = L/\sqrt{2}$ was used for the clamped-clamped end conditions present in the grips of the mechanical testing system.

Cyclic Stress-Strain Curves and Data Processing

Typical uniaxial tension/compression cyclic stress-strain testing results for a SAC solder uniaxial specimen are shown in Figure 5. In this case, the specimen was tested in strain controlled cycling for 10 cycles with strain limits of ±0.0018. The sample was initially relaxed (unloaded, point A), and then a tensile loading was applied to generate the initial portion of the stress-strain curve (A to B). Tensile loading was continued until the strain reached +0.0018 (point B). At that time, the loading direction and the direction of cross-head motion were reversed. The sample was then compressed until the strain reached -0.0018 (point C), when the direction of loading and extension was reversed again (to be tensile). Testing continued in tension until the strain again reached +0.0018 (point B) and the first hysteresis loop was completed. Further loops (between points B and C) were generated by continuing the alternating compressive and tensile loadings in a similar manner.

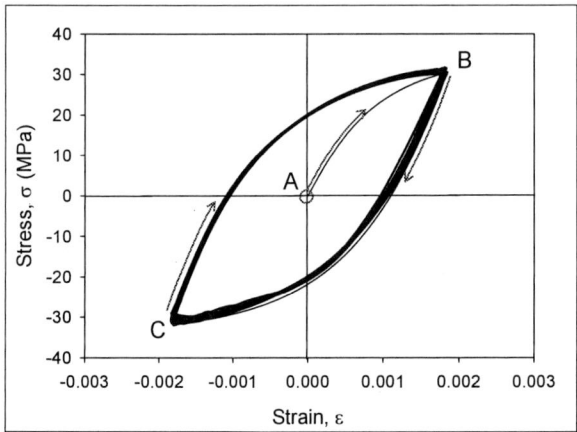

Figure 5 - Typical Cyclic Stress-Strain Test Results for SAC Solder (Strain Controlled)

Although, several different empirical models can be used to represent the observed stress-strain data for solder, we have chosen to use four parameter hyperbolic tangent model [43-46]

$$\sigma = C_1 \tanh(C_2 \varepsilon) + C_3 \tanh(C_4 \varepsilon) \quad (3)$$

where initial effective elastic modulus can be calculated as

$$E = C_1 C_2 + C_3 C_4 \quad (4)$$

To process the cyclic stress-strain data and calculate the areas of the associated hysteresis loops, a pair of empirical models have been used to represent the tension portion (C to B, top) and the compression portion (B to C, bottom) of the stress-strain behavior in each cycle. Figure 6 illustrates the definitions of the two portions of the stress-strain cycle and the empirical fits $f_1(\varepsilon)$ for the compression loading region (bottom of the hysteresis loop) and $f_2(\varepsilon)$ for the tensile region (top of the hysteresis loop). Extending the hyperbolic tangent expression in Eq. (3), the following empirical models were found to fit our experimental cyclic stress-strain data well:

$$f_1(\varepsilon) = -P_1 \tanh(P_2(-\varepsilon + \varepsilon_2)) - P_3 \tanh(P_4(-\varepsilon + \varepsilon_2)) + \sigma_1$$

$$f_2(\varepsilon) = R_1 \tanh(R_2(\varepsilon - \varepsilon_1)) + R_3 \tanh(R_4(\varepsilon - \varepsilon_1)) + \sigma_2 \quad (5)$$

$$\varepsilon_1 \le \varepsilon \le \varepsilon_2$$

From the stress-strain data for each cycle, the constants in the empirical models in Eq. (5) can be determined through a nonlinear regression analysis. The hysteresis loop area can then be evaluated using

$$\Delta W = \int_{\varepsilon_1}^{\varepsilon_2} [f_2(\varepsilon) - f_1(\varepsilon)] \, d\varepsilon \quad (6)$$

This area represents the energy density dissipated per cycle (as heat) during the cyclic loading.

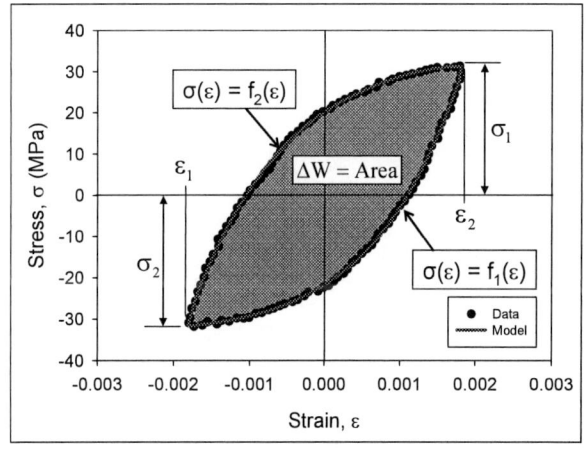

Figure 6 - Typical Hysteresis Loop and Area Calculation

Cyclic Stress-Strain Testing Results

In previous papers [43-44], we have investigated the effects of aging on the cyclic stress-strain behavior of SAC solders. Testing was performed for up to 10 loading and unloading cycles. The major observations of these studies are summarized below.

Effects of Aging for Strain Controlled Cyclic Testing

Uniaxial tension/compression strain controlled cyclic tests were performed on a SAC105 samples subjected to aging at 125 C. After solidification, the samples were subjected to aging at 125 C for various durations including 0, 5, 10, 30, 60, 90, 120, 150, 180, 270 and 360 days (0-12 months). The 0-day aging specimens represented non-aged samples, which were tested within a few minutes after solidification. For each aging time, there were 5 specimens tested. The strain controlled testing was performed with strain limits of ±0.0018, a strain rate of $\dot{\epsilon} = 0.0001$ sec^{-1}, and a testing temperature of 25 C.

Figure 7 illustrates example plots of the cyclic stress-strain behavior for different aging times. Each curve is the plot of the 10th stress-strain cycle for one of the five tests performed for a particular aging duration. Results for the other 4 samples were remarkably similar, and the data showed a high level of consistency and repeatability from sample to sample. It is observed that the hysteresis loops rotate and become smaller in height for samples with increased aging. From the recorded data, the hysteresis loop areas were calculated using the procedure outlined above. A plot of the evolution of hysteresis loop area with aging time is shown in Figure 8. In this graph, each data point represents the average value of the 5 tests performed for a particular aging duration. The area of hysteresis loop denotes the energy density (volumetric) dissipated per cycle, and is related to the damage accumulated per cycle. The units are in J/m^3 (energy per unit volume). The data in Figure 8 illustrate a smooth reduction of the hysteresis loop size with aging. It is observed that the most dramatic changes in the hysteresis loop area occur within the first 20-30 days of aging. The rate of change of the area with aging time then becomes much smaller and the loop size decreases linearly with long aging times.

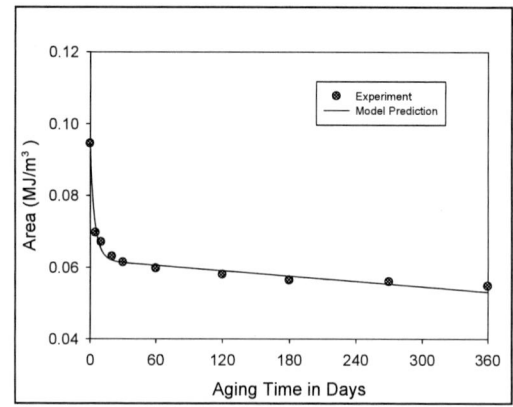

Figure 8 - Variation of the Hysteresis Loop Area with Aging Time (SAC105, Strain Controlled)

Morrow [47] suggested that on the microscopic level, cyclic plastic strain is related to the movement of dislocations and the cyclic stress is related to the resistance to their motion. Repeated stress without accompanying plastic strain (no movement of dislocations) will not cause fatigue damage nor will repeated slip without repeated stress (resistance to dislocation movement). Both stress and plastic strain are required for the dissipation of mechanical energy (a hysteresis loop exists). Thus, the plastic strain energy per cycle may be regarded as a composite measure of the amount of fatigue damage per cycle, and the fatigue resistance of a material may be characterized in terms of its capacity to absorb and dissipate plastic strain energy. From the test results, it is evident that for strain controlled cyclic testing, the movement of dislocations does not change significantly for different aging conditions. However, the height of a hysteresis loop decreases significantly as the resistance to dislocation movement decreases due to microstructural changes occurring during aging.

Uniaxial Fatigue Testing

Effects of Aging on Fatigue Life

Uniaxial tension/compression fatigue testing was performed using both SAC105 and SAC305 test specimens subjected to prior aging. After solidification, the samples were aged at 125 C for various durations up to 1 year as shown in the test matrix in Figure 9. The 0-day aging specimens represented non-aged samples, which were tested within a few minutes after solidification. Strain controlled fatigue testing was performed with various strain limits ranging from ±0.0005 to ±0.006 (e.g. ±0.0015, ±0.0025, ±0.0035, ±0.0045, etc.). For each set of aging conditions and strain range, at least 5 samples were tested to failure. The cyclic stress-strain response and the number of cycles to failure were recorded for each test, and the average and standard deviation of the number of cycles to failure were calculated for each set of aging conditions and strain range. A constant strain rate of $\dot{\epsilon} = 0.001$ sec^{-1} was maintained during the uniaxial fatigue tests.

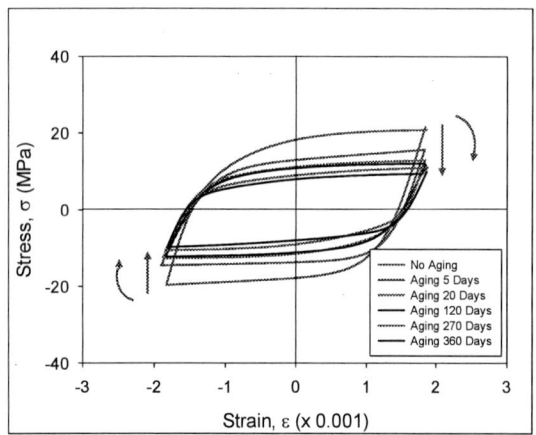

Figure 7 - Hysteresis Loops for Cyclic Stress-Strain Testing of SAC105 Solder (Strain Controlled, Aging for 0-12 Months at 125 C)

The reduction of the loop size with aging suggests that damage accumulation is mitigated with aging. However, this trend is mainly due to the fact that there are large drops in yield stress and ultimate strength of the solder material with aging. These changes reduce the height of the hysteresis loop. Since the cyclic testing was performed using a fixed set of strain limits, the hysteresis loop width stays constant. Thus, the area must decrease because of the material strength reductions associated with aging.

978-1-4799-2408-0/14 $31.00 © 2014 IEEE

Aging Time (Days)	SAC105	SAC305
0	X	X
30	X	X
120	X	X
210	X	X
300	X	X
360	X	X

Figure 9 - Aging Test Matrix for Fatigue Testing
(X = Completed, O = In Progress)

Figure 10 illustrates a typical example of the hysteresis loops generated during a single fatigue test for a SAC105 sample performed until specimen failure (2000+ cycles). In this plot, the hysteresis loops are shown in increments of 100 cycles (i.e. the cycles shown include N = 1, 100, 200, 300, etc.). It is evident that during the cyclic loading the peak stress drops continuously due to damage accumulation. Eventually crack initiation occurs, followed by crack growth, and fracture. In this work, the fatigue life (cycles to failure) was defined to be the point in the cyclic test where a 50% load drop occurred. Park and Lee [48] showed that there were abrupt increases in the resistance of BGA solder joints at 50% load drop during fatigue testing.

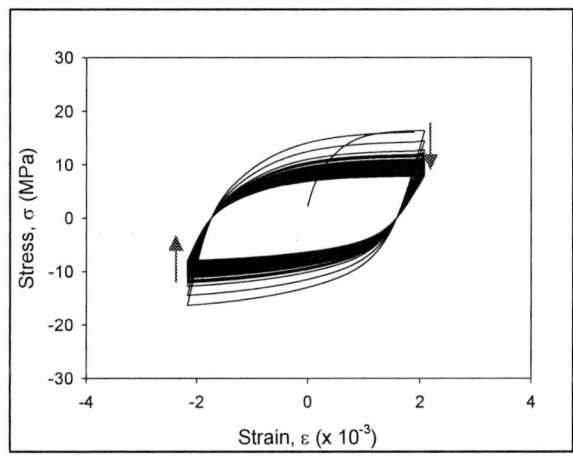

Figure 10 - Hysteresis Loop Evolution during a Fatigue Test

Figure 11 illustrates the recorded fatigue data (plastic strain range vs. number of cycles to failure) for SAC105 lead free solder samples subjected to different aging conditions (aging at 125 C for 0, 30, 120, 210, 300 and 360 days) prior to cyclic testing. Each data point represents an average of at least 5 fatigue tests for a particular set of aging/strain conditions, and the error bars indicated the standard deviations in the data. For each set of aging conditions, the $\Delta\varepsilon_p$ vs. N_f data demonstrated a nearly linear variation when graphed on a log-log scale. This suggests that the Coffin-Manson fatigue law [49-50] can accurately represent the data:

$$N_f^n \Delta\varepsilon_p = C$$
$$\log\Delta\varepsilon_p = \log C - n \log N_f \tag{7}$$

where $\Delta\varepsilon_p$ is the inelastic strain range, N_f is the fatigue life (number of cycles to failure), n is the fatigue exponent, and C is the fatigue ductility coefficient. For each set of aging conditions, Eq. (7) has been regression fit to the data, and the colored straight lines in Figure 16 are the Coffin-Manson fatigue curves that result from the best fits to the data for the various aging conditions. The Coffin-Manson coefficients C and n calculated by least-squares regression analysis are tabulated in Figure 12.

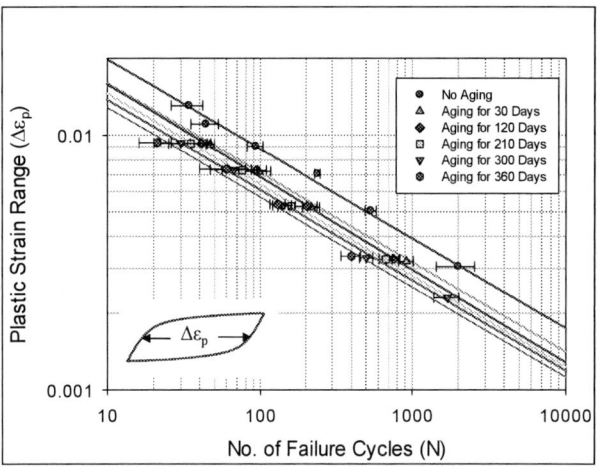

Figure 11 - Fatigue Data (Plastic Strain Range vs. Cycles to Failure) for SAC105 Solder Subjected to Aging

Aging Conditions	Fatigue Ductility Coefficient, C	Fatigue Exponent, n
No Aging	0.04455	0.3514
Aging for 30 Days	0.03598	0.3515
Aging for 120 Days	0.03655	0.3636
Aging for 210 Days	0.03321	0.3575
Aging for 300 Days	0.03120	0.3545
Aging for 360 Days	0.02900	0.3529

Figure 12 - Coffin-Manson Coefficients for SAC105

From the plots in Figure 11, it can be seen that the Coffin-Manson curves/fits are nearly parallel. This is reflected by the small changes in coefficient n (see Figure 12) that occur with aging. The largest shift in the fatigue curves occurs in the first 30 days of aging. This is reflected by the relatively large (19.2%) change in coefficient C that occurs between no aging and 30 days of aging. Microscopy of the fracture surfaces has shown that transgranular cracking was predominant in the failed SAC solder samples with no prior aging. With aging, there is significant micro-structural evolution and grain size coarsening. These effects promoted intergranular crack propagation and contributed to the reduction in the fatigue life with aging, as crack propagation rates at grain boundaries are faster than those in the grain interior. These observations agree with those made by Kariya, et al. [51], who measured the fatigue life of SAC solders at room temperature and elevated temperature (100 C).

The fatigue data in Figure 11 has been recast to be in terms of the strain energy density (ΔW) dissipated per cycle, and the results are shown in Figure 13. The value of ΔW for each fatigue test was calculated from the area of the first cycle hysteresis loop. As before, each data point represents the average of the fatigue tests for the particular set of aging/strain conditions, and the error bars indicated the standard deviations in the data. For each set of aging conditions, the recast ΔW vs. N_f data also demonstrated a nearly linear variation when graphed on a log-log scale. This suggests that the Morrow energy-based fatigue law [47] can accurately represent the data:

$$N_f^m \Delta W = K$$
$$\log \Delta W = \log K - m \log N_f \tag{8}$$

where m is the fatigue exponent, and K is the material ductility coefficient. For each set of aging conditions, Eq. (8) has been regression fit to the data. The colored straight lines in Figure 13 are the Morrow model fatigue curves that result from the best fits to the data for the various aging conditions. The coefficients K and m calculated by least-square regression analysis are tabulated in Figure 14.

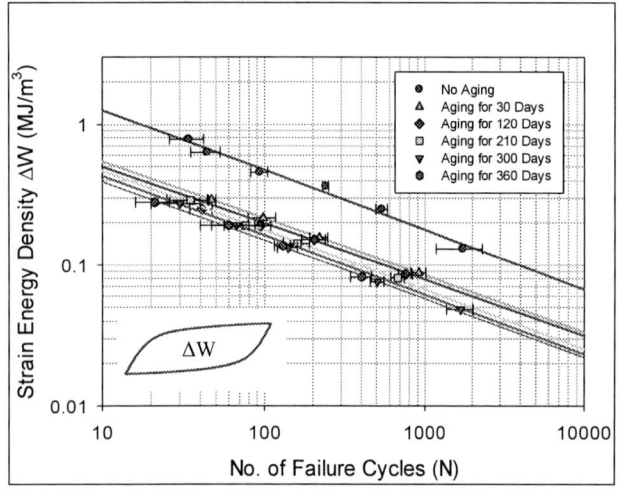

Figure 13 - Fatigue Data (Energy Dissipation vs. Cycles to Failure) for SAC105 Solder Subjected to Aging

Aging Conditions	Fatigue Ductility Coefficient, K	Fatigue Exponent, m
No Aging	3.3510	0.4246
Aging for 30 Days	1.3816	0.4038
Aging for 120 Days	1.2534	0.4000
Aging for 210 Days	1.2370	0.4222
Aging for 300 Days	1.1347	0.4218
Aging for 360 Days	1.0162	0.4156

Figure 14 - Morrow Model Coefficients for SAC105

Similar to Coffin-Manson model prediction, it can be seen that the Morrow model curves/fits are nearly parallel. This is reflected by the small changes in coefficient m (see

Figure 14) that occur with aging. The largest shift in the fatigue curves again occurred in the first 30 days of aging. This is reflected by the relatively large (58.8%) change in coefficient K that occurs between no aging and 30 days of aging. As mentioned previously, the value of ΔW used for each data point in the fatigue plots was found from the area of the first cycle during mechanical cycling. In actuality, the value of ΔW evolves dramatically during each fatigue test. This is shown pictorially in Figure 15 and quantitatively in Figure 16 for the example fatigue test cyclic stress-strain curves in Figure 10.

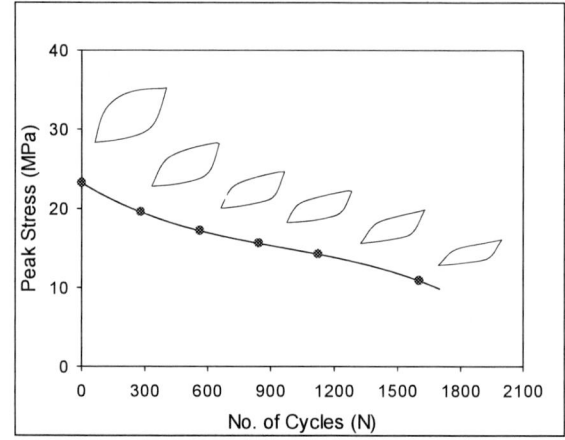

Figure 15 - Evolution of Hysteresis Loop Shape with Number of Cycles during a Fatigue Test

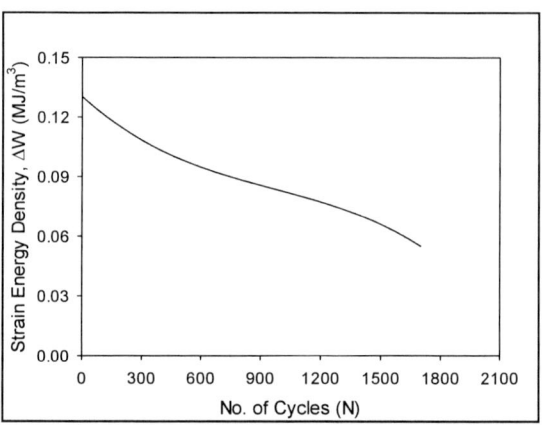

Figure 16 - Evolution of Hysteresis Loop Area with Number of Cycles during a Fatigue Test

Figure 17 contains the analogous fatigue data for SAC305 along with the Coffin-Manson fatigue curves, and Figure 18 contains the calculated Coffin-Manson coefficients C and n for various aging conditions. Likewise, the SAC305 fatigue data in terms of the strain energy density (ΔW) dissipated per cycle are shown in Figure 19, while Figure 20 contains the calculated Morrow model coefficients K and m for the various aging conditions. The fatigue curves again show a progression with duration of aging, with the largest changes occurring with the first 30 days of aging.

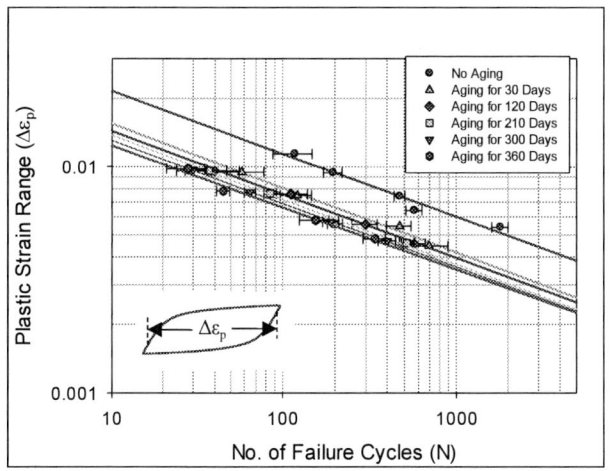

Figure 17 - Fatigue Data (Plastic Strain Range vs. Cycles to Failure) for SAC305 Solder Subjected to Aging

Aging Conditions	Fatigue Ductility Coefficient, C	Fatigue Exponent, n
No Aging	0.04106	0.2782
Aging for 30 Days	0.02994	0.2850
Aging for 120 Days	0.02752	0.2809
Aging for 210 Days	0.02670	0.2856
Aging for 300 Days	0.02490	0.2799
Aging for 360 Days	0.02340	0.2743

Figure 18 - Coffin-Manson Coefficients for SAC305

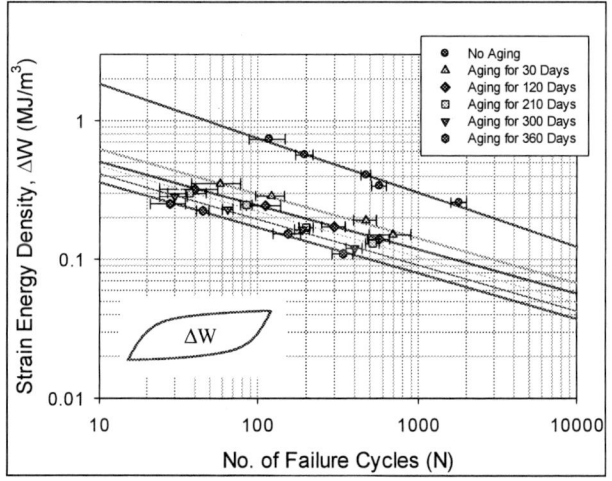

Figure 19 - Fatigue Data (Energy Dissipation vs. Cycles to Failure) for SAC305 Solder Subjected to Aging

In addition, the fatigue curves are nearly parallel for the various aging conditions (m and n are nearly independent of aging). However, while the slopes of the fatigue curves are nearly constant for each SAC alloy taken individually, they are different for the 2 alloy systems. For example, Figure 21 shows the fatigue data for no aging for both SAC105 and SAC305. The convergence of the curves for very high strain levels is clearly evident.

Aging Conditions	Fatigue Ductility Coefficient, K	Fatigue Exponent, m
No Aging	4.5040	0.3906
Aging for 30 Days	1.3028	0.3214
Aging for 120 Days	1.0490	0.3165
Aging for 210 Days	1.0055	0.3286
Aging for 300 Days	0.8828	0.3292
Aging for 360 Days	0.7674	0.3282

Figure 20 - Morrow Model Coefficients for SAC305

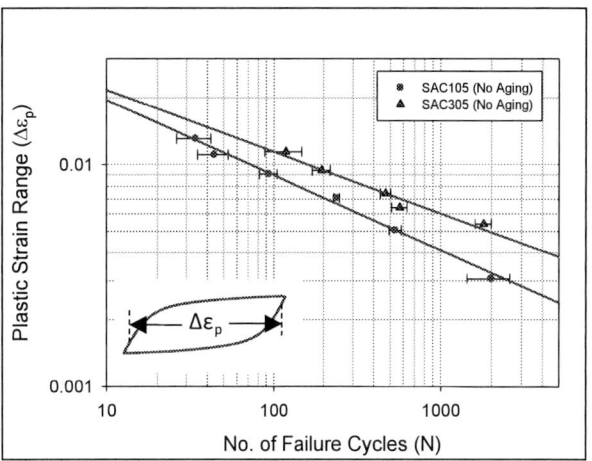

Figure 21 - Fatigue Data (Plastic Strain Range vs. Cycles to Failure) for SAC105 and SAC305 Solders (No Aging)

Shear Fatigue Testing

Shear Testing Techniques

Due to differences in the coefficients of thermal expansion in the materials making up electronic assemblies, solder joints are often subjected to cyclic shear deformations during thermal cycling. Similar to the tension/compression fatigue described above, the effects of aging on the shear fatigue behavior of lead free solder joints has also been explored. Several different shear loading methods have been proposed for solder materials including the Iosipescu technique [19-20, 52-55], ring-and-plug shear [56-57], single lap shear [58-60], double lap shear [61], and solder ball shear [62]. Although none of these techniques provides a perfectly uniform state of pure shear stress in the test sample, the Iosipescu method has been popular due to its widespread use for shear testing of fiber-reinforced composite materials. Zhang, et al. [19-20, 54] have used an Iosipescu-type specimen to perform cyclic testing and creep testing of solders in shear. The uniformity of the shear stress/strain state in the utilized test samples has been explored by Mukherjee and Dasgupta [55]. In the present work, we have utilized a similar modified lap shear technique using an Iosipescu shear specimen.

Iosipescu Shear Specimen Preparation

Details of the developed Iosipescu shear specimen and loading fixture have been presented in reference [44]. The specimen assembly utilizes copper end pieces of dimensions

13 x 5 x 3 mm, with 45° angular cuts (see Figure 22). They are joined together by the 3 x 3 x 0.25 mm solder layer to be tested. As discussed above and in reference [44], the angled cuts in the copper pieces help reduce the effects of stress concentrations and establish a fairly uniform shear stress distribution throughout the solder layer. Good wetting of the solder to copper is required, so that all of the surfaces must be clean and free of all forms of contamination.

Figure 22 - Iosipescu Shear Specimen Assembly

Before assembly, each copper part is placed in a holding fixture and polished using multiple grades of waterproof metallographic sandpaper (600 and 1200 grit), a 0.3 μm diamond paste, and a 0.05 μm colloidal silica suspension. The copper surface is then cleaned with acetone and isopropyl alcohol, and RMA flux is used to reduce oxidation and increase wettability of the solder to copper surface. During assembly, the two copper end pieces are placed into an aluminum mold with the wedge shaped ends facing each other. The mold is sized so that there is a nominal 0.25 mm clearance between the copper pieces, which is just big enough to fit a thin solder rectangular preform (3 x 3 x 0.25 mm). A spring-loaded lid is used so that the solder perform is maintained under a small pressure during reflow and solidification. The reflow temperature profile shown in Figure 2 has been used for forming the Iosipescu shear specimens (same profile as used for the uniaxial fatigue specimens discussed earlier). The microtester system described previously was used with special specimen grips to apply shear loading to the thin solder layers in the Iosipescu assemblies (see Figure 23).

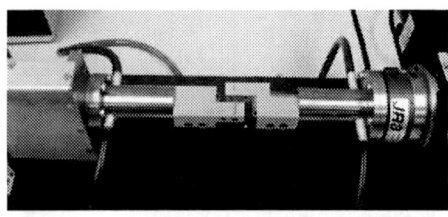

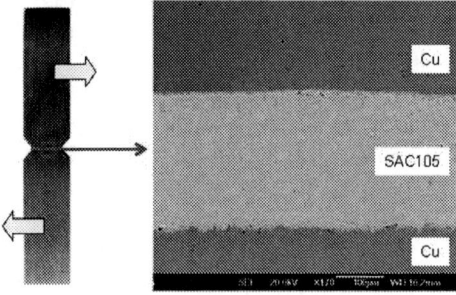

Figure 23 - Testing System with Iosipescu Shear Specimen

Effects of Aging on Shear Fatigue Life

Shear fatigue testing was performed using SAC105 Iosipescu shear specimen assemblies. After solidification, the samples were aged at 125 C for various durations up to 1 year. The samples in each group were cooled to room temperature (25 C) after aging. The 0-day aging specimens represented non-aged samples, which were tested within a few minutes after solidification. Strain controlled fatigue testing was performed with various shear strain limits. For each set of aging conditions and strain range, at least 5 samples were tested to failure. The cyclic stress-strain response and the number of cycles to failure were recorded for each test, and the average and standard deviation of the number of cycles to failure were calculated for the specimens tested for each set of aging conditions and strain range. The shear strain rate was maintained at $\dot{\gamma} = 0.001$ sec^{-1} during the fatigue tests. The fatigue life (cycles to failure) was defined to be the point in the cyclic test where a 50% load drop occurred.

Figure 29 illustrates the recorded fatigue data (plastic shear strain range vs. number of cycles to failure) for SAC105 lead free solder layers subjected to different aging conditions (aging at 125 C for 0, 30, 120, 300 and 360 days) prior to cyclic testing. Each data point represents an average of at least 5 fatigue tests for a particular set of aging/strain conditions, and the error bars indicate the standard deviations in the data. For each set of aging conditions, the $\Delta\gamma_p$ vs. N$_f$ data demonstrated a nearly linear variation when graphed on a log-log scale. This suggests that the Coffin-Manson fatigue law [49-50] can accurately represent the data:

$$N_f^{\alpha}\Delta\gamma_p = \theta$$
$$\log\Delta\gamma_p = \log\theta - \alpha\log N_f \tag{9}$$

where $\Delta\gamma_p$ is the inelastic strain range, N_f is the fatigue life (cycles to failure), α is the fatigue exponent, and θ is the fatigue ductility coefficient. For each set of aging conditions, Eq. (9) has been regression fit to the data, and the colored straight lines in Figure 24 are the Coffin-Manson fatigue curves that result from the best fits to the data for the various aging conditions.

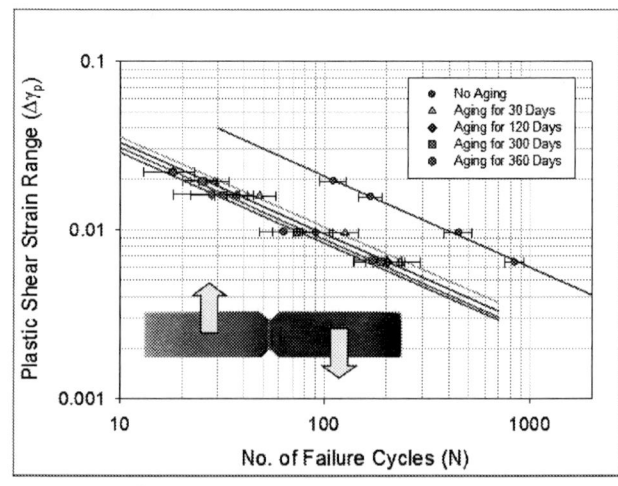

Figure 24 - Shear Fatigue Data (Shear Plastic Strain Range vs. Cycles to Failure) for SAC105 Solder Subjected to Aging at 125 C

The Coffin-Manson coefficients θ and α calculated by least-squares regression analysis are tabulated in Figure 25. From the plots in Figure 24, it can be seen that the Coffin-Manson curves/fits are nearly parallel. This is reflected by the small changes in coefficient α (see Figure 25) that occur with aging. The largest shift in the fatigue curves occurs in the first 30 days of aging. This is reflected by the relatively large (52.3%) change in coefficient θ that occurs between no aging and 30 days of aging.

Aging Conditions	Fatigue Ductility Coefficient, θ	Fatigue Exponent, α
No Aging	0.2516	0.5413
Aging for 30 Days	0.1200	0.5310
Aging for 120 Days	0.1144	0.5415
Aging for 300 Days	0.1073	0.5444
Aging for 360 Days	0.0988	0.5379

Figure 25 - Coffin-Manson Coefficients for Shear Fatigue of SAC105 Solder

The fatigue data in Figure 24 has been recast to be in terms of the shear strain energy density (ΔW) dissipated per cycle, and the results are shown in Figure 26. The value of ΔW for each fatigue test was calculated from the area of the first cycle shear stress-strain hysteresis loop. As before, each data point represents the average of the fatigue tests for the particular set of aging/strain conditions, and the error bars indicated the standard deviations in the data. For each set of aging conditions, the recast ΔW vs. N_f data also demonstrated a nearly linear variation when graphed on a log-log scale. This suggests that the Morrow energy-based fatigue law [47] can accurately represent the data:

$$N_f^{\beta} \Delta W = \varphi \qquad (10)$$
$$\log \Delta W = \log \varphi - \beta \log N_f$$

where β is the fatigue exponent, and φ is the material ductility coefficient. For each set of aging conditions, Eq. (10) has been regression fit to the data, and the colored straight lines in Figure 26 are the Morrow model fatigue curves that result from the best fits to the data for the various aging conditions. The coefficients φ and β calculated by least-square regression analysis are tabulated in Figure 27.

Similar to Coffin-Manson model prediction, it can be seen that the Morrow model curves/fits are nearly parallel. This is reflected by the small changes in coefficient β (see Figure 27) that occur with aging. The largest shift in the fatigue curves again occurred in the first 30 days of aging. This is reflected by the relatively large (63.8%) change in coefficient φ that occurs between no aging and 30 days of aging.

As mentioned previously, the value of ΔW used for each data point in the fatigue plots was found from the area of the first cycle during mechanical cycling. In actuality, the value of ΔW evolves dramatically during each fatigue test .

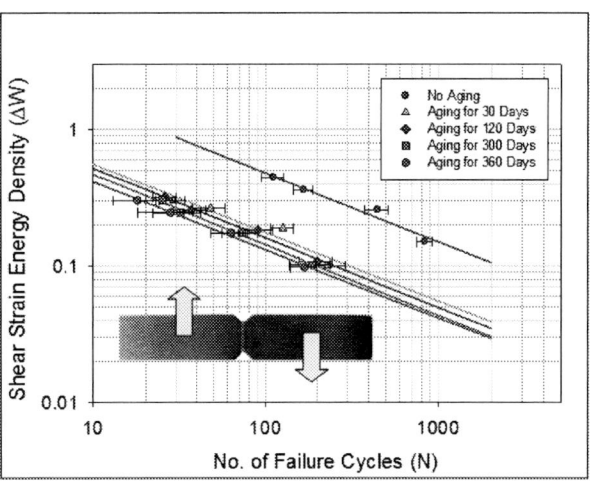

Figure 26 - Shear Fatigue Data (Energy Dissipation vs. Cycles to Failure) for SAC105 Solder Subjected to Aging at 125 C

Aging Conditions	Fatigue Ductility Coefficient, φ	Fatigue Exponent, β
No Aging	4.896	0.5047
Aging for 30 Days	1.769	0.5017
Aging for 120 Days	1.683	0.5175
Aging for 300 Days	1.547	0.5175
Aging for 360 Days	1.310	0.5000

Figure 27 - Morrow Model Coefficients for Shear Fatigue of SAC105 Solder

<u>Equivalency of Uniaxial and Shear Fatigue Data</u>

A comparison has been made between the SAC105 lead free solder fatigue data measured using uniaxial specimens (tension/compression) and shear test specimens. To perform the comparison, the cyclic stress-strain data in each set of experiments have been converted to equivalent stress vs. equivalent strain data using the Von Mises failure/yield criterion. The equivalent (Von Mises) stress is defined as

$$\sigma_{eq} = \sigma_{VM} = \left(\frac{1}{2} \left[(\sigma_1 - \sigma_2)^2 + (\sigma_1 - \sigma_3)^2 + (\sigma_2 - \sigma_3)^2 \right] \right)^{\frac{1}{2}} \quad (11)$$

where σ_1, σ_2, σ_3 are the principal stresses. For the case of plastic strains, the equivalent (Von Mises) strain can be shown to be

$$\varepsilon_{eq} = \varepsilon_{VM} = \frac{\sqrt{2}}{3} \left[(\varepsilon_1 - \varepsilon_2)^2 + (\varepsilon_1 - \varepsilon_3)^2 + (\varepsilon_2 - \varepsilon_3)^2 \right]^{\frac{1}{2}} \quad (12)$$

where ε_1, ε_2, ε_3 are the principal strains. For uniaxial and pure shear loading cases, the equivalent stresses and strains become:

Uniaxial: $\sigma_{eq} = \sigma$ and $\varepsilon_{eq} = \varepsilon$ $\qquad (13)$

Pure Shear: $\sigma_{eq} = \sqrt{3}\,\tau$ and $\varepsilon_{eq} = \gamma / \sqrt{3}$

978-1-4799-2408-0/14 $31.00 © 2014 IEEE

where σ and ε are the stress and strain in a uniaxial (tension/compression) cyclic test, and τ and γ are the shear stress and shear strain in a shear cyclic test.

Using the relationships in Eq. (13), the plastic strain range vs. cycles to failure fatigue data in Figure 11 (uniaxial tension/compression) and Figure 24 (shear) have been recast in terms of equivalent plastic strains. The combined results are shown in Figure 28 for the various aging conditions. In this plot, the different colors again represent different aging conditions. Both uniaxial and shear test data are included, and different symbols are used to differentiate the data coming from uniaxial and shear tests. It can be seen that the mixed uniaxial/shear data for each aging condition demonstrate a fairly linear variation when graphed on a log-log scale. This suggests that the Coffin-Manson type relationship in Eq. (7) is applicable for general strain states if the equivalent plastic strain is used.

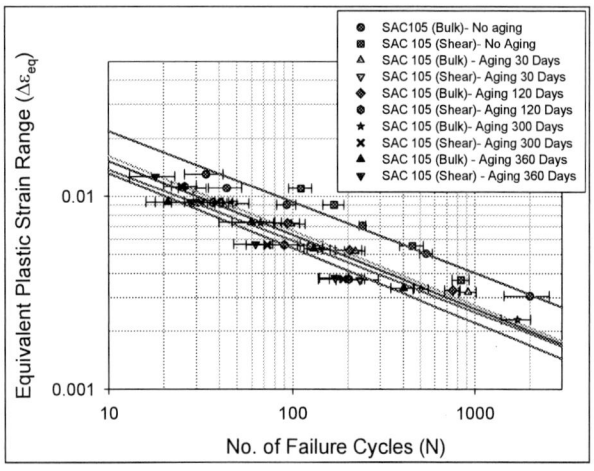

Figure 28 - Uniaxial and Shear Fatigue Data for SAC105 Solder Using Equivalent Plastic Strain

Similarly, the energy dissipation fatigue data in Figure 13 (uniaxial tension/compression) and Figure 26 (shear) have been recast in terms of hysteresis loops calculated from plots of equivalent stress vs. equivalent plastic strain found using the relationships in Eq. (13). The combined results are shown in Figure 29 for the various aging conditions. In this plot, the different colors again represent different aging conditions. Both uniaxial and shear test data are included, and different symbols are used to differentiate the data coming from uniaxial and shear tests. It can be seen that the mixed uniaxial/shear data for each aging condition demonstrate a fairly linear variation when graphed on a log-log scale. This suggests that the Morrow type relationship in Eq. (8) is applicable for general stress and strain states if the energy dissipation is calculated from the equivalent stress and equivalent plastic strain.

Microstructural Damage Evolution

The fatigue and reliability behavior of lead free solder materials are affected by their microstructures, which are inherently unstable because of their high homologous temperatures at or above room temperature. Intermetallics

and grains will tend to grow in size over time to achieve an energetically more favorable morphology. During cyclic loading, grain growth and/or recrystallization processes can be an indication of the accumulation of fatigue damage. Highly strained zones can become coarsened and recrystallized, and crack initiation and microvoids can occur along weakened grain boundaries and grain boundary intersections. These can grow into microfissures, which in turn will grow into cracks and eventually lead to total fracture.

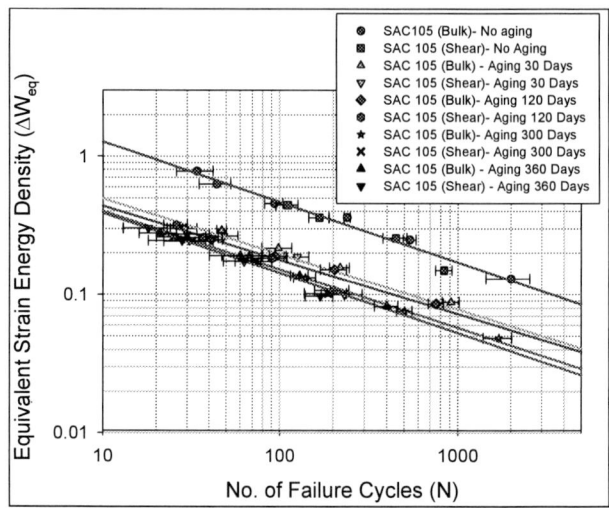

Figure 29 - Uniaxial and Shear Fatigue Data for SAC105 Solder Using Energy Dissipation Calculated Using Plots of Equivalent Stress and Equivalent Plastic Strain

Failure analysis has been performed on the uniaxial fatigue specimens in this work, and typical SEM microstructures are shown in Figures 30 and 31 for non-aged and aged SAC105 specimens, respectively. The non-aged samples (Figure 30) were tested immediately after reflow solidification, while the aged samples (Figure 31) experienced an exposure of 210 days at 125 C prior to fatigue testing. In both cases, samples were cross-sectioned in their (a) initial (unloaded) state, as well as after (b) 30% load drop, and (c) 50% load drop (fatigue failure). Figures 32 and 33 contain analogous optical microscopy images after 50% load drop (fatigue failure) of non-aged and aged samples, respectively.

Prior to aging or fatigue testing (Figure 30a), the microstructure for the ternary Sn-Ag-Cu (SAC105) solder alloy illustrates the typical composition of primary dendrites of β-Sn surrounded by a honeycomb structure of densely distributed eutectic micro-constituents comprised of Ag_3Sn and Cu_6Sn_5 intermetallic particles. When the non-aged samples are cyclically loaded, the IMC particles coalesce and become more sparsely distributed, and microvoids and microcracks are found around the particles (see Figure 30b). After 50% of peak cyclic stress drop, it is observed in Figure 30(c) that micro-cracks grow and coalesce into macro-cracks, which will lead to total fracture. The fatigue microcracks and macro-cracks in the non-aged sample are often transgranular, growing through subgrains (see Figure 32, and Figures 30b and 30c).

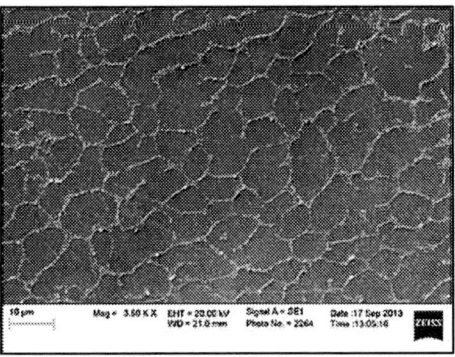

(a) Initial Microstructure (No Cycling)

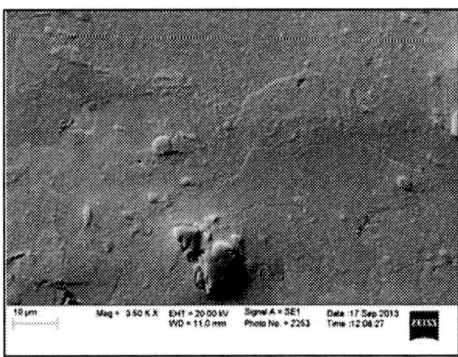

(a) Initial Microstructure (No Cycling)

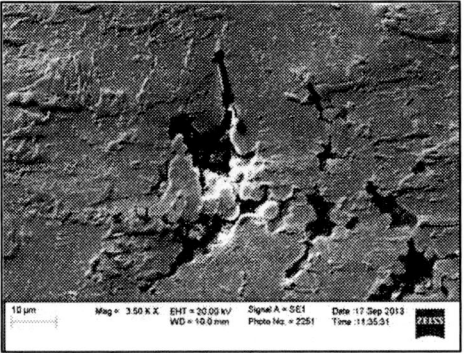

(b) 30% Load Drop

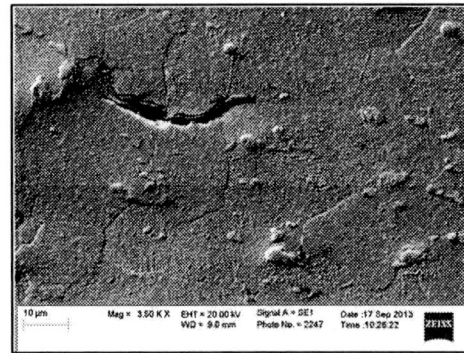

(b) 30% Load Drop

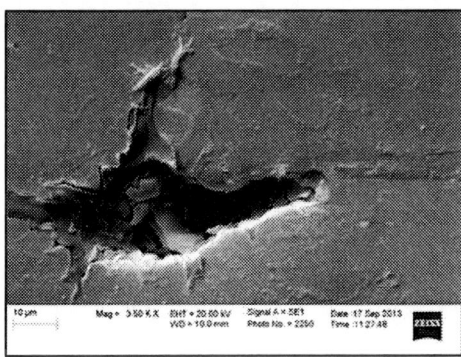

(c) 50% Load Drop

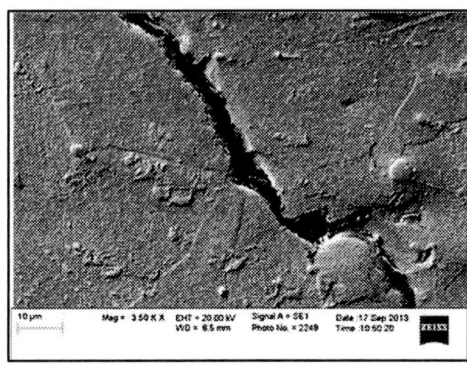

(c) 50% Load Drop

Figure 30 - Microstructural Damage Evolution
(SAC105, No Aging)

Figure 31 - Microstructural Damage Evolution
(SAC105, 210 Days Aging at 125 C)

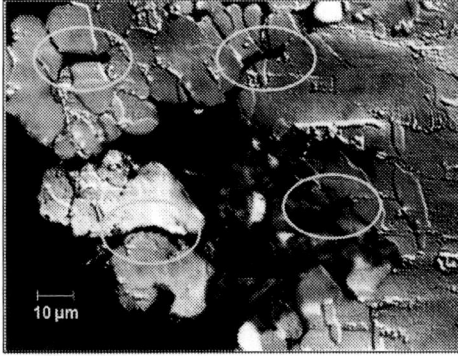

Figure 32 - Transgranular Cracking in Fatigue Specimen
(SAC105, No Aging)

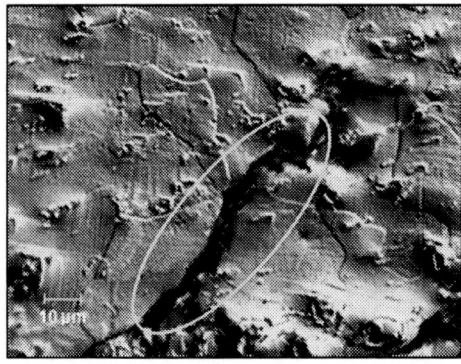

Figure 33 - Intergranular Cracking in Fatigue Specimen
(SAC105, 210 Days Aging at 125 C)

978-1-4799-2408-0/14 $31.00 © 2014 IEEE

This is especially true when the stress state, and orientation and geometry of the grains are not in favor of cracking along grain boundaries. Occasionally, the cracks may be found propagating intergranularly when the recrystallized grain size is small. Similar phenomena were also reported by Chen, et al. [63] for cyclic loading of lead free solders

The basic microstructure of the SAC105 alloy changes dramatically with aging. Figure 31a illustrates the aged alloy before cyclic loading, and larger β-Sn dendrites with conglomerated and coarser IMC particles are observed (relative to Figure 30a). For the aged samples, microfissures were seen to form at recrystallized subgrain boundaries after 30% load drop (Figure 31b), and large cracks were observed at the subgrain boundaries after 50% load drop (Figure 31c and Figure 33).

From this discussion and Figures 32-33, it is evident that two different primary crack propagation modes exist in non-aged and aged samples. In the non-aged samples, the crack propagation mode is mostly trangranular; whereas intergranular crack propagation occurs at the subgrain boundaries in the aged samples. Cracking along recrystallized subgrain boundaries need less energy (smaller hysteresis loops) than transgranular cracking [64].

Fatigue Model Including Aging Effects

The Coffin-Manson fatigue curves present in each of Figures 11, 17, 24, and 28 form sets of parallel lines (constant exponent n) for different aging conditions. Similarly, the Morrow fatigue curves present in each of Figures 13, 19, 26, and 29 form sets of parallel lines (constant exponent m) for different aging conditions. In attempt to build aging into the fatigue criteria, it is proposed to incorporate a shift factor into the conventional equations, and to explicitly use the same fatigue exponent for all aging conditions. With this approach, the modified Coffin-Manson fatigue criterion becomes

$$(N_t a)^n \Delta\varepsilon_p = C \qquad (14)$$

where N_t is number of cycles to failure for solder with aging time t, a is the shift factor, n is the fatigue exponent and C is the fatigue ductility coefficient. Likewise, the modified Morrow model becomes

$$(N_t a)^m \Delta W = K \qquad (15)$$

where N_t is number of cycles to failure for solder with aging time t, a is the shift factor, m is the fatigue exponent and K is the material ductility coefficient. In each of these expressions, the shift factor is assumed to be a shift relative to the non-aged solder material (t = 0), so that

$$a = \frac{N_0}{N_t} \qquad (16)$$

where N_0 is the number of cycles to failure for the solder material when no prior aging has occurred. With this assumption, a = 1 for t = 0.

The shift factor "a" is a function of the microstructural evolution occurring during aging and accelerated damage occurring during fatigue at different testing conditions. It will depend on the prior aging conditions (aging temperature and aging time) of the specimens, as well as the testing temperature and strain rate during fatigue loading, and possibly other factors. If the testing conditions (temperature and strain rate) are constant (as in this study), the shift factor can be assumed to be related only to the effects of aging. In this work, we have assumed aging manifests itself through microstructural changes (phase and grain/sub-grain coarsening) that are detrimental to the fatigue life. The variation of grain size was assumed to depend on the aging time using the combined effects of the model for grain/subgrain coarsening during static annealing proposed by Speight [65] and Ardell [66], and the model for dynamic grain/subgrain coarsening during plastic deformation proposed by Senkov and Myshlyaev [67]:

$$\frac{d}{d_0} = A_0 + A_1 t + A_2 e^{-A_3 t} \qquad (17)$$

where d is the average grain diameter after aging, d_0 is the average grain diameter prior to aging, and t is the aging time. Coefficients A_0, A_1, A_2, and A_3 are functions of the alloy composition, aging temperature, and possibly other factors.

In this work, it is proposed that the degradation of the fatigue life of solder materials with aging occurs primarily due to the grain/subgrain coarsening during aging. Therefore, the shift function was assumed to vary with aging time using the same functional form as the expression in Eq. (17):

$$a(t) = A_0 + A_1 t + A_2 e^{-A_3 t} \qquad (18)$$

where t is the aging time; A_0, A_1, A_2, and A_3 are coefficients to be determined by fitting the fatigue data; and a(0) = A_0 + A_2 = 1 from Eq. (16). By using Eq. (18) within Eq. (14), the Coffin-Manson fatigue criterion has been adapted to include aging time and the microstructural evolution caused by aging. Similarly, by using Eq. (18) within Eq. (15), the Morrow model has been adapted to include aging time and the microstructural evolution caused by aging.

A regression fitting technique has been developed to determine the four coefficients in the shift function for each of the two microstructure based fatigue models. For example, the uniaxial fatigue data in Figure 17 (SAC305, plastic strain vs. failure cycles) was used to calculate the shift factor for several aging times by using Eq. (16) at various strain levels. These results are plotted in Figure 34, and Eq. (18) was then fit to the extracted shift factor vs. aging time data points using a nonlinear regression analysis to determine the optimum values for coefficients A_0, A_1, A_2, and A_3. Finally the fatigue data were fit with Eq. (14) and the determined shift function to calculate the single values of the Coffin-Manson fatigue exponent n and fatigue ductility constant C that are valid for all aging conditions. The fit of Eq. (14) to the SAC305 fatigue data is shown in Figure 35, while the corresponding fit for the SAC105 fatigue data is shown in Figure 36.

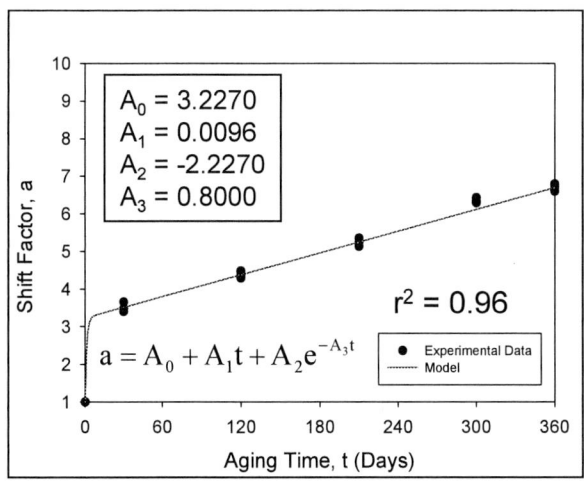

Figure 34 - Shift Factor vs. Aging Time Data
(SAC305, from Figure 17)

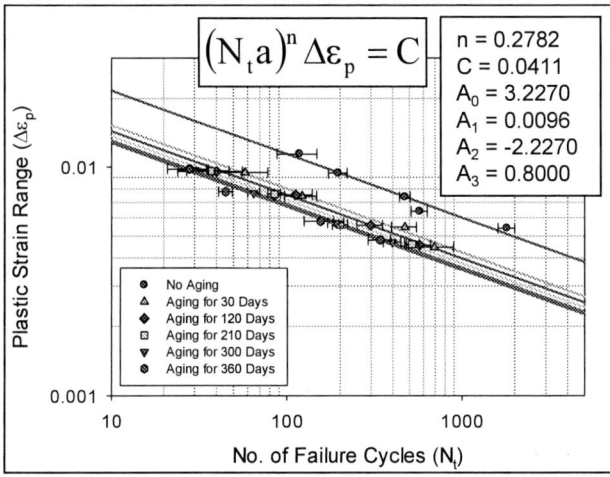

Figure 35 - Modified Coffin-Manson Model with Shift
Function for SAC305

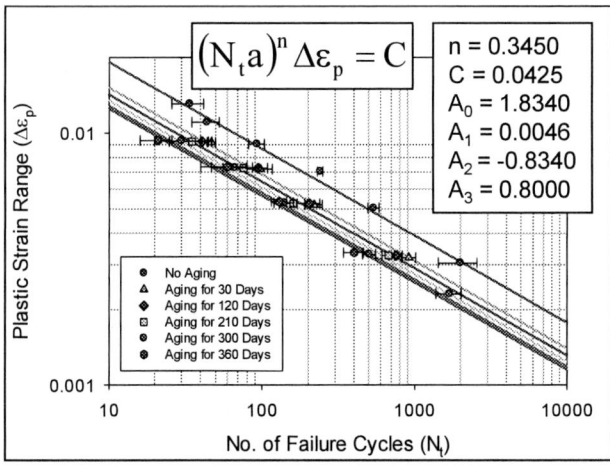

Figure 36 - Modified Coffin-Manson Model with Shift
Function for SAC105

A similar approach has been used to fit the modified Morrow model in Eq. (15) to the fatigue data for SAC105 (Figure 13) and SAC 305 (Figure 19), and the results are shown in Figures 37 and 38, respectively. In all cases, the modified fatigue criteria are able to fit the experimental data well for all aging times.

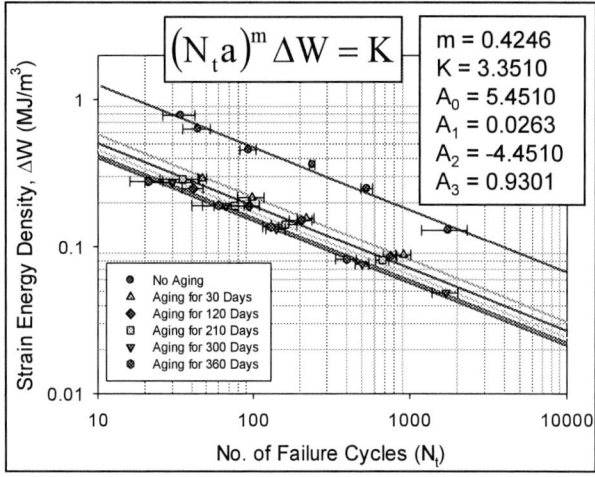

Figure 37 - Modified Morrow Model with Shift
Function for SAC105

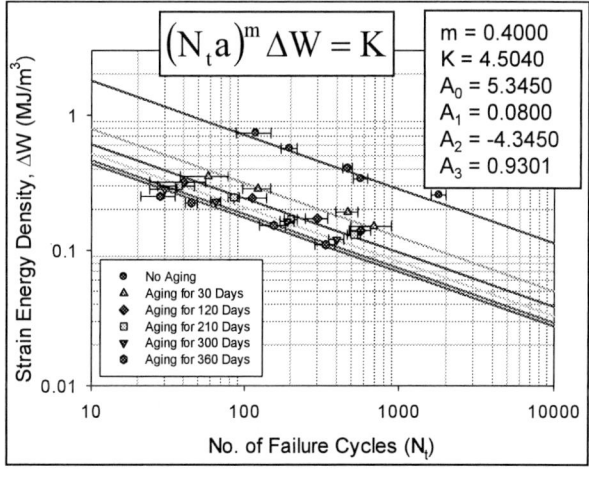

Figure 38 - Modified Morrow Model with Shift
Function for SAC305

Fatigue Life Regions

The data presented above for extreme aging (one year at 125 C) can be taken to approximately represent a lower bound on the fatigue life for a particular alloy. This is because the heavy aging at high temperature has caused extreme coarsening of the microstructure, and additional aging will cause only relatively small further coarsening and fatigue life reduction. The upper bound on the fatigue life in the data presented so far is found from the data for no aging (measured immediately after solder reflow). To push this upper bound even higher, we have performed uniaxial tension/compression fatigue tests on water quenched samples immediately after quenching. These samples have finer microstructure than the

978-1-4799-2408-0/14 $31.00 © 2014 IEEE

reflowed samples tested previously, and represent an extreme case of high modulus, strength, and fatigue resistance for a solder alloy.

The new water quenched results have been added to the uniaxial fatigue data for SAC105 (Figures 11 and 13) and SAC305 (Figures 17 and 19) shown earlier. The new fatigue life plots are shown in Figures 39-40 for SAC105, and Figures 41-42 for SAC305. In these plots, the purple shaded regions outline the possible fatigue life regions for the alloys for all possible microstructures. The tops of these regions represent the water quenched upper limits, while the bottoms of these regions represent the lower limits for the extremely coarse microstructures caused by one year of aging at high temperature.

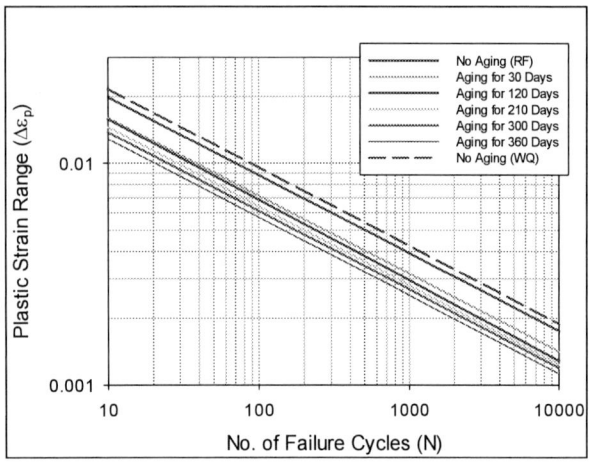

Figure 39 - Fatigue Life Region for SAC105
Based on Plastic Strain Range

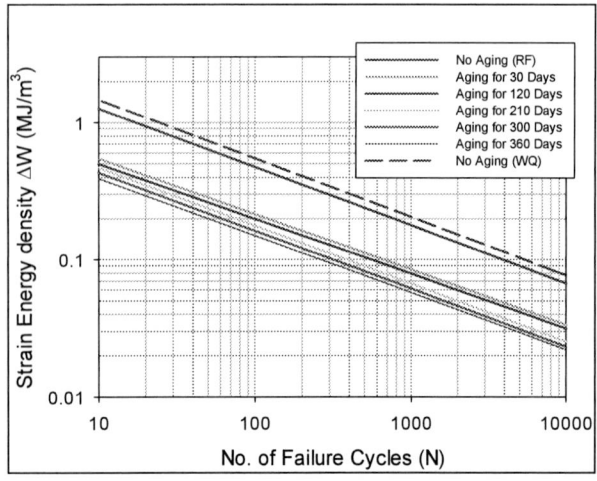

Figure 40 - Fatigue Life Region for SAC105
Based on Plastic Strain Energy

Summary and Conclusions

In this investigation, we have examined the effects of aging on the cyclic stress-strain and fatigue behavior of lead free solders. Uniaxial SAC105 and SAC305 lead free solder specimens were subjected to cyclic (tension/compression)

mechanical loading under strain control (constant positive and negative strain limits). The hysteresis loop size (area) was calculated from the measured cyclic stress-strain curves for a given solder alloy and temperature. This area represents the strain energy density dissipated per cycle, which can be typically correlated to the damage accumulation in the solder material. It was observed that the hysteresis loop area changes significantly with aging, and mathematical models have been found to fit the aging induced evolution. For strain controlled cyclic testing, the hysteresis loop size decreases with aging time due to the large reductions in elastic modulus and ultimate strength that occur with aging.

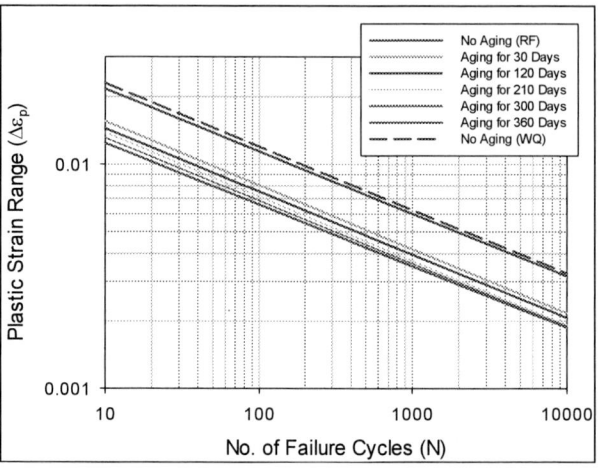

Figure 41 - Fatigue Life Region for SAC305
Based on Plastic Strain Range

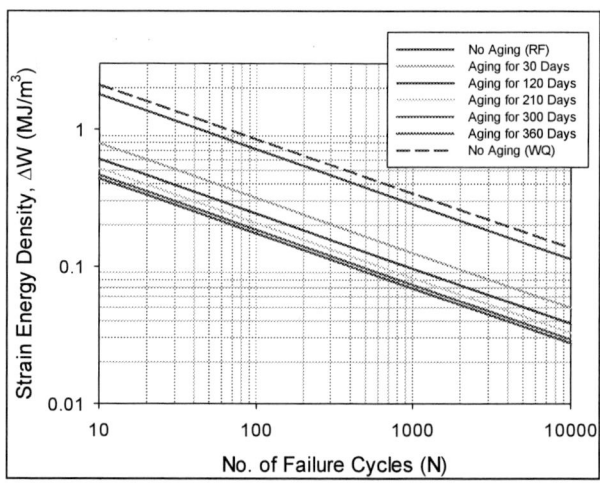

Figure 42 - Fatigue Life Region for SAC305
Based on Plastic Strain Energy

For both alloys, the fatigue data expressed as plastic strain change vs. cycles to failure demonstrated a nearly linear variation for all aging conditions when graphed on a log-log scale. This suggests that the Coffin-Manson fatigue law can accurately represent the data for the various aging conditions, and the largest shift in the fatigue curves occurred in the first 30 days of aging. While the slopes of the fatigue curves are nearly constant for each SAC alloy taken individually, they are different for the 2 alloy systems. A clear progression of the fatigue curves was observed with longer aging times prior to mechanical cycling. The fatigue data was also recast in terms of the energy dissipated per cycle (ΔW) in the first cycle hysteresis loop. The fatigue curves in this case were found to be well fit by the Morrow energy-based fatigue law for all aging conditions.

A quick, repeatable and novel method was developed to prepare Iosipescu solder specimens for shear fatigue testing. Similar to the uniaxial specimen data, aging caused degradations of the cyclic shear fatigue life of the lead free solders, and the Coffin-Manson and Morrow fatigue criteria fit the data well for all aging conditions. Comparisons of the low-cycle fatigue results for uniaxial tension/compression and shear loadings were made using the equivalent stresses and strains calculated using the von Mises failure criterion. It has been shown that uniaxial fatigue data and shear fatigue data for SAC105 lead free solder were in good agreement.

Microstructural damage evolution has also been investigated for both aged and non-aged SAC105 fatigue specimens. Transgranular cracking was mainly observed in the non-aged samples, while intergranular cracking was predominant in the aged samples. The Coffin-Manson model and Morrow model were modified using a microstructurally adaptive shift function to include aging effects. The new models were found to fit all of the fatigue data well of all aging conditions.

Acknowledgments

This work was supported by the US Army and the NSF Center for Advanced Vehicle and Extreme Environment Electronics (CAVE[3]).

References

1. Solomon, H. D., "Creep Strain Rate Sensitivity and Low Cyclic Fatigue of 60/40 Solder," *Brazing and Soldering*, Vol. 11, pp. 68-75. 1986,
2. Guo, Q., Cutiongco, E. C., Keer, L. M., and Fine, M. E., "Thermomechanical Fatigue Life Prediction of 63Sn/37Pb Solder," *Journal of Electronic Packaging*, Vol. 114, pp.145-151, 1992.
3. Busso, E. P., Kitano, M., and Kumazawa, T., "A Visco-plastic Constitutive Model for 60/40 Tin-lead Solder used in IC Package Joints," *Journal of Engineering Materials Technology*, Vol. 114, pp. 333-337, 1992.
4. Frear, D. R., Burchett, S. N., Nielsen, M. K., and Stephens, J. J., "Microstructurally based Finite Element Simulation of Solder Joint Behavior," *Soldering and Surface Mount Technology*, Vol. 2, pp. 39-42, 1997.
5. McDowell, D. L., Miller, M. P., and Brooks, D.C., "A Unified Creep-plasticity Theory for Solder Alloys," *Fatigue of Electronic Materials,* ASTM STP, Vol. 1153, pp. 42-59, 1994.
6. Lau, J. H. (Ed.), *Solder Joint Reliability, Theory and Applications*, Van Nostrand Reinhold, New York, 1991.
7. Shi, X. Q., Zhou, W., Pang, H. L. J., and Wang, Z. P., "Effect of Temperature and Strain Rate on Mechanical Properties of 63Sn/37Pb Solder Alloy." *Journal of Electronic Packaging*, Vol. 121, pp. 179-185, 1999.
8. Yang, X. J., Chow, C. L., and Lau, K. J., "Time Dependent Cyclic Deformation and Failure of 63Sn/37Pb Solder Alloy," *International Journal of Fatigue*, Vol. 25, pp. 533-546, 2003.
9. Hall, P. M., "Forces, Moments, and Displacements During Thermal Chamber Cycling of Leadless Ceramic Chip Carriers Soldered to Printed Boards," *IEEE Transactions on Components, Hybrids, and Manufacturing Technology,* Vol. 7(4), pp. 314-327. 1984.
10. Hall, P. M., "Creep and Stress Relaxation in Solder Joints of Surface Mounted Chip Carriers," *IEEE Transactions on Components, Hybrids and Manufacturing Technology*, Vol. 12(4), pp. 556-565, 1987
11. Pao, Y.-H., Badgley, S., Govila, R. and Jih, E., "An experimental and modeling study of thermal cyclic behavior of Sn–Cu and Sn–Pb solder joints," Electronic Packaging Materials Science VII, (Materials Research Society Symposium, Number 323), pp. 153-158, 1994.
12. Pao, Y.-H., Chen, K. L. and Kuo, A. Y., "A Nonlinear and Time Dependent Finite Element Analysis of Solder Joints in Surface Mounted Components under Thermal Cycling," in *Proceedings of the Materials Research Society Symposium*, Vol. 226, pp. 23-28, 1991.
13. Pao, Y.-H., Govila, R., Badgley, S. and Jih, E., "An Experimental and Finite Element Study of Thermal Fatigue Fracture of Pb-Sn Solder Joints," *Journal of Electronic Packaging*, Vol. 115, pp. 1-8, 1993.
14. Haacke, P. Sprecher, A. F. and Conrad, H., "Computer Simulation of Thermomechanical Fatigue of Solder Joints Including Microstructural Coarsening," *Journal of Electronic Packaging*, Vol. 115, pp. 153-158, 1993.
15. Ma, H. and Suhling, J. C., "A Review of Mechanical Properties of Lead Free Solders for Electronic Packaging," *Journal of Material Science*, Vol. 44, pp. 1141-1158, 2009.
16. Raeder, C. H., Felton, L. E., Messler, R. W., and Coffin, L. F., "Thermomechanical Stress-Strain Hysteresis of Sn-Bi Eutectic Solder Alloy," *Proceedings of the IEEE International Electronics Manufacturing Technology Symposium*, pp. 263-268, 1995.
17. Dusek, M., and Hunt, C., "Low Cycle Isothermal Fatigue Properties of Lead-free Solders," *Soldering and Surface Mount Technology*, pp. 25-32, 2007.
18. Herkommer, D., Punch, J. and Reid, M., "A Reliability Model for SAC Solder Covering Isothermal Mechanical Cycling and Thermal Cycling Conditions," *Journal of Microelectronics Reliability*, Vol. 50, pp. 116-126, 2010.
19. Zhang, Q. and Dasgupta, A., "Constitutive and Durability Properties of Selected Lead-Free Solders," *Pb-Free Electronics*, Chapter 6, Edited by M. Pecht and S. Ganesan, 2nd Edition, 2006, Wiley.
20. Zhang, Q., Dasgupta, A. and Haswell, P., "Isothermal Mechanical Durability of Three Selected Pb-Free Solders: Sn3.9Ag0.6Cu, Sn3.5Ag, and Sn0.7Cu," *Journal of Electronic Packaging*, Vol. 127(4), pp. 512-522, 2005.

21. Whitelaw, R. S., Neu, R. W. and Scott, D. T., "Deformation Behavior of Two Lead-Free Solders: Indalloy 227 and Castin Alloy," *Journal of Electronic Packaging*, Vol. 121(2), pp. 99-107, 1999.

22. Korhonen, T. K., Lahman, P. L., Korhonen, M. A. and Henderson, W. D., "Isothermal Fatigue Behavior of the Near Eutectic Sn-Ag-Cu Alloy Between -25 ℃ and 125 ℃," *Journal of Electronic Materials*, Vol. 36(2), pp. 173-178, 2007.

23. Kanchanomai, C., Miyashita, Y., Mouth, Y. and Mannan, S. L., "Low Cycle Fatigue and Fatigue Crack Growth Behaviour of Sn-Ag Eutectic Solder," *Soldering and Surface Mount Technology*, Vol. 14(3), pp. 30-36, 2002.

24. Shang, J. K., Zeng, Q. L., Zhang, L. and Zhu, Q. S., "Mechanical Fatigue of Sn-rich Pb-Free Solder Alloys," *Journal of Material Science*, Vol. 18, pp. 211-227, 2007.

25. Andersson, C., Lai, Z., Liu, J., Jiang, H. and Yu, Y., "Comparison of Isothermal Mechanical Fatigue Properties of Lead-Free Solder Joints and Bulk Solders," *Material Science and Engineering A*, Vol. 394, pp. 20-27, 2005.

26. Pang, J. H. L., Xiong, B. S. and Low, T. H., "Low Cycle Fatigue Models for Lead-Free Solders," *Journal of Thin Solid Films*, Vol. 462-463, pp. 408-412, 2004.

27. Chou, G. J. S., "Microstructure Evolution of SnPb and SnAgCu BGA Solder Joints During Thermal Aging," *Proceedings of the 8th Symposium on Advanced Packaging Materials*, pp. 39-46, 2002.

28. Hasegawa, K., Noudou, T., Takahashi, A., and Nakaso, A., "Thermal Aging Reliability of Solder Ball Joint for Semiconductor Package Substrate," *Proceedings of the 2001 SMTA International*, pp. 1-8, 2001.

29. Chiu, T. C., Zeng, K., Stierman, R., Edwards, D., and Ano, K., "Effect of Thermal Aging on Board Level Drop Reliability for Pb-free BGA Packages," *Proceedings of the 54th Electronic Components and Technology Conference*, pp. 1256-1262, 2004.

30. Ding, Y., Wang, C., Li, M., and Bang, H. S., "Aging Effects on Fracture Behavior of 63Sn37Pb Eutectic Solder During Tensile Tests Under the SEM," *Materials Science and Engineering*, Vol. A384, 314-323, 2004.

31. Pang, J. H. L., Low, T. H., Xiong, B. S., Xu, L., and Neo, C. C., "Thermal cycling aging effects on Sn–Ag–Cu solder joint microstructure," *Thin Solid Films*, Vol. 462-463, pp. 370-375, 2004.

32. Ma, H., Suhling, J. C., Lall P., Bozack, M. J., "Reliability of the Aging Lead-free Solder Joint," *Proceeding of the 56th Electronic Components and Technology Conference*, San Diego, California, pp. 849-864, 2006

33. Ma, H., Suhling, J. C., Zhang, Y., Lall, P., and Bozack, M. J., "The Influence of Elevated Temperature Aging on Reliability of Lead Free Solder Joints," *Proceedings of the 57th IEEE Electronic Components and Technology Conference*, pp. 653-668, Reno, NV, May 29-June 1, 2007.

34. Zhang, Y., Cai, Z., Suhling, J. C., Lall, P., and Bozack, M. J., "The Effects of Aging Temperature on SAC Solder Joint Material Behavior and Reliability," *Proceedings of the 58th IEEE Electronic Components and Technology Conference*, pp. 99-112, Orlando, FL, May 27-30, 2008.

35. Zhang, Y., Cai, Z., Suhling, J. C., Lall, P., and Bozack, M. J., "The Effects of SAC Alloy Composition on Aging Resistance and Reliability," *Proceedings of the 59th IEEE Electronic Components and Technology Conference*, pp. 370-389, San Diego, CA, May 27-29, 2009.

36. Cai, Z., Zhang, Y., Suhling, J. C., Lall, P., Johnson, R. W., Bozack, M. J., "Reduction of Lead Free Solder Aging Effects using Doped SAC Alloys," *Proceedings of the 60th Electronic Components and Technology Conference*, pp. 1493-1511, 2010.

37. Zhang, J., Hai, Z., Thirugnanasambandam, S., Evans, J. L., Bozack, M. J., Sesek, R., Zhang, Y., Suhling, J. C., "Correlation of Aging Effects on Creep Rate and Reliability in Lead Free Solder Joints," *SMTA Journal*, Volume 25(3), pp. 19-28, 2012.

38. Zhang, J., Hai, Z., Thirugnanasambandam, S., Evans, J. L., Bozack, M. J., Zhang, Y., Suhling, J. C., "Thermal Aging Effects on Thermal Cycling Reliability of Lead-Free Fine Pitch Packages," *IEEE Transactions on Components and Packaging Technologies*, Vol. 3(8), pp. 1348-1357, 2013.

39. Motalab, M., Cai, Z., Suhling, J. C., Zhang, J., Evans, J. L., Bozack, M. J., Lall, P., "Improved Predictions of Lead Free Solder Joint Reliability that Include Aging Effects," *Proceedings of the 62nd IEEE Electronic Components and Technology Conference*, pp. 513-531, San Diego, CA, May 30 - June 1, 2012.

40. Motalab, M., Cai, Z., Suhling, J. C., Zhang, J., Evans, J. L., Bozack, M. J., Lall, P., "Correlation of Reliability Models Including Aging Effects with Thermal Cycling Reliability Data," *Proceedings of the 63rd IEEE Electronic Components and Technology Conference*, pp. 986-1004, Las Vegas, NV, May 28-31, 2013.

41. Hasnine, M., Mustafa, M., Suhling, J. C., Prorok, B. C., Bozack, M. J., Lall, P., "Characterization of Aging Effects in Lead Free Solder Joints Using Nanoindentation," *Proceedings of the 63rd IEEE Electronic Components and Technology Conference*, pp. 166-178, Las Vegas, NV, May 28-31, 2013.

42. Lall, P., Shantaram, S., Suhling, J., Locker, D., "Effect of Aging on the High Strain Rate Mechanical Properties of SAC105 and SAC305 Leadfree Alloys," *Proceedings of the 63rd IEEE Electronic Components and Technology Conference*, pp. 1277-1293, Las Vegas, NV, May 28-31, 2013.

43. Mustafa M., Cai Z., Suhling J. C., Lall P., "The Effects of Aging on the Cyclic Stress-Strain Behavior and Hysteresis Loop Evolution of Lead Free Solders," *Proceedings of the 61st Electronic Components and Technology Conference*, pp. 927-939, 2011.

44. Mustafa, M., Cai, Z., Roberts, J. R., Suhling, J. C., Lall, P., "Evolution of the Tension/Compression and Shear Cyclic Stress-Strain Behavior of Lead-Free Solder Subjected to Isothermal Aging," *Proceedings of ITherm 2012*, pp. 765-780, San Diego, CA, May 30 - June 1, 2012.

45. Chhanda N. J., Suhling J. C., Lall P., "Experimental Characterization and Viscoelastic Modeling of the Temperature Dependent Material Behavior of Underfill Encapsulants," *Proceedings of InterPACK 2011*, pp. 749-761, Portland, OR, 2011.

46. Chhanda N. J., Suhling J., Lall P., "Implementation of A Viscoelastic Model for the Temperature Dependent Material Behavior of Underfill Encapsulants," *Proceedings of ITherm 2012*, pp. 269-281, San Diego, CA, 2012.

47. Morrow, J., "Cyclic Plastic Strain Energy and Fatigue of Metals," *Internal Friction, Damping and Cyclic Plasticity*, ASTM STP 378 ASTM, West Conshohocken, PA, pp. 45, 1965.

48. Park, S. T., and Lee, S. B., "Low Cycle Fatigue Testing of Ball Grid Array Solder Joints under Mixed-Mode Loading Conditions," *Journal of Electronic Packaging*, Vol. 125, pp. 237-244, 2005.

49. Coffin, L. F., "A Study of the Effects of Cyclic Thermal Stresses on Ductile Metal," *ASME Transactions*, Vol. 76, pp. 931-950, 1954.

50. Manson, S. S., "Fatigue: A Complex Subject - Some SPL Approximations," *Experimental Mechanics*, Vol. 5, pp. 193-226, 1965.

51. Kariya Y., Niimi T., Otsuka M., "Isothermal Fatigue Properties of Sn-Ag-Cu Alloy Evaluated by Micro Size Specimen," *Materials Transactions*, Vol. 46(11), pp. 2309-2315, 2005.

52. Unal, O., Barnard, D. J., and Anderson, I. E., "A Shear Test Method to Measure Shear Strength of Metallic Materials and Solder Joints Using Small Specimens," *Scripta Materialia*, Vol. 40(3), pp. 271-276, 1999.

53. Cook, B. A., Anderson, I. E., Harringa, J. L., Terpstra, R. L., Foley, J. C., and Laabs, F. C., "Shear Deformation in Sn-3.5Ag and Sn-3.6 Ag-1.0Cu Solder Joints Subjected to Asymmetric Four Point Bend Tests," *Journal of Electronic Materials*, Vol. 30(9), pp.1214-1221, 2001.

54. Zhang, Q., Dasgupta, A., and Haswell, P., "Partitioned Viscoplastic-Constitutive Properties of the Pb-Free Sn3.9Ag0.6Cu Solder," *Journal of Electronic Materials*, Vol. 33(11), pp. 1338-1349, 2004.

55. Mukherjee, S., and Dasgupta, A., "An Evaluation of a Modified Iosipescu Specimen For Measurement of Elastic-Plastic Properties of Solder Materials," *Proceedings of the ASME 2010 International Mechanical Engineering Congress & Exposition*, Paper IMECE2010-39309, Vancouver, Canada, 2010.

56. Dirnfeld, S. F, and Ramon, J. J., "Microstructure Investigation of Copper-Tin Intermetallics and the Influence of Layer Thickness on Shear Strength," *Welding Journal*, pp. 373-377, 1990.

57. Vianco, P. T., and Rejent, J. A. "Properties of Ternary Sn-Ag-Bi Solder Alloys. Part II. Wettability and Mechanical Properties Analyses," *Journal of Electronic Materials*, Vol. 28(10), pp. 1138-1143, 1999.

58. Wright, C., "Effect of Solid-State Reactions upon Solder Lap Shear Strength," *IEEE Transactions of Components, Hybrids, and Packaging Technologies*, Volume PHP-13, Issue 3, pp. 202-207, 1977.

59. Reader, C. H., Felton, L. E., Tanzi, V. A., and Knorr, D. B., "The Effect of Aging on Microstructure, Room Temperature Deformation, and Fracture of Sn-Bi/Cu Solder Joints," *Journal of Electronic Materials*, Vol. 23(7), pp. 611-617, 1994.

60. Shieu, F. S., Chang, Z. C, Sheen, J. G and Chen, C. F., "Microstructure and Shear Strength of a Au-In Microjoint," *Intermetallics*, Vol. 8, pp. 623-627, 2000.

61. Frear, D., Grivas, D., and Morris, J. W., "Parameters Affecting Thermal Fatigue Behavior of 60Sn-40Pb Solder Joints," *Journal of Electronic Materials*, Vol. 18, pp. 671-680, 1989.

62. Darveaux, R., "Shear Deformation of Lead Free Solder Joints," *Proceedings of the 55th Electronic Components and Technology Conference*, pp. 882-892, 2005.

63. Chen, H. T., Mattila, T., Li, J., Liu, X. W., Li, M. Y., Kivilahti, J. K., "Localized Recrystallization and Cracking Behavior of Lead-free Solder Interconnections Under Thermal Cycling," *Proceedings of the International Conference on Electronic Packaging Technology and High Density Packaging (ICEPT-HDP)*, pp. 562-568, 2009.

64. Mattila, T. T., Vuorinen, V., Kivilahti, J. K., "Impact of Printed Wiring Board Coatings on the Reliability of Lead-free Chip-scale Package Interconnections," *Journal of Materials Research*, Vol. 19(11), pp. 3214-3223, 2004.

65. Speight, M. V., "Growth Kinetics of Grain-Boundary Precipitates," *Acta Metallurgica*, Vol. 16, pp. 133-135, 1968.

66. Ardell, A. J., "On the Coarsening of Grain Boundary Precipitates," *Acta Metallurgica*, Vol. 20, pp. 601-609, 1972.

67. Senkov, O., and Myshlyaev, M., "Grain Growth in a Superplastic Zn-22%Al Alloy," *Acta Metallurgica*, Vol. 34(1), pp. 97-106, 1986.

Solder Joint Height Impact on Temperature Cycle Reliability of BGA Components with Thermal Enabling Load

Yun Ge, Jeffrey Cook, Min Pei, Milena Vujosevic, Bite Zhou, Suddhasattwa Nad
Intel Corporation
5000 W. Chandler Blvd., CH5-263, Chandler, AZ 85226
yun.ge@intel.com

Abstract

Thinner organic BGA packages typically lead to higher warpage before surface mount which, after reflow, causes package corner solder joints (SJ) to have higher stand-off compared to solder joints in the package center. Hence the solder joints at package corners can change from typical barrel shape to an elongated hour-glass shape. Conventional understanding of the impact of SJ height on temperature cycling (TC) reliability states that taller SJs results in longer life. Since that understanding was derived based on the evaluations of BGAs without an enabling load applied, the question was if the same conclusions can be extended to BGAs with an enabling load, where critical SJs are in the package corner. To answer that question, a comprehensive experimental-modeling study was conducted. To investigate the solder joint height impact on temperature cycle reliability of BGA components with thermal enabling load, empirical data was collected on identical BGA packages that were intentionally deformed to create different incoming warpage and as a result different SJ height. The experimental results showed the solder joint shape impacts reliability under temperature cycle condition and that reliability is highly correlated with SJ height. Finite Element Method (FEM) modeling studies was conducted to provide physical explanation of observed behaviors.   The modeling results correlate very well with experimental data. The key result of the study is that incoming warpage driven SJ height variations will impact reliability of BGAs with enabling load, that taller SJs will results in shorter life then "normal" barrel shaped joint which is contrary to the understanding held in the past.

Introduction

The ball grid array (BGA) package development roadmap is rapidly moving towards finer pitch, thinner die, thinner packages, and thinner printed circuit boards due to demands in consumer devices, such as hand-held segments. All these changes result in significant package and board deformation during BGA surface mount, which impacts the shape and quality of the solder joints. It is important to evaluate the impact that different SJ shapes can have on reliability.

One of the main factors controlling the solder joint shape is the standoff height of the solder joint. Thin plastic packages have high incoming warpage before surface mount which, after reflow, causes package corner solder joints to have higher standoff than solder joints in the package center. Hence the solder joints at package corners can assume an hour-glass shape.

Past studies suggested that the taller solder joints offer better reliability performance under thermal cycling. The explanation for the improvement has been based on the higher shear compliance of taller solder joints [1-6]. These studies were focused on BGA components without thermal enabling load; under these conditions the critical solder joints are in the die shadow, where the mismatch in coefficient of thermal expansion is the highest. The question was whether the same conclusions would hold true for the cases where thermal enabling load is present.

This is relevant, because in many applications of practical interest (from servers to laptops to smartphones) enabling load is applied to dissipate the heat generated during the operation of the device [7]. The presence of a thermal enabling load impacts the deformation of the assembly. In addition to high stresses being present in the die shadow solder joints (as in the case without load), package corner solder joints can also become critical. There had been no studies (to the knowledge of these authors) that addressed the effect of solder joint shape on TC reliability of components with enabling load.

This study focuses on the impact that solder joint shape has on temperature cycling reliability performance of BGA components with thermal enabling load. It is the first study of this kind, and is based on both experimental evaluations and modeling analysis. Empirical data was collected on identical BGA packages that were deformed to create different incoming warpage. The experimental work was complemented by computational mechanics analysis to provide a physical explanation of the experimental findings.

The paper is organized as follows. The first section explains the experimental procedure used in the study, including the description of the specimens created, and setup of the assembly put through reliability testing. The second section describes the modeling approach, key features of the model, and minimization of energy approach used to define various SJ shapes. Results of the experimental and modeling work and their correlation are discussed in the details in the third section. Key conclusions are summarized at the end.

Experimental Methodology

In order to investigate the thermal cycling reliability of BGA packages with different warpage-induced solder joint shapes, empirical data was collected on identical BGA packages that were deformed to create different incoming warpage. These BGAs were then assembled to 32mil printed circuit boards.

Because of the package warpage, the stand-off heights of solder joints at different locations are different. By increasing the incoming package warpage, the solder joint stand-off height was highly impacted at the package corners. Therefore, the solder joint shape can be controlled to vary from barrel shaped to hour-glass shaped. Samples were grouped into 4 different levels of warpage. Figure 1 is images of the cross-

978-1-4799-2408-0/14 $31.00 © 2014 IEEE

session of assembled package with different incoming warpages. There were 39 samples prepared for each warpage.

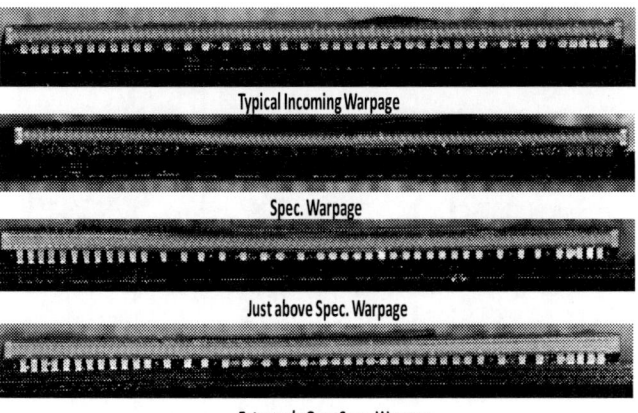

Figure 1. Cross-session of assembled package with different incoming warpage

The assembled packages were submitted to a temperature cycle test with load fixture specially designed to simulate heat solution compressive load impact on solder joint reliability (SJR) [8]. The package assembled on PCB was sandwiched between a load fixture and a 0.9mm backplate. A compressive load of 15lbs was applied on the CPU die, and fastened with 4 bolts through fixture to the backplate on the back side of PCB, as shown in Figure 2.

Figure 2. Load Fixture setup for BGA enabled temperature cycle risk assessment test [8].

During the test, three readouts were taken at 600, 900, and 1200 temperature cycles. Samples were taken out of the temperature cycle chamber at each readout level and sent for failure analysis. Dye and Pull (DnP) and cross-sections were performed to check the percentage of crack area/length at each BGA location.

Modeling Methodology

A finite element thermo-mechanics model was developed to model the BGA package with different SJ heights and with different enabling loads under TC conditions. The global model used a coarse mesh for BGA and it output substrate and board displacement to the local model which had a fine mesh of local solder joint. The global-local finite element approach was used to save computational time while maintaining numerical method accuracy [7][9][10].

BGA package modeled in this study has 29x29 mm2 substrate size, 13.5x13.5 mm2 die size, 0.65mm BGA pitch and 14x14 mm2 BGA depopulated area in the die shadow region. The package was assembled to a 32mil PCB board. The elements of the model geometry are shown in Figure 3. The BGA solder joints (Figure 3b) were modeled with a solder mask defined pad on the substrate side. On the board side, because there are typically solder mask defined pads (SMD) at package corner, all solder joints were modeled as solder mask defined pads, which are worst case for TC. This was also consistent with the empirical data in this study, since all corner solder joints on the BGA package had solder mask defined pad on the board side.

A uniform compression load of 15 lbs was applied on the die top surface to simulate the fixture compressive load. A 0.9mm thick backplate was placed under the PCB board. Contact boundary condition was defined between PCB and backplate. In the local-model, the displacement of the nodes highlighted in Figure 3.c was from the global model. The package was investigated under temperature cycle condition from -40 to 100 °C. Because of the symmetry of the BGA package geometry and boundary condition, only a quarter-symmetric model was built for simulation in this study.

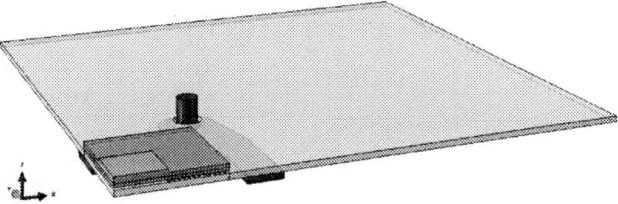

a) Geometry of the global model: BGA component, board, with enabling load and back plate

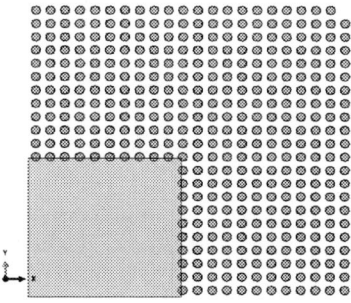

b) Global finite element model geometry: BGA map

c) Local finite element model of BGA joint

Figure 3. Model geometry: key features of the global and local model.

Because of the incoming warpage, there is different stand-off height at different locations. To investigate the solder joint height impact at the package corner, the dependence of stand-off height on location was ignored and the BGA height was uniformly fixed at a certain value to make uniform elongated BGA joints. Three levels of joint heights were used: normal height h, and two elongated heights 1.25h and 1.5h. Joints with normal height have 'normal' barrel shape solder joint height for a flat package after surface mount. Joints with height of 1.25h and 1.5h were generated by increasing the distance between substrate and board, so a uniformly elongated BGA was achieved.

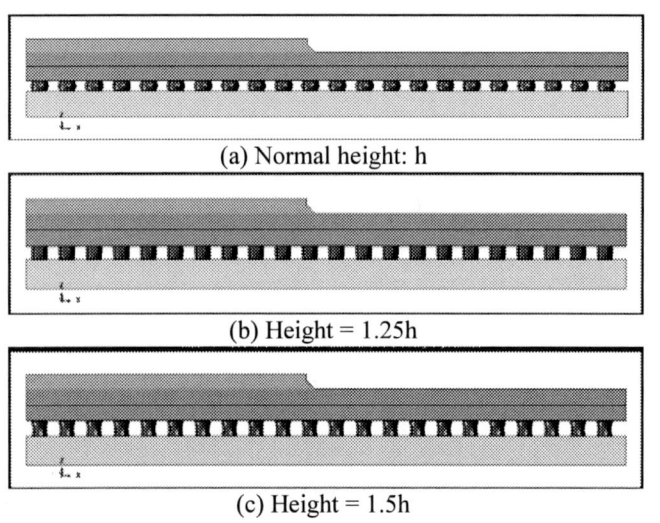

(a) Normal height: h

(b) Height = 1.25h

(c) Height = 1.5h

Figure 4. Solder Joint Shape vs. Solder Joint Standoff Height

A key modeling issue was to decide how BGA stand-off height will impact solder joint shape. The shape of solder joint can be determined by minimizing the total energy of the liquid solder drop. Because gravity and internal pressure effects can be ignored during solder joint shaping [2], with a constant solder ball volume, the solder joint shape is determined by minimizing its surface area, yielding minimum surface energy. When other geometric factors are fixed, the solder shape is only dependent on stand-off height. When the stand-off height increases, the solder joint shape changes from a barrel shape to an hour-glass shape. Global models built with different stand-off-heights was shown in Figure 4. Figure 5 shows the detailed local models of the solder shapes with different stand-off-heights.

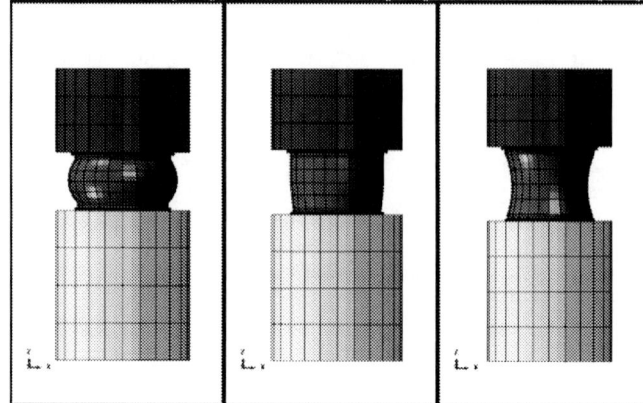

Figure 5. Solder Joint Shape vs. Solder Joint Standoff Height

Results and Discussion

This section summarizes key results. It quantifies the impact of package warpage on SJ height and on SJ shape, describes type of failure observed, and discusses impact of SJ height on SJ reliability. It then describes modeling results, their correlation with test data and uses them to provide physics based explanations of the observed behaviors.

1. Solder Joint Shape vs. Stand-off Height

As expected, the incoming package warpage impacts the standoff height. Figure 6 quantifies that impact. Each data point in Figure 6 came from one corner most solder ball in one sample; each sample thus contributes to 4 data points (four corner joints). The average value of solder joint stand-off height increased with increasing warpage. The package corner joint height varied from h to 1.5h which means with the largest considered incoming warpage, the solder joint height can be up to 150% of the designed value. Because the SJ volume was kept the same, the shape of the solder joint changed from a barrel shape to an hour-glass shape, as shown in Figure 7.

978-1-4799-2408-0/14 $31.00 © 2014 IEEE

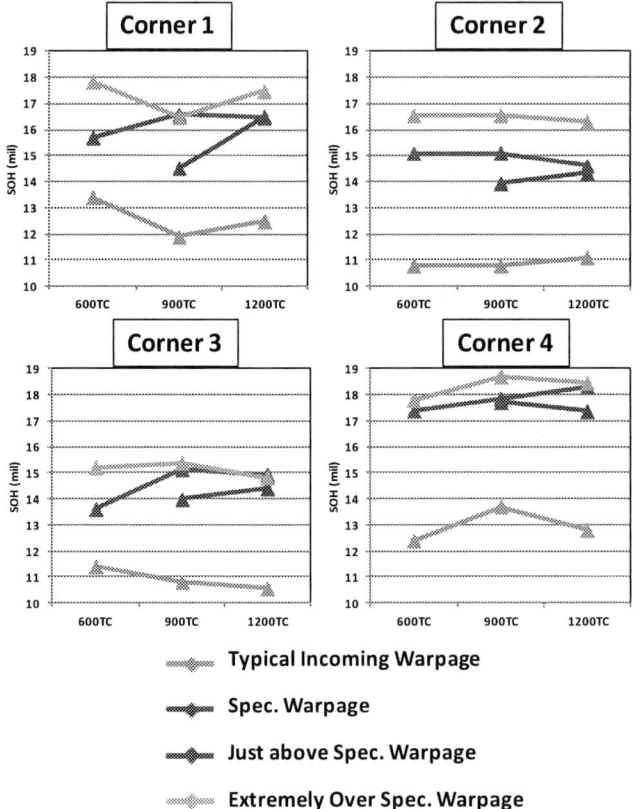

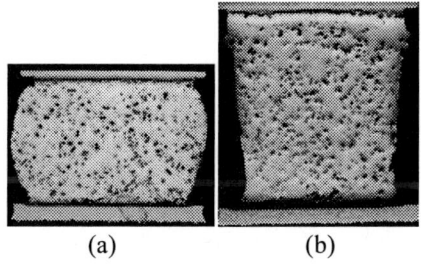

Figure 6. Height of package corner most joints with different incoming warpages measured at different readouts.

Figure 7. Solder Joint Cross-sections at 600 TCT. (a) Nominal height h, (b) Height=1.5h

2. Crack Propagation

FEA results (Figure 8.b,c) showed the z-direction stress distribution in the package corner solder joint during high temperature dwell and low temperature dwell. The color contour indicates positive z-direction stress and the grey area represents the area under compression stress. The center of the package is toward the right in the image, and the outer corner of the package is toward the left. The portion of solder under stress was larger at high temperature dwell than at low temperature dwell. This indicates that if peel stress drives crack propagation, cracks should propagate from the direction of the package corner towards the center of the package. The FEA prediction was confirmed on actual joint failures by failure analysis showing that most of the time failure occurred

at the joints in the corners. Crack was observed at near the interface between solder joint material and the copper pad on the PCB side. As shown in Figure 9, from 600 to 1200TC, crack propagation was observed from the package corner outside towards the center of the package.

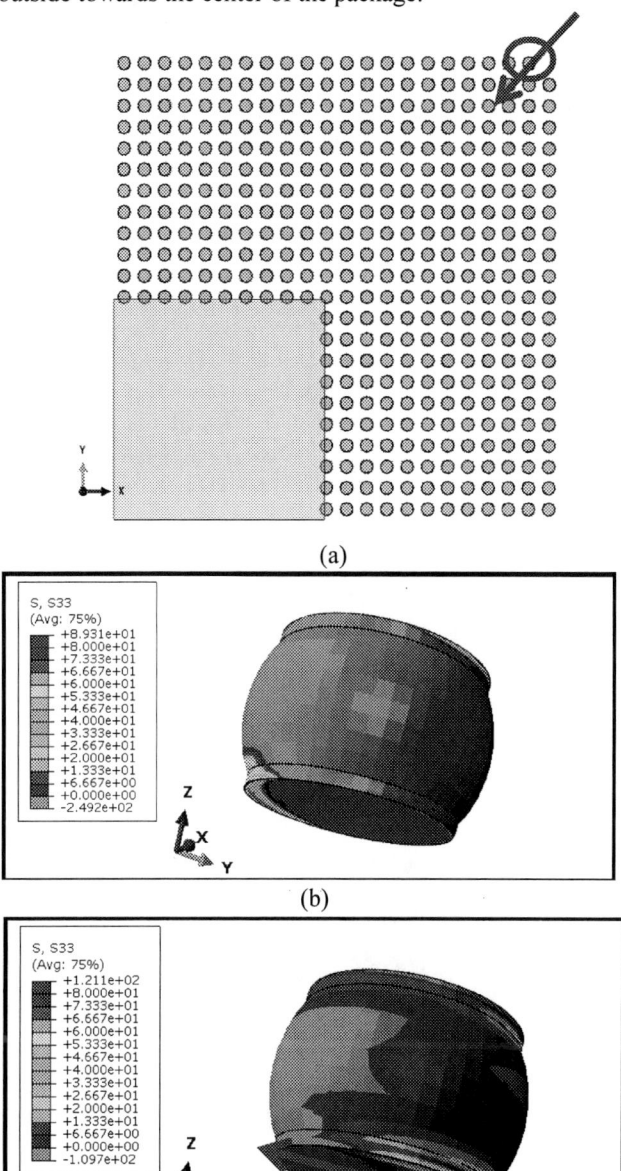

Figure 8. Package corner solder joint (a): Joint location in BGA map and Crack Propagating Direction) z-direction tensile stress distribution at low (b) and high (c) temperature dwells.

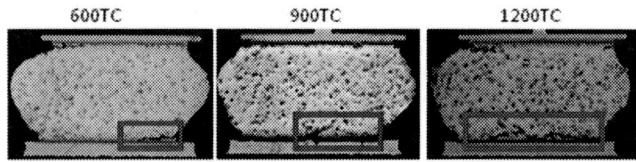

Figure 9. Package corner solder joint at corner 2 crack propagation at readouts: 600TC, 900TC and 1200TC.

3. Solder Join Height vs. Solder Joint Reliability

Figure 10 shows cross-sectional failure analysis of packages with different incoming warpage at each readout (600, 900, and 1200TC). The empirical data indicated that cracks were most commonly observed in the region of solder joint material near the copper pad on the PCB side. From the first column in Figure 10, readout 600 TC, solder joint crack length at the PCB side increased with increasing SJ height. The crack length in the hour-glass shaped joint was much higher than in barrel-shaped joints. This indicates that the solder joint failure is strongly correlated with solder joint height.

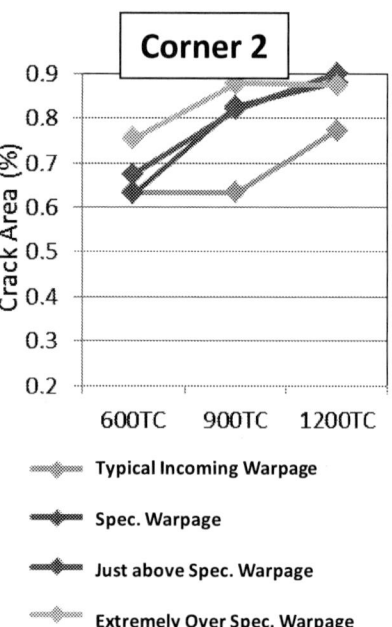

Typical Incoming Warpage

Spec. Warpage

Just above Spec. Warpage

Extremely Over Spec. Warpage

Figure 11. Crack area (in percentage) obtained by DnP showing the correlation between the incoming package warpage and different TC readouts for four package corner joints.

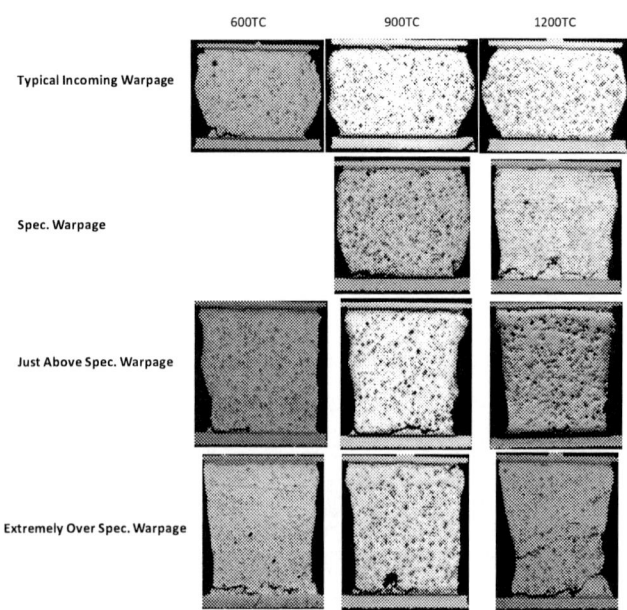

Figure 10. Cross-sections micrographs showing crack lengths grew from units at 600TC, 900TC and 1200TC readouts in package corner solder joints at corner 2 with different stand-off-heights.

Crack area measurements (CAM) were collected by Dye and Pull (DnP) for the corner solder joints, as shown in Figure 11. In most cases, CAM values increased with increasing number of TC cycles, which is consistent with expectations and the cross-section results in Figure 10. In general, CAM values increased with increasing SJ height (i.e. increasing package warpage).

Both DnP CAM and cross-section crack length empirical results indicate that for BGAs with enabling load packages with taller corner solder joints have worse temperature cycling performance than packages with shorter joints. This observation was contrary to the conventional understanding of solder joint height impact on BGA reliability, which held that the shear loading on die area joints during temperature cycling decreases with increasing solder joint height.

From the FEA global-local modeling, we were able to compute package corner solder joint stress-distributions at high temperature, as shown in Figure 12. The contour plot illustrated that the region of tensile stress was larger in elongated hour-glass shaped solder joints than in barrel shaped solder joints. (The colors of positive tensile peel stress amplitude in all figures are the same. Grey regions indicate compression stress.)

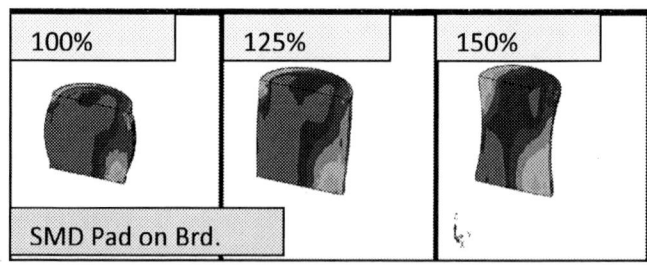

Figure 12. Package corner solder joint z-direction tensile stress distribution at high temperature as determined by modeling.

To quantitatively measure the peel stress for each solder joint shape, the peel stress was averaged by volume to avoid stress concentrations and mesh density issues. As shown in Figure 13, a 25um thin layer of solder material from the solder/copper interface was pre-defined. This approach has been commonly used in solder fatigue analysis [7-10]

$$\overline{\sigma_{33}} = \frac{\int \sigma_{33} dv}{V} \qquad \text{Eq.(1)}$$

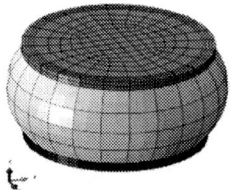

Figure 13. 25um pre-defined layers (in brown color) in solder material

Volume averaged tensile stress on the whole solder joint, on a 25um layer close to substrate, and on a 25um layer close to board within different solder joint stand-off heights are shown in Figure 14. The results indicate that peel stress in the solder near board side was higher than the peel stress near the substrate side, hence SJ failure near the board side is more likely.

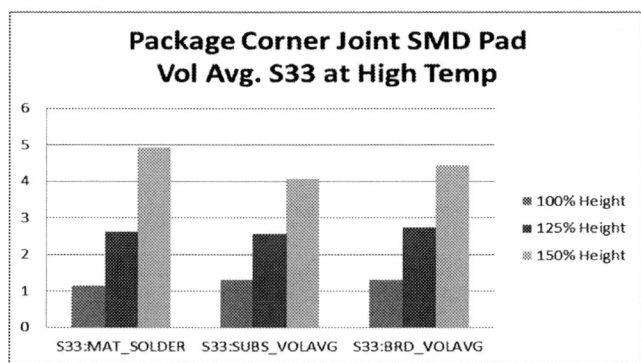

Figure 14. Volume Averaged Peel Stress in Package Corner Solder Joint vs. Different SOH: in whole solder, in 25um near substrate, in 25um near board side

The physical explanation of the observed experimental results can be summarized as follows. For BGA components with thermal enabling load, package corner SJs are usually the highest risk SJs. In the case of higher incoming warpage these corner SJ will assume elongated hour-glass shape. The hour-glass shaped joints at the package corner undergo higher out-of-plane peel stress at high temperature dwell than barrel shaped joints do. These joints experience higher peel stress concentrated in the lower part of the joint towards the outside of package on the board side. Cracks initiate at peel stress concentrations and then propagate toward the inside of the package. Because taller SJs fail earlier than shorter SJs joints, loaded BGA components with elongated corner SJs will have shorter life in TC.

Conclusion

This is the first study addressing the impact of solder joint height/shape on temperature cycling reliability of BGA components with thermal enabling load. The results indicate that for loaded BGA configurations the packages with taller SJ in the corner have shorter life then packages with "normal" barrel shaped SJs. The conclusions of this study indicate that commonly held beliefs that taller SJ have longer life cannot be extended to BGAs with enabling load. This is due to the fact that the underlining physics driving the failure of loaded BGAs is different and dominated by significant bending as opposed to shear. The results of the study have important relevance for comprehending thermo-mechanical performance of new BGAs characterized by very thin substrates prone to high deformations that can impact SJ shape and height, and as shown here, significantly impact the reliability.

References

1. Lim, S.S., etc "Reliability Performance of Stretch Solder Interconnections", 2006, IEMT
2. Qin, H., etc "Solder Volume Effects on Fatigue Life of BGA Strucutre Cu/Sn-3.0Ag-0.5Cu/Cu Interconnects"
3. Darveaux, R. , "Effect of Simulation Methodology on Solder Joint Crack Growth Correlation", 2000, ECTC.
4. Yu,.Q., etc, "A Study of the Effects of BGA Solder Geometry on Fatigue Life and Reliability Assessment', 1998, ISCTP
5. Xu, H., etc, "Manufacture of Hourglass-shaped Solder Joint by Induction Heating Reflow", 2008 ICEPT-HDP.
6. Yang, S., etc, " Optimization of Solder Height and Shape to Improve the Thermo-mechanical Reliability of Wafer-Level Chip Scale Packages", 2013, ECTC.
7. Han.R., etc "New Temperature Cycle (TC) Enabling Load Testing and Modeling Method from Use Condition (UC) and Teardown Analysis and New Spectrum to TC Requirements" ESTC 2013.
8. Min,etc. "Define Electrical Packaging Temperature Cycling Requirement with Field Measured User Behavior Data, ECTC 2013.
9. Syed, A., "Updated Life Prediction Models for Solder Joints with Removal of Modeling Assumptions and Effect of Constitutive Equations", 7th. Int. Conf. on Thermal, Mechanical and Multiphysics Simulation and Experiments in Micro-Electronics and Micro-Systems, EuroSimE 2006, pp.1-9.
10. Han, R., etc. "Solder Fatigue Life Prediction Model by FEA Simulation and Experiment for New Physical Damage Based Use Condition to TC Requirement Approach", ECTC 2013.

Controlling the Sn Grain Morphology of SnAg C4 Solder Bumps

Gregory Parks, *Minhua Lu, #Eric Perfecto, and Eric Cotts
Physics Department and Materials Science Program
Binghamton University; Binghamton, NY
*IBM T. J. Watson Research Center
Yorktown Heights, NY 10598
#IBM Microelectronics
Hopewell Junction, NY

Abstract

The effects of solder composition, under bump metallurgy (UBM), and solder joint geometry on Sn grain morphology in Pb free, C4 solder joints were examined. The solder joints examined had compositions of either Sn-1%Ag or Sn-2.4%Ag, and were on either Cu or Ni pads. The geometries of the solder joints were varied: diameters ranged from 11 to 100 microns and heights from 11 to 30 microns. To study the effects of aspect ratio on the Sn grain morphology, pairs of microbumps with similar pad diameters, solder height and composition were bonded together. Samples were reflowed by means of differential scanning calorimetry (DSC); thus the solidification temperatures were carefully monitored. Sn grain morphologies were characterized by optical microscopy with crossed polarizers (XP) and by scanning electron microscopy with secondary electrons (SE) and an electron backscattered diffraction detector (EBSD). Sn-2.4%Ag solder bumps exhibited smaller Sn grains than Sn-1%Ag alloys. The nucleation rates of Sn from the melt in Pb free solder joints at a number of different temperatures were also measured. An expression for the nucleation of Sn in Sn-2.4%Ag and Sn-1%Ag solder joints on Cu or Ni pads was formulated, providing a predictive capability for Sn solidification temperatures. This study has shown that variation in the temperature of solidification of Sn from the melt results in significant differences in Sn grain morphology.

Introduction

The Sn grain morphology and precipitate microstructure of SnAgCu alloys plays a large role in determining the lifetime of Pb-free solder joints. The size, number, and spacing of Ag_3Sn and Cu_6Sn_5 precipitates in a SAC solder are important factors in determining the mechanical strength and reliability of SAC solder alloys [1,2]. The number and orientation of the β-Sn grains in these solder joints heavily influence the lifetime of these solder joints. For instance during thermal cycling tests, single grained solder joints in which the c-axis of the Sn was parallel to the substrate failed much earlier than solder joints with different Sn grain orientations or multiple Sn grains [3]. This is due to the fact that the coefficient of thermal expansion and Young's modulus along the c-axis is about two to three times larger than it is along the a-axis.

During electromigration (EM) tests the number and orientation of solder joint has a dramatic effect on the lifetime of SAC solders. The interstitial diffusion of Ni and Cu atoms

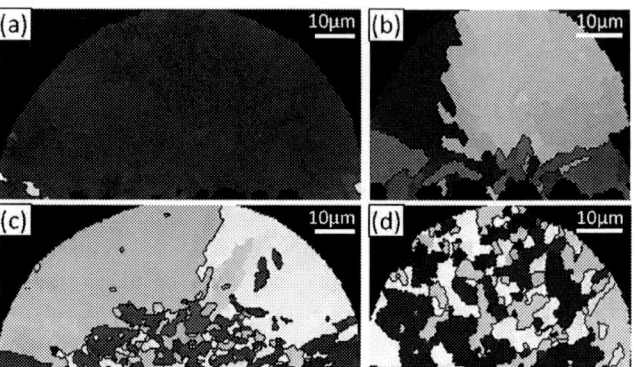

Figure 1: EBSD orientation maps showing examples of (a) single grain, (b) beach ball, (c) mixed, and (d) interlaced Sn grain morphologies in Sn-2.4%Ag solder bumps. Solid lines are drawn on the orientation maps along grain boundaries misoriented by more than 55°.

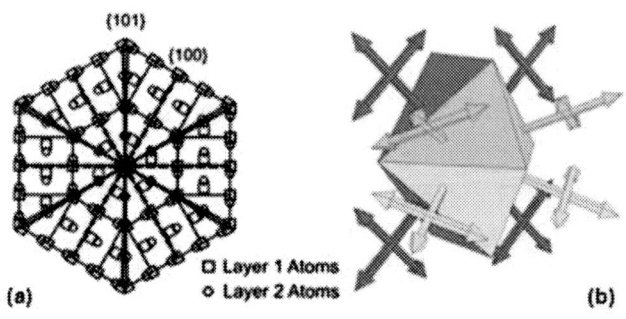

Figure 2: (a) In the model of ref. 7 a hexagonal Sn cluster forms around a Ag or Cu impurity atom forming a six-fold, cyclically twinned nucleus in an undercooled SnAgCu melt. (b) A sketch of growth in the [110] and [1-10] directions from the facetted surface of the cyclically twinned nucleus of (a).

through the Sn matrix is about 500 times faster along the c-axis than it is along the a-axis. This disparity in diffusion times causes early failures in solder joints that have single Sn grains that span the entire sample with its c-axis parallel to the current direction. The work done by several groups show lifetime of solder joints with large grains that subtend the joint and have their c-axis parallel to the current direction fail about 6 times sooner than similar joints with different c-axis orientations [4-6]. In a study by Wang et. al, reflowing solder joints several

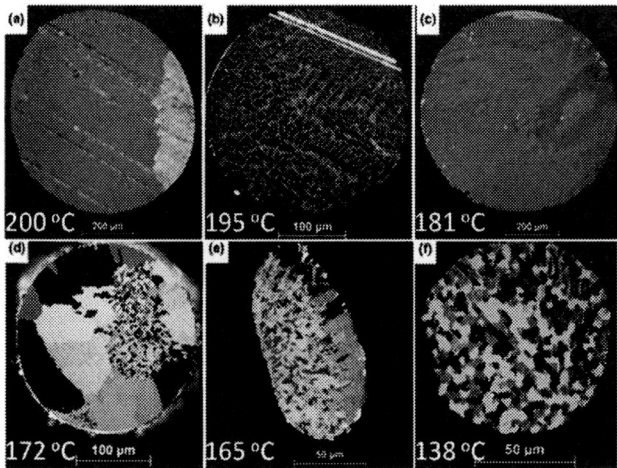

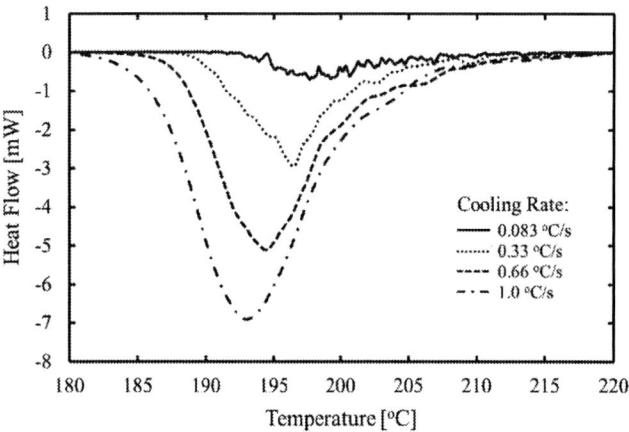

Figure 3: The Sn grain morphology of SAC305 samples is observed to change from a single grain or beach ball morphology to a completely interlaced morphology as the diameter and solidification temperature decreases. (from [21])

Figure 4: Heat flow versus temperature curves for Sn-2.4%Ag alloys on 100μm diameter Ni pads during cooling rates of (solid line) 0.083°C/s, (dotted line) 0.33°C/s, (dashed line) 0.66°C, and (dot-dashed line) 1.0°C/s.

times increased the average lifetime and predictability of solder joints during EM tests [6]. With electron backscattered diffraction (EBSD) they discovered that the solder joints that underwent multiple reflows typically had a region with many small Sn grains (an interlaced Sn grain morphology), rather than the single grain or beach ball morphologies that are typically observed in solder joints that had only one reflow. While many small grains were observed, it was evident from the EBSD maps that only main three Sn grain orientations were present.

In general, only three unique Sn grain orientations are present in SAC microbumps, although the Sn grain morphology may vary significantly. In Fig. 1 typical Sn grain morphologies observed in SAC alloys are illustrated for Sn-2.4%Ag alloy solder bumps on Cu pads. The four commonly observed Sn grain morphologies are: Fig. 1a, single grain; Fig. 1b, beach ball; Fig. 1c, mixed; and Fig. 1d, interlaced. In each case (excluding the single grained solder bump) there are three primary Sn grain orientations present, and the Sn grain orientations are rotated by approximately 60° around their [100] axes relative to each other, a natural twin axis in the β-Sn crystal structure. The origin of this morphology in SAC solders is believed to be a result of the nucleation and growth process that occurs in undercooled melts [7-14].

These twinned microstructures are produced during the nucleation and growth of Sn from undercooled SnAgCu melts. A simple model hypothesized that the nucleus is a six-fold, multiply twinned β-Sn structure [10], grown from a small hexagonal cluster of Sn atoms centered on a Ag or Cu atom forms (Fig. 2a). For temperatures near the melting point, β-Sn dendrites grow from the facetted surface in their preferred [110] and [1-10] directions (figure 2b) [10, 15-20]. As the undercooling of the Sn liquid increases, the growth of Sn dendrites moves away from the [110] and [1-10] directions creating an interpenetrating network of dendrites, also known as an interlaced structure [19]. In previous studies, the interlaced Sn grain structure has in fact been correlated with low solidification temperatures [7, 21]. In Fig. 3, free standing SAC305 solder bumps that solidified at progressively lower temperatures show the progression from single grain or beach ball to a mixed Sn grain morphology to a fully interlaced structure.

The present work examines the nature of the Sn grain morphology in different arrays of SnAgCu solder bumps. Measurements of the distribution of solidification temperatures in such arrays, as well as Sn grain morphologies in solidified solder bumps, are correlated with changes in pad metallurgy, pad diameter, and solder composition. The variation of the Sn grain morphology with solidification temperature is carefully examined.

Experiment

The solidification temperatures and Sn grain morphologies of an array of Sn-Ag, C4 solder bumps were systematically examined. The surface of a Si chip was populated with Sn-2.4%Ag solder bumps with either Cu or Ni metallizations. Solder bumps had pad diameters in the range of 14μm to 100μm. Each chip contained between 22,078 (14μm pad mask diameter) and 644 (100μm pad mask diameter) solder bumps. The height of the 14μm to 50μm diameter solder bumps was 11μm and the height of the 63μm and 100μm diameter solder bumps was 30μm. The solidification temperatures of C4 solder bump assemblies were also monitored. For this, two chips with 30 μm thick (total of 60μm of solder) Sn-2.4%Ag alloys on 100μm diameter pads were aligned and reflowed with a FineTech optical aligner. Each combination of pad metallurgy was produced, i.e. Cu/Cu, Ni/Ni, and Ni/Cu pad combinations.

The solidification temperatures of the array of solder bumps and sandwiched assemblies were precisely measured in a differential scanning calorimeter (DSC) during constant cooling, and during isothermal holds. The Si chips containing

978-1-4799-2408-0/14 $31.00 © 2014 IEEE

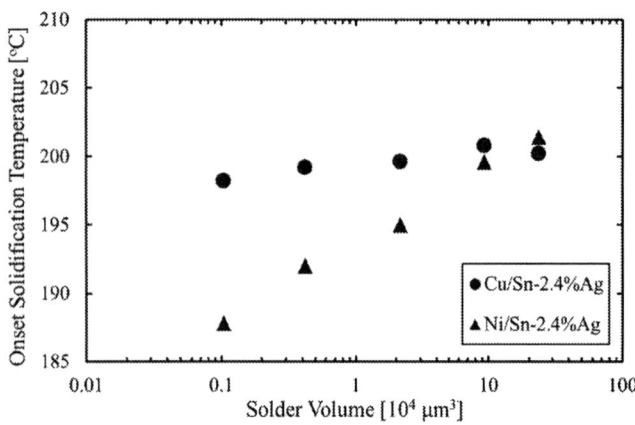

Figure 5: Plots of the onset solidification temperature versus solder volume are shown for Sn-2.4%Ag solder bumps on different diameter (circles) Cu and (triangles) Ni pads.

the array of solder bumps were placed directly into the DSC calorimeter, without encapsulation. In the DSC the chips were heated to 240°C, held there for 2 minutes then cooled to room temperature at 60°C/min. Each chip was reflowed only one time. For the isothermal experiments, the chips were heated to 240°C, held for 2 minutes then cooled at 20°C/min to the desired isothermal temperature and held there for 30 minutes. After the isotherm, the chips were cooled at 2°C/min to room temperature. During all of the experiments in the DSC Ar gas

was flowing throughout the sample chamber to ensure similar conditions for each chip.

The Sn grain morphology of the solder bumps was characterized using optical microscopy (OM, Zeiss mlm Optical Microscope). Optical microscopy with crossed polarizers (XP) was used to examine the Sn grain morphology in the solder bumps. The different colors in a XP micrograph qualitatively represent the Sn grain orientations in sample. The Sn grain morphologies of several solder bumps were examined and it was determined whether an individual bump had an interlaced Sn grain morphology, or not.

Results and Discussion

In order to provide more reliable C4 solder joints, consistent production of an interlaced Sn grain morphology is advantageous. Past work has correlated both solder alloy composition, and Sn solidification temperature, with the determination of Sn grain morphology in a Sn based solder joint [7, 22]. In this study, such influences are examined.

Solidification Temperature and Sn grain Morphology in C4 Solder Bumps

Cooling an array of solder bumps in the DSC produces a curve reflecting the distribution of solidification temperatures. Fig. 4 shows heat flow versus temperature plots for 4 different chips with Sn-2.4%Ag solder bumps on 100µm diameter Ni pads cooled at rates of 1.0°C/s (dot-dashed line), 0.66°C/s

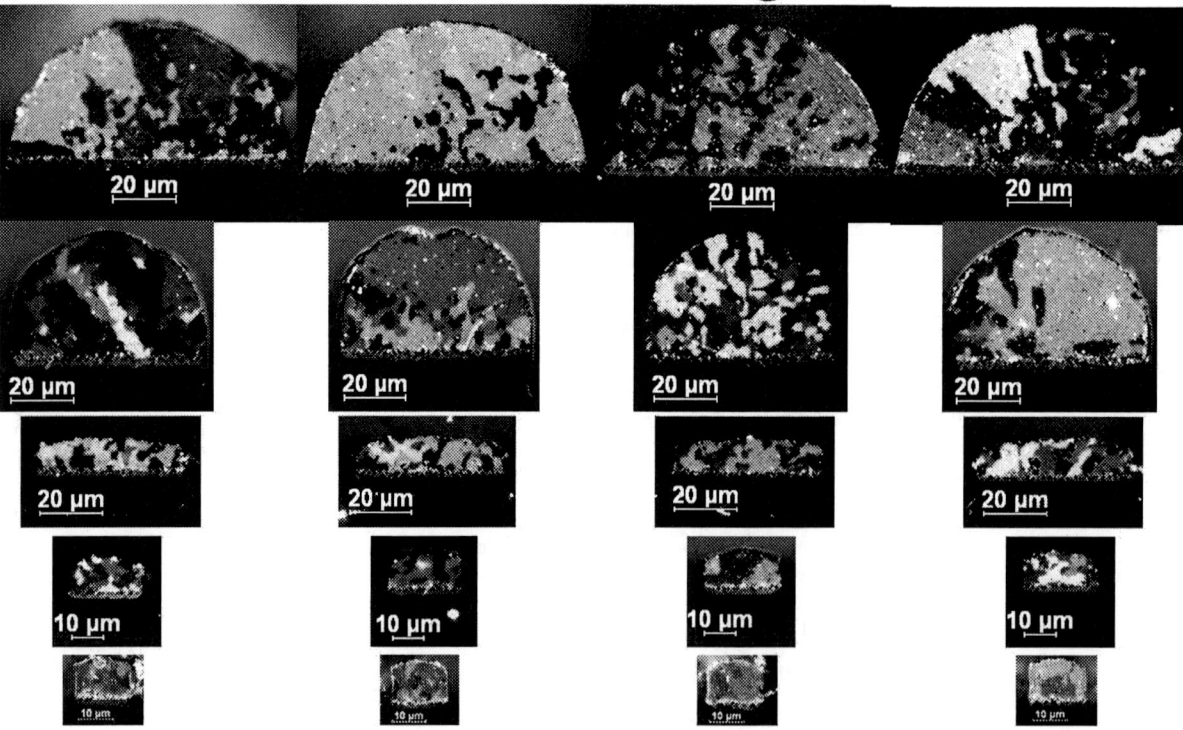

Figure 6: Optical microscopy with cross polarizers of representative Sn-2.4%Ag alloys on (row 1) 100µm diameter, (row 2) 63µm diameter, (row 3) 50µm diameter, (row 4) 24µm, and (row 5) 14µm diameter Ni pads.

Cu/Sn-2.4%Ag

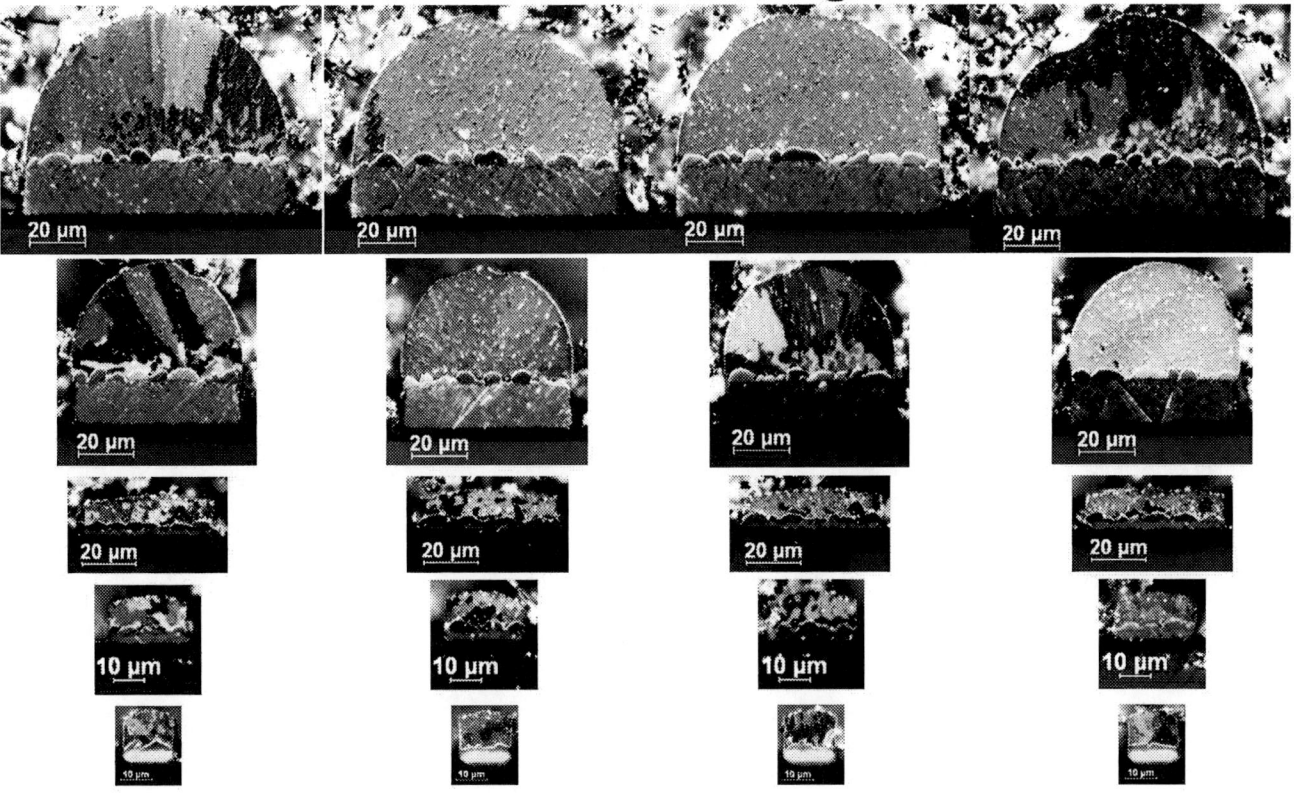

Figure 7: Optical microscopy with cross polarizers of representative Sn-2.4%Ag alloys on (row 1) 100µm diameter, (row 2) 63µm diameter, (row 3) 50µm diameter, (row 4) 24µm, and (row 5) 14µm diameter Cu pads.

(dashed line), 0.33°C/s (dotted line), and 0.083°C/s (solid line). At lower cooling rates, the solid curve in the heat flow versus temperature graph of Fig. 1 reveals significant detail. The small peaks at the beginning and end of the distribution are representative of single solidification events, while at intermediate temperatures simultaneous solidification of individual solder bumps is observed. These solder bumps solidify over a 25°C range of temperatures, taking a total time of about 1200 seconds. As the cooling rate is increased, the range of solidification increases to about 35°C at 1°C/s and the total time of solidification is reduced to 35 seconds. The shortened amount of time it takes for all 644 solder bumps to solidify (about 35 seconds) coupled with the thermal response time of the DSC (about 5 seconds) results in the DSC curve that represent the superposition of many solidification events in a smooth distribution. The onset of these solidification curves, which represent the highest possible solidification temperature of such solder joints, is used to characterize the effects of solder volume and pad metallization on Sn solidification.

The onset solidification temperature of the distribution of solidification temperatures were used to characterize the solidification behavior of the solder bumps. In Fig. 5, the onset solidification temperatures for different Sn-2.4%Ag solder bumps on either Cu and Ni pads is plotted as a function of solder volume. For both Cu and Ni substrates, the onset

solidification temperature decreases with decreasing solder volume, although this effect is more prominent for solder bumps on Ni pads. The onset solidification temperature is decreased by 14°C by decreasing the Ni pad diameter from 100µm to 14µm, while the same decrease in Cu pad diameter results in a decrease in the onset solidification temperature of about 4°C. The Cu_6Sn_5 intermetallic compound is most likely the reason a small dependence of the onset solidification temperature on the pad diameter is observed in this study. It is thought that the vertex between two scalloped shaped Cu_6Sn_5 IMC's provides a potent nucleation site for β-Sn.

The Sn grain morphology of the solder bumps was examined with optical microscopy with crossed polarized light. After reflow in the DSC, the chips were carefully cross-sectioned to expose an entire row containing between 28 (100µm diameter) to 128 (14µm diameter) solder bumps. Representative micrographs showing the Sn grain morphology on a given chip is shown in cross polarized micrographs in Fig. 6 for solder bumps on Ni pads, and Fig. 7 for solder bumps on Cu pads. In these micrographs, different colors represent different Sn grain orientations. As shown in Fig. 1, there are four typical Sn grain morphologies that appear in Sn-Ag solder bumps, thus the frequency of solder bumps with an interlaced or mixed morphology (Figures 1c, 1d) was tabulated. In general, the solder bumps on Ni pads with diameters of 100µm and 63µm had a higher tendency to form the interlaced

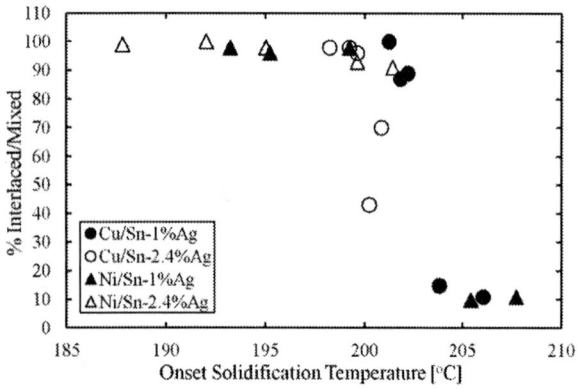

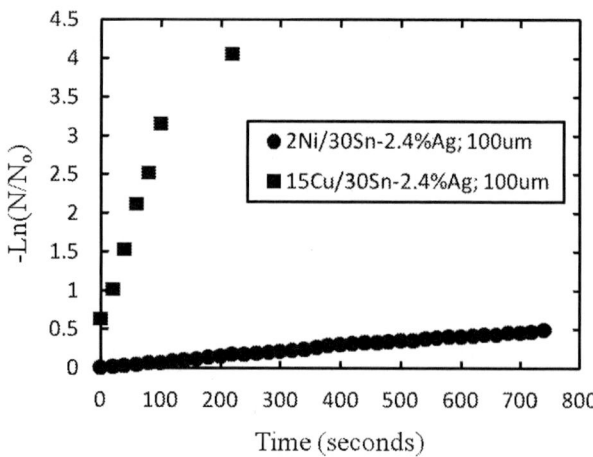

Figure 8: Plots of the percentage of solder bumps with an interlaced or mixed morphology versus pad diameter for Sn-2.4%Ag alloys (open symbols) or Sn-1.0%Ag alloys (solid symbols) on pads with either Cu metallizations (circles) or Ni metallizations (triangles).

Figure 9: Plots of $-Ln(N/N_o)$ versus Time for Sn-2.4%Ag solder bumps on 100μm diameter (squares) Cu and (circles) Ni pads during a 201°C isotherm in the DSC.

structure than similar bumps on Cu pads. The interlaced Sn grain morphology was almost always observed in the smaller 50μm to 14μm solder bumps with either Cu or Ni pad metallizations.

The interlaced Sn grain morphology has been previously associated with lower solidification temperatures, therefore correlations between Sn grain morphology and solidification temperature were examined in the present study. The percentage of solder bumps with an interlaced or mixed structure versus onset solidification temperature was plotted in Fig. 8. In this figure, a clear correlation is observed between interlaced percentage and solidification temperature. At temperatures above 203°C, the percentage of solder bumps with an interlaced or mixed morphology is low, not exceeding 15%. Below solidification temperatures of 199°C, the percentage of solder bumps with an interlaced or mixed morphology increases to greater than 90%. This high percentage of interlaced samples on Ni pads is correlated with lower solidification temperatures experienced while cooling from the melt.

Nucleation Rates of β-Sn in C4 Solder Bumps

The Sn grain morphology is strongly correlated with the solidification temperature of the Sn-Ag solder bumps (c.f. Fig. 8), thus quantifying the temperature dependence of the nucleation rate could lead to an expression for the degree of interlacing in these solder bumps. Having hundreds of individual solder bumps in the DSC at once allows combinatorial measurements, and more precise determination of nucleation rates. During isothermal holds, the solidification of an individual solder bump was clearly discernable. At a given temperature, the rate of nucleation R, in N similar solder balls in the liquid state at time t can be written in terms of the volume, v, of a single sample and the nucleation I_{het}:

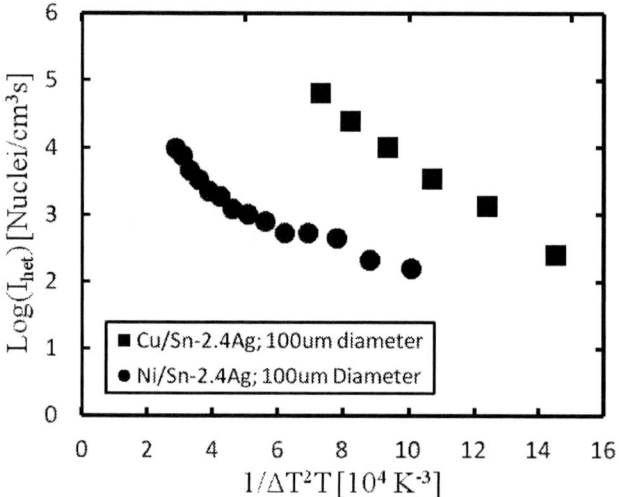

Figure 10: Plots of $Log(I_{het})$ versus $1/\Delta T^2 T$ for Sn-2.4%Ag solder bumps on 100μm diameter (squares) Cu and (circles) Ni pads.

$$R = N\, I_{het}\, v = N\, \alpha$$

where $\alpha = I_{het}\, v$

The growth rate of β-Sn in a solder bump is very high and the recalescent heat given off during solidification creates conditions in which only one critical nucleus forms in each sample. The measured rate of solidification of solder balls, dN/dt, can be assumed to be proportional to I_{het} and to N:

$$dN/dt = -\alpha\, N$$

A simple calculation leads to:

$$Ln(N/N_o) = -\alpha\, t$$

Where N_o is the total number of solder balls in the liquid state at $t = 0$.

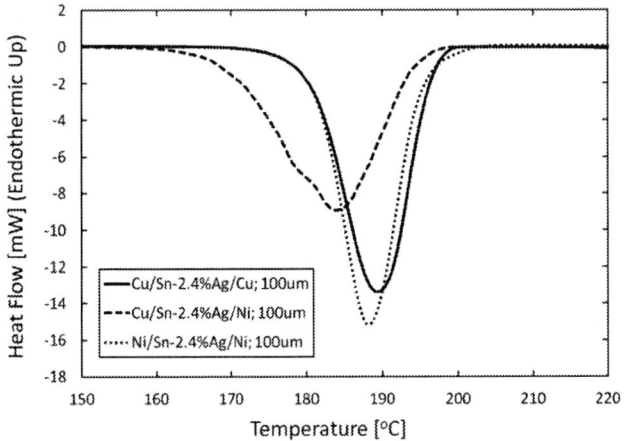

Figure 11: Plots of Heat Flow versus Temperature for Sn-2.4%Ag solder bumps sandwiched between (solid line) two Cu pads, (dotted) two Ni pads, and (dashed) a Cu and Ni pad while cooling at a rate of 1°C/s in the DSC.

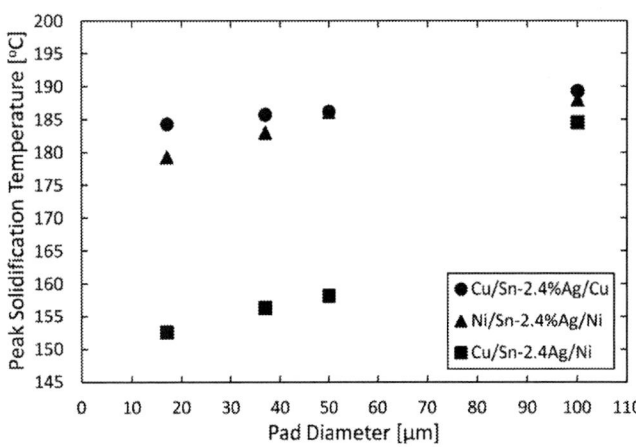

Figure 12: Plots of Peak Solidification Temperature versus Pad Diameter for Sn-2.4%Ag solder bumps sandwiched between (circles) two Cu pads, (triangles) two Ni pads, and (squares) a Cu and Ni pad.

Plots of $-\ln(N/N_o)$ versus time in Fig. 9 for an isothermal temperature of 201°C were constructed with solidification data from Sn-2.4%Ag solder bumps on 100μm diameter Cu and Ni pads. These plots clearly show the difference in nucleation kinetics for similar solder bumps on Cu pads versus Ni pads. The slope of a straight line plot, α, divided by the volume yields a value for the nucleation rate at a given temperature. It is clear that the nucleation rate of Sn-Ag solder bumps on Ni is much lower than similar bumps on Cu pads. For instance, at a temperature of 201°C, the nucleation rate of solder bumps on Ni pads was calculated to be 2.25×10^9 nuclei/m^3 s, while on Cu pads the nucleation rate was 6.70×10^{10} nuclei/m^3 s.

The nucleation rates of the solder bumps on Ni and Cu pads were measured at several different temperatures. In Fig. 10, plots the $\log(I_{het})$ versus $1/\Delta T^2 T$ plot are presented. The straight line plots in Fig. 10 are used to determine the activation energy for forming a critical nucleus in the undercooled melt at different temperatures. A distinctly lower nucleation barrier is found for solder bumps on Cu pads.

Solidification Temperatures of Sandwiched C4 Solder Bumps

The solidification temperatures of assemblies of solder were measured in the DSC during constant cooling rates of 1°C/s. The heat flow versus temperature plots in Fig. 11 show the distributions of solidification temperatures for 60μm thick Sn-2.4%Ag solder bumps sandwiched between two 100μm diameter pads with metallurgies of (solid line) Cu/Cu, (dotted line) Ni/Ni, and (dashed line) Cu/Ni. The solder joints with Cu/Cu and Ni/Ni pads have peak shapes similar to the single sided chips during a 1°C/s cooling rate (figure 4). The onset solidification temperatures of the Ni solder bumps are slightly higher than those on Cu, 203°C on Ni versus 200°C on Cu. The onset solidification temperature for the solder joints sandwiched between Cu and Ni pads occurs at 198°C. A small shoulder is observed at the tail end of the solidification distribution of solder bumps with Cu/Ni pads, whereas the peak shapes for Cu/Cu, Ni/Ni, and single chips are nearly symmetrical. Along with the small shoulder, the peak of the temperature distribution is lower for the Cu/Ni solder bumps than both the Cu/Cu and Ni/Ni solder bumps.

A large decrease in peak temperature of the solder bumps sandwiched between a Ni and Cu pads with diameters of 19μm, 30μm, 50μm and 100μm is observed upon constant cooling in the DSC from the melt. The peak temperature in the temperature distributions seen in heat flow versus temperature plots (Figures 4 and 10) are representative of an average solidification temperature for the solder bumps in the DSC. In Fig. 12 the peak solidification temperatures of the different sandwiches are plotted as a function of pad diameter. For the 19μm to 50μm diameter solder bumps between a Cu and Ni pad, the peak temperature is observed to be about 25°C to 30°C lower than solder bumps between two Cu pads and two Ni pads. The mechanism behind this dramatic decrease in peak solidification temperature observed for solder bumps between Cu/Ni pads is under further investigation.

Conclusions

Variations in solder composition, under bump metallurgy (UBM), and solder joint geometry were each observed to vary the Sn grain morphology of Pb free, C4 solder joints. While correlations between the degree of interlacing and each of these parameters were observed, the clearest correlation was between degree of interlacing and solidification temperature. It was observed that most microbumps solidified below199°C were interlaced. The nucleation rates of Sn from the melt in Pb free solder joints at a number of different temperatures were measured.

Acknowledgments

SERDP and IEEC are gratefully acknowledged for support of this project.

References

1. M. Kerr, N. Chawla, Creep Deformation behavior of Sn-3.5 Ag solder/Cu couple at small length scales, Acta Materialla, vol. 52, 4527, July, 2004

2. P. Kumar, Z Huang, S. Chavali, D. Chan, I. Dutta, G. Subbarayan, V. Gupta, Microstructurally adaptive model for primary and secondary creep of Sn-Ag based solders, IEEE Transactions on Components, Packaging and Manufacturing Technology, Vol. 2, No. 2, 256-265, February, 2012.

3. T. Bieler, H. Jiang, L. Lehman, T. Kirkpatrick, E. Cotts, Influence of Sn grain size and orientation on the thermomechanical response and reliability of Pb-Free solder joints, 56th IEEE Transactions on Components and Packaging Technology Conference, vol. 31, 370, 2006

4. M. Lu, D. Shih, P. Lauro, C. Goldsmith, D. Henderson, Effect of Sn grain orientation on electromigration degradation mechanism in high Sn-based Pb-free solders, Applied Physics Letters, vol. 92, 211909, May, 2008.

5. K. Lee, K. Kim, Y. Tsukada, K. Suganuma, K. Yamanaka, S. Kuritani, M. Ueshima, Effects of the crystallographic orientation of Sn on the electromigration of Cu/Sn-Ag-Cu/Cu ball joints, Materials Research Society, Vol. 26, No. 3, 467-474, February, 2011.

6. Y. Wang, K. Lu, V. Gupta, L. Stiborek, D. Shirley, S. Chae, J. Im, P. Ho, Effect of Sn grain structure on the electromigration of Sn-Ag solder joints, Materials Research Society, Vol. 27, No. 28, 1131-1141, April, 2012.

7. G. Parks, B. Arfaei, M. Benedict, E. Cotts, M. Lu, E. Perfecto, The dependence of the Sn grain structure of Pb-free solder joints on composition and geometry, Electronics Components and Technology Conference (ECTC), 2012 IEEE 62nd, 703-709, May 2012.

8. D. Porter, K. Easterling, M. Sherif, Phase transformations in metals and alloys, CRC Press, 3rd Edition, 2009.

9. K. Kelton, A. Greer, Nucleation in condensed matter, Elsevier Ltd., 2010.

10. L. Lehman, Y. Xing, T. Bieler, E. Cotts, Cyclic twin nucleation in Sn-based solder alloys, Acta Materialia, Vol. 58, 3546-3556, March, 2010.

11. B. Arfaei, E. Cotts, Correlations between the microstructure and fatigue life of near-eutectic Sn-Ag-Cu Pb-free solders, Journal of Electronic Materials, Vol. 38, No. 12, September, 2009.

12. P. Borgesen, T. Bieler, L. Lehman, E. Cotts, Pb-free solder: New materials considerations for microelectronics processing, MRS Bulletin, Vol. 32, 360-365, April, 2007.

13. B. Arfaei, M. Benedict, E. Cotts, Nucleation rates of Sn in undercooled Sn-Ag-Cu flip-chip solder joints, Journal of Applied Physics, Vol. 114, Novermber 2013.

14. L. Lehman, S. Athavale, T. Fullem, A. Giamis, R. Kinyanjui, M. Lowenstein, K. Mather, R. Patel, D. Rae, J. Wang, Y. Xing, L. Zavalij, P. Borgesen, and E. Cotts, Growth of Sn and intermetallic compounds in Sn-Ag-Cu Solder, Journal of Electronic Materials, Vol. 33, No. 12, 1429-1439, June, 2004.

15. G. Powell, G. Colligan, V. Surprenant, A Urquhart, The growth rate of dendrites in undercooled Tin, Metallurgical Transactions A, Vol 8A, 971-973, June, 1977.

16. D. Lee, K. Kim, Y. Lee, C. Choi, Factors for determining crystal orientation of dendritic growth during solidification, Materials Chemistry and Physics, Vol. 47, 154-158, January, 1996.

17. J. Warner, J. Verhoeven, Morphology of Sn dendrites in near-eutectic alloys, Metallurgical Transactions, Vol. 3, 1001-1002, April, 1972.

18. S. O'Hara, Controlled growth of Sn dendrites, Acta Maetallurgica, Vol. 15, 231-236, Febuary, 1967.

19. V. Esin, G. Pankin, R. Nasyrov, Crystallographic orientation of dendrites of pure Sn, Soviet Physics: Crystallography, Vol. 18, No. 2, 246-249, 1973.

20. A. Rosenberg, W. Winegard, The rate of growth of dendrites in supercooled Sn, Acta Metallurgica, Vol. 2, No. 2, 342, March, 1954.

21. B. Arfaei, N. Kim, E. Cotts, Dependence of Sn grain morphology of Sn-Ag-Cu solder on solidification temperature, Journal of Electronic Materials, Vol. 41, No. 2, 362-374, October, 2011.

The Impact of Microstructure Evolution, Localized Recrystallization and Board Thickness on Sn-Ag-Cu Interconnect Board Level Shock Performance

[1]Tae-Kyu Lee, [1]Weidong Xie, [2]Thomas R. Bieler and [3]Choong-Un Kim

[1]Component Quality and Technology Group, Cisco Systems, Inc., San Jose, CA 95134

[2]Michigan State University, East Lansing, MI, [3]University of Texas, Arlington, TX

taeklee@cisco.com

ABSTRACT

The mechanical stability of solder joints with SnAgCu alloy on various board thicknesses were investigated in a high G level shock environment. A test vehicle with 31mil, 62mil and 93mil board thickness, which has three different strain and shock level condition combination per board, was used to identify the joint stability and failure modes. The results revealed that joint stability is sensitive to board thickness and that the first failure location shift from the corner location near the stand off to the center with increased board thickness. Also the impact of isothermal aging and fine grain structure transformation on mechanical shock performance of solder joints were investigated. The results revealed that joint stability during shock loading is sensitive to the level of shock that can be absorbed during each shock cycle based on the capability of single to multi grain transformation. The localized fine grain structure distributions were analyzed to identify correlations between the microstructure evolution and shock performance.

INTRODUCTION

With more wider end use conditions and applications, the mechanical stability of solder interconnects during dynamic shock and loading conditions are crucial for high reliability electronic devices. [1-4] To have a stable interconnect in a mechanical shock environment, the interconnections need to endure mechanical strain, which develops in a way that is dependent upon the reaction between the component location and applied shock. [5-8] Failure modes vary from the package side interface intermetallics to the laminate region right below the Cu pad. Mitigation of crack initiation or crack propagation can be achieved if the shock energy is consumed or absorbed in the interconnection system, in other words by the solder joint bulk solder material. [9,10] Figure 1 shows cross section polarized light microstructures of the outermost 5 joints on each side of the outside edge of the shock-tested samples. Since the pad design of these samples were NSMD (non-solder mask defined), the crack locations were typically at the upper interface or in the laminate below the board side copper pad, as indicated by an outlined box in the figure. [10] As shown in Figure 1(a), most of the unaged solder joints have single, bi or tri-crystal structure with a small number of grains per joint. However, after shock testing, the SAC105 microstructure revealed development of a fine grain microstructure that is much finer for isothermally aged (Figure 1(c)) than the unaged samples (Figure 1(b)). This transformation from single or dual grains to multi grain or fine localized grain structure may account for better absorption of the shock induced strain energy. Shown in this example, the transformation from single to multi grains during shock cycling produced additional grain boundaries. With this microstructural transformation, the shock induced strain can be absorbed in two ways by motion of dislocations nucleated at grain boundaries, and by grain boundary sliding. [10] Thus, the shock that progress to the interface region is reduced resulting in an improved shock performance. This shock induced strain consumption can increase by the number increase of new grain orientations and/or grain boundary area, thus can reduce or delay damage accumulation at the high stress concentration locations at the intermetallic interface layer or at the laminate and Cu pad interface. [10] But given the fact that these samples in this example are NSMD pad design samples, the possibility of crack development at the IMC at the board side interface is minimal. Shock induced

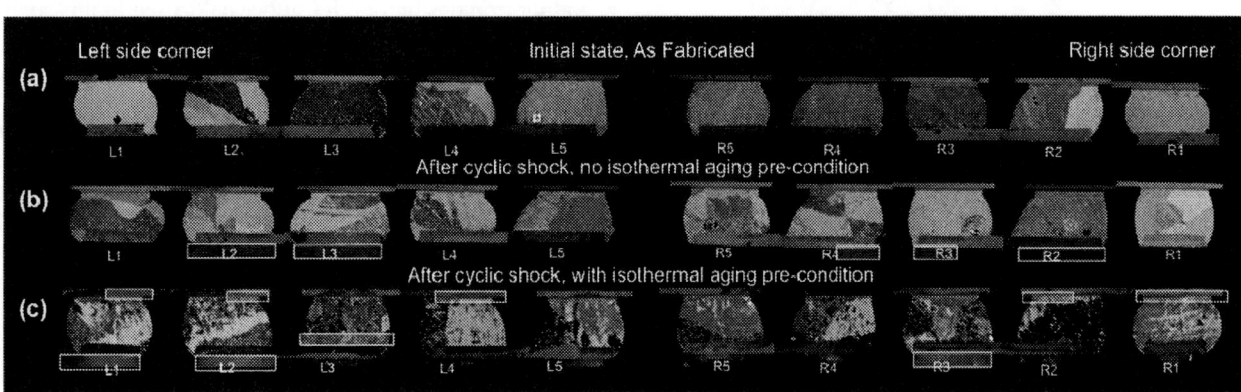

Figure1. Polarized light microstructure cross sections of the 10 joints (location indicated in Figure 2) before and after shock testing. SAC105 (a) As assembled, (b) after shock test and (c) after 150°C/500h aging and shock test The crack locations are indicated by boxes. [10]

strain can be absorbed by the whole solder joint body, since the bonding at the board side is secured by the pad design itself. If the pad design is SMD (solder mask defined), which allows the crack to be initiated at the corner of the board side interface where the IMC is exposed to the stress concentration point, the shock performance is expected to be vastly different than the results with NSMD pad design samples. In this study, the impact of the isothermal aging and board thickness on SMD pad design samples was investigated and the potential mechanisms are discussed. The impact of isothermal aging and recrystallization on high G board level shock performance are analyzed. By following the evolution of the grain structure and distribution, interrelationships between the fine grain distributions, mechanical shock are identified.

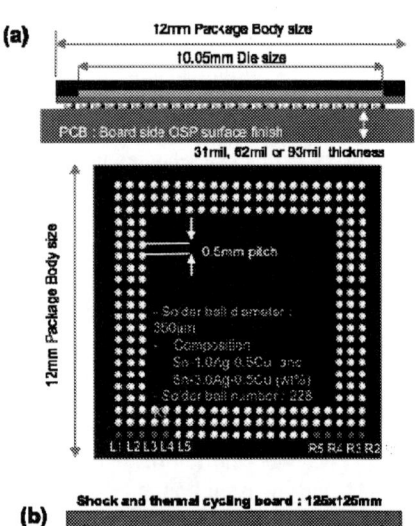

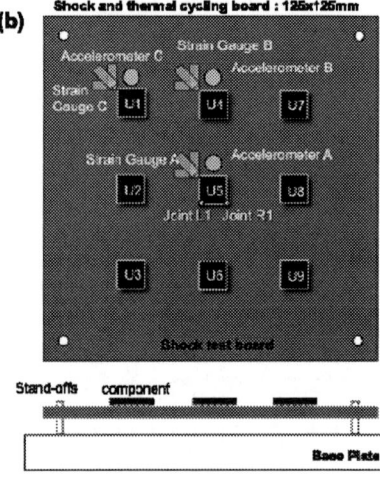

Figure 2. Sample component (a) and Shock test board (b) schematics. Location of packages U1-U9, strain gauges and accelerometers are indicated on the board. [10]

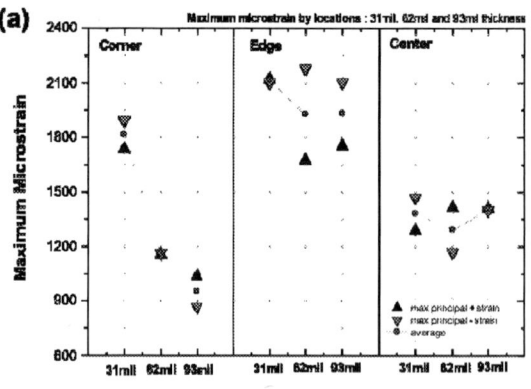

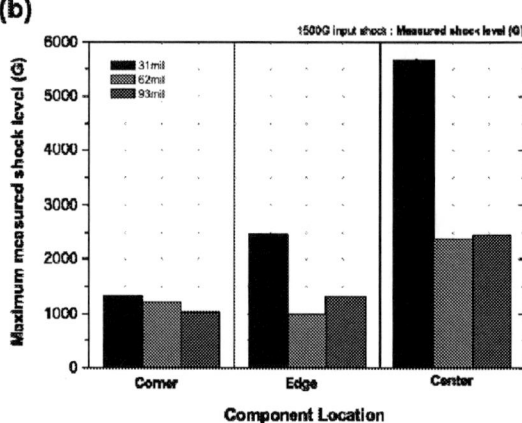

Figure 3. Strain and shock level (G) during 1500G table input shock with 0.5millisecond half sine pulse duration. (a) Maximum microstrain by location for each board thickness samples. (b) Measured shock level (G) at each location per board thickness.

EXPERIMENTAL PROCEDURE

Shown in Figure 2(a), the samples used in this study were 12x12 mm body size ball grid arrays (BGAs) with a three-row perimeter array, 0.5mm pitch and a total of 228 solder joints with solder ball diameter of 350 μm. The package side substrates had electrolytic Ni/Au surface finish and the composition of the solder balls used in this study was Sn-3.0Ag-0.5Cu (wt%) (SAC305). The packages were assembled onto 31mil, 62mil or 93mil high Tg FR4 boards with organic surface preservative (OSP) surface finish. All test samples board pad design were SMD (solder mask defined) and the reflow condition was 1 minute above the liquidus with a 245°C peak temperature profile. As shown schematically there are 9 components per board and for a given input shock event, each location experienced a different shock level and strain level. The test vehicle was designed to provide three different shock and strain conditions and the shock and strain response was explained in detail in earlier publications. [9,10] A fixed input shock of 1500G with 0.5 millisecond half-sine pulse duration at the table was imposed, which is based on the JEDEC22-B111 standard. [11] The connections between the component daisy chains were continuously monitored and a 20% increase in resistance

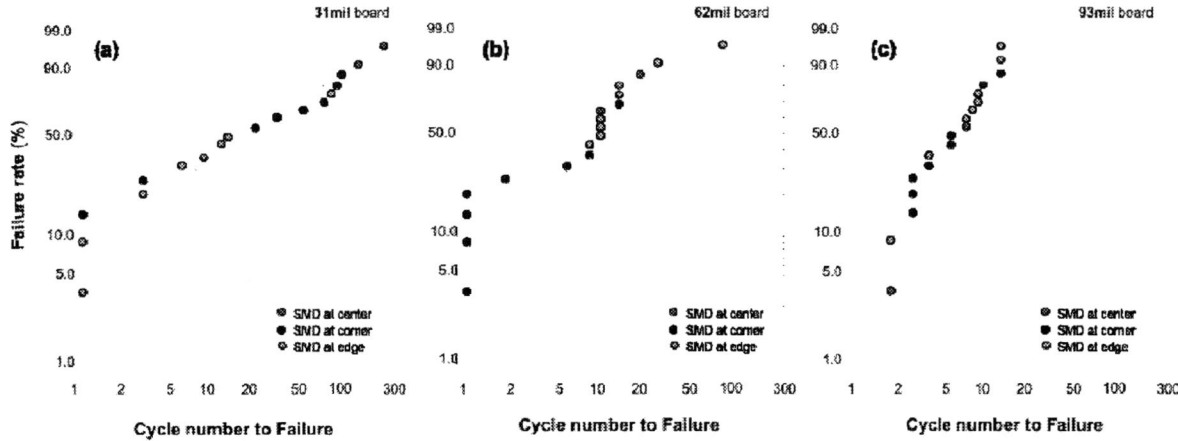

Figure 4. Weibull plot for shock test to failure. (a) 31mil thickness boards, (b) 62mil thickness boards and (c) 93mil thickness boards.

during the shock test, compared to the initial resistivity value, was considered to be a failure. To study the effect of isothermal aging, samples were isothermally aged at either 100°C or 150°C for 500 hours in ambient air. To identify the crack locations and their distribution in the sample, cross section and dye-and-pry methods were used. Both bright field and polarized optical and SEM microscopy were used to identify the crack propagation path. Polarized light images were used to identify initial and localized fine grain structures.

RESULTS AND DISCUSSION

Figure 3 shows the strain and shock level per board thickness and location of the measurement point. As shown in Figure 3(a), the maximum strain level per location of the board show similar value at the edge and center for all three board thicknesses at around 2000 microstrain for edge location and 1400 microstrain for the center location. But the corner location strain show different levels with a high strain value for 31mil boards and low value for thicker 93mil boards. Compared to the variation of the microstrain level at the corner per board thickness, the measured G level at the corner show similar levels compared to a larger difference at the center, where a higher G level was detected for 31mil boards than 62mil or 93mil boards. Through these series of observation, we can see that the combinations of strain and shock level are varied by board thickness and thus a variety of shock performance can be expected.

Impact of the board thickness on shock performance

Figure 4 shows the shock test results based on different board thicknesses. The locations of the samples are indicated on the weibull plot so that a shift of the early failure location can be seen. Figure 4(a) shows the shock test results on the 31mil thickness board. Compared to shock test results on thicker boards, the beta slope is smaller compared to other board thickness test results with the first failure at the first cycle. The characteristic life cycle number is around 50 cycles. The first failures occurred at the edge and corner samples as shown with a dominance at the edge samples. With the higher cycle number to failure, the corner samples

begin to show as the primary failure location followed by the center located samples. It is interesting to see the trend of having the center samples failed last in 31mil boards changed to failing first with thicker board thickness. As shown in Figure 4(b) and (c), with thicker board thickness, the center location samples shifted to an earlier failure cycle among the tested samples within the same thickness board. In 93mil boards, the center location samples failed first indicating a shift of failure mechanism among different board thickness test samples. A summarized plot is shown in Figure 5. The failure cycle for the edge, corner and center samples are plotted with a characteristic life cycle number indicated for each board thickness test conditions. Overall, the thinner board (31mil) shows a better shock performance, especially for the center location samples. Comparing the corner and center strain and shock level measurement shown in Figure 3(a) and (b), a few observations can be listed as follows; (1) shock level (G) at the center location showed the highest value for 31mil compared to 62mil and 93mil boards. But the cycle number to failure was the highest for 31mil center location samples. This indicates that the higher shock level (G) is not directly playing as a sole failure driving factor; (2) the microstrain value at the corner location for each board

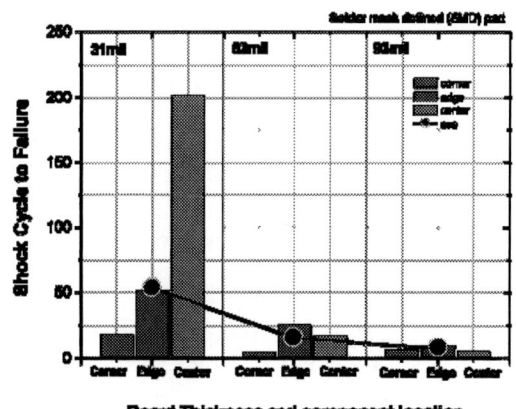

Figure 5. Summary of plot per board thickness and sample location.

978-1-4799-2408-0/14 $31.00 © 2014 IEEE

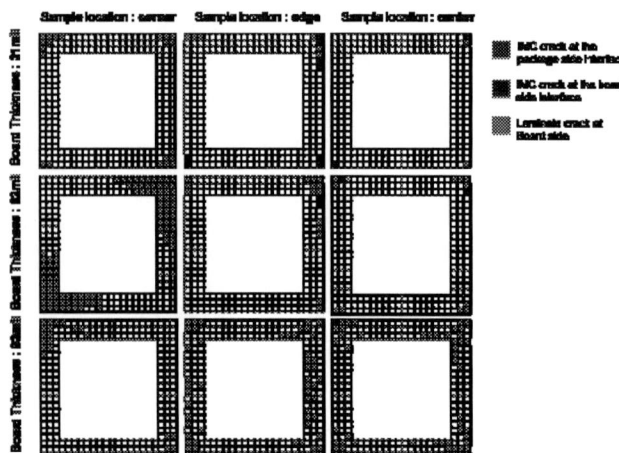

Figure 6. Dye and pry fracture distribution map per shock tested board thickness

thickness samples show a variation with 31mil board as highest and 93mil as lowest, but the shock performance at the corner did not follow the strain value and show relatively similar performance with even a slightly higher performance for 31mil boards, which actually showed the highest maximum strain value; (3) The strain value at the center location show similar level of microstrain but the life cycle at the center sample location show the most variation in number cycles between the three board thickness samples. Based on these three observations, we can see that the strain and G level measurement are not the sole factors, which defines the stability of the solder joints. It is important to note and identify that additional factors are influencing and contributing the cycle number to failure per each sample location and per board thicknesses. An investigation using FEAM is under evaluation to analyze the board reflection per location of the board. It is expected to have a higher fluctuation at the corner location for thinner boards, which was already observed and showed in Figure 3. The fluctuation is expected to be with lower frequency compared to the thicker boards. The results provide potential explanation that the thinner boards experience fewer bending cycles compared to thicker boards even with experiencing the maximum strain value. The thicker boards with higher frequency have higher potential to have the failure earlier and that is probably the phenomena we can see in these series of results. Figure 6 is the dye and pry results on the shock tested samples for 31mil, 62mil and 93mil board samples. The package side IMC interface is shown as the dominant failure site with a few failure locations at the board side IMC interface.

Impact of Isothermal aging on shock performance

The impact of isothermal aging on shock performance were published earlier on 1.0mm pitch samples [12] but comparing to finer pitch samples used in this study, a different trend is observed with a shift of failure mode. The weibull plot of the shock test results with 0.5mm pitch samples on 62mil thickness boards were presented in Figure 7(a) and the summary plot is shown in Figure 7(b). The clear separation per isothermal aging condition for the 1.0mm pitch samples [12] was not so predominant in the 0.5mm pitch SMD

samples. As shown in Figure 7(b), the 100°C aged samples actually show an improved shock performance compared to the as-assembled samples, and the first failure for both aged and non-aged samples were similar. The optical microscope cross section view of corner solder joints from center, edge and corner components after shock test per isothermal aging condition are shown in Figure 8. The regions where a crack developed are indicated in the outlined box. As shown in the figure, the failure mode shifted from a mixed failure mode of package side and board side IMC crack to a package side dominant crack formation and propagation after isothermal aging. This trend can be also seen in the dye and pry result shown in Figure 9.

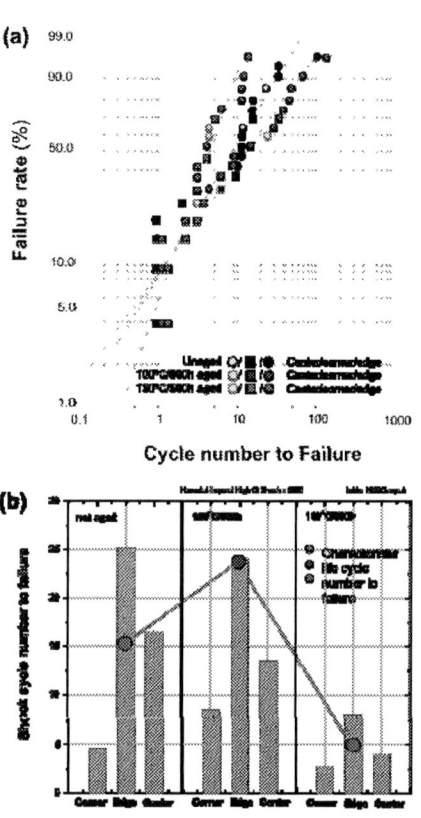

Figure 7. Weibull plot of number of shock cycles to failure with and without isothermal aging using SMD pad test board (a) and Number of shocks to failure before and after isothermal aging based on package locations (b)

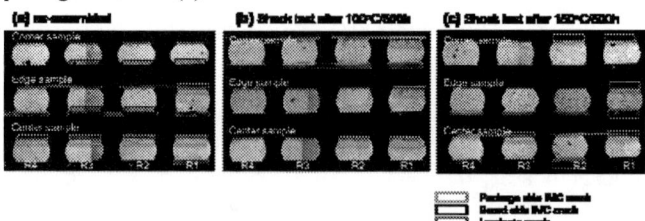

Figure 8. Optical microscope cross section view after shock test on (a) as-assembled (b) shock tested with 100°C/500h isothermal aging and (c) shock tested with 150°C/500h isothermal aging

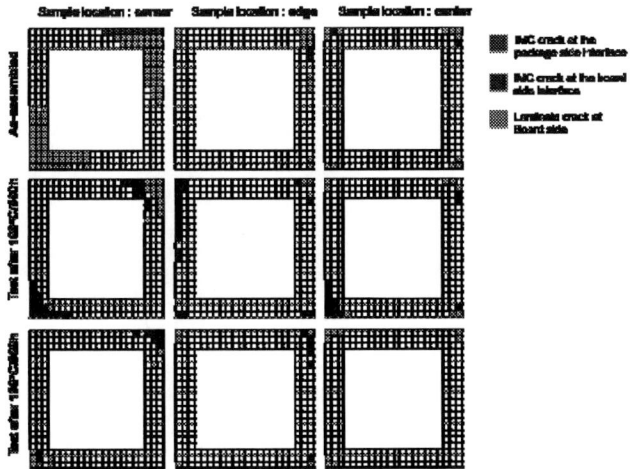

Figure 9. Dye and pry fracture distribution map with and without isothermal aging and shock test.

With the representing SEM pictures on each aging condition samples, the package side crack propagation happened through the interface between the Ni layer and the root of the $(Ni,Cu)_6Sn_5$ layer in all three aging condition samples. The difference in the solder bulk structure after aging can be also seen in Figure 10 (d)-(f), where finer Ag_3Sn precipitate network become coarser and fewer in numbers after aging at 150°C for 500 hours. But it is difficult to explain the increase of the shock performance after 100°C aging with the crack propagation path and the solder bulk IMC precipitation change observed and described above. A possible explanation can be found in the bulk microstructure transformation during shock test. As already explained in an earlier publication, the bulk solder can transform into a multi grained structure from a single/dual grain structure during shock test. [9,10] With this transformation, the shock induced strain can be absorbed and result in an improved shock performance. A similar single to multi grain transformation is also observed in the isothermally aged samples after shock. Figure 11 shows selected solder joints at the corner of the shock-tested samples per aging condition. The solder joint in Figure 11(a) is a joint after shock with no isothermal aging. A localized area shows a small faction of fine-grained area development (indicated with an arrow) but the grain structure at the package and board side near the interface remained relatively unchanged. Compared to the no aged samples solder joint, the joint after shock with 100°C aging condition shows a relatively large amount of fine grain structure development. It is hard to tell that the fine grain structure was already developed before the shock test or during the shock, but with the given fine grain structure, the ability to absorb the shock induced strain is higher, thus it has a higher potential to show a good shock performance. The fine grain structure near at the board side IMC and solder bulk interface show also a fine grain structure which might be an interlace structure. These fine grain structure can potentially absorb the shock energy and the shock induced strain. The 150°C aged solder joint after shock is shown in Figure 11(c). It is expected to have more capability to absorb the shock induced strain, but at the same time the IMC layer at the package and board side are constantly growing, and this brings a competition between the capability of absorbing the shock induced strain and the IMC bonding at both interfaces. The cycle number to failure show that the IMC interface at the package side is getting weaker after aging and resulting in an earlier solder fracture compared to the no aged samples.

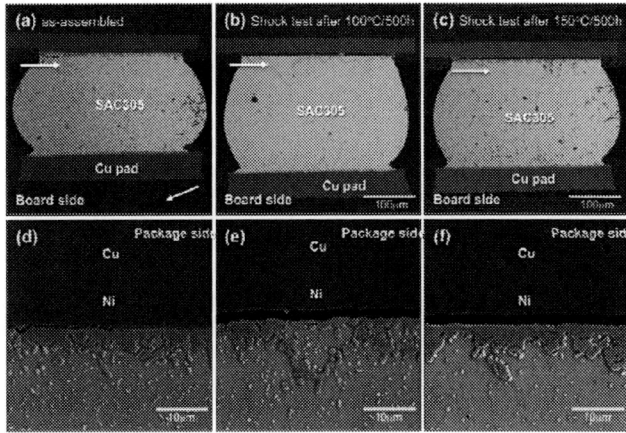

Figure 10. Selected solder joint SEM images after shock test per isothermal pre-condition.

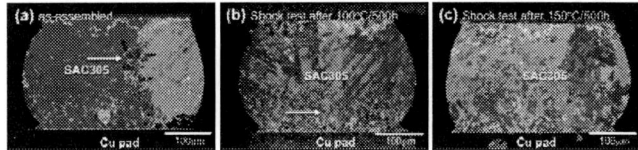

Figure 11. Selected solder joint optical polarized images after shock test per isothermal pre-condition. (a) as-assembled (b) shock tested with 100°C/500h isothermal aging and (c) shock tested with 150°C/500h isothermal aging

CONCLUSIONS

The mechanical stability of solder joints with SnAgCu alloy on various board thicknesses was investigated in a high G level shock environment. Test vehicles with 31mil, 62mil and 93mil board thicknesses, which have three different strain and shock level condition couples per board, were used to identify the joint stability and failure modes. The results revealed that joint stability is sensitive to board thickness and that the first failure location shift from the corner location near the stand off to the center by increased board thickness. Also the impact of isothermal aging of solder joints reveal that joint stability during shock loading is sensitive to the level of shock that can be absorbed during each shock cycle. Isothermal aging not only increase the IMC thickness, but also increase the capability of absorbing the shock-induced strain. The improvement of shock performance for samples aged at 100°C show an example of the solder bulk shock absorption mechanism. It is too early to confirm the mechanism in this alloy systems but an earlier publication saw this phenomenon in low silver alloy solder joints, which provide potential explanation that there are competition between the bulk shock absorption and the IMC structure stability against mechanical shock.

978-1-4799-2408-0/14 $31.00 © 2014 IEEE

REFERENCES

1. W. Xie, T.-K. Lee, K.-C. Liu, and J. Xue, IEEE 60th Electronic Components and Technology Conference (ECTC) (Las Vegas, NV, June 2010).

2. H.G. Song, J.W. Morris Jr., and F. Hua, JOM 56, 30 (2002)

3. D.R. Frear and P.T. Vianco, Metall. Trans. A 25A, 1509 (1994).

4. H.K. Kim and K.N. Tu, Phys. Rev. B 53, 16027 (1996).

5. P. Lall, R. Lowe, and K. Goebel, IEEE 60th Electronic Components and Technology Conference (ECTC) (Las Vegas, NV, June 2010), p. 889.

6. W.Peng and M.E.Marques, J.Electron.Mater. 36, 1679(2007).

7. B. Noh, J. Yoon, S. Ha, and S. Jung, J. Electron. Mater. 40, 224 (2011).

8. G. Godbole, B. Roggeman, P. Borgesen, and K. Srihari, IEEE 59th Electronic Components and Technology Conference (ECTC) (San Diego, CA, June 2010).

9. T.-K. Lee, B. Zhou, T. Bieler, C.Tseng, and J. Duh, J. Electron. Mater. 42 (2), 215 (2013).

10. Tae-Kyu Lee, C. Kim and T. Bieler, J. Electron. Mater. 43 (1), 69 (2014).

11. JEDEC Standard JESD22-B111, July 2003.

12. T.-K. Lee, B. Zhou, T. Bieler, and K.-C. Liu, J. Electron. Mater. 41 (2), 273 (2012).

Thermal Cycle Fatigue Life Prediction for Flip Chip Solder Joints

Robert Darveaux
Skyworks Solutions, Inc., Irvine, CA
Robert.Darveaux@skyworksinc.com

Abstract

Finite element analysis was used to predict the thermal cycle fatigue life of flip chip solder joints. Three different data sets of measured thermal cycle fatigue life were simulated to assess model correlation. Variables in the measured data included solder resist opening diameter, UBM diameter, joint height, solder volume, underfill material, package structure, and temperature range.

Viscoplastic strain energy density (SED) per cycle was used as the damage indicator. Three different post processing schemes were compared for extracting the damage indicator: 1) maximum elemental SED value, 2) maximum SED^2 / average SED, and 3) layer averaged SED. It was found that all three methods could achieve some degree of correlation between measured and predicted life for a given data set. The best and worst legs of an experiment could be predicted well, but the ranking of all legs in a given experiment could not be perfectly predicted. For the experiment on joint design, max SED^2 / avg SED gave the best prediction, followed by max SED, and worst was layer averaged SED. For the experiments on power cycle conditions and package structure, all methods gave similar results.

Finally, the three data sets were plotted together with additional predictions from previous work on WLCSPs [1]. There was a large discrepancy in the *absolute* prediction from one data set to the next. However, the ability to predict the *relative* fatigue life within a data set was consistent, and the results did match the previous work on WLCSP solder joints.

Introduction

Analytical models to predict solder joint fatigue have several practical uses: 1) rapid design optimization during the development phase of a product, 2) predicting field use limits, and 3) failure analysis of product returned from the field or failed in a qualification test.

Many fatigue life models have been published over the last several decades. They generally require an indicator that represents the amount of damage inflicted on the solder joint per cycle. Common damage indicators are strain range, strain energy density (aka plastic work per unit volume), stress range, etc. The damage indicator might also be partitioned out into components, such as elastic, plastic, and creep.

The solder joint fatigue life can be calculated from the damage indicator by using a set of constants. This calculation can be done 1) directly, or 2) through a crack initiation and growth law. The important difference between these two variants is in how the size of the joint is accounted for. In the direct calculation method, it is best if the damage indicator scales with the size of the joint. In the crack initiation and growth method, the joint size is accounted for explicitly.

There are many methods and nuances in how to calculate the damage indicator. The basic inputs include the dimensions of the assembly, material properties, and the thermal cycle profile. The damage indicator is then typically calculated using closed form methods or finite element analysis (FEA). The advantage of FEA is that complex configurations can be represented. The disadvantage of FEA is that mathematical singularities cause the calculated damage indicator to be a function of the element size chosen by the analyst.

The two most common solutions employed to deal with the issue of element size dependence in FEA are 1) use the same minimum element size across all simulations, or 2) volumetrically average the output results over a layer of elements of fixed thickness that is used across all simulations. These solutions work well for a given range of solder joint size, but they require modification for some flip chip applications where the entire joint might be smaller than the minimum element size used in a BGA assembly simulation.

In the present paper, three different data sets are analyzed to determine correlation between calculated damage indicators and measured thermal cycle fatigue life for underfilled flip chip solder joints. A wide range of design and material variables was considered. Two experiments used thermal cycling, and one used power cycling to fatigue the solder joints. The data and correlations from the current study were then compared to predictions based on previous work using non-underfilled WLCSPs [1].

Experimental Data Sets

Three experimental data sets were used in this study as summarized in Table 1. For data set #1, both lidded and bare die FCBGA packages were used as depicted in Figures 1 and 2. Two different underfill materials and two different core materials were evaluated. A non-underfilled assembly was also tested. For data set #2, an fcCSP package configuration was used as depicted in Figure 3. The primary variables were solder resist opening diameter, UBM diameter, joint height, and solder volume. For data set #3, an fcCSP package configuration was used, but it was mounted to a motherboard, which was subsequently attached to a water cooled cold plate. The primary variable in this experiment was power dissipation on the die, which in turn drove the temperature range per cycle. The range of bump dimensions used across all three data sets is shown schematically in Figure 4.

Table 1
Summary of experimental data sets

Data Set	#1	#2	#3
Package Configuration	Fig 1 and Fig 2	Fig 3	Fig 3 mounted
Applied Stress	Thermal Cycling	Thermal Cycling	Power Cycling
Primary Variable(s)	Package configuration and material set	Solder joint design	Power level

Figure 1. Lidded FCBGA.

Figure 2. Bare die FCBGA.

Figure 3. fcCSP.

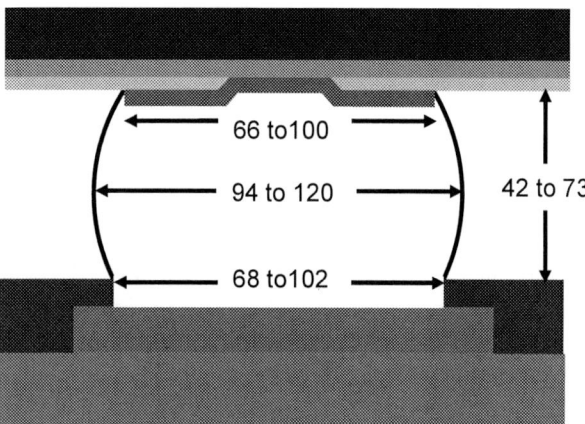

Figure 4. Range of bump dimensions (um).

Material Properties

Three different capillary underfill (CUF) materials were used in the experimental studies. Samples were extracted from actual packages and measured to determine the material properties [2]. CUF1 and CUF 2 were used in data set #1. CUF3 was used in data sets #2 and #3. Shown in Figure 5 is the expansion behavior of the underfill materials. A range of CTE values and glass transition temperatures is represented. Shown in Figure 6 is the temperature dependent elastic modulus for the underfill materials. A significant range in properties is represented across the test vehicles.

The measured temperature dependent elastic-plastic behavior of the underfill materials is shown in Figure 7. The data were fit to a sinh creep equation. The modeled response is shown in Figure 8, and it is seen that there is good agreement to the measured response in Figure 7.

The linear elastic material properties for all other materials are given in Table 2. The silicon, copper, solder mask, and lead free solder properties were taken from the literature or provided by suppliers. All other properties were measured from samples extracted from the test vehicle hardware. Creep properties for the lead free flip chip solder joints are given in Ref [3].

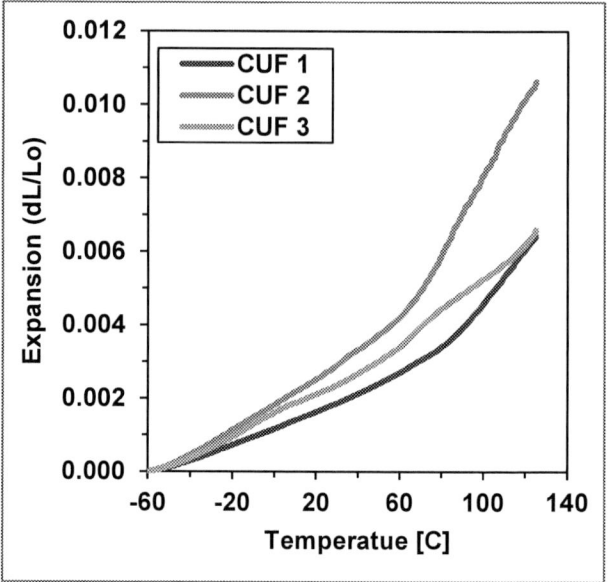

Figure 5. Expansion behavior of underfill materials.

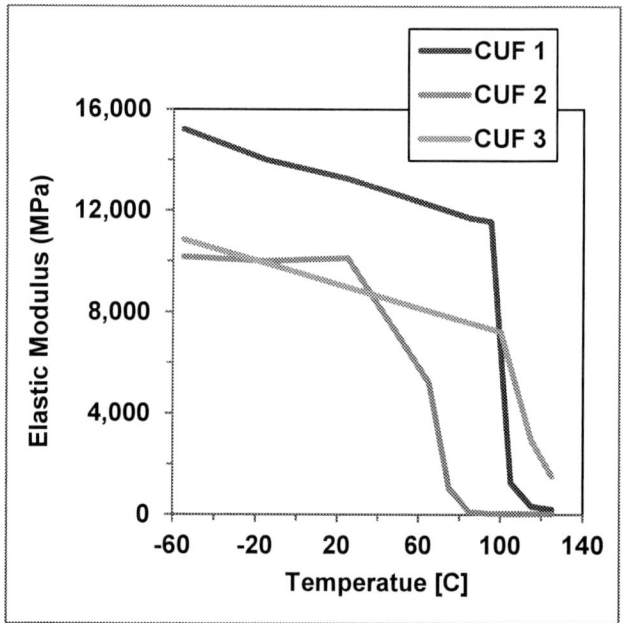

Figure 6. Elastic modulus of underfill materials.

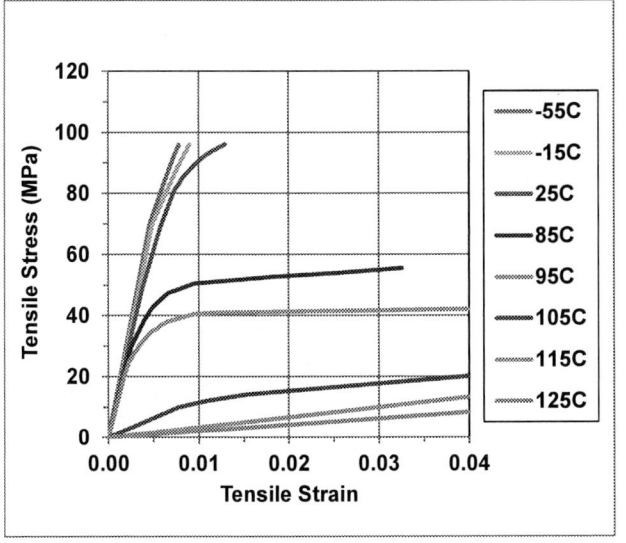

a) CUF 1

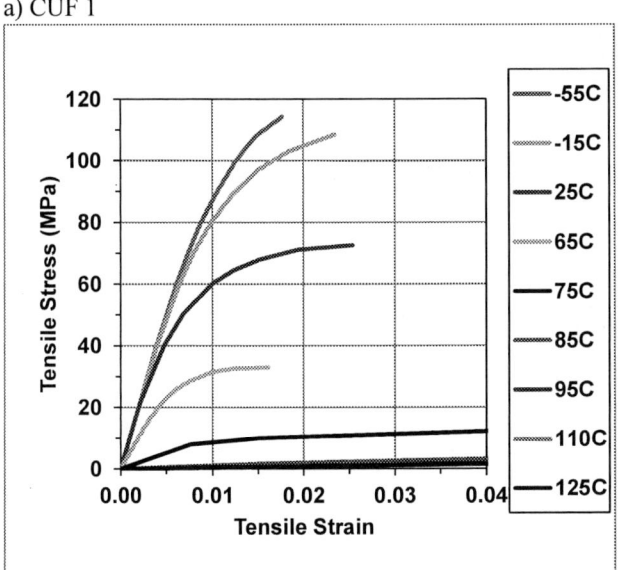

b) CUF 2

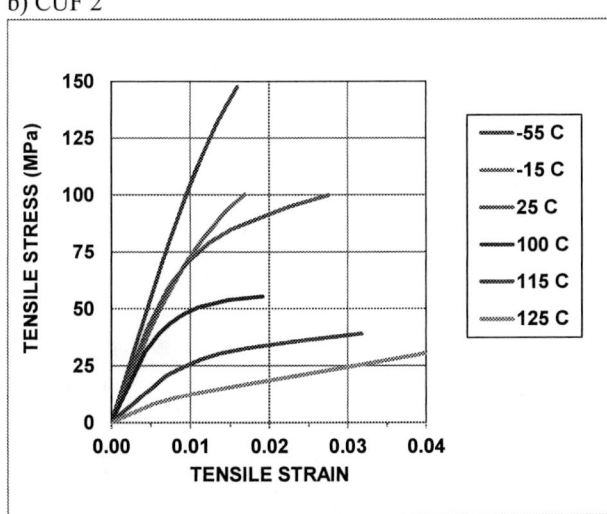

c) CUF 3

Figure 7. Measured elastic-plastic behavior of underfill materials.

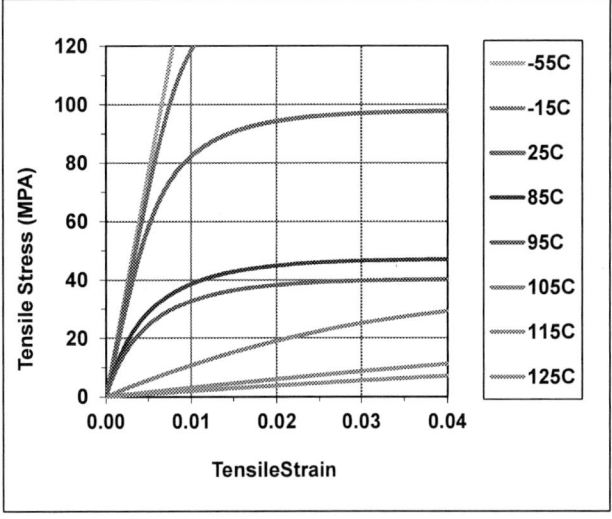

a) CUF 1

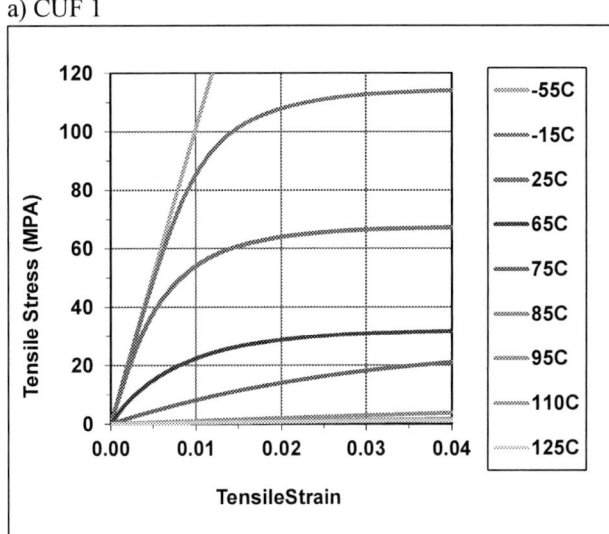

b) CUF 2

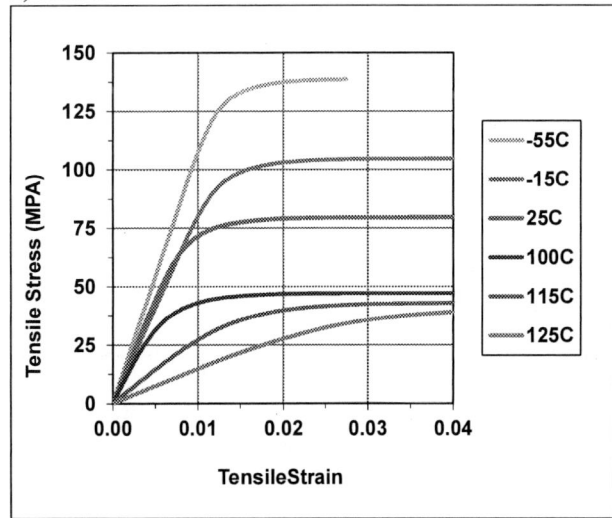

c) CUF 3

Figure 8. Plastic response of underfill materials modeled using sinh creep equation.

978-1-4799-2408-0/14 $31.00 © 2014 IEEE

Table 2
Linear elastic material properties.

	Elastic Modulus @-55C (GPa)	Elastic Modulus @125C (GPa)	Poisson's Ratio	CTE (ppm/C)
Silicon	130.3	128.9	0.28	2.7
Pb Free Solder	62.4	28.9	0.40	22.0
Copper	132.2	124.5	0.34	16.7
Core 1	24.6	21.0	0.30	14.4
Core 2	21.8	16.8	0.30	14.4
Core 3	19.9	15.3	0.30	17.1
Build Up Layers	22.7	17.3	0.30	24.5
Solder Mask 1	3.3	0.5	0.20	72.5
Solder Mask 2	3.2	3.2	0.32	40.0
Mold Compound 1	30.0	23.4	0.20	12.1
Mold Compound 2	24.7	18.0	0.20	10.4
Lid Attach Adhesive	0.003	0.003	0.40	143
TIM 1	0.005	0.005	0.40	280
TIM 2	0.013	0.013	0.40	280
Heat Sink	73.7	64.3	0.33	20.6
Motherboard x	21.0	17.8	0.11 / 0.39	12.8
Motherboard y	17.8	14.6	0.11 / 0.39	15.0
Motherboard z	8.4	7.0	0.39	64.1

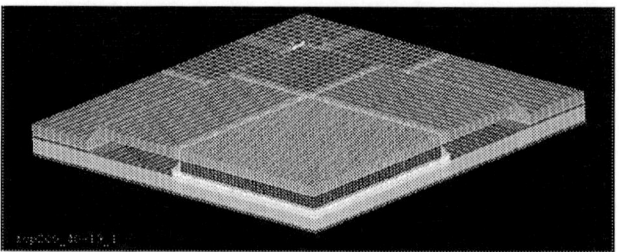

Figure 9. Lidded FCBGA model for data set #1. Quarter symmetry, 240,375 elements

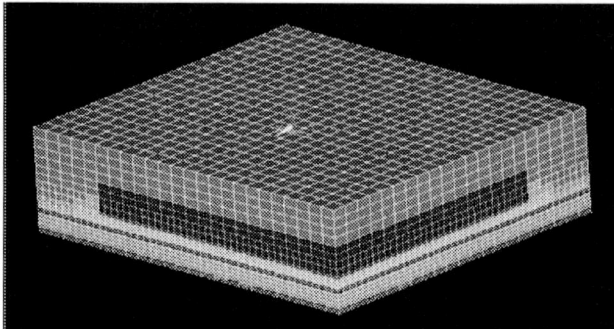

a) Quarter symmetry. 54,776 elements

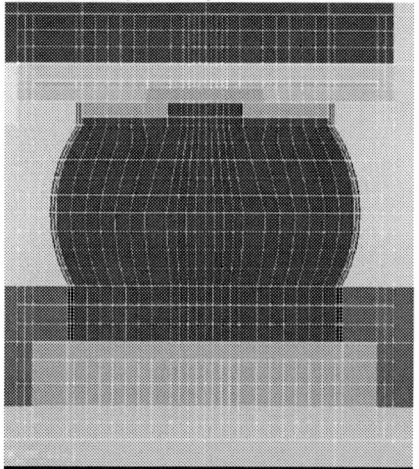

b) Fine mesh local model of bump.
Figure 10. fcCSP model for data set #2.

Finite Element Analysis

FEA was conducted to simulate the thermomechanical response of each assembly. The models used for data sets #1, #2, and #3 are shown in Figures 9, 10, and 11, respectively. Shown in Figure 10b is the fine mesh local model of an individual bump. All the models had two or three fine mesh models incorporated into the global model. Constraint equations were used at the interface between fine mesh and coarse mesh regions. These mesh interfaces were always positioned within a given material, not at the interface between two materials. The model in Figure 11 is more complex because a power cycling test was simulated with the package mounted to a motherboard that was subsequently attached to a water cooled heat sink.

For the thermal cycle evaluations in data sets #1 and #2, two complete thermal cycles were simulated. The viscoplastic strain energy density (SED) (a.k.a. plastic work per unit volume) increment in the solder joints was taken from the second cycle. For the power cycle evaluation in data set #3, the assembly was cooled down from motherboard attach, held at room temperature, then two power cycles were simulated. The temperature distribution was calculated as a function of time, and the nodal temperatures were input to a

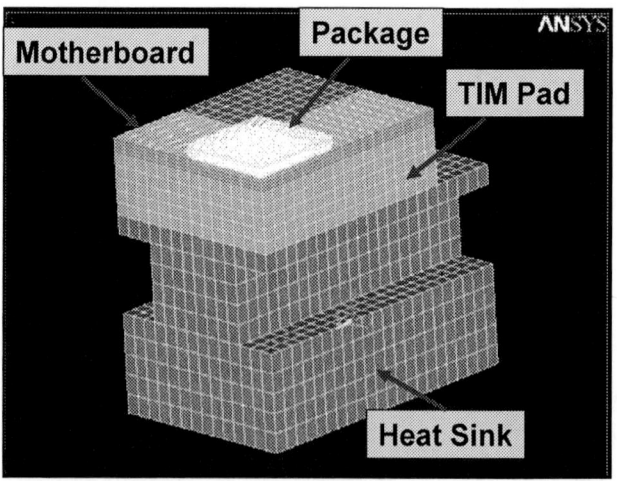

Figure 11. fcCSP model for data set #3. Quarter symmetry. Power cycling evaluation. 295,038 elements.

structural model to calculate the stress and strain distribution as a function of time. The SED increment per cycle in the solder joints was taken from the second power cycle.

In post processing, the elemental SED values were extracted for each solder element in the fine meshed joints. The worst case joint was determined, and volumetric averages were calculated for the entire joint, and for each 5um thick element layer in the joint. Three different damage indicators were ultimately extracted: 1) maximum elemental SED per cycle, 2) maximum SED^2 / average SED per cycle, and 3) layer averaged SED per cycle.

Finite Element Analysis Results

Shown in Figure 12 are calculated strain energy density distributions for the worst case flip chip solder joint in data set #2. This experiment had 8 legs, and the variables were solder resist opening (SRO) diameter, UBM diameter, joint height, and solder volume. The images represent a diagonal cross section through the 3-dimensional joint, viewed perpendicular to the diagonal. It is seen that the highest SED location can be either on the UBM side (top) or SRO side of the joint. If the UBM is larger than the SRO, the SED concentrates more on the SRO side. Conversely, if the SRO is larger than the UBM, the SED concentrates more on the UBM side. It should be recognized that there are both local and global expansion mismatches driving the SED distribution in the solder joints. There are global expansion mismatches between the die, substrate, and mold compound. There are local mismatches between the solder, underfill, solder mask, etc. The combination of these expansion mismatches and the shape of the flip chip solder joints drive the SED distribution, and ultimately the fatigue life.

The measured fatigue life performance ranking for data set #2 is also shown in Figure 12. The general observation is that more uniform SED distributions have better performance. Also, it is expected that the lowest maximum SED values should have better performance. Hence, a new damage indicator is proposed as maximum SED^2 / average SED. This damage indicator incorporates the effects of both the level of SED per cycle, and the SED distribution in the solder joint. It is easy to calculate, and does not require integrating around a crack tip in the model.

Shown in Table 3 is performance ranking for the legs of the experiment in data set #2. The measured ranking is given alongside the predicted ranking using three different damage indicators. All damage indicators correctly predicted leg 8 as the best and leg 7 as the worst performing joint designs. However, there was some significant variation in the measured vs. predicted ranking for the other legs. The most accurate prediction was made by maximum SED^2 / average SED, followed by maximum SED, and the worst was layer averaged SED.

Shown in Figure 13 is a comparison of results using viscoplastic versus elastic underfill properties. The maximum elemental SED per cycle is plotted for the outermost six joints along the die diagonal for an FCBGA package. Joint 1 represents the corner joint in Figure 13. It is seen that incorporating non-linear (viscoplastic) behavior in the underfill material dramatically increases the calculated damage to the solder joints. The SED values increased by

nearly 2X. Clearly, accurate properties for the underfill material are vitally important to flip chip solder joint fatigue life prediction.

Shown in Figure 14 is a comparison of two FCBGA models with the same bump pattern, but slightly different die sizes. Hence, the distance from outmost bump center to the die edge was varied from 178um to 278um. It is seen that even this seemingly small change impacted the calculated maximum SED by 15%. This highlights the importance of measuring actual dimensions on test vehicles in order to get the best possible simulation accuracy.

The results shown in Figures 13 and 14 underscore the difficulty in making *absolute* fatigue life predictions. Here, absolute is defined as a blind prediction of fatigue life with no measured life data as a baseline on a particular assembly. There are a massive number of dimensional and material property variables that are inputs to any given finite element model. Even when the engineer is very diligent, it is difficult to accurately measure all the required input values. This fact, coupled with highly non-linear properties and mathematical singularities at material interfaces, make absolute life prediction very hard over any significant range of configurations.

Table 3

Predicted vs. measured performance ranking for data set #2
Listed from best joint design to worst joint design

Measured	Predicted Layer Avg SED	Predicted Max SED	Predicted Max SED^2 / Avg SED
8	8	8	8
4	3	1	1
1	2	2	2
2	1	3	4
6	6	4	3
5	5	5	5
3	4	6	6
7	7	7	7

Fatigue Life Correlations

One of the primary goals of the present paper is to establish correlation between the *measured* flip chip solder joint fatigue life and the *calculated* damage indicator. The equation describing this correlation can then be used to predict the fatigue life as a function of changes in material set, design, stress conditions, etc. For measured fatigue life, the N05, or cycles to 5% cumulative failures was chosen for correlation. N50, or mean life is also commonly used for such correlations because there is typically less scatter in the data, and the correlation looks better. However, there were relatively few legs of the data sets that were taken to N50, so N05 was used here instead. Also, N05 is more important from a practical point of view since most engineers need to make decisions based on the lower end of the fatigue life distribution, not the middle of it.

978-1-4799-2408-0/14 $31.00 © 2014 IEEE

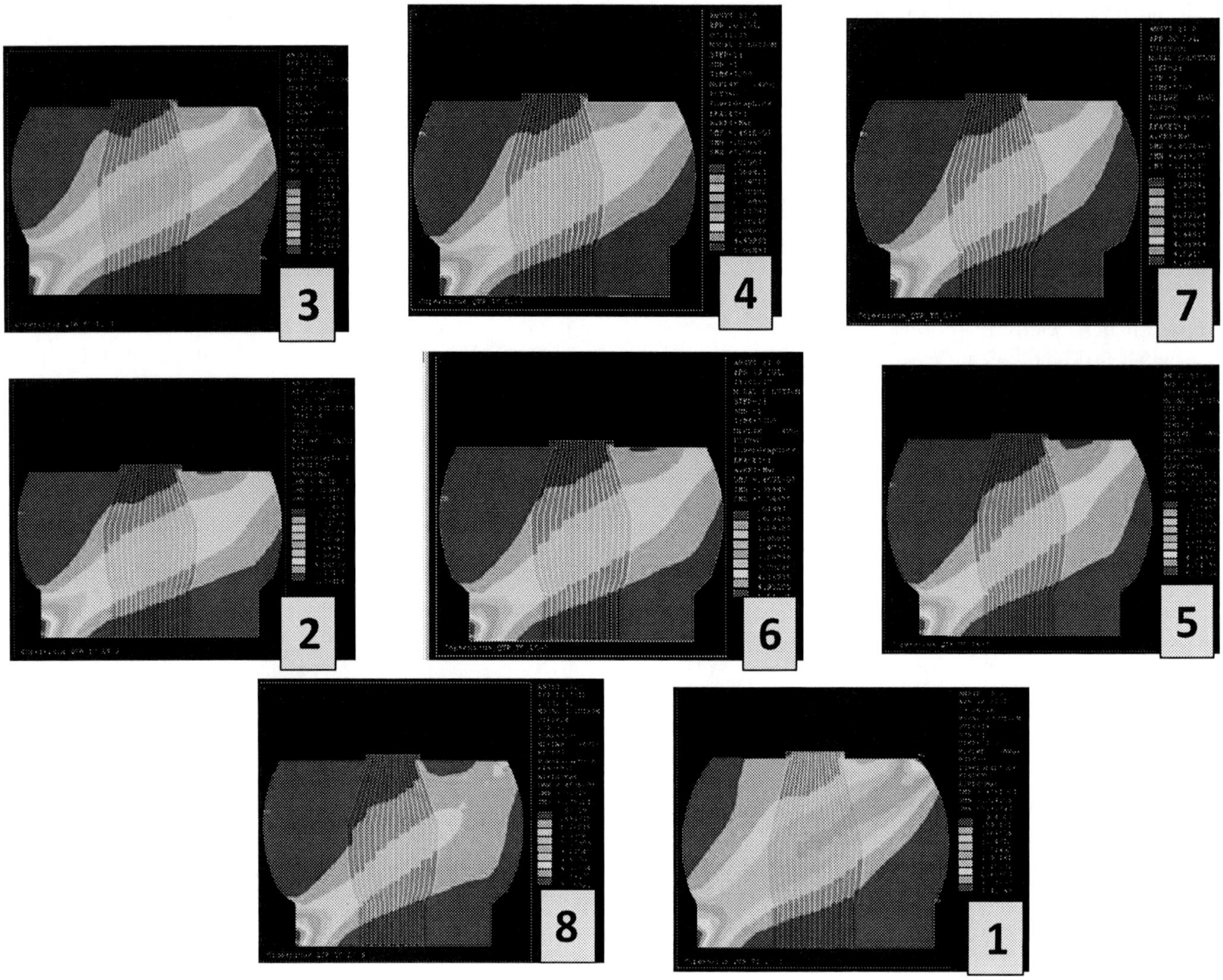

Figure 12. Strain energy density distribution after 2 thermal cycles compared with reliability performance ranking for fcCSP flip chip joints in data set #2.

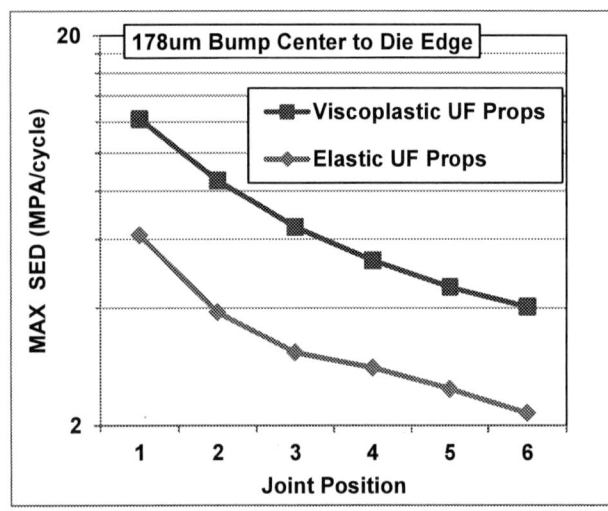

Figure 13. Effect of underfill material property model on maximum elemental strain energy density per cycle for outermost solder joints along diagonal for FCBGA.

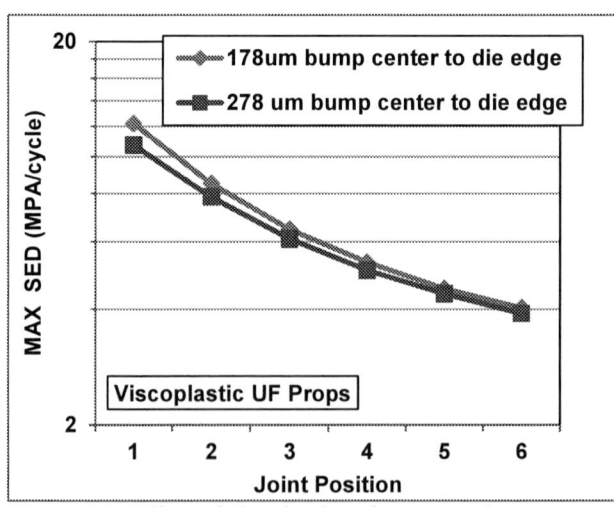

Figure 14. Effect of die edge location on maximum elemental strain energy density per cycle for outermost solder joints along diagonal for FCBGA.

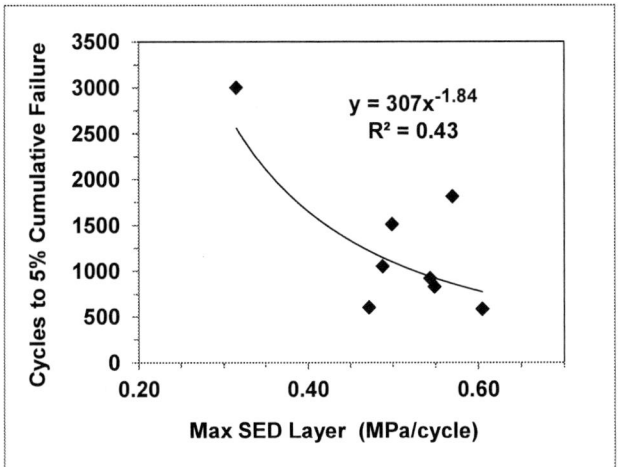

Figure 15. Fatigue life correlation for fcCSP under thermal cycling conditions (data set #2) using layer averaged strain energy density per cycle.

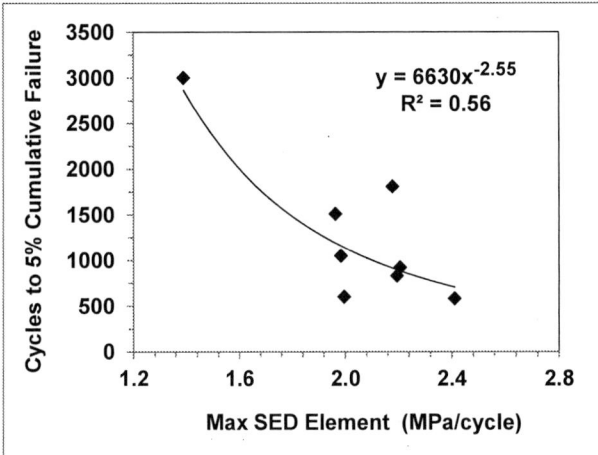

Figure 16. Fatigue life correlation for fcCSP under thermal cycling conditions (data set #2) using maximum elemental strain energy density per cycle.

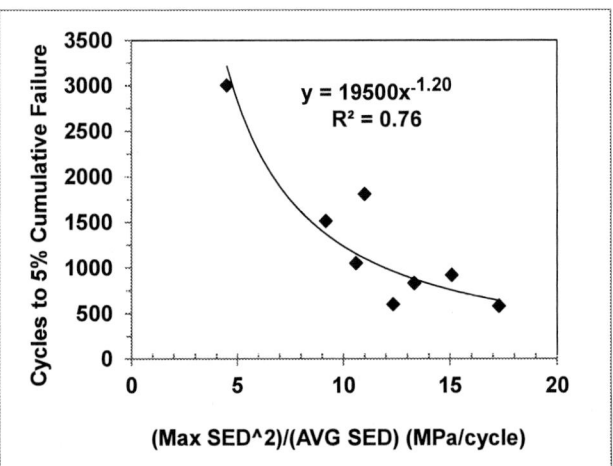

Figure 17. Fatigue life correlation for fcCSP under thermal cycling conditions (data set #2) using maximum SED^2 / average SED per cycle.

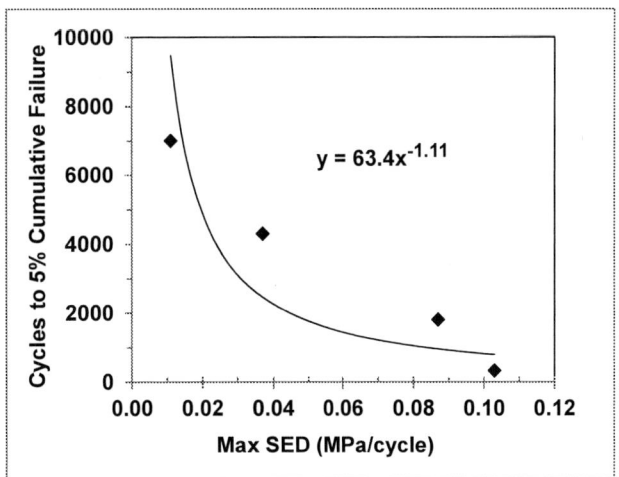

Figure 18. Fatigue life correlation for fcCSP under power cycling conditions (data set #3) using maximum elemental strain energy density per cycle.

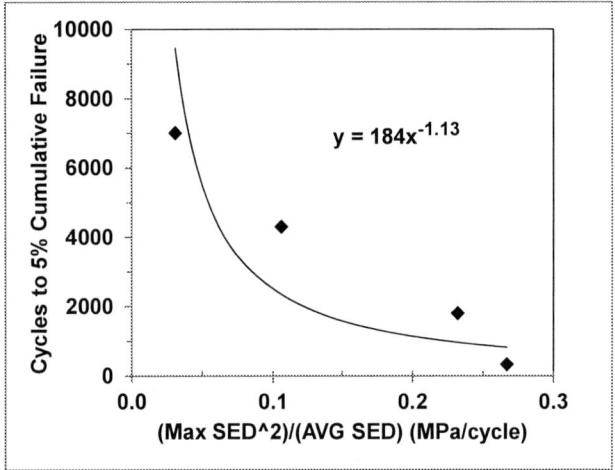

Figure 19. Fatigue life correlation for fcCSP under power cycling conditions (data set #3) using maximum SED^2 / average SED per cycle.

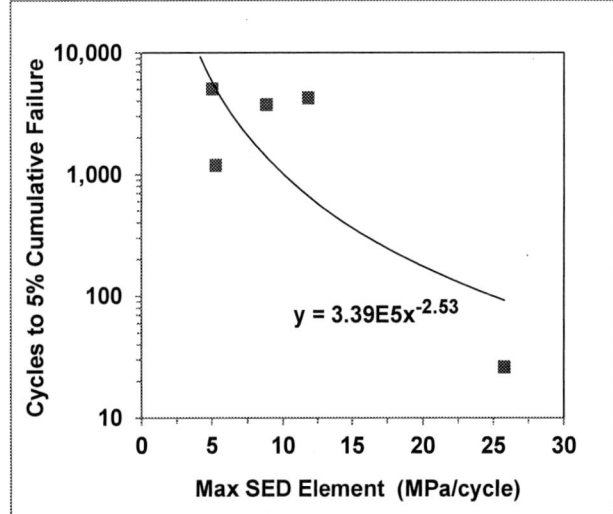

Figure 20. Fatigue life correlation for FCBGA under thermal cycling conditions (data set #1) using maximum elemental strain energy density per cycle.

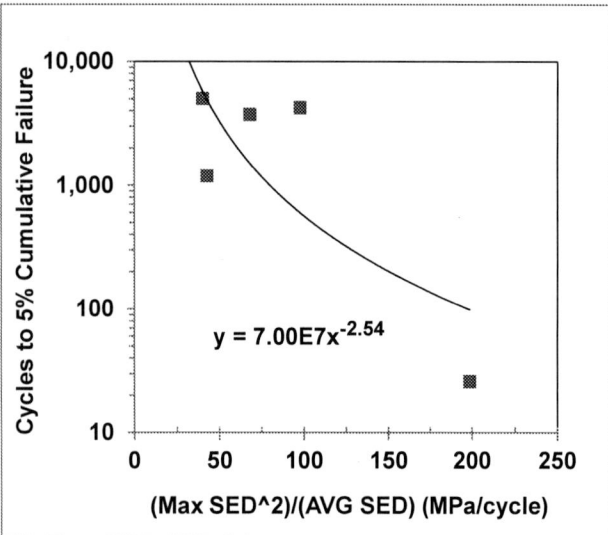

Figure 21. Fatigue life correlation for FCBGA under thermal cycling conditions (data set #1) using maximum SED^2 / average SED per cycle.

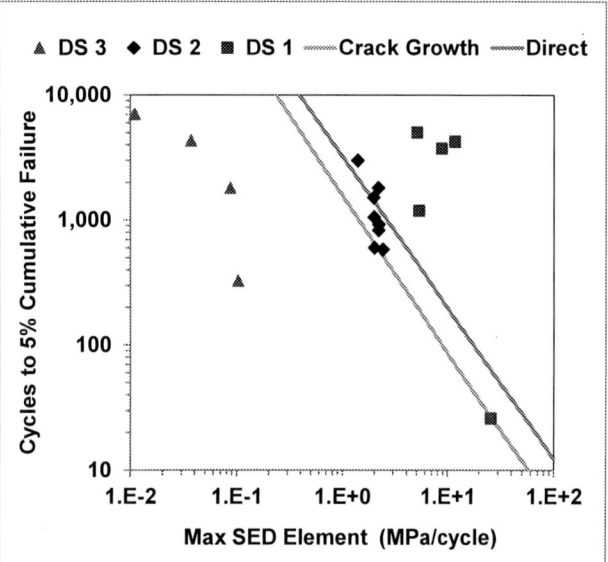

Figure 22. Comparison of fatigue life correlations using maximum elemental strain energy density per cycle as the damage indicator. "Direct" and "Crack Growth" use WLCSP data from Ref [1].

Shown in Figures 15 to 17 are fatigue life correlations for data set #2. The damage indicator used in Figs 15, 16, and 17 were layer averaged SED, maximum SED, and max SED^2 / avg SED, respectively. Based on the correlation coefficients (R^2 values), it is seen that layer averaged SED gave the weakest correlation, and max SED^2 / avg SED gave the strongest correlation. The primary variable in this data set was the flip chip solder joint design (solder resist opening diameter, UBM diameter, joint height, and solder volume).

Shown in Figures 18 and 19 are fatigue life correlations for data set #3. Both max SED and max SED^2 / avg SED gave similar results for the limited data set of four points. The variable in this evaluation was temperature range (as driven by the power dissipation on the test die).

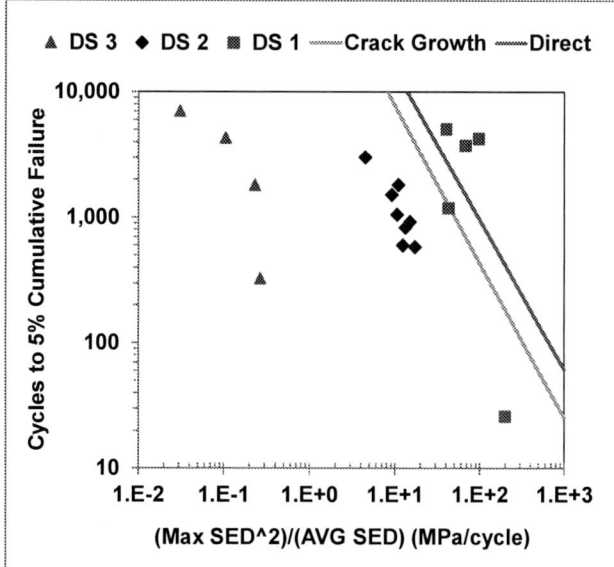

Figure 23. Comparison of fatigue life correlations using maximum SED^2 / average SED per cycle as the damage indicator. "Direct" and "Crack Growth" use WLCSP data from Ref [1].

Shown in Figures 20 and 21 are fatigue life correlations for data set #1. The variables in this data set were package structure, underfill material, and substrate core material. The lowest fatigue life data point was for non-underfilled packages (only 26 cycles to N05). Both max SED and max SED^2 / avg SED gave similar results for the limited data set of five points.

Discussion

Figures 15 to 21 demonstrate that it is possible to correlate measured solder joint fatigue life with calculated damage indicators for the three data sets, individually. In order to have a universal model that makes *absolute* fatigue life predictions, one needs to compare the three data sets on a single plot. Such comparisons are shown in Figures 22 and 23. Maximum SED was used as the damage indicator in Figure 22, and max SED^2 / avg SED was used in Figure 23. It is seen that no universal correlation fits all three data sets. The general slope of the correlation is similar between data sets, but each group of data points is offset from the others.

Also shown in Figures 22 and 23 are predicted fatigue life versus damage indicator plots based on correlations from non-underfilled WLCSP solder joints in Ref [1]. The alloy from Ref [1] selected for this comparison was SAC125Ni, since it has the closest measured creep behavior to the SnAg flip chip solder joints. The slopes of the WLCSP correlations are similar to that found in the current study. There is a slightly better match in slope for the five correlations when using the max SED^2 / avg SED as the damage indicator. The slope of the correlation is the most important feature, because this allows for *relative* fatigue life predictions (i.e. making an estimate of fatigue life for new conditions when there is at least one measured data point available).

In comparing the positions of the three data sets in Figures 22 and 23, it is seen that data set #3 is the furthest left on the x-axis. In other words, the simulation model produced the lowest calculated damage for a given measured fatigue life performance. There are at least two theories to explain this: 1) the higher complexity of this particular simulation resulted in more error in the calculated damage indicator, or 2) thermomigration and/or electromigration effects reduced the measured solder joint fatigue life.

Data set #3 was a power cycling experiment which required the most complex simulation model. Both thermal and structural simulations were needed to estimate the damage indicator. It had the largest number of materials, and some interfaces that were not actually bonded together (only clamped). Hence, there were more material and dimensional variables, and more uncertainty in some of the boundary conditions. It is likely that this results in more error in estimating the effective stiffness and exact expansion mismatches in the assembly.

Since data set #3 was a power cycling experiment, it is possible that more than just thermal cycle fatigue was damaging the solder joints. Thermal gradients across the flip chip solder joints can cause thermomigration of the solder atoms, which could accelerate the growth of voids and cracks. Likewise, electromigration can occur in the solder joints carrying high current used to heat the daisy chain test vehicle. This effect would also accelerate the growth of voids and cracks in the solder.

As a general question, why does the measured life not match up perfectly with the predicted life? There are several possible reasons.

On the experimental side, there are lot-to-lot variations in the material properties and solder joint quality. Simulation models typically predict the corner joint in an array will fail first. This is not always observed experimentally, so there must be some local variations between solder joints. Such effects are especially likely in flip chip joints that might only have a few grains per joint. Excessive warpage in packages can cause joints to be already cracked after assembly. Sometimes interface failures are observed. Measuring the material properties of packaging materials is not an easy task. There is often quite a large scatter in both elastic modulus and thermal expansion coefficient data. Test chambers are often stopped during a test that last several weeks or months. An excessive number of stops in the test could affect the results. Control of the chamber to the specified thermal cycle test profile can be a problem. The event detector that monitors joint failure can malfunction, or the actual definition of electrical failure is not always consistent.

On the simulation side, there are several simplifying assumptions made. Individual material layers are often omitted, and detailed features of the metal traces are rarely incorporated. There are singularities at the edges of the joint, so element size selection and damage indicator definition become important. A crack free joint is typically modeled, even though it is known that most of the life of a joint is spent in propagating a crack. Hence, the relative stiffness of the assembly increases as the joint cracks. The effect of this change in stiffness would also depend on the initial assembly stiffness and the degree of cracking in neighboring joints. One would expect this effect to be more pronounced for non-underfilled assemblies compared to the underfilled flip chip packages studied here.

All of the above factors can impact the accuracy of *absolute* fatigue life predictions. However, these factors have less impact when just making *relative* predictions. Solder joint fatigue life prediction models are useful for: 1) rapid design optimization during the development phase of a product, 2) predicting field use limits, and 3) failure analysis of product returned from the field or failed in a qualification test. In all of these scenarios, the analyst is typically making relative predictions of an unknown case from measured data on a known case. The important guideline in making relative fatigue life predictions is to be consistent in the methodology.

Conclusions

1) Finite element analysis was used to predict the thermal cycle fatigue life of flip chip solder joints. Three different data sets of measured thermal cycle fatigue life were simulated to assess model correlation. Variables in the measured data included solder resist opening diameter, UBM diameter, joint height, solder volume, underfill material, package structure, and temperature range.

2) Viscoplastic strain energy density (SED) per cycle was used as the damage indicator for the solder joints. Using a non-linear material property model for the underfill increased SED in the worst case solder joint by 2X compared to an elastic underfill model. Increasing the distance from outmost bump center to the die edge from 178um to 278um reduced the calculated maximum SED by 15% for the worst case solder joint.

3) Three different post processing schemes were compared for extracting the damage indicator: 1) maximum elemental SED value, 2) maximum SED^2 / average SED, and 3) layer averaged SED. It was found that all three methods could achieve some degree of correlation between measured and predicted life for a given data set. The best and worst legs of an experiment could be predicted well, but the ranking of all legs in a given experiment could not be perfectly predicted. For the experiment on joint design, max SED^2 / avg SED gave the best prediction, followed by max SED, and worst was layer averaged SED. For the experiments on power cycle conditions and package structure, all methods gave similar results.

4) The three data sets were plotted together with additional predictions from previous work on WLCSPs. There was a large discrepancy in the *absolute* prediction from one data set to the next. However, the ability to predict the *relative* fatigue life within a data set was consistent, and the results did match the previous work on WLCSP solder joints.

References

1] R. Darveaux, "Thermal Cycle Fatigue Life Models for WLCSP Solder Joints," Proc. SMTAI, 2013.

2] A, Syed, G. Sharon, and R. Darveaux, "Factors Affecting Pb-free Flip Chip Bump Reliability Modeling for Life Prediction," Proc. ECTC 2012.

3] R. Darveaux and C. Reichman, "Solder Alloy Creep Constants for Use in Thermal Stress Analysis," Proc. SMTAI, 2012.

High Thermo-Mechanical Fatigue and Drop Impact Resistant Ni-Bi Doped Lead Free Solder

Jae Hong Lee, Santosh Kumar, Hui Joong Kim, Young Woo Lee, Jeong Tak Moon
MK electron., LTD
405, Geumeo-ro, Pogok-eup, Cheoin-gu Youngin, Gyeonggi, Korea
leejh@mke.co.kr

Abstract

Presently, Sn-Ag-Cu (SAC) solders are most commonly used as the interconnect materials in the semiconductor package. However, their thermal fatigue and drop impact resistant properties depends on the Ag content and therefore, most semiconductor package assemblers are forced to implement multiple SAC alloys depending on intended performance. Sn–xAg–Cu solders with high Ag content (x>3 mass%) give good temperature cycling (TC) reliability but poor drop impact reliability whereas Sn–xAg–Cu solders with low Ag content (x<2 mass%) show poor temperature cycling reliability but good drop impact reliability. So, there is need to develop solders having both good TC and drop impact reliability. For the present study, we developed a new SAC solder by micro-alloying it with Ni and Bi. It was found that the thermal fatigue and drop impact resistant is improved dramatically simultaneously. The improvement in the drop impact resistance is attributed to the decrease in the bulk and joint IMC thickness and grain refining. The high TC reliability is due to the unique network like structure of Ag3Sn in the bulk solder microstructure. The newly developed solder show high and stable shear and pull strength as compared to SAC solders and the dominant fracture mode in the high speed shear test is ductile. This new solder has potential to become interconnect material for all types of semiconductor packages.

Introduction

Sn-Ag-Cu solder is most common interconnect material used in electronic devices. However, their thermal fatigue and drop shock reliability performance depends on the Ag content in the solder. High Ag content ($\geq$ 3wt%) solder shows better thermal cycle reliability performance and poor drop performance whereas low Ag content solder ($\leq$ 2wt%) shows just opposite performance [1, 2].

For devices such as computer, fax, printers etc., which requires high thermal cycling performance, high Ag content solders Sn-3.0Ag-0.5/0.7Cu are used. For handheld devices (mobile, digital camera etc.), low Ag content solder Sn-1.0Ag-0.5Cu is used.

The present and future electronic devices are getting miniaturized while their functionality is continuously increasing. These devices require solders having both good TC and drop impact reliability. So, there is clear need to develop solders which satisfy reliability requirements in all conditions. It is well known that adding minor dopants in the normal SAC solders can enhance their mechanical properties and reliability. Following extensive research and experimentation, we developed the Ni and Bi doped novel SAC solder. The solder composition chosen for this study was Sn-2.5Ag-0.8Cu-Ni-Bi. In the present study, we evaluated the wetting, mechanical and bonding properties of our novel solder. Thermal cycling and drop test was done to evaluate the reliability of the solder joints. Similar experiments are also conducted for the Sn-3.0Ag-0.5Cu solder and their results are compared and discussed.

Experiment method

2.1 Sample preparation

All the elements used to make solder alloy is of 4N purity. Appropriate amounts of elements are taken and melted in the vacuum induction furnace under the protection of high purity inert gas atmosphere to prepare the alloy of desired composition. The composition of the alloy was continuously monitored by the spark-emission equipment till the desired composition was obtained. Finally the alloy was atomized to obtain the solder balls of 300um, having composition Sn-3.0Ag-0.5Cu alloy and Sn-2.5Ag-0.8Cu-Ni-Bi.

The solder ball were attached to the FR-4 substrate having Cu OSP pad finish (pad opening 260um) and reflowed in the multi-zone oven (HELLER 1706-EXL) to form the solder joint. The peak reflow temperature was 518K and the dwell time was 45s. The reflow was carried according to JEDEC_J_STD-020C. The test coupons (dimension10mm x 10mm x 1mm) have daisy chains for the reliability test. The board were fabricated by the NSMD (non-solder mask defined) type, the components were fabricated by the SMD (solder mask defined) type.

2.2 Property Evaluation

The melting point of solders was measured using the DSC (Differential Scanning Calorimeter) equipment. The maximum temperature was 523K and the raising temperature was 278 K/min. The DSC curves for the solder alloys are shown in Fig. 1. The wettability of the solder alloys were evaluated using the solder checker (RHESCA STA-5100). The hardness of the as manufactured solder ball and reflowed bump was measured using the Fischer Scope equipment, in accordance with JIS-Z-3198 standards.

2.3 Bonding evaluation & reliability

The solder joint integrity was evaluated using normal and high speed shear test. Normal shear test was carried out in accordance with JESD22-B117 using the DAGE-4000 equipment (Test condition: 300um/sec shear speed, 20um shear height). High speed shear test was carried out in accordance with JESD2-B117A using the DAGE-4000HS equipment (Test condition: 500mm/sec shear speed, 20um shear height).

2.4 Thermal cycle test

The dimension of the test vehicles used for thermal cycling test was 132 x 72 x 1.0 mm. The test boards were subjected to a cycling loading between -40 to 125°C with a dwell time of 15min. The solder joint resistance during test was continuously monitored using the daisy chain.

2.5 Drop test

The dimensions of test vehicles were same as used in the TC test. There are total 15 packages mounted in board. The drop shock pulse had the peak acceleration of 1500G with pulse duration of 0.5milliseconds. Event detector, with the daisy-chain links, was used to measure the resistance after each drop cycle. Resistance with more than 1000ohm was considered as a failure.

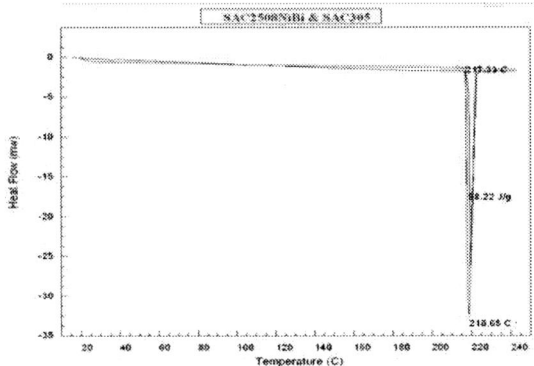

Figure 1. DSC curve of the solder alloys and hardness

Result and discussion

1. Mechanical properties

   1-1. Melting point

The melting points of Sn-3.0Ag-0.5Cu and Sn-2.5Ag-0.8Cu-Ni-Bi solders are shown in Fig 3. The melting point of Sn-2.5Ag-0.8Cu-Ni-Bi solder was in the range 217~219, which is same as that of Sn-3.0Ag-0.5Cu. Owing to the low Ag content, it was expected that Sn-2.5Ag-0.8Cu-Ni-Bi should have higher melting point as compared to SAC305, but the melting point of both alloys were almost similar. This is probably attributed to the Bi, which has the effect of lowering the melting point of SAC solder.

   1-2. Wettability

The wettability curves of Sn-3.0Ag-0.5Cu and Sn-2.5Ag-0.8Cu-Ni-Bi solders are shown in Fig.2. The zero-cross time and wetting force is the measure of the wettability of solder. Lower zero-cross time and higher wetting force means higher wettability. Table 1 shows the wettability of solders decreases in the order: SAC2508NiBi> SAC2508Bi> SAC305> SAC2508Ni> SAC2508.

Fig.3 shows the hardness value of Sn-3.0Ag-0.5Cu and Sn-2.5Ag-0.8Cu-Ni-Bi solder alloys in as prepared (raw) condition and after reflow state. The hardness values of Sn-3.0Ag-0.5Cu and Sn-2.5Ag-0.8Cu-Ni-Bi in as prepared state is 18.3HV and 16.6HV while in the reflow condition, it is 16.1HV and 16.8HV respectively. The hardness property is related to the microstructure of solder. After the reflow and high temperature condition the hardness value, in general,

decreases due to change in solder microstructure and it will affect the bonding strength and reliability of the solder joints.

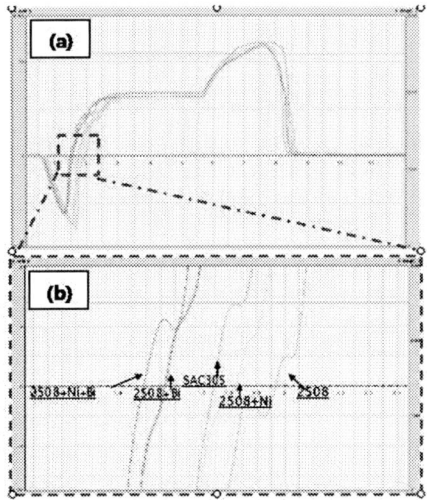

Figure 2. Result of wetting test
(a) Whole test graph
(b) Magnified image of blue square part

Table 1. Wetting force and zero-cross time from graph of Fig. 2

Wetting time (sec)	Zero cross time (sec)	Maximum force (uN)
2508(ref)	1.30	6.41
2508-Ni	1.19	6.16
2508-Bi	0.96	6.36
SAC305	1.12	6.16
2508-Ni-Bi	0.89	6.77

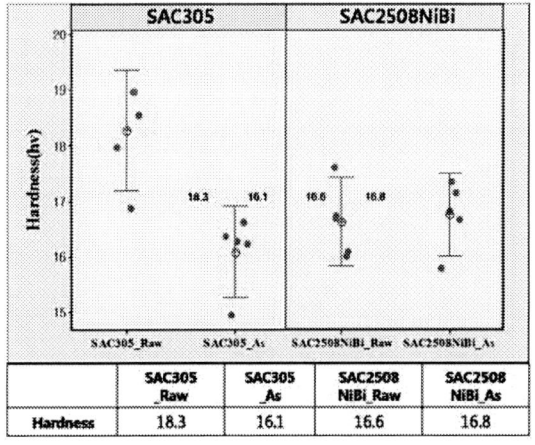

	SAC305 _Raw	SAC305 _As	SAC2508 NiBi_Raw	SAC2508 NiBi_As
Hardness	18.3	16.1	16.6	16.8

Figure 3. Hardness of Sn-3.0Ag-0.5Cu and Sn-2.5Ag-0.8Cu-Ni-Bi solder in raw condition and after reflow

   1-3. Hardness

The low hardness of Sn-2.5Ag-0.8Cu-Ni-Bi as compared to Sn-3.0Ag-0.5Cu is attributed to the low Ag content and presence of Bi in the former. It should be noted that the hardness value of Sn-2.5Ag-0.8Cu-Ni-Bi is almost same even after the reflow. Addition of Bi results in the significant grain

978-1-4799-2408-0/14 $31.00 © 2014 IEEE   713

refining in the SAC solder and there is less degradation and coarsening of the phases occurs in Bi doped solder materials relative to non-doped solders after reflow and severe aging [3].

2. Bonding properties

2-1. Shear test, high speed shear test

Fig.4 shows the shear strength of Sn-3.0Ag-0.5Cu and Sn-2.5Ag-0.8Cu-Ni-Bi solders according to the number of reflows. It can be seen that the shear strength of SAC305 solders decreases gradually with the number of reflows whereas, in the case of Sn-2.5Ag-0.8Cu-Ni-Bi solder, the shear strength remains constant even after the multiple reflows.

Even though the Sn-2.5Ag-0.8Cu-Ni-Bi solder has lower Ag content as compared to Sn-3.0Ag-0.5Cu, the shear strength of both solder alloys is almost similar. This is probably because of the higher Cu content and presence of Ni and Bi in the former. The addition of Bi leads to significant grain refining and Ni suppresses the growth of intermetallic compound formed between the solder and Cu-OSP pad at the solder joint. This explains the almost constant shear strength in case of Sn-2.5Ag-0.8Cu-Ni-Bi even after multiple reflows.

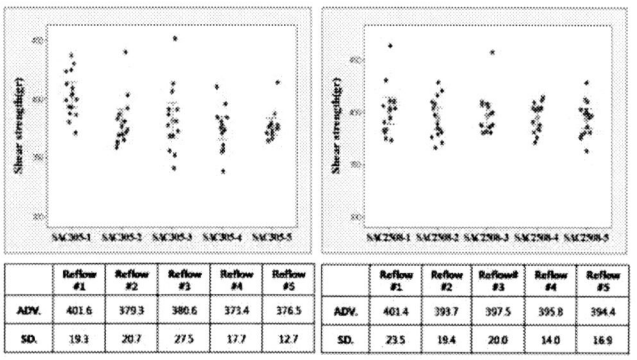

	Reflow #1	Reflow #2	Reflow #3	Reflow #4	Reflow #5			Reflow #1	Reflow #2	Reflow #3	Reflow #4	Reflow #5
ADV.	401.6	379.3	380.6	373.4	376.5		ADV.	401.4	393.7	397.5	395.8	394.4
S.D.	19.3	20.7	27.5	17.7	12.7		S.D.	23.5	19.4	20.0	14.0	16.9

Figure 4. Shear strength of solder alloys after different reflow times (Normal speed shear test).

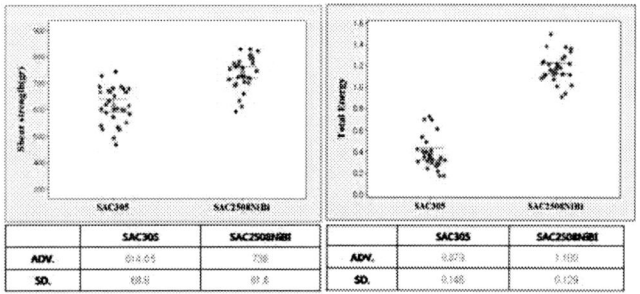

	SAC305	SAC250ENiBi			SAC305	SAC250ENiBi
ADV.	614.05	739		ADV.	0.379	1.180
S.D.	68.8	61.8		S.D.	0.148	0.129

Figure 5. Shear strength and total energy of Sn-3.0Ag-0.5Cu and Sn-2.5Ag-0.8Cu-Ni-Bi solders in high speed shear test.

Fig. 5 shows the results of high speed shear test in Sn-3.0Ag-0.5Cu and Sn-2.5Ag-0.8Cu-Ni-Bi in as-reflowed state. High speed shear test stimulates the condition of high strain rate (and also impact) and in useful test to measure the reliability of solder joints, particularly in mobile devices. It is very interesting observation that the average shear strength of

Sn-2.5Ag-0.8Cu-Ni-Bi solder (739gf) is 125gf higher than Sn-3.0Ag-0.5Cu solder (614gf) in the high speed shear test, even though they have almost similar average shear strength value in the normal speed shear test. In the high strain rate condition, Sn-2.5Ag-0.8Cu-Ni-Bi solder joints have clearly higher bonding strength as compared to Sn-3.0Ag-0.5Cu solder and the former is ideal replacement of later in the extreme high strain rate and impact environment condition. Also, we see that the average energy value of Sn-2.5Ag-0.8Cu-Ni-Bi solder (1.180mJ) in high speed shear test is 0.801mJ higher than in case of Sn-3.0Ag-0.5Cu solder (0.379mJ). The area under the shear strength-displacement curve is gives the energy value and is measure of fracture toughness of the solder joint. The measure of energy value is more meaningful criteria to evaluate the solder joint integrity in high speed shear test than the maximum shear force [4, 5]. It can be clearly seen from Fig.6 that the area under the shear strength-displacement curve (hence, energy) for Sn-2.5Ag-0.8Cu-Ni-Bi is significantly higher than Sn-3.0Ag-0.5Cu and thus the fracture toughness of solder joints of former is higher than later.

The difference in the fracture toughness (energy) of both solders can be explained by observing the fracture mode in high speed shear test. The fracture mode can be divided in four types: ductile, quasi-ductile, quasi-brittle, brittle. Ductile fracture mode is fracture in the bulk solder; Brittle mode is fracture in the IMC or between IMC and solder. Quasi fracture depends on the degree of ductility or brittleness in the fracture surface. The fracture mode having > 50% ductile fracture is quasi ductile and vice-versa. Fig.7 shows the fracture mode and fracture ratio in both solders. It can be clearly see that the Sn-2.5Ag-0.8Cu-Ni-Bi has 97% ductile and 3% quasi-ductile fracture and has no brittle or quasi-brittle fracture whereas, in case of Sn-3.0Ag-0.5Cu, almost all fracture is brittle or quasi-brittle. In the case of ductile fracture the area under the shear strength-displacement curve is higher as compared to brittle fracture. This explains why the fracture toughness of Sn-2.5Ag-0.8Cu-Ni-Bi solder is higher as compared to Sn-3.0Ag-0.5Cu solder.

Intermetallic compounds (IMCs) are formed between the solder and the pad metal as a result of reaction between them. Formation of IMCs is necessary for the sound joint strength but if the IMCs are too thicker, it will reduce the joint strength and reliability. IMCs are the main source of void and crack generation and being brittle it will lead to easy failure of solder joints during reliability test. In the case of Cu-OSP pad finish, the main IMC formed with solder at the joint interface is Cu6Sn5.

Fig.8 shows the SEM image of the bulk microstructure and the IMCs in the Sn-3.0Ag-0.5Cu and Sn-2.5Ag-0.8Cu-Ni-Bi solders for different reflow times. It can be observed that in the Sn-3.0Ag-0.5Cu solder, the IMC thickness is increased with the number of reflow and Ag3Sn plates are also elongated and thickened and even encompass the whole solder ball in some case. Also the spalling of the Ag3Sn occurs.

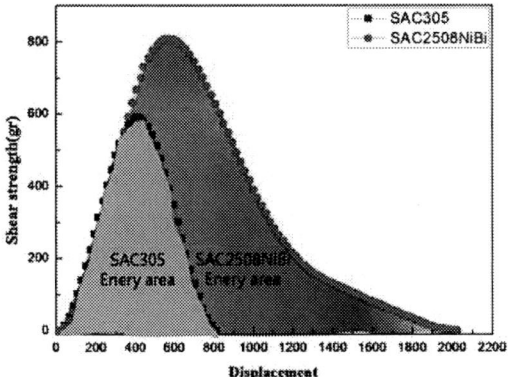

Figure 6. Shear strength-displacement curve of Sn-3.0Ag-0.5Cu and Sn-2.5Ag-0.8Cu-Ni-Bi solders (High speed shear test)

However, in the Sn-2.5Ag-0.8Cu-Ni-Bi solder, the interfacial and bulk IMCs growth is very slow as compared to Sn-3.0Ag-0.5Cu solder even after multiple reflows. The bulk solder microstructure is also not changed after the multiple reflows. The growth of the interfacial IMC is retarded due to Ni effect. Ni atom incorporates in the Cu6Sn5 crystal structure by substituting Cu atoms, resulting in the formation of (Cu, Ni)6Sn5. Cu6Sn5 has monoclinic crystal structure. Addition of Ni atoms in Cu6Sn5 results in the formation of (Cu, Ni)6Sn5 having more stabilized η-hexagonal crystal [6]. The addition of Ni stabilizes the hexagonal allotrope of η-(Cu, Ni)6Sn5 at room temperature minimizing stresses associated with phase transformation. In this way, Ni additions reduce the cracking of Cu6Sn5 layers and improve reliability [7, 8]. Also the Ni atom has smaller radius than Cu and thus (Cu, Ni)6Sn5 has smaller lattice constant as compared to (Cu)6Sn5. Thus the residual stress generated during formation of (Cu, Ni)6Sn5 is smaller as compared to (Cu)6Sn5. All these factors helps in improving the bonding strength.

The solder bulk grain structure refining and forming of dense Ag3Sn network is attributed to the presence of Bi. It has been known that the addition of Bi helps in the grain refining and change in microstructure that improves the thermal fatigue resistance of low silver alloys [9]. Even after the multiple reflows, the Ag3Sn IMC is not coarsened and their network like structure remains intact. The exact mechanism regarding the Bi effect to maintain the solder bulk microstructure (dense Ag3Sn network) even after multiple reflow is not known. However, it can be postulated that the Bi atoms have some precipitate toughening effect. The Bi atoms precipitate in the bulk solder on the grain boundary of Ag3Sn-β-Sn network and it pins the grain boundaries during the ageing and helps maintain the dense Ag3Sn network. More research can be needed to elucidate the exact role of Bi in this regard.

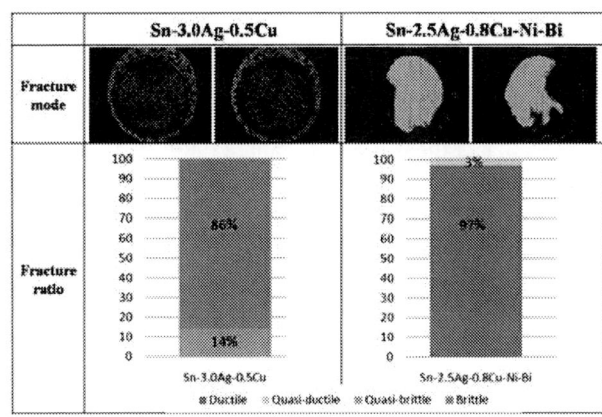

Figure 7. Result of high speed shear test (Fracture mode and fracture ratio)

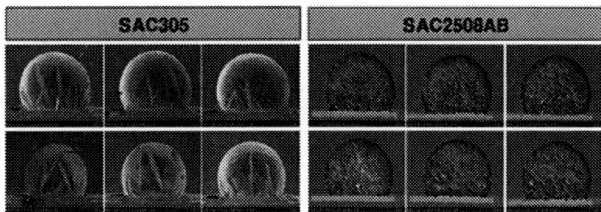

Figure 8. SEM image of solders after multiple reflows (Samples half etched)

Fig.9 shows the SEM image of Sn-3.0Ag-0.5Cu solder after multiple reflowing by half etching. It can be seen that the big IMCs are formed inside the Sn-3.0Ag-0.5Cu solder while in the case of Sn-2.5Ag-0.8Cu-Ni-Bi solder, IMCs are small and uniformly distributed. The EDX analysis (Fig. 9(b)), shows the distribution of Sn, Cu and Ag element in the huge IMC in the Sn-3.0Ag-0.5Cu solder. These huge IMCs are detrimental to the mechanical strength and reliability of solder joint. It acts as the site of crack generation and propagation which ultimately leads to the failure of the bond during the reliability test.

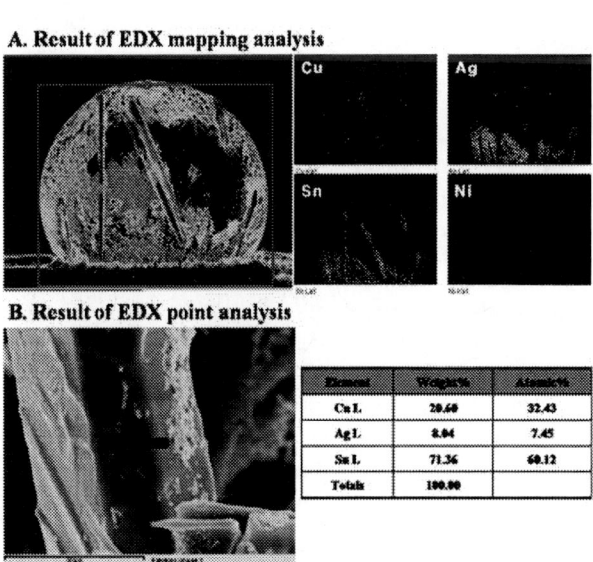

Figure 9. (a)SEM image of the SAC 305 solder (etched) showing IMC morphology; (b) EDX analysis of bulk plate like IMC.

978-1-4799-2408-0/14 $31.00 © 2014 IEEE

Fig. 10 shows the result of thermal cycling reliability test. It can be clearly seen that the Sn-2.5Ag-0.8Cu-Ni-Bi solder shows much better performance than Sn-3.0Ag-0.5Cu solder. In the Sn-2.5Ag-0.8Cu-Ni-Bi solder, the initial failure in 3100cycle and there was no failure until 3500 cycle. As already mentioned, this can be attributed to the grain refining effect of Bi and the Ag3Sn dense network which does not degrade even after the multiple reflows. Fig. 11 shows the drop test results of both solders. The Sn-2.5Ag-0.8Cu-Ni-Bi solder performed has significantly higher drop performance than Sn-3.0Ag-0.5Cu solder. This is because of small interfacial IMC thickness during joint formation and its slow growth during multiple reflows in the Sn-2.5Ag-0.8Cu-Ni-Bi solder as compared to Sn-3.0Ag-0.5Cu solder and as already explained, it can be attributed to the presence of Ni.

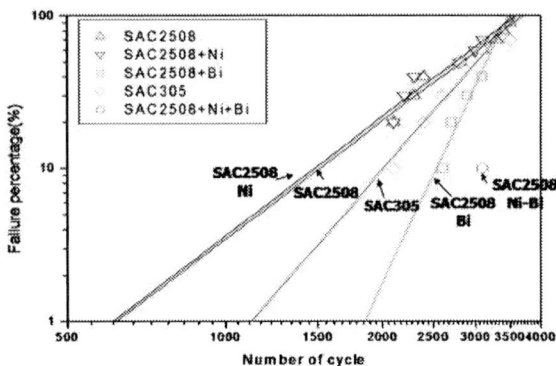

Figure 10. Weibull distribution curve showing Thermal cycling reliability (number of cycles to failure) of solder alloys

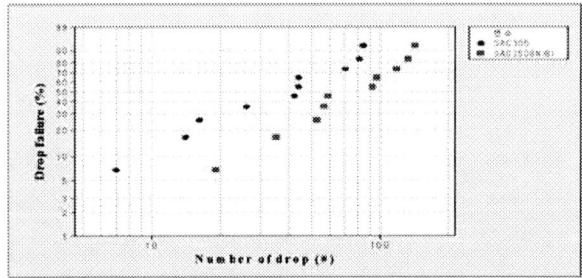

Figure 11. Weibull distribution curve showing number of drops to failure of solder alloys

Conclusions

Following results are confirmed during the development of high drop and thermal cycling performance solder having composition Sn-2.5Ag-0.8Cu-Ni-Bi.

1) The effect of addition of Ni is to reduce the IMC thickness during the formation of solder joint and ageing.

2) The Bi effect is to refine the bulk microstructure and the formation of dense Ag3Sn network even after multiple reflows.

3) The shear strength and fracture toughness of Sn-2.5Ag-0.8Cu-Ni-Bi solder is higher than Sn-3.0Ag-0.5Cu solder. This confirms the higher ductility of Sn-2.5Ag-0.8Cu-Ni-Bi solder.

4) The hardness of Sn-2.5Ag-0.8Cu-Ni-Bi solder is lower than Sn-3.0Ag-0.5Cu solder.

5) Sn-2.5Ag-0.8Cu-Ni-Bi solder shows much higher drop and thermal cycling performance as compared to Sn-3.0Ag-0.5Cu solder.

References

1. Zhu, W.H.; Luhua Xu; Pang, J.H.L.; Zhang, X.R.; Poh, E.; Sun, Y.F.; Sun, A.Y.S.; Wang, C.K.; Tan, H. B., "Drop reliability study of PBGA assemblies with SAC305, SAC105 and SAC105-Ni solder ball on Cu-OSP and ENIG surface finish," *Electronic Components and Technology Conference, 2008. ECTC 2008. 58th*, vol., no., pp.1667,1672, 27-30 May 2008

2. Desmond Y.R. Chong, F.X. Che, John H.L. Pang, L.H. Xu, B.S. Xiong, H.J. Tohl, B.K. Lim. Evaluation on Influencing Factors of Board-level Drop Reliability for Chip Scale Packages (Fine-pitch Ball Grid Array). IEEE Transaction on Advanced Packaging, vol. 31, issue 1, February 2008.

3. Zijie Cai; Suhling, J.C.; Lall, P.; Bozack, M.J., "Mitigation of lead free solder aging effects using doped SAC-X alloys," *Thermal and Thermomechanical Phenomena in Electronic Systems (ITherm), 2012 13th IEEE Intersociety Conference on*, vol., no., pp.896,909, May 30 2012-June 1 2012.

4. Santosh Kumar, Jae Yong Park, Jae Pil Jung., "Analysis on high speed shear characteristics of Sn-Ag-Cu solder joint" Electronic Materials Letters, Vol. 7, pp. 365-373, 2011.

5. Kumar, S.; Dohyun Jung; Jaepil Jung, "High-Speed Shear Test for Low Alpha Sn-1.0%Ag-0.5%Cu (SAC-105) Solder Ball of Sub-100- \mu{\rm m} Dimension for Wafer Level Packaging," Components, Packaging and Manufacturing Technology, IEEE Transactions on , vol.3, no.3, pp.441,451, March 2013.

6. K. Nogita, C. M. Gourlay, T. Nishimura., "Cracking and phase stability in reaction layers between Sn-Cu-Ni solders and Cu substrates" JOM, June 2009, Volume 61, Issue 6, pp 45-51.

7. K. Nogita, D. Mu, S. D. McDonald, J. Read and Y. Q. Wu, "Effect of Ni on phase stability and thermal expansion of Cu6-xNixSn5 (X=0, 0.5, 1, 1.5 and 2)", Intermetallics, 26 (2012) 78-85.

8. K. Nogita, "Stabilisation of Cu6Sn5 by Ni in Sn-0.7Cu 0.05Ni lead-free solder alloys", Intermetallics 18 (2010) 145-149.

9. R. Pandher and T. Lawlor, "Effect of Silver in Common Lead-free alloys", Proceedings of the International Conference on Soldering and Reliabilty, 2009, Toronto

Optimal Relaxation of I/O Electrical Requirements under Packaging Uncertainty by Stochastic Methods

Xu Chen, Juan S. Ochoa, José E. Schutt-Ainé, Andreas C. Cangellaris
Department of Electrical and Computer Engineering
University of Illinois at Urbana-Champaign
1406 W. Green St., Urbana, Illinois 61801
xuchen1@illinois.edu, jochoa@qti.qualcomm.com, jesa@illinois.edu, cangella@illinois.edu

Abstract

Fast and accurate evaluation of system failure rate is performed using stochastic collocation methods. First, we demonstrate that variability in I/O performance, such as driven voltage and slew rate, will impact failure probability of the link. For instance, higher slew rates will lead to increased levels of crosstalk between signals. Crosstalk above a certain threshold can be an indicator of system failure. Then, we demonstrate that by defining an upper bound for failure probability, an optimally relaxed set of I/O performance metrics can be generated corresponding to a stochastic interconnect model. This can be achieved by minimizing a cost function formulated with some performance metric, if these metrics are random variables. Lastly, we demonstrate that, if more information about the packaging design becomes available, the randomness of the interconnect model can be reduced, leading to a more relaxed set of I/O performance metrics. The method proposed in [1] is then be used to verify that the packaging design will meet the tolerated failure probability.

Introduction

As part of the iterative design cycle for high-speed and high-density systems, a set of acceptable I/O performance metrics is provided to circuit designers based on preliminary electrical simulations [2], [3]. These metrics are often defined such that the packaging engineers are highly confident that a satisfactory interconnect can be realized within packaging design constraints and manufacturing tolerances. Hence, the preliminary simulations that lead to these I/O performance metrics often make conservative assumptions regarding the impact of packaging on link performance [4]. This tendency is exacerbated by the fact that during initial design stages, a fully realized packaging design is usually unavailable. Thus, these I/O performance metrics, while robust, are often overly restrictive, leading to unnecessarily tight design constraints for circuit designers, increased design time, and lowered yield of fabricated chips. A more relaxed set of I/O requirements will shorten design-to-market time, and improve yield.

The premise of this paper is that the overall system design objective is to keep system failure rate below a pre-determined maximum. We recognize that, in the absence of a realized physical design, the interconnect can be modeled as a stochastic system, the uncertainties of which are induced both by lack of information regarding the final design and by tolerances of typical manufacturing processes. Furthermore, as the packaging design progresses, more information becomes available, thus reducing the randomness of the interconnect model. With this in mind, the objective of this

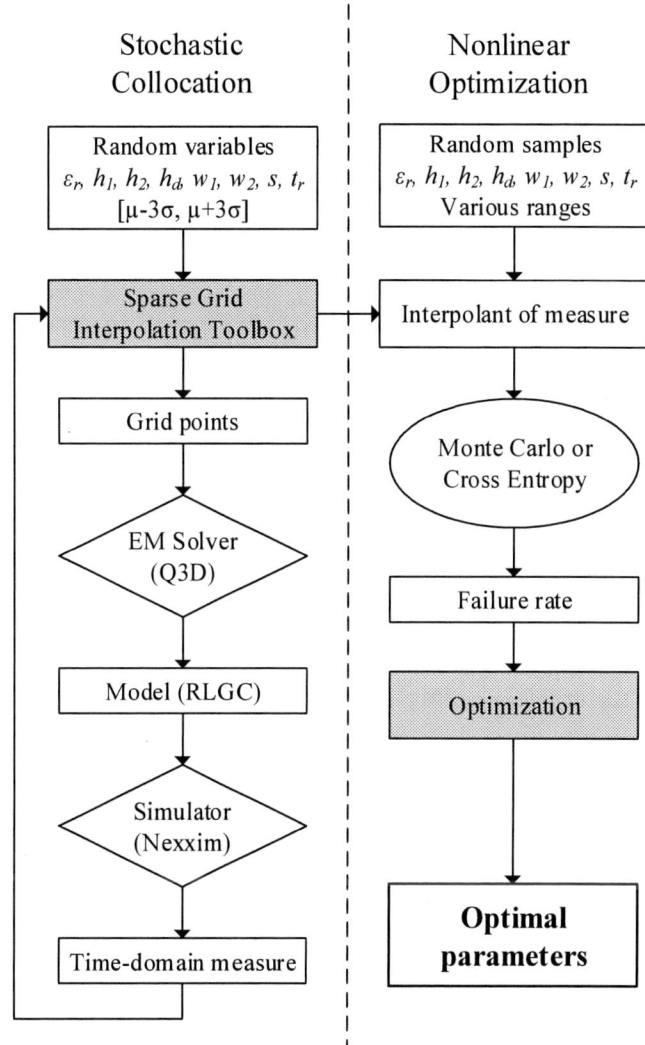

Figure 1. Flowchart for process of optimizing I/O parameters for specific failure rates.

work is to develop a framework for providing an optimally relaxed set of I/O performance metrics given a system yield target and interconnect uncertainties.

Process Flow

The process presented in this paper is separated into two major portions: uncertainty qualification (UQ) using stochastic collocation [5], and failure rate estimation and constraint relaxation using nonlinear optimization. The process is summarized in a flowchart in Figure 1.

The random space is first defined, which includes both variables in the interconnect model, as well as system variables. At this stage, it is very important to ensure that all random dimensions of interest are captured, and their ranges well-defined. The distribution of the random variables should include uncertainty from both manufacturing variations and lack of information regarding design. Since the range of variables will be used to construct a sparse grid for interpolating the system response, and evaluation of the response at each grid node can be expensive, it is beneficial to be generous when choosing ranges to avoid having to construct the interpolant again and run additional simulations later. For this paper, the same interpolant is used to perform stochastic collocation, failure rate estimation, and optimization.

For the optimization portion of the process, a new set of probability distributions is defined for the random space, and the interpolant from before is used to determine system failure by comparing a measure of interest against a hard threshold. We discretize the optimization variables and estimate the failure rate of the system at each node using sampling techniques. Finally, nonlinear optimization routines are employed to obtain the optimal design parameters. This portion can be repeated using different probability distributions for the model and system variables. When more information is known about the design, the randomness of the system and consequently the variance of these random variables can be reduced, which will often lead to a more relaxed optimum.

Stochastic Collocation

While Monte Carlo method is the most natural way to treat a stochastic problem, its convergence goes as the inverse of the square root of the number of iterations, requiring a large number of electromagnetic simulations; therefore, making its use computationally unfeasible. In this paper we employ an efficient alternative, called the stochastic collocation method that, roughly speaking, is aimed at the interpolation of the output in terms of the random input parameters by appropriately sampling the random space. Then, the interpolation is employed in the statistical assessment of our output of interest. This technique is a non-intrusive methodology in the sense that its application does not require modification of the EM solver.

The selection of the interpolation grid nodes is dictated by the Smolyak algorithm [5]-[7]. The grid of nodes required by this method is *sparse* and its number of nodes, considerably less than the number of points required by the tensor product scheme to achieve the same accuracy of interpolation. The number of nodes in the sparse grid method grows polynomially with the dimensions and it is given by an accuracy parameter known as the level of the interpolation, k [5]. The higher the level k, the more accurate the interpolation is and the more nodes the grid utilizes.

For our purposes we employ the interpolation tool developed by Klimke [8]-[10] to propagate the input uncertainty to the output response, x. The key attributes of the interpolation are summarized below. The reader is referred to [8] for more details.

The Smolyak scheme approximates the general tensor product multivariate interpolation formula as follows:

$$x_{k,D} = \sum_{|\mathbf{i}|<k} (\Delta^{i_1} \otimes \dots \otimes \Delta^{i_D}), \qquad (1)$$

where we use the notation $|\mathbf{i}|=i_1+\dots+i_D$ and $\Delta^i = U^i - U^{i-1}$, and the operant U represents the interpolation in each dimension,

$$U^i(x) = \sum_{j=1}^{m_i} a_j^i(\xi^i)\psi(\xi_j^i). \qquad (2)$$

In the above equation a_j are the basis functions associated with the j-th grid value of the parameter ξ^i, and $U^0 = 0$. The number of sampling nodes for each dimension, m_i, depends on the particular grid choice and the value of i. For the Clenshaw-Curtis-type sparse grid it is [8], [9]:

$$m_i = \begin{cases} 1, & \text{if } i = 1, \\ 2^{i-1} + 1, & \text{if } i > 1, \end{cases} \qquad (3)$$

$$\xi_j^i = \begin{cases} \dfrac{2(j-1)}{(m_i-1)} - 1, & \text{for } j = 1, \dots, m_i \text{ if } m_i > 1, \\ 0, & \text{for } j = 1 \text{ if } m_i = 1, \end{cases} \qquad (4)$$

where values of variable ξ are assumed to be in the range $[-1, 1]$. Such values can be scaled to the actual range of the variable of interest. The key advantage of the Smolyak formula is that we only need to find the value of the output with the typically computationally expensive field solver at the sparse grid nodes,

$$H_{k,D} = \bigcup_{k-D+1\le|\mathbf{i}|\le D} (X^{i_1} \times \dots \times X^{i_D}), \qquad (5)$$

where the set X^{i_j} contains the samples of the j-th variable, corresponding to the interpolation parameter i.

Nonlinear Optimization

Numerical optimization is a well-developed field, and many algorithms are available to solve unconstrained and constrained problems fairly efficiently [11]-[13]. Optimization problems always take the following form:

$$\underset{x\in S}{\text{minimize}} \ f(x) \qquad (6)$$

where S is the constraint set over which we are searching for the minimum, $f(x)$ is the objective or cost function, and x is the optimization variable. For the example in this paper, the optimization variable x is t_r. In general, we define x to be the design parameters we wish to optimize, such as rise time and source voltage, and it can have many dimensions. The cost function $f(x)$ is defined in (11), which is constructed to have a clear minimum. It is most ideal for $f(x)$ to be convex to avoid the algorithm finding local minima instead of global minima.

978-1-4799-2408-0/14 $31.00 © 2014 IEEE

If $f(x)$ is continuously differentiable, then any minimum x^* must satisfy the necessary first-order condition for optimality:

$$\nabla f(x^*) = 0 \qquad (7)$$

All minima over a set can be classified as local and global minima. If the set S and cost function $f(x)$ are convex, then all minima are global minima. Furthermore if $f(x)$ is strictly convex, then the minimum is unique.

If $f(x)$ is twice continuously differentiable, then x^* satisfying (7) and the following second-order condition:

$$\nabla^2 f(x^*) > 0 \qquad (8)$$

are sufficient for x^* to be a minimum. We note here that if the cost function $f(x)$ is interpolated using a piecewise-linear basis function, then it is only once continuously differentiable, and thus we cannot apply condition (8) in our search for minima.

Since in most realistic engineering applications, sets S over which we optimize are closed and bounded, the Weierstrass extreme-value theorem [13] guarantees that if $f(x)$ is a continuous function, a global minimum will be achieved over the constraint set. Hence it is very important in our applications to formulate the optimization problem such that the cost function is continuous.

For this paper, the coordinate descent method [12] is used when the cost function $f(x)$ is constructed using piecewise-linear basis functions. This is a non-derivative optimization algorithm which searches through the constraint set one direction at a time in a cyclical fashion using line search. Although effective, this method may converge slowly when the optimization variable has high dimensions. If the cost function is constructed using polynomial basis functions, then it is sufficiently smooth and a gradient method such as conjugate gradient [12] can be used which offers better convergence.

Numerical Example: Coupled Stripline

To test our optimization methodology, we consider a 2GHz bus using single-ended signaling scheme over a lossless coupled stripline on printed circuit board (PCB) substrate. We consider many cross-sectional geometry variables, such as line widths, line spacing, and dielectric thicknesses as random variables. The cross-section is shown in Figure 1. The source of the randomness is due to both manufacturing tolerances and lack of information regarding the final design. For instance, if the manufacturer and part number of the raw PCB substrate is still unknown, then we must model the dielectric constant with more variance to account for all the possible options in selecting the part. When a final part number is decided, the analysis can be repeated while modeling the dielectric constant with a different mean and a much smaller variance to account for only the statistical variability in the samples.

First, the stripline is modeled using a commercial 2-D quasi-static solver [14]. Any EM solvers can be used for this purpose, but it is important for the geometric variables to be parameterized in the model. The dielectric constant of the

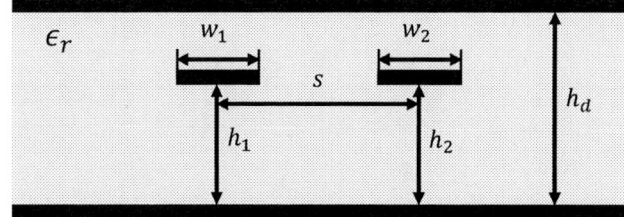

Figure 2. Cross-section of coupled stripline example with the variables $w_1, w_2, s, h_1, h_2, h_d$ labeled. The conductors are in a homogenous dielectric with permittivity ε_r.

PCB substrate and other material properties where variability is present should also be parameterized. In this case, the model variables are listed in Table 1. There is a total of 7 random model dimensions. Variations in these random parameters will influence the magnitude of crosstalk present on the victim line, and the influence of each parameter on crosstalk will be captured in the statistical model.

Table 1. Coupled stripline model variables and their mean μ and standard deviation σ.

Variable	Mean μ	Std. Deviation σ
ε_r	4.4	0.147
h_1	0.2 mm	0.007 mm
h_2	0.2 mm	0.007 mm
h_d	0.3 mm	0.01 mm
w_1	0.1 mm	0.003 mm
w_2	0.1 mm	0.003 mm
s	0.3 mm	0.01 mm

In addition to variability in the EM model, engineers must also consider variability in the operating environment, such as circuit excitations. For this example, we perform transient simulations [15] of the coupled line system by exciting the aggressor line with a continuously-switching clock signal V_{in} whose rise time, t_r, is defined as an additional system random dimension. As shown in Figure 3, the peak value of V_{in} is set to be 1.2 V in this example, but this value can also be a source of uncertainty. It is well known that the aggressor slew rate is a strong influence in the magnitude of crosstalk on victim lines. We would expect a voltage source with smaller rise time to induce stronger crosstalk compared to a voltage source with larger rise time, if the same instantiation of transmission line model is used.

The seven random parameters in the stripline model combined with t_r yield a random space of dimension 8. The next step is to create a high-dimensional interpolant for a measure of interest. A hard threshold will be defined on the measure, which becomes an indicator function for failure. In the first case, we define the maximum crosstalk magnitude on a quiet victim as the failure indicator. The transient simulation topology for this is shown in Figure 4. The aggressor signal drives into a line terminated with 50 Ohms, and the victim line is terminated at both ends by 50 Ohms. The far end voltage V_{vic} on the victim line is simulated for 100 ns, and the maximum value over this time is measured.

978-1-4799-2408-0/14 $31.00 © 2014 IEEE   719

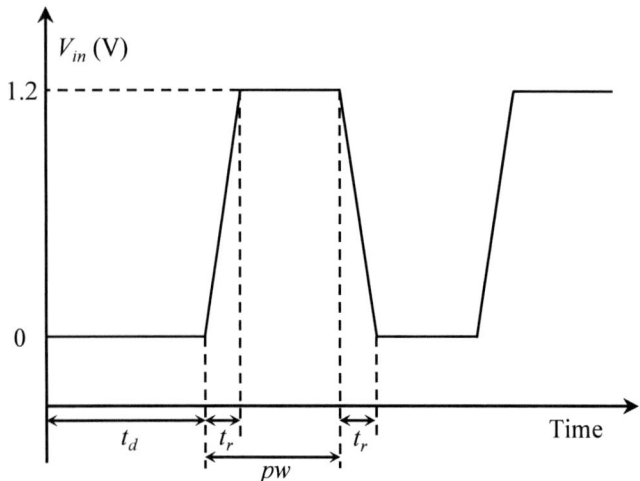

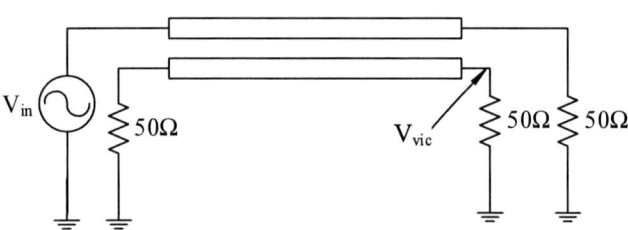

Figure 3. Source voltage connected to the aggressor line. The signal is running at 1 GHz with $pw = 1$ ns, $t_d = 20$ ns, and the rise time t_r is a random variable with mean 175 ps and standard deviation of 50 ps.

Figure 4. Transient simulation circuit for measuring far-end crosstalk (FEXT) on victim line.

It is quickly apparent that the amount of simulations required to construct the interpolant is insurmountable. If we take only 10 samples of each random dimension, we will need to perform $10^7 = 10,000,000$ EM simulations and $10^8 = 100,000,000$ circuit simulations to interpolate. To overcome this limitation, we utilize sparse grid based on the Smolyak algorithm. Using a Clenshaw-Curtis type grid with 4 levels, the total number of nodes needed to construct the 8-dimensional interpolant is 3937, of which only 2467 unique EM simulations are required. With each instantiation of the model taking approximately 2 minutes to extract, the simulation time is around 4 days. Although the computational resource is still significant, this effort only needs to be undertaken once per design if the range of the each dimension is carefully chosen. Stochastic collocation can still be performed on the interpolant if the random samples fall within a subset of the initial range. Additionally, simulation time can be significantly reduced using distributed computing.

After the interpolant is constructed, we can estimate the failure rate by performing Monte Carlo sampling on the collocated function, and comparing the interpolated value against a hard threshold α to create a failure indicator function:

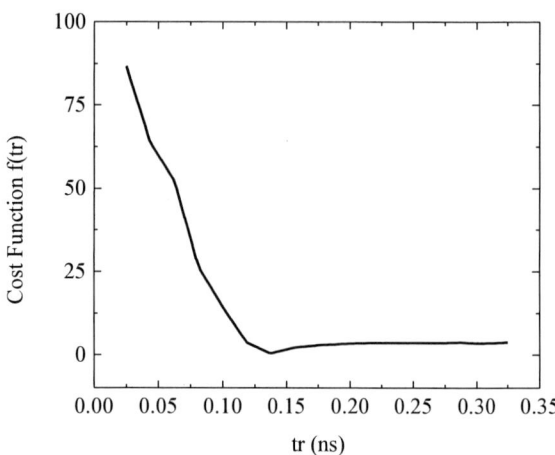

Figure 5. Cost function for finding optimal lower bound for t_r, for failure rate of 5%. All 7 other random variables are set to mean and variances listed in Table 1.

$$\text{Failure Rate} = \frac{\sum_{i=1}^{N} \mathbf{1}_F(x_i)}{N} \times 100 \ (\%) \qquad (9)$$

$$\mathbf{1}_F(x) = \begin{cases} 1 & if \ x \geq \alpha \\ 0 & if \ x < \alpha \end{cases} \qquad (10)$$

In this example, we use $\alpha = 6$ mV. In other words, if a random sample shows crosstalk of greater than α on the victim line, the sample is considered a failure. We take a total of $N = 10000$ samples. For ordinary Monte Carlo analysis, the accuracy of the result converges with order $1/\sqrt{N}$ according to the Law of Large Numbers. Consequently, to increase the accuracy of the estimated failure rate by an order of magnitude, 100 times more samples need to be taken. Although stochastic collocation speeds up the evaluation of sample points significantly, this process can still become very time consuming if very high accuracy is desired. Depending on the failure threshold being used, failure could be a rare event. It is recommended that at a minimum, N should be large enough so that its inverse is less than 1/10 of the resolution desired in error rate estimation. For this example, N is chosen such that the resolution for failure rate is 0.1%. More samples will always produce more accurate results. Additionally, techniques such as Cross Entropy can be utilized to increase the efficiency of the sampling scheme when the failure rate is expected to be very low.

To construct a cost function for optimization, we discretize the optimization variable, in this case t_r, into 17 points and perform Monte Carlo analysis at each point to obtain the failure rate. Furthermore, we need to fix a desired failure rate β in order to perform optimization. The cost function for optimization is then:

$$f(t_r) = |\text{Failure Rate}(t_r) - \beta| \qquad (11)$$

When more details is known about the design, we are able to redefine the random model parameters. The uncertainty corresponding to lack of design and manufacturing information is now reduced, and the random variables will have reduced variance and possibly different mean than what was assumed in the initial analysis. As long as the new ranges of the parameters fall within those used in the initial interpolation of the model, the original interpolant can still be used and no new simulation needs to be performed. We change the mean values of ε_r and s, and reduced their standard deviations. The new values are listed in Table 2. With the randomness reduced, the lower bound for t_r is now 81.3 ps, corresponding to a slew rate of 14.76 V/ns. This shows that when more information is known about the design, the I/O requirements can be relaxed.

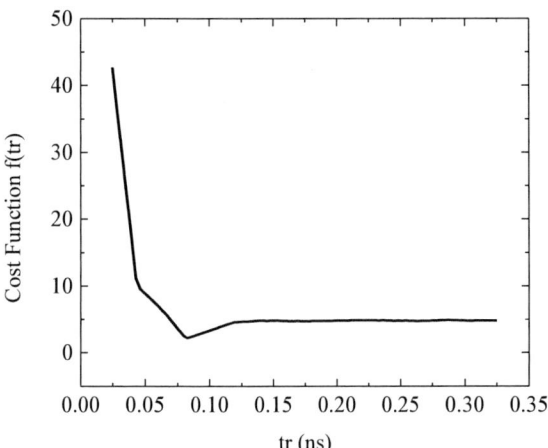

Figure 6. Cost function for finding optimal lower bound for t_r for failure of 5%. The line spacing s and dielectric constant ε_r have different mean and reduced variance compared to the nominal values. The values are listed in Table 2.

Table 2. Mean and standard deviation for Nominal and Reduced random variable ranges with their respective optimal lower bound for t_r and the corresponding slew rate (SR).

Variable	Nominal (μ, σ)	Reduced (μ, σ)
ε_r	(4.4, 0.147)	**(4.2, 0.014)**
h_1	(0.2, 0.007) mm	(0.2, 0.007) mm
h_2	(0.2, 0.007) mm	(0.2, 0.007) mm
h_d	(0.3, 0.01) mm	(0.3, 0.01) mm
w_1	(0.1, 0.003) mm	(0.1, 0.003) mm
w_2	(0.1, 0.003) mm	(0.1, 0.003) mm
s	(0.3, 0.01) mm	**(0.32, 0.0011) mm**
t_r^*	**137.5 ps**	**81.3 ps**
SR	**8.72 V/ns**	**14.76 V/ns**

Conclusions

This paper presented a method to calculate optimal I/O design requirements using stochastic interconnect models. By leveraging stochastic collocation techniques, we demonstrated that all the costly EM simulations for design analysis can be performed ahead of time, and the resulting interpolant be used for subsequent parameter sampling and failure rate estimation processes. Furthermore, due to the non-intrusive nature of the numerical methods used in this methodology, any existing EM and circuit solvers can be used to perform this analysis.

Through a stripline example, we demonstrated that for a desired failure rate, the I/O requirements can be relaxed if the randomness of the packaging model is reduced. Although our example is limited to the optimization of rise time, the designer may wish to relax requirements for multiple I/O parameters. In these cases, optimization will need to be performed over high-dimensional constraint sets.

Acknowledgments

This material is based upon work supported by, or in part by, the U.S. Army Research and the U.S. Army Research Office under grant number W911NF-10-1-0269.

The authors would like to thank Maxim Raginsky and Rishi Ratan at the University of Illinois at Urbana-Champaign for their helpful discussions.

References

[1] Ochoa, J.S.; Cangellaris, A.C., "Random-Space Dimensionality Reduction for Expedient Yield Estimation of Passive Microwave Structures," *Microwave Theory and Techniques, IEEE Transactions on* , vol.61, no.12, pp.4313,4321, Dec. 2013.

[2] S. H. Hall and H. L. Heck, *Advanced Signal Integrity for High-Speed Digital Designs*. Hoboken, N.J.: Wiley : IEEE, 2009.

[3] K.S. Oh, X. Yuan, *High-Speed Signaling: Jitter Modeling, Analysis, and Budgeting*. Boston, MA: Pearson Education, 2012.

[4] E. Matoglu, N. Pham, D. N. D. Araujo, M. Cases, and M. Swaminathan, "Statistical Signal Integrity Analysis and Diagnosis Methodology for High-Speed Systems," *Advanced Packaging, IEEE Transactions on*, vol. 27, no. 4, pp. 611–629, 2004.

[5] D. Xiu and J. S. Hesthaven, "High-Order Collocation Methods for Differential Equations with Random Inputs," *SIAM Journal on Scientific Computing*, vol. 27, no. 3, pp. 357–78, 2005.

[6] D. Xiu, "Fast Numerical Methods for Stochastic Computations: a Review," *Communications in Computational Physics*, vol. 5, no. 2-4, pp. 242–72, 02 2009.

[7] D. Xiu, *Numerical Methods for Stochastic Computations: a Spectral Method Approach*. Princeton, N.J.: Princeton University Press, 2010.

[8] A. Klimke and B. Wohlmuth, "Algorithm 847: spinterp: piecewise multilinear hierarchical sparse grid interpolation in matlab," *ACM Transactions on Mathematical Software*, vol. 31, no. 4, pp. 561–579, Dec. 2005.

[9] A. Klimke, "Sparse Grid Interpolation Toolbox – User's Guide," University of Stuttgart, Germany, IANS report 2007/017, 2007.

[10] A. Klimke, "Sparse grid inerpolation toolbox," January 2006. [Online]. Available: http://www.ians.uni-stuttgart.de/spinterp/

[11] S. P. Boyd, L. Vandenberghe, *Convex Optimization*. New York: Cambridge, 2004.

[12] D. G. Luenberger and Y. Ye, *Linear and Nonlinear Programming*. New York: Springer, 2004.

[13] D. P. Bertsekas, *Nonlinear Programming*. Belmont, Mass.: Athena Scientific, 1999.

[14] ANSYS Q3D Extractor, Release 12.0.3.

[15] ANSYS Designer, Release 8.0.3.

An Accurate and Convenient Lumped/Discrete Port De-Embedding Method for the 3D Integration and Packaging Full-Wave Modeling by Splitting and Absorbing the Error-Cancelling Network

Zhaoqing Chen
IBM Corporation
2455 South Rd, B002, Poughkeepsie, NY 12601
zhaoqing@us.ibm.com, (845)435-5595

Abstract

In this paper, an accurate and convenient lumped/discrete port de-embedding method is proposed for full-wave modeling interconnect and packaging components with bump/pin array interfaces in the 3D integration and packaging systems. This method is based on splitting and absorbing the Error-Cancelling Network which was inserted in between two connected S-parameter models to cancel any error by the port parasitic parameters and uncertainty in port current and voltage definitions. In the proposed approach, the Error-Cancelling Network is split into two identical Half-Error-Cancelling networks by making use of the square root of the corresponding T-matrix of the error-cancelling network. Each half-error-cancelling network can be attached to and combined with the original S-parameter models to generate de-embedded S-parameters models without any additional and individual Error-Cancelling Network in appearance. This de-embedding method is a wide-band approach valid on every frequency point in the S-parameter model.

1. Introduction

In 3D integration and packaging system design, vertical connection through arrays of pins, balls, C4 bumps, and micro-C4 bumps (Fig.1) becomes a major type of interconnect [1][2] because of not only on the high-speed but also on the high-density requirements on the system packaging and interconnect. As the system signal bit rate increases dramatically in recent years, the S-parameter models derived by full-wave electromagnetic simulation tools are widely used in system signal integrity (SI) simulations for better high frequency accuracy[3]-[6]. In addition, the full-wave approach can accurately cover some particular effects of inhomogeneous semiconductor and dielectric materials in the 3D integration and packaging components.

The best way to generate the S-parameter model of a whole net group for system SI simulation is to combine individual S-parameter models of single chip and single package by port connection. The ports defined in the model can be either signal ports only or general ports including signal and power/ground ports [7]. The port definitions should be consistent for both ports of any connected port pair so that not only the signal nodes but also the reference notes (local grounds) are really connected in the circuit schematic.

An important issue in the port definition is the electromagnetic field distribution at the port locations. A good port definition needs the electromagnetic field concentrating in the port area with the port voltage and current well defined. However, the situations for the port voltage and current definitions might be very bad at the component natural boundaries in many applications such as the chip stack case. In this situation, the accurate port de-embedding is required to minimize any port connection error caused by the port parasitic parameters and the uncertainty in port voltage and current definitions.

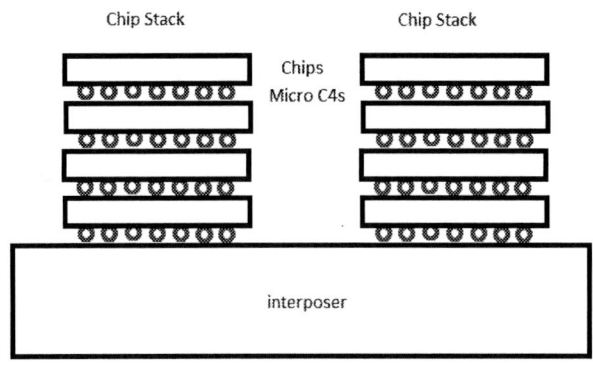

Fig.1 Illustration of the vertical interconnects in the 3D integration

In most commercial electromagnetic simulation tools, there are two different types of port definition, namely the waveguide port (or wave port) and the lumped port (or discrete port). For the vertical array interface, the waveguide port is very difficult to apply. Some special lumped port with realistic, not artificial, parasitic parameters like the circular lumped port is suitable only in some particular cases such as the port defined at via clearance anti-pad location, for example. In this paper, the gap lumped/discrete port as defined at the natural boundaries of different chip/package components in [7][18] is used for good port connectivity since not only the port signal node pair but also the port reference node pair of the two connected ports are connected perfectly.

There are several other published de-embedding methods for electromagnetic simulation tools on the lumped/discrete ports [9]-[14]. However, most of them need to be implemented into an electromagnetic simulation tool which is usually not accessible to general tool users.

A lumped/discrete port de-embedding method by the error-cancelling network was proposed [18] for broadband application targeting better accuracy and passivity/causality properties than previous approaches. In stead of deriving individual lumped parasitic capacitance and inductance, the method in Reference [18] is a full solution using S-parameter matrix operation based on full wave electromagnetic simulations of three calibration structures as briefly reviewed in the following section.

The error-cancelling network can be applied in the system SI simulation by inserting it in between two connected original component models which are derived from the electromagnetic simulation directly without port de-embedding. However, the above de-embedding form is totally different from conventional de-embedding which simply uses a de-embedded model to replace the original model.

In order to make the de-embedding in a traditional form without any additional network, a new method is proposed in this paper for application of the Error-Cancelling Network. The Error-Cancelling Network will be split into two identical parts. Each of them will be absorbed into one of the two original models to form two de-embedded models. The port errors will be eliminated in the de-embedded models with the same accuracy of the method in Reference [18] which needs a additional network.

2. Methodology

A. Brief Review of the Error-Cancelling Network [18]

In [18], a de-embedding method was proposed by inserting an Error-Cancelling Network S_{err_c} in between two connected original networks, S_a and S_b, for cancelling the errors caused by parasitic effects and the uncertainties in current and voltage definitions at the lumped/discrete ports. The Error-Canceling Network S_{err_c} in Fig.2 can be derived by Equation (1) as shown below from three calibration S-parameter models $S_{a'}$, $S_{b'}$, and $S_{a'b'}$ as shown in Fig.3.

$$S_{err\_c} = \left(S_c^{bb} + S_c^{ba} \left(S_{a'b'} - S_c^{aa} \right)^{-1} S_c^{ab} \right)^{-1} \tag{1}$$

where

$$S_c^{aa} = \begin{bmatrix} S_{a'}^{\alpha\alpha} & 0 \\ 0 & S_{b'}^{\alpha\alpha} \end{bmatrix} \quad S_c^{ab} = \begin{bmatrix} S_{a'}^{\alpha\beta} & 0 \\ 0 & S_{b'}^{\alpha\beta} \end{bmatrix}$$

$$S_c^{ba} = \begin{bmatrix} S_{a'}^{\beta\alpha} & 0 \\ 0 & S_{b'}^{\beta\alpha} \end{bmatrix} \quad S_c^{bb} = \begin{bmatrix} S_{a'}^{\beta\beta} & 0 \\ 0 & S_{b'}^{\beta\beta} \end{bmatrix} \tag{2}$$

$$S_{a'} = \begin{bmatrix} S_{a'}^{\alpha\alpha} & S_{a'}^{\alpha\beta} \\ S_{a'}^{\beta\alpha} & S_{a'}^{\beta\beta} \end{bmatrix} \quad S_{b'} = \begin{bmatrix} S_{b'}^{\alpha\alpha} & S_{b'}^{\alpha\beta} \\ S_{b'}^{\beta\alpha} & S_{b'}^{\beta\beta} \end{bmatrix} \tag{3}$$

The lumped/discrete port de-embedding by Equation (1) is based on S-parameter matrix operation so that it is a broadband approach which is valid on every frequency point in the S-parameter model. The accuracy is determined by the broadband electromagnetic simulations for each component modeling including the original structures and the calibration structures.

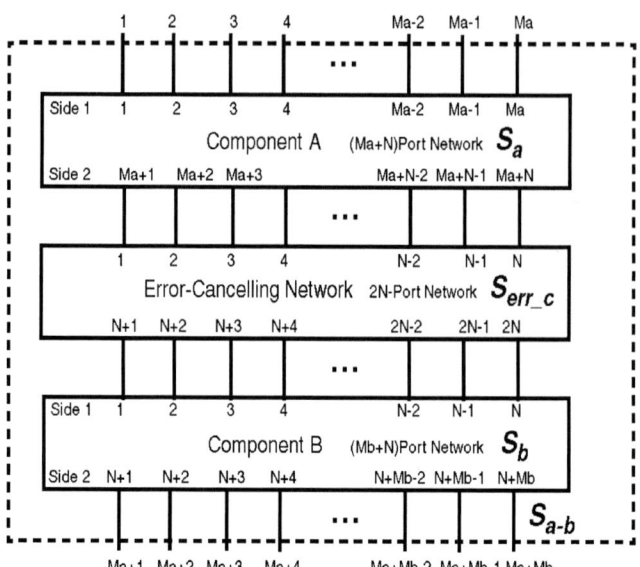

Fig.2 Original component models S_a and S_b with the Error-Cancelling Network S_{err_c} inserted in between

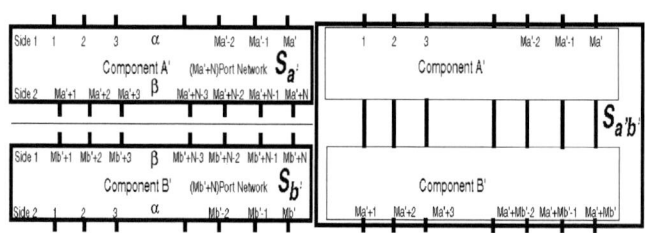

Fig.3 Illustration of three calibration S-matrixes $S_{a'}$, $S_{b'}$, and $S_{a'b'}$

B. Splitting the Error-Cancelling Network into Two Identical Half-Error-Canceling Networks

The error-cancelling network S_{err-c} derived by Equations (1)-(3) will completely cancel any error by the port parasitic parameters and the uncertainty in port voltage and current definitions if there is no meshing and other numerical errors in electromagnetic simulations. Some direct interactions between the two networks, S_a and S_b, other than those by the port current and voltage are taken into account in S_{err-c} by Equations (1)-(3) already. Actually, the Error-Cancelling Network S_{err-c} is customized for connecting two particular networks, S_a and S_b, as shown in Fig.2.

The direct application of the de-embedding by introducing the Error-Cancelling Network is not straightforward for the model users to run system SI simulations since an additional network, the Error-Canceling Network, is required for port de-embedding. Most users would prefer two de-embedded networks other than two original networks plus an Error-Cancelling Network inserted in between.

In order to derive individual de-embedded model, we need to split the Error-Cancelling Network into two parts, or two Half-Error-Cancelling Networks. In general, splitting a network into two parts is not a unique solution. There are multiple solutions. In this paper, splitting the error-cancelling network into two identical parts is assumed as shown in Fig.4. It may not be the best way but the procedure is relatively simple. It should be mentioned that the divide as shown as the dash line in Fig.4 is just a mathematically conceptual boundary and may not have a corresponding real physical location.

For easily splitting, the original multi-port S-parameter model of the Error-Cancelling Network S_{err_c} is transferred into the multi-port T-parameters T_{err_c} by Equations (4)-(5) from References [16] and [17].

The square-root of T_{err_c} is derived through the matrix diagonal transformation using the eigenvector method. After perform the square-root of each diagonal element, the diagonal matrix is transformed back to get the square-root matrix T_{herr_c} which is the Half-Error-Cancelling Matrix in T-parameter format.

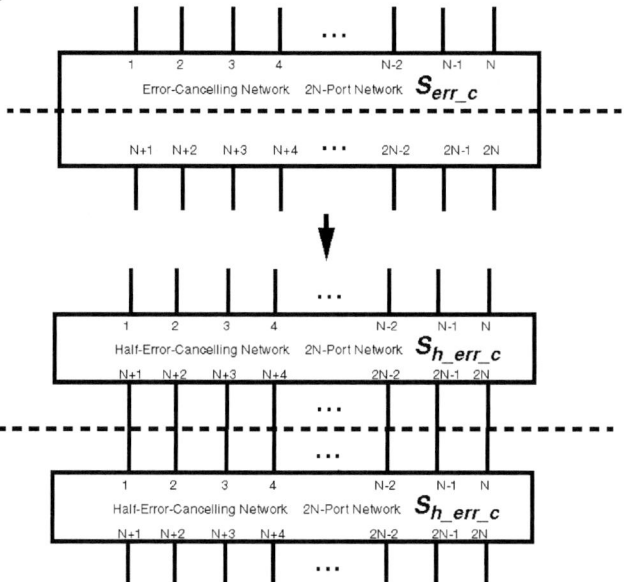

Fig.4 Splitting the Error-Cancelling Network into two identical parts.

By Equations (7) and (8) which come from References [16] and [17], the T-parameter matrix T_{herr_c} can be transformed back into the S-parameter matrix S_{herr_c}.

$$T_{err\_c} = \begin{bmatrix} S_{err\_cI,II} - S_{err\_cI,I}\, S_{err\_cII,I}^{-1}\, S_{err\_cII,II} & S_{err\_cI,I}\, S_{err\_cII,I}^{-1} \\ -S_{err\_cII,I}^{-1}\, S_{err\_cII,Ii} & S_{err\_cII,I}^{-1} \end{bmatrix} \quad (4)$$

$$S_{err\_c} = \begin{bmatrix} S_{err\_cI,I} & S_{err\_cI,II} \\ S_{err\_cII,I} & S_{err\_cII,II} \end{bmatrix} \quad (5)$$

$$T_{h\_err\_c} = T_{err\_c}^{1/2} = V\,(V^{-1}\, T_{err\_c}\, V)^{(1/2)}\, V^{-1} \quad (6)$$

where V is the eigenvector matrix of T_{err_c}. $(.)^{1/2}$ is the operator of matrix square root, $(.)^{(1/2)}$ is the operator of matrix individual element square root.

$$S_{h\_err\_c} = \begin{bmatrix} T_{h\_err\_cI,II}\, T_{h\_err\_cII,II}^{-1} & T_{h\_err\_cI,I} - T_{h\_err\_cI,II}\, T_{h\_err\_cII,II}^{-1}\, T_{h\_err\_cII,I} \\ T_{herr\_cII,II}^{-1} & -T_{h\_err\_cII,II}^{-1}\, T_{h\_err\_cII,I} \end{bmatrix} \quad (7)$$

$$T_{h\_err\_c} = \begin{bmatrix} T_{h\_err\_cI,I} & T_{h\_err\_cI,II} \\ T_{h\_err\_cII,I} & T_{h\_err\_cII,Ii} \end{bmatrix} \quad (8)$$

C. De-Embedded S-Parameter Model for Each Component by Using the Half-Error-Cancelling Networks

The Half-Error-Cancelling network S_{herr_c} can be attached to and combined with one of the two original models by using the S-parameter model combination methods such as in [18].

For each original model in the vertical connect applications, there are up to two Half-Error-Cancelling Networks, one on each side, as S_{herr_c1} and S_{herr_c2} shown in Fig.5. The network S_{a_de} consisting of S_{herr_c1}, S_a, and S_{herr_c2} is the S-parameter model after de-embedding by the proposed method.

The Error-Cancelling Network and Half-Error-Cancelling network are introduced to cancel the port error, so they are usually not passive and causal themselves by nature. There should not be any passivity and causality violations in the network after de-embedding, like S_{a_de}, assuming there are no numerical errors in the electromagnetic simulations and the matrix operations. However, in reality there may be still small passivity and causality violations in the de-embedded network S_{a_de}. These small violations can be eliminated easily by using passivity enforcement in some commercial tools like *IdEM Plus*[20]. The passive and causal models made this way are ready for transient simulation applications.

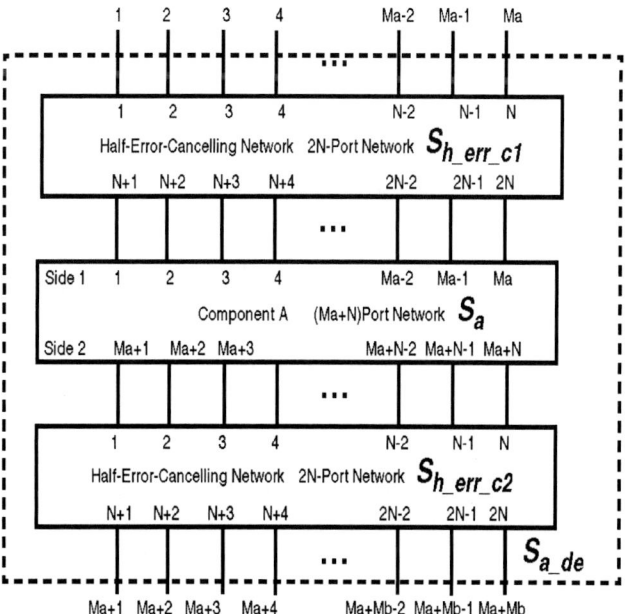

Fig.5 Model combination schematic for the de-embedded model S_{a_de}.

978-1-4799-2408-0/14 $31.00 © 2014 IEEE

3. Verification and Application

For verification purpose, a test case of a 9-chip stack will be given here. In the test case, the simulation result by the proposed method is compared with a known-good reference by running electromagnetic simulation on the corresponding whole structure as one piece and with the result without port de-embedding. It should be mentioned in the test case the outer ports (at Side 1 of Component A and Side 2 of Component B) are directly from the electromagnetic simulations without any de-embedding. It does not affect the verification of the proposed de-embedding method which is applied to all inner ports (at Side 2 of Component A and Side 1 of Component B) here. In real application, we may need to use other de-embedding methods on the outer ports separately since the proposed de-embedding method only applies to the inner ports of the whole stacked structure.

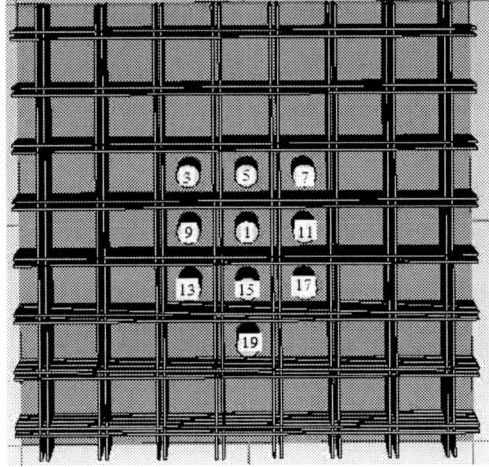

Fig.6 A 9-Chip stack in 3D integration by micro-C4 bump arrays (with power/ground grids shown, Silicon substrate not shown)

The test case is a 20-port 9-chip stack as shown in Fig.6 (Silicon substrate not shown). The diameter of the Through-Silicon-Via (TSV) is 0.010mm. The diameter of the micro-C4 bump is 0.020mm with a height of 0.0144mm. The discrete port gap is 0.002mm. The Silicon properties used in modeling are ε_r=11.9, σ=10.0S/m coated by a SiO$_2$ round layer with ε_r=3.9 and thickness of 0.002mm. For simplicity the x-y size of each chip in Case 1 is set to 0.4mm×0.4mm which is smaller than most real practical applications. The Silicon thickness in the z-direction is 0.05mm. The power and ground grid are similar to those used in [15], with a width of 0.002mm and a thickness of 0.001mm, a grid pitch of 0.050mm, and a power-to-ground pitch of 0.006mm. The x- and y-grid are connected by a 0.002mm×0.002mm×0.001mm Cu block at the power-to-power or ground-to-ground crossover points. The distance between the top of Silicon and top of the grid is 0.001mm. The TSV terminated with Ports 5 and 6 is connected to the power grid in the chip. The TSV terminated with Ports 15 and 16 is connected to the ground grid in the chip.

The electromagnetic simulation tool CST Microwave Studio [4] was used for this test case. The calibration structures Components A' and B' are the same as the original structures Components A and B as shown in Fig.7 although we may use simpler calibration structures as in Cases 2 and 3 in [18]. The calibration structure by combining Components A' and B' as a one-piece structure Component A'B' is shown in Fig.8. It still makes sense to use the original component structures for the calibration structures in this 9-chip stack case, because the maximum number of chips included in electromagnetic calibration structure simulation is only two instead of nine. We can avoid running electromagnetic simulation on stacked nine chips as a whole structure.

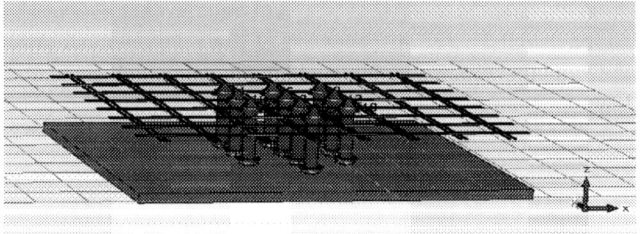

Fig.7 Component A' (the same structure as Component A)

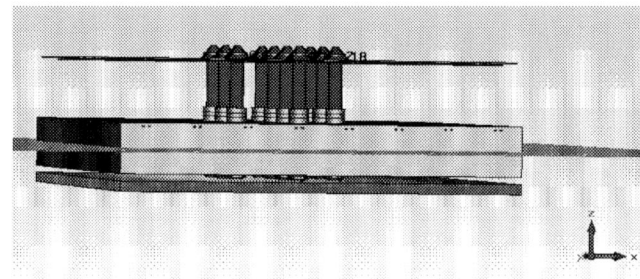

Fig.8 Component A'B' consisting of connected Components A' and B' (Silicon substrate in Component A' not shown)

One of the S-parameter components, S_{21}, of the Error-Cancelling Network and Half-Error-Cancelling Network are shown in Fig. 9. It is easy to see $|S_{21}| > 1.0$ at some frequency points. That means they are non-passive.

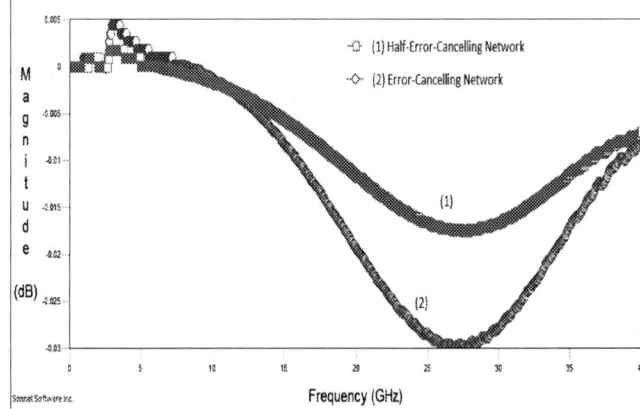

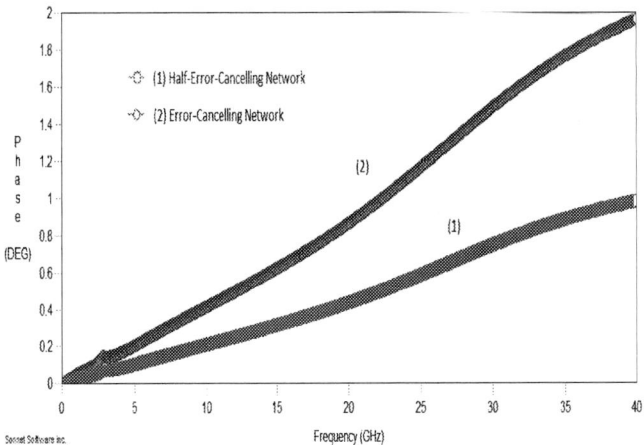

Fig.9 Simulated S_{21} of the Error-Cancelling Network and the Half-Error-Cancelling Network

Figs. 10-11 show two of the two-side de-embedded S-Parameter components, S_{11} and S_{21}, of an inner chip (one of Chips 2-8) with $|S_{21}| < 1.0$ in the whole frequency range of 0-40GHz. That means after de-embedding, the network cascaded by three networks as shown in Fig.7 can be passive although two Half-Error-Cancelling Networks are non-passive.

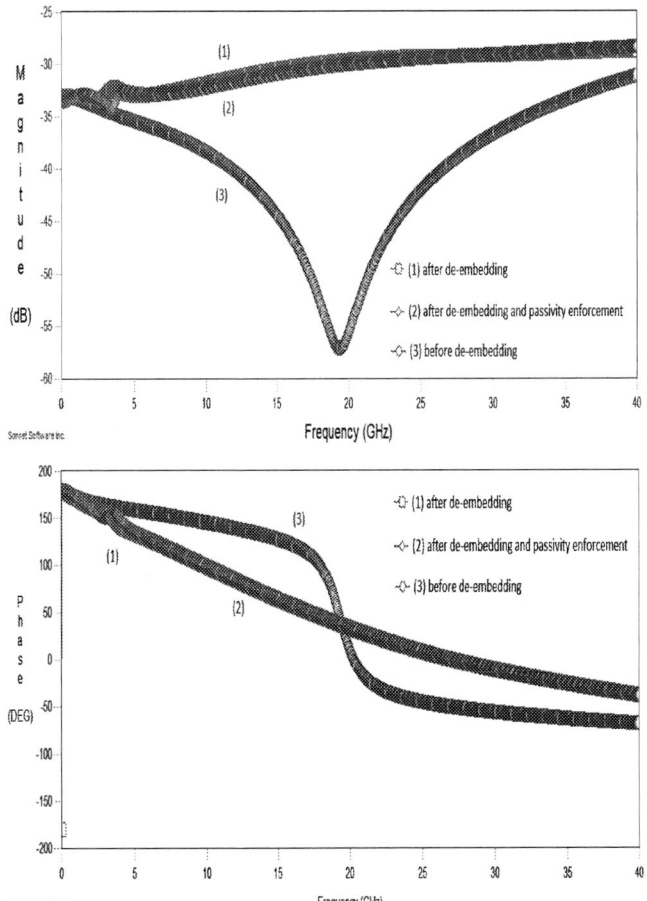

Fig.10 Simulated S_{11} of an inner chip in the 9-chip stack

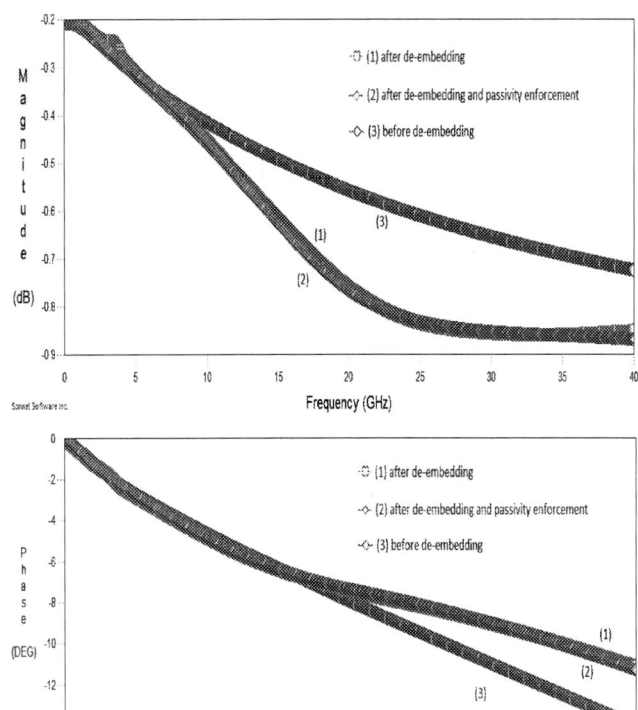

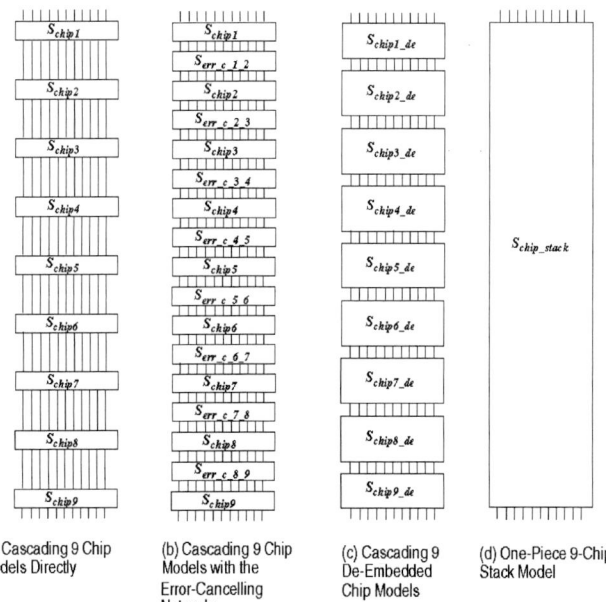

Fig.11 Simulated S_{21} of an inner chip in the 9-chip stack

(a) Cascading 9 Chip Models Directly

(b) Cascading 9 Chip Models with the Error-Cancelling Networks

(c) Cascading 9 De-Embedded Chip Models

(d) One-Piece 9-Chip Stack Model

Fig.12 Application of the proposed method by cascading 9 de-embedded chip models(c) compared with the direct connection (a), directly using the Error-Canceling Networks (b), and the one-piece structure modeling (d).

By using the proposed method as shown in Fig.12(c), the S-Parameters of the 9-chip stack are derived and compared with those by reference(Fig.12(d)) and by other methods (Figs. 12 (a) and (b)).

978-1-4799-2408-0/14 $31.00 © 2014 IEEE   727

The simulated curves by the proposed method are very close to those by the one-piece reference (Figs.13 and 14). The figures also show that the curves without lumped/discrete port de-embedding have too large error especially for amplitude of S_{11} and phase of S_{21}. That means de-embedding in this application is very effective, accurate, and absolutely necessary.

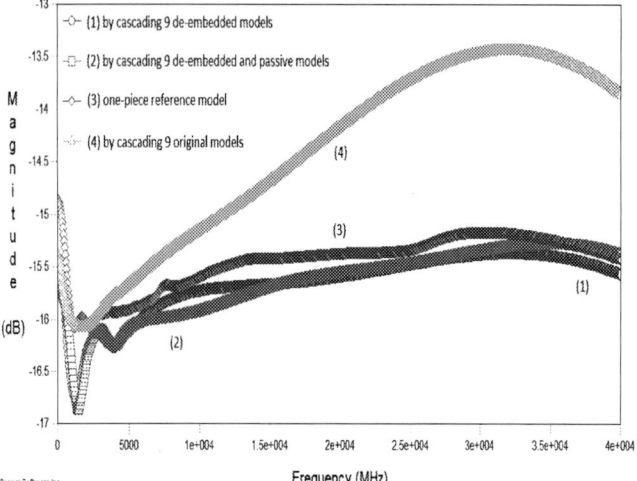

Fig. 13 Simulated $|S_{11}|$ of the 9-chip stack

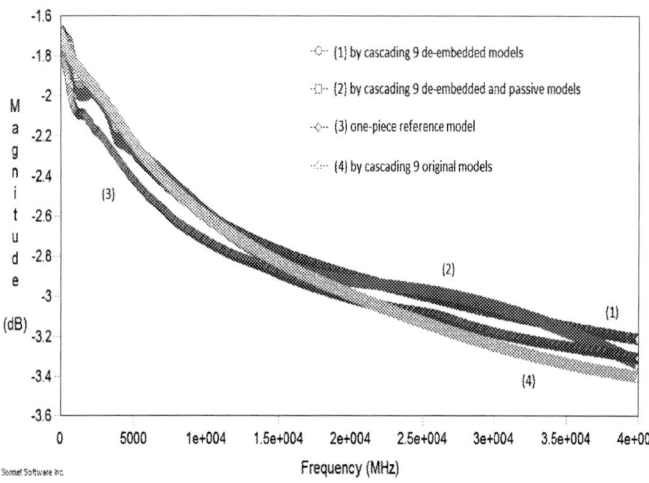

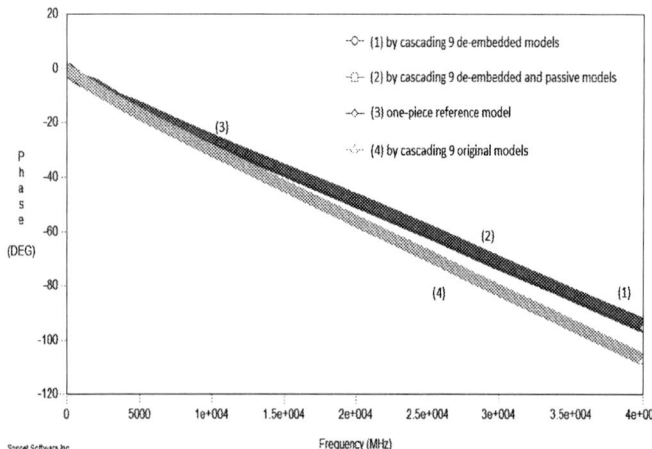

Fig.14 Simulated S_{21} of the 9-Chip Stack

4. Conclusions

In this paper, a general lumped/discrete port de-embedding method for full-wave electromagnetic modeling of 3D integration and packaging with vertical connection interfaces has been proposed. This method is based on splitting and absorbing the Error-Cancelling Network which was inserted in between two connected S-parameter models to cancel any error by the port parasitic parameters and uncertainty in port current and voltage definitions. In the proposed approach, the Error-Cancelling Network is split into two identical Half-Error-Cancelling Networks by making use of the square root of the corresponding T-matrix of the error-cancelling network. Each Half-Error-Cancelling Network is attached to and combined with the original S-parameter models to generate de-embedded S-parameters models without any additional and individual error-cancelling network in appearance. This de-embedding method is a wide-band approach valid on every frequency point in the S-parameter model.

A test case of a 20-port 9-chip stack with signal, power, and ground ports is shown for verification and application of the proposed method. The simulation results by the proposed approach have good agreement with the references. Although the test case covers only a small portion of the 3D chip in the real product, the feasibility and correctness of the proposed lumped/discrete port de-embedding method have been verified.

References

[1] Muhannad S. Bakir and James D. Meindl, Editors, *Integrated Interconnect Technologies for 3D Nanoelectronic Systems*, Artech House, Norwood, MA, 2009.

[2] J.U.Knickerbocker, P.S.Andry, E.Colgan, B.Dang, T.Dickson, X.Gu, C.Haymes, C.Jahnes, Y.Liu, J.Maria, R.J.Polastre, C.K.Tsang, L.Turlapati, B.C.Webb, L.Wiggins, and S.L.Wringht, "2.5D and 3D technology challenges and test vehicle demonstration," in *Proc. 62nd Electronic Components & Technology Conference*, San Diego, CA, May 29-June 1, 2012, pp.1068-1076.

[3] ANSYS, *HFSS 3D Full-Wave Electromagnetic Field Simulation,* http://www.ansoft.com/products/hf/hfss

[4] CST, *CST Studio Suite 2011.*

[5] Soon Wee Ho, Seung Wook Yoon, Qiaoer Zhou, Krishnamachar Pasad, Vaidyanathan Kripesh, and John H. Lau, "High RF performance TSV silicon carrier for high frequency application," in *Proc. 58th Electronic Components & Technology Conference*, Lake Buena Vista, FL, May 27-30, 2008, pp.1946-1952.

[6] Darryl Kostka, *TSV and Interposer Modeling, Design and Characterization*, CST Web Seminar, October 4, 2012.

[7] Zhaoqing Chen, "A general co-design approach to multi-level package modeling based on individual single-level package full-wave S-parameter modeling including signal and power/ground ports," in *Proc. 62nd Electronic Components & Technology Conference*, San Diego, CA, May 29-June 1, 2012, pp.1687-1694.

[8] Zhaoqing Chen, "A simple lumped port S-parameter de-embedding method for interconnect and packaging component high-frequency modeling, " in *Proc. 12th*

IEEE Workshop on Signal Propagation on Interconnects, Avignon, France, May 12-15, 2008.

[9] Barry Rubin and Shahrokh Daijavad, "Calculation of multi-port parameters of electric packages using a general purpose electromagnetic code," in *Proc. IEEE 2nd Topical Meeting on Electrical Performance of Electronic Packaging*, Monterey, CA, Oct. 1993, pp.37-39.

[10] James C. Rautio, "A de-embedding algorithm for electromagnetics," *International Journal of Microwave and Millimeter-Wave Computer-Aided Engineering*, vol.1, no. 3, pp282-287, 1991.

[11] Lei Zhu and Ke Wu, "Unified equivalent-circuit model of planar discontinuities suitable for field theory-based CAD and optimization of M(H)MIC's," *IEEE Trans. on Microwave Theory and Tech.*, vol.47, no.9, 1999, pp.1589-1602.

[12] Marco Farina and Tullio Rozzi, "A short-open deembedding technique for Method-of-Moments-based electromagnetic analysis," *IEEE Trans. on Microwave Theory and Tech.*, vol.49, no.4, 2001, pp.624-628.

[13] Vladimir I. Okhmatocski, Jason Morsey, and Andress C.Cangellaris, "On deembedding of port discontinuities in full-wave CAD models of multiport circuits," *IEEE Trans. on Microwave Theory and Tech.*, vol.51, no.12, 2003, pp.2355-2365.

[14] Zhaoqing Chen and Sungjun Chun, "Per-unit-length *RLGC* extraction using a lumped port de-embedding method for application on periodically loaded transmission lines," in *Proc. 56th Electronic Components & Technology Conference*, San Diego, CA, 2006, pp.1770-1775.

[15] Zheng Xu, Xiaoxiong Gu, Michael Scheuermann, Kenneth Rose, Buckwell Webb, John U. Knickerbocker, and Jian-Qiang Lu, "Modeling of power delivery into 3D chips on Silicon interposer," in *Proc. 62nd Electronic Components & Technology Conference*, San Diego, CA, May 29-June 1, 2012, pp.683-689.

[16] Christophe Seguinot, Patrick Kennis, Jean-Franicos Legier, Fabrice Huret, Erick Paleczny, and Leonard Hayden, "Multimode TRL-a new concept in microwave measurements: theory and experimental verification," *IEEE Trans. on Microwave Theory and Tech.* vol.46, no.5, pp.536-542, May 1998.

[17] James Frei, Xiao-Dong Cai, and Stephen Muller, "Multiport *S*-parameter and T-parameter conversion with symmetry extension," *IEEE Trans. on Microwave Theory and Tech.* vol.56, no.11, pp.2493-2504, Nov. 2008.

[18] Zhaoqing Chen, "A lumped/discrete port de-embedding method by port connection error-cancelling network in full-wave electromagnetic modeling of 3D integration and packaging with vertical interconnects," To be published in *Proc. 63th Electronic Components & Technology Conference*, Las Vegas, NV, May 2013.

[19] K.C.Gupta, Ramesh Garg, and Rakesh Chadha, *Computer-Aided Design of Microwave Circuits*, Artech House, Dedham MA, 1981.

[20] IdemWorks, *IdEM Plus*, www.idemworks.com

Design, Modeling, and Characterization of Passive Channels for Data Rates of 50 Gbps and Beyond

Wendemagegnehu Beyene, Yeon-Chang Hahm, Dave Secker, Don Mullen, Yuriy Shlepnev[1]
Rambus Inc.
1050 Enterprise Way, Suite 700
Sunnyvale, CA 94089

Abstract

The design of interconnects for links operating at *50 Gbps* and beyond is very challenging. The loss, dispersion, and discontinuities along the signaling path have to be minimized over a wide frequency range. Frequency dependent material properties and surface roughness has to be accurately considered. The impacts of short via stubs that are ignored at lower data rates can severely degrade the signals when operating at higher data rates. In order to provide ways to mitigate these effects and optimize the performance of the system, it is primarily essential to correctly model and characterize the passive channel. In this paper, the modeling and characterization techniques that guarantee successful designs of passive channels for data rates of *50 Gbps* and beyond will be presented. Detailed studies and measurement results on the effects of short via stubs are also presented.

Introduction

Memory bandwidth requirements for high performance computing, data centers, servers and storages, driven by internet in general and multi-core memory and processors architectures in particular, are rapidly increasing to meet the performance demands of today's applications [1]-[2]. The high bandwidth necessitates a large increase in the interface data rate and width.

The design of high-speed links that operate at data rate exceeding *50 Gbps* is necessary to support Terabit backplane systems. In the past, the increase the performance of the input/output (I/O) circuits and the use of more complex equalization, complicated coding and modulation and other signal processing techniques have been able to sustain the growth of data rate. However, the electronic and I/O power consumption significantly increases with increasing the interface speed. Thus, the link data rate increase cannot only come from circuit design and process improvements. To improve and extend the reach of copper-based interconnects, several improvement have been suggested including high-speed channel design using low-loss dielectric, smooth copper surfaces, improved connectors and packages [3]-[4].

The printed circuit boards (PCB) where the longest signal traces commonly found and where the signal is significantly attenuated impose limitation on the supported speed. In order to predict and optimize the performance of high-speed links operating at *50 Gbps* and beyond, it is essential to accurately model and characterize the interconnect systems, such as long traces in backplanes. The models of interconnects have to be broadband and include high frequency effects that were not critical at lower data rates [5]. For higher data rates, proper identification of the frequency-dependent properties of the

dielectric and conductor over extremely wide frequency band is required to accurately model signal propagation in PCB and package traces. In addition, 3D modeling and characterization are essential to understand and optimize the wave propagation and minimize the mismatch across various 3D transition structures such as connectors and via and BGA at the interface between boards and packages.

Low-loss laminates such, FR408HR from Isola Group, Nelco N4000-13 EPSI from Park Electrochemical Corp. and Megtron 6 from Panasonic are expected to be key enablers to design boards to operate at higher data rates. These laminates have considerably low loss at high frequencies and offer much more stable dielectric characteristics. To investigate the effect of low-loss laminates and see the impact of surface roughness, dielectric properties, glass weave effects, several boards with Isola FR408HR with Reverse-Treated Foil (RTF) finish copper foil and standard glass weave, Nelco 4000-13 EPSI with RTF copper foil and standard glass weave, and Megtron 6 with RTF finish and Hyper Very Low Profile (HVLP) finish are designed and characterized. Table I shows typical electric properties of the four low-pass laminates that are studied in this paper and the property of a typical FR-4 board for comparisons. Figure 1 shows the photos of the boards designed to characterize the traces and the dielectrics.

Table I : The board types studied and their electrical properties.

Laminate Types	Dielectric Constant	Dissipation Factor	Amplitude of Surface Roughness
Megtron 6 HVLP	3.6	0.004	1.5 – 2.0 um
Megtron 6 RTF	3.6	0.004	7.0 – 8.0 um
Nelco N4000-13 EPSI	3.2	0.008	7.0 – 8.0 um
FR408HR	3.65	0.0095	7.0 – 8.0 um
Typical FR-4	4.3	0.02	7.0 – 8.0 um

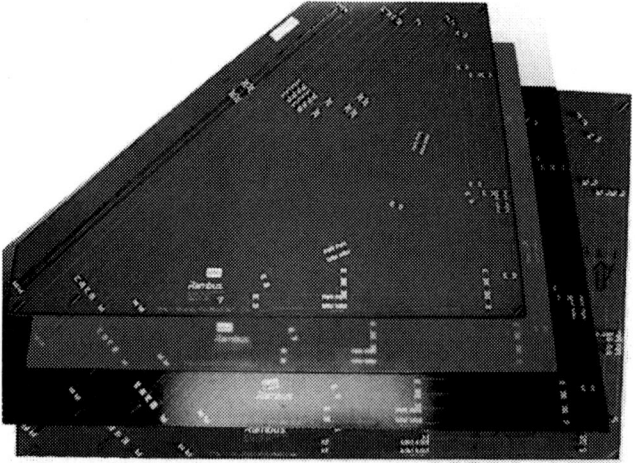

Figure 1: The photo of the four boards that are designed to extract trace models for low-loss boards.

---

[1] Yuriy Shlepnev is with Simberian Inc.

978-1-4799-2408-0/14 $31.00 © 2014 IEEE

For material characterization, several microstrips and striplines of various lengths, as well as de-embedding structures with probe pads and connectors are designed. The pad to pad links of two set of differential nets of *6-in.* and *12-in.* long striplines, shown in Figure 2, are used to characterize the low-loss materials and traces.

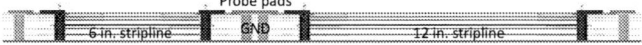

Figure 2: The cross section of the 8-layer board showing the *6 in.* and *12 in.* long striplines on the 4th layer.

First, the manufactured boards are cross-sectioned to accurately verify all the dimensions of the transmission lines. Obtaining the accurate dimensions of the cross-sections is very critical to generate accurate models of the striplines. Figure 3 (a), (b), (c) and (d) show the cross-section of the boards: Isola's FR408HR, Nelco N4000-13 EPSI, Megtron 6 with RTF and HVLP finishes, respectively. The dimensions for the conductor thickness, width, the trace spacing, and the top and bottom layer heights are all marked in microns (μm).

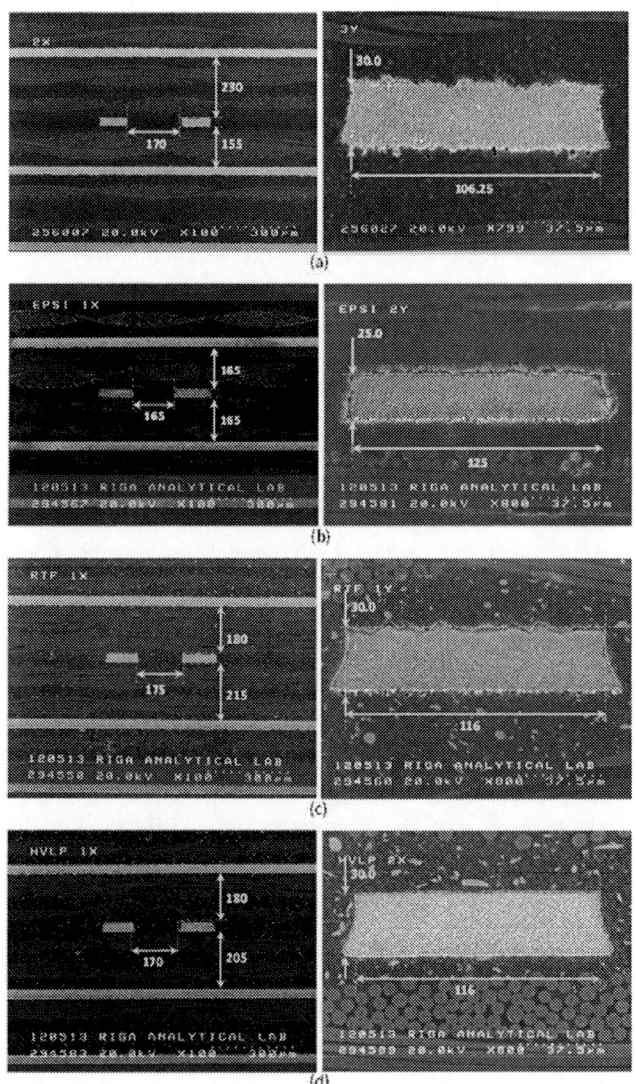

Figure 3: The Micro-photographs of (a) FR408H, (b) Nelco N4000-13EPSI, (c) Megtron 6 RTF, (d) Megtron 6 HVLP boards showing cross-sections of the traces and copper roughness (dimensions are in μm).

Using high-frequency probes with *200*-μm pitch and GSSG configuration and connectors, several scattering parameter measurements are performed using a *4*-port *67-GHz* vector network analyzer (VNA). The measured S parameters, using GSSG probes, of the two sets of differential nets with *6-in.* and *12-in.* long traces are used to characterize the FR408HR, Nelco N4000-13 EPSI, Megtron 6 with RTF, and HVLP finish board traces. Figure 4 (a) and (b) show the measured differential and common mode insertion loss for the *12 in.* traces of the four boards, respectively. The simulated insertion losses of similar structures in FR-4 boards are also plotted for comparisons. The plots show attenuations that agree well with the electrical properties of these laminates given in Table I. The measured differential insertion loss of the Megtron 6 with HVLP finish shows about *2 dB* improvement over that of the Megtron 6 with RTF finish at *25 GHz*. The Megtron 6 with HVLP finish also shows about 4 dB and *6 dB* improvements over Nelco N4000-13 EPSI and FR408HR, respectively. The *12-in.* trace in Megtron 6 with HVLP laminate shows about *20 dB* improvement in attenuation when compared to similar trace in FR-4 board.

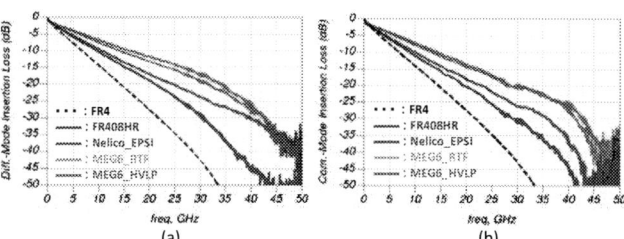

Figure 4: Measured insertion losses, (a) differential and (b) common-mode for the four boards: Megtron 6 HVLP, Megtron 6 RTF, Nelco N4000-13 EPSI, and FR408HR, and typical FR-4 board from simulation.

The measured four-port S-parameters are used to calculate the differential group delays of the *12-in.* traces. The simulated group delay for FR-4 board is also included in the plots. The delays per inch of the four boards are plotted as functions of frequency in Figure 5 (a). The typical FR-4 shows the longest delay as predicted from its higher dielectric constant. The Nelco N4000-13 EPSI exhibits the smallest delay as expected from the dielectric constant value of this laminate given in Table I.

The measured S-parameters are also used to obtain time-domain responses using the time-domain engine of Advanced Design System (ADS) [6]. In order to calculate the single-bit response of the link, an excitation of a pulse with amplitude of *1 V* and width of *20 ps* (corresponding to a data rate of *50 Gbps*) and rise and fall time of *8 ps* is used at the launch pads. Figure 5(b) shows that the single-bit response of FR-4 suffered the larger attenuation and edge degradation closely followed by the single response of FR408HR when compared to the Megtron 6 boards. On the other hand, the single-bit responses of the Megtron 6 board experienced the smallest attenuations as predicted by glancing at the differential insertion loss shown in Figure 4(a). Although the single-bit response of the Nelco N4000-13 EPSI suffered similar attenuation and dispersion as FR408HR, it experienced the smallest delay due to its low dielectric constant.

978-1-4799-2408-0/14 $31.00 © 2014 IEEE   731

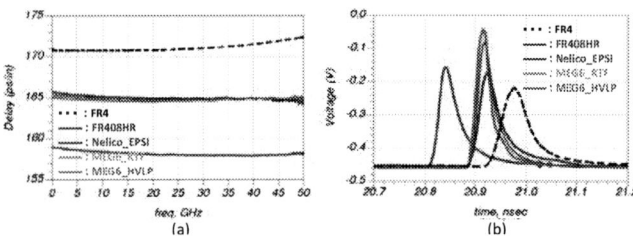

Figure 5: The group delays (delays per inch) and the single-bit responses (at *50 Gbps*) of the *12-in.* long traces generated using the measured scattering parameters of the four boards: Megtron 6 HVLP, Megtron 6 RTF, Nelco N4000-13 EPSI, and FR408HR and FR-4 (from simulation).

Next, the broadband dielectric and conductor roughness models are extracted using the generalized modal S-parameters obtained from the measured S-parameters of the two stripline of length of *6 in.* and *12 in.* [7]-[8]. The detail of the procedure is described in the next section. The ability to properly identify conductor and dielectric losses for a wide range of cross-section and low-loss dielectrics is very critical to construct models that correlate with measured data for wide-range of cross-sections often used in interconnect design. The accurate transmission line and 3D transition models are then used to perform analysis and optimization of high-speed interconnects operating at *50 Gbps.*

Broadband Transmission Line Models

The long striplines and microstrips, commonly found in backplanes and daughter cards, can be modeled and simulated as transmission line segments. Models for transmission lines are usually generated using 2D or 3D electromagnetic field solvers. Transmission lines, such as striplines, with homogeneous dielectrics can be accurately analyzed with quasi-static field solvers whereas lines with inhomogeneous dielectric may require analysis with a full-wave electromagnetic solver to account for the high-frequency dispersion. The broadband dielectric and conductor roughness models are necessary to accurately model transmission lines in low-loss laminates for high data operations. Wideband Debye (Djordjevic-Sarkar) and multi-pole Debye models are successfully used to model the dispersive dielectric properties in PCB and packaging interconnects [9]. These models are casual and passive. Frequency dependent complex dielectric constant of the multi-pole Debye model is defined as,

$$\varepsilon(f) = \varepsilon(\infty) + \sum_{n=1}^{N} \frac{\Delta\varepsilon_n}{1 + i\frac{f}{f_{rn}}},$$

where f_{rn} and $\Delta\varepsilon_n$ are real poles and residues, and $\varepsilon(\infty)$ is the dielectric constant value at infinite frequency. All parameters for such models are not readily available from manufacturers and have to be identified. Manufacturers of dielectric laminates usually provide dielectric parameters at few frequency points at the most. These frequency points can be used as initial parameters in the wideband Debye model.

To simulate the effect of conductor roughness of PCB traces, several models are available. Huray's snowball [10] and modified Hammerstad [11] are used to successfully model conductor roughness at high data rates. The modified Hammerstad formula is defined as,

$$k_{sr} = 1 + \left(\frac{2}{\pi} \cdot \arctan\left[1.4\frac{\Delta}{\delta_s}\right]\right) \cdot (RF - 1),$$

where δ_s is the skin depth, Δ is the surface roughness RMS peak-to-valley height and RF is the surface roughness factor. Parameters for these electrical roughness models are not usually available either from the PCB manufacturers or from manufacturers of the copper foils.

Thus, the material characterization and trace modeling must first start with the identification of dielectric constant using the values given by the manufacturer and then generating the dielectric and conductor roughness models over the frequency band of interest. Successfully generating accurate broadband material models is very critical to design high-speed links in general and passive channels in particular. Extraction and validation of dielectric and conductor models are performed by means of generalized modal scattering-parameters (GMS-parameters) using the Simbeor Electromagnetic Signal Integrity Software [12]-[14]. The basic identification procedure for the dielectric and conductors surface roughness models is illustrated in Figure 6.

The following are the key steps in using the procedure in Simbeor software successfully:

- Measure good quality scattering parameters for two different lengths of line segments.
- Validate the quality of the scattering parameters in terms of reciprocity, passivity, causality, and symmetry
- Verifying the cross-sections of the board, thickness of the conductor and laminates, and length of the line segments
- Accurately de-embed or model the probe launches and via transitions (including the via stubs)
- Extract the generalized S parameters of the line segment difference, L, from the two measured S parameters.
- Compute the generalized modal S-parameters of line segment difference, L, using electromagnetic field solver.

The procedure begins by accurately measuring the S-parameters of for *6 in.* and *12 in.* differential stripline segments with the GSSG differential probes for all four boards. The quality of the measurement and calibration are verified for passivity and reciprocity. We observed that most of the scattering parameters have non-symmetrical properties even though the test structures are designed to be symmetrical. The fiber weave effect caused by difference in dielectric properties of glass fiber and fill resin may be possible source of the asymmetry. The differences in the geometry of the launches, due to unequal depths of back drilling that result in different lengths of stub vias at the ends of the striplines, can be another source of asymmetry in the measured S-parameters. In addition, the inhomogeneity of line segments and non-idealities of the launches can reduce the frequency range of the extracted GMS-parameters.

Two models have been identified for all four boards, one with all losses included into dielectric model and shown in Table II. To construct another model we followed procedure of the loss effect separation suggested in [13]. Loss tangent in dielectric model was set equal to the value provided in the material specifications. Then, modified Hammerstad model for conductor roughness [11] was identified by matching generalized insertion loss. The results are listed in Table III.

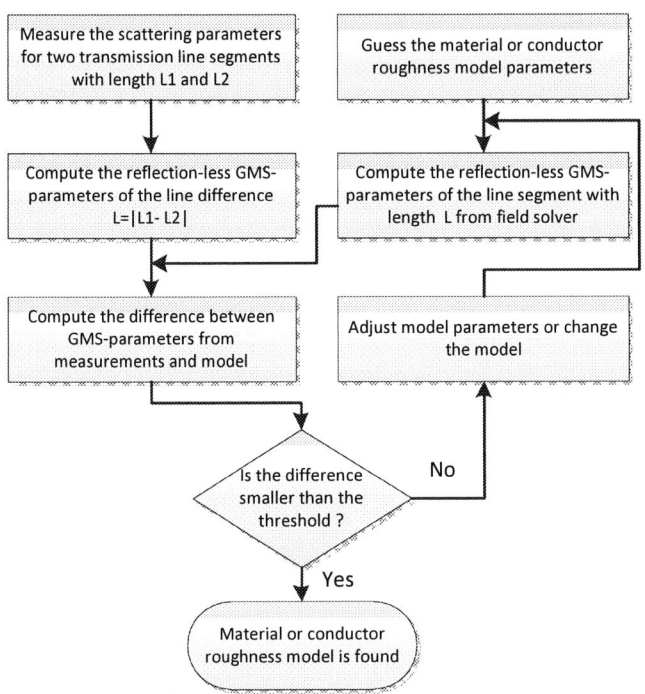

Figure 6: Model Extraction procedure for dielectric material and surface roughness models.

Table II: Wideband Debye dielectric model with conductor roughness effect included.

Board Types	Dielectric Constant @ 1 GHz	Loss Tangent @ 1 GHz
FR408HR with RTF copper	3.76	0.012
FR408HR with RTF copper, inhomogeneous	3.95, 3.5	0.01, 0.012
Megtron-6 with HVLP copper	3.69	0.0065
Megtron-6 with RTF copper	3.75	0.0083
Nelco N4000-13EPSI with RTF copper	3.425	0.011

Table III: Wideband Debye dielectric model and modified Hammerstad model for conductor roughness.

Board Types	Dielectric Constant @ 1 GHz	Loss Tangent @ 1 GHz	Surface Roughness, rms (um)	Surface Roughness Factor
Megtron-6 with HVLP copper	3.64	0.002	0.38	3.15
Megtron-6 with RTF copper	3.72	0.002	0.37	4
Nelco N4000-13EPSI with RTF copper	3.425	0.008	0.49	2.3

Finally, the probe to probe link path is closely examined to study some of the discrepancies that are discussed in the previous sections. The launch structures at both ends of the traces that include via, via stubs and pads are examined carefully. In order to include the impact of the launch structures, the scattering parameters of the via and pad structures are generated using full wave electromagnetic solver. The side view of the via and pad structures with four different stub length are shown in Figure 7. The frequency responses of these structures are analyzed as a function of the stub lengths. The corresponding differential insertion loss and return loss are shown in Figure 8 (a) and (b), respectively. It is important to notice that in order to extend the bandwidth of the launch structure to *50 GHz* range, the lengths of the via stubs need to be short. Commonly accepted tolerance of

minimum stub lengths of *7±3 mil*, after back drilling, can be limiting factor in accurately measuring the stripline characteristics at higher frequencies.

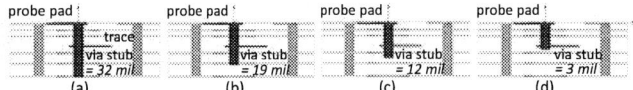

Figure 7: The launch structures and the stub vias from 3D EM solver.

After careful inspection of the board under microscope, it is found that there are signifcant variations in the length of the via stubs. Digital microscope images of the back-drilled vias are taken to accurately determine the length of the via stubs. Figure 9 shows the top view and 3D images from a digital microscope of the back-drilled vias (holes).

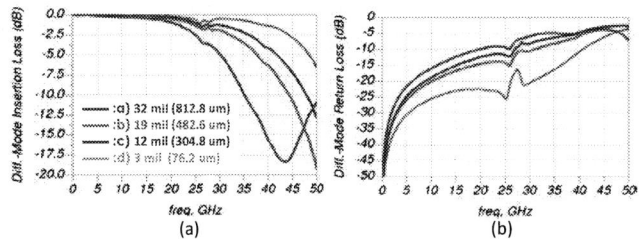

Figure 8: Scattering parameters of the lunch structures as a function of via stub length: (a) Differential insertion and (b) differential return loss.

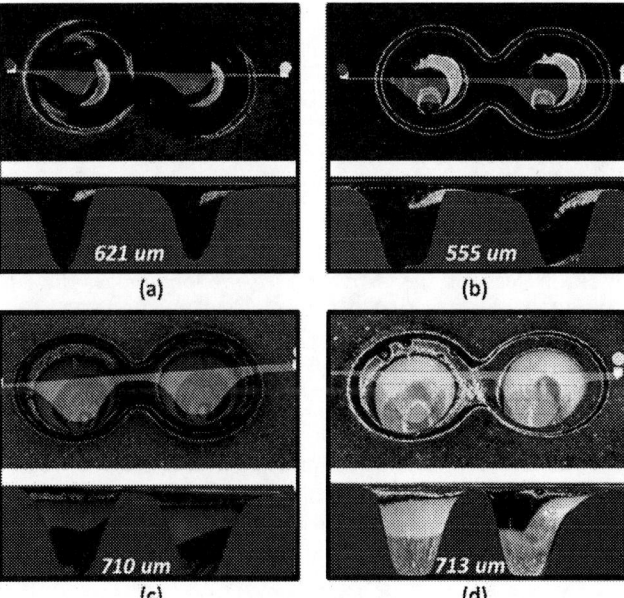

Figure 9: 3D images from a digital microscope of the back-drilled vias (holes) showing the top view and cross-sections for (a) FR-408HR, (b) Nelco-N4000 13EPSI, (c) Megtron 6 RTF, and (d) Megtron 6 HVLP.

The lengths of via stubs are estimated from the board cross-section of Figure 2 and the back-drilled via profile of Figure 9, for the four boards as shown in Table IV. Notice that the back-drilled holes are significantly different from board to board. We also noticed that the back-drilled holes show different depth even on the same board for vias very close to each other, as shown on Figure 9 (a). For example, the Megtron 6 RTF board shows an average back-drilled depth of 713 um, the minimum and maximum depths of

681um and 748 um, respectively. In general, there are some via stubs that are successfully back drilled, some are partially back-drilled and some are not back-drilled or entirely missed.

Table IV: The board thickness, via and via stub length in microns (μm).

Board Types	Board Thickness	Pad to trace Via length	Back drill depth	Via stub length
FR408 HR	1470	600	620	250
Nelco N4000-13 EPSI	1360	540	555	265
Megatron 6 RTF	1400	560	710	130
Megtron 6 HVLP	1400	560	713	127

Based on the via stub length, the S-parameters of for the via and pad structure obtained from full-wave electromagnetic solver are cascaded to both ends of the trace model as shown in Figure 10. The trace model consists of transmission line models based on the parameters extracted for the four boards. The frequency-domain simulations of the complete links are carried out in ADS using the transmission line models with extracted parameters listed in Table III [6]. Then, the measured and simulated S-parameters of the overall S-parameters of both *6 in.* and *12 in.* long traces are compared for multiple trace lengths and boards to verify the quality of the extracted trace models.

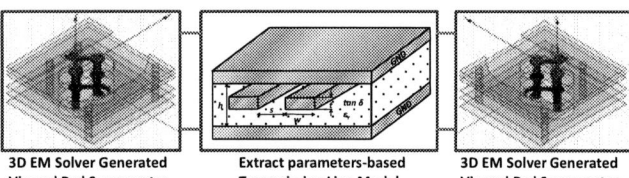

3D EM Solver Generated Via and Pad S parameter   Extract parameters-based Transmission Line Model   3D EM Solver Generated Via and Pad S parameter

Figure 10: Cascade of the S-parameters of the via and pad structures at both ends and the traces and the extracted transmission line model.

The magnitude and phase (unwrap) of the differential insertion loss of the *6-in.* and *12-in.* long traces for the FR408HR, Nelco N4000-13EPSI, Megtron 6 RTF, and Megtron 6 HLVP boards are shown in Figures 11 (a), (b), (c), and (b), respectively. As shown in the plots, the measurements and the extracted model parameters for the four boards show good agreements over a wide frequency range. The minor divergence of the measurement and model results at very high frequency can be attributed to the uncertainty in the length of the remaining via stubs.

Complete Chip-to-Chip Channel

To further validate the trace models, a chip-to-chip interconnect system with several channels of different lengths and structures are designed in the boards and packages. The differential nets consisting of PCB and package traces and transitions are measured using a VNA up to *50 GHz*. The transmitter and receiver packages (i.e. without the silicon devices mounted) are soldered down onto the PCB as shown in Figure 12. Two different-length differential pairs are analyzed. The shorter pair consists of PCB trace of *109-mm* long and package trace of *9-mm* on both ends. The longer pair consists of PCB trace of *293-mm* long and package trace of *13-mm* on both ends. The four-port scattering parameters of

the differential nets are measured at the transmitter and receiver package pads using GSSG probes.

When the back-drilled vias of the high-speed nets are careful investgated, signifcant variations in the lengths of the via stubs are also observed in the chip-to-chip board. In order to accurately determine the length of the via stubs, digital microscope images of the back-drilled vias are also taken. Figure 13 shows the top view and 3D images from a digital microscope of the back-drilled vias holes. Most nets consists of one or more incorrectly processed back-drilled vias. Figure 13(b) shows the carefully removed back-drilled vias stubs under a microscope. Repeated measurements are performed before and after manually removing the remaining stub vias after the back drill.

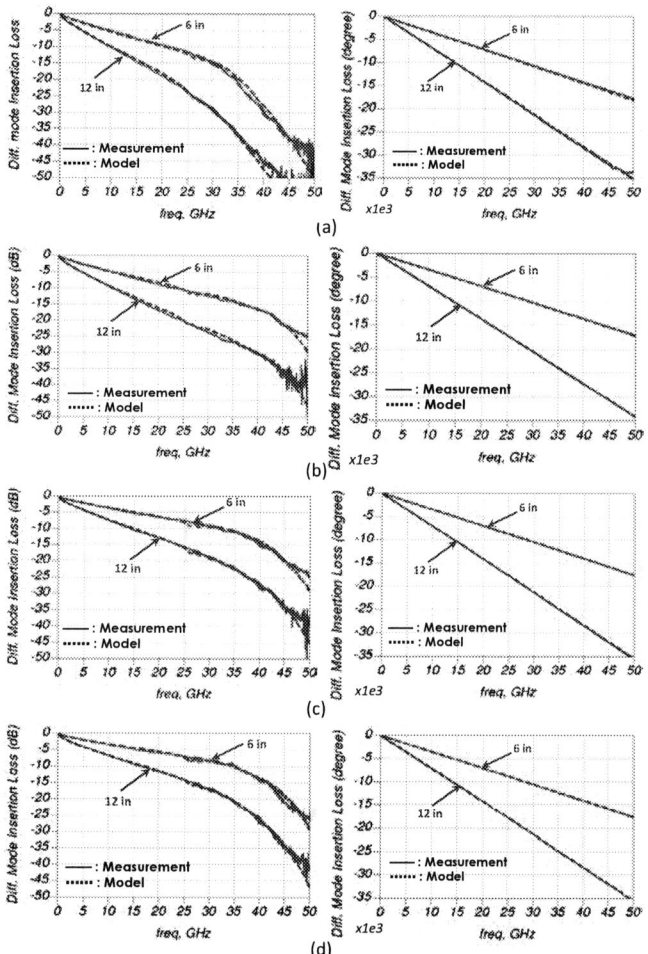

Figure 11: Magnitude and phase of the measured and simulated differential insertion loss for the four boards: (a) FR408HR, (b) Nelco N4000-13EPSI, (c) Megtron 6 with RTF, and (d) Megtron 6 with HVLP finishes.

The channel insertion losses before and after removing the reaming stub vias are shown in Figure 14 for the short and long nets. The via stubs severely degrade the signals by creating resonances and reflections at frequencies 18 GHz and 20 GHz, shorter and longer nets, respectively. Removing the remaining via stubs significantly improves the insertion loss of both short and long nets.

978-1-4799-2408-0/14 $31.00 © 2014 IEEE

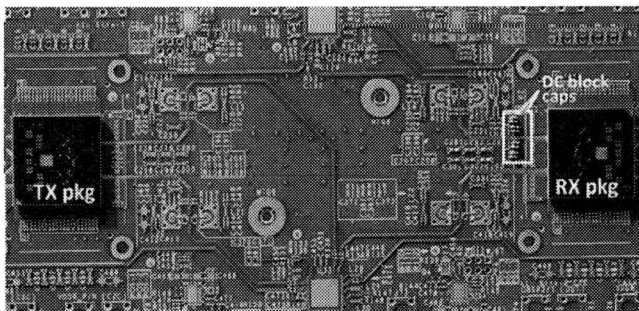

Figure 12: A passive chip-to-chip interconnect system consisting of several different pairs of different length.

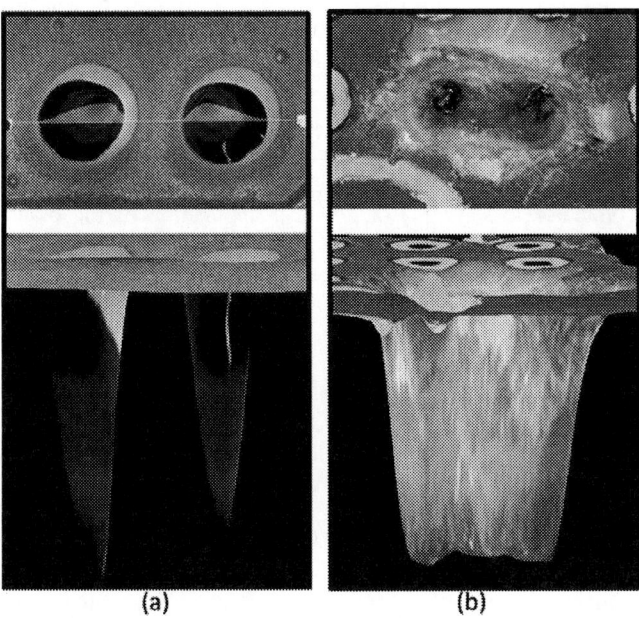

Figure 13: 3D images from a digital microscope of the back-drilled vias (holes) of the chip-to-chip interconnect board shown in Figure 10. (a) An example of the back-drilled via holes and (b) manually back drilled to remove stub vias on a differential net.

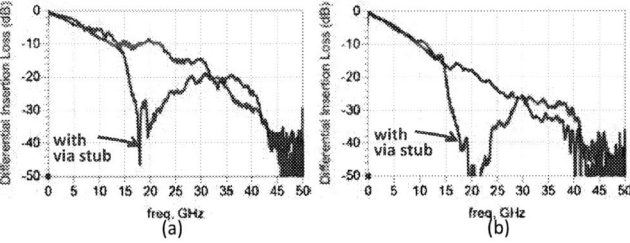

Figure 14: Insertion loss of differential nets with and without via stubs: (a) 109 mm and (b) 293 mm PCB traces

Conclusions

The design of passive interconnect systems for links operating at *50 Gbps* and beyond is very challenging; as the loss, dispersion, and discontinuities along the signaling path have to be minimized over a wide frequency range. Several low-loss PCB laminates are characterized over wide frequencies. It is shown that the low-loss boards can make it possible to extend the copper-based backplane links to *50 Gbps*. Using the low-loss board, for example, the loss of a *12-in.* PCB trace is improved by over *20 dB* when compared to FR-4 board. It is also shown that even short via stubs can significantly degrade high-speed signals. The problem can be reduced either by eliminating those stubs or reducing stub length The short stubs due to typical back-drilling tolerance can impact channel performance at *50 Gbps* and higher data rates. The practice of minimization of via stubs by use of blind and buried vias needs to be carefully considered as the resulting resonances and discontinuities can adversely impact the link performance.

References

1. A. F. Benner, *et al.*, "A roadmap to 100G Ethernet at the enterprise data center," *IEEE. Communications Magazine,* vol. 45, no. 11, pp. 10-17, Nov. 2007.

2. B. Casper, *et al.,* "Future microprocessor interfaces: analysis, design and optimization," *IEEE Custom Integrated Circuit Conf.,* pp. 479-486, 2007.

3. R. Kollipara, et al., "Practical design considerations for 10 to 25 Gbps copper backplane serial links," in *DesignCon,* Santa Clara, CA, Feb. 6-9, 2006.

4. H. Braunisch, *et al.,* "High-speed flex-circuit chip-to-chip interconnects," *IEEE Trans. on Advanced Packaging Technologies*, vol. 31, no.1, pp. 82-90, Feb. 2008.

5. W. T. Beyene, *et al.*, "Advanced modeling and accurate characterization of a 16 Gb/s memory interface," *IEEE Transactions on Advanced Packaging Technologies*, vol. 32, no.2, pp. 306-327, May 2009.

6. Agilent Technologies, "Advanced Design System, ADS" http://www.home.agilent.com/en/pc1297113/advanced-design-system-ads?&cc=DZ&lc=eng.Inc.

7. Shlepnev, Modeling frequency-dependent conductor losses and dispersion in serial data channel interconnects, Simberian Application Note #2007_02, http://www.simberian.com/AppNotes.php

8. Y. Shlepnev, Modeling frequency-dependent dielectric loss and dispersion for multi-gigabit data channels (with experimental validation), Simberian Application Note #2008_06, http://www.simberian.com/AppNotes.php

9. Djordjevic, R. M. Biljic, V. D. Likar-Smiljanic, T. K. Sarkar, Wideband frequency domain characterization of FR-4 and time-domain causality, *IEEE Trans. on EMC*, vol. 43, N4, 2001, p. 662-667.

10. P. G. Huray, O. Oluwafemi, J. Loyer, E. Bogatin, X. Ye, Impact of Copper Surface Texture on Loss: A Model that Works, DesignCon 2010.

11. Y. Shlepnev, C. Nwachukwu, Roughness characterization for interconnect analysis., *Proc. of the 2011 IEEE International Symposium on Electromagnetic Compatibility*, Long Beach, CA, USA, August, 2011, p. 518-523.

12. Y. Shlepnev, System and method for identification of complex permittivity of transmission line dielectric, US Patent #8577632, Nov. 5, 2013, Provisional App. #61/296237 filed on Jan. 19, 2010.

13. Y. Shlepnev, System and method for identification of conductor surface roughness model for transmission lines, Patent Pending, App. #14/045,392 filed on Oct. 3, 2013.

14. Simbeor Electromagnetic Signal Integrity Software, www.simberian.com

Low Loss Conductors for CMOS and Through Glass/Silicon Via (TGV/TSV) Structures Using Eddy Current Cancelling Superlattice Structure

Arian Rahimi and Yong-Kyu 'YK' Yoon
Department of Electrical and Computer Engineering, University of Florida
Gainesville, FL 32611, USA
arahimi@ufl.edu, ykyoon@ece.ufl.edu

Abstract

Superlattice structures consisting of non-ferromagnetic/ferromagnetic metals have been used to create high performance conductors for radio frequency (RF) transmission lines and low loss vias in CMOS and through silicon/glass via (TSV/TGV) structures, whose ohmic resistance and RC delays have been greatly reduced. Two permalloys of $Ni_{80}Fe_{20}$ and FeCo are studied as the ferromagnetic materials with low and high magnetization saturation that can be used for designing superlattice structures with low and high GHz frequency ranges, respectively. The effects of design parameters including the number of layers and thickness ratio of the superlattice structures have been studied. Full wave simulations have been used to verify them. Finally, a radial superlattice structure consisting of Cu/NiFe layers has been implemented and its resistance has been compared with the control solid-core devices made of copper; proving the effectiveness of the proposed radial superlattice structure for reducing the RF loss.

Introduction

The operation frequency of monolithic integrated circuits has been growing up and reaching the GHz range in modern communication and consumer electronic applications and the clock frequency of today's microprocessors has reached 3 GHz and tends to move to higher frequencies [1]. One of the limiting factors of the high frequency operation is the radio frequency (RF) loss including the dielectric and conductor loss which are associated with devices and circuits operating in the RF range. The mentioned losses together with the inherent parasitic capacitance in those circuits result in the so called RC time delay, preventing the operating frequencies from going higher. The dielectric loss would be reduced by locally removing the dielectric materials forming air-lifted architectures ([2-3]) or using very low loss dielectric materials [4]. Meantime, most electrodes or interconnectors are adopting copper as a conducting material because of its low electrical resistivity, ease of deposition and moderate cost. However, in higher frequencies, its higher conductivity is not as effective and beneficial as it is in the low frequency and DC operation because of the skin effect, where most current is confined in the outermost surface of the conductor and therefore the volume of the conductor is underutilized. As a result, the effective cross section is reduced in RF frequencies and the resistance and conductor loss are increased. Therefore, the high frequency interconnects, transmission lines and vias in a standard CMOS process and in through silicon/glass via (TSV/TGV) structures suffer from large conductor loss. The larger conductor loss will also be significant in high speed digital circuits including analog-to-digital/digital-to-analog converters and processors which will lead to a substantially large RC delays and limit the maximum operation frequency [5].

The theory of superlattice structure is proposed in [6] to reduce the conductor loss by forcing the current to flow inside the volume of the conductor instead of its edges. In the planar superlattice structure, the multiple layers of ferromagnetic/non-ferromagnetic metals are used as the conductor where negative permeability of the ferromagnetic material has the effect of cancelling out the eddy currents inside the conductor and allows the current to flow inside the volume to reduce the conductor loss. In [7], a practical application of the planar superlattice structure has been showed where RF inductors using a planar multilayered superlattice structure are fabricated and skin effect suppression and increased quality factors have been demonstrated. In [8] and [9], multilayer interconnects with reduced loss have been reported where the loss and quality factor of a coplanar waveguide (CPW) transmission line has been improved. Recently, we have reported a skin effect suppression using a cylindrical radial superlattice (CRS) architecture in the microwave coaxial cable setup [10], where the round shape conductors benefiting from the skin effect suppression have been demonstrated. In this work, low conductor loss cylindrical via architectures are demonstrated using the CRS scheme consisting of alternating nanoscopic non-ferromagnetic/ferromagnetic conductors and the design procedure is detailed for the high frequency TSV/TGV and CMOS via usage.

Theory and Analysis

At higher frequencies, the ohmic loss and consequently the RC delays of conductors are totally governed by the conductivity (σ) and the skin depth (δ),

$$\delta = \sqrt{\frac{2}{\omega\mu\sigma}} \qquad (1)$$

where ω is the angular frequency, μ is the permeability of the conductor and σ is its conductivity. Therefore, by setting the μ inside the conductor to a value very close to zero, it would be possible to enlarge the skin depth. By assuming the multiple non-ferromagnetic/ferromagnetic layers as the conductor carrying the high frequency signal, the effective permeability of the stack layers is,

$$\mu_{eff} = \frac{\mu_N t_N + \mu_F t_F}{t_N + t_F} \qquad (2)$$

where μ_N and μ_F are the permeability of the non-ferromagnetic and ferromagnetic metals, and t_N and t_F are their thicknesses, respectively. The permeability of a thin film ferromagnetic material is given by:

$$\mu = \mu' + j\mu'' \qquad (3)$$

where μ' is the real part and μ" is the imaginary part of the permeability. The dynamic response of magnetic thin film has been investigated theoretically [11] and experimentally [12]. The Landau-Lifshitz-Gilbert (LLG) equation is used to estimate the dynamic response of the ferromagnetic thin film by which the complex permeability of the thin film can be given by:

$$\mu = \left\{ 1 + \gamma^2 4\pi M_s + \frac{[H_{Kp} - H_{Ku} + 4\pi M_s + j\omega\alpha/\gamma]}{\gamma^2 H_{Kp}[H_{Kp} - H_{Ku} + 4\pi M_s] - \omega^2 + j\omega\alpha\gamma[2H_{Kp} - H_{Ku} + 4\pi M_s]} \right\} \times \frac{\tanh[(1+j)t/(2\delta)]}{(1+j)t/(2\delta)} \quad (4)$$

where M_S is the magnetic saturation of the thin film, H_{Kp} is the in-plane anisotropy field, H_{Ku} is the out-of-plane anisotropic field, γ is the gyromagnetic ratio, α is the Gilbert damping parameter, and t is the thickness of the thin film. The skin depth, δ, is calculated by using the electrical conductivity (σ).

The device is operating between the ferromagnetic resonance frequency (f_{MR}) and the anti-resonance frequency (f_{AR}) where the real part of the permeability, μ', is negative. Therefore, the t_N/t_F is an important design parameter and by properly choosing the thickness ratio based on (2), it would be possible to make the effective magnetic permeability close to zero and enlarge the skin depth, resulting in the reduction of the conduction loss in the frequency range of interest.

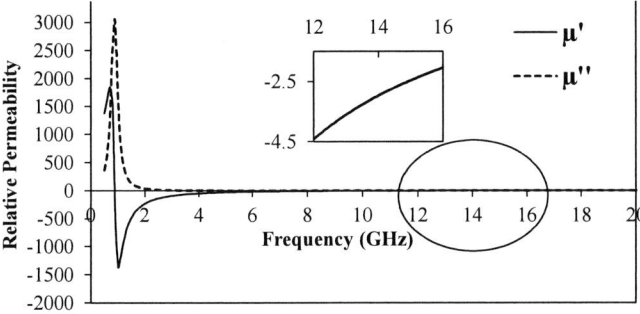

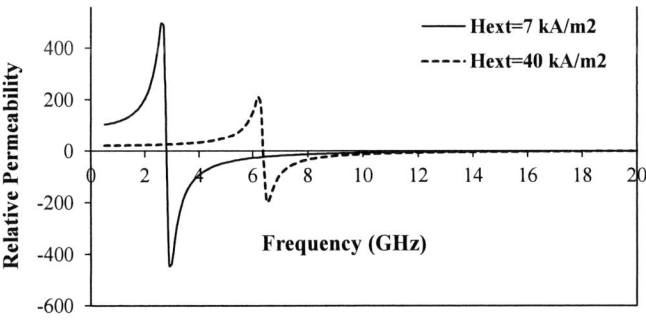

Figure 1. (a) The calculated permeability of the Permalloy assuming no external magnetic field. The inset shows the real part μ' for the frequency between 12 to 16 GHz, (b) the calculated real part of the permeability for different external magnetic fields.

Permalloy ($Ni_{80}Fe_{20}$) thin films are one of the most commonly studied soft magnetic materials. Fig. 1 (a) shows the calculated permeability of the $Ni_{80}Fe_{20}$ using the LLG equation where $M_s = 10000\ G = 1\ T$ for permalloy is the magnetic

saturation, $H_{Kp} = 10\ Oe$ the in-plane anisotropy field, $H_{Ku} = 10\ Oe$ the out-of-plane anisotropic field, $\gamma = 1.75 \times 10^7\ s^{-1}Oe^{-1}$ the gyromagnetic ratio, $\alpha = 0.01$ the Gilbert damping parameter, and t the typical thickness of the thin film. The skin depth, δ, is calculated by using the electrical conductivity of permalloy and copper where $\sigma_{permalloy} = 6 \times 10^6\ S/m$ and $\sigma_{Cu} = 5.8 \times 10^7\ S/m$, respectively. The resonance frequency of permalloy thin film, $f_{MR} \approx 900\ MHz$ and the anti-resonance frequency is $f_{AR} \approx 28\ GHz$.

Fig. 1 (b) shows the tenability of the permeability of the ferromagnetic materials by using an external magnetic field. The f_{MR} frequency shifts to higher frequencies as the external magnetic field increases. Therefore, the μ_F in (2) also shifts and it would be possible to design the conductor for multiple target frequencies.

Fig. 2 (a) shows the conductor using a planar superlattice structure where the high frequency current is flowing in the longitudinal y direction with transvers magnetic field flowing along each layer in z direction. The negative permeability of the ferromagnetic layers operating in the design frequency results in a longitudinal direction eddy current opposite to the one generated by metal layers with positive permeability. Since the forward and reverse eddy currents could be manipulated to the extent that they could be evenly cancelled in each layer, the skin depth and the effective area will be enlarged, the alternating current flow will be more uniform, and the microwave resistance at the designed frequency range will be reduced.

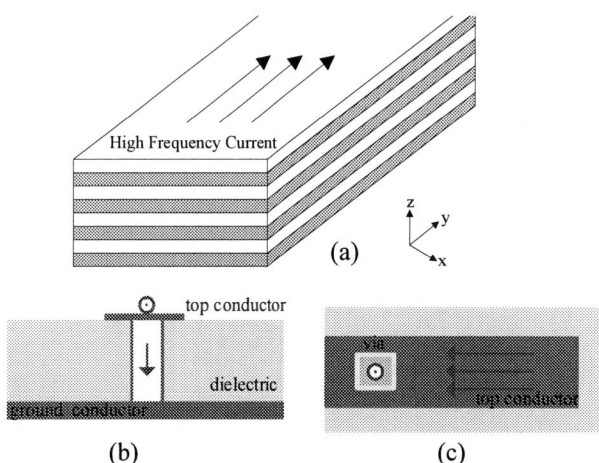

Figure 2. (a) The planar-type superlattice conductor, (b) the cross section view of a microstrip line, and (c) the top view of the microstrip line transmission line. The arrows show the direction of the high frequency current to be carried by the horizontal conductors and vertical vias.

Figs. 2 (b) and (c) show a microstrip transmission line with a via connecting the top conductor to the bottom ground plate. The vias in a standard CMOS or TSV/TGV process are vertically connecting the horizontal conductors. Therefore, in order to benefit from the eddy current cancelling structure of the conductors for the vias, one should vertically grow the stack layers inside the via opening. Deposition of the metal layers vertically on the walls of the via might involve some uniformity issues and not be compatible with some microfabrication processes. Also, those planar superlattice conductors have

978-1-4799-2408-0/14 $31.00 © 2014 IEEE

electromagnetic discontinuity at the edge of the conductor and therefore large fringing effects, and the eddy current suppression of the planar superlattice conductor and vias might be limited in practice.

The Low Loss Radial Superlattice Via

To overcome such drawbacks of the planar superlattice conductors, the Cylindrical Radial Superlattice (CRS) structure has been proposed in [10], which inherently has a closed boundary condition in a radial direction and therefore is considered more appropriate for eddy current suppression and low conductor loss. In this work, the Radial Superlattice Via (RSV) structure has been proposed where the CRS structure has been employed to fabricate low loss vias to replace the vias that are known to produce high loss in a CMOS or TSV/TGV process. Fig. 3 (a) depicts the cross section of the CRS conductor where the arrows show the generated eddy currents when an alternating current is flowing in z direction. Fig. 3 (b) shows the typical TSV/TGV vias where the RSV structure has been employed to reduce the ohmic loss of the vias.

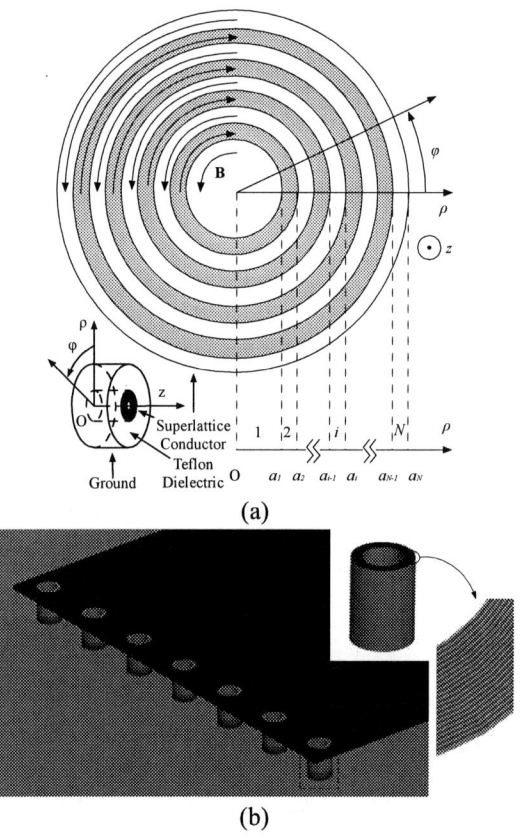

Figure 3. (a) A cylindrical radial superlattice (CRS) conductor composed of metal (white region) and ferromagnetic thin film (grey region) with N layers and total radius a_N in the cylindrical coordinate system. The arrows show the direction of the induced eddy currents where opposing each other in the design frequency region due to negative permeability of the ferromagnetic thin film, (b) the radial superlattice vias (RSV) constructed using the CRS structure in a typical RF MEMS process and it's cross section view.

The analysis of the superlattice vias is followed by the full-wave simulations using the high frequency structure simulator (HFSS, ANSYS Inc.) that proves the effectiveness of the proposed method for conductor loss reduction of the vias. The CRS conductor consists of a solid core conductor covered by laminated superlattice structures (Fig. 3 (a)) where a ground plane is used underneath the structure, forming a coaxial structure, to allow propagation of transverse electromagnetic (TEM) waves. The lumped element equivalent circuit model, as shown in Fig. 4, will be used to extract the conduction resistance of the conductor. By analyzing the Y-parameters, the value of the equivalent circuit elements could be extracted,

$$\mathrm{Re}[Y21] = \frac{-R_1}{R_1^2 + \omega^2 L_1^2} \qquad (5.\,a)$$

$$\mathrm{Im}[Y21] = \frac{\omega L_1}{R_1^2 + \omega^2 L_1^2} \qquad (5.\,b)$$

where the resistance and inductance can be found by solving the system of equations in (5),

$$L_1 = \frac{\mathrm{Im}[Y21]}{\omega\left[\left(\mathrm{Re}[Y21]\right)^2 + \left(\mathrm{Im}(Y21)\right)^2\right]} \qquad (6.\,a)$$

$$R_1 = \frac{\left(-\mathrm{Re}[Y21]\right)^{-1} \pm \sqrt{\left(\mathrm{Re}[Y21]\right)^{-2} - 4\omega^2 L_1^2}}{2} \qquad (6.\,b)$$

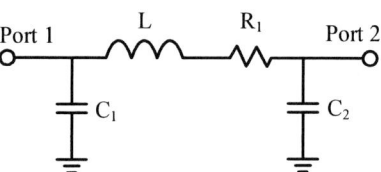

Figure 4. The lumped element circuit model of the CRS conductor.

The RSV structure using Cu/NiFe

Copper is widely used as the conductor in standard manufacturing processes that has high conductivity and can be deposited using standard microfabrication processes. Fig. 5 shows the simulation results of the RSV structure where Cu/NiFe are used as the nonmagnetic/magnetic metal layers. Relatively low magnetization saturation of the NiFe [13] allows designing the RSV to operate in the lower frequency region (up to minimum 5 GHz). The extracted resistance of the radial superlattice conductor using equation (6) has been compared to that of the solid conductor with the same thickness where no laminated structure is used. In Fig. 5, the resistance of the Cu/NiFe conductor has been simulated where the effect of different thickness ratios $r = t_{Cu}/t_{NiFe}$ has been studied to show the capability of tuning the operation frequency (frequency where the resistance spectra is minimum) while the number of layers (N=12) and the thickness of the NiFe (t_p=50 nm) are kept constant. The frequency where the effective permeability is close to zero which makes the ohmic resistance minimum, is tuned based on the given thickness ratio. By increasing the thickness ratio, the minimum frequency shifts to lower frequencies. The minimum resistance in Fig. 5 is for the conductor with the thickness ratio of 4 because the overall conductor thickness is larger than those of other conductors with a ratio of 2.5 and 1.6.

978-1-4799-2408-0/14 $31.00 © 2014 IEEE

In Fig. 6, three different ratios of the superlattice structure $r = t_{Cu}/t_{NiFe}$ with a constant $t_{NiFe} = 50nm$ and a constant total conductor thickness has been investigated. The overall diameter of the RSV structure is kept the same (7 µm thick) in order to investigate the conductor resistance while the minimum values of the resistances are close to each other due to the same total conductor thickness for different Cu/NiFe ratios.

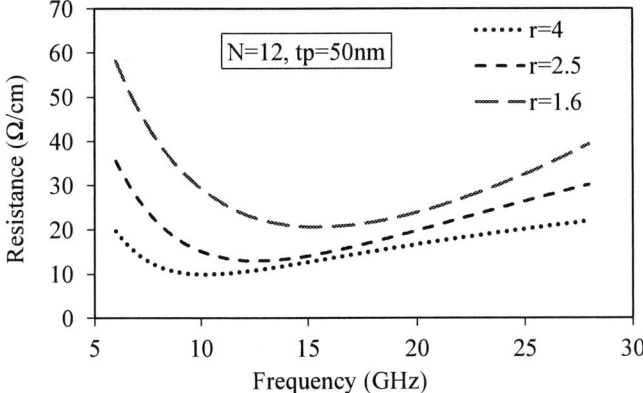

Figure 5. The simulation results of the Cu/NiFe RSV structures with $N = 12$ and $t_{NiFe} = 50nm$ (*i*) r = 1.6, $t_{Cu} = 80nm$, (*ii*) r = 2.5, $t_{Cu} = 125nm$, and (*iii*) r = 4.0, $t_{Cu} = 200nm$.

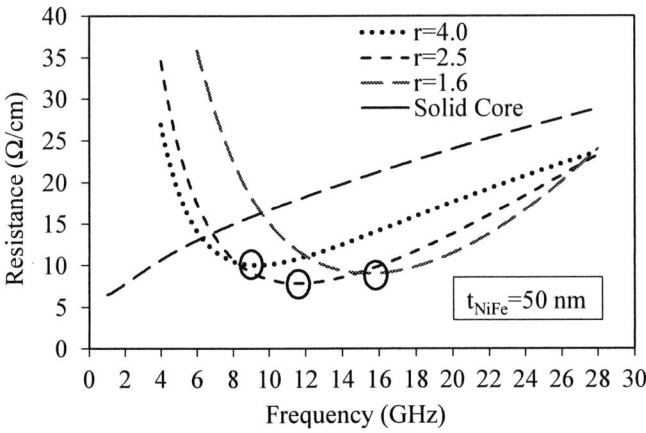

Figure 6. The simulation results of the 7-µm thick Cu/NiFe RSV structure with $t_{NiFe} = 50nm$, (*i*) r = 1.6, N = 21, $t_{Cu} = 80nm$, (*ii*) r = 2.5, N = 16, $t_{Cu} = 125nm$, and (*iii*) r = 4.0, N = 11, $t_{Cu} = 200nm$.

By properly selecting the thickness ratio of the conductors, it would be possible to place the minimum resistance area in the frequency of operation range. In Fig. 6, the frequencies of 9, 12 and 16 GHz are obtained by using the $r = t_{Cu}/t_{NiFe}$ ratio of 4, 2.5, and 1.6, respectively. At the lower frequency range (below 5 GHz), the resistance of the superlattice structure remains high because of the high permeability of the ferromagnetic layers that make the skin depth very thin (based on (1)) and increases the ohmic loss. The eddy current canceling effect begins in the frequency range where the effective permeability approaches zero, and so the resistance does not go up with the frequency rise. After the operation frequency range, the resistance starts increasing normally with the order of $\sqrt{f}$. Fig. 7 shows the simulation results where the

current distribution throughout the CRS conductor is showed (a) and compared with that of the regular conductor made of copper only (b) where the same minimum and maximum current boundaries are set for those graphs.

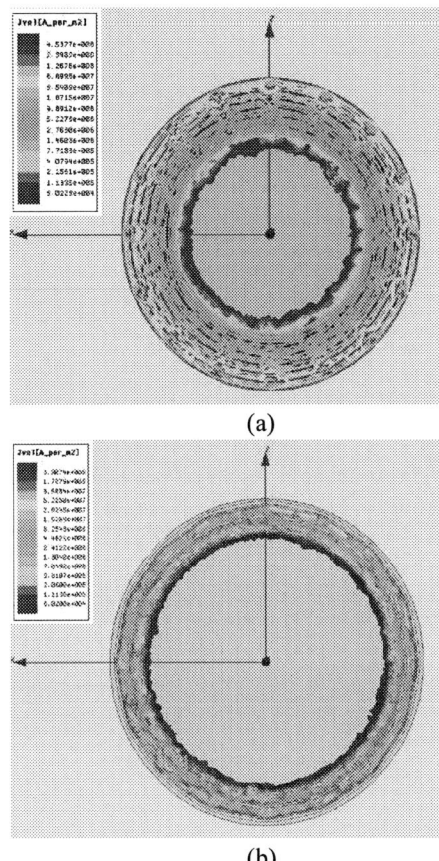

Figure 7. Current distribution throughout the volume of (a) the CRS conductor consisting of 21 layers at the design frequency of 16 GHz where the resistance has been minimized, (b) the solid core conductor where the current has been confined in the outermost region of the conductor due to the skin effect.

Another set of simulations have been performed using a 13 µm thick RSV (Fig. 8) where the $r = t_{Cu}/t_{NiFe}$ ratio has been kept as a constant while the number of layers are changed to study the effect of different number of layers on the resistance spectra. The extracted resistance of the control solid-core Cu conductor has also been showed in Fig. 8 for the sake of comparison. By increasing the number of layers while keeping the same ratio, the frequency where the minimum resistance occurs, doesn't shift much while the higher number of layers with thinner each layer, shows a more effective macroscopic eddy current cancelling and the resultant resistance is smaller in the frequency range of operation. Its fabrication would be more difficult.

The RSV structure using Cu/FeCo

In order to utilize the RSV structures in higher frequencies, ferromagnetic materials with higher saturation magnetization and higher f_{MR} frequency are required. Therefore, the eddy current cancelling will occur at the higher frequency and the overall system could be designed at higher frequency ranges.

FeCo is considered as a good candidate magnetic material [14] that has a high f_{MR} and can be deposited using standard processes allowing the operation in higher frequencies (above 30 GHz). Fig. 9 depicts the simulation results of the resistance spectra of the RSV's consisting of Cu/FeCo layers that are performed up to 80 GHz.

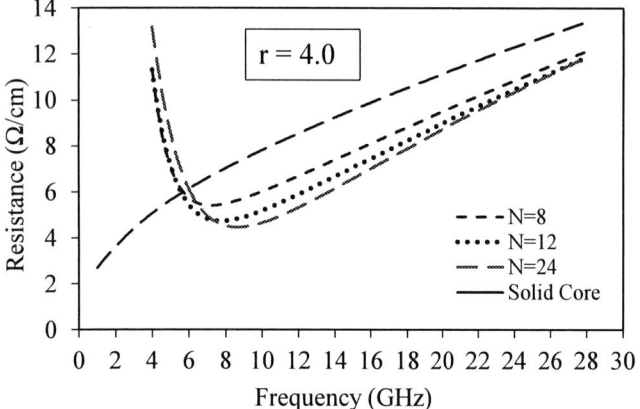

Figure 8. The simulation results of the 13-µm thick Cu/NiFe RSV structure with $r = t_{Cu}/t_{NiFe} = 4.0$, (*i*) N = 8, $t_{NiFe} = 150nm$, $t_{Cu} = 600nm$ (*ii*) N = 12, $t_{NiFe} = 100nm$, $t_{Cu} = 400nm$, (*iii*) N = 24, $t_{NiFe} = 50nm$, $t_{Cu} = 200nm$.

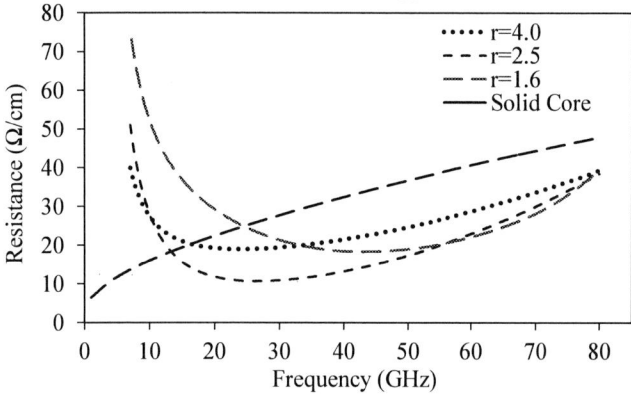

Figure 9. The simulation results of the 7-µm thick Cu/FeCo RSV structure with $t_{FeCo} = 50nm$, (*i*) r = 1.6, N = 21, $t_{Cu} = 80nm$, (*ii*) r = 2.5, N = 16, $t_{Cu} = 125nm$, and (*iii*) r = 4.0, N = 11, $t_{Cu} = 200nm$.

The fabricated Cu/NiFe CRS conductors

In order to verify the performance of the proposed CRS and RSV structures, the CRS superlattice structure has been fabricated on a radial conductor. Fig. 10 shows the schematic of the CRS conductor where a radial gold wire has been implemented using a commercial wire bonder and one-port coplanar waveguide (CPW) input. The equivalent circuit of the conductor in Fig. 10 is the one given in Fig. 4 while port 2 is grounded. The CRS conductors are fabricated on a glass substrate using a gold wire as the structural core followed by multiple-step electroplating of Cu/NiFe. Due to the radial shape of the structures, electroplating has been selected as the thin film deposition technique to ensure uniformity as other methods like DC sputtering would not work best for these devices [7-9]. Arrays of conductors with different dimensions are fabricated (Fig. 11).

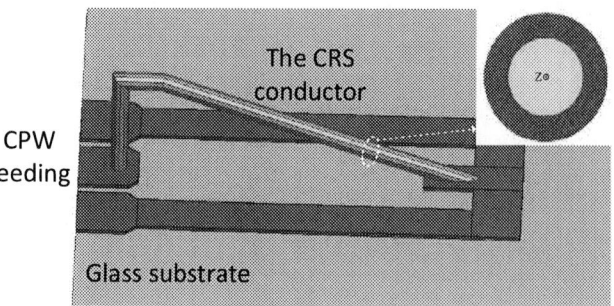

Figure 10. The schematic of the CRS using a radial wire bonded conductor core and CPW feeding.

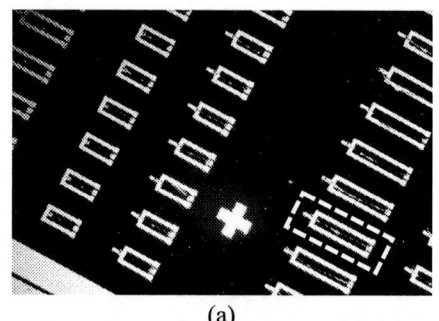

(a)

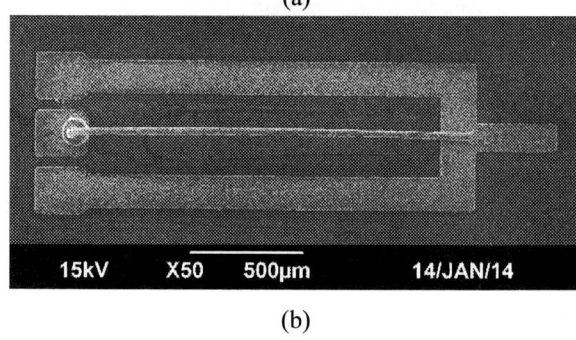

(b)

Figure 11. (a) The glass substrate holding the array of CRS conductors, and (b) the fabricated CRS conductor of 1-mm length after electroplating. The backbone wire has been implemented using a gold wire bonder.

Fig. 12 shows a comparison of the measured resistance of the conductors of 1-mm and 2-mm length. The conductors have higher resistance in low frequency range due to the high permeability of NiFe thin films in that frequency range which makes the skin depth smaller and increases the ohmic loss. However, as the frequency increases, the total CRS resistance is kept below that of solid core devices and a wide-band of operation is achieved.

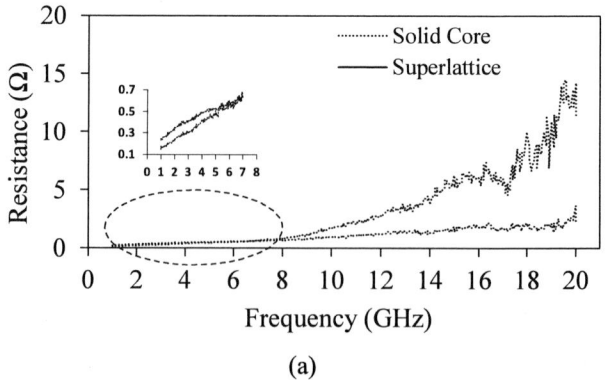

(a)

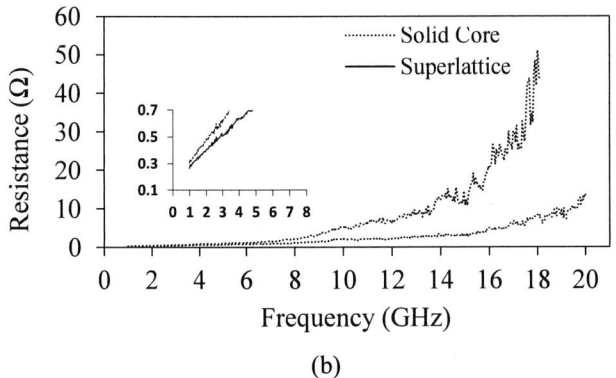

(b)

Figure 12. A comparison of the measured resistance of the conductors of (a) 1 mm length, and (b) 2 mm length.

Conclusions

The theoretical and simulation results show up to 100% reduction of the conductor loss of the vias in microelectronic structures including TSVs/TGVs. It is shown that magnetic materials with smaller magnetic saturation (M_s) such as NiFe have lower f_{MR} and are suitable for lower frequency ranges, which finds applications such as in digital microprocessors. NiFe and FeCo are used as the ferromagnetic materials for low and high frequency operation, respectively, while electroplated Cu is used as a non-ferromagnetic conductor material. The effect of the number of layers for the superlattice structure and the characteristics of the vias using different materials for a number of frequency ranges have been reported. The fabrication of the superlattice structure is fully compatible with the standard MEMS and CMOS processes. Therefore, the integration of the proposed method with the current microfabrication processes is further extended to a long-term solution for maximally reducing the loss in high-speed and radio frequency (RF) devices ranging from analog, digital and mixed signal applications in digital microprocessors, analog-to-digital converters to RF inductors, antennas and metamaterials.

Acknowledgments

This work is supported in part by NSF ECCS (1132413). Microfabrication and magnetic characterization have been performed in Nanoscale Research Facilities (NRF) cleanroom and Interdisciplinary Microsystems Group (IMG), at the University of Florida.

References

1. "The evolution of a revolution", an overview of Intel processor history, http://www.intel.com/pressroom/kits/quickreffam.htm

2. B. Pan, Y. Li, M. M. Tentzeris, and J.Papapolymerou, "A novel low-loss integrated 60 GHz cavity filter with source-load coupling using surface micromachining technology," *Microwave Symposium Digest, IEEE MTT-S International*, pp.639-642, Jun. 2008.

3. Y.K. Yoon, J.W. Park, and M. G. Allen, "Polymer-core conductor approaches for RF MEMS," *IEEE Journal of Microelectromechanical Systems*, vol. 14, no. 5, pp. 886-894, Oct. 2005.

4. D. E. Senior, X. Cheng, and Y.K. Yoon, "Highly compact surface micromachined metamaterial circuits using multilayers of low-loss Benzocyclobutene for microwave

and millimeter wave applications," *IEEE Electronic Components and Technology Conference (ECTC)*, pp. 2062-2069, Jun. 2012.

5. S. H. Hall, G. W. Hall, and J. A. McCall, "High-speed digital system design: a handbook of interconnect theory and design practices" New York: Wiley, 2000.

6. B. Rejaei and M. Vroubel, "Suppression of skin effect in metal/ferromagnet superlattice conductors," *Journal of Applied Physics,* 96.11 (2004): 6863-6868.

7. I. Iramnaaz, T. Sandovala, Y. Zhuanga, H. Schellevisb, and B. Rejaei, "High quality factor RF inductors using low loss conductor featured with skin effect suppression for standard CMOS/BiCMOS," *IEEE Electronic Components and Technology Conference (ECTC)*, Jun. 2011.

8. Y. Zhuang, B. Rejaei, H. Schellevis, M. Vroubel, and J. N. Burghartz, "Magnetic-multilayered interconnects featuring skin effect suppression," *IEEE Electron Device Letters*, vol. 29, no. 4, 2008.

9. M. Yamaguchi, Y. Endo, N. Sato, and A. Ludwig, "Overview of RF high-permeability ferromagnetic thin films and its application to a new ferromagnetic/ conductive multilayer to suppress skin effect in RF on-chip conductors," *IEEE 6th European Microwave Integrated Circuits Conf. (EuMIC)*, 2011.

10. J. Wu and Y.K. Yoon, "A Low Ohmic Loss Radial Superlattice Conductor at 15 GHz using Eddy Current Canceling Effect," *IEEE Antenna and Propagation Society International Symp. (APS/URSI)*, pp. 1552 – 1553, Jul. 2013.

11. J. B. Youssef, N. Vukadinovic, D. Billet, and M. Labrune "Thickness-dependent magnetic excitations in permalloy films with nonuniform magnetization," *Physical Review B* 69.17, 2004.

12. Y. Liu, L. Chen, C. Y. Tan, H. J. Liu, and C. K. Ong, "Broadband complex permeability characterization of magnetic thin films using shorted microstrip transmission-line perturbation," *Review of Scientific Instruments*, 76.6, 063911, 2005.

13. I. Iramnaaz, H. Schellevis , B. Rejaei , R. Fitch , and Y. Zhuang, "Self-Biased Low Loss Conductor Featured With Skin Effect Suppression for High Quality RF Passives," *IEEE Transactions on Magnetics*, vol. 48, no.11, pp. 4139-4142, Nov. 2012.

14. Y. Liu, C. Y. Tan, Z. W. Liu, and C. K. Ong "FeCoSiN film with ordered FeCo nanoparticles embedded in a Si-rich matrix," *Applied physics letters*, 90, 112506, 2007.

Modeling, Design, Fabrication and Characterization of First Large 2.5D Glass Interposer as a Superior Alternative to Silicon and Organic Interposers at 50 micron bump pitch

Brett Sawyer, Hao Lu, Yuya Suzuki[†], Yutaka Takagi[‡], Makoto Kobayashi[±], Vanessa Smet, Taiji Sakai[*], Venky Sundaram and Rao Tummala

3D Systems Packaging Research Center, Georgia Institute of Technology,
813 Ferst Dr. N.W., Atlanta, GA 30332.
[†] Zeon Corporation, Kawasaki, Kanagawa, Japan
[‡] NGK Spark plug Co., Ltd., Komaki, Aichi, Japan
[*] Fujitsu Laboratories Ltd., Atsugi, Kanagawa, Japan
[±] Namics Corporation, Niigata, Japan
Email: bsawyer@gatech.edu

Abstract

This paper describes the first design and fabrication of a large 2.5D glass interposer with 50 μm pitch chip-level interconnections made of 6 layers of 3 μm re-distribution (RDL) wiring. Many applications including high-performance networking and cloud computing data centers require ultra-high-bandwidth of the magnitude of 512 GB/s. Silicon-based 2.5D interposers are the only approaches being pursued by the industry to meet this need, enabled by sub-micron BEOL wiring in the wafer fabs. Such interposers, however, are too expensive for most applications. Glass interposers are superior to silicon interposers due to their high dimensional stability, low loss tangent, and large panel processing ultimately leading to lower cost. This paper presents the design, fabrication and electrical characterization, leading to the first fabrication of 2.5D glass interposers with 50 μm I/O pitch with 3 μm lines. Double-sided panel processing utilizing thin, low-loss dryfilm polymer dielectrics and SAP copper plating, with differential spray etching techniques, was used to fabricate 3 μm wide transmission lines on 25mm x 30mm glass interposers processed on a 300 μm thick 150mm x 150mm glass panels. A six-metal layer test vehicle with two daisy chain, 10mm x 10mm test chips at 100 μm spacing, was fabricated and assembled by thermo-compression bonding of Cu microbumps and SnAg solder caps. Ultra-fine 3 μm escape routing was demonstrated on a two-metal layer test vehicle. High frequency characterization of 3 μm lines showed low loss of 0.12 dB/mm at 2 GHz.

Introduction

System bandwidth for high-performance applications are expected to require 512 GB/s to 1 TB/s in the near future. The only way to meet this need is by ultra-fine pitch interconnections between logic and memory devices either in 3D with vertical TSV interconnections or by horizontal 2.5D interconnections, as shown in Fig. 1a and 1b. The industry has also been aggressively developing through silicon via (TSV) technology for high-bandwidth memory (HBM) [1] to support high channel density. To achieve both, the industry began to pursue back-end-of-line (BEOL) silicon wafers in both 2.5D and 3D silicon interposers [2, 3] capable of sub-micron wiring dimensions. However, electrical loss associated with silicon is a limiting factor. High permittivity and high loss tangent of silicon limit silicon interposer technology to single-sided signal routing with short, wide I/O interconnections. Furthermore, silicon interposers are limited by high cost,

resulting from small number of interposers from each 200 or 300 mm wafers. Low-TCE and coreless organic substrates have improved electrical performance [4], but are limited in pitch due to their dimensional instability affecting layer-to-layer via registration; in body size, and thinness due to high warpage during substrate fabrication and assembly.

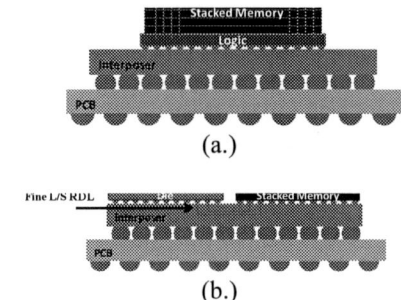

Figure 1. High bandwidth (a.) 3D and (b) 2.5D package architectures.

Glass interposers have been demonstrated with superior electrical properties and with 5um wiring lines [5]. This paper goes beyond the prior work in glass to fabricate the first 2.5D glass interposer in large body size with 3 μm RDL wiring at 50 μm I/O pitch, as shown in Figure 2. The main contributions of this work include: (a) the first analysis of double-sided RDL signal routing for wide I/O applications, (b) 3 μm RDL routing demonstration on glass using low cost, panel-based fabrication processes, (c) simultaneous thermo-compression bonding (TCB) of chips with side-by-side assembly processes, and (d) characterization of 3 μm width transmission line structures up to 5 mm signal length on glass.

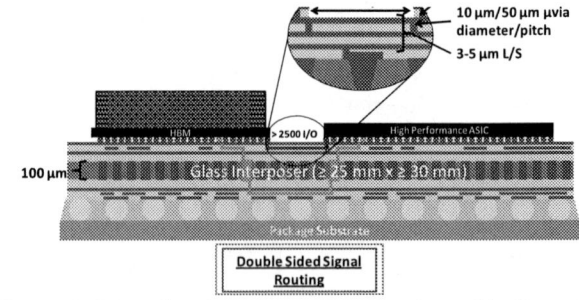

Figure 2. 2.5D glass interposer cross section with 50 um I/O Pitch.

The second section of the paper describes the electrical modeling and design of re-distribution layers (RDL) which includes an investigation of escape routing capacity and signal integrity of high density die-to-die interconnections. The third section discusses the details of a 2.5D glass interposer test vehicle designed to study the substrate fabrication process and assembly. The fourth section presents the details of the glass interposer fabrication and multiple-chip assembly processes. Such a test vehicle demonstrates 3 μm RDL wiring as well as side-by-side die chip assembly. The final section presents high frequency characterization, performed on 3 μm CPW transmission line structures, demonstrating the superior electrical performance of glass over silicon and organic interposers.

Design and Modeling

High-performance applications require interposers and packages that can support a wide, high-density, and high aggregate bandwidth data bus (signal speeds up to 1 Gbps). The first step in the design of a 2.5D glass interposer was the modeling and analysis of a wide I/O channel that uses double-sided signal routing, to accurately predict the transmission line insertion loss and line-to-line coupling noise. The objective was to establish a set of design rule targets for RDL and chip-level interconnections in 2.5D glass interposers.

High-density Die-to-Die Routing on Glass

An initial set of design rules was assumed as shown in Table 1, based on previous processing experience with through vias (TPV) and multi-layer, double-sided RDL on glass interposers. The routing layer design was based on, but not limited to a 2.5D interposer with side-by-side HBM and logic ICs (see Figure 2)

Table 1. Initial Design Rule Targets for 2.5D glass interposers.

Design Parameter	Target
Minimum L/S	3/3 μm
μ-via pad/size/pitch	16/10/50 μm
TPV pad[1]/size/pitch	50/30/50 μm
RDL Stack-up[2]	3/0/3
Bump pitch/die spacing	20-50 μm/ ≤ 100 μm

[1] 40 μm bottom pad size
[2] ultra-fine feature size on all metal layers

A routing study was performed based on these design rules. The three goals for this routing study were to (a) reduce layer count (and consequently cost), (b) use GSSG and GS transmission structures for better signal integrity, and (c) demonstrate escape routing for up to 24 signal rows on a die with 96/55 μm face-centered rectangular (FCR) stagger ball pattern.

Figure 3 shows an isometric view of the escape routing study performed in Cadence APD [7]. Using GSSG and GS transmission line structures, 2492 die-to-die interconnections are required between logic and memory—indicating the need for a logic die with 20 μm FLI pitch. Assuming a 15mm x 15mm logic die with 250 μm die edge keep out and 20 μm FLI pitch, four exterior in-line signal rows are needed.

Routing at 20 μm FLI pitch is possible using a fan-in/fan-out escape technique at target design rules in Table 1.

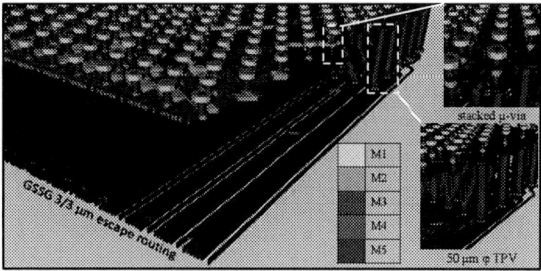

Figure 3. HBM escape routing study on a six-metal layer 2.5D glass interposer.

Insertion Loss and Crosstalk in High-density Channels

Minimum line and space requirements identified in Table 1 have two consequences on signal integrity for high-density die-to-die interconnects. First, insertion loss increases with decreasing line width. Second, the number of ground and supply lines that can be routed from a die limits the line-to-line shielding. Therefore, worst-case insertion loss, far-end crosstalk (FEXT) and near-end crosstalk (NEXT) were analyzed using 3D-EM simulation [7]. The stack-up in this analysis was based on the routing study results in Figure 3 and is shown in Figure 4 below.

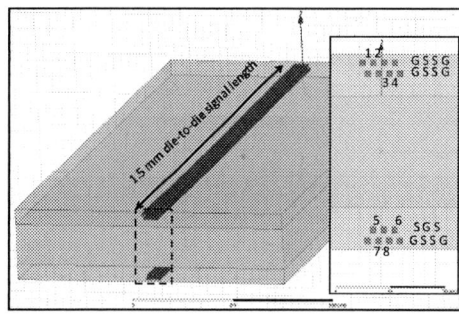

Figure 4. Routing structure in four-metal layer, double-sided glass interposer used for electrical modeling.

The routing density is lower on TPV layers due to the increased via pad size. Decreased signal density on these signal layers is represented by port 5 and 6 in the data bus cross-section in Figure 4. The signal integrity analysis for insertion loss and crosstalk did not include a direct comparison between glass and silicon since through interposer lines were not included in the model and signal length was only 1.5 mm.

The worst-case insertion loss for an 8-bit data bus section is shown in Figure 5. The insertion loss plot is for a signal line on the TPV signal layer (port 5). Up to f = 18.5 GHz, insertion losses less than 0.67 dB/mm are observed. The worst case line-to-line coupling for the 8-bit data bus section is shown in Figure 6. The NEXT and FEXT plots shown correspond to a signal line on the TPV signal layer (port 5). At low frequency up to f = 6 GHz crosstalk between data lines is less than 20 dB.

The simulated channel performance is adequate for wide I/O, low data rate per channel applications such as HBM.

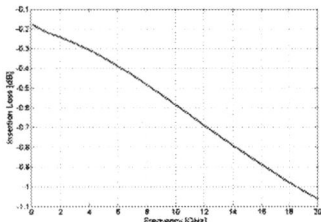

Figure 5. Worst-case insertion loss for an 8-bit high-density die-to-die data bus.

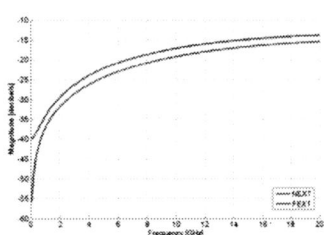

Figure 6. Worst-case crosstalk on high-density die-to-die data bus.

Time Domain Analysis of Double-side High-density Channel

A five-metal layer GSSG structure was modeled to better understand the effect of multiple RDL microvia and TPV transitions on line performance. Insertion loss simulations and time domain analyses were performed to observe the performance of double-side signal routing scheme for a 2.5D interposer architecture. In this analysis, a direct comparison with wafer silicon interposer technology is beneficial to quantify the performance benefits of glass over silicon. The design rules shown in Table 1 were again used for the glass interposer. The design rules for silicon were based on published literature [8, 9].

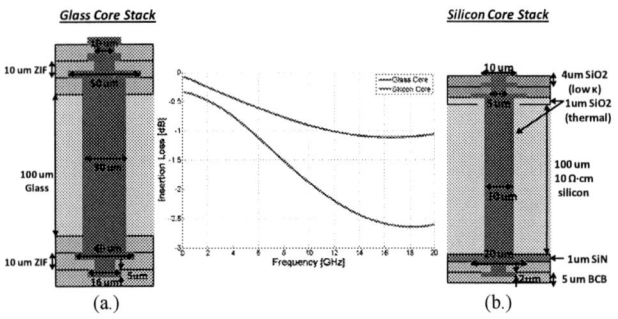

Figure 7. Insertion loss of through package lines in high-density die-to-die data bus

Double-sided signal routing performance was compared for the glass and silicon interposer structures as shown in Figure 7. Superior electrical performance was demonstrated for glass over the range of frequencies analyzed. Highest insertion loss for glass occurred at f = 16 GHz, with S21 of

1.11 dB, while highest insertion loss for silicon occurred at f = 18 GHz, with S21 of 2.65 dB.

To further assess whether double-sided signal routing was feasible for glass interposers, a time domain analysis was conducted to simulate full channel performance for those signal lines routed through the interposer and bottom most signal layer (M5 in Figure 2).

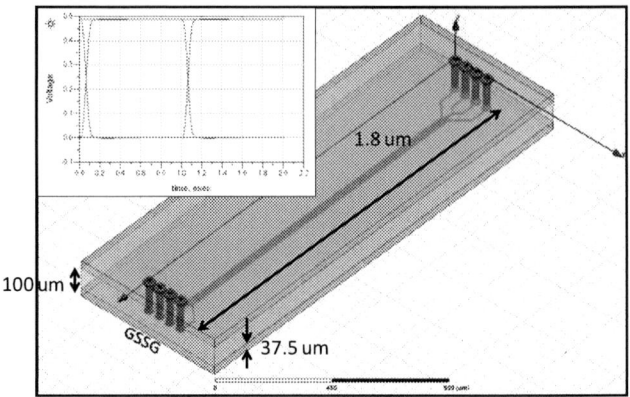

Figure 8. Eye diagram of backside GSSG structure driven with 1 Gbps data signal (rise time t_r, fall time t_f = 100 ps).

A GSSG structure was chosen for this analysis to capture the effect of line-to-line coupling on channel performance. Signal routing density is expected to be adequate to support this transmission line structure, as the M5 signal layer is not limited by TPV pad size. The results of this time domain analysis for a 1 Gbps pseudo-random bit sequence (PRBS) input signal are shown in Figure 8. Large eye opening for these signals lines demonstrates the feasibility of double-sided signal routing on glass interposers.

Test Vehicle Design

Routing studies as well as the modeling and simulation analysis above justify the need for 3 μm wiring technology on 2.5D glass interposers, for escape routing at 20-50 μm bump pitch. Similar feature sizes have been demonstrated on organic [10] and silicon [2] interposers using high-cost, single-side thin film processes such as vacuum deposition and chemical mechanical polishing (CMP). The double-side process used in this study can be scaled to large panels, using lamination of dry film polymer dielectrics and wet chemical plating technologies, leading to much lower cost than wafer processes [11].

Three types of structures were included in the test vehicle layout: (a) ultra-fine line and space escape routing, (b) transmission lines to characterize 3 μm signal widths at high frequencies, and (c) fine bump pitch daisy chains for 2.5D assembly process optimization and reliability testing.

Test Vehicle Structure A - Escape routing

The escape routing coupon was a 25mm x 30mm body size 2.5D glass interposer with 879 die-to-die interconnections. The test vehicle was designed for two 10 mm x 10 mm dies with a die-to-die spacing of 100 μm, and chip-level interconnection pitch of 50 μm. A bump landing pad size of 20 μm on the top metal layer of the glass

978-1-4799-2408-0/14 $31.00 © 2014 IEEE

interposer enables the routing of 4 interior signal rows in one layer with 3 μm wiring technology. Outside the die shadow, the traces fan-out to 5 μm lines, with 500 μm length between the dies. Consequently, the largest line length at 3/3 μm was approximately 300 μm.

Test Vehicle Structure B- High-frequency Structures

High-frequency test structures include 24 co-planar waveguide (CPW) variants. Six different signal widths were implemented in the design: (i) 3 μm, (ii) 5 μm, (iii) 10 μm, (iv) 15 μm, (v) 20 μm, and (vi) 25 μm. For each signal width, four different signal lengths were included: (a) 660 μm, (b) 5 mm, (c) 15 mm, and (d) 25 mm.

Test Vehicle Structure C - 2.5D Two Chip Assembly

The 2.5D assembly coupon also used a 25mm x 30mm body size glass interposer with 87 die-to-die interconnections. A coarser line and space of 40 μm is used for this coupon to implement a serpentine daisy chain connection between die. The 2.5D assembly coupon uses the same die design described above for the escape routing coupon, namely, two 10mm x 10mm dies with 50 μm FLI pitch (5 row in-line design) and central area array at 150 μm pitch with 100 μm die-to-die spacing. The micro-bumps were 15 μm in diameter, with a 5 μm Cu height and a 10 μm high SnAg cap.

Fabrication and Assembly

The test vehicles were fabricated on a 150mm x 150mm square glass panel that is 300 μm thick. For the initial fine line routing process optimization and electrical characterization test structures, a simple two-metal layer glass panel (1/0/1 stack up) was fabricated. The chip-level interconnect reliability test structures were fabricated on a six-metal layer glass panel (3/0/3 stack up). The interior layers (M2-M5) in the six-metal layer panel included a copper mesh pattern with approximately 55% copper coverage. The fabrication process details are described in another publication [11], but a brief summary is shown in Table 2.

Table 2. Process for two-metal layer (double-sided) panel fabrication. Six-metal layer panel, Test Vehicle Structure C, fabrication sequence repeats steps 1-7 three times. Two-metal layer panel, Test Vehicle Structures A and B, fabrication sequence follows steps 1-6.

#	Process
1	17.5 μm dry film polymer lamination
2	Electroless copper plating
3	Photolithography[1,2] (dry film resist)
4	Electrolytic copper plating
5	Dry film resist removal
6	Seed layer etching[3]
7	Surface treatment (adhesion promoter)

[1] Two-metal layer fabrication sequence uses Ushio UX-44101 projection mask aligner [12]
[2] Six-metal layer fabrication sequence uses contact photolithography
[3] Differential spray etching used for two-metal layer fabrication sequence

Two-metal Layer Panel Fabrication

The fabricated structures with 3 μm lines and spaces in the escape region, and 5 μm lines and spaces in the die-to-die space are shown in Figure 9. Optical and scanning electron

microscope (SEM) inspections confirmed good yield of the ultra-fine lines and spaces on glass for all 879 die-to-die interconnections without delamination or bridging. The measured line widths were slightly smaller than the designed dimensions mainly due to over etch during copper seed layer removal. Future designs will include an etch compensation factor to account for this deviation.

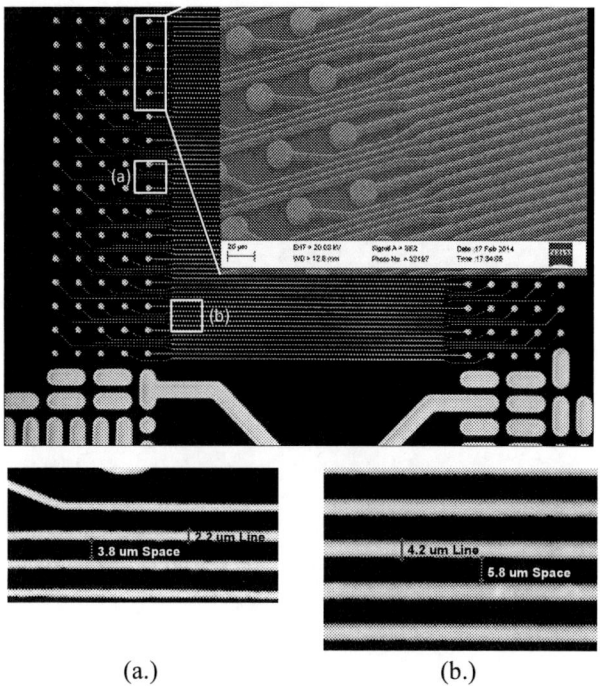

(a.)　　　　　　　　(b.)

Figure 9. Demonstration of (a) 3/3 μm escape routing and (b) 5/5 μm fan-out on 1/0/1 glass interposer at 50 μm bump pitch.

Six-metal Layer Panel Fabrication and Assembly

Results for six-metal layer panel fabrication using the process flow described above are shown in Figure 10. Topside metal layers are shown in the detailed cross-section of Figure 10c. Similar results were observed for bottom side metallization.

The 2.5D assembly was performed using a semi-automatic Finetech Matrix Flip-Chip Bonder [13] with a placement accuracy of 3 μm. Pre-applied underfill (B-Stageable No-Flow Underfill by Namics Corporation [14]) was used to confine the solder, prevent bridging at 50 μm pitch, and to control the shape of the solder joint during assembly by viscosity control. The silicon dies were sequentially picked and placed onto the interposer, using a 10mm x 10mm vacuum-locked spring gimbal tool. The interposer was maintained at 70°C to reduce the viscosity of the underfill material for adequate bump-to-pad contact. Underfill volume was optimized to prevent any movement of the first die during the placement of the second die due to flow of the material in the die gap. After sequential placement, both dies were simultaneously bonded by low-pressure thermo-compression at 260°C peak temperature, using a 20mm x 20mm pre-leveled gimbal tool. The placement pressure was reduced upon reaching the low viscosity point of the underfill material to allow excess material to flow.

Figure 10a shows the top view of the six-metal layer interposer after 2.5D assembly. X-ray characterization of the glass assembled structure was performed after placement of the first and second die, respectively, and after thermo-compression bonding, confirming that there was no displacement of either the first or second die during the assembly sequence. After the side-by-side chip assembly, sample cross-sections indicated acceptable joint quality and interconnection yield, despite a slight misalignment, as seen in Figure 10b and 10c.

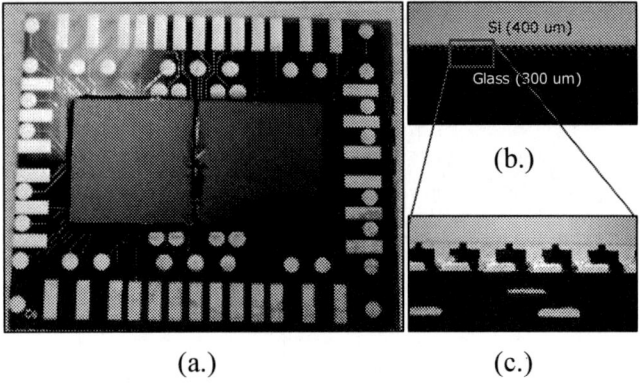

(a.)   (c.)

Figure 10. Demonstration of 2.5D FLI assembly at 100 μm die-to-die spacing (a.) top view of 25 mm x 30 mm six-metal layer glass interposer (b) cross-section (c) FLI detailed cross-section.

Electrical Characterization

Ultra-fine lines, while required for high-density die-to-die interconnects, pose electrical design challenges for signal integrity especially at increased signal length. As a first step in the high frequency characterization of 3 μm signal lines on glass, the insertion loss of CPW transmission lines was measured up to 20 GHz after performing a SOLT calibration. Measurement results shown in Figure 11 indicate good correlation between 3D-EM simulations and VNA measurements at signal lengths of 660 μm and 5 mm, and, more importantly, low insertion loss of approximately 0.12 dB/mm at 2 GHz.

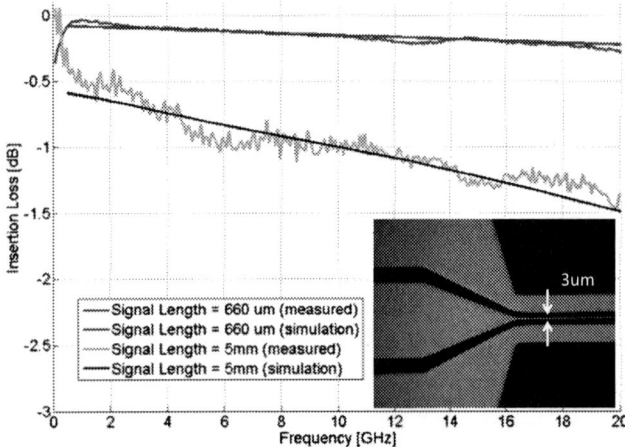

Figure 11. Simulated and measured insertion loss of a 3 μm signal CPW transmission line up to 20 GHz.

Conclusions

Results shown in the above analysis indicate the advantages of glass interposers compared to silicon and organic technology. This paper, for the first time, shows the design, fabrication, and assembly of a panel-based, low-cost, six-metal layer, 2.5D, 300 μm thick glass interposer at 50 μm bump pitch. Side-by-side chip assembly with sequential pick-and-place, and simultaneous TCB was demonstrated at 100 μm die spacing. Ultra-fine escape routing structures down to 3 μm line lithography were fabricated at high yield, verified by optical and SEM inspections. Modeling, simulation, design, and characterization of 3 μm, single-layer transmission line structures confirmed low signal insertion loss up to 20 GHz. Furthermore, double-sided signal routing on glass was analyzed, and these simulated results show good signal integrity for frequencies up to 6 GHz. Low loss TPVs resulted in lower insertion loss in glass, compared to wafer silicon.

In summary, this paper describes the first panel-based, 2.5D glass interposer fabrication, demonstrating 3 μm RDL wiring technology at 50 μm bump pitch. Such a technology, when implemented in large panel manufacturing is expected to result in about 5X cost reduction compared to silicon from wafer fabs [15]. The electrical simulation and characterization results demonstrate the superiority of glass for low-data rate double-sided signal routing as well as low insertion loss for 3 μm lines fabricated using the aforementioned low-cost processes.

Acknowledgements

Research results described above are part of the Low-cost 3D Glass Interposers and Packages (LGIP) program at Georgia Tech PRC (GT-PRC). The authors acknowledge those LGIP member companies and supply chain partners in supporting this research effort. Additionally, the authors would like to thank Jialing Tong from GT-PRC, and Ryuta Furuya, a visiting engineer to GT-PRC from USHIO, for his help in operating the Ushio UX-44101 photolithography tool with panel stepper lens (model "Square 70") used in fabrication.

References

1. *High-bandwidth Memory (HBM) DRAM*, JESD 235, Oct. 2013

2. Chaware, R.; Nagarajan, K.; Ramalingam, S., "Assembly and reliability challenges in 3D integration of 28nm FPGA die on a large high density 65nm passive interposer," *Electronic Components and Technology Conference (ECTC), 2012 IEEE 62nd*, vol., no., pp.279,283, May 29 2012-June 1 2012 doi: 10.1109/ECTC.2012.6248841

3. Li Li; Peng Su; Jie Xue; Brillhart, M.; Lau, J.; Tzeng, P. -J; Lee, C. K.; Zhan, C. J.; Dai, M.J.; Chien, H. C.; Wu, S.T., "Addressing bandwidth challenges in next generation high-performance network systems with 3D IC integration," Electronic Components and Technology Conference (ECTC), 2012 IEEE 62nd, vol., no., pp.1040,1046, May 29 2012-June 1 2012

4. Savic, J.; Aria, P.; Priest, J.; Dugbartey, N.; Pomerleau, R.; Shanker, B. J.; Nagar, M.; Lim, J.; Sue Teng; Li Li; Jie Xue, "Electrical performance assessment of advanced substrate technologies for high speed networking

applications," *Electronic Components and Technology Conference, 2009. ECTC 2009. 59th* , vol., no., pp.1193,1199, 26-29 May 2009

5. Sukumaran, V.; Bandyopadhyay, T.; Chen, Q.; Kumbhat, N.; Fuhan Liu; Pucha, R.; Sato, Y.; Watanabe, M.; Kitaoka, Kenji; Ono, M.; Suzuki, Y.; Karoui, C.; Nopper, C.; Swaminathan, M.; Sundaram, V.; Tummala, R., "Design, fabrication and characterization of low-cost glass interposers with fine-pitch through-package-vias," *Electronic Components and Technology Conference (ECTC), 2011 IEEE 61st* , vol., no., pp.583,588, May 31 2011-June 3 2011

6. Cadence Design Systems, Inc. (2011). *Cadence IC Package Design* [Online]. Available: http://www.cadence.com/rl/Resources/datasheets/7429_Al legro_IC_PKG_DS_FINAL.pdf

7. ANSYS, Inc. (2014). *ANSYS HFSS* [Online]. Available: http://www.ansys.com/Products/Simulation+Technology/E lectromagnetics/Signal+Integrity/ANSYS+HFSS

8. Dickson, T.O.; Yong Liu; Rylov, S.V.; Dang, B.; Tsang, C.K.; Andry, P.S.; Bulzacchelli, J.F.; Ainspan, H.A.; Xiaoxiong Gu; Turlapati, L.; Beakes, M.P.; Parker, B.D.; Knickerbocker, J.U.; Friedman, D.J., "An 8x 10-Gb/s Source-Synchronous I/O System Based on High-Density Silicon Carrier Interconnects," *Solid-State Circuits, IEEE Journal of* , vol.47, no.4, pp.884,896, April 2012

9. Xiaoxiong Gu; Turlapati, L.; Dang, B.; Tsang, C.K.; Andry, P.S.; Dickson, T.O.; Beakes, M.P.; Knickerbocker, J.U.; Friedman, D.J., "High-density silicon carrier transmission line design for chip-to-chip interconnects," *Electrical Performance of Electronic Packaging and Systems (EPEPS), 2011 IEEE 20th Conference on* , vol., no., pp.27,30, 23-26 Oct. 2011

10. Shimizu, N.; Kaneda, W.; Arisaka, H; Koizumi, N.; Sunohara, S.; Rokugawa, A.; Koyama, T., "Development of Organic Multi Chip Package for High-performance Application," *International Microelectronics Assembly and Packaging* (IMAP), 2013

11. Lu, H.; Takagi, Y.; Suzuki, Y.; Sawyer, B.; Sundaram, V.; Tummala, R.,"Demonstration of low cost 3-5 um RDL line lithography on glass interposers," *Electronic Components and Technology Conference (ECTC), 2014 IEEE 64th*, May 27 2014-May 30 2014

12. USHIO. (2013). *Lighting Edge Technologies* [Online]. Available: http://www.ushio.co.jp/en/index.html

13. Finetech. (2014). *FINEPLACER matrix ma Semi-automatic Die Bonder* [Online]. http://www.finetechusa. com/bonders/products/fineplacerr-matrix-ma.html

14. Namics Corporation. (2013). *Flip Chip Underfill (UF)* [Online]. http://www.namics.co.jp/e/product/chipcoat01. Html

15. Sukumaran, V.; Bandyopadhyay, T.; Sundaram, V.; Tummala, R., "Low-Cost Thin Glass Interposers as a Superior Alternative to Silicon and Organic Interposers for Packaging of 3-D ICs," *Components, Packaging and Manufacturing Technology, IEEE Transactions on* , vol.2, no.9, pp.1426,1433, Sept. 2012

978-1-4799-2408-0/14 $31.00 © 2014 IEEE

Coupling Impact of Single Ended Signals to LVDS Interface

June Feng, Chooi Ian Loh, Edward Lin, Ellen Du, Guang Chen and Dan Oh
Altera Corporation
101 Innovation Dr, San Jose, CA 95134
jfeng@altera.com, 408-544-7000

Abstract

The speed of general purpose input output (GPIO) continues to increase as more consumer applications utilize "smart" devices. The low-voltage differential signaling (LVDS) is often times the highest speed that GPIO interface needs to support in the mixture of different single ended signaling pins. Although LVDS is differential and somewhat immune to direct signal coupling from other signals, it is still subject to coupling through a shared power supply noise. A thorough SSN analysis between single ended to differential signals is presented in this paper. To help designing GPIO systems, we have considered different single ended signaling types such as SSTL, LVTTL,etc.

Signal integrity issues for multiprotocol FPGA GPIO interfaces

Due to advance in consumer electronics, more consumer devices require higher speed general purpose input output interfaces. For instance, emerging 4K LCD/LED panels require significant speed boost for LVDS to equivalent differential link. GPIO interface is often designed to support both differential and single ended signals. At high-end LVDS operations, the coupling impact of single ended signal to differential signal contributes a significant reduction in channel margin. Hence, accurate performance analysis is highly desirable for reliable system operation or to avoid over designing the system. This is particularly true for FPGA designs since FPGA devices provide lots flexibility in design including configuring I/O settings.

Typically, single ended signal used in GPIO interface operates at relatively low speed below a few hundreds megahertz and often do not have performance issues. Although these low speed signals do not have any issues of their own, it can be quite detrimental to high speed differential signals. Due to the generality of GPIO, both signal ended and differential signals often share the same power supply rails and supply noise from single ended signals due to simultaneous switching noise (SSN) can be translated to significant timing degradation for differential signal. Furthermore, wirebond package designs, often used in consumer applications, can also add significant package coupling noise.

GPIO system level simulation methodology

In this paper, we present a thorough coupling analysis between single ended and differential signals using a mid-end FPGA device. SSN analysis is carried out using a well-established SPICE simulation flow [1][2]. To find the optimum configuration, various single ended driver settings are considered depending on the performance requirements. The proposal is validated with extensive measurement data.

System level channel analysis is used to characterize the coupling effects from single ended signaling to LVDS channels. Signal xtalk between different IOs is easy to simulate. The results are very accurate using commercial EDA tools. But accurate signal and power integrity co-simulation is not easy and straight forward.

In order to simulate SSN effect accurately, full transistor models were used instead of behavior models. All the PDN models for pre-drivers and output drivers were modeled using lumped RLC circuits includes on-die PDN, package, PCB and VRM PDN models. Horizontal and vertical signal coupling were modeled using 3D structure models. Power and signal xtalk is dominated in package wirebond section. In order to capture all xtalk and SSN effects among signal and power domains, full bank wirebond and via models are a must.

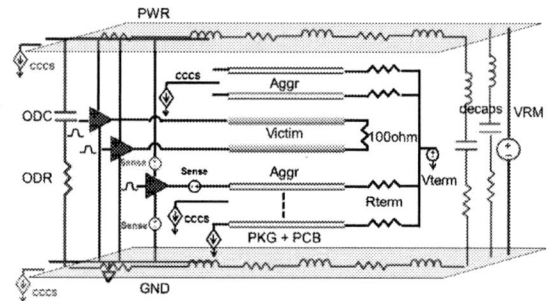

Figure 1. GPIO channel model scheme of SI/PI co-simulation analysis

Figure 1 illustrates a write channel SI/PI co-simulation model scheme for a GPIO environment. The whole channel models start with full transistor models including pre-driver and last stage output drivers for LVDS and single ended signaling. For system channel SI/PI co-simulation, accuracy and fast simulation time are equally important. One of the reasons to use FPGA is its flexibility. Therefore, FPGA provides a lot of different GPIO signalings in a single bank. Normally, Xtalk and PDN noise simulations within the same signaling are analyzed. The SSN impact between single ended signaling and LVDS is often ignored. We will demonstrate the signaling coupling and power delivery noise among different protocols.

With more than 70 IOs' full transistor models and full bank wirebonds, vias, transmission line trances, and solder balls, the simulation time is extremely long and not practical. Current control current source (CCCS) was used to represent the current profile and to simulate SSO effects [2].

In the next few sections, we will discuss the xtalk and PDN effects from single ended signaling to LVDS. Also model to hardware correlation will be presented.

Single ended signal xtalk impact on LVDS channels

In a multi-protocol system environment like Altera FPGA package, the routing real estate is very limited due to the complexity of the design. Xtalk is hard to eliminate while cost is always a concern.

For a FPGA package especially a wirebond package, signal xtalk between single ended and differential IOs often causes problems. Most of the failures appear within VIH/VIL region at rise and falling edges by coupling glitches.

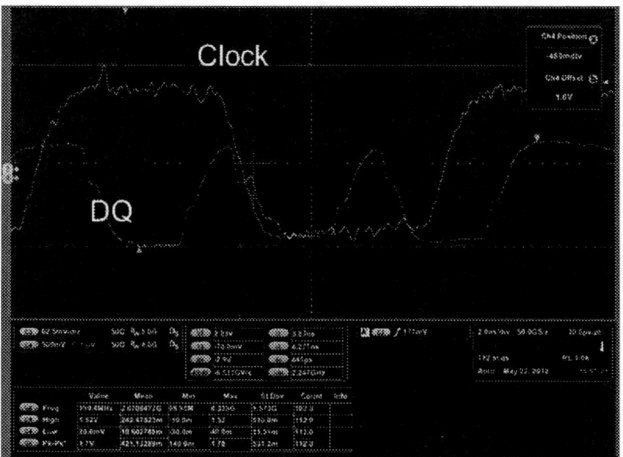

Figure 2. Noise coupling between single ended signals to differential clock

Sometimes the differential clock drivers are not true differential. Coupling noise affects P and N pins differently. Therefore, it is hard to cancel out the coupling noise between differential pairs. Figure 2 shows the signal coupling noise from DDR3 DQ pins (aggressors) to differential clocks (victims). Wirebond inductive coupling from single ended aggressors causes ripples at victim high/low status and VIH/VIL transition regions. Since clock pins have enough margins at high/low stages, the glitches will not be a problem. Rise and falling edges, on the other hand, are sensitive with all the coupled noise which triggers the false input for clocks.

Some workarounds of how to avoid clock failures are presented. First, we will take a look at signal slew rates.

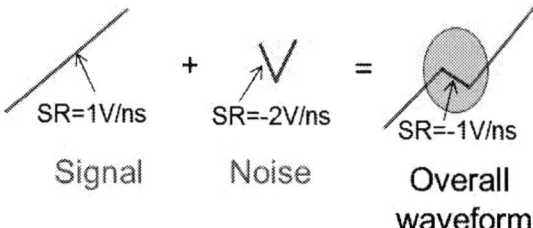

Figure 3. Slow edge victim with coupling noise

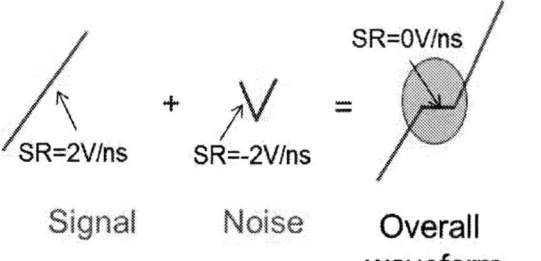

Figure 4. Fast edge victim with coupling noise

Figure3 and Figure4 illustrate the noise effects on slow and fast slew rate edge victim pins. Fast slew provides less transition time within VIL/VIH region and less chance to have glitches by coupling noise. In order to achieve small noise impact at the rise/fall region on the victim pins, the rule of thumb is to increase victim lines slew rate while decreasing aggressor lines slew rate.

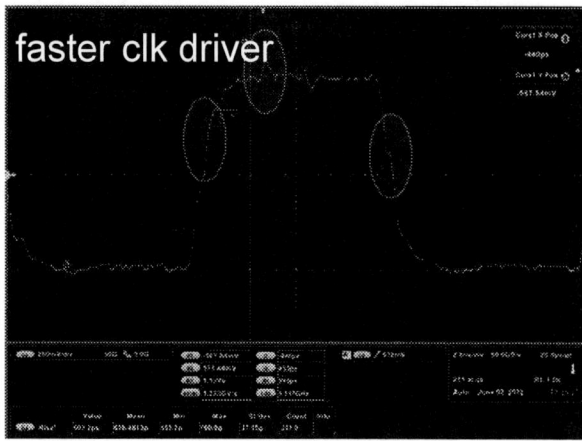

Figure 5. Rise/falling edge transition with faster clock driver

Using a faster slew rate clk driver, glitches during rise/fall transition region almost disappeared in the measured waveform shown in Figure 5. Coupling noise is unchanged at high/low stage. Slower slew rate aggressors create less noise on the victim lines. We should select slowest aggressor slew rate that meets the system speed requirement.

Another way to eliminate differential clock failure which caused by coupling noise from the SE signals is to control the buffer delay. So the data and clock output timing has some slew.

Having a good ground return path and adding more ground isolation between different signaling is always an ideal way to improve system signal integrity.

Aggressor IO signaling has different effects on victim pins. Unlike SSTL signaling, LVTTL/LVCMOS has open termination channel topology. A large part of noise will bounce back and forth between aggressors and victim pins which increases xtalk noise.

There are some general recommendations for reducing xtalk between signal ended signals to differential signals. First of all, we need to control the noise source, like adding ground guards and using less number of aggressor signals. Tightly coupled differential lines can greatly reduce the coupling noise from single ended aggressors. Increasing slew rate and drive strength of victim signals is also an effective way to mitigate xtalk. Sometimes, it is necessary to have timing delay between the aggressors and victim lines (differential clock) to avoid having voltage glitches on clock rising and falling transition periods.

Single ended signal SSN noise impact on LVDS channels

Simultaneous Switching Noise is one of the major margin loss contributors for single ended signaling environment. As we have known, FPGA provides variety of signaling

978-1-4799-2408-0/14 $31.00 © 2014 IEEE   749

standards in a same bank. SSN induced by single ended signaling can easily couple to differential channels. Designing a complex multi-protocol system with low cost and high performance, signal integrity and power integrity issues become extremely challenging. It is essential to analyze all noise effects. Single ended signaling with higher data rate, DDR3/4 channel noise over 1Gbps, has been simulated thoroughly [2]. In the next few sections, SSN noise impact from SE to LVDS will be demonstrated followed by model to hardware correlation.

Beside the routing complexity of GPIO signal traces in a FPGA package, the numbers of power rails are very high as well. Table 1 lists some popular I/O standards that Altera mid-end FPGA products support [3]. Due to the flexibility and large IO account of FPGA, different kind of I/O standards will appear in the same bank which causes the xtalk between single ended signaling and differential signaling.

I/O Standard	Vccio(V)	Vccpd(V)
SSTL	2.5/1.8/1.5	2.5/1.8/1.5
LVDS	2.5	2.5
LVCMOS/LVTTL	3.3/3.0	3.3/3.0

Table 1. I/O standards Voltage Levels in Cyclone V Devices

Because of the generality of FPGA IOs, there are so many power rails for each IO standards, including core power, pre-driver power, and output IO driver power. A practical power design is to share the same power for different I/O standards. In this study case shown in Figure 6, LVDS pre-driver and output driver power rails are tied together with the rest of single ended IO output drivers which could have more than 70 IO drivers. Normally, differential signaling is not sensitive to SSO noise. But this LVDS pre-driver circuits are single ended and output driver circuits are pseudo differential. Therefore, SSO noise voltage at pseudo differential output drivers can only be canceled partially. The timing push out comparing to ideal clock at pre-driver stage will accumulate and pass it on output driver.

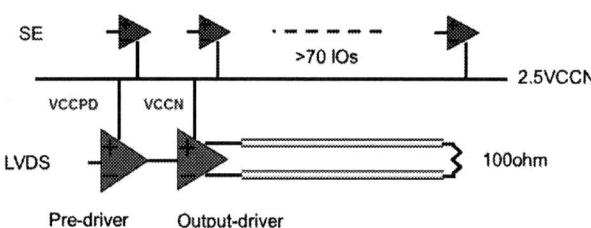

Figure 6. GPIO channel topology

The total power induced noise is a function of system PDN impedance and the current loop shown in equation 1.

$$V_{noise}(f) = Z_{pdn}(f) * I_{noise}(f) \qquad \text{Equation 1}$$

In order to predict single ended IO PDN effect on LVDS channels, pre-driver (VCCPD) and output driver (VCCN) PDN impedance curves were studied. Figure 7 shows only the VCCN PDN impedance over the frequencies for three

different wirebond package designs. From the Zpdn plot, package inductance and one die capacitance resonant peaks are varied from 100MHz and 200MHz. Naturally SE IO PDN noise at 170MHz has most impact on LVDS channel. For comparison, PDN noise at 20MHz was measured as well.

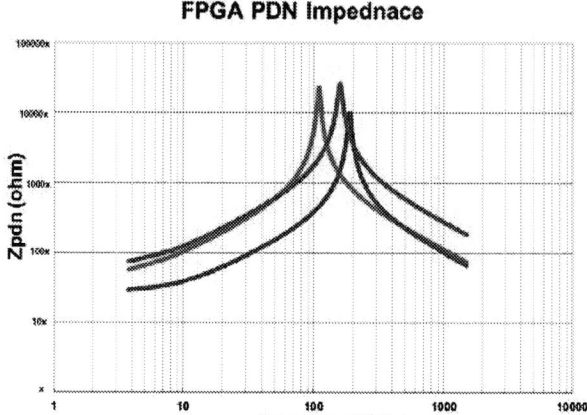

Figure 7. FPGA PDN impedance curves
(All numbers are normalized)

In addition to PDN frequencies, IO channel topology plays a big role on LVDS channels in term of total noise. SSN noise couples back and forth between power rails and signal channels. SSTL is well terminated at both controller and DRAM side. But LVTTL and LVCMOS have open termination channels which amplify signal overall noise due to reflection.

The PDN noise measurement setup and results from single ended signals to LVDS were described in details [4].

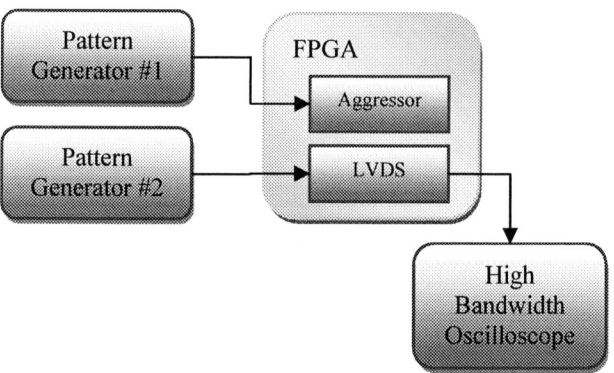

Figure 8. Altera test chip PDN noise measurement setup

In the hardware measurement, two different pattern generators are used to provide clock input to the single ended signals and LVDS showing in Figure 8. Asynchronous SSO noise is created under this condition.

The aggressors are multiple single ended signals that toggle synchronously and simultaneously with a 1010 pattern. Single ended signals' arrangement on die-pad location is shown in Figure 9. In the measurement, the aggressors will be turned on based on the sequence 1, 2, 4, 8, 16, 32, 64 and 74 as assigned in the pin map. The PDN noise versus the number of aggressor is measured using the high bandwidth oscilloscope.

978-1-4799-2408-0/14 $31.00 © 2014 IEEE

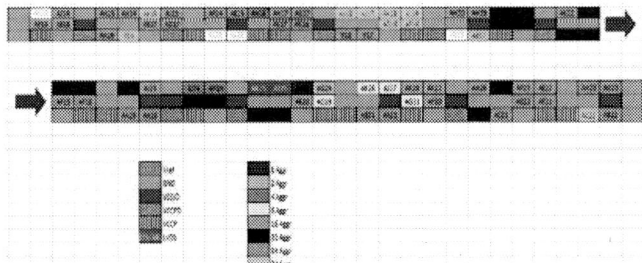

Figure 9. Single ended signals arrangement in Altera test chip (die-pad view)

The measurement procedure is listed as below:

1) Provide clock input to the LVDS signal
2) Provide desired clock input to the aggressors' clock in, it is 20 or 170-MHz in this paper
3) Turn on the single ended signal
4) Measure the PDN noise and eye diagram
5) Repeat step 3 and 4 with different number of aggressors

It is a well-known fact that PDN noise increases with flowing factors: 1) the number of aggressors switching around victim lines and 2) aggressors' slew rate, the higher number of IO switching and faster the slew rate the bigger the noise.

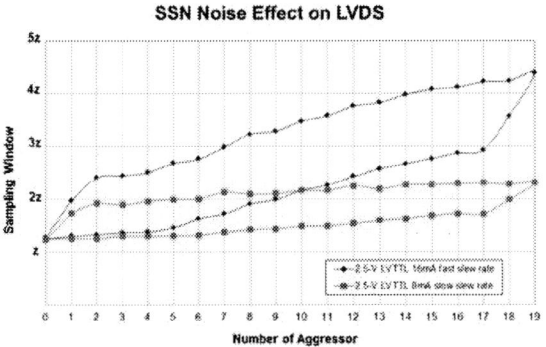

Figure 10. SSN noise effect on LVDS (All numbers are normalized)

Figure10 shows SE induced SSN noise effect on LVDS as number of switching IO increases. The blue curve and magenta curve represent the relationship of SSN noise and number of aggressor switching for 2.5V LVTTL 16mA and 8mA driver, respectively. The sequence of aggressor switching is called spiral and reverse spiral. The slopes of the curves indicate the inductance coupling and PDN effects. Mark the victim pin the center of the spiral and turn on adjacent aggressor one by one till the farthest location. For reverse process, turn the furthest pins on first and reverse spiral to the victim pin. Let's use 2.5V LVTTL 16mA, the blue curve, as an example. When closely coupled pins turned on, the noise increases sharply. As the number of IO switching reaching higher, the noise slope slows down. In the other hand, when reverse spiral happened, noise slope increases slowly until the last few pins. Clearly, 16mA driver has larger SSN effect on the victim line than 8mA driver.

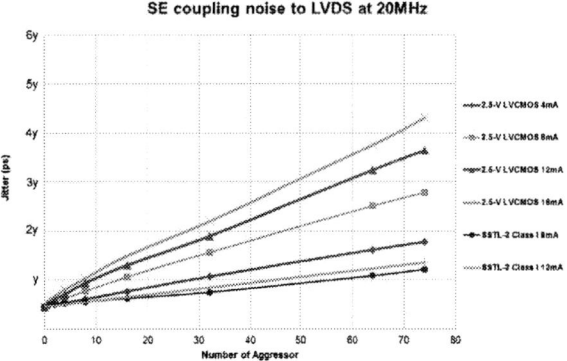

Figure 11. SE coupling noise to LVDS at 20MHz (All numbers are normalized)

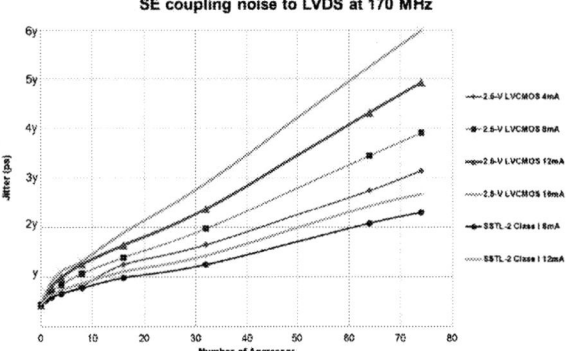

Figure 12. SE coupling noise to LVDS at 170MHz (All numbers are normalized)

In this section, more single ended signaling induced SSN noise will be studied.

Figure 11 and Figure 12 illustrate SSN noise from SE signaling coupled to LVDS pins by increasing the number of aggressors and by changing data rate, signaling IO standards, and drive strength of the aggressors.

Firstly, the outcome of these noise measurements shows the noise increases with number of aggressors switching. Secondly, open terminated IOs have higher noise impact than well terminated IOs due to reflection. Thirdly, bigger driver strength or faster slew rate causes more noise. Lastly, aggressors toggling at PDN resonant frequency introduce the highest PDN noise.

Model to hardware correlation

In this section, model to hardware correlation of single ended channel PDN effects on LVDS pins is demonstrated. The measurement setup is the same as shown in Figure 8. The whole channel includes all bank Altera LVDS/SE driver models, FPGA package models and PCB models which capture IO to channel ISI, xtalk and SSO effects.

The input data for the victim lines is fixed to be PRBS15 at 840Mbps. The first simulation tests on victim lines ISI performance with PRBS15 data pattern while all the aggressors being quiet. Plots A and B show the measurement and simulation data, respectively. Then all single ended pins were turned on in the same bank as long as the victim pins to demonstrate PDN effects. Different signal topologies,

978-1-4799-2408-0/14 $31.00 © 2014 IEEE   751

different drive strength, different data rates were performed on the aggressors.

All the eye diagrams of victim pins, measured at the receiver sides of the channel for measurement and simulation correlation, are shown in Figure 13. All cases here are with LVDS transmitter equalization on. Reasonable model and simulation correlation was achieved. The simulated noise only includes pre-driver, output driver, and channel effects. It has an ideal condition which the real measurements have core noise, random noise and other circuit effects. LVDS voltage noise has been partially canceled. But timing error has been canceled and varied with different aggressors' signal topology, IO driver strength, and data rate.

From plot A to H in Figure 13, the overall PDN noise on LVDS pins increases with aggressor output driver drive strength, open termination IOs, and PDN resonant frequency. Based on the system PDN effects on LVDS signals discussed on this section, single ended IOs need to be carefully designed around LVDS to guarantee channel margin.

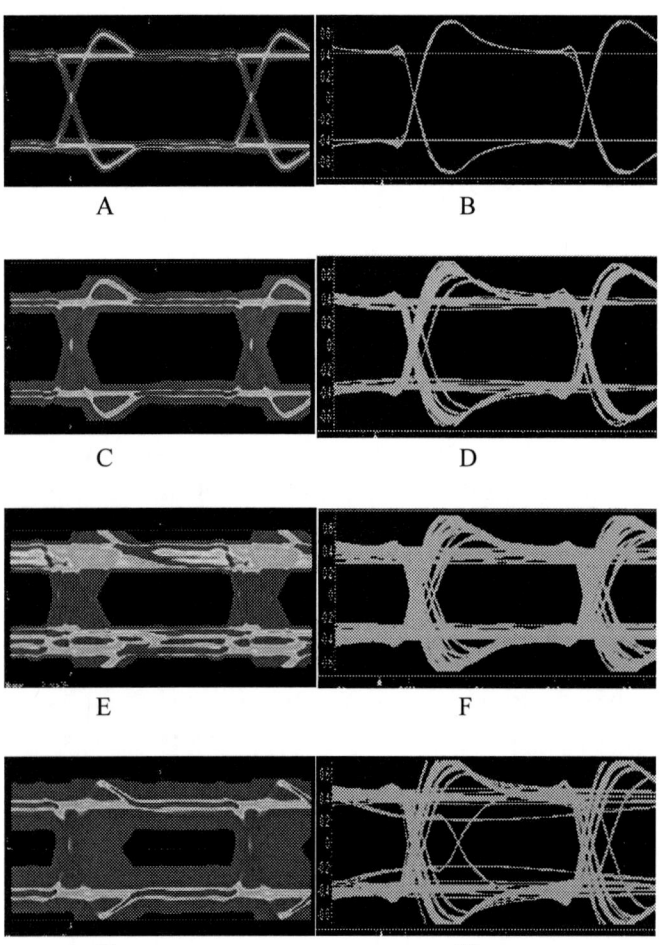

Figure 13. Measured vs. Simulated channel eye diagram

(A): Measured eye at quiet condition
(B): Simulated eye at quiet condition
(C): Measured eye with SSTL 8mA driver at 20MHz
(D): Simulated eye with SSTL 8mA driver at 20MHz
(E): Measured eye with SSTL 8mA driver at 167MHz

(F): Simulated eye with SSTL 8mA driver at 167MHz
(G): Measured eye with LVCMOS/LVTTL 16mA driver at 20MHz
(H): Simulated eye with LVCMOS/LVTTL 16mA driver at 20MHz

Conclusions

In a complex multi-protocol GPIO environment, it is hard to balance the cost and the performance. To design a cost effective and high performance FPGA system, signal integrity and power integrity issues become more and more challenging. Coupling impact of single ended signals to LVDS interfaces has been demonstrated.

The system level SI/PI co-simulation methodology was presented to analysis coupling noise. Two examples, about coupling noise between single ended IO to differential IO, were discussed. Aggressor slew rate (drive strength), channel topology, number of IO switching, system PDN and aggressor data rate are all affect overall noise to LVDS pins.

It is important and yet difficult to design a robust system to achieve a low cost and high performance criteria. The last part of this paper shows the model and measurement correlation based on the SI/PI co-simulation methodology. Good correlation was achieved.

Acknowledgments

Special thanks to Hazlina Ramly, Yujeong Shim, Yin Mei Yap, Yee Huan Yew, Chit Zhung Tan, Lian Nee Soh, Siong Hee Lim for discussion and support.

References

1. D. Oh and C. Yuan, High-Speed Signaling: Jitter Modeling, Analysis, and Budgeting, Prentice Hall, 2011
2. J. Feng, B. Dhavale, J. Chandrasekhar, Y. Tretiakov, D. Oh, "System Level Signal and Power Integrity Analysis for 3200Mbps DDR4 Interface", ECTC 2013
3. Altera Cyclone V Device Handbook
4. J.Kho, CI. Loh, B. Krsnik, Z. Li, CS. Fong, P. Boyle, and MO Wong, "Effect of SSN-induced PDN Noise on a LVDS Output Buffer", APACE 2007

Analysis on Interference between Multi-Giga bit Display Serial Link and RF Components in Smart Mobile Device

Youchul Jeong, Jaemin Kim, and Baegin Sung
Silicon Image Inc.
1140 E. Arques Ave. Sunnyvale CA 94085 USA
Tel: 408-616-1579, Fax: 408-616-6399, E-mail: yjeong@siliconimage.com

Abstract

In Smart Mobile Devices, many components are integrated in a small space while each of them is operating at higher frequency ever. In the case of RF components, they are working at a specific single frequency, and it is relatively easy to avoid the interference by allocating the frequency band adequately. However the digital signal which is used in cutting-edge application processor and Multi-Giga bit display serial link does not have a single frequency but has a broad frequency spectrum, and it is very hard to avoid a frequency conflict between digital components or between digital and RF components. RF components are especially vulnerable to external noise, so the high-speed serial link usually becomes a noise source and RF components become its victim. In this paper, we analyzed noise emissions from various components on a Multi-Giga bit display serial link channel, and we found out how the interference between the link and RF components can be minimized, and signal integrity of the link can be improved by proper design.

Introduction

As demand for smart mobile device ecosystems increases, a display interface between mobile device and TV/monitor becomes a must-have feature. To transfer high quality video and audio through a very limited number of pins, a Multi-Giga bit serial link has been adopted in display interface, and its data rate becomes even higher than that of the application processor. Because of the ultra-high-speed data rate, system design considering signal integrity and EMI gets harder and harder. Especially, minimizing noise interference between high-speed digital signal and RF components is very important because RF components are very sensitive to digital noise and even a very small amount of noise coupling degrades RF sensitivity of the mobile device significantly.

In this paper, we studied how the components on the display serial link channel (e.g. copper etch, ESD, common mode choke, and connector) generate noise, and how this noise interference is minimized by proper design. There is a myth that differential signaling solves all radiation-related issues by canceling out its emission. However, in mobile devices, noise coupling happens in a very small space because many components are integrated on a small PCB, so differential signaling does not guarantee the inference free channel as expected. So, even though most high-speed serial links use differential signaling, all copper etches and discrete components on the PCB need to be considered as a kind of RF antenna in noise analysis.

This paper consists of three sections. First we simulated near field emission from copper etches on PCB using 3D full-wave simulator. We verified the simulation data with test results using a near field scanner, and we optimized design for copper traces, via, and pads based on the simulation results. Second we analyzed discrete components on the serial link channel as noise emission sources. Since it is not possible to model the components accurately in full wave simulator, we made many test boards with different kinds of components, and we measured radiated emission using near field scanner. We compared two different sizes of ESD diodes, two different types of common mode chokes, and two different types of uUSB connectors. After this noise emission analysis based on simulation and measurement, we checked the signal integrity of serial link channel using these components by simulation. After all, we concluded how we can resolve link performance and emission problems together from these simulation and measurement results.

Simulation method

To calculate the near field emission from copper etches, we simulated various PCB patterns using HFSS 3D full wave simulator [1]. The simulation conditions are summarized in Table 1.

Frequency	870MHz ~ 895Mhz, 5MHz step
Calculated field	Maximum Hz
Vertical distance	1mm
Mesh size	0.35mm

Table 1 3D full wave simulation condition

In this paper, we focused on the GSM850 band (Downlink: 869.2MHz–894.2MHz) and we swept the frequency from 870MHz to 895MHz with 5MHz frequency steps. We calculated the maximum magnitude of H field because PCB etches and discrete components radiate H field very well, rather than E field and RF antenna is also very sensitive to H field. The Hz field is calculated over copper etches at 1mm height from the PCB, considering real smart phone design in which components are placed close to each other as short as 1mm.

We set the mesh size for field calculation based on physical length ($0.35mm = \lambda/1000@850MHz$) manually instead of using auto-generated mesh by the simulation tool. The recent 3D full-wave simulator generates its mesh automatically and the size and the number of mesh are decided by algorithm for high simulation efficiency. However the auto-generated mesh is optimized for accurate S-parameter results and simulation time to extract the parameters, and it is not optimized for near field simulation. Figure 1 shows the auto-generated mesh by field solver at 10GHz and the manual mesh based on 0.35mm physical length. To reduce calculation time, we only applied the fine mesh to the area around traces.

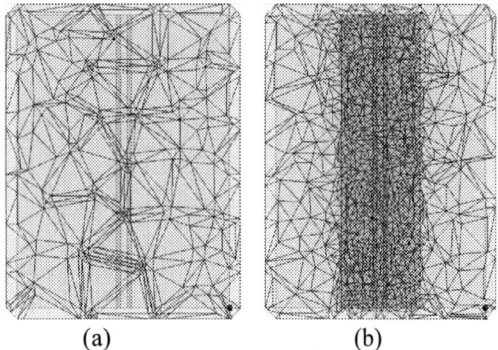

Figure 1 Mesh plots (a) Auto-generated mesh by simulation tool at 10GHz (b) Manual mesh

Figure 2 shows the field heat map of Hz field depending on mesh style. With auto-generated mesh, even though it is solved at high enough frequency (10GHz) comparing to the target frequency (850MHz), it is difficult to see the field distribution and to find the location where the field strength is the strongest. On the other hand, manual mesh gives a clear picture of Hz field distribution and it is easy to find out where the noise coupling will be maximized.

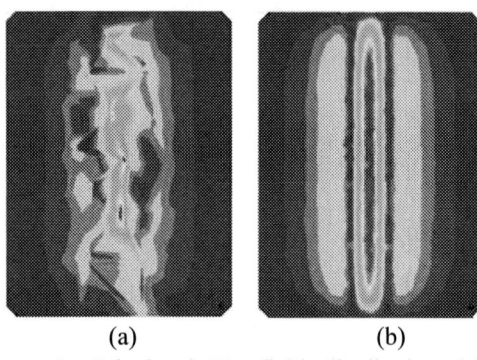

Figure 2 Calculated Hz field distribution (a) Auto-generated mesh at 10GHz (b) Manual mesh

With 0.5V differential signal excitation, the maximum Hz field is captured and summarized in Table 2.

H field (dBA/m)		Auto mesh	Manual mesh
Frequency (MHz)	870	-11.89	-15.82
	875	-11.89	-15.82
	880	-11.89	-15.82
	885	-11.89	-15.82
	890	-11.89	-15.82
	895	-11.89	-15.82

Table 2 Maximum field from auto mesh and manual mesh

As shown in the table, the maximum Hz field values are different depending on mesh style, and auto-generated mesh gives 4dB bigger values than manual mesh. In frequency sweep, Hz field does not change in GSM850 band regardless of mesh type. So we fixed the frequency to 885MHz instead of sweeping the frequency for the analysis on PCB etches and components based on simulation and measurement.

Noise emission from copper etches on PCB

First we compared the near field emission from 50ohm single-ended trace and four kinds of differential traces of which width and space are different while keeping 100ohm differential mode impedance. Figure 3 shows the dimension of the single-ended trace and differential traces.

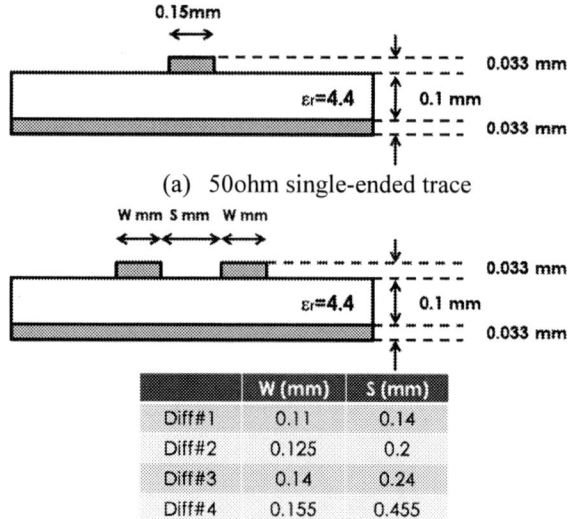

(a) 50ohm single-ended trace

	W (mm)	S (mm)
Diff#1	0.11	0.14
Diff#2	0.125	0.2
Diff#3	0.14	0.24
Diff#4	0.155	0.455

(b) 100ohm differential trace

Figure 3 Dimension of single-ended and differential traces

To keep 100ohm differential impedance, we increased the trace width and adjusted the space between two traces accordingly. Figure 4 shows the Hz field distribution at 885MHz and Figure 5 shows the maximum field values over several cases. As expected, single-ended trace emits more Hz field than differential traces and its field distribution is broader than that of differential traces. On the other hand, the Hz field is confined between two traces in differential traces and the maximum field level is lower than that from single-ended trace. As the space between two differential traces increases, two traces becomes uncoupled and the Hz field from differential traces becomes bigger. The difference between Diff#1 and Diff#4 is about 6dB and the noise emission from Diff#4 is comparable to that from single-ended trace.

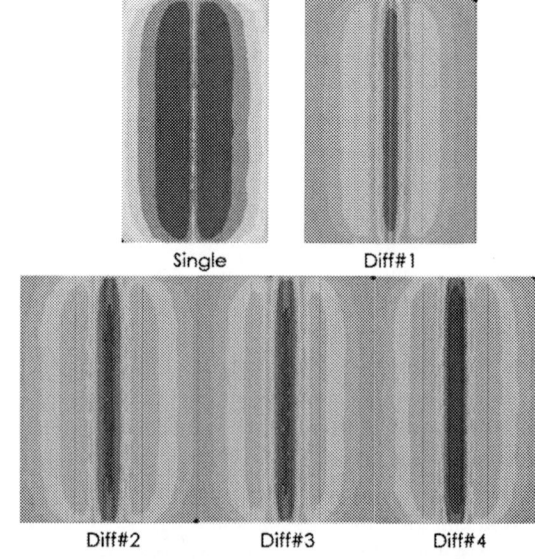

Figure 4 Hz field distribution depending on copper etches

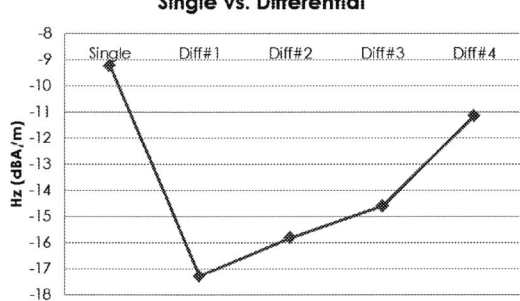

Figure 5 Maximum Hz field value from various copper etches on PCB

From these results, we can conclude that improper design of differential traces could be bad as much as a single-ended trace, and the single-ended trace needs to be shielded very well to reduce the radiated emission in a mobile device. The best way to shield the single-ended trace is to use strip line instead of u-strip line. The Hz field can be lower than that from the best differential traces using strip line.

Next we simulated the Hz field from pads which are used for component mounting or for signal monitoring. Figure 6 shows the dimensions of the circular pads we analyzed.

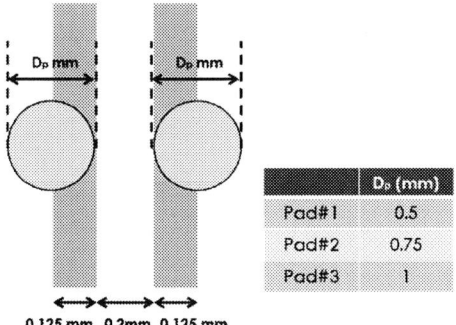

Figure 6 Dimensions for pads on differential traces

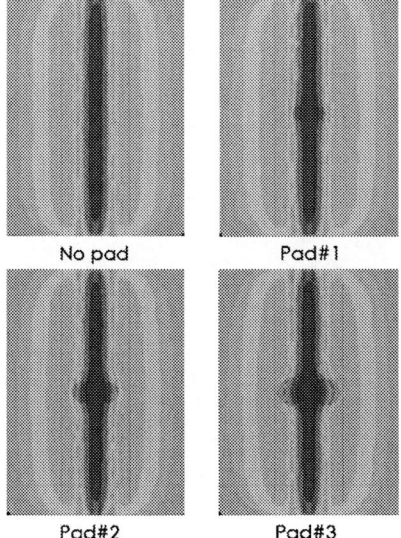

Figure 7 Hz field distribution depending on pad size

Figure 7 shows the Hz field distribution at 885MHz and Figure 8 shows the maximum field values per each case. In these two figures, the magnitude of Hz field increases in proportional to pad size, and the area of Hz field increases as well.

Basically, any pad on the PCB with a size wider than the differential traces increases parasitic capacitance and it reduces the characteristic impedance consequently. As a result, this impedance discontinuity corrupts current flowing on the differential traces and it causes more emission as well as it degrades signal integrity.

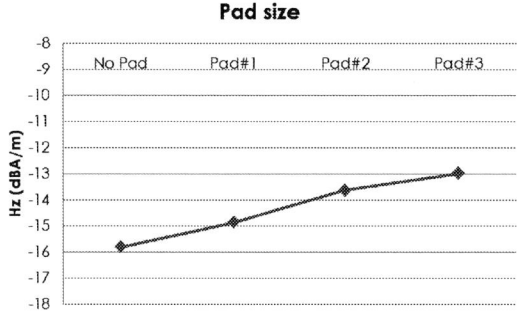

Figure 8 Maximum Hz field depending on pad size

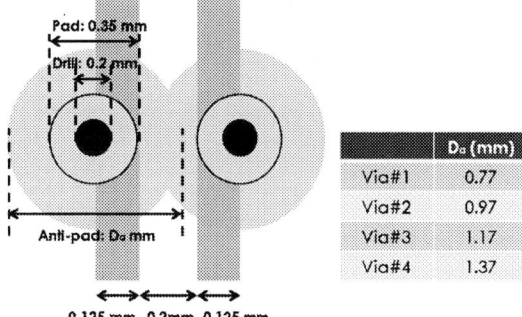

Figure 9 Dimensions for vias on differential traces

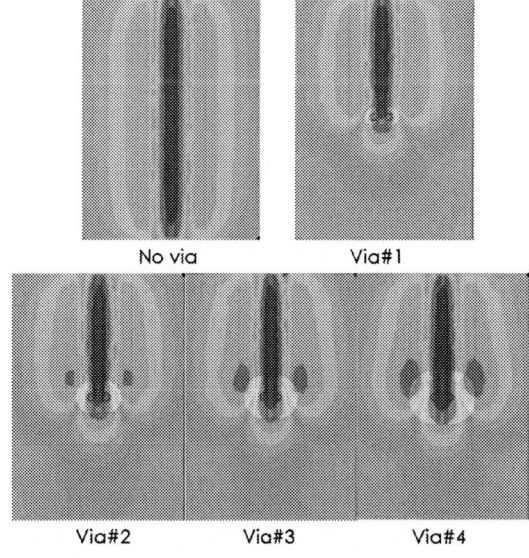

Figure 10 Hz field distribution depending on anti-pad size for via

The next structure is the signal via for layer change. Figure 9 shows the dimension of the signal via. To simplify the analysis, drill and pad sizes were fixed and anti-pad size was varied. Figure 10 and Figure 11 show Hz field distribution and maximum Hz field value depending on anti-pad size. In this case, the anti-pad increases parasitic inductance of the differential trace and decreases parasitic capacitance of via, and it causes an impedance discontinuity by increasing characteristic impedance. Therefore, the larger anti-pad causes the bigger emission as shown in Figure 11. However the dependency on the anti-pad size is less compared to the dependency on the pad size.

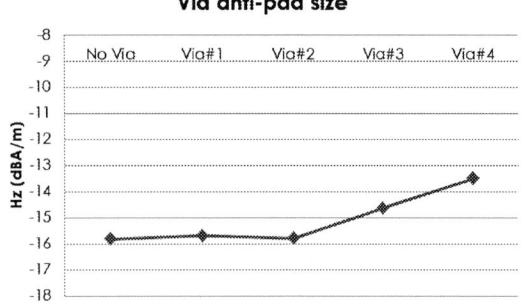

Figure 11 Maximum Hz field depending on anti-pad size

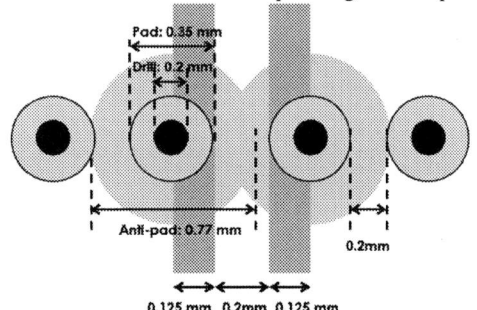

Figure 12 Differential signal vias with two ground vias and its dimensions

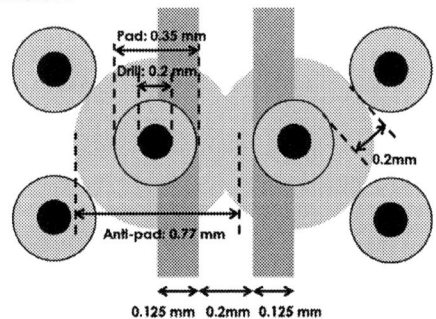

Figure 13 Differential signal vias with four ground vias and its dimensions

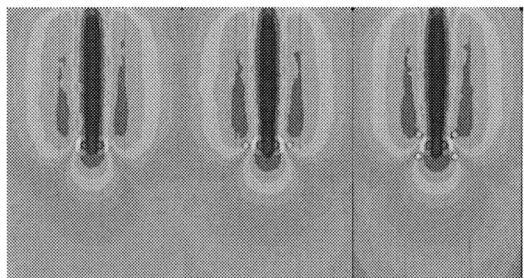

Figure 14 Hz field distribution depending on number of ground vias

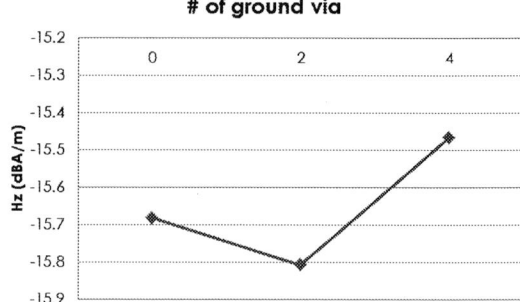

Figure 15 Maximum Hz field depending on the number of ground vias.

When a signal trace changes layers through a via, a ground via is usually placed next to the signal via to reduce noise coupling and to keep its characteristic impedance as designed.

In this paper, we simulated Hz field when two ground vias and four ground vias are placed next to differential signal vias, and Figure 12 and Figure 13 show their dimensions. Figure 14 and Figure 15 show the Hz field distribution and the maximum Hz field depending on the number of ground vias. As shown in the results, there is no dependency on the number of ground via. This is because we captured the Hz field and the ground vias do not help reduce Hz field due to its orientation.

Near field measurement method

For H field measurement, we used a spectrum analyzer, a near field probe, and a near field scanner. Table 3 summarizes the measurement conditions. The frequency was fixed at 885MHz and 10KHz resolution bandwidth was applied in the spectrum analyzer considering SNR. We used two kinds of near field probes for Hz and Hy field measurement. The Hy field is measured for connector evaluation.

Frequency	885MHz
Resolution BW	10KHz
Probe	Langer RF-B 0,3-3 for Hz field Langer RF-R 0,3-3 for Hy field
Step	0.2mm

Table 3 Near field measurement condition

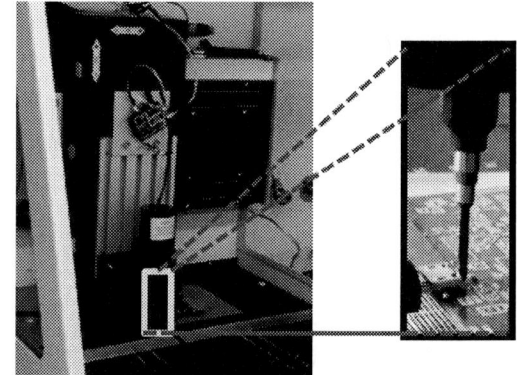

Figure 16 Near field scanner and probe

Figure 16 shows the photo of the near field scanner and the near field probe. To correlate the measurement and simulation results, the Hz field was captured at 1mm height from the PCB etches and the two-dimensional field distribution was scanned with 0.2mm step.

To check the validity of full wave simulation and near field test, we picked one of the differential traces (Diff#2) and compared the Hz field distribution from simulation and measurement. Figure 17 shows the results.

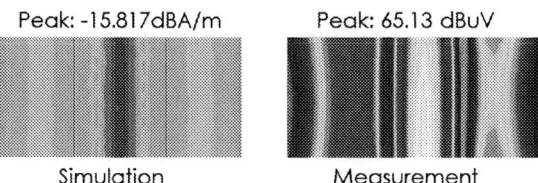

Peak: -15.817dBA/m — Simulation

Peak: 65.13 dBuV — Measurement

Figure 17 Hz field distribution obtained from simulation and measurement

Even though distribution is somewhat asymmetric in the test results, the peak Hz field was well captured in between two traces and it matched well with simulation result. By comparing between two maximum Hz field values, we calculated the correction coefficient, and this number was used in components evaluation.

Near field emission from components on PCB

In high-speed serial link, ESD diode and common mode choke (CMC) are used for ESD protection and EMI reduction, and they are adopted in many smart mobile devices as well.

When an ESD diode is selected for high-speed serial link, the key value in the datasheet is the parasitic capacitance which affects signal bandwidth directly. So, in the first step, ESD diodes which do not meet the target capacitance need to be fileted out. After this step, near field emission must be scanned and analyzed especially in smart mobile device. In this paper, we compared two different sizes of ESD diodes and Table 4 shows their dimensions.

ESD#1	Diode type	1.3mm x 0.8mm, 0.4mm height
ESD#2	Diode type	1.6mm x 1.0mm, 0.58mm height

Table 4 Dimensions of two different ESD diodes - ESD#1 and ESD#2

Figure 18 shows the measured near field emission from two kinds of ESD diodes and, in this test, it was turned out that the bigger size of the ESD diode gave more emission than the smaller part. However, it is not always true because ESD diode is active device and emission mechanism is complicated. Therefore ESD diode needs to be tested before it is used in smart mobile device.

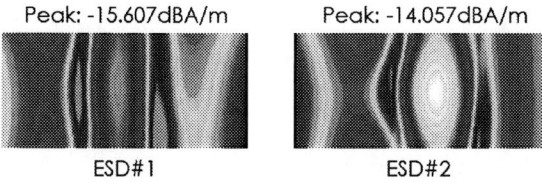

Peak: -15.607dBA/m — ESD#1

Peak: -14.057dBA/m — ESD#2

Figure 18 Hz field distribution depending on ESD diode

In similar to ESD diode, 3dB bandwidth of Common Mode Choke (CMC) is the most critical factor to consider in high-speed serial link, CMC components which do not meet the bandwidth requirement must be screened out first. We selected two types of high-bandwidth CMCs depending on their internal structure, which are film type and wire wound type as shown in Table 5. Because the height of each CMC is close to 1mm, we scanned Hz field at 2mm height instead.

Figure 19 shows the Hz field distribution. Depending on the CMC internal implementation, Hz field distributions are different and there is about 0.9dB difference between two CMCs.

CMC#1	Film type	1.25mm x 0.5mm, 0.82mm height
CMC#2	Wire wound type	2.0mm x 1.2mm, 0.58mm height

Table 5 Comparison between CMC#1 and CMC#2

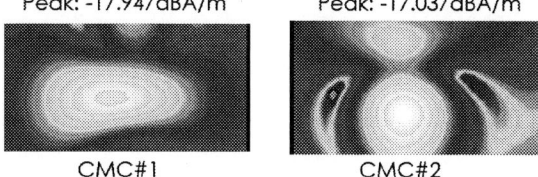

Peak: -17.947dBA/m — CMC#1

Peak: -17.037dBA/m — CMC#2

Figure 19 Hz field distribution depending on ESD diode

The last components we tested are uUSB connectors which are used at smart mobile device for charging, USB, and display link connection. In the case of connectors, there is not much freedom to change its structure because of the specification to which the connector belongs. So we just picked two kinds of connectors in the market and the structures are shown in Figure 20. The big difference between two connectors is the location of the vertical pin connection between receptacle connector and PCB board. To see the effectiveness of design change, we measured Hy field instead of Hz field.

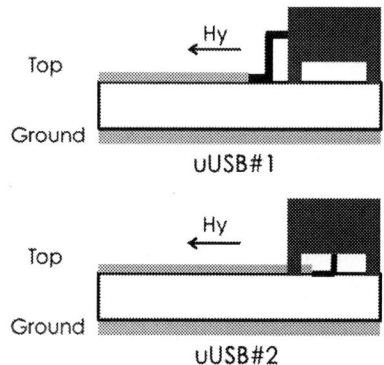

Figure 20 Two different uUSB connector - side view

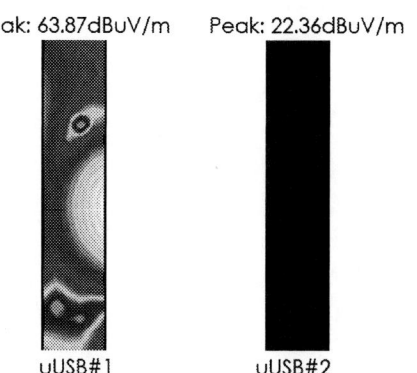

Peak: 63.87dBuV/m — uUSB#1

Peak: 22.36dBuV/m — uUSB#2

Figure 21 Hy field distribution depending on uUSB connector

Because of the difficulty in the design of reference test fixture for Hy field measurement, the correlation coefficient was not calculated and the measurement values from connector were not converted to dBA/m. So it was not possible to compare the emission amount from connectors and that from other components, but it is still meaningful to compare the measured values from two structures as is. Figure 21 shows the Hy field distribution and the peak magnitude of Hy field, and we can see that the Hy field is reduced significantly in uUSB#2 by hiding vertical pins under the metal enclosure.

Signal integrity analysis

For signal integrity analysis, we ran full-wave simulation to extract S-parameter from PCB etches and we measured S-parameter using vector network analyzer to get S-parameter for ESD diode, CMC and uUSB components.

Figure 22 shows the serial link channel from transmitter chip to connector used in this simulation and we had the total channel S-parameter by combining all S-parameters for PCB etches and components.

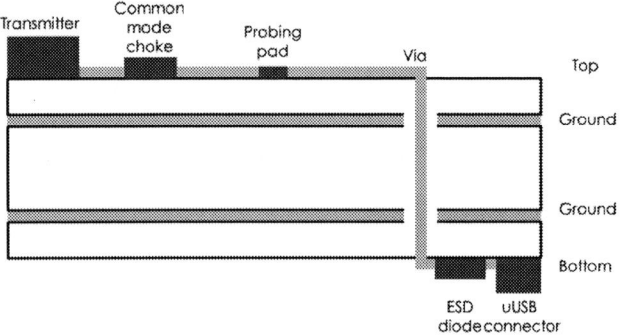

Figure 22 Multi-Giga bit display serial link channel

As shown in the previous section, PCB etches give less near field emission when the impedance discontinuity is minimized and it is also good for signal integrity. So we selected the best combination and the worst combination for PCB etches including differential trace, probing pad and via, while CMC#1, ESD#1, and uUSB#2 are used for both cases. The PCB etches for the best combination are Diff#1, Pad#1, and Via#1, and the worst combination consists of Diff#4, Pad#3, and Via#4. Figure 23 shows the differential insertion loss for both cases, and it is verified that the best case shows less insertion loss overall.

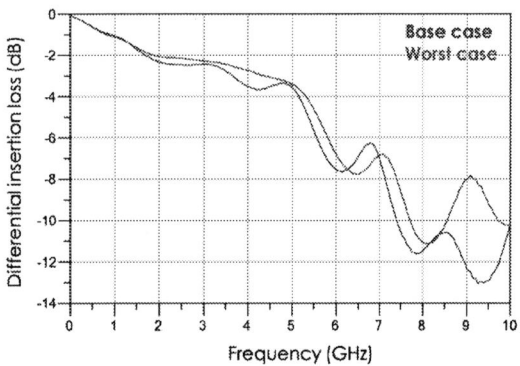

Figure 23 Differential insertion loss for the best and the worst combination

For eye diagram simulation, a 6Gbps differential transmitter model was applied and the results are shown in

Figure 24. By changing PCB etches, there was a 12% improvement in eye height.

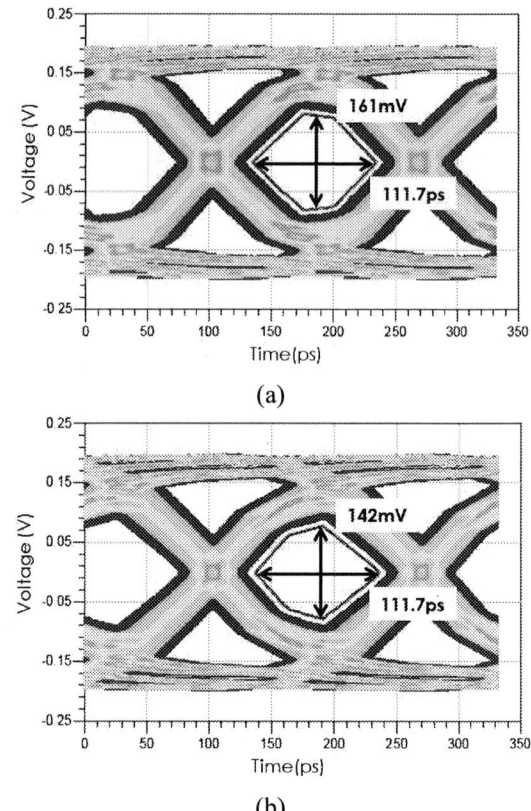

(a)

(b)

Figure 24 6Gbps eye diagram (a) best case (b) worst case

Conclusions

We analyzed near field emission from various PCB etches and components based on full-wave simulation and near field measurement. By doing adequate full-wave simulation and near field measurement, the noise interference between display serial link and RF components can be minimized and signal integrity can be improved together.

Acknowledgments

We acknowledge Dr. Hoo Kim for near field data collection, Duc Nguyen for layout design of test PCB patterns, and William Altmann for final document review.

References

1. Ansoft, "Ansoft HFSS Field Calculator Cookbook"

AUTHOR INDEX

A

Aasmundtveit, Knut 139, 498
Abe, Hidenori 1523
Agarwal, Rahul 590
Agrawal, Akash 2014
Agrawal, Jai P. 1838
Ahn, Billy 2027
Ahn, Byungki 1616
Ahn, Yesul 1361
Aizawa, Mitsuhiro 1166
Akaike, Masakate 1143
Akin, Meriem Ben-Salah 1595
Alatorre, Roseann 215
Alcira, Cecille 47
Allen, Aileen M. 1433
Allen, Craig 878
Almeida, Rodrigo 935
Althoff, Simon 1549
Alvanos, Tyson 452
Alvarado, Rey 100, 925, 1173
Amara, Karima 1198
Aminov, Parvizso 1838
Amirkhany, Amir 560
Anderson, R. 1641
Andreassen, Erik 139
Andry, P. 576, 883, 1372, 1880
Angyal, Matthew 647
Anselm, Martin 119
Anzai, Nobuhiro 829
Aoki, Issei 1776
Aoyagi, Masahiro 62, 1915, 1926, 2003
Araga, Shinji 2194
Arai, Yoshiyuki 913
Arakere, Guruprasad 395
Arfaei, Babak 425, 655, 1769
Arik, Mehmet 209
Arkalgud, Sitaram 862
Armutlulu, Andac 1098
Arnold, Kim 894
Arvin, Charles 1940

Aryasomayajula, Lavanya	348
Asahi, Noboru	913
Asai, Kosuke	186
Asai, Osamu	1885
Asai, Satoshi	1028
Asano, Tanemasa	2303
Aschenbrenner, R.	940
Atsumi, Koichiro	2227
Audoin, M.	1714
Auerswald, E.	1134
Augustine, Anne E.	528
Aung, Kyaw Oo	596
Awatsuji, Yasuhiro	186

B

Baba, Mikio	763
Baba, Shunji	68
Bader, V.	940
Bae, Byeong-Hyun	1635
Bae, Byung-Hyun	1735
Bae, Hyun-Cheol	1154
Bae, In-Tae	1562
Bae, J.-C.	304
Bae, Jangyong	973
Baek, Hyunho	554
Bailey, Chris	1342
Bailey, Susan	894
Baillin, X.	1714
Bajwa, A.A.	1707, 2181
Bakir, Muhannad S.	13
Baks, Christian	1016, 1272, 1955
Ball, Sabrina	2035
Baloglu, Bora	1231, 1401
Balucani, M.	194, 1571, 1842, 2137
Bandarenka, H.	194
Bao, Andy	47
Barros, Isabel	935
Barth, Holly D.	1457
Barwicz, Tymon	179
Bauer, J.	940
Bchir, Omar	1396
Bea, JiChel	636
Becker, K.-F.	940
Beer, Gottfried	1183
Belardini, A.	194
Beleran, John D.	490

Benedetti, A. 194
Benedetto, Elizabeth 425, 1433
Benjamin, Shuki 1021
Bennett, N.S. 1517
Berg, J. 1071
Berger, Daniel 647
Berger, François 478
Berghuvud, A. 2220
Bernardi, D. 1571, 1842
Berry, C.J. 1965
Beyene, Wendemagegnehu 730
Beyer, G. 33, 309, 572, 850, 894
Beyne, E. 26, 33, 309, 572, 613, 850, 894
Bezuk, Steve 47, 925, 1173, 1396
Bian, Yuan 1902
Bieler, Thomas R. 697
Björnängen, T. 2220
Blair, Justin 609
Böck, J. 956
Bock, K. 1482
Bohm, Johannes 1509
Bollmann, D. 1482
Bondarenko, V. 194
Borgesen, P. 371
Bösch, Wolfgang 1183
Böttcher, M. 625
Bowen, Terry 1054
Boyer, Nicolas 179
Boyne, Dan 1539
Bozack, Michael J. 379
Bozack, Mike 242
Brady, D. 74
Brand, Sebastian 850
Brandt, Lutz 279, 1652
Braun, T. 940
Bravin, Julian 888
Breach, Christopher 1803
Brofman, Peter 255, 1308
Brusberg, Lars 1033, 1498, 1759
Budd, R. 883
Burggraf, Jürgen 888

C

Caffey, Kevin 1173
Caggiano, Michael 2159
Cai, Jian 1378

Calvert, Jeffrey	342
Campos, José	935
Cangellaris, Andreas C.	717, 2094
Cao, Andrew	862, 1810
Cao, Liqiang	1116, 1932, 2008, 2131
Cao, Zhihua	464
Caplet, S.	1714
Capuz, G.	572
Cardoso, Paulo	935
Castagné, Laetitia	1198
Castro, José	935
Cate, S.	74, 952
Cha, Seungyong	354
Chada, Arun Reddy	2081
Chahal, Premjeet	775, 2144, 2168
Chainer, Timothy	1955
Chakraborti, Parthasarathi	541, 1492
Chan, M.	952
Chan, Yan Cheong	1342
Chang, Chih-Wei	512
Chang, David	947
Chang, Gee-Kung	1054
Chang, Hong-Da	947
Chang, Hsiang-Hung	1891
Chang, Hung-Hsein	868, 1628
Chang, Jenny	1562
Chang, Nistec	81
Chang, Pai-Cheng	290
Chang, Po-Chih	1853
Chang, Shih-Chieh	316
Chang, Xin	2200
Chang, Yiu-Hsiang	1853
Chao, Chun-Chieh	868
Chao, Shu-Han	1908
Chao, Yu-Lin	290
Chau, Ellis	1810
Chen, Agassi	1750
Chen, Cheng-Fu	963, 2020
Chen, Chien-Chou	1853
Chen, Chih	1908
Chen, Chunwei	26
Chen, Dong	2299
Chen, Eason	81
Chen, Erh-Hao	1853
Chen, Gang	1080
Chen, George	470

Chen, Guang	566, 748
Chen, Hsien-Wen	1
Chen, Jing	1902, 2099
Chen, Jui-Chin	1853
Chen, K.M.	297
Chen, Kim-Hui	2299
Chen, Kuan-Neng	512
Chen, Kuo-Hua	512
Chen, M.T.	572
Chen, Qi	1992
Chen, Qianwen	1372
Chen, Quan	1992
Chen, Scott	1562
Chen, Shang-Chun	1853
Chen, Shi-Ching	868
Chen, Shih-Ching	1628
Chen, Stephen	1
Chen, Su-Mei	1908
Chen, Tai-Yu	2069
Chen, Xu	717, 1080
Chen, Yan-Heng	1750
Chen, Yan-Siang	1584
Chen, Yen-Chi	1258
Chen, Yi	1584
Chen, Yi-Ling	1529
Chen, Yin-Fa	419
Chen, Youpeng	1488
Chen, Yu	1378
Chen, Yu Hua	360, 1590
Chen, Zhaoqing	723, 2104
Cheng, Hung-Hsiang	2112
Cheng, Lianxi	1335
Cheng, M.D.	572
Cheng, Ren-Shin	290, 1908
Cheng, Xiaojin	1926
Cheng, Yu-Mei	290
Chéramy, S.	906
Cherman, V.	309, 1517
Chia, Pierre	1122
Chiang, Tzu-Hsing	419
Chien, Chun-Hsien	290, 1891
Chien, F.L.	56
Chien, Heng-Chieh	290, 963
Chin, Wai Foo	1790
Chiou, Jin-Chern	512
Chiu, Chi-Tsung	512, 2112

Chiu, Steve 1, 947
Chiu, Ying-Ta 419
Chlieh, Outmane Lemtiri 2293
Cho, Byoungwoo 1361
Cho, Jounghyun 541
Cho, Junghyun 1328
Cho, Sangbeom 1247
Choi, Alan 925
Choi, BaekKyu 2075
Choi, Eun-Kuk 1735
Choi, Hoi Wai 919
Choi, Hyeseon 1004
Choi, Joonyoung 836
Choi, Kwang-Seong 1154
Choi, Won Kyung 596
Choi, Yongwon 1128, 1661, 1755, 2241
Choi, Youjoung 836
Choi, Young Won 1765
Choki, Koji 1148
Chong, Chan Kai 490
Choo, Joung-Hoon 1735
Chou, Bruce C. 1054
Chou, Lei-Chun 512
Chua, S.L. 324
Chuang, Ching-Te 512
Chuang, Shih-Keng 1944
Chung, Ming-Feng 2112
Chung, YoungSuk 582, 1816
Chylak, Bob 1523
Ciarniello, Danilo 1571
Clauberg, Horst 1523
Co, Rey 215, 1389
Cochet, Philippe 20, 523
Collado, Ana 796
Collins, Sian 878
Conti, Fosca 1464
Cook, Jeffery 684
Cordes, Steven A. 1782
Cotts, Eric 655, 690, 1769
Coutier, C. 1714
Cowley, A. 1517
Coyle, Richard 425, 655
Cremaldi, Joseph 255
Crescenzi, Rocco 1571
Croes, Kristof 613
Crouthamel, D. 74

Cui, Tong 100
Cunningham, G. 8
Czurratis, Peter 850

D

Daerhan, Daerhan 759
Dai, Ming-Ji 290
Daily, Derek 1308
Daily, R. 165, 309, 572
Dal Molin, Renzo 1198
Dalal, Mitul 145
Dan, Qiang 1010, 1690, 2245
Dang, B. 576, 883, 1372, 1880
Dang, J. 74
Danilewsky, A.N. 1517
Darveaux, Robert 703
Daubenspeck, T. 1949
Davis, Roy I. 1996
Davis, Taryn 236
De Messemaeker, J. 33, 613
De Vos, J. 309
De Wolf, I. 309, 613, 850, 1517
Defay, E. 1296
Dehag, G. 1714
Dej, Sebastian 590
Demir, Kaya 1098
Demosthenous, Andreas 2213
Derix, Robert 1464
Deshpande, Shantanu 242, 1449, 1973
Desmaris, V. 1071
Detalle, Mikael 33
D'hiver, Philippe 1198
Dieng, K. 1296
Dinan, Thomas 862
Ding, L. 1212
Ding, Wen 2236
Djuric, Tatjana 850
Doany, Fuad E. 1016
Dobashi, Masahiro 763
Dobritz, Stephan 873
Donaldson, Nick 2213
Dong, Jianwei 342
Dong, Mingzhi 1192
Dornala, Kalyan 85
Dragoi, Viorel 888
Dressler, C. 1714

Drewniak, James L. 2081
Drost, A. 1482
Du, Ellen 748
Duffy, Daniel 1803
Durfee, Loren 236
Dzarnoski, John 157

E

Eastep, Brian 279
Eggen, Trym 498
Ehrhardt, Christian 1321
Eichstadt, David 1803
Eisenstadt, William R. 554
Elger, Gordon 1464
Ellis, Charles D. 1086
Elmer, John W. 1457
Elsherbini, Adel A. 1610
Emerich, S. 74, 952
Endo, Yoshinori 1523
Engelmann, Sebastian 179
Eom, Yong-Sung 1154
Esfahani, Zahra Kolahdouz 1556
Eto, Michiyuki 2003

F

Fan, Chia-Wen 1908
Fan, Chuanhong 1885
Fan, Jun 2081
Fan, Xuejun 967
Fang, Runiu 641, 1902, 1986, 2099
Fanget, S. 1714
Farahmand, Farid 2159
Farcy, A. 1296
Farrugia, Mark-Luke 411
Fay, Owen 2014
Feng, June 748
Feng, Wei 1915
Fiedler, C. 625
Fischer, Thorsten 1421
Fischer, U.H.P. 1862
Flack, Warren 26
Fortier, Paul 179
Franiatte, R. 906
Frank, Torrey W. 1610
Frye, Robert 1284, 2027
Fu, Hailuo 1652

Fu, Huan-Chun 290, 1891
Fu, Shancan 1080
Fu, Xianzhu 464
Fu, Xingming 170, 1189
Fu, Yifeng 459
Fujii, Atsushi 829
Fujimaru, Koichi 913
Fujimoto, Koji 1853
Fujino, Masahisa 1504
Fujita, Mitsuru 829
Fukuoka, Yoshitaka 1673
Fukushima, T. 304, 636, 856, 1110, 1148
Fukuzono, Kenji 68
Furuya, Ryosuke 1885

G

Gaherty, Lee 279
Gandhi, Saumya 541, 1492
Gao, Guilian 862, 1810
Garant, John 452
Gardner, Donald S. 1290
Garnier, A. 906
Gaynes, Michael 255, 1372, 1955
Ge, Y.P. 1212
Ge, Yun 684
Gebara, Edward 2293
Gelorme, Jeffrey 1308
Georgakopoulos, Stavros V. 759
Georgiadis, Apostolos 796
Ghaffari, Roozbeh 145
Gharaibeh, Mohammad 119
Giagka, Vasiliki 2213
Gieser, H. 1482
Gilham, David 504
Gill, Harpreet 354
Gissila, T. 2220
Glodde, M. 883
Goldsmith, Charles 1940
Golick, L. 952
Gonzalez, M. 33, 309, 1517
Goodwin, S. 8
Gottfried, Knut 1218
Goumans, Leon 411
Goyal, Deepak 1457
Grafe, Juergen 873
Grams, A. 625

Graves-Abe, Troy 647
Greco, F. 1714
Gregory, C. 8
Greve, Hannes. 1314
Grosinger, Jasmin. 1183
Grymyr, Ole Johannes 139
Gu, Ping 464
Gu, Sam. 609
Gu, Xiaoxiong 548, 1272
Gu, Yingke 1378
Guan, Yong. 1902
Guerin, Luc 647
Guerrero, Alice. 894
Guevara, Gabe 215
Guidotti, Daniel 1932
Guiller, O. 1296
Gundurao, Anil. 354
Günther, Wolfgang. 1218
Guo, F.M. 1212
Guo, Maofeng 2236
Guo, W. 309
Guo, Xing 1010, 2283
Gurrum, Siva P. 821
Guthrie, William 647
Guven, Ibrahim 2035

H

Hagelauer, A. 956
Hahm, Yeon-Chang 730
Hale, Cassandra 236
Halonen, Eerik 151
Halvorsen, Per Steinar. 139
Hamasha, S. 371
Hamilton, Michael C. 441, 1086
Han, C.J. 1641
Han, Gyuwan 1361
Han, KiJin 2124
Han, Kyu 782
Han, Michael 47
Han, Minghui 560
Han, SeungChul 582, 1816
Han, Sungwon 973
Hanna, Carlton 1278
Hao, Jifa 1241
Harada, Takeshi 1857
Harel, Stephane 179

Hartner, W.	956
Hasharoni, Kobi	1021
Hashiguchi, H.	856, 1110
Hashimoto, H.	304, 636
Hashimoto, Masakazu	1955
Hashimoto, Tomoaki	596
Hasnine, Mohammad	379
Haupt, M.	1862
Hau-Riege, Christine	1173
He, Hongwen	1116
He, Huanyu	548
He, Yi	2008, 2131
Heinrich, T.	108
Hell, W.	1482
Hemsel, Tobias	1549
Henriques, Vitor	935
Herbold, Christian	203
Herbst, Christian	1033
Hernandez, George A.	1086
Heuer, Henning	1509
Heumann, Wolfgang	1595
Higashi, Mitsutoshi	1166, 1366
Higgins, Leo	1539
Hill, Michael J.	528
Hilton, A.	8
Hiner, Dave	590
Hirai, Hiroyuki	1729
Ho, Heidi S.Y.	1965
Ho, Lung-Hua	316
Ho, Paul S.	1122
Hoff, Lars	139
Hoffrogge, Peter	850
Hoivik, Nils	139
Hölck, O.	625, 1134
Höll, S.	1862
Holmes, Pat	47
Holweg, Gerald	1183
Hong, J.B.	1533
Hong, Shengping	1622
Honrao, Chinmay	1160
Horibe, A.	803
Hoshino, Manabu	1955
Hoshiyama, M.	803
Hou, Fei	1690, 2048
Hou, Fengze	2008
Howell, Keith	425

Hsiao, Zhi-Cheng 290
Hsieh, Robert 26
Hsieh, Wan-Lin 1908
Hsu, Chao-Kai 1891
Hsu, Chih-Chung 1258
Hsu, H.S. 947
Hsu, Yung-Yu 145
Hu, Dyi-Chung 360, 1590
Hu, Hao 759
Hu, Je-Ping 1944
Hu, Y.H. 572
Hu, Yating 1189
Huang, Chen-Yu 1
Huang, Cui 1722
Huang, Hsiao-Chun 868, 947
Huang, Joseph 1533
Huang, Kuo-Hsin 620
Huang, Louie 419
Huang, Po-Tsang 512
Huang, Rui 1122
Huang, Shin-Yi 1908
Huang, Timothy 2266
Huang, Ting-Chia 1160
Huang, Yaping 2236
Huang, Yu-Wei 290, 1891, 1908
Huffman, A. 8, 20
Hui, Ho-Yee 1492
Hummler, Klaus 1648
Hunegnaw, Sara 1652
Hung, Chih-Pin 2112
Hung, Yin-Po 360
Huppert, Gil 145
Hutter, Matthias 1321
Hutzler, Aaron 1700
Huylenbroeck, S.V. 572
Hwang, Lih-Tyng 1303
Hwang, Seung Min 74
Hwang, Wei 512
Hwang, Yuchul 973

I

Iijima, Yu 452
Ikari, Gary 1366
Im, Jay 1122
Imai, H. 899
Imajyo, Nobuhiko 2271

Imanari, Masaaki 342
Imenes, Kristin 139, 498
Imura, Fumito 1915
Indyk, Richard 452
Inomata, Katsumi 1796
Iruvanti, Sushumna 236, 1253
Ishida, Hiroya 404
Ishigure, Takaaki 1042
Ishiguro, Kazuya 1308
Ishino, H. 1179
Islam, Md. R. 1272
Islam, Nokibul 50, 836
Ito, Yuka 1148
Itoh, Toshihiro 1857
Ivankovic, A. 309, 1517
Iwanabe, Keiichiro 2303
Iwata, Yoshiharu 2227
Iyer, Subramanian 647
Izzi, M. 194, 1842

J

Jang, BoRa 582, 1816
Jang, Myong-Gi 447
Jao, Pitfee 789
Jeong, Min-Su 1735
Jeong, Yonghyuk 836
Jeong, Youchul 753
Ji, Jason 1750
Ji, Liang 1116
Ji, Yu 1488, 2299
Jiang, D.S. 1533
Jiang, Feng 1740
Jiang, Hanqing 470
Jiang, Tengfei 1122
Jimbo, Toshihiko 1955
Jin, Howard (Hwa II) 1790
Jin, Yufeng 641, 1902, 1986, 2099
Jin, Zhenrong 535
Jing, Xiangmeng 1116, 1740
Jinka, Oblesh 1782
Jo, Chanmin 2075
Joblot, S. 1296
Johansson, A. 1071
Johansson, Susie 157
John, P. 625
Johnson, Christopher 1577

Johnson, Mark 2014
Johnson, Michael 1965
Jomaa, Houssam 1396
Jonah, Olutola 759
Joshi, Gaurang 119
Joshi, Yogendra 1247
Jourdain, Anne 894
Jouve, A. 906
Juang, Jing-Ye 1908
Jung, Dae Young 1562
Jung, Seong-Yoon 2255
Jürgensen, Nils 1498
Juskey, Frank 1264

K

Kabir, M.S. 1071
Kahle, R. 940
Kaletta, K. 1204
Kamgaing, Telesphor 1264, 1610
Kamlapurkar, Swetha 179
Kanagavel, Senthil 1790
Kandaswamy, Shri Vishnu 1464
Kang, Dongchul 1523
Kang, Kuiwon 1396
Kang, Sung K. 1782
Kang, SungGeun 590
Kang, Un Byung 1128, 1755
Kannan, Sukeshwar 590
Kao, C.R. 2277
Kao, Ming-Jer 290, 1853
Kao, Nicholas 1750
Karikalan, Sam 2119
Karlicek, Jr., Robert F. 2207
Karsli, Kivanc 209
Kashyap, Anirudh 279
Kata, Keiichirou 596
Katkar, Rajesh 1389
Kato, Fumiki 62, 1926
Katoh, Y. 1179
Kaur, Amanpreet 2144, 2168
Kaushik, B.K. 1091, 1921
Kawamoto, S. 803
Kawanami, Satoshi 186
Kawano, Kenji 2003
Kaynak, M. 1204
Ke, Chang-Bo 2249

Keech, John	20
Keezer, David	2260, 2287
Kenyon, Gareth	26
Keser, Beth	100, 925, 1173
Khan, Wasif T.	2293
Khim, JooHyun	582
Kholostov, K.	194, 1571, 1842
Ki, Wing-Hung	919
Kida, Tsuyoshi	596
Kijkanjanapaiboon, Kasemsak	967
Kikuchi, Katsuya	62, 1915, 1926
Kim, Cheolbok	1103
Kim, Cheolgyu	1004
Kim, Chimyung	1616
Kim, Chin Kwan	1396
Kim, Choong-Un	133, 697
Kim, Chunho	263
Kim, DoHyeong	582, 590, 1816
Kim, Dong Wook	609
Kim, Dongsu	41
Kim, Ga Won	354
Kim, Gwang	50, 2027
Kim, Hakgu	1616
Kim, Henry	2119
Kim, Hui Joong	712
Kim, HyunTai	2027
Kim, Hyup Jong	1103
Kim, Il	1661
Kim, Jaemin	753
Kim, Jinseong	1361
Kim, JooHyun	1816
Kim, Joungho	541
Kim, Jun Chul	41
Kim, June-Bum	1635
Kim, Kwonil	279
Kim, KyungOe	50
Kim, Min Sung	1004
Kim, Namhoon	2119
Kim, Nam-Seog	1122, 1635
Kim, Samuel	1471
Kim, Seung-Ho	841, 2241
Kim, Seung-Hyun	1635
Kim, Sung Jin	1264, 1384
Kim, Sungjoo	2075
Kim, Tae Wan	1060
Kim, Taek-Soo	1004

Kim, Taewan	271, 1616
Kim, YongSang	1816
Kim, Yoosun	841, 2241
Kim, Young Soon	1128, 1661, 1755, 1765
Kim, Young Suk	1853
Kim, Younghoon	354
Kimura, Kazushi	230
Kimura, Michitaka	596
King, A.	601
Kino, H.	856, 1110
Kintaka, Kenji	186
Kitada, Hideki	1853
Klink, G.	1482
Klyshko, A.	194, 1571
Knickerbocker, J.	576, 647, 883, 1372, 1880
Ko, Cheng-Ta	1891
Ko, Wen-Ching	1944
Kobayashi, Makoto	284, 365, 484, 742, 1160, 1384, 1959
Kobayashi, Yuta	913
Kodama, Shoichi	1853
Kodani, K.	1077
Kodemura, Junji	1955
Kohara, S.	647, 803
Koide, Masateru	68
Koiwa, Sumio	1673
Kono, Tsutomu	2042
Kotake, Tomohiko	1407
Koyama, Toshinori	348
Koyanagi, M.	304, 636, 856, 1110, 1148
Kraft, Silke	1700
Krieger, William E.R.	983
Krüger, Michael	114
Ku, Tzu-Kun	1853
Kubo, A.	899
Kuchta, Daniel M.	1016
Kumar, Gokul	541
Kumar, Santosh	712
Kumar, Vobulapuram Ramesh	1091
Kumari, Archana	1921
Kumpulainen, Tero	1685
Kunimoto, Yuji	348
Kuo, An-Yu	1303
Kuo, C.L.	297
Kuo, Chih-Ming	316
Kuo, H.J.	572
Kuo, Kuei Hsiao (Frank)	56

Kuramochi, Satoru 1673
Kurihara, Takashi 348
Kurokawa, Tsuyoshi 1776
Kutlu, Zafer 348
Kutter, C. 1482
Kwark, Young 1955
Kwatra, Abhishek 983

L

La Manna, A. 33, 309
Lachner, R. 956
Laflamme, Simon 179
Lagae, Liesbet 165
Lai, Chiming 2299
Lai, J.Y. 1
Lai, Yi-Che 947, 1750
Lall, Pradeep 85, 242, 379, 666, 990, 1449, 1973, 2053
LaManna, A. 572
Lambert, William J. 528
Lamy, Y. 1296
Lan, Albert 81, 1533
Lan, Jia-Shen 1235
Landesberger, C. 1482
Lang, K.-D. 114, 625, 873, 940, 1033, 1204, 1218, 1321, 1421, 1498, 1759
Langlois, Richard 236, 647
Lannon, J. 8, 1641
Larson, Lyndon 236
Lau, John 56, 290, 1908
Lau, Kei May 919
Lauro, Paul 1308
LaVoie, Annique 236
Law, Edward 518
Lazerand, Thierry 1577
Le, Fuliang 919
Le, Taoran 769
Lee, Bong-Sub 862, 1810
Lee, Byeong Rok 1635, 1735
Lee, Chang-Chi 2112
Lee, Chin C. 1335, 1529
Lee, Ching-Kuan 290, 1891
Lee, ChoonHeung 582, 1361, 1816
Lee, Chu-Chung (Stephen) 1539
Lee, Chung-Fa 2069
Lee, DongHoon 1816
Lee, DongHun 582
Lee, Fred C. 504

Lee, Haksun..1154

Lee, Heng..815

Lee, Inho...342

Lee, Jae Hong...712

Lee, Jason..56

Lee, K.W...304, 856

Lee, Kangwook...636, 1110, 1148

Lee, Kenny...47

Lee, KiWook..590

Lee, Minkyu...1616

Lee, Ning-Cheng..655

Lee, Rick...56

Lee, S.W. Ricky...919

Lee, Sang Hoon..271, 1060

Lee, Sangmin...2075

Lee, Seungbae...354

Lee, Seungyong..1616

Lee, Shih-Wei...512

Lee, Taeik...1004

Lee, Tae-Kyu...133, 697

Lee, Yil-Hak..342

Lee, YongTaek..2027

Lee, Young Woo...712

Lee, Yuan-Chang...290

Lee, Yuh-Lang...1584

Levine, Lyle E..1648

Lewandowski, Eric...255

Leyrer, Benjamin...203

Li, Bin..2249

Li, C..1641

Li, Guangfeng..1231

Li, Heng...1411

Li, Jun...2131

Li, K.H..324

Li, Lang..1876

Li, Leo Hongyu..2119

Li, Li...1122, 1366

Li, Liyi..631, 1492, 1745, 1848

Li, Long...1690

Li, Menglu...609

Li, Qiang...504

Li, Shidong...1253

Li, Xin...1080

Li, Y.Q...1212

Li, Yan...1457

Li, Yuan...338

Li, Yuefa	170
Li, Zhe	338
Li, Zhuo	1745, 1828
Liang, Hanshuang	470
Liang, Jiang-Long	1584
Liao, Anmou	2131
Liao, Jash	1750
Liao, Li-Ling	290
Liao, Mark	81
Liao, Sue-Chen	1853
Libsch, Frank	1016
Lii, M.J.	572
Liimatta, Toni	151
Lin, C.F.	297
Lin, Cha-Hsin	1853
Lin, Chun-Tang	947
Lin, Chun-Te	1891
Lin, Chun-Ting	1590
Lin, Edward	748
Lin, Frank M.-S.	620
Lin, Kung-An	316
Lin, Kwang-Lung	419, 1584
Lin, M.J.	297
Lin, Puru	1590
Lin, Wei	647, 1401
Lin, Y.C.	297
Lin, Y.J.	952
Lin, Yang-Kai	1258
Lin, Yaojian	931
Lin, Yi-Fan	1944
Lin, Yu-Min	1891, 1908
Lin, Ziyin	447, 769, 1828, 2231
Lindner, Paul	888
Ling, Feng	2094
Liou, Chang-Ho	1944
Liu, C.S.	572
Liu, Changqing	1348
Liu, Chaojun	1010, 1189
Liu, Cheng-Hsiang	868, 1628
Liu, Chengxun	165
Liu, Duixian	1272
Liu, Fengman	2008, 2131
Liu, Fuhan	1384
Liu, Hsichang	647
Liu, Hung-Wen	1750
Liu, Johan	459

Liu, Kai	1284, 2027
Liu, Kenny	947
Liu, Li	1348
Liu, Sheng	170, 1010, 1189, 1690, 1876, 2048, 2245, 2283
Liu, X.Y.	1212
Liu, Y.	1880
Liu, Yan	447, 2231
Liu, Yingxia	609
Liu, Yi-Wei	1750
Liu, Yong	808, 1241, 1601, 2063
Liu, Yumin	808, 1601
Liu, Zhiming	1652
Lo, Wei-Chung	290, 360, 1891
Lofrano, M.	309
Loh, Chooi Ian	748
Longgood, Stuart	425
Longworth, Hai	236
Lopez-Montesinos, Pedro	342
Lu, Chang-Lun	868, 947, 1628
Lu, Guo-Quan	1080
Lu, Hao	742, 1416
Lu, Hsin-Chia	1944
Lu, Hua	1342
Lu, James J.-Q.	2207
Lu, Jian	1857
Lu, Jian-Qiang	548
Lu, Minhua	690, 1940
Lu, PingHung	26
Lu, Terren	1
Lu, Wengao	2099
Lu, Yongqiang	878
Lu, Yuan	1354, 2008
Lu, Yung Jean (Rachel)	1891
Luckenbach, Thomas	1606
Lueck, M.	8, 20
Luesebrink, Helge	1021
Luo, L.	815, 1212, 1411
Luo, Xiaobing	170, 1189, 1876, 1992
Luo, Yihua	242
Luo, Yu	1782
Luo, Zhang	1690, 2048, 2245
Lv, Cheng	470, 1290
Lwo, Ben-Je	620

M

Ma, Shenglin 1902
Ma, Teng 470
Mackenzie, Kenneth D. 1577
Madenci, Erdogan 973
Maeda, Nobuhide 1853
Maeda, Ryutaro 1857
Maekawa, Katsuhiro 2194
Magaya, Tafadzwa 1652
Maikowske, Stefan 203
Majeed, Bivragh 165
Majumder, M.K. 1921
Maldonado, Alberto 2159
Mallampati, Sandeep 1328
Malta, D. 8
Maman, Avi 1021
Manhas, S.K. 1921
Maniatty, Antoinette 1241
Manier, C.-A. 1204
Mäntysalo, Matti 151, 1685
Mao, Cindy 56
Marcoux, Phil 1071
Maria, Joana 1372
Mariappan, Murugesan 636
Martin, Bill 2124
Martinez, Linnell 1577
Marzano, Frank Silvio 2137
Mason, Maurice 1782
Mathewson, Alan 1064
Matsubara, Takahiro 1028
Matsumae, Takashi 1504
Matsumoto, Rie 1870
Matsushita, Kiyoto 404
Matsuura, Ikkei 1955
Matthias, Thorsten 888
Mauer, Laura B. 878
Mavinkurve, Amar 411
Mawer, Andrew 1539
Mayer, Michael 1471
McCann, Scott R. 2189
McCarson, Brian 2035
McCleary, Roger 523
McCluskey, F. Patrick 1314
Mclaughlin, Kevin 878
McLeod, Mark 1308

McMullen, Tom 126
McNally, P.J. 1517
Mehr, M. Yazdan 1477
Mehta, Gaurav 490
Mei, Yunhui 1080
Meindl, Manfred 1183
Meinecke, Christoph 1218
Melville, Robert 1284
Mendoza, Nestor 1523
Menezes, Gary 1959
Mesh, Michael 1021
Miao, Min 641, 1902, 1986, 2099
Middendorf, Andreas 114
Milanes II, Ninoy 490
Miller, A. 33
Miller, Allen 1033
Miller, Andy 26, 894
Milton, Basil 1523
Min, Max (Sungwan) 354
Minami, Shigenobu 1729
Mino, Masayuki 2042
Mirza, Kazi 990
Misra, E. 1949
Mita, Mamoru 2194
Mitachi, Seiko 230
Miura, Hideo 1885
Miura, Testunosuke 186
Miyamoto, Yoshinori 913
Miyatake, Masato 1407
Miyazaki, Tomokazu 165
Mizuno, Hikaru 1796
Mizusawa, Ryuma 1870
Mizutani, Daisuke 68
Moeini, S. Ali 1314
Mok, Philip K.T. 919
Möller, Eike 1707
Moon, Jeong Tak 712
Moon, Kyoung-Sik 447, 1828, 2231
Moon, Seongjae 2075
Morales, Jorge Mario Herrera 478
Morey, Briana 145
Mori, H. 803
Morikawa, Y. 846, 1697
Morris, Jeffrey 221
Mu, Mingkai 504
Mukai, Kenichiroh 279

Mullen, Don......................730
Murai, Hikari......................1407
Murayama, Kei......................1166
Murayama, T......................846, 1697
Murtuza, Masood......................242, 1973
Murugesan, M......................304, 1148
Mustafa, Muhannad......................666
Mutnury, Bhyrav......................2081
Mutuku, Francis......................425, 655, 1769

N

Na, Duk Ju......................596
Na, Nanju......................2151
Nad, Suddhasattwa......................684
Nagao, S......................1077, 1179
Nah, Jae-Woong......................1308, 1372
Nair, Vijay......................782, 1264, 1278
Nakagawa, Hiroshi......................1915
Nakamoto, Mark......................1226
Nakamura, Shinya......................1523
Nakamuta, Y......................1697
Nakazuru, Kazumi......................1028
Nam, Kiyoung......................1616
Natarajan, Arun......................1272
Nauchi, Takashi......................1308
Ndip, Ivan......................1498
Neitz, Marcel......................1759
Nemoto, Shunsuke......................1915
Nenzi, Paolo......................1571, 2137
Neo, Chong-Wei......................518
Neuhaus, Jan......................1549
Newman, Keith......................1433
Nguyen, Anh Tuan Thai......................139
Nguyen, Hoa......................470
Nguyen, Hoang-Vu......................498
Nguyen, Luu......................242, 1973
Ni, Chih-Hsien......................316
Ni, Jiamin......................1241
Nicholls, Lou......................1361
Nicolas, S......................1714
Niittynen, Juha......................151
Niizeki, Shoichi......................913
Nimura, Masatsugu......................913
Ning, Wenguo......................815, 1411
Niotaki, Kyriaki......................796
Nishio, Kenzo......................186

Nishizono, Shinji 763
Niwa, Hiroyuki 913
Nolmans, P. 33
Nonaka, Toshihisa 913
Noriki, A. 856
Numata, Hidetoshi 179, 1870
Nuss, M. 625

O

Obeng, Yaw 1648
Ochiai, Toshihiko 596
Ochoa, Juan S. 717
O'Connell, Barry 1241
Oh, Dan 566, 748, 1896
Oh, Tac-Keun 1635
O'Halloran, G.M. 411
Ohba, Takayuki 1853
Oi, Kiyoshi 348
Okada, Hironao 1857
Okamoto, Kazuya 2227
Okamoto, Keishiro 68
Okamoto, Kenji 1796
Okoro, Chukwudi 1648
Onishi, M. 304
Ooshima, Nobuhiro 1729
Oppermann, H. 1204, 1321
Orii, Y. 803, 1308
Osaka, Masahiko 1523
Ose, Masahisa 1407
Osenbach, J. 74, 952
Oster, Sasha N. 1610
Ostrowicki, Gregory T. 821
Otaka, S. 348, 899
Oterkus, Erkan 973
Oterkus, Selda 973
O'Toole, Eoin 935
Ou, Jack 2159
Ouyang, Eric 836
Oya, S. 899

P

Pacheco, Mario 1457
Pachler, Walther 1183
Paek, JongSik 590
Pagliucci, Roberto 1571
Paik, Kyung-Wook 271, 841, 1060, 1128, 1661, 1755, 1765, 2241, 2255

Pakbaz, Faraydon	535
Palma, Fabrizio	2137
Pan, Jie	2163
Pan, Po-Chih	2112
Pan, Shiji	2088
Pan, Yi	332, 1253
Pang, Cheng	2163
Papakyrikos, Cole	145
Papapolymerou, John	2293
Parat, Guy	1198
Park, Dongjoo	1361
Park, John	1667
Park, JungSoo	582, 1816
Park, Kyoung Youl	775
Park, S.W.	1077, 1179
Park, Susan	2027
Park, Yongsun	1616
Park, Young-Bae	1635, 1735
Parker, Ben	1272
Parker, Richard	425
Parkinson, Dilworth Y.	609
Parkinson, Dula	1457
Parks, Gregory	690
Patnaik, Amalendu	1091
Paul, Jens	590
Pavlidis, Spyridon	2293
Pays-Volard, David	1577
Pedreira, Olalla Varela	613
Pei, Min	395, 684
Peng, Jr-Wei	1584
Peng, Shih-Liang	1
Perfecto, Eric	647, 690, 1308, 1606, 1940
Pesika, Noshir	255
Peterson, Andrew F.	2271
Peterson, K.	1641
Pham, Nam	535
Philipsen, Harold	613, 850
Phommahaxay, Alain	850, 894
Pinho, Nelson	935
Pitarresi, James	119
Pitwon, Richard	1033
Plante, David	236
Pleyer, Wolfgang	1021
Polastre, R.	883, 1880
Pollard, Scott	20
Prange, Jonathan	342

Pressel, Klaus 1183
Prewitz, T. 625
Privett, Mark 894
Prorok, Barton C. 379
Pucha, Raghuram 1098
Pufall, R. 1134

Q

Qasaimeh, A. 371
Qian, Jiangyuan 2088
Qian, Richard 2063
Qin, Hong-Bo 2249
Qin, Ivy 1523
Qin, Y. 342, 2181
Qin, Zheng 2163
Qiu, Delong 1932
Qiu, Jinghao 1690
Qu, Shichun 808
Quay, R. 2181
Queisser, Marco 1759
Questad, D. 332, 1253, 1949

R

Radhakrishnan, Kaladhar 528
Radojcic, Riko 1226
Raghavan, Sathyanarayanan 983
Rahimi, Arian 736, 789
Raj, Milan 145
Raj, P. Markondeya 284, 484, 541, 782, 1160, 1264, 1384, 1492, 1959, 2266
Rajendran, H. 1134
Rajoo, Ranjan 490
Ralph, W. Carter 1433
Ramachandran, Koushik 647
Ramachandran, Vidhya 1226
Ramm, P. 1134
Ranjan, Manish 26
Rao, Valluri R. 1610
Rastjagaev, Eugen 1707
Rastogi, Rajiv 2035
Ratel, David 478
Razdan, Sandeep 1054
Razeeb, Kafil M. 1064
Razzaq, A. 324
Rebibis, K.J. 572
Rebibis, Kenneth 894
Reiner, R. 2181

Ren, Xiaoli 2163

Reynolds, Scott K. 1272

Rimolo-Donadio, Renato 1955

Rissing, Lutz 1595

Roa, Fernando 1656

Robertazzi, Raphael 1372

Roberts, Jordan C. 666

Röder, Julia 1033

Rodriguez, Daniel 2151

Rogoff, Rich 523

Romankiw, Lubomyr T. 1782

Rong, Aosheng 2094

Rongen, René 411

Rosenthal, Christopher 452

Rouhana, Layal 1396

Roy, Rajiv 523

Ruhmer, Klaus 20, 523

Rylyakov, Alexander V. 1016

Ryu, Woonghwan 354, 2075

S

Sabuncuoglu, Deniz 165

Saeidi, Nooshin 2213

Saephan, Chio 2159

Sakai, Taiji 742

Sakalaukus, Peter 2053

Sakamoto, S. 1077

Sakuishi, T. 846, 1697

Sakuma, Katsuyuki 647, 1372

Sakurai, Tomohiko 1796

Sakuyama, Seiki 68

Saleem, A.M. 1071

Sammakia, Bahgat 2175

Samoilov, Arkadii 1679

Sandhu, Javed 518

Santagata, Fabio 1192

Sasaki, Hideki 763

Satio, Yukihiro 2227

Sato, Osamu 452

Sato, Ryohei 2227

Sato, Y. 304

Sato, Yasuhiro 1729

Sato, Yoichiro 365, 1247, 2271

Sato, Yutaka 636

Sawyer, Brett 742, 1416

Scheuermann, Michael 1372

Schletz, Andreas 1700
Schmitz, D. 371
Schmitz, Stefan 114
Schneider, Marc 203
Schoeller, Harry 1328
Schow, Clint L. 1016
Schröder, Henning 1033, 1759
Schultz, Mark 1016
Schutt-Ainé, José E. 717
Schwarz, Mark 100, 925
Secker, Dave 730
Seki, Toshitake 365
Seler, E. 956
Senior, David E. 789, 1103
Seo, Hochul 1616
Seo, SeongMin 582, 1816
Seo, Sun Kyung 1128, 1755
Seo, YoungChul 582
Serenelli, L. 1842
Sextro, Walter 1549
Shaba, Hala 215
Shaddock, David 1328
Shafiee, S. 1071
Shah, Milind 1396
Shan, Lei 1955
Shang, Jintang 1488, 1833, 2299
Shang, Wenya 2163
Shangguan, Dongkai 1116, 1740, 2008, 2131, 2163
Sharifi, Reza 2119
Sharma, Himani 782, 1492, 2266
Shen, Hong 862
Shen, J.H. 1212
Shen, Wen-Wei 1891
Shi, Shawn 263
Shibuya, Hiroki 763
Shigemoto, Takumi 2227
Shih, W.L. 2277
Shim, Yujeong 1896
Shimizu, Kozo 68
Shimizu, Noriyoshi 348
Shimizu, Tadashi 763
Shin, Jaemin 2075
Shin, Jiwon 841, 1128, 1661, 1755, 1765
Shiraishi, Takashi 1048
Shirangi, M.H. 108
Shirazi, S. 371

Shlepnev, Yuriy	730
Shorey, Aric	20, 1103
Shoukrun, Aki	1606
Shuto, Takanori	2303
Sibilia, C.	194
Sikka, Kamal K.	332
Sillanpää, Hannu	151
Simon, Gilles	478, 1198
Simons, V.	309
Sin, Johnny K.O.	919
Sinha, Pankaj	2035
Sitaraman, Srikrishna	1264
Sitaraman, Suresh K.	983, 2189
Skidmore, G.	1641
Skordas, Spyridon	647
Slabbekoorn, John	26
Sleeper, Scott	263
Smet, Vanessa	284, 365, 484, 742, 1054, 1160, 1384, 1959
Smith, Dan	590
Snyder, S.	601
Soler, Vilmarie	1606
Soltani, Ayat	1685
Soma, Kazutomo	1042
Son, Ho-Young	1122, 1635
Son, Yong	590
Sone, H.	803
Song, Chongshen	1116
Sorce, Peter	1308
Souchon, F.	1714
Souriau, Jean-Charles	478, 1198
Spreitzer, Ronald	1278
Sprenger, Daniel J.	1996
Steffek, Roland	1021
Steller, Wolfram	1218
Sten-Nilsen, Bjørnar	498
Stepanov, Stanislav	1021
Stewart, Aaron	119
Stömmer, Christian	1021
Stratton, Ken	1401
Strothmann, Tom	931
Struyf, Herbert	850
Su, Meiying	1116
Su, Peng	1122
Su, Quang	119
Su, Yipeng	504
Suaoke, Kuniaki	1870, 647

Suga, Tadatomo 1143, 1504
Sugahara, T. 1077, 1179
Suganuma, Daisuke 1042
Suganuma, K. 1077, 1179
Sugase, Naoki 452
Sugawara, Mariko 1048
Sugawara, Yohei 1110
Sugiura, K. 1179
Suh, Min-Suk 1122, 1635
Suhling, Jeff 379, 666, 990, 1449
Suk, Kyung-Lim 1661
Sun, Jibin 1054
Sun, Minggang 641, 1986
Sun, Rong 464
Sun, Xiaofeng 1354, 1378
Sun, Xin 641, 1902, 1986, 2099
Sun, Yangyang 47
Sun, Zhuowen 862
Sundaram, Venkatesh 1098, 2189
Sundaram, Venky 284, 365, 541, 742, 1054, 1247, 1264,
.... 1384, 1416, 1427, 1959, 2266, 2271
Sung, Baegin 753
Suthau, Eike 173
Suthiwongshunthorn, Nathapong 490
Suu, K. 846, 1697
Suzuki, A. 1697
Suzuki, Hiroko 763
Suzuki, Ken 1885
Suzuki, Kousuke 1673
Suzuki, Naoya 2042
Suzuki, Ryoichi 1308
Suzuki, Takuji 1729
Suzuki, Yuya 742, 1264, 1416, 1427
Swaminathan, Madhavan 782, 2124, 2260, 2287
Swan, Johanna 1278
Sweatman, Keith 425, 655
Syed, Ahmer 100, 1173
Sze, Henry 518

T

Taddei, John 878
Tai, Rui-Feng 947
Tain, Ra-Min 290, 360
Taira, Yoichi 179, 1870
Takagi, Yutaka 365, 742, 1416, 1427
Takaguchi, Akira 1308

Takahashi, Hiroshi	2042
Takahashi, Naoki	1028
Takahashi, Shintaro	2271
Takanezawa, Shin	1407
Takano, Akihito	1366
Takasugi, Hiroshi	1776
Takegami, Toshifumi	913
Takekoshi, Masaaki	1407
Takenobu, Shotaro	179
Takizawa, Hideo	452
Tamura, K.	899
Tan, C.S.	324
Tan, Keith	518
Tanaka, Kazuhiro	1048
Tanaka, Masato	348
Tanaka, T.	304, 636, 856, 1110, 1148
Taner, Ozgur	967
Tang, Rui	470
Tang, Yin	236
Tanikawa, Seiya	1110
Tao, Jing	1064
Taylor, Robin	1416
Telikepalli, Satyanarayana	2260, 2287
Temple, D.	8, 1641
Tentzeris, M.M.	759, 769
Terajima, Katsushi	763
Teraki, Shin	1776
Thangaraju, Sara	590
Thomas, T.	940
Thomas, Windsor	1103
Thompson, J.	601
Tian, Shurong	1016
Tillack, B.	1204
Tjulkins, Fjodors	139
Tkachenko, Anton	2207
Toepper, Michael	1421
Tokarski, Adam	1700
Tong, Ho-Ming	512
Tong, Jialing	1098, 2271
Töpper, Michael	1498
Toriyama, Kazushige	1308
Toukhy, Medhat	26
Trampert, Stefan	114
Tran, TuAnh	1539
Tsai, Jensen	1533
Tsai, Jyun-Ling	868, 1628

Tsai, Wen-Li..............290
Tsang, C. 576, 883, 1372
Tsang, Leung..........2200
Tsebo, Simo G. 108
Tseng, Stephen 1
Tsukuda, Tatsuaki..........763
Tsunoda, Masatoshi..........1028
Tsuruta, K. 1179
Tu, Chia-Jung 316
Tu, K.N. 609
Tuan, Chia-Chi..........447, 2231
Tucci, M. 194, 1842
Tummala, Rao 284, 365, 484, 541, 742, 782, 1054, 1098, 1160, 1247, 1264, 1384, 1416, 1427, 1492, 1959, 2189, 2266, 2271
Tung, Bui Thanh..........62, 1926
Tunga, K. 1949
Tyberg, Christy 1372
Tyler, P. 601

U

Uhrmann, Thomas 888
Ulusoy, A. Cagri..........2293
Unnikrishnan, R. 108
Ura, Shogo..........186
Uwataki, R. 1179
Uzoh, Cyprian 862

V

Valdes-Garcia, Alberto..........1272
Van Acker, Lut 165
Van der Donck, Tom..........613
Van der Plas, G. 309
Van Driel, W.D. 1477
van Zeijl, Henk..........1556
Vandevelde, B. 33, 1517
Vanstreels, K. 309
Vardakas, John..........796
Varlamava, Volha 2137
Velenis, D. 572
Verbinnen, Greet 894
Vick, E...........8, 1641
Vlasov, Yurii..........179
Vogel, D. 1134
Voges, S. 940
Voitsekhivska, Tetiana 173
von Kouwen, Maarten 1464
Vujosevic, Milena 395, 684

W

Wagner, Rebecca	236
Wahrmund, Wieland	873
Walls, Lloyd	535
Walter, H.	625, 1421
Waltereit, P.	2181
Walters, E.	601
Wan, Lixi	1354, 1378, 1932
Wang, Chen-Chao	2112
Wang, Daijiao	1965
Wang, Guanjiang	641, 1986, 2099
Wang, Hanguo	221
Wang, Howard	2175
Wang, Huijuan	2163
Wang, Liang	215, 862
Wang, M.J.	1212
Wang, Nan	459
Wang, Qian	1378
Wang, Qibin	1740
Wang, Qidong	1932
Wang, Shiqiang	1086
Wang, Shuling	1932
Wang, T.	309, 572
Wang, Tao	284, 484
Wang, Tzu-Chang	1258
Wang, W.	1212, 2048, 2245
Wang, Xianyan	145
Wang, Xugang	1932
Wang, Yiman	1876
Wang, Z.	108
Wang, Zheyao	1722
Wang, Zhihua	1378
Wassick, T.	1940, 1949
Watanabe, Manabu	68
Watanabe, Naoya	62, 1915, 1926, 2003
Watanabe, Shoji	348
Weatherspoon, M.R.	601
Webb, Bucknell	576
Weber, Daniel	1033
Wee, K.H.	324
Wegner, M.	1204
Wei, Frank	452
Wei, Jia	1192
Wei, Junchao	1449, 2053
Wei, Pinghung	145

Weigel, R. 956
Wen, Ming 1010, 2048, 2245
Wen, Shengmin 1231, 1361
Wentlent, L. 371
Werner, Randall J. 1253
Westerman, Russell 1577
Whalley, Simon 1033
White, Christopher 1264
Wietstruck, M. 1204
Wilde, Jürgen 1707, 2181
Wilke, M. 1204
Wimplinger, Markus 888
Win, Si T. 2151
Winstel, Kevin 647
Wittler, O. 625
Woertink, Julia 342
Wöhrmann, Markus 1421, 1498
Wojewoda, Leigh 528
Wojnowski, M. 956
Woldt, Gregor 1218
Wolf, H. 1482
Wolf, M.J. 625, 873, 1218
Wolter, Klaus-Juergen 173, 1509
Wong, C.P. 447, 464, 631, 769, 1488, 1492,
.... 1745, 1828, 1833, 1848, 2231, 2299
Wong, C.S. 1517
Wong, K. 952
Woo, Min 126
Wood, James 1606
Wright, S.L. 1880
Wu, C.Y. 297
Wu, Chenglin 1122
Wu, Chuan-Yu 316
Wu, Chung-Hsi 512
Wu, Dong 1722
Wu, Fei-Jain 316
Wu, Hao 470, 1290
Wu, Li Wei 1533
Wu, Mei-Ling 1235
Wu, Peng 2131
Wu, Shang-Lin 512
Wu, Sheng-Tsai 290, 963
Wu, Wei-Hsin 316
Wu, Xiaomeng 2008, 2131
Wu, Xin 1010, 1189, 2048, 2245, 2283
Wu, Yuan-Yun 1529

Wu, Yung Shen 316
Wu, Zhongming 126
Wu, Zihan 1384
Wunderle, B. 1134

X

Xia, Deyang 2236
Xie, Dongji 85, 126
Xie, John 338
Xie, Weidong 697
Xie, Xiang 1378
Xiong, Wei 560
Xu, Cheng 1116
Xu, Chunlin 2283
Xu, Gaowei 815, 1411
Xu, Hui 1523
Xu, Huili 133
Xu, Jiafeng 1833
Xu, P. 1086, 1212
Xu, Ruqing 1648
Xu, Sha 1342
Xu, Steven 100
Xu, Yichao 641, 1986
Xu, Yong 170, 1010, 1189, 2048, 2245, 2283
Xue, Jie 1366
Xue, Kai 1116, 1740

Y

Yacoub-George, E. 1482
Yagisawa, Takatoshi 1048
Yamada, Hiroshi 1729
Yamaguchi, Mitsugu 2194
Yamamoto, Tsuyoshi 68
Yamasaki, Kazuhiko 2194
Yan, Xiaolong 1622
Yang, Hyung Suk 13
Yang, Melinda (Ling) 354
Yang, Ming-Hsien 1
Yang, Ping-Feng 419, 2112
Yang, T.L. 2277
Yang, Wenhua 1143
Yang, Xianbo 2144, 2168
Yau, YouWen 1173
Yazdani, Farhang 1667
Ye, J.T. 1212
Ye, Jiaotuo 815

Ye, Lilei 459

Ye, Tiachun 1932

Yeap, Geoffrey 47

Yeh, C.T. 297

Yeung, Tak-Sang 518

Yin, L. 371, 1328

Yoneta, Kiyoto 2227

Yong, Andy Chang Bum 596

Yoo, Sehoon 447

Yook, Jong-Min 41

Yoon, Jong-Sun 1735

Yoon, Juhoon 1361

Yoon, S.W. 596, 931, 952

Yoon, Yong-Kyu 736, 789, 1103

Yoshikawa, Tomoyasu 1308

Yoshioka, T. 899

You, Eileen 354

You, Se-Ho 354

Youssef, Ramey 878

Yu, Daquan 1116, 1740, 1932, 2163

Yu, Doug C.H. 572

Yu, H. 324, 470, 1290

Yu, J.J. 2277

Yu, Man 2245

Yuan, Cadmus 1192

Yue, C. Patrick 919

Yue, Wu 2249

Yuen, Matthew M.F. 464, 1822

Yun, Feng 2236

Z

Zandén, Carl 459

Zhan, Chau-Jie 290, 1891, 1908

Zhang, Andy 85, 1441

Zhang, Chaoqi 13

Zhang, David C. 2260

Zhang, Di 85

Zhang, G.Q. 1477, 1556

Zhang, Gaugping 464

Zhang, Guoqi 1192

Zhang, Haipeng 1054

Zhang, Jinshen 170

Zhang, Kai 464

Zhang, Li 2299

Zhang, Mingchuan 2163

Zhang, Minyan 2236

Zhang, Pengtu 459
Zhang, Ron 215
Zhang, S.H. 1212
Zhang, Shengzhi 1010, 1690, 2283
Zhang, Tingting 2175
Zhang, Tonglong 1622
Zhang, Weihan 2236
Zhang, Weihong 1622
Zhang, Wenli 504
Zhang, Wenqi 1116
Zhang, Xia 2008
Zhang, Xin-Ping 2249
Zhang, Xuefeng 47
Zhang, Ye 2236
Zhao, Lily 47
Zhao, Wei 1226
Zhao, Xueying 1745
Zhao, Yukun 2236
Zhao, Zhili 1876
Zheng, H.Z. 1212
Zheng, Huai 1876
Zheng, Min 2236
Zhong, Jie 1086
Zhong, Xiao 1902
Zhong, Yinsheng 1822
Zhou, Bite 684
Zhou, Eric 221
Zhou, Feng 1622
Zhou, Longzao 1348
Zhou, Min-Bo 2249
Zhou, Tiao 1679
Zhu, Chunsheng 815, 1411
Zhu, Xiaoxin 1342
Zhu, Xunxun 1378
Zhu, Yunhui 641, 1902, 1986, 2099
Zilch, Christian 1183
Zischler, Sigrid 1700
Zitz, Jeffrey 332, 647
Zohni, Wael 1389
Zoschke, K. 1204
Zou, Simin 441
Zou, Yu 2299
Zou, Zhihua 1290
Zschenderlein, U. 1134

2014 IEEE 64th Electronic Components and Technology Conference

(ECTC 2014)

Lake Buena Vista, Florida, USA
27-30 May 2014

Pages 759-1516

IEEE Catalog Number: CFP14ECT-POD
ISBN: 978-1-4799-2408-0

**Copyright © 2014 by the Institute of Electrical and Electronic Engineers, Inc
All Rights Reserved**

Copyright and Reprint Permissions: Abstracting is permitted with credit to the source. Libraries are permitted to photocopy beyond the limit of U.S. copyright law for private use of patrons those articles in this volume that carry a code at the bottom of the first page, provided the per-copy fee indicated in the code is paid through Copyright Clearance Center, 222 Rosewood Drive, Danvers, MA 01923.

For other copying, reprint or republication permission, write to IEEE Copyrights Manager, IEEE Service Center, 445 Hoes Lane, Piscataway, NJ 08854. All rights reserved.

***This publication is a representation of what appears in the IEEE Digital Libraries. Some format issues inherent in the e-media version may also appear in this print version.

IEEE Catalog Number: CFP14ECT-POD
ISBN 13:   978-1-4799-2408-0

Additional Copies of This Publication Are Available From:

Curran Associates, Inc
57 Morehouse Lane
Red Hook, NY 12571 USA
Phone: (845) 758-0400
Fax: (845) 758-2633
E-mail: curran@proceedings.com
Web: www.proceedings.com

TABLE OF CONTENTS

1: Interposer Technologies
Chairs: Subhash L. Shinde, *Sandia National Laboratory*
John Knickerbocker, *IBM Corporation*

Integration Study of Die Strength and Various Bumping Volume and Reliability Performance on 2.5D Silicon Interposer Assembly 1
Shih-Liang Peng, *Siliconware Precision Industries Co., Ltd.*; Chen-Yu Huang, *Siliconware Precision Industries Co., Ltd.*; Ming-Hsien Yang, *Siliconware Precision Industries Co., Ltd.*; Stephen Tseng, *Siliconware Precision Industries Co., Ltd.*; J.Y. Lai, *Siliconware Precision Industries Co., Ltd.*; Terren Lu, *Siliconware Precision Industries Co., Ltd.*; Hsien-Wen Chen, *Siliconware Precision Industries Co., Ltd.*; Steve Chiu, *Siliconware Precision Industries Co., Ltd.*; Stephen Chen, *Siliconware Precision Industries Co., Ltd.*

Process Integration, Improvements, and Testing of Si Interposers for Embedded Computing Applications 8
S. Goodwin, *RTI International*; J. Lannon, Jr., *RTI International*; A. Hilton, *RTI International*; A. Huffman, *RTI International*; M. Lueck, *RTI International*; E. Vick, *RTI International*; G. Cunningham, *RTI International*; D. Malta, *RTI International*; C. Gregory, *RTI International*; D. Temple, *RTI International*

Mechanically Flexible Interconnects with Highly Scalable Pitch and Large Stand-off Height for Silicon Interposer Tile and Bridge Interconnection 13
Chaoqi Zhang, *Georgia Institute of Technology*; Hyung Suk Yang, *Georgia Institute of Technology*; Muhannad S. Bakir, *Georgia Institute of Technology*

Advancements in Fabrication of Glass Interposers 20
Aric Shorey, *Corning Incorporated*; Philippe Cochet, *Rudolph Technologies*; Alan Huffman, *RTI International*; John Keech, *Corning Incorporated*; Matt Lueck, *RTI International*; Scott Pollard, *Corning Incorporated*; Klaus Ruhmer, *Rudolph Technologies*

Large Area Interposer Lithography 26
Warren Flack, *Ultratech, Inc.*; Robert Hsieh, *Ultratech, Inc.*; Gareth Kenyon, *Ultratech, Inc.*; Manish Ranjan, *Ultratech, Inc.*; John Slabbekoorn, *IMEC*; Andy Miller, *IMEC*; Eric Beyne, *IMEC*; Medhat Toukhy, *AZ Electronics Materials USA Corporation*; PingHung Lu, *AZ Electronics Materials USA Corporation*; Chunwei Chen, *AZ Electronics Materials USA Corporation*

Minimizing Interposer Warpage by Process Control and Design Optimization 33
Mikael Detalle, *IMEC*; B. Vandevelde, *IMEC*; P. Nolmans, *IMEC*; J. De Messemaeker, *IMEC*; M. Gonzalez, *IMEC*; A. Miller, *IMEC*; A. La Manna, *IMEC*; G. Beyer, *IMEC*; E. Beyne, *IMEC*

High Performance IPDs (Integrated Passive Devices) and TGV (Through Glass Via) Interposer Technology Using the Photosensitive Glass 41
Jong-Min Yook, *Korea Electronics Technology Institute*; Dongsu Kim, *Korea Electronics Technology Institute*; Jun Chul Kim, *Korea Electronics Technology Institute*

2: Advances in Copper Pillar & Solder Based Flip Chip Technologies

Chairs: Tom Gregorich, *Micron*
Bernd Ebersberger, *Intel Corporation*

Challenges and Opportunities of Chip Package Interaction with Fine Pitch Cu Pillar for 28nm 47
Andy Bao, *Qualcomm, Inc.*; Lily Zhao, *Qualcomm, Inc.*; Yangyang Sun, *Qualcomm, Inc.*; Michael Han, *Qualcomm, Inc.*; Geoffrey Yeap, *Qualcomm, Inc.*; Steve Bezuk, *Qualcomm, Inc.*; Pat Holmes, *Qualcomm, Inc.*; Cecille Alcira, *Qualcomm, Inc.*; Xuefeng Zhang, *Qualcomm, Inc.*; Kenny Lee, *Qualcomm, Inc.*

Electromigration for Advanced Cu Interconnect and the Challenges with Reduced Pitch Bumps 50
Nokibul Islam, *STATS ChipPAC, Inc.*; Gwang Kim, *STATS ChipPAC, Inc.*;
KyungOe Kim, *STATS ChipPAC, Inc.*

Electromigration Performance of Cu Pillar Bump for Flip Chip Packaging with Bump on Trace by Using Thermal Compression Bonding 56
Kuei Hsiao (Frank) Kuo, *Siliconware Precision Industries Co., Ltd.*; Jason Lee, *Siliconware Precision Industries Co., Ltd.*; F.L. Chien, *Siliconware Precision Industries Co., Ltd.*; Rick Lee, *Siliconware Precision Industries Co., Ltd.*; Cindy Mao, *Siliconware Precision Industries Co., Ltd.*; John Lau, *ITRI*

Flip-Chip Bonding Alignment Accuracy Enhancement Using Self-Aligned Interconnection Elements to Realize Low-Temperature Construction of Ultrafine-Pitch Copper Bump Interconnections 62
Bui Thanh Tung, *Nanoelectronics Research Institute*; *Institute for Photonics-Electronics Convergence System Technology*; Naoya Watanabe, *Nanoelectronics Research Institute*; Fumiki Kato, *Nanoelectronics Research Institute*; Katsuya Kikuchi, *Nanoelectronics Research Institute*; Masahiro Aoyagi, *Nanoelectronics Research Institute*; *Institute for Photonics-Electronics Convergence System Technology*

Development of Second-Level Connection Method for Large-Size CPU Package 68
Shunji Baba, *Fujitsu Advanced Technologies, Ltd.*; Masateru Koide, *Fujitsu Advanced Technologies, Ltd.*; Manabu Watanabe, *Fujitsu Advanced Technologies, Ltd.*; Kenji Fukuzono, *Fujitsu Advanced Technologies, Ltd.*; Tsuyoshi Yamamoto, *Fujitsu Advanced Technologies, Ltd.*; Seiki Sakuyama, *Fujitsu Laboratories, Ltd.*; Kozo Shimizu, *Fujitsu Laboratories, Ltd.*; Keishiro Okamoto, *Fujitsu Laboratories, Ltd.*; Daisuke Mizutani, *Fujitsu Laboratories, Ltd.*

Development of Fine Pitch Area Array Cu Pillar/Lead Free Solder Bumps for Large 28nm Die in Large Organic Flip Chip Packages 74
John Osenbach, *LSI Corporation*; Sue Emerich, *LSI Corporation*; S. Cate, *LSI Corporation*; D. Brady, *Amkor Technology, Inc.*; Seung Min Hwang, *Amkor Technology, Inc.*; J. Dang, *Kyocera America Inc.*; D. Crouthamel, *LSI Corporation*

ELK Delaminate Improvement Methodology on Cu Pillar Interconnect BOP Structure 81
Nistec Chang, *Siliconware Precision Industries Co., Ltd.*; Albert Lan, *Siliconware Precision Industries Co., Ltd.*; Mark Liao, *Siliconware Precision Industries Co., Ltd.*; Eason Chen, *Siliconware Precision Industries Co., Ltd.*

3: Dynamic Mechanical Characterization

Chairs: Darvin R. Edwards, *Edwards Enterprises*
Tim Chaudhry, *Broadcom Corporation*

Transient Dynamics Model and 3D-DIC Analysis of New-Candidate for JEDEC JESD22-B111 Test Board 85
Pradeep Lall, *Auburn University*; Kalyan Dornala, *Auburn University*; Di Zhang, *Auburn University*; Dongji Xie, *Nvidia Corporation*; Andy Zhang, *Texas Instruments, Inc.*

Interconnect Reliability Prediction for Wafer Level Packages (WLP) for Temperature Cycle and Drop Load Conditions 100
Tong Cui, *Qualcomm Technologies, Inc.*; Ahmer Syed, *Qualcomm Technologies, Inc.*; Beth Keser, *Qualcomm Technologies, Inc.*; Rey Alvarado, *Qualcomm Technologies, Inc.*; Steven Xu, *Qualcomm Technologies, Inc.*; Mark Schwarz, *Qualcomm Technologies, Inc.*

A Novel Drop Test Methodology for Highly Stressed Interconnects in Automotive Electronic Control Units 108

M.H. Shirangi, *Robert Bosch GmbH*; Simo G. Tsebo, *Robert Bosch GmbH*; Z. Wang, *Bosch Automotive Products (Suzhou) Co., Ltd.*; R. Unnikrishnan, *RWTH Aachen University*; T. Heinrich, *Robert Bosch GmbH*

Early-State Crack Detection Method for Heel-Cracks in Wire Bond Interconnects 114

Michael Krüger, *Technical University Berlin*; Stefan Trampert, *Fraunhofer IZM*; Andreas Middendorf, *Technical University Berlin*; Stefan Schmitz, *Fraunhofer IZM*; Klaus-Dieter Lang, *Technical University Berlin*

Accelerated Vibration Reliability Testing of Electronic Assemblies Using Sine Dwell with Resonance Tracking 119

Quang Su, *Binghamton University*; James Pitarresi, *Binghamton University*; Mohammad Gharaibeh, *Binghamton University*; Aaron Stewart, *Binghamton University*; Gaurang Joshi, *Binghamton University*; Martin Anselm, *Universal Instruments Corporation*

Crack Monitoring and Life Modeling Technique Towards High Thermal Cyclic and Mechanical Reliability of fcBGA Solder Joint 126

Dongji Xie, *Nvidia Corporation*; Zhongming Wu, *Nvidia Corporation*; Min Woo, *Nvidia Corporation*; Tom McMullen, *Nvidia Corporation*

Fatigue Properties of Lead-Free Solder Joints in Electronic Packaging Assembly Investigated by Isothermal Cyclic Shear Fatigue 133

Huili Xu, *University of Texas, Arlington; Intel Corporation*; Tae-Kyu Lee, *Cisco Systems, Inc.*; Choong-Un Kim, *University of Texas, Arlington*

4: Bio & Flexible Electronics

Chairs: Joana Maria, *IBM Corporation*
C.S. Premachandran, *GLOBALFOUNDRIES*

MEMS-Based Implantable Heart Monitoring System with Integrated Pacing Function 139

Fjodors Tjulkins, *Buskerud and Vestfold University College*; Anh Tuan Thai Nguyen, *Buskerud and Vestfold University College*; Erik Andreassen, *Buskerud and Vestfold University College; SINTEF Materials and Chemistry*; Nils Hoivik, *Buskerud and Vestfold University College*; Knut Aasmundtveit, *Buskerud and Vestfold University College*; Lars Hoff, *Buskerud and Vestfold University College*; Ole Johannes Grymyr, *Oslo University Hospital Intervention Centre*; Per Steinar Halvorsen, *Oslo University Hospital Intervention Centre*; Kristin Imenes, *Buskerud and Vestfold University College*

Archipelago Platform for Skin-Mounted Wearable and Stretchable Electronics 145

Yung-Yu Hsu, *MC10, Inc.*; Cole Papakyrikos, *MC10, Inc.*; Milan Raj, *MC10, Inc.*; Mitul Dalal, *MC10, Inc.*; Pinghung Wei, *MC10, Inc.*; Xianyan Wang, *MC10, Inc.*; Gil Huppert, *MC10, Inc.*; Briana Morey, *MC10, Inc.*; Roozbeh Ghaffari, *MC10, Inc.*

Inkjet Printing in Manufacturing of Stretchable Interconnects 151

Toni Liimatta, *Tampere University of Technology*; Eerik Halonen, *Tampere University of Technology*; Hannu Sillanpää, *Tampere University of Technology*; Juha Niittynen, *Tampere University of Technology*; Matti Mäntysalo, *Tampere University of Technology*

Ultra Small Hearing Aid Electronic Packaging Enabled by Chip-in-Flex 157

John Dzarnoski, *Starkey Hearing Technologies*; Susie Johansson, *Starkey Hearing Technologies*

Fabrication of Silicon Based Microfluidics Device for Cell Sorting Application 165

Bivragh Majeed, *IMEC*; Chengxun Liu, *IMEC*; Lut Van Acker, *IMEC*; Robert Daily, *IMEC*; Tomokazu Miyazaki, *JSR Micro NV*; Deniz Sabuncuoglu, *IMEC*; Liesbet Lagae, *IMEC*

A Novel 3D Neural Probe with Integrated Channel and Its Package 170

Xingming Fu, *Wuhan University*; Yong Xu, *Wayne State University*; Yuefa Li, *Wayne State University*; Jinshen Zhang, *Wayne State University*; Xiaobing Luo, *Huazhong University of Science & Technology*; Sheng Liu, *Wuhan University*

CMOS Multiplexer for Portable Biosensing System with Integrated Microfluidic Interface 173

Tetiana Voitsekhivska, *Technical University, Dresden*; Eike Suthau, *Technical University, Dresden*; Klaus-Juergen Wolter, *Technical University, Dresden*

5: Silicon Photonics & LEDs

Chairs: Fuad Doany, *IBM Corporation*

Stefan Weiss, *II-VI Laser Enterprise GmbH*

Assembly of Mechanically Compliant Interfaces between Optical Fibers and Nanophotonic Chips 179

Tymon Barwicz, *IBM Corporation*; Yoichi Taira, *IBM Corporation*; Hidetoshi Numata, *IBM Corporation*; Nicolas Boyer, *IBM Corporation*; Stephane Harel, *IBM Corporation*; Swetha Kamlapurkar, *IBM Corporation*; Shotaro Takenobu, *Asahi Glass Corporation*; Simon Laflamme, *IBM Corporation*; Sebastian Engelmann, *IBM Corporation*; Yurii Vlasov, *IBM Corporation*; Paul Fortier, *IBM Corporation*

Proposal of Integrated-Optic Wavelength-Selective Modulator Based on Coupling-Efficiency Control of Distributed Bragg Reflector in Straight Waveguide 186

Shogo Ura, *Kyoto Institute of Technology*; Testunosuke Miura, *Kyoto Institute of Technology*; Satoshi Kawanami, *Kyoto Institute of Technology*; Kenji Kintaka, *National Institute of Advanced Industrial Science and Technology*; Kosuke Asai, *Kyoto Institute of Technology*; Kenzo Nishio, *Kyoto Institute of Technology*; Yasuhiro Awatsuji, *Kyoto Institute of Technology*

Porous Silicon Technology, a Breakthrough for Silicon Photonics: From Packaging to Monolithic Integration 194

M. Balucani, *Sapienza University of Rome*; A. Klyshko, *Sapienza University of Rome*; K. Kholostov, *Sapienza University of Rome*; A. Benedetti, *Sapienza University of Rome*; A. Belardini, *Sapienza University of Rome*; C. Sibilia, *Sapienza University of Rome*; M. Izzi, *Enea Casaccia Research Centre Rome*; M. Tucci, *Enea Casaccia Research Centre Rome*; H. Bandarenka, *Belarusian State University of Informatics and Radioelectronics*; V. Bondarenko, *Belarusian State University of Informatics and Radioelectronics*

High Power Density LED Modules with Silver Sintering Die Attach on Aluminum Nitride Substrates 203

Marc Schneider, *Karlsruhe Institute of Technology*; Benjamin Leyrer, *Karlsruhe Institute of Technology*; Christian Herbold, *Karlsruhe Institute of Technology*; Stefan Maikowske, *Karlsruhe Institute of Technology*

Effect of Optical Design on the Thermal Management for the Smart TV LED Backlight Systems 209

Kivanc Karsli, *Vestel AS*; Mehmet Arik, *Ozyegin University*

Wafer Level LED Packaging with Optimal Light Output and Thermal Dissipation for High-Brightness Lighting 215

Liang Wang, *Invensas Corporation*; Gabe Guevara, *Invensas Corporation*; Hala Shaba, *Invensas Corporation*; Roseann Alatorre, *Invensas Corporation*; Rey Co, *Invensas Corporation*; Ron Zhang, *Invensas Corporation*

High Power Laser Packaging Challenges and Standardization 221

Eric Zhou, *LDX Optronics, Inc.*; Jeffrey Morris, *LDX Optronics, Inc.*; Hanguo Wang, *University of California, Los Angeles*

6: Adhesives, Underfills, and Thermal Interface Materials

Chairs: Don Frye, *ATMI*
C. Robert Kao, *National Taiwan University*

Novel Highly Moisture Resistant Optical Adhesives and Their High Power Resistivity 230
Seiko Mitachi, *Tokyo University of Technology*; Kazushi Kimura, *Yokohama Rubber Co. Ltd.*

Engineered Thermal Interface Material 236
Lyndon Larson, *Dow Corning Corporation*; Yin Tang, *Dow Corning Corporation*; Loren Durfee, *Dow Corning Corporation*; Cassandra Hale, *Dow Corning Corporation*; David Plante, *Dow Corning Corporation*; Sushumna Iruvanti, *IBM Corporation*; Rebecca Wagner, *IBM Corporation*; Taryn Davis, *IBM Corporation*; Hai Longworth, *IBM Corporation*; Annique LaVoie, *IBM Corporation*; Richard Langlois, *IBM Corporation*

Degradation Mechanisms in Electronic Mold Compounds Subjected to High Temperature in Neighborhood of 200°C 242
Pradeep Lall, *Auburn University*; Shantanu Deshpande, *Auburn University*; Yihua Luo, *Auburn University*; Mike Bozack, *Auburn University*; Luu Nguyen, *Texas Instruments*; Masood Murtuza, *Texas Instruments*

Time, Temperature, and Mechanical Fatigue Dependence on Underfill Adhesion 255
Joseph Cremaldi, *Tulane University*; Michael Gaynes, *IBM Corporation*; Peter Brofman, *IBM Corporation*; Noshir Pesika, *Tulane University*; Eric Lewandowski, *IBM Corporation*

Study on Isotropic Electrically Conductive Adhesive for Medical Device Applications 263
Shawn Shi, *Medtronic, Inc.*; Scott Sleeper, *Medtronic, Inc.*; Chunho Kim, *Medtronic, Inc.*

Effect of Aligned Nanofiber in Nanofiber Solder Anisotropic Conductive Films (ACFs) on the Solder Ball Movement for Flex-on-Flex (FOF) Assembly 271
Tae-Wan Kim, *Korea Advanced Institute of Science and Technology (KAIST)*; Sang-Hoon Lee, *Korea Advanced Institute of Science and Technology (KAIST)*; Kyung-Wook Paik, *Korea Advanced Institute of Science and Technology (KAIST)*

Adhesive Enabling Technology for Directly Plating Metal on Molding Resin 279
Kwonil Kim, *Atotech USA, Inc.*; Kenichiroh Mukai, *Atotech USA, Inc.*; Brian Eastep, *Atotech USA, Inc.*; Lee Gaherty, *Atotech USA, Inc.*; Anirudh Kashyap, *Atotech USA, Inc.*; Lutz Brandt, *Atotech USA, Inc.*

7: Interposers & 3D Integration

Chairs: Katsuyuki Sakuma, *IBM Corporation*
Lou Nicholls, *Amkor Technology, Inc*

Modeling, Design, and Demonstration of Low-Temperature Cu Interconnections to Ultra-Thin Glass Interposers at 20 μm Pitch 284
Tao Wang, *Georgia Institute of Technology*; Vanessa Smet, *Georgia Institute of Technology*; Makoto Kobayashi, *Namics Corporation*; Venky Sundaram, *Georgia Institute of Technology*; P. Mardkondeya Raj, *Georgia Institute of Technology*; Rao Tummala, *Georgia Institute of Technology*

Low-Cost TSH (Through-Silicon Hole) Interposers for 3D IC Integration 290

John H. Lau, *Industrial Technology Research Institute (ITRI)*; Ching-Kuan Lee, *Industrial Technology Research Institute (ITRI)*; Chau-Jie Zhan, *Industrial Technology Research Institute (ITRI)*; Sheng-Tsai Wu, *Industrial Technology Research Institute (ITRI)*; Yu-Lin Chao, *Industrial Technology Research Institute (ITRI)*; Ming-Ji Dai, *Industrial Technology Research Institute (ITRI)*; Ra-Min Tain, *Industrial Technology Research Institute (ITRI)*; Heng-Chieh Chien, *Industrial Technology Research Institute (ITRI)*; Chun-Hsien Chien, *Industrial Technology Research Institute (ITRI)*; Ren-Shin Cheng, *Industrial Technology Research Institute (ITRI)*; Yu-Wei Huang, *Industrial Technology Research Institute (ITRI)*; Yuan-Chang Lee, *Industrial Technology Research Institute (ITRI)*; Zhi-Cheng Hsiao, *Industrial Technology Research Institute (ITRI)*; Wen-Li Tsai, *Industrial Technology Research Institute (ITRI)*; Pai-Cheng Chang, *Industrial Technology Research Institute (ITRI)*; Huan-Chun Fu, *Industrial Technology Research Institute (ITRI)*; Yu-Mei Cheng, *Industrial Technology Research Institute (ITRI)*; Li-Ling Liao, *Industrial Technology Research Institute (ITRI)*; Wei-Chung Lo, Industrial Technology Research Institute (ITRI); Ming-Jer Kao, *Industrial Technology Research Institute (ITRI)*

Cu Pattern Density Impacts on 2.5D TSI Warpage Using Experimental and FEM Analysis 297

C.T. Yeh, *United Microelectronics Corporation*; C.Y. Wu, *United Microelectronics Corporation*; C.F. Lin, *United Microelectronics Corporation*; K.M. Chen, *United Microelectronics Corporation*; M.J. Lin, *United Microelectronics Corporation*; Y.C. Lin, *United Microelectronics Corporation*; C.L. Kuo, *United Microelectronics Corporation*

A Resilient 3-D Stacked Multicore Processor Fabricated Using Die-Level 3-D Integration and Backside TSV Technologies 304

K.W. Lee, *Tohoku University*; H. Hashimoto, *Tohoku University*; M. Onishi, *Tohoku University*; Y. Sato, *Tohoku University*; M. Murugesan, *Tohoku University*; J.-C. Bae, *Tohoku University*; T. Fukushima, *Tohoku University*; T. Tanaka, *Tohoku University*; M. Koyanagi, *Tohoku University*

3D Stacking Induced Mechanical Stress Effects 309

V. Cherman, *IMEC*; G. Van der Plas, *IMEC*; J. De Vos, *IMEC*; A. Ivankovic, *IMEC; KU Leuven*; M. Lofrano, *IMEC*; V. Simons, *IMEC*; M. Gonzalez, *IMEC*; K. Vanstreels, *IMEC*; T. Wang, *IMEC*; R. Daily, *IMEC*; W. Guo, *IMEC*; G. Beyer, *IMEC*; A. La Manna, *IMEC*; I. De Wolf, *IMEC*; E. Beyne, *IMEC*

Six-Die Stacking: Three-Dimensional Interconnects Using Au and Pillar Bumps 316

Fei-Jain Wu, *Chipbond Technology Corporation*; Lung-Hua Ho, *Chipbond Technology Corporation*; Chih-Ming Kuo, *Chipbond Technology Corporation*; Chia-Jung Tu, *Chipbond Technology Corporation*; Chih-Hsien Ni, *Chipbond Technology Corporation*; Shih-Chieh Chang, *Chipbond Technology Corporation*; Chuan-Yu Wu, *Chipbond Technology Corporation*; Kung-An Lin, *Chipbond Technology Corporation*; Wei-Hsin Wu, *Chipbond Technology Corporation*; Yung Shen Wu, *Chipbond Technology Corporation*

TSV-Less 3D Stacking of MEMS and CMOS via Low Temperature Al-Au Direct Bonding with Simultaneous Formation of Hermetic Seal 324

S.L. Chua, *Nanyang Technological University*; A. Razzaq, *Nanyang Technological University*; K.H. Wee, *DSO National Laboratory*; K.H. Li, *Nanyang Technological University*; H. Yu, *Nanyang Technological University*; C.S. Tan, *Nanyang Technological University*

8: Flip Chip Packaging & Advanced Substrate

Chairs: Young-Gon Kim, *IDT*
Omar Bchir, *Qualcomm, Inc.*

Chip Package Interaction: An Experiment Study on White Bump Mitigation Using Flat Laminates 332

Yi Pan, *IBM Corporation*; Jeffrey A. Zitz, *IBM Corporation*; David L. Questad, *IBM Corporation*; Kamal K. Sikka, *IBM Corporation*

Design and Package Technology Development of Face-to-Face Die Stacking as a Low Cost Alternative for 3D IC Integration 338

Zhe Li, *Altera Corporation*; Yuan Li, *Altera Corporation*; John Xie, *Altera Corporation*

From C4 to Micro-Bump: Adapting Lead Free Solder Electroplating Processes to Next-Gen Advanced Packaging Applications 342

Julia Woertink, *Dow Electronic Materials*; Yi Qin, *Dow Electronic Materials*; Jonathan Prange, *Dow Electronic Materials*; Pedro Lopez-Montesinos, *Dow Electronic Materials*; Inho Lee, *Dow Electronic Materials*; Yil-Hak Lee, *Dow Electronic Materials*; Masaaki Imanari, *Dow Electronic Materials*; Jianwei Dong, *Dow Electronic Materials*; Jeffrey Calvert, *Dow Electronic Materials*

Development of New 2.5D Package with Novel Integrated Organic Interposer Substrate with Ultra-Fine Wiring and High Density Bumps 348

Kiyoshi Oi, *Shinko Electric Industries Company, Ltd.*; Satoshi Otake, *Shinko Electric Industries Company, Ltd.*; Noriyoshi Shimizu, *Shinko Electric Industries Company, Ltd.*; Shoji Watanabe, *Shinko Electric Industries Company, Ltd.*; Yuji Kunimoto, *Shinko Electric Industries Company, Ltd.*; Takashi Kurihara, *Shinko Electric Industries Company, Ltd.*; Toshinori Koyama, *Shinko Electric Industries Company, Ltd.*; Masato Tanaka, *Shinko Electric Industries Company, Ltd.*; Lavanya Aryasomayajula, *GLOBALFOUNDRIES, Inc.*; Zafer Kutlu, *GLOBALFOUNDRIES, Inc.*

Package Embedded Decoupling Capacitor Impact on Core Power Delivery Network for ARM SoC Application 354

Ga Won Kim, *Samsung Semiconductor Inc.*; Max (Sungwan) Min, *Samsung Semiconductor Inc.*; Melinda (Ling) Yang, *Samsung Semiconductor Inc.*; Anil Gundurao, *Samsung Semiconductor Inc.*; Eileen You, *Samsung Semiconductor Inc.*; Harpreet Gill, *Samsung Semiconductor Inc.*; Seungyong Cha, *Samsung Electronics Corporation*; Younghoon Kim, *Samsung Electronics Corporation*; Se-Ho You, *Samsung Electronics Corporation*; Seungbae Lee, *Samsung Electronics Corporation*; Woonghwan Ryu, *Samsung Electronics Corporation*

Embed Glass Interposer to Substrate for High Density Interconnection 360

Dyi-Chung Hu, *Unimicron Technology Corporation*; Yin-Po Hung, *Unimicron Technology Corporation*; Yu-Hua Chen, *Unimicron Technology Corporation*; Ra-Min Tain, *Unimicron Technology Corporation*; Wei-Chung Lo, *Industrial Technology Research Institute (ITRI)*

First Demonstration of a Surface Mountable, Ultra-Thin Glass BGA Package for Smart Mobile Logic Devices 365

Venky Sundaram, *Georgia Institute of Technology*; Yoichiro Sato, *Asahi Glass Company*; Toshitake Seki, *NGK Spark Plug Co., Ltd.*; Yutaka Takagi, *NGK Spark Plug Co., Ltd.*; Vanessa Smet, *Georgia Institute of Technology*; Makoto Kobayashi, *Namics Corporation*; Rao Tummala, *Georgia Institute of Technology*

9: Interconnect Reliability

Chairs: Tz-Cheng Chiu, *National Cheng Kung University*
Vikas Gupta, *Texas Instruments*

Towards a Quantitative Mechanistic Understanding of the Thermal Cycling of SnAgCu Solder Joints 371

D. Schmitz, *Binghamton University*; S. Shirazi, *Binghamton University*; L. Wentlent, *Binghamton University*; S. Hamasha, *Binghamton University*; L. Yin, *GE Global Research*; A. Qasaimeh, *Tennessee Tech University*; P. Borgesen, *Binghamton University*

Exploration of Aging Induced Evolution of Solder Joints Using Nanoindentation and Microdiffraction 379

Mohammad Hasnine, *Auburn University*; Jeffrey C. Suhling, *Auburn University*; Barton C. Prorok, *Auburn University*; Michael J. Bozack, *Auburn University*; Pradeep Lall, *Auburn University*

Accessing Adhesive Induced Risk for BGAs in Temperature Cycling 395

Guruprasad Arakere, *Intel Corporation*; Milena Vujosevic, *Intel Corporation*; Min Pei, *Intel Corporation*

Characteristics of Ceramic BGA Using Polymer Core Solder Balls 404

Hiroya Ishida, *Sekisui Chemical Co., Ltd.*; Kiyoto Matsushita, *Sekisui Chemical Co., Ltd.*

Lifetime Prediction of Cu-Al Wire Bonded Contacts for Different Mould Compounds 411

René Rongen, *NXP Semiconductors*; G.M. O'Halloran, *NXP Semiconductors*; Amar Mavinkurve, *NXP Semiconductors*; Leon Goumans, *NXP Semiconductors*; Mark-Luke Farrugia, *NXP Semiconductors*

The Corrosion Performance of Cu Alloy Wire Bond on Al Pad in Molding Compounds of Various Chlorine Contents under Biased-HAST 419

Ying-Ta Chiu, *ASE Group*; Tzu-Hsing Chiang, *ASE Group*; Yin-Fa Chen, *ASE Group*; Ping-Feng Yang, *ASE Group*; Louie Huang, *ASE Group*; Kwang-Lung Lin, *National Cheng Kung University*

The Effect of Nickel Microalloying on Thermal Fatigue Reliability and Microstructure of SAC105 and SAC205 Solders 425

Richard Coyle, *Alcatel-Lucent*; Richard Parker, *iNEMI*; Babak Arfaei, *Universal Instruments*; Francis Mutuku, *Binghamton University*; Keith Sweatman, *Nihon Superior Co., Ltd.*; Keith Howell, *Nihon Superior Co., Ltd.*; Stuart Longgood, *Delphi*; Elizabeth Benedetto, *Hewlett Packard Company*

10: Novel Materials & Processes
Chairs: Ivan Shubin, *Oracle*
Bing Dang, *IBM Corporation*

Flexible Non-Volatile Cu/CuxO/Ag ReRAM Memory Devices Fabricated Using Ink-Jet Printing Technology 441

Simin Zou, *Auburn University*; Michael C. Hamilton, *Auburn University*

Ultra-High Refractive Index LED Encapsulant 447

Chia-Chi Tuan, *Georgia Institute of Technology*; Ziyin Lin, *Georgia Institute of Technology*; Yan Liu, *Georgia Institute of Technology*; Kyoung-Sik Moon, *Georgia Institute of Technology*; Sehoon Yoo, *Korea Institute of Industrial Technology*; Myong-Gi Jang, *EI Lighting Co. Ltd.*; Ching-Ping Wong, *Georgia Institute of Technology; Chinese University of Hong Kong*

A Novel Methodology for Wafer-Specific Feed-Forward Management of Backside Silicon Removal by Wafer Grinding for Optimized Through Silicon Via Reveal 452

Tyson Alvanos, *Disco Hi Tec America, Inc.*; John Garant, *IBM Corporation*; Yu Iijima, *Disco Hi Tec America, Inc.*; Richard Indyk, *IBM Corporation*; Christopher Rosenthal, *Lasertec USA, Inc.*; Osamu Sato, *Lasertec Corporation*; Naoki Sugase, *Lasertec Corporation*; Hideo Takizawa, *Lasertec Corporation*; Frank Wei, *Disco Hi Tec America, Inc.*

Thermal Characterization of Power Devices Using Graphene-Based Film 459

Pengtu Zhang, *Chalmers University of Technology; East China University of Science and Technology*; Nan Wang, *Chalmers University of Technology*; Carl Zandén, *Chalmers University of Technology*; Lilei Ye, *Smart High Tech AB*; Yifeng Fu, *Smart High Tech AB*; Johan Liu, *Chalmers University of Technology*

High Performance Phase Change Thermal Interface Materials Based on Porous Graphitic Carbon Spheres-Paraffin Wax Composite 464

Zhihua Cao, *Shenzhen Institutes of Advanced Technology, Chinese Academy of Sciences; University of Science and Technology of China*; Kai Zhang, *Hong Kong University of Science and Technology*; Gaugping Zhang, *Shenzhen Institutes of Advanced Technology, Chinese Academy of Sciences*; Matthew M.F. Yuen, *Hong Kong University of Science and Technology*; Ping Gu, *University of Science and Technology of China*; Xianzhu Fu, *Shenzhen Institutes of Advanced Technology, Chinese Academy of Sciences*; Rong Sun, *Shenzhen Institutes of Advanced Technology, Chinese Academy of Sciences*; C.P. Wong, *Chinese University of Hong Kong*

High Sensitivity In-Plane Strain Measurement Using a Laser Scanning Technique 470

Hanshuang Liang, *Arizona State University*; Teng Ma, *Arizona State University*; Cheng Lv, *Arizona State University*; Hoa Nguyen, *Arizona State University*; George Chen, *Arizona State University*; Hao Wu, *Arizona State University*; Rui Tang, *Arizona State University*; Hanqing Jiang, *Arizona State University*; Hongbin Yu, *Arizona State University*

Biophysicochemical Evaluation of Passivation Layers for the Packaging of Silicon Microsystems in Medical Devices 478

Jorge Mario Herrera Morales, *CEA-LETI*; Jean-Charles Souriau, *CEA-LETI*; David Ratel, *CEA-LETI*; François Berger, *CEA-LETI*; Gilles Simon, *CEA-LETI*

11: Innovative Packaging Technologies
Chairs: Paul Tiner, *Texas Instruments*
Shichun Qu, *Fairchild Semiconductor*

A New Era in Manufacturable, Low-Temperature and Ultra-Fine Pitch Cu Interconnections and Assembly without Solders 484

Vanessa Smet, *Georgia Institute of Technology*; Makoto Kobayashi, *Namics Corporation*; Tao Wang, *Georgia Institute of Technology*; Pulugurtha Markondeya Raj, *Georgia Institute of Technology*; Rao Tummala, *Georgia Institute of Technology*

Enabling Fine Pitch Cu & Ag Alloy Wire Bond Assessment for 28nm Ultra Low-k Structure 490

John D. Beleran, *United Test and Assembly Center, Ltd.*; Ninoy Milanes II, *United Test and Assembly Center, Ltd.*; Gaurav Mehta, *United Test and Assembly Center, Ltd.*; Nathapong Suthiwongshunthorn, *United Test and Assembly Center, Ltd.*; Ranjan Rajoo, *GLOBALFOUNDRIES, Inc.*; Chan Kai Chong, *GLOBALFOUNDRIES, Inc.*

Assembly of Multiple Chips on Flexible Substrate Using Anisotropic Conductive Film for Medical Imaging Applications 498

Hoang-Vu Nguyen, *Buskerud and Vestfold University College*; Trym Eggen, *GE Vingmed Ultrasound AS*; Bjørnar Sten-Nilsen, *GE Vingmed Ultrasound AS*; Kristin Imenes, *Buskerud and Vestfold University College*; Knut E. Aasmundtveit, *Buskerud and Vestfold University College*

High Frequency High Current Point of Load Modules with Integrated Planar Inductors 504

Wenli Zhang, *Virginia Polytechnic Institute and State University*; Yipeng Su, *Virginia Polytechnic Institute and State University*; David Gilham, *Virginia Polytechnic Institute and State University*; Mingkai Mu, *Virginia Polytechnic Institute and State University*; Qiang Li, *Virginia Polytechnic Institute and State University*; Fred C. Lee, *Virginia Polytechnic Institute and State University*

Integrated Microprobe Array and CMOS MEMS by TSV Technology for Bio-Signal Recording Application 512

Lei-Chun Chou, *National Chiao Tung University*; Shih-Wei Lee, *National Chiao Tung University*; Po-Tsang Huang, *National Chiao Tung University*; Chih-Wei Chang, *University of California, Los Angeles*; Shang-Lin Wu, *National Chiao Tung University*; Jin-Chern Chiou, *National Chiao Tung University*; *China Medical University*; Ching-Te Chuang, *National Chiao Tung University*; Wei Hwang, *National Chiao Tung University*; *Advanced Semiconductor Engineering, Inc.*; Chung-Hsi Wu, *Advanced Semiconductor Engineering, Inc.*; Kuo-Hua Chen, *Advanced Semiconductor Engineering, Inc.*; Chi-Tsung Chiu, *Advanced Semiconductor Engineering, Inc.*; Ho-Ming Tong, *Advanced Semiconductor Engineering, Inc.*; Kuan-Neng Chen, *National Chiao Tung University*

Material Characterization of a Novel Lead-Free Solder Material - SACQ 518

Tak-Sang Yeung, *Broadcom Corporation*; Henry Sze, *Broadcom Corporation*; Keith Tan, *Broadcom Corporation*; Javed Sandhu, *Broadcom Corporation*; Chong-Wei Neo, *Broadcom Corporation*; Edward Law, *Broadcom Corporation*

Lithography Challenges for 2.5D Interposer Manufacturing 523

Klaus Ruhmer, *Rudolph Technologies, Inc.*; Philippe Cochet, *Rudolph Technologies, Inc.*; Roger McCleary, *Rudolph Technologies, Inc.*; Rich Rogoff, *Rudolph Technologies, Inc.*; Rajiv Roy, *Rudolph Technologies, Inc.*

12: Power Integrity & Passive Component Modeling
Chairs: Wendem Beyene, *Rambus Inc.*
Daniel de Araujo, *Nimbic, Inc.*

Package Embedded Inductors for Integrated Voltage Regulators 528
William J. Lambert, *Intel Corporation*; Michael J. Hill, *Intel Corporation*; Kaladhar Radhakrishnan, *Intel Corporation*; Leigh Wojewoda, *Intel Corporation*; Anne E. Augustine, *Intel Corporation*

Power Supply Filter for PLL Circuit in Digital Systems 535
Nam Pham, *IBM Corporation*; Faraydon Pakbaz, *IBM Corporation*; Zhenrong Jin, *IBM Corporation*; Lloyd Walls, *IBM Corporation*

Coaxial Through-Package-Vias (TPVs) for Enhancing Power Integrity in 3D Double-Side Glass Interposers 541
Gokul Kumar, *Georgia Institute of Technology*; P. Markondeya Raj, *Georgia Institute of Technology*; Jounghyun Cho, *Korea Advanced Institute of Science and Technology (KAIST)*; Saumya Gandhi, *Georgia Institute of Technology*; Parthasarathi Chakraborti, *Georgia Institute of Technology*; Venky Sundaram, *Georgia Institute of Technology*; Joungho Kim, *Korea Advanced Institute of Science and Technology (KAIST)*; Rao Tummala, *Georgia Institute of Technology*

Modeling of Switching Noise and Coupling in Multiple Chips of 3D TSV-Based Systems 548
Huanyu He, *Rensselaer Polytechnic Institute*; Xiaoxiong Gu, *IBM Corporation*; Jian-Qiang Lu, *Rensselaer Polytechnic Institute*

Characterization of On-Die Power Supply Noise in FCBGA (Flip-Chip Ball Grid Array) Packages 554
Hyunho Baek, *University of Florida*; William R. Eisenstadt, *University of Florida*

An Enhanced Power Integrity Analysis Flow Based on the Interdependence between Simultaneous Switching Output Noise and Static IR Drop 560
Minghui Han, *Samsung Display*; Amir Amirkhany, *Samsung Display*; Wei Xiong, *Samsung Display*

Improving the Target Impedance Method for PCB Decoupling of Core Power 566
Guang Chen, *Altera Corporation*; Dan Oh, *Altera Corporation*

13: 3D Process Integration & Die Stacking
Chairs: Rozalia Beica, *Yole Developpement*
Jianwei Dong, *Dow Electronic Materials*

Process Development to Enable 3D IC Multi-Tier Die Bond for 20μm Pitch and Beyond 572
Y.H. Hu, *TSMC*; C.S. Liu, *TSMC*; M.T. Chen, *TSMC*; M.D. Cheng, *TSMC*; H.J. Kuo, *TSMC*; M.J. Lii, *TSMC*; A. LaManna, *IMEC*; K.J. Rebibis, *IMEC*; T. Wang, *IMEC*; S.V. Huylenbroeck, *IMEC*; R. Daily, *IMEC*; G. Capuz, *IMEC*; D. Velenis, *IMEC*; G. Beyer, *IMEC*; E. Beyne, *IMEC*; Doug C.H. Yu, *TSMC*

Factors in the Selection of Temporary Wafer Handlers for 3D/2.5D Integration 576
Bing Dang, *IBM Corporation*; Bucknell Webb, *IBM Corporation*; Cornelia Tsang, *IBM Corporation*; Paul Andry, *IBM Corporation*; John Knickerbocker, *IBM Corporation*

Optimization and Challenges on TSV MEOL Integration 582
DoHyeong Kim, *Amkor Technology Korea, Inc.*; DongHun Lee, *Amkor Technology Korea, Inc.*; YoungChul Seo, *Amkor Technology Korea, Inc.*; JungSoo Park, *Amkor Technology Korea, Inc.*; SeungChul Han, *Amkor Technology Korea, Inc.*; BoRa Jang, *Amkor Technology Korea, Inc.*; JooHyun Khim, *Amkor Technology Korea, Inc.*; YoungSuk Chung, *Amkor Technology Korea, Inc.*; SeongMin Seo, *Amkor Technology Korea, Inc.*; ChoonHeung Lee, *Amkor Technology Korea, Inc.*

TSV Integration on 20nm Logic Si: 3D Assembly and Reliability Results 590
Rahul Agarwal, *GLOBALFOUNDRIES, Inc.*; Dave Hiner, *Amkor Technology, Inc.*; Sukeshwar Kannan, *GLOBALFOUNDRIES, Inc.*; KiWook Lee, *Amkor Technology, Inc.*; DoHyeong Kim, *Amkor Technology, Inc.*; JongSik Paek, *Amkor Technology, Inc.*; SungGeun Kang, *Amkor Technology, Inc.*; Yong Son, *Amkor Technology, Inc.*; Sebastian Dej, *GLOBALFOUNDRIES, Inc.*; Dan Smith, *GLOBALFOUNDRIES, Inc.*; Sara Thangaraju, *GLOBALFOUNDRIES, Inc.*; Jens Paul, *GLOBALFOUNDRIES, Inc.*

TSV MEOL (Mid End of Line) and Packaging Technology of Mobile 3D-IC Stacking 596
Duk Ju Na, *STATS ChipPAC, Ltd.*; Kyaw Oo Aung, *STATS ChipPAC, Ltd.*; Won Kyung Choi, *STATS ChipPAC, Ltd.*; Tsuyoshi Kida, *Renesas Electronics Company*; Toshihiko Ochiai, *Renesas Electronics Company*; Tomoaki Hashimoto, *Renesas Electronics Company*; Michitaka Kimura, *Renesas Electronics Company*; Keiichirou Kata, *Renesas Electronics Company*; Seung Wook Yoon, *STATS ChipPAC, Ltd.*; Andy Chang Bum Yong, *STATS ChipPAC, Ltd.*

Thermally Enhanced 3 Dimensional Integrated Circuit (TE3DIC) Packaging 601
S. Snyder, *Harris Corporation GCSD*; J. Thompson, *Harris Corporation GCSD*; A. King, *Harris Corporation GCSD*; E. Walters, *Harris Corporation GCSD*; P. Tyler, *Harris Corporation GCSD*; M.R. Weatherspoon, *Harris Corporation GCSD*

Filler Trap and Solder Extrusion in 3D IC Thermo-Compression Bonded Microbumps 609
Yingxia Liu, *University of California, Los Angeles*; Menglu Li, *University of California, Los Angeles*; Dong Wook Kim, *Qualcomm, Inc.*; Sam Gu, *Qualcomm, Inc.*; Dilworth Y. Parkinson, *Lawrence Berkeley National Laboratory*; Justin Blair, *Lawrence Berkeley National Laboratory*; K.N. Tu, *University of California, Los Angeles*

14: TSV Fabrication & Its Reliability Impact
Chairs: Li Li, *Cisco Systems, Inc.*
Wei-Chung Lo, *ITRI*

Correlation between Cu Microstructure and TSV Cu Pumping 613
Joke De Messemaeker, *IMEC*; Olalla Varela Pedreira, *IMEC*; Harold Philipsen, *IMEC*; Eric Beyne, *IMEC*; Ingrid De Wolf, *IMEC*; Tom Van der Donck, *KU Leuven*; Kristof Croes, *IMEC*

TSV Reliability Model under Various Stress Tests 620
Ben-Je Lwo, *National Defense University*; Frank M.-S. Lin, *National Defense University*; Kuo-Hsin Huang, *National Defense University*

Development of Process and Design Criteria for Stress Management in Through Silicon Vias 625
O. Hölck, *Fraunhofer IZM*; M. Nuss, *Fraunhofer IZM*; A. Grams, *Fraunhofer IZM*; T. Prewitz, *Fraunhofer IZM*; P. John, *Fraunhofer IZM*; C. Fiedler, *Fraunhofer IZM*; M. Böttcher, *Fraunhofer IZM*; H. Walter, *Fraunhofer IZM*; M.J. Wolf, *Fraunhofer IZM*; O. Wittler, *Fraunhofer IZM*; K.-D. Lang, *Technical University Berlin*

High-Speed Wet Etching of Through Silicon Vias (TSVs) in Micro- and Nanoscale 631
Liyi Li, *Georgia Institute of Technology*; Ching-Ping Wong, *Georgia Institute of Technology*; *Chinese University of Hong Kong*

Replacing the PECVD-SiO$_2$ in the Through-Silicon Via of High-Density 3D LSIs with Highly Scalable Low Cost Organic Liner: Merits and Demerits 636
Murugesan Mariappan, *NICHe, Tohoku University*; Takafumi Fukushima, *NICHe, Tohoku University*; JiChel Beatrix, *NICHe, Tohoku University*; Hiroyuki Hashimoto, *NICHe, Tohoku University*; Yutaka Sato, *NICHe, Tohoku University*; Kangwook Lee, *NICHe, Tohoku University*; Tetsu Tanaka, *NICHe, Tohoku University*; Mitsumasa Koyanagi, *NICHe, Tohoku University*

Investigation of a TSV-RDL In-line Fault-Diagnosis System and Test Methodology for Wafer-level Commercial Production 641
Runiu Fang, *Peking University*; Min Miao, *Beijing Information Science and Technology University*; Xin Sun, *Peking University*; Yunhui Zhu, *Peking University*; Guanjiang Wang, *Peking University Shenzhen Graduate School*; Yichao Xu, *Peking University Shenzhen Graduate School*; Minggang Sun, *Beijing Information Science and Technology University*; Yufeng Jin, *Peking University*

Bonding Technologies for Chip Level and Wafer Level 3D Integration 647

Katsuyuki Sakuma, *IBM Corporation*; Spyridon Skordas, *IBM Corporation*; Jeffrey Zitz, *IBM Corporation*; Eric Perfecto, *IBM Corporation*; William Guthrie, *IBM Corporation*; Luc Guerin, *IBM Corporation*; Richard Langlois, *IBM Corporation*; Hsichang Liu, *IBM Corporation*; Koushik Ramachandran, *IBM Corporation*; Wei Lin, *IBM Corporation*; Kevin Winstel, *IBM Corporation*; Sayuri Kohara, *IBM Corporation*; Kuniaki Sueoka, *IBM Corporation*; Matthew Angyal, *IBM Corporation*; Troy Graves-Abe, *IBM Corporation*; Daniel Berger, *IBM Corporation*; John Knickerbocker, *IBM Corporation*; Subramanian Iyer, *IBM Corporation*

15: Solder Joint Reliability
Chairs: Keith Newman, *Hewlett-Packard Company*
Toni Mattila, *Aalto University*

Dependence of Solder Joint Reliability on Solder Volume, Composition and Printed Circuit Board Surface Finish 655

Babak Arfaei, *Universal Instruments Corporation*; Francis Mutuku, *Binghamton University*; Keith Sweatman, *Nihon-Superior*; Ning-Cheng Lee, *Indium Corporation*; Eric Cotts, *Binghamton University*; Richard Coyle, *Alcatel-Lucent*

The Effects of Aging on the Fatigue Life of Lead Free Solders 666

Muhannad Mustafa, *Auburn University*; Jordan C. Roberts, *Auburn University*; Jeffrey C. Suhling, *Auburn University*; Pradeep Lall, *Auburn University*

Solder Joint Height Impact on Temperature Cycle Reliability of BGA Components with Thermal Enabling Load 684

Yun Ge, *Intel Corporation*; Jeffery Cook, *Intel Corporation*; Min Pei, *Intel Corporation*; Milena Vujosevic, *Intel Corporation*; Bite Zhou, *Intel Corporation*; Suddhasattwa Nad, *Intel Corporation*

Controlling the Sn Grain Morphology of SnAg C4 Solder Bumps 690

Gregory Parks, *Binghamton University*; Minhua Lu, *IBM Corporation*; Eric Perfecto, *IBM Corporation*; Eric Cotts, *Binghamton University*

The Impact of Microstructure Evolution, Localized Recrystallization and Board Thickness on Sn-Ag-Cu Interconnect Board Level Shock Performance 697

Tae-Kyu Lee, *Cisco Systems, Inc.*; Weidong Xie, *Cisco Systems, Inc.*; Thomas R. Bieler, *Michigan State University*; Choong-Un Kim, *University of Texas, Arlington*

Thermal Cycle Fatigue Life Prediction for Flip Chip Solder Joints 703

Robert Darveaux, *Skyworks Solutions, Inc.*

High Thermo-Mechanical Fatigue and Drop Impact Resistant Ni-Bi Doped Lead Free Solder 712

Jae Hong Lee, *MK Electron, Ltd.*; Santosh Kumar, *MK Electron, Ltd.*; Hui Joong Kim, *MK Electron, Ltd.*; Young Woo Lee, *MK Electron, Ltd.*; Jeong Tak Moon, *MK Electron, Ltd.*

16: Advances in Signal Integrity & High-Speed System Design
Chairs: Xiaoxiong (Kevin) Gu, *IBM Corporation*
Kemal Aygun, *Intel Corporation*

Optimal Relaxation of I/O Electrical Requirements under Packaging Uncertainty by Stochastic Methods 717

Xu Chen, *University of Illinois, Urbana-Champaign*; Juan S. Ochoa, *University of Illinois, Urbana-Champaign*; José E. Schutt-Ainé, *University of Illinois, Urbana-Champaign*; Andreas C. Cangellaris, *University of Illinois, Urbana-Champaign*

An Accurate and Convenient Lumped/Discrete Port De-Embedding Method for the 3D Integration and Packaging Full-Wave Modeling by Splitting and Absorbing the Error-Cancelling Network 723

Zhaoqing Chen, *IBM Corporation*

Design, Modeling, and Characterization of Passive Channels for Data Rates of 50 Gbps and Beyond 730
Wendemagegnehu Beyene, *Rambus, Inc.*; Yeon-Chang Hahm, *Rambus, Inc.*; Dave Secker, *Rambus, Inc.*; Don Mullen, *Rambus, Inc.*; Yuriy Shlepnev, *Simberian Inc.*

Low Loss Conductors for CMOS and Through Glass/Silicon Via (TGV/TSV) Structures Using Eddy Current Cancelling Superlattice Structure 736
Arian Rahimi, *University of Florida*; Yong-Kyu Yoon, *University of Florida*

Modeling, Design, Fabrication and Characterization of First Large 2.5D Glass Interposer as a Superior Alternative to Silicon and Organic Interposers at 50 Micron Bump Pitch 742
Brett Sawyer, *Georgia Institute of Technology*; Hao Lu, *Georgia Institute of Technology*; Yuya Suzuki, *Zeon Corporation*; Yutaka Takagi, *NGK Spark Plug Co. Ltd.*; Makoto Kobayashi, *Namics Corporation*; Vanessa Smet, *Georgia Institute of Technology*; Taiji Sakai, *Fujitsu Laboratories Ltd.*; Venky Sundaram, *Georgia Institute of Technology*; Rao Tummala, *Georgia Institute of Technology*

Coupling Impact of Single Ended Signals to LVDS Interface 748
June Feng, *Altera Corporation*; Chooi Ian Loh, *Altera Corporation*; Edward Lin, *Altera Corporation*; Ellen Du, *Altera Corporation*; Guang Chen, *Altera Corporation*; Dan Oh, *Altera Corporation*

Analysis on Interference between Multi-Giga Bit Display Serial Link and RF Components in Smart Mobile Device 753
Youchul Jeong, *Silicon Image Inc.*; Jaemin Kim, *Silicon Image Inc.*; Baegin Sung, *Silicon Image Inc.*

17: Emerging Wireless Technologies & Design
Chairs: Amit P. Agrawal, *Cisco Systems, Inc.*
Lih-Tyng Hwang, *National Sun Yat-Sen University*

Novel Highly-Efficient and Misalignment Insensitive Wireless Power Transfer Systems Utilizing Strongly Coupled Magnetic Resonance Principles 759
Daerhan Daerhan, *Florida International University*; Olutola Jonah, *Florida International University*; Hao Hu, *Florida International University*; Stavros V. Georgakopoulos, *Florida International University*; Manos M. Tentzeris, *Georgia Institute of Technology*

A Wireless Charging and Near-field Communication Combination Module for Mobile Applications 763
Hiroki Shibuya, *Renesas Electronics Corporation*; Tatsuaki Tsukuda, *Renesas Electronics Corporation*; Hiroko Suzuki, *Renesas Electronics Corporation*; Tadashi Shimizu, *Renesas Electronics Corporation*; Masahiro Dobashi, *Renesas Electronics Corporation*; Shinji Nishizono, *Renesas Electronics Corporation*; Mikio Baba, *Renesas Electronics Corporation*; Hideki Sasaki, *Renesas Electronics Corporation*; Katsushi Terajima, *Renesas Electronics Corporation*

Enhanced-Performance Wireless Conformal "Smart Skins" Utilizing Inkjet-Printed Carbon-Nanostructures 769
Taoran Le, *Georgia Institute of Technology*; Ziyin Lin, *Georgia Institute of Technology*; C.P. Wong, *Georgia Institute of Technology*; M.M. Tentzeris, *Georgia Institute of Technology*

Novel THz Imaging Array Using High Resistivity Metasurfaces 775
Kyoung Youl Park, *Michigan State University*; Premjeet Chahal, *Michigan State University*

Magneto-Dielectric Characterization and Antenna Design 782
Kyu Han, *Georgia Institute of Technology*; Madhavan Swaminathan, *Georgia Institute of Technology*; P. Markondeya Raj, *Georgia Institute of Technology*; Himani Sharma, *Georgia Institute of Technology*; Rao Tummala, *Georgia Institute of Technology*; Vijay Nair, *Intel Corporation*

Flexible Liquid Crystal Polymer Based Complementary Split Ring Resonator Loaded Quarter Mode Substrate Integrated Waveguide Filters for Compact and Wearable Broadband RF Applications 789

David E. Senior, *Universidad Tecnológica de Bolívar; University of Florida*; Arian Rahimi, *University of Florida*; Pitfee Jao, *University of Florida*; Yong-Kyu Yoon, *University of Florida*

A Dual-Band Power Amplifier Based on Composite Right/Left-Handed Matching Networks 796

Kyriaki Niotaki, *Centre Tecnologic de Telecomunicacions de Catalunya*; Ana Collado, *Centre Tecnologic de Telecomunicacions de Catalunya*; Apostolos Georgiadis, *Centre Tecnologic de Telecomunicacions de Catalunya*; John Vardakas, *Iquadrat S. L.*

18: WLCSP, Flip Chip, and PoP
Chairs: Valerie Oberson, *IBM Corporation*
Sa Huang, *Medtronic Corporation*

Wafer-Level Non Conductive Films for Exascale Servers 803

A. Horibe, *IBM Corporation*; S. Kohara, *IBM Corporation*; H. Mori, *IBM Corporation*; Y. Orii, *IBM Corporation*; S. Kawamoto, *Namics Corporation*; H. Sone, *Namics Corporation*; M. Hoshiyama, *Namics Corporation*

Bump Geometric Deviation on the Reliability of BOR WLCSP 808

Yumin Liu, *Fairchild Semiconductor Corporation*; Yong Liu, *Fairchild Semiconductor Corporation*; Shichun Qu, *Fairchild Semiconductor Corporation*

Experimental Identification of Warpage Origination During the Wafer Level Packaging Process 815

Chunsheng Zhu, *Chinese Academy of Sciences*; Wenguo Ning, *Chinese Academy of Sciences*; Heng Lee, *Chinese Academy of Sciences*; Jiaotuo Ye, *Chinese Academy of Sciences*; Gaowei Xu, *Chinese Academy of Sciences*; Le Luo, *Chinese Academy of Sciences*

A Stress-Based Effective Film Technique for Wafer Warpage Prediction of Arbitrarily Patterned Films 821

Gregory T. Ostrowicki, *Texas Instruments, Inc.*; Siva P. Gurrum, *Texas Instruments, Inc.*

Drop Test and TCT Reliability of Buffer Coating Material for WLCSP 829

Nobuhiro Anzai, *Asahi Kasei E-Materials Corporation*; Mitsuru Fujita, *Asahi Kasei E-Materials Corporation*; Atsushi Fujii, *Asahi Kasei E-Materials Corporation*

Optimization of Compression Bonding Processing Temperature for Fine Pitch Cu-Column Flip Chip Devices 836

Yonghyuk Jeong, *STATS ChipPAC, Inc.*; Joonyoung Choi, *STATS ChipPAC, Inc.*; Youjoung Choi, *STATS ChipPAC, Inc.*; Nokibul Islam, *STATS ChipPAC, Inc.*; Eric Ouyang, *STATS ChipPAC, Inc.*

Reliability Improvement Methods of Solder Anisotropic Conductive Film (ACF) Joints Using Morphology Control of Solder ACF Joints 841

Yoo-Sun Kim, *Korea Advanced Institute of Science and Technology (KAIST)*; Seung-Ho Kim, *Korea Advanced Institute of Science and Technology (KAIST)*; Jiwon Shin, *Korea Advanced Institute of Science and Technology (KAIST)*; Kyung-Wook Paik, *Korea Advanced Institute of Science and Technology (KAIST)*

19: Progress in 3D Integration
Chairs: Shawn Shi, *Medtronic Corporation*
Mark Gerber, *Texas Instruments*

Development of the Technology to Control the Spatial Distribution of Plasma Using Double ICP Coil 846

T. Sakuishi, *ULVAC, Inc.; NMEMS Technology Research Organization*; T. Murayama, *ULVAC, Inc.; NMEMS Technology Research Organization*; Y. Morikawa, *ULVAC, Inc.; NMEMS Technology Research Organization*; K. Suu, *ULVAC, Inc.; NMEMS Technology Research Organization*

Defect Detection in Through Silicon Vias by GHz Scanning Acoustic Microscopy: Key Ultrasonic Characteristics 850

Alain Phommahaxay, *IMEC*; Ingrid De Wolf, *IMEC; KU Leuven*; Tatjana Djuric, *PVA TePla Analytical Systems GmbH*; Peter Hoffrogge, *PVA TePla Analytical Systems GmbH*; Sebastian Brand, *Fraunhofer IWM*; Peter Czurratis, *PVA TePla Analytical Systems GmbH*; Harold Philipsen, *IMEC*; Gerald Beyer, *IMEC*; Herbert Struyf, *IMEC*; Eric Beyne, *IMEC*

Temporary Spin-on Glass Bonding Technologies for Via-Last/Backside-Via 3D Integration Using Multichip Self-Assembly 856

H. Hashiguchi, *Tohoku University*; T. Fukushima, *Tohoku University*; A. Noriki, *Tohoku University*; H. Kino, *Tohoku University*; K.-W. Lee, *Tohoku University*; T. Tanaka, *Tohoku University*; M. Koyanagi, *Tohoku University*

TSV Module Optimization for High Performance Silicon Interposer 862

Andrew Cao, *Invensas Corporation*; Thomas Dinan, *Invensas Corporation*; Zhuowen Sun, *Invensas Corporation*; Guilian Gao, *Invensas Corporation*; Cyprian Uzoh, *Invensas Corporation*; Bong-Sub Lee, *Invensas Corporation*; Liang Wang, *Invensas Corporation*; Hong Shen, *Invensas Corporation*; Sitaram Arkalgud, *Invensas Corporation*

Study of TSV Thinning Wafer Strength Enhancement for 3DIC Package 868

Jyun-Ling Tsai, *Siliconware Precision Industries Co., Ltd.*; Chun-Chieh Chao, *Siliconware Precision Industries Co., Ltd.*; Hsiao-Chun Huang, *Siliconware Precision Industries Co., Ltd.*; Cheng-Hsiang Liu, *Siliconware Precision Industries Co., Ltd.*; Hung-Hsein Chang, *Siliconware Precision Industries Co., Ltd.*; Chang-Lun Lu, *Siliconware Precision Industries Co., Ltd.*; Shi-Ching Chen, *Siliconware Precision Industries Co., Ltd.*

Challenges in 3D Die Stacking 873

Juergen Grafe, *Fraunhofer IZM*; Wieland Wahrmund, *Fraunhofer IZM*; Stephan Dobritz, *Fraunhofer IZM*; Juergen Wolf, *Fraunhofer IZM*; Klaus-Dieter Lang, *Fraunhofer IZM*

Wet Silicon Etch Process for TSV Reveal 878

Laura B. Mauer, *Solid State Equipment, LLC*; John Taddei, *Solid State Equipment, LLC*; Ramey Youssef, *Solid State Equipment, LLC*; Yongqiang Lu, *SACHEM, Inc.*; Sian Collins, *SACHEM, Inc.*; Kevin Mclaughlin, *SACHEM, Inc.*; Craig Allen, *SACHEM, Inc.*

20: 3D Materials & Processing

Chairs: Myung Jin Yim, *Intel Corporation*
Daniel D. Lu, *Henkel Corporation*

Advanced Wafer Bonding and Laser Debonding 883

P. Andry, *IBM Corporation*; R. Budd, *IBM Corporation*; R. Polastre, *IBM Corporation*; C. Tsang, *IBM Corporation*; B. Dang, *IBM Corporation*; J. Knickerbocker, *IBM Corporation*; M. Glodde, *IBM Corporation*

Versatile Thin Wafer Stacking Technology for Monolithic Integration of Temporary Bonded Thin Wafers 888

Thomas Uhrmann, *EV Group*; Jürgen Burggraf, *EV Group*; Julian Bravin, *EV Group*; Viorel Dragoi, *EV Group*; Markus Wimplinger, *EV Group*; Thorsten Matthias, *EV Group*; Paul Lindner, *EV Group*

Temporary Bonding for High-Topography Applications: Spin-on Material versus Dry Film 894

Anne Jourdain, *IMEC*; Alain Phommahaxay, *IMEC*; Greet Verbinnen, *IMEC*; Alice Guerrero, *Brewer Science, Inc.*; Susan Bailey, *Brewer Science, Inc.*; Mark Privett, *Brewer Science, Inc.*; Kim Arnold, *Brewer Science, Inc.*; Andy Miller, *IMEC*; Kenneth Rebibis, *IMEC*; Gerald Beyer, *IMEC*; Eric Beyne, *IMEC*

Development of New Concept Thermoplastic Temporary Adhesive for 3D-IC Integration 899

A. Kubo, *Tokyo Ohka Kogyo Co., Ltd.*; K. Tamura, *Tokyo Ohka Kogyo Co., Ltd.*; H. Imai, *Tokyo Ohka Kogyo Co., Ltd.*; T. Yoshioka, *Tokyo Ohka Kogyo Co., Ltd.*; S. Oya, *Tokyo Ohka Kogyo Co., Ltd.*; S. Otaka, *Tokyo Ohka Kogyo Co., Ltd.*

Underfilling Techniques Comparison in 3D CtW Stacking Approach 906
A. Garnier, *CEA-LETI*; A. Jouve, *CEA-LETI*; R. Franiatte, *CEA-LETI*; S. Chéramy, *CEA-LETI*

High Throughput Thermal Compression NCF Bonding 913
Toshihisa Nonaka, *Toray Industries, Inc.*; Yuta Kobayashi, *Toray Industries, Inc.*; Noboru Asahi, *Toray Industries, Inc.*; Shoichi Niizeki, *Toray Industries, Inc.*; Koichi Fujimaru, *Toray Industries, Inc.*; Yoshiyuki Arai, *Toray Engineering Co., Ltd.*; Toshifumi Takegami, *Toray Engineering Co., Ltd.*; Yoshinori Miyamoto, *Toray Engineering Co., Ltd.*; Masatsugu Nimura, *Toray Engineering Co., Ltd.*; Hiroyuki Niwa, *Toray International America Inc.*

Through Silicon Underfill Dispensing for 3D Die/Interposer Stacking 919
Fuliang Le, *Hong Kong University of Science and Technology*; S.W. Ricky Lee, *Hong Kong University of Science and Technology*; Kei May Lau, *Hong Kong University of Science and Technology*; C. Patrick Yue, *Hong Kong University of Science and Technology*; Johnny K.O. Sin, *Hong Kong University of Science and Technology*; Philip K.T. Mok, *Hong Kong University of Science and Technology*; Wing-Hung Ki, *Hong Kong University of Science and Technology*; Hoi Wai Choi, *University of Hong Kong*

21: Wafer-Level & Fan-Out Packages
Chairs: Christopher Bower, *X-Celeprint Ltd.*
E. Jan Vardaman, *TechSearch International, Inc.*

Board Level Reliability and Surface Mount Assembly of 0.35mm and 0.3mm Pitch Wafer Level Packages 925
Beth Keser, *Qualcomm Technologies, Inc.*; Rey Alvarado, *Qualcomm Technologies, Inc.*; Alan Choi, *Qualcomm Technologies, Inc.*; Mark Schwarz, *Qualcomm Technologies, Inc.*; Steve Bezuk, *Qualcomm Technologies, Inc.*

Encapsulated Wafer Level Package Technology (eWLCS) 931
Tom Strothmann, *STATS ChipPAC, Inc.*; Seung Wook Yoon, *STATS ChipPAC, Ltd.*; Yaojian Lin, *STATS ChipPAC, Ltd.*

Enabling of Fan-Out WLP for More Demanding Applications by Introduction of Enhanced Dielectric Material for Higher Reliability 935
Rodrigo Almeida, *Namium, S.A.*; Isabel Barros, *Namium, S.A.*; José Campos, *Namium, S.A.*; Paulo Cardoso, *Namium, S.A.*; José Castro, *Namium, S.A.*; Vitor Henriques, *Namium, S.A.*; Eoin O'Toole, *Namium, S.A.*; Nelson Pinho, *Namium, S.A.*

24" x 18" Fan-Out Panel Level Packaging 940
T. Braun, *Fraunhofer IZM*; K.-F. Becker, *Fraunhofer IZM*; S. Voges, *Technical University Berlin*; J. Bauer, *Fraunhofer IZM*; R. Kahle, *Technical University Berlin*; V. Bader, *Fraunhofer IZM*; T. Thomas, *Technical University Berlin*; R. Aschenbrenner, *Fraunhofer IZM*; K.-D. Lang, *Technical University Berlin*

Development and Characterization of New Generation Panel Fan-Out (P-FO) Packaging Technology 947
Hong-Da Chang, *Siliconware Precision Industries Co., Ltd.*; David Chang, *Siliconware Precision Industries Co., Ltd.*; Kenny Liu, *Siliconware Precision Industries Co., Ltd.*; H.S. Hsu, *Siliconware Precision Industries Co., Ltd.*; Rui-Feng Tai, *Siliconware Precision Industries Co., Ltd.*; Hsiao-Chun Huang, *Siliconware Precision Industries Co., Ltd.*; Yi-Che Lai, *Siliconware Precision Industries Co., Ltd.*; Chang-Lun Lu, *Siliconware Precision Industries Co., Ltd.*; Chun-Tang Lin, *Siliconware Precision Industries Co., Ltd.*; Steve Chiu, *Siliconware Precision Industries Co., Ltd.*

Development of Exposed Die Large Body to Die Size Ratio Wafer Level Package Technology 952
J. Osenbach, *LSI Corporation*; S. Emerich, *LSI Corporation*; L. Golick, *LSI Corporation*; S. Cate, *LSI Corporation*; M. Chan, *STATS ChipPAC, Ltd.*; S.W. Yoon, *STATS ChipPAC, Ltd.*; Y.J. Lin, *STATS ChipPAC, Ltd.*; K. Wong, *STATS ChipPAC, Inc.*

3D Rectangular Waveguide Integrated in Embedded Wafer Level Ball Grid Array (eWLB) Package 956

E. Seler, *Friedrich-Alexander University Erlangen-Nuremberg*; M. Wojnowski, *Infineon Technologies AG*; W. Hartner, *Infineon Technologies AG*; J. Böck, *Infineon Technologies AG*; R. Lachner, *Infineon Technologies AG*; R. Weigel, *University of Erlangen-Nuremberg*; A. Hagelauer, *Friedrich-Alexander University Erlangen-Nuremberg*

22: System-Level Thermal & Mechanical Models I

Chairs: Yong Liu, *Fairchild Semiconductor Corporation*
Sandeep Sane, *Intel Corporation*

Interplay and Influence of Thermomechanical Stress in Copper-Filled TSV Interposers 963

Sheng-Tsai Wu, *Industrial Technology Research Institute (ITRI)*; Cheng-Fu Chen, *University of Alaska, Fairbanks*; Heng-Chieh Chien, *Industrial Technology Research Institute (ITRI)*

Does Current Crowding Induce Vacancy Concentration Singularity in Electromigration? 967

Ozgur Taner, *Lamar University*; Kasemsak Kijkanjanapaiboon, *Lamar University*; Xuejun Fan, *Lamar University*

Hygro-Thermo-Mechanical Analysis and Failure Prediction in Electronic Packages by Using Peridynamics 973

Selda Oterkus, *University of Arizona*; Erdogan Madenci, *University of Arizona*; Erkan Oterkus, *University of Strathclyde*; Yuchul Hwang, *Samsung Electronics Company, Ltd.*; Jangyong Bae, *Samsung Electronics Company, Ltd.*; Sungwon Han, *Samsung Electronics Company, Ltd.*

Cohesive Zone Experiments for Copper/Mold Compound Delamination 983

William E.R. Krieger, *Georgia Institute of Technology*; Sathyanarayanan Raghavan, *Georgia Institute of Technology*; Abhishek Kwatra, *Georgia Institute of Technology*; Suresh K. Sitaraman, *Georgia Institute of Technology*

Damage Pre-Cursor Based Life Prediction of the Effect of Mean Temperature of Thermal Cycle on the SnAgCu Solder Joint Reliability 990

Pradeep Lall, *Auburn University*; Kazi Mirza, *Auburn University*; Jeff Suhling, *Auburn University*

Methodology Development of Warpage Analysis of Polymer Based Packaging Substrate 1004

Cheolgyu Kim, *Korea Advanced Institute of Science and Technology (KAIST)*; Taeik Lee, *Korea Advanced Institute of Science and Technology (KAIST)*; Hyeseon Choi, *Korea Advanced Institute of Science and Technology (KAIST)*; Min Sung Kim, *Samsung Electro-Mechanics*; Taek-Soo Kim, *Korea Advanced Institute of Science and Technology (KAIST)*

Simulations for the Impact of Warpage on the Accuracy of Attitude and Heading Reference System 1010

Shengzhi Zhang, *Huazhong University of Science & Technology; Wuhan National Laboratory for Optoelectronics*; Qiang Dan, *Huazhong University of Science & Technology; Wuhan National Laboratory for Optoelectronics*; Chaojun Liu, *Huazhong University of Science & Technology; Wuhan National Laboratory for Optoelectronics*; Yong Xu, *Wayne State University*; Xin Wu, *Wayne State University*; Sheng Liu, *Wuhan University*; Xing Guo, *Huazhong University of Science & Technology; Wuhan National Laboratory for Optoelectronics*; Ming Wen, *Huazhong University of Science & Technology; Wuhan National Laboratory for Optoelectronics*

23: Optical Interconnects

Chairs: Hiren Thacker, *Oracle*
Ping Zhou, *LDX Optronics, Inc.*

Multicore Fiber 4 TX + 4 RX Optical Transceiver Based on Holey SiGe IC 1016

Fuad E. Doany, *IBM Corporation*; Daniel M. Kuchta, *IBM Corporation*; Alexander V. Rylyakov, *IBM Corporation*; Christian Baks, *IBM Corporation*; Shurong Tian, *IBM Corporation*; Mark Schultz, *IBM Corporation*; Frank Libsch, *IBM Corporation*; Clint L. Schow, *IBM Corporation*

336-Channel Electro-Optical Interconnect: Underfill Process Improvement, Fiber Bundle and Reliability Results 1021

Shuki Benjamin, *Compass-EOS*; Kobi Hasharoni, *Compass-EOS*; Avi Maman, *Compass-EOS*; Stanislav Stepanov, *Compass-EOS*; Michael Mesh, *Compass-EOS*; Helge Luesebrink, *PVA TePla AG*; Roland Steffek, *PVA TePla AG*; Wolfgang Pleyer, *PVA TePla AG*; Christian Stömmer, *PVA TePla AG*

Development of Optical Multi-Channel Connector for Rigid Waveguide – Fiber Optical Interconnection 1028

Kazumi Nakazuru, *Kyocera Corporation*; Satoshi Asai, *Kyocera Corporation*; Masatoshi Tsunoda, *Kyocera Corporation*; Naoki Takahashi, *Kyocera Corporation*; Takahiro Matsubara, *Kyocera Corporation*

Electro-Optical Backplane Demonstrator with Gradient-Index Multimode Glass Waveguides for Board-to-Board Interconnection 1033

Lars Brusberg, *Fraunhofer Institute IZM*; Henning Schröder, *Fraunhofer Institute IZM*; Richard Pitwon, *Xyratex Technology Ltd.*; Simon Whalley, *ILFA Feinstleitertechnik GmbH*; Allen Miller, *Xyratex Technology Ltd.*; Christian Herbst, *Technical University of Berlin*; Julia Röder, *Fraunhofer Institute IZM*; Daniel Weber, *Fraunhofer Institute IZM*; Klaus-Dieter Lang, *Technical University of Berlin*

Three-Dimensional High-Density Channel Integration of Polymer Optical Waveguide Using the Mosquito Method 1042

Takaaki Ishigure, *Keio University*; Daisuke Suganuma, *Keio University*; Kazutomo Soma, *Keio University*

Novel Trace Design for High Data-Rate, Multi-Channel Optical Transceiver Assembled Using Flip-Chip Bonding 1048

Takatoshi Yagisawa, *Fujitsu Laboratories, Ltd.*; Takashi Shiraishi, *Fujitsu Laboratories, Ltd.*; Mariko Sugawara, *Fujitsu Laboratories, Ltd.*; Kazuhiro Tanaka, *Fujitsu Laboratories, Ltd.*

Modeling, Design, and Demonstration of Ultra-Miniaturized and High Efficiency 3D Glass Photonics Modules 1054

Bruce C. Chou, *Georgia Institute of Technology*; Sandeep Razdan, *TE Connectivity*; Haipeng Zhang, *TE Connectivity*; Jibin Sun, *TE Connectivity*; Terry Bowen, *TE Connectivity*; Vanessa Smet, *Georgia Institute of Technology*; Gee-Kung Chang, *Georgia Institute of Technology*; Venky Sundaram, *Georgia Institute of Technology*; Rao Tummala, *Georgia Institute of Technology*

24: Innovative Interconnections

Chairs: James E. Morris, *Portland State University*
            Nathan Lower, *Rockwell Collins, Inc.*

A Study on Nanofiber Anisotropic Conductive Films (ACFs) for Fine Pitch Chip-on-Glass (COG) Interconnections 1060

Sang Hoon Lee, *Korea Advanced Institute of Science and Technology (KAIST)*; Tae Wan Kim, *Korea Advanced Institute of Science and Technology (KAIST)*; Kyung-Wook Paik, *Korea Advanced Institute of Science and Technology (KAIST)*

Study of Fine Pitch Micro-Interconnections Formed by Low Temperature Bonded Copper Nanowires Based Anisotropic Conductive Film 1064

Jing Tao, *University College Cork*; Alan Mathewson, *University College Cork*; Kafil M. Razeeb, *University College Cork*

Carbon Nanofibers (CNF) for Enhanced Solder-Based Nano-Scale Integration and On-Chip Interconnect Solutions 1071

V. Desmaris, *Smoltek AB*; A.M. Saleem, *Smoltek AB*; S. Shafiee, *Smoltek AB*; J. Berg, *Smoltek AB*; M.S. Kabir, *Smoltek AB*; A. Johansson, *Smoltek AB*; Phil Marcoux, *PPM Associates*

Pressure-Less Plasma Sintering of Cu Paste for SiC Die-Attach of High-Temperature Power Device Manufacturing 1077

S. Nagao, *Osaka University*; K. Kodani, *Nissin, Inc.*; S. Sakamoto, *Osaka University*; S.-W. Park, *Osaka University*; T. Sugahara, *Osaka University*; K. Suganuma, *Osaka University*

Bonding 1200 V, 150 A IGBT Chips (13.5 mm x 13.5 mm) with DBC Substrate by Pressureless Sintering Nanosilver Paste for Power Electronic Packaging .... N/A

Shancan Fu, *Tianjin University*; Yunhui Mei, *Tianjin University*; Guo-Quan Lu, *Tianjin University, Virginia Tech*; Xin Li, *Tianjin University*; Gang Chen, *Tianjin University*; Xu Chen, *Tianjin University*

Flip Chip Based on Compliant Double Helix Interconnect for High Frequency Applications 1086

Pingye Xu, *Auburn University*; George A. Hernandez, *Auburn University*; Shiqiang Wang, *Auburn University*; Jie Zhong, *Auburn University*; Charles D. Ellis, *Auburn University*; Michael C. Hamilton, *Auburn University*

Modeling of Crosstalk Effects in Coupled MLGNR Interconnects Based on FDTD Method 1091

Vobulapuram Ramesh Kumar, *Indian Institute of Technology Roorkee*; Brajesh Kumar Kaushik, *Indian Institute of Technology Roorkee*; Amalendu Patnaik, *Indian Institute of Technology Roorkee*

25: Recent Advances in 3D Package Reliability
Chairs: Deepak Goyal, *Intel Corporation*
Jeffrey Suhling, *Auburn University*

First Demonstration of Reliable Copper-Plated 30µm Diameter Through-Package-Vias in Ultra-Thin Bare Glass Interposers 1098

Kaya Demir, *Georgia Institute of Technology*; Andac Armutlulu, *Georgia Institute of Technology*; Jialing Tong, *Georgia Institute of Technology*; Raghuram Pucha, *Georgia Institute of Technology*; Venkatesh Sundaram, *Georgia Institute of Technology*; Rao Tummala, *Georgia Institute of Technology*

Through-Glass Interposer Integrated High Quality RF Components 1103

Cheolbok Kim, *University of Florida*; David E. Senior, *University of Florida; Universidad Tecnológica de Bolívar*; Aric Shorey, *Corning, Inc.*; Hyup Jong Kim, *University of Florida*; Windsor Thomas, *Corning, Inc.*; Yong-Kyu Yoon, *University of Florida*

Minimization of Keep-Out Zone (KOZ) in 3D IC by Local Bending Stress Suppression with Low Temperature Curing Adhesive 1110

Hisashi Kino, *Tohoku University*; Hideto Hashiguchi, *Tohoku University*; Yohei Sugawara, *Tohoku University*; Seiya Tanikawa, *Tohoku University*; Takafumi Fukushima, *Tohoku University*; Kangwook Lee, *Tohoku University*; Mitsumasa Koyanagi, *Tohoku University*; Tetsu Tanaka, *Tohoku University*

Effect of Thermal Annealing on TSV Cu Protrusion and Local Stress 1116

Xiangmeng Jing, *National Center for Advanced Packaging; Chinese Academy of Sciences*; Hongwen He, *National Center for Advanced Packaging; Chinese Academy of Sciences*; Liang Ji, *National Center for Advanced Packaging*; Cheng Xu, *National Center for Advanced Packaging*; Kai Xue, *National Center for Advanced Packaging*; Meiying Su, *National Center for Advanced Packaging; Chinese Academy of Sciences*; Chongshen Song, *National Center for Advanced Packaging; Chinese Academy of Sciences*; Daquan Yu, *National Center for Advanced Packaging; Chinese Academy of Sciences*; Liqiang Cao, *National Center for Advanced Packaging; Chinese Academy of Sciences*; Wenqi Zhang, *National Center for Advanced Packaging*; Dongkai Shangguan, *National Center for Advanced Packaging; Chinese Academy of Sciences*

Effect of High Temperature Storage on the Stress and Reliability of 3D Stacked Chips 1122

Tengfei Jiang, *University of Texas, Austin*; Chenglin Wu, *University of Texas, Austin*; Peng Su, *Cisco Systems, Inc.*; Pierre Chia, *Cisco Systems, Inc.*; Li Li, *Cisco Systems, Inc.*; Ho-Young Son, *SK Hynix, Inc.*; Min-Suk Suh, *SK Hynix, Inc.*; Nam-Seog Kim, *SK Hynix, Inc.*; Jay Im, *University of Texas, Austin*; Rui Huang, *University of Texas, Austin*; Paul S. Ho, *University of Texas, Austin*

A Novel Fine Pitch TSV Interconnection Method Using NCF with Zn Nano-Particles 1128

Ji-Won Shin, *Korea Advanced Institute of Science and Technology (KAIST)*; Yong-Won Choi, *Korea Advanced Institute of Science and Technology (KAIST)*; Young Soon Kim, *Korea Advanced Institute of Science and Technology (KAIST)*; Un Byung Kang, *Samsung Electronics Company, Ltd.*; Sun Kyung Seo, *Samsung Electronics Company, Ltd.*; Kyung-Wook Paik, *Korea Advanced Institute of Science and Technology (KAIST)*

Residual Stress Investigations at TSVs in 3D Micro Structures by HR-XRD, Raman Spectroscopy and fibDAC 1134

U. Zschenderlein, *Technical University Chemnitz*; D. Vogel, *Fraunhofer ENAS*; E. Auerswald, *Fraunhofer ENAS*; O. Hölck, *Technical University Chemnitz*; H. Rajendran, *Technical University Chemnitz*; P. Ramm, *Fraunhofer EMFT*; R. Pufall, *Infineon Technologies*; B. Wunderle, *Technical University Chemnitz*; *Fraunhofer ENAS*

26: 3D Microbumps

Chairs: Kathy Cook, *Ziptronix*
Lei Shan, *IBM Corporation*

Formic Acid Treatment with Pt Catalyst for Cu Direct and Hybrid Bonding at Low Temperature 1143

Tadatomo Suga, *University of Tokyo*; Masakate Akaike, *University of Tokyo*; Wenhua Yang, *University of Tokyo*

Direct Multichip-to-Wafer 3D Integration Technology Using Flip-Chip Self-Assembly of NCF-Covered Known Good Dies 1148

Yuka Ito, *Tohoku University; Sumitomo Bakelite Co., Ltd.*; Mariappan Murugesan, *Tohoku University*; Takafumi Fukushima, *Tohoku University*; Kang-Wook Lee, *Tohoku University*; Koji Choki, *Sumitomo Bakelite Co., Ltd.*; Tetsu Tanaka, *Tohoku University*; Mitsumasa Koyanagi, *Tohoku University*

Maskless Screen Printing Technology for 20μm-Pitch, 52InSn Solder Interconnections in Display Applications 1154

Kwang-Seong Choi, *ETRI*; Haksun Lee, *ETRI*; Hyun-Cheol Bae, *ETRI*; Yong-Sung Eom, *ETRI*

Accelerated SLID Bonding Using Thin Multi-Layer Copper-Solder Stack for Fine-Pitch Interconnections 1160

Chinmay Honrao, *Georgia Institute of Technology*; Ting-Chia Huang, *Georgia Institute of Technology*; Makoto Kobayashi, *Namics Corporation*; Vanessa Smet, *Georgia Institute of Technology*; P. Markondeya Raj, *Georgia Institute of Technology*; Rao Tummala, *Georgia Institute of Technology*

Study of Electro-Migration Resistivity of Micro Bump Using SnBi Solder 1166

Kei Murayama, *Shinko Electric Industries Company, Ltd.*; Mitsuhiro Aizawa, *Shinko Electric Industries Company, Ltd.*; Mitsutoshi Higashi, *Shinko Electric Industries Company, Ltd.*

The Impact of Different Under Bump Metallurgies and Redistribution Layers on the Electromigration of Solder Balls for Wafer-Level Packaging 1173

Christine Hau-Riege, *Qualcomm Technologies, Inc.*; Beth Keser, *Qualcomm Technologies, Inc.*; Rey Alvarado, *Qualcomm Technologies, Inc.*; Ahmer Syed, *Qualcomm Technologies, Inc.*; YouWen Yau, *Qualcomm Technologies, Inc.*; Steve Bezuk, *Qualcomm Technologies, Inc.*; Kevin Caffey, *Qualcomm Technologies, Inc.*

Low-Pressure Sintering Bonding with Cu and CuO Flake Paste for Power Devices 1179

S.W. Park, *Osaka University*; R. Uwataki, *Osaka University*; S. Nagao, *Osaka University*; T. Sugahara, *Osaka University*; Y. Katoh, *Denso Corporation*; H. Ishino, *Denso Corporation*; K. Sugiura, *Denso Corporation*; K. Tsuruta, *Denso Corporation*; K. Suganuma, *Osaka University*

27: Sensors & MEMS Technologies

Chairs: Joseph W. Soucy, *Draper Laboratory*
Daniel Baldwin, *Engent, Inc.*

A Novel 3D Packaging Concept for RF Powered Sensor Grains 1183

Walther Pachler, *Graz University of Technology*; Klaus Pressel, *Infineon Technologies AG*; Jasmin Grosinger, *Graz University of Technology*; Gottfried Beer, *Infineon Technologies AG*; Wolfgang Bösch, *Graz University of Technology*; Gerald Holweg, *Infineon Technologies AG*; Christian Zilch, *Magna Diagnostics GmbH*; Manfred Meindl, *Danube Mobile Communications Engineering GmbH & Co. KG*

A Novel Sound Sensor and Its Package Used in Lung Sound Diagnosis 1189

Xingming Fu, *Wuhan University*; Chaojun Liu, *Wuhan University*; Yong Xu, *Wuhan University; Wayne State University*; Yating Hu, *Wayne State University*; Xiaobing Luo, *Huazhong University of Science & Technology*; Xin Wu, *Wayne State University*; Sheng Liu, *Wuhan University*

Novel System-in-Package Design and Packaging Solution for Solid State Lighting Systems 1192

Mingzhi Dong, *Delft University of Technology; State Key Laboratory of Solid State Lighting*; Fabio Santagata, *Delft University of Technology; State Key Laboratory of Solid State Lighting*; Jia Wei, *Delft University of Technology; State Key Laboratory of Solid State Lighting*; Cadmus Yuan, *Chinese Academy of Sciences; State Key Laboratory of Solid State Lighting*; Guoqi Zhang, *Chinese Academy of Sciences; Delft University of Technology*

Implantable Device Including a MEMS Accelerometer and an ASIC Chip Encapsulated in a Hermetic Silicon Box for Measurement of Cardiac Physiological Parameter 1198

Jean-Charles Souriau, *CEA-LETI*; Laetitia Castagné, *CEA-LETI*; Guy Parat, *CEA-LETI*; Gilles Simon, *CEA-LETI*; Karima Amara, *Sorin CRM SAS*; Philippe D'hiver, *Sorin CRM SAS*; Renzo Dal Molin, *Sorin CRM SAS*

Capping Technologies for Wafer Level MEMS Packaging Based on Permanent and Temporary Wafer Bonding 1204

K. Zoschke, *Fraunhofer IZM*; M. Wilke, *Fraunhofer IZM*; M. Wegner, *Fraunhofer IZM*; K. Kaletta, *Fraunhofer IZM*; C.-A. Manier, *Fraunhofer IZM*; H. Oppermann, *Fraunhofer IZM*; M. Wietstruck, *IHP GmbH*; B. Tillack, *IHP GmbH*; M. Kaynak, *IHP GmbH*; K.-D. Lang, *Technical University Berlin*

The Novel Assembly Method of a Field Deployable Biosensor Unit 1212

P. Xu, *East China Normal University*; F.M. Guo, *East China Normal University*; X.Y. Liu, *East China Normal University*; J.H. Shen, *East China Normal University*; L. Ding, *East China Normal University*; W. Wang, *East China Normal University*; Y.Q. Li, *East China Normal University*; Y.P. Ge, *East China Normal University*; S.H. Zhang, *East China Normal University*; M.J. Wang, *East China Normal University*; H.Z. Zheng, *East China Normal University*; J.T. Ye, *Chinese Academy of Sciences*; L.; Luo Chinese Academy of Sciences

SIMEIT-Project: High Precision Inertial Sensor Integration on a Modular 3D-Interposer Platform 1218

Wolfram Steller, *Fraunhofer IZM*; Christoph Meinecke, *Technical University Chemnitz*; Knut Gottfried, *Fraunhofer ENAS*; Gregor Woldt, *Microelectronic Packaging Dresden GmbH*; Wolfgang Günther, *GEMAC*; M. Juergen Wolf, *Fraunhofer IZM*; K. Dieter Lang, *Fraunhofer IZM*

28: System-Level Thermal & Mechanical Models II

Chairs: Pradeep Lall, *Auburn University*
Xuejun Fan, *Lamar University*

Mechanical Stress Management for Electrical Chip-Package Interaction (e-CPI) 1226

Wei Zhao, *Qualcomm Technologies, Inc.*; Mark Nakamoto, *Qualcomm Technologies, Inc.*; Vidhya Ramachandran, *Qualcomm Technologies, Inc.*; Riko Radojcic, *Qualcomm Technologies, Inc.*

Cu Pillar Flip Chip Assembly: Chip Attach Process Failure Mode Study 1231

Shengmin Wen, *Amkor Technology*; Bora Baloglu, *Amkor Technology*; Guangfeng Li, *Amkor Assembly and Test (Shanghai) Co., Ltd.*

Mechanical and Thermo-Mechanical Stress Considerations in Applying 3D ICs to a Design 1235

Jia-Shen Lan, *National Sun Yat-Sen University*; Mei-Ling Wu, *National Sun Yat-Sen University*

Modeling Microstructure Effects on Electromigration in Lead-Free Solder Joints 1241

Jiamin Ni, *Rensselaer Polytechnic Institute*; Yong Liu, *Fairchild Semiconductor*; Jifa Hao, *Fairchild Semiconductor*; Antoinette Maniatty, *Rensselaer Polytechnic Institute*; Barry O'Connell, *Fairchild Semiconductor*

Experimental Demonstration of the Effect of Copper TPVs (Through Package Vias) on Thermal Performance of Glass Interposers 1247

Sangbeom Cho, *Georgia Institute of Technology*; Yoichiro Sato, *Asahi Glass*; Venky Sundaram, *Georgia Institute of Technology*; Yogendra Joshi, *Georgia Institute of Technology*; Rao Tummala, *Georgia Institute of Technology*

Failure Mechanism Investigation of Stacked Via Cracking in Organic Chip Carrier 1253

Shidong Li, *IBM Corporation*; Yi Pan, *IBM Corporation*; Sushumna Iruvanti, *IBM Corporation*; David L. Questad, *IBM Corporation*; Randall J. Werner, *IBM Corporation*

A Novel Method to Predict Fluid/Structure Interaction in IC Packaging 1258

Chih-Chung Hsu, *National Tsing Hua University*; Tzu-Chang Wang, *CoreTech System (Moldex3D) Co., Ltd.*; Yen-Chi Chen, *CoreTech System (Moldex3D) Co., Ltd.*; Yang-Kai Lin, *CoreTech System (Moldex3D) Co., Ltd.*

29: Integrated RF & Power Modules
Chairs: Rockwell Hsu, *Cisco Systems, Inc.*
           P. Markondeya Raj, *Georgia Institute of Technology*

Modeling, Design and Demonstration of Multi-Die Embedded WLAN RF Front-End Module with Ultra-Miniaturized and High-Performance Passives 1264

Srikrishna Sitaraman, *Georgia Institute of Technology*; Yuya Suzuki, *Zeon Corporation*; Christopher White, *Georgia Institute of Technology*; Vijay Nair, *Intel Corporation*; Telesphor Kamgaing, *Intel Corporation*; Frank Juskey, *TriQuint Semiconductor*; Sung Jin Kim, *Georgia Institute of Technology*; P. Markondeya Raj, *Georgia Institute of Technology*; Venky Sundaram, *Georgia Institute of Technology*; Rao Tummala, *Georgia Institute of Technology*

A Compact 4-Chip Package with 64 Embedded Dual-Polarization Antennas for W-Band Phased-Array Transceivers 1272

Xiaoxiong Gu, *IBM Corporation*; Duixian Liu, *IBM Corporation*; Christian Baks, *IBM Corporation*; Alberto Valdes-Garcia, *IBM Corporation*; Ben Parker, *IBM Corporation*; Md. R. Islam, *IBM Corporation*; Arun Natarajan, *IBM Corporation; Oregon State University*; Scott K. Reynolds, *IBM Corporation*

Active Die Embedded Small Form Factor RF Packages for Ultrabooks and Smartphones 1278

Vijay K. Nair, *Intel Corporation*; Carlton Hanna, *Intel Corporation*; Ronald Spreitzer, *Intel Corporation*; Johanna Swan, *Intel Corporation*

Design and Material Contributions to Second-Harmonic Nonlinearities in RF Silicon Integrated Passive Devices 1284

Robert Frye, *RF Design Consulting, LLC*; Robert Melville, *Emecon, LLC*; Kai Liu, *STATS ChipPAC, Inc.*

Integration of Magnetic Materials into Package RF and Power Inductors on Organic Substrates for System in Package (SiP) Applications 1290

Hao Wu, *Arizona State University*; Donald S. Gardner, *Intel Corporation*; Cheng Lv, *Arizona State University*; Zhihua Zou, *Intel Corporation*; Hongbin Yu, *Arizona State University*

Through Silicon Capacitor Co-Integrated with TSV as an Efficient 3D Decoupling Capacitor Solution for Power Management on Silicon Interposer 1296

O. Guiller, *STMicroelectronics*; S. Joblot, *STMicroelectronics*; Y. Lamy, *CEA-LETI*; A. Farcy, *STMicroelectronics*; E. Defay, *CEA-LETI*; K. Dieng, *Université de Savoie*

Design of RF and Thermal Pads of CMOS PAs Using Copper to Copper Bonding Technology 1303

Lih-Tyng Hwang, *National Sun Yat-Sen University*; An-Yu Kuo, *Cadence Design Systems, Inc.*

30: Solders & Bonding

Chairs: Mikel Miller, *Draper Laboratory*
Grace Yi Li, *Intel Corporation*

Wafer IMS (Injection Molded Solder) – A New Fine Pitch Solder Bumping Technology on Wafers with Solder Alloy Composition Flexibility 1308
Jae-Woong Nah, *IBM Corporation*; Jeffrey Gelorme, *IBM Corporation*; Peter Sorce, *IBM Corporation*; Paul Lauro, *IBM Corporation*; Eric Perfecto, *IBM Corporation*; Mark McLeod, *IBM Corporation*; Kazushige Toriyama, *IBM Corporation*; Yasumitsu Orii, *IBM Corporation*; Peter Brofman, *IBM Corporation*; Takashi Nauchi, *Senju Metal Industry Co., Ltd.*; Akira Takaguchi, *Senju System Technology Co., Ltd.*; Kazuya Ishiguro, *Senju System Technology Co., Ltd.*; Tomoyasu Yoshikawa, *Senju Comtek Corporation*; Derek Daily, *Senju Comtek Corporation*; Ryoichi Suzuki, *Senju Metal Industry Co., Ltd.*

Reliability of Paste Based Transient Liquid Phase Sintered Interconnects 1314
Hannes Greve, *University of Maryland*; S. Ali Moeini, *University of Maryland*; F. Patrick McCluskey, *University of Maryland*

A Lead Free Joining Technology for High Temperature Interconnects Using Transient Liquid Phase Soldering (TLPS) 1321
Christian Ehrhardt, *Technical University Berlin*; Matthias Hutter, *Fraunhofer IZM*; Hermann Oppermann, *Fraunhofer IZM*; Klaus-Dieter Lang, *Technical University Berlin*

Developments of High-Bi Alloys as a High Temperature Pb-Free Solder 1328
Sandeep Mallampati, *Binghamton University*; Harry Schoeller, *Universal Instruments Corporation*; Liang Yin, *GE Global Research*; David Shaddock, *GE Global Research*; Junghyun Cho, *Binghamton University*

The Quantum Theory of Solid-State Atomic Bonding 1335
Chin C. Lee, *University of California, Irvine*; Lianxi Cheng, *University of California, Irvine*

Effective Method to Disperse and Incorporate Carbon Nanotubes in Electroless Ni-P Deposits 1342
Sha Xu, *City University of Hong Kong*; Yan Cheong Chan, *City University of Hong Kong*; Xiaoxin Zhu, *University of Greenwich*; Hua Lu, *University of Greenwich*; Chris Bailey, *University of Greenwich*

Electroless Ni-W-P Alloy as a Barrier Layer between Zn-Based High Temperature Solders and Cu Substrates 1348
Li Liu, *Loughborough University*; Longzao Zhou, *Huazhong University of Science & Technology*; Changqing Liu, *Loughborough University*

31: PoP, SiP, and Die Stacking

Chairs: Raj N. Master, *Microsoft Corporation*
Deborah Patterson, *Amkor Technology, Inc.*

Fabrication and Reliability Evaluation of a Novel Package-on-Package (PoP) Structure Based on Organic Substrate 1354
Xiaofeng Sun, *National Center for Advanced Packaging*; *Chinese Academy of Sciences*; Lixi Wan, *Chinese Academy of Sciences*; Yuan Lu, *National Center for Advanced Packaging*; *Chinese Academy of Sciences*

Strip Grinding Introduction for Thin PoP 1361
Jinseong Kim, *Amkor Technology Korea, Inc.*; Yesul Ahn, *Amkor Technology Korea, Inc.*; Gyuwan Han, *Amkor Technology Korea, Inc.*; Byoungwoo Cho, *Amkor Technology Korea, Inc.*; Dongjoo Park, *Amkor Technology Korea, Inc.*; Juhoon Yoon, *Amkor Technology Korea, Inc.*; Choonheung Lee, *Amkor Technology Korea, Inc.*; Lou Nicholls, *Amkor Technology Inc.*; Shengmin Wen, *Amkor Technology Inc.*

Cost and Performance Effective Silicon Interposer and Vertical Interconnect for 3D ASIC and Memory Integration 1366
Li Li, *Cisco Systems, Inc.*; Mitsutoshi Higashi, *Shinko Electric Industries Company, Ltd.*; Akihito Takano, *Shinko Electric Industries Company, Ltd.*; Jie Xue, *Cisco Systems, Inc.*; Gary Ikari, *Shinko Electric Industries Company, Ltd.*

Assembly and Packaging of Non-Bumped 3D Chip Stacks on Bumped Substrates 1372
Bing Dang, *IBM Corporation*; Joana Maria, *IBM Corporation*; Qianwen Chen, *IBM Corporation*; Jae-Woong Nah, *IBM Corporation*; Paul Andry, *IBM Corporation*; Cornelia Tsang, *IBM Corporation*; Katsuyuki Sakuma, *IBM Corporation*; Christy Tyberg, *IBM Corporation*; Raphael Robertazzi, *IBM Corporation*; Michael Scheuermann, *IBM Corporation*; Michael Gaynes, *IBM Corporation*; John Knickerbocker, *IBM Corporation*

The Miniaturization of a Micro-Ball Endoscope by SiP Approach 1378
Xunxun Zhu, *Tsinghua University*; Jian Cai, *Tsinghua University*; Yu Chen, *Tsinghua University*; Yingke Gu, *Tsinghua University*; Xiang Xie, *Tsinghua University*; Qian Wang, *Tsinghua University*; Zhihua Wang, *Tsinghua University*; Xiaofeng Sun, *Chinese Academy of Sciences*; Lixi Wan, *Chinese Academy of Sciences*

Design and Demonstration of Paper-Thin and Low-Warpage Single and 3D Organic Packages with Chip-Last Embedding Technology for Smart Mobile Applications 1384
Sung Jin Kim, *Georgia Institute of Technology*; Zihan Wu, *Georgia Institute of Technology*; Makoto Kobayashi, *Namics Corporation*; Fuhan Liu, *Georgia Institute of Technology*; Vanessa Smet, *Georgia Institute of Technology*; P. Markondeya Raj, *Georgia Institute of Technology*; Venky Sundaram, *Georgia Institute of Technology*; Rao Tummala, *Georgia Institute of Technology*

Manufacturing Readiness of BVA Technology for Ultra-High Bandwidth Package-on-Package 1389
Rajesh Katkar, *Invensas Corporation*; Rey Co, *Invensas Corporation*; Wael Zohni, *Invensas Corporation*

32: Substrates
Chairs: Yu-Hua Chen, *Unimicron*
Dong Wook Kim, *Qualcomm, Inc.*

Improvement of Substrate and Package Warpage by Copper Plating Process Optimization 1396
Omar Bchir, *Qualcomm Technologies, Inc.*; Houssam Jomaa, *Qualcomm Technologies, Inc.*; Chin Kwan Kim, *Qualcomm Technologies, Inc.*; Layal Rouhana, *Qualcomm Technologies, Inc.*; Kuiwon Kang, *Qualcomm Technologies, Inc.*; Milind Shah, *Qualcomm Technologies, Inc.*; Steve Bezuk, *Qualcomm Technologies, Inc.*

Coreless Substrate with Asymmetric Design to Improve Package Warpage 1401
Wei Lin, *Amkor Technology*; Bora Baloglu, *Amkor Technology*; Ken Stratton, *Amkor Technology*

Ultra Low CTE (1.8 ppm/°C) Core Material for Next Generation Thin CSP 1407
Tomohiko Kotake, *Hitachi Chemical Co., Ltd.*; Hikari Murai, *Hitachi Chemical Co., Ltd.*; Shin Takanezawa, *Hitachi Chemical Co., Ltd.*; Masato Miyatake, *Hitachi Chemical Co., Ltd.*; Masaaki Takekoshi, *Hitachi Chemical Co., Ltd.*; Masahisa Ose, *Hitachi Chemical Co., Ltd.*

A Novel Redistribution Layer Tailored by Nanotwinned Copper Decreases Warpage in Wafer Level Packaging 1411
Heng Li, *Shanghai Institute of Microsystem and Information Technology, Chinese Academy of Sciences*; Wenguo Ning, *Shanghai Institute of Microsystem and Information Technology, Chinese Academy of Sciences*; Chunsheng Zhu, *Shanghai Institute of Microsystem and Information Technology, Chinese Academy of Sciences*; Gaowei Xu, *Shanghai Institute of Microsystem and Information Technology, Chinese Academy of Sciences*; Le Luo, *Shanghai Institute of Microsystem and Information Technology, Chinese Academy of Sciences*

Demonstration of 3–5 μm RDL Line Lithography on Panel-Based Glass Interposers 1416
Hao Lu, *Georgia Institute of Technology*; Yutaka Takagi, *NGK Spark Plug Co., Ltd.*; Yuya Suzuki, *Georgia Institute of Technology*; Brett Sawyer, *Georgia Institute of Technology*; Robin Taylor, *Atotech GmbH*; Venky Sundaram, *Georgia Institute of Technology*; Rao Tummala, *Georgia Institute of Technology*

Characterization of Thin Polymer Films with the Focus on Lateral Stress and Mechanical Properties and Their Relevance to Microelectronics 1421
Markus Woehrmann, *Technical University Berlin*; Thorsten Fischer, *Fraunhofer IZM*; Hans Walter, *Fraunhofer IZM*; Michael Toepper, *Fraunhofer IZM*; Klaus-Dieter Lang, *Technical University Berlin*

Thin Polymer Dry-Film Dielectric Material and a Process for 10 μm Interlayer Vias in High Density Organic and Glass Interposers 1427

Yuya Suzuki, *Zeon Corporation; Georgia Institute of Technology*; Yutaka Takagi, *NGK Spark Plug Co., Ltd.*; Venky Sundaram, *Georgia Institute of Technology*; Rao Tummala, *Georgia Institute of Technology*

33: Novel Test Methods

Chairs: Lakshmi N. Ramanathan, *Microsoft Corporation*
Sridhar Canumalla, *Microsoft Corporation*

Pad Crater Detection Using Acoustic Waveform Analysis 1433

W. Carter Ralph, *Southern Research Institute*; Elizabeth E. Benedetto, *Hewlett Packard*; Aileen M. Allen, *Hewlett Packard*; Keith Newman, *Hewlett Packard*

High Acceleration Board Level Reliability Drop Test Using Dual Mass Shock Amplifier 1441

Andy Zhang, *Texas Instruments, Inc.*

Non-Destructive Crack and Defect Detection in SAC Solder Interconnects Using Cross-Sectioning and X-Ray Micro-CT Using Cross-Sectioning and X-Ray Micro-CT 1449

Pradeep Lall, *Auburn University*; Shantanu Deshpande, *Auburn University*; Junchao Wei, *Auburn University*; Jeff Suhling, *Auburn University*

High Resolution and Fast Throughput-Time X-Ray Computed Tomography for Semiconductor Packaging Applications 1457

Yan Li, *Intel Corporation*; Mario Pacheco, *Intel Corporation*; Deepak Goyal, *Intel Corporation*; John W. Elmer, *Lawrence Livermore National Laboratory*; Holly D. Barth, *Lawrence Livermore National Laboratory*; Dula Parkinson, *Lawrence Berkeley National Laboratory*

In-Situ Measurements of the Relative Thermal Resistance: Highly Sensitive Method to Detect Crack Propagation in Solder Joints 1464

Gordon Elger, *Technische Hochschule Ingolstadt*; Shri Vishnu Kandaswamy, *Technische Hochschule Ingolstadt*; Maarten von Kouwen, *Philips Technology GmbH*; Robert Derix, *Philips Technology GmbH*; Fosca Conti, *University of Padova*

Reliability Testing of Wire Bonds Using Pad Resistance with van der Pauw Method 1471

Michael Mayer, *University of Waterloo*; Samuel Kim, *University of Waterloo*

Colour Shift in Remote Phosphor Based LED Products 1477

M. Yazdan Mehr, *Materials Innovation Institute; Delft University of Technology*; W.D. Van Driel, *Philips Lighting; Delft University of Technology*; G.Q. Zhang, *Delft University of Technology*

34: Novel Packaging

Chairs: Vasudeva P. Atluri, *Renavitas Technologies*
Jai Agrawal, *Purdue University*

Multifunctional System Integration in Flexible Substrates 1482

K. Bock, *Fraunhofer EMFT*; E. Yacoub-George, *Fraunhofer EMFT*; W. Hell, *Fraunhofer EMFT*; A. Drost, *Fraunhofer EMFT*; H. Wolf, *Fraunhofer EMFT*; D. Bollmann, *Fraunhofer EMFT*; C. Landesberger, *Fraunhofer EMFT*; G. Klink, *Fraunhofer EMFT*; H. Gieser, *Fraunhofer EMFT*; C. Kutter, *Fraunhofer EMFT*

Preparation of a Micro Rubidium Vapor Cell and Its Integration in a Chip-Scale Atomic Magnetometer 1488

Yu Ji, *Southeast University*; Jintang Shang, *Southeast University*; Youpeng Chen, *Southeast University*; Ching-Ping Wong, *Chinese University of Hong Kong*

Nanowires-Based High-Density Capacitors and Thinfilm Power Sources in Ultra-Thin 3D Glass Modules 1492

Saumya Gandhi, *Georgia Institute of Technology*; Liyi Li, *Georgia Institute of Technology*; Ho-Yee Hui, *Georgia Institute of Technology*; Parthasarathi Chakraborti, *Georgia Institute of Technology*; Himani Sharma, *Georgia Institute of Technology*; P. Markondeya Raj, *Georgia Institute of Technology*; C.P. Wong, *Georgia Institute of Technology*; Rao Tummala, *Georgia Institute of Technology*

Development of a High Density Glass Interposer Based on Wafer Level Packaging Technologies 1498

Michael Töpper, *Fraunhofer IZM*; Markus Wöhrmann, *Technical University Berlin*; Lars Brusberg, *Fraunhofer IZM*; Nils Jürgensen, *Fraunhofer IZM*; Ivan Ndip, *Fraunhofer IZM*; Klaus-Dieter Lang, *Technical University Berlin*

Novel Sealing Technology for Organic EL Display and Lighting by Means of Modified Surface Activated Bonding Method 1504

Takashi Matsumae, *University of Tokyo*; Masahisa Fujino, *University of Tokyo*; Tadatomo Suga, *University of Tokyo*

Solder Joint Inspection with Induction Thermography 1509

Johannes Bohm, *Technical University Dresden*; Klaus-Juergen Wolter, *Technical University Dresden*; Henning Heuer, *Technical University Dresden*

Development of B-Spline X-Ray Diffraction Imaging Techniques for Die Warpage and Stress Monitoring inside Fully Encapsulated Packaged Chips 1517

C.S. Wong, *Dublin City University*; A. Ivankovic, *IMEC; KU Leuven*; A. Cowley, *Dublin City University*; N.S. Bennett, *Dublin City University*; A.N. Danilewsky, *Albert-Ludwigs-Universität*; M. Gonzalez, *IMEC*; V. Cherman, *IMEC*; B. Vandevelde, *IMEC*; I. De Wolf, *IMEC; KU Leuven*; P.J. McNally, *Dublin City University*

35: Innovations in Wirebond Technology

Chairs: William Chen, *Advanced Semiconductor Engineering, Inc.*
Gilles Poupon, *CEA-LETI*

Process Optimization and Reliability Study for Cu Wire Bonding Advanced Nodes 1523

Ivy Qin, *Kulicke and Soffa Industries, Inc.*; Hui Xu, *Kulicke and Soffa Industries, Inc.*; Basil Milton, *Kulicke and Soffa Industries, Inc.*; Nestor Mendoza, *Kulicke and Soffa Industries, Inc.*; Horst Clauberg, *Kulicke and Soffa Industries, Inc.*; Bob Chylak, *Kulicke and Soffa Industries, Inc.*; Hidenori Abe, *Hitachi Chemical Co., Ltd.*; Dongchul Kang, *Hitachi Chemical Co., Ltd.*; Yoshinori Endo, *Hitachi Chemical Co., Ltd.*; Masahiko Osaka, *Hitachi Chemical Co., Ltd.*; Shinya Nakamura, *Hitachi Chemical Co., Ltd.*

Silver-Assisted Copper Wire Bonding Using Solid-State Processes 1529

Yi-Ling Chen, *University of California, Irvine*; Yuan-Yun Wu, *University of California, Irvine*; Chin C. Lee, *University of California, Irvine*

Ag Alloy Wire Characteristic and Benefits 1533

Jensen Tsai, *Siliconware Precision Industries Co., Ltd.*; Albert Lan, *Siliconware Precision Industries Co., Ltd.*; D.S. Jiang, *Siliconware Precision Industries Co., Ltd.*; Li Wei Wu, *Siliconware Precision Industries Co., Ltd.*; Joseph Huang, *Siliconware Precision Industries Co., Ltd.*; J.B. Hong, *Siliconware Precision Industries Co., Ltd.*

Copper versus Palladium Coated Copper Wire Process and Reliability Differences 1539

Chu-Chung (Stephen) Lee, *Freescale Semiconductor, Inc.*; TuAnh Tran, *Freescale Semiconductor, Inc.*; Dan Boyne, *Freescale Semiconductor, Inc.*; Leo Higgins, *Freescale Semiconductor, Inc.*; Andrew Mawer, *Freescale Semiconductor, Inc.*

Improving the Bond Quality of Copper Wire Bonds Using a Friction Model Approach 1549

Simon Althoff, *University of Paderborn*; Jan Neuhaus, *University of Paderborn*; Tobias Hemsel, *University of Paderborn*; Walter Sextro, *University of Paderborn*

High Aspect Ratio Lithography for Litho-Defined Wire Bonding 1556

Zahra Kolahdouz Esfahani, *Delft University of Technology*; Henk van Zeijl, *Delft University of Technology*; G.Q. Zhang, *Delft University of Technology*

Comprehensive Intermetallic Compound Phase Analysis and Its Thermal Evolution at Cu Wirebond Interface 1562

In-Tae Bae, *Binghamton University*; Dae Young Jung, *Binghamton University*; Jenny Chang, *Advanced Semiconductor Engineering, Inc.*; Scott Chen, *Advanced Semiconductor Engineering, Inc.*

36: Recent Advancement in Manufacturing Technology

Chairs: Paul Houston, *Engent*
Hirofumi Nakajima, *Consultant*

High Uniformity and High Speed Copper Pillar Plating Technique 1571

Konstantin Kholostov, *Sapienza University of Rome*; Aliaksei Klyshko, *Sapienza University of Rome*; Danilo Ciarniello, *Rise Technology S.r.l.*; Paolo Nenzi, *Rise Technology S.r.l.*; Roberto Pagliucci, *Rise Technology S.r.l.*; Rocco Crescenzi, *Sapienza University of Rome*; Dario Bernardi, *2BG*; Marco Balucani, *Sapienza University of Rome, Rise Technology S.r.l.*

Plasma-Based Die Singulation Processing Technology 1577

Kenneth D. Mackenzie, *Plasma-Therm LLC*; David Pays-Volard, *Plasma-Therm LLC*; Linnell Martinez, *Plasma-Therm LLC*; Christopher Johnson, *Plasma-Therm LLC*; Thierry Lazerand, *Plasma-Therm LLC*; Russell Westerman, *Plasma-Therm LLC*

Removed Organic Solderability Preservative (OSP) by Ar/O2 Microwave Plasma to Improve Solder Joint in Thermal Compression Flip Chip Bonding 1584

Jr-Wei Peng, *ASE Group*; Yan-Siang Chen, *ASE Group*; Yi Chen, *ASE Group*; Jiang-Long Liang, *National Cheng Kung University*; Kwang-Lung Lin, *National Cheng Kung University*; Yuh-Lang Lee, *National Cheng Kung University*

A PoP Structure to Support I/O over 2000 1590

Dyi-Chung Hu, *Unimicron Technology Corporation*; Puru Lin, *Unimicron Technology Corporation*; Yu Hua Chen, *Unimicron Technology Corporation*; Chun-Ting Lin, *Unimicron Technology Corporation*

Enabling Eutectic Soldering of 3D Opto-Electronics onto Low Tg Flexible Interposers 1595

Meriem Ben-Salah Akin, *Leibniz University of Hanover*; Lutz Rissing, *Leibniz University of Hanover*; Wolfgang Heumann, *Leibniz University of Hanover*

Parameter Optimization in Assembly Manufacturing Process for a Power Module 1601

Yumin Liu, *Fairchild Semiconductor Corporation*; Yong Liu, *Fairchild Semiconductor Corporation*

Automated Inspection and Metrology for 2.5D and 3D/TSV Process Assurance 1606

James Wood, *IBM Corporation*; Vilmarie Soler, *IBM Corporation*; Eric Perfecto, *IBM Corporation*; Thomas Luckenbach, *Camtek USA*; Aki Shoukrun, *Camtek Ltd.*

37: Interactive Presentations 1

Chairs: Mark Poliks, *i3 Electronics, Inc.*
Ibrahim Guven, *University of Arizona*

Investigation of a Photodefinable Glass Substrate for Millimeter-Wave Radios on Package 1610

Telesphor Kamgaing, *Intel Corporation*; Adel A. Elsherbini, *Intel Corporation*; Torrey W. Frank, *Intel Corporation*; Sasha N. Oster, *Intel Corporation*; Valluri R. Rao, *Intel Corporation*

Design and Fabrication of Low-Pressure Piezoresistive MEMS Sensor for Fuel Cell Electric Vehicles 1616

Minkyu Lee, *Hyundai Motor Company*; Kiyoung Nam, *Hyundai Motor Company*; Seungyong Lee, *Hyundai Motor Company*; Hakgu Kim, *Hyundai Motor Company*; Chimyung Kim, *Hyundai Motor Company*; Yongsun Park, *Hyundai Motor Company*; Byungki Ahn, *Hyundai Motor Company*; Taewan Kim, *Sejong Industrial Company, Ltd.*; Hochul Seo, *Sejong Industrial Company, Ltd.*

Demonstration of TCNCP Flip Chip Reliability with 30µm Pitch Cu Bump and Substrate with Thin Ni and Thick Au Surface Finish 1622

Weihong Zhang, *Nantong Fujitsu Microelectronics Co., Ltd.*; Shengping Hong, *Nantong Fujitsu Microelectronics Co., Ltd.*; Xiaolong Yan, *Nantong Fujitsu Microelectronics Co., Ltd.*; Feng Zhou, *Nantong Fujitsu Microelectronics Co., Ltd.*; Tonglong Zhang, *Nantong Fujitsu Microelectronics Co., Ltd.*

Integrated Process Characterization and Fabrication Challenges for 2.5D IC Packaging Utilizing Silicon Interposer with Backside Via Reveal Process 1628

Cheng-Hsiang Liu, *Siliconware Precision Industries Co., Ltd.*; Jyun-Ling Tsai, *Siliconware Precision Industries Co., Ltd.*; Hung-Hsien Chang, *Siliconware Precision Industries Co., Ltd.*; Chang-Lun Lu, *Siliconware Precision Industries Co., Ltd.*; Shih-Ching Chen, *Siliconware Precision Industries Co., Ltd.*

Structure Effects on the Electrical Reliability of Fine-Pitch Cu Micro-Bumps for 3D Integration 1635

Byeong-Rok Lee, *Andong National University*; June-Bum Kim, *Andong National University*; Seung-Hyun Kim, *Andong National University*; Byeong-Hyun Bae, *Andong National University*; Ho-Young Son, *SK Hynix Inc.*; Tac-Keun Oh, *SK Hynix Inc.*; Min-Suk Suh, *SK Hynix Inc.*; Nam-Seog Kim, *SK Hynix Inc.*; Young-Bae Park, *Andong National University*

Demonstration of Low Cost TSV Fabrication in Thick Silicon Wafers 1641

E. Vick, *RTI International*; D.S. Temple, *RTI International*; R. Anderson, *RTI International*; J. Lannon, *RTI International*; C. Li, *DRS RSTA, Inc.*; K. Peterson, *DRS RSTA, Inc.*; G. Skidmore, *DRS RSTA, Inc.*; C.J. Han, *DRS RSTA, Inc.*

X-Ray Micro-Beam Diffraction Measurement of the Effect of Thermal Cycling on Stress in Cu TSV: A Comparative Study 1648

Chukwudi Okoro, *NIST*; Lyle E. Levine, *NIST*; Ruqing Xu, *Argonne National Laboratory*; Klaus Hummler, *SEMATECH*; Yaw Obeng, *NIST*

Adhesive Enabling Technology for Directly Plating Copper onto Glass/Ceramic Substrates 1652

Hailuo Fu, *Atotech USA Inc.*; Sara Hunegnaw, *Atotech USA Inc.*; Zhiming Liu, *Atotech USA Inc.*; Lutz Brandt, *Atotech USA Inc.*; Tafadzwa Magaya, *Atotech USA Inc.*

Very Thin POP and SIP Packaging Approaches to Achieve Functionality Integration Prior to TSV Implementation 1656

Fernando Roa, *Amkor Technology, Inc.*

A Study on the Fine Pitch Chip Interconnection Using Cu/SnAg Bumps and B-Stage Non-Conductive Films (NCFs) for 3D-TSV Vertical Interconnection 1661

Yongwon Choi, *Korea Advanced Institute of Science and Technology (KAIST)*; Jiwon Shin, *Korea Advanced Institute of Science and Technology (KAIST)*; Young Soon Kim, *Korea Advanced Institute of Science and Technology (KAIST)*; Kyung-Lim Suk, *Korea Advanced Institute of Science and Technology (KAIST)*; Il Kim, *Korea Advanced Institute of Science and Technology (KAIST)*; Kyung-Wook Paik, *Korea Advanced Institute of Science and Technology (KAIST)*

Pathfinding Methodology for Optimal Design and Integration of 2.5D/3D Interconnects 1667

Farhang Yazdani, *BroadPak Corporation*; John Park, *Mentor Graphics Corporation*

Cost Effective Interposer for Advanced Electronic Packages 1673

Satoru Kuramochi, *Dai Nippon Printing Co., Ltd.*; Sumio Koiwa, *Dai Nippon Printing Co., Ltd.*; Kousuke Suzuki, *Dai Nippon Printing Co., Ltd.*; Yoshitaka Fukuoka, *WEISTI*

Thermal Management for Wafer Level Packaging (WLP) 1679

Tiao Zhou, *Maxim Integrated*; Arkadii Samoilov, *Maxim Integrated*

Inkjet Printed Nano-Particle Cu Process for Fabrication of Re-Distribution Layers on Silicon Wafer 1685

Ayat Soltani, *Tampere University of Technology*; Tero Kumpulainen, *Tampere University of Technology*; Matti Mäntysalo, *Tampere University of Technology*

Design of Multi-Sensor for Safety Monitoring of Heavy Machinery 1690

Long Li, *Huazhong University of Science & Technology*; Fei Hou, *Dongfeng Automobile Electronics Co., Ltd.*; Jinghao Qiu, *Nanjing University of Aeronautics and Astronautics*; Zhang Luo, *Huazhong University of Science & Technology*; Shengzhi Zhang, *Huazhong University of Science & Technology*; Qiang Dan, *Huazhong University of Science & Technology*; Sheng Liu, *Wuhan University*

Novel TSV Process Technologies for 2.5D/3D Packaging 1697

Y. Morikawa, *ULVAC, Inc.; NMEMS Technology Research Organization*; T. Murayama, *ULVAC, Inc.; NMEMS Technology Research Organization*; T. Sakuishi, *ULVAC, Inc.; NMEMS Technology Research Organization*; A. Suzuki, *ULVAC, Inc.*; Y. Nakamuta, *ULVAC, Inc.*; K. Suu, *ULVAC, Inc.; NMEMS Technology Research Organization*

Increasing the Lifetime of Electronic Packaging by Higher Temperatures: Solders vs. Silver Sintering 1700

Aaron Hutzler, *Fraunhofer IISB*; Adam Tokarski, *Fraunhofer IISB*; Silke Kraft, *Fraunhofer IISB*; Sigrid Zischler, *Fraunhofer IISB*; Andreas Schletz, *Fraunhofer IISB*

Comparison of New Die-Attachment Technologies for Power Electronic Assemblies 1707

Eike Möller, *University of Freiburg*; Adeel Ahmad Bajwa, *University of Freiburg*; Eugen Rastjagaev, *Infineon Technologies AG*; Jürgen Wilde, *University of Freiburg*

High Vacuum Wafer Level Packaging for High-Value MEMS Applications 1714

S. Nicolas, *CEA-LETI*; F. Greco, *CEA-LETI*; S. Caplet, *CEA-LETI*; C. Coutier, *CEA-LETI*; C. Dressler, *CEA-LETI*; M. Audoin, *CEA-LETI*; X. Baillin, *CEA-LETI*; G. Dehag, *CEA-LETI*; F. Souchon, *CEA-LETI*; S. Fanget, *CEA-LETI*

Thermal and Electrical Tests of Air-Gap TSV 1722

Cui Huang, *Tsinghua University*; Dong Wu, *Tsinghua University*; Zheyao Wang, *Tsinghua University*

Heterogeneous System Integration Pseudo-SoC Technology for Smart-Health-Care Intelligent Life Monitor Engine & Eco-System (SILMEE) 1729

Hiroshi Yamada, *Toshiba Corporation*; Yasuhiro Sato, *Toshiba Corporation*; Nobuhiro Ooshima, *Toshiba Corporation*; Hiroyuki Hirai, *Toshiba Corporation*; Takuji Suzuki, *Toshiba Corporation*; Shigenobu Minami, *Toshiba Corporation*

Effects of Various Environmental Conditions on the Electrical Properties and Interfacial Reliability of Printed Ag/Polyimide System 1735

Byung-Hyun Bae, *Andong National University*; Min-Su Jeong, *Andong National University*; Byeong Rok Lee, *Andong National University*; Joung-Hoon Choo, *HICEL*; Eun-Kuk Choi, *HICEL*; Jong-Sun Yoon, *HICEL*; Young-Bae Park, *Andong National University*

Wafer Level Warpage Characterization for Backside Manufacturing Processes of TSV Interposers 1740

Feng Jiang, *National Center for Advanced Packaging*; Qibin Wang, *National Center for Advanced Packaging; Chinese Academy of Sciences*; Kai Xue, *National Center for Advanced Packaging; Chinese Academy of Sciences*; Xiangmeng Jing, *National Center for Advanced Packaging; Chinese Academy of Sciences*; Daquan Yu, *National Center for Advanced Packaging; Chinese Academy of Sciences*; Dongkai Shangguan, *National Center for Advanced Packaging; Chinese Academy of Sciences*

Stretchable and Transparent Silicone/Zinc Oxide Nanocomposite for Advanced LED Packaging 1745

Xueying Zhao, *Georgia Institute of Technology*; Liyi Li, *Georgia Institute of Technology*; Zhuo Li, *Georgia Institute of Technology*; Ching-Ping Wong, *Georgia Institute of Technology; Chinese University of Hong Kong*

Warpage Characterization of Panel Fan-Out (P-FO) Package 1750

Hung-Wen Liu, *Siliconware Precision Industries Co., Ltd.*; Yi-Wei Liu, *Siliconware Precision Industries Co., Ltd.*; Jason Ji, *Siliconware Precision Industries Co., Ltd.*; Jash Liao, *Siliconware Precision Industries Co., Ltd.*; Agassi Chen, *Siliconware Precision Industries Co., Ltd.*; Yan-Heng Chen, *Siliconware Precision Industries Co., Ltd.*; Nicholas Kao, *Siliconware Precision Industries Co., Ltd.*; Yi-Che Lai, *Siliconware Precision Industries Co., Ltd.*

38: Interactive Presentations 2
Chairs: Mark Eblen, *Kyocera America, Inc.*
Michael Mayer, *University of Waterloo*

A Novel Double Layer NCF for Highly Reliable Micro-Bump Interconnection 1755

Ji-Won Shin, *Korea Advanced Institute of Science and Technology (KAIST)*; Yong-Won Choi, *Korea Advanced Institute of Science and Technology (KAIST)*; Young Soon Kim, *Korea Advanced Institute of Science and Technology (KAIST)*; Un Byung Kang, *Samsung Electronics Company, Ltd.*; Sun Kyung Seo, *Samsung Electronics Company, Ltd.*; Kyung-Wook Paik, *Korea Advanced Institute of Science and Technology (KAIST)*

CO2-Laser Drilling of TGVs for Glass Interposer Applications 1759

Lars Brusberg, *Fraunhofer IZM*; Marco Queisser, *Technical University Berlin*; Marcel Neitz, *Technical University Berlin*; Henning Schröder, *Fraunhofer IZM*; Klaus-Dieter Lang, *Technical University Berlin*

Effects of Pad Surface Finish on Interfacial Reliabilities of Cu-Pillar/Sn-Ag Bumps of 2.5D TSV-Interposer on PCB Applications 1765

Youngsoon Kim, *Samsung Electro-Mechanics Company, Ltd.*; Ji-Won Shin, *Korea Advanced Institute of Science and Technology (KAIST)*; Young Won Choi, *Korea Advanced Institute of Science and Technology (KAIST)*; Kyung-Wook Paik, *Korea Advanced Institute of Science and Technology (KAIST)*

Effect of Variation in the Reflow Profile on the Microstructure of Near Eutectic SnAgCu Alloys 1769

Francis Mutuku, *Binghamton University*; Babak Arfaei, *Binghamton University; Universal Instruments Corporation*; Eric J. Cotts, *Binghamton University*

Development of the Thin Film with High Thermal Conductivity for Power Devices 1776

Hiroshi Takasugi, *Namics Corporation*; Shin Teraki, *Namics Corporation*; Tsuyoshi Kurokawa, *Namics Corporation*; Issei Aoki, *Namics Corporation*

Development of Electroless Nickel-Iron Plating Process for Microelectronic Applications 1782

Yu Luo, *IBM Corporation*; Sung K. Kang, *IBM Corporation*; Oblesh Jinka, *IBM Corporation*; Maurice Mason, *IBM Corporation*; Steven A. Cordes, *IBM Corporation*; Lubomyr T. Romankiw, *IBM Corporation*

Novel Conductive Paste Using Hybrid Silver Sintering Technology for High Reliability Power Semiconductor Packaging 1790

Howard (Hwa II) Jin, *Alpha Advanced Materials*; Senthil Kanagavel, *Alpha Advanced Materials*; Wai Foo Chin, *Alpha Advanced Materials*

Novel Low Temperature Curable Photo-Sensitive Insulator 1796

Kenji Okamoto, *JSR Corporation*; Hikaru Mizuno, *JSR Corporation*; Tomohiko Sakurai, *JSR Corporation*; Katsumi Inomata, *JSR Corporation*

3D and 2.5D Packaging Assembly with Highly Silica Filled One Step Chip Attach Materials for Both Thermal Compression Bonding and Mass Reflow Processes 1803

Christopher Breach, *Kester Inc.*; Daniel Duffy, *Kester Inc.*; David Eichstadt, *Kester Inc.*

Process Compatibility of Conventional and Low-Temperature Curable Organic Insulation Materials for 2.5D and 3D IC Packaging – A User's Perspective 1810

Guilian Gao, *Invensas Corporation*; Bong-Sub Lee, *Invensas Corporation*; Andrew Cao, *Invensas Corporation*; Ellis Chau, *Invensas Corporation*

Optimization of CMP Process for TSV Reveal in Consideration of Critical Defect 1816

DongHoon Lee, *Amkor Technology Korea, Inc.; Sungkyunkwan University*; DoHyeong Kim, *Amkor Technology Korea, Inc.*; SeungChul Han, *Amkor Technology Korea, Inc.*; JooHyun Kim, *Amkor Technology Korea, Inc.*; JungSoo Park, *Amkor Technology Korea, Inc.*; BoRa Jang, *Amkor Technology Korea, Inc.*; YoungSuk Chung, *Amkor Technology Korea, Inc.*; SeongMin Seo, *Amkor Technology Korea, Inc.*; YongSang Kim, *Sungkyunkwan University*; ChoonHeung Lee, *Amkor Technology Korea, Inc.*

High Throughput Roller Type Nano-Pattern Transfer Technique on Both Rigid Flexible Substrates and Mold Deformation Analysis under Atmospheric Imprint Environment 1822

Yinsheng Zhong, *Hong Kong University of Science and Technology*; Matthew M.F. Yuen, *Hong Kong University of Science and Technology*

Capacitive Deionization of Water Coolant Using Hybrid Carbon Electrodes for High Power Electronic Applications 1828

Ziyin Lin, *Georgia Institute of Technology*; Zhuo Li, *Georgia Institute of Technology*; Kyoung-Sik Moon, *Georgia Institute of Technology*; Ching-Ping Wong, *Georgia Institute of Technology; Chinese University of Hong Kong*

A Microfluidic Chip Integrated with a Sono-Transducer Using Combined Resonance between Oscillations of Hemispherical Micro Glass Shell and Enclosed Microfluid .... N/A

Jiafeng Xu, *Southeast University*; Jintang Shang, *Southeast University*; Ching-Ping Wong, *Chinese University of Hong Kong*

RF Energy Harvesting 1838

Parvizso Aminov, *Purdue University*; Jai P. Agrawal, *Purdue University*

Localized Metal Plating on Aluminum Back Side PV Cells 1842

M. Balucani, *Sapienza University of Rome; Rise Technology S.r.l.*; K. Kholostov, *Sapienza University of Rome*; L. Serenelli, *ENEA Casaccia Research Centre*; M. Izzi, *ENEA Casaccia Research Centre*; D. Bernardi, *2BG S.r.l.*; M. Tucci, *ENEA Casaccia Research Centre*

Wet Etching of Deep Trenches on Silicon with Three-Dimensional (3D) Controllability 1848

Liyi Li, *Georgia Institute of Technology*; Ching-Ping Wong, *Georgia Institute of Technology; Chinese University of Hong Kong*

An Innovative Bumpless Stacking with Through Silicon Via for 3D Wafer-On-Wafer (WOW) Integration 1853

Sue-Chen Liao, *Industrial Technology Research Institute (ITRI)*; Erh-Hao Chen, *Industrial Technology Research Institute (ITRI)*; Chien-Chou Chen, *Industrial Technology Research Institute (ITRI)*; Shang-Chun Chen, *Industrial Technology Research Institute (ITRI)*; Jui-Chin Chen, *Industrial Technology Research Institute (ITRI)*; Po-Chih Chang, *Industrial Technology Research Institute (ITRI)*; Yiu-Hsiang Chang, *Industrial Technology Research Institute (ITRI)*; Cha-Hsin Lin, *Industrial Technology Research Institute (ITRI)*; Tzu-Kun Ku, *Industrial Technology Research Institute (ITRI)*; Ming-Jer Kao, *Industrial Technology Research Institute (ITRI)*; Young Suk Kim, *Tokyo Institute of Technology*; Nobuhide Maeda, *Tokyo Institute of Technology*; Shoichi Kodama, *Tokyo Institute of Technology*; Hideki Kitada, *Tokyo Institute of Technology*; Koji Fujimoto, *Tokyo Institute of Technology*; Takayuki Ohba, *Tokyo Institute of Technology*

3D Integration and Assembly of Wireless Sensor Nodes for 'Green' Sensor Networks 1857

Jian Lu, *National Institute of AIST; NMEMS Technology Research Organization*; Hironao Okada, *National Institute of AIST; NMEMS Technology Research Organization*; Toshihiro Itoh, *National Institute of AIST; NMEMS Technology Research Organization*; Takeshi Harada, *NMEMS Technology Research Organization*; Ryutaro Maeda, *National Institute of AIST; NMEMS Technology Research Organization*

New Demultiplexer Component for Optical Polymer Fiber Communication Systems 1862

S. Höll, *Harz University of Applied Sciences*; M. Haupt, *Harz University of Applied Sciences*; U.H.P. Fischer, *Harz University of Applied Sciences*

Nanofiller Based Spin-on Materials for Negligible Reflection of Silicon Photonic External Coupling 1870

Yoichi Taira, *IBM Corporation*; Ryuma Mizusawa, *Tokyo Ohka Kogyo Co., Ltd.*; Rie Matsumoto, *Tokyo Ohka Kogyo Co., Ltd.*; Kuniaki Suaoke, *IBM Corporation*; Hidetoshi Numata, *IBM Corporation*

Effect of Patterned Substrate on Light Extraction Efficiency of Chip-on-Board Packaging LEDs 1876

Huai Zheng, *Huazhong University of Science & Technology*; Zhili Zhao, *Huazhong University of Science & Technology*; Yiman Wang, *Huazhong University of Science & Technology*; Lang Li, *Huazhong University of Science & Technology*; Sheng Liu, *Huazhong University of Science & Technology*; Xiaobing Luo, *Huazhong University of Science & Technology*

39: Interactive Presentations 3

Chairs: Patrick Thompson, *Texas Instruments, Inc.*
Rao Bonda, *Amkor Technology, Inc.*

Transferrable Fine Pitch Probe Technology 1880

Y. Liu, *IBM Corporation*; S.L. Wright, *IBM Corporation*; B. Dang, *IBM Corporation*; P. Andry, *IBM Corporation*; R. Polastre, *IBM Corporation*; J. Knickerbocker, *IBM Corporation*

Improvement of the Crystallinity of Electroplated Copper Thin Films for Highly Reliable 3D Interconnections 1885

Chuanhong Fan, *Tohoku University*; Osamu Asai, *Tohoku University*; Ryosuke Furuya, *Tohoku University*; Ken Suzuki, *Tohoku University*; Hideo Miura, *Tohoku University*

Process, Assembly and Electromigration Characteristics of Glass Interposer for 3D Integration 1891

Chun-Hsien Chien, *Industrial Technology Research Institute (ITRI)*; Ching-Kuan Lee, *Industrial Technology Research Institute (ITRI)*; Chun-Te Lin, *Industrial Technology Research Institute (ITRI)*; Yu-Min Lin, *Industrial Technology Research Institute (ITRI)*; Chau-Jie Zhan, *Industrial Technology Research Institute (ITRI)*; Hsiang-Hung Chang, *Industrial Technology Research Institute (ITRI)*; Chao-Kai Hsu, *Industrial Technology Research Institute (ITRI)*; Huan-Chun Fu, *Industrial Technology Research Institute (ITRI)*; Wen-Wei Shen, *Industrial Technology Research Institute (ITRI)*; Yu-Wei Huang, *Industrial Technology Research Institute (ITRI)*; Cheng-Ta Ko, *Industrial Technology Research Institute (ITRI)*; Wei-Chung Lo, *Industrial Technology Research Institute (ITRI)*; Yung Jean (Rachel) Lu, *Corning Inc.*

Improved PCB Via Pattern to Reduce Crosstalk at Package BGA Region for High Speed Serial Interface 1896

Yujeong Shim, *Altera Corporation*; Dan Oh, *Altera Corporation*

A Wafer Level Through-Stack-Via Integration Process with One-Time Bottom-up Copper Filling 1902

Yunhui Zhu, *Peking University*; Shenglin Ma, *Xiamen University, Peking University*; Xin Sun, *Peking University*; Runiu Fang, *Peking University*; Xiao Zhong, *Peking University*; Yuan Bian, *Peking University*; Yong Guan, *Peking University*; Jing Chen, *Peking University*; Min Miao, *Peking University, Beijing Information Science and Technology University*; Yufeng Jin, *Peking University*

Effect of Joint Shape Controlled by Thermocompression Bonding on the Reliability Performance of 60μm-Pitch Solder Micro Bump Interconnections 1908

Yu-Wei Huang, *Industrial Technology Research Institute (ITRI)*; Chau-Jie Zhan, *Industrial Technology Research Institute (ITRI)*; Jing-Ye Juang, *Industrial Technology Research Institute (ITRI)*; Yu-Min Lin, *Industrial Technology Research Institute (ITRI)*; Shin-Yi Huang, *Industrial Technology Research Institute (ITRI)*; Su-Mei Chen, *Industrial Technology Research Institute (ITRI)*; Chia-Wen Fan, *Industrial Technology Research Institute (ITRI)*; Ren-Shin Cheng, *Industrial Technology Research Institute (ITRI)*; Shu-Han Chao, *National Chiao Tung University*; Wan-Lin Hsieh, *National Chiao Tung University*; Chih Chen, *National Chiao Tung University*; John H. Lau, *Industrial Technology Research Institute (ITRI)*

Development of Micro Bump Joints Fabrication Process Using Cone Shape Au Bumps for 3D LSI Chip Stacking 1915

Fumito Imura, *National Institute of AIST*; Naoya Watanabe, *National Institute of AIST*; Shunsuke Nemoto, *National Institute of AIST*; Wei Feng, *National Institute of AIST*; Katsuya Kikuchi, *National Institute of AIST*; Hiroshi Nakagawa, *National Institute of AIST*; Masahiro Aoyagi, *National Institute of AIST*

Effect of Polymer Liners in CNT Based Through Silicon Vias 1921

Archana Kumari, *Indian Institute of Technology Roorkee*; M.K. Majumder, *Indian Institute of Technology Roorkee*; B.K. Kaushik, *Indian Institute of Technology Roorkee*; S.K. Manhas, *Indian Institute of Technology Roorkee*

Investigation of Low-Temperature Deposition High-Uniformity Coverage Parylene-HT as a Dielectric Layer for 3D Interconnection 1926

Bui Thanh Tung, *National Institute of AIST*; Xiaojin Cheng, *National Institute of AIST; Loughborough University*; Naoya Watanabe, *National Institute of AIST*; Fumiki Kato, *National Institute of AIST*; Katsuya Kikuchi, *National Institute of AIST*; Masahiro Aoyagi, *National Institute of AIST*

Arrays of Millimeter-Wave Silicon Waveguides for Interchip Communication on Glass Interposer 1932

Qidong Wang, *Chinese Academy of Sciences; National Center for Advanced Packaging*; Daniel Guidotti, *Chinese Academy of Sciences; National Center for Advanced Packaging*; Liqiang Cao, *Chinese Academy of Sciences; National Center for Advanced Packaging*; Delong Qiu, *National Center for Advanced Packaging*; Daquan Yu, *Chinese Academy of Sciences; National Center for Advanced Packaging*; Shuling Wang, *Chinese Academy of Sciences; National Center for Advanced Packaging*; Xugang Wang, *Chinese Academy of Sciences; National Center for Advanced Packaging*; Tiachun Ye, *Chinese Academy of Sciences; National Center for Advanced Packaging*; Lixi Wan, *National Center for Advanced Packaging*

Effect of Ag and Cu Content in Sn Based Pb-Free Solder on Electromigration 1940

Minhua Lu, *IBM Corporation*; Charles Goldsmith, *IBM Corporation*; Thomas Wassick, *IBM Corporation*; Eric Perfecto, *IBM Corporation*; Charles Arvin, *IBM Corporation*

Low Loss Transmission Lines on Flexible COP Substrate by Standard Lamination Process 1944

Chang-Ho Liou, *Industrial Technology Research Institute (ITRI)*; Hsin-Chia Lu, *National Taiwan University*; Yi-Fan Lin, *National Taiwan University*; Shih-Keng Chuang, *National Taiwan University*; Wen-Ching Ko, *Industrial Technology Research Institute (ITRI)*; Je-Ping Hu, *Industrial Technology Research Institute (ITRI)*

FBEOL No-Aluminum Pad Integration in Pb-Free C4 Products for Environmental, Cost and Reliability Benefits 1949

E. Misra, *IBM Corporation*; T. Daubenspeck, *IBM Corporation*; T. Wassick, *IBM Corporation*; K. Tunga, *IBM Corporation*; D. Questad, *IBM Corporation*

Preparing 25Gbps Electrical I/O for Exascale Computing Systems 1955

Lei Shan, *IBM Corporation*; Young Kwark, *IBM Corporation*; Renato Rimolo-Donadio, *IBM Corporation*; Christian Baks, *IBM Corporation*; Michael Gaynes, *IBM Corporation*; Timothy Chainer, *IBM Corporation*; Manabu Hoshino, *Zeon Corporation*; Masakazu Hashimoto, *Zeon Corporation*; Toshihiko Jimbo, *Zeon Corporation*; Junji Kodemura, *Zeon Corporation*; Ikkei Matsuura, *Zeon Corporation*

Large Low-CTE Glass Package-to-PCB Interconnections with Solder Strain-Relief Using Polymer Collars 1959

Gary Menezes, *Georgia Institute of Technology*; Vanessa Smet, *Georgia Institute of Technology*; Makoto Kobayashi, *Namics Corporation*; Venky Sundaram, *Georgia Institute of Technology*; Pulugurtha Markondeya Raj, *Georgia Institute of Technology*; Rao Tummala, *Georgia Institute of Technology*

The Study of Bare-Die FCBGA Die Damage in Response to Applied Mechanical Stress During Heat Sink Assembly 1965

Heidi S.Y. Ho, *Broadcom Corporation*; Daijiao Wang, *Broadcom Corporation*; Michael Johnson, *Amkor Technology, Inc.*; C.J. Berry, *Amkor Technology, Inc.*

Prognostication of Copper-Aluminum Wirebond Reliability under High Temperature Storage and Temperature-Humidity 1973

Pradeep Lall, *Auburn University*; Shantanu Deshpande, *Auburn University*; Luu Nguyen, *Texas Instruments*; Masood Murtuza, *Texas Instruments*

Low-Frequency Testing of Through Silicon Vias for Defect Diagnosis in Three-Dimensional Integration Circuit Stacking Technology 1986

Yichao Xu, *Peking University*; Min Miao, *Beijing Information Science & Technology University; Peking University*; Runiu Fang, *Peking University*; Xin Sun, *Peking University*; Yunhui Zhu, *Peking University*; Minggang Sun, *Beijing Information Science & Technology University*; Guanjiang Wang, *Peking University*; Yufeng Jin, *Peking University*

Fast Estimation of LED's Accelerated Lifetime by Online Test Method 1992

Qi Chen, *Huazhong University of Science & Technology*; Quan Chen, *Huazhong University of Science & Technology*; Xiaobing Luo, *Huazhong University of Science & Technology*

Methodology and Apparatus for Rapid Power Cycle Accumulation and In-Situ Incipient Failure Monitoring for Power Electronic Modules 1996

Roy I. Davis, *Fairchild Semiconductor Corporation*; Daniel J. Sprenger, *Fairchild Semiconductor Corporation*

Fine-Pitch Probing on TSVs and Microbumps Using a Chip Prober Having a Transparent Membrane Probe Card 2003

Naoya Watanabe, *National Institute of AIST*; Michiyuki Eto, *STK Technology Co., Ltd.*; Kenji Kawano, *STK Technology Co., Ltd.*; Masahiro Aoyagi, *National Institute of AIST*

40: Interactive Presentations 4

Chairs: Nam Pham, *IBM Corporation*
Rabindra N. Das, *MIT Lincoln Labs*

Thermal Management of 3D RF PoP Based on Ceramic Substrate 2008

Fengze Hou, *National Center for Advanced Packaging; Chinese Academy of Sciences*; Fengman Liu, *National Center for Advanced Packaging; Chinese Academy of Sciences*; Yi He, *Chinese Academy of Sciences*; Xiaomeng Wu, *Chinese Academy of Sciences*; Xia Zhang, *National Center for Advanced Packaging; Chinese Academy of Sciences*; Liqiang Cao, *National Center for Advanced Packaging; Chinese Academy of Sciences*; Yuan Lu, *National Center for Advanced Packaging; Chinese Academy of Sciences*; Dongkai Shangguan, *National Center for Advanced Packaging; Chinese Academy of Sciences*

Bump Pattern Optimization and Stress Comparison Study for DCA Packages 2014

Akash Agrawal, *Micron Technology Inc.*; Owen Fay, *Micron Technology Inc.*; Mark Johnson, *Micron Technology Inc.*

Characterization of In-Plane Stress in TSV Array – A Unit Model Approach 2020

Cheng-Fu Chen, *University of Alaska, Fairbanks*

Electrical-Thermal Characterization of Wires in Packages 2027

Kai Liu, *STATS ChipPAC, Inc.*; Robert Frye, *STATS ChipPAC, Inc.*; HyunTai Kim, *STATS ChipPAC, Inc.*; YongTaek Lee, *STATS ChipPAC, Inc.*; Gwang Kim, *STATS ChipPAC, Inc.*; Susan Park, *STATS ChipPAC, Inc.*; Billy Ahn, *STATS ChipPAC, Inc.*

Computational Investigation of Failure in Anodized Aluminum 2035

Sabrina Ball, *University of Arizona*; Ibrahim Guven, *University of Arizona*; Pankaj Sinha, *Intel Corporation*; Rajiv Rastogi, *Intel Corporation*; Brian McCarson, *Intel Corporation*

Study on Prediction about Residual Position of Void Generated by Resin Flow 2042

Masayuki Mino, *Hitachi, Ltd.*; Naoya Suzuki, *Hitachi Chemical Co., Ltd.*; Hiroshi Takahashi, *Hitachi Chemical Co., Ltd.*; Tsutomu Kono, *Hitachi, Ltd.*

Modeling and Analysis of Temperature Effect on MEMS Gyroscope 2048

Ming Wen, *Huazhong University of Science & Technology*; Weihui Wang, *Huazhong University of Science & Technology*; Zhang Luo, *Huazhong University of Science & Technology*; Yong Xu, *Wuhan University; Wayne State University*; Xin Wu, *Wayne State University*; Fei Hou, *Dongfeng Automobile Electronics Co., Ltd.*; Sheng Liu, *Wuhan University*

Life Prediction and Classification of Failure Modes in Solid State Luminaires Using Bayesian Probabilistic Models 2053
Pradeep Lall, *Auburn University*; Junchao Wei, *Auburn University*; Peter Sakalaukus, *Auburn University*

Modeling for Reliability of Ultra Thin Chips in a System in Package 2063
Richard Qian, *Fairchild Semiconductor Corporation*; Yong Liu, *Fairchild Semiconductor Corporation*

Development of Effective Thermal Characterization on Handheld Devices by Matrix Method 2069
Tai-Yu Chen, *MediaTek Inc.*; Chung-Fa Lee, *MediaTek Inc.*

Comprehensive Design Optimization for 2.133 Gbps LPDDR3 Extension for Mobile Platform System 2075
Chanmin Jo, *Samsung Electronics*; Jaemin Shin, *Samsung Electronics*; BaekKyu Choi, *Samsung Electronics*; Sangmin Lee, *Samsung Electronics*; Seongjae Moon, *Samsung Electronics*; Sungjoo Kim, *Samsung Electronics*; Woong Hwan Ryu, *Samsung Electronics*

Estimation of Mode Conversion and Crosstalk Impact from a Single-Ended Aggressor to a Differential Victim Using Statistical BER Analysis 2081
Arun Reddy Chada, *Missouri S&T EMC Laboratory*; Jun Fan, *Missouri S&T EMC Laboratory*; James L. Drewniak, *Missouri S&T EMC Laboratory*; Bhyrav Mutnury, *Dell, Inc.*

Power Distribution Network Worst-Case Power Noise and an Efficient Estimation Method 2088
Jiangyuan Qian, *Broadcom Corporation*; Shiji Pan, *University of California, Irvine*

Fast Calculation of Electromagnetic Interference by Through-Silicon Vias 2094
Aosheng Rong, *University of Illinois, Urbana-Champaign*; Andreas C. Cangellaris, *University of Illinois, Urbana-Champaign*; Feng Ling, *Nanjing University of Science and Technology*

Electrical Simulation and Analysis of Si Interposer for 3D IC Integration 2099
Xin Sun, *Peking University*; Min Miao, *Beijing Information Science & Technology University*; Yunhui Zhu, *Peking University*; Runiu Fang, *Peking University*; Guanjiang Wang, *Peking University*; Wengao Lu, *Peking University*; Jing Chen, *Peking University*; Yufeng Jin, *Peking University*

A SPICE Model of Multi-Mode Optical Fiber in Mid-Channel Link for Package System SI Transient Simulations 2104
Zhaoqing Chen, *IBM Corporation*

Next Generation Package-on-Package Solution to Support Wide IO and High Bandwidth Interface 2112
Hung-Hsiang Cheng, *Advanced Semiconductor Engineering, Inc.*; Chang-Chi Lee, *Advanced Semiconductor Engineering, Inc.*; Ming-Feng Chung, *Advanced Semiconductor Engineering, Inc.*; Po-Chih Pan, *Advanced Semiconductor Engineering, Inc.*; Ping-Feng Yang, *Advanced Semiconductor Engineering, Inc.*; Chi-Tsung Chiu, *Advanced Semiconductor Engineering, Inc.*; Chih-Pin Hung, *Advanced Semiconductor Engineering, Inc.*; Chen-Chao Wang, *Advanced Semiconductor Engineering, Inc.*

Package-Level Electromagnetic Interference Analysis 2119
Namhoon Kim, *Broadcom Corporation*; Leo Hongyu Li, *Broadcom Corporation*; Sam Karikalan, *Broadcom Corporation*; Reza Sharifi, *Broadcom Corporation*; Henry Kim, *Broadcom Corporation*

A Path Finding Based SI Design Methodology for 3D Integration 2124
Bill Martin, *E-System Design*; KiJin Han, *UNIST*; Madhavan Swaminathan, *Georgia Institute of Technology*

Design and Implementation of a 700-2600 MHz RF SiP for Micro Base Station .... N/A
Yi He, *National Center for Advanced Packaging*; *Chinese Academy of Sciences*; Fengman Liu, *National Center for Advanced Packaging*; *Chinese Academy of Sciences*; Anmou Liao, *National Center for Advanced Packaging*; *Chinese Academy of Sciences*; Jun Li, *National Center for Advanced Packaging*; *Chinese Academy of Sciences*; Xiaomeng Wu, *National Center for Advanced Packaging*; Peng Wu, *National Center for Advanced Packaging*; *Chinese Academy of Sciences*; Liqiang Cao, *National Center for Advanced Packaging*; *Chinese Academy of Sciences*; Dongkai Shangguan, *National Center for Advanced Packaging*

Dielectric Lens Optimization for Conical Helix THz Antennas 2137

Paolo Nenzi, *ENEA Frascati Research Center*; Volha Varlamava, *Sapienza University of Rome*; Frank Silvio Marzano, *Sapienza University of Rome*; Fabrizio Palma, *Sapienza University of Rome*; Marco Balucani, *Sapienza University of Rome*

Embedded Diodes for Microwave and Millimeter Wave Circuits 2144

Xianbo Yang, *Michigan State University*; Amanpreet Kaur, *Michigan State University*; Premjeet Chahal, *Michigan State University*

PCIe Gen3 Link Design and Tuning in Server Systems with End Devices from Multiple IP Suppliers 2151

Si T. Win, *IBM Corporation*; Daniel Rodriguez, *IBM Corporation*; Nanju Na, *IBM Corporation*

A Low-Cost PCB Fabrication Process 2159

Jack Ou, *Sonoma State University*; Alberto Maldonado, *Sonoma State University*; Chio Saephan, *Sonoma State University*; Farid Farahmand, *Sonoma State University*; Michael Caggiano, *Rutgers University*

Novel Band-Pass Filters on Thin Glass Substrate with Through Glass Vias (TGVs) .... N/A

Cheng Pang, *National Center for Advanced Packaging*; *Chinese Academy of Sciences*; Wenya Shang, *National Center for Advanced Packaging*; *Chinese Academy of Sciences*; Mingchuan Zhang, *Chinese Academy of Sciences*; Zheng Qin, *National Center for Advanced Packaging*; *Chinese Academy of Sciences*; Huijuan Wang, *National Center for Advanced Packaging*; *Chinese Academy of Sciences*; Xiaoli Ren, *National Center for Advanced Packaging*; Jie Pan, *Chinese Academy of Sciences*; Daquan Yu, *National Center for Advanced Packaging*; *Chinese Academy of Sciences*; Dongkai Shangguan, *National Center for Advanced Packaging*; *Chinese Academy of Sciences*

Study of Microwave Circuits Based on Metal-Insulator-Metal (MIM) Diodes on Flex Substrates 2168

Amanpreet Kaur, *Michigan State University*; Xianbo Yang, *Michigan State University*; Premjeet Chahal, *Michigan State University*

41: Student Interactive Presentations
Chairs: Mark Poliks, *i3 Electronics, Inc.*
Ibrahim Guven, *University of Arizona*

Nanocomposite Pastes for Thermal and Mechanical Bonding 2175

Tingting Zhang, *Binghamton University*; Bahgat Sammakia, *Binghamton University*; Howard Wang, *Binghamton University*

Assembly and Packaging Technologies for High-Temperature and High-Power GaN HEMTs 2181

A.A. Bajwa, *University of Freiburg*; Y. Qin, *University of Freiburg*; J. Wilde, *University of Freiburg*; R. Reiner, *Fraunhofer Institute IAF*; P. Waltereit, *Fraunhofer Institute IAF*; R. Quay, *Fraunhofer Institute IAF*

Flip-Chip on Glass (FCOG) Package for Low Warpage 2189

Scott R. McCann, *Georgia Institute of Technology*; Venkatesh Sundaram, *Georgia Institute of Technology*; Rao R. Tummala, *Georgia Institute of Technology*; Suresh K. Sitaraman, *Georgia Institute of Technology*

Laser-Based Conductive Film Forming with Gold Nanoparticles for Electrical Contacts 2194

Mitsugu Yamaguchi, *Ibaraki University*; Shinji Araga, *Ibaraki Giken Ltd.*; Mamoru Mita, *M&M Research Laboratory*; Kazuhiko Yamasaki, *Ibaraki University*; Katsuhiro Maekawa, *Ibaraki University*

Analysis of Modes Effect on Signal/Power Integrity in Finite Cavity for Chip and Die Level Packaging Based on a Hybrid Full Wave Method 2200

Xin Chang, *University of Washington*; Leung Tsang, *University of Washington*

Directed Self-Assembly of Mesoscopic Dies Using Magnetic Force and Shape Recognition 2207

Anton Tkachenko, *Rensselaer Polytechnic Institute*; Robert F. Karlicek, Jr., *Rensselaer Polytechnic Institute*; James J.-Q. Lu, *Rensselaer Polytechnic Institute*

Controlled Silicon IC Thinning on Individual Die Level for Active Implant Integration Using a Purely Mechanical Process 2213

Vasiliki Giagka, *University College London*; Nooshin Saeidi, *University College London*; Andreas Demosthenous, *University College London*; Nick Donaldson, *University College London*

Connectors and Vibrations – Damages in Different Electrical Environments .... N/A

A. Berghuvud, *Blekinge Institute of Technology*; T. Björnängen, *Blekinge Institute of Technology*; T. Gissila, *Blekinge Institute of Technology*

Study of Extreme Low Temperature and Load Solid-Phase Sn-Ag System Bonding Mechanism for 3D ICs 2227

Kiyoto Yoneta, *Osaka University*; Ryohei Sato, *Osaka University*; Yoshiharu Iwata, *Osaka University*; Koichiro Atsumi, *Osaka University*; Kazuya Okamoto, *Osaka University*; Yukihiro Satio, *Osaka University*; Takumi Shigemoto, *Osaka University*

Self-Patterning, Pre-Applied Underfilling Technology for Stack-Die Packaging 2231

Chia-Chi Tuan, *Georgia Institute of Technology*; Ziyin Lin, *Georgia Institute of Technology*; Yan Liu, *Georgia Institute of Technology*; Kyoung-Sik Moon, *Georgia Institute of Technology*; Ching-Ping Wong, *Georgia Institute of Technology; Chinese University of Hong Kong*

Study of High CRI White Light-Emitting Diode Devices with Multi-Chromatic Phosphor 2236

Min Zheng, *Xi'an Jiaotong University*; Wen Ding, *Xi'an Jiaotong University*; Feng Yun, *Xi'an Jiaotong University*; Deyang Xia, *Xi'an Jiaotong University*; Yaping Huang, *Xi'an Jiaotong University*; Yukun Zhao, *Xi'an Jiaotong University*; Weihan Zhang, *Xi'an Jiaotong University*; Minyan Zhang, *Xi'an Jiaotong University*; Maofeng Guo, *Xi'an Jiaotong University*; Ye Zhang, *Xi'an Jiaotong University*

The Effects of Self-Fluxing Additives in Solder Anisotropic Conductive Films (ACFs) on Solder Wettability and Joint Reliability of Flex-on-Board (FOB) Assemblies 2241

Seung-Ho Kim, *Korea Advanced Institute of Science and Technology (KAIST)*; Yongwon Choi, *Korea Advanced Institute of Science and Technology (KAIST)*; Yoosun Kim, *Korea Advanced Institute of Science and Technology (KAIST)*; Kyung-Wook Paik, *Korea Advanced Institute of Science and Technology (KAIST)*

Modeling and Analysis of Frequency Shift of MEMS Gyroscope Subjected to Temperature Change 2245

Weihui Wang, *Huazhong University of Science & Technology*; Sheng Liu, *Wuhan University*; Zhang Luo, *Huazhong University of Science & Technology*; Ming Wen, *Huazhong University of Science & Technology*; Qiang Dan, *Huazhong University of Science & Technology*; Man Yu, *Huazhong University of Science & Technology*; Yong Xu, *Wuhan University, Wayne State University*; Xin Wu, *Wayne State University*

Interaction Effect between Electromigration and Microstructure Evolution in Cu/Sn-58Bi/Cu Solder Interconnect 2249

Hong-Bo Qin, *South China University of Technology*; Bin Li, *Southern Methodist University*; Wu Yue, *South China University of Technology*; Chang-Bo Ke, *South China University of Technology*; Min-Bo Zhou, *South China University of Technology*; Xin-Ping Zhang, *South China University of Technology*

Effects of Alignment of Graphene Flakes on Water Permeability of Graphene-Epoxy Composite Film 2255

Seong-Yoon Jung, *Korea Advanced Institute of Science and Technology (KAIST)*; Kyung-Wook Paik, *Korea Advanced Institute of Science and Technology (KAIST)*

Characterization of Alternate Power Distribution Methods for 3D Integration 2260

David C. Zhang, *Georgia Institute of Technology*; Madhavan Swaminathan, *Georgia Institute of Technology*; David Keezer, *Georgia Institute of Technology*; Satyanarayana Telikepalli, *Georgia Institute of Technology*

Adhesion and Reliability of Direct Cu Metallization of Through-Package Vias in Glass Interposers 2266

Timothy Huang, *Georgia Institute of Technology*; Venky Sundaram, *Georgia Institute of Technology*; P. Markondeya Raj, *Georgia Institute of Technology*; Himani Sharma, *Georgia Institute of Technology*; Rao Tummala, *Georgia Institute of Technology*

High-Frequency Characterization of Through-Package Vias Formed by Focused Electrical-Discharge in Thin Glass Interposers 2271

Jialing Tong, *Georgia Institute of Technology*; Yoichiro Sato, *Asahi Glass Company*; Shintaro Takahashi, *Asahi Glass Company*; Nobuhiko Imajyo, *Asahi Glass Company*; Andrew F. Peterson, *Georgia Institute of Technology*; Venky Sundaram, *Georgia Institute of Technology*; Rao Tummala, *Georgia Institute of Technology*

Interfacial Reactions between Cu and Sn, Sn-Ag, Sn-Bi, Sn-Zn Solder under Space Confinement for 3D IC Micro Joint Applications 2277

T.L. Yang, *National Taiwan University*; W.L. Shih, *National Taiwan University*; J.J. Yu, *National Taiwan University*; C.R. Kao, *National Taiwan University*

Simulation and Optimization of a Micro Flow Sensor 2283

Xing Guo, *Huazhong University of Science & Technology*; Chunlin Xu, *Huazhong University of Science & Technology*; Shengzhi Zhang, *Huazhong University of Science & Technology*; Yong Xu, *Wayne State University*; Xin Wu, *Wayne State University*; Sheng Liu, *Wuhan University*

Minimizing Coupling of Power Supply Noise between Digital and RF Circuit Blocks in Mixed Signal Systems 2287

Satyanarayana Telikepalli, *Georgia Institute of Technology*; Madhavan Swaminathan, *Georgia Institute of Technology*; David Keezer, *Georgia Institute of Technology*

A Feasibility Study of Flip-Chip Packaged Gallium Nitride HEMTs on Organic Substrates for Wideband RF Amplifier Applications 2293

Spyridon Pavlidis, *Georgia Institute of Technology*; A. Cagri Ulusoy, *Georgia Institute of Technology*; Wasif T. Khan, *Georgia Institute of Technology*; Outmane Lemtiri Chlieh, *Georgia Institute of Technology*; Edward Gebara, *I2R Nanowave Inc.*; John Papapolymerou, *Georgia Institute of Technology*

A Novel Molding Process for Wafer Level LED Packaging Using Uniform Micro Glass Bubble Arrays 2299

Yu Zou, *Southeast University*; Jintang Shang, *Southeast University*; Yu Ji, *Southeast University*; Li Zhang, *Jiangyin Changdian Advanced Packaging Co. Ltd.*; Chiming Lai, *Jiangyin Changdian Advanced Packaging Co. Ltd.*; Dong Chen, *Jiangyin Changdian Advanced Packaging Co. Ltd.*; Kim-Hui Chen, *Jiangyin Changdian Advanced Packaging Co. Ltd.*; Ching-Ping Wong, *Chinese University of Hong Kong*

Analysis of Room-Temperature Bonded Compliant Bump with Ultrasonic Bonding 2303

Keiichiro Iwanabe, *Kyushu University*; Takanori Shuto, *Kyushu University*; Tanemasa Asano, *Kyushu University*

Novel Highly-Efficient and Misalignment Insensitive Wireless Power Transfer Systems Utilizing Strongly Coupled Magnetic Resonance Principles

Daerhan Daerhan[1], Olutola Jonah[1], Hao Hu[1], Stavros V. Georgakopoulos[1], and Manos M. Tentzeris[2]

[1]Florida International University, 10555 W Flagler St, Miami, FL, 33174
[2]Georgia Institute of Technology, 777 Atlantic Drive NW, Atlanta, GA 30332-0250

Abstract

The wireless powering efficiency of traditional Strongly Coupled Magnetic Resonance (SCMR) systems is highly sensitive to the alignment between the transmitter and receiver elements; an issue that has limited their applicability in practical wireless power transfer systems. This paper proposes a novel set of SCMR-based topologies that are insensitive to misalignment and isotropic while providing large wireless powering efficiencies. The systems, which are presented here, achieve power transfer efficiencies above 50% over the complete misalignment range of 0−90° with performance that is significantly better than typical SCMR elements.

Introduction

Wireless **power** transfer methods have been utilized in the past for various applications. The Strongly Coupled Magnetic Resonance SCMR method is a highly efficient wireless power transfer technique that has been recently developed [1]. In order for SCMR to achieve high efficiency, it requires that the transmitting and receiving elements (typically loops or coils) are designed so that they resonate at the desired operating frequency that must coincide with the frequency where the elements naturally exhibit maximum Q-factor. A main disadvantage of conventional SCMR systems is that they are highly sensitive to the alignment between transmitter and receiver. An optimization technique for improving the efficiency of SCMR systems under lateral misalignment was presented in [2]. Specifically, 48.4% efficiency was achieved by using an adaptive matching network. However, no solution for angular misalignment was proposed in [2]. Analytical formulations for the power transfer efficiencies of inductive links under lateral and angular coil misalignment were presented in [3]. In addition, the effects of angular and lateral misalignment in inductively coupled systems, which simultaneously achieve WPT and data communication, were examined in [4]. Furthermore, analytical models for SCMR that incorporate misalignment effects were presented in [5].

SCMR's radial and angular misalignment sensitivity were examined by [6] and [7], respectively. Only [7], attempted to alleviate SCMR's angular misalignment sensitivity by using tuning circuits, which were not able to maintain high efficiency above 60° of misalignment. Also, it should be pointed out that tuning circuits add to the complexity of SCMR RX systems and cannot compensate for large angular and radial misalignments as they cannot recover the lost flux density between TX and RX. However, tuning circuits can be used for compensating the effects of variable axial distance between TX and RX [8], [9]. It can be seen that previous research efforts have investigated the effects of both lateral and angular misalignment of SCMR systems. However, no solutions, which are based on new geometries of SCMR elements, have been proposed to address the sensitivity of SCMR systems to lateral and angular misalignment. Also, no paper has attempted to design an SCMR system that exhibits isotropic wireless powering performance. This paper proposes a novel set of SCMR-based topologies that are insensitive to misalignment and isotropic while providing large wireless powering efficiencies. This type of system was first presented in [10]. The systems, which are presented here, achieve power transfer efficiencies above 50% over the complete misalignment range of 0−90° with performance that is significantly better than typical SCMR elements.

Standard SCMR

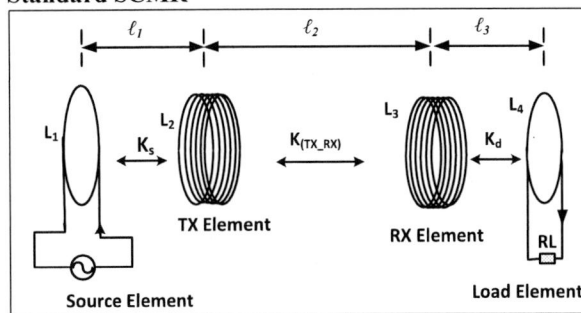

Figure 1. Standard SCMR System

Standard SCMR systems consist of four elements (see Fig. 1) and require that TX and RX systems are aligned in order to achieve high efficiency. In fact, such standard SCMR systems suffer a significant decrease of their wireless powering efficiency when they are misaligned. This disadvantage greatly limits the application of standard SCMR systems since mobile, wearable or implantable, devices cannot be designed so that they are always aligned.

The main purpose of this paper is to provide a novel way to perform WPT that is misalignment insensitive. The following types of misalignment will be examined:

a. Angular azimuth misalignment - In this case, the RX system rotates around the Z axis from 0° to 90°, while the TX system is fixed.

b. Angular elevation misalignment - In this case, the RX system (RX resonator and load element) is rotated in the YZ (elevation) plane from 0° to 90° from the center between the two devices on their common axis, while the TX system (TX resonator and source element) is fixed.

c. Isotropy azimuth misalignment – In this case, we examine the performance when we rotate the RX system around the TX system by keeping its distance from TX system constant. Therefore the RX system moves along an azimuth sphere with a radius of the distance of these two.

d. Isotropy elevation misalignment – Much like previous case, we move the RX system along elevation sphere with a radius of the distance between RX and TX.

The performance of standard SCMR is examined here. All the simulations in this paper are performed in ANSYS HFSS and Designer. All load loops in the designs examined here are connected to 50 ohm loads. All SCMR resonators examined below were designed to resonate at the frequency where their Q-factor is maximum, which was accomplished using analytical equations for the standard loops and simulations for the proposed 3-D loops.

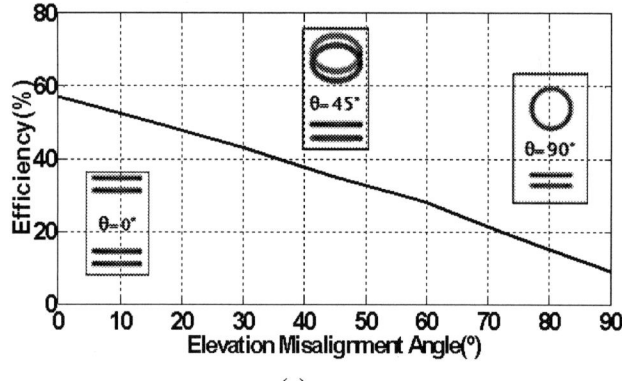

(a)

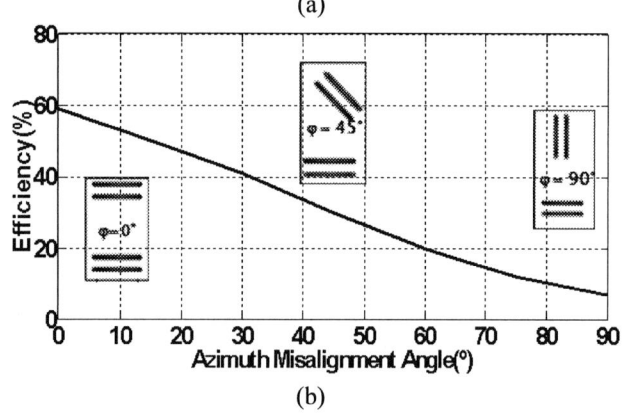

(b)

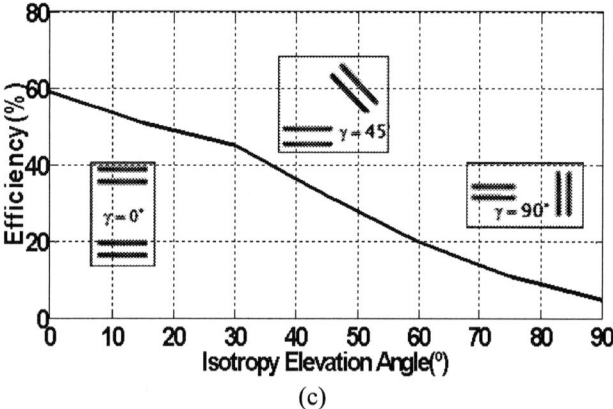

(c)

Figure 2. Standard SCMR under different misalignment conditions. (a) angular elevation, (b) angular azimuth, and (c) isotropy elevation.

The specifications of this SCMR system are as follows: the radius of the four loops is 50 mm, the thickness of the wire is 2.2 mm, the distance between RX and TX is 150 mm. A 78

pF capacitor is needed to achieve maximum efficiency, and the operating frequency is 40.8 MHz.

Fig. 2 shows the simulation results of the SCMR system for the different misalignment conditions. Specifically, Figs. 2(a) and 2(b) show the variation of SCMR's efficiency for elevation and angular misalignment conditions. The results clearly demonstrate the gradual decrease of efficiency and thus the high sensitivity of conventional SCMR systems to misalignment. Also, the isotropy of the SCMR system was examined in the elevation plane and the results are shown in Fig. 2(c). These results illustrate that the efficiency of the standard SCMR system does not exhibit isotropic performance in the elevation plane. Similarly, due to its geometric symmetry, this standard SCMR system does not exhibit isotropic WPT in the azimuth plane either.

Misalignment Insensitive SCMR System (3- loop structure)

In an effort to develop an SCMR system that is misalignment insensitive and isotropic, we designed the system shown in Fig. 3. The difference between this system and the standard SCMR is that each element (TX resonator, RX resonator, source and load) is a 3D loop that comprises of three continuous connected orthogonal loops. Due to its geometrical symmetry, this design is expected to provide misalignment insensitive performance. The 3-D RX and TX resonator loops are designed so that they exhibit their highest Q-factor at the frequency of operation and capacitors are connected at their feed point to resonate them at the operating frequency. The designs of this system are done using ANSYS HFSS simulation analysis and the specifications are discussed as follows.

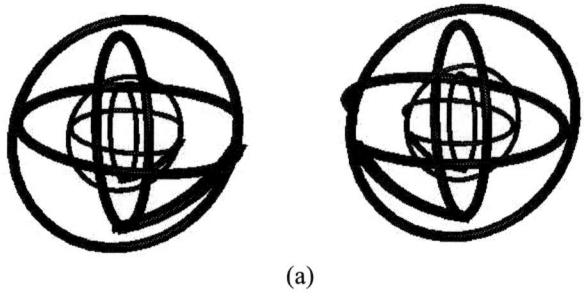

(a)

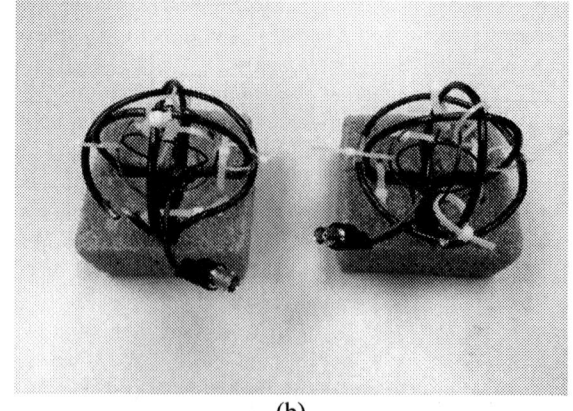

(b)

Figure 3. (a) Simulation Model, (b) Prototype

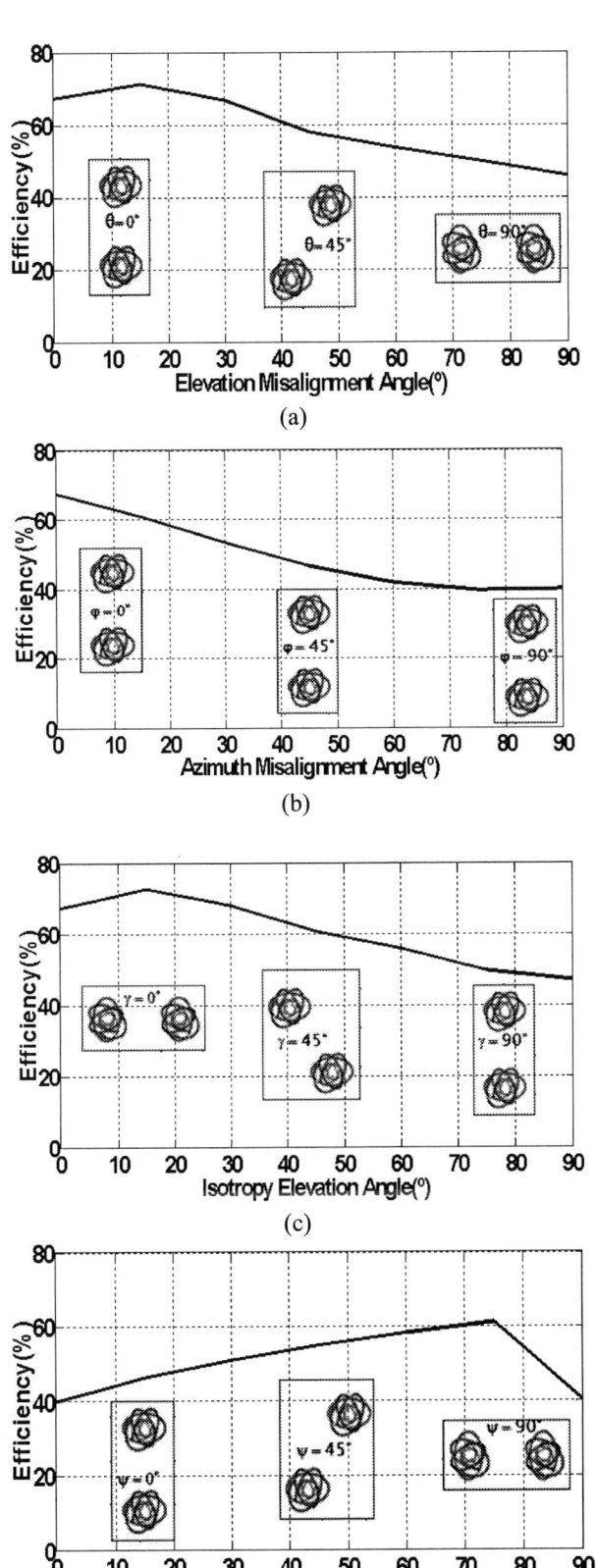

Figure 4. SCMR system of Fig. 3 under different misalignment conditions. (a) angular elevation, (b) angular azimuth, (c) isotropy elevation, and (d) isotropy azimuth.

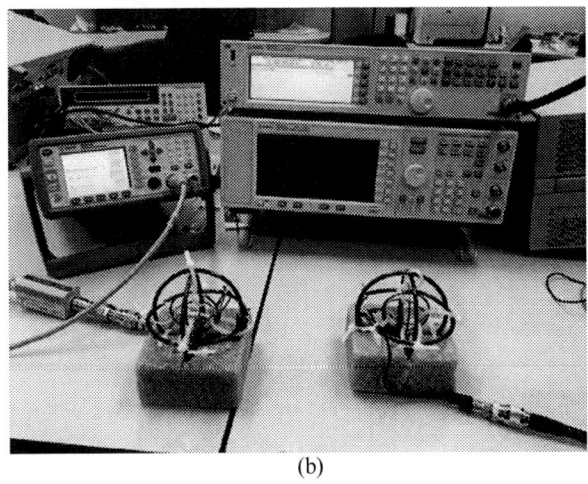

(b)

Figure 5. Measurement setup.

The geometric specifications of the system, which is examined here, are described in what follows. The TX and RX resonators are comprised of three connected orthogonal loops with the following radii: the radius of the outermost loop is 50 mm, the radius of the inner loop is 45 mm, the radius of the innermost loop is 40 mm, and thickness of the wire is 2.2 mm. The load and source elements are comprised of three connected orthogonal loops with the following radii: the radius of the outermost loop is 25 mm, the radius of the inner loop is 22.5 mm, the radius of the innermost loop is 20 mm, and the thickness of the wire is 1.1 mm. The distance between the center of TX and RX systems is 150 mm. Each of the TX and RX resonators are connected to a 100 pF capacitor. This capacitance value was found by simulation analysis so that maximum WPT efficiency is achieved by the system. The operating frequency is 51.6 MHz.

Figs. 4(a) and 4(b) illustrate through simulation results that this system provides efficiency that is insensitive to angular (azimuth and elevation) misalignment. Also, Figs. 4(c) and 4(d) show that the system exhibits nearly isotropic wireless power transfer in both elevation and azimuth planes. The measurement setup is shown in Fig. 5

Conclusion

The paper proposed a novel SCMR system that exhibit misalignment-insensitive wireless power transfer. The system uses 3-D elements (source, load, TX and RX resonators) comprising of three continuous and connected orthogonal loops. Also, in this system the source and load elements are concentrically embedded into the TX and RX resonators, respectively. This system achieves power transfer efficiencies above 50% over the complete misalignment range of 0 - 90° with performance dramatically better than typical SCMR elements that are highly sensitive to misalignment. Also, this system is the first WPT system that exhibits isotropic performance in both elevation and azimuth planes thereby providing an efficiency that only depends on the distance between the TX and RX subsystems. The misalignment insensitive and isotropic performance of the proposed SCMR system is due to its geometrical spherical symmetry.

978-1-4799-2408-0/14 $31.00 © 2014 IEEE

References

[1] A. Kurs, A. Karalis, R. Moffatt, J. D. Joannopoulos, P. Fisher, and M. Soljacic, "Wireless Energy Transfer via Strongly Coupled Magnetic Resonances, " *Science*, vol. 317, pp. 83-85, 2007.

[2] S. G. Lee, H. Hoang, Y. H. Choi, F. Bien, "Efficiency improvement for magnetic resonance based wireless power transfer with axial-misalignment," *Electronics Letters* , vol.48, no.6, pp.339-340, March 15 2012.

[3] K. Fotopoulou and B. W. Flynn, "Wireless Power Transfer in Loosely Coupled Links: Coil Misalignment Model," Magnetics, *IEEE Transactions on Magnetics*, vol.47, no.2, pp.416-430, Feb. 2011.

[4] M. A. Adeeb, A. B. Islam, M. R. Haider, F. S. Tulip, M. N. Ericson, and S. K. Islam, "An Inductive Link-Based Wireless Power Transfer System for Biomedical Applications," *Active and Passive Electronic Components*, vol. 2012, Article ID 879294, 11 pages, 2012.

[5] Wang Junhua, S. L Ho, W. N. Fu and Sun Mingui, "Analytical Design Study of a Novel Witricity Charger With Lateral and Angular Misalignments for Efficient Wireless Energy Transmission," *IEEE Transactions on Magnetics*, vol.47, no.10, pp.2616-2619, Oct. 2011.

[6] Fei Zhang, S. A. Hackwoth, Xiaoyu Liu, Li Chengliu, Sun Mingui, "Wireless power delivery for wearable sensors and implants in Body Sensor Networks," *2010 IEEE Annual International Conference of Engineering in Medicine and Biology Society (EMBC)*, vol., no., pp.692-695, Aug. 31 2010-Sept. 4 2010.

[7] A. P. Sample, D. A. Meyer, J. R. Smith, "Analysis, Experimental Results, and Range Adaptation of Magnetically Coupled Resonators for Wireless Power Transfer," *IEEE Transactions on Industrial Electronics*, vol.58, no.2, pp.544-554, Feb. 2011.

[8] Hao Qiang, XueLiang Huang, LinLin Tan, QingJing Ji, JiaMing Zhao, " Achieving maximum power transfer of inductively coupled wireless power transfer system based on dynamic tuning control", *Science China Technological Sciences,* Science China Press, July 2012, Volume 55, Issue 7, pp. 1886-1893.

[9] Y. Endo, Y. Furukawa,"Proposal for a new resonance adjustment method in magnetically coupled resonance type wireless power transmission," *2012 IEEE MTT-S International Microwave Workshop Series on Innovative Wireless Power Transmission: Technologies, Systems, and Applications*, vol., no., pp.263,266, 10-11 May 2012.

[10] O. Jonah, S. V. Georgakopoulos, and M. M. Tentzeris, "Orientation insensitive power transfer by magnetic resonance for mobile devices," *IEEE Wireless Power Transfer Conference*, Perugia, Italy, May 15-16, 2013.

A Wireless Charging and Near-field Communication Combination Module for Mobile Applications

Hiroki Shibuya, Tatsuaki Tsukuda, Hiroko Suzuki,
Tadashi Shimizu, Masahiro Dobashi, Shinji Nishizono, Mikio Baba, Hideki Sasaki and Katsushi Terajima
Package and Test Technology Division, Renesas Electronics Corporation
1753 Shimonumabe, Nakahara-ku, Kawasaki-shi, Kanagawa 211-8668, Japan
E-mail: hiroki.shibuya.ak@renesas.com

Abstract

This paper presents the first demonstration of an ultra-miniature module combining with 13.56/6.78-MHz wireless charging receiving functions and types-A/B/F near-field communication (NFC) functions. In order to be able to embed this module into mobile terminals, the electrical and thermal designs are optimized and then the size is 14 x 26 x 1.86 mm. A simulation bench for verifying the efficiency of the wireless charging, performance of NFC by a wireless charging antenna, thermal design for embedding the module into a mobile terminal and EMI reduction of the wireless charging system are described.

Keywords: wireless power transfer, wireless charging, near-field communication, NFC, module

Introduction

Wireless power transfer or wireless charging is one of hot topics in the technical fields of both RF and power electronics. Existing wireless charging systems, such as Qi standardized by Wireless Power Consortium (WPC), for mobile applications uses inductive coupling with the frequency range from 100 kHz to 200 kHz. Since the frequency of the wireless power transfer is lower than 200 kHz, the antenna is surely bigger than the one for ex. wireless LAN and Bluetooth operated at 2.45 GHz. It generates a negative impact for downsizing mobile terminals. Also, heat generation by several hundred-kHz wireless charging in mobile terminals is typically larger than heat generation of wire charging such as USB, because the coil-type antenna generates heavy heat as well as power management IC and the peripheral components generate heat. The heat has the performance of Lithium ion batteries degraded. Furthermore, strong magnetic field used by wireless power transfer causes Electromagnetic Interference (EMI) problems. EMI noise interferes with TV and broadcast and decreases the performances of wireless communications in the terminals.

In order to solve these problems, 13.56-MHz[1][2] or 6.78-MHz[3][4] wireless charging has been investigated. This choice can downsize the antenna for wireless charging and can decrease heat generation of the antenna. In addition, to realize a safe wireless charging system, wireless communication function suitable for wireless charging has been investigated.

This paper presents the first demonstration of an ultra-miniature module combining with 13.56/6.78-MHz wireless charging receiving functions and types-A/B/F near-field communication (NFC) functions. The biggest advantage of the system in the packaging point of view is to use the same antenna for NFC and 13.56/6.78-MHz wireless power transfer. This wireless communication can exchange information on the temperature of the system for realizing a safe system. This module includes a power management IC for wireless charging and a microcontroller for NFC. The module is also capable of 5-watt power transfer with small and thin size (14 x 26 x 1.86 mm).

In order to be able to embed this module into mobile terminals, a simulation bench for verifying the efficiency of the wireless charging, performance of NFC by a wireless charging antenna, thermal design for embedding the module into a mobile terminal and EMI reduction of the wireless charging system are described in this paper.

System block diagram and receiver module

Figure 1 shows a block diagram of a system combining with wireless charging and NFC. The left side a transmitter (Tx) module and the right side a receiver (Rx) one. The Tx module is not necessary to miniaturize the size, because the module with an antenna is implemented into a stand or box putting a mobile terminal on. On the other hand, the Rx module is strongly requested to miniaturize the size, because the module with an antenna is embedded into a mobile terminal.

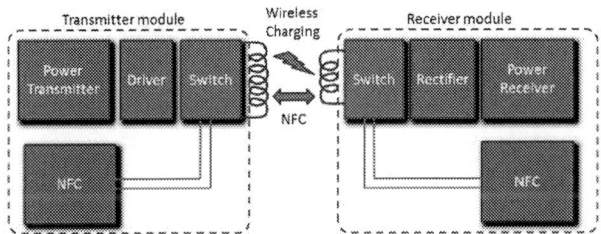

Fig 1. A block diagram of a wireless charging system

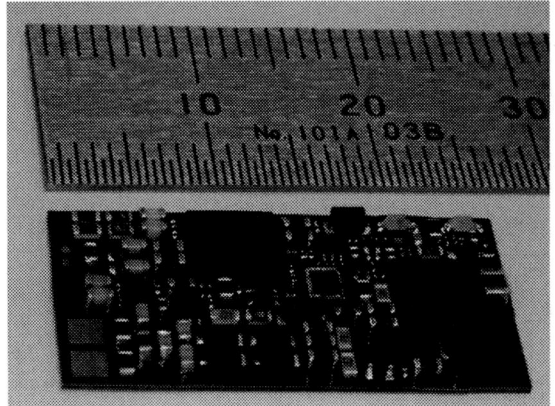

Fig 2. NFC wireless charging receiver module

Figure 2 shows the prototype of wireless charging receiver module we designed and fabricated. The module has both of 13.56/6.78-MHz wireless charging receiver function and types-A/B/F NFC card mode function. The module is composed of a power management IC, a choke coil, a rectifier, ripple filters, a microcontroller, a crystal unit, passive elements and some others. The size of the module is 14 x 26 x 1.86 mm. This size was designed for a mobile terminal such as a smart phone. For example, the module can be embedded into a back cover of a smart phone. This module can charge a battery by connecting an antenna and a battery to the module. Interface pads to a system board of a mobile terminal are allocated backside of the module for mounting the module on the system board as LGA.

Simulation bench for verifying power transfer efficiency

In order to design a high efficient wireless charging system, at first, a simulation bench was set. Figure 3 shows a rough sketch of a wireless charging system we investigated. After extracting the impedance models of power transmitter IC output and power receiver IC input, mainly, the simulation models of driver circuit, Tx antenna, Rx antenna and rectifier circuit were optimized for improving the efficiency of the wireless power transfer.

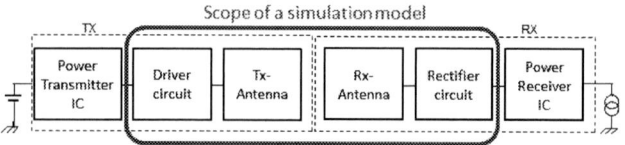

Fig 3. A rough sketch of a wireless charging system

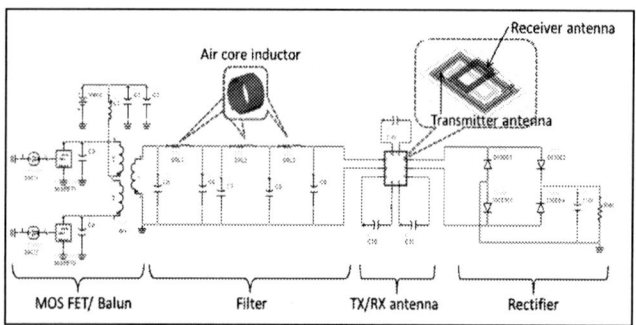

Fig 4. A simulation bench of a wireless charging system

Figure 4 shows a detailed simulation bench of the wireless charging system we investigated. The simulation model of a MOSFET is a behavior model. The balun model is a passive element. The models of the filter, air-core coils, are consisted of passive elements, but in order to improve the accuracy of the models, the parameters of the models was extracted from S parameters by using 3D electromagnetic simulator. The models of Tx and Rx antenna were expressed as a black box model of S parameters. The model was extracted by a 3D electromagnetic field simulator. The rectifier model is consisted of behavior models of diodes and a RC circuit expressing to input impedance of the power receiver IC.

In order to realize a high efficient wireless charging system, Tx and Rx antennas are optimized by using an electromagnetic field simulator. The antennas were constructed by three layers; an antenna board, a magnetic sheet and a metal plate. The antenna board is a both-side printed circuit board. The antennas are basically loop antennas resonated at the frequency of wireless charging and NFC. The magnetic sheet was utilized as the path of magnetic field flux in between the loop antenna and the metal plate. The metal plate was applied for stabilizing the antenna's Q ($\omega L/R$) value regardless of the surrounding environment. The Rx antenna is a small one-coil antenna in order to be able to embed this antenna into mobile terminal. On the other hand, a Tx antennas have a larger size and a three-coil configuration. One of them is a booster antenna for the resonance.

A higher Q antenna is preferable for a wireless charging. However, it is not suitable for NFC. Table 1 shows efficiencies of wireless charging in two cases. The size of Tx antenna is 116 x 66 mm, and the one of Rx antenna is 44 x 31 mm. The Q factors of Tx antennas were the same, however, the Q factors of the Rx antennas were different. The Q factor of the Rx antenna in the case 1 was 39.1. It is an example preferable for NFC, however, the efficiency was 36.0%. The Q factor of the Rx antenna in the case 2 was 76.1 by optimizing the shape of the antenna pattern. The efficiency was 61.2%. It was reasonable performance for wireless charging system. However, it was not good for NFC, because the distance of the communication shortened.

Table 1. Q factors vs. power transfer efficiency@6.78MHz

	Q of Tx Antenna	Q of Rx Antenna	Measured Efficiency (%)
Case 1	161.2	39.6	36.0
Case 2	161.2	76.1	61.2

The accuracy of the simulation bench shown in Fig 4 was confirmed by the correlation to the measurement results. In this paper, the only data for 6.78-MHz wireless charging was showed, because there are not clear differences from data for 13.56-MHz wireless charging. In the case 2, the simulated efficiency was 64.7%, while the measured efficiency was 61.2%.

Performance of NFC with a wireless charging antenna

Figures 5, 6 and 7 show the measurement results of three types of NFC, types A, B and F using the Rx antenna in the case 1. The operational frequency of NFC is 13.56 MHz. The Q factor of the Rx antenna at 13.56 MHz was approximately 58.3. The Q factor is approximately 5 times higher than that of a typical NFC antenna. These measured results of Figures 5-7 shows that all types of NFC succeed in the distance from 0 to 24 mm. This distance is enough for general use cases of NFC. Therefore, the results show that NFC functions of this module can be achieved by using a wireless charging antenna.

There are no data of the measurement results using the Rx antenna of the case 2. The Q factor of the antenna is approximately two times higher than that of the antenna in case 1. Through our experience, the NFC performance using the antenna of the case 2 is predicted to degrade. The remaining work is to confirm the trade-off between performance of NFC and efficiency of wireless charging and to clarify the design rule of the antenna suitable for both of wireless charging and NFC.

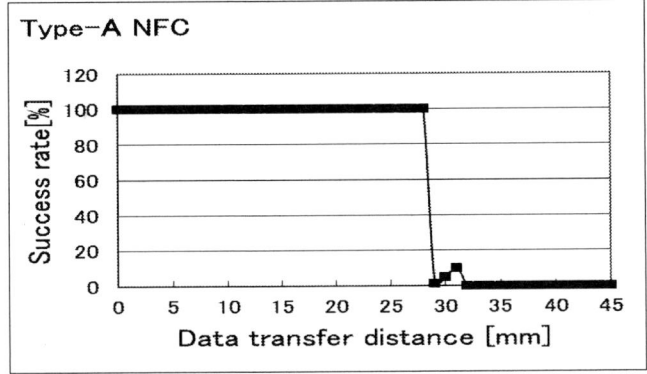

Fig 5. NFC performance for Type-A

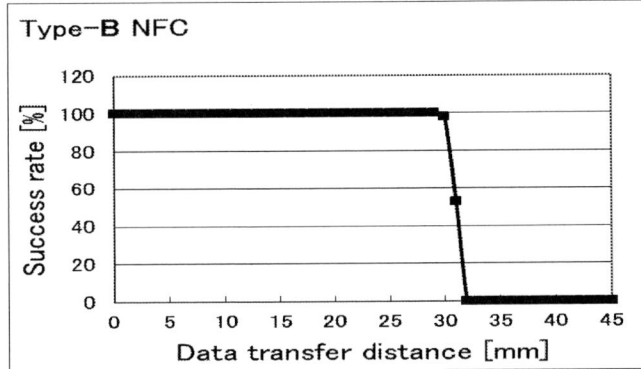

Fig 6. NFC performance for Type-B

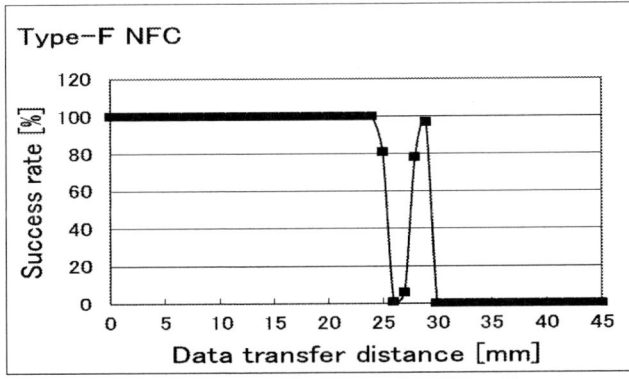

Fig 7. NFC performance for Type-F

Thermal design for embedding the module into a mobile terminal

In order to reduce the module dimentions, in general, small and thin components are densely mounted on the module. Smaller and thiner components however generate larger heat due to increase of the thermal resistance. In addition, the smaller module has the smaller diffusion area of the heat. Therefore, the reduction of the demensions raises the module temperature. On the other hand, almost all mobile terminals have a battery. A battery hates high temperature because high temprature causes degradation of the battery performance. In addition, the surface of a housing covering a mobile terminal

should keep lower temprature because of avoiding skin burn. Thus, it is important to suppress increase of the temperature caused by downsizing of the module. In this demonstration, the target temperatures of the prototype module were lower than 60 degrees Celsius at the surface of components and around 45 degrees Celsius at the surface of the housing.

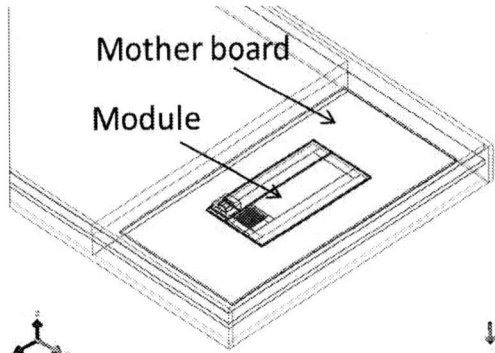

Fig 8. Thermal simulation model of the module embedded into a mobile terminal

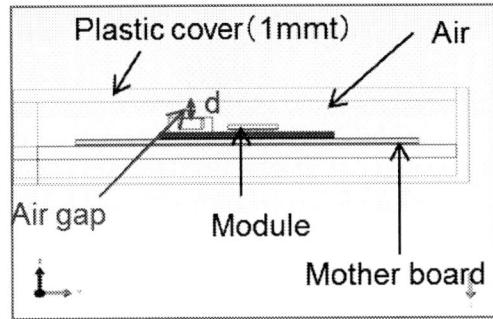

Fig 9. A cross section of the thermal simulation model

Figures 8 and 9 show thermal simulation models of the module embedded into a mobile terminal. The module was mounted on a mother board in the housing made of plastic. To control the temperature on the surface of plastic housing cover, the air gap between heating components and the plastic cover was changed. The heating components of this simulation were a power receiver IC and an inductor of the DC-DC converter. The height of the inductor is higher than that of a power receiver IC. Then, the definition of the air gap is the distance d between the top of the inductor and the bottom of the plastic cover. In figure 9, the left component the inductor and the right the power receiver IC.

Detailed parameters of the thermal simulation model are followings. The module substrate has six metal layers. The dimension is 14 x 26 x 0.46 mm. The mother board has six metal layers. The dimention is 28 x 54 x 0.40 mm. Ambient temperature is 25 degree Celsius. Thermal conductivity of the plastic is 0.3 W/m·K. The estimated charging power is 5 watts, however, it is not the total power to provide the heating components. The total power consumed at the heating components was estimated by correlation with masurement data.

Figure 10 shows a thermal simulation result for the surface of the heating components. Heat from the inductor and the IC

978-1-4799-2408-0/14 $31.00 © 2014 IEEE

spreads into the mother board and the body of the mobile terminal through the module. The maximum temperature in this situation was 73.7 degree Celsius at the surface of the inductor. On the other hand, Figure 11 shows a thermal simulation result for the surface of the plastic housing cover. The maximum temperature on the surface of the housing cover was 45.2 degree Celsius. It almost accepted the target temperature of 45 degree Celsius.

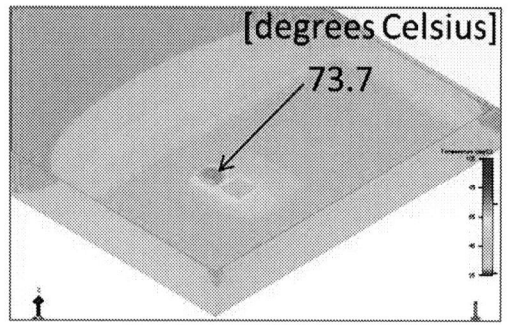

Fig 10. Thermal simulation result on the components

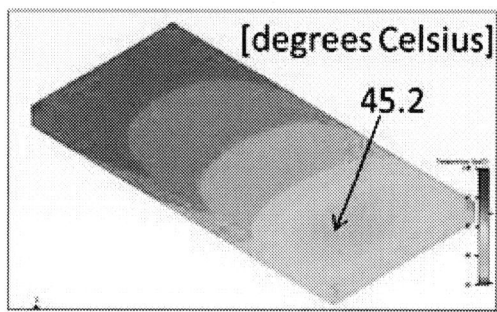

Fig 11. Thermal simulation result on the housing

In order to confirm the simulation results, the temperature of the heating components was measured. Figure 12 shows a picture of the module mounting on the board assumed as a mother board. The temperature was measured with a thermistor by mounting it on the heating components.

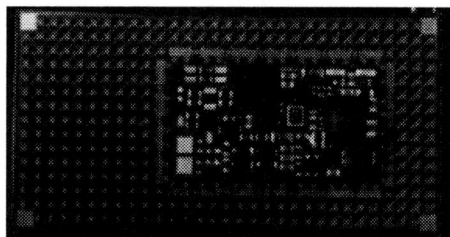

Fig 12. The module for thermal measurement

Table 2 Temperature on the surface of the inductor

Conditions	Temperature (degree C)
Simulation	73.7
Measurement	70.6

As mentioned before, in this investigation, the inductor generated the highest heat. Table 2 shows the temperatures of simulation and measurement on the surface of the inductor. The measurement result, 70.6 degree C, was lower than

simulation result, 73.7 degree C. This comparison data shows that the thermal simulation provides reasonable data for thermal design.

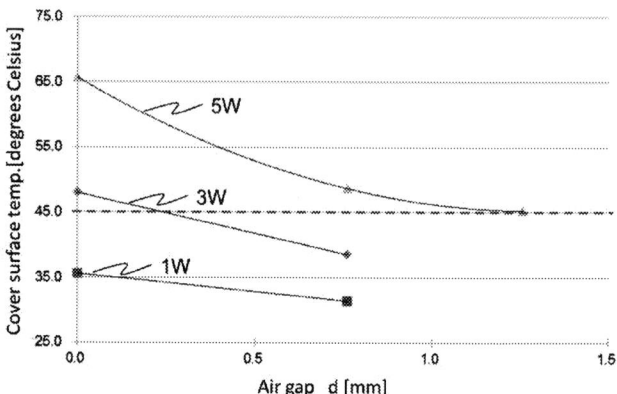

Fig 13. Thermal simulation results of the cover temp.

Figure 13 shows thermal simulation results of the cover temperature. In order to realize that the temperature of the cover surface is 45 degree C and less, these simulations were achieved. The estimated receiving powers by wireless charging were 5, 3 and 1 watt. In the case of 1 watt, it was unnecessary to keep an air gap between the top of the heating components and the bottom of the plastic cover. In the cases of 3 and 5 watts, however, it was necessary to keep the air gap. 0.75-mm air gap was enough for 3-watt wireless charging, however, it was not enough for 5-watt wireless charging. In the case of 5-watt wireless charging, 1.26-mm air gap was necessary for realizing that the surface temperature of the cover was 45 degree C as shown in Fig 11.

Table 3 Meausred temperatures depending on module board condition @5-watt wireless charging

	# of Layer	# of Via	Max. Temp.(degree C).
Case 1	6	Typical	70.6
Case 2	8	Rich	52.6

Table 3 shows measured maximum temperatures of the heating components depending on the module board conditions. The case 1 is the module board condition described before. On the other hand, the case 2 is the different board condition for aggressively decreasing the temperature. Although the thickness of the module boards are almost the same, the numbers of the layer and the via in the module boards are different from each other. Larger numbers of the layer and the via decreases the maximum temperature of the heating components mounting on the module board. This result shows improvement of heat spread for the board module is one of good options for thermal design.

EMI reduction of the wireless charging system

Wireless charging systems necessarily generate strong electromagnetic radiation and cause EMI (Electro-Magnetic Interference) problem with other wireless communications, because power transfer from a charger to a terminal utilizes

978-1-4799-2408-0/14 $31.00 © 2014 IEEE

electromagnetic coupling with Tx and Rx coil antennas. In order to reduce radiated emissions, sources and mechanism of EMI on wireless charging system we designed were investigated.

In order to find the source of EMI, near magnetic field distributions were measured by using a small magnetic field probe. Figure 14 shows measured results of near magnetic field distribution over module boards of the wireless charging receiver. These modules are not exactly the same as the module shown in Figure 2. The basic circuit of the wireless charging receiver module is the same as that of the previous module. The sizes of the modules are however larger than the previous one for inserting several EMI filters into the circuit. The measured frequency of near field is 142.38 MHz. It is twenty first harmonics of 6.78 MHz. In the frequency, radiated emission from the system was larger than the other frequencies.

Figure 14 (a) shows measured near magnetic field distribution of the module without EMI filters and Figure 14 (b) shows measured near field distribution with EMI filters. The arrows show direction of the magnetic field. The EMI filters inserted between the antenna pad and the rectifier shown in Fig 14(b). Strength of the magnetic field above the module was reduced. This experiment demonstrates that a rectifier is one of large EMI sources in a wireless charging system.

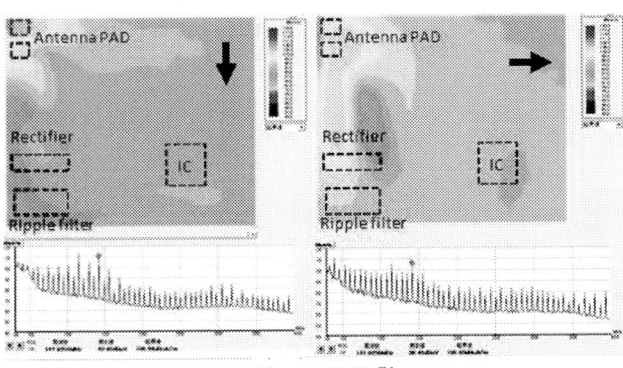

(a) without EMI filters

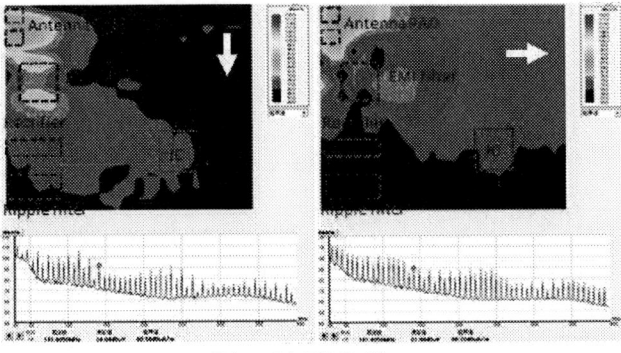

(b) with EMI filters

Fig 14. Measured near field distributions@142.38 MHz

Figure 15 shows measured results of far-field emissions radiated from the wireless charging system we designed. The distance of the measurement is 10 m accorrding to CISPR 32.

At approximetely 150 MHz of the horizontal field, the emission slightly exceeded the limitation of CISPR32. However, at the wide range of both holizontal and vertical fields, the emissions were lower than the limitations. This measurement demonstrates that the wireless charging system we designed has highly potential to pass EMI regulation of CISPR32.

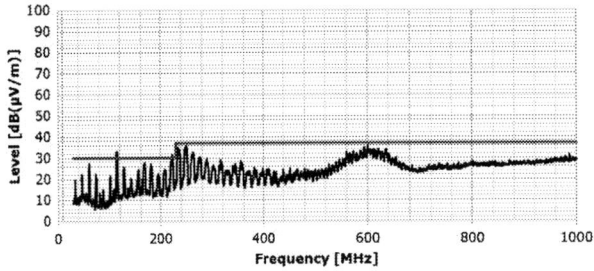

(a) Horizontal far field

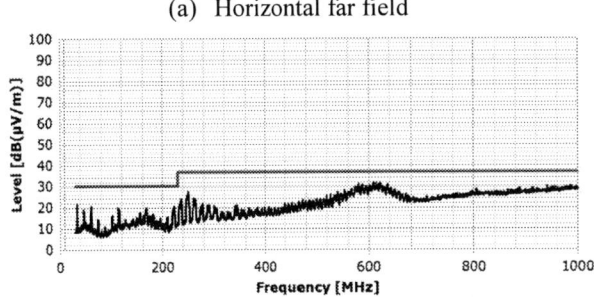

(b) Vertical far field

Fig. 15 measured results of 10-m radiated emissions

Conclusions

This paper presented an ultra-miniature module combining with 13.56/6.78-MHz wireless charging receiving functions and types-A/B/F near-field communication (NFC) functions. In order to be able to embed this module into mobile terminals, the electrical and thermal designs were optimized and then the size was 14 x 26 x 1.86 mm. The simulation bench we set could verify the efficiency of a wireless charging. The antenna we designed for wireless charging could function as NFC antenna. Thermal simulations and measurements showed that the temperature of surface on a mobile terminal could control by designing an air gap between the cover and components. Finally, measurement results of near fields and far fields demonstrated that the wireless charging system we designed had a potential to pass EMI regulation of CISPR32.

References

1. R.Goncalves, et al., "Increasing the RFID Readability Range Using Wireless Power Transmission Enhancements," in *Proc. IEEE Wireless Power Transfer Conf. (WPTC)*, Perugia, Italy, May 15-16, 2013, pp.135-138.

2. M. Fu, et al., "A 13.56 MHz Wireless Power Transfer System Without Impedance Matching Networks," in *Proc. IEEE Wireless Power Transfer Conf. (WPTC)*, Perugia, Italy, May 15-16, 2013, pp.222-225.

978-1-4799-2408-0/14 $31.00 © 2014 IEEE

3. J. Kim, et al., "Impedance Matching Considering Cross Coupling for Wireless Power Transfer to Multiple Receivers," in *Proc. IEEE Wireless Power Transfer Conf. (WPTC)*, Perugia, Italy, May 15-16, 2013, pp.226-229.

4. J. Nadakuduti, et al., "Operating Frequency Selection for Loosely Coupled Wireless Power Transfer Systems with Respect to RF Emissions and RF Exposure Requirements," in *Proc. IEEE Wireless Power Transfer Conf. (WPTC)*, Perugia, Italy, May 15-16, 2013, pp.234-237.

Enhanced-Performance Wireless Conformal "Smart Skins" Utilizing Inkjet-Printed Carbon-Nanostructures

Taoran Le[1], Ziyin Lin[2], C. P. Wong[2], and M. M. Tentzeris[1]

[1]School of Electrical and Computer Engineering, [2]School of Materials Science and Engineering
Georgia Institute of Technology
Atlanta, GA 30332
taoran.le@ece.gatech.edu

Abstract

This paper introduces for the first time the integration of a UHF radio frequency identification (RFID) antenna with reduced graphene oxide (rGO), developed using direct-write techniques and utilizing an RFID chip for chemical gas detection. The module is realized by inkjet printing on a low-cost paper-based substrate, and the RFID tag is designed for the North America UHF RFID band. The electrical impedance of the rGO thin film changes in the presence of very small quantities of certain toxic gases, resulting in a variation of the backscattered power level which is easily detected by the RFID reader to realize reliable wireless toxic gas sensing. The inkjet printed RFID tag demonstrated a change in backscattered power of 9.18% upon exposure of 40 ppm NO_2 for 5 minutes.

Index Terms — *Gas sensor, graphene, inkjet printing, UHF RFID, wireless, T-match.*

I. Introduction

The strict demands to both business and process in today's industry require robust, accurate sensors developed using low cost, reliable fabrication methods. In addition, uses such as environmental cognition require a robust sensor with extremely low power consumption. Therefore, methods for development of accurate, low power sensors using proven, low cost fabrication systems and off the shelf (OTC) components are extremely desirable. This paper provides detail and design of a sensor developed using conventional inkjet printing methods and OTC RFID components to create an accurate, robust, low power wireless gas detection system.

Carbon materials, such as carbon nanotube (CNT) and graphene, have been investigated previously as candidates for the early and accurate detection of various chemicals [1-3]. These materials alter their properties in the presence of a given substance due to their ability to absorb the chemicals on their surface [4-5]. Chemical absorption produces changes in material properties such as real and imaginary impedance, DC resistance, and effective dielectric constant [1]. These changes can be exploited to determine the presence of various chemical compounds by translating the material effects into measureable electrical quantities such as changes in voltage, current, resonant frequency, and backscattered power amplitude. Excellent electrical conductivity and the ability to be functionalized for a wide range of chemicals make these novel materials ideal candidates for the development of a broad spectrum of portable and wearable sensors. Moreover, the ability to deposit these materials via inkjet printing of aqueous solutions on low cost, flexible, environmentally friendly substrates opens the possibility of mass producing such devices and taking advantage of economies of scale.

In our previous research efforts, we have demonstrated excellent gas detection sensitivity for nitrogen-based gases [1,6]. At 10ppm NO_2 we achieved 21% sensitivity, and at 4ppm NH_3 a 9% sensitivity was obtained at 864 MHz using functionalized multi walled CNT (MWCNT) dissolved in water and inkjet printed on paper substrate. The incorporation of the previous graphene-based sensing thin films with micro-strip lines served as proof of concept [2, 6]. Of course these transmission lines can be thought of as parts of a wide range of antenna and microwave structures. Under this analog transmission approach, the wireless remote sensing mechanism is actually the detection of shifts in resonant frequency and / or magnitude of the backscattered power as a direct result of the change in impedance of the graphene stripes and, thus, also of the whole structure. Calibration may be needed in order to map particular amounts of shift in the resonant frequency or change in backscattered power levels to the concentration of toxic gases under test.

This paper will establish for the first time the principles of wireless gas sensor design utilizing rGO based thin films integrated into RFID platforms, and featuring the first ever inkjet-printing of aqueous graphene oxide solution on environmentally friendly paper. Use of water based carbon inks and paper substrates can set the standard for the development of a class of low cost, environmentally friendly rGO based sensors utilizing RFID principles.

II. Passive UHF RFID Tag Design

A. *General Design Consideration for rGO based Inkjet-Printed Gas Sensor Tags*

In the previous work on SWCNT based wireless gas sensors [7], a passive RFID system was utilized. This work uses a similar methodology. The RFID reader sends an interrogating RF signal to the RFID tag, which contains an antenna and an IC chip. The IC responds to the reader by varying its input impedance, thereby modulating the backscattered radiation levels. The modulation scheme often used in RFID applications is amplitude shift keying (ASK). The IC impedance switches between the matched state and the mismatched state, altering levels of backscattered radiation [8]. As illustrated in Fig.1, the rGO thin film is integrated with the printed antenna by direct-write methods, and acts as a tunable part of the antenna with an impedance value determined by the concentration of the target gas. The RFID reader monitors the backscattered power level. Once the power level changes, it shows that there is a variation of the rGO film impedance, therefore, the wireless sensor detects the existence and the concentration of the target gas.

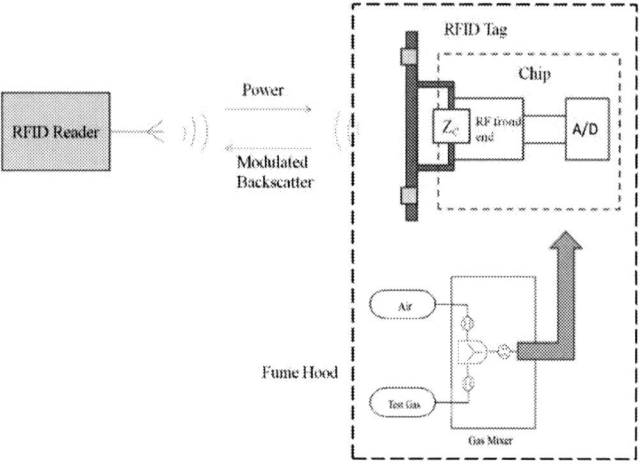

Figure 1. Conceptual diagram of the RFID-enabled wireless gas sensor module.

The power level of the received signal can be calculated using the Friis free-space equation as:

$$P_{tag} = P_t G_t G_r \left(\frac{\lambda}{4\pi d}\right)^2 \qquad (1)$$

where P_t is the power transmitted from the reader antenna, G_t and G_r are the gains of the reader antenna and the tag antenna respectively, and d is the distance between the reader and the tag.

B. Tag Antenna Design Parameters

The design goals for the tag antenna were to achieve an omnidirectional read pattern and maximum read range. Two types of known planar tag antennas which can provide an omnidirectional pattern are a slot antennas and dipole antennas. However, a slot type tag antenna usually requires significantly more metallization and a larger footprint than a dipole. Since a goal of the project was to provide a low cost system, a dipole design was chosen for the tag antenna design.

The maximum read range of a passive UHF RFID tag depends on the gain and quality of impedance matching between the tag antenna and IC [9]. The read range can be optimized by selecting an IC with low-power consumption, maximizing the antenna gain, and arranging a complex-conjugate impedance matching between the IC and tag antenna for maximal power transfer [10]. Two IC chips (SOT 1040AB2 and Higgs-4 SOT) were chosen for design in the North America UHF RFID band. The parameters are listed in the Table I. The power transfer between the antenna and IC is described by a power transmission coefficient (PTC) as:

$$\tau = P_C/P_A \qquad (2)$$

where P_C is the power absorbed by the chip from the antenna, and P_A is the maximum available power from the antenna.

The maximum power can be transferred when Z_C is the conjugate value of Z_A [11].

TABLE I.
IC Chip Parameters at 915MHz

Symbol	Quantity	Value [Ω]
Z_{C1}	SOT 1040Ab2 IC impedance	13.3- j122
Z_{A1}	Tag 1 antenna design impedance	13.1+ j122
Z_{C2}	Higgs-4 SOT IC impedance	18.5- j181.3
Z_{A2}	Tag 2 antenna design impedance	18.5 +j 181.3

The T-match is an effective way modify the input impedance of a dipole by introducing a centered short-circuit stub, as shown in Fig.2 [12].

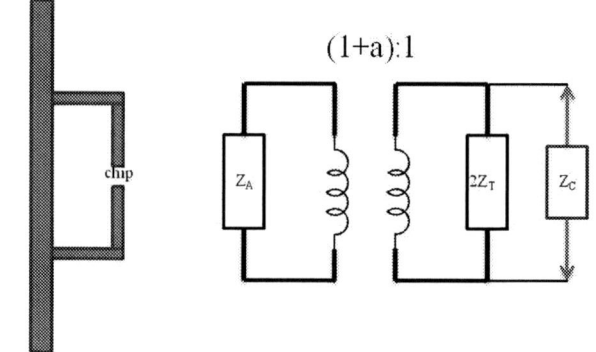

Figure 2. T-match configuration for dipoles and equivalent circuit.

The terminal impedance is given by [13],

$$Z_{in} = \frac{2Z_T(1+\alpha)^2 Z_A}{2Z_T + (1+\alpha)^2 Z_A} \qquad (3)$$

where ($Z_T = jZotank\alpha/2$) is the input impedance of the short-circuit stub by the T-match circuit and the middle part of the dipole. Preliminary results demonstrate that the surface area of the inkjet-printed graphene rather than the number of its inkjet-printed layers determine the sensitivity to the presence of small concentrations of poisonous gases [14]. Therefore, as long as the surface area of the rGO exposed to the air remains constant, the gas sensitivity remains the same. Two antenna tags with two different surface area rGO film patterns were designed in Computer Simulation Technolgy (CST) STUDIO SUITE® 2013 for validating and optimizing the gas sensor sensitivity. The parameters used to model the inkjet-printed conductor are listed in the Table II. The Kodak photopaper substrate parameters are the same as used in the previous work [15]. The Cabot silver ink was also used, as in [16].

978-1-4799-2408-0/14 $31.00 © 2014 IEEE

TABLE II.
Design Specifications

Symbol	Quantity	Value
f_0	Operating frequency	915 MHz
εr	Dielectric constant	2.9
H	Paper thickness	230 um
Tan δ	Loss tangent of paper	0.05
T	Carbot silver ink thickness	1.6 um
σ	Cabot silver ink conductivity	$5e^6$ S/m

The tag antenna design for SOT 1040AB2 IC chip is shown in Fig.3. The antenna tag design for Higgs-4 SOT IC chip is in Fig.4.

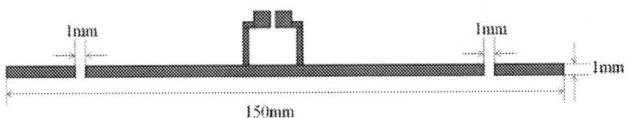

Figure 3. Antenna design for SOT 1040AB2.

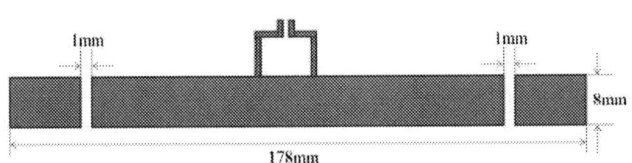

Figure 4. Antenna design for Higgs-4 SOT.

III. Inkjet Printing of Gas Sensor Prototype

The core of the wireless gas sensor is a prototype board made from graphene deposited onto a Kodak photo paper. Graphene was chosen for several reasons. First, it exhibits remarkable electronic and mechanical properties [3, 5], and graphene oxide materials can be easily dispersed in water. In our previous work on CNT based wireless gas sensors, the short lifetime and poor dispersion of the CNT ink limited the obtainable material performance [1, 7].

Cabot's CCI-300 conductive nanoparticle silver ink was used to create the traces connecting the graphene pad to the external circuitry. Microstrip topology was chosen because the microstrip elements can be easily integrated with inkjet printed rGO thin films.

The tag creation process, shown in Fig. 5, includes:

1) Deposition of the GO ink on paper substrate
2) Curing and reduction of the GO thin film
3) Alignment of the GO film and printing of the antenna
4) Sintering of the silver ink
5) Integration of the target RFID IC chip on the tag

The following sections will provide a more in depth discussion of these steps.

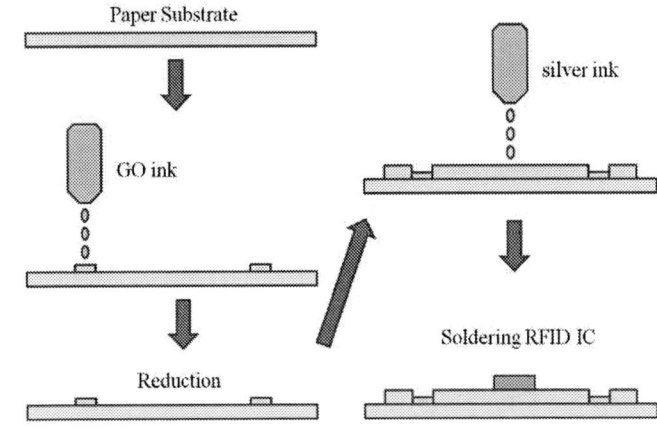

Figure 5. The fabrication process of the gas sensor prototype.

A. Creation of Graphene Oxide

The first step in the sensor development process was the creation of stable, long-life, inkjet-printable graphene-based inks. This was accomplished by first converting the graphene into graphene oxide powder. Unlike pristine graphene, which has very poor dispersion in common solvents, graphene oxide (GO) exhibits excellent solubility in water due to the existence of hydrophilic functional groups on the surface [17], rendering it an excellent candidate for development of environmental friendly water-based inks. After deposition, the graphene was restored by the reduction of GO, which reverts the conjugated basal plane and restores the electrical properties of the material. The reduction of GO is considered as one of most promising methods for low-cost, high-yield and scalable preparation of graphene materials [18].

The GO was produced by chemical oxidation of graphite, which introduces oxygen-containing functional groups to exfoliate pristine graphite into individual GO sheets [2, 19]. To prepare the GO ink, dry GO powder was dispersed in a water/glycerol solution and sonicated to form a homogenous dispersion.

B. Fabrication via Inkjet Printing

A Dimatix Materials Printer (DMP-2800) Series material deposition system, as shown in Fig. 6, was used to print both the silver and GO inks.

Figure 6. The Dimatix materials printer.

Based on the previous experimentation, we determined that the application of additional 5 layers of silver applied on top of the graphene guarantees better connectivity between the graphene and silver traces and helps to de-embed any additional impedance which would offset the measured change of impedance unevenly among different samples of the similar dimensions [2]. The fabrication process involved deposition 5 layers of GO ink followed by 5 layers of conductive silver onto the paper substrate. First, the 5 layers of GO ink were deposited and cured at 80°C overnight. After the decution of GO thin film, 5 layers silver GO were deposited and cured at 120°C for 30 minutes. To insure optimum contact between the rGO thin film and the silver ink, a 0.5 mm overlap of the two surfaces was chosen.

C. Reduction of Graphene Oxide

After curing GO ink, chemical oxidation was used to remove some of oxygen-containing groups and restore the conjugated system and the electrical conductivity. Typical methods for GO reduction include thermal reduction, N_2H_4 reduction and $NaBH_4$ reduction etc. The final structure of chemically reduced graphene is highly dependent on the reduction method, and it is possible to control the final structure by selective removal of specific types of oxygen-containing groups for optimized material properties. Due to the paper substrate thermal tolerances, the N_2H_4 reduction was chosen. The GO thin film was reduced by N_2H_4 vapor at 90 °C overnight.

D. Final Packaging Steps

The final step in the fabrication was to solder the IC chips to their respective antennas, as shown in Figs. 7 and 8 below

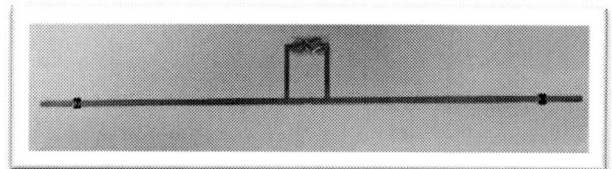

Figure 7. The antenna tag with SOT 1040AB2 .

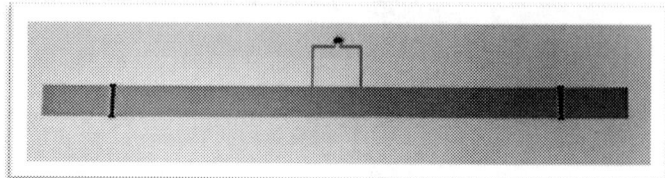

Figure 8. The antenna tag with Higgs-4 SOT.

E. Optimization of the Reduced Graphene Oxide Thin Film

Optimization of the RGO thin films was done in order to obtain the proper sensor area to ensure maximum sensitivity. To determine the optimum sensor area, two patterns of different lengths of RGO material were produced as shown in Fig. 3 and Fig. 4. From the previous work in [14], it was expected that the larger surface area of the rGO pattern in Fig. 4 would yield the higher sensitivity.

IV. Gas Sensor Experimentation

A. Experiment Setup

The gas test setup is shown in Fig. 9 below. The tags were tested using a KIN-TEK FlexStream gas standards generator to provide a stable gas source with accurate concentrations. The setup was capable of producing reliable mixtures up to 50 ppm of nitrogen dioxide gas diluted in nitrogen gas. A Tagformance Lite RFID reader (Voyantic Inc. 2011) is adopted for interrogation power threshold measurements. At each frequency point, the reader varies the interrogation power until the power is just enough to activate the RFID chip. After the reader sweeps through the entire target frequency range from 800MHz to 1GHz, the interrogation power threshold versus frequency curve can be obtained. When the antenna sensor deforms due to gas, the antenna resonance frequency changes accordingly. Therefore, gas sensing is achieved utilizing the relationship between antenna power threshold and gas concentration.

To begin the test, the reader was first calibrated using a reference tag at a distance of 0.3m away from the reader antenna. Then, the antenna tag was placed at the exact interrogation position as the reference tag. The reader recorded the sweep from 800MHz to 1GHz. Nitrogen was then flowed through the system for 5 minutes to establish a system baseline. Next, the 40 ppm nitrogen dioxide was introduced into the system and the reader swept the entire frequency range again. Finally, the gas source was removed from the device while measurements continued to be taken for another 5 minute interval in order to measure the recovery time of the sensor.

978-1-4799-2408-0/14 $31.00 © 2014 IEEE

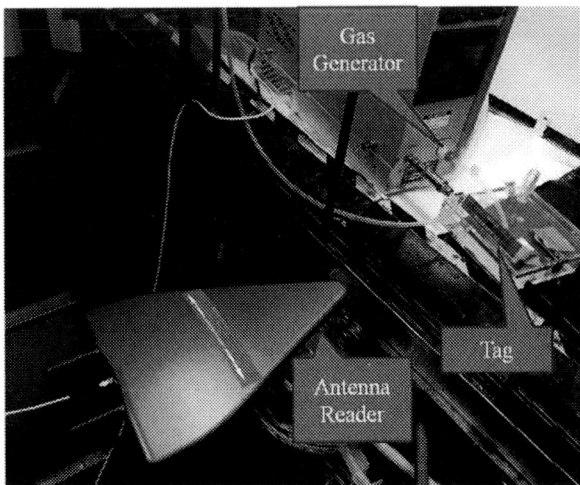

Figure 9. Gas test setup for sensor measurements.

B. Experimental Results

The results of 40 ppm NO_2 gas testing are shown in Fig. 10. The inkjet-printed rGO thin films of different dimensions demonstrated similar responses to 40 ppm NH_3. Due to the poor connection of the Higgs-4 SOT IC with the antenna, the result was not able to be validated. Fig. 10 shows the power threshold measurement of the antenna tag with SOT 1040AB2 IC. After exposure to the target gas, the mismatching of the impedance between antenna and IC chip caused the backscattered power level to be reduced as compared to the original power level around 915MHz target frequency. The power level of 9.18% difference was observed at the target frequency. After 5 minutes period of pure nitrogen purging, full recovery of the system was observed. We are presently continuing investigations into further improving the recovery time.

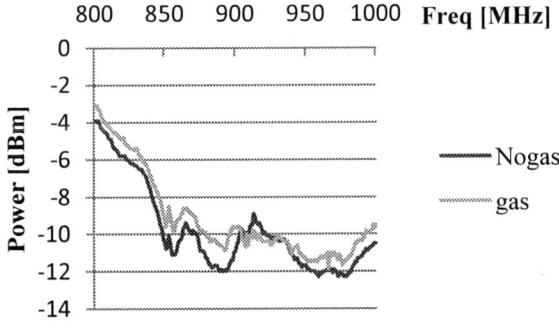

Figure 10. Measured response of RGO thin films with and without NO_2.

VI. Conclusion

We have demonstrated for the first time integration of rGO thin films onto an inkjet printed UHF RFID module on paper to form a wireless gas sensor. We also provide detailed results of our efforts in the development of environmentally friendly, stable, low cost, inkjet-printable GO inks. The design demonstrated the viability of inkjet-printed rGO for the realization of fully integrated "green" wireless RFID-enabled flexible sensors and the ultrasensitive impedance properties of the rGO materials utilized.

The prototype device exceeded our expectations for initial tests, producing a 9.18% change in backscattered power level at 40 ppm concentration of nitrogen dioxide gas. Moreover, the sensor demonstrated fast recovery time in comparison to the current state of technology without the use of heat or UV treatments to assist in the material recovery, providing an advantageous result for sensing in natural environmental conditions.

V. Future work

Future work includes several areas of improvement on the current design, now that the concept has been proven. The dipole antenna design can be optimized to minimize its size for accommodation of mobile and wearable sensing applications. A meander line design is currently being considered for the generation II prototype. The RFID tag range can also be improved in the next generation, and more directional designs may be considered. Finally, based on the findings here, the sensor sensitivity can be improved by incorporating larger surface area rGO thin films into the design. This is also under consideration in choosing the design parameters of the generation III prototype.

Acknowledgments

The authors would like to thank NSF-ECS for their support of the research, and the helpful discussion with coworkers John Kimionis, Sangkil Kim and James Cooper.

References

1. Lakafosis, Vasileios, Xiaohua Yi, Taoran Le, Edward Gebara, Yang Wang, and Manos M. Tentzeris. "Wireless sensing with smart skins." In Sensors, 2011 IEEE, pp. 623-626. IEEE, 2011.
2. Le, Taoran, Vasileios Lakafosis, Ziyin Lin, C. P. Wong, and M. M. Tentzeris. "Inkjet-printed graphene-based wireless gas sensor modules." In Electronic Components and Technology Conference (ECTC), 2012 IEEE 62nd, pp. 1003-1008. IEEE, 2012.
3. Geim, Andre K., and Konstantin S. Novoselov. "The rise of graphene." Nature materials 6, no. 3 (2007): 183-191.
4. Schedin, F., A. K. Geim, S. V. Morozov, E. W. Hill, P. Blake, M. I. Katsnelson, and K. S. Novoselov. "Detection of individual gas molecules adsorbed on graphene." Nature materials 6, no. 9 (2007): 652-655.
5. Li, Dan, and Richard B. Kaner. "Graphene-based materials." Nat Nanotechnol 3 (2008): 101.
6. Le, Taoran, Ziyin Lin, C. P. Wong, and M. M. Tentzeris. "Novel enhancement techniques for ultra-high-performance conformal wireless sensors and" smart skins" utilizing inkjet-printed graphene." In Electronic Components and Technology Conference (ECTC), 2013 IEEE 63rd, pp. 1640-1643. IEEE, 2013.
7. Yang, Li, Rongwei Zhang, Daniela Staiculescu, C. P. Wong, and Manos M. Tentzeris. "A novel conformal RFID-enabled module utilizing inkjet-printed antennas and carbon nanotubes for gas-detection applications." Antennas and Wireless Propagation Letters, IEEE 8 (2009): 653-656.

8. Nikitin, Pavel V., and K. V. S. Rao. "Performance limitations of passive UHF RFID systems." In IEEE Antennas and Propagation Society International Symposium, vol. 1011. 2006.

9. Rao, KV Seshagiri, Pavel V. Nikitin, and Sander F. Lam. "Antenna design for UHF RFID tags: A review and a practical application." Antennas and Propagation, IEEE Transactions on 53, no. 12 (2005): 3870-3876.

10. Kurokawa, Kaneyuki. "Power waves and the scattering matrix." Microwave Theory and Techniques, IEEE Transactions on 13, no. 2 (1965): 194-202.

11. Rao, KV Seshagiri, Pavel V. Nikitin, and Sander F. Lam. "Impedance matching concepts in RFID transponder design." In Automatic Identification Advanced Technologies, 2005. Fourth IEEE Workshop on, pp. 39-42. IEEE, 2005.

12. Balanis, Constantine A. Antenna theory: analysis and design. John Wiley & Sons, 2012.

13. Marrocco, Gaetano. "The art of UHF RFID antenna design: impedance-matching and size-reduction techniques." Antennas and Propagation Magazine, IEEE 50, no. 1 (2008): 66-79.

14. Le, Taoran, Vasileios Lakafosis, Sangkil Kim, Benjamin Cook, Manos M. Tentzeris, Ziyin Lin, and Ching-ping Wong. "A novel graphene-based inkjet-printed WISP-enabled wireless gas sensor." In Microwave Conference (EuMC), 2012 42nd European, pp. 412-415. IEEE, 2012.

15. Kim, Sangkil, Benjamin Cook, Taoran Le, James Cooper, Hoseon Lee, Vasileios Lakafosis, Rushi Vyas et al. "Inkjet-printed antennas, sensors and circuits on paper substrate." IET Microwaves, Antennas & Propagation 7, no. 10 (2013): 858-868.

16. http://www.cabot-corp.com/ CCI-300 Data Sheet.

17. Lin, Ziyin, Yagang Yao, Zhuo Li, Yan Liu, Zhou Li, and Ching-Ping Wong. "Solvent-assisted thermal reduction of graphite oxide." The Journal of Physical Chemistry C 114, no. 35 (2010): 14819-14825.

18. Li, Zhuo, Yagang Yao, Ziyin Lin, Kyoung-Sik Moon, Wei Lin, and Chingping Wong. "Ultrafast, dry microwave synthesis of graphene sheets." Journal of Materials Chemistry 20, no. 23 (2010): 4781-4783.

19. Hummers Jr, William S., and Richard E. Offeman. "Preparation of graphitic oxide." Journal of the American Chemical Society 80, no. 6 (1958): 1339-1339.

20. Leenaerts, O., B. Partoens, and F. M. Peeters. "Adsorption of H_2O, NH_3, CO, NO_2, and NO on graphene: A first-principles study." Physical Review B77, no. 12 (2008): 125416.

Novel THz Imaging Array Using High Resistivity Metasurfaces

Kyoung Youl Park and Premjeet Chahal
Michigan State University
428 S. Shaw Lane, East Lansing, MI, 48824, USA
chahal@egr.msu.edu

Abstract

In this paper, lossy metasurfaces are employed in the design of uncooled terahertz (THz) bolometers on polymer substrate. The metasurface bolometer is composed of a metal backed low loss dielectric substrate and thin patterned lossy metal layer. Numerical simulations using finite element method (FEM) are carried out to design optimal unit cells with high absorptivity. An array of these unit cells are combined together to form a pixel element of an imaging array. Unit cells with center frequency of 0.1 and 0.4 THz are designed. A 3×3 imaging array operating at 0.1THz is fabricated on a plastic substrate and measured results are presented. An image of a concealed object inside a paper envelope is also demonstrated in transmission mode using this array. Results show that high sensitivity THz bolometers can be fabricated on low-cost substrates using strongly absorbing metasurfaces.

Introduction

Over the last two decades significant interest has grown in the area of terahertz (THz). Many THz systems are being investigated for a range of applications [1, 2]. Among the many THz systems, imagers are one of the most critical systems that form the basis for many applications including biomedical imaging, security adverse weather landing and driving, and non-destructive evaluation (NDE) [3]. Many types of detection methods exist including heterodyne detectors using Schottky diode, superconductor or hot electron bolometer mixer, and direct detectors using bolometers (e.g., Bi, VOx based devices) or wide-band antenna coupled Schottky diode rectifiers [4-9]. Among these, bolometers are attractive as they provide a low-cost, low-power avenue to imaging and they are simple to design and implement. A bolometer converts the incoming radiation to heat and the heat induces a change in the resistivity of the film which is directly measured using a readout circuit. Use of bolometers in THz detection is in its infancy and the most of the existing approaches rely on technologies borrowed from infrared imagers [2]. For a high sensitivity bolometer design, key requirements are: i) small area, ii) low-thermal loading, iii) high coupling efficiency and iv) high temperature coefficient resistance (TCR) [10]. Area of a typical infrared bolometer is in the range of $10\mu m \times 10\mu m$ to $50\mu m \times 50\mu m$. For efficient THz detection an antenna element is coupled to these small bolometer structures [11]. However, impedance matching between the bolometer element and the antenna is very difficult while maintaining low thermal loading.

In this paper, we demonstrate novel detector elements based on metasurface structures. In the proposed design, detectors based on sub-wavelength size resonant structures made from high resistivity thin films are utilized to achieve strong absorption leading to high sensitivity [12-14]. In place of an antenna element, an array of resonant structures are grouped together to form a pixel element of an imaging array. Frequency selective surfaces (FSS) have been utilized in the design of strong microwave and millimeter wave absorbers [15]. However, to design a high sensitivity bolometer ultra small unit cells are desirable. Recently, metamaterials composed of sub-wavelength resonant structures have been studied in order to improve absorption [13, 16]. For the design of bolometer, beyond small size, structures having high loss properties are required. This is because the absorbed energy should be converted to heat in a localized area. Furthermore, the absorbing structure should be made of films having high temperature of coefficient of resistance (TCR) so that the change is resistance is strongly dependent on absorbed power. The substrates on which these structures are fabricated should have low-loss characteristics so that all of the power is absorbed by the lossy metal. Also, the substrate should poses low thermal conductivity to minimize heat loss.

Here thin film metamaterials (metasurfaces) are utilized in the design of THz bolometers. These structures are made from lossy metal film on low-loss dielectric substrate with a solid ground plane. Metamaterial unit cells are designed that are compact, simple to fabricate and provide strong absorption. Noise equivalent power (NEP), which is a measure of sensitivity, is determined of a pixel element. A 3×3 imaging array is demonstrated by capturing an image of a concealed object in a paper envelope in the transmission mode.

Design and simulation

One of the key advantages of using metamaterials is that the frequency of operation can be changed arbitrarily through manipulation of geometry of a unit cell. Here a modified slit ring resonator (SRR), [12], is adopted in the design of unit cells operating at 0.1 and 0.4 THz. These cells are scaled version of each other. Geometry of each unit cell is optimized to achieve a compact sub-wavelength size structure. Figure 1 shows the design layout of the unit cells. Figure 1 (a) is the basic model of the SRR, it is disjointed from its neighboring cells. Figure 1 (b) and (c) shows the unit cell which is composed of interconnected metamaterial unit cells. The unit cells are connected together so that the resistance change can be directly measured across these group of cells. These unit cells are connected together using two approaches: straight and zigzag. In the straight connection approach, the top sections of the unit cells are connected together. In the zigzag approach, the diagonals of the unit cells are connected together. The zigzag approach provides higher resistance path. Table 1 summarizes the physical dimensions of the unit cell and array. The periodicity of the unit cells is approximately one sixth of the free space wavelength at these design frequencies.

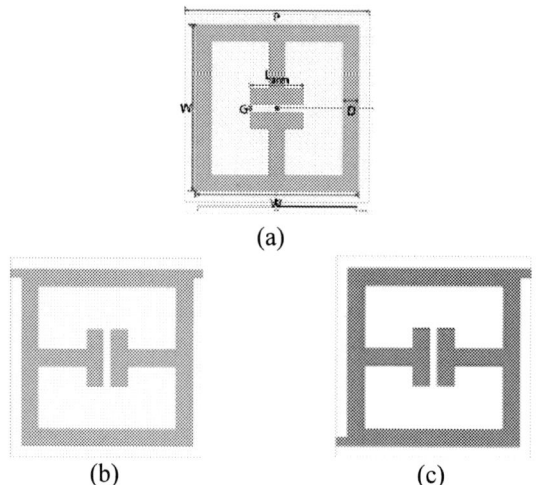

(a)

(b)   (c)

Figure 1. The geometry of SRR structure, (a) a modified SRR unit cell, (b) a straight connection unit cell, and (c) a zigzag connection array of unit cells.

Table 1. Dimensions of a 0.1 and 0.4 THz unit cell design.

Parameters	Dimensions		Remarks
	0.1 THz	0.4 THz	
Periodicity (P)	0.47 mm	0.115 mm	$\lambda_{0.1THz} = 3mm$ ($\lambda/p = 6.38$) $\lambda_{0.4THz} = 0.75mm$ ($\lambda/p = 6.52$)
Width (W)	0.415 mm	0.105 mm	
length of arm (L_{arm})	0.135 mm	0.035 mm	
Width of arm (D)	40 µm	10 µm	
Gap (G)	20 µm	5 µm	

Ansoft HFSS v.14, a full wave EM simulator based on finite element method (FEM), was utilized in the design of unit cell structures. A low loss polymer substrate, PEEK (polyether ether ketone), is used as a substrate material. The substrate thickness is 250 µm and its electrical properties in the THz spectral region are presented in [17]. The dielectric properties of PEEK are: dielectric constant, ε_r=3 and loss tangent, tan δ = 0.05 over a frequency range of 0.1 to 0.6 THz [17]. PEEK is inert to most chemicals used in microfabrication, has low moisture absorption and low thermal conductivity. A solid ground plane (Cu, 1 µm thick with thin Ti adhesive layer) is e-beam deposited on one side of the substrate. A 500 nm thick e-beam deposited nickel (Ni) is used in the patterning of the absorbing structures. The thickness of this conductor is less than the wavelength at 0.1 THz and it is a few times thicker than its skin depth.

Figures 2 (a) and (b) shows the simulation results of unit cells of microbolometer operating 0.1 and 0.4 THz, respectively. The proposed SRR is dependent on the polarization of incident waves. Simulations are carried out for both vertical (V-pol) and horizontal polarization (H-pol) incident waves. For the analysis, the array is assumed to be infinite in periodicity and thus Floquet's theorem analysis is carried out on HFSS.

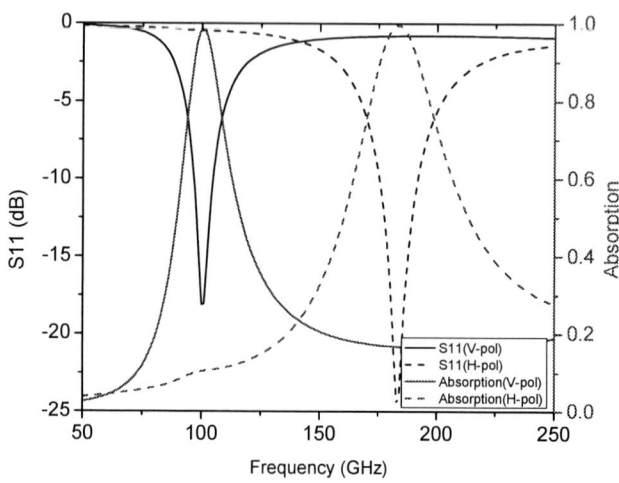

(a)

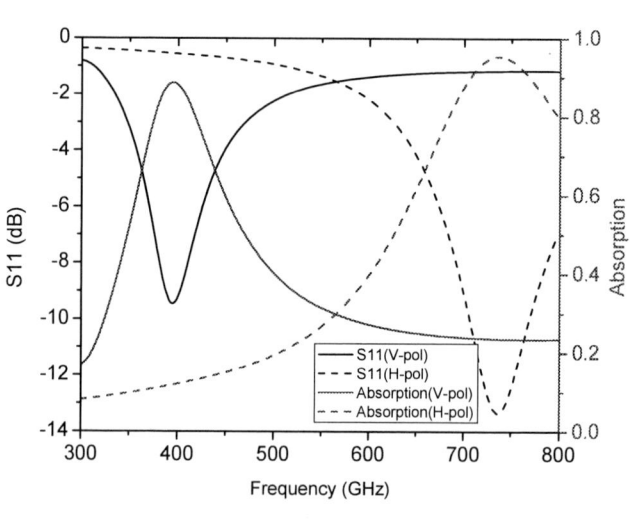

(b)

Figure 2. Simulation results (reflectivity and absorption) of a proposed unit cell of imaging array, (a) V- and H-polarization of 0.1 THz unit cell and (b) V- and H-polarization of 0.4 THz unitcell.

In case of vertical incident wave, the resonant frequency of the structure is excited at 0.1 THz. The reflected signal (S_{11}) is approximately -18 dB and this translates to 98.5 % absorption. The absorption is simply calculated by subtracting the reflected power from unitary incident power. The power transmitted through the sample is negligible as there is a solid ground plane present on the opposite side. The usable bandwidth (above 50 % of absorption) of the unit cell (V-pol) is 26 GHz (90 to 116 GHz). In the case of H-pol incident wave, the resonant frequency is excited at 0.183 THz. This is different from the excitement of V-pol resonance frequency because of difference in electrical lengths (shown in Figure 3 of surface current distributions). The minimum S_{11} was calculated to be -24 dB at resonant frequency of 0.183 THz.

978-1-4799-2408-0/14 $31.00 © 2014 IEEE

The maximum calculated absorption is 99.5 % at this resonance frequency. The usable half power absorption bandwidth was calculated to be approximately 54 GHz (between 0.16 to 0.214 THz).

Similar design and analysis was carried out for a scaled 0.4 THz unit cell structure. For the physical parameters tabulated in Table 1, a 0.394 THz resonance frequency is calculated for V-pol mode excitation. The reflected signal, S_{11}, is determined to be -9.5 dB. This translates to absorption of 88.7 % at the resonant frequency. The bandwidth of 0.4 THz unit cell (V-pol) is 127 GHz (between 0.350 to 0.47 THz). For H-pol mode excitation, the resonant frequency was calculated to be 0.736 THz and the reflected signal (S_{11}) is approximately -13.4 dB.

Imaging array design

In the above analysis, an infinite array of unit cells working in tandem is assumed. In a lossy metamaterial structure the field decays rapidly around the resonator. In other words, the fields are weakly coupled between the resonant structures and thus finite numbers of unit cells can be used. Also, in the design of an imaging array each pixel element should be large so that it is not diffraction limited. For high sensitivity, small bolometer structures are desirable. Based on these key conditions a pixel element design composed of 4×4 unit cells was chosen. No other optics is used in tandem with this pixel element design. In the future, a thorough analysis will be carried out to design an optimum detector pixel element that considers all of these aspects.

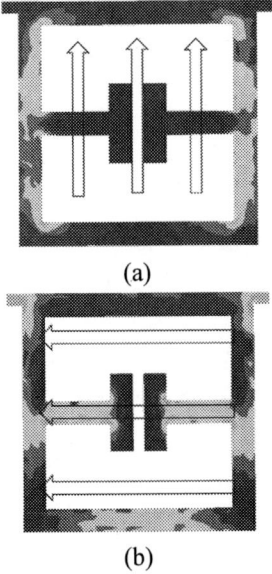

(a)

(b)

Figure 3. Surface current distributions of each excited mode, Magnitude of surface currents, (a) vertical polarization and (b) horizontal polarization.

Figure 4 (a) and (b) show two design layouts of 3×3 pixels imaging arrays. In both of these designs, each pixel element is composed of 4×4 unit cells. The key difference between these two designs is the interconnection scheme of the unit cells. In order to detect the incoming signal, a change in resistance of the pixel element is measured. Each pixel is connected with Cu traces for measurement of series resistance. One side of the pixel element is connected to a common (ground) and the other end is interrogated independently. Figure 4 (a) and (b) shows the layout of proposed THz imaging arrays. The straight connection array consists of nine microbolometers, nine signal pads, and two common ground pads as shown Figure 4 (a). The zigzag connected imaging array, shown in Figure 4 (b) has a similar layout. Pads shown in Figure 4 are formed from copper that is known to have good electrical conductivity. These pads on the edge are connected to readout circuits. Total size of the 3 × 3 THz imaging array is 2 × 2 cm² (including test pads), the active detection area is approximately 7.5 × 7.5 mm². Fabrication of this structure requires two mask layers. The first mask layer is to pattern the Ni layer and the second mask is needed to pattern the Cu pads and traces.

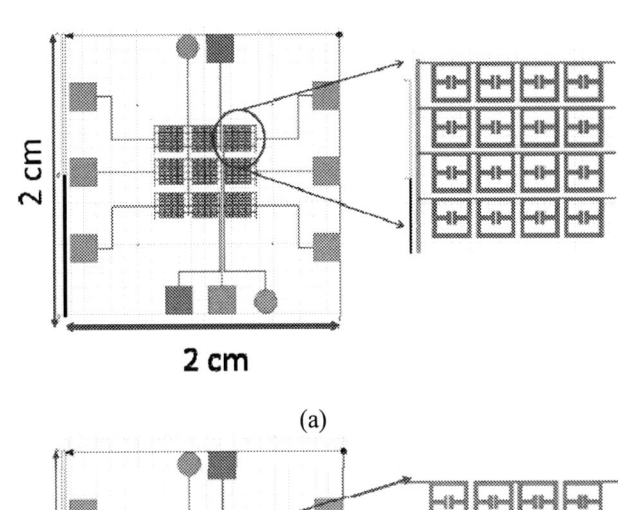

(a)

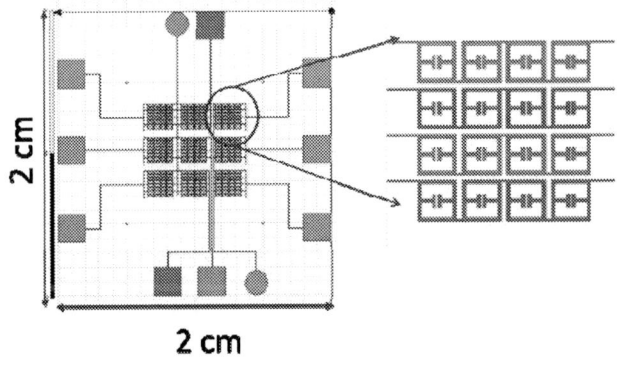

(b)

Figure 4. Layouts of 0.1 THz imaging array pixels, (a) straight connected unit cells pixel element, and (b) the zigzag connected unit cells pixel element.

Fabrication

First step of the fabrication is to clean the PEEK substrate and this was done using acetone, methanol, isopropanol rinse in DI-water. After drying in an oven, photoresist, Shipley 1813 positive resist, of approximately 1.5 μm is deposited on the substrate using a spin coater. Following prebake, exposure, post-bake and develop, "descum" using oxygen plasma in reactive ion etcher (RIE) was carried out to remove photoresist residue from the surface of PEEK. A 120W, 30 second 100% oxygen plasma was used to descum resist. Titanium (t = 300 Å) and nickel (5,000 Å) were deposited using e-beam. Titanium is used as an adhesive layer. The metal liftoff was performed in acetone bath using a sonicator.

978-1-4799-2408-0/14 $31.00 © 2014 IEEE

The sonicator helps decrease the lift-off time. After lift-off, the substrate is cleaned in isopropanol followed by DI water and N$_2$ blow dry. Using the same microlithography process as above, the second metal layer (Cu) was patterned. Process steps are shown in Figure 5.

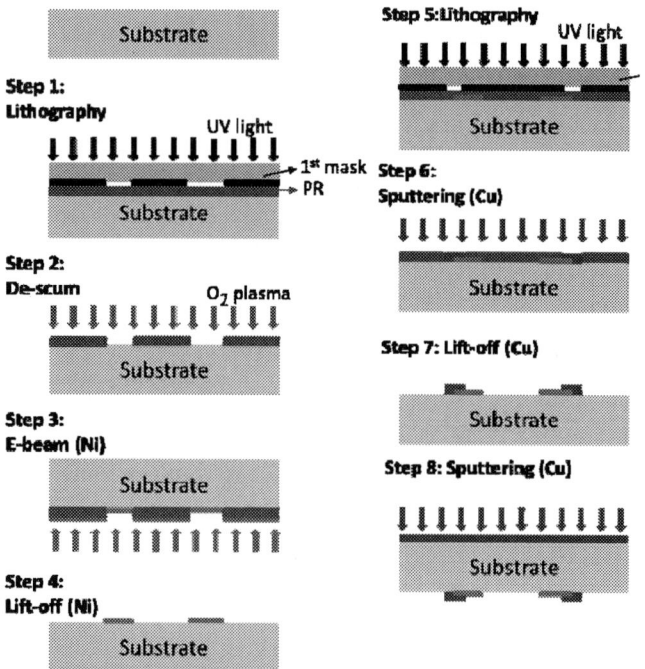

Figure 5. Microfabrication process of a proposed THz imaging array.

Figure 6 shows example photographs of the fabricated THz imaging array. The physical dimension of the fabricated nickel microbolometer pixel array was inspected using a microscope. As shown in pictures, the metal patterns and interconnections were well aligned.

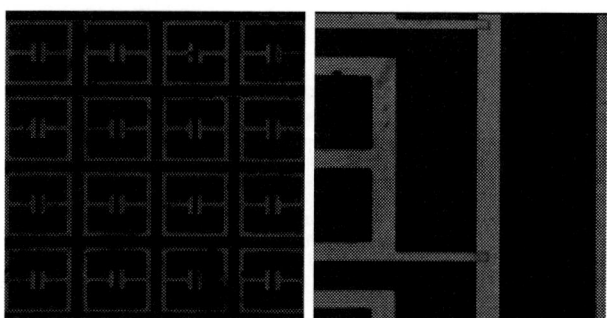

Figure 6. Microphotographs of the fabricated imaging pixel element (Zigzag connected unit cells).

Measurement setup

Two measurement setups are used in the characterization of the pixel elements and the imaging array. First setup is made to determine the absorbed power by the pixel elements and the second setup is made to characterize imaging characteristics of the fabricated devices.

The setup for the measurement of power absorbed depicted in Figure 7 (a). In this setup, an array of absorbing elements is placed in front of the horn antenna. A W-band

directional coupler is attached between the horn antenna and the backward wave oscillator (BWO, W-band source). A W-band receiver (a calibrated harmonic mixer) attached to the directional coupler was used to measure the reflected power. For calibration, a flat metal sheet was used that was placed at the same distance as the absorbing elements. The reflected signal from the reference and the array elements was measured and the difference in signal is used to calculate the absorbed power.

An open-ended W-band waveguide was used to illuminate the object to be imaged. Measurement was setup for capturing the image in the transmission mode depicted in Figure 7 (b). Since the array is composed of 3×3 array, a 2-axis robotic manipulator was used with step resolution of 200 μm on each axis to raster the sample to be able to capture a complete image of the sample. A multimeter along with data acquisition was used to measure the resistance across the individual pixel elements. The THz imaging array was fixed, whereas the object under scan was attached to the robotic arm. Through this setup, change of surface resistance of each pixel across the array was measured to capture the image.

Generally, the noise equivalent power (NEP) of a bolometer is defined as the power needed to achieve a signal to noise ratio of unit at 1 Hz bandwidth [10]. To indirectly measure the intrinsic NEP of the bolometer, a bias circuit with a DC voltage and a load resistor was used. In this configuration [18], a voltage is applied across the series connected bolometer and fixed temperature insensitive resistor. A voltage drop is measured across the bolometer as a function of applied current.

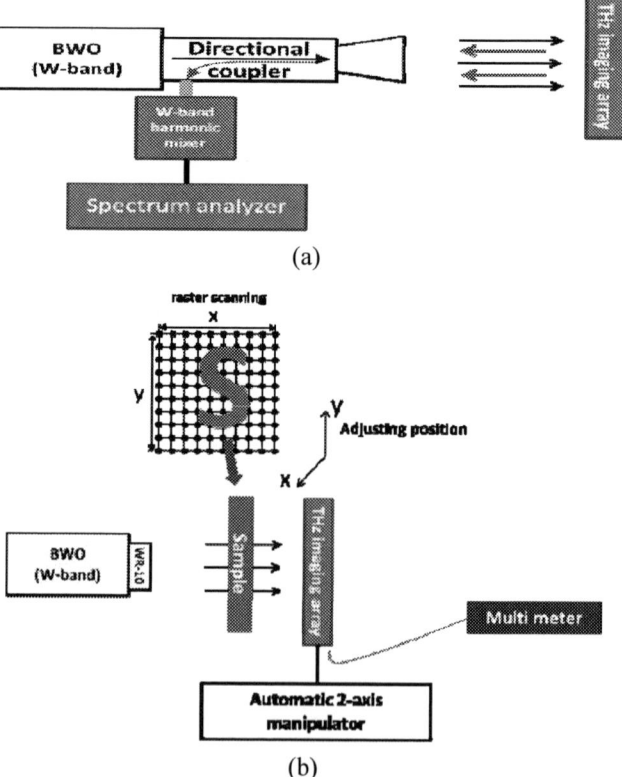

Figure 7. Measurement setup block diagrams, (a) Absorption measurement setup, and (b) Imaging measurement setup.

978-1-4799-2408-0/14 $31.00 © 2014 IEEE

Measurements

Absorption measurements were carried out from 85 to 105 GHz with step resolution of 1 GHz. Figure 8 shows the measured result of the proposed THz microbolometer array. In order to characterize the absorption of the bolometer array, the main orientation of polarization, vertical polarization was excited. The measured results matches closely with the simulation results (V-pol: 0.1 THz at Figure 2 (a)). The results show strong absorption for V-pol from 87 GHz to 107 GHz, Figure 8. The absorption is determined to be 9.8 dB at 102 GHz. It is lower than the simulated absorption of 18 dB. Rest of frequency range could not be measured as the source frequency is limited to 110 GHz. However, we can estimate that the performance of rest of frequencies will have similar results as the simulation results shown in Figure 2 (a).

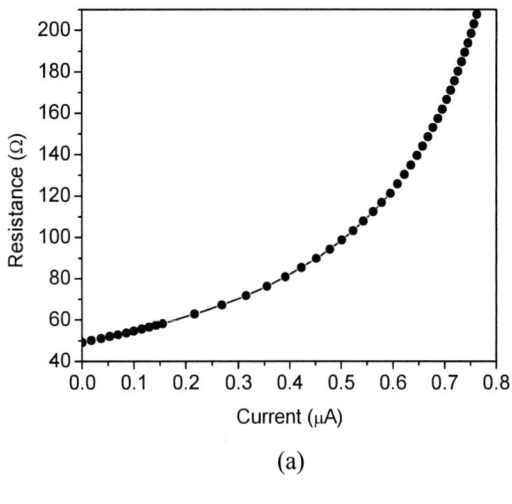

(a)

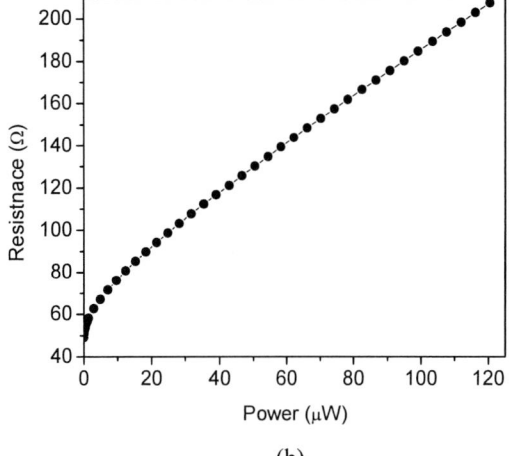

(b)

Figure 9. Measured resistance versus applied (a) current and (b) power.

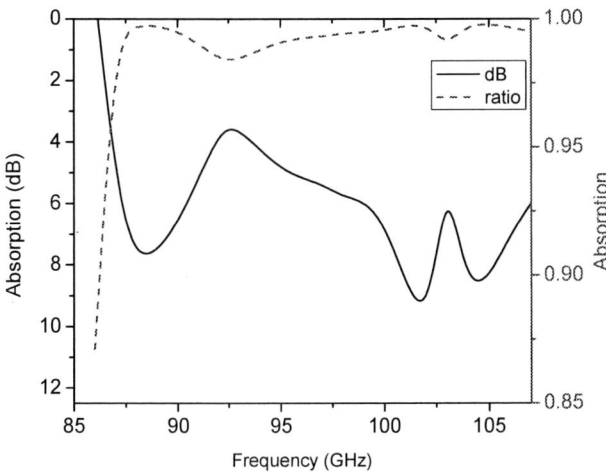

Figure 8. Measured result of a far-field test of THz imaging array.

Figure 9 (a) show the measured resistance of a bolometer versus applied current while the environment temperature is fixed (room temperature, T=296 °K). The power drop across the bolometer converts to heat and which in turn increases the resistance of the bolometer as seen in Figure 9 (a). This can be plotted as a function resistance versus power drop across a bolometer as shown in Figure 9 (b). The slope of Figure 9 (b) gives the sensitivity of the bolometer. NEP of the bolometer can be calculated using the slope (sensitivity), TCR of the film and Johnson noise [10, 19]. The TCR of Ni film with similar thickness processed using e-beam is $\alpha=2.7\times10^{-3}$ grad^{-1} [14]. Using these values and supposing the Johnson noise is the only parameter defining the radiation equivalent to the power of the NEP, the calculated intrinsic NEP of a bolometer is approximately 1.6×10^{-9} W/Hz$^{1/2}$. This value does not account for reflected signal at the measurement frequency. However, it does provide a simple approach to characterize the sensor elements.

In order to demonstrate the viability of the proposed imaging array, an imaging scan was carried using the fabricated structures. Figure 10 (a) shows a photograph of a sample with punched holes in a metallic sheet. The minimum size of holes is 0.9 cm × 0.9 cm which is approximately twice the size of a pixel element on each side. Figure 10 (b) shows the transmitted signal as a function of position across the sample. The measurement step of X- and Y-axis robotic arm set at 6mm in each direction. A total of 11 × 11 = 121 points are measured over the XY plane and the spacing between the source and the detector was set to approximately 7cm.

Figure 10 (b) shows that such setup can be used in transmission measurements and shapes of structures can be resolved using a bolometer pixel element of this paper. Scattering at the edge of punched holes makes the holes appear larger than their actual size. For the measurement of an actual sample hidden in a paper envelope the setup shown in Figure 11 was used. Here, to minimize scattering and to attain a close to a plane wave impinging on a sample an open ended WR-10 waveguide at the source end was used in place of the horn antenna. Furthermore, the raster steps were decreased to improve resolution. Measured image of a metalized sample

978-1-4799-2408-0/14 $31.00 © 2014 IEEE   779

placed in an optically opaque envelope between the source and the imaging array is shown in Figure 11. In this picture, colors represent the density of the signal after passing through the sample. Blue color represents low-transmission and red color represents regions with high transmitted signal. The physical size of the sample is approximately 2.8 cm × 2.5 cm. In this measurement, the axial step size is set as 2mm per step. A total of 21 × 21 = 441 points are collected in the XY plane. The image scanned clearly shows the hidden object. This clearly shows that proposed imaging array of this paper can be used in high sensitivity imaging applications.

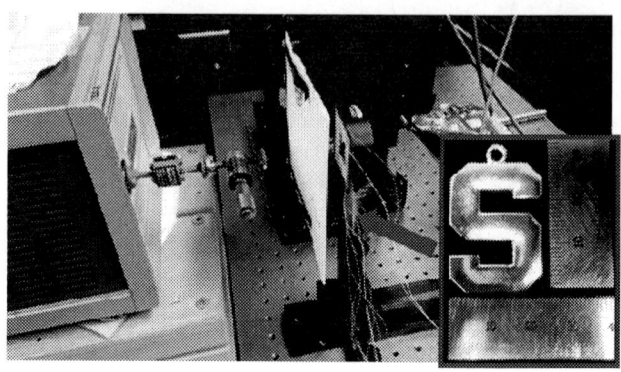

(a)

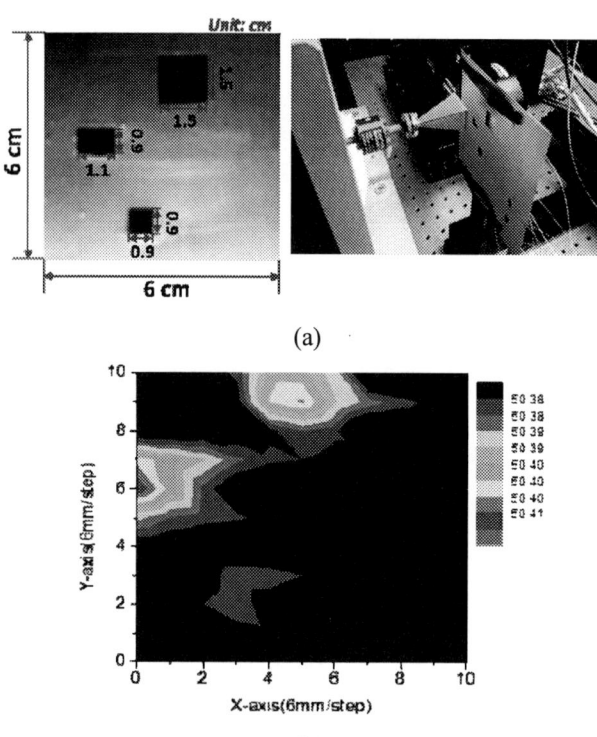

(a)

(b)

Figure 10. Scanning result of punched holes.

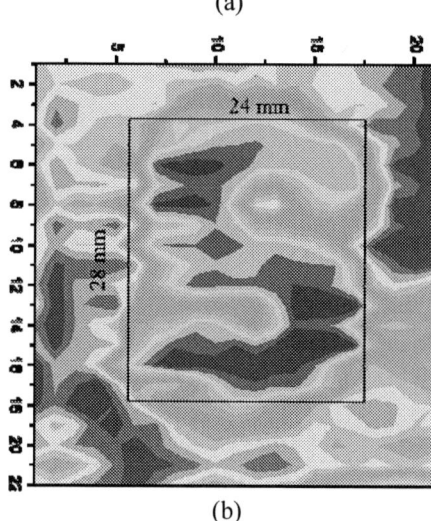

(b)

Figure 41. Scanned results of a concealed S-shaped metallized plastic object in a paper envelope, (a) measurement setup, and optical picture of the S-shaped pendant, (b) transmission mode THz scanned results (stepper resolution: 2mm/step).

Conclusions

In this paper, a new THz imaging array is proposed and demonstrated. The proposed imaging array is based on high resistivity metasurface that provides high absorption of incident signal. The imaging pixel element is composed of an array of sub-wavelength metamaterial unit cells. The unit cells can be scaled to design detector element in the desired band of operation. Furthermore, the size of the unit cells can be reduced down into subwavelength size. These subwavelength structures directly absorb THz radiation and this can be utilized as a detector through a change in the resistance across the pixel element. An intrinsic NEP of 1.6×10^{-9} W/Hz$^{1/2}$ is demonstrated at 0.1 THz. This can further be improved by optimizing thermal isolation, TCR of the film and effective resistance of the film. The results of this paper clearly demonstrate that lossy metasurfaces can be utilized in the design of highly sensitive bolometer for THz imager applications.

Acknowledgments

The authors would like to thank members of TeSLa research lab for helpful discussion. This work was supported in part by the DARPA YFA program (Grant Number: N66001-12-1-4238).

References

1. Ferguson, B. and Zhang, "Material for terahertz science and technology", *nature material*, Vol 1. pp.26-33, Sep, 2002.

2. Siegel, P. H, "Terahertz Technology", *IEEE Transactions on Microwave Theory and Techniques*, Vol. 50, No.3, pp.910-928, March, 2002.

3. D. M. Sheen and D. L. McMakin, 'Three-dimensional millimeter-wave imaging for concealed weapon detection', *IEEE Transactions on Microwave Theory and Techniques* 49(9) (2001), pp. 1581–1592.

4. Weinreb, S., Kerr, A.R., 'Cryogenic cooling of mixers for millimeter and centimeter wavelengths', *IEEE J. Solid-State Circuits* 8, pp.58–63, 1973.

978-1-4799-2408-0/14 $31.00 © 2014 IEEE

5. McColl, M., Millea, M.F., Silver, A.H., 'The superconductor–semiconductor Schottky barrier diode detector', *Appl. Phys. Lett.* 23, pp. 263–264,1973.

6. McGrath, W.R., "Hot-electron bolometer mixers for submillimeter wavelengths: an overview of recent developments," *6th Int. Space Terahertz Technol. Symp.,* Pasadena, CA, pp. 216–228 ,1995.

7. E. M. Gershenzon, G. N. Goltsman, I. G. Gogidze, Y. P. Gusev, A. I. Elantev, B. S. Karasik, and A. D. Semenov, "Millimeter and submillimeter range mixer based on electronic heating of superconductive films in the resistive state," *Superconductivity,* vol. 3, no. 10, pp. 1582–1597, 1990.

8. W. R. McGrath, B. S. Karasik, A. Skalare, R. Wyss, B. Bumble, and H. G. LeDuc, "Hot-electron superconductive mixers for THz frequencies," in *SPIE Terahertz Spectroscopy Applicat. Conf.,* vol. 3617, pp. 80–88, Jan. 1999.

9. F. Niklaus, C. Vieider, H. Jakobsen, 'MEMS-Based Uncooled Infrared Bolometer Arrays –A Review', *Proc. of SPIE*, vol.6836, 2007.

10. W.Holland, "Bolometers for Submillimeter and Millimeter Astronomy", *Single-Dish Astronomy: Techniques and Applications, ASP Conference proceedings,* vol.278, 2002

11. S.Cherednichenko, et al., "A Room Temperature Bolometer for Terahertz Coherent and Incoherent Detection", *IEEE Transactions on Terahertz Science and Technology,* vol.1, No.2, Nov.2011.

12. H.T, Chen, W.J.Padilla, *et al*, 'Active terahertz metamaterial devices', *nature letters*, vo.444, pp.597-600, Nov. 2006.

13. K. Y. Park, *et al.*, "Metamaterial-inspired absorbers for terahertz packaging application", *61st Electronic Components and Technology Conference (ECTC)*, pp. 2107-2113, May, 2011.

14. K. Y. Park, *et al*, "A novel terahertz power meter using metamaterial-inspired thin-filme absorber," *36th International conference on Infrared, Millimeter and Terahertz Waves (IRMMW-THz)*, Oct, 2011.

15. Mei Li, et al., "An Ultrathin and Broadband Radar Absorber Using Resistive FSS", *IEEE Antennas and Wireless Propagation Letters*, vol.11, pp.748-751, 2012.

16. L. Sun, et al., "Broadband metamaterial absorber based on coupling resistive freqeuncy selective surface", *Optics Express*, vol.20, No.4, pp.4675-4680, Feb.2012

17. J. Hejase, *et al.*, "Terahertz characterization of dielectric substrates for component design and nondestructive evaluation of packages," *IEEE Transactions on Components, Packaging and Manufacturing Technology*, vol. 1(11), pp.1685-1694, 2011.

18. R.K. Bhan, R.S.Saxena, C.R.Jalwania, and S.K.Lomash, 'Uncooled Infrared Microbolometer Arrays and their Characterization Techiniques', *Defense Science Journal*, vol.59, No.6, pp.580-589, Nov.2009.

19. R. Novak and J.Hrbek, "Nickel bolometers", *Czech Journal of Physics*, B.18, No.11, pp.1423-1432, 1968.

Magneto-dielectric Characterization and Antenna Design

Kyu Han[1,2], Madhavan Swaminathan[1,2], P. Markondeya Raj[3], Himani Sharma[3], Rao Tummala[3] and Vijay Nair[4]

[1]Interconnect and Packaging Center (IPC), SRC Center of Excellence @ GT

[2]School of Electrical and Computer Engineering, Georgia Institute of Technology

266 Ferst Drive, Atlanta, GA 30332 USA

Email: khan7@gatech.edu

[3]Packaging Research Center, Georgia Institute of Technology

[4]Intel Corporation, Chandler, Arizona

Abstract

Antenna size has fundamental limits based on the frequency of operation and performance required. In the past, various methods have been developed to miniaturize antennas with limited success. Magneto-dielectric materials, however, have been reported as providing new opportunities for effective antenna size reduction in many recent studies. In this paper, a novel material characterization method which is a cavity perturbation technique (CPT) with substrate integrated waveguide (SIW) cavity resonator is presented for measuring electric and magnetic properties of magneto-dielectric material. CPT formulas for extracting complex permittivity and complex permeability are explained and modification process using 3D EM simulation tool is discussed. Design and fabrication of SIW cavity resonators is presented. The frequency dependent properties of permittivity and permeability for synthesized magneto-dielectric material are extracted in the frequency range of 1-4 GHz. Planar inverted F antenna (PIFA) working at 1GHz on magneto-dielectric substrate has been designed and simulated in this paper.

Introduction

In wireless communication systems, size of mobile device is a key specification. Since the size of the antenna is determined by its electrical length, miniaturization of the antenna can be very challenging. One method for decreasing the antenna size is by using high permittivity materials [1]. However, using the high dielectric constant material as the antenna substrate leads to narrow bandwidth and low efficiency [2]. Recently, magneto-dielectric materials, which have both permittivity and permeability greater than 1, have stirred the interest of antenna designers since the material can reduce antenna size without deteriorating antenna performance [3].

Magneto-dielectric materials are not available readily in nature and have to be realized using material synthesis where magnetic metal particles are mixed with low loss dielectric materials. Since the antenna response is affected by the frequency dependent permeability and permittivity of the material, an accurate method is required to extract the frequency dependent properties (ε', ε'', μ' and μ'') of the magneto-dielectric material. This can be challenging since the electric and magnetic properties need to be separated through measurements. This has been achieved by using two different structures, as described in [4] and [5]. One structure is sensitive to change in permittivity and electric loss tangent while the other structure is sensitive to change in permeability and magnetic loss tangent. In this paper, a cavity perturbation technique (CPT) is used to measure both electric and magnetic properties using a single substrate integrated waveguide (SIW) structure. Simulation and measurements have been used to design the SIW and demonstrate the characterization process on a magneto-dielectric material synthesized using nano-cobalt magnetic particles in a polymer dielectric material. The frequency dependent permittivity and permeability have been extracted for this material in the frequency range 1-4 GHz. The second part of the paper uses the magneto-dielectric material for PIFA design and presents improvement in size reduction, bandwidth and radiation efficiency. The sensitivity of the material parameters for fine tuning the antenna is discussed, to determine the parameters that have the largest effect on antenna performance.

CPT for SIW

CPT is a well-known method for extracting electromagnetic properties of dielectrics, semiconductors, magnetic materials, and composite materials [6]. Permittivity and permeability of the magneto-dielectric sample can be calculated from changes in the resonant frequency and quality factor by introducing the sample at positions where the electric field and magnetic field are maximum in the cavity, respectively. For complex permittivity measurements, modified CPT formulae from [7] can be used which are:

$$\varepsilon'_s = \frac{A\varepsilon'_r V_c}{V_s}\left(\frac{f_o - f_s}{f_s}\right) + \varepsilon'_r. \tag{1}$$

$$\varepsilon''_s = \frac{B V_c}{V_s}\left(\frac{\varepsilon'^2_r + \varepsilon''^2_r}{\varepsilon'_r}\right)\left(\frac{Q_o - Q_s}{Q_s Q_o}\right) + \frac{\varepsilon'_s \varepsilon''_r}{\varepsilon'_r}. \tag{2}$$

where ε'_s and ε''_s correspond to real and imaginary permittivity of the sample, respectively, ε'_r and ε''_r are real and imaginary part of relative permittivity of cavity substrate. Q_o and Q_s are the quality factors of the empty and loaded with sample cavity. f_o and f_s are the resonant frequencies before and after the sample perturbation, respectively. V_c is a volume of the cavity and V_s is a volume of the sample. In the above equations constants A and B are obtained experimentally by using standard samples with known dielectric properties. Similarly, for complex permeability measurement, equation (3) and (4) can be used [6].

$$\mu'_s = \frac{C V_c}{V_s}\left(\frac{f_o - f_s}{f_s}\right) + 1. \tag{3}$$

$$\mu''_s = \frac{D V_c}{V_s}\left(\frac{Q_o - Q_s}{Q_s Q_o}\right). \tag{4}$$

978-1-4799-2408-0/14 $31.00 © 2014 IEEE

where μ'_s and μ''_s are real and imaginary part of permeability of the sample where constants C and D in equations (3) and (4) are also obtained from the measurement of standard sample.

Design of SIW Cavities and CPT Analysis

SIW technology has been used to implement cavities for CPT in this paper, as shown in Fig. 1, since they have high Q, are highly sensitive to material properties and have minimum radiation effect. This technology has been used to measure complex permittivity of dielectric materials in [7] and [8]. The resonant frequency of these resonators for TE_{m0k} mode is related to the width W and the length L of the cavity as follows [7]:

$$f_r = \frac{c}{2\sqrt{\varepsilon'_r \mu'_r}} \sqrt{\left(\frac{m}{W}\right)^2 + \left(\frac{k}{L}\right)^2} \qquad (5)$$

where c is the speed of light in free space, ε'_r and μ'_r are the relative permittivity and permeability of SIW substrate respectively and m, k are mode numbers.

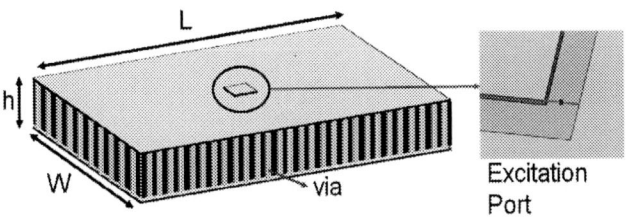

Figure 1. TE_{103} mode SIW cavity resonator with GSG probe excitation.

The SIW cavities are designed for the dominant TE_{102} or TE_{103} mode. TE_{102} mode can be used to measure material properties in the low frequency range since it can reduce the SIW cavity dimension. Fig. 2 shows the electric and magnetic field distribution in the SIW with TE_{103} mode. The cavity is excited at one of the maximum positions of the electric field where a GSG probe is used for the excitation, with the signal probe on the center patch and ground probes on the SIW on either side (corner-to-corner probing). In the CPT, placing the sample at either the E or H-field maximum position as shown in Fig. 2 with dashed circles, changes the resonance frequency of the SIW cavity based on the sample's permittivity and permeability, respectively. During measurements it is required that, the permeability of the sample does not perturb the permittivity measurement and also that the permittivity of the sample does not affect the permeability measurement. Separating the permittivity and permeability parameters can be a very challenging process during characterization of magneto-dielectric materials. 3D EM simulations with CST microwave studio have been used in this paper to design the SIW cavity with appropriate locations for excitation and sample placement such that the permittivity and permeability measurements have minimum effect on each other. The results of simulations for the SIW cavity resonating at ~2GHz is shown in Fig. 3 and 4 where the complex permittivity and

permeability parameters are varied to demonstrate that these parameters are isolated from each other. In these figures, ε_r and μ_r of the sample are varied between 6-8 and 1-3 respectively while both electric and magnetic loss tangent of the sample are varied from 0.01 to 0.03. As shown in Fig. 3, magnetic properties are not sensitive to the resonant frequency and quality factor when the sample is located at the E-field maximum position. Similarly electric properties of the sample are not sensitive to changes in the cavity response when the sample is located at the H-field maximum position as shown in Fig. 4.

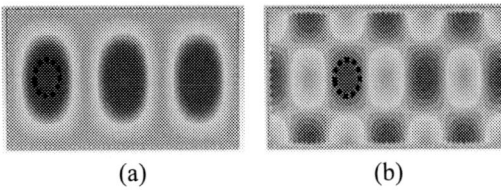

Figure 2. Field distribution of TE_{103} mode SIW cavity (a) E-Field, (b) H-Field (Circle shows E and H-Field maximum).

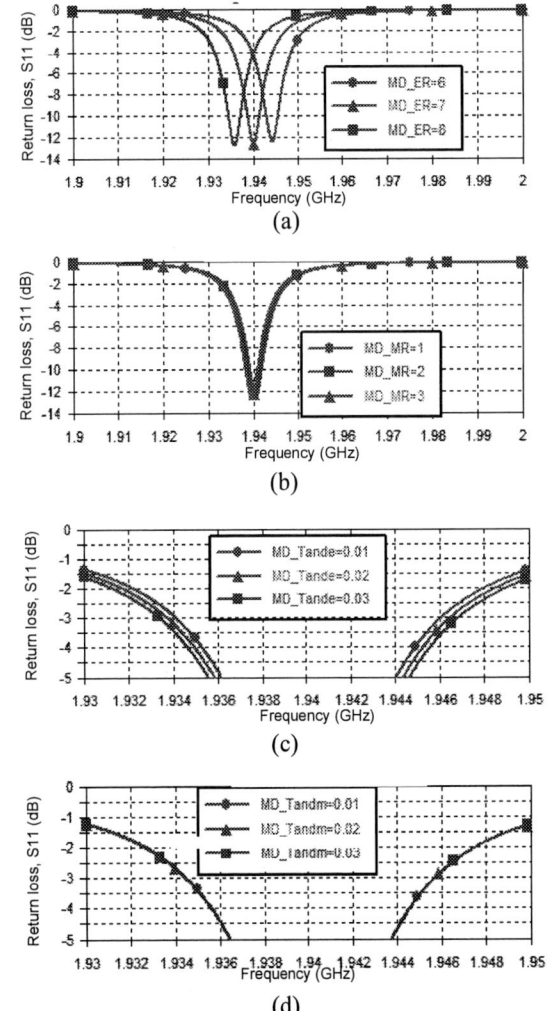

Figure 3. Parameter sweep when the sample is at E-field maximum position: (a) ε_r, (b) μ_r, (c) electric loss tangent, (d) magnetic loss tangent.

978-1-4799-2408-0/14 $31.00 © 2014 IEEE

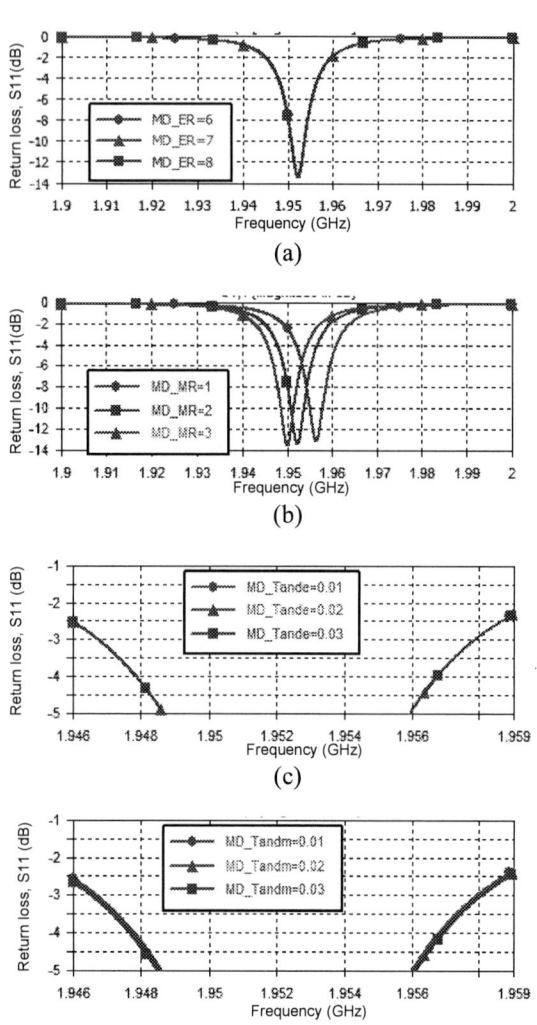

(a)

(b)

(c)

(d)

Figure 4. Parameter sweep when the sample is at H-field maximum position: (a) ε_r, (b) μ_r, (c) electric loss tangent, (d) magnetic loss tangent.

Fabrication of SIW Cavities

In order to validate the simulation, FR4 material was used to fabricate SIW cavities in [9], Rogers 3003 material which has parameters ε_r=3.0, tanδ=0.0013 (electric loss tangent) has been used instead in this study. Thickness of the cavity is 1.524mm. Since Rogers 3003 has less dielectric loss than FR4 material, cavity with Rogers 3003 has higher Q factor than cavity with FR4. This advantage can give better accuracy for loss tangent extraction. Seven SIW cavities were designed and fabricated to characterize the magneto-dielectric sample in the frequency range 1-4 GHz, as shown in Fig. 5. Table 1 shows the SIW cavity operating mode, resonant frequency and dimension. Fig. 6 shows the fabricated SIW cavity which is resonating at 3 GHz with drilled hole for sample insertion. Dimension of the hole is 6 x 6mm². SIW cavity in Fig. 6 (a) has a hole at the E-field maximum position and this cavity was used for permittivity measurement. Another SIW cavity in Fig. 6 (b) has a hole at the H-field maximum position and it was used for permeability measurement.

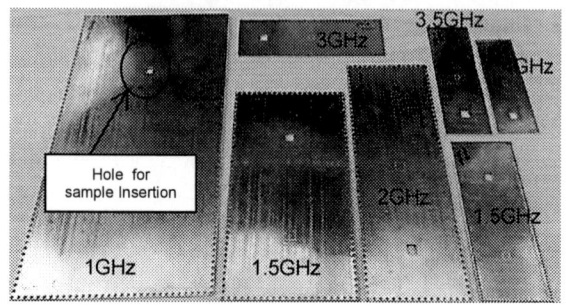

Figure 5. Fabricated SIW cavities with different resonant frequency

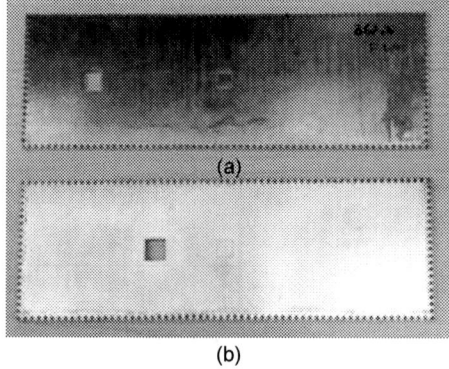

(a)

(b)

Figure 6. SIW cavity with different holes location for (a) permittivity (b) permeability measurements.

Table 1. SIW cavity specification

Resonance (GHz)	Operation Mode	W (mm)	L (mm)
1	TE$_{102}$	242	126
1.5	TE$_{102}$	162	86
2	TE$_{103}$	190	66
2.5	TE$_{102}$	117	45
3	TE$_{103}$	123	41
3.5	TE$_{103}$	109	35
4	TE$_{103}$	93	31

Magneto-dielectric Material Synthesis

The magneto-dielectric material used in this study was synthesized using high volume loading (50-70 vol %) of cobalt nano-particles in a dielectric polymer matrix. [10], where it shown in Fig 7 (a). Higher permeability at high frequencies can be achieved by reducing the metal particle size and the separation between adjacent metal particles down to the nano-scale [11]. The partially-oxide-passivated cobalt nanoparticles were commercially obtained from US Nanomaterial as powders while the polymer was obtained from Asahi Inc. The hard metal aggregates of cobalt were broken down to their primary particle sizes of ~20-30 nm using ball-milling process. As-received metal powders were suspended in anhydrous toluene solvent and milled for 10-15 hours with zirconia balls to break the aggregates. The dielectric polymer was then added to the suspension and

978-1-4799-2408-0/14 $31.00 © 2014 IEEE   784

milled again for 4-6 hours to ensure complete homogenization of the polymer and the metal particles. The final polymer-metal slurry was dried into a powder at 80°C for 30 minutes in a nitrogen atmosphere.

The dried metal-polymer composite powder was compacted using a mechanical hydraulic press in varied shapes and sizes. For the high-frequency measurements, the compacts were made with 6 x 6 mm² stainless steel mold (shown in Fig. 7(b)), while the thicknesses were maintained close to 1.5 mm to match the SIW cavity substrate height. A high mechanical load of 2 Tons/cm² was applied on the 6x6 mm² mold, to ensure good packing density is achieved in the pressed compacts. The compacts were then thermally treated in inert atmosphere at 250°C for 90 minutes to cure the fluoropolymer matrix. A picture of the compacted pellets is shown in Fig. 7(a).

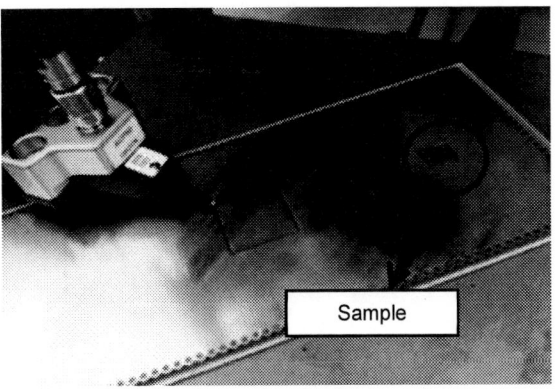

Figure 8. Corner-to-corner probing method.

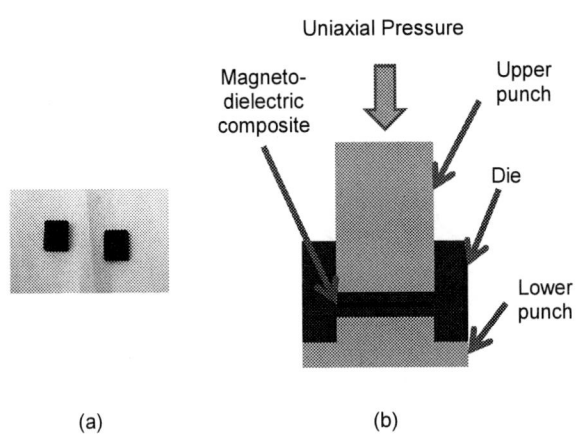

Figure 7. (a) Magneto-dielectric composite samples, (b) Schematic of the compaction set-up.

Magneto-dielectric Material Characterization

The magneto-dielectric composite samples were characterized in the range 1-4 GHz. Rogers dielectric material RO4360, TMM 6 and TMM 10 were chosen as the standard sample to obtain the constants A and B in equations (1) and (2) for each frequency. Similarly, Cuming Microwave FLX-10 magnetic absorber material was used as the standard sample for obtaining the constants C and D in equations (3) – (4) for each frequency. The samples were prepared in a hexahedron form with dimension of 6x6x1.524mm³ and were inserted into the hole machined in the cavity. The cavity response was measured with a VNA using SOLT calibration. Corner-to-corner probing method is shown in Fig. 8. GSG 500 probe was used to excite the SIW cavity. As shown in Fig. 8, samples were inserted in the hole and copper tape was used to cover the top and bottom of the hole. Fig. 9 shows the measured response of the 2GHz SIW cavity, with the various samples inserted in the cavity. From Fig. 9, the samples with higher permittivity or permeability values have lower resonant frequencies and samples with higher loss tangent show wider 3dB bandwidth corresponding to a lower Q factor.

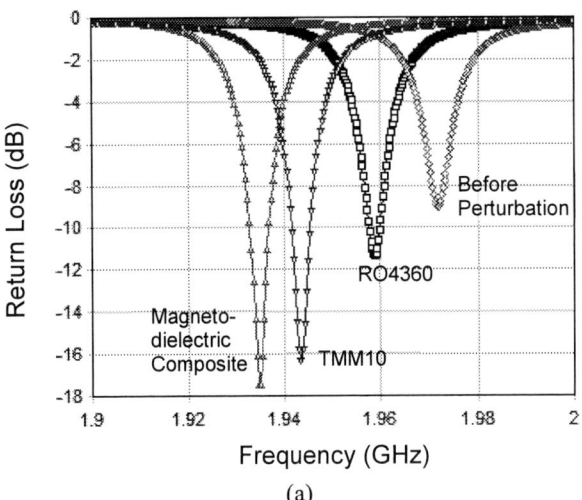

(a)

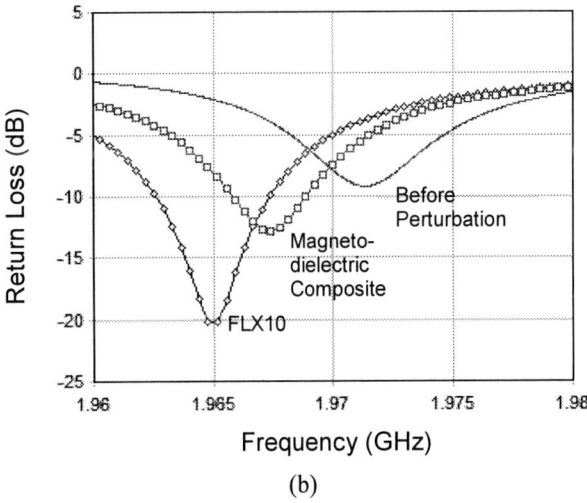

(b)

Figure 9. SIW cavity measurement with various sample materials (a) @ E-field maximum and (b) @ H-field maximum.

Fig. 10 shows the extracted properties of the magneto-dielectric composite material using the seven SIW cavities provided in Table I. In Fig. 10 (a) and (b), the relative permeability μ_r' gradually decreases as frequency increases

978-1-4799-2408-0/14 $31.00 © 2014 IEEE

while ε_r' is fairly constant. The value of the extracted relative permittivity (ε_r') is 12±0.5 and relative permeability (μ_r') was 1.9±0.2 in the frequency range 1-4 GHz. Electric loss tangent is 0.0035±0.001. Magnetic loss tangent increases with frequency increases. It is 0.0614 at 1GHz and increases to 0.48 at 4GHz.

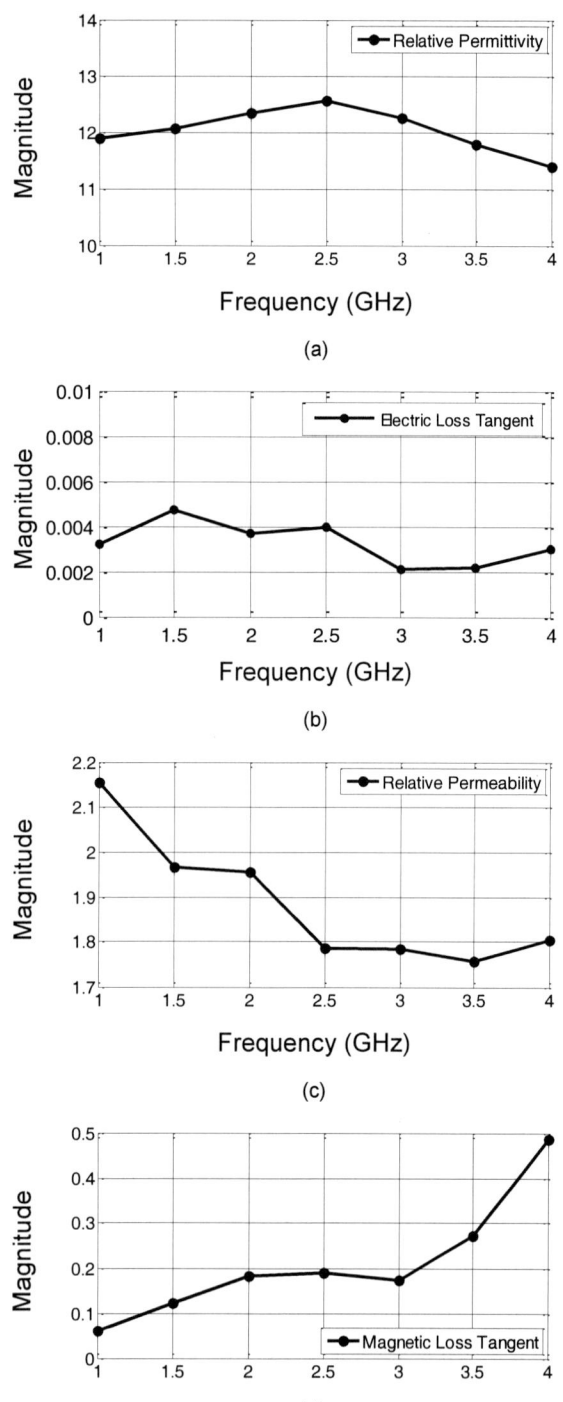

Figure 10. Extracted Cobalt nano particle composite material properties (a) Permittivity, (b) Electric loss tangent, (c) Permeability and (d) Magnetic loss tangent.

PIFA on magneto-dielectric substrate

A 1 GHz Planar Inverted-F Antenna (PIFA) on magneto-dielectric substrate was designed using CST and the material properties from previous section was used for the design, as shown in Fig. 11. The relative permittivity and permeability of the magneto-dielectric used was 11.9 and 2.153 at 1GHz respectively. Also electric loss tangent and magnetic loss tangent used was 0.0032 and 0.0614 at 1GHz respectively. These values are extracted material properties at 1 GHz. As shown in Fig. 11, ground plane of the antenna is supported by FR4 material and magneto-dielectric material was used as substrate for PIFA. Size of the magneto-dielectric substrate is 20 x 20mm². Height of FR4 and magneto-dielectric substrate is 1 mm and 1.5 mm respectively. Size of the FR4 material is 60 x 120mm². Actual dimension of two patterned conductors is shown in Fig. 12. U shaped top plane is connected to the ground plane with 2mm width shorting pin at top left corner of the antenna. Distance between the shorting pin and port is 4.8mm and it is optimized for good matching. This antenna showed 9.73% bandwidth and 72.11% efficiency as shown in Fig. 13.

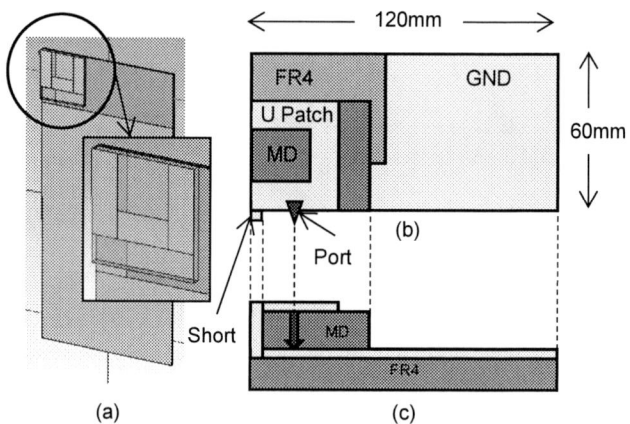

Figure 11. Planar inverted-F antenna with magneto-dielectric material (a) Perspective view, (b) top view, (c) side view.

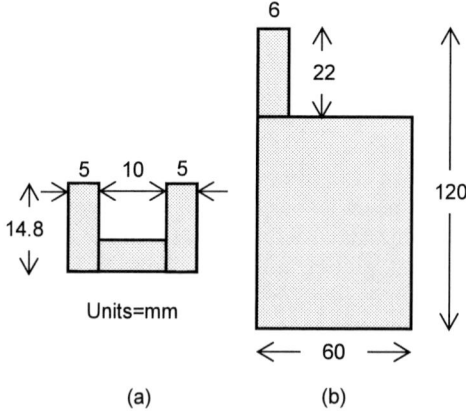

Figure 12. Dimension of patterned conductor (a) U-shaped patch, (b) Ground plane.

Dielectric substrate with ε_r=21, tan δ=0.0646 and the same thickness h=1.5 mm was also designed and simulated to

978-1-4799-2408-0/14 $31.00 © 2014 IEEE

compare antenna performance between the magneto-dielectric substrate and the high dielectric constant substrate. Dielectric constant of 21 was chosen to maintain the same antenna size and resonant frequency. For comparison, electric loss tangent of 0.0646 was chosen for the high dielectric constant material to equal the total loss of magneto-dielectric material. The high dielectric constant substrate antenna showed 8.62% bandwidth and lower efficiency of 51.6% as shown in Fig. 13. PIFA with the magneto-dielectric material substrate shows better performance for both bandwidth and efficiency than the antenna with high dielectric constant material substrate. It shows that the magneto-dielectric material is an effective material for antenna miniaturization. Losses of magneto-dielectric substrate effect on antenna performance have also been analyzed using EM simulation. Without changing the values of real permittivity and real permeability, if the magnetic loss tangent was reduced to 0.0307, the antenna efficiency increased to 82.5% and the bandwidth decreased a little to 8.82% as shown in Fig. 13. This narrow bandwidth can be compensated using patch design optimization. Therefore, if the loss of the magneto-dielectric material is reduced and patch design optimized, antenna performance can be improved further.

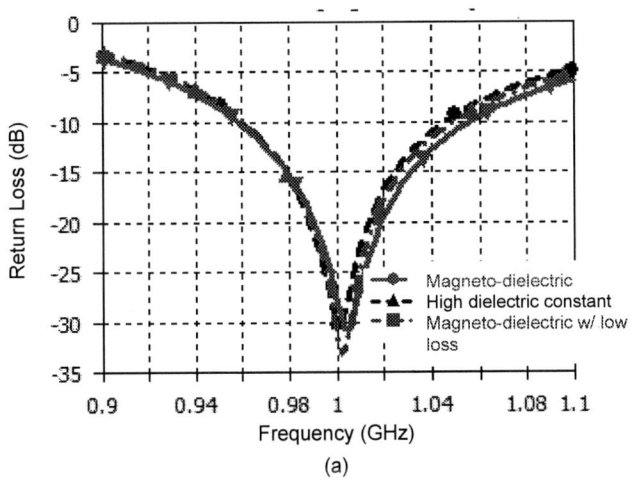

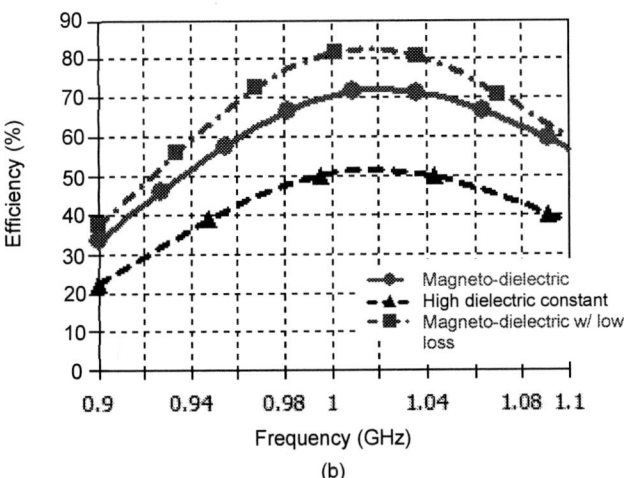

Figure 13. Antenna performance of PIFA with different substrates (a) Return loss, (b) Efficiency.

Conclusions

In this paper, CPT with SIW cavity resonators for magneto-dielectric material is discussed and demonstrated with simulation and measurement. The novelty arises in the use of a single SIW cavity structure to extract both properties, which otherwise would require two separate structures. Simulation and experiment shows that magnetic properties of the sample do not affect electric properties extraction and vice versa. Magneto-dielectric composite material which is synthesized with cobalt metal particles has been measured with this method. PIFA on magneto-dielectric material with extracted properties has been designed using CST. The magneto-dielectric substrate antenna shows better bandwidth and efficiency as compared to using the high dielectric constant material substrate. If the total loss of magneto-dielectric material is reduced using enhanced material synthesis technique, antenna performance can be improved further. In conclusion, effective antenna miniaturization can be achieved by using the magneto-dielectric material.

References

1. C. A. Balanis, *Antenna Theory: Analysis and Design*, Wiley, New York, 1997, pp. 812-813.
2. J. S. Colburn and Y. Rahmat-Samii, "Patch antennas on externally perforated high dielectric constant substrates," *IEEE Trans. Antennas Propag.*, vol. 47, no. 12, pp.1785-1784, 1999.
3. P. M. T. Ikonen and S. A. Tretyakov, "On the advantages of magnetic materials in microstrip antenna miniaturization," *Microwave Opt. Technol. Lett.*, vol. 52, pp. 3131-3134, 2008.
4. N. Altunyult, M. Swaminathan, P. M. Raj, V. Nair, "Antenna miniaturization using magneto-dielectric substrates," *in Proc. IEEE Electronic Components and Technol. Conf. (ECTC)*, May 2009, pp. 801-808.
5. K. Han, M. Swaminathna, P. M. Raj, H. Sharma, K. P. Murali, R. Tummala, V. Nair, "Extraction of Electricl Properties of Nanomagnetic Materials through Meander-Shaped Inductor and Inverted-F Antenna Structures," *in Proc. IEEE Electronic components and Technol. Comf. (ECTC)*, 2012, pp.1808-1813.
6. L. F. Chen, C. K. Ong, C. P. Neo, V. V. Varadan, V. K. Varada, *Microwave Electronics: Measurement and Materials Characterization*, Wiley, New York, 2004, pp. 250-286.
7. H. Lobato-Morales, A. Corona-Chavex, D. V. B. Murthy, J. L. Olvera-Cervantes, "Complex permittivity measurements using cavity perturbation technique with substrate integrated waveguide cavities," *Review of Scientific Instruments*, 81.6, 2010.
8. K. Saeed, R. D. Pollard, I. C. Hunter, "Substrate integrated waveguide cavity resonators for complex permittivity characterization of materials," *IEEE Trans. Microwave Theory and Techniques*, vol. 56, no. 10, pp.2340-2347 2008.
9. K. Han, M. Swaminathan, P. M. Raj, H. Sharma, R. Tummala, V. Nair, "Magneto-dielectric material characterization and antenna design for RF applications," *European Conference on Antenna Propagation (EuCAP)*, 2014, paper accepted.

10. P. M. Raj, H. Sharma, D. Mishra, K. P. Murali, K. Han, M. Swaminathan, R. Tummala, "Nanomagnetics for high-performance, miniaturized power, and RF components," *IEEE Nanotechnology magazine*, vol. 6, no. 3, pp.18-23 2012.

11. N. Tang, W. Zhong, X. Wu, H. Jiang, W. Liu, and Y. Du, "Synthesis and complex permeability of Co/SiO_2 nanocomposites," *Matter. Lett.*, vol. 59, no. 14-15, pp.1723-1726, 2005.

Flexible Liquid Crystal Polymer Based Complementary Split Ring Resonator Loaded Quarter Mode Substrate Integrated Waveguide Filters for Compact and Wearable Broadband RF Applications

David E. Senior[1,2], Arian Rahimi[2], Pitfee Jao[2], Yong-Kyu Yoon[2]

[1] Department of Electrical and Electronics Engineering, Universidad Tecnológica de Bolívar

Parque Industrial y Tecnológico Carlos Vélez Pombo, Ternera, Km. 1, Cartagena, Colombia

[2]Department of Electrical and Computer Engineering, University of Florida

225 Larsen Hall, Gainesville, FL, USA

E-mail :dsenior@unitecnologica.edu.co, ykyoon@ece.ufl.edu

Abstract

In this paper the flexible Liquid Crystal Polymer (LCP) substrate is used to implement broadband wearable/foldable conformal bandpass filters that use compact cavity resonators working under the principle of quarter mode substrate integrated waveguide (QMSIW), which features a 75% size reduction with respect to the conventional substrate integrated waveguide (SIW) counterpart. Further size reduction is realized with the use of a complementary split ring resonator (CSRR) metamaterial unit cell integrated with the QMSIW architecture. The resulting CSRR-loaded QMSIW cavity has its main resonance frequency below the quasi-$TE_{0.5,0,0.5}$ resonance mode of the original QMSIW cavity due to the evanescent wave amplification phenomenon with CSRR loading. A low temperature surface micromachining process on the LCP and mechanical drilling of via holes are used for fabrication. The realized CSRR-loaded QMSIW cavity features a moderate quality factor (Q) that makes it useful for the design of bandpass filters with much broader fractional bandwidth (FBW) when compared to those using conventional SIW cavities. A 2nd order and a 3rd order surface micromachined Chebyshev BPFs are demonstrated for operation at a center frequency of 25.5 GHz. More than 11% FBW with an in-band return loss of better than 20 dB and an insertion loss of less than 1.5 dB, including transitions, are obtained for both filters. Theoretical analysis of the working principle is explained. Measured results are in good agreement with the 3D full wave structure simulations.

Introduction

Demand for compact wearable/foldable and conformal wireless devices continuously grows as the health, sports, fitness, industry and automotive products equipped with wireless connectivity at the GHz range become prevailing in the market [1-3]. The last decade has witnessed a great demand for miniaturized high performance wireless communication devices that can be easily integrated in systems on package (SoP), systems on chip (SoC), wireless sensor networks technologies and wearable/conformal applications [4-7]. Broadband operation and size reduction for SoC or SoP platforms have become the main targets for these applications. After the authorization in 2002 by the Federal Communications Commission (FCC) to use the 3.1-10.6 GHz band for ultrawideband (UWB) communications and the 22-29 GHz band for automotive radar systems [8] under no license requirements, many works have demonstrated broadband bandpass filters (BPFs) and antennas operating within the entire frequency range from 3 to 10.6 GHz, or

having a bandwidth within that range [9-11]. In contrast, few quasi-millimeter or millimeter wave filters have been reported for broadband operation, mainly due to fabrication constraints and bandwidth limitations of the selected electromagnetic structures [2, 12, 13]. On the other hand, modern wireless systems, such as ground penetrating radar (GPR), automotive radar systems, UWB body area networks, high-resolution sensing and imaging devices, have a great demand for broader bandwidth to support multiple band communication systems, including those working at K-Ku frequency bands [2, 13-16]. Bandpass filters with low insertion loss and compact sizes are the key components for those modern wireless communication systems [17]. Conventionally, planar broadband BPFs have been implemented with microstrip or coplanar waveguide (CPW) technology using a great variety of resonators such as stub loaded ring resonators, multimode resonators and stepped impedance resonators [17,19]. Special attention has been given to the use of standard rigid printed circuit board (PCB) and low temperature co-fired ceramic (LTCC) processes [17, 20, 21] for BPFs designs. However, LTCC is expensive and does not offer the targeted flexibility many wearable/conformal applications of SoC or SoP must have. During the last decade, the organic Liquid Crystal Polymer (LCP) has emerged as a low cost, low processing temperature alternative substrate for implementing highly integrated planar, flexible and multilayer radiofrequency (RF) devices; mainly due its unique electrical (ε_r = 2.9 to 3, tan δ < 0.004), thermal and mechanical properties [22,23,24]. Microstrip and coplanar waveguide transmission lines technologies have been used as the preferred waveguiding structures for such implementations. However, conformal devices are subject to the deformation along the non flat surface of the wearers or the particular application, and thus the flexibility of the devices and associated components is required for wearer's comfort and satisfaction [1]. Meanwhile, such deforming activities result in frequent system/device failure due to disassembly and disconnect between the components and the substrate. To minimize such a failure mode, substrate integrated device architectures for RF devices are highly desirable [25]. In this work, the substrate integrated waveguide (SIW) architecture in/on a flexible liquid crystal polymer (LCP) substrate is explored for wearable RF component applications [26]. For compact device realization, an advanced complementary split ring resonator (CSRR) metamaterial unit cell is integrated with the SIW architecture [27]. Furthermore, quarter mode SIW (QMSIW) is exploited instead of a conventional SIW or half mode substrate integrated waveguide (HMSIW) [28, 29]. It is noticed that the

978-1-4799-2408-0/14 $31.00 © 2014 IEEE

2014 Electronic Components & Technology Conference

QMSIW based filters offer much broader bandwidths compared with the SIW and HMSIW counterparts because of the nature of their lower Q-factors and the higher values of magnetic coupling between neighboring resonators, which is easily controlled by changing the opening window size of the coupling via walls [28].

As a test vehicle, CSRR-loaded QMSIW broadband bandpass filters for compact and wearable RF applications are demonstrated. A surface micromachining process on the flexible LCP substrate is used for fabrication. The working principle and frequency characterization of the proposed CSRR-loaded cavity are also studied.

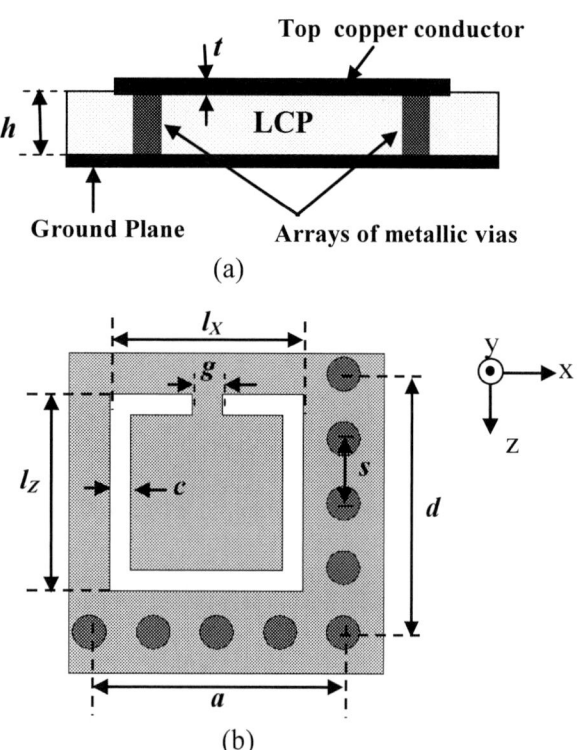

Figure 1. (a) Cross section of SIW devices on LCP substrate. (b) The layout of the proposed single-ring CSRR-loaded quarter mode substrate integrated waveguide (QMSIW) cavity on LCP. Metallization is shown in gray.

Topology and the CSRR-loaded QMSIW Cavity

Double clad liquid crystal polymer (LCP) laminate Ultralam 3850 (ε_r = 2.9, tan δ = 0.0025 at 10 GHz) from Rogers Corporation with a thickness of 100 μm and a metal thickness of 18 μm (½ oz. of Copper) is selected as the substrate to implement the CSRR-loaded QMSIW bandpass filters. One of the copper (Cu) layers of the laminate is etched off as the first step to implement the proposed devices. The remaining copper layer is used as the ground plane. Then, the surface micromachining process is used to implement the signal layer for the fabrication of the BPFs on the LCP substrate. Figure 1(a) shows the general cross section of substrate integrated waveguide devices on the LCP substrate. To keep the original geometry of the LCP layers, the thickness of the deposited top copper layer is kept to be 18 μm as well.

Next, Figure 1(b) shows the layout of the proposed quarter mode SIW cavity loaded with a complementary split ring resonator (CSRR-loaded QMSIW). A single-ring CSRR is selected for this work; however, double ring CSRR can also be used. Arrays of metalized vias with a diameter of 0.2 mm and a center-to-center spacing s of 0.3125 mm are used to create the metallic walls for the QMSIW cavity.

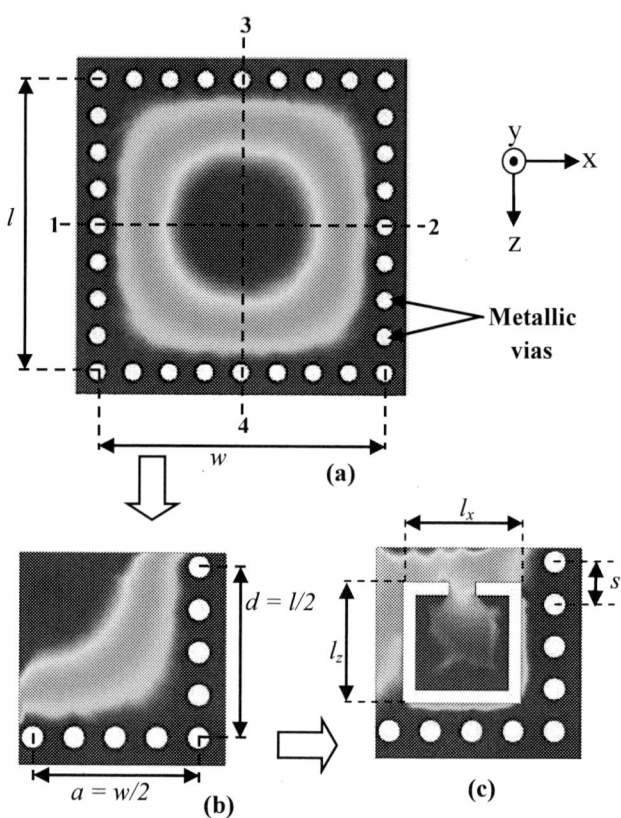

Figure 2. Electric field distributions for SIW cavities on the LCP substrate. (a) E-Field for the quasi-TE_{101} mode in a conventional SIW cavity (b) E-Field for the Quasi-$TE_{0.5,0,0.5}$ mode in the resulting QMSIW cavity at 48 GHz, (c) E-Field for the main resonance mode at 25.5 GHz in the QMSIW cavity loaded with a single ring CSRR. Geometrical parameters are w = l = 2.7 mm, l_X = l_Z = 0.95 mm, s = 0.3125 mm, c = 0.1 mm, g = 0.18 mm.

Resonator Study

Figure 2 presents the comparison of the electric field distribution for the conventional SIW cavity, the original QMSIW cavity and the proposed CSRR-loaded QMSIW cavity in this work. Eigenmode simulation using High Frequency Structure Simulator (HFSS 13, Ansys. Inc.) is performed here to obtain the resonance modes of the cavities. In Figure 2(a), the electric field distribution of the quasi-TE_{101} mode in a conventional substrate integrated waveguide cavity is presented. The cavity is designed and optimized on the LCP substrate to have a resonance frequency of 48 GHz. As observed, the resonance mode resembles the TE_{101} mode of a conventional rectangular waveguide cavity [30]. Arrays of metalized via holes with a spacing s of 0.3125 mm are used as the side walls of the waveguide cavity. The resonance frequency of the quasi-TE_{101} mode can be approximated with the formula for the resonance frequency of the TE_{101} mode in a conventional rectangular waveguide cavity [30].

Next, taking into account that for the SIW architecture to be modified to a reduced mode version, such as the half mode or the quarter mode SIW structures, the thickness of the substrate h should be much smaller than the width w of the waveguide [28, 29]. For this particular case, the height h of the substrate is much smaller than the width w of the cavity ($h \ll w$). Then, two fictitious magnetic walls 1-2 and 3-4 allow splitting the SIW into four independent square cavity resonators [26, 29]. Each cavity represents one-quarter of the original SIW cavity, or a 75% size reduction, and then it is named the quarter-mode substrate integrated waveguide cavity (QMSIW) [28]. As observed in Figure 2(b), the QMSIW propagates one-quarter of the quasi-TE_{101} mode in the original SIW cavity, called here the QMSIW quasi-$TE_{0.5,0,0.5}$ mode.

In this work, the size of the cavity is further reduced by loading a single ring CSRR on the QMSIW structure. Figure 2(c) shows the E-field distribution of the first propagating mode in the CSRR-loaded QMSIW cavity. As observed, the propagating mode in the new cavity is a result of the effect of loading the CSRR in the waveguide structure. The electric field is more confined around the CSRR than at the corner of the QMSIW cavity, indicating that the CSRR interacts with the waveguide structure to create a new propagating mode.

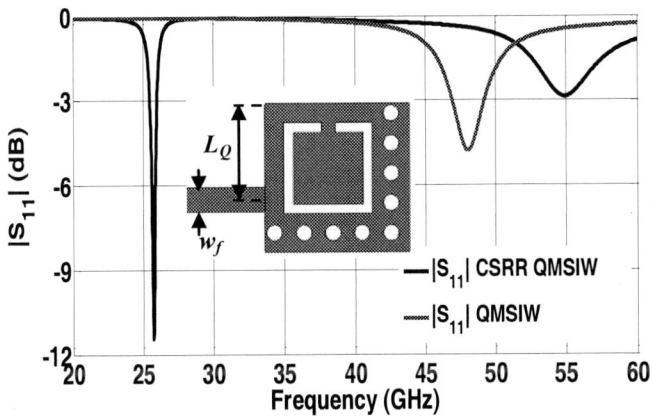

Figure 3. Frequency responses of the conventional QMSIW and the CSRR-loaded QMSIW cavities. Geometrical parameters are $L_Q = 1$ mm, $w_f = 0.26$ mm. A microstrip 50 Ω feeding line is used to excite the cavity.

To confirm the effect the CSRR loading has in the resonance frequency, Figure 3 shows the comparison of simulated frequency responses of the original QMSIW and the new CSRR-loaded QMSIW cavities. A direct connection of a 50 Ω microstrip line is used to excite both the QMSIW and the CSRR-loaded QMSIW cavities, as observed in the inset of Figure 3. Full wave structure simulations are performed in HFSS 13. As observed, the original QMSIW cavity resonates around 48 GHz, as expected. On the other hand, it is clear that the CSRR-loaded QMSIW cavity has two resonant modes with completely different frequency values when compared with the previous cavity. The first resonance frequency of the CSRR-loaded QMSIW cavity is around 25.5 GHz, which is due to the interaction of the CSRR with the QMSIW cavity,

while the second mode is observed at 54 GHz. We suspect the quasi-$TE_{0.5,0,0.5}$ mode of the original QMSIW cavity has been pushed up to 54 GHz as a result of the CSRR loading, however, more studies should be done in order to confirm it.

Further, as observed in Figure 3, the resonance frequency of the CSRR-loaded QMSIW cavity has been reduced from 48 GHz to 25.5 GHz, which represents 47% reduction. Then, the CSRR-loaded QMSIW cavity offers an additional size reduction with respect to its conventional SIW counterpart, which in this case is nearly half of the size of the original QMSIW cavity and around 13% of the size of the conventional SIW cavity. In terms of the guide wavelength λ_g at 25.5 GHz, the size of the proposed CSRR-loaded QMSIW cavity on LCP is $0.1954\lambda_g \times 0.1954\lambda_g$.

From the previous analysis, it is concluded that the CSRR-loaded QMSIW cavity represents a new step toward the miniaturization of bandpass filters. The working principle is based on the effect the negative permittivity CSRR has on the waveguiding structure, which in this case allows evanescent wave amplification below the quasi-$TE_{0.5,0,0.5}$ mode resonance frequency of the original QMSIW [26, 27]. Further analysis is suggested to better explain the interaction of the CSRR with the QMSIW cavity and the generation of a forward-wave propagation band below the main resonant mode of the original QMSIW cavity.

Bandpass Filter Design

As test vehicles to demonstrate the use of the CSRR-loaded QMSIW cavity, a two-pole and a three-pole Chebyshev BPFs are designed for operation at a center frequency of 25.5 GHz with a 20 dB return loss FBW of more than 11%. It is important to follow the classical coupled resonator design methodology to achieve the desired results. The required external quality factor Q_e of the CSRR-loaded QMSIW cavity and the coupling coefficient K_{ij} between neighboring cavities for the design of a bandpass filter are first obtained from the low-pass prototype filter [17]. Next, the sizes of the resonators are optimized in order to fulfill the frequency and bandwidth requirements. The variation of the external quality factor and the internal coupling coefficients are next investigated.

External Quality Factor Study

The variation of the external or loaded quality factor (Q_e) of the CSRR-loaded QMSIW cavity is obtained by exciting the cavity with one of the open sides, as shown in the inset of Figure 3. Since the first propagating mode, in which we are interested, is completely different from a quasi-TE_{101} mode in an SIW cavity, direct connection of a microstrip 50 Ω transmission line is used to excite the cavity. The total length of the single ring CSRR is optimized to be 3.65 mm, which is initially selected as half guided wavelength on the LCP substrate at the resonance frequency of 25.5 GHz (3.45 mm).

The distance L_Q is used to vary the external quality factor, which represents the distance from the top corner of the cavity to the center point of the feeding line. Figure 4 shows the available external quality factors Q_e for a variation of the distance L_Q from 0.2 mm to 0.9 mm. In comparison with a conventional SIW cavity, the obtained external quality factors Q_e are smaller since the size of the CSRR-loaded QMSIW cavity is also smaller [28]. A cavity with small quality factor is useful for the design of bandpass filters with broad bandwidth [28]. In this case, a Q_e as low as 5.2 has been obtained, which is not easy to get in conventional SIW or even HMSIW cavities [26, 28]. On the other hand, the unloaded quality factor Q_u obtained from the Eigenmode simulations is 141.62, which is in the same range of those obtained for conventional QMSIW cavities [28].

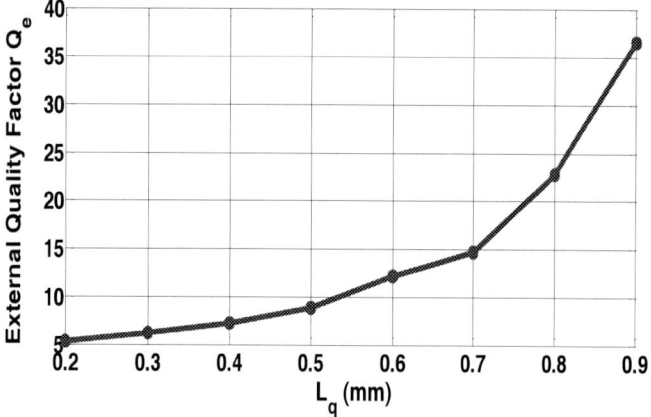

Figure 4. External quality factor Q_e variation with the offset distance L_Q for the CSRR-loaded QMSIW cavity.

Coupling Coefficients Study

Next, the coupling coefficient (K_{ij}) variation between magnetically coupled CSRR-loaded QMSIW cavities is studied. Figure 5(a) shows the configuration that is used for this purpose. Here, instead of using extensive Eigenmode simulations to obtain the coupling coefficient [27], the cavities are excited with a high external quality factor Q_e in order to have a weak coupling from the source [17]. An opening of width W_c in the metalized via row is used to efficiently adjust the internal coupling coefficient, as shown in Figure 5(a). Figure 5(b) shows the obtained coupling coefficient K_{ij} for a variation of the opening W_c up to 1 mm. As observed, coupling coefficient values higher than 0.2 are possible, which is not easy to obtain with conventional SIW cavities. The achievable high values of the coupling coefficient indicate that the CSRR-loaded QMSIW cavity is also useful for broadband filter applications as the conventional QMSIW cavity is [28].

Filter Designs

For demonstration purposes, two pole and three pole Chevyshev bandpass filters are designed for a center frequency of 25.5 GHz. The classic methodology for coupled resonator filter design is followed [17]. Table 1 summarizes the calculated design parameters following the methodology. Figure 6 shows the layouts of the two and three pole filters. The previous resonator study gives the initial geometrical parameters, which are further optimized through full wave structure simulations in HFSS 13. Grounded coplanar waveguide (GCPW) launches are included to test the filters with a Cascade Microtech probe station with GSG (ground-signal-ground) probes and a pitch distance of 150 μm. In Figure 6(b), the CSRR of the second cavity resonator (middle) of the three pole filter is scaled up with a factor of 1.122 in order to optimize the frequency response.

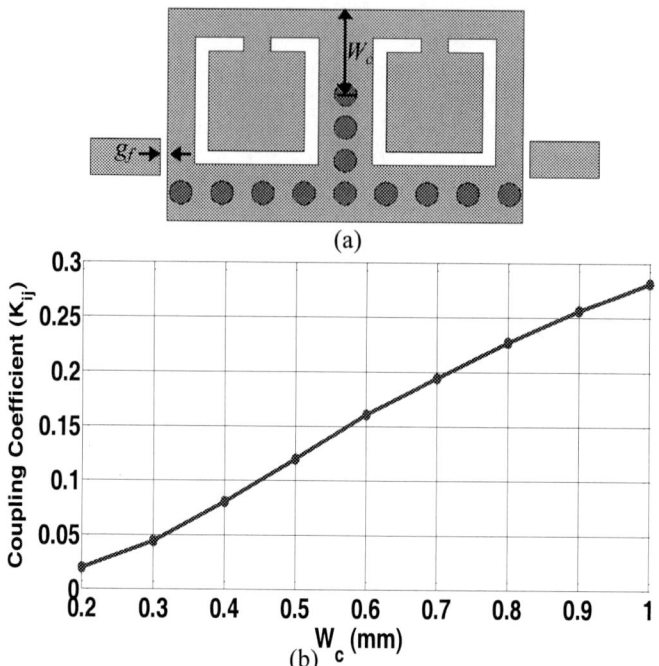

Figure 5. Variation of the internal coupling coefficient (K_{ij}) of coupled cavities. (a) Configuration for the simulation. The opening W_c in the via row varies K_{ij}. High Q_e is achieved with a feeding gap g_f of 0.05mm. (b) Simulated results.

TABLE 1. Design parameters of the filters on LCP

Filter	f_0(GHz)	20 dB RL FBW	Q_e	K_{12}	K_{23}
Two poles	25.5	11%	6.044	0.1828	-
Three poles	25.5	16%	5.320	0.1650	0.165

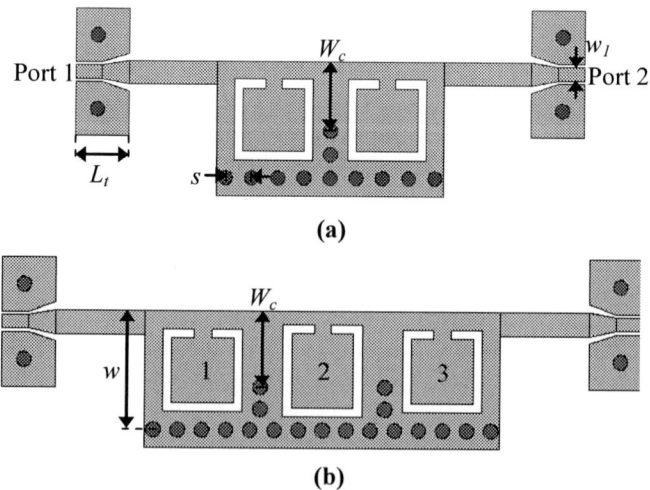

Figure 6. Layouts of the CSRR-loaded QMSIW BPFs. (a) Two-pole filter. Dimensions are W_c = 0.51 mm, w_1 = 35 μm, w = 1.3 mm, $l_x = l_z$ = 0.9 mm, L_q = 0.13 mm, L_f = 1 mm, s = 0.3 mm. (b) Three pole filter. Dimensions are the same except for W_c = 0.6 mm. Dimension l_z of the CSRR in resonator 2 is 1.01 mm.

978-1-4799-2408-0/14 $31.00 © 2014 IEEE

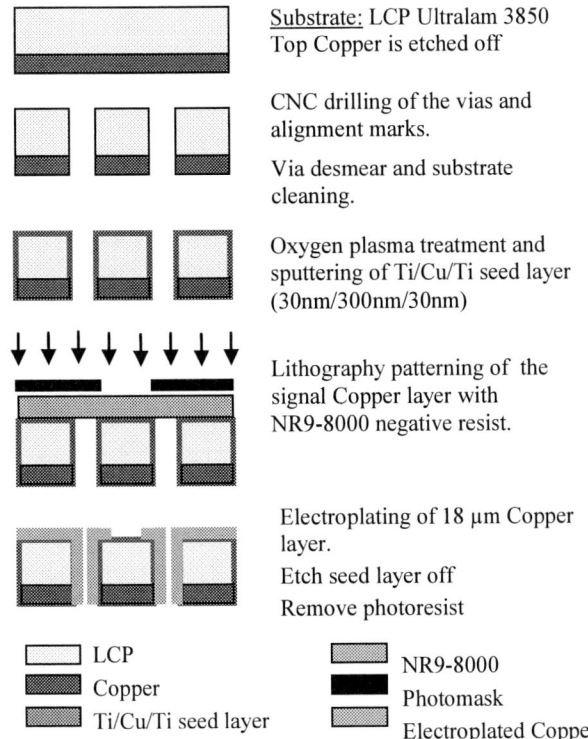

Substrate: LCP Ultralam 3850
Top Copper is etched off

CNC drilling of the vias and
alignment marks.

Via desmear and substrate
cleaning.

Oxygen plasma treatment and
sputtering of Ti/Cu/Ti seed layer
(30nm/300nm/30nm)

Lithography patterning of the
signal Copper layer with
NR9-8000 negative resist.

Electroplating of 18 μm Copper
layer.
Etch seed layer off
Remove photoresist

▢ LCP
▣ Copper
▣ Ti/Cu/Ti seed layer

▣ NR9-8000
▣ Photomask
▣ Electroplated Copper

Figure 7. Fabrication procedure.

Fabrication of the Filters on the LCP

In this work, a surface micromachined process is used for the fabrication of the filters. Standard PCB fabrication is not selected due to the small geometrical parameters in the design, for instance, the gap of the CPW launches (35 μm). Fabrication based on wet chemical etching is avoided to eliminate possible errors due to copper undercut. Standard UV lithography and copper electroplating are used to fabricate the patterns of the filters.

Figure 7 shows the fabrication procedure, which starts with the etching of the top copper layer of a double clad Ultralam 3850 LCP sheet by using standard wet chemical etching procedures. For this purpose, one layer of the LCP substrate is covered with photoresist AZ5214 and heated up to 100°C for one minute on the flat top surface of a hot plate. Then, the exposed copper layer is removed using H_2SO_4:H_2O_2:DI-water wet chemical etching (1:1:5 ratio mixture). Photoresist is removed, the substrate is cleaned and the via holes of 200 μm diameter are mechanically drilled using an automatic CNC milling machine. Further, cleaning of the substrate is performed in a heated ultrasonic bath using Acetone, Isopropanol and DI-water in a sequence, prior to the deposition of any metallic layer. Next, desmear of the vias is performed with an oxygen plasma treatment prior to metal deposition, which also improves the adhesion of metal layers to the LCP substrate, as recommended in [22]. Further, DC-sputtering of Ti/Cu/Ti (30 nm/300 nm/ 30 nm) seed layer is performed, which improves adhesion between the LCP and the electroplated layers with fine features. Next, the lithographical pattern of the top layer signal is realized with the use of a 20 μm thick negative resist NR9-8000 with recommended processing procedures (Futurrex Inc.). Further, copper electroplating is performed to deposit the signal layer

of the devices with a thickness of 18 μm. Finally, seed layer is removed and the devices are ready for measurements.

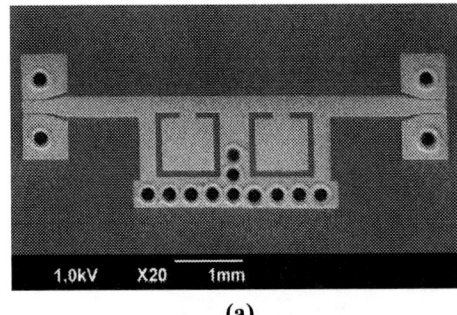

(a)

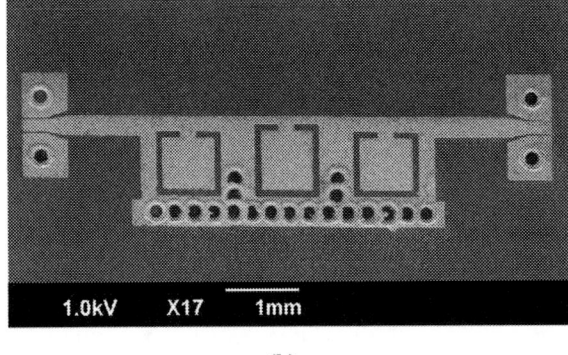

(b)

Figure 8. SEM images of the fabricated filters. (a) Two-pole filter. (b) Three-pole filter.

Fabricated Devices and Measurements

The scanning electron microscope (SEM) images of the fabricated filters are shown in Figure 8. The size of the two pole and thee pole filters are $0.22\lambda_0 \times 0.11\lambda_0$ and $0.33\lambda_0 \times 0.11\lambda_0$, respectively; where λ_0 is the free space wavelength at 25.5 GHz. Measurements are performed with a vector network analyzer Agilent E8361A (Agilent, Inc.) and a Cascade Microtech probe station with a probe pitch of 150 μm. Conventional SOLT (Short, Open, Load, Thru) calibration is done from 10 GHz to 40 GHz. No bending effects on the performance of the devices are reported at this time. The substrate is kept flat at all times.

Figure 9 presents the measured and simulated results. Also, the insets of Figure 9 show the microphotograph of the realized filters. Less than 1.5 dB insertion loss with an in-band return loss of better than 20 dB, including the transitions and feeding lines, are achieved in both cases. More than 11.6% 20-dB return loss FBW is obtained for the two-pole filter and a FBW of 16% for a return loss of more than 18 dB is obtained for the three pole BPF. Measured results are in good agreement with the 3D structure simulations in HFSS. A maximum shift in the frequency of 4% is observed for the three pole filter, which is attributed to the tolerance in the fabrication process, especially in the lithography and electroplating procedures, as well as the variation in the electrical properties of the LCP substrate. The results show that QMSIW cavities can be successfully fabricated with a repeatable surface micromachining process on low loss flexible/bendable organic substrates for wireless conformal applications. The demonstration of filters with basic configurations open the possibility of using the CSRR-loaded QMSIW cavity resonator and the same fabrication process for the implementation of higher order filters, which also include transmission zeros and sharper frequency responses [17]. Further demonstrations are left for a future work.

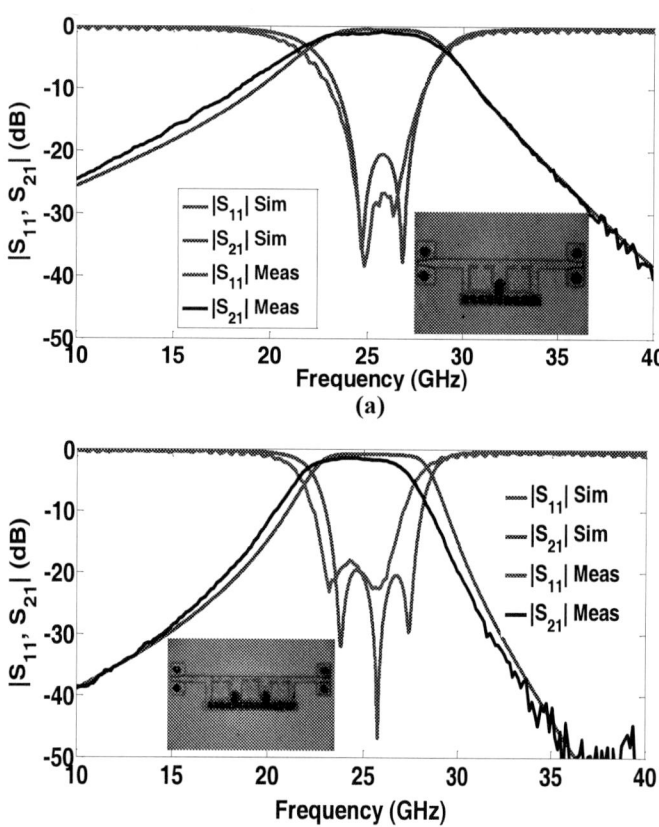

Figure 9. Simulated and measured results. (a) Two-pole filter. (b) Three-pole filter.

Conclusions

This work has proposed the use of the flexible liquid crystal polymer (LCP) substrate to implement quasi-millimeter wave broadband bandpass filters with SIW architecture for flexible/foldable conformal applications. For compact device realization and broadband operation, the QMSIW architecture has been combined with the negative permittivity metamaterial particle CSRR, which creates a new resonance frequency below the original resonant mode of a conventional QMSIW cavity. The variations of the external quality factor of the proposed resonator and the internal coupling coefficient of magnetically coupled cavities have been presented. Surface micromachined Chevyshev BPFs are demonstrated for 25.5 GHz with more than 11 % 20 dB RL FBW. Measurement results are obtained up to 40 GHz and show good agreement with simulated ones. The demonstrated devices feature great size reduction and substrate flexibility.

Acknowledgments

This work was supported in part by the National Science Foundation under Grant No. 1132413. David E. Senior is supported by Universidad Tecnológica de Bolívar in Cartagena, Colombia.

References

1. M.R. Yuce, C.K. Ho, S.C. Moo, "Wideband Communication for Implantable and Wearable Systems," *IEEE Trans. Microwave Theory and Techniques,* , vol.57, no.10, pp.2597,2604, Oct. 2009
2. M. Zhewang, M. Ohira, C.P. Chen, T. Anada, "A novel compact high-performance microstrip 26 GHz ultra-wideband (UWB) bandpass filter for vehicle radar systems," *in Proc. IEEE MTT-S International Microwave Workshop Series on Millimeter Wave Wireless Technology and Applications (IMWS),* vol., no., pp.1,4, 18-20 Sept. 2012
3. V. Jain, S. Sundararaman, P. Heydari, "A 22–29-GHz UWB Pulse-Radar Receiver Front-End in 0.18-μm CMOS," *IEEE Trans. Microwave Theory and Techniques,* vol.57, no.8, pp.1903,1914, Aug. 2009
4. T.H. Teo, X. Qian, K.P. Gopalakrishnan, Y.S. Hwan, K. Haridas, C.Y. Pang; H-K. Cha, M. Je, "A 700-μ W Wireless Sensor Node SoC for Continuous Real-Time Health Monitoring," *IEEE Journal of Solid-State Circuits,* vol.45, no.11, pp.2292,2299, Nov. 2010
5. M. Alhawari, A. Khandoker, B. Mohammad, H. Saleh, K. Khalaf, M. Al-Qutayri, M.K. Yapici, S. Singh, M. Ismail, "Energy efficient system-on-chip architecture for non-invasive mobile monitoring of diabetics" *in Proc. Interational Conference on Design & Technology of Integrated Systems in Nanoscale Era (DTIS), 2013 8th,* vol., no., pp.180,181, 26-28 March 2013
6. T. Torfs, T. Sterken, S. Brebels, J. Santana, R. van den Hoven, V. Spiering, N. Bertsch, D. Trapani, D. Zonta, "Low Power Wireless Sensor Network for Building Monitoring," *IEEE Sensors Journal,* vol.13, no.3, pp.909,915, March 2013.
7. L. Brown, B. Grundlehner, J. van de Molengraft, J. Penders, B. Gyselinckx, "Body area network for monitoring autonomic nervous system responses," *in Proc. 3rd International Conference on Pervasive Computing Technologies for Healthcare, 2009. PervasiveHealth 2009.* , vol., no., pp.1,3, 1-3 April 2009.
8. Cheolbok Kim, *Chapter 2. Ultra-Wideband Antenna in Microwave and Millimeter Wave Technologies Modern UWB antennas and equipment,* IN-TECH, March 2010
9. Z-C. Hao, J-S. Hong, J.P. Parry, , D.P. Hand, "Ultra-Wideband Bandpass Filter With Multiple Notch Bands Using Nonuniform Periodical Slotted Ground Structure," *IEEE Trans. Microwave Theory and Techniques,* vol.57, no.12, pp.3080,3088, Dec. 2009
10. G-M. Yang, R. Jin, C. Vittoria, V.G. Harris, N.X. Sun, "Small Ultra-Wideband (UWB) Bandpass Filter With Notched Band," *IEEE Microwave and Wireless Components Letters,* vol.18, no.3, pp.176,178, March 2008
11. K.-R Chen, C.-Y.-D. Sim, J.-S. Row, "A compact monopole antenna for super wideband applications," *IEEE Antennas Wireless Propag. Lett.,* vol. 10, pp. 488-491, 2011.
12. X. Zhang, C. Karnfelt, J. Liu, S. Ma, W. Xu, L. Meng, H. Zirath, "Realization of Ultra Wideband Bandpass Filter based on LCP Substrate for Wireless Application," *in Proc. International Symposium on High Density*

packaging and Microsystem Integration, 2007. HDP '07., vol., no., pp.1,5, 26-28 June 2007.

13. C. Kim, J. K. Kim, K. T. Kim, Y-K. Yoon, "Micromachined wearable/foldable super wideband (SWA) monopole antenna based on a flexible liquid crystal polymer (LCP) substrate toward imaging/sensing/health monitoring systems," *in Proce. IEEE Electronic Components and Technology Conference (ECTC)*, Las Vegas, NV. vol., no., pp.1926,1932, 28-31 May 2013.

14. A. Alomainy, A. Sani, A. Rahman, J.G. Santas, Y. Hao, "Transient Characteristics of Wearable Antennas and Radio Propagation Channels for Ultrawideband Body-Centric Wireless Communications," *IEEE Trans. Antennas and Propagation*, vol.57, no.4, pp.875,884, April 2009.

15. A. Yarovoy, "Ultra-wideband radars for high-resolution imaging and target classification," in *Proc. 4th European Radar Conference (EuRAD)*, Munich, Germany, Oct. 10-12, 2007.

16. S. Ye, B. Zhou, G. Fang, "Design of a novel ultrawideband digital receiver for pulse ground penetrating radar," *IEEE Geosci. Remote Sens. Lett.*, vol. 8, no. 4, pp. 656-660, July 2011.

17. J. S. Hong and M. J. Lancaster, *Microstrip Filters for RF/Microwave Applications*. New York: Wiley, 2001, ch. 8.

18. Z.-C. Hao and J.-S. Hong, "Ultrawideband Filter Technologies", *IEEE, Microwave Magazine*, Volume: 11 , Issue: 4, pp. 56-68, 2010.

19. L. Zhu, S. Sun and W. Menzel, "Ultra-wideband (UWB) bandpass filter using multiple-mode resonator," *IEEE Microw. Wireless Compon. Lett.*,vol.15, no.11, pp. 796-798, Nov. 2005.

20. H. Chien, T. Shen, T. Huang, W. Wang, R. Wu, "Miniaturized Bandpass Filters With Double-Folded Substrate Integrated Waveguide Resonators in LTCC," *IEEE Trans. Microwave Theory & Tech* , vol.57, no.7, pp.1774,1782, July 2009.

21. T. H. Duong and I. S. Kim, "New Elliptic Funtion Type UWB BPF Based on Capacitively Coupled λ/4 Open T Resonator," *IEEE Trans. Microw. Theory Tech.*, vol., 57, no. 12, pp. 3089 – 3098, Dec. 2009.

22. T T. Zhang, W. Johnson, B. Farrell, and M. St. Lawrence, "The processing and assembly of liquid crystalline polymer printed circuits," 2002 *in Proc. Int. Symp. on Microelectronics, 2002.*

23. D. C. Thompson, O. Tantot, H. Jallageas, G. E. Ponchak, M. M. Tentzeris, and J. Papapolymerou,"Characterization of liquid crystal polymer (LCP) material and transmission lines on LCP substrates from 30–110 GHz," *IEEE Trans. Microw. Theory Tech.*, vol. 52, no. 4, pp. 1343–1352, Apr. 2004.

24. M. F. Davis, S.-W. Yoon, S. Pinel, K. Lim, and J. Laskar, "Liquid crystal polymer—based integrated passive development for RF applications," in *Proc. IEEE Int. Microwave Symp.*, vol. 2, 2003, pp. 1155–1158.

25. K. Wu, D. Deslandes, and Y. Cassivi, "The substrate integrated circuits- A new concept for high frequency electronics and optoelectronics," *in Proc. Conference on Telecommunications in Modern Satellite Cable and Broadcasting Service.* Oct. 2003, vol. I, pp. P-III-IX..

26. D. E. Senior, *Advanced metamaterial circuits for microwave and millimeter wave applications*. PhD Dissertation. Gainesville, FL. University of Florida. 2012.

27. Y. Dong, and T. Itoh, "Promising Future of Metamaterials," *IEEE Microwave Magazine,* vol.13, no.2, pp.39,56, March-April 2012.

28. Z. Zhang, N. Yang and K. Wu, "5-GHz bandpass filter demonstration using quarter-mode substrate integrated waveguide cavity for wireless systems," in *Proc. IEEE Radio and Wireless Symposium,* vol., no., pp.95-98, Jan. 2009.

29. S. Zhang, T. Bian, Y. Zhai, W. Liu, G. Yang, and F. Liu, "Quarter substrate integrated waveguide resonator applied to fractal-shaped BPFs," *Microw. Journal,* vol. 55, no. 5, pp. 200-208, May 2012.

30. D.M. Pozar, Microwave Engineering, 3d, ed. New York, Wiley & Sons, 2005.

A Dual-Band Power Amplifier Based on Composite Right/Left-Handed Matching Networks

Kyriaki Niotaki[1], Ana Collado[1], Apostolos Georgiadis[1], John Vardakas[2]

[1]Centre Tecnologic de Telecomunicacions de Catalunya (CTTC), Castelldefels, 08860, Spain

[2]Iquadrat S.L., 08009 Barcelona (Spain)

Corresponding author: K. Niotaki; email: kniotaki@cttc.es; phone: +34 936452900

Abstract

This paper presents the design of a dual-band power amplifier operating at 2.4 GHz and 3.35 GHz. The proposed topology is based on the use of composite right/left-handed unit cells. A composite right/left-handed cell exhibits a dual-band frequency response at an arbitrary pair of frequencies because of its phase characteristics. A power amplifier based on an enhancement mode pseudomorphic HEMT transistor is simulated and manufactured. The fabricated prototype leads to a dual-band amplification and is characterized in terms of measurements. A maximum drain efficiency of 65% and 52% is achieved for an output level of 28.7 dBm and 27.5 dBm at 2.4 GHz and 3.35 GHz, respectively. The presented approach can be applied for the design of dual-band matching networks for microwave circuits operating at two arbitrary frequencies.

Introduction

Multiband transceivers are receiving special attention due to the emergence of new communication standards [1]. One of the most challenging tasks for the design of transceivers capable of operating at multiple frequency bands is the implementation of multiband RF power amplifier circuits. The design of efficient and linear multiband/broadband power amplifiers will allow the operation of different standards in the same equipment and thus will lead to cheaper and smaller devices.

Power amplifiers are fundamental blocks in any transmitter as they dominate the power consumption and thus the cost of the device. High efficiency amplification is mainly desired for longer battery lifetime and simple heat management configurations. Simultaneously, linearity requirements are increased due to the emergence of new communication standards and high peak to average power ratio (PAPR) signals. At the same time, the concurrent operation of these circuits in different frequency bands is an additional requirement that comes from the need for compact devices operating with multiple communication standards.

The increasing need for broadband and multiband microwave circuits is mainly focused on the design of multiband and broadband impedance matching networks. The implementation of multiband matching networks is a complex task in which the desired tradeoff between linearity and efficiency at the desired frequencies should be taken into account. The difficulty in the design of multiband components comes from the fact that the same behavior should be met at arbitrary frequencies.

The design of any multiband/broadband microwave circuits, such as power amplifiers, focus on the selection of the input and matching networks that have to be optimized for operation in a couple of frequency bands or for operation over a wide frequency range. In the literature many efforts have been made to realize impedance matching networks operating in more than one frequency bands, including lumped element networks and transmission line implementations [2] [3] [4].

Various approaches for designing power amplifiers with multiband capabilities already have already been proposed and considered in the literature [4]. One approach that stands for the design of a multiband power amplifier is by cascading multiple frequency matching networks. In [5], the design and the implementation of a dual-band power amplifier with two frequency matching networks based on the low-pass Chebyshev form impedance transformer design is presented. A dual-band matching network can also be implemented based on a π-topology matching network. A power amplifier circuit based on π-topology matching network and operating at 1.9 GHz and 3.5 GHz is shown in [6]. A 1.81 GHz and 2.65 GHz dual-band power amplifier based on multiharmonic load transformation network is analyzed in [4].

Among the efforts for dual-band operation is the implementation of power amplifier circuits with composite right/left-handed cells as the impedance matching networks. Composite right/left-handed (CRLH) transmission lines can be applied for the design of dual-band components for an arbitrary pair of frequencies because of its phase characteristics [7].

In this work, the properties of composite right/left-handed unit cells are applied to the implementation of dual-band matching networks for a power amplifier. The power amplifier should operate at ISM band and 3.35 GHz WiMAX band. A brief analysis of the networks is presented and the dual-band properties are validated with a fabricated prototype.

This paper is organized as follows. Initially, the maximization of power transfer at two frequencies is briefly discussed as part of the design of power amplifier circuits. Then, the properties and equivalent topologies of the composite right/left-handed cells are presented. Two approaches have been considered and simulated. The simulation results show that by using a single unit cell as the input and output matching network, the desired operation is achieved. The simulation process for the design of the power amplifier is described in detail. Finally, the experimental characterization of the fabricated device is presented. The simulation and measurement results exhibit a dual-band frequency response. The paper concludes with the achieved results.

Dual-Band Impedance Matching Networks

The selection of the impedance matching networks is an important design challenge for the implementation of an efficient power amplifier. The design of a power amplifier circuit implies the design of two matching networks as it is shown in the simplified power amplifier topology of Fig. 1. The input matching network matches the signal source to the gate of the device, while the output matching network offers an impedance matching from the drain of the transistor to the 50 Ω output load.

The selection of the matching networks depends on the system specifications and the operating frequency or frequencies. A variety of matching network topologies can be applied for the matching networks when operating at a single frequency. The simplest structures for an impedance transformation network are the L-type and π-type networks [8]. Their implementation varies according to the device specifications and the available technology. Implementations with commercial lumped elements or transmission lines (TL) have been presented in the literature [8].

The design of dual-band matching networks is a more complex task as the same behavior should be met at arbitrary frequencies. Composite right/left-handed cells can be applied as the impedance matching networks of broadband and multiband microwave circuits [9] [10] [11].

The composite right/left-handed TL is the combination of series connected left-handed (LH) and right-handed (RH) transmission lines. The circuit model of a right-handed TL is shown in Fig. 2(a) [8] [9]. The dual of the right-handed model is called left-handed TL and is presented in Fig. 2(b).

The CRLH network has been introduced as a consequence of the presence of parasitic elements during the operation of left-handed and right-handed unit cells [9]. The topology shown in Fig. 2(c) shows the circuit model of a CRLH transmission line unit cell (series connected LH and RH unit cells) when only one unit cell is used. Additional unit cells (N) can be put in series according to the application scenario and the available technology. The total phase shift depends on the number of cells along a CRLH line and is given by:

$$\varphi = N * \Delta\varphi$$

where N is the number of unit cells and $\Delta\varphi$ is the phase shift for each unit cell.

The CRLH unit cell operates as follows: it shows a left-handed response at the low operating frequency as C_R and L_R tend to be open and short at f_1 [9]. Likewise, the CRLH unit cell shows right-handed behavior at the high frequency f_2 as C_L and L_L tend to be short and open at f_2.

A CRLH transmission line can be implemented with asymmetric or symmetric unit cells. The asymmetric structure of a CRLH TL (Fig. 2(c)) has unequal input and output impedance. On the other hand, the symmetric CRLH unit cell (Fig. 3(a)) is characterized with equal input and output impedance.

In the literature, alternative circuit models have been proposed for the implementation of the CRLH concept in different technologies [9]. A π-shaped CRLH transmission line has been proposed for substrate integrated waveguide (SIW) technology and is shown in Fig. 3(b). The proposed structure has found successful implementation in the design of substrate integrated waveguide matching structures [12] [13] [14].

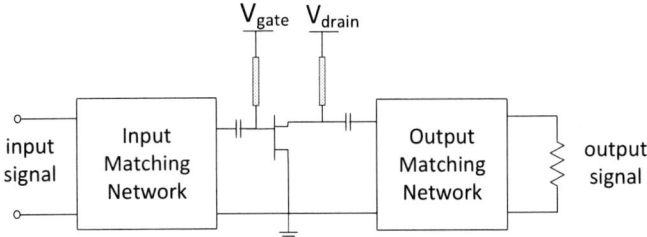

Figure 1. Simplified power amplifier circuit topology that consists of the input and output impedance matching networks, two DC blocking capacitors and the bias of the transistor. The active device is an enhancement mode pseudomorphic HEMT transistor (ATF-50189) from Avago.

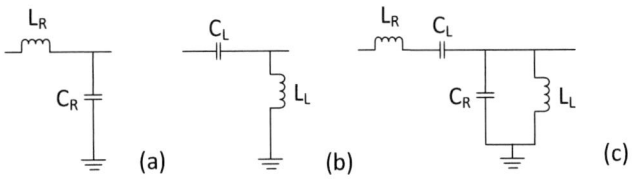

Figure 2. Equivalent circuit model for: (a) one unit cell of an deal lumped element right-handed TL, (b) one unit cell of an ideal lumped element left-handed TL and (c) the unit cell of an ideal asymmetric lumped element composite right/left-handed transmission line (T-shaped topology).

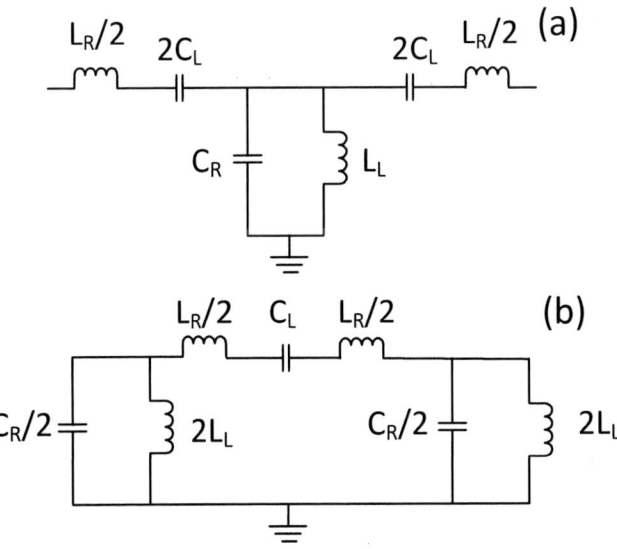

Figure 3. Equivalent circuit model for conventional symmetric composite right/left-handed unit cell (T-shaped topology) and (b) equivalent composite right/left handed unit cell topology (π-shaped topology).

Simulation and Experimental Results

A dual-band power amplifier is designed based on the ATF-50189 E-pHEMT transistor from Avago (operating frequency from 400 MHz to 3.9 GHz) [15]. We investigated in simulation the power amplifier performance using one and two unit cells. Fig. 4 and Fig. 5 show the simulated results in terms of the S_{11} parameter and the drain efficiency, which demonstrate that the use of a single cell can lead to a good performance, while additionally it minimizes the size and number of components used in the amplifier. As a result the amplifier with N=1 was selected for the implementation of the input and output impedance networks.

Harmonic balance simulation in combination with non-linear optimization procedure is used for an efficient and stable operation at the selected frequencies. Optimization goals are carefully selected to match the system requirements as far as the output power and the efficiency of the system.

The simulation set up includes the following optimization goals for the output power and the drain efficiency of the topology:

$$P_{out[f1]} > P_{min} \quad and \quad n_{[f1]} > n_{min} \; at \; f_1 = 2.4 \; GHz$$

and

$$P_{out[f2]} > P_{min} \quad and \quad n_{[f2]} > n_{min} \; at \; f_2 = 3.35 \; GHz$$

where $P_{out[f1]}$ and $n_{[f1]}$ is the output power and the drain efficiency at the low operating frequency ($f_1 = 2.4 \; GHz$). $P_{out[f2]}$ and $n_{[f2]}$ is the output power and the drain efficiency at the high operating frequency ($f_2 = 3.35 \; GHz$). The lumped component values are optimized to fulfill these goals when the device operates with an input power level of 19 dBm.

Small S-parameters and large signal scattering parameter (LSSP) analysis is also used to ensure the impedance matching of the circuit at the selected frequencies. Both the input and output matching network are optimized for maximum power transfer from the RF input signal to the gate of the transistor and from the drain of the transistor to the 50 Ω output load.

The simulation data from the S-parameter analysis is also used for a stability analysis of the topology. The stability factors K and B are calculated based on the analysis results to predict the operation of the system under different operating conditions. In particular, the design is optimized in order to meet the necessary and sufficient conditions for unconditional stability $B > 0$ and $K > 1$ [16] [17].

The simulation result shows that by placing a series resistance of 22 Ω at the gate of the transistor, the power amplifier exhibits a stable operation for all the expected operating conditions in terms of input power and biasing.

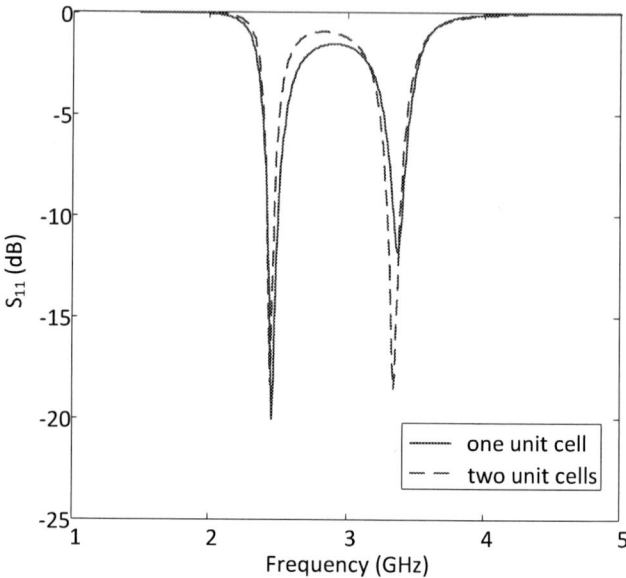

Figure 4. Simulated S_{11} versus operating frequency. The power amplifier performance is evaluated when the input and output impedance matching networks consist of one and two CRLH unit cells.

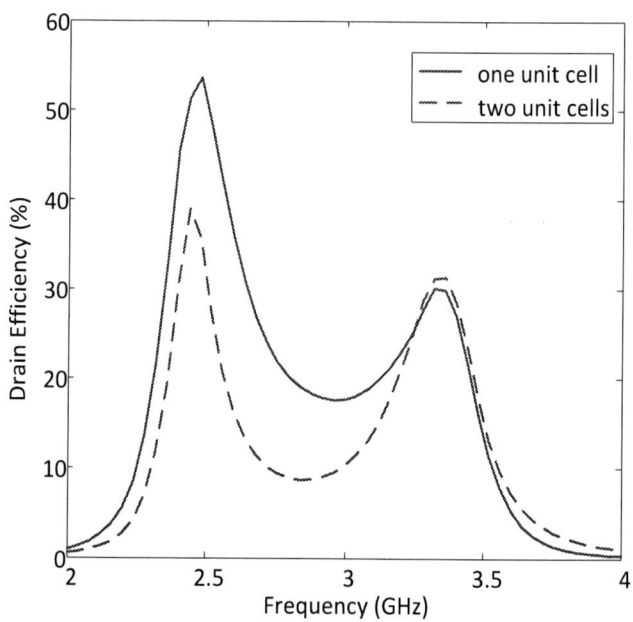

Figure 5. Simulated drain efficiency versus operating frequency. The input power level is 19 dBm. The power amplifier performance is evaluated when the input and output impedance matching networks consist of one and two CRLH unit cells.

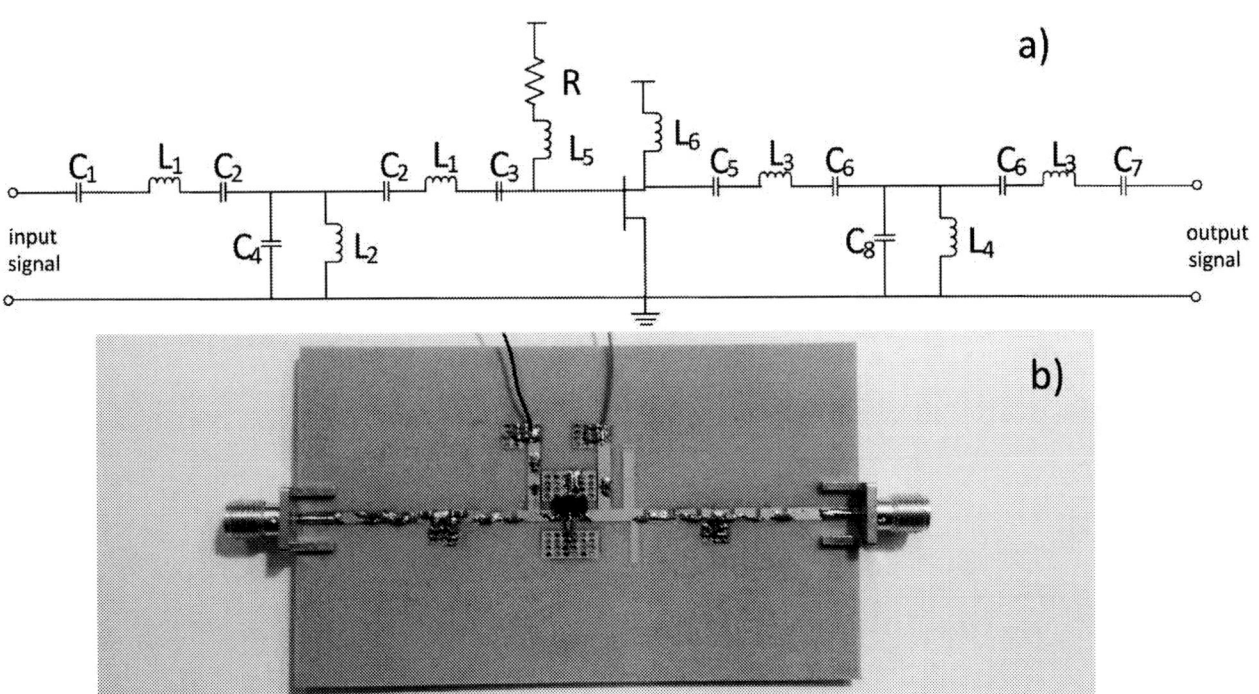

Figure 6. Dual-band power amplifier: a) Schematic of the circuit topology and b) fabricated prototype. The fabricated prototype has a total size of 7 cm x 5 cm.

Component	Value	Component	Value
C_1	1.8 pF	R	22 Ω
C_2	3.9 pF	L_1	0.9 nH
C_3	0.45 pF	L_2	1 nH
C_4	0.25 pF	L_3	1 nH
C_5	5.6 pF	L_4	1.8 nH
C_6	12 pF	L_5	16 nH
C_7	10 pF	L_6	100 nH
C_8	1 pF		

Table 1. Power amplifier circuit component values.

The final circuit schematic and the component values are shown in Fig. 6(a) and Table 1, respectively. Commercial capacitors from Murata and AVX and inductors from Coilcraft have been used for both the simulation and the implementation of the current design.

The power amplifier is fabricated in Arlon 25N substrate with 3.38 dielectric constant and 30 mil thickness utilizing a milling machine. A photo of the implemented prototype is shown in Fig. 6(b). The fabricated prototype has a total size of 7 cm x 5 cm.

The performance of the power amplifier is evaluated in terms of measurements. Initially, the S-parameters of the power amplifier are measured using a signal with an input power of 0 dBm. The S_{11} and S_{22} parameters are shown in Fig. 7 showing a good matching at the two operating frequencies (2.4 GHz and 3.35 GHz).

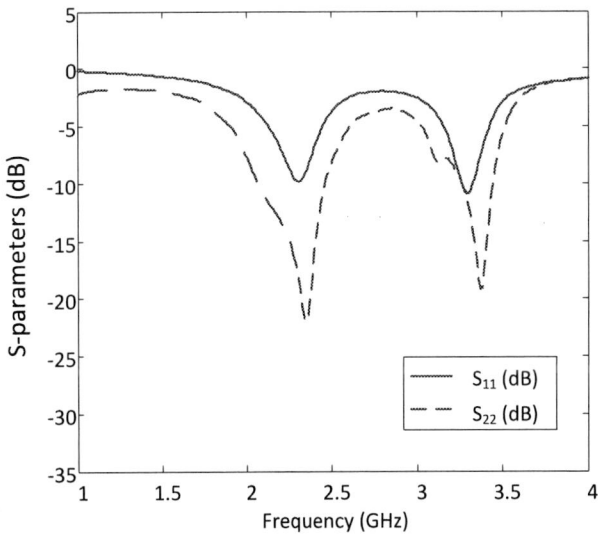

Figure 7. Measured small signal S-parameters (S_{11} and S_{22}) versus operating frequency for an input power of 0 dBm.

The S_{21} and S_{12} parameters are shown in Fig. 8.The circuit exhibits a small signal gain of 11 dB and 7.8 dB for an input power of 0 dBm at 2.4 GHz 3.35 GHz, respectively.

The small signal gain of the circuit is tested for different gate to source voltages as Fig. 9 shows. As it is can be observed from the measurements, the device achieves a maximum gain for V_{gs}=0.54 V. Thus, the characterization of the circuit versus operating frequency and input power level is made for V_{gs}=0.54 V.

978-1-4799-2408-0/14 $31.00 © 2014 IEEE

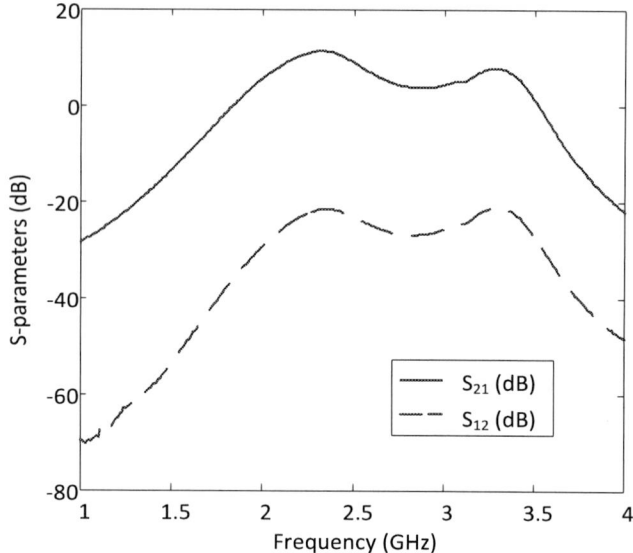

Figure 8. Measured small signal S-parameters (S_{21} and S_{12}) versus operating frequency for an input power of 0 dBm.

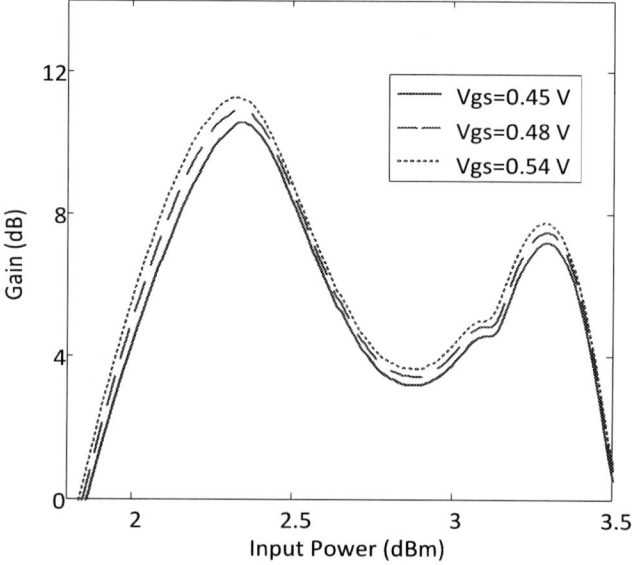

Figure 9. Measured S_{21} for different gate to source voltages (V_{gs}= 0.45V, V_{gs}= 0.48V and V_{gs}= 0. 54V). The input power level is 0 dBm.

The circuit is characterized for an input power level of 19 dBm versus operating frequency. The measured output power and drain efficiency are shown in Fig. 10 and Fig. 11, respectively. A measured efficiency of 58 % and 38 % was obtained for an input power level of 19 dBm at 2.4 GHz and 3.35 GHz, respectively. Additionally, Fig. 12 shows the gain of the circuit over frequency.

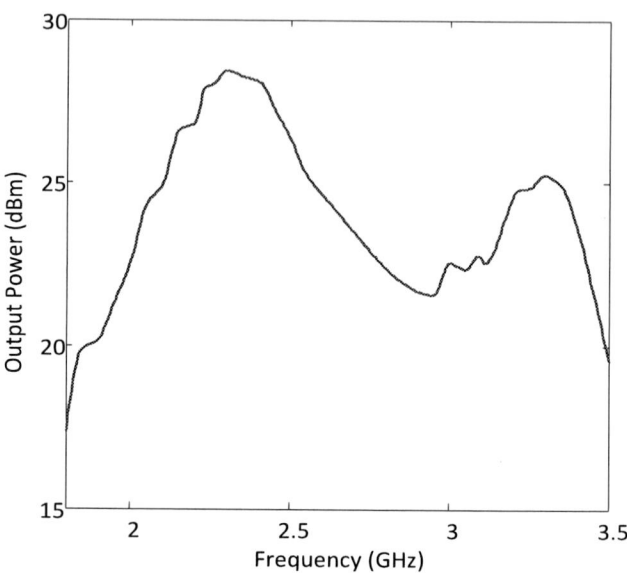

Figure 10. Measured output power level versus operating frequency for an input power of 19 dBm.

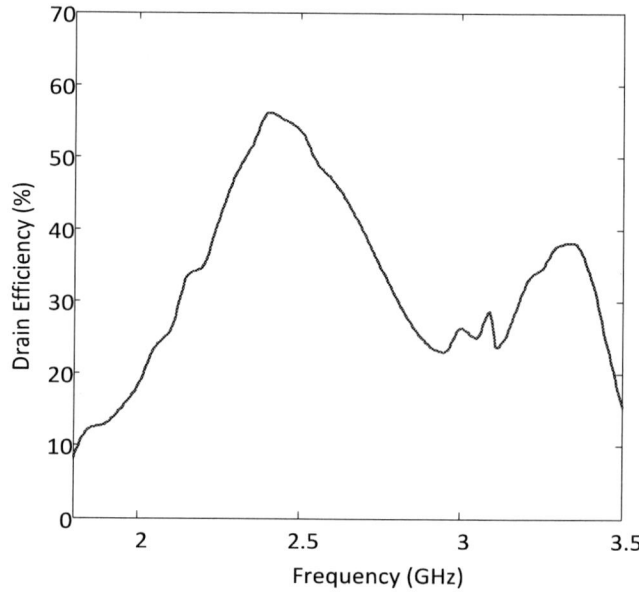

Figure 11. Measured drain efficiency versus operating frequency for an input power of 19 dBm.

The circuit is also tested versus input power level at the two operating frequencies. Fig. 13 shows the measured drain efficiency of the device when the input power level varies from 0 to 24 dBm. A maximum drain efficiency of 65% and 52% is achieved for an output level of 28.7 dBm and 27.5 dBm at 2.4 GHz and 3.35 GHz, respectively. Finally, Fig. 14 depicts the measured gain of the circuit versus input power.

978-1-4799-2408-0/14 $31.00 © 2014 IEEE

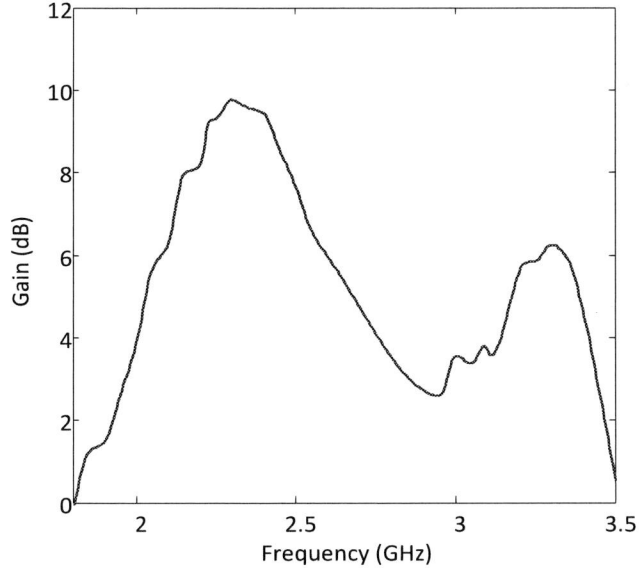

Figure 12. Measured gain versus operating frequency for an input power of 19 dBm.

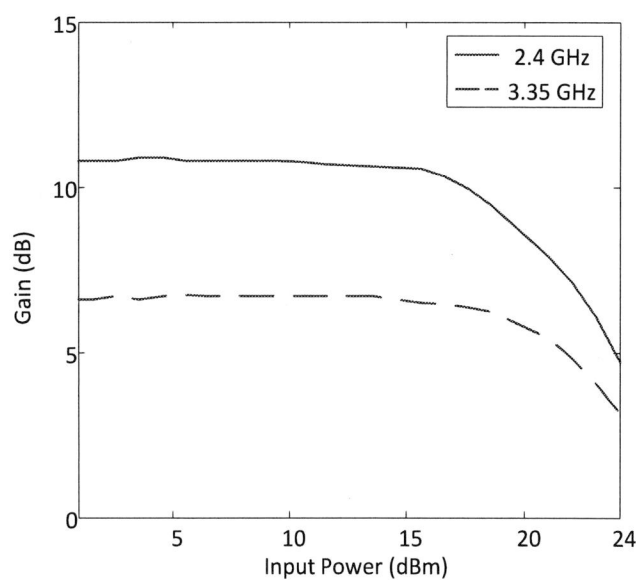

Figure 14. Measured gain versus input power level for the two operating frequencies (2.4 GHz and 3.35 GHz).

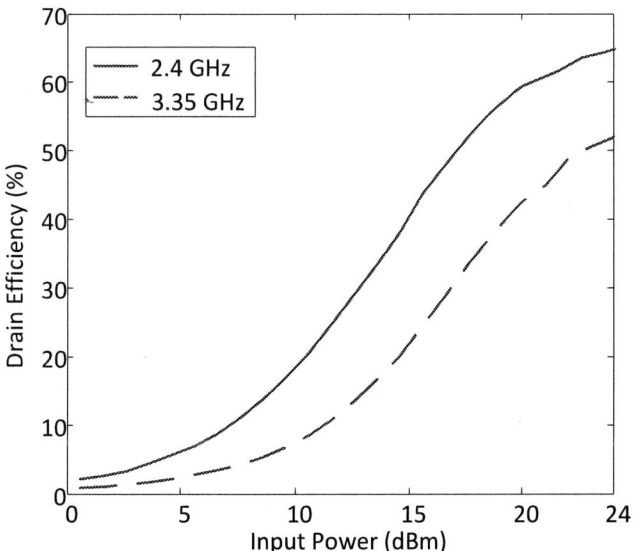

Figure 13. Measured drain efficiency versus input power level for the two operating frequencies (2.4 GHz and 3.35 GHz).

Conclusions

A detailed discussion for the design of impedance matching networks for microwave circuits and especially power amplifiers is presented. The design of a dual-band power amplifier based on composite right/left-handed unit cells is described and validated by measurements.

The design technique is successfully applied to the design of a dual-band power amplifier operating at 2.4 GHz and 3.35 GHz. The resulting power amplifier topology demonstrates a promising solution for the design of compact dual-band power amplifiers operating at two arbitrary frequencies.

Future work includes the design of a power amplifier circuit with improved performance in terms of efficiency and gain. Additionally, special attention will be paid to extend the bandwidth of the second operating frequency of 3.35 GHz.

Acknowledgments

This work was supported by the Spanish Ministry of Economy and Competitiveness project TEC2012-39143, the ENIAC JU project ARTEMOS and the Marie Curie project SWAP, FP7-PEOPLE-2009-IAPP 251557.

The authors would like to acknowledge Coilcraft, Inc. for providing the inductors used in the development of the power amplifier prototype.

References

1. K. Rawat, M.S. Hashmi, F.D. Ghannouchi, "Double the Band and Optimize," *IEEE Microwave Magazine,* vol.13, no.2, pp.69-82, March-April 2012.
2. Fu Xin, D.T. Bespalko, S. Boumaiza, "Novel Dual-Band Matching Network for Effective Design of Concurrent Dual-Band Power Amplifiers," IEEE Transactions Circuits and Systems, vol.61, no.1, pp.293-301, Jan. 2014.
3. Pei Shifeng, Nan Jingchang, Qu Yun, "Two kinds of dual-band power amplifier matching networks using three-section microstrip line," *in Proc Antennas, Propagation & EM Theory (ISAPE),* Oct. 22-26, 2012, pp. 702-705.
4. D. Kalim, R. Negra, "Concurrent planar multiharmonic dual-band load coupling network for switching-mode power amplifiers," *in Proc. 2011 IEEE MTT-S International Microwave Symposium Digest*, Baltimore, MD, June 5-10, 2011, pp.1-4.

5. K. Uchida, Y. Takayama, T. Fujita, and K. Maenaka, "Dual-band GaAs FET power amplifier with two-frequency matching circuits," in *Proc Asia–Pacific Microw. Conf.*, Dec. 2005, vol. 1, pp. 4–7

6. A. Cidronali, N. Giovannelli, I. Magrini, G. Manes, "Compact Concurrent Dual-Band Power Amplifier for 1.9GHz WCDMA and 3.5GHz OFDM Wireless Systems," *in Proc 38th European Microwave Conference*, Oct. 27-31, 2008, pp.1545-1548.

7. Seung Hun Ji, Gyu Seok Hwang, Choon Sik Cho, Jae W. Lee, and Jaeheung Kim, "836 MHz/1.95GHz Dual-Band Class-E Power Amplifier Using Composite Right/Left-Handed Transmission Lines," in *Proc. 36th European Microwave Conference*, Manchester, Sept., 2006, pp.356-359.

8. D. M. Pozar, *Microwave Engineering*, John Wiley & Sons, 1993.

9. C. Caloz, T. Itoh, *Electromagnetic Metamaterials: Transmission line theory and microwave applications*, Hoboken, New Jersey, John Wiley & Sons, 2006.

10. G.V. Eleftheriades, EM transmission-line metamaterials, Materials Today, vol. 12, Issue 3, pp. 30-41, March 2009.

11. I-Hsiang Lin, M. DeVincentis, C. Caloz, and T. Itoh, "Arbitrary dual-band components using composite right/left-handed transmission lines," *IEEE Trans. Microw. Theory Tech.*, vol. 52, no. 4, pp. 1142-1149, April 2004.

12. K. Niotaki, F. Giuppi, A. Collado, A. Georgiadis and J. Vardakas, "A Broadband Power Amplifier Based on Composite Right/Left-Handed Half-Mode Substrate Integrated Waveguide," *in Proc. Design of Circuits and Integrated Systems*, San Sebastian, Spain. Nov. 27-29, 2013.

13. Y. Dong, T. Itoh, "Composite Right/Left-Handed Substrate Integrated Waveguide and Half Mode Substrate Integrated Waveguide Leaky-Wave Structures," *IEEE Trans. on Antennas and Propagation*, vol. 59, no. 3, pp. 767-775, Mar. 2011.

14. Y. Dong, T. Itoh, "Promising Future of Metamaterials," *IEEE Microwave Magazine*, vol. 13, no. 2, pp. 39-56, Mar.-Apr. 2012.

15. Avago technologies, Data Sheets & Technical Specifications, http://www.avagotech.com/pages/en/rf_microwave/transistors/fet/atf-50189/

16. S. Kassim, F. Malek, "Microwave FET amplifier stability analysis using Geometrically-Derived Stability Factors," *in Proc. 2010 Intelligent and Advanced Systems*, June, 2010, pp.1-5.

17. G. Gonzalez, Microwave Transistor Amplifiers, second edition, Prentice-Hall, 1997.

Wafer-Level Non Conductive Films for Exascale Servers

A. Horibe, S. Kohara, H. Mori, Y. Orii

IBM Research – Tokyo

S. Kawamoto, H. Sone, M. Hoshiyama

Namics Corporation

NANOBIC 1001, 7-7 Shin-kawasaki, Saiwai-ku, Kawasaki, 212-0032 Japan

hory@jp.ibm.com

Abstract

Wafer-Level Non Conductive Films (WLNCFs) were evaluated as a potential underfill solution for future exascale server packages. The fundamental chip-joining ability and reliability on a small die package were tested. The fillet shapes, fillet overcoating, adhesion strength, and simplifications of the chip joining conditions, and the thermomechanical stresses in relation to the lower CTE substrate were evaluated and assessed for the future advanced packages.

1. Introduction

Semiconductor scaling has strongly driven improved performance, higher power efficiency, and cost reductions for electronic devices, but now the scaling technologies are reaching their physical limits. High-density packaging technologies may extend the life of Moore's Law. Future server systems will integrate (1) denser MCMs (Multiple Chip Modules), which may be placed in a plane or stacked with TSVs (Through Silicon Vias), (2) optical devices for higher bandwidth I/O between modules, and (3) more efficient and more integrated cooling devices. Such densely integrated devices such as in-plane MCMs or 3D/2.5D integrated modules will require fine pitch bumps which makes capillary flow of underfill difficult and also require chip-scale packaging without large side-drops of the underfill to minimize the wiring lengths between the chips for faster and lower power data transmissions. For such narrow gaps and tight fillets, pre-applied underfill technology are a potential solution.

Conventional capillary underfills are drawn into the gap under the chips by capillary action. With narrow gaps, filling the underfill without voids is a critical concern. At the same time, very thin dies are easily contaminated on the back sides if there are excess amounts of underfill at the chip edges. In contrast, a film underfill can be laminated onto a bumped wafer as a pre-applied underfill before dicing. This provides mechanical protection for thinned silicon wafers and ultra low-k layers of chips, without bump-location-dependent material flows that can occur with capillary or paste underfill flow.

We are now studying WLNCFs to apply to the future advanced packages. The film underfill is laminated onto the wafer, diced into dies, and then bonded by a thermal compression bonder. This can also support a back-grinding process to make thin wafers with simple processing steps.

2. Exascale servers

Exascale servers [1] will integrate CPUs and stacked memories on a limited-footprint low-CTE (Coefficient of Temperature Expansion) substrate. The stacked memories with TSVs should consist of thinned memory dies. The substrate will be made of lower CTE materials to reduce the thermomechanical stress between the chips and the substrate, and high bandwidth interconnections will require many small joints between dies or die and substrate and fine-pitch data lines between CPU and memory. The reduction of the CTE of the substrate will enables the use of larger dies and stacked dies.

In the coming cognitive era, the exascale servers should work on processing enormous data which are increasing significantly. Concurrently, neurosynaptic chips and systems [2] are anticipated for self-learning and flexible recognition of images and sounds. These brain-like devices are expecting to have higher efficiency for certain processes with lower power consumption than conventional data processing systems. Such brainmorphic devices should need enormous interconnections to connect between large numbers of neurosynaptic elements corresponding to neurons, axons, and synapses. This implies that both exascale and neurosynaptic devices will consist of fine pitch interconnections integrated in high-density packages.

3. WLNCF materials

Table 1 presents the material properties of the WLNCF materials that were evaluated in this study. The filler size and content have to be optimized for tackiness of the raw material, alignment mark visibility, load pressure, and reliability. Using raw materials with low tackiness reduces the risks of chip handling errors or contamination by foreign materials. The viscosity was measured with a rheometer (temperature ramping speed: 10°C/min.). This value is only a reference because the actual temperature ramping speed of the chip joining step is usually more than 100 times faster than this measurement condition. The glass transition temperature (T_g), the (CTE), and the tensile modulus affect the long term reliability of the modules. The impact of these parameters onto exascale servers will be discussed in a later section.

4. Experiments for material evaluations

To understand the basic characteristics for a flip chip package with fine pitch bumps, we tested the chip-join ability and the reliability of the modules with WLNCF. As shown in Table 2, the test vehicle consists of 125-μm-thick die with 50-μm pitch peripheral Cu pillar bumps and a thin-core laminate with Cu + OSP pad. A image of the joints and C-SAM (acoustic microscope) and a planar section image for void analysis are shown in Fig. 1 and 2, respectively. Table 3 shows the reliability test results. There were no failed modules after the test.

Table 1. Relevant material properties of WLNCFs.

ITEM		UNIT	WLNCF-1	WLNCF-2
Filler Content		wt%	40	40
Filler Size		μm	0.1	0.1
Tackiness @ 25°C		N	0.4	0.1
Minimum Viscosity	Viscosity	Pa·s	2500	5200
	Temperature	°C	104	113
Tg		°C	105	99
CTE	<Tg	ppm/K	35	35
	>Tg	ppm/K	142	140
Tensile Modulus		GPa	2.4	3.2

Table 2. Test vehicle specifications.

Chip	Size	7.3 × 7.3 × 0.125 mm
	Passivation	SiN
	Bump	Cu (30 μm-t) + solder (15 μm-t) (50 μm pitch, peripheral)
Laminate	Thickness	360 μm
	Pad finish	Cu + OSP

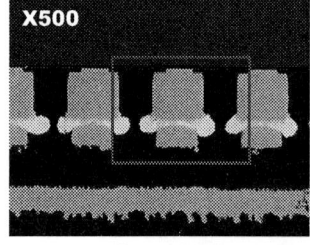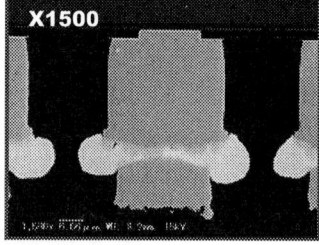

Figure 1. Cross-section images of joints.

(a) C-SAM

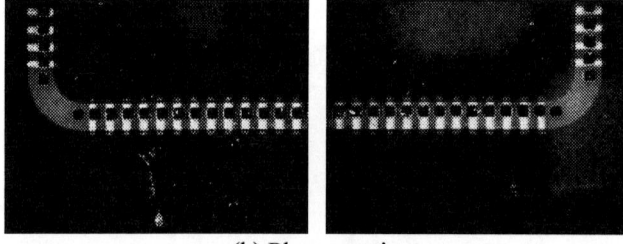

(b) Planar section

Figure 2. C-SAM and planer section images.

Table 3. Reliability test result. (-55/125°C)

	Time=0	After MRT	500 cycles	1000 cycles
CHIP/NCF				
NCF/Sub				
Passed modules	3/3	3/3	3/3	3/3

5. WLNCF challenges for exascale server applications

As explained above, we confirmed that the WLNCF meets the basic requirements for the module with fine-pitch bumps without voids and with good reliability. As mentioned above, we anticipate that exascale server packages will consist of CPU chips with many cores and caches and 3D stacked high bandwidth memories adjacent to the CPUs. The bandwidth between CPUs and memories will be high, and shortening the distance between the dies will contribute to the high bandwidth, reduced power consumption, and minimizing the size of the interposer. Also, these dies on the interposer have fine-pitch joints which make it difficult to clean the flux or to encourage capillary flow of the underfill.

In addition, the capillary underfill needs lower viscosity to fill these narrow gaps, but a low viscosity underfill can easily spread to surrounding area of the dies. On the other hand, the WLNCF process offers some of the highest potential to fill with the underfill into the narrow gaps between die-die and die-substrate and minimize the fillet width for exascale-server packages.

A typical fillet shape for a WLNCF module with a 100-μm thick chip is shown in Fig. 3. The shape is tapered along the chip edge. The center of the edge is thick and corners are thin. The shape of the cross-section of the fillet is like a droplet or an extruded film, which differs from a typical underfill fillet that becomes thinner like slope along the distance from the chip. The thin corner fillet leads to lower reliability in cyclic thermomechanical stress environment.

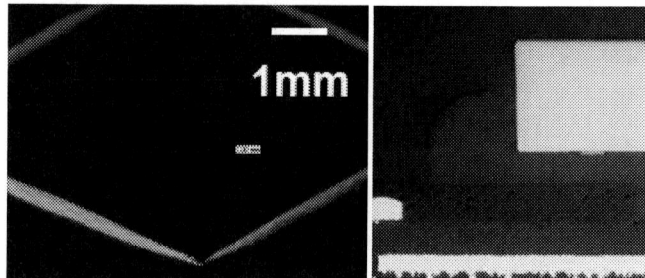

Figure 3. Fillet shape of a WLNCF module.

With WLNCF, one of the advantages is the precisely-controlled thickness of the laminated material on the chip before chip-join step, resulting in enabling underfill volume minimization for small foot-print packages. However, although the thickness of the raw material on the wafer can be controlled, the thickness in the chip area can't be controlled (= uniform). As a result, the control of the fillet volume at each location along the chip edge is difficult. The material viscosity must be low enough so that the material will spread to the corners. However, if the volume of the filler is too low, then the reliability will be threatened.

In contrast, non-conductive paste (NCP) materials that can be dispensed directly on a substrate or a chip may be also considered as another option for fine-pitch joints. To control the volume of the underfill material, jet dispensing technology has good precision (+/- 1% for each die) and enables flexible dispensing pattern of underfill materials which leads to uniform fillet volume along the chip edge. However, the narrower the gap, the more difficult the optimization of the dispensing pattern to prevent voids, because the dispensing volume per unit length from the tool has a limited minimum value (which means the minimum dot volume of the jet dispenser, and is not small enough for the small underfill volume for the narrow gap.). Thus, when the gap is relatively small compared to the size of the chip, the total dispensing length becomes short. (Fig. 4) In this case, the material has to spread to a wider area in the gap from the originally-dispensed location. This would result in enlarging a risk of entrapping voids. This is one of the concerns of this method for the advanced fine pitch packages.

(a) for 30-μm gap   (b) for 20-μm gap

Figure 4 Examples of NCP dispensing patterns (line width: 0.5 mm) for 7.3 mm chip modules with different height gaps.

For WLNCF, the film thickness uniformity in the wider area of the rolled sheet affects the required volume of the material for each chip in the production. Table 4 shows the relationship between the film thickness and calculated fillet width variations. The result shows that +/-2um thickness variation in a WLNCF film continuously fabricated by a roll-to-roll processing system causes relatively large fillet size variation for large die. This means, WLNCF will also faces challenges for maximizing the yield for large die packages

Table 4. WLNCD thickness vs. calculated fillet width variation.*

Fillet width comparison		WLNCF thickness variation (+/- 2 μm)	
		30 μm-t	34 μm-t
Chip size	7.3 mm	(Ref.: No fillet)	111 μm
	20 mm	(Ref.: No fiilet)	305 μm

* Chip is 100 μm thick with a triangular fillet shape at each location.

To improve the reliability of the WLNCF modules, the adhesion strength of the WLNCF is a key parameter. Table 5 shows the adhesion strengths of two WLNCFs adhering to the SiN surfaces of Si chips. WLNCF-2 has especially strong adhesion strength even if exposed in high humidity.

Table 5. Shear strength (kg/mm^2) of WLNCFs.

Condition	WLNCF-1	WLNCF-2
Initial	2.8	**9.2**
MRT L3	7.1	**8.7**
MRT L2	3.8	**8.1**

Another approach to improve the reliability of WLNCF is an overcoating on the fillet, especially at the chip corners. The chip joining time using the thermal compression bonder also has to be minimized to reduce manufacturing costs. As a result, the WLNCF material is only partially cured. Because the material is only half-cured, another underfill material can be applied as an overcoating on the WLNCF fillet. These two materials can be bonded chemically and form appropriate fillet shapes even at the chip corners. (Fig. 5)

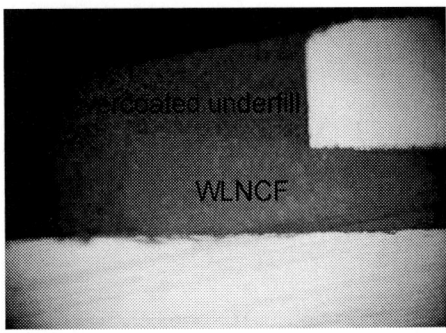

Figure 5. Overcoated fillet structure near the chip corner.

Another important property of the WLNCF is that it can withstand loading pressure, which allows for a simpler chip joining process without controlling the gap precisely in the thermal compression bonding step, even if there is a solder layer in the joints. The viscosity of the WLNCF has strong dependencies on time and temperature. A complex bonder tool sequence including timing of loading pressure control makes yield optimization difficult. With a WLNCF, the parameters of the bonding profile can be reduced to temperature with a constant loading pressure. (Fig.6)

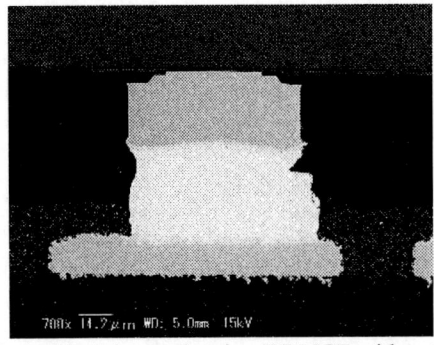

Figure 6. A chip-join result using WLNCF without loading pressure control.

The bump metals and structure are the keys for high joint yield, stress reduction, and improved electro-migration resistance of the packages. To apply a solder cap on the Cu post, the solder volume in each joint has to be optimized to prevent solder extrusion by the WLNCF flow from underneath the chip. A thicker solder cap in the joints improves the joint yield, but can easily be extruded by the WLNCF into the fillet. Fig. 7 shows the schematic diagram of the WLNCF flow directions under the chip, and the flow speeds at each location in the chip. The likelihood of solder extrusion is affected not only by the rheology of the WLNCF material but also by the solder shape and the surface tension in the resin.

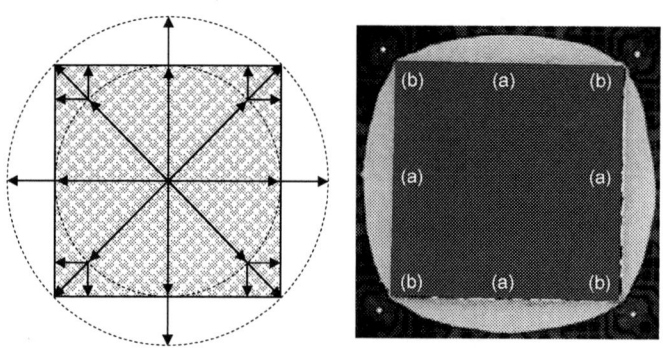

Figure 7. WLNCF flow under chip.
(a) fast flow speed, (b) slow flow speed

Table 6. Parameters of WLNCF modules for FEM.

WLNCF	CTE (ppm/K)	35
	Modulus (GPa)	3.2
Normal UF (Ref.)	CTE (ppm/K)	25
	Modulus (GPa)	9
Laminate	CTE (ppm/K)	17(Ref.) → 6
	Modulus (GPa)	30

6. WLNCF modeling for exascale packages

The WLNCFs evaluated in this study include small-size and small amount of fillers to improve the visibility of the alignment marks. Generally, finer fillers increase the material viscosity due to the wider contact area between the fillers, which increases the loading pressure if the same amount of filler is used as in a standard capillary underfill. In contrast, lower filler content leads to lower modulus and higher CTE than the standard underfills. Therefore, the thermomechanical analysis for the WLNCFs is done with a finite element method (FEM) to understand the failure risk of the WLNCF packages under thermomechanical stress. As shown in Table 6, the key parameter in this model (Fig. 8) is the CTE of the laminate, which is expected to reduce the stress in the future packages. Fig. 9 and Fig. 10 show the maximum elastic energy density in the underfills at chip corner. In case of the same laminate's CTE, the energy density of WLNCF is higher than that of a normal underfill. However, the energy density of the WLNCF on the low-CTE laminate is much lower than that of the normal underfill on the normal

laminate. These results show that a low-CTE laminate clearly contributes to reducing the energy density in the WLNCF material at chip corners, where there is often high stress. At the same time, this suggests that the WLNCF can be applicable for the future exascale packages with a low-CTE substrate, even if the filler content is relatively lower than normal underfills.

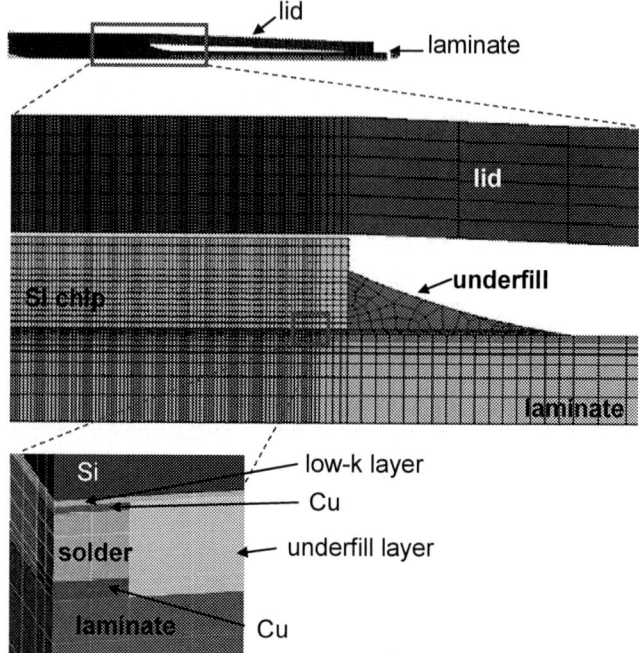

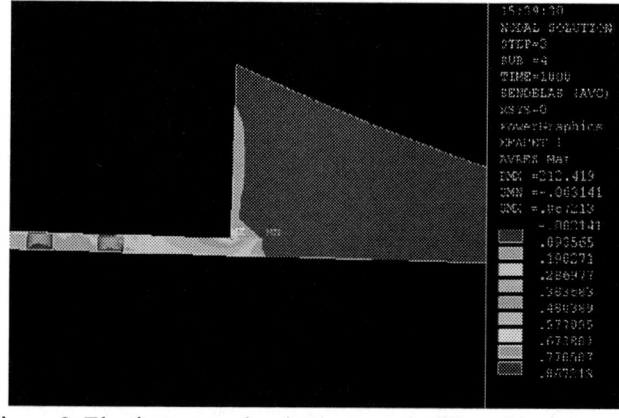

Figure 8. Simulation model.

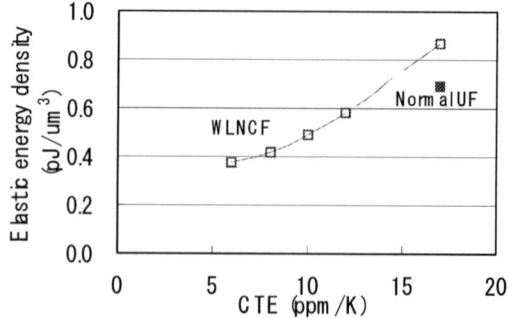

Figure 9. Elastic energy density in an underfill at chip corner.

Figure 10. Relationship of the elastic energy density in underfills to the substrate's CTE.

7. Results and discussion

In this study, first we measured the fundamental properties of WLNCF as a pre-applied wafer level underfill. The material produced good solder joints without voids and a highly reliability components. For exascale servers, some of the special expectations and concerns were discussed. WLNCF has a small fillet at the corners of the chips, which can receive an overcoat using additional underfill dispensed just after chip join (and before the final cure step), resulting in high mechanical strength enhancement and few chip-corner defects during the reliability testing. Regarding the underfill application methods, NCP dispensing and WLNCF coating methods were considered for the exascale packages with larger die and finer pitch bumps. NCP has been studied for industrial applications, and allows for process simplification with minimum process changes in the production line. However, there are concerns about the uniform spreading under the chips with narrow gap or large chips. In contrast, WLNCF can minimize the material flow under the chip, resulting in higher repeatability, but the fillet sizes may not be uniform, especially on large die, depending on variations in the original film's thickness.

Some potential advantages of WLNCF were also discussed. Basically WLNCFs have higher viscosity than NCP resulting in flat films and stable WLNCF wafers or chip handling in such stages as dicing the wafer, picking up the chip, etc. These characteristics also have side effects. For example, the WLNCFs with optimized viscosity can insure uniform chip height under constant loading pressure during the chip-joining process. This makes it easier to optimize the process conditions for a various size of chips in the production line.

For future exascale server packages, thermomechanical modeling was used to investigate the stress in the packages in relation to differing CTEs of the substrates. This is expected to be a trend for future packages. The results show that the maximum elastic energy density of the WLNCF with 40wt% silica filler (0.1 μm in diam.) on the low-CTE substrate is lower than that of a common capillary underfill material with 60 wt% filler (several micron in diam.) on the current commercial substrate.

These characteristics and analysis results will be helpful in optimizing WLNCF parameters and chip joining conditions for reliable exascale server packages. And, another challenge of the NCF will be the thermal conductivity enhancement, whose requirement depends on the power consumption of the actual chips and the number of stacked chips.

References

1. P.W. Coteus, J.U. Knickerbocker, C.H. Lam, Y.A. Vlasov, "Technologies for exascale systems," in *IBM J. Res. & Dev.*, Vol. 55, no. 5, pp. 14:1 - 14:12 , 2011.

2. J. Seo, et al., "A 45-nm CMOS neuromorphic chip with a scalable architechture for learning in networks of spiking neurons," in IEEE Custom Integrated Circuits Conference (CICC), 2011

3. A. Horibe, et al., "Thermally Enhanced Pre-applied Underfills for 3D Integration," in *Proc. IEEE Electronic Components and Technol. Conf. (ECTC)*, Las Vegas, NV, May 28–31, 2013, pp. 909–914.

4. S. Kawamoto, et al., "The Optimization of the Composition of Non-Conductive Film and the Lamination to Wafer," in *Proc. IEEE Electronic Components and Technol. Conf. (ECTC)*, Las Vegas, NV, May 28–31, 2013, pp. 778–784.

5. S. Kawamoto, et al., "Effect of NCF design for the assembly of Flip Chip and reliability", in *Proc. IEEE Electronic Components and Technol. Conf. (ECTC)*, San Diego, CA, May 29–June 1, 2012, pp. 399–405.

Bump Geometric Deviation on the Reliability of BOR WLCSP

Yumin Liu, Yong Liu and Shichun Qu
Fairchild Semiconductor Corp.
82 Running Hill Rd, South Portland, ME 04074
Tel: (207) 761-3155, Fax: (207) 761-6339, Email: yong.liu@fairhildsemi.com

Abstract

To solve the mystery of early thermal cycling failures on a relatively small size (1.6 mm × 2.4 mm) wafer level chip scale package (WLCSP) with bump on re-passivation (BOR) structure, details of bump geometry parameters are looked into. While the under bump metal layer stack is first confirmed within specifications, UBM offset from center point is found greater than typical on the failed units. To fully understand the impact of this bump parameter shift that is defined by lithographic process, additional geometric configurations of the bump structure and materials, as well as bump parameter deviations are studied on the stress in solder, UBM and on chip SiN passivations. The study takes into consideration the thickness/size of UBM, aluminum metal pad, SiN passivation layer and PI re-passivation layer. UBM materials, such as nickel vs copper, are also included in the study. Board-level thermal cycling test (TMCL) is conducted on a WLCSP with 8x8 bump array at 0.4mm pitch. Simulation reveals that size ratio and bump offset indeed change the stress level in the critical passivation layer, and the results well explain the early failures recorded in a device qualification reliability test and follow up experimental verifications. Conclusion of this work should be applied when defining design rules for robust reliability performance and manufacturability of this traditional WLCSP bumping technology.

Introduction

WLCSP has become the package technology of choice in mobile applications, largely because of the smallest possible package size, high efficiency manufacturing process and low cost associated with it. Early WLCSP has an under ball metallization layer (UBM) deposited and patterned over the SiN passivation openings, which subsequently has a solder ball dropped on and reflowed [1& 2]. Board-level temperature cycling test is often tough for this kind of bumping technology with one common failure mode of SiN passivation and underlying metal layer crack. It is known that thin film SiN is brittle and is prone to crack under tensile stress. This issue is significantly improved by polymer re-passivation, i.e., depositing a polyimide (PI) layer on top of the SiN passivation layer before the deposition of UBM layer. However, SiN passivation crack may still occur under certain circumstances, even with the protection and cushioning of PI re-passivation layer, and that is to be investigated.

Intensive studies on the reliability of WLCSP have been conducted in recent years. In [3], the impact of silicon thickness, back side laminate (BSL), PCB with/without buried copper planes, PCB material, and the design rule of depopulated solder array, was thoroughly studied during the board-level thermal cycling test and drop test. The failure of WLCSP solder joints was investigated in [4], it found that incorrectly placed through vias in PCB board could indeed induce high solder stress at non-corner locations and cause premature component failures in the reliability tests. The effect of UBM on the thermo-mechanical reliability of WLCSP was reported in [5 & 6], in which the UBM thickness and UBM layer materials was investigated. From the literature, very few researches have been conducted on the geometric configurations of the bump structure for WLCSP, especially the size ratio and bump offset.

During the qualification of WLCSP BOR bumping, a 4×5 array, 0.4 mm pitch WLCSP failed unexpectedly in the component TMCL test. Failure mode was quickly identified as crack through chip passivation and the metal layer (Figure 1). After elimination of other potential causes, bumping layer shift came up as a potential cause for the unusual failure mode on this regular size WLCSP.

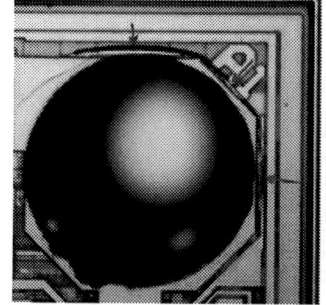

Figure 1. Optical images show cracks (red arrows) in the SiN passivation after the chip removal from the test PCB (left) and bump removal (right). The bottom hollow arrow highlights the gap due to off-center UBM.

To fully understand the impact of the bumping layer shift to the reliability of BOR WLCSP, board level TMCL simulation is conducted on an 8×8 array 0.4 mm pitch WLCSP, with variations in UBM metal stack, thickness, size and offset, etc. (Table 1).

Table 1 Simulation cases

Sims Case	UBM Stack	UBM Size	PI* Open	Al Pad/ Passivation	UBM offset
Ref	Ni: 2 μm	205 μm	170 μm	225/215 μm	0
1a	Ni: 2 μm	215 μm	180 μm	225/215 μm	8
1b	Cu: 8 μm	215 μm	180 μm	225/215 μm	8
2a	Ni: 2 μm	235 μm	180 μm	225/215 μm	0
2b	Cu: 8 μm	235 μm	180 μm	225/215 μm	0
3a	Ni: 2 μm	215 μm	180 μm	225/215 μm	0
3b	Ni: 2 μm	215 μm	180 μm	235/225 μm	0
3c	Ni: 2 μm	235 μm	195 μm	250/240 μm	0

978-1-4799-2408-0/14 $31.00 © 2014 IEEE

* PI thickness is 10 μm for all simulation cases.

Stresses in the solder, UBM and chip passivation are particularly looked at to identify root cause of the failure mode observed in the component reliability test.

Figure 2 below provides top view of relative sizes of the simulation reference and simulation case 1a and 1b, both featuring 8 μm off center shift which exposes UBM over the edge of under bump aluminum pad. WLCSP with purposely oversized UBM (case 2a) was manufactured and tested in TMCL for the verification of FEA simulation results.

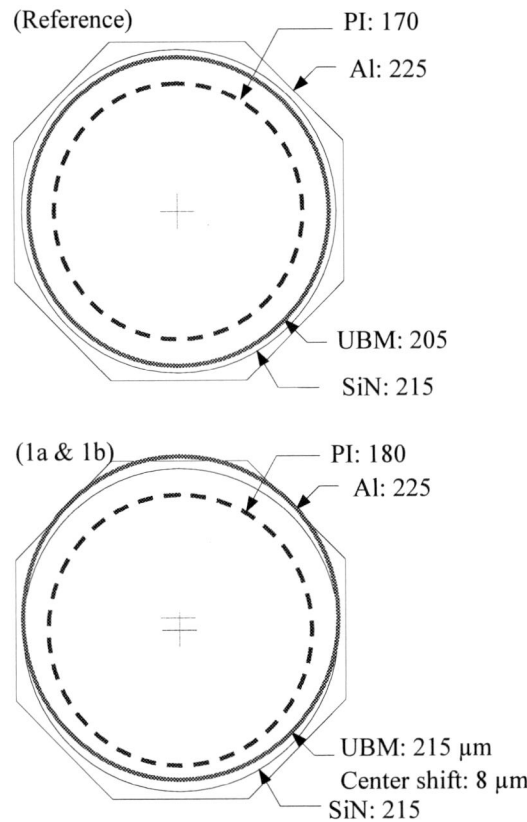

Figure 2. Relative sizes of simulation reference case and simulation case 1a and 1b – both cases featuring 8 μm UBM shift from center point, which extend UBM edge beyond the aluminum pad edge.

FEA Model Setup

An 8×8 array, 0.4 mm pitch WLCSP is modeled in the investigation of the influence of relative size and offset in board-level thermal cycling test. Solder bumps, UBM, the Al pads, SiN passivation layer, PI re-passivation layer, and silicon bulk were all factored in the FEA model. A total of 8 simulation cases were studied (Table 1).

Figure 2 above reveals the size and position relationships in the top view of the centered and shifted UBM cases. Figure 3 in this section shows the detailed cross-section views of all the FEA models in the simulation study. Case 1a and 1b have 8 μm UBM center offset, which is quite obvious in the cross sections.

For TMCL simulation, the assembled PCB with mounted WLCSPs follows the JEDEC Condition G (-40 to 125 °C) [7], with 15 minute dwell time for a typical frequency of two cycles per hour. Standard JEDEC drop board, as shown in Figure 4, is also referenced in the simulations.

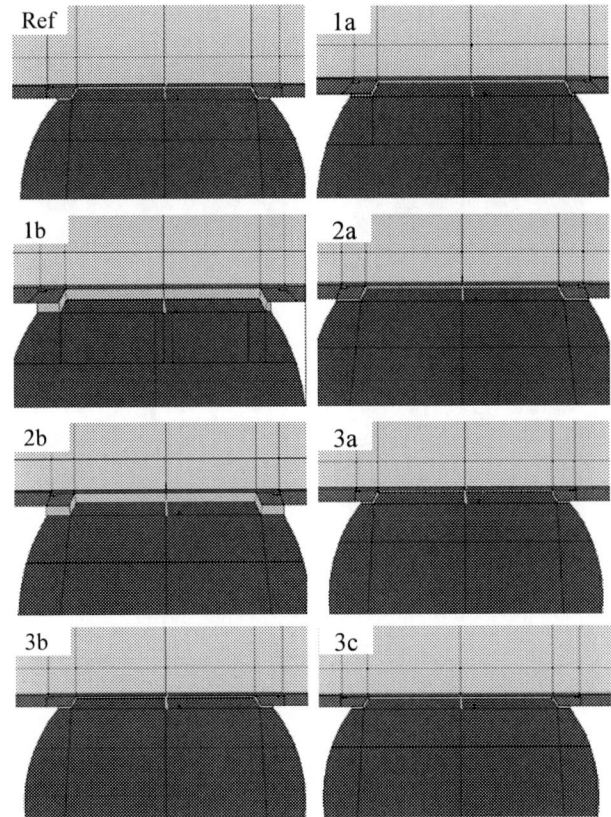

Figure 3. Cross sectional views of all simulation cases (including the reference). Case 1a and 1b feature 8 μm UBM shift from center point, which extend UBM edge beyond the aluminum pad edge.

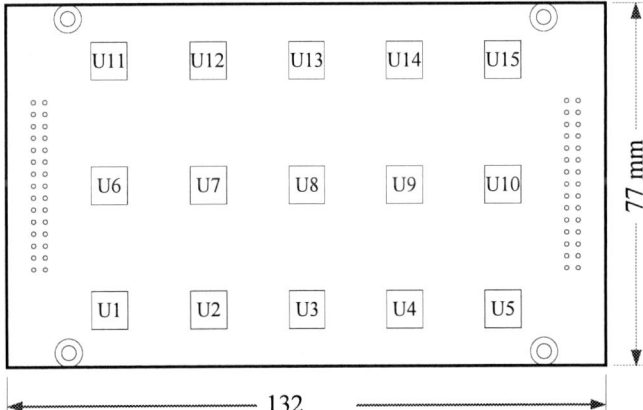

Figure 4. JEDEC PCB adopted in the simulation and 15 surface mounted components.

From previous experience, the inter-impact between different units on the PCB board is negligible. Therefore, a simplified symmetrical model is built as shown in Figure 5, with a quarter of the WLCSP and part of the PCB board. In the model, only the corner solder joint is meshed with fine elements, and all the others have coarse meshes. The loading condition in the FEA follows the JEDEC condition, and the stress-free temperature is supposed to be the high soak temperature 125 °C. It is assumed that the temperature is

978-1-4799-2408-0/14 $31.00 © 2014 IEEE

uniform throughout the whole model at a specific time. The basic material property data of the WLCSP package are listed in Table 2a. The effect of visco-plasticity of the solder joint material is considered, and Anand model is used in the simulations. The parameters of Anand model of the solder material are selected from [8].

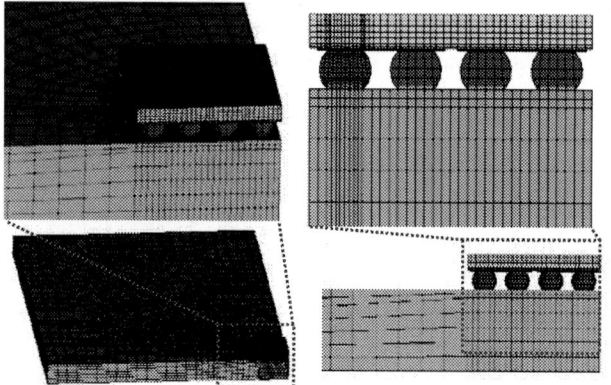

Figure 5. FEA model details.

Table 2a: simulation basic materials properties

Materials	Modulus (GPa)	Poisson Ratio	CTE ($\times 10^{-6}$ ppm/°C)
Silicon	131	0.278	2.4
Solder joint	see table 4	0.4	21.9
Passivation	314	0.33	4
Polyimide	3.5	0.35	35
FR4	Ex = 25.42 Ey = 25.42 Ez = 11 Gxz = 4.97 Gyz = 4.97 Gxy=11.45	Nuxy=0.11 Nuxz=0.39 Nuyz=0.39	Alpx = 16 Alpy = 16 Alpz = 60
Cu	117	0.33	16.12
Al Pad	68.9	0.33	20
Ni	124.5	0.299	15

Pad aluminum is considered a bilinear material. Table 2b lists its yield stress and tangent modulus at different temperatures. Cu and Ni are also bilinear materials. Table 2c shows their yield stress and tangent modulus.

Table 2b: Yield stress and tangent modulus of Al pad

Temperature (°C)	25	125
Yield stress (MPa)	200	164.7
Tangent Modulus (MPa)	300	150

Table 2c Yield stress and tangent modulus of Cu and Ni

	Cu	Ni
Yield stress (MPa)	70	59
Tangent Modulus (MPa)	700	590

Table 3a and Table 3b list the non-linear properties for the solder joint. Table 3a gives temperature dependent elastic modulus of solder materials. Table 3b gives the parameters of a rate-dependent Anand model for the solder joint.

Table 3a. Temperature dependent elastic modulus of solder

Temperature (°C)	35	70	100	140
Modulus (GPa)	26.38	25.8	25.01	24.15

Table 3b: Anand model constants for solder alloy

Description	Symbol	Constant
Initial value of s	s_o	1.3 MPa
Activation energy	Q/R	9000 K
Pre-exponential factor	A	500/s
Stress multiplier	ζ	7.1
Strain rate sensitivity of stress	m	0.3
Hardening coefficient	h_o	5900 MPa
Coefficient for deformation resistance saturation value	$\hat{s}$	39.4 MPa
Strain rate sensitivity of saturation value	n	0.03
Strain rate sensitivity of hardening coefficient	a	1.4

Simulation Results

The commercial FE code Ansys® is applied for all the simulations. During the thermal cycling test, the deformation of solder joints is accumulated for each cycle due to the CTE mismatch of different materials. The max stress of the solder joints occurs at the corner solder joint with fine mesh, which has the longest distance from the WLCSP center point. The von Mises stress contours of the solder joints at -40C are shown in Figure 6. It can be seen that the max von Mises stress occurs at the interface of solder joint and UBM layer. The von Mises stress of solder joints can be reduced by changing the UBM material from Ni to Cu (case 1b, 2b). Comparing cases 1a and 3a, the offset of UBM from center point (case 1a) increases the von Mises stress of solder joints.

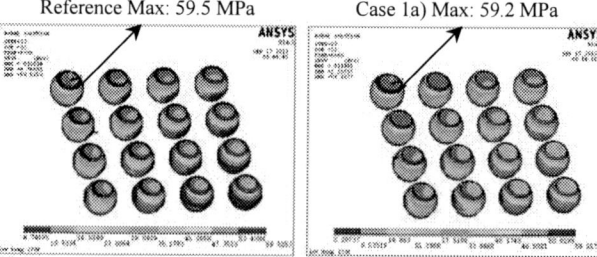

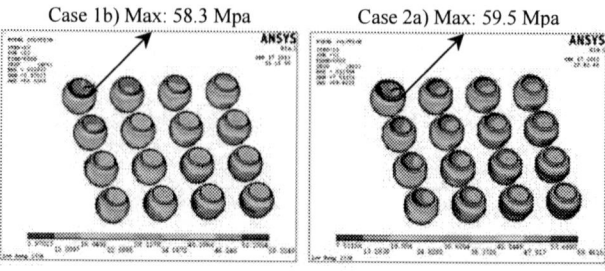

978-1-4799-2408-0/14 $31.00 © 2014 IEEE   810

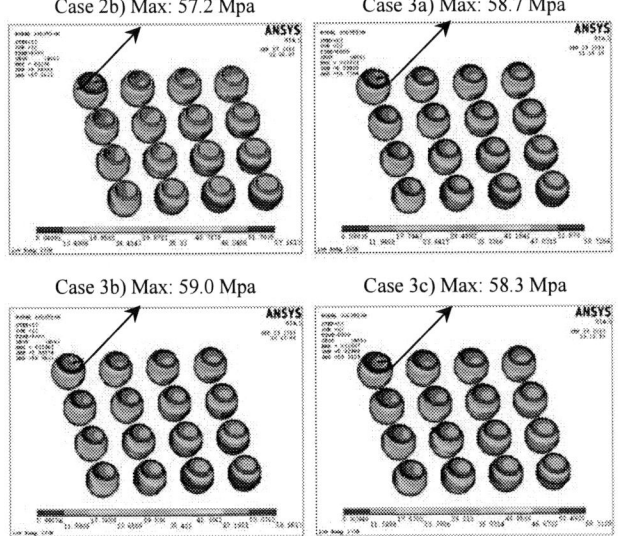

Figure 6. Von Mises stress of the solder for all 8 simulation cases.

The von Mises stress contours of UBM at -40 °C for all 8 simulation cases are shown in Figure 7. The copper UBM is found to reduce the von Mises stress in UBM layer (case 1a vs.1b; 2a vs. 2b), regardless of alignment ship to the center. Larger UBM size is also found to reduce the von Mises stress in UBM layer (case 3b vs. 3c). The offset of Ni UBM from center point (case 1a) increases the von Mises stress of the UBM layer by 21.3%, comparing with the normal no-offset UBM layer (case 3a).

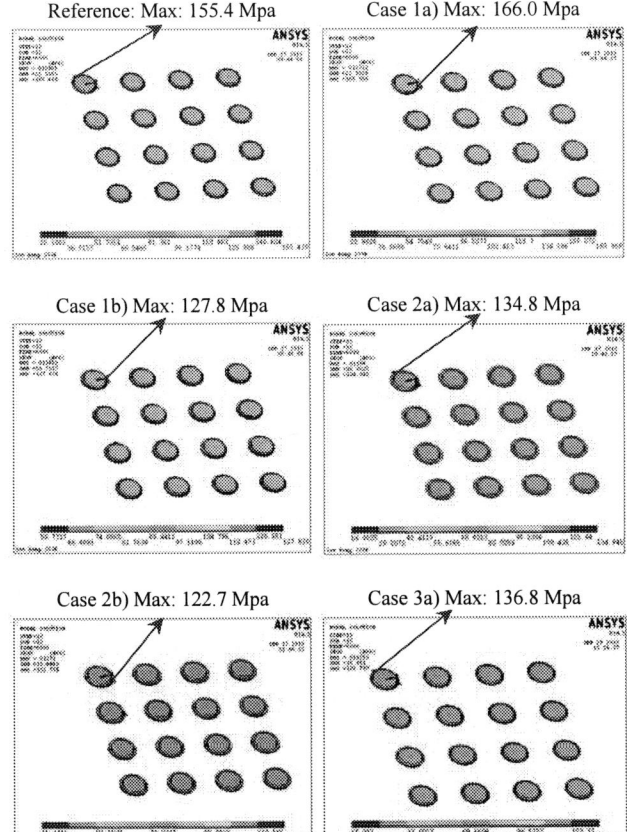

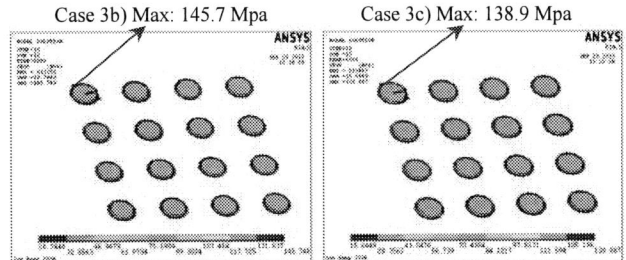

Figure 7. Von Mises stress of the UBM for all 8 simulation cases.

The SiN passivation is a brittle material of interest since identified failures are cracks in passivation and underlying metal layers. Tensile stress is used as the judging criterion. The 1st principal stress (S1) contours of the SiN passivation layer at -40 °C are shown in Figure 8. Comparing cases 1a and 3a, the offset of UBM from center point (case 1a) indeed produces higher tensile stress (+4.2%) in SiN passivation layer. On the other hand, over sizing UBM by just 10 μm with the same size aluminum pad seems to produce much significant (17.1%) tensile stress increase (reference vs. case 3a). Further extending the UBM size over the edge of aluminum pad only makes the situation much worse in the SiN passivation (case 2a vs. reference), which sees a 61.7% increase in tensile stress. It is worth noting that the tensile stress of SiN passivation layer can be slightly reduced by changing the UBM material from Ni to Cu. Size increase of Al metal pad/SiN passivation can help to reduce the tensile stress in SiN passivation layer (case 3a, 3b, 3c). So maintaining the proper size of aluminum pad/SiN passivation and UBM is important for not over stressing the SiN passivation.

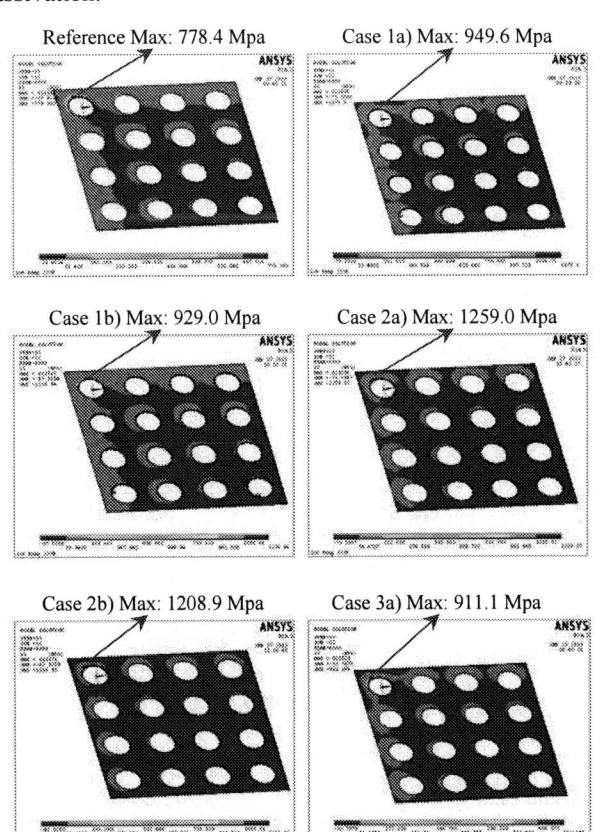

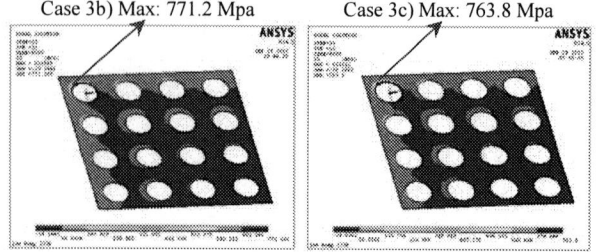

Figure 8. The S1 stress of SiN passivation for all 8 simulation cases

Table 4a & 4b summarize the stresses in Solder joints, UBM layer, Al metal pad, and SiN passivation layer at -40 °C for all 8 simulation cases.

Table 4a. Stress comparison of simulation case 1a, 1b, 2a and 2b

Sim Case	1a	1b	2a	2b
UBM size [μm]	215	215	235	235
UBM shift [μm]	8	8	0	0
Metal stack	2μm Ni	8μm Cu	2μm Ni	8μm Cu
PI open [μm]	180	180	180	180
Al pad [μm]	225	225	225	225
Pass open [μm]	215	215	215	215
S1 of solder* [MPa]	89.1	81.2	72.6	64.1
Seqv of solder* [MPa]	59.2	58.3	59.5	57.2
Seqv of UBM* [MPa]	166.0	127.8	134.8	122.7
Seqv of Al* [MPa]	200.7	200.5	200.8	200.8
S1 of Passi* [MPa]	949.6	929.0	1259.0	1206.9

* All maximum values

Table 4b. Stress comparison of simulation case 3a, 3b, 3c and Ref

Sim Case	3a	3b	3c	Ref
UBM size [μm]	215	215	230	205
UBM shift [μm]	0	0	0	0
Metal stack	2μm Ni	2μm Ni	2μm Ni	2μm Ni
PI open [μm]	180	180	195	170
Al pad [μm]	225	235	250	225
Pass open [μm]	215	225	240	215
S1 of solder* [MPa]	82.0	81.0	77.5	86.3
Seqv of solder* [MPa]	58.7	59.0	58.3	59.5
Seqv of UBM* [MPa]	136.8	145.7	138.9	155.4
Seqv of Al* [MPa]	200.8	200.9	200.9	200.8
S1 of Passi* [MPa]	911.1	771.2	763.8	778.4

* All maximum values

The fatigue life of the solder joints is predicted based on the Darveaux methodology [9], which applies both energy and damage accumulation-based theories. Based on previous tests and investigations [3], [4], [10], during the thermal cycling test the solder crack may firstly occur at the package side rather than at the PCB side. Therefore, the slice of solder joint at the package side is taken to calculate the fatigue life. The plastic energy density contours of the corner solder joint at the die side at 125 °C are shown in Figure 9. It can be seen that the copper UBM can help to decrease the plastic energy

density of the solder joint. Comparing case 1a and 3a, the offset of UBM from center point (case 1a) produces higher plastic energy density of the solder joints. The fatigue life of solder joints is calculated based on the assumed crack length and average increase of the plastic energy density between 2 successive thermal cycles. The calculated fatigue life of the corner solder joint for all the 8 cases are illustrated in Figure 10. Cases 2a, 2b, and 3c have relatively longer fatigue life among all the 8 cases. With the increase of the UBM size, the solder joint crack length becomes bigger, so the solder joints have longer fatigue life. The WLCSPs with copper UBM have longer fatigue life than those with Ni UBM layers. The offset of UBM from the center point has negative effect on the fatigue life of solder joints (See case 1a vs. case 3a). The reference case has the shortest fatigue life during the thermal cycling.

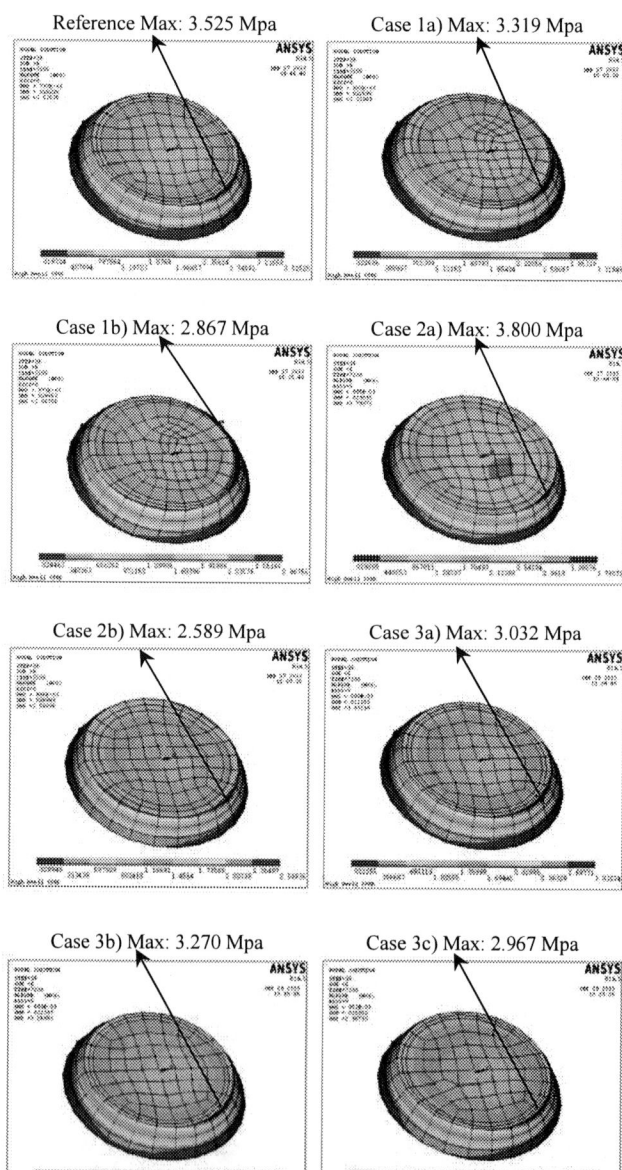

Figure 9. The plastic energy density of the corner solder joint at the die side for all 8 simulation cases.

In order to have better understanding of the effect of offset/size/material of UBM layer, size ratio of UBM vs Al metal pad, size of SiN passivation layer, thickness/opening of PI re-passivation layer, the tensile stress in SiN passivation

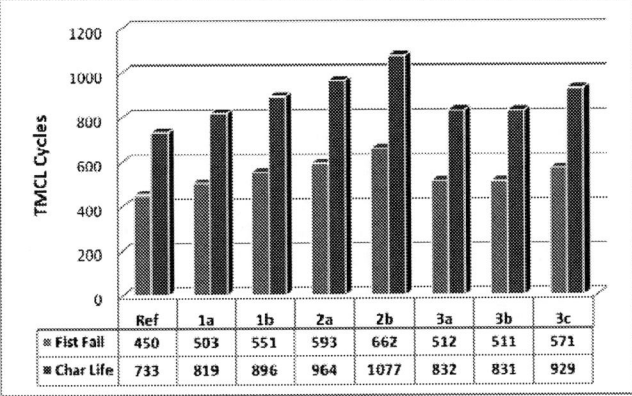

	Ref	1a	1b	2a	2b	3a	3b	3c
※ Fist Fail	450	503	551	593	662	512	511	571
※ Char Life	733	819	896	964	1077	832	831	929

Figure 10. Fatigue life of the corner solder joint at the die side for all 8 simulation cases.

layer and fatigue life of solder joints are taken as judging criteria, which are illustrated in Figure 11. It can be seen that the copper UBM, size of Al metal pad / SiN passivation, have positive impact on the thermal cycling performance of the WLCSP, since it produces lower tensile stress in SiN passivation layer and longer fatigue life of solder joints. Larger size ratio of UBM vs Al metal pad produces longer fatigue life of solder joints, but brings more risk in the SiN passivation layer. The offset of UBM from the center point has negative effect on the thermal cycling performance of the WLCSP, by increasing the tensile stress in SiN passivation layer and reducing the fatigue life of the solder joints.

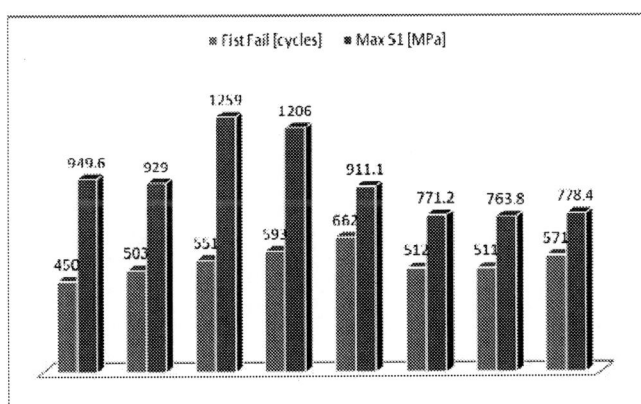

Figure 11. Simulation results comparison of the 8 cases.

Experimental Verification

The same 4×5, 0.4 mm WLCSP observed early TMCL failures with off-center UBM previously are fabricated with oversized UBM (case 2a) for board-level thermal cycling test. Data is compared to the reference. For the reference (subgroup A), UBM diameter is 10 μm smaller than the Al metal pad. For test group (subgroup B), the UBM diameter is 10 μm greater than the Al metal pad. At 850-cycle read point, there are 3 failures observed for subgroup B but zero failure for subgroup A. thorough failure analysis is performed on the 3 failed units, with combination of non-destructive and deprocessing techniques. Curve tracer is first performed to confirm the failures post sample removal from test PCB. Then Back side TIVA/NIR located the potential cracks in the silicon (Figure 12). After selectively removal of the solder and UBM metal layers, it is confirmed that the failure mode is indeed cracking in passivation/metal layers (Figure 13), which is typically seen under normal condition for the bumping technology at this particular bump count.

Figure 12. Back Side NIR image showing the silicon crack on one TMCL failure unit.

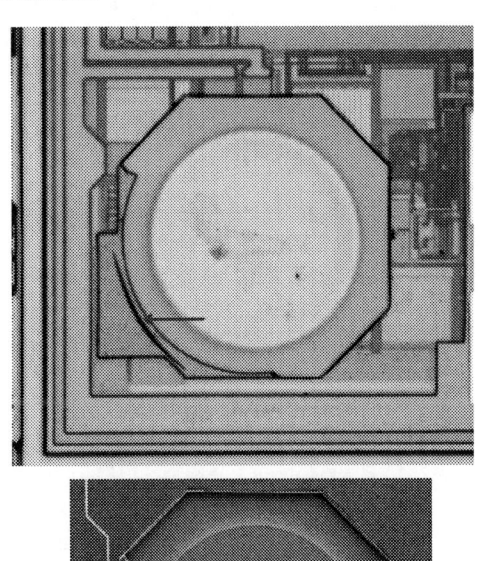

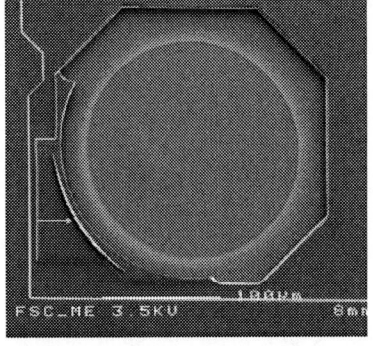

Figure 13. Optical photo and SEM image showing SiN passivation crack post solder bumps/polyimide removal.

Discussion & Conclusions

In this paper, the board-level thermal cycling simulations of a WLCSP with 8×8 bump array at 0.4 mm pitch are

978-1-4799-2408-0/14 $31.00 © 2014 IEEE

thoroughly investigated to understand the impact of the geometric configurations of the bump structure, the bump parameter shift, and UBM materials. The board-level thermal cycling reliability tests are conducted in the reliability lab, and the failed devices are carefully analyzed by a combination of FA techniques. From the FE simulation results and experimental results, it can be concluded:

(1) The offset of UBM from the center point has negative effect on the thermal cycling performance of the WLCSPs, by increasing the tensile stress in SiN passivation layer and reducing the fatigue life of the solder joints. In the extreme case, the off-center UBM can result early TMCL failures by inducing passivation/metal crack.

(2) Larger size ratio of UBM vs Al metal pad produces longer fatigue life of solder joints, but brings more risk in the SiN passivation layer during the thermal cycling test.

(3) The copper UBM, larger size of Al metal pad / SiN passivation, can improve the board-level thermal cycling performance of the WLCSPs.

Based on the above conclusions, design rules governing UBM and aluminum pad size rule should thoughtfully set up to extend the solder joint TMCL life while avoiding excessive stress in the SiN passivation. Bumping process capability should also be well characterized to limit the misalignment of UBM to the chip Al/passivations. Well centered bump structure is essential for the robust TMCL performance of the BoR WLCSP bumping technology.

Acknowledgments

The authors wish to thank the support from Fairchild package development group, product line, and analytical group. In particular, Etan Shacham, Fabio Principi and Yun Chow are acknowledged for their unique contributions in initial technical discussions and follow up experimentations.

References

1. J. Hao, Y. Liu, J. Hunt, T. Tessier, H. Kuisma and T. Wakabayashi, "Demand for wafer-level chip-scale package accelerates," 3D Packaging, No.22, Feb, 2012.

2. Application Note, "Wafer Level Chip Scale Packaging", Freescale Semiconductor, Rev2.0, 8/2009.

3. Y.M. Liu, Y. Liu, "Prediction of board-level performance of WLCSP," in Proc. IEEE Electronic Components and Technol. Conf. (ECTC), 2013, pp. 840-845.

4. Y. Liu, Q. Qian, M. Ring, J. Kim & D. Kinzer, "Modeling for Critical Design of Wafer Level Chip Scale Package," in Proc. IEEE Electronic Components and Technol. Conf. (ECTC), 2012, pp. 959-964.

5. Y.S. Chan, S.W.R. Lee, F. Song, C.C.J. Lo, T. Jiang, "Effect of UBM and BCB layers on the thermo-mechanical reliability of wafer level chip scale package (WLCSP)," in Proc. Microsystems, Packaging, Assembly and Circuits Technology Conf. (IMPACT), 2009, pp. 407-410.

6. K. Ruhmer, E. Laine, K. O'Donnell, J. Kostetsky, et al, "Alternative UBM for Lead Free Solder Bumping using C4NP," in Proc. IEEE Electronic Components and Technol. Conf. (ECTC), 2007, pp. 15-21.

7. JEDEC JESD-A104D, "Temperature Cycling", Mar 2009.

8. T.O., Reinikainen, P., Marjamäki, and J.K., Kivilahti, "Deformation Characteristics and Microstructural Evolution of SnAgCu Solder Joints," EuroSime Conference Proc., Germany, Apr. 2005, pp. 91-98.

9. R. Darveaux, K. Banerji, A. Mawer, and G. Dody, Reliability of Plastic Ball Grid Array Assembly, Ball Grid Array Technology, McGraw-Hill, New York, 1995, pp. 379-442.

10. Q. Qian, Y. Liu, "Reliability Analysis of Next Generation Wafer Level Chip Scale," EuroSime Conference Proc., 2013, pp. 1-7.

Experimental Identification of Warpage Origination During the Wafer Level Packaging Process

Chunsheng Zhu[1,2], Wenguo Ning[1,2], Heng Lee[1,2], Jiaotuo Ye[1,2], Gaowei Xu[1], Le Luo[1,3]

[1]State Key Laboratory of Transducer Technology, Shanghai Institute of Microsystem and Information Technology,
Chinese Academy of Sciences (CAS), 200050, Shanghai, China
[2]University of Chinese Academy of Sciences, 100049, Beijing, China
[3]Contact E-mail: leluo@mail.sim.ac.cn

Abstract

Redistribution layer (RDL) composing of polyimide (PI) dielectric layer and electro-chemical deposited (ECD) Cu trace is a critical part for wafer level packaging (WLP). One concern of this multi-layered film structure is the wafer warpage induced during the process, which poses threats to automatic handling, 3-D integration and device reliability. In this paper, the warpage origination during the WLP process was identified and analyzed by experiments and simulations. The wafer warpage evolution during the WLP process was measured by a Multi-beam Optical Sensor system. We found that the cure shrinkage of PI has little effect on the warpage, however, it is mainly caused by the coefficient of thermal expansion (CTE) mismatch between the deposited materials. The ECD Cu trace in RDL accounted for a substantial proportion to the total wafer warpage and lead to a hysteresis response during the thermal processes indicating plastic deformation has taken place. For in-depth understanding, the plastic behavior of ECD Cu film was investigated and the kinematic hardening plastic model was established. Finally, the stresses distribution in RDL structure was simulated by numerical method and the influence of ECD Cu trace pattern on the wafer warpage was evaluated.

Introduction

Novel materials, new designs, and new packaging technologies are introduced frequently to meet the demands for smaller, faster, and cheaper electronic components. Wafer level packaging (WLP) is one of the emerging package technologies that have the key advantages of reduced cost and smaller form factor [1-3]. In recent years, a variety of WLP technologies have been developed and the Standard WLP, which is similar to a typical flip chip technology, has evolved with the incorporation of redistribution layer (RDL) process, copper post process, and other compliant layer process [1, 4].

The RDL is often used to re-route the peripheral I/O layout into a new area array footprint for solder bumping. In the RDL process, thick polyimide (PI) is widely adopted as the dielectric layer and electro-chemical deposited (ECD) Cu serving as the trace layer. To fabricate the RDL and the following solder bump, wafer will underwent several high temperature thermal processes. For example, the temperature of PI curing process could be as high as 375 ℃. Due to the cure shrinkage of PI and the coefficient of thermal expansion (CTE) mismatch between the deposited materials and Si substrate, large wafer warpage will generate after thermal processes and it poses a threat to handling of the wafer and affects the quality of sequential process steps such as lithography [5-7]. The warpage problem becomes seriously with the adoption of larger size wafer and thinner substrate for 3-dimensional integration. Moreover, the induced large thermal stresses might cause cracking or interfacial delamination in the layered structure [8, 9]. It is thus important to understand the evolution of the thermal stresses as well as its potential impact on the wafer warpage and the reliability of WLP. Experimental methods to measure the wafer warpage evolution during the thermal process and to be validated with numerical models are urgently needed.

In this paper, wafer warpage evolution during a typical WLP process with RDL has been characterized and analyzed. The thermal behaviors of PI film and ECD Cu film were studied for in-depth understanding their effects on the wafer warpage. Plastic material model of ECD Cu was established and adopted by finite element analysis (FEA) to elucidate the stresses distribution in the RDL structure. In the study, it is found that the ECD Cu trace has a great influence on the warpage, therefore, the relationship between the ECD Cu trace pattern and the wafer warpage was also investigated.

Experimental

Figure 1 shows the WLP structure adopted in our experiment and the corresponding process flow. It was done on a 4-inch silicon wafer with a thickness of 420 um and each chip unit had a size of 4.3 mm × 4.3 mm.

1) First, Al pad with a thickness of 0.5 μm was fabricated on the Si wafer using sputtering deposition and wet etching method.

2) A 0.5 um thick SiN$_x$ passivation was deposited on the wafer surface, and contact window was etched upon each pad to form the interconnection with the RDL.

3) The 1st PI layer was span coated and patterned. Then the wafer underwent the first thermal process to curing the 1st PI layer. After curing, the thickness was 9.8 ± 0.18 μm.

4) A 5 μm thick ECD Cu trace was deposited using through-mask plating method with sputtered Ti/Cu serving as the adhesion layer and seed layer, respectively.

5) The 2nd PI layer was span on the wafer surface, and then the wafer went through the second thermal process to cure the 2nd PI layer. After curing, the thickness of 2nd PI layer was similar to that of the 1st one.

6) The under bump metallization (UBM) was deposited, which consisted of ECD Cu/Sputtered Cu/TiW. The thickness of ECD Cu, sputtered Cu and TiW were 5, 0.4, and 0.2 μm, respectively.

7) Finally, solder paste was screened on the UBM and, then, reflow was processed at a peak temperature of 260 ℃ in the nitrogen ambience with an oxygen concentration of less than 100ppm.

In the experiment, PMDA-ODA based photosensitive PI was adopted and the curing temperature was 375 °C for an hour with a heating and cooling rate of 5 °C/min. The curing process was carried out in nitrogen atmosphere. The glass temperature of the PI is around 290 °C. The ECD Cu trace pattern adopted in our experiment is also plotted in figure 1. It occupies certain area on the PI and can be separated in two parts, i.e., the lines and the pads. Generally, the pattern and the coverage of Cu trace could vary greatly depending on design requirements. The Cu trace area ratio might change from 10% to 80% of the total wafer area. In this work, the Cu trace area ratio is about 40%.

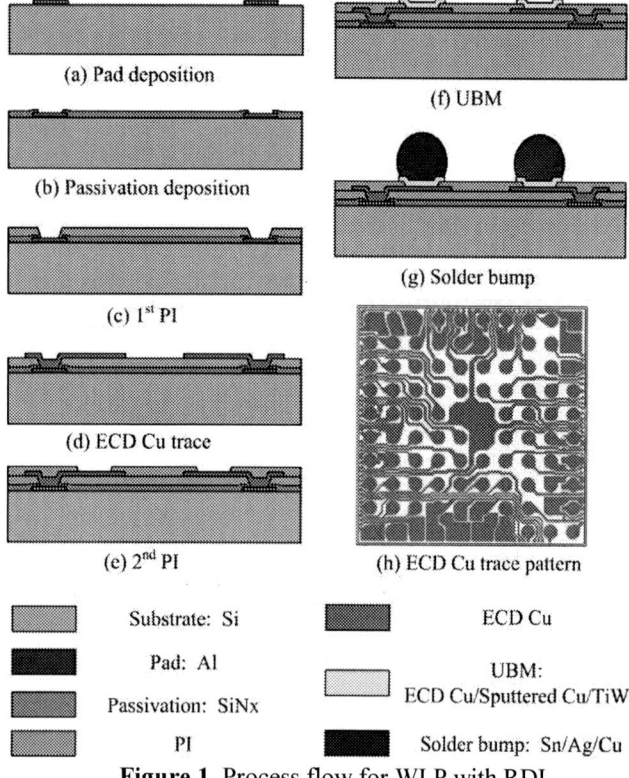

(a) Pad deposition

(b) Passivation deposition

(c) 1st PI

(d) ECD Cu trace

(e) 2nd PI

(f) UBM

(g) Solder bump

(h) ECD Cu trace pattern

Substrate: Si

Pad: Al

Passivation: SiNx

PI

ECD Cu

UBM:
ECD Cu/Sputtered Cu/TiW

Solder bump: Sn/Ag/Cu

Figure 1. Process flow for WLP with RDL

Wafer warpage after each process was measured by a Multi-beam Optical Sensor (MOS) system from k-Space Associate Inc., which also allows measuring the curvature evolution during the thermal process in suit. The principle of the MOS system is similar to the normal wafer curvature measurement method for characterizing stress in the thin film, which is based on the laser reflection theory [10]. Since the laser reflection of the front surface of the wafer could be affected by the presence of the RDL and solder bump structures, double-side polished wafer was used and all the measurements were conducted at the polished back surface [11].

Typically, Stoney's equation is used to determine the average stress in a thin film structure based on the measured wafer curvature:

$$\sigma_f = \frac{Et_s^2}{6(1-\upsilon)t}(K_f - K_0)$$

where E is the Young's Modulus, t_s is the wafer thickness, t is the film thickness, υ is Poisson's ratio, K_f is the final wafer curvature and K_0 is the initial wafer curvature [12]. A positive sign for stress indicates tensile stress.

Results and discussions

The final wafer deformation after solder reflow process is shown in figure 2, which is like a bow. The bow height is about 154.2 um. The wafer curvature after each process is plotted in figure 3. It has two large increments, which corresponding to the 1st PI layer and the 2nd PI layer deposition. The curvature induced by Al pad and SiN$_x$ is small due to their thin thickness. The electro-chemical deposited process also has little effect on the curvature since it was carried out at room temperature. The figure also indicates that the formation of solder bumps and the reflow process have little contribution to the wafer curvature because the solder bumps are under-constraint and have limited contact area with the wafer. To further investigate the warpage origination, curvature evolutions during the two PI curing processes were characterized.

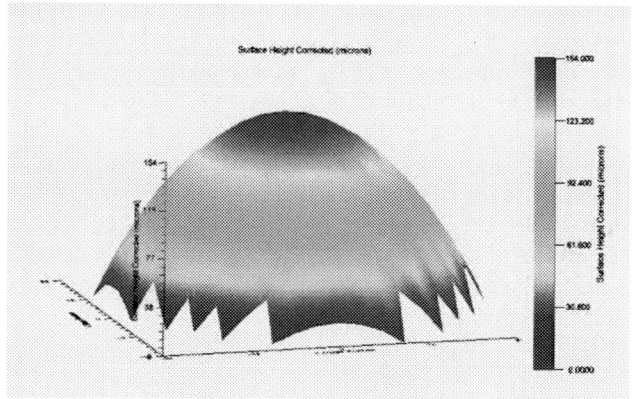

Figure 2. Final wafer deformation

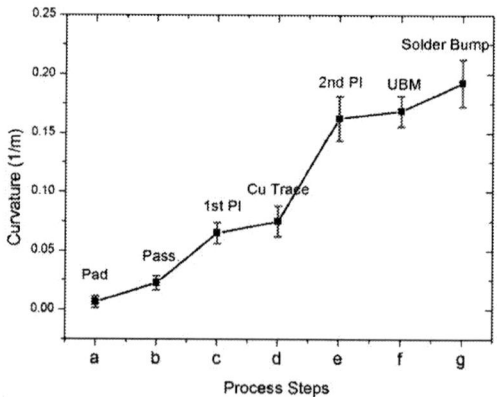

Figure 3. Wafer Curvature after each processes

Curvature Evolution during the Curing Process of the 1st PI Layer

As for the 1st PI layer during its curing process, the wafer curvature was small in the heating stage and a tensile stress was built in the cooling stage from 290 °C to room temperature, which is shown in figure 4. Since the

978-1-4799-2408-0/14 $31.00 © 2014 IEEE

polymerization of PI was carried out at 375 °C and the curvature changed little at this stage, therefore, it can be concluded that the cure shrinkage of PI has small contribution to the warpage and the stress was mainly caused by the CTE mismatch between the PI film and the substrate. Generally, the thermal stress induced by CTE mismatch can be calculated as

$$\sigma_{thermal} = \int_{T_1}^{T_2} (\frac{E_f(T)}{1-\upsilon_f})[\alpha_s - \alpha_f(T)]dT$$

where α_s and $\alpha_f(T)$ are the CTE of substrate and film, respectively, $E_f(T)$ and υ_f are the Young's modulus and Poisson's ratio of the film. Previous research indicates that PI owns viscoelastic properties and its modulus increases monotonically with decreasing temperature while the CTE decreases monotonically [13]. The linear behavior in figure 4 implies that the product of the modulus and the CTE of PI is approximately a constant between room temperature and 290 °C. To investigate the stability of the cured PI, an additional thermal process was proceeded on this sample, which only consisted the 1st PI layer. As shown in figure 4, the additional heating and cooling results overlap each other without hysteresis loop, which indicates the cured PI has stable thermal-mechanical properties.

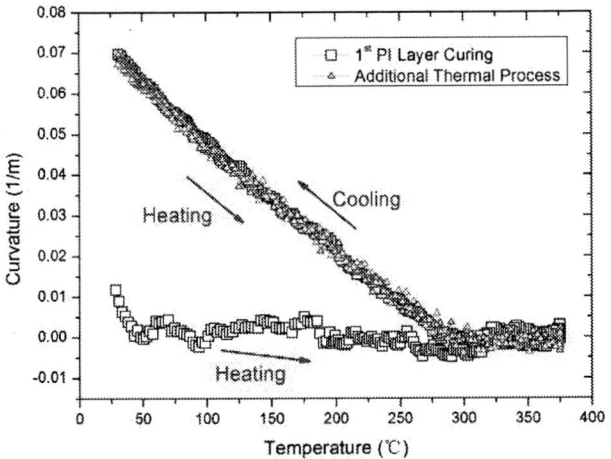

Figure 4. Wafer curvature evolution during the curing process of the 1st PI layer

Curvature Evolution during the Curing Process of the 2nd PI Layer

The result of the 2nd PI layer curing, plotted in figure 5 with solid square lines, is very different from the 1st one since the structure contains ECD Cu trace. Upon initial heating, the curvature decreased linearly with the temperature as ECD Cu and PI have a larger CTE than Si substrate. Further heating above 100 °C, the slope of the curve decreased and became much smaller. This phenomenon is caused by the microstructural change of the ECD Cu trace in the sample [10, 14, 15]. The Cu after electroplating was unstable, therefore, during the first thermal process, microstructural evolving accompanied by stress relaxation and grain growth will occur under the influence of induced thermal stress at high temperature. Two successive thermal cycling processes were carried out on this sample for further investigation, which is plotted in figure 5 with hollowed triangular symbols. As can

be seen, the microstructure of the Cu was stabilized and a hysteresis response formed, which indicates plastic deformation has taken place and this phenomenon is completely different from the response of the ECD Cu in through silicon via (TSV) of 3D packaging reported by S. K. Ryu [10]. As for the Cu in TSV, no hysteresis loop was observed. This might originate from the different shapes of the Cu structure. The geometry of the Cu vias is uniform in all directions and the stress state is like triaxial, however, the stress state of the Cu trace in RDL is more like a biaxial stress because the thickness of the trace is much smaller compared to its in-plane feature size, especially at the pad region. The biaxial stress will produce a relatively large effective shear stress required for plastic deformation, which has been widely investigated with blank film structure, and also for this reason an obvious hysteresis loop was observed.

Figure 5 also shows that the wafer curvature held around zero above 290 °C. This phenomenon is caused by the stress relaxation of PI, since the PI is in rubbery state above its glass transition temperature.

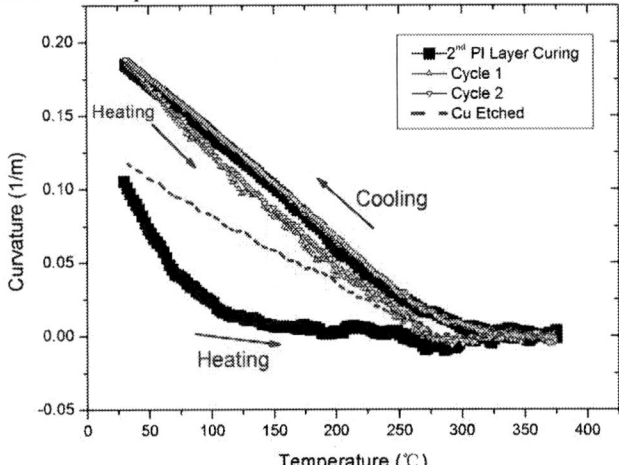

Figure 5. Wafer curvature evolution during the curing process of the 2nd PI layer

Contribution of ECD Cu Trace to the Wafer Curvature

To evaluate the influence of Cu on the total wafer curvature, ECD Cu trace in the RDL sample wafer was etched off using ammonium persulfate solution. Then the wafer (only contains two PI layers) underwent a thermal process between room temperature and 375 °C, and its curvature evolution was plotted with dash lines in figure 5. The curvature of the etched sample at room temperature is about 0.12 m^{-1}, therefore, the contribution of ECD Cu to the total curvature, about 0.08 m^{-1}, can be calculated. In this typical configuration, the ratio of ECD Cu and PI to the wafer warpage is roughly about 2:3. The Cu also plays a key role in influencing the warpage since it owns large Young's modulus and relative large CTE. To precisely calculate the wafer warpage, the effect of Cu trace should be taken into consideration and studied carefully.

Plastic Deformation of ECD Cu Film

Since the ECD Cu also has relative large influence on the wafer warpage, its properties were fully studied. A 5 μm blank ECD Cu film was deposited on a 4-inch Si substrate and

978-1-4799-2408-0/14 $31.00 © 2014 IEEE

underwent thermal cycling. Figure 3 shows the measured curvature-temperature relationship after stabilizing and a distinct plastic hysteresis loop can be observed. The right Y-axis is the corresponding stress calculated from the Stoney's equation. The curve exhibits distinct strain hardening and Bauschinger-like effect, which is similar to the result of the sputtered Cu film with silicon oxide passivation layer reported by Y. Shen [16]. The film can sustain a compression stress up to 100 MPa even at 375 °C and this characteristic can be attributed to the impurity incorporated during the electroplating process. This characteristic also verifies the previous conclusion that stress relaxation in RDL structure above 290 °C was caused by the softness and stress relaxation of PI. After cooling to room temperature, the curvature caused by the ECD Cu film is about 0.25 m^{-1}.

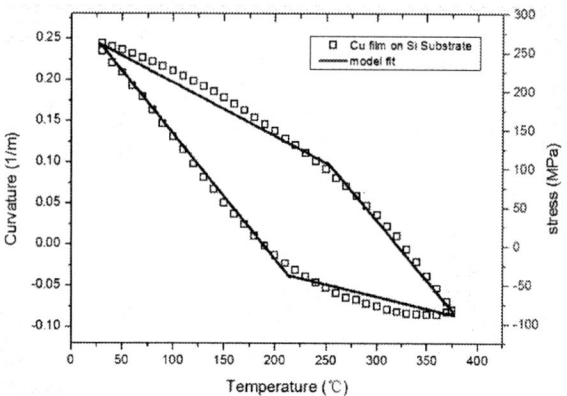

Figure 6. Experimental data of 5 μm ECD Cu film in the thermal cycling

Stresses Distribution in RDL Structure

FEA was performed to determine the stresses distribution and plastic deformation in the sample with RDL. Firstly, the kinematic hardening plasticity material model of ECD Cu was established based on the measured result of ECD Cu film in the previous section using the method provided by Y. Shen [16]. The fitted parameters are listed in table I, in which σ_0 and T_0 are reference constants, σ_y is the yield strength, H_{heat} and H_{cool} are the hardening rate for heating and cooling, respectively. The blue curve in figure 6 is the result of the fitted model. From the fitted parameters, we can get the yield strength of the ECD Cu film at room temperature is 172.6 MPa. All the other materials used in the FEA are modeled as linear elastic materials and are also listed in table I [17]. Due to the high aspect ratio between the thickness of the RDL and the diameter of the wafer, only a small section of the wafer has been modeled. The warpage of this equivalent cell was fitted by a quadratic function and the actual warpage of the wafer could be acquired by extrapolating the fitting curve, therefore, the wafer curvature can be calculated [18]. Since the solder bumps have little contribution to the wafer warpage, they were ignored in the simulation.

The wafer curvature calculated from the simulation is 0.17 m^{-1}, which is in well agreement with the measured result. The calculated von Mises stress and plastic strain distribution in

the RDL structure are shown in figure 7. The stress in Cu is much larger than that of PI and silicon substrate, which is due to its large CTE and Young's modulus. It is noted that the von Mises stress as the effective shear stress for plastic deformation reaches the yield strength, 172.6MPa, in the whole Cu structure. Both of the maximum stress and plastic strain locate at the center of the trace pad with a corresponding value of 257.26 MPa and 0.49%, respectively.

As the UBM and solder bump will be directly deposited on the surface of the trace, large plastic strain in the ECD Cu trace not only influences the wafer warpage, but also has potential impact on the void formation and reliability of the solder bump.

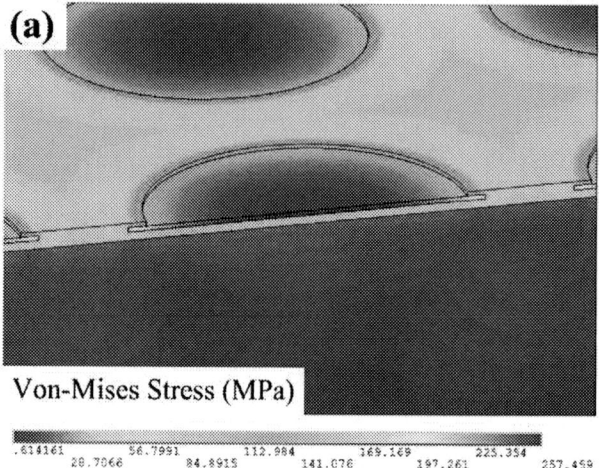

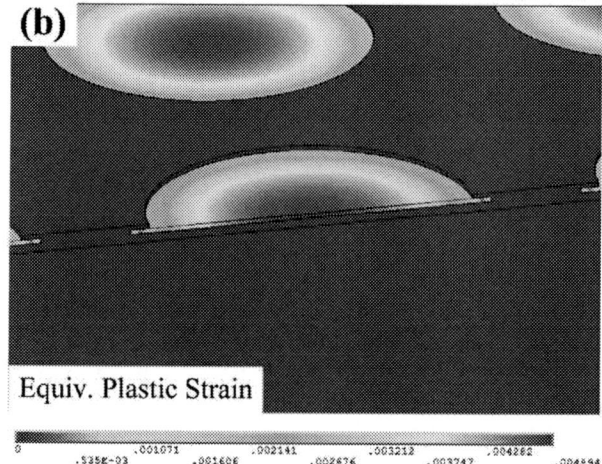

Figure 7. Simulation results: (a) von-Mises stress distribution; (b) Equivalent plastic strain distribution.

Effect of Cu Trace Area Fraction on the Wafer Curvature

The ratio of ECD Cu and PI to the wafer warpage is related to their area fraction in RDL when the thickness keeps unchanged. In engineering applications, the ECD Cu trace pattern may vary greatly dependent on design requirements. Therefore, the influence of Cu trace area fraction on the wafer curvature was studied by numerical method. The simulation model was consistent with the experiment sample, except the ECD Cu trace area fraction varied. As shown in figure 8, the relationship between the Cu trace area fraction in RDL and the wafer curvature is approximately quadratic polynomial. With increment of the Cu trace area fraction, its contribution to the warpage becomes much more important.

Table I. Material properties used in FEA

	Young's modulus (GPa)		CTE (ppm/°C)		Poisson's ratio	
Si	130		2.6		0.28	
PI	3.4		0.35		35	
Cu	91		16.5		0.36	
	σ_0 (MPa)	T_0 (K)	σ_y at 298K (MPa)		H_{heat} (GPa)	H_{cool} (GPa)
	252.9	938.9	172.6		32.6	32.6

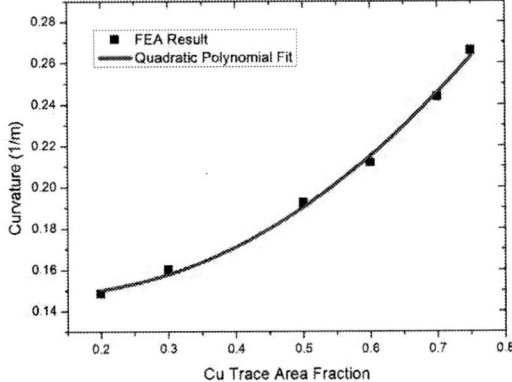

Figure 8. Simulated relationship between the Cu trace area fraction and the wafer curvatures

Conclusions

In summary, the curvature evolution during the WLP processes was measured by a MOS system. The characteristic of thermal stresses in the RDL were analyzed based on the measured results and FEA.

People generally think that PI plays the major role on the wafer warpage since it has significant volume shrinkage during the curing process. However, our experiment reveals that the cure shrinkage of PI has little effect and the warpage is primarily induced by CTE mismatch between different materials in the structure during the thermal process. The ECD Cu trace in the RDL also accounted for a substantial proportion in the total wafer warpage and lead to a hysteresis response during the thermal cycling, indicating an induced plastic deformation. The thermal behavior of ECD Cu film was studied and its kinematic hardening plasticity material model has been established. Finally, the relationship of the Cu trace area fraction and the wafer curvature has been investigated.

Acknowledgments

This work was supported by National Science and Technology Major Project (NO.2009ZX02025,

NO.2011ZX02602) and Natural Science Foundation of Shanghai (NO.13ZR1447300).

References

1. X. J. Fan, B. Varia, and Q. Han, "Design and optimization of thermo-mechanical reliability in wafer level packaging," *Microelectronics Reliability,* vol. 50, no. 4, pp. 536-546, Apr, 2010.
2. Y. S. Lai, C. L. Kao, Y. T. Chiu *et al.*, "Electromigration Reliability of Redistribution Lines in Wafer-level Chip-Scale Packages," *2011 IEEE 61st Electronic Components and Technology Conference (ECTC)*, pp. 326-331, 2011.
3. C. S. Zhu, W. G. Ning, J. T. Ye *et al.*, "FEA Study of the Evolution of Wafer Warpage During Reflow Process in WLP," *2012 13th International Conference on Electronic Packaging Technology & High Density Packaging (ICEPT-HDP 2012)*, pp. 660-664, 2012.
4. P. Tumne, V. Venkatadri, S. Kudtarkar *et al.*, "Effect of Design Parameters on Drop Test Performance of Wafer Level Chip Scale Packages," *Journal of Electronic Packaging,* vol. 134, no. 2, Jun, 2012.
5. S. S. Deng, S. J. Hwang, and H. H. Lee, "Warpage Prediction and Experiments of Fan-Out Waferlevel Package During Encapsulation Process," *IEEE Transactions on Components Packaging and Manufacturing Technology,* vol. 3, no. 3, pp. 452-458, Mar, 2013.
6. N. P. Pham, M. Rosmeulen, G. Bryce *et al.*, "Wafer bow of substrate transfer process for GaNLED on Si 8 inch," *Proceedings of the 2012 IEEE 14th Electronics Packaging Technology Conference*, pp. 202-205, 2012.
7. A. Tay, W. K. Ho, N. Hu *et al.*, "Estimation of wafer warpage profile during thermal processing in microlithography," *Review of Scientific Instruments,* vol. 76, no. 7, Jul, 2005.
8. F. Brunner, A. Mogilatenko, A. Knauer *et al.*, "Analysis of doping induced wafer bow during GaN:Si growth on sapphire," *Journal of Applied Physics,* vol. 112, no. 3, Aug 1, 2012.
9. H. J. Kim, S. C. Chong, D. S. W. Ho *et al.*, "Process and Reliability Assessment of 200 mu m-Thin Embedded Wafer Level Packages (EMWLPs)," *2011 IEEE 61st Electronic Components and Technology Conference (ECTC)*, pp. 78-83, 2011.
10. S. K. Ryu, T. F. Jiang, K. H. Lu *et al.*, "Characterization of thermal stresses in through-silicon vias for three-dimensional interconnects by bending beam technique," *Applied Physics Letters,* vol. 100, no. 4, Jan 23, 2012.
11. S. R. Oh, K. Yao, C. L. Chow *et al.*, "Residual stress in piezoelectric poly(vinylidene-fluoride-co-trifluoroethylene) thin films deposited on silicon substrates," *Thin Solid Films,* vol. 519, no. 4, pp. 1441-1444, Dec 1, 2010.
12. W. D. Nix, "Mechanical-Properties of Thin-Films," *Metallurgical Transactions a-Physical Metallurgy and Materials Science,* vol. 20, no. 11, pp. 2217-2245, Nov, 1989.

13. J. N. Antonakakis, P. Bhargava, K. C. Chuang *et al.*, "Linear viscoelastic properties of HFPE-II-52 polyimide," *Journal of Applied Polymer Science,* vol. 100, no. 4, pp. 3255-3263, May 15, 2006.

14. S. P. Baker, A. Kretschmann, and E. Arzt, "Thermomechanical behavior of different texture components in Cu thin films," *Acta Materialia,* vol. 49, no. 12, pp. 2145-2160, Jul 17, 2001.

15. K. Mirpuri, H. Wendrock, S. Menzel *et al.*, "Texture evolution in Copper film at high temperature studied in situ by electron back-scatter diffraction," *Thin Solid Films,* vol. 496, no. 2, pp. 703-717, Feb 21, 2006.

16. Y. L. Shen, and U. Ramamurty, "Constitutive response of passivated copper films to thermal cycling," *Journal of Applied Physics,* vol. 93, no. 3, pp. 1806-1812, Feb 1, 2003.

17. J. M. G. Ong, A. A. O. Tay, X. Zhang *et al.*, "Optimization of the Thermomechanical Reliability of a 65 nm Cu/low-k Large-Die Flip Chip Package," *IEEE Transactions on Components and Packaging Technologies,* vol. 32, no. 4, pp. 838-848, Dec, 2009.

18. M. Lofrano, N. Pham, M. Rosmeulen *et al.*, "Stress and wafer warpage analysis of GaN thin film induced by transfer bonding process on 200mm Si substrate," *2013 14th International Conference on Thermal, Mechanical and Multi-Physics Simulation and Experiments in Microelectronics and Microsystems (Eurosime)*, 2013.

A Stress-Based Effective Film Technique for Wafer Warpage Prediction of Arbitrarily Patterned Films

Gregory T. Ostrowicki, Siva P. Gurrum
Texas Instruments, Inc.
12500 T I Blvd, Dallas, TX 75243
gtostrowicki@ti.com

Abstract

Initially flat silicon wafers are prone to warp due to the high levels of intrinsic stress of deposited films, particularly metallic films. Processing and handling of warped wafers in the fab is a challenge. One of the ways to control the degree of warpage is by limiting the amount of metallization allowed on the wafer. However, this imposes a constraint on the silicon designers, and can lead to decreased performance of the IC. Therefore, there is a need to accurately predict the amount of wafer warpage caused by a proposed layout in order to give designers the most freedom to develop IC solutions while ensuring that the processed wafers meet the manufacturing equipment requirements.

The metal artwork (in addition to other materials, layer thicknesses, processing parameters, etc.) is an important factor in determining wafer curvature. Simple analytical methods, such as Stoney's Formula, cannot capture the non-uniform warpage due to these patterned films. On the other hand, numerical methods which require detailed modeling of the film patterns across the whole wafer are computationally expensive. Thus, a new finite element modeling technique was developed in which the entire patterned film stack is represented as a uniform effective orthotropic film bonded onto a silicon substrate. The orthotropic properties are determined from a small set of virtual experiments using a unit-cell model that is characteristic of the actual pattern. The resultant effective film, despite using a very course mesh, is able to capture the non-uniform surface stress induced by a patterned multi-layer film stack, and thus results in very similar wafer warpage as in the conventional detailed model. Several example film patterns will be presented here, where the warpage difference between the detailed model and the effective film model are less than a few percent across the whole wafer.

1. Introduction

Wafers warp from the accumulation of stress within the deposited films due to CTE mismatch, lattice mismatch, impurities, recrystallization, creep, cure shrinkage, and other phenomena [1]. It is a challenge to characterize all the individual mechanisms that can contribute to the film stress for each material in the stack. However, many of these mechanisms can be lumped together into an "effective CTE" (e.g. [2]). Therefore, this work focuses on thermal expansion as the driving warpage mechanism.

As a silicon wafer goes through the IC manufacturing flow, a series of patterned films are deposited across the wafer to create a rectangular array of identical dies. Each die within the wafer interior is essentially a unit-cell, with a characteristic stress distribution through the film stack. Much of the film stress is concentrated within the metal artwork, due to the relatively high modulus and significant CTE mismatch of metals with respect to the silicon substrate. Thus, the patterning of the metal circuitry typically results in a non-uniform biaxial stress depending on both the volume of metal within the film as well as the design of the circuitry itself. Any technique used to predict the global warpage of the wafer after processing must therefore capture this non-uniform stress in the film stack at the die level. For finite element methods, this can require modeling the patterns exactly, using element sizes that are on the order of the minimum feature size in the film. Since the metal line width and thickness can be in the micron to sub-micron scale, the necessary element count can approach the millions for a single die depending on the artwork, number of layers, and the die size. It can quickly become computationally unreasonable to include this level of detail across a whole wafer. For example, a 300 mm wafer composed of 1x1 mm dies contains about 70,000 dies. To simplify matters, one can model only a small portion of the wafer and extrapolate the local curvature to estimate the global warpage. However, this approach alone cannot capture the effects of gravity, mechanical fixturing, and other global body forces and constraints that can impact the warpage.

Previous numerical approaches have been used to model global wafer warpage based on a film patterns of parallel lines of a single material [3]. However, actual IC wafers can have very complex film stacks. In this work, a finite element approach is outlined to simplify the detailed nature of the composite film stack into a uniform material with effective properties, as shown in Figure 1. This representative film can be modeled with a very coarse mesh, but still develop an equivalent biaxial stress (and thus, equivalent warpage) resulting from its CTE mismatch with the silicon substrate. A detailed unit-cell model with representative film pattern (typically one die) is first modeled in order to calculate the effective film properties. Then, a second model of the wafer with effective film can be used to determine the global warpage by using a very coarse mesh.

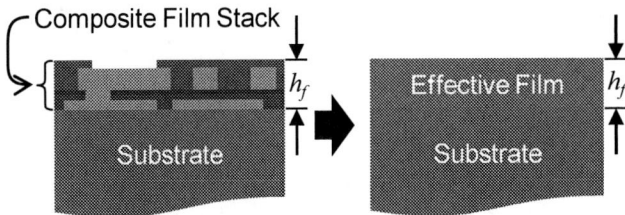

Figure 1. Simplification of composite film stackup into an effective film with uniform properties.

2. Theory

Stress and Warpage

Stoney famously related the curvature of a plate to the film stress through

$$\sigma_f = \frac{h_s^2 E_s}{6 R h_f (1 - v_s)},$$ (2.1)

where R is the radius of curvature, E is the Young's modulus, h is the thickness, v is the Poisson's ratio, and the substripts f and s correspond to the film and substrate, respectively [4]. This relationship holds for the case where the substrate and film materials are uniform, isotropic, the film is much thinner than the substrate, and the radius of curvature is much greater than the substrate thickness (i.e. $h_f \ll h_s \ll R$). The film stack on an IC wafer, however, is a mixture of different layers, materials, and patterns. These heterogeneities can create a non-uniform film stress, and thus non-uniform warpage along the different radial directions of the wafer. This work aims to simplify the composite film stack into a homogeneous effective film for warpage prediction. Since any warpage model must capture the directional nature of the film stress, an anisotropic material model is used for the effective film.

Orthotropic Thin Film

A linear orthotropic material model is used to represent the effective properties of an arbitrary composite film stack. Since the film is thin and at a free surface, it can be assumed to be in a state of plane stress, for which the conventions of the stress components are shown in Figure 2.

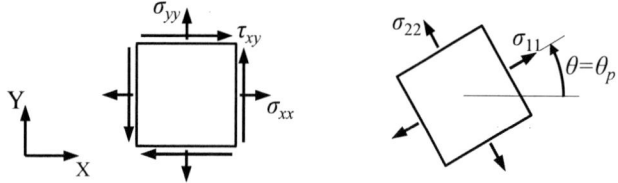

Figure 2. Conventions for element in plane stress.

Hooke's law for a thin orthotropic material can be expressed in matrix form as

$$\begin{bmatrix} \varepsilon_{xx} \\ \varepsilon_{yy} \\ \gamma_{xy} \end{bmatrix} = \begin{bmatrix} \frac{1}{E_x} & -\frac{v_{yx}}{E_y} & 0 \\ -\frac{v_{xy}}{E_x} & \frac{1}{E_y} & 0 \\ 0 & 0 & \frac{1}{G_{xy}} \end{bmatrix} \begin{bmatrix} \sigma_{xx} \\ \sigma_{yy} \\ \tau_{xy} \end{bmatrix},$$ (2.2)

where E is the Young's modulus, G is the shear modulus, and v is Poisson's ratio [5]. The following constraint is also required to make the stiffness matrix symmetric:

$$\frac{v_{xy}}{E_x} = \frac{v_{yx}}{E_y}.$$ (2.3)

In addition, there is an orientation θ_p at which the biaxial stresses are aligned with the principal stresses and there is no shear stress (i.e. $\tau = 0$). The principal stresses σ_{11} and σ_{22} can be determined by

$$\sigma_{11}, \sigma_{22} = \frac{(\sigma_{xx} + \sigma_{yy})}{2} \pm \sqrt{\left(\frac{\sigma_{xx} + \sigma_{yy}}{2}\right)^2 + \tau_{xy}^2},$$ (2.4)

and the principal orientation angle θ_p is determined by

$$\theta_p = \frac{1}{2} \tan^{-1}\left(\frac{2\tau_{xy}}{\sigma_{xx} - \sigma_{yy}}\right).$$ (2.5)

Therefore, for an orthotropic material in plane stress, and oriented along the principal stress directions (i.e. $\theta = \theta_p$), Hooke's law can be further simplified to just the following relationships

$$\begin{cases} \varepsilon_{11} = \dfrac{\sigma_{11}}{E_1} - \sigma_{22}\dfrac{v_{21}}{E_2} \\ \varepsilon_{22} = \dfrac{\sigma_{22}}{E_2} - \sigma_{11}\dfrac{v_{12}}{E_1}, \end{cases}$$ (2.6)

where the subscripts 1 and 2 refer to the principal directions, and the corresponding stiffness symmetry constraint is

$$\frac{v_{12}}{E_1} = \frac{v_{21}}{E_2}.$$ (2.7)

This result indicates that only the biaxial Young's moduli and Poisson's ratios are required to fully define the stiffness of the effective film, provided the film stress is oriented in the principal stress directions. In addition, the film can be assumed to have biaxial thermal expansion coefficients α_1 and α_2. Thus for thermo-mechanical stress analysis, the effective orthotropic film can be fully defined by only six material parameters: E_1, E_2, v_{12}, v_{21}, α_1, and α_2.

Thermally Induced Stress-Based Effective Film

In the modeling approaches, a stress is induced in the film by subjecting the wafer to a uniform change in temperature. A 1-D approximation of such a bi-material strip undergoing thermal loading is shown in Figure 3. Here the total strain of each material must be equivalent as described by

$$\alpha_f \Delta T + \varepsilon_f = \alpha_s \Delta T + \varepsilon_s,$$ (2.8)

where ΔT is the change in temperature from an initial stress-free condition, and the subscripts f and s refer to film and substrate, respectively. When the film is very thin (and relatively soft) with respect to the substrate, the mechanical strain in the substrate can be neglected (i.e. $\varepsilon_s = 0$) and (2.8) can be expressed as

$$\frac{\varepsilon_f}{\Delta T} = \alpha_s - \alpha_f.$$ (2.9)

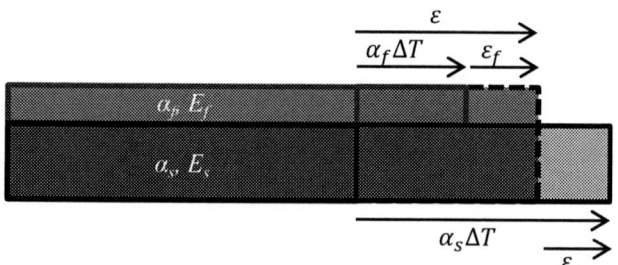

Figure 3. 1-D illustration of bi-material strip under thermal expansion

Although crystalline silicon is an anisotropic material, it has been shown that a typical (100) oriented wafer has a symmetric biaxial modulus [6], and its thermal expansion can also be treated as isotropic [7]. Thus assuming an orthotropic film on an isotropic substrate, (2.9) can be expanded along the film's principal directions to result in

978-1-4799-2408-0/14 $31.00 © 2014 IEEE

$$\begin{cases} \dfrac{\varepsilon_{11}}{\Delta T} = \alpha_s - \alpha_1 \\ \dfrac{\varepsilon_{22}}{\Delta T} = \alpha_s - \alpha_2 \end{cases}, \qquad (2.10)$$

where the subscripts *1* and *2* correspond to the film along the respective principal directions. Substituting (2.6) into (2.10), after some algebra results in

$$\begin{cases} \dfrac{\sigma_{11}}{\Delta T} = m_1 \alpha_s + b_1 \\ \dfrac{\sigma_{22}}{\Delta T} = m_2 \alpha_s + b_2 \end{cases}, \qquad (2.11)$$

where

$$\begin{aligned} m_1 &= E_1\left(\frac{1+\nu_{21}}{1-\nu_{12}\nu_{21}}\right) & b_1 &= E_1\left(\frac{\alpha_1+\alpha_2\nu_{21}}{1-\nu_{12}\nu_{21}}\right) \\ m_2 &= E_2\left(\frac{1+\nu_{12}}{1-\nu_{12}\nu_{21}}\right) & b_2 &= E_2\left(\frac{\alpha_2+\alpha_1\nu_{12}}{1-\nu_{12}\nu_{21}}\right) \end{aligned} \qquad (2.12)$$

Equation (2.11) shows that for a given temperature difference, the film's principal stress is simply a linear function of the substrate CTE, where the constants m and b are purely functions of the film material properties. This implies that effective orthotropic properties of a composite film stack could potentially be determined if the average thermally induced stress was measured for at least two samples, on substrates with different CTEs. While this can be impractical to exercise in a real experiment, it is straightforward to implement in a numerical experiment. The proposed process to determine the effective film properties is thus outlined below:

1. Create detailed unit-cell model of the wafer with representative film pattern (e.g. a single die).

2. Subject unit-cell to a characteristic temperature change, and calculate the resulting volume-averaged film stress along the principal directions.

3. Modify the CTE of the substrate in the unit-cell model and repeat Step 2.

4. Solve for the constants m and b using (2.11) and the results from Steps 2-3.

5. Find an appropriate property set for the film which satisfies (2.7) and (2.12).

The final step in the process outlined above presents a potential caveat with this approach. As mentioned before, the film is assumed to have six independent material parameters (E_1, E_2, ν_{12}, ν_{21}, α_1, α_2). However, (2.7) and (2.12) provide for only five constraint equations from which to solve for the six properties. This in an unconstrained problem for which there can be an infinite number of property sets which satisfy the conditions. In the absence of a physical constraint, it is possible to define an additional condition in an effort to obtain a unique solution, such as assuming an isotropic CTE (i.e. $\alpha_1 = \alpha_2$). However, the authors feel that such a constraint may be arbitrary, and in their experience may result in a poorly defined material property set. Instead, an evolutionary algorithm was used in EXCEL® to find a suitable material property set, where the bounds of the effective modulus and CTE were set to be within the maximum values of the component materials in the film stack (i.e. $0 \le E_1$, $E_2 \le E_{max}$ and $0 \le \alpha_1$, $\alpha_2 \le \alpha_{max}$), and the Poisson's ratios were set to be realistic values (i.e. $0 \le \nu_{12}$, $\nu_{21} \le 0.5$).

3. Numerical Models

A hypothetical wafer design was devised for use as a baseline in order to demonstrate and compare the warpage modeling approaches. A unit-cell of this wafer is shown in Figure 4, and is representative of the patterning on a single die. This die features alternating parallel strips of metal and polymer film over a silicon substrate. Representative material properties for the film and substrate are shown in Table 1. Here $h_f = 10$ um, $h_s = 725$ um, $b = d = 1$ mm, $w_m = 71.4$ um, and $v_m = 0.5$, where h_f is the film thickness, h_s is the substrate thickness, b is the die width, d is the die depth, w_m is the metal line width, and v_m is the volume fraction of metal in the film, respectively.

For warpage modeling, only a small section of the wafer is modeled due to element restrictions, but which is sufficient to capture the wafer curvature in the absence of external forces. As shown in Figure 5, the warpage model is a disk with radius of 5 mm and is composed of a rectangular array of the baseline die. Symmetry is utilized so that only ¼ of the model is represented, where appropriate symmetry conditions are imposed on the cut faces. Body forces such as gravity are neglected such that the resulting curvature is purely a function of the CTE mismatch between the film and substrate.

For the initial case, the entire film/substrate system is assumed to be stress-free at elevated temperature and then subjected to a drop in temperature to ambient conditions, where $\Delta T = -100°C$. In a more realistic scenario, however, each individual film material is likely to have a unique stress-free temperature depending on the material's processing history. This situation is addressed later as described in Section 4, and is only briefly neglected here in order to demonstrate the methodology of the approach. It should also be noted that the film pattern was specifically chosen so that the resulting principal stress directions in the film are aligned with the coordinate system (i.e. σ_{11}, $\sigma_{22} = \sigma_{xx}$, σ_{yy}). Die designs where this is not the case are also discussed in Section 4.

Finite element models were developed using ANSYS® 14.5. First, a detailed model was constructed to establish a benchmark of the wafer warpage. Then, a new approach was pursued which first determined the effective properties of the film based on stress within a unit-cell, and then applied the effective properties to a simplified global warpage model.

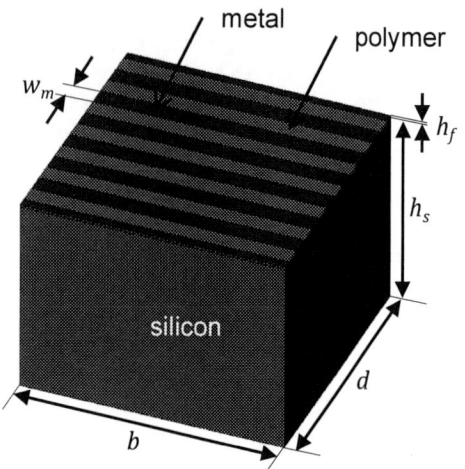

Figure 4. Baseline die design with patterned film of alternating metal and polymer strips on a silicon substrate.

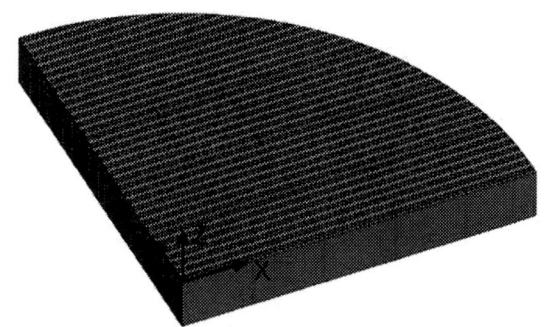

Figure 5. Baseline warpage model domain.

Table 1. Material properties of film and substrate.

	Metal film	Polymer film	Si substrate
E	100 GPa	5 GPa	131 GPa
v	0.33	0.33	0.28
CTE	20 ppm/°C	40 ppm/°C	2.6 ppm/°C

Benchmark Approach: Detailed Model

A detailed finite element warpage model was constructed using solid quadratic elements (SOLID186) with element sizes of ~5x5x1 um in the film and ~5x5x145 um in the substrate (see Figure 8). A sufficiently fine mesh was used to capture the stress distribution through the film. The resulting warpage is shown in Figure 6 for $\Delta T = -100°C$. The radius of curvature R was calculated along the X- and Y- axis by fitting a circular arc to the out-of-plane displacement profiles shown in Figure 7, which was found to be $R_x = 11.492$ m and $R_y = 20.401$ m. As a side note, if this curvature was extrapolated to a 200 mm wafer, that would translate to a maximum warpage of 435 um and 245 um along the X- and Y- axis, respectively.

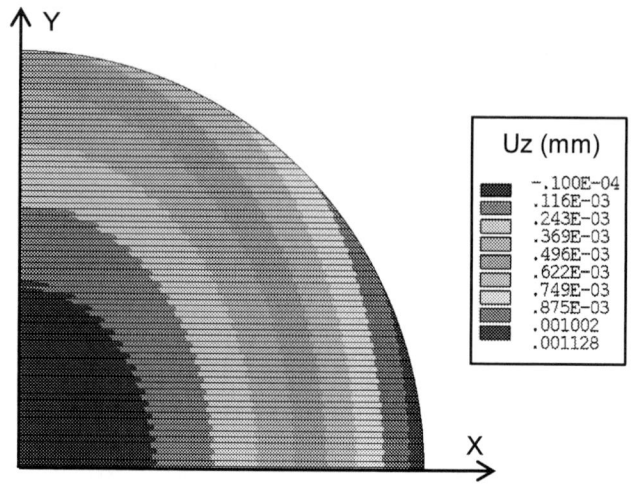

Figure 6. Benchmark warpage contours of detailed model.

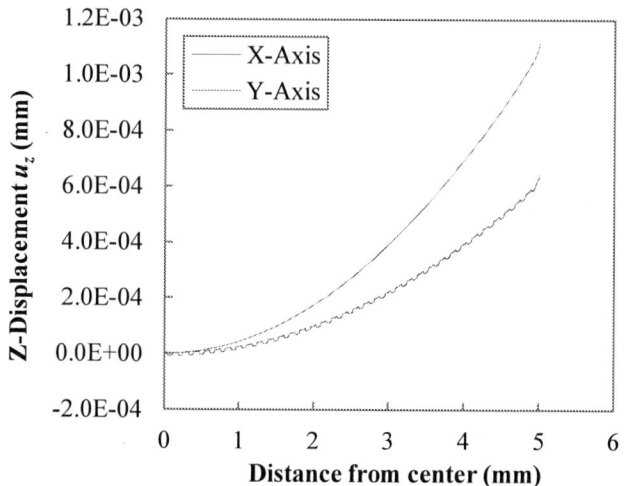

Figure 7. Warpage profile of benchmark model.

The disparity of the warpage along the two axes can be explained by examining the stress within the composite film as shown in Figure 8. Here the stress is concentrated in the metal, and is much higher in the direction of the strips than perpendicular to them. This is due to the fact that the metal is less constrained near the sidewall adjacent to the low stress polymer. In order to determine the effective film stress across the wafer, the volume-averaged stress $\sigma^{(avg)}$ in the film was calculated to be $\sigma_{xx}^{(avg)} = 123$ MPa and $\sigma_{yy}^{(avg)} = 90$ MPa. It is interesting to note that although the film stress in the X-direction is only about 1.4X higher than in the Y-direction, the curvature $1/R$ is almost 1.8X higher. Only a relatively small non-uniformity in the biaxial stress is necessary to create a significant difference in the warpage along the two perpendicular axes.

978-1-4799-2408-0/14 $31.00 © 2014 IEEE

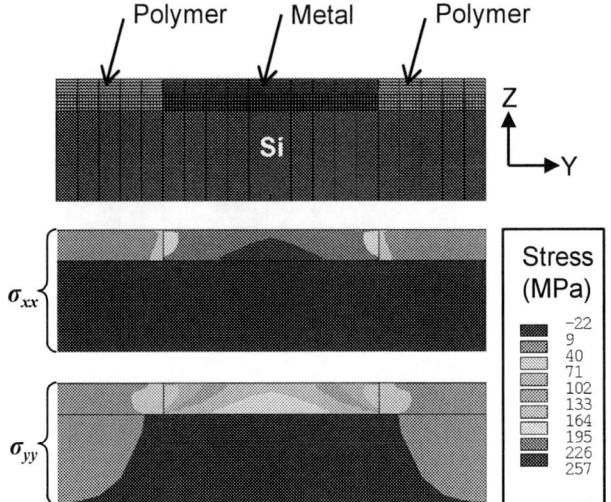

Figure 8. Cross section of the detailed model near the film/substrate interface. The corresponding contour plots of σ_{xx} and σ_{yy} indicate that the overall film stress is more tensile in the direction of the strips than perpendicular to them.

New Approach: Stress-Based Effective Orthotropic Film

Whereas the benchmark approach involves using a detailed model for the entire warpage domain, the new approach employs a two-step process. First, a detailed model is constructed only for one representative unit-cell (or die), from which the smeared effective film properties are calculated as described in Section 2. Then, a separate simplified model, composed of just the substrate and effective film, is used to determine the warpage over the entire domain. Since the effective film has uniform properties across the wafer, a relatively coarse mesh can be used for the warpage model, which enables a much more efficient solution than the benchmark approach.

The detailed unit-cell model was created for the die shown in Figure 4, again using SOLID186 elements and a fine mesh. Conventional unit-cell boundary conditions were applied on the cut faces (i.e. out-of-plane displacements were coupled across each of the four respective cut faces). The model was subjected to ΔT = -100°C, and the resulting volume-averaged biaxial film stress was determined. Then, a second identical unit-cell model was made except that the CTE of the Si was artificially changed to a different value, and again the resulting volume-averaged biaxial stress was determined. Although these two unit-cell models are sufficient to determine m and b, a total of four models were created with Si substrate CTEs of 1, 2.61, 5, and 10 ppm/°C in order to clearly demonstrate that the trend is indeed linear as predicted by (2.11) and shown in Figure 9. A suitable set of effective film properties was found using the strategy described in Section 2 with values listed in Table 1.

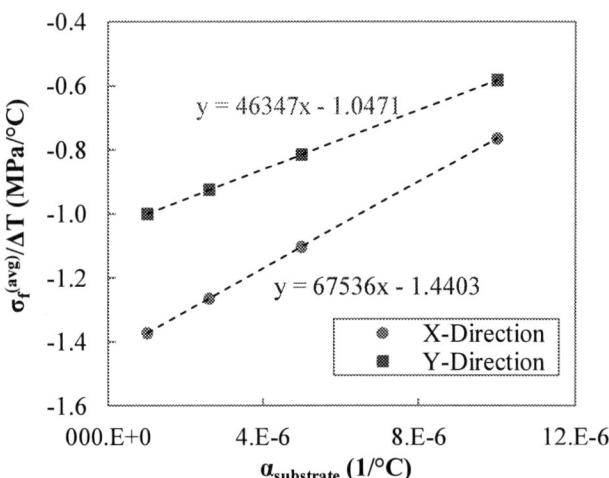

Figure 9. Biaxial film stress using detailed unit-cell model as a function of substrate CTE. This data is from four separate models in order to demonstrate that the trends are linear. In general, two models are sufficient.

Table 2. Calculated effective film material property set.

	Effective Film
E_x	56378 MPa
E_y	35973 MPa
v_{xy}	0.2408
v_{yx}	0.1536
α_x	21.08E-6 °C^{-1}
α_y	22.96E-6 °C^{-1}

Next, a warpage model was constructed using SOLID186 elements for the Si substrate and quadratic shell elements (SHELL281) for the film. Unlike heterogeneous films which can have a large variation in biaxial stress through the thickness of the film (e.g. see Figure 8), the stress in a homogeneous film is uniformly distributed through the film thickness. Thus it is appropriate to use shell elements for the effective film and this further reduces the model complexity. Using the calculated effective film properties in Table 2, the resulting warpage for this model is shown in Figure 10, and can be seen to match very closely with the results of the benchmark model in Figure 6. A comparison of the two models is shown in Table 3, where the difference in the total film stress and warpage is only 0-1.3%.

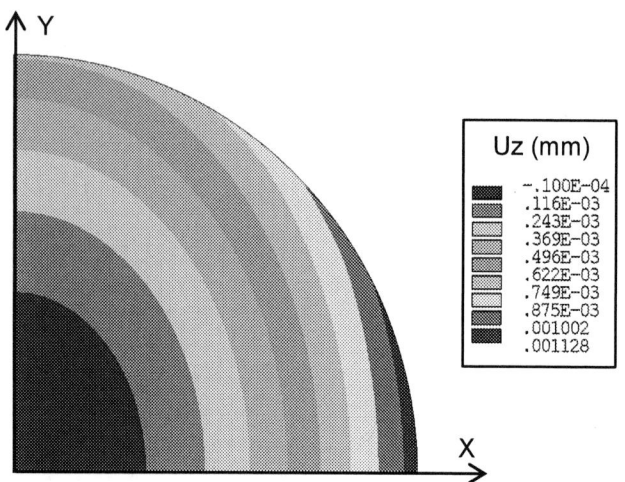

Figure 10. Warpage contours of simplified model using effective film properties.

Table 3. Comparison of baseline warpage models.

	Detailed Model	Effective Model	% Difference
$\sigma_{xx}^{(avg)}$	123.14 MPa	123.25 MPa	0.09%
$\sigma_{yy}^{(avg)}$	90.27 MPa	91.05 MPa	0.86%
R_x	11.49 m	11.63 m	1.21%
R_y	20.40 m	20.13 m	1.32%

4. Additional Model Considerations

Arbitrary Patterns

For the example die design used in the previous analysis, the principal stress in the film is naturally oriented with the global coordinate system since the patterning is symmetric with respect to the imposed unit-cell boundary conditions. Therefore, it was sufficient to calculate the effective film properties along the defined X- and Y- axes. However, for an arbitrary die layout, the principal film stress may not be aligned with the defined coordinate system. Since the maximum warpage is typically in the direction of the maximum principal stress, it is then necessary to rotate the coordinate system so that the effective properties are determined in the proper orientation.

To illustrate this procedure, consider the die layouts in Figure 11. Design (a) is the same unit-cell as shown in Figure 4, with horizontally aligned strips where $\varphi = 0°$. Design (b) has the same strip pattern except that the strips are at an angle of $\varphi = 45°$. In fact, these two designs are really just alternative unit-cells for the same repeating pattern of parallel film strips. Therefore, both dies ought to have equivalent principal stress magnitudes, and thus equivalent warpage. The resulting volume-averaged film stress for each of these unit-cell designs is shown in Figure 12, where each of the three planar stress components were calculated in the rotated X'-Y' coordinate system, for $0°\leq\theta\leq90°$. Indeed, the principal stresses for design (a) were found at $\theta_p^{(a)} = 0°$, and are in close agreement with those of design (b) where $\theta_p^{(b)} = 45°$.

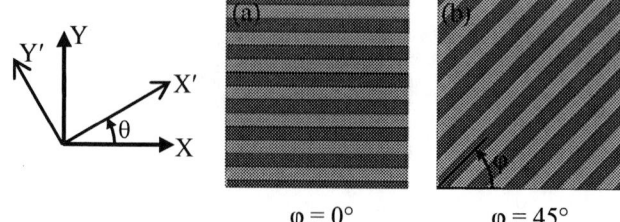

Figure 11. Two alternative unit-cells that result in the same global wafer design.

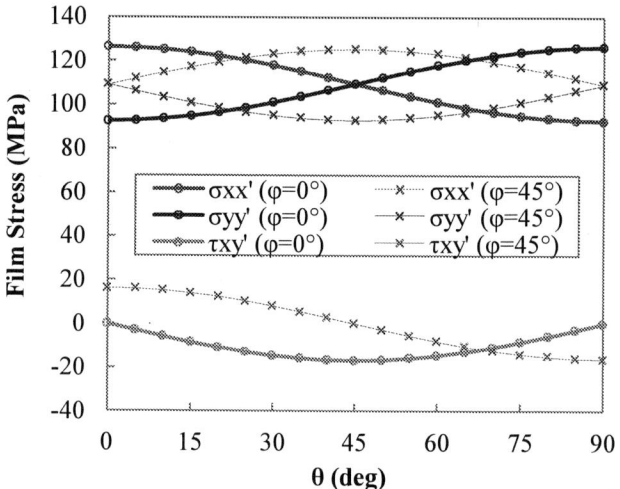

Figure 12. Volume-averaged plane stress of the film in the rotated X'Y' coordinate system, where θ=0-90°. Two cases are shown, for unit-cells (a) and (b).

The von Mises stress for the two designs is shown in Figure 13. While the stress distribution is very similar for both cases, there is some distortion near the cut boundary of the die with angled strips. This is due to the fact that the imposed unit-cell boundary conditions do not allow for global shear deformation, and this enforces symmetry constraints along the bounding faces even when the geometry is not symmetric. Fortunately, this stress artifact is isolated to the periphery of the die, and does not significantly affect the volume-averaged stress calculations.

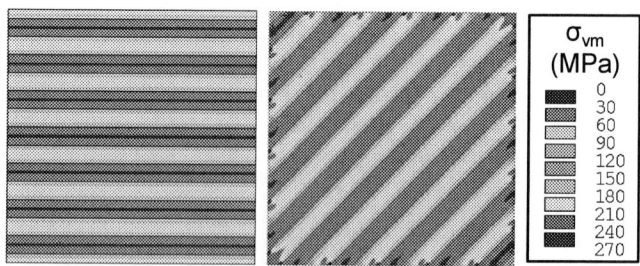

Figure 13. Von Mises film stress for the two unit-cell designs. There is some artificial distortion in the stress near the cut boundary of the angled strips due to the asymmetric nature of the design with respect to the imposed unit-cell constraints.

Multiple Stress-Free Temperatures

In all the previous models, the wafer was considered to be stress-free at an initial temperature and then subjected to a temperature ramp. In an actual wafer processing flow, however, film materials are serially deposited, layer-by-layer, at a wide range of temperatures and environmental conditions. It is unlikely the wafer is stress-free at any given temperature. In order to properly capture the component stresses of such a heterogeneous film, the thermal assembly process can be mimicked so that the appropriate film material is activated at its respective stress-free temperature. This can be accomplished using the element birth capability within the finite element model.

Using the unit-cell from Figure 4 as an example, the stress-free temperature T_{ref} is assumed to be 120°C for the metal film and 220°C for the polymer film, respectively. The resulting stress distribution for the die at an ambient temperature T_{amb} of 20°C can be determined using the strategy shown in Figure 14, where the results of each load step from the analysis are shown in Figure 15. In order to find the effective film properties for this case, the model was repeated using a modified Si substrate CTE as was done in Section 3. Again the trends are clearly linear as shown in Figure 16, from which appropriate effective film properties can be calculated.

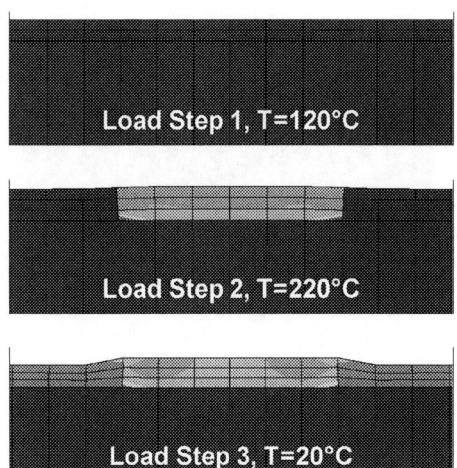

Figure 15. Progression of stress as the temperature is ramped according to the film deposition sequence, and the appropriate elements in the film are activated upon reaching the stress-free temperatures of metal and polymer, respectively.

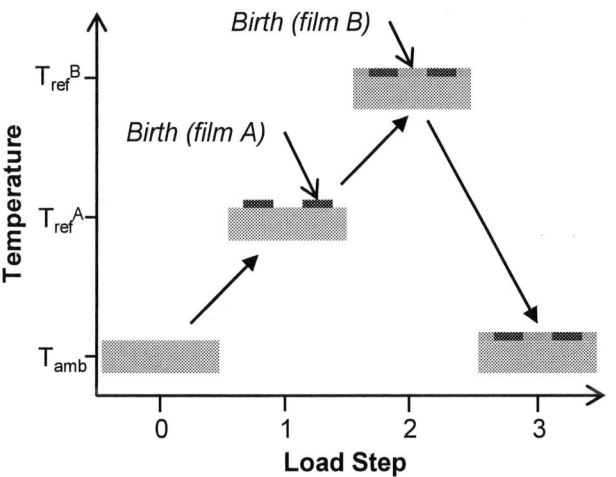

Figure 14. Example illustration of a multi-step film deposition sequence, where each component film material is activated through element birth at its respective stress-free temperature.

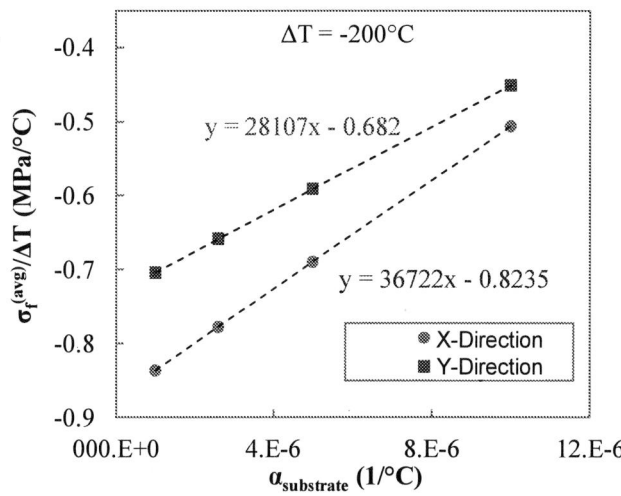

Figure 16. Biaxial film stress for unit-cell design with different stress-free temperatures for metal and polymer film.

5. Realistic Wafer

The methods outlined above were used to model the warpage of a patterned wafer with a realistic die layout shown in Figure 17. Here the metal artwork is partially covered by a polymer film except for the circular regions that define an exposed metal pad. The wafer thickness is 15 mil, and the material properties used were the same as specified in Table 1, with stress-free temperatures of 120°C and 220°C for the metal and polymer film, respectively.

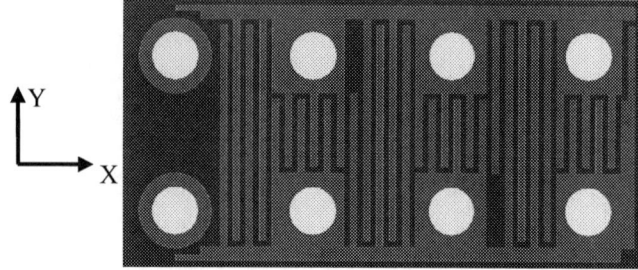

Figure 17. Example die layout, with metal artwork covered by polymer film.

Both detailed and effective-film warpage models were constructed for this wafer pattern in the same manner as above, and subjected to the temperature history as shown in Figure 14. The detailed model was composed of only a 4x6 die array to limit the computation expense, whereas the effective film could be modeled for the entire wafer domain due to its very low element count requirement. The results for both models are compared in Table 4, where they are in excellent agreement with a difference of just a few percent. The principal stress orientation for this pattern resulted in the maximum curvature to be in the Y-direction due to the pattern of thin metal strips which run in parallel along the same direction.

Table 4. Comparison of warpage models with realistic pattern.

	Detailed Model	Effective Model	% Difference
$\sigma_{xx}^{(avg)}$	112.03 MPa	110.04 MPa	1.77%
$\sigma_{yy}^{(avg)}$	118.85 MPa	116.98 MPa	1.58%
R_x	2.762 m	2.699 m	2.29%
R_y	2.399 m	2.421 m	0.91%

6. Conclusions

Wafer warpage is due not only to the amount of highly-stressed metal content in the patterned film, but also the design of the artwork. These patterns can have tendencies to cause greater warpage in certain directions more so than others, and cause handling issues during the manufacturing flow. Conventional modeling of the warpage thus requires modeling the fine details of the patterns, which is computationally prohibitive for large areas of the wafer. A better approach is to use an effective film with properties that represent the cumulative effect of all the individual layers and materials that make up the film stack on a patterned wafer.

A method for modeling the warpage of arbitrarily patterned wafers was thus developed which uses an effective orthotropic film to represent the composite film stack. A detailed model of a representative unit-cell of the wafer was first used to extract the material properties for an effective film, based on thermal mismatch with the silicon substrate and the resulting volume-averaged stress within the film stack. Then, a global warpage model was constructed using the effective film without having to capture the details of the pattern. The stress-based effective film approach was also

demonstrated to be capable of capturing the unique stress-free temperatures of each of the patterned film components.

The difference between the resulting biaxial warpage of the simplified model with effective properties with that of the detailed model was less than a few percent. Furthermore, the proposed approach can be used to model the global warpage across the entire wafer domain, which would be virtually impossible to do with a detailed model. This approach therefore allows for including the effects of external forces such as gravity and mechanical fixturing which can impact the wafer curvature.

References

1. P. A. Flinn *et al.*, "Measurement and interpretation of stress in aluminum-based metallization as a functino of thermal history," *IEEE Trans. Electron Devices*, vol. 34, no. 3, pp. 689-699, Mar. 1987.

2. C.-H. Hsueh *et al.*, "Residual stresses in thermal barrier coatings: effects of interface asperity curvature/height and oxide thickness," *Mater. Sci. Eng. A*, vol. 283, pp. 46-55, 2000.

3. Y.-L. Shen *et al.*, "Stresses, curvatures, and shape changes arising from patterned lines on silicon wafers," *J. Appl. Phys.*, vol. 80, no. 3, pp. 1388-1398, 1996.

4. G. G. Stoney, "The tension of metallic films deposited by electrolysis," *Proc. R. Soc. Lond. A*, vol. 82, no. 553, pp. 172-175, 1909.

5. N. E. Dowling, *Mechanical Behavior of Materials 2ⁿᵈ Ed.*, Prentice Hall, Upper Saddle River, 1999, pp. 190-192.

6. M. A. Hopcroft, *et al.*, "What is the Young's modulus of silicon?," *JMEMS*, vol. 19, no. 2, pp. 229-238, Apr. 2010.

7. C. A. Swenson, "Recommended values for the thermal expansivity of silicon from 0 to 1000 K," *J. Phys. Chem. Ref. Data*, vol. 12, no. 2, 1983.

Drop Test and TCT Reliability of Buffer Coating Material for WLCSP

Nobuhiro Anzai[1*], Mitsuru Fujita [1], Atsushi Fujii[1]

1) Asahi Kasei E-materials Corporation
2-1, Samejima, Fuji-city, Shizuoka, Japan
*E-mail: anzai.nb@om.asahi-kasei.co.jp

Abstract

Portable electronic products has been extended for high-performance and high-density LSI applications in recent years. One of the major concerns of portable electronic products is performance of package during drop impact and temperature cycle test (TCT) because fracture and/or delamination of the fine joint parts are caused. In addition, in terms of process limitation or device yields, process temperature limitation is required below 200deg.C in some packaging applications.

Recently, the organic passivation as polyimide, polybenzoxazole and BCB are widely used in top layer of IC chips. In terms of material development, it is interesting to clarify which material properties are most important to release the stress during TCT, drop test, and other reliability tests.

In this study, TCT and drop test of WLCSP with various kinds of buffer coating material were carried out. As the results of TCT and drop test, some polymers show good reliability performance. Above all, photosensitive polyimide PIMEL[TM] BL-series was lower fail rate by TCT and drop test regardless of the chip size. PIMEL[TM] BL-series is 200deg.C curable material which is expected to show good reliability.

Introduction

In case of WLCSP, underfill is generally filled in between IC chip and interposer, and the underfill and passivation layer release the stress concentration at bumps. However, underfill is not filled in some of package designs as small chips. In that case, the stress concentration at bump is only released by passivation layer. The motivation of underfill elimination is cutting cost in production. In other words, if the stress release function of passivation is sufficient, cost down products by underfill elimination may increase.

It is expected that the stress of drop impact is absorbed by solder bump, underfill (UF) and organic passivation as polyimide (PI) at top layer of IC chips. In case of Wafer Level CSP (WLCSP), because of small chip size and large bump, the stress on chip is not so large as Flipchip for CPU/GPU. Therefore, UF could be eliminated from process. As a result of UF elimination, drop impact should be absorbed by bump and PI layer. However, solder bump material have been changed from eutectic to fragile lead-free, it is difficult to absorb drop impact by bump. Therefore, it is expected that PI layer has good stress absorption function.

In general, it is told that low modulus material has more shock absorbance function because the material is deformed, and released the concentrated stress at impact position.

In our previous study, the drop impact of WLCSP package without UF was calculated with FEA using a zooming technique. Thicker PI thickness and lower PI modulus contribute release the drop stress at bump edge. On the other hand, thinner PI thickness and higher PI modulus contribute release the drop impact at pad area. It is necessary to choose appropriate thickness and modulus under high strain rate [1].

Judging from our previous study by Finite Element Analysis (FEA), it was found that appropriate modulus and thickness combination shall be selected to minimize the stress concentration. The simulation result indicates the stress release function is defined by not only modulus but also other parameters.

On the other hand, the stress buffer effect by thermal stress is also important for reliability. For latest FC packages with a 28nm node and beyond, ELK (extra low-k)/Cu interconnect with fragile and lower mechanical strength is implemented [2], fracture and/or delamination of ELK induced by thermal stresses have been observed [3]. In general, significant ELK damaged failure mode has been expected when the cooling step during reflow for flipchip bonding, and during TCT.

In our member's previous study, the stress buffer effect by polyimide was investigated and clarified by utilizing FEM analysis [4]. Thicker PI with higher modulus is effective to protect ELK under the edge of UBM, and smaller PI opening size is preferable to reduce stress under PI opening edge in the cooling step during reflow process. However, TCT simulation analysis with UF was not discussed in the study.

In this study, board level test were done by simple designed WLCSP to clarify the influence of material properties. The reliability test contents are TCT and drop test. In order to verify the influence of material property difference, some kinds of polymers as polyimide, polybenzoxazole and others are prepared. To compare the buffer function under severe condition, some sizes of chips are prepared.

Dielectric materials used for WLCSP packaging

There are some kinds of material which could be used as a dielectric layer for Wafer Level Packaging (WLP), for example, polyimide, polybenzoxazole, phenolic polymer and benzocyclobutene etc. In this study, 2 kinds of polyimide and a polybenzoxazole and a phenolic polymer are prepared.

Polyimide is one of most famous material which is used for WLP. The features of polyimide are high thermal properties, good mechanical properties, good chemical resistance for bumping process, and good adhesion with substrate and metal layers (Cu, Al, Ti, TiW etc.). One of typical PI structure is shown in Figure 1. Asahi Kasei E-materials Corporation is supplier of the photosensitive polyimide precursor PIMEL[TM] which are developed for buffer coating material or dielectric material in semiconductor field. The photosensitive precursor becomes polyimide structure through curing process. Curing temperature of typical photosensitive polyimide precursor is over 300deg.C due to finalize the imidization reaction. However, imidizaion of PIMEL™ BL-series and BM-series are completely finished at 200deg.C by polymer structure and components modification. These materials are applied to many kinds of WLP. In this study, PIMEL™ BL-300 and BM-300 is selected to confirm the reliability for bumping application.

PI precursor structure before curing

↓ Thermal curing

PI structure after curing

Figure 1. Example of a PI structure

Polybenzoxazole have been also widely applied for buffer coating material of WLP. One of typical PBO structure is shown in Figure 2. Curing temperature of typical photosensitive polybenzoxazole precursor is also over 300deg.C due to finalize the cyclic reaction.

PBO precursor structure before curing

↓ Thermal curing

PBO structure after curing

Figure 2. Example of a PBO structure

Phenolic polymer which does not cause cyclic reaction has been used for photoresist in the semiconductor field. In general, thermal resistance and mechanical properties of phenolic polymers are inferior to that of PI. Especially, concerning point of phenolic material is reliability for TCT because its film is brittle and fragile. Elongation and tensile strength are one of key parameters to define fragile nature. Elongation of typical phenolic polymer is around 5%. Therefore, in this study, high elongation phenolic material is selected to confirm the reliability for bumping application.

The film properties of PI, PBO and phenolic material are shown in Table 1.

Material		BL-s	BM-s	PBO	Phenolic polymer
Cure temp. [deg.C]		200	200	350	220
5% wt loss temp	deg.C	335	335	480	330
Tg	deg.C	200	220	300	230
CTE	ppm/K	60	50	60	35
Modulus	GPa	3.5	4.8	2.7	3.8
Tensile strength	MPa	130	150	140	135
Elongation	%	50	30	45	25
Residual stress	MPa	19	19	35	15
Adhesion (Si)	MPa	> 70	> 70	>70	>70

Table 1. Typical properties of dielectric material

Description of the Test chip

The reliability of WLCSP depends on the chip structure, chip size and with/without UF. Three different types of chips 3 mm x 3 mm size, 5 mm x 5 mm size and 9 mm x 9 mm size with 400 µm pitch were prepared for the evaluation. Numbers of SAC 305 solder balls are 36 pieces for 3 mm x 3 mm size, 100 pieces for 5 mm x 5 mm size and 256 pieces for 9 mm x 9 mm size. As the drop test chips, 3mm x 3mm size and 5mm x 5mm size were selected in accordance with standard WLCSP size without UF. As the TCT chips, 3mm x 3mm size and 9mm x 9mm size were selected to clarify the chip size influence. The chips were prepared by Fraunhofer IZM. The cross section image of chip is shown in Figure 3.

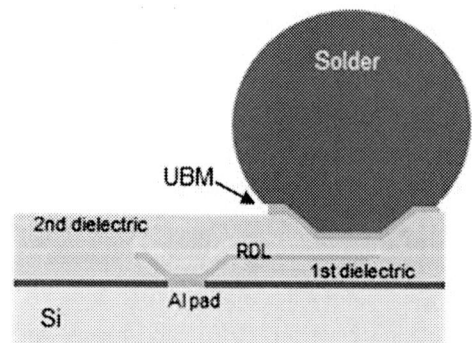

Figure 3. Cross section image of chip

The thickness of different layers are as below.
- Al pad: 1 µm
- Oxide passivation (PE-CVD): 1 µm
- 1st dielectric polymer : 7 µm
- RDL (TiW/Cu sputtering + Cu plating): 3 µm
- 2nd dielectric polymer : 7 µm
- UBM: Ni 5 µm, Au 100 nm
- UBM diameter: 235 µm
- Solder ball size: 250 µm (SAC 305)

978-1-4799-2408-0/14 $31.00 © 2014 IEEE

The chips are assembled on test boards which are based on standard FR4. Different board designs are adopted for drop test and TCT.

Description of the Test board for drop test

In case of drop test, the structure is based on JEDEC standard JESD22-B111. The composition of the board layer is shown in Table 2. The properties of each layer are shown in Table 3.

Board Layer	Thickness (microns)	Copper Coverage (%)	Material
Solder Mask	20		LPI
Layer 1	35	Pads + traces	Copper
Dielectric 1-2	65		RCC*
Layer 2	35	40% including daisy chain links	Copper
Dielectric 2-3	130		FR4†
Layer 3	18	70%	Copper
Dielectric 3-4	130		FR4†
Layer 4	18	70%	Copper
Dielectric 4-5	130		FR4†
Layer 5	18	70%	Copper
Dielectric 5-6	130		FR4†
Layer 6	18	70%	Copper
Dielectric 6-7	130		FR4†
Layer7	35	40%	Copper
Dielectric 7-8	65		RCC*
Layer 8	35	Pads + Traces + daisy chain links	Copper
Solder Mask	20		LPI

* Suggested RCC Material: Polyclad PCL-CF-400 12/35/35
† Suggested FR4 Material: NELCO N-4000-6 or equivalent

Table 2. Substrate layout for drop test

Property	Unit	FR4	RCC
Tensile Strength	MPa	>100	>50
Tensile Modulus	GPa	20 ± 2	2 ± 1
Tensile Elongation	%	>3	>3
In-plane CTE (below Tg)	ppm/°C	15 ± 2	60-80
Tg	°C	>130	>130
Cu Peel	kgf/cm	>1	>1

Table 3. Properties of board layers

The PCB was 8 layers with dimensions of 132mm x 77mm x 1mm and used a standard FR4 material. A total of 5 chips are mounted in a line per PCB on a single side in a 3 x 5 matrix which is defined by JEDEC (Figure 4). Connections to the test board were confirmed by the resistance measurement at the end of the board. The positions of resistance measurement area were connected at most outer side of bumps in the chip where the drop impact stresses concentrate.

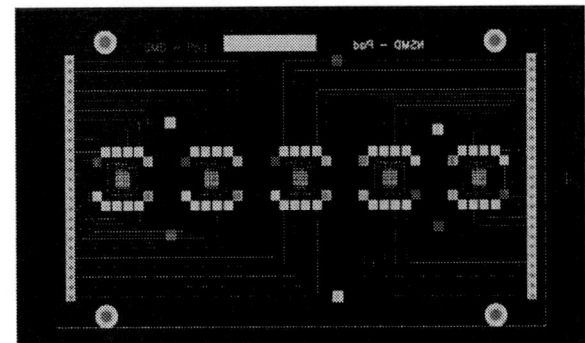

Figure 4. Test board image of a drop test

In this study, SAC305 solder balls were used. The reflow peak temperature was 254deg.C. The reflow profile during chips and a board connection is shown in Figure 5.

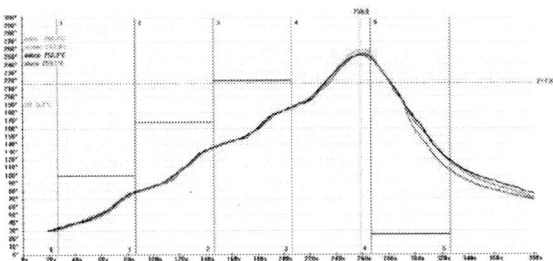

Heating rate: 1.8K/s
Cooling rate: 2.6K/s
Time over Ts: 46.3 sec

Figure 5. Reflow profile

The cross section images of a chip mounted on a board is shown in Figure 6. After reflow, all solder joints were good wetting in cross section.

Whole chip image

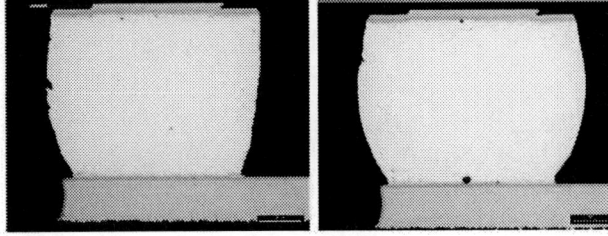

Figure 6. Cross section images of a drop test board

Drop test and data collection were conducted in accordance with JESD22-B111. The test apparatus setting image and shock event are shown in Figure 7. The board is fixed on the drop table by using pillars at 4 points of corners. Initial height of the drop table before drop test should be determined by acceleration when the table hit on the strike surface. The acceleration would be controlled by the material of strike surface and the height of drop table.

The parameters of drop test is shown as below,

- Equivalent drop height: 112 cm
- Velocity change: 467 cm/sec
- Peak acceleration: 1500 G
- Pulse duration: 0.5 msec

The board were dropped a maximum of 30 times. The failure criteria were judged whether intermittent discontinuities of resistance were over 1000 ohms and lasting for 1msec or longer.

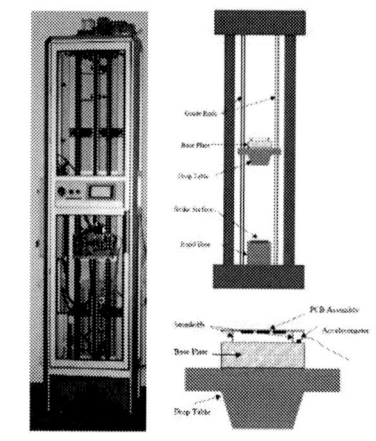

(a)

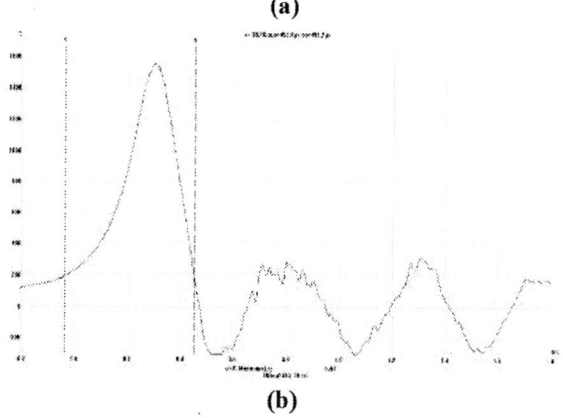

(b)

Figure 7. Drop test configuration (a) and sample shock event (b)

Description of the Test board for TCT

In case of TCT, all daisy chain structures of the chips were manually measured. Solder resist covered on top layer of the PCB, and design is briefly shown as below.

- Pad size: 300 μm
- Solder mask opening: 220 μm
- Solder mask thickness 15 μm

For example, maximum measurement channel numbers are 34 for 9mm x 9mm size chip with 400μm pitch (Figure 8).

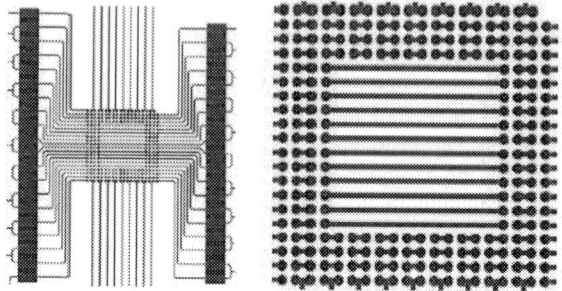

Figure 8. Test board design image of TCT 9mm x 9mm size with 400μm pitch

After the chips mounted on a board, TCT data was collected using an ESPEC TSD 100 (Figure 9). The test conditions for the AATC (air to air thermal cycling) are shown as below.

- 2 Chambers (ESPEC TSD 100)
- Lower Temp: -55deg.C for 30min
- Higher Temp: +125deg.C for 30min

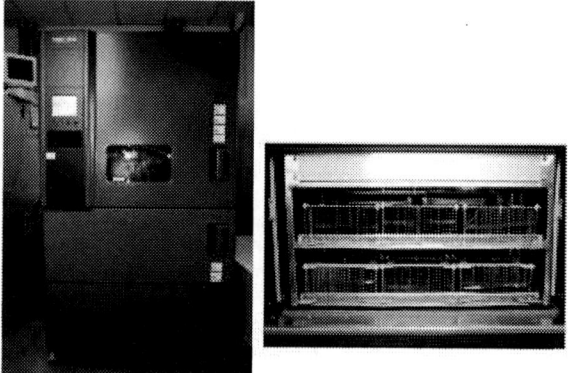

Figure 9. Equipment of AATC

Drop test results and discussion

Drop test results are listed in Table 4. Comparing the results for 3mm x 3mm chips without UF after 30 times drops, there was no obvious difference seen on all materials. As the results for 5mm x 5mm chips without UF after 30 times drops, total two chips of electric resistance were increased. Judging from cross section of the position, conductive layer crack were found at PCB side. However, the film breakage or RDL breakage at chip side did not occur. Since the purpose of this test is to clarify whether the film properties of dielectric materials affects the electrical failures, the influence of conductive layer cracks at PCB side are omitted from data collection. The conductive layer crack at PCB side was not observed when UF was filling between a chip and a board for 5mm x 5mm size.

Chip size	UF	Fail rate by different dielectric [%]			
		BL-300 PI	BM-300 PI	PBO	Phenolic Polymer
3x3mm	No	0	0	0	0
5x5mm	No	0	0	0	0
5x5mm	Yes	0	0	0	0

(a)

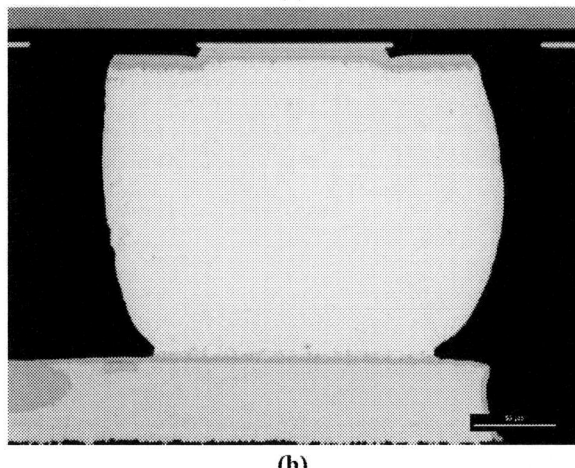

(b)

Table 4. Failure rate of drop test (a)
Cross section example after 30 times drops (b)

As the result of FEM analysis in our previous study [1], film thickness and modulus are key factors to reduce stress under bump area and solder bump.

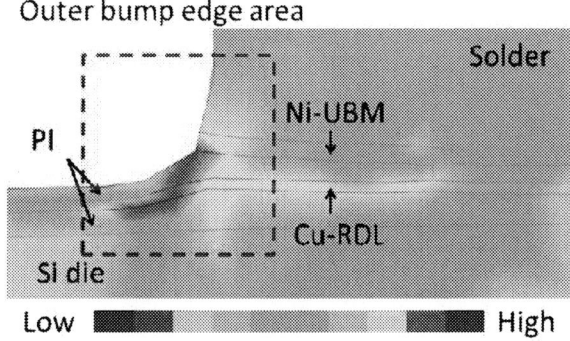

(a)

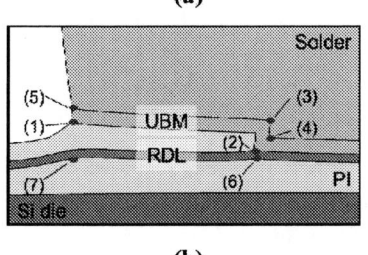

(b)

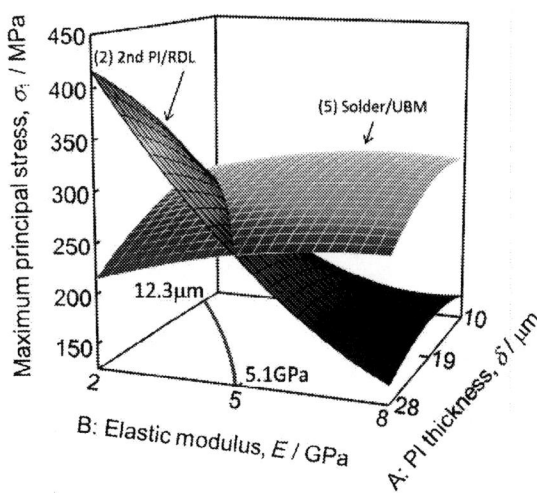

*Thickness: total thickness of 1st and 2nd dielectric layer

(c)

Figure 10. Image of deformation and stress concentration (a), stress simulated position in a chip side (b), Response surface of interaction at position 2 and 5 (c)

As the response surface in Figure 10 (c), thicker thickness and low modulus contribute stress increase at PI opening edge of pad area. On the other hand, thicker thickness and low modulus contribute strass reduction at bump edge area. In this time experiments, the dielectric layer thickness is total 14 μm, and modulus is different in between 2.7 and 4.8GPa. It is expected that the stress at PI opening edge of pad area (Figure 10(c), position 2) and the stress of interface between UBM and solder at bump edge area (Figure 10(c), position 5) are not so high. As the FEM results, stress at bump edge area (Figure 10(b), position 1 and 7) are similar tendency as Figure 10(b), position 5. Therefore, it is supposed that stress at pad area and bump edge area are not so high.

Actually, there was no film breakage, RDL breakage and solder breakage for all samples at chip side in drop test.

TCT results and discussion

TCT results of 3mm x 3mm chip without UF are listed in Table 5. PIMEL™ BL-300 was tested as a representative polyimide, and typical PBO was tested as a reference. There was no electrical failure for both samples till TCT 250 cycle, however, PBO was failed after TCT 500 cycle. The representative typical failure mode is seen by cross section in Table 5 (b). Solder breakage occurred at interface between a solder and a PCB. Dielectric material film breakage was also found under a bump edge. RDL metal layer breakage was also found on some of chips.

978-1-4799-2408-0/14 $31.00 © 2014 IEEE   833

Chip size & UF	Cycle times	Fail rate by different dielectric [%]	
		BL-300 PI	PBO
3x3mm without UF	100	0	0
	250	0	0
	500	0	40

(a)

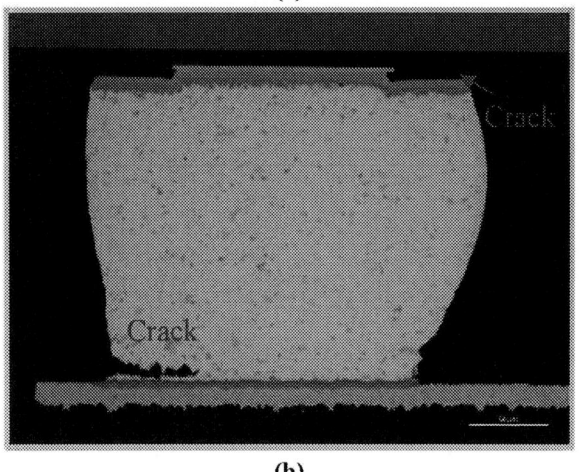

(b)

Table 5. TCT results without UF (a)
Cross section example image after TCT (b)

As the result of FEM analysis in our previous study [4], the principal stress shall be most concentrated under bump edge in the cooling step during thermal change. In the cooling step, tensile strength of a general organic material is getting higher, and the elongation is getting smaller because the material becomes brittle and fragile. Multiple thermal deformations are induced to the dielectric film during TCT. When the deformation becomes larger than the film elongation, the dielectric material breakage might be caused. Moreover, the film breakage becomes a trigger of RDL breakage and/or delamination between RDL and dielectric material. Therefore, the mechanical property at -55deg.C is very important to discuss electrical failure in chip side. Relationships of tensile strength vs. elongation of BL-series are shown in Figure 11(a). PIMEL[TM] BL-300 has very stable mechanical property at -55deg.C. The stability might make good TCT resistance.

On the other hand, PBO elongation at -55deg.C was much lower than the elongation at 25deg.C (Figure 11(b)). It is supposed that the PBO material tends to cause breakage during cooling step. The failure rate difference between PIMEL[TM] BL-300 and PBO might be caused by the film breakability at low temperature.

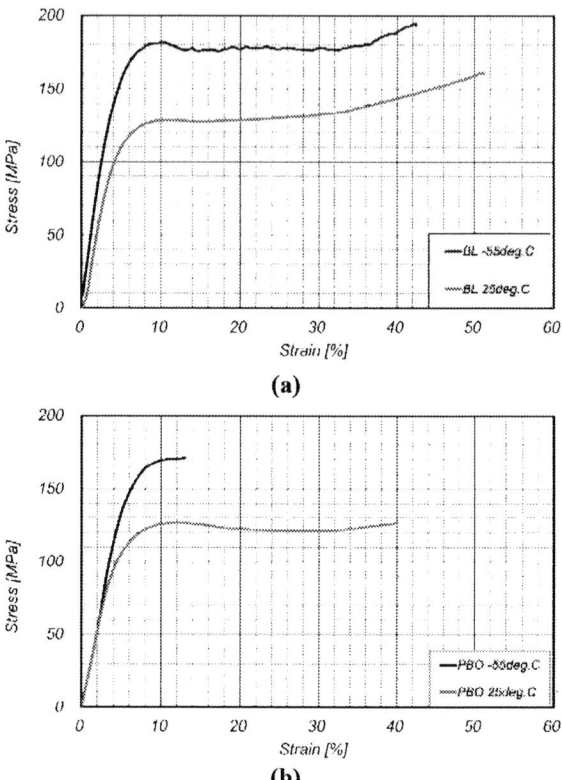

(a)

(b)

Figure 11. S-S curve of dielectric material under different ambient temperature (PIMEL[TM] BL-series (a), PBO (b))

TCT results of 9mm x 9mm chip with UF are listed in Table 6. There was no electrical failure for PIMEL[TM] BL-300 after TCT 1000 cycle. First failure was found on PBO after TCT 1000 cycle. Comparing the result of without UF, electrical failure is much lower due to buffer effect of UF.

There was no obvious solder breakage by cross section of failure position (Table 6 (b)). However, PCB breakage and dielectric film breakage were found on PBO sample.

In case of applying UF, concentrated stress at solder bump might be released to PCB side, and the stress concentrated to the solder resist which is top surface of PCB. When the stress is higher than the breakage energy of solder resist which is defined as the integration of tensile strength and elongation, solder resist breakage would be caused. The solder resist breakage is also a trigger of PCB breakage and cause electrical failure. Although the mechanical property data of solder resist at -55deg.C is not measured, it is supposed that the elongation is low because the elongation at room temperature is around 3%. The low elongation might be a cause of PCB breakage.

As for the chip side, PBO film breakage was also found even though applying UF. The mechanism of film breakage would be same as 3mm x 3mm chip without UF, film breakage at low temperature. Therefore, elongation at low temperature is important.

Chip size & UF	Cycle times	Fail rate by different dielectric [%]	
		BL-300 PI	PBO
9x9mm with UF	100	0	0
	250	0	0
	500	0	0
	750	0	0
	1000	0	5

(a)

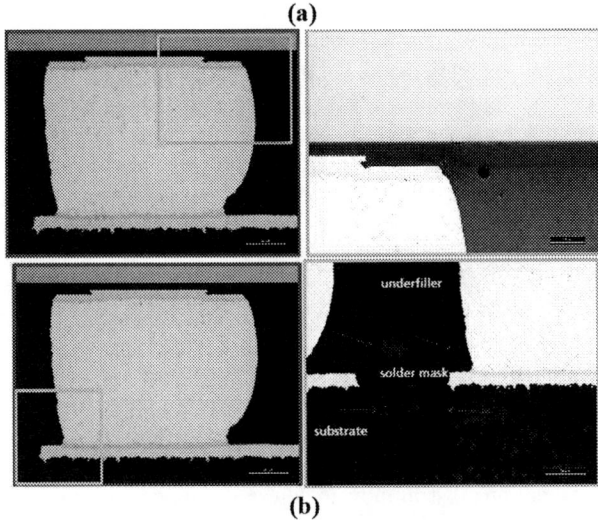

(b)

Table 6. TCT results with UF (a)
Cross section example image after TCT (b)

Conclusions

There are various kinds of photosensitive dielectric materials in the semiconductor filed. In terms of process limitation or device yields, process temperature limitation is required below 200deg.C in some packaging application. In this study, reliability performance of 200deg.C curable PIMEL™ BL-300 was verified.

PIMEL™ BL-300 showed good drop test and TCT performance comparing with PBO which is widely used in WLCSP. The key parameter of TCT resistance might be high elongation at -55deg.C. PIMEL™ BL-300 is a suitable photosensitive dielectric material to apply WLCSP for 200deg.C cure.

Acknowledgments

The Authors would like to extend their appreciation to Dr. Michael Töpper with Fraunhofer IZM for their cooperation.

References

1. N.Anzai, "Analysis of Stress Buffer Effect of Polyimide for Board Level Drop Test by a Finite Element Analysis," in *Proc. IEEE International Microsystems Packaging Assembly and Circuits Technology conference (IMPACT-IAAC),* Taipei, Taiwan, Oct. 22-25, 2013, pp. 64.

2. W.Volksen, *et al.,* "Low Dielectric Constant Materials," *Chem. Rev.,* 2010, Vol.110, pp.56-109.

3. G. Wang, *et al.,* "Chip-packaging interaction: a critical concern for Cu/low k packaging," *Microelectronics Reliability,* 2005, Vol.45, pp.1079-1093.

4. M.Niwa, "Analysis of stress buffer effect of polyimide on thermo-mechanical stress of ELK layer in flip-chip packages," in *Proc. IEEE International Conference on Electronics Packaging (ICEP), Japan, April 18-20, 2012.*

Optimization of Compression Bonding Processing Temperature for Fine Pitch Cu-Column Flip Chip Devices

Yonghyuk Jeong, Joonyoung Choi, Youjoung Choi, Nokibul Islam, Eric Ouyang
STATS ChipPAC Inc
yonghyuk.jeong@statschippac.com,
eric.ouyang@statschippac.com

Abstract

For the demand of high density input/output (I/O), fine-pitch, and low-k materials in copper column bump flip chip packages, Thermal Compression Bonding (TCB) with pre-applied Non Conductive Paste (NCP) has been developed in order to ensure manufacturing reliability. The narrow bonding process window of pre-applied NCP, short bonding time, and high bonding head temperature can cause low yield issues such as NCP voiding in solder and no solder wetting on substrate. For this reason, the bonding parameters, such as bonding temperature profiles and dwell times, have to be controlled and optimized to achieve good solder wettability.

In this paper, the optimized maximum bonding temperatures and timing of the TCB process for fine pitch copper column flip chip package are examined. A thermal simulation is also conducted to correlate with experimental data. In the experiment, the bump temperature is measured with a thermocouple while the bonding head temperature and time are controlled with a heat controller. In the thermal simulation, a transient approach is used to consider the bonding temperature profiles and boundary conditions.

The paper concludes with an approach and methodology to obtain optimized bonding temperature profiles which is crucial for the development of next generation fine pitch flip chip devices.

Introduction

As the number of I/Os continues to increase in many advanced devices, semiconductor package sizes and bump pitches become smaller and silicon (Si) nodes get narrower. For the conventional flip chip bonding process, the copper and solder bumps are self-aligned in between the die and substrate in a standard reflow in the manufacturing process to achieve high production yields. However, there is a high risk of die and bump cracking in the manufacturing process, especially for advanced packaging requirements. The typical failures are bump and die cracking, and bump bridging in the underfill fill steps. One of root causes of the failures is that the traditional reflow process introduces high thermal stresses [1]. To address these kinds of issues and to simplify the manufacturing processes for the next generation of packaging, TCB provides a good solution.

Thermal Compression with Non-conductive Paste (TCNCP) is an alternative solution to improve the quality of interconnection between die and substrate. This process basically combines both capillary underfill process (CUF) and molded underfill process (MUF) into one single process. TCNCP minimizes the number of process such as fluxing, die attach, reflow, deflux, and baking, which are typically used in the conventional CUF or MUF processes. The TCNCP process is able to prevent bump bridging and die/bump cracking as well as minimizing the voiding issue which is commonly seen in the CUF and MUF processes, especially for very small gaps in between the die and substrate.

However, the application of TCNCP requires careful thermal management. The temperature controls for the heating sources and the heat transfer from heating source to bumps must be tightly controlled in order to achieve the optimized bump temperature for the purposes of melting and soldering. To minimize voiding issues such as air entraps, a very short time window for curing has to be used in the NCP process.

The goal of this paper is to predict the temperature distribution of the copper/solder interconnection location. The temperature distribution is the key factor determining the NCP filling quality, solder shapes, adhesion strength of interconnections, and thermal stresses. Both experimental and simulation approaches are used in this study. The numerical approach provides an economic means to predict the temperature distribution to lower the manufacturing cost [2, 3].

TCNCP test vehicle

In this study, the test vehicle is a flip chip very fine Ball Grid Array (fcVFBGA). The device has 14x14x0.8mm body size and flip chip die 11x11x0.1mm. The test die is built with daisy-chain circuits for the open/short test and there are over three thousand bumps populated in the center and at the peripheral of the die. The pitch of bumps is around 80um for the central bumps and 40um for the peripheral bumps. The height of copper/solder interconnection is 42um. The substrate has 4 metal layers and Figure 1 is the schematic drawing of the test vehicle. Table 1 lists the package and substrate information.

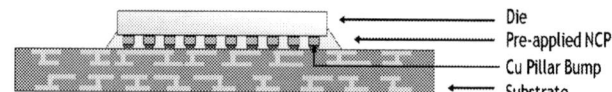

Figure 1. Cross sectional view of the package

	Item	Description
PKG Type	PKG type	FCVFBGA
	PKG size	14x14mm
Device	Die size	11x11mm
	Die thickness	100um
Bump	Bump Count	Over 3000
	Bump pitch	40/80um
	Bump height	42um
Substrate	Type	Strip type
	Substrate layer	4-layer
	Core Thickness	150um
	Bond Pad Finish	OSP
NCP Material		Material A

Table 1. Package and substrate

Advantages of TCNCP

There are a number of advantages with the TCNCP process. For the conventional flip chip manufacturing process, there are several process steps involved such as die preparation (DP), chip attach and reflow, flux cleaning or deflux, baking, underfill and underfill curing, and the molding. With the application of TCNCP, two to three processing steps may be removed as shown in the Figure 2. Before the thermal compression bonding with pre-applied NCP, the TCNCP goes through the die preparation (DP) which is the same as a conventional flip chip manufacturing process. The DP involves wafer back grinding and laser grooving which separates each die from the wafer. After the DP, the substrate in the TCNCP process is then baked to remove the moisture and prevent the voiding risks. Following the substrate baking, the substrate was cleaned with a plasma cleaning process to remove surface impurities and achieve a better surface adhesion for compression bonding and better flow of NCP materials. Figure 2 shows that with TCNCP process, the reflow and deflux steps are removed, thus the simplification and lower manufacturing cost.

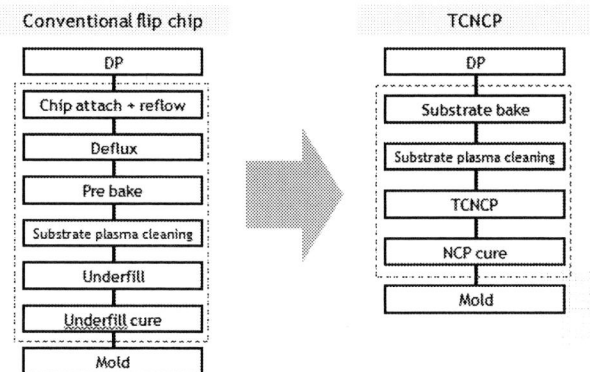

Figure 2. Conventional flip chip vs. TCNCP manufacturing process

Consideration of processing parameters

From previous section, the TCNCP couples the chip attach step with underfill step to lower the manufacturing cost. The simplified process involves detailed procedures and considerations such as the bonding force, NCP dispensing, bonding tool setting, and the temperature profile as shown in the Figure 3. These four key process parameters have to be carefully controlled and monitored to obtain the best optimization. Various conditions and parameters were carefully tested at STATS ChipPAC to ensure the reliability of the final packages.

Bonding force was adjusted and selected to have a good quality of bump shape after soldering and stand-off height. If the bonding force is too high or too low, the solder cap may be squeezed too much or may have non-wetting issues. The dispensing volume, rate, and pattern of NCP were also carefully controlled to minimize the void and overflow. The bonding tools were calibrated and adjusted to ensure the test parts and machine components were accurately aligned. This step is crucial, especially for the fine pitch devices, to achieve short process time and to avoid the contamination of NCP

overflow or foreign material on the bonding heads and machine tools [4, 5].

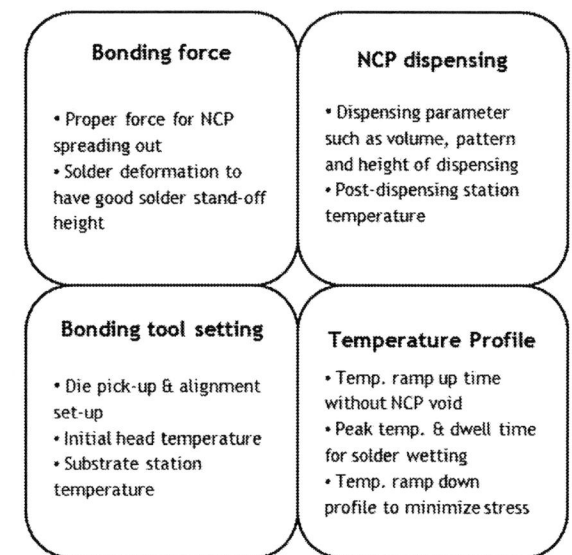

Figure 3. Key parameters for TCNCP

To have a successful bond, one of the most important keys is to obtain an optimized temperature profile which considers the ramp up/down speeds and times. The ramp up/down speeds and times affects the NCP flow behavior, void creation, and residual stress of the final product. In this regard, the peak and dwell time shall be precisely controlled to provide enough time for melting and soldering of bumps with substrate pads.

In this study, the target bonding peak temperature of the test vehicle surface was set to be 250C. In order to achieve this temperature within a short time, the temperature of bonding heads has to be heated to over 400C using rapid ramp-up speed. For the production concern, the entire process flow, which includes the NCP spreading and melting of solder, has to be controlled to within 5 seconds. Figure 4 demonstrates a schematic drawing of the temperature profile of the compression bonding steps.

(Step 1) Pick and deliver a die from wafer to the substrate and align the die with substrate which is already being dispensed with NCP material.

(Step 2) Place and mount the die on the substrate. The bottom substrate and top die have different initial temperatures, but all parts are heated rapidly as shown in the figure.

(Step 3) In this step, the parts are heated rapidly to reduce the process or cycle time. The peak temperature and the dwell time, which are for the melting of solder and the curing of NCP, have to be carefully monitored and controlled. The curing rate and time will determine if voids or air trappings occur inside the NCP. The improper temperature or time may also cause solder bleedings, abnormal bump shapes, bridging bumps, or non-wetting bumps. After this step, the NCP is partially cured.

(Step 4) After the die and substrate are bonded, the heating bonding head separates from the silicon and the temperatures in the parts start to drop rapidly. The

978-1-4799-2408-0/14 $31.00 © 2014 IEEE

temperature in this step must be carefully controlled and monitored as well to avoid the bump or die cracking.

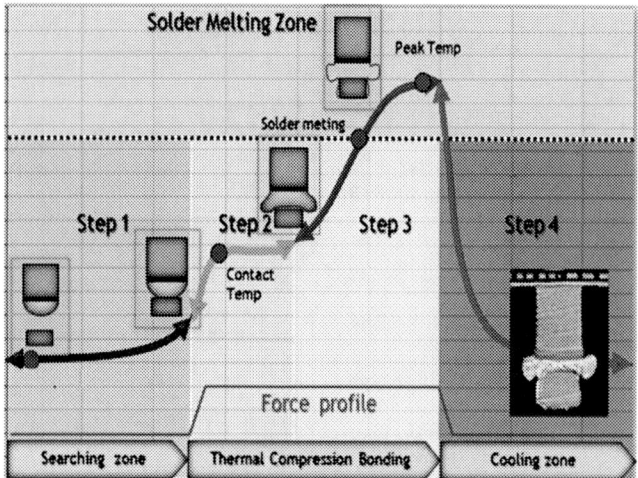

Figure 4. Thermal compression bonding steps

As mentioned in the previous four steps, the temperature profile has to be carefully monitored and controlled, otherwise, defects may occur.

Experimental setup

To monitor the temperature distribution of copper/solder interconnections, a thermocouple is used. The thermocouple is designated to be placed in positions where high density bumps are located. In this study, the thermocouple is placed inside the NCP material and at the peripheral bumping area, as shown in the Figure 5. The procedures for the preparation and measurement of thermocouple were as follows: (a) dispense NCP material on the substrate; (b) place and attach a thermocouple near the edge of die, and ensure the device is in the region where high density copper/solder interconnections are located. Figure 6 illustrates one sample of NCP coverage after the process.

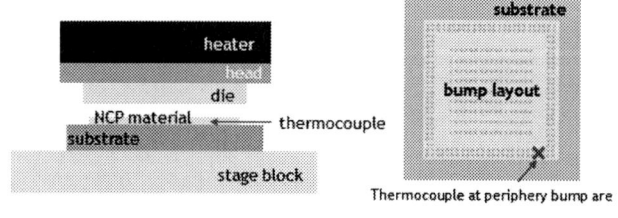

Figure 5. The placement of thermocouple to measure the temperature

After the preparation of thermocouple, the device goes through a standard TCNCP process flow which includes baking the substrate, plasma cleaning of the surfaces with Argon gas and beginning the pattern dispensing of flux and NCP materials. After this the device is moved to the bonding stage and the substrate was pre-heated on the stage.

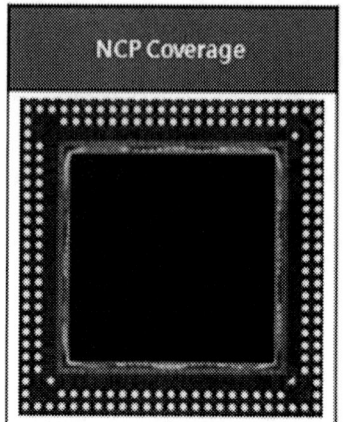

Figure 6. NCP coverage after process

Figure 7 below illustrates an example of the measured temperature profile near the copper/solder interconnections. The optimized temperature profile is obtained through various experiments. The performance of each temperature profile is verified by the quality of final processed devices.

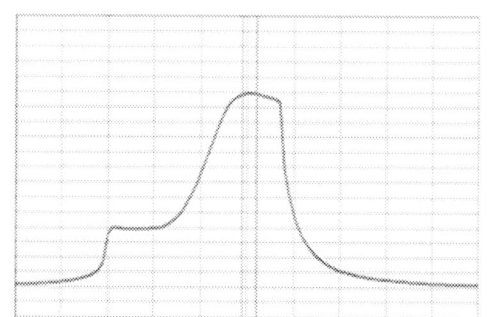

Figure 7. The measured temperature profile from thermocouple

Optimization of temperature profile with simulation

Besides the experimental approach, a numerical simulation is conducted to correlate with experimental data. The simulation provides an advantage to obtain numerous data points under different conditions without the need to do all experimental legs. In this study, the focus was on the temperature prediction near the die edge in order to have an optimized temperature profile. A transient thermal simulation was conducted to study the temperature response of interconnection and the detailed simulation approaches are introduced below.

(i) Configuration of the system and boundary conditions

For this system level thermal simulation, the ambient temperature was set to be 25C and a convective heat transfer coefficient of 6W/mK was applied on the outer surfaces of the bonding machine tools. The bonding machine tools comprise the components of heater, head, and stage, as shown in Figure 8. The heater and head are made of ceramic material to sustain the high temperature application. The detailed structures and dimensions of the components are considered in the simulation.

978-1-4799-2408-0/14 $31.00 © 2014 IEEE 838

(ii) Configuration of the package and material properties of the components

In order to accurately predict the temperatures of the copper/solder interconnections, the detailed features of the substrate were considered. The effective material properties were considered in order to determine the temperature responses of the device. Table 2 shows the material properties of the components used in the simulation.

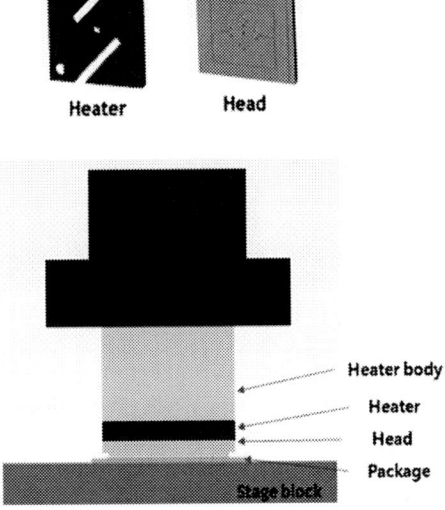

Figure 8. System level thermal modeling

	Item	Material name	Thermal conductivity (W/mK)	Density (g/cm3)	Specific heat (J/Kg-K)
TC Bonding	Heater & Stage	Ceramic A	150	3.31	880
	Head	Ceramic B	90	3.2	670
Package	Substrate	Dielectric	0.3	1.2	880
	NCP	Material A	0.85	1.57	930
	Cu pillar	Copper	401	8.94	385
	Solder	SnAg	73.52	7.34	227
	Die	Silicon	117.5	2.34	700

Table 2. Material properties of the components

The focus of the current simulation was to predict the temperature responses of the copper/solder interconnections. Thus the distribution of copper/solder interconnection was calculated. In this study, the test vehicle package had more than 3000 copper/solder interconnections and the copper/solder interconnections with four different scenarios was considered, as shown in Figure 9.

(Option 1)
Smear all components and all substrate layers into one single material and one single layer. For this case, there was only one effective material property.

(Option 2)
Smear top copper column and top NCP as one layer, and smear bottom solder and bottom NCP as another layer. For this case, there were two effective material properties corresponding to these separate layers.

(Option 3)
Smear the internal copper/solder interconnections and internal NCP as one region, and smear the outer square of copper/solder interconnections and NCP material as another region. Outside of these two regions was the pure NCP material. For this case, the localized effective material properties were used inside each region.

(Option 4)
Simulation software was then used to import the detailed distribution of the copper/solder interconnections. For this approach, all the layers and components were captured into the simulation, thus, the material property will be position dependent. The drawback of this approach was the computational resources and time required because of the significant amount of numerical meshes.

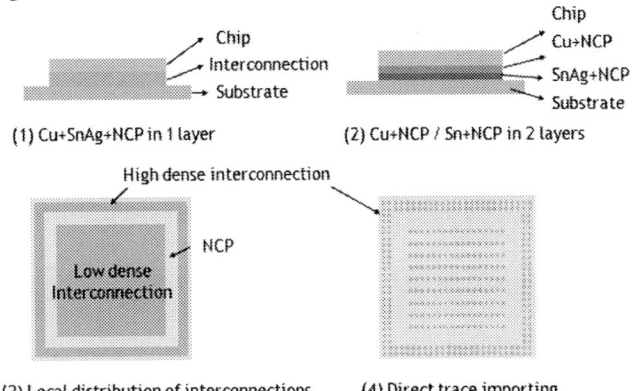

Figure 9. Consideration of copper/solder interconnections with four scenarios

Table 3 shows the calculated effective thermal conductivities of the scenarios used in the simulation. From this table it is evident that the peripheral region of option C has a much higher thermal conductivity due to the high density distribution of copper/solder interconnections.

Option	Composition	Thermal conductivity (W/mK)	Density (g/cm3)	Specific heat (J/kgK)
1	Cu+SnAg+NCP	4.18	1.59	986
2	Cu+NCP	5.83	1.56	1030
	SnAg+NCP	1.75	1.64	921
3	Center area	3.41	1.58	973
	Peripheral area	21.87	1.70	1282
	No bumps(NCP)	0.85	1.57	930
4	Position dependent	Position dependent		

Table 3. Effective material properties

Correlation of simulation and experimental data

Figure 10 illustrates the temperature setting at the heater, the experimental measured temperatures at the peripheral NCP and copper/solder interconnections, and the simulated temperature data using four different options as mentioned in previous section. The temperature profile from the heater was calculated based on heater's ramp-up heating rate, dwell time, and the ramp-down cooling rate. The temperature profile was obtained through various optimization tests.

978-1-4799-2408-0/14 $31.00 © 2014 IEEE

The measured and simulated temperature distribution may be roughly divided into three zones, namely contact zone, rise zone, and drop zone. The contact zone refers to the heater and die that are in contact with the substrate. The rise zone is the temperature rising after the parts are in contact with each other and the heater is rapidly powered. The drop zone refers to the separation of heater from the device. In the simulation, the transient boundary conditions of the heater and head were considered, and a temperature monitoring point was set at the corner peripheral ring position which is the same location as the actual thermocouple.

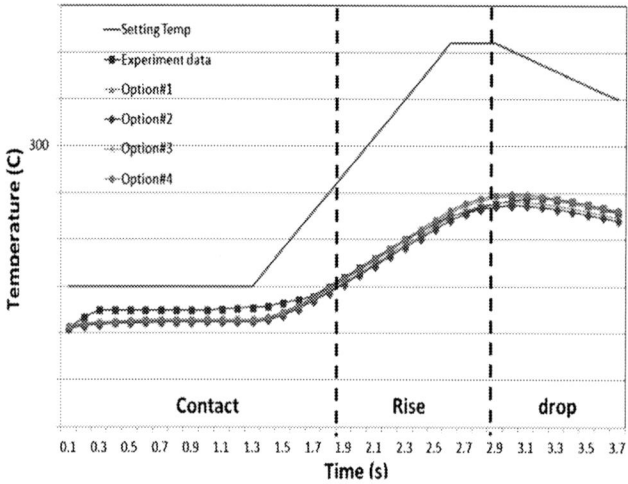

Figure 10. Simulation result and experimental data

Figure 11 shows the experimental and simulated temperatures of the peripheral copper/solder interconnections as a function of time near their peak temperatures. Options 1 and 2, which do not consider the localized high density copper/solder distribution, have lower peak temperatures. If the simulated temperatures and the times are used in the real process, they may not be enough to melt the solder. For options 3 and 4, the peak temperatures are closer to experimental data and this is due to the consideration of localized high density copper/solder distributions.

Figures 10 and 11 predict the temperature distribution of interconnections as a function of time during the TCNCP process. The combination of the experimental and simulation approaches provide a methodology to optimize the TCNCP volume manufacturing process.

Conclusions

Following are the conclusions of this study:

(1) Through experiment, which includes various adjustments of the manufacturing parameters and processing optimization, end-of-line fcVFBGA devices have very high yields and the defects are minimized.

(2) The temperature control to bond the silicon, NCP material, and substrate is one of the key parameters to obtain defect-free devices.

(3) A thermocouple was used to monitor the temperatures of NCP and copper/solder interconnections, and the temperature was compared

with numerical simulation. The experimental and simulation data were very close to each other.

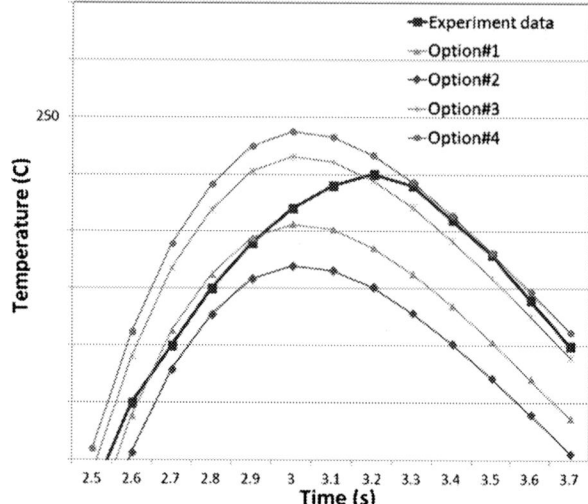

Figure 11. Experimental and simulated peak temperatures

(4) A transient thermal simulation was conducted to predict the transient temperature distribution near copper and solder interconnections. The copper trace and via on the substrate were modeled with four different scenarios. The simulation scenarios which do not consider the localized copper/solder density lead to lower peak temperatures when compared with real experiment.

(5) The combination of the experimental and simulation approaches provide a methodology for the optimization of NCTCP process.

References

1. Julien Sylvestre et al, "The Impact of Process Parameters on the Fracture of Device Structures During Chip Joining on Organic Laminates", in *Proc.58th Electronics Components and Technology Conf.(ECTC)*, Lake Buena Vista, FL, May 2008, pp.82-88

2. Mark Gerber "Next Generation Fine Pitch Cu Pillar Technology-Enabling Next Generation Silicon Nodes", in *Proc. 61st Electronics Components and Technology Conf.(ECTC)*, Lake Buena Vista, FL, May 2011, pp.612-618

3. Lee, Minjae, et al, "Study of Interconnection Process for Fine Pitch Flip Chip",in *Proc 59th Electronic Components and Technology Conf.(ECTC)*, San Diego, May 2009, CA, pp.720-723

4. Y.M.Cheung, "Process considerations of TC-NCP Fine-pitch Copper Column FC Bonding" *14th Electronics Materials and Packaging*, Dec 2012, pp.1-7

5. Tan, S.C., et al, "Process optimization to overcome void formation in nonconductive paste interconnections for fine-pitch applications", *Journal of Electronic Matierlas*, Vol.34, Issue 8, pp 1143-1149

Reliability Improvement Methods of Solder Anisotropic Conductive Film (ACF) Joints Using Morphology Control of Solder ACF Joints

Yoo-Sun Kim, Seung-Ho Kim, Jiwon Shin and Kyung-Wook Paik

Nano Packaging and Interconnect Lab. (NPIL) Korea Advanced Institute of Science and Technology (KAIST)
Guseong-dong 373-1, Yuseong-gu, Daejeon, 305-701, Republic of Korea
phone: +82-42-350-3375, fax:+82-42-350-3310

email: kys870505@kaist.ac.kr

Abstract

As the use of ACFs increased in various areas of electronic packaging such as semiconductors and flexible devices, demands for highly reliable ACFs have been also increased. For this demand, solder ACFs which use solder particles as conductive particles were introduced. In the solder ACF bonding, the solder ACF joint morphology should be controlled because the failures of solder ACF joints by crack propagation have close relationship with morphology of solder ACF joints. In this study, in order to improve the reliability of solder ACF joints in electronic packaging, the morphologies of solder ACF joints were controlled and the reliability depending on the morphologies was investigated. For the investigation of ACF joint reliability, unbiased autoclave tests were performed at 121℃, 2atm, and 100% relative humidity. According to the results, as bonding pressure increased from 2 MPa to 6 MPa, aspect ratio (Joint area/joint gap) increased by increased joint area and decreased joint gap. In unbiased autoclave tests, some of solder ACF joints with bonding pressure of 2MPa showed electrical open failures after 60 hours because tensile stress was applied to solders due to polymer resin expansion by water absorption. On the other hand, solder ACF joints with bonding pressure of 6 MPa showed no open failures for 60 hours due to higher tensile strength by higher aspect ratio compared with those with bonding pressure of 2 MPa. Solder ACF joints with bonding temperature of 250℃ showed hourglass shape by large spreading of solders and 5 times higher radius of curvature of stress concentration region than that of solder joints with bonding temperature of 200℃ which showed barrel shape. Solder ACF joints with bonding temperature of 250℃ showed higher reliability than those with bonding temperature of 200℃ due to smaller amount of concentrated stress by hourglass shape. In terms of ACF resin materials, despite of bonding temperature of 200℃, low curing rate acrylate ACFs showed hourglass shape and higher reliability than that of solder joints with conventional acrylate ACFs. The reason of that was low curing rate acrylate ACFs showed lower degree of cure of resin around solder at solder melting point than that of conventional ACFs. These results indicate that solder ACF joint morphology can be controlled by adjusting bonding conditions and ACF materials. Furthermore, the morphologies of solder ACF joints can be significantly important factors for highly reliable ACF joints in high temperature and high humidity.

1. Introduction

Anisotropic conductive film (ACF) is a film-type interconnection adhesive material which consists of polymer adhesive resins and randomly dispersed conductive particles. ACFs have been widely used due to low bonding temperature, fine-pitch capability, and cost effectiveness [1]. Recently, as the use of ACFs has increased in electronic packaging such as flexible devices and semiconductors, demands for highly reliable ACFs has been also increased. For this demand, solder ACFs which use solder particles as conductive particles of ACFs has been introduced combined with an ultrasonic bonding method before. The electrical properties of solder ACFs such as lower contact resistance and higher current handling capability can be obtained by solder particle melted joints compared with physical contact metal conductive particles of conventional ACFs [2]. However, even with these advantages of solder ACFs, further investigations are needed to enhance the reliability of solder joint during high temperature and high humidity conditions [3].

In the solder ACF bonding, solders are melted simultaneously with polymer resin curing reaction and the solders between electrodes form various joint morphologies depending on the ACF resin flow and curing characteristics. Furthermore, the solder joint morphologies between electrodes can affect the ACF joint reliability in high temperature and high humidity conditions. Therefore, the study on the effects of solder joint morphology on the reliability of solder ACF joints is needed to obtain highly reliable solder ACF bonding.

In this study, solder joint morphologies with different wetting shape and aspect ratio were obtained by controlling bonding conditions and ACF resin materials, and the reliabilities of solder ACF joints were evaluated using unbiased autoclave tests.

2. Experiment

2.1 Materials

In this study, 20 wt% Sn58Bi solder balls with diameter of 25 um were used in the ACFs. The thickness of ACF films was 35 μm and 5 wt% Ni balls with diameter of 8 μm were also used as spacers of solder ACFs to maintain uniform gaps between electrodes.

Test vehicle Flex-On-Board (FOB) was performed using solder ACFs. 1 mm-thick rigid printed circuit boards (PCBs) and 250 μm-thick flexible printed circuit boards (FPCBs) were used. Figure 1 shows the FOB structure used in the experiments. Both substrates have 500 µm pitch Cu patterns with electroplated Ni/Au surface finish.

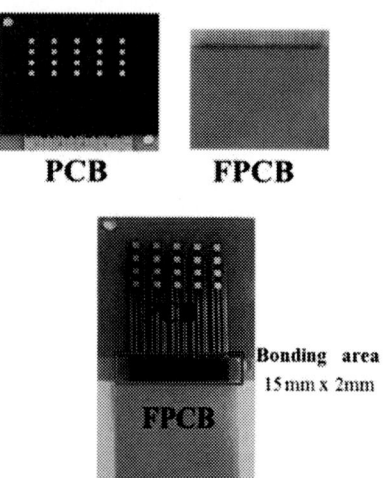

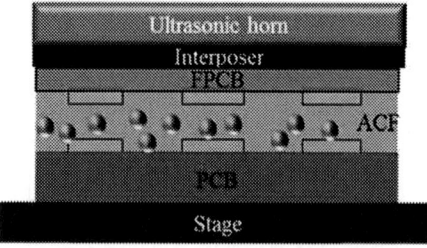

Figure 1. Schematic of FOB samples used in this study

2.2 Experimental parameter conditions
2.2.1 Bonding parameters

In this study, ultrasonic bonding method was used for the stable solder ACF joint formation. Figure 2 shows the experimental set up in ultrasonic bonding. In this bonding method, solder oxide layers were broken by ultrasonic vibration and solders between electrodes can show good wettability. In this bonding process, the ACF bonding temperatures were measured using the control of the ultrasonic amplitude and the bonding pressures were applied simultaneously with the ultrasonic vibration [4].

In order to investigate the effects of bonding pressures on solder ACF joint morphologies, bonding pressures were changed from 2 MPa to 6 MPa. As amount of resin flow between electrodes increase by large bonding pressures, the solder joint gap decrease and the reliability can be changed. In terms of bonding temperature, the ACF joints were bonded at 200℃ and 250℃. As different bonding temperatures are used, degree of cure of resin around solders at solder melting point (MP) was different by different heating rate and this factor can have effect on the joint morphologies and joint reliability. In these experiments, conventional acrylate resins were used and bonding time was 9 seconds.

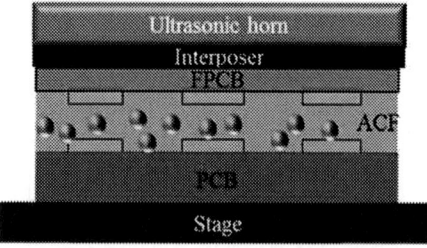

Figure 2. Schematic of the experimental set up in the ultrasonic bonding

2.2.2 ACF material property

Two acrylate-based resins with different curing rates were used. In this experiment, bonding temperature of 200℃ and

bonding pressure of 2 MPa were used. Figure 3 shows curing behaviors of used resins which were measured by differential scanning calorimetry (DSC). Low curing rate acrylate ACFs showed about 20℃ higher on-set curing temperature than that of conventional acrylate ACFs.

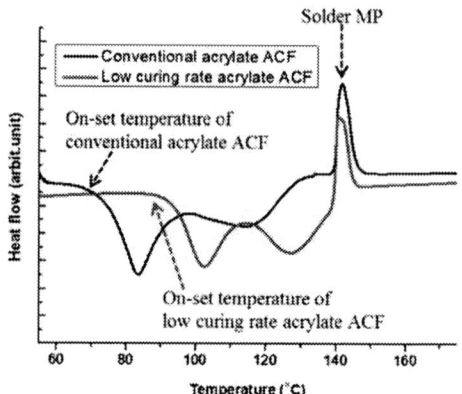

Figure 3. Curing behaviors of used resins

2.3 Observation of solder ACF joints

Solder ACF joints were observed after bonding in terms of aspect ratio and wetting shape of joints. For the calculation of aspect ratio (Joint area/joint gap) of ACF joints, joint gaps and joint areas were measured as shown in figure 4. Joint gaps between electrodes were measured by cross-sectional analysis and joint areas were measured by observing remaining solders on substrates after peeling off FPCBs and PCBs. In terms of solder wetting shape, radius of curvature of stress concentration regions was calculated as shown in figure 5.

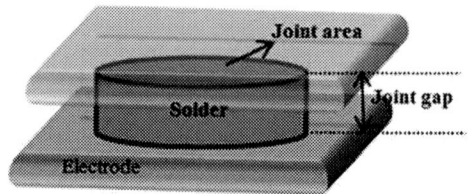

Figure 4. Structure of solder ACF joints after bonding

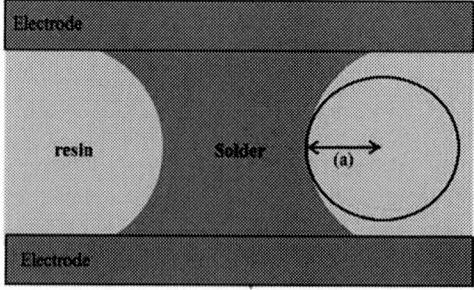

Figure 5. Schematic of solder ACF joint morphology and (a) represents radius of curvature of stress concentration region

2.4 Electrical test & unbiased autoclave test

Joint resistances were measured to examine the electrical continuity after bonding. Figure 6 shows the 4-point kelvin method used for a single joint resistance measurement. Joint resistances were measured at the section where the single through a FPCB and the single through of a PCB overlap each other. For the reliability evaluation of the solder ACF joints, unbiased autoclave tests were performed. Unbiased autoclave tests were performed at 121°C, 2 atm and 100% relative humidity and joint resistances were observed at every 12 hours during the test.

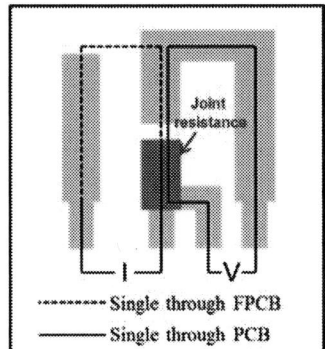

Figure 6. 4-point kelvin method used for joint resistance measurement

3. Results & discussions

3.1 Effects of bonding parameters on joint morphologies & reliability

Figure 7 showed solder ACF joints depending on the bonding pressures and figure 8 showed measured joint gap (a) and joint area (b) and calculated aspect ratio (c). As bonding pressure increased from 2 MPa to 6 MPa at bonding temperature of 200 ℃, aspect ratio increased from 318 to 1320 by decreased joint gap and increased joint area because large amount of resin flow out by large force applied to ACF joints in high bonding pressure. As shown in figure 9, in unbiased autoclave tests, average of joint resistances of solder ACF joints with bonding pressure of 2 MPa showed about 400 mOhm after 60 hours which was 20 times higher than initial resistances because tensile stress was applied to solders due to polymer resin expansion by water absorption. On the other hand, solder ACF joints with bonding pressure of 6 MPa showed stable joint resistance after reliability tests due to higher tensile strength by high aspect ratio compared with those with bonding pressure of 2 MPa [5].

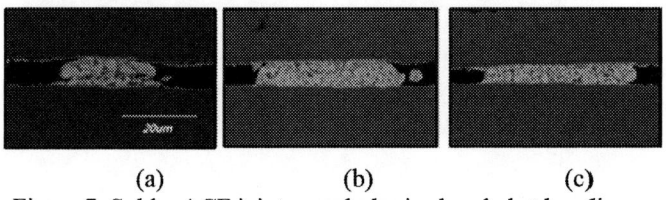

(a)   (b)   (c)

Figure 7. Solder ACF joint morphologies bonded at bonding pressure of 2 MPa (a), 4 MPa (b) and 6 MPa (c).

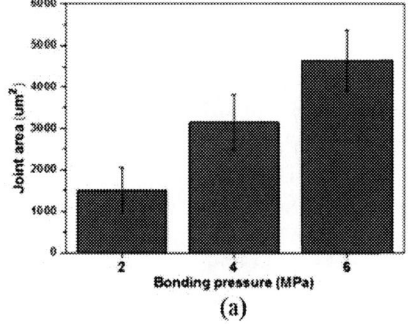

(a)

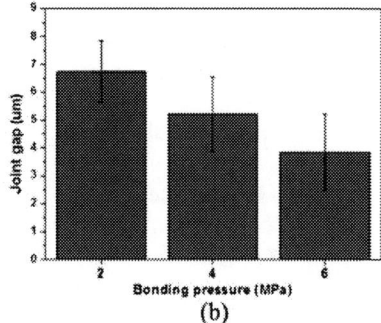

(b)

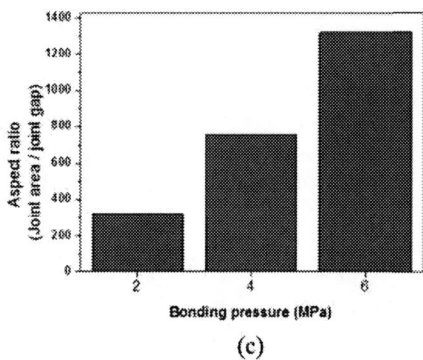

(c)

Figure 8. Joint areas (a), joint gaps (b) and aspect ratios (c) of solder ACF joints depending on the bonding pressures

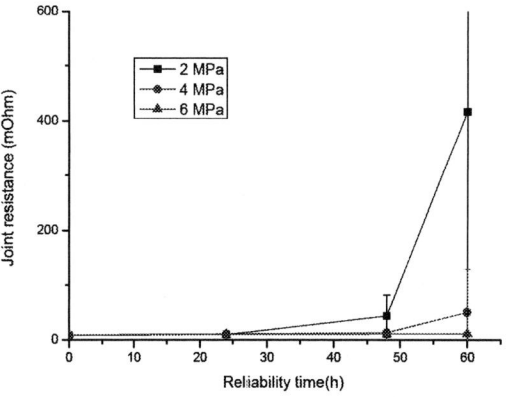

Figure 9 Joint resistances after unbiased autoclave tests

Figure 10 shows solder ACF joints bonded at bonding temperature of 200℃ and 250℃. The joints bonded at 200℃

showed barrel shape and the joints bonded at 250℃ showed hourglass shape. Moreover, the average radius of curvature of stress concentration region was 3.56 in ten solder ACF joints with bonding temperature of 250℃, which was 5 times larger compared with those bonded with bonding temperature of 200℃. These results are related to the degree of cure at solder melting point. Figure 11 shows the ACF temperatures measured during the bonding process and degree of cure of resin at the time when ACF temperature reached at solder melting point. The degree of cure of the resin was measured by fourier transform infrared spectroscopy (FTIR). The resin around solder with bonding temperature of 250℃ showed 13% degree of cure. On the other hand, the resin with bonding temperature of 200℃ showed two times higher degree of cure because ACF temperature with higher bonding temperature reached to solder melting temperature in shorter time. Therefore, the solder joints showed large spreading at bonding temperature of 250℃. In the concept of stress concentration in the material fracture, when nominal tensile stress is applied, the higher stress is applied to the smaller radius of curvature of stress concentration region [6]. According to the concept, in unbiased autoclave test, solder ACF joints with bonding temperature of 250℃ showed stable joint resistances after 60 hours as shown in figure 12. On the other hand, some solder ACF joints with bonding temperature of 200℃ showed open failures after 60 hours in unbiased autoclave tests. The result was due to lower stress applied to solder by larger radius of curvature of stress concentration region with higher bonding temperature.

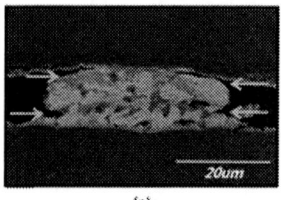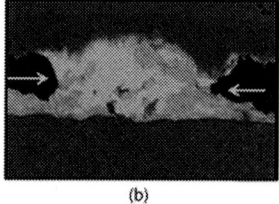

(a)   (b)

Figure 10. Solder ACF joint morphologies with bonding temperature of 200℃ (a) and 250℃ (b) and arrows represent stress concentration regions

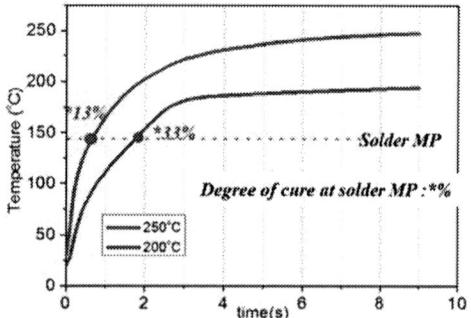

Figure 11. ACF temperature during bonding process and degree of cure of resin at solder MP

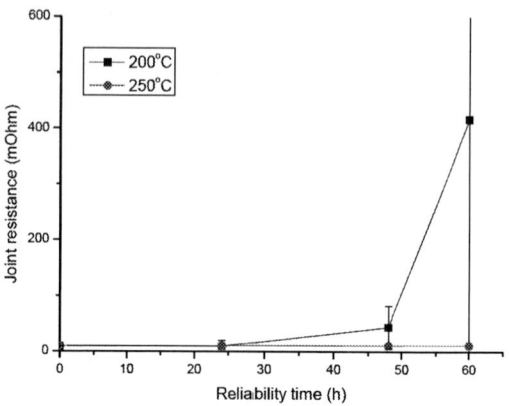

Figure 12. Joint resistances depending on the bonding temperature

3.2 Effects of ACF material property on joint morphologies & reliability

Figure 13 shows solder ACF joints bonded with two type of acrylate rate ACFs. The ACF joints were bonded at bonding temperature of 200℃ and bonding pressure of 2 MPa. Solder ACF joints with low curing rate acrylate ACFs showed larger spreading and hourglass shape despite of bonding temperature of 200℃. The reason of that was low curing rate acrylate ACFs showed lower degree of cure at solder melting point than that of conventional acrylate ACFs. In the results of the effects of bonding temperature, hourglass-shaped solder joints showed higher reliability than barrel-shaped solder joints because the amount of concentrated stress was small in hourglass-shaped solder joints by larger curvature of radius at stress concentration regions. Therefore, low curing rate acrylate ACFs showed more stable joint resistance after 60 hours in unbiased autoclave tests than that of conventional acrylate ACFs due to hourglass-shaped solder joints as shown in figure 14.

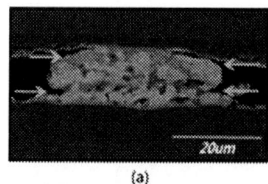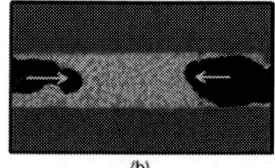

(a)   (b)

Figure 13. Solder ACF joint morphologies bonded with conventional acrylate ACF (a) and low curing rate acrylate ACF (b) and arrows represent stress concentration regions

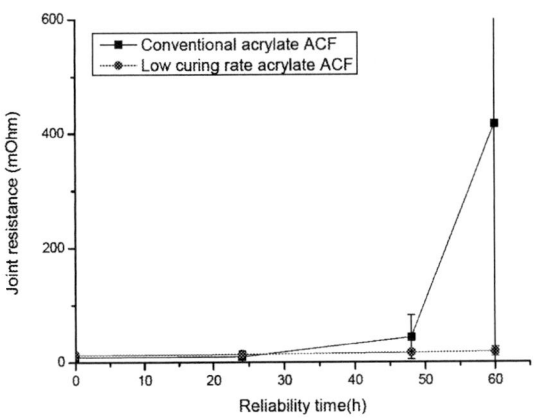

Figure 14. Joint resistances depending on the ACF material property

4. Conclusions

In this study, the reliabilities of solder ACF joints depending on the bonding parameters and ACF material properties were investigated. The results showed that as bonding pressures increased from 2 MPa to 6 MPa, solder aspect ratio (Joint area / joint gap) in the ACF joints increased from 320 to 1320 by decreased joint gaps and increased joint areas. It was mainly due to larger amount of resin flow and higher force applied to solders at higher bonding pressures. In unbiased autoclave tests, some of solder ACF joints bonded at 2 MPa showed open failures in 60 hours because tensile stress was applied to solders due to polymer resin expansion by water absorption. On the other hand, solder ACF joints bonded at 6 MPa showed no open failures for 60 hours due to higher tensile strength by high aspect ratio compared with those bonded at 2 MPa. In terms of bonding temperature, the solder ACF joints bonded with bonding temperature of 250℃ showed hourglass shape and the solder ACF joints with bonding temperature of 200℃ showed barrel shape. Moreover, the joints bonded at 250℃ showed 5 times larger radius of curvature of stress concentration region in solder joints compared with those bonded with bonding temperature of 200℃. The reason of that was larger solder spreading with bonding temperature of 250℃ by lower degree of cure of resin around solders at solder melting temperature, 138 °C. In the concept of stress concentration in the material fracture, when nominal tensile stress is applied, the greater stress is applied to the smaller radius of curvature of stress concentration region. According to the concept, in unbiased autoclave test, Solder ACF joints with bonding temperature of 250℃ showed open failure rate in 60 hours later than those with bonding temperature of 200℃. The result was due to lower stress applied to solder by larger radius of curvature at stress concentration region. However, despite of bonding temperature of 200℃, low curing rate acrylate ACFs showed hourglass shape and higher reliability than that of solder joints with conventional acrylate ACFs. The reason of that was low curing rate acrylate ACFs showed lower degree of cure at solder melting point than that of conventional acrylate ACFs. The significance of the results was the reliability of the ACF joints can be improved by the control of ACF joint morphology. Moreover, the morphology is related to the bonding conditions and ACF material properties.

References

[1] Myung-Jin Yim and Kyung-Wook Paik, "Design and understanding of anisotropic conductive films (ACF's) for LCD packaging, IEEE Transactions on Components, Packaging, and Manufacturing Technology-part A, Vol. 21, No. 2, pp.226-234 (1998.06)

[2] Kiwon Lee et al., "High Power and Fine Pitch Assembly Method using Solder Anisotropic Conductive Films (ACFs) combined with Ultrasonic Bonding Technique", 59th Electronic Components and Technology Conference, San Diego, California, USA, 2009

[3] Lei L. Mercado et al., "Failure Mechanism Study of Anisotropic Conductive Film (ACF) Packages", Vol.26, No.3, PP 509-516 (2003. 09)

[4] Kiwon Lee et al., "High-speed Flex-On-Board Assembly Method using Anisotropic Conductive Films (ACFs) Combined with Room Temperature Ultrasonic (US) Bonding for High-density Module Interconnection in Mobile Phones" 61st Electronic Components and Technology Conference, 2011

[5] JOHN P. RANIERI et al., "Plastic Constraint of Large Aspect Ratio Solder Joints", Journal of Electronic Materials, Vol. 24, No. 10, 1995

[6] Callister, William D. Materials Science and Engineering: An Introduction - 3rd Edition. John Wiley & Sons, Inc.: New York, 1994

Development of the Technology to Control the Spatial Distribution of Plasma using Double ICP Coil

T.Sakuishi[1,2], T. Murayama[1,2], Y. Morikawa[1,2], K.Suu[1,2]

[1]ULVAC Inc., [2]NMEMS technology research organization

1220-1 Suyama, Susono, Shizuoka, 410-1231 Japan

Tel:+81-55-998-1592 , E-mail:toshiyuki_sakuishi@ulvac.com

Abstract

With the shrink of TSV diameter, smoothness of side wall and taper angle control become more and more important. We have developed scallop free etching by direct etching method. Direct etching method is continuous, not cyclical etching and deposition. Scallop does not occur in principle. Double ICP antenna newly developed has good controllability of side wall taper angle.

Introduction

Data traffic is increasing explosively with the spread of mobile devices such as mobile phone and tablet PC. Data server required high-speed processing and low power consumption. Devices used in these equipments are required high-speed processing, high density packaging and low power consumption. One of the most powerful way to achieve these requirements are 2.5D and 3D packaging using TSV. Via diameter becomes 10um or less, subsequent deposition process becomes difficult, because of high aspect ratio. Sidewall smoothness and controllability of taper angle have significant influence on the reliability and production costs. We have developed scallop-free etching using double ICP antenna. In this paper, we focused on the controllability of taper angle. Scallop-free and taper etching is expected to improve the reliability and production costs of liner/barrier seed layer deposition. Figure1. shows the etched profile by conventional method and our countermeasures.

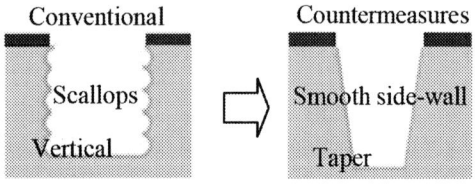

Figure1. Scallop-free tapered etching and conventional method.

Technical issues

In direct etching, sidewall passivation and etching occur simultaneously and, taper angle can be controlled by the ratio of deposition and etching. With the increase of deposition ratio, etched profile changes positive taper from vertical. On the contrary, deposition ratio decrease, etched profile changes negative taper or bowing. (Figure2.) However, in the case of conventional ICP plasma, etched profile of center and edge are quite different. (Figure3.) In the case of vertical etching, stage bias covers the difference of deposition ratio. But, in taper etching, surface reaction itself needs to be uniform. For this, next two things are very important. First, each of the particles contributes to etching or deposition need to be uniform within a wafer. Second, surface reaction needs to uniform within a wafer. For first purpose, we developed double ICP antenna and tried to control the spatial distribution of the plasma. For second purpose, we used multi-zone control stage for precise wafer temperature control.

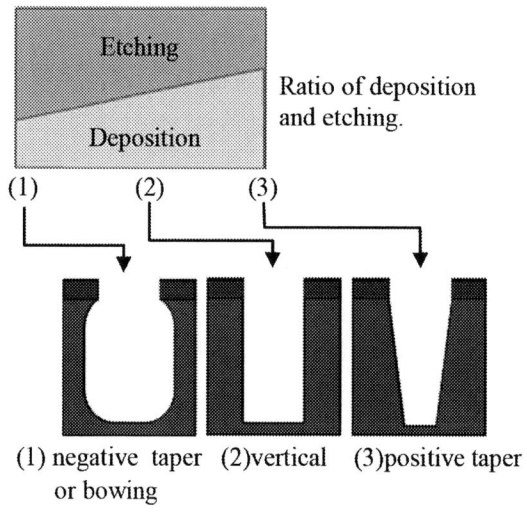

(1) negative taper (2)vertical (3)positive taper or bowing

Figure2. Taper control by deposition ratio

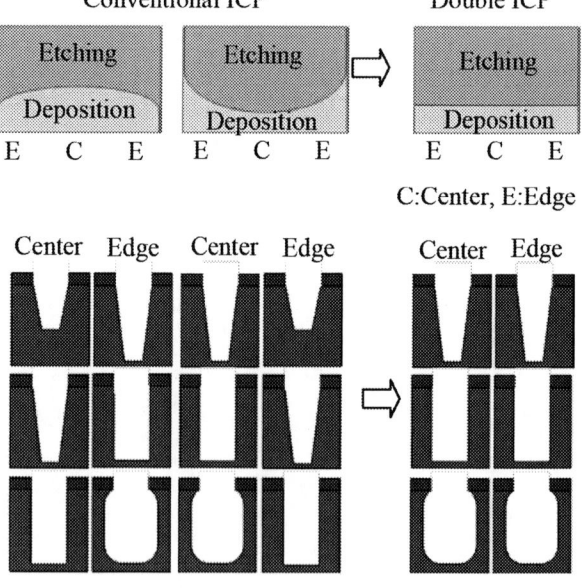

Figure3. Etched profile in a wafer

Results and Discussions

Etching gas is SF6/O2 mixture. Silicon etching by fluorine radical generated from SF6 is originally isotropic. Anisotropic etching is realized because the sidewall is passivated with oxygen. First, we show the results of conventional single ICP antenna. Figure4. shows the schematic diagram of conventional ICP source. Upper ICP antenna is applied RF and the stage is applied RF bias independently. Gas inlets are center and side. Figure5. show the cross-sectional SEM image of TSV etched by conventional ICP antenna. Figure5(a) is the result of vertical etching, etched profile of center and edge is almost same. Figure5(b) is the result of taper etching, etched profile of center and edge is quite different. Figure6. shows the model of vertical etching. Even if the deposition ratio is different, passivation of the bottom is prevented by ion bombardment. Therefore, it is possible to obtain the same profile by adjusting the bias power. On the other hand, at the case of taper etching, ions hit the sidewall and break the passivation layer. Where the passivation layer is broken, isotropic etching will occur and result in bowing. (Figure7.) If the taper etching, surface reaction itself need to be controlled precisely. For this, distribution of fluorine radical and oxygen radical will need to be very uniform.

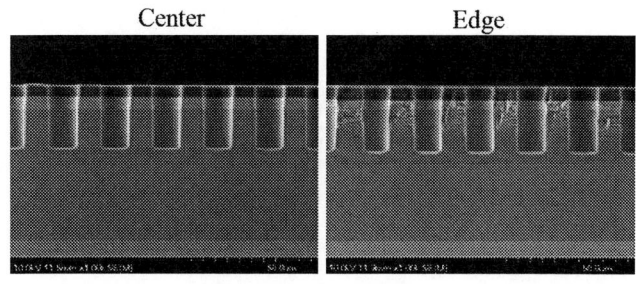

(a) Vertical Etching

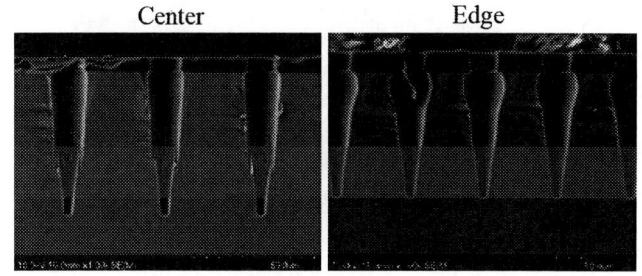

(b) Taper Etching

Figure5. Etched profile by conventional ICP source

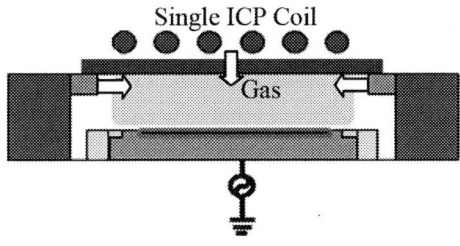

Figure 4. Conventional Single ICP source

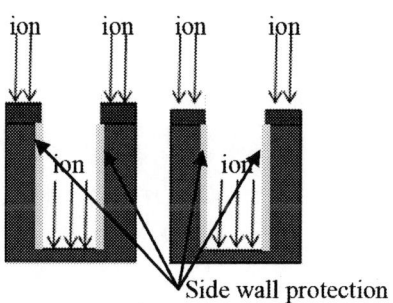

Figure 6. Model of vertical Etching

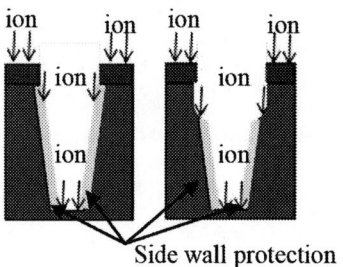

Figure 7. Model of taper Etching

Next, in order to confirm the role of the ICP antenna, we divided the conventional antenna. Figure8(a) shows the plasma source of only inner antenna and (b) shows the only outer antenna. Figure9(a) shows the etched profile of only inner antenna and (b) shows the profile of only outer antenna. Difference of the two suggests the possibility of controlling the special distribution of the plasma by two separated antenna.

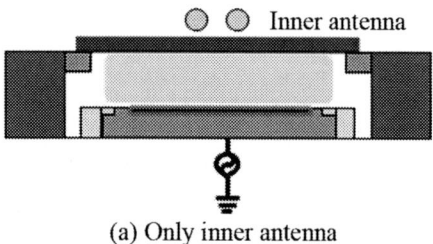

(a) Only inner antenna

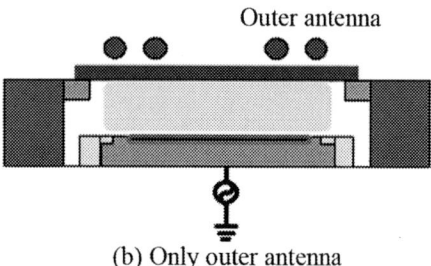

(b) Only outer antenna

Figure8. Plasma source

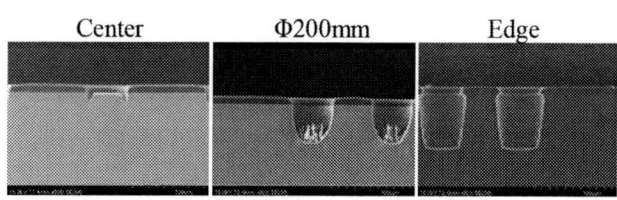

Figure 9 (a) Etched profile of only inner antenna

Figure 9 (b) Etched profile of only outer antenna

Figure10. shows the schematic diagram of double ICP source which is newly developed. RF power is applied in parallel to the inner and outer antenna. Self-inductance of double ICP antenna was reduced to 0.8uH from 4.1uH, inductive coupling was improved. Antenna Vpp reduced about 1/2 and the sputtering of the dielectric window was reduced. These improvements are also important in mass production equipment. Figure11. Show the distribution of SiO2 etching by single antenna and double ICP antenna. In double ICP antenna, uniformity was improved ±4.1% from ±14.6%. This indicates that the uniformity of plasma is improved by double ICP antenna.

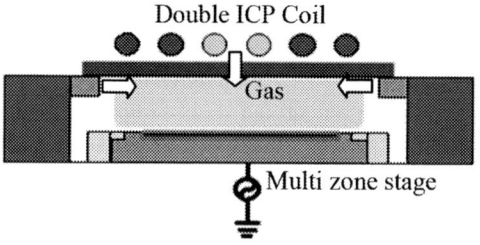

Figure 10. Double ICP source newly developed

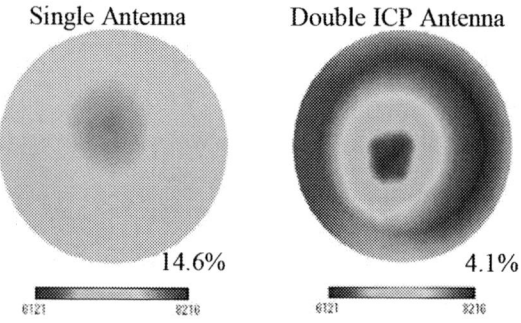

Figure 11. Uniformity of SiO2 etching

Figure12. shows the etched profile of double ICP antenna. Etching gas is SF6/O2 mixture. Test wafer is Φ300mm Si wafer and the mask is photo-resist. Diameter of the via are Φ 20um. ICP power is applied in parallel to the inner and outer antenna, stage bias is also applied. Wafer temperature is controlled by multi zone control ESC (Electrostatic chuck). Observed point is shown right figure. (center/Φ100mm/Φ 200mm/Edge) Taper angle is 82±1degree in Φ300mm wafer. (Figure13.) This result suggests good taper angle controllability.

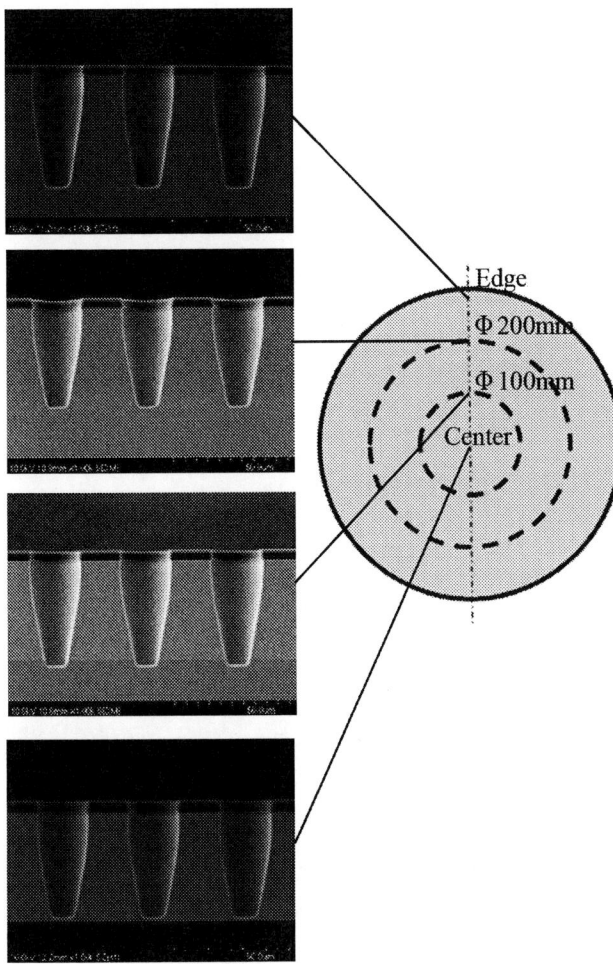

Figure 12. Etched profile of Φ20um Via

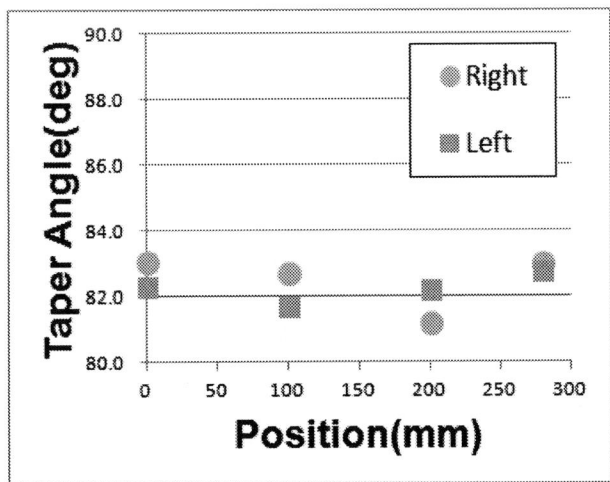

Figure 13. Taper angle of Φ20um Via

Figure14. shows the etched profile of Φ2um small via. Etching gas is SF6/O2 mixture. Etched depth is 20um and the aspect ratio is 10. Etching rate is 4.3um/min. Double ICP source has good uniformity to the small via etching.

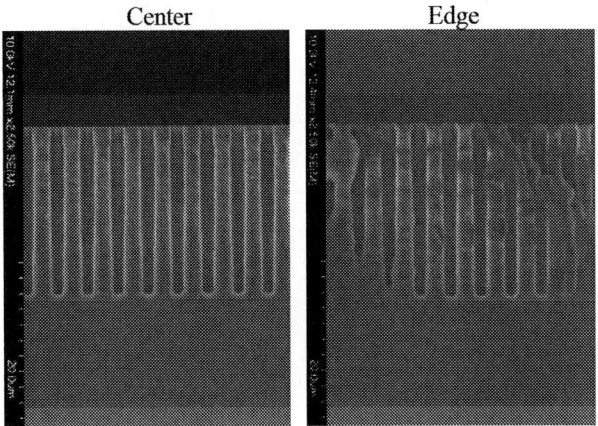

Figure 14. Etching profile of Φ2um Via

Conclusions

Biggest challenges in TSV are reliability and cost. Taper etching is very effective in improving two problems. Direct etching method using newly developed double ICP antenna enables scallop-free and taper controlled etching. Taper angle uniformity is significantly improved by double ICP antenna. Taper angle of Φ20um via etched by double ICP antenna is 82 ± 1degree within Φ300mm wafer. Small via etching also improved by double ICP antenna.

Acknowledgments

This work of Si etch process was partly supported by NEDO (New Energy and Industrial Technology Development Organization of Japan).

References

1. M. Wilke et al., "Process Modeling of Dry Etching for the 3D-Integration with Tapered TSVs," in *Proc. IEEE Electronic Components and Technol. Conf. (ECTC)*, San Diego, CA, May 29–June1, 2012, pp. 83–89.

2. T.Nakamura, H.Kitada et al., "Comparative sutudy of side-wall roughness effects on leakage currents in through-silicon via interconnects," in *Proc. 3D Systems Integration Conf. (3DIC)*, Osaka, Japan, Jan. 31–Feb.2, 2011, pp. 1–4

3. K-W Lee et al., "High reliability and fine seze 5-um diameter backside Cu through-silicon Via(TSV) for high reliability and high-end 3-D LSIs." in *Proc. 3D Systems Integration Conf. (3DIC)*, Osaka, Japan, Jan. 31–Feb.2, 2012, pp. 1–4

4. Pejman Monajemi et al., "DESIGN AND PROCESS OPTIMIZATION OF THROUGH SILICON VIA INTERPOSER FOR 3D-IC INTEGRATION." in *Proc. (IMAPS)*, San Diego, CA, Sep.9–13, 2012

5. Y. Morikawa et al., "A novel scallop free TSV etching method in magnetic neutral loop discharge plasma." in *Proc. IEEE Electronic Components and Technol. Conf. (ECTC)*, San Diego, CA, May 29–June1, 2012, pp. 794–795

Defect Detection in Through Silicon Vias by GHz Scanning Acoustic Microscopy: Key Ultrasonic Characteristics

Alain Phommahaxay[*1], Ingrid De Wolf[1,2], Tatjana Djuric[3], Peter Hoffrogge[3], Sebastian Brand[4], Peter Czurratis[3], Harold Philipsen[1], Gerald Beyer[1], Herbert Struyf[1] and Eric Beyne[1]

[1]imec, Kapeldreef 75, 3001 Leuven, Belgium
[2]Dept. Material science, KU Leuven, 3001 Leuven, Belgium
[3]PVA TePla Analytical Systems GmbH, Deutschordenstrasse 38, 73463 Westhausen, Germany
[4]Fraunhofer Institute for Mechanics of Materials IWM, 06120 Halle, Germany

*Phone: +32-16-28-79-05
E-mail: phomma@imec.be

Abstract

Among the technological developments pushed by the emergence of 3D-ICs, Through Silicon Via (TSV) technology has become a standard element in device processing over the past years. As volume increases, defect detection in the overall TSV formation sequence is becoming a major element of focus nowadays. Robust methods for in-line void detection during TSV processing are therefore needed especially for scaled down dimensions. Within this framework, the current contribution describes the application field of GHz Scanning Acoustic Microscopy (SAM) to TSV void detection.

Introduction

Scanning Acoustic Microscopy (SAM) has been widely used in the past for failure analysis of packaged micro-components and is now commonly used for wafer to carrier bonding and stacking nondestructive in-line inspection [1]. With the development of novel ultrasonic transducers operating beyond the typical 10 MHz-150 MHz range, new applications are now achievable. Indeed recent progress in thin film technology has enabled the fabrication of acoustic resonators reaching GHz frequencies, hence increasing the spatial resolution of Scanning Acoustic Microscopy.

In parallel, the emergence of 3D-IC technology has brought new requirements in terms of in-line metrology. One of the vital element of 3D processes being the formation of Through Silicon Vias, robust methods for in-line process control are becoming key for the adoption of this technology but are still awaited.

Thanks to the progresses made by ultrasonic inspection techniques, we have previously demonstrated the feasibility of micron-size defect detection within a through-silicon via. [2] In the frame of this paper, we will further investigate such technique on various TSV dimensions and defects.

Via-middle process description

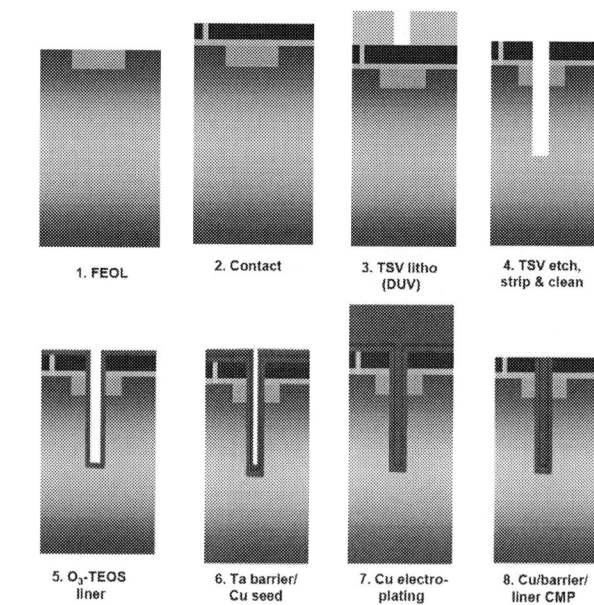

Figure 1. Schematic process flow followed at imec for via middle formation.

The conceptual via-middle process flow followed at imec is depicted in Fig. 1. All process steps are performed in the imec 300mm pilot production line [3].

After formation of the Front End of Line and contact formation, TSVs are formed by Reactive Ion Etching (RIE) with a Bosch process. An insulating layer consisting of O3-TEOS is then deposited by Chemical Vapor Deposition.

Barrier and seed layers consisting of TaN/Cu are subsequently deposited by PVD prior to TSV filling by copper electrochemical plating. Finally excessing copper is removed by CMP.

978-1-4799-2408-0/14 $31.00 © 2014 IEEE

a) Correctly filled TSV

b) TSVs with incomplete fill

Figure 2. Example of FIB analysis of TSVs after Cu plating

Typical Focused Ion Beam (FIB) pictures of TSVs after Cu filling (Fig. 1 step 7) are depicted in Fig. 2. Although FIB can provide information on both correctly filled TSV (Fig. 2 a)) and TSV presenting voids (Fig. 2 b)), it remains a destructive failure analysis technique. Hence we will focus on the use of Scanning Acoustic Microscopy as an alternative method enabling in-line inspection.

Within this contribution, we are exploring various types of defects (size and position) in different TSV dimensions: 5x50 μm, 3x50 μm and 10x100 μm.

Acoustic signal properties at very high frequency

Given the acoustic propagation velocity v summarized in Table I, one can derive the ultrasound wavelengths by the relation described in Equation 1.

$$\lambda = \frac{v}{f} \qquad (1)$$

Material	v	λ (100 MHz)	λ (1 GHz)
Air	343 ms^{-1}	3.43 μm	0.34 μm
H$_2$O	1495 ms^{-1}	14.95 μm	1.49 μm
Cu	4759 ms^{-1}	47.59 μm	4.75 μm
Si	8430 ms^{-1}	84.30 μm	8.43 μm

Table I: Longitudinal acoustic properties at high frequencies

When considering the TSV depth, ranging from typically 50 to 100 μm, one can notice that the ultrasonic wavelength in copper is much smaller than the TSV depth. Hence we will use a 1D wave propagation theory to model the ultrasonic interaction with defects in TSV.

Scanning acoustic microscopy in TSV

The typical electro-acoustic emitter-receiver configuration used in scanning acoustic microscopy is depicted in Fig. 3. In a pulse-echo detection mode, small acoustic bursts are sent through an ultrasonic transducer. The sound wave is propagating through the sample of interest until it encounters a change in acoustic impedance Z.

At a certain boundary between materials 1 and 2, the reflection coefficient Γ is directly linked to the difference between $Z1$ and $Z2$ as indicated in Equation 2.

$$\Gamma = \left| \frac{Z2 - Z1}{Z2 + Z1} \right| \qquad (2)$$

The reflected acoustic signal from the sample is then recorded back by the acoustic transducer and additionally amplified typically using a low-noise amplifier. The information is finally digitalized through an analog digital converter before being computed.

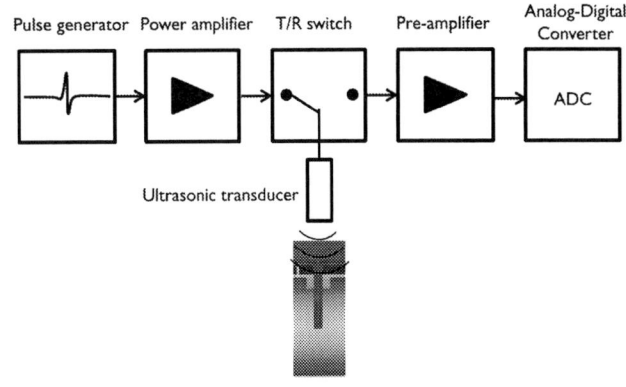

Figure 3. Principle for acoustic signal generation and detection

Interface	Γ
Cu-Air	100%
Cu-H$_2$O	93%
Cu-Si	35%

Table II: Acoustic reflection coefficient for different boundary interface with Cu

978-1-4799-2408-0/14 $31.00 © 2014 IEEE

Considering various possible scenarios for copper TSVs, incomplete filling during electrochemical deposition will lead to the formation of liquid pocket within the structure or gaseous voids after annealing. In both cases the acoustic reflection coefficient Γ indicates strong signal bounce back on the copper-defect interface, as more than 90% of the signal will be reflected as indicated in Table II.

Hence, the acoustic signal recorded on top of a TSV containing voids would present key characteristics:

- Earlier signal reflection as described in Fig. 5. vs Fig. 4.

  Given the acoustic velocity in Cu (4700 ms^{-1}) and TSV depth of 50 μm, any reflection recorded <20 ns would indicate the presence of defects within the TSV.

- Stronger signal reflection compared to filled TSVs: Γ factor above >90% for voids vs 35% for completely filled vias (equivalent to 8 dB in signal strength difference).

In order to demonstrate the feasibility of TSV void detection using SAM, various kind of TSV samples taken right after Cu plating and proper FIB characterization have been submitted to GHz SAM analysis.

The early results obtained with a 1 GHz transducer shown in Fig. 6 clearly indicate a leap in spatial resolution. Each single TSV can now be clearly imaged. Although all process steps performed in the TSV formation are of parallel nature, the GHz acoustic inspection has indicated the presence of singular vias, which was later confirmed by FIB cross section on the singular structures [2].

Time domain analysis: key limiting parameters

Similar analysis with other types of samples has been furthermore attempted. At first sight one could imagine that defects located just beneath the surface of a TSV would be detected much more easily, but it is not the case. Indeed time-domain based analysis of the sample depicted in Fig. 7 did not reveal the singularities between the TSVs that was indicated by FIB cross section.

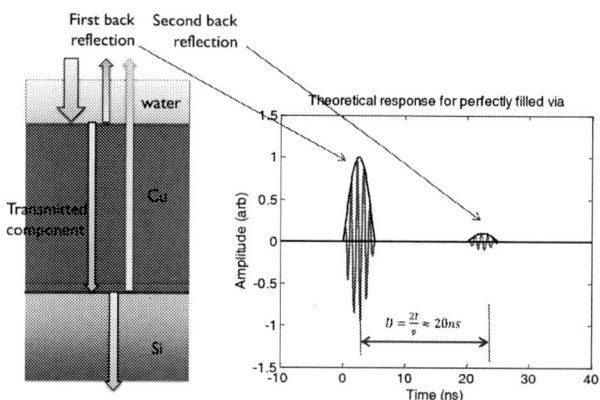

Figure 4. Theoretical acoustic signal in perfectly filled TSV in time-domain

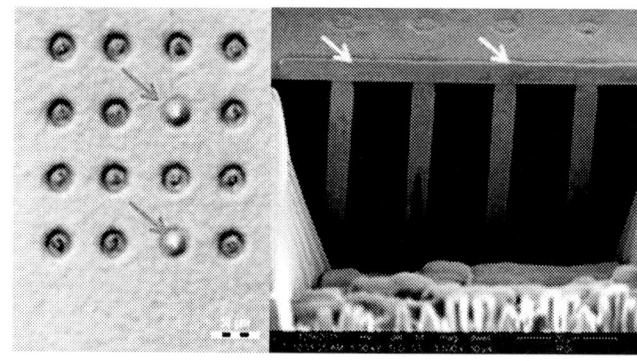

Figure 6. 1 GHz SAM vs FIB correlation results on 5x50 μm TSV [2]

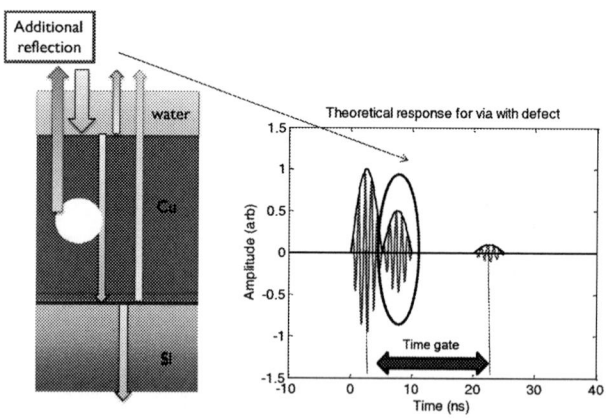

Figure 5. Theoretical acoustic signal in TSV with defect in time-domain

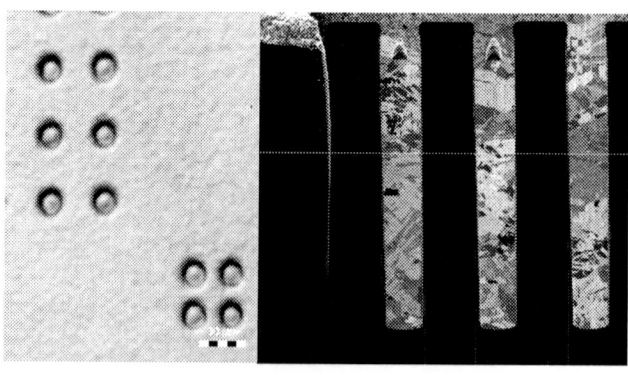

Figure 7. 1 GHz SAM vs FIB correlation results on sub-surface voids in a 5x50 μm TSV

Two main phenomenons occur with sub-surface defects in time domain as depicted in Fig. 8:

- Overlapping pulses
- High dynamic signal range

Indeed, the reflected wave coming from sub-surface defect interferes with the incident wave. Based on the schematic shown in Fig. 3, the resolution of this overlap is mainly dependent on:

- The acoustic pulse width generated
- The ADC sampling rate
- The pre-amplifier SNR

For instance, with a signal sampling rate of 1 giga-samples per second, one can distinguish 2 ns signal difference, corresponding to approximately a depth variation of 5 µm in copper. With such setup, time domain analysis become challenging for voids located down to 10 µm below the surface. The signal amplitude difference between the incident and reflected wave is evaluated at 38 dB, thus requiring very low noise amplifier.

While some limitation may be overcome by the continuous performance improvements in ADC and LNA, we are exploring further aspects of acoustic wave propagation and interaction in order to solve these challenges.

V(z) analysis principle

Up to now, we have only considered the longitudinal component of the acoustic wave. When dealing with very high acoustic frequencies, other propagation mode can be excited as illustrated in Fig. 9.

Both longitudinal and transverse components can co-exist. Longitudinal wave tends to propagate through the bulk of the TSV while Rayleigh waves are likely to be present at the surface.

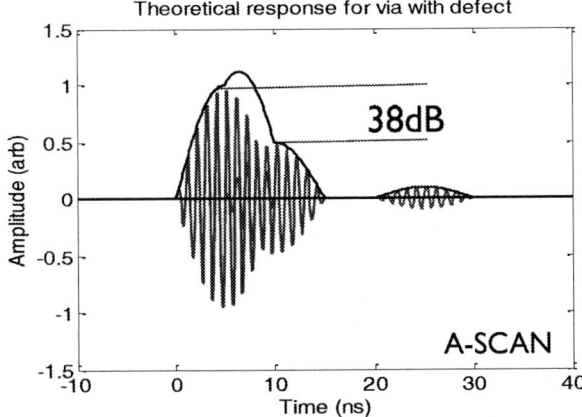

Figure 8. Impact of wide acoustic pulse or close surface defect on time domain signal

By tuning the acoustic transducer physical position, one can modulate the intensity of the signal recorded back, which is directly linked to the interaction between the different acoustic wave component. This type of analysis is called V(z) curve, where z is the acoustic transducer defocus amount. The plotted curve can be used to characterize the propagation medium mechanical properties [4]. A typical example of V(z) curve can be seen on Fig. 10.

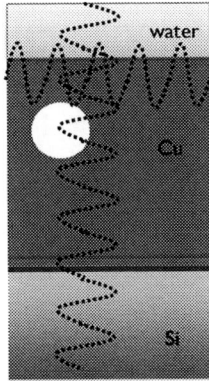

Figure 9. Illustration of acoustic wave component: longitudinal and transverse

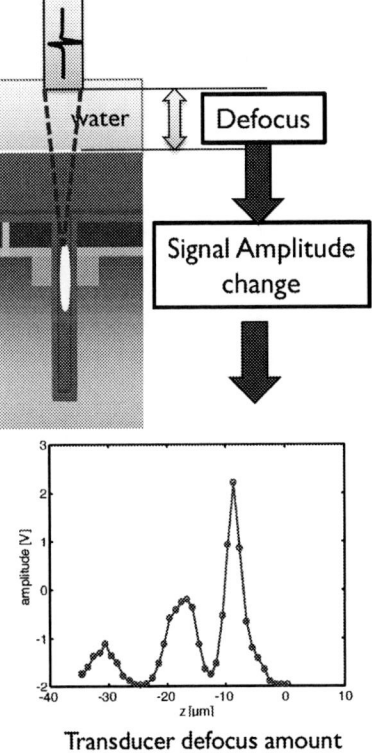

Transducer defocus amount

Figure 10. Example of V(z) curve

V(z) analysis applied to TSV void categorization

Various kind of TSV samples taken right after Cu plating and proper FIB characterization have been submitted to SAM analysis. Different void types could therefore been investigated using ultrasonic inspection. A non-exhaustive example of TSV defects is shown in Fig. 11, 12 and 13.

On top of the 5x50 µm TSV previously discussed in [2], other TSV dimensions have been investigated: 3x50 µm and 10x100 µm to assess the lateral spatial resolution limit and penetration depth of acoustic signal.

By comparing Fig. 11 to 12, one can notice the impact of sub-surface void on 10x100 µm TSV in the V(z) curve. Indeed ripples appear in the V(z) curve by defocussing the acoustic transducer on top of the defective TSV, while the signal remains close to zero for the perfectly filled TSV. This is probably explained by the lower amount of reflected acoustic energy for good TSVs.

In addition, smaller diameter TSVs have been also characterized as shown in Fig. 13. In this example, delamination of the Cu from the oxide liner occurred at the bottom of the TSV. This phenomenon has been revealed in the V(z) curve for a single TSV. A more comprehensive surface scan (C(z)) analysis of a TSV array shows singularities between the vias as illustrated in Fig. 14.

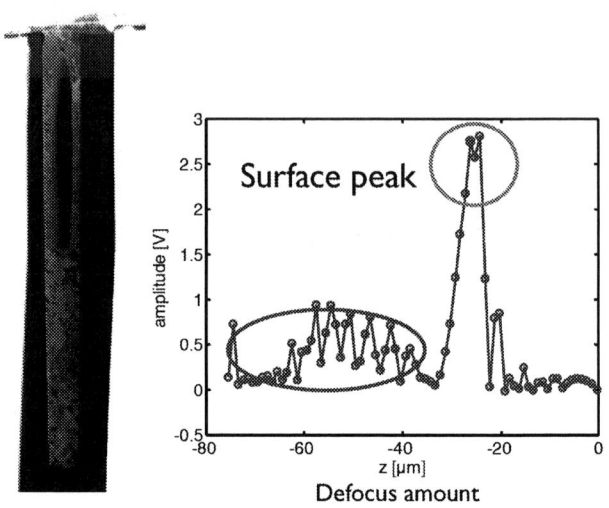

Figure 12. V(z) characteristic of a 10x100 µm TSV with subsurface void

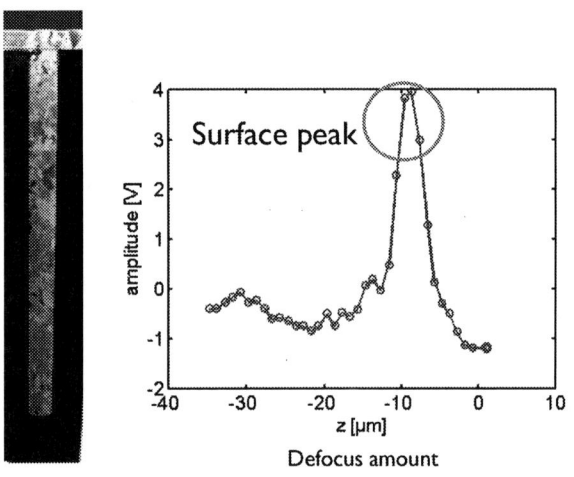

Figure 11. V(z) characteristic of a reference 10x100 µm TSV

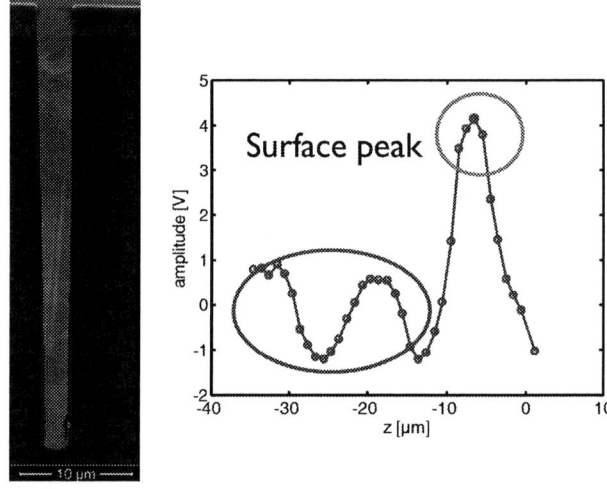

Figure 13. V(z) characteristic of a 3x50 µm TSV with void at the bottom of the via

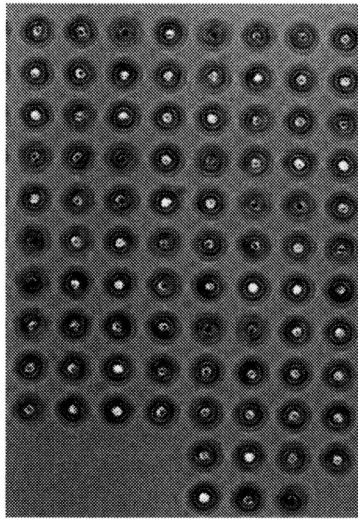

Figure 14. C(z) inspection of 3x50 μm TSVs with voids at the bottom of the vias

Conclusions

The detection of voids in Through Silicon Vias by Scanning Acoustic Microscopy has been demonstrated for the first time using a V(z) approach. While deep defects can be detected by time domain analysis, sub-surface voids require the use of novel analysis techniques allowed by the interaction of specular wave with surface Rayleigh wave.

First results indicate the feasibility of such method to a wide range of TSV dimensions up to 100 μm deep and down to 3 μm in diameter. Further dimensions with be explored in the future.

Moreover, V(z) curve characteristics such as oscillation periodicity and amplitude may provide more information related to the mechanical properties of the TSV. The exploitation of such characteristic curve will be further investigated, potentially providing a first step towards automatic binning of structures.

As such, these results open new range of applications for SAM Inspection. It furthermore provides a potential leap forward in 3D process control. It is indeed a non-destructive inspection technique and may be implemented for in-line process monitoring. It also enables early defect detection right after TSV filling by electroplating, thus avoiding unnecessary and costly CMP steps on faulty samples.

Acknowledgments

The authors would like to thank the imec partners of the Industrial Affiliation Programs on 3D integration. This work is performed in the frame of a Joint-Development-Project between imec and PVA Tepla Analytical Systems.

References

1. S. Brand, M. Petzold, P. Czurratis, J.D. Reed, M. Lueck, C. Gregory, A. Huffman, J.M. Lennon, D.S. Temple, "Acoustic inspection of high-density-interconnects for 3D-integration," in *Proc. IEEE Ultrasonics Symposium (IUS)*, Orlando, FL, October 18-21, 2011, pp. 1076-1079.

2. A. Phommahaxay, I. De Wolf, P. Hoffrogge, S. Brand, P. Czurratis, H. Philipsen, Y. Civale, K. Vandersmissen, S. Halder, G. Beyer, B. Swinnen, A. Miller, E. Beyne, "High frequency scanning acoustic microscopy applied to 3D integrated process: Void detection in Through Silicon Vias," *Proc. IEEE Electronic Components and Technology Conference (ECTC)*, Las Vegas, NV, 28-31 May 2013, pp.227-231.

3. A. Redolfi, D. Velenis, S. Thangaraju, P. Nolmans, P. Jaenen, M. Kostermans, U. Baier, E. Van Besien, H. Dekkers, T. Witters, N. Jourdan, A. Van Ammel, K Vandersmissen, S. Rodet, H.G.G. Philipsen, A. Radisic, N. Heylen, Y. Travaly, B. Swinnen, E. Beyne, "Implementation of an industry compliant, 5×50μm, via-middle TSV technology on 300mm wafers", in *Proc. IEEE Electronic Components and Technol. Conf. (ECTC)*, Orlando, FL, 31 May-3 June, 2011, pp. 1384-1388.

4. Z. Yu, "Scanning Acoustic Microscopy and its Applications to Materials Characterization", *Reviews of Modern Physics*, Vol. 67, No. 4, October 1995.

Temporary Spin-on Glass Bonding Technologies for Via-Last/Backside-Via 3D Integration Using Multichip Self-Assembly

H. Hashiguchi[1], T. Fukushima[2], A. Noriki[3], H. Kino[3], K.-W. Lee[2], T. Tanaka[1,3], M. Koyanagi[2]

[1] Dept. of Bioengineering and Robotics, Graduate School of Engineering, Tohoku University
[2] New Industry Creation Hatchery Center (NICHe), Tohoku University
[3] Dept. of Biomedical Engineering, Graduate School of Biomedical Engineering, Tohoku University
6-6-01 Aza-Aoba, Aramaki, Aoba-ku, Sendai 980-8579, Japan
Phone: +81-22-795-6909 E-mail: link@lbc.mech.tohoku.ac.jp

Abstract

In this study, we proposed and demonstrated self-assembly-based via-last/backside-via 3D integration using a temporary spin-on glass (SOG) bonding technology. A hydrogenated amorphous silicon (a-Si:H) was employed as a debonding layer. Known good dies (KGDs) were precisely self-assembled right side up on an electrostatic carrier wafer by surface tension of water, and then, the KGDs were fixed by applying DC voltage to the carrier. After that, the KGDs were temporarily bonded and transferred to another support glass wafer on which the a-Si:H and SOG layers were deposited. After multichip thinning, Cu-TSVs were formed on the KGDs. The resulting TSV daisy chains showed good electrical characteristics. The KGDs can be debonded with a 308-nm laser and transferred again to target interposer wafers.

Introduction

Since the late 1980s, we have studied 3D integration technologies using vertical buried interconnections to be later called through-silicon vias (TSVs) and pioneered the fabrication of innovative device chips such as a 3D image sensor, a 3D shared memory, a 3D retina chip, and a 3D microprocessor [1]-[12]. In standard 3D integration processes, large scale integration (LSI) chips are uniformly thinned, vertically stacked, and mutually interconnected with huge numbers of TSVs. 3D LSI chips have great advantages in smaller chip size, shorter interconnect lengths, higher processing speed, and lower power consumption, compared to ordinary planar LSI chips. To further increase production throughput and yield of 3D LSIs, we have proposed and developed self-assembly-based multichip-to-wafer 3D integration with a reconfigured wafer as shown in Fig. 1 [13][14]. The reconfigured wafer consists of a support wafer and various kinds of many KGDs. These KGDs are precisely and instantly self-assembled on the support wafer with alignment errors of within 1 μm and with alignment times of within 0.1 second at once. Then, the KGDs are processed and finally transferred to a target interposer wafer in batch processing. In the advanced chip-to-wafer 3D integration processes, multichip bonding and debonding are the most important key technologies. In general, KGDs are temporarily bonded on wafers with organic thermoplastic or thermosetting adhesives. However, these temporary adhesives have a serious disadvantage in low stability in high temperature processes. In order to overcome this problem, we propose a new temporary bonding technology using SOG as a bonding layer and a-Si:H as a debonding layer.

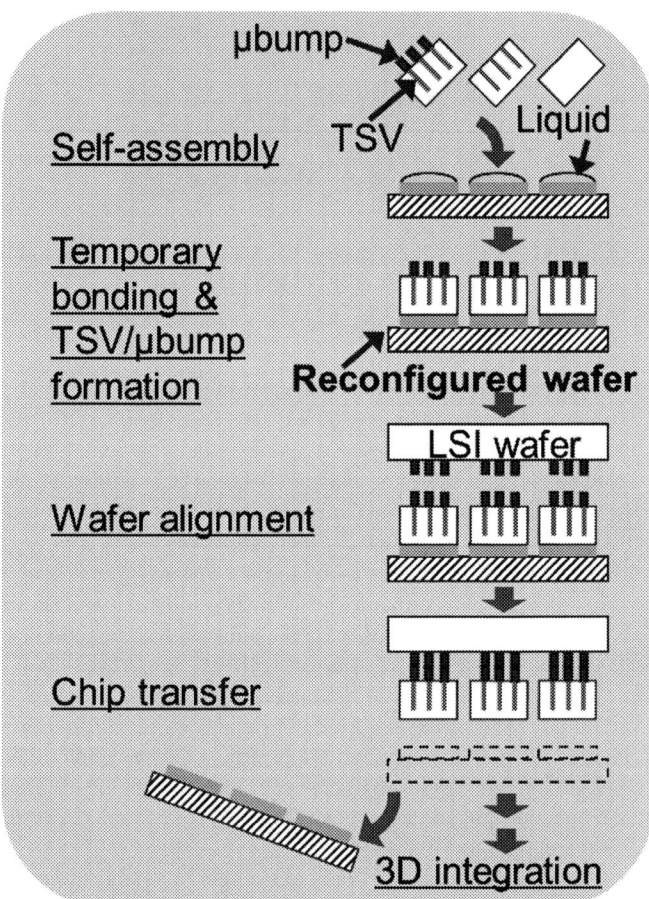

Figure 1. A reconfigured wafer-to-wafer 3D integration scheme using multichip self-assembly.

Temporary SOG bonding technology

Direct oxide-oxide bonding generally requires highly smooth and clean surfaces resulted from chemical mechanical polishing (CMP) and severe particle control. In contrast, adhesive bonding with the organic thermoplastic or thermosetting polymers is restricted by process temperatures although they easily deform and planarize the surface of patterned and bumping wafers to be bonded. Chemical vapor deposition (CVD) for TSV liner formation and annealing for electroplated Cu are higher temperature processes in via-last/backside-via 3D integration approaches. Lin *et al.*, have reported a SOG technology for the permanent bonding of VCSEL (vertical cavity surface emitting laser) chips on silicon wafers without CMP [15]. SOG is a representative spin-on inorganic dielectric used for semiconductor industries, and thus, the heat resistance and electrical/mechanical

978-1-4799-2408-0/14 $31.00 © 2014 IEEE
2014 Electronic Components & Technology Conference

reliability are high. In this work, we develop new via-last/backside-via 3D integration combining temporary SOG bonding with a debonding technique using excimer laser irradiation to a-Si:H. Sameshima have reported that a-Si:H formed by CVD acts as a debonding layer to remove a thin Al film from a substrate by laser ablation [16]. As shown in Fig. 2, KGDs self-assembled and electrostatically fixed on a carrier wafer are simultaneously flip-chip bonded to another glass support wafer by thermal compression through SOG coated on a a-Si:H layer deposited on the glass wafer. Finally, the KGDs are debonded from the glass wafer by irradiation with a 308-nm excimer laser after thinning and subsequent TSV formation and so forth.

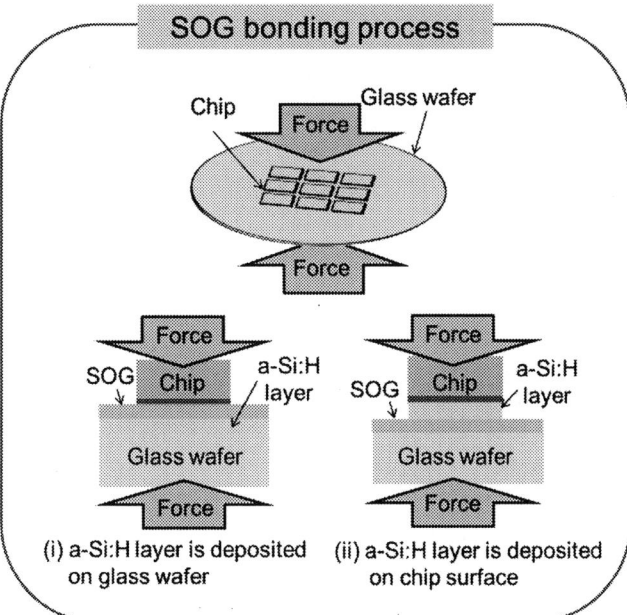

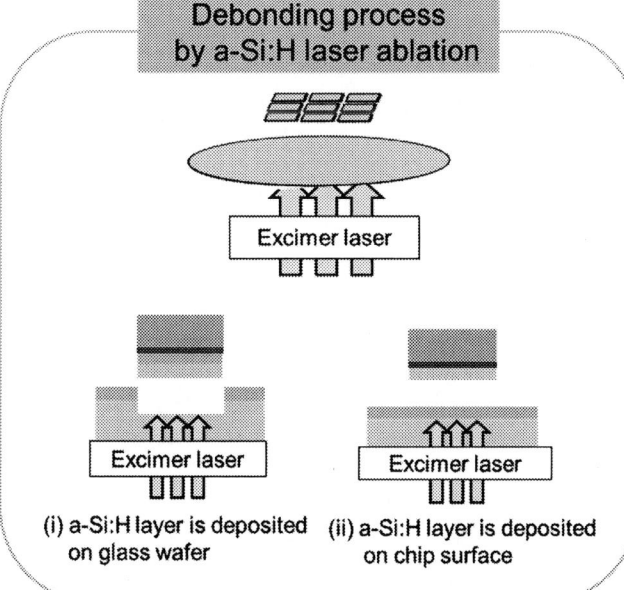

Figure 2. Concepts of temporary SOG bonding and a-Si:H debonding.

Firstly, we evaluated bonding strengths between chips and wafers through SOG. In the experiments, we prepared 100-μm-thick square chips (4.2 mm by 5.2 mm), a self-assembly and electrostatic temporary bonding carrier (SAE carrier) [17], and a glass support wafer on which SOG was coated and a a-Si:H layer was deposited. At first, the chips were self-assembled in a face-up fashion to the SAE carrier by self-assembly using water droplets. After that, these chips were transferred to the glass support wafer. Then, the chips were thermally compressed to the wafer under the bonding conditions described in Table 1 and the SOG was cured in vacuum. After cooling, we measured the alignment accuracies using vernier scale patterns photolithographically formed on the chips and the wafer. The alignment errors of the self-assembled and transferred chips were found to be within 2 μm. Next, bonding strengths between transferred chips and the glass support wafer were measured against pull stress as shown in Table 2. The maximum, minimum, and average chip bonding strengths were 3.89MPa, 1.89 MPa, and 2.72 MPa. These results show that SOG bonding keeps high-precision alignment and provides high bonding strengths. By using the SOG temporary bonding technology, TSV liner oxide can be deposited at higher temperature above 350°C by plasma enhanced chemical vapor deposition (PECVD) or thermal CVD with ozone, resulting in high quality of the TSV liner that acts as highly-reliable insulator showing high resistivity, low leak current, and high endurance for high voltages.

Table 1. Bonding conditions

Bonding Temperature	200 °C
Bonding Pressure	50 N/chip

We evaluated compatibility of chip debonding processes with a a-Si:H layer by laser ablation to via-last/backside-via 3D integration. In these experiments, the chips (5.0 mm by 5.0 mm) having a 100-nm-thick a-Si:H layer deposited by PECVD at 320°C on their top surface were employed. These chips were bonded on a glass wafer by SOG bonding in a face-down fashion. Thus, the a-Si:H layer exists between the chips and the glass wafer. After SOG bonding, 20-nanosecond-pulsed XeCl excimer laser was irradiated to the a-Si:H layer with energies of 100, 138, 200, or 280 mJ/cm^2 through the glass wafers.

Table 2. Measured pull bonding strengths between chips and wafers through SOG

MAX	3.89 MPa
MIN	1.89 MPa
AVE	2.72 MPa

In addition, we evaluated the relationship between chip debonding ability and laser energy density. In this debonding experiment, we used quartz and non-alkali glass as chip bonded wafers. The laser condition was shown in Table 3. As shown in Table 4, All chips were successfully debonded at above 200 mJ/cm^2. The chips bonded on quartz wafers were debonded from the wafers, whereas the chips bonded on non-alkali glass wafers were not debonded. No chips were debonded from each glass wafer below 100 mJ/cm2.

The debonded interfaces between the chips and the wafer were covered with the resulting decomposed a-Si:H residues as shown in Fig. 3. The residual a-Si:H layer was fully etched with 2.38 % tetramethylammonium hydroxide (TMAH) solution. These results show that the chips bonded through SOG are easily debonded from the glass wafer by pulsed-excimer-laser ablation of the a-Si:H layer.

Table 3. Experimental conditions for XeCl laser irradiation

Laser wavelength	308 nm
Pulse width	20 nm
Pulse frequency	1 Hz
Scan speed	1 mm/sec

Table 4. Impact of laser energy density on debonding performance for two types of glass wafers

Laser energy density (mJ/cm²)	Quartz	Non-alkali glass
100	Not debonded	Not debonded
138	Debonded	Not debonded
200	Debonded	Debonded
280	Debonded	Debonded

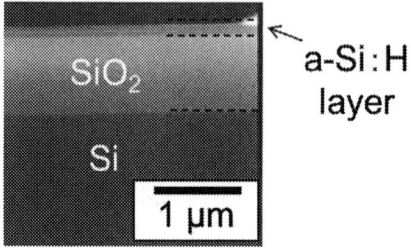

(a) After a-Si:H deposition

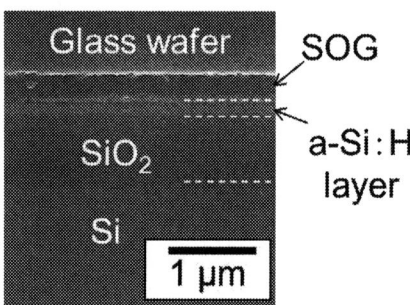

(b) After SOG bonding

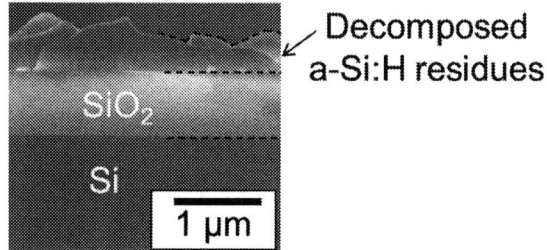

(c) After debonding from the glass wafer

Figure 3. Cross-sectional SEM images of bonding interfaces at each process.

Via-last/backside-via 3D process integration

Here, we introduce a new via-last/backside-via 3D integration process using multichip self-assembly, temporary SOG bonding, and a-Si:H debonding technologies as shown in Fig. 4. Firstly, we assembled KGDs (4.2 mm by 5.2 mm) on a SAE carrier in a face-up manner. The SAE carrier has hydrophilic bonding areas and the other hydrophobic areas for self-alignment using water droplets. After water droplets were dropped on the hydrophilic areas, the KGDs were released on the droplets. Then, these KGDs were driven by water surface tension and precisely aligned to the hydrophilic areas in a moment. After water evaporation, the KDGs were electrostatically clamped on the SAE carrier by applying DC voltage to bipolar electrodes embedded in the SAE carrier as shown in Fig. 5 (a). Secondly, these self-assembled KGDs were temporarily bonded to the corresponding support glass wafer by SOG bonding. An 100-nm-thick a-Si:H layer was deposited on the support wafer by PECVD at 320 ºC using SiH$_4$ and H$_2$ gases, followed by coating and the subsequent partial curing of SOG. The wafer bow was small enough to proceed with wafer-level 3D integration as shown in Fig. 6.

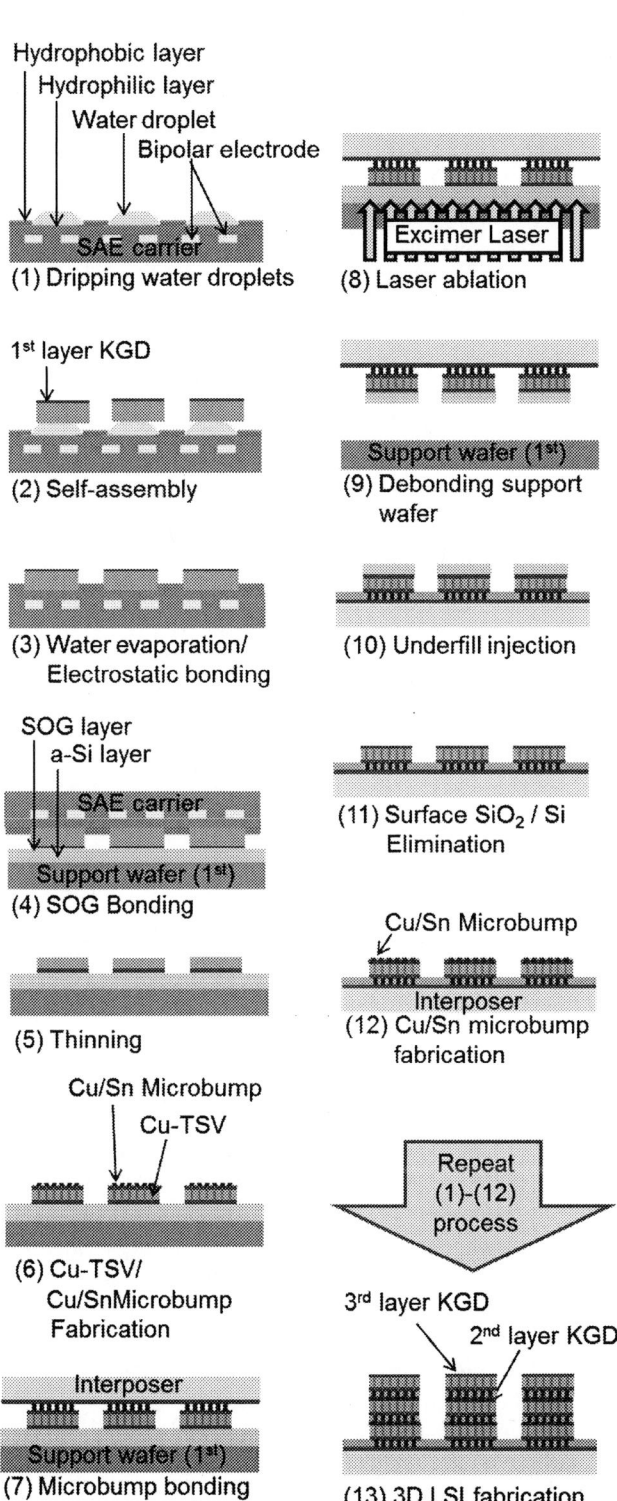

Figure 4. Via-last/backside-via 3D integration process using multichip self-assembly and temporary SOG bonding.

The KGDs were transferred from the SAE carrier to the support wafer in a face-down manner. After that, these KGDs were thermally compressed to the support wafer at 50N/chip. Subsequently, all KGDs were tightly bonded on the support wafer by the SOG bonding technique.

These KGDs were simultaneously thinned down to 50 μm by grinding and CMP. These KGDs are tightly bonded to the support wafer in order to adequately withstand the mechanical stress during the grinding and polishing processes. Next, 1-μm-thick SiO_2 layer was deposited on the polished backside surface of the KGDs by PECVD with TEOS (Tetraethyl orthosilicate) at 350°C as shown in Fig. 5 (b). After 15-μm-square Si via holes were formed by Bosch process, 300-nm-thick SiO_2 was conformably deposited along the deep Si vias by O_3-TEOS-CVD, followed by reactive ion etching to completely remove the thin SiO_2 layer on the bottom of the via holes. After barrier Ta and seed Cu layer deposition by high-aspect-ratio sputtering, the following process was bottom-up Cu electroplating. The electroplated Cu was fully filled into the via holes, and consequently, Cu-TSVs were formed in the thinned KGDs. Finally, 20-μm-square Cu/Sn microbumps were formed by electroplating on the Cu-TSVs. Fig. 7 shows a cross-sectional view of the resulting Cu-TSVs and Cu/Sn microbumps. We measured electrical characteristics of the resulting Cu-TSV daisy chain as shown in Fig. 8, the chain resistance was approximately 97 Ω including front-side and backside wirings and contact resistance between the microbumps and the wirings: a Cu-TSV resistance was nearly 18 mΩ. Therefore, both chip front-side and backside wirings were electrically connected to Cu-TSVs without failure.

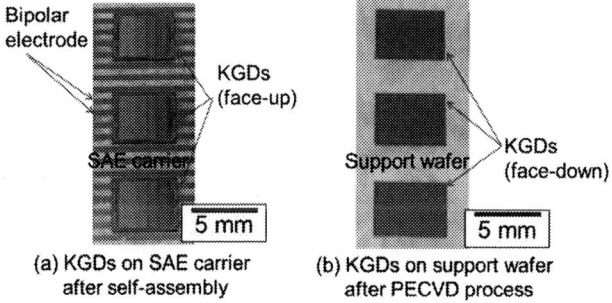

Figure 5. Top views of the array of KGDs self-assembled and fixed on wafers in via-last/backside-via 3D integration processes.

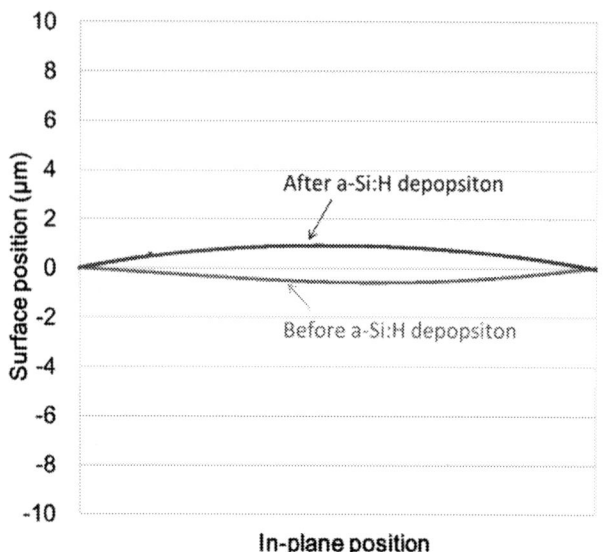

Figure 6. Wafer bow before and after a-Si:H deposition.

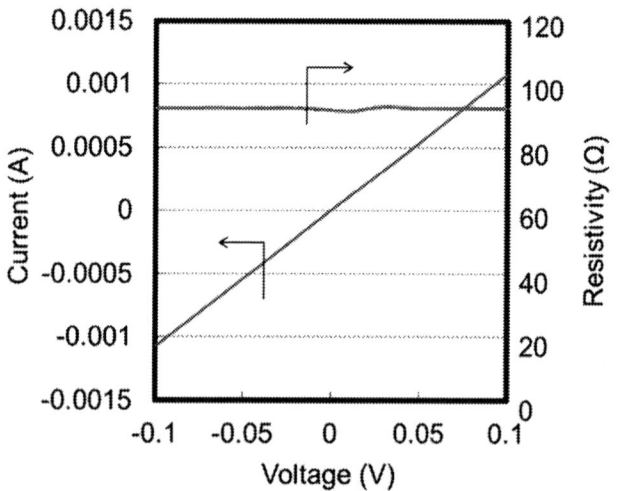

Figure 7. Cross-sectional view of Cu-TSVs and Cu/Sn microbumps in thinned KGDs.

Figure 8. I-V characteristics of a daisy chain with Cu-TSVs and wirings.

Conclusions

A temporary SOG bonding technology was developed for via-last/backside-via 3D integration using multichip self-assembly with water. The first topic was SOG temporary bonding. Surface tension-driven chips were temporarily fixed on an SAE carrier, and subsequently, the chips were temporarily bonded to another glass wafer through SOG. The average chip bonding strengths against pull test was 2.72 MPa. The second topic was a-Si:H debonding. The temporarily bonded chips were easily debonded from the quartz and non-alkali glass wafers by 20-nanosecond-pulsed XeCl excimer laser irradiation over 200 mJ/cm². The temporary SOG bonding and a-Si:H debonding technologies were applied for multichip-to-wafer via-last/backside-via 3D integration. As a consequence, we successfully fabricated KGDs having Cu-TSVs and Cu/Sn microbumps, and obtained good I-V characteristics of the resulting Cu-TSV daisy chains.

Acknowledgments

Laser ablation experiment was performed in Beams Inc., Tokyo, Japan. This work was performed in the Micro/Nano-machining research and education Center (MNC) and Jun-ichi Nishizawa Research Center at Tohoku University. This research was supported by Tohoku University International Advanced Research and Education Organization (IAREO), and Japan Society for the Promotion of Science (JSPS), Grant-in-Aid for Scientific Research "Grant-in-Aid for Scientific Research (S)", No. 21226009.

References

[1] M. Koyanagi, "Roadblocks in Achieving Three-Dimensional LSI," Proc. 8th Symposium on Future Electron Devices, pp. 55-60 (1989).

[2] T. Matsumoto, Y. Kudoh, M. Tahara, K.-H. Yu, N.Miyakawa, H. Itani, T. Ichikizaki, A. Fujiwara, H. Tsukamoto, and M. Koyanagi, "Three-dimensional integration technology based on wafer bonding techniqueusing micro-bumps," in Proc. Int. Conf. Solid State Devices and Mater. (SSDM), pp. 1073–1074 (1995).

[3] T. Matsumoto, M. Satoh, K. Sakuma, H. Kurino, N. Miyakawa, H. Itani, and M. Koyanagi, "New three-dimensional wafer bonding technology using adhesive injection method," in Proc. 29th Int. Conf. Solid State Devices and Mater. (SSDM), pp. 460-461 (1997).

[4] T. Matsumoto, M. Satoh, K. Sakuma, H. Kurino, N. Miyakawa, H. Itani, and M. Koyanagi, "New threedimensional wafer bonding technology using the adhesive injection method," Jap. J. Appl. Phys. Part 1, vol. 37, pp. 1217-1221 (1998).

[5] K. Kurino, K.-W. Lee, K. Sakuma, T. Nakamura, and M. Koyanagi, "A new wafer scale chip-on-chip (WCoC) packaging technology using adhesive injection method," Jap. J. Appl. Phys. Part 1, vol. 38, pp. 2406-2410 (1999).

[6] H. Kurino, K. W. Lee, T. Nakamura, K. Sakuma, H. Hashimoto, K. T. Park, N. Miyakawa, H. Shimazutsu, K. Y. Kim, K. Inamura, and M. Koyanagi, "Intelligent image sensor chip with three dimensional structure," in Tech.

Dig. Int. Electron Devices Meeting (IEDM), pp. 879–882 (1999).

[7] K. W. Lee, T. Nakamura, K. Sakuma, K. T. Park, H. Shimazutsu, N. Miyakawa, K.-Y. Kim, H. Kurino, and M. Koyanagi, "Development of Three-dimensiona Integration technology highly parallel image-processing chips," Jap. J. Appl. Phys. Part 1, vol. 39, pp. 2473-2477 (2000).

[8] K. W. Lee, T. Nakamura, T. Ono, Y. Yamada, T. Mizukusa, H. Hashimoto, K. T. Park, H. Kurino, and M. Koyanagi, "Three-dimensional-shared memory fabricated using wafer stacking technology," in Tech. Dig. Int. Electron Devices Meeting (IEDM), pp. 165–168 (2000).

[9] M. Koyanagi, Y. Nakagawa, K.-W. Lee, T. Nakamura, Y. Yamada, K. Inamura, K.-T. Park, and H. Kurino, "Neuromorphic vision chip fabricated using three dimensional integration tech-nology," in Proc. IEEE Int. Solid State Circuits Conf. (ISSCC), pp. 270–271 (2001).

[10] T. Ono, T. Mizukusa, T. Nakamura, Y. Yamada, Y. Iga-rashi, T. Morooka, H. Kurino, and M. Koyanagi, "Three-dimensional processor system fabricated by wafer stacking technology," in Proc. Int. Symp. Low-Power and High-Speed Chips (COOL Chips V), pp. 186–193 (2002).

[11] M. Koyanagi, H. Kurino, K. W. Lee, K. Sakuma, N. Miyakawa, and H. Itani, "Future system-on-silicon LSI chips," IEEE Micro, vol. 18, pp. 17–22 (1998).

[12] M. Koyanagi, T. Nakamura, Y. Yamada, H. Kikuchi, T. Fukushima, T. Tanaka, and H. Kurino, "Three-dimensional integration technology based on wafer bonding with vertical buried interconnections," IEEE Trans. Electron Devices, vol. 53, pp. 2799–2808 (2006).

[13] Takafumi Fukushima, Hirokazu Kikuchi, Yusuke Yamada, Takayuki Konno, Jun Liang, Keiichi Sasaki, Kiyoshi Inamura, Tetsu Tanaka, and Mitsumasa Koyanagi, "New Three-Dimensional Integration Technology Based on Reconfigured Wafer-on-Wafer Bonding Technique," IEEE International Electron Devices Meeting (IEDM) Technical Digest, pp. 985-988 (2007).

[14] T. Fukushima, E. Iwata, K.-W. Lee, T. Tanaka, and M. Koyanagi, "Self-Assembly Technology for Reconfigured Wafer-to-Wafer 3D Integration," the 60th Electronic Components and Technology Conference (ECTC), pp.1050-1053 (2010).

[15] H. C. Lin, K. L. Chang, G. W. Pickrell, K. C. Hsieh, and K. Y. Cheng, "Low temperature wafer bonding by spin on glass," J. Vac. Sci. Technol. B, vol. 20, No. 2, pp. 752-754 (2002).

[16] T. Sameshima, "Laser beam application to thin film transistors," Applied Surface Sci., vol. 96–98, pp. 352–358, (1996).

[17] T. Fukushima, H. Hashiguchi, J. Bea, Y. Ohara, M. Murugesan, K.-W. Lee, T. Tanaka, and M. Koyanagi, "New Chip-to-wafer 3D integration technology using hybrid self-assembly and electrostatic temporary bonding", in Tech. Dig. Int. Electron Devices Meeting (IEDM), pp. 789-792 (2012).

TSV Module Optimization for High Performance Silicon Interposer

Andrew Cao, Thomas Dinan, Zhuowen Sun, Guilian Gao, Cyprian Uzoh, Bong-Sub Lee, Liang Wang,
Hong Shen, Sitaram Arkalgud.
Invensas, Inc.
3025 Orchard Parkway, San Jose, CA 95134, U.S.A.
acao@invensas.com

Abstract

This paper presents Invensas' silicon interposer technology for heterogeneous chip integration. Various process module and integrated blocks were optimized for yield and high performance in the interposer. The modules under evaluation include TSV etch, barrier deposition, electrochemical plating, chemical mechanical polishing (CMP), temporary bonding, low temperature oxide (LTO) and low temperature polyimide (LTPI) passivation.

Introduction

Each generation of semiconductor products from cellphones to servers is expected to be faster and more powerful than the previous one. Along with the ever-expanding internet and widespread availability of portable devices, more computing power and functionality need to be available in small packages. The number of diverse chips inside these products is increasing rapidly and their integration within a small form factor is driving dense chip packaging technology, such as multi-chip modules (MCM), package on package (POP), package in package (PIP) and 3D integration.

2.5D integration using a silicon interposer offers a way to achieve dense chip packaging with high performance interconnects for these applications. The benefits include smaller foot print, lower power consumption, and shorter delays. Silicon interposer can also reduce stress and warpage caused by CTE mismatch between chips and package substrate. Chips built with different process flows, such as logic memory, CMOS image sensor (CIS) and microelectro mechanical systems (MEMS) can be closely connected using the top routing layer of an interposer. Since chips are placed side by side on an interposer, this assembly method reduces the thermal issues of high-power 3D stacked dies.

The market for 2.5D technology includes FPGA, ASIC, CPU, GPU, APU, high end memory and mobile applications. A number of companies are working on TSV interposer technology [1-9], and close to 80 million interposer based products are expected to be fabricated in 2014 [10]. Invensas is also developing technology for this growing market. There are many technical challenges in TSV interposer manufacturing. This paper describes our interposer test vehicle and our process optimization results.

Electrical Simulation

We conducted numerical simulations to understand the impact of various TSV process parameters to the TSV's electrical performance. While there have been many studies on this topic [11-14], our analysis includes more details in order to fine-tune our TSV process development and optimization. For example, we included in our model the often-ignored barrier layer in the TSV metallization. To assess its impact on electrical performance, we used High Frequency Structural Simulation (HFSS), a finite element method (FEM) based numerical tool that calculates the 3D electromagnetic fields. The model parameters are typical for a TSV process.

Figure 1 shows the insertion loss of a 10x100um TSV pair (one signal and one ground) with different barrier layer thickness and conductivity. As seen in the plot, the insertion loss is strongly affected by the choice of the barrier layer. Especially in the high frequency range, the insertion loss gets worse for a TSV pair with less conductive barrier. This is understandable, as more current flows near the TSV surface at higher frequency due to the skin effect. The barrier material should be highly conductive and be kept as thin as possible to prevent excessive loss at high frequency. We evaluated PVD Ti and ALD TiN as the barrier layer materials in this study.

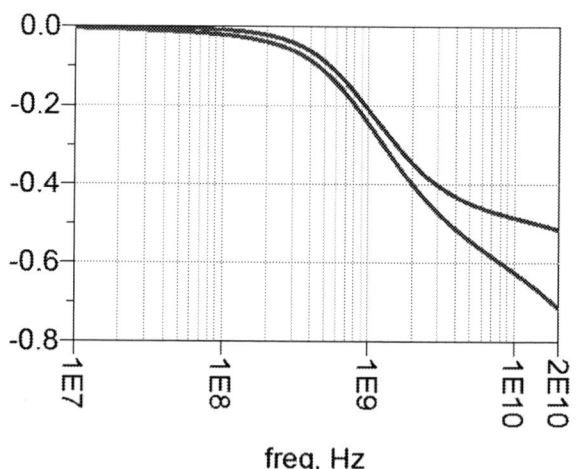

Figure. 1 Insertion loss of a signal-ground TSV pair using different barrier conductivity (Blue: 6.7E6 S/m, Red: 4E5 S/m)

Test Vehicle Design

The test vehicle consists of two dies stacked on a TSV interposer. This stack is bonded to an organic substrate as shown in Figure 2.

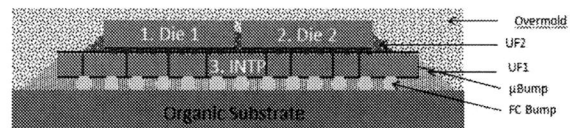

Figure 2. Invensas test vehicle with top die, interposer and organic substrate

The top dies are meant to simulate different dies, such as memory, logic, sensors or other chips that are connected to each other via the top RDL layer on the interposer and routed to the package substrate. The interposer is 19x27x0.1mm with approximately 7700 TSVs. The top dies are 10x12 mm and are connected to the TSV by more than 7200 microbumps on each die with 45um pitch.

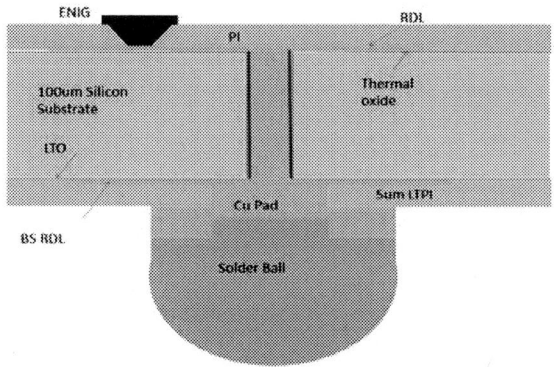

Figure 3. Cross section schematic of TSV interposer

The signal and power connections to the backside of the interposer are made using TSVs and C4 bumps on the interposer backside. The C4 bumps are bonded to an organic substrate, which distributes the power and signal to the bottom side of the package. The cross section of the interposer is shown in Figure 3.

Process Flow

The TSV are first patterned and etched using a photo resist mask and Bosch deep reactive ion etching (DRIE) process. Thermal oxide is grown to electrically isolate the vias. Barrier and seed layers are deposited, and then copper is electroplated into the vias. The Cu overburden, seed, and barrier layer is removed by CMP leaving the Cu in the TSVs, flush with top of the isolation oxide. The top Cu RDL layer is deposited and patterned. The RDL layer is passivated using polyimide, which also serves as a stress buffer during assembly. Microbump pads are formed by pattering openings in the polyimide to reach the RDL. ENIG pads are plated into the polyimide openings until their top surface is above the polyimide passivation. 10x12mm top dies simulating logic or memory dies are bonded to the interposer top surface using lead-free microbumps on Cu pillars at 45um pitch.

The interposer wafer with top surface completed is temporarily bonded to a carrier wafer and ground down to 120um, and the TSVs are revealed by dry etching.

After temporary bonding, all subsequent processing is limited to 200°C or below, and so low temperature isolation oxide is necessary for RDL isolation. TSVs exposed at the reveal process are covered with LTO and polished by CMP to make the backside of the via flush with the bottom isolation oxide. A copper RDL layer is deposited, patterned, and then passivated using photosensitive low temperature polyimide. Under bump metallization (UBM) is plated on top of the backside RDL through openings in the LTPI and bumped with 120um diameter C4 solder balls. The process flow is shown in Figure 4. This optimization of specific process modules is discussed in the following sections of this paper.

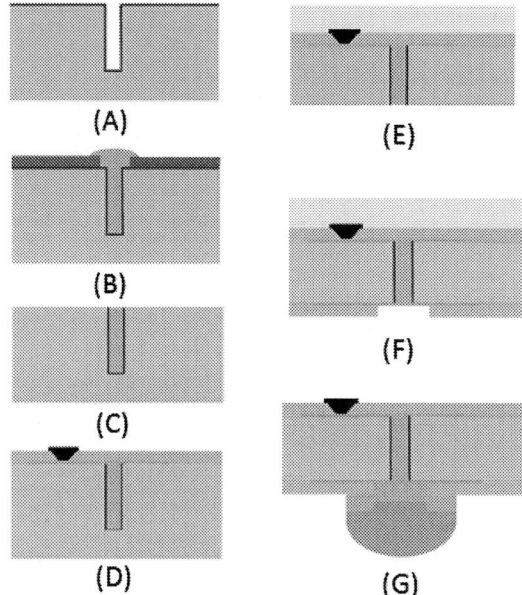

Figure 4. Interposer process flow. A) TSV etched, grow thermal oxide and deposit barrier and seed layers. B) Define resist pattern on top of TSV and plate through resist. C) CMP overburden until Cu in TSV is flush with oxide. D) Define top RDL and top ENIG pads. E) Temporary bond to carrier, wafer thin, TSV reveal and deposit LTO isolation. F) Define bottom RDL and low temperature polyimide isolation. G) Electroplate UBM and solder ball.

TSV Etch

TSVs are Bosch etched using a photo resist mask, which is simpler and less expensive than a hard mask process. TSV etch using hard mask can leave overhang or flaring at the TSV entrance, which can cause the sputtered barrier and seed layers to be discontinuous, leading to poor plating quality. The Bosch etch process commonly used for TSV drilling has alternating etch and passivation steps that leave scalloping on the side wall. Faster etch rate causes larger scalloping where the overhanging portions of the scallop can shadow the recessed regions during barrier and seed deposition; this can cause discontinuity of barrier and seed films as shown in Figure 5. Barrier and seed materials deposited by sputtering are much thicker

978-1-4799-2408-0/14 $31.00 © 2014 IEEE

on the wafer surface than inside the TSV. The thinnest region of the films is the deepest straight section of the TSV just prior to reaching the bottom. The barrier and seed thickness at this portion of the TSV is only a few percent of the film thickness at the wafer surface; depending on TSV size, aspect ratio and film materials. Shadowing caused by sidewall scalloping can make the sputtered films completely discontinuous in this region. Then the barrier layer cannot serve as diffusion barrier and copper seed cannot provide a continuous current path for plating. Depositing very thick layers of barrier and seed material will lead to excessive film-stress and possible delamination. It can also warp the wafer and require longer CMP process times.

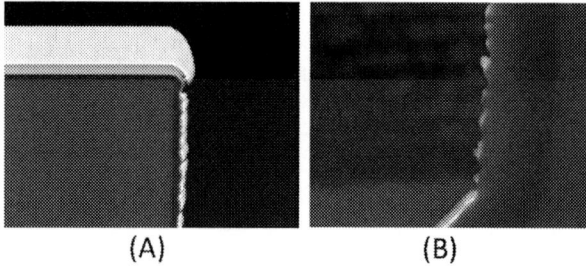

(A)   (B)

Figure 5. TSV cross section showing barrier and seed discontinuity A) TSV entrance where even a small overhang can cause barrier or seed continuity B) Bottom straight portion of TSV where the barrier and seed is the thinnest and discontinuous

Barrier and seed thickness requirements can be reduced by having straight TSV entrance profile, and flaring or rounding at the top corner. It is also desirable to remove scalloping from Bosch etch and its overhang shadowing effect by smoothing the TSV sidewall. Scalloping and surface roughness caused by DRIE etch can also act as crack initation sites, where stress in assembly and TCT might cause crack to initiate from these weak spots. Removing scallops can be accomplished by growing a thermal oxide on the wafer after TSV etch, and then wet etching the oxide away. A second thermal oxide is then grown to provide electrical isolation and serve as top CMP stop. The cross sectional profile after wet etch and thermal oxidation is much smoother than the as-etched surface. Scalloping is not detectable even at 100K magnification and roughness is well under 100A as shown in Figure 6.

(A)   (B)

Figure 6. TSV cross section (A) Sidewall is rough after etch (B) TSV sidewall is very smooth after thermal oxidation then oxide removal.

Barrier Deposition

We used both sputtered Ti and atomic layer deposition (ALD) TiN as barrier materials. Ti is an inexpensive barrier material commonly used for TSV fabrication. In this study, we deposited 1um of Ti on the wafer surface resulting in 10nm of Ti at the bottom TSV sidewall. In comparison, we also used ALD TiN as the barrier layer material. ALD TiN is not as common as Ti, and the theoretical film resistivity of TiN is about three times that of Ti (43 vs. 130$\mu\Omega$-cm). However, ALD TiN film is smooth and insures a continuous barrier layer with little thickness variation from the wafer surface (50nm) to the bottom TSV side wall (30nm) and TSV bottom (30nm). The sputtered Cu seed layer is 2.5um thick on the wafer surface but is only 20nm thick at the bottom TSV sidewall. TSV cross section with continuous ALD TiN barrier and Cu seed layer is shown in Figure 7.

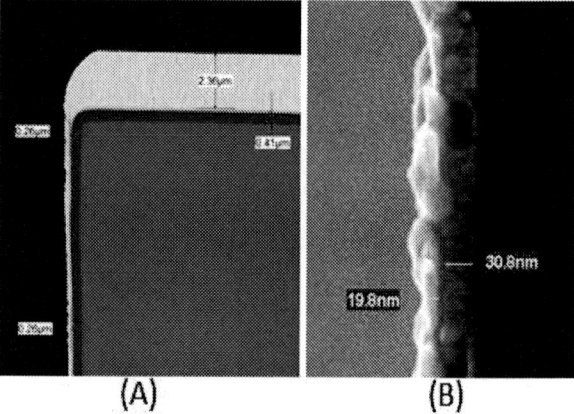

(A)   (B)

Figure 7. Continuous TiN barrier and Cu seed coverage at A) TSV top and B) TSV sidewall closest to bottom

Electrochemical Plating:

Via filling by electroplating is an enabling technology for TSV formation [15]. High aspect ratio vias must be filled without voids or seams that could jeopardize the wafer during grinding, CMP or reliability testing. The plated Cu must adhere well to the TSV sidewall and RDL layers to ensure interposer reliability. Some plating chemistries can cause a thick overburden on the wafer surface, resulting in high stress and warpage of the wafer. The warpage and overburden thickness make subsequent CMP to wafer surface long and expensive. This can lead to excessive isolation oxide removal and non-uniform dishing in the TSV across the wafer. Dishing in the vias can make the RDL layer contacting the TSV to be non-flat, making subsequent RDL layers difficult to fabricate. This type of electroplating can be extremely slow and requires extra process steps to integrate. A typical cross section of TSV after plating is shown in Figure 8.

978-1-4799-2408-0/14 $31.00 © 2014 IEEE

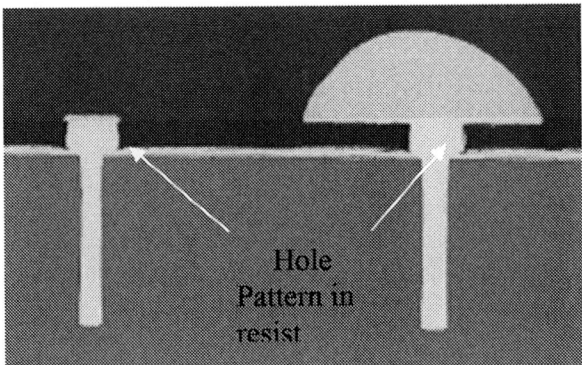

Figure 8. Conformal plating through resist with non-uniform excessive overburden on top of TSV.

We have demonstrated bottom up electroplating with low impurity content to address voids, seams and excessive overburden in the TSV. An acidic copper sulfate bath was used for the TSV copper fill. This bath contained two proprietary organic additives, an accelerator and a leveler. Agitation was provided by a paddle which oscillated in close proximity to the TSV wafer. Total electroplating time was 6X faster than the previous example. Figure 9 is an x-ray image of 20um filled vias with 50um, 100um, and 180um pitch. It can be seen that the vias are free of voids and seams at all pitches. Figure 10 is an SEM image of a cross-section from the same wafer. The overburden on the TSV is 9um thick including 2um seed and 1um Ti barrier. The overburden is relatively flat and improves CMP uniformity.

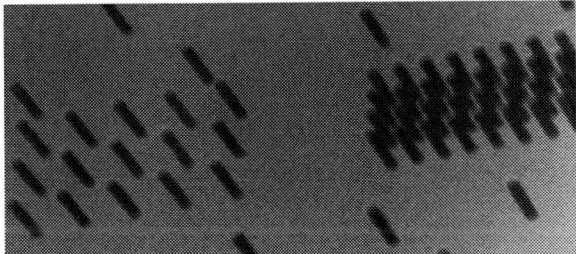

Figure 9. An X-ray image of Cu electroplated TSVs with 50um, 100um, and 180um pitches.

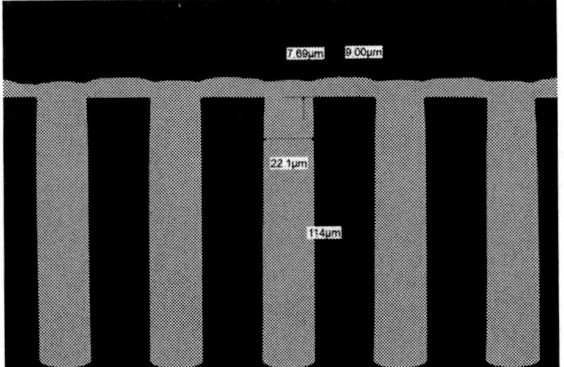

Figure 10. An SEM cross-section from the wafer in Figure 9 showing via dimensions of 114μm length and 22.1μm diameter. The copper overburden is 9.0μm including seed and barrier.

CMP:

Plated copper TSVs are annealed to promote grain growth, followed by CMP to polish away overburden from plating, copper seed and barrier. The choice of barrier material determines the slurries needed for CMP. In some cases, a separate etch process is needed to remove the barrier. A wrong choice of processes and materials to remove overburden and barrier on top of the isolation oxide can attack the TSV as shown in Figure 11, where the Cu near the TSV top has been etched away and shows truncated cones inside the vias. This will have an adverse effect during reliability testing. An optimized integrated process shows that the TSVs after polish are flush with the wafer surface with no sign of delamination or Cu attack.

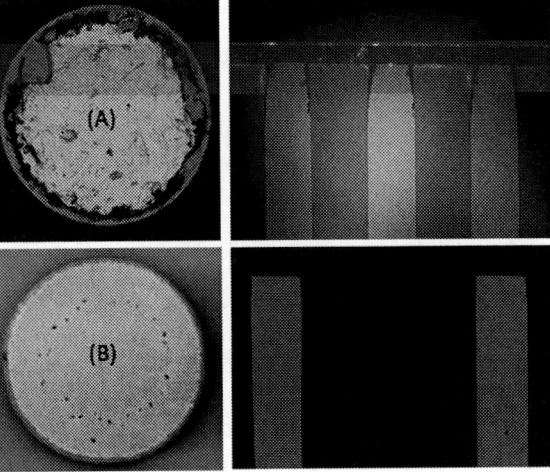

Figure 11. TSV top and side view of A) Bad Cu CMP process showing damage to top of TSV Cu and barrier layer B) Good copper CMP process showing no sign of attack

RDL and Microbump Pads

After TSVs are finished and CMPed flush with wafer surface, the next process block forms the top copper RDL above the isolation oxide, which routes connections between the dies bonded on top of the interposer. The top RDL layer is passivated with polyimide. Openings are etched in the polyimide to form connections to the RDL layers. These openings are plated with ENIG pads that protrude about 1um on top of the polyimide. These ENIG pads will be bonded to lead free microbumps on the top dies. At this point, front side wafer processing is complete.

Temporary Bonding

Temporary adds extra non value added steps to interposer processing. This leads to defects and yield loss. In addition, temporary bonding puts both temperature and chemical limitations on wafer processing. Although there are now more than 10 bonding material suppliers on the market, temporary bond and de-bond remain a major challenge to high volume manufacturing. Some manufacturers claim their bond and de-bond process can tolerate temperature up to

250°C or even 300°C, but these results depend on the specific materials bonded to the temporary bond adhesive, and must be evaluated on a case by case basis.

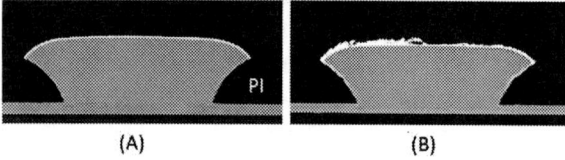

(A)   (B)

Figure 12. ENIG PAD on top of RDL and PI passivation. (A) Good ENIG Pad (B) Pad damaged after de-bond

In some cases, polyimide passivation might stick so well to the temporary bond adhesive that the bond becomes permanent. Figure 12 shows such an example where gold on top of ENIG bump surface was damaged during the debond process.

Temporary bond material can also contaminate the wafer during backside CMP processing. In one temporary bond process, a thin bead of temporary bond material was found between the carrier wafer and the device wafer. During CMP, the temporary bond material can be transferred to the wafer surface and cause contamination. Figure 13 shows temporary bond material that has accumulated in the densely populated TSV areas during backside CMP. The temporary bond material slows down CMP in dense areas, causing high CMP non-uniformity. The temporary bond material cannot be easily removed, as using aggressive chemicals can attack the temporary bond interface. The bead of temporary bond material supports the wafer and makes the bond interface strong. Trimming off the edge bead can make the wafer more susceptible to edge delamination and wafer breakage during subsequent processing. We are investigating alternative approaches that do not require temporary bonding and de-bonding.

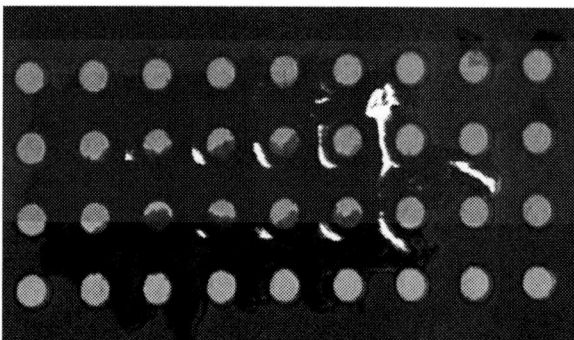

Figure 13. Temporary bond adhesive accumulated between dense TSV area during backside CMP process, causing high TTV.

Low Temperature Oxide and Polyimide

Temporary bonding also limits the choice of passivation materials that can used on backside wafer processing. Typical backside isolation uses low temperature oxide (LTO), or low temperature nitride

(LTN) with a typical deposition temperature limit of 150°C to 200°C. LTO and LTN tend to be weaker and more porous than oxide and nitrides deposited at higher temperatures. These weaker passivation materials are more likely to cracks or delamination during temperature cycling. This problem can be minimized by first choosing the right deposition process and then optimizing the deposition parameters and LTO thickness. We have developed a simple but elegant method to detect LTO cracking and delamination by temperature cycling and warpage measurements. Hysteresis in a warpage vs. temperature curve suggests a film has either delaminated or cracked. A stable LTO film will have a repeatable warpage vs. temperature curve as shown in Figure 14.

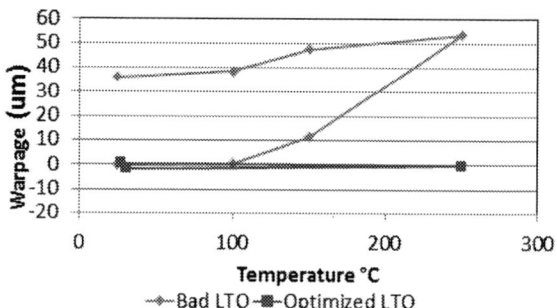

Figure 14. Die warpage hysteresis during temperature cycling indicates bad LTO was damaged. Optimized LTO shows low warpage and no hysteresis within measurement error of the tool.

Temporary bond also restricts backside polyimide to a curing temperature around 200°C. Low temperature polyimide (LTPI) lacks the chemical resistance, temperature resistance, and mechanical strength of conventional polyimides that are typically cured around 300°C to 400°C [16]. LTPI cured at low temperature tend to outgas above the curing temperature, which can lead to delamination between LTPI and underfill. UBM and solder balls are typically processed after LTPI passivation, therefore several high temperature process steps and exposure to various chemicals can threaten the LTPI integrity. The high temperature steps include pre-sputter etch, seed layer sputtering, and solder ball reflow. Chemicals exposure includes photo resist, developer, resist stripper, plating baths for copper, tin and silver, flux and flux remover. We have experienced LTPI wrinkling during sputtering and LTPI cracking during solder ball reflow. Figure 15 A shows cracked LTPI after solder ball reflow, where the same process did not have issues with normal polyimide cured at higher temperature. We experimented with different flux materials, flux cleaning methods and reflow profiles and conditions in order to avoid LTPI cracking. We successfully resolved the issue by reflowing solder balls in optimized conditions, as shown in Figure 15B.

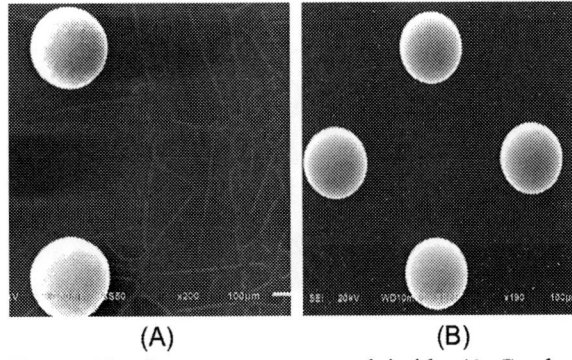

(A)   (B)

Figure 15. Low temperature polyimide A) Cracked after reflow. B) Optimized reflow process

Conclusion

2.5D and 3D technology is a growth area in high density electronic packaging. This work describes our work in optimizing individual process blocks in 3D packaging technology to deliver well integrated process blocks for dense, high performance interposers.

Acknowledgments

The authors wish to thank Invensas management for support and all our partners for wafer and die processing services.

References

1. Saban, Kirk. "Xilinx stacked silicon interconnect technology delivers breakthrough FPGA capacity, bandwith, and power efficiency. Xylinx white paper: Virtex-7 FPGAs

2. Banijamali B. et al. "Advanced Reliability Study of TSV Interposer and Interconnects for the 28nm Technology FPGA". ECTC conference proceeding 2011

3. Zhang X. et al. "Development of Through Silicon Via (TSV) Interposer Technology for Large Die (21x21mm) Fine-Pitch Cu/low-K CFBGA package". ECTC conference proceeding 2009.

4. Baez F. M. et al. "Electrical Design and Performance of a Multichip Module on a Silicon Interposer". Electrical Performance of Electronic Packaging and Systems (EPEPS), 2012 IEEE 21st Conference on

5. Andry et al. "Fabrication and characterization of robust through-silicon vias for silicon carrier application" IBM J. Res & Dev. Vol. 56, No. 6 (2008)

6. Lin et al. "Reliability Characterization of Chip-on-Wafer-on-Substrate (CoWoS) 3D IC Integration Technology. -TSMC

7. Vodrahalli N. "Silicon TSV Interposers with embedded capacitors for high performance VLSI packaging". CMPT symposium Japan, 2010 IEEE

8. Detalle M. Manna A. L., Vos De, et al. "Interposer Technology for High Band Width Interconnect Applications". ECTC conference proceeding 2013

9. Murayama, K. ; Aizawa, M. ; Hara, K. ; Sunohara, M. et al. "Warpage control of silicon interposer for 2.5D package application. Electronic Components and Technology Conference (ECTC), 2013 IEEE 63rd

10. The multi-component IC packaging market, 2014 edition. New venture research corp.

11. P. Muthana, et al "The Road to 1TBps Bandwidth Systems: A case study", DesignCon, Santa Clara, CA 2014

12. A. Martwick and J. Drew, "3D STACKED SIGNALING, JITTER AND MEASURMENT", DesignCon, Santa Clara, CA 2014

13. Z. Xu and J. Q. Lu, "Through-Silicon-Via Fabrication Technologies, Passives Extraction, and Electrical Modeling for 3-D Integration/Packaging", Trans. Semi Manufacturing, pp23-34, Feb 2013

14. J. S. Pak et al, "Slow Wave and Dielectric Quasi-TEM Modes of Metal-Insulator-Semiconductor (MIS) Structure Through Silicon Via (TSV) in Signal Propagation and Power Delivery in 3D Chip Packag", IEEE 60th Electronic Components and Technology Conference (ECTC 2010), Las Vegas, NV 2010

15. Rozalia Beica, Charles Sharbono, and Tom Ritzdorf. "Through silicon via copper electrodeposition for 3D integration," in *Proc. IEEE Electronic Components and Technol. Conf. (ECTC)*, Lake Buena Vista, FL, May 27-20, 2008, pp. 577-583.

16. Gao, G, Lee B., Cao A, Chau. E "Process Compatibility of Conventional and Low-Temperature Curable Organic Insulation Materials for 2.5D and 3D IC Packaging – A User's Perspective" ECTC Conference proceeding 2014

Study of TSV Thinning Wafer Strength Enhancement for 3DIC Package

Jyun-Ling Tsai, Chun-Chieh Chao, Hsiao-Chun Huang, Cheng-Hsiang Liu, Hung-Hsien Chang,
Chang-Lun Lu, Shih-Ching Chen
Siliconware Precision Industries Co., Ltd. (SPIL)
No.153, Sec.3, Chung Shan Rd., Tantzu Dist., Taichung 42756, Taiwan, R.O.C.
genietsai@spil.com.tw +886-4-2534-1525 ext.4618

Abstract

Interfacial delamination between backside of TSV thin wafer silicon, low temperature PECVD silicon nitride and UBM (under bump metallurgy) layer under room temperature and thermal cycling or processing have been investigated in this paper. FEA (Finite element analysis) was used to assessment the thermal stresses and the driving force of thin wafer bump UBM delamaination. The 3D modeling results were validated by process improve solutions which are including add a dielectric materials PBO (polybenzoxazole) buffer layer to release stresses and adjust UBM metal structure thickness to reduce UBM corner stresses. In the other hand, seeking enhancement of interfacial adhesion between thin wafer silicon and PECVD SiNx is important work in parallel.

Integrate PECVD SiNx adhesion enhanced on TSV backside via reveal process and addition PBO buffer layer, reduce UBM metal thickness during C4 bump formation process was effectively to release die stresses (increase die strength around 8 times) for delamination free.

Eliminate die crack issue during Back-End stacking process is another good improve process yield achievement. At the same time, Interposer die warpage has 70% improved from crying direction to close to fully flat interposer die which is good achievement for Back-End chip to chip stacking process. The Mid-End process improvement was proved and validated consistent with thermal stress simulation.

Introduction

3DIC stacking technology is performed Z-direction interconnect through TSV (through silicon via) which can provide the shortest and most effective interconnections between chip and chip or between chip and substrate. The reason are demand for miniaturization trend and homogeneous, hetero-integration device between high capacity memory and logic chips are increase performance and functionality, TSV is one of the most promising technologies in extensive 3D chip stacking technology.

In order to develop 3DIC package, one of model has been discussing and realize in semiconductor business chain. The TSV/BEoL process fabricate by foundry houses and Mid-End/Back-End process fabricate by OSAT (Outsourced Semiconductor Assembly and Test), in this paper is focusing on Mid-End/Back-End process development and challenges, the full Mid-End process can be categorized into bumping stage process and TSV backside via reveal stage process, the bumping stage may include top die uBump (microbump), interposer front-side uPad (micropad) and interposer back-side RDL (Re-Distribution Layer), C4 bump process; the TSV backside via reveal stage may include thin wafer handling process, wafer thinning, DRIE (Deep Reactive Ion Etching), dielectric passivation deposition, and CMP (Chemical Mechanical Polishing). The full Back-End process can be categorized into interposer die flatness & CoC (Chip stack on Chip) stage process and CoS (Chip module stack on Substrate) stage process, the CoS is same as standard flip chip BGA assembly.

Although 3DIC package have great electrical benefit, such kind of long process flow and thermal loading are corresponding thermal mechanical problem which build on TSV thin wafer are big challenges. This consist of interposer/chip module problems of material CTE (coefficient of thermal expansion) mismatch, induce stresses, poor adhesion, interfacial delamination, die crack and so on. Consider balancing wafer/die warpage and stresses are major challenge within Mid-End/Back-End process integration.

In this paper, die crack during back-end process and interfacial delamination under room temperature, TCT (thermal cycling test) or Back-End processing challenges and solutions in Med-End process have been investigated. Figure 1 shows the cross section pictures of process failure mode: interfacial delamination between interposer silicon/dielectric passivation and die crack into interposer silicon.

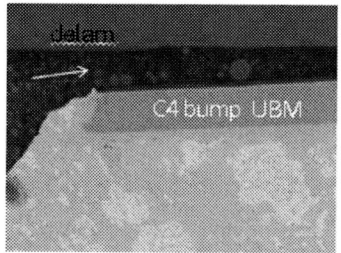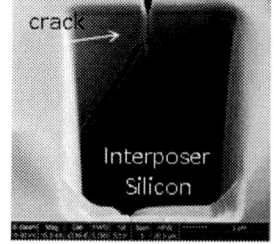

Figure 1. Cross section pictures of process failure mode: interfacial delamination and die crack at Back-End process.

Photosensitive (PI) polyimide and (PBO) Polybenzoxazole products are not only widely used for passivation layer for chip surface protection, but also for dielectrics passivation of bump (RDL) redistribution structure design. PI and PBO products are specialty stress relief coatings which have good heat, chemical resistances and electronic, mechanical properties. However, even PI is standard production dielectric passivation for stress release buffer layer, but curing temperature is excess of 350℃. To demand thin wafer handing using adhesive for temporary bonding to glass carrier, low temperature process has been required for interposer backside process. Low temperature curing also brings up lower interposer warpage effect on future Med-End process. Base on this temperature limitation concern, the passivation buffer layer must select low temperature curing (<200℃) material. Low temperature PBO has used in this investigation. [1-2]

978-1-4799-2408-0/14 $31.00 © 2014 IEEE 868 2014 Electronic Components & Technology Conference

Finite Element Analysis Model

FEA model is an interposer silicon die stack on organic substrate, FEA simulate thermal stress effect to different material and different opening size design. The interposer schemes have PECVD (SiN) Nitride on interposer backside, backside Cu UBM and connect C4 eutectic or lead-free bump material. Interposer front-side micro-pad and top die Cu pillar bump joined will be ignoring in this study, we focus on interposer backside and organic substrate solder tip pad joined. The scheme was described in figure 2.

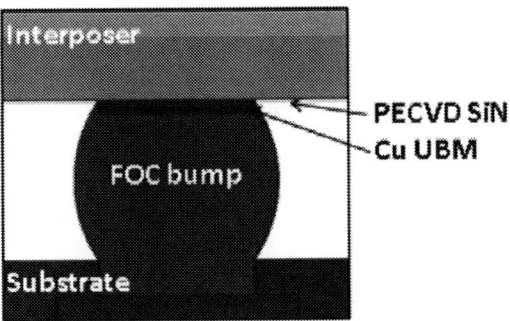

Figure 2. Scheme of flip chip package model for thermal mechanical finite element simulation, interposer frontside and top die has been ignore in this FEA.

The flip chip package thermal loading is from stress free condition in 187°C for eutectic bump melting temperature cooling drop to 25°C temperature. Material properties were listed in Table 1.

Table 1 Material properties of each component.

Material	Modulus(Kg/mm²)	CTE(ppm/°C)
PBO	180	80
Cu	12100	16.3
Ni	20500	12.3
Eutectic Bump	3388	22
Lead-Free Bump	4082	20

The FEA stress simulation condition split factor includes passivation layer PBO-free, PBO thickness, PBO open size, and UBM thickness, and solder bump height (solder volume). The split condition table was described in table 2.

Table 2 DOE split for Finite Element Analysis Model.

Split factor	Condition 1	Condition 2	Condition 3
PBO-free (control)	--	--	--
PBO thickness	3 um	6 um	10 um
PBO open size	70 um	50 um	30 um
UBM thickness	5 um	3 um	--
Solder bump height	90 um	80 um	--

Die Strength Test Methodology

To demand miniaturization device and increase electric performance and functionality, Z-direction integration by TSV is one optimal solution for 3D IC stacking technology. Z-direction decrease for thin wafer/die is current industry mainstream which has been driven by high electrical performance and better functionality devices of high capacity memory and logic chips. 3DIC is one of technology which be required the maximum wafer thickness to 100um. However several studies have conducted wafer surface damage during wafer thinning process from 775um to 100um or thinner thickness, the chip surface damage was caused by backside grind and wafer dicing process. The saw mark (roughness) or chip size is inversely trend with die strength have been investigated. Meanwhile a study indicated an increasing trend in the breaking load when die thickness increase form 150um to 394um. [3-4] Interposer is different scheme with general flip chip die; the flip chip die is exposed silicon post backside grinding process, the interposer has thin dielectric passivation film protect the silicon surface. Even through interposer has well control wafer saw mark and roughness during wafer thinning process, the dielectric film still influence die strength to make die crack during Back-End chip stacking process.

Die strength test was used for verify dielectric passivation film property adjustment to improve die strength, and also verified reducing interposer die UBM corner stress is success achieve die strength increase to prevent die crack during Back-End process. The die strength test can calculate the failure strength of material is maximum stress loading at material breakage moment; the test conditions are keep load speed 1mm/min and die size 100um thickness. Figure 3 is described die strength test methodology. We put a thin interposer die on a plate fixture, the plate fixture have one diameter 4mm hole for push pin through, calculate maximum stress loading to come out die strength value.

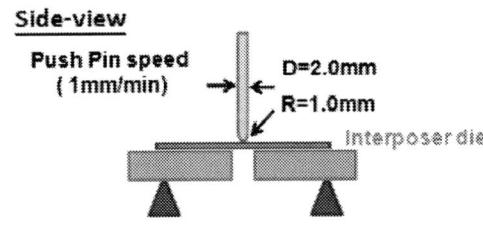

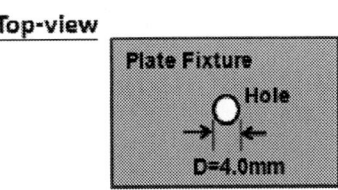

Test scheme

Figure 3. is described die strength test methodology.

In order to enhance thin wafer die strength and interfacial adhesion between interposer backside silicon and backside low temperature dielectric film, we adjustment parameter of RF power (W) to create dielectric film surface densification characteristic to deposition on interposer backside surface, the split condition described as table 3.

Table 3 Interposer dielectric film split for TSI die strength and TSI die warpage collect

Split factor	Condition
Baseline (control)	Low RF (W)
New dielectric film	High RF (W)

Result and Discussion

General FOC (directly bump on cupper) structure has observed the maximum stress distribute on the outer edge of UBM at room temperature 25℃, which location is focused of stress release in below following investigations. The UBM layer stress oscillation from positive to negative values. Figure 4 is show interposer die stacked on substrate, the interposer backside C4 bump is typical directly bump on pad structure stress distribution scheme.

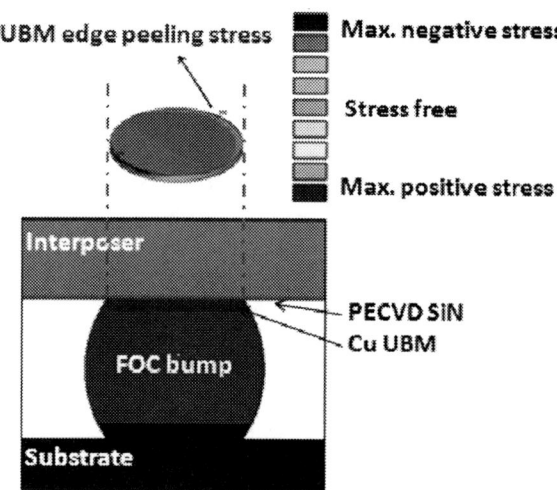

Figure 4 is described directly bump on pad structure stress distribution scheme, the maximum stress distribute on outer edge of UBM.

First achievement of this study is PBO coatings relief over 60% stresses decreasing which compared with original directly bump on pad structure. Figure 5 result shows PBO coating layer not only stress decrease substantially, but also PBO thickness reduction in direct proportion to stress relief. 3um and 6um thickness PBO was appeared larger stress release condition. Keep same condition: 6um PBO thickness to split different opening size to understand UBM edge stress release behavior. Apparently, the result shows best condition is 70um open size which decrease stress over 90%; 50um and 30um open CD was appeared same stress decreased level which around 80%. Figure 6. is showed same PBO thickness condition to evaluate stress relief ratio by different PBO opening size.

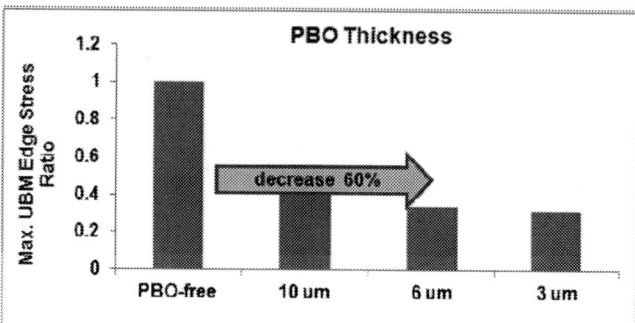

Figure 5. is described UBM edge stress release performance by different PBO thickness design.

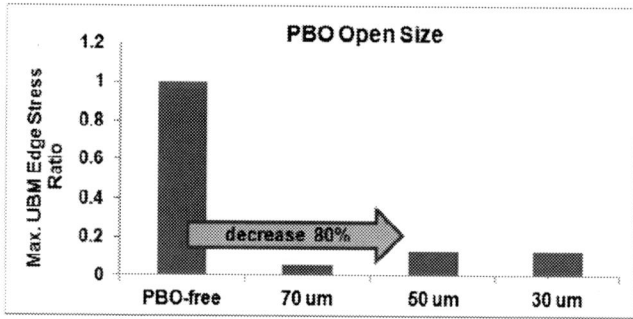

Figure 6. is showed UBM edge stress trend by different PBO open size with same 6um PBO thickness.

UBM thickness is one of stress reduction study item since Cu UBM has larger young's modules (E: 12100 Kg/mm2). Apparently, PBO coating condition has reduce around 40% UBM edge stress. Also, compare to original 5um Cu UBM thickness and 3um Cu thickness with PBO coating layer result was observed there is no significant between 5um and 3um Cu. However 3um is less stress effect design then 5um thickness. Figure 7. is showed Cu thickness effect of UBM edge stress. Meanwhile, reduce solder volume from solder bump height 90um to 80um comparison was appeared no significant on these two condition split. Stress release trend by solder bump volume change condition is showed at Figure 8.

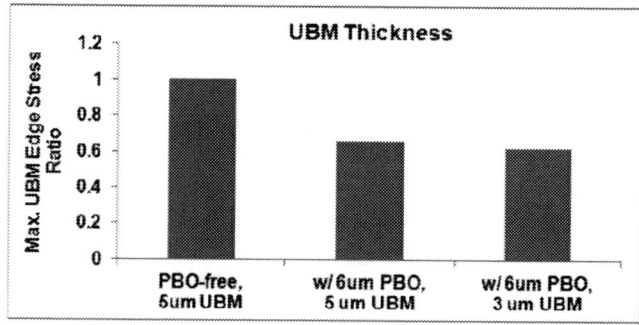

Figure 7. is showed UBM edge stress effect by different UBM thickness with and without PBO buffer layer.

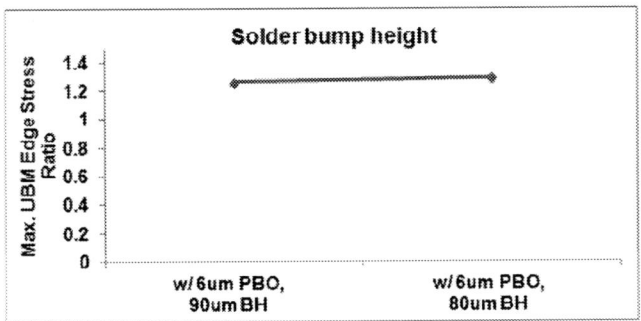

Figure 8. is described solder bump volume effect of UBM edge stress.

To avoid UBM corner delamination due to peeling stress concentrates on UBM edge, the PBO coatings has effectively release major stresses. In addition, increase die strength is another work to prevent die crack phenomenon during Back-End chip stacking process. We are adjust interposer backside dielectric passivation film parameter to create more surface densification characteristics to increase die strength and interfacial adhesion of interposer backside silicon/dielectric film [5-6], using die strength tester to compare die strength for original PBO-free structure and PBO coatings structure with LRF dielectric film scheme, the result is apparently, PBO coating layer was increase almost achieve 5 times die strength level. Also more surface densification dielectric film (HRF condition) has additional 3 times die strength increased with same PBO coated layer. This is a big jump to enhance die strength of thin wafer to eliminated die crack phenomenon at Back-End process handling now. The die strength enhancement trend was described as Figure 9.

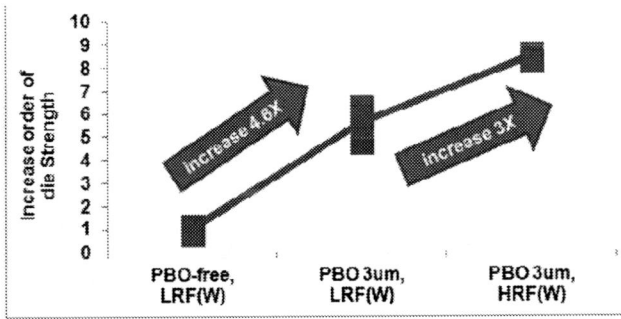

Figure 9. is showed PBO coating layer and dielectric film adjustment resulting larger enhancement of die strength.

Considering Back-End chip on chip micro-joining process is susceptible to die warpage at room temperature and reflow temperature behavior, which is because die center and die edge are unbalance formation of micro-joining by die warpage. Bump Non-wet or bump bridge at die center or die edge are major failure concern by TSI die warpage. Multi-layer material on thin interposer die will influence warpage behavior, especially for addition PBO coating layer on thin wafer/die. Compare die warpage between PBO-free and PBO coating condition by original (un-adjust) dielectric film could observe 30% of die warpage reduction at room temperature. Furthermore, adjust and un-adjust dielectric film of die warpage performance appearance substantial improve 40%

die warpage to most flatness behavior at room temperature. Also, 30% improved at reflow temperature. This is big challenging to improve die warpage to approach flatness die handling before die bonding and smooth warpage change behavior under thermal profile from room temperature to micro-joining temperature. Attached Figure 10 is showed interposer die warpage behavior improve to approach flatness interposer die from root temperature to micro-joining temperature.

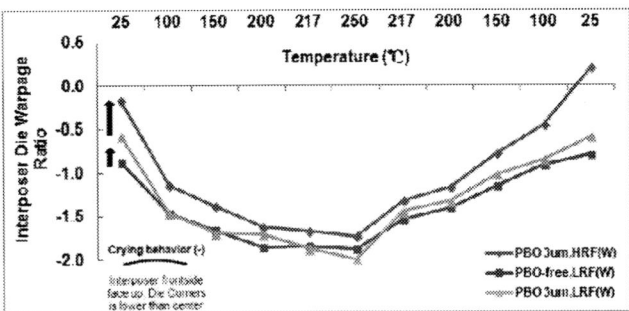

Figure 10. is showed addition PBO coating layer and combine dielectric film adjustment resulting substantially improve die warpage to most flatness behavior at room temperature.

In addition, UBM corner stress release and die strength enhance and die warpage reduce was implement and validated, QTC (quick temperature cycle) test is one effectively evaluation methodology for bump crack/bump interface delaminate. Therefore, QTC test ($-40°C \sim 60°C$) was implemented to verified the robustness of interposer which after additional PBO layer and dielectric film improvement at this study, the chip on chip stack module and chip module stack on substrate without underfill dispensing sample build to quickly evaluate C4 bump robustness. The inspection gating was using C-SAM technology to monitor white bump or interfacial delamination, also cross-section analysis was using for future confirm bump crack interface layer which is judged by C-SAM failure location. In this evaluation, we perform QTC cycle times are 0X, 5X, 10X, 20X, 30X, 40X and the result are all pass C-SAM inspection and pass bump cross-section structure check. The QTC test pass is successes development by PBO coating layer and dielectric film adjustment which were improve UBM corner delaminate free and die crack free.

Conclusions

In this paper, we investigated several approach to reduce UBM corner stress which are including low temperature PBO coating layer thickness, PBO opening size, UBM Cu thickness, solder volume and so on. As well known, addition of PBO coating layer can substantially release 80% of UBM edge stresses to improve interposer silicon and C4 bump UBM interfacial delamination free. Moreover, backside dielectric CVD film adhesion enhancement and film surface densification characteristics to increase die strength to around 8 times, it is success prevent thin interposer die crack at chip stacking handling process and achieved inline process yield improvement. Surface densification dielectric film not only that eliminate die crack but also reduce 40% of interposer die warpage to approach die flatness, which is increase chip to chip stacking process window to prevent bump bridge or bump non-wetting failure after high reflow temperature. In addition, quick temperature cycle test (-40℃ ~60℃) was validated by 40 cycles pass, therefore we obtained the robustness interposer after PBO coating layer and surface densification dielectric film implemented for the improvement at this study.

Acknowledgments

The authors would like to thanks SPIL TSV/MEoL/BEoL team for valuable study data on this development.

References

1. M. Niwa, "Analysis of stress buffer effect of polymide on thermal-mechanical stress of ELK layer in flip chip package," in Proc. IEEE International Conference on Electronics Package (ICEP), India, November 4-9, 2012

2. R.L.Hubbard,"Low warpage and improved 2.5/3DIC process capability with a low stress polyimide dielectric" in Proc. Internal Wafer Level Package Conference (IWLPC), November 5-8, 2012

3. S. Schonfelder, "Investigation of strengh properties of Ultra-Thin silicon,"IEEE, 2005

4. Y.R. Chong, "Mechanical characterization in failure strength of silicon dice," in Proc. IEEE InterSociety Conference on thermal Phenomena, 2004

5. M. Martyniuk, "Stress response of low temperature PECVD silicon nitride thin film to cryogenic thermal cycling," IEEE, 2005

6. J. Liao,"Study of the effect on die backside stress from coating of a nitride layer," IEEE, 2012

Challenges in 3D Die Stacking

Juergen Grafe, Wieland Wahrmund, Stephan Dobritz, Juergen Wolf, Klaus-Dieter Lang
Fraunhofer IZM - ASSID
Ringstrasse 12, 01468 Moritzburg, Germany
juergen.grafe@assid.izm.fraunhofer.de

Abstract

Many semiconductor companies are currently engaged in 3D system integration. The assembly of 3D compliant chips becomes a vital factor of the 3D application success and reliability. Major challenges are provided by very low chip thickness, large die size, small interconnect diameter and pitch. Diverse 3D assembly technologies and methods are currently under investigations which address these specific technical challenges. Stable and volume capable assembly processes must be developed in order to manufacture such products in future with reasonable cost. Wafer-to-wafer (W2W) assembly is not yet recommended for most of the advanced 3D applications since it still suffering from too high yield losses what would translate into unacceptable W2W stack yield. For that reason the die-to-die (D2D) assembly is considered as the more efficient way for the time being. For that reason we're developing integrated assembly and test concepts on 300 mm wafer size to evaluate and validate various assembly technologies regarding to their capabilities with respect to interconnect materials, dimension, pitch and I/O density.

Introduction

This study is focusing on 3D assembly and test of high-density interconnect arrays related design rules, wafer process integration schemes and results obtained from 3D die stacking.

Assembly and test technologies applied to 3D applications must be considered to a much higher extend in wafer design and process integration as usually done for single sided standard wafers. Since the trend goes towards through silicon vias (TSVs) with ≤ 5μm diameter (AR ≥ 1:10) the silicon wafer thickness is continuously decreasing towards 50 - 30μm. For that reason it is important to determine 3D process integration schemes in such a manner that ultra-thin and fragile device wafers will be exposed to the wafer bonding and debonding process one time only. In order to meet this requirement the different interconnect elements (flip chip pads and micro-bumps) are to be placed on the appropriate wafer side and outlined according to the assembly technology and reliability requirements dedicated to the application. All process steps in wafer back side manufacturing, singulation and assembly should be executed with the lowest level of difficulty regarding to topologies impacts, thermal and vacuum budgets applied in DRIE, CVD and PVD as well as cross contamination or wafer edge damages during wafer handling. This is also in accordance with the demand to further reduce manufacturing cost for 3D applications what can be also achieved by the simplification of process integrations schemes and the related yield improvements.

Experimental Procedure

The ATC2 integrated assembly and test concept is an advancement of our former ATC1 approach /1/ and was especially established to primarily support assembly and tests investigations. For that reason we defined for the ATC2 concept a minimum set of mask layers needed for manufacturing of TSVs, interconnect elements at both wafer sides and very limited metallization layers providing signal traces for daisy chain measurements. The ATC2 addresses high-density interconnect arrays with ~36.000 I/O per die in order to allow an extraction of electrical TSV and interconnect parameter on a high statistical base and to challenge the capabilities of assembly technologies and processes applied.

The ATC2 concept uses three different chips types: (1) mother wafer/die, (2) stackable daughter die and (3) top die. Figure 1 shows an example of a 3D die stack construction. The 3D stack height is scalable via the number of stackable dies assembled. The die size is (10.2 x 10.2 x 0.1) mm.

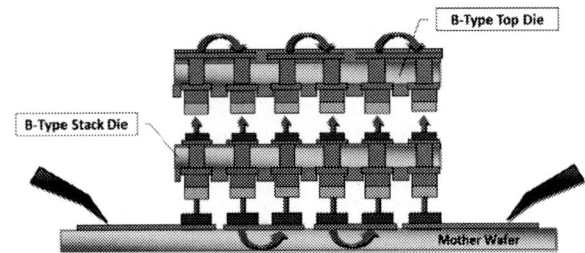

Figure 1: 3D die stack construction (ATC2).

Die stacks are to be assembled on the mother wafer/die and must be completed by a top die. The design of all three chip types is arranged in such a manner that all existing interconnects and TSVs are integrated in one electrical daisy chain regardless the number of chips assembled. Figure 2 shows an optical image and a cross section of a triple die stack assembled according this approach.

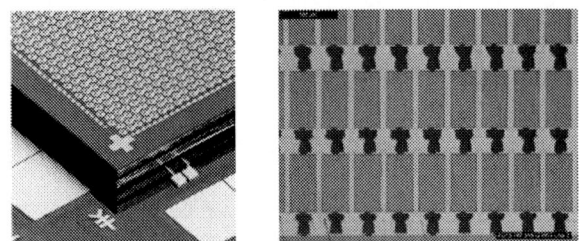

Figure 2: Triple die stack (ATC2).

The micro-bump diameter of the ATC2 is defined with 25μm and the pitch with 55μm. The micro-bump construction

978-1-4799-2408-0/14 $31.00 © 2014 IEEE

2014 Electronic Components & Technology Conference

used in this investigation consists of a 15μm tall Cu pillar with a 12μm SnAg cap determined for reflow soldering as assembly technology. Figure 3 shows an image of the high-density micro-bump array at the ATC2 chip back side.

Figure 3: Flip chip bump array at ATC2 chip back side

The flip chip pad is designed as "pedestal" pad in order to generate together with the flip chip micro-bump sufficient underfill gap for capillary underfill (Figure 4). The "pedestal" pad has a diameter of 35μm. It consists of a 25μm tall Cu pillar with Ni/Au finish (3/0.5μm). This construction provides a total underfill gap of ~50μm after reflow soldering.

Figure 4: Flip chip pad array at ATC2 mother chip top side

The wafer process flow for both chip types was arranged as described in Figure 5. The process flow starts with the TSV module containing all required process steps for TSV patterning, etch, clean, isolation, barrier/seed deposition, electro-chemical copper deposition (TSV fill) and CMP. The second process module includes the front side isolation deposition and patterning followed by the micro-bump plating and reflow (Figure 3).

The first process module in wafer back side processing contains the process steps temporary wafer bonding and wafer thinning. Within the next process module the TSV contact gets established which requires TSV recess etch, isolation and opening. The third process module covers the deposition and patterning of the flip chip pad metallization. Finally the device

wafers will be debonded from the temporary carrier and singulated in wafer front side-up position.

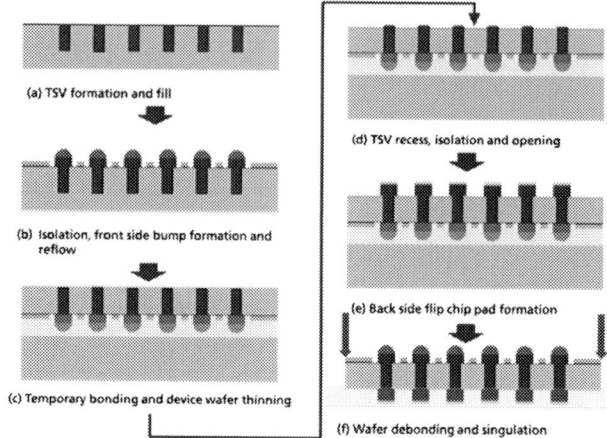

Figure 5: ATC2 basic wafer process flow

The 3D assembly was done by flip chip bonding. This process was split in two-step sequences: (1) pick&place and (2) external reflow soldering. The pick&place process of the flip chips were performed on a Panasonic FCB3. Fluxing of the flip chip bumps has been executed before each pick&place step. Finally all 3D die stacks got underfilled and electrically tested. The process flow applied in 3D die stacking is illustrated in Figure 6.

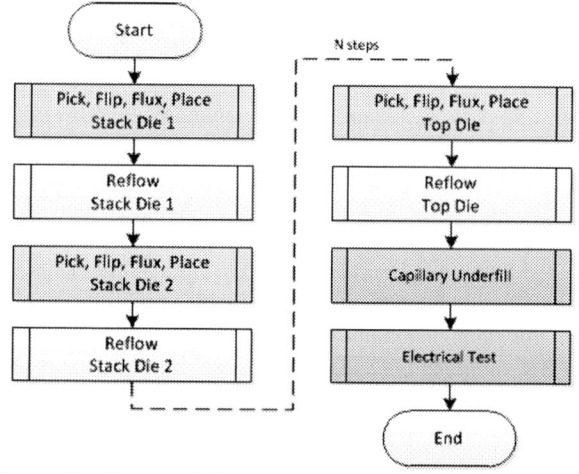

Figure 6: 3D assembly process flow

Discussion

A high singulation quality of double-sided patterned wafer with high-density micro-bump arrays on both wafer sides is of very high importance for the following assembly processes. Mechanical blade dicing of such wafers is very challenging since always one of the micro-bump arrays is per default positioned on dicing tape and can't be efficiently sealed against waste water penetration during the dicing process. Any particles, water stains or other contamination will have serious impact on the assembly yield later on. For that reason laser stealth dicing using a Disco DFL 7360 laser dicing saw has been chosen for 3D wafer singulation since this is a dry and almost particle-free process. The chip edge and corner quality out of laser stealth dicing is superior compared to mechanical blade dicing. The significantly reduced edge and corner chipping makes the final 3D stack also more robust against stress and strains when it gets for example molded in plastic packages.

The assembly of high-density interconnect arrays requires very low chip warpage and high uniformities in interconnect height, shape and surface conditions. The chip warpage is mainly influenced by the layer stack design and the materials used at both wafer sides. It is well known that assembly yield losses will be obtained in reflow soldering if the interconnects suffering from insufficient wetting caused by excessive chip bending /2/. Therefore the chip warpage has been initially measured under room temperature using a µScan laser profilometer from Nanofocus. Both chip types (top die and stackable die) revealed concave chip warpage (micro-bumps in bottom-up position). Higher chip warpage was measured for the stackable die. This can be explained by a higher compression of the stack die micro-bump side compared to the top die caused by a different front side layer stack. The inherent difference is a thicker polymer layer at the top die front side surface which results in a better stress compensation for this chip type (Figure 7).

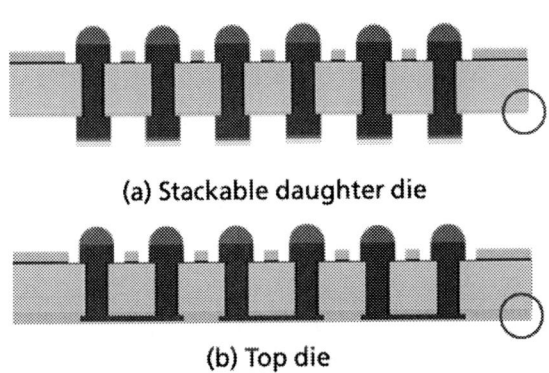

(a) Stackable daughter die

(b) Top die

Figure 7: Layer stack design of stackable vs. top die

Typical measurement results taken from both chip types are shown in Figure 8 and Figure 9. The highest chip warpage was measured in diagonal chip direction as expected. Even the diagonal chip warpage of the stackable die was judged critical for external mass reflow a rather low impact on the assembly result has been experienced. This is in accordance with theoretical evaluations published in /2/. The large solder volume in combination with the low silicon die bulk thickness results in a high solder force which is flattening the assembled die during the time the solder is liquid. The higher expansion of the polymer material during temperature ramp-up contributes also to the die warpage reduction. This warpage state will be more or less frozen during solder solidification. This behaviour has been confirmed by replicated chip warpage measurements after each reflow step at room temperature. It revealed that the chip warpage was always below 5µm (Figure 10) what is uncritical for the next die to be

stacked. We have also conducted evaluations where we assembled 3D die stacks by applying only one final reflow step. This was still working for quad die stacks but the flux must be very tacky to hold the stacked dies in position during handling and assembly of the remaining dies.

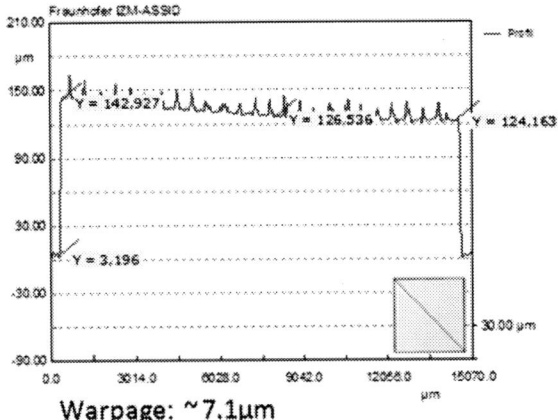

Figure 8: Chip warpage of ATC2 top die

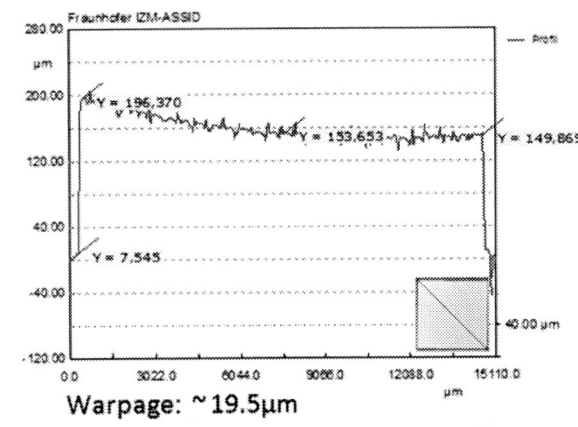

Figure 9: Chip warpage of ATC2 stackable die

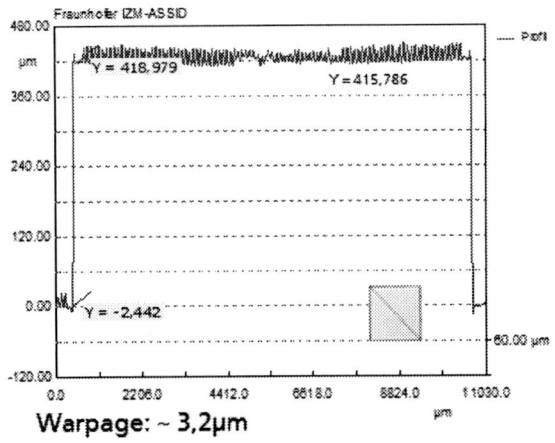

Figure 10: Chip warpage measurement on top of 3rd stacked die after reflow soldering

Different 3D die stack constructions (single, dual, triple and quad stacks) have been assembled, underfilled and electrically characterized. All dies have been placed with a bond force of 100 N and the placement time was set to 1 sec. The temperature profile applied in external mass reflow using a Rehm V8/Nitro 2.1 belt furnace is shown in Figure 11.

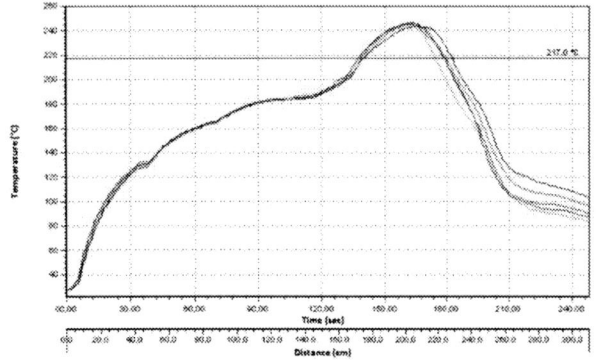

Figure 11: Temperature profile used in mass reflow

The process integration scheme applied on wafer back side takes effect on the flip chip pad outline. It is well known that any difference in height below interconnect will be transferred to its top side in electro-plating. The most precise shaped interconnects will be obtained out of a dual damascene process as published in /3/. Both chip types assembled in this investigation have been manufactured with an integration scheme which doesn't use CMP in TSV recess. We developed this integration scheme to reduce manufacturing cost in wafer backside processing and evaluated it first time in this study. The approach was to deposit a 5µm thick polymer layer onto the wafer back side directly after recess etching of the TSVs. This polymer layer has been reduced in thickness via a mask-less G-line exposure process until the TSV tips were slightly exposed (~ 2µm) at the polymer surface. Finally a short oxide dry etch step was applied to remove the oxide isolation at the TSV tips. It has been experienced that the TSV tip protrusion was unfortunately transferred with high extend to the pad surface and created in combination with a slight alignment failure an uneven pad surface shape (Figure 12a) at the back side of the stackable dies. It was difficult to achieve a correct die placement on such pads during flip chip bonding but the final reflow result was acceptable due to the given chip self-alignment capability.

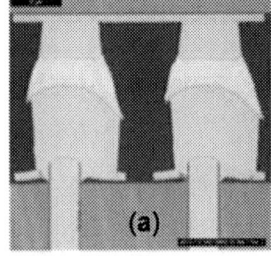 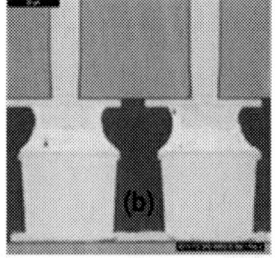

Figure 12: Flip chip interconnect on stackable die (a) and on mother wafer/die (b)

Electrical measurements have been conducted after assembly and underfill for all 3D stack constructions (10 pcs each). The measured electrical resistance values are shown in Figure 13. A comparative study using top dies with and without TSVs allowed us to determine the TSV resistance for one die stack layer containing 35.904 Cu TSVs (10x100µm) with 737.5 Ohm. This results in single TSV resistance of 0.020 Ohm. Considering the area resistance value of the plated Cu RDL metallization and the total trace length given we were also able to extract the single interconnect resistance with 0.003 Ohm. The statistical base of the TSV and interconnect resistances is very high due to the high number of such elements within the daisy chain which is further increasing with each additional die stack layer. This allows also a good estimation of second order effects like the TSV seed layer and other impacts. The extracted values are comparable with similar investigations results published /4/ where the resistance value for a TSV with 10µm diameter and 40µm depth was extracted with 0,0087 Ohm and the interconnect resistance of Cu/Sn micro-contact with 10µm diameter after solid liquid interdiffusion w/ 0,002 Ohm.

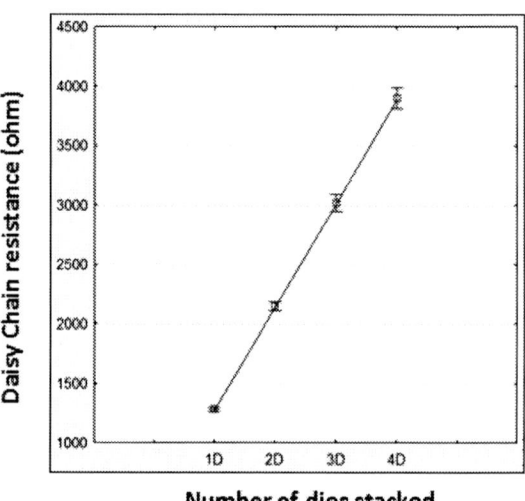

Figure 13: Daisy chain resistance plot for different die stack constructions

Conclusion

This investigation has shown that high-density interconnect arrays can be assembled by solder reflow. Due to the high number of interconnects included within one electrical daisy chain (e.g. 143.616 I/O for a 4L die stack) a good statistical base for the extraction of single resistance values for the TSV and interconnect was given. A critical issue was the decreasing assembly yield with increasing number of dies stacked. This was mainly caused by placement failures during flip chip bonding due to the realized pad shape degradation of the flip chip pad at the back side of the stackable die. For that reason we further optimized the back side integration scheme what has significantly improved the assembly yield.

Based on the results of this study we established some recommendations for 3D process integration schemes, wafer design and wafer processing:

1. The layer stack construction on wafer/chip front and back side should support stress compensation between both wafer sides to reduce the chip warpage. This can be done via an implementation of metal stress compensation elements and/or thickness optimization of the isolation layers on both wafer sides (especially for polymer layers). Thermo-mechanical simulation and Thermoiré measurements are effective methods to predict and proof the chip warpage under temperature exposure.

2. All dicing streets should be designed in such a manner that a small part of the kerf (15-20µm) is free of metal structures on wafer front and back side and free of oxide and nitride at least at one wafer side. This is to support laser stealth dicing in achieving smooth cutting lines.

3. Flip chip micro-bumps should be preferably allocated at the wafer back side of 3D wafers what prevents wafer flipping during carrier debonding and provides the debonded wafer in micro-bump bottom-up position to the wafer singulation. Micro-bumps on wafer back side allow also the application of pre-applied underfill on carrier level before debonding.

4. Flip chip pads should be preferably allocated to the wafer front side. The goal is to manufacture low-topology pads with optimized surface shape pending on the assembly technology applied. Low-topology pads reduce also the waferbond adhesive thickness in temporary bonding. This has a positive impact on the heat dissipation during DRIE, CVD and PVD and improves also the outgoing quality in waferbond adhesive cleaning after debonding.

5. The definition of the under-bump metallization depends on the application requirements (e.g. electro-migration, intermetallic growth etc.) as well as the assembly technology to be used. Good investigation results and recommendations were published in /5/.

6. The flip chip pad finish becomes very important when thermo-compression bonding (TCB) gets applied. It has to be ensured that the contact of the heated bond tool with the flip chip pads at the chip backside doesn't result into contamination, stains or discoloration what may prevent solder wetting during reflow or TCB at the flip chip pads located on the back side of the die to be assembled. Unprotected Cu pads should be avoided.

This investigation will be continued with focus on the interconnect resistance after TCB. This requires a modification of the micro-bump and flip chip pad in such a manner that both interconnect parts will be furnished with a thinner Sn cap to reduce the solder excess during TCB. Goal is to establish a pure Cu_3Sn intermetallic compound after bonding either as direct result of the TCB process or after additional anneal. This will allow a comparison of the interconnect resistivity's established in solder reflow vs. transient liquid phase bonding. Reliability tests will be conducted to study the behaviour of the daisy chain resistivity.

Acknowledgments

Special acknowledgement is given to several scientists and technical assistants of the IZM-ASSID as there are Frank Windrich, Mario Schima, Alexander Wollanke, Steffen Mimietz and Michael Lorenz for their contribution to these investigations as well as to Juliane Krause for her excellent support in physical analysis.

References

1. S. Dobritz et al. "3D Integration - Technology and Test Strategy", *oral presentation at Smart System Integration International Conf, (SSI)*, Zurich, Swizerland, March 21 – 22, 2012

2. C. L. Yeh et al. "Micro-bump Bondability Design Guidelines for High Throughput 2.5D & 3D IC Assemblies" in *Proc. IEEE Electronic Components and Technol. Conf. (ECTC)*, Las Vegas, NV, May 28–31, 2013, pp. 897-903.

3. P. J. Tzeng et al. "Process Integration of 3D Si Interposer with Double-Sided Active Chip Attachments" in *Proc. IEEE Electronic Components and Technol. Conf. (ECTC)*, Las Vegas, NV, May 28–31, 2013, pp. 86 -93.

4. Y. J. Chang et al. "Electrical investigation and Reliability of 3D Integration Platform using Cu TSVs and Micro-Bumps with Cu/Sn-BCB Hybrid Bonding" in *Proc. IEEE Electronic Components and Technol. Conf. (ECTC)*, Las Vegas, NV, May 28–31, 2013, pp. 64 -70.

5. H. Y. Son et al. "Reliability Studies on Micro-Bumps for 3-D TSV Integration" in *Proc. IEEE Electronic Components and Technol. Conf. (ECTC)*, Las Vegas, NV, May 28–31, 2013, pp. 29 - 34.

Wet Silicon Etch Process for TSV Reveal

Laura B. Mauer, John Taddei, Ramey Youssef
Solid State Equipment, LLC
185 Gibraltar Road
Horsham, PA 19044 USA
Yongqiang Lu, Sian Collins, Kevin McLaughlin, Craig Allen
SACHEM, Inc.
821 E. Woodward St, Austin, TX 78704, USA

Abstract

This paper presents a wet process as a simple and cost-effective alternative to the polish/plasma etch TSV reveal process. By combining silicon thickness measurement, wet etch, and cleaning in a single-wafer process system, this platform provides a low cost-of-ownership solution for TSV reveal. The process uses a wet etch chemistry with a fast etch rate and high selectivity, in a single-wafer process tool. The new selective etch chemistry improves the etch rate by 50% or more over traditional Si etchants currently used in the industry, such as tetramethylammonium hydroxide (TMAH). This new etch chemistry also has high silicon-etch selectivity over the oxide liner and Cu, with etch rate (ER) ratios greater than 10,000 and 1000, respectively. TMAH is not a component in the chosen chemistry because of safety concerns specifically related to TMAH toxicity.

Variations in the depth of the Si overburden occur due to non-uniformities in post-grind thickness, via depth, and bonding. To compensate, an algorithm is used to control etch profiles. Integration of wafer thickness measurements before and after etching–within the single-wafer equipment–provides the high-accuracy process control needed for high-volume manufacturing. Improvement in surface roughness and etch uniformity are achieved with this wet process through the combination of chemistry performance and process optimization.

Introduction

Mobility and performance demands from semiconductor end users have continually driven the semiconductor device geometry to smaller dimensions. The same pressure has also resulted in many innovations from the semiconductor packaging industry. One of these innovations is the Through Silicon Via (TSV) 3D packaging technology. Through Silicon Via has become the key enabling technology in 3D packaging by reducing interconnect length to increase device speed, and by increasing interconnect density to reduce the package form factor. There are three different integration schemes with the TSV process: via-first, via-middle, and via-last. The via-first process forms the TSV in the substrate silicon before the front-end process. The via-middle process forms the TSV at the front end or interconnect steps with the regular wafer process flow. The via-last process makes the TSV from the backside of the wafer after completing the BEOL processing. In via-first and via-middle TSV integration flows, Si wafers must be thinned from the backside to reveal the Cu vias for the wafer to make contact with another wafer or chips. Typically, this thinning is accomplished by grinding the back side of the wafer, CMP polishing to remove the subsurface damage and to eliminate stress in the wafer, then etching with a plasma or wet process to reveal the Cu vias [1]. The CMP process involves using expensive slurries and critical post cleaning steps to remove the slurry particles and other introduced contaminants. Dry etch processes usually require expensive equipment and etching gases. On the other hand, wet chemicals such as KOH and TMAH have been used as cost-effective wet etch alternatives for plasma etching to reveal the TSV [1, 2]. The issue with the KOH process is that KOH adds metal ion (K^+) contamination on wafer surfaces. Typically, a cleaning process is required after the KOH etch, to remove the residual K^+. The additional cleaning process reduces the tool throughput and, therefore, is not desired in mass production. Tetramethylammonium hydroxide (TMAH) has been used to replace the KOH in TSV reveal wet processing to eliminate metal contamination. However, TMAH is toxic. Some semiconductor fabs try to avoid it whenever possible. This paper discusses the results of TSV reveal etch using SACHEM's proprietary Si etchant, SMC6-42-1, that does not contain TMAH or metal-containing (inorganic) hydroxide.

Experimental
Coupon Tests

Coupons of about 20x20 mm made from P-type single-crystal Si wafers with [100] orientation were used in all lab tests. A coupon was premeasured for surface area, pretreated with 2% HF to remove native oxide, and preweighed. The coupon was then submerged in an etch solution in a PTFE beaker for a specified time. Temperature was controlled at 75°C for all tests unless otherwise indicated. After the specified etch period, the coupon was removed from the etchant and immediately rinsed with DI water and then IPA. The coupon was then dried using N_2, and weighed to calculate the total etch amount and etch rate.

Wafer Tests

Ground wafers and Si test wafers (all 300mm) were etched on a commercial-grade SSEC 3300ML single-wafer process tool to verify the lab results. Process and equipment parameters were developed for optimum etch rate, surface roughness, and surface defects. The etch amount was determined using an ISIS StraDex f2-300 IR sensor by measuring the pre- and post-etch wafer thickness. Surface roughness was measured using a KLA P16 surface profilometer, and reviewed with a Veeco Icon AFM. Surface defects and conditions were reviewed using a Hitachi S-3700N SEM. Finally, production TSV wafers were processed under the optimized chemical and equipment conditions determined using test wafers.

Results and Discussion
Etch Rates and Selectivities

For a successful wet process, a high silicon etch rate is essential because it determines the throughput, one of the key cost factors for the process. Table 1 compares the [100] Si wafer, thermal oxide (Tox), and Cu (sputtered film) etch rates from beaker tests in the SACHEM R&D lab. Hydroxide ion, [OH⁻] is believed to be the dominant active silicon etch species in strong base solutions through the following general silicon etch reaction [3-6]

$$Si + 2OH^- + 2H_2O \bullet \rightarrow Si(OH)_2O_2^{2-} + 2H_2$$

The [OH⁻] concentration was kept at the same level (<2M). Temperature was maintained at 75-80°C. As the table indicates, the SMC6-42-1 etch rate is about 2x and 4x faster than TMAH and KOH, respectively, at the same molar concentration. It should be mentioned that KOH etches Si (100) faster at higher concentrations, such as 2.2~3.6M (12.5-20%), as reported in the literature [3-5]. Detailed discussion of anisotropic Si etch using strong bases is outside the scope of this paper and can be found in other references [3-6].

It is critical that, during the reveal etch, the etchant does not significantly etch the oxide liners that insulates plated copper vias from the bulk silicon in the TSV. This requirement is the main reason that the industry prefers strong base etchants to acidic etchants based on HF, such as HF/HNO₃.

The data from Table 1 suggest that SMC6-42-1 has about the same Tox etch rate as TMAH and KOH, but 2x the selectivity due to a high Si etch rate. Cu etch rates among these etchants are comparable and at a low level of 1nm/min. Under normal circumstances Cu vias are not exposed during the reveal etch since the vias are covered with oxide. Therefore the Cu etch rate is not as critical as the oxide etch rate.

Table 1. Comparison of Si, Tox, and Cu etch rates in beaker tests

Etchant	Si (100) ER (nm/min)	Tox ER (nm/min)	Cu ER (nm/min)
SMC6-42-1	1600	0.11	1.05
TMAH	800	0.1	0.9
KOH	363	0.1	1.0

With the promising beaker test results, the new Si etchant was tested using an SSEC single-wafer processor on regular test wafers and ground wafers. The ground wafers were thinned by Strasbaugh using the same grinding parameters as for real TSV device wafers. Therefore, the surface roughness, waviness, texture, and thickness variations are the same as they would be for TSV wafers. Table 2 summarizes the etch data generated in the SSEC single-wafer tool.

Table 2. Ground wafer ER in a commercial-grade SSEC single-wafer tool

Etchant	Ground Wafer ER (nm/min)	Tox ER (nm/min)	OH Conc (M)	Temp (°C)
SMC6-42-1	800	0.12	1	95
TMAH	260	0.1	2.8	80
KOH	600	0.1	4.2	72

The data in table 2 indicate that the SMC6-42-1 etches Si significantly higher than TMAH and KOH at the conditions tested. It is especially interesting that the higher silicon etch rate of SMC6-42-1 was achieved with only 1/3 and 1/4 respectively of the TMAH and KOH molar concentrations.

Significant ER differences were observed between SMC6-42-1 in the commercial tool and in a beaker test. This is probably due to the temperature drop of the etching solution when spreading to a thin film on the wafer surface. In a lab setup, a small coupon is submerged in a relatively large amount of solution kept at constant temperature.

Surface Roughness

Surface roughness affects film deposition following the TSV reveal process and should be tested and tightly controlled in high-volume production. Therefore surface roughness (pre and post etch process) was measured for all wafers tested on the commercial grade SSEC single-wafer processor using an AFM and/or a profilometer. Figure 1 shows the surface of a TSV wafer after grinding and after SMC6-42-1 etch.

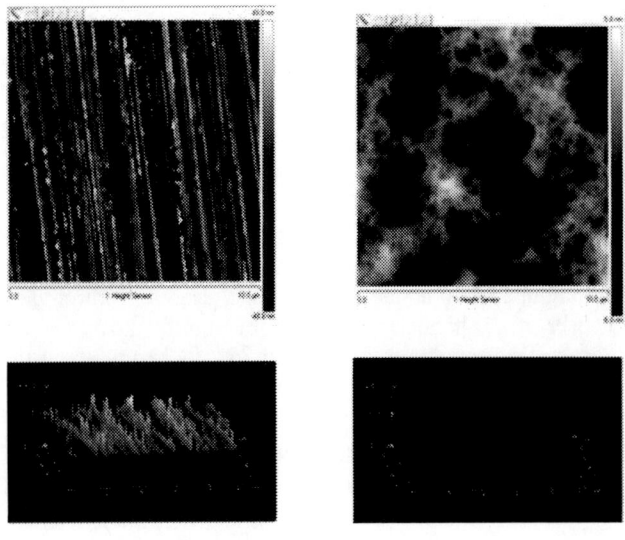

Figure 1. AFM images of TSV wafer surface post grind (*left*) and post 10-μm etch (*right*)

As indicated in Figure 1, post-grind TSV wafers appear rough, with visible grinding marks of maximum peak height up to 50nm. After a 10 µm chemical etch the maximum peak is below 15nm. Most of this roughness is removed by the chemical etch. Here, the chemical etch process applied was a two-step approach. First, the wafer was etched using HF/HNO_3 that etches silicon isotropically at a very fast etch rate, up to 10-µm/min., depending on the amount of silicon to be removed. The HF-based chemical cannot be used to reveal the TSV because it would etch the SiO_2 liner and the Cu in the vias. The isotropic etch is used to quickly smooth out the peaks and valleys created during the grinding process. An anisotropic etchant has to be used to finish etching the remaining silicon to reveal the TSV vias. As is well documented [4-6], anisotropic etching leaves well defined pyramids and pits on single crystalline silicon surfaces. Figure 2 shows an example of such defects. SMC6-42-1 is specially formulated to prevent pit and pyramid formation.

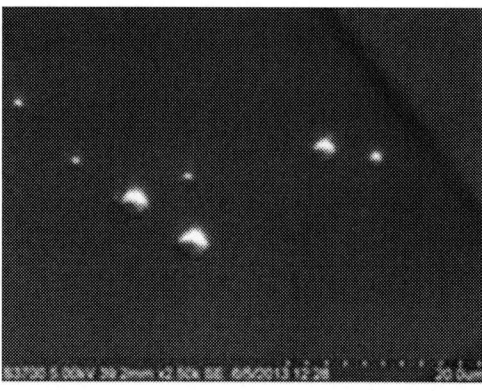

Figure 2. Surface pits and pyramids on a TSV wafer after anisotropic etch

Two wet chemical TSV reveal schemes were tested with the special formulation of SMC6-42-1. First, wafers were ground then CMP polished, and etched using SMC6-42-1. Second, wafers were ground, then etched using HF/HNO_3 (isotropic) and followed by the anisotropic chemical etching process using SMC6-42-1. Both the isotropic and anisotropic etches were conducted in the SSEC single wafer tool. Pre and post etch wafer surface roughness was measured using a profilometer and reported in Table 3. As the data demonstrated, for the Grinding/CMP/Etch process, the chemical etch increased the wafer surface roughness (Ra) from about 15Å to about 23Å. For the grinding/etch process however, the chemical etch process significantly reduced wafer surface roughness from about 75Å to 22Å, about the same with that obtained from the grinding-CMP-etch process.

Surface Damage and Stress Removal

As mentioned in the introduction section, one function of CMP or wet chemical process is to remove the mechanical subsurface damage and release subsurface stress caused from the grinding process. TEM was used to image the cross section of the wafers post grinding and post two step wet chemical etch. The TEM results are presented in Figure 3a

and b. Figure 3a is the TEM image of a wafer cross section close to the surface. The image clearly shows the subsurface damages up to about 100nm deep. The subtle contrast change from the surface to the bulk suggests stress existing at the surface about 300nm into the bulk. After the two step wet chemical etch the subsurface damages and stress are removed as the TEM image shows in Figure 3b. This TEM and the roughness data demonstrate the CMP process step can be replaced with a wet etch process.

Table 3. Surface roughness using grind/CMP/Sachem etch compared with the grind followed by two-step wet etch process (HF/HNO_3 + SMC6-42-1)

Wafer #	Process	Ra Pre (A)	Ra Post (A)
1	Grind/CMP/Etch	19	23.55
2	Grind/CMP/Etch	12.5	24.85
3	Grind/CMP/Etch	14.9	26.25
4	Grind/CMP/Etch	11.55	18.05
5	Grind/Etch	86.9	22.25
6	Grind/Etch	62.4	21.65
7	Grind/Etch	73	27.5
8	Grind/Etch	77.5	16.45

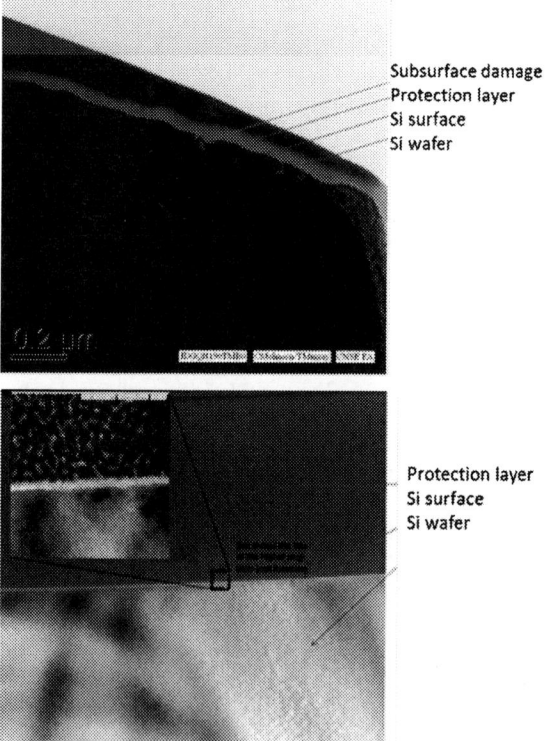

Figure 3 a. TEM image of a ground wafer cross section close to the surface. **b.** Cross section TEM image of a ground wafer after the two-steps wet etch (~10µm Si removal)

TSV Wafer Results

TSV production wafers were received after the grinding process. The wafers were processed using the two step chemical etch using HF/HNO3 and SACHEM SMC6-42-1. Wafers were successfully processed on the SSEC single-wafer tool with the optimum known parameters. The TSVs are revealed cleanly and the wafer surface is smooth, as shown in Figure 4.

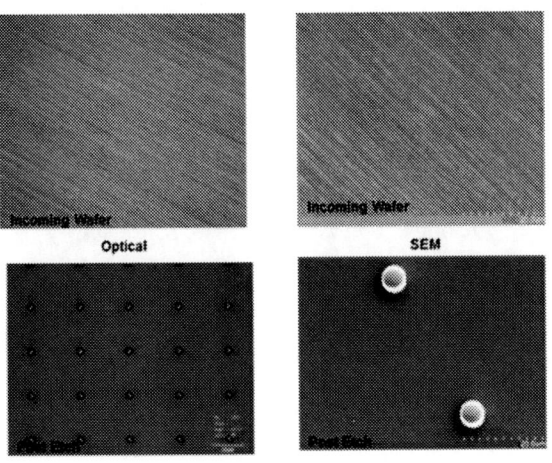

Figure 4. TSV production wafers pre- and post-reveal-etch

Toxicity of the Chemicals

SMC6-42-1 is formulated for fast etch rate and smooth surface finishing without the use of TMAH. The composition is listed in Table 4. Its toxicity data for rat dermal exposure, using LD50 as the indicator, is presented in Table 5.

Table 4. SMC6-42-1 Components

Component	Conc (wt %)	Purpose
Organic Base (Non-TMAH)	8-25	etchant
Organic additive	<1	Etch enhancement
DI Water	74-91	

Table 5. Comparison of LD50 for SMC6-42-1 and TMAH

Component	LD50 (mg/kg)
SMC6-42-1	1000
TMAH	157

LD50 is the dosage for 50% or more of test rats to survive after skin contact with the test chemicals in a given time. Generally, the higher the LD50, the less toxic the chemical. As indicated in Table 5, SMC6-42-1 LD50 for rat dermal exposure is 1000mg/kg and more than 5 times higher than that of TMAH.

Etch Profile Control

Post-grind TSV wafers have significant thickness variations. These variations or non-uniformities may come from the grinders used and/or from thickness variations of the adhesive layer used for mounting the device wafer to a carrier. These non-uniformities are often radial in nature, due to the process that caused them, such as the top curve shown in Figure 5. In this case, the wafer has a center and edge thicker profile. The incoming thickness variations could cause serious issues such as unrevealed vias. To address this non-uniformity issue, the SSEC single-wafer process system integrates a wafer-thickness measurement sensor in the system–the ISIS StraDex f2-300. The sensor is incorporated in a separate chamber in the system design, thereby eliminating the need for off-line thickness metrology. The ISIS StraDex f2-300 sensor uses spectral coherence interferometry at a 1300-nm wavelength to obtain thickness measurements. The control system utilizes measurements taken across the diameter of the wafer. A feed-forward control system uses pre-processing thickness measurements to adjust the radial etch depth to compensate for incoming thickness variation. In addition, a feed-back control mechanism utilizes thickness measurements taken post processing to adjust etch times for the subsequent wafers, thus addressing variations in etch rate. This closed-loop process control is especially important for high-volume manufacturing. Figure 5 (bottom curve) shows the actual measurement results on a TSV wafer.

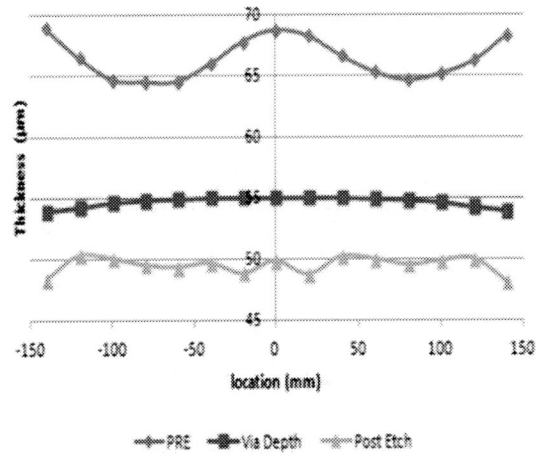

Figure 5. On-line-measured pre- and post-etch TSV wafer thickness

As shown in Figure 5, by measuring the incoming wafer thickness (PRE) and forward feeding this data to the control algorithm the single-wafer process tool automatically adjusted the process parameters to compensate. The result is a reduction in the post etch thickness variation and uniformly revealed TSVs to within ±1.0μm of the desired reveal height. The capability of measuring incoming, and controlling post-etch, wafer thickness and radial profile provides a great advantage in high-volume production. More-detailed discussion can be found in a previous publication by Mauer et al [7].

It is important that the etchant does not preferentially attack along the sidewall of the vias as the etch reveals the TSV. Figure 6 shows a fractured cross-section of a TSV that has been revealed. The oxide liner and barrier metal are intact. No dishing is observed at the intersection between the silicon and TSV.

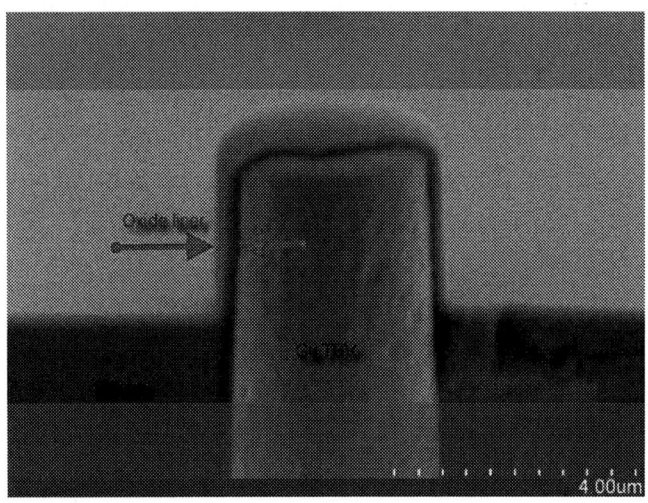

Figure 6. SEM cross section of revealed TSV

Conclusions

We have demonstrated a wet etch process as a simple and cost-effective alternative to the CMP/plasma etch TSV reveal process. The alternative process uses a wet etch chemistry with a fast etch rate and high selectivity, within a single-wafer process tool. The new selective etch chemistry improves the etch rate by 50% or more over traditional Si etchants currently used in the industry. SMC6-42-1 offers the added benefits of either reducing toxicity or eliminating metal ion contamination, depending on the alternative chemistry. By combining silicon-thickness measurement and wet etch in a single-wafer process system, this platform provides the lowest cost of ownership and excellent process control capability for high-volume production.

References

1. Olson, S. and K. Hummler. 2012. "TSV Reveal Etch for 3D Integration." Paper presented at Proceedings of 3D Systems Integration Conference, 2011 IEEE International, Osaka, Japan.

2. Mauer, L. et al. 2012. "Silicon Wafer Thinning to Reveal Cu TSV." Paper presented at Proceedings of IMAPS Device Packaging Conference, Fountain Hills, AZ.

3. Sato, K. et al. 1988. "Characterization of Orientation-Dependent Etching Properties of Single-Crystal Silicon: Effects of KOH Concentration," *Sensors and Actuators* A 64, 87-93.

4. Wind, R. A. and M. A. Hines. 2000. "Macroscopic Etch Anisotropies and Microscopic Reaction Mechanisms: a Micromachined Structure for the Rapid Assay of Etchant Anisotropy." *Surface Science* 460 21–38.

5. Alvi, P. A. et al. 2008. "A Study on Anisotropic Etching of (100) Silicon in Aqueous KOH Solution." *Int. J. Chem. Sci.*: 6(3), 1168-1176.

6. Wind, R. A. et al. 2002. "Orientation-Resolved Chemical Kinetics: Using Microfabrication to Unravel the Complicated Chemistry of KOH/Si Etching." *Journal of Physical Chemistry B* 106, 1557-1569.

7. Mauer, L. et al. 2013. "Silicon Etch with Integrated Metrology for Through Silicon Via (TSV) Reveal." Paper presented at 3D Systems Integration Conference (3DIC), 2013 IEEE International, San Francisco, CA.

Advanced Wafer Bonding and Laser Debonding

P. Andry, R. Budd, R. Polastre, C. Tsang, B. Dang, J. Knickerbocker and M. Glodde
IBM T.J. Watson Research Center
1101 Kitchawan Road, Yorktown Heights, NY 10598, USA

Abstract

This paper describes the development of a wafer debonding process and tooling based on 355 nm UV laser ablation. While laser–assisted debonding of polyimide-based materials at shorter UV wavelengths has been described previously, this work describes a method having two major advantages over earlier methods: 1) a significantly more compact and affordable diode-pumped solid state 355 nm laser source is combined with a high-speed optical scanner to create a rapid debonding module with a small footprint, and 2) the addition of a very thin UV ablation layer to the glass handler ensures that release will occur, independently of the adhesive used to bond the wafer. The first enhancement is designed to enable high-throughput debonding at lower cost than earlier laser tools, while the second enhancement is designed to greatly expand the adhesive choices available to device manufacturers. Flexibility in material choice for both the release layer and the adhesive layer permits the manufacturer to meet post-wafer thinning process and temperature compatibility needs. In this research paper, the many benefits of this novel room-temperature debonding technology are reported along with examples of successfully demonstrated adhesives and release layers.

Introduction

Temporary wafer bonding and debonding are among a handful of key enabling processes in the world of 2.5D and 3D technology. There exist a number of approaches to the debonding of thinned device wafers: they may be released by exposure to chemical solvents delivered through perforations in the handler, by mechanical peeling from an edge-initiated separation point, or by heating the adhesive to the point where the silicon device wafer may be removed by sheering or peeling. Room-temperature debonding techniques, which include laser-assisted debonding as well as mechanical peeling [1] appear to be gaining wider acceptance than other methods due to their compatibility with standard dicing tape frame mounting and materials. Ultraviolet (UV) laser ablation using an excimer source in combination with an x-y scanning stage has previously been demonstrated to effectively debond wafers that have been bonded using polyimide-based adhesives [2]. At the same time, a number of mechanical debonding approaches have arisen that rely on controlled peeling of a handler from a thinned wafer, for example by engineering the handler to have a central low-adhesion zone, or by the application of a special release layer to the wafer before bonding [3].

In spite of the advances that have been made in temporary bonding / debonding, both in materials and tooling, there exists at this time a general consensus that this area of 3D enablement is lagging other key areas (e.g. TSV etch, fill and integration) and contributing to delays in the rollout of 2.5D /

3D products due to insufficient performance in throughput, yield and total cost [4]. Thus, there remains opportunity for improvement in debonding methods and tooling which address these issues. Herein we describe such a process based on 355 nm laser ablation of a thin interfacial layer placed between the bonding adhesive and a glass handler.

Overview of the Process

The debonding process is illustrated schematically in Fig.1 which shows the bonded interface between a glass handler and a 3D silicon device wafer. A thin ablation layer directly adjacent to the glass is rapidly decomposed by the rapidly scanned laser spot, and effects release of the glass leaving the bonding adhesive on the 3D device wafer. The glass, the ablation layer and the adhesive each have important roles to play. The silicon-CTE matched glass must easily transmit the UV wavelength chosen for maximum efficiency of release.

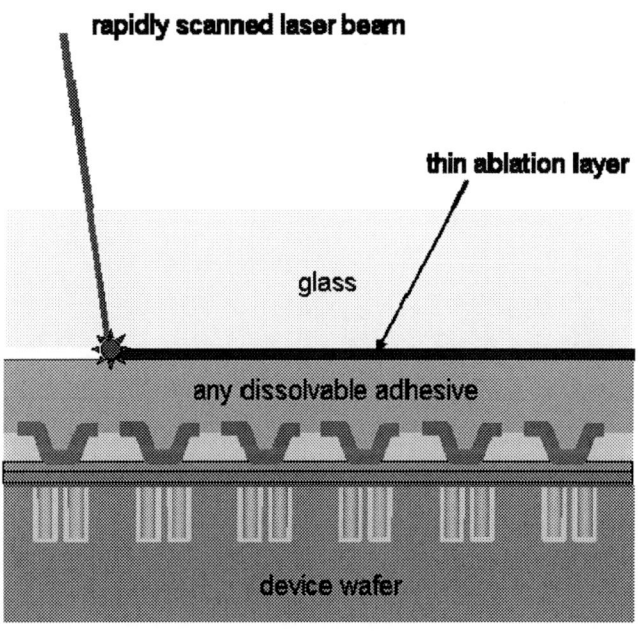

Figure 1. Schematic of a bonded wafer interface undergoing debonding by UV laser ablation.

The ablation layer must bond well to both glass and adhesive, it must have good chemical and thermal stability to withstand bonding and downstream processing, and it should be very sensitive to the UV laser wavelength so that it will cleanly and rapidly decompose. For ease of 3D process development, especially in the area of defect inspection, it is highly advantageous for the ablation material to be thin enough for direct optical viewing of the bonded interface throughout 3D processing. The bonding adhesive can be any suitable bonding adhesive the application requires, subject to the condition that it must be able to be cleanly removed form the device wafer after ablation. This condition favors adhesives that are typically dissolved in one of a number of cleaning solvents.

Wavelength Selection

Three primary criteria were considered in determining the best wavelength for this application. The first was transmission of the glass handler as mentioned above; the second was availability, cost and characteristics of the laser source; the third was availability and sensitivity of suitable ablation materials. Fig. 2 shows the transmission characteristics of a 0.7 mm sheet of typical silicon CTE-matched wafer glass. Three common UV laser wavelengths are marked on the graph at 355 nm, 308 nm and 266 nm. Of these, only the 355nm wavelength shows no absorption loss, having the same 92% transmission as the visible spectrum (4% Fresnel reflective loss at front and back surfaces). Transmission at 308 nm is about 83% indicating very slight absorption, while transmission at 266 nm is barely 20% indicating very strong absorption. Since the goal is to transmit as much laser radiation as possible through the glass to the ablation layer, the results show that 266 nm is a poor choice.

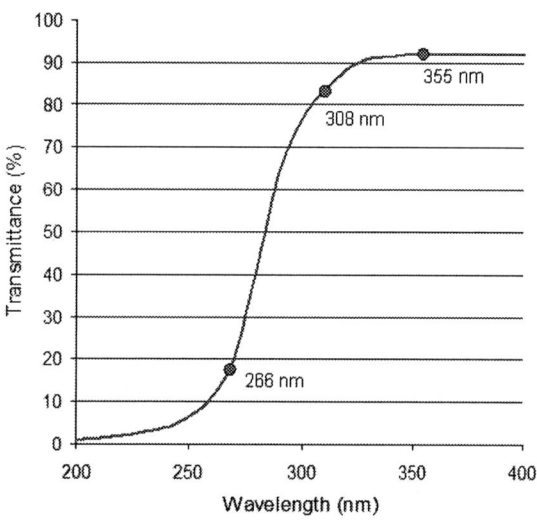

Figure 2. UV transmission of typical glass handler.

We have previously reported good debonding of glass coated with a polyimide-based adhesive using the 308 nm wavelength from a XeCl excimer laser [2]. In that work, a portion of the adhesive itself served as the ablation layer, and a relatively high-power line beam (~50 mm wide) was scanned across the wafer using a moving stage. In the present work, we have decoupled the absorption layer from the adhesive, giving the process integrator much more latitude in choosing the proper adhesive for a specific application. Furthermore, we have chosen a much more compact diode-pumped solid state (DPSS) 355 nm laser source. While excimer lasers have large output powers, they are also physically quite large, have complicated waveguides for steering of the beam, and require gas cabinets to house the

halogen gas mixtures with which they are charged. In contrast, DPSS sources are physically much smaller, with simple rack-mounted power supplies, and fiber optic delivery of the fundamental 1064 nm laser output to a compact laser head containing the third harmonic generation crystal which "triples down" frequency to 355 nm. While the maximum output power is much lower than an excimer laser, the output beam produces a slightly elliptical Gaussian spot which can be quickly rastered across the wafer surface using an optical scanner.

Laser Debonding System

The experimental laser ablation system developed for debonding is illustrated in Fig. 3. The system is comprised of a number of critical components including: the 355 nm DPSS Q-switched laser, a high-speed optical scanner, an F-theta lens and a PC containing appropriate control software and an interface card. The laser chosen for this work was puts out a maximum power of just under 6 Watts at a 50 kHz repetition rate and a ~12 ns pulse width. The 1 mm nominal width beam enters a beam expander, and the variably expanded beam is fed into a high performance optical scanner. The scanning output beam is focused by an F-theta lens onto the wafer placed ~800 mm below. The laser and the scanner are PC-controlled by means of an interface card and SCANLAB's laserDESK software. The system was designed to be very flexible, and can raster scan spot sizes on the order of one to several hundred microns at speeds in excess of 10 m/s.

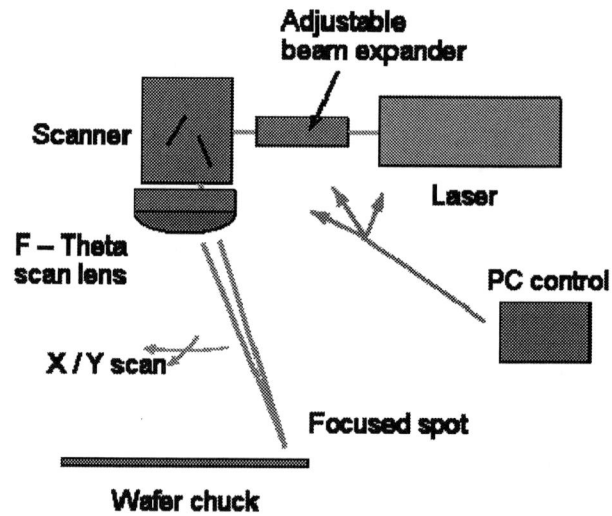

Figure 3. Schematic of the laser debonding system.

Knowledge of the effective spot dimension at the wafer surface is important in selecting the proper scanning parameters for uniform ablation and release of the handler. However, unlike other laser systems having line beams with a uniform "top hat" cross-section, the system described here produces a slightly elliptical spot having a Gaussian cross-section. We have chosen to define the effective spot size as the Full Width Half Maximum (FWHM) point because it can be measured in a fairly straightforward way. Assuming a Gaussian distribution, a spot size defined by its FWHM captures ~50% of the total pulse power within the area of the spot. Figure 4 illustrates schematically how knowledge of the spot size along with judicious choice of laser power and scanning parameters allow a uniform ablation fluence to be delivered to the bonded interface. For illustrative purposes a normalized Gaussian of sigma = 1 is used. The scanning direction is left to right and back in a serpentine pattern.

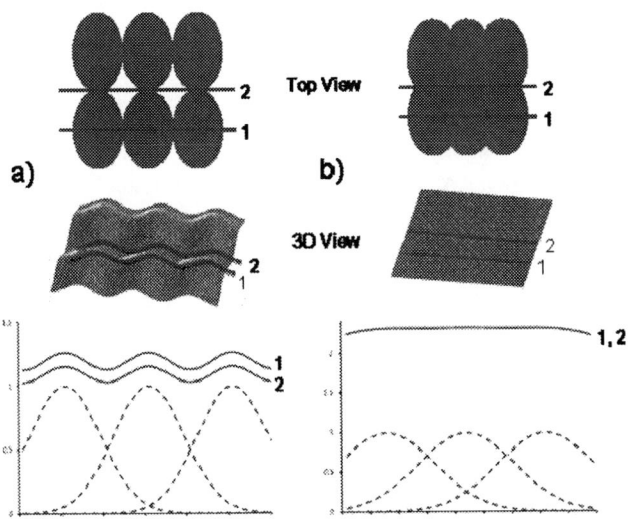

Figure 4. Schematic of energy delivered by scanning laser system under two different conditions: a) setting scan speed and row pitch for zero overlap of laser pulses, b) setting the same parameters for ~30% overlap in both axes.

Figure 4a) shows an elliptical spot with a height to width ratio of 3:2, a row pitch equal to the spot height, and dot pitch equal to the spot width. Consider a FWHM spot width = 100 um. For a 50 kHz laser repetition rate, the dot pitch of 100 um shown in Fig. 4a) would result from a scanning speed selection of 5 m/s; row pitch would be set at the spot height of 150 um. The additive pulse energy impinging on the wafer surface along lines connecting the horizontal spot centers and between spot rows is illustrated in the dot pattern above and shows that the energy ripples up and down by about 14% of peak pulse power along a line and by about 26% across the wafer overall. With laser power set such that the ablation material threshold is just met along the centerline of each scanned row, it is quite obvious that this pattern is sub-optimal and may result in complete ablation of the interface. The diagonal areas between the pulses will see a notable reduction in energy, and will remain intact. Note: jogging the spot position has a negligible effect.

Fig. 4b) illustrates an approach to improve exposure uniformity by adjusting the scanning parameters to achieve a certain degree of overlap between the pulses along both scan axes. A spot overlap of ~30% is illustrated in the figure. This corresponds to a scanning speed of ~3.5 m/s and a row pitch of ~105 um. The net additive energy is more than twice as high as the peak pulse energy at any spot center, but moreover, the ripple is nearly eliminated and a uniform

fluence is achieved. As long as the minimum ablation threshold fluence set by the laser peak power is met at all locations, the handler will release. The larger the output power of the laser, the larger the spot size can be made, the larger the row pitch can be set, and the faster the beam can be scanned (up to the speed limit of the scanner) resulting in maximum throughput of the debonding system. Apart from the system, there is really only one other noteworthy parameter: the ablation threshold fluence of the release material itself.

Choice of UV Ablation Materials

Based on previous experience using a 308 nm laser and UV-absorbing materials on glass handlers [2] it is known that debonding by UV ablation begins at a minimum threshold fluence controlled primarily by the absorption and thickness of the material being ablated. As mentioned above, the material must adhere well both to the glass handler and the bonding adhesive, it must be thermally stable up to at least the maximum bonding temperature, it must be highly absorbing at 355 nm, and for ease of inspection of the bonded interface throughout 3D wafer processing, it should be as transparent as possible at optical wavelengths. At first this might seem to be a fairly stringent set of requirements, but it so happens that there exists a class of materials used in advanced deep-UV lithography known variously as inert underlayer (UL) films, spin-on carbon (SOC) films and organic planarizing layers (OPL), and while originally designed for an entirely different purpose [5], many of these do meet the requirements.

OPL materials are produced by a number of suppliers, and they typically are used to planarize pre-existing patterns to enable lithography of the subsequent level [5]. Their resistance to reactive ion etching is used to transfer very fine patterns in photoresist to the substrate. They are typically spin-applied to a thickness on the order of 200 nm, and simply baked on a hotplate, yielding optically transparent films. Most are thermally stable up to 300C and several are stable above this temperature. In order to be useful in this debonding application, however, it was necessary to gauge if these films are likely to rapidly decompose under laser irradiation at 355 nm, and a good indicator of this is their UV absorption. Thin films of several OPLs were spin-applied to handler glass, cured at 250C and 300C, and measured using an Ocean Optics fiber optic spectrometer. The transmission results are shown in Table I.

Table I. 355 nm transmission characteristics for OPL ablation materials on glass.

	355 nm Transmission		Film thickness
	250 C cure	300 C cure	(nm)
air	100%	100%	n/a
0.7 mm glass	92%	92%	n/a
glass/OPL_A1	85%	86%	200
glass/OPL_A2	90%	88%	200
glass/OPL_A3	82%	77%	200
glass/OPL_A4	16%	15%	260
glass/OPL_B1	68%	66%	120
glass/OPL_B2	31%	29%	200

As noted previously, bare glass has 92% transmission relative to air. From the data, it can be seen there is a wide variation in the 355 nm absorption of the OPL materials studied. With the exception of one material (OPL_A3) the absorption did not appreciably change with an increase in curing temperature from 250C to 300C, owing to the inherent thermal stability of these materials. One material from each supplier stood out as top candidates based on their high level of UV absorption: OPL_A4, where a 260 nm film on glass transmitted only ~15%, and OPL_B2, where a 200 nm film transmitted ~30% of the incident beam.

Bonding & Debonding Study

In order to verify compatibility with a range of bonding processes, samples of all ablation layer candidates were spin applied to glass samples and cured at 300C. Silicon wafers containing simple BEOL Cu damascene wiring patterns and a top PECVD passivation bilayer of silicon nitride / SiO_2 were coated with either an acrylic-based low-temperature adhesive (LTA), or a polyimide-based high-temperature adhesive (HTA). The wafers coated with the LTA were baked for 3 minutes each at 100C, 150C and 200C after which the samples of glass coated with ablation layer were bonded using a low bonding force at a temperature of 200C. Selected samples were also bonded to silicon wafers coated with the HTA which was cured in tube oven for 30 minutes at 350C. These samples were bonded using a fairly high force at a temperature of 280C. Following bonding, the bonded interface was inspected through the glass for any signs of voids or defects which might indicate instability / gassing of the thin ablation layer.

All samples were taken to the laser ablation lab for 355 nm debonding. Before beginning the ablation threshold tests, the scanned laser spot size was measured using a custom-designed spot size detector. The FWHM was measured both parallel and perpendicular to the scanning axis, and the values obtained were compared directly with the ablation spot imprints created in several of the ablation materials scanned at laser powers between about 0.7 Watts and 4 Watts. The elliptical spot size for all of these experiments was measured to be ~80 um wide and ~120 um tall.

Following the method outlined above for creating a uniform fluence across the sample surface, a scanning speed of 3 m/s was chosen giving a dot pitch of ~60 um for good horizontal overlap of each pulse. Row pitch was set to ~90um ensuring good vertical overlap. For each ablation material, a small area sample was used to first determine a minimum threshold energy at which a faint spot was barely visible. The laser power level was noted, and the level was then reduced and a bonded sample of that material was then completely scanned. If the glass could be easily separated, the threshold power was noted. If not, the power was incremented upwards until the glass could be easily removed. Figure 5 shows laser power setting required for glass release plotted against the % transmission for each ablation layer. As can be seen, the higher the transmission, the higher the laser power required to ablate the material, which makes sense since materials which are highly absorbing at 355 nm concentrate a much higher portion of the impinging UV radiation into the thin volume occupied by the ablation layer, thus enabling rapid decomposition. Materials such as OPL_A1 and OPL_A2 required ~2 Watts for release. Due to their fairly low UV absorption at 355 nm, they are not ideal candidates because a large proportion of the impinging radiation is transmitted through to the adhesive below, and if that adhesive is highly transmissive, that radiation can penetrate to the materials typically found at the surface of semiconductor device wafers, including polyimide and silicon nitride. Both of these materials will absorb 355 nm radiation, and can be damaged if the fluence is too high.

In contrast, the four other ablation layers tested all released at laser power levels proportional to their absorption. The highest performing ablation materials for each supplier were OPL_A4 which released cleanly at 0.76 Watts, and OPL_B2 which released at 0.97 Watts. For our laboratory laser system, we could easily double the spot size (allowing us to double both the scan rate and row pitch) to quadruple the throughput with these two materials, and still have room for further improvement before running out of available power. In a commercial debonding system, one would choose to run at a laser power on the order 80% of the maximum power level at all times to improve pulse uniformity and maximize throughput.

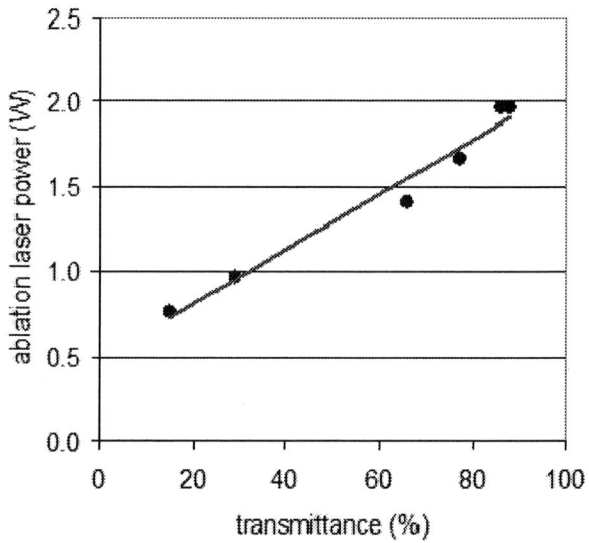

Figure 5. Ablation threshold for various ablation materials as a function of their 355nm transmission (elliptical spot size ~80 um x ~120 um, scan speed = 3 m/s, row pitch ~90 um)

Following the ablation trials, each of the debonded silicon samples was immersed in a cleaning solvent to remove thin ablation residue and the adhesive below. Each sample was dipped in room-temperature solvent for 3 minutes, and then rinsed with IPA and water, and blown dry with a nitrogen gun. Figure 6 shows the results before and after solvent cleaning for the two most absorbing ablation materials from each supplier. The additive spot pattern is clearly visible for each sample before cleaning, while the post-clean images are basically identical. No damage to the surface nitride/oxide layer was noted for any of the samples shown in Figure 6. Slight damage to the surface nitride layers was noted for samples bonded using the acrylic-based LTA after cleaning for the OPL_A1 and OPL_A2 materials only, due to the high UV transmission of both these ablation layers and the adhesive. No such damage was noted for these same two ablation layers when they were used with the polyimide-based HTA, due to absorption of UV in the HTA itself.

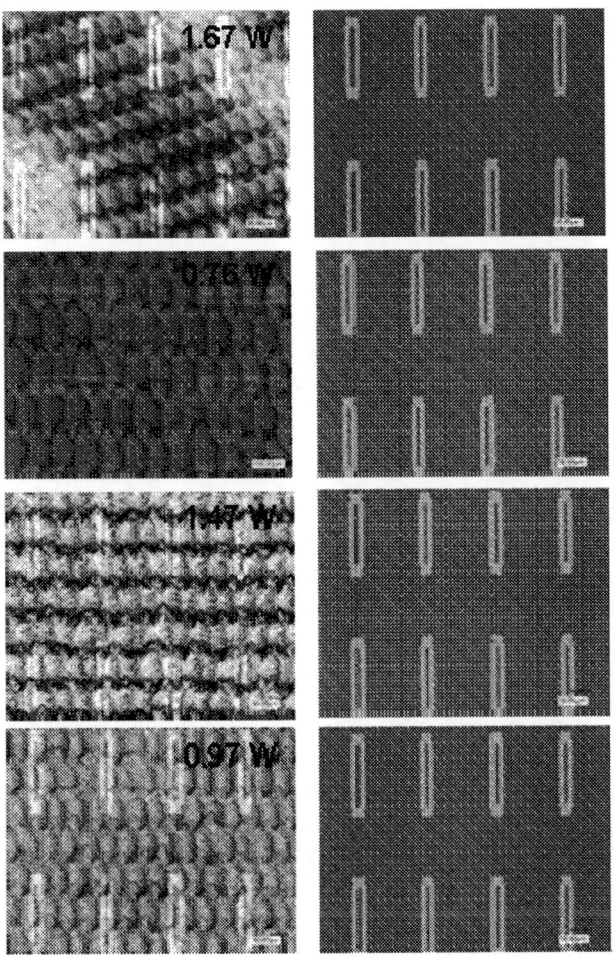

Figure 6. Images of laser debonded samples before (left) and after (right) solvent cleaning for four ablation materials. From top to bottom the materials are OPL_A3, OPL_A4, OPL_B1 and OPL_B2. The laser power threshold for release is shown for each material.

Conclusions

Advanced laser debonding at 355 nm is made possible by a rapidly scanning laser system operating in combination with a thin UV ablation layer applied to a glass handler before bonding. A compact diode-pumped solid state laser source combined with a high-speed optical scanner creates a rapid debonding module with a small footprint. Proper choice of system components and scanning parameters leads to a system capable of delivering a very uniform UV fluence at the bonded interface. Organic planarizing layer materials, commonly used in deep UV lithography, have interesting thermal and UV absorption properties which make several of

them excellent choices as ablation layers, even though they are very thin. The overall process is designed to decrease tool cost and enhance debonding throughput while expanding the adhesive choices available to device manufacturers.

Acknowledgments

The authors would like to thank the support of engineers from the Microelectronics Research Lab (MRL) and Central Scientific Services (CSS) labs at T.J. Watson Research.

References

1. M. Zorberbier and S. Lutter, "Latest Insights in Thin Wafer Handling Technologies," *Chip Scale Reveiw*, vol. 17, no. 4, pp. 54–57, July/August 2013.

2. B. Dang, P. Andry, C. Tsang, J. Maria, R. Polastre, R. Trzcinski, A. Prabhakar and J. Knickerbocker, "CMOS compatible thin wafer processing using temporary mechanical wafer, adhesive and laser release of thin chips/wafers for 3D integration," *Electronic Components and Technology Conference (ECTC), 2010 Proceedings 60th* , pp.1393–1398, 1-4 June 2010.

3. Brewer Science, Products, Temporary Bonding Materials "ZoneBOND Thin-Wafer Handling Technology" http://www.brewescience.com/products/temporary-bonding-materials/zonebond.

4. Jan Vardaman, "Progress on the Road to 3D and 2.5D: What's Next?" in *3D Architectures for Semiconductor Integration and Packaging (3D ASIP)*, Burlingame, CA, December 11–13, 2013, session 2 paper 4.

5. M. Glodde, S, Engelmann, M. Guillorn et al. "Systematic Studies on Reactive-Ion-Etched-Induced Defrmations of Organic Underlayers," in *Proc. of SPIE* Vol. 7972, pp. h797216-1 to 797216-1, 2011.

Versatile Thin Wafer Stacking Technology for Monolithic Integration of Temporary Bonded Thin Wafers

Thomas Uhrmann, Jürgen Burggraf, Julian Bravin, Viorel Dragoi, Markus Wimplinger, Thorsten Matthias, and Paul Lindner

EV Group

DI Erich Thallner Str. 1, 4782 St. Florian am Inn, Austria

E-mail: t.uhrmann@evgroup.com Phone: +43-7712-5311-0

Abstract

This paper will focus on recent results for wafer stacking of temporary bonded wafers for the integration in a monolithic device process. For ease of process integration, this process enables the face-to-back stacking of several device layers. Plasma activated fusion bonding could be shown to be an enabling step to lower annealing temperatures into a CMOS compatible range. Furthermore, plasma activation enables to use thermoplastic adhesives. Two types of test vehicles have been fabricated, showing on the one hand a successful stacking of a 11µm thin device wafer onto another thick substrate wafers. On the other hand, a triple stack of thick substrate wafer and two 20µm thin devices is shown as well. Bonding results have been measured using state-of-the-art measurement techniques, such as infrared scanning, scanning acoustic microscopy and scanning white light interferometry, to detect interface defects, bond integrity and temporary adhesive properties, respectively.

Introduction

3D-IC stacking utilizing through silicon via (TSV) technology is on progress to high volume production for few applications. Depending on the target application and functionality of the stacked IC layers, multiple TSV processing flows are considered. Hybrid bonding process flows are gaining considerable attention, as wafer bonding besides the mechanical connection simultaneously enables electrical connection between individual chip layers. On the other hand, via-last processes are frequently being used, where layer stacking by direct bonding is followed by a subsequent via processing. Here, restrictions apply in regards to chip arrangement, where face-to-face transistor bonding is not the most favorable chip layout [1,2].

A versatile 3D-IC thin wafer stacking technology for temporary bonded thin wafers will be presented. Following to front-side processing, passivation and several metal redistribution layers, wafer stacking by plasma-activated fusion bonding offers a permanent connection between individual layers with maximum temperature ranging typically between 200°C – 400°C. To enable face-to-back connection of the individual thin layers, which is usually being required for via-last process integration, an additional temporary transfer handling step is required. Temporary bonding to another silicon carrier is used to attach the device wafer for subsequent thinning. The lack of topography on the wafer surface enables very thin bond lines for lowest total thickness variation (TTV) as well as reduced temporary adhesive consumption for considerable process cost reduction. A low temperature PECVD layer has been deposited, leading to high yield fusion bonds. As a final step, the carrier mounted wafer has been plasma-activated and fusion bonded onto a full

thickness device wafer. A final annealing process is needed to convert hydrogen-bridge to permanent covalent bonds. Multiple thin wafers can be stacked on one device wafer by reiteration of the above process flow. All experimental results are backed by state-of-the-art measurement techniques, such as infrared scanning, scanning acoustic microscopy and scanning white light interferometry, to detect interface defects, bond integrity and temporary adhesive properties, respectively.

According to the final device process flow depicted in Fig.1 – temporary bonding, wafer stacking and debonding – this paper is structured in the same way. Each used process technology is shortly introduced before central experimental results are reviewed. Conclusively, we will present two different device results, transferring and stacking thinned device layers of 20µm as well as 10µm onto a base substrate.

Temporary Wafer Bonding

Industry consensus has been developed for temporary wafer bonding to be an inevitable step for 3D device manufacturing. On the one hand, temporary bonding enables reliable backside processing of devices, being essential for stacking more than two devices onto each other. On the other hand, handling of device wafers as thin as 11µm, which has been the minimum device thickness in this study, is not being possible without a rigid carrier solution.

In our experiments a carrier wafer has been coated with Brewer Science HT10.10 temporary bonding adhesive [3] using spin coating. The choice for this adhesive has been made due to its compatibility with slide-off as well as ZoneBOND® debonding. The difference between the two for the bonding is only related to the carrier. While standard 200mm silicon wafers have been utilized for wafer stacks using slide-off debonding, the carrier wafers have been pre-treated with a release layer coating in the case of ZoneBOND [4].

In the case of thermoplastic adhesives curing carries major importance to drive out all containing solvents from the thermoplastic adhesive. In our case, three different curing temperatures of 120°C, 150°C and 180°C have been held for 3min at each. Fig. 2(a) shows a thickness mapping, obtained by IR interferometry using EVG's inline metrology equipment [8], revealing a total thickness variation (TTV) after coating and baking of 300nm for an adhesive thickness of 20µm. In a following step, wafers have been aligned mechanically onto the wafer carrier.

978-1-4799-2408-0/14 $31.00 © 2014 IEEE

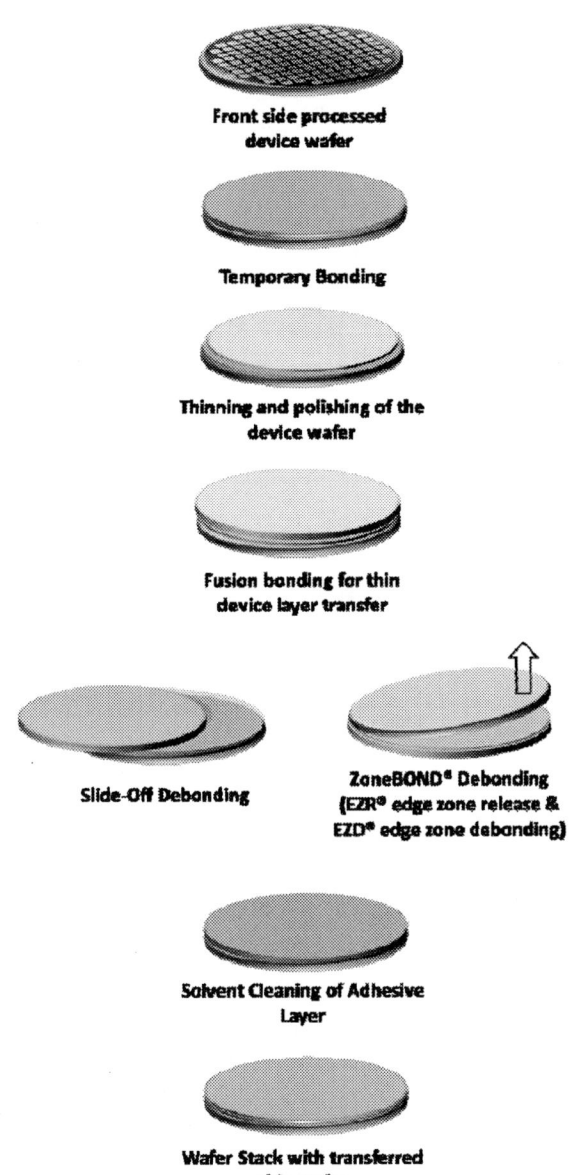

Figure 1. Schematic process flow of the different thin wafer debonding techniques used in this study: (left) thermal slide-off debonding; (right) ZoneBOND LowTemp™ Debonding.

Alignment of carrier and device wafer is critical, in the sense that support of the valuable device wafer needs to be provided at any time during grinding, polishing and backside processing.

We found mechanical alignment inside the bond chamber, generating wafer to wafer alignment around ±50μm, is fine for the presented process flow. Please note alternative process flows may require a more accurate optical wafer-to-wafer alignment. Especially, as more advanced design nodes are applied for backside processing in combination with standard edge trimming of the device wafer, accurate alignment is needed for yield and cost reasons.

Wafers are subsequently being bonded under vacuum applying standard conditions for the used thermoplastic adhesive [3]. After bonding the TTV increased slightly to 1.3μm (3 sigma), as shown in Fig. 2(b).

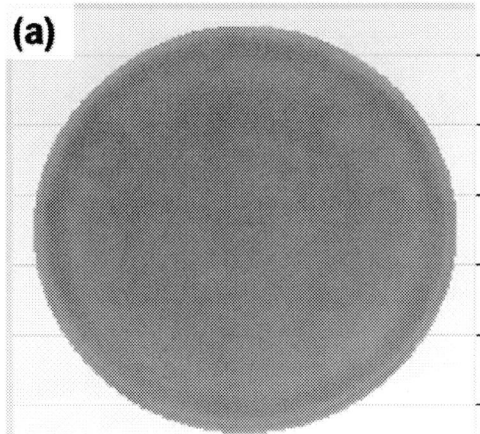

Post coat TTV: 300nm

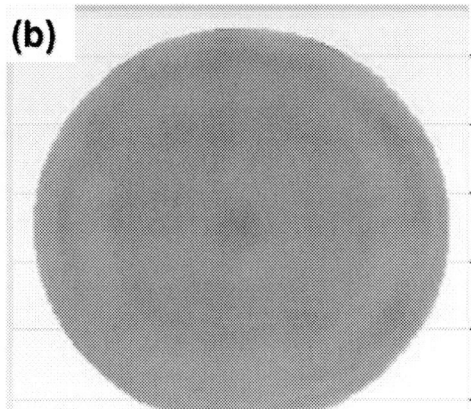

Post bond TTV: 1,3μm

Figure 2. Adhesive thickness map of 200mm carrier wafer coated with 10μm Brewer Science HT10.10 adhesive, qualified with EVG's in-line metrology module: (a) adhesive thickness map after coating and baking, revealing a total thickness variation (TTV) of 300nm; (b) post bond thickness map of carrier and device wafer showing a stack TTV of 1.3μm.

After bonding we have conducted backgrinding of the device wafer. An industry standard multi spindle grinder has been applied, where most of the wafer thickness is subtracted using coarse grinding. In a subsequent fine grind and concluding chemical mechanical polishing (CMP) step, the wafer backside shows no more defects as well as low enough surface roughness suitable for direct wafer bonding. Figure 3 depicts a thickness map of the device wafer after the full backgrinding process, including a CMP touch up. For this experiment the TTV of the device wafer is 2.0μm for a device thickness of 10μm on 20μm thin thermoplastic adhesive. We note TTV is evolving during processing due to several influence factors. Typically, standard silicon wafers, as being used in this experiment, have standard TTV values of up to 2μm.

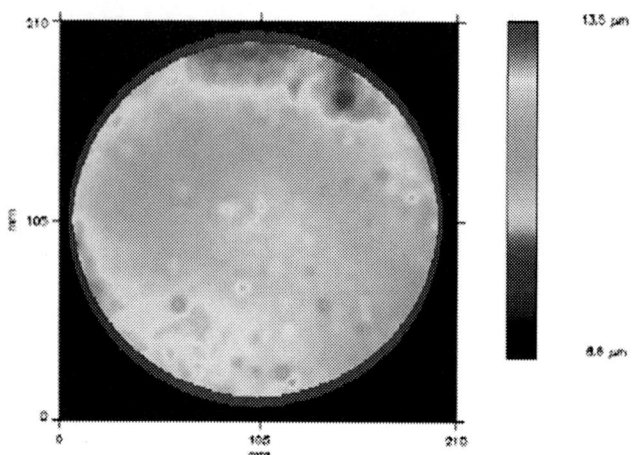

Figure 3. Thickness map of the thinned device wafer using IR interferometry.

Wafer Stacking

Oxide direct or fusion wafer bonding is one key process for 3D stacking, enabling permanent transfer of thin silicon layers to an acceptor substrate. Fusion bonding is based on the formation of molecular bonds between two joint wafer surfaces. Annealing is essential since final bond strength can only be reached by transferring instantaneously formed hydrogen-bridges into permanent covalent bonds. This transfer normally requires annealing temperature between 700°C and 1100°C, eliminating from any use in CMOS processing where temperature restricted at most to 400°C in logic devices or even less than 300°C for memory wafers. Such high annealing temperatures have substantially been reduced using low temperature plasma activated direct wafer bonding [5, 6]. Fusion wafer bonding requires flat wafer surfaces with a roughness of well below 1nm. As already being pre-mentioned, this makes CMP processing after backgrinding an essential step, to have flat enough surfaces for a high bonding yield.

As being depicted in Fig. 4, successful wafer stacking involves several preprocessing steps. As previously described, the thinned, carrier-mounted device wafer received a CMP touch-up in order to have suitable surface roughness for bonding. The landing/acceptor wafer has 0.55µm PECVD silicon oxide on top, also being polished by CMP. Plasma activation of the landing wafer has been carried out using the latest generation EVG810 plasma module, specifically designed for wafer bonding applications. The wafer is placed on the lower electrode and the device interface is subjected to the plasma for about 30s. The gas atmosphere and hence plasma radicals are recipe controlled and depend on the interfaces to be bonded. In our specific experiment O_2 plasma has been used to activate the wafers with a power of 100W.

Prior to bonding both wafer are rinsed with DI water in order to remove eventually remaining particles from the wafer surface. Besides cleaning, rinsing carries also major impact for a successful bonding result, where water is essential for the formation of hydrogen bridges between both wafers.

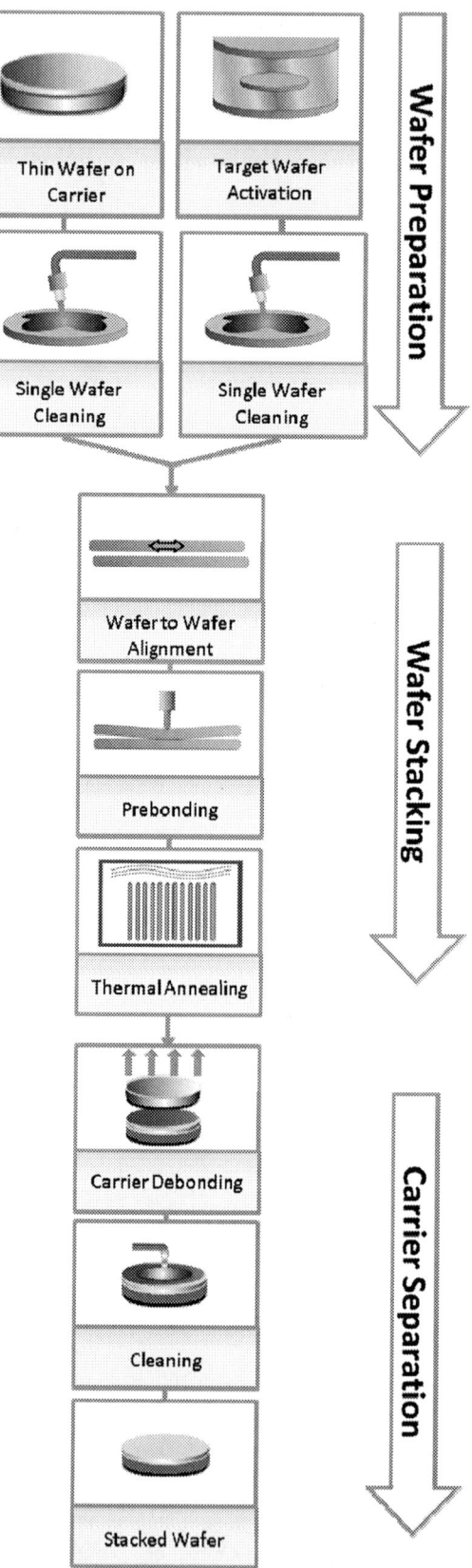

Figure 4. Schematic process flow showing the performed process steps for wafer stacking and subsequent carrier separation.

This best known method recipe as well as further insight into room temperature wafer bonding has been discussed previously by *Plach et al.* [7].

High accuracy alignment of both wafers is another essential step for monolithic device integration. In order to contact the acceptor/landing wafer and stacked device wafer, via-last processing is generally being conducted. Final metallization on top of state-of-the-art device wafers is typically being done using dual-damascene processes, where dimension of the metal layers for interconnection are in the couple of micrometer range. Etching through silicon vias through the stacked wafer and finding the landing metal pads on the lower layer is the main reason for high accurate wafer to wafer alignment. Using an industry-leading EVG SmartView®NT face-to-face bond aligner, alignment accuracies of well below 500nm (3 sigma) are feasible. Alignment accuracies of 250nm (3 sigma) have been shown elsewhere [6,9,10].

Fusion bonding is a two-step process where after the alignment routing, wafers are prebonded before they undergo an annealing step to strengthen bonding. Previous experiments have shown, a suitable plasma activation step results in a initial bond strength between 0.2 and 0.6 J/m^2 [7]. Such high bond strength is essential to reliably the stacked wafers onto each other and having no die shift by any means during detachment of the temporary carrier wafer.

Carrier Detachment

In our experiments two different debonding/detachment schemes of the carrier wafer have been applied, namely thermal slide-off as well as ZoneBOND debonding. Both these processes flows require the use of thermoplastic adhesives, which offer the advantage of having the option to perform solvent cleaning after debonding to remove adhesive residues, prior to stacking further thin wafers. In fact, cleaning on the temporary bonding adhesive is an essential step, since fusion bonding is highly sensitive with regards to contamination, leading to decreased bonding yield.

Slide-off debonding is making use of one inherent property of thermoplastic adhesive, their decreasing rheology for increasing temperatures. Carrier detachment is achieved by heating the wafer stack to 190°C on the thermal slide debonding module, where the carrier is separated by applying shear force.

ZoneBOND is more recent debonding process, where the debonding properties are incorporated into the carrier. Such ZoneBOND carrier has an inner zone with reduce adhesion, whereas the outside rim of the carrier is untreated and provides full adhesion properties. For debonding the outside, high adhesion zone is chemically dissolved using the EVG EZR® module (Edge Zone Release). Subsequently, the carrier wafer can simply being lifted of the functional wafer stack using EVG's EZD module (Edge Zone Debond) [4].

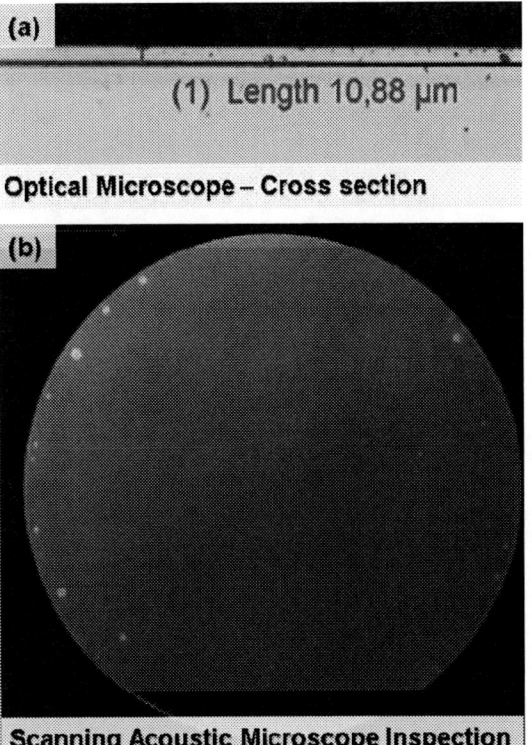

Figure 5. (a) Microscopic cross section image providing some insight of the bond interface; (b) Surface Acoustic Measurement (SAM) of a 11µm thin device wafer being fusion bonded onto a full thickness landing wafer, both on 200mm.

Experimental Details

We have carried out several experiments transferring temporary bonded and backgrinded wafers onto landing wafers using direct fusion bonding. All the results have been investigated for different success criteria. Initially, direct wafer bonding yield has been investigated using surface acoustic microscopy (SAM), enabling the inspection of the bonding interface in a non-destructive way. Furthermore, several wafer stacks have been diced and cross-section has been further analyzed using standard optical microscopy and scanning electron microscopy (SEM) techniques.

In a first integration run feasibility of a single layer transfer has been investigated using the discussed process flow on 200mm silicon substrates (Fig. 1). The device wafer has been mounted to a temporary carrier using 20µm adhesive, followed by backgrinding to a device thickness of 11µm and stacked to a full thickness substrate. The carrier wafer has been detached using slide-off debonding. The final device thickness of the stacked thin wafer can be seen in Fig. 5(a). In order to investigate bonding yield we have carried out several SAM measurements from the different test vehicles, where Fig. 5(b) shows a characteristic measurement. All investigated samples show a high bonding yield, with a minimum amount of dispersed voids at the bonding interface. The root cause of these few voids has been found in the non-optimized processing conditions in this set up phase, where wafer are processed in semi-automated modules, making them prone to increase particle exposure. A detailed investigation for fusion bond quality and potential improvements has been already carried out earlier by V. Dragoi *et al.* [11].

In a second sample test we have investigated the subsequent stacking of two 20μm thin wafers onto a thick landing wafer, as being shown in Fig. 6(a). The integration process remains unchanged to the previous process flow. It should be pointed that the first plasma activation and bonding has been again carried out on a landing wafer with 0.55μm PECVD oxide. The second stacking process has been carried on two native oxide interfaces. Also here, a high bonding yield could be achieved, where the SAM image is being shown in Fig. 7.

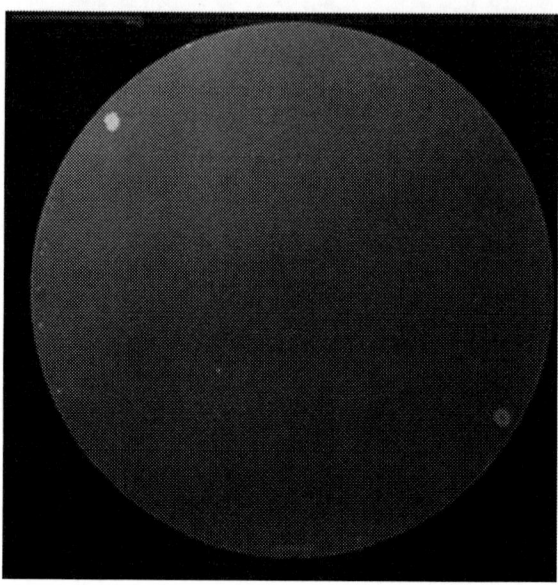

Figure 7. SAM image of the bonded triple stack of 20μm thin wafers

Conclusions

In conclusion we have investigated the bonding performance of temporary bonded wafers. In this way we could achieve the successful face-to-back stacking of several thinned devices wafers, increasing process freedom for monolithic device manufacturing.

Combining low thermoplastic adhesive layer thickness we have managed to achieve a minimum device layer thickness for transfer of 10μm with a TTV of 2μm. Furthermore, we have shown the stacking of two thinned wafers with thicknesses of 20μm onto a thick landing wafer, forming a triple stack.

Plasma activated low temperature wafer direct bonding enables to use thermoplastic adhesives, even for the thermal annealing needed to strengthen the prebond state and keep the high alignment accuracy of fusion bonded wafers. As debonding occurs while the thin wafer is only held by hydrogen-bridges to the device wafer, these pre-bond energies are of essential interest. We could show, that typical tacking energies of $0.2 - 0.6$ J/m^2 is adequate for high process yields.

Furthermore, we showed that different debonding solutions such as thermal slide-off as well as LowTemp ZoneBOND debonding are suitable for layer transfer of thin wafer for monolithic integration.

Acknowledgments

ZoneBOND® is a registered trademark of Brewer Science, Inc.

LowTemp™ is a trademark of EV Group.

EZR® and EZD® are registered trademarks of EV Group.

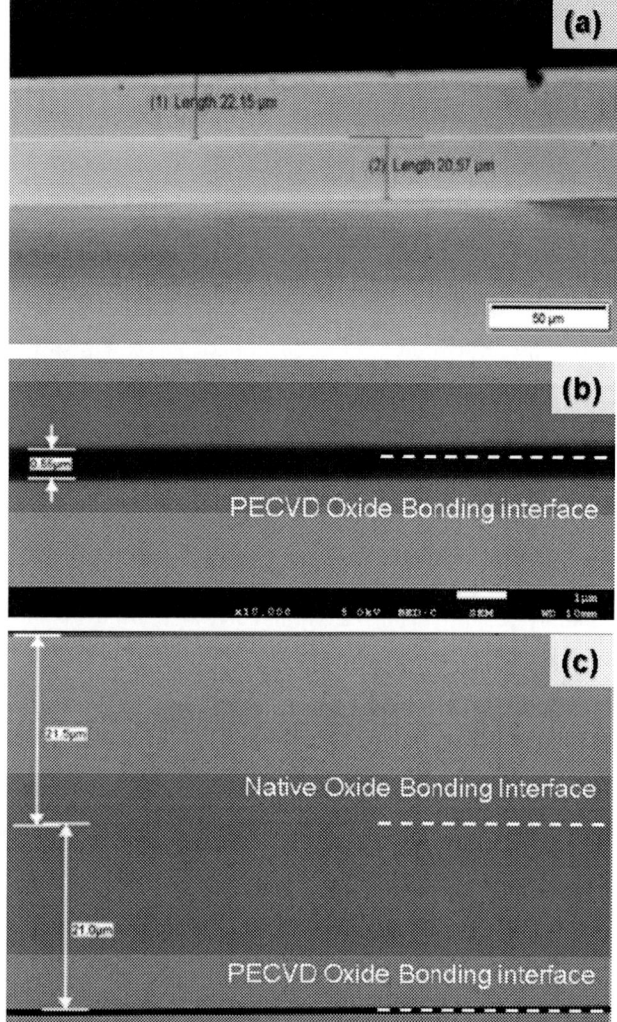

Figure 6. (a) Optical microscope cross section through a bonded triple stack; (b) high magnification SEM of the bonding interface between thick acceptor wafer with 0.55μm PECVD oxide and the first stacked thin wafer; (c) low magnification SEM image of the whole stack

References

1. P. Ramm, Handbook of WaferBonding, Wiley, 2012
2. P. Garrou, Handbook of 3D Integration, Wiley, 2012
3. J. Charbonnier, et.al., "Integration of a Temporary Carrier in a TSV Process Flow", Proceedings of IEEE Electronic Components and Technology Conference (ECTC), 2009.
4. T. Matthias, J. Burggraf, D. Burgstaller, M. Wimplinger, P. Lindner, "Thin die stacking for wide I/O memory-on-logic", Solid State Technology. June 2012, pp.17-20.
5. Q.-Y. Tong and U. Gösele, Semiconductor Wafer Bonding: Science and Technology, Wiley Interscience, New York (1998)
6. G. Gaudin et al., "Low temperature direct wafer to wafer bonding for 3D integration", in Proc. IEEE 3D System Integration Conference (3DIC) 2010.
7. T. Plach, K. Hingerl, S. Tollabimazraehno, G. Hesser, V. Dragoi, and M. Wimplinger, "Mechanisms for room temperature direct wafer bonding," J.Appl.Phys., vol. 113, 094905, 2013.
8. T. Matthias, D. Burgstaller, J. Burggraf, P. Kettner, M. Wimplinger, P. Lindner, "Thin wafer processing – yield enhancement through integrated metrology", Proceedings of IEEE Electronics Packaging Technology Conference (EPTC), 2011.
9. W.H. The, C. Deeb, J. Burggraf, D. Arazi, R. Young, C. Senowitz, and A. Buxbaum, "Post-Bond Sub-500 nm Alignment in 300 mm Integrated Face-To-Face Wafer-To-Wafer Cu-Cu Thermocompression, Si-Si Fusion and Oxide-Oxide Fusion Bonding", in Proc. IEEE 3D System Integration Conference (3DIC) 2010.
10. T. Matthias, T. Uhrmann, V. Dragoi, T. Wagenleitner, P. Lindner, "Wafer Bonding for Backside Illuminated Image Sensors", ECS Transactions, 44 (1) 1269-1274 (2012)
11. V. Dragoi, P. Czurratis, S. Brand, J. Beyersdorfer, C. Patzig, J.P. Krugers, F. Schrank, J. Siegert, and M. Petzold, "Low Temperature Fusion Wafer Bonding Quality Investigation for Failure Mode Analysis", ECS Transactions, 50 (7) 227-239 (2012)

Temporary Bonding for High-topography Applications: Spin-on Material Versus Dry Film

Anne Jourdain, Alain Phommahaxay, Greet Verbinnen, Alice Guerrero[1], Susan Bailey[1], Mark Privett[1], Kim Arnold[1],
Andy Miller, Kenneth Rebibis, Gerald Beyer and Eric Beyne

imec vzw, Kapeldreef 75, B-3001 Leuven, Belgium
[1]Brewer Science, Inc., 2401 Brewer Drive, Rolla, MO 65401, USA
Tel: +32 16 281 909 Fax: +32 16 281 097 Email: jourdain@imec.be

Abstract

Handling wafers with sub–100 μm thicknesses requires a support or carrier wafer during handling, transport and processing in a semiconductor process line. This thin wafer support system should be very stable during these operations but should also allow for easy wafer debonding. Therefore, room temperature debonding methods are the favored solutions, in particular, room temperature peel debonding, which is one the industry focuses on today. We have estimated that a TTV of less than 2μm is required to enable a 'soft' backside via reveal process for temporary bonding material thicknesses less than 20μm. For high frontside topography applications involving Cu pillars or C4 types of bumps, the challenge is to maintain a low post-grinding TTV value while processing the device wafer with thick temporary bonding materials (typically thicker than 50μm). Reducing the post-thinning TTV can be done either by reducing the edge bead effect of the spin-on temporary bonding material, or by looking into novel types of material applications such as thick film lamination. In this paper, we compare the thinning performance in terms of post-grinding TTV of two materials from Brewer Science: the thick spin-on ZoneBOND®5150 material and the experimental BrewerBOND™ dry film.

Introduction

One of the key steps in 3D technology as of today is the bonding of the device wafer to a carrier wafer prior to wafer thinning and subsequent backside processing. Such wafer bonding allows the use of standard semiconductor equipment during post-thinning processing. The most critical aspect of this process flow is the choice of the interfacial bonding material: it must combine a variety of properties and functions that are often contradictory. The bonding must be physically, chemically and thermally stable during the thinning and subsequent backside processing steps. At the end of the process flow, however, the material must be easily debondable from the thinned wafer, without leaving residues and risking wafer damage.

A wide variety of approaches have been proposed and are being developed today. First-generation materials, including thermoplastics and laser-degradable or chemically dissolvable bonding materials, are readily available on the market, and they already enable early 3D demonstration processing. Yet, when it comes to debonding the thin wafer after backside processing, some of these materials can present some integration limitations, especially if low-melting-point solders or high-topography structures are present on the backside of the wafers. Therefore, a room temperature carrier-peel debonding technique is the preferred approach and the focus of most of the semiconductor players today [1-2]. The ease of the debonding is just one part of the overall process. The ability of the bonding material to be very uniform in thickness across the wafer before and after bonding enables a very low total thickness variation (TTV) after thinning and a uniform TSV reveal process. This becomes extremely critical when the device wafer contains high-topography structures on its frontside, such as Cu pillars or C4 bumps, which can range in thickness typically between 50 and 80 μm. The challenge here is to find a temporary bonding material that can maintain a low TTV after material coating and temporary bonding. Reducing the post-thinning TTV can be done either by reducing the edge bead effect of the spin-on temporary bonding material, or by looking into novel types of material applications such as film lamination.

We have already compared the thinning performance on blanket wafers (no frontside topography) of two materials from Brewer Science: the spin-on ZoneBOND®5150 material [3,4] and the BrewerBOND™ dry film that is laminated onto the carrier wafers prior to bonding [5]. In this paper, we build further on these 2 material performance and we demonstrate their ability to comply with high-topography wafers containing 50μm Cu pillars frontside.

Spin-on ZoneBOND®5150 process optimization and results

We used 300mm device wafers processed with 50μm high Cu pillars frontside (pitch: 100μm) and 5x50 μm TSVs in a via-middle approach. We focused on a 55μm thick ZoneBOND®5150 material spin-on layer coated on the carrier wafers prior to bonding as depicted in Figure 1. The coating and bonding are performed on a SUSS XBC300 platform.

978-1-4799-2408-0/14 $31.00 © 2014 IEEE

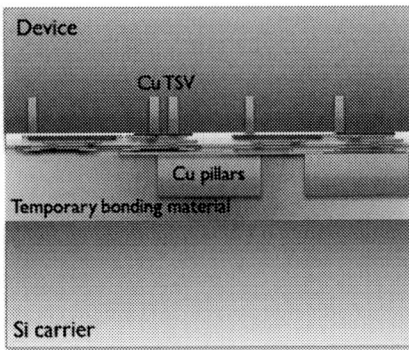

Figure 1. Temporary wafer bonding of a high frontside topography device wafer containing frontside Cu pillars and Cu TSVs.

The carrier coating approach presents the advantage to become independent from the device wafer layout and pattern density, and one single coat recipe is required that can cover a wider range of applications. But as all spin-on materials, the thicker the material, the larger the edge effect as shown in Figure 2 that compares a 20μm thick and a 55μm thick layer of ZoneBOND®5150 material. The 20μm thick material profile was measured on an ellipsometer while the thick material profile was measured on an ISIS SEMDEX301-2 metrology tool. The use of two metrology tools was justified in order to cover the wide material thickness range. For the thick material layer, an edge bead of more than 20μm has been measured on blanket carrier wafers while a thickness variation of 0.5μm is obtained for the thin material after optimization.

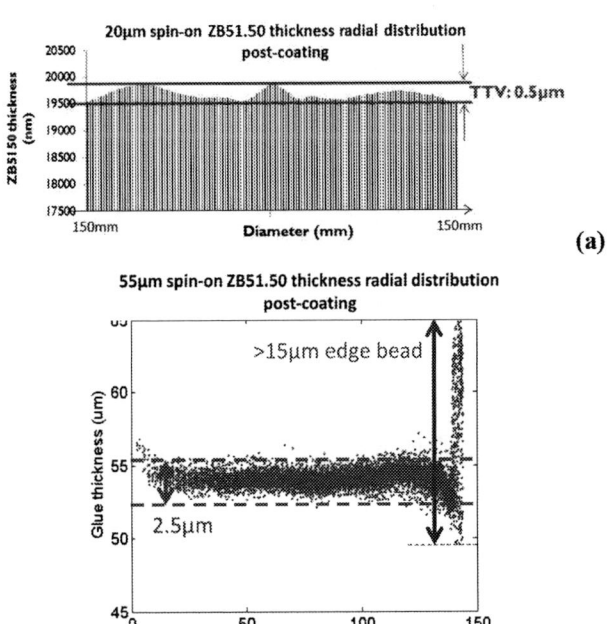

(a)

(b)

Figure 2. Thickness variation across a blanket wafer for a 20μm thick (a) and a 55μm thick ZoneBOND®5150 layer (b) respectively.

Two process steps can be optimized to reduce the edge bead effect before exploring the thinning performance of the material: the coating and the bonding steps respectively.

The edge bead effect related to the thick film spin-coating can be minimized by reducing the baking ramp rate after coating as shown in Figure 3: up to 60% edge bead reduction is achieved if the temperature ramp rate is extended from 3 to 9min. A defect inspection of the carrier wafers in a NANDA SPARK system after coating also shows a significant defect reduction in terms of adhesive shrinkage at the edge of the wafers and number of bubbles trapped into the film. No residual solvent was detected by chemical analysis after bake.

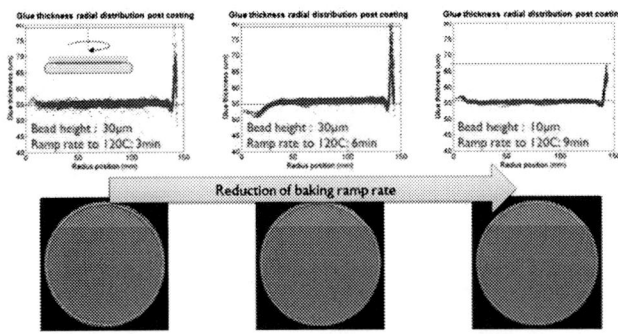

Figure 3. Impact of post-coating bake on edge bead reduction for a 55μm thick ZoneBOND®5150 film.

An initial slow temperature ramp-up is recommended for thick spin-on materials:

[1] Reduction of bubble formation

[2] Reduction of overall bead height

For throughput optimization, a multiple hotplate option might be recommended for thick adhesive processing.

The edge effect can also be reduced during the bonding process by reducing the bond temperature as shown in Figure 4**Error! Reference source not found.**: the material reflows and the edge bead is flattened during the bonding process. An initial 4.3μm material TTV measured after bonding on blanket wafers is reduced to 1.5μm by reducing the bond temperature from 200°C to 170°C. On the other hand, the temperature should not be too low: at lower temperature, the adhesive does not reflow sufficiently, nor stick well, inducing some tiny voids at the bond interface. The edge bead after coat is a secondary element of the full process and the major part of TTV comes from the bond process itself.

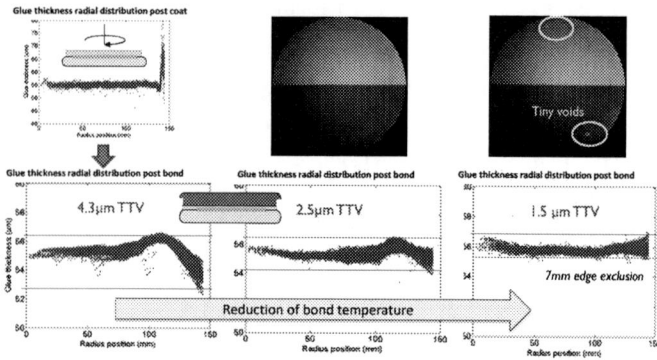

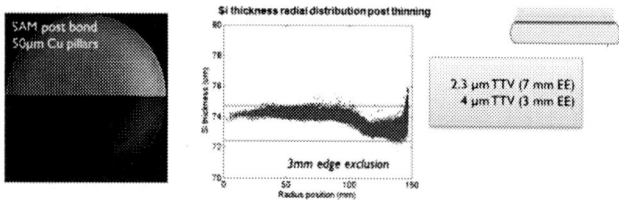

Figure 4. Impact of the bonding temperature on the thick adhesive profile and TTV.

This optimized coating and bonding processes have been applied to device wafers containing 50μm high Cu pillars frontside (100μm pitch), and a 4μm post-grinding TTV has been achieved for the device wafer as shown in Figure 5. A uniform, void free bond was obtained, without any adhesive squeeze out.

Figure 5. Optimized bonding process transferred to 50μm high Cu pillar wafers. A 4μm post-grinding TTV is achieved on the device wafer.

This shows that the bonding recipe optimization is the key process to reduce the post-coat edge bead and to achieve low post-grinding TTVs. Further TTV reduction will involve fundamental understanding of dynamics in the bond chamber, and further performance gain will require strong tool and material vendor interaction.

BrewerBOND™ process development and results

Maintaining a low post-thinning TTV can be done either by reducing the edge bead effect of the spin-on temporary bonding material (as described in previous section), or by looking into novel types of material applications such as film lamination. The BrewerBOND™ dry film concept offers the possibility to laminate on the carrier wafers a constant material thickness across the entire wafer. The advantages of dry films as compared to spin on materials are multiple:

- They reduce the edge bead effect and therefore offer a potentially better TTV after thinning

- They improve the wafer coverage on high topography wafers

- They remove the pattern density and layout issues seen with highly viscous materials

- They negate the drainage compatibility issues related to incompatible solvent systems

Two films thicknesses have been investigated: 55μm and 110μm respectively. Thicknesses above 55μm are achieved by two sequential film laminations onto the same carrier wafer. Thickness and TTV measurements performed in the ISIS tool did not show a significant TTV increase for the double film thickness as shown in Figure 6.

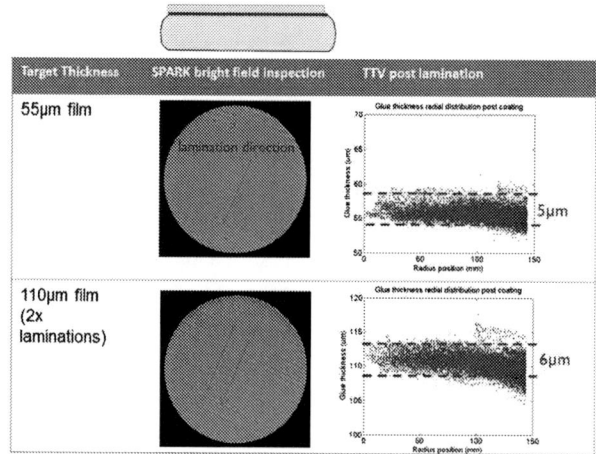

Figure 6. TTV evolution for increasing BrewerBOND™ dry film thicknesses.

Note that the lamination has been performed in a manual mode, and a TTV improvement is expected by using a fully automated lamination process.

When comparing the post-application performance of a 55μm thick dry film after lamination on 300mm blanket wafers and a 55μm thick spin-on ZoneBOND®5150 layer, we see that one big advantage of the dry bonding film is the reduced edge bead effect compared to the spin-on material as seen in Figures 7 and 8. Indeed, most of the spin-on material thickness variation comes from the edge bead (which can exceed 20μm), while the dry film shows a relatively uniform thickness variation across the wafer diameter.

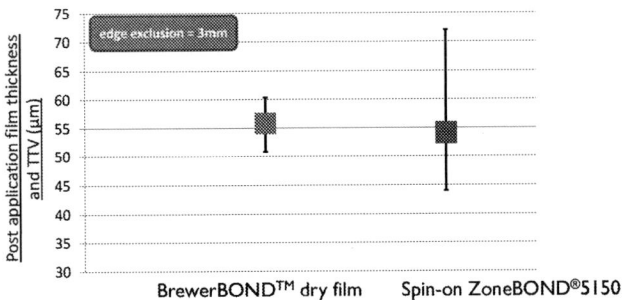

Figure 7. Thick spin-on ZoneBOND®5150 and BrewerBOND™ dry film thickness direct comparison after material application on a blanket carrier wafer.

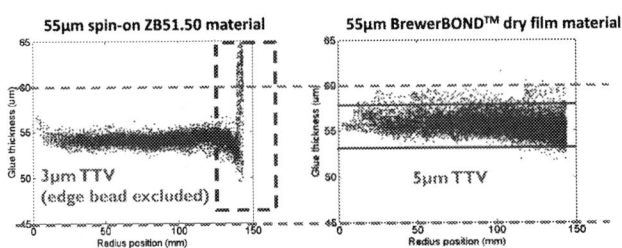

Figure 8. Thickness radial distribution of the ZoneBOND®51.50 spin-on material versus the BrewerBOND™ dry film.

Temporary wafer bonding and grinding has been performed on blanket and topography wafers (containing 50μm Cu pillars, 100μm pitch) using the 55μm and the 110μm thick films respectively. Results are shown in Figure 9. Both approaches show a uniform, void free bond, indicating that the material provides a good reflow during bonding around the 50μm high Cu pillar structures. After back grinding to 55μm final Si thickness, a Si TTV between 3-4μm has been measured for the 2 cases (with 0mm edge exclusion!). However, a slight edge effect is observed for the 110μm thick film that slightly increases the TTV to 8μm. It is to note that the thicker adhesive on the edge is not resulting in a thinner Si, but rather the bonding is compressing the material at the edge. This is further evidence that this is a bonding phenomenon and indicates that the bonding mechanism is the main contributor to the final post-thinning TTV. A bonding optimization is still possible for the 110μm thick film.

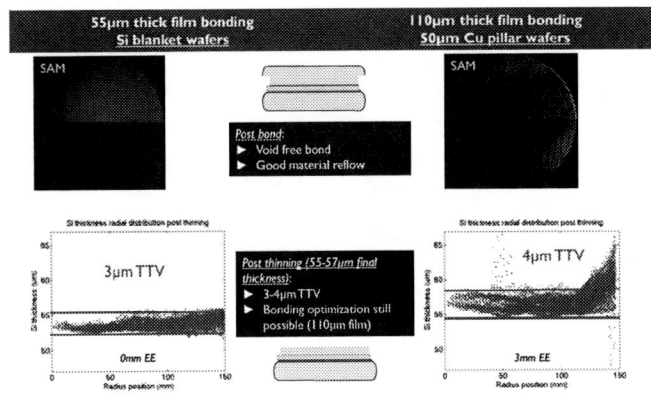

Figure 9. Bonding and thinning results using 55μm and 110μm thick BrewerBOND™ dry films on blanket and topography wafers respectively.

Room-temperature thin wafer debonding

After completion of backside processing (that can include interconnects layers and microbumps that are not described in this paper), the process is further completed by debonding the thin wafer from the carrier wafer and transferring to a dicing tape on frame. This process was already described [3] and implemented for the first time on blanket wafers [3,4]. This step is performed at room temperature in a peel-off type of process in a SUSS DB12T debonder as described in Figure 10. In a first step, a wet edge preparation is required in order to eliminate the high adhesion area of the adhesive layer between the 2 substrates. Then, the thin device wafer still bonded to the carrier is laminated onto a dicing tape on frame. Next, the room temperature peel off debonding separates the thin wafer from the carrier while the thin wafer is still on tape and on frame. A final cleaning step on tape is performed to remove adhesive residues from the device wafer. The device wafer is then ready for dicing and stacking while the carrier wafer can be recycled for a new temporary bonding. At this stage, tape selection is critical as it should be chemically compatible with the cleaning solvents used to strip the remaining adhesive from the thin device wafer. Furthermore, customized tooling was used to remove the adhesive. A dedicated tool, the SUSS AR12, was adapted to handle dicing frames and was used to clean wafers on tape without degradation. This process flow was successfully implemented on 300mm wafers, with and without frontside topography as shown in Figure 11.

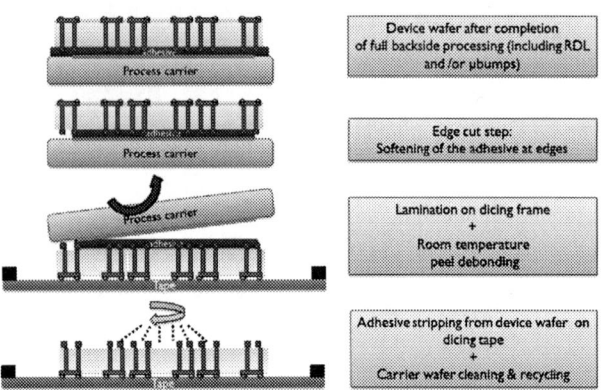

Figure 10. ZoneBOND®51.50 typical debonding process.

55µm thick spin-on ZoneBOND®5150 on 50µm Cu pillars   *110µm BrewerBOND™ dry film on blanket wafer*

Figure 11. 55µm thin wafers on tape after room-temperature peel debonding using the thick ZoneBOND®5150 material and the BrewerBOND™ dry film concepts.

Conclusions

For high frontside topography applications, the challenge is to maintain a low post-thinning TTV (typically less than 2µm) while processing the device wafers with very thick temporary bonding materials (typically above 50µm thicknesses). Reducing the post-thinning TTV can be done either by reducing the edge bead effect of the spin-on temporary bonding material, which may require quite some process optimization, or by looking into novel types of material applications such as dry film lamination. The proof of concept of a novel type of dry film material has been successfully demonstrated, and the good material reflow properties showed to be fully compatible with high topography wafers. The absence of edge bead effect potentially reduces the post thinning TTV, and the 3µm TTV obtained with the 55µm thick film is already comparable to the 20µm thin spin-on ZoneBOND®5150 process. The performance optimization of thicker films (>100µm) would require further material and process optimization (like automated lamination and bonding optimization). The experimental BrewerBOND™ dry bonding film is a very good alternative to thick spin-on materials to achieve ultralow TTV after thinning. This novel concept is very promising for adhesive thickness scale-up & future large substrate size.

Acknowledgments

Authors would like to thank Walter Spiess, Erika Demke and Stefan Lutter from SUSS MicroTec as well as Pieter Bex and Ingrid Demonie from imec for their strong support in wafer bonding and debonding.

This work was carried out in the frame of Joint Development Programs between Suss MicroTec - Brewer Science – imec.

References

1. J. McCutcheon and D. Bai, "Advanced thin wafer support processes for temporary wafer bonding," *IMAPS 2010 - 43rd International Symposium on Microelectronics*, 2010, pp. 361-363.

2. J. McCutcheon, R. Brown, and J. Dachsteiner, "ZoneBOND thin wafer support process for wafer bonding applications," *Journal of Microelectronics and Electronic Packaging*, vol. 7, no. 3, 2010, pp. 138-142.

3. A. Phommahaxay, A. Jourdain, G. Verbinnen, T. Woitke, P. Bisson, M. Gabriel, W. Spiess, A. Guerrero, J. McCutcheon, R. Puligadda, P. Bex, A. Van den Eede, B. Swinnen, G. Beyer, A. Miller, and E. Beyne, "Ultrathin wafer handling in 3D stacked IC manufacturing combining a novel ZoneBOND™ temporary bonding process with room temperature peel debonding," *IEEE International 3D Systems Integration Conference (IEEE 3DIC)*, January 31-February 2, 2012, pp. 1-4

4. A. Jourdain, A. Phommahaxay, G. Verbinnen, S. Suhard, A. Miller, A. La Manna, B. Swinnen, G. Beyer, and E. Beyne, "Integration of the ZoneBOND™ temporary bonding material in backside processing for 3D applications", *Proc. ESTC2012*, September 17-20, 2012.

5. A Jourdain, A. Phommahaxay, A. Guerrero, S. Bailey, K. Arnold, K. Rebibis, A. Miller, G. Beyer and E. Beyne, "temporary bonding for high-topography applications: spin-on versus dry bonding film processes", *Proc. WaferBOND'13 Conference*, December 5-7 2013.

Development of New Concept Thermoplastic Temporary Adhesive for 3D-IC Integration

A. Kubo, K. Tamura, H. Imai, T. Yoshioka, S. Oya, S. Otaka

TOKYO OHKA KOGYO CO., LTD.,

7-8-16, Ichinomiya, Samukawa, Koza, Kanagawa 253-0111, Japan

a-kubo@tok.co.jp

Abstract

2.5D and 3D integration technology using temporary bonding has become main stream in the semiconductor industry in recent years. However, thermal stability, low damage, and low temperature debonding are still areas of challenge. Meanwhile, the demand for a solution has been rapidly increasing to enable the high volume manufacturing of 3D-IC products. In this paper, we report the development of a new type thermoplastic temporary adhesive.

In general, thermoplastic material is considered difficult to overcome the critical issues that it faces during the back side processing in 3D-IC production due to its thermal behavior. In a previous study, the glass transition point (Tg) of adhesive was considered as a key parameter to improve thermal resistance. However, the high Tg material also induced large warpage from its internal stress, and had difficulty in obtaining sufficient coverage on high topography wafer surface.

The latest adhesive by TOK provides solutions to those problems. The new thermoplastic adhesive, which has optimized rheological property and elastic modulus, can reduce warpage and maintain stability through thermal processes such as PECVD and solder reflow. In addition, it can reduce void issues found during PECVD passivation process. The new thermoplastic adhesive can also coat over 100 μm in thickness with good total thickness variation (TTV) post bonding, which can be adapted to the C4 bump application. This study on adhesive was demonstrated with the thin wafer handling system by TOK called Zero Newton®. The temporary adhesive is compatible with both perforated and non-perforated glass carriers and is separated with each corresponding scheme. The adhesive is removed by solvent dissolving from the substrate, so low residue and damage are expected as a result. The details of this study on thermoplastic material and demonstration of wafer handling process are fully described in this report.

Introduction

Recent interest in 3D integration technology may be attributed to both its technological potential and its possible insertion to high volume manufacturing for numerous applications in various fields. The temporary bonding system is recognized as a key step for 3D integration approach and therefore several kinds of temporary bonding system have been introduced and available today. However, there are still room for improvement when enabling high volume manufacturing [1, 2].

Thermoplastic materials have been considered to be most suitable for use in temporary bonding adhesive for its ease in removal from bonded substrate during solvent clean with low damage to structures on the wafer surface, while it is understood that those materials show unstable behavior under high temperatures condition. This poor thermal stability imposes limitation on temperature range for critical backside processing condition such as PECVD process [3-5].

Previously we showed a thermally stable thermoplastic temporary adhesive which can be adapted to CMOS type 3D integration [6]. However, these types of adhesive are not sufficient to meet the various requirements for temporary adhesive process in recent condition. Major requirements for the adhesive are summarized in Table 1.

The biggest problem of thermoplastic materials for temporary bonding application is that there are tradeoffs in thermal bonding temperature and thermal stability. In other words, if an adhesive demonstrates bonding at lower temperature, then the thermal stability will decrease. Meanwhile, increase in Tg and softening point of the adhesive for thermal stability also cause bonding difficulty, especially in non-perforated support carrier. Our internal testing showed that an adhesive of high Tg exhibited the trend of inducing difficulty of high thickness coating and higher warpage.

In an attempt to improve those properties of thermoplastic adhesive and while maintaining the advantage with respect to adhesive removal at the final step of temporary bonding, we achieved to development of new type of adhesive.

Table 1. Required adhesive property

Property	Description
High Thickness coating	Sufficient coverage for C4 bump
Bonding capability	No void, Good TTV after bonding, High throughput
Low temperature bonding	Below usual solder bump melting point
BG/CMP resistance	No delamination and chipping during backside grind and polish processing
Thermal stability	Stability to thermal backside processing
Chemical stability	Stability to chemicals used in TSV processing
Cleaning speed	High dissolution speed
Good cleaning property	No residue after debonding and cleaning

Temporary bonding process flow

The wafer bonding flow includes spin coating on wafer, pre-bake steps, and thermal bonding. The debonding flow includes carrier separation, adhesive dissolution, and cleans as presented in Figure 1.

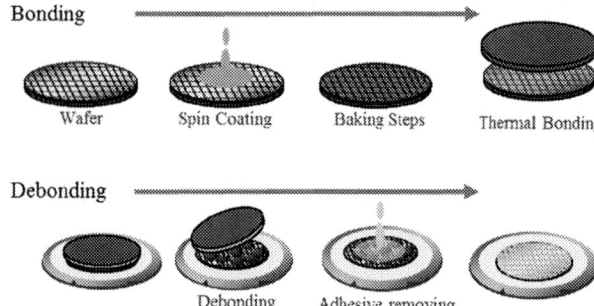

Figure 1. Temporary bonding and debonding process

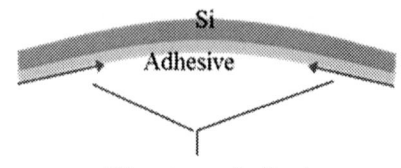

Figure 3. Schematic of adhesive film stress measurement of coated Si wafer by laser method

Table 2. Tg and softening point determined by TMA and film stress measurement

Adhesive	Tg (°C)	Softening point (°C)	Adhesive film stress (MPa)
Sample-A	55	183	6.0
Sample-B	168	270	19.8
Sample-C	68	240	0.5
Sample-D	70	260	0.5

Order of experimental topics presented

First, the fundamental property of conventional (Sample-A and Sample-B) and new adhesive (Sample-C and Sample-D) was examined. Then the temporary bonding process of the new adhesive with blanket wafers was demonstrated. Finally, bumped wafers were used for evaluation for the new adhesive.

Adhesive glass transition and softening point

The temperature dependence of several elastic property of adhesive was examined as the initial approach of evaluation.

Thermal mechanical analysis (TMA) was used to determine the Tg and softening point of conventional thermoplastic type of Sample-A as shown in Figure 2. The data shows the adhesive thickness plotted as a function of temperature under a fixed load condition. The curve indicates that the adhesive remains stable and at a firm state in the temperature range between the Tg and the softening point. The adhesive becomes unstable due to rapidly viscosity decrease above the softening point.

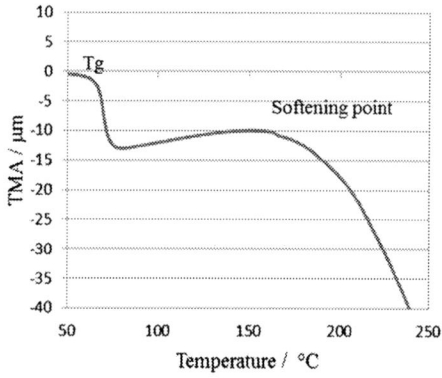

Figure 2. TMA data of Sample-A

The newly developed adhesive Sample-C and Sample-D were prepared and compared to the conventional type, Sample-A and Sample-B, which was originally designed for perforated support glass carrier as following.

The glass transition and softening point were obtained by using TMA as well. Then the adhesive film stress was measured by laser method as described in Figure 3 and those results are summarized in Table 2.

The data show that there are no significant difference between Tg of Sample-A, Sample-C, and Sample-D. Sample-B was the only adhesive that showed much higher Tg temperature. Adhesives other than the Sample-A exhibited softening point of over 240°C. Both conventional types, Sample-A and Sample-B, indicate larger film stress and this may cause high wafer warpage, especially in the case of large diameter wafer. In contrast, new adhesives, Sample-C and Sample-D, indicate low stress. Although their softening point are close to that of Sample-D.

The softening point was considered as the dominant factor to determine the thermal bonding temperature in the previous study. In fact Sample-B showed poor bonding result due to its high thermal resistance. It was successfully bonded to perforated glass carrier but not bonded well to non-perforated glass carrier under an appropriate bonding temperature (< 220°C)

Rheological study

Rheological property was measured by using rheometer (UBM, Reogel-E4000) to confirm more detailed thermal elastic behavior. The shearing test by rheometer provides several rheological elastic properties such as storage modulus (G'), loss modulus (G''), and viscosity of adhesive as function of temperature and measurement frequency. In particular, temporary adhesive must to be softer during bonding process, while in a solid state to prevent adhesive bleed from wafer edge during backside thermal processing. The measurement in changing frequency with the rheometer provides modulus data of static area to dynamic area of adhesive, which is possibly useful for understanding the adhesive behavior of actual temporary bonding process if there is correlation between those results.

Sample adhesives were tested with changing frequency condition (1~20 Hz) by rheometer. Figure 4 shows the adhesive viscosity plotted as a function of frequency at 180°C. The curves indicate that adhesives Sample-C and

Sample-D have a clear effect by frequency change. On the other hand, Sample-A exhibits very little dependence on frequency.

If there is a correlation between frequency dependence of adhesive to bonding properties, Sample-C and Sample-D should bond at moderate condition. Bonding test was performed using a TOK bonding unit, TWM12000 series, with 300 mm blanket wafer and non-perforated glass carrier to confirm this relationship. (The conditions are described in detail in the latter part of this paper.) Sample-A, Sample-C, and Sample-D were bonded successfully at 215°C, 180°C, and 215°C respectively. These bonding temperatures are much lower than that of what is expected from its softening point. Hence it is reasonable to assume that the bonding process with pressure is similar to dynamic rheological property.

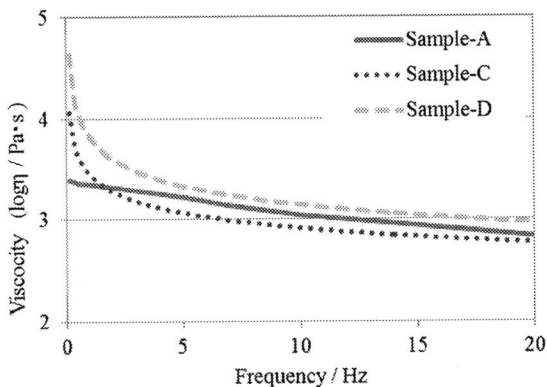

Figure 4. Dependence of viscosity on frequency at 180°C

Moreover, the elastic modulus on temperature dependence was obtained by shearing test at 1 Hz measurement frequency conditions to study adhesive stability similar to static condition with high temperature (Figure 5). This can be used to predict adhesive behavior under static high temperature condition of backside process. Figure 5 indicates Sample-D is the most stable and followed by Sample-C. Sample-A perfomed worse than the new adhesives at the general temperature range of temporary bonding process.

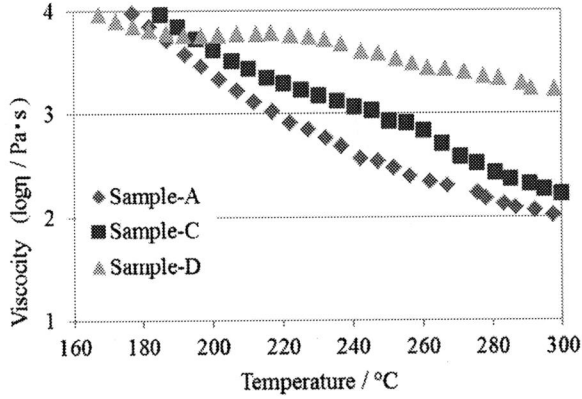

Figure 5. The dependence of viscosity on temperature

Actual thermal test was performed to comfirm the preceding results for viscosity behavior. All bonded wafers were proceeded to wafer thin down process by using a generic back grinding equipment with a target silicon thickness of 50 μm and subsiquently treated by solder reflow process simulation with a curing furnace at condition of 260°C for 10 minutes under N_2 flow. The wafer was then examined with a microscope around the edge area from the silicon side to confirm if there was any adhesive bleed issue (Figure 6).

Table 3 summarizes the bonding capability and thermal stability of each adhesive. Conventional type a Sample-A can be used for temporary bonding process but there is limitation due to the lack of thermal stability. Sample-C and Sample-D can be used under much higher thermal condition. Sample-C can be applied at low temperature bonding for low melting point bump material in particular.

Measurement of the dependence of elastic modulus on temperature and frequency is important in terms of evaluating bonding properties and thermal stability.

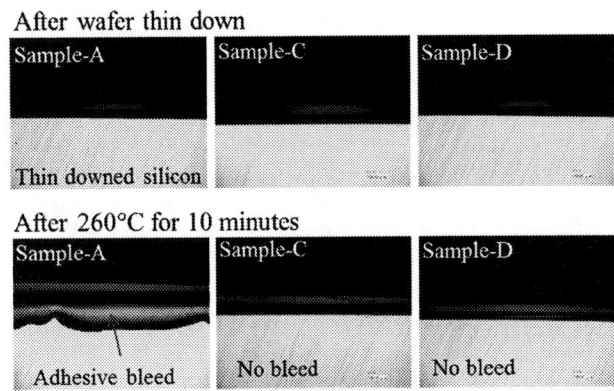

Figure 6. Adhesive bleed check after wafer thin down and thermal process

Table 3. Adhesive bonding property and compatibility

Adhesive	Possible bonding temperature	Stability around 260°C	Compatibility for temporary bonding process
Sample-A	215°C	Poor	Limited use
Sample-C	180°C	Good	Wide range and low temperature bonding
Sample-D	215°C	Good	Wide range

Thermal resistance of Adhesive

A thermogravimetric analysis (TGA) and thermal desorption analysis (TDS) are general approaches for evaluating the thermal resistance of the adhesives described. The thermal decomposition of the adhesives is shown as TGA data in Figure 7. These adhesive were very stable up to temperatures in excess of 300°C, while the weight loss at 300°C was measured to be less than 1.0%.

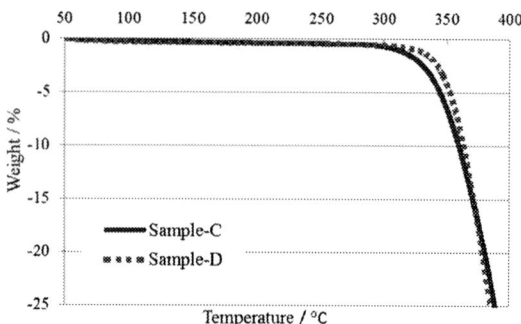

Figure 7. TGA data under inert condition (N_2)

Out-gassing from the adhesive poses a serious risk for void or delamination issues during the temporary bonding process, especially under high vacuum condition. The out-gassing source is possibly from adhesive decomposition or residual solvent in the adhesive film. The residual solvent in the adhesive film may cause tool contamination, so it is critical that all the solvent be removed after the wafer coat step. Coated adhesive on blanket wafer was cut into individual coupons (1 cm^2) and measured by TDS with pressure range of 10^{-7} Torr. Figure 8 shows an out-gassing TDS measurement of the Sample-C and Sample-D following the standard prebake condition and blanket silicon for comparison. As shown in Figure 8, the intensity of out-gassing is relatively low at the temperature range below 300°C. Absorption of water at the adhesive surface was slightly detected. This result confirms that the solvent is completely removed by the standard prebake and there is no decomposition up to 300°C even under such high vacuum condition.

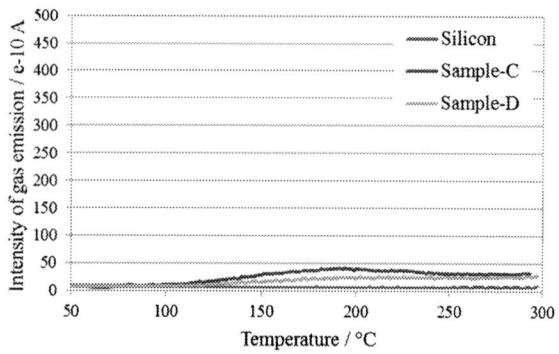

Figure 8. TDS data of adhesive and blanket silicon

Chemical stability test

Chemical stability is also an important requirement item for temporary bonding process because poor chemical stability or adhesion with substrate can possibly induce delamination. Adhesive resistance against chemical solutions generally used in wet process was examined and shown in Table 4. Evaluation includes bonded and blanket adhesive coated wafer soaked in chemicals in various condition as described on the Table 4. After soaking, weight change and visual delamination was confirmed and then evaluated.

Table 4. Chemical resistance property

Major wet process chemicals	Test condition soaking time at 25°C	Results
1% HF	5 min	Good
31% H_2O_2	5 min	Good
2.38% TMAH	10 min	Good
PGME	5 min	Good
PGMEA	5 min	Good
IPA	10 min	Good
Acetone	5 min	Good

Spin coating and bonding evaluation with blanket wafer

TOK TWM12000 series was utilized for actual bonding test with 300 mm blanket wafer. Bonding test includes adhesive spin coating, subsequent prebaking and bonding to carrier under vacuum condition with pressure. The two previously mentioned adhesives were coated on the wafer. Coating thickness ranging from 30 to 120 μm can be obtained by varying the spin rotation in a single coat. We chose a thickness of 50 μm for our testing. After the bake step, the wafer were brought into the vacuum bonding chamber and placed on bonder plate, followed by a mount by prepared non-perforated support carrier then bonded at each corresponding temperature (180°C for Sample-C and 215°C for Sample-D). After bonding TTV thickness was measured using a 57 point measurement probe with 3 mm edge exclusion. The measurement employ laser type optical micro thickness gauge. The bonding results provide the data of uniform bonding interface with no voids. The TTV after bonding was confirmed to achieve 2.12 μm (Sample-C) and 2.19 μm (Sample-D).

Wafer back grinding

The bonded wafers were further processed to wafer thin down process. Generic back grinding (BG) equipment was used for the test. Silicon thickness of approximately 50 μm was obtained after a two-step rough and fine grinding. Figure 9 shows thinned thickness mapping of blanket wafer with Sample-D, which indicates silicon TTV of approximately 2 μm as reflected bonding TTV. The evaluation items of wafer edge chipping, cracking, and adhesive delamination were not observed through optical microscope observation.

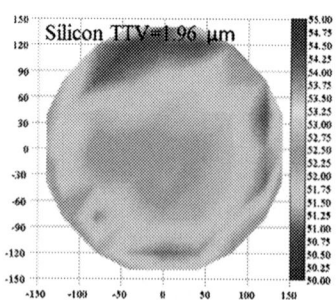

Figure 9. Silicon thickness distribution after thin down

Heat treatment test as a simulation for PECVD and reflow process

As next step of evaluation, the thinned bonded wafers were examined under vacuum and high temperatures condition for confirmation of backside process. Vacuum N_2 plasma chamber was used at a temperature of 220°C for 10 minutes with pressure of 1 Torr and power of 1000 W radio frequency (RF) to simulate PECVD process condition. On the other hand, curing furnace was used at 260°C for 10 minutes as a simulation of reflow process. Thermal treated wafer were examined after each process to confirm void generation, delamination and adhesive bleed from edge side by using microscope. No negative results were observed on this evaluation.

TTV trend

TTV change during temporary bonding process is regarded as the highest concern for thermoplastic material. Figure 10 shows TTV trend after each process step and good TTV trend through backside process simulation was observed.

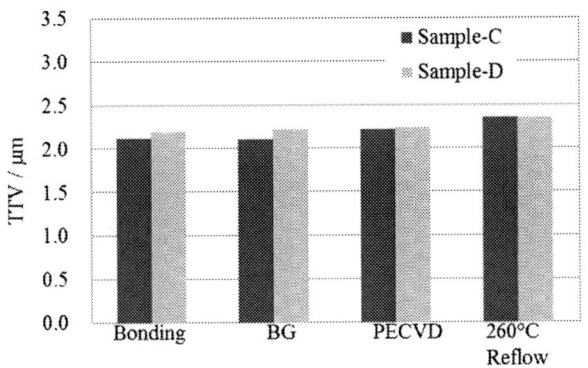

Figure 10. TTV trend after each process

Debonding and cleaning

Heat treated bonded wafers were proceeded to carrier debonding process. TOK TWR12000 series was used for debonding and adhesive removal from blanket wafer.

In general, temporary debonding processes are susceptible to cracking or chipping issues of the thinned wafers during carrier removal due to physical stress. In the TOK system, however, once the carrier is removed, the wafer is sent to adhesive dissolution and through a final rinse process to completely remove any remaining adhesive and then the thinned wafer on the tape frame in spin dried.

Cleaning result

After debonding carrier of the Sample-D wafer, particle counter was used to confirm residue and particle on the blanket wafer by using wafer surface inspection system LS6600 (Hitachi High-Technologies Corporation). Particle count of 24 (5 μm up) and 11 (10 μm up) were detected. This result looks quite few number of particle and confirms no residue after cleaning.

Evaluation of adhesive compatibility with bumped wafer

In the final portion of the evaluation, wafers with different bump heights were prepared for further study of the adhesive compatibility with device wafer. Wafer profile and test condition are summarized in Table 5.

Table 5. Bump height and adhesive treatment condition

Wafer Type	Bump height (μm)	Adhesive	Coating adhesive thickness (μm)	Bonding temperature (°C)
Wafer A	80	Sample-C	120	180
Wafer B	25	Sample-D	50	215

TOK TWM12000 was used for adhesive coating and bonding with 300 mm bumped wafer as described previously.

Sample-C was first coated on wafer A with sufficient coverage of 120 μm thickness and bonded to a prepared non-perforated carrier support glass at 180°C with pressure. On the other hand, the Sample-D was coated on wafer B with 50 μm thickness and bonded to the carrier at 215°C with pressure. Bonding defects such as void and bubble was not observed in the adhesive and at the interface of glass carrier with microscopy. The bonding results for wafer A with Sample-C and wafer B with Sample-D are shown in Figure 11. The wafer map shows good TTV even over topographies of 25 and 80 μm bump at low bonding temperatures.

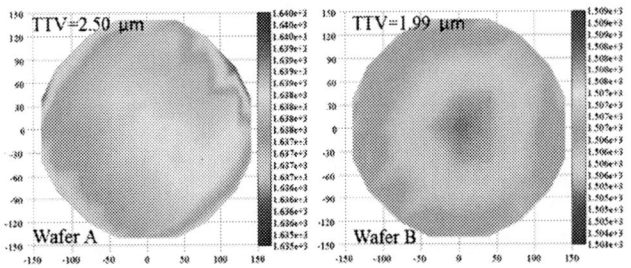

Figure 11. Thickness distribution after bonding of wafer A and wafer B

Figure 12 and 13 shows wafer warpage profile of the original bumped wafer and after bonding. Both bumped wafer show high warpage in its original state, but after bonding low warpage were maintained.

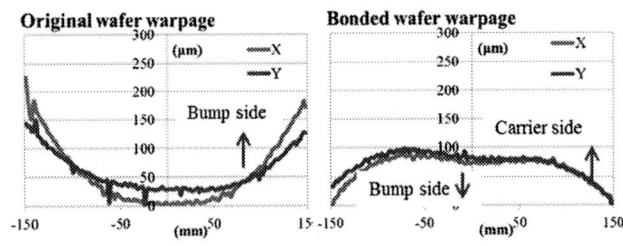

Figure 12. Warpage profile change before and after bonding of wafer A (80 μm height bump)

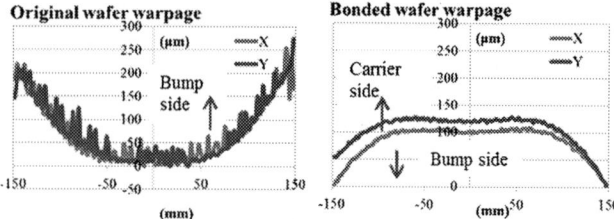

Figure 13. Warpage profile change before and after bonding of wafer B (25 µm height bump)

Heat treatment test for simulation of PECVD and reflow

Bonded bumped wafer also treated with thermal simulation test as described previously. The bumped wafers did not show any void and delamination.

Debonding and cleaning

Heat treated bonded wafers were proceeded to carrier debonding process. TOK TWR12000 series was used for carrier debond and adhesive removal from bumped wafer as well. The debonded result of wafer A is shown in Figure 14. The post debonding wafer observation did not indicate wafer damage such as cracking and chipping. Residue around the bump and bump deformation were not observed with SEM as shown in Figure 15.

Figure 14. Observation of wafer A after debonding

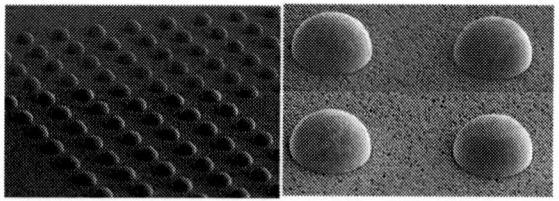

Figure 15. SEM images of wafer A after debonding

Additional heat treatment test of Sample-D

Additional thermal test was demonstrated with Sample-D by heat cycle test of PECVD simulation and thermal curing

process in order to confirm thermal stability limitation. For PECVD simulation, the test blanket wafer was bonded to the carrier glass with the adhesive and thin downed and then treated with a heat cycle step. The bonded wafer was treated in vacuum plasma chamber at temperatures of 200°C, 230°C, and 250°C each for 60 minutes. After each step, thickness variation (glass carrier plus adhesive thickness) was measured by optical micro gauge (Hamamatsu, C8125) and plotted as shown in Figure 16. There was almost no change observed.

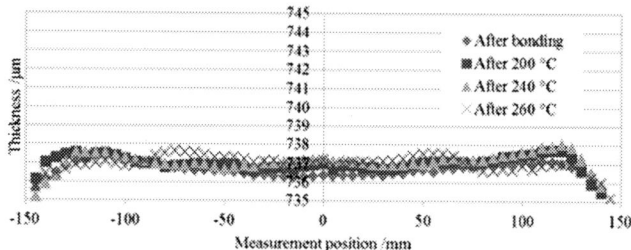

Figure 16. Thickness variation after thermal treatment (PECVD simulation for 60 min at each step)

Another bonded wafer was prepared for curing test as same as mentioned above, and processed in curing furnace at temperature of 220°C, 240°C, and 260°C each for 3 hours. Thickness variation was measurement after each step and plotted as shown in Figure 17. The result confirms that such extreme thermal condition does not affect thickness variation.

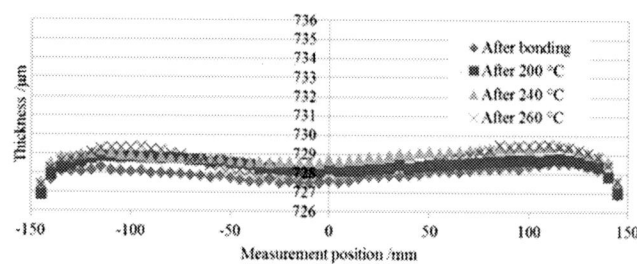

Figure 17. Thickness variation after thermal treatment (Curing for 3 hours at each step)

Conclusions

We have successfully developed a new concept thermoplastic temporary adhesive with temperature tolerance of up to 260°C with good removal property. The new adhesive also enables thermal bonding at moderate bonding condition (180°C to 215°C) with adequate thermal stability. Sample-D in particular exhibits thermal stability of low TTV change after excess heating cycle test of PECVD simulated process at temperatures of 200°C to 250°C for 60 minutes and curing process at temperatures of 220°C to 260°C for 3 hours.

We have also established a method for characterizing the adhesive behavior under thermal condition by measuring rheological properties dependence on frequency and found its correlation with thermal bonding property, which can be useful for evaluating thermoplastic temporary bonding adhesive.

Lastly, we have demonstrated temporary bonding process using 300 mm blanket and bumped wafers with the new adhesive. Good bonding TTV of 2 to 3 μm was obtained, and there was no significant issue such as edge void observed throughout the wafer processing. Consequently, debonding and cleaning also resulted in no residue and low particle on the wafer. Fundamental study for TGA, TDS, and chemical resistance of the adhesive also indicate sufficient stability against backside process condition.

References

1. Y.H.Chen et al., "Low Temperature Process Evaluation for 3DIC Integrated Thin Wafer Handling," in *Proc. 3rd IEEE International Workshop on low temperature bonding for 3D Integration,* Tokyo, May 2012,pp.235-240.

2. J.A. Sharpe et al., "Analyzing the Behaviour and Shear Strength of Common Adhesives used in Temporary Wafer Bonding" in *Proc. IEEE Electronic Components and Technol. Conf. (ECTC),* Las Vegas, NV, May, 2013, pp. 94–99.

3. Kath Crook et al., "Dielectric Stack Engineering for Via-Reveal Passivation" in *Proc. IEEE Electronic Components and Technol. Conf. (ECTC),* Las Vegas, NV, May, 2013, pp. 576–580.

4. Anne Jourdain et al., "Integration of TSVs, wafer thining and backeside passivation on full 300mm CMOS wafers for 3D applications" in *Proc. IEEE Electronic Components and Technol. Conf. (ECTC),* Orland, FL, May, 2011, pp. 1122–1125.

5. Erh-Hao Chen et al., "Fine-pitch Backside Via-last TSV Process with Optimization on Temporary Glue and Bonding Conditions" in *Proc. IEEE Electronic Components and Technol. Conf. (ECTC),* Las Vegas, NV, May, 2013, pp. 1811–1814.

6. K. Tamura et al., "Novel Adhesive Development for CMOS-Compatible Thin Wafer Handling" in *Proc. IEEE Electronic Components and Technol. Conf. (ECTC),* Las Vegas, NV, June, 2010, pp.1239–1244.

Underfilling Techniques Comparison In 3D CtW Stacking Approach

A. Garnier, A. Jouve, R. Franiatte, S. Cheramy
CEA, LETI, MINATEC Campus
17 rue des Martyrs, F38054 GRENOBLE Cedex 9, France
arnaud.garnier@cea.fr

Abstract

Die stacking in 3D integration increasingly deals with smaller soldered joints on flip chips which have to meet reliability requirements especially thermal cycling, vibrations, shocks... Adding an underfill between stacked chips is a solution to improve the structural integrity of those joints. In this work, different underfilling techniques are compared in chip to wafer (CtW) approach: one capillary underfill (CUF) and three pre-applied underfills (PAUF) including one non conductive paste (NCP) and two wafer level underfills (WLUF). These underfilling solutions are assessed using a test vehicle including daisy chains for electrical tests. Preconditioning and temperature cycling tests were carried out to monitor reliability. CUF and NCP enable to get good interconnections electrical resistance after 500 cycles. On the other side, WLUF process currently appears to be harder to implement because of lack of reproducibility and polymer entrapment at the bonding interface, preventing a reliable electrical contact. Advantages and drawbacks of each underfilling processes are also discussed regarding for instance maturity, easiness of process, throughput, creeping risks, entrapment risks, fine pitch and fine gap compatibility. It is obvious that PAUF are inevitable for 3D high density involving gap between stacked chips. However, related process speed is still low for PAUF. Further work on products and processes is thus needed to get reliability performances and cost-effectiveness suitable with high volume manufacturing.

Introduction

Microelectronic performances improvement is currently following Moore's law which aims at scaling down transistor size. This trend was always followed by a decrease in cost per transistor. However, the move from 28nm to 20nm node is expected to raise this cost [1] for the first time (requiring longer design cycles and substantial investments) which will likely slow down Moore's law application. In parallel, 3D integration is considered as a way to carry on enhancing performances. In this approach, engineering is focused at the package scale instead of integrated circuit scale. 3D integration based on chip stacking and through silicon vias leads to shorter interconnections between assembled chips. Accordingly, 3D integration enables smaller package size, higher bandwidth and smaller power consumption.

Chip stacking is usually implemented with underfill integration. This extra material located between both stacked chips is used for different purposes. It acts as a physical barrier to moisture therefore limiting potential interconnections corrosion. It is required in case of overmolding to avoid air trapping between the chips that could subsequently burst when heating. And the major reason for integrating the underfill is its ability to lower strains and stresses in the interconnections when subjected to thermomechanical fatigue. Actually, it mechanically couples the chip to the wafer or substrate to reduce lateral strain between the two interfaces. Even in the case of silicon chip to silicon chip bonding where no global coefficient of thermal expansion (CTE) mismatch exist, fatigue life can be improved by as much as 2.5 times compared with no underfill configuration [2]. In this case, the underfill enables to decrease high local stresses originating from CTE mismatch between metallic layers forming the interconnections (Al, Cu, Ni, SnAg). Thus, although it involves an additional plastic strain located along the solder boundary [3], the underfill integration finally shows significant benefits.

Traditional underfilling technique is capillary underfill (CUF) process consisting in firstly die stacking meaning pick and place followed by mass reflow as a soldering process, and secondly underfill dispensing. The UF actually flows in between both parts to be assembled throughout the interconnection array. This technique was born in mid 60s and has been improved, mainly by decreasing the cure time and enhancing its ability to flow, in order to meet industrial requirements [4]. This technique is suitable for rather large interconnection pitch and die to die gap (>50µm in standard). It reaches its limits for high interconnects density and large chips which are scheduled in 2.5D and 3D applications. To overcome these challenges, pre-applied underfill (PAUF) were developed. Among them, non conductive paste (NCP) and wafer level underfill (WLUF) processes stand for promising techniques compatible with high density 3D interconnection. NCP consists in dispensing the underfill with a defined pattern on the bottom wafer before stacking the top chip and completing the assembly with a thermal-compression (TC) step. On the other hand, WLUF process starts with WLUF lamination on top wafer before the dicing step. The stacking is carried out by TC. In both PAUF techniques, the UF has to be driven out of the interconnections to create a good electrical contact [5].

The purpose of this study is to compare different underfilling techniques in order to select the one suitable for the targeted application. A chip to wafer configuration is implemented to get higher interconnection density than a chip to substrate would allow. However, the conclusion of this work may also be applied to chip to substrate case.

Process and layout descriptions

The first step was to process bottom and top wafers in order to create the pillars for interconnections and associated daisy chains. Bottom wafers are made of Cu/Ni/Au pillar with ~5µm as total height. These pillars were electroplated above aluminum pads lying on top of Cu damascene redistribution layer and mineral passivation. Top wafers are made of Cu/Sn3.5Ag pillars with a total height of ~25µm after reflow.

These pillars were electroplated above aluminum redistribution layer and mineral passivation.

Selected test vehicle consists in a 6x5mm² chip that includes more than 2000 interconnections with 25µm in diameter and 50µm in pitch. Figure 1 shows the pillar layout.

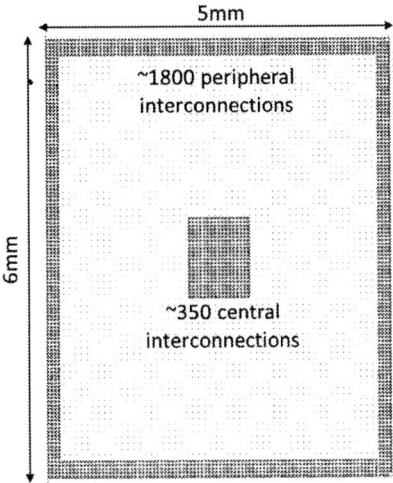

Figure 1. Test vehicle pillar layout

Bottom wafer can host ~180 top chips (figure 2). After assembly, the test vehicle enables to check electrical continuity between the chip and the wafer with daisy chains ranging from 1 interconnection (Kelvin measurement) to 340 interconnections.

Figure 2. Chip to wafer test vehicle with ~180 chips assembled to bottom wafer

Underfill introduction

For this study, 4 underfills were assessed: one CUF, one NCP and two WLUF (WLUF1 and WLUF2).

In the case of CUF, top wafer is thinned down to 200µm and diced. Then, chips are dipped in a flux and accurately placed on bottom wafer. Afterwards, the populated wafer undergoes reflow to bond each chip. Finally, CUF process takes place with a head jetting the underfill from the chip edge.

In the case of NCP, top wafer is also thinned down to 200µm and diced. Then, the underfill is dispensed according to an appropriate pattern on the bottom wafer, and chips are thermocompressed to spread the NCP and bond each top pillar to its bottom counterpart.

In the case of WLUF1, the product is a classical 30µm thick dry film that is stocked at low temperature ~5°C. Top

wafer is first thinned down to 200µm and then laminated by the dry film. Afterwards, the thinned wafer coated with WLUF1 is diced. The chips are eventually bonded to the bottom wafer by TC to make the WLUF flow and complete solder bonding.

In the case of WLUF2, top wafer is laminated with a 2 layers dry film comprising a 30µm thick WLUF layer in contact with the wafer front side and a 100µm thick underlying grinding tape. The laminated wafer is then thinned down to 200µm. The grinding tape is removed after the usual flip flop on dicing tape. The thinned wafer coated with WLUF2 is then diced. The end of process is the same as for WLUF1 i. e. TC bonding. It is worth mentioning that the 2 layers dry film has to be stocked at low temperature ~5°C.

The final step for all these underfills is curing at ~150-180°C.

Reliability tests and characterizations

After stacking, the wafers underwent preconditioning test according to JEDEC J-STD-020D Level 1. It includes 3 steps: firstly a 24h at 125°C thermal treatment, secondly a soak of 168h at 85°C under 85% relative humidity, and thirdly 3 reflows with maximum temperature between 255 and 260°C. After preconditioning, temperature cycling test (TCT) were carried out according to standard JESD22-A104D: wafers with bonded chips were subjected up to 500 cycles between -55°C and 150°C, with 10min soak time and 14°C/min temperature ramp.

Underfill fillet which corresponds to the polymer part extending on bottom wafer beyond the chip was observed using optical microscopy. Chip to wafer alignment and interfaces between the underfill and chip or wafer were monitored using transmission infra-red (IR) microscopy. Interconnections quality was assessed using electrical and physical characterizations. Electrical tests were carried out using four wires method to measure only the daisy chains resistance contribution and not access resistances. Kelvin resistance corresponds to one interconnection resistance and is located at the chip edge whereas the 340 daisy chain is related to the interconnections at the chip center. Kelvin resistance is the most accurate indicator related to the solder interconnection whereas 340 interconnections resistance is more a statistical indicator revealing the possibility to form long daisy chains without short cuts or opens. For physical characterization, interconnections cross sections were obtained after mechanical polishing and a focused ion beam (FIB) finishing. This FIB sputtering process is absolutely required to remove mechanical polishing induced defects and to reveal voids that could have been filled during the previous process. Inspection was completed by cross sectional scanning electron microscopy. FIB imaging was also used to enhance metallurgical phases contrast. Finally, chip surface deformation was assessed using confocal optical profilometry.

Results and discussion

Figure 3 shows top view of the chips after stacking with their underfill. The fillet is the black area bordering each chip. Figure 4 displays chip to wafer optical cross sections after stacking with a focus on the underfill fillet. Cross sections are located close to a chip corner to avoid dicing through the stack and a too large mechanical polishing distance.

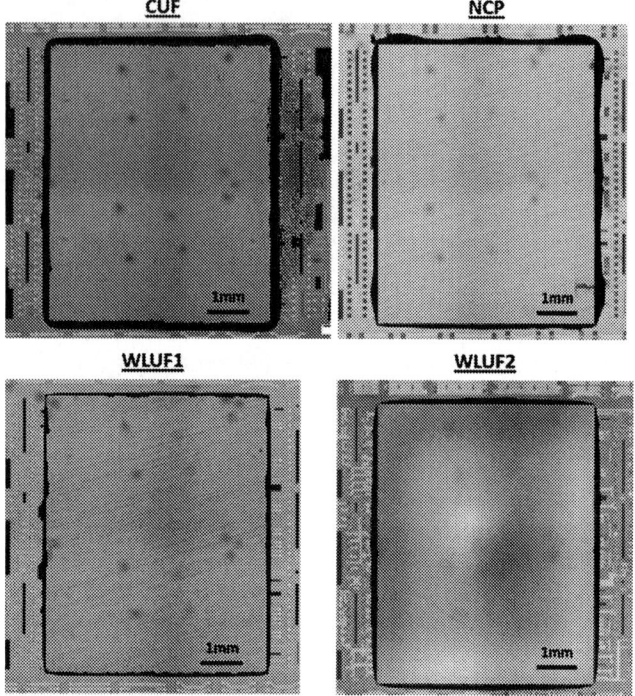

Figure 3. Chips top view after stacking with different underfills (CUF, NCP, WLUF1 and WLUF2).

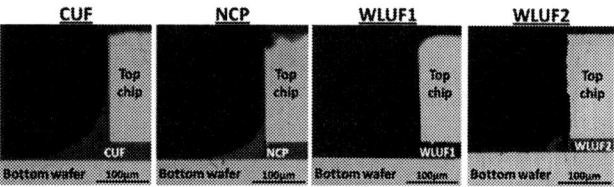

Figure 4. Chip to wafer optical cross sections after stacking with different underfills (CUF, NCP, WLUF1, WLUF2). Cross sections are located close to a chip corner.

CUF fillet is rather uniform and corners are well surrounded which is known as a favorable feature for reliability. The fillet extends as far as 200µm on the dispensed side and ~150µm on the opposite side. Moreover, the dispensed side features some underfill residues stretching up to 1.2mm. Regarding cross section image, CUF fillet has a typical liquid to solid good wetting shape (low wetting angle). In the case of NCP, the fillet surrounds well the entire chip despite a non-uniform extension ranging from 50 to 200µm. This uniformity can be improved by customizing the dispensed polymer pattern shape and size. Cross section image shows fillet with almost the same shape as CUF one but less perfect (larger wetting angle and less uniform slope curve). This is consistent with a polymerization beginning while flowing during the TC step, unlike CUF which completely flows until perfect wetting before cross-linking stage. It is worth mentioning that NCP is prone to creeping with the underfill contaminating chip backside and the TC head. Adjusting chip thickness, NCP volume and pattern, and TC force is the key to prevent creeping. In both WLUF configurations, WLUF fillet is visible on the chip sides but not around its corners. Fillet extends up to ~100µm for WLUF1 and ~200µm for WLUF2. This difficulty to spread the polymer around the corner means that WLUF ability to flow is limited. This is consistent with cross section images that show typical WLUF fillet shape with polymer squeeze out without wetting phenomenon. Thus the dry film viscosity moves to a minimum during TC but not as low as for NCP or CUF.

Figure 5 are IR microscopy images of the chips after stacking. Black patterns correspond to metals that are opaque to IR. Black squares are aluminum pad on top of which copper pillars are growth.

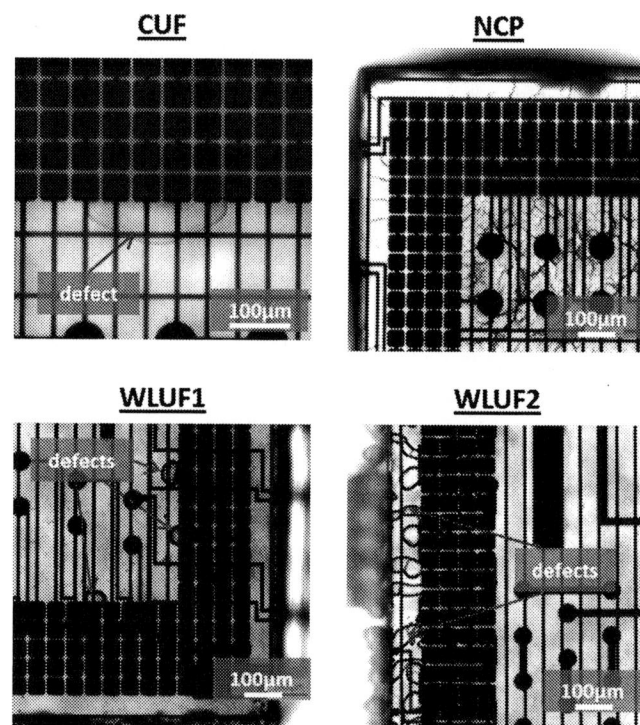

Figure 5. IR microscopy images of the chips after stacking with different underfills (CUF, NCP, WLUF1 and WLUF2).

In the case of CUF, some small isolated stretched bubble-like defects are detected. These defects are fixed to some array of interconnections. NCP case is different with cracks-like defects in the underfill, mainly located close to the chip corner but inside the peripheral interconnections rings. For both WLUF configurations, bubble-like defects occur but with different features. WLUF1 includes few isolated defects each fixed to one or two interconnections inside the peripheral interconnections rings, whereas WLUF2 features some areas with high bubble density located on chip border outside the interconnections ring.

After aging, NCP and WLUF1 keep the same features regarding these IR detected defects. In the case of CUF, preconditioning lead to the formation of new defects with stretched bubble shape (Figure 6). They are located mainly around the peripheral interconnections ring. It is worth mentioning that their contrast in IR microscopy is very weak. After TCT, no new defect occurs.

978-1-4799-2408-0/14 $31.00 © 2014 IEEE

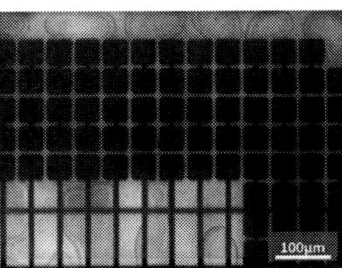

Figure 6. IR microscopy image after preconditioning of the chip stacked with CUF.

Regarding WLUF2, no aging test was carried out because no good reproducible electrical resistances were obtained. This particular case may be explained by the fact that WLUF2 had to withstand several steps before bonding including wafer thinning and dicing at room temperature and in water environment. The time spent out of storing specifications could lead to polymer alterations and subsequent bonding issues. For the other underfills, electrical tests were achieved before and after aging tests: figures 7 and 8 respectively give Kelvin and 340 interconnections resistance measurements.

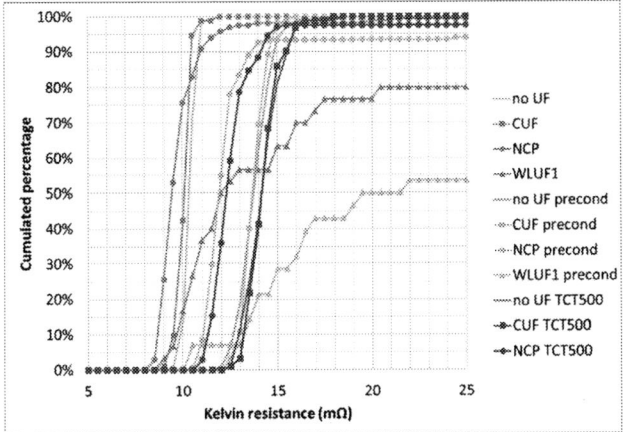

Figure 7. Plot of chips cumulated percentage as a function of Kelvin resistance: after stacking (blue lines), after preconditioning (green lines), after thermal cycling (red lines).

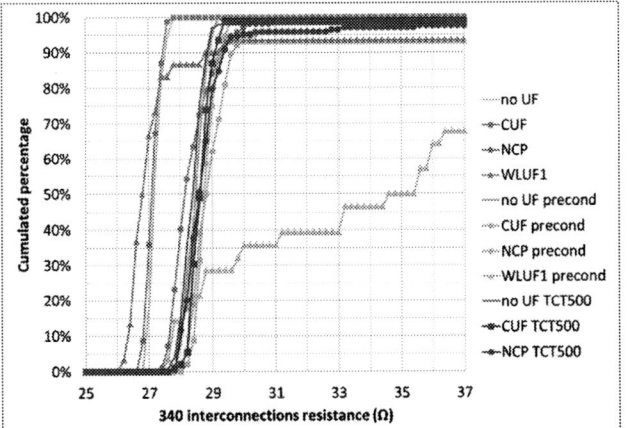

Figure 8. Plot of chips cumulated percentage as a function of 340 interconnections resistance: after stacking (blue lines), after preconditioning (green lines), after thermal cycling (red lines).

For no underfill, CUF and NCP configurations, resistances distributions are normal without significant broadening after preconditioning and 500 TCT. With a resistance criteria <30Ω for 340 daisy chains and <16mΩ for Kelvin, the yield is above 95% for these three configurations. However, preconditioning results in 2 to 3mΩ increase for Kelvin resistance and 0.7 to 1.8Ω rise for 340 daisy chain. The resistance evolution after 500 TCT is very slight: a little increase of less than 0.5mΩ for Kelvin resistance and quasi-steadiness for 340 daisy chains. This resistance increase mainly visible after preconditioning probably originates form interconnections evolution during the subjected thermal budget: intermetallic compounds (IMC) and remaining solder keep on reacting by interdiffusion leading to chemical changes (phases composition) and physical changes (phase thickness, voids occurrence). No degradation was highlighted for the case without underfill probably because cycle number is too low for the selected tests vehicle. For instance, discrimination was found for around 2000 to 4000 cycle in [4].

On the other hand, WLUF1 electrical behavior features worse results. Kelvin resistance distribution is broader with a 80% yield after stacking and 50% after preconditioning for criteria <20mΩ. 340 interconnections resistance distribution is rather good with almost 100% yield after stacking but drops down to 35% after preconditioning with criteria <30Ω. The results after stacking show that interconnection resistances are closer to the specifications for 340 interconnections than Kelvin are. This probably reflects an edge/center effect with a better interconnection quality at the center. Generally, chips well aligned but out of specifications have trapped underfill at the bonding interface. So it appears that WLUF1 is probably more easily compressed at the center of the chip than on edges. This may be ascribed to the combination of the polymer rheology, chip size, process cycle (time, temperature, force) and TC head pressure and temperature uniformity. Regarding after preconditioning results, both chip center and edge have a large part of abnormal interconnections resistance increases with eventually a lower yield. This probably means that polymer was initially trapped after stacking, even at the chip center.

To check if trapping occurred, SEM cross sections were achieved with results for CUF and NCP after 500 TCT on figure 9.

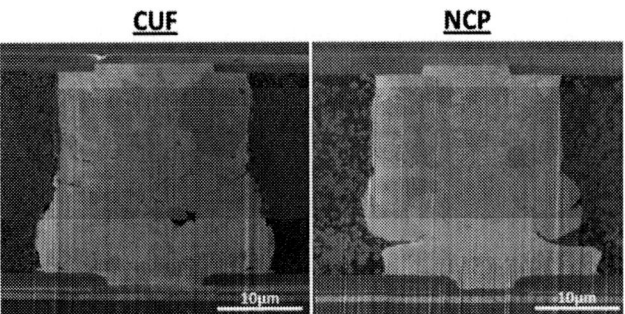

Figure 9. SEM cross section images for CUF and NCP after 500 TCT

In the case of CUF, the interconnection has a well-defined shape. The underfill correctly fills the gap between chip and wafer until the pillars borders despite little delamination located at the interface between CUF and the pillar. This delamination probably originates from cross section preparation (mechanical polishing and FIB) according to our experience. In the case of NCP, the interconnection shape is not perfect because bonding did not occur on the bottom pillar borders. Nevertheless, the underfill still properly fills the gap between chip and wafer. It is worth mentioning that filler entrapment can occur as depicted on figure 10 after stacking. Proper TC bonding force and ramp can prevent this issue.

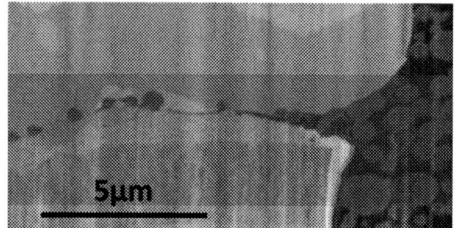

Figure 10. SEM cross section images for NCP after stacking with too low force

Figure 11 gives two different WLUF configurations after stacking.

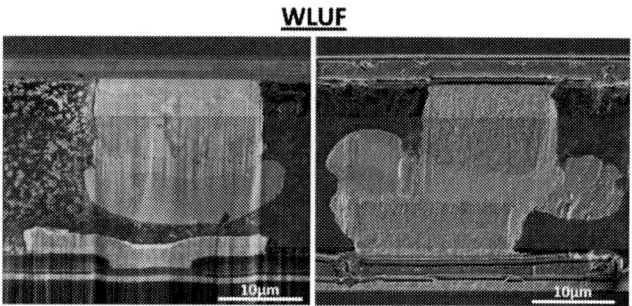

Figure 11. SEM cross section images for WLUF with two extreme configurations

On these cross sections, two cases are depicted: one where polymer entrapment is well visible and one where solder squeezes out without short cuts. Both are extreme configurations and are not especially related to WLUF1 or WLUF2. In the first case, electrical continuity is not secured. In the second case, electrical continuity is OK for some daisy chains and not for others on the same chip. This means that interconnection shape uniformity across the chip has to be improved. More generally, poor repeatability was obtained for WLUF regarding interconnection shape and electrical properties across one chip and between different chips with same TC bonding conditions.

Figure 12 depicts chip surface 3D images for no UF and CUF after 500 TCT.

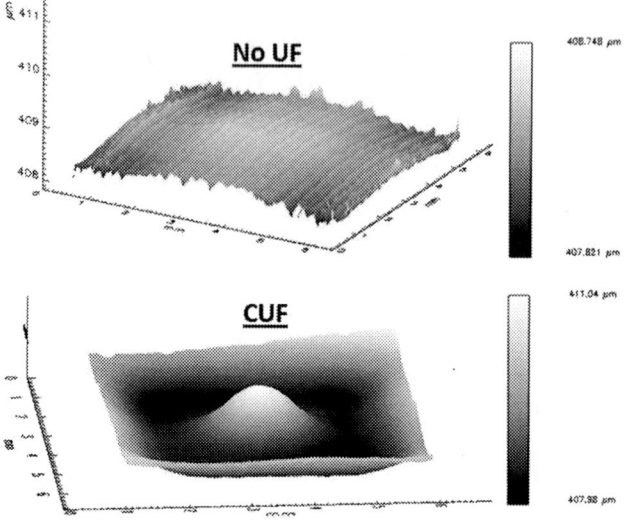

Figure 12. Chip surface 3D images for no UF and CUF after 500 TCT

A focus on quantitative deformation along one diagonal is displayed on figure 13.

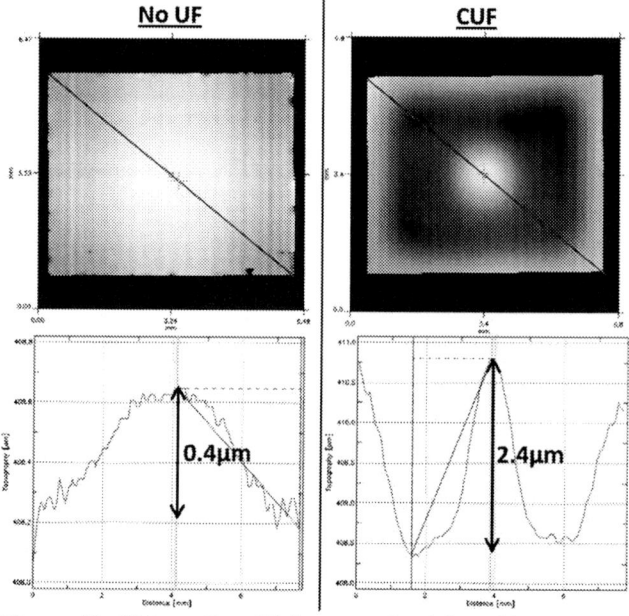

Figure 13. Chip surface 2D image and axial profilometry for WLUF2

Without underfill, chip surface is slightly convex with a 0.4µm altitude difference between the maximum at the center and the minimum on edges. In the case of CUF, chip surface has a shape that reflects the underlying interconnections. Maximum altitude points are located opposite to the interconnections, i. e. at the chip center and on the edges. For NCP and the other two WLUF configurations, the shape is roughly the same: the maximum altitude difference is around 2µm (+/-1µm). The general deformation shape obtained here can be explained as follows. After bonding, interconnections act as standoffs. Then, during the polymer cure ramp up to 150-180°C, the underfill should not impose much deformation because it is not fully cross-linked. After dwell time at cure temperature (typically 1h) the underfill turns to its final form with larger Young's modulus and CTE than before cure. After curing, because the underfill CTE (ranging from 30 to 140ppm/°C) is larger than the ones of solder (22ppm/°C), copper (17ppm/°C), Ni (13ppm/°C)s or IMCs (16ppm/°C) [3,4], the underfill size shrinks more than the interconnection pillars. So the underfill pulls down the thinned silicon membrane especially as it is far from the pillars standoffs, which leads to the obtained deformation shapes.

The same deformation shape and range suggest that the induced stress is approximately the same whatever the underfill.

Discussion

In this section, we will try to compare CUF, NCP and WLUF techniques regarding their potential throughput and application field. Of course, these features will largely depend on the following parameters: chip size, chip to wafer gap, interconnection pitch, interconnection density, interconnection design, underfill chemistry, tool configuration... So we will give magnitude orders and highlight bottlenecks in each process. Figure 14 displays process flows comparison between each underfilling technique.

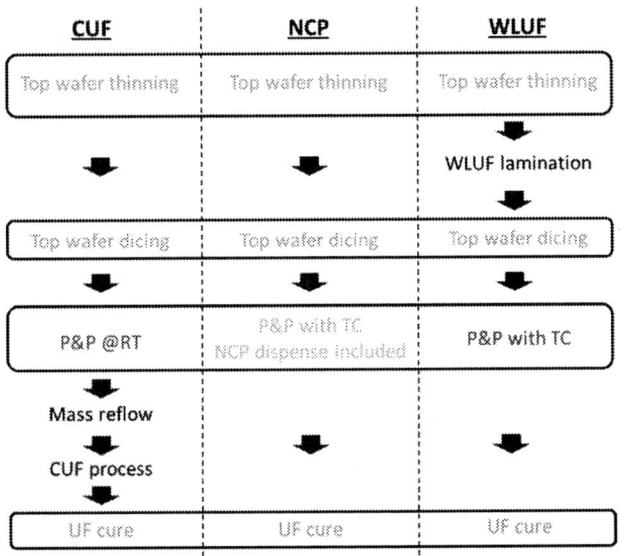

Figure 14. Process flows comparison between CUF, NCP and WLUF approaches (equivalent steps between each technique are highlighted in green)

The first step deals with top wafer thinning which is a common point for each technique. It usually includes grinding tape lamination, wafer grinding, and flip flop on wafer dicing tape. In the case of WLUF2, the grinding tape also includes WLUF film which enables to avoid the subsequent WLUF lamination step carried out on thinned wafer. Although this latter step would last around 1min per wafer, the advantage of WLUF2 technique would be to spare a lamination tool and an additional process step that extends time cycle. The following step is wafer dicing which is similar for CUF and NCP approaches. For WLUF technique, dicing is carried out through WLUF on wafer stack, involving potential delamination and dust issues. However, the throughput should be roughly the same regarding dicing for these three

techniques. Then, pick and place process is probably the most differentiating step. For CUF technique, stacking at room temperature enables to reach around 1000-3000 unit per hour (UPH) for a single head tool. This is followed by mass reflow lasting about 10min and CUF dispensing process. This latter step has a throughput ranging from 100 to 1000UPH that strongly depends on chip size, chip to wafer gap, interconnection pitch and design. For instance, larger chips lead to lower throughput. Another way for CUF process flow is to achieve the whole assembly in the pick and place tool by adding a thermal cycle with a small pressure to bond each chip individually. This configuration is considered for example if one needs to flatten a thinned chip prone to warpage during the assembly at 260°C, or if one need to do vertical multiple chips stacking with a localized reflow (in order not to melt underlying interconnections). For NCP technique, stacking is completed after NCP dispense and TC step. The wafer must be held at low temperature (<100°C) in order not to initiate too much change on the NCP before TC. Moreover, chip pick-up can be done at standard temperatures around 150°C. In these conditions, the limiting factor is the cooling rate after TC, leading to throughput in the range of 200 to 500UPH. It is possible to perform NCP dispense on the whole wafer on a dedicated tool. This alternative does not significantly lower the throughput and may enable to rationalize the tools and also to outgas NCP if needed to ensure void-free underfilling. Regarding WLUF approach, stacking is carried out by a TC step. The wafer can be hold at standard temperatures around 150°C because it includes no underfill. Because WLUF is on the chip, pick-up must be done at low temperature (<100°C) to avoid sticking issues. In these conditions, cooling rate is still the limiting factor, leading to throughput ranging from 50 to 150UPH. During pick and place step, optical detection may be a challenge with WLUF material because patterns have to be detected through the normally transparent polymer that also can hold back dicing dust. Finally, the last step which is common for each technique is underfill cure which has similar throughput for the three processes.

To summarize, for CUF technique, the bottleneck is CUF dispensing process whereas it is TC process for NCP and WLUF, especially cooling times.

Figure 15 lists advantages and drawbacks for the 3 techniques.

	CUF	NCP	WLUF
Process maturity	+	+/-	-
High chip density compatibility (2.5D and 3D)	-	+	+
High interconnection density compatibility	-	+	+
Fine gap compatibility (<30µm)	-	+	+
Large chip compatibility	-	+	+
Underfill entrapment	+	-	-
Fillet monitoring and shape	+	+/-	+/-
Underfill creeping	+	+	+
Backgrinding and sawing potential issues	+	+	-
Underfill coating througput	-	+	+
Assembly throughput (P&P tool)	+ 1000-3000UPH	+/- 200-500UPH	- 50-150UPH

Figure 15. List of advantages and drawbacks for CUF, NCP and WLUF approaches.

This comparison shows that CUF technique is a mature process with relatively high throughput as long as chips size is not too large, chip to wafer gap is large enough and interconnections density is rather low. Actually this technique is not suitable for 3D high density (fine interconnection pitch, fine gap and 2.5D multi chips stacking on interposers). NCP and WLUF are more suitable for this application field but are not currently as mature as CUF. Both these PAUF have to deal with polymer entrapment issues. According to this study, these problems are currently more significant for WLUF process certainly because WLUF has to turn from a solid state at ambient temperature (to withstand dicing step) into fluid state at medium temperature (to easily spread and allow electrical contact) before cross-linking at higher temperature. This situation is more complex compared to NCP where the product has to turn from a fluid state at low temperature into a solid state during the TC step. Regarding underfill fillet, WLUF enables to limit its extension but in turn fails to properly cover chip corners. NCP fillet extension can be monitored to get the proper shape but its fluid state at the beginning of TC can involve creeping if chip thickness is too low (<80μm for instance). Finally, WLUF which has to be laminated before dicing and even before grinding in the case of WLUF2 has to deal with polymer potential alteration before pick and place step, but also sawing issues as previously mentioned.

According to this analysis, each considered underfilling technique has its application range and its maturity level. PAUF are expected for 3D high density but need product and process/tool optimizations to meet high volume manufacturing requirements.

Conclusions

Different underfilling techniques including CUF, NCP and WLUF have been compared using a dedicated test vehicle in chip to wafer bonding configuration. Electrical and physical characterizations were carried out after stacking, preconditioning and temperature cycling tests. CUF and NCP enable to get good interconnections electrical resistance after 500 cycles. On the other side, WLUF process appears to be harder to implement because of lack of reproducibility and polymer entrapment at the bonding interface, preventing a reliable electrical contact. Moreover, these techniques were compared regarding their advantages and drawbacks. CUF is the most mature solution but is not suitable with 3D high density. NCP and WLUF are promising techniques to overcome CUF limitations. However, they would currently both have lower throughput than CUF mainly because of long cooling time just after TC step. NCP is more mature than WLUF but this latter technique seem to better fit for very thin chip (~50μm) stacking. Finally, PAUF are inevitable for 3D high density involving gap between stacked chips. But products and processes have to be improved to meet high volume manufacturing requirements.

Acknowledgments

The authors would like to express their grateful thanks to Lionel Vignoud for 3D chip profiles, Catherine Brunet-Manquat, David Bouchu and Perceval Coudrain for cross section preparation and observation, and finally to STMicroelectronics Corporate Packaging and Automation team for their support on backgrinding, dicing, CUF and NCP processes.

References

1. H. Jones, "Why the Technology Road Maps for the Semiconductor Industry Will Be Different", *Industry Strategy Symposium*, Half Moon Bay, CA, January 12-15, 2014.
2. K. Gilleo, D. Blumel, "The Ultimate Flip Chip-Integrated Flux/Underfill," in *Proc. NEPCON West*, Anaheim, CA., 1999, pp. 1477-1488.
3. Y.-L. Shen and R. W. Johnson, "Misalignment induced shear deformation in 3D chip stacking: A parametric numerical assessment", Microelectronic Reliability, vol. 53, no. 1, pp. 79-89, Jan. 2013.
4. H.-C. Cheng *et al.* "Interconnect Reliability Characterization of a High-Density 3-D Chip-on-Chip Interconnect Technology", *IEEE Trans. CPMT*, vol. 3, no. 12, pp. 2037–2047, Dec. 2013.
5. A. Taluy *et al.* "Wafer Level Underfill entrapment in solder joint during thermocompression: simulation and experimental validation", in *Proc. IEEE Electronic Components and Technol. Conf. (ECTC)*, Las Vegas, NV, May 28–31, 2013, pp. 768–772.

High Throughput Thermal Compression NCF Bonding

Toshihisa Nonaka, Yuta Kobayashi, Noboru Asahi, Shoichi Niizeki and Koichi Fujimaru,
Electronic & Imaging Materials Res. Labs.
Toray Industries Inc.
1-2 Sonoyama 3-chome Otsu, Shiga, Japan, 520-0842
Email: Toshihisa_Nonaka@nts.toray.co.jp

Yoshiyuki Arai, Toshifumi Takegami, Yoshinori Miyamoto and Masatsugu Nimura
Solution Center, Research & Development Div.
Toray Engineering Co., Ltd.
1-45 Oe 1-chome, Otsu Shiga, Japan, 520-2141

Hiroyuki Niwa
Toray International America Inc.
411 Borel Avenue, Suite 520, San Mateo, CA 94402
Email: h.niwa@toray-intl.com

Abstract

High through put thermal compression NCF bonding was studied and the new process consisting of dividing pre and main bonding, and the multi die gang main bonding has been developed. The dividing could change the process from serial to parallel and enabled to use the constant heated bonder head, which eliminated the time consuming head cooling process of the conventional serial thermal compression bonding. The die of 7.3 x 7.3 x 0.1 mm size with bumps of 38 x 38 μm^2 square Cu pillar covered by Sn-Ag cap, which had the pitches of 80 μm at peripheral and 300 μm at corer area, and the organic laminated substrate with Cu/OSP trace were used as the test vehicle in this study. Firstly, the dividing of pre and main bonding process in the case of single die was investigated. The pre bonding was the die placement to the NCF on the substrate, which was carried out at 150°C for 0.5 second. The substrate was kept at 80°C during the process. After the pre bonding the test vehicle was removed out from the equipment and cooled down to room temperature. And then it was mounted back to the equipment again and main bonding was carried out at 240°C for 20 seconds. The same substrate temperature as the pre bonding process was kept. Solder joint formation and NCF curing was made at the process. The assembled test vehicle was evaluated. The cross sectional observation results showed that the bump solder wetted the Cu trace on the substrate and no void was detected in the NCF by C-SAM observation. Secondly, the multi die main gang bonding was studied. The equipment was newly designed and built. 15 dies were pre bonded on the substrate with the same condition as that of the single die experiment. After the pre bonding was finished, the substrate was moved to the main gang bonder. During the transportation the substrate was cooled down to room temperature. The 15 dies were bonded at one time at 240°C for 10 seconds. The substrate was heated at 240°C during the process. The evaluation of the assembled dies revealed that the solder wettability of the joints and void detection in the NCF was almost the same as those of the single die pre and main divided bonding. This main bonding process time corresponded to 2700 UPH.

Introduction

Thermal compression bonding (TCB) technology is thought to become a major production process of chip by chip bonding in TSV-3D die stacking [1-2] and TSV-2.5D assembly [3]. In conventional chip on board flip chip assembly area TCB has already been used. Substituting for conventional mass reflow it is often used with pre applied adhesive especially in the case of thin and/or large die, and/or the die with fine pitch bumps, which are built up of Cu pillar and solder cap or Au stud. TCB combined with pre applied adhesive has various advantages which are, for example, the applicability of narrow gap bonding, the high accuracy placement of die on substrate, the stress relief just after cool down from solder joining temperature regarding avoiding white bump issue. There are two types of pre applied adhesive, one is a paste material which is called NCP (Non Conductive Paste) [4] and the other one is NCF (Non Conductive Film) which is a b-stage film [5-7]. NCF has some advantages. One is the higher controllability of fillet flow out. It is very important for thin die and/or narrow space adjacent dies bonding, which can be critical point for 3D die stacking and 2.5D assembly. Another one is wafer level process compatibility. NCF can be applied on wafer, and then singulated chip having the same size NCF on the surface was obtained by dicing process. Such chip can be assembled on substrate or wafer by TCB process without flux process nor capillary flow under fill applying process.

However, the throughput of current TCB process is lower than that of mass reflow process in general. Because TCB process is composed of die pick and place, and bump solder melt and the joint formation processes, which are carried out one chip by one chip on the same stage as a serial process. Improvement of the low throughput is important to meet the various demands of 2.5 D and 3D multi die stacking, and thin die flip chip bonding in consumer products.

In this paper the high throughput TCB bonding with the NCF was studied and demonstrated. The process is divided into the two steps. The first one is pre bonding which is pick and placement of a die, and the second one is main bonding which completes the bonding of the multi dies on the substrate at one time. This pre and main divided (PMD) bonding process has a potential of improving the throughput dramatically.

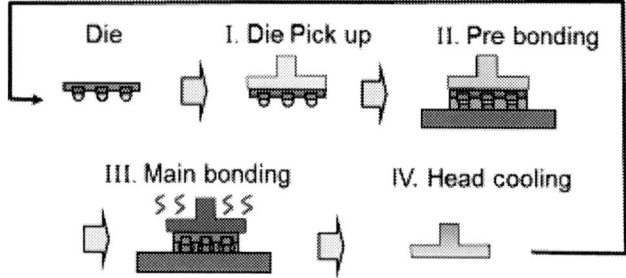

Figure 1. A conventional TCB process flow

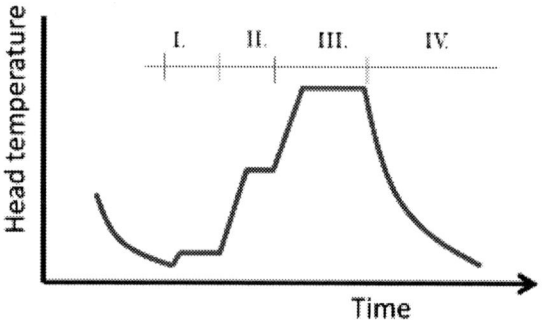

Figure 2. A typical head temperature profile of conventional TCB process.

A conventional TCB process flow is illustrated in Fig. 1, which composed of 4 major steps of die pick up (I), pre bonding (II), main bonding (III) and head cooling (IV). Figure 2 shows a typical head temperature profile of the 4 steps. It must be changed rapidly under precise control in the bonding cycle. A heater which can be heated up very rapidly is usually used to meet such requirement. At the process the head temperature should be initialized before the first step of die pick up starts, which needs the head temperature pushing down after the main bonding of the previous cycle. That is the fourth step of the head cooling and it takes a certain time, which influences the bonding throughput significantly.

PMD process concept is drawn in Fig. 3. It consists of 2 parallel processes of A and B. Process A is for pre bonding process, which is composed of die pick up, transportation, alignment and placement steps. The process is repeated until the substrate is filled by the dies. The die filled substrate is sent to process B and the next process A starts for a new substrate immediately. The head temperature of process A isn't needed to be changed during the whole process and a simple heater can be used, which only keeps the head temperature constant. Process B, which is a multi-die main bonding, starts with the plural dies pre bonded substrate. It is a simple process of just press with heat, which doesn't need an alignment procedure. The constant head temperature also can be adopted and it makes the head structure simple too.

Process A and B can be operated parallel. Typical head temperature profiles of process B is shown in Fig. 4. Pre applied NCF is used in process A. It works to prevent the moving of the die after the placement at the substrate transportation step from process A to B.

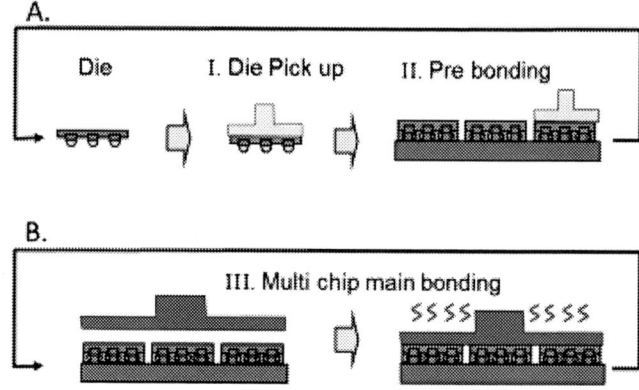

Figure 3. PMD process flow.

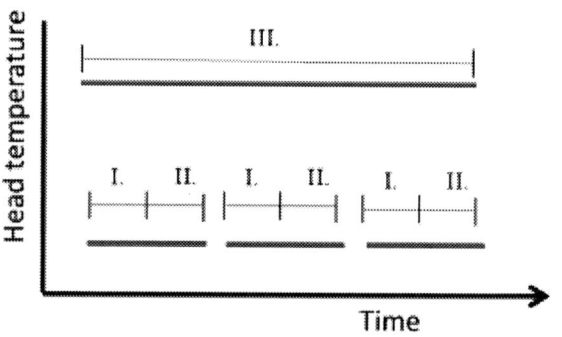

Figure 4. Typical head temperature profiles of PMD TCB process

Experimental Procedure

This research has been performed by 2 steps. The first one was concept proof of dividing pre and main NCF TCB bonding with getting the same quality of a conventional NCF-TCB process in which pre and main bonding were carried out serially. A commercialized TCB equipment of FC-3000 (Toray Engineering Co., Ltd) was used for both of pre and main bonding in this experiment. And the development of gang main bonding equipment and process and the demonstration were implemented as the second step. At the process all joints of plural pre bonded dies and the corresponded electrodes of the substrate were interconnected simultaneously.

7.3 x 7.3 mm² die with Cu pillar and Sn-Ag solder cap bumps which locate at both of peripheral and core area was used as a test vehicle in this study. The Cu pillar height was 30 μm and the thickenss of the Sn-Ag cap was 15 μm. The bump size and the pitch at peripheral area was 38 x 38 μm² square and 80 μm, respectively. Those bumps are aligned staggered in two rows. The same dimensions bumps were distributed with 300 μm pitch at the core area. The Cu trace of the test vehicle substrate was 20 μm thick, which had an OSP (Organic Solderability Preservative) treatment. The thicknesses of the die and the substrate were 100 and 356 μm, respectively. The 40 μm thick NCF was pre applied on the substrate by vacuum lamination. Each dimension of the test vehicle is indicated in Fig. 5.

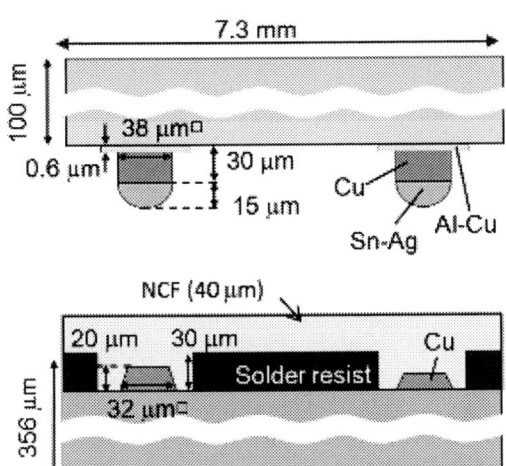

Figure 5. Test vehicle dimensions

Investigation of the solder wettability at the joints of the bump and the substrate electrode were performed with cross sectional observation by lapping. Void in the NCF after main boding was investigated by C-SAM (Constant-depth mode Scaning Acoustic Microscorpe). And microscope observation was also carried out after the specimens were parallel lapped.

Results & discussions
-Concept proof of dividing pre and main bonding

After some initial optimization work of each condition the concept proof of dividing pre and main bonding was demonstrated.

Firstly, the pre bonding was carried out. The TCB head temperature was kept constant and it made the die temperature at 150°C at the whole pre bonding process which consisted of die pick up, transportation, alignment and placement. The stage temperature was also kept constant, which was 80°C. The placement process took 0.5 second.

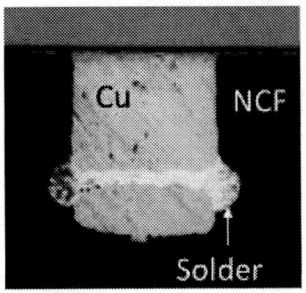 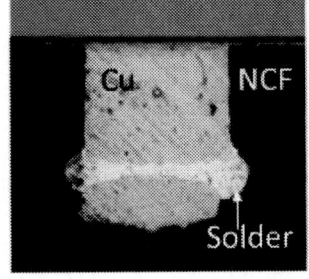

Figure 6. Typical cross sectional microscopic images of the joints at peripheral by the NCF-PMD process.

After the pre bonding process the die pre bonded substrate was removed from the TCB stage and kept at room temperature. After the TCB head temperature was raised up, the substrate was placed back to the stage again. The main bonding was performed at 240°C for 20 seconds on the stage kept constantly at 80°C.

The typical cross sectional microscopic images of the joints at peripheral area are shown in Fig. 6. The joints of the chip bumps and the substrate electrodes were welded well by the solder which was preformed on bump top.

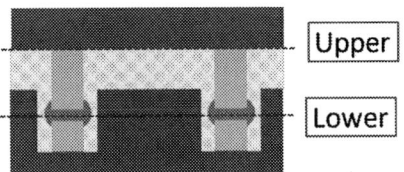

Figure 7. Void observation level.

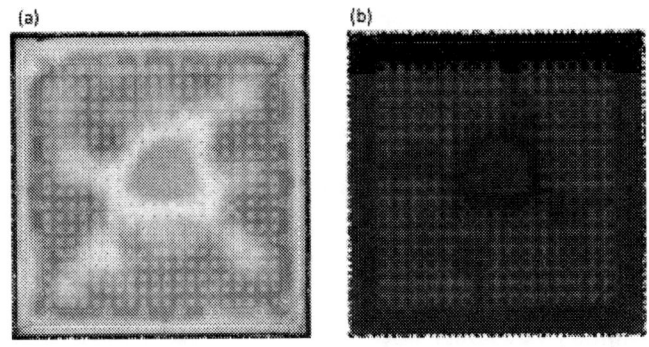

Figure 8. C-SAM observation results at the upper level (a) and the lower level (b), respectively.

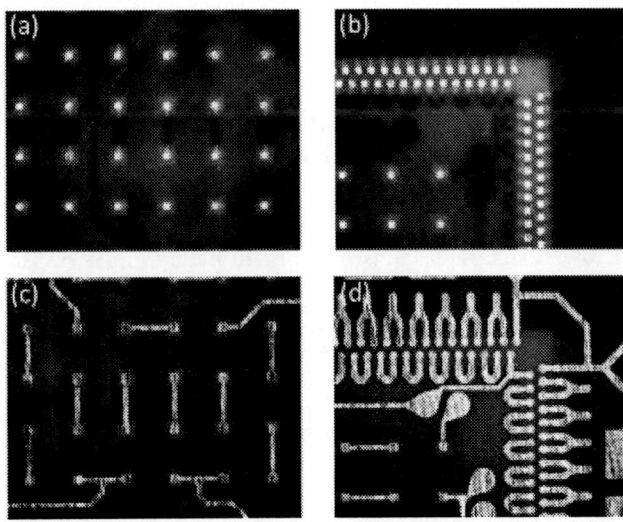

Figure 9. Microscopic observation results after parallel lapping of the package. (a), (b), (c) and (d) are images at the area of the upper level central, the upper level peripheral, the lower level central and the lower level peripheral, respectively.

Void in NCF after main bonding was investigated at 2 different levels normal to the substrate surface. Upper one was just underneath the die and the other lower was around solder joint level, which are illustrated in Fig. 7. The C-SAM observation results at the 2 levels are shown in Fig. 8. No void was detected in either the images of upper level (a) or lower level (b). Large shades in both images may reflected the swell of the package. After the C-SAM observation, the package was lapped parallel to the surface to the 2 levels step by step and observed by microscope and Figure 10 shows the typical results of the central and peripheral area. (a), (b), (c) and (d) in Figure 9 are images at the area of the upper level central, the upper level peripheral, the lower level central and the lower level peripheral, respectively. Void wasn't detected in each level at the whole area.

It is fundamentally confirmed that the NCF-PMD process can prepare the same quality level bonding by a conventional NCF-TCB process from the joint and the void detection investigation which were shown above. The process time balance of pre bonding and main bonding are described the formula below.

$$(ts + tp) \times N \approx tm$$

ts, tp, N and tm are pick up & transportation time, pre bonding time, die numbers per main bonding and main bonding time, respectively. Table 1 shows the estimated examples of each process time and corresponding main bonding time. As "tp" of above study is 0.5 second, assuming "ts" is 2.0 seconds, the pre bonding full process time becomes 2.5 seconds. As 2.5 (ts + tp) x 8 (N) = 20.0, which is case 2 in Table 1, if 8 dies can be bonded simultaneously, the main bonding process time of 20 seconds, which is adopted condition above, is balanced to the pre bonding process. In case of the die numbers per main bonding "N" is large enough which is like a case 1 in Table1, a long main bonding time "tm" can be applicable from the process time balance of pre and main bonding. However, when N is small, "tm" should become shorter, which is like case 3. Speeding up of pick &transportation and pre bonding time and enlarging of numbers of main bonding dies at one time enhance the productivity of NCF-TCB.

Table 1: Examples of pre and main bonding process time estimation and the balance

Case	Pick up & transportation time :ts (s)	Pre bonding time :tp (s)	Die numbers per main bonding :N	Main bonding time :tm (s)
1	2.0	0.5	15	37.5
2	2.0	0.5	8	20
3	2.0	0.5	4	10
4	2.5	0.5	8	24
5	1.5	0.5	8	16

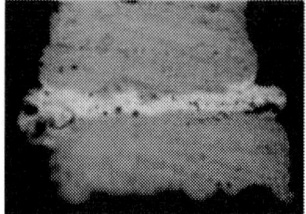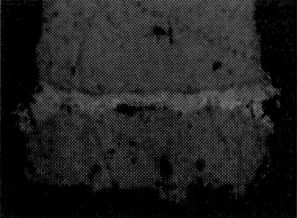

Figure 10. Typical cross sectional microscopic images of the joints at peripheral by the pre and main bonding divided process with NCF. Main boding time was 1 and 3 seconds, and shown in (a) and (b), respectively.

The impact of the shorter main bonding time was investigated focusing on the joint formation of the bump and the substrate electrode. Main bonding time of 1 and 3 seconds were carried out, in which the same pre bonding condition as the 20 seconds main bonding experiment were applied. The microscopic cross sectional observation results are shown in Fig. 10. Those revealed that both of the conditions could make the solder joints and the solder wetting to the substrate electrode from the bump. It didn't look significant difference in the joints among bonding time of 1, 3 and 20 seconds.

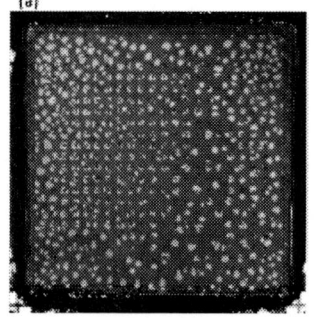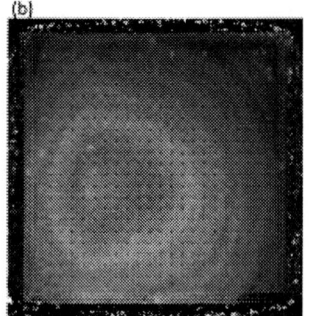

Figure 11. Upper level C-SAM observation results at of 1 (a) and 5 (b) seconds samples, respectively.

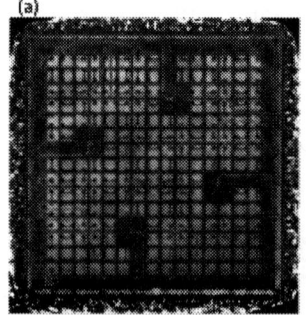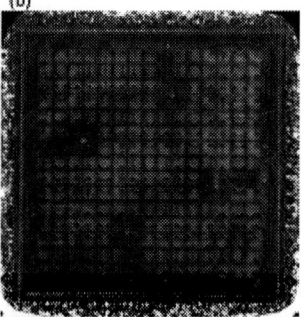

Figure 12. Lower level C-SAM observation results at of 5 (a) and 10 (b) seconds samples, respectively.

Bonding void of the fabricated samples by 1 and 5 second main bonding were also investigated by C-SAM observation at the upper level. The results were shown in Fig. 11. There were a lot of void in 1 second main bonding sample (a), but no void was detected from 5 seconds main bonding sample (b).

Figure 12 (a) and (b) show the lower level C-SAM observation results of 5 and 10 seconds main bonding samples, respectively. There were some void at peripheral area in Fig. 12 (a). Void wasn't detected from the 10 seconds main bonding sample shown in Fig. 12 (b). These results indicate the void at the upper area was disappeared between 1 to 5 seconds, but void still existed at the lower area at 5 seconds. The lower area void disappeared from 5 to 10 seconds.

-Gang main bonding

To perform the gang main bonding successfully the specially designed new equipment was developed. 15 dies were pre bonded on the substrate one by one with the same condition as that of the single die experiment above. After the pre bonding was finished, the substrate was moved to the new main gang bonder. During the transportation the substrate was cooled down to room temperature. The gang bonding was performed by a unit of block on the substrate, which includes 15 dies. The dies were simultaneously pressed with heat by the single large head. Its temperature was kept constant. Those processes images were shown in Fig. 13 (a) and (b). The substrate temperature of the gang main bonding was 240°C. The large main bonding head was also heated to 240°C.

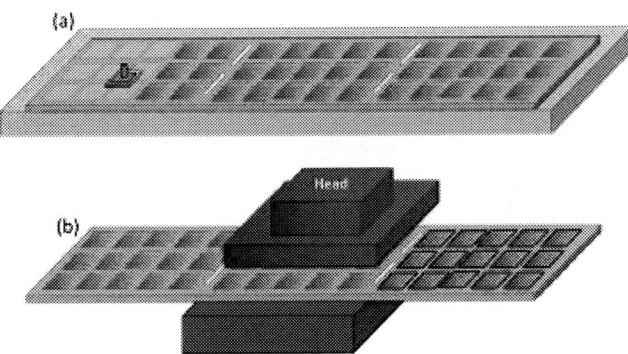

Figure 13. The image of the pre bonding and the main bonding processes are shown in (a) and (b), respectively.

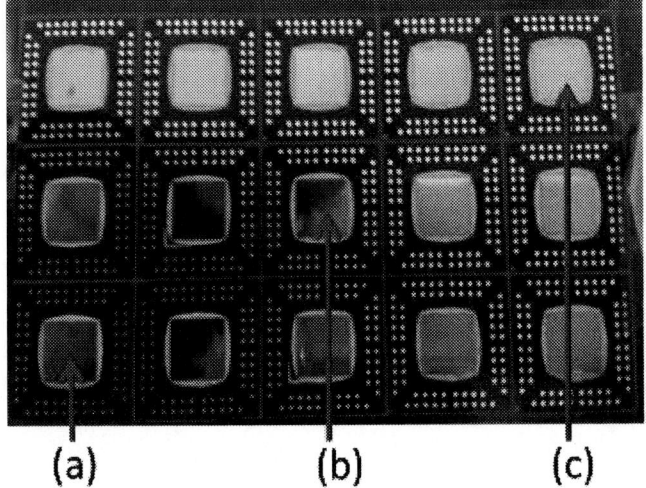

Figure 14. 15 dies were gang main bonded on a substrate simultaneously.

The gang bonded sample photo image is shown in Fig. 14. The 15 dies were main bonded simultaneously for 10 second, which is comparable to 2700 UPH (Unit per hour). The UHP number isn't behind that of a conventional mass reflow process.

One of the major concerns of the gang main bonding process was the shift of the dies which had been princely aligned, placed and fixed by NCF at the pre bonding process. The shift was evaluated by cross sectional observation of the dies. The results were show in Fig. 15 (a), (b) and (c) which were the cross sections of the dies of left-hand side bottom, center and right-hand side top in Fig. 14, respectively. The

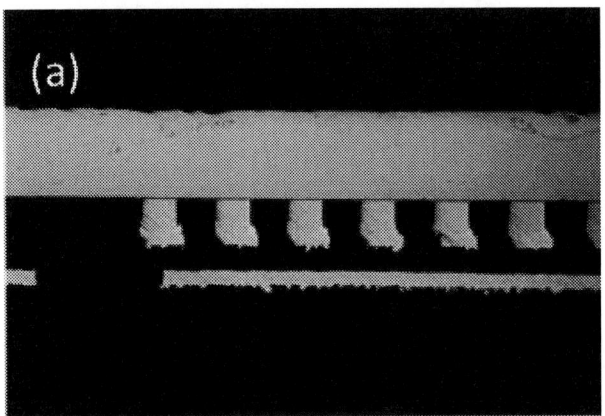

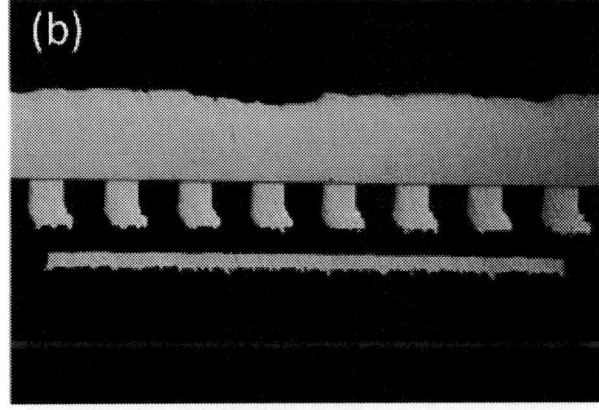

Figure 15. Typical microscopic cross sectional observation results of the main gang bonded die.

shifts of the corner dies of the block which were (a) and (c) were the same as that of the center die.

Bonding void evaluation was also implemented by C-SAM observation. Figure 16 shows the typical image. No void was detected.

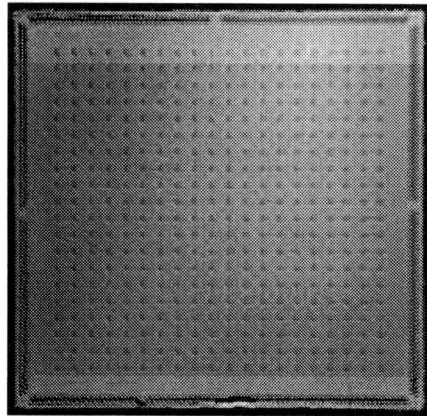

Figure 16. Typical C-SAM observation result of the main gang bonded die.

Conclusions

High through put NCF-TCB process was studied by the pre and main bonding divided (PMD) process, in which the main bonding was performed with multi die simultaneously.

As the first step the dividing of pre and main bonding with single die was investigated. The test vehicle was taken off from the bonding equipment after the pre bonding. Its temperature went down to room temperature. And then it was set back into the equipment again and the main bonding was carried out at 240°C for 20 seconds. The head temperature of these 2 processes was kept constant at each temperature. It meant that head cooling process could be eliminated. The test vehicle die and substrate were 7.3 x 7.3 x 0.1 mm size with bumps of 38 x 38 μm^2 square Cu pillar and Sn-Ag cap and the organic laminated substrate with Cu/OSP trace, respectively.

The assembled test vehicle was evaluated. The cross sectional observation results showed that the bump solder wetted the Cu trace on the substrate and no void was detected in the NCF by C-SAM observation.

As the second step the multi dies main gang bonding was performed with the newly designed and built equipment for the PMD process. After the one by one pre bonding of 15 dies on the substrate with the same condition as the first step. The substrate was moved to the main bonder after it passed the room temperature. The 15 dies were bonded at one time at 240°C for 10 seconds. The substrate was heated at 240°C. The evaluation of the assembled dies revealed that the solder wettability of the joints and void detection in the NCF was the same level as those of the first step PMD bonding die.

This main bonding process time corresponded to 2700 UPH. It was demonstrated that the potential of the productivity of the PMD process was comparable well to a conventional mass reflow process.

Acknowledgments

The authors thank to Mr. Sakabe for his work in NCF preparation.

References

1. Dong Wook Kim et al, "Development of 3D Through Silicon Stack (TSS) Assembly for Wide IO Memory to Logic Devices Integration," in *Proc. IEEE Electronic Components and Technol. Conf. (ECTC)*, Las Vegas, NV, May 28 - 31, 2013, pp. 77–80.
2. A. La Manna et al, "Challenges and Improvements for 3D-IC Integration Using Ultra Thin (25μm) Devices," in *Proc. IEEE Electronic Components and Technol. Conf. (ECTC)*, San Diego, CA, May 29 – June 1, 2012, pp. 532–536.
3. Zhe Li et al, "Development of an Optimized Power Delivery System for 3D IC Integration with TSV Silicon Interposer," in *Proc. IEEE Electronic Components and Technol. Conf. (ECTC)*, San Diego, CA, May 29 – June 1, 2012, pp. 678–682.
4. Yanggyoo Jung et al, "Development of Large Die Fine Pitch Flip Chip BGA using TCNCP Technology," in *Proc. IEEE Electronic Components and Technol. Conf. (ECTC)*, San Diego, CA, May 29 – June 1, 2012, pp. 439–443.
5. Nonaka, T. et al, "Development of Wafer Level NCF (Non Conductive Film)," *Proc 58th Electronic Components and Technology Conf*, Orlando, FL, May. 2008, pp. 1550-1555.
6. Nonaka, T. et al, "Wafer and/or chip bonding adhesives for 3D package," Proc IEEE CPMT Symposium Japan, Aug. 2010, 10-4.
7. Nonaka, T. et al, "Low temperature touch down and suppressing filler trapping bonding process with wafer level pre applied underfilling film adhesive," *Proc 62th Electronic Components and Technology Conf*, San Diego, CA, May. 2012, pp. 444-449.

Through Silicon Underfill Dispensing for 3D Die/Interposer Stacking

Fuliang Le[1], S. W. Ricky Lee[1,*], Kei May Lau[2], C. Patrick Yue[2]
Johnny K. O. Sin[2], Philip K. T. Mok[2], Wing-Hung Ki[2], Hoi Wai Choi[3]
[1]Department of Mechanical and Aerospace Engineering, Hong Kong University of Science and Technology
[2]Department of Electronic and Computer Engineering, Hong Kong University of Science and Technology
[3]Department of Electrical and Electronic Engineering, University of Hong Kong
*(Tel) +852-23587203, (E-mail) rickylee@ust.hk

Abstract

This study describes a through-silicon-underfill dispensing that the encapsulant is dispensed through through-silicon-vias (TSVs). The TSVs function as entrances for encapsulant dispensing or paths for fluid flow. Typically, the inflow for TSV dispensing may be flow with a constant speed or free droplets. A model was developed to investigate the filling time and the pressure distribution for the quasi-steady, radial and laminar flow between parallel plates. Compared with free droplets, a constant inflow has shorter filling time at the expense of increasing the fluid pressure. 3D stacking with the same-size interposers forms several planar sidewalls. Encapsulant may flow out from the edges of the sidewalls and form an edge flood failure if the edge flow of an underfill can overcome the surface force. An optimized pattern of TSVs was designed for the underfill of a 3D package to identify the trade-off between the filling efficiency and the lower risk of the edge flood. In each interposer, the TSVs are classified into two groups: the central group is dispensed by a constant inflow whereas the outer group is dispensed by free droplets; the inflow of free droplets eliminates the risk of an edge flood. A four-stack 3D package with the optimized TSV pattern was developed for validation. Edge dispensing was used to fill the gaps at the bottom levels as much as possible. Subsequently, the remaining gaps were encapsulated by TSV dispensing (with both a constant inflow and an inflow of free droplets). Inspections by scanning acoustic microscopy and cross-sectioning verified that the combined underfill could result in a void-free encapsulation and suitable fillets.

Introduction

Over the years, 3D die/interposer stacking with TSVs has emerged as a solution for a smaller footprint and higher performance. Multiple dies and interposers are vertically stacked to allow the use of the third dimension for electrical interconnections. This leads to a significant improvement in silicon efficiency [1]. As the packaging technology moves to the 3D stacking with TSVs, the demand for the underfill of the stacked gaps among vertical dies or interposers also arises. The stacked dies or interposers usually have the same sizes; therefore it is almost impossible to underfill the stacked gaps with conventional edge dispensing since there is no space to form "encapsulant reservoirs" for capillary action. In current products, edge dispensing is only partially applied between the substrate and the flip chip at the bottom level, but the stacked gaps are still not filled.

To address this challenge, multiple pre-applied underfills (PAUF) have been developed by various research teams. The PAUF is generally pre-applied either onto a bumped wafer or a wafer without solder bumps, using an appropriate technique such as printing or coating. One typical product is a solid "underfilm" tape [2]. The solid underfilm, which consists of a thermoset/thermoplastic composite, is laminated onto the bumped substrate in vacuum. Heat is applied under vacuum to ensure the complete wetting of the underfilm and an optional patterning process can be carried out to expose the solder bumps. The subsequent assembly is a process like no-flow underfill in which a curable flux adhesive is applied on the board and then the package is reflowed. Another popular pre-applied process developed by Aguila Technologies makes use of coating and laser drilling techniques. The highly filled underfill is screen printed onto an unbumped wafer and then cured. This material is laser ablated to form vias that expose the bond pads. The vias are filled with solder paste and then reflowed. Pre-applied underfill has the common challenges of non-uniform deposition, difficult vision recognition of chip alignment, high investment of the wafer-level facilities, short shelf-life and poor wetting of the underfilm [3].

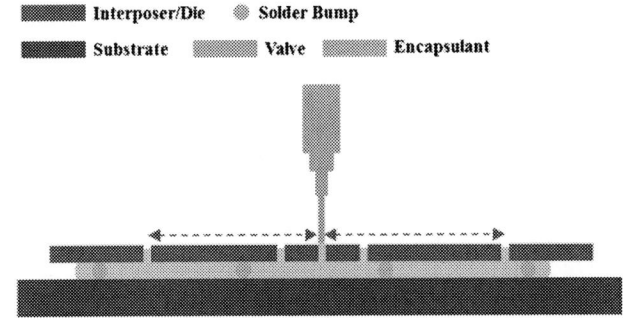

Figure 1. Schematic diagram of TSV underfill dispensing

A dispensing approach through TSVs, which is described in our previous researches [4] [5], has also been developed to solve the underfill application. A schematic diagram of TSV dispensing is shown in Figure 1. The TSVs function as entrances for fluid dispensing or paths for fluid flow. TSV dispensing is applied after the flip chip interconnects are formed, and the encapsulant is dispensed through TSVs other than edges. TSV dispensing is completely compatible with the current underfill industry, thus it is unnecessary to spend a vast amount of money to upgrade current equipment.

Analysis of through silicon underfill dispensing

Each application in the underfill dispensing has its own specific fluid volume to be dispensed to assure the proper fillet size that finally impacts on the stress of solder interconnections [6]. Typically, an auger valve or a jet is used in the underfill to control the dispensing volume and to form acceptable fillets to decrease the shear stress at solder joints or the stress around the bottom corner of a flip chip. The flow behavior during the underfill is strongly related to the boundary condition of the inlet that is exerted by a dispensing valve. The inflow dispensed by an auger valve can be a constant volumetric flow rate (Figure 2) if the motor speed is constant, or it can be free droplets (Figure 3) if the valve dispenses encapsulant discretely [5]. The jet valve is a non-contact dispenser with only the output of free droplets as the encapsulant is punched out instead of squeezed.

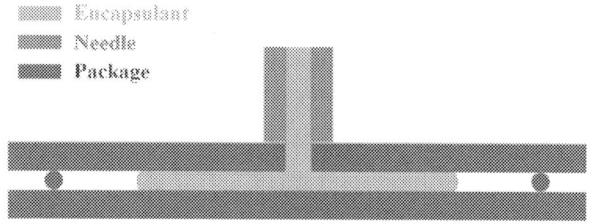

Figure 2. Inlet boundary: constant flow rate

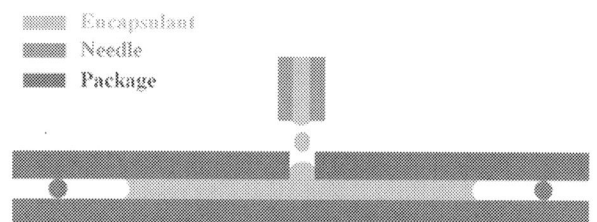

Figure 3. Inlet boundary: free droplets

The fluid flow under TSV dispensing can be modeled as a two-dimensional, laminar and creeping flow between two parallel plates as the Reynolds number of the fluid flow is vanishingly small ($Re \ll 1$). Considering a bidirectional flow on the rz-plane (direction z is perpendicular to the substrate as shown in Figure 4), the continuity equation is automatically satisfied by introducing Lagrange's stream function $\psi (r, z)$:

$$v_r = \frac{1}{r}\frac{\partial \psi}{\partial z}; \quad v_z = -\frac{1}{r}\frac{\partial \psi}{\partial r} \quad (1)$$

where v_r and v_z are respectively the velocity component of r, z. The pressure p can be eliminated by differentiating Navier-Stokes equation with respect to r and z respectively, and by subtracting one component from the other. Substituting v_r and v_z, in terms of ψ, into the resulting equation leads to [7]

$$E^4 \psi = E^2 (E^2 \psi) = 0; \quad (2)$$

where the differential operator E^2 is defined by:

$$E^2 \equiv \frac{\partial^2}{\partial r^2} - \frac{1}{r}\frac{\partial}{\partial r} + \frac{\partial^2}{\partial z^2} \quad (3)$$

In the TSV dispensing, the stream function ψ is stipulated to satisfy the power-law function to separate the axial from the radial dependence with a constant λ:

$$\psi = r^\lambda f(z) \quad (4)$$

A. Solutions for a constant inflow

Introducing the cylindrical coordinates in a through silicon underfill dispensing as shown in Figure 4 and employing the defined stream function, the general solutions of the control volume (CV) 1 and 2, irrespective of the boundary conditions, are given by

$$\psi(r,z) = \begin{cases} r^2 (c_2 z^2 + c_3 z^3) & \text{(in CV 1)} \\ c_4 - \frac{3}{2}c_7 hz^2 + c_7 z^3 & \text{(in CV 2)} \end{cases} \quad (5)$$

$$p(r,z) = \begin{cases} -3\mu c_3 (2z^2 - r^2) - 4\mu c_2 z + C_1 & \text{(in CV 1)} \\ 6c_7 \mu ln r + C_2 & \text{(in CV 2)} \end{cases}$$

where μ is the viscosity of the encapsulant and six constants c_2, c_3, c_4, c_7, C_1 and C_2 are governed by boundary conditions. Assuming all the walls are in the no-slip condition and the volumetric flow rate of the inlet boundary is constant Q, then the analytical solutions for the constant inflow are

$$t_c = \frac{\pi h r^2}{Q} \quad (h \gg h^*, h^* \to 0) \quad (6)$$

$$p_c(r) = \overline{p_1} - \frac{6Q\mu}{\pi h^3}ln\frac{2r}{\varepsilon h} \quad (r > \frac{d}{2}) \quad (7)$$

$$\overline{p_1} = \overline{p_0} + \frac{4Q\mu}{\pi h^3}\left(\frac{7}{5\varepsilon^2} + \frac{1}{10}\right) \quad (8)$$

where t_c is the filling time when the flow radius is r (the subscript c means a constant inflow), p_c is the relative pressure of the fluid when the radius is r, p_i is the relative pressure of plane i (or surface i), d is the diameter of the TSV, h is the gap, and ε is a ratio ($\varepsilon = d/h$). Actually, d is determined by the size of the dispensing needle [5]. Currently, the minimum inner diameter of the needle supplied by EFD Company is 250µm.

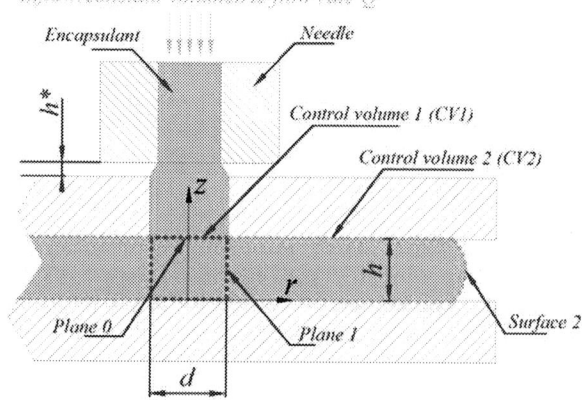

Figure 4. Diagram of TSV dispensing with a constant inflow

B. Solutions for an inflow of free droplets

The filling time and the pressure of the TSV dispensing with an inflow of free droplets can be obtained by applying equation 5 on the control volume 2 in Figure 5. As the inflow is free droplets, the relative pressure inside the control volume 1 is equal to zero. Assuming the diameter of the TSV is far smaller than the flow radius, the radial velocity v_{rd} (the subscript d means an inflow of free droplets) of the fluid is

$$v_{rd}(r,z) = \frac{z(h-z)p_{II}}{2\mu r ln\frac{2r}{d}} \quad (r > \frac{d}{2}) \quad (9)$$

where p_{II} is the relative pressure of surface II, μ is viscosity. Assuming contact angle α between the encapsulant and the flip chip is the same as that between the encapsulant and the substrate, the relative pressure p_{II}, which is determined by the Laplace-Young equation, is described as

978-1-4799-2408-0/14 $31.00 © 2014 IEEE

$$p_{II} = -\frac{2\sigma\cos\alpha}{h} \tag{10}$$

where σ is the surface tension coefficient of the encapsulant.

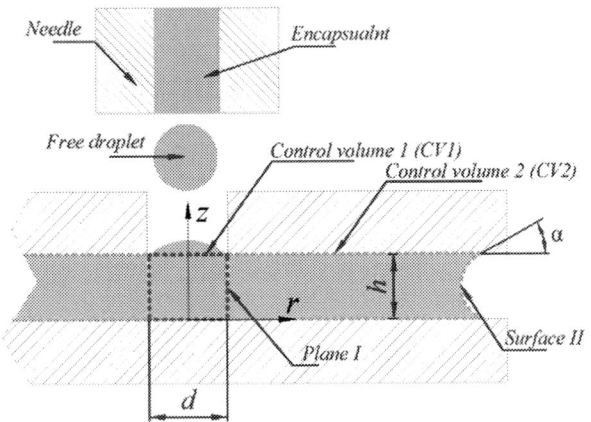

Figure 5. Diagram of TSV dispensing with free droplets

Assuming all the walls are in the no-slip condition and contact angle α is equivalent to static contact angle α_c during underfill dispensing, the filling time t_d (dispensed by free droplets) is

$$t_d = \frac{\mu}{h\sigma\cos\alpha_c}\left[3r^2 ln\left(\frac{2r}{d}\right) - \frac{3}{2}r^2 + \frac{3}{8}d^2\right] \tag{11}$$

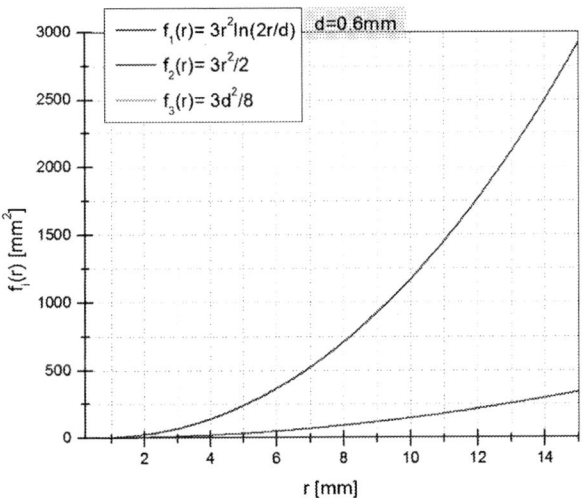

Figure 6. Comparison of three terms in the equation 11

Notice that equation 11 consists of three terms. The variation of the three terms versus flow radius r is plotted in Figure 6. The comparison of the three terms shows that it is sufficient to only take the first term $f_1(r)$ for considering the approximate solution of the filling time. That is, by ignoring $f_2(r)$ and $f_3(r)$, the equation 12 or the filling time under the inflow of free droplets becomes

$$t_d \approx \frac{3\mu r^2}{h\sigma\cos\alpha_c}ln\left(\frac{2r}{d}\right) \tag{12}$$

Model verification and comparison of the two inflows

The correlations between the filling time and the flow radius under a constant flow rate and free droplets are verified by experiments. The configuration of the experimental sample is shown in Figure 7. A silicon flip chip is bonded to a silicon substrate, and the dimension of the flip chip is 30mm×30mm. The average gap between the flip chip and the substrate is 55μm, 90μm or 230μm. The temperature of the substrate and the dispensing valve are the same, 60°C. The experiment only considers bump-free area. A shear viscosity of 0.27Pa.s was measured at an average shear rate of 20s^{-1} and a temperature of 60°C. A static contact angle of 23° was measured on a silicon wafer at a temperature of 60°C. A surface tension of 13.08mN/m was determined by the drop weight method and the Guggenheim-Katayama equation.

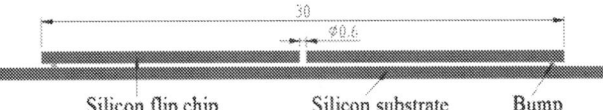

Figure 7. Sample structure for the model verification

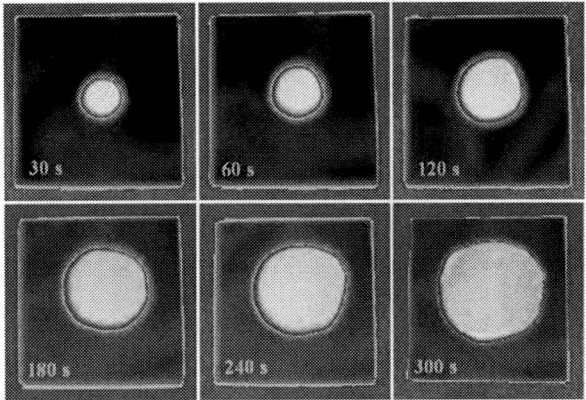

Figure 8. Experimental result of TSV dispensing with a constant volumetric inflow of 0.05μL/s under a 55μm gap

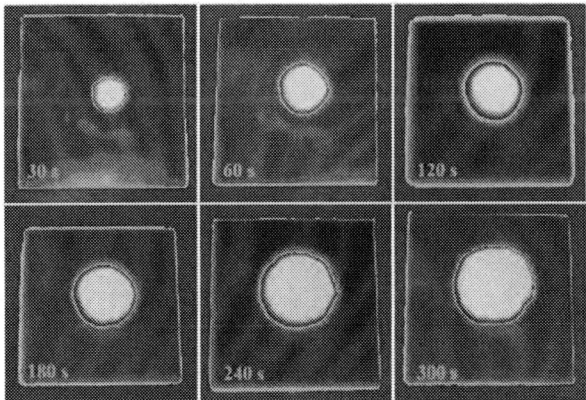

Figure 9. Experimental result of TSV dispensing with an inflow of free droplets under a 55μm gap

The actual flow radius in the experiment was observed by an infrared camera. The front surface of the fluid flow can be observed clearly since both the silicon flip chip and the silicon substrate are transparent in the infrared camera's images. Figure 8 shows the experimental result of TSV dispensing with a constant volumetric inflow of 0.05μL/s under a 55μm gap and Figure 9 shows the result with an inflow of free droplets under the same condition. The average flow radius r_e in the experiment is evaluated by

$$r_e = \sqrt{\frac{P_F S_C}{\pi P_C}} \qquad (13)$$

where P_F donates the pixels of the filled area in the infrared image, P_C donates the pixels of the flip chip and S_C is the area of the flip chip. For the TSV dispensing with constant inflows, comparisons of the average flow radius r_e with the analytical solution under the gap of 55μm, 90μm or 230μm are shown in Figure 10. It should be noted that the maximum actuator power of an auger valve is constant; therefore the volumetric flow rate will increase with the gap as shown in equation 7. The volumetric flow rate Q in Figure 10 is the value when the power of the actuator is set as maximum.

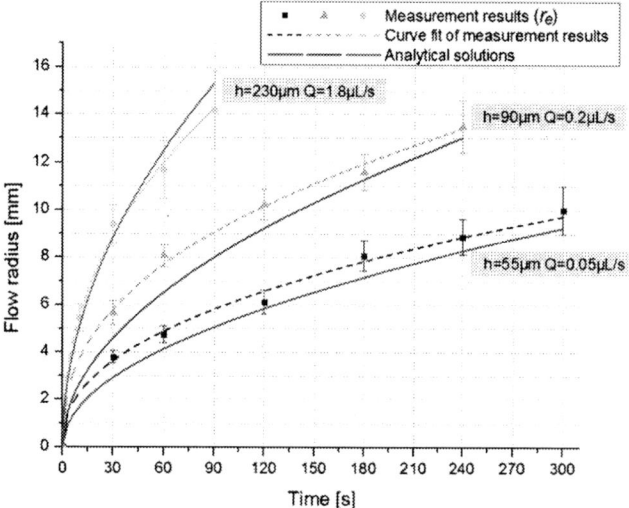

Figure 10. Comparison of the analytical solutions (equation 6) with the experimental results at the gap of 55μm, 90μm or 230μm (inflow = constant volumetric flow rate)

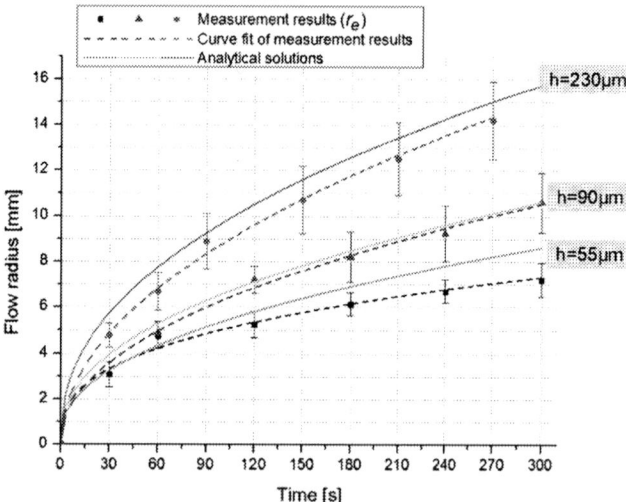

Figure 11. Comparison of the analytical solutions (equation 12) with the experimental results at the gap of 55μm, 90μm or 230μm (inflow = free droplets, shear rate = 20 s⁻¹)

Figure 11 shows the comparisons of TSV dispensing with an inflow of free droplets under the gap of 55μm, 90μm or

230μm. By comparing the filling time under the same gap, it can be observed that TSV dispensing with a constant inflow has much shorter filling time than the dispensing with free droplets inflow. As the gap is increased, the distinction of the filling time between the two inflows is significantly enlarged.

Edge flood when dispensing stacked interposers

A typical design to encapsulate a package with multiple layers is shown in Figure 12. The TSVs in each level have the same size and they are aligned vertically. The above model is still valid in each level. The flow radius decreases from top to bottom due to the pressure loss inside the TSVs. In theory, each interposer with one coaxial TSV is enough for TSV dispensing with a constant inflow (to obtain a fast filling effect). However, the boundaries of the stacked gaps, which are formed by the edges of the interposers, actually can function as "outlets" if edge flow can overcome surface force or as "walls" if edge flow is constrained in the stacked gaps. If the edge flow exceeds the surface force, the encapsulant will flow down along the sidewalls. Then, an edge flood failure occurs as shown in Figure 13.

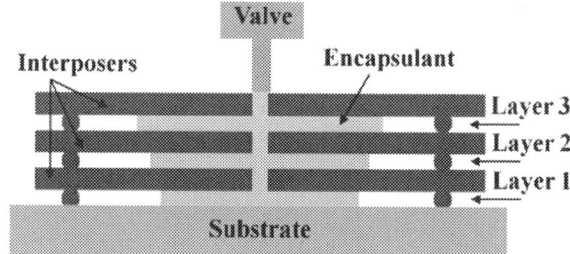

Figure 12. TSVs are vertically aligned when underfilling a 3D package with multiple layers

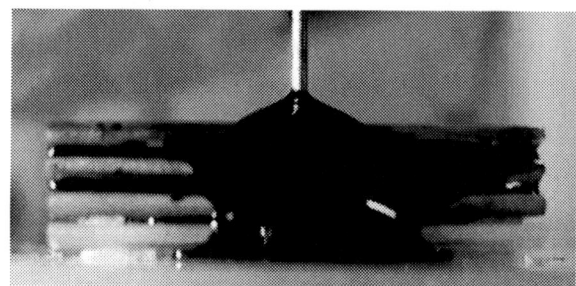

Figure 13. Edge flood failure

Figure 14 shows a schematic diagram of an experiment to investigate the transition moment from a wall to an outlet and the development of an edge flood. An underfill encapsulant was dispensed by a time-pressure valve in which the pressure of the dispensed fluid is gradually increased with the time. Figure 15 shows multiple key screenshots of the progression of the edge flood in two video recordings. The edge fluid initially just expands over the edges of the stacked interposers until reaching the limit equilibrium position. Then, the upper contact point (the red point in the figure) separates from the bottom-left corner of the upper interposer and the encapsulant begins to flow down and finally an edge flood forms. Thus, the relative pressure of the edge flow should be lower than the limit equilibrium pressure p_{eq} to avoid an edge flood failure. The pressure p_{eq} can be calculated by the Laplace-Young equation and the curvature equation after constructing the curve fitting of the front surface at the limit equilibrium position. Table 1 shows the calculation results of p_{eq} at the gaps of 50μm~300μm for the silicon material.

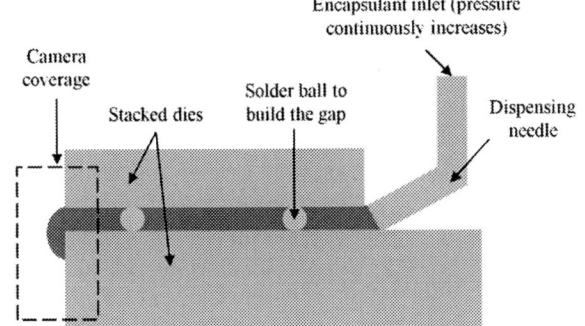

Figure 14. Configurations for the investigation of edge flood

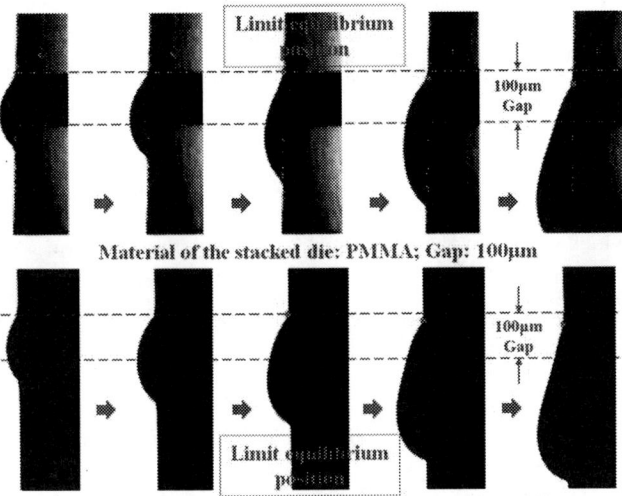

Figure 15. Development of edge flood and the limit equilibrium position

Table 1. Results of the limit equilibrium pressure p_{eq}

Gap clearance [μm]	Temperature [°C]	Material of stacked dies	Encapsulant	Advancing contact angle θ_A [°]	Receding contact angle θ_R [°]	Limit equilibrium pressure p_{eq} [Pa]
50	60	Silicon die	FP3549	37.3	19.0	220
100	60	Silicon die	FP3549	42.1	21.1	122
200	60	Silicon die	FP3549	48.7	29.2	72
300	60	Silicon die	FP3549	52.6	32.9	53

In the TSV dispensing with an inflow of free droplets, the relative pressure of the encapsulant varies gradually from zero to a negative value (equation 10), so it completely eliminates the occurrence of an edge flood. However, in the dispensing with a constant inflow, the encapsulant is squeezed forward by a positive pressure-difference between the encapsulant and the air. Therefore, the relative pressure of the front surface, which is governed by equation 7, is always positive. As the value of p_{eq} is about several hundred Pascal, the dispensing will be extremely difficult (to avoid an edge flood failure) if the underfill is only achieved by a constant inflow. Figure 16 shows an optimized TSV pattern design that finds a trade-off

between the shorter filling time and the lower risk of edge flood. Two groups of TSVs are set in each interposer. The group of central red TSVs is set for a constant inflow; therefore these TSVs are set as far away from all edges as possible to avoid the potential edge flood. The group of outer green TSVs is dispensed by an inflow of free droplets, and therefore it eliminates the risk of an edge flood.

Underfill dispensing on multiple layers

A 3D package with stacked dies/interposers, as shown in Figure 17, was developed for concept validation. The package has a stacked flip chip-on-interposer structure and eight silicon flip chips are arranged in four vertical silicon interposers with multiple TSVs. All the flip chips are 9.8mm × 5.7mm and all the silicon interposers are 18.8mm × 10.8mm. The diameter of all TSVs is about 300μm, and each interposer has the same optimized TSV pattern as described above. The average gap between two adjacent interposers is about 300μm, and the average gap between a flip chip and an interposer is about 50μm.

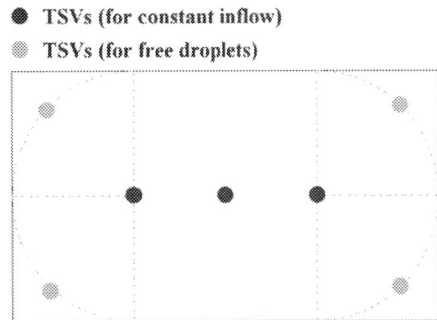

Figure 16. An optimized TSV pattern for TSV dispensing on stacked interposers

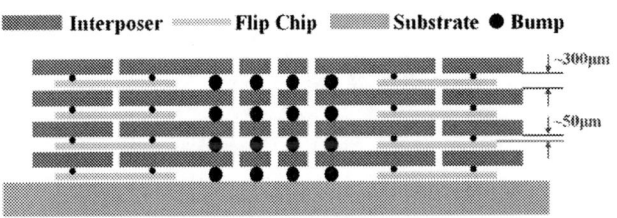

Figure 17. Package structure for concept validation

Figure 18 shows the steps for the underfill dispensing on the 3D package. Edge dispensing is still used to fill the gaps at the bottom levels as much as possible if the fillet size can satisfy requirements [6] (as shown in step 1 and 2). The TSV dispensing with a constant inflow is 3.5 times quicker than an inflow of free droplets at a gap of 230μm as shown in Figures 10 and 11. As the average gap between each two interposers is even wider (~300μm), in step 3, an inflow of constant volumetric flow rate is applied on the central TSVs to encapsulate most areas until the fluid almost reaches the edges. Then in step 4, the remaining unfilled area and the fillets are compensated by dispensing free droplets into the outer TSVs (to avoid the edge flood). The total time of the combined underfill dispensing for the 3D package is about 3mins. It should be noted that the formation of edge reservoirs (in step 1 and 2) usually need to dispense the edge several times, therefore the underfill process can be accelerated by properly alternating the dispensing between the edges (steps 1 and 2) and the outer TSVs (step 4) .

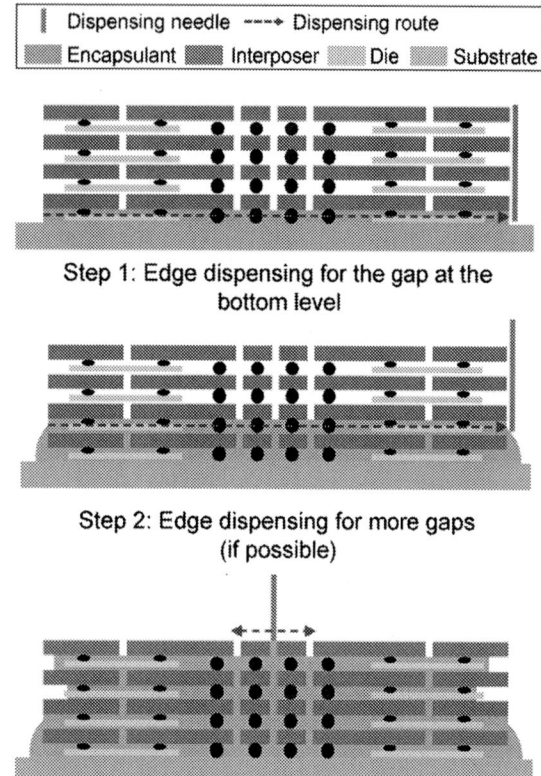

Step 1: Edge dispensing for the gap at the bottom level

Step 2: Edge dispensing for more gaps (if possible)

Step 3: TSV dispensing with a constant inflow for the remaining gaps

Step 4: TSV dispensing with an inflow of free droplets (for compensation)

Figure 18. Underfill dispensing on multiple layers

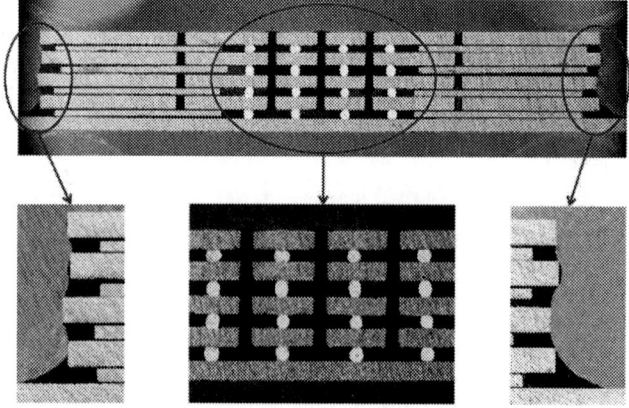

Figure 19. A cross-section cut of the cured 3D package

An inspection of acoustic scanning was conducted on the cured 3D package, and no voids could be found in each level. Moreover, a central cross-section cut (as shown in Figure 19) was performed to further verify the effect of underfill.

Conclusions

In this paper, TSV underfill dispensing was discussed for the encapsulation of 3D die/interposer stacking. The inflow of TSV dispensing may be flow with a constant rate or free droplets. The filling time under a constant inflow is the function of the filled area, the flow rate and the gap clearance, whereas the filling time under an inflow of free droplets is governed by the effect of capillary action, the encapsulant viscosity, the filled area and the gap clearance. The analytical model agreed well with the experimental results. An edge flood failure might occur around the edges of stacked interposers if the pressure of edge flow exceeded the limiting equilibrium pressure. Therefore, an optimized TSV pattern was designed to identify the trade-off between the filling efficiency and the lower risk of edge flood. In the TSV pattern, central TSVs are dispensed by a constant inflow to obtain a fast filling effect whereas outer TSVs are dispensed by free droplets to avoid the edge flood and to compensate the fillets. A four-stack flip-chip-on-interposer package was developed for concept validation. The average gap between a flip chip and an interposer is ~50μm whereas the average gap between two adjacent interposers is ~300μm. An underfill, which combines the dispensing of edges with TSVs, was used to encapsulate the package. Inspections by scanning acoustic microscopy and cross-sectioning indicated that the encapsulation was void-free and the fillets were suitable.

Acknowledgments

The authors acknowledge the support of a grant (T23-612/12-R) from the Research Grants Council (RGC) of the Hong Kong Special Administrative Region Government under the Theme-based Research Scheme.

References

1. John H. Lau, "TSV interposer: the most cost-effective integrator for 3D IC integration," *Sematech Symposium*, Seoul, Korea, September, 2012.
2. Zenner, et al., "Wafer-applied underfill film laminating," *8th International Symposium Advanced Packaging Materials*, Stone Mountian, GA, 2002, pp. 317–325.
3. Daniel Lu and C. P. Wong, *Materials for Advanced Packaging*, Springer, Atlanta, GA, 2009, pp. 326–332.
4. Yat Kit Tsui, S. W. Ricky Lee, "Design and fabrication of a flip-chip-on-chip 3D packaging structure and through-silicon-via for underfill dispensing," *IEEE Transactions on Advanced Packaging*, Vol. 28, No. 3, 2005, pp. 413-420.
5. Fuliang Le and S. W. Ricky Lee, "Underfill dispensing for 3D die stacking with through silicon vias," *International Microelectronics Assembly and Packaging Society Conference*, San Diego, CA, 2012.
6. Karan Kacker and Sidharth, "Impact of underfill fillet geometry on interfacial delamination in organic flip chip packages," *6th Electronic Components and Technology Conference*, Atlanta, GA, 2006, pp.1604-1610.
7. Tasos C. P., et al, *Viscous Fluid Flow*, CRC Press, Florida, 2000, pp. 373-401.

978-1-4799-2408-0/14 $31.00 © 2014 IEEE

Board Level Reliability and Surface Mount Assembly of 0.35mm and 0.3mm Pitch Wafer Level Packages

Beth Keser, Rey Alvarado, Alan Choi, Mark Schwarz, Steve Bezuk
Qualcomm Technologies, Inc.
5775 Morehouse Drive, San Diego, CA 92121
reya@qti.qualcomm.com, 858-845-6865

Abstract

Board level reliability studies have been performed on wafer level packages (WLP) on various die sizes with 0.35mm and 0.3mm ball pitches. The 0.35mm pitch test vehicles included 4mm x 4mm, 5mm x 5mm, and 6mm x 6mm package sizes. The 0.3mm pitch test vehicles were 3mm x 3mm and 4mm x 4mm. All test vehicles were fully populated ball arrays. The parts were assembled at 2 different suppliers. All of the WLPs studied passed drop shock. All of the test vehicles passed board level temperature cycle initially except for the 6mm x 6mm. The SMT process optimization included variations of stencil aperture ratios. These modifications impacted temperature cycling reliability. The reliability of the largest package size was improved from this optimization.

Introduction

Mobile applications are consuming wafer level packages at a rapid pace. The cumulative annual growth rate of WLP from 2012 to 2017 is predicted to be 11% growing from 21.8B units to 36.8B.[1] WLP is used for power management, RFIC, bluetooth, WiFi, and other mobile applications. Image sensors, microcontrollers, memory, discretes, and integrated passive devices also use WLP packages. WLP's are widespread in consumer mobile wireless devices due to their small size and low cost. WLP eliminates the interconnects found in flip chip and wirebond packages as well as the leadframe or package substrate. The limitation of a WLP is its small real estate. All the power, ground, and signal pins must fit within the die area at the pitch of a BGA. Currently, 0.5mm and 0.4mm pitch wafer level packages are the most common in the industry. [2-5] The WLP die real estate cannot grow beyond 6mm x 6mm due to the coefficient of thermal expansion (CTE) mismatch between the silicon device and the PCB. Large WLP's suffer solder joint failure during board level temperature cycle test due to this CTE mismatch. In order to increase the number of BGA per WLP the pitch must shrink. Alternatively, to save on silicon real estate cost, a smaller pitch is required to shrink the die.

Although finer pitches may shrink the die to save cost or increase the number of pins for increased functionality, there are drawbacks. The ability to surface mount finer pitches needs to be studied with the goal to use existing equipment at mobile phone assembly sites. The PCB board design rules and features to support finer pitches need to be evaluated to insure finer pitches are not driving up board costs. Electromigration limits of the smaller solder joints needs to be characterized. Lastly, the board level reliability of the smaller solder joints need to be studied.

0.35mm and 0.3mm pitches drive new design rules for geometries such as the under-bump metallurgy (UBM), the polymer passivation opening (PM2), and the redistribution layer (RDL) capture pad size as well as the ball size. Also the maximum die size for finer pitches needs to be established with these new design rules. This paper describes the impact of 0.35mm and 0.3mm pitch on board level reliability including drop shock and temperature cycle as well as the surface mount assembly of these wafer level packages.

Background

Although pitch finer and finer ball pitch progression from 0.5 to 0.4 to 0.35 and 0.3mm is natural, there are no references in the literature to 0.35 and 0.3mm pitch wafer level package reliability studies. Currently, all references discuss the study and adoption of finer pitches for WLP, but there is no specific data given. [6,7] Only recently have others started publishing on ball grid array (BGA) and chip scale packages (CSP) at finer pitches. [8-10]

The wafer level packages created for this study use a Cu RDL metal structure to connect the die pad to the solder ball (Figure 1) rather than a ball on pad or ball on repassivation structure where the solder ball sits directly on the die pad or a polymer passivation opening over the die pad, respectively. The RDL type WLP process flow is shown in Figure 2. The electromigration (EM) performance of the baseline WLP package construction with a 5um polyimide passivation, 4um Cu RDL metal, a second 5um polyimide passivation, 8.6um thick Cu underbump metallurgy (UBM) and a SAC405 solder ball was reported previously for 0.4mm pitch devices[11] EM performance of various UBM types and RDL thicknesses were also reported previously [12]. The board level reliability of a similar constructions have been published [13-14]. EM results on the baseline construction at 0.3mm ball pitch have been studied and reported. [15]

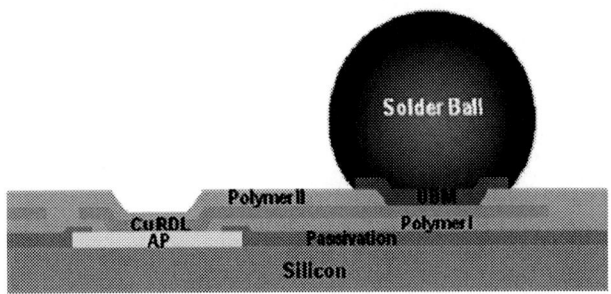

Figure 1. RDL type Wafer Level Package.

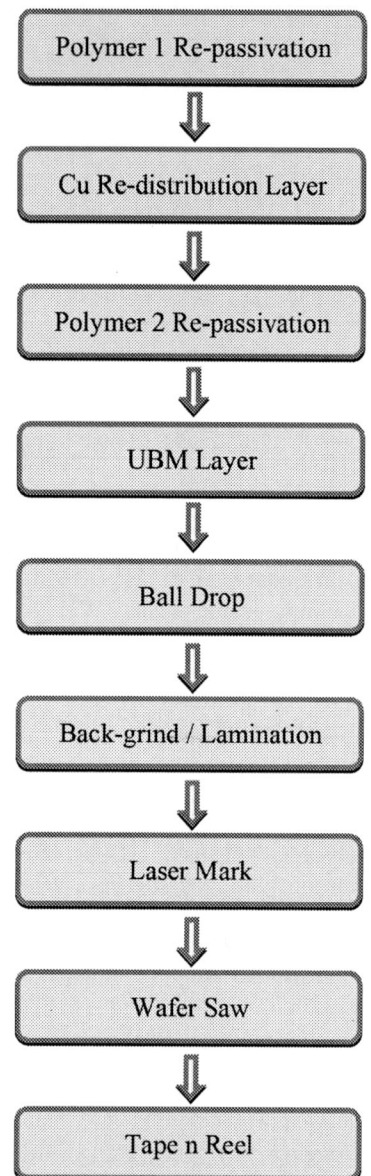

Figure 2. Process flow for RDL type Wafer Level Package.

Experimental

Daisy chain wafer level package test vehicles were used for the board level reliability and SMT yield evaluations. The daisy chain is connected in the top Al metal of the device. The devices evaluated are listed in Table 1.

Name	Package Size (mm)	Pin Count	Pitch (mm)
98WLPSP	3.14x3.14	98	0.3
181WLPSP	4.16x4.16	181	0.3
200WLPSP	4.14x4.14	128	0.35
200WLPSP	5.06x5.06	200	0.35
313WLPSP	6.04x6.04	313	0.35

Table 1. Daisy chain devices used in this study.

The top metal is 54um in diameter with a passivation opening of 49um. The top passivation on the die is nitride. The daisy chain connections made in the PCB test board include 2 separate chains. One chain isolates the corner balls and the other chain includes all non-corner balls such that the corner ball board level reliability performance can be assessed separately from the non-corner balls.

In order to accommodate for the finer pitch, new ball sizes, UBM sizes, polymer2 passivation openings (PM2), and RDL capture pad sizes had to be used. The dimensions used for these daisy chains are listed in Table 2. All of the devices are fully populated with solder balls. Also, all of the daisy chains use a 45 degree ball layout except for the largest 313WLPSP, which uses a 60 degree ball layout to test the limits of SMT at 0.35mm pitch by using dense solder ball packing.

Design Rule	0.3mm Pitch	0.35mm Pitch
Ball Diameter (μm)	200	220
UBM Diameter (μm)	200	220
PM2 Opening Diameter (μm)	170	190
RDL Capture Pad Diameter (μm)	220	240

Table 2. Design rules for 0.3mm pitch and 0.35mm pitch daisy chain devices.

JEDEC JESD22-B111 criteria specifies the fabrication of the test boards used in this study. 8-layer (1-6-1) PCB boards use a standard core material with dimensions of 132mm x 77mm are shown in Figure 3. The Cu pads on the test board have organic solderability preservative (OSP) finish. All tests were run with non-soldermask defined (NSMD) pads. The NSMD opening requires a minimum 25um clearance from the Cu PCB pad, and the nominal Cu PCB pad size is 0.220 for 0.35mm pitch PCBs and 0.200mm for 0.30mm pitch. 15 units were mounted in a 3 x 5 matrix on every board. Probe locations were added to the board for failure isolation.

PCBs were inspected before SMT to verify if PCB manufactures could meet PCB specifications. The 0.35mm pitch PCBs fell within spec for both Cu pad diameter and soldermask clearance. However, the 0.30mm pitch PCBs did not meet the Cu pad specification (see Table 3). The Cu pads for both test vehicles were undersized. Typically, smaller PCB pads would raise a concern for BLR, but since these were NSMD pads and they PCBs met the SM clearance spec, there was no concern the smaller pads would impact BLR.

The SMT reflow peak temperature is 245C and a sample profile is shown in Figure 4. A SAC305 solder paste was printed on the board using a 100um thick stencil. Visual and X-ray inspection is conducted after assembly to monitor placement, voiding, bridging quality, etc.

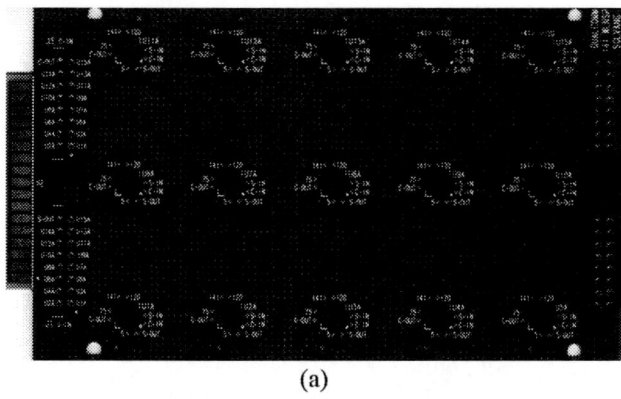

(a)

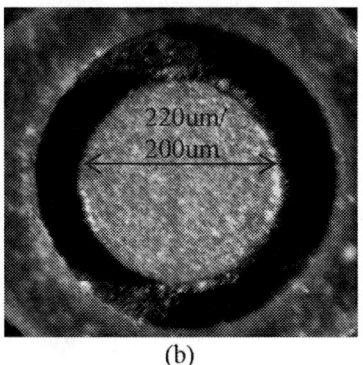

(b)

Figure 3. Test board (a) and pad design (b).

Pitch	0.30mm		0.35mm		
Package Size	3x3	4x4	4x4	5x5	6x6
I/O	98	181	128	200	313
Cu Pad Spec	200 +/- 25um		220 +/- 25um		
Cu Pad Actual	165um	170um	210um	210um	210um
SM Clearance	25um	25um	35um	30um	30um

Table 3. PCB Inspection Measurements

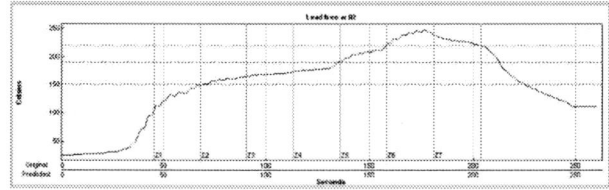

Figure 4. SMT reflow profile.

Temperature cycle (TC) on board experiments were run using JESD22-A104C criteria [16]. For each leg, 2 fully populated boards (30 samples) were tested up to 1000 cycles. Condition G, mode III was used with in-situ monitoring of the daisy chain devices. The mounting set-up and temperature cycle profile are shown in Figure 5.

(a)

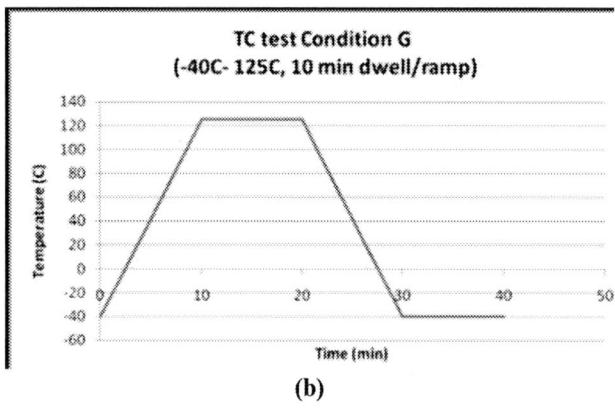

(b)

Figure 5. Board mounting set-up (a) and temperature cycle profile (b).

JESD22-B111 criteria was used for drop shock (DS) testing [17]. For each leg, 4 fully populated boards (60 samples) were tested to 200 drops. The desired shock wave was 1500g 0.5ms half sine event. A picture of the shock event is shown in Figure 6(a) along with a picture of the set-up in Fig. 8(b). For each test run, 2 PCB's were mounted with the packages facing down with an accelerometer mounted in the center between the 2 boards. Acceleration was recorded for every drop.

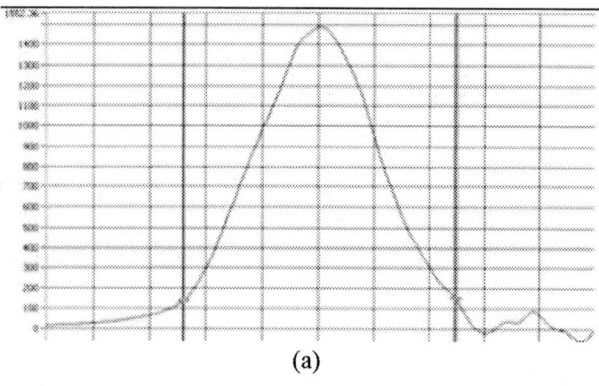

(a)

(b)

Figure 6. Sample shock event (a) and drop shock mounting configuration (b).

Results

All of the daisy chains passed drop shock. Most daisy chains passed with no failures. The units that did fail were cross-sectioned to confirm failure mode. The drop shock results are shown in Table 4. The 181WLPSP had one corner ball failure at 139 drops. This drop was confirmed to fail in the bulk solder on the device side of the solder ball. See figure 7 for cross-section. This is consistent with a drop shock failure mode typical of WLP. Another failure occurred at 74 drops for the 181WLPSP. This failure occurred on a unit built by another supplier and was a non-corner ball failure. It is unusual to have non-corner balls fail before corner balls, so this failure was cross-sectioned. The cross-section, shown in Figure 8, found that the failure was due cracking in the board underneath the Cu pad. These board types of failures can be discounted as they do not measure the robustness of the package.

The non-corner balls of the 3mm x 3mm, 4mm x 4mm, and 5mm x 5mm daisy chain devices at 0.3 and 0.35mm ball pitch passed temperature cycle without failures below 500 cycles. However, the 6mm x 6mm 313 WLPSP with 0.35mm pitch at 60 degree ball layout did have non-corner ball failures below 500 cycles at one supplier. Both fails were confirmed to be bulk solder fails as shown in Figures 9 and 10. This required SMT stencil print optimization before it could pass temperature cycle. The temperature cycle results are show in Table 5.

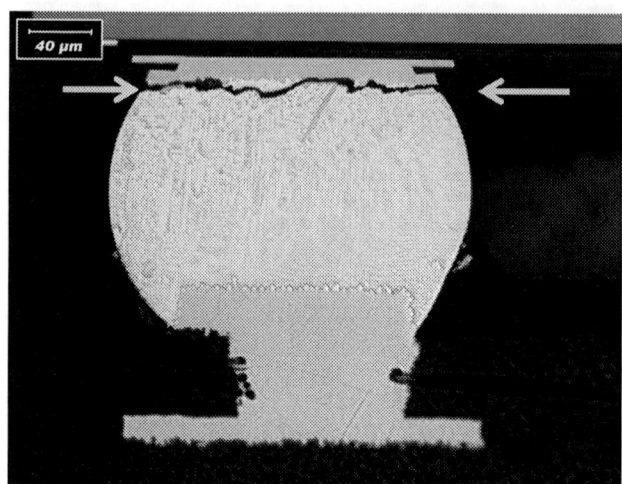

Figure 7. Cross-section of 181WLPSP corner ball failure after 139 drops.

Name	Pitch (mm)	Supplier	Corner First Fail	Non-Corner First Fails
98 WLPSP	0.3	A	none	none
		B	none	none
181 WLPSP	0.3	A	139	none
		B	none	74
128 WLPSP	0.35	A	none	none
		B	none	none
200 WLPSP	0.35	A	none	none
		B	none	none
313 WLPSP	0.35	A	none	none
		B	none	none

Table 4. Drop shock results for 0.3mm pitch and 0.35mm pitch daisy chain devices.

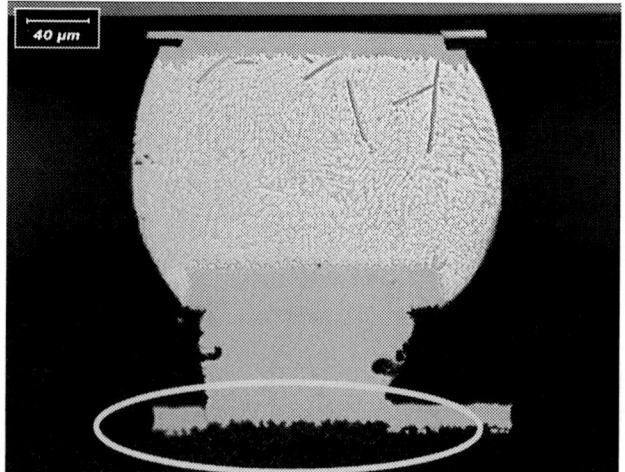

Figure 8. Cross-section of 181WLPSP non-corner ball failure after 74 drops.

Name	Pitch (mm)	Supplier	Corner First Fail	Non-Corner First Fails
98 WLPSP	0.3	A	none	none
		B	708	923
181 WLPSP	0.3	A	508	763
		B	607	718
128 WLPSP	0.35	A	480	739
		B	464	652
200 WLPSP	0.35	A	428	750
		B	471	694
313 WLPSP	0.35	A	407	446
		B	383	566

Table 5. Temperature cycle results for 0.3mm pitch and 0.35mm pitch daisy chain devices.

SMT assembly yields were also monitored during the assembly of units to boards for board level reliability. The goal was to assess mobile wireless device vendor's ability to successfully place finer pitch units on their boards. The results are shown in Table 6. Overall, acceptable yields were obtained for 0.35mm pitch units with 45 degree ball arrays; however, the 0.3mm pitch TVs and the 0.35mm pitch 313WLPSP TV with the 60 degree ball array had some yield issues caused from the excessive paste during the stencil printing process. All yield loss for the 128WLPSP was due to solder bridging that was caused by the placement misalignment. Once the placement tool was fixed, the remaining boards had 100% SMT yield. The stencil thickness of the 313WLPSP was increased from 80μm to 100μm to evaluate the current standard 0.4mm pitch SMT process using the most challenging test vehicle to achieve 100% SMT yield. The results were promising seeing minimal bridging. In addition increasing the stencil thickness improved temperature cycling reliability. More tuning is needed to find a solution with optimal yield and board level reliability for 60 degree arrays if needed for product designs.

Mobile wireless device vendors will need to modify their current SMT process to accommodate 0.3mm pitch. No significant changes are necessary for customer SMT process when moving from WLP devices with 0.4mm pitch to 0.35mm pitch for 45 degree array ball layouts or other less dense ball arrays.

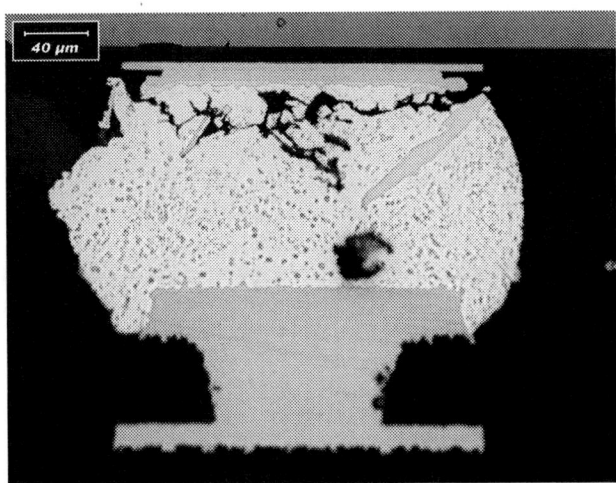

Figure 9. Cross-section of 313WLPSP unit that had corner ball failure at 407 temperature cycles.

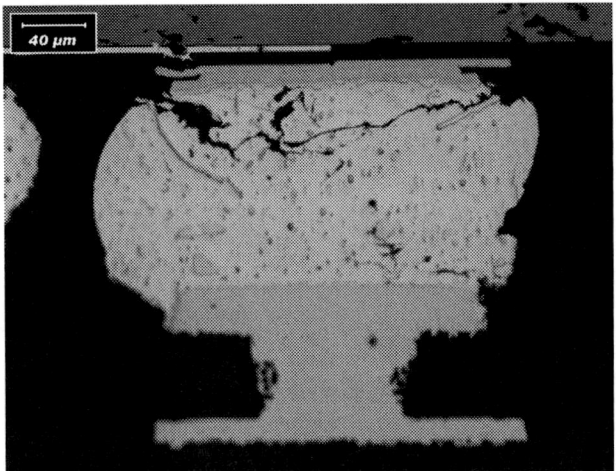

Figure 10. Cross-section of 313WLPSP unit that had a non-corner ball failure at 446 temperature cycles.

Pin	Pitch (mm)	Stencil Thickness (μm)	Yield (%)	Comment
98	0.3	80	100	
181	0.3	80	99.49	Bridging
128	0.35	80	94.33	Bridging, PnP misalign
200	0.35	80	100	
313	0.35	80	100	
		100	99.6	Bridging

Table 6. Summary of SMT Yield results.

Conclusions

In this study, WLP board level drop shock and temperature cycle were evaluated using 5 different daisy chain WLP devices ranging from 3mm x 3mm at 0.3mm pitch to 6mm x 6mm at 0.35mm pitch and varying the ball array layout from 45 degrees to 60 degrees. The goal was to find the die size, pitch, pin count, and array types that will pass JEDEC board level reliability criteria and with good SMT yield. The results show that all of the daisy chains evaluated passed drop shock for corner and non-corner balls. All daisy chains passed temperature cycle for non-corner balls once solder print thickness on the board was optimized; however, excess solder print on the board also caused some yield loss due to bridging. Also, SMT yield showed that mobile wireless device vendors would have to adopt processes to improve 0.3mm pitch WLP devices. Changing device pitch from 0.4mm to 0.35mm is transparent, but 0.3mm pitch would not.

Acknowledgments

The authors would like to thank Qualcomm's Packaging Lab for support of the board design, procurement, and execution of the drop shock and temperature cycle tests and failure confirmation. Also, the authors would like to thank the Qualcomm Failure Analysis Lab for the cross-sections completed for this study.

References

1. TechSearch International, Inc., "2013 Flip Chip and WLP: Recent Developments and Market Forecasts", 2013, p. 9. Prismark Semiconductor Package Report Q1 2012, p16.
2. Tong Yan Tee, Long Bin Tan, Rex Anderson, Hun Shen Ng, Jim Hee Low, Choong Peng Khoo, Robert Moody, Boyd Rogers "Advanced Analysis of WLCSP Copper Interconnect Reliability under Board Level Drop Test," *Proc. IEEE Electronics Packaging Technology Conference,* 2008, pp. 1086-1095.
3. Xuejan Fan and Qiang Han, "Design and Reliability in Wafer Level Packaging," *Proc. IEEE Electronics Packaging Technology Conference,* 2008, pp. 834-841.
4. L. Cergel, L. Wetz, B. Keser, J. White, "Chip Size Packages with Wafer Level Ball Attach and their Reliability," *Proc. of 4th International Conference on Advanced Semiconductor Devices and Microsystems,* 2002, pp. 27-30.
5. D. Yang. X. Ye, F. Xiao, D. Chen, L. Zhang, "Reliability of Fine Pitch Wafer Level Packages," *International Conference on Electronic Packaging Technology and High Density Packaging*, 2012, pp. 1097-1101.
6. B. Prior, "IC Package Miniaturization and System in Package (SiP) Trends," *Burn-in and Test Socket Workshop*, 2011.
7. J. Hung, "Value Engineered Wafer Level Packages for Mobile Devices," *Semicon West*, 2013.
8. A. Choi, B. Roggeman, M. Schwarz, "Demonstrated Process and Reliability of 0.35mm Pitch BGA Devices for Mobile Environment," *Proc. of SMTA International*, 2013.
9. J. Sjoberg, J. Lee, S. Alam, R. Aranda, D. Geiger "0.3mm Pitch Chip Scale Package (CSP) Process Development and Printed Circuit Board (PCB) Design", *Proc. of SMTA International,* 2011.
10. D. Barbini, M. Meilunas, "Investigations into 0.3mm Pitch Assembly and Reliability", *Proc. of SMTA International*, 2013, pp147-158.

11. C. Hau-Riege, R. Zang, Y. W. Yau, P. Yadav, B. Keser, J. K. Lin, "Electromigration Studies of Lead-Free Solder Balls used for Wafer-Level Packaging," *Proc. IEEE Electronic Components and Technol. Conf.* (ECTC), 2011, pp. 717-721.
12. C. Hau-Riege, B. Keser, S. Bezuk, Y. W. Yau, "Electromigration of Solder Balls for Wafer-Level Packaging with Different Under Bump Metallurgy and Redistribution Layer Thicknesses," *Proc. IEEE Electronic Components and Technol. Conf.* (ECTC), 2013, pp. 707-713.
13. P. Yadav, S. Kalchuri, B. Keser, R. Zang, M. Schwarz, B. Stone, "Reliability Evaluation on Low k Wafer Level Packages," *Proc. IEEE Electronic Components and Technol. Conf. (ECTC)*, 2011, pp. 71-77.
14. S. Xu, B. Keser, C. Hau-Riege, S. Bezuk, Y. W. Yau, "A Study of Wafer Level Package Board Level Reliability," *Proc. IEEE Electronic Components and Technol. Conf.* (ECTC), 2013, pp. 1204-1209.
15. C. Hau-Riege, B. Keser, R. Alvarado, A. Syed, S. Bezuk, Y. W. Yau, "Electromigration of Solder Balls for Wafer-Level Packaging with Different Under Bump Metallurgies and Redistribution Layers," *Proc. IEEE Electronic Components and Technol. Conf.* (ECTC), 2014.
16. JESD22-A104C, JEDEC Standard, "Temperature Cycling", May 2005.
17. JESD22-B111, JEDEC Standard, "Board Level Drop Test Method of Components for Handheld Electronic Products", July 2003.

Encapsulated Wafer Level Package Technology (eWLCS)

Tom Strothmann, *Seung Wook Yoon, *Yaojian Lin
STATS ChipPAC Inc.1711 W Greentree Drive Tempe, AZ 85284 USA
* STATS ChipPAC Ltd. 5 Yishun Street 23, Singapore 768442
tom.strothmann@statschppac.com

Abstract

This paper introduces a new encapsulated WLCSP product (eWLCS). The new product has a thin protective coating applied to all exposed silicon surfaces on the die. The applied coating protects the silicon and fragile dielectrics and prevents handling damage during dicing and assembly operations, effectively providing a durable packaged part in the form factor of a WLCSP. The manufacturing process leverages existing high volume manufacturing methods with exceptionally high process yields. In this process the silicon wafer is diced prior to the wafer level packaging process. The dice are then reconstituted into a new wafer form with adequate distance between the die to allow for a thin layer of protective coating to remain after final singulation. Standard methods are used to apply dielectrics, thin film metals, and solder bumps. The resulting structure is identical to a conventional WLCSP product with the addition of the protective sidewall coating. This paper discusses the key attributes of the new package as well as the manufacturing process used to create it. Reliability data will be presented and compared to conventional WLCSP products and improvements in package durability will be discussed and compared to conventional WLCSP.

Improving the Conventional WLCSP Structure

The wafer level chip scale package was introduced in 1998 as a semiconductor package wherein all packaging operations were done in wafer form [1]. The resultant package has dielectrics, thin film metals, and solder bumps directly on the surface of the die with no additional packaging. The basic structure of the WLCSP has an active surface with polymer coatings and bumps with bare silicon exposed on the remaining sides and back of the die. The WLCSP is the smallest possible package size since the final package is no larger than the required circuit area. Based on the small form factor and low cost, the number of Wafer Level Packages used in semiconductor packaging has experienced significant growth since its introduction. The growth has been driven aggressively by mobile consumer products because of the small form factor and high performance required in the package design. Although WLCSP is now a widely accepted package option, the initial acceptance of WLCSP was limited by concerns with the SMT assembly process and the fragile nature of the exposed silicon inherent in the package design. Assembly skills and methods have improved since the introduction of the package, however damage to the exposed silicon remains a concern. This is particularly true for advanced node products with fragile dielectric layers. One method commonly used to improve die strength and reduce silicon chipping during assembly is lamination of an epoxy film on the back of the die. The film is laminated and cured on the back of the wafer prior to singulation to strengthen the die, in spite of the fact it adds cost to the package. By the nature of the backside lamination process, the uncoated sides of the die continue to be exposed after dicing the wafer and the silicon continues to be at risk for chipping, cracking, and other handling damage during the assembly process.

A process has been developed to provide five sided protection for the exposed silicon surfaces in a WLCSP. The process starts with a high volume manufacturing flow developed by STATS ChipPAC for fan-out products. The implementation of this process flow into 300mm diameter reconstituted wafers has been described in detail in previous presentations [2]. In this manufacturing method the wafer is diced at the start of the process and then reconstituted into a standardized wafer (or panel) shape for the subsequent process steps. The basic process flow for creating the reconstituted wafer is shown in Figure 1. 1) The reconstitution process starts by laminating an adhesive foil onto a carrier. 2) The singulated die are accurately placed face down onto the carrier with a pick and place tool. 3) A compression molding process is used to encapsulate the die with mold compound while the active face of the die is protected. 4) After curing the mold compound, the carrier and adhesive foil are removed in a de-bonding process resulting in a reconstituted wafer where the mold compound encapsulates all exposed silicon die surfaces. The eWLB process is unique since the reconstituted wafer does not require a carrier during the subsequent wafer level packaging processes.

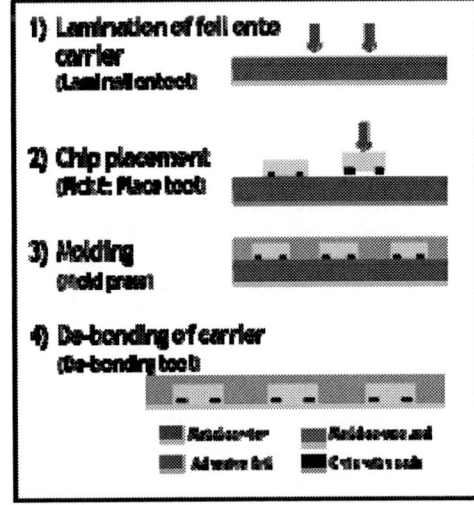

Figure 1. The Reconstitution Process Flow

After the reconstitution process, the reconstituted wafer is processed with conventional wafer level packaging techniques for the application and patterning of dielectric layers, thin film metals for redistribution and under bump metal, and solder bumps. In the final dicing operation a thin layer of mold compound, typically < 70um, is left on the side of the die as a protective layer. The back of the die is also protected with mold compound, although with a greater thickness. A schematic drawing of a typical structure is shown in Figure 2 for greater clarity. Alternatively, the backside mold compound can be removed and the body made thinner with an optional back grind operation without damaging the protective sidewall layer. The remaining sidewall coating will continue to protect the fragile silicon sides of the die during the assembly operation.

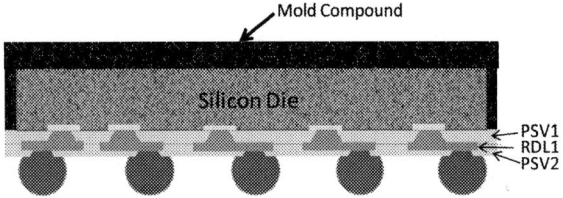

Figure 2. eWLCSP™ Structure

The Encapsulated WLCSP Process

The FO-WLP process has been discussed in many venues and it is recognized as an industry standard process. In the FO-WLP process the area of the package is increased to allow for placement of RDL layers and solder balls outside of the silicon die area [3]. This packaging method allows the die to shrink to a minimum size independent of the required area for an array of solder balls at industry standard BGA ball pitches [4]. It also allows for novel multi-die structures, 2.5D structures and 3D structures [5]. The Fan-out process has been qualified to a 28nm process node with the same dielectrics and Cu plating as are used in the eWLCSP process described here [5]. The eWLCSP process data presented in this paper was generated with a 300mm round reconstituted panel [2]. In the case of conventional FO-WLP the die are typically widely spaced to allow for the expanded RDL and bump area and the conventional saw street. In the case of eWLCSP the die are closely spaced allowing for only the sidewall thickness in addition to a street area of 80um. The die size used this evaluation was 4.5x4.5mm similar to the construction shown in Figure 2. The final structure had 2 layers of polymer and 1 layer of plated Cu RDL with the solder ball mounted directly on the RDL without the use of a separate UBM layer. The process flow used is shown in Figure 3 and the details of the structure are shown in Table 1.

PSV 1 (um)	7.0 - 11.0
RDL 1 (um)	7.0 - 10.0
PSV 2 (um)	7.0 - 11.0
Ball pitch (mm)	0.4
Ball size (um)	250
Solder alloy	SAC 405

Table 1. Layer Thickness

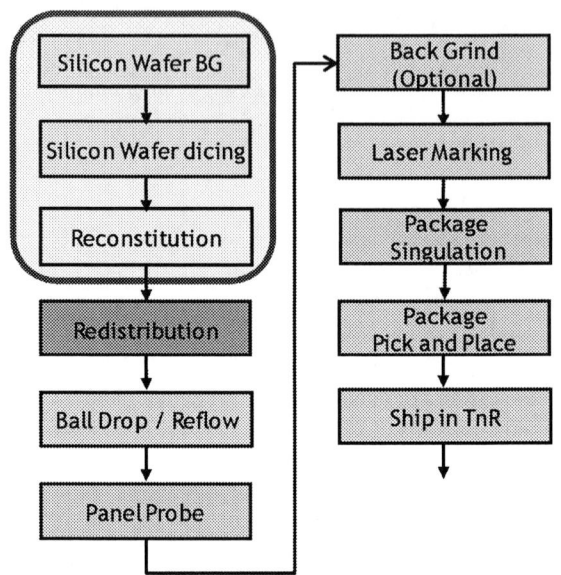

Figure 3. eWLCSP Process Flow

Intuitively the process flow shown in Figure 3 would have higher cost since there are additional steps required for reconstitution at the start of the flow. There are two key factors that offset the cost of the additional steps required for the reconstitution to make this a commercially viable process. 1) Panel size scaling reduces the unit cost if the source silicon wafer is smaller than the reconstituted panel size. In the case of the 300mm reconstituted panel used here the cost is very competitive for silicon wafers with a diameter of 200mm and below. The cost of processing a 300mm reconstituted panel for WLCSP is approximately 1.7x the cost of processing a conventional 200mm silicon wafer in WLCSP, however the units processed per panel increases by a factor of 2.3x, effectively offsetting the cost of reconstitution. 2) Since known good die can be selected at the start of the process, advanced devices that have a lower electrical yield can be tested in wafer form prior to the process. If the incoming wafer has a probe yield of 85%, then 15% more units per reconstituted panel can be processed to offset the cost of the reconstitution process. Since the reconstituted panel size is no longer linked to the incoming silicon wafers size, the panel size can be increased over time and change from a round to a much larger rectangular format. This scaling to a larger panel size will provide a compelling cost reduction when compared to conventional WLCSP packaging methods where the round silicon format is maintained throughout the wafer level packaging process.

One difference in processing panels in the reconstitution flow is found in the attributes of the polymers that are used. In conventional WLCSP either polyimide (PI) or polybenzoxazole (PBO) are used as the dielectrics for planarization, stress buffering and RDL insulation. In the case of the reconstituted panel, the mold compound has a lower temperature threshold than silicon and sustained temperatures over 200°C can cause degradation of the material. PI typically has a cure temperature of 380°C and PBO has a typical cure temperature of 300°C and therefore cannot be used in the

978-1-4799-2408-0/14 $31.00 © 2014 IEEE

process. A new low temperature polymer has been developed for this application that has a cure temperature compatible with the 200°C threshold temperature of the mold compound.

A SEM cross-section of a eWLCSP part created in the process is shown in Figure 4. In this case a thicker sidewall protection layer was used to demonstrate the process on an existing production device running in the conventional 200mm WLCSP production line. The device demonstrated equivalent electrical yield, Component Level Reliability and Board Level Reliability performance to the conventional WLCSP.

Figure 4. Cross-section of eWLCSP

A second SEM cross-section is shown in Figure 5 showing the finished package with a thin protective sidewall coating and the use of the optional back grind to thin the body thickness.

Figure 5. Cross-section of thin body eWLCSP

A WLCSP product that is currently in production using a conventional WLCSP process can be converted to a eWLCSP product without any design change required, regardless of the current silicon wafer diameter. If a reduced thickness is required for the specific application, an optional back grind step can be added to the process flow to reduce the body

thickness while retaining the protective sidewall coating. Since the dice are singulated at the start of the process, the manufacturing equipment and bill of materials are the same for any incoming wafer size. The initial back grind and dicing tools are the only wafer size dedicated equipment required for the process. Very little process development and very little additional capital will be required to package 450mm silicon wafers as eWLCSP.

eWLCSP Product Assessment

The unique attribute of the eWLCSP package is the protective sidewall coating. The protective layer is durable and will prevent silicon chipping on the side of the package. This protective layer has the ability to protect the silicon during socket insertion for test. This has been demonstrated through multiple insertion test on completed products with no observed damage to the protective coating.

The eWLCSP process has passed standard reliability tests used in wafer level packaging including Component Level Reliability (CLR), Temperature Cycle on Board (TCoB), and Drop Test.

Component Level Reliability was completed with the test conditions shown in Table 2. The evaluation results were confirmed by visual inspection and electrical test. No delamination of the protective coating was detected during the CLR evaluation.

Component Level Test	Condition		Status
MSL1	MSL1, 260C Reflow (3x)	-	Pass
Temperature Cycling (TC) after Precon	-55°C to 125°C	1000 x	Pass
HAST (w/o bias) after Precon	130°C / 85% RH	192 hrs	Pass
High Temperature Storage (HTS)	150°C	1000 hrs	Pass

Table 2. Component Level Reliability Results

Thermal Cycle Reliability Test (TCoB) was completed and passed 500 cycles with the results shown in Table 3 and the Weibull plot in Figure 6. Results were obtained from electrical measurement of daisy chain bump structures. Results are comparable to conventional WLCSP product produced with polyimide dielectrics.

TCoB (Cond B)	Failure Rate	Characteristic life (η)	Weibull slope β	First Failure
-40°C to 125°C	0.635	1219.4	10.13	864x

Table 3. TCoB Reliability Test Results

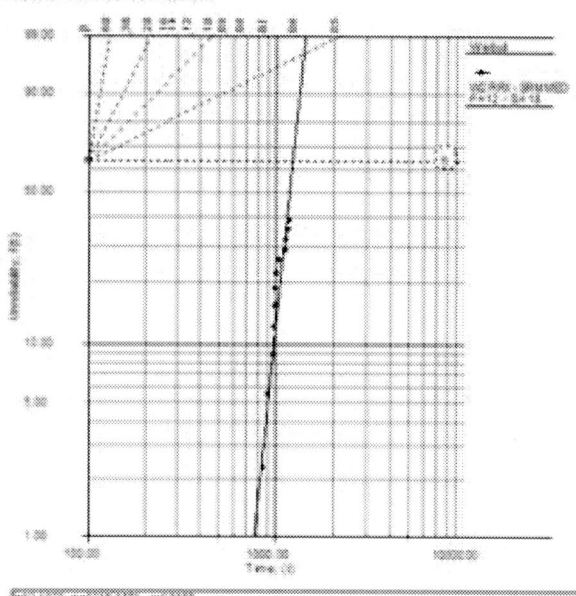

Figure 6. TCoB Weibull Plot for eWLCSP

Drop Test was completed and passed the JEDEC requirement of 30 drops with the results shown in Table 4 and the Weibull plot in Figure 7. Results were obtained from electrical measurement of daisy chain bump structures. Results are comparable to conventional WLCSP product produced with polyimide dielectrics.

Drop Test	Failure Rate	Characteristic life (η)	Weibull slope (β)	First Failure
JEDEC	0.635	1553.5	5.97	772x

Table 4. Drop Test Results for eWLCSP

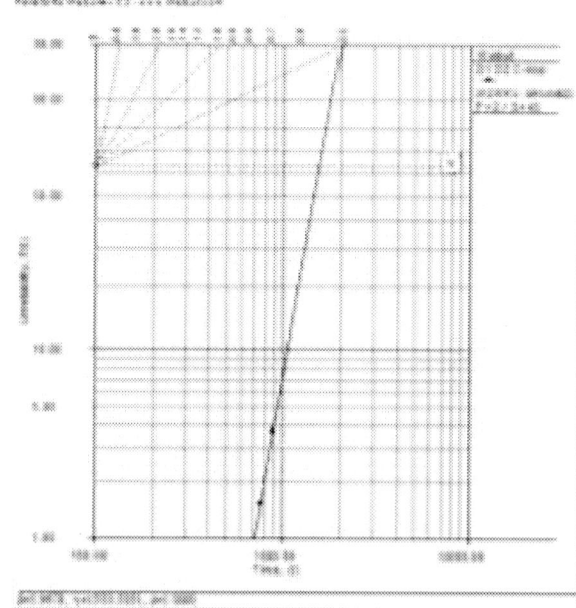

Figure 7. Drop Test Weibull Plot for eWLCSP

Conclusions

A new encapsulated WLCSP process has been developed and verified with reliability testing. The process provides mechanical sidewall protection to WLCSP parts with an increase in package size of less than 100um in X and Y dimensions. The sidewall protection resolves the problem of silicon damage during the assembly process and provides a path to significant cost savings for the customers as the panel size is increased. The eWLCSP process described is wafer size agnostic, so the same manufacturing line can process the eWLCSP products regardless of the incoming wafers size. 450mm wafers can easily be accommodated for the encapsulated WLCSP process once the service is required by the customers.

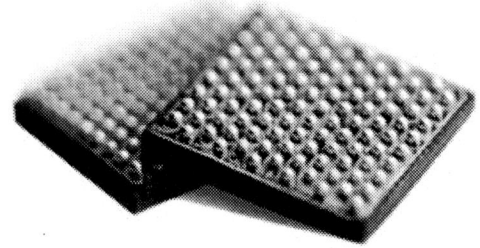

eWLCSP Product with Protective Sidewall Coating

Acknowledgments

STATS ChipPAC WP/eWLB development team supporting WLP Research and Development.

References

1. P. Elenius, "The Ultra CSP Wafer Scale Package", Electronics Packaging Technology Conference, 1998.
2. M. Prashant, S.W. Yoon, Y.J. Lin, and P.C. Marimuthu, "Cost effective 300mm large scale eWLB (embedded Wafer Level BGA) Technology", 2011 13th Electronics Packaging Technology Conference.
3. M. Brunnbauer, et al., "Embedded Wafer Level Ball Grid Array (eWLB)," Proceedings of 8th Electronic Packaging Technology Conference, 2009, Singapore (2006)
4. M. Brunnbauer, T Meyer, "Embedded Wafer Level Ball Grid Array (eWLB)," IMAPS Device Packaging Conference 2008, Arizona, US
5. SW Yoon, P Tang, R Emigh, YJ Lin, PC Marimuthu, R Pendse, "Fanout Flipchip eWLB (embedded Wafer Level Ball Grid Array) Technology as 2.5D Packaging Solutions", 2013 Electronic Components & Technology Conference.

Enabling of Fan-Out WLP for More Demanding Applications by Introduction of Enhanced Dielectric Material for Higher Reliability

Almeida Rodrigo, Barros Isabel, Campos José, Cardoso Paulo, Castro José, Henriques Vítor, O'Toole Eoin, Pinho Nelson
NANIUM, S.A., Av.1º de Maio 801, 4485-629 Vila do Conde, PORTUGAL
Telephone: +351 252 24 78 39

Abstract

Market is requesting more and more IC packages capable to cope with more demanding reliability requirements, in order to meet customer increased expectations regarding functionality, size, power dissipation and cost. Miniaturization needs, in an increasing number of applications, accelerate the implementation of WLP (Wafer Level Packaging) even in market segments not using WLP by today. This requires more robust and more reliable WLP solution, where Fan-Out versions like eWLB (embedded Wafer Level Ball Grid Array) will play an emphasized role in terms of system integration and high density packaging.

To address those expectations, the packaging construction, especially RDL (redistribution layer) and dielectric materials of eWLB, has to be further enhanced.

This paper will present the development activities performed regarding dielectric material characterization and selection, process development, integration, validation, qualification and transfer into volume production. The eWLB products qualified at NANIUM high volume manufacturing line cover a wide range of package configurations, mainly in terms of package size and thickness, proving the capability of the improved material/process for eWLB technology platform.

As result of this development, NANIUM enabled the eWLB technology to fulfill also the needs of more demanding applications and market segments. The new eWLB packages are exceeding 1.000 cycles in component level based Temperature Cycling Test (TCT -55°C to 125°C) according to JEDEC JESD47 (condition B) and 500 to 1.000 cycles, in board level based Temperature Cycling on Board Test (TCoB -40°C to 125 °C) according IPC-9701 (condition TC3). The paper will also show results from Drop Test according JEDEC JESD22-B111.

Due to its nature eWLB FO-WLP implies specific properties and processing conditions to dielectric materials candidates, like curing temperature, shrinkage level and stress buffer capability as examples. A full material characterization of main related properties was initially performed to support the selection of the best candidate.

The introduction of this enhanced dielectric material required deep development of several process steps and optimization of its integration. New surface cleaning steps were tested, completely new lithography processing conditions were introduced, curing and drying steps were adapted and new AOI configuration was developed.

Finally, specific test vehicles were designed and produced to test and validate the complete technology. This included all package and reliability tests according to JEDEC more stringent standards, like uHAST, THB, TCT, HTS, TCoB and Drop Test.

Material selection

As referred in [1], due to the nature of eWLB, the reliability of the package is a balance of the capability of the different layers that constitutes the package in absorb shocks and mechanical stress from the different materials CTEs. In IC packaging interfaces, the dielectric material, plays a significant role absorbing thermal stress and mechanical shocks slowing down cracks propagation.

A wide range of material classes has been considered as polymeric dielectric, such as Polybenzoxazole (PBO), Polyimide (PI), nano-filled Phenol resin, Benzocyclobutene (BCB), Silicone, Epoxy-based, Fluorinated Polymer, Siloxane-based, Polynorbornene and Acrylate. Fifty six polymeric dielectrics from seventeen different manufactures were compared concerning physical, mechanical, thermal, electrical and chemical properties. Data are from technical papers [2, 3], internet and material suppliers.

Most significant material properties considered for the dielectric selection were the elongation to break, the tensile strength and the Young's modulus as they are typically used as an initial indicator of how a polymer will perform under mechanical stress caused by CTE mismatch between the die and the molding compound in thermal cycling and in mechanical shock drop tests.

The best candidates among the polymers suitable for microelectronic applications, were polyimides that have received increased attention due to their thermal and chemical stability, low dielectric constant and high electrical resistivity [4].

The curing temperature condition was another restriction considered due to FO-WLP packaging temperature limitations. Imidization reactions of polyimide precursors requires typically activation energy levels with temperatures above 300ºC which makes technically impossible to use it due to the highly negative impact in package fan-out molded area and in wafer warpage. One of the PIs precursors formulations was selected based on its low curing temperature compatible with eWLB FO-WLP products and processing temperature restrictions. The selected PI precursor formulation is NMP/NEP solvents free and complies with environmental legislation like RoHS and Reach directives, and also with big IDMs environmental requirements standards. It is also compatible with copper and all the other chemicals used in production process like solvents, bases and acids.

Thus, the PI precursor formulation was selected to be used as buffer layer and also as RDL top layer. A full material characterization was carried out in terms of mechanical (as function of different curing conditions), thermal and electrical parameters. It shows larger elongation to brake value even at negative temperatures, the highest stress condition. Electrically shows a good compromise between low dielectric constant and dielectric loss at high frequencies, and is very suited for RF antenna applications that requires high frequencies working.

eWLB FO-WLP process integration

The integration of this new dielectric material in NANIUM 12" wafers eWLB FO-WLP production process flow required several changes in different steps.

Photolithography

On top of the utilization of different equipments to process this new photosensitive PI-type material developable in organic solvents, Lithography process recipes development was very challenging due to the different behavior showed by reconstituted wafers in terms of shape, warpage and stiffness. After the spin curve building and the set point defined to achieve component topography coverage with a high uniformity (range <0.5μm), soft bake conditions were tested obtaining a wide process window.

To define process window, resolution capabilities and optimal exposure settings for the new dielectric material a FEM (Focus Exposure Matrix) was printed followed by optical inspection showing good resolution in vias with dimensions between 5μm and 30μm, as shown in Figure 1.

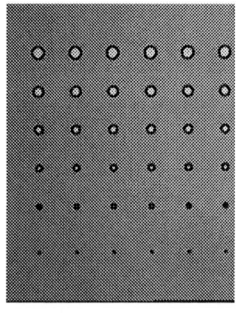

Figure 1. Microscope images of via opening on polyimide after RIE plasma descum.

The exposing process exhibits a large process latitude as can be observed on the Bossung Plot (Figure 2) with the best isofocal dose in yellow.

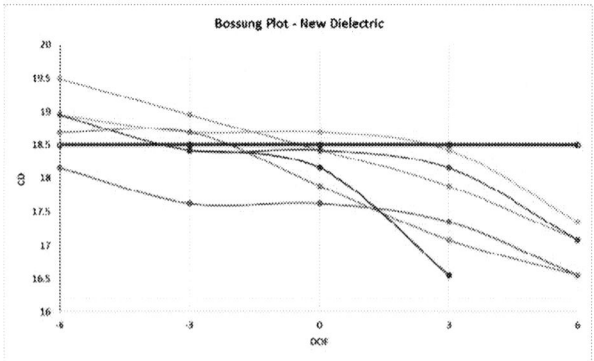

Figure 2. Bossung plot.

Curing

It was also proved that is possible to achieve a high imidization reaction extent, at low temperatures maintaining at the same time a very high performance in terms of elongation to brake values.

Descum

Due to UV light reflection on die pads during Exposing, a PI-precursor residual film is left after Developing at vias bottom. After PI formation during Curing, this footing has to be removed because reduces the available area for electrical contact (Figure 3).

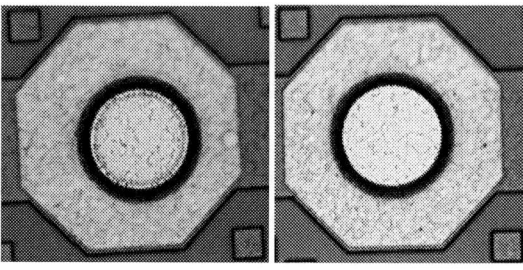

Figure 3. Microscope images of via opening on polyimide, before and after RIE plasma descum.

A reactive ion etching (RIE) reactor is used for anisotropic descum process improving vias openings profile and cleaning metal pads surface enabling a low contact resistance with the RDL metal patterning deposition (Figure 4).

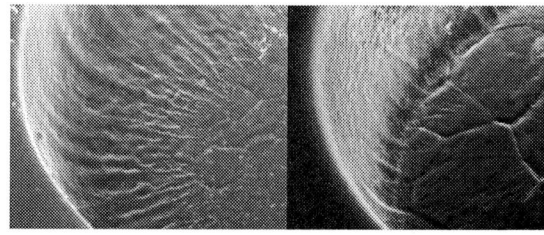

Figure 4. SEM images of via opening on polyimide, before and after RIE plasma descum.

Automatic Optical Inspection

The selected dielectric presented challenges for post RDL Automated Inspection due to its high level of transparency.

In Figure 5(a) we have the reflectance for the basic materials and the chosen CCD gain function (sensitivity) as a function of wavelength. Ideally the setup of an illumination strategy should attempt to maximize the contrast between each material group, simultaneously guaranteeing detectability of the desired defect groups. However, to achieve contrast of dielectric over copper, one loses contrast in the remaining materials.

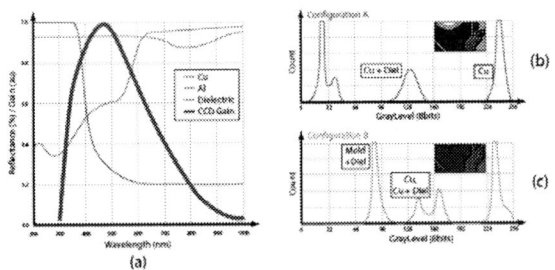

(a)

(b)

(c)

Figure 5 (a) Material reflectance in function of wavelength of perpendicular incidence. (b) GrayLevel Histogram for an image of the full package with illumination condition A. Superimposed is an example of a selected package zone. (c) GrayLevel Histogram for an image of the full package with illumination condition B. Superimposed is an example of a selected package zone.

The selected illumination strategy was therefore to have two complementary light conditions

Results

For reliability characterization purposes, stress tests according to JEDEC more stringent standards were carried out under Pb free solder conditions. Tables 2 and 3 summarize the test conditions applied and results for component and board level reliability.

Samples from two different test vehicles (TV) were used for these tests with dimensions listed in Table 1.

Table 1 – Details of Test Vehicles used for Reliability Characterization

Test Vehicle	Package Type	Package size	Die Size	Bump / Pad Pitch
TV A	BGA	9.25 x 8.8 x 0.8mm	5.6mm x 5.3mm	0.5mm
TV B	BGA	7.5 x 7.5mm x 0.8mm	5.0 x 4.96mm	0.4mm

Table 2 - First level (components) reliability test conditions and results summary

Stress (Standard)	Condition	Criteria	Status
PRECON (JESD22-A113/J-STD-020)	MSL1	Level 1 (T_{peak}: 260ºC)	PASS
PRECON + TC (JESD22-A104)	Condition B -55ºC↔125ºC 2 cycle/h	0 fail 1000/1500 cycles	PASS
PRECON + uHAST (JESD22-A118)	Condition A: 130ºC / 85% rH	0 fail 96/188h	PASS
HTS (JESD22-A103)	Condition B: 150ºC	0 fail 1000h	PASS
THB (JESD22-A101)	85ºC / 85% RH; V_{cc} :5V	0 fail 1000h	PASS

Table 3 - Second level (board) reliability test conditions and results summary

Stress	Condition	Criteria	Status
TCoB (IPC-97-01)	Condition TC3 -40ºC↔125ºC 1 cycle/h	FF > 500/850 cycles	PASS
Drop Test (JESD22-B11)	B 1500 Gs; 0.5 millisecond duration; half-sine pulse	< 10% fails @ 20 drops	PASS

Thermal Cycling

The mechanical stability and reliability under TC stress of the stack (dielectric layers and RDL copper layer) was largely improved by the new dielectric material.

For the new dielectric material, a comprehensive characterization was carried out, with read-outs (physical characterization) at 700, 1000 and 1500 cycles. This characterization included a top down analysis through selective delayering, layer-by-layer, at 700 and 1000 cycles and cross-sections at 1000 and 1500 cycles.

In the case of PI, crack propagation process slows down once the crack reaches the bottom dielectric layer.

A sample of 3 lots x 25 units was tested electrically after 1000 cycles with a yield of 100%. All units were pass under the pass criteria. Therefore, the electrical functionally of the unit is preserved at least until 1000 cycles and the cross-sections carried out during characterization seems to indicate that this limit can be even exceeded.

Thermal Cycling on Board

TCoB with continuous in situ electrical monitoring was carried out, which allowed estimating the parameters of the Weibull distribution. For this purpose it was considered the number of cycles until the first fail in a given unit. Stop criteria was 80%.

Figure 6 depicts the Weibull plot for one of the test vehicle (TV A). The estimated scale parameter, η and shape parameter β, are the following:

β=8.87 η= 1283.82 cycles ρ=0.87

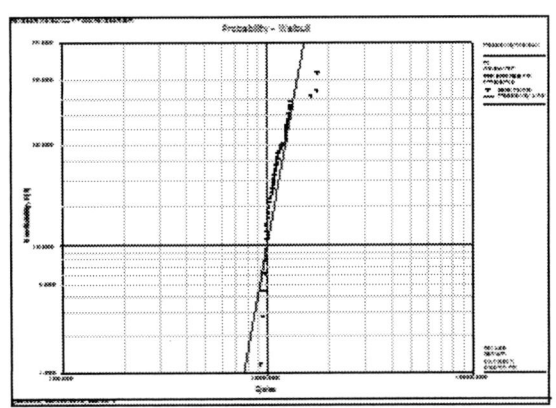

Figure 6. Weibull plot for TCoB.

Based on the correlation coefficient, the Weibull distribution is suitable to model the data. Using the estimated parameters in Weibull cdf, the unreliability is 0% until 500 cycles. It will be 1% for 766 cycles. The MTTF is 1215 cycles. These results fit with the expected performance for packages with this size and pitch and considering that it uses a 3 Mask process (no-UBM).

Physical characterization of the solder ball in the failing net revealed a damage due to solder fatigue with cracks developing from the corner of IMC formed between solder ball and copper pads of the PCB and propagating through the solder bulk, as shown in Figure 7. This failure mode is more likely to be related to SMT process.

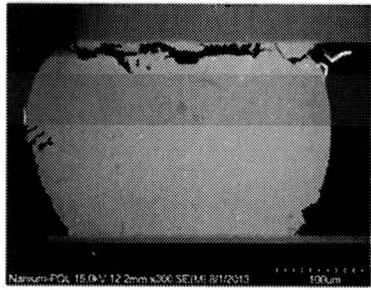

Figure 7. Solder ball bulk crack, PCB side.

Drop Test

As in TCoB, continuous in situ electrical monitoring during the stress was carried out. Again, the test continued until 80% of the board fails.

Figure 8 depicts the Weibull plots. The estimated parameters are the following:

β=1.40 η= 1161.81 drops ρ=0.97

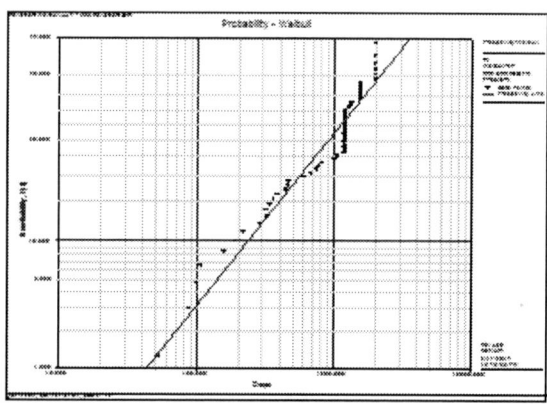

Figure 8. Weibull plot for Drop Test.

Considering a more strict pass criteria of less than 5% fails at 30 drops, based on the cdf it would require 140 drops, in order to achieve the 5% of fails. The MTTF is 1059 drops.

The physical failure characterization on the first fail also revealed that the only failure mechanism observed was solder joint crack at the component side, specifically in the

intermetallic layer which became the more fragile interlayer connection (Figure 9).

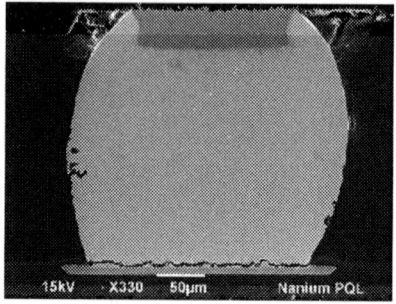

Figure 9. Intermetallic crack at component side.

Another evaluation made on this new PI-type dielectric material was using a QFN-type test vehicle.

The QFN-type test vehicle was designed as a plug and play replacement for QFN packages whilst offering a significant advantage in terms of package height. This configuration was envisaged to simulate a very small thin package ranging from a relatively low to relatively high silicon occupation as shown in Table 4. The redistribution layer design was such that a simple daisy chain structure is created between the package and the test PCB. LGA pads were designed above the second dielectric, and were finished in a Copper-Nickel-Gold stack.

Table 4 – QFN-type test vehicle configurations

Die SOR	Package XY	Package height
25%	1mm X 1mm	0.225mm / 0.33mm
73%	1mm X 1mm	0.225mm / 0.33mm
5%	3mm X 3mm	0.225mm / 0.33mm
95%	3mm X 3mm	0.225mm / 0.33mm
Dual Die 15%	3mm X 3mm	0.225mm / 0.33mm

The final package appearance can be seen in Figures 10 and 11 illustrating two examples of different SOR – Silicon Occupation Ratio.

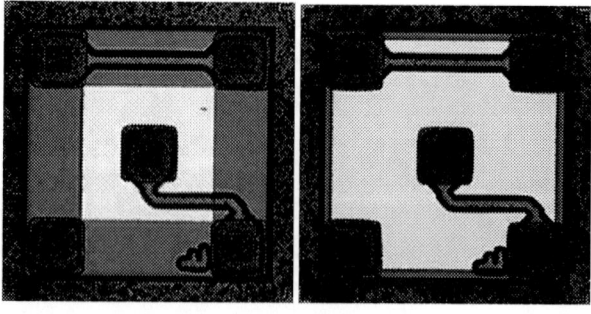

Figure 10. QFN-like Package with 25% SOR.

Figure 11. QFN-like Package with 73% SOR.

A cross section of the LGA pads is shown in Figure 12.

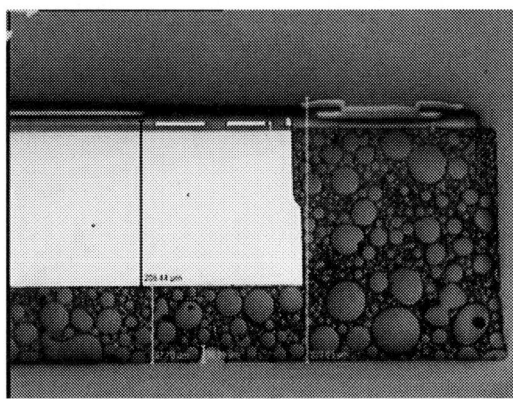

Figure 12. QFN-like Package cross section with LGA detail and dimensions.

A summary of the reliability results for the 1mm X 1mm package is shown in Tables 5 and 6.

Table 5 – First Level reliability for 1mm X 1mm, 0.33mm Low and High SOR packages

Stress (Standard)	Condition	Criteria	Status
PRECON (JESD22-A113/J-STD-020)	MSL1	Level 1 (T_{peak}: 260ºC)	PASS
PRECON + TC (JESD22-A104)	Condition B -55ºC↔125ºC 2 cycle/h	0 fail 2000 cycles	PASS
PRECON + uHAST (JESD22-A118)	Condition A: 130ºC / 85% RH; 96h	0 fail	PASS
HTS (JESD22-A103)	Condition B: 150ºC 1000h	0 fail	PASS

Table 6 - Second level reliability results summary for 1mm X 1mm, 0.33mm Low and High SOR packages

Stress	Condition	Criteria	Status
TCoB (IPC-97-01)	Condition TC3 -40ºC↔125ºC 1 cycle/h	0 fails @2000 cycles	PASS
Drop Test (JESD22-B11)	B 1500 Gs; 0.5 millisecond duration; half-sine pulse	0 fails @\500 drops	PASS

Conclusions

The introduction of PI-based new dielectric material in eWLB packages clearly results in an improvement of the mechanical properties of the entire package with no set back regarding other properties.

Component and board level thermal cycling clearly indicates an improvement in the life of the units, now increased above 1000 cycles, even under a much more aggressive stress, with lower minimum temperatures. Drop test also revealed a considerable increase in the lifetime of the package, with a MTTF above 1000 drops.

The packaging of eWLB FO-WLP units using this new dielectric PI-type material was qualified in different products that successfully passed all industry standard reliability tests and are running actually in high volume production.

Acknowledgments

This paper and the development work behind it was only possible due to the contribution and support from customer and supplier and NANIUM organization people from all departments.

References

1. Development of Early Process Control Indicators for Reliability Drop Test Performance of eWLB Products; Azevedo Alexandre, Cardoso André, Teixeira Jorge, Tavares Oriza, Marques Rui; 2012 IEEE.
2. "A Comparison of Thin Film Polymers for Wafer Level Packaging", M. Töpper, T. Fischer, T. Baumgartner and H. Reichl, Electronic Components and Technology Conference, 2010.
3. "Characterization of Material Properties of Low Temperature Curing Polymer Dielectrics", A. Huffman, M. Butler, J. Piascik, P. Garrou and J. Im, Electronic Components and Technology Conference, 2011.
4. Soane DS, Martynenko Z. Polymers in Electronics: Fundamentals and Applications, Elsevier, Amsterdam, 1989.

24"x18" Fan-out Panel Level Packing

T. Braun ([1]), K.-F. Becker ([1]), S. Voges ([2]), J. Bauer ([1]), , R. Kahle ([2]), V. Bader ([1]) T. Thomas ([2]),
R. Aschenbrenner ([1]), K.-D. Lang ([2])
([1]) Fraunhofer Institute for Reliability and Microintegration
Gustav-Meyer-Allee 25, 13355 Berlin, Germany
phone: +49-30/464 03 244 fax.: +49-30/464 03 254
e-mail: tanja.braun@izm.fraunhofer.de
([2]) Technical University Berlin, Microperipheric Center

Abstract

Fan-out Wafer Level Packaging (FOWLP) is one of the latest packaging trends in microelectronics. Mold embedding for this technology is currently done on wafer level up to 12"/300 mm size. For higher productivity and therewith lower costs larger mold embedding form factors are forecasted for the near future. Following the wafer level approach then the next step will be a reconfigured wafer diameter of 450 mm. An alternative option would be leaving the wafer shape and moving to panel sizes leading to Fan-out Panel Level Packaging (FOPLP). Sizes for the panel could range up to 18"x24" or even larger. For reconfigured mold embedding, compression mold processes are used in combination with liquid or granular compound. As an alternative process, lamination can be considered where also sheet compounds can be used.

Using maskless laser direct imaging technologies (LDI) instead of photolithography has a high potential for further cost reduction with intrinsic process advantages. The LDI cost advantage is backed by LDI availability for large panel sizes, also including 450 mm wafer form factors.

Already today PCB technologies offer the potential for large area panel packaging up to 24"x18"/610 x 457 mm² and can be applied to form a redistribution layer [RDL] for large area reconfigured wafers or panels, replacing thin film redistribution. For PCB based RDLs a resin coated copper sheet (RCC) is laminated on the reconfigured wafer or panel, respectively. Micro vias are drilled through the RCC layer to the die pads and electrically connected by Cu plating. Final process step is the etching of Cu lines using LDI techniques for maskless patterning. State of the art equipment and materials allow the manufacturing of structures down to 20 μm lines and spaces with a clear development trend to 10 μm lines and spaces and hence getting close to photolithography thin film structure sizes.

Based on the technology described above the Fan-out Panel Level Packaging approach will be demonstrated on full 24"x18"/610x457 mm² format including large area embedding and redistribution. Related technology challenges as die shift, warpage, panel handling or yield will be discussed in detail. Using maskless LDI technology real die positions can be automatically adapted to the redistribution and hence less accurate die placement can be compensated and higher die shift could be tolerated which is a big advantage when moving towards large area with acceptable yield.

In summary this paper describes the technological path from wafer level embedding to 24"x18" fan-out panel level packaging technology in combination with low cost PCB based RDL processes. The technology described offers a cost effective packaging solution for various scenarios e.g. as packages for handheld consumer applications or bio-medical applications as sensor integration into microfluidics.

Introduction

Drivers for 3D packaging solutions are manifold and each requirement calls for different answers and technologies. Main goal is miniaturization, but component density and performance, simplification of design and assembly, flexibility, functionality and finally, cost and time-to-market have been found to be the core drivers for going 3D as well. Besides die and package stacking and folded packages, embedding dies is a key technology for heterogeneous system integration [1].

There are two main approaches for embedded die technologies: Fan Out Wafer level integration, where dies are embedded into polymer encapsulants and 3D vertical integration, where dies are embedded into the substrate. A lot of activities are running worldwide dealing with wafer level integration. Main drivers are here the Embedded Wafer Level Ball Grid Array (eWLB) by Infineon [2] and the Redistributed Chip Package (RCP) by Freescale [3]. Singulated dies are assembled on an intermediate carrier and encapsulated by compression molding, forming a polymer wafer with embedded silicon dies. This "reconfigured" wafer is then released from the carrier. Using thin film technology, an electrical redistribution layer is routed on the wafer. Finally, the wafer is singulated by sawing into single packages. Current trend in eWLB technology is a double sided eWLB packaging with integration of vias through the encapsulant by integration of preformed PCB based vias allowing the stacking of eWLB packages [4].

Another concept for 3D integration of active components is the Chip in Polymer (CiP) technology, introduced by Fraunhofer IZM, TU Berlin and Wuerth. It is based on embedding of ultra-thin dies into build-up layers of printed circuit boards (PCBs). The dies are bonded onto a core substrate using an adhesive; a resin coated copper (RCC) layer with thin Cu is used for the subsequent lamination. Interconnects are established by laser drilled micro vias followed by a PCB-compatible Cu plating [5].

The combination of both concepts, embedding into polymer by compression molding or mold compound lamination and redistribution using PCB technologies, which allows the production on panel size up to 515x610 mm² (Figure 1) has the potential for highly integrated low cost

978-1-4799-2408-0/14 $31.00 © 2014 IEEE

packages. The proof of concept for this technology combination was successfully done for a 2-chip LGA package [6], for a stackable BGA package [7], functional sensor-ASIC package [8] as well as for a CMOS sensor integration into a microfluidic application [9].

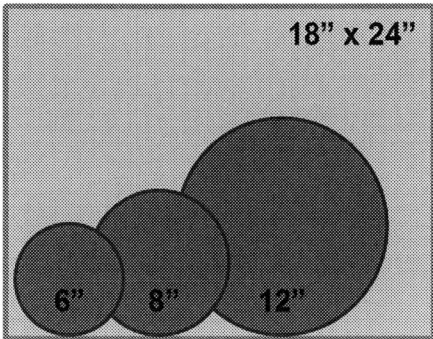

Figure 1:   PCB panels size versus wafer sizes.

PCB based process flow

The process flow for the combination of embedding into polymer by molding and redistribution using PCB technologies is quite similar to wafer level embedded packages with thin film redistribution. For a LGA/BGA package with through mold vias (TMV) for 3D routing (see Figure 2) the processing starts with the lamination of an adhesive film to a carrier. This special adhesive film has one pressure adhesive side and one thermo-release side, i.e. by heating up the tape above a certain temperature, the thermo-release side of the tape loses its adhesion strength.

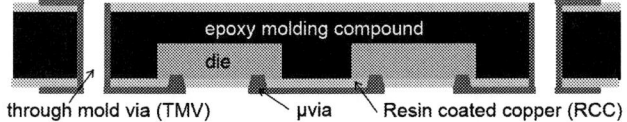

Figure 2:   Schematic of a stackable panel level embedded package with PCB based redistribution technology.

Dies are precisely placed on the adhesive film attached to the carrier, the active side facing down towards the carrier. High accuracy is needed as die pads have to match with the redistribution layer. Encapsulation is done by large area compression molding or mold compound lamination. For chip redistribution, low cost PCB based technology with RCC has been selected. After lamination of the RCC film on both panel sides in one step, µvias to the die pads and through mold vias are drilled in the same process step to connect top and bottom side. Next process steps are cleaning, palladium activation and copper plating. By plating, vias are filled, die pads are connected to the copper layer and connection of top copper layer to bottom copper layer is achieved. Conductor line formation is done by laser direct imaging (LDI) in combination with a dry film resist and copper etching. Finally, a solder mask and solderable surface finish as NiAu and solder balls can be applied.

Materials and processes for panel mold embedding

Within the next section materials as well as processes for large area embedding will be discussed in detail. Focus for this investigation was put on the large area assembly and embedding process steps. A 2-chip package of 8x8 mm^2 with BGA pin-out has been designed for technology demonstration. The package consists of two chips with a size of 2x3 mm^2 (see Figure 4).

Assembly

For die placement an ASM Siplace CA3 was used that allows die placement on the entire panel in one step. Dies and fiducials were directly picked from the diced wafer and placed face down on the carrier. Additionally, spacers for panel thickness control were assembled as for embedding lamination without thickness control was used [10]. The overall layout is shown in Figure 3 and a detailed view in Figure 4. Altogether 5528 chips have been placed on the panel with an assembly speed of around 6500 chips/h using one collect and place 20-nozzle revolver head. Process might be accelerated up to 8000 chips/h and with the option of using four heads maximum assembly speed could be around 32000 chips/h.

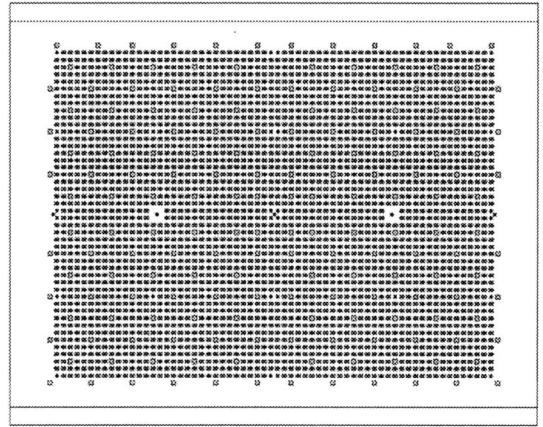

Figure 3:   Assembly panel layout.

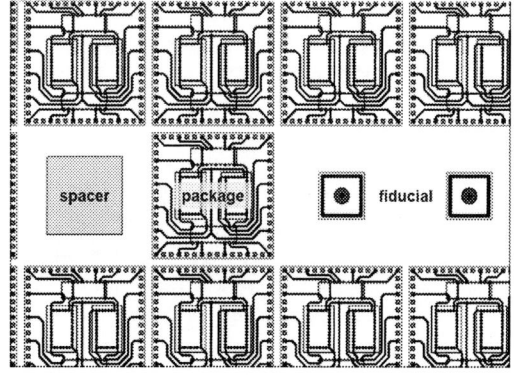

Figure 4:   Panel layout detail.

A photo of the assembled 18"x24" carrier with dies, fiducials and distance pieces is depicted in Figure 5.

978-1-4799-2408-0/14 $31.00 © 2014 IEEE   941

Figure 5:   Assembled reconfigured 18"x24" panel.

Embedding

There is a variety of compression molding compounds for embedded wafer level molding from different suppliers on the market available. For mold embedding technology materials should have low chemical shrinkage, low cure temperature and match thermo-mechanical properties for low warpage of the molded panel and low die shift after molding. Flow properties should allow homogeneous filling of large cavities. Basically, state of the art materials can be divided in liquid, granular and sheet compounds (Figure 6).

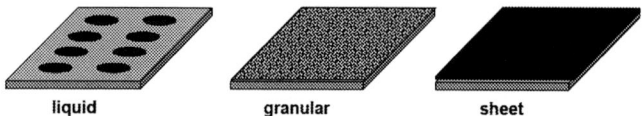

Figure 6:   Epoxy molding compound types and application.

From the processing point of view, the liquid materials are dispensed in the middle of the cavity and flow during closing and compression of the tooling to fill the entire cavity. In opposition to the liquid, the granular compound is distributed nearly homogeneously all over the cavity. The compound melts and the droplets have to fuse during closing and compression of the tool (Figure 7). State of the art production equipment for e.g. eWLB wafer molding is currently available up to 300 mm wafer or panel molding. But first compression mold machines for panel encapsulation up to 18"x24" have been already announced by several machine suppliers and will be available for mold evaluations within the next months.

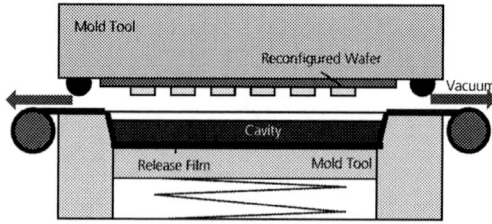

Figure 7:   Principle sketch of compression molding.

Sheet or roll lamination under vacuum are alternative process options for large area encapsulation allowing embedding up to 18"x24" and have been used for this study (Figure 8). Lamination process is slightly different compared to compression molding as lamination presses used for PCB manufacturing allow the application of a temperature profile

over time where compression molding works at constant temperatures. It has been demonstrated that same materials for encapsulation by lamination can be used for compression molding [10] but for thickness control of the molded sheet a height control is need. This can be done by assembling spacers on the panel and hence defining the panel thickness.

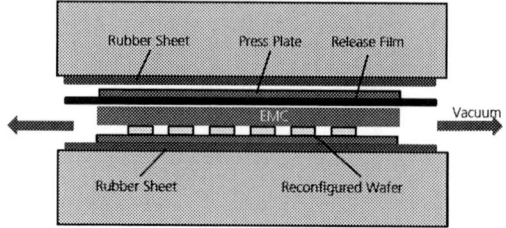

Figure 8:   Schematic diagram of the lamination process.

For this study a commercially available sheet molding compound has been selected for panel mold embedding. Table **1** summarizes data sheet material properties of this epoxy sheet molding compound.

Table 1:   Data sheet properties of sheet molding compound used for lamination.

properties	sheet compound
filler content	88 wt.-%
filler cut size	52 µm
CTE_1	7 ppm/K
CTE_1	40 ppm/K
T_g	110°C
flexural modulus @ RT	4,5 GPa
mold temperature	90 – 120°C
inmold cure time	120 – 180 s
PMC temperature	130 – 140°C
PMC time	2 – 3 h

Differential Scanning Calorimetry (DSC) using a DSC Q2000 (TA Instruments) and rheological characterization using a rheometer AR-G2 from TA Instruments of the sheet compound have been performed to study melting and curing of the material to adapt and optimize lamination parameters.

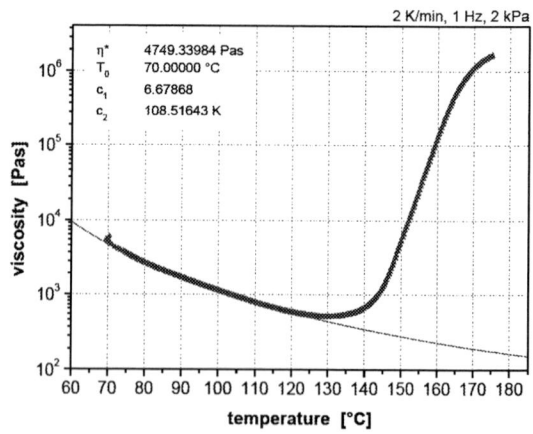

Figure 9:   Rheological analysis of sheet compound used for lamination. (dynamic scan, heating rate 2 K/min)

The rheological characterization of the temperature depending viscosity gave a minimum viscosity for the sheet compound between 120 °C and 140 °C where flow behavior should be best and hence void-free encapsulation feasible (see Figure 9). Above 140 °C curing and therewith gelation starts quickly. This is in good alignment with the DSC results where reaction peak can be observed between 140 °C and 150 °C (see Figure 10).

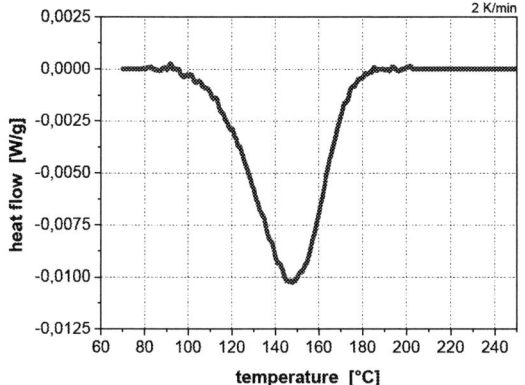

Figure 10: DSC analysis of sheet compound used for lamination. (dynamic scan, heating rate 2 K/min)

A lamination profile has been developed based on these results on viscosity and curing behavior. The thermal release step of the mold panel from the carrier is a critical process step. Here, a homogeneous heating at a defined temperature of the carrier and mold is needed. On the one hand the panel will be not released from the carrier if the temperature is too low. On the other hand the panel will permanently stick to film if the temperature budget (time and/or temperature) is too high. In consequence the temperature profile was adapted to allow encapsulation, curing as well as release from the carrier in one process step. Figure 11 depicts the lamination profile used for panel embedding.

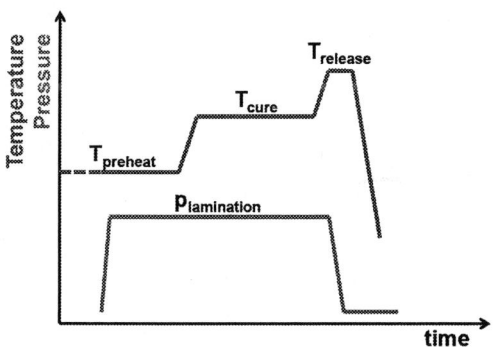

Figure 11: Schematic lamination profile; blue: pressure, red: temperature.

For lamination experiments a standard PCB lamination press from Lauffer was used. Lamination was done with the profile described above under vacuum. The molded panel with embedded dies is shown in Figure 12 and demonstrates successfully the overall proof of concept. Dies were fully embedded in the molding compound without air entrapments and the molded panel could be released from the carrier without tape residues.

Figure 12: Lamination of an embedded reconfigured 18"x24" panel.

Results and discussion

One important requirement on the mold embedding process on large panels is the position accuracy and tolerance of the embedded dies after molding/lamination and cure. Looking at the process steps, the precision that can be achieved is determined by the capability of the die placement equipment used and the possibility of compensating the shift of the dies during molding and cure. Therefore, die positions have been measured before and after lamination and die shift was subsequently calculated for x- and y-direction (see Figure 13 and Figure 14).

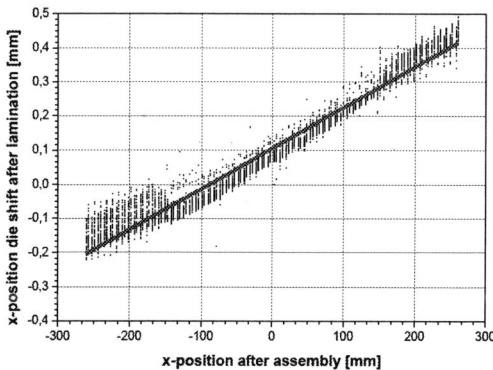

Figure 13: Die shift in x-direction after lamination on 18"x24" panel.

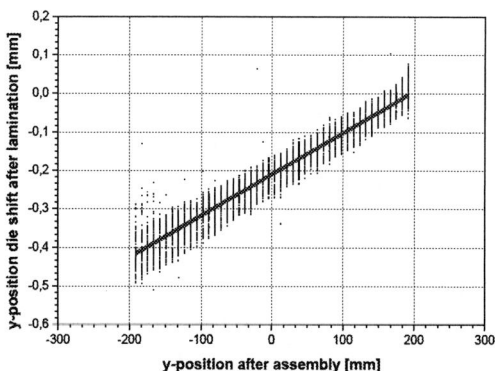

Figure 14: Die shift in y-direction after lamination on 18"x24" panel.

978-1-4799-2408-0/14 $31.00 © 2014 IEEE

Die placement can be done getting a constant precision and statistical variation over the whole panel as guaranteed by the pick and place equipment manufacturer and independently measured.

On the other hand, our investigations clearly showed that the shift of the dies from their original position caused by molding, shrinkage and thermal expansion increases nearly equivalent to the distance of the geometrical center. (see Figure 13 and Figure 14). Additionally, not only the shift increases for larger distances but also its variation, leading to larger inaccuracy and thus to larger compensation.

An easy approach leading to a lower compensation level of die placement and, therefore, a higher accuracy after molding and cure, is the introduction of four independent coordinate systems as basis for placement and compensation (see Figure 15).

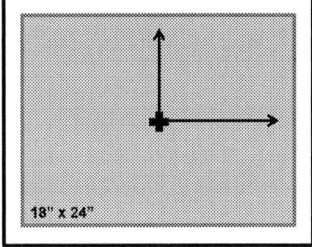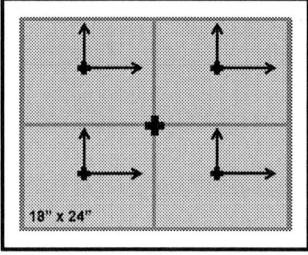

Figure 15: Position of Cartesian coordinate systems used for assembly of dies on a large panel; left – single coordinate system with central position of origin, right – segmentation of panel in quadrants with local centered coordinate systems.

The idea of this approach is to virtually divide the whole panel into parts, e.g. quadrants and to introduce an independent coordinate system with an own reference point for each part. These new reference points act as basis for assembling the dies within their quadrant respectively. Thus, the overall shift of a die during the process can be divided into the shift relative to its corresponding reference point and the shift of this reference relative to the center of the panel i.e. the origin of the overall coordinate system.

The advantage of this approach is the prevention of large values of compensating shifts during placement and their associated large variations (see Figure 13 and Figure 14). Furthermore, more accurate placement processes can be achieved when working with additional reference points. In that way, a higher precision and a lower variation of the position of the dies in each quadrant after molding and curing can be achieved. Such a higher accuracy is needed to connect fine pitch dies with small pads or to guarantee free lines for the saw streets also for panels with a high density of dies. Moreover, mask-based processes such as photolithography also need low tolerances of die positions and their connection pads.

Using maskless processes for die connection and rewiring gives the opportunity to tolerate larger die misplacement by adapting the layout to the real die position. This adaptation can be done in addition to a die shift compensation if higher accuracy is required. The dedicated layouts for maskless μvia connection and redistribution structuring can be automatically adapted according to measured die positions. Even if these two process steps are maskless they can be highly productive. State of the art laser drill equipment as e.g. Schmoll Picodrill can manufacture more than 500 μvias/s and the writing of the wiring with e.g. an Orbotech Paragon will take between 1 and 2 min for a full format panel as described here.

Automated layout adaption has been already implemented for x-/y-package shift. For this case die edges are measured and die centers and consequently package center is calculated. The virtual center of μvia and wiring layout is then aligned to the real package center (see Figure 16). By implication adaption can be only done within package outline limits. Drawback of this alignment is that no relative movement between both dies or a rotation is taken into account.

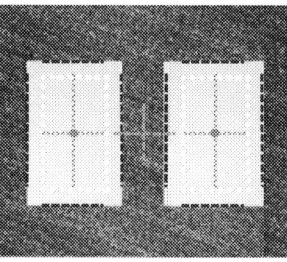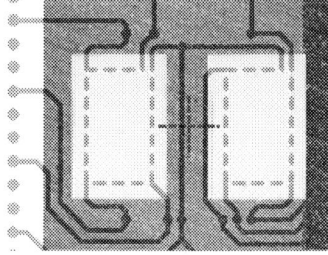

Figure 16: x-/y-package shift adaptation for maskless redistribution; left: die position measurement, red: measured Si edges, green: calculated chip center, yellow: calculated package center; right: adapted layout, blue: layout registration.

For the package described in this work this method resulted in good alignment to connect all dies on an 8" reconfigured wafer. Figure 17 depicts an x-ray image of package with PCB based redistribution and x-/y-package shift adaptation.

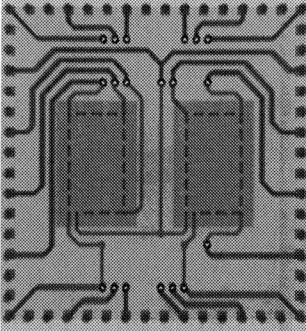

Figure 17: X-ray images of 2-chip package showing good alignment of μvias and routing after x-/y-package shift adaptation.

For packages with higher accuracy demands also the rotation of the package could be adapted. With the same measurement procedure as described above but with higher data processing effort also the package rotation could be calculated beneath the package x-/y-shift (see Figure 18).

Full layout adaptation as shown in Figure 19 including die and package shift as well as rotation should be basically feasible but will need profound developments of software algorithms as it is not sufficient to shift and rotate the layout.

978-1-4799-2408-0/14 $31.00 © 2014 IEEE

In this case also the layout has to be stretched or shrinked. And this is even more complicated if the package outline and connecting pads need fixed positions.

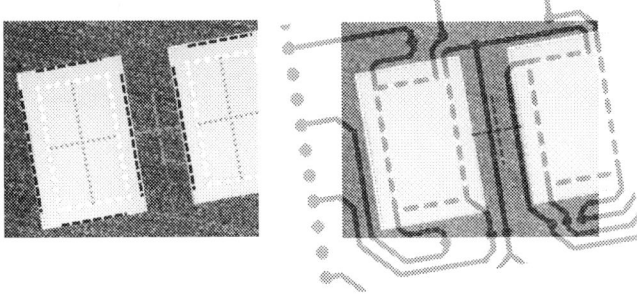

Figure 18: x-/y-package and rotation shift adaptation for maskless redistribution; left: die position measurement, red: measured Si edges, green: calculated chip center, yellow: calculated package center; right: adapted layout, blue: layout registration.

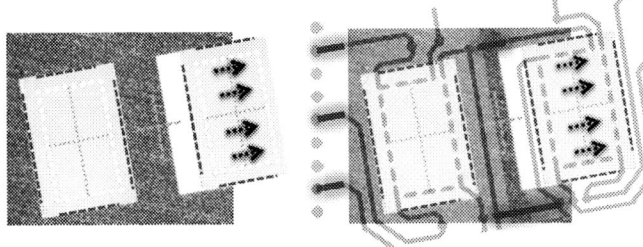

Figure 19: Full package and die shift adaptation for maskless redistribution; left: die position measurement, red: measured Si edges, green: calculated chip center, yellow: calculated package center; right: adapted layout, blue: layout registration.

Next steps

Based on the results shown above reconfigured panels will be used for further processing. Redistribution layers will be applied by resin coated copper (RCC) lamination to the reconfigured sheets and evaluated on their adhesion and resulting warpage after RCC application. Finally, packages will be manufactured demonstrating the overall process flow on 18"x24". In addition to lamination compression molding will be evaluated for these large reconfigured panel embedding processes.

Conclusions

The combination of large area embedding technology with PCB based redistribution has the potential for low cost highly miniaturized packages. Within this study large area assembly and lamination processes has been evaluated for fan-out panel level packaging (FOPLP) for panels up to 18"x24".

Lamination is definitely an option for panel embedding. One advantage might be here the option for temperature profiles with defined heating and cooling rates during lamination. This allows embedding, compound curing and carrier release without adhesive film residues and low warpage in one step. Nevertheless, thickness accuracy is one major point for improvement as lamination presses for PCB

manufacturing as used in this study have typically a slightly lower accuracy as the tooling for compression molding.

Concerning die placement accuracy and die shift, materials and process parameters have to be carefully selected to keep the die shift over the entire area tolerable and feasible to compensate within the package limits. Here, also intelligent assembly strategies can help to optimize die placement accuracy. For packages where even higher accuracy is needed an automatic layout adaptation to real die position can be implemented into the proposed maskless process flow based on PCB technologies.

Acknowledgments

Part of this work has been supported by the German ministry for education and research [BMBF] and by VDI/VDE Innovation + Technik GmbH in the MST LowCostTMV Project, (Contract No. W40043).

Additionally the authors want to thank M. Minkus for analytical support, M. Töpper and D. Jäger for test die manufacturing and L. Böttcher for electroless Ni bumping.

References

[1] T. Thomas, F. Yuwen Lin, K.-F. Becker, T. Braun, E. Jung, R. Aschenbrenner, H. Reichl; State-of-the-art of 3D SiP Technology; Proceedings of IMAPS Poland 2009.

[2] T. Meyer, G. Ofner, S. Bradl, M. Brunnbauer, R. Hagen; Embedded Wafer Level Ball Grid Array (eWLB); Proceedings of EPTC 2008, Singapore.

[3] B. Keser, C. Amrine, T. Duong, O. Fay, S. Hayes, G. Leal, W. Lytle, D. Mitchell, R. Wenzel; The Redistributed Chip Package: A Breakthrough for Advanced Packaging, Proceedings of ECTC 2007, Reno/Nevada, USA.

[4] Y. Jin, X.r Baraton, S. W. Yoon, Y. Lin, P. C. Marimuthu, V. P. Ganesh, T. Meyer, A. Bahr; Next Generation eWLB (embedded Wafer Level BGA) Packaging; Proceedings of EPTC 2010, Singapore.

[5] L. Boettcher, D. Manessis, S. Karaszkiewicz, A. Ostmann, R. Aschenbrenner, H. Reichl; Innovative embedded-chip QFN package realization; Proceedings of SMTA International Wafer Level Packaging Conference 2009, Santa Clara, California, USA.

[6] T. Braun, K.-F. Becker, L. Böttcher, J. Bauer, T. Thomas, M. Koch, R. Kahle, A. Ostmann, R. Aschenbrenner, H. Reichl, M. Bründel, J.F. Haag, U. Scholz; Large Area Embedding for Heterogeneous System Integration; Proceedings of ECTC 2010, Las Vegas, USA.

[7] T. Braun, K.-F. Becker, S. Voges, T. Thomas, R. Kahle, V. Bader, J. Bauer, K. Piefke, R. Krüger, R. Aschenbrenner, K.-D. Lang; Through Mold Vias for Stacking of Mold Embedded Packages; Proc. of ECTC 2011, Orlando, USA.

[8] T. Braun, M. Bründel, K.-F. Becker, R. Kahle, K. Piefke, U. Scholz, F. Haag, V. Bader, S. Voges, T. Thomas, R. Aschenbrenner, K.-D. Lang; Through Mold

Via Technology for Multi-Sensor Stacking; Proc. of EPTC 2012; Singapore.

[9] T. Ebefors, J. Fredlund, E. Jung, T. Braun; Recent Results using Met-Via TSV Interposer Technology as TMV Element in Wafer Level Through Mold Via Packaging of CMOS Biosensors; Proc. of IWLPC 2013, San Jose, CA, USA.

[10] T. Braun, K.-F. Becker, S. Voges, T. Thomas, R. Kahle, J. Bauer, R. Aschenbrenner, K.-D. Lang; From Wafer Level to Panel Level Mold Embedding; Proc. of ECTC 2013; Las Vegas, USA.

Development and Characterization of New Generation Panel Fan-out (P-FO) Packaging Technology

Hong-Da Chang, David Chang, Kenny Liu, H.S Hsu, Rui-Feng Tai, Hsiao-Chun Huang,
Yi-Che Lai, Chang-Lun Lu, Chun-Tang Lin, Steve Chiu
Siliconware Precision Industries Co., Ltd., Taichung, Taiwan, R.O.C.
No.153, Sec.3, Chung Shan Rd., Tantzu Dist., Taichung 42756, Taiwan, R.O.C.
Hungdachang@spil.com.tw +886-4-2534-1525 ext.7807

Abstract

Wafer Level Packaging (WLP) is a packaging technology focusing on integrated circuit (IC) packaging at wafer level instead of die level. WLP essentially consists of IC foundry fabrication process and subsequent device interconnection and back-end passivation process. The general wafer level packages (WLPs) are designed for fan-in chip scale packaging but the shrinkage of pad pitch and size at the chip to package interface is much faster than the shrinkage at the package to board interface. The embedded package technologies are developed to provide larger package size in order to offer a sufficient area to accommodate the redistribution interconnection with standard pitches at the packaging.

Embedded technology (FO-WLP / eWLB and embedded die) are unique technological breakthroughs enabling further growth of next-generation packaging modules and brings several benefits such as small form factor of packages, electrical and thermal performance improvement and cost reduction for IDMs. Compare to FO-WLP, panel level packaging (PLP) has the advantage of using large panel processing compare with FO-WLP while 12" is commonly used. Hence panel fan-out package (P- FO) has a better cost down potential than FO-WLP.

To demonstrate the new generation panel fan-out package conception, the several hetero-technologies are integrated into the packaging which composed of PCB, LCD, Bumping and FOWLP has substantially broadened the spectrum of the well established wafer level packaging (WLP) technologies. The new generation panel fan-out packaging combines the different infrastructures from several fields to realize the conception. The achieved results for the package, such as the large panel size process and precise die alignment capability point to the outstanding potential of this novel hetero-system. In this paper, mixing PCB, semiconductor back-end, semiconductor WLP and LCD Gen 2.5 (370X470mm) size processing technologies combined with innovation as well as integration of P-FO techniques are proposed, including high accuracy die bonding and die shift compensation at film lamination, lower warpage sheet form film lamination, good copper trace plating uniformity control at large panel area and also precise photolithographic technique. In addition to technology elaboration and process depiction, relevant experimental data and final reliability test results on board level have also been thoroughly demonstrated and discussed.

Introduction

Highly integrated and microminiaturized semiconductor packaging devices for portable devices had strong and increasing demand from end user at mobile phone. Current portable electronic products are driving component packaging towards several packaging technologies for integrating memory die and application processors (AP), including digital, analog and radio frequency (RF) into one multiple function system. Base on different application and Si nodes, TSV, 2.5D interposers, eWLB (embedded Wafer Level Ball Grid Array) / FO-WLP (Fan-out Wafer Level Package) and coreless substrate are developed to meet different needs. Driven by consumer electronics, smaller, thinner and cost effective package had demanded the use of economic panel size substrate as a carrier to reduce overall cost. Fig. 1 shows the package cost reduction with substrate size and the evolution of panel scale packaging platforms.

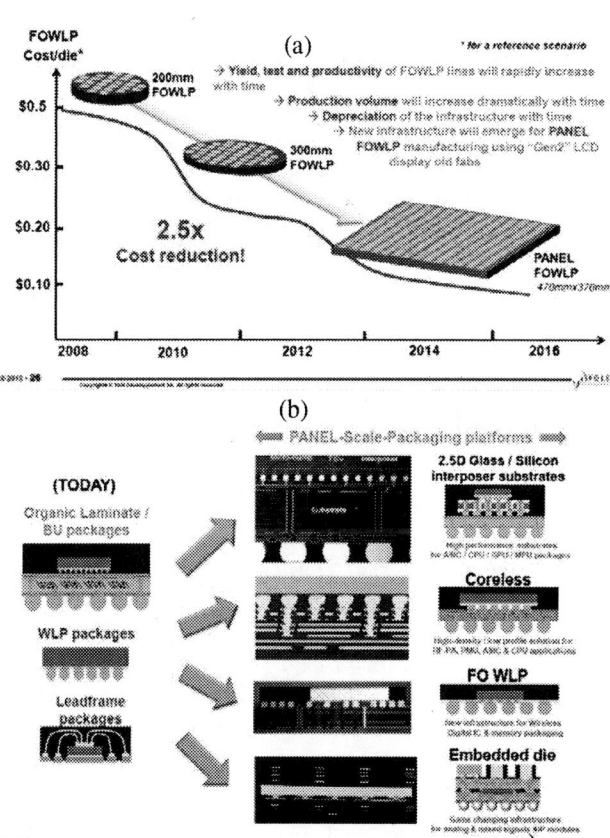

Figure 1. (a) Cost/die development based on substrate size for Fan-Out packages. [1] (b) Evolution of PANEL-Scale-Packaging platforms. [2]

With the advancement of new CMOS technologies, high I/O density and highly integrated high volume packages such as copper pillars, micro-bumps or RDL (redistribution layers) continue to require smaller Lines and Spaces (L/S). The

---

978-1-4799-2408-0/14 $31.00 © 2014 IEEE   947   2014 Electronic Components & Technology Conference

increased sophistication of interconnection from die to die is pushing the combination of typical packaging technical art and new high accuracy manufacturing techniques into one. Panel fan-out (P-FO) package can judiciously this purpose – form a high volume, smaller package footprint, medium to high I/O count, as well as good electrical and thermal performance. In this article packaging evolution is special focus on the challenge and solution of panel size fan-out package.

Fan-out-WLP Technology

FO-WLP technology combines conventional front- and back-end manufacturing techniques with parallel processing of all chips. Tested known good dies are reconstructed and surrounded by a suitable material which spreads the package footprint outside the die.FO-WLP combines four major stages to fabricate package including KGD reconstruction, surrounding film lamination, RDL and backend process as shown in figure 2. [3]

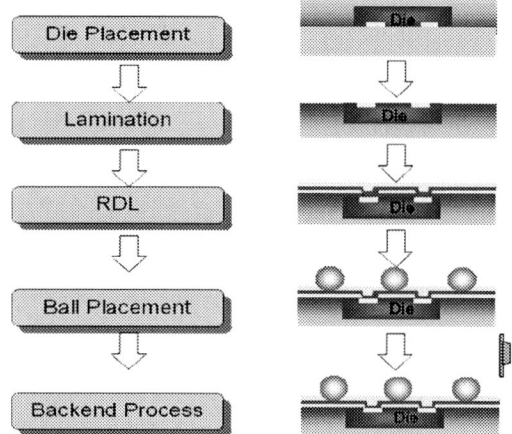

Figure 2. Process flow of FO-WLP

As the Si nodes move to 16nm, there are more challenges for flip chip assembly to deal with micro-bump structure and material, underfill process & material and substrate materials to secure good solder joint reliability. FO-WLP provides a spread area to extended RDL line and enlarger both pad pitch and size for solder joint that provide a connection between device and motherboard without substrate, underfill and sophisticated UBM/micro-bump designs. Hence FO-WLP is a cost effective package which compete with standard flip chip assembly. [4]

Panel Fan-out (P-FO) Package Technology

In order to reduce overall cost further, panel form fan-out package technology which provides more effective die-setting and provides high volume manufacture is demonstrated. The known good die is reconstructed on the LCD Gen 2.5 (370X470mm) size glass carrier with adhesive temporary bonding material directly without any small strip fan-out cube. [5]

Although highly integrated panel form FO package provide high volume manufacture possibility, there are still many challenges such as warpage, die shift, coordinates compensation at lithography and Cu plating uniformity.

I. Design Simulation

FO-WLP technology provides a fan-out area to reconfigure I/O with larger bump pad size and coarse pad pitch. Fan-out area which is constructed from thermosetting compound has different material properties than silicon. Stronger and reliable Cu trace design of RDL is needed to stretch the electronic connection from device to fan-out area. The stress distribution contour of Cu trace is showed at Fig.3. Trace at fan-in and fan-out boundary and bump pad shape should be well defined for stress reduction.

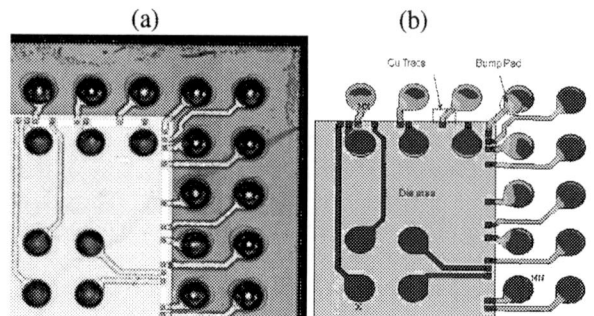

Figure 3. (a) RDL trace on FO-WLP (b) Stress contour of RDL during reliability test.

II. Warpage

Larger footprint of panel also induces higher warpage and decrease the package yield. The mismatch of material properties will cause structure bending. The simulation modeling of panel size fan-out structure is established with several possible schemes which including surrounding material, carrier, adhesive glue, thickness and structure of package.

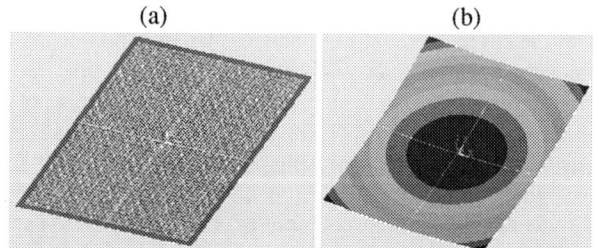

Figure 4. (a) Panel FO simulation modeling (b) Warpage contour of panel FO

Base on the quickly prediction from simulation model, the warpage of panel FO structure can be controlled within +/- 0.5mm after carrier de-bond. The key parameter of material properties can be defined from the model and use for the process improvement. Fig. 5 shows the warpage trend chart of each assembly step and feature of P-FO package panel after lamination and reconstruction process.

978-1-4799-2408-0/14 $31.00 © 2014 IEEE   948

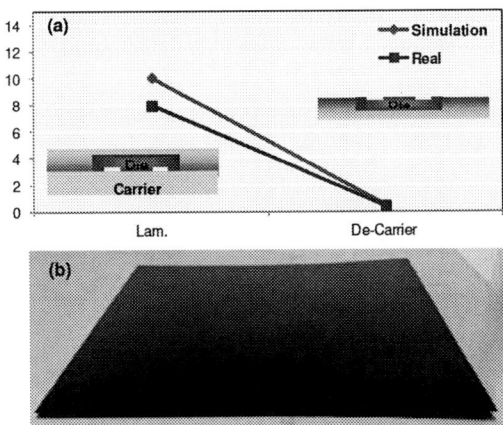

Figure 5. (a)Warpage trend chart of simulation and real process data (b) Side view of P-FO package panel after lamination and reconstruction process

III. Die Shift Compensation

The sheet form lamination film is pressed with heat on the reconstructed die. The melted film will push the components on the carrier at high down force pressure. Die shift can be overcome by enlarge CD opening at RDL design on 8 inch or 12 inch wafer but hard to achieve at panel size which has larger die shift at panel edge. Figure 6 shows schema of die shift on different size carrier.

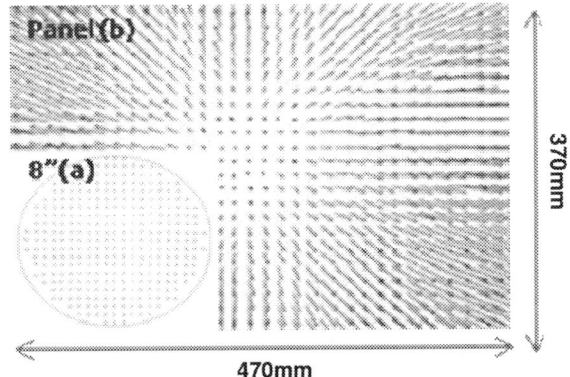

Figure 6. Die shift vector diagram after lamination (a) 8" wafer (b) Gen 2.5 Panel (370mm x 470mm)

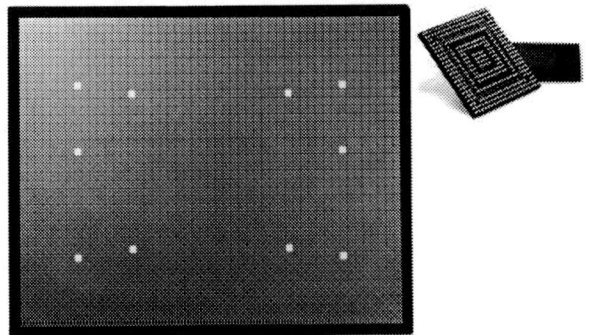

Figure 7. Panel Fan out package of 370mm x 470mm footprint (9X9mm package size)

The compensable patterning method for photo lithography technique is applied on the panel fan-out technology to achieve high yield and high volume application. The alignment method can compensate the die coordinates to get good overlay capability.

Combine with existing panel size plating infrastructures of PCB industry and assembly technique arts. The fan out package with 370mm X 470mm panel size sample was demonstrated which shows in fig.7.

Board Level Reliability

Two different kinds solder ball was tested to enhance the reliability during drop test. These solder balls were jointed on Cu trace directly without UBM which called BOT (Ball on Trace) bump structure. The test condition follows the JEDEC (Joint Electron Device Engineering Council) standard to accelerate 1500G force on test board with 1msec pulse duration. SACX which doped with 0.1% Bi shows improvement at drop shock and pass 500 drop cycles. Figure 8 shows the Wei-bull distribution of SACX solder ball. The element Bi can lower the solidus temperature, improve the wetting and alloy spreading, refine the Sn matrix through precipitation hardening, and suppress the formation of large Ag3Sn IMCs in the bulk solder. [6]

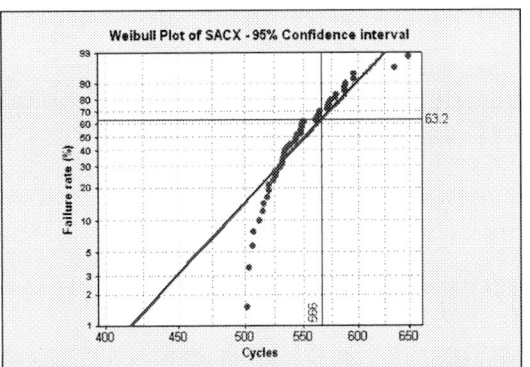

Figure 8. Wei-bull plot of board level drop reliable test

Figure 9 shows the different failure mode of SAC405 and SACX without underfill.

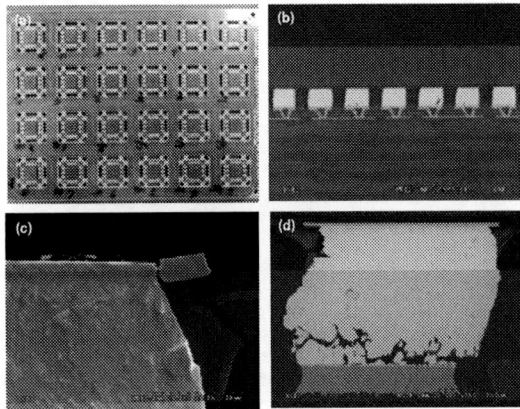

Figure 9. (a) Test board outline of drop (b) X-section of solder joint between package and board (c) Failure mode of SAC405: bump pad crack (d) Failure mode of SACX (w/ Bi): solder joint crack

978-1-4799-2408-0/14 $31.00 © 2014 IEEE

For solder joint crack of SACX, underfill encapsulation between package and board will increase drop cycles of reliability test but depend on requirement of applications.

Temperature-dependent Warpage

The temperature-dependent warpage of package was measured from shadow Moiré. The result is shown in Figure 10. The warpage variation of panel fan-out package performs within +/-30um and shows lower material properties mismatch between die and surrounding compound at high temperature.

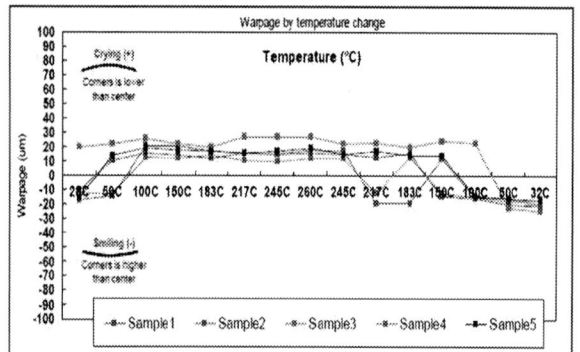

Figure 10. Temperature-dependent warpage measurement from shadow Moiré

Next Generation FO Package Challenges

The use of fan out package in a package-on-package (POP) configuration to function as the base for a 2.5D TSV interposer configuration is critical to enable a more cost effective package solution. [7] The package allows for vertical integration of the memory package and the logic package into one stacked package. Fig. 11 shows the packaging market evolution roadmap.

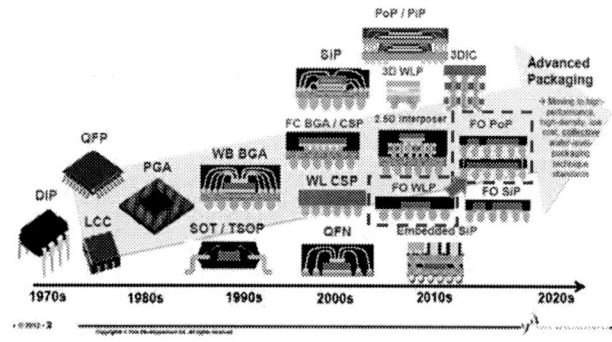

Figure 11. Semiconductor chip packaging market evolution [8]

Combining logic device with the wide I/O memory interfaces or high bandwidth memory (HBM) with the 3D FO packaging capability can provide an optimum solution for achieving the highly integrated multi-die stacks for high volume manufacturing. There is several technology of 3D FO packaging technology demanded for the next generation applications including:

1. Multi- layer interconnection for higher logic I/O.

2. Fine line/space Cu trace of RDL.
3. Fine pitch / opening through package via (TPV)

Figure 12 shows relationship between RDL Line/Space and end–markets application. The RDL Line/ Space (L/S) is pushing to lower than 5/5um from the requirement at mobile application at FO technology.

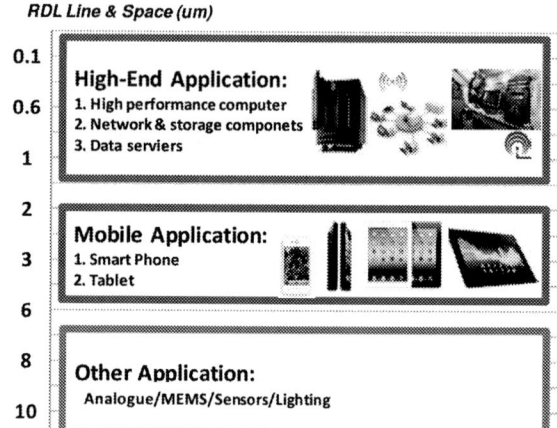

Figure 12. Relationship between RDL Line/Space and end–markets application

Three layer RDL and fine line Cu trace was demonstrated for process limitation and design rule check which shows in Fig. 13. From the individual module study results, the concept of multi layer RDL and ultra fine Line / Space (L/S) Cu trace feasibility was demonstrated. There are still more challenges to apply the package experience to panel form working area and infrastructures integration. Panel FO package technology shows the possibility of well established new platform with constitution of PCB, LCD, bumping and assembly technologies.

Figure 13. X-section (a) Multi-layer RDL (b) Fine line RDL

Conclusions

Panel fan-out package is a cost effective manufacturing technology and enhance the standard FO-WLP, allowing the next generation of panel platform for multitude of complex and highly integrated solutions that portable and mobile applications require. The advantage of panel fan-out package include a predictable stress simulation model, lower mismatch material combined structure and compensable patterning method of lithography technique for lower cost and high volume manufacturing solution. Panel FO provides enhanced bump solder joint reliability with a BOT (Ball on Trace) bump structure approach for advanced Si node devices. And it shows good temperature-dependent warpage of package from shadow Moiré measurement.

Panel FO technology will provide more exciting developments in the future as the scalability of LCD 2.5 generation panel size (370mm X 470mm) carrier with multi-layer RDL, ultra fine Cu trace Line/ Space (L/S) and fine pitch through package via in order to drive more cost effective and highly integrated semiconductor package.

Acknowledgments

The authors would like to thank our colleagues Yan-Heng Chen, Benson Lai, Minging Chang, Carina jim, JP Huang, Agassi Chen, Yi-Wei Lu, Po-I Wu, Bruce Chiu, Steven Fan, Alex Chen, Jay Chao, Jash Lai, and Ken Chang of SPIL Corporate R&D.

References

1. Yole Development, "Convergence of FE, BE and FPD manufacturing technologies: a closer look," Jan 7th, 2012 http://www.i-micronews.com/news.

2. Yole Development, "An overview of recent panel-scale packaging developments throughout the industry," in MiNaPAD Forum 2012, Grenoble, France, April 25-26, 2012
http://www.semi.org/eu/sites/semi.org.

3. H.S Hsu, "Innovative Fan-Out Wafer Level Package using Lamination Process and Adhered Si Wafer on the Backside," in Proc. IEEE Electronic Components and Technol. Conf. (ECTC) 62nd, May 29 -June 1, 2012

4. Seung Wook Yoon, "Fanout Flipchip eWLB (embedded Wafer Level Ball Grid Array) Technology as 2.5D Packaging Solutions," in Proc. IEEE Electronic Components and Technol. Conf. (ECTC) 63rd, May 28-31, 2013

5. John Hunt, "A Hybrid Panel Embedding Process for Fanout," *in Proc. IEEE Electronics Packaging Technology Conference (EPTC) 14th*, Dec 5-7, 2012

6. Zijie Cai, "Reduction of Lead Free Solder Aging Effects Using Doped SAC Alloys," thesis of doctor degree of Philosophy, Auburn University, Alabama, December 8, 2012

7. Seung Wook Yoon, "Advanced low profile PoP solution with embedded wafer level PoP (eWLB-PoP) technology," in Proc. IEEE Electronic Components and Technology Conference (ECTC) 62nd, May 29 -June 1, 2012

8. Yole Development, "3DIC & TSV interconnects," Semicon Taiwan, Taipei, Taiwan, Sep 5-7, 2012
http:// www.semicontaiwan.org/en/sites.

Development of Exposed Die Large Body to Die Size Ratio Wafer Level Package Technology

J. Osenbach[1], S. Emerich[1], L. Golick[1], S. Cate[2], M. Chan[3], S.W. Yoon[3], Y.J. Lin[4] & K. Wong[5],

[1]LSI Corporation (USA), 1110 American Parkway NE Lehigh Valley Central, Allentown, PA 18109, USA
[2]LSI Corporation (USA), 1320 Ridder Park Drive, San Jose, CA 95131, USA
[3]STATSChipPAC Ltd, 5 Yishun Street 23, Singapore 768442
[4]STATSChipPAC Ltd, 2 Woodlands Sector 1 #01-20, Woodlands Spectrum 1, Singapore 738068
[5]STATSChipPAC, Inc, 47400 Kato Road, Fremont, CA 94538, USA

Abstract

Traditionally fan out wafer level package technology has been associated with lower power, smaller body sizes (typically < 8mmx8mm), small body-to-die size area ratios (<2) and fine pitch BGAs (0.4mm or less). This work extends this technology to larger body sizes up to 13mm x 13mm, higher powers, > 5W, and larger body-to-die size area ratios up to 10.5. It is shown that such packages can be readily manufactured in a 300mm wafer format with yields exceeding 99% and final package warpage < 75um. Further, data is presented showing that 10mm x 10mm packages with a body to die area ratio of 6.25 are compatible with moisture sensitivity level 1, and easily pass 2000 temperature cycle (-55C to 125C air to air) and 288 hr uHAST. That is to say they have reliability that is compatible with that required for all storage and communications applications. Larger package sizes, up to 13mm x 13mm, and body-to-die area ratios, > 10, have also been demonstrated. However, failures in extended temperature cycle were found in these larger packages. All of the failures were due to pre-identified package design flaws that violated well established rules. This indicates if such packages were designed with no rule violations then they would meet the reliability requirements needed for communications and storage applications.

Introduction

Although there have been significant challenges for silicon technology to follow Moore's law, leading some to predict its demise, the industry has been able to overcome these challenges and appears ready to continue developing solutions for the challenges at least for the foreseeable future. As such integrated circuits (ICs) continue to move towards finer feature sizes and increased functionality. This drives more die and package I/O with higherpower densities than packages designed andmanufactured in previous generation technology nodes. Further, mobile devices are driving smaller form factor ICs with renewed pressure to reducepackage thickness. The combination of these forces has driven the development of a varietyof chip scale packagesolutionsincluding wafer level packages (WLP), dual row quad flat pack packages (DRQFN), flip chip -chip scale packages (fcCSP), and fan out wafer level packages (FO-WLP).

FO-WLP is an emerging package technology platformthatprovides opportunities for extremely thin, small foot print, high I/O density packages with electrical performance rivaling many of the low-to mid-range fcCSPs and exceeding that for the large majority of lead frame wire bond package technologies [1]. FO-WLP technology has traditionally been associated with lower power, smaller body sizes (typically < 8mmx8mm), small body-to-die size ratios (<2) and fine pitch BGAs (0.4mm or less). However, recent advancements of this technology include demonstrations of side by side multi-chip module (MCM), Package-on-Package (POP), 2.5D, and 3D configurationswhich seem to indicate such limitations may not be fundamental technologybarriers, but rather manufacturing and materials challenges that are waiting to be solved. [2]

To increase the breadth and type of markets that this technology can be applied to, improvements in thermal performance and larger body-to-die area ratios are required. These market segments impose requirements such as: BGA pitchesas large as 0.65mm and in some cases 0.8mm; > 5watts power dissipation; body sizes of 15mm or larger; and body-to-die size ratios of >10. Furthermore, some of these markets segments also require a minimum lifetime of 10 years. In this paper, our results on the development and qualification of an exposed die fan-out wafer level package technology with body-to-die aspect ratios of >10 and body sizes up to 13mm x 13mm are presented.

Process Development and Control

A schematic of the package technology developed in this work is shown in Figure 1. The package contains a 250um thick, 16mm^2 die embedded in a mold compound with 2 layers of package routing connected to 0.25mm diameter, 305SAC BGA solder balls. The die are designed into a 40nm silicon node technology. The maximum package height is 0.68mm and the number of I/O was > 200. Two sizes of packages were used for the work, i) 10mm x10mm with a BGA pitch of 0.5mm ; and ii) 13mm x 13mm with a BGA pitch of 0.65mm. Thus, the body-to-die area ratios were 6.25 and 10.5 respectively. The same die was used for both packages; however, the package routing was different. The large body-to-die ratio tends to drive to higher package warpage making it difficult to meet JEDEC and IPC requirements. [3]

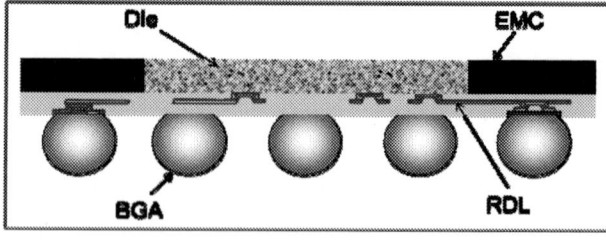

Figure 1: Schematic of an exposed die FO-WLP

Although back side mold compound on the die helps to mitigate the difficulty in meeting the warpage specifications, the 5W power dissipation requirement for these market applications prevents, as discussed next, the use of this warpage mitigation approach from being implemented. The thermal simulation done for both package sizes using the design constraints and thermal boundary conditions of the market segment this device is targeted for indicate a mold compound thickness greater than 35um will prevent the package from meeting the 5W requirements, Figure 2. In this simulation, the system configuration and restriction on air flow were modeled with 3-D finite element model, 3w/mK TIM material, a die size 4mm x 4mm in a 10mm x 10mm package.After assessment of variation in die performance at all silicon and FO-WLP process corners, as well as the system corners, it was concluded that an exposed die format was needed to meet the thermal requirements.

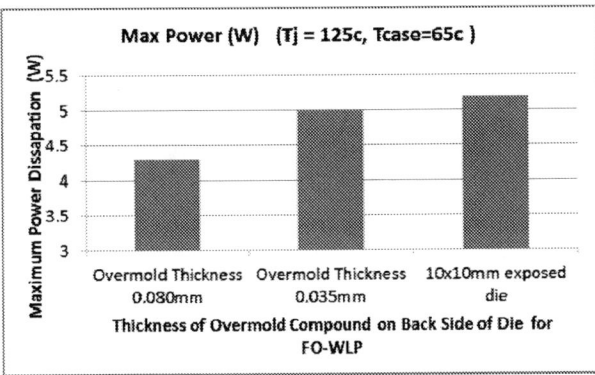

Figure 2: Dependence of max power dissipation on mold thickness on back side of die in a 10mm x 10mm FO-WLP

In this work the FO-WLP were made in 12" diameter format. An overview of the process flow is given in Figure 3. Known good die are picked from the diced wafer and placed active side face down onto a carrier array at predefined x, y positions. The gap between each die is used to create the FO-WLP routing and BGA connections. There were approximately 40% fewer die and thus packages for the 13mm x 13mm package than for the 10mm x 10mm package, driven completely by area considerations. The carrier wafer has a commercially available adhesive film on it which is used to hold the die in place at pick and place as well as during the molding process. After the carrier wafer is fully populated with die, the entire carrier wafer is overmolded with a commercially available mold compound. Following mold compound cure, the carrier is removed from the composite wafer, die plus mold compound, exposing the active side and bond pads for further processing. The package routing is then completedusing standard spin deposition of interlevel dielectrics, lithography, and Cu plating technology. First, dielectric layer 1 is spun on the composite wafer and patterned, opening up the bond pad regions of the die. Cu is deposited and patterned, followed by a second layer of dielectric and metal. The final passivation layer is then deposited and patterned. Note the final passivation layer in the FO-WLP is equivalent to a solder mask in a standard substrate based technology. The wafer is then fluxed after which BGA balls are placed on each pad and reflowed. Because the product thermal requirement can only be met with an exposed die back side configuration, the back side mold compound is then removed.The fully processed carrier wafer, which is now a wafer array of packages, is then diced creating individual packages. The packages are then tested and packed into dry bags ready for shipment.

Figure 4 shows a cross section of a 13x13mm FO-WLP test vehicle. The die, mold compound (EMC), BGA, Cu RDL layers and dielectric RDL layers are identified in this figure. As shown all interfaces are intact with no signs of delamination, separation or cracking of the different materials or interfaces. Figure 5 is a micrograph of the BGA side of both the 10mm x 10mm and 13mm x 13mm packages. Also show is the approximate position and size of the die.

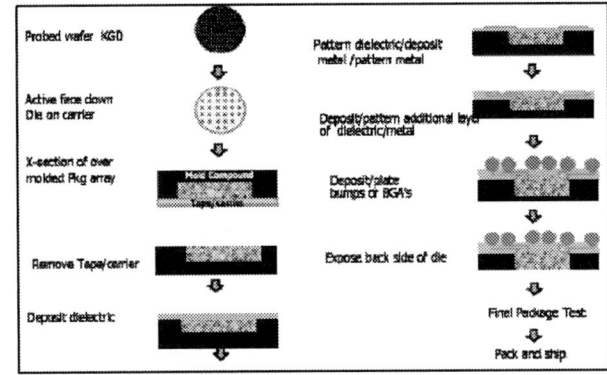

Figure 3: Overview of process flow for the devices used in this work

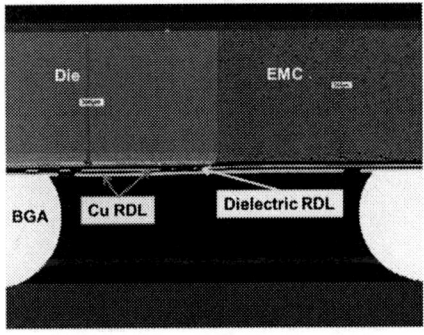

Figure 4: Cross section of a 13x13mm FO-WLP test vehicle

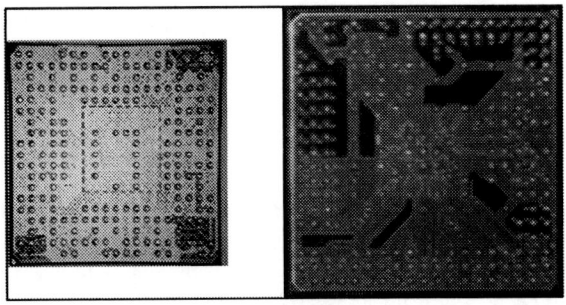

Figure 5: Micrograph of 10mm x10mm and 13mm x 13mm packages; outline of approximate position of the die are shown as the red dotted box.

978-1-4799-2408-0/14 $31.00 © 2014 IEEE

The body-to-die area ratios, 6.25-10.5, used in this development along with the exposed back side construction and 2 layers of routing were substantially outside of the established design and process window for FO-WLP. As such significant development was needed to establish a process with a well-controlled-, reproducible-, and predictable-die shift and composite wafer warpage, as these two physical attributes largely influence interconnect patterning and wafer handling both of which ultimately control the overall package yield. Initial builds made with the previously "established" processes and materials led to package yields that were less than 90%, clearly not an acceptable yield for low cost, high volume manufacturing. Subsequent process and materials development and optimization led to yields approaching or exceeding 99%. It was thus concluded that this type of package technology is capable of competing on a yield basis with well-established traditional wire bond and flip chip package technology even for the packages developed in this work which have somewhat extreme body-to-die ratios in an exposed die configuration.

In addition, the yield as measured by electrical testing, control of the final package dimensions (X, Y and Z) and warpage is critical. Figure 6 is a plot of the dimensions (X, Y and Z) and warpage of the 10mm x 10mm packages. Similar behavior was observed for the 13mm x 13mm packages, however, the number of 13mm x 13mm FO-wafers and packages made was significantly smaller than made the number of 10mm x 10mm FO-wafer and packages. Thus, conclusions about dimensional (X, Y and Z) and warpage control are somewhat less well-founded in data for the 13mm x 13mm package than for the 10mm x 10mm package. As a result we restricted this discussion to the 10mm x10mm package. As shown, for the 10mm x 10mm packages, the package dimensions and wapage are well controlled with warpage being < 50um on average with a three sigma maximum of ~75um which is well below the JEDEC and IPC requirements for this package and BGA size.

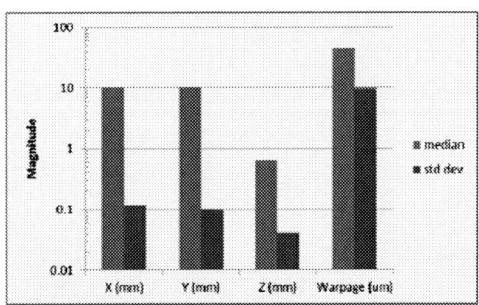

Figure 6: Median and standard deviation of X, Y, and Z dimensional control and of final package warpage at room temperature for the 10mm x 10mm package

Reliability

The reliability work is summarized in Table 1. All devices were tested then subjected to Moisture Sensitivity Level (MSL) exposure followed by 3x-260C reflow and re-tested. The 10mm x 10mm packages were subjected to MSL1 and the 13mm x 13mm to MSL3. The 10mm x 10mm packages used in cell 1 were taken from two different FO-wafer builds whereas those in cell 2 were taken from 4 wafer builds made

some 3-4 months after those in cell 1. All 13mm x 13mm devices were taken from one wafer build. Following MSL/reflow exposure, devices from each package type were then subjected to either temperature cycle JEDEC condition B (-55C to + 125C air to air) up to 2000 cycles or to unbiased Hast (uHAST- 130C/85%RH) up to 288hrs [4]. Periodically the packages were removed from the stress tests and electrically characterized. All electrically good devices were then placed back into the stress test chamber for additional aging. The temperature cycling and uHAST testing duration was extended with the intent of insuring the packages were not close to end of life after exposure to the more conventional JEDEC tests. The thesis being, exposure to extended temperature cycles and/or time in uHAST conditions would either show where the ultimate failure lifetime would be or if no failures were found indicate the packages are significantly more robust than required for use in high reliability environments with no failures .

Cell	Package Type	MSL	TCB (Cycles-#fail/#tested)				uHAST (hr-#fail/#tested)		
			500	1000	1500	2000	96	192	288
1	10mm x 10mm	0/320	0/160	0/160	----	----	0/160	0/160	----
2	10mm x 10mm	0/660	0/330	0/330	0/330	0/330	0/330	0/330	0/330
3	13mm x 13mm	0/94	0/47	1/47	----	32/47	0/47	0/47	----

Table 1: Summary of qualification testing done for 10mm x 10mm and 13mm x 13mm packages(10x10mm device has 0.5mm BGA pitch, 13x13mm device has 0.65mm BGA pitch)

As shown, failures were not observed in either cell in any of the stress tests independent of duration for the 10mm x 10mm packages. Given that the devices used for this test were taken from a total of 6 wafer builds over a multiple month time period, the sample sizes used in each test and the fact that TCB and uHAST stress tests were extended 2x and 3x respectively beyond standard JEDEC recomendations, it is conclude that such packages are very reliable and thus compatible with all storage and communications applications.

Although the 13mm x 13mm package easily passed MSL3 and extended uHAST, ~2% (1/47) of the devices failed after exposure to 1000 temperature cycles, and ~ 70% (32/47) of the devices failed after 2000 temperature cycles. The failure site and mechanism was similar for all 32 failures. Figure 7 shows two micrographs of the failure site. The failure is a result of metal fatigue of a metal 1 trace near the die edge at the die to mold compound interface. This particular trace is also under the BGA and BGA pad in metal 2. Placement of the BGA in this location is and was a design rule violation which was identified in the design stage. It was flagged as a potential weakness in the design that could be susceptible to temperature cycling induced failure. In addition the RDL stack used in this work was designed for solder balls compatible with BGA pitch ≤ 0.5mm, note for 0.65mm pitch as was used for the 13mm x 13mm package used here. Given larger diameter BGA solder balls drive higher localized stresses in lower level structures, this was also flagged as a potential reliability issues in the design stage. However, because there was a need to make the 13mm x 13mm FO-WLP pin to pin compatible with a 2 layer flip chip version of the same device, the design rule violationsand potential reliability risks/failure modes were noted, but the design proceeded with no changes to accommodate the rule violations. Thus it was understood at the outset of the program that the 13mm x 13mm package could be susceptible to failure in temperature cycling and/or other qualification stress test. Given all temperature failures were related to the design violations, it is concluded that FO-WLP designs with body-to die ratios$\leq$10.5, that follow the existing design rules, have a high probability of meeting and, in fact, exceeding the JEDEC temperature cycling requirements.

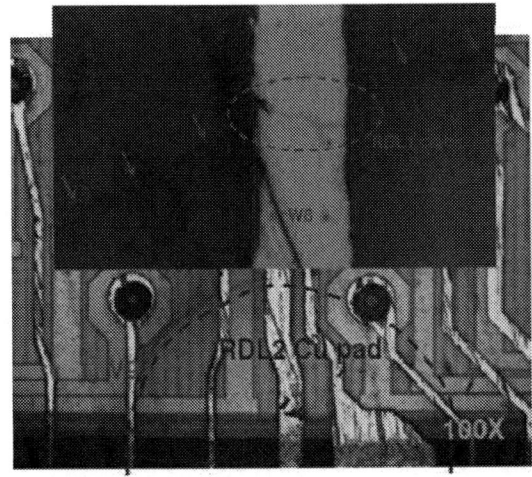

Figure 7: Micrographs representative of all 13mm x 13mm TCB induced failures.

Conclusions:

Traditional fan out wafer level package technology has been associated with lower power, smaller body sizes (typically < 8mmx8mm), small body-to-die size area ratios (<2) and fine pitch BGAs (0.4mm or less) package designs. The work presented here extends this technology to larger body sizes, up to 13mm x 13mm, higher powers, > 5W, and larger body-to-die size area ratios,up to 10.5. It is shown that such packages can be readily manufactured in a 300mm wafer format with yields exceeding 99% and final package warpage < 75um. Further, data is presented showing that 10mm x 10mm packages with a body to die area ratio of 6.25 are compatible with moisture sensitivity level 1, easily pass 2000 temperature cycle (-55C to 125C air to air) and 288 hr uHAST. That is to say they have reliability that is compatible with or exceeds that required for all storage and communications applications. Larger package sizes, up to 13mm x 13mm, with body-to-die area ratios in excess of 10, have also been demonstrated, however, failures in extended temperature cycle were found. All of the failures were due to pre-identified package design flaws that violated well established rules. This indicates if such packages were designed with no rule violations then they would meet the reliability requirements needed for communications and storage applications

Acknowledgments

The authors would like to express their gratitude to all the people who were involved in this program. Particularly, thanks to Jar Ray Estela from the STATS ChipPAC Singapore Test team and Law Siew Chin from the STATS ChipPAC Singapore Process Engineering team for their support and contributions.

References

1. M. Brunnbauer, et al., "Embedded Wafer Level Ball Grid Array (eWLB)," Proceedings of 8th Electronic Packaging Technology Conference, 2009, Singapore.

2. Seung Wook Yoon, Patrick Tang, Roger Emigh, Yaojian Lin, Pandi C. Marimuthu, and Raj Pendse, "Fanout Flipchip eWLB (embedded Wafer Level Ball Grid Array) Technology as 2.5D Packaging Solutions", Proceedigns of 63[rd]Electronic Components and Packaging Technology, 2013, Las Vegas

3. Peter Chen, M. Bachman, John Osenbach, Feng Kao , Erik So , Eason Chen , Jun Min Liao , and Jui Tsung, "2L OMEDFC Development for Low Cost & High Performance Application", Proceedings of 15th Electronic Packaging Technology Conference, 2013, Singapore

4. JEDEC Standards JESD51-2 Integrated Circuits Thermal Test Method Environment Conditions Natural Convection (December 1995) and JESD51-9 Test Boards for Area Array Surface Mount Package Thermal Measurements (July 2000).

3D Rectangular Waveguide Integrated in embedded Wafer Level Ball Grid Array (eWLB) Package

E. Seler[1], M. Wojnowski[2], W. Hartner[3], J. Böck[2], R. Lachner[2], R. Weigel[1], A. Hagelauer[1]

[1]Friedrich-Alexander University Erlangen-Nuremberg, Cauerstr. 9, 91058 Erlangen, Germany
[2]Infineon Technologies AG, Am Campeon 1-12, 85579 Neubiberg, Germany
[3]Infineon Technologies AG, Wernerwerkstraße 2, 93049 Regensburg, Germany
E-Mail: Ernst.Seler@Infineon.com

Abstract

In this paper, we present for the first time the realization of a 3D rectangular waveguide in the fan-out area of an embedded Wafer Level Ball Grid Array (eWLB) Package using laminate inserts. To obtain the waveguide side walls in eWLB, a RF laminate with micro-vias is inserted in the fan-out area. The classical redistribution layer (RDL) on the one surface and an additional back side metallization on the other surface of the package are used to realize the top and bottom walls of the waveguide. Furthermore, we focus on a single-ended coplanar waveguide (CPW) to rectangular waveguide transition. We investigate and compare two concepts for the transformation of the transverse electromagnetic (TEM) mode of a CPW to the transverse electric (TE) mode of a rectangular waveguide. The first concept is using a via in the waveguide structure to excite the waveguide TE mode. The second concept is a planar realization using a modified shape of the RDL to realize the mode transformation. We show by the means of measurements the characteristics of each version. We show that the mode change is done with almost no additional losses to those related to physical length of the transmission line. Furthermore we present several on-wafer measurements which agree very well with the RF behavior predicted by the simulations. This 3D waveguide integration in eWLB enables the realization of many new RF features in chip embedding technologies in the future.

Introduction

Radio-frequency (RF) components and systems operating above 60 GHz have received increased interest in recent years, as new applications are being introduced and developed [1]. A variety of applications have been recently proposed working at millimeter-waves like wireless networks, various radars, high-resolution radio imaging and biomedical devices.

Due to the increase of millimeter-wave applications a lot of research and development in the field of new technologies and system concepts is ongoing. The success of such new technologies and system concepts depends mainly on the availability of a cost-effective production process suitable for mass production. In addition, millimeter-wave technologies require high-resolution techniques due to the small wavelength. Even small variations of layout dimensions due to production tolerances can have a significant influence on the system performance.

One innovative packaging technology for millimeter-wave applications is the embedded wafer level ball grid array (eWLB) technology [2], [3]. eWLB is a promising candidate for system in package (SiP) and 3D eWLB applications [4], [5].

Another promising candidate for future millimeter-wave applications is the substrate integrated waveguide (SIW) technology [6], [7]. SIW structures are generally fabricated by using two rows of conducting cylinders, e.g. micro-vias, or slots that are embedded in a dielectric substrate that connects two parallel metal plates. Classical rectangular components like filters or antennas can be realized in planar form using SIW technology [7].

In this paper we present the combination of both technologies. We show the realization of a rectangular waveguide in the fan out area of the eWLB package. This paper is organized as follows. First, we briefly describe the eWLB package concept and the rectangular waveguide theory with respect to SIW. Then, we present the technical solution of the realization of a rectangular waveguide in eWLB. We present the possibility of using through encapsulant vias (TEV) and pre-fabricated via bars for the production of the rectangular waveguide in the eWLB fan-out area. In the following part we present two different electromagnetic concepts to transform the transverse electromagnetic (TEM) mode of the CPW line to the transverse electric (TE) mode of the rectangular waveguide. The first concept uses a via realized in the rectangular waveguide to excite the TE mode in the rectangular waveguide. The second electromagnetic concept is a planar approach, using the RDL of the eWLB to achieve the mode transformation. Finally we present several measurement results. We show the measurement structures and the rectangular waveguide characterization. We also show the measured influence of the production tolerances on both electromagnetic concepts and we show that the measurements go very well with the simulation results.

Embedded Wafer Level BGA (eWLB) Package

The eWLB is one of the most advantageous packaging technologies with respect to high input/output (I/O) density, integration flexibility and electrical performance [2]. When it comes to RF, the latter has to be considered particularly. Therefore much research is ongoing to minimize the losses at the chip-to-periphery (e.g. board, antenna) RF transitions. Additional aims are higher level of integration, reduced size and low production costs. Due to low losses at millimeter-wave frequencies and high design flexibility, integration of a rectangular waveguide in a chip-embedding package is a promising approach to achieve the mentioned goals.

The excellent electrical properties of the mold compound, which acts as a carrier for the fan-out area, enables the realization of low low-loss transmission lines and high-Q passives [8]. The eWLB technology has demonstrated

978-1-4799-2408-0/14 $31.00 © 2014 IEEE

outstanding electrical capabilities for RF and millimeter-wave applications [3].

Substrate Integrated Waveguide (SIW)

The propagation characteristics of travelling waves in SIW structures are similar to those of rectangular waveguides, provided that the metallic vias are closely spaced and therefore the radiation leakage can be neglected (Fig. 1) [9]. More precisely, SIW modes only coincide with a subset of the guided modes of the rectangular waveguide. In SIW only TE_{n0} modes, with $n = 1, 2, \ldots$ can propagate (Fig. 2). Transverse magnetic (TM) modes are not supported by SIW. TM modes require longitudinal surface currents, which are suppressed by the gaps between the vias. Analytical descriptions of the dimensions of SIW components can be found in [10], [11]. All in this paper presented SIW components have been dimensioned using commercial full-wave analysis tools.

Rectangular Waveguide realized in eWLB

Fig. 3a and Fig. 3b show two possible ways to realize the via bars for the rectangular waveguide. In both approaches, the two RDLs on both surfaces of the package are used to form the upper and lower metal layer of the rectangular waveguide.

The first approach uses TEVs to realize the sidewalls. TEVs are realized in the mold compound of the fan-out area (Fig. 3a). Using laser drilling, TEVs can be positioned freely over the mold compound area. The filling of the laser drilled holes can be done by application of an electro less plated seed layer and electroplating and/or it can be plugged with conductive or non conductive paste [5].

The second way to realize the via bars in the fan-out area is the application of pre-fabricated printed circuit board (PCB) structures (Fig. 3b). Fig. 4 shows the cross section of a via and parts of the RDL using pre-fabricated via bars. Fig. 5 illustrates the process flow of this variant. Pre-fabricated PCB pieces with via bars are placed on the mold carrier prior to molding. Then the reconstitution wafer is generated via compression molding. This artificial layer is then grinded down to make the via bars accessible for the connection with a

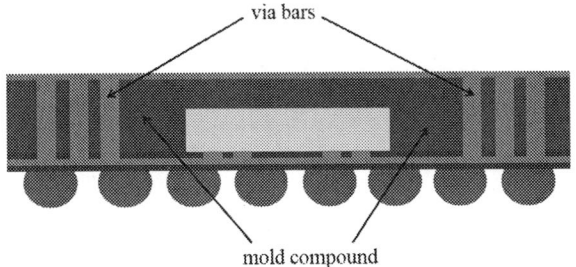

Figure 3a. Via bars realized in the fan-out area of the eWLB using TEVs.

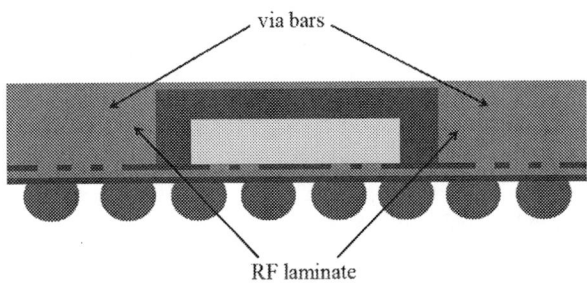

Figure 3b. Via bars realized in the fan-out area of the eWLB using pre-fabricated via bars.

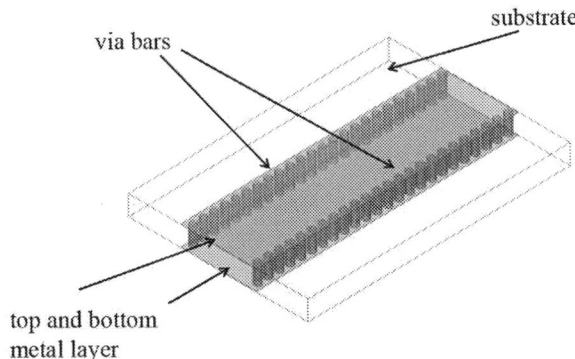

Figure 1. General structure of a SIW.

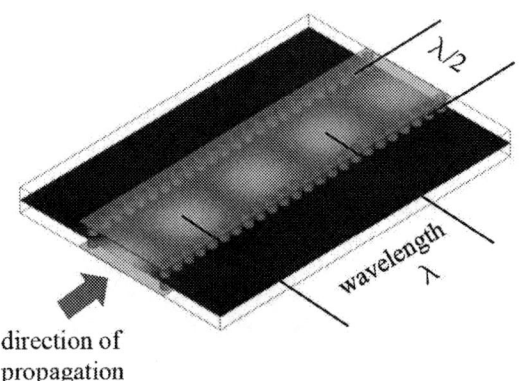

Figure 2. The fundamental SIW mode (amplitude of the electric field).

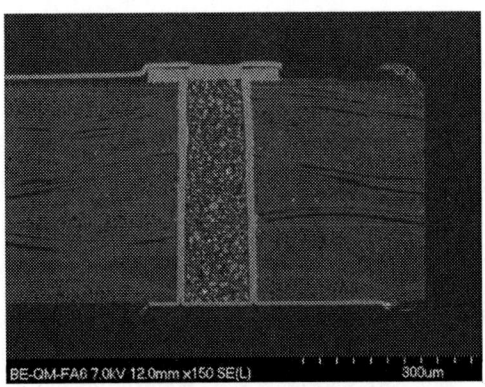

Figure 6. Process flow with pre-fabricated PCB.

Figure 4. Cross section of a via realized in the fan-out area of an eWLB package using pre-fabricated via bars.

1) pick and place

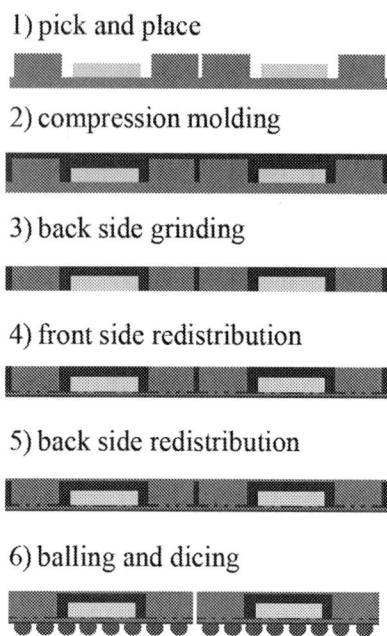

2) compression molding

3) back side grinding

4) front side redistribution

5) back side redistribution

6) balling and dicing

Figure 5. Process flow with pre-fabricated via bars.

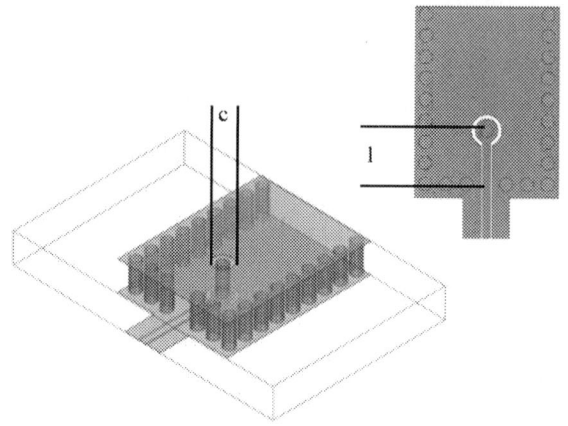

Figure 6a. Electromagnetic concept for mode transformation based on via.

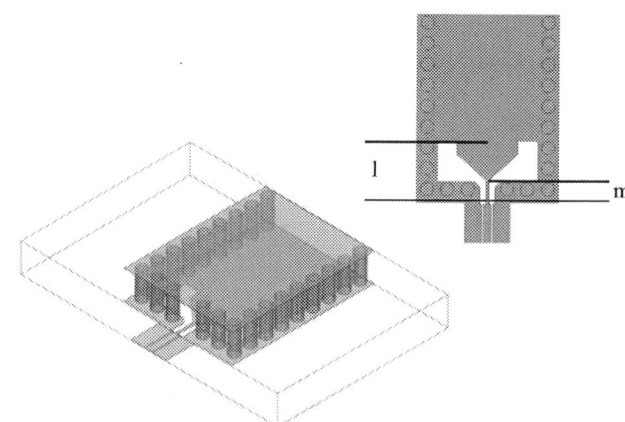

Figure 6b. Electromagnetic concept for mode transformation based on a tapered line.

redistribution layer. The further steps are like the normal process flow of an eWLB package, but is executed on both surfaces. Many process steps can be applied to front and back side simultaneously [5]. We chose the application of pre-fabricated PCB structures for the realization of the rectangular waveguide presented in this paper.

Electromagnetic Concepts for Transition

The transitions between planar transmission lines and SIW structures represent an important element related to SIW components. In [7] several broadband transitions between microstrip line (MSL) and CPW to SIW structures are presented. Usually a CPW line is used in eWLB packages for single-ended signals. Therefore a CPW to rectangular waveguide transition is required. In this paper we investigate two different concepts for the realization of the CPW to rectangular waveguide transition. The first concept uses a via in the rectangular waveguide to excite the travelling wave (Fig. 6a). The distance between this via and the back wall of the rectangular waveguide is a quarter wavelength. Currents in this via excite the magnetic field of the TE_{10} mode. The second concept is based on a tapered line to match the transition (Fig. 6b). In this concept, the layout of this transition section can be modified in a wide range to obtain required characteristics like narrow-band or broad-band behavior.

Realization of Transitions

The analytical description and optimization of the two presented electromagnetic concepts in Fig. 6a and Fig. 6b would be very difficult and time consuming. Therefore commercial software using a finite element method (FEM) is used to simulate and optimize the structures. The CPW used

for both concepts is designed for a 50 Ω characteristic impedance. Both transitions have been optimized for a frequency of 77 GHz. The distance l is in both concepts about a quarter wavelength. The distance between the two via bars is a half wavelength, respectively. In Fig. 6a the distance c is used for the optimization. The dimension of this opening has a strong influence on the capacitive characteristics in this area. Therefore it is a good instrument to match the transition.

For the matching of the transition in Fig. 6b there exist more degrees of freedom. First, there is a short line section with a higher inductance inserted. The length is about m = 250 μm and the width is about 30 μm. After that section a taper with an angle of 45° is located. This taper works as transition from the smaller CPW line dimension to the bigger rectangular waveguide dimensions. The planar layout (Fig. 6b) for the transition offers a more broad band characteristic. By decreasing the length l the transition can be designed more small band. The matching of this transition has been done by shifting the taper. Again there has been used an optimization algorithm of a commercial 3D-FEM software.

Measurement Results

The measured structures are shown in Fig. 7a and Fig. 7b. The transition with the electromagnetic concept using a via is shown in Fig. 7a. For the characterization of the rectangular waveguide three back-to-back connections with a different length of the rectangular waveguide in-between have been realized. Also a short has been realized. In Fig. 7b the same composition for the planar electromagnetic concept using a tapered line is shown. All layouts are designed for a ground-signal-ground (GSG) RF contact tip with a pitch of 100 μm.

Insertion and Return Loss

The measurement results of the shortest through connection of the electromagnetic concept with the via are shown in Fig. 8a. All presented measurement results are performed on a back-to-back configuration. The length of the back-to-back configuration between contact tip to contact tip is 3.4 mm. The insertion loss at 77 GHz is 1.9 dB. The insertion loss decreases slightly to a value of 2.3 dB at 100 GHz. The insertion loss decreases strongly above 100 GHz. The cut-off frequency of the rectangular waveguide is about 65 GHz. The measured return loss is about 15 dB at 77 GHz (Fig. 8b). The return losses from both sides are identical ($S_{11}=S_{22}$). This underlines the excellent symmetry of the realized structures. The observable ripples result from the system drift. The measurement results of the planar electromagnetic concept are shown in Fig. 9a. Again, the measured length is about 3.4 mm from contact tip to contact tip. The insertion loss is slightly better with 1.6 dB for the

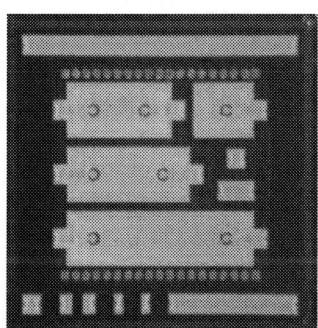

Figure 7a. Photograph of the electromagnetic concept based on a via.

Figure 7b. Photograph of the electromagnetic concept based on a taper.

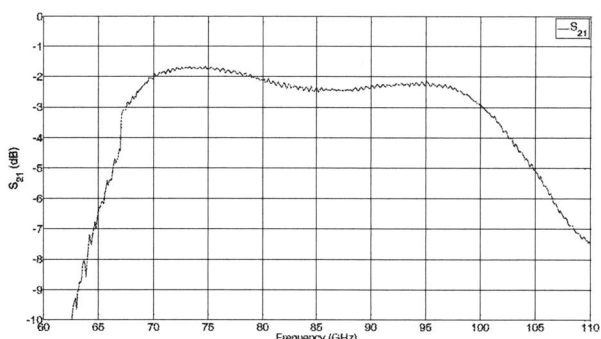

Figure 8a. Measured transmission coefficient of the electromagnetic concept using a via (back-to-back).

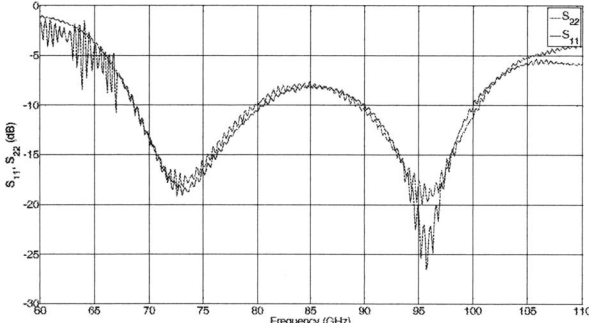

Figure 8b. Measured reflection coefficient of the electromagnetic concept using a via (back-to-back).

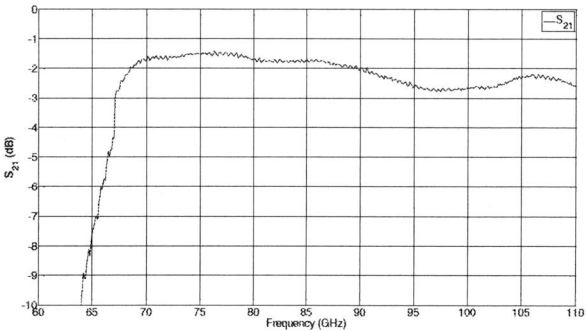

Figure 9a. Measured transmission coefficient of the electromagnetic concept based on a taper (back-to-back).

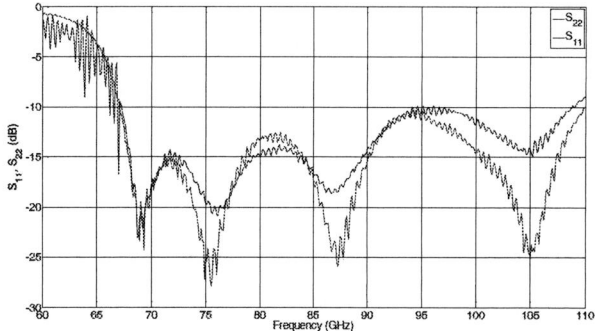

Figure 9b. Measured reflection coefficient of the electromagnetic concept based on a taper (back-to-back).

back-to-back transition at 77 GHz. For this concept, there is only a slight decrease of the insertion loss up to 110 GHz recognizable. The transition has been optimized for 77 GHz. Considering the measurement results, the optimization of the transition for the complete E band (60 GHz to 90 GHz) or W band (75 GHz to 110 GHz) should be possible.

Characterization of the Rectangular Waveguide

The measurement results of the back-to-back transition of the planar electromagnetic concept with different length of the rectangular waveguide are shown in Fig. 10. The length of the short line is again 3.4 mm. The length of the long line is 5.7 mm. The difference of the insertion loss between the measurements with different length is about 1.0 dB. Therefore

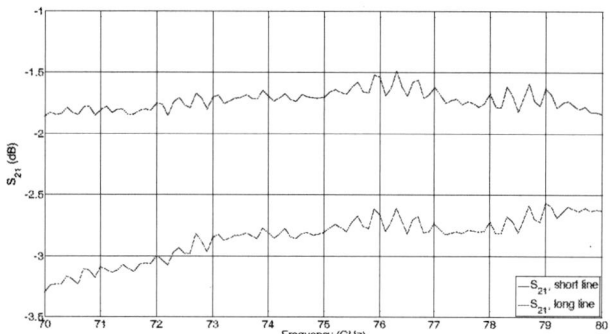

Figure 10. Transmission coefficient of a back-to-back transition with 3.4 mm length and 5.7 mm length.

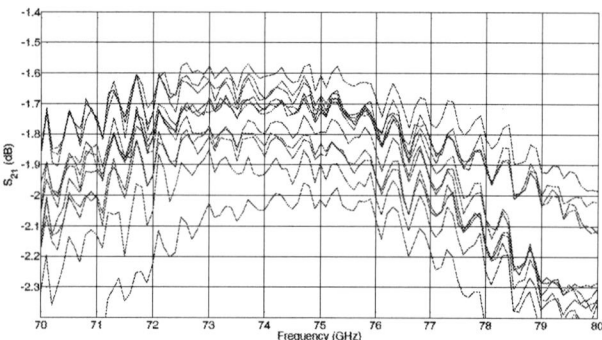

Figure 11a. Reproducibility of the CPW to rectangular waveguide transition based on a via.

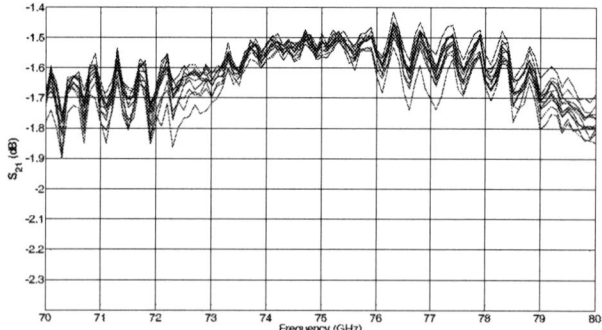

Figure 11b. Reproducibility of the CPW to rectangular waveguide transition based on a taper.

the rectangular waveguide shows attenuation of only about 0.4 dB/mm at 77 GHz. The attenuation of a CPW line in eWLB is almost the same with about 0.3 dB/mm at 77 GHz [3]. A rectangular waveguide line with a length of 3.4 mm has an calculated insertion loss of about 1.4 dB. The measured back-to-back transition with the same length has an insertion loss of 1.6 dB. This approximate calculation shows that the mode transformation is done with almost no additional losses to those related to physical length of the transmission line.

Reproducibility

In Fig. 11a several measurements of the electromagnetic concept based on a via are shown. Each on-wafer measurement took place on a different location of the eWLB wafer. All measured values at 77 GHz have a deviation of 0.2 dB. The mean value of the performed measurements of the insertion loss is 1.9 dB. Simulations have shown that the location of the via, placed in the rectangular waveguide in the fan-out area of the package is very sensitive with respect to production tolerances. This simulated behavior can be seen very clear in the measurement results. In Fig. 11b several measurements of the electromagnetic concept based on the tapered line are shown. Again, each on-wafer measurement took place on a different location of the eWLB wafer. The deviation of all measurements at 77 GHz is only ± 0.05 dB. The mean value of the performed measurements of the insertion loss is 1.6 dB. The observed differences are in the range of the measurement reproducibility. The placement

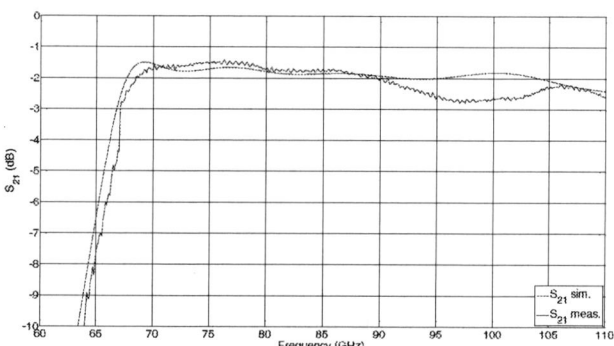

Figure 12a. Comparison of the simulated and measured transmission coefficient of the planar concept.

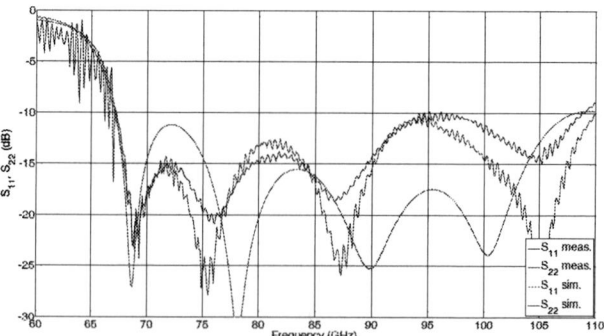

Figure 12b. Comparison of the simulated and measured reflection coefficient of the planar concept.

accuracy of the RF laminate insert with the micro-vias is not critical for the electromagnetic concept with the planar tapered line. In this concept, the electromagnetic fields are mostly guided by the RDL of the standard eWLB package and the thin-film RDL has a high metal pattern resolution. This confirms the very good suitability of the used technologies for millimeter-wave applications.

Comparison to Simulation

Fig. 12a and Fig. 12b show the comparison of simulation results with measurement results. Especially for 77 GHz the insertion loss is almost the same. The larger discrepancies observed for return loss can be explained by simplifications of the simulation model and limited measurement accuracy for low reflected signal levels. In particular, the shift of the resonance positions is due to uncertainty of electrical permittivity of the insert material used in the simulation.

Single Transition

The manufactured back-to-back structures allow one to determine the S-parameters of the single CPW to rectangular waveguide using the thru-reflect-line (TRL) method [12]. Fig. 13a and Fig. 13b show the measurement and simulation results of a single transition using the planar concept. The insertion loss of a single transition is 0.8 dB which agrees perfect with the insertion loss of 1.6 dB measured for a back-to-back version. A good agreement between measurement and simulation is observed.

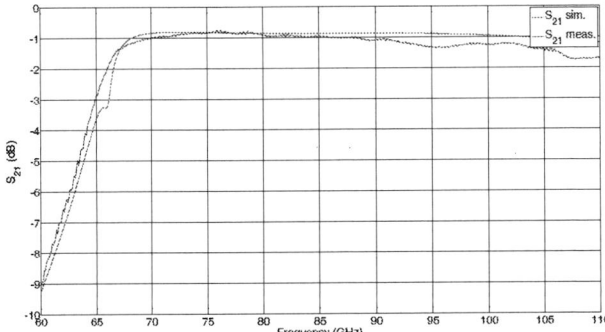

Figure 13a. Simulated and measured transmission coefficient of a single transition based on the planar concept.

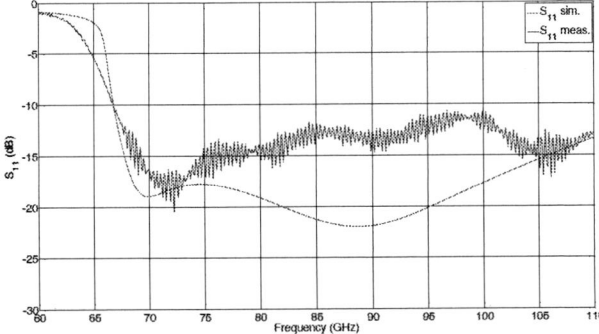

Figure 13b. Simulated and measured reflection coefficient of a single transition based on the planar concept.

Conclusions

We presented for the first time the realization of a 3D rectangular waveguide in the fan-out area of an eWLB package using pre-fabricated RF laminate. Micro-vias realized in the RF laminate are used to form the sidewalls of a rectangular waveguide structure. RDL structures on the top and bottom surface of the package are used to realize top and bottom walls of the waveguide. We presented two electromagnetic concepts to transform the TEM mode of a CPW line to the TE mode of the rectangular waveguide. The first concept uses a via in the waveguide structure for mode transformation. Measurements of a 3.4 mm back-to-back transition show an insertion loss of 1.9 dB at 77 GHz. The measured return loss is better than 10 dB. The second concept uses a tapered line for mode transformation. The measurements of a 3.4 mm back-to-back transition show an insertion loss of 1.6 dB at 77 GHz. The measured return loss is again better than 10 dB. The second concept is clearly less sensitive with respect to placement accuracy of the pre-fabricated RF laminate. This is verified by several measurements of different packages using the same concept. The rectangular waveguide shows attenuation of about 0.4 dB/mm at 77 GHz. The agreement between measurement results and simulation results is very good. This 3D waveguide integration in eWLB will enable the realization of many new RF features in chip embedding technologies in the future.

Acknowledgments

This work has been performed in the project EFA2014 II, hich is funded by the German Bundesministerium für ildung und Forschung (BMBF) under contract 16N11947. he authors also want to acknowledge the very good ooperation with the Robert Bosch GmbH within the project FA2014.

eferences

"The International Technology Roadmap for Semiconductors (ITRS), 2012 Update Overview," Available on-line: https://www.itrs.net.

M. Brunnbauer, T. Meyer, G. Ofner, K. Müller, R. Hagen, "Empedded Wafer Level Ball Grid Array (eWLB)," *in Proc. 33rd Eloctronic Manufacturing Technology Symposium (IEMT 2008),* Penang, Nov. 2008

3. M. Wojnowski, R. Lachner, J. Böck, C. Wagner, F. Starzer, G. Sommer, K. Pressel, R. Weigel, "Embedded Wafer Level Ball Grid Array (eWLB) Technology for Millimeter-Wave Applications," *in Proc. 13th Electronic Packaging Technology Conference (EPTC 2011),* Singapore, Dec. 2011

K. Pressel, G. Beer, T. Meyer, M. Wojnowski, M. Fink, G. Ofner, B. Römer, "Embedded Wafer Level Ball Grid Array (eWLB) Technology for System Integration," *in Proc. International Symposium on Components, Packaging, and Manufacturing Technology (ICCSJ 2010),* Tokyo, Aug. 2010

T. Meyer, K. Pressel, G. Ofner, B. Römer, "System integration with eWLB," *in Proc. IEEE 3rd Electronic System-Integration Technology Conference (ESTC 2010),* Berlin, Sept. 2010

6. U. Hiroshi, T. Takeshi, M. Fujii, "Development of a laminated waveguide," *IEEE Trans. on Microwave Theory and Techniques*. 1998, vol 46, no. 12, pp. 2438-2443

7. M. Bozzi, A. Georgiadis, K. Wu, "Review of substrate-integrated waveguide circuits and antennas," *in IET Microwaves, Antennas & Propagation,* 2010, Vol. 5, Iss. 8, pp 909-920

8. M. Wojnowski, V. Issakov, G. Knoblinger, K. Pressel, G. Sommer, R. Weigel, "High-Q Inductors Embedded in the Fan-Out Area of an eWLB," *IEEE Trans. on Components, Packaging and Manufacturing Technology,* 2012, vol 2, Iss. 8, pp. 1280-1292

9. F. Xu, K.Wu, "Guided-wave and leakage characteristics of substrate integrated waveguide," *IEEE Trans. Microwave Theory and Techniques,* 2005, vol 53 no. 1, pp. 66-73

10. Cassivi, Y., Perregrini, L., Arcioni, P., Bressan, M., Wu, K., Conciauro, G.: 'Dispersion characteristics of substrate integrated rectangular waveguide', *IEEE Microw. Wirel. Compon. Lett.*, 2002, 12, (9), pp. 333–335

11. Che, W., Deng, K., Wang, D., Chow, Y.L.: 'Analytical equivalence between substrate-integrated waveguide and rectangular waveguide', *IET Microw. Antennas Propag.,* 2008, vol 2, no. 1, pp. 35–41

12. M.Wojnowski, G.Sommer, A.Klumpp, and W.Weber, "Electrical Characterization of 3D Interconnection Structures up to Millimeter Wave Frequencies," *in Proc. 10th Electronic Packaging Technology Conference (EPTC 2008)*, Singapore, Dec. 2008

Interplay and Influence of Thermomechanical Stress in Copper-Filled TSV Interposers

Sheng-Tsai Wu,[1] Cheng-fu Chen,[2*] and Heng-Chieh Chien[1]

1. Electronic and Optoelectronic Research Lab, Industrial Technology Research Institute
No.195, Sec. 4, Chung Hsing Rd., Chutung, Hsinchu, Taiwan 31040, R.O.C.
2. Department of Mechanical Engineering, University of Alaska Fairbanks, Fairbanks
PO Box 755905, Fairbanks, AK 99775-5905, US
*Correspondent: cf.chen@alaska.edu

Abstract

In this paper we use finite element simulations to determine the influence of two key design parameters on the thermomechanical stress and the stress interplay in copper-filled TSV arrays. The two parameters are the via's pitch-to-diameter ratio and thickness-to-radius (aspect) ratio. Our analytical results has suggested that the in-plane stress interplay becomes insignificant when the TSV pitch is at least five times of the via's radius. This criterion will be numerically verified in this paper. This work also suggests that it would be inadequate to approximate the thermomechanical stress into a simplified 2D models even for a small H/D ratio (which resembles a think, 2D-like structure).

Introduction

Through-silicon via (TSV) plays a key role in accomplishing 3D IC integration. Copper-filled TSV interposers enable vertical stacking of heterogeneous components, in which device dies stack on one another for a small form factor in packaging. Multifunctionality and better performance can thus be achieved, yet at reduced power consumption because vias channel electrons at the shortest distance for less joule heating. TSV interposers also allowdie-to-die stacking, which facilitates the IC industry for fabricating Moore's law chips at reasonable costs [1-3].

Stress interplay is a phenomenon that describes the overlap of thermomechanical stress induced from one TSV over the stress from another. The thermomechanical stress is caused by the great thermal mismatch among Si (2.8 ppm/°C), SiO$_2$ (0.5 ppm/°C), and Cu (18 ppm/°C) in Cu-filled TSV silicon interposers. The stress interplay can be apparent in the in-plane direction, out-of-plane direction, or both. Locally, thermomechanical stresses around each TSV yields copper protrusion (a.k.a. copper pumping [4]) and causes stress concentration along the Si/SiO2/Cu interface. Globally, the stress interplay causes additional stress concentration that will affect electron mobility. The high stress among the vias, silicon wafer, and the re-distribution layers causes warpage in the interposer, which makes bonding of device dies difficult.

The industry has been looking into the thermomechanical issue and searches for an optimal design of balancing the needs for required pin counts and thermomechanical reliability. In general, adjusting the design parameters such as the via size, pitch, and the TSV array configuration can mitigate the thermomechanical stresses in TSV interposers. In particular, the interplay of thermomechanical stresses among TSVs in the silicon interposer should be minimized for reducing the "kept-away zone". However, due to the great number of design parameters and the order of magnitude difference in the dimensions of components in 3D IC, the tasks of analyses and design optimization are curbed. These factors make experimental and modeling work challenging; sometimes, prohibitive.

In this investigation, we use finite element modeling to study the influence of a few key design parameters of TSV on the stress interplay phenomenon. The analytical calculation results suggest that the in-plane stress interplay becomes insignificant if the TSV pitch is at least five times of the via's radius [5-6]. This criterion will be verified by numerical analysis in this paper. Besides, the range of the via's aspect ratio (TSV thickness over via's diameter) will also be determined for understanding when a 2D TSV model will suffice to capture the features of thermomechanical stress in TSV interposers. These results will facilitate global modeling of 3D IC integration by replacing the TSV interposer with a equivalent, homogenized model by the analytical calculations.

Finite Element Modeling and Analysis

Finite element (FE) software ANASYS Mechanical (version 14.5) was used for simulating the thermomechanical stress in a single unit TSV shown in Figure 1. This unit TSV model was characterized by its thickness (H), via's radius (r), and the TSV pitch (2R). The SiO$_2$ layer is 0.5 μm thick, which is a fixed value in all our FE models.

Figure 2 shows an octahedral element- finite element model with quarter symmetry. The boundary conditions include the plane-symmetry set up at the yz and xz planes, and fix the origin to avoid any rigid-body motion in simulations. Here we only considered elastic deformation. The material properties are listed in the Table 1. The element type SOLID 186 was used for meshing.

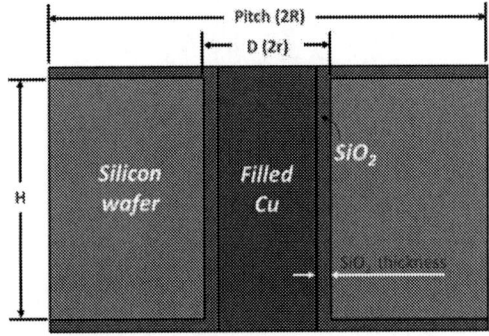

Figure. 1 Schematic of a unit TSV structure.

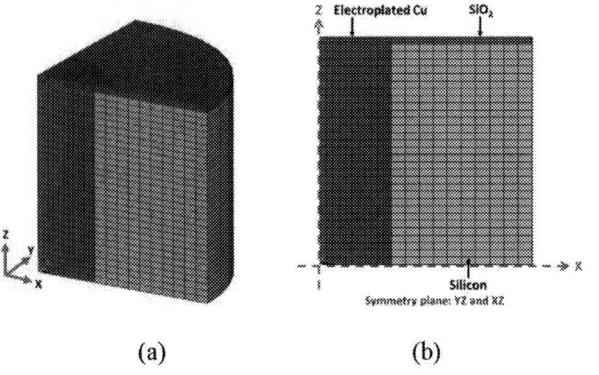

(a)   (b)

Figure. 2 (a) a quarter-symmetric finite element model. (b) Boundary conditions.

Table 1. Material properties of TSV model

Material	E (GPa)	Poisson's ratio	CTE (ppm/°C)
Silicon	129	0.28	2.8
Electroplated Cu [5]	70	0.34	17.0
SiO$_2$ [4]	70	0.16	0.6

Figure 3 shows a contour plot of the von Mises strain for a unit TSV model with r = 5um, R = 25um, and H = 50um; and subject to a field temperature change of 100 °C. Figure 3 shows that the stress (or strain) concentrates around the interfaces of the SiO$_2$ passivation to Cu via. Figure 4 shows the contours of deformation in the radial and axial directions. Figure 4(a) shows that the in-plane radial deformation is uniform around the outside area. Figure 4(b) shows that the axial deformation is uniform in the region away from the boundary, thank to the Saint-Venant's principle.

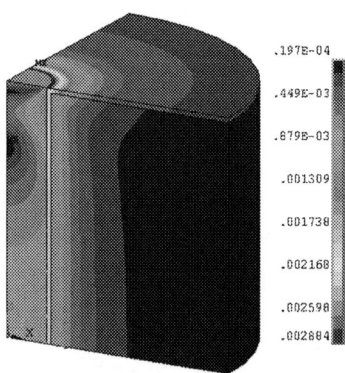

Figure. 3 Contour of von Mises strain.

The Key Parameters Effect of TSV structure

In order to determine the stress interplay phenomenon of TSV structure, the in-plane equivalent strain and axial strain are respectively used to indicate the in-plane and out-of-plane

thermomechanical features. The in-plane equivalent strain is calculated by taking the square root of squared sum of the following strain components: normal radial and tangential strains and the in-plane shearing strain.

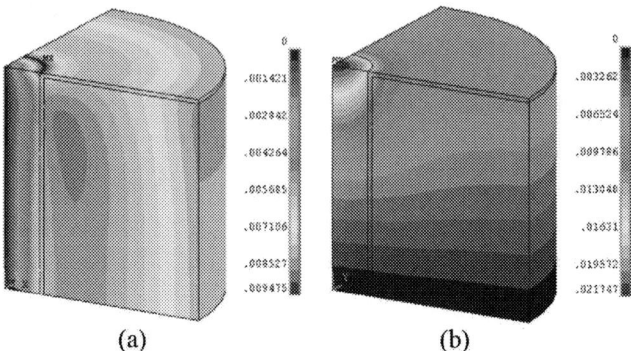

(a)   (b)

Figure. 4 Contour of deformation field in two directions: (a) radial, and (b) axial.

In the following, we conduct a parametric study on the influence of two key design parameters on the thermomechanical behavior of the TSV: the pitch of via and the aspect thickness-to-radius ratio of the via. In the radial direction, we take the in-plane equivalent strain at the radial position *r=R*. In the axial direction, we choose an averaged axial strain at a section that locates at 10% of the TSV thickness from the bottom of the TSV quarter-symmetry model.

In Figure 5 and Figure 6, the H/D ratio is 10. Figure 5 shows the change in the in-plane equivalent strain by varying the ratio of the via's pitch and size (R/r), which is an indicator to the volume ratio of vias in an interposer. It shows that the in-plane strain converges to its limit value at about 400 µ. It also apparent that when the pitch is greater than 5 times of the via diameter, the strain is barely influenced by the parameter R/r. It implies an insignificant stress interplay. In the low R/r regime, which indicate a large volume fraction of vias, it should expects a large in-plane strain.

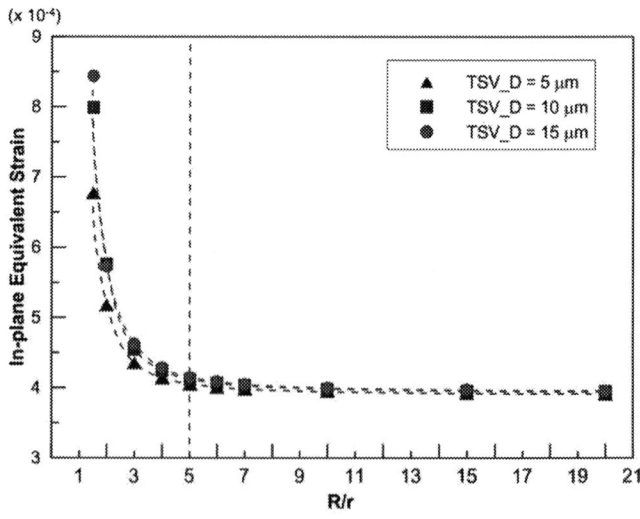

Figure. 5 The in-plane strain versus R/r.

Figure 6 shows the change in the axial as a function of R/r. The trend on the change in the axial strain is similar to that in the in-plane equivalent strain. Again, from the curves in Figure 6 we choose 3R/r as a threshold to small stress interplay. In combination of the results in Figure 5, we can see that the thermomechanical stress (or strain) becomes barely changed in a TSV interposer as long as the R/r ratio is greater than 5. Therefore, it is evident that an TSV array at a large pitch (i.e., five times of the via's diameter) or larger, it will mitigate the thermomechanical stress.

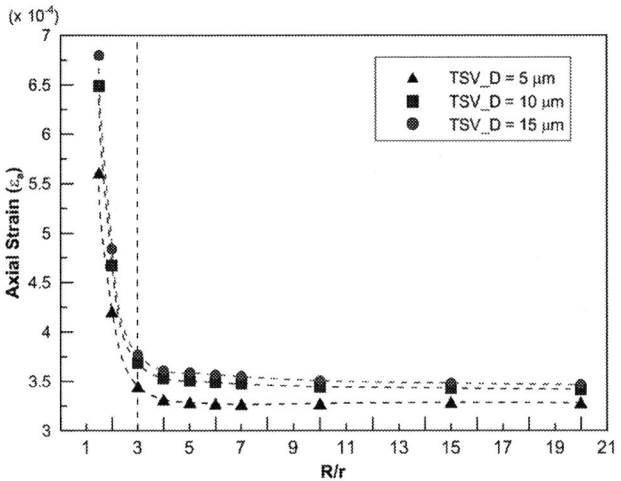

Figure. 6 The axial strain versus R/r.

Figure 7 and Figure 8 show the change in the in-plane strain and axial strain by varying the aspect ratio H/D of the TSV array. Here the R/r ratio is 10 for the results presented in both of the figures. The in-plane equivalent strain converges to 395 µ as H/D becomes larger. At a low H/D ratio, the in-plane strain is little smaller, yet still remains in the range of 360~400 µ. It implies that the aspect ratio H/D has little influence on the build-up of the in-plain strain.

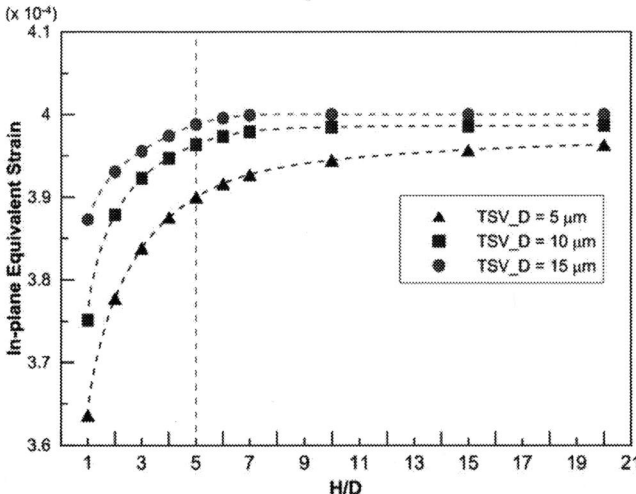

Figure. 7 The in-plane strain versus aspect ratio H/D.

Figure 8 shows the influence of the aspect ratio H/D on the axial strain. In the smaller H/D regime there shows a larger axial strain. It speculates that the high strain value at smaller H/D ratios because of the stress concentration due to cupper pumping, as shown in Figure 9.

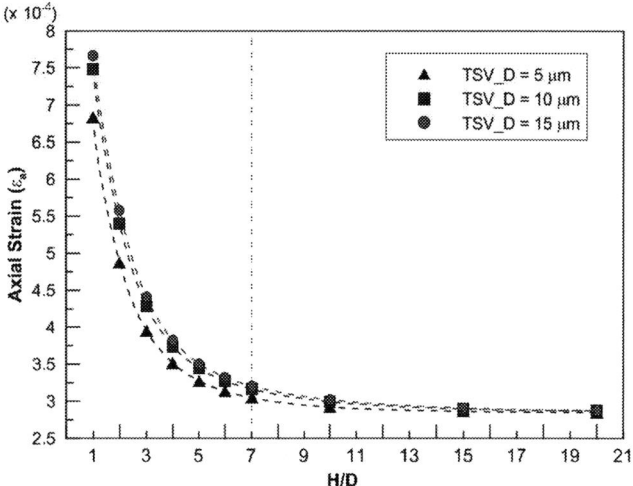

Figure. 8 The axial strain versus aspect ratio H/D.

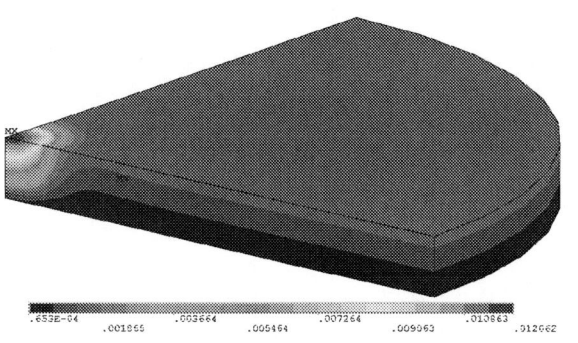

Figure. 9 The axial strain for H/D = 1, r = 5 µm.

Discussion

In this work, we used numerical means to discuss the conditions under which the stress interplay is minimal, such that the analysis of a single unit TSV model can be representative of the entire TSV array in interposers. The first finding is that, when the via's pitch-to-diameter ratio R/r is greater than five, analysis on a single unit TSV model suffices to capture the majority of the thermomechanical features in a TSV interposer. This verifies the observation from an analytical model published [5-6]. We also found that the in-plane strain (or stress) is little influenced by the via's thickness-to-diameter (aspect) ratio H/D.

In Figures 5-8, it becomes clear that the in-plane strain and axial strain are compatible under any combination of the design parameters. For a small H/D ratio (which resembles a think, 2D-like structure), our numerical results suggest that it would be insufficient to approximate the thermomechanical stress into a simplified 2D models.

978-1-4799-2408-0/14 $31.00 © 2014 IEEE

Conclusion

In this paper, the finite element models were used to study the influence of two design parameters of TSV interposers on the thermomechanical stress: the via's pitch-to-diameter ratio R/r and the thickness-to-diameter (aspect) ratio H/D. The numerical results suggest that the in-plane strain (or stress) can be minimal and unchanged when R/r is greater than five. The in-plane strain is little influenced by the aspect ratio H/D. With a numerical example we also speculate that the stress concentration (e.g., cupper pumping) has an influence on the analysis of the axial strain.

Acknowledgements

The authors would like to thank the financial support of Ministry of Economic Affairs (MOEA), Taiwan. The strong support of the 3D IC Integration program by Dr. W. C. Lo, director of advanced package technology division is greatly appreciated.

References

[1] Lau, J. H., *Through-Silicon Via for 3D Integration*, McGraw-Hill Book Company, New York, NY, 2013.

[2] Lau, J. H., *Reliability of ROHS-Compliant 2D and 3D IC Integration*, McGraw-Hill Book Company, New York, NY, 2011.

[3] Lau, J. H., C. K. Lee, C. S. Premachandran, and A. Yu, *Advanced MEMS Packaging*, McGraw-Hill Book Company, New York, NY, 2010.

[4] Cheryl S. Selvanayagam, John H. Lau, Xiaowu Zhang S. K.W. Seah, Kripesh Vaidyanathan and T. C. Chai, "*Nonlinear Thermal Stress/Strain Analyses of Copper Filled TSV (Through Silicon Via) and their Flip-Chip Microbumps*", Proceedings of IEEE/ECTC, Lake Buena Vista, FL, May 2008, pp. 1073-1081.

[5] Cheng-fu Chen, "*Homogenization of TSV interposer and quick assessment of its thermomechanical influence on 3D packages*", Proceedings of IEEE/ECTC, Las Vegas, NV, May 2013, pp. 2249-2254.

[6] Cheng-fu Chen, "*Characterization of in-plane stress in TSV array - a unit model approach*", Proceedings of IEEE/ECTC, Lake Buena Vista, FL, May 27 - 30, 2014, pp. TBD.

978-1-4799-2408-0/14 $31.00 © 2014 IEEE

Does Current Crowding Induce Vacancy Concentration Singularity in Electromigration?

Ozgur Taner, Kasemsak Kijkanjanapaiboon, and Xuejun Fan
Department of Mechanical Engineering
Lamar University
PO Box 10028, Beaumont, TX 77710, USA
Tel: 409-880-7792; xuejun.fan@lamar.edu

Abstract

Mathematical model of electromigration in terms of vacancy concentration is studied analytically and numerically in this paper with the combined effect of vacancy gradient (Fickian term) and electric flow. A 2-D, L-shaped, homogeneous material model with perfect blocking boundary condition ($J = 0$) is chosen as the problem of interest. Dandu and Fan have shown that current density singularity exists at the tip of the wedges when the angle $\theta_0 < 90°$ [13]. This study investigates the effect of current density singularity at the tip of the wedges towards the vacancy concentration at the same location. The results of the study, both analytically and numerically, show that the location of maximum vacancy concentration occurs at cathode side, but not at the location of current density singularity.

Introduction

Electromigration is a phenomenon of mass transport in electrical conductor under the driving force of electrical current. Open and/or short circuit in electronic devices are typical failures caused by electromigration due to voids nucleation near cathode side and hillock development near anode side. Several mathematical models have been proposed in order to understand and/or predict the electromigration [1-10]. They can also be utilized to improve the design of electronic devices in order to increase their reliability against electromigration failure [11]. Many studies employ the divergence as a metric for the failure, with four driving forces considered as the sources of electromigration failure. These four driving forces are vacancy concentration gradient (Fickian term), electric field or current, mechanical stress, and temperature gradient. In this paper, we consider electric field as the main driving force of electromigration failure, and the mathematical model of vacancy concentration coupling with electric potential is studied, analytically and numerically.

Dandu and Fan [12, 13] have shown that current density singularity, i.e. current crowding, exists at the tip of the wedges when the angles $\theta_0 < 90°$. Black has shown that the median time to failure due to electromigration is inversely proportional to the square of current density through experiments [1]. Based on this information, one might conclude that the location of maximum vacancy concentration would be at the location of current density singularity. In order to verify this statement, the vacancy concentration coupling with electric potential electromigration model is studied analytically and numerically. A 2-D, L-shaped, homogeneous material model with perfect blocking boundary condition ($J = 0$) is chosen to be the problem of interest because several

electromigration experiments and/or observations are conducted under this configuration.

Mathematical Model

Electromigration is a phenomenon of mass transport, which can be described by Fick's diffusion equation as follows.

$$\frac{\partial C_v}{\partial t} = -\nabla \cdot J + G \tag{1}$$

where C_v is the vacancy concentration, J is the total vacancy flux, G is a generation or annihilation term. The total vacancy flux, J, is the sum of electromigration driving forces. The two driving forces, vacancy concentration gradient (J_1) and electric field (J_2), considered in this study can be expressed as follows.

$$J_1 = -D_v \nabla C_v \tag{2}$$

$$J_2 = -\frac{D_v C_v}{kT} Z^* e \nabla V \tag{3}$$

where D_v is the vacancy diffusivity, Z^* is the effective charge number, e is the elementary charge, V is the electric potential, k is the Boltzmann constant, T is the absolute temperature. Considering that the electrostatic potential must be equal to zero ($\nabla^2 V = 0$), then the governing equation for the two dimensional (2-D) vacancy concentration coupling with electric potential electromigration model in Cartesian coordinate system can be written as

$$\frac{\partial C_v}{\partial t} = D_v \left[\frac{\partial^2 C_v}{\partial x^2} + \frac{\partial^2 C_v}{\partial y^2} + \frac{Z^* e}{kT} \left(\frac{\partial C_v}{\partial x} \frac{\partial V}{\partial x} + \frac{\partial C_v}{\partial y} \frac{\partial V}{\partial y} \right) \right] \tag{4}$$

where the sink/source term $G = 0$ is used, which means grain boundary is not considered in this study, i.e. homogeneous material. In polar coordinate system, this governing equation can be written as

$$\frac{\partial C_v}{\partial t} = D_v \left[\frac{\partial^2 C_v}{\partial r^2} + \frac{1}{r} \frac{\partial C_v}{\partial r} + \frac{1}{r^2} \frac{\partial^2 C_v}{\partial \theta^2} \right.$$
$$\left. + \frac{Z^* e}{kT} \left(\frac{\partial C_v}{\partial r} \frac{\partial V}{\partial r} + \frac{1}{r^2} \frac{\partial C_v}{\partial \theta} \frac{\partial V}{\partial \theta} \right) \right] \tag{5}$$

The boundary condition studied in this paper is perfect blocking boundary condition, where total flux equals to zero ($J = 0$). Clement and Lloyd [3] call this boundary condition constant-volume boundary condition; it "corresponds to a situation where vacancies are conserved which could be maintained in a system where a thick strong passivation layer would preclude changes in the volume of the conductor."

$$J = J_1 + J_2 = 0 \tag{6}$$

The initial condition studied in this paper is

$$C_v(x, y, 0) = C_{v0} \tag{7}$$

where C_{v0} is the initial vacancy concentration.

Analytical Solutions

In this section, the analytical solution to vacancy concentration (C_v) in polar coordinate system at the location of current density singularity is derived. Since our interest is to study vacancy concentration at the location of current density singularity, steady state solution $(\partial C_v / \partial t = 0)$ is considered. In polar coordinate system, after applying the steady state condition, the governing equation (5) becomes

$$\frac{\partial^2 C_v}{\partial r^2} + \frac{1}{r}\frac{\partial C_v}{\partial r} + \frac{1}{r^2}\frac{\partial^2 C_v}{\partial \theta^2} + \frac{Z^* e}{kT}\left(\frac{\partial C_v}{\partial r}\frac{\partial V}{\partial r} + \frac{1}{r^2}\frac{\partial C_v}{\partial \theta}\frac{\partial V}{\partial \theta}\right) = 0 \tag{8}$$

For asymptotic solution, let electric potential (V) and vacancy concentration (C_v) be functions of r and θ as follows.

$$V(r,\theta) = r^{\lambda_1} \cdot f(\theta) \tag{9}$$

$$C_v(r,\theta) = r^{\lambda_2} \cdot g(\theta) \tag{10}$$

The schematic diagram of the analytical problem is illustrated in (Fig. 1).

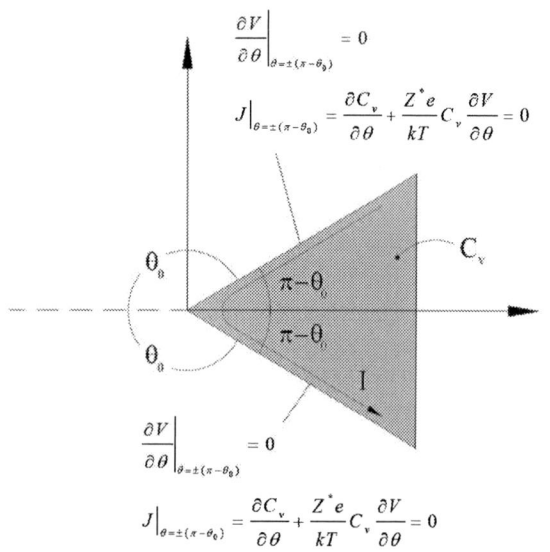

Figure 1. Schematic diagram of the analytical problem

According to Dandu and Fan [13], current density singularity occurs at the locations where $r \to 0$ and $\theta_0 < 90°$. Their analytical solution to voltage function is

$$V(r,\theta) = f_0(\theta) + r^{\left[\frac{\pi}{2(\pi-\theta_0)}\right]} \cdot f_1(\theta) + r^{\left[\frac{\pi}{(\pi-\theta_0)}\right]} \cdot f_2(\theta)$$
$$+ r^{\left[\frac{3\pi}{2(\pi-\theta_0)}\right]} \cdot f_3(\theta) + \ldots \tag{11}$$

$$f(\theta) = A\cos\lambda(\theta) + B\sin\lambda(\theta) \tag{12}$$

Note that voltage function is expressed in series form, however, only the second term, which is $r^{\left[\frac{\pi}{2(\pi-\theta_0)}\right]} \cdot f_1(\theta)$, contributes to current density singularity $(\partial V / \partial r \to \infty)$. So,

in order to study its effect towards vacancy concentration, we can simplify the voltage function to

$$V(r,\theta) \cong r^{\left[\frac{\pi}{2(\pi-\theta_0)}\right]} \cdot f_1(\theta) \tag{12}$$

Apply equations (10) and (12) into the governing equation (8), we have

$$r^{\lambda_2-2}\left[\lambda_2\left\{\lambda_2 + \frac{\pi}{2(\pi-\theta_0)}\frac{Z^* e}{kT}r^{\left[\frac{\pi}{2(\pi-\theta_0)}\right]}f(\theta)\right\}g(\theta)\right.$$
$$\left.+ \frac{Z^* e}{kT}r^{\left[\frac{\pi}{2(\pi-\theta_0)}\right]}f'(\theta)g'(\theta) + g''(\theta)\right] = 0 \tag{13}$$

Since the location of current density singularity is where $r \to 0$ and $\theta_0 < 90°$, we find the term $r^{\left[\frac{\pi}{2(\pi-\theta_0)}\right]} \to 0$. Thus, the governing equation (13) can be reduced to

$$r^{\lambda_2-2}[g''(\theta) + \lambda_2^2 g(\theta)] = 0 \tag{14}$$

Next, we consider the self-diffusion of vacancy concentration under singularity analysis. Its governing equation in polar coordinate can be written as

$$\frac{\partial^2 C_v}{\partial r^2} + \frac{1}{r}\frac{\partial C_v}{\partial r} + \frac{1}{r^2}\frac{\partial^2 C_v}{\partial \theta^2} = 0 \tag{15}$$

By applying equation (10) into equation (15), we have the same result as equation (14), which indicating that current density distribution does not induce vacancy concentration singularity. We find agreement to this finding in the numerical study, as is discussed in the following section.

Numerical Solutions

To further investigate the 2-D electric field coupled vacancy concentration problem, the vacancy concentration equation is numerically solved using finite element method. Simulations are done in COMSOL Multiphysics software, with Free Equation modeling options COMSOL offers such as the "PDE Weak Form Subdomain" module. The problems are implemented by providing COMSOL, the domain and boundary integrals that construct the variational form equations. In this study, two problems are examined. Both of the problems are defined by identical governing equations including the boundary conditions imposed on the systems. However, the geometries of the problem domains are different; the first problem, a rectangular plane, and the second problem, an L-shaped line, respectively. In the first model, owing to the rectangular domain geometry, and the boundary conditions employed, an electric field occurs with the existence of only longitudinal electric potential gradient. Next, the second problem is considered, where current crowding is induced due to the L-shaped geometry.

Both numerical solutions belong to the vacancy transport equation with perfect blocking boundary conditions only with the contributions of two driving forces, vacancy gradient and electromigration force due to electric potential gradients. Diffusion coefficients in both problems are taken as constant.

In this work, normalized equations are taken into consideration. Therefore non-dimensional equations for the

Eq. (4), and electrostatic potential equation are derived. Normalized vacancy transport equation and Laplacian electric potential equations are written respectively as follows.

$$\frac{\partial C^*}{\partial \tau} = \frac{\partial^2 C^*}{\partial x^{*2}} + \frac{\partial^2 C^*}{\partial y^{*2}} + \frac{Z^* e}{kT} V_0 \left(\frac{\partial C^*}{\partial x^*} \frac{\partial V^*}{\partial x^*} + \frac{\partial C^*}{\partial y^*} \frac{\partial V^*}{\partial y^*} \right) \quad (16)$$

With normalized boundary conditions

$$J^* = -\left(\frac{\partial C^*}{\partial x^*} + \frac{\partial C^*}{\partial y^*} \right) - \frac{Z^* e}{kT} V_0 \left(\frac{\partial V^*}{\partial x^*} + \frac{\partial V^*}{\partial y^*} \right) C^* \quad (17)$$

And normalized electrostatic potential equation is written as:

$$\nabla^2 V^* = 0 \quad (18)$$

With Dirichlet boundary conditions, $V^* = 1$ and $V^* = 0$, at anode and cathode ends, respectively, and Neumann boundary conditions on the remaining boundaries. Normalized parameters are given as:

$$\tau = \frac{t}{t_c}, \; t_c = \frac{D_v}{L_0^2}, \; x^* = \frac{x}{L_0}, \; y^* = \frac{y}{L_0}, \; V^* = \frac{V}{V_0},$$

$$V_0 = \rho_0 j L_0, \text{ and } C^* = \frac{C_v}{C_{v0}}$$

Here, electromigration parameters are chosen in a similar manner that is introduced in Lloyd and Clement's paper [3]. In their work, coefficient in front of the electromigration force is suggested to be in a range of 2-8. In our simulation, a similar procedure is followed and the coefficient in front of the coupling terms of normalized vacancy and gradient of normalized electric potential is taken as, $\dfrac{Z^* e}{kT} V_0 = 1$.

Weak form equations of the normalized vacancy concentration and electrostatic potential equations derived to implement our model in COMSOL, which are used for both of the simulations, are written as:

$$\int_\Omega \frac{\partial C^*}{\partial \tau} \delta C^* d\tau = \int_\Omega \frac{\partial C^*}{\partial x^*} \delta \left(\frac{\partial C^*}{\partial x^*} \right) dA + \int_\Omega \frac{\partial C^*}{\partial y^*} \delta \left(\frac{\partial C^*}{\partial y^*} \right) dA$$

$$+ \int_\Omega \left(\frac{\partial C^*}{\partial x^*} \frac{\partial V^*}{\partial x^*} + \frac{\partial C^*}{\partial y^*} \frac{\partial C^*}{\partial y^*} \right) \delta C^* dA - \oint_\Gamma \frac{\partial C^*}{\partial x^*} \vec{n} \delta C^* ds$$

$$- \oint_\Gamma \frac{\partial C^*}{\partial y^*} \vec{n} \delta C^* ds \quad (19)$$

1. Solution of the First Problem

In this problem, vacancy concentration equation is coupled with electric potential equation with perfect blocking boundary conditions for both ends of the line, corresponding to anode and cathode. Other driving forces forged by mechanical stress gradient and temperature gradient are neglected. Simulation of this model basically yields the same results with the solution for the one dimensional vacancy transport equation driven by electromigration and diffusion with perfect blocking boundary conditions. Electromigration force is calculated coupling the electric potential gradients. Considering the Neumann boundary conditions and the shape of the line as illustrated in (Fig. 2), vertically constant distance between two boundary lines, electric potential gradients of

only one direction will be effective on the electromigration force. Problem domain is modeled in COMSOL with a mesh consisting of 122 rectangular elements.

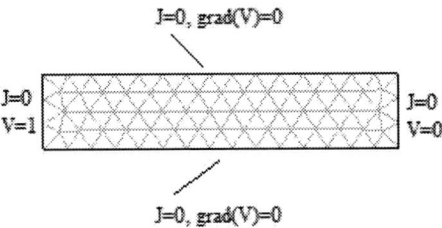

Figure 2. Mesh structure and boundary conditions

Results obtained by solving the Laplacian equation are shown in (Fig. 3). Gradients in vertical direction are not generated, thus reducing the problem to be 1-D in nature.

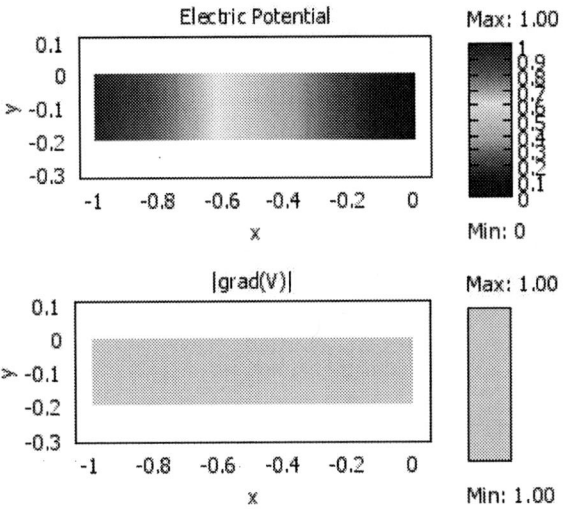

Figure 3. Electric Potential and Electric Field

This property enables the comparison of the simulation results with similar problems from literature. Additionally, an analytical solution is available in several published work for 1-D vacancy transport problems [9, 14], thus numerical results can be verified with exact solutions. In the end, a confirmation of the correct implementation of the equation is made before moving on to the more complex problem.

As expected, vacancy concentration increases at cathode end, referring to our model right end as shown in (Fig. 4). On the other hand, vacancy concentration decreases at left end, i.e. anode, and causes depletion in the vacancy. This is a natural result of the problem, mainly led by the electromigration force. Neumann boundary conditions employed for the solution of the electric potential equation on all the boundaries except at cathode and anode ends represent electric insulation, restriction of electron flow in this direction; thus, electric potential gradients perpendicular to the insulation boundaries will be zero. In the end, one dimensional electromigration force shifts the vacancy concentration balance in anode-cathode end direction, causing a higher vacancy concentration on the cathode end, and the opposite on the other end.

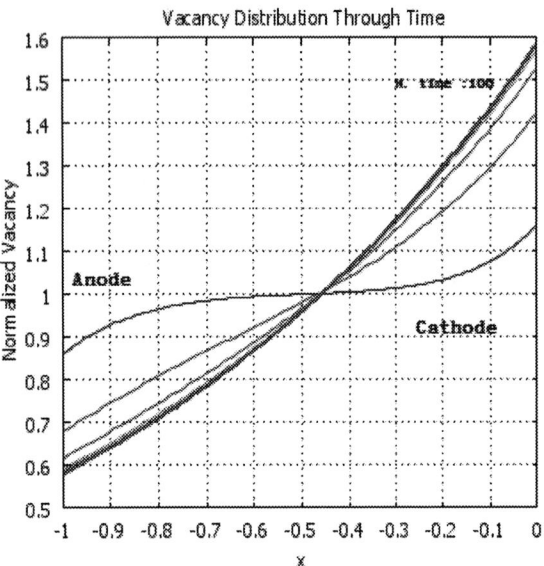

Figure 4. Vacancy concentration of the line at different time

2. Solution of the 2-D L-Shape Problem

In this problem, normalized vacancy concentration equation on an L-shaped geometry is studied. The main goal is to investigate the vacancy concentration, coupling with electric potential, which already shows singularity in its gradients due to current crowding effect, as discussed in the Analytical Solutions section. Here again, we take diffusion coefficient to be constant for entire domain, thus grain boundaries and vacancy sink/source terms are neglected. A schematic presentation of the problem with boundary conditions is given in (Fig. 5).

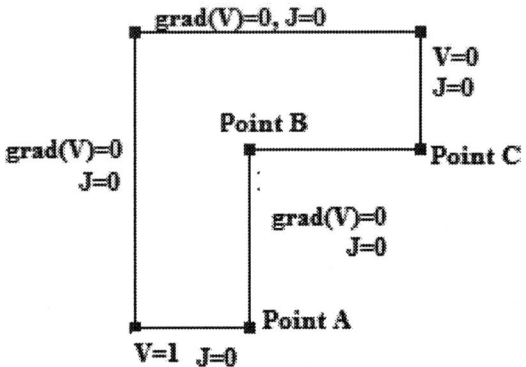

Figure 5. Problem geometry and boundary conditions

As illustrated in (Fig. 5), several locations are denoted as Point A, B, C, in order to be referred to when we investigate the vacancy concentration and electric field values.

Electric potential and electric field on the L-shaped geometry is illustrated in (Fig. 6). Electric field that is needed to compute electromigration force and the divergence of this force is obtained by solving electric potential equation with two Dirichlet boundary conditions at anode and cathode, and

Neumann boundary conditions to maintain electric insulation on remaining boundaries.

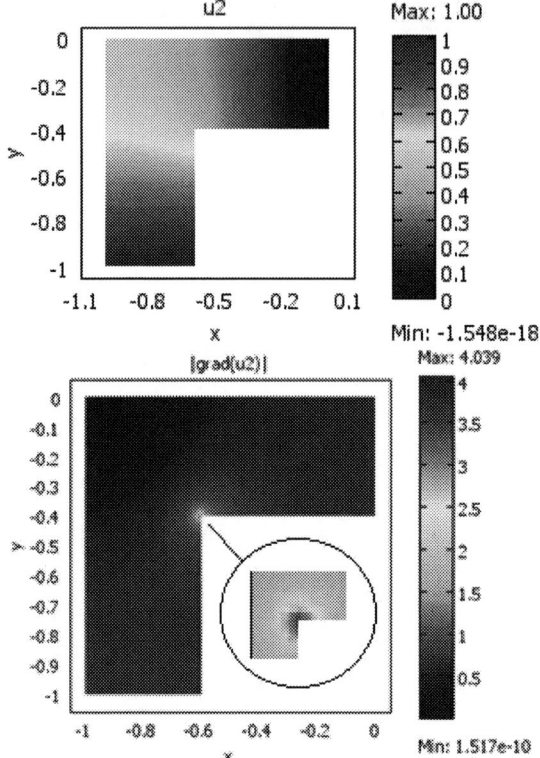

Figure 6. Electric potential and gradient

Electric gradient profile, on the L-shaped line, is found to be in agreement with analytical solution of Dandu and Fan [13], where singularity is observed for gradient of the voltage. Electric potential decreases from normalized value of one to zero, from anode end to the cathode end. Current crowding occurs at the corner on the path which electron flow follows.

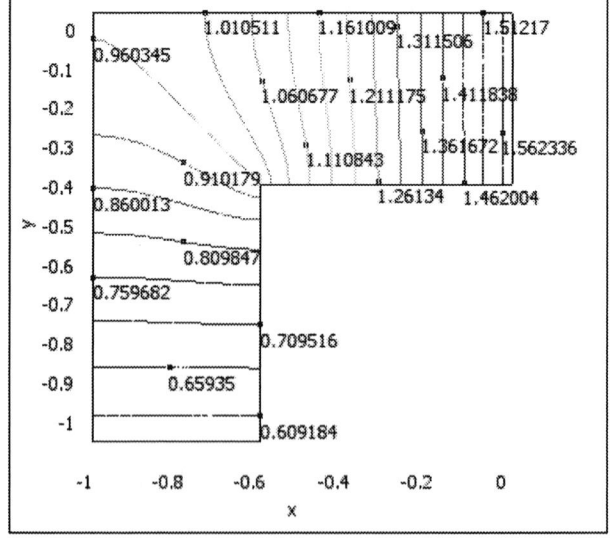

Figure 7. Vacancy concentration profile on the domain

Steady state solution of normalized vacancy concentration distribution along the L-shaped line is figured in (Fig. 7). Maximum vacancy concentration increases at Point C, located on the cathode end, where vacancy concentration decreases at Point A, i.e. anode, causing vacancy depletion. At Point B, where current singularity can be easily observed, vacancy concentration exhibits a lower value than the initial value. Vacancy concentration on the region surrounding this point remains mostly isometric, getting to the same steady-state vacancy concentration values.

Vacancy concentration values with the increasing time at the points of interest are figured in (Fig. 8). The results of our simulation comply with our analytical finding obtained by applying William's method [15] for solving singular stress field to our problem. It is observed that vacancy concentration does not show singularity where current singularity exists, nor is there an increased vacancy concentration on the surrounding region.

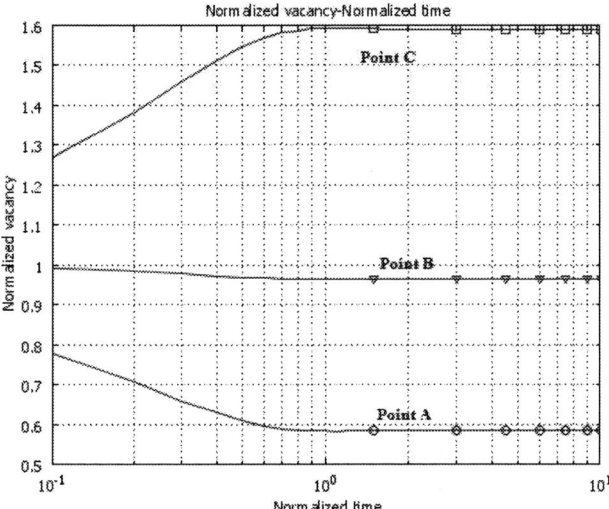

Figure 8. Vacancy progress with time at accumulation, depletion, and current crowding points

In the end, simulations show that, maximum vacancy concentration is located at the cathode end while vacancy depletion is seen at the anode side of the L-shaped line. With the increase in time, vacancy concentration increases at Point C, and decreases at points A and B. Vacancy concentration value is in a descendent regime even at Point B, where the magnitude of electric potential gradient becomes maximum, until problem enters the steady state.

Discussion

This study considers only two electromigration driving forces, vacancy concentration gradient and electric field. All other possible electromigration driving forces are excluded from this study. This is because electric field is considered the main driving force of electromigration failure, that is, without electric current there would be no electromigration failure. The solitary effect of electric field towards vacancy concentration, without distortion from other possible electromigration diving forces, is also a subject of interest. The results from this study can be used as a baseline to understand the effects of other electromigration driving forces should they be implemented into the model.

For homogeneous material with constant vacancy diffusivity and perfect blocking boundary condition, the results from both analytical and numerical study show that there is no indication of vacancy concentration singularity at the location of current density singularity. Instead, the location of maximum vacancy concentration is found to be at cathode side as shown in the numerical results. The experiment performed by Nemoto et al. [16] shows that, for conductor with passivation film, which is considered perfect blocking boundary condition in this study, voids accumulated at cathode side, but, for conductor without passivation film, generation and accumulation of voids was observed at the tip of the wedge. The study done by Zhang et al. shows that voids in solder joint generated at the location of current crowding [17]. At first, their results seem to be in contradiction to the results of this study, however, should the effects of two different conducting materials contacting each other and their interfacial boundary be considered, then the location of current crowding is also the location of cathode of the solder joint, where maximum vacancy concentration occurs.

Despite the agreement we find with other studies, voids accumulation due to electromigration do occur at the corner of L-shape interconnect structure as shown in the study published by He et al [6]. From our study, maximum vacancy concentration does not occur at the location of current density singularity, which means some other forces are likely in effect. The model proposed by Nemoto et al. shows that mechanical stress can shift the location of maximum vacancy concentration [16]. However, whether or not mechanical stress has any influence to the location of maximum vacancy concentration, this topic is to be discussed in another study.

Conclusion

The mathematical model of vacancy concentration coupling with electric potential is studied analytically and numerically. A 2-D, L-shaped, homogeneous material with constant vacancy diffusivity and perfect boundary condition is chosen as the problem of interest. Current density singularity is known to present at the tip of the wedge; however, the results from the study, both analytically and numerically, show that there is no vacancy concentration singularity at this location. The maximum vacancy concentration is found to be at cathode side, instead of the location of current density singularity.

References

1. J. R. Black, "Electromigration – A Brief Survey and Some Recent Results," *IEEE Transactions on Electron Devices*, vol. ED-16, no. 4, pp. 338–347, Apr. 1969.
2. I. A. Blech, "Electromigration in thin aluminum films on titanium nitride," *Journal of Applied Physics*, vol. 47, no. 4, pp. 1203–1208, Apr. 1976.
3. Clement and Lloyd, "Numerical investigations of the electromigration boundary value problem," *Journal of Applied Physics*, 71 (4), pp. 1729–1731, Feb. 1992.

978-1-4799-2408-0/14 $31.00 © 2014 IEEE

4. Korhonen et al., "Stress evolution due to electromigration in confined metal lines," *Journal of Applied Physics*, 73 (8), pp. 3790–3799, Apr. 1993.

5. Filippi et al., "The effect of current density, stripe length, stripe width, and temperature on resistance saturation during electromigration testing," *Journal of Applied Physics*, vol. 91, no. 9, pp. 5787–5795, May 2002.

6. He et al., "Electromigration lifetime and critical void volume," *Applied Physics Letters*, vol. 85, no. 20, pp. 4639–4641, Nov. 2004.

7. Cacho et al., "Modeling of Electromigration Induced Failure Mechanism in Semiconductor Devices," *COMSOL Users Conference 2007 Grenoble*, Oct. 2007.

8. Zheng et al., "A 2-D mesoscopic model coupling mechanical and diffusion for electromigration in thin films," *Computational Materials Science*, 46, pp. 443–446, 2009.

9. Orio et al., "Physically based models of electromigration: From Black's equation to modern TCAD models," *Microelectronics Reliability*, 50, pp. 775–789, 2010.

10. Y. Zhang, Y. Liu, L. Liang, and X. J. Fan, " The effect of atomic density gradient in electromigration," *Int. J. Materials and Structural Integrity*, vol. 6, no. 1, pp. 36–53, 2012.

11. P. Dandu, X. J. Fan, Y. Liu, and C. Diao, "Finite element modeling on electromigration of solder joints in wafer lelvel packages," *Microelectronics Reliability*, 50, pp. 547–555, 2010.

12. P. Dandu, X. J. Fan, and Y. Liu, "Some Remarks on Finite Element Modeling of Electromigration in Solder Joints," *2010 Electronic Components and Technology Conference*, 2010.

13. P. Dandu and X. J. Fan, "Assessment of Current Density Singularity in Electromigration of Solder Bumps," *2011 Electronic Components and Technology Conference*, pp. 2192–2196, Jun. 2011.

14. S. R. De Groot, "THEORIE PHENOMENOLOGIQUE DE L'EFFT SORET," *Physica IX*, no. 7, pp. 699–708, Jul. 1942.

15. Williams M., "On the stress distribution at the base of a stationary crack," *Journal of Applied Mechanics*, 24, pp. 109–114, 1957.

16. Nemoto et al., "In-situ observation of electromigration failure in AlCuSi interconnects with a passivation film," *Strength, Fracture and Complexity 5*, pp. 97–107, Sep. 2007.

17. Zhang et al., "Effect of current crowding on void propagation at the interface between intermetallic compound and solder in flip chip solder joints," *Applied Physics Letters*, 88, pp. 012106-1–3, 2006.

Hygro-Thermo-Mechanical Analysis and Failure Prediction in Electronic Packages by Using Peridynamics

Selda Oterkus and Erdogan Madenci
University of Arizona, Tucson, AZ

Erkan Oterkus
University of Strathclyde, Glasgow, UK

Yuchul Hwang, Jangyong Bae and Sungwon Han
Samsung Electronics Co., LTD, Korea.

Abstract

This study presents an integrated approach for the simulation of hygro-thermo-vapor-deformation analysis of electronic packages by using peridynamics. This theory is suitable for such analysis because of its mathematical structure. Its governing equation is an integro-differential equation and it is valid regardless of the existence of material and geometric discontinuities in the structure. It permits the specification of distinct properties of interfaces between dissimilar materials in the direct modeling of thermal and moisture diffusion, and deformation. Therefore, it enables progressive damage analysis in materials or layered material systems such as the electronic packages. It describes the validation procedure by considering a particular package for each thermomechanical, hygromechanical deformation as well as vapor pressure predictions. Also, it presents results concerning failure sites and mechanisms due to hygro-thermo-vapor-deformation.

Introduction

During packaging, transportation and storage, IC packages may absorb moisture leading to differential swelling between the polymeric and nonpolymeric materials, and among the polymeric materials. This differential swelling exacerbates the thermal deformation during the solder reflow process. In order to minimize the mismatch in swelling of the dissimilar materials of the IC packages during the solder reflow process, the packages are subjected to moisture conditioning (baking) for a period of time prior to this process. The estimate of baking period necessary to minimize the differential swelling and prevent possible cracking during the solder reflow process was investigated by Tay and Lin [1].

Although the baking process is essential in reducing the thermo-mechanical deformation during the solder reflow, it influences the moisture concentration distribution, and induces significant hygro-mechanical deformation. Also, the distribution of moisture concentration dictates the vapor pressure in micro voids while reducing the interfacial adhesion strength. The decrease in adhesion strength is caused by moisture absorption. Furthermore, the package cracking is not controlled by the absolute moisture content but its concentration at the critical interface, Kitano et al. [2]. An extensive discussion of moisture induced failure mechanisms in IC packages can be found in a study by Tee and Ng [3].

Coupled with the vapor pressure in micro voids, hygro-mechanical and thermo-mechanical deformation may cause interfacial delamination, and subsequent cracking at the pad-encapsulant interface, die-attach layer, and the die-encapsulant interface. Delamination and/or cracking at the die-attach layer is one of the primary failure mechanisms in plastic IC packages and often lowers the threshold for other mechanical, and electrical failures, Suhir [4] and Wong et al. [5].

The vapor pressure, dictated by the moisture concentration after baking, saturates much faster than the moisture diffusion, and that a near uniform vapor pressure is reached in the package, Tee and Ng [3]. The vapor pressure introduces an additional strain of the same order as that of the hygro-mechanical strains to the package. The hygro-mechanical stresses induced through moisture conditioning (baking) are significant compared to the thermo-mechanical stresses induced during the solder flow. Combination of these stresses can be detrimental to the reliability of the IC packages.

Therefore, the determination of the moisture concentration and temperature distributions is essential in order to determine the vapor pressure in micro voids, hygro-mechanical and thermo-mechanical stresses. There exists no known technique for the measurement of moisture distribution inside the package. Therefore, the predictive methods become unavoidable for investigating the effect of moisture conditioning.

Traditionally, moisture diffusion analysis is performed by using thermal-moisture analogy. Since the moisture concentration is not continuous along interfaces, a new parameter called "wetness" is introduced to render it continuous. Wetness is the ratio of the moisture concentration to its value at the saturated state, and it is continuous along interfaces. Although this approach is commonly accepted, it is not always valid because the saturated moisture concentration is not constant during the reflow process. A direct concentration approach (DCA) (Fan et al. [6]) should be employed to address this issue by imposing continuity condition along the interface between dissimilar materials.

A new continuum mechanics theory referred to as peridynamics (Silling [7]), removes this requirement because it is not necessary to impose continuity conditions. This feature of peridynamics emerges because the governing equations are based on integro-differential equations rather than partial differential equations of classical theory.

Furthermore, peridynamics is also very suitable for failure prediction which allows cracks to initiate and grow naturally in the structure without resorting to any external crack growth law.

This study presents an integrated hygro-thermo-vapor-deformation analysis using peridynamics to predict failure in electronic packages.

Peridynamic (PD) Theory

The peridynamic theory is a nonlocal continuum theory, and its continuum mechanics formulation was introduced by Silling [7] to overcome the difficulties arising due to the existence of discontinuities in the structure. The theory depends on integration rather than the spatial differentiation of PDEs as in classical continuum mechanics. Hence, it can be easily applicable to problems with discontinuities. As opposed to classical continuum mechanics, a material point inside the body can interact with other material points within its domain of influence called *horizon*, $\Re$ as shown in Fig. 1. The interaction (bond) between two material points **x** and **x'** are expressed by using a response function, **f**. Although PD formulation is originally given for mechanical field, it is applicable in other fields as well. The detailed derivation and capability of PD theory is given in the book by Madenci and Oterkus [8].

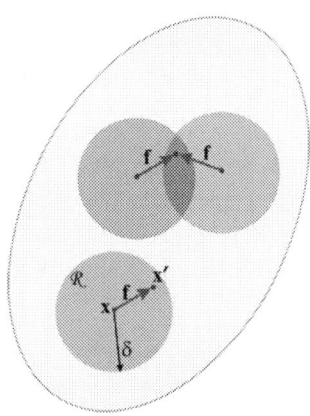

Fig. 1. Interaction of a material point with its neighboring points.

Basics - According to the PD theory, the field is analyzed by considering the interaction of a PD material point, **x**, with the other, possibly infinitely many, material points in the body. Therefore, an infinite number of interactions may exist between the material point at location **x** and other material points. Hence, the PD state may contain particular information on an infinite number of interactions. However, the influence of the material points interacting with **x** is assumed to vanish beyond a local region (horizon), denoted by $\Re$ shown in Fig. 1. The range of material point, **x** is defined by δ referred to as the "horizon". Also, the material points within a distance δ of **x** is called the family of **x**, $\Re$. The interaction of material points is prescribed through the response function which contains all of the constitutive information associated with the material. The response function also includes a

length parameter (horizon), δ. The locality of interactions depends on the horizon, and the interactions become more local with a decreasing horizon.

Hygrothermomechanics with vapor pressure - For known temperature and moisture concentration, the equation of motion of a material point can be expressed as

$$\rho(\mathbf{x})\ddot{\mathbf{u}}(\mathbf{x},t) = \int_{\Re} \mathbf{f}(\mathbf{u}'-\mathbf{u},\mathbf{x}'-\mathbf{x})dV_{\mathbf{x}'} + \mathbf{b}(\mathbf{x},t) \tag{1}$$

where the response function is defined as

$$\mathbf{f}(\mathbf{u}'-\mathbf{u},\mathbf{x}'-\mathbf{x}) = c\left(s - \alpha T_{avg} - \beta C_{avg} - \gamma p\right)\frac{(\mathbf{x}'+\mathbf{u}')-(\mathbf{x}+\mathbf{u})}{\left|(\mathbf{x}'+\mathbf{u}')-(\mathbf{x}+\mathbf{u})\right|} \tag{2}$$

in which c is the bond constant and can be expressed in terms Young's modulus, E

$$c = \frac{9E}{\pi h \delta^3} \text{ for 2D}, \quad c = \frac{12E}{\pi \delta^4} \text{ for 3D} \tag{3}$$

with h denoting the thickness.

The parameter s represents the stretch between material points and given by

$$s = \frac{\left|(\mathbf{x}'+\mathbf{u}')-(\mathbf{x}+\mathbf{u})\right| - \left|\mathbf{x}'-\mathbf{x}\right|}{\left|\mathbf{x}'-\mathbf{x}\right|} \tag{4}$$

where **u** and **u'** are the displacements of material points, **x** and **x'**. Thermal related parameter α is the coefficient of thermal expansion and T_{avg} is defined as

$$T_{avg} = \frac{(T-T_0)+(T'-T_0)}{2} \tag{5}$$

in which T and T' are the temperature of material points **x** and **x'** with T_0 being the reference temperature. Similarly, moisture related parameter β is the coefficient of moisture expansion and C_{avg} is defined as

$$C_{avg} = \frac{(C-C_0)+(C'-C_0)}{2} \tag{6}$$

in which C and C' are the moisture concentration of material points **x** and **x'** with C_0 being the reference moisture concentration. Finally, vapor pressure related parameter γ is the coefficient of vapor pressure expansion, i.e.

$$\gamma = \frac{1-\nu}{E} \text{ for 2D}, \quad \gamma = \frac{1-2\nu}{E} \text{ for 3D} \tag{7}$$

with ν being the Poisson's ratio, and p_{avg} is defined as

$$p_{avg} = \frac{p + p'}{2} \tag{8}$$

in which p and p' are the vapor pressure of material points $\mathbf{x}$ and $\mathbf{x}'$.

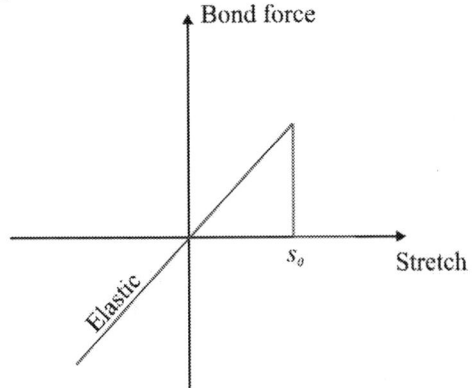

Fig. 2. Constitutive relation between material points in an elastic material.

In peridynamics, failure is introduced through a history-dependent scalar-valued function, μ, which is defined as

$$\mu(t, \mathbf{x}'\text{-}\mathbf{x}) = \begin{cases} 1 & \text{if } s(t', \mathbf{x}'\text{-}\mathbf{x}) < s_0 \text{ for all } 0 \le t' \le t \\ 0 & \text{otherwise} \end{cases} \tag{9}$$

in which s_0 is the critical stretch where failure occurs, as shown in Fig. 2. Its value is calculated by the relationship

$$s_0 = \sqrt{4\pi G_f / (9E\delta)} \text{ for 2D}, \; s_0 = \sqrt{5G_f / (6E\delta)} \text{ for 3D} \tag{10}$$

where G_f denotes the energy release rate. As suggested by Silling and Askari [9], an unambiguous notion of local damage at a point can be defined as

$$\varphi(\mathbf{x}, t) = 1 - \left(\int_{\Re} dV_{\mathbf{x}'} \mu(t, \mathbf{x}'\text{-}\mathbf{x}) \Big/ \int_{\Re} dV_{\mathbf{x}'} \right) \tag{11}$$

Since the response function is nonlinear, the peridynamic equation of motion is solved numerically. Therefore, in order to carry out the numerical integration, the region of interest is first discretized into sub-domains in which the displacement and velocity fields are assumed to be constant. Hence, each sub-domain can be represented as a single collocation point located at the mass center of the sub-domain. After discretization, the peridynamic equation of motion can be cast as

$$\rho \ddot{\mathbf{u}}_i = \sum_j \mathbf{f}(\mathbf{u}, \mathbf{u}', \mathbf{x}, \mathbf{x}', t) V_j + \mathbf{b}(\mathbf{x}_i, t) \tag{12}$$

in which V_j is the volume of the sub-domain that is represented by the collocation point located at $\mathbf{x}_j$.

Thermal diffusion - As opposed to the deformation field, the peridynamic formulation for the thermal field concerns the interaction due to heat exchange between material points. The material points are connected through thermal bonds. The peridynamic heat conduction equation is derived as (Oterkus et al. [10])

$$\rho c_v \dot{T}(\mathbf{x}, t) = \int_{\Re} f_q\left(T(\mathbf{x}, t), T'(\mathbf{x}', t), \mathbf{x}', \mathbf{x}, t\right) dV_{\mathbf{x}'} + q_E(\mathbf{x}, t) \tag{13}$$

where f_q is the thermal response function and q_E is the joule heating term. The pairwise response function for isotropic materials can be written as

$$f_q(\mathbf{x}', \mathbf{x}, t) = \kappa \frac{T(\mathbf{x}', t) - T(\mathbf{x}, t)}{|\mathbf{x}' - \mathbf{x}|} \tag{14}$$

in which the thermal bond constant, κ is defined in terms of the thermal conductivity, k as

$$\kappa = \frac{6k}{\pi h \delta^3} \text{ for 2D}, \quad \kappa = \frac{6k}{\pi \delta^4} \text{ for 3D} \tag{15}$$

Moisture diffusion - Direct approach for moisture diffusion analysis is essential (Fan et al. [6]). Similar to the heat conduction equation, the peridynamic formulation for the moisture concentration field concerns the interaction due to moisture exchange between material points. The material points are connected through hygro bonds. The peridynamic moisture concentration equation can be expressed as

$$\dot{C}(\mathbf{x}, t) = \int_{\Re} f_c\left(C(\mathbf{x}, t), C'(\mathbf{x}', t), \mathbf{x}', \mathbf{x}, t\right) dV_{\mathbf{x}'} \tag{16}$$

where f_c is the moisture concentration response function. The pairwise response function for isotropic materials can be written as

$$f_c(\mathbf{x}', \mathbf{x}, t) = d \frac{C(\mathbf{x}', t) - C(\mathbf{x}, t)}{|\mathbf{x}' - \mathbf{x}|} \tag{17}$$

in which the hygro bond constant, d is defined in terms of the moisture diffusivity, D as

$$d = \frac{6D}{\pi h \delta^3} \text{ for 2D}, \; d = \frac{6D}{\pi \delta^4} \text{ for 3D} \tag{18}$$

Vapor pressure - Simulations of vapor pressure field are performed by adopting the procedure suggested by Fan et al. [11], in which they proposed a micro-mechanics approach. It is assumed that the voids in polymer materials are uniformly distributed, thus a representative volume element (RVE) can be used for a continuum representation of vapor pressure within the material. Fig. 3 shows a typical RVE with the moisture in both liquid and vapor state within the void. There exist three possibilities for the moisture state within the voids:

Case 1: The moisture within the void is in single vapor phase at pre-conditioning temperature (T_0). Thus the moisture within the void is also in single vapor phase at reflow temperature (T_r). In this case, the transition temperature (T_1) at which all the moisture is transformed into vapor phase is less than the pre-conditioning temperature ($T_1 < T_0$).

Case 2: The moisture within the void is in mixed liquid and vapor phase at pre-conditioning temperature (T_0) and in single vapor phase at reflow temperature (T_r). In this case, the transition temperature (T_1) is greater than the pre-conditioning temperature and less than the reflow temperature ($T_0 < T_1 < T_r$).

Case 3: The moisture within the void is in mixed liquid and vapor phase at both pre-conditioning temperature (T_0) and reflow temperature (T_r). In this case, the transition temperature (T_1) is greater than the reflow temperature ($T_r < T_1$) thus the vapor pressure in the void is the saturated vapor pressure at reflow temperature ($p_g(T_r)$) as given in steam tables.

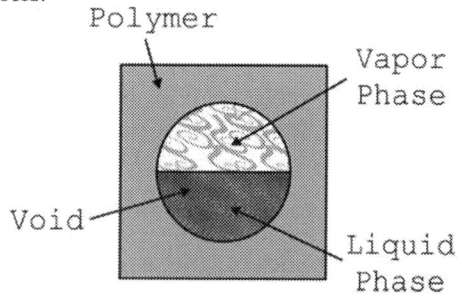

Fig. 3. Representative volume element (RVE) with a void.

Fig. 4 illustrates these three distinct cases. The vapor pressure, p at each node is calculated by following the flowchart given in Fig. 5. Its saturated value is denoted by $p_g(T)$. In the computations, the following formulae are used for:

Case 1: $p = \dfrac{C \cdot p_g(T_0) \cdot T_r}{f \cdot \rho_g(T_0) \cdot T_0}$ (19a)

Case 2: $p = \dfrac{p_g(T_1) \cdot T_r}{T_1}$ (19b)

Case 3: $p = p_g(T_r)$ (19c)

in which, f is the void volume fraction with its initial value being f_0. The moisture density and saturated moisture density are defined by $\rho_m = C / f_0$ and $\rho_g(T)$, respectively.

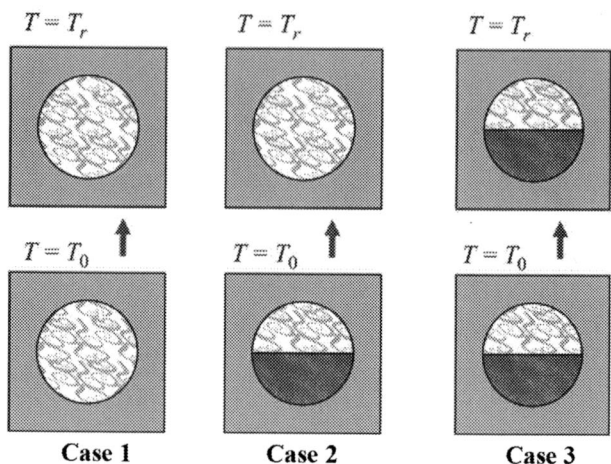

Fig. 4. Three distinct moisture states in RVE.

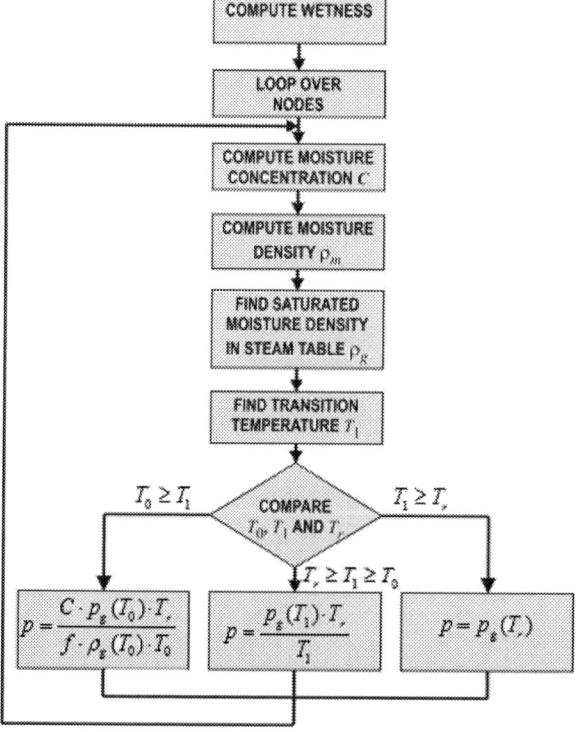

Fig. 5. Flowchart of vapor pressure simulations.

Saturated pressure $p_g(T)$ and saturated moisture density, $\rho_g(T)$ can be obtained from the steam tables. Their dependence on temperature is shown in Figs. 6 and 7. Furthermore, the variation of vapor pressure as a function of void volume fraction is depicted in Fig. 8.

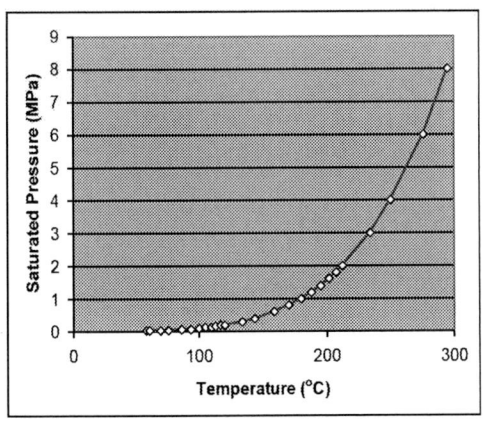

Fig. 6. Saturated pressure variation.

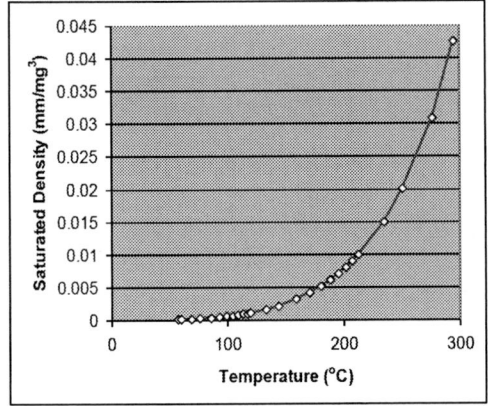

Fig. 7. Saturated moisture density variation.

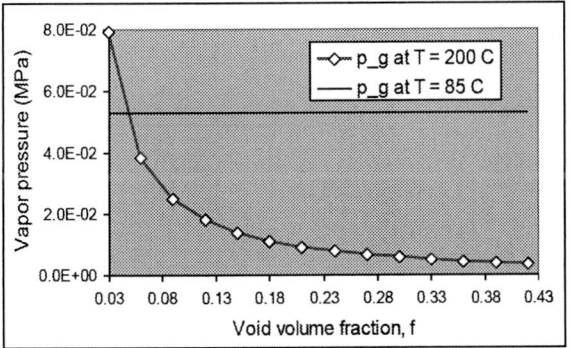

Fig. 8. Vapor pressure variation as a function of void volume fraction.

Numerical results

The validity of the peridynamic simulations is established by comparing the solutions for a particular package considered previously by Tee and Ng [3] against finite element predictions. The package geometry is described in Fig. 9.

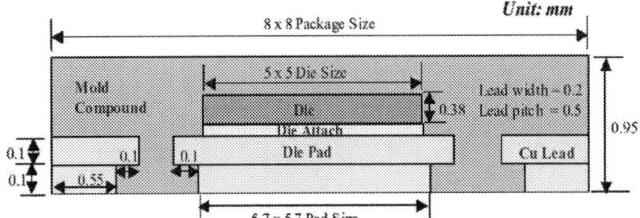

Fig. 9. Geometric and material configuration for the Quad Flat Non-Lead (QFN) package (figure taken from (Tee and Ng, 2000)).

The geometric parameters are specified as shown in Fig. 10:

$$L_1 = 2.5\,mm, \quad L_2 = 0.45\,mm, \quad L_3 = 0.4\,mm, \quad L_4 = 0.65\,mm$$
$$H_1 = 0.2\,mm, H_2 = 0.05\,mm, \quad H_3 = 0.4\,mm, \quad H_4 = 0.35\,mm$$
$$h = 0.01\,mm$$

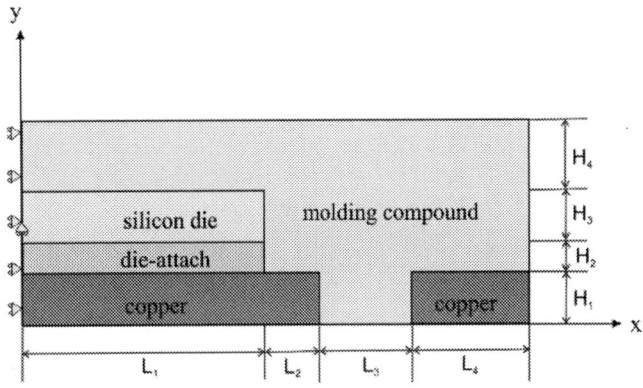

Fig. 10. Geometric parameters and symmetry conditions.

The material properties are specified as

$$E_1 = 15 \times 10^9\,Pa, \quad E_2 = 7.4 \times 10^9\,Pa,$$
$$E_3 = 163 \times 10^9\,Pa, \quad E_4 = 129 \times 10^9\,Pa$$

$$\nu_1 = 0.25, \quad \nu_2 = 0.4, \quad \nu_3 = 0.278, \quad \nu_4 = 0.355$$

$$\beta C_{sat1} = 1.57 \times 10^{-3}, \quad \beta C_{sat2} = 3.22 \times 10^{-3},$$
$$\beta C_{sat3} = 0, \quad \beta C_{sat4} = 0$$
$$C_{sat1} = 7.06\,kg\,/\,m^3, C_{sat2} = 6.20\,kg\,/\,m^3,$$
$$C_{sat3} = 0, \quad C_{sat4} = 0$$

$$\alpha_1 = 16 \times 10^{-6}\,1/^\circ K, \alpha_2 = 52 \times 10^{-6}\,1/^\circ K,$$
$$\alpha_3 = 2.6 \times 10^{-6}\,1/^\circ K, \alpha_4 = 14.3 \times 10^{-6}\,1/^\circ K$$

$$\rho_1 = 1180\,kg\,/\,m^3, \rho_2 = 6450\,kg\,/\,m^3,$$
$$\rho_3 = 2330\,kg\,/\,m^3, \rho_2 = 8940\,kg\,/\,m^3$$

During absorption

$D_1 = 2.6748 \times 10^{-9} \, m^2 / hr$

$D_2 = 45 \times 10^{-9} \, m^2 / hr$

$D_3 = 1.0 \times 10^{-11} \, m^2 / hr$

$D_4 = 1.0 \times 10^{-11} \, m^2 / hr$

During desorption

$D_1 = 6.0 \times 10^{-7} \, m^2 / hr$

$D_2 = 1.5 \times 10^{-6} \, m^2 / hr$

$D_3 = 1.0 \times 10^{-11} \, m^2 / hr$

$D_4 = 1.0 \times 10^{-11} \, m^2 / hr$

in which subscript 1, 2, 3 and 4 represent the molding compound, die attach, die and copper, respectively.

Thermomechanics - For thermomechanical deformation, the boundary conditions are specified as

$$u_x(x=0,y,t) = 0, \quad u_y(x=0,y=H_1+H_2,t) = 0$$

and a uniform temperature change of

$$\Delta T = 1.0^\circ K$$

is assumed throughout the domain.

As shown in Figs. 11 and 12, the peridynamic and ANSYS predictions are in excellent agreement for thermomechanic deformations.

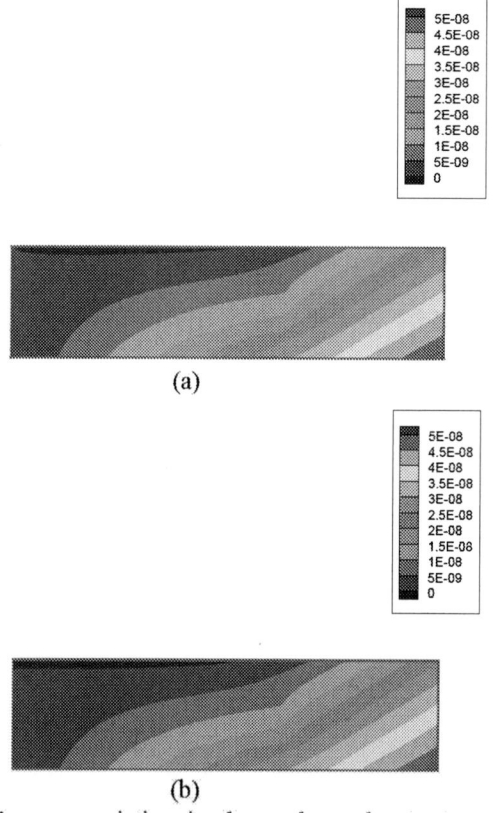

Fig. 11. u_x variation in the package due to temperature change (a) PD results (b) ANSYS results

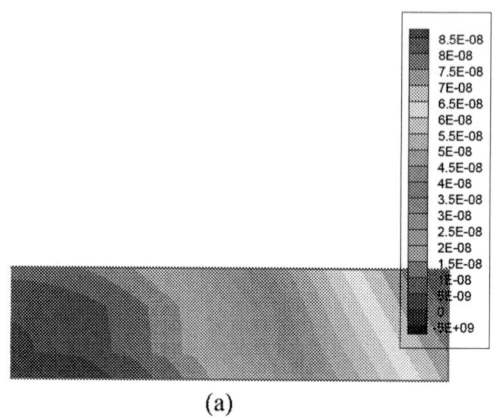

(a)

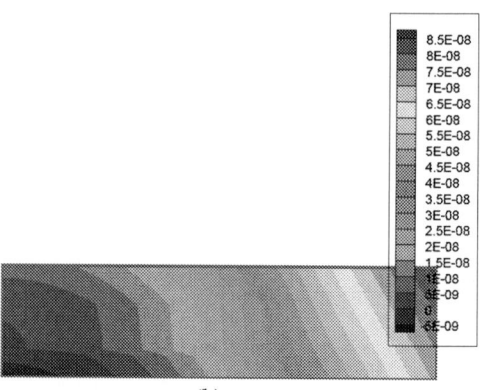

(b)

Fig. 12. u_y variation in the package due to temperature change (a) PD results (b) ANSYS results

Hygromechanics - For hygromechanical deformation, the boundary conditions are specified as

$$u_x(x=0,y,t) = 0, \quad u_y(x=0,y=H_1+H_2,t) = 0$$

and the moisture concentration values in each material region are assumed as

$$C_1 = C_{sat1}, C_2 = C_{sat2}, C_3 = 0, C_4 = 0.$$

As shown in Figs. 13 and 14, the peridynamic and ANSYS predictions are also in excellent agreement for hygromechanic deformations.

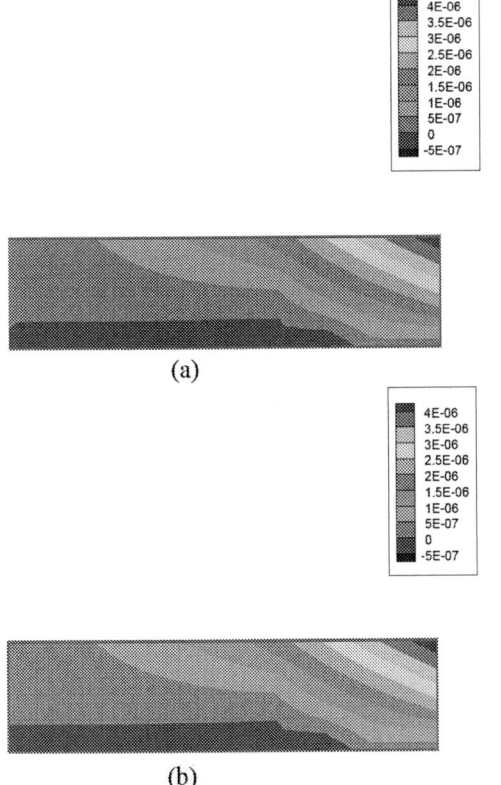

Fig. 13. u_x variation in the package due to moisture concentration (a) PD results (b) ANSYS results

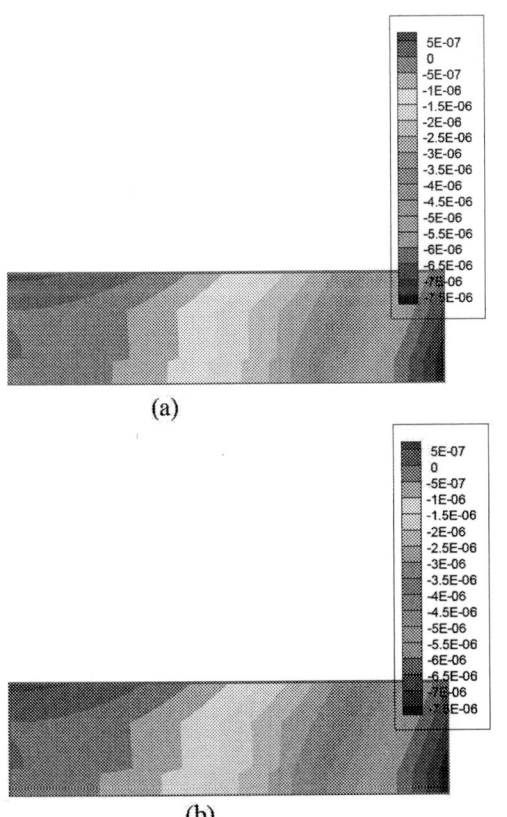

Fig. 14. u_y variation in the package due to moisture concentration (a) PD results (b) ANSYS results

Vapor pressure distribution - In order to determine the effect of vapor pressure, first the package is subjected to 168hr. absorption and 5 min desorption at 220°C (reflow temperature). The wetness distribution from peridynamic analysis is compared against ANSYS predictions as shown in Fig. 15 to validate the wetness results. Then, the desorption process is extended to 25 min and new wetness and corresponding concentration distributions are obtained as depicted in Fig. 16. At this concentration and temperature state, a maximum vapor pressure of 2.2 MPa is obtained as demonstrated in Fig. 17. Corresponding deformations from both peridynamics and ANSYS predictions match very well as shown in Fig. 18 and 19.

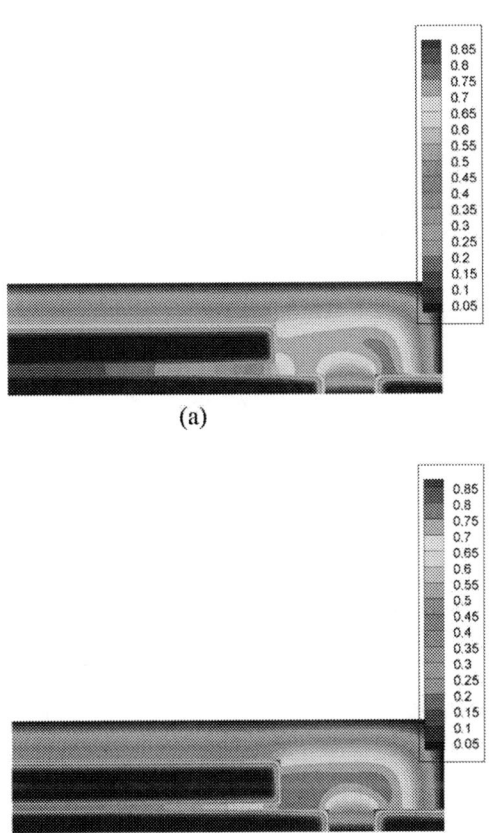

Fig. 15. Wetness distribution after absorption and desorption a) PD results b) ANSYS results (168hr. absorption and 5 min desorption)

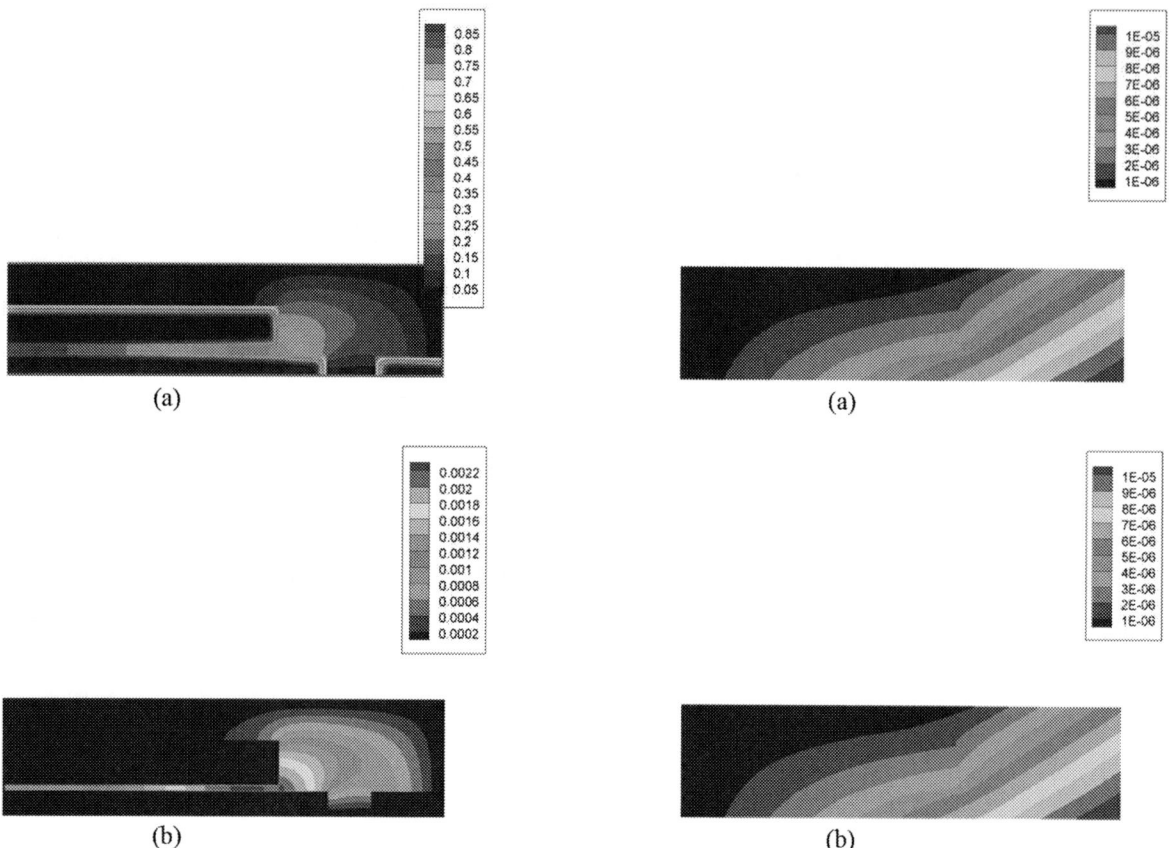

Fig. 16. PD results for a) wetness and b) corresponding concentration distribution after absorption and desorption (168hr. absorption and 25 min desorption)

Fig. 18. u_x variation in the package due to temperature change, concentration change and vapor pressure (a) PD results (b) ANSYS results

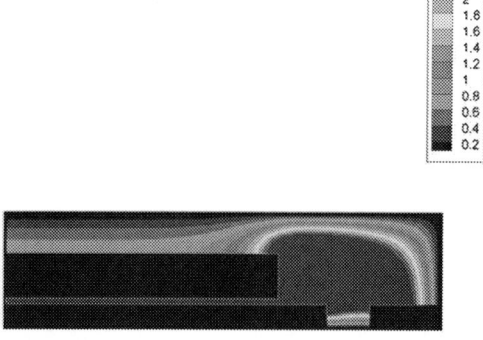

Fig. 17. Pressure distribution after absorption and desorption (MPa) (168hr. absorption and 25 min desorption)

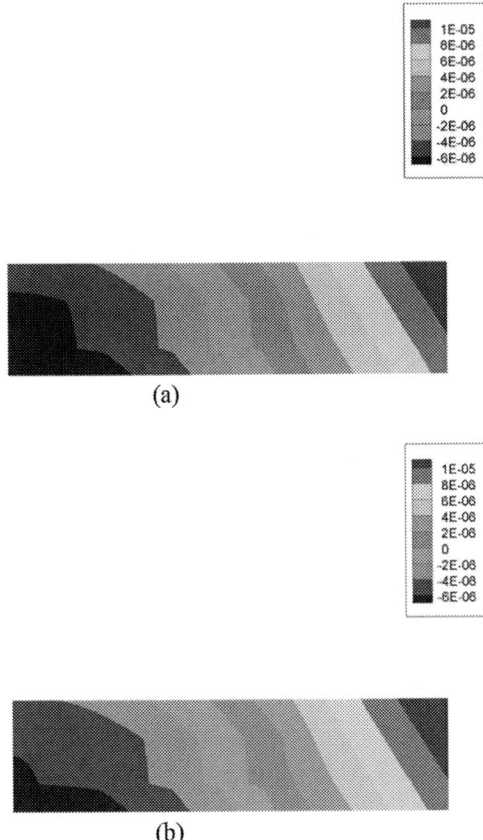

(a)

(b)

Fig. 19. u_y variation in the package due to temperature change, concentration change and vapor pressure (a) PD results (b) ANSYS results

Failure prediction – In order to demonstrate the failure prediction capability of peridynamics, a critical stretch value of $s_0 = 0.015$ is assumed for all bonds inside the package. For this critical stretch value, it is observed that delamination/cracking occurs along the die-attach region and partial cracking occurs inside the molding compound as shown in Fig. 20. Extensive demonstration cases are left for a future study by considering different critical stretch values based on the individual material component and interface properties.

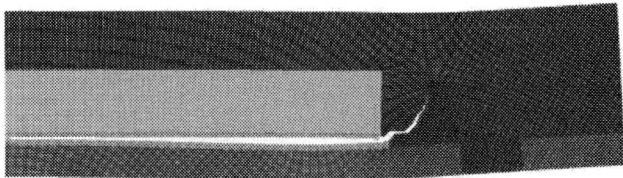

Fig. 20. PD failure prediction inside the package.

Conclusions

In this study, a new hygro-thermo-vapor-deformation analysis of electronic packages is presented by using peridynamics. As a result of the integro-differential formulation of peridynamics, it is straightforward to obtain moisture concentration without using "wetness" parameter by direct concentration approach (DCA). Furthermore, the formulation is also very suitable for failure analysis and for the demonstration case chosen, delamination/cracking failure along the die-attach region is observed which is a common failure mechanism in electronic packages.

References

1. Tay, A. A. O. and Lin , T. Y, 1998, "Quantifying the Effect of Prebaking on Delamination in Plastic IC Packages," EEP-Vol. 25, Workshop on Mechanical Reliability of Polymeric Materials and Plastic Packages of IC Devices, ASME, pp. 215-221.

2. Kitano, M., Nishimura, A., Kawai, S. and Nishi, K., 1988, "Analysis of Package Cracking During Reflow Soldering Process," IEEE/IRPS, pp. 90-95.

3. Tee, T. Y. and Ng, H. S., 2002, "Whole Field Vapor Pressure Modeling of QFN During Reflow with Coupled Hygro-mechanical and Thermo-mechanical Stresses," Proceedings, 52nd Electronic Components and Technology Conference, San Diego, California.

4. Suhir, E., 1997, " Failure Criterion for Moisture-Sensitive Plastic Packages of Integrated Circuit Devices: Application of Von-Karman's Equations with Consideration of Thermoelastic Strains," International Journal of Solids and Structures, Vol. 34, pp. 2991-3019.

5. Wong, E. H., Chan, K. C., Rajoo, R. and Lim, T. B., 2000, "The Mechanics and Impact of Hygroscopic Swelling of Polymeric Materials in Electronic Packaging," Proceedings, 50nd Electronic Components and Technology Conference, Las Vegas, Nevada.

6. Fan, X. J., Tee, T. Y., Shi, X. Q. and Xie, B., 2010, "Modeling of Moisture Diffusion and Whole-Field Vapor Pressure in Plastic Packages of IC Devices," in Moisture Sensitivity of Plastic Packages of IC Devices, X. J. Fan and E. Suhir (eds.) , Springer, New York.

7. Silling S. A., 2000, "Reformulation of Elasticity Theory for Discontinuities and Long-range Forces," Journal of the Mechanics and Physics of Solids, Vol. 48, pp. 175-209.

8. Madenci, E. and Oterkus, E., 2013, Peridynamic Theory and Its Applications, Springer, New York.

9. Silling, S. A. and Askari, E., 2005, "A Meshfree Method Based on the Peridynamic Model of Solid Mechanics," Computers & Structures, Vol. 83, pp. 1526-1535.

10. Oterkus, S., Madenci, E. and Agwai, A., 2014, "Peridynamic Thermal Diffusion, " Journal of Computational Physics, Vol. 265, pp.71-96.

11. Fan, X. J., Zhang, G. Q. and Ernst, L. J., 2002, "A Micro-mechanics Approach for Polymeric Material Failures in Microelectronic Devices," Proceedings of the conference EuroSimE 2002, Paris, France, pp. 154-164.

Cohesive Zone Experiments for Copper/Mold Compound Delamination

William E. R. Krieger, Sathyanarayanan Raghavan, Abhishek Kwatra, and Suresh K. Sitaraman*
The George W. Woodruff School of Mechanical Engineering
Georgia Institute of Technology, Atlanta, GA 30332
*contact: suresh.sitaraman@me.gatech.edu

Abstract

As complex multi-layered packaging becomes more common in microelectronic design, delamination remains a prominent failure mechanism due to coefficient of thermal expansion mismatch. Numerous studies have investigated interfacial cracking in microelectronic packages. These studies commonly use classical interfacial fracture mechanics analyses, but such analyses require knowledge of starter crack size, locations, and propagation paths. Cohesive zone theory has been identified as an alternative method for modeling crack propagation and delamination without the need for a pre-existing crack. This paper presents a framework to determine mixed-mode cohesive zone parameters using experimental methods. We demonstrate this method by characterizing cohesive zone parameters for a copper/epoxy molding compound interface. Fully characterized cohesive zone elements can be placed at interfaces in finite-element models of microelectronic systems to simulate loading and failure in mixed-mode conditions.

Background

Miniaturization and rising performance demands have led to the introduction of multilayered structures in modern microelectronic packages. During fabrication and assembly processes, these multilayered systems are subjected to several thermal excursions. During such thermal excursions, thermo-mechanical stresses develop due to coefficient of thermal expansion (CTE) mismatch among different material layers in the package, and these stresses can be high enough to result in interfacial delamination.

Multi-material interfaces are common points of CTE mismatch failure. Directly bonded interfaces and adhesively bonded interfaces are generally weaker than cohesive materials, and this has prompted numerous experimental [1, 2] and theoretical studies [3, 4] of interfacial strength. To avoid singularity issues, fracture mechanics based approaches are preferred over stress-based approaches in studying interfacial delamination.

An energy-based fracture-mechanics approach uses available strain energy release rate G to study interfacial crack propagation. G depends on many parameters including magnitude and orientation of applied loading, package dimensions, and materials. A crack is expected to grow when G reaches a critical value, defined as critical energy release rate G_C. Unlike cohesive materials, G_C is not a constant value for a given interface. Rather, it depends on the ratio of shear to normal loading applied. This ratio is quantified as mode-mixity ψ in angular units, where 0° signifies pure normal (mode I) loading and 90° signifies pure shear (mode II) loading.

Since G_C is a function of mode-mixity, G_C must be characterized over a range of ψ for any particular interface.

Researchers have demonstrated several techniques to characterize G_C at varying ψ. These include button shear test, mixed-mode bend, tab pull test, end-notched flexure, four-point bend (FPB), superlayer, and magnetic actuation [e.g. 5, 6-13]. For metal/polymer interfaces, G_C under pure shear loading is approximately ten times G_C under pure normal loading conditions [5, 9]. In this work we accept this as a valid assumption.

Critical strain energy release rate measurements have been used to perform classical fracture mechanics analyses of interfacial delamination in microelectronic packaging [13-16]. Finite element modeling (FEM) is used to create a model of a package such as small outline integrated circuit (SOIC) package or flip-chip package, and a starter crack of known geometry and size is modeled at a probable failure region. Appropriate loading is applied. G and ψ are obtained from the model using well-known methods like virtual crack closure technique (VCCT), virtual crack extension technique, J-integral, etc. [e.g. 14, 15, 17, 18, 19]. The modeled G is compared to G_C from experiments, and the crack is expected to grow if G exceeds G_C.

Fracture mechanics is adequate for analyzing geometries with well-known starter crack locations. Crack propagation can be simulated with FEM using nodal release techniques, but several iterations are required to re-check failure criteria as the crack grows. Also, fracture mechanics does not describe crack initiation, and analyses must be repeated several times to simulate different starter crack geometries. In any case, fracture mechanics involves several unknowns since starter crack size and geometry are rarely known *a priori*.

Cohesive Zone Modeling for Interfaces

Cohesive zone (CZ) modeling is an emerging technology capable of simulating crack initiation and crack propagation. CZ models have been used to simulate several types of interface, including integrated thin-film structures, adhesively bonded polymers, glass/elastomer, and on-chip interfaces [20-23]. Here we apply a CZ technique to model delamination between copper leadframe and epoxy molding compound (EMC).

In a CZ model, interfacial separation occurs within a cohesive damage zone when the damage exceeds a pre-set limit. Within the cohesive zone, there are active traction stresses between the cohesive surfaces, and the interaction is governed by the traction-separation law. Many proposed shapes of traction-separation law are available, including exponential, bilinear, and trapezoidal. Area beneath the traction-separation law is equivalent to G_C.

We select a bilinear traction-separation law proposed by Alfano and Crisfield [24]. Six parameters are required to fully define the law in FEM software. The bilinear law appears in Fig. 1.

978-1-4799-2408-0/14 $31.00 © 2014 IEEE

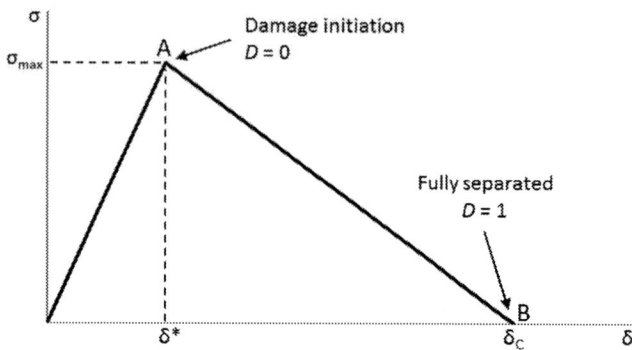

Figure 1: Bilinear traction-separation law for cohesive zone interfacial elements.

The bilinear law shows interfacial traction σ vs. interfacial separation δ. As interfacial elements undergo deformation, they exhibit elastic loading for $\delta < \delta^*$. In this region, no damage is accumulated in the interface, and unloading returns interfacial elements to their initial configuration. At point A, a critical traction σ_{max} is reached and damage is initiated. Delamination is tracked by a damage parameter D calculated by (1). When $\delta > \delta^*$, D increases, and when $\delta \geq \delta_C$, D reaches a maximum value of 1.

$$D = \begin{cases} 0 & \text{if } \delta \leq \delta^* \\ \left(\frac{\delta - \delta^*}{\delta}\right)\left(\frac{\delta_c}{\delta_c - \delta^*}\right) & \text{if } \delta^* < \delta < \delta_c \\ 1 & \text{if } \delta \geq \delta_c \end{cases} \quad (1)$$

Regardless of the current magnitude of δ, the damage value D can never decrease. In other words, unloading will not reduce the damage that has accumulated. When the damage parameter $D = 1$, the interfacial element is said to be fully damaged, and the stiffness of the cohesive zone element is zero, as seen at point B. Thus, a fully damaged element has been completely separated and will not produce interactions between layers. Throughout separation, traction is a function of interfacial separation given by (2).

$$\sigma = \frac{\sigma_{max}}{\delta^*}(1 - D)\delta \quad (2)$$

The area under the traction-separation profile is the critical energy release rate, and thus, for the bilinear law, $G_C = 0.5\,\delta_C\,\sigma_{max}$.

In applications, interfacial cracking always propagates in mixed-mode conditions [25]. Therefore, six independent parameters are required to fully characterize CZ models for mixed-mode loading conditions. These parameters define two bilinear traction-separation laws corresponding to pure mode I and pure mode II delamination. In this work, we present a methodology to determine CZ parameters for mixed-mode loading conditions.

Critical Strain Energy Release Rate Characterization

Critical strain energy release rate of the copper/EMC interface is determined analytically from experiments. FEM is used to calculate ψ and validate the measured value of G_C.

For experimental characterization, an FPB test is selected due to the simplicity of determining G_C. The advantage of the FPB test is that the loading produces a constant bending moment between the inner loading pins. As a result, steady-state delamination occurs at a constant critical load, and G is independent of crack length. Load-displacement data is expected to show steady-state delamination at a constant force P_{crit}.

Pre-molded bimaterial strip specimens are used for experimental testing. *Sumitomo Sumikon® EME-G630AY* molding compound is applied via transfer mold to a CDA194 copper alloy. The assembled specimen is cured at 175°C and cooled to room temperature. To prepare the specimen for FPB testing, a notch is placed in the mold compound at the middle of the beam using a *DISCO* automatic dicing saw. The depth of cut is approximately 95% of the mold compound thickness. An untested specimen is shown in Fig. 2. Specimen dimensions include EMC thickness $t_{EMC} = 1.524$ *mm*, copper thickness $t_{Cu} = 0.254$ *mm*, and width $b = 6$ *mm*.

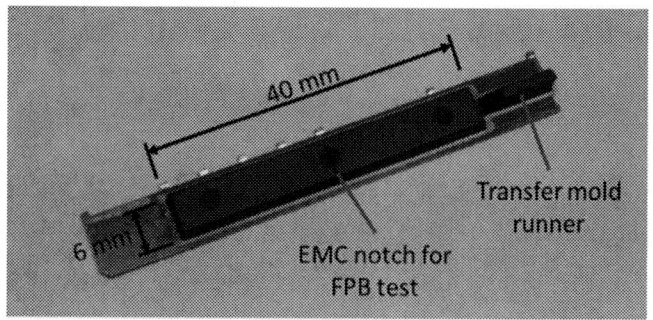

Figure 2: Bimaterial specimen for FPB testing.

The FPB setup is shown in Fig. 3a. The notched bimaterial specimen is placed on two fixed support pins with the copper side down. The notch is centered between the support pins, and the loading pins are lowered to contact the specimen. All pins have a diameter of 1 *mm*. Displacement-controlled loading is applied at a rate of 0.50 *mm/min*. As loading increases, a crack is expected to propagate from the notch to the copper/EMC interface and continue along the interface. Fig. 3b shows a test in progress. Tests are performed on a *TestResources* tensile test machine. Load and displacement data are recorded throughout by a *TestResources* force gauge and *Epsilon Technology* extensometer.

Fig. 4 shows a load-displacement response from an experiment at room temperature. After some initial slack in the system, the data exhibits elastic bending with stiffness determined by the depth of the EMC notch. When the load reaches approximately 8.2 *N*, one load drop is observed indicating a crack has probably propagated from the notch through mold compound and then the load begins to increase momentarily. A second, larger load drop is observed, indicating a crack has started to propagate at the EMC and copper interface. The load stabilizes as displacement increases, indicating the interfacial crack continues to propagate. The test is repeated with several other samples, and the steady-state delamination is observed at an average critical load of $P_{crit} = 6.526$ *N*.

978-1-4799-2408-0/14 $31.00 © 2014 IEEE

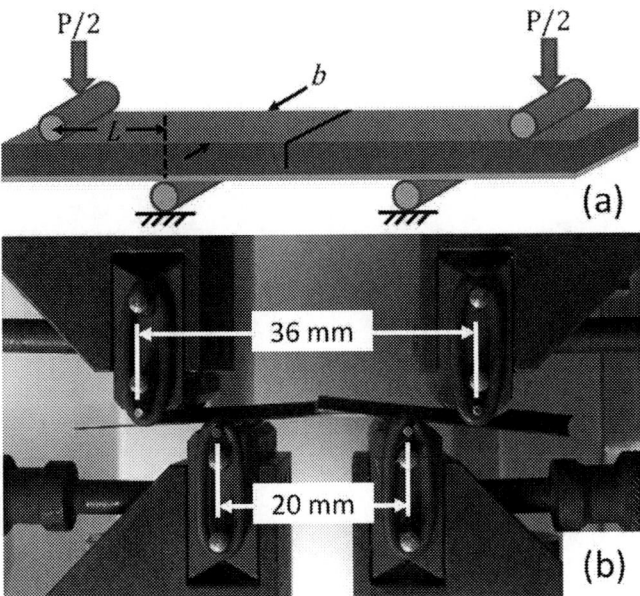

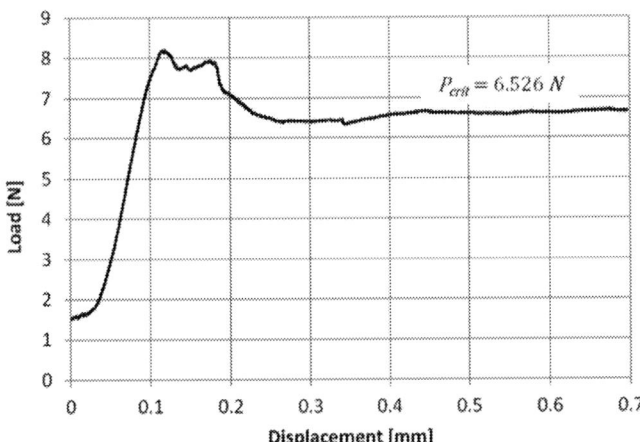

Figure 3: Four-point bend test (a) experimental schematic and (b) experiment in progress.

Figure 4: Load-displacement data obtained from an FPB experiment.

For a bimaterial interface, strain energy release rate may be computed by (3) from Charalambides *et al.* [26].

$$G = \frac{(1-v_{Cu}^2)P^2L^2}{8E_{Cu}b^2}\left(\frac{1}{I_{Cu}}-\frac{\lambda}{I_C}\right) \quad (3)$$

$$I_{Cu} = \frac{t_{Cu}^3}{12} \qquad \lambda = \frac{E_{Cu}(1-v_{EMC}^2)}{E_{EMC}(1-v_{Cu}^2)}$$

$$I_C = \lambda I_{Cu} + \frac{t_{EMC}^3}{12} + \frac{\lambda t_{Cu}t_{EMC}(t_{Cu}+t_{EMC})^2}{4(\lambda t_{Cu}+t_{EMC})}$$

G is the strain energy release rate produced by the FPB loading configuration from applied load P. From Fig. 3a, L is the moment arm and b is specimen width. Layer thicknesses are given previously as t_{Cu} and t_{EMC}. I_{Cu} is the area moment of inertia per unit width for the copper layer, and I_C represents the same for the composite beam as a whole. λ is a non-

dimensional parameter that gives the ratio of copper to EMC plane strain moduli. Material properties E and v are obtained from material datasheets and appear in Table 1.

Table 1. Material properties at room temperature for bimaterial specimens.

	Copper	EMC
E [GPa]	121	25.0
v	0.33	0.30

Critical strain energy release rate G_C is obtained by calculating G during delamination. Therefore, P_{crit} observed in experiments can be substituted into (3) to determine G_C. For the purpose of calculating G_C, perfectly symmetric delamination is a valid assumption and any effects of one-sided delamination may be ignored [27]. Using (3), we compute a critical strain energy release rate $G_C = 50.63\ J/m^2$.

To calculate the associated mode-mixity, a 2D plane-strain finite element model of the specimen is created using *ANSYS Mechanical APDL 14.5®*. As shown in Fig. 5, only the right half of the specimen is modeled to save computation time. The purpose of this model is to determine ψ and G_C by simulating an instant during interfacial crack propagation, so a pre-crack is constructed at the interface. G is independent of crack length, so a pre-crack length of 2 *mm* from the symmetry boundary is chosen arbitrarily. To simulate the vertical EMC notch, nodes on the left side of the EMC are unconstrained and a symmetry boundary condition is placed on the left side of the copper layer. The node at the interior loading pin is constrained in the vertical direction. A vertical downward force $P_{crit}/(2b)$ is applied at the exterior loading pin, with P_{crit} obtained from experiments. For small deformations, both copper and EMC are assumed linear elastic at room temperature.

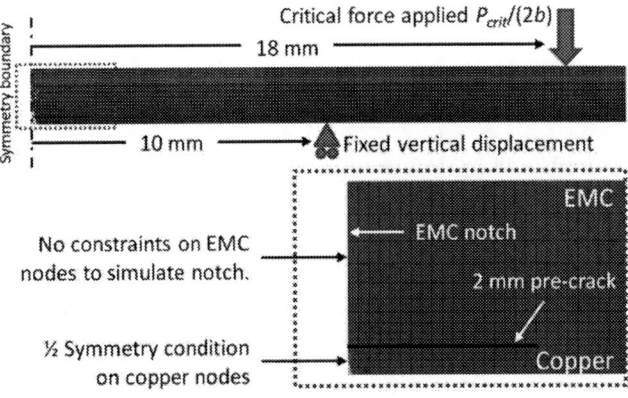

Figure 5: 2D FEM model of FPB test with pre-crack.

The VCCT method is applied to the deformed model in Fig. 6 to calculate the mode I and mode II energy release rates, and ψ and G_C are obtained using (4). The VCCT method returns $G_I = 41.54\ J/m^2$ and $G_{II} = 9.09\ J/m^2$. VCCT calculations were repeated with varying crack extension length Δa to verify that results have converged. Δa also represents the element size at the crack tip. Fig. 7 shows that

978-1-4799-2408-0/14 $31.00 © 2014 IEEE

mode-mixity varies by less than 5° as Δa changes. When G_I and G_{II} are added to obtain the total energy release rate, it is seen from Fig. 7 that the resulting magnitude agrees well with the analytical value obtained from equation (3). For subsequent calculations, we use $\psi = 22.6°$.

$$\tan^2 \psi = \frac{G_{II}}{G_I} \qquad G = G_I + G_{II} \qquad (4)$$

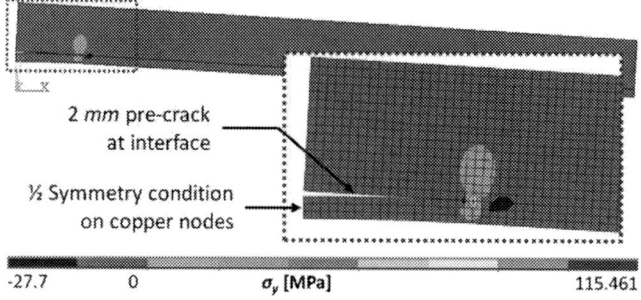

Figure 6: Deformed FEM model for calculating ψ and G_C via the VCCT method.

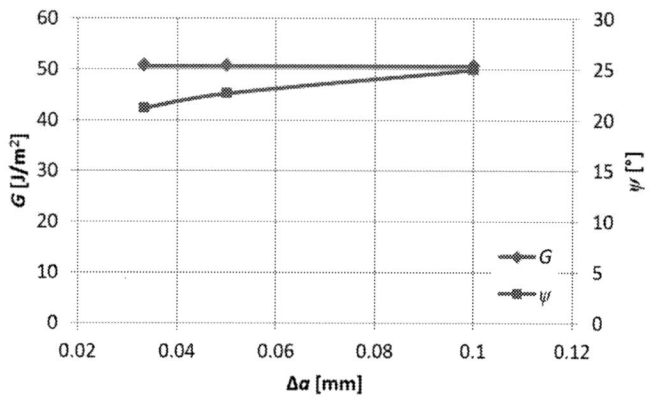

Figure 7: Dependence of G and ψ on Δa in VCCT calculations.

Cohesive Zone Modeling

To obtain properties for cohesive zone elements, the FPB experimental test is modeled and simulated with *ANSYS*. Material models are again assumed to be linear elastic.

A half-symmetry 2D plane-strain model of the FPB experiment is constructed (Fig. 8). No pre-crack is modeled, and cohesive zone elements are inserted along the entire interface. A symmetry boundary condition is applied only to the copper layer, as done in the fracture-mechanics model discussed earlier. Crack propagation through the EMC to the interface is not simulated. The interior pin is modeled as a node fixed in the vertical direction. Displacement-controlled loading is applied to one node at the exterior loading pin. The model is used to simulate force-displacement data, and the cohesive zone properties are modified to replicate experimental data.

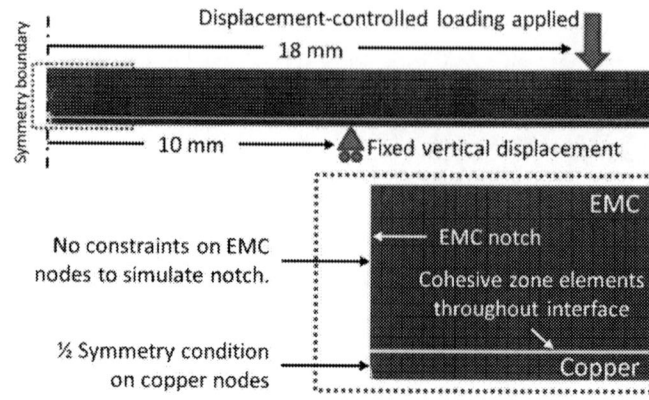

Figure 8: Cohesive zone finite element model of FPB test.

Other interfacial fracture experiments are currently ongoing at *Georgia Institute of Technology* to obtain G_C as a function of mode mixity. In the meantime, for this publication, since experimental data is obtained for one mode-mixity, we must assume values of G_C for 0° and 90° mode-mixity. Hutchinson and Suo give (5) as one equation for characterizing critical strain energy release rate with regard to mode-mixity [28].

$$G_C = G_{I,C}[1 + \tan^2(\psi(1 - \lambda))] \qquad (5)$$

$G_{I,C}$ is the mode I G_C at $\psi = 0°$. Based on results from literature, we assume a condition that $G_C = 10G_{I,C}$ at $\psi = 90°$ [5, 9]. By applying (5) for $\psi = 22.6°$ and $\psi = 90°$, we obtain $\lambda = 0.2048$ and $G_{I,C} = 45.8$ J/m^2. This produces the relationship shown in Fig. 9.

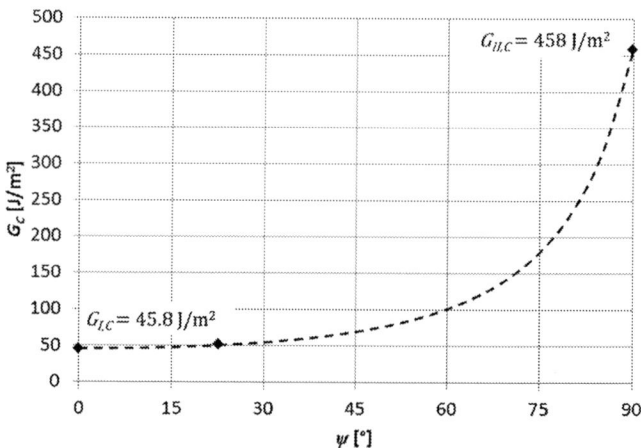

Figure 9: The Hutchinson and Suo model of G_C vs. ψ is used for determining cohesive zone parameters.

To determine cohesive zone parameters, a general design procedure follows. Initial guesses are selected for opening and shear traction-separation laws such that the areas are equivalent to $G_{I,C}$ and $G_{II,C}$ respectively. σ_{max} is raised or lowered accordingly to adjust the critical force at which the interface delaminates. As σ_{max} changes, δ_C is updated to maintain a constant area.

Results

Simulated load-displacement data is plotted with experimental data in Fig. 10. As crack propagation through the EMC is not modeled, so no peak appears before delamination initiates. In other words, the assumed notch depth is through 100% of the EMC. Fig. 11 shows normal and tangential traction-separation laws used for the simulation. All cohesive zone parameters are listed in Table 2, where $\alpha = \delta^*/\delta_C$.

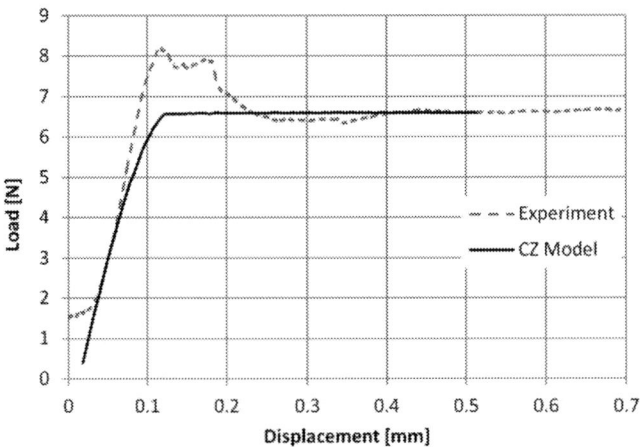

Figure 10: Simulated FPB load-displacement data compared to results from experiments.

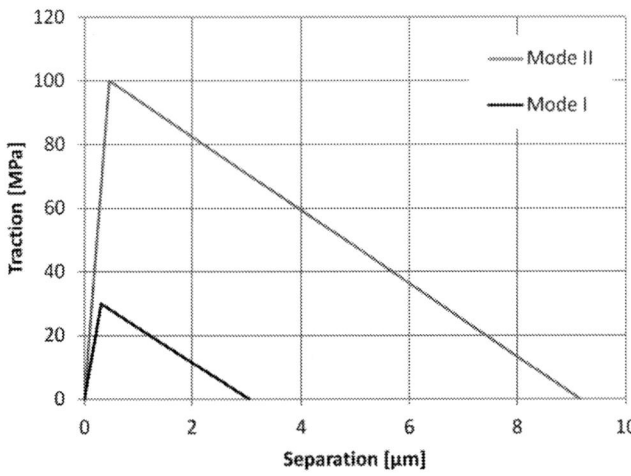

Figure 11: Bilinear traction-separation laws defined for modeling of the copper/EMC interface.

Table 2: Copper/EMC interfacial cohesive zone parameters for bilinear traction-separation laws.

	Normal	Tangential
σ_{max} [MPa]	30	100
δ_C [μm]	3.053	9.160
α	0.1	0.05

With cohesive zone elements, crack propagation and interfacial separation can be modeled directly. Fig. 12 shows specimen deformation and peel stress σ_y at several instances of delamination. Crack tip movement along the interface is observed in the movement of the concentrated peel stresses. Once the damage D of a CZ element reaches a value of 1, the CZ element stiffness becomes zero, and the surfaces have fully cracked. Images in Fig. 12 show the interface (i) before delamination has initiated, (ii) just after separation has begun, (iii) after the delamination has propagated some distance, and (iv) after significant delamination.

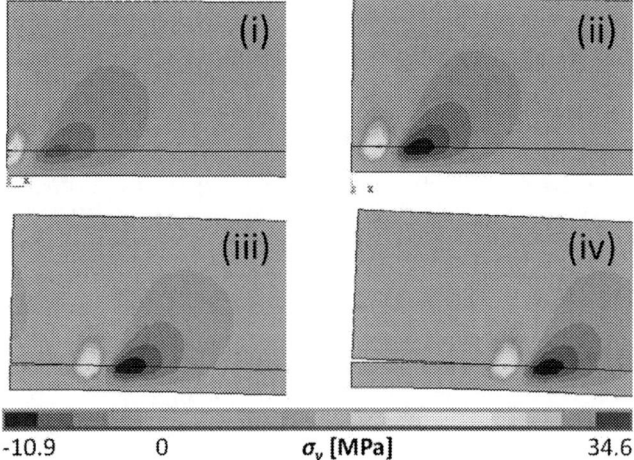

Figure 12: Peel stresses near the crack tip at several stages of interfacial delamination propagation.

Conclusion

A methodology for determining cohesive zone parameters for modeling of a metal/polymer interface has been demonstrated. First FPB experiments are performed to measure G_C of a particular interface and load-displacement data is recorded. By comparing results to literature, G_C values for opening and shear mode-mixities are estimated. Critical strain energy release rate under pure mode I and pure mode II loading are applied to a cohesive zone model, and the model is used to simulate load-displacement experimental data. Cohesive zone parameters are adjusted via a design procedure to closely mimic the experimental data. The result is a set of parameters which define separation behavior of the copper/EMC interface. These cohesive zone parameters are necessary for a predictive model for interfacial delamination.

Results of this study may be fortified by inclusion of an additional interfacial strength experiment. A double cantilever beam test will provide a quantitative measurement of critical strain energy release rate close to $\psi \approx 0°$ as well as another set of load-displacement data to mimic. Once fully characterized, the cohesive zone elements can be inserted along copper/EMC interfaces in models of SOICs, flip-chips, or similar packaging assemblies to simulate interfacial cracking under mechanical loading, thermal cycling, or fabrication process thermal excursions.

Cohesive zone modeling is still an emerging technique and more work is needed to streamline the procedure and examine other effects on the cohesive zone model. An exhaustive study may investigate how to manipulate the cohesive zone model to account for moisture, temperature, or surface roughness, all of which are known to affect critical strain energy release rate of copper/EMC interfaces.

Acknowledgments

This work is supported by the *Semiconductor Research Corporation*. The authors would like to thank industry liaisons Torsten Hauck, Vijay Sarijan, and Ilko Schmadlak from *Freescale Semiconductor* for providing valuable input and guidance throughout the course of this project. The authors also thank Nishant Lakhera of *Freescale Semiconductor* for the bimaterial strip specimens used in the experiments. Thanks are also due to Christina Weiler of *Georgia Institute of Technology* for starting initial exploratory experiments.

References

1. Nishimura, A., I. Hirose, and N. Tanaka, *A New Method for Measuring Adhesion Strength of IC Molding Compounds*. Journal of Electronic Packaging, 1992. **114**(4): p. 407.

2. Sheng, L., M. Yuhai, and T.Y. Wu, *Bimaterial interfacial crack growth as a function of mode-mixity*. Components, Packaging, and Manufacturing Technology, Part A, IEEE Transactions on, 1995. **18**(3): p. 618-626.

3. O'Dowd, N.P., C.F. Shih, and M.G. Stout, *Test geometries for measuring interfacial fracture toughness*. International Journal of Solids and Structures, 1992. **29**(5): p. 571-589.

4. Davidson, B.D. and V. Sundararaman, *A single leg bending test for interfacial fracture toughness determination*. International Journal of Fracture, 1996. **78**(2): p. 193-210.

5. Durix, L., et al., *On the development of a modified button shear specimen to characterize the mixed mode delamination toughness*. Engineering Fracture Mechanics, 2012. **84**: p. 25-40.

6. Xiao, A., et al. *Interfacial fracture properties and failure modeling for microelectronics*. in *Electronic Components and Technology Conference, 2008. ECTC 2008. 58th*. 2008. IEEE.

7. Xu, L., et al. *Adhesion behavior between epoxy molding compound and different leadframes in plastic packaging*. in *Electronic Packaging Technology & High Density Packaging, 2009. ICEPT-HDP'09. International Conference on*. 2009. IEEE.

8. Shirangi, M., W. Müller, and B. Michel. *Determination of Copper/EMC interface fracture toughness during manufacturing, moisture preconditioning and solder reflow process of semiconductor packages*. in *ICF12, Ottawa 2009*. 2013.

9. Tran, H.T., et al., *Temperature, moisture and mode-mixity effects on copper leadframe/EMC interfacial fracture toughness*. International Journal of Fracture, 2013. **185**(1-2): p. 115-127.

10. Lam, W.K., et al. *A method for evaluating delamination between epoxy moulding compounds and different plated leadframes*. in *Electronic Materials and Packaging, 2000. (EMAP 2000). International Symposium on*. 2000.

11. Ostrowicki, G.T. and S.K. Sitaraman, *Magnetically actuated peel test for thin films*. Thin Solid Films, 2012. **520**(11): p. 3987-3993.

12. Modi, M.B. and S.K. Sitaraman, *Interfacial fracture toughness measurement for thin film interfaces*. Engineering Fracture Mechanics, 2004. **71**(9–10): p. 1219-1234.

13. Xie, W. and S.K. Sitaraman, *Investigation of interfacial delamination of a copper-epoxy interface under monotonic and cyclic loading: experimental characterization*. Advanced Packaging, IEEE Transactions on, 2003. **26**(4): p. 447-452.

14. Van Driel, W., et al., *Prediction of delamination related IC & packaging reliability problems*. Microelectronics Reliability, 2005. **45**(9): p. 1633-1638.

15. Guofeng, X., et al. *Interfacial delamination and reliability design of exposed pad packages*. in *Electronic Packaging Technology and High Density Packaging (ICEPT-HDP), 2012 13th International Conference on*. 2012.

16. Liu, X., et al., *Failure analysis of through-silicon vias in free-standing wafer under thermal-shock test*. Microelectronics Reliability, 2013. **53**(1): p. 70-78.

17. Harries, R.J. and S.K. Sitaraman, *Numerical modeling of interfacial delamination propagation in a novel peripheral array package*. Components and Packaging Technologies, IEEE Transactions on, 2001. **24**(2): p. 256-264.

18. Sundararaman, V. and S.K. Sitaraman, *Interfacial fracture toughness for delamination growth prediction in a novel peripheral array package*. Components and Packaging Technologies, IEEE Transactions on, 2001. **24**(2): p. 265-270.

19. Xie, W. and S.K. Sitaraman, *Investigation of interfacial delamination of a copper-epoxy interface under monotonic and cyclic loading: modeling and evaluation*. Advanced Packaging, IEEE Transactions on, 2003. **26**(4): p. 441-446.

20. Mei, H., et al. *Initiation and propagation of interfacial delamination in integrated thin-film structures*. in *Thermal and Thermomechanical Phenomena in Electronic Systems (ITherm), 2010 12th IEEE Intersociety Conference on*. 2010. IEEE.

21. Li, S., et al., *Mixed-mode cohesive-zone models for fracture of an adhesively bonded polymer–matrix composite*. Engineering fracture mechanics, 2006. **73**(1): p. 64-78.

22. Rahul-Kumar, P., et al., *Polymer interfacial fracture simulations using cohesive elements*. Acta materialia, 1999. **47**(15): p. 4161-4169.

23. Raghavan, S., et al. *Framework to Extract Cohesive Zone Parameters Using Double Cantilever Beam and Four-Point Bend Fracture Tests*. in *Thermal, Mechanical and Multi-Physics Simulation and Experiments in Microelectronics and Microsystems (EuroSimE), 2014 15th International Conference on*. 2014. IEEE.

24. Alfano, G. and M. Crisfield, *Finite element interface models for the delamination analysis of laminated composites: mechanical and computational issues*. International journal for numerical methods in engineering, 2001. **50**(7): p. 1701-1736.

25. Rice, J.R., *Elastic Fracture Mechanics Concepts for Interfacial Cracks*. Journal of Applied Mechanics, 1988. **55**(1): p. 98.

26. Charalambides, P.G., et al., *A Test Specimen for Determining the Fracture Resistance of Bimaterial Interfaces*. Journal of Applied Mechanics, 1989. **56**(1): p. 77.

27. Noijen, S., O. van der Sluis, and P. Timmermans. *An extensive investigation of the four point bending test for interface characterization*. in *Thermal, Mechanical and Multi-Physics Simulation and Experiments in Microelectronics and Microsystems (EuroSimE), 2012 13th International Conference on*. 2012. IEEE.

28. Hutchinson, J. and Z. Suo, *Mixed mode cracking in layered materials*. Advances in applied mechanics, 1992. **29**(63): p. 191.

Damage Pre-Cursor Based Life Prediction of the Effect of Mean Temperature of Thermal Cycle on the SnAgCu Solder Joint Reliability

Pradeep Lall, Kazi Mirza, Jeff Suhling
Auburn University
NSF-CAVE3 Electronic Research Center
Department of Mechanical Engineering
Auburn, AL. 36849
Tele: (334) 844-3424
E-mail: lall@auburn.edu

Abstract

Electronics in automotive underhood applications may be subjected to temperatures in the neighborhood of 150°C to 175°C. Several of the electronics functions such as lane departure warning systems, collision avoidance systems are critical to vehicle operation. Prior studies have shown that low silver leadfree SnAgCu alloys exhibit pronounced deterioration in mechanical properties even after short exposure to high temperatures. Current life prediction models for second level interconnects do not provide a method for quick-turn assessment of the effect of mean temperature on cyclic life. In this paper, a method has been developed for assessment of the effect of mean cyclic temperature on the thermal fatigue reliability based on physics based leading damage indicators including phase-growth rate and the intermetallic thickness. Since the quantification of the thermal profile in the field applications may be often very difficult, the proposed method does not require the acquisition of the thermal profile history. Three environments of -50°C to +50°C, 0°C to 100°C, 50°C to 150°C with identical thermal excursion and different mean temperatures have been studied. Test assemblies with three different packages including CABGA 144, PBGA 324, and PBGA 676 have been used for the study. Damage-proxy based damage-equivalency relationships have been derived for the three thermal cycles. Weibull distributions have been developed for the three test assemblies to evaluate the effect of the mean cyclic temperature on the thermal fatigue life. Data indicates that the thermal fatigue lie drops with the increase in mean temperature of the thermal cycle even if the thermal excursion magnitude is kept constant. Damage equivalency model predictions of the effect of mean temperature of the thermal cycle have been validated versus weibull life distributions. The damage proxy based damage equivalency methodology shows good correlation with experimental data.

Introduction

Electronics in a variety of applications such as automotive underhood on-engine, on-transmission, high-performance computing, military, and defense applications may be subjected to prolonged exposure to high temperature in addition to wide cyclic temperature extremes. Furthermore, the mean temperature of the thermal excursion may vary based on application. Prior, studies have revealed that leadfree material properties degrade with prolonged exposure to high temperature. Detrimental effects on properties include the degradation in the yield strength and ultimate tensile strength of the materials. [Chou 2002; Hasegawa 2001; Zhang 2009]. Furthermore, prior exposure to high temperature aging has been shown to reduce mechanical integrity by as much as 50-percent. Evolution of mechanical properties has been verified in the solder alloys even at high strain rates in the neighborhood of 1-to-100 sec[-1] typical of shock and vibration [Lall 2013]. The effects are most pronounced in the widely used SnAgCu based alloys including SAC105, SAC205, SAC305 and SAC405 solders. Lower silver solders such as the SAC105, often touted for their resistance to transient dynamic shock and vibration, are the most susceptible to thermal aging amongst the SAC solders. Thus, prior data suggests that the cyclic life for leadfree assemblies cannot be considered without accounting for mean cyclic temperature.

The evolution of mechanical properties of leadfree solders alloys may pose a potential reliability problem in long-life systems such as automotive applications in which electronics often resides underhood of the car at temperatures in the neighborhood of 150°C-175°C. Several of the electronic modules in automotive electronics applications may perform critical functions such as lane departure warning, collision avoidance, adaptive cruise control and antilock braking. There is a need for tools and techniques to enable damage mapping between different thermal cycle conditions and for evaluating the effect of the mean temperature of the thermal cycle. However, the current, closed form life prediction models for leadfree second-level interconnects do not provide any method for quick-assessment of effect of mean temperature on the expected life under thermal cycling. In this paper, a new model has been developed for the assessment of the effect of the thermal cycle's mean temperature on the cyclic life of a leadfree assembly. Three ball-grid arrays including CABGA 144, PBGA 324, and PBGA 676 have been subjected to three thermal cycles including -50°C to +50°C, 0°C to 100°C, 50°C to 150°C. The thermal cyclic magnitude has been kept the same while the mean temperature has been varied in the thermal cycle. Previously, leading indicators of damage have been used to quantify the accrued thermo-mechanical damage under steady-state and cyclic temperature exposure in leadfree solders. [Lall 2011[a-b]; 2012[a-b]]. in this paper, microstructural indicators including phase growth in solder interconnects, intermetallic thickness, intermetallic composition has been measured. In addition, a separate population of the parts has been cycled to failure under each of the three conditions. Predictive model has been developed for mapping the cyclic damage for leadfree electronics subjected to mean cyclic temperatures. Model predictions have been correlated with experimental weibull failure distributions in order to quantify the model accuracy and precision. The ability to assess the effect of mean cyclic temperature on the thermal fatigue reliability will allow reliability assessment of leadfree electronics using prior accelerated test data for any mean temperature and thermal cycle magnitude.

Test Vehicle

In this study, three different leadfree assemblies with CABGA 144, PBGA-676 and PBGA-256 packages have been used. The packages are full-array configuration and Sn3Ag0.5Cu solder interconnects. The ball diameter for the 144 I/O, 256 I/O and 676 I/O BGA is 0.3mm, 0.32mm and 0.48mm respectively. Package attributes are shown in Table 1. The printed circuit board for all assemblies was a double-sided FR4-06 material. The printed circuit board pads were solder mask defined (SMD) with immersion silver finish. Figure 1 shows the packages. All test vehicles were subjected to three thermal cycling environments including TC1: -50°C to 50°C, TC2: 0°C to 100°C and TC3: 50°C to 150°C for various numbers of cycles. The test board is a JEDEC form-factor test board with corner holes. Each test package has four daisy chain patterns corresponding to the four quadrants. Board assemblies were assembled at in-house surface mount facility of CAVE3. The reflow profile used for assembly is shown in Figure 2.

Table 1: Package Attributes

	Board A	Board B	Board C
Solder	Sn3Ag0.5Cu	Sn3Ag0.5Cu	Sn3Ag0.5Cu
Body Size (mm)	27 mm	17 mm	13mm
Package Type	PBGA	PBGA	CABGA
I/O Count	676	256	144
I/O Pitch (mm)	1	1	1
Ball Diameter (mm)	0.48	0.32	0.3
Matrix	26 x 26	16 x 16	12 x 12
Pad (board)	NSMD	NSMD	NSMD
Pad (package)	SMD	SMD	SMD
Board Finish	ImAg	ImAg	ImAg

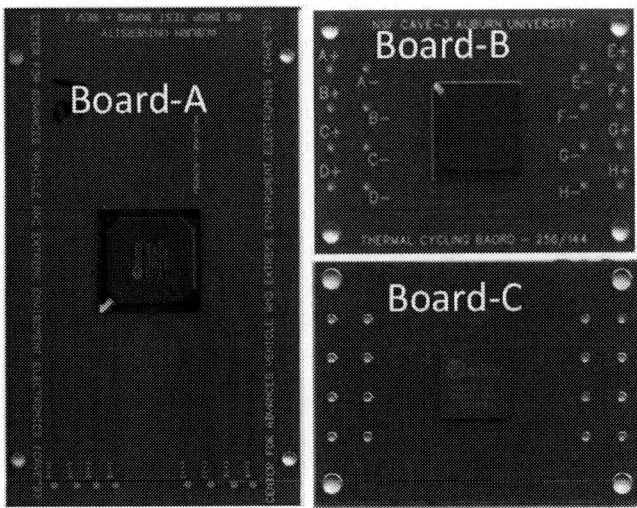

Figure 1: Test Vehicle

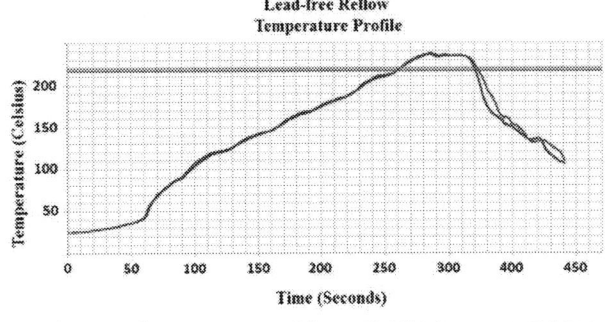

Figure 2: Reflow profile

Approach For Interrogation of Accrued Damage

In operational environments, electronics may be subjected to a variety of thermal loads including environmental thermal excursions, thermal storage, and power on-off cycling. The wide array of possible thermal environments makes the job of correlating the accelerated test results with the expected performance in operational environments somewhat difficult. In this paper, a damage proxy based damage mapping method has been presented using the microstructural evolution of damage in the solder joints in addition to the intermetallic growth at the interfaces. The methodology has been used to study the effect of the mean temperature of the thermal cycle (T_{mean}), keeping the temperature-cycle magnitude (ΔT) constant in SnAgCu leadfree interconnects. The approach has been developed in three steps:

Micro-structural Evolution of Damage

In this step, three sets of board assemblies have been subjected to thermal cycling. The package architecture used for the experiment has been kept constant between the groups in order to minimize the effects of changes in dimensions, processing or manufacturing variances on the differences in the measured value of damage proxies between groups. Measurement of the damage proxies is a destructive test involving cross-sectioning of the samples to access the corner solder joint in the die shadow region. Samples were withdrawn at periodic intervals and cross-sectioned. Damage proxies including the phase growth and the intermetallic thickness were measured using optical confocal microscopy. Phase growth has been quantified using a parameter 'S' which is the defined as the power difference in the phase size at the measurement time w.r.t to the phase size at the time of the board assembly manufacture, mathematically represented by:

$$S = g^4 - g_0^{\ 4} = aN^b \tag{1}$$

$$\ln S = \ln(g^4 - g_0^{\ 4}) = \ln a + b \ln N \tag{2}$$

$$\frac{d \ln S}{d \ln N} = b \tag{3}$$

where, g is the average grain size at time of prognostication, g_0 is the average grain size of solder after reflow, N is the number of cycle, S is the phase growth parameter, parameters a and b are the coefficient and exponent respectively. Previously, the rate of change in phase growth parameter [d(lnS)/d(lnN)] or the slope of the curve between the phase growth parameter versus number of cycles has been shown to be a valid damage proxy for prognostication of thermo-mechanical damage in solder interconnects and for assessment of residual life. Prior research has shown a positive correlation between the rate of change in phase growth parameter [d(lnS)/d(lnN)] and plastic work or the inelastic strain energy density, a parameter often used to quantify damage in finite element models and used as an input to life-prediction relationships [Lall 2004a, 2005a, 2006c, 2006d, 2007c, 2007e, 2008c, 2008d, 2009c, 2009d].

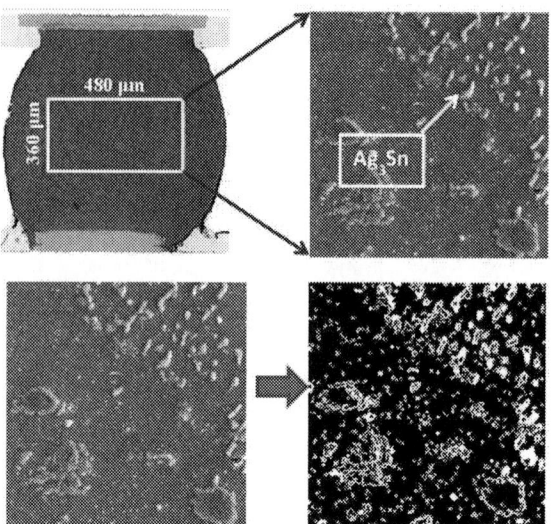

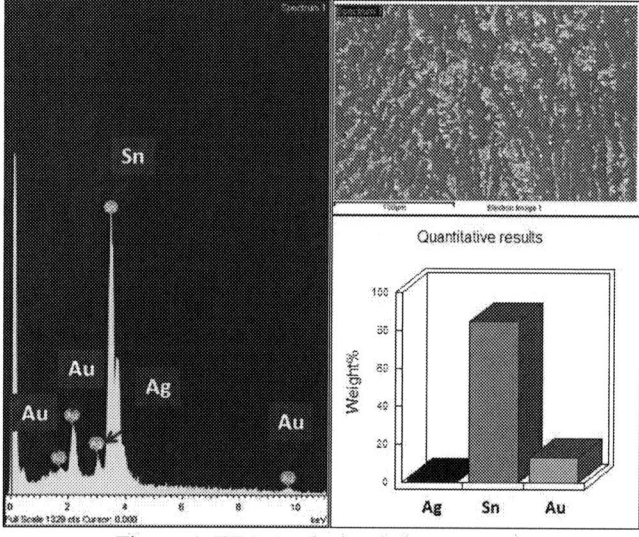

Figure 3: Micrograph and Gray scale mapping of image using image analysis software.

Figure 4: EDS Analysis of phase growth

The rate of change in phase growth parameter [d(lnS)/d(lnN)] damage proxy has been studied in the present effort to assess its potential for mapping damage between different operating conditions including between different accelerated test conditions or between accelerated test and operating conditions. Images of polished samples were taken for the critical solder joint in the chip-shadow region. The growth rate of Ag3Sn phases was optically measured using image analysis software NI-MAQ. Measurements of Ag3Sn phase size were taken from a rectangular region 480μm x 360μm in size from the confocal optical images. The gray scale images

were then mapped to black-and-white images. The mapped image was overlapped with the original image in each case to ensure the accuracy of the mapping. Figure 3 shows intermediate steps in mapping of the gray scale image. The location and appearance of the phases was confirmed using EDS in a scanning electron microscope (Figure 4).

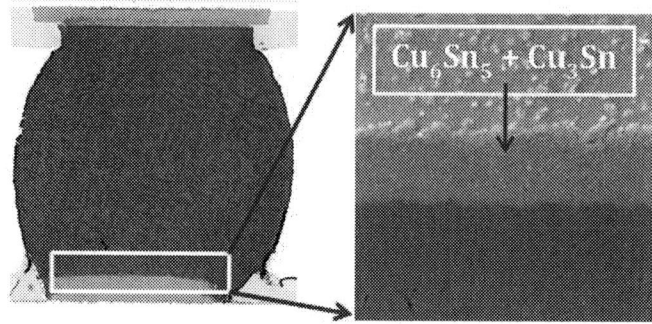

Figure 5: Image of IMC growth

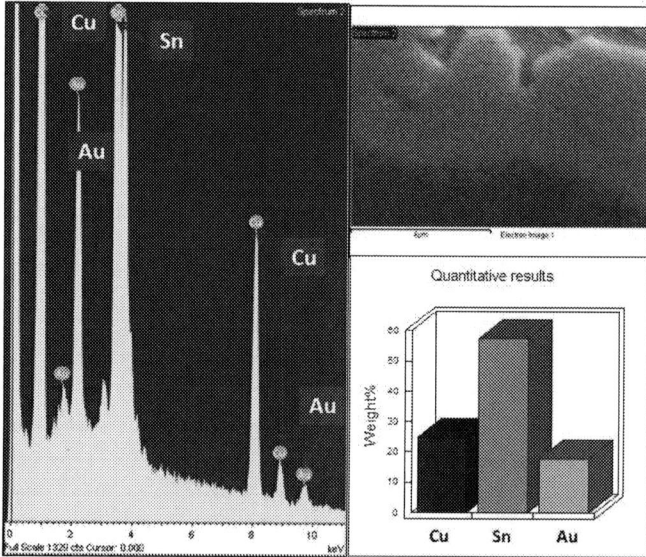

Figure 6: EDS Analysis of IMC at the solder joint interface.

The residual Au peak is because of the gold-sputter used in sample preparation. The potential use of intermetallic thickness for damage mapping between thermal cycling environments has been studied as a second leading indicator of damage progression in solder interconnects. The use of IMC thickness of damage mapping has been inspired by the prior studies which have shown IMC thickness to be a damage precursor for computation of remaining useful life [Lall 2005a, 2006c, 2006d, 2007c, 2007e, 2008c, 2008d, 2009c, 2009d]. The interfacial intermetallic layers are formed between solder and copper, and some precipitates appear near the interface of the IMCs/solder as shown in Figure 5. These intermetallic layers have been identified to consist of Cu3Sn and Cu6Sn5 phases [Lall 2005a]. Location and thickness of the intermetallic layer has been identified using EDS analysis in the SEM as shown in Figure 6. Equation (4) shows the relationship between IMC thickness versus the number of cycles:

$$y(t) = y_0 + kN^n \exp\left(\frac{-E_A}{K_B T}\right) \tag{4}$$

where $y(t)$ is IMC growth thickness during cycling, y_0 is the initial thickness of intermetallic compounds, k is the coefficient of IMC growth, N is number of cycles, E_A is the activation energy, K_B is Boltzmann's Constant (8.617×10^{-5} ev/K) and T is mean temperature of cycling range in Kelvin.

Damage Mapping Relationships for Phase Growth

For the purpose of damage mapping, the relation between phase growth and number of cycles has been normalized with respect to the initial phase size, as follows,

$$S_n = \left(\frac{g_p}{g_0}\right)^4 - 1 = a_0 N^{b_0} \tag{5}$$

$$S_n = a_1 \exp\left(\frac{-E_A}{K_B T}\right) N^{\left[b_1 \exp\left(\frac{-E_B}{K_B T}\right)\right]} \tag{6}$$

Where S_n is the normalized phase growth parameter, 'a_1' is the coefficient for phase growth, 'b_1' is the phase-growth exponent, E_A and E_B is the activation energy for phase growth coefficient and exponent respectively, K_B is the Boltzmann Constant, T is the mean temperature of cycling range in Kelvin. The normalized phase growth expression from Equation (5) has been rearranged as follows:

$$S_n = a_0 N^{b_0} \tag{7}$$

$$\ln S_n = \ln a_0 + b_0 \ln N \tag{8}$$

$$a_0 = a_1 \exp\left(\frac{-E_A}{K_B T}\right) \tag{9}$$

$$b_0 = b_1 \exp\left(\frac{-E_B}{K_B T}\right) \tag{10}$$

Where a_0 is the temperature dependent phase growth coefficient and b_0 is temperature dependent phase growth exponent. Taking a natural logarithm of Equation (9) and Equation (10), the relationship has been reduced to that of a straight line,

$$\ln a_0 = \ln a_1 - \left(\frac{E_A}{K_B T}\right) \tag{11}$$

$$\ln b_0 = \ln b_1 - \left(\frac{E_B}{K_B T}\right) \tag{12}$$

By plotting experimental values of normalized phase growth (S_n) with respect to number of cycles (N) the values of coefficient and exponent in Equation (7) have been computed. The intercept ($\ln a_0$) and slope (b_0) of Equation (11) and Equation (12) have been plotted with respect to mean temperature to calculate the Activation energy of phase growth for coefficient (E_A) and Activation energy of phase growth for exponent (E_B) respectively.

Damage Mapping Relationships for IMC Growth

The intermetallic thickness based damage proxy has been related to mean temperature of cycling range and number of cycles using the following normalized IMC thickness equation,

$$Y_n = \frac{y_p}{y_0} - 1 = k_0 N^{b_0} \tag{13}$$

$$\ln Y_n = \ln k_0 + b_0 \ln N$$

Where

$$k_0 = k_1 \exp\left(\frac{-E_A}{K_B T}\right) \tag{14}$$

$$\ln b_0 = \ln b_1 - \left(\frac{E_B}{K_B T}\right) \tag{15}$$

Here k_0 is the temperature dependent IMC growth coefficient and b_0 is temperature dependent IMC growth exponent. The experimental values of normalized IMC thickness (Y_n) have been plotted with respect to number of cycles (N) to compute the coefficient and exponent for Equation (13). The intercept ($\ln k_0$) and slope (b_0) of Equation (13) have been plotted with respect to mean temperature to calculate the Activation energy of IMC growth for coefficient (E_A) and Activation energy of IMC growth for exponent (E_B) respectively.

Leading Indicators for Thermal Cycling

Three separate sets of board assemblies A, B and C were subjected to temperature cycling at TC1: -50°C to 50°C, TC2: 0°C to 100°C and TC3: 50°C to 150°C. The cyclic magnitude of the temperature cycle is 100°C in each case but the mean temperature is 0°C for TC1, 50°C for TC2 and 100°C for TC3. The board assemblies were withdrawn after periodic time-intervals of 250 cycles. The samples were cross sectioned, potted and polished. The same joint, generally the corner joint in the die shadow region was examined in each cross-section. The two damage proxies of phase growth rate and intermetallic thickness were studied using images taken by SEM at periodic intervals (Figure 7, Figure 8, Figure 9).

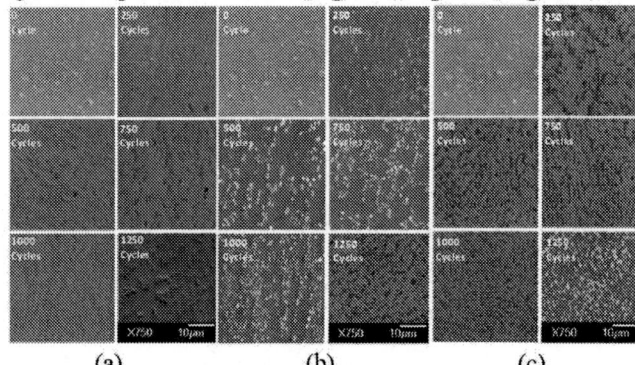

(a)　　　　　(b)　　　　　(c)

Figure 7: SEM images of Phase Growth in CABGA144 at different cycles at 750X magnification (a) for -50°C to 50°C (b) 0°C to 100°C and (c) 50°C to 150°C.

Phase-Growth Damage Proxy

Micrographs of phase structure are shown in Figure 7 for CABGA 144, in Figure 8 for PBGA 256, and in Figure 9 for

PBGA 676. The image analysis software has been used to measure the average phase size. In each case, the phase growth rate has been measured and plotted to calculate the slope of the phase growth curve versus number of thermal cycles for Equation (7).

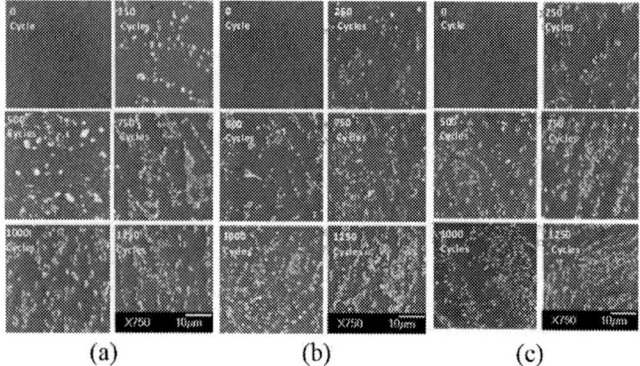

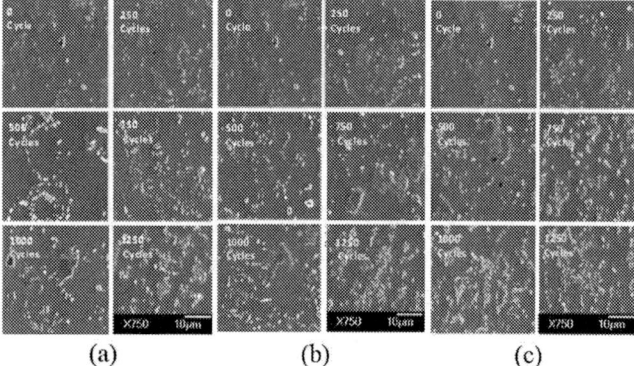

Figure 8: SEM images of Phase Growth in PBGA256 at different cycles at 750X magnification (a) for -50°C to 50°C (b) 0°C to 100°C and (c) 50°C to 150°C.

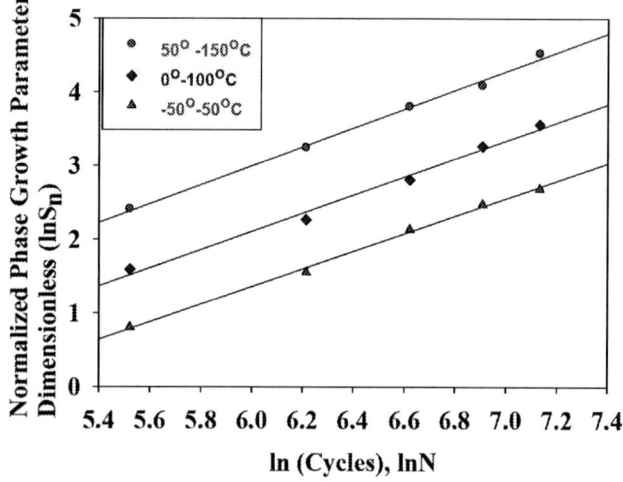

Figure 9: SEM images of Phase Growth in PBGA676 at different cycles at 750X magnification (a) for -50°C to 50°C (b) 0°C to 100°C and (c) 50°C to 150°C.

PBGA 676 respectively, which clearly reveals that phase growth is higher in smaller packages with higher value of phase growth accumulation rate ($dlnS_n/dlnN$). Both the coefficient and exponent term change with mean temperature of cycling range because the underlying agglomeration of phases proceeds at a faster pace at a higher temperature. Table 2, Table 3 and Table 4 show the value of phase growth coefficient (a_0) and phase growth exponent or phase growth accumulation rate (b_0) for CABGA144, PBGA 256 and PBGA 676 respectively. The values have been calculated from the regression fit of the experimental data. Both phase growth coefficient (a_0) and phase growth exponent or phase growth accumulation rate (b_0) show an increase with mean temperature of the thermal cycle.

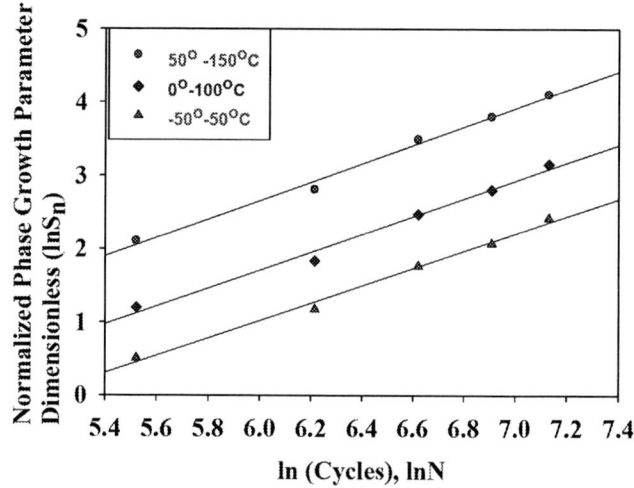

Figure 11: Relationship between Normalized Phase Growth (lnSn) and Number of Cycles (lnN) for PBGA 256 with SAC305 interconnects

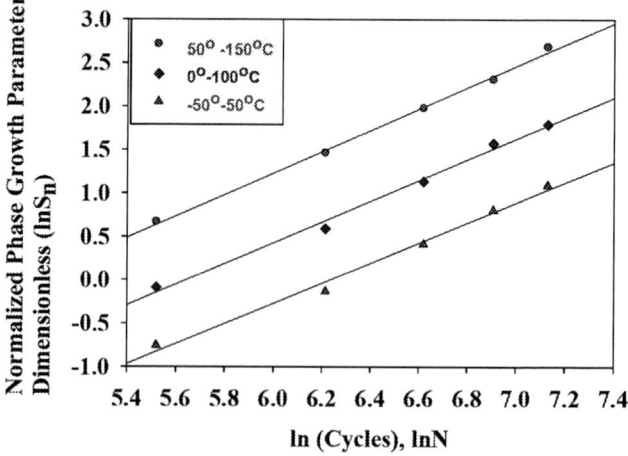

Figure 10: Relationship between Normalized Phase Growth (lnSn) and Number of Cycles (lnN) for CABGA 144 with SAC305 interconnects

Figure 12: Relationship between Normalized Phase Growth (lnSn) and Number of Cycles (lnN) for PBGA 676 with SAC305 interconnects

Figure 10, Figure 11 and Figure 12 show the plot of normalized phase growth parameter (S_n) with number of cycles (N) on a log-log scale for CABGA144, PBGA 256 and

Table 2: Normalized Phase Growth Coefficients and Exponents for the CABGA 144 Package

Temp. range	T_{mean}	$S_n = a_0 N^{b_0}$ $\ln S_n = \ln a_0 + b_0 \ln N$	
		$\ln a_0$	b_0 (d$\ln S_n$/d$\ln N$)
+50°C to +150°C	373K	-4.71	1.28
0°C to +100°C	323K	-5.3	1.23
-50°C to +50°C	273K	-5.81	1.19

Table 3: Normalized Phase Growth Coefficients and Exponents for the PBGA 256 Package

Temp. range	T_{mean}	$S_n = a_0 N^{b_0}$ $\ln S_n = \ln a_0 + b_0 \ln N$	
		$\ln a_0$	b_0 (d$\ln S_n$/d$\ln N$)
+50°C to +150°C	373K	-4.92	1.27
0°C to +100°C	323K	-5.64	1.22
-50°C to +50°C	273K	-6.1	1.18

Table 4: Normalized Phase Growth Coefficients and Exponents for the PBGA676 Package

Temp. range	T_{mean}	$S_n = a_0 N^{b_0}$ $\ln S_n = \ln a_0 + b_0 \ln N$	
		$\ln a_0$	b_0 (d$\ln S_n$/d$\ln N$)
+50°C to +150°C	373K	-6.18	1.23
0°C to +100°C	323K	-6.74	1.19
-50°C to +50°C	273K	-7.21	1.16

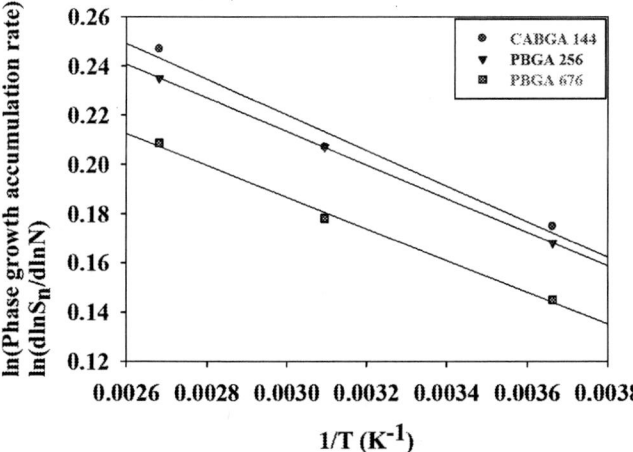

Figure 13: Plot of phase growth accumulation rate (d$\ln S_n$/d$\ln N$) vs. inverse of mean temperature of cycling range (1/T)

The activation energy of the exponent term has been computed by fitting the data in Equation (12). Figure 13 shows the relationship between $\ln(b_0)$ and (1/T), where T is the mean temperature in Kelvin scale. Since the parameter d$\ln S_n$/d$\ln N$ is proportional to damage accumulated per cycle, lower value of mean temperature corresponds to lower damage accumulated per cycle. Thus, thermal cycles with higher mean temperatures will result in higher plastic work per cycle and a lower cyclic thermal fatigue life.

Table 5: Activation energy for phase growth rate

	Activation energy for phase growth accumulation rate (E_b)
CABGA 144	6.24E-03 ev
PBGA 256	5.83E-03 ev
PBGA 676	5.56E-03 ev

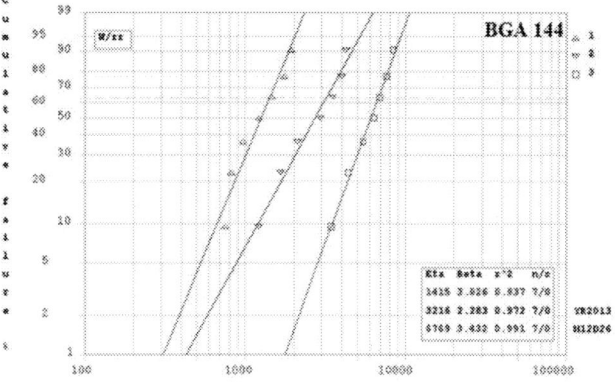

Figure 14: Weibull Plot for CABGA 144

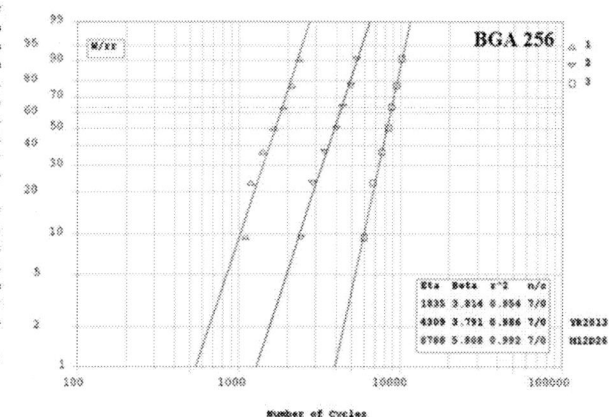

Figure 15: Weibull Plot for PBGA 256

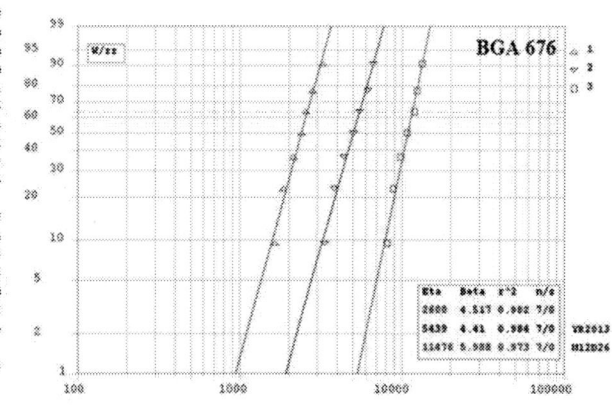

Figure 16: Weibull Plot for PBGA 676

Slope of the fit of $\ln b_0$ or $\ln(d\ln S_n/d\ln N)$ versus $(1/T)$ is E_b/K_B, where E_b is the activation energy; K_B is Boltzmann's Constant $(8.617 \times 10^{-5} \text{ ev/K})$. Table 5 shows the Activation Energy (E_b). A separate set of boards A, B and C test assemblies were subjected to the TC1, TC2 and TC3 cyclic environments and the Weibull characteristic life (63.2%) was calculated. Figure 14, Figure 15, and Figure 16 show the Weibull Plot for CABGA 144, PBGA 256 and PBGA 676 respectively.

Table 6: Characteristic life

	CABGA 144 (Cycles)	PBGA 256 (Cycles)	PBGA 676 (Cycles)
+50°C to +150°C	1431	1835	2600
0°C to +100°C	3216	4309	5439
-50°C to +50°C	6769	8788	11476

Table 6 contains the Characteristic life of all the test vehicles we used for three different cyclic environments. The Phase Growth Accumulation Rate $(d\ln S_n/d\ln N)$ vs Characteristic life has been plotted as shown in Figure 17 to derive the relationship between solder joint accrued damage (considering Phase Growth Accumulation Rate as a proxy to solder joint accrued damage) and Characteristic life (N_f).

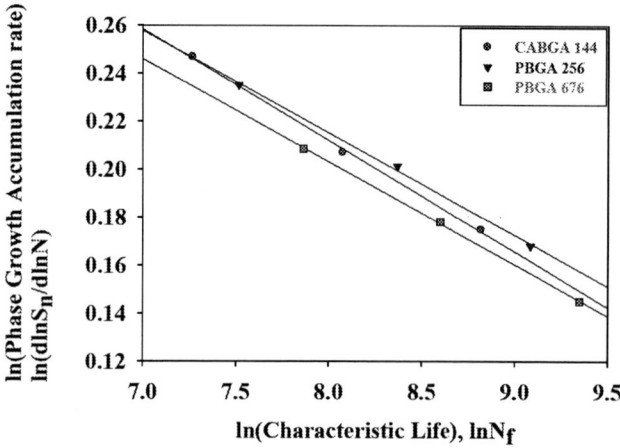

Figure 17: Plot of phase growth accumulation rate (dlnSn/dlnN) vs. Characteristic Life (Nf)

Table 7: Relation between Phase Growth Accumulation Rate and Characteristic life

Package	Damage Relationship
CABGA 144	$N_f = 288447 \, (d\ln S_n/d\ln N)^{-21.55}$
PBGA 256	$N_f = 461337.5 \, (d\ln S_n/d\ln N)^{-23.42}$
PBGA 676	$N_f = 340379 \, (d\ln S_n/d\ln N)^{-23.31}$

The relation between Characteristic Life and Phase growth Accumulation Rate as derived from the data fit are shown in Table 7. In Figure 18, the measurements of the Phase Growth Accumulation Rate $(d\ln S_n/d\ln N)$ from CABGA 144, PBGA 256 and PBGA 676 have been plotted versus the characteristic life to develop a predictive model.

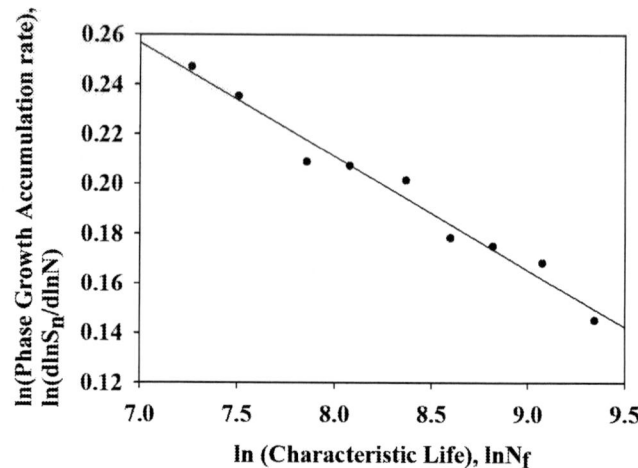

Figure 18: Predictive model using Phase Growth as a damage proxy

The equation derived from the least square fit of the Phase Growth Accumulation Rate $(d\ln S_n/d\ln N)$ versus characteristic life of all tested board assemblies under the thermal cycles of TC1, TC2, TC3 is shown in Figure 18 is as follows;

$$\ln\left(\frac{d\ln S_n}{d\ln N_f}\right) = -0.0457 \ln N_f + 0.5766 \qquad (16)$$

Equation (16) has then been used in predictive mode to assess the error of the model predictions with experimentally measured life distributions. The Characteristic Life for all three test vehicles has been calculated using Equation (16). The results using this predictive model along with results derived experimentally are shown in Table 8.

Table 8: Comparison of experimental results and results from predictive model using Phase Growth as a damage proxy

	CABGA 144		
	Exp.	Predict.	Error (%)
+50°C to +150°C	1431	1360	4.95
0°C to +100°C	3216	3253	-1.14
-50°C to +50°C	6769	6583	2.75
	PBGA 256		
	Exp.	Predict.	Error (%)
+50°C to +150°C	1835	1761	4.07
0°C to +100°C	4309	3685	14.47
-50°C to +50°C	8788	7629	13.19
	PBGA 676		
	Exp.	Predict.	Error (%)
+50°C to +150°C	2600	3139	-20.73
0°C to +100°C	5439	6118	-12.47
-50°C to +50°C	11476	12644	-10.18

Inter-Metallic Thickness Growth Damage Proxy

Equation (13) represents the evolution of IMC thickness growth in thermal cycling based on experimental data. The test data has been used to derive the parameters for normalized IMC growth of Equation (13). Micrographs of IMC thickness are shown in Figure 19 for CABGA 144, in

Figure 20 for PBGA 256, and in Figure 21 for PBGA 676. Figure 22, Figure 23 and Figure 24 show the plot of normalized IMC growth parameter (Y_n) with number of cycles (N) in log-log scale for CABGA144, PBGA 256 and PBGA 676 respectively, which clearly reveals that IMC growth is higher in smaller packages and at a higher IMC growth accumulation rate ($d \ln Y_n / d \ln N$). The larger ball-grid array packages (e.g. PBGA676) in this study have larger solder balls compared to smaller ball-grid array packages (e.g. CABGA144), which gives larger package a higher ability to resist shear strain under similar temperature loading condition because of the higher package stand-off and the higher solder volume. Both the coefficient and exponent term change with mean temperature of cycling range because the underlying agglomeration of IMC proceeds at a faster pace at a higher temperature. Table 9, Table 10 and Table 11 show the value of IMC growth coefficient (k_0) and IMC growth exponent or IMC accumulation rate (b_0) for CABGA144, PBGA 256 and PBGA 676 respectively. Again these three tables clearly show that both IMC growth coefficient (k_0) and IMC growth exponent or IMC accumulation rate (b_0) are increasing with mean temperature.

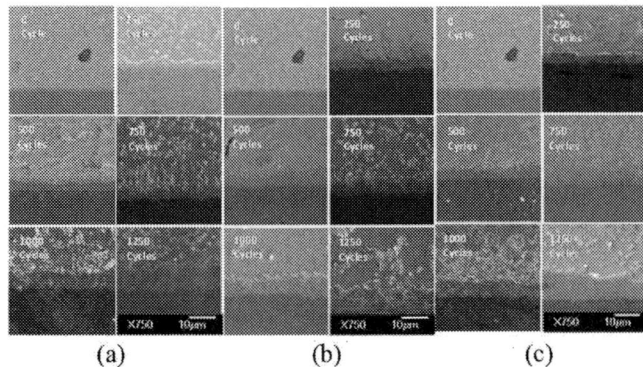

(a)　　　　(b)　　　　(c)

Figure 19:SEM images of IMC Growth in CABGA144 at different cycles at 750X magnification (a) for -50°C to 50°C (b) 0°C to 100°C and (c) 50°C to 150°C.

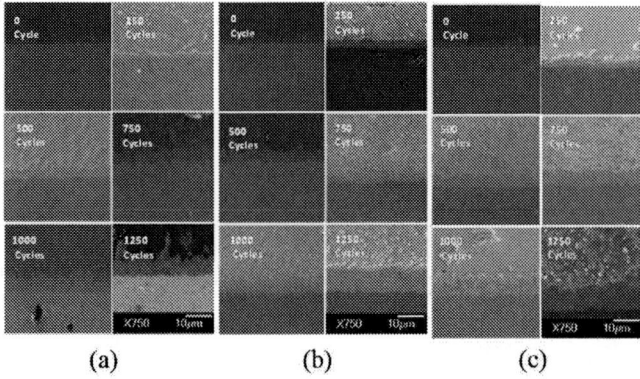

(a)　　　　(b)　　　　(c)

Figure 20:SEM images of IMC Growth in PBGA256 at different cycles at 750X magnification (a) for -50°C to 50°C (b) 0°C to 100°C and (c) 50°C to 150°C.

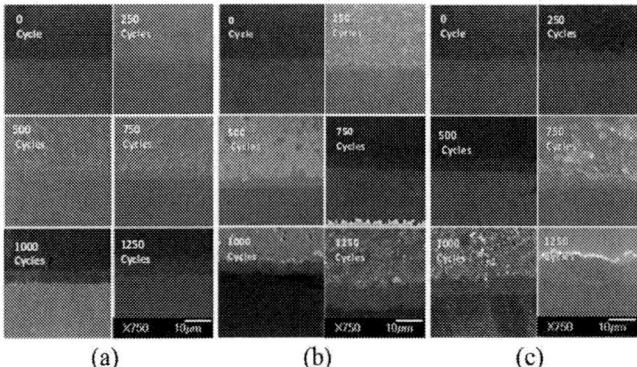

(a)　　　　(b)　　　　(c)

Figure 21:SEM images of IMC Growth in PBGA676 at different cycles at 750X magnification (a) for -50°C to 50°C (b) 0°C to 100°C and (c) 50°C to 150°C.

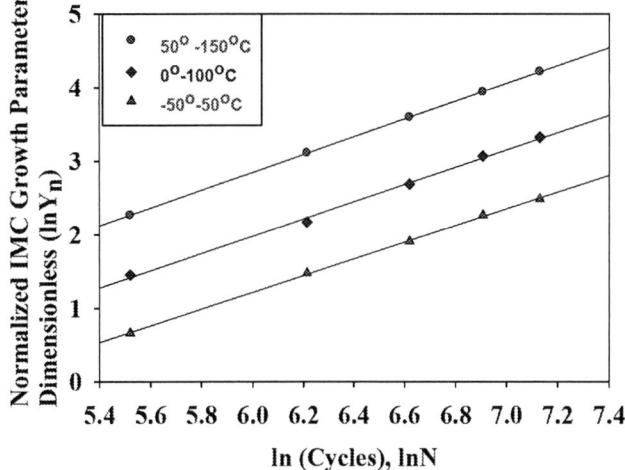

Figure 22: Relation between Normalized IMC Growth (lnYn) and Number of Cycles (lnN) for CABGA 144.

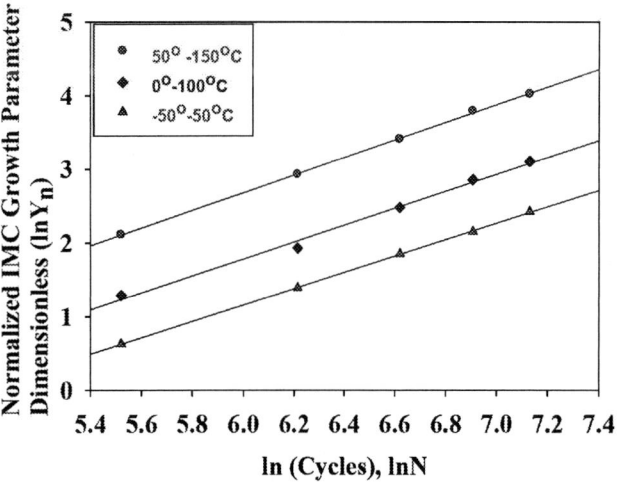

Figure 23: Relation between Normalized IMC Growth (lnYn) and Number of Cycles (lnN) for PBGA 256.

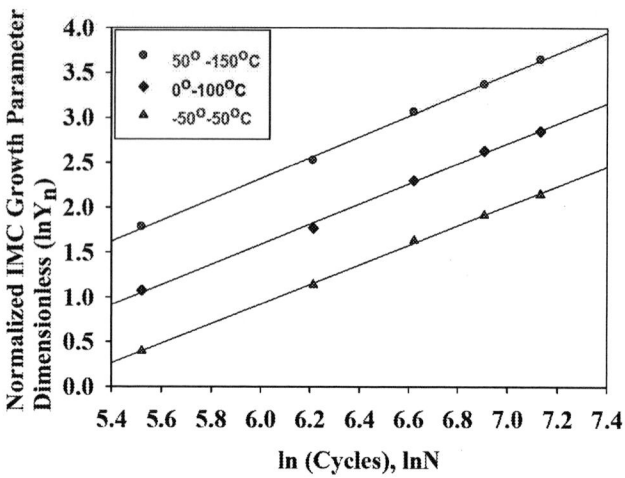

Figure 24: Relation between Normalized IMC Growth (lnYn) and Number of Cycles (lnN) for PBGA 676.

Table 9: Normalized IMC Growth Coefficients and Exponents for the CABGA 144 Package

Temp. range	T_{mean}	$Y_n = k_0 N^{b_0}$ $\ln Y_n = \ln k_0 + b_0 \ln N$	
		$\ln k_0$	$b_0 (d\ln Y_n/d\ln N)$
+50°C to +150°C	373K	-4.41	1.21
0°C to +100°C	323K	-5.07	1.173
-50°C to +50°C	273K	-5.6	1.14

Table 10: Normalized IMC Growth Coefficients and Exponents for the PBGA 256 Package

Temp. range	T_{mean}	$Y_n = k_0 N^{b_0}$ $\ln Y_n = \ln k_0 + b_0 \ln N$	
		$\ln k_0$	$b_0 (d\ln Y_n/d\ln N)$
+50°C to +150°C	373K	-4.49	1.19
0°C to +100°C	323K	-5.11	1.147
-50°C to +50°C	273K	-5.5	1.11

Table 11: Normalized IMC Growth Coefficients and Exponents for the PBGA 676 Package

Temp. range	T_{mean}	$Y_n = k_0 N^{b_0}$ $\ln Y_n = \ln k_0 + b_0 \ln N$	
		$\ln k_0$	$b_0 (d\ln Y_n/d\ln N)$
+50°C to +150°C	373K	-4.66	1.164
0°C to +100°C	323K	-5.16	1.12
-50°C to +50°C	273K	-5.65	1.09

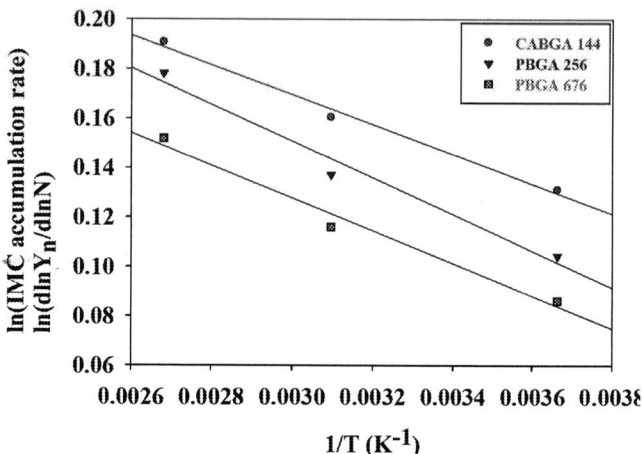

Figure 25: Plot of IMC accumulation rate ($d\ln Y_n/d\ln N$) vs. inverse of mean temperature of cycling range (1/T)

The activation energy for the exponent term has been computed by fitting the data to Equation (15). Figure 25 shows the relationship between $\ln(b_0)$ and $(1/T)$, where T is the mean temperature in Kelvin scale. Table 12 shows the computed values of Activation Energy for IMC accumulation rate for the test board assemblies.

Table 12: Activation energy for IMC growth rate per cycle

	Activation energy for IMC accumulation rate (E_b)
CABGA 144	5.19E-03 ev
PBGA 256	6.4E-03 ev
PBGA 676	5.69E-03 ev

Slope of the fit of $\ln b_0$ or $\ln(d\ln Y_n/d\ln N)$ versus $(1/T)$ is E_b/K_B, where E_b is the activation energy; K_B is Boltzmann's Constant (8.617×10^{-5} ev/K). The IMC Accumulation Rate ($d\ln Y_n/d\ln N$) vs Characteristic life (Table 6) has been plotted in Figure 26 to get a relation between solder joint accrued damage (considering IMC Accumulation Rate as a proxy to solder joint accrued damage) and Characteristic life (N_f).

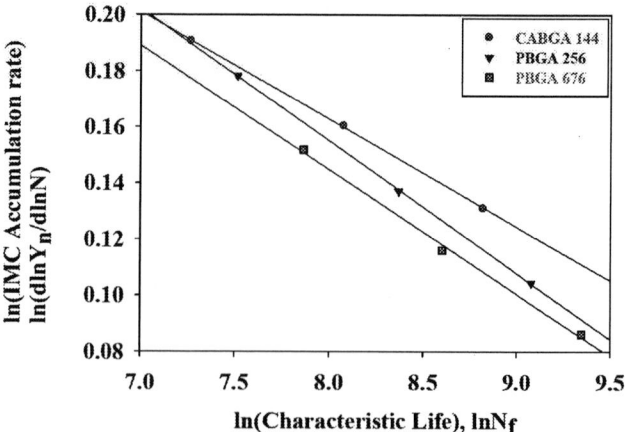

Figure 26: Plot of IMC accumulation rate ($d\ln Y_n/d\ln N$) vs. Characteristic Life (N_f)

Table 13: Relation between IMC Accumulation Rate and Characteristic life

Package	Damage Accumulation Rate
CABGA 144	$N_f = 213586 \, (dlnY_n/dlnN)^{-26.11}$
PBGA 256	$N_f = 55911.6 \, (dlnY_n/dlnN)^{-18.97}$
PBGA 676	$N_f = 80218 \, (dlnY_n/dlnN)^{-22.62}$

The relation between Characteristic Life and IMC Accumulation Rate as derived from the least squares of test data fit is shown in Table 13. In Figure 27, all data on IMC Accumulation Rate (dlnY$_n$/dlnN) from CABGA 144, PBGA 256 and PBGA 676 has been plotted in one single plot to develop a predictive model.

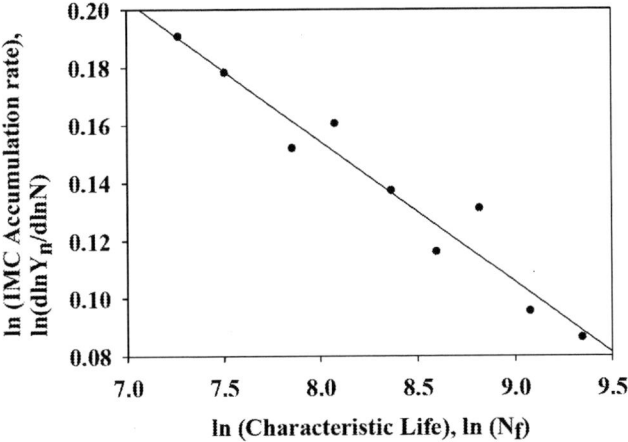

Figure 27: Predicting model using IMC as a damage proxy

Table 14: Comparison of experimental results and results from predictive model using IMC as a damage proxy

	CABGA 144		
	Exp.	Predict.	Error (%)
50°-150°C	1431	1422	2.4
0°-100°C	3216	2641	17.55
-50°-50°C	6769	4822	26.89
	PBGA 256		
	Exp.	Predict.	Error (%)
50°-150°C	1835	1836	0.8314585
0°-100°C	4309	4254	-0.8435428
-50°-50°C	8788	10026	0.8073895
	PBGA 676		
	Exp.	Predict.	Error (%)
50°-150°C	2600	3146.5	-22.2902
0°-100°C	5439	6560.74	-25.1664
-50°-50°C	11476	12088.95	-11.7461

The equation derived from the least square fit of the IMC Accumulation Rate (dlnY$_n$/dlnN) versus characteristic life of all tested board assemblies under the thermal cycles of TC1, TC2, TC3, shown in Figure 27 is as follows:

$$\ln\left(\frac{d\ln Y_n}{d\ln N_f}\right) = -0.047 \ln N_f + 0.532 \qquad (17)$$

Equation (17) has been used in predictive mode to assess the error of the model predictions with experimentally measured life distributions. The Characteristic Life for all three test vehicles was calculated using Equation (17). The results using this predictive model along with results derived experimentally are shown in Table 14.

Damage Equivalency Relationships

Damage accrued from the two different thermal cycling environments has been equivalenced based on two damage proxies including normalized intermetallic thickness and normalized phase growth to get Damage Equivalency Relationships shown in Table 16, Table 18 and Table 20.Table 15, Table 17 and Table 19 show the damage mapping relationship for CABGA 144, PBGA 256 and PBGA 676 respectively obtained by plotting Normalized Phase Growth or Normalized IMC Growth parameter with respect to number of cycles. The damage mapping relationships for any two test conditions have been equivalenced to get damage equivalency relationships to enable computation of the number of cycles or equivalent time required under any two test condition to reach a nearly state of accrued damage.

Table 15: Damage mapping relationship for CABGA 144

	Phase Growth	IMC
+50°C to 150°C (TC3)	$S_{nTC3}=$ $0.009N_{TC3}^{1.28}$	$Y_{nTC3}=$ $0.012N_{TC3}^{1.21}$
0°C to +100°C (TC2)	$S_{nTC2}=$ $0.005N_{TC2}^{1.23}$	$Y_{nTC2}=$ $0.0063N_{TC2}^{1.174}$
-50°C to +50°C (TC1)	$S_{nTC1}=$ $0.003N_{TC1}^{1.191}$	$Y_{nTC1}=$ $0.0037N_{TC1}^{1.14}$

Table 16: Damage Equivalency relationship for CABGA 144

Phase Growth	IMC
$N_{TC3}=0.63N_{TC2}^{0.96}$	$N_{TC3}=0.59N_{TC2}^{0.97}$
$N_{TC2}=0.66N_{TC1}^{0.968}$	$N_{TC2}=0.63N_{TC1}^{0.97}$
$N_{TC3}=0.42N_{TC1}^{0.93}$	$N_{TC3}=0.38N_{TC1}^{0.942}$

Table 17: Damage mapping relationship for PBGA 256

	Phase Growth	IMC
+50°C to 150°C (TC3)	$S_{nTC3}=$ $0.0073N_{TC3}^{1.265}$	$Y_{nTC3}=$ $0.011N_{TC3}^{1.195}$
0°C to +100°C (TC2)	$S_{nTC2}=$ $0.00355N_{TC2}^{1.223}$	$Y_{nTC2}=$ $0.00603N_{TC2}^{1.147}$
-50°C to +50°C (TC1)	$S_{nTC1}=$ $0.00225N_{TC1}^{1.183}$	$Y_{nTC1}=$ $0.00401N_{TC1}^{1.1}$

Table 18: Damage Equivalency relationship for PBGA 256

Phase Growth	IMC
$N_{TC3}=0.565N_{TC2}^{0.967}$	$N_{TC3}=0.6N_{TC2}^{0.96}$
$N_{TC2}=0.69N_{TC1}^{0.967}$	$N_{TC2}=0.7N_{TC1}^{0.96}$
$N_{TC3}=0.39N_{TC1}^{0.935}$	$N_{TC3}=0.43N_{TC1}^{0.92}$

Table 19: Damage mapping relationship for PBGA 676

	Phase Growth	IMC
+50°C to 150°C (TC3)	$S_{nTC3}=$ $0.00207N_{TC3}^{1.232}$	$Y_{nTC3}=$ $0.00946N_{TC3}^{1.164}$
0°C to +100°C (TC2)	$S_{nTC2}=$ $0.00117N_{TC2}^{1.195}$	$Y_{nTC2}=$ $0.00574N_{TC2}^{1.12}$
-50°C to +50°C (TC1)	$S_{nTC1}=$ $0.00074N_{TC1}^{1.156}$	$Y_{nTC1}=$ $0.00352N_{TC1}^{1.09}$

Table 20: Damage Equivalency relationship for PBGA 676

Phase Growth	IMC
$N_{TC3}=0.63N_{TC2}^{0.97}$	$N_{TC3}=0.65N_{TC2}^{0.965}$
$N_{TC2}=0.68N_{TC1}^{0.967}$	$N_{TC2}=0.65N_{TC1}^{0.971}$
$N_{TC3}=0.43N_{TC1}^{0.938}$	$N_{TC3}=0.43N_{TC1}^{0.936}$

Since, both phase growth rate and IMC growth rate have been shown to be valid damage proxies; therefore the damage equivalency relationships obtained using both these damage proxies should be similar for any component under a pair of equivalenced test conditions. Table 16, Table 18 and Table 20 reveals the convergence of the damage equivalency relationships from two separate damage proxies which supports the validity of the correlation.

Prognostication of Accrued Damage

The damage equivalency relationships developed in this paper have been used to prognosticate the accrued damage in assemblies that have been subjected to thermal cycling and determine the expected performance of the test assemblies in thermal cycle conditions prior to development of complete failure distributions. In addition, the ability of the prognostic indicators to predict performance of the test assemblies in thermal cycle conditions even not tested, has been assessed.

Phase-Growth Rate as Damage Proxy

In order to prognosticate the accrued damage, test assemblies have been withdrawn at periodic intervals of 250 cycles to measure the phase growth and intermetallic growth. The withdrawn samples have been cross-sectioned. The phase growth rate and intermetallic thickness growth rate has been studied under SEM and measured by using NIMAQ software. The measured values of phase growth and intermetallic thickness have been used as data input in Levenberg-Marquardt Algorithm for prognostication. For prognostication using Phase growth as a damage proxy, a simplified version of Equation (1) has been used, as shown in Equation (18). In this equation g_0 (Initial Phase Growth), a (Phase Growth Coefficient), b (Phase Growth Exponent) and N (Number of Cycles) are four unknowns; therefore four equations are needed for solution as shown in Equation (18) to (21).

$$g_p = \sqrt[4]{g_0^4 + aN^b} \tag{18}$$

$$g_{p+\Delta N} = \sqrt[4]{g_0^4 + a(N + \Delta N)^b} \tag{19}$$

$$g_{p+2\Delta N} = \sqrt[4]{g_0^4 + a(N + 2\Delta N)^b} \tag{20}$$

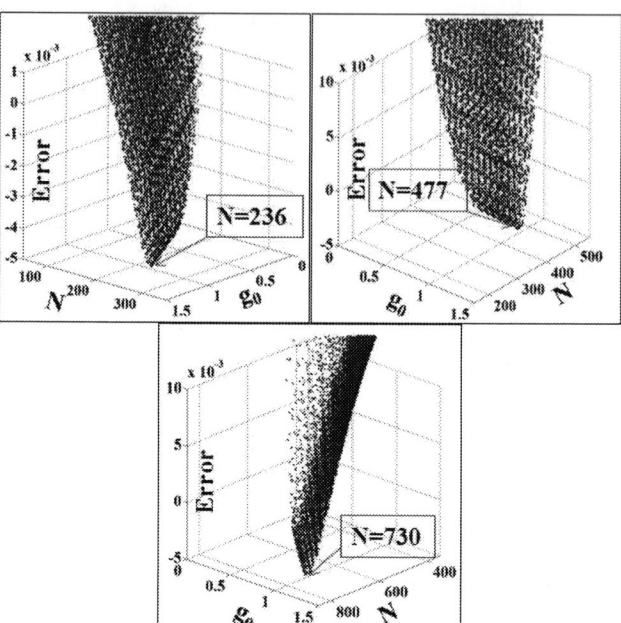

Figure 28: 3D plot of Error vs Number of Cycles and Normalized Phase Growth for CABGA 144 for temperature cycle +50°C to +150°C

Table 21: Comparison of results obtained from experiment and prognostication for +50°C to +150°C

Phase Growth	
N (Exp.)	N(Prog.)
250	236
500	521
750	730

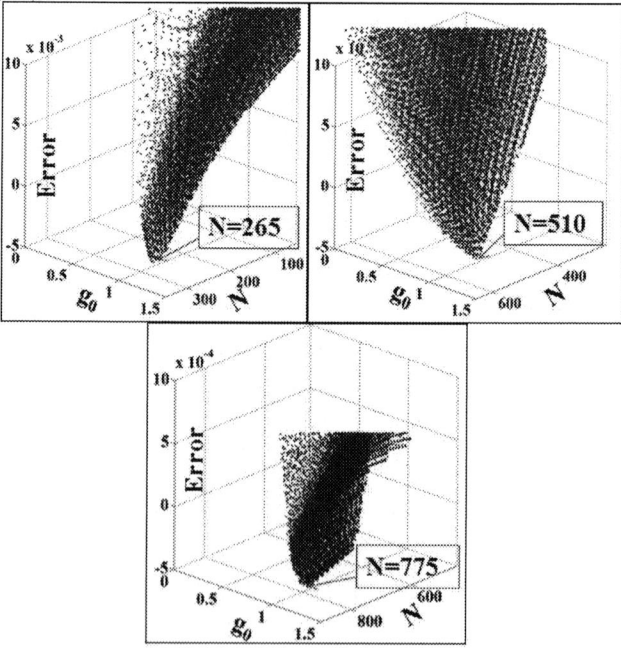

Figure 29: 3D plot of Error vs Number of Cycles and Normalized Phase Growth for CABGA 144 for temperature cycle 0°C to +100°C

$$g_{p+3\Delta N} = \sqrt[4]{g_0^4 + a(N+3\Delta N)^b} \tag{21}$$

Where $\Delta N = 250$ cycles. Figure 28 and Table 21 show the prognosticated values of prior accrued damage using phase growth rate as a damage indicator for the CABGA144 test assemblies exposed to thermal cycle of +50°C to +150°C. Board assemblies have been prognosticated after exposure to 250 cycles, 500 cycles and 750 cycles. Prognosticated values of damage are 236 cycles, 477 cycles, and 730 cycles which correlate well with the actual accrued damage.

Table 22: Comparison of results obtained from experiment and prognostication for 0°C to +100°C

Phase Growth	
N (Exp.)	N(Prog.)
250	265
500	510
750	775

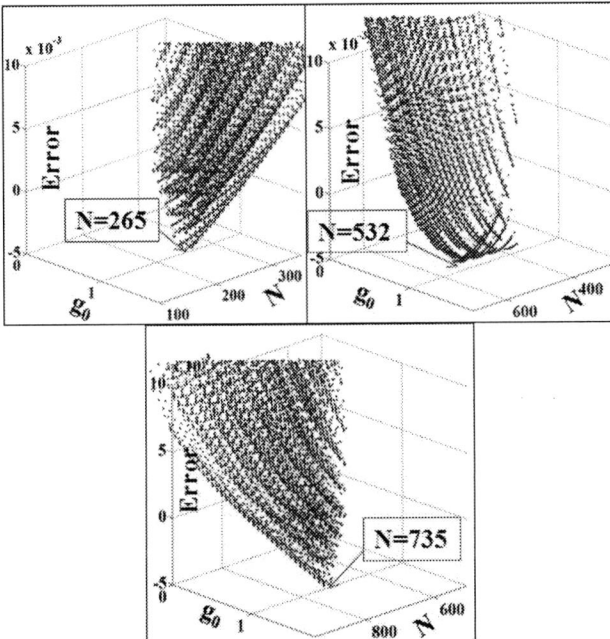

Figure 30: 3D plot of Error vs No. of Cycles and Normalized Phase Growth for CABGA 144 for temperature cycle -50°C to +50°C

Table 23: Comparison of results obtained from experiment and prognostication for -50°C to +50°C

Phase Growth	
N (Exp.)	N(Prog.)
250	265
500	532
750	735

Figure 29 and Table 22 shows the prognosticated values of prior accrued damage using phase growth rate as a damage indicator for the CABGA144 test assemblies exposed to thermal cycle of 0°C to +100°C. Board assemblies have been prognosticated after exposure to 250 cycles, 500 cycles and

750 cycles. Prognosticated values of damage are 265 cycles, 510 cycles, and 775 cycles which correlate well with the actual accrued damage. Figure 30 and shows the prognosticated values of prior accrued damage using phase growth rate as a damage indicator for the CABGA144 test assemblies exposed to thermal cycle of -50°C to +50°C. Board assemblies have been prognosticated after exposure to 250 cycles, 500 cycles and 750 cycles. Prognosticated values of damage are 265 cycles, 532 cycles, and 735 cycles which correlate well with the actual accrued damage.

IMC Thickness as Damage Proxy

For prognostication of prior accrued damage using IMC growth as a damage proxy, Equation (22)which is obtained after simplifying Equation (13) has been used. In this equation y_0 (Initial IMC Growth), k (IMC Growth Coefficient), b (IMC Growth Exponent) and N (Number of Cycles) are four unknowns; therefore four equations are needed for solution as shown in Equation (22) to (25).

$$y_p = y_0 + kN^b \tag{22}$$

$$y_{p+\Delta N} = y_0 + k(N+\Delta N)^b \tag{23}$$

$$y_{p+2\Delta N} = y_0 + k(N+2\Delta N)^b \tag{24}$$

$$y_{p+3\Delta N} = y_0 + k(N+3\Delta N)^b \tag{25}$$

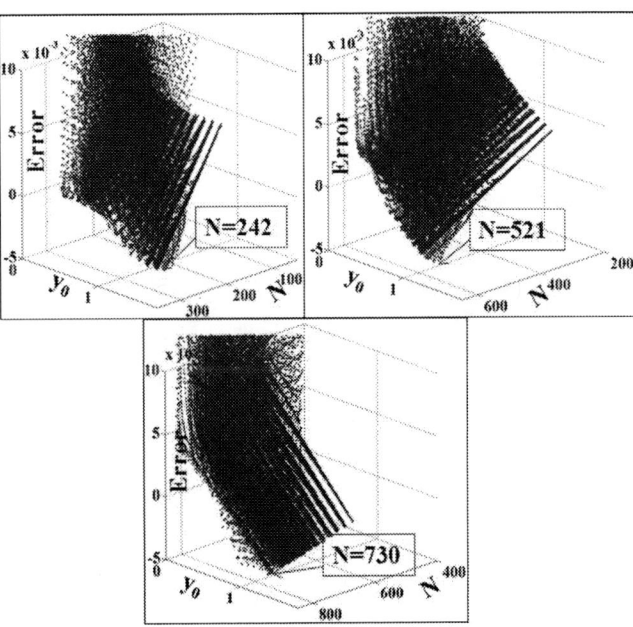

Figure 31: 3D plot of Error vs No. of Cycles and Normalized IMC Growth for CABGA 144 for temperature cycle +50°C to +150°C

Table 24: Comparison of results obtained from experiment and prognostication for +50°C to +150°C

IMC growth	
N (Exp.)	N(Prog.)
250	242
500	521
750	730

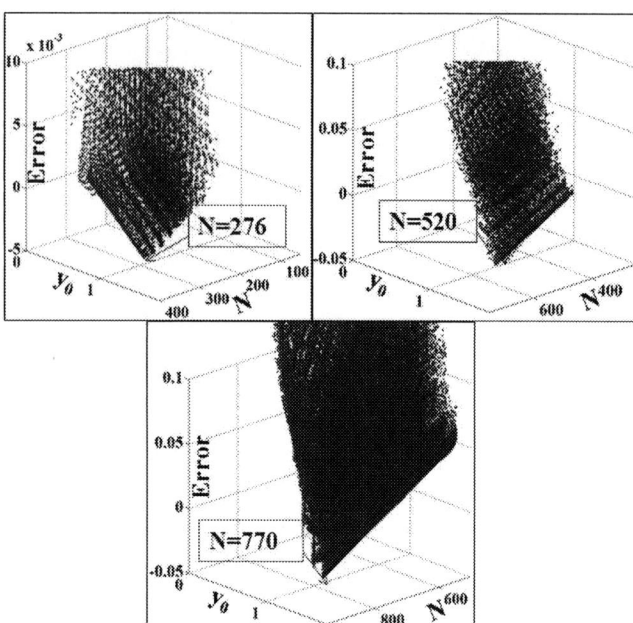

Figure 32: 3D plot of Error vs No. of Cycles and Normalized IMC Growth for CABGA 144 for temperature cycle 0°C to +100°C

Table 25: Comparison of results obtained from experiment and prognostication for 0°C to +100°C

IMC growth	
N (Exp.)	N(Prog.)
250	276
500	520
750	770

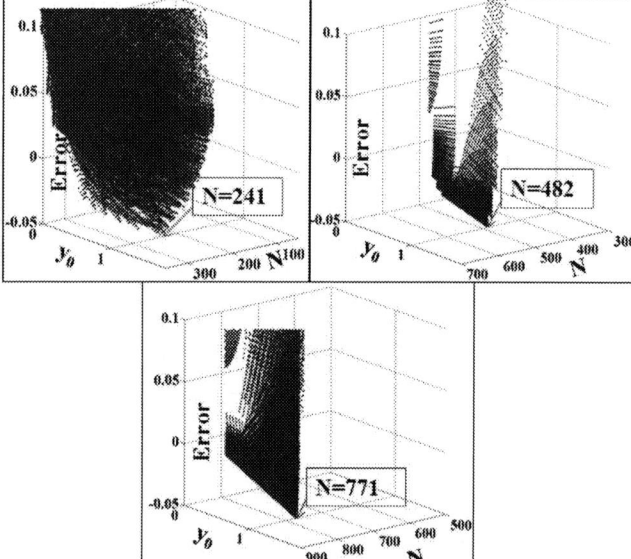

Figure 33: 3D plot of Error vs No. of Cycles and Normalized IMC Growth for CABGA 144 for temperature cycle -50°C to +50°C

Figure 31 and Table 24 show the prognosticated values of prior accrued damage using IMC growth rate as a damage indicator for the CABGA144 test assemblies exposed to

thermal cycle of +50°C to +150°C. Board assemblies have been prognosticated after exposure to 250 cycles, 500 cycles and 750 cycles. Prognosticated values of damage are 242 cycles, 521 cycles, and 730 cycles which correlate well with the actual accrued damage.

Table 26: Comparison of results obtained from experiment and prognostication for -50°C to +50°C

IMC growth	
N (Exp.)	N(Prog.)
250	241
500	482
750	771

Figure 32 and Table 25 show the prognosticated values of prior accrued damage using IMC growth rate as a damage indicator for the CABGA144 test assemblies exposed to thermal cycle of 0°C to +100°C. Board assemblies have been prognosticated after exposure to 250 cycles, 500 cycles and 750 cycles. Prognosticated values of damage are 276 cycles, 520 cycles, and 770 cycles which correlate well with the actual accrued damage. Figure 33 and Table 26 show the prognosticated values of prior accrued damage using IMC growth rate as a damage indicator for the CABGA144 test assemblies exposed to thermal cycle of -50°C to +50°C. Board assemblies have been prognosticated after exposure to 250 cycles, 500 cycles and 750 cycles. Prognosticated values of damage are 241 cycles, 482 cycles, and 771 cycles which correlate well with the actual accrued damage.

Conclusions

Using three different packages, effect of mean temperature of temperature cycling on package reliability was investigated. Thermal cycle conditions examined in the study include, -50°C to +50°C, 0°C to 100°C, 50°C to 150°C. A damage mapping method has been developed based on the underlying physics-based leading indicators to relate the damage accrual rate to life consumed under three different temperature cyclic environments. A Predictive Model has also been developed using the failure data of all three packages analyzed. Convergence of both Damage Mapping relationships and Predictive Model using two damage proxies and convergence of Prognosticated Values and Experimental Values of damage indicators established the validity of our approach. Experimental results indicate that the higher mean temperature reduces the thermal fatigue life even if the cyclic temperature range of the thermal excursion is kept constant. Damage equivalency relationships derived based on damage proxies of normalized phase growth parameter and normalized IMC thickness can be used for damage mapping between thermal cycle conditions.

Acknowledgments

The research results presented in this paper are based on projects supported by industrial members of the NSF-CAVE3 Electronics Research Center at Auburn University.

References

Allen, D., Probabilities Associated with a Built-in-Test System, Focus on False Alarms, Proceedings of IEEE AUTOTESTCON, pp. 643-645, Sept 22-25, 2003.

Anderson, N., and Wilcoxon, R., Framework for Prognostics of Electronic Systems, Proceedings of International Military and Aerospace Avionics COTS Conference, Seattle, WA, Aug 3-5, 2004.

Chou, G. J. S., "Microstructure Evolution of SnPb and SnAgCu BGA Solder Joints During Thermal Aging," Proceedings of the 8th Symposium on Advanced Packaging Materials, pp. 39-46, 2002.

Drees, R., and Young, N., Role of BIT in Support System Maintenance and Availability, IEEE A&E Systems Magazine, pp. 3-7, August 2004.

Gao, R. X., Suryavanshi, A., BIT for Intelligent System Design and Condition Monitoring, IEEE Transactions on Instrumentation and Measurement, Vol. 51, Issue: 5, pp. 1061-1067, October 2002.

Hasegawa, K., Noudou, T., Takahashi, A., and Nakaso, A., "Thermal Aging Reliability of Solder Ball Joint for Semiconductor Package Substrate," Proceedings of the 2001 SMTA International, pp.1-8, 2001.

Hassan, A., Agarwal, V. K., Nadeau-Dostie, B., Rajski, J., BIST of PCB Interconnects Using Boundary- Scan Architecture, IEEE Transactions on Computer-Aided Design, Vol. 11, No. 10, pp. 1278-1288, October 1992.

Jarrell, D., Sisk, D., Bond, L., Prognostics and Condition Based Maintenance (CBM) A Scientific Crystal Ball, Pacific Northwest National Laboratory, Richland, WA, International Congress on Advanced Nuclear Power Plants (ICAPP), paper number 194 June 2002.

Lall, P., Islam, N., Suhling, J., Prognostication and Health Monitoring of Leaded and Lead Free Electronic and MEMS Packages in Harsh Environments, Proceedings of the 55th ECTC, pp. 1305-1313, Orlando, FL, 2005.

Lall, P., Harsha, M., Kumar, K., Goebel, K., Jones, J., Suhling, J., Interrogation of Accrued Damage and Remaining Life in Field-Deployed Electronics Subjected to Multiple Thermal Environments of Thermal Aging and Thermal Cycling, Proceedings of the 61st ECTC, pp.775-789, 2011a.

Lall, P., Lowe, R., Goebel, K., Keynote Presentation: Prognostics and Health Monitoring of Electronic Systems, IEEE 12th. Int. Conf. on Thermal, Mechanical and Multiphysics Simulation and Experiments in Microelectronics and Microsystems, EuroSimE, Linz, Austria, pp. 1-17, 2011b.

Lall, P., Harsha, M., Suhling, J., Goebel, K., Sustained Damage and Remaining Useful Life Assessment in Leadfree Electronics Subjected to Sequential Multiple Thermal Environments, Proceedings of the 62nd ECTC, pp.1695-1708, May 29-June 1, 2012a.

Lall, P., Lowe, R., Goebel, K., Prognostication of Accrued Damage in Board Assemblies Under Thermal and Mechanical Stresses, Proceedings of the 62nd ECTC, pp.1475 - 1487 , May 29-June 1, 2012b.

Lourakis, M., I., A., A brief Description of the Levenberg-Marquardt algorithm implemented by Levmar, Foundation of Research and Technology – Hellas (Forth), Greece, pp. 1- 6, Feb 11, 2005.

McCann, R. S., L. Spirkovska, Human Factors of Integrated Systems Health Management on Next-Generation Spacecraft, First International Forum on Integrated System Health Engineering and Management in Aerospace, Napa, CA, pp. 1-18, November 7-10, 2005.

Madsen, K., Nielsen, H., B., Tingleff, O., Methods for Non-Linear Least Squares Problems, Technical University of Denmark, Lecture notes, available at http://www.imm.dtu.dk/courses/02611/nllsq.pdf, 2nd Edition, pp. 1-30, 2004.

Marko, K.A., J.V. James, T.M. Feldkamp, C.V. Puskorius, J.A. Feldkamp, and D. Roller, Applications of Neural Networks to the Construction of "Virtual" Sensors and Model-Based Diagnostics, Proceedings of ISATA 29th International Symposium on Automotive Technology and Automation, pp.133-138, June 3-6, 1996.

Mishra, S., Pecht, M., In-situ Sensors for Product Reliability Monitoring, Proceedings of SPIE, vol. 4755, pp. 10-19, 2002.

Nielsen, H., B, Damping Parameter in Marquardt's Method, Technical Report, IMM-REP-1999-05, Technical University of Denmark, Available at http://www.imm.dtu.dk/˜hbn, pp. 1-16, 1999.

Rosenthal, D., and Wadell, B., Predicting and Eliminating Built-in Test False Alarms, IEEE Transactions on Reliability, Vol. 39, No 4, pp. 500-505, October 1990.

Saxena, A., J. Celaya, B. Saha, S. Saha, and K. Goebel, Evaluating Algorithm Performance Metrics Tailored for Prognostics, IEEE Aerospace Conference, Big Sky, MT, pp. 1-11, March 2008a.

Saxena, A., J. Celaya, E. Balaban, K. Goebel, B. Saha, S Saha, and M. Schwabacher, Metrics for Evaluating Performance of Prognostic Techniques, Intl. Conf. on Prognostics and Health Management, Denver, Colorado, pp. 1-17, October 2008b.

Schauz, J. R., Wavelet Neural Networks for EEG Modeling and Classification, PhD Thesis, Georgia Institute of Technology, 1996.

Shiroishi, J., Y. Li, S. Liang, T. Kurfess, and S. Danyluk, Bearing Condition Diagnostics via Vibration and Acoustic Emission Measurements, Mechanical Systems and Signal Processing, Vol.11, No.5, pp.693-705, Sept. 1997.

Williams, T. W., Parker, K. P., Design for Testability- Survey, Proceedings of the IEEE, Vol. 71, No. 1, pp. 98- 112, January 1983.

Zhang, Y., Cai, Z., Suhling, J., Lall, P., Bozack M., The Effects of SAC Alloy Composition on Aging Resistance and Reliability, 59th Electronic Component and Technology Conference, San Diego, CA, pp. 370 – 389, May 26-29, 2009.

Zorian, Y., A Structured Testability Approach for Multi Chip Boards Based on BIST and Boundary Scan, IEEE Transactions on Components, Packaging, and Manufacturing Technology-Part B, Vol. 17, No. 3, pp. 283-290, August 1994.

Methodology Development of Warpage Analysis of Polymer Based Packaging Substrate

Cheolgyu Kim[1], Taeik Lee[1], Hyeseon Choi[1], Min Sung Kim[2], Taek-Soo Kim[1]

[1]Department of Mechanical Engineering, KAIST. 373- 291 Daehak-ro (373-1 Guseong-dong), Yuseong-gu, Daejeon 305-701, Korea.

[2]Samsung Electro-Mechanics, 25 Samsung-gil,Yeandong-myun,Saejong 339-702,Korea

E-mail: tskim1@kaist.ac.kr

Abstract

Warpage of packaging substrate has been at issue due to thin and flexible substrate. It is occurred during manufacturing processes. Although warpage research based on thin metal film and silicon substrate was actively studied, it has difficulty about research of polymer based electronic due to its flexibility and low stiffness compared to metal and silicon substrate. We suggest a new methodology of warpage analysis to predict the warpage behavior of polymer composite substrate based bilayer specimen during temperature rising. The warpage analysis is performed in the sequence of scanning 3D surface, calculating curvature and built-in stress of film and verifying the result using FEM simulation. Two main factors of warpage behavior are built-in stress in film layer and stress induced by misfit of coefficient of thermal expansion between film and substrate. Built-in stress, arisen from built-in strain, is generated during film lay-up process such as electro-plating of copper, curing process of polymer. It is computed from the curvature at room temperature using the strain-curvature relation. Though this analysis method, the predicted curvature through temperature cycle showed good agreement with the experiment.

Introduction

Recently, electronic devices are developed to be multi-functional, of high performance. Since electronic device trends to be thinner and smaller than before, Packaging structure in the various devices could not be avoided to get thinner and more complex to be asymmetric design. Serious warpage problem is occurred due to asymmetry design and flexibility of thin film structure [1, 2]. Thermal processes heat and/or cool the substrate to be deflected and deformed. Thermal residual stress is accumulated in the multilayered structure. Packaging device is disqualified for severe deflection, cannot conduct its own work. In addition, deformed substrate is frequently stuck in the manufacturing lines. It cause rate of production to decrease dramatically. Therefore, it needs to analysis warpage behavior of packaging substrate.

Displacement between maximum point and minimum point of specimens, which is called maximum deflection measurement, were frequently used as warpage standard. Although, it is easy to measure, it cannot contain any information about warpage mode and orientation. Therefore, we adopt curvature, which shows overall shape of specimens, as warpage standard. Comparison of two methods is shown Figure 1. Edge of specimen is used for measuring maximum deflection. It is easily affected by routing condition. Due to this boundary effect, deflection could not represent accurate warpage standard. On the contrary, we could gain not only amount of deflection but also warpage direction and orientation form the curvature. Coefficient of thermal

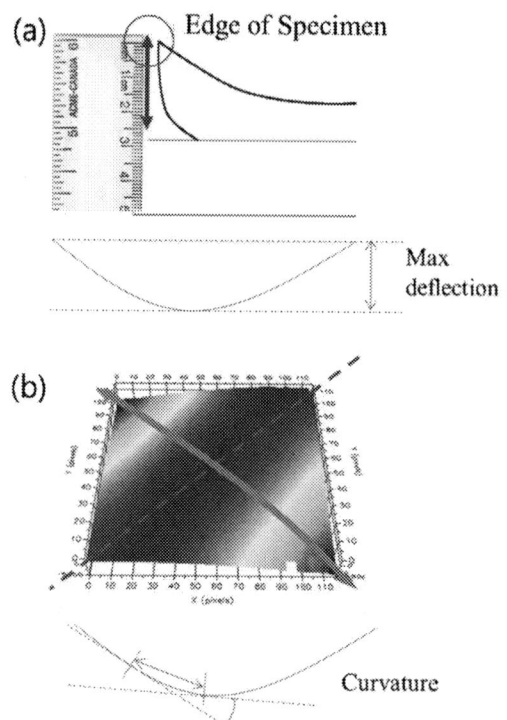

Figure 1. Warpage calculation methods (a) Maximum deflection measurement (b) Curvature measurement

expansion (CTE) difference between film and substrate was previously considered as dominant factor of deformation of packaging structure [3]. However, deflection caused by room temperature process was not attributed to CTE misfit. Film/substrate structure was deformed at room temperature, although it does not pass through the thermal process. Therefore, Non-mechanical stress should be studied to understand the warpage behavior of packaging structure at room temperature. A number of manufacturing processes made various intrinsic stresses such as plating stress and curing stress, thermal stress by CTE. These intrinsic stresses are the main factors of warpage behavior of room temperature.

To analyze the in-plane stress distribution, proper analytical solution should consider characteristics of electronic packaging substrate structure. Stoney's equation [4] shows basic relation between curvature and misfit strain with simply equation. However, commercialized packaging structures have complex multilayer structure. In the case of PCB, films on the substrate such as copper circuit, solder resist are not thin compare to the substrate. Moreover stresses were high enough to make packaging structure to be severely deformed. Hence, classical Stoney's equation [4] is no longer valid to analyze warpage behavior of packaging substrate.

In this study, we present new analysis method that predicts curvature of packaging substrate structure by comparing

978-1-4799-2408-0/14 $31.00 © 2014 IEEE

2014 Electronic Components & Technology Conference

calculated built-in stress on the various films and simulation result. Built-in stresses were calculated by equation discussed by Gleskova [5] which is applicable for large deformation case.

Procedure of Warpage Analysis

Figure 2 shows the schematic of procedure of warpage analysis. To analyze the warpage deformation, curvature should be calculated in advance. It is necessary to measure the deformed contour of the warpage. This is a process for extracting deformation shape of the specimens, the data for calculating the direction of the warpage (warpage orientation) and mode of warpage (convex, concave). Contour data is used for equation fitting. It is possible to calculate the curvatures through fitted equation. After calculating the curvatures of specimen, built-in stress is directly calculated by putting the curvature into analytic model, which is suitable for large deformation case. Next step is running FEM simulation for predicting warpage behavior with temperature variation. Built-in stress and temperature field are applied in simulation model. Finally, simulation result is verified through the comparing the actual deformation mode and orientation of experiment result.

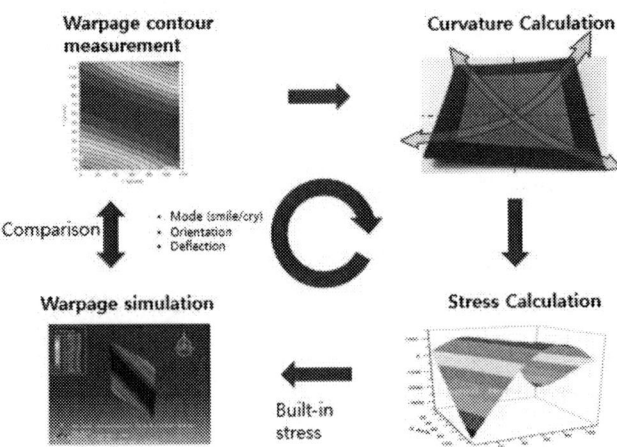

Figure 2. Flow diagram of warpage anaylsis

Theoretical Background of Curvature Analysis

Stress measurement in thin film structure based on the observation of substrate curvature has been developed with classical Stoney formula and various applications. Stoney's theoretical model [4] is well known as most common approach of getting biaxial stress state from the substrate curvature. The curvature of Stoney's equation with thin film in equibiaxial stress is given as

$$k_{stoney} = \frac{6\sigma_f h_f}{E'_s h_s^2} \qquad (1)$$

Where σ_f is equibiaxial stress, h_f, h_s are thickness film and substrate and of E' is equibiaxial young's modulus $E/(1-v_s)$. However this model is applicable only for limited case. There are several assumptions to make Stoney's equation simple. Substrate has to have uniform thickness and substrate and films are linear elastic and isotropic. Effects of deposited film are negligible. The out-of –plane displacement is at least an order of magnitude less than the substrate thickness. Therefore, it was impossible to apply Stoney's equation directly to our method. If substrate is relatively compliance and film is highly stressed, film/substrate behaves toward minimize

of the total strain energy. As film intrinsic stress increasing, the curvature is proportionally increasing with spherical shape of curvature within linear range. Once film strain becomes large and overcome the linear range, it is no long maintain its own shape. It rapidly turns into cylindrical shape with one dominant curve direction as shown in Figure 3. This phenomenon is called curvature of bifurcation. The schematic of curvature bifurcation phenomenon is shown in Figure 4. After curvature bifurcation, polymer based substrate structure could be analyzed using analytic model, while the Stoney equation could not be applicable anymore.

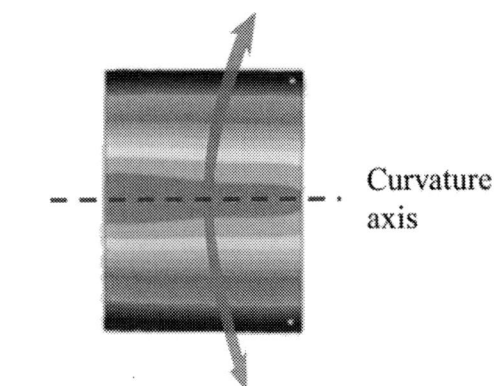

Figure 3. Shapes of specimen with principal curvatures

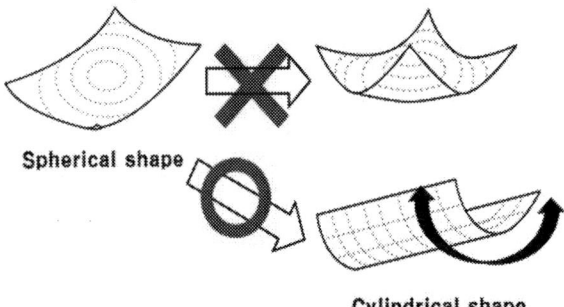

Figure 4. Curvature bifurcation

Curvature calculation

The equations of the deformed contour from initial state at the flat surface to the out-of- surface were measured by 3D scanner. The out-of plane deflections were fitted to an equation of the 2D polynomial form. Shapes of specimen are categorized using curvature. Curvature of specimen was calculated by equation below

$$k_{\alpha\beta} = \frac{\partial^2 f}{\partial x_\alpha \partial x_\beta} \cdot \frac{1}{\sqrt{1+\left(\frac{\partial f}{\partial x_\alpha}\right)^2+\left(\frac{\partial f}{\partial x_\beta}\right)^2}} \qquad (2)$$

For displacement surfaces where the squares of the slopes are small compared to unity, the curvatures of each axis could be defined as

$$k_x = \frac{\partial^2 f}{\partial x^2}, k_y = \frac{\partial^2 f}{\partial y^2}, k_{xy} = \frac{\partial^2 f}{\partial x \partial y}, k_{yx} = \frac{\partial^2 f}{\partial y \partial x} \qquad (3)$$
$$(x_\alpha = x, x_\beta = y)$$

Where z=f(x, y) is specimen contour, the equation of measured deformed displacement from the flat plane state. k_x and k_y are curvatures along two mutually orthogonal x,y directions. k_{xy} and k_{yx} are defined as twist curvatures which are the rate of change in slope perpendicular to the direction of travel. Due to the symmetry of curvature tensor, k_{xy} and k_{yx} are theoretically same each other. Principal curvatures could get using coordinate transformation and removing k_{xy} as same steps to get a maximum and minimum stresses using Mohr circle.

$$k_1 = \frac{1}{2}\left[k_x + k_y + \sqrt{(k_x - k_y)^2 + 4k_{xy}^2}\right] \qquad (4)$$

$$k_2 = \frac{1}{2}\left[k_x + k_y - \sqrt{(k_x - k_y)^2 + 4k_{xy}^2}\right] \qquad (5)$$

Where k_1 is principal maximum curvature and k_2 is principal minimum curvature. Figure 5 shows different shapes along with principal curvature values.

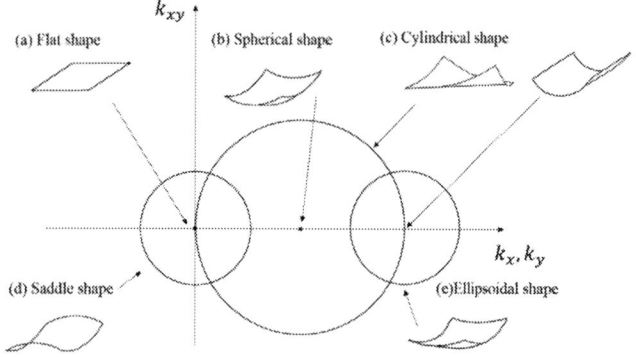

Figure 5. Image of test specimen

(a)   Flat shape: $k_x = 0,\ k_y = 0,\ k_{xy} = 0$

(b)   Spherical shape: $k_x = k_y,\ k_{xy} = 0$

(c)   Cylindrical shape: $k_x \neq 0,\ k_y = 0,\ k_{xy} = 0$

       (or) $k_x = 0,\ k_y \neq 0,\ k_{xy} = 0$

(d)   Saddle shape: $k_x > 0,\ k_y < 0,\ k_{xy} = 0$

       (or) $k_y > 0,\ k_x < 0,\ k_{xy} = 0$

(e)   Ellipsoidal shape: $k_x > k_y > 0$

Stress – Curvature relation

Curvature induced by intrinsic stress increase, system does not go along with linear model. Harper and Wu [6], Masters and Salamon [7, 8] and other researchers [9-12] presented models for predicting geometrically nonlinear stress – curvature relation. Film and substrate greatly deformed into cylindrical shape after bifurcation due to large deformation induced by intrinsic stress. Gleskova *et al.* also derived a theoretical model for nonlinearly behaving bilayer structure [5,13]. This model is applicable when substrate is compliant and film is relatively stiff than substrate. It is valid when both have similar products of elastic modulus and thickness, $E_f t_f \approx E_s t_s$.

Film/structure easily deform into cylindrical shape with relatively small stress in film. Bilayer stress state was calculated based on specimen curvature as Stoney equation works. In contrast with Stoney's equation, theoretical model of this research considers not only film mechanical properties but also substrate mechanical properties. The radius of curvature is given by following equation [5].

$$R = \frac{d_s}{6 \cdot \frac{E_f'}{E_s'} \cdot \frac{d_f}{d_s} \cdot (e_f - e_s)} \cdot$$

$$\left\{ \frac{\left\{\left[1 + \frac{E_f'}{E_s'}\left(\frac{d_f}{d_s}\right)^2\right] + 4\frac{E_f'}{E_s'}\frac{d_f}{d_s}\left(1 + \frac{d_f}{d_s}\right)^2\right]\left[(1 - \nu_s^2) + \left(\frac{E_f'}{E_s'}\frac{d_f}{d_s}\right)^2(1 - \nu_f^2)\right]}{\left(1 + \frac{d_f}{d_s}\right)\cdot\left(1 + \frac{E_f'}{E_s'}\frac{d_f}{d_s}\right)\left[(1 - \nu_s^2)(1 + \nu_f) + \frac{E_f'}{E_s'}\frac{d_f}{d_s}(1 - \nu_f^2)(1 + \nu_s)\right]} \right.$$
$$\left. + \frac{3\cdot\left(\frac{E_f'}{E_s'}\frac{d_f}{d_s}\right)^2\left(1 + \frac{d_f}{d_s}\right)^2\left[(1 - \nu_s^2) + (1 - \nu_f^2)\right] + 2\frac{E_f'}{E_s'}\frac{d_f}{d_s}(1 - \nu_s\nu_f)\cdot\left(1 + \frac{E_f'}{E_s'}\frac{d_f}{d_s}\right)\left(1 + \frac{E_f'}{E_s'}\left(\frac{d_f}{d_s}\right)^3\right)}{\left(1 + \frac{d_f}{d_s}\right)\left(1 + \frac{E_f'}{E_s'}\frac{d_f}{d_s}\right)\left[(1 - \nu_s^2)(1 + \nu_f) + \frac{E_f'}{E_s'}\frac{d_f}{d_s}(1 - \nu_f^2)(1 + \nu_s)\right]} \right\}$$

(6)

R is radius of curvature of specimen and ε is the mismatch strain between the film and the substrate. E', d and ν are the equibiaxial Young's modulus, thickness and Poisson ratio respectively. Subscript f and s stand for film and substrate respectively. R presents specimen radius of curvature of cylindrical deformed shape after bifurcation occurs. Mechanical properties and thickness of film and substrate are known and radius of curvature is calculated from the fitted equation. Therefore, one can pick out mismatch strain. Mismatch strain is constituted of two parts. One is thermal mismatch strain caused by CTE mismatch during temperature change. Another is built- in strain during film deposition process.

$$e_f - e_s = (\alpha_f - \alpha_s)\cdot(T_r - T_d) + \varepsilon_{bi} \qquad (7)$$

$$\sigma_{built\_in} = E'_{film} \cdot \epsilon_{built\_in} \qquad (8)$$

Specimen curvature caused without thermal process is thoroughly depends on built-in strain on the film. For example, copper electrodepositing process is manufactured in room temperature plating bath, therefore, specimens go though no thermal process. One could easily get built-in stress with cross product of built-in strain and Young's modulus of film. We know the built-in stress and strain, film and substrate strain caused by temperature change could be achieved through equation.

Bilayer specimen preparation

Bilayer specimens were prepared with structure of film on packaging substrate. Base copper was attached on both sides of substrate. Test specimens were made by deposition method of various types of layer on polymer substrate as bilayer form. It is a thin layer about 1μm for copper electrochemical plating, which is deposited through the rolling process. Since state of packaging substrate is stress free, it is flat. Copper plating test pieces was made by electroplating on the base copper about 20~30μm thickness. Both sides of base copper were electroplated in the plating bath symmetrically. It was a flat while electroplating process due to stress balance. However, it soon deformed with etching of one side of copper and plating stress from remaining side of copper made bilayer specimen to curve. All specimens were cut down using routing machine.

Characteristics of polymer based packaging substrate

Polymer based packaging composites are materials composed of resin and bundles of fiber. Resin content in the polymer substrate typically varies range from 35 to 44 percent. Reinforcement comes in the form of bundles of individual fiber filaments. Thousands of filaments are woven together to make fibers. Fiber composites are categorized into unidirectional reinforcement and fabric reinforcement depending on fiber type. Unidirectional fibers is a bundles of fibers oriented in one direction. Another is fabric that bundles of fiber are orthogonally woven. Fabrics make substrate to have orthotropic property. There are two directions, warp and weft which are same as the direction of fibers. The warp is the longitudinal fibers that are under tension of roll, while the weft is the transverse fibers drawn through the warp fibers. Cross section and schematic of glass fiber composite is shown in Figure 6.

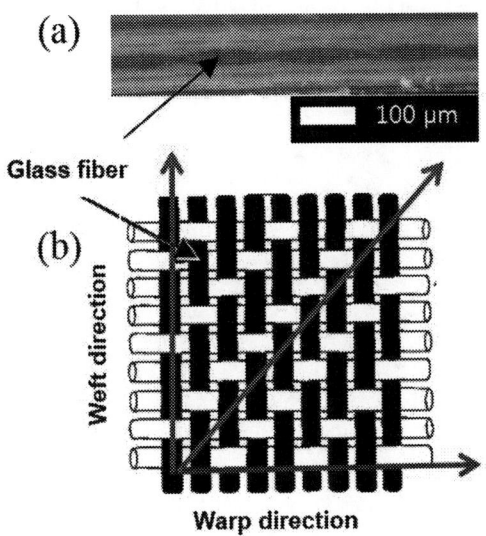

Figure 6. (a) Cross section of glass fiber composite (b) Schematic of woven glass fiber

Fibers reinforce the woven insulator and strengthen along the direction of fiber's normal direction. However, fibers do not work in the compressive or transversal force. Therefore, deformation behavior of system depends on properties of resin rather than those of fibers. Properties such as young's modulus and coefficient of thermal expansion of packaging substrate vary depending of not only the temperature but also direction.

FEM simulation

In order to verify the experimental result, we performed finite element method (FEM) simulations. ABAQUS (version 6.12) was used as simulation tool. We compared the curvature of simulation result with the curvature of experiment result with temperature variation. Mesh element type was S4R shell composite. The geometry was drawn in 2D shell model. Polymer composite substrate was positioned bottom side. For accurate observation of overall warpage behavior, we used full geometry model instead of half or quarter model. Concerning mechanical boundary condition, three displacement and rotation conditions were restricted at the center point. This condition prevents model from rigid body rotation. The initial thermal conditions for the whole layers were set to the ambient temperature of 25°C. Temperature increased up to 150~180°C through whole region (upper limitation of temperature is changed depend on specimens). In actual experiments, specimens were heated from the bottom first. However, thermocouples attached top and bottom sides indicate almost same temperature. Thickness of structure was thin enough to conduct heat rapidly thorough the specimen region. Therefore,

temperature was applied as predefined field through whole region.

Result and discussion

We used the curvature analysis method to calculate stress and compared specimens. Curvature comparison between measured data and simulation result is shown Figure 7. Specimen conditions are shown in Table 1 and Table 2.

Table 1. Condition of copper bilayer specimen

Specimen	Copper	Polymer composite
Size	90 mm X 90 mm	
Film thickness	20 μm	150 μm
Young's modulus	90 GPa	23 GPa
Shear modulus	33.5 GPa	6.3 GPa
Poisson	0.34	0.2
CTE	17 ppm	7 ppm

Table 2. Condition of prepreg bilayer specimen

Specimen	Prepreg	Polymer composite
Size	75mm X 60mm	
Film thickness	40 μm	150 μm
Young's modulus	30 GPa	23 GPa
Shear modulus	11.5 GPa	6.3 GPa
Poisson	0.3	0.2
CTE	13 ppm	7 ppm

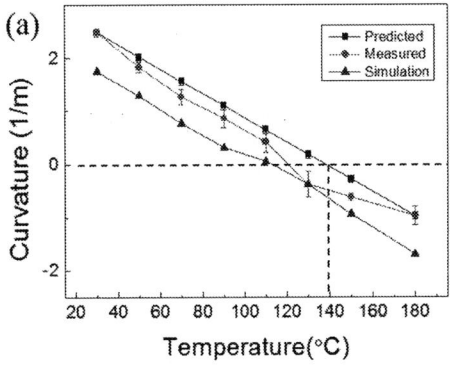

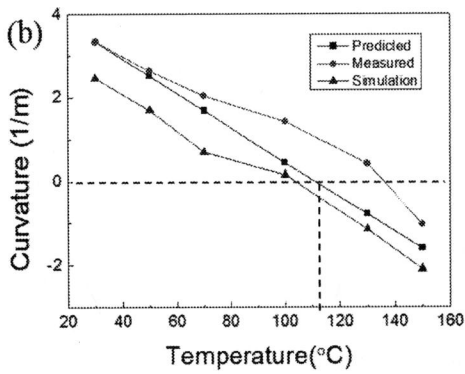

Figure 7. Curvature Comparison graph among measured curvature, predicted curvature and simulation result (a) Cu bilayer (b) Prepreg bilayer

At the first room temperature point, built in stress was calculated. Warpage of specimen at room temperature was induced by built-in stress. Warpage changes as temperature rises because CTE misfit between layers causes thermal stress.

Built-in strain is calculated from specimen's curvature at room temperature. Built-in strain induces tensile or compressible built-in stress. While room temperature curvature was thoroughly depends on built-in stress, additional curvature during the temperature change is due to CTE mismatch that leads to thermal stress. Electroplated copper had tensile built-in stress and the CTE of copper is larger than that of core substrate. Subsequently, we could expect that original concave shape specimen turned to the shape of convex. We predict curvature change through built-in strain. CTE mismatch strain of film and substrate calculated from analytic model.

In FEM simulation, built-in stress in room temperature is applied to the simulation model and heating history was applied as next step. We verified that warpage decreases as temperature increases due to CTE misfit. We used the curvature analysis method to calculate stress and compared specimens that had experienced different process. Curvature was decreased due to the influence of thermal stress induced by CTE misfit between layers during the heating cycle. Compressive stress was applied to the film layer and tensile stress was applied to core substrate relatively.

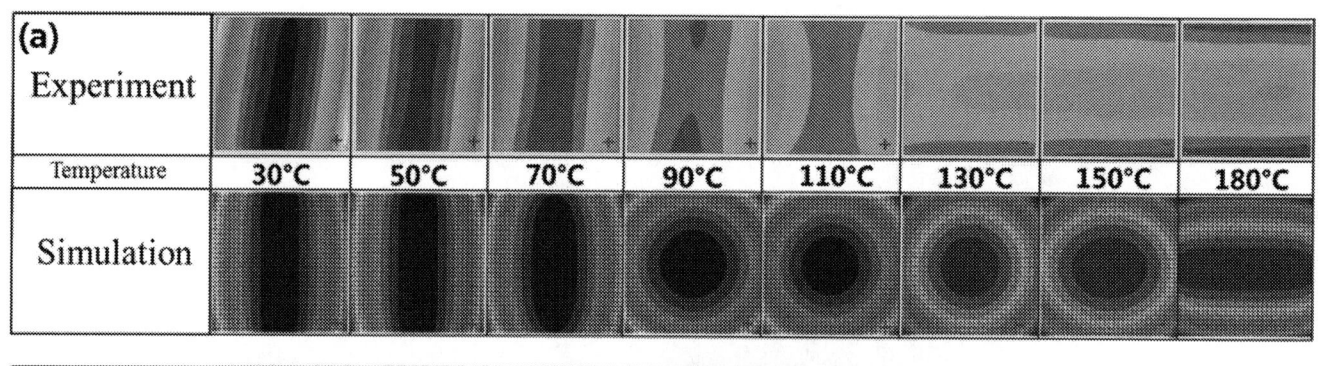

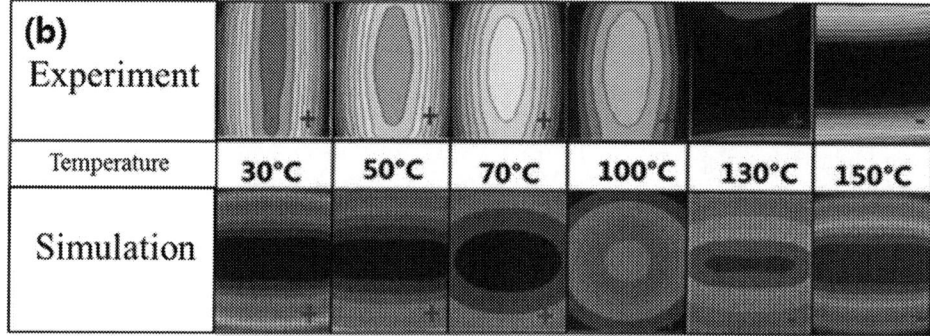

Figure 8. (a) Copper bilayer and (b) prepreg bilayer specimen's curvature comparison experiment image and simulation result

Conclusions

A new method of predicting the behavior of packaging substrate structure based on analytic solution and its verification using FEM simulation have been presented. Thin film structure with compliant substrate has shown large deformation even under small stress. Due to this large deformation, thin film structure was unable to maintain axially symmetric shape, and thus bifurcation of curvature occurred. Using the shape of the specimens that have gone through bifurcation and with the aid of appropriate theoretical solutions, we have analyzed their warpages. Contour measurement and curvature calculation have been developed for the characterization of built-in strain and stress. Warpages during heating process were predicted with analytic solution, and were compared to the results of FEM simulation; the results were in good agreement. We have identified built-in stress at room temperature and thermal stress caused by CTE mismatch as the main factors that cause the warpage. For the accurate measurement and analysis, further study is required about warpage orientation and warpage of patterned structures.

Acknowledgments

This work was supported by Samsung Electro-Mechanics, Basic Science Research Program (2012R1A1A1006072) and the Global Frontier R&D Program on Center for Multiscale Energy System (2011-0031569) and the Global Frontier R&D Program on Center for Multiscale Energy System (2011-0031569) funded by the National Research Foundation under the Ministry of Science, ICT & Future Planning, Korea, and also by Graphene Materials and Components Development Program of MOTIE/KEIT (10044412, Development of basic and applied technologies for OLEDs with graphene).

References

1. V. Z. Rizzo and R. D. Mansano, "Electro-optically sensitive diamond-like carbon thin films deposited by reactive magnetron sputtering for electronic device applications," *Progress in Organic Coatings,* vol. 70, pp. 365-368, 2011.

2. S. Miyajima, S. Nagamatsu, S. S. Pandey, S. Hayase, K. Kaneto, and W. Takashima, "Electrophoretic deposition onto an insulator for thin film preparation toward electronic device fabrication," *Applied Physics Letters,* vol. 101, pp. 193305-193305-4, 2012.

3. C. Hsueh, "Thermal stresses in elastic multilayer systems," *Thin Solid Films,* vol. 418, pp. 182-188, 2002.

4. G. G. Stoney, "The tension of metallic films deposited by electrolysis," *Proceedings of the Royal Society of London. Series A, Containing Papers of a Mathematical and Physical Character,* pp. 172-175, 1909.

5. H. Gleskova, I. C. Cheng, S. Wagner, J. C. Sturm, and Z. Suo, "Mechanics of thin-film transistors and solar cells on flexible substrates," *Solar Energy,* vol. 80, pp. 687-693, 2006.

6. B. D. Harper and W. Chih-Ping, "A geometrically nonlinear model for predicting the intrinsic film stress by the bending-plate method," *International journal of solids and structures,* vol. 26, pp. 511-525, 1990.

7. C. B. Masters and N. Salamon, "Geometrically nonlinear stress-deflection relations for thin film/substrate systems," *International journal of engineering science,* vol. 31, pp. 915-925, 1993.

8. C. Masters and N. J. Salamon, "Geometrically nonlinear stress-deflection relations for thin film/substrate systems with a finite element comparison," *Journal of applied mechanics,* vol. 61, p. 872, 1994.

9. M. Finot and S. Suresh, "Small and large deformation of thick and thin-film multi-layers: effects of layer geometry, plasticity and compositional gradients," *Journal of the Mechanics and Physics of Solids,* vol. 44, pp. 683-721, 1996.

10. M. Finot, I. Blech, S. Suresh, and H. Fujimoto, "Large deformation and geometric instability of substrates with thin-film deposits," *Journal of Applied Physics,* vol. 81, pp. 3457-3464, 1997.

11. W. Jun and C. Hong, "Effect of residual shear strain on the cured shape of unsymmetric cross-ply thin laminates," *Composites science and technology,* vol. 38, pp. 55-67, 1990.

12. L. Freund, "Substrate curvature due to thin film mismatch strain in the nonlinear deformation range," *Journal of the Mechanics and Physics of Solids,* vol. 48, pp. 1159-1174, 2000.

13. William S. Wong and Alberto Salleo, *Flexible Electronics : Materials and Applications,* Springer,New York, 2009, pp. 29-51.

Simulations for the Impact of Warpage on the Accuracy of Attitude and Heading Reference System

Shengzhi Zhang[1,2], Qiang Dan[1,2], Chaojun Liu[1,2], Yong Xu[4], Xin Wu[5], Sheng Liu[3*], Xing Guo[1,2], and Ming Wen[1,2]

[1] Institute of Microsystems, State Key Laboratory for Digital Manufacturing Equipment & Technology,
School of Mechanical & Engineering, Huazhong University of Science & Technology, Wuhan, Hubei, 430074, China

[2] Division of MOEMS, Wuhan National Laboratory for Optoelectronics, Wuhan, Hubei, 430074, China

[3] Cross-disciplinary Institute of Engineering Sciences, School of Power and Mechanical Engineering, Wuhan University,
Wuhan, Hubei, 430072, China

[4] Department of Electrical and Computer Engineering, Wayne State University, Detroit, MI 48202, USA

[5] Department of Mechanical Engineering, Wayne State University, Detroit, MI 48202, USA

* Corresponding author: victor_liu63@126.com

Abstract

Design and implementation is critically important for micro electro mechanical systems (MEMS) based devices, modules. As a typical instance, modern attitude and heading reference systems (AHRS) generally use Kalman Filter to integrate gyroscopes with some other augmenting sensors, such as accelerometers and magnetometers, so as to provide a long term stable orientation solution. While warpage of printed circuit board (PCB) induced by initial installation error or thermal deflection may result in non-orthogonality and misalignment of MEMS devices, similar to inner inter-axis alignment error of multi-axis MEMS device, consequently output values of MEMS devices are unable to reflect the actual angular velocity, acceleration and magnetic field intensity, etc. These physical quantities obtained in the body frame cannot be transformed into the local level frame exactly, so that the accuracy of attitude and heading reference system cannot be guaranteed during the long time Kalman filtering process.

This paper deals with the warpage of a micro AHRS composed of the low-cost inertial and magnetic sensors. Experimental results show that thermal cycling has a significant impact on the accuracy of AHRS. A warpage model of PCB based on thermomechanics is developed. Then the non-orthogonality and misalignment matrices induced by thermal deflection are obtained. A body/local level frame transformation matrix is used to study that mechanical parameters (orthogonalization and alignment parameters) are not independent of temperature and time during operation. The most important concern is that this accumulated error is unbounded in time. We analyze the interactions between MEMS device and its application to system so as to guarantee the accuracy of AHRS. Experiments and simulation results are consistent with what we envisage before.

1 Introduction

Gyroscopes have been successfully used to determine system orientation by numerically integrating the angular rates (pitch, roll, and yaw rates) [1]. Due to the drift of the null bias point and the presence of the noise in the gyro output signal, there is a considerable amount of the error accumulation. The most important issue we concern is that this accumulated error is unbounded in time, so that only the short-term accuracy can be achieved using the rate gyro measurements. Accelerometers and magnetic sensors are also used for attitude and heading calculating [2]. Because of the fact that gravity acceleration cannot be separated accurately from the accelerometer measurements, especially obvious in long-term high dynamic operation, accelerometers are usually used as augmenting sensor [3]. Simultaneously, geomagnetic field is very susceptible to the interference of additional magnetic field in the space [4]. Therefore, in many applications of attitude and heading reference system (AHRS), combinations of these three kinds of sensors have been applied to provide a robust orientation solution.

Nowadays, a typical AHRS usually consists of MEMS rated-gyroscopes, electronic compass composed of accelerometers and magnetometers, with Kalman Filter program embedded so as to provide a long-term stable orientation solution. Attitude information for moving objects is currently obtained by an AHRS, such as unmanned aerial [5][6] and underwater vehicles, handheld navigation devices, human motion tracking and many more, where the attitude is essential factor for control or monitoring purposes.

Many researchers have studied deflection of printed circuit board (PCB) induced by surface mounting process, thermal shock, or assembly error, mechanical shock, etc. [7]. Most of them merely concentrate on stress and deformation, or fatigue and fracture [8]. However, sometimes interactions of devices mounted in PCB and PCB itself may play a more important role in integrated systems, and may even involve the resonance [9]. After all, the accuracy of the system (e.g. AHRS) is determined by precise output values of sensors, any non-orthogonality and misalignment of MEMS devices (like gyroscopes, accelerometers and magnetometers) will lead to invalid measurements.

The scope of this paper is to present methods on how the warpage can have a considerable effect on the accuracy for such a micro AHRS composed of the low-cost inertial and magnetic sensors. A warpage model of PCB based on thermo-mechanics has been developed. Subsequently, transformation matrix is used to study that mechanical parameters (orthogonalization and alignment parameters) are not independent of temperature and time during operation. The most important concern is that this accumulated error is unbounded in time. Compared with the inherent temperature properties of sensors, thermal deflection induced by temperature change plays the same important role on the accuracy of AHRS at an extreme high temperature.

2 Orientation Determination

2.1. Orientation representation

As is well known, attitude and heading (the orientation) of a rigid body expressed in the inertial coordinate frame can be represented in different ways such as direction cosine matrix (DCM), Euler angles (roll, pitch, yaw) or quaternions (q_0, q_1, q_2 and q_3). Each method comes up with its inherent advantages and disadvantages. According to our requirements, for the orientation representation in our case the quaternions are used. The transformation between the Euler angles and the four quaternion components is quite formulaic [10]. Additional motive was the simplified data fusion algorithm design, since the orientation transformation matrix provided by the tri-axis accelerometer and electronic compass is represented in a simple way.

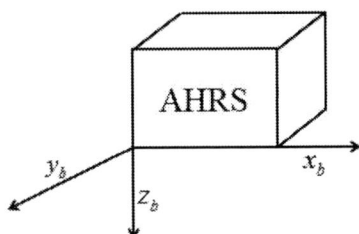

Figure 1. Body coordinate frame attached to the casing of AHRS

The body coordinate frame (**B** frame) is attached to the body of AHRS's casing. The x_b-axis points to the forward direction and the z_b-axis points to the bottom of AHRS, with the y_b-axis formed right-handed orthogonal coordinate system (Fig. 1). The local level frame, so called East, North, Up (**ENU**) frame, is set as inertial frame or navigation frame (**I** frame) for attitude calculation. The axes x_i and y_i lie on the local level tangent plane. The x_i-axis points to the east and the y_i-axis to the north. The z_i-axis completes **I** frame by pointing to the straight up.

2.2 Rate gyro approach

The transformation between **I** frame and **B** frame of a column vector $\vec{x}(t)$, whose components are generally functions of time t, is expressed as:

$$\vec{x}^n(t) = C_b^n \left[q(t) \right] \vec{x}^b(t) \tag{1}$$

Henceforward, parameter t will be omitted for the sake of simplicity. Regarded as body/local level frame transformation matrix, $C_b^n \left[q(t) \right]$ represents the transformation matrix form **B** frame to **I** frame. A rigid body angular motion must comply with the differential equation:

$$\dot{q} = \frac{1}{2} A q \tag{2}$$

where

$$A = \begin{bmatrix} 0 & -\omega_x & -\omega_y & -\omega_z \\ \omega_x & 0 & \omega_z & -\omega_y \\ \omega_y & -\omega_z & 0 & \omega_x \\ \omega_z & \omega_y & -\omega_x & 0 \end{bmatrix} \tag{3}$$

$$q = \begin{bmatrix} q_0 & q_1 & q_2 & q_3 \end{bmatrix}^T$$

The state equation of extended Kalman Filter (EKF) could be formulated based on equation (2).

$\omega = \begin{bmatrix} \omega_x & \omega_y & \omega_z \end{bmatrix}^T$ is the angular velocity of **B** frame relative to **I** frame, which can be obtained by gyroscopes integrated in AHRS. The discrete-time equation model corresponding to (2) is

$$\begin{cases} q_{k+1} = \exp\left(A t_\Delta \right) q_k, k = 0, 1, \dots \\ q_0 = q(0) \end{cases} \tag{4}$$

where t_Δ is the sampling interval of the AHRS. On the assumption that the angular velocity ω measured at time instants $k t_\Delta$ is constant in the interval $\left[k t_\Delta \left(k+1 \right) t_\Delta \right]$. Once equation (4) has been solved in time, attitude information will be updated immediately.

2.3 Accelerometer and electronic compass approach

Outputs of accelerometer and magnetometer are written in vector form as:

$$\begin{cases} \vec{a}_b = C_n^b(q)[\vec{g} + \vec{a}_m] \\ \vec{m}_b = C_n^b(q)\vec{h} \end{cases} \tag{5}$$

$\vec{a}_b$ and $\vec{m}_b$ represent the output values of the accelerometer and magnetometer in the body frame, respectively. $\vec{a}_m$ represents the linear acceleration along each axis in **B** frame with its origin at the center of gravity. $\vec{g}$ and $\vec{h}$ represent the vector of local gravity acceleration and local geomagnetic field in the **I** frame. It is not easy to obtain true $\vec{a}_m$ using accelerometer. We assume that the measured object is moving in a low dynamic state, $\vec{a}_m$ can be neglected. Equation (5) will be simplified to:

$$\begin{bmatrix} \vec{a}_b \\ \vec{m}_b \end{bmatrix} = \begin{bmatrix} C_n^b(q) & \mathbf{0}_{3 \times 3} \\ \mathbf{0}_{3 \times 3} & C_n^b(q) \end{bmatrix} \begin{bmatrix} \vec{g} \\ \vec{h} \end{bmatrix} \tag{6}$$

Attitude transformation matrix $C_n^b(q)$ can be obtained by solving equation (6). And the observation equation of EKF could be formulated based on equation (6).

3 Data Fusion Algorithm

Now that we have acquired redundant attitude data, how to deal with redundant data is urgent. As the accumulating error of attitude computed from gyroscopes is unbounded in time, meanwhile, accelerometer and magnetometer provides long-term accuracy with high noise contents. Hence, an extended

Kalman Filter process is implemented to integrate the attitude information from gyroscopes, accelerometer and magnetometer. The dynamic equation of the AHRS can be expressed as:

$$\begin{cases} \dot{x} = Ax + w \\ z = h(x) + v \end{cases} \quad (7)$$

where $x = \begin{bmatrix} q_0 & q_1 & q_2 & q_3 \end{bmatrix}^T$ is regarded as the state vector of the AHRS, $z = \begin{bmatrix} a & m \end{bmatrix}^T$ is regarded as the measurement vector, and w, v are the process noise and measurement noise of system, respectively. The discrete-time equation model corresponding to equation (7) is:

$$\begin{cases} x_{n+1} = F(t_\Delta, \omega_k)(x_n) + w(n) \\ z_n = h(x_n) + v(n) \end{cases}, n = 0, 1, \dots \quad (8)$$

For the sake of reader's convenience, the EKF equations are summarized below.

Update of discrete Kalman filter time equations:

$$x_{k+1}^- = F(t_\Delta, \omega_k) x_k \quad (9)$$

$$P_{k+1}^- = F(t_\Delta, \omega_k) P_k F(t_\Delta, \omega_k)^T + Q_k \quad (10)$$

Update of discrete Kalman filter measurement equations:

$$K_{k+1} = P_{k+1}^- F_{k+1}^T \left(F_{k+1} P_{k+1}^- F_{k+1}^T + R_{k+1} \right)^{-1} \quad (11)$$

$$x_{k+1} = x_{k+1}^- + K_{k+1} \left[z_{k+1} - f_{k+1}(x_{k+1}^-) \right] \quad (12)$$

$$P_{k+1} = P_{k+1}^- - K_{k+1} F_{k+1} P_{k+1}^- \quad (13)$$

With continuous iterations, the state vector of x will be updated in real time. The validity of the proposed EKF is verified by programing using MATLAB V.7.8 software [11] according equations above. In order to let the results be more visualized, four quaternion components are transformed into three Euler angles. The raw data of sensors were collected from MTi-300 (produced by the Xsens), which is a miniature AHRS comprised of three single-axis MEMS gyroscopes, two dual-axis MEMS accelerometers and a low-cost tri-axis magnetometer, at a sampling rate of 400 Hz. Two different test scenarios were implemented: stationary test and in-motion test. Then the raw data were taken into EKF program for iterations. Figures 2, 3 illustrate the roll and pitch differences between the results of the proposed EKF algorithm (black dotted line) and the reference outputs of MTi-300 (red solid line) for stationary test and in-motion test, respectively.

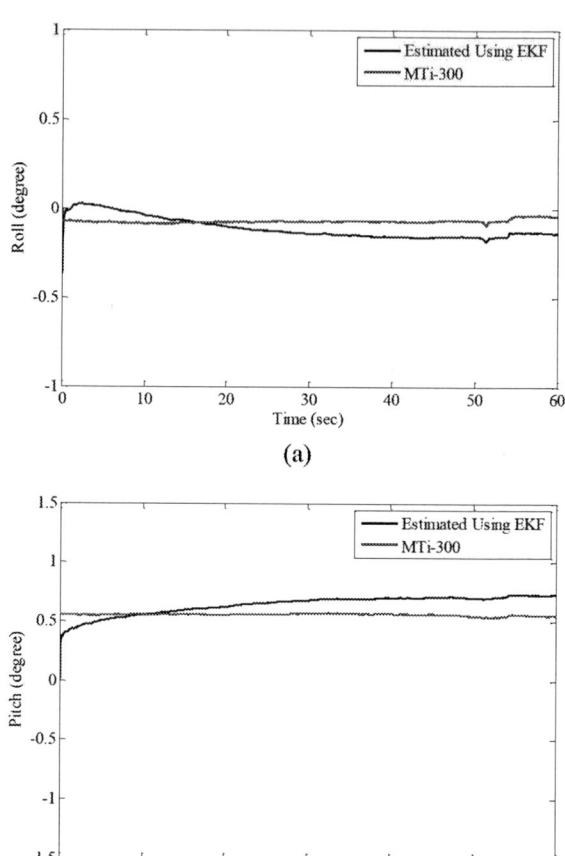

Figure 2. The roll (a) and pitch (b) angles for stationary test.

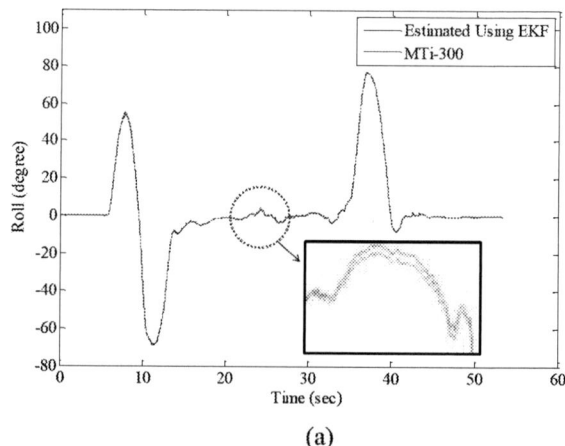

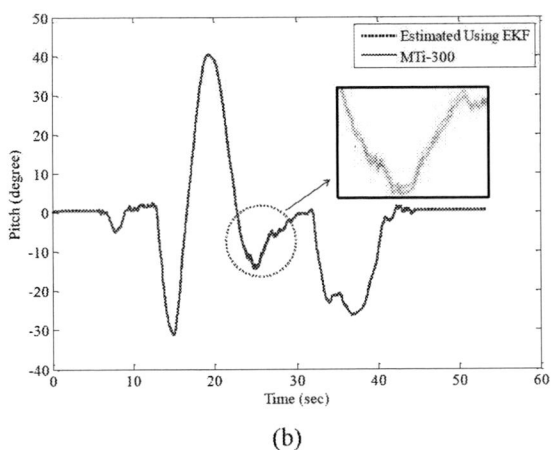

(b)

Figure 3. The roll (a) and pitch (b) angles for in-motion test.

To obtain additional insight into the orientation errors, the residuals of the orientation solution of the Kalman filtering are calculated. The results demonstrate that the proposed EKF is able to estimate the orientation effectively as the root mean square (RMS) Euler angle errors remained within the limits of $0.3°$: roll error, $0.19°$ RMS; pitch error, $0.24°$ RMS. Good dynamic response demonstrates that the proposed EKF program is robust enough without any latency. A prototype of the AHRS named FIS100 (Fig.4) has been developed successfully with on-line running EKF program codes embedded in.

Figure 4. Prototype of AHRS named FIS100

4 Simulations of the Impact of Warpage

In the previous sections we have discussed the issue how to deal with the outputs of various sensors for attitude determination. It is obvious that there is no discussion involved with the temperature. Hereon, the emphasis that we will focus on is not the temperature properties of sensors themselves, but the analysis of non-orthogonality and misalignment induced by the warpage.

As we know, thermal deflection generated by temperature gradient will lead to the warpage of PCB. In our design of the AHRS, three single-axis gyroscopes, a tri-axis accelerometer and a tri-axis magnetometer are mounted in PCB with perpendicular sensitivity axes. Each sensor must be aligned to the AHRS's casing. It is difficult to achieve this aim in reality and is impossible under the condition with thermal cycling. When the warpage of PCB occurs, non-orthogonality and misalignment both exist in the gyroscopes triad. But for the accelerometer and magnetometer, we will just take the issue of misalignment into consideration.

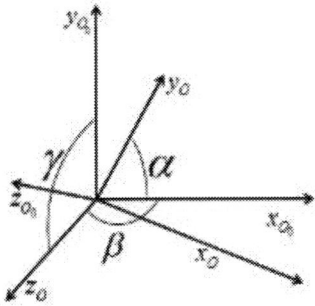

Figure 5. Orthogonalization of the sensor frame O.

First of all, two matrices need to be introduced: the orthogonalization matrix T_k and the alignment matrix M_k, regarded as a body/orthogonal coordinate transformation matrix. Index k represents the type of sensor. Other effects like sensitivities and nonlinearities of the sensors are neglected. The orthogonalization matrix T_k transforms the outputs vector expressed in the non-orthogonal reference frame O into the vector expressed in the orthogonal reference frame O_0. Alignment matrix M_k is a parameterized rotation matrix expressed in an aerospace sequence Euler angles, which rotates (alignment) the body reference frame **B** to the orthogonal reference frame O_0. Because of the wapage of PCB is still confined to the category of small deformation, non-orthogonality angles (α, β, γ) are approximately $90°$, and misalignment angles (ϕ, θ, φ) are approximately $0°$. To make an approximation without any significant loss of accuracy, orthogonalization and alignment matrices can be simplified as:

$$T_k \approx \begin{bmatrix} 1 & 0 & 0 \\ \cos\alpha_k & 1 & 0 \\ \cos\beta_k & \cos\gamma_k & 1 \end{bmatrix} \quad (14)$$

$$M_k = \begin{bmatrix} 1 & 0 & 0 \\ 0 & \cos\phi_k & \sin\phi_k \\ 0 & -\sin\phi_k & \cos\phi_k \end{bmatrix} \begin{bmatrix} \cos\theta_k & 0 & -\sin\theta_k \\ 0 & 1 & 0 \\ \sin\theta_k & 0 & \cos\theta_k \end{bmatrix} \begin{bmatrix} \cos\varphi_k & \sin\varphi_k & 0 \\ -\sin\varphi_k & \cos\varphi_k & 0 \\ 0 & 0 & 1 \end{bmatrix} \quad (15)$$

Define the column vector $\vec{u}_k$ as measured quantities, $\vec{y}_k^B$ as the outputs of sensors in orthogonal coordinate frame. The simulation trials are assumed to start with perfectly calibrated sensors, namely unit scale factor and null bias. Thus, $\vec{y}_k^B$ can be written as:

$$\vec{y}_k^B = T_k M_k \vec{u}_k \quad (16)$$

From equations (14), (15) and (16) it is obvious that 6 arguments must be determined. Different with predecessor using experiments, we developed a warpage model of PCB based on thermomechanics to acquire these two matrices. A thermal mechanical model of sensor elements and PCB based finite element method, is established using the commercial software called ABAQUS (v.6.10) [12]. Three cubes lain in the upper left are the gyroscopes triad, the tri-axis accelerometer is in lower left, the tri-axis magnetometer is in the lower right. All sensors are just regarded as display bodies

978-1-4799-2408-0/14 $31.00 © 2014 IEEE

attached to the PCB, while other components like resistors and capacitors are neglected.

Considering the working temperature ranges (-40℃~85℃, room temperature 20℃) of the AHRS, high temperature was set at 85℃ when running the simulation for thermomechanics. The warpage deformation contour of PCB induced by thermal deflection in Z direction is shown in Fig.6. Compared with the length of PCB (57mm), the max displacement 0.17mm is really considerable.

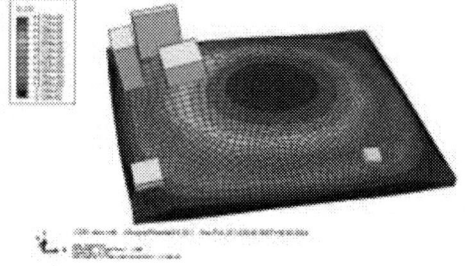

Figure 6. Displacement contour of PCB in Z direction at high temperature.

The PCB can be regarded as a reference plane in the view of math, while becoming a surface after deformation. Fortunately, ABAQUS supplies an access to obtain the displacement of the finite element nodes directly. Next, orthogonalization matrix T_k and the alignment matrix M_k are both obtained by means of mathematical relationship between before and after deformation of PCB expediently. According to equation (16), available outputs of sensors $\vec{y}_k^B$ can be modified properly. Omitting unnecessary derivations, matrix T_k and matrix M_k are determined as:

$$T_g = \begin{bmatrix} 1 & 0 & 0 \\ 0.0052 & 1 & 0 \\ -0.0035 & -0.0070 & 1 \end{bmatrix}, \; M_g = \begin{bmatrix} 0.9998 & -0.0052 & -0.0175 \\ 0.0051 & 0.9999 & -0.0087 \\ 0.0175 & 0.0086 & 0.9998 \end{bmatrix},$$

$$T_a = \begin{bmatrix} 1 & 0 & 0 \\ 0.0140 & 1 & 0 \\ -0.0052 & -0.0017 & 1 \end{bmatrix}, \; T_m = \begin{bmatrix} 1 & 0 & 0 \\ 0.0136 & 1 & 0 \\ -0.0061 & -0.0026 & 1 \end{bmatrix} \quad (17)$$

The raw data of sensors at 85℃ were collected from FIS 100 at a sampling rate of 50 Hz. Figures 7, 8 show the roll and pitch angles outputted by FIS100 directly (black line), compared with the results calculated after measurements of sensors modified (red line) during the stationary test at 85℃ .

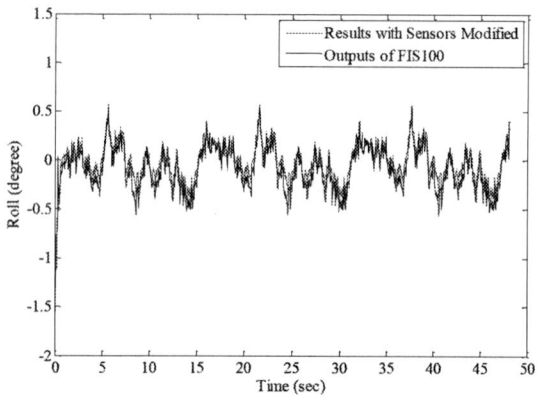

Figure 7. The roll angle for stationary test.

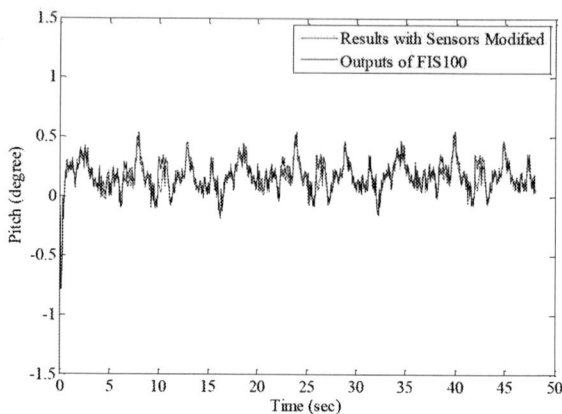

Figure 8. The pitch angle for stationary test.

Conclusions and Discussions

An extended Kalman Filter model for the quaternion vector measurements is introduced and implemented in this article. A warpage model of PCB based on thermomechanics has been developed. According to the results of finite element analysis, non-orthogonality matrix and misalignment matrix induced by thermal deflection are obtained. It is observed that the wapage deflection has a considerable impact on the accuracy of the AHRS. The validity of proposed method is verified by a stationary test. Restricted by the experimental conditions, we haven't acquired the performance of AHRS under the in-motion condition. Future works will be focused on optimizing simulation and dynamic test.

Acknowledgments

The support of National High Tech Program of Ministry of Science and Technology (863) with a contract number of 2012AA040501 and MEMS-volume Manufacture Technology of National Basic Research Project (973) with a contract number of 2011CB309504 of Ministry of Science and Technology of PR China is highly appreciated.

References

1. D. H. Titterton and J. L. Weston, *Strapdown Inertial Navigation Technology.* 2-nd Edition. The Institution of Electronical Engineers, Reston, USA, 2004.

2. D. Gebre-Egziabher , G. H. Elkaim , J. D. Powell and B. W. Parkinson, "A gyro-free quaternion-based attitude determination system suitable for implementation using low cost sensors", *Proc. IEEE Position, Location and Navigation Symp.*, San Diego, CA, USA, Mar., pp.185 - 192, 2000.

3. S. Han , J. Wang, "A novel method to integrate IMU and magnetometers in attitude and heading reference systems", *J. Navig.*, vol. 64, no. 4, pp.727 -738, 2011.

4. A. M. Sabatini, "Quaternion-based extended Kalman filter for determining orientation by inertial and magnetic sensing", *IEEE Trans. Biomed. Eng.*, vol. 53, no. 7, pp.1346 -1356, 2006.

5. D.-M. Ma, J.-K. Shiau, I.-C. Wang, and Y.-H. Lin, "Attitude Determination Using a MEMS-Based Flight

Information Measurement Unit." *Sensors*, vol. 12, no. 1, pp. 1-23, Jan. 2012.

6. S. K. Hong, "Fuzzy logic based closed-loop strapdown attitude system for unmanned aerial vehicle (uav)," *Sens. Actuators A, Phys.*, vol. 107, no. 2, pp. 109 - 118, 2003.

7. S. Liu, Y. Liu, *Modeling and Simulation for Microelectronic Packaging Assembly: Manufacturing, Reliability and Testing*, John Wiley & Sons, 2011.

8. G. Li and A. A. Tseng, "Low stress packaging of a micromachined accelerometer," *IEEE Trans. Electronics Packaging Manufacturing*, vol. 24, no. 1, pp. 18-25, Jan. 2001.

9. D. Jurman , M. Jankovec , R. Kamnik and M. Topic, "Calibration and data fusion solution for the miniature attitude and heading reference system", *Sens. Actuators A, Phys.*, vol. 138, no. 2, pp.411-420, 2007.

10. J. B. Kuipers, *Quaternions and Rotation Sequences*, Princeton University Press, Princeton, NJ, 1999.

11. MATLAB, http://www.mathworks.com/.

12. ABAQUS, http://www.3ds.com/products-ervices/simulia/ portfolio/abaqus/latest-release/.

Multicore Fiber 4 TX + 4 RX Optical Transceiver Based on Holey SiGe IC

Fuad E. Doany, Daniel M. Kuchta, Alexander V. Rylyakov, Christian Baks,
Shurong Tian, Mark Schultz, Frank Libsch, Clint L. Schow
IBM T. J. Watson Research Center
1101 Kitchawan Road, Yorktown Heights, NY 10698
Email: doany@us.ibm.com, Tel: (914) 945-2831, Fax: 914-945-4219

Abstract

A novel optical transceiver with 4 transmitter plus 4 receiver channels designed for coupling to multicore multimode fiber has been fabricated and characterized. The transceiver is based on the holey Optochip concept where 4-channel VCSEL and photodiode arrays are flip-chip attached to a single-chip SiGe IC using AuSn solder. Optical vias (holes) are fabricated into the SiGe IC to enable optical access to the conventional topside emitting 850-nm optoelectronic arrays. The optoelectronic arrays are arranged in a quad-VCSEL and quad-photodiode configuration where the 4 devices are on a 2 x 2 array on a dense 50-μm pitch. The transceiver module is completed by flip-chip soldering the Optochip onto a 8 mm x 8 mm high-speed high-density organic carrier. Optical access through the backside of the IC is provided through 2 optical vias. Electrical I/O is supplied through BGA pads on 0.8 mm pitch at the bottom of the module.

High-speed characterization was carried out between 2 modules soldered to test cards, a transmitter (TX) and a receiver (RX) module. Each of the 4 optical outputs from the TX Optochip was coupled into a MMF and directed to individual photodiodes in the RX module. Eye-diagrams were measured for TX outputs as well as TX-to-RX links at data rates 20 Gb/s to 42 Gb/s. The 4 optical links operate error free up to 40 Gb/s, achieving a record data rate for multimode parallel optical transceivers.

Introduction

VCSEL-based optical transceivers are ubiquitous today in short-reach (<100 m) datacom and computercom applications. A notable example is IBM's *Sequoia* BlueGene/Q supercomputer at the Lawrence Livermore National Laboratory, which employs over 620,000 optical links [1]. As the bandwidth requirements and data rates continue to increase, future systems will require dense parallel optical transceivers which are more tightly integrated with the chip packaging. Over the past few years we have demonstrated dense optical transceivers based on flip-chip soldering of optoelectronic arrays directly onto a transceiver IC to form the transceiver "Optochip." The Optochip packaging concept provides optimum density and high-speed performance through the close integration of the optoelectronic (OE) devices with the IC amplifier circuits. To facilitate use of industry-standard surface emitting/detecting 850 nm VCSEL and photodiopde (PD) arrays, a novel "holey" Optochip transceiver was developed which incorporates optical vias (holes) fabricated into the IC [2,3]. The optical vias enable optical I/O from flip-chip soldered 850-nm OEs through the otherwise absorbing bulk silicon substrate.

A schematic representation of the holey Optochip is depicted in Figure 1. At the heart of the assembly is the transceiver Optochip with the integrated optical vias and flip-chip soldered VCSEL and PD arrays. To complete the transceiver packaging, the Optochip is flip-chip soldered to an organic carrier. The transceiver module can be soldered directly to a printed circuit board (PCB) for characterization. Two versions of the holey transceivers have been previously demonstrated, the first based on a CMOS IC and the second based on a SiGe IC.

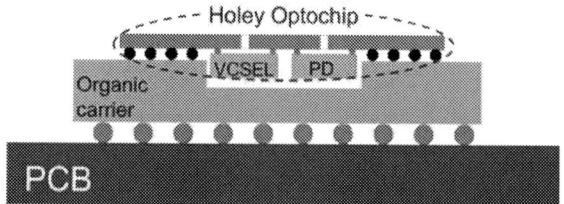

Figure 1. Cross-section of holey Optochip transceiver module with VCSEL and PD arrays.

The CMOS holey Optochip is based on 24 TX + 24 RX IC fabricated in IBM 90-nm CMOS process and incorporates flip-chip attached 2 x 12 VCSEL and PD arrays. A wafer-scale process is used to etch 48 optical vias at the center of IC. The CMOS holey Optochip transceiver operates up to 20 Gb/s/ch, providing a record aggregate data rate of 480 Gb/s TX + 480 Gb/s RX [2,3].

The SiGe holey Optochip is based on 4 TX + 4 RX IC fabricated in IBM 130nm BiCMOS8HP. Last year we reported on a transceiver Optochip consisting of this SiGe IC with flip-chip attached 4-channel VCSEL and photodiode arrays arranged in the typical configuration of 4 devices on a linear 250 μm pitch. Optical access to the conventional topside emitting 850-nm optoelectronic arrays is achieved through 8 optical vias fabricated into the transceiver IC [4]. High speed characterization showed all 8 optical links between the 2 transceiver modules operate error free (BER $< 10^{-12}$) up to 36 Gb/s [4]. At 36 Gb/s, the module achieves a bidirectional aggregate bandwidth of 144 Gb/s.

Here we report a new 4 TX + 4 RX transceiver based on the same holey SiGe IC but designed for coupling to a multicore MMF. The primary modification is the use of novel 4-channel VCSELs and photodiodes arranged in a multicore configuration, where the 4 OE devices are grouped together into a 2 x 2 array on a 50-μm pitch. This quad-configuration is designed to match a multicore, multimode fiber (MC-MMF), which has four cores in a single glass fiber. In addition, since only a single fiber is used for TX and a second

978-1-4799-2408-0/14 $31.00 © 2014 IEEE 2014 Electronic Components & Technology Conference

fiber for RX, the multicore holey Optochip requires only 2 optical vias.

Transceiver Design and Assembly

The single-chip SiGe IC with 4 transmitter circuits plus 4 receiver circuits was described in detail elsewhere [4]. Briefly, the IC is fabricated in standard IBM 130 nm SiGe8HP technology and measures 6.0 mm x 4.25 mm. The circuit areas for the 4-channel TX and RX blocks each occupy only about 0.6 mm² with a similar area reserved for optical via fabrication. Typical C4 bond pads are used for packaging of the IC while the majority of the chip surface area can be used for future thermal management solutions. To enable high-speed operation beyond 30 Gb/s, the circuit design included implementation of feed-forward equalizer (FFE) output blocks into both the TX and RX designs. Optical via fabrication was carried out using a laser ablation process [4]. Only a single optical via is required for the quad-VCSEL array and a second for the quad-photodiode array.

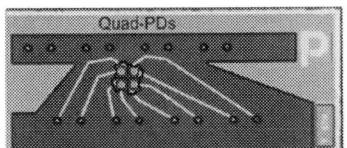 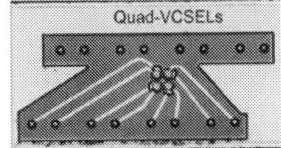

Figure 2. Quad-PD (left) and quad-VCSEL (right) arrays.

Figure 2 shows the quad-VCSEL and quad-PD arrays, which are custom 25 Gb/s-class high-speed optoelectronic arrays fabricated by Sumitomo Electric [5]. The OE devices are arranged in a 2 x 2 array on 50-µm pitch, matching prototype multimode multicore fiber fabricated by OFS having four 26-µm graded index cores within a single 125-µm glass fiber [6]. The PD has an approximately square active area of 18 µm dimensions, which maximizes the optical coupling to devices on 50-µm pitch. The VCSEL has a ~6.5 µm aperture and sub-mA threshold. The pairs of OE bond pads are arranged on a 250-µm pitch to match the identical pads on the SiGe IC that were originally designed for 1 x 4 parallel OE arrays. There are an additional 8 mechanical bond pads (top of array) for improved mechanical stability. AuSn solder was pre-deposited on the bond pads of the OE arrays.

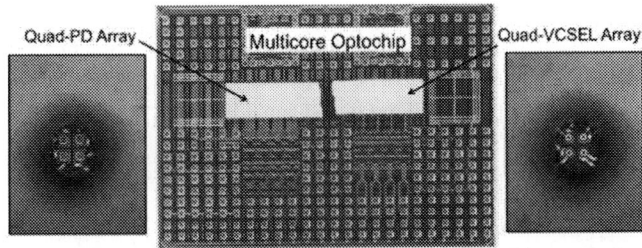

Figure 3. Multicore Optochip with flip-chip attached quad-PD (left) and quad-VCSEL arrays (right).

The OE arrays were flip-chip soldered to the SiGe IC using a Karl Suss FC-150 bonder at a temperature of about 300 °C to form the multicore transceiver Optochip. The Optochip seen in Figure 3 shows the SiGe IC with the attached

multicore OE arrays. The insets (left and right) show the OE devices seen through the two optical vias, with the individual quad-VCSELs and quad-photodiodes clearly evident.

The Optochip is then assembled onto the high-speed, high-density organic carrier. The carrier is fabricated using Endicott Interconnects CoreEZ technology and measures 8 mm x 8 mm with C4 pads on the top surface for attachment of the Optochip and BGA pads on 0.8 mm pitch on the bottom surface for mounting to a test card. Prior to Optochip attachment, a cavity is milled out at the center of the carrier to accommodate the OEs. The Optochip is flip-chip attached to the organic carrier using PbSn solder pre-deposited on the C4 pads of the carrier. The module is reflowed at ~200 °C to complete the assembly shown in Figure 4. The 2 optical vias are seen in the backside of the SiGe IC.

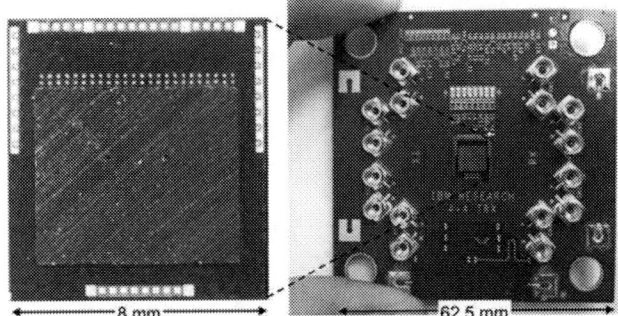

Figure 4. Fully assembled multicore transceiver (left) and transceiver module soldered to high-speed test card (right).

The complete module can be soldered to a test card for high-speed characterization, as shown in Fig. 4 (right). On the Nelco 4000 test card, low speed bias and control signal are supplied though a ribbon connector at bottom of card (not shown). The differential high-speed TX inputs and RX outputs are routed to pairs of high-speed connectors on the test card. Finally, the test card can be used for DC and high-speed characterization of the complete multicore transceiver module.

Thermal Modeling and Management

Initial high-speed characterization showed that the multicore transceiver operated at <30 Gb/s, whereas the previous linear 4 TX and 4 RX transceiver operated at >36 Gb/s [4]. In particular, the multicore laser performance was degraded compared to the linear VCSEL layout under similar operating parameters. These initial measurements and the previous linear array results were carried out under forced air cooling applied to the bare die transceiver module. Additional studies identified the primary limitation to high-speed performance as the thermal environment of the VCSEL device. Specifically, we observed (1) a higher device temperature for the dense quad-VCSEL configuration (50-µm 2-D pitch) relative to that of the linear 250-µm pitch configuration, and (2) optimization of the air cooling conditions was critical to minimize device temperature and improve performance.

The thermal environment of the VCSEL was assessed from the shift in the laser wavelength under a variety of VCSEL and IC drive conditions. The temperature increase was calculated

978-1-4799-2408-0/14 $31.00 © 2014 IEEE

from the shift in wavelength compared to the minimum lasing conditions (negligible IC power and near-threshold laser output) using the typical VCSEL temperature dependence factor of 0.06 nm/°C.

With minimal airflow, the multicore VCSEL temperature increases by >100 °C under nominal operating conditions. That is, the VCSEL junction temperature (T_j) is estimated at >130 °C. The junction temperature can be reduced by 10 – 20 °C by using moderate (few m/s) to aggressive airflow. VCSEL temperature >100 °C limits the high-speed performance as well as the long-term reliability.

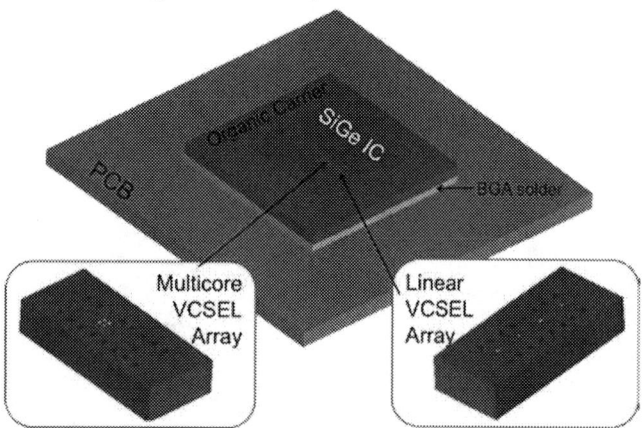

Figure 5. Transceiver model used for thermal modeling. Airflow is applied to the carrier from one side.

Modeling was carried out using the Ansys Icepak software package to further understand the thermal effects and to identify thermal management solutions. The 3-D model shown in Figure 5 includes all the components of the transceiver shown in Fig. 4: the VCSEL chip, the holey SiGe IC (transparent in Fig. 5), the 8 mm organic carrier and the test PCB. The organic carrier and PCB are modeled in detail with the applicable Cu/dielectric layer structure and multiple Cu planes interconnected through internal via fields within the dielectric layers. The components are interconnected through the bond pads: AuSn micro solder pads for GaAs to SiGe IC, PbSn C4 pads for SiGe IC to organic laminate and typical ball grid array (BGA) PbSn solder connections to the test PCB.

The VCSEL array is modeled as a GaAs substrate with dimensions 0.65 mm x 1.5 mm x 0.3 mm and 4 heat sources of 28-μm diameter by 5-μm height representing the mesa structure of the 4 devices. For the multicore version, the heat sources are on a 2 x 2 array on 50-μm center-to-center pitch while the linear model used a linear array of sources on 250-μm pitch.

The experimental condition of applying high-velocity air directly to the transceiver carrier through a 6 mm tube was simulated. For this simulation, typical transmitter operating conditions were used, namely 1.25 W SiGe driver power and 15 mW/laser. Only the transmitter power of the SiGe IC was included for these simulations.

Table I gives the VCSEL junction temperature under airflow conditions of 3.9 and 8.9 m/s. At low to moderate airflow, the VCSEL junction temperature rises beyond recommended operating parameters, with the multicore

configuration showing even higher temperatures. This is in agreement with experimental results of degraded high-speed performance. Only at the airflow of 8.9 m/s does the VCSEL temperature approach the desired range. Although these extreme airflow velocities are possible in a laboratory environment, they are not practical for typical operation.

Airflow	Multicore T_j (°C)	Linear T_j (°C)
3.9 m/s	122	118
8.9 m/s	90	87

Table I. VCSEL temperature for Multicore and Linear arrangement at various airflow conditions.

The temperature contour of the VCSEL die under the aggressive airflow of 8.9 m/s is shown in Figure 6(a). The temperature of the die itself is about 75 °C. Only the area adjacent to the devices is significantly higher, with a maximum temperature of 90 °C at the 4 devices.

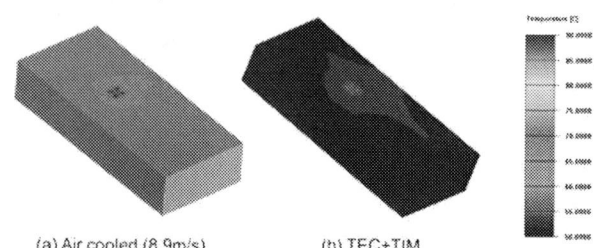

(a) Air cooled (8.9m/s) (b) TEC+TIM

Figure 6. Thermal modeling results: Temperature contour of VCSEL chip using (a) 8.9 m/s airflow and (b) TEC in thermal contact of PCB.

For improved high-speed performance, the experimental setup was modified to include a thermo-electric cooler (TEC) attached to the bottom of the PCB. This introduces a conductive thermal path through the package compared to the initial primarily convective heat transfer through the limited IC surface area.

The thermal model was modified to include a 30 mm x 30 mm TEC interface at the bottom of the PCB which is held at 25 °C. The model also includes the 0.5 mm thermal interface material (TIM) between the PCB and TEC. Table II gives the VCSEL temperature reached when this more efficient heat transfer path is simulated. As evident in Table II, the provision of a thermally conductive path through the package provides a vast improvement to thermal environment of the VCSEL devices. In both the multicore and linear VCSEL arrays, the junction temperature is greatly reduced to only 70 °C and 67 °C, respectively.

	Multicore T_j (°C)	Linear T_j (°C)
Bare GaAs	70	67
GaAs w/ TIM	65	62

Table II. VCSEL junction temperature for Multicore and Linear arrangements using TEC at bottom of PCB.

An additional concern is the limited heat transfer path from the GaAs laser chip, which is provided through the 16 micro solder pads between the GaAs and SiGe chips. The GaAs chip is otherwise surrounded by air in the organic carrier cavity. Improved thermal performance can be achieved by including a conductive path from the back of the GaAs chip. This can be achieved by exposing a Cu ground plane in the cavity of the organic carrier along with the application of a thermal interface material between the GaAs chip and the Cu plane. The effect of incorporating this TIM interface into the model is also given in Table II. The simulations show an additional 5 °C temperature reduction is achieved by the addition of a TIM at the backside of the GaAs chip. The simulated temperature contour of the VCSEL die under this optimized thermal condition is shown in Fig. 6(b). A substantial temperature reduction is seen compared to the temperature profile of the air cooling simulation seen in Fig. 6(a).

The TEC solution does provide effective thermal management for the VCSEL arrays and is suitable for the laboratory measurements. For general applications, additional modeling was carried out which verified the use of a finned heat sink in thermal contact with the perimeter of the SiGe IC combined with moderate airflow can provide adequate thermal management and limit the VCSEL temperature to suitable range.

High Speed Characterization

The high-speed measurements presented in this section were carried out using a TEC module attached to the bottom of the test card for both the transmit module and the receive module. In addition, the transmit module includes a TIM material added to the interface between the backside of the GaAs chip and a ground plane within the organic carrier. To further reduce the VCSEL temperature, the VCSEL drive current was decreased to 6 mA per device. This setting provides the optimum high-speed performance with a power consumption of about 12.5 mW per device.

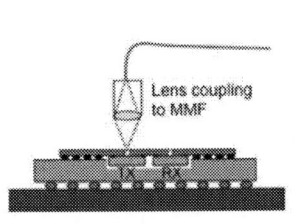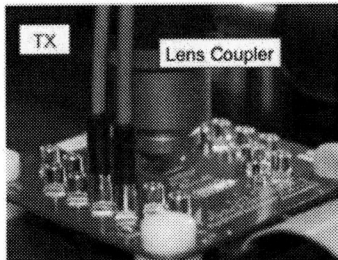

Figure 7. Measurement setup for high-speed TX characterization.

For high speed characterization, electrical input/output to the module is accessed using the high-speed SMP surface mount connectors on the test card, while optical input/output is coupled to MMF. For the TX, optical output is accessed using a lens-coupled MMF as shown in Figure 7. A 2-lens optical coupler was developed to couple light from an individual VCSEL from the quad array into a MMF. The lens system shown in Fig. 7 provides >20 dB isolation of the coupled VCSEL light from the neighboring 3 VCSELs in the quad array. Power and control signals are provided to the module through ribbon cables connected to the test circuit board.

The blue cables in Fig. 7 provide the high-speed electrical I/O to one channel at a time. An 8 m MMF fiber link is used to couple TX output to the RX module, one channel at a time. The test pattern used is 2^7-1 PRBS. Measurements are carried out using a 56 Gb/s SHF pattern generator and error analyzer. Eye diagrams are collected on a Tektronix DSA 8200 sampling oscilloscope with 60 GHz bandwidth.

Figure 8 presents the eye-diagrams measured for one TX channel at various data rates from 20 Gb/s to 42 Gb/s. The TX output eye diagrams are measured as a function of data rate using a Newport D25xr photodiode with 17 GHz bandwidth, which contributes marked eye closure at >30 Gb/s. Note that the TX FFE circuit is enabled for all measurements, with the resulting characteristic over/undershoot visible in the VCSEL outputs.

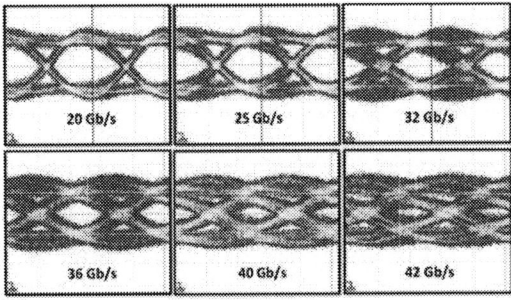

Figure 8. Optical eye diagrams for a typical TX channel at data rates 20 Gb/s to 42 Gb/s.

On the receive side, a lensed MMF is used to couple light into individual photodiodes. Since the PD device locations are offset from the center of the optical via, standard MMF with 125-μm diameter does not allow sufficient travel within the approximately 150 μm diameter optical via to access the individual PDs. Instead, we used a custom lensed MMF probes fabricated using 50/80 MMF; that is, 50-μm core and 80-μm clad.

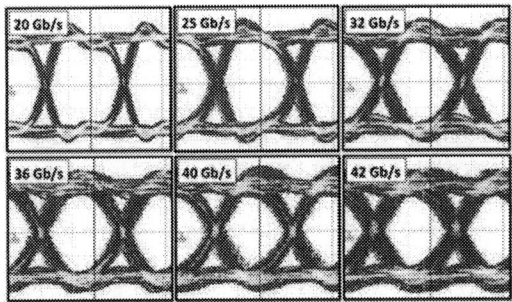

Figure 9. Electrical eye diagrams for a typical TX-to-RX link at data rates 20 Gb/s to 42 Gb/s.

Figure 9 shows eye diagrams for a complete TX-to-RX link for one channel at data rates form 20 Gb/s to 42 Gb/s. Best performance is achieved by enabling the RX FFE to drive through the organic carrier, few cm transmission lines in the test card, and the ~8" cables connecting the RX outputs to the error detector and oscilloscope. As evident in Fig. 9, open eye diagrams are seen for data rates up to 42 Gb/s.

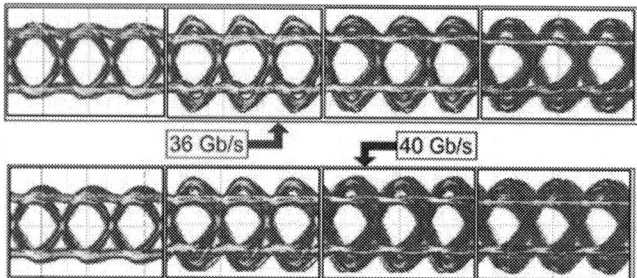

Figure 10. Electrical eye diagrams for all 4 TX-to-RX links at data rates 36 Gb/s (top) and 40 Gb/s (bottom).

Figure 10 shows eye-diagrams obtained from all 4 channels between the TX module shown in Fig. 7 and a separate RX module with the same IC settings. The data represents 4 unique links from each VCSEL in the TX module to 4 different photodiodes in the RX module. Fig. 10 (top) presents the RX electrical output eye-diagrams at 36 Gb/s, while Fig. 10 (bottom) shows the eye-diagrams produced at 40 Gb/s. All 4 links show open eye diagrams up to 40 Gb/s.

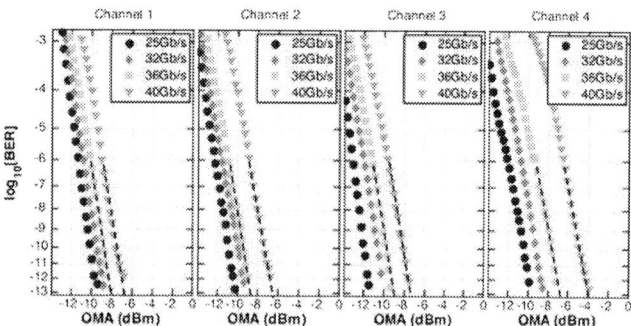

Figure 11. Sensitivity characteristics measured for all 4 links (channels 1 - 4) between TX module and RX module at data rates 25, 32, 36 and 38 Gb/s.

Receiver sensitivity characteristics were measured for all 4 links over 8 m of MMF. Figure 11 shows the RX sensitivity results for the 4 TX-to-RX links (channels 1 – 4) at various data rates from 25 Gb/s to 40 Gb/s. The bit error ratio (BER) is plotted as a function of optical modulation amplitude (OMA). At the lowest data rates of 25 Gb/s, the sensitivity at BER = 10^{-12} is < -10 dBm OMA for all 4 links. As the data rate is increased to 32 Gb/s, a small penalty of 1 dB is incurred. Additional 0.5 -1.5 dB penalty is incurred in moving to the 36 Gb/s. At 40 Gb/s rates, all channels operate error-free although a significant variation in sensitivity is observed due to link bandwidth limitations.

Conclusions

We have successfully demonstrated the design, fabrication and high-speed characterization of a 4 TX + 4 RX optical transceiver designed for coupling to multicore MMF. Based on the highly integrated holey Optochip, the transceiver is a single-chip SiGe IC with flip-chip soldered 4-channel VCSEL and photodiode arrays. The novel aspect of the transceiver is the use of quad-VCSEL and quad-PD devices on a 2 x 2 array on 50-μm pitch. This dense arrangement is designed for direct coupling to multimode MMF, where 4 graded-index cores are fabricated within a single 125-μm glass fiber. The SiGe Optochip also incorporates two optical vias, one for the TX and one for the RX, for optical access from the substrate surface using 2 multicore fibers. The Optochip is further flip-chip soldered to an 8 mm x 8 mm high density organic laminate to complete the transceiver module.

High-speed characterization was carried out between 2 modules, a transmitter module and a receiver module, that were soldered to test circuit boards. Measurements on the 4 optical links between the 2 modules showed all 4 links operate error free (BER < 10^{-12}) up to 40 Gb/s, achieving a record data rate for multimode parallel optical transceivers. The 40 Gb/s data rate, achieved using 25 Gb/s class VCSELs and photodiodes, is enabled using innovative transmitter pre-distortion circuitry incorporated into conventional IBM Silicon Germanium technology.

References

1. Lawrence Livermore National Laboratory, Advanced Simulation and Computing, *Sequoia*, https://asc.llnl.gov/computing_resources/sequoia/
2. C.L.Schow et al., "A 24-channel 300 Gb/s 8.2 pJ/bit full duplex fiber-coupled optical transceiver module based on single Holey CMOS IC," IEEE J. Lightwave Technol., vol. 29, no. 4, pp. 542-553, Feb 2011.
3. F.E.Doany, et al., "Terabit/sec VCSEL-Based 48-Channel Optical Module Based on Holey CMOS Transceiver IC," J. Lightwave Technol., vol. 31, no. 4, pp. 672-680, 2013.
4. F. E.Doany et al., "Single-chip 4 TX + 4 RX Optical Module Based on Holey SiGe Transceiver IC," Electronic Components and Technology Conference (ECTC), 2013 IEEE 63st, pp.268-274, May 28 -June 1 2013.
5. N.Y. Li, C. L. Schow, D. M. Kuchta, F. E. Doany, B. G. Lee, W. Luo, C. Xie, X. Sun, K. P. Jackson, and C. Lei, "High-Performance 850 nm VCSEL and Photodetector Arrays for 25 Gb/s Parallel Optical Interconnects," Optical Fiber Communication Conference (OFC), 2010.
6. OFS Fitel, http://www.ofsoptics.com/

336-Channel Electro-Optical Interconnect: Underfill Process Improvement, Fiber Bundle and Reliability Results

Shuki Benjamin[1], Kobi Hasharoni[1], Avi Maman[1], Stanislav Stepanov[1], Michael Mesh[1], Helge Luesebrink[2], Roland Steffek[2], Wolfgang Pleyer[2] and Christian Stömmer[2]

[1] Compass-EOS, Netanya, Israel

shuki@compass-eos.com

[2] PVA TePla AG

Abstract

The icPhotonics™ optical module assembly development was presented in [1]. The module allows for aggregate full duplex data rate of better than 1.30 Tb/s with Bit Error Rate (BER) < 1E-12 into and out of a single ASIC die for distances up to 300m. The module assembly uses standard semiconductor processing equipment and materials. The icPhotonics™ components – an organic interposer (substrate), ASIC (standard TSMC 65nm node technology), two 168-channel 1000nm optical dies (VCSELs and photodiodes) and two Micro Lens Arrays (MLA's) dies of 168 pixels each – are all assembled using flip-chip technology, allowing for a very high density data rate transfer.

The 1.344 Tb/s data transfer rate is enabled from the 168 VCSELs and 168 PD's through two sets of 168 multi-mode fiber bundles. The fiber-bundle's design and special production sequence enable high efficiency and low-loss optical data transfer to a standard MT-Ferrule connector. Several icPhotonics™ modules can thus be connected and allow applications which require a very high BW data transferred between different ASIC dies, with bandwidth density of 64Gb/s/mm^2.

A major production challenge in the icPhotonics™ production was associated with the unusual underfill application for an optical module with 2 optical dies in the center of the ASIC die and a hole in the center of the organic substrate. The overall yield of the Optical Interconnect (OI) was enhanced using MW-plasma treatment before applying underfill. In this paper the general failure modes associated with the underfill process are examined and a comparison of the yield with and without the plasma activation is presented.

Various reliability tests were performed on the icPhotonics™ module, its components and the accompanied fiber-bundle assembly. The entire assembly was exposed to thermal cycles and rapid aging procedures, testing the soldering performance. The optical components, especially the VCSELs, were also exposed to prolonged rapid aging procedures while testing their optical performances.

Reliability tests and results are presented, as well as additional thermal cycles performed on the large format fiber-bundle, testing BER results changes after exposure to environmental conditions.

Introduction

The exponential increase in bandwidth requirement, and the crucial role played out by the optical interconnect was discussed and reviewed in depth [2]-[4]. The icPhotonics™ module technology was presented as an enabler for very high density data rate transfer solution ([1], [5]-[6]). The schematic structure of the icPhotonics™ module can be seen in Figure 1, and the actual module in Figure 2.

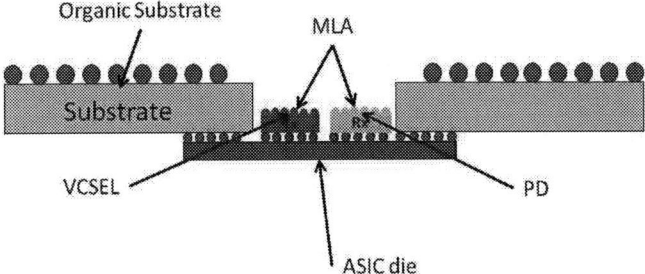

Figure 1: A schematic cross-section of the icPhotonics™ module

This technology, when utilizing 8Gb/s VCSEL and PD, had demonstrated a total of 1.344 Tb/s full duplex optical connectivity measured with bit-error-rate (BER) < 10^{-12} and power efficiency of 10pJ/bit, with data rate density of 64Gb/s/mm^2. The first application for this technology had also been demonstrated, when integrated with a core router traffic manager. It is very easy to extrapolate for the total BW one can achieve when moving to the already established 25-32Gb/s VCSEL and PD technology, without any major technology changes in the icPhotonics™ module structure, or when using larger optical dies without adding to the module assembly complexity.

This paper would further detail several issues concerning the icPhotonics™ technology and the complete OI assembly.

Figure 2: The icPhotonics™ module where the hole in the middle of the substrate can be clearly seen.

978-1-4799-2408-0/14 $31.00 © 2014 IEEE

2014 Electronic Components & Technology Conference

The first issue presented regards the unique underfill application problem encountered in this assembly design. Underfill application is now a well-established practice in flip-chip assembled devices ([7]-[9]). Current practice had developed some comprehensive as well as empirical models for the flow between ASIC die and substrate, and common application practices for different sizes and shapes of devices were published therein. However, in the icPhotonics™ module a unique situation presents itself, since the large-format optical dies are attached to the center of the ASIC die – which forced us to create a hole in the substrate (Figure 2). There is now a new limitation on the underfill process, where the underfill adhesive cannot be allowed to flow into the optical dies area. This limitation in addition to the large die (more than 400 mm^2) and high pin-count (~7000 bumps) made the otherwise common underfill process into an engineering challenge.

The second issue regards the complementary part of the icPhotonics™ module – the high density large-format fiber-bundle which is required in order to transmit and receive the data to other modules and complete the OI structure. The bundle general description was given in ([10], [11]). Here some more details explaining the assembly procedure will be given.

The third and last issue concerns the reliability data of the entire OI as was measured in Compass-EOS. Reliability data is presented for the new features presented in this module, both in connectivity, optical power loss and optical dies performances.

Underfill application and MW Plasma for surface state control

There were two main issues required to overcome when applying the underfill to the icPhotonics™ module: (i) preventing the underfill glue from flowing over to the optical dies and (ii) controlling the surface state within the relative large die size and high pin-count with different pitches.

(i) Around the hole

After deliberating several options it became evident that the hole should be used as an escape route for the trapped air between the ASIC and the substrate. The underfill adhesive has to be injected from all sides around the die, creating a uniform front which moves inward, driving the air to the hole.

The Asymtec S-820 system with DJ-9000 is used to control underfill dispensing position and weight. By controlling the weight dispensed it was possible to control the underfill adhesive and prevent it from overflowing out over the optical dies (Figure 3). This method also helped in preventing air voids from forming between the ASIC die and the substrate.

Figure 3: A close-up view of the underfill fillet inside the hole. No overflow on optical die – the flow is well controlled and stops at the edge of the substrate.

(ii) Controlling surface state

The developments and trends in Flip Chip die size increase and standoff reduction pose continuous challenges for void free packaging in flip chip underfill processes in the industry. Current conditioning techniques prior to underfill or molding, are typical surface treatments using low pressure plasma for cleaning and activation [12 - 14]. The main goal of the plasma treatment is to prevent delamination or voids formation in the polymer substrate / chip interface resulting in yield loss (Figure 4).

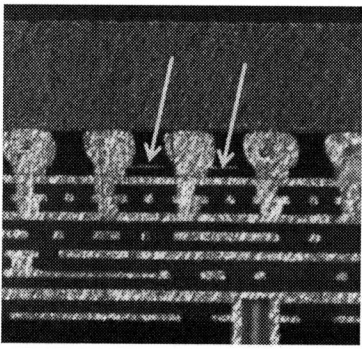

Figure 4: Solder flowing into delaminated voids which were created due to bad surface preparation

Typically low pressure plasma systems are used with different plasma excitation frequencies such as 13.56 MHz for RF or 2.45 GHz for microwave (MW). The effects of the plasma treatment are twofold: There is the change in surface chemistry by creating polar groups (typically implanting H, or O), and a removal of surface contamination. This can be achieved either physically by ion bombardment (RF), where surface and subsurface bulk atoms are physically removed /sputtered from the substrate or chemically (MW), where free radicals modify the substrate surface composition for activation and subsequently cleaning by forming new volatile molecules that will be removed by the evacuation system. In addition there is a directional, anisotropic, physical effect in RF plasma, attracting the positively charged ions towards the electrode, resulting in a perpendicular impact on the sample surface. When using MW plasma, there is a more isotropic, chemical effect taking advantage of the radical's free mean path to reach the center under the die and react with all exposed surfaces.

When examining these effects on Flip Chip applications treated with RF plasma it can be expected to see a shadowing effect of the Flip Chip itself preventing a sufficient surface treatment in the center under the die to ensure good adhesion of the underfill compound. This has been examined by comparing the surface tension of Flip Chip samples with different die sizes. A contact angle (CA) measurement with goniometer and DI water has been conducted to determine the starting condition before plasma. The CA before plasma was 80° on the substrate. Similar gas mixtures of Argon (Ar) and Oxygen (O_2), and a plasma time of 30sec have been applied to evaluate both physical and chemical effects with each plasma excitation frequency. As seen in Figure 5, the surface activation efficiency decreases for RF plasma treated samples with increasing die size. CA measurements were performed in the center position under the flip chip . Gasses used were Ar / O_2 with power, flow and time optimized.

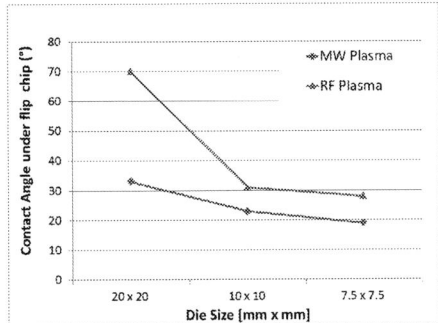

Figure 5: CA comparison for different flip chip die size. Gap size was 100µm, 40µm and 70µm and pitch 130µm, 60 µm and 90µm for die sizes 20X20, 10X10 and 7.5X7.5 mm respectively

Another factor influencing the plasma system choice for the icPhotonics[TM] treatment was the effect it might have on the optical dies. With RF plasma there is the concern that ion bombardment might cause electrical charge accumulation on the surface of the dies, where plasma charging induced damage may occur.

Taking these effects and results into consideration a further process development using MW plasma has been initiated using a batch system GIGA 690 MW plasma.

The particular challenge for the optical module was its unusual structure that required absolute control on surface condition and underfill flow pattern in order to prevent the underfill from flowing onto the optical dies. To accomplish this, MW plasma treatment was selected also due to the fact that the surface modification is mainly chemical with consistent uniformity along the edge and across the surface under the die. The activated substrate and Flip Chip show uniform flow patterns of the fillet and no creeping of underfill material were detected onto the die (Figure 3, Figure 6).

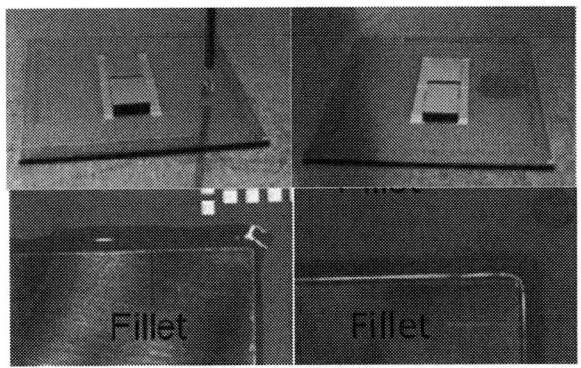

Figure 6: On the left, CA and fillet shape without applying MW plasma. On the right a much better CA can be seen and the resulted fillet shape after applying MW plasma.

The MW plasma process was gradually added to modules in production and resulted in a significant decrease of x4 in failures related to underfill delamination when assembling the module on the OI (Figure 7).

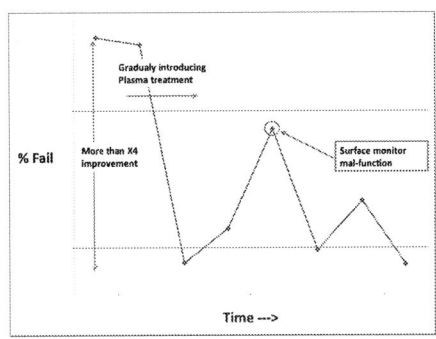

Figure 7: icPhotonics[TM] module Failure [%] related to bad surface preparation.

High-density Fiber-Bundle

The icPhotonics[TM] module will always be conjugated with a similar large format fiber-bundle which will be used to transmit and receive high density data rate from one module to the other, or alternatively to and from other systems. The ability to assemble this fiber-bundle with high yield rates and excellent reliability is one of the key elements of the OI features presented here.

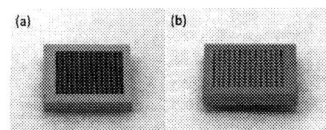

Figure 8 Si die with 168 through via's where (a) the back side and (b) the front side with alignment-marks and die ID can be seen.

The general structure and building blocks of the large-format fiber-bundle are further described in [10] and [11]. The basic building block is a Silicon die with 168 through-holes via's (Figure 8). These dies are fabricated using standard Si processes, which allows achieving high yield and inexpensive processes. After the Si wafers are fabricated and diced, the dies are tested for internal diameter and wall angle and then moved to fiber insertion.

Figure 9: fiber ribbons insertion. The ribbons are inserted one at a time: (a) the insertion of the first ribbon, and consequently (b) the 5th and (c) the 14th.

The fibers are prepared in advance in groups of standard 12 fibers in each ribbon, where one side is stripped, and the other is tested with a standard MT-Ferrule connector. The ribbons are then inserted using specially designed fixtures in groups of one ribbon at a time. Each ribbon is aligned to the back-side of the Si die and inserted through the holes (Figure 9) until all of the 168 fibers are through.

When all 168 fibers are inserted, glue is applied and the cover is placed to protect the fibers (Figure 10). The front side of the Si die is then treated for fibers planarity and tested for damage, continuity and optical power-loss in transmittance.

Figure 10: Adjusting the cover on the fibers.

After gluing and adjusting the bundle cover and housing, testing the bundles and marking them, the assembly is ready for a micro-lens-array die to be glued on the front side of the Si die of the bundle (Figure 11a).

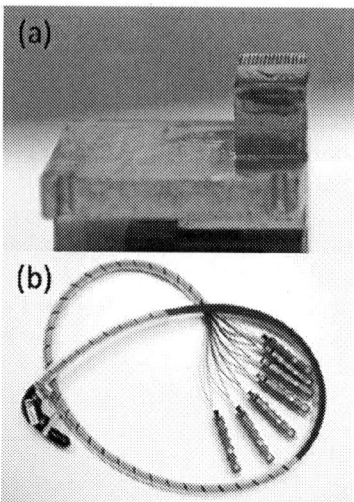

Figure 11: a complet assembled and tested large-format fiber-bundle. At the top (11a) – a close-up view of the MLA glued to the Si die in the bundle. At the bottom (11b) – the complete bundle assembly with the fibers protected and ready to be installed with the icPhotonics™ module.

The optical design for the OI is based on a 2-lens relay system. The first lens is used to collimate the optical beam and the second then focuses the beam back into the fiber or the detector pixel (depending on transmit direction). This design allows for a very relaxed alignment tolerances of the fiber-bundle to the optical dies, as shown in Figure 12.

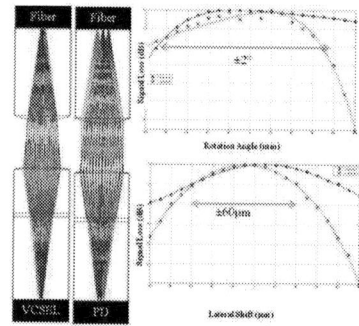

Figure 12: The OI optical system design. Using a two-lens system allows for a very-relaxed alignment tolerances as shown on the right.

The final bundle is now ready to be attached (Figure 11b) and complement the full OI assembly (Figure 13), allowing for a very powerful parallel optical data transmitting to and from the ASIC die using 168 channels.

This type of OI was first applied for core-router application as described in [1], [5]-[6].

Figure 13: Two large-format fiber-bundles complement the icPhotonics™ module and create a powerful 1.3Tb/s parallel Optical-Interconnect used in the r10004 router application.

Reliability tests results

This section will focus on the special reliability considerations involving some of the unique features of the OI when compared to regular chip packages. While there are some regular features, such as bumps soldering of the ASIC to the substrate or BGA soldering of the substrate to the PCB, there are also many unique features, for example: soldering the optical dies to the ASIC die in the icPhotonics™ module, using the large-format fiber-bundle and assembling it onto the PCB and even the endurance of the optical dies themselves – all are unique features that do not fall into any formal category for device qualification. However, since the OI main usages are in communication systems, where the chassis usually undergo through NEBS qualification, it was decided to run a custom reliability program for the OI following GR-63, GR-1221 and GR-468 standards for environmental criteria, passive optics and electro-optics devices.

Three main failure scenarios were tested here, and are illustrated in Figure 14: (i) the soldering of the optical dies to the ASIC die, (ii) the optical dies reliability while working in the OI environment and (iii) the bundle assembly reliability.

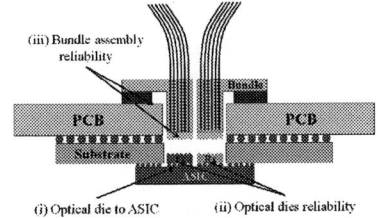

Figure 14: A schematic drawing of the OI with the three failure scenarios presented here for reliability study.

(i)   Optical dies to ASIC die reliability

The main failure mechanism at this scenario is the disconnection of the solder due to creep formation arising from the Coefficient of Thermal Expansion (CTE) differences between the Si and the III-V materials of the optical dies. In order to overcome this failure mechanism, gold pillars are used with Au/Sn eutectic as described in [1]. For reliability study, a custom Si and III-V dies were designed with alternate daisy-chain connections. The design allows testing continuity of the connection at several intermediate test-points along the die (Figure 15)

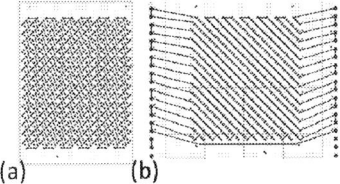

Figure 15: The optical (a) and Si (b) dies daisy chain design. There are 16 seperate chains which can be tested seperatly to locate a disconnection if required.

The dies were assembled with the standard production recipe, tested for continuity and then performed both thermal-cycles (-40^{0}C / +80^{0}C, 1000 cycles) and accelerated aging (170^{0}C, 1300hrs – equivalent to 10 years normal operation) processes. Sample size was 25 units. No failure was found on any of the tested assemblies.

(ii)   Optical dies reliability

The VCSEL and PD arrays used in the OI are standard commercial products, and as such passed full qualification and reliability tests by their perspective vendors (Philips photonics for the VCSEL dies and Albis optoelectronics for the PD). Nevertheless it was decided to repeat these tests using the assembly procedures used in the OI system. Particular emphasize was given to infant mortality rate, with the goal of deciding whether to install a burn-in process for incoming wafers.

The optical dies were flip-chip bonded to customized Si dies, which were bonded to a specially designed PCB (Figure 16). The device temperature was controlled and logged. The number of working channels was recorded after assembly, and then tested again every couple of hundreds of hours. Devices were tested for optical power (VCSELs) and dark-current (PDs).

Figure 16: Reliability study fixtures for optical dies, where (a) presents the custom Si die with optical dies flip-chip attached and temperature measurements leads and (b)+(c) is an overview of all the devices being tested.

The results indicate an infant mortality rate which is less than 1%. This rate is acceptable with the OI specification sheet, and thus it was concluded that no burn-in process was required with these lasers. The stressed optical power and dark-current trends are presented in (Figure 17). Optical power fail criteria for the VCSEL is 2dB loss, and for the PD a 2x increase of dark-current from initial dark-current values.

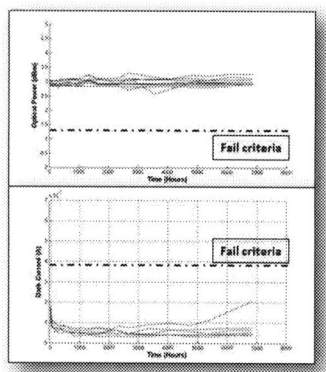

Figure 17: VCSEL optical power degradation (top) and PD dark-current increase (bottom) after 7000hrs. No failures are seen and the test is on-going.

(iii) Bundle assembly reliability

A lot of effort was invested in testing the bundle attachment reliability, since this is a key feature and a crucial factor in the OI architecture. The assembly itself follows NEBS guidelines. There were two tested objects in this study – the MLA attachment to the bundle and the entire bundle assembly.

The MLA gluing procedure was tested after alignment and glue, with an aggressive stress test, applying 120^0C in 1.5Atm at 100% humidity for 300 hours. The samples were then tested for alignment again and went through shear test. The results show no alignment failures at all. The shear force measured was > 2 kg ($x10^5$ of the MLA weight). This is much higher than the typical ~10g acceleration present during linecard plugging in the r10004 chassis.

The bundle assembly test method was to expose the entire OI (with the PCB and the icPhotonics^{TM} module) to harsh conditions and then test for BER using a sampling scope. Three tests were performed: Fatigue test with 1000 thermal cycles of $-20^0C/+60^0C$, humidity cycles with 5 cycles of exposure to 24hrs 95% humidity and vibration tests with 1g acceleration along 3-axis at frequencies of 5-100 Hz.

After 1000 thermal cycles, no change could be seen in the eye-scope. This implies that there was no creep movement creating fibers misalignment, and thus increasing error count. The results after the humidity cycles were plotted as a distribution of maximal changes in the optical power (Figure 18)

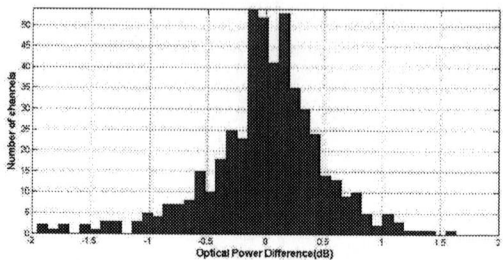

Figure 18: Optical power difference before and after humidity cycles for all devices tested.

The same method is used for presenting the vibration tests results (Figure 19)

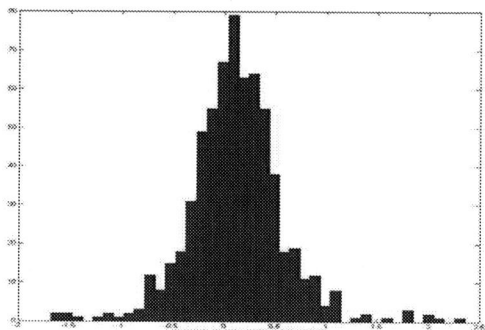

Figure 19: Optical power differences before and after vibration tests for all devices tested.

All of the results are well within the system spec for optical power degradation and demonstrate the robustness and stability of the OI at any system operation environment.

Conclusions

This work illustrates several new aspects of the icPhotonics^{TM} module underfill application in particular with the introduction of MW plasma controlling the surface state, as well as details on the large-format fiber-bundle assembly. There are also some new reliability results of the entire OI assembly demonstrating high reliability as required by communication systems.

The OI presented demonstrates inherent flexible design which allows for a very high density band-width transmission between ASIC modules of various applications. Ramping up for higher band-width densities is easily extrapolated while using faster optical dies or larger formats, all this without changing the basic assembly technology.

Acknowledgments

We gratefully acknowledge the help and technical expertise of our colleagues at the Fraunhofer Institute for reliability and micro integration, in particular Dr. H. Oppermann, Dr. M. Hutter, Karl-Friedrich Becker and their groups, in the development of the OM assembly process.

References

1. Benjamin, S. et al., "Assembly development of 1.3 Tb/s full duplex optical module", in *Proc. IEEE Electronic Components and Technol. Conf. (ECTC)*, Las-Vegas, NV, May 28-31, 2013, pp. 292-296

2. Berthold J., "Optical Networking for Data Centers Across Wide Area Networks," in Proc. OFC, paper OW1J.1 (2012).

3. Lam, C. F. et. al., "Fiber Optic Communication Technologies: What's Needed for Datacenter Network Operations", IEEE Communications Magazine, 48, 32-39 (2010).

4. Taubenblatt, M. A., "Optical Interconnects for High-Performance Computing," J. Lightwave Technology, 21:5, 1242–1255 (2003).

5. Hasharoni, K. et al., "A High End Routing Platform for Core and Edge Applications Based on Chip To Chip Optical Interconnect," in Optical Fiber Communication Conference/National Fiber Optic Engineers Conference 2013, OSA Technical Digest (online) (Optical Society of America, 2013), paper OTu3H.2.

6. Hasharoni K. et. al., "A 1.3 Tb.s Parallel Optics VCSEL Link", in Proceedings of SPIE vol. 8991 (2014) (to be published)

7. Li P. C. et al., "Underflow Process for Direct-Chip-Attachment Packaging", in Proc. IEEE Symp. Polymeric Electronics Packaging, 1997, pp. 273-283

8. Davidson D. et al., "Effect of Geometric Surface Features on Void Formation: Application to the Underfill Process", in *Proc. IEEE Electronic Components and Technol. Conf. (ECTC)*, pp.1411-1418, (2002)

9. Wang J. et al., "The Effect of Flux Residue and Substrate Wettability on Underfill Flow Process in Flip Chip Packages", in *Proc. IEEE Electronic Components and Technol. Conf. (ECTC), pp. 467-473, (2006).*

10. A. Geron and G. Katz, US D675,996 S

11. A. Geron and G. Katz, US 2012/0106898 A1

12. Lin, T. S. et al., "Impact of low-K wire bond stacked flip chip CSP package material on reliability test", in *Proc. IEEE Microsystem, Packaging, Assembly, Circuits Technology International Conf. (IMPACT), pp. 22-25, (2007)*

13. Yeh, C. T.et al., "Underfill Adhesion Enhancement via Optimization of Plasma Treatment for 3D IC", in *Proc. IEEE Microsystem, Packaging, Assembly, Circuits Technology International Conf. (IMPACT), pp. 327-329, (2013)*

14. Teo, M. et al., "Plasma Surface Modification and Impact on MSL Performance for Flip Chip Packaging", in *IEEE Electronics Packaging Technology Conference, 2007 (EPTC)*, pp. 657-663 (2007)

Development of Optical Multi-channel Connector for Rigid Waveguide - Fiber Optical Interconnection

Kazumi Nakazuru[1], Satoshi Asai[2], Masatoshi Tsunoda[1], Naoki Takahashi[1] and Takahiro Matsubara[2]

[1] R&D Division, KYOCERA Connector Products Corporation
402-1 Nakayama, Midori, Yokohama, Kanagawa, 226-8512 Japan
kazumi_nakazuru@kyocera-connector.jp

[2] R&D Center, Keihanna, KYOCERA Corporation
3-5-3 Hikaridai, Seika, Souraku, Kyoto, 619-0237 Japan

Abstract

We propose a newly developed optical multi-channel connector to connect a rigid optical waveguide array and an optical fiber array for an on-board optical interconnection.

The prototype of a 24-channel optical card edge connector to connect between a rigid multimode waveguide array fabricated onto a substrate and a multi-mode fiber array is manufactured, and two types of optical connection losses are defined and evaluated by the direction of the optical signal. The optical connection losses from the waveguide to the fiber as transmitting function and from the fiber to the waveguide as receiving function, are less than 0.7 dB and 0.6 dB, respectively.

Introduction

An on-board optical interconnection to connect high-performance chips in advanced electrical equipment for computing and networking is strongly demanded. The limitation of electrical signaling among large-scale integration chips using metal lines in such equipment is recognized in speed, density, and mainly power consumption. Many discussions and developments have been reported for on-board optical interconnections to connect chips with respect to optical waveguides, optical coupling and connection, assembly techniques, and module-level integration of these elements [1] [2] [3]. These developments describe the application of optical waveguides and/or optical fibers as optical path lines to transmit high-speed signals inside equipment.

With respect to the wiring of the waveguides and the fibers, optical fibers formed using silica glass has a strong advantage in terms of optical transmission loss. Current optical waveguides that are mainly formed from transparent polymer material exhibit a hundred times more loss. On the other hand, optical waveguides have the advantages of line density, and the size of the input and output (I/O) ports when fabricated onto rigid substrate. Optical I/O ports of both optical modules and signal lines on a substrate can be designed to have a smaller size when modules are directly attached onto a waveguide without a larger-sized optical connector to terminate the optical fiber array.

Therefore, we believe that with an optical interconnection structure to combine the waveguide and fiber can achieve good signaling for a required link distance and higher-density packaging in a circuit substrate. It is possible to apply rigid optical waveguides on substrates for higher density optical I/O to attach the optical module and couple optical signals, and to apply optical fiber to transmit signals over longer distances

[4]. Figure 1 shows the block diagram of the on-board optical interconnection.

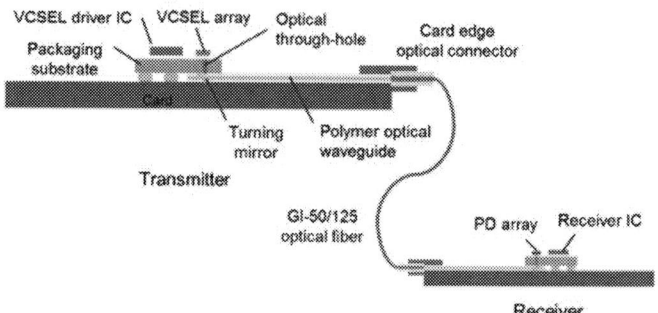

Figure 1. Block diagram of the on-board optical interconnection

In addition, the structure shown in Figure 1 also has advantages for both the heat dissipation of optical modules and the handling of fiber ribbons. In the case of fiber ribbons connected to the top the of photo-semiconductor chips (VCSEL array, PIN-PD array) using an optical connector and coupling lens, the heat sink cannot be attached onto the driver/receiver ICs. Furthermore, it is also time consuming to bundle and fix fiber ribbons on the card surface.

Thus, the use of optical multi-channel connectors to connect rigid optical waveguides which are fabricated onto substrates, and optical fibers at the edge of the card is effective and required.

Optical Design of Card-Edge Connector

Optical connector losses

Optical power and loss designs that include the optical connection between the optical waveguide and fiber are required to establish an on-board, high-speed optical link. In the case of a 10 Gbit/s transmission between chip-to-chip on the board, and card-to-card interconnection, as shown in Figure 1, the link distance is almost 1 m in rack-type equipment such as servers and routers, the total loss should be less than 10 dB, including the power penalty caused by mode dispersion and crosstalk is estimated. The maximum loss estimation as the link budget is; VCSEL to waveguide coupling: 2 dB, waveguide: 0.5 dB (0.1 dB/cm × 5 cm), waveguide to PD coupling: 2 dB, power penalty: 2.5 dB, then a 1.0 dB loss is allowed on each optical connector to connect the waveguide and fiber at the card edge. For higher link speed such as 25 Gbit/s or more, lower loss is required according to link budget design.

978-1-4799-2408-0/14 $31.00 © 2014 IEEE   1028   2014 Electronic Components & Technology Conference

For the multi-channel connection of the optical waveguide to fiber, several methods have been developed. However, a lower-loss and easy-handle optical connector is still required. Therefore, we developed and reported the new type card-edge optical connector that achieved the lowest loss, it connects the end face of waveguide and fiber at the card edge without either index matching or any lenses [5].

Polymer optical waveguide

The cross-sectional shape of the polymer waveguide core is typically a rectangle. The important design parameter is the core size of the optical waveguide for lower loss in an optical connector. Figure 2 shows the calculated result of the connection loss, which was obtained using the ray-trace method on a rectangular core optical waveguide and graded-index-type multimode optical fiber, a 50 μm diameter core, and a 125 μm diameter cladding (GI-50/125) connection without index matching.

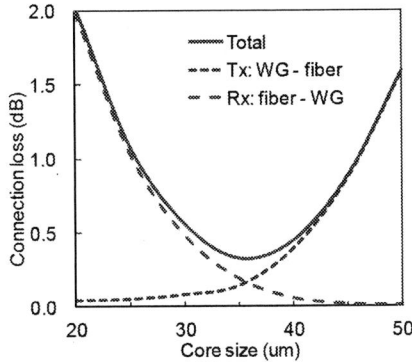

Figure 2. Calculated results for the relation between the core size of the optical waveguide and connection loss

As shown in Figure 2, in the waveguide to the fiber direction (called the transmitter function, Tx), a larger size of waveguide core has a larger loss. On the other hand, a larger core has a lower loss in the fiber to the waveguide direction (called the receiver function, Rx). With regards to the total loss, the minimum loss is for a 35 μm square core.

(a) Cross-section of a 24-channel polymer optical waveguide

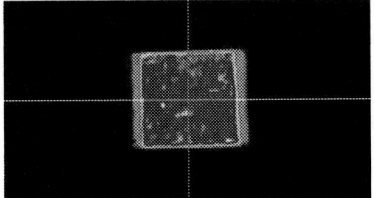

(b) Near-field pattern of the optical waveguide

Figure 3. Polymer optical waveguide

Figure 3(a) shows the cross-section of a 24-channel optical waveguide. The core size is a 35 μm^2, and the pitch of the core is 125 μm. The near-field pattern of the optical waveguide is shown in Figure 3(b). It was measured with 850 nm light emitted from VCSEL and, then inputted via the GI-50/125 optical fiber. The refractive index of the core is 1.58, and the index ratio is 2.5%.

The waveguide consist of three layers: the under cladding, core, and over cladding. The waveguide using transparent polymer material is directly fabricated onto the organic substrate. Conventional UV photolithography process is applied to form the core and each cladding of the waveguide.

Figure 4 shows the top view of the optical card. The 95 x 55 mm size card is designed to interconnect a large size LSI and an optical fiber. It features; i) The BGA pad to mount LSI of 1000 pad classes are designed at left-side area, ii) The area to attach two of 9 mm size and two of 5 mm size optical modules are formed in the center of the card. There are small size BGA pads, and reflective mirrors to couple optical modules and waveguides. iii) The curved optical waveguide to convert the pitch into 125um at the card edge from 250 um at the module area is fabricated in the right-side area. The alignment studs are fabricated both the module area and the card edge portion, respectively.

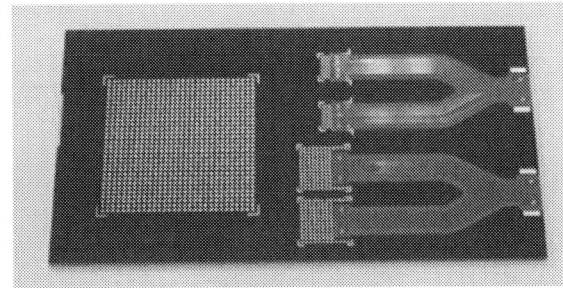

(a) Optical card

(b) Part of the curved optical waveguide

Figure 4. Curved optical waveguide fabricated onto the card

Optical characteristics of the waveguide and fiber joint

Figure 5 shows the increase of insertion loss with a lateral offset of GI-50/125 fiber to the waveguide without index matching. According to this experiment, the tolerance between the waveguide and fiber is within +/− 8 μm for less than 0.5 dB loss increase.

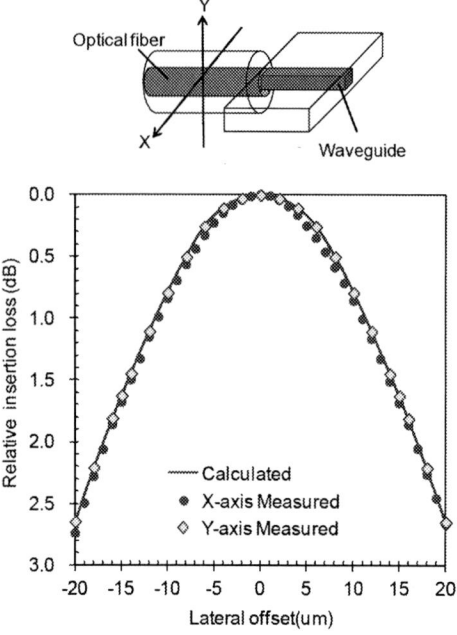

(a) Waveguide – Fiber direction (Tx)

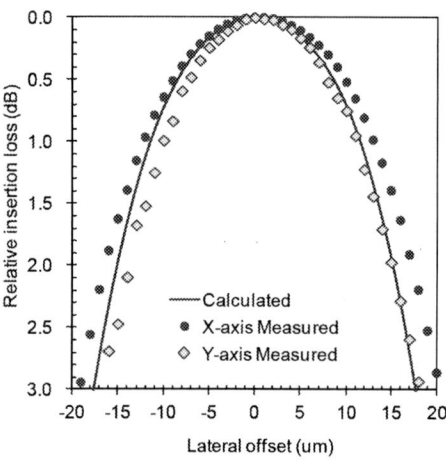

(b) Fiber – Waveguide direction (Rx)

Figure 5. Insertion loss of the optical waveguide and fiber with lateral offset

Specified connector design

The connection loss is also caused at a gap between the waveguide edge and fiber as an optical discontinuous point. The end face of the optical waveguide is formed flat shape only with saw dicing. On the other hand, the physical contact (PC) structure is formed on the edge of the ferrule; an optical fiber was made to project minutely from an end face of a ferrule, and was ground in the shape of a surface of a sphere. Then the fiber is pushed to the waveguide using a force of 700–800 gf to achieve the optically continuous connection structure.

We apply the structure the optical fiber is arranged in a single line. In the optical connector exceeding 12 channels,

the ferrule arranged the optical fiber in the shape of matrix is developed [6] [7]. These shapes are good for minimizing the width of an optical connector and a fiber ribbons, however, in the case of a rigid waveguide fabricated onto a substrate, a waveguide must be laminated and formed in two or more layers. Therefore, there is the problem of the optical signal levels are differ in each layer. Because an inner layer has a longer coupling length from multi-channel optical module mounted onto a surface of a waveguide, so a coupling loss must be larger than a surface layer. As the result, the optical coupling levels are varied between each layer. To decrease this difference of an optical coupling level in multi-layer optical waveguide, some structures are developed [8] [9]. These structures require additional components and/or additional process. We realized this point, to require a special structure is problematic, thus the single line type optical fiber arrangement in a ferrule is designed to connect single-layer type optical waveguide.

Mechanical Design of Card-Edge Connector

Regarding the development of the optical connecter which connects fiber to optical waveguide, some development results have been already reported [10] [11] [12] [13]. Most of these connectors are designed on the basis of well-known MT connector, and require highly precise structure using a special process such as a slot, a notch in a card, and additional core without covered by a cladding to align the part of connector to the optical waveguide.

Figure 6 shows the fundamental design of the newly developed optical card-edge connector.

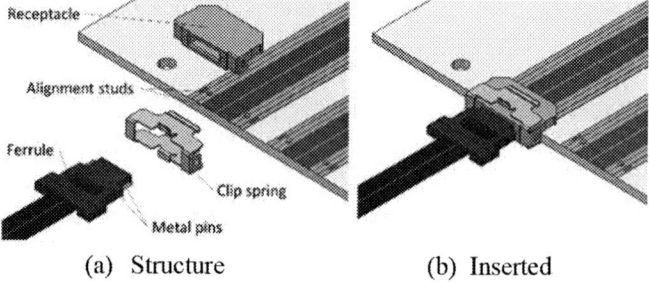

(a) Structure   (b) Inserted

Figure 6. Fundamental design of card edge connector

On the optical waveguide side, the alignment studs are directly formed onto the surface of the optical waveguides by the conventional UV photolithography process using photosensitive materials. The receptacle, which is a precision part, is positioned and aligned at the card edge with the alignment studs onto the optical waveguide.

On the fiber side, the ferrule terminates the optical fiber. The edge of the optical fiber is polished in a PC shape to achieve physical contact with the edge of the optical waveguide core.

The receptacle has a rectangular parallelepiped hole to insert the ferrule, and two highly precise pins are attached to both sides of the ferrule tip. These pins are inserted into the highly precise cylindrical holes at both sides of the receptacle.

Figure 7 shows the structure of the optical card-edge connector.

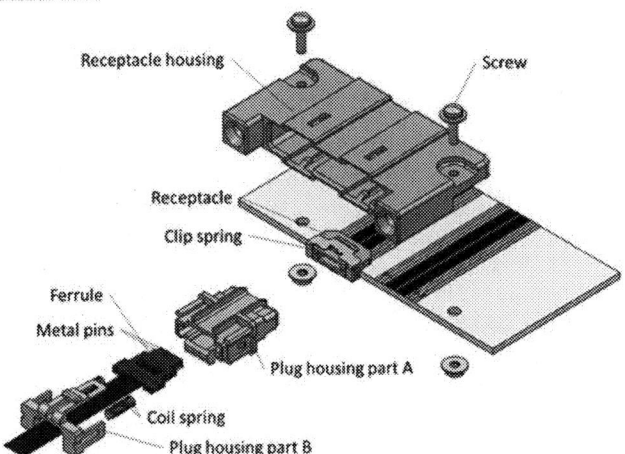

Figure 7. Structure of the optical card-edge connector

The receptacle is fixed to the waveguide on the card by a clip spring. The receptacle housing placed onto the optical waveguide is fixed by the small screws without any adhesives. Therefore, the receptacle assembly can be easily detached, and does not need to be scrapped if a problem occurs.

The ferrule, terminates optical fibers, is held inside the two parts of the plug housing with two coil springs, generating a force that pushes the ferrule to the edge of waveguide. The two parts of the plug housing can be joined without additional materials such as an adhesive and can be removed if a problem occurs.

Performance of the 24-channel Optical Connector

Figure 8 shows the prototype of the two-set 24-channel optical card-edge connector that connects to the rigid multimode waveguide array fabricated onto the substrate, and the GI-50/125 fiber array with 250μm pitch. The receptacle housing has two slots, it achieves 48-channel connection in the one housing on the waveguide and two plugs attached to the edge of the fiber array.

The optical connector after insertion in the two slots is 56 mm in width, 10 mm in thickness, and 38 mm in length outside of card. These values are designed considering the operation of the human hand. Furthermore, the receptacle housing is designed to apply it in backplane applications in the future. The plug can be easily inserted and extracted by one action.

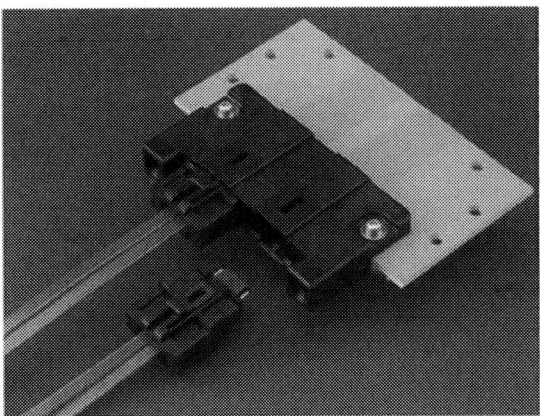

Figure 8. Prototype of the two-set 24-channel optical card-edge connector

We evaluate the connection losses of 16 of the 24-channel optical card-edge connector. Two types of losses are defined by the direction of the optical signal: one goes from the waveguide to the fiber as the transmitting function (Tx), while the other goes from the fiber to the waveguide as the receiving function (Rx). The wavelength is 850 nm, and index matching gel is not applied.

To measure the connection loss in the Tx direction, the output power from the core of the optical waveguide is first measured as P_0, and then the output power from the connected plug and the fiber is measured as P_1. The connection loss of the connector (L, dB) is calculated by $L = -10 \log_{10}(P_1/P_0)$.

In addition, to measure the connection loss in the Rx direction, from the fiber to the waveguide, the output power from the plug is measured as P_0, and then the plug is inserted to the receptacle, and the output power from the optical waveguide is measured as P_1. The insertion loss of the optical connector and waveguide (L_1) is calculated with the same formula as the loss in the Tx direction. Finally, the insertion loss of the optical waveguide (L_w) is subtracted from L_1 to quantify the loss caused by the card-edge connector (L).

The connection losses of the optical card-edge connector are shown in Figure 9. The connection losses in the Tx direction are from 0.12 dB to 0.64 dB, with an average of 0.40 dB. The losses in the Rx direction are from 0.16 dB to 0.52 dB, with an average of 0.30 dB.

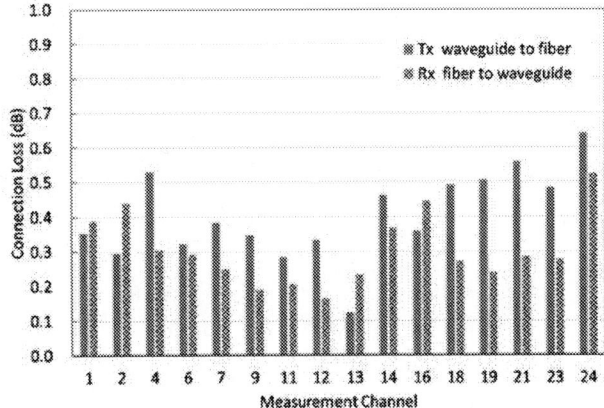

Figure 9. Connection losses of the 24-channel optical card-edge connector

978-1-4799-2408-0/14 $31.00 © 2014 IEEE

The measured losses in Figure 9 is appropriate to the theoretical loss calculated with ray-trace method as shown in Figure 2 and the excess loss caused by the size difference of parts.

Conclusions

We propose a newly developed optical multi-channel connector to connect a rigid optical waveguide array and an optical fiber array for an on-board optical interconnection. The connection losses of the prototype of 24-channel optical card-edge connector are 0.7 dB on the waveguide to fiber direction, and 0.6 dB in the fiber to waveguide direction.

Acknowledgments

The authors would like to thank Keizo Honda, Toshikazu Fuji, and the Engineers at the Products Technology Center, KYOCERA Connector Products Corp. for their support during development. The authors also would like to thank Maraki Maetani, KYOCERA Corp. for designing the optical card.

References

1. E.Mohammed, J.Liao, A.Kern, D.Lu, H.Braunisch, T.Thomas, S.Hyvonen, S.Palermo, and I.A.Young, "Optical hybrid package with an 8-channel 18GT/s CMOS transceiver for chip-to-chip optical interconnect", *Proc. of SPIE (Photonics West)*, Vol. 6899, 68990Z, January, 2008

2. M.Tokunari, Y.Tsukada, K.Toroyama, H.Numatqa, and S.Nakagawa, "High-Bandwidth Density Optical I/O for High-Speed Logic Chip on Waveguide-Integrated Organic Carrier", *Proc of 61st Electronic Components and Technology (ECTC)*, pp819-822, May-June 2011.

3. T.Matsubara, K.Oda, K.Watanabe, M.Maetani, K.Tanaka, and S.Tanahashi, "Three-Dimensional dimensional Optical Lines fabricated onto Substrate for On-Board Interconnection", *Proc. of SPIE (Photonics West)*, Vol. 7221, 72210A, January 2009.

4. S.Nakagawa, Y.Taira, H.Numata, "High-Bandwidth, Chip-Based Optical Interconnects on Waveguide-Integrated SLC for Optical Off-Chip I/O", *Proc. of 59th Electronic Components and Technology Conference (ECTC)*, pp2086-2091, May-June 2009.

5. K.Nakazuru, N.Takahashi, M.Tsunoda, S.Asai, and T.Matsubara, "Optical Multi-channel Connector for Rigid Waveguide and Fiber Connection", *IEEE CPMT Symposium Japan 2013*, pp107-110, November, 2013

6. T.Ohta, S.Shida, K.Takizawa, A.Nishimura, T.Arikawa and Y.Tamaki, "Two Dimensional Array Optical Fiber Connector", *Fujikura Technical Review*, No.29, 2000

7. R.Nagase, "Technical Trends in Optical Fiber Connectors for Telecommunication Systems", *IEICE Trans. Electron*, Vol.E86-C, No.6, pp968-974, June 2003

8. F.Libsch, R.Budd, P.Chiniwalla, P.Hobbs, M.Mastro, J. Sanford, J. Xu, "MCM LGA Package with Optical I/O Passively Aligned to Dual Layer Polymer Waveguides in PCB", *Proc of 56th Electronic Components and Technology (ECTC)*, pp1693-1699, May-June 2006.

9. M. Shishikura, M. Matsuoka, T. Ban, T. Shibata, and A. Takahashi, "A High-Coupling-Efficiency Multilayer Optical Printed Wiring Board with a Cube-Core Structure for High-Density Optical Interconnections", *Proc of 57th Electronic Components and Technology (ECTC)*, pp1275-1280, May-June 2007.

10. L. Dellmann, C. Berger, R. Beyeler, R. Dangel, M. Gmür, R. Hamelin, F. Horst, T. Lamprecht, N. Meier, T. Morf, S. Oggioni, M. Spreafico, R. Stevens, B. J. Offrein, "120 Gb/s Optical Card-to-Card Interconnect Link Demonstrator with Embedded Waveguides", *Proc of 57th Electronic Components and Technology (ECTC)*, pp1288-1293. May-June 2007.

11. Y. Hatakeyama, S. Imamura, J. Kobayashi, H. Takahara, "PMT connectors for multi-channel film waveguides", *Proc. of SPIE (Photonics West)*, Vol. 7213 72130V, January 2009.

12. Y. Taira, H. Numata, F. Yamada, Y. Katayama, S. Nakagawa, M. Hasegawa, "OE Device device Integration integration for Optically optically Enabled enabled MCM", *Proc. of 57th Electronic Components and Technology Conference (ECTC)*, pp1262-1267, May-June 2007

13. M,Takaya, S.Nagasawa, "An Easily-Assembled Opttical Device for Coupling Single-Mode Planar Waveguides to aFiber Array Employing Plastic Plug Components", *IEICE Trans. Electron*, Vol.E85-C, No.4, pp921-926, April 2002

978-1-4799-2408-0/14 $31.00 © 2014 IEEE

Electro-optical Backplane Demonstrator with Gradient-index Multimode Glass Waveguides for Board-to-board Interconnection

Lars Brusberg[1], Henning Schröder[1], Richard Pitwon[2], Simon Whalley[3], Allen Miller[2], Christian Herbst[4], Julia Röder[1], Daniel Weber[1], Klaus-Dieter Lang[4]

1) Fraunhofer Institute for Reliability and Microintegration, Gustav-Meyer-Allee 25, 13355 Berlin, Germany,

2) Xyratex Technology Ltd., Langstone Road, Havant, Hampshire, PO9 1SA, United Kingdom

3) ILFA Feinstleitertechnik GmbH, Lohweg 3, 30559 Hannover, Germany

4) Technical University of Berlin, Gustav-Meyer-Allee 25, 13355 Berlin, Germany,

lars.brusberg@izm.fraunhofer.de

Abstract

First time an electro-optical circuit board (EOCB) is demonstrated with integrated planar glass multimode waveguides and with optical pluggable line card connectors. The waveguides are patterned inside commercially available thin-glass panels by performing a two-step thermal ion-exchange process. The resulting low-loss multimode waveguides possess a gradient-index profile. The glass waveguide panel is embedded within the layer stack-up of a printed circuit board (PCB) using proven industrial processes. Cut-outs inside the PCB are structured for assembling a pluggable optical connector and receptacle system for connecting optical fiber based waveguides on the line cards to integrated optical waveguides in the backplane. The demonstration platform comprises a standardized sub-rack chassis and five pluggable test cards with pluggable optical connectors and designed for housing optical engines. The test cards support a variety of different data interfaces for bidirectional signal integrity measurements. The evaluated demonstrator system performed with bit error free data transmission at 10.3 Gb/s for the tested wavelengths of 850 and 1310 nm.

Introduction

Optical interconnects for data transmission at board level offer significant reduction in power consumption, increased energy efficiency, system density and bandwidth scalability compared to purely copper driven systems. So far embedded optical architectures do not exist in data center and network systems but the goal is to replace the electrical signal lines with optical interconnects for high-speed data transmission. The system enclosure consists of different peripheral line cards that are plugged into an electro-optical backplane where signals are routed across. The bandwidth by length limitation of copper signal lines can be overcome by deployment of embedded optical interconnect structures in the backplane and pluggable optical connections the line cards. On the line cards, optical engines can be closely located to application-specific integrated circuits (ASIC), central processing units (CPU) or memory ICs, in order to convert high speed electrical signals into optical signals for data transmission through the system enclosure. Such a configuration is shown in Figure 1 with fiber based optical signal links on the line cards and integrated waveguides based optical interconnections in the backplane. The benefit is the simple line card configuration based on deployment of commercially

available optical mid-board engines (e.g. MicroPOD™ [1], BOA [2]) and low loss optical fiber patch cords above the PCB surface. Pure electrical PCBs without any embedded optics are used for the line cards. On the line cards the optical engine is closely positioned and electrically interconnected to the ASIC, CPU or memory IC, which require bit error free high-speed data transfer within the system enclosure. Also, in the event of multiple optical engines assembled onto a single line card, the fiber layout on the line-card will remain manageable while the resulting increased complexity in optical routing can be displaced to the electro-optical backplane. Then the motivation for an electro-optical backplane with integrated waveguides is very high because of the high number of optical interconnects and routing possibilities. A key enabling requirement for enclosure based optical board-to-board interconnections are pluggable optical connectors. Such optical board-to-board connector technologies would be subject to tight alignment tolerances due to the inherently small sizes of optical channels, and, in order to be deployed within a predominantly forced air cooled environment, would certainly require the means to protect against dust contamination and prevent operator exposure to hazardous levels of laser radiation.

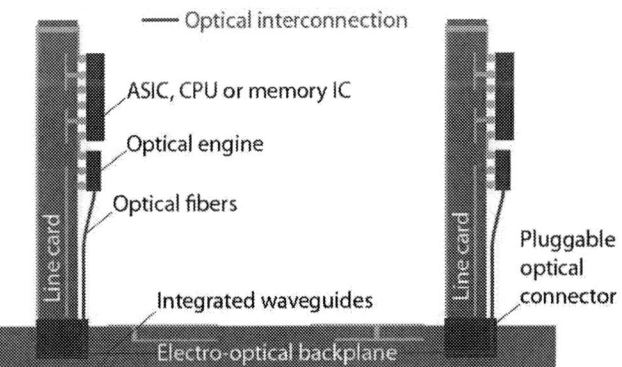

Figure 1. Optical engines on different line-cards are optically interconnected (red line) over optical fiber links on the line card, pluggable optical board-to-board connectors and an electro-optical backplane with integrated waveguides.

Electro-optical backplane design

The work presented here was carried out as part of the EU Piano+ project "SEPIANet", the aim of which was to develop enabling technologies for system embedded optical infrastructures in the 1310 nm wavelength range, including

978-1-4799-2408-0/14 $31.00 © 2014 IEEE

the development of board mounted optical transceivers [3]. This paper focuses on the design, fabrication and characterization of the electro-optical backplane and connector demonstrator.

The electro-optical backplane in question was created by laminating a glass panel with integrated multimode waveguides into a central layer of a conventional multi-layer electronic PCB stack-up. The waveguide properties are not altered by the thermal or mechanical stresses inherent to the lamination process.

Afterwards the optical board-to-board connectors were assembled onto the backplane. The complete electro-optical backplane was mounted inside a chassis. A schematic view of the demonstrator platform is shown in Figure 2.

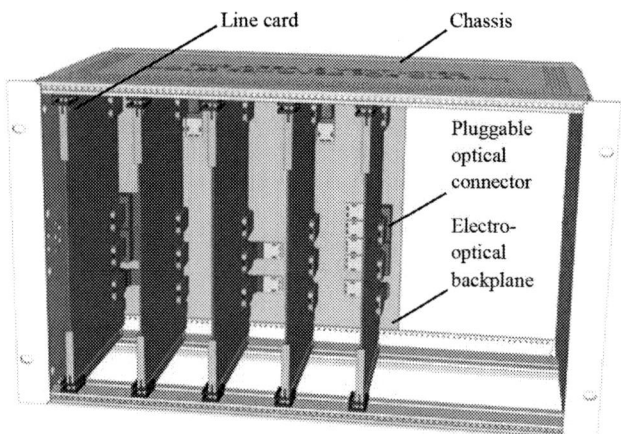

Figure 2. Design of the demonstrator platform.

The paper discusses the design, fabrication and characterization of two demonstration platforms which are proposed with the designations SEPDEM1 and SEPDEM2. Figure 2 shows the design of SEPDEM2 containing a sub-rack chassis, 5 line cards and an electro-optical backplane with pluggable optical board-to-board connectors.

The electro-optical backplane (EOCB) itself comprises 4 conventional electrical PCB layers with an outer board area of 281 x 233 mm² and board thickness of 3.5 mm, encasing a smaller 199 x 160 mm² glass waveguide panel with a thickness of 500 µm. The optical waveguide layout contains a variety of waveguide groups with point-to-point geometries, which would be suitable for the key system enclosures in data centers and access networks such as storage server and head-end system enclosures. As seen in Figure 3, these geometries include groups of straight waveguides, groups of waveguides with varied bend radii and waveguide crossovers. Two EOCB variants were designed: SEPPLANE1 is an EOCB with one embedded planar glass waveguide panel, while SEPPLANE2 has 2 smaller embedded planar glass waveguide panels. The purpose of the latter was to demonstrate the ability to laminate multiple glass panels into a single backplane, which would be a critical means of providing optical interconnect across larger high density backplanes if individual glass waveguide panel sizes are constrained in future.

The EOCB includes exposed cavities in the PCB prepreg material showing sections of the embedded glass waveguide

panel encased within. This allows observers to see certain waveguide features as well as six designated windows, in which the aluminum mask covering has been selectively retained with the logos of the partner companies.

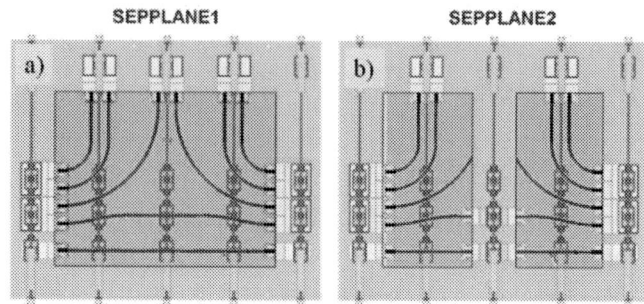

Figure 3. Electro-optical backplane variants SEPPLANE1 and SEPPLANE2: a) Schematic view of SEPPLANE1 with single glass waveguide layer and connector layout and b) Schematic view of SEPPLANE2 with dual single glass waveguide and connector layout.

Glass waveguide panel technology

For that work, the Fraunhofer IZM glass panel waveguide process line was used to produce glass panels with sizes of 210 x 297 mm² and 500 µm glass thicknesses. The waveguide fabrication in the glass panels was based on an ion-exchange process between a molten salt melt and the glass causing an increase in refractive index near the surface where the glass and salt melt interact.

A glass containing adequate amount of ions that are exchangeable with a counterpart in the salt melt is required for the waveguide process. We chose Schott D263Teco, a borosilicate glass containing monovalent sodium ions.

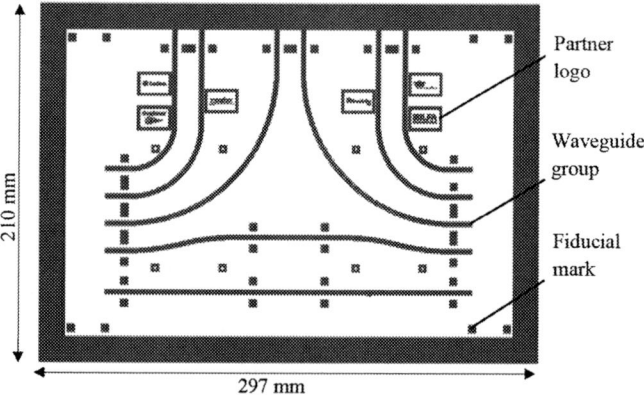

Figure 4. Central layout of 14 inch chrome-on-glass mask which was transferred to the aluminum mask on the glass panel needed as diffusion barrier during waveguide fabrication.

For waveguide integration, a diffusion process was enabled for the defined demonstrator waveguide pattern. This first required an aluminum diffusion barrier to be deposited on top of the glass surface, which was then patterned with the demonstrator waveguide layout using lithography and wet-

978-1-4799-2408-0/14 $31.00 © 2014 IEEE

chemical processes to open the aluminum layer where diffusion had to occur. For lithography we used dip coating for photoresist deposition in combination with a mask contact exposure. The exposure was done on an Orbotec Paragon 9000 laser direct imaging (LDI) system. The 14 inch chrome-on-glass mask designed for the demonstrator contained the waveguides, fiducial marks and partner logos as shown in Figure 4. After exposure the photoresist was developed and the underlying aluminum layer was structured with acid treatment before removal of the photoresist. The aluminum diffusion mask layer on the glass panel is shown in Figure 5.

Figure 5. Aluminum diffusion mask layer on glass panel (297 x 210 mm²).

Multiple glass panel samples with the aluminum mask layer were inserted vertically into the hot salt melt for the ion-exchange process. During this process, the sodium ions in the glass are exchanged with silver ions, which increases the refractive index of the glass. The diffusion process forms an isotropic refractive index profile in the glass with the highest refractive index change occurring at the mask openings on the glass surface.

After this diffusion process, the aluminum mask was removed from the glass surface with the exception of fiducial marks needed for post alignment processes or signs for identification needs. In a second diffusion step, the concentration of the silver ions in the surface area was reduced until the refraction index maximum (waveguide core center) was shifted to a point at a certain depth below the glass surface.

In the final step, the glass waveguide panel was trimmed with a CO_2-laser cutting process [4] to the panel size required for the electro-optical backplanes. The surface quality of the processed glass edges was sufficiently high to ensure low-loss optical coupling to the waveguides.

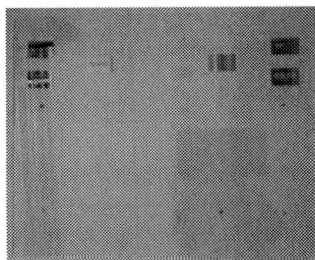

Figure 6. Glass waveguide panel (199 x 160 mm) for SEPPLANE1.

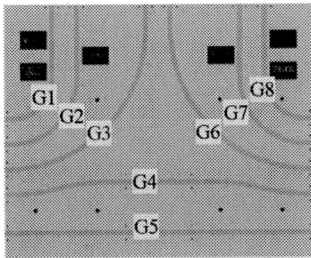

Figure 7. Waveguide layout of glass panel for SEPPLANE1.

The 297 x 210 mm processed glass panels were cut to a size of 199 x 160 mm for SEPPLANE1 (Figure 6) and two smaller panels of 79.25 x 160 mm for SEPPLANE2, in which they would be embedded with a separation of 40.5 mm.

Each waveguide core was characterized by an elliptic graded index profile, with the long axis along the waveguide width due to the isotropic diffusion characteristics of the ion-exchange process. In previous work [5] the waveguide characteristics were extensively characterized. A typical refractive index profile for multimode waveguides can be seen in Figure 8. The process parameters applied in that case caused the refractive index maximum to form at a depth of 18 μm below the glass surface with a maximum refractive index difference in the range $\Delta n = 0.014 \dots 0.016$.

The propagation and coupling losses on the fabricated glass waveguide panel were characterized by the cut-back method at a wavelength of 1310 nm. The propagation loss was measured to be 0.07 dB/cm and coupling loss was 2.15 dB for a GI MM fiber launch. The results are comparable with those of previous work [5].

The developed process works reliably and can fabricate waveguides with reproducible waveguide characteristics. The process parameters have been selected to manufacture low loss waveguides in the 1310 nm wavelength range to meet the project targets. In addition, the waveguides were characterized at a wavelength of 850 nm with propagation loss measured to be 0.41 dB/cm and coupling loss 2.7 dB for a GI MM fiber launch. It is expected that, through process improvement, the propagation losses can be further reduced by a factor of $2 - 4$. That will be studied in much more detail in subsequent work.

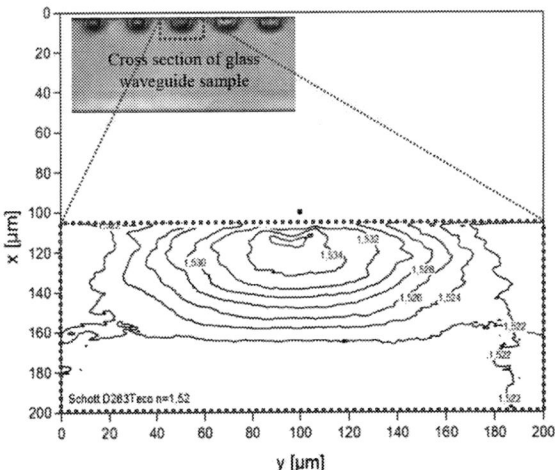

Figure 8. Refractive index profile at measurement wavelength of 678 nm.

The SEPPLANE1 layout as shown in Figure 7 consists of 8 waveguide groups. Each group comprises a row of 12 waveguides with a center to center channel separation of 250 μm, which is compliant with conventional parallel optical arrays such as those based on MT interfaces (MTP, MPO). On SEPPLANE1 waveguide layout groups G1-G3 are curved over a 90° arc with varied concentric bend radii. Group G3 and G6 are complete arcs, whilst groups G1, G2, G7 and G8 include an additional straight vertical section to maintain the minimum radial distance between input and output waveguide. Group G4 includes a gentle S-bend, while group G5 is completely straight. Group G8 however includes a series of stepped waveguide stubs crossing the group orthogonally. Within this group, the top 2 waveguides have no waveguides stubs crossing them to serve as a non-crossing reference. The next two have two waveguide stubs crossing them and then four over the next two waveguides. The trend continues until the final two waveguides which have 10 waveguides stubs crossing them. The insertion loss measurement results for SM and MM launching condition for all waveguide groups and related bend radii for all waveguide groups G1-G8 are summarized in Table 1. Insertion loss for a 90° arc is in the range between 2.6 dB and 2.9 dB for MM launch condition with slight variations. That means that a waveguide bend with a small radius like 25 mm suffers from higher propagation loss compared to a waveguide bend with a much larger radius like 90 mm. The waveguide crossing with maximum number of 10 intersections does not show significant reduction of insertion loss. These results confirm our estimations based on our previous study [5].

Table 1. Insertion loss of different waveguide groups of a SEPPLANE1 glass panel dependent on bend radii and launching condition.

Waveguide group	Bend radii	SM launch condition	MM launch condition
G1	25–27.75 mm	1.21 ± 0.24 dB	2.62 ± 0.10 dB
G2	41–43.75 mm	1.10 ± 0.14 dB	2.75 ± 0.10 dB
G3	88.125 – 90.875 mm	1.15 ± 0.06 dB	2.92 ± 0.27 dB
G4	134.0 mm	1.76 ± 0.21 dB	2.98 ± 0.09 dB
G5	without bend	1.72 ± 0.17 dB	3.08 ± 0.21 dB
G6	88.125 – 90.875 mm	1.22 ± 0.26 dB	2.83 ± 0.20 dB
G7	41 – 43.75 mm	0.96 ± 0.14 dB	2.74 ± 0.13 dB
G8	25 – 27.75 mm	1.04 ± 0.11 dB	2.90 ± 0.20 dB

On SEPPLANE2 the glass panel with the same waveguide layout as in SEPPLANE1 has a vertical strip cut out of it. The former groups G4 and G5 in SEPPLANE1 are each split into two separate groups in SEPPLANE2. G3 and G6 were not used because the portion inside of the vertical strip that was cut out.

Electro-optical backplane fabrication

At ILFA, a novel glass substrate EOCB bonding technology was developed; utilizing low temperature adhesive foils to negate difficulties resulting from the disparity in the dimensional behavior of the different materials by applying thermal load during lamination. Here, we introduced an alternative procedure for laminating the optical and electrical cores using 50 μm thick adhesive foils that require only minimal thermal energy for an adequate bonding. With process temperatures far below 100°C during the lamination process, the thermal expansion of the individual layers in the stack-up does not result in critical amounts of built-in stress within the manufactured PCB. No issues resulted from the horizontal coefficient of thermal expansion (CTE) mismatch of the different materials within the stack-up during lamination. Typical CTE values are:

- Copper: $\alpha = 16.5 \cdot 10^{-6}$ K^{-1}
- Glass D263Teco: $\alpha = 7.2 \cdot 10^{-6}$ K^{-1}
- FR4 pregreg: $\alpha = 12..14 \cdot 10^{-6}$ K^{-1}

Furthermore, being able to neglect the CTE broadly extends the overall selection of suitable PCB materials for EOCBs and thus allowing cost efficient and highly specialized products. The bonding strength of the adhesive foil was evaluated by IPC standardized tests. A FR4-stack-up with multiple adhesive foil layers was thermal shock tested for 10 seconds with peak temperature of 288°C. A thermal transition test with 100 cycles in temperature range of +125°C and -40°C was also performed without impact.

Figure 9. Electrical package with two copper layers.

The electrical package made of multiple copper and FR4 prepreg layers can be fabricated with standard PCB technologies and thus allowing advanced high-density electrical PCB designs with multiple layers. Figure 9 shows such a two layer electrical package.

The cold lamination technique was only applied for bonding of the electrical cores to the inner optical cores. Applying the adhesive foils to the core, one glass waveguide panel (SEPPLANE1) or two glass waveguide panels (SEPPLANE2) were set within a FR4 prepreg rigid protective core in the same layer and enclosed with adhesive foil by two additional prepreg layers on top and bottom side with milled cut-outs for optical connector openings or cavities e.g. for partner logos. Such an optical package is shown in Figure 10 for the SEPPLANE2 backplane with two separate glass waveguide panels in the core layer.

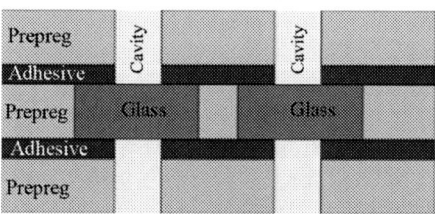

Figure 10. Optical package with two embedded glass waveguide panels for SEPPLANE2 in the core layer between prepreg layers with cavities.

After initial laminate manufacturing, each optical or electrical package was subsequently bonded together utilizing the low temperature adhesive foil lamination. The final EOCB stack-up is shown in Figure 11 with a total thickness of 3.5 mm.

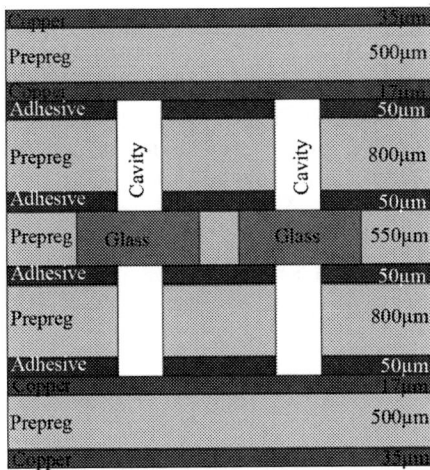

Figure 11. EOCB stack-up which was made of one optical and two electrical packages.

Through holes were only drilled outside the glass panel area and plated to interconnect the electrical packages. After processing the outer layers (metallization, soldering stop layer), the cavities of the optical connector openings are exposed by milling through surrounding top layers in the area of the cut-outs of the optical package. The fabricated backplane is shown in Figure 12. The backplane is electrically fully functional and the glass is undamaged after fabrication.

Figure 12. Electro-optical backplane after fabrication.

Manufacturing the different packages separately and utilizing low temperature adhesive foils for embedding glass foils within an optical package or bonding packages containing glass foils to other packages allows the number of electrical and optical layers to be scaled freely while allowing a vast range of PCB technologies without limiting the material selection in order to match the CTE of the glass sheet.

Optical board-to-board connector design and fabrication

Xyratex designed and developed a passive optical connector system to enable pluggable optical connectivity between peripheral test cards and the electro-optical backplane. The design was tailored to meet the opto-mechanical waveguide requirements of the glass foils with embedded planar waveguides as deployed within a multilayer electronic PCB stack-up.

The optical connector system comprises three parts: 1) connector plug, 2) backplane receptacles and 3) MT receptacle mount.

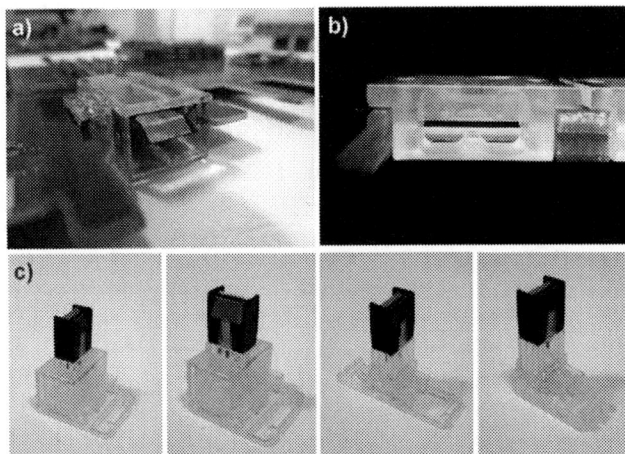

Figure 13. a) Connector plug, b) MT receptacle mount, c) Backplane receptacle varieties.

Connector plug - The connector plug section is mounted on the edge of a connecting test card. The plug housing is designed to support an MT type ferrule and contains engagement features to allow the MT ferrule to couple accurately to another compliant MT ferrule housed in the receptacle section mounted on the backplane. The plug housing also includes a shutter to prevent both dust contamination of the internal parallel optical engagement interface and inadvertent user exposure to propagating laser light (Figure 13a).

Backplane receptacle - The receptacle section includes a head section (black section) designed to support a compliant MT ferrule and its own shutter designed to interlock with that of the engaging plug. The receptacle housing (white section) incorporates a short fiber-optic patchcord bent by 90° to enable the required deflection of the optical signals from the vertical plug interface axis to the horizontal embedded waveguide interface axis. The backplane receptacle housing provides a cover for the fiber jumper and holds the receptacle head at the required position. Four varieties of backplane receptacle were designed to accommodate different board positions, connector configurations and orientations (Figure 13b).

MT receptacle mount - Each connector assembly requires a high precision MT receptacle mount that must be actively aligned over the waveguides in the glass foil and assembled. The receptacle will allow a fiber-optic MT patchcord to be connected directly to waveguides within the EOCB, however, in order to avoid cutting notches into the glass panel either side of the waveguide group, the datum between the MT pins of the connecting ferrule must be vertically offset from the fibre array. This can be achieved by using a standard 6x12 MT ferrule of which only the lowest row is populated with 12 fibers (Figure 13c).

Board assembly of MT receptacle mounts

Board assembly of MT receptacle mounts was done at Fraunhofer IZM. An active alignment and assembling routine was selected for decreasing misalignment and achieving highest coupling efficiency between the fiber-optic MT patchcord and 12 channel waveguide array. Central to the process is the 5-axis pick-and-place assembly equipment by ficonTEC with three translational axes and two rotary axes. A vacuum tool was designed for holding the MT receptacle mount with plugged fiber-optic MT patchcord. Additionally the ficonTEC equipment provides adhesive dispensing and UV curing as well as top and bottom vision control cameras for alignment purpose. For our assembly task the equipment was upgraded with a separate 3-axis translation stage for positioning a second fiber-optic MT patchcord in front of the same waveguide group for launching purpose for the active alignment routine. In Figure 14 a SEPPLANE1 EOCB is shown on the working stage of the ficonTEC equipment during an active alignment process. For that a two channel laser source and photodetector with operating wavelength of 1310 nm were connected to the two fiber-optic MT patchcords for in-situ insertion loss measurement.

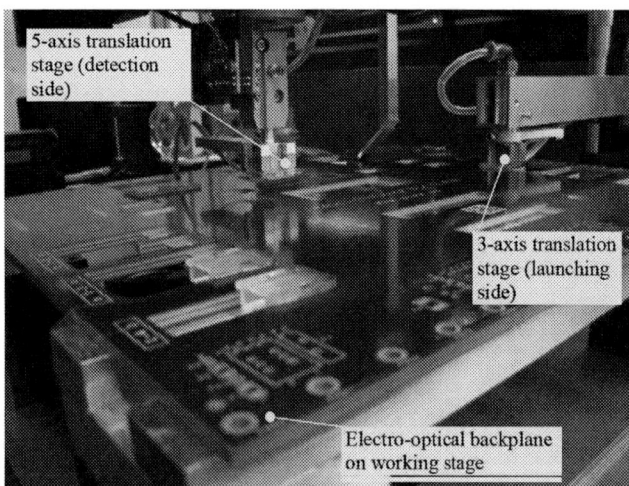

Figure 14. View into the ficonTEC assembling equipment during active alignment of MT receptacle mount.

The developed semi-automated assembling routine consists of the following process steps:

(1) **Backplane placing** - Electro-optical backplane is placed on the working stage.
(2) **Alignment launching side** - Position of fiber-optic MT patchcord is aligned by 3-axis translation stage in front of the selected waveguide group for light in-coupling. The light coupling of the two outer waveguide channels (CH1 & CH12) is monitored on the opposite waveguide facet by a vision control camera. The alignment is continued until two light peaks are detected in case of waveguide beam propagation.
(3) **Alignment detection side** - The MT receptacle mount with plugged fiber-optic MT patchcord is positioned in the PCB cut-out. Pre-positioning is achieved by scanning fiducial marks on the glass

panel and calculating the related position. Then the active alignment routine starts using the two launched outer waveguide channels (CH1 & CH12) for active optical alignment by monitoring the insertion loss values. Alignment will be done until insertion loss of both channels is reduced to minimum.

(4) **Assembly detection side** - Having the receptacle mount in the correct position UV-curable glue is applied to the glass surface in the PCB cut-out. The mount is permanently held during the curing process to guarantee stability of its position. A permanent bond of MT receptacle mount to the EOCB is achieved after UV curing (Figure 15).

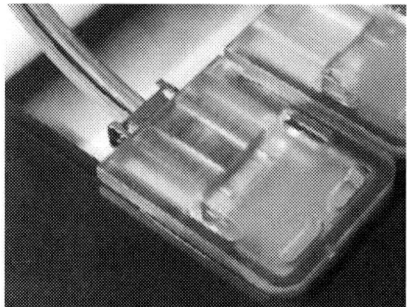

Figure 15. MT receptacle mount with plugged MT fiber ribbon patch cord.

(5) **Assembly launching side** - Process step (3) and (4) are repeated for the same waveguide group using the 5-axis translation stage. Afterwards assembling of MT receptacle mounts is completed for the selected waveguide group.

The foregoing assembly process was performed for every single waveguide group on the electro-optical backplane. A fully assembled SEPPLANE1 with 8 waveguide groups and 16 MT receptacle mounts is shown in Figure 16.

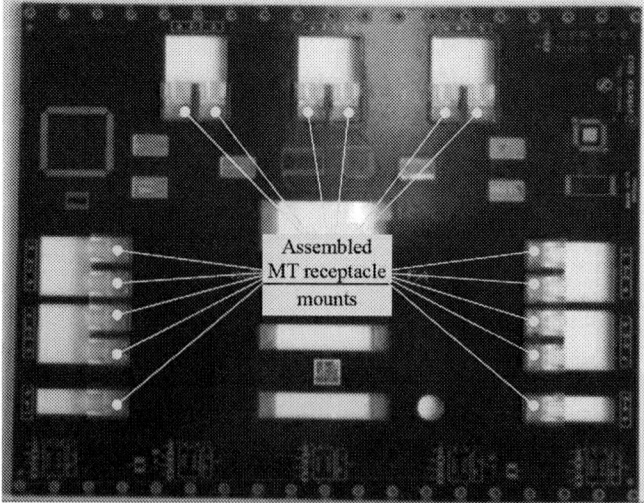

Figure 16. SEPPLANE1 after assembling of all MT receptacle mounts.

Board assembly of optical pluggable connectors

To provide a low-loss high density reliable 90° deflection of optical signals from the horizontal EOCB waveguides to the vertical test card fibres and vice versa, custom short fibre jumpers were designed and developed. These jumpers split out 2 groups of 12 fibres from a single vertical 2x12 MT ferrule to the lowest rows of 2 separate 6x12 MT ferrules (Figure 17). This arrangement allows one backplane receptacle to accommodate 2 adjacent MT receptacle mounts.

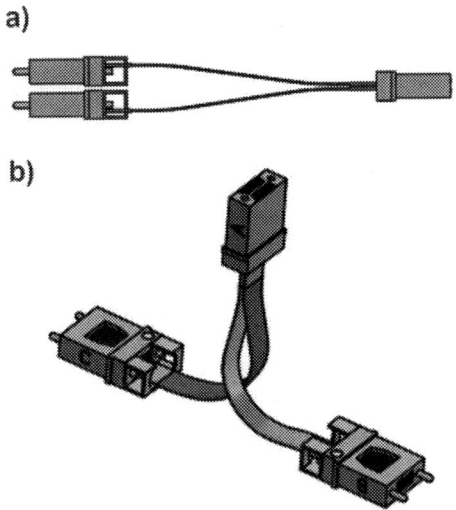

Figure 17. Custom fibre jumpers splitting out 2 groups of 12 fibres from a single vertical 2x12 MT ferrule to 2 separate 6x12 MT ferrules a) Side view of flattened fibre jumper, b) Oblique view of fibre jumper arranged within a midboard connector.

The 6x12 MT ferrule ends of special fiber jumpers were then passively connected into the horizontal MT ferrule receptacle mounts assembled to the electro-optical backplane by Fraunhofer IZM and the vertical 2x12 MT ferrule end mounted into the backplane receptacle head section . A spring clip is provided to ensure that the horizontal MT ferrules are held in place against the glass waveguide interface. The backplane receptacles are clamped into the PCB cut-outs with screws pulling the top and bottom sections together.

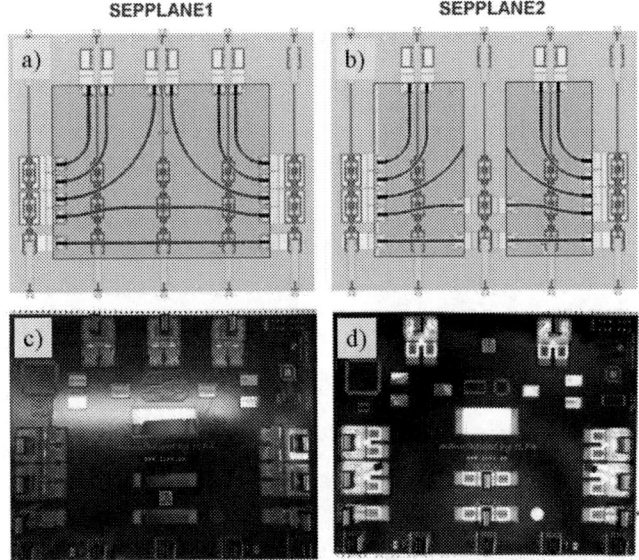

Figure 18. Electro-optical backplane variants SEPPLANE1 and SEPPLANE2: a) Schematic view of SEPPLANE1 with single glass waveguide layer and connector layout, c) photo of SEPPLANE1, b) Schematic view of SEPPLANE2 with dual with single glass waveguide and connector layout, d) Photo of SEPPLANE2.

Demonstrator and optical link performance

As shown in Figure 19 two demonstration platforms SEPDEM1 and SEPDEM2 were designed by and assembled by Xyratex to demonstrate the full technology eco-system for system embedded photonics interconnect including: planar glass waveguide based electro-optical circuit boards, pluggable connectors and optical engines. To this end, Xyratex designed the two types of EOCB and a suite of test cards to accommodate different optical engines and a variety of data interfaces.

Each demonstration platform comprises:

- 7U (311.15 mm) high 84 HP (426.72 mm) sub-rack chassis with an integrated fan tray and PSU
- 5 pluggable test cards based on the Euro-card form factor supporting 850 nm optical engines and pluggable optical connectors
- Pluggable connector system comprising connector plug, different backplane receptacles and MT receptacle mount for direct passive optical connectivity
- EOCB with integrated planar glass waveguides

Figure 19. Two demonstration platforms SEPDEM1 and SEPDEM2.

Test and measurement regime

A comprehensive test and measurement regime was carried out on the two SEPIANet demonstrator platforms SEPDEM1 and SEPDEM2. In all cases, optical data was conveyed along the EOCB embedded optical waveguides through the pluggable connector technologies from various optical sources.

Bidirectional insertion loss measurements

Insertion loss measurements were carried out at 850 nm and 1310 nm for all functional waveguides in each EOCB as mounted in a demonstration platform. Each waveguide was measured twice with the optical signal conveyed in opposite directions across the waveguide link, in order to evaluate reciprocity of the measurements. This highlighted differences in the quality of waveguide interfaces.

The insertion loss was measured over the complete waveguide, connector and patchcord link between the points 1 shown in Figure 20.

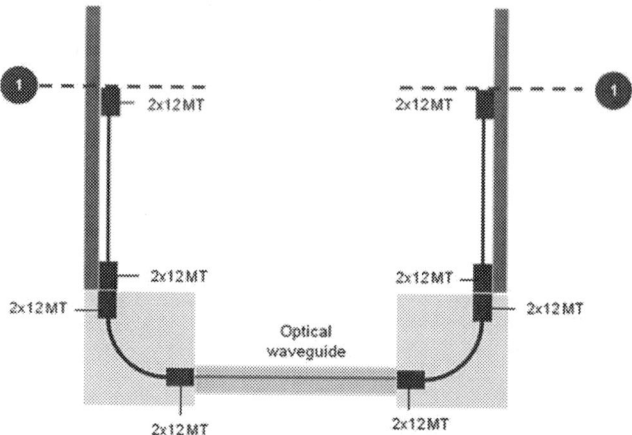

Figure 20. Schematic view of complete optical connector and waveguide link under test including an additional short fibre patchcord connected to the backplane receptacles. Insertion loss measurements were made from point 1.

The average insertion loss over the complete link at 1310 nm for the best performing EOCB was 3.9 dB. The average insertion loss over the complete link at 850 nm for the best performing EOCB was 10.7 dB.

Bidirectional signal integrity measurements

A 10.3125 Gb/s PRBS 2^{15-1} data pattern was generated by an Anritsu MP1800A pattern generator and BERT system as shown for all 12 channels of a waveguide group in Figure 21. Bidirectional signal integrity measurements were carried out on all functional waveguides with 850 nm and 1310 nm optical signals.

Figure 21. Eye diagrams for 10.3 Gb/s with PRBS 2^{15-1} test signal conveyed at 1310 nm over all 12 waveguides on an electro-optical backplane type SEPPLANE1.

The average peak to peak jitter measured over the complete connector and waveguide link at 1310 nm was 27.9 ps compared to a reference source of 25 ps. The average peak to peak jitter measured over the same link at 850 nm was 84.3 ps compared to a reference source of 24 ps.

BERT measurements

In addition, two 3rd party 850 nm optical engines were mounted on two test cards connected to each other across an EOCB type SEPPLANE1 within a demonstration platform and bit error free operation of BER $<10^{-13}$ for a 10.3 Gb/s test data stream was measured consistently over multiple 2 hour long test periods.

Summary

We developed a planar glass based electro-optical backplane technology for data center environments. Waveguide glass panels were fabricated and laminated into the PCB. A pluggable optical board-to-board connector was assembled for optical interconnection between fiber based peripheral line cards and the integrated glass waveguides of the electro-optical backplane. The backplane and line cards were mounted in a sub-rack chassis and data transmission for 10.3 Gb/s data rate was successfully demonstrated with average insertion loss over the complete link at 850 nm and 1310 nm.

Acknowledgments

Some of the research leading to these results has received funding from German Ministry of Education and Research (BMBF) and from British Technology Strategy Board (TSB) within the European piano+ project SEPIANet. The authors would like to thank Anritsu for supporting the research with the provision of an Anritsu MP1800A pattern generator and BERT system. Furthermore the authors would like to thank all the colleagues who have supported this work.

References

1. http://www.avagotech.com/pages/minipod_micropod
2. http://www.finisar.com/products/optical-engines
3. Dorward, R.M.; Symington, K.; Brusberg, L.; Kropp, J.R.; Miller, A.; Pitwon, R.; Whalley, S., "Market drivers and architectural requirements for backplane inter-connect capacities in Next Generation PON Head-End equipment in the Access Network," *15th International Conference on Transparent Optical Networks (ICTON),* 23-27 June 2013, pp.1-4.
4. Hermanns, C.; Middleton, J. "Laser separation of flat glass in electronic-, optic-, display-, and bio-industry (Invited Paper)", Proc. of SPIE, Vol. 571312 (2005)
5. Brusberg, L.; Schroder, H.; Pitwon, R.; Whalley, S.; Herbst, C.; Miller, A.; Neitz, M.; Roder, J.; Lang, K.-D., "Optical backplane for board-to-board interconnection based on a glass panel gradient-index multimode waveguide technology," *63rd Electronic Components and Technology Conference (ECTC),* 28-31 May 2013, pp. 260-267.

Three-Dimensional High Density Channel Integration of Polymer Optical Waveguide Using the Mosquito Method

Takaaki Ishigure, Daisuke Suganuma, and Kazutomo Soma
Faculty of Science and Technology, Keio University
3-14-1, Hiyoshi, Kohoku-ku, Yokohama, 223-8522, JAPAN
E-mail: ishigure@appi.keio.ac.jp   URL: www.//ishigure.appi.keio.ac.jp /index-e.html

Abstract

We experimentally demonstrate that the Mosquito method is capable of fabricating polymer optical waveguides with extremely narrow pitch. The Mosquito method we developed recently is a technique for directly writing polymer waveguides with circular cores on a substrate using a microdispenser. Actually, even when parallel cores are formed by repetitive horizontal scans of a syringe needle with a 230-μm outer diameter, a pitch of 40 μm is achieved. It is noteworthy that the pitch is almost one third of the outer needle radius. For the waveguide fabrication, a silicone resin (FX-712 form ADEKA Corp.) is used for the core, which is a sufficiently durable polymer material. Furthermore, we also represent in this paper that a three-dimensional channel integration is also realized with the Mosquito method by varying the vertical position of the needle tip in the cladding layer. To the best of our knowledge, this paper is the first report of three-dimensionally aligned fan-out polymer waveguide realized with GI-circular-core including the pitch and core-height conversions.

Introduction

Over the last couple of years, optical interconnects have been studied extensively as an interesting technology for high-performance computers and high-end servers and routers, because they have the potential to contribute not only to the data processing speed but also to the power dissipation of the whole computing system. In particular, optical printed circuit boards (OPCBs) on which multimode polymer optical waveguides are incorporated have been regarded as a promising component, and actually several reports have already been published [1, 2]. Over the past decades, OPCB links have aimed to transmit a data rate up to 10 Gbps/ch. However, due to the technical advances in the electrical counterparts, optical channels are now required to transmit at least 25 Gbps. Hence, in order to encourage the introduction of optical wirings into on-board interconnects, the key advantage of optics, the ability to support high-bandwidth-density wiring, should be fully utilized. However, high interchannel crosstalk could be a concern in such densely-aligned arrays, particularly when the core has a conventional step-index (SI) structure.

In order to realize high-bandwidth-density OPCB links, we have proposed a graded-index (GI) core even into polymer optical waveguides, by demonstrating superior optical characteristics such as low propagation loss, low coupling loss with GI multimode fibers (MMFs), and low interchannel crosstalk. In this paper, a very simple fabrication methodology for GI-core polymer waveguides, the Mosquito method is introduced [3]. GI polymer waveguides with a core diameter smaller than 40 μm are fabricated using the Mosquito method for high-density wiring, and finally, a single-mode polymer waveguide whose core diameter is less than 10 μm is successfully fabricated. Then, in order to achieve even higher density, we investigate how we can decrease the interchannel pitch.

After specifying how the core diameter and pitch could be reduced, we investigate three-dimensional channel alignment. There are several reports on three-dimensionally aligned polymer waveguides, but most of them are fabricated using the conventional photo-lithography method [4, 5]. Hence, the primary subject is waveguides with SI-square cores, and for stacking the 2-D waveguide arrays, very high position accuracy is required, which makes the fabrication process very complicated. Meanwhile, the Mosquito method has a large advantage in writing three-dimensional waveguides, because the Mosquito method resembles to popular 3-D printing technologies. Finally, the obtained waveguides are characterized.

Inter- and intra-board interconnect

In some peta-flop scale super computers, MMF links are already deployed mainly for connecting the racks and even boards in them. In those applications, optical data are generated at the edge of boards [6], and then transmitted through MMF ribbons with a data rate of 10 to 14 Gbps/ch. In the next advanced systems, the MMF links would be extended to allow electrical and optical (EO/OE) data conversions near the processors on the boards. Such OE/EO conversions on a chip is already realized in a commercially available IP router from Compass EOS Inc. [7]. In this system, 168 (12×14) bi-directional links are implemented, and the data rate is 8 Gbps/ch., which allows a 1.340 Tbps total throughput. However, with increasing the data throughput, more MMF links would be required for interconnects, and the wiring of MMFs would make the system integration more complicated. Therefore, polymer waveguides are anticipated to replace the electrical circuits on current PCBs, potentially making optical data routing simpler. Hence, in this paper, we focus on the optical interconnects from between boards to on-board as schematically shown in Fig. 1. As shown in Fig. 1, the MMF links remain used between board edges, while polymer optical waveguides on boards connect the processor chips and MMFs.

Figure 1. Configuration of inter O-PCB link model composed of two polymer waveguides and a multimode fiber.

Here, one of the advantages of optical interconnects over the electrical contenders is the possibility of high-bandwidth-density wiring, as mentioned before. For high-density wiring, channel alignment density of current fiber ribbons would not be high enough, so that multi-core fibers (MCFs: multiple cores in one fiber) are drawing much attention [8]. For a high connectivity with such MCFs, polymer waveguides on boards should also have high density in the channel alignment. Hence, in this paper, we show how high-density alignment is achieved by polymer optical waveguides utilizing the Mosquito method: three-dimensional wiring is also investigated.

The Mosquito Method

Most of the polymer waveguides previously reported have been of SI type with square or rectangular core shapes. This is attributed to the fabrication method of the polymer optical waveguides: the photo-lithography and other similar techniques using photo-masks have generally been adopted. However, for the applications shown in Fig.1, the waveguides need to be connected with MMFs with a GI-circular core. So, the connectivity between MMFs and the SI-square-core polymer waveguides could be a problem particularly for realizing high-speed optical links [9], because the link power budget is quite limited. In order to address the connectivity issue, the same structure as GI-MMFs: GI-circular core in polymer waveguides is a promising solution, although conventional photo-lithography would be unable to form the circular cores. Therefore, we have focused on how to fabricate GI-circular-core polymer optical waveguides.

Recently, we proposed a new methodology for fabricating GI-circular core polymer waveguides: the Mosquito method [3, 10]. The Mosquito method utilizing a microdispenser is a very simple fabrication technique for polymer waveguides. A viscous core monomer is dispensed within a cladding monomer from a thin-needle of a syringe that is connected to a dispenser. Here, if the core monomer is dispensed on a hard-cured cladding, the bottom of the core transcribes the flat upper surface of the cladding, and thus it is impossible to maintain the original circular cross-sectional shape of the core just after dispensed from the needle. Meanwhile, in the Mosquito method, the needle tip remains inserted into the "liquid state" cladding monomer while dispensing the core monomer. The needle horizontally scans within the cladding monomer, resulting in forming a waveguide structure. Parallel arrays of circular cores are fabricated by repetitive parallel scans of a single needle, as schematically shown in Fig. 2.

For the waveguide material, we use silicone resins supplied by ADEKA Corp. (FX-W712 and FX-W713 for the core and the cladding, respectively). Since these two monomers are miscible, after the core monomer is dispensed, the two monomers diffuse slightly into each other to form a concentration distribution. After dispensing multiple cores, both the cores and cladding are exposed to UV light emitted from a UV LED for curing. As the obtained polymer is a three-dimensionally cross-linked copolymer, the fabricated waveguides have a high durability compared to the dopant-based GI-core waveguides we previously fabricated using the preform method [11].

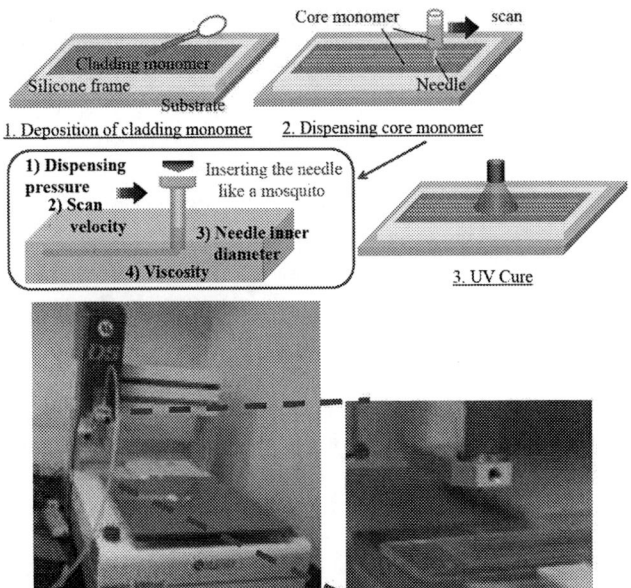

Figure 2. Fabrication process of polymer waveguides using the Mosquito method. Photo (left): desktop robot with a syringe (right): needle dispensing core into cladding.

Core diameter control for GI-circular-core waveguide

The MMFs used for the inter-board connections shown in Fig. 1 generally have a core diameter of 50 µm, so the waveguides connected with the MMFs should have similar dimensions. Meanwhile, in MCFs, since the cladding diameter is fixed to the same size as the standard MMFs, the core diameter should be smaller than 50 µm. In particular, a core diameter of 25 µm is chosen for a 7-core multimode MCF [8], while the interchannel pitch is as narrow as 40µm. Hence, for the applications of intra- and inter-board interconnects shown in Fig. 2 with the seven-core MCFs, the core diameter and interchannel pitch of the polymer waveguides should be controlled very accurately to fit to the MCF.

We already found that the dispensing pressure, the needle scan velocity, the needle inner diameter, and the viscosity of monomers were the key parameters in the Mosquito method for controlling the core diameter and interchannel pitch [3, 10]. By adjusting those parameters, we succeeded in fabricating a GI circular core polymer waveguide with a 40-µm core diameter and 125-µm pitch [3].

In this paper, much smaller core diameter is formed by optimizing the fabrication conditions. The results are summarized in Table 1. From Table 1, it is easily understood that the inner diameter of the needle directly influences the core diameter. On the other hand, it should be noted that even if a small-diameter needle is not used, high scan velocity or low dispensing pressure makes it possible to form a core diameter as small as 10 µm, as shown in Table I. Here, the needle scan is completely automated: we use a desktop robot for the dispenser (Musashi Engineering, Inc., SHOT Master 300DS), as shown in Fig. 2.

In Table 1, the measured insertion losses are also indicated. In the insertion loss measurement, a 4-Ch. waveguide sample to be tested is butt-coupled to launch and detection probes. For the launch probe, a single-mode fiber (SMF) is used, while two

Table 1 Core diameter control in the Mosquito method and the measured insertion losses of fabricated 5-cm long waveguides

Core Cross Section					
	50 μm	50 μm	50 μm	50 μm	50 μm
Needle I.D Pressure Scan Velocity	Large High High	Large High Low	Large High Medium	Medium High High	Small Low Low
Core Diameter (μm)	30	20	18	12	9
Insertion Loss at 850-nm Average (lowest) — Detection Probe 105-SI MMF	1.20 dB (1.17 dB)	1.31 dB (1.11 dB)	1.64 dB (1.37 dB)	2.11 dB (1.72 dB)	2.70 dB (2.17 dB)
Insertion Loss at 850-nm Average (lowest) — Detection Probe SMF	11.88 dB (10.39 dB)	7.08 dB (6.52 dB)	6.72 dB (6.13 dB)	5.31 dB (4.81 dB)	6.79 dB (6.46 dB)

different probes: 105-μm SI MMF and SMF are used for the detection, as indicated in Table 1. When an SMF and 105-μm SI-MMF are used for the probes, the coupling losses at the connections with the waveguide could be minimized, although the Fresnel reflection losses are still present. On the other hand, the losses while using two SMFs for the probes indicate the possibility of the single-mode operation of the waveguides: Even if the waveguide satisfies the single-mode condition, the difference of the mode fields between the probe and waveguide leads to large coupling losses.

From Table 1, it is found that the insertion loss measured using the SMF and SI-MMF probes gradually increases with decreasing the core diameter, but they are almost the same values of the insertion loss we previously measured for the waveguide with 40-μm cores [3]. It should be emphasized that even if the core diameter is 9 μm, the insertion loss of 2.17 dB (the lowest in all four channels) is still low enough. However, when the waveguide is inserted in two SMF probes, the insertion loss increases to 6 dB and higher even in the case of the smallest core waveguide, which is due to the mode field mismatch. Hence, for the single-mode operation, the waveguides should have the same mode field as the SMFs.

Narrow pitch waveguide

We already succeeded in fabricating a waveguide with a 40-μm core and 125-μm pitch [3]. In this paper, we focus on how we can decrease the interchannel pitch. First, we fabricate a waveguide with a 40-μm core and 62.5-μm pitch. A cross-section of the fabricated waveguide is shown in Fig. 3. The pitch of this waveguide is measured to be 61.7±3.4μm (average value is for the twelve channels). From Fig. 3, we confirm high pitch uniformity even for a pitch as narrow as 62.5 μm. We succeeded in decreasing the interchannel pitch of GI-core polymer waveguides to 62.5 μm for the first time to the best of our knowledge.

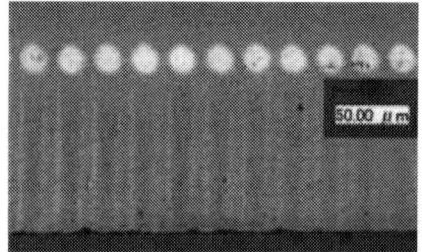

Figure 3. Cross-section of waveguide with 40-μmø cores and a 62.5-μm pitch.

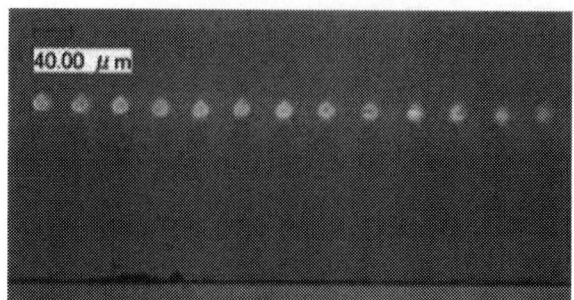

Figure 4. Cross-section of waveguides with 25-μmø cores and a 40-μm pitch.

Three-dimensional wiring

Since the needle tip is inserted into the cladding, if the needle scans in the vertical directions as well as in the horizontal directions, three-dimensional waveguides can be formed. In our previous investigations for multimode (large core) waveguides 10], we found that the dispensed core monomer sank slightly in the cladding monomer probably due to the dispensing pressure, as schematically illustrated in Fig. 5. If the needle tip is inserted deep into the cladding to reach very close to the substrate surface, the core collapses because the dispensed core cannot sink into the cladding but is stuck between the needle tip and the substrate, as shown in Fig. 6(a). Therefore, for forming 3-D waveguides, the needle tip needs to scan in an appropriate range of height from the substrate surface, specifically at the bottom of the cladding. In order to find the appropriate needle-tip height, cores are dispensed into a cladding monomer by deliberately varying the needle-tip height, similar to the one shown in Fig. 6(a). The cores shown in Fig. 6(a) was dispensed to have a core diameter of 50 μm, while in this paper, the dispensing conditions for a 10-μm core diameter are selected. A cross-section of the obtained waveguides is shown in Fig. 6(b).

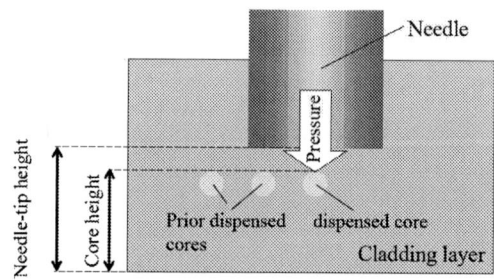

Figure 5. Height of dispensed cores in the cladding.

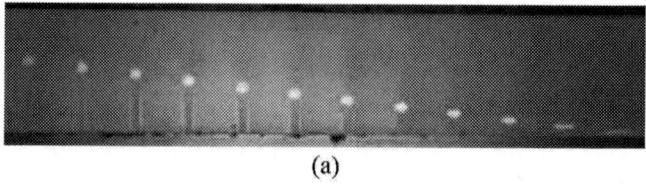

(a)

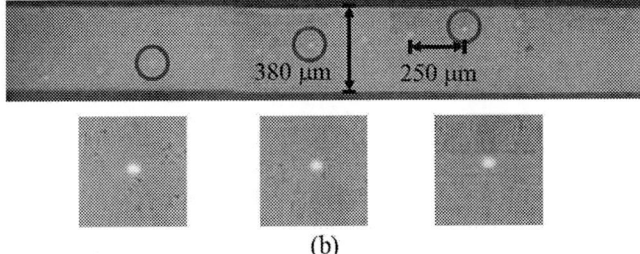

(b)

Figure 6. Cross-sections of waveguides in which the core height is deliberately varied. (a): Core diameter is preset to be 50 μm.[9] (b): Core diameter is preset to be 10 μm. The three photos at the bottom are enlarged images of the cores in the corresponding red circles.

As shown in Fig. 6 (b), if the cladding thickness is 380 μm, the height of the needle should be within 120 to 240 μm in order to form the cores with a high circularity. If we zoom in the cores in three red circles in the cross-section, it is confirmed that the cores in the middle has the highest circularity, while the two at the both sides have elliptical shapes.

Applying the results in Fig. 6(a), we design a three-dimensional fan-out waveguide in which three cores are aligned three dimensionally. Here, the core diameter is preset to be 25 μm, same as that of the multimode MCF aforementioned. The fan-out structure is formed in a 10-cm long waveguide: at one end facet (A) of the waveguide, three cores are obliquely aligned with a 40-μm pitch, as shown in Fig. 7. The height difference between the neighbor cores is set to be 10 μm. The three-core alignment at the waveguide end facet (A) is maintained for the first 2-cm long section, followed by a 5-cm long fan-out structure to expand the pitch only in the horizontal direction to 250 μm. In the next 1-cm long section, the heights of the two edge cores are conformed to that of the middle core (height conversion). Then, at last, the three cores horizontally aligned are extended to the other end facet (B) for 2 cm.

The designed fan-out device is fabricated utilizing the Mosquito method. The cross-sections of both ends (A) and (B) are shown in Fig. 8. Three circular cores are successfully aligned obliquely, while those cores are perfectly fanned-out and horizontally aligned in line with a 250-μm pitch.

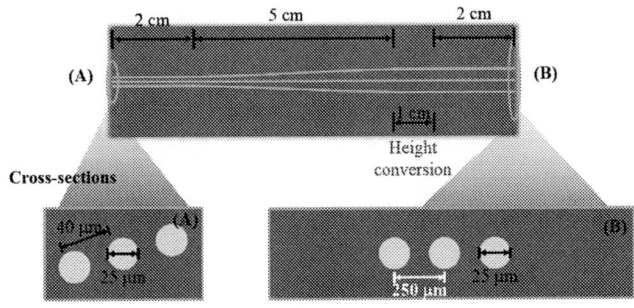

Figure 7. Design of a 3-D waveguide.

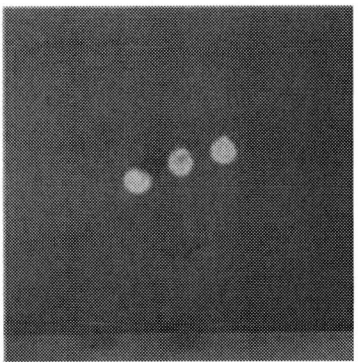

Facet (A)

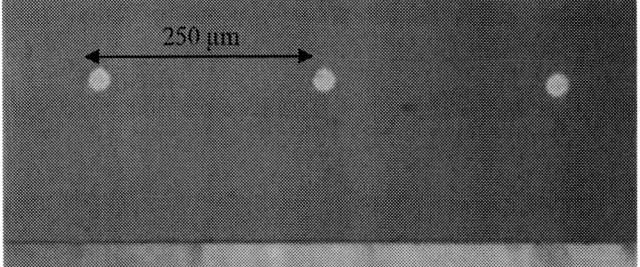

Facet (B)

Figure 8. Cross-sections of a three-core fan-out waveguide.

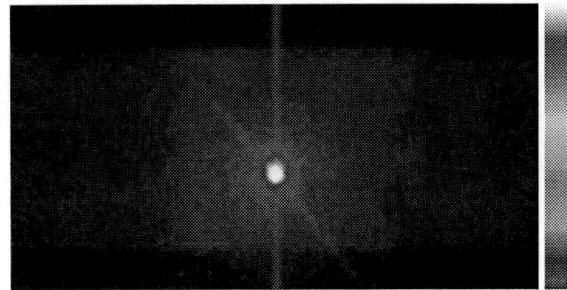

Figure 9. Near-field pattern form the middle core at End (B) shown in Fig. 8.

Near-field pattern (NFP) from the middle of three cores is measured, as shown in Fig. 9. For measuring the NFP, an 850-nm VCSEL is used and the middle core at End (A) in the fan-out waveguide is launched via an SMF probe. Although three cores are aligned in parallel at End (B), as shown in Fig. 8, the optical output is observed only from one core in the NFP, which means that, the crosstalk to the other cores is remarkably low. The low crosstalk is attributed to the tight optical confinement effect of GI-core [3, 10, 11].

The insertion losses of the three cores in the fan-out waveguide are also evaluated. The same launch condition as the NFP measurement is adopted for the insertion loss measurement, while a 50-μm core GI MMF is used for guiding the output light from the core at End (B) to an optical power meter. At the connection points between the launch/detection probes and the waveguide, the cores are butt-coupled without using matching oil. Therefore, Fresnel reflection losses are part of the observed insertion loss results. The insertion loss of the middle core is 2.95 dB, while the other two edge cores show 3.05 and 2.85 dB. Although the edge cores include curved structures for realizing the fan-out design, one of the edge cores shows the lowest loss. Only the middle core is straight, but the length variation of the three cores due to the fan-out structure is negligible.

For comparison, we fabricated a straight waveguide with the same length (10 cm), independently. The fabrication conditions are the same as those for the fan-out waveguide, so the core diameter is controlled to be 25 μm. The insertion losses evaluated for eight cores are 1.70 dB on average, and 1.18 and 2.70 dB are the lowest and highest values, respectively. The losses show a 1.5-dB variation, which could be because the fabrication condition is not necessarily optimized. However, the straight waveguide shows approximately 1 dB lower average loss than the fan-out waveguide. Meanwhile, the lowest loss in the three cores in the fan-out waveguide is comparable with the highest loss from the straight waveguide. Therefore, we still have the possibility to reduce the loss of the fan-out waveguide by optimizing the fabrication condition.

The facet (A) of the fan-out device was originally designed for fitting to the structure of an MCF (7 cores), and only three cores in seven were extracted in Figs. 7 to 9. However, the final structural design should be hexagonally stacking the 7 cores with a pitch as narrow as possible, as shown in Fig. 10 (a).

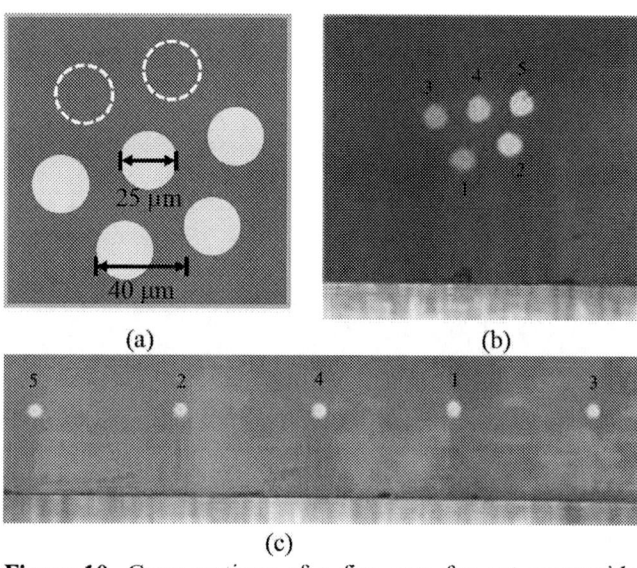

(a)   (b)

(c)

Figure 10. Cross-sections of a five-core fan-out waveguide (a): Structural design   (b): End facet (A), (c): End facet (B)

Since it is technically difficult to directly increase the core numbers from 3 to 7, a 5-core fan-out waveguide is fabricated for the second step. The cross-section of the end facet (A) is shown in Fig. 10 (b). As designed in Fig. 10(a), five cores are accurately aligned, while at the other end facet (B) shown in Fig. 10(c), the pitch of the five cores are extended to 250 μm, and all the cores are well aligned in line. The channel numbers from 1 to 5 shown in Fig. 10(b) correspond to the same channel numbers in Fig. 10(c), hence channels 1, 2, 3 and 5 need to be bent to satisfy the fan-out design, and channel 1 needs the largest height change of all the five cores. Although a slight increase in the insertion loss is observed, an insertion loss as low as 2.56 dB for 9.5-cm long fan-out waveguide is observed from channel 4 (straight channel) and even 5 (including a curved structure). Therefore, it would be possible to fabricate seven-core fan-out waveguide that is connectable to a seven-

core MCF. We are now investigating how the needle should scan to obrain the seven-core waveguide. The results will soon be published elsewhere.

Conclusions

By optimizing the dispensing and needle scan conditions, we succeeded in fabricating a GI-circular core waveguide with a core diameter of 10 μm and less, which allows us to fabricate single-mode polymer waveguides with such a simple fabrication methodology. For high-density wiring, not only the core diameter is reduced but also the interchannel pitch is decreased to 40 μm, despite the needle outer diameter larger than 300 μm. Such a narrow pitch alignment is possible because the dispensed core from the needle sinks slightly into the cladding: repetitive needle scans show little effect on the core shapes and alignment.

Furthermore, needle scans both in the horizontal and vertical directions in the Mosquito method could create three-dimensional waveguide circuit: low-loss three-core and five-core fan-out waveguides were successfully fabricated. If we align seven cores with a hexagonal stacking, a fan-out waveguide for multimode MCF could be fabricated. The Mosquito method that is very similar to a 3-D printing technology, would be a promising way to create polymer optical components not only for high-bandwidth-density on-board optical interconnects but also for wide variety of optical applications. To the best of our knowledge, this paper is the first report of three-dimensionally aligned fan-out polymer waveguide realized with GI-circular-core including the pitch and core-height conversions.

This work is partially supported by NEDO.

Acknowledgments

The authors would like to acknowledge Y. Ishikawa of ADEKA Corporation for supplying silicone polymer materials and for his continuous technical support to this research

. References

1. F. E. Doany, C. L. Schow, B. G. Lee, R. A. Budd, C. W. Baks, C. K. Tsang, J. U. Knickerbocker, R. Dangel, B. Chan, H. Lin, C. Carver, J. Huang, J. Berry, D. Bajkowski, F. Libsch, and J. A. Kash, "Terabit/s-class optical PCB links incorporating 360-Gb/s bidirectional 850 nm parallel optical transceivers," *J. Lightw. Technol.*, vol. 30, no.4, pp. 560-571, Feb. 2012

2. N. Bamiedakis, J. Beals, IV, R. V. Penty, I. H. White, J. V. Degroot, Jr., and T. V. Clapp, "Cost-effective multimode polymer waveguides for high-speed on-board optical interconnects," *J. Quant. Electron.*, vol. 45, no. 4, pp. 415-424, Apr. 2009

3. K. Soma and T. Ishigure, "Fabrication of a graded-index circular-core polymer parallel optical waveguide using a microdispenser for a high-density optical printed circuit board," *IEEE J. Sel. Top. Quant. Electron.*, vol. 19, no. 2 p.3600310, March/April. 2013

4. Y. Matsuoka, D. Kawamura, K. Adachi, Y. Lee, S. Hamamura, T. Takai, T. Shibata, H. Masuda, N. Chujo, and T. Sugawara, "20-Gb/s/ch high-speed low-power 1-Tb/s

multilayer optical printed circuit board with lens-integrated optical devises and CMOS IC," *IEEE Photon. Technol. Lett.*, vol. 23, no. 18, pp. 1352-1354, Sep. 2011.

5. R. Dangel, F. Horst, D. Jubin, N. Meier, J. Weiss, B. J. Offrein, B. W. Swatowski, C. M. Amb, D. J. DeShazer, and W. K. Weidner, "Development of versatile polymer waveguide flex technology for use in optical interconnects," *J. Lightw. Technol.*, vol. 31, no. 24, pp. 3915-3926, Dec. 2013

6. M. A. Taubenblatt, "Optical interconnects for high performance computing," in Technical Digest of Opt. Fiber Commun. Conf., OThH3 (Los Angeles, California, USA., 2011)

7. K. Hasharoni, S. Benjamin, A. Geron, G. Katz, S. Stepanov, N. Margalit and M. Mesh, "A high end routing platform for core and edge applications based on chip to chip optical interconnect," in Technical Digest of Opt. Fiber Commun. Conf., OTh3H. 2 (Los Angeles, California, USA., 2013)

8. B. G. Lee, D. M. Kuchta, F. E. Doany, C. L. Schow, P. Pepeljugoski, C. Baks, T. F. Taunay, B. Zhu, M. F. Yan, G. E. Oulundsen, D. S. Vaidya, W. Luo, and N. Li, "End-to-end multicore multimode fiber optic link operating up to 120 Gb/s," *J. Lightw. Technol.*, vol. 30, no. 6, pp. 886–892, Mar. 2012

9. S. Yakabe, T. Ishigure, and S. Nakamura, "Link power budget advantage in GI-core polymer optical waveguide link for optical printed circuit boards," in *Proc. SPIE*, 8267, 82670J, 2012

10. R. Kinoshita, D. Suganuma, and T. Ishigure, "Accurate interchannel pitch control in graded-index circular-core polymer parallel optical waveguide using the Mosquito method," Opt. Express, Submitted.

11. Y. Takeyoshi and T. Ishigure, "High-density 2x4 channel polymer optical waveguide with graded-index circular cores," *J. Lightw. Technol.*, vol. 27, no.14, pp. 2852 – 2861, Jul. 2009

Novel Trace Design for High Data-rate Multi-channel Optical Transceiver Assembled using Flip-chip Bonding

Takatoshi Yagisawa, Takashi Shiraishi, Mariko Sugawara, and Kazuhiro Tanaka
Fujitsu Laboratories Ltd.
1-1, Kamikodanaka, 4-chome, Nakahara-ku, Kawasaki 211-8588, Japan
t-yagisawa@jp.fujitsu.com

Abstract

We propose a trace design for high data-rate multi-channel optical transceiver assembled using flip-chip bonding technology. The trace, which is formed on a transceiver substrate, compensates for bandwidth degradation due to the parasitic capacitance of the flip-chip bonding pad for optical and electrical devices. The trace was adopted to the receiver side of a flexible-printed-circuit-based optical engine that consists of a 25-μm-thick flexible printed circuit substrate on which a vertical-cavity surface-emitting laser array, driver, photodiode array, and transimpedance amplifier are mounted by flip-chip bonding. The polymer waveguide is laminated from the backside of the substrate via microlens-imprinted film for optical coupling. The sensitivity degradation caused by the parasitic capacitance was successfully suppressed to 0.6 dB at 40 Gb/s by using our trace design compared to that with a minimum length wire bonding connection.

Introduction

Optical interconnection is a key technology for overcoming bandwidth bottlenecks in high-performance computing systems and high-end servers [1], [2]. Recently, several studies have been carried out on high-speed, multi-channel optical transceivers for interconnects assembled using flip-chip bonding technologies such as 3-dimensional (3-D) stacking (i.e. optics on chip) [3], [4], chip on glass carrier [5], and chip on film structures [6], [7]. This is because flip-chip bonding technology enables simultaneous connection of a large number of channels of an arrayed optical device to a substrate.

We previously proposed an optical transceiver based on optical engines (OEs) that use polymer waveguides and flexible-printed-circuits (FPCs) [8]. A FPC-based optical engine (FPC-OE) includes electrical interfaces connected to a printed circuit board, a vertical-cavity surface-emitting laser (VCSEL) array, driver integrated circuit (IC), a photodiode (PD) array, and a transimpedance amplifier (TIA) mounted by flip-chip bonding on the FPC. A thin microlens-imprinted film is attached to the bottom of the FPC. The film is then pasted on a polymer waveguide that has a 45° mirror. The FPC has through-holes just beneath the VCSEL and PD so that the FPC does not interfere with the lens structure. These simple passive-alignment processes reduce the assembly cost [9]. The FPC-OE is placed on a transceiver module substrate with an FPC connector as shown in Figure 1.

However, a serious issue occurs when this technology is applied to high-speed, multi-channel optical transceivers. The impedance of a VCSEL or a PD is vastly different from that of a driver IC or a TIA. Moreover, the high-speed electrical signals between electrical and optical devices have to be connected with traces formed on a substrate. Usually, the length of the traces should be several millimeters, and longer traces may be required for 2-D aligned optical arrayed devices. Therefore, when we use flip-chip technology for the connection between a VCSEL and driver IC or a PD and TIA, it is necessary to suppress the degradation caused by a several-millimeter impedance mismatch connection. A maximum 3-dB degradation of link sensitivity introduced by a 3-mm transmission line formed on a glass substrate was observed at 30 Gb/s, was reported [10].

For this study, we investigated the degradation factors of the connections between a VCSEL and driver IC and between a PD and TIA mounted using flip-chip bonding to an FPC substrate. We propose a trace design that suppresses sensitivity degradation by compensating the parasitic capacitance with an inductive line. The sensitivity degradation is suppressed to less than 0.6 dB at 40 Gb/s compared with a minimum length wire bonding connection.

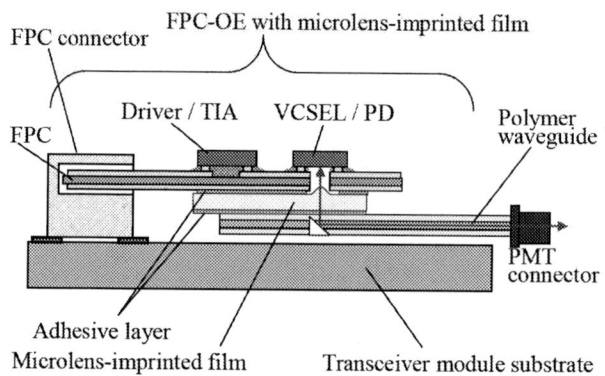

Figure 1. Structure of FPC-OE with microlens-imprinted film and optical polymer waveguide

Issues on connection between driver and VCSEL

We first clarified the degradation factors of flip-chip bonding for a multi-channel transmitter by comparing the equivalent circuits with a conventional wire bonding structure. For wire bonding, the VCSEL array is mounted near the driver IC to shorten the length of the bonding wire, as shown in Figure 2. For the flip-chip bonding, the VCSEL array and the driver are mounted at a distance of several millimeters to take a space for the fillet of an underfilling that fixes the chips to the substrate, as shown in Figure 3. When applying flip-chip bonding to the transmitter, the following two issues with the driver circuits and transmission line to be addressed.

978-1-4799-2408-0/14 $31.00 © 2014 IEEE   1048   2014 Electronic Components & Technology Conference

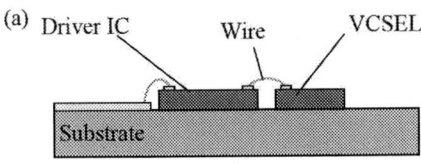

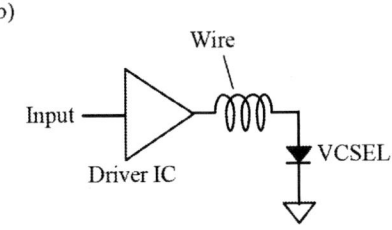

Figure 2. (a) Structure of wire bonding connection between driver IC and VCSEL, (b) and its schematic

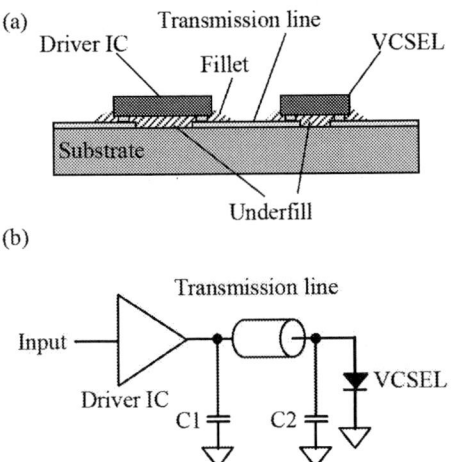

Figure 3. (a) Structure of flip-chip bonding connection between driver IC and VCSEL, and (b) its schematic

First, back termination has to be implemented in the driver circuit accompanied with an impedance controlled transmission line to reduce the signal reflection via the transmission line. Since the VCSEL has high capacitive impedance, back termination dose not result in multiple reflections in the transmission line. Impedance matching of the output buffer of the driver with load VCSEL is one of the most important issues.

The second issue is the VCSEL driving polarity, which relates to the VCSEL array structure for high-density interconnect. In general, the cathode-driving type [11], [12] is easier to design due to its high-speed characteristics. However, the cathode of array VCSELs should be commonly connected to ground (GND) for this application, and cannot be modulated independently. This is because high-density transmission lines for array VCSELs need to be designed on a stable GND reference plane (see [13] for further details of the transmitter circuits).

Issues on connection between PD and TIA

The receiver side shows more noticeable degradation due to the instability of a bias voltage for the PD cathode in addition to the several-millimeter high-speed impedance mismatched connection. A schematic of the receiver side for wire bonding connection and flip-chip bonding connection are shown in Figures 4 (a), (b), respectively. Since the structure of the receiver side is the same as the transmitter side, they are omitted. When applying flip-chip bonding to the receiver, the following issues with the PD bias and transmission line need to be addressed.

The first issue is the parasitic capacitance. In case of flip-chip bonding, additional parasitic capacitance of the bonding pads C1 and C2 are introduced. The bandwidth of the receiver is determined by the total capacitance between the PD anode and TIA and the impedance of the frontend of TIA. The second is the instability of the PD cathode bias that supplied only from the TIA mounted several millimeters from the PD. The instability causes sensitivity degradation.

To solve these issues, we propose an improved circuits shown in Figure 4 (c). We propose an inductive line on the substrate to compensate for the bandwidth degradation caused by the parasitic capacitance C1 and C2. We also propose the capacitor C3 that stabilizes the cathode bias and suppresses the resonance that occurs with the cathode line length between the PD and TIA.

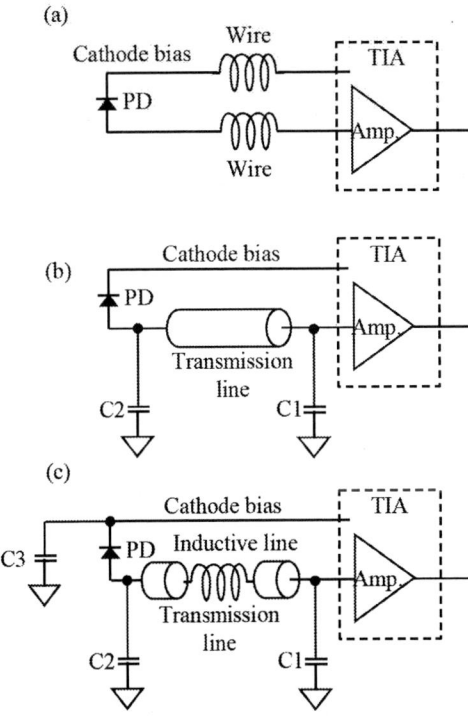

Figure 4. Schematic of (a) wire bonding connection, (b) flip-chip bonding connection, and (c) proposed improved circuits for flip-chip bonding

Characteristics of transmission line

We fabricated and evaluated a trace for the transmission line for the transmitter and receiver. The line was fabricated on a 25-μm-thick FPC substrate. The material of the substrate was polyimide. The trace was designed for a 125-μm-pitch GSGS (G: GND, S: signal) to connect the anode and cathode pads of the 250-μm-pitch optical arrayed device. The trace

978-1-4799-2408-0/14 $31.00 © 2014 IEEE   1049

was evaluated using a vector network analyzer. Broadband probes that had a bandwidth of 65 GHz were used by measuring at ports 1 and 2, as shown in Figure 5.

The measured small signal frequency responses of S-parameters for the fabricated FPC trace for the transmitter are shown in Figure 6. These results include the 1.5-mm transmission line formed on the FPC. The reflection of SDD_{11} remained below −10 dB for frequencies up to 61 GHz, and the −3 dB bandwidth was more than 60 GHz. This is a sufficient frequency response of the fabricated trace for the high-speed optical transceivers.

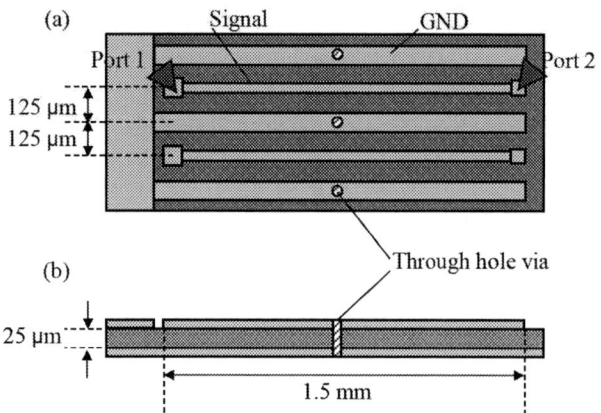

Figure 5. (a) Structure of transmission line fabricated on FPC, and (b) cross-section of structure

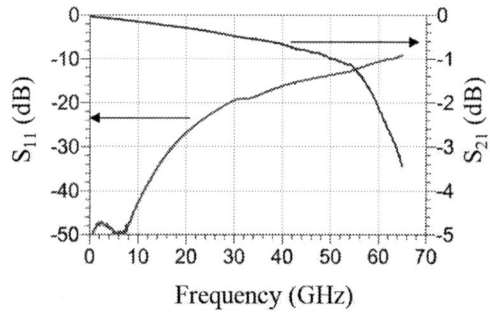

Figure 6. Measured S-parameters of transmission line fabricated on FPC

Proposed trace design for receiver

For the receiver, it is necessary to implement the inductive peaking line in the transmission line as described in the previous paragraph. The inductive line was created by cutting off the reference plane located on the back surface of the substrate, as shown in Figure 7. The inductive line was formed at the middle of the transmission line. The impedance of the inductive region of the line was calculated to be 119 Ω by using the results of electro-magnetic analysis.

The connection between the PD and TIA is composed of cathode lines supplied from the TIA and anode lines that transmit the high-speed signals into the TIA. The lines are aligned at a 125-μm pitch; therefore, the transmission lines

with inductive region causes instability of the cathode line. Therefore, we used the cathode as a reference plane.

We evaluated the effectiveness of the cathode reference plane thorough simulation. Figure 8 shows the simulation model. The lines were formed on a 25-μm-thick FPC substrate. The material of the substrate was polyimide. The lines were aligned at a 125-μm pitch to connect the anode and cathode pad of the 250-μm pitch PD model. Figure 9 shows the differences in the eye diagram between the lines that use the GND and cathode references. The eye diagram of the cathode reference (Figure 9 (b)) was much clearer than for the GND reference (Figure 9 (a)), because the stability of the cathode improved by using the cathode as the reference plane.

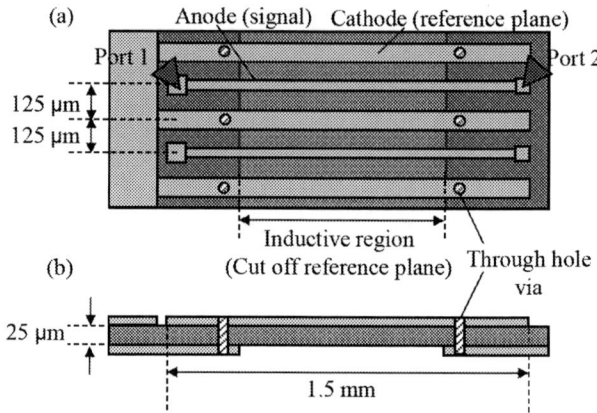

Figure 7. (a) Structure of trace for transmission line between PD and TIA, and (b) cross-section of structure

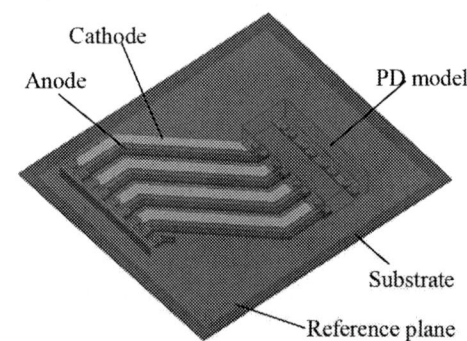

Figure 8. Implantation model for reference plane

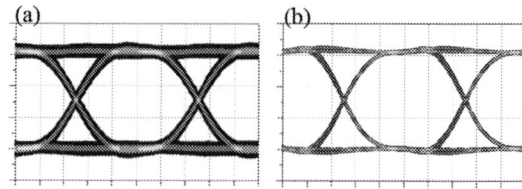

Figure 9. Simulated eye diagram of (a) GND reference and (b) cathode reference

Fabricated FPC-based optical engine

We fabricated an FPC-OE by using our proposed traces. A driver IC with dual peak-tunable pre-emphasis [13], [14]

978-1-4799-2408-0/14 $31.00 © 2014 IEEE   1050

which uses 0.13-µm SiGe BiCMOS technology, was used with a GaAs-based 850-nm VCSEL array that has a −3-dB bandwidth of 16 GHz with an 8-mA bias current for the transmitter side. The virtual back termination techniques enable the output impedance of 50 Ω of the driver to be connected to the VCSEL by flip-chip mounting on the impedance-controlled transmission line formed on the FPC, mitigating the degradation of the high-speed signals.

A GaAs-based PD array with a 25-µm active area diameter and a TIA that uses 0.13-µm SiGe BiCMOS technology were used for the receiver side. The inductive peaking lines are formed on the FPC substrate that connects the PD and TIA to compensate for the bandwidth limitation caused by the large parasitic capacitance of the flip-chip bonding pads. A thin microlens-imprinted film is attached to the bottom of the FPC. The film is then pasted on a polymer waveguide that has a 45° mirror. The FPC has through-holes just beneath the VCSEL and PD so that the FPC does not interfere with the lens structure. The fabricated FPC-OE of 13.5 × 10.6 mm with a 27.8-mm polymer waveguide is shown in Figure 10.

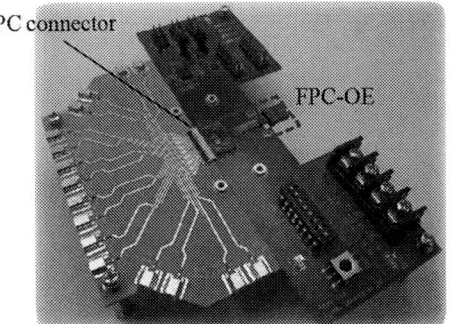

Figure 11. Structure of EVB

The measurement results of the output waveform of the fabricated FPC-OE are shown in Figure 12. Clear eye-openings of the optical output with extinction ratios of 4.1, 3.8, 3.4, 2.5 dB were obtained while operating at 25, 30, 35, 40 Gb/s, respectively. Clear eye-openings of the electrical output waveform were also obtained.

The receiver sensitivities of the fabricated FPC-OE are shown in Fig. 8. The sensitivity at a BER = 10-12 measured at 25, 30, 35, 40 Gb/s were −5.0, −4.1, −3.4, −1.6 dBm, respectively. The sensitivity of the PD and TIA mounted on a minimum length wire bonding evaluation board was −2.2 dBm at 40 Gb/s. Therefore, the degradation of the flip-chip bonding was suppressed to 0.6 dB by using our trace design.

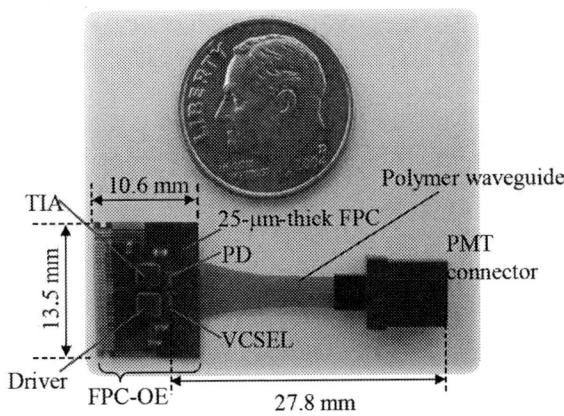

Figure 10. Structure of fabricated 4-channel FPC-OE

Measurement results

The fabricated FPC-OE was mounted on an evaluation board (EVB) by using a high-speed FPC connector [15]. The EVB had a coaxial connector interface for connecting the high-speed signals from the measurement equipment as shown in Figure 11. The input signal was 25 to 40 Gb/s non-return to zero (NRZ) with a 2^7-1 pseudo-random bit sequence (PRBS). The VCSEL bias current was controlled at about 10 mA. The operating ambient temperature was 25°C. A sampling oscilloscope with a bandwidth of 20 GHz was used with a precision time base module to monitor the optical waveforms. The output optical signals of the transmitter were connected to the receiver side of another FPC-OE, which converted the optical signals into electrical ones in order to test the bit error rate (BER). The measured results of the optical characteristics were obtained at the output of a multi-mode fiber (MMF) which connected to the PMT connector of the FPC-OE.

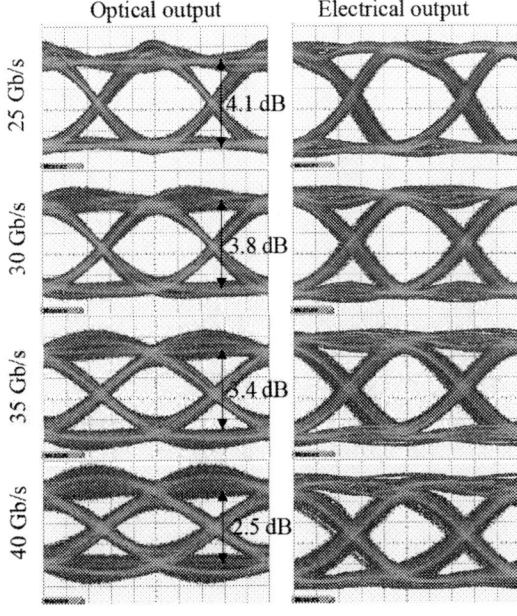

Figure 12. Output optical and electrical waveform of fabricated FPC-OE

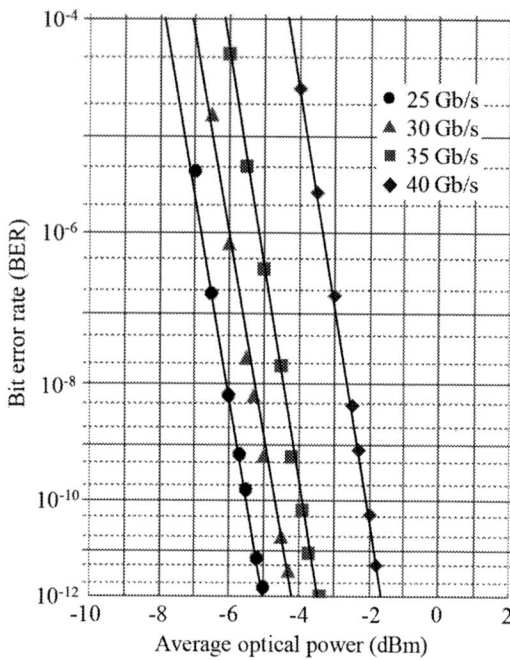

Figure 13. Receiver sensitivity of fabricated FPC-OE.

Conclusions

We proposed a trace design for high data-rate multi-channel optical transceivers assembled using flip-chip bonding technology. The trace, which is formed on a transceiver substrate, compensated for bandwidth degradation due to the parasitic capacitance of the flip-chip bonding pad for optical and electrical devices. The trace was adopted to the receiver side of a fabricated FPC-OE that consists of a 25-μm-thick FPC substrate on which a VCSEL array, driver, PD array, and a TIA are mounted by flip-chip bonding. Clear eye openings and error-free operation up to 40 Gb/s were successfully demonstrated. The sensitivity degradation caused by the parasitic capacitance was suppressed to 0.6 dB at 40 Gb/s compared with a minimum length wire bonding connection by using our trace design.

References

1. Alan Benner, "Optical Interconnect Opportunities in Supercomputers and High End Computing," Proc. *The Optical Fiber Commun. Conf. (OFC)*, Los Angeles, CA, March, 2012, OTu2B.4.

2. Jun Matsui, Tomohiro Ishihara, Tsuyoshi Yamamoto, Kazuhiro Tanaka, Satoshi Ide, Shigenori Aoki, Tsuyoshi Aoki, Mitsuhiro Iwaya, Kenji Kamoto, Katsuki Suematsu, and Masato Shiino, "High Bandwidth Optical Interconnection for Density Integrated Server," Proc. *The Optical Fiber Commun. Conf. (OFC)*, Anaheim, CA, March, 2013, OW4A.4.

3. Laurent Schares, Jeffrey A. Kash, Fuad E. Doany, Clint L. Schow, Christian Schuster, Daniel M. Kuchta, Petar K. Pepeljugoski, Jean M. Trewhella, Christian W. Baks, Richard A. John, Lei Shan, Young H. Kwark, Russell A. Budd, Punit Chiniwalla, Frank R. Libsch, Joanna Rosner,

Cornelia K. Tsang, Chirag S. Patel, Jeremy D. Schaub, Roger Dangel, Folkert Horst, Bert J. Offrein, Daniel Kucharski, Drew Guckenberger, Shashikant Hegde, Harold Nyikal, Chao-Kun Lin, Ashish Tandon, Gary R. Trott, Michael Nystrom, David P. Bour, Michael R. T. Tan, and David W. Dolfi, "Terabus: Terabit/Second-Class Card-Level Optical Interconnect Technologies," *IEEE J. of Selected Topics in Quantum Electronics*, Vol. 12, No. 5, September/October, 2006.

4. Yehoshua Benjamin, Kobi Hasharoni, and Michael Mesh, "Assembly Development of 1.3 Tb/s Full Duplex Optical Module," Proc. *IEEE Electronic Components and Technol. Conf. (ECTC)*, Las Vegas, NV, May, 2013, pp. 292–296.

5. Takashi Shiraishi, Takatoshi Yagisawa, Tadashi Ikeuchi, Satoshi Ide, and Kazuhiro Tanaka, "Cost-effective Optical Transceiver Subassembly with Lens-integrated High-k, Low-Tg Glass for Optical Interconnection," Proc. *IEEE Electronic Components and Technol. Conf. (ECTC)*, Lake Buena Vista, FL, May 2011, pp.798-804.

6. Tomoaki Shibata and Atsushi Takahashi, "Flexible Opto-Electronic Circuit Board for In-device Interconnection," Proc. *IEEE Electronic Components and Technol. Conf. (ECTC)*, Lake Buena Vista, FL, May, 2008, pp. 261–267.

7. Yuka Ito, Shinsuke Terada, Mayank Kumar Singh, Shinya Arai, and Koji Choki, "Demonstration of High-Bandwidth Data Transmission above 240 Gbps for Optoelectronic Module with Low-Loss and Low-Crosstalk Polynorbornene Waveguides," Proc. *IEEE Electronic Components and Technol. Conf. (ECTC)*, San Diego, CA, May 2012, pp. 1526–1531.

8. Takashi Shiraishi, Takatoshi Yagisawa, Tadashi ikeuchi, Satoshi Ide, and Kazuhiro Tanaka, "Cost-effective On-board Optical Interconnection using Waveguide Sheet with Flexible Printed Circuit Optical Engine," Proc. *The Optical Fiber Commun. Conf. (OFC)*, Los Angeles, CA, March, 2011, OTuQ5.

9. Takashi Shiraishi, Takatoshi Yagisawa, Tadashi Ikeuchi, Satoshi Ide, and Kazuhiro Tanaka, "Cost-effective Low-Loss Flexible Optical Engine with Microlens-imprinted Film for High-speed On-board optical Interconnection," Proc. *IEEE Electronic Components and Technol. Conf. (ECTC)*, San Diego, CA, May 2012, pp. 1505–1510.

10. Xiaoxiong Gu, Renato Rimolo-Donadio, Russell Budd, Christian W. Baks, Lavanya Turlapati, Christopher Jahnes, Daniel M. Kuchta, Clint L. Schow, and Frank Libsch, "High-Speed Signaling Performance of Multilevel Wiring on Glass Substrates for 2.5-D Integrated Circuit and Optoelectronic Integration," Proc. *IEEE Electronic Components and Technol. Conf. (ECTC)*, Las Vegas, NV, May 2013, pp. 846–851.

11. Shigeru Nakagawa, Daniel Kuchta, Clint Schow, and Richard John, "1.5mW/Gbps Low Power Optical Interconnect Transmitter Exploiting High-Efficiency VCSEL and CMOS Driver," Proc. *The Optical Fiber Commun. Conf. (OFC)*, Anaheim, CA, March, 2008, OThS3.

12. Samuel Palermo, Azita Emami-Neyestanak, and Mark Horowitz, "A 90 nm CMOS 16 Gb/s Transceiver for Optical Interconnects," *IEEE J. of Solid-state Circuits*, VOL. 43, NO. 5, 1235-1246, May, 2008.

13. Mariko Sugawara, Yukito Tsunoda, Hideki Oku, Satoshi Ide, and Kazuhiro Tanaka, "Novel VCSEL driving technique with virtual back termination for high-speed optical interconnection," Proc. *The International Society for Optics and Photonics. (SPIE)*, Vol. 8267, 826713, February, 2012.

14. Yukito Tsunoda, Mariko Sugawara, Hideki Oku, Satoshi Ide, and Kazuhiro Tanaka, "A 40Gb/s VCSEL Over-Driving IC with Group-Delay-Tunable Pre-Emphasis for Optical Interconnection," Proc. *IEEE International Solid-State Circuits Conf. (ISSCC)*, San Fransisco, CA, Feb. 2014, 8.9.

15. Takatoshi Yagisawa, Takashi Shiraishi, Tadashi Ikeuchi, and Kazuhiro Tanaka, "FPC-Based Compact 25-Gb/s Optical Transceiver Module for Optical Interconnect Utilizing Novel High-Speed FPC Connector," Proc. *IEEE Electronic Components and Technol. Conf. (ECTC)*, Las vegas, NV, May 2013, pp. 274-279.

Modeling, Design, and Demonstration of Ultra-miniaturized and High Efficiency 3D Glass Photonic Modules

Bruce C. Chou^, Sandeep Razdan*, Haipeng Zhang*, Jibin Sun*, Terry Bowen*, Vanessa Smet, Gee-Kung Chang, Venky Sundaram, and Rao Tummala

3D Systems Packaging Research Center,
813 Ferst Drive NW, Georgia Institute of Technology, Atlanta, GA, USA
*TE Connectivity, Menlo Park, CA, USA
^cchou36@gatech.edu, (352)-359-3182

Abstract

This paper presents the modeling, design, and demonstration of an ultra-miniaturized 2.5D optical transceiver module using ultra-thin glass interposers with electrical and optical through vias. The 3D Glass Photonics (3DGP) technology with double sided attach of electrical and photonics ICs can achieve ultra-high bandwidth with improved power efficiency at lower cost than other photonic integration such as silicon photonics and organic boards. Thin glass substrates with 60um diameter through vias were fabricated with copper plated electrical vias and polymer-filled optical vias, formed simultaneously. Re-distribution layers were fabricated on top of these integrated vias for electrical interconnections. The 2.5D optical module produced this way features flip-chip bonded VCSEL and driver chips. Initial measurements of the optical vias showed 1.2 dB of loss.

Keywords: 2.5 D glass interposer, optical vias, optical transceiver.

I. Introduction

Due to intrinsic lower propagation loss and higher channel capacity of photons versus electrons, optical interconnects have gradually replaced electrical interconnects on bandwidth-critical nodes at shorter distances. Since the early 2000s, intense efforts in optical interconnects, beyond board-level, achieved impressive research results, including IBM's Terabus, Fraunhofer IZM's EOCB (Electro-Optical Circuit Board), and Intel's silicon photonics research that spearheaded many groundbreaking accomplishments [1 – 3]. Silicon photonics, which leverages high density CMOS technology and high index contrast between silicon and silicon dioxide, is no doubt the most suitable for intra-chip optical communications. However, silicon photonics cannot realistically extend beyond a single die due to fabrication cost. Furthermore, the huge refractive index mismatch between silicon-based optical waveguides and glass-based optical fibers introduces high loss and requires costly sub-micron alignment. The fiber-to-waveguide transition happens at board and chip level, as shown in Figure 1. In fact, packaging cost and fiber-to-waveguide loss are two of the biggest factors why optical communications have yet to replace electrical communications at board and chip levels in spite of decades of research efforts [4]. To go beyond board-level optical communications, the three metrics of energy efficiency (Joules per bit), density, and cost of packaging optoelectronics systems must be optimized simultaneously [5].

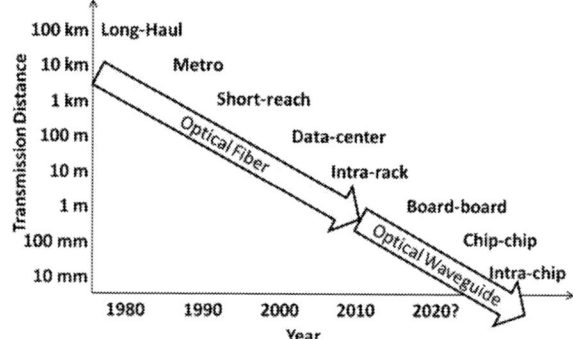

Figure 1. Evolution of optical communications

The 3D Glass Photonics (3DGP) concept aims to provide a simple and low cost photonics packaging solution utilizing ultra-thin glass interposer technology [6,7]. Glass offers several advantages over silicon and organic substrates for photonics packaging: optical transparency, reflective index matching to glass fiber, low-loss electrical signaling capability, good thermal isolation, excellent dimensional stability, and large panel processing. The 3DGP research aims to demonstrate the advantages of glass substrate in the following four research areas, as illustrated in Figure 2:

1. Fine-pitch and low-loss optical vias in glass.
2. Ultra-thin glass interposer with both optical and electrical vias.
3. 3D assembly of dies such as Photonic Integrated Circuit (PIC) on glass interposers.
4. Fiber-to-die transition.

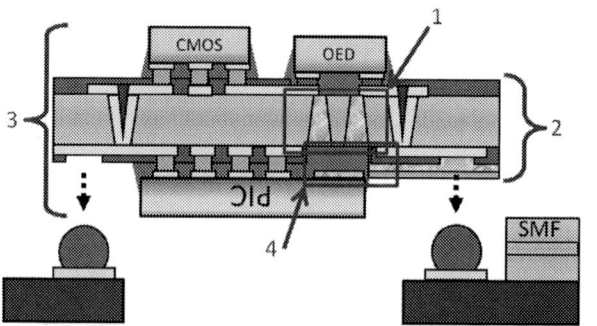

Figure 2. The four focus areas of 3D Glass Photonics (3DGP)

Initial modeling and fabrication of passive photonic devices in a glass interposer, namely polymer-based optical vias and waveguides, were presented earlier [8]. The current paper goes beyond this prior work in higher level of module integration. A VCSEL-based direct modulation optical transceiver was chosen to showcase the capability of 3DGP technology. Prior work has demonstrated the feasibility of a 2D optical transceiver module on glass substrate, but the thickness and dimension of such a module can be further miniaturized [9]. An ultra-thin 2.5D optical transceiver module is fabricated, featuring high density optical and electrical vias using a simplified process that can be modified for panel level processing. The 2.5D optical transceiver module is targeted at size reduction, electrical/optical loss reduction and lower cost enabled by large panel fabrication processes, precisely as shown on the top side of Figure **2**.

The rest of the paper is organized as follows. Section II covers the design of optical transceiver module and Beam Propagation Model (BPM) of optical wave propagation through glass interposer with through package vias (TPV). Section III describes the simplified fabrication and assembly steps used to build the 2.5 D transceiver module. Section IV shows the optical characterization results, with a summary and conclusion in Section V.

II. Modeling & Design

A typical 40 G short-reach optical transceiver module consists of a 4-channel laser source made of either VCSELs or EELs driven by a CMOS driver on the transmit side, and a 4-channel Photo-Detector (PD) array feeding into an amplifier such as a Trans-Impedance Amplifier (TIA) on the receiver side [10]. The optical modulation scheme is the simple On-Off Keying (OOK) scheme and the lasing wavelength used is typically 850 nm rather than 1550 nm, which uses multi-mode fibers (MMF) with 62.5/125 um core/clad diameter rather than single-mode fibers. The connection from the optoelectronic devices to MMF is through a 45 degree mirror and then to a standard MT-connector mounted on the PCB.

The electrical design of the 2.5 D transceiver module is completed in collaboration with TE Connectivity, and features impedance matched differential signal lines at minimized distances between the dies. The optoelectronic dies and CMOS dies used are provided by TE specifically for 40 G application. The optimized dimension of the interposer is 5 x 5 mm^2, which achieves 4x area reduction comparing to 10x10 mm^2 reported in [9]. Shown in Figure 3 is a simplified (to protect proprietary design details) side-by-side comparison of the 3DGP transceiver module versus the previously reported module. The reduction in area is achieved by 1) shorter electrical signal traces from the CMOS dies to the bumps through the use of electrical vias and 2) possibility of dies on top of bumps by utilizing both layers. In addition, thermal vias are deployed to help heat dissipation on a smaller area. Further reduction in area is achievable through 3D assembly.

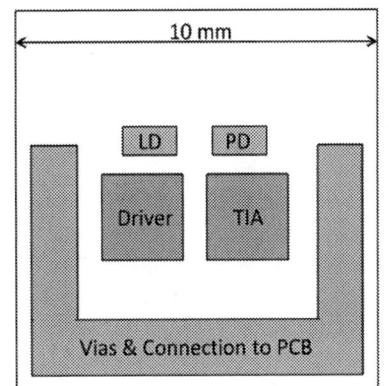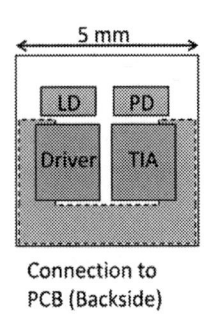

Figure 3. Simplified diagram showing area reduction of the 3DGP module (right) versus previously published module (left) [9].

One of 3DGP's research areas is the fabrication of fine-pitch optical vias. Depending on the thickness of the interposer, required laser/photo detector pitch, and fiber core diameter, an optical via might not be necessary.

Beam Propagation Model is used to determine the need for optical via. BPM is used in favor of the more accurate but more computationally intensive Finite-Difference Time-Domain (FDTD) method. BPM is sufficient because the critical dimensions (>20 um) are much larger than the wavelength of interest (0.85 um), and the lack of sharp angles along the path of light ensures paraxial model is reliable. BPM is implemented using MATLAB. Light propagation is simulated for three cases: 1) bare glass, 2) bare glass with polymer lens, 3) bare glass with fully filled polymer vias. In our model, the light wave travels through VCSEL-underfill-interposer-gel-fiber interface. The VCSEL is assumed to be 30 um away from the surface of glass, the glass thickness, t, varies from 100 um to 200 um, and the fiber is assumed to be 30 um from the exit side. The via has an entrance diameter of 60 um and a taper angle of 2.8°. The three cases are illustrated in Figure 4.

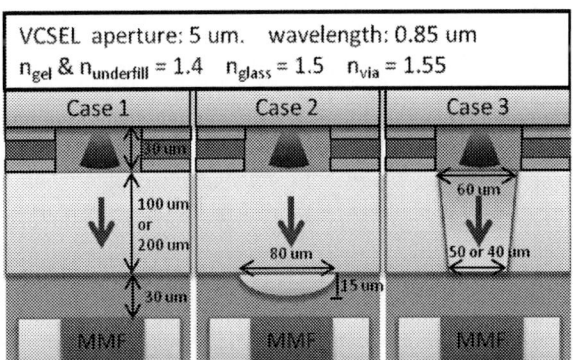

Figure 4. BPM simulation setup for the three cases.

The modeling results for t = 100 um is shown in Figure 5. The spreading of the light wave within the distance between VCSEL and fiber is short enough that bare glass is the

978-1-4799-2408-0/14 $31.00 © 2014 IEEE

simplest and best approach. In fact, a comparison of the optical power measured at the fiber core showed the fiber core will be able to capture close to 100% of the optical power in the bare glass case. While the addition of via did reduce the loss in the simulation, the added fabrication steps and optical loss due to surface and sidewall roughness could rendered the use of optical via excessive.

On the other hand, the modeling results for $t = 200$ um showed that the optical loss in the bare glass case is no longer negligible due to the dispersion of light, as shown in Figure 6. As shown in simulation, an optical via can help guide the beam to the fiber core through a 200 um thick interposer.

Similar simulation had been completed for $t = 300$ um and the loss at the fiber end was measured. The optical loss, calculated in dB as 10*log(Pout/Pin), for the three cases from 100 to 300 um has been plotted in Figure 7. The optical loss is effectively controlled to less than -0.5 dB at 300 um while the optical loss for the bare glass and polymer lens case exceeded -2 dB due to dispersion. The interposer built by GT-PRC features thin glass between $100 \sim 130$ um thick. In this range via is not needed to guide light into a MMF as the dispersion has not spread beyond the core region.

However, if a thicker interposer is used or SMF is needed for higher wavelength, an optical via will be the best choice with the most consistent dispersion control among the three.

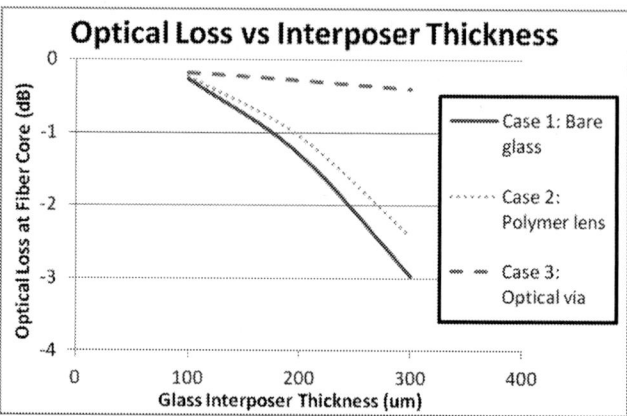

Figure 7. Calculated optical loss from BPM simulation.

III. Fabrication

Two fabrication processes were explored: electroless plated seed layer on dielectric laminated glass and sputtered seed layer on bare glass. The detailed process steps are illustrated in Figure 8. Both processes are implemented to compare the advantages and disadvantages of each. In the processed below, steps 1 ~ 5 are panel level processes while step 6 is done at the coupon level currently. Panel level processing on a 150 mm x 150 mm glass panel with 5 mm square coupons can yield 576 coupons per panel; therefore, potentially reducing the cost per coupon drastically.

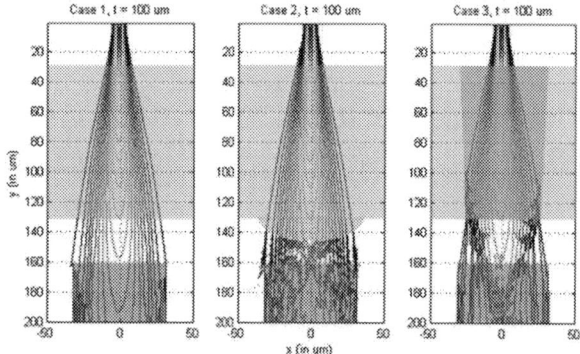

Figure 5. BPM simulation for t = 100 um.

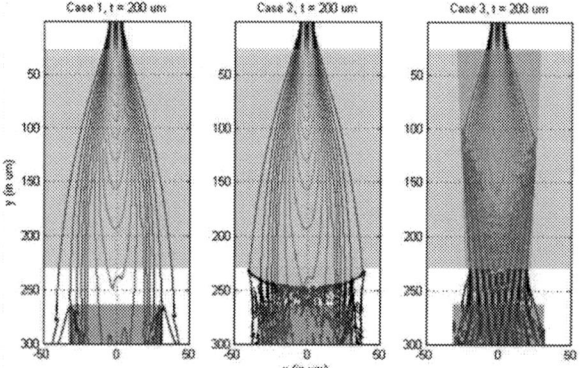

Figure 6. BPM simulation for t = 200 um.

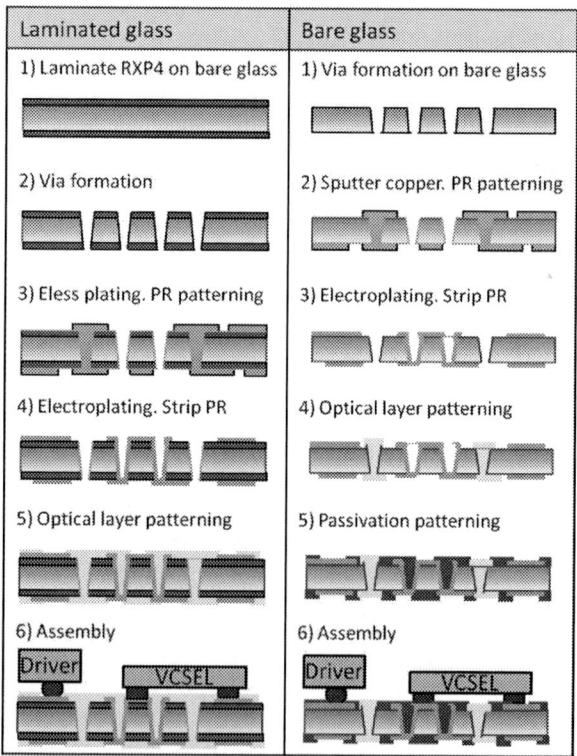

Figure 8. The two 3DGP substrate process sequences explored.

III.1 Laminated Glass Process

The laminated glass approach has been developed and optimized by GT-PRC [7]; however, the optically opaque dielectric laminate restricts optical waveguide formation on glass. The laminated glass approach nevertheless presents a viable option when only optical vias are needed. For the demonstrator, a 6" x 6" 100 um thick glass panel from Asahi Glass Company (AGC) is laminated on both sides by RXP-4 dielectric from Rogers. RXP-4 is chosen for its higher temperature tolerance (>350 C), which is important for assembly. 60 um vias are drilled by AGC at 250 um pitch, to match typical VCSEL pitch. Electroless process by Atotech is used to plate the copper seed layer. Electrical vias are exposed after photoresist patterning and plated to desired thickness, while optical vias are left bare.

The optical material used is the photo-definable Cyclotene 4024 from Dow Chemical, which is based on bisbenzocyclobutene (BCB) chemistry [11]. BCB exhibits higher refractive index comparing to glass and a high Tg; therefore making it suitable for the assembly process involving AuSn solder. Processing of optical waveguides and fully filled optical vias on glass has been reported in our previous work and will not be discussed in detail [8]. In current work, an attempt was undertaken to use BCB as passivation in addition to optical structures. If successful, the number of process steps can be reduced. Unfortunately, the process had not been optimized at writing and severe warpage made assembly unfeasible. On the other hand, the optical vias were filled completely, while the electrical vias were filled conformally. Figure 9 shows the panel-view of finished demonstrator containing 289 interposers, and Figure 10 shows the cross-sectional view of the electrical and optical vias from one of the coupons.

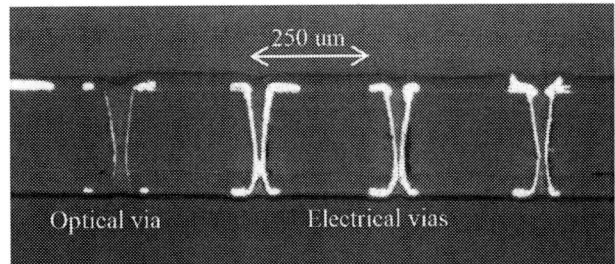

Figure 10. Cross-section of optical and electrical vias using laminated glass process.

III.2 Bare Glass Process

The bare glass process utilizes Corning's bare glass via formation technology to drill 60 um vias on 130 um thick 6"x 6" glass panel, also provided by Corning [12]. Titanium and copper seed layers were sputtered on bare glass by Tango Systems Inc. Based on the results from laminated glass, BCB was only used for optical via filling to reduce warpage, while solder resist from Hitachi Chemical was used for the passivation layer. ENEPIG (Electroless Nickel, Electroless Palladium, Immersion Gold) process by Atotech was applied to create gold finish on the surface. The completed panel featuring 289 interposers is shown in Figure 11. The panel was diced after surface finish to obtain the singulated interposers for assembly. VCSEL die with AuSn solder joints was flip-chip bonded first, followed by flip-chip bonding of driver die with tin-silver solder joints. The receiver dies were not assembled in the current demonstrator. Underfill was dispensed after bonding and cured in oven in air. The cross-section of the glass interposer with bonded VCSEL and driver dies is shown in Figure 12.

Figure 9. 3DGP demonstrator panel using laminated glass process on 100 um glass provided by AGC.

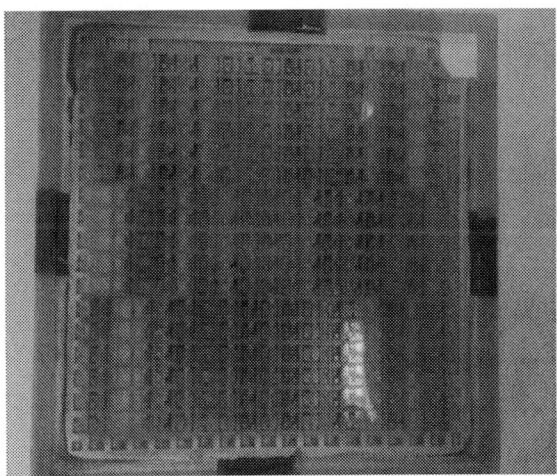

Figure 11. 3DGP demonstrator panel using bare glass process on 130 um glass provided by Corning.

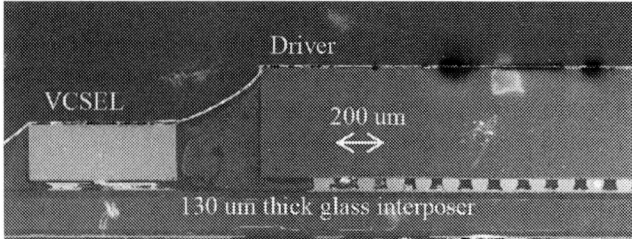

Figure 12. SEM of cross-sectioned interposer with VCSEL and driver.

IV. Characterization:

Optical loss characterization had been performed on via level, where the loss through the optical via was measured using direct abutment method. The measurement setup is shown in Figure 13. A standalone 850 nm VCSEL was used to drive a single mode fiber (SMF), which was fastened to a xyz micro-positioner. The DUT, which in this case is the glass interposer, was mounted on a second xyz micro-positioner. A photodetector, mounted on top of an FR4 board, was attached to the third xyz micro-positioner. The schematic of the sectup is shown in Figure 13a), while a picture of the actual setup is shown in Figure 13b).

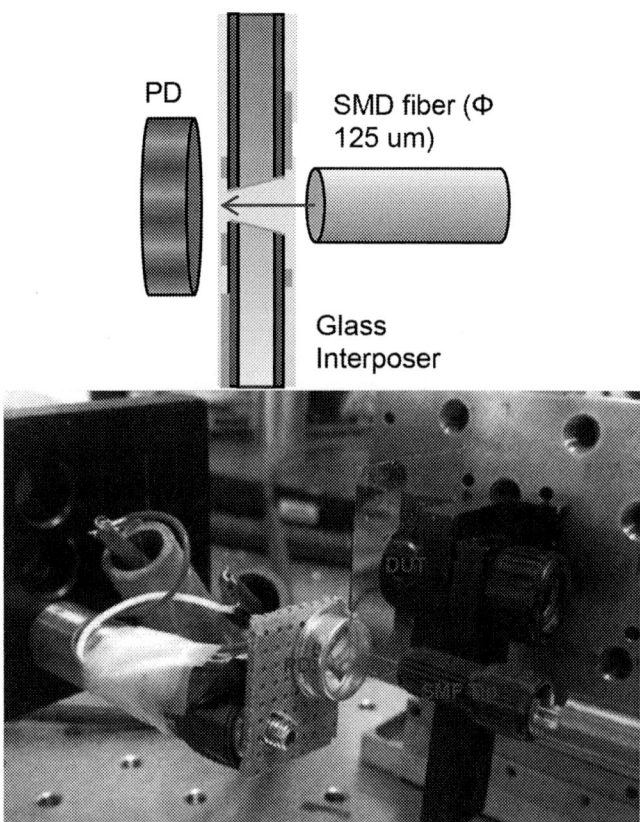

Figure 13. a) Schematic of optical loss measurement setup. b) Picture of setup showing DUT.

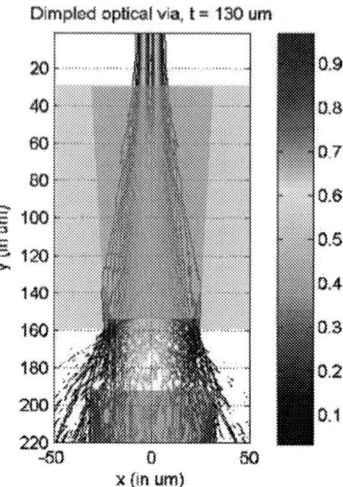

Figure 14. BPM of dimpled via showing light dispersion at the exit side of the via.

Relative optical loss can be measured by normalizing the receiving power calculated using PD voltage and current reading with respect to VCSEL power with no DUT (through air). Measurements performed on the 100 um laminated glass samples were recorded in Table I. The loss is higher in the optical vias than BPM simulation predicted, while the loss is comparable for the bare glass case. The higher loss through the optical via could be caused by either the surface or sidewall roughness of the optical via. In the BPM simulation of the filled via with surface dimples, as shown in Figure 14, visible optical loss due to dispersion could be observed, while simulation showed 0.48 dB loss. Since BPM could not capture the loss due to micron level surface roughness, the lower loss in simulation was expected.

Table I. Optical loss measurement of optical vias

Interface	Measured Loss	Simulated loss
Air	Normalized to 0 dB	0 dB
Bare Glass	0.25 dB	0.23 dB
Optical via	1.2 dB	0.14 dB / 0.48 dB

V. Conclusion:

A 2.5D short-reach optical transceiver module was demonstrated using GT-PRC's 3D glass photonics technology using both the laminated glass and the bare glass approach. The bare-glass approach was used for assembly trials due to lower warpage. The optical transceiver interposer featured reduction in dimensions in the x, y, and z direction, thus achieving more than 10x improvements in density compared to previously- published results. Both optical and electrical vias were integrated in the interposers. Optical measurements and simulations confirmed 0.48 to 1.2 dB loss in the optical via at 100 um thickness which could be reduced with process improvements.

Acknowledgement:

The authors would like to thank William Vis, Jialing Tong, Timothy Huang, and Vijay Sukumaran for their generous help in fabrication. Furthermore, the authors would like to thank Chris White and Jason Bishop for lab support. The authors would also like to thank Meg Gerstner and Jim Toth from TE Connectivity for their leadership and guidance.

REFRENCES

[1] F. Doany, "Power-Efficient, High-Bandwidth Optical Interconnects for High Performance Computing," conference presentation, Hot Interconnects, 2012.

[2] H. Schroder et al., "Advanced Thin Glass Based Photonic PCB Integration," in Electronic Components and Technology Conference, 2012.

[3] http://newsroom.intel.com/community/intel_newsroom/blo g/2013/04/11/chip-shot-intel-silicon-photonics-demonstrated-at-100-gbps

[4] D. Miller, "Device Requirements for Optical Interconnects to Silicon Chips", Proc. of the IEEE , Vol. 97, No.7, pp. 1166 - 1185 (2009)

[5] M. Taubenblatt, "Optical Interconnects for High-Performance Computing," in IEEE Journal of Lightwave Technology, Vol. 30, No. 4, 2012.

[6] V. Sukumaran et al., "Low-Cost Thin Glass Interposers as a Superior Alternative to Silicon and Organic Interposers for Packaging of 3-D IC, " in IEEE Transactions on Components, Packaging, and Manufacturing Technology, Vol. 2, No. 9, Sept. 2012.

[7] V. Sukumaran et al., "Design, Fabrication and Characterization of Low-Cost Glass Interposers with Fine-Pitch Through-Package-Vias," in Electronic Components and Technology Conference, 2011.

[8] B. Chou et al., "Modeling, Design, and Fabrication of Ultra-high Bandwidth 3D Glass Photonics (3DGP) in Glass Interposers," in Electronic Components and Technology Conference, 2013.

[9] L. Brusberg et al., "Glass Carrier Based Packaging Approach Demonstrated on a Parallel Optoelectronic Transceiver Module for PCB Assembling," in Electronic Components and Technology Conference, 2010.

[10] http://www.optcore.net/html_news/40G-&-100G-Optical-Transceivers-Basics-12.html

[11] Cyclotene 4000 Series Advanced Electronics Resins (Photo BCB), Dow Chemical Company, 2012

[12] A. Shorey et al., "Development of Substrates for Through Glass Vias (TGV) for 3DS-IC Integration," in Electronic Components and Technology Conference, 2012.

A Study on Nanofiber Anisotropic Conductive Films (ACFs) for Fine Pitch Chip-on-Glass (COG) Interconnections

Sang Hoon Lee, Tae Wan Kim, and Kyung-Wook Paik
Department of Materials Science and Engineering
Korea Advanced Institute of Science and Technology
291 Daehak-ro, Yuseong-gu, Daejeon, 305-701, South Korea
Corresponding author: kwpaik@kaist.ac.kr
Phone) +82-42-350-3375 Fax) +82-42-869-8124

Abstract

As pitch and space get finer in COG package, interconnection problems such as short circuit or open open circuit have been raised. In order to solve these problems, a novel concept of ACFs called nanofiber ACFs, a combination of nanofiber technology and electronic packaging, was introduced as one promising method in fine pitch assembly. In this study, nanofiber ACF technology was first applied to COG assembly, and nanofiber ACFs were investigated in terms of particle movement and electrical properties. For the nanofiber ACFs, polyvinylidene fluoride (PVDF) and polybutylene succinate (PBS) were used as nanofiber polymer materials. For the conductive particles, insulated conductive particles (ICPs) were used and incorporated into each nanofibers by electrospinning technology and adhesive films were laminated to fabricate nanofiber ACFs. The free movement of conductive particles was successfully suppressed in COG assembly by using PVDF and PBS nanofiber ACFs, resulting in excellent particle capture rate and electrical properties.

Keywords: chip-on-glass, fine pitch, interconnection, nanofiber anisotropic conductive film

1. Introduction

The demand of smart devices such as smart phones and tablets has risen dramatically over the last decades. As a result, display market also has grown and display technology has been continuously evolved. Industry wants more competitive technology like ultra high definition (UHD), and the current electronic packaging trend requires high performance, multi-functionalization, and miniaturization. As more I/O pins are required within the same space, the pitch has become finer, and packaging and interconnection technology in fine pitch has become very important in assembly processes [1].

In electronic packaging industry, anisotropic conductive films (ACFs), well known adhesive interconnecting materials, consist of thermosetting resin and conductive particles in a film format, have been widely used [2]. However, as the size of devices has been miniaturized and the pitch has become finer, interconnection problems such as short circuit issue and open circuit issue have been raised in fine pitch assembly. As a result, a novel concept of ACFs called "nanofiber ACFs" was introduced to solve these problems [3]-[5]. It has been proposed that combining nanofiber technology and electronic packaging is one promising method in fine pitch assembly. Nanofiber suppresses the free movement of conductive particles to prevent electrical interconnection issues such as short or open circuits.

In this study, nanofiber ACFs were further investigated for Chip-on-Glass (COG) assembly where interconnection structure is established by a flip-chip bonding process using ACFs. In this process, an electrical path is formed through captured conductive particles between a gold bump and a thin film electrode of the glass substrate. Nanofiber ACFs containing insulated conductive particles (ICPs) were investigated in terms of particle capture rate and electrical properties at 20 μm fine pitch COG assembly whose gap was only 8 μm. For nanofiber polymer material, polyvinylidene fluoride (PVDF) and polybutylene succinate (PBS) were selected for low temperature bonding process in order to avoid any heat damage during the bonding process. Then, ICPs coated by insulating polymer beads were incorporated into PVDF and PBS nanofiber by electro-spinning technology to maximize the insulation effect of nanofiber ACF.

Nanofiber ACFs are expected to suppress the free movement of conductive particles and prevent them from flowing out during bonding process. In other words, it can increase the number of captured particles and decrease the number of particles agglomerated between COG bumps which eventually prevents electrical issues like open and short circuits in fine pitch COG assembly. Moreover, polymer beads of ICPs are expected to provide higher electrical insulation along with nanofiber coating layers.

2. Experiment

2-1. Matierals

For PVDF nanofiber ACF, a mixture of Polyvinylidene fluoride (PVDF), Dimethylacetamide, Acetone, and insulated conductive particles (ICPs) of 3.5 μm diameter were used for electrospinning. For PBS nanofiber ACF, Polybutylene succinate (PBS), Chloroform, 3-chloro-1-propanol, Tetrabutylammonium bromide (TBAB) and the same ICPs were used [5]-[6].

2-2. Fabrication of nanofiber ACFs

Nanofiber ACFs where ICPs are incorporated into nanofiber were fabricated by an "electrospinning" method shown in Figure 1. When a high voltage is applied to a droplet of a polymer solution of a syringe needle and a target, the body of the polymer solution becomes charged and electrostatic repulsion counteracts the surface tension and stretches the droplet results in a shape of Taylor cone. At above the critical voltage, a charged liquid jet is formed. As the jet dries in flight, it is elongated and starts to form nanofibers [7]

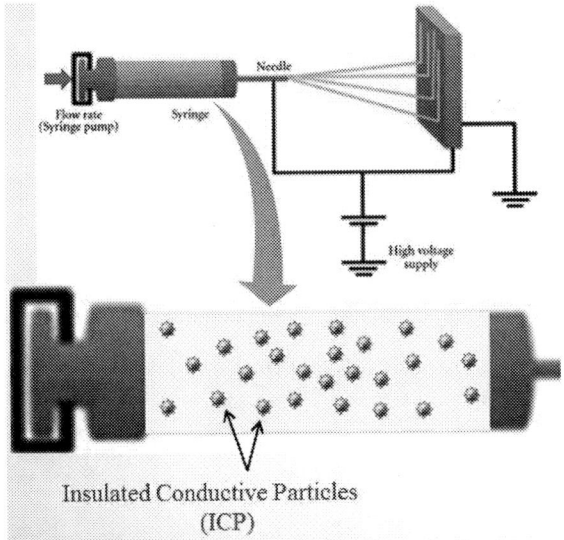

Figure 1. Electrospinning technology

For the nanofiber polymer solutions, Polyvinylidene fluoride (PVDF) nanofiber solution and Polybutyelene succinate (PBS) nanofiber solution were used at optimized solution contents shown in Table 1. In order to increase the conductivity of the PBS solution, 0.1 wt% of TBAB was added.

	PVDF	PBS
Concentration (wt%)	18	12
ICP content (polymer : ICP)	1:0.3	1:0.3
Solvent	DMAC Acetone	Chloroform 3-chloro-1-propanol
ICP diameter (μm)	3.5	
Applied voltage (kV)	8	12.5
Spinning rate (μL/min)	10	20
Working distance (cm)	15	12.5

Table 1. Electrospinning condition

For the conductive particles, ICPs coated by insulating polymer beads with diameters of 3.5 μm were used. The ICPs have core-shell structure. Poly beads were used as core polymer balls and nickel/gold were coated as 1st and 2nd coating layers. Then, polymer beads were coated at the outer layer as shown in Figure 2 whose thickness is 50 to 250 nm.

Finally, two epoxy nonconductive films (NCFs) were laminated on top and bottom of the electrospun nanofiber containing ICPs by a roll-laminator as shown in Figure 3. The bottom NCF with the thickness of 2 μm and the top NCF with the thickness of 12 μm were used. The minimum viscosity of the bottom NCF was 10 times higher than that of the top NCF. Two layer ACF has been considered to be favorable for capturing ICPs in COG structure, and the same concept has been applied to fabricate the nanofiber ACFs. Finally, the laminated ACF was laminated once again by a vacuum laminator to remove voids at increasing temperature to fabricate the nanofiber ACFs.

Figure 2. SEM image of insulated conductive particles (ICPs)

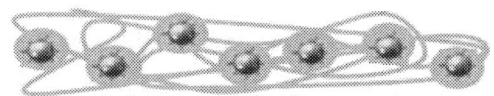

nanofiber containing ICP

Adhesive film

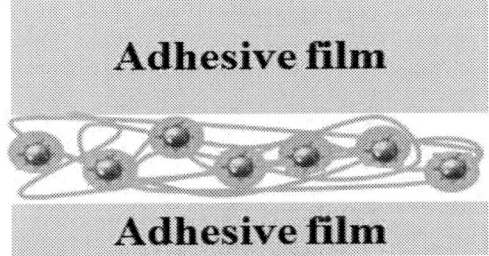

Adhesive film

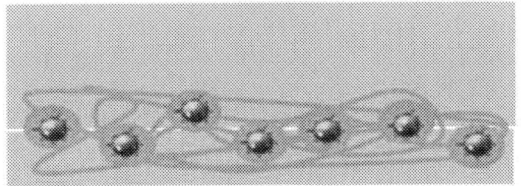

nanofiber ACF

Figure 3. Manufacturing procedures of nanofiber ACFs

2-3. Particle movement of nanofiber ACFs

In this experiment, bonding condition was divided into prebonding condition and main bonding condition. During the prebonding condition, 70 MPa was applied at 80 °C for 15 s to flow out the resin and capture the conductive particles between bumps and thin-film electrodes. In this process, nanofiber layers suppress the free movement of conductive particles and prevent them from flowing out. This process is a key to increase the capture rate of conductive particles. For the main bonding condition of PVDF nanofiber ACF, 30 MPa was applied at 190 °C for 5s to melt the nanofibers and cure the resin. For the PBS nanofiber ACF, 30 MPa was applied, but the bonding was performed at 150 °C for 5 s to melt the nanofibers and cure the resin. The bonding temperature conditions were decided based on the melting point of each nanofiber materials. Once the fine pitch COG assembly was bonded using the nanofiber ACFs, conductive particle movements of each nanofiber ACFs were analyzed and compared with conventional ACF by optical microscope see how the nanofibers contribute to controlling of conductive particle movement.

2-4. Characterization of ACF joint properties

The electrical properties of nanofiber ACFs were analyzed at 20 μm pitch COG packages whose pattern gap is only 8 μm. The COG packages have Kelvin structures for the bump contact resistance measurement and inspection of any open circuit issue caused by insufficient trapped conductive particles or unmelted nanofiber layer. They also contain insulation pattern for insulation resistance measurement and observation of any short circuit issue caused by agglomerated conductive particles between the neighboring bumps.

3. Results and Discussion

3-1. Fabrication of nanofiber ACFs

PVDF and PBS nanofibers containing ICPs electrospun at optimized electrospinning conditions are shown in Figure 4. Both nanofibers were ejected uniformly without forming any beads. The diameter of nanofibers was around 500 nm.

3-2 Particle Movement of nanofiber ACFs

Movement of the ICPs of nanofiber ACFs were analyzed by counting the number of ICPs on the bumps and converted into capture rate as shown in Figure 5. After main bonding, the number of ICPs of conventional ACF dropped gradually down to 35% while that of PVDF and PBS nanofiber ACFs dropped down to 60% and 65% each. It seemed that the ICPs in conventional ACF were free to move during the bonding. However, nanofiber layers in nanofiber ACFs suppressed the free movement of ICPs during the prebonding process and increased the capture rate. Higher capture rate means fewer ICPs are needed in the ACFs for the same performance. If ACFs contain fewer ICPs, there would be a smaller number of ICPs captured between bumps resulting less chance of having short circuits. In other words, nanofiber ACFs could capture more ICPs after bonding compared with conventional ACF which would eventually improve interconnection by reducing short circuit issue.

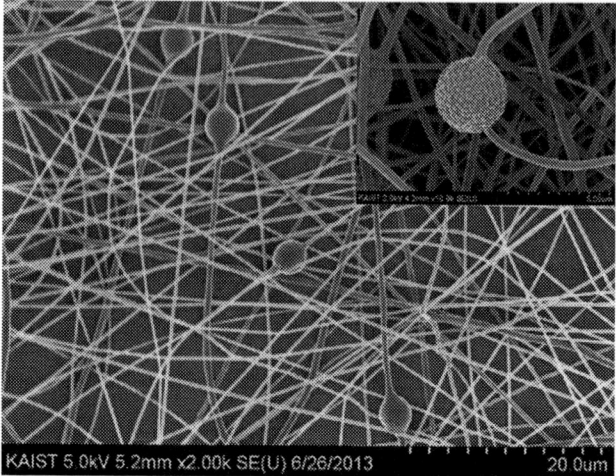

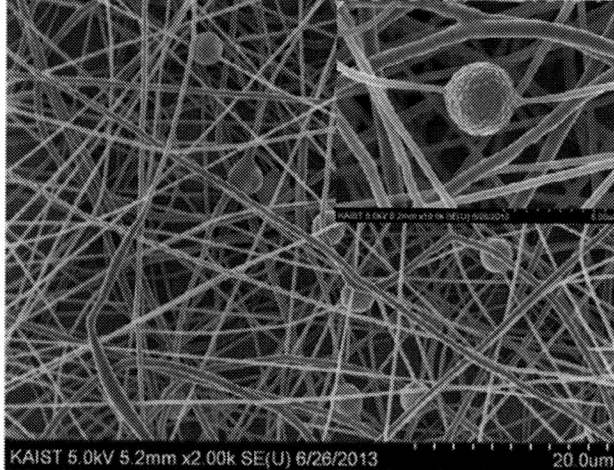

Figure 4. ICPs incorporated into electrospun PVDF nanofiber (top) and PBS nanofiber (bottom)

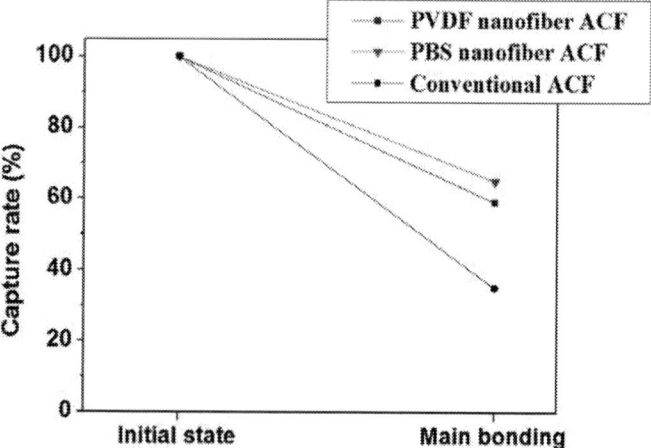

Figure 5. Analysis of the movement of conductive particles after main bonding process

3-3. Characterization of ACF joint properties

3-3-1. Contact resistance

Contact resistance of fine pitch COG assembly using PVDF and PBS nanofiber ACFs was measured and compared with the conventional ACF as shown in Figure 6. All three

ACFs showed relatively stable contact resistances below 500 mΩ. Both nanofiber ACFs showed similar contact resistance value compared to that of the conventional ACF despite the existence of the nanofiber layers inside the ACFs and different capture rate.

If nanofiber layers were unmelted and remained after the main bonding process, contact resistance value could be very high and even open circuit could be observed. However, both nanofibers in ACFs were melted completely at each main bonding process and stable contact resistance was able to be obtained.

Although the capture rates of conventional ACF and nanofiber ACFs were different, the initial number of ICP in conventional ACF was two times more than nanofiber ACFs. As a result, number of captured particles after main bonding was similar and total contact area of three ACFs was also similar.

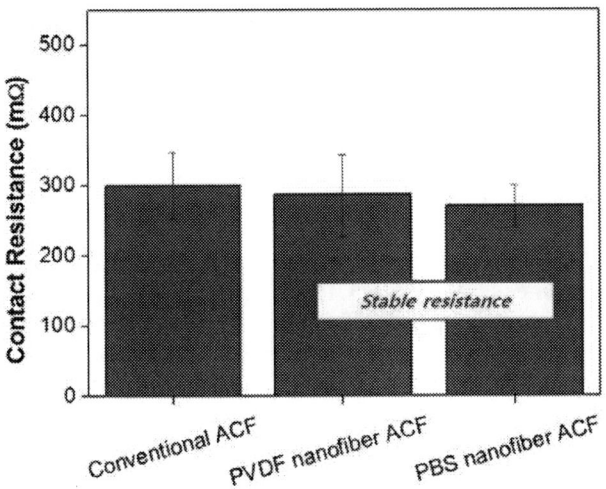

Figure 6. Contact resistance of ACF joints

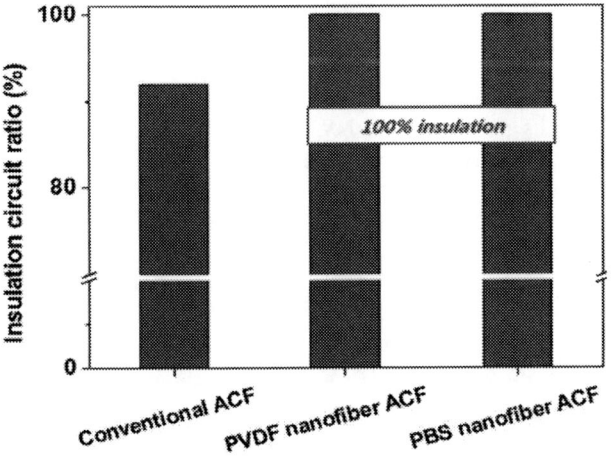

Figure 7. Insulation resistance of ACF joints

3-2-2. Insulation resistance

Insulation resistance at 20 μm pitch was measured when bonded by three different ACFs: conventional ACF, PVDF and PBS nanofiber ACFs. Both nanofiber ACFs showed perfect insulation while conventional ACF showed 91% of

insulation circuit rate as shown in Figure 7. Since nanofiber ACFs contain only half of ICPs compared with conventional ACF, they have less chance of having short circuit than conventional ACF. Moreover, nanofiber coating layer and insulating beads of ICP maximized the insulation effect of nanofiber ACFs.

4. Conclusion

20 μm fine pitch COG assembly was successfully demonstrated using PVDF and PBS nanofiber ACFs. Nanofibers contributed to suppressing the free movement of conductive particles and over 60% of initial conductive particles were captured after main bonding process. It was remarkable result since conventional ACF could only capture 35% of initial conductive particles after main bonding process. Nanofiber ACFs also showed excellent electrical properties such as stable contact resistance and no open or short circuit. This study has shown that nanofiber technology is a promising technology in ACF and COG packaging field which can contribute to the advancement in fine pitch package.

References

1. M. J. Yim, "Anisotropic conductive films (ACFs) for ultra-fine pitch chip-on-glass (COG) applications, *International Journal of Adhesion and Adhesives,* vol.27, pp. 77-84, Jan. 2007
2. M. J. Yim, "Design and understanding of anisotropic conductive films (ACFs) for LCD packaging", *IEEE Polymeric Electronics Packaging,* pp. 233-242, Oct. 1997
3. K. L. Suk, "Nanofiber anisotropic conductive adhesives (ACAs) for ultra fine pitch chip-on-film (COF) packaging," *Proc. 61st Electronic Components and Technology Conf.,* Lake Beuna Vista, FL, 2011, pp. 656-660
4. S. H. Lee, "Study on fine pitch flex-on-flex assembly using nanofiber/solder anisotropic conductive film and ultrasonic bonding method", *IEEE Trans. Comp. Packag. Manuf. Technol.,* vol. 2, pp. 2108-2114, Dec. 2012
5. T. W. Kim, "Low temperature flex-on-flex assembly using Polyvinylidene Fluoride nanofiber incorporated Sn58Bi solder anisotropic conductive films and vertical ultrasonic bonding", *Journal of Nanomaterials,* vol. 2013, Nov. 2013
6. E. H. Jeong, "Electrospinning and structural characterization of ultrafine poly(butylene succinate) fibers", *Polymer,* vol. 46, pp. 9538-9543, Nov. 2005
7. Teylor, G., "Electrically Driven Jets," *Proc. R. Soc. Lond. A,* Dec. 1969, Vol. 313, pp. 453-475

Study of Fine Pitch Micro-Interconnections Formed by Low Temperature Bonded Copper Nanowires Based Anisotropic Conductive Film

Jing Tao, Alan Mathewson and Kafil M. Razeeb*
Tyndall National Institute, University College Cork
Lee Maltings, Dyke Parade, Cork, Ireland
*kafil.mahmood@tyndall.ie

Abstract

In this paper, Cu nanowire based anisotropic conductive film (ACF) has been systematically investigated in terms of its fine-pitch potential, electrical/ mechanical properties, process related issues and failure mechanisms. A test chip module including daisy-chain and 4-point structures was fabricated by electroplating and photolithography process, which has 160, 80, 40 and 30 μm bond pad pitches. The selection of membrane templates and the template based open/short failures have been discussed. Bond pads were finished with electroplated In to lower the bonding temperature and contact resistance. The electrical and mechanical performance of the interconnections has been studied in term of bonding force. The interconnect resistance of various pad sizes was 0.3–0.6 Ω per pad at the low bonding force of 1.5 N and dropped to 0.02-0.04 Ω per pad at the high bonding force of 10 N. The influence of bonding pressure on such nanowires formed interconnects was studied by micro-sectional analysis. For all pitch sizes, the electrical insulation was maintained for an applied voltage of 20 V. A shear strength of 1-5 MPa was achieved as a function of the bonding force and the fractural surface analysis verifies the Cu-In joint formed in such interconnections.

Introduction

The fabrication of stacked and vertically interconnected device layers (3D integration) with through Si via (TSV) technology [1] has provided a path to system scaling and performance enhancement. One of the key challenges in 3D integration is the fine pitch interconnection scheme, which has to accommodate various thermal and mechanical constraints, especially for the non-Si elements in the future 3D integrated package [2]. For advanced 3D stacking, various interconnection schemes and process technology have been proposed and most of them require a temperature higher than 250 °C and the underfill process during the bonding process [3, 4]. As an alternative, anisotropic conductive film (ACF) can be a low-temperature, low-cost fine-pitch interconnection solution for future 3D integration. The conventional ACF, composed of randomly dispersed conductive micro particles in an adhesive polymer matrix, faces agglomeration problem when the size is reduced, resulting in potentially unstable electrical properties [5]. Moreover, the micro particles in the film have to be deformed to create electrical contact with high pressure, which constrains the minimum pitch size to ~ 100 μm for most commercial ACFs [6]. In order to further relax the thermal and mechanical constraints of interconnection process, nanowire based ACF (NW-ACF), containing metallic nanowire arrays vertically distributed in a porous polymer template, has been proposed as a low-temperature

process with a finer pitch potential. It avoids the particle agglomeration issue compared to conventional ACF, ease the co-planarity issue and remove underfill requirements compared to solder micro bump technology.

The synthesis of nanowires is based on electrodeposition method with nanoporous templates [7, 8]. Two kinds of templates are commonly used, anodic aluminum oxide (AAO) and ion-track etched polymer membranes (TEM). AAO templates have the advantage of the high pore density and ordered pore structure, but not suitable for direct bonding due to the brittleness of the template. Cheng et al. developed a novel Ag/Co-nanowire/polymer nanocomposite-film typed flip-chip technology, by fabricating the nanowires through AAO template, removing the AAO by chemical etching and refilling the nanowire arrays under a magnetic field with polyimide [9]. On the other hand, investigation on NW-ACF formed by TEM templates shows preliminary electrical and mechanical results in terms of bonding conditions with the unpatterned chips [10, 11]. Unlike nanowire arrays formed in AAO template, the nanowires grown in TEM template has a lower density (up to $10^9/cm^2$ depending on pore diameter, compared to AAO $\sim 10^{11}/cm^2$) and less ordered. The common problems with commercially available TEM are the overlapping pores at a relatively high porosity and the angular distribution of the pore channels with the surface normal [12, 13]. Overlapping pores can results in cross-linking of the grown nanowires and the distribution angle may affect the contact formed between the nanowires and the bond pads. Despite the statistically distributed pores of TEM template, it is still a very promising candidate to form an ACF material due to its relatively simple and cheap fabrication process.

In this paper, the Cu NW-ACF based on different TEM templates were investigated for its electrical performance at fine pitches. Three types of NW-ACFs were electrically screened for the open/short failures. The functional NW-ACF was studied at all pitch sizes for its contact resistance and insulation properties. Also, the influence of bonding pressure on the formed interconnects was evaluated. Finally, the preliminary shear strength results were demonstrated with the fractural surface analysis of the sheared samples.

Experiments

Fabrication of NW-ACF

The NW-ACF was fabricated based on TEM templates which are commercially available. The material variables include the membrane thickness, pore size and pore density. Table 1 lists the material properties of different membranes with similar pore size, purchased from different companies.

978-1-4799-2408-0/14 $31.00 © 2014 IEEE

Table 1 Material property of polycarbonate membranes.

NW-ACF type	Template Company/ Trade Name	Thickness (μm)	Pore size (μm)	Pore Density (pores/cm^2)
1	Isopore™	25	0.22	4.5×10^8
2	Nuclepore®	10	0.2	3×10^8
3	Sterlitech	10	0.2	3×10^8

The Cu NW-ACF was fabricated according to the process described in ref [10]. All templates were electrodeposited galvanostatically under the same condition by applying a current density of 3 mAcm^{-2}. Fig. 1 shows the SEM images of the morphology of the three NW-ACF films based on each template. The films were laterally torn to view the pore shapes and the distribution of the nanowires.

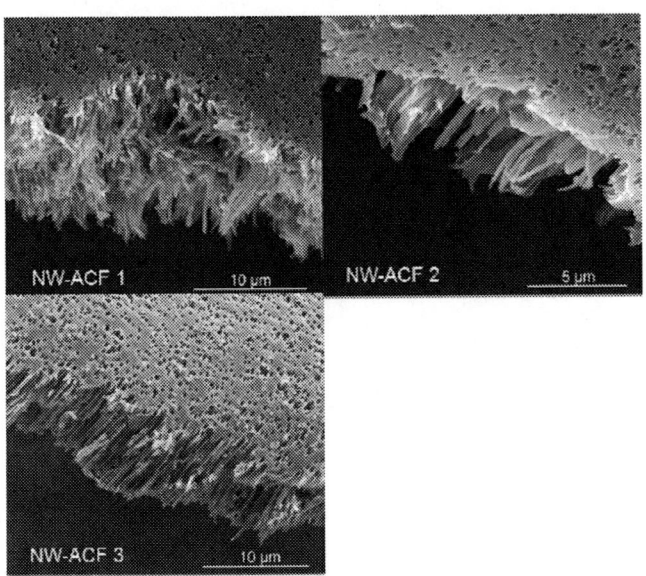

Fig. 1 SEM image showing both the surface and side view of three types of NW-ACF by laterally tearing the film.

Test Chips

The test chips include bottom and top chip with the dimension of 5 mm × 5 mm × 0.5 mm and 2.5 mm × 2.5 mm × 0.5 mm (or 1 mm × 1 mm × 0.5 mm), respectively. Four groups of chips are designed with the different pitch sizes. A typical layout of the test chip pair is shown in Fig. 2. The test chips include 100 daisy chain structure, two 4-point structure and four insulating pads. The resistance of daisy chain was probed in the bottom chip at the interconnection number of 2, 6, 8, 10, 30, 60, 96, 98 and 100. The track resistance was minimized by designing Kelvin probes (the voltage probing point being placed very close to the bond pads). The purpose of 4-point structure is to measure the resistance of single interconnect by eliminating the track resistance. The specifications of the pads design are shown in Table 2. A multilayer metallization composed of 3 μm thick Cu and 2 μm thick In was electroplated as bond pad finish. The metal tracks connecting between pads were covered with a thin

layer (200 nm) of sputtered SiO$_2$ for insulation. The test wafer fabrication process was described in [14].

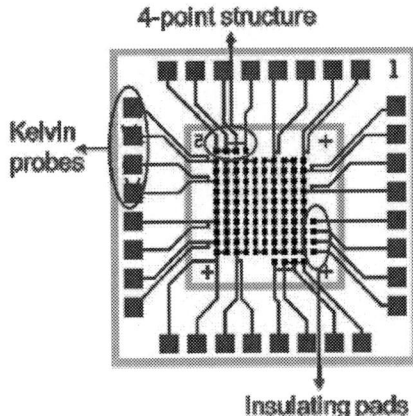

Fig. 2 Test chip layout (image demonstrating the alignment of the bottom chip and the flipped top chip).

Table 2 Specifications of the test chips.

	No. of daisy chain	No. of 4-point structure	No. of insulating pads	Bond pad size (μm)	Bond pad pitch
Chip 1	100	2	4	80	160
Chip 2	100	2	4	40	80
Chip 3	100	2	4	20	40
Chip 4	100	2	4	10	30

Bonding Process

Flip chip bonding was carried out with FINETECH lambda bonding machine. Firstly, the top chip was picked up by the tool head and bottom chip was held by the vacuum on heating plate. The top and bottom chip was manually aligned and adjust for co-planarity by camera system as shown in Fig. 3(a). Then the NW-ACF film was placed on top of the bottom chip. The top chip was then brought down and the temperature and force was applied. In the experiment, the bonding temperature was fixed at 220 °C and the peak time was 60 s [10]. A typical bonding profile is shown in Fig. 3(b). The bonding force was adjusted from 1.5 N to 20 N for each pitch size.

Results and Discussion

Template based Open/Short Failure

NW-ACF fabricated with different templates (Table 1) was evaluated for the electrical properties by bonding with the test chips of 80 μm pad size. All samples were bonded under the same condition (220 °C, 15N, 60 s). Electrical test with the multimeter was conducted after bonding. Bonded samples using NW-ACF 1 failed in short-circuit when probing between two insulating pads. Whereas, NW-ACF 2 and 3 did not show any short-circuit problem. The electrical continuity was verified with daisy chain structure. NW-ACF 2 failed in this test with only conducting (< 100 Ω) between some neighboring pads. To understand the electrical failures caused by the templates, micro-sectional analysis was carried out.

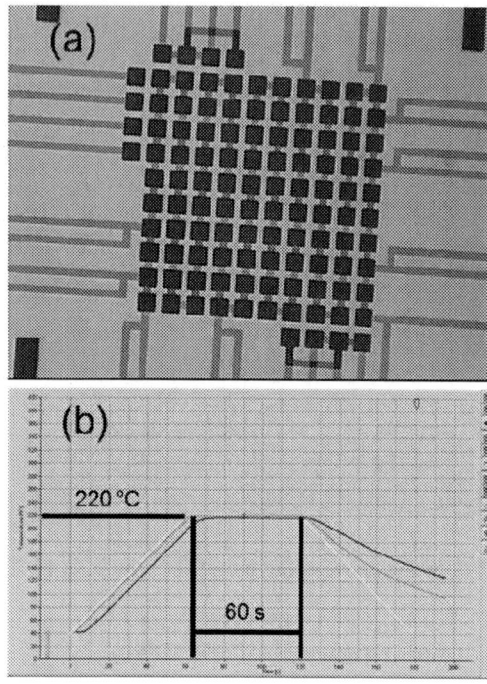

Fig. 3 Bonding alignment at 40 μm pitch in (a) and a typical bonding profile in (b).

Fig. 4 shows the SEM images of the bonding interface of different NW-ACF films with the 80 μm bond pads. It can be seen from the images that the high density and the angular distribution of nanowires in NW-ACF 1 may attribute to the shorting problem. For NW-ACF 2, the gap between nanowires and the pads formed by polymer can hinder the electrical connection. It is to be noted that NW-ACF 3 had the same wire density as NW-ACF 2. However, most nanowires are seen to contact both pad surfaces. The formed polymer gap in case of NW-ACF 2 is probably due to the higher roughness on one side of the template as observed by SEM shown in Fig. 4 (b'). The preliminary electrical screening results are summarized in Table 3, where it can be seen that the NW-ACF 3 passed all tests and thereby applied in the subsequent experiments.

Electrical Performance at Fine Pitch

A fine pitch capability down to 30 μm was assessed with NW-ACF 3. Electrical test with Kelvin probing was carried out with Cascade manual probe station and Agilent B1500 semiconductor parameter analyzer. For nanowire interconnects, bonding pressure below 100 MPa would reach the sufficient proximity to form joint [15]. The minimum applicable bonding force is 1.5 N with the lambda flip-chip bonder. By dividing the pad area, the equivalent bonding pressure to force data are presented in Table 4. For 1.5 N force, the bonding pressures equal to 2, 8, 32 and 128 MPa with the decreasing pad size. Fig. 5 shows the resistance as a function of number of interconnects, measured with 4-point and daisy chain structure for each pitch size bonded applying 1.5 N force. The data was linear fitted to obtain the average

resistance of the interconnects, where the slope of each line represents the average resistance. The results show the average resistance of 0.2-0.3 Ω for 10 and 20 μm pad sizes and 0.4-0.6 Ω for 40 and 80 μm pad sizes, respectively. Despite the smaller contacting area at the finer pitch, the equivalent bonding pressure was increased by 4 times at the same bonding force for subsequent pads of 80, 40, 20 and 10 μm size. This caused a lower resistance at smaller bond pads compared to the larger ones.

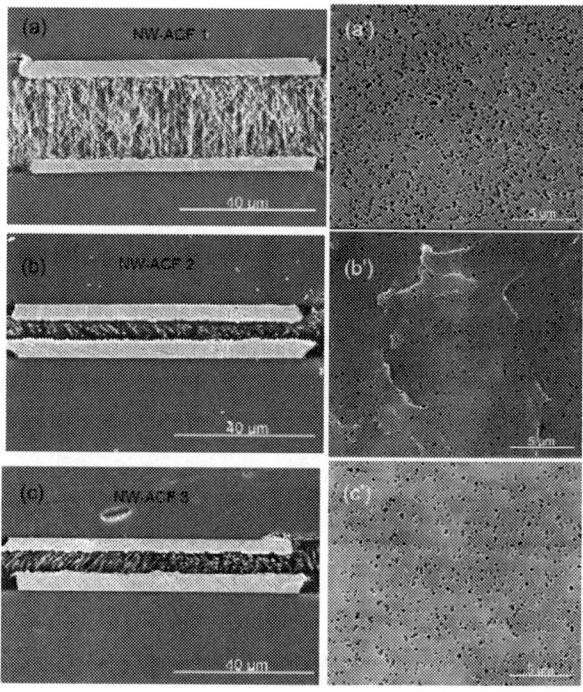

Fig. 4 Bonding Interface of three types of NW-ACF interconnects and the surface morphology of the corresponding templates.

Table 3 Electrical screening of NW-ACF based on templates.

Tests	NW-ACF 1	NW-ACF 2	NW-ACF 3
Short-circuit	FAIL	PASS	PASS
Electrical Continuity	PASS	FAIL	PASS

Table 4 Equivalent bonding pressure to bonding force.

Pad size (μm)	Equivalent Pressure (MPa)					
	1.5 N	**3 N**	**5 N**	**10 N**	**15 N**	**20 N**
80	2	4	7	14	21	28
40	8	16	28	56	84	112
20	32	64	112	224	336	N. A.
10	128	256	448	896	1344	N. A.

By changing the bonding force, the average resistance for each pad size was obtained and plotted in Fig. 6. It was found that the resistance generally decreased with the increasing force for all pad sizes. The resistances became stable at 5 N force for 10 and 20 μm pads and 10 N for 40 and 80 μm pads. It is worth noting that the resistance change is not only related to equivalent pressure but also the contact area of the corresponding pitch size.   Some fluctuating values were found in the curves, e.g., the 40 μm pad at 5 N and the 80 μm pad at 15 N. The possible cause could be the bonding errors like a bad alignment. Table 5 represents the value of the resistances of the corresponding data points in Fig. 6. It was observed that the resistance could drop 10-30 times from ~0.6 Ω to ~0.02 Ω with the increasing bonding force.

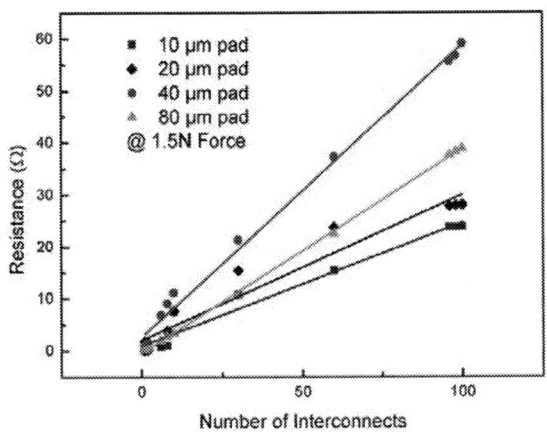

Fig. 5 4-point and daisy chain resistance for various pad sizes bonded using smallest bonding force (the average resistance per pad size is the slope of each curve).

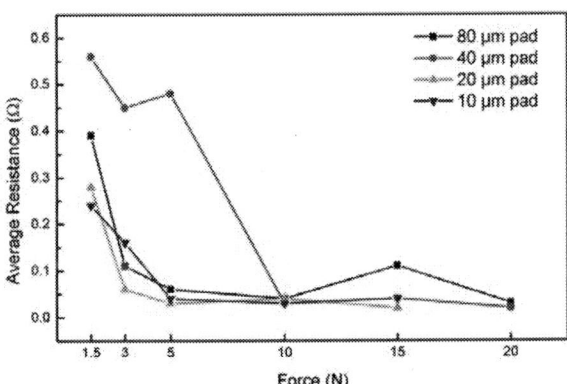

Fig. 6 Average interconnect resistance in terms of bonding force for each pad size.

Table 5 Resistance data corresponding to Fig. 6.

Pad size (μm)	Resistance (Ω) per bonding force					
	1.5 N	3 N	5 N	10 N	15 N	20 N
80	0.39	0.11	0.06	0.04	0.11	0.03
40	0.56	0.45	0.48	0.03	0.04	0.02
20	0.28	0.06	0.03	0.04	0.02	N.A.
10	0.24	0.16	0.04	0.03	0.04	N.A.

The insulation property of NW-ACF formed interconnects were evaluated with the same chips by the four insulating pads. The failure criteria of leakage current up to 1 nA [16]

was applied. The voltage was applied at each pitch size up to a value of 20 V. Table 6 shows the results of the withstanding voltage in terms of pad pitch and bonding force. It was found that for 160 and 80 μm pitch size, samples bonded at all forces (except for the 80 μm pitch size bonded under 5 N), the insulation can be maintained for an applied voltage up to 20 V. For smaller pitch size, only the interconnection formed at a bonding force above 10 N can withstand an applied voltage of 20 V. However, only one sample failed for the 30 μm pitch, bonded using 3 N force. It is suspected that the failure may be resulted from the bad joints formed during bonding. However, further experiments are required to verify this assumption.

Table 6 Voltage tolerance of NW-ACF interconnects.

Pitch size (μm)	Insulation Voltage (V) (Failure criteria: I_{leak} >1nA)					
	1.5 N	3 N	5 N	10 N	15 N	20 N
160	20	20	20	20	20	20
80	20	20	6	20	20	20
40	9	7.5	18	20	20	N.A.
30	20	Fail	14	20	20	N.A.

Bonding Pressure on NW-ACF Interconnection

The interconnects formed by NW-ACF was analyzed by micro-sectional analysis. Fig. 7 shows a group of bonding interfaces at different pad size and bonding pressures. Two images (a) and (b) at low bonding pressure (< 10 MPa) shows the bondline thickness of ~ 5 μm for 80 μm pad and ~3 μm for 40 μm pad. At the cross section, the nanowires were found to be bended and making an angle with the surface normal. For the same pad size as (a) and (b), the bonding interface at a higher bonding pressure (> 30 MPa) was compared by (c) and (d). The nanowires were nearly 90° bended with the wire tips visible in the SEM images. The mechanical bending of the nanowires, pressed between two pads under a certain pressure was comparable to the finding in [15]. The bondline thickness was found to be much decreased for the samples (c) and (d) compared to (a) and (b), respectively. Thus, the shrinking bondline thickness can be attributed to the increase of the normalized pressures. For the smaller pad sizes, the high pressure was reached at even smaller bonding force. In the SEM images of (e) and (f), the nanowires are hardly visible and the indium surfaces were brought to contact. The bondline thickness became nearly zero at such pressure. Due to the increased contact area formed by the bond pads at a sufficiently high bonding pressure, a resultant resistance drop of 10-30 times (Table 5) was observed. Further experiments will be carried out to quantify the relation of the bonding pressure and the bondline thickness for each pad size. An EDX line scan was performed for the bonding interface of 80 μm pad to verify the elements as shown in Fig. 8.

978-1-4799-2408-0/14 $31.00 © 2014 IEEE

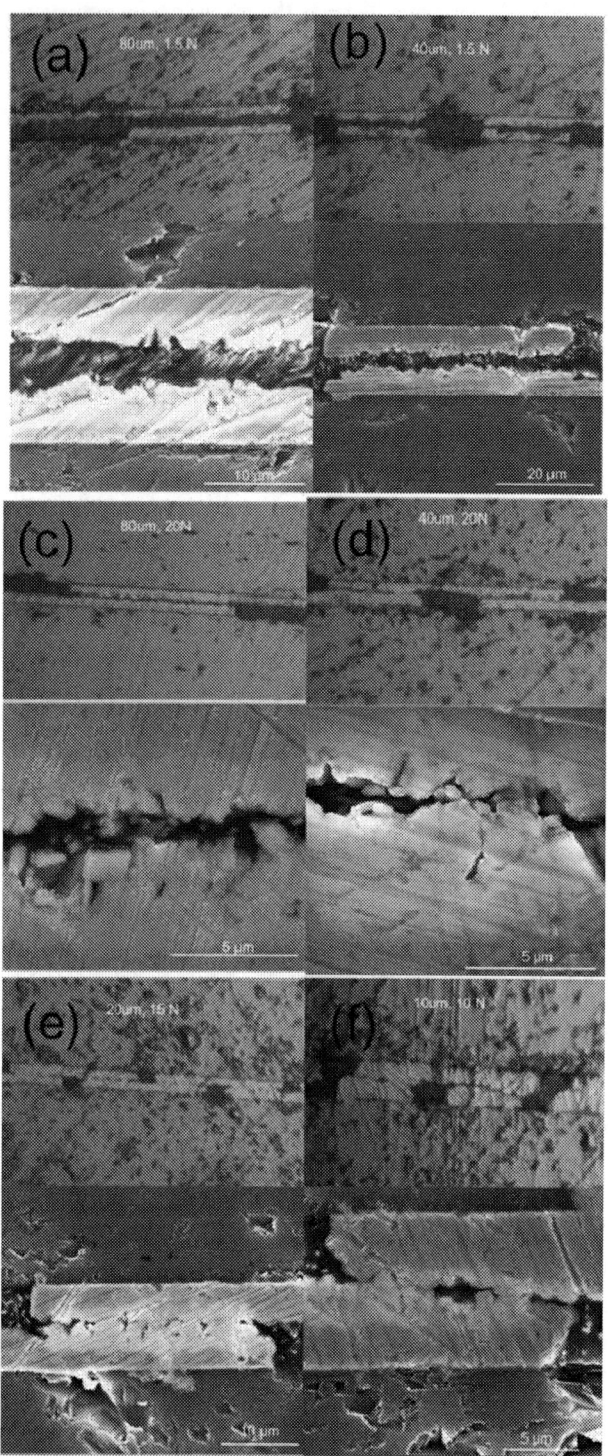

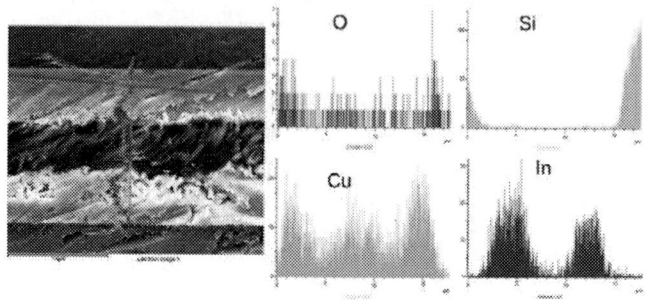

Fig. 8 The EDX line scan of the bonding interface of 80 μm pad at 1.5 N force.

Shear Strength and Fractural Surface Analysis

The shear test was carried out with ROYCE 550 system. Fig. 9 shows the preliminary results of shear strength for pad sizes of 10, 20 and 40 μm in terms of bonding force from 5 to 20 N (the data for the 80 μm pad size and other applied bonding force was not completed during the reporting of this work). The shear strength was around 1 to 5 MPa, which is comparable to the data in [10]. It was found that the shear strength reached its peak at 10 N force for 10 and 20 μm pad size and 15 N force for 40 μm pad size. The higher strength of 10 μm pad compared to 20 μm pad could be due to the higher equivalent bonding pressure at 10 μm pads. The fractural surface analysis was performed with the sheared samples. The typical fractural surfaces of different pad sizes are shown in Fig. 10. The swollen surfaces demonstrate a good polymer adhesion. On 10 μm pad, a considerable amount of nanowires are visible underneath the polymer. Some broken nanowire joints can also be found in the indium surface as shown in Fig. 11. The EDX spectrum was performed to verify the present of such Cu-In joint as shown in Fig. 12.

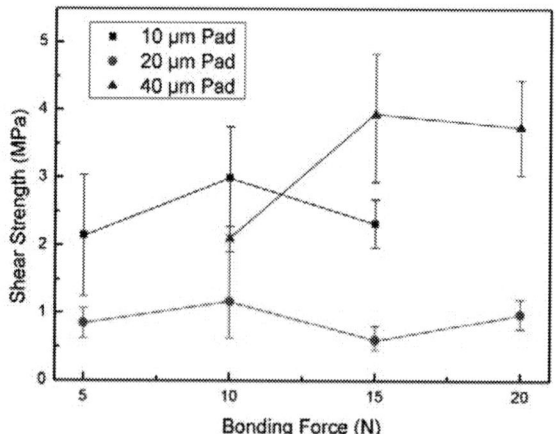

Fig. 9 The shear strength of different pad sizes per bonding force.

Fig. 7 Optical and magnified SEM images of the bonding interfaces of the different pad size and bond force; (a) 80 μm pad at 2 MPa (1.5 N), (b) 40 μm at 8 MPa (1.5 N), (C) 80 μm at 28 MPa (20 N), (d) 40 μm at 112 MPa (20 N), (e) 20 μm at 336 MPa (15 N) and (f) 10 μm at 896 MPa (10 N).

Fig. 10 The morphology of the fractural surface of different pad size.

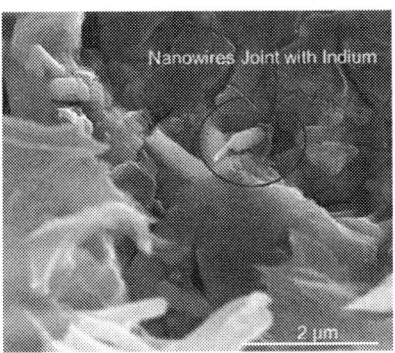

Fig. 11 Site of broken nanowires joint in the fractured indium surface.

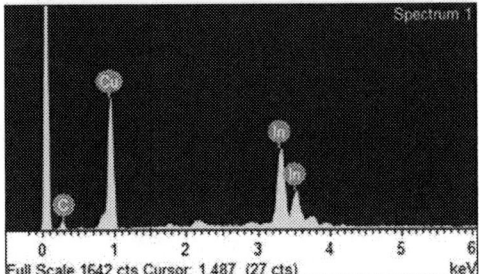

Fig. 12 SEM image and the EDX spectrum performed at fractural surface.

Conclusions

In conclusion, Cu nanowire based ACF showed satisfactory electrical and mechanical performance at a fine pitch size down to 30 μm. The specifications of the TEM template, e.g., porosity, pore shape and surface roughness is critical to fabricate a reliable NW-ACF. It was found that with the increase of the bonding pressure, the nanowires were mechanically bended and the bondline thickness was decreased to nearly zero. This may result in the corresponding interconnect resistance drop from 0.3-0.6 Ω per pad to 0.02-0.04 Ω per pad under certain bonding forces for each pad size. Despite the bending of nanowires at high bonding force, the electrical insulation was maintained for an applied voltage up to 20 V for each pitch size. The shear strength was 1-5 MPa for various pitch sizes. By fractural surface analysis, a considerable amount of nanowires were found at the smallest pad size and the Cu-In joints were verified by the EDX analysis. The promising results of NW-ACF showed a great potential to apply this nanocomposite film in ultra-fine pitch 3D interconnections.

Acknowledgments

The authors would like to acknowledge financial support from the European Union through the Integrated Project e-BRAINS under Grant FP7-ICT-257488 and their colleagues in Tyndall National Institute for their technical support in wafer fabrication.

References

1. P. Ramm, M. Wolf, A. Klumpp, R. Wieland, B. Wunderle, B. Michel, and H. Reichl, "Through silicon via technology—processes and reliability for wafer-level 3D system integration," in *Electronic Components and Technology Conference, 2008. ECTC 2008. 58th*, 2008, pp. 841-846.
2. R. S. Pai and K. M. Walsh, "The viability of anisotropic conductive film as a flip chip interconnect technology for MEMS devices," *Journal of Micromechanics and Microengineering*, vol. 15, p. 1131, 2005.
3. R. Agarwal, W. Zhang, P. Limaye, R. Labie, B. Dimcic, A. Phommahaxay, and P. Soussan, "Cu/Sn microbumps interconnect for 3D TSV chip stacking," in *Electronic Components and Technology Conference (ECTC), 2010 Proceedings 60th*, 2010, pp. 858-863.
4. Y. Ohara, A. Noriki, K. Sakuma, K.-W. Lee, M. Murugesan, J. Bea, F. Yamada, T. Fukushima, T. Tanaka, and M. Koyanagi, "10 μm fine pitch Cu/Sn micro-bumps for 3-D super-chip stack," in *3D System Integration, 2009. 3DIC 2009. IEEE International Conference on*, 2009, pp. 1-6.
5. Y. Chiu, Y. Chan, and S. Lui, "Study of short-circuiting between adjacent joints under electric field effects in fine pitch anisotropic conductive adhesive interconnects," *Microelectronics Reliability*, vol. 42, pp. 1945-1951, 2002.
6. K.-W. Jang, C.-K. Chung, W.-S. Lee, and K.-W. Paik, "Material properties of anisotropic conductive films (ACFs) and their flip chip assembly reliability in NAND flash memory applications," *Microelectronics Reliability*, vol. 48, pp. 1052-1061, 2008.

7. C. Schönenberger, B. Van der Zande, L. Fokkink, M. Henny, C. Schmid, M. Krüger, A. Bachtold, R. Huber, H. Birk, and U. Staufer, "Template synthesis of nanowires in porous polycarbonate membranes: electrochemistry and morphology," *The Journal of Physical Chemistry B,* vol. 101, pp. 5497-5505, 1997.

8. J. Xu, A. Munari, E. Dalton, A. Mathewson, and K. M. Razeeb, "Silver nanowire array-polymer composite as thermal interface material," *Journal of Applied Physics,* vol. 106, p. 124310, 2009.

9. H.-C. Cheng, K.-Y. Hsieh, and K.-M. Chen, "Thermal–mechanical optimization of a novel nanocomposite-film typed flip chip technology," *Microelectronics Reliability,* vol. 51, pp. 826-836, 2011.

10. J. Tao, M. Hasan, J. Xu, A. Mathewson, and K. M. Razeeb, "Investigation of Process Parameters and Characterization of Nanowire Anisotropic Conductive Film for Interconnection Applications," *IEEE Trans.Compon., Packag. Manuf. Technol.,* in press, 2014 (DOI: 10.1109/TCPMT.2013.2297734).

11. F. Stam, K. Razeeb, S. Salwa, and A. Mathewson, "Micro-nano interconnect between gold bond pads and copper nano-wires embedded in a polymer template," in *Electronic Components and Technology Conference, 2009. ECTC 2009. 59th,* 2009, pp. 1470-1474.

12. M. E. Toimil-Molares, "Characterization and properties of micro- and nanowires of controlled size, composition, and geometry fabricated by electrodeposition and ion-track technology," *Beilstein Journal of Nanotechnology 3,* pp. 860-883, 2012.

13. H. He and N. J. Tao, "Electrochemical fabrication of metal nanowires," *Encyclopedia of Nanoscience and Nanotechnology,* vol. 2, pp. 755-772, 2003.

14. J. Tao, M. Hasan, J. Xu, A. Mathewson, and K. M. Razeeb, "Test Structure for Electrical Characterization of Copper Nanowire Anisotropic Conductive Film (NW-ACF) for 3D Stacking Applications," *Microelectronic Test Structures (ICMTS), 2010 IEEE International Conference on,* submitted, 2014.

15. S. Fiedler, M. Zwanzig, R. Schmidt, E. Auerswald, M. Klein, W. Scheel, and H. Reichl, "Evaluation of metallic nano-lawn structures for application in microelectronic packaging," in *Electronics Systemintegration Technology Conference, 2006. 1st,* 2006, pp. 886-891.

16. M. Stucchi, D. Perry, G. Katti, and W. Dehaene, "Test structures for characterization of through silicon vias," in *Microelectronic Test Structures (ICMTS), 2010 IEEE International Conference on,* 2010, pp. 130-134.

Carbon Nanofibers (CNF) for Enhanced Solder-based Nano-Scale Integration and on-chip Interconnect Solutions.

V. Desmaris, A. M. Saleem, S. Shafiee, J. Berg, M. S. Kabir, A. Johansson

Smoltek AB,

Regnbagsgatan 3, 41755 Gothenburg, Sweden

Vincent@smoltek.com

Phil Marcoux

PPM Associates

Mountain View, CA, USA.

Abstract

While the density of chip-to-chip and chip-to-package component interconnections increases and their size decreases the ease of manufacture and the interconnection reliability are being dangerously reduced.

This paper introduces the use of Carbon Nanofibers (CNF) grown on chip as an embedded reinforcing material for nano-solder interconnections and as bonding material (adhesive) for chip-to-package solutions.

Interconnections are realized by means of microbumps which can be less than 10 μm in diameter and up to 20 μm high. Such micro-bumps are shown to be solderable using conventional thermal-compression and micro-bumps.

Using CNF embedded in polymer is shown to provide a robust solution for chip-to-package interconnections.

Introduction

During the last two decades, the eventual ban of toxic Sn-Pb solder for chip soldering lead to surge of alternative solder, which are nowadays well standardized and established [1-3]. However the continuous increase of the density of the chip-to-chip and chip-to-package, as consequence of Moore's law [4] make lead-free solders more challenging to implement, since the interconnect pitch get smaller and the current and power densities, that they should handle, larger. In addition, the interconnect reliability is dangerously being reduced as well because of warpage and heat.

Beside the potential toxicity, solders such as SnAgCu or Sn-Pb are still limited in by their relatively high temperature processing, excess growth of IMC and/or cost [5-6]. To mitigate these problems, composite solders have been suggested as prospective future solder material [7]. Nano materials such as nanoparticles [8], Carbon nanotubes (CNT) [9] or Carbon Nanofibers (CNF) [10] have been considered for forming alloyed interconnects. CNT-or CNF- reinforced solders have shown promising results, yet their implementation required complex processing including solder material transfer from one substrate to another or high temperature processing

Because of the more controllable electrical DC andf RF behavior [11] of CNF as compare to CNT and their high resistance to corrosion, we present, in this paper, the use of CNF in the forms of solderable micro-bumps grown directly on chips, providing prospectively high aspect ratio bumps with diameter in the order of 10-20 micrometers. In addition we show that CNF can be used as adhesive in combination with polymers and provide good mechanical connection to packages such as lead-frame or PCB.

CNF on chip

Using Smoltek's patented technology, CNFs are produced using DC Plasma Enhanced Chemical Vapor Deposition (Fig.1.) at a temperature of 390 C, which is considered at present to be a CMOS compatible process temperature. The growth process itself is of catalytic nature, which allows full control of the placement of the CNF over chips. Therefore The CNF formation process includes the deposition and patterning of a catalyst before growth an environment containing ammonia and acetylene as a Carbon precursor. The use of a DC-plasma yield vertically aligned CNF of typically 50-100 nm in diameter and 2-150 micrometers in lengths (Fig.2-Fig3.).

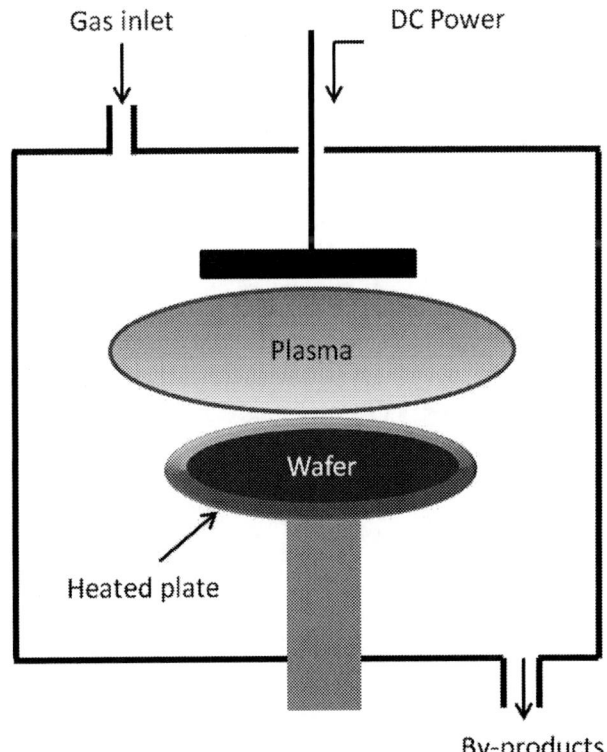

Figure 1. A high resolution of image of the carbon nanofibers reveals the diameter ranging from 20-100nm

Figure 2. A high resolution of image of the carbon nanofibers reveals the diameter ranging from 20-100nm

Figure 3. A forest of CNFs grown onto a metal IC bond pad. The pad size is approximately 20μm x 20 μm.

Wetability of CNF

The CNF were first grown as a film on Al-coated Si-chip to verify their wetability with low temperature solders such as Ind290 [12]. Regardless of an eventual coating the CNF were shown to wet very well the solder. As shown on Fig. 4.,the TEM analysis of a cross section of the Solder-CNF composite did not reveal the presence of voids, which would be compromising the reliability of the interconnects made out of it. In addition, the wetability of the CNF to the solder could prevent the solder from wicking out and away from the pads, thus reducing the number of solder shorts that occur as the solder pad-pad pitch gets smaller.

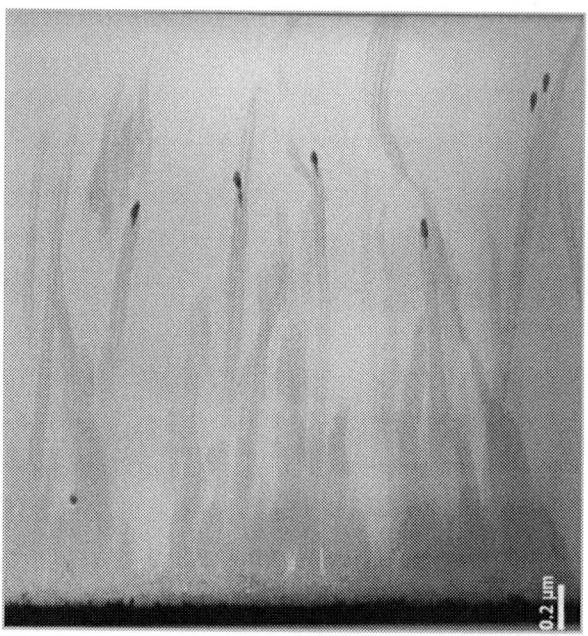

Figure 4. TEM Cross sectional view of CNFs rooted in a metal pad and wetted with Ind290 solder.

CNF as solderable micro-bumps

The considerable Young Modulus of the CNFs along their growth axis confer them remarkable piercing properties, e.g. for nano-imprinting [13]. In addition, the CNF remarkable compressive strength would imply that they could provide considerable reinforcement of a solder bump, preventing it to deform and dimple during testing [14].

The demonstration of solderable micro-bumps was conducted using standard thermal compression flip-chip technique and test vehicle made of a readout wafer with 40 micrometer In-Sn solder bumps.

Two set of samples were used for these experiments: uniformly Al-coated Si chips and Al-patterned Si chips. The CNF were grown at 390 C for both type of samples yielding either CNF films or CNF micro-bumps. The CNF, were later coated with Au, using sputtering.

The Au-coated CNF films were flip-chipped to the read out using a NILT Nanoimprint CNI system under vacuum, whereas the Si Chips with micro CNF-bumps were flip-chipped at 180 C using a conventional SET FC 150 flip-chip bonder and later reflowed at 170 C for 7 min as illustrated on Fig.5.

978-1-4799-2408-0/14 $31.00 © 2014 IEEE

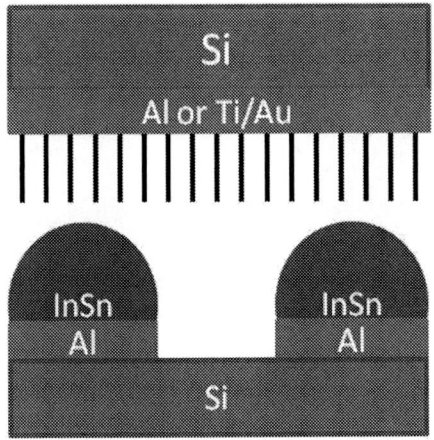

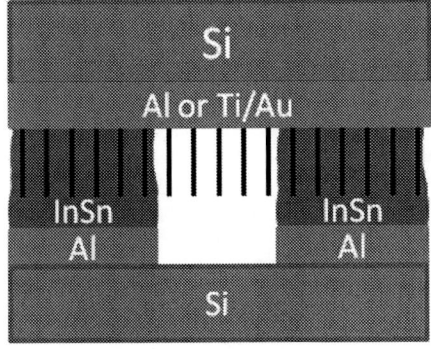

Figure 5. Schematic view of the micro-bump soldering.

Upon disassembly, the wetting of the fibers to the In-Sn solder was investigated using SEM. As shown on Fig 6, fibers inclusions in the solder are visible indicating the solder did wet the Au-coated fibers and formed an interconnection.

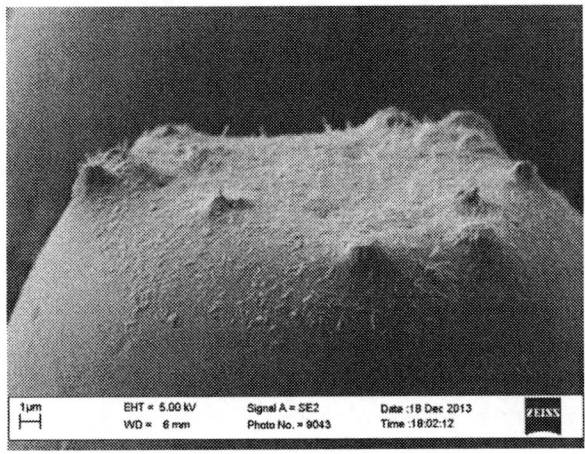

Figure 6. SEM picture of the top of the In-Sn bumps after disassembly.

The transfer of CNFs from their original substrate to micro-solder bump confirmed by inspecting the original substrate in SEM after disassembling (Fig. 7.). Clearly, the original substrate show large areas, from which fibers are missing, which support the idea that the fibers are embedded into the solder.

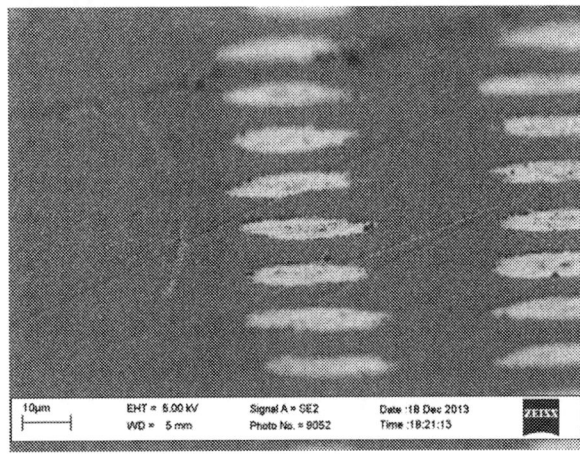

Figure 7. SEM picture of the original substrate after disassembly, covered with a film of CNF, except where it was in contact with the solder

The electrical behavior of the interconnection was tested using the chips with CNF films and compared to the electrical behavior of chips without any CNF (Au coated Si). The results of 2 point resistance measurements performed with a Keithley 4200-SCS parameter analyser and a DC-probe station are summarized on Fig 8 and Fig 9. The measured resistances for chips with Au-coated CNF or solely Au are similar. This indicate that it is possible to make solderable CNF micro-bumps, without compromising the electrical quality of the interconnects.

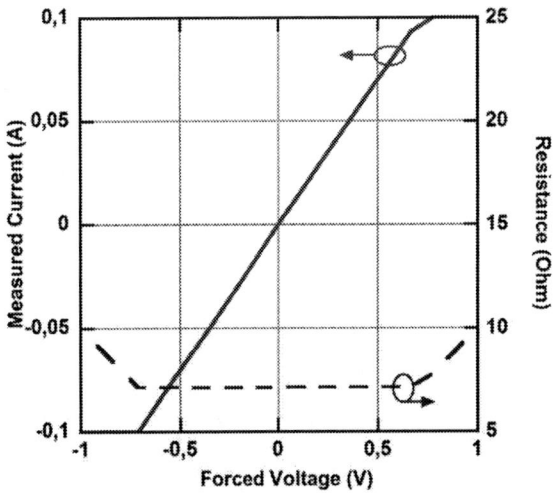

Figure 8. Electrical characteristics of the solder bumps on bare metal (i.e. without CNF)

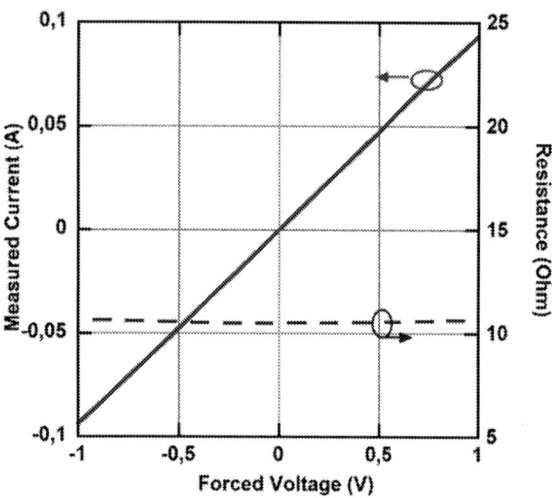

Figure 9. Electrical characteristics of the solder bumps on with embedded CNF

CNF-based adhesives interconnects

CNF films can also be used as performance enhancer of flip-chip bonding, when soldering processes could not be used.

For this purpose, CNF were first grown at the backside of SiC power transistors. Then, a low-temperature polymer was used to bond the fiber-covered transistor (back sides) to substrates. A double-layer of polymer was spun onto the fiber-covered transistor chip, and the chip was bonded to different substrates, namely a Si-chip with measurement lines made of electroplated Gold (Fig. 10.) and direct-copper-bonding (DCB) substrates (Fig. 11.). The SiC chips were bonded to the substrates for 5 min at 160 C under a bonding pressure of 50 bar.

Figure 10. SiC power transistor with nanofibers bonded to a silicon substrate

Figure 11. SiC power transistor with nanofibers bonded to a DCB substrate

Electrical measurements were carried out using Keithley instrument both for two and four probe measurements. Representative results from four probe measurements are presented in Figure 12.

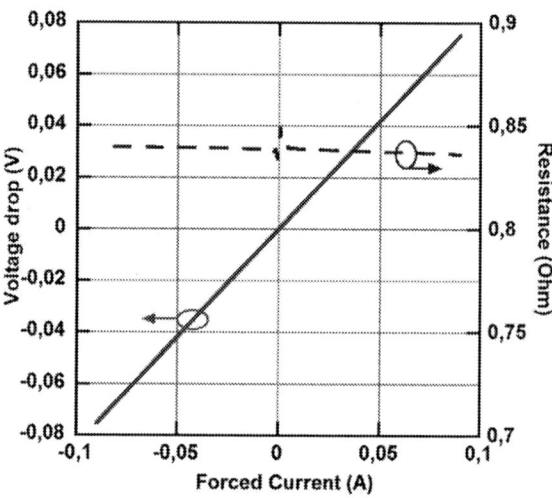

Figure 12. IV characteristics of the interconnect between SiC power transistor with nanofibers bonded and the silicon substrate

The interconnection between the SiC chip and the Si-substrate exhibit good ohmic behavior, whereas the bonding to the DCB substrate shows some non-linearity. This behavior can be ascribed to the large RMS surface roughness of the substrates (~12 micrometers) as compared to the fiber length (2 um).

Taking advantage of the Transmission Line Measurements [15] configuration of the Si-chip, and using a simple resistance to model the interconnect behavior (Fig. 13.), the resistance value showed 21 mΩ per contact points with linear IV proving the fact that it is possible to utilize CNF on SiC substrate for Flip Chip bonding with good ohmic contacts. Specifically, the resistance per contact point was found to be of the order of

$9 \cdot 10^{-5}$ $\Omega \cdot cm^2$ which is within the frame of electrical requirements for FC bumping purposes.

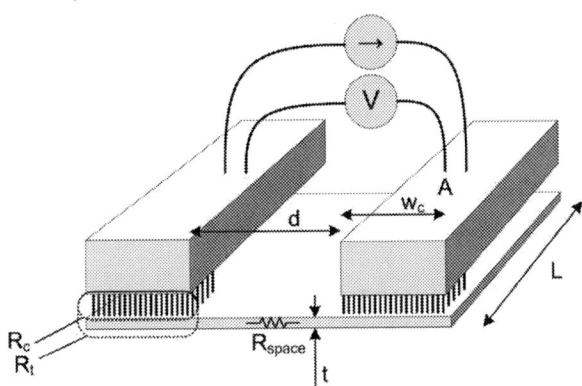

Figure 13. Schematic of the model used for specific resistance extraction

The mechanical strength of the CNF-based adhesive was investigated using small Si transistor chips bonded to Ag plated leadframe (Fig. 14.). The Si mechanical dies were 1.45x1.45 mm. CNF were grown at the backside of the chips before being attached to the leadframe using double layer of. A double-layer of polymer was spun onto the fiber-covered transistor chip. The Si chips were bonded to the substrates for 5 min at 160 C under a bonding pressure of 50 bar.

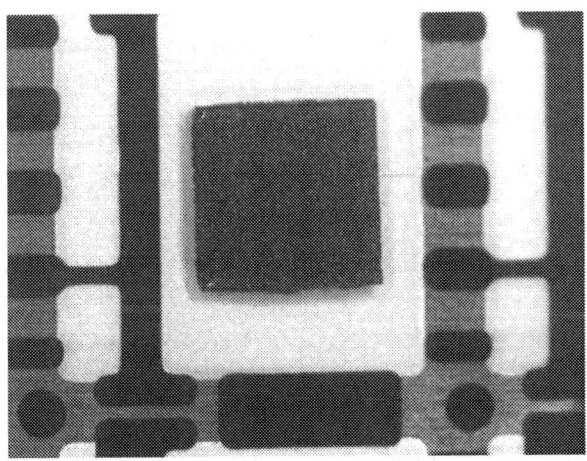

Figure 14. Si chips on Ag plated leadframe

Statistics of the minimum shear force required to detach the chips from the leadframe were taken on 10 samples and lead to an average of 2300g and sample standard deviation of 0,497g. The obtained results qualify the assembly as MIL x1 according to the MIL-STD-883 standard for microelectronic devices.

Conclusions

Prospective CNF-based solutions for next generation of interconnects have been presented.

Solderable CNF Micro-bumps with high aspect ratio have been demonstrated and resulted in interconnects made of CNF reinforced In-Sn solder composite. We believe this approach is very suitable for the formation of short pitch chip-to-chip interconnects.

In addition, the use of embedded CNF into polymer to enhance the performance of the Chip-to-package interconnect has been demonstrate in terms of electrical performance. Furthermore, prototype assemblies made of Si power devices attached to Ag coated leadframe was confirmed to perform beyond the MILx1 standards in terms of mechanical strength.

Acknowledgments

The authors would like to acknowledge the Swedish Governmental Agency for Innovation Systems (VINNOVA) for its financial support regarding the CNF-based Adhesives for SiC devices.

References

1. J. Glazer, "Microstructure and mechanical properties of Pb-free solder alloys for low cost electronic assembly," *Journal of Electronic Materials*, vol. 23, no. 8, pp. 693–700, Aug. 1994.
2. Y. Lin, L. Yin, X. Wei, "Recent progress in the studies of low melting Sn-based Pb-free solders", *International Conference on Electronic Packaging Technology and High Density Packaging (ICEPT-HDP)*, Shanghai, PR. China, Aug 8-11, 2011, pp1-4.
3. S. A. Musa, M. A. A. M. Salleh, S. Norainiza, *"Zn-Sn Based High Temperature Solder – A short Review"*, *Advanced Material Research*, vol 795, pp. 518-521, 2013
4. G. E. Moore, "Cramming more components onto integrated circuits", *Electronics*, pp. 114-117. Apr.19, 1965.
5. G. Zeng, S. Xue, L. Zhang, L. Gao, W. Dai, J. Luo, "A review of interfacial intermetallic compompounds between Sn-Ag-Cu based solders and substrates", *Journal of Materials Science: Materials in Electronics*, vol. 21, no. 5, pp. 421–440, May. 2010.
6. A. Koupa, D. Andersson, N. Pearce, A. Watson, A. Dinsdale, S. Mucklejohn, "Current Problems and Possible Solutions in High-Temperature Lead-Free Soldering", *Journal of Materials Engineering and Performance*, Vol. 21, no 5, pp.629-637, May 2012.
7. F. Gao, "Composite lead-free electronics", *J. Mater. Sci. Mater. El.* Vol 18, pp. 129-145, 2007.
8. E. E. M. Noor, A. Snigh, Y. T. Chuan, "A review: influence of nanoparticles reinforced on solder alloy", *Soldering & Surface Mount Technology*, vol. 25, no 4. Pp. 229-241, 2013.
9. K. M. Kumar, V. Kripesh, A. A. O. Tay, "Influence of single-wall Carbon nanotube(SWCNT) functionalized Sn-Ag-Cu lead-free composite solder", *J. Alloys Compd*, vol. 455, no 1-2, pp. 148-158, 2008.
10. S. Chen, "Ultra-short vertically aligned carbon nanofibers transfer and application as bonding material", *Soldering & Surface Mount Technology*, vol. 25, no 4. pp. 229-241, 2013
11. M. Kabir, V. Desmaris, A. M. Saleem, J. Berg, P. Enoksson, L-G Huss, R. Jonsson, S. Rudner, M. Hoijer, M. S. Sarto, A. Tamburrano, "A test Vehicle for RF/DC Evaluation and Destructive testing of Vertically Grown Nanostructures (VGCNS), *International conference on the*

Science and Application of Nanotubes, Cambridge, UK, Jul. 10-16, 2011.

12. Indium Corp. http://www.indium.com

13. A. M. Saleem, J. Berg, V. Desmaris, M. Kabir, "Nanoimprint lithography using vertically aligned carbon nanostructures as stamp", *Nanotechnology,* vol. 20, No. 27, pp375302-375306, 2009.

14. K.Smith, "Probing 25μm-diameter micro-bumps for Wide-I/O 3D SICs", *Chipscale Review,* Vol 18. No. 1, pp20-23, 2014.

15. G. Reeves, H. B. Harrison, "Obtaining the specific contact resistance from transmision line model measurements", *IEEE. Electronc Device Letters,*vol 3. no. 5, pp.111-113, 1982.

Pressure-less Plasma Sintering of Cu Paste for SiC Die-Attach of High-Temperature Power Device Manufacturing

S. Nagao[1], K. Kodani[2], S. Sakamoto[1], S.-W. Park[1], T. Sugahara[1], and K. Suganuma[1]

[1] The institute of Scientific and Industry Research, Osaka University, Mihogaoka 8-1, Ibaraki, Japan
[2] Nissin Inc., Kameichou 10-7, Takarazuka, Japan

shijo.nagao@sanken.osaka-u.ac.jp, Tel: +81-6-6879-8521

Abstract

Pressure-less sintering of Cu flake paste is achieved assisted by hydrogen plasma process toward die-attach technique of next-generation high-temperature power semiconductor devices. The sintered paste shows high bond strength aver 50 MPa, showing large abnormal grain growth at the bonding interface, with homogeneously distributed void of porous interconnection layer. Our results indicate that still the reduction of the surface oxide of Cu flake powders, and thus both the metal powder synthesis and the bonding process must be optimized to achieve a sound bonding interface.

Introduction

The performance of electronic power devices are crucial for the total energy consumption in future human society, and thus wide band-gap (WBG) semiconductors like SiC or GaN with their excellent physics properties are emerging to replace well-matured silicon-based power devices widely used everywhere in the word nowadays. During the last few years, both the device and process designs suitable for these WBG semiconductors have been significantly improved, and actual power devices products using WBG chips are appearing in the market [1]. However, there still exist many steps to realize global mass-productions of such devices, and critical issues mostly remain in the reliability of the packaging technology applied. In particular, die-attach technique is a challenge for the existing packaging materials to assure the sufficient reliability under the expected high operation-temperatures over 200 °C of WBG power devices [1]. Various high-temperature resistant die-attach materials and processing have thus been attracted much attention, namely Pb-free high-temperature soldering [2], temporary liquid phase (TLP) bonding [3], and Ag paste sintering [4]. Pb-free solders needs more improvement to assure their high-temperature resistance because there is basically no way to reduce the bonding temperature lower than the melting point that determines the upper limit of the operating temperature. TLP bonding appears quite promising, but the searching for the appropriate materials with suitable bonding reactions is on the way, and may needs more time to confirm their mechanical and electrical reliabilities toward the market-level productions. Ag paste sintering is indeed one of the most promising die-attach technology, at this moment. However, the cost of Ag paste is still not negligible for the mass production, and serious questions are imposed on the reliability risk of the material's reliability due to their easy sulfidation.

Therefore, we would here propose a pressure-less Cu-paste sintering method assisted by hydrogen plasma for die-attach of high-temperature WBG devices. Because of the almost inevitable surface oxidation of Cu particles, typical Cu pastes need a reduction process for the efficient sintering, e. g. H_2 or formic acid gas atmosphere [5]. In our study, hydrogen plasma is selected for Cu flake paste sintering process [6-7], resulting in a high bonding strength with obvious abnormal growths of Cu grains. Our test samples have indeed exhibited over 50 MPa of die-shear strength, which may be sufficient for SiC die-attach. The major advantage of hydrogen plasma compared to other reduction gases may come from the energetic hydrogen ion – i.e. proton – bombarded into the Cu paste, accelerating the atom diffusions without elevating the temperature of the whole sample body commonly heated from the outside. In addition to the reduction by hydrogen gas, the concentrated heat in Cu paste produced by the energy directly transferred from the hydrogen plasma accelerates the sintering process, and shortens the bonding time into several minutes. Our plasma sintering method suggests that metal-paste die-attach methods instead of solders are promising bonding technology suitable for high-temperature power device manufacturing in future.

Experimental

Several types of copper flake pastes are prepared from commercially available Cu flakes (Mitsui Mining & Smelting Co. Ltd., Japan). Three different size of the particles of 1000YP series, namely d_{Ave} = 1.3, 3.0, and 6.9 μm are adopted (see **Fig. 1a**). The other two types of Cu powders are made by conventional atomized method, and then mechanically comminuted to the size of d_{Ave} = 11.8, and MA-C: d_{Ave} = 17.1 μm (see **Fig. 1b**). These Cu powders are mixed with a solvent of decanol, and the viscosity of the paste is adjusted to about 200 Pa·s, suitable for metal mask printing used in this study.

Cu metal plates with t = 0.8 mm is selected for dummy chip and substrate, where the dimensions of chip and substrate are 12×12 mm and 3×3 mm, respectively. The dummy parts are cleaned by a short plasma process: 30 sec of pre-heat at 110 °C followed by 30 sec of hydrogen plasma exposure. Cu paste is then printed on the cleaned substrate by a squeezee on the 50 μm thick metal-mask with a 3 x 3 mm aperture. Then the dummy chip is put on the printed paste by tweezers, and pressurized to maintain a better contacts by a slight load about 5 N for 2 sec.

978-1-4799-2408-0/14 $31.00 © 2014 IEEE

2014 Electronic Components & Technology Conference

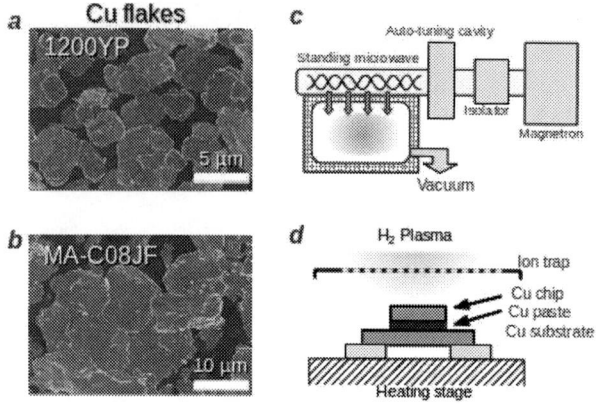

Figure 1. Schematic of the plasma sintering process of Cu paste; (*a*) typical SEM image of 1200 YP Cu flake powders, and (*b*) MA-C08JF; (*c*) the microwave plasma chamber, and (*d*) the sample setting in the plasma chamber.

The plasma sintering process starts with pre-heating at 110 °C for 3 min in air, to make sure the solvent is well evaporated before the vacuum process, as schematically show in **Fig. 1c-d**. The air in the chamber is purged for 1 min, and then the pressure reduced to less than 10 Pa. After this air evacuation, 100% H_2 gas is introduced into the chamber, where the flow rate was 10 sccm. The chamber pressure becomes stable about 15 Pa after 30 sec of gas flow. Microwave is then applied on this stable atmosphere to produce hydrogen plasma with using 1.5 kW power for 5 min. The sample temperature is elevated because of the heat generated by the energetic plasma gas for the sintering process. The sintered samples are immediately cooled down by introducing a 97% He + 3% H_2 gas, and then the introduce gas purged for 90 sec. Finally the chamber door is open, and the samples are take out for further cooling in air at RT.

To evaluate the reduction and sintering effect of both the H_2 gas and proton plasma, an ion-trap grid is introduce to cut out the hydrogen plasma (see **Fig. 1d**). The ion-trap grid is supposed to shield both the microwave and energetic plasma gas, and thus the H_2 gas is the major contributor to the sintering process.

The mechanical strength of the bonded samples are evaluated by using Dage400 (Nordson Corporation, USA) die-shear tester. In addition, the cross-sectional microstructure of the sintered Cu pasted is observed by scanning electron microscopy (SEM).

Results and discussion

The die-shear strength obtained for the bonded samples by the H_2 plasma sintering process without ion-trap are summarized in **Fig. 2**, where the averaged strength for each type of Cu paste is determined from 5 samples tested by Dage4000 die-shear tester. The paste made from YP series Cu powders result in surprisingly high bonding strength about 50 MPa. The strength appears independent to the diameter of these Cu flakes, even though the standard deviation of 1400YP is slightly larger than the others. This may be explained by the less homogeneous printing quality due to the larger size of the 1400YP flake. On the other hand, MA-C series give relatively lower strength, but the shear strength of the both pastes exceed 20 MPa that is usually required for industrial standard for die-attach process. These results indicate that the size of the flakes used here is not critical, but the method of metal powder creation and the printing process of Cu paste may be major issues.

The sample temperature profiles are recorded with using a thermocouple sensor placed between the dummy chip and substrate. As plotted in **Fig. 3**, lower the gas flow basically produces higher energy of proton plasma, and thus the sample temperature increases with decreasing H_2 gas flow. When the ion trap is inserted with 10 sccm of H_2 gas flow, the sample temperature is increased up to 600 °C, which is similar to that with the 100 sccm gas flow without the ion-trap. This means proton plasma contribute more to sample heating than the original H_2 gas.

To investigate the source of the high shear-strength obtained for the bonding, cross-section microstructures of the sintered Cu paste are observed by SEM observations (see **Fig. 4**). The SEM image of **Fig. 4a** shows the YP flakes sintered with the ion-trap conditions. The macrostructure indicates rather incomplete sintering of the paste with the ion-trap. Other samples of **Fig. 4c-d** show well-sintered microstructure by the plasma process; the necking of the flakes appears almost perfect, and no bonding boundaries can be recognized from the cross-sectional SEM images. One of the clear characters of the microstructures is the many voids homogeneously distributed over the whole bonding layer. This voids can be a source of reliability risks of the bonding, but at the same time, the homogeneous porosity can be an advantage to reduce thermal stresses caused by the high operating-temperature of the WBG power devices. Further research may be needed to evaluate the microstructure of the sintered paste suitable for the die-attach of the final power device products.

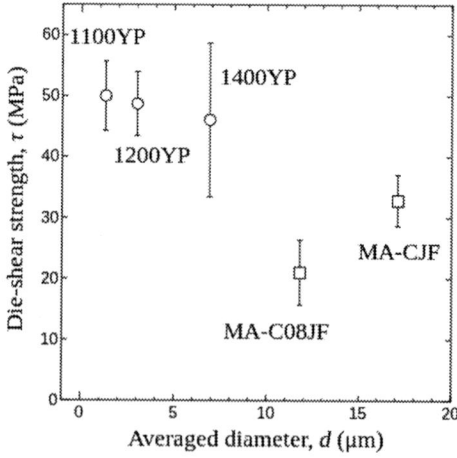

Figure 2. Die-shear strength of the Cu flake paste bonding by the hydrogen plasma sintering, plotted as a function of the diameter size of the original Cu flake powders.

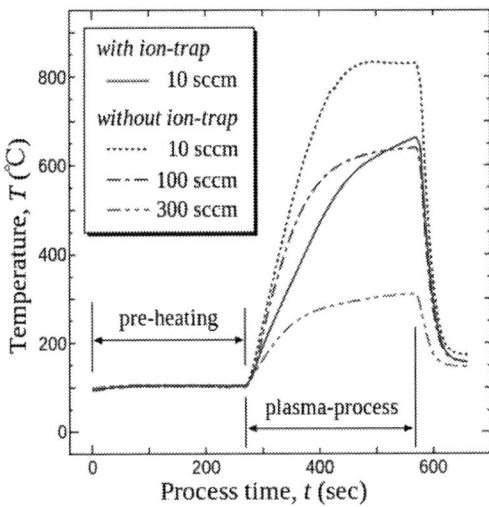

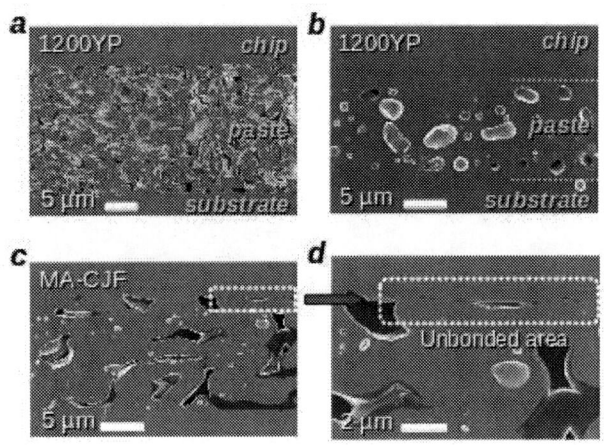

Figure 3. Temperature profiles of Cu samples recorded under the various H_2 gas flows with or without using the ion-trap.

Figure 4. Cross-section SEM images of the sintered Cu flake pastes; (*a*) 1200YP paste sintered with using the ion-trap, (*b*) without the ion-trap. (*c*) MA-CJF powder paste, and (*d*) typical uncompleted sintering due to the surface oxidation.

Looking at the difference of the microstructures between the YP and MA-C series flake pastes, the YP series give the excellent sintering microstructure consisting of large Cu grains as usually found in a bulk Cu material (see **Fig. 4*b***). The abnormal grain growth is indeed the origin of the high shear-strength of the bonding. MA-C series flakes also displays well-sintered bonding interfaces, but the microstructure is more porous than that in YP series pastes as shown in **Fig. 4*c***. Moreover, there are a few places where the bonding is incomplete particularly at the interface to the Cu substrate or chip (**Fig. 4*d***). Such bonding flaws are possibly due to the surface oxide remained through the plasma sintering process, and may cause an initial fracture under shear stress. Hence the reduction of the oxide is still the key to increase the strength of the bonding. In addition, this observation suggests that the different strength confirmed between YP and MA-C series pastes is due to the different amount of the surface oxide originally created at the synthesis process of the Cu particles.

Conclusions

In this study, we have explored the H_2 plasma sintering process of Cu flake paste to evaluate the possible applications for die-attach used in WBG power semiconductor devices. The obtained high strengths appear promising for the future applications. The high strength is rooted in the well-sintered Cu paste microstructure with homogeneously distributed porosity. Surface oxidation of the Cu flakes is still the key factor for the successful sintering process, and thus the synthesis method of the metal powders as well. Further evaluation and research shall be needed on both the Cu flake paste and the plasma sintering process before the practical die-attach applications for WBG power semiconductor devices.

Acknowledgments

The study is partially supported by Grant-in-Aid for Scientific Research (S) Grant Number 24226017.

References

1. F. Xu, T. J. Han, D. Jiang, L. M. Tolbert, F. Wang, J. Nagashima, S. J. Kim, S. Kulkarni, and F. Barlow, "Development of a SiC JFET-based six-pack power module for a fully integrated inverter," IEEE Trans. Power Electron., vol. 28, no. 3, pp. 1464–1478, 2013.

2. K. Suganuma, S.-J. Kim, and K.-S. Kim, "High-temperature lead-free solders: Properties and possibilities," JOM, vol. 61, no. 1, pp. 64–71, 2009.

3. H. A. Mustain, W. D. Brown, and S. S. Ang, "Transient liquid phase die attach for high-temperature silicon carbide power devices,"IEEE Trans. Compon. Packag. Technol., vol. 33, no. 3, pp. 563–570, 2010

4. S. Sakamoto, S. Nagao, and K. Suganuma, "Thermal fatigue of Ag flake sintering die-attachment for Si/SiC power devices," J. Mater. Sci.: Mater. Electron, vol. 24, no. 7, pp. 2593–2601, 2013

5. N. Terada, K. Shiokawa, and T. Yamakawa, "Influence of joining conditions on bonding strength of joints: Efficacy of low-temperature bonding using Cu nanoparticle paste," J. Electron. Mater., vol. 42, no. 6, pp. 1260–1267, 2013.

6. S. Q. Xiao, S. Xu, H. P. Zhou, D. Y. Wei, S. Y. Huang, L. X. Xu, C. C. Sern, Y. N. Guo, S. Khan, and Y. Xu, "Silicon homojunction solar cells via a hydrogen plasma etching process," J. Phys. D: Appl. Phys., vol. 46, 105103 (6pp), 2013.

7. A. Descoeudres, L. Barraud, S. De Wolf, B. Strahm, D. Lachenal, C. Gue'rin, Z. C. Holman, F. Zicarelli, B. Demaurex, J. Seif, J. Holovsky, and C. Ballif, "Improved amorphous/crystalline silicon interface passivation by hydrogen plasma treatment," Appl. Phys. Lett., vol. 99, 123506, 2011.

Gap in pagination due to withheld paper.

Pages 1080-1085

Flip Chip Based on Compliant Double Helix Interconnect for High Frequency Applications

Pingye Xu[1], George A. Hernandez[1], Shiqiang Wang[1], Jie Zhong[2], Charles D. Ellis[1], Michael C. Hamilton[1]*
[1]Electrical and Computer Engineering, [2]Chemical Engineering
Auburn University, Auburn, Alabama 36849-5201
Email: mch0021@auburn.edu

Abstract

Compliant interconnects are a viable solution for coefficient of thermal expansion (CTE) mismatch failures and it is therefore important to understand their performance at high frequency. In this work, we present double helix shaped compliant interconnects that are designed and fabricated on top of a coplanar waveguide (CPW) test structure followed by flip chip bonding. The high frequency performance of compliant double helix interconnect is simulated and measured. The measured insertion and reflection loss were less than -0.6 dB and -15 dB up to 50 GHz, respectively.

Introduction

Due to the development of wireless applications in recent years, many researchers have focused on millimeter-wave frequency range systems. One important aspect of these systems is module packaging. The proper strategy to assemble and interconnect monolithic microwave integrated circuits (MMICs) is critical to the performance of the system. Flip-chip is an approach is considered to be the most promising to make interconnects and assemble chips due to its low reflection and insertion loss as well as low fabrication cost [1]. One issue with the conventional flip-chip solder interconnect is the increasing risk of failure as the number of I/O interconnects is scaled down and the density of interconnects is increased. Coefficient of thermal expansion (CTE) mismatch is a common cause of failure in electronic components [2]. A solution to mitigate CTE mismatch is the use of underfill, but the underfill process is slow and time-consuming process that can have detrimental impact to the performance of high frequency signal interconnects [3].

Another solution to CTE mismatch issue is the use of compliant interconnects. Compliant interconnects are off-chip mechanically flexible structure fabricated between the carrier and the chip. Due to the compliance of the structure, the stress caused by CTE mismatch can be mitigated. Various compliant interconnect structures have been researched in the last few decades, including micro-spring [4]-[6], micro-helix [7], sea of leads [8], and sea of polymer pillars [9]. Additionally, compliant interconnects are advantageous for mounting chips onto non-planar surface.

Despite the works on various structures of the compliant interconnect, to the best of our knowledge, none of them have studied the high frequency performance of their compliant interconnect. This work presents the design and fabrication a flip chip bonded coplanar waveguide (CPW) structure using compliant double helix structures as interconnects [10]. The CPW serves as a test structure for characterization of the double helix interconnects. A repetitive photolithography and copper electroplating process is used to fabricate the structure. To facilitate flip chip, indium bumps were deposited by an electroplating process, which serves as the bonding adhesive. The fabrication process is compatible with wafer-level packaging and can be an extension of the standard back-end-of-line (BEOL) with high I/O density. The compliant double helix has a width of 100 μm and a height of 25 μm (including the height of indium bumps for bonding). The performance of the structure was simulated and measured up to 50 GHz. The insertion loss and reflection loss were simulated and measured to be reasonably low.

Design and Fabrication

1. Design of flip chip bonded CPW

The coplanar waveguides were used as test structure of the double helix interconnects. The schematic of the flip chip bonded CPW with double helix interconnects is shown in Figure 1 and the geometric parameters are listed in Table I. Glass substrates were used due to their low dielectric loss. The structure was designed to have a characteristic impedance of 50 Ω. In this design, the high frequency signal transits from one CPW on the carrier to the other through 2 flip chip bonded double helix interconnects transitions and a CPW on the die. Since the signal loss through the CPW is low, the test structure provides insight on the high frequency electrical performance of the double helix interconnects.

To reduce the reflection at the interconnects, a staggered interconnects design is adopted. The center double helix interconnects were designed further away from the double helixes on the ground lines, resulting in characteristic line capacitance decreases and less reflection due to the field concentration in the air region [1].

2. Design of the double helix interconnect

The double helix interconnect is illustrated in Figure 2. One of the helixes is semi-transparent for clarity. The structure is composed of two identical half-turn helixes and fabricated by stacking several layers of electroplated copper. The design goal of the structure is to provide compliance in both the vertical and horizontal directions while keeping the electrical inductance and resistance to be sufficiently low. The geometry of the overhang is critical since it determines the mechanical and electrical properties of the interconnect structure. The increase of thickness (t) and width (w_1) deteriorates the compliance of structure but improve its electrical conductance. The compliance of the structure also benefits from taller pillars, but taller pillars also induce higher electrical resistance and inductance. As described in our previous work [10], the design of this double helix interconnect has the advantage of low electrical resistance compared to other work, primarily due to its high volumetric density of the conductor.

978-1-4799-2408-0/14 $31.00 © 2014 IEEE

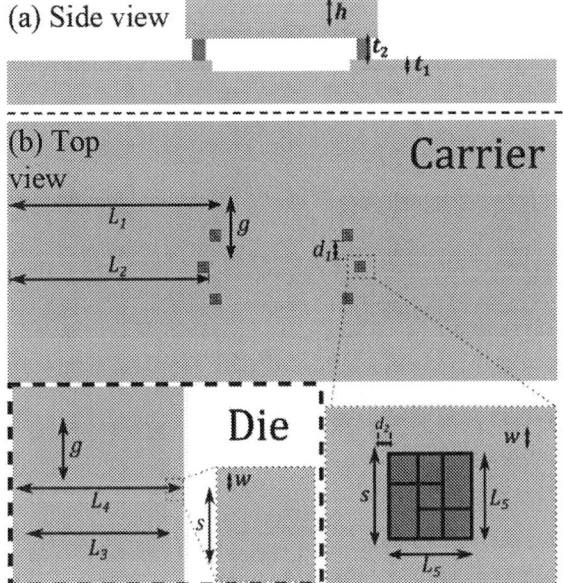

Figure 1. Geometry of the flip chip bonded CPW. The glass substrate, CPW lines and double helix interconnects are in blue, light orange and dark orange.

TABLE I
GEOMETRICAL PARAMETERS OF FLIP CHIP BONDED CPW DESIGN.

Parameter (mm)	Value	Parameter (μm)	Value
L_1	1.9	s	120
L_2	1.78	w	20
L_3	9.60	d_1	175
L_4	1.2	d_2	10
L_5	0.1	t_1	2.5
h	0.5	t_2	25
g	0.6		

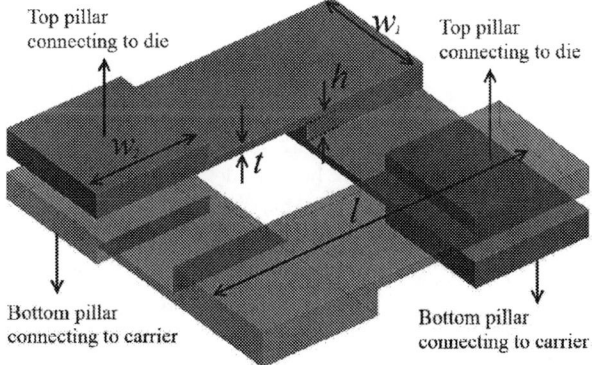

Figure 2. Geometric design of double helix interconnect. Overhang geometry: $w_1 \times l \times t = 35\ \mu m \times 100\ \mu m \times 2.5\ \mu m$. Pillar geometry: $w_1 \times w_2 \times h = 35\ \mu m \times 35\ \mu m \times 5\ \mu m$.

3. Fabrication of the double helix interconnect

Conventional photolithography, electron beam deposition and electroplating were used to fabricate the double helix interconnect. A detailed fabrication process of a similar structure is described in reference [10]. The process is illustrated in Figure 3. In brief, a Ti/Cu seed layer was first

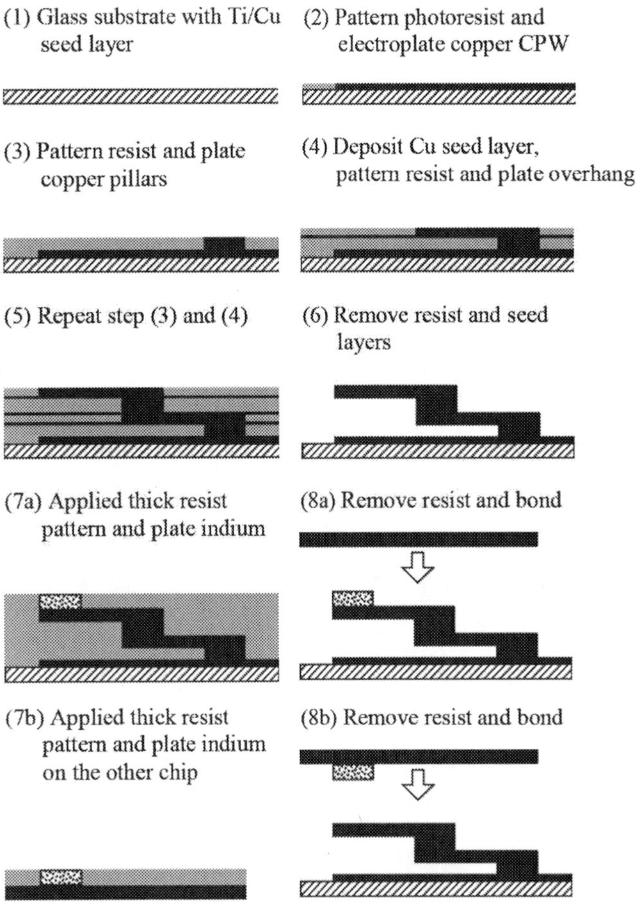

(1) Glass substrate with Ti/Cu seed layer

(2) Pattern photoresist and electroplate copper CPW

(3) Pattern resist and plate copper pillars

(4) Deposit Cu seed layer, pattern resist and plate overhang

(5) Repeat step (3) and (4)

(6) Remove resist and seed layers

(7a) Applied thick resist pattern and plate indium

(8a) Remove resist and bond

(7b) Applied thick resist pattern and plate indium on the other chip

(8b) Remove resist and bond

Figure 3. Schematic fabrication flow for double helix compliant interconnect. The photoresist, copper, indium and substrate are represented by grey, black, dotted and hatched patterns.

deposited on a glass substrate and photoresist was applied and patterned (Figure 3(1)). Copper was electroplated to form a 2.5 μm thick CPW (Figure 3(2)), which was used as a test structure for high frequency characterization. The photoresist layer was then stripped. Another photolithography step was used to form the pattern of the pillars of the double helix, followed by another copper plating process (Figure 3(3)). Without removing the resist, another copper seed layer was deposited. Sequentially, the overhang of the double helix was patterned and electroplated (Figure 3(4)). The photolithography/electroplating process were repeated until a complete double helix structure was completed (Figure 3(5)). After removing the photoresist and seed layers, the interconnect is ready for indium deposition and flip chip bonding (Figure 3(6)). We note that the applied photoresist should approximate the thickness of the copper layer to be plated, since it is critical to allow a planar and uniform surface and facilitate the subsequent fabrication steps.

To fabricate indium bumps on the double helix structure, a thick photoresist layer needs to be carefully deposited onto the substrate and patterned (Figure 3(7a)). Since spin coating process is used in this work to deposit photoresist, it is critical to keep the spinning speed and acceleration low to avoid damaging the double helix structures. The indium bumps can

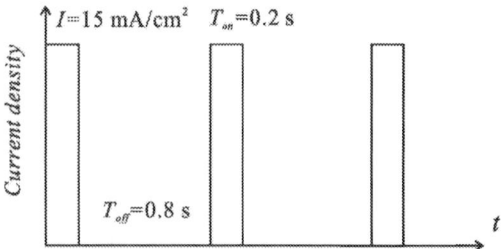

Figure 4. Pulsed current indium plating waveform.

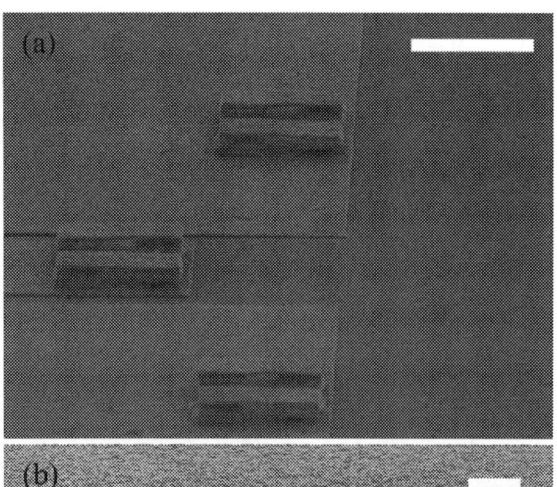

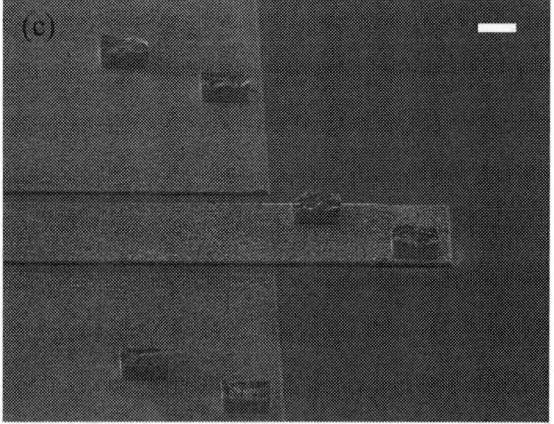

Figure 5. (a) Double helix interconnects on CPW with no indium bumps. Scale bar: 100 μm. (b) Double helix interconnects with indium bumps on top. Scale bar: 20 μm. (c) Indium bumps electroplated on the other chip for flip-chip bonding. Scale bar: 20 μm.

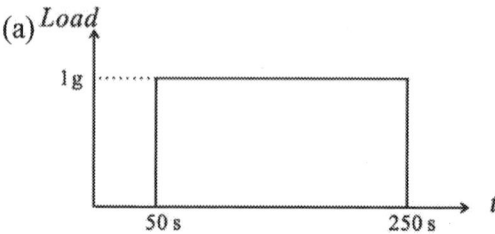

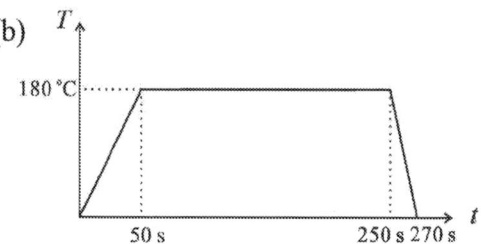

Figure 6. Load and temperature profile used for flip chip bonding.

also be fabricated on the other chip prior to flip chip bonding, as shown in Figure 3(7b). The indium sulfamate plating bath is purchased from Indium Corp. A pulsed current plating process was used for more refined grain and better uniformity [11]. The indium plating waveform is shown in Figure 4. Since large current density may induce hydrogen gas bubbles near the surface of the plated substrate and undermine the coverage of solution into the micro-scale patterns [11], the average current density was kept low (3 mA/cm²).

The scanning electron microscopy (SEM) images of the fabricated double helix structures are shown in Figure 5. The fabricated structures have a high yield and uniformity. Figure 5(b) and 5(c) show the fabricated double helix interconnects with indium bumps and indium bumps directly deposited on the other chip, respectively. The seed layer in Figure 5(c) was kept while in Figure 5(b) the seed layer was removed and therefore the indium bumps were corroded during the seed layer removal process. Since the process is based on conventional photolithography and electroplating, it can be used to fabricate structures with finer pitch size. Also, it can be performed at the wafer level and be an extension of the standard back-end-of-line (BEOL).

4. Flip chip bonding of the double helix interconnects

The flip chip bond process was performed using a FC-150 flip chip bonder. The profile of the process is shown in Figure 6. Since the bonding load is required to be lower than the yield strength of the electroplated copper of the double helix interconnects, it is important to estimate the yield strength of the structure. Using a finite element analysis software ANSYS Workbench 14.5, the yield strength of each double helix is determined to be approximately 2 mN and therefore the applied load should be less than 1.2 g because 6 double helix interconnects were pressured. Thus, a 1 g load was applied during flip chip bonding. In the ANSYS simulation, the yield strength of electroplated copper was assumed to be 280 MPa. The Young's modulus and the tangent modulus were assumed to be 140 GPa and 4.67 GPa, respectively [12]. The arm and the chuck of the FC-150 were heated to slightly higher than

978-1-4799-2408-0/14 $31.00 © 2014 IEEE

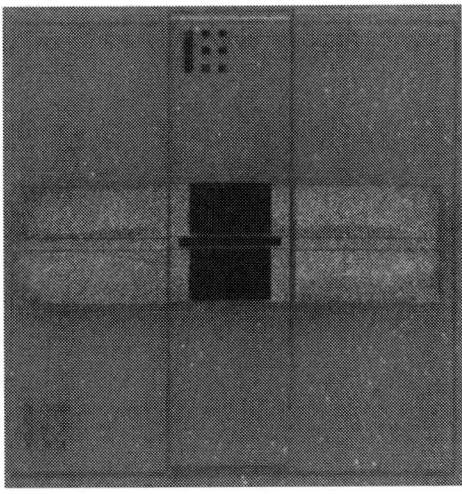

Figure 7. Flip chip bonded CPW with double helix interconnects.

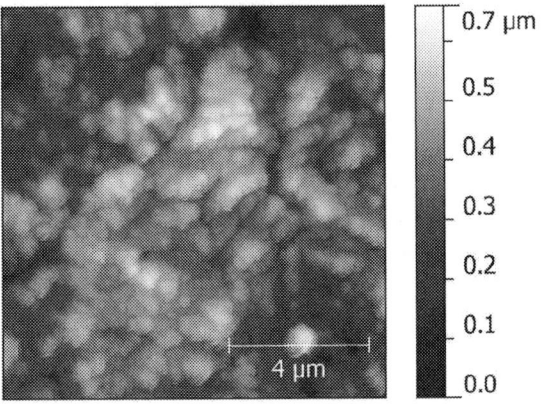

Figure 8. AFM image of the electroplated copper surface.

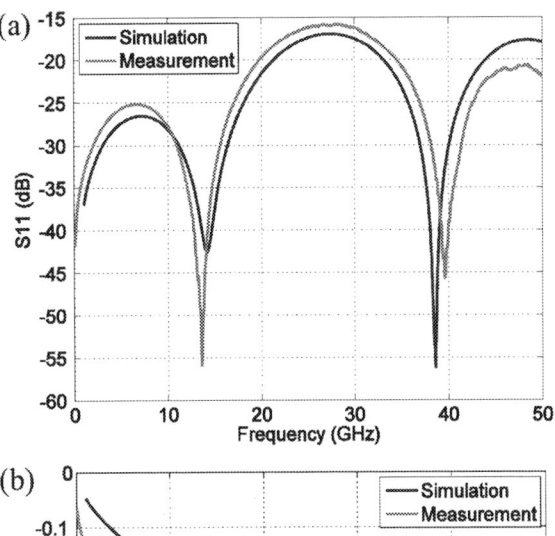

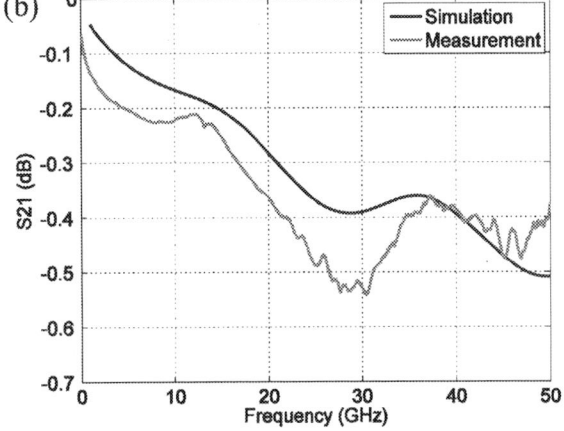

Figure 9. S-parameters of the Flip chip bonded CPW with double helix interconnects. (a) Reflection loss S11. (b) Insertion loss S21.

the melting temperature of indium at 180 °C during bonding. The bonded CPW with double helix interconnects is shown in Figure 7.

High frequency characterization

The structure was modeled using HFSS 14.0. The resistivity of electroplated copper and indium was assumed to be 2.4 μΩ·cm and 15.5 μΩ·cm, respectively [13]. Since the skin effect plays an important role at high frequency, it should be modeled. In this work, Groisse surface roughness model was used. Atomic force microscopy (AFM) was used to determine the root-mean-square (rms) surface roughness of the electroplated copper surface to be approximately 91.6 nm, as shown in Figure 8.

The high frequency characterization was performed using an Agilent N5227A PNA. On-wafer probing measurement was carried out using 150 μm-pitch Cascade ACP65 GSG probes. A full two-port SOLT (Short-Open-Load-Through) calibration was performed in order to move the reference plane close to the measured structure. The measured S-parameters were shown in Figure 9. The reflection loss was measured to be less than -15 dB up to 50 GHz and the insertion loss is less than -0.6 dB, including the influence of

the CPW test structure. The simulation results match reasonably well with the measurement results. The difference might be due to the fabrication tolerance and underestimation of the contact resistance of the indium bump. Since the flip chip bonding process was performed at an elevated temperature in air, the copper and indium structures are easily oxidized, leading to higher resistance. Furthermore, the indium bumps were modeled in HFSS to be perfectly shaped rectangular cuboid, which is inaccurate since the indium bumps deformed during the bonding process and there might be voids remaining inside the indium bumps after bonding. With better process control, reduced loss of the structure is expected.

Conclusion

Compliant double helix interconnects were designed and fabricated on top of a CPW test structure. Indium was plated to facilitate flip chip bonding. Since the fabrication process is based on conventional photolithography and electroplating, it has the potential to make structures with ultra-fine pitch size. Additionally, it can be performed at the wafer level and be an extension of the BEOL. The high frequency performance of compliant double helix interconnect is simulated and measured. The measured insertion and reflection loss were

978-1-4799-2408-0/14 $31.00 © 2014 IEEE   1089

less than -0.6 dB and -15 dB, respectively. Future work includes further polishing of the fabrication process. Reworkable double helix interconnects using sliding contact mode will be fabricated and characterized.

References

1. A. Jentzsch and W. Heinrich, "Theory and measurements of flip-chip interconnects for frequencies up to 100 GHz," *Microwave Theory and Techniques, IEEE Transactions on*, vol. 49, pp. 871-878, 2001.

2. J. H. Pang, D. Chong, and T. Low, "Thermal cycling analysis of flip-chip solder joint reliability," *Components and Packaging Technologies, IEEE Transactions on*, vol. 24, pp. 705-712, 2001.

3. T. Chen, J. Wang, and D. Lu, "Emerging challenges of underfill for flip chip application," in *Proc. Electron. Compon. Technol. Conf.*, 54th, 2004, pp. 175-179.

4. B. Cheng, D. De Bruyker, C. Chua, K. Sahasrabuddhe, I. Shubin, J. E. Cunningham, et al., "Microspring Characterization and Flip-Chip Assembly Reliability," *Components, Packaging and Manufacturing Technology, IEEE Transactions on*, vol. 3, pp. 187-196, 2013.

5. L. Ma, Q. Zhu, S. K. Sitaraman, C. Chua, and D. K. Fork, "Compliant cantilevered spring interconnects for flip-chip packaging," in *Proc. Electron. Compon. Technol. Conf.*, 51st, 2001, pp. 761-766.

6. R. B. Marcus, "A new coiled microspring contact technology," in *Proc. Electron. Compon. Technol. Conf.*, 51st, 2001, pp. 1227-1232.

7. G. Lo and S. K. Sitaraman, "G-helix: Lithography-based wafer-level compliant chip-to-substrate interconnects," in *Proc. Electron. Compon. Technol. Conf.*, 54th, 2004, pp. 320-325.

8. M. S. Bakir, B. Dang, R. Emery, G. Vandentop, P. A. Kohl, and J. D. Meindl, "Sea of Leads Compliant I/O Interconnect Process Integration for the Ultimate Enabling of Chips With Low-Interlayer Dielectrics," *Advanced Packaging, IEEE Transactions on*, vol. 28, pp. 488-494, 2005.

9. M. S. Bakir and J. D. Meindl, "Sea of polymer pillars electrical and optical chip I/O interconnections for gigascale integration," *Electron Devices, IEEE Transactions on*, vol. 51, pp. 1069-1077, 2004.

10. P. Xu, A. H. Pfeiffenberger, C. D. Ellis and M. C. Hamilton, "Fabrication and characterization of double helix structures for compliant and reworkable electrical interconneccts," *Microelectromechanical Systems, Journal of*, 2014. (Accepted)

11. Y. Tian, C. Liu, D. Hutt, and B. Stevens, "Electrodeposition of indium for bump bonding," in *Proc. Electron. Compon. Technol. Conf.*, 58th, 2008, pp. 2096-2100.

12. Y. Xiang, T. Tsui, and J. J. Vlassak, "The mechanical properties of freestanding electroplated Cu thin films," *Journal of materials research*, vol. 21, pp. 1607-1618, 2006.

13. P. Xu and M. C. Hamilton, "Reduced-Loss Ink-Jet Printed Flexible CPW With Copper Coating," 2013.

IEEE Microw. Wireless Compon. Lett., vol. 23, no. 4, pp. 178-180, Apr. 2012.

Modeling of Crosstalk Effects in Coupled MLGNR Interconnects Based on FDTD Method

Vobulapuram Ramesh Kumar, Brajesh Kumar Kaushik, and Amalendu Patnaik
Department of Electronics and Communication Engineering,
Indian Institute of Technology, Roorkee, India.
kumardec@iitr.ac.in, bkk23fec@iitr.ac.in, apfecfec@iitr.ac.in, +91-1332-285662

Abstract

This paper presents an accurate and efficient model for the crosstalk analysis of coupled multilayer graphene nanoribbon (MLGNR) interconnects using finite-difference time-domain (FDTD) technique. The proposed model can be used for accurately estimating the voltage and current at any particular point on the interconnect line. The model is further extended to coupled-n interconnect lines with a low computational cost. Crosstalk induced performance parameters are measured using the proposed model and validated by comparing it to the HSPICE simulations. It is observed that the average error in estimating the noise peak voltage and propagation delay is less than 2%.

1. Introduction

The conventional copper interconnect material is unable to meet the requirements of future technology. In the era of nanotechnology, the copper material demonstrates lower reliability with down scaling of interconnect dimensions. Moreover, the resistivity of copper increases due to electron-surface scattering and grain-boundary scattering with smaller dimensions. Therefore, researchers are forced to find an alternative solution for global VLSI interconnects. Graphene nanoribbon (GNR) has been proposed as a promising interconnect material [1], [2]. The current density of GNR is much higher than the conventional interconnect materials, provides a large mean free path, and high thermal stability [1]. These properties create lots of interest among researchers to use GNR as a VLSI interconnect material.

Graphene nanoribbon can be classified into single-layer GNR (SLGNR) and multi-layer GNR (MLGNR). The most promising interconnect solution for VLSI interconnect is MLGNR due to their high current carrying capabilities than SLGNR. Additionally, the easier fabrication process of MLGNR makes it as promising candidate for VLSI interconnect material [3].

The performance of an MLGNR interconnect line is generally evaluated by means of an equivalent transmission line model. The equivalent model considers all the parasitic parameters based on the quantum mechanical behavior of the nanowire, and its electrostatic and magnetic characteristics. Li *et al.* proposed a multi-conductor transmission line model (MTL) to represent the MLGNR interconnect [2]. However, the analysis of MLGNR with N number of layers leads to the solution of differential equation with the dimensional system of $2N$, which can be computationally expensive. For this reason, an equivalent single conductor (ESC) model was proposed in [1], using the assumption that voltage at an arbitrary cross-section along MLGNR are the same such that all layers are connected in parallel at the both ends. The accuracy of the ESC model in comparison to the MTL model has been reported by several researchers [4], [5]. It was observed that the transient responses to a pulsed input of MTL model and ESC model are in good agreement.

For the first time this research work presents an accurate numerical model for comprehensive crosstalk analysis of coupled MLGNR interconnect lines based on FDTD method. Using the FDTD method, the voltage and current can be accurately estimated at any particular point on the interconnect line. Moreover, the proposed model can be extended to coupled-n interconnect lines with low computational cost. Since the proposed model requires a lesser number of assumptions, the accuracy is very high. Using the proposed FDTD method the transient response of coupled MLGNR lines are analyzed. Furthermore, the functional and dynamic crosstalk analysis is also carried out. The results demonstrate that the proposed model has high accuracy that matches closely with the HSPICE results. In addition to this, the proposed model is highly time efficient than the HSPICE.

The rest of the paper is organized as follows: Section 2 describes the ESC model of an MLGNR. In Section 3, the FDTD technique is developed for coupled-two MLGNR interconnect lines and these expressions can be extended two coupled-n interconnect lines. Section 4 is devoted to the validation of proposed model for coupled-two lines and Section 5 confirms the validation of model for extended coupled-three lines. Finally, Section 6 concludes this paper.

2. Equivalent Single Conductor (ESC) model of MLGNR

This section presents an equivalent *RLC* model of an MLGNR interconnect line. Consider a horizontal MLGNR interconnect line positioned over a ground plane at a distance d and placed in a dielectric medium with dielectric constant ε. The geometry of an MLGNR is shown in Figure 1. The MLGNR interconnect consists of N layers with interlayer distances (=0.34nm), width w, and thickness t. The cross-sectional view of coupled-two lines is shown in Figure 2. The crosstalk induced noise peak voltage and propagation delay is observed on victim line 2. The number of conducting channels per layer can be approximately expressed by [6]

$$N_{ch} = \sum_{i=1}^{n_c} \left[1+e^{(E_i-E_F)/k_BT}\right]^{-1} + \sum_{i=1}^{n_v} \left[1+e^{(E_i+E_F)/k_BT}\right]^{-1} \quad (1)$$

where T is the temperature, k_B is the Boltzmann constant, E_F is the Fermi level.

The ESC model of an MLGNR has been presented in [1], [5] the main features of the model are discussed in this section. The ESC model of an MLGNR is shown in Figure 3.

The number of layers in an MLGNR can be expressed as

$$N_{layer} = 1 + Integer\left(\frac{t}{\delta}\right) \tag{2}$$

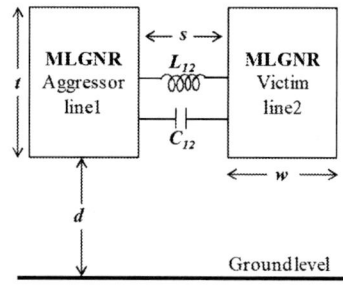

Figure 1. Geometry of an MLGNR with N layers above a ground plane.

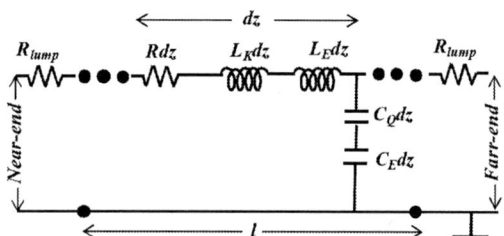

Figure 2. Cross-sectional view of coupled MLGNR interconnects.

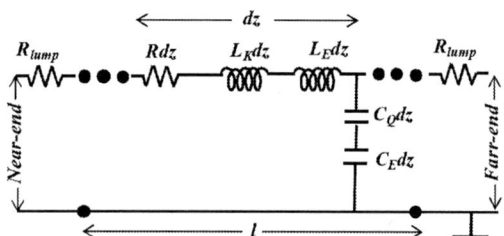

Figure 3. Circuit representation of MLGNR, ESC model.

Each layer in the MLGNR primarily demonstrates three different types of resistances: 1) quantum resistance (R_q) due to the finite conductance value of quantum wire if there is no scatterings along the length; 2) imperfect metal-nanowire contact resistance (R_{mc}) that exhibits a value of few kilo-ohms depending on the fabrication process; 3) scattering resistance (R_S) that is mainly due to acoustic phonon scattering, and optical phonon scattering occurs when the quantum wire lengths exceeding mean free path (mfp) of electrons.

$$R_{lump} = \frac{1}{2}\left[\frac{h/2e^2}{N_{ch}N_{layer}} + \frac{R_{mc}}{N_{layer}}\right] \tag{3}$$

$$R_S = \frac{h/2e^2}{N_{ch}N_{layer}\lambda_{mfp}} \tag{4}$$

where h and e represent the Planck's constant and the charge of an electron, respectively. The value of R_{mc} is highly process dependent, its typical value is about 20k per layer [7]. The scattering resistance is a distributed $p.u.l$ resistance of GNR, and it is primarily depends on the mean free path,

λ_{mfp}. The value of λ_{mfp} is considered as 419nm for MLGNR, which is experimentally determined by measuring the conductance of bulk graphite [1].

The capacitance of the MLGNR arises from two sources: 1) the classical electrostatic capacitance ($C_{E,ESC}$), occurs due to the electric field coupling to neighboring interconnects that can be extracted using electrostatic field solver Ansoft Maxwell [8]; 2) the quantum capacitance (C_Q), occurs due to the fact that, it is only possible to add the electrons into the quantum wire at an available quantum state above Fermi-level, that accounts for the quantum electrostatic energy stored in a quantum wire. By equating this energy to effective capacitance energy, the C_Q is expressed as [9]

$$C_Q = \frac{2q^2}{hv_F} \tag{5-a}$$

For an MLGNR interconnect of N conducting layers the equivalent quantum capacitance ($C_{Q,ESC}$) is expressed as

$$C_{Q,ESC} = N_{ch}N_{layer}\frac{2\times 2q^2}{hv_F} \tag{5-b}$$

Similar to the capacitance the GNR inductance is also arises from two sources: 1) the classical magnetic inductance (L_M), occurs due to magnetic field coupling to neighboring interconnects that can be extracted using magnetic field solver Fast Henry [10]; 2) the kinetic inductance (L_K), can be expressed by equating kinetic energy stored in a channel to effective inductance energy [9]

$$L_K = \frac{h}{2q^2 v_F} \tag{6-a}$$

For an MLGNR of N conducting layers the equivalent kinetic inductance ($L_{K,ESC}$) is expressed as

$$L_{K,ESC} = \frac{1}{N_{ch}N_{layer}}\frac{h}{2\times 2q^2 v_F} \tag{6-b}$$

where v_F is the Fermi velocity considered as 8×10^5 m/s for graphene. The kinetic inductance value per channel is 16 nH/ m, which is verified by the experimental observations as well [11]. Since the GNR layers in a bundle are assumed to be at the same potential, therefore, the effect of coupling capacitance between layer to layer can be neglected. Moreover, the mutual inductance between the layers is observed to be four orders less than the kinetic inductance and therefore, this factor can be safely ignored in the ESC model [12].

3. Proposed Model Formulation

The FDTD technique is used to model the coupled MLGNR interconnect lines. The coupled-two interconnect lines are analyzed in this section; however, the model can be extended to coupled-n lines with a low computational cost.

3.1. MLGNR interconnect line

The Coupled-two MLGNR interconnect line structure is shown in Figure 4, where R_1, R_2 are the scattering resistance, L_{K1}, L_{K2} are the kinetic inductance, L_{E1}, L_{E2} are the magnetic inductance, C_{Q1}, C_{Q2} are the quantum capacitance, C_{E1}, C_{E2} are the electrostatic capacitance, and C_{L1}, C_{L2} are the load capacitance of line 1 and line 2, respectively. All these values are mentioned in per unit length (p.u.l). C_{12} and L_{12} are the p.u.l coupling capacitance and mutual inductance, respectively. The position along the interconnect line, and time are denoted as z and t, respectively [13], [14].

For uniform coupled-two transmission lines the telegrapher's equations in the transverse electro-magnetic (TEM) mode is embedded as [15]

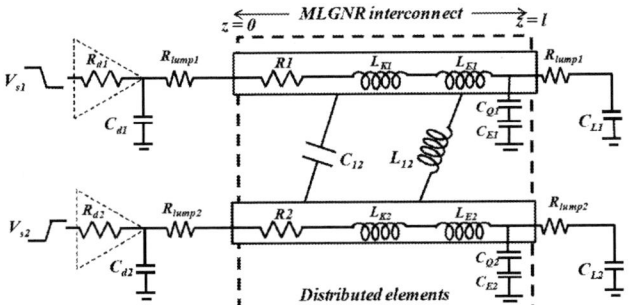

Figure 4. Coupled-two MLGNR interconnect lines.

$$\frac{d}{dz} V(z,t) + R I(z,t) + L \frac{d}{dt} I(z,t) = 0 \qquad (7\text{-}a)$$

$$\frac{d}{dz} I(z,t) + \frac{d}{dt} C V(z,t) = 0 \qquad (7\text{-}b)$$

where V and I are 2×1 column vectors of line voltages and currents, respectively. The line parasitic elements are obtained in 2×2 per unit length matrix form i.e.,

$$V = \begin{bmatrix} V_1 \\ V_2 \end{bmatrix}, \quad I = \begin{bmatrix} I_1 \\ I_2 \end{bmatrix}$$

$$R = \begin{bmatrix} R_1 & 0 \\ 0 & R_2 \end{bmatrix}, \quad L = \begin{bmatrix} L_{K1}+L_{E1} & L_{12} \\ L_{12} & L_{K2}+L_{E2} \end{bmatrix} \text{ and}$$

$$C = \begin{bmatrix} \left(1/C_{Q1}+1/C_{E1}\right)^{-1}+C_{12} & -C_{12} \\ -C_{12} & \left(1/C_{Q2}+1/C_{E2}\right)^{-1}+C_{12} \end{bmatrix}$$

Center difference approximation is used to analyze the first-order differential equations (7-a) and (7-b). Using the FDTD method the analysis of telegrapher's equations has shown better accuracy if the voltage and current points are chosen as alternate in space location and separated by one-half of the position discretization, i.e., $\Delta z/2$. In the same manner, the solution time for V and I should also be separated by $\Delta t/2$ as shown in Figure 5.

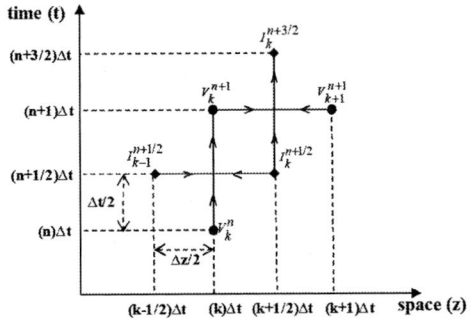

Figure 5. The relation between spatial and time discretization to achieve second order accuracy.

The interconnect line of length l is driven by a resistive driver at $z = 0$ and terminated by a capacitive load at $z = l$. The line is discretized into NDZ uniform segments of length $\Delta z = l/NDZ$. The voltage and current solution points are discretized along the line as shown in Figure 6.

Applying finite difference approximations to (7-a) and (7-b) results in

$$\frac{V_{k+1}^{n+1}-V_k^{n+1}}{\Delta z} + L \frac{I_k^{n+3/2}-I_k^{n+1/2}}{\Delta t} + R \frac{I_k^{n+3/2}+I_k^{n+1/2}}{2} = 0 \qquad (8\text{-}a)$$

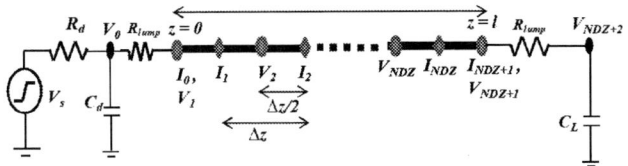

Figure 6. Illustration of space discretization of line for FDTD implementation.

$$I_k^{n+3/2} = BDI_k^{n+1/2} + B\left[V_k^{n+1}-V_{k+1}^{n+1}\right] \qquad (8\text{-}b)$$

for $k = 1, 2, \ldots\ldots, NDZ$

where $B = \left[\frac{\Delta z}{\Delta t} L + \frac{\Delta z}{2} R\right]^{-1}$, $D = \left[\frac{\Delta z}{\Delta t} L - \frac{\Delta z}{2} R\right]$

$$\frac{I_k^{n+1/2}-I_{k-1}^{n+1/2}}{\Delta z} + C \frac{V_k^{n+1}-V_k^n}{\Delta t} = 0 \qquad (9\text{-}a)$$

$$V_k^{n+1} = V_k^n + A\left[I_{k-1}^{n+1/2}-I_k^{n+1/2}\right] \qquad (9\text{-}b)$$

for $k = 2, 3, \ldots\ldots, NDZ$

where $A = \left[\frac{\Delta z}{\Delta t} C\right]^{-1}$

Notice that the calculations are interleaved in both space and time. For example, in (9-b) the new value of V is calculated from the previous value of V and the most recent values of I. For integer values of i and j, V and I vectors are denoted as

$$V_i^j = V\left[i\Delta z, j\Delta t\right], \quad I_i^j = I\left[(i+1/2)\Delta z, j\Delta t\right] \qquad (10)$$

3.2. Near-end terminal

The voltage and current points at the near end terminal are represented by V_1 and I_0, respectively. As indicated in Figure 6, it is observed that to apply the boundary conditions in (9-b), z is replaced by $\Delta z/2$. Therefore, at $k = 1$ equation (9-b) becomes

$$V_1^{n+1} = V_1^n + 2A\left[I_0^{n+1/2}-I_1^{n+1/2}\right] \qquad (11\text{-}a)$$

The source current I_0 at $(n+1/2)$ time interval is obtained by averaging the source current at (n) and $(n+1)$ time intervals then the equation (11-a) becomes

$$V_1^{n+1} = V_1^n + 2A\left[\frac{I_0^{n+1}+I_0^n}{2}-I_1^{n+1/2}\right] \qquad (11\text{-}b)$$

here I_0 is the driver current. Applying Kirchhoff's current law (KCL) at near-end terminal I_0 is expressed as

$$V_0 - V_1 = R_{lump}I_0 \qquad (11\text{-}c)$$

In order to use the equation (11-c) in (11-b), the source current I_0 must be represented in discretized form

$$I_0^{n+1} = \frac{1}{R_{lump}}\left[V_0^{n+1} - V_1^{n+1}\right] \quad (11\text{-}d)$$

using (11-b) and (11-d)

$$V_1^{n+1} = EV_1^n + 2EA\left[\frac{V_0^{n+1}}{2R_{lump}} + \frac{I_0^n}{2} - I_1^{n+1/2}\right] \quad (11\text{-}e)$$

where $E = \left[U + \dfrac{A}{R_{lump}}\right]^{-1}$

Applying KCL at near-end terminal V_0 is expressed as

$$V_0^{n+1} = F\left[\frac{C_d}{\Delta t}V_0^n + \frac{U}{R_d}V_S^{n+1} - I_0^{n+1}\right] \text{ where } F = \left[\frac{C_d}{\Delta t} + \frac{1}{R_d}\right]^{-1} \quad (12)$$

3.3. Far-end terminal

The object is to derive the voltage expression at $k = NDZ+1$, and $NDZ+2$. At $k = NDZ+1$ the equation (9-b) becomes

$$V_{NDZ+1}^{n+1} = V_{NDZ+1}^n + 2A\left(I_{NDZ}^{n+1/2} - \frac{I_{NDZ+1}^{n+1} + I_{NDZ+1}^n}{2}\div\right) \quad (13\text{-}a)$$

Applying KCL at far-end terminal the output current (I_{NDZ+1}) is expressed as

$$V_{NDZ+1} - V_{NDZ+2} = R_{lump}I_{NDZ+1} \quad (13\text{-}b)$$

The discretization form of (13-b) is expressed as

$$I_{NDZ+1}^{n+1} = \frac{1}{R_{lump}}\left[V_{NDZ+1}^{n+1} - V_{NDZ+2}^{n+1}\right] \quad (13\text{-}c)$$

using (13-c) and (13-a) the far-end voltage V_{NDZ+1} is expressed as

$$V_{NDZ+1}^{n+1} = EV_{NDZ+1}^n + 2EA\left[\frac{V_{NDZ+2}^{n+1}}{2R_{lump}} + I_{NDZ}^{n+1/2} - \frac{I_{NDZ+1}^n}{2}\right] \quad (13\text{-}d)$$

and the load voltage V_{NDZ+2} is expressed as

$$V_{NDZ+2}^{n+1} = V_{NDZ+2}^n + \frac{\Delta t}{C_L}I_{NDZ+1}^n \quad (13\text{-}e)$$

These equations are evaluated in a bootstrapping fashion. Initially, the voltages along the line are evaluated for a specific time from equations (12), (11-e), (9-b), (13-e), and (13-d) in terms of the previous values of voltage and current values. Thereafter, the currents are evaluated from (11-d), (8-b), and (13-c) in terms of these voltages and previous current values. The analysis starts with a quiescent line having zero voltage and current values. Nevertheless, the FDTD method provides an exact solution if the following two conditions are satisfied: 1) the spatial increment steps (Δz) must be small enough in comparison to the wavelength (generally 10-20 steps per wave length) to obtain a good spatial resolution; 2) time step (Δt) must be small enough to satisfy the Courant stability condition $\Delta t \leq \Delta z/v$. However, since the boundary conditions have the explicit forms derived from the implicit

equations, there is no stability problem at the two boundaries. Therefore, the stability of the system is only determined by the transmission line portion of the system. For a coupled-multiple interconnect lines these equations remain valid except the few manipulations in the matrix notation. For instance, in a coupled-three interconnect line system the parasitics of interconnect line, resistive driver and load capacitance elements become 3×3 matrices, and the voltage and current values are calculated in a 3×1 column vector form.

4. Validation

The proposed model is tested by comparing its results with that from HSPICE simulations. The coupled-two MLGNR interconnect line structure is considered for the validation of the proposed model as shown in Figure 4.

The length of the interconnect line is chosen as 100μm. The load capacitance is chosen as 14fF. Based on the ITRS data [16] the interconnect dimensions and design parameters of the driver-interconnect system are mentioned in Table 1.

Table 1. 32nm technology node design parameters for global interconnects based on ITRS data [16]

Width (w) (nm)	Thickness (t) (nm)	Distance (d) (nm)	R_d (KΩ)	C_d (fF)	Vdd (V)	ε_r
48	144	86.4	13.85*	0.07*	0.9	2.25

* To drive the large interconnect load the driver strength is considered as 100× times the minimum size.

Using the above mentioned dimensions and the procedure mentioned in Section 2, the corresponding line parasitics are found as

$$R = \begin{bmatrix} 29 & 0 \\ 0 & 29 \end{bmatrix}\frac{\Omega}{\mu m}, \; L = \begin{bmatrix} 8.34 & 0.49 \\ 0.49 & 8.34 \end{bmatrix}\frac{H}{\mu m}, \text{ and}$$

$$C = \begin{bmatrix} 110.3 & -80.7 \\ -80.7 & 110.3 \end{bmatrix}\frac{aF}{\mu m}$$

In the coupled-two line system, line 1 and line 2 are considered as aggressor and victim lines respectively. The crosstalk effects studied in two different cases. First case considers functional crosstalk by switching the aggressor line while keeping the victim in quiescent mode. A second case considers the dynamic crosstalk effect by switching the both lines simultaneously, either in-phase or out-phase. The input signal rise and fall transition times are chosen as 50ps. For the above-mentioned conditions, transient waveforms are compared at the far end terminal on the victim line 2. The functional, in-phase, and out-phase transient waveforms are shown in Figure 7, 8, and 9, respectively. During the out-phase switching, signal transition takes more time than the in-phase transition due to the effect of Miller coupling capacitance (C_{12}) as shown in Figure 9. It is found that in all cases of input switching the proposed model matches accurately with the simulation results. In addition to that, the proposed model is time efficient than the HSPICE.

978-1-4799-2408-0/14 $31.00 © 2014 IEEE

The propagation delay comparison during the in-phase and out-phase transition on victim line 2 is shown in Figure 10. To test the robustness of the model, we examine the propagation delay at different input transition times. The propagation delay during the out-phase transition is found to be high due to Miller capacitance effect. It is observed that the proposed model accurately predicts the propagation delay in both in-phase and out-phase transitions.

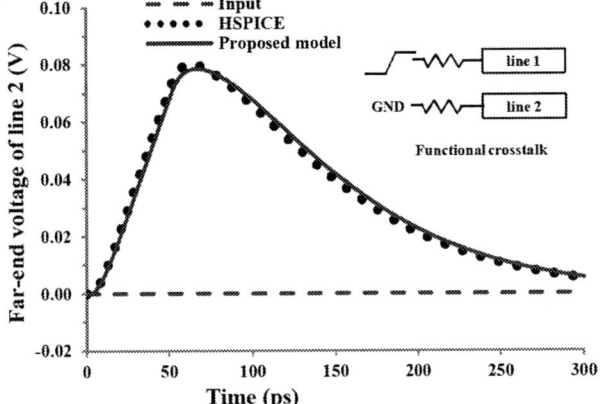

Figure 7. Transient response of quiescent line 2.

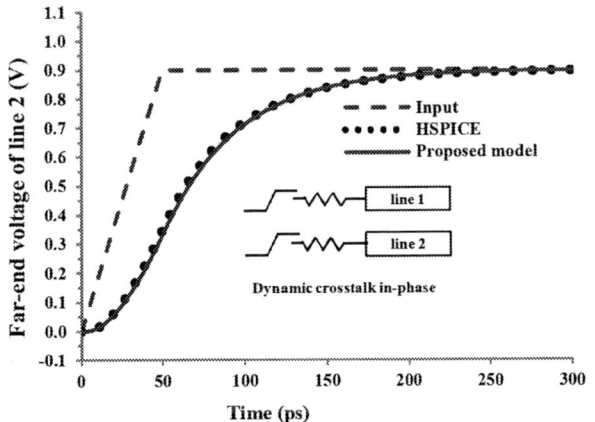

Figure 8. In-phase transient response of line 2.

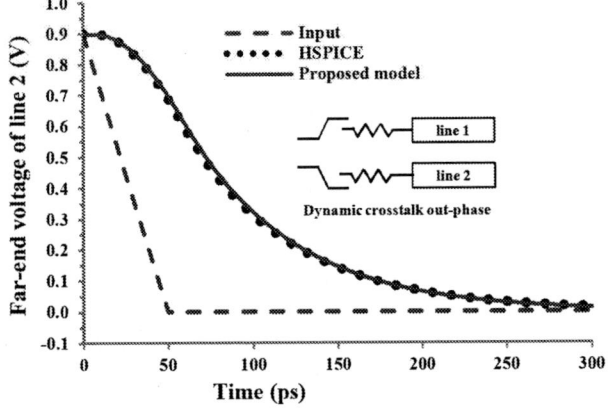

Figure 9. Out-phase transient response of line 2.

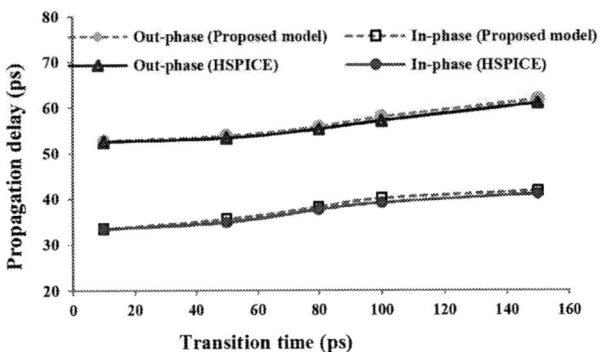

Figure 10. Variation of propagation delay with respect to input transition time.

5. Extensions and Observations

In the previous section, it was demonstrated that the proposed model accurately analyze the coupled-two interconnects. In this section, the interconnect line model is extended to three lines as shown in Figure 11, and it is validated with HSPICE simulations. The input transition time and the load capacitance are chosen as 50 ps and 14 fF, respectively. The signal integrity analysis is carried out at 32 nm technology node. Using the dimensions described in Section 4, the following interconnect parameters are used in the crosstalk analysis of coupled-three lines. The coupling capacitance between 1 and 3 can be safely neglected because of large spacing distance [17].

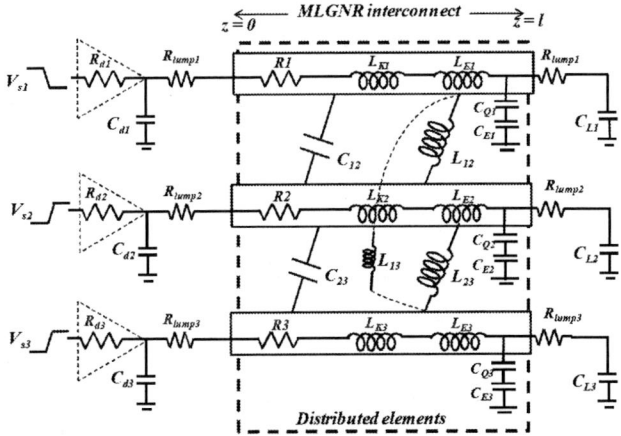

Figure 11. Coupled-three MLGNR interconnect lines.

$$R = \begin{bmatrix} 29 & 0 & 0 \\ 0 & 29 & 0 \\ 0 & 0 & 29 \end{bmatrix} \frac{\Omega}{\mu m} \ , \quad L = \begin{bmatrix} 8.34 & 0.49 & 0.38 \\ 0.49 & 8.34 & 0.49 \\ 0.38 & 0.49 & 8.34 \end{bmatrix} \frac{H}{\mu m} \ , \text{and}$$

$$C = \begin{bmatrix} 110.3 & -80.7 & 0 \\ -80.7 & 191 & -80.7 \\ 0 & -80.7 & 110.3 \end{bmatrix} \frac{aF}{\mu m}$$

Table 2 provides the computational error involved in the estimation of crosstalk induced propagation delay with respect to HSPICE. It is observed that the average error for prediction of propagation delay on victim line 2 is less than 2%. It is also observed that the crosstalk induced delay increases with the increasing switching type mode. This fact can be understood by the Miller capacitance effect. The Miller capacitance factor highly influences the signal propagation when two wires (aggressor and victim) are switching in the opposite direction. Consequently, for a victim line 2, the Type-5 transition produces worst-case delay in high-speed on-chip interconnects.

Table 2. Computational error involved for propagation delay on victim line 2

Type_mode	Input switching modes			Propagation delay on victim line 2		
	Line 1 (Agg.)	Line 2 (Vic.)	Line 3 (Agg.)	HSPICE (ps)	Proposed model (ps)	% error
Type-1	0 1	0 1	0 1	34.8	35	0.57
Type-2	0 1	0 1	Gnd	42.7	43.45	1.76
Type-3	0 1	0 1	1 0	52.1	52.9	1.54
Type-4	1 0	0 1	Gnd	62.69	63.65	1.53
Type-5	1 0	0 1	1 0	73.8	74.9	1.49

Finally, to make a simple decision about the quality of the comparison the datasets of Figure 8 are converted into natural language descriptor (excellent, very good, good, fair, poor, and very poor). Using the Feature Selective Validation (FSV) tool [18] the histogram of the global difference measure (GDM) is generated and shown in Figure 12. The results convincingly demonstrate that using the proposed model the data under an excellently matched region is 88% with respect to HSPICE simulations.

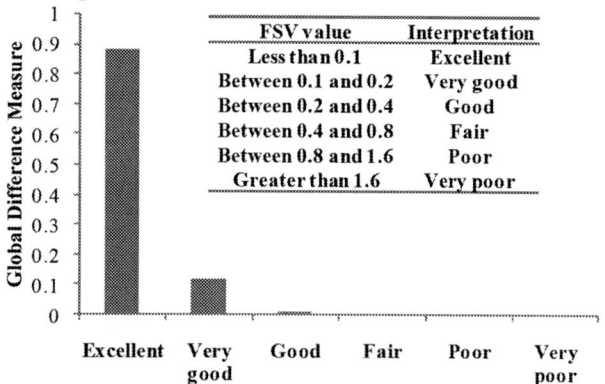

Figure 12. Histogram from the datasets of Figure 8 using FSV tool.

Modified nodal analysis (MNA) is the core analysis approach used in SPICE to formulate the system equations. Applying the Kirchhoff's current law and following the energy conversion principle the MNA generates the set of matrix equations. The order of the matrix is determined by the number of nodes and unknown variables in the circuit. The unknown variables are solved after the inversion of the matrix and therefore, it requires more computation time. However, the FDTD operator is matrix free and therefore, fast and memory efficient as compared to HSPICE simulations. For more clarification, we compared the CPU runtime between proposed FDTD model and HSPICE. An average reduction of 90% is achieved using the proposed model. For instance, using the interconnect length of 100μm with 100-space and 1000-time segments, the runtime using proposed model is 0.58sec against 7.26sec using HSPICE, while maintaining excellent accuracy.

6. Conclusion

This paper presented an accurate model to analyze the crosstalk effects in coupled MLGNR interconnect lines using the FDTD technique. Crosstalk effects, including the functional and dynamic crosstalk analysis in coupled two and three interconnect lines are presented. Furthermore, the proposed model can be extended to *n*-coupled lines with minimum computational effort. It has been observed that the results exhibit a good agreement with HSPICE simulations. Over the random number of test cases, the average error in the propagation delay measurement is found to be within 2%. The analysis suggests that, with continuous advancements in FDTD technique the proposed model will occupy a significant role in performance analysis of MLGNR on-chip interconnects, and it can be incorporated in the TCAD simulation tools.

References

1. C. Xu, H. Li, and K. Banerjee, Modeling, analysis, and design of graphene nano-ribbon interconnects, *IEEE Trans. on Electron Devices*, vol. 56, no. 8, pp. 1567 1578, 2009.

2. H. Li, C. Xu, N. Srivastava, K. Banerjee, Carbon nanomaterials for next-generation interconnects and passives : physics, status, and prospects, *IEEE Trans. on Electron Devices,* vol. 56, no. 9, pp. 1799 1821, 2009.

3. Q. Yu, J. Lian, S. Siriponglert, H. Li, Y. P. Chen, and S.-S. Pei, Graphene segregated on Ni surfaces and transferred to insulators, *Appl. Phys. Lett.*, vol. 93, no. 11, p. 113103, Nov. 2008.

4. J. Cui, W. Zhao, W. Yin, and J. Hu, Signal Transmission Analysis of multilayer graphene nano-ribbon (MLGNR) interconnects, *IEEE Trans. on Electromagnetic Compatibility*, vol. 54, no. 1, pp. 126 132, 2012.

5. M. S. Sarto and A. Tamburrano, Comparative analysis of TL models for multilayer graphene nanoribbon and multiwall carbon nanotube interconnects, *in Proc. of IEEE International Symposium on Electromagnetic Compatibility*, 2010.

6. S. H. Nasiri, M.K.M.Moravvej-Farshi, and R. Faez, Stability analysis in graphene nanoribbon interconnects, *IEEE Electron. Device Lett.,* vol. 31, no. 12, pp. 1458 1460, Dec. 2010.

7. D. Fathi, B. Forouzandeh, S. Mohajerzadeh, and R. Sarvari, Accurate analysis of carbon nanotube interconnects using transmission line model, *Micro & Nano Lett.,* vol. 4, no. 2, pp. 116 121, 2009.

8. Maxwell 2D student version, Ansoft Corp., Pittsburgh, PA., 2005.

9. P. J. Burke, Lüttinger liquid theory as a model of the gigahertz electrical properties of carbon nanotubes, *IEEE Transactions on Nanotechnology*, vol. 1, no. 3, pp. 129 144, 2002.

10. FastHenry Version 3.32. [Online]. Available at: http://www.fastfieldsolvers.com., 2011.

978-1-4799-2408-0/14 $31.00 © 2014 IEEE

11. J. J. Plombon, High-frequency electrical properties of individual and bundled carbon nanotubes, *Applied Physics Lett.,* vol. 90, p. 063106, 2007.

12. A. Naeemi, R. Sarvari, and J. D. Meindl, Performance comparison between carbon nanotube and copper interconnect for GSI, *In Proc. IEEE Int. Electron Devices Meeting,* pp. 699 702, 2004.

13. V. R. Kumar, B. K. Kaushik, and A. Patnaik, "An accurate model for dynamic crosstalk analysis of CMOS gate driven on-chip interconnects using FDTD method," *Microelectronics Journal, Elsevier,* (*in press*).

14. V. R. Kumar, B. K. Kaushik, and A. Patnaik, "An accurate FDTD model for crosstalk analysis of CMOS-gate-driven coupled RLC interconnects," *IEEE Trans. on Electromagnetic Compatibility,* (*in press*).

15. C. R. Paul, Incorporation of terminal constraints in the FDTD analysis of transmission lines, *IEEE Trans. Electromagnetic Compatibility.* vol. 36, no. 2, pp. 85-91, 1994.

16. International Technology Roadmap for Semiconductors [Online]. Available at: http://public.itrs.net, 2009.

17. J. A. Davis, and J. D. Meindl, Compact distributed RLC interconnect models-Part II: Coupled line transient expressions and peak crosstalk in multilevel networks, *IEEE Trans. on Electron Devices.,* vol. 47, no. 11, pp. 2078 2087, 2000.

18. A. P. Duffy, A. J. M. Martin, A. Orlandi, G. Antonini, T.M. Benson, and M.S. Woolfson, Feature Selective Validation (FSV) for Validation of Computational Electromagnetics (CEM). Part I   The FSV Method, *IEEE Trans. on Electromagnetic Compatibility.* vol. 48, no. 3, pp. 449 459, 2006.

978-1-4799-2408-0/14 $31.00 © 2014 IEEE

First Demonstration of Reliable Copper-plated 30μm Diameter Through-Package-Vias in Ultra-thin Bare Glass Interposers

Kaya Demir, Andac Armutlulu, Jialing Tong, Raghuram Pucha, Venkatesh Sundaram and Rao Tummala
3D Systems Packaging Research Center, Georgia Institute of Technology, Atlanta, GA, USA

Abstract

This paper reports the first demonstration of the reliability of through package copper vias (TPV) with 30μm diameter at 120μm pitch in ultra-thin100μm thick glass to achieve high-density vertical interconnections in 2.5D and 3D interposers and packages. Bare glass with 100μm thickness was used to demonstrate the via formation, metallization and reliability of small through-package-vias. The reliability concerns at this fine pitch were addressed through modeling and design from first principles, followed by experimental validation with test-vehicle fabrication and reliability characterization. Thermo mechanical reliability of TPV was analyzed through finite element modeling to estimate stresses inside TPV during thermal cycling and provide design guidelines. For experimental validation, test samples with daisy chains of glass TPVs were fabricated and subjected to accelerated thermal cycling tests between -55°C and 125°C to assess the thermomechanical reliability. Resistance of each daisy chain was measured periodically as a method to detect failure initiation. Majority of the TPV chains passed the reliability test without significant change in resistance. TPV daisy chains that showed changes in resistance were cross-sectioned and failure analysis indicated that the early failures of TPV in glass were related to process defects coming from via-hole formation and metallization.

Key Words

TPV (Through Package Via), Reliability, Glass Interposer, Thermal Cycling Test (TCT), Mechanical Modeling

Introduction

Glass interposers are emerging as an attractive alternative to organic and silicon interposers [1-2]. One of these reasons for the superiority of glass is its ability to achieving high density I/Os enabled by fine-pitch TPVs and RDLs [3]. Moreover, the CTE of glass package can be designed to match that of silicon, leading to lower thermomechanical stresses in ICs [4]. Previous works focused on utilizing attractive electrical properties of glass in order to demonstrate superior RF performance of passive devices, RF filters and transmission lines on glass interposers with TPVs [5, 6]. However, the brittleness of glass raises reliability concerns especially at lower thickness and with fine-pitch TPVs. Glass is an inherently strong material but its strength is degraded by defects created during various fabrication steps [7]. Forming TPV holes for vertical electrical interconnections is another step that is shown to decrease the mechanical strength of glass [8].

Reliability studies of TPV in glass interposers have been very limited [9, 10]. Previous work [10] at GT-PRC modeled, fabricated and characterized reliability of through-package vias in 180μm thick glass interposers with 60μm diameter at 120μm pitch. However, reliability of TPV becomes a major concern as the TPV diameter and interposer thickness are decreased due to the following reasons. Firstly, ultra-thin glass interposer creates additional handling challenges due to the brittleness of glass panels. Secondly, CTE mismatch between materials elevates the TPV stresses at smaller diameters. Moreover, at small TPV diameters, the hole can get fully-filled with copper (Cu) which further increases stress on brittle glass TPV wall. As a result, TPV formation with minimal defects becomes even more critical. This paper focuses on addressing these reliability concerns which have not been investigated yet in detail.

Another key focus of the paper is to use bare glass as the interposer substrate instead of polymer-laminated glass [11]. Polymer lamination of glass makes it easier to handle thin glass. Polymer protects and fills ultra-small defects on glass surfaces, thus increasing its strength during via-formation and processing. Another advantage is that the polymer acts as a buffer layer, reducing the stress induced by the metal layers on glass. Additionally, polymer lamination also helps with metallization. However, there are many advantages of using bare glass as interposer material: 1) there are more via formation options on bare glass compared to polymer laminated glass. 2) Polymer layers on both sides of glass increase RF losses due to higher dielectric loss of most polymers compared to glass. 3) Omitting polymer lamination decreases number of fabrication steps and also the total thickness of interposer. 4) High-CTE polymer layers increase the interconnect strains, which might decrease the fatigue life.

This paper provides a comprehensive modeling, design and fabrication study to demonstrate fine-pitch TPV reliability in bare glass. The first part of the paper focuses on design guidelines for reliable through-package-vias in glass interposers using thermomechanical modeling concepts. The fabrication of test structures with 30μm diameter through-package vias is then described. The last part of the paper presents results from accelerated reliability tests investigating thermomechanical reliability of ultra-small TPVs in thin glass interposer.

Thermo mechanical reliability of TPVs in bare Glass

Thermomechanical reliability of TPVs in glass is studied by finite element modeling to provide design guidelines for subsequent test-vehicle fabrication and reliability characterization. Metallization of TPVs on bare glass is achieved in two steps: 1) Seed-layer formation by Ti/Cu sputtering 2) Cu electroplating. Titanium (Ti) is used as the adhesion enhancement layer between Cu and glass. The TPV cross-section schematically shown in the Figure 1 is used for finite element modeling.

Thermomechanical FEM simulations were run to obtain the stress and strain contours of different TPVs with changing glass and metallization thickness. Simulation results at extreme cold temperatures are shown in Figure 2. The simulations results suggest that material junctions are critical parts for failure.

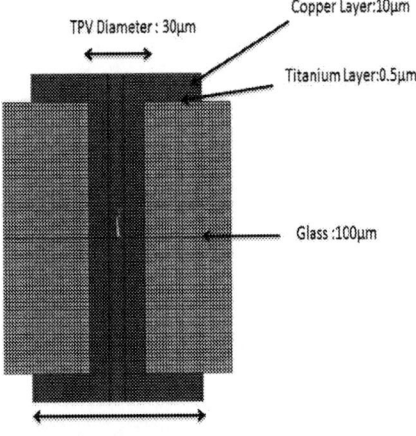

Figure 1. Schematical cross-section of the analyzed TPV

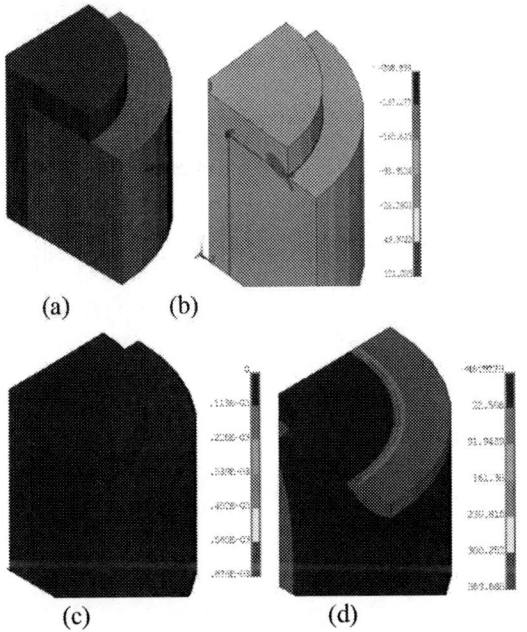

Figure 2. (a) Quarter TPV model b) Shear stress between materials interfaces c) Plastic strain in Cu

d) Axial stress in glass

CTE mismatch between the materials constituting the TPV structure results in stresses during thermal cycling. In order to assess the effect of thermal cycles on stresses inside the TPV, an axisymmetric finite element model was created with Ansys™ . Due to the symmetry of the structure, only quarter of a TPV is modeled to reduce simulation time while maintaining accuracy. Standard temperature cycling between -55°C and 125°C was applied as the source for thermomechanical loading. For copper, bilinear kinematic hardening model was used. Ti and glass were considered as linear elastic materials. Glass with through-via holes is supplied by Corning with 100µm thickness. TPV diameter is 30µm with 120µm pitch. Thermal and mechanical properties of these materials are given in Table 1.

Table 1. Material Properties

	Elastic Modulus(GPa)	Poisson's Ratio	CTE (ppm/°C)
Glass	63.6	0.23	3.17
Ti	116	0.32	8.6
Cu	121	0.3	17.3

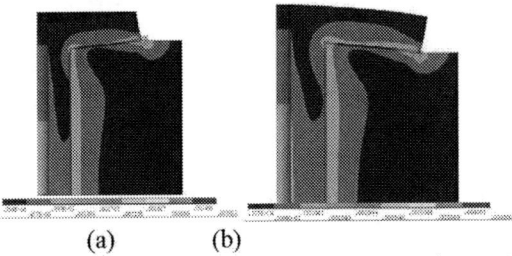

(a)   (b)

Figure 3. a) TPV stresses at extreme cold and b) hot temperatures during cycling.

Due to the brittle nature of glass, axial tensile stresses in the glass are considered as the failure metrics. Plastic strain in copper is also considered as a failure metric due to its influence on via fatigue lifetime. Figure 3 shows the TPV stresses at extreme temperatures. Stress distribution in glass is basically tensile at the low temperatures and compressive at extreme hot temperatures. Parametric simulations are run to assess the effect of various geometries on failure metrics. Results are normalized by the maximum value in that set.

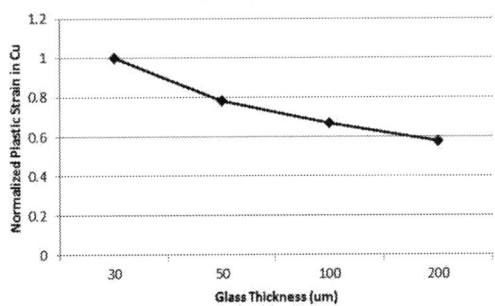

Figure 4. Plastic strain in Cu vs glass thickness

It is shown that plastic strain in copper increases with decreasing glass thickness, keeping other parameters constant. This is one of the challenges of using thinner glass.

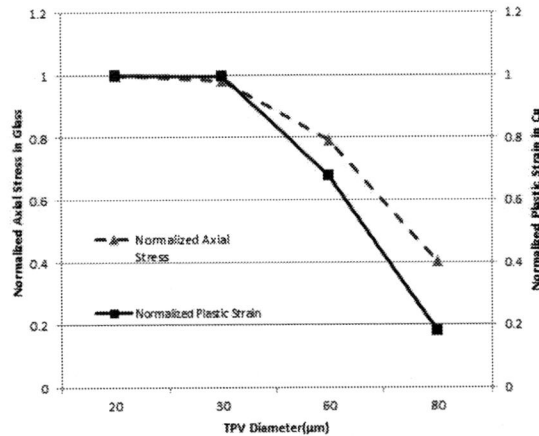

Figure 5. Plastic strain in Cu and axial tensile stress in glass with repect to diameter

Both plastic strain in Cu and tension in glass increases with decreasing via diameter. This trend suggests that more design optimization is needed for high I/O densities.

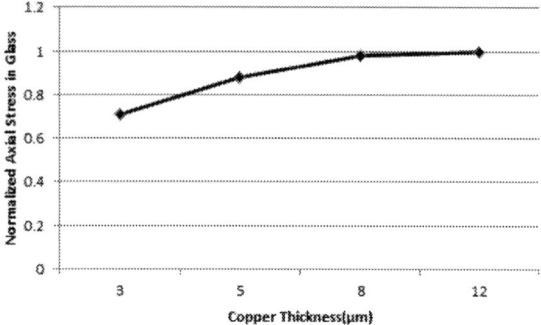

Figure 6.Axial tensile stress in glass vs Cu thickness at extreme cold

Increasing metallization thickness results in higher tensile stresses in glass. This is an expected result as the load on glass arises due to due to its CTE mismatch with glass, as shown in Figure 6. This fact presents a challenge if high copper thickness is required.

The expected lifetime of the fabricated TPV was calculated based on Cu fatigue failure. As indicated by simulations, CTE mismatch between Cu, glass and titanium results in plastic deformation in copper during every cycle. The plastic deformation leads to fatigue failure of TPVs. Accumulated plastic strain per cycle can be calculated from simulations. Using a Coffin-Manson type equation, lifetime of TPVs for different materials and geometries can be estimated.

$$N_f{}^c \times \varepsilon_f{}^{0.75} = \Delta\varepsilon_p$$

c : Fatigue ductility exponent of Cu (-0.6)

ε'_f : Fatigue ductility coefficient of Cu (0.3)

N_f : Number of cycles to failure

$\Delta\varepsilon_p$: Accumulated Plastic strain in each cycle

Number of cycles to failure can be solved to approximately obtain the fatigue life. The fatigue life of TPV is estimated to be over 1500 thermal cycles for the geometry shown in Figure 2. If TPVs in Figure 1 are fabricated without defects, they are expected to meet 1000 thermal cycles reliability.

Test Vehicle Design and Fabrication
The test vehicles design used for investigating thermomechanical reliability of TPVs is shown in Figure 7.

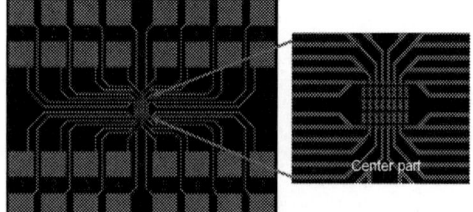

Figure 7.CAD design of test vehicle

Each test coupon consists of 64 vias arranged in a 8x8 array and metalized to form daisy-chains of 8 TPVs. The diameter of the TPV is 30µm and pitch is 100µm. Each array is connected to 4 ports in order to accurately measure the resistance with four-probe technique.

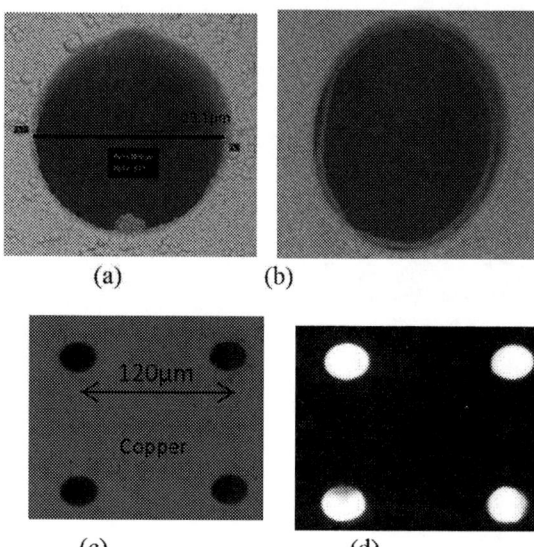

Figure 8.a,b) SEM images of TPVs after Cu sputtering c) TPV array picture d) Transmission mode microscope picture of the same TPV array

Fabrication steps of the test samples are shown in Figure 9. Basically, this low-cost and high-throughput fabrication process consists of two major steps: 1) TPV hole formation 2) TPV and surface metallization through a semi-additive process

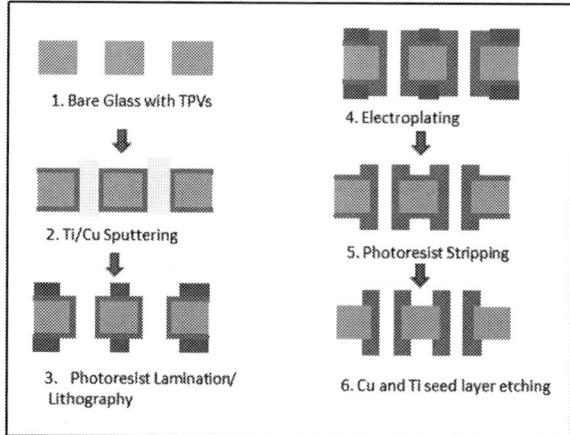

Figure 9. Fabrication process of glass interposer

Bare glass with through vias is provided by Corning. The glass samples were metalized through sputtering of Ti and Cu layers from their high-purity targets with diameters of 3 and 8 inches, respectively (DC sputterer, CVC Inc.). Sputtering of high aspect ratio TPV with small diameter presented a challenge, which was solved by applying longer sputtering times at higher power. The Ti layer serving as the adhesion promoter between the glass substrate and the Cu layer was sputtered for 20 minutes at a power of 350 W, yielding an approximate thickness of 500 nm. Following the formation of the Ti film, approximately 2μm thick Cu layer was sputtered onto the substrate at 1000 W. The reported thickness values for the Ti and Cu layers were determined for the top surface of the substrates via surface profiler (Tencor P15). Hence, the thickness of the sputtered films inside the vias is expected to be significantly lower than the measured values. To prevent the oxidation of the deposits, as well as the targets, the sputtering process for both layers was carried out in 100% Argon environment at a pressure of 6 mTorr. Fairly good adhesion strength of Ti/Cu to bare glass was confirmed by tape test.

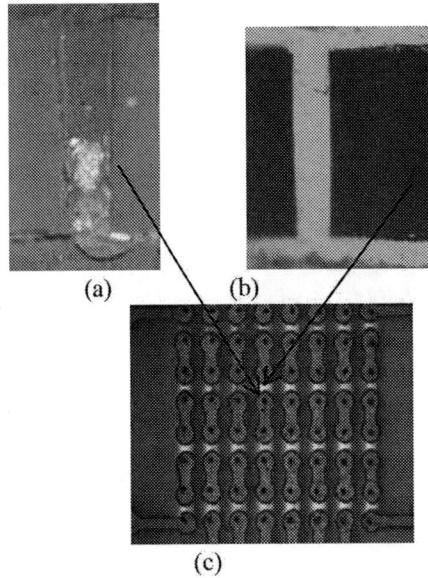

Figure 10. a) TPV hole in glass b) Electroplated TPV c) Fabricated daisy chains

Following the metallization step, glass surfaces were patterned by photoresist deposition and lithography. Another fabrication challenge is the possible clogging of the vias during photoresist development. This was avoided by applying longer development times to ensure the complete dissolution of unexposed photoresist regions, particularly inside the vias. Before proceeding with metallization, through-via holes are checked for openness using transmission light microscope. Plasma ashing was applied for 5 minutes on both sides of the samples to descum any photoresist residue. Samples were then dipped into S-2 cleaner (Atotech) solution for increasing the wettability.

Upon completion of the lithography steps, electroplating of Cu was carried out in a two-electrode-cell configuration. Prior to electrodeposition process, the sample was briefly dipped in diluted HCl solution (DI: HCl, 4:1) to remove the oxide layer on the Cu surface. An aqueous commercial Cu electroplating bath (Clean Earth Cu-Mirror Solution, Grobet) was utilized where a high purity Cu sheet of the same size and geometry as the substrate was used as the anode. The current density was set to 10mA/cm^2 and the solution was stirred at a moderate speed while the sample was slightly moved back and forth to ensure fresh electrolyte access to vias and thus, a conformal deposition within. To further improve the conformal deposition, the sample was flipped after 30 minutes of plating and plated for another 30 minutes

while the back side is facing the Cu anode. At the end of the plating process, the Cu thickness was measured to be approximately 10μm on both sides of the substrate.

Electroplating is followed by removing the photoresist and remaining Cu and Ti seed layers.

Figure 11. Completed test vehicle

Figure 12. TPV after 250 thermal cycles

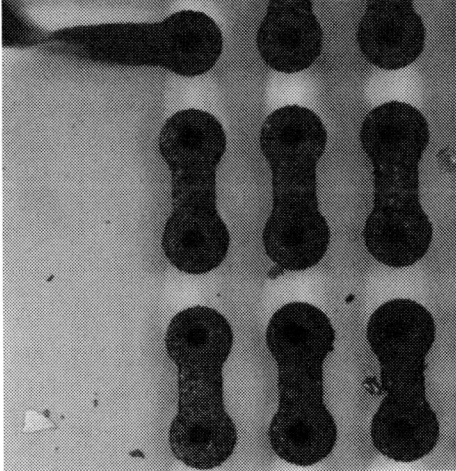

Figure 13. Delamination of copper line

Test Details, Results and Discussion

The test vehicles were first subjected to a 24h bake at 125°C to remove any remaining moisture, and then kept in humidity chamber for preconditioning (60°C, 60% RH and 40hrs) following moisture sensitivity level-3(MSL-3) standards. This is followed by 3-times reflow at a peak temperature of 260°C to simulate lead-free boards assembly process and to investigate the effect of moisture absorption on reliability. The test vehicles were then subjected to thermal cycles between -55°C and 125°C with a dwell time of 15minutes at each extreme as described in JEDEC JESD22-A104 condition B test standards.

Failures were identified based on changes in electrical resistance. Electrical resistance of the TPV daisy chains was measured by using four-point probe after fabrication, after reflow, after 50 cycles and after 250 cycles. More than 10% change in resistance of the TPV daisy chain is chosen as the failure criterion. The daisy-chains were cross-sectioned and analyzed to understand the TPV failure mechanism. Twenty daisy chains corresponding to 160 TPVs were subjected to TCT. 20 out of 20 of the daisy chains survived thermal cycling test without any significant change in the electrical behavior, as was expected from the simulations.

A passed TPV is shown in Figure 12. There were no detected failures related to brittle nature of glass, such as cracks in glass. We did not observe any failure related to TPV however, we observed delamination of copper from smooth glass surface in 3 out of 20 RDL lines connecting the TPV array to contact pads for measurement as shown in figure 13. Methods to increase the adhesion of copper to glass are currently being explored to overcome this issue. Warpage after reflow and thermal cycling was not visibly significant, as expected from the high dimensional stability of glass.

Summary

This paper presents the first demonstration of reliable copper-plated 30μm through package copper vias (TPVs) at fine-pitch in ultra-thin glass interposers. Bare glass with 100mm thickness was used to demonstrate 100mm via-pitch for high-density interconnections .Thermomechanical reliability of TPVs is studied through finite element modeling of various TPV geometries. Modeling results are validated by thermal cycling tests. Copper-plated 30mm diameter TPV holes at 100mm pitch, without any process-related defects survived 250 thermal cycles between -55°C and 125°C. Results demonstrate that 30μm diameter copper plated TPV in ultra-thin bare glass is highly reliable if TPV hole-formation creates minimal defects. TPV in glass, therefore, meets reliability specifications even with bare glass without polymer lamination. Glass is shown to be strong enough to resist against stress, arising from CTE mismatch, without cracking. However, adhesion of copper to glass can be another issue which needs to be further explored.

Acknowledgments

This research was supported by the Low-cost Glass Interposer (LGIP) program in the PRC industry consortium. The authors would also like to thank Corning for providing ultra-thin glasses with through-vias.

References

1. Garrou, Philip, Christopher Bower, and Peter Ramm, eds. *Handbook of 3D Integration: Volume 1-Technology and Applications of 3D Integrated Circuits*. Wiley. com, 2011.

2. Sukumaran, Vijay, et al. "Low-Cost Thin Glass Interposers as a Superior Alternative to Silicon and Organic Interposers for Packaging of 3-D ICs." (2012): 1-1.

3. Sukumaran, Vijay, et al. "Through-package-via formation and metallization of glass interposers." *Electronic Components and Technology Conference (ECTC), 2010 Proceedings 60th*. IEEE, 2010.

4. Sukumaran, Vijay, et al. "Design, fabrication and characterization of low-cost glass interposers with fine-pitch through-package-vias." *Electronic Components and Technology Conference (ECTC), 2011 IEEE 61st*. IEEE, 2011.

5. Anderson, O. L. "The Griffith criterion for glass fracture." *ICF0, Swampscott-MA (USA) 1959*. 2012.

6. Lin, Y. J., et al. "Study of the thermo-mechanical behavior of glass interposer for flip chip packaging applications." *Electronic Components and Technology Conference (ECTC), 2011 IEEE 61st*. IEEE, 2011.

7. Sridharan, Vivek, et al. "Design and fabrication of bandpass filters in glass interposer with through-package-vias (TPV)." *Electronic Components and Technology Conference (ECTC), 2010 Proceedings 60th*. IEEE, 2010.

8. Sato, Yoichiro, et al. "Ultra-miniaturized and surface-mountable glass-based 3D IPAC packages for RF modules." *Electronic Components and Technology Conference (ECTC), 2013 IEEE 63rd*. IEEE, 2013.

9. Topper, Michael, et al. "3-D thin film interposer based on TGV (Through Glass Vias): An alternative to Si-interposer." *Electronic Components and Technology Conference (ECTC), 2010 Proceedings 60th*. IEEE, 2010.

10. Demir, Kaya, et al. "Thermomechanical and electrochemical reliability of fine-pitch through-package-copper vias (TPV) in thin glass interposers and packages." Electronic Components and Technology Conference (ECTC), 2013 IEEE 63rd. IEEE, 2013.

11. Shorey, Aric, et al. "Development of substrates for through glass vias (TGV) for 3DS-IC integration." *Electronic Components and Technology Conference (ECTC), 2012 IEEE 62nd*. IEEE, 2012.

Through-Glass Interposer Integrated High Quality RF Components

Cheolbok Kim[1], David E. Senior[1,2], Aric Shorey[3], Hyup Jong Kim[1], Windsor Thomas[3], and Yong-Kyu Yoon[1]

[1]University of Florida, Gainesville, Florida, USA
[2]Universidad Tecnológica de Bolívar, Cartagena, Colombia
[3]Corning Inc., Corning, New York, USA
ykyoon@ece.ufl.edu

Abstract

High quality and compact RF devices, using the half mode substrate integrated waveguide (HMSIW) architecture loaded with a complementary split ring resonator (CSRR), are implemented on a glass interposer layer, which therefore serves as an interconnection layer and as a host medium for integrated passive RF components. Compared with the silicon interposer approach, which suffers from large electrical conductivity and therefore substrate loss, the glass interposer has advantages of low substrate loss, allowing high quality interconnection and passive circuits, and low material and manufacturing costs. Corning fusion glass is selected as the substrate to realize the compact CSRR-loaded HMSIW resonators and bandpass filters (BPFs) working under the principle of evanescent wave amplification. Two and three pole bandpass filters are designed for broadband operation at 5.8 GHz. Thru glass vias (TGVs) are used to define the side-wall of the substrate integrated waveguiding structure. Surface micromachining techniques are used to fabricate the proposed devices. The variations of the external quality factor (Q_e) of the resonator and the internal coupling coefficient (M) of the coupled resonators are studied for filter design. Operation of the filters at 5.8 GHz with a fractional bandwidth (FBW) of more than 10% for an in-band return loss of better than 20 dB and an low insertion loss of less than 1.35 dB has been obtained, which is not feasible with a usual Si interposer approach. Measurement results are presented from 2 to 10 GHz and show good agreement with simulated ones.

Introduction

The through-glass interposer (TGI) technology rapidly grows up as a promising alternative to the through-silicon interposer (TSI) because of its low substrate loss in the RF/microwave range, the mechanical robustness and low material and manufacturing cost [1-3]. Also, recent advancement on the corning fusion process for pristine surface glass substrates and through glass via (TGV) processes including wet and dry etching, laser drilling, and W-plug [1], have made the glass interposer much viable in the market. Previously, we have reported the high frequency characterization of Corning glass using a ring resonator, as well as the modeling of high frequency TGV using Corning glasses [4]. High quality factor (Q-factor) radiofrequency (RF) performance of the resonator has implicated the glass interposer can be a good hosting medium for high quality RF circuits, i.e. bandpass filters (BPF), supporting modern devices required for system on package (SoP) and system on chip (SoC) technologies [5,6].

It is known that during the years BPFs for wireless systems have used the waveguide, microstrip and coplanar waveguide technologies for their unit resonator implementation in a great variety of planar and non-planar forms [7-9]. On the other hand, integrated passive devices technologies (IPD) on glass and silicon substrates have been recently exploited for the design of high quality RF devices, which use conventional transmission line resonators or high quality semi-lumped components [10,11]. To achieve a higher level of integration, substrate integrated devices are desirable [12-13]. During the last decade, the substrate integrated waveguide (SIW) technology has become a good candidate to implement wireless devices for SoP and SoC platforms [13-14]. Substrate integrated waveguide (SIW) is a planar-type waveguide architecture with two metallic plates on the upper and bottom sides of the substrate connected by side-walls made of arrays of metalized vias. This architecture was exploited for microwave and millimeter wave components with simple design, high quality factor with low loss substrates, high power handling capability, high integrability with other planar circuits and low manufacturing cost [12]. However, many implementations of SIW devices have been done on the conventional printed circuit board (PCB) technology or with low-temperature co-fired ceramic (LTCC) processes [13-16]. To achieve higher integrability, silicon or glass substrates are desirable, for which one of the main challenges is to create via holes in those substrates. TSV technology has been explored during the years, which however, has demonstrated poor RF performance [17]. Alternatively, glass interposer technology shows great promise with low substrate loss, which is a very appealing property for RF devices.

As the size of electronic systems is getting smaller, the miniaturization of component size is highly demanded. In order to reduce the size of SIW devices, reduced mode versions of the SIW technique have been studied, i.e. the half mode substrate integrated waveguide (HMSIW) and the quarter mode substrate integrated waveguide (QMSIW) [18-19]. Such higher order SIW devices, however, being still large in size compared with their microstrip or CPW counterparts [19], further size reduction could be realized with the inclusion of metamaterial resonators, such as the complementary split-ring resonators (CSRRs) [20] in them. The CSRR designed to have a negative effective permittivity at its resonance frequency, is combined with the SIW and HMSIW structures, to generate a forward-wave transmission band below the waveguide original cutoff frequency. The operation mode is completely different from the conventional TE_{10} waveguide mode, thanks to the principle of evanescent wave amplification [21-23].

In this work, we demonstrate the glass interposer working as a hosting medium for the high quality compact RF components as well as the conventional interconnecting layer. The implementation of high quality substrate integrated waveguide RF components using glass substrates is explored. Also broadband bandpass filter operation, instead of narrowband one, is selected to demonstrate the use of high quality glass substrate with advanced waveguiding structures. The merits of this work include the combination of the superior glass processing technology [1] and the advanced RF passive and metamaterial circuit architectures [21-23]. The Corning glass process used in this work offers important attributes such as pristine surface, low thickness variation and excellent flatness, with the ability to scale to large wafer and panel formats as well as via forming in various thicknesses in a few tens μm and millimeters, which offers great opportunities to achieve truly cost effective processes [1],[4]. Direct integration of TGVs and other passive RF components is demonstrated, whose approach produces unique values and features such as high integrability, low component loss, small device form factor, manufacturability and low cost. The low RF loss and low parasitic capacitance of the high-quality fusion glass greatly enhance the operating frequency range to a level that might not be achievable by the silicon counterpart. Also, the half mode substrate integrated waveguide (HMSIW) architecture is adopted here to show the superiority of the low loss substrate, where the TGVs are designed to form side-walls for the electromagnetic waveguide in the glass interposer substrate. As a test vehicle, an evanescent wave HMSIW resonator, a two-pole and a three-pole bandpass filters are fully demonstrated by using the half mode substrate integrated waveguide (HMSIW) loaded with a single ring CSRR (CSRR-loaded HMSIW) [22] for broadband operation covering the ISM band of 5.8 GHz. A single ring CSRR is used instead of a double ring CSRR in order to achieve a wider bandwidth. Full wave structure simulations are compared with measured results.

Proposed Resonator and Filters on Glass

In this work, Corning fusion glass substrates are used to implement the CSRR-loaded HMSIW resonator and filters. The Corning in-house fabrication process allows to create via holes in the glass substrate by using different fabrication methods, including mechanical and laser drilling [1]. The glass substrate with a thickness h of 300 μm has been previously characterized to have an electrical permittivity of 5.8 and a loss tangent of 0.0042 at 5 GHz [4]. Metallization of the ground plane, thru-hole vias and the signal layer for the proposed devices is performed with DC-sputtering and subsequent Copper (Cu) electroplating with a Cu thickness of 10 μm, which is selected to be more than five times the skin depth of copper at 5.8 GHz in order to reduce RF conductor losses [24]. Figure 1(a) shows the cross section of the substrate integrated waveguide devices on the glass substrate, where the metalized via holes define the metallic side-walls.

Figure 1(b) shows the layout of a CSRR-loaded HMSIW resonator, as previously proposed in [22]. Here, a single ring complementary split ring resonator is used. Although the use of a single ring CSRR offers a higher resonance frequency and thus a larger size than those offered by the use of a double ring CSRR, the out-of-band frequency response of the resonator is better with higher order harmonics suppressed. Metalized vias are used as the side-walls for the cavity. Via diameter is 80 μm with a center-to-center pitch of s = 160 μm.

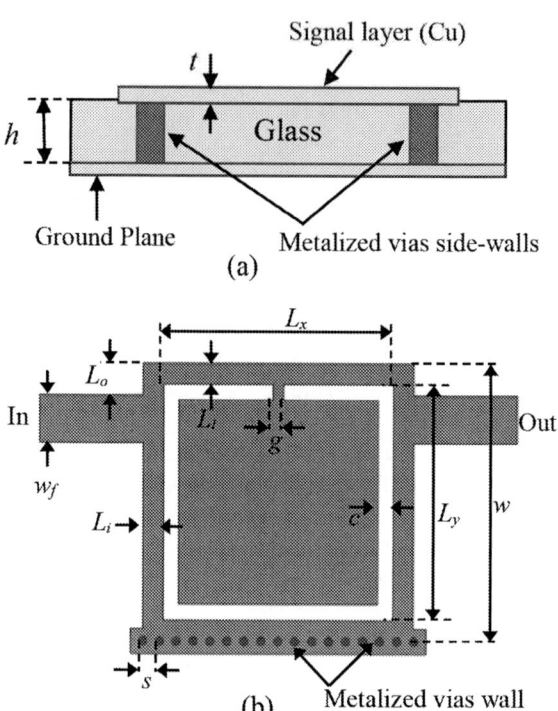

Figure 1. CSRR loaded HMSIW resonator: (a) Cross section view of the SIW architecture on glass substrate and (b) Layout of the resonator. Metal layer is shown in dark color. The diameter of the vias is 80 μm.

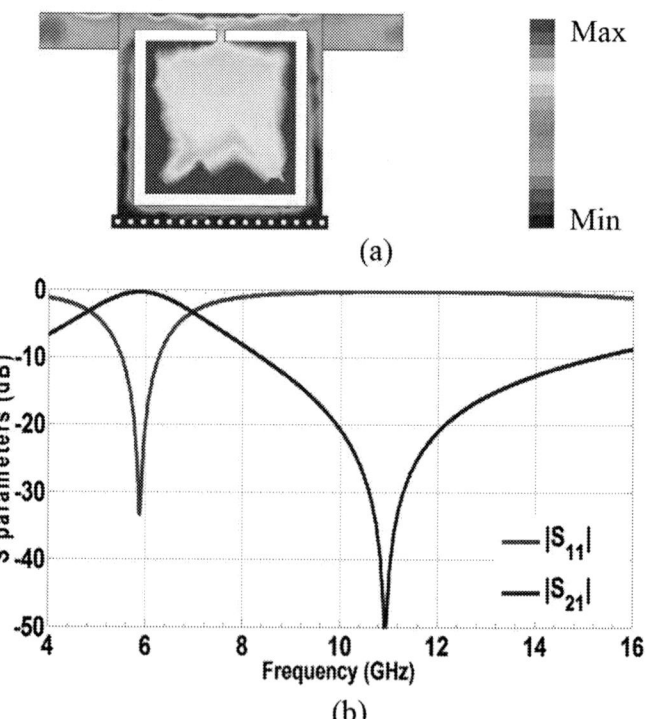

Figure 2. Simulation of the CSRR-loaded HMSIW cavity on a glass substrate designed for 5.8 GHz. (a) Electric field distribution at the resonance frequency and (b) Frequency response. Dimmensions are w = 2.62 mm, w_f = 0.45 mm, L_x = L_y = 2.22 mm, s = 160 μm, c = 0.15 mm, g = 0.1 mm, L_o = 0 mm, L_i = 0.2 mm.

978-1-4799-2408-0/14 $31.00 © 2014 IEEE

Resonator Characterization

It is important to follow the classic methodology for coupled resonator bandpass filters design [8], in which the variations of the external quality factor Q_e and the coupling coefficient of coupled resonators M are first obtained. To briefly explain the working principle of a CSRR-loaded HMSIW resonator, the layout of the resonator in Figure 1(b) is used. The effective width w of the HMSIW controls the waveguide cutoff frequency f_c [22]. As in previous work [22,23], a direct connection of a 50 Ω microstrip line is used to excite the CSRR-loaded HMSIW resonator, as observed in Figure 1(b). The external quality factor Q_e can be controlled by either the offset distance L_o or the internal offset distance L_i [22]. The geometrical parameters of the CSRR control the main resonance frequency of the resonator. The total length of the CSRR is initially selected as one half of the guided wavelength λ_g at the center frequency f_o. Since this is an evanescent wave resonator, the cutoff frequency of the waveguide is higher than the resonance frequency of the CSRR-loaded HMSIW resonator [21,22]. The CSRR loading allows having a compact resonator.

Figure 2(a) shows the electric field distribution at the resonance frequency for the single-ring CSRR-loaded HMSIW resonator used in this work. The cavity is designed and optimized to resonate at 5.8 GHz. A double loaded cavity allows the easy calculation of the external quality factor Q_e as follows [24]:

$$Q_e = \frac{2f_o}{BW_{3dB}} \qquad (1)$$

,where f_o is the resonance frequency and BW_{3dB} is the 3dB bandwidth of the resonator. Here, an external quality factor Q_e of 5.4 has been obtained from full wave 3D structure simulations, shown in Figure 2(b). Similar to the results in [22], the resonant mode shows that the electric field is confined around the CSRR. The obtained resonance frequency is 5.8 GHz, which is below the TE_{10} cutoff frequency of the original HMSIW structure (12 GHz), indicating that the working principle follows the evanescent mode amplification [20,21]. The size of the CSRR-loaded HMSIW resonator on the glass substrate is $0.122\lambda_g \times 0.122\lambda_g$, where λ_g is the guided wavelength at 5.8 GHz.

Bandpass Filter Design

As test vehicles to demonstrate the use of the CSRR-loaded HMSIW cavity, a two-pole and a three-pole Chebyshev BPFs are designed for operation at a center frequency of 5.8 GHz with a 20 dB return loss FBW of more than 10%. The coupled resonator design methodology is following to achieve the specifications [8]. The external quality factor Q_e of the double loaded resonator and the coupling coefficient M of the coupled resonators are obtained through electromagnetic simulations.

Variation of the External Quality Factor

The variation of the loaded quality factor (Q_e) of the CSRR-loaded HMSIW cavity is obtained for a doubly loaded resonator [8]. Microstrip feeding lines with a characteristic impedance of 50 Ω are used to excite the resonator, as previously shown in Figure 1(b). The distance L_o is used to vary the external quality factor, which represents the distance from

the top open side of the cavity to the top point of the feeding line. Figure 3(a) shows the available external quality factors Q_e for a variation of the distance L_o from 0 mm to 1 mm. In comparison with a conventional SIW cavity, the obtained external quality factors Q_e are smaller since the size of the CSRR-loaded HMSIW cavity is also smaller [15, 19, 21]. In this case, a Q_e as low as 5.4 has been obtained, which is not easy to get in conventional SIW and HMSIW cavities. The unloaded quality factor Q_u obtained from the simulation of a double loaded cavity is 150, which is comparable to that obtained for low Q factor cavities for broadband filters [19]. The formula used to calculate the Q_u is as follows:

$$Q_u = \frac{Q_e}{1 - |S_{21}|} \qquad (2)$$

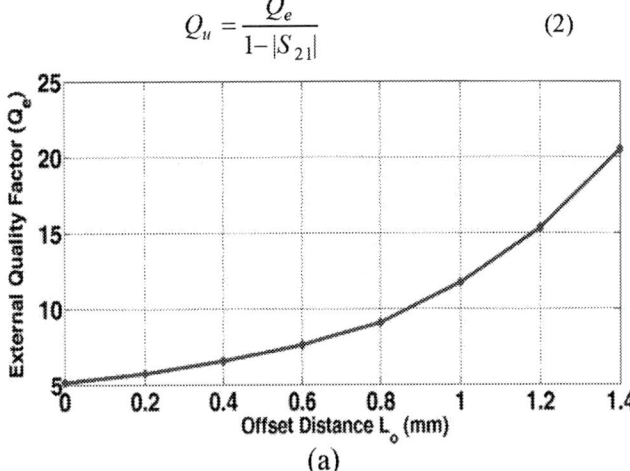

(a)

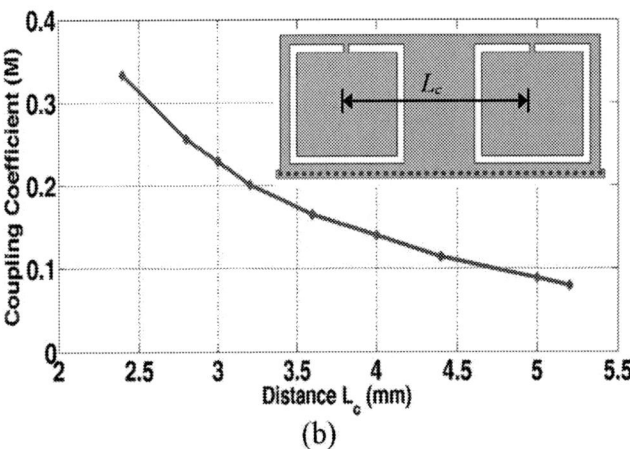

(b)

Figure 3. Characterization of the CSRR-loaded HMSIW resonator on glass: (a) Variation of the external quality factor Q_e and (b) Variation of the coupling coefficient M.

Variation of the Coupling Coefficient

The variation of the coupling coefficient M of coupled resonators is shown in Figure 3(b). The configuration in the inset of Figure 3(b) is used for simulation purposes. Two coupled resonators are excited with a high Q_e. The inter-resonator distance L_c is used to control the coupling coefficient as shown in the inset of Figure 3(b). The coupling coefficient is obtained for a variation of the distance L_c from 2.4 mm up to 5.2 mm. Further, due to the small size of the cavity, higher coupling coefficients than those obtained with conventional SIW and HMSIW cavities are possible. Then, the single ring CSRR-loaded HMSIW cavity is useful for the design of bandpass filter with broader bandwidth than those obtained with conventional SIW and HMSIW cavities [15, 29, 21].

Filter Designs

The design parameters of the two pole and three pole Chevyshev bandpass filters are shown in Table 1. Figure 4 shows the configurations of the proposed filters. Initial dimensions are obtained from the previous analysis, which are further optimized through full wave structure simulations. For testing purposes, the grounded coplanar waveguide (GCPW) with a ground-signal-ground pitch of 150 μm is included in the design. The length L_x of the CSRR of the second resonator in the three-pole filter is optimized to a value of 2.618 mm to fulfill frequency requirements, as shown in Figure 4(b). It is observed that the final dimensions of the resonators are not the same as the original dimensions used in the resonator characterization, since optimizations have been done in order to compensate frequency shift due to the interaction of the resonators.

TABLE 1. Design parameters of the filters on glass

Filter	f_0 (GHz)	20 dB RL FBW	Q_e	M_{12}	M_{23}
Two poles	5.8	12%	5.8	0.1994	-
Three poles	5.8	14%	6.083	0.1444	0.1444

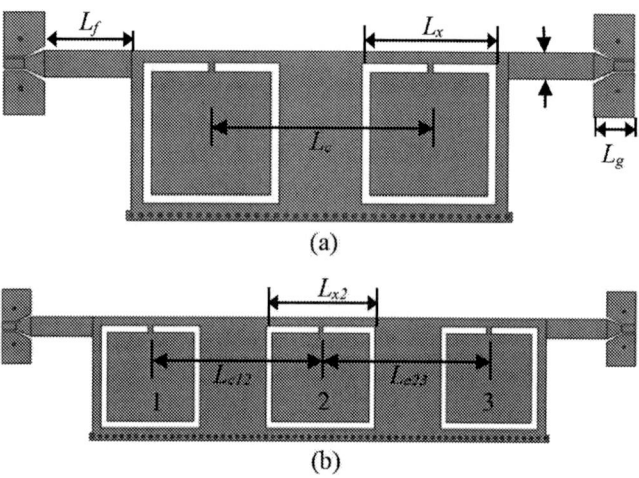

(a)

(b)

Figure 4. Detailed layouts of the realized filters: (a) Two-pole filter. Dimensions are $w = 2.76$ mm, $L_c = 3.76$ mm, $L_x = L_y = 2.36$ mm, $L_o = 0$ mm, $L_f = 1.5$ mm, $s = 160$ μm, $L_g = 0.7$ mm and (b) Three pole filter. Dimensions are $w = 2.78$ mm, $L_x = L_y = 2.38$ mm, $L_{x2} = 2.618$ mm, $L_{c12} = L_{c23} = 4.1$ mm, $L_o = 0$ mm.

Fabrication

The glass substrate is fabricated by the Corning fusion process which allows a pristine surface, a thin and strong glass [25]. A via is formed in glass using optimal fabrication methods such as mechanical, chemical electrical discharge and various types of laser drilling [1,25]. Figure 5 shows the front and back side SEM image of the fabricated vias. The average diameters

of the front and back side are 80.79 μm and 75.60 μm, respectively. Also, average circularity of the front and back side are 2.11 μm and 1.13 μm, respectively.

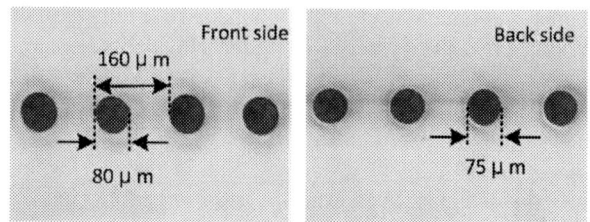

Figure 5. SEM image of the fabricated holes on the glass substrate.

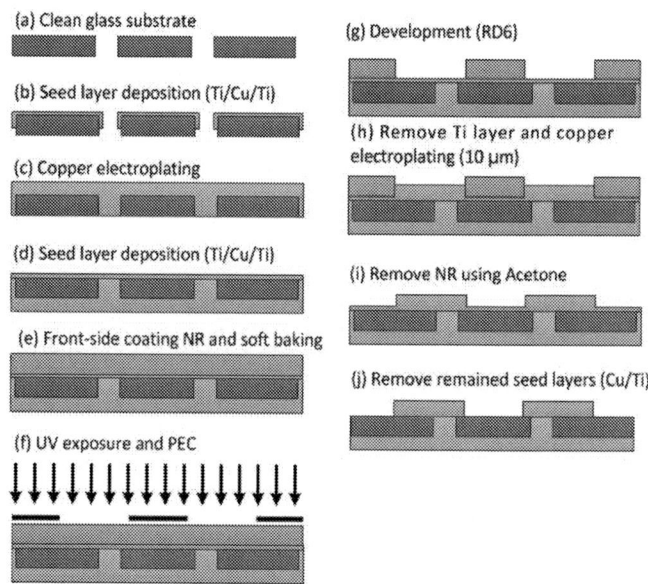

Figure 6. Fabrication process of the HMSIW bandpass filters.

The designed CSRR-loaded HMSIW resonator and bandpass filters are fabricated on a 300 μm thick glass substrate. Figure 6 shows the detailed fabrication process. On the glass substrate with 100 μm diameter holes, a thin film of titanium (Ti)/copper (Cu)/ titanium (Ti) (50 nm/500 nm/50 nm) is deposited on one side of the glass substrate as a seed layer using a sputtering system (Kurt J. Lesker CMS-18) after cleaning (b). After removing the first Ti layer, copper electroplating is performed to fill the holes up with copper (c). The thick copper layer on top of the substrate would be a ground plane. A seed layer of titanium (Ti)/copper (Cu)/ titanium (Ti) (50 nm/500 nm/50 nm) is deposited on the other side of the glass substrate for patterning the top filter layer (d). A negative ton photoresist, NR2-8000P, is spincoated on the front-side of the glass, followed by soft baking at 120 °C for 10 minutes (e). *i*-line light (365 nm) is used for exposure, and the post-exposure bake is performed at 80 °C for 5 minutes (f). After development into the development solution, RD6, the crosslinked areas remain on the substrate (g). One more copper electroplating is performed to increase the copper thickness to 17 μm (h). After removing the remained photoresist (i) and seed layers (j), the designed HMSIW bandpass filter is fabricated as shown in Figure 7

Results

Measurements are performed using a vector network analyzer (E5071C, Agilent, Inc.) after a standard two ports short-open-load-thru (SOLT) calibration between 2 GHz and 10 GHz. A Cascade Microtech probe station with a ground-signal-ground (GSG) pitch of 150 μm is used for testing. The measured and simulated results of the fabricated devices are shown in Figure 8. Figure 8(a) shows the results for the resonator. It is observed that there is a slight frequency shift. A resonance frequency of 5.63 GHz is measured with an insertion loss of 0.35 dB and a return loss of 28.7 dB. Differences with simulated results might be due to tolerance in the fabrication process and the variation of the dielectric constant of the glass substrate. A 3 dB bandwidth and external Q factor, Q_e of 1.95 GHz (4.49-6.64 GHz) and 5.76, respectively, are obtained for the resonator. An unloaded quality factor Q_u of 144 is obtained from measurements.

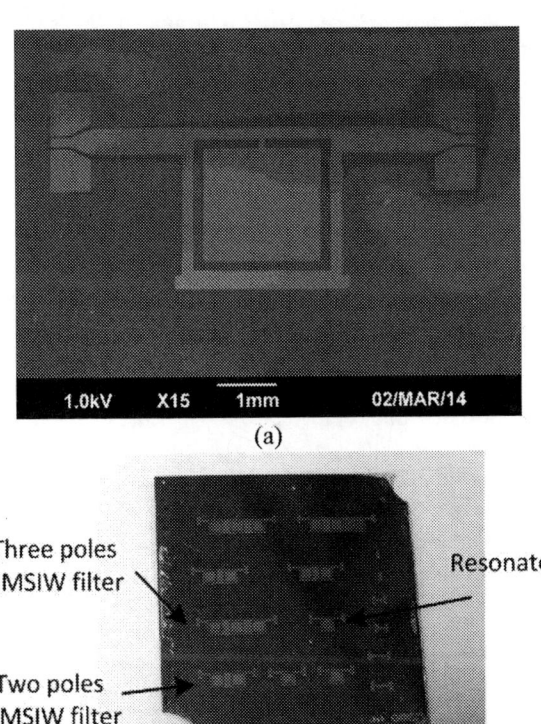

(a)

Figure 7. Fabricated CSRR-loaded HMSIW devices on glass: (a) SEM image of the resonator and (b) Optical image of the fabricated devices.

Next, the simulated and measured S-parameters of the two and three pole filters are shown in Figure 8 (b) and (c), respectively. The measured resonance frequency of the two pole filter is 5.76 GHz with an in-band insertion loss of 0.79. A 3dB bandwidth of 1.22 GHz is observed (5.15-6.37 GHz), which represents a 3 dB FBW of 21.18%. In band return loss better than 15 dB is obtained. The measured center frequency of the three pole filter is 5.81 GHz with an insertion loss of 1.35 dB. A measured 3 dB bandwidth of 1.1 GHz (5.2-6.3 GHz) is obtained, which gives a 3 dB FBW of 18.9%. The details are summarized in Table 2.

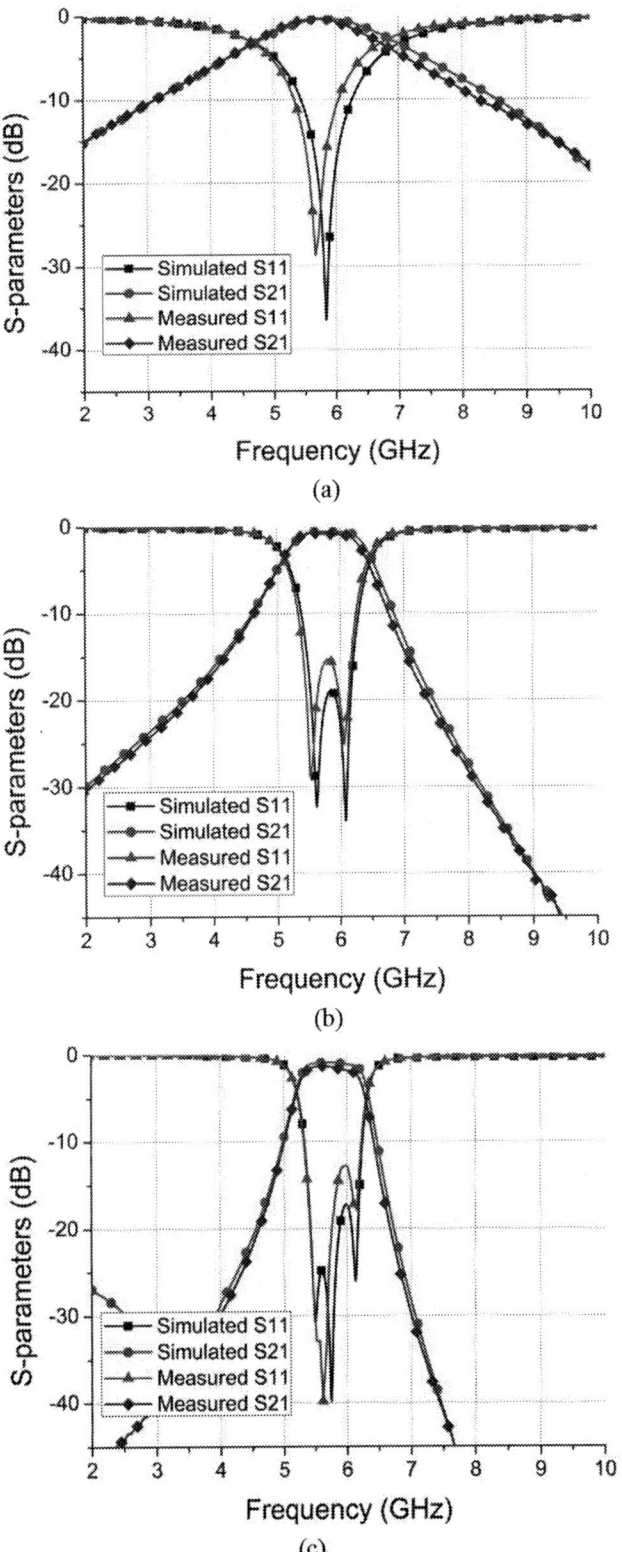

Figure 8. Simulated and measured results: (a) CSRR-loaded HMSIW resonator. Size is $0.122\lambda_g \times 0.122\ \lambda_g$, (b) Two pole filter. Size is $0.1285\lambda_g \times 0.285\ \lambda_g$ and (c) Three pole filter. Size is $0.1294\lambda_g \times 0.496\ \lambda_g$ at 5.8 GHz.

978-1-4799-2408-0/14 $31.00 © 2014 IEEE

Table 2. Summary of the results for CSRR-loaded HMSIW devices

	Resonator		Two poles		Three poles	
	Sim.	Mea.	Sim.	Mea.	Sim.	Mea.
f_0 (GHz)	5.84	5.63	5.81	5.76	5.77	5.79
BW (GHz)	2.16	1.95	1.29	1.22	1.03	0.98
Q_e	5.4	5.76				
IL (dB)	0.34	0.35	0.60	0.79	0.94	1.35
RL (dB)	36.4	28.7				

Conclusions

This work has explored the implementation of compact high quality RF devices on Corning glass substrates. Substrate integrated waveguide architectures in combination with metamaterial particles have been selected to realize compact resonators and bandpass filters working under the principle of the evanescent wave amplification. Compact devices with broadband operation are achieved with the used of the HMSIW architecture loaded with a single ring complementary split ring resonator. Thru Glass Vias (TGV) process and a surface micromachined Chevyshev BPF are demonstrated for the implementation of BPFs operating at 5.8 GHz with more than 10 % FBW 20 dB RL. Measurement results are presented from 2 to 10 GHz and show good agreement with simulated ones.

Acknowledgments

This work was supported in part by the National Science Foundation under Grant No. 1132413. David E. Senior is supported by Universidad Tecnológica de Bolívar in Cartagena, Colombia.

References

1. A. Shorey, S. Pollard, A. Streltsov, G. Piech and R. Wagner, "Development of substrates featuring through glass vias (TGV) for 3D-IC integration," *in 61st Electron. Compon. Technol. Conf. (ECTC)*, San Diego, CA, May 29-June 1, 2012.
2. V. Sukumaran, T. Bandyopadhyay, Q. Chen, N. Kumbhat, F. Liu, R. Pucha, Y. Sato, M. Watanabe, M. Kitaoka, M. ono, Y. Suzuki, C. Karoui, C. Nopper, M. Swaminathan, V. Suzuki and R. Tummala, "Design, Fabrication and characterization of low-cost glass interposers with fin-pitch through-package-vias," in *61st Electron. Compon. Technol. Conf. (ECTC)*, Lake Buena Vista, FL, May 31-June 3, 2011.
3. M. Topper, I. Ndip, R. Erxleben, L. Brusberg, N. Nissen, H. Schroder, H. Yamamoto, G. Todt and H. Reichl, "3-D thin film interposer based on TGV (Through Glass Vias): An alternative to Si-interposer," *in 60th Electron. Compon. Technol. Conf. (ECTC)*, Las Vegas, NV, June 1-4, 2010.
4. C. Kim, Y-K. Yoon, "High Frequency Characterization and Analytical Modeling of Through Glass Via (TGV) for 3D Thin-film Interposer and MEMS packaging," *the 63rd Electron. Compon. Technol. Conf. (ECTC) 2013*, Las Vegas, Nevada, USA, May 28 – May 31, 2013.
5. T.H. Teo, X. Qian, K.P. Gopalakrishnan, Y.S. Hwan, K. Haridas, C.Y. Pang; H-K. Cha, M. Je, "A 700- μW Wireless

Sensor Node SoC for Continuous Real-Time Health Monitoring," *IEEE J. of Solid-State Circuits*, vol.45, no.11, pp.2292,2299, Nov. 2010.
6. M. Alhawari, A. Khandoker, B. Mohammad, H. Saleh, K. Khalaf, M. Al-Qutayri, M.K. Yapici, S. Singh, M. Ismail, "Energy efficient system-on-chip architecture for non-invasive mobile monitoring of diabetics" *in Proc. Int. Con. on Design & Techno. of Integrated Systems in Nanoscale Era (DTIS)*, 2013 8th, vol., no., pp.180,181, 26-28 March 2013.
7. V.E. Boria and B. Gimeno, "Waveguide filters for satellites," *IEEE Microw. Mag.*, vol.8, no.5, pp.60-70, October 2007.
8. J. S. Hong and M. J. Lancaster, *Microstrip Filters for RF/Microwave Applications*. New York: Wiley, 2001, ch. 8.
9. Z.-C. Hao and J.-S. Hong, "Ultrawideband Filter Technologies", *IEEE, Microw. Mag.*, vol. 11 , no. 4, pp. 56-68, 2010.
10. Z. Wu, Y. Shim, M. Rais-Zadeh, "Miniaturized UWB Filters Integrated With Tunable Notch Filters Using a Silicon-Based Integrated Passive Device Technology," *IEEE Trans. Microw. Theory and Tech.*, vol. 60, no. 3, pp.518-527, March 2012.
11. C-Y. Hsiao, S.S.H. Hsu, D-C. Chang, "A Compact V-Band Bandpass Filter in IPD Technology," *IEEE Microw. Wireless Compon. Lett.*. vol. 21, no. 10, pp.531-533, Oct. 2011.
12. K. Wu, D. Deslandes, and Y. Cassivi, "The substrate integrated circuits- A new concept for high frequency electronics and optoelectronics," *Telecommunications in Modern Satellite Cable and Broadcasting Service Conf.*, Oct. 2003, vol. I, pp. P-III-IX.
13. W. Shen, W-Y. Yin, X-W. Sun, L-S. Wu, "Substrate-Integrated Waveguide Bandpass Filters With Planar Resonators for System-on-Package," *IEEE Trans. Comp., Packaging and Manufacturing Tech.*, vol. 3, no. 2, pp.253-261, Feb. 2013.
14. K.S. Chin, C-C. Chang, C-H. Chen, Z. Guo, D. Wang, W. Che W, "LTCC Multilayered Substrate-Integrated Waveguide Filter With Enhanced Frequency Selectivity for System-in-Package Applications," *IEEE Trans. Comp., Packaging and Manufacturing Tech.*, vol.PP, no.99, pp.1,2014.
15. X-P. Chen, K. Wu, "Substrate Integrated Waveguide Cross-Coupled Filter With Negative Coupling Structure," *IEEE Trans. Microw. Theory and Tech.*, vol. 56, no. 1, pp.142-149, Jan. 2008.
16. X.-P. Chen, K. Wu, "Self-Packaged Millimeter-Wave Substrate Integrated Waveguide Filter With Asymmetric Frequency Response," *IEEE Trans. Comp., Packaging and Manufacturing Tech.*, vol. 2, no. 5, pp.775-782, May 2012.
17. Y.P.R. Lamy, K.B. Jinesh, F. Roozeboom, D.J. Gravesteijn, W.F.A. Besling, "RF Characterization and Analytical Modelling of Through Silicon Vias and Coplanar Waveguides for 3D Integration," *IEEE Trans. Advanced Packaging,*, vol. 33, no. 4, pp.1072-1079, Nov. 2010.
18. Y. Wang, W. Hong, Y. Dong, B. Liu, H.J. Tang, J. Chen, X. Yin and K. Wu, "Half mode substrate integrated

978-1-4799-2408-0/14 $31.00 © 2014 IEEE

waveguide (HMSIW) bandpass filter," *IEEE Microw. Wireless Compon. Lett.*, vol. 17, no. 4, pp. 265-267. April 2007.

19. Z. Zhang, N. Yang and K. Wu, "5-GHz bandpass filter demonstration using quarter-mode substrate integrated waveguide cavity for wireless systems," *in Proc. IEEE Radio and Wireless Sym.*, pp.95-98, Jan. 2009.

20. F. Falcone, T. Lopetegi, J.D. Baena, R. Marques, F. Martin and M. Sorolla, "Effective negative-epsilon stopband microstrip lines based on complementary split ring resonators," *IEEE Microw. Wireless Compon. Lett.*, vol. 14, no. 14, pp.280-282, June. 2004.

21. Y. D. Dong, T. Yang, and T. Itoh, "Substrate integrated waveguide loaded by complementary split-ring resonators and its applications to miniaturized waveguide filters," *IEEE Trans. Microw. Theory Tech.*, vol. 57, no. 9, pp. 2211-2222, Sep. 2009.

22. D.E. Senior, X. Cheng, M. Machado, and Y.-K Yoon, "Single and Dual Band Bandpass Filters Using Complementary Split Ring Resonator Loaded Half Mode Substrate Integrated Waveguide," *2010 IEEE Antenna Propagation Symposium*, Toronto, Canada, July 2010.

23. D.E Senior, X. Cheng, Y.-K. Yoon, "Electrically Tunable Evanescent Mode Half-Mode Substrate-Integrated-Waveguide Resonators," IEEE *Microw. and Wireless Components Lett.*, vol. 22, no. 3, pp.123-125, March 2012.

24. D.M. Pozar, *Microwave Engineering*, 3rd ed. New York, Wiley & Sons, 2005.

25. B. K. Wang, Y.-A. Chen, A. Shorey and G. Piech, "Thin glass substrate development and integration for through glass vias (TGV) with copper (Cu) interconnections," *7th Int. Microsystem, Packaging, Assembly and Circuit Tech. Conf.*, Taipei, Thailand, 24-26 Oct. 2012.

Minimization of Keep-Out-Zone (KOZ) in 3D IC by Local Bending Stress Suppression with Low Temperature Curing Adhesive

Hisashi Kino[1], Hideto Hashiguchi[2], Yohei Sugawara[2], Seiya Tanikawa[2],
Takafumi Fukushima[3], Kangwook Lee[3], Mitsumasa Koyanagi[3], and Tetsu Tanaka[1,2]

1. Graduate School of Biomedical Engineering Tohoku University
2. Graduate School of Engineering Tohoku University
3. New Industry Creation Hatchery Center (NICHe), Tohoku University
6-6-01 Aza-Aoba, Aramaki, Aoba-ku, Sendai 980-8579, Japan
E-mail: kino@lbc.mech.tohoku.ac.jp Phone: +81-22-795-6909

Abstract

Three dimensional IC (3D IC) has lots of through-Si vias (TSVs) and metal microbumps for electrical connection between stacked IC chips, and also has organic adhesives to enhance the mechanical strength of 3D IC. However, the coefficient of thermal expansion (CTE) mismatch between microbumps and organic adhesives generate the local bending stress in thinned IC chips. Therefore, Keep-Out-Zone (KOZ) for transistors must be considered in 3D IC design to eliminate characteristic fluctuations and degradations due to the local bending stress. In this study, for the first time, we evaluated the effects of low temperature curing adhesive on both the local bending stress and the resultant transistor characteristics for decrease in KOZ of 3D IC.

Introduction

Three dimensional IC (3D IC) has attracted much attention as a promising method to enhance performance of IC [1]. 3D IC has great advantages, such as parallel processing, high packaging density, low power consumption, short global wiring length, and high-speed operation [2]-[6]. However, it has some problems to be solved for practical applications. Recently, great interests in electrical and mechanical reliability issues are increasing for production of 3D IC. Conventional 3D IC consist of five key technologies, as shown in Fig. 1 [7]-[14]. 3D ICs consist of vertically stacked several thin IC chips with lots of through-Si vias (TSVs) and metal microbumps that electrically connect each IC chips. Metal microbumps are surrounded by organic adhesive called underfill material. Chip alignment technology is required to have alignment accuracy less than 1 μm to realize high-density TSV interconnects. The thinning of the IC chips leads to a thinner IC package and a shorter TSV. A shorter TSV makes it possible to decrease parasitic capacitance, electric resistance, and process cost. However, the thinning of the IC chips lead to the low flexural rigidity of IC chips. In additionally, coefficient of thermal expansion (CTE) of the underfill material is larger than that of metal microbumps. This CTE mismatch induces local bending stress in thinned Si chips, because the underfil material shrink more than metal microbumps and the IC chips is bended by this shrinkage after 3D integration process. In addition, this local bending stress would affect MOSFET performance in thinned Si chips. One of the causes of this issue is temperature change for adhesive curing. Therefore, we proposed to use low temperature curing adhesive to suppress the temperature change.

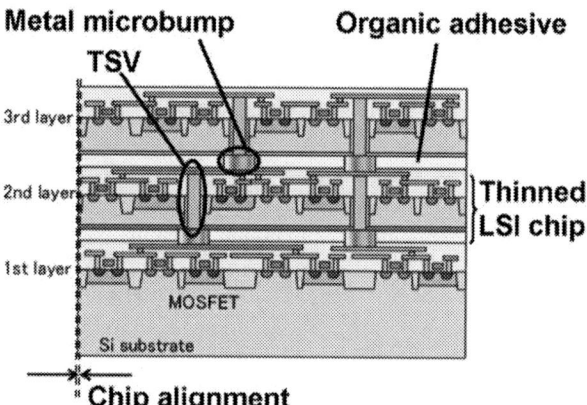

Figure 1. Cross-sectional schematic structure of 3D IC.

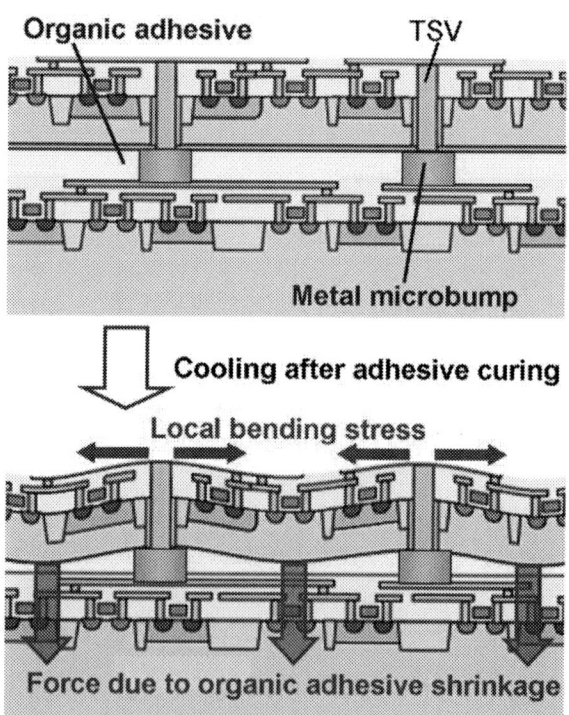

Figure. 2. Schematic cross-section of 3D IC expressing local bending stress due to the CTE mismatch between organic adhesive and metal microbumps.

In this paper, we evaluated the effect of low temperature curing adhesive on the local bending stress by measuring surface profile and change of MOSFETs characteristics including electron mobility characteristics before and after underfill process with our unique test structure.

Experimental

Figurer 3 shows a schematic cross-section of the test structure to evaluate the effects of local bending stress. As 3D IC has several issues, that affect the MOSFET characteristics, such as metal diffusion from TSVs and metal microbumps, disappearance of intrinsic gettering (IG) layers, and mechanical stress/strain caused by TSVs, it is very difficult to evaluate effects elicited only by local bending of the thinned chip on MOSFET characteristics using actual 3D IC [15]-[19]. The test structure is composed of Si (dummy) microbumps, an organic adhesive, and a thinned Si chip stacked on Si substrate. This test structure can generate the controlled local bending stress in a thinned chip on Si microbumps, and it enables us to evaluate only local bending stress due to the CTE mismatch between organic adhesive and Si microbumps, because there is no cause of other several issues. In previous works, we have demonstrated several characteristic changes of MOSFETs on thinned IC stacked on microbumps and adhesive chip with this test structure [20][21]. In this study, we proposed a method to suppress the local bending stress and evaluated the effect of the method using our unique test structure.

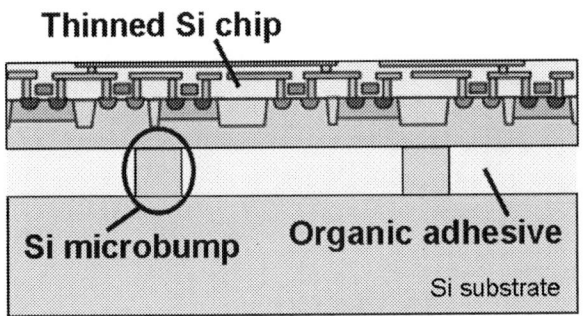

Figure. 3. Schematic image of the test structure to evaluate the effect of controlled local bending stress due to CTE mismatch between microbumps and organic adhesive.

The test structure was fabricated with following processes, as shown in Fig.4. First, Si microbumps were formed on the Si substrate by inductively coupled plasma reactive ion etching (ICP-RIE) with SF_6 as etching gas and C_4F_8 as deposition gas. The bump size and height were 20 μm by 20 μm and 20 μm. The bump pitches were 50, 100, 300, and 500 μm. Figure 5 shows a birds-eye SEM image of fabricated Si microbumps with 50-μm pitch. We could observe the Si microbus has fabricated as designed. Then, the thinned Si chips were bonded on the Si substrate which was coated with an adhesive A. In this paper, the thinned chip thickness was 35 μm. After that, two kinds of epoxy resin were additionally coated around the thinned chips depicted as the adhesive B in Fig. 4. Here, one resin was a heat curing type, and the other was a low temperature curing type. The chip and the Si substrate were temporarily exposed in vacuum atmosphere and opened to the air to completely fill gaps between the thinned chip and the substrates by the epoxy. Finally, the heat curing epoxy resin was cured at 190 degree C for 1 hour, and the low temperature epoxy resin was held at room temperature for curing. Figure 6 shows a cross-sectional SEM image of fabricated test structure after adhesive curing. It is clearly observed that no voids exits in the organic adhesive region. This voidless injection is achieved by vacuum injection shown in fig. 4. It have been demonstrated that this structure can give a tensile stress of more than 1 GPa to thinned Si chip using heat curing resin [22]. In general, the procedures after fourth step of fig. 4 were called underfill process. After underfill process, we measured the surface profile and the current-voltage and carrier mobility characteristics of nMOSFET fabricated on the locally-bended thin Si chip.

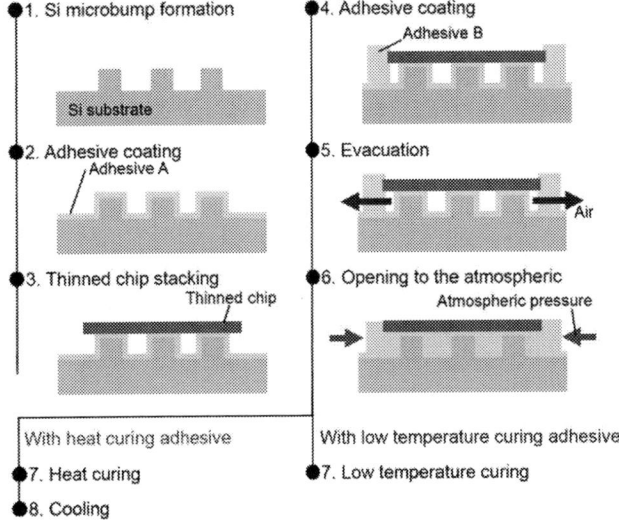

Figure. 4. Process flow of the test structure fabrication. (Underfill process consists of procedures after fourth step.)

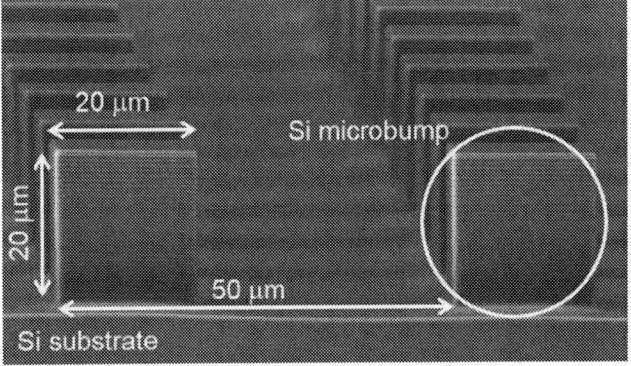

Figure. 5. Birds-eye SEM image of fabricated Si microbumps with 50-μm pitch.

978-1-4799-2408-0/14 $31.00 © 2014 IEEE

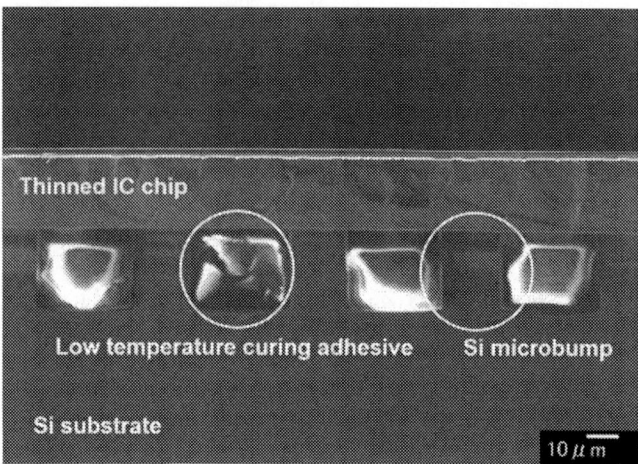

Figure. 6. Cross sectional SEM image of thinned IC chip stacked on the substrate with the Si microbumps.

Results and discussion

First, we measured the surface profile on Si chips stacked on the substrate with the Si microbumps. Figure 7 shows the schematic image and results of contact-type surface-profile measurements of thinned Si chips stacked on the substrates with the 300-μm-pitch Si microbumps. First, we could observe the 300-μm-cycle waveform. This fact indicated that Si chip is bended by cooling shrinkage of the adhesive surrounding 300-μm-pitch Si microbumps. Therefore, we should suppress this adhesive shrinkage for reduction of the local bending of Si chip. In this measurement, we used two kinds of adhesive materials. One is heat curing adhesive, and another is low temperature curing adhesive. It was clearly observed that the deflection using the low temperature curing epoxy was approximately 37% of the deflection using the heart curing adhesive, as shown in Fig. 7. Here, a deflection is a height difference between top and bottom of the surface of bended Si chip. This result demonstrated that low temperature curing adhesive reduced the local bending by adhesive shrinkage during cooling process, as shown in Fig. 8, because no temperature change exited in the underfill process with low temperature curing adhesive.

Therfore using low temperature curing adhesive can reduce a keep-out-zone (KOZ) that transistors should not be placed, as shown in Fig. 9. KOZ is defined by the area where Si strain is more than specified value. It is well known in mechanics of materials that the strain is proportional to d^2Y/dX^2. Here, Y is a displacement perpendicular to Si chip surface, and X is a displacement in parallel with Si chip surface. By using low temperature curing adhesive, d^2Y/dX^2 totally decreased in accordance with decreases of the Si-chip deflection, as shown in Fig. 9. In this case, low temperature curing adhesive reduced the local bending stress by 63%.

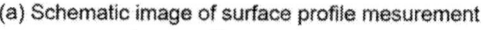
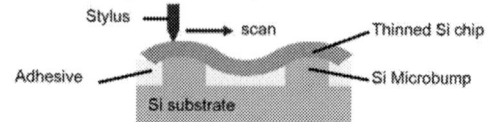
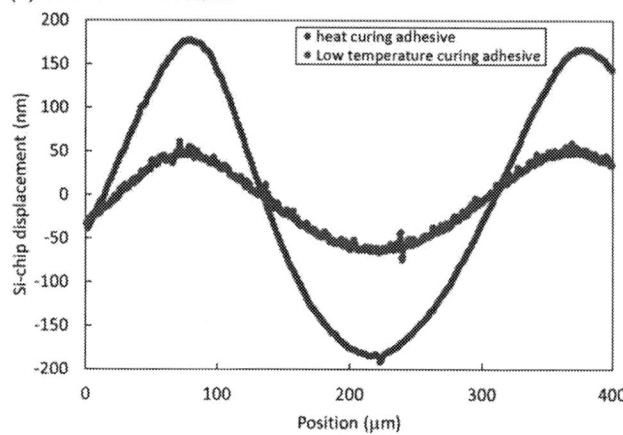

Figure. 7. Surface profile of thinned Si chip on 300-μm-pitch Si microbumps with two types of adhesives.

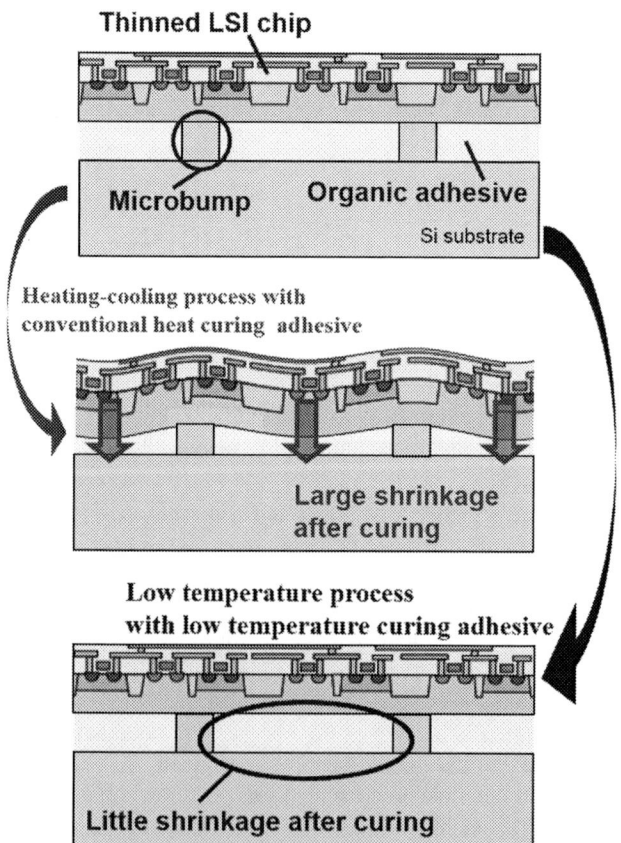

Figure. 8. Schematic image of local-bending reduction with low temperature curing adhesive.

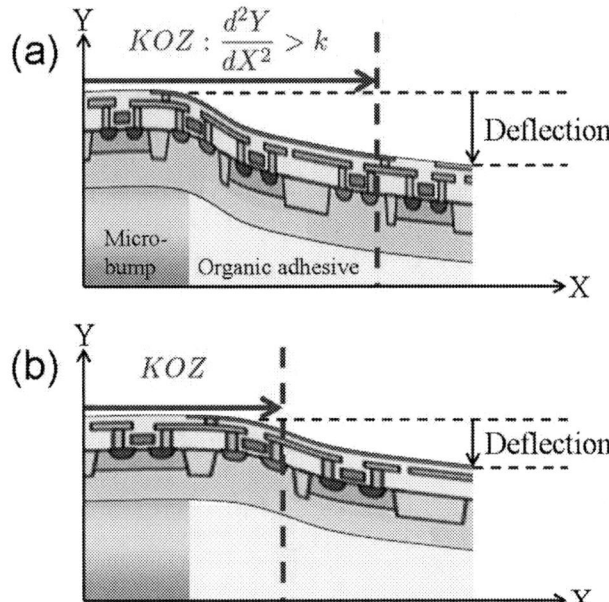

Figure. 9. KOZ around microbump with (a) heat curing adhesive and (b) low temperature curing adhesive.

Next, we evaluated the MOSFETs characteristics fabricated in the different Si chips with different epoxies. Figure 10 shows drain current (Id) – drain voltage (Vd) characteristics of nMOSFETs with Lg/Wg of 0.5 μm/10 μm in thinned Si chips before and after underfill process. Id-Vd characteristics are measured at gate voltage (Vg) of 0, 0.6, 1.2, and 1.8 V with Agilent B1500. The nMOSFET was located at 15-μm away from the center of the nearest Si microbump, as shown in Fig. 10(a). With the heat curing adhesive, the drain current increased by 6.7% at Vd and Vg of 1.8 V by the local bending stress. The point nMOSFET was placed at was impressed tensile stress. It is well known that electron mobility increases with tensile strains in the Si (100) plane [23][24]. These results agree well with theoretical predictions. We could observe that drive current change of MOSFET was suppressed by using low temperature curing adhesive. Meanwhile, the drain current increased by 0.1% at Vd and Vg of 1.8V with the low temperature curing adhesive. These results demonstrated that low temperature curing adhesive suppressed the effect of local bending stress on MOSFET. Therefore low temperature adhesive can reduce the KOZ.

Finally, we measured the electron mobility characteristics of nMOSFETs in thinned Si chip stacked on Si microbumps and adhesive. The nMOSFET was located at 15-μm away from the center of the nearest Si microbump in common with Id-Vd measurement. Figure 11 shows the results of electrom mobility characteristics of nMOSFET. Electron mobility is measured by split-CV method [25]. Especially at an effective electrical field of 0.5 MV/cm in nMOSFETs, electron mobility increased by 13.4% and 0.14% with heat curing epoxy and low temperature curing epoxy, respectively. This increase is caused by tensile stress due to local bending of thinned Si chip. These results also indicated that the low

temperature curing adhesive was one of the most effective to suppress the chip local bending.

Therefore, the low temperature curing adhesive can reduce the KOZ in 3D IC, and enlarges the area where the transistors can be freely placed.

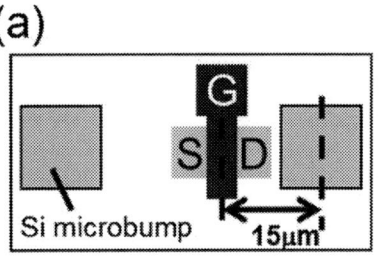

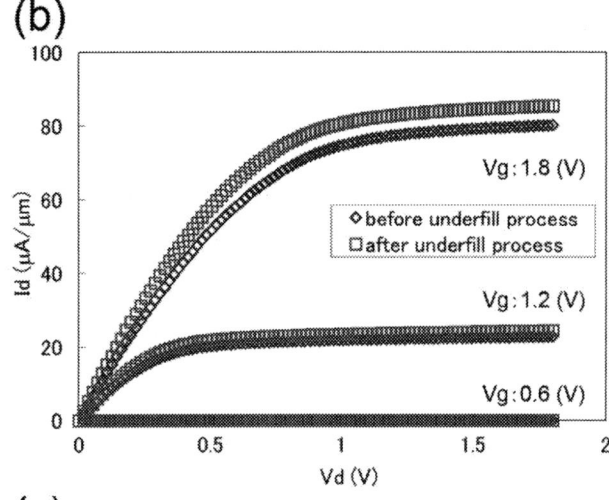

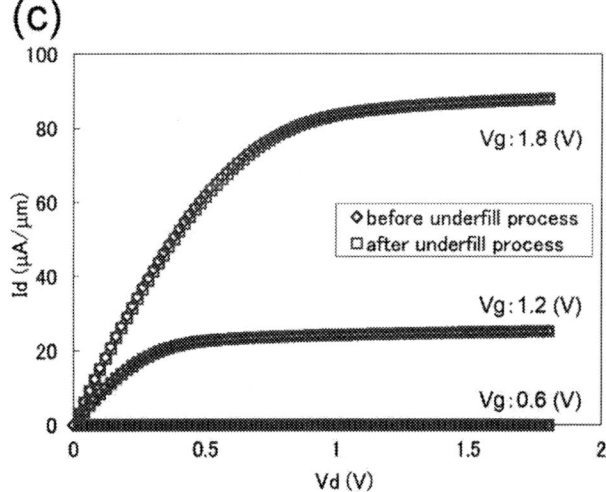

Figure. 10. (a) Layout of Si microbumps and MOSFET in thinned Si chip, and Id-Vd characteristics of nMOSFET on thinned Si chip with (b) heat curing adhesive, and (c) low temperature curing adhesive.

(a) With heat curing adhesive

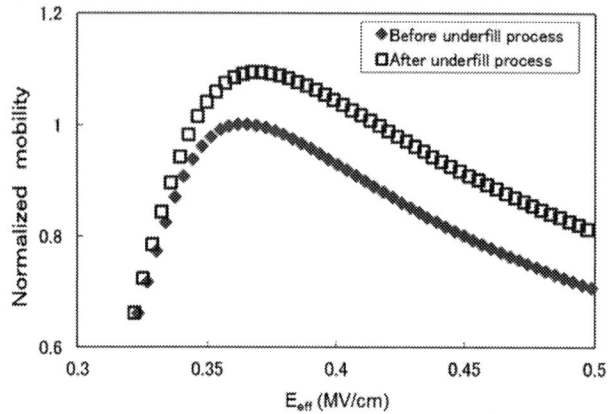

(b) With low temperature curing adhesive

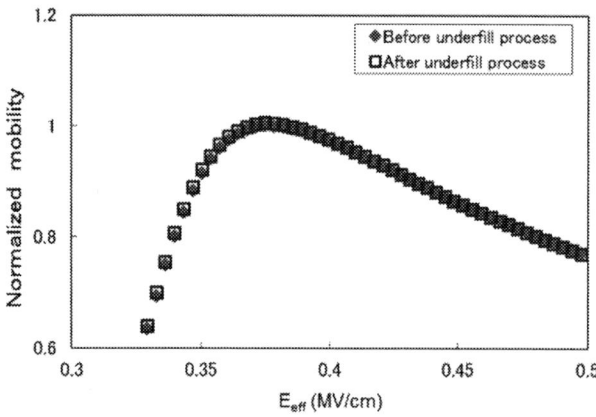

Figure. 11. Electron mobility characteristics of nMOSFET on thinned Si chip with (a) heat curing adhesive, and (b) low temperature curing adhesive.

Conclusions

The effects of local bending stress applied to the thinned Si chips were investigated in detail with the unique test structure that can generate controlled local bending stress by the CTE mismatch between organic adhesive and Si microbumps. The local bending stress affected the I-V characteristics of MOSFET. It was obviously clarified that the low temperature curing adhesive can suppress the local bending of the thinned Si chip. In order to realize higher performance 3D IC with diminished circuit performance degradation and fluctuation, careful transistor layout and process design including choice of the underfill material were strongly required.

Acknowledgments

This work was supported by JSPS Grants-in-Aid for Scientific Research Grant Number 25820133. The work is also supported by VLSI Design and Education Center (VDEC), the University of Tokyo, in collaboration with Cadence Design Systems. This work was performed in the Micro/Nano-Machining Research and Education Center at Tohoku University.

References

1. M. Koyanagi, T. Nakamura, Y. Yamada, H. Kikuchi, T. Fukushima, T. Tanaka, and H. Kurino, *IEEE Trans. Electron Devices* vol. 53 pp. 2799-2808 2006.
2. K. W. Lee, A. Noriki, K. Kiyoyama, S. Kannno, R. Kobayashi, W. C. Jeong, J. C. Bea, T. Fukushima, T. Tanaka, and M. Koyanagi, *IEDM Tech. Dig.*, p. 531-534 2009.
3. K. W. Lee, A. Noriki, K. Kiyoyama, T. Fukushima, T. Tanaka, and M. Koyanagi, *IEEE Trans. Electron Devices* vol. 58 pp. 748-757 2011
4. M. Koyanag: Proc. 8th Symp. *Future Electron Devices*, p. 55-60 1989.
5. M. Sekikawa, K. Kiyoyama, H. Hasegawa, K. Miura, T. Fukushima, S. Ikeda, T. Tanaka, H. Ohno, and M. Koyanagi, *IEDM Tech. Dig.*, pp. 935-938 2008.
6. T. Tanaka, H. Kino, R. Nakazawa, K. Kiyoyama, H. Ohno, and M. Koyanagi, *Symp. VLSI Technology Dig. Tech. Pap.*, pp. 169-170 2012.
7. H. Kino M. Murugesan, T. Kojima, T. Fukushima, T. Tanaka, and M. Koyanagi, *IEICE Trans. Electronics Jpn. Edition,* vol. J94-C pp. 411-418 2011.
8. T. Matsumoto, M. Satoh, K. Sakuma, H. Kurino, N. Miyakawa, H. Itani, and M. Koyanagi, *Ext. Abstr. Solid State Devices and Materials*, pp. 460-461 1997.
9. M. Motoyoshi, K. Kamibayashi, M. Koyanagi, and M. Bonkohara, *Tech. Dig. Int. 3D System Integration Conf.*, p. 8.1 2007.
10. T. Tanaka, and H. Kurino, *IEEE Trans. Electron Devices* vol. 53 pp. 2799-2808 2006.
11. T. Fukushima, E. Iwata, Y. Ohara, A. Noriki, K. Inamura, K. W. Lee, J. Bea, T. Tanaka, and M. Koyanagi, *IEDM Tech. Dig.*, p. 349-352 2009.
12. A. Yu, J. H. Lau, S. W. Ho, A. Kumar, W. Y. Hnin, D. Q. Yu, M. C. Jong, V. Kripesh, D. Pinjala, and D. –L. Kwong, *Proc. Electric Components and Technology Conf.*, pp. 6-10 2009.
13. Y. Ohara, A. Noriki, K. Sakuma, K.W. Lee, M. Murugesan, J. Bea, F. Yamada, T. Fukushima, T. Tanaka, and M. Koyanagi, *Tech. Dig. Int. 3D System Integration Conf.*, 2009.
14. T. Fukushima, Y. Yamada, H. Kikuchi, and M. Koyanagi: *Jpn. J. Apl. Phys.*, vol. 45 pp. 3030-3035 2006
15. M. Murugesan, H. Kino, J. C. Bea, A. Horibe, F. Yamada, C. Miyasaki, H. Kobayashi, T. Fukushima, T. Tanaka, and M. Koyanagi, *IEDM Tech. Dig.*, pp. 30-33 2010.
16. M. Murugesan, J. C. Bea, H. Kino, Y. Ohara, T. Kojima, A. Noriki, K. W. Lee, T. Fukushima, H. Nohira, T. Hattori, E. Ikenaga, T. Tanaka, and M. Koyanagi, *IEDM Tech. Dig.*, pp. 361-364 2009.
17. A. D. Trigg, L. H. Yu, C. K. Cheng, R. Kumar, D. L. Kwong, T. Ueda, T. Ishigaki, K. kang, and W. S. Yoo, *Appl. Phys. Express* vol. 3 pp. 086601-1-086601-3, 2010.
18. C. S. Selvanayagam, J. H. Lau, X. Zhang, S. Seah, K. Vidyanathan, and T. C. Chai, *IEEE Trans. Adv. Packag.* vol. 32 pp. 720-728 2009.

19. C. S. Selvanayagam, X. Zhang, R. Rajoo, and D. Pinjala, *Proc. Electric Components and Technology Conf.*, pp. 612-618 2009.

20. H. Kino, J.-C. Bea, M. Murugesan, K.-W. Lee, T. Fukushima, M. Koyanagi, and T. Tanaka, Proc. *Electric Componets and Technology Conf.*, pp. 360-365 2013.

21. H. Kino, J.-C. Bea, M. Murugesan, K.-W. Lee, T. Fukusima, M. Koyanagi, and T. Tanaka, *Jpn. J. Apl. Phys.*, vol. 52 pp. 04CB11-1-04CB11-6 2013.

22. H. Kino, J.-C. Bea, M. Murugesan, K.-W. Lee, T. Fukusima, T. Tanaka, and M. Koyanagi, *Ext. Abstr. Solid State Devices and Materials*, pp. 52-53 2011.

23. S. E. Thompson, S. Suthram, Y. Sun, G.. Sun, S. Parthasarathy, M. Chu, and T, Nishida, *IEDM Tech. Dig.*, pp. 1-4 2006.

24. M. Emam, S. Houri, D. Vanhoenacker-Janvier, and J. P. Raskin, *J. Telecommun. Inf. Tech.*, vol. 10 pp. 18-24 2009.

25. C. G. Sodini, T. W. Ekstedt, and J. L. Moll, *Solid State Electron*, Vol. 25, pp. 833-841 1982.

Effect of Thermal Annealing on TSV Cu Protrusion and Local Stress

Xiangmeng Jing[1,2], Hongwen He[1,2], Liang Ji[1], Cheng Xu[1], Kai Xue[1], Meiying Su[1,2], Chongshen Song[1,2],
Daquan Yu[1,2], Liqiang Cao[1,2], Wenqi Zhang[1], Dongkai Shangguan[1,2]
1 National Center for Advanced Packaging, 214315 Wuxi, China
2 Institute of Microelectronics, Chinese Academy of Sciences, 100029 Beijing, China
E-mail: xiangmengjing@ncap-cn.com

Abstract

Through silicon vias (TSVs) are regarded as one of the key enabling component to achieve three-dimensional (3D) integrated circuit (IC) functionality. In this paper, we present the investigation on TSV protrusion and stress at different annealing conditions tested by means of optical profiler and high efficiency micro-Raman microscopy. Finite element method is utilized to model and simulate the thermo-mechanical behavior of the TSV having a diameter of 20 μm and a depth of 120 μm under different annealing temperatures. The measured protrusion increases with annealing temperature below 400°C, and then decreases when being further annealed. The maximum measured silicon stress as a function of annealing temperature has shown similar trend to the protrusion. The pre-annealing has limited effect on protrusion, but is helpful to reduce the silicon stress.

Introduction

Stacking of ICs using three-dimensional (3D) integration technology helps in significantly reducing wiring lengths, interconnection latency, and power dissipation while enhancing performance. Through silicon vias (TSVs) are a key enabling technology for 3D IC integration which assists in the realization of highly miniaturized and complex next-generation systems [1]. Current efforts are mostly focused on the development and improvement of TSV fabrication process, efforts to study in-depth the associated reliability issues in TSVs are still limited [2-3]. Thermo-mechanical reliability is one of the main reliability concerns, because process conditions during subsequent bumping and die-stacking processes subject the wafers to repeated thermal loadings. Due to the mismatch in coefficient of thermal expansion (CTE) between copper, surrounding dielectric, and silicon substrate, thermal stresses can be developed in the silicon and interconnect structures, resulting in several reliability problems such as cracking, delamination, and voiding [4]. It is reported that the performance of transistor devices is degraded due to its sensitivity to stresses. A stress of 100 MPa can change the carrier mobility by 7% in a metal-oxide-semiconductor field-effect transistor (MOSFET) devices [5]. The area near the copper via is not preferred to put transistors and, therefore, named keep off zone or keep out zone (KOZ). In addition, the interface between the copper and the silicon substrate can be delaminated due to the shear stress concentration at the edge of the TSV. Another reliability concern is the TSV extrusion (or called TSV pumping) problem. During TSV annealing and subsequent fabrication processes, Cu TSVs were found to extrude out of the Si wafer surface, causing fracture of the overlying dielectric material. This is a potential threat to the IC interconnection layer, particularly for low-k materials, since it can lead to cracking of the dielectric layer in the BEOL structure. It was observed that, if the TSVs have not received proper heat treatment prior to backside processing, it will bulge upwards and deform the metal/dielectric stack on top. Previous studies have reported that initial heat treatment at 420°C for 20 min could solve the TSV protrusion [6]. It is analyzed that, during the first thermal cycling, the Cu via undergoes material transformation, such as grain growth and recrystallization, thus resulting in an open hysteresis loop. The succeeding cycles after the first cycle are found to have the loops that are similar in shape, size, and position. This means that the material behavior of Cu becomes stable after the first cycle to the maximum thermal cycling condition [7-8]. Therefore, suitable heat treatment and non-destructive detecting the TSV deformation and stress are of great importance for TSV based 3D integration.

Many efforts have been paid on detecting the TSV stresses in recent years. A.S. Budiman et al has reported the stress measurement around Cu by synchrotron X-ray micro-diffraction method [9]. As the high brilliance synchrotron-sourced X-rays can penetrate silicon and copper with high resolution, strain or stress measurements can be achieved during and after the TSVs being filled. The power of this technique as a local stress probe for micro and nano scale devices stems from its two features. First, the submicron-sized focused X-ray beam that enables measurement of stresses at submicron resolution. Secondly, the continuous range of wavelengths in the white X-ray beam allows measurements of the unpredicted components of the stresses in addition to the standard ability of an X-ray diffraction technique to measure hydrostatic components of the stress tensors. This technique has been proven to be useful in the study of mechanical stresses in advanced micro-devices. However, the difficulty is that the synchrotron light source is not easily available. Then many other researchers move to intensively investigate the alternatives that measure the stresses by micro-Raman spectroscopy [10-13]. Micro-Raman is a spectroscopic technique where the stress magnitude is deduced from the frequency shift of the impinging laser light as a result of inelastic scattering by Si lattice. The lateral resolution of micro-Raman spectroscopy is in the order of 0.5 μm, depending on the laser wavelength and the substrate lattice. For silicon, the Raman penetration depth ranges up to 2 μm, again depending on the laser wavelength. Moreover, the Raman technique can be used to measure the near-surface stresses in Si around TSVs even with an oxide layer covering the wafer surface because the laser can penetrate the oxide layer with nearly 95% transparency [14]. In this paper, we present our latest results on thermal mechanical study of TSV

978-1-4799-2408-0/14 $31.00 © 2014 IEEE

from simulation approach, test vehicle preparation and characterization by white light optical profiler and high efficiency micro-Raman microscopy.

Thermo-mechanical modeling and simulation

The finite element method (FEM) is utilized to model and simulate the thermal mechanical behavior of TSV under the annealing condition. By using ANSYS software, a single TSV with 20μm in diameter and 120μm in depth, surrounded by 300 nm SiO_2 liner and (100) silicon substrate is 3D modeled. SOLID 185 is adopted as the element type and the model is sweep meshed. The materials property used in this simulation is listed in table 1.

Table 1 Materials parameters used in the simulation

Material	Young's Modulus, E (GPa)	Poisson's Ratio	Coefficient of Thermal Expansion, (ppm/ k)	Plastic Curve- stress （Mpa）vs strain
Silicon	Ex=Ey=169 Ez=130 Gyz=Gzx=79.6 Gxy=50.9	Vyz=0.36 Vzx=0.28 Vxy=0.064	2.8	-
Copper	121	0.34	17.3	121@0.001ε 186@0.004ε 217@0.01ε 234@0.02ε 248@0.04ε
SiO₂	131	0.27	3.3	-

In Fig.1 is shown the simulation result of the cross section of TSV deformation after being annealed. Considering the plasticity of the electroplated copper, the copper protrudes out and results in permanent deformation at elevated temperature due to the CTE mismatch between copper and silicon. The maximum strain occurs at the center of the copper via, and the strain decreases with the distance to the center. The minimum strain happens at the edge of the TSV.

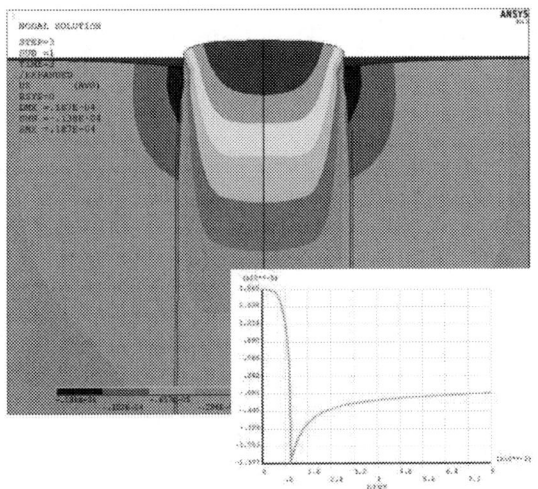

Fig.1 TSV deformation after annealing

In Fig.2 is shown the copper protrusion height as a function of annealing temperature. From room temperature to 450°C, with increasing of the annealing temperature, the copper tend to protrude more volume. The protrusion stabilizes in the range of 350°C to 400°C, and then continues increasing. It means that, if there is no proper heat treatment, the chips with TSVs working at higher temperature are more likely to be failure. Note the annealing in this simulation is after the CMP process, and the protrusion will be different when being annealed before the CMP process.

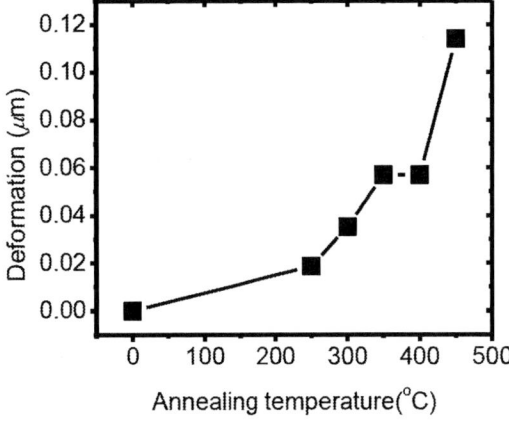

Fig.2 Cu protrusion as a function of annealing temperature

In Fig.3 is shown the top view of stress distribution on silicon surface around the copper via, considering the anisotropic property of silicon material. The maximum stress locates at the boundary between copper and silicon. The stress decreases with the distance to the copper. The area suffered by the stress is called the KOZ.

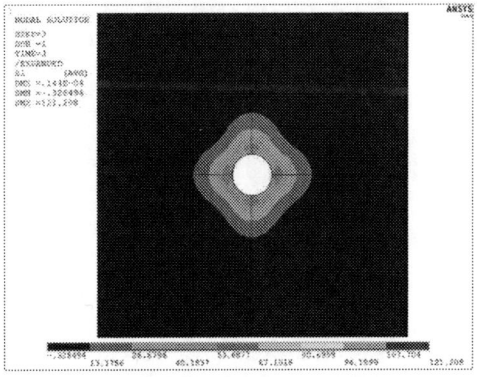

Fig.3 stress distribution on silicon surface around the TSV

In Fig. 4 is shown the maximum silicon stress as a function of annealing temperature. In the annealing temperature range from room temperature to 450°C, the maximum silicon stress is proportional to the annealing temperature. At the higher annealing temperature, the copper expanses larger and presses more heavily to the surrounding silicon. Therefore, higher annealing temperature tends to lead to larger KOZ area.

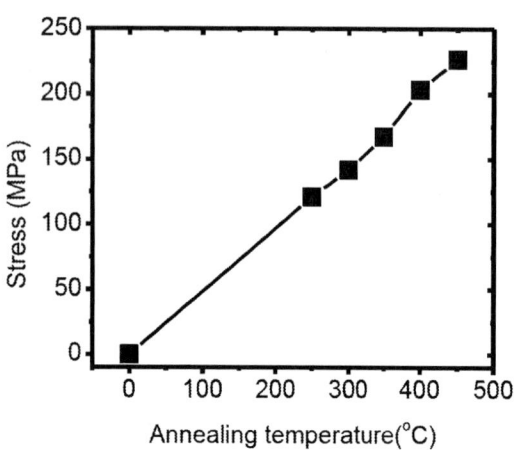

Fig. 4 maximum silicon stress around TSV as a function of annealing temperature

TSV Test Vehicle Fabrication

The test chip is fabricated by the typical TSV process flow. A 200 mm (100) silicon wafer with 670 μm in thickness is used. Firstly, a layer of 10 μm positive photoresist is spin-coated, exposed, and developed to form the via pattern on the wafer. The via diameter is defined as 20 μm with array layout in a varied pitch. Then, the exposed area is deep-etched into 120 μm by Bosch deep reactive ion etching (DRIE) process. The sidewall scallop is controlled in a range of several tens of nanometer at the fairly fast etch rate, as shown in Fig.5.

Fig.5 Deep reactive ion etching of silicon vias and their sidewall scallops

After the steps of photoresist stripping and via residue cleaning, a layer of silicon dioxide (SiO_2) is deposited by Tetraethyl orthosilicate (TEOS) chemical vapor deposition (CVD). In Fig.6 is shown the oxide layer at the via's upper corner and bottom. The step coverage for this case is 12.5%.

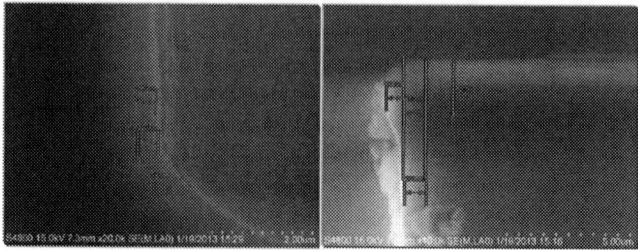

Fig.6 oxide layer at the via's upper corner and bottom

Next, a composite barrier and seed layer of Ti/Cu is coated on SiO_2 by physical vapor deposition (PVD). The via conductor is copper, which is deposited by electroplating in a copper sulfate with organic additives to facilitate complete via filling. In Fig.7 is shown the cross section of bottom up filled vias and the 3D X-ray image. All the vias are void free filled successfully.

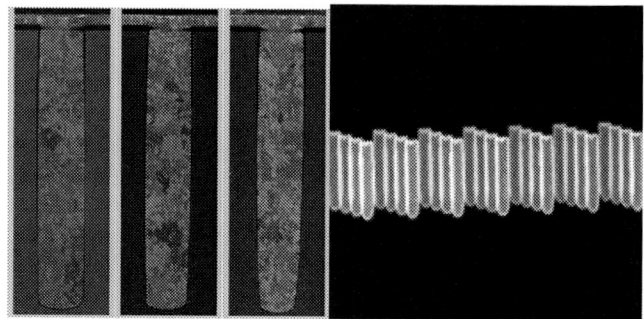

Fig. 7 cross section of filled vias and their 3D X-ray image

The overburden is removed by chemical mechanical polishing (CMP). With the slurry provided by Anji Microelectronic Inc., the TSV wafer is polished at 5 PSI down force and 50 rpm rotation speed for 30 minutes. In Fig.8 is shown an optical picture of the polished TSV array.

Fig. 8 optical picture of the polished TSV array

The wafers after CMP polish are annealed in a N_2 furnace at 250°C, 300°C, 350°C, 400°C, 450°C and 500°C, respectively, for 30 minutes. The furnace ramps up at 5 °C/min and naturally cools down. For a comparison, another pair of wafers have been pre-annealed at 300°C for 30 minutes to evaluate the effect of pre-annealing on the TSV stress and deformation. The copper protrusion and stress distribution on silicon surface are then characterized.

978-1-4799-2408-0/14 $31.00 © 2014 IEEE

Characterization and Discussion

The copper protrusion is measured by a white light optical profiler from Bruker. Cu protrusion volume and shape are changed with annealing conditions. In Fig.9 is shown a typical 3D Cu protrusion image annealed at 250°C for 30 minutes. The Cu protrusion height and shape are then measured based on the optical profiler. It also shows that the surface silicon around the copper via is deformed by the copper expansion.

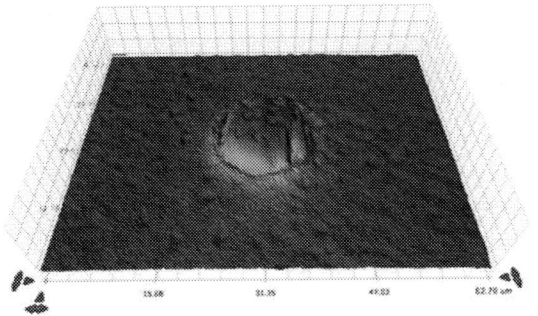

Fig.9 typical Cu protrusion annealed at 250°C for 30 min

In Fig.10 is shown the Cu protrusion height as a function of annealing temperature. The initial protrusion at room temperature is due to the deviation of planarization in the CMP process. When there is no pre-annealing treatment, the protrusion increases with annealing temperature below 400°C, and then decreases with annealing temperature. The pre-annealed wafers show the similar trend but the highest protrusion happens at 450°C. The measured protrusion results do not fully agree with the simulation, especially after 450°C. It could be explained by the grain structure of the plated copper re-constructing at higher temperature, which results in the shrinking of protrusion volume. For different copper plating conditions, the maximum protrusion temperature could be different.

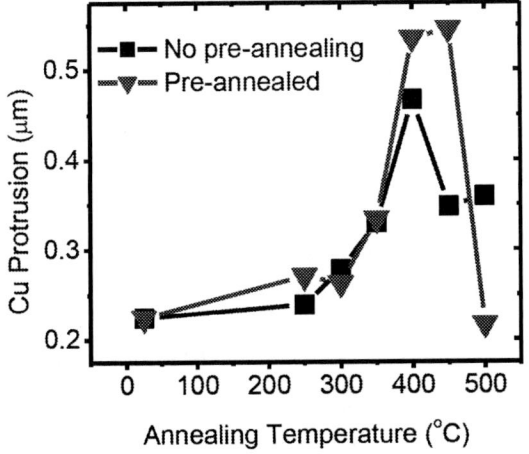

Fig.10 Cu protrusion as a function of annealing temperature

Fig.10 also reveals the pre-annealing effects on the total protrusion. Even though being annealed at 300°C for 30 minutes, TSV wafers exhibit similar protrusion no more than 350°C. It can be understand that the pre-annealing at 300°C does not influence the total protrusion for the annealing temperature lower than 350°C. When the temperature continuously heats up, the pre-annealed TSVs changes greatly. It could be explained by the grain growth and internal stress release of the plated copper at the higher temperature are more drastic when experiencing double heating treatment.

The surface silicon stress around TSV is characterized by micro-Raman microscopy. Materials suffered pressure or stress have internal strain in the crystal structure. The strain modulates frequency of Raman scattering light that is related with crystal structure. This is a feature of Raman scattering or called inelastic scattering. The Raman spectroscopy is widely employed for stress-strain study.

The peak of unstrained (100) silicon locates at $520cm^{-1}$ and the peak shift is linearly increased with the amount of strain. By measuring the amount of wavenumber shift precisely, the amount of strain can be achieved. In the (100) plane, the relation of Raman frequency shift and subjected stress (compressive or tensile) follows equation (1). Compressive uniaxial or biaxial stress will result in an increase of the Raman frequency, while tensile stress will cause a decrease.

$$\Delta\omega(cm^{-1}) = -4 \times 10^{-9}\left(\frac{\sigma_{xx} + \sigma_{yy}}{2}\right)(Pa) \qquad (1)$$

The literatures have report their measurement using conventional Raman tools. The micro-Raman system for the stress distribution imaging is highly demanded on the spectroscopy repeatability and stability. In this paper, we conduct the stress characterization by a newly developed Raman microscopy named *Nanophoton* 11, which has the advantages of higher efficiency (400 times faster) and higher spatial resolution (with smaller laser spot size), making it more suitable to be utilized in the high volume manufacturing. In Fig.11 is shown a typical stress distribution image annealed at 250°C for 30 minutes. It shows similar distribution with the simulation result that the silicon surface close to the copper via exhibits highly stressful character. Each point of the Raman image reflects the strain magnitude at the position. In Fig.12 is shown the precise wavenumber spectra at the position A and B in Fig.11. The spectra have a high signal-to-noise ratio, and the wavenumber shift can be easily detected.

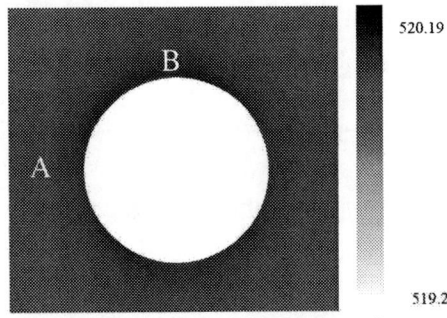

Fig.11 stress distribution by micro-Raman microscopy

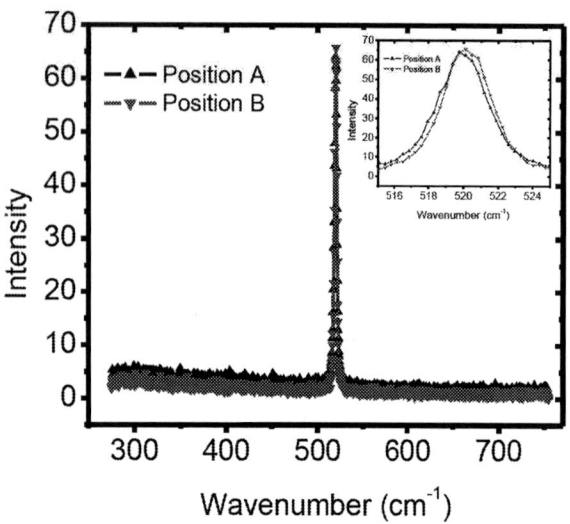

Fig.12 Raman spectra at specific positions

In Fig.13 is shown the maximum stress around TSV, measured by micro-Raman microscopy, as a function of annealing temperature. The silicon stress of the initial state after CMP is 25 MPa, which is a small stress level proving good stress management during the fabrication, mainly by the low temperature electroplating process. The overall stress level of Raman measurement is higher than, but in the same order of magnitude with, the simulation results. The maximum silicon stress increases with annealing temperature below 400°C, which is 562.5MPa, and then decreases. It is found that the pre-annealing has a positive effect on the silicon stress control. The TSVs experienced the pre-annealing show the lower stresses at all the annealing temperatures. It proves double annealing helps reducing the silicon stress around TSV.

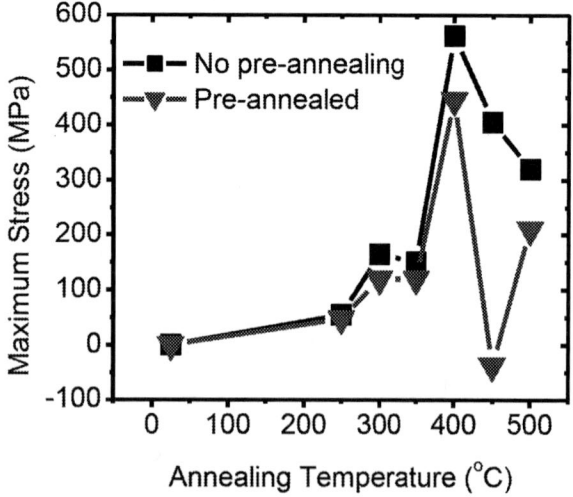

Fig.13 silicon stress as a function of annealing temperature

The silicon maximum stress curves are in accordance with the copper protrusion curves (Fig.10) that both stress and protrusion increase with annealing temperature below 400°C, and drop after that. Higher copper protrusion volume is accompany with larger compressive load to the surrounding silicon, resulting in higher stress level. Again, the measured stress trend does not fully agree with simulation result, but agrees well with the protrusion measurement. The initial stress state of the plated copper and microstructure reconstruction could be the reason for the difference.

Conclusions

The effect of thermal annealing on TSV Cu protrusion and local stress is investigated in this paper. The TSVs with 20μm in diameter and 120μm in depth on 200mm silicon wafer are annealed at 250°C, 300°C, 350°C, 400°C, 450°C and 500°C, respectively, for 30 minutes. Finite element method is utilized to model and simulate the thermo-mechanical behavior of TSV under different annealing temperatures. The copper protrusion is measured by white light optical profiler. It reveals that the protrusion increases with annealing temperature below 400°C and then decreases when being further annealed. Below 350°C, the pre-annealing at 300°C has limited effect on the copper protrusion. The silicon surface stress around TSV is characterized by a newly developed high-efficiency micro Raman tool. The maximum measured silicon stress as a function of annealing temperature has shown similar trend to the measured protrusion. The largest stress, 562.5MPa, happens at 400°C. The pre-annealing at 300°C for 30 minutes is helpful to reduce the silicon stress.

Acknowledgments

The research funding from China National Science Foundation with contract No. 61204115 and National S&T Major Project with the contract No. 2013ZX02051 is greatly acknowledged. The authors would like to thank Dr. Ruiqun Chen from *Nanophoton* for micro Raman characterization.

References

1. Suk-Kyu Ryu, Qiu Zhao, Michael Hecker, et al., "Micro-Raman spectroscopy and analysis of near-surface stresses in silicon around through-silicon vias for three-dimensional interconnects," J. Appl. Phys., 20 12, 111, 063513

2. W. S. Kwon, D. T. Alastair, K. H. Teo, et al. "Stress evolution in surrounding silicon of Cu-filled through-silicon via undergoing thermal annealing by multiwavelength micro-Raman spectroscopy," Appl. Phys. Lett., 2011, 98, 232106

3. Holger Roth, Zhenhui He, Thomas Mayer, "Inspection of Through Silicon Vias (TSV) and other Interconnections in IC packages by Computed Tomography," 11th Electronics Packaging Technology Conference, Singapore 2009

4. Sven Niese, Peter Krueger, Ehrenfrid Zscheh, "NanoXCT-A High-Resolution Technique for TSV characterization," AIP Conf. Proc. 2011, 1378, pp. 168-173

5. K. Sueoka, F. Yamada, A. Horibe, et al., "TSV Diagnostics by X-ray Microscopy," 13th Electronics

978-1-4799-2408-0/14 $31.00 © 2014 IEEE

Packaging Technology Conference, Singapore 2011, pp. 695-698

6. Holger Roth, Tobias Neubrand, "Non-destructive Inspection of through Mould vias in Stacked Embedded Packages by Micro-CT," 13th Electronics Packaging Technology Conference, Singapore 2011, pp. 710-713

7. L. W. Kong, J. R. Lloyd, K. B Yeap, et al, "Applying x-ray microscopy and finite element modeling to identify the mechanism of stress-assisted void growth in through-silicon vias," JOURNAL OF APPLIED PHYSICS, 110, (2011), pp. 053502

8. Holger Roth, Tobias Neubrand, Thomas Mayer, "Improved Inspection of Miniaturised Interconnections by Digital X-ray Inspection and Computed Tomography," 12th Electronics Packaging Technology Conference, Singapore 2010, pp.441-444

9. V. N. Sekhar1, Sam Neo, Li Hong Yu, et al, "Non-Destructive Testing of a High Dense Small Dimension Through Silicon Via (TSV) Array Structures by Using 3D X-ray Computed Tomography Method (CT scan)," 12th Electronics Packaging Technology Conference, Singapore 2010, pp.462-466

10. LayWai Kong, Peter Krueger, Ehrenfried Zschech, et al, "Sub-imaging Techniques For 3D-Interconnects On Bonded Wafer Pairs," 11th International Workshop on Stress-Induced Phenomena in Metallization, Dresden/Bad Schandau 2010, pp. 221-228

11. Heim, S.; Friedrich, D.; Guttmann, P. et al, "Dynamical X-ray Microscopy Study of Stress-Induced Voiding in Cu Interconnects," 10th International Workshop on Stress-Induced Phenomena in Metallization, Austin 2008, pp. 20-30

12. Christian Weinekoetter, "X-ray Nanofocus CT: Visualising of Internal 3D-Structures with Submicromiter Resolution," An International Conference on the Applications of Computerized Tomography," AIP Conference Proceedings, Vol. 1050, pp. 3-14 (2008)

13. Banqiu Wu, Ajay Kumar, Sesh Ramaswami, 3D IC Stacking Technology, McGraw-Hill (2011), pp.309-349

14. Philip Garrou, Christopher Bower, Peter Ramm, Handbook of 3D Integration: Technology and Applications of 3D Integrated Circuits, WILEY-CVH (2007), pp. 133-15

Effect of High Temperature Storage on the Stress and Reliability of 3D Stacked Chip

Tengfei Jiang[1], Chenglin Wu[2], Peng Su[3], Pierre Chia[3], Li Li[3], Ho-Young Son[4], Min-Suk Suh[4], Nam-Seog Kim[4], Jay Im[1], Rui Huang[2], and Paul S. Ho[1]

[1]*Microelectronics Research Center, The University of Texas, Austin, TX 78712*
[2]*Dept. of Aerospace Engineering and Engineering Mechanics, The University of Texas, Austin, TX 78712*
[3]*Cisco Systems Inc., San Jose, CA 95134*
[4]*SK Hynix Inc., Icheon-si, Gyeonggi-do, Korea*
Email: jiangt@mail.utexas.edu

Abstract

In this work, the effect of high temperature storage (HTS) on the stress in and around Cu TSVs in 3D stacked chips is studied by scanning white beam x-ray microdiffraction. The x-ray microdiffraction measurements were conducted on different die levels in the stacked chips before and after HTS test. High resolution mappings of stress distribution were obtained and compared between pre-HTS and post-HTS for both the Cu via and the surrounding Si. The x-ray microdiffraction technique provides a means for non-destructive, direct stress measurement in a 3D die stack structure. Finite element analysis (FEA) was carried out for the test structure to interpret the measurement results and to discuss the thermal aging effect on the 3D chip. Overall, the results show reduced stress in both Cu and Si after HTS, which can be explained by stress relaxation occurred during HTS. The implication of the HTS results on long term reliability of 3D die stacks is discussed.

Introduction

Three-dimensional (3D) integration with multiple dies stacked and interconnected by Cu through-silicon vias (TSVs) has the potential to overcome the wiring limit of interconnect structures for better performance with smaller form factors [1,2]. One challenge for 3D integration is the thermo-mechanical reliability of the die stacks, where significant thermal stress can be induced due to the thermal expansion mismatch between the Cu TSV and Si, as well as the mismatch between the packaging materials and the thin chips. The thermal stress can cause reliability issues such as via pop-up and degradation of device performance [3-6]. For development of 3D process integration, it is important to characterize and understand the effects of thermal stress in the 3D stacked chips. In addition, for 3D product qualifications, it is important to study the response of residual stresses after high temperature storage (HTS) for long-term reliability assessment.

Experimentally, a number of techniques have been used for stress measurement in TSV structures, including micro-Raman spectroscopy and wafer curvature method [7-9]. More recently, synchrotron x-ray based microdiffraction technique has emerged as a useful method for direct, non-invasive stress measurement in 3D structures. In the x-ray microdiffraction technique, highly focused x-ray microbeam is used to perform high resolution scanning in the area of interest. This technique has the unique capability to probe the stress characteristics of the Cu vias and Si in 3D structures with submicron resolution and minimal sample preparation is required [10].

In this study, the thermal stress behaviors of Cu and Si in and around the TSVs for a multi-level die stack structure are characterized using the x-ray microdiffraction technique. The measurement was conducted for different die levels in the 3D package before and after HTS tests. Together with FEA modeling, the effect of HTS on stress in the die stacks is discussed.

Test Vehicle

The test structure used in this study was fully processed 3D chips containing 5 thin dies. On each die, the Cu TSVs were formed using standard via-middle scheme following the formation of FEOL devices. The fabrication of the TSVs involved deep etching of Si, barrier and seed layer deposition, via filling by electroplating, post-electroplating annealing, and CMP removal of overburden. Afterwards, the BEOL layers were deposited, followed by front side bumping, TSV reveal and back side bumping. Then thin wafer was diced and stacked on top of each other by thermo-compression bonding. The via diameter was 8μm and thickness of each die was 50μm. The thin dies were connected by Sn based microbumps which were spaced 50μm apart. Finally, the chip stack was packaged by encapsulating with epoxy molding compound (EMC) and mounting on a PCB substrate. In this paper, the die immediately connected to the substrate will be referred to as the "bottom die", or "tier 1". The die furthest away from the substrate will be called the "top die", or "tier 5". The test structure is schematically illustrated in Figure 1.

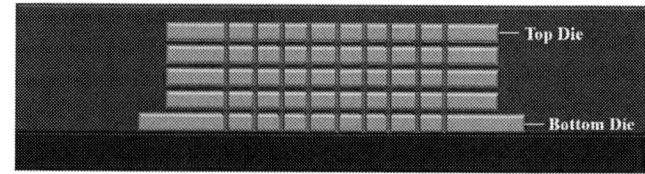

Figure 1. Schematic of the test structure.

High Temperature Storage (HTS) Test

In order to study the effect of long-time storage at elevated temperatures on the stress and reliability of the 3D package, HTS test was conducted on the samples in a chamber held at 150°C for 450 hours. The x-ray microdiffraction of the test vehicle was conducted in its as-received condition and after HTS.

978-1-4799-2408-0/14 $31.00 © 2014 IEEE

X-ray Microdiffraction Measurements

The synchrotron-based x-ray microdiffraction measurements were carried out at beamline 12.3.2 at the Advanced Light Source (ALS), the Lawrence Berkeley National Laboratory (LBNL). The sample was diced and polished, with the polished surface parallel to a row of TSV/microbump which was aligned to the <110> direction in Si. The polishing was stopped very close to, but not cutting into, the Cu vias to avoid polishing-induced stresses in Cu.

For the measurement, the sample was mounted on a high precision stage in a 45° reflection geometry. The white beam contained x-ray with energies between 5keV and 22keV, and the beam was focused to 1 μm in size. The sample was raster scanned by the focused beam at 1 μm/step with 1 second exposure time. The Laue reflection from each point in the scan matrix was collected by a large area DECTRIS Pilatus pixel array detector. Details of the measurement setup can be found elsewhere [10,11].

Since the vias were buried in Si underneath the polished surface, the location of the vias was identified from the intensity map by synchrotron-based scanning x-ray microfluoresence (μSXRF), as shown in Figure 2. Guided by the XRF intensity maps, white beam scanning was conducted around vias in the top die and the bottom die. The scanned area covered not only the TSVs but the surrounding Si and the microbumps.

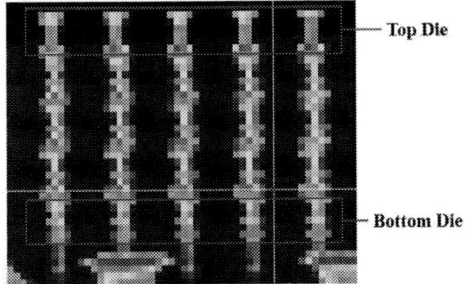

Figure 2. XRF intensity map showing five die levels in the 3D die stacks.

Results from the White Beam Measurement

The diffraction of the white beam x-ray generated Laue patterns from each point in the scan matrix. The Laue patterns were indexed with respect to Cu, Si, and Sn, using an analysis routine developed at the Beamline [11]. Several important material properties can be obtained from the analysis of the Laue patterns. By indexation of the Laue patterns, the out-of-plane grain orientation, i.e. the angle between the [001] crystal direction and the normal of the sample surface can be obtained. The deviatoric strain tensor of the material can be calculated from the shift of the reflected spots in the Laue pattern [11]. The deviatoric strain tensor, ε'_{ij} is defined as

$$\varepsilon'_{ij} = \varepsilon_{ij} - \Delta,$$ where ε_{ij} is the full strain tensor and Δ is the dilatational strain tensor. For a material with cubic symmetry such as Si, the deviatoric strain was converted to deviatoric

stress by Hooke's law of linear elasticity. The shapes of the diffraction peak can be analyzed to reveal the presence of plasticity [12].

In the following discussion, the microdiffractrion results before and after HTS for the Cu vias, the Si surrounding the TSVs, and the Sn in the microbump area will be presented.

Cu TSV

In Figure 3, the grain orientation of Cu is plotted from measurements of vias in the top and bottom dies before and after HTS. Note that, because of the large number of vias in the test structure, the measurement before and after HTS was not conducted on the same via. The results showed that the grain orientation in the Cu via was random, consistent with microstructures reported in previous studies [9]. Before and after HTS, there seemed to be no significant change in the grain structure, both in terms of the grain size and the randomness of the grain orientation. This suggests that the grain structure of Cu was rather stable, which is to be expected since the grain structure was stabilized after post-plating annealing which is normally carried out at a much higher temperature than that used in the HTS test.

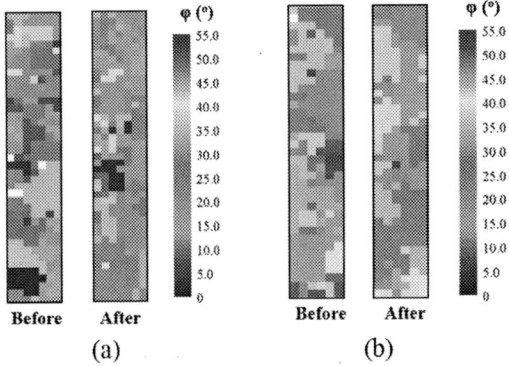

Figure 3. Out-of-plane Cu grain orientation in TSV before and after HTS for (a) top die and (b) bottom die.

The presence of geometrically necessary dislocation (GND) in Cu can cause peak broadening of the Laue reflections [12,13]. The average peak width (APW) obtained in the data analysis provides a measure of the dislocation density and also serves as an indication of plasticity [12,13]. In Figure 4, the APW is plotted for vias in the top die and bottom die before and after HTS. Before HTS, a large APW can be seen near the end of vias, indicating local plasticity in that area. The location of the large APW corresponded to location of stress concentration which caused plasticity in Cu. However, after HTS, the APW in the Cu via was noticeably decreased in both the top and bottom dies. This suggests that qualitatively, the dislocation density in Cu was considerably reduced after the sample being held at an elevated temperature for a prolonged period of time. In comparison, the APW in the bottom die after HTS was slightly larger near the upper end where the TSV is connected to the microbump.

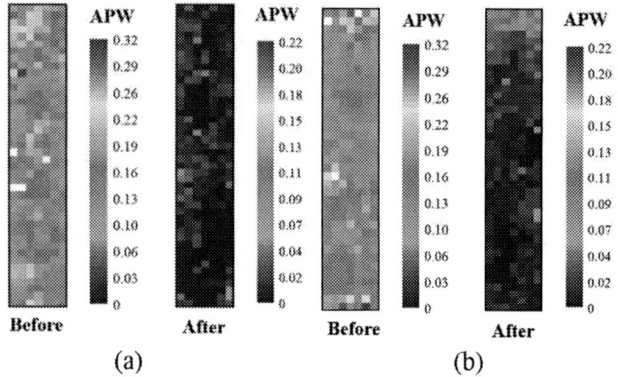

Figure 4. Average peak width (APW) of Cu in the (a) top die and (b) bottom die before and after HTS.

Silicon

For Si, a change of the out-of-plane orientation around the top of the via was observed for the top die. In Figure 5, the relative change of the lattice orientation, $\Delta\varphi$, is plotted, which was obtained by subtracting the Si orientation far away from the via where the stress was negligible. The change of Si lattice orientation indicated deformation of Si around the TSV due to bending of the Si lattice planes. Because of the geometry of the TSV structure, the bending was symmetric around the via axial direction, resulting in the opposite sign of $\Delta\varphi$ on the two sides of the via. The bending was more pronounced near the top of the via in the upper die where the chip was in contact with the molding compound. After HTS, the degree of Si lattice bending became less in the top die, suggesting a reduction in deformation. In comparison, the bending was much smaller in the bottom die both before and after HTS.

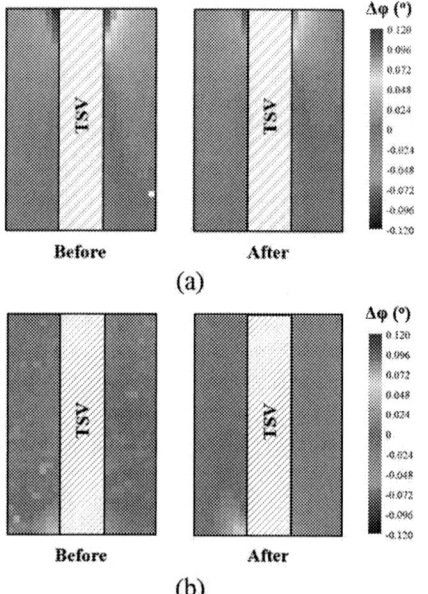

Figure 5. Relative change of the out-of-plane orientation of Si in the (a) top die and (b) bottom die before and after HTS.

The six deviatoric stress components in Si are plotted in Figure 6 for both the top and bottom dies. Overall, the Si in the top die showed higher stresses than those in the bottom die. The stress distribution was not uniform showing concentration near the Cu via. After HTS test, the sample showed a similar stress behavior, although the magnitude of stress in the top die became smaller. The deviatoric stress components were small to begin with in the bottom die before HTS, and showed no significant change after the HTS test.

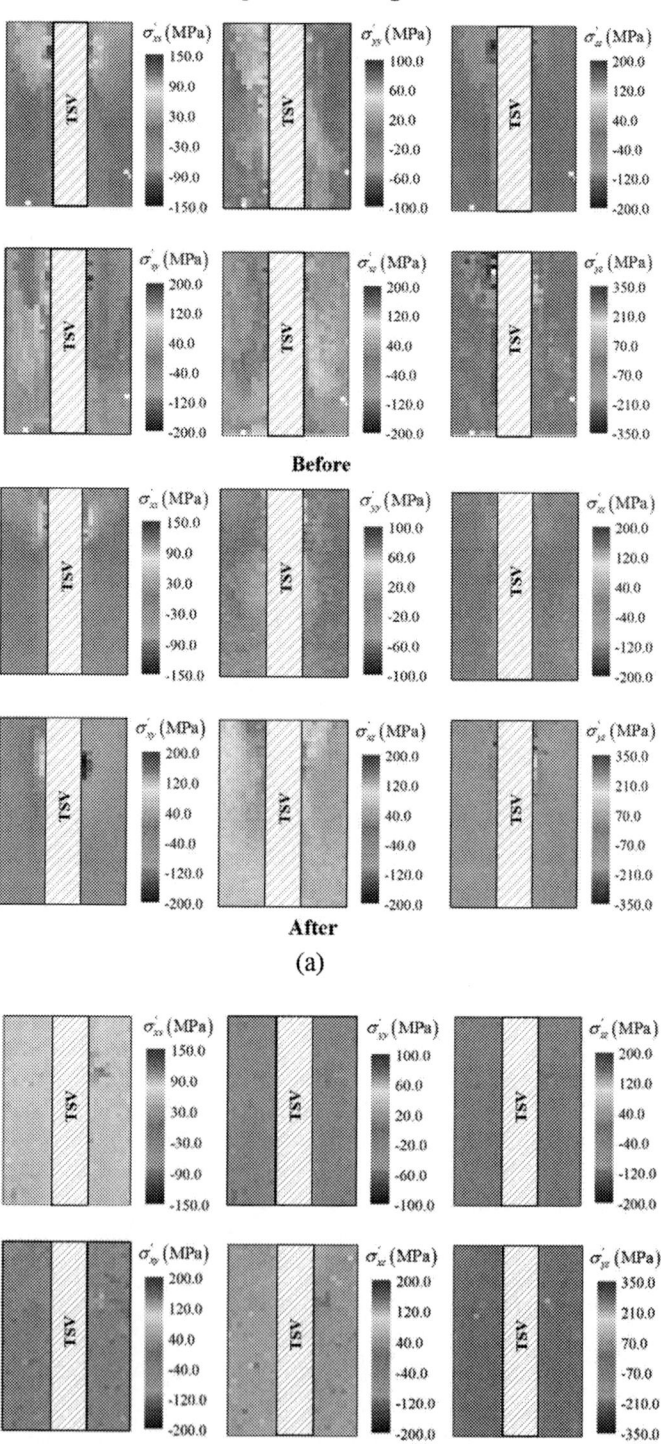

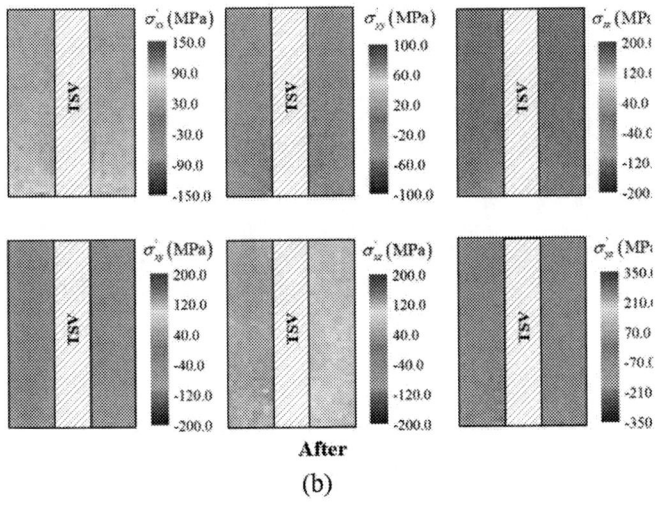

Figure 6. Deviatoric stress of Si for the (a) top die and (b) bottom die before and after HTS.

Microbump

Sn-based fine-pitch microbumps were used in the test structure to connect the die stacks. The white beam scans were indexed for Sn and shown in Figure 7. Before HTS, the microbumps in the top die and bottom die contained two large Sn grains. After HTS, part of the microbump region could not be indexed with respect to Sn. This suggests that Sn in the microbump has been consumed, most likely by conversion to intermetallic compound (IMC) during HTS.

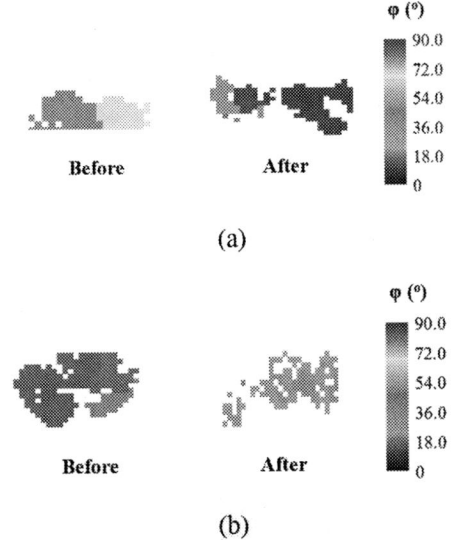

Figure 7. Out-of-plane grain orientation of Sn in (a) the top die and (b) the bottom die before and after HTS.

Discussion

Experimental Observations

The results from the x-ray microdiffraction measurements can be summarized as follows for the Cu TSV, the surrounding Si, and the microbump area before and after

HTS. (1) The grain sizes in the Cu vias for both the top die and the bottom die showed no obvious change after HTS, and the randomness of Cu grain orientation was not affected by HTS. (2) Peak broadening of the Laue diffractions, i.e. an increase in APW, was observed in the Cu via, but much reduced after HTS, both for the top die and bottom die. (3) In Si, a lattice bending due to deformation was observed in the top die near the region where the chip was in contact with the EMC. After HTS, the lattice bending became smaller. In the bottom die, the deformation was small both before and after HTS. (4) The deviatoric stress components were larger in the top die than in the bottom die, both before and after HTS. After HTS test, a reduction of stress could be observed for the top die. (5) In its initial state, the microbump contained a few large Sn grains. After HTS, the indexation of Sn became difficult, and the Sn area became much smaller, especially near the middle of the microbump, where the microbump was aligned with the TSV.

Effect of HTS

When the 3D package was held for a prolonged time for 450 hour at an elevated temperature of 150°C, the following processes were possible. First, stress relaxation in Cu occurred after the via-middle process and subsequently through several thermal processes involving heating and cooling from elevated temperatures, such as the BEOL, the die stacking and packaging processes. The CTE mismatch between Cu and Si, as well as the overall stress in the 3D package could induce stress and deformation in Cu and Si surrounding the TSV. The stress could cause plastic deformation and increase the stored energy in Cu. Microscopically, the stress which the via was subjected to resulted in an increased dislocation density and defect concentration. This was observed as an increase in APW in Cu in the as-received sample. During HTS test, Cu atoms could undergo diffusional creep to reduce the defect density and dislocation density, where a rearrangement of dislocation configurations could also happen [14]. Such effects will result in a reduction of the APW observed in the Cu via after HTS. Grain growth was also a possible mechanism to reduce stress in Cu, but grain growth typically requires a higher temperature than that in the HTS test. No obvious change in grain size was observed.

The deformation of Si was manifested as lattice bending, indicating the presence of large residual stress near the Cu/Si interface. The residual stress might be caused by the thermal mismatch between Si and Cu, as well as by the interaction between the chip and the packaging materials. In the as-received sample, the deformation was larger in the top die than in the bottom die. This shows that dies at different level of the 3D stacks had different stress levels, dependent on die stacking and packaging processes. During HTS, as stress relaxation occurred in Cu, the residual stress in Si would correspondingly be reduced. In addition, the temperature of the HTS test was near or slightly above the glass transition temperature (Tg) of the organic materials used in the 3D structure, such as the molding compound and underfill. Under such conditions, the organic materials would become compliant to induce stress relaxation in the 3D package. This leads to an overall stress reduction in Si at room temperature after the HTS test.

Although limited results were available for microbumps, the consumption of Sn to form the Cu-Sn IMC such as Cu_3Sn was observed during the annealing process. The formation of IMC was accompanied by volume shrinkage. Also, the modulus and hardness of IMCs are much larger than those of Sn, which could cause higher stress concentration and potential failure in the microbump [15,16]. This could result in the relatively large APW observed near the top of the via in the bottom die where the TSV is connected to the microbump.

Finite Element Analysis

To qualitatively show the effect of HTS on the overall stress in the 3D package, finite element analysis (FEA) was conducted. A two-dimensional (2D) FEA model was constructed for half the package with a symmetric boundary condition on the right end of the model as shown Figure 8. Furthermore, the bottom right corner was fixed and a 2D plane strain condition was assumed. Key components considered for modeling were the Si chip, TSV arrays, Sn solderbumps, underfill layers, molding compound, and organic substrate, while the barrier layers were ignored. For Cu, Si and Sn, the following material properties were used: Young's modulus, $E_{Cu} = 110$ GPa, $E_{Si} = 130$ GPa, and $E_{Sn} = 70$ GPa; Poisson's ratio, $\nu_{Cu} = 0.35$, $\nu_{Si} = 0.28$, and $\nu_{Sn} = 0.35$. The CTEs, $\alpha_{Cu} = 17$ ppm/°C, $\alpha_{Si} = 2.3$ ppm/°C and $\alpha_{Sn} = 22$ ppm/°C. For material plasticity, the yield strength of Sn was assumed $\sigma_{y,Sn} = 25$ MPa. For Cu, before HTS, $\sigma_{y,Cu} = 250$ MPa was used, and after HTS, a reduced Cu yield strength of $\sigma_{y,Cu} = 200$ MPa was assumed, in line with the reduced dislocation population observed. Temperature-dependent material properties as provided by the material suppliers were used for the underfill and molding compound.

Figure 8. Half model of the die stacks for FEA.

For the as-received sample, a simplified thermal process was used for qualitative analysis. This consisted of heating from 180°C to 260°C to simulate solder reflow and cooling to room temperature (RT) with zero stress assumed to occur at 180°C based on the processing conditions of the test structure. For the sample subjected to HTS test, the thermal process was taken to be cooling from 150°C, where zero stress was assumed. Thus, the thermal stress induced by the maximum thermal load was computed to qualitatively show the effect of HTS.

In Figure 9, the overall deformation shape and the von-Mises stress distribution are shown for of the package before and after HTS. Within the package, the von-Mises stress was higher in the upper portion of the package and decreased towards the bottom of the package. After HTS (Figure 9b), the degree of warpage and stress levels were reduced in the structure, although the stress was still higher in the top die.

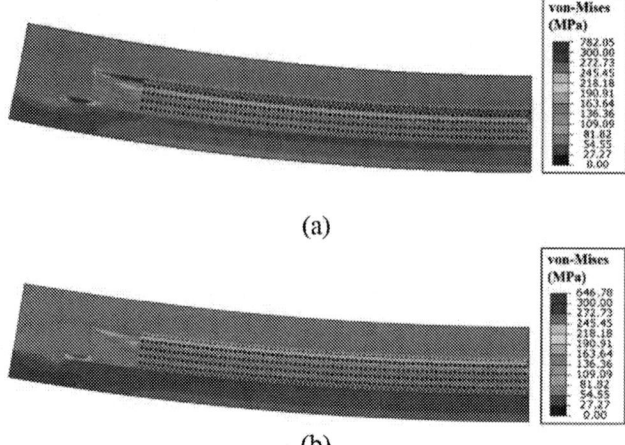

(a)

(b)

Figure 9. Distribution of von-Mises stress in the package: (a) before and (b) after HTS. Deformation scale is 1:5.

The difference in the stress levels between the top die and the bottom die was further compared in Figure 10, where the von-Mises stress is plotted for Si before and after HTS. The von-Mises stress remained larger in the top die than in the bottom die, but the reduction of the stress was more significant for the top die. This trend of stress distribution was consistent with the x-ray microdiffraction measurement.

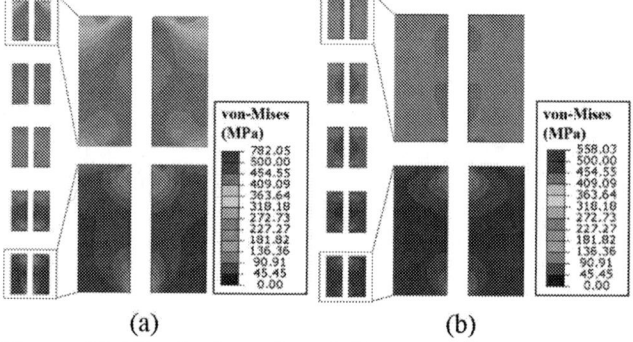

(a)   (b)

Figure 10. Distribution of von-Mises stress in the package: (a) before and (b) after HTS. Deformation scale is 1:5.

The equivalent plastic strain in Cu, which is a measure of the plastic deformation, is plotted in Figure 11 for both before and after HTS. Reduction of the plastic strain can be seen after HTS, which was also observed in the x-ray microdiffraction measurement.

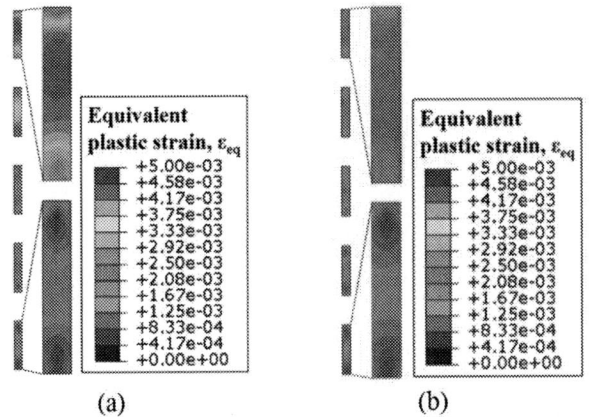

Figure 11. Distribution of Plastic strain in Cu vias (a) before and (b) after HTS.

Quantitatively, the reduction of stress and strain in the 3D die stack structure after HTS was observed in the FEA modeling, which was consistent with the results from the x-ray microdiffraction measurement. The reduced stress and strain were the results of stress relaxation occurred during the HTS test, where the sample was held at elevated temperature for a prolonged time. Under such a condition, atomic diffusion of Cu and relaxation in the package could happen to cause the reduced stress in Si and less plasticity in Cu.

Quantitatively, there was still discrepancy between the measured stress from the x-ray microdiffraction and FEA. This could be due to the penetration effect of the high intensity x-ray as well as the simplified conditions used in the FEA model. The actual fabrication of the 3D die stacks involves many more steps and materials. More detailed modeling analysis will be performed in the future to establish better correlation between the measurement and modeling.

In terms of reliability implications, the reduced stress in Si and Cu by HTS would be beneficial. However, HTS can promote a complete conversion of Sn to Cu-Sn IMC in the microbumps, which might create an adverse effect of stress concentration and potential cracking if unattended.

Summary

The stress in a multi-level die stack structure was characterized using an x-ray microdiffraction before and after HTS test. The unique capability of the x-ray microdiffraction enabled direct measurements of the Cu via, the surrounding Si and the microbump in the 3D chip stacks. The stress and deformation in Si and plasticity in Cu TSVs have been directly observed. Overall, the experimental results showed that HTS could result in stress relaxation in the 3D package and conversion of IMC in the microbumps. FEA has been performed to verify the stress reduction in the package after HTS. While the stress relief can be beneficial, one must be mindful of the accompanying effect such as IMC formation, which might lead to potential cracking of the microbumps..

Acknowledgments

The authors gratefully acknowledge financial support of this work by Semiconductor Research Corporation. We are thankful to SK Hynix Inc. for providing the test structure. The authors also thank Drs. N. Tamura and M. Kunz at ALS for helpful discussions. The Advanced Light Source is supported by the Director, Office of Science, Office of Basic Energy Sciences, Materials Sciences Division, of the U.S. Department of Energy under Contract No. DE-AC02-05CH11231 at Lawrence Berkeley National Laboratory and University of California, Berkeley, California. The move of the micro-diffraction program from ALS beamline 7.3.3 onto to the ALS superbend source 12.3.2 was enabled through the NSF grant #0416243.

References

1. J.U. Knickerbocker et al., IEEE J Solid-St Circ, 41(8), pp.1718- 1725 (2006).
2. J. Lau, El. Packag. Tech. Conf., pp.560-570 (2010).
3. J. Van Olmen et al., Microelectron Eng., 88(5), pp.745-748 (2010).
4. S. Kang et al., IEEE International 3D Systems Integration Conference (3DIC), pp.1-3 (2012).
5. A. Mercha et al., Proc. IEEE Electron Device Meeting (IEDM), pp. 2.2.1-2.2.4 (2010).
6. S.K. Ryu et al., IEEE Trans. Device and Materials Reliability, 12(2), pp. 255-262 (2012).
7. W.S. Kwon et al., Appl. Phys. Lett., 98, 232106 (2011).
8. T. Jiang et al., Microelectron. Reliab., 53, pp. 53–62 (2013).
9. S.K. Ryu et al., Appl. Phys. Lett., 100, 041901 (2012).
10. A.S. Budiman et al., Microelectron. Reliab., 52, pp.530-533 (2012).
11. N. Tamura et al., J. Synchrotron Rad., 10, pp.137-143 (2003).
12. B. C. Valek et al., Appl. Phys. Lett., 81, 4168 (2002).
13. T. Jiang et al., Appl. Phys. Lett. 103, 211906 (2013).
14. R. Abbaschian, R. E. Reed-Hill, Physical Metallurgy Principles, 4th edition, CL-Engineering (2009).
15. Y. Wang et al., Proc. IEEE Electronic Components and Technology Conference (ECTC), pp.1953- 1958 (2013).
16. H.-Y. Son et al., Proc. IEEE Electronic Components and Technology Conference (ECTC), pp. 29 - 34 (2013).

A Novel Fine Pitch TSV Interconnection Method Using NCF with Zn Nano-particles

Ji-Won Shin[1], Yong-Won Choi[1], Young Soon Kim[1], Un Byung Kang[2], Sun Kyung Seo[2], and Kyung-Wook Paik[1]

1) Dept. of Materials Science and Engineering
Korea Advanced Institute of Science and Technology (KAIST)
2) Package Development team
Semiconductor R&D Center
Semiconductor Business
Samsung Electronics Co.,LTD

Abstract

Non-conductive film (NCF) with Zn nano-particles is an effective solution for fine-pitch Cu-pillar/Sn-Ag hybrid bump interconnection in terms of manufacturing process and interfacial reliability. In this study, NCFs with Zn nano-particles of different acidity, viscosity, and curing speed were formulated, and diffused Zn contents in the Cu pillar/Sn-Ag hybrid bumps were measured after 3D TSV chip-stack bonding. Amount of Zn diffusion into the Cu pillar/Sn-Ag bumps increased as the acidity of resin increased, as the viscosity of resin decreased, as the curing speed of resin decreased, and as the bonding temperature increased. Diffusion of Zn nano-particles into the Cu pillar/Sn-Ag bumps are maximized when the resin viscosity became lowered and the solder oxide layer was removed. To analyze the effect of Zn on IMC reduction, NCFs with 0 wt%, 1 wt%, 5 wt%, and 10 wt% of Zn nano-particles were bonded on the test vehicles, and aged at 150°C up to 500 hours. NCF with 10wt% Zn nano-particle showed remarkable suppression in Cu_6Sn_5 and $(Cu,Ni)_6Sn_5$ IMC compared to NCFs with 0 wt%, 1 wt%, and 5 wt% of Zn nano-particles. However, in terms of Cu_3Sn IMC suppression, which is the most critical goal of this experiment NCFs with 1 wt%, 5wt%, and 10wt% showed an equal amount of IMC suppression. As a result, it was successfully demonstrated that the suppression of Cu-Sn IMCs was achieved by the addition of Zn nano-particles in the NCFs resulting an enhanced reliability performance in the Cu/Sn-Ag hybrid bumps bonding in 3D TSV interconnection.

1. Introduction

As demands for smaller and faster electronic products increased rapidly, 3D-packaging became inevitable in electro-packaging industry. Among various 3D packaging methods, 3D chip-stacking using through silicon via (TSV) technology becomes one of the best candidates to further reduce the size and improve electrical performance [1]. Due to tremendous effort of many research organizations and industries, state-of-art TSV forming and copper via filling can be successfully developed. As TSV vertical interconnection methods, Cu pillar/Sn-Ag hybrid bump structure is being widely investigated for 3D TSV chip stacking [2-8]. Recently, 40 um pitch TSV-chip stacking is being applied in real products in IBM, Xillinx, Samsung, etc., however, there are two major

problems in Cu pillar/Sn-Ag hybrid bonding. The first problem is the voids formation within underfill materials and flux residue due to the limitation of capillary flow at the narrow gap of 3D-TSV stacked chips [9]. The second problem is the faster-growth of Cu-Sn IMCs due to reduced size of Cu/Sn-Ag bumps and limited amount of Sn [10,11]. These two problems can cause serious reliability problems. Therefore, the new idea of non-conductive film (NCF) with Zn nano-particles was suggested as a potential solution for solving both problems. Wafer-level applied NCF is one of the most promising candidates to replace conventional underfill process due to its voidless underfill ability and easy processing, and it is being studied in many research organization [12-15]. Figure 1 shows the advantage of wafer-level NCF compared conventional reflow and underfill process. Wafer-level NCF process provides process reduction by performing fluxing/bonding/underfill process at once and it also provides higher reliability due to void-less underfill. Zn has been previously reported to reduce IMC growth in Sn solders, when Zn was added as a dopant in Sn bulk [16] or as an ingredient in flux [17] or as a plated material in bonding pad [18,19]. However, the first method has difficulty to be adopted in bumping process of Sn-Ag-Zn since it is three phase electroplating process. The second method requires additional fluxing process which is unrequired process in NCF bonding process and the third method has problems of Zn oxidation and bad plating behavior of Zn. Compared to those methods, adding Zn nano-particles to NCF material doesn't change any chip/substrate structure or take additional flux removal process to apply Zn into the solder or the wettability problem of solders on pads due to oxidation of Zn. Therefore, NCFs with Zn nano-particles for IMC growth reduction is a promising method in terms of costs and process-ability.

Zn nano-particles were used as diffusing material to the solder in the NCFs since they wouldn't form volatile species and would have good mobility in NCFs to diffuse into the solder bumps. Zn nano-particles in NCF are expected to diffuse into the solder bumps and suppress IMC growth as shown in Figure 2. In this research, diffusion mechanism of Zn nano-particle into the solder bumps in NCFs will be investigated and effects of Zn nano-particles on interfacial reactions of Cu/Sn-Ag hybrid bumps would be also investigated.

978-1-4799-2408-0/14 $31.00 © 2014 IEEE

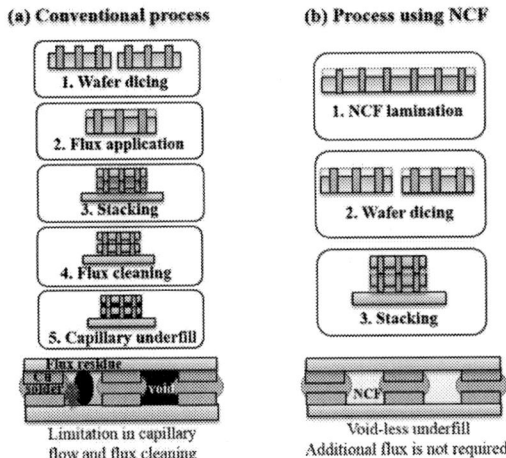

Figure 1. Schematic diagram of (a) conventional reflow and underfill process and (b) wafer-level NCF process

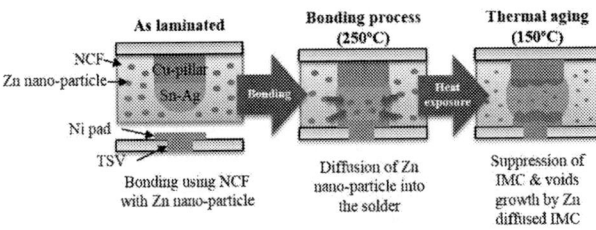

Figure 2. Schematic diagram of the Zn nano-particle diffusion from the NCFs to solder bumps

2. Experiments

2.1. Test vehicle

As a test vehicle for the experiment, a top chip with dimensions of 8 mm x 8 mm x 0.2 mm and a TSV Si-interposer with dimension of 12 mm x 12 mm x 0.06 mm were used. The TSV Si-interposer was already bonded on the PCB substrate with a conventional SMT process, and the top chip was bonded on TSV Si-interposer using NCFs laminated on top chip as shown in Figure 3. The top chip consists of Cu pillar/Sn-Ag hybrid bumps with height of 20 μm, and the TSV Si-interposer has Cu pads with height of 5 μm as shown on Figure 4.

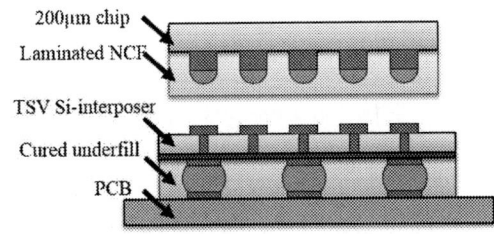

Figure 3. Schematic structures of top chip, TSV Si-interposer, and PCB substrates

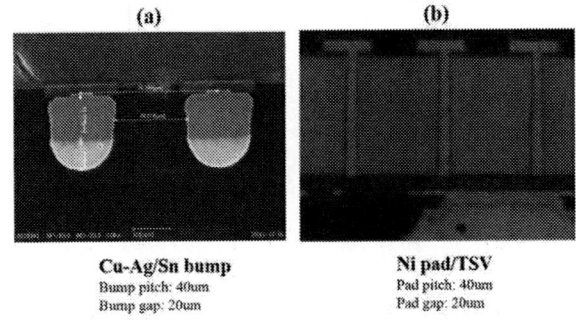

Cu-Ag/Sn bump	Ni pad/TSV
Bump pitch: 40um	Pad pitch: 40um
Bump gap: 20um	Pad gap: 20um

Figure 4. Shape of (a) Cu pillar/Sn-Ag hybrid bump and (b) TSV Si-interposer with Ni pad

2.2. NCF materials

NCFs used in this experiment basically consist of Zn nano-particle, resin, curing agent, curing accelerator, fluxing agents, and silica filler. 40nm size and 99% purity unoxidized Zn nano-particles were used in the experiment. Zn nano-particles were first sonicated in the methyl ethyl ketone (MEK), and they were homogeneously mixed with the resin which consists of Bisphenol-F, Bisphenol-A, phenoxy, and thermo-plastic resins. The resin mixture was subsequently coated into 30μm-thick film using film coater at temperature of 60°C. By this method, Zn nano-particles were kept unoxidized within the NCFs. Two types of resin curing systems were used in the experiment. The first one is epoxy-anhydride system, which basically forms acids during curing reaction, and the second one is epoxy-DICY system with no acidic reaction. For each type, curing accelerators were used to control the curing speed. To grant acidity to epoxy-DICY NCF, thermal acid generator (TAG) which emits acid at certain temperature was additionally added as a fluxing agent. Acids formed by anhydride and TAG take place in resin curing reaction after solder oxide removal, and hence it has been reported to not to degrade package reliability if suitable amounts are added [21,22]. Silica filler with average size of 7 nm was added to control the viscosity of NCFs.

2.3. Reflow & bonding condition

In this experiment, both conventional oven reflow and thermo-compression were performed. As shown in figure 5, conventional oven reflow was performed to analyze mechanisms of Zn nano-particle diffusion in the Cu/Sn-Ag hybrid bump. Thermo-compression bonding was performed of chip on TSV Si-interposer was performed using flip-chip bonder and general procedures for the bonding are shown on figure 6. To relax and minimize warpage stress caused by CTE mismatch between PCB, underfill material, and Si-chip, pre-heating was performed on TSV Si-interposer/PCB substrate with temperature of 120°C. At 120°C, top chip and TSV chip were aligned, and pre-bonded so that mechanical contact between bumps and pads can be made. After the pre-bonding, main bonding was performed with heating rate of 6°C/s and peak temperature of 250°C for 10s so that sufficient wetting of solder and curing of NCFs could be made. All the bonding was performed with bonding pressure of 40 N (0.28 MPa).

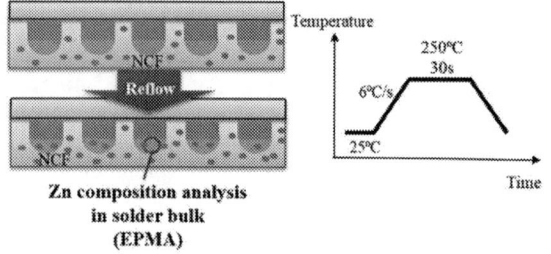

Figure 5. Schematic procedures of oven reflow process for Zn nano-particle diffusion

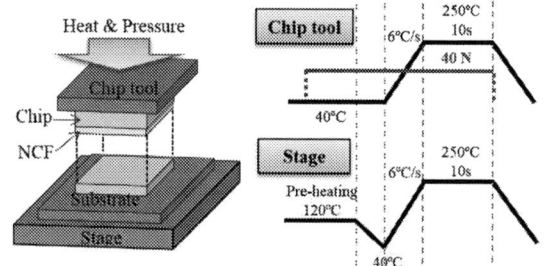

Figure 6. Procedures and profile of the thermo-compression bonding

3. Results and discussion

3.5. Effect of reflow temperature on Zn diffusion

The solder bumps laminated with NCF-A2 were taken through reflow with various peak temperatures of 100°C, 150°C, 200°C, and 250°C to investigate effect of reflow temperatures on Zn nano-particle diffusion. Figure 7 shows the DSC curve and viscosity of NCF-A2 with heating rate of 6°C/s and corresponding Zn contents in solder bumps at reflow temperature. Figure 8 shows Zn-EPMA mapping of the solder bumps at various temperature. At 100°C, no diffusion of Zn nano-particles into the solder bumps and no movement of Zn nano-particles in NCF was observed. At 150°C, 0.51 wt% of Zn was detected in the solder and active movement of Zn nano-particles to the solder surface was observed. At 200°C, Zn contents in the solder bumps significantly increased to 1.44 wt%, and 1.52 wt% of Zn was detected in the solder bumps reflowed at 250°C. This behavior can be interpreted in terms of heat flow and resin viscosity depending on temperatures. Variation of viscosity depending on temperature caused the variation of Zn nano-particle mobility. Movement of Zn nano-particles to the solder surface at 150°C is mainly due to reduction of viscosity. Peak in the heat flow indicates removal of zinc oxide and tin oxide by acids which are formed from curing reaction of epoxy-anhydride. The significant increase in Zn contents at 200°C is due to the removal of zinc oxide and tin oxide by the acids.

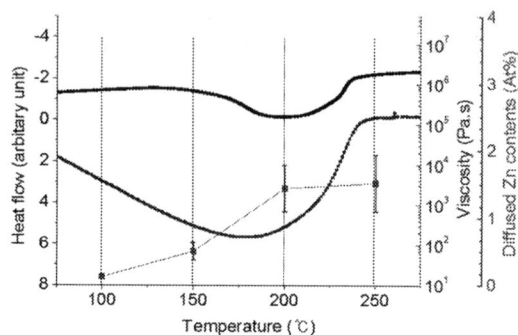

Figure 7. DSC curve and viscosity of NCF-A2 with heating rate of 6°C/s and Zn contents of solder bumps depending on reflow temperature

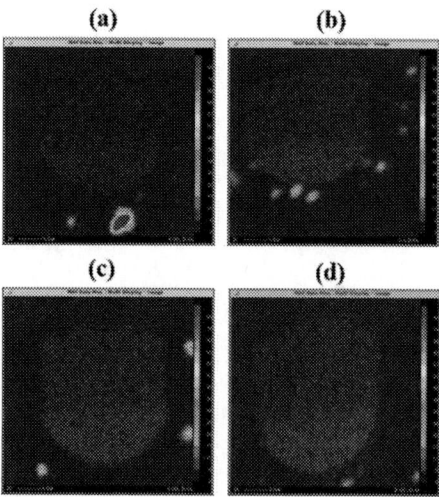

Figure 8. Zn-EPMA mapping of solder bumps after reflow at (a) 100°C, (b) 150°C, (c) 200°C, (d) 250°C

3.6. Mechanism of Zn nano-particle diffusion in NCF

The effects of resin acidity, resin viscosity, resin curing speed, and reflow temperature can be organized as shown in Table 1. Considering such effects and thermo-mechanical behavior of NCF-A2, the mechanism of Zn nano-particle diffusion into the solder bumps can be explained as shown in figure 9. At 100°C Zn nano-particles are dispersed in the NCFs and movement is restricted due to high viscosity. At 150°C, resin between chip and substrate gains flow under flow-able resin viscosity. The resin trapped between parallel plates gains horizontal flow when the bonding force is applied vertically. Under the horizontal resin flow, Zn nano-particles gain movement to be attached on solder surface. At 200°C which is temperature higher than eutectic temperature of Sn-Zn (198.5°C), tin oxide is effectively removed due to acids from the epoxy-anhydride curing reaction, and then Zn nano-particles are actively diffused into the solder bumps. At 250°C, the diffusion of Zn nano-particles into the solder bumps is relatively decreased due to an increase in resin viscosity by curing reaction of NCFs and depletion of Zn in the NCFs.

Table 1. Effects of NCF material properties and temperature on Zn diffusion

Variables	Effects
Acidity↑	Oxide removal↑
Viscosity↓	Zn mobility↑
Curing speed↓	Zn flow time↑
Temperature↑	Zn mobility↑

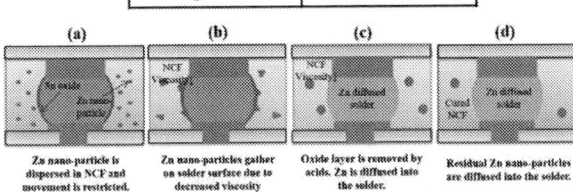

Figure 9. Mechanism of Zn nano-particle diffusion into the solder bump laminated by NCF-A2 at (a) 100°C, (b) 150°C, (c) 200°C, (d) 250°C

3.7. Joint structure using NCFs with Zn nano-particle

Based on the properties of NCF-A2, NCFs with 0 wt%, 1 wt%, 5 wt%, and 10 wt% of Zn nano-particles were fabricated. The fabricated NCFs were bonded using test vehicles with Cu pillar/Sn-Ag solder/Ni pad structure. Figure 10 shows Zn contents in the solder bumps of NCFs with 0wt%, 1wt%, 5wt%, and 10wt% of Zn nano-particles. Contents of Zn in the solder bumps increased as Zn nano-particle contents in NCF increased as shown in Figure 11. However, amount of Zn contents in the solder were only in the range of 1.24~1.84 wt% showing smaller values compared with the contents of Zn nano-particles added in NCFs (1~10 wt%). EPMA mapping of the joints indicates that NCFs with 0wt%, 1wt%, 5wt%, and 10wt% of Zn nano-particles made clear diffusion of Zn nano-particles into the solder bumps.

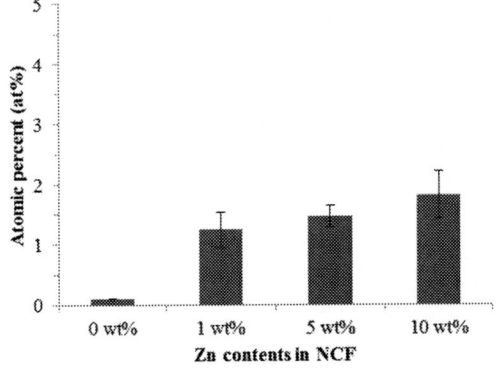

Figure 10. Zn contents of the solder bumps in NCF with 0wt%, 1wt%, 5wt%, and 10wt% Zn nano-particles

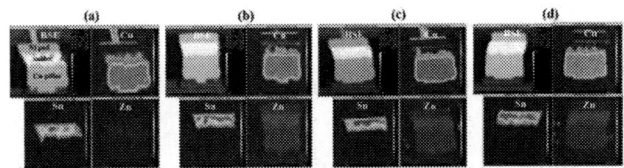

Figure 11. EPMA mapping of NCF with (a) 0wt%, (b) 1wt%, (c) 5wt%, and (d) 10wt% Zn nano-particles

3.8. Analysis on IMC height

To analyze the effect of Zn addition on IMC reduction, test vehicles bonded using NCFs with 0wt%, 1wt%, 5wt%, and 10wt% Zn nano-particles were taken under high temperature storage test (HTST) at 150°C. Figure 12 shows cross-sectional images of the joints after aging for 0, 24, and 48 hours. IMC heights of Cu_6Sn_5 at the Cu-pillar and $(Cu,Ni)_6Sn_5$ at Ni pad is shown on Figure 13. Compared to NCF without Zn nano-particles, NCFs with 1wt%, 5wt%, and 10wt% of Zn nano-particles showed -11.2%, -9.2%, and -14.4% of Cu_6Sn_5 IMC reduction. In case of $(Cu,Ni)_6Sn_5$ IMC, NCFs with 1wt%, 5wt%, and 10wt% of Zn nano-particles showed -13.9%, -14.8%, and -20.4% of IMC reduction compared to NCF-0wt%. In terms of the IMCs at short aging times, the amount of Cu_6Sn_5 and $(Cu,Ni)_6Sn_5$ IMC growth decreased as Zn nano-particles contents in NCFs increased. This phenomena occurs due to increased Zn diffusion in solder bumps as Zn contents in NCFs increased. Diffused Zn in solder bumps primarily reacts with Cu meaning that it relatively decrease reaction between Cu and Sn [16]. Therefore, the growth of Cu_6Sn_5 which is primary Cu-Sn IMC is reduced as the contents of Zn nano-particles in NCFs increased. Figure 14 shows cross-sectional images of the joints after aging for 125, 250, and 500 hours. Since the whole area of Sn-Ag solder has reacted with Cu to form Cu_6Sn_5 IMC, analysis on the height of Cu_3Sn which is secondary IMC was performed and is shown on Figure 15. Compared to NCFs without Zn nano-particles, 1wt%, 5wt%, and 10wt% Zn showed -18.6%, -19.2%, and -19.8% of Cu_3Sn IMC reduction. In terms of IMCs at long aging time, NCFs with 1wt%, 5wt%, and 10wt% showed similar decrease in Cu_3Sn IMC growth. Minor difference in Cu_3Sn IMC height depending on Zn nano-particle contents can be interpreted as follows. As explained, Zn has a role of reducing diffusion the between Cu and Sn by primarily reacting with Cu. During the long term aging, all the Zn has participated to react with Cu due to limited Zn source in small bump volume, and consequently the role of Zn depending on its contents has diminished.

978-1-4799-2408-0/14 $31.00 © 2014 IEEE

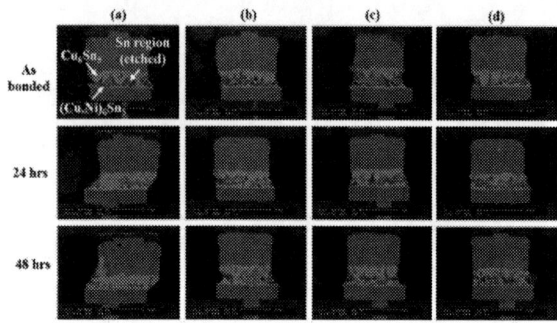

Figure 12. Cross-sectional images of the joints bonded using NCF with (a) 0wt%, (b) 1wt%, (c) 5wt%, and (d) 10wt% Zn nano-particles after aging for 0, 24, and 48 hours (Sn wet etched to reveal Cu-Sn IMCs.)

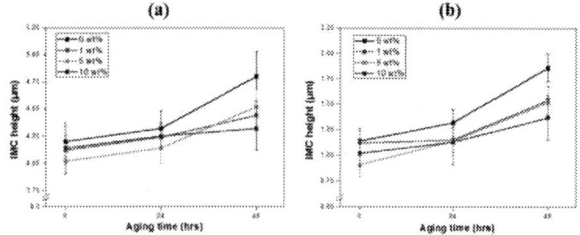

Figure 13. IMC heights of (a) Cu_6Sn_5 formed at Cu-pillar and (b) $(Cu,Ni)_6Sn_5$ formed at Ni pad

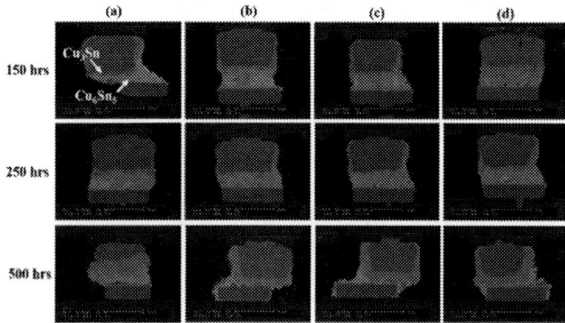

Figure 14. Cross-sectional images of the joints bonded using NCF with (a) 0wt%, (b) 1wt%, (c) 5wt%, and (d) 10wt% Zn nano-particles after aging for 125, 250, and 500 hours

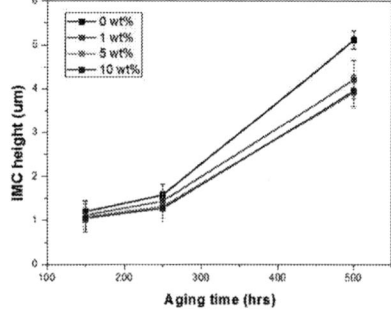

Figure 15. IMC height of Cu_3Sn formed at Cu pillar side

4. Conclusions

In this study, a practical method of alloying Zn into fine-pitch Cu-pillar/Sn-Ag hybrid bump using NCFs with Zn nano-particles was demonstrated. Additionally, the diffusion mechanism of Zn nano-particles to the solder bumps in NCF was proposed and the effects of Zn nano-particles on IMC growth were investigated. The amount of Zn nano-particle diffusion into the solder bumps increased as resin acidity increased, resin viscosity decreased, resin curing speed decreased, and reflow temperature increased. The results indicate that ability to remove Sn oxides, mobility of Zn nano-particles, and sufficient flow time are required for the enhanced Zn diffusion. The amount of Zn diffusion to the solder bumps is maximized when Zn nano-particles move to the surface of the solder bumps at flow-able resin viscosity and diffuse into the solder bumps with the removal of zinc oxide and tin oxide. By the flux function NCFs with Zn nano-particles showed significant suppression of Cu_6Sn_5, and $(Cu,Ni)_6Sn_5$ IMC in short aging time and Cu_3Sn in long aging time. The actual bonding using test vehicle was performed using NCFs with Zn nano-particles. The diffused Zn contents in solder bumps increased as Zn nano-particles contents increased. Higher the Zn contents in solder bumps, the growth rate of Cu_6Sn_5 and $(Cu,Ni)_6Sn_5$ IMCs decreased due to priority reactions of Cu-Zn. However, in the long term aging, all the NCFs with Zn similarly showed about 20% of Cu_3Sn IMC suppression compared to NCFs without Zn.

Acknowledgments

The authors would like to give special thanks to Un Byung Kang and Young Kun Jee, Samsung Electronics Co., LTD for test vehicles and helpful discussions.

References

1. M.J. Wang, C.Y. Hung, C.L. Kao, P.N. Lee, C.H. Chen, C.P. Hung, H.M. Tong, TSV technology for 2.5D IC solution, Proceedings - Electronic Components and Technology Conference, (2012) 284-288.

2. H.Y. Son, G.J. Jung, B.J. Park, K.W. Paik, A study on the thermal reliability of Cu/SnAg double-bump flip-chip assemblies on organic substrates, Journal of Electronic Materials, 37 (2008) 1832-1842.

3. R. Dunne, Y. Takahashi, K. Mawatari, M. Matsuura, T. Bonifield, P. Steinmann, D. Stepniak, Development of a stacked WCSP package platform using TSV (through silicon via) technology, Proceedings - Electronic Components and Technology Conference, (2012) 1062-1067.

4. J. Hwang, J. Kim, W. Kwon, U. Kang, T. Cho, S. Kang, Fine pitch chip interconnection technology for 3D integration, Proceedings - Electronic Components and Technology Conference, (2010) 1399-1403.

5. K. Takahashi, M. Umemoto, N. Tanaka, K. Tanida, Y. Nemoto, Y. Tomita, M. Tago, M. Bonkohara, Ultra-high-

density interconnection technology of three-dimensional packaging, Microelectronics Reliability, 43 (2003) 1267-1279.

6. B. Ebersberger, C. Lee, Cu pillar bumps as a lead-free drop-in replacement for solder-bumped, flip-chip interconnects, Proceedings - Electronic Components and Technology Conference, (2008) 59-66.

7. A. Yu, J.H. Lau, S.W. Ho, A. Kumar, W.Y. Hnin, W.S. Lee, M.C. Jong, V.N. Sekhar, V. Kripesh, D. Pinjala, S. Chen, C.F. Chan, C.C. Chao, C.H. Chiu, C.M. Huang, C. Chen, Fabrication of high aspect ratio TSV and assembly with fine-pitch low-cost solder microbump for Si interposer technology with high-density interconnects, IEEE Transactions on Components, Packaging and Manufacturing Technology, 1 (2011) 1336-1344.

8. H. Gan, S.L. Wright, R. Polastre, L.P. Buchwalter, R. Horton, P.S. Andry, C. Patel, C. Tsang, J. Knickerbocker, E. Sprogis, A. Pavlova, S.K. Kang, K.W. Lee, Pb-free micro-joints (50 μm pitch) for the next generation micro-systems: The fabrication, assembly and characterization, Proceedings - Electronic Components and Technology Conference, 2006 (2006) 1210-1215.

9. M.K. Schwiebert, W.H. Leong, Underfill flow as viscous flow between parallel plates driven by capillary action, IEEE transactions on components, packaging and manufacturing technology. Part C. Manufacturing, 19 (1996) 133-137.

10. K.N. Tu, Reliability challenges in 3D IC packaging technology, Microelectronics Reliability, 51 (2011) 517-523.

11. K.N. Tu, H.Y. Hsiao, C. Chen, Transition from flip chip solder joint to 3D IC microbump: Its effect on microstructure anisotropy, Microelectronics Reliability, 53 (2013) 2-6.

12. C. Feger, N. LaBianca, M. Gaynes, S. Steen, Z. Liu, R. Peddi, M. Francis, The over-bump applied resin wafer-level underfill process: Process, material and reliability, Proceedings - Electronic Components and Technology Conference, (2009) 1502-1505.

13. K. Honda, T. Enomoto, A. Nagai, N. Takano, NCF for wafer lamination process in higher density electronic packages, Proceedings - Electronic Components and Technology Conference, (2010) 1853-1860.

14. T.F. Yang, K.S. Kao, R.S. Cheng, J.Y. Chang, C.J. Zhan, Development of wafer-level underfill bonding process for 3D chip stacking, Proc. IEEE ICEP, (2011) 74-78.

15. S. Kawamoto, M. Yoshida, S. Teraki, H. Iida, Effect of NCF design for the assembly of Flip Chip and reliability, Electronic Components and Technology Conference (ECTC), (2012) 399-405.

16. M.G. Cho, S.K. Kang, D.-Y. Shih, H.M. Lee, Effects of Minor Additions of Zn on Interfacial Reactions of Sn-Ag-Cu and Sn-Cu Solders with Various Cu Substrates during Thermal Aging, Journal of Electronic Materials, 36 (2007) 1501-1509.

17. H. Sakurai, A. Baated, K. Lee, S. Kim, K.-S. Kim, Y. Kukimoto, S. Kumamoto, K. Suganuma, Effects of Zn-Containing Flux on Sn-3.5Ag Soldering with an Electroless Ni-P/Au Surface Finish: Microstructure and Wettability, Journal of Electronic Materials, 39 (2010) 2598-2604.

18. C.-Y. Yu, W.-Y. Chen, J.-G. Duh, Suppressing the growth of Cu–Sn intermetallic compounds in Ni/Sn–Ag–Cu/Cu–Zn solder joints during thermal aging, Intermetallics, 26 (2012) 11-17.

19. J.H. Lee, Y.M. Kim, J.H. Hwang, Y.-H. Kim, Wetting characteristics of Cu–xZn layers for Sn–3.0Ag–0.5Cu solders, Journal of Alloys and Compounds, 567 (2013) 10-14.

20. K. Gilleo, M. Witt, D. Blumel, P. Ongley, Towards a better understanding of underfill encapsulation for flip chip technology: proposed developments for the future, Microelectronics international, 16 (1999) 39-43.

21. S.-H. Kim, Y. Choi, Y. Kim, K.-W. Paik, Flux function added solder anisotropic conductive films (ACFs) for high power and fine pitch assemblies, Electronic Components and Technology Conference (ECTC), 2013 IEEE 63rd, IEEE2013, pp. 1713-1716.

978-1-4799-2408-0/14 $31.00 © 2014 IEEE

Residual Stress Investigations at TSVs in 3D Micro Structures by HR-XRD, Raman Spectroscopy and fibDAC

U. Zschenderlein[1], D. Vogel[2], E. Auerswald[2], O. Hölck[1], H. Rajendran[1], P. Ramm[3], R. Pufall[4], B. Wunderle[1,2]

[1]TU Chemnitz, Chair Materials and Reliability of Micro Systems, Chemnitz, Germany
[2]Fraunhofer ENAS, Micro Materials Center (MMC), Chemnitz, Germany
[3]Fraunhofer EMFT, Munich, Germany
[4]Infineon Technologies AG, Munich, Germany
uwe.zschenderlein@etit.tu-chemnitz.de, +49 371 531 35616

Abstract

In this paper the residual stress in single-crystalline Si around W-filled TSVs was determined experimentally by three methods with high spatial resolution and compared to one another. In contrast to Cu as TSV filler, W has the potential advantage of a lower CTE mismatch to Si resulting in lower thermally induced stress at the TSV-interface. As test layout a cross-sectioned double-die stack was used consisting of a top die with TSVs which is bonded by Cu-Sn Solid Liquid Interdiffusion Bonding (SLID) to the bottom die. Three different experimental methods have been used to determine mechanical stresses in silicon nearby tungsten TSVs - HR-XRD performed at a synchrotron beamline, microRaman spectroscopy and stress relief techniques put into effect by FIB milling. All methods possess, to a different extend, high spatial resolution capabilities. However they differ in their sensitivity and response to the particular stress tensor components relevant for the residual stress state nearby TSV structures. Stress measurements were performed on test samples with W-TSVs in thinned dies, which were SLID bonded to a thicker Si substrate die. The measurements captured stresses introduced by the W-TSV as well as by the wafer bonding process. A stress range from several MPa to hundreds of MPa could have been covered with a spatial resolution ranging from 100 nm to tens of microns. Measurement results were compared to one another and to simulated stresses from finite element analysis (FEA).

All experimental methods show the influence of W and Cu-Sn-Bond in Si. The very high stress sensitivity for HR-XRD below 1 MPa could be shown. For small stress gradients the analysis of the peak position gives reasonable results and for larger stress gradients a profile analysis of the diffraction peak is more accurate. The results show that in intrinsic stress in W may have to be considered in FEA and more attention should be directed to the accuracy of the FE-modelled Cu-Sn SLID bond with respect to shrinkage during phase formation of Cu_3Sn.

1. Introduction

The heterogeneous integration of 3D system-in-package (SiP) is a challenge for reliability. At Through Silicon Vias (TSVs) the different thermo-mechanical behavior of the used dissimilar materials may lead to residual stress with related failures. Therefore the determination of localized residual stresses is a key focus for industry [1-5].

Scanning X-ray micro diffraction (μSXRD) [9] and Raman spectroscopy are among the most often used methods to analyze residual stress around TSVs. Although μSXRD is fast with high lateral resolution, its stress resolution is lower compared to the classical high resolution X-ray diffraction (HR-XRD) scans in reciprocal space. HR-XRD provides appropriate diffraction patterns for a profile analysis. Further, Raman spectroscopy is easy to perform, highly local and surface sensitive. But for non-uniaxial stress states it is very complicated to extract stress tensor components. Finite Element Analysis (FEA) is widely used. However, complex stress built-up processes have to be modeled adequately, which is cumbersome and sensitive to plenty of potential input errors.

In order to ensure reliable stress data, in this paper the residual stress in single-crystalline Si around W-filled TSVs was determined experimentally by three independent methods with high spatial resolution and compared to one another. Besides HR-XRD and microRaman, a recently developed new method base on stress relief measurements (fibDAC) was extended to measure high stress gradient fields. Finally, all measured stress data was compared to respective FEA results. In contrast to Cu as TSV filler, W has the potential advantage of a lower CTE mismatch to Si resulting in lower thermal induced stress at the TSV-interface at room temperature, despite its larger elastic modulus. As test layout a cross-sectioned double-die stack was used consisting of a top die with TSVs which was bonded by Cu-Sn Solid Liquid Interdiffusion Bonding (SLID) to the bottom die, exploiting fast diffusion in the emerging transient liquid phase.

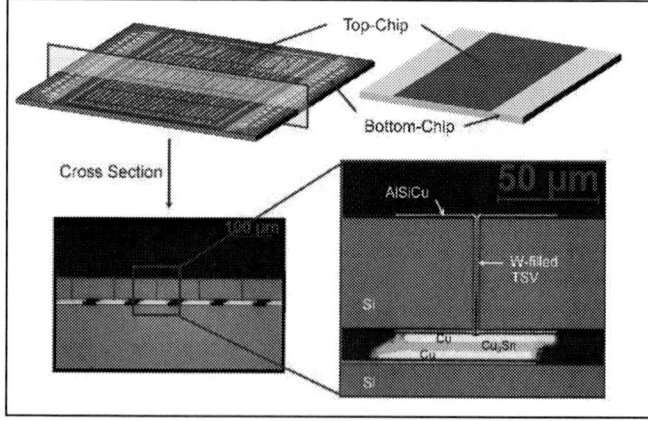

Figure 1: Cross section of double die (light microscope). Top die and bottom die are bonded by solid liquid interdiffusion (SLID) of Cu and Sn in chip to wafer process.

Analyzed samples. All experiments were carried out on cross sections consisting of two stacked dies (Figure 1). The layout is schematically pictured in Figure 2. It consists of a thin top die with W-TSV and contact pads which is bonded to a thick bottom die by a Cu-Sn-SLID in a chip to wafer process. The cross sectioned W-TSV as well as the SLID bond are shown in Figure 1. The cross section of a W-filled TSV is 10 x 3 µm² (Figure 2, top view). Its height is 50 µm, what equals the thickness of the top die. The via metallisation was made by TiN-CVD-seed layer and W-CVD-process at 400°C. The SLID-bonds were formed at 240°C.

Finite element analysis. An FE model of the double stack was built up in ANSYS to simulate its manufacturing steps obtaining the stress state. The model and the processing parameters are shown in Figure 3. The applied materials, their parameters and the mechanical models are listed in Table 1. The volume shrinkage during the formation of theCuSn phases are neglected.

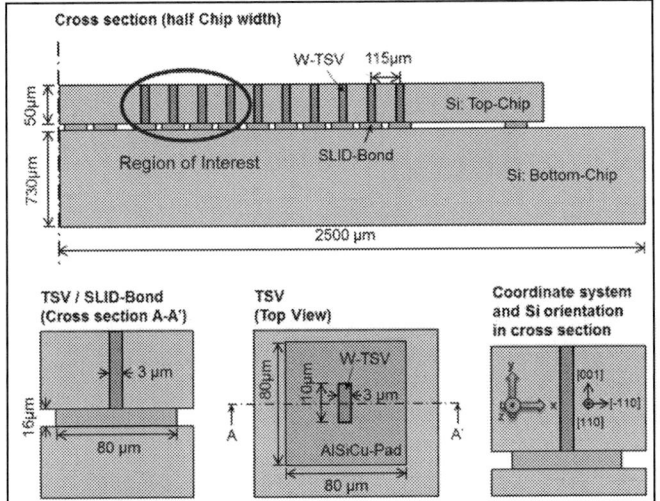

Figure 2: Double die sample layout with TSVs and SLID-Bonds (schematically).

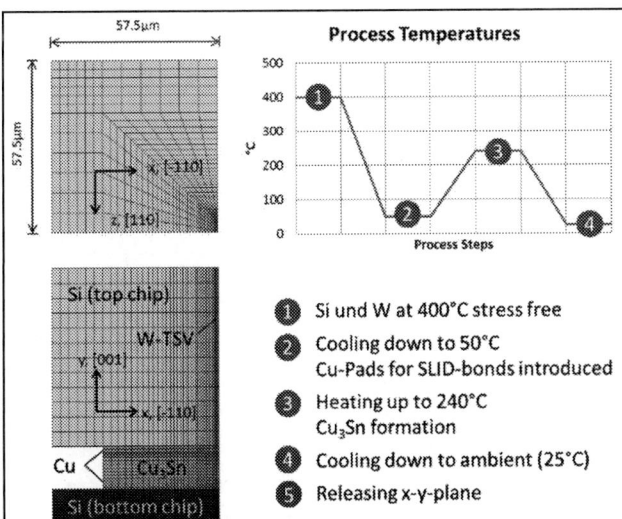

Figure 3: FE model und applied process parameters.

The double die sample exhibits TSV-areas, so the FE model is a quarter model with periodic boundary conditions. The symmetry planes are defined by the symmetry axes of a TSV cross section. The cutting along the x-y-symmetry plane is considered by allowing relaxation of nodes in z-direction after the last processing step. The orientation of the Si single crystal is indicated by the direction indices in Figure 2 and Figure 3.

Table 1: Material properties and mechanical models used in FEA. [6-9]. Youngs Modulus of W, Cu and Cu₃Sn were checked by Nanoindentation experiments.

Material	CTE (ppm/K)	E (GPa), sij (1/TPa)	Poissons ratio v	σ_y (MPa)	Mechanical Model
Si	2.8	s11: 7.74 s12: -2.14 s44: 12.56			anisotropic, elastic
CVD-W	4.4	E: 420	0.28		elastic
CVD/EPD-Cu	16.6	E: 90	0.36	200	elastic-plastic
Cu₃Sn	18	E: 100	0.30	480	elastic

An overview of the stress distribution is given by the contour plots of normal (σ_x, σ_y, σ_z) and principal stress components (σ_1, σ_2, σ_3) in Figure 4. Looking at the Cu₃Sn-Phase of the SLID-bond, we find for σ_x large tensile stresses up to 480 MPa due to the large CTE-Mismatch with Si. Releasing the x-y-plane will result in redistribution of stress. Higher stress than that is not observed since the yield stress is 480 MPa (Table 1). The normal stress of that new surface should be generally zero, which is approved by the σ_z – distribution. As a result of this the σ_y - component the Cu₃Sn-Phase shows high compressive stress at the surface, which rapidly drops down within a few microns below surface.

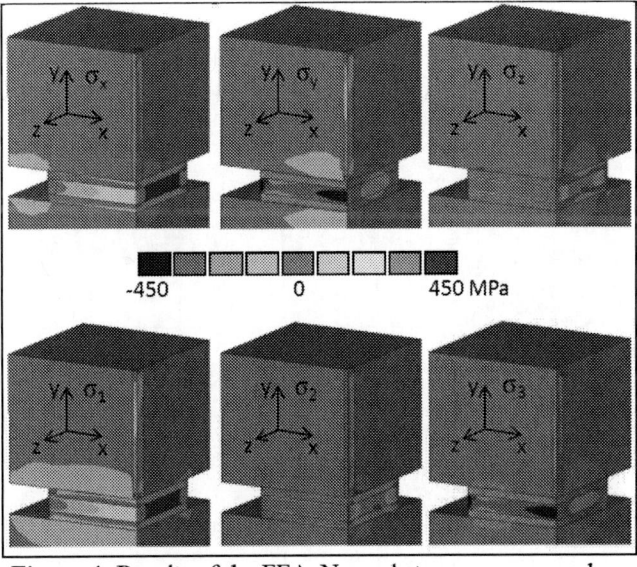

Figure 4: Results of the FEA. Normal stresses: σ_x, σ_y and σ_z; principal stresses: σ_1, σ_2 and σ_3.

The stress inside the Cu of the SLID-Bond is generally lower. One reason for that is the lower yield stress of 200 MPa. The Cu is than like a stress interposer between Cu₃Sn and Si, suffering shear stress. Those shear stresses are indicated by the principle stress σ_1 at the interface between Si and Cu, at both, top and bottom die. At the W-TSV stresses in

Si and W developed due to their high CTE mismatch. In W the in-plane stress is dominated by distinctive tensile stress in y-direction. The stress is in Si is mostly low. We will later see that strong stress gradients are observed right at the interface between Si and W. This is discussed inside the HR-XRD section. In general we can draw two conclusions: At first the stress inside the Si within some microns distance to the TSV edge is influenced by the interaction with the W metallisation. And secondly the influence of the SLID bond increases strongly by increasing the distance to the TSV edge. This is especially seen for the contour plots of σ_1 and σ_3.

2. Methods for Stress Determination

XSA by HR-XRD. The XSA make use of the fact that the Bragg angle ϑ_0 is correlated to the lattice plane spacing in a crystal by the Bragg-Equation. Its partial derivative gives an expression of the deviation of the Bragg angle $\Delta\vartheta$ as strain $\varepsilon_{(hkl)}$ of the normal **r** to the lattice planes (hkl) caused by mechanical stress:

$$\varepsilon_{(hkl)} = \Delta d / d_0 = -(\Delta\vartheta)/\vartheta_0$$

The strain $\varepsilon_{(hkl)}$ is as projection of the strain tensor **ε** on to **r**:

$$\varepsilon_{(hkl)} = r_i \cdot r_j \cdot \varepsilon_{ij}$$

This equation permits access to a full strain tensor by obtaining the strain $\varepsilon_{(hkl)}$ for a minimum of six linearly independent (hkl). Applying the linear theory of elasticity determines than the stress tensor if elastic constants are known.

XSA was reported at Cu-TSVs by means of µSXRD [10]. This technique bases on Laue-Backscatter-Patterns with a high level of automation [11]. µSXRD was successfully employed by different groups [12, 13].
XSA at Cu-TVSs obtained where strain were obtained by HR-XRD a higher angle resolution from a 2ϑ-ω-scan [14]. For use as a matter of routine only a few beamlines at synchrotron sources are available which provides a sufficiently small spot.

Table 2: Properties of detectors used for HR-XRD

Detector	Material	Pixel Number	Detector Size	Pixel Depth	Read out Time
Mythen (SLS detector group)	Si	1 x 1280	50 µm x 8 mm	18 bit	250 µs
Cyberstar (FMB Oxford / Oxford Danfysik)	NaI	1 x 1	Ø 30 mm	-	-

Experiments in this paper were performed for determination of residual stress inside Si of the top die nearby W-TSVs and close to SLID-bonds. All HR-XRD experiments were carried out at the beamline P08 at PETRA III (DESY). That beamline is described in detail elsewhere [10]. The beamline was run in *Micro Mode*. That mode provides a beam with fine line focus of a nominal cross section of 30 µm x 2 µm. Its divergence is in the order of 1/100°. Thus the width of a rocking curve obtained in a Si single crystal is in the same order. In Si wafers the typical misorientation of crystals are in the order of 1/10°. The die-to-wafer bonding process is usually even less accurate. At our samples we measured misorientations between top and bottom die of a few degrees.

Since we do not perform diffraction on higher indexed lattice planes we can therefore assume that in our samples diffraction peaks of both dies can easily be separated.

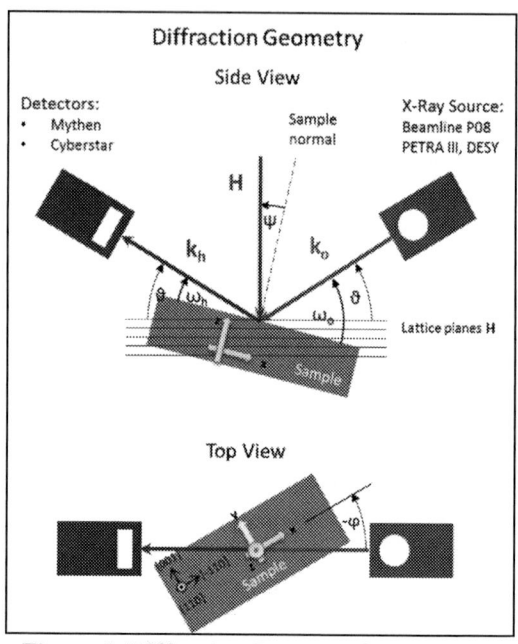

Figure 5: Diffraction Geometry for HR-XRD.

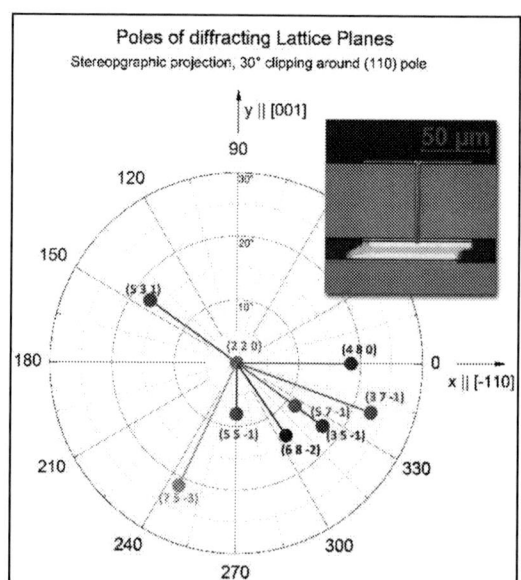

Figure 6: Reflex normals in stereographic projection around (1 1 0)-pole (30° clipping).

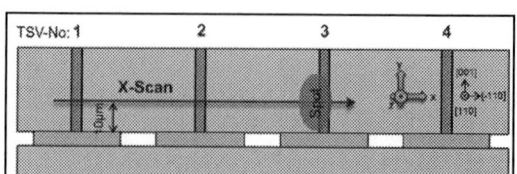

Figure 7: Path of line scan for HR-XRD.

We used a (6+2)-Circle-Diffractometer from Kohzu Precision, Kawasaki, Japan. The main axes provide a precision of von 2×10^{-5} deg. For registration of angle-

978-1-4799-2408-0/14 $31.00 © 2014 IEEE

dispersed X-rays a Mythen line detector was used. Some of its properties are shown in Table 2. The line was aligned to the tangential at the θ-circle on the detector side Figure 5, side view), scanning the 2θ-space. Its resolution was determined to 1.63×10^{-3} deg per pixel. For easier peak search a Cyberstar Counter supported the adjustments and alignments (Table 2).

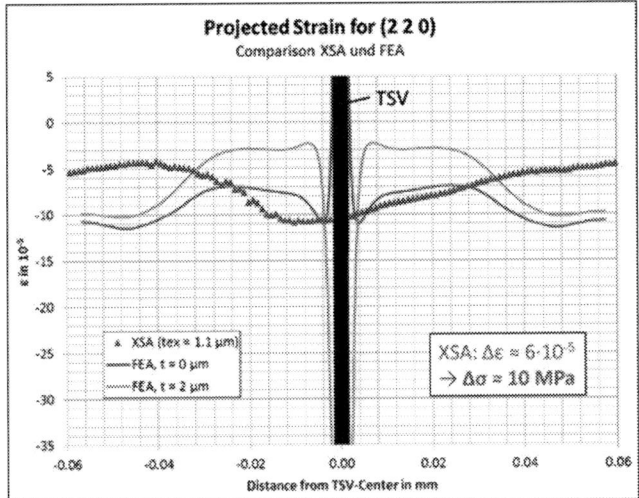

Figure 8: Projected strain for (2 2 0). Comparison between XSA and FEA for a scan along the x-direction (10 μm distance to the SLID-Bond).

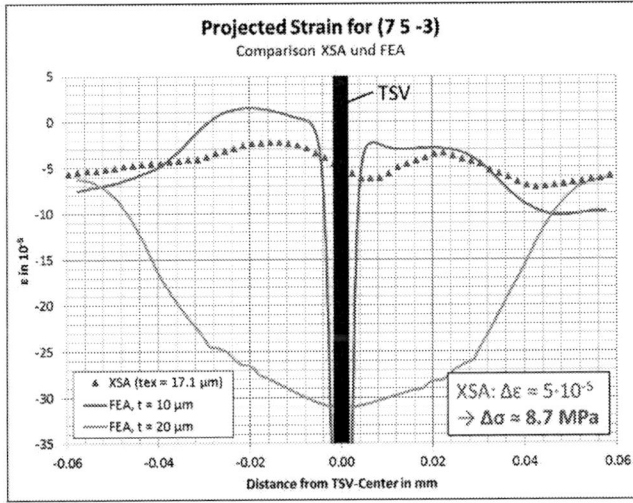

Figure 9: Projected strain onto (2 2 0)-plane (a) and (7 5 -3)-plane (b). Comparison between XSA and FEA for a scan along the x-direction at 10 μm distance to the SLID-Bond.

We determined an energy of 17.486±0.003 keV using a stress free Si analyser crystal. The diffraction geometry with relevant angles is shown in Figure 5. The poles of the lattice planes for the diffraction experiments are shown at Figure 6 in stereographic projection around the (1 1 0). These poles represent measuring directions and it is seen that all of them have a strong out-of-plane component. For every plane the reflexes were obtained from line scans on a path over four TSVs along the x-direction with 10 μm distance to the SLID edge (Figure 7). The positions of the TSVs were determined by intensity minimums of the detector signal. These

minimums are due to shadowing of the W in TSVs. Analysing the diffraction peaks of different lattice planes an X-ray stress analysis (XSA) can be done to obtain strain or stress tensor [16].

Two of the measured reflexes are to be discussed in that paper. Figure 8 illustrates the influence of the W-TSV and the SLID bond on the spacing of the lattice planes (2 2 0) and Figure 9 the influence on (7 5 -3). The results of the XSA are obtained from a scan along the x-direction at 10 μm distance to the SLID-Bond (Figure 7). They are compared to two plots obtained from FEA at different depths with respect to the extinction depth of diffraction at specific lattice planes.

(2 2 0): The lattice plane for the (2 2 0) X-ray reflex lies in the surface of the (1 1 0) cross section. The measuring direction [1 1 0] lies in the TSV-centre (x=0), a structural mirror plane. The diffraction geometry and the strain distribution are therefore symmetrical to x=0. The projected spot size into the x-direction was 23.2 μm, which is relatively large due to the low incident angle and produces a strong convolution of the diffraction signal. The extinction depth (tex) is 1.1 μm. From FEA thus x-paths at surface and in depth of 2μm were chosen in Figure 8. They show a projection of strain onto the (2 2 0) normal. Both FEA graphs reveil small compression of some 10^{-5} and look similar at larger distance from TSV. The compression at the surface is slightly higher than just below the surface. The difference between both paths increases with decreasing distance to the TSV. Within 20 μm distance from the TSV centre the strains differ strongly and show a kind of reverse behaviour. This is due to rearrangement of stress during cross sectioning as well as the slight influence of stress around the 5 μm deep corners of the TSV. Due to the higher CTE of Cu and Cu_3Sn, the Si is under tension in x-direction and the surface is bent at material interfaces. Due to lateral contraction Si is under compression normal to (220). If the strong gradients of both FEA paths are neglected, one finds ε varying within 10^{-4}. The values of the XSA cover a slightly smaller range of 6×10^{-5}. This is plausible in consideration of convolution and corresponds to stress of 10 MPa, if shear is neglected. The global strain is satisfyingly matched. The strong gradients near the TSV couldn't be resolved with the XSA. The decreasing strain towards the TSV correspond with the FEA distribution below surface (t =2 μm), but seems a little bit stronger than predicted. The relative compression can be explained by the higher Cu-CTE as well as the shrinkage of Cu_3Sn after the SLID formation. The XSA distribution doesn't show the maximum at distance of 30 μm and the decrease of strains at larger distance from TSV.

(7 5 -3): For the (7 5 -3) X-ray reflex the measuring direction is off any symmetry axis or plane, so that asymmetrical strain distributions were expected Figure 9). The diffraction at (7 5 -3) yields an extinction depth of 17.1 μm. For comparison FEA-paths at depths of 10 and 20 μm were chosen. The W-TSV reaches only 5 μm deep so that its influence onto the strain distribution should be weaker with increasing depths. Both FEA-plots differ strongly and cross each other near the SLID edges at x ≈ ±40 μm. The curve at 20 μm only shows influence of the SLID-bond. The projected spot size of the X-ray beam is 16.4 μm in x-direction, hence the convolution is weaker than for the (2 2 0) reflex resulting in less blurring of the strain distribution profile. The XSA results match the FEA results of the path in 10 μm depth well. Since the extinction is nonlinear with depth and the majority of X-rays are coming from depths below 16.4 μm, it is reasonable that the XSA results should match the FE-result in 10 μm depth better than in 20 μm. The XSA results cover a strain range of 5×10^{-5} which resembles about 9 MPa, if shear is neglected. If we keep in mind that the measuring signal is strongly convolved we can expect actually higher strain variations than predicted by the FEA. This is an indication that the FEA underestimates stresses in its model.

(6 8 -2): The effect of convolution on the peak shift will become more obvious, if we have a look onto the diffraction peak profile. Figure 10 shows the Intensity over the Bragg angle deviation $\vartheta - \vartheta_0$ for the (6 8 -2)-reflex (for that reflex see also Figure 6). The three graphs depict diffraction peaks at different distances from the TSV-center. Far away from the TSV (x = 50 μm) a narrow diffraction peak forms because only small stress gradients inside the excited diffraction volume exists. If we assume only minor influence from the SLID-Bond the intensity maximum indicates global background stress in sample. Closer to the TSV (x = 15 μm) we find a small broadening of the peak – especially at the right high intensity flank. That indicates a small stress gradient inside the diffraction volume. This gradient is represented by a peak shift compared to the first peak. The third peak is taken right over the TSV-center. The projected spot size is large enough to excite Si on both sides of the TSV. The resulting diffraction peak has a distinctive diffuse diffraction profile covering more than 0.05° at its right side. This reflects a strong stress gradient. The extinction depth for the (6 8 -2)-reflex in our experiments was is 20.1 μm. With this we cannot distinguish, whether the gradient lies in lateral or vertical direction, it's just located somewhere in the diffraction volume. The peak position is determined by the center of mass. Therefore it does not represent those strong gradients, because the diffuse part has only marginal impact on the center of mass.

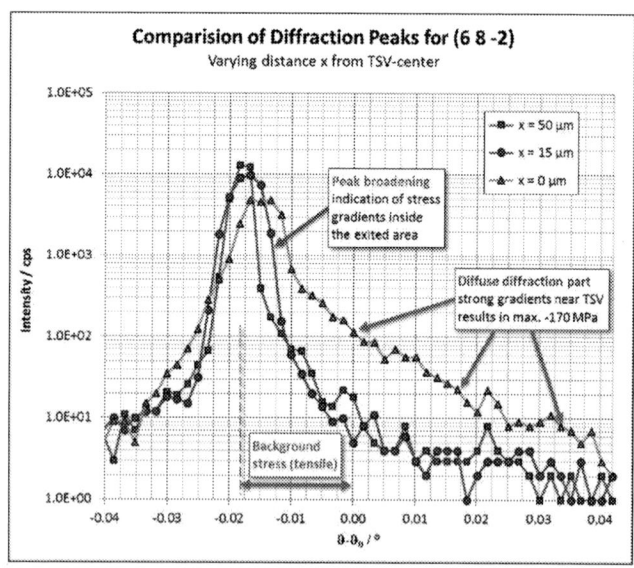

Figure 10: Comparison of diffraction peak profiles for the (6 8 -2) for varying distances to TSV-center. ϑ is the diffraction angle and ϑ_0 the Bragg angle.

The actual distributions of the projected strain for the (6 8 -2)-plane looks similar to the one of (7 5 -3) in Figure 9. For the FEA we find right at the TSV maximum strain gradients covering 3×10^{-4}, resulting in approximately 60 MPa, if shear stress is neglected. This is in contrast to our total peak shift of for (6 8 -2), that gives a maximum of 170 MPa compressive stress. Consequently, we assume that at the measured location the out-of-plane component of the stress tensor is larger than predicted by the FEA.

MicroRaman measurements. Measurements on electronics devices have been applied already in the 1990s [17]. Although the method exhibits clear advantages like fast stress mapping, stress extraction from Raman line shift features challenges for many objects of interest. As far as Raman shifts commonly depend on more than one stress tensor component, straightforward stress extraction is not possible from a measured single Raman shift [18]. Attempts have been made in recent years to overcome this handicap [19], basically incorporating the excitation of more than one phonon. In some cases combined data analysis from both Raman measurement and Finite Element Simulation can be applied [20]. Utilizing this approach, Raman shift calculation from stress tensor values obtained by FEA has been used in this paper to compare FEA and measurement stress data.

Measuring stress in Si nearby TSVs, one has to mind the following circumstances:

- Measuring on cross sectioned samples, as in this paper, implies a severe strain/stress re-distribution at the new free surface. Stresses in the volume of the original untouched specimen may derived by adapting appropriate FEA.

- Near surface stresses originating from strain mismatches at material interfaces often reveal a fast strain/stress re-distribution with increasing measurement depth. For this reason all measurements on TSVs reported here have been excited with a He-Cd laser wavelength of 325 nm. The penetration depth in Si [21] is limited to approximately 10 nm only.

- Approaching the material interface with the scanning laser spot, different kind of artefacts can appear. Corresponding artificial peak shifts in the vicinity of the interface may distort the shift value over a final approach distance of ~ 1 μm. These errors cannot be avoided by corrections with laser plasma lines, which behave not the same way (see [20]).

Our Raman measurements in Si surrounding tungsten TSVs were carried out on cross sectioned specimen, as line scans perpendicular to the trench direction with different distances to the SLID bond. The cross section has a (1 1 0) orientation.

Figure 11 shows the measured Raman shifts vs. the distance to the TSV center. In comparison, computed Raman shifts for different phonon excitation are given, which have been calculated from stress values obtained by FEA.

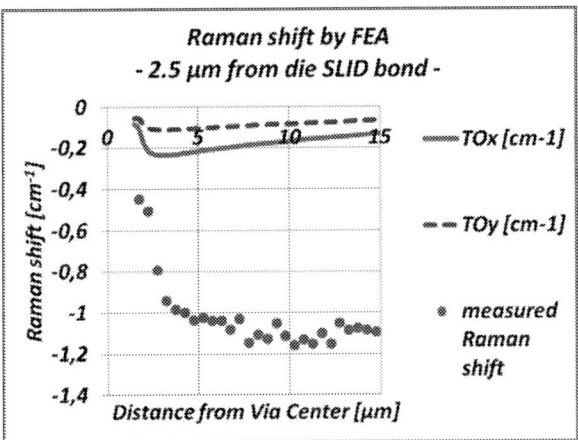

Figure 11: Raman shift along a line perpendicular to the TSV direction. Measurement on cross sectioned sample (110). Besides the measured data computed shifts are given, which would be expected for different phonon excitation, if simulated stresses from FEA are used.

The comparison of the measured Raman shift with the simulated shifts (Figure 11) reveals much higher measured shifts than simulated, independent from the selected phonon for the calculated peak shift. Namely, the steep shift jump for the 2..3 μm gap adjacent to the TSV is not reflected by the FEA data. The measured stress change over this distance is at least five times higher than the expected from FEA. It indicates normal stress levels not less than at least 500 MPa. For this estimation, a possible mutual compensation of stress tensor components in the resultant Raman shift is neglected and could lead to even higher values.

fibDAC stress relief measurements. Stress determination measuring deformation due to stress relaxation after local material removal is a classic experimental method. In the past years this method has been adopted on high spatial resolution treatment and imaging tools. So, focused ion beam (FIB) milling in combination with stress relief measurements by Digital Image Correlation (DIC) techniques were used, starting from a first publication in 2003 [22]. Several kinds of this approach have been reported later, e.g. [23-25].

The type of measurement presented in this paper uses the focused ion beam (FIB) to mill trenches of approximately 100 nm width into the surface of the specimen to trigger stress relief in its vicinity. Capturing the corresponding deformation on high resolution SEM micrographs by local digital image correlation (DIC), the original stress can be determined by simulating the elastic stress relief process with the help of finite element analysis. This simulation starts from an arbitrarily chosen residual stress value. Then, stress relief displacements are scaled linearly until they fit the experimental values obtained by DIC. Because of the linear elastic material relaxation after ion milling, the same scaling factor can be used to determine the true stress at the place of ion milling. This trench technique has been established under the name "fibDAC stress relief technique" [26, 23]. In the present work a new fibDAC approach is introduced allowing stress determination in gradient fields with a dedicated spatial resolution by using a single trench milling only.

The fibDAC method has been used to measure the stress gradient nearby the TSV. The stress relief trench was placed perpendicular to the TSV direction on the cross sectioned TSVs (see Figure 12), i.e. stress sensitivity is maximised for the normal stress component in TSV direction.

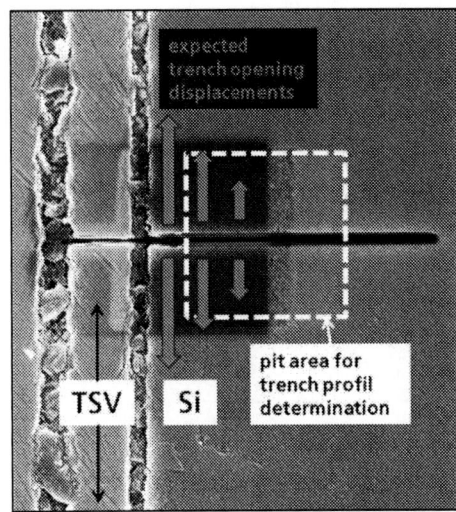

Figure 12: fibDAC trench cutting through the W-TSV and the adjacent Si. Bold arrows schematically depict trench opening displacements. After fibDAC measurement a small pit is ion milled in order to get imaging access to the real trench profile.

After the images for DIC analysis have been picked up, a pit is milled into silicon to capture the trench profile, which is introduced in the finite element stress relief model.

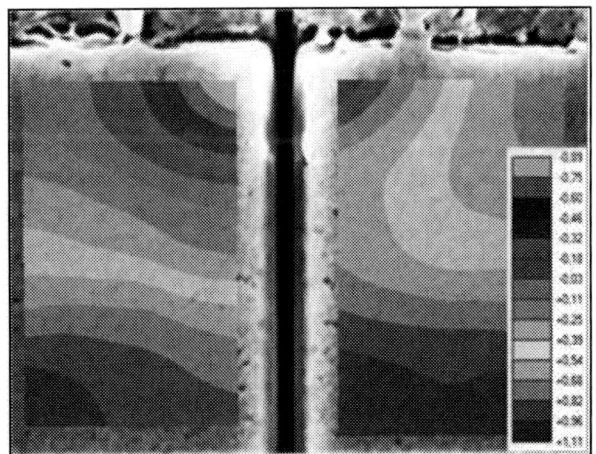

Figure 13: Isoline plot of the trench opening displacement fields in pixel units. Decreasing displacement discontinuity between the trench edges with increasing distance from the TSV (at the top if the image). In comparison to Figure 12 this top view has been clockwise rotated over 90°.

Figure 13 depicts the experimental DIC displacement field, respectively. The found displacement field indicates a trench opening after ion milling which refers to tensile stress. The trench opening amount rapidly decreases with increasing distance from the W-Si interface, i.e. the measured stress is caused mostly by the strains from TSV processing. The trench opening is caused by Si stress (and not by W stress), because similar opening takes place, if the trench is not extended into the W-TSV.

Stress extraction from the DIC displacement field based on the described above fitting procedure between scaled FEA and measured DIC displacements. Standard fibDAC procedures commonly look for stress relaxation displacements in the middle of a longer trench [26], postulating relaxation displacements independent of the position along the trench. Because stresses change in our case within the evaluated area with increasing distance to the TSV, thin horizontal slices of the displacement field in Figure 13 have been analysed separately under the assumption of approximately constant relaxation displacement in each of the slices.

Resultant stress values over distance to the TSV center is given in Figure 14. For comparison respective data from FEA for the same stress tensor component s_{yy} has been added.

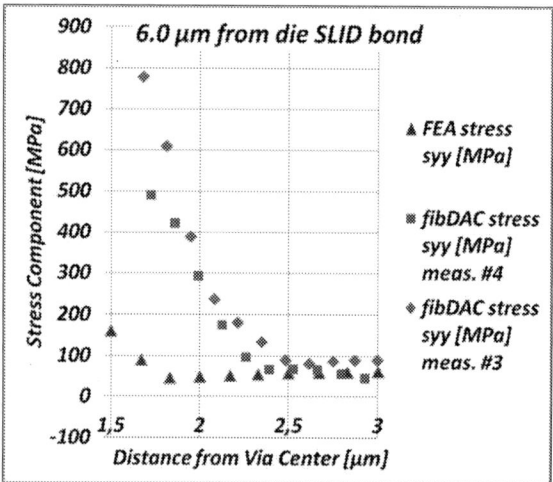

Figure 14: Normal stress along TSV direction (Figure 12) in silicon approaching a W-TSV of 3 µm width. Most stress disappears over a 1 µm distance away from the TSV rim. The remaining far reaching stress of approx. 100 MPa (in experiment) probably is due to the Cu-Sn-Cu bond underneath the TSV. Measurement #3 is the resultant data referring to the displacement relaxation field in Figure 13 and the trench top view of Figure 12.

Measured stresses exceed significantly the simulated ones: The large increase of experimentally found stress approaching the TSV interface is not reflected by simulation data. Both measurements exhibit similar stress behavior and values.

Analyzing the results of the fibDAC measurements (Figure 14) a similar conclusion is to be made as for Raman. The measured stress increase approaching the TSV interface exceeds the simulated at minimum by a factor five. This large stress gain appears over a 1 µm distance, which is somewhat faster than shown by Raman. Considering the lower lateral spatial resolution of Raman (~ 1 µm vs. 100 nm for fibDAC), it seems reasonable to assume, to a certain extent, a data smoothing of the steep increase by Raman scanning. fibDAC

measurements are in well agreement with FEA for distances larger 2.5 µm from the center of the TSV.

Probably stresses for larger distances from the TSV are induced rather by the SLID bonding underneath the TSV. In Raman shifts as well as in fibDAC stress they keep constant over a larger detected distance aside the narrow TSV and are of the magnitude of about 50..100 MPa for the s_{yy} stress component. Concluding, the mutual comparison between Raman, fibDAC and FEA data indicates, that a severe increase of stresses in Si over the final 1 µm to the TSV exists. It reaches at least some hundreds of MPa and seems to originate from the TSV mismatch to the Si. It is not reflected by the FEA, which still is not well understood. I.e. the performed FEA modeling obviously does not describe the very near interface behavior sufficiently, but explains quite well the influence of the SLID bonding between the dies.

3. Discussion

Stress values have been determined in Si around tungsten TSVs within an area of maximum 60 µm. The applied methods possess different spatial resolutions and realize dissimilar access to particular strain / stress components as well. As a matter of fact, comparison between results from different techniques has to take into account the distinguished capabilities of methods applied. For that reason Table 3 summarizes them, indicating Cartesian directions as shown in Figure 2. Listed available data refers to the particular experiments carried out.

Table 3: Overview on the conditions of stress determination

Measurement method	Lateral spatial resolution [µm]	Available data
Finite element analysis (FEA)	optional, depending on meshing	any of the strain or stress tensor components
HR-XRD w/ XSA	10 ... 25 µm	Generally full strain or stress tensor, in this paper strain projected on (2 2 0) and (7 5 -3) directions in the unit cell system
microRaman	0.7 ... 1 µm	Raman wavelength shift, being an arithmetic function of normal stress tensor components
fibDAC stress relief technique	100 nm	normal stress s_{yy}

A first look at Table 3 reveals that two of the measurement techniques allowed a rather coarse lateral spatial resolution in the micrometer region, whereas the fibDAC method provides a significant higher resolution of 100 nm. As a result the fibDAC tool only gives a realistic insight view to stress built-up due to the Si substrate/W-TSV mismatch, as far as this stress degrades over a distance ≤ 1 µm.

For HR-XRD measured changes of strain may cover the range of some 10^{-5} (Figure 8). This results in stress variation between 5 and 10 MPa if shear is neglected. This high stress resolution is achieved by HR-XRD only. In the opposite, microRaman and fibDAC measurements provide resolutions smaller than 10 MPa. XRD results show the high precision of the 2θ-Sans in combination with the XSA. The XSA unfolds its full potential in out-of-plane measuring direction. Strong gradients have not been resolved by standard XSA due to the relatively large projected spot size. Here a profile analysis of the diffraction peaks gives better results.

One drawback of the HR-XRD by scanning single reflexes sequentially is its long adjusting time. Here the µSXRD method is faster and provides therefore an alternative to map strains around TSVs [11]. But due to its smaller angular resolution the precision is somewhat worse. Another issue is that large in-plane components in XRD-measuring directions require low incident or exit angles. This results in very large projected spot sizes and therefore in low spatial resolution. Here XRD beamlines with a sufficiently small spot size below 1 µm are required.

In case of small strain gradients the match between FEA and XSA is satisfying. The results give an indication that there is more stress in systems than predicted by FEA. Koseski showed that intrinsic (athermal) stress in W may be responsible for higher stress values in W-TSVs [27]. For W-CVD at 400°C, diffusion, crystal regeneration, recrystallization or creep are blocked due to high melting temperature of W, thus giving a significant contribution to intrinsic stress [28].

However, judging the graphs from Figure 8 till Figure 10 it has to be stated that the XSA measurements average over the penetration depth. I.e. they will not coincide with FEA data representing a particular depth under the surface. As far as both strain projections in XSA are equal or nearby to the surface normal, stress in the captured surface layer is small and changes rapidly starting from a plane stress state at the very surface to moderate stresses in projection direction, when the penetration depth is reached.

The comparison of the measured Raman shift with the simulated shifts Figure 11 reveals much higher measured shift than simulated, independent on the selected phonon for the emulated peak shift. Namely, the steep shift jump for the 2..3 µm gap adjacent to the TSV is not reflected by the FEA data. The measured stress change over this distance is at least 5 times higher than the expected from FEA. It indicates normal stress levels not less than 500 MPa. For this estimation, a possible mutual compensation of stress tensor components in the resultant Raman shift is neglected and could lead to even higher values.

Comparing HR-XRD stress levels with that of Raman and fibDAC, it has to be taken into account that HR-XRD presents basically out-of-plane stresses with highest precision, which are rather low nearby the surface, whereas the other measurements correspond to the higher in-plane stresses.

4. Conclusions

Four different methods have been utilized to obtain a comprehensive view on stress development in silicon surrounding tungsten TSVs after via processing and SLID bonding of dies. Measurements were performed on cross sectioned specimens. All three experimental methods show the influence of W and Cu-Sn-Bond in Si. The influence of W ranges only about some µm and produces a very strong stress-gradient in Si. In larger distances the influence of SLID-Bond dominates.

HR-XRD was used to study stresses in a nearby surface layer with emphasis on out-of-plane components. For small stress gradients the analysis of the peak position gives reasonable results. For stronger gradients and larger spot sizes of the X-ray beam, a profile analysis of the diffraction peak gives more accurate results. Those results indicate higher stresses in Si around the W-TSV than predicted by FEA.

Raman and fibDAC measurements gave access to in-plane components. A severe stress increase in a 1 µm thick layer adjacent to the TSV was found, as well as a stress offset level above the SLID bonding width. The latter could have been well described by Finite Element Analysis, whereas the TSV induced stresses were not reflected accordingly by simulations.

The results show that intrinsic (athermal) stress in W may have to be considered. Additionally more attention should be directed to the accuracy of the FE-modelled Cu-Sn SLID bond causing large background stresses around the TSV, to account for shrinkage during phase transition as well as the thermo-mechanical properties of Cu_3Sn and Cu_6Sn_5

We are convinced that this approach provides further insight into the full 3D stress distribution around TSVs during processing and fatigue testing, thus serving as corrective for life time model assumptions, and is transferable to Cu-TSVs and liners.

Acknowledgments

The work was part of the project "Best-Reliable Ambient Intelligent Nano Sensor Systems" (eBRAINS). The project is funded under the Seventh Framework Programme for Research of the European Commission (ICT-257488).

We thank Oliver Seek and his team from beamline P08 at PETRA III (DESY) for the support during our HR-XRD experiments in Hamburg.

Raman measurements have been performed at Fraunhofer IZM. We acknowledge the assistance by A. Gollhardt.

References

1. B. Wunderle, R. Mrossko, O. Wittler, E. Kaulfersch, P. Ramm, B. Michel und H. Reichl, „Thermo-Mechanical Reliability of 3D-integrated Microstructures in Stacked Silicon" MRS Online Proceedings Library, Bd. 970, 2006.

2. B. Wunderle und B. Michel, „Lifetime modelling for microsystems integration: from nano to systems" Microsystem Technologies, Bd. 15, Nr. 6, pp. 799-812, 2009.

3. I. De Wolf, V. Simons, V. Cherman, R. Labie, B. Vandevelde und E. Beyne, „In-depth Raman spectroscopy analysis of various" in Proc. of 62th IEEE ECTC, San Diego, 2012.

4. J. V. Olmen, C. Huyghebaert, J. Coenen, J. V. Aelst, E. Sleeckx, A. V. Ammel, S. Armini, G. Katti, J. Vaes, E. Beyne und Y. Travaly, „Integration challenges of copper Through Silicon Via (TSV) metallization for 3D-stacked IC integration" Microelectronic Engineering, Bd. 88, Nr. 5, pp. 745-748, 2011.

5. U. Zschenderlein, PhD-Thesis:"Zerstörungsfreie Eigenspannungsbestimmung für die Zuverlässigkeitsbewertung 3D-integrierter Kontaktstrukturen in Silizium", Chemnitz, 2013.

6. P. R. Raffo, „Yielding and fracture in tungsten and tungsten-rhenium alloys", Journal of the Less Common Metals, Bd. 17, Nr. 2, pp. 133-149, 1969.

7. P. Ramm, M. J. Wolf und B. Wunderle, „Wafer-Level 3D System Integration" in Handbook of 3D Integration, Wiley-VCH Verlag GmbH & Co. KGaA KGaA (ISBN 978-3-527-32034-0), 2008, p. 289–318.

8. P. Garrou und C. Bower, „Overview of 3D Integration Process Technology" in Handbook of 3D Integration, Wiley-VCH Verlag GmbH & Co. KGaA KGaA (ISBN 978-3-527-32034-0), 2008, p. 25–44.

9. J. R. Davis, ASM handbook. Vol. 2, Properties and selection : nonferrous alloys and special- purpose materials, Materials Park: ASM International, 1992.

10. A. S. Budiman, H. A. S. Shin, B. J. Kim, S. H. Hwang, H. Y. Son, M. S. Suh, Q. H. Chung, K. Y. Byun, N. Tamura, M. Kunz und Y. C. Joo, „Measurement of stresses in Cu and Si around through-silicon via by synchrotron X-ray microdiffraction for 3-dimensional integrated circuits", *Microelectronics Reliability,* Bd. 52, Nr. 3, pp. 530-533, 2012.

11. N. Tamura, A. A. MacDowell, R. Spolenak, B. C. Valek, J. C. Bravman, W. L. Brown, R. S. Celestre, H. A. Padmore, B. W. Batterman und J. R. Patel, „Scanning X-ray microdiffraction with submicrometer white beam for strain/stress and orientation mapping in thin films," Journal of synchrotron radiation, Bd. 10, Nr. Pt 2, pp. 137-143, 2003.

12. C. Okoro, L. E. Levine, R. Xu, J. Tischler, W. Liu, O. Kirillov, K. Hummler und Y. S. Obeng, „X-Ray Micro-Beam Diffraction Determination of Full Stress Tensors in Cu TSVs" in Proc. 63rd IEEE ECTC, Las Vegas, 2013.

13. T. Jiang, C. Wu, P. Su, X. Liu, P. Chia, L. Li, H. Y. Son, J. S. Oh, K. Y. Byun, N. S. Kim, J. Im, R. Huang und P. S. Ho, „Characterization of Plasticity and Stresses in TSV Structures in Stacked Dies Using Synchrotron X-ray Microdiffraction" in Proc. 63rd IEEE ECTC, Las Vegas, 2013.

14. O. Nakatsuka, H. Kitada, Y. S. Kim, Y. Mizushima, T. Nakamura, T. Ohba und S. Zaima, „Characterization of Local Strain around Through-Silicon Via Interconnects by using X-ray Microdiffraction" *Japanese Journal of Applied Physics,* Bd. 50, Nr. 5, 2011.

15. O. H. Seeck, C. Deiter, K. Pflaum, F. Bertam, A. Beerlink, H. Franz, J. Horbach, H. Schulte-Schrepping, B. M. Murphy, M. Greve und O. Magnussen, „The high-resolution diffraction beamline P08 at PETRA III" Journal of Synchrotron Radiation, Bd. 19, Nr. 1, pp. 30-38, 2012.

16. L. Spieß, G. Teichert, R. Schwarzer, H. Behnken and C. Genzel, Moderne Röntgenbeugung: Röntgendiffraktometrie für Materialwissenschaftler, Physiker und Chemiker, Wiesbaden: Teubner, 2008.

17. I. d. Wolf, "Micro-Raman spectroscopy to study local mechanical stress in silicon integrated circuits" Semicond. Sci. Technol., vol. 11, pp. 139-154, 1996.

18. Qiu Zhao, J. Im, R. Huang and P. S. Ho, "Extension of Micro-Raman Spectroscopy for Full-Component Stress Characterization of TSV Structures" in Proc. of ECTC, Las Vegas, 2013.

19. I. de Wolf, V. Simons, V. Cherman, R. Labie, B. Vandevelde and E. Beyne, „In-depth Raman Spectroscopy Analysis of Various Parameters Affecting the Mechanical Stress" in Proc. of IEEE, San Diego, 2012.

20. D. Vogel, E. Auerswald, J. Auersperg, S. Rzepka, B. Michel, "Measuring Techniques for Deformation and Stress Analysis in Micro-Dimensions" in Proc. of EuroSimE, Wroclaw, 2013.

21. D.E. Aspnes, A.A. Studna, „Dielectric functions and optical parameters of Si, Ge, GaP, GaAs, GaSb,InP, InAs, and InSb from 1.5 to 6.0 eV" Phys. Rev., Bd. B 27, p. 985, 1983.

22. K.J. Kang, N. Yao, M.Y. He, A.G. Evans, "A method for in-situ measurement of the residual stress in thin films by using the focused ion beam" Thin Solid Films, vol. 443, pp. 71-77, 2003.

23. N. Sabaté, D. Vogel, A. Gollhardt, J. Keller, B. Michel, "Measurement of residual stresses in micromachined structures in a micro region" Appl. Physics Letters, vol. 07910, 2006.

24. A.M. Korsunsky, M. Sebastiani, E. Bemporad, "Focused ion beam ring drilling for residual stress evaluation" Materials Letters, vol. 63, pp. 1961-1963, 2009.

25. D. Vogel, S. Rzepka, B. Michel, "Local stress measurement on metal lines and dielectrics of BEoL pattern by stress relief technique" in Proc. of Semiconductor Conference, Dresden, 2011.

26. D. Vogel,A. Gollhardt, N. Sabate, J. Keller, B. Michel, H. Reichl, "Localized Stress Measurements – A New Approach Covering Needs for Advanced Micro and Nanoscale System Development" in Proc. of 57th IEEE ECTC, Reno, 2007.

27. R. P. Koseski, W. A. Osborn, S. J. Stranick, F. W. DelRio, M. D. Vaudin, T. Dao, V. H. Adams und R. F. Cock, "Micro-scale measurement and modeling of stress in silicon surrounding a tungsten-filled through-silicon via" J. of Appl. Phys., 110 (7), 2011.

28. G. J. Leusink, T. G. M. Oosterlaken, G. C. a. M. Janssen, S. Radelaar, "The evolution of growth stresses in chemical vapor deposited tungsten films studied by insitu wafer curvature measurements", J. of Appl. Phys., 74(6), 1993

Formic Acid Treatment with Pt Catalyst for Cu Direct and Hybrid Bonding at Low Temperature

Tadatomo Suga, Masakate Akaike, Wenhua Yang

The Department of Precision Engineering, The School of Engineering, The University of Tokyo.
Hongo 7-3-1, Bunkyo-ku, 113-8656 Tokyo, Japan.
Mail: ssuga@pe.t.u-tokyo.ac.jp Phone: +81-3-5841-6491

Abstract

A new process of Cu-Cu direct bonding at lower than 200°C in atmospheric atmosphere was developed. The method is composed of formic acid treatment with Pt catalyst. To demonstrate the feasibility of the method, Cu electrodes samples were designed for characterizing the bonding strength and the contact resistance for the Cu-Cu direct bonding. The bonding strength is about 40MPa, and the contact resistance at the bonding interface is about 0.17mΩ. Both of them are better than those without Pt catalyst. The effects of the Pt catalyst on bonding were investigated and it is concluded that the surface oxide of Cu is reduced effectively by the hydrogen radical generated by formic acid decomposition with Pt catalyst and the Cu formats resulted by the formic acid treatment is decomposed to precipitate very fine Cu grains at the bonded interface.

Introduction

Cu electrodes direct bonding is expected for TSV stack integration or low resistance bonding for power devices because cooper is low resistivity material. Usually, the temperature above 350°C is required for Cu/Cu direct bonding [1, 2]. For Cu low temperature bonding process, surface oxide layer should be removed before bonding. Surface activated bonding (SAB) [3-6] using ion bombardment is an effective technology for strong bonding, but treatment and bonding should be performed in high vacuum chamber to keep activated surface. Formic acid also is used to reduce surface oxide for some metals such as Cu, Pb, Sn, and solders for bonding [7-9]. For Cu, however, a treatment temperature is required as higher than 250°C for effective reduction [10]. Using Pt as catalyst, formic acid decomposes easy and produces more H radicals, which may reduce Cu oxide effectively at low temperature. Therefore, formic acid treatment process with Pt catalyst was suggested to reduce Cu surface oxide at 200°C for Cu low temperature bonding [11]. However, the effect of Pt catalyst on the bonding performance and bonding mechanism was not explained clearly.

In this study, we designed Cu electrodes samples for evaluation of bonding performance and study of the bonding mechanism using formic acid treatment with Pt catalyst. Designed samples were treated by formic acid treatment without Pt catalyst and with Pt catalyst respectively for low temperature bonding. After treatment, Cu surfaces were analyzed by X-ray photoelectron spectroscopy (XPS). The bond strength was evaluated by die shear tester, and the contact resistance of the bonded interface was measured using 4-point resistance measurement. The bonded interface was observed using scanning electron microscopy (SEM) and transmission electron microscopy (TEM). Finally, the mechanism of Cu low temperature bonding using formic acid treatment process with Pt catalyst was explored.

Experimental

Figure 1 shows the bonding equipment used to perform the treatment and bonding which combines a formic acid generation system and a bonding chamber. Two different square nozzles were prepared to introduce formic acid vapor to the surface of samples. One nozzle is empty for the treatment without Pt catalyst. Another nozzle is filled with Pt foil and a heater is inserted in the nozzle to heat Pt foil for the treatment with Pt catalyst. N_2 as carrier gas was applied to generate mixed gas (1% formic acid and 99% N_2) from formic acid solution. Samples were treated at 200°C for 20min using formic acid without/with Pt catalyst. After the treatment, the bonding was performed immediately at 200°C under a contact load of 1500N.

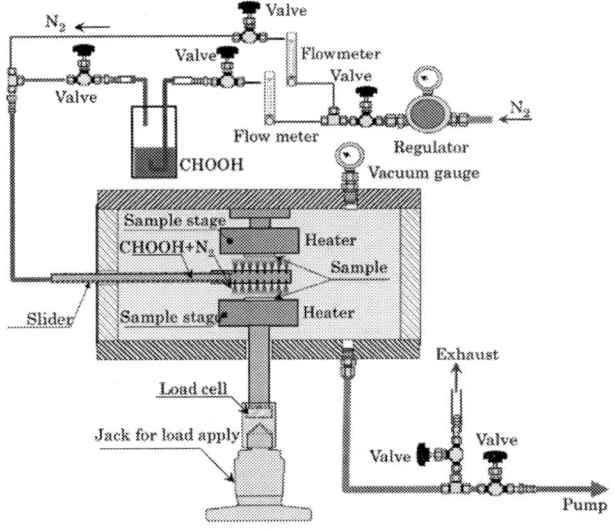

Figure 1. Schematic diagram of bonding equipment.

To measure the contact resistance of Cu bonded interconnects and the bond strength, chip samples were prepared: They consist of Cu electrodes and wiring layers fabricated on 10mm×10mm Si chip. The wiring layers are fabricated as Cu(400nm)/Ti(20nm) on SiO_2, and Cu electrodes Cu(600nm)/Ti(20nm) on the wiring layers using electro-beam deposition. The size of the electrodes is 30μm × 30μm. An example of the samples is shown in Figure 2. The contact resistance of the bonded interface was measured by the four-point probe method.

GC analysis for formic acid decomposition

For the formic acid treatment without Pt and with Pt catalyst, the decomposition of formic acid and H radicals generation was confirmed using a GC analysis. The formic acid flowing through the nozzle is decomposed and H radicals are generated under the catalysis effect of Pt, which was reported previously by the authors [11]. The amount of the H radicals was estimated by H_2 concentration measured by GC analysis for the treatment both without Pt catalyst and with Pt catalyst. Figure 3(a) and (b) show the GC analysis spectra for the treatment both without Pt and with Pt catalyst, respectively. In GC analysis spectra, usually, the peak at 2.3 min indicates H_2. For the formic acid treatment without Pt, no peak appears at 2.3 min, which suggests that H radical is hardly generated during the treatment without Pt catalyst. For the formic acid treatment with Pt catalyst, one peak appears at 2.3 min, which shows H radicals generation during the treatment. The concentration of H_2 is about 0.27%, which was measured through the comparison with standard H_2 GC spectra.

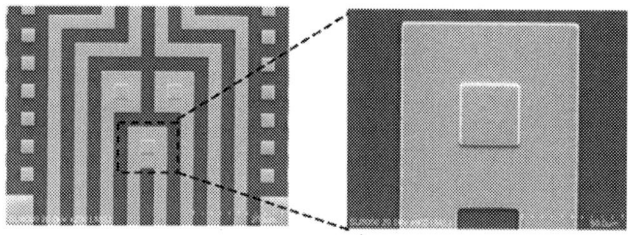

Figure 2. A part of Cu electrode samples.

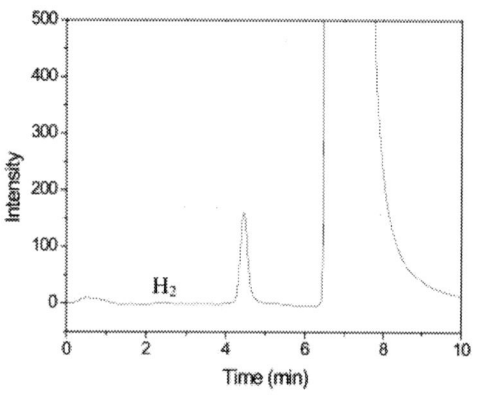

(a) Without Pt catalyst

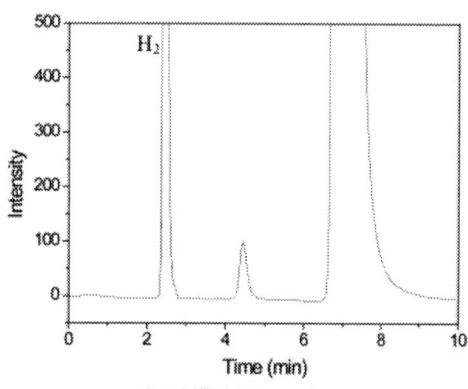

(b) With Pt catalyst

Figure 3. GC analysis for formic acid (a) without and (b) with Pt catalyst.

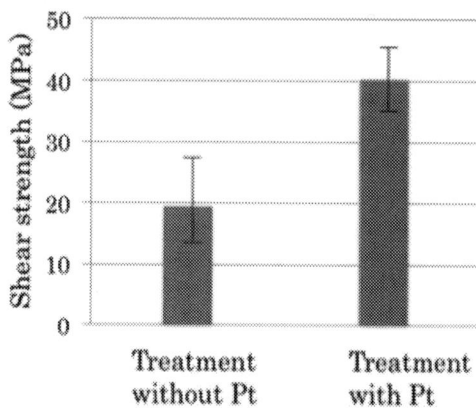

Figure 4. The shear strength of the bonded samples using formic acid treatment with/without Pt catalyst.

Characterization of the bonding of Cu

The bond strength was evaluated using die shear tester. Two samples were bonded at 200°C under a contact load of 1500N for 5min after formic acid treatment without/with Pt catalyst for 20min. Figure 4 shows shear strength for the two formic acid treatment processes with and without Pt catalyst. For the formic acid treatment with Pt catalyst, the shear strength is about 40MPa and 19.5MPa without Pt catalyst, which is half of that with Pt catalyst.

Figure 5 is the contact resistance of the interfaces bonded both without and with Pt catalyst. For the formic acid treatment with Pt catalyst, the contact resistance is about 0.17mΩ, which is lower than that without Pt catalyst. For the Cu bonding well treated with Pt catalyst, the failure occurs in the Cu part whereas only a small area was bonded on electrode surface in case of the formic acid treatment without Pt catalyst. It also suggests the strong bonding and the low contact resistivity using formic acid treatment with Pt catalyst.

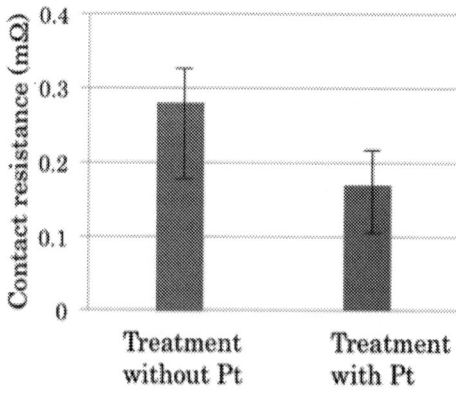

Figure 5. Contact resistance of bonded samples using formic acid treatment with/without Pt catalyst

Observation of the bonded interface

The bonded interfaces for the samples obtained by the formic acid treatment without Pt catalyst and with Pt catalyst were tried to be compared. However, with the standard condition for the bonding: at 200°C under a contact load of 1500N for 5min after formic acid treatment, there were too many gaps and voids found at the bonded interface in case without Pt catalysis to prepare a sample for cross sectional observation of the bonded interface. Therefore an additional annealing time of 120min was applied to the samples bonded without Pt catalysis. The cross sectional images thus obtained for the bonded interfaces for different conditions are shown in Figure 6. Figure 6(a) shows the cross-section image of the bonded interface using the treatment without Pt catalyst. The bonded interface was observed clearly between Cu electrodes, showing an interface layer with a thickness of 10~20nm, which is formed through thermal diffusion between Cu electrodes during the thermal annealing at 200°C for 120min. Figure 6(b) is the bonded interface by the treatment with Pt catalyst. The interface is not a clear straight line, but an uneven wide intermediate layer of a thickness between 100nm and 200nm is visible. The interface layer is composed of Cu fine grains with the size of several tens nm.

XPS analysis for Cu surfaces

In order to investigate the effect Pt catalyst in the formic acid treatment on Cu surface, XPS analysis was performed for the Cu surfaces before treatment and after treated without Pt, and with Pt, respectively. Figure 7 (a), (b) are Cu 2p3/2 spectra and O 1s spectra, respectively. For the Cu 2p3/2 spectra, Cu/Cu_2O peak intensity becomes higher after treatment, and it is highest for the treatment with Pt catalyst probably due to the reduction of Cu_2O by formic acid. For the O 1s spectra, O peak is divided into two peaks of CuO and $Cu_2O/Cu_2(OH)$ and the CuO is reduced effectively by the formic acid treatment.

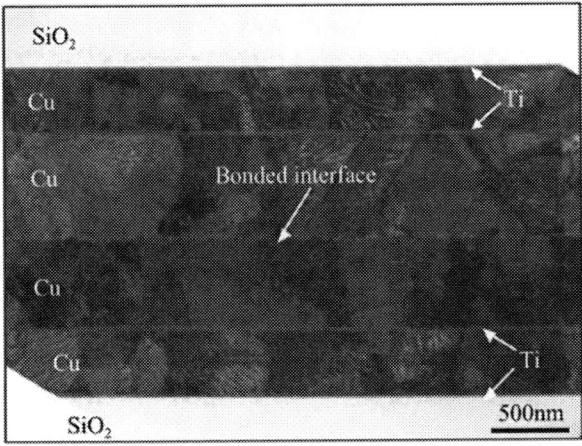

(a) Without Pt catalyst

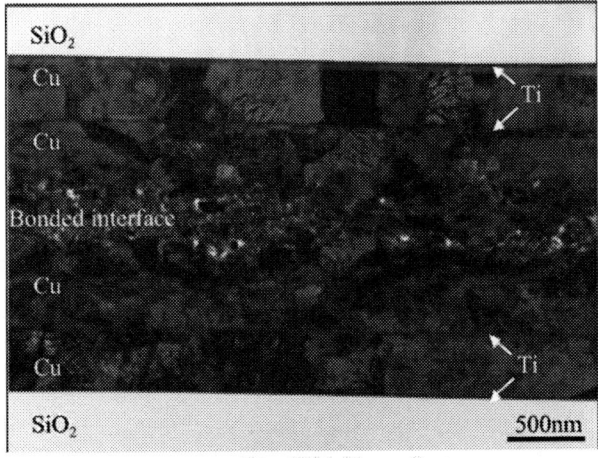

(b) With Pt catalyst

Figure 6. Cross sectional TEM image of the bonded interfaces by formic acid treatment (a) without Pt catalyst, bonded at 200°C under 1500N for120min, and (b) Treatment with Pt catalyst, bonded at 200°C under 1500N for 5min.

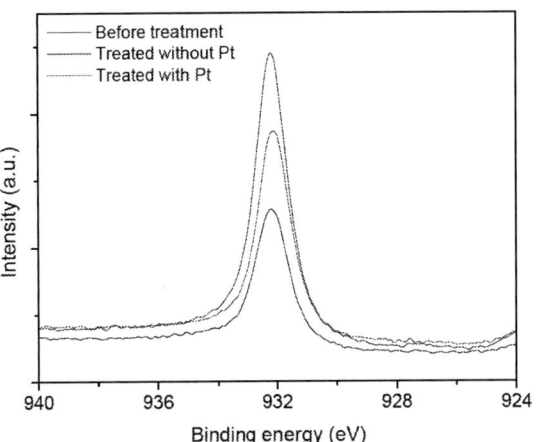

Cu 2p3/2 spectra

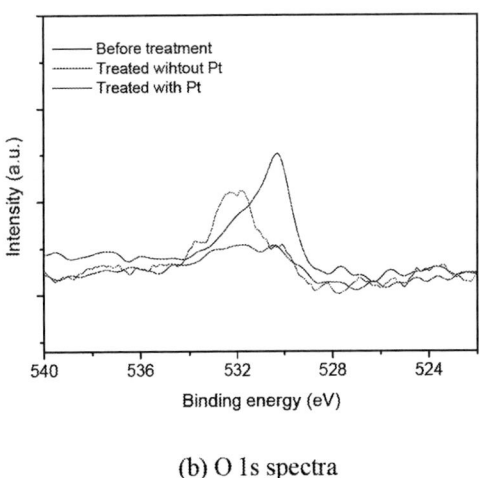

(b) O 1s spectra

Figure 7. XPS Cu 2p3/2 spectra spectra of Cu surfaces before treatment and after treatment without and with Pt catalyst.

Bonding mechanism

It is known that the Cu surface is oxidized at first as CuO, and further oxidation will grow Cu_2O layer under the CuO surface oxide. The thickness of the oxide layers are very thin, such as several nanometer, in case of a clean CMP surface. However the present study showed that the Pt catalyst is much effective to reduce the Cu oxide layer that the treatment without Pt catalyst. Based on the above observations, therefore, a bonding mechanism of Cu at low temperature using the formic acid treatment with Pt catalyst can be assumed as following.

Firstly, Cu surface oxide, probably CuO at the top of the Cu surface, is reduced effectively by H radicals which are generated from the formic acid by its decomposition using Pt as catalyst. Then, the formic acid reacts with further Cu oxide, preferentially Cu_2O, and Cu formate is formed at the surface of Cu. At the same time, Cu formate is decomposed, and Cu nano-particles are precipitated and sintered together to form Cu nano-particles during cooling down. For formic acid treatment without Pt catalyst, the surface oxide cannot be reduced effectively by the formic acid. As the result, only few fewer Cu nano-particles are produced on surface, and Cu grains are not formed at the bonded interface, which causes the weak bonding under same conditions with that with Pt catalyst.

To enhance such effects, the Cu surface was oxidized by annealing at 200°C for 20min in air before the formic acid treatment. Figure 8 shows the SEM image of Cu surfaces before the formic acid treatment, after treatment without Pt, and after treatment with Pt. Few Cu grains were generated on the surface after the treatment without Pt catalyst, whereas more Cu nano particles are generated on the surface after the treatment with Pt catalyst.

Feasibility of the proposed method on hybrid bonding

The proposed method is now being applied to hybrid bonding of TSV, Cu-Cu direct bonding below 200°C. Since the method generate hydrogen radicals without usual plasma process, cleaning of Cu surface can be carried out with very simple manner and there is no contamination from the plasma. Therefore it will be implemented also easily to any bonding equipment. Combining oxidation process with oxygen plasma, or hydrophilic process using water plasma treatment, the proposed method provides more variety of the surface activation method. The oxygen plasma and water vapor treatment were combined using a linear ion source of hollow-cathode type and make the treated surface hydrophilic. The –OH terminated surfaces of silicon oxide or metal oxide are stable, and therefore easily be reduced by subsequently applied process of the formic acid treatment. The process was installed to bonding equipment using ion beam source for combining the surface activation process and the formic acid treatment.

Conclusions

The effect of Pt catalyst on Cu low temperature bonding using formic acid treatment was studied. For the treatment with Pt catalyst, both of bonding strength and electrical performance are better than those without Pt catalyst. Based on the observation of the bonding interfaces for the treatments with/without Pt catalyst, bonding mechanism with Pt catalyst was explained. The bonding interface was composed of Cu grains generated by formate decomposition at Cu surface. It is

easy to realize high quality bonding using Pt catalyst at low temperature because Cu grains are formed easily using Pt catalyst. Therefore, it can be concluded that, using formic acid treatment with Pt catalyst, higher bonding quality can be achieved for Cu low temperature bonding due to Cu grains formation on Cu surface easily.

Acknowledgments

This work was supported by Grant-in-Aid for Scientific Research (A)(22360025)

References

1. Eun-jung, Seungmin Hyun, Hak-Joo Lee, and Young-Bae Park, "Effect of Wet Pretreatment on Interfacial Adhesion Energy of Cu-Cu Thermocompression Bond for 3D IC Packages", Journal of Electronic materials, Vol.38, No.12, 2009, pp. 2449-2454.

2. Rajappa Tadepalli, Carl V Thompson, "Quantitative Characterization and Process Optimization of Low-Temperature Bonded Copper Interconnects for 3-ID Integrated Circuits", 2003 Proceedings of the IEEE Interconnect Technology Conference, 2003, pp.36-38.

3. T. H. Kim, M. M. R. Howlader, T. Itoh, T. Suga, "Room temperature Cu–Cu direct bonding using surface activated bonding method," Journal of Vacuum Science & Technology A, Vol. 21, No. 2, 2003, pp.449-453.

4. A. Shigetou, T. Itoh, T. Suga, "Direct Bonding of CMP-Cu Films by Surface Activated Bonding (SAB) Method," J. Mater. Sci., Vol. 40, 2005, pp.3149-3154.

5. Akitsu Shigetou, Toshihiro Itoh, Mie Matsuo, "Bumpless Interconnect Through Ultrafine Cu Electrodes by Means of Surface-Activated Bonding (SAB) Method," IEEE Transactions on Advanced Packaging, Vol.29, No.2, (2006), pp. 218-226.

6. T. Suga, "Cu-Cu Room Temperature Bonding-Current Status of Surface Activated Bonding(SAB)", ECS Trans., Vol.3, No.6, 2006, pp.155-163.

7. Wei Lin and Y. C. Lee, "Study of Fluxless Soldering Using Formic Acid Vapor", IEEE Transaction of Advanced Packaging, Vol. 22, No. 40, 1999, pp. 592-601.

8. I. Kotoulas, A. Schizodimou and G. Kyriacou, "Electrochemical Reduction of Formic Acid on a Copper-Tin-Lead Cathode", The Open Electrochemistry Journal, Vol. 5, 5, 2013, pp.8-12.

9. Shinji Koyama, Yukinari Aoki and Ikuo Shohji, "Effect of Formic Acid Surface Modification on Bond Strength of Solid-State Bonded Interface of Tin and Copper", Materials Transactions, Vol. 51, No. 10, 2010, pp.1759-1763.

10. S. Poulston, E. Rowbotham, P. Stone, P. Parlett and M. Bowker, "Temperature-programmed desorption studies of methanol and formic acid decomposition on copper oxide surfaces," Catalysis Letters, Vol. 52, 1998, pp.63-67.

11. Wenhua Yang, Masatake Akaike, Masahisa Fujino and Tadatomo Suga, "A Combined Process of Formic Acid Pretreatment for Low-temperature Bonding of Copper Electrodes", ECS Journal of Solid State Science and Technology, 2 (6), 2013, pp.271-274.

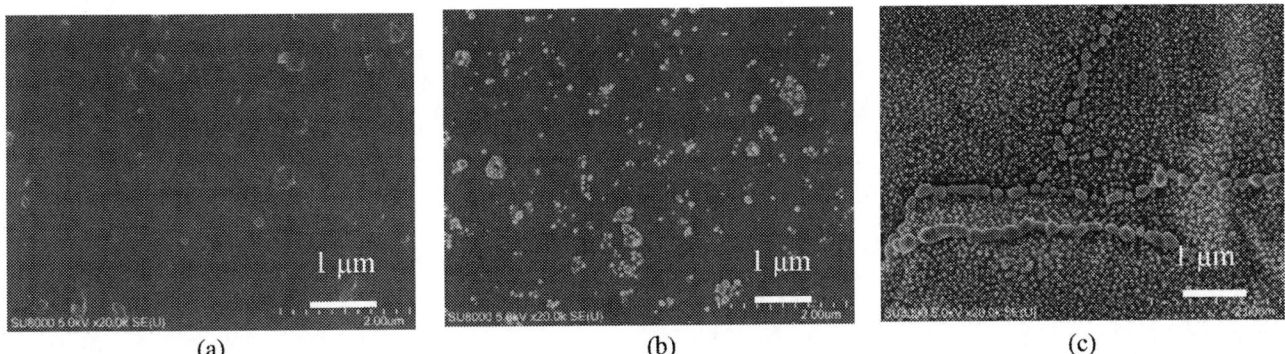

(a)   (b)   (c)

Figure 8. Cu surfaces (a) before and after formic acid treatments (b) without and (c) with Pt catalyst. Beforehand, Cu samples are heated in air at 200°C for 20min to enhance oxidation of the surface in order to see the effects of the Pt catalyst more clearly.

Direct Multichip-to-Wafer 3D Integration Technology
Using Flip-Chip Self-Assembly of NCF-Covered Known Good Dies

Yuka Ito[1,2,*], Mariappan Murugesan[3], Takafumi Fukushima[3,*], Kang-Wook Lee[3],
Koji Choki[2], Tetsu Tanaka[1,4], and Mitsumasa Koyanagi[3]

[1]Department of Bioengineering and Robotics, Graduate School of Engineering, Tohoku University
6-6-01 Aza-Aoba, Aramaki, Aoba-ku, Sendai 980-8579, Japan
E-mail: link@lbc.mech.tohoku.ac.jp, Phone: +81-22-795-6909,
[2]Circuitry with Optical Interconnection Business Development Dept., Sumitomo Bakelite Co., Ltd.
[3]New Industry Creation Hatchery Center, Tohoku University
[4]Deptartment of Biomedical Engineering, Graduate School of Biomedical Engineering Tohoku University

Abstract

We demonstrated surface tension-driven self-assembly and microbump bonding using NCF (non-conductive film)-covered chips with Cu/Sn-Ag microbumps for high-throughput and high-yield direct multichip-to-wafer 3D integration. The NCF is a promising candidate to completely fill gaps between fine-pitch microbumps, and is essential for realizing highly-reliable microbump-to-microbump interconnections. Here, by applying the self-assembly method with strong water surface tension, the NCF-covered chips were precisely aligned to hydrophilic assembly sites defined on host Si substrates in a face-down manner with alignment accuracies of approximately 1 μm. The self-assembled chips having Cu/Sn-Ag microbumps covered with NCF were thermally compressed to obtain electrical joints between the chips and substrate after the self-assembly process. The resulting daisy chains showed good electrical characteristics with contact resistance of 53 mΩ/joint.

Introduction

Chip-to-wafer stacking is a promising candidate to realize high-production yield for 3D and hetero system integration owing to the use of known good dies (KGDs) [1]-[7]. In the 3D and hetero system integration, different-sized chips and various materials/devices are highly integrated. Therefore, high-precision, high-throughput, and high-yield 3D chip stacking technologies are essential for creating advanced 3D and heterogeneous systems with TSVs. Conventional robotic pick-and-place machineries such as a flip chip bonder are still mainstream method for chip assembly. However, the one-by-one sequential methods have serious trade-off between assembly throughput and alignment accuracy. Thus, to realize the advanced 3D and hetero system integration, we have developed multichip self-assembly techniques using liquid surface tension that drives KGDs to precisely predetermined positions with higher alignment accuracies [8]-[16]. In the series of our previous works, we proposed and developed "Direct multichip-to-wafer 3D integration technology" based on the self-assembly techniques that are essential to improve throughput of the whole integration processes, as shown in Figure 1. The 3D integration is completed by repeating the stacking process of chips. This technology is free from above-mentioned trade-off between throughput and alignment accuracy lying in conventional mechanical pick-and-place

assembly methods. In addition, no support wafers are required for 3D integration processes.

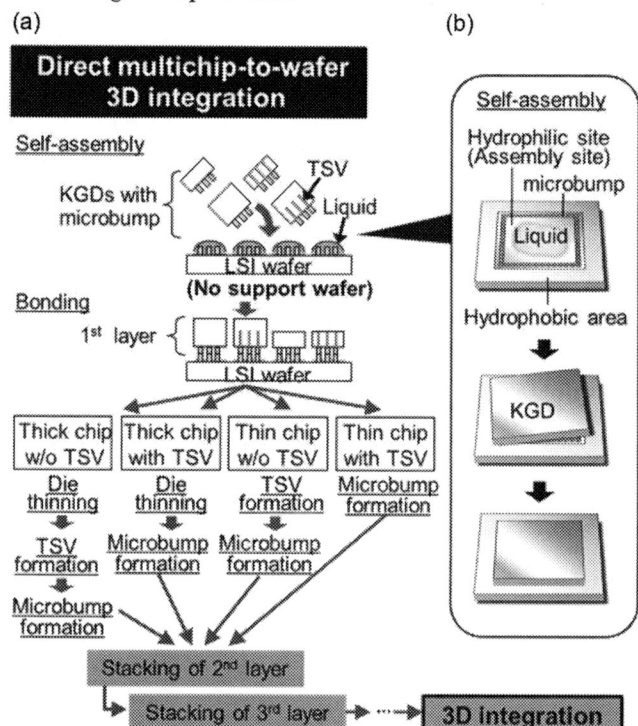

Figure 1. Conceptual drawings of "Direct multichip-to-wafer 3D integration using flip-chip self-assembly" : (a) stacking process of 1st layer and 3D integration, and (b) schematic drawing for self-assembly technique.

We have previously reported precise flip-chip self-assembly and microbump interconnections using KGDs with fine-pitch In/Au bumps [12-14] and with Cu/Sn bumps [15, 16], respectively. This paper proposes a new flip-chip self-assembly method using KGDs covered with NCFs to realize high-throughput and highly-reliable microbump interconnections for advanced direct multichip-to-wafer 3D integration.

Compared to conventional die-level capillary underfills with a one-by-one injection process, in recent years, wafer-level underfill using pre-underfill materials called "non-conductive film (NCF)" will become further necessary for the chip-to-wafer 3D integration to fully fill narrow gaps between KGDs and a substrate wafer with fine-pitch (< 10 μm)

978-1-4799-2408-0/14 $31.00 © 2014 IEEE

microbumps. In this paper, to realize high-throughput and highly-reliable multichip-to-wafer 3D integration, we demonstrate a new multichip self-assembly with KGDs having fine-pitch Cu/Sn-Ag microbumps covered with a NCF. Figure 2 show a conceptual drawings of self-assembly and bonding process with the NCF-covered KGDs. The KGDs with the NCF are directly self-assembled on a substrate wafer in a flip-chip bonding manner. And then, the subsequent wafer-level thermal compression provides rigid microbump interconnections. We also describe the evaluation of alignment accuracies using the KGDs covered with the NCF and electrical characteristics of microbump interconnections after the KGD assembly and the NCF bonding. Furthermore, we discuss throughput of the direct multichip-to-wafer 3D integration technology.

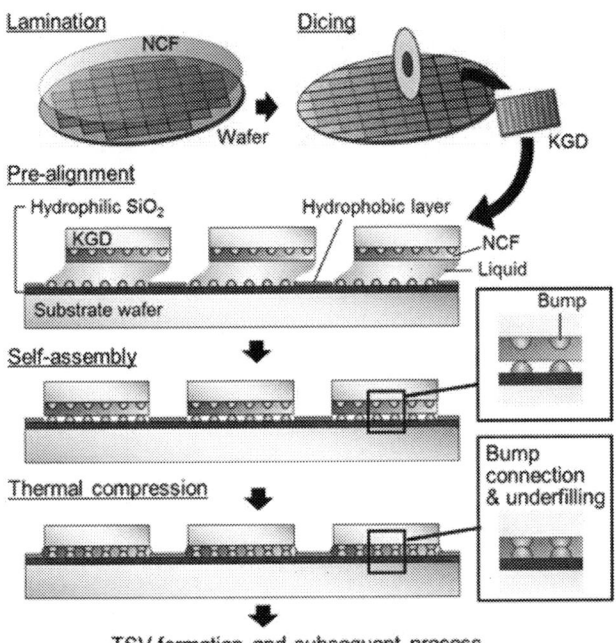

Figure 2. Self-assembly and bonding processes with NCF-covered KGDs in batch processing.

Fabrication of chips and substrates with microbumps

8-inch wafers were utilized for the fabrication of interposer substrates without the NCF and 3-mm-square chips covered with 17-μm-thick NCF. Cu/Sn-Ag microbumps with the size of 10-μm-square, 30-μm-pitch, and 6.5-μm-height were formed on both wafers. Firstly, Ti barrier and Cu seed layers were sputtered onto a thermal silicon dioxide layer on the wafers. 2.5-μm-thick Cu wirings were fabricated by photolithography and electroplating. After Cu seed and Ti barrier etching, a silicon dioxide layer as an insulation layer and assembly sites was deposited by plasma-enhanced chemical vapor deposition on the Cu wirings, and contact holes were formed by reactive ion etching (RIE). Cu/Sn-Ag microbumps were fabricated by photolithography, electroplating, and vapor deposition. 1.5-μm-thick Cu microbumps and the following 2.5-μm-thick Sn-Ag microbumps were deposited on the thin Ti and Cu layers. The exposed thin Cu and Ti were etched out from the wafers. On the wafer for the substrates, a hydrophobic fluorocarbon layer

was formed outside the assembly sites in order to define precise assembly sites with the same size of 3-mm square to the designed chip outer size. In contrast, the NCF was laminated on the wafers for the chips with vacuum laminator. Finally, the wafers were diced with a standard blade dicer. Figures 3(a) and 3(b) show photographs of the resulting chips and substrates with Cu/Sn-Ag microbumps and Cu wirings. An SEM image of the representative microbumps formed on the substrates and the chips is shown in Figure 3(c).

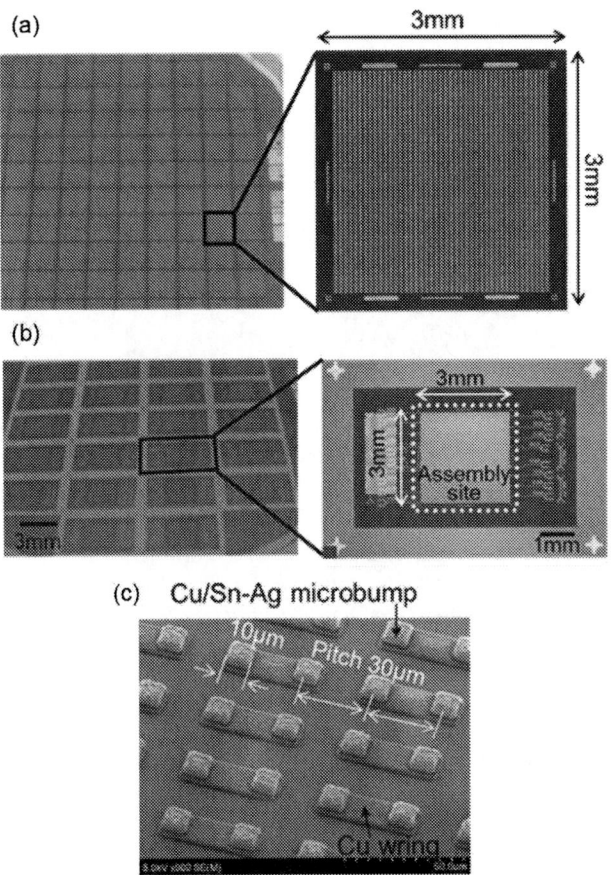

Figure 3. (a) Photographs of the wafer covered with the NCF after chip dicing, (b) photographs of the wafer for the substrate, (c) an SEM image of Cu/Sn-Ag microbumps.

Capability of precise chip self-assembly with water surface tension

Liquid wetting properties on the assembly site and the surrounding area on substrates are essential to precisely align chips on the assembly sites. Static contact angles of water droplets were evaluated on the both surfaces. Contact angles on the hydrophilic assembly sites were less than 30° as shown in Figure 4(a). In contrast, the surrounding areas of the assembly sites rendered hydrophobic by fluorocarbon had contact angle of more than 110° as shown in Figure 4(b). These results indicate that the wetting properties of the resulting substrate are appropriate to precise self-assembly [11].

Differences of size and size variation between the chip and the assembly site also have an impact on alignment accuracies. Correlation with chip outer size and alignment accuracies in self-assembly were compared between chips dicing with plasma etching and with standard blade dicer. High alignment accuracies are expected with the chips and substrates fabricated with a high-precision process. As shown in Figure 5(a)-(b), both the resulting chips had similar chip outer size, and were precisely aligned toward the assembly sites within 2.0 μm. The alignment accuracies of 2.0 μm are eventually consistent with chip-size accuracies. In addition, Figure 5 indicates that wafer dicing with a standard saw dicer have little effect on alignment accuracy in these self-assembly experiments. Therefore, the chips with NCF/microbumps were also singulated from wafers with the standard dicer. In the blade dicing of the chips with the NCF/microbumps, the chip outer size was 2.999 mm ±2.5 μm, and the standard deviation of chip size was 1.1 μm respectively, shown as in Figure 6 (a).

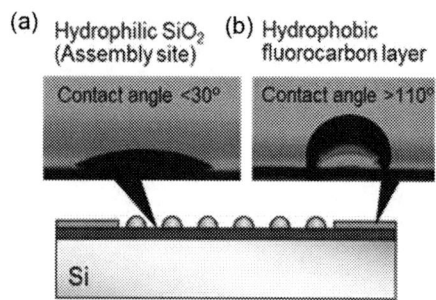

Figure 4. Contact angles of water droplets (a) on the hydrophilic assembly site, and (b) on the surrounding hydrophobic area.

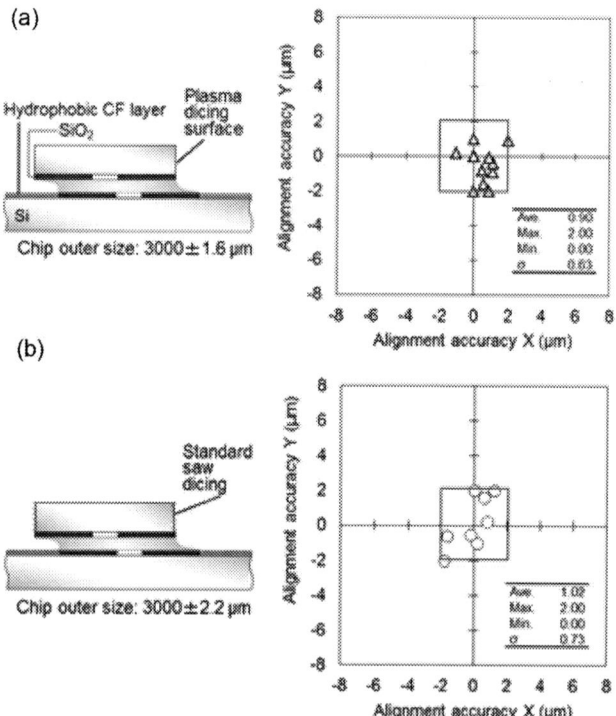

Figure 5. Alignment accuracies of Si chips without NCF: (a) chips singulated by plasma etching, and (b) chips singulated by standard blade dicing.

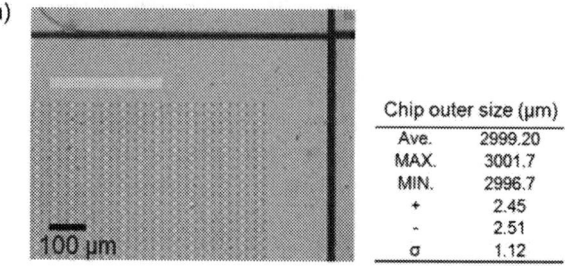

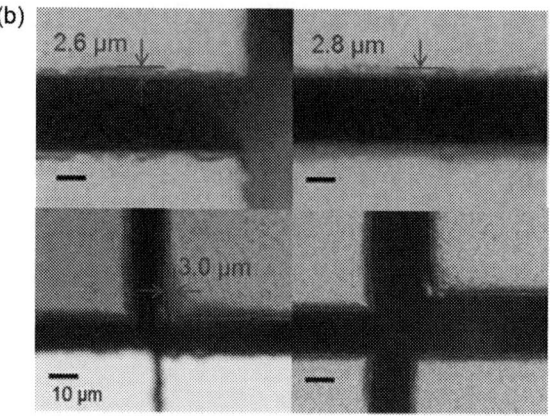

Figure 6. (a) A photograph of the chip laminated with the NCF after dicing, and (b) the magnified images of the chip surface.

Self-assembly evaluation using the chip with NCF and Cu/Sn-Ag microbumps

Shortly before chip self-assembly, the bonding surfaces of the substrates were treated with 172-nm-wavelength excimer lamp irradiation for 4.5 sec to remove contamination and enhance water wettability on the assembly sites. A self-assembly scholar robot developed for the multichip-to-wafer 3D integration was utilized to supply liquid droplets and to sort and supply chips onto the assembly sites. The positioning reproducibility of the robotic was approximately 100 μm, which was high enough to precisely align the chips on assembly sites formed on the corresponding substrates by water surface tension.

Figure 7 shows self-assembly flow and photographs for chips in each step. A 0.4 μl droplet of ultrapure water was supplied onto hydrophilic assembly sites on the substrates. The chips were pre-aligned to the assembly sites in a face-down manner using an image-recognition function of the robot, and then the chips were kept in contact with the droplet. After that, the chips were released out from the vacuum tweezer, and finally, the chips were immediately self-aligned to the assembly sites by water surface tension. In Figure 7, the chip was released from 500 μm initial offset position along x-axis, and was self-aligned within 0.3 sec. The alignment accuracies were evaluated with vernier scales on the self-assembled chip and the substrate with an IR microscope, as shown in Figure 8.

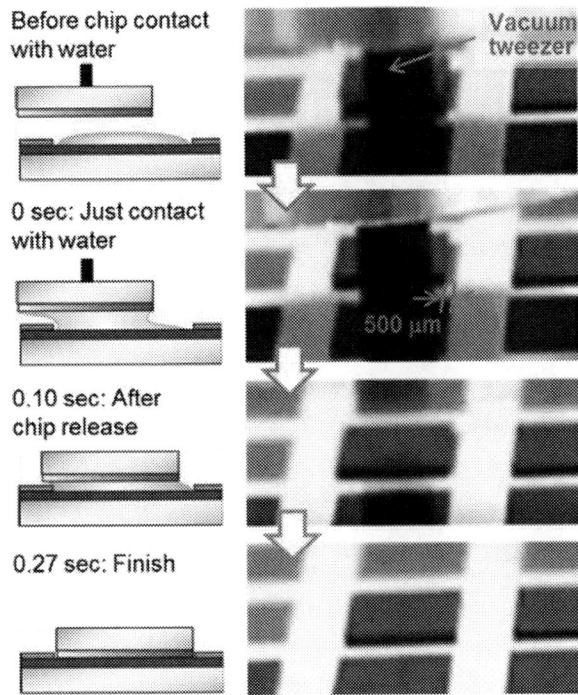

Figure 7. Self-assembly behavior of a NCF-covered chip released with an initial offset of 500 μm.

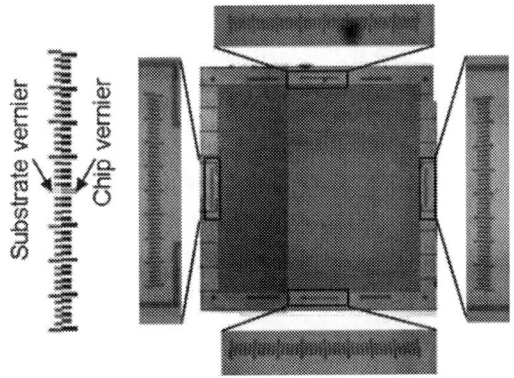

Figure 8. IR microscopic overlap images of the 3-mm square chip and the substrate after self-assembly.

The impact of initial offset along x-axis prior to chip release on alignment accuracies was investigated in this work. Figure 9 shows self-alignment behaviors of the chips released from initial offsets in 100 μm, 500 μm, and 1500 μm. The NCF-covered chips with the initial offsets were also spontaneously self-aligned by water surface tension. Figure 10 shows alignment accuracies of chips released from initial offset in 0 μm to 1500 μm. The chips with the NCFs can be precisely assembled toward the corresponding assembly sites. In all conditions, most chips can be precisely self-aligned with alignment accuracies within 3.0 μm. With even 1.5-mm initial offset, the alignment accuracies were equivalent to that of chips released from other smaller initial offsets, as shown in Figure 9(c), 10(a) and (e). In Figure 10, the assembly accuracies were 1.5 μm on average and 2.9 μm at worst, and the total accuracy variation were below 0.7 μm.

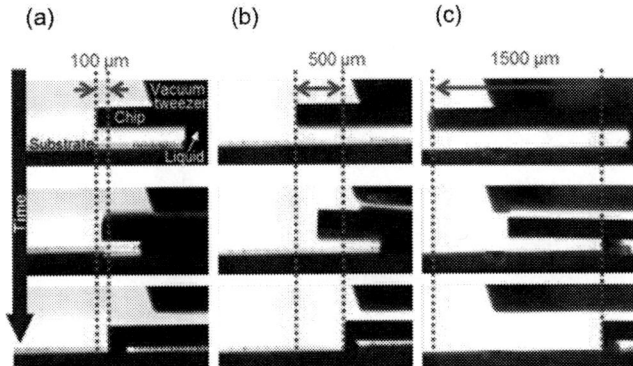

Figure 9. Side views of self-assembling chips to substrates with initial offsets of (a) 100 μm, (b) 500 μm, and (c) 1500 μm.

There were some differences between alignment accuracies of chips with NCF in Figure 10(a) and without the NCF in Figure 5(b). The difference was derived mainly from chip-size variations, NCF delamination, and small deformation of the NCF due to its viscoelastic property and stress derived from mechanical dicing. Figure 6(b) shows magnified images of the chip edges after blade dicing. The deformations of 3.0 μm width were observed on the edges of the NCF. Therefore, the alignment accuracy of the worst 3.0 μm would just result from NCF deformation and delamination. However, NCF deformation and delamination protruding inward from the chip edge would have little influence on the chip alignment, because the chip outer sizes were not changed.

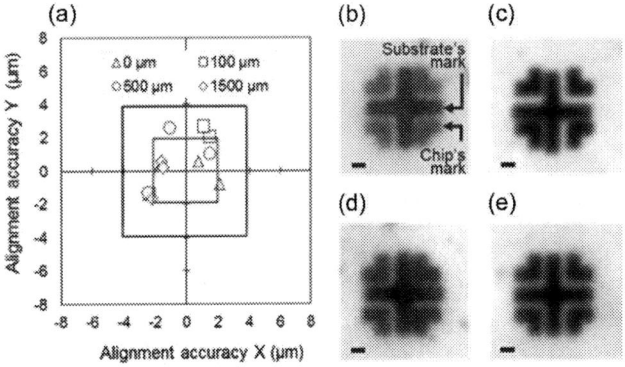

Figure 10. (a) Alignment accuracies of NCF-covered chips with microbumps, and the resulting IR images of alignment marks on the self-assembled chip with initial offsets of (b) 0 μm, (c) 100 μm, (d) 500 μm, and (e) 1500 μm. Scale bars in (b)-(e): 10 μm.

Thermal compression bonding of Cu/Sn-Ag microbumps and underfilling after self-assembly

After the chips are self-aligned to assembly sites on the substrates by water surface tension and the subsequent evaporation of water, Cu/Sn-Ag microbump interconnections between the chips with the NCFs and the substrates were obtained by thermal compression. The chips with alignment accuracies below 2.0 μm were evaluated in this section. Here, the self-assembled chips were thermally compressed in the chip-level processing using flip-chip bonder (FC3000, Toray). It is noted that chip alignment function of the flip-chip bonder was not employed. After optimization of bonding conditions such as temperature and load, the chips self-assembled on the substrate were bonded under the temperature and load profiles, as described in Figure 11.

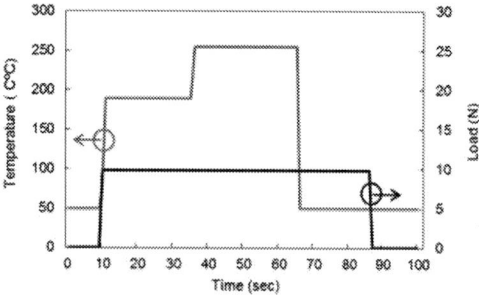

Figure 11. Bonding temperature and load profiles during thermal compression for chips after self-assembly the flip-chip bonder (chips are self-assembled on the substrates).

Figure 12 (a) and (b) show the plots of alignment accuracies and IR images of alignment marks before and after NCF bonding, respectively. After NCF bonding, chip alignment accuracy were 1.5 μm on average and 3.4 μm at worst. And, such a small decrease in alignment accuracies after thermal compression would be attribute to chip sliding on the substrate owing some tiny tilts, which would be derived from variations of the NCF thickness, the microbump height, and the bonder head planarity.

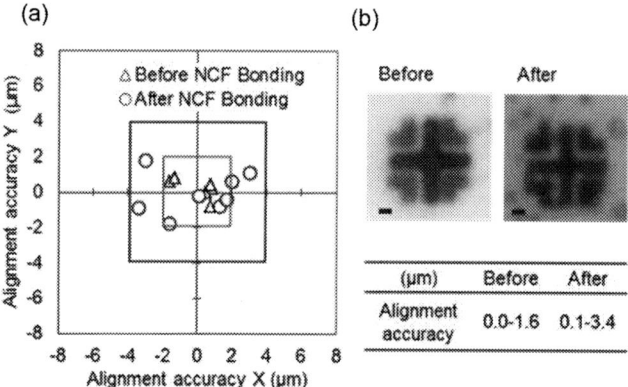

Figure 12. (a) Alignment accuracies of chips, and (b) IR images of alignment marks before and after NCF bonding. Scale bar in (b): 10 μm.

Figure 13 shows cross-sectional images of–microbumps before bonding and bonding interfaces after bonding. The gaps retained by Cu microbumps between the chip and the substrate were above 10 μm without electrically shorted solder bridges and bonding failures. The gaps were filled with the NCF material. *I-V* characteristics were measured using the daisy chain patterns interconnected through Cu/Sn-Ag microbump joints. The electrical characteristics were measured using a 4-terminal probe on substrates. Figure 14 shows the obtained *I-V* curves. The daisy chains including 5,000 joints of microbumps showed good electrical

characteristics with contact resistances of 57 mΩ/joints or below without bridge short and open failures. Good electrical connection of Cu/Sn-Ag microbumps through the NCF was obtained.

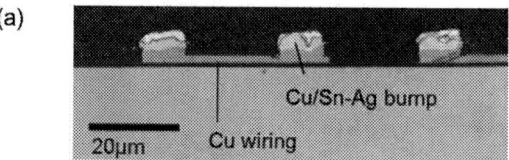

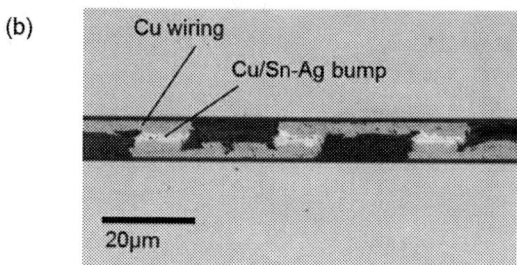

Figure 13. Cross-sectional images of (a) microbumps before self-assembly, and (b) microbump joints between the upper chip and lower substrate.

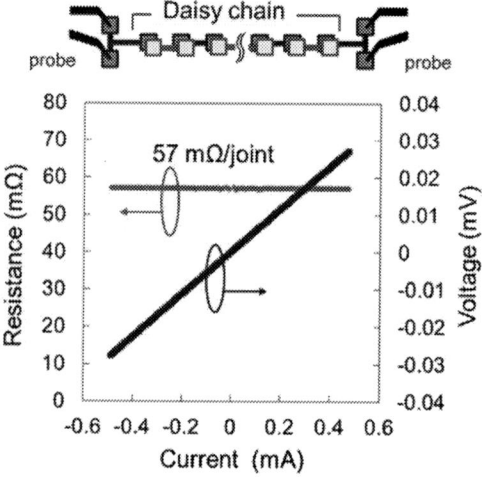

Figure 14. *I-V* characteristics of daisy chain with Cu/Sn-Ag microbump through NCF.

Conclusions

Chips with Cu/Sn-Ag microbumps covered by the NCF were successfully self-assembled onto substrates with the corresponding Cu/Sn-Ag microbumps. The alignment accuracies were found to be 1.4 μm on average. Although the chips were fabricated with a standard blade dicing, the chip outer size and the size variation had less influence on the alignment accuracy in self-assembly using chips having Cu/Sn-Ag microbumps with a size/pitch of 10/30 μm. In the subsequent thermal compression process, Cu/Sn-Ag microbumps on the chips and the substrates were electrically connected through the NCF, and good electrical properties

978-1-4799-2408-0/14 $31.00 © 2014 IEEE   1152

with contact resistances of 57 mΩ/joints were obtained. By applying a series of direct multichip-on-wafer 3D stacking including pre-underfilling, self-assembly of NCF-covered chips, and thermal compression in wafer-level batch processing, highly-integrated 3D and hetero systems would be realized.

Acknowledgments

This work was performed at Micro/Nano-Machining Research and Education Center (MNC) and Jun-ichi Nishizawa Research Center at Tohoku University. This work was also performed at Global INTegration Initiative (GINTI) in Tohoku University, Japan. We would like to acknowledge DISCO CORPORATION for supporting implementation of wafer dicing. This research was supported by Japan Society for the Promotion of Science (JSPS), Grant-in-Aid for Scientific Research "Grant-in-Aid for Scientific Research (S)", No. 21226009. We would like to acknowledge Sumitomo Bakelite Co.,Ltd., for their material support in this work.

References

1. M. Koyanagi, T. Nakamura, Y. Yamada, H. Kikuchi, T. Fukushima, T. Tanaka, and H. Kurino: IEEE Trans. Electron Devices 53 (2006) 2799.
2. Kunio, K. Oyama, Y. Hayashi, and M. Morimoto: IEDM Tech. Dig., 1989, p. 837.
3. M. Koyanagi, H. Kurino, K.-W. Lee, K. Sakuma, N. Miyakawa, and H. Itani: IEEE Micro 18 [4] (1998) 17.
4. S. J. Souri, K. Banerjee, A. Mehrotra, and K. C. Saraswat: Proc. 37th ACM Design Automation Conf., 2000, p. 873.
5. K. Banerjee, S. J. Souri, P. Kapur, and K. C. Saraswat: Proc. IEEE 89 (2001) 602.
6. P. Ramm, D. Bonfert, H. Gieser, J. Haufe, F. Iberl, A. Klumpp, A. Kux, and R.Wieland: Proc. IEEE Interconnect Technology Conf. (IITC), 2001, p. 160.
7. K.-W. Lee, A. Noriki, K. Kiyoyama, T. Fukushima, T. Tanaka, and M. Koyanagi: IEEE Trans. Electrons Devices 58 (2011) 748.
8. T. Fukushima, Y. Yamada, H. Kikuchi, and M. Koyanagi: IEDM Tech. Dig., 2005, p. 359.
9. T. Fukushima, H. Kikuchi, Y. Yamada, T. Konno, L. Jun, K. Sasaki, K. Inamura, T. Tanaka, and M. Koyanagi: IEDM Tech. Dig., 2007, p. 985.
10. T. Fukushima, T. Konno, K. Kiyoyama, M. Murugesan, K. Sato, W.-C. Jeong, Y. Ohara, A. Noriki, S. Kanno, Y. Kaiho, H. Kino, K. Makita, R. Kobayashi, C.-K. Yin, K. Inamura, K.-L. Lee, J.-C. Bea, T. Tanaka, and M. Koyanagi: IEDM Tech. Dig., 2008, p. 499.
11. T. Fukushima, E. Iwata, Y. Ohara, M. Murugesan, J.-C. Bea, K.-W. Lee, T. Tanaka, and M. Koyanagi: IEEE Trans. Trans Compon Packag Manuf Tech. 1 (2011) 1873
12. T. Fukushima, E. Iwata, Y. Ohara, A. Noriki, K. Inamura, K.-W. Lee, J.-C. Bea, T. Tanaka, and M. Koyanagi: IEDM Tech. Dig., 2009, p. 349.
13. T. Fukushima, Y. Ohara, M. Murugesan, J.-C. Bea, K.-W. Lee, T. Tanaka, and M. Koyanagi: Proc. 61st Electron Components Technology Conf. (ECTC), 2011, p. 205
14. T. Fukushima, E. Iwata, Y. Ohara, M. Murugesan, J.-C. Bea, K.-W. Lee, T. Tanaka, and M. Koyanagi: IEEE Trans. Electron Devices 59 (2012) 2956
15. Y. Ito, T. Fukushima, K.-W. Lee, K. Choki, T. Tanaka, and M. Koyanagi, Jpn. J. Appl. Phys. 52 (2013) 04CB09.
16. Y. Ito, T. Fukushima, K.-W. Lee, K. Choki, T. Tanaka, and M. Koyanagi, Proc. 63rd Electron Components Technology Conf. (ECTC), 2013, p. 891

Maskless Screen Printing Technology for 20μm-Pitch, 52InSn Solder Interconnections in Display Applications

Kwang-Seong Choi, Haksun Lee, Hyun-Cheol Bae, and Yong-Sung Eom
IT Materials and Components Laboratory, ETRI
138, Gajeong-ro, Yuseong-gu Daejeon, 305-700, Korea
E-mail: kschoi@etri.re.kr, Tel: 82-42-860-6033, Fax: 82-42-860-5077

Abstract

Traditionally, ACF (Anisotropic Conductive Adhesive) technology has been used for CoG (Chip-on-Glass) and FoG (Flex-on-Glass) interconnections in display packaging area. The electrical contacts of ACF technology are based on the mechanical contacts between the electrodes on substrates and conductive particles in ACF. As pitches of these interconnections tend to get finer than 30 μm and bonding temperature needs to be decreased because of warpage concerns during the bonding process, a novel interconnection technology for the advanced display systems is necessary.

In this paper, a maskless screen printing technology is proposed to form and bond 20 μm-pitch, 52InSn solder interconnections for advanced display systems. InSn solder is selected to decrease the bonding temperature because its melting point is 118 °C. A novel material, called as solder bump maker (SBM) is developed to have InSn solder powder in SBM used for InSn bumping process. The polymer matrix and deoxidizing agent in SBM are carefully designed to make InSn solder powder in SBM wet on Cu or Au electrodes on a substrate during the bumping process. Since InSn solder powder resides only on electrodes on a substrate with temperature variations because of the surface tensions between the solder powder and metal electrodes, a maskless screen printing process can be adopted for the InSn, fine-pitch bumping process. Using a maskless screen printing process with SBM, 20 μm-pitch, InSn solder interconnections on a glass substrate are successfully formed. We, also, developed a no-flow underfill material, name as fluxing underfill, for a bonding material of InSn interconnections. It plays roles of flux and underfill at the same time during the bonding process. The bonding process for 20 μm-pitch, InSn solder interconnections is successfully achieved using fluxing underfill. Its peak temperature of the bonding process is 130 °C.

Introduction

Nowadays, flexible display and electronics get to be highlighted because of the market demand. As a solution, transparent plastic films such as polyethylene terephthalate (PET) and polyethersulfone (PES) are developed for the flexible applications. Since the glass transition temperatures of these materials are lower than 200 °C, the process temperature to implement displays or electronics on these substances should be lowered. Conventionally, anisotropic conductive films (ACF) have been widely used as interconnection materials in both flat panel displays and semiconductor packaging applications [1]. Since the electrical contacts of ACF interconnections rely on the mechanical contacts between the conductive particles in ACF and metal electrodes as shown in

Fig 1 (a), they inherently exhibit high electrical contact resistances and low adhesion. Additionally, the usual bonding temperature of ACF is higher than 150 °C, which may lead to the permanent deformation of flexible substrates during the bonding process.

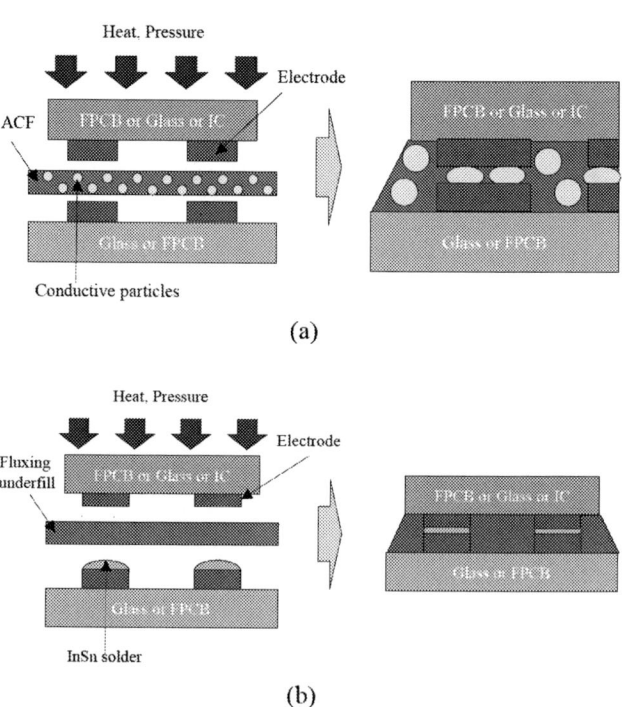

Fig. 1. Bonding process (a) using ACF and
(b) using InSn solder and fluxing underfill.

To reduce the bonding temperature of an ACF display interconnection, several technologies have been proposed. Conductive nano-scale films with nano-silver powder at a bonding temperature of 180°C were formulated [2]. Instead of using conductive particles in an ACF, SnBi solder powder and nanofiber were used through an ultrasonic bonding method to generate a bonding temperature of 200°C [3]. Combinations of acrylic-based film, SnBi solder powder and an ultrasonic-assisted thermocompression bonding method were proposed to reduce the ACF bonding temperature down to 150°C [4]. UV curing of an ACF using a photo-active curing agent (PA-ACFs) was introduced to decrease the bonding temperature to 110°C [5]. These studies show a similar approach in using conductive particles as a conductive medium so that the electrical resistance between electrodes may increase especially for a fine-pitch applications of less than 40μm because of a limited

978-1-4799-2408-0/14 $31.00 © 2014 IEEE   1154   2014 Electronic Components & Technology Conference

number of conductive particles in a small electrode area. Another approach to obtain low bonding temperature was using nanoparticles of solder materials [6]-[8]. The melting temperature of about 120 °C was achieved using nanoparticles of SnAgCu solder. However, a broad size distribution of the nanoparticles tends to melt step by step, so that their differential scanning calorimetry (DSC) curves show broad endothermic peaks, which means the bonding temperature cannot be easily determined for the practical applications.

In this paper, we propose a novel interconnection technology using 52InSn solder-on-pad (SoP) and fluxing underfill technology for display applications. Figure 1 (b) shows the novel process; InSn SoP on the metal electrodes on a substrate is formed, and two substrates are bonded using a fluxing underfill. Since the melting point of InSn solder is 118°C, low-temperature bonding below 150°C can be achievable. Since the whole electrode area is used for the electrical contacts, the joint can show a lower electrical contact resistance than the joint with ACF. To implement such a novel technology, novel materials, a solder bump maker (SBM) for the bumping process and a fluxing underfill for bonding process are developed. An SBM is based on the rheological behavior of the solder in a resin [9]-[20]. The resin used is distinguishable as having low viscosity around the melting point of the solder, a deoxidizing capacity of the oxide layer on the surface of the solder, and no out-gassing related with the solvents during the bumping process. A maskless screen printing process is developed using an SBM for implementing a fine-pitch solder-on-pad (SoP) on a substrate. The fluxing underfill is designed to have the characteristics of a flux and underfill at the same time without forming voids inside underfill during the bonding process. Fluxing underfill can have the bonding process done in only three steps: dispensing a fluxing underfill, the alignment of two substrates, and thermocompression bonding. An InSn SoP array with a 20μm pitch is made using a maskless screen printing process with an SBM. After that, low temperature bonding below 150°C for two glass substrates using a fluxing underfill are successfully performed, and the microstructure of the bonded joints is observed.

Motivations

To achieve lower bonding temperature than 150 °C and low electrical contact resistance of the bonded joints for the flexible displays and electronics, we considered the bonding using solder alloy appropriate because the infrastructures related with solder alloy are well established and the process temperature can be easily chosen by selecting a proper solder. Table 1 shows the solder alloys we considered as candidate materials. We thought the eutectic composition suitable for the bonding process because the solder can melt at once at the composition. Pb and Cd among them cannot be used because of the environment effects. Bi was not considered because of its well-known brittleness. Ternary alloy cannot be a strong candidate because it is not available commercially yet. Therefore, 52In48Sn solder alloy was determined as a solder material for the process. Its melting point is 118 °C and thus the process temperature lower than 150 °C can be achievable.

To implement the bonding procedure as shown in Fig. 1 (b), first, InSn solder bump array should be formed on the metal electrodes on a substrate and second, the formed bump array needs to be contacted mechanically and electrically using the bonding process. We, already, reported the solder-on-pad (SoP) technology using a SBM for solder bumping on a PCB [9]-[16]. Solder alloys used were SnBi, Sn3.0Cu0.5Cu. In a previous paper, we compared the electrical contact resistance of the joints formed using ACF with that of SoP technology using an SBM and fluxing underfill. From the literatures, the contact resistances of the ACF joints were reported ranged from about 35 mΩ/2400 μm² to about 47 mΩ/2400μm², while those of the joints made of SBM and fluxing underfill was 0.32 mΩ/6400μm², which means the latter exhibited lower resistance by 2.7 times than the former [22].

Table 1. Solder alloys and their melting points.

No.	Solder compositions	Melting point
1	50Bi28Pb22Sn	100 °C
2	52.2In46Sn1.8Zn	108 °C
3	67Bi33In	109 °C
4	52In48Sn	118 °C
5	74In26Cd	123 °C

Figure 2 shows the SEM images of (a) a joint bonded using ACF and (b) a joint bonded using fluxing underfill. In Fig. 2 (a), a limited number of conductive particles was trapped between the electrodes and substrate. The number of the trapped conductive particles have the major effect on the electrical resistance. The number of the trapped conductive is determined according to probability. There is no adhesion between the electrodes and the conductive particles in ACF after the bonding process, and thus the polymer matrix in ACF contributes mainly the adhesion strength of the joint.

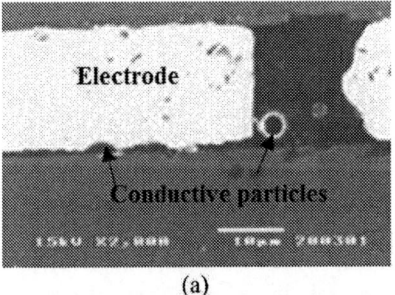

(a)

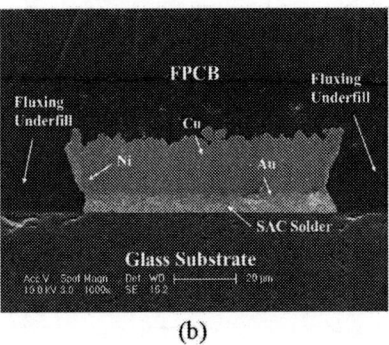

(b)

Fig. 2. SEM images of (a) joint bonded using ACF [21] and (b) joint bonded using fluxing underfill [22].

The joint bonded using a fluxing underfill shows the whole area in the electrode is used for the mechanical and electrical contact, which results in lower electrical contact resistance and higher adhesion of the joint. In addition, there is no conductive medium in the fluxing underfill as observed in Fig. 2 (a), since the solder formed on the electrode reacts only with the metal electrode on the glass substrate. No conductive medium between electrodes is considered crucial, especially for the fine-pitch applications to prevent short faults between the electrodes. The solder joint is considered to contribute the adhesion strength as well as the underfill, and the underfill can also increase the toughness of the joint, relieving the brittleness of the intermetallic compounds formed between the solder and metal electrodes. The properties of fluxing underfill, such as the coefficient of thermal expansion and modulus, can be optimized by adding filler to underfill.

Bumping using SBM

To develop the SBM with InSn solder powder, we fabricated InSn solder powder from an InSn solder lump. Generally, the fabrication of the solder powder can be done in two ways: the physical way and chemical one. The atomization process is a typical example of the physical ways. Usually, the physical methods cannot prevent the formation of the oxide layers on the solder powder during the process even though hermetic or nitrogen environments are applied. However, it has a merit of high throughput. On the other hand, the chemical methods has a low throughput, conventionally, and the oxidation of the formed solder powder can be easily disturbed, because the solvents in the bath suppress the interactions between the oxygen and formed solder powder. In this study, we developed a physical method without the oxidation of InSn solder powder. The solder powder is fabricated physically in a polymer matrix, which refrains the reactions between the oxygen and InSn solder powder. The polymer marix is one of the major components of the SBM, and thus the SBM with SnIn solder powder is easily formulated by adding other chemical components.

Figure 3 (a) and (b) shows a photography of the fabricated InSn solder powder and the particle size distribution of InSn solder powder, respectively. The dispersion behavior of the solder powder in a polymer matrix is good, with the result that the solder powder do not aggregate each other. The average size of the solder is 2.43 μm and its range is from 0.8 to 9 μm. The volumetric mixing ratio between the matrix and solder powder is 8:2.

After fabricating the solder powder in the polymer matrix, a deoxidizing agent and additives are mixed with the polymer matrix. The roles and requirements of the constituents of the resin of the SBM were reported in previous studies [9]-[16]. To control the surface tensions in the bumping process, we investigated several candidate materials for the polymer matrix and additives of the SBM. After choosing the proper materials, the effects of the mixing ratio between the polymer matrix, deoxidizing agent, and additives were investigated to obtain the proper interactions between the solder powder and the metal pads on a substrate during the bumping process. As in previous reports, the solvents are not mixed in the resin to minimize the out-gassing from the resin during the bumping process.

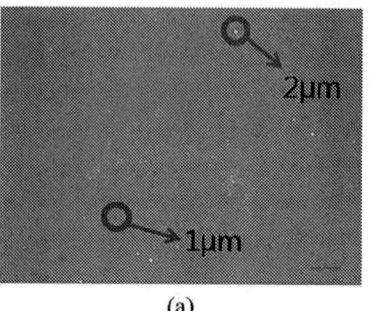

(a)

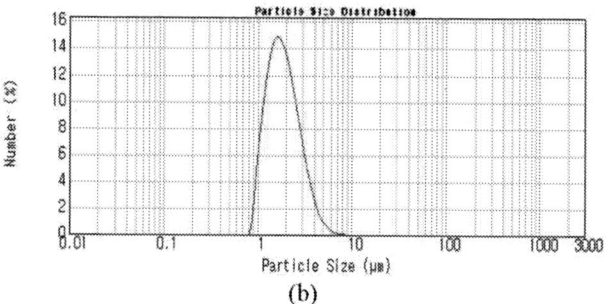

(b)

Fig. 3. (a) Photography of fabricated InSn solder powder and (b) particle size distribution of InSn solder powder

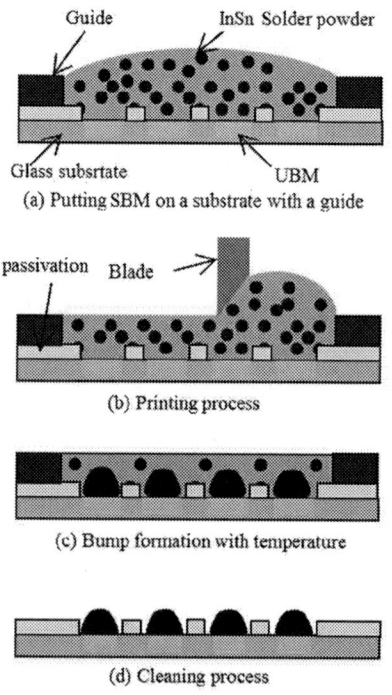

(a) Putting SBM on a substrate with a guide

(b) Printing process

(c) Bump formation with temperature

(d) Cleaning process

Fig. 4. Process flow of SoP using SBM

The whole bumping process is developed as shown in Fig. 4. First, the SBM is placed on a glass substrate. Its working area and thickness are defined using a guide. The thickness of the guide is 35 μm. The guide is not a conventional mask in that it does not isolate each metal pad on the substrate. We do not use a mask owing to its unsuitability for fine-pitch applications of less than 130 μm because of the process faults related with the miniaturized apertures. For bumping using the SBM an under bump metallization (UBM) pad array a pitch of 20 μm on a glass substrate is prepared. The UBM structure is made of T/Ni/Cu/Au, and its length is 0.8 mm. As in the screen process, a blade is used to make the thickness of the SBM on the substrate uniform. The temperature of the substrate is increased to 150 ºC in an oven. The dwell time at the peak temperature is 1 min. The oxygen amount in the oven is controlled at 1000 ppm. A cleaning process is then performed to remove the remaining SBM on the substrate.

To enable such a process, the design of the SBM was crucial. Ordinary solder pastes cannot be applied the bumping process as shown in Fig. 4, because of the solvents in the solder pastes. As temperature increases, the solvents tends to evaporate, resulting in the increase of the viscosity of the solder pastes. The flux agents in the solder pastes concurrently reduce the oxide layers on the whole solder powder. Then, the remaining solder powder contribute to make connections each other so that short faults cannot be avoided. On the other hands, the SBM consists mainly of the polymer matrix and does not have any solvents. Therefore, the viscosity change of the SBM during the bumping process is not abrupt. To maintain the viscosity uniformly, we designed the chemical reactions between the constituents in the SBM. If the amount of the heat of reactions are high, then the viscosity of the SBM gets to changes, while those are small, then the phase separations between the components in the SBM can appear, which may lead to the inhomogeneous bumping performance.

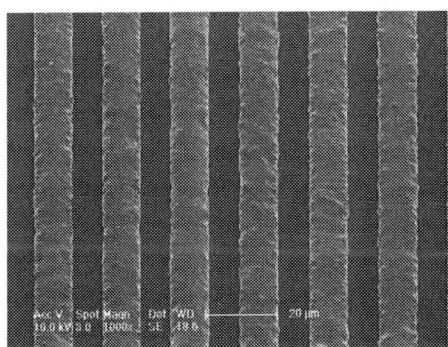

Fig. 5. SEM image of InSn SoP array formed
with a pitch of 20 μm.

An SEM image of the InSn SoP array formed using the novel SBM is shown in Fig. 5. A quite uniform SoP array was obtained using a maskless screen printing process, which features a low-cost process. A single guide can be used may times because there is no aperture in the guide for this process because there is no aperture in the guide for the process, and thus there are no process faults such as a skipping or slim observed during the screen printing process with a mask. The diameter of the solder powder for the bumping process is generally necessary to be less than one-fifth of the width of the UBM for good solder bumping using screen printing. Therefore, the diameter distribution shown in Fig. 2 may be considered inappropriate for the UBM with a 20 μm pitch on a glass substrate. Although the pitch is as small as 20 μm, the length of the UBM is as large as 0.8 mm, and thus the ratio between the maximal diameter of the solder powder and the longitudinal length of the UBM is about 0.1, which is small enough to obtain a uniform SoP array. The peak temperature of the bumping process needs to be lower than 150 ºC, which is the on-going topic of this study.

Bonding using Fluxing Underfill

The resin of the fluxing underfill consists of a polymer matrix, a deoxidizing agent, a hardener, and additives. A hardener reacts with a polymer matrix, which leads to curing. The cured fluxing underfill enhances the mechanical and environmental reliability of the bonded joints. The material design of the fluxing approach is quite similar with that of the SBM. However, careful material design is necessary to prevent a chemical reaction between a deoxidizing agent and a hardener. This may disturb the function of the fluxing underfill. Additionally, the reaction order between the constituents of the fluxing underfill is important. A deoxidizing agent reduces the oxide layers of the solder powder and metal pads on a substrate with temperature, and chemical reactions between a polymer matrix and hardener then occurs such that the underfill can be cured.

We observe dynamically the behavior of an InSn solder lump in the fluxing underfill with temperature to see if we can set the peak temperature of the bonding process to 130 ºC. The heating rate is 2 ºC/sec. When the temperature reaches 126 ºC, the InSn solder lump started to melt. When the temperature reached above 128 ºC, the InSn solder is fully wet on the metal which is made of an electro-plated Au on a laminated FR-4. This, therefore, proves that solder can wet, and the deoxidizing agent in the fluxing underfill can effectively remove the oxide layers on the InSn solder. From the result, it can be concluded that the peak temperature of the bonding process for InSn solder using a fluxing underfill is 130 ºC.

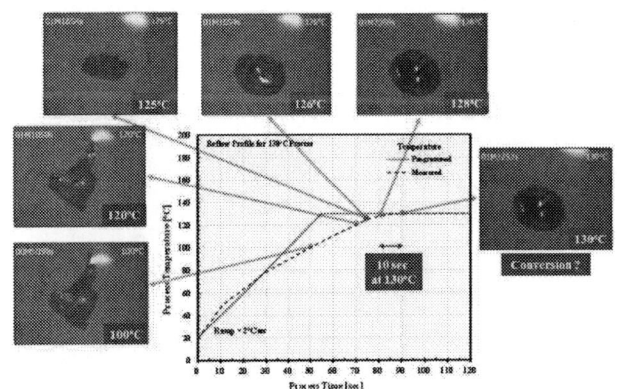

Fig. 6. Dynamic visual observation of InSn solder lump behavior in the fluxing underfill with temperature.

Fig. 7 shows the bonding process of two glass substrates. An InSn solder bump array is formed on a glass substrate and only an UBM is on the other substrate. The fluxing underfill is dispensed using a syringe. The, the glass substrate is aligned using a flip chip bonding machine. Thermocompression bonding is done using this machine. After bonding, a cross-section SEM image of the bonded joints is observed to investigate its microstructure.

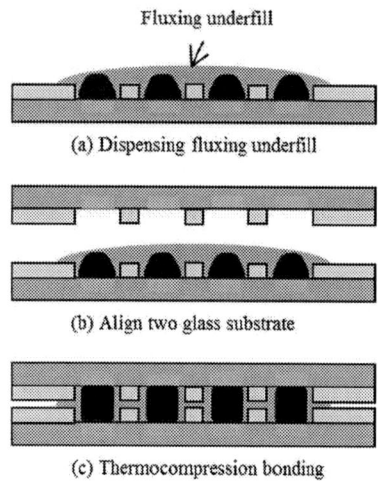

Fluxing underfill

(a) Dispensing fluxing underfill

(b) Align two glass substrate

(c) Thermocompression bonding

Fig. 7. Bonding process using fluxing underfill.

The thermocompression bonding of a glass substrate with an InSn SoP array, and a glass substate with an UBM array, was performed under a process temperature of 130 °C. Since the fluxing underfill has a capability of reducing the oxide layers on the InSn SoP array, any surface treatment to deoxidize the oxide layers of the solder bump array on the glass substrates is not necessary. Figure 8(a) shows an optical photograph of the top surface of the bonded two-glass substrate. InSn solder is partly observed next to the UBM, because of the misalignment during the bonding process. Figure 8(b) shows the InSn solder joints and interface between solder and UBM layer. The misalignment observed was about 2 μm. It can be concluded that the bonded InSn solder joints are mechanically reliable with the UBM layer because of the presence of the intermetallic compounds at the interface.

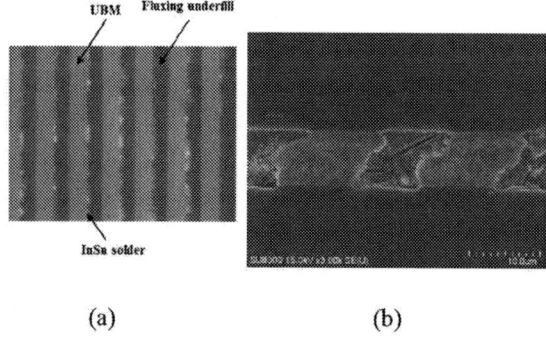

(a)   (b)

Fig. 8. (a) Optical photograph of the top surface of the bonded two-glass substrates and (b) a cross-sectional SEM image of the InSn joints.

Conclusions

To achieve lower bonding temperature than 150 °C and low electrical contact resistance of the bonded joints for the flexible displays and electronics, a novel interconnection technology based on 52InSn solder was developed. Using a novel bumping material, SBM, with InSn solder powder and a maskless screen printing process, an InSn SoP array with a 20 μm pitch on a glass substrate was successfully formed. Using the fluxing underfill, the bonding temperature for an InSn SoP array on a glass substrate was decreased to 130 °C.

Acknowledgments

This work was, partly, supported by ETRI, and the Components and the R&D Convergence Program of MSIP (Ministry of Science, ICT and Future Planning) and ISTK (Korea Research Council for Industrial Science and Technology) of Republic of Korea (Grant B551179-12-04-00, [Development of an image-based, real-time inspection and isolation system for hyperfine faults).

References

1. M.-J. Yim, et al, "Design and Understanding of Anisotropic Conductive Films (ACF's) for LCD Packaging," *IEEE Trans. CPMT Part-A*, vol. 21, no. 2, pp. 226-234, June 1998,

2. Y. Li, et al, "Novel Nano-Scale Conductive Films With Enhanced Electrical Performance and Reliability for High Performance Fine Picth Interconnect," *IEEE Trans. Advanced Packaging*, vol. 32, no. 1. pp. 123-129, Feb. 2009.

3. T. W. Kim, et al, "Low Temperature Fine Pitch Flex-on-Flex (FOF) Assembly using Nanofiber Sn58Bi Solder Anisotropic Conductive Films (ACFs) and Ultrasonic Bonding Method," in *Proc. Electronic Components and Technology Conf. (ECTC)*, 2013, pp. 461-467.

4. Y.-S. Kim, et al, "Low-Temperature Camera Module Assembly Using Acrylic-based Solder ACFs with Ultrasonic-Assisted Thermo-Compression Bonding Method," in *Proc. Electronic Components and Technology Conf. (ECTC)*, 2013, pp. 1613-1616.

5. I. Kim, et al, "Low Temperature Curable Anisotropic Conductive Films (ACFs) with Photo-active Curing Agent (PA-ACFs)," in *Proc. Electronic Components and Technology Conf. (ECTC)*, 2012, pp. 412-415.

6. C. Zou, et al., "Melting and Solidification Properties of the Nanoparticles of Sn3.0Ag0.5Cu Lead-free Solder Alloy," *Materials Charcterization*, pp. 474-480. 2010.

7. Y. Gao, et al., "Nanoparticles of SnAgCu Lead-free Solder Alloy with an Equivalent Melting Temperature of SnPb Solder Alloy," *J. Alloys and Compounds*, pp. 777-781, 2009.

8. C.-D. Zou, et al., "Size-dependent Melting Properties of Sn Nanoparticles by Chemical Reduction Synthesis," *Trans. Nonferrous Met. Soc. China*, pp. 248-253, 2010.

9. K.-S. Choi et al., "Novel Maskless Bumping for 3D Integration," *ETRI J.*, vol. 32, no. 2, pp. 342-344, Apr. 2010.

10. K.-S. Choi et al., "Novel Bumping Material for Solder-on-Pad Technology," *ETRI J.*, vol. 33, no. 4, pp. 637-640, Aug. 2011.

11. K.-J. Sung et al, "Novel Bumping and Underfill Technologies for 3D IC Integration," *ETRI J.*, vol. 34, no. 5, pp. 706-712, Oct. 2012.

12. Y.-S. Eom et al., "Characterization of Polymer Matrix and Low Melting Point Solder for Anisotropic Conductive Film," *Microelectron. Eng.*, vol. 85, pp. 327-331, 2008.

13. Y.-S. Eom et al., "Electrical Interconnection with a Smart ACA Composed of Fluxing Polymer and Solder Powder," *ETRI J.*, vol. 32, no. 3, pp. 414-421, June 2010.

14. K.-S. Choi et al, "Novel Bumping Process for Solder on Pad Technology," *ETRI J.*, vol. 35, no. 2, pp. 340-343, April 2013.

15. Y.-S. Eom, et al, "Optimization of Material and Process for Fine Pitch LVSoP Technology," ETRI J., vol. 35, no. 4, pp. 625-631, Aug 2013.

16. H.-C. Bae, et al, "Fine-Pitch Solder on Pad Process for Microbump Interconnection," *ETRI J.*, vol. 35, no. 6, pp. 1152-1155, Dec. 2013.

17. J.-W. Baek, et al., "Chemo-rheological Characteristic of a Self-assembling Anisotropic Conductive Adhesive System Containing a Low-Melting Point Solder," *Microelectron. Eng.*, vol. 87, pp. 1968-1972 , 2010.

18. K.-S. Jang, Y.-S. Eom, J.-T. Moon, et al., "Catalytic Behavior of Sn/Bi Metal Powder in Anhydride-Based Epoxy Curing," *J. Nanosci. Nanotechnol.*, vol. 9, no. 12, pp. 7461-7466, 2009.

19. Y.-S. Eom, K.-S. Choi, S.-H. Moon, et al., "Characterization of a Hybrid Cu Paste as an Isotropic Conductive Adhesive," *ETRI J.*, vol. 33, no. 6, pp. 864-870, 2011.

20. Y.-S. Eom, J.-H. Son, K.S. Jang, et al., "Characterization of Fluxing and Hybrid Underfills with Micro-encapsulated Catalyst for a Long Pot Life," *ETRI J.*, accepted, 2014.

21. M. J. Yim, et al., "Anisotropic Conductive Film (ACFs) for Ultra-fine Pitch Chip-On-Glass (COG) Applications," *Int'l. J. Adhesion & Adhesives*, pp. 77-84, 2007.

22. H. Lee, et al., "Novel Interconnection Technology for Flex-on-Glass (FOG) Applications," in *Proc. Microelectronics Packaging Conf. (EMPC)*, Sept. 2013, pp. 1-5.

Accelerated SLID Bonding Using Thin Multi-layer Copper-Solder Stack for Fine-pitch Interconnections

Chinmay Honrao, Ting-Chia Huang, Makoto Kobayashi[#], Vanessa Smet, P. Markondeya Raj,
and Rao Tummala
3D Systems Packaging Research Center, Georgia Institute of Technology
813 Ferst Dr NW, Atlanta, GA 30332
Namics Corporation, 3993 Nigorikawa, Kita-ku, Niigata City, Niigata Prefecture 950-313, Japan
Email: chinmay.honrao@gatech.edu / Phone: 734-604-4216

Abstract

Emerging 2.5D and 3D package-integration technologies for mobile and high-performance applications are primarily limited by advances in ultra-short and fine-pitch off-chip interconnections. A range of technologies are being pursued to advance interconnections, most notably with direct Cu-Cu interconnections or Cu pillars with solder caps. While manufacturability is still a major concern for the Cu-Cu interconnections technologies, the copper-solder approaches face limitations due to solder-bridging at fine-pitch, electromigration, and reliability issues. Thus, novel low-temperature, low-pressure, high-throughput, cost-effective and manufacturable technologies are needed to enable interconnections with pitches finer than 15 microns.

This paper focuses on an innovative multi-layered copper-solder stack approach to achieve fine-pitch off-chip interconnections with no residual solders after assembly. Interconnections using this new technology enable higher current-handling because of the stable intermetallics, high-throughput assembly, and high yield even at low stand-off heights. The elimination of solder-intermetallic (IMC) interfaces is also expected to enhance the joint strength. This paper describes the design, fabrication, assembly and characterization of such stacked copper-solder interconnections. A detailed study of the effect of bonding parameters such as temperature and time on the rate of formation of stable Cu-IMC-Cu structures is presented. Test-vehicles were designed and fabricated as the first demonstration of this technology.

Introduction

The I/O density, speed and bandwidth requirements for emerging mobile and high-performance systems are projected to drive the off-chip interconnection pitch to less than 20μm by 2015 [1]. Various flip-chip interconnection materials and processes have been developed over the past two decades to meet the need for higher I/Os and enhanced electrical performance. Lead-free solder bumps have been serving the industry for the past 10 years, but face shortcomings for emerging fine-pitch applications because of issues such as solder-bridging and electromigration.

Direct copper-copper interconnections without solders have the highest current-handling and lowest pitch capabilities. However, there are fundamental challenges associated with direct copper-copper bonding that include a high temperature solid-state bonding process, inability to accommodate non-planarity and non-uniformity of interconnection bumps, and complex processes that are required for the removal of residual oxides on the copper surfaces prior to bonding [2]. The Georgia Tech-Packaging Research Center recently made pioneering advances and patented a low-temperature copper-copper thermo-compression bonding process at less than 200°C, and demonstrated HAST, TCT and electro-migration reliability at 30μm pitch using 10-15μm copper bumps [3]. The 30μm pitch copper interconnections showed stable resistance for more than 1000 hours even at $10^6 A/cm^2$, proving the high current-carrying capability [4].

The copper pillar and solder-cap approach combines some of the advantages of both copper and solder bump technologies, and is the preferred option from the manufacturability standpoint. Paik et al. recently demonstrated reliability of 40μm pitch Cu-SnAg interconnections with a stand-off height of 20μm using anhydride-based NCFs [5]. However, the current interconnection approaches using this technology are not scalable to finer pitches as a result of electromigration and reliability issues arising with decreased solder content. Being a low-strength and low-fatigue resistance material, the solder strains increase with decreased solder height. The formation of copper-tin intermetallics leads to stresses at the IMC (brittle)-solder (ductile) interfaces, which get further aggravated at smaller stand-off heights and lower solder volumes.

Solid Liquid Inter-Diffusion (SLID) bonding is being explored as a promising approach to overcome some of the challenges previously faced by copper-solder interconnections. This technology is based on the rapid formation of intermetallics between a high melting component (in this case, Cu) and a low melting component (in this case, Sn solder) at a temperature above the melting point of the latter [6]. At this temperature, the copper diffuses into the liquid tin at a very high rate, leading to much faster IMC growth as compared to that in case of solid tin.

The reliability performance of SLID bonding has also been investigated and reported. Labie et al. demonstrated electromigration testing of 20μm diameter, Cu-Sn SLID-bonded chip-chip interconnections at a current density of $6.3 \times 10^4 A/cm^2$ at 150°C [7]. No failures were observed till 1000 hours of current stressing. Chang et al. demonstrated pressure-assisted SLID bonding of 20μm pitch micro-bumps consisting of a 4μm copper-pillar and a 4μm tin-cap structure, using a post-curing step at a temperature of 150°C for 30 minutes [8]. These interconnections were shown to be reliable over more than 1000 cycles of thermal cycling. SLID bonding, however, faces certain process challenges. IMC formation, being a diffusion driven process, requires long assembly or post-annealing processes for a complete conversion of solders to stable intermetallics. Infineon Technologies developed a chip-stacking process based on SLID bonding called SOLID-F2F, in which two chips are bonded in the F2F (face-to-face) orientation [9]. The bonding was completed in two steps, an initial soldering step at 260°C for 1 minute, during which the Sn solder was completely converted to Cu_6Sn_5 intermetallics. This was followed by a 20-minute-anneal at 300°C to convert the Cu_6Sn_5 into Cu_3Sn. For 30μm- pitch interconnections formed using this method, they successfully demonstrated 1000 hours of temperature cycling between -65°C and 150°C without any significant increase in daisy-chain resistance [10].

A novel approach, based on a combination of SLID bonding and alternate stacking of copper-solder layers, for faster conversion of copper-solder to thermally-stable and electromigration-resistant intermetallics is proposed. Figure 1 schematically shows the proposed interconnection approach as compared to the current approach.

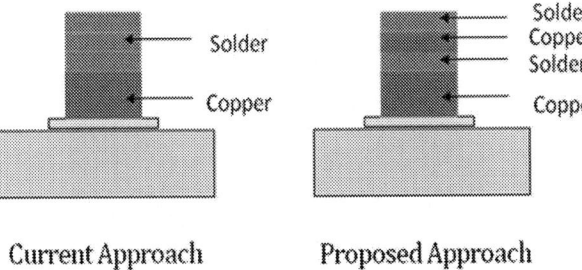

Figure 1. Current and proposed approaches for copper-solder interconnections

The key advantages of such a technology are: (i) higher electromigration resistance compared to traditional copper-solder approaches, (ii) high throughput assembly at ultra-fine pitch and low stand-off height without facing challenges such as solder-bridging and solder-cracking, (iii) lower bonding temperatures and pressures as compared to that used for Cu-Cu interconnections, and (iv) enhanced thermal and mechanical stability due to the elimination of solder/IMC interfaces.

A first demonstration of this technology is presented along with design, fabrication, assembly and characterization results.

Modeling and Design

The Cu/Sn stacked structure should enable SLID bonding with stable intermetallics using a short assembly time. This section models the Cu/Sn structure to accomplish this. Presence of silver in the solder inhibits the formation of intermetallics. Since this approach requires faster formation of intermetallics, pure tin was used as the solder. Based on the atomic weights and densities of copper and tin, the minimum thickness ratio of the copper and tin layers was calculated to be 1.3 for conversion of tin to Cu_3Sn.

Ideally, the thickness of individual copper and tin layers should be as small as possible for lowest diffusion distances. This, however, is restricted by the process capability for copper and tin electroplating. A very thin layer of tin results in insufficient wetting due to instant solidification of the tin upon melting. From previous literature, it was observed that the tin thickness used for SLID bonding is usually in the range from 1-4μm. Based on the process capability of the available copper and tin electroplating setup, a 1.5μm thick tin layer was chosen for the bump structure. Based on the thickness ratio previously calculated, the corresponding thickness of the copper layer was 2μm.

The thickness of the initial copper layer was chosen to be 5μm as this layer is responsible for providing adhesion to the seed layer, and as such, cannot be completely consumed to form IMCs. For the final tin layer, the thickness was chosen to be 3μm to ensure all bumps land on the substrate. A total of three layers each of copper and tin were chosen to be stacked alternately, resulting in a final bump height of 15μm. Figure 2 shows the final configuration of the bump for the copper-solder stacked interconnections approach.

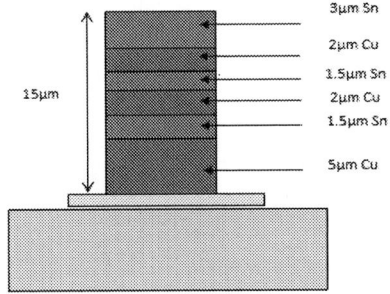

Figure 2. Bump configuration for copper-solder stacked interconnections

Intermetallic formation between copper and tin is a diffusion-limited process. During the bonding, Cu initially reacts with Sn to form Cu_6Sn_5. Due to the high diffusion rate of Cu into liquid Sn, complete conversion of Sn into Cu_6Sn_5 is achieved in a few seconds [6]. Cu_3Sn formation requires solid-state interdiffusion between Cu and the previously formed Cu_6Sn_5. As such, the Cu_3Sn IMC formation can be modeled using a parabolic law which is based on Fick's first law of diffusion, where the interdiffusion coefficient can be calculated using the Arrhenius relationship. The parabolic law and the Arrhenius relationship used are shown in Equation (I).

$$\Delta x^2 = k \cdot t$$

$$k = k_0 \exp\left(-\frac{Q}{RT}\right)$$

Equation (I)

Δx = thickness of Cu_3Sn after time t

k = inter-diffusion coefficient

k_0 = intrinsic diffusivity

Q = activation energy

T = temperature

The assumptions made while simulating the intermetallic growth were (i) constant concentration of the diffusing species at the inter-layer boundaries, and (ii) constant concentration-gradient along the inter-layer. Values for the activation energy for the formation of Cu_3Sn (Q) and its intrinsic diffusivity (k_0) were taken from previous studies based on the growth of copper-tin intermetallics. These values differ with processing techniques, and the values selected for this study were applicable to thin films of copper and tin. The 'Q' and 'k_0' values used were 66.1kJ/mol and 5.3E-8m^2/s respectively [6].

Modeling the IMC formation is of importance as it provides an estimation of the bonding temperature and time needed for complete conversion of solder to Cu_3Sn using the above bump configuration. Based on the parabolic law, Arrhenius relationship and the values for 'Q' and 'k_0' it was calculated that 2µm of Cu_3Sn could be formed in 5 minutes at a temperature of 250°C.

Fabrication of Copper-Solder Multi-layer Stack

As a proof-of-concept, 80µm pitch interconnections with multi-layered copper-tin stack were fabricated. The process was completed in two photo-lithography steps, one for the routing layer and the other for the bumps. Figure 3 gives an overview of the fabrication process.

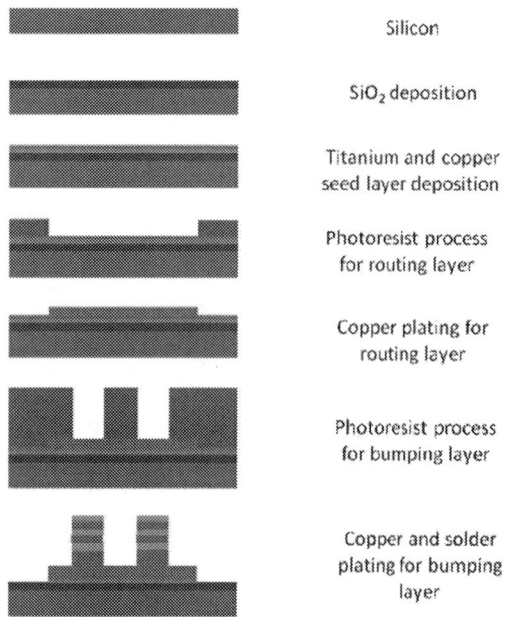

Figure 3. Process for fabrication of copper-solder multi-layer stacked structure

Hitachi RY-5315EB dry-film photoresist having a thickness of 15µm was used for patterning the routing layer. Karl-Suss MA6 Mask Aligner was used to expose the wafers with a dose of 95mJ/cm^2 using hard contact, after which they were developed in 3% Na_2CO_3 solution at 85°C for about 2 minutes. Once the photoresist development was complete, the wafers were plasma-cleaned to remove organic residue, if any, from the openings in the photoresist. Copper was then electroplated through these openings to form a 2-3µm thick

routing layer. Finally, Enthone PC 4025 was used to strip the photoresist.

The same photoresist was used for patterning the bumping layer. Photoresist lamination, exposure and development steps were similar to the ones followed for patterning the routing layer. After patterning, the wafers were plasma-cleaned, before continuing with the plating process. Copper and tin were plated alternately to obtain the copper-tin stacked structure. Cupracid TP chemistry was used for plating copper while tin was plated using the Stannobond FC chemistry, both provided by Atotech. The wafers were thoroughly rinsed and dried after electroplating each layer of the bump, so as not to contaminate the two plating baths. The current densities used for copper and tin electroplating were 15mA/cm^2 and 20mA/cm^2 respectively, and the resultant plating rates were 0.33µm/min and 1µm/min respectively. The plating time for each layer was determined based on their respective target thicknesses. The plated thickness was measured after each plating step using the Dektak Profilometer.

After completing the plating, the photoresist was removed using the Enthone PC 4025 stripper solution. This was followed by seed-layer etching to remove the underlying copper and tin seed layers. The alternate layers of copper and tin in the fabricated interconnect structure are clearly visible, as can be seen in Figure 4.

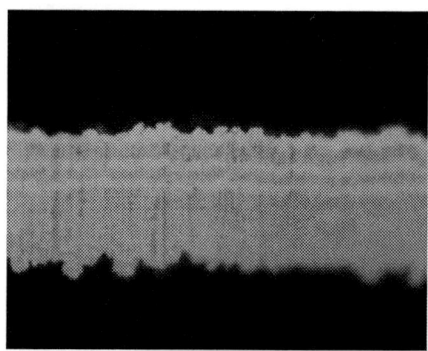

Figure 4. Copper-solder multi-layer stacked structure

Assembly and Characterization of Copper-Solder Stacked Interconnections

Interconnections with 80µm pitch with the traditional copper pillar-solder cap structure were assembled using SLID bonding. SLID bonding assembly has to be performed at a temperature above the melting point of tin, so as to enhance the formation of intermetallics through diffusion of copper in liquid tin. In this study, the aim was to convert the tin to the intermetallics during the assembly process itself.

A FINETECH Lambda flip-chip bonder was used to perform for assembly. Pre-applied BNUF was used to minimize the process defects and further improve the reliability of these interconnections. As determined by the diffusion modeling described previously, the bonding temperature and dwell-time used were 250°C and 300-900 seconds respectively. Figure 5 shows the temperature profile used for this assembly process. The force applied during bonding was 7.5N, which resulted in an equivalent pressure of 15MPa.

978-1-4799-2408-0/14 $31.00 © 2014 IEEE

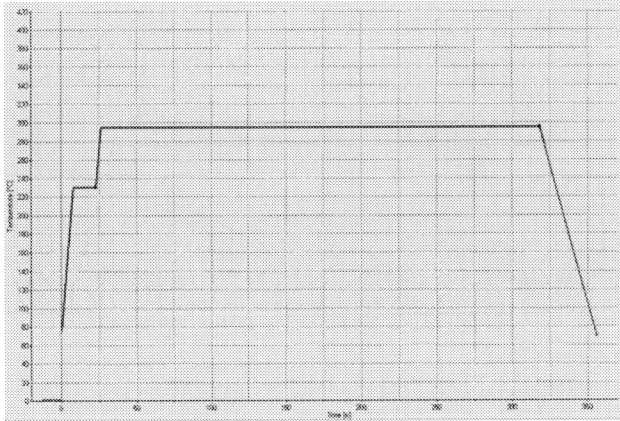

Figure 5. Temperature profile used for SLID bonding of copper-solder stacked interconnections

Assembled samples were cross-sectioned and characterized using Energy Dispersive X-ray Spectroscopy (EDS) to determine the presence of Cu_6Sn_5 and Cu_3Sn, and to study the thickness of the intermetallics.

Results and Discussion

1] IMC Formation Study

The diffusion model predicts the formation of 2μm of Cu_3Sn in 5 minutes at a bonding temperature of 250°C in the ideal case. Referring to the Scanning Electron Microscope (SEM) image shown in Figure 6(a), it can be seen that in 5 minutes, Cu_3Sn was formed at the top-most and bottom-most interfaces. Cu_3Sn thickness was about 1μm at both interfaces while the joint was mainly composed by Cu_6Sn_5. Moreover, the intermediate Cu layer originally plated between the solder layers was no longer observed. With extended assembly time of 15 minutes, larger amounts of Cu_3Sn were observed, while significant amounts of Cu_6Sn_5 still remained, as seen in Figure 6(b). The different formation mechanisms for these two intermetallics can help in explaining both the absence of intermediate Cu layer, and the lower formation rate of Cu_3Sn, as described below.

The melting temperatures of Sn, Cu_6Sn_5 and Cu_3Sn are 231.9°C, 415°C and 670°C respectively. The bonding temperature used in this research is 250°C, which is slightly higher than the melting temperature of the tin solder. By exposing the copper-tin stacked structure to this temperature for a sufficient amount of time, all of the tin is converted to Cu_6Sn_5. During this liquid-phase reaction, the growth of Cu_6Sn_5 consists of Cu dissolution and Cu_6Sn_5 precipitation from molten solder. The dissolution of Cu into molten Sn can be described by Dybkov's analysis [11], where 'Cs' is the solubility of Cu in molten solder at the reaction temperature, 'C' is the current concentration of Cu in molten solder, 'k' is the dissolution rate constant, 'S' is the surface area of Cu pad and 'V' is the volume of molten solder.

$$\frac{dc}{dt} = k\frac{S}{V}(c_s - c) \qquad \textbf{Equation (II)}$$

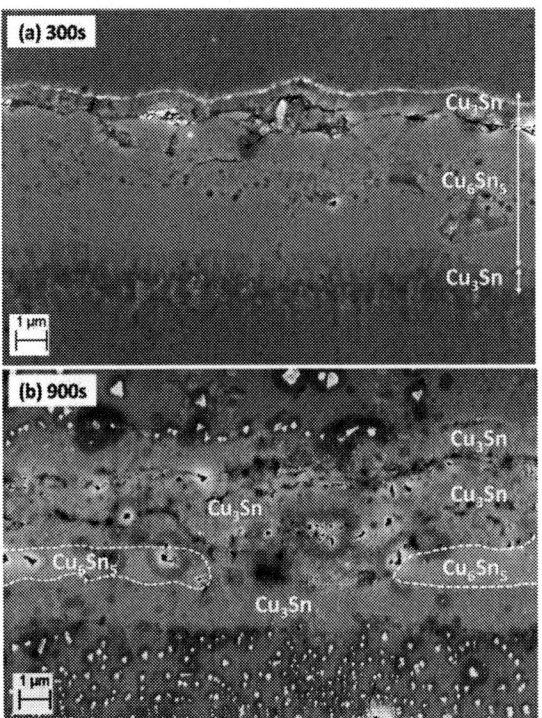

Figure 6. Cross-sections of stack-plated Cu-Sn interconnections

At the early stage of liquid-phase reaction, the term (Cs-C) will dominate the dissolution rate. In this study, pure tin was used within the stack-plated structure. This implies that the dissolution rate of Cu will be high during the beginning of the liquid-phase reaction. This high dissolution rate of Cu has been demonstrated in previous studies [12-13]. Since the thickness of Cu layer within the stack-plated structure was only 2μm, this dissolution mechanism can explain the absence of the intermediate Cu layer that was originally a part of the stack.

As the bonding temperature is below the melting point of the formed Cu_6Sn_5, it is converted to Cu_3Sn only through solid-state diffusion. As a result, the conversion of Cu_6Sn_5 to Cu_3Sn requires much more time than that needed for the formation of Cu_6Sn_5. Referring to the SEM image and EDS characterization, the joint was mainly composed of Cu_6Sn_5 even after 900 seconds of bonding with a maximum temperature of 250°C. The solder-based interconnection had not fully transformed into Cu_3Sn, but only a mixture of $Cu_3Sn + Cu_6Sn_5$. This result agrees with previous studies focusing on Cu/Sn/Cu structure. Li et al. [14] found that reflow at 350°C for 90 minutes was required to transform a 25μm Sn layer into Cu_6Sn_5, but it required another 390 minutes to achieve complete transformation from Cu_6Sn_5 to Cu_3Sn. For a 10μm Sn layer, Cu_6Sn_5 remained as the main part of the solder joints even after 20 minutes reflow at 250°C [15]. Although the designs varied from each other, all these results indicate that Cu_6Sn_5 is the main product of liquid-phase reaction. The growth rate of Cu_3Sn by consuming Cu_6Sn_5 has been found to be much lower than the rate of formation of Cu_6Sn_5.

Therefore, 5-15 min of bonding time at 250°C is sufficient to completely eliminate the residual solders, though the most stable Cu_3Sn is not yet achieved. Increasing either the temperature or time can lead to the stable Cu_3Sn intermetallics. However, due to the lower diffusion distances resulting from the stacked structure of copper and tin, the Cu_3Sn formation time will at least be 2-3X smaller with the current approach than what could be achieved with the traditional solder cap structures.

2] Reliability of Cu/IMC joints

Solder and copper-tin intermetallics formed during assembly significantly differ in their mechanical properties. IMCs are inherently brittle while solders are ductile in nature. Thus IMC formation is a known reason for stress generation in the interconnection bumps during the cooling down of solder. These stresses are usually concentrated at the interface between the solder and the intermetallics [16]. This has an adverse effect on the interconnection reliability, and this issue needs to be addressed in order to achieve good-quality joints at fine pitches and low stand-off heights.

The SLID bonding approach minimizes such stresses at the solder-IMC interfaces by eliminating the solder-IMC interface itself during an isothermal heat-treatment step. This is achieved by converting all of the solder to IMCs, so as to have uniform composition across the joint. Copper and tin form two intermetallics, Cu_6Sn_5 and Cu_3Sn, the latter being the stable intermetallic. The residual solder after assembly reflow in traditional copper-solder interconnections is susceptible to electromigration and thermal-migration. Cu_6Sn_5 and Cu_3Sn, on the other hand, have been shown to have a higher electromigration resistance and better stability as compared to solder [17]. By completely converting the Sn solder to Cu_3Sn, thermodynamic and metallurgical stability in the joints can be achieved. The interfacial shear strength for joints consisting of IMCs has been found to be higher than Sn-dominated joints [18]. Thus, a Cu-IMC-Cu structure is not only highly electromigration-resistant, but also mechanically stable as compared to the Cu-IMC-SnAg-IMC-Cu structure found in traditional copper-solder joints. The solder-free all-intermetallic interconnections with Cu-solder SLID bonding are shown to have good electromigration resistance and thermal cycling reliability, as reported earlier [7-9]. The reliability of SLID bonding with the present Cu/Sn stack structures after complete elimination of residual solders is currently being investigated as the next phase of this work.

Conclusions

An innovative bumping process with alternating copper and tin plating layers to pre-designed thicknesses was developed to fabricate ultra-short, fine-pitch interconnections for 2.5 and 3D interposers and packages. Alternate layers of copper and tin were electroplated on a blanket wafer and at 80 micron pitch, as a first demonstration of this stack-technology. Formation of the intermetallics Cu_6Sn_5 and Cu_3Sn was investigated by SLID-bonding these stack-plated dies with test substrates. The resulting interconnection structures showed a mixture of Cu_6Sn_5 and Cu_3Sn, and no presence of any residual solders, potentially resulting in benefits such as enhanced electromigration resistance and higher joint strength

with shorter processing times. With further process development and optimization, this novel copper-solder stacked approach can potentially achieve ultra-short fine-pitch interconnections capable of handling current densities of $10^5 A/cm^2$ or higher.

Acknowledgments

The authors are grateful to the industry sponsors and mentors for their funding and technical guidance. The authors would also like to thank the staff at the Packaging Research Center for their help in this research project.

References

1. David McCann, Global Foundries.
2. Radu, I. et al., "Recent developments of Cu-Cu non-thermocompression bonding for wafer-to-wafer 3D stacking," 3DIC, 2010.
3. N. Kumbhat, et al., "Highly-reliable, 30m pitch copper interconnects using nano-ACF/NCF," in Electronic Components and Technology Conf, May 2009, pp. 1479-1485.
4. Khan, S. et al., "High Current-Carrying and Highly-Reliable 30μm Diameter Cu-Cu Area-Array Interconnections Without Solder," Electronic Components and Technology Conf, May 2009, pp. 1479-1485.
5. Paik, K. et al, "3D-TSV vertical interconnection method using Cu/SnAg double bumps and B-stage non-conductive adhesives (NCAs)," Electronic Components and Technology Conference, May 2012, pp. 1077-1080
6. Bader, S. et al, "Rapid formation of intermetallic compounds interdiffusion in the Cu–Sn and Ni–Sn systems," Acta Metallurgica et. Materialia, Jan. 1995, vol. 43, no. 1, pp. 329–337
7. Labie, R. et al, "Resistance to electromigration of purely intermetallic micro-bump interconnections for 3D-device stacking," Interconnect Technology Conference, June 2008, pp 19-21
8. Chang, T.C, et al, "Reliable Microjoints for Chip Stacking Formed by Solid-Liquid Interdiffusion (SLID) Bonding," Components, Packaging and Manufacturing Technology, IEEE Transactions, June 2012, pp. 979-984
9. Hubner H, Ehrmann O, Eigner M, Gruber W, Klumpp A, Merkel R, Ramm P, Roth M, Weber J, Wieland R (2002) Face-to-face chip integration with full metal interface. In: Melnick BM, Cale TS, Zaima S, Ohta T (eds) Advanced Metallization Conference, San Diego, V18:53–58
10. Hubner H, Penka S, Eigner M, GruberW, Nobis M, Kristen G, Schneegans M, Barchmann B, Janka S (2006) Micro contacts with sub-30 μm pitch for 3D chip-on-chip integration. MAM, Genoble
11. V. I. Dybkov, "Growth Kinetics of Chemical Compound Layers", Cambridge International Science, Cambridge, MA, 1889
12. M. O. Alam et al.," Cu addition in Sn-3.5%Ag solder on the dissolution rate of Cu metallization," Journal of Applied Physics 94, 2003.
13. M. L. Huang, T. Loeher, A. Ostmann and H. Reichl, "Role of Cu in dissolution kinetics of Cu metallization in

molten Sn-based solders," Applied Physics Letter 86, 2005.

14. J. F. Li, P. A. Agyakwa, C. M. Johnson, "Interfacial reaction in Cu/Sn/Cu system during the transient liquid phase soldering process," Acta Materialia 59, 2011

15. H. Y. Chuang, T. L. Yang, M. S. Kuo, Y. J. Chen, J. J. Yu, C. C. Li, and C.R. Kao, "Critical Concerns in Soldering Reactions Arising from Space Confinement in 3D IC Packages," Transactions on Device and Materials Reliability 12, 2012.

16. Honrao, C. "Fine-pitch Cu-SnAg die-to-die and die-to-interposer interconnections using advanced SLID bonding," Smartech, December 2013

17. Munding, A. et al, "Cu/Sn Solid–Liquid Interdiffusion Bonding," Wafer Level 3-D ICs Process Technology Integrated Circuits and Systems, 2008, pp 1-39

18. Lee et al.,"Chip to Chip Bonding using Micro-Cu Bumps with Sn Capping Layers," Microelectronics and Packaging Conf, June 2009

Study of Electro-migration Resistivity of Micro Bump Using SnBi Solder

Kei Murayama, Mitsuhiro Aizawa, and Mitsutoshi Higashi
Interconnect Technology Development Dept. Research & Development Div.
SHINKO ELECTRIC INDUSTRIES CO., LTD.
36, Kita Owaribe Nagano-shi, 381-0014, Japan
Phone: +81 26 263-4594/ Fax: +81 26 263-4562 / E-mail: kei_murayama@shinko.co.jp

Abstract

There has been a great discussion about electro-migration behavior in semiconductor area. And it has been often discussed that electro-migration behavior of the flip chip package using Sn-Ag bump. However, little study has been done to explore the electro-migration behavior of low temperature solder such as a Sn-Bi solder. In this report, we investigated electro-migration behaviors of micro pillar bump (100 μm diameter) and fine pitch micro bump (25 μm diameter) using Sn57wt%Bi solder. In the case of micro pillar bump, Bi quickly migrated and accumulated on the anode side (Cu pillar) and Sn migrated to the cathode side (substrate pad). And interconnect resistance was quickly increased 80 % from initial during about 150 hours. There was no electrically break failure and it was stabilized at 80% of initial resistance for more than 2800 hours. On the other hand, in the case of fine pitch micro bump, almost of Sn atoms were consumed to form Cu-Sn or Ni-Sn intermetallic compounds (IMCs) after bonding process. The resistance increase was less than 9 %, it is stabilized even for more than 2200 hours and there were no electrically break failure. Additionally, it is evident from electromagnetic field simulation that the maximum current density of the fine pitch micro bump are less than half compared with that of Cu-pillar bumps. Fine pitch micro bump using Sn57 Bi solder is promising candidates for the bonding technology of high performance packages.

Introduction

In order to achieve high speed transmission and large volume data processing, large size silicon die and fine pitch connecting bump for semiconductor package has been required. The bonding technique for high density Flip Chip (F.C.) packages requires a low temperature and a low stress process to achieve high reliability of the micro joining.

Moreover, over the past years, many researchers have interested in electro-migration behavior of the flip chip package. The current density at the micro solder joint is expected to be in the order of $10kA/cm^2$ [1, 2]. Many studies have reported the electro-migration mechanism and the metallurgical reaction between solder and under bump metallurgy (UBM) such as a Sn-Ag solder and Cu pad [3-5]. And several studies have reported that the bump geometry of fine pitch micro joining using Sn-Ag solder after electro-migration test [6, 7]. But fine pitch micro bump using a Sn-Bi solder has been little investigated [8].

Sn-Bi solder joining has been noted as a low temperature bonding methods.

In our previous study, we reported that electro-migration behavior of Sn57Bi solder at 125 degree C. Bi quickly migrated and accumulated on the anode side. Due to the electrical resistivity of Bi is higher than that of eutectic Sn-Bi solder, the resistance was increased to 80 % from initial at first stage. However, it is stabilized more than 3300 hours and there were no electrically break failure [9, 10].

We also investigated electro-migration behavior of Sn3.0wt%Ag0.5wt%Cu (SAC305) solder with electroless Ni/Au on Cu pad. In the case of SAC305, open failure was observed at solder /Ni/Au pad interface (cathode side) [9, 11].

In this paper we will discuss electro-migration phenomena for two types of bump structures connected to two types of surface finish pads. One was micro bump (100 μm diameter) and the other was fine pitch micro bump (25 μm diameter).

Experimental

Appearance of an electro-migration test vehicle and schematic of cross-sections of the bump connections used in this study are shown in Figure 1 and 2, respectively. Table 1 summarizes the package specification in this study.

Two types of bump structure were studied. One was 176 μm pitch Cu pillar with Sn57Bi solder ball (Cu pillar bump). Other was 50 μm pitch Cu pillar with Sn57Bi solder plating (fine pitch micro bump). Additionally, two types of surface finish of substrate were studied, that is, Cu pad and Ni/Au on Cu pad.

In the case of Cu pillar bump, the size of the TEG chip was 9 mm × 9 mm × 0.55 (t) mm. The daisy chain lines were routed using a wafer-level-packaging redistribution process, that yielded 5μm thick Cu. Cu pillars with 50 μm height and 100 μm diameter were fabricated by electroplating on the pads of Si chip. The organic substrate was 35 mm × 35 mm × 1.1 (t) mm with Cu pads and with electroless Ni(P) (6 μm)/Au (0.5 μm) on Cu pad. Sn57Bi solder balls with 75μm diameter were used to fabricate the bumps for connecting Si chip to organic substrate. A chip was mounted on a organic substrate and was reflowed at 180 degree C for 5 minutes. After reflow, underfill resin was supplied and was cured at 165 degree C for 30minutes.

In the case of fine pitch micro bump, the size of the TEG chip was 9 mm × 9 mm × 0.725 (t) mm. The daisy chain lines were routed using a wafer-level-packaging redistribution process, that yielded 3 μm thick Cu. The Cu, Ni and Sn57Bi layers were electroplated and the thickness were 5, 2 and 8 μm, respectively. The Si substrate was 35 mm × 35 mm × 0.725 (t) with Cu pads and with electroplated Ni (2μm) and Au (0.05μm) on Cu pads. A chip was mounted on a silicon substrate and was reflowed at 180 degree C for 5 minutes. After reflow, underfill resin was supplied and was cured at 165 degree C for 30 minutes.

Table 1. Package specification.

	Item	Cu-pillar TEG	Fine pitch TEG
Die	size	9x9x0.55mm	9x9x0.725mm
	Pillar Diameter	100 μm	25 μm
	Bump pitch	176 μm	50 μm
	Solder	None	Sn57wt% Bi Plating (8μm thickness)
	UBM	None	Ni
Substrate	Material	Organic	Silicon
	Pad diameter	74 μm	25 μm
	Solder	Sn57wt% Bi Micro ball (Φ75 μm)	None
	Pad surface finish	Cu / electroless Ni/Au	Cu / electrolytic Ni/Au

Cross-sectional views of 50 μm pitch Cu-pillar are shown in Figure 3.

Focusing on after Sn-Bi plating, large crystal grains were observed. On the other hand, after reflow, the grain was refined and lamellar structure was observed.

The electro-migration test was carried out by reference to the JEDEC standard JEP 154 [12]. The schematic illustration of current supply bump circuit used in this study is shown in Figure 4. Electro-migration is the movement of metal atoms in the direction of electron flow, leads to the formation of cracks at cathode side or some extrusions at other points. In the case of Cu pillar connected to an organic substrate, the electro-migration damage occurred in bumps with electron flow from the Cu pad side (organic substrate) to the Cu pillar side (chip side) [11]. In order to occur failures only one polarity of bump, multiple bumps were provided and current was focused to the bump of interest. A daisy chain of ten interest bumps was constructed in a test vehicle. Therefore the direction of electron flow to induce the electro-migration was decided from substrate side to chip side with current density of 40k A/cm^2 at 150 degree C. In these cases the current density were determined by the smallest cross-section area of the Cu pad on the solder mask opening diameter. To minimize the effect of Joule heating, heat sink was attached on the back side of chip. The chip temperature was measured to be in the range of 2 to 4 degree C higher than the atmospheric temperature of 150 degree C by measuring the resistance patterned on chip.

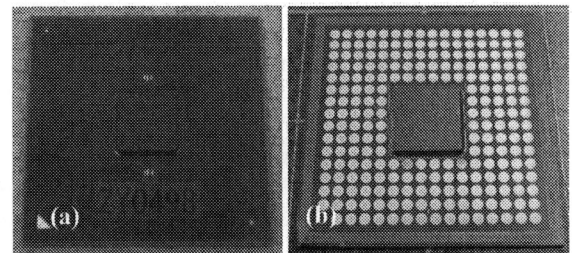

Figure 1. Appearance of electro-migration test vehicles.
(a) for Cu pillar bump   (b) for fine pitch micro bump

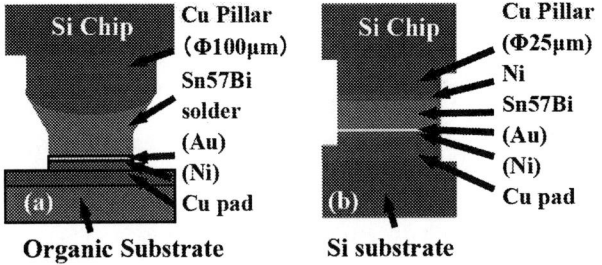

Figure 2. Schematic of cross sections of the each bump connection.
(a) Cu pillar bump   (b) Fine pitch micro bump

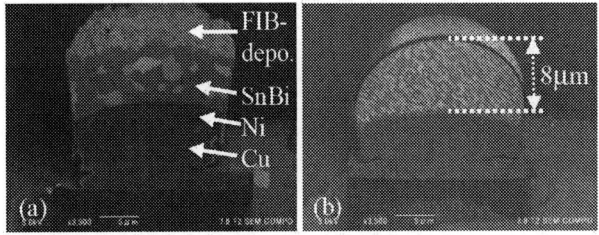

Figure 3. Cross–sectional view of 50 μm pitch Cu-pillar.
(a) As plating   (b) As reflow

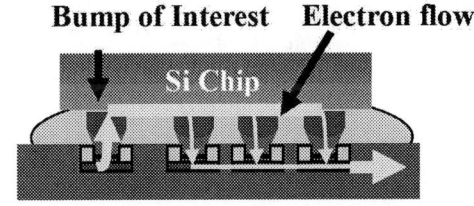

Figure 4. Schematic illustration of current supply bumps circuit.

The resistance of the daisy chain of the ten interest bumps was measured using automatic measuring system. Four-wire method was employed for canceling the resistance of the organic and Si substrate wiring.

Microstructure analysis were performed on both of before and after electro-migration test. The interest bumps were cross-sectioned and were observed by Scanning Electron Microscope (SEM). After that, the bumps were treated with focused ion beam (FIB) and were analyzed by Electron probe micro analyzer (EPMA). We investigated phase analysis from X-ray maps and scatter diagrams that was used to identify the composition of IMCs.

978-1-4799-2408-0/14 $31.00 © 2014 IEEE

Comparison of IMC growth in Sn57Bi Bump

With miniaturization of a bump size, solder volume in bump is reduced and the thickness of IMCs layers increases in bump connections. Cross-sectional views of several size of Sn57Bi bump after bonding process are shown in Figure 5 and 6. In the cases of BGA bump and Cu pillar bump, the solder structure show lamellar and eutectic structure after bonding process. On the other hand, in the case of fine pitch micro bump, all of Sn atoms were consumed to form Cu_3Sn and large Cu_6Sn_5. Bi atoms was observed in large area of Cu_6Sn_5. It should be noted that, after mass reflow and underfill curing, almost of Sn atoms were transformed into Sn-Cu IMCs. Figure 7 shows Backscattered Electron (BSE) images and phase maps determined from X-ray maps and scatter diagrams of Sn3.0wt%Ag0.5wt%Cu (SAC305) bump (25 μm diameter) after bonding process. In this case, un-reacted Sn layer was observed. SAC305 have a composition of more than 90 wt% of Sn but Sn57Bi have only 43 wt% of Sn. It is less than half content of Sn compared with SAC305 joints. Additionally, in the case of fine pitch micro bump using Sn57Bi solder, solder bump height is only 8 μm (Figure 3(b)). Due to the finite Sn atoms content, all of Sn atoms were consumed to form Cu-Sn IMCs during bonding process.

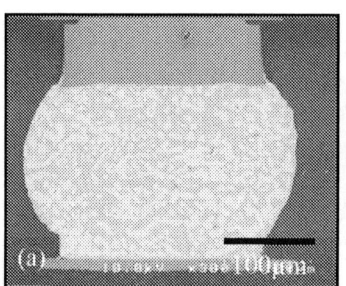 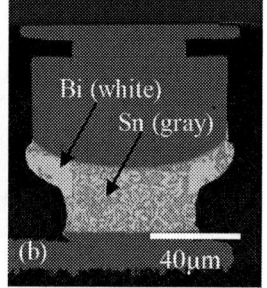

Figure 5. Cross-sectional view of Sn57Bi solder bump after bonding process.

(a) BGA bump (Φ250 μm)   (b) Cu pillar bump (Φ100 μm)

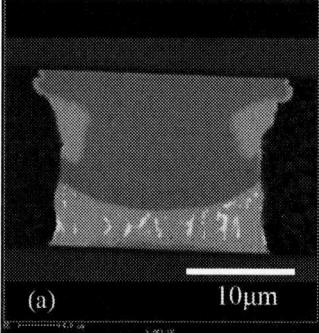 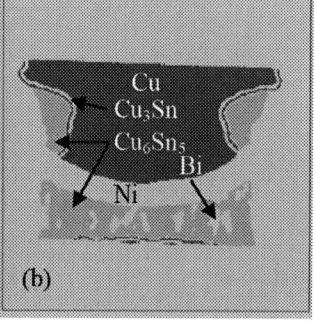

Figure 6. Cross-sectional view of Sn57Bi fine pitch micro bump on Cu pad after bonding (Φ25 μm).

(a) BSE image (b) Phase maps determined from X-ray maps and scatter diagrams of the bump.

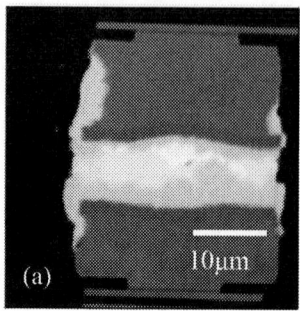 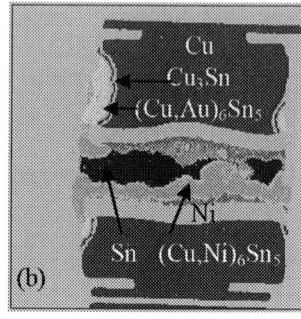

Figure 7. Cross-sectional view of SAC305 bump after bonding process (Φ25 μm).

(a) BSE image (b) Phase maps determined from X-ray maps and scatter diagrams of the bump.

Electro-migration behavior of Sn57Bi micro bump

Figure 8 shows typical resistance increase of using the Cu pillar bump with the current density of 40k A/cm² at 150 degree C.

As connected to Ni/Au pad, the resistance was rapidly increased to about 80 % during 80 hours. And after, the resistance was very slowly decreased to about 77 % during 2800 hours. On the other hand, as connected to Cu pad, the resistance was quickly increased to about 81 % during 130 hours. And after, the resistance was fast decreased to 55 % during 990 hours. The fast resistance decrease is assumed due to the transformation of Cu_6Sn_5 at Cu-pillar side and Cu pad side to Cu_3Sn [9, 13]. And then, the resistance was stabilized until at least 2800 hours. In both cases electrically failures were not observed.

Figure 9 and 10 show BSE images and phase maps determined from X-ray maps and scatter diagrams of the Cu pillar bump after current stressed by 40k A/cm² at 150 degree C. for 2800 hours.

Focusing on connecting to Ni/Au pad (Figure 9), all of Bi atoms migrated to the anode side (Cu-pillar side) and Bi atoms formed an uniform layer. Cu_3Sn / thin Cu_6Sn_5 IMCs were formed at the Cu-pillar side. Under the Bi layer thick $(Cu,Ni)_6Sn_5$ / thin $(Ni,Cu)_3Sn_4$ IMCs were formed. In our previous study, we investigated the electro-migration behavior of Sn57Bi with electroless Ni/Au on Cu pad at 125 degree C. In initial reaction stage, an increment in electrical resistance strongly depended on the growth of Bi-layer at anode side. In second reaction stage, Cu-Sn IMCs were formed and the resistance was slowly increased until at least 3300 hours [8, 9]. In first stage of this case, the behavior is similar to the previous results at 125 degree C. In second stage, it can be presumed that Cu-Sn IMCs were more quickly formed than the previous results at 125 degree C.

As connected to Cu pad (Figure 10), all of Bi atoms migrated to the anode side (Cu-pillar side). However, Bi atoms did not form an uniform layer. Cu_3Sn IMCs were formed at around center location in the connection. Cu_6Sn_5 IMCs were slightly formed at the Cu pad side. In both cases, all of Sn atoms were consumed to form Cu-Sn IMCs during second stage.

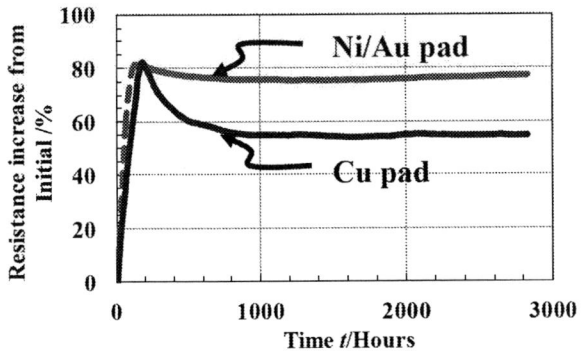

Figure 8. Typical resistance increase of Cu pillar bump with 40k A/cm² at 150 degree C.

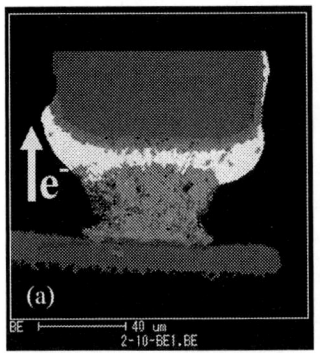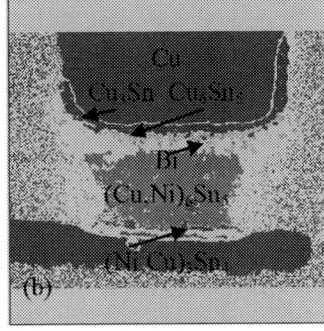

Figure 9. Cross-sectional view of Cu pillar bump on Ni/Au pad after current stressed by 40k A/cm² at 150 degree C for 2800 hours

(a) BSE image (b) Phase maps determined from X-ray maps and scatter diagrams of the bump.

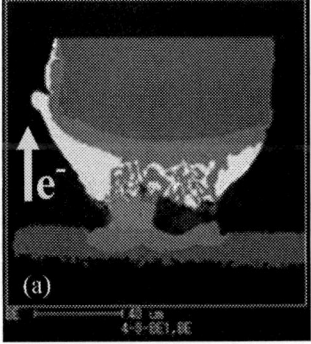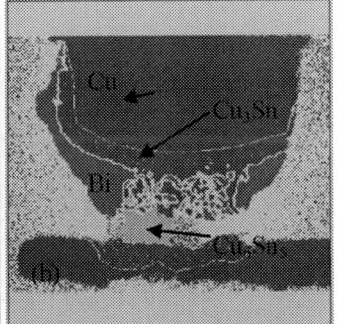

Figure 10. Cross-sectional view of Cu pillar bump on Cu pad after current stressed by 40k A/cm² at 150 degree C for 2800 hours

(a) BSE image (b) Phase maps determined from X-ray maps and scatter diagrams of the bump.

Figure 11 shows typical resistance increase of the fine pitch micro bump with the current density of 40k A/cm² at 150 degree C.

As connected Ni/Au pad, the resistance was rapidly increased to about 8.8 %, during 120 hours. And after, the resistance was slowly decreased to about 7.6% during 460 hours. And then, the resistance was varied between 7.2 to

7.8 % during 2200 hours. On the other hand, as connected to Cu pad, the resistance was increased to about 4.2 % during 310 hours. And after, the resistance was very slowly decreased to 3.8 % during 2200 hours. In both cases electrically failures were not observed.

The resistance increase in the fine pitch micro bumps were reduced less than 1/8 compared with that in the Cu-pillar bumps. The cause of Bi thickness of bump center is smaller than that in the Cu-pillar bump and Bi atoms does not form an uniform layer.

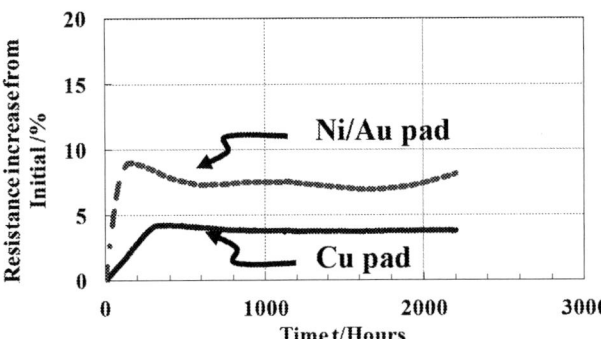

Figure 11. Typical resistance increase of fine pitch micro bump on Ni/Au pad with 40k A/cm².

Figure 12 and 13 show BSE images and phase maps determined from X-ray maps and scatter diagrams of the fine pitch micro bump after current stressed by 40k A/cm² at 150 degree C. for 263 hours.

As connected to Ni/Au pad (Figure 12), Bi atoms was transported from the inside of solder to outside of bump connection. Large $(Ni,Cu)_3Sn_4$ were formed at both of the micro bump side and the Ni/Au pad side. Sn atoms was not existed.

On the other hand, as connected to Cu pad (Figure 13), Bi atoms was transported from inside of solder to outside of bump area as same as observed at Ni/Au pad. However, thin $(Cu,Ni)_6Sn_5$ and large Cu_6Sn_5 were formed at the micro bump side. Sn atoms was not observed.

The IMCs formed were different between Ni/Au pad and Cu pad. It was reported that the kind of IMC formed in Sn, Ni, Cu system was determined by the Cu concentration, that is, when the Cu concentration was low (x = 0.2wt %), the reaction product was $(Ni_{1-x}Cu_x)_3Sn_4$, at high Cu concentrations (x = 0.7 and 1 wt %), the reaction product was $(Cu_{1-y}Ni_y)_6Sn_5$, when the Cu concentration was in-between (x = 0.4 wt %), both $(Ni_{1-x}Cu_x)_3Sn_4$ and $(Cu_{1-y}Ni_y)_6Sn_5$, were formed [15, 16]. It can be presumed that Ni atoms in solder to Ni/Au pad is higher than that to Cu pad [14-16].

Additionally, the resistance increase on Ni/Au pad is larger than that on Cu pad. As connected to Ni/Au pad, Bi layer was existed between Cu pillar and pad after electro-migration test. The electrical resistivity of Bi is higher than that of eutectic Sn-Bi solder and IMCs formed in bump [8-10, 17]. The diffusion rate of Ni atoms in Sn is lower than that of Cu atoms in Sn [18]. As connected to Ni/Au pad, it can be presumed that un-reacted Sn-Bi solder was existed in solder after bonding process. On the other hand, as connected to Cu pad, all of Sn atoms were consumed Cu–Sn IMCs.

As connected to Ni/Au pad, two types of reaction mode are occurred simultaneity. One is Bi transportation from the inside of solder to outside of bump connection. The other is Ni-Sn IMCs formation. On the other hand, as connected to Cu pad, in initial reaction stage, Cu-Sn IMCs were formed. In second reaction stage, Bi atoms was transported from the inside of solder to outside of bump connection.

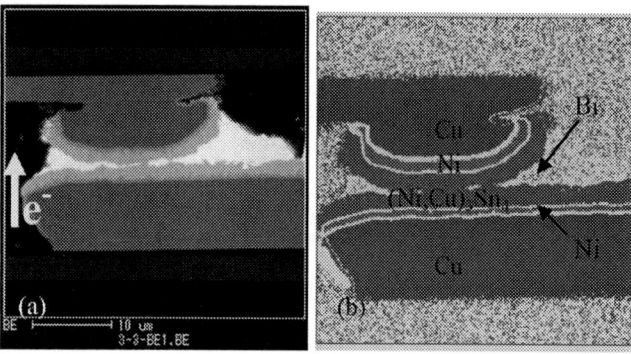

Figure 12. Cross-sectional view of fine pitch micro bump on Ni/Au pad after current stressed by 40k A/cm² at 150 degree C for 263 hours
(a) BSE image (b) Phase maps determined from X-ray maps and scatter diagrams of the bump.

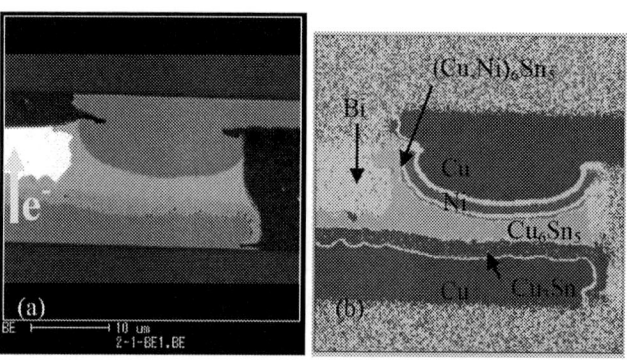

Figure 13. Cross-sectional view of fine pitch micro bump on Cu pad after current stressed by 40k A/cm² at 150 degree C for 263 hours
(a) BSE image (b) Phase maps determined from X-ray maps and scatter diagrams of the bump.

Current density Distribution

A current density calculated from only bump area is deferent from the real value. The current distribution is influenced on the conductor structure and its resistivity.

Therefore, it is important that a current density of bump is simulated with its consideration of bump structure.

An electromagnetic field simulation was employed to analyze current density distributions. The solver was Ansys Q3D Extractor Ver.12, using the finite Element Method with DC analysis.

The current value were determined by the smallest cross-section area of solder so as to be the current density 40k A/cm². Analyze model of micro bump and fine pitch micro bump are shown in Fig. 14 and 15, respectively.

Results of simulated current distribution of micro bump and that of fine pitch micro bump are shown in Figure16 and Figure17, respectively.

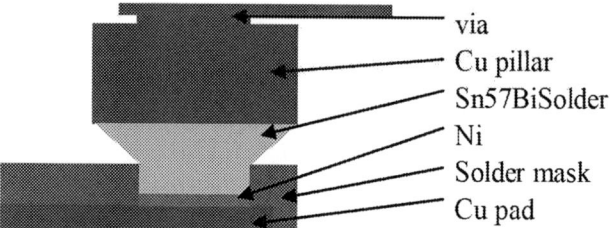

Figure 14. Analyze model of pine pitch micro bump.

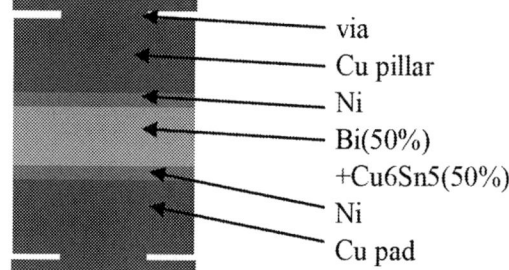

Figure 15. Analyze model of pine pitch micro bump.

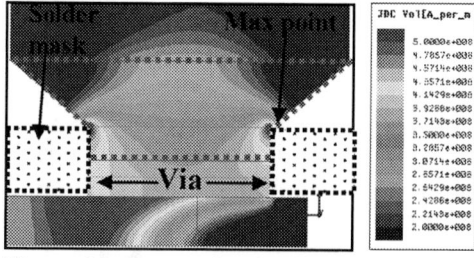

Figure 16. Current density distribution of micro bump.

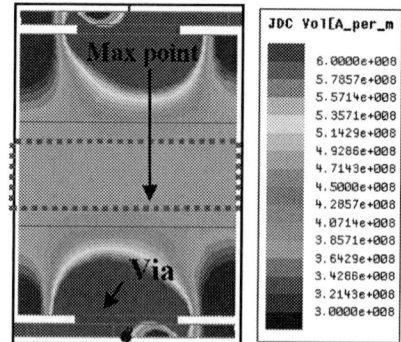

Figure 17. Current density distribution of fine pitch micro bump.

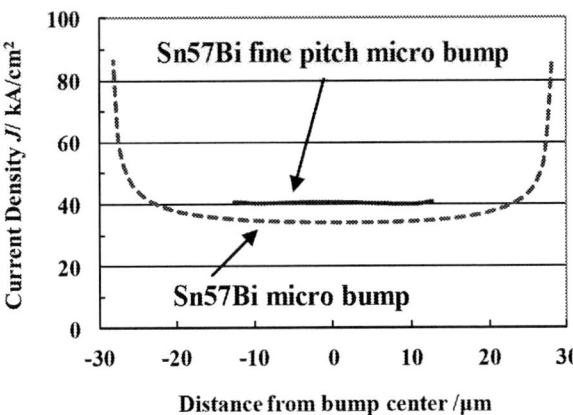

Figure 18. Current density distribution of Solder /land interface.

The pink dash lines indicate the solder area. Since the electro-migration resistance of solder materials is lower than that of Cu, current density distribution of solder area was focused. In Cu micro bump, on the edge of solder mask opening, the maximum current density was 86.1 kA/cm². This value is as large as twice over than calculated value by the smallest cross-section area. In contrast, in Cu fine pitch micro bump, the maximum current density is 40.5 kA/cm². This value is almost the same as the defined value.

The current density of solder / land interface is plotted in Fig.18 as a function of distance from bump center.

In the micro bump, the current density was rapidly increased at the land edge. On the other hand in the fine pitch micro bump, the distribution of current density was uniform in all of solder area. This results indicates that applying a fine pitch micro bump facilitate better electro-migration resistance.

Conclusions

The electro-migration test is performed on Sn57Bi with Cu-pillar and Sn57Bi with fine pitch micro bump.

In the case of Cu-pillar, the resistance was quickly increased to 80 % from initial, it is stabilized more than 2800 hours and there were no electrically break failure.

In the case of fine pitch micro bump, the resistance increase was less than 8.8% from initial and it is stabilized more than 2200 hours and there were no electrically break failure.

The maximum current density of the fine pitch micro bumps are less than half that of Cu-pillar bumps.

Acknowledgments

The authors would like to thank Shinko analysis team and characterization team. We would like also to thank co-workers at Shinko who supported us on preparing test die, substrate and assembly.

References

1. H. Gan, K. N. Tu, "Effect of Electromigration on Intermetallic Compound Formation in Pb-free Solder–Cu Interfaces", Proceedings of 52thECTC, San Diego, CA USA, May 2002.

2. W. J. Choi, E. C. C. Yeh, K. N. Tu, "Meen-time-to-failuer study of flip chip solder joints on Cu/Ni(V)/Al thin-film under-bump-metallization", Journal of Applid Physics, vol. 94, number9, pp.5665-5671, 2003.

3. Kimihiro Yamanaka, Yutaka Tsukada and Katsuaki Suganuma, "Electromigration effect on solder bump in Cu/Sn-3Ag-0.5Cu/Cu system", Spectra materialia, Vol.55, pp. 55867-870, Aug. 2006.

4. Minhau Lu, Da-Yuan Shih, Paul Lauro, "Electromigration in Pb-free Solders", IEEE International Conference on Electronic Packaging Technology & High Density Packaging, 2008.

5. Kiju Lee, Keun-Soo Kim ,Kimihiro Yamanaka ,Yutaka Tsukada, Soichi Kuritani, Minoru Ueshima and Katsuaki Suganuma, "Effects of crystallographic orientation of Sn on electromigration behavior", Proceedings of IMAPS2010, Raleigh, NC, USA, October 2010, pp. 792-797.

6. Yasumitsu Orii, "Microstructure Observation of Electromigration Behavior in Peripheral C2 Flip Chip Interconnection with Solder Capped Cu Pillar Bump", Proceedings of IMAPS2011, Long Beach, CA, USA, October 2011, pp.828-836.

7. Hsiao-Yun Chen et al., "Generic Rules to Achieve Bump Electromigration Immortality for 3D IC Integration", Proceedings of 63thECTC, Las Vegas, Nevada, USA, May 2013, pp.49-57.

8. Kenichi Yasaka, Yasuhisa Ohtake, Toshiya Akamatsu, Nobuhiro Imaizumi, Seiki Sakuyama, Keisuke Uenishi, "Microstructural Changes in Micro-joints between Sn-58Bi Solders and Copper by Electro-migration", Proceedings of ICEP2010, Sapporo, Hokkaido, JAPAN, May 2010, pp.475-478.

9. Kei Murayama, Taiji sakai, Nobuaki Imaizumi and Mitsutoshi Higashi, "Electro-migration Behavior in Micro-joints of Sn-57Bi solder and Cu Post Bumps ", Proceedings of IMAPS2011, Long Beach, CA, USA, October 2011, pp. 997-1006.

10. Kei Murayama, Takashi Kurihara, Taiji sakai, Nobuaki Imaizumi Kozo Shimizu, Seili Sakuyama and Mitsutoshi Higashi, "Electro-migration Behavior in Eutectic Sn-Bi Flip Chip Solder Joints with Cu-Pillar Electrodes" Journal of smart processing, Vol. 2, No. 4, pp. 178-185, 2013.

11. Shigeaki Suganuma, Toshio Gomyo, Yuya Yamagishi, Kei Imafuji, Masaki Sanada, Yuko Karasawa, Kei Murayama, Kurihara Takashi, Yukiharu Takeuchi, " Break Down Failure Process of the Flip Chip Bumps Caused by Current Stressing ", Proceedings of IMAPS2009, San Jose, CA, USA, November 2009, pp. 346-353.

12. JEDEC Standard, JEP154, "Guideline for Characterizing Solder Bump Electromigration under Constant Current and Temperature Stress", JEDEC solid State Technology Association, Arlington, VA, USA, 2008.

13. Riet Labie, Wouter Ruythooren, Kris Baert, Eric Beyne and Bart Swinnen, "Resistance to electromigration of purely intermetallic micro-bump interconnections for 3D-device stacking", Proccedings of IITC2008, Burlingame, CA, USA, June 2008, pp.19-21.

14. Kei Murayama, Taiji sakai, Nobuhiro Imaizumi and Mitsutoshi Higashi, "Electro-migration behavior in Low Temperature Flip Chip bonding", Proceedings of ECTC2012, San Diego, CA, USA (2012), pp. 608-614.

15. W. T. Chena, C. E. Hoal and C. R. Kaoa," Effect of Cu concentration on the interfacial reactions between Ni and Sn–Cu solders ", Journal of Materials Research, Vol.17, pp. 263-266, 2002.

16. Y.D. Jeon et al.,"Comarison of Interfacial Reactions and Reliabilities of Sn3.5Ag, Sn4.0Ag0.5Cu, and Sn0.7Cu Solder Bumps on Electroless Ni-P UBMs", Proceedings of 53thECTC, New Orleans, LA USA, May 2003,pp.1203-1208.

17. Hsiao-Yun Chen, Min-Feng Ku and Chih Chen, "Effect of under-bump-metallization structure on electromigration of Sn-Ag solder joints", Advances in Materials Research, Vol. 1, No. 1 (2012) 83-92.

18. Daisuke Toyoshima, Kenichi Yasaka, Toru Sakai, Toshiya Akamatsu, Nobuhiro Imaizumi, Seki Sakuyama and Keisuke Uenishi, "Influence of UBM Layers on Electro-migration Behavior of Micro-joints using Sn-Ag Solders", Proceedings of ICEP2011, Nara, Nara, JAPAN, April 2011, pp.548-552.

The Impact of Different Under Bump Metallurgies and Redistribution Layers on the Electromigration of Solder Balls for Wafer-Level Packaging

Christine Hau-Riege[1], Beth Keser[2], Rey Alvarado[2], Ahmer Syed[2], YouWen Yau[2], Steve Bezuk[2], and Kevin Caffey[2]

Qualcomm Technologies, Inc

[1] 3165 Kifer Road, Santa Clara, CA 95051

[2] 5775 Morehouse Drive, San Diego, CA 92121

chaurieg@qti.qualcomm.com, 408-533-9647

Abstract

Electromigration performance has been characterized for lead-free solder balls in wafer-level packaging for different solder metallurgy, under bump metallurgy thickness, and redistribution layer thickness and composition. The electromigration lifetimes in this study were found to strongly correlate with the thickness of under bump metallurgy as well as redistribution layer, spanning more than an order-of-magnitude in median time to failure. Also, a redistribution layer comprising of a Ni/Cu bilayer led to a significant lifetime improvement over its Cu-only counterpart, while a change in solder composition did not affect lifetime. Through extensive failure analysis, the differences in lifetimes can be linked to the amount CuSn formation as determined by the under bump metallurgy thickness as well as the location of the CuSn formation as determined by the redistribution layer thickness. Finally, activation energy has been characterized for a process leg with Cu redistribution layer and under bump metallurgy to be 1.34eV, and a current density exponent to be 3.8.

Introduction

Wafer-level packaging (WLP) achieves a small-form factor and lower cost by eliminating the package substrate, and connecting the die directly to the printed circuit board (PCB) via solder balls. As such, different materials and structures are being investigated for WLP, which require an in-depth understanding of the associated reliability [1-2]. High currents are often required, as in the case for power management integrated circuits, so that electromigration is a key reliability concern.

Fundamentally, EM is the self-diffusion of metal(s) induced by electric current, which can lead to voiding and interconnect failure, and has been an area of major reliability concern for decades. Electromigration lifetime quickly degrades with increasing temperature and current density, as described by the well-known Black's Law established in 1969 [3]:

$$MTF \propto j^{-n} \exp\left(\frac{E_A}{kT}\right) \quad \text{Equation 1.}$$

where MTF is the median time to failure, j is the current density, n is the current density exponent, Ea is the activation energy, k is the Boltzmann's constant and T is the temperature. The kinetic parameters Ea and n are critical for the accurate extrapolation of EM lifetime from test to use conditions and the setting the maximum current specification. In surveying the recent literature, a wide range of parameters have been reported for similar lead-free interconnect systems [4-8], which may be a result of the differences in process details as well as testing methodology. Therefore it is prudent to explicitly characterize these parameters for each significant technology change.

In the case of lead-free solder-based systems which use Cu metallization, the formation of CuSn intermetallic (IMC) is observed. IMC plays an important role for multiple reliability mechanisms, including thermal cycling [9-11], power cycling [12], and electromigration [13-14]. In most cases, failure voids develop at the IMC interface, thereby linking the IMC formation rate and location to the reliability lifetime, where electromigration is no exception [13-14].

This work systematically investigates the EM reliability of the die-to-solder ball connection for different UBM and RDL thicknesses, RDL materials, and lead-free solder alloys. Kinetics parameters and failure modes are also reported. This work builds upon previously published results [15], and will be considered together.

Experiment

The WLP package used in this study consists of a polyimide passivation, Cu RDL metal, a second polyimide passivation, a UBM of Cu, and a lead-free solder ball (Figure 1), where the top passivation on the die is nitride. The test vehicle is 0.3mm ball pitch with 145 pins total, and the Si die is 5mm by 5mm. The solder ball is connected to a high-Tg printed circuit board (PCB). Figures 2 and 3 show schematics of the test structures from plan- and cross-sectional views, respectively, where the left-most ball is the tested ball.

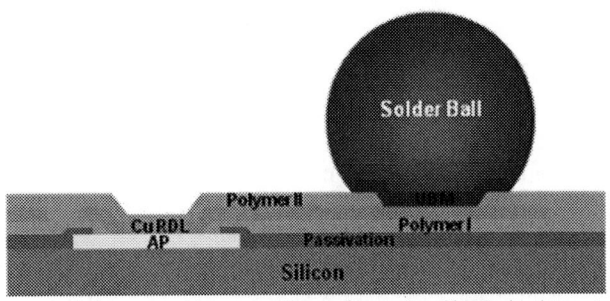

Figure 1. Cross-section schematic of a Wafer Level Package.

978-1-4799-2408-0/14 $31.00 © 2014 IEEE

2014 Electronic Components & Technology Conference

Figure 2. Schematic of the EM test structure from plan view.

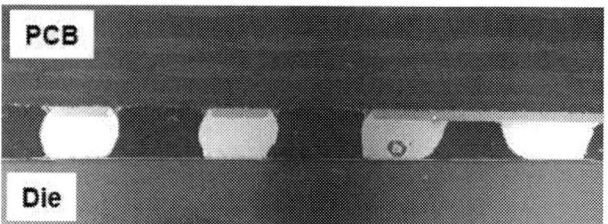

Figure 3. A cross-sectional SEM of the solder ball EM test structures. The die-side was the electron-source side, and the left-most ball is the tested ball, while the other three are designed for current supply and voltage sensing.

Six process legs were tested with different Cu UBM thicknesses (i.e., 0, 8, 12 and 17µm thick) and RDL schemes (i.e., 4 and 9µm thick Cu RDL, and 2µmNi/7umCu bilayer RDL). These legs are designated with an "a" in the experiment identification (ID) as seen on Table 1. For comparison purposes, this paper also refers to results from a previous related work [15], which used the same test structure layout and test methodology, and is designated with a "b" in the ID column in Table 1. It is noted that legs 1a and 7b are identical and have statistically similar lifetime distributions, thereby validating a direct comparison of the two experiments sets. Also, "LF35" (2a) is Sn, 1.2% Ag, 0.5% Cu, 0.05% Ni. Otherwise, all solder balls were SAC405 alloy unless noted.

ID	Description	RDL [um]	UBM [um]	RDL+UBM [um]
1a	8.6um UBM	4	8.6	12.6
2a	LF35	4	8.6	12.6
3a	12um UBM	4	12	16
4a	17um UBM	4	17	21
5a	9um RDL	9	8.6	17.6
6a	2umNi/7umCu RDL	9	0	9
7b	8.6um UBM	4	8.6	12.6
8b	SAC405NiGe	4	8.6	12.6
9b	SAC305NiGe	4	8.6	12.6
10b	no UBM	9	0	9

Table 1. A description of the process legs discussed in this paper.

The electromigration tests of each process leg were conducted at a current density of 0.32×10^4 A/cm² based on the UBM or SRO opening, and a temperature of 162C. Also, the die-side was the electron source. Three additional test conditions were utilized for kinetics studies on leg 1a at a temperature range of 153C to 162C and a current density of 0.38×10^4 A/cm² (Table 2). The test temperatures cited in this study include Joule heating, which was characterized prior to each electromigration experiment, and is less than 5 degrees Celsius in each case.

ID	temperature [°C]	current density x 10^4 [MA/cm²]
1a.1	153	0.38
1a.2	162	0.38
1a.3	162	0.32
1a.4	162.5	0.32

Table 2. A description of test conditions used for kinetics studies on leg 1a.

EM failure was defined as 10% rise in initial resistance, though tests were not necessarily stopped immediately after reaching this criterion. Statistical analysis is based on right-censoring since not all samples failed at this time.

Scanning electron microscopy (SEM) and energy dispersive X-ray (EDX) were routinely conducted for failure site location and IMC formation using.

Results and Discussions: Lifetime distributions

The lifetime distributions for legs 1a through 6a are shown in Figure 4. All distributions are monomodal, and cover a wide range of failure times. Figure 5 shows the corresponding median time to failures (MTFs) as well as their 90% upper and lower confidence intervals, with the addition of leg 10b. All legs have distinctly different lifetimes relative to leg 1a, except for leg 2a. Further, it can be seen that leg 10b (Cu RDL) is inferior to leg 6a (Cu/Ni RDL) even though both had the same overall thickness in RDL and no UBM.

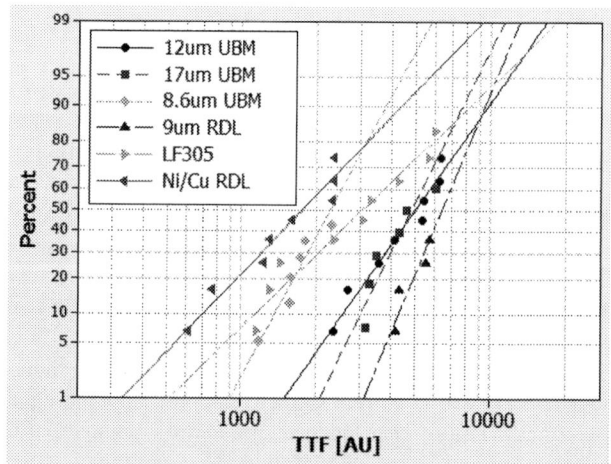

Figure 4. Electromigration lifetime distributions for process leg 1a through 6a as shown in Table 1.

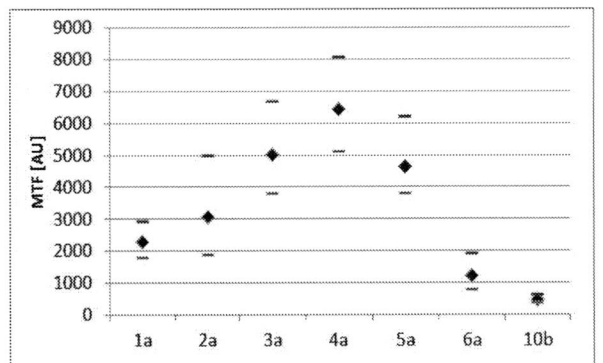

Figure 5. Electromigration lifetime distributions for process leg 1a through 6a as shown in Table 1.

The results in Figure 5 are also represented as MTF versus combined thickness of RDL and UBM (Figure 6). Here, a very strong correlation is observed leading to a regression fit of $R^2 = 0.94$. It has already been established that a thicker UBM enhances EM performance by lowering Joule heating and current crowding [15-17]. This paper seeks to explain how the increase of UBM and RDL thickness lead to MTF improvements for fundamentally different reasons, related to CuSn formation and location, respectively, through failure analysis, and will be discussed in the following sections.

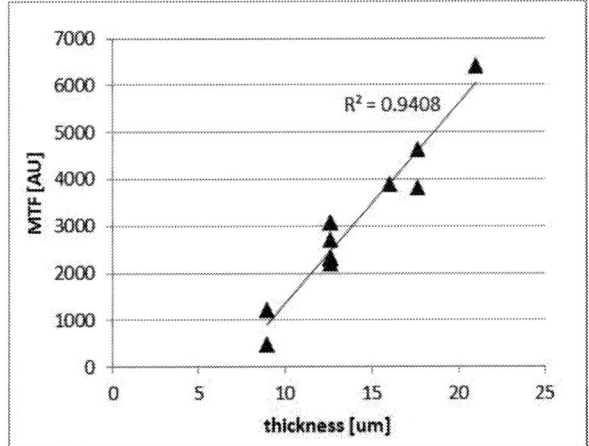

Figure 6. The MTFs of all process legs shown on Table 1 were found to correlate directly to the combined thickness of RDL and UBM layers.

It was also observed that the lifetime distribution of the SAC305 alloy (leg 2a) is statistically similar to the other lead-free solders SAC405 (leg 1a), SAC405NiGe (leg8b) and SAC305NiGe (leg 9b), where all legs had 4um Cu RDL and 8.6um Cu UBM (Figure 7), which indicates the insensitivity of these solder compositions on electromigration lifetime. That is, the change in Ag content as well as Ni and Ge dopant did not impact the IMC formation and electromigration voiding rates. Other papers have reported that a larger amount of dopant can change electromigration lifetime [18-19].

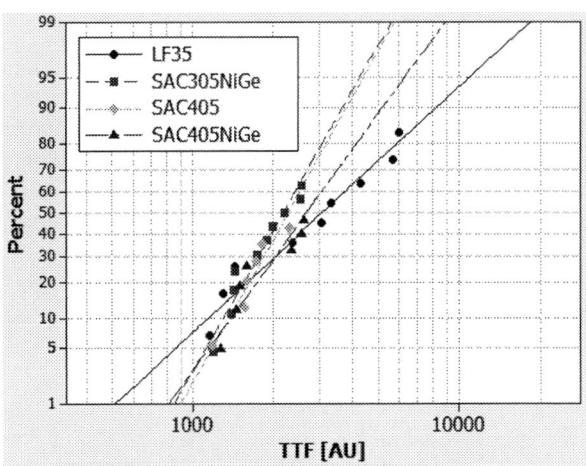

Figure 7. The lifetime distributions for legs 1a, 2a, 8b and 9b which had different solder alloys were statistically similar. It should be noted that all samples had identical RDL and UBM layers.

Results and Discussions: Kinetics Study

Leg 1a was characterized for kinetics parameters Ea and n. Figure 8 shows the lifetime distributions for the four test conditions used in the kinetics study as described in Table 2, resulting in an Ea = 1.34eV and n = 3.8. While there are not many kinetics values published in literature for WLP solder ball, these values fall into the range published for analogous lead-free flip-chip bumps [4-8]. Also, the authors note that the n-value is higher than the typical range of 1 to 2, and could be a consequence of local Joule heating [20], which amplifies the effect of increasing currents on lifetime (equation 1).

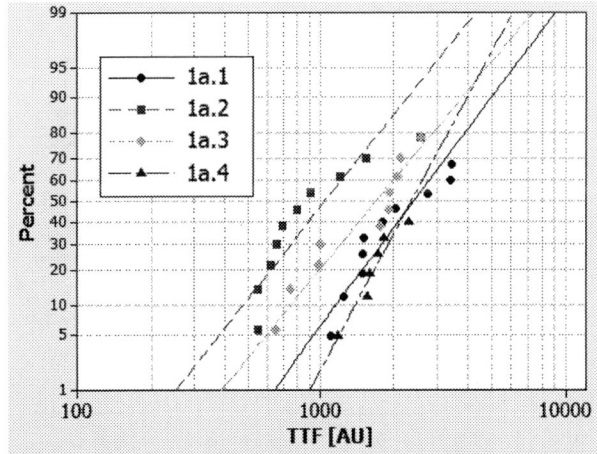

Figure 8. EM lifetime distributions of the 1a process leg at 4 different test conditions.

Results and Discussions: Failure modes

We report the failure analysis for legs 1a, 4a, 5a and 6a in this section. It should be noted that all images are oriented with the die at the top, and electron-source from the right.

978-1-4799-2408-0/14 $31.00 © 2014 IEEE   1175

LF305 solder, 4um Cu RDL and 8.6um UBM

Figure 9 is a set of SEM cross-sections from different samples from leg 1a, arranged in order of increasing final resistance. In each sample, the Cu UBM is no longer present, but has either formed CuSn or been swept to the anode side of the ball. Sn is also observed in the once-UBM area, which is the result of CuSn formation followed by Cu dissolution, which has been described in detail elsewhere [21].

The RDL next to the ball is also no longer pure Cu, but has completely transformed into CuSn. The resulting failure mode in each sample is a pancake void along the CuSn/solder interface in or near the RDL. Clearly, breakage in the RDL connection due to EM leads to significant resistance increase, so it is not surprising that the failure times of this leg were amongst the shortest in this paper (Figure 5). Also this failure mode was also observed in other samples with the same UBM (i.e., legs 7b, 8b, and 9b in Table 1) [15].

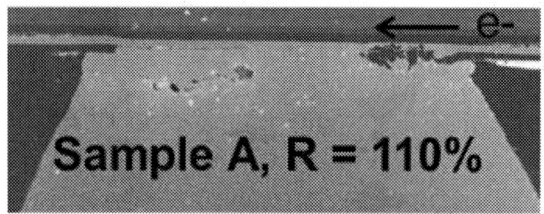

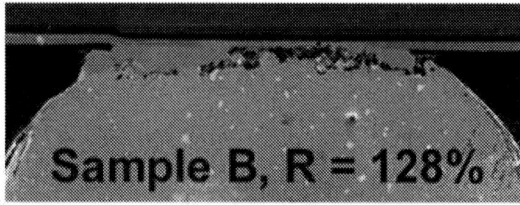

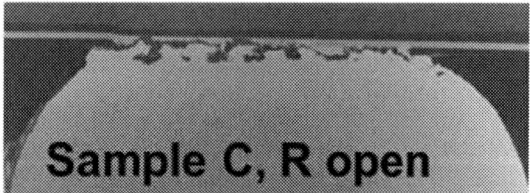

Figure 9. Each of these samples from leg 1a failed by pancake voiding in or near the RDL. The RDL has transformed into CuSn and the Cu UBM is no longer present.

Figure 10 shows the corresponding resistance traces for each sample shown in Figure 9. The final relative resistances of each sample are as follows: Sample A at 110%, Sample B at 128%, and Sample C as open failure. Evidently, the EM void initiates at the interface of the RDL/ball on the cathode side where the current density is highest, and progressively elongates, or "unzips" at this interface towards the left, eventually leading to open circuit failure (see Sample C). Incidentally, this finding differs from other works which report failure by the growth of pre-existing voids which grow and coalesce to form an open circuit [22]. However, time zero analysis of these samples did not show pre-existing voids.

Therefore, the light degree of voiding seen in Sample A away from the cathode-side occurred by EM at the IMC interface.

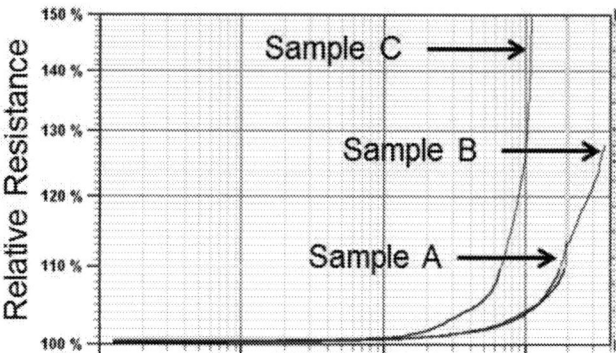

Figure 10. The relative resistance traces show rapid rise for leg 2a, indicating the unzipping of the ball/RDL interface due to electromigration.

LF405 solder, 4um Cu RDL and 17um UBM

With 4um of Cu RDL and 17um of Cu UBM, leg 4a had the largest Cu supply of this study, which resulted in the greatest amount of intermetallic formation in the tested ball. Figure 11 shows a close-up of the cathode side of the ball. Unlike the previous section, the RDL and part of the UBM region remained largely as untransformed Cu in these samples, especially in the region away from the electron-source (i.e., the left-side of the figure). However, Sn did migrate into the RDL near the electron-source where current crowding was stronger to form intermetallic, so that the failure mode is still at the RDL, and resembles the failure analysis of sample A from leg 1a (Figure 9).

Unlike leg 1a, this leg exhibited very long lifetimes, which is due to the massive amount of Cu that must either be swept to the anode side or transformed into CuSn. The authors believe that the "unzipping" of this sample may trace along the IMC which is larger, as suggested by the arrows in Figure 11.

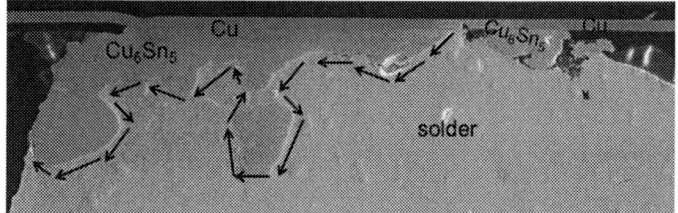

Figure 11. A large amount of IMC formed in leg 4a. Also, the EM void traces the interface between the CuSn and the solder, which is highlighted with arrows. The relative resistance of this sample was 110%.

LF405 solder, 9um Cu RDL and 8.6um UBM

The failure mode in leg 5a with thick RDL was also analyzed (Figure 12). A new failure mode is observed here: the CuSn did not breach the RDL, but transformed mainly in the UBM region, leading to a uniform block of intermetallic at about 8um-thick. The EM void then formed at the CuSn/solder interface, initiating on the cathode-side, in the ball and far from the RDL. Therefore, MTF of this leg (5a) is more than 2x higher than that of the thinner RDL leg (1a) (Figure 5); i.e., the latter failed inside the RDL, so that a smaller void is required for a given resistance rise, while the former failed outside of the RDL so that a larger void is required for a given resistance rise. The MTF of this leg (5a) is also less than the 17um thick UBM leg (4a), suggesting that CuSn formation was quicker owing to the smaller supply of Cu from the UBM. It is supposed, though not proven, that a process leg with thick RDL and thicker UBM would result in even higher lifetimes than that which was examined in this study.

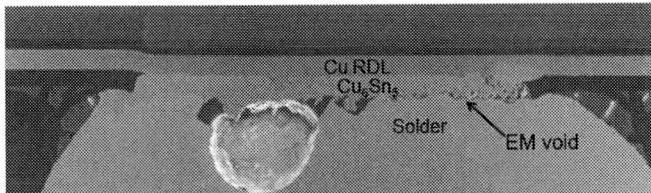

Figure 12. The intermetallic region of leg 5a is quite thick, resulting from the thick Cu RDL layer. Further, the EM void formed at the intermetallic/solder interface. The relative resistance of this sample was 110%.

LF405 solder, 2um Cu/7um Ni RDL, and no UBM

The Ni/Cu RDL samples of leg 6a were also analyzed for failure. Here, the Ni layer is still intact, though it has depleted to a thickness of less than 1um, forming Cu(Ni,Sn) intermetallic in the RDL and ball regions (Figure 13). Consequently, the EM void has formed inside the RDL along the IMC/solder region in the "pancake" mode. Without the Ni layer, the EM void cuts through the thickness of the RDL for Cu RDL samples that have no UBM (leg 10b) where the entire RDL transformed into IMC (Figure 14) [15], and had a statistically shorter lifetime. Once again, the relative lifetime of the process leg can be connected to the location and size of the IMC and the resulting void formation.

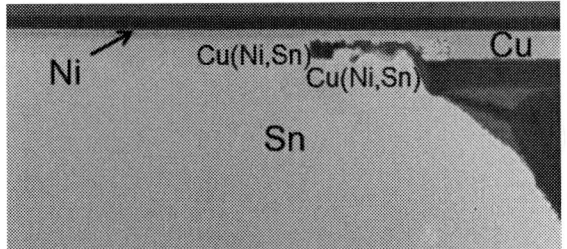

Figure 13. The EM void traces the interface between the Ni layer and the transformed CuSn region with the RDL. The relative resistance of this sample was 113%.

Figure 14. The EM void cut through the thickness of the Cu RDL for samples without UBM [15].

Conclusions

The electromigration performance of solder balls for WLP is critical, especially for high-power applications which require large currents. In this work, we have investigated multiple process legs with various solder alloys, Cu RDL and UBM thicknesses, and RDL types (pure Cu and Ni/Cu bilayer). Increasing the Cu supply in the RDL and UBM layers increased lifetime, but for different reasons: A thicker UBM promoted greater CuSn formation at lower local Joule heating and current crowding, thus prolonging life and failed by voiding in or near the RDL, while the thicker RDL pushed the CuSn region away from the RDL thereby requiring a larger void for a given resistance increase and is a different failure mode. Additionally, a Ni/Cu bilayer (with no UBM) leads to enhanced EM performance relative to Cu-only RDL (with no UBM). The presence of the Ni layer changed the failure mode from a though-RDL-thickness void to a pancake void, requiring longer void growth times to reach failure. Further, we have measured the kinetics parameters of Ea and n to be 1.34eV and 3.8, respectively, which is necessary for proper lifetime extrapolation, and is also within the range reported in literature.

Acknowledgements

The authors would like to thank Hosain Farr from Qualitau, Inc. for his electromigration test support, and Ron Cantrell from ICE for his failure analysis work.

References

1. X. Fan, Q. Han, "Reliability Challenges and Design Consideration for Wafer-Level Packages", International Electronic Packaging Technology & High Density Packaging (2008).

2. X. Fan, "Wafer Level Packaging (WLP): Fan-in, Fan-out and Three-Dimensional Integration", 11th Int. Conf. on Thermal, Mechanical and Multiphysics Simulation and Experiments in Micro-Electronics and Micro-Systems (2010).

3. J.R. Black, "Electromigration—A brief survey and some recent results", *IEEE Transaction on Electron Devices* (IEEE) **ED-16** (4): 338, 1969.

4. J.-H. Lee, Y.-K. Lee, Y.-Bae Park, S.-T. Yang, M.-S. Suh, Q.-H. Chung, K.-Y. Byun, "Joule Heating Effect on the Electromigration Lifetimes and Failure Mechanisms of Sn-3.5Ag Solder Bump", Electronic Components and Technology Conference 2007.

5. M. Lu, D.-Y. Shih, C. Goldsmith, T. Wassick, "Comparison of electromigration behaviors of SnAg and

SnCu solders", 47th Annual International Reliability Physics Symposium 2009.

6. L. Nicholls, R. Darveaux, A. Syed, S. Loo, T. Y. Tee, T. A. Wassick, B. Batchelor, Electronic Components and Technology Conference 2009.

7. M. Ding. G. Want, B. Chao, and P.S. Ho, International Reliability Physics Symposium 2005.

8. C. Hau-Riege, Y.-W. Yau, and N. Yu, International Reliability Physics Symposium 2011.

9. P. Yadav; Kalchuri, S.; Keser, B.; Zang, R.; Schwarz, M.; Stone, B., "Reliability evaluation on low k wafer level packages" Electronic Components and Technology Conference, 2011.

10. S. Xu, B. Keser, C. Hau-Riege, S. Bezuk, and Y.-W. Yau, "A Study of Wafer Level Package Board Level Reliability", Electronic Components and Technology Conference, 2013.

11. Graver Chuan-Chun Chang, Chi-Ko Yu, Tina Shao, Cherie Chen, Jeffrey Lee, "The Relationship of Life Prediction Between Cyclic Bending and Thermal Cycle Testing on CSP Package", Electronic Packaging & High Density Packaging, 2011.

12. S.Y. Yang, I. Kim, S.-B. Lee, "A Study on the Thermal Fatigue Behavior of Solder Joints Under Power Cycling Conditions", Components and Packaging Technologies, Vol. 31, No. 1, 2008.

13. C. Hau-Riege, R. Zang, Y.-W. Yau, P. Yadav, B. Keser, J.-K. Lin, "Electromigration Studies of Lead-Free Solder Balls used for Wafer-Level Packaging", Electronic Components and Technology Conference 2011.

14. C. Hau-Riege, "Tutorial on Package-level Reliability", International Reliability Physics Symposium (2013).

15. C. Hau-Riege, B. Keser, S. Bezuk, and Y.-W. Yau, "Electromigration of Solder Balls for Wafer-Level Packaging with Different Under Bump Metallurgy and Redistribution Layer Thickness", Electronic Components and Technology Conference 2013.

16. R. Bauer, A. Fischer, C. Birzer, L. Alexa, "Electromigration behavior of interconnects between chip and board for embedded wafer level ball grid array (eWLB)", Electronic Components and Technology Conference, 2011.

17. Syed, A. ; Dhandapani, K. ; Berry, C. ; Moody, R. ; Whiting, R., "Electromigration reliability and current carrying capacity of various WLCSP interconnect structures", Electronic Components and Technology Conference (ECTC), 2013

18. X. Zhao, M. Saka, M. Yamashita, "The effect of adding Ni and Ge microelements on the electromigration resistance of low-Ag based SnAgCu", Microsyst Technol (2012) 18:2077-2084.

19. M. Lu, D.-Y. Shih, P. Lauro, S. Kang, C. Goldsmith, S.-Y. Seo, "The effects of Ag, Cu compositions and Zn doping on the electromigration performance of Pb-free solders", Electronic Components and Technology Conference 2009.

20. J. Lloyd and J. J. Clement, Appl. Phys. Lett., "Electromigration damage due to copper depletion in Al/Cu alloy conductors", 69, 2486 (1996).

21. Frank, T.; Chappaz, C.; Arnaud, L.; Federspiel, X.; Colella, F.; Petitprez, E.; Anghel, L., "Electromigration degradation mechanism analysis of SnAgCu interconnects for eWLB package", International Reliability Physics Symposium 2011.

22. T Tian, K Chen, AA MacDowell, D Parkinson, YS Lai, KN Tu, "Quantitative X-ray microtomography study of 3-D void growth induced by electromigration in eutectic SnPb flip-chip solder joints", Scripta Materialia 65 (7), 646-649 2011.

Low-Pressure Sintering Bonding with Cu and CuO Flake Paste for Power Devices

S.W. Park[1],[*], R. Uwataki[1], S. Nagao[1], T. Sugahara[1], Y. Katoh[2], H. Ishino[2], K. Sugiura[2], K. Tsuruta[2], and K. Suganuma[1]

1) Institute of Scientific and Industrial Research, Osaka University, Mihogaoka 8-1, Ibaraki, Osaka 567-0047, Japan
2) DENSO CORPORATION, Materials R&D Div., Komenoki-cho Minamiyama 500-1, Nissin, Aichi 470-0111, Japan
*E-mail: swpark@eco.sanken.osaka-u.ac.jp / Phone: +81-6-6879-8521

Abstract

Low-temperature sintering bonding has been proposed as an alternative technique for the soldering to overcome such high operating temperature in wide-gap semiconductor power devices. Ag nanoparticle sintering is one of the candidates in die-attach bonding, but there are certain obstacles for mass production mainly due to the high cost of silver. In addition, metal nano-particle paste including Ag nanoparticle paste bonding needs to apply certain high pressure of MPa order. For mass productions, it is necessary to decrease the applying pressure during the bonding process. In the present study, the authors make flake-shaped Cu based particles by using mechanical milling for improving the contact area between the particles to decrease the required pressure. The die-bonding with Cu flake pastes was carried out at 300 °C with a formic acid. Resulting die-shear strength exceeds 15 MPa for bonded at 300 °C for 60 minutes low pressure (0.4 MPa). Moreover, Cu flake pastes with polyethylene glycol (PEG) solvent showed solid interface layer like bulk Cu. Thus, the Cu flake PEG paste is one of the most promising bonding materials with the remarkably high strength of the sintered bonding.

1. Introduction

Until recently, the research and development of power semiconductor devices had been focused on Si-based semiconductor because of its excellent availability and affordable production cost. For the recent application, however, under high frequency, high temperature and high voltage condition, wide band-gap compound materials (e.g. SiC, GaN, or InP) are demanded as replacements of Si [1–5]. These wide band- gap semiconductors offer a higher thermal conductivity, higher breakdown electric field, larger band-gap, and higher saturation velocity than Si. Particularly, because of the excellent heat resistant with high breakdown voltage, wide-gap semiconductor is expected to be applied in a severe condition of high temperature (>250 °C) [3, 5]. The increasing operating temperature beyond 250 °C, the electronics packages should be withstand intrinsic harsh environment.

The packaging technologies to fabricate the wide-gap semiconductor devices, die-attach materials play an important role to maintain both electric and mechanical connection between die and electrode in the extreme conditions. Therefore, the development of advanced lead-free technologies should be required. In fact, low-temperature sintering Ag particle pastes have been an increasing focus on interconnect technologies as an alternative technique for the soldering [6–10]. However, Ag pastes have certain obstacles for mass production due to high cost of silver particles. Moreover, Ag is vulnerable to the electrochemical migration (called ion migration) [11, 12]. In contrast, Cu paste is a very promising low-cost alternative to silver-based pastes [13, 14]. It offers proper coefficient of thermal expansion (CTE), high electrical and thermal conductivities [13, 14].

However, Cu nano-particle paste as well as Ag nano-particle paste bonding needs very high applying pressure around MPa orders [6–8, 13, 14], which can cause serious damage of devices. The low-pressure process enables it to be applied to a wide range of applications in power semiconductor industries. In recent papers, we proposed low pressure bonding using the Ag flakes / Ag sub-micron particles hybrid paste [9] and Ag flake paste [10] as a die-bonding material. These flake based pastes could be used in a achieved low applying pressure of 0.07 MPa [9] and of 0.4 MPa [10], respectively.

To decrease the applying pressure, in the present work, the authors make Cu based flake shape particles using mechanical milling for improving the contact area of the each filler. Therefore, the authors try to optimize the low temperature and low applying pressure sintering bonding of flake type Cu paste as high performance die attach materials.

2. Experimental procedure

2.1. Cu flake filler and paste

We used the sub-micron Cu filler as the starting materials: CU1020Y (MITSUI MINING & SMELTING CO., LTD., Japan) with an average particle diameter of 0.5μm. The Cu flake shape particles were fabricated mechanically using a wetting process type rotation attritor mill (NIPPON COKE & ENGINEERING CO., LTD., Japan). The Cu particles were mixed with ethanol and stearic acid as a process control agent (PCA) and a dispersant, respectively. The mechanical milling was operated at the same shaft speed of 300 rpm for 12 hours. After the mechanical milling, the Cu fillers were heat up to 70 °C for 30min for completely evaporating the solvents. The mechanical-milled Cu flake filler was determined by X-ray diffraction (XRD) (Rigaku, Rint-2500).

Cu flake pastes are prepared with an alcohol-base solvent 1-decanol (DN), and a series of glycol solvents, namely, diethylene glycol (DEG), triethylene glycol (TEG), and polyethylene glycol (PEG). These glycol-base solvent have various carbon chain length and molar weight [8]. The Cu powders were mixed with these solvents to ensure a suitable viscosity for screen printing around 300 Pa·s.

2.2. Bonding with Cu flake paste

Polished Cu dummy chips (4 mm × 4 mm, $t = 0.8$ mm) and substrates (10 mm × 10 mm, $t = 0.8$ mm) were used. The paste was screen printed on a Cu substrate, on which a Cu dummy chip was place as shown in Fig. 1(a). Either approximately

0.05 or 0.4 MPa pressure was applied during the die bonding, which was carried out with a formic acid at 300 °C for 60 min.

The shear strength of the joints was evaluated by using a bond tester (DAGE-series4000) at head speed of 50 μm/s, where the fly height was 100μm from the bottom of the specimen as shown in Fig. 1(b). Microstructural observations were carried out using a scanning electron microscope (SEM) operated at 15 kV (JEOL JSM-6700F).

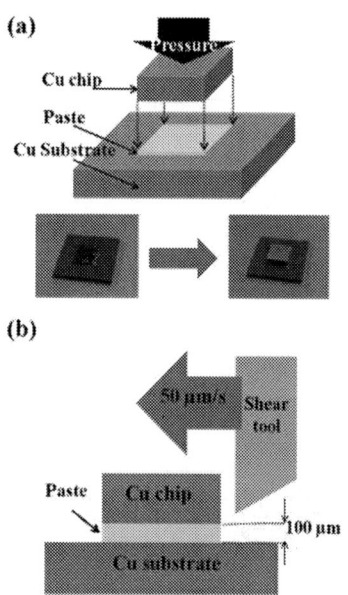

Figure 1. Schematic representation of die-bonding samples used in the bonding experiment.

3. Results discussion

3.1. Properties of Cu flake filler and paste

Changes in powder particle morphology and structure during the milling process of sub-micron Cu powder are illustrated in Fig. 2. The original particles are spherical (see Fig. 2(a)), but after the mechanical milling, they become equiaxed flakes as show in Fig. 2(b). The observations confirm that we can control the flake shape of Cu particles to increase the contact area of filler during die-bonding.

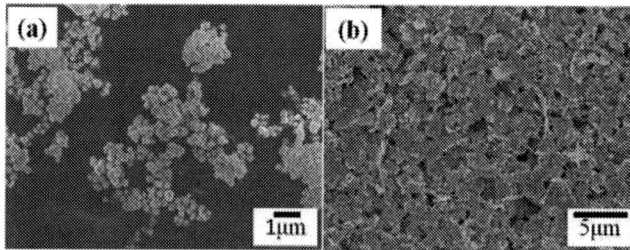

Figure 2. SEM images of the Cu fillers. (a) Sub-micron particle before mechanical milling (b) Cu flake after mechanical milling.

Figure 3 shows the typical diffraction patterns of Cu flakes. The XRD results indicate that Cu particles were oxidized during the milling, as both Cu_2O and CuO peaks appear on the XRD profile in Fig. 3. The oxidized Cu may be useful for sintering process because reduced Cu atoms may recrystallize in the formic acid gas. This reduction and recrystallization processes are the driving force for the well-sintered bonding interface.

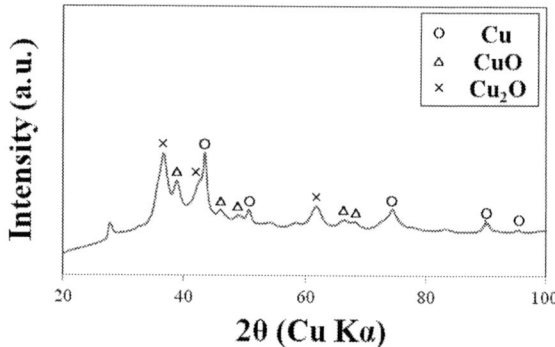

Figure 3. X-ray diffraction pattern of Cu flake filler.

To optimize the bonding conditions and to investigate the influence of the reducing solvent on the decomposition behavior of Cu flake paste, the thermal reaction of the Cu paste evaluated using *in-situ* observation heating furnace. The results of *in-situ* observation are shown in Fig. 4. All the pastes show the oxidized non-metallic color below 250 °C. Over 300 °C the paste color changes to a bright color of Cu metal. We hence suppose that these Cu pastes require about 300 °C to be sintered by the reduction effect of these solvents.

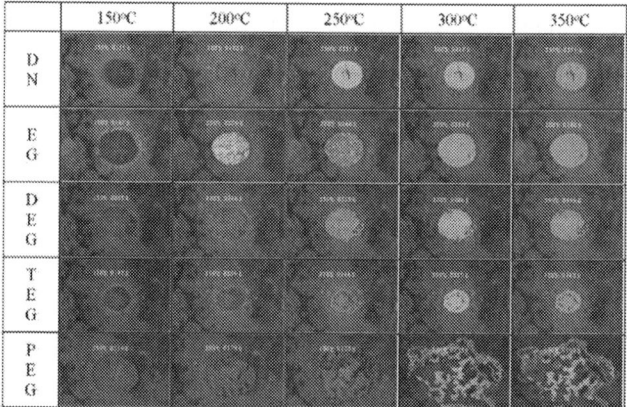

Figure 4. *In-situ* observation of each paste under heating condition.

3.2. Bonding with Cu flake paste

The die-attachment is conducted at 300 °C, which is higher than the boiling points of the solvents as obtained in the paste heating tests in Fig. 4. To evaluate the bonding quality, we have carried out die-shear tests of the bonded samples. Figure 5 shows the result of the die-shear strength test for the joints prepared with each type of Cu flake paste and different applying pressure.

978-1-4799-2408-0/14 $31.00 © 2014 IEEE   1180

At the applying pressure of 0.05 MPa, the shear strength is very low below 10 MPa. Among them, the bonding samples with DN paste are even separated holding with tweezers. To investigate the reason of low bonding, we observed the cross-section of joints. Figure 6 shows that Cu flakes almost maintain their original shape after sintering at 300 °C for 60 min. The rate of the bonded interface area for Cu flake with DN is too small, and is negligible under 0.05 MPa applying pressure.

In case of glycol based paste, the bonding strength increased with the molecular weight of the solvent. Among the glycol series paste, paste with PEG shows the best bonding strength over 10 MPa. We suppose that PEG assists the bonding reaction as a reducing agent (See Fig. 4).

According to some literature about semiconductor standard, the die-shear strength is considered as 12.5 MPa [15]. The applying pressure of 0.05 MPa is very low to satisfy the bonding strength of semiconductor standard. Therefore, the authors increased the applying pressure up to 0.4 MPa.

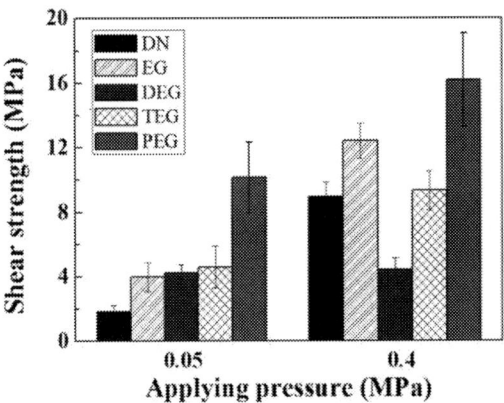

Figure 5. Comparison of die-shear strength using Cu flake paste at different applying pressure.

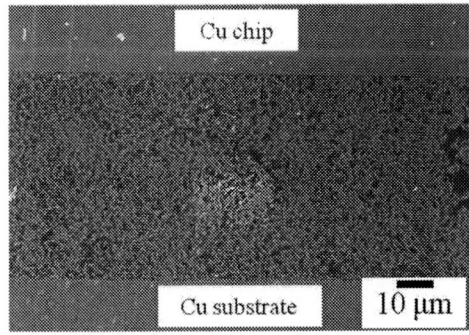

Figure 6. SEM micrographs of jointing interface with DN paste under applying pressure 0.05 MPa.

With an increase in applying pressure at 0.4 MPa, the shear strength of the joint with paste increases quite. Average shear strength of specimens using each paste, except TEG paste 9.6, 13.2, 9.8, 16.4 MPa, respectively. Especially, the shear strength of joints with PEG paste about 16.4 MPa, which is lower than that of reported pure Zn solder (about 60 MPa) [16], but slightly higher than that of semiconductor standard (12.5 MPa) [15]. Therefore, we consider that the best

Cu flake bonding paste of present study is PEG paste with 0.4 MPa pressure.

To clarify the relation between shear strength and applying pressure, we also observe the cross-section microstructure of bonding interface. Figures 7 and 8 show the cross-section microstructure of bonding interface with PEG paste. The interface with applying pressure of 0.05 MPa was observed the partially sintered Cu paste layer. Although the sintering is progressed, some void and not sintered layer is also observed as shown in Fig 7(b). In contrast, applying pressure of 0.4 MPa shows completely sintered layer, it could not distinguish between Cu substrate from Cu flake paste interface (see Fig. 7 (c)).

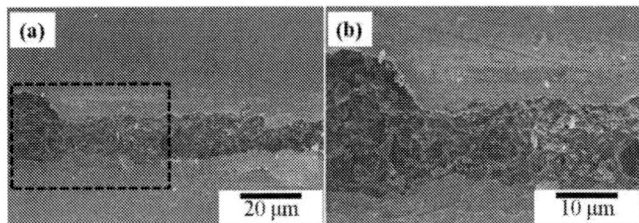

Figure 7. SEM micrographs of jointing interface with 0.05 MPa applying pressure using a PEG paste. (a) low magnification image, (b) enlargement image of (a).

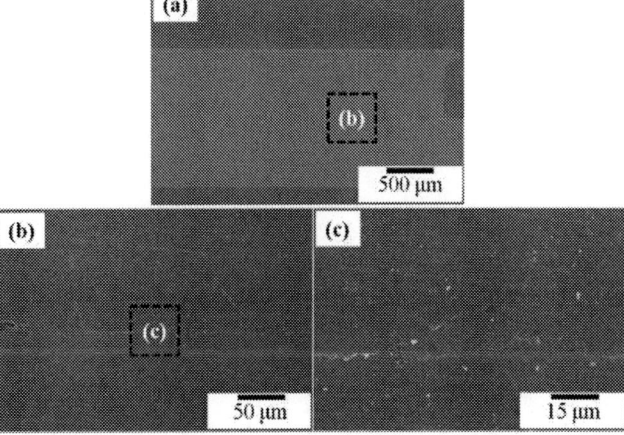

Figure 8. SEM micrographs of jointing interface with 0.4 MPa applying pressure using PEG paste. (a) low magnification image, (b) the part shown in dotted line (a), (c) high magnification image of (b).

Summary

In this work, Cu sintering bonding with flake shape fillers was investigated their bonding strength and interface for use in a high temperature power devices. The results could be summarized as follows:

(1) The present Cu flake was investigated with various solvent such as 1-decanol (DN), diethylene glycol (DEG), triethylene glycol (TEG), and polyethylene glycol (PEG). We found that the molecular weight or chain length affect to the

bonding strength. The present work, the heaviest molecular weight of PEG paste shows the best bonding strength.

(2) In case of PEG solvent Cu flake paste, the bonding strength was obtained with a low applying pressure of 0.4 MPa. The strength was indicates reasonable bonding strength about 16.4 MPa.

(3) Cu flake pastes with PEG solvent show solid interface layer like bulk Cu under when sintered under a slight pressure of 0.4 MPa.

Thus, it can be conclude that the present Cu flake paste with PEG has great potential as die-attach materials for wide-gap semiconductor devices. A further study on the reliability of the die-attachment with Cu flake paste during thermal cycling should be conducted.

Acknowledgments

This work was partly supported by Grant-in-Aid for Scientific Research (S), Grant No. 24226017.

References

1. A. Katz, C. H. Lee, and K. L. Tai, "Advanced metallization schemes for bonding of InP-based laser devices to CVD-diamond heatsinks," *Mater. Chem. Phys.*, vol. 37, no. 4, pp. 303–328, 1994.

2. M. Asif Khan *et al.*, "GaN based heterostructure for high power devices," *Solid-State Electron.*, vol. 41, no. 10, pp. 1555–1559, 1997.

3. W. Wondrak *et al.*, "SiC devices for advanced power and high temperature applications," *IEEE Trans. Ind. Electron.*, vol. 48, no.2, pp. 307–308, 2001.

4. Y.Sugawara, "SiC devices for high voltage high power applications," *Mater. Sci. Forum*, vols. 457–460, pp. 963–968, 2004.

5. H.S. Chin, K.Y. Cheong, and A.B. Ismail, "A Review on Die Attach Materials for SiC-Based High Temperature Power Devices," *Metall. Mater.Trans. B*, vol. 41, no. 4, pp.824–832, 2010.

6. Andrew A. Wereszczak *et al.*, "Sintered Silver Joint Strength Dependence on Substrate Topography and Attachment Pad Geometry," in *Proc. of the 7th IEEE International Conference on Intergrated Power Electronics Systems (CIPS 2012)*, Nuremberg, Germany, March 2012, pp. 451-456.

7. Gang Chen *et al.*, " Pressure-Assisted Low-Temperature Sintering of Nanosilver Paste for 5 × 5-mm² Chip Attachment," *IEEE Trans. Comp. Packag. Technol.*, vol. 2, no. 11, pp. 1759–1767, 2012.

8. S. Takata, *et al.*, "Effects of Solvents in the Polyethylene Glycol Series on the Bonding of Copper Joints Using Ag_2O Paste," *J. Eletron. Mater.*, vol. 42, No.3, pp. 507–515, 2013

9. K. Suganuma *et al.*, "Low-temperature low-pressure die attach with hybrid silver particle paste," *Microelectron. Reliab.*, vol. 52, no. 2, pp. 375– 380, 2012.

10. Sakamoto Soichi and Katsuaki Suganuma, "Low Temperature and Low-Pressure Die Bonding Using Thin Ag-Flake and Ag-Particle Pastes for Power Devices," *IEEE Trans. Comp. Packag. Technol.*, vol. 3, no. 6, pp. 923–929, 2013.

11. Bo-in Noh *et al.*, " Microstructure, Electrical Properties, and Electrochemical Migration of a Directly Printed Ag Pattern,"*J. Eletron. Mater.*, vol. 40, No.1, pp. 35–41, 2011.

12. Kwang-Seok Kim, Jae-Oh Bang, and Seung-Boo Jung, "Electrochemical migration behavior of silver nanopaste screen-printed for flexible and printable electronics," *Curr. Appl. Phys.* vol. 13, pp. S190–S194, 2013.

13. Shutesh Krishnan, A. S. M. A. Haseeb, and Mohd Rafie Johan, "Preparation and Low-Temperature Sintering of Cu Nanoparticles for High-Power Devices," *IEEE Trans. Comp. Packag. Technol.*, vol. 2, no. 4, pp. 587–592, 2012.

14. Julian Kähler *et al.*, "Sintering of Copper Particles for Die Attach," *IEEE Trans. Comp. Packag. Technol.*, vol.2, no. 10, pp. 1587–1591, 2012

15. A. Haque *et al.*, "Die attach properties of Zn–Al–Mg–Ga based high-temperature lead-free solder on Cu lead-frame," *J. Mater. Sci.-Mater. Electron.*, vol. 23, no. 1, pp.115– 123.

16. K. Suganuma, and S. Kim, "Ultra Heat-Shock Resistant Die Attachment for Silicon Carbide With Pure Zinc," *IEEE Electr. Device. L.*, vol. 31, no. 12, pp. 1467– 1469, 2010.

A Novel 3D Packaging Concept for RF Powered Sensor Grains

Walther Pachler[1], Klaus Pressel[2], Jasmin Grosinger[1], Gottfried Beer[2],
Wolfgang Bösch[1], Gerald Holweg[3], Christian Zilch[4], Manfred Meindl[5]

[1]Institute of Microwave and Photonic Engineering, Graz University of Technology, Graz, Austria
[2]Infineon Technologies AG, Wernerwerksstraße 2, 93049 Regensburg, Germany
[3]Infineon Technologies AG, Babenbergerstraße 10, 8010 Graz, Austria
[4]Magna Diagnostics GmbH, Leipzig, Germany
[5]Danube Mobile Communications Engineering GmbH & Co KG, Linz, Austria

Abstract

We present a novel three-dimensional (3D) embedded wafer-level ball grid array (eWLB) system in package (SiP) solution for biochips and micro labs. This 3D SiP includes three major components, a complementary metal oxide semiconductor (CMOS)-tunnel magneto resistance (TMR) sensor biochip for magnetic bead-sensing stacked on a radio frequency identification (RFID) microchip and a 13.56 MHz coil antenna for wireless energy and data transfer. The power supply and the serial peripheral interface (SPI) chip interconnections between the CMOS-TMR sensor biochip (slave) and the RFID microchip (master) are implemented with a novel embedded Z-line (EZL) vertical contact technology through the mold compound. The 13.56 MHz antenna is embedded into the fan-out area of the bottom redistribution layer of the eWLB. With this setup we are able to maximize the RFID reading distance and to ensure a displacement to the TMR sensor surface. We achieve an overall volume of the 3D SiP of only 5.6 mm x 3.6 mm x 0.7 mm applying the eWLB technology. Due to the RFID technology the developed 3D SiP does not need any external contacts and cabling. Therefore it can be encapsulated into harsh environments. In addition the top fan-out surface of the eWLB can be used for adhesive bonding to higher level analyzing setups. The results demonstrate that innovative SiP technology using the eWLB technology combined with chip and antenna design allow to realize modern subsystems e.g. for medical applications.

Introduction

The point-of-care (POC) technology for health care already makes inroads into our life [1]. Various simple biological rapid tests are available on the market, such as blood glucose testing, pregnancy testing, and hemoglobin diagnostics. These tests are fast, low-cost, effective and simple. However the diagnostic of life-threatening infections and complex illnesses such as sepsis and cancer is still done by laboratories. If there is a serious illness suspicion, specialists need to analyze blood or tissue samples with time-consuming and expensive procedures.

Recently, the number of reported highly miniaturized sensor grains or micro labs has increased significantly. Amperometric measurements with biochips [2], impedance spectroscopy [3], or even deoxyribonucleic acid (DNA) detection using complementary metal oxide semiconductor (CMOS) sensor arrays [4] [5] were published in the last years. Indeed these prototypes promise affordable POC diagnostics

for complex diseases. Nevertheless the custom made analyzing hardware and data interfaces are far away from commercial viability. In particular the biochip packages are connected with bond wires [2-5]. The bond wires and the non-active/sensing chip surface are encapsulated with silicone or epoxy. The mechanical robustness and commercial viability of this approach are debatable. Furthermore the analyzing hardware is often realized with external microcontroller- or field-programmable gate array (FPGA)-boards. The dimensions of such assemblies lead to various drawbacks for POC diagnostic systems.

To introduce the POC technology into the consumer market, the work presented here, demonstrates a novel approach for the design of heterogeneous three-dimensional (3D) biochip packaging. All biochip components are assembled within a minimum volume utilizing the embedded wafer-level ball grid array (eWLB) technology. The presented system in package (SiP) includes a CMOS-tunnel magneto resistance (TMR) sensor biochip for magnetic bead-sensing, which is stacked on a radio frequency identification (RFID) microchip, and a 13.56 MHz coil antenna for wireless energy and data transfer. The RFID chip enables the SiP to work autonomously. Other benefits of the presented SiP are that it can easily be attached to higher level systems and that it assures an unprecedented prototype in the field of biochip and in general fluidic micro labs due to planar surfaces simplifying sealing. This application demonstrates that the eWLB technology has outstanding capabilities to design innovative SiP solutions.

The paper is organized as follows. First we provide a brief description of the CMOS-TMR biochip and its application. Then we present the design of the eWLB 3D SiP. This section focuses on all the SiP components and the benefits achieved by applying the eWLB packaging technology using for the first time the vertical contact embedded Z-line (EZL) technology. Next, we show the results on the contactless RFID communication interface of the presented eWLB SiP and the 13.56 MHz antenna. We discuss electromagnetic field simulations of the antenna in the package and verify them by measurements. Finally, we draw conclusions that point out the unique concept which enables new capabilities and leads to affordable POC diagnostic systems in the field of complex medical applications.

TMR biochip application

The presented CMOS-TMR biochip was developed within the EU-funded project "eBrains" [6]. The main aim of the biochip is to detect magnetically marked bio-molecules from

978-1-4799-2408-0/14 $31.00 © 2014 IEEE   1183   2014 Electronic Components & Technology Conference

test substances, such as a blood sample. After a preparation procedure and the immobilization of these DNA molecules (see Figure 1), magnetic beads are hybridized on to the sensor array of the biochip. Special bound magnetic beads can be detected by the use of the TMR biochip surface. TMR cells change the resistance with the magnetic field [7]. Therefore the biochip also comprises analog-to-digital converter (ADC) and memory banks to detect and save the change of these resistances. Different pathogens, which are detected, can lead to a complex diagnosis, such as to diagnose sepsis.

The chip features 8 x 16 sensor spots with 32 x 32 TMR cells each. In contrast to recently reported biochips, we packaged the TMR biochip with a novel eWLB technology concept. The eWLB 3D SiP, which is the main focus of the paper, can be easily attached to a micro-fluidic card as shown in Figure 1. Furthermore the SiP could also be applied to realize intelligent measurement tapes.

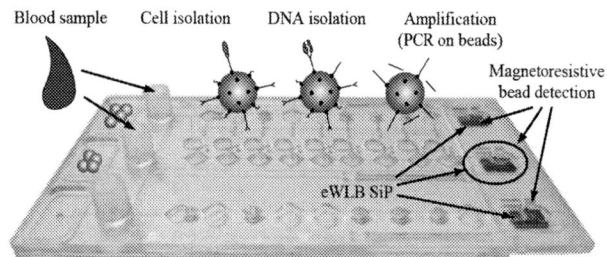

Figure 1. POC micro-fluidic card: A blood sample is inserted to a micro-fluidic cartridge and binds to paramagnetic beads. Three steps are fulfilled until the presented eWLB 3D SiP detects the magnetic beads: cell isolation, DNA isolation, and amplification. Circle: one of the three eWLB SiPs.

3D eWLB biochip packaging concept

The main focus of the paper is an eWLB 3D SiP. The eWLB technology is an innovative packaging technology developed during recent years [8]. The technology was developed to cross the interconnect gap, to achieve further miniaturization, and to reduce parasitic effects. The package has excellent capabilities for high frequency applications and allows outstanding capabilities to design innovative SiP solutions. During recent years Infineon developed a magnitude of toolbox elements for this technology [9]. In this paper we show novel toolbox elements that were developed and apply them to the example of a novel medical application. Figure 2 shows the example of a basic eWLB package [10].

Figure 3 shows a schematic of the process flow. This is provided here to demonstrate the packaging of the medical SiP demonstrator setup. The process flow can be divided into three main parts: reconstitution, redistribution, and ball apply with singulation [11]. In case of wireless energy/data transfer the ball apply for external connections is simple skipped.

The eWLB assembly technology offers outstanding capability for 3D system integration. On the one hand chips can be advantageously placed side by side. On the other hand the chips can be stacked in the package and additional passive components can be integrated. The fan-out area provides more space for additional solder balls and routing. Furthermore it can be used to add passive components to the package, as reported in [12]. Due to the material characteristics of the

mold compound (e.g. dielectric constant $\varepsilon_r = 4.1$) [12], the fan-out area can also be used for antennas. The reported eWLB packages are typically mounted to higher level carrier boards connected via solder balls. Figure 4 shows various toolbox elements to generate SiP devices. The cross-section on the left shows two dies stacked and a redistribution layer (RDL) on top and bottom of the eWLB. In addition passive components were soldered on top of the package. Furthermore a through encapsulated via (TEV) vertical contact was designed. Two choices were investigated in the past: (i) via bars that are based on printed circuit board (PCB) technology (shown in the middle); (ii) TEV filled with copper by electroplating (applied in the left SiP).

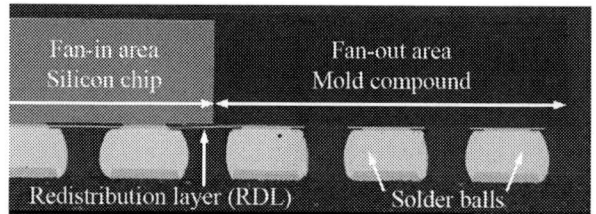

Figure 2. Basic eWLB package: The fan-out area can for example be used for the design of inductors, antennas, or vertical contacts [10].

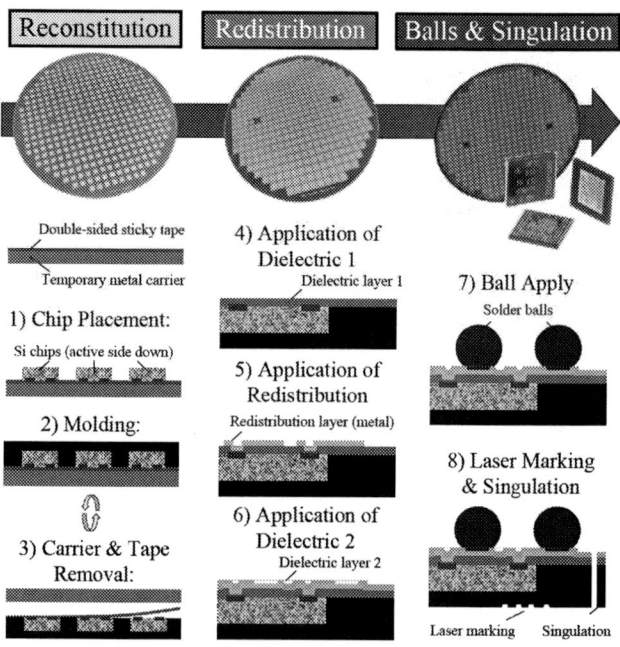

Figure 3. The three main parts of the eWLB process flow: reconstitution, redistribution, and balling and singulation

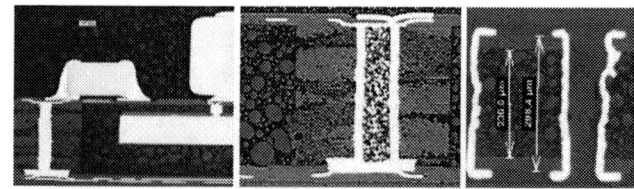

Figure 4. Various toolbox elements to build SiP devices: The figure shows a 3D-SiP with two dice stacked and passives on top (picture on the left), a TEV made of via bars (picture in the middle), and a TEV made with laser drilling and filled with copper by electroplating (picture on the right).

978-1-4799-2408-0/14 $31.00 © 2014 IEEE

For this work two chips, the power management RFID chip, and the TMR biochip with an array of 8 x 16 independent sensor cells are required to be integrated. This work disclaims the use of any external contact. An RFID microcontroller chip provides the data and energy interface between the biochip and a reader device. In order to miniaturize the size of the biochip package, the chips were stacked above each other. The area of the overall package was designed to meet 5.6 mm x 3.6 mm. Due to the smaller size of the TMR biochip (4.3 mm x 2 mm), 43 % of the resulting top surface was fabricated as fan-out mold compound. While the top fan-out surface is used for adhesive bonding and fluidic sealing to the already described micro-fluidic plastic card, the bottom fan-out area provides enough area to embed the high frequency (HF) RFID antenna. The RFID antenna itself, as well as the routing is realized with two RDLs. Each coil antenna with more than one winding needs an underpass. To realize this underpass and to interconnect the bottom RDL with the top RDL through the mold compound, a novel vertical contact, the so-called EZL technology is used.

The EZLs are prepared components based on the eWLB RDL technology. Contact lines are generated by the eWLB thin film technique just on mold compound. This mold compound is then cut into pieces. The components are then turned by 90° and placed as independent components on the artificial wafer (see Figure 3: eWLB redistribution). The EZL has the advantage that much finer structures can be generated than for via bars and laser drilled TEVs.

Figure 5 shows a top view of the EZL component (black) together with the power management chip, which is stacked onto the TMR sensor chip before molding (compare Figure 3 left). Figure 5 shows the two semiconductor chips stacked on each other together with the EZL component, which includes the vertical contacts, next to them. Only a limited number of the two active devices were available due to early silicon on shared reticle. Thus, we had to use quadratic silicon dummy chips to compensate voids. Identical silicon content is needed during molding to achieve homogeneous moldflow.

Fig 6a) shows a computer tomographic picture of the produced SiP. The chips and all metallic materials are illustrated. The size of the bottom chip (blue) is 3000 μm x 2100 μm. It is directly connected to the HF coil antenna. As it can be seen the coil antenna, realized on the bottom RDL, surrounds the chip at its fan-out area. Due to the fact, that no chip metal coats the antenna, the parasitic capacitance of the antenna could be minimized. The size of the top chip (red) is 2075 μm x 4300 μm. The computer tomography image shows the chip metals and the TMR sensor array. The TMR biochip is connected with the top RDL. There are five interconnections. While the first two connections (left) provide the power supply (VDD and GND), the other connections realize the data interface to the RF chip. The data interface is designed as serial peripheral interface (SPI). Therefore there are three more connections: MISO (master in slave out), MOSI (master out slave in), and CLK (clock).

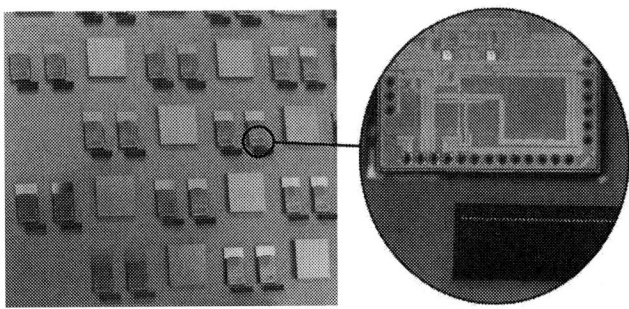

Figure 5. (left) The two stacked semiconductor chips next to the black EZL device with a quadratic dummy silicon chip before molding; (right) the top view of an EZL device (black) with the power management chip (yellow) is highlighted

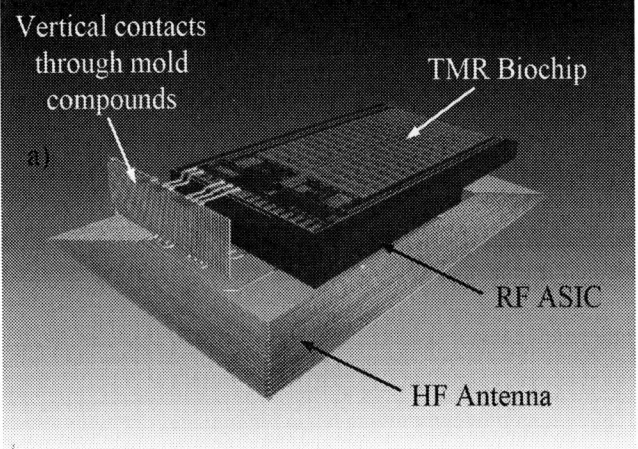

Figure 6. Top: a) Computer tomography of the produced eWLB package. The biochip and the radio frequency communication chip are arranged above each other. The interconnection is realized with vertical contacts through the mold compounds.
Bottom: Produced eWLB package b) top side c) bottom side The overall size of the eWLB package is 5.6 mm x 3.6 mm.

978-1-4799-2408-0/14 $31.00 © 2014 IEEE   1185

Contactless communication interface

In general near field RFID transponder (tag) antennas in the HF band at 13.56 MHz are resonant loop antennas [14]. The geometrical dimensions and the antenna environment determine the electrical parameters of the antenna, i.e. the value of the inductance, resistance, and parasitic capacitance. Due to the magnetic coupling between two loop antennas of the RFID reader and the tag, the RFID reader is able to transfer energy and data by modulating the magnetic field [14].

Loop antenna design

In order to design a small resonant loop antenna for the tag, the input impedance of the tag microchip has to be considered. A simplified equivalent circuit of the presented SiP HF RFID tag is illustrated in Figure 7. It is important to note that this equivalent circuit is only precisely correct at the designed operating frequency of 13.56 MHz.

The analog frontend of conventional HF RFID chips is usually equipped with an additional parallel on-chip capacitor. Typically the capacitor values are in the range of several picofarad (pF). With a good resonant antenna design no additional passive tuning elements are needed. Due to the fact that the size of the presented SiP and therefore the size of the loop antenna (smaller value of inductance) are miniaturized to single digits of millimeters, this additional chip capacitance is increased to 56 pF.

The values of C_{Chip} and R_{Chip} are voltage dependent as already mentioned and illustrated in Figure 7. Due to the well-defined application of the presented eWLB SiP application a chip capacitance of $C_{Chip} = 56$ pF is chosen. The HF antenna is characterized by the frequency dependent capacitance C_A, resistance R_A, and inductance L_A circuit. The coil inductance L_A depends on the size of the antenna, especially on the length of the wire-track and the number of windings (due to mutual coupling from one winding to each other) [14]. The resistance R_A describes the losses in the antenna. The resistance includes not only the frequency independent material specific resistive losses of the coil metal; it also describes frequency dependent losses due to eddy currents and the skin effect [14]. The proximity effects of metallic environments influence the antenna [15]. Due to the small eWLB 3D SiP, the metals of the two chips, the EZL and the RDL routing influence the electrical parameters of the small antenna. These effects have to be considered in the antenna design. As already mention before, the capacitance C_A describes the parasitic capacitance of the antenna. As it can be seen in Figure 6, the coil antenna is embedded in the fan-out area of the eWLB package and surrounds the RFID chip. As a result, no chip metal is in close vicinity of the antenna and the electric coupling. Thus, the parasitic capacitance is minimized. The remaining capacitance C_A is due to the electric coupling between the coil windings. This coupling is affected by the dielectric constant of the mold compound ($\varepsilon_r = 4.1$). Figure 8 shows the bottom side of the package and the RDL of the eWLB SiP. The fabricated bottom RDL can also be seen in Figure 6b. This RDL is used to realize the coil antenna and to connect to the RFID chip. Measurement pads and connections for possible additional passive capacitors were added to the design (Figure 8). The overall size of the coil spans to 3570 µm x 5775 µm. The track width and the gap width between the coil windings is

20 µm. In order to design a resonant loop antenna, the electric parameters of the coil have to fulfill Equation (1):

$$f_r = \frac{1}{2 \cdot \Pi} \cdot \sqrt{\frac{1}{L_A \cdot (C_{Chip} + C_A)} - \frac{R_A{}^2}{L_A{}^2}} \quad (1)$$

Equation 1 describes the damped parallel LC resonant circuit of the antenna, shown in Figure 7 ($f_r = 13.56$ MHz).

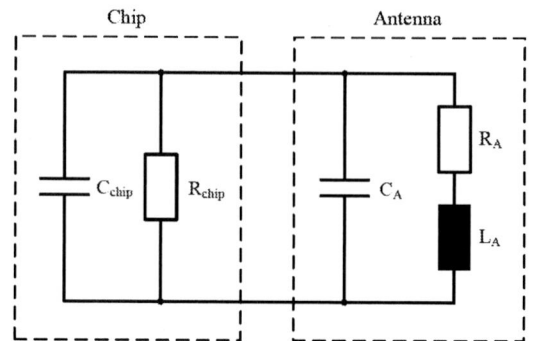

Figure 7. Equivalent circuit of the tag: The voltage dependent elements, the capacitance C_{Chip} and the resistance R_{Chip}, represent the tag chip. The connected HF antenna is characterized by the frequency dependent capacitance C_A, resistance R_A, and inductance L_A circuit.

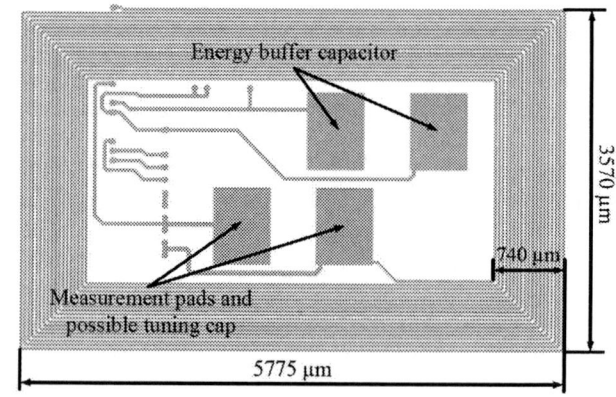

Figure 8. Geometric dimensions of the coil antenna (bottom side of eWLB package - redistribution layer): The HF RFID coil antenna consist of 19 windings with track and gap widths of 20 µm.

Antenna simulation and measurements

The electromagnetic simulation of the antenna is done with the 3D electromagnetic field simulator CST Microwave Studio [16]. The simulation shows that 19 coil windings generate the required inductance of L_A. The unknown parasitic capacitance C_A and the resistance R_A of the 13.56 MHz antenna are also determined by simulation. The inductance and the parasitic capacitance of the simulated 19 turn loop antenna is $L_A = 2.33$ µH and $C_A = 2.45$ pF. The magnetic field and thus the loop antenna is slightly influenced by the eWLB packaged CMOS chips as shown in Figure 9. We observe that the magnetic field is damped to the top side. At the bottom side the magnetic field is able to propagate unhindered.

Figure 10 shows the simulated and measured reflection coefficient S11 at the antenna input. The measurement was done with a vector network analyzer. The illustrated Smith Chart is normalized to 50 Ω. Figure 10 also considers the the parallel capacitor C_{Chip} of the chip at the simulation and measurement ports. The material resistance of the RDL coil is displayed at the direct current (DC) or low frequencies (e.g. 500 kHz). The simulation fits to the measured $R_A = 34\ \Omega$ @500 kHz. As already mentioned this part of R_A describes the real part losses due to the coil metal. The inductance and the parasitic capacitance of the simulated 19 turn loop antenna is $L_A = 2.33\ \mu H$ and $C_A = 2.45\ pF$. These values were verified by measurements, using an appropriate measurement bridge configuration. As it is illustrated in the Smith Chart (see Figure 10) the imaginary parts of the simulated coil is zero at the resonance frequency (markers 1, 2 = 13.56 MHz), meaning the antenna design is well matched to the RFID chip capacitance. The measured operation frequency (marker 2) is slightly shifted to the capacitive area due to measurement tolerances. Thus the measured resonance frequency of the produced SiP is 13.43 MHz.

Contactless communication measurement

Additionally a contactless measurement setup verified the communication between the SiP tag and the reader. At the reader side a two winding 6x6 mm coil antenna was connected to the vector network analyzer. This reader antenna has a self-resonance frequency far beyond 13.56 MHz. In order to analyze the influence of the already described microfluidic card to the SiP antenna, the overall system with this card was tested (see Figure 1). Therefore the SiP eWLB was attached to the microfluidic card and moved towards the 6x6 mm couple antenna. Figure 11 shows the measured real part of the reader coil antenna. The maximum of the real part shows the SiP tag resonance frequency that is $f_r = 13.43$ MHz. This result shows that the card does not influence the eWLB SiP and the value of the resonance frequency verifies the contact based measurement results of Figure 10.

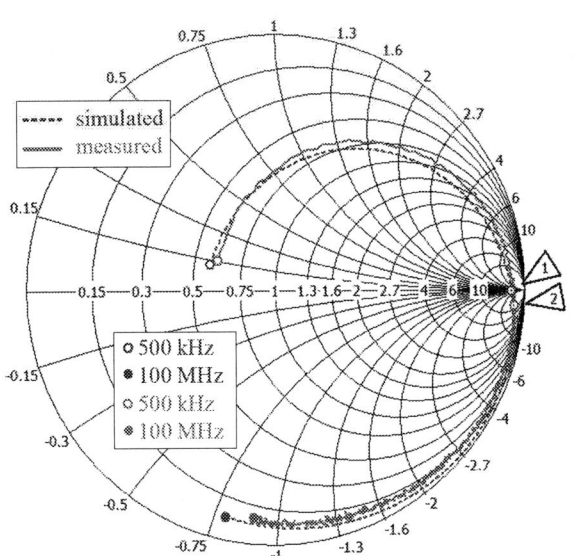

Figure 10. Simulated and measured reflection coefficient S11. We observe that the imaginary part of the presented SiP coil antenna is precisely matched to 13.56 MHz (marker 1,2). Note that the Smith chart is normalized to 50 Ω

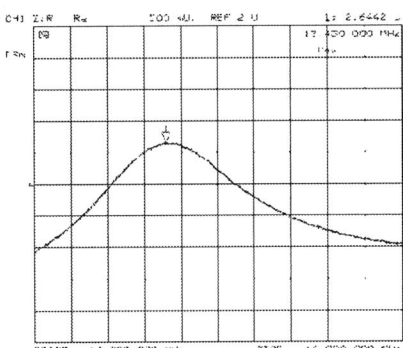

Figure 11. Real part of a coupling coil: The resonance frequency of the SiP coil-antenna is $f_r = 13.43$ MHz.

Next generation test setup of eWLB 3D SiP

Figure 12 illustrates a future point-of-care (POC) application. A test substance is analyzed by the eWLB packaged micro lab. The eWLB SiP itself is mounted onto a conventional measurement tape. In contrast to Figure 1 this test setup presents an example to use a eWLB 3D SiP autonomously.

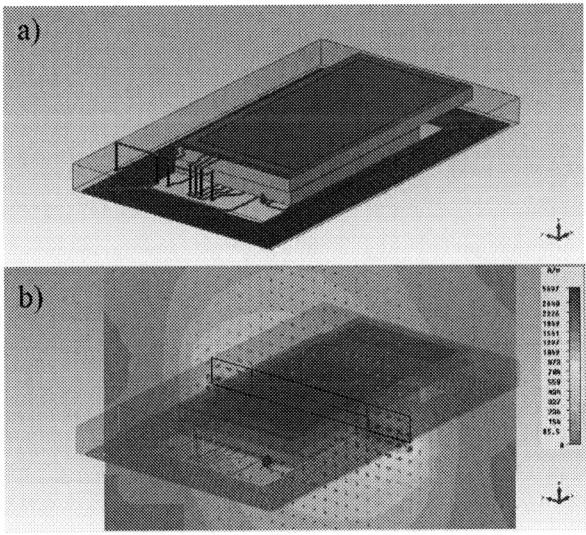

Figure 9. a) 3D image of the eWLB SiP
b) Electromagnetic simulation with the finite element simulator CST [16]: The magnetic vector field is displayed. We observe that the magnetic field is slightly damped to the top side. At the bottom side the magnetic field is able to propagate unhindered.

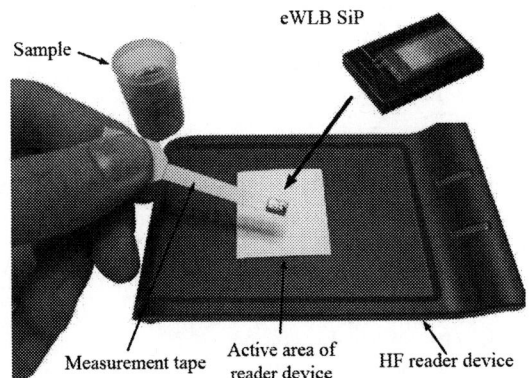

Figure 12. Measurement tape and modified reader device. Due to the contactless interface the presented SiP is able to work autonomously.

Conclusion and Outlook

The presented 3D SiP based on the eWLB technology provides a model for next generation biochip and micro lab packaging. It is shown that the fan-out area of the eWLB can be perfectly used to attach the sensitive biochip sensor-surface to higher level analytic hardware, such as microfluidic cartridges. Furthermore it is shown that the fan-out area allows outstanding capabilities to realize antennas. A 13.56 MHz coil antenna and an RFID power management chip, enables the presented SiP to work autonomously through the contactless communication interface. Energy as well as data is transferred due to the RFID technology. Due to chip-stacking in the eWLB package, the 3D SiP could be minimized to a volume of only 5.6 mm x 3.6 mm x 0.7 mm. This work demonstrates that the eWLB technology has outstanding capabilities to build complex and innovative packaging solutions. To the authors' vision the presented biochip 3D SiP packaging concept accelerates the progress in complex POC developments.

Acknowledgments

This work was funded by the European EU FP7 project "eBrains" (Project-no FP7-ICT-257488). The development of the eWLB technology was also financially supported by the ENIAC Joint Undertaking and the "Bundesministerium für Bildung und Forschung (BMBF)" in the project ESiP. The authors are indebted to the eWLB team of Infineon Regensburg who made the assembly and packaging developments possible.

References

1. Beyette, F.R.; Kost, G.J.; Gaydos, C.A.; Weigl, B.H., "Point-of-Care Technologies for Health Care," *Biomedical Engineering*, vol.58, no.3, pp.732,735, March 2011.
2. Lin Li; Xiaowen Liu; Qureshi, W.A.; Mason, A.J., "CMOS Amperometric Instrumentation and Packaging for Biosensor Array Applications," *Biomedical Circuits and Systems, IEEE*, vol.5, no.5, pp.439,448, Oct. 2011.
3. Manickam, A.; Chevalier, A.; McDermott, M.; Ellington, A.D.; Hassibi, A., "A CMOS electrochemical impedance spectroscopy biosensor array for label-free biomolecular detection," *Solid-State Circuits Conference Digest of Technical Papers (ISSCC), 2010 IEEE International*, pp.130,131, 7-11 Feb. 2010.
4. Musayev, J.; Adiguzel, Y.; Kulah, H.; Eminoglu, S.; Akin, T., "Label-free DNA Detection Using a Charge Sensitive CMOS Microarray Sensor Chip," *Sensors Journal, IEEE* , Januar 2014.
5. Stagni, C.; Guiducci, C.; Benini, L.; Ricco, B.; Carrara, S.; Paulus, C.; Schienle, M.; Thewes, R., "A Fully Electronic Label-Free DNA Sensor Chip," *Sensors Journal, IEEE* , vol.7, no.4, pp.577,585.
6. "The e-BRAINS project", a European Commission Framework Programme 7 (Project-no FP7-ICT-257488). Online: http://www.e-brains.org.
7. Bruckl, H.; Panhorst, M.; Schotter, J.; Kamp, P. B.; Becker, A., "Magnetic particles as markers and carriers of biomolecules," *Nanobiotechnology, IEEE Proceedings -* , vol.152, no.1, pp.41,46, 4 Feb. 2005.

8. Meyer, T.; Pressel, K.; Ofner, G.; Römer, B., "System integration with eWLB," Electronic System-Integration Technology Conference (ESTC), 2010 3rd, pp.1,9, 13-16 Sept. 2010.
9. Pressel, K.; Beer, G.; Wojnowski, M.; „Assembly and Packaging Enabling System Integration" 46th Internation Symposium on Microelectronics (IMAPS) October 2013.
10. Wojnowski M., Wagner C., Lachner R., Böck J., Sommer G., and Pressel K., "A 77-GHz SiGe Single-Chip Four-Channel Transceiver Module with Integrated Antennas in Embedded Wafer-Level BGA Package," in Proc. 62nd Electronic Components and Technology Conference (ECTC 2012), San Diego, USA, May 2012.
11. Brunnbauer, M.; Fuergut, E.; Beer, G.; Meyer, T., "Embedded wafer level ball grid array (eWLB)," Electronics Packaging Technology Conference, 2006. EPTC '06. 8th, pp.1,5, 6-8 Dec. 2006.
12. Wojnowski, M.; Sommer, G.; Pressel, K.; Beer, G., "3D eWLB — Horizontal and vertical interconnects for integration of passive components," *Electronic Components and Technology Conference (ECTC), 2013 IEEE 63rd* , pp.2121,2125, 28-31 May 2013.
13. PourMousavi, M.; Wojnowski, M.; Agethen, R.; Weigel, R.; Hagelauer, A., "The impact of embedded wafer level BGA package on the antenna performance," *Antennas and Propagation in Wireless Communications (APWC), 2013 IEEE-APS Topical Conference on* , pp.828,831, 9-13 Sept. 2013.
14. Finkenzeller, RFID Handbook: Fundamentals and Applications in Contactless Smart Cards, Radio Frequency Identification and Near-Field Communication", third revised edition, 2010.
15. Xianming Qing; Zhi Ning Chen, "Proximity Effects of Metallic Environments on High Frequency RFID Reader Antenna: Study and Applications," Antennas and Propagation, IEEE, vol.55, no.11, pp.3105,3111, Nov. 2007.
16. CST STUDIO SUITE TM, 3D Electromagnetic simulation software, http://www.cst.com.

A Novel Sound Sensor and Its Package Used in Lung Sound Diagnosis

Xingming Fu[1], Chaojun Liu[1], Yong Xu[1,3], Yating Hu[3], Xiaobing Luo[2], Xin Wu[4], Sheng Liu[1*]

1 Cross-disciplinary Institute of Engineering Sciences, School of Power and Mechanical Engineering, Wuhan University, Wuhan, Hubei, 430072, China
2 Huazhong University of Science and Tech, Wuhan, Hubei, 430074, China
3 Department of Electrical and Computer Engineering, Wayne State University, Detroit, MI 48202 USA
4 Department of Mechanical Engineering, Wayne State University, Detroit, MI 48202 USA
Victor_liu63@126.com

Abstract

Auscultation method has been used since the early 19th century in diagnostics of heart and lung diseases. Traditional stethoscope coupled with ears of physicians detects the heart and lung sound, and it is formed by microphone, air chamber and other infrastructures. However, some disadvantages exist in this kind of detecting tools, such as bulky size, friction noise due to hand-held, and inconvenience in long-term continuing monitoring, making it impossible to be integrated with clothing and hard to be interfaced with chest or inner cloth. A novel sound sensor used in lung sound detection with small size and high sensitivity will be presented in this paper. The sound sensor and its packaging are based on asymmetrically-gapped cantilever, consisting of a movable proof mass and three connection beams with two parallel beams that integrate one piezoelectric element at the top side and one mechanical beam at the bottom of sensor. As an example of lung sound diagnosis, a sound sensor with 5g weight, 86V/g sensitivity and 40ng/√Hz minimum detection acceleration is used in this paper. Due to the advantage of small size, high sensitivity and low cost, this sensor can be used in subtle sound and acceleration detection by a long term continuous monitoring mode. The application of this sound sensor will be presented in details in the lung sound detection, with particular focus on the deep breath induced sound. The monitoring experimental results show that this new sensor makes the detection of breathing clearer and easier, and heart sound can be sensed during breath holding period. The test data for regular breathing, deep breathing, and breathing & holding cases will be reported. It has been observed that lung sound can be measured and consistent results have been achieved. In addition, heart sound was tried to be extracted from lung sound, and the heart rate for the measured person is found to be 62 pulses per minute for an individual tested.

Introduction

Since lung sounds were first recorded and analyzed in the 1950s in addition to the heart sound [1-6]. The stethoscope has been used by clinicians around the world since the 19th century to help the diagnosis of heart and lung (cardio-respiratory) disorders. The application of the stethoscope in research and clinical trials has been limited due to the inherent physician variability and subjectivity in the interpretation of heart and lung sounds. With increasing needs of the aging population, needs for non-cooperative patients such as those in intensive care unit (ICU), patients in sleeping mode, babies, an objective assessment of lung and heart sound is essential. Most efforts have been limited by commercial off-the-shelf microphones and accelerometers, with the latter ones without enough sensitivity and bulky, not wearable [7]. The advances of the bio-acoustical methods are evident. However, with the disadvantages of more complex, expensive, and far less harmless procedures (computed X-ray tomography, magnetic resonance tomography), lung auscultation should not be treated as an assistance tool only in medical practices. However, with the increasing shortage of nurses, doctors, and aging population, and the wide use of intelligent cell phones, wearable electronics based bio-acoustic images seem to be an important part of electronic medical records. As a new application case, this system can also be used for the monitoring tool for those who are trained for daily breathing for willpower in physiology. Kelly McGonigal pointed out in her book [8] that the willpower is able to help prevent some bad habits, such as the delaying addiction, and internet addiction, etc. It is therefore critical to develop cheat compatible, wearable device and system, which are sensitive enough to be able to detect both heard and lung sound. The approach based on accelerometer is a different way from the traditional stethoscope. Many efforts have been made in the development of accelerometer based acoustic sensors for chest sound monitoring [9-11], and to optimize the packaging size and improve their reliability, much works need to be further done [12]. Yong Xu et al. [13-16] researched the asymmetrical gapped cantilever structure and developed a novel accelerometer with high sensitivity for heart and lung sound detection, and obtained better detection results compared with the traditional stethoscope.

Sensor Description [13]

The sound sensor is based on a novel asymmetrical gapped cantilever structure which is composed of a bottom mechanical layer and a top piezoelectric layer separated by a gap. The sensitivity can be increased significantly by the asymmetrical gapped structure. The overall energy conversion efficiency can reach about 90% for asymmetrical gapped cantilevers and only below 39% for conventional cantilevers.

The size of the prototype sensor is 35mm×18mm×7.8mm (length×width×thickness), shown as Fig.1. A charge amplifier was built in. The sensitivity, resonate frequency and noise floor are 86V/g, 1100Hz and 40ng/√Hz above 100 Hz, respectively. The sensitivity of this sound sensor is orders of magnitude higher than the conventional one.

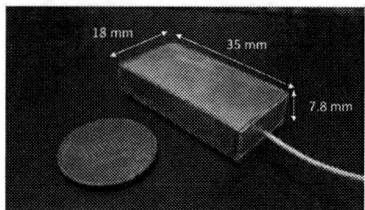

Figure 1. The prototype sensor

Lung Sound Monitoring

Lung sound detection are carried out on healthy volunteers in a regular laboratory environment. 20 kHz of sample rate is adopted. A filter with a bandwidth from 200 Hz to 2000 Hz is applied to extract the lung sound, and a filter with a bandwidth from 20 Hz to 500 Hz is applied to extract the heart sound. The device location is chosen to be at right anterior intercostal space above the level of the 3rd rib for breathing signal detection.

A comparison has been made between the asymmetrical gapped accelerometer and a high-end electronic stethoscope in detecting heart and lung sounds[13]. It was found that The signal-to-noise ratio of the new sensor is about two times higher in heart sound detection. lung sounds are much weaker than heart sounds and thus are more difficult to detect by stethoscope, especially for a gentle breathing. While the new sound sensor was able to detect the weak lung sounds clearly.

Fig. 2(a) plots the lung sounds recorded by sound sensor for 5 cycles of regular breathing; Fig. 2(b) plots the lung sounds recorded for 6 cycles of regular breathing by the new sensor. It is observed that when testing in lung location, the heart sounds are so weaker than lung sounds that it is difficult to distinguish it because of the large distance away from heart. The sensor is not very sensitive to air-borne noise due to the regular laboratory environment full of air-borne noises.

Fig. 3 plots the lung sounds recorded by the new sound sensor for deep breathing with 4 breathing cycles per minute. It can be seen that good consistence are got among the four deep breathing. It is worth noting that during deep breathing the lung sounds are many times stronger than that of regular breathing as shown in Fig. 2. This large acceleration happens when the lung changes from inhale to exhale. The lung sound of deep breathing is about twice stronger than that of heart sound when the sensor is positioned on the heart location.

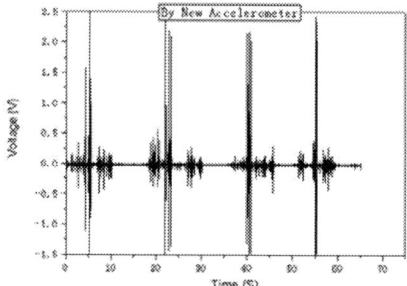

Figure 3. Sample waveforms of lung sound during deep breathing

Fig. 4 plots the lung sounds recording of one regular breathing and then breath holding. It can be seen that there is a strong waveform in forepart which is the breathing time and a weaker waveform in the later part which is the breathing holding time. It is interesting to note that in the breath holding part of this waveform some weak peaks are found which is not the air-born noise. In fact, these peaks are produced by heart impact which can be seen in Fig. 5. Therefore, the sounds measured in lung position can be considered as the superposition of lung sound and heart sound. It is easily to extract the heart sounds from lung sounds during breath holding period.

Fig. 5 plots the sound waveform extracted from Fig. 4 between the time of 22s and 25s by a filter with a bandwidth from 20 Hz to 500 Hz. Good consistence and regular intervals are shown in Fig. 5. Take these 3 cycles as an example, average interval time of heart beat is 0.96s, and the interval time between heart contract and diastole is 0.32s. That is, the heart rate for the measured person is 62 pulses per minute.

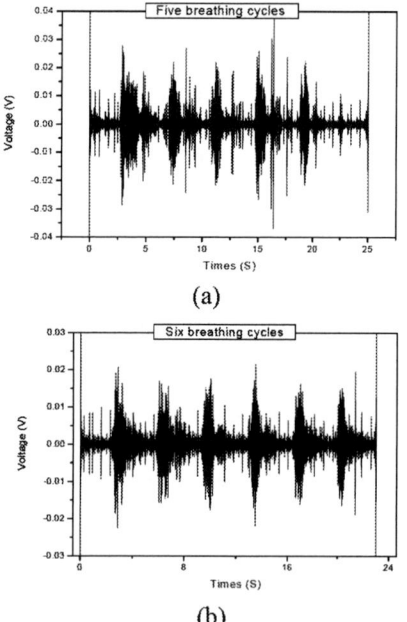

Figure 2. Regular breathing of the (a) five cycles regular breathing and (b) six cycles regular breating.

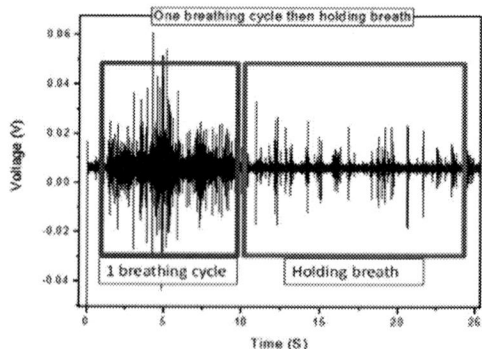

Figure 4. Sample waveforms of lung sound during one regular breathing and breath holding.

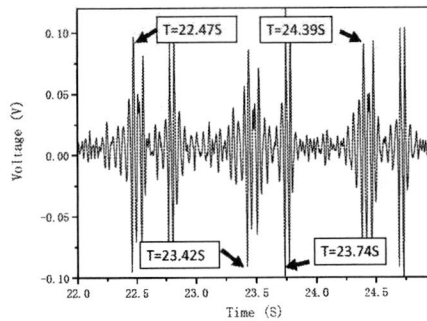

Figure 5. Heart sound waveforms of 3 intervals of heart contract and diastole.

Conclusions

A small size novel sound sensor with 5g weight, 86V/g sensitivity and 40ng/ √ Hz minimum detection acceleration used in lung sound monitoring was demonstrated in this paper. The sound sensor and its packaging are based on asymmetrically-gapped cantilever, and piezoelectric elements were integrated.

The sound sensor was used in the regular breathing, deep breathing induced sound and breathing & holding cases. The monitoring experimental results showed that this sensor made the detection of breathing clearer and easier. Monitoring results obviously showed that the lung sound of regular breathing was much weaker than heart sound. While lung sound of deep breathing was much stronger than heart sound.

It has been observed that lung sound could be measured and consistent results have been achieved. It was found that the sounds measured in lung position is the superposition of lung sound and heart sound. Heart sounds were extracted from holding breathing part, and the heart rate for the measured person was found to be 62 pulses per minute.

Due to the advantage of small size, high sensitivity and low cost, this sensor can be used in subtle sound and acceleration detection by a long term continuous monitoring mode. This system can also be used for the monitoring tool for those who are trained for daily breathing for willpower in physiology, which is said to be able to help prevent the delaying addiction, and internet addiction.

Acknowledgments

The support of Hightech Proram (863) with a contract number of 2012AA040501 and Basic Research (973) with a contract number of 2011CB309504 of Ministry of Science and Technology of PR China is highly appreciated. Research is also partially supported by National Science Foundation under Grant No. 0747620.

References

1. V. McKusick, J. T. Jenkins, and G. N. Webb, "The acoustic basis of the chest examination; studies by means of sound spectrography," Am. Rev. Tuberc., vol. 72, pp. 12–34, 1955.

2. A. Jalil and M. Aronovitch, "Recording and analysis of breath sounds in normal individuals," Indian. J. Chest Dis., vol. 8, pp. 158–166, 1966,concl.

3. P. Forgacs, "Crackles and wheezes," Lancet, vol. 2, pp. 203–205, 1967

4. "Lung sounds," Br. J. Dis. Chest, vol. 63, pp. 1–12, 1969.

5. S. J. Slodki and P. M. Shah, "The Q-II interval. II. Astudy of the second heart sound in pulmonary hypertension," Chic. Med. Sch. Quart., vol.25, pp. 108–113, 1965.

6. B. W. Cobbs, Jr., R. B. Logue, and E. R. Dorney, "The second heart sound in pulmonary embolism and pulmonary hypertension," Am. Heart J., vol. 71, pp. 843–844, 1966.

7. Kraman S S, Pressler G A, Pasterkamp H, et al. "Design, construction, and evaluation of a bioacoustic transducer testing (BATT) system for respiratory sounds". Biomedical Engineering, IEEE Transactions on, 53(8): 1711-1715, 2006.

8. McGonigal K. "The Willpower Instinct: How Self-Control Works, Why It Matters, and What You Can Doto Get More of It'. Penguin, 2011.

9. V. Padmanabhan, J. L. Semmlow, and W. Welkowitz, "Accelerometer type cardiac transducer for detection of low-level heart sounds," IEEE Transactions on Biomedical Engineering, vol. 40, pp. 21-28, Jan 1993.

10. K. J. Cho and H. H. Asada, "Wireless, battery-less stethoscope for wearable health monitoring," in IEEE 28th Annual Northeast Bioengineering Conference, April 20, 2002 - April 21, 2002, Philadelphia, PA, United states, pp. 187-188 , 2002

11. S. Henneberg, B. Hok, L. Wiklund, and G. Sjodin, "Remote auscultatory patient monitoring during magnetic resonance imaging," Journal of clinical monitoring, vol. 8, pp. 37-43, Jan 1992.

12. Liu S, Liu Y. "Modeling and Simulation for Microelectronic Packaging Assembly: Manufacturing, Reliability and Testing". John Wiley & Sons, 2011.

13. Yating Hu, Yong Xu. "An ultra-sensitive wearable accelerometer for continuous heart and lung sound monitoring," 34th Annual International Conference of the IEEE EMBSSan Diego, California USA, pp.694-697, September, 2012

14. Q. L. Zheng and Y. Xu, "Asymmetric air-spaced cantilevers for vibration energy harvesting," Smart Materials & Structures, vol. 17, Oct 2008.

15. Y. F. Li, Q. L. Zheng, Y. T. Hu, and Y. Xu, "Micromachined piezoresistive accelerometers bBased on an asymmetrically gapped cantilever," Journal of Microelectromechanical Systems, vol. 20, pp. 83-94, Feb, 2011.

16. Q. Zheng and Y. Xu, "Vibration energy harvesting device based on asymmetric air-spaced cantilevers for tire pressure monitoring system," in PowerMEMS Workshp, Washington DC, USA, December 1-4, 2009.

Novel System-in-Package Design and Packaging Solution for Solid State Lighting Systems

Mingzhi Dong[1,4*], Fabio Santagata[1,4], Jia Wei[1,4], Cadmus Yuan[2,4], Guoqi Zhang[2,3]

[1]Beijing Research Center, Delft University of Technology, Beijing, China
[2]Institute of Semiconductors, Chinese Academy of Sciences, Beijing, China
[3]DIMES, Delft University of Technology, Delft, the Netherlands
[4]State Key Laboratory of Solid State Lighting, China
*mzdong85@gmail.com

Abstract

With more implementation of LED in lighting and communication applications, increasing demand for function enrichment and miniaturization has emerged. Existed technologies are highly challenged. System-in-package (SiP) technology is very promising in terms of function integration and cost reduction. In this paper, a novel SiP platform for LED system with integrated driver and wireless control function has been developed for the first time and will be presented here. This platform consists of silicon submount design and fabrication, module packaging, system assembling, and testing and analyzing. As a demonstrator, this paper presents a bulb fabrication process and the testing results. The proposed SiP module includes driver and wireless control function besides light source and optics. The process can be achieved by regular packaging technology and existed standard infrastructures. According to the testing results, the assembled system shows satisfying performance.

Key words: LED, System-in-package (SiP), Silicon submount, 3D integration

Introduction

Solid state lighting, represented by light emitting diodes (LEDs), has been rapidly replacing the traditional luminaires due to longer lifetime and lower energy consumption. On one hand, in the past decades effort has been put into increasing light output of LEDs and meanwhile reducing the cost to make the LED bulbs competitive with traditional luminaires. On the other hand, increasing need for smart lighting has emerged in daily life. Thanks to the semiconductor nature, the LEDs provide more colorful and tunable light compared with other light sources. Moreover, controlling the light output by electronics is made possible. Smart lighting, like any other integrated electronic system, is facing the challenge of function enrichment and miniaturization. From the aspect of electronic packaging, new supporting technologies need developing.

Current approach for lighting system integration, like Philips hue bulbs, is based on standard printed circuit board (PCB) assembly [1]. The intrinsic bulkiness of this approach obstructs further function enrichment and miniaturization. Besides, thermal management of such system is extremely challenging. SiP technology has recently emerged and already shown potential in applications of RF and networking. So far, SiP design for lighting system is quite new. Gielen proposed a design of intelligent integration of LED system, in which standard PCB was used as packaging substrate [2]. However, for LED applications, the main barriers for the implementation of PCB or other polymer-based substrates are insufficient thermal dissipation and coefficient of thermal expansion (CTE) mismatch-induced issues, such as unexpected warpage.

Silicon has been used as packaging substrate of LED packages for some time due to its compatibility with IC process and high thermal conductivity. Tsou et al. first developed a silicon-based packaging platform for LED packaging [3]. Recently, researchers have tried to achieve more function out of silicon substrates. Kang et al. tried to integrate zener diode into silicon submount for multi-chip LED module [4]. In general, all the above endeavors aim to obtain discrete silicon-based LED packages other than functional systems.

In this paper, a novel SiP design for LED modules is proposed and presented. The proposed SiP module includes driver and wireless control function besides light source and optics. As a demonstrator, this paper shows great potential of silicon-based SiP solution for smart lighting system development.

Design

I. Electrical circuit description

Figure 1 shows a typical application circuit of LED lamps. As indicated, two driver ICs limit the current of each chain of LEDs, respectively. The two chains of LEDs are designed with different color temperatures, which are warm white and cold white, so that the color temperature of the lamp can be tuned. The programmable MCU can adjust the brightness of each chain of LEDs by controlling the switches (MOSFETs). Connected with an infrared receiver, the lamp can be wirelessly controlled by a remote controller.

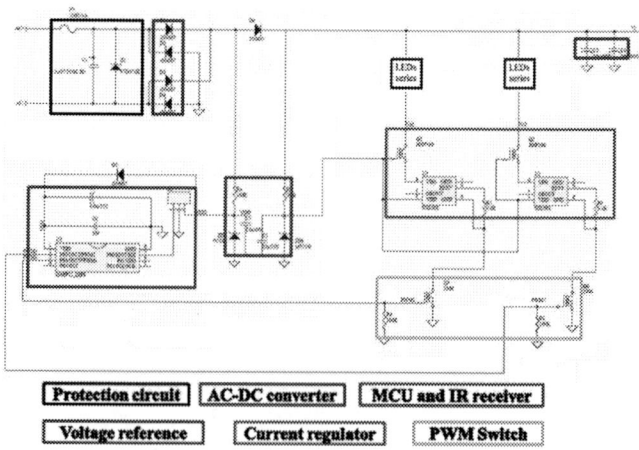

Figure 1. Circuit of LED bulb.

II. Module description

Based on the functions, the aforementioned circuit was divided into three parts: protection circuit, light engine, and driver and control circuit. Figure 2 illustrates a schematic of the design. Due to the bulkiness of the components, the protection circuit was kept out of the integrated module. The light engine, containing all the LEDs (bare die), was made onto one silicon submount (top). The driver and control circuit was assembled onto another silicon submount (bottom), with interconnections to the light engine submount. In this way, either of the two submounts can work independently so that either of them is changeable and replaceable, providing more freedom for the design.

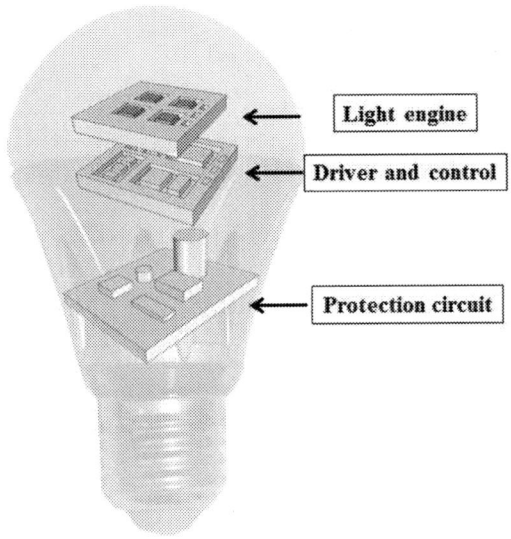

Figure 2. Schematic of SiP design of LED bulb.

All the bare dice and passives were embedded into cavities on the submounts. The top submount was bonded on top of the bottom submount by thermal interface material (TIM). The through-silicon-vias (TSVs) were fabricated on top submount to connect the bottom submount and external protection circuit. Drawing of the module (cross-sectional view) is shown in Figure 3.

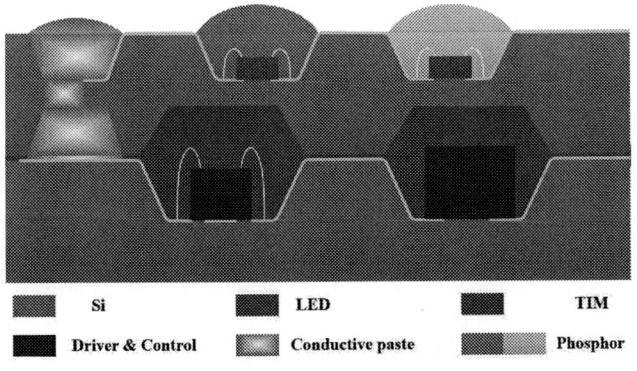

Figure 3. Cross section of stacked Si submounts module.

Thermal simulation

Thermal management is one of the crucial issues for LED system because LEDs generate quite a lot of heat when they are working and the performance of LED is highly dependent on the temperature [5]. Therefore, thermal management is a key design aspect for LED modules. In this section, the thermal performance of stacked silicon submounts is investigated through finite element analysis (FEA). The model used in the simulation is the same as shown in Figure 3. The silicon module was glued onto heatsink using TIM. The thickness of TIM layer was set as 100 μm while the thermal conductivity varied. The material property and module geometry used in the thermal simulation are listed in Table 1 and Table 2.

Table 1. Material properties used in the simulation.

Material	Thermal Conductivity [W/m·K]	Density [kg/m³]
LED	35	3965
Silicon	130	2329
TIM	1~50	1120
Heat sink	100	3000
Phosphor	0.22	980

Table 2. Simulation model geometry.

Function	Geometry	Diameter / sides	Height/thickness
Heat sink	hollow cylinder	50 mm	50 mm
TIM_1 (between module and heatsink)	cylinder	25 mm	100 μm
TIM_2 (between submounts)	cuboid	15 mm	100 μm
Si submount	cuboid	15 mm	0.5 mm
LED	cuboid	1 mm	0.2 mm
IC	cuboid	1×2 mm	0.3 mm
Phosphor	cuboid	15 mm	0.5 mm

Eight LEDs were mounted on the top submount; each of those generates 0.6 W of heat. On the bottom submount, six chips were simulated as driver and control ICs, generating 0.4 W of heat in total. Above the top submount, a layer of phosphor is coated. All the outer surfaces of the structure were convection cooled by the environmental temperature of 20°C.

As interface material, TIM is the most critical link on the path of heat dissipation from LEDs to heatsink. In the simulation, only the thermal conductivity of TIM is changed to evaluate the thermal performance of the whole module. Figure 4 shows the temperature distribution images using different TIMs. The hotspots appear on the LEDs covered by the phosphor. Compared with phosphor, the heatsink has much higher thermal conductivity so that it provides dominant heat dissipation path. The junction temperature of LEDs and maximum temperature of the phosphor layer are shown in Fig. 5. Higher thermal conductivity usually leads to lower module temperature. This trend is obvious until the thermal conductivity is up to 10 W/m·K. After the thermal conductivity of TIM is higher than 10 W/m·K, the temperature shows little drop. For most thermal greases, the thermal conductivity is lower than 10 W/m·K, meaning that the interface thermal resistance is always the bottleneck.

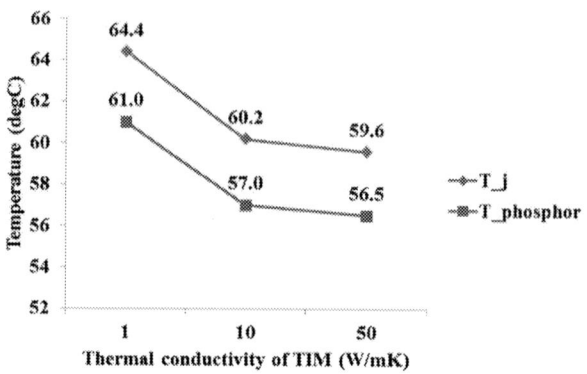

Figure 5. Change of junction temperature with TIM properties.

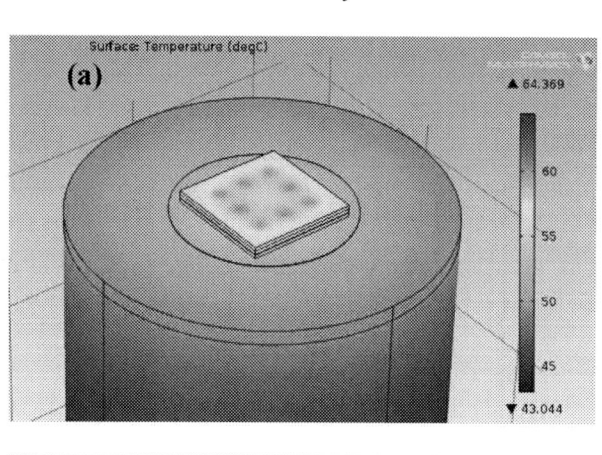

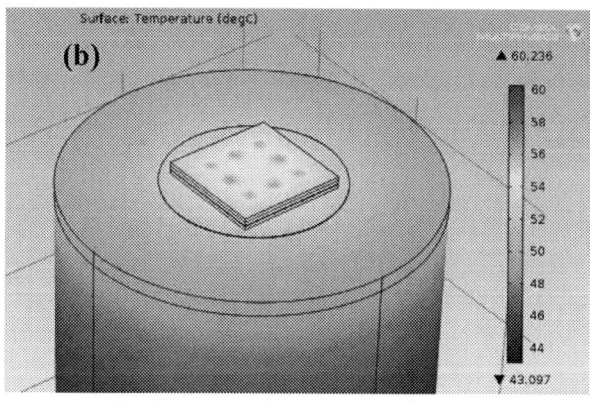

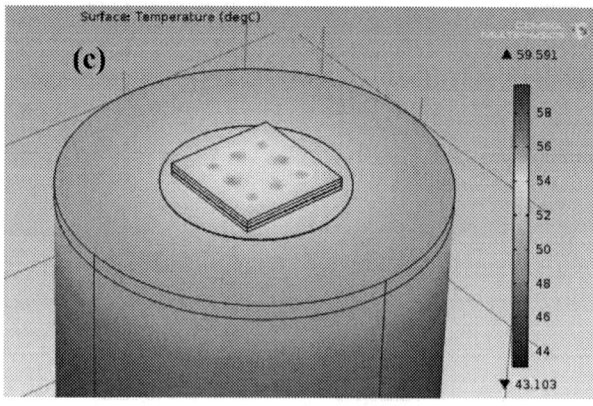

Figure 4. Simulation result with varied TIM: (a) k=1 W/m·K, (b) k=10 W/m·K), (c) k=50 W/m·K.

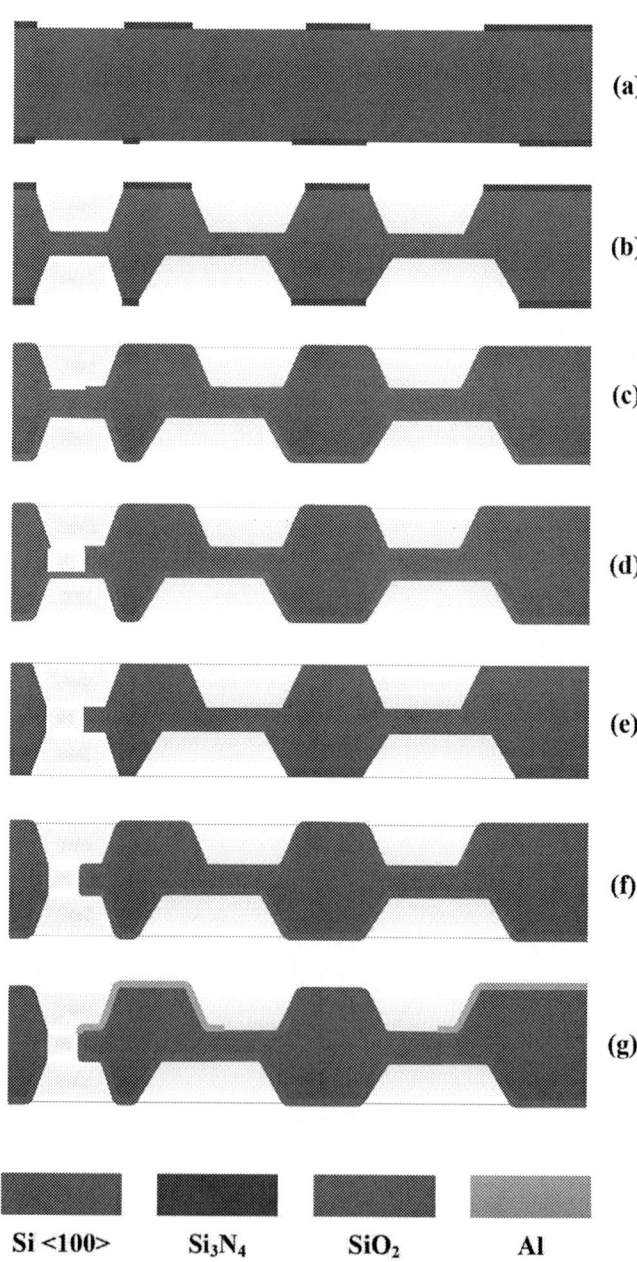

Figure 6. Process flow of silicon submount fabrication.

Fabrication of Si submount

The presented module consists of two stacked Si submounts; the top silicon submount contains the light engine, whereas the driver and control circuit was assembled onto the bottom silicon submount. The fabrication of top and bottom submount is very similar. The only differences are that the depth for the KOH cavity is larger on the bottom submount and also that the TSVs are only made on the top submount.

Figure 6 outlines the fabrication process to realize a double-side silicon submount with TSVs. The fabrication starts with a <100> silicon wafer coated with a layer of silicon nitride on both sides (thickness: 200 nm). The first step is anisotropic wet etching of Si using KOH solution (a-b). The etching depth is 250 μm and silicon nitride is removed by RIE after etching. A 2 μm thick oxide layer is coated on etched wafer using thermal oxidation process and used as mask layer when the TSVs are formed by DRIE (c-e). A new layer of oxide is coated as insulation layer (f). A layer of 2 μm Al is sputtered on the front side of the wafer and then patterned by photolithography (g). However, because of the high topography, spray-coating is used instead of standard photoresist spinning in order to successfully pattern metal into the KOH cavities.

The KOH cavities on the front side of the submount act as reflectors of LEDs while the cavities on the backside provide space for the components on bottom submount when the submounts are stacked. The metal Al acts as both conductive trace of the circuit and light reflective layer. As said, the bottom submount (driver and control) fabrication process is quite similar. The only difference is the KOH cavities were only made on the front side with the depth of 450μm and no TSVs were fabricated.

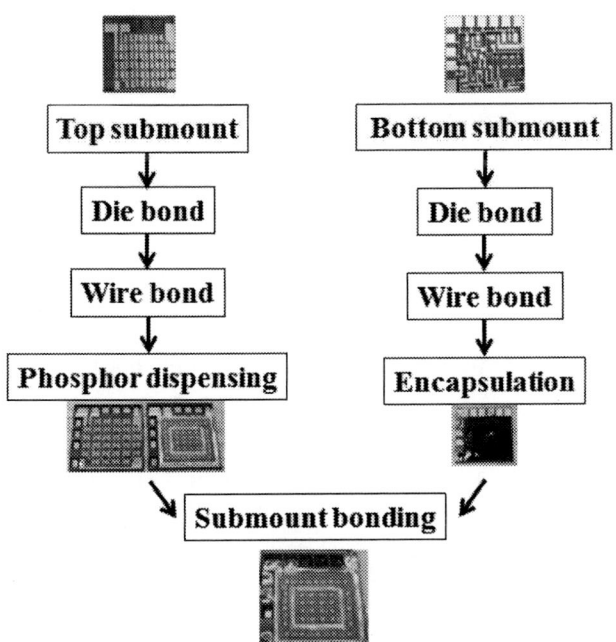

Figure 7. Module packaging process flow.

Packaging and assembling

The module was packaged by standard packaging process, as shown in Figure 7. All the bare dice were mounted on the submount using die attach glue and connected into the circuit by wire bonding. The passives used in this module are all surface mount devices (SMDs). Conductive silver glue was used as interconnection material instead of reflowed solder paste. Normally, all the above packaging process can be done at wafer-level, including the stacking step.

Figure 8 shows a packaged bottom submount (top view), including the driver and control circuit components. Table 3 lists the dimensions and components information of the module. The packaged submounts were stacked and bonded by a thin layer of TIM. The TSVs were filled with silver glue connecting the I/O pads on the bottom submount with the external protection PCB.

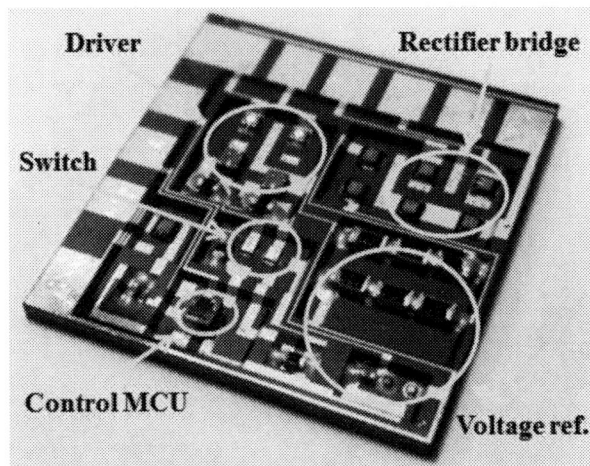

Figure 8. Driver and control module (bottom submount, top view).

Table 3. Dimensions and components of SiP module.

Submount	Dimension	Amount of components	Component type
Top (light engine)	17×17×0.8 mm³	32	Bare die (LED)
Bottom (driver and control)	17×17×0.8 mm³	28	Bare die ICs SMD passives

Testing

The stacked module was assembled with external protection PCB for testing, as shown in Figure 9.

I. Function testing

Firstly all the designed functions were checked. Controlled by an infrared remote controller, the brightness of each LED chain can be adjusted independently. Since the two LED chains emit light with different color temperature, both the brightness and color temperature of the bulb is tunable. The controller can also switch on / off the bulb from distance (Figure 10).

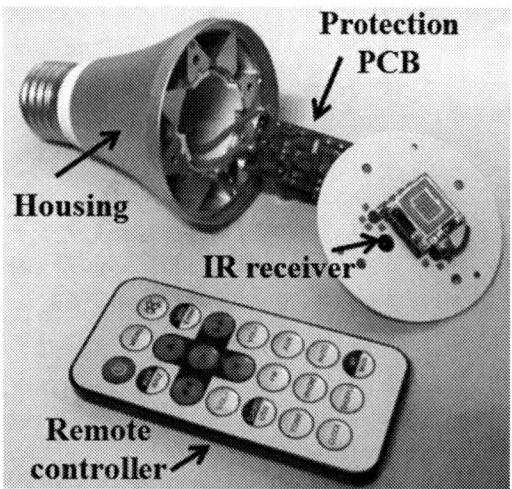

Figure 9. Bulb assembly and remote control device.

Figure 10. Functional testing (left: bulb off; right: bulb on).

II. Thermal testing

The temperature of the working bulb (without the cap) was measured by an infrared imager. Figure 11 shows the temperature distribution at normal working condition. According to the infrared image, the hotspots locate at the LEDs with the highest temperature of 52~54 °C (on the phosphor layer). This result agrees with the simulation. The slight difference can be explained by that when tested the bulb was connected with the socket which drained away some heat by means of thermal conduction.

Since the infrared image does not indicate the junction temperature, the simulation result can be taken as reference. From the simulation, the junction temperature is 3~4 °C higher than the phosphor temperature. Reasonably, the junction temperature of the working bulb can be estimated as 55~58 °C. Consider the environmental temperature as 20 °C and the input power as 6 W, the total thermal resistance from LED junction to ambient is calculated as ~ 6 K/W. For indoor lighting bulbs, this result is fully acceptable, which strongly proves that the silicon is suitable for SiP module in terms of thermal management.

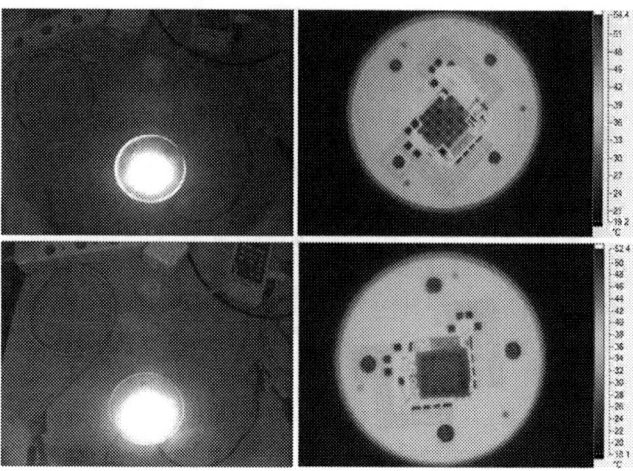

Figure 11. Infrared image of the working bulb.

Conclusion

A novel SiP design platform for LED lighting systems with integrated driver and wireless control function has been developed and outlined in this study. This platform consists of silicon submount design and fabrication, module packaging and system assembling, and testing and analyzing. As a demonstrator, this paper presents a bulb fabrication process and testing results. The process can be achieved by regular packaging technology and existed standard infrastructures. According to the testing results, the assembled bulb shows satisfying performance. Further evaluation of this SiP module is currently underway to enhance the reliability of and to optimize the design.

Acknowledgment

This research was carried out within the project "3D integration technology of multi-functional LED systems", cooperated with Unilumin Group (China), and funded by the government of Guangdong province of China. The authors would like to thank the APSI group at the State Key Laboratory of Solid State Lighting, China for the technical support. In addition, the authors are grateful to the Chinese Academy of Sciences for providing support to this project.

References

1. Philips product *Hue*, http://www.meethue.com/en-CN
2. A.W.J. Gielen, P. Hesen, F. Swartjes, H. van Zeijl, F. Boschman, J-E. Bullema, R.J.Werkhoven, S. Koh, "Development of an intelligent integrated LED system-in-package," in *Proc. European Microelectronics and Packaging Conference (EMPC)*, Brighton, UK, Sept. 12–15, 2011, pp.1–7.
3. Chingfu Tsou, Yu-Sheng Huang, "Silicon-Based Packaging Platform for Light-Emitting Diode," *IEEE Transactions on Advanced Packaging*, vol. 29, no. 3, pp. 607–614, Aug. 2006.

4. Jeung-Mo Kang, Jeong-Hyeon Choi, Du-Hyun Kim, Jae-Wook Kim, Yong-Seon Song, Geun-Ho Kim, and Sang-Kook Han, "Fabrication and Thermal Analysis of Wafer-Level Light-Emitting Diode Packages," *IEEE Electron Device Letters*, vol. 29, no. 10, pp. 1118–1120, Oct. 2008.

5. Mingzhi Dong, Jia Wei, Huaiyu Ye, Cadmus Yuan, Kouchi Zhang, "Thermal analysis of remote phosphor in LED modules," *Journal of Semiconductors*, vol. 34 no. 5, pp. 053007, May 2013.

Implantable Device Including a MEMS Accelerometer and an ASIC Chip Encapsulated in a Hermetic Silicon Box for Measurement of Cardiac Physiological Parameter

Jean-Charles Souriau[1], Laetitia Castagné[1], Guy Parat[1], Gilles Simon[1],
Karima Amara[2], Philippe D'hiver[2], Renzo Dal Molin[2]
[1] Léti, MINATEC Campus
17 rue des Martyrs ; 38054 Grenoble - France
[2] Sorin CRM SAS
Parc d'affaires NOVEOS; 4 avenue Réaumur; 92143 Clamart - France

Abstract

The objective of this paper is to present a new technology that integrates a Micro Electro-Mechanical Systems (MEMS) accelerometer and an Application-Specific Integrated Circuit (ASIC) chip encapsulated in a hermetic silicon box that could be embedded in a transvenous cardiac lead in order to sense the endocardial acceleration signal. The originality of the approach consists of using an interposer and a lid, both made of conductive doped silicon to connect the device. The MEMS and the ASIC are attached on the silicon interposer and the silicon lid is bonded using eutectic AuSi ring. The electrical interconnection to the two conductor wires is obtained through the interposer and the lid using the conductivity of doped silicon. The system integration was performed at the wafer level. A test vehicle which allows characterizing the required technologies was designed and manufactured. A technical focus on the most important process steps for the integration is presented and discussed in this paper. This includes interconnection on doped silicon, dies on wafer bonding and wafer to wafer bonding. The hermeticity and biocompatibility encapsulation of the device is also addressed and a prototype which has been designed is described.

Introduction

The aging population and health are strong drivers for research and the miniaturization of microsystem offers a tremendous opportunity for medical application [1]. Gradually the electronic devices that were previously external invade the human body in order to improve therapy or physiological parameter measurements. However, human body is a new environment for such devices and two main issues have to be considered [2]. First of all the body has to be preserved from toxic elements of the device and secondly, electronics parts have to be preserved from corrosive substances which constitute the major part of the body. The focus on this paper is a cardiac implantable device which measures endocardial acceleration signal. A micro electro-mechanical systems (MEMS) accelerometer and an application-specific integrated circuit (ASIC) chip are placed inside a cardiac lead. A packaging architecture which consists of a small form factor silicon box is presented in this paper. The MEMS and the ASIC are attached on the silicon interposer and the silicon lid is bonded using eutectic AuSi ring. The electrical interconnection to the two external conductor wires is obtained through the interposer and the lid using the conductivity of doped silicon (0.1 to 5 mOhm.cm). A schematic view of the package is presented in Figure 1.

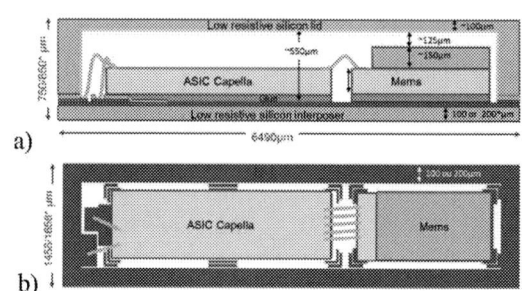

Figure 1. Hermetic Silicon Package schematic: a) cross section view; b) top view.

The ASIC and custom MEMS were designed by Sorin CRM.

Wafer level device integration

The system integration was performed at the wafer level. The process flow, as illustrated in figure 2, is outlined as follows:

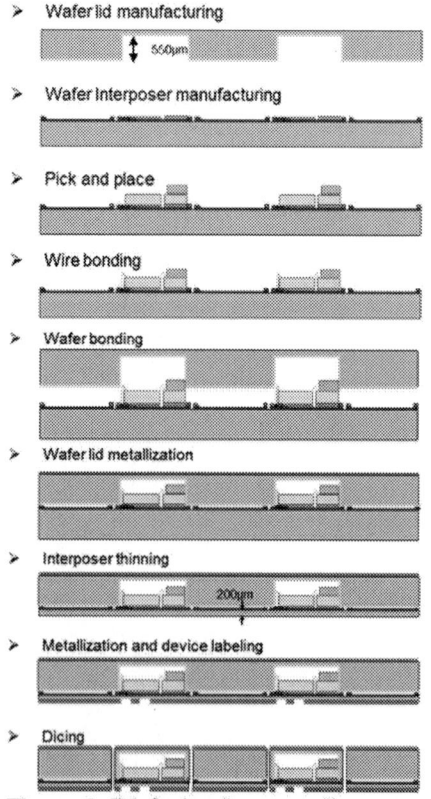

- Wafer lid manufacturing
- Wafer interposer manufacturing
- Pick and place
- Wire bonding
- Wafer bonding
- Wafer lid metallization
- Interposer thinning
- Metallization and device labeling
- Dicing

Figure 2. Wafer level process flow.

Lid wafer manufacturing:

Lid wafer was made from 200mm doped silicon wafer. SiO2 layer was deposited and patterned to be used as hard mask. Cavities of 550μm depth were etched by Deep Reactive Ion Etching. By considering 13 deep measurements, the typical standard deviation on a wafer was around 9μm. The figure 3 shows the mapping of the deviation with the average value on one wafer.

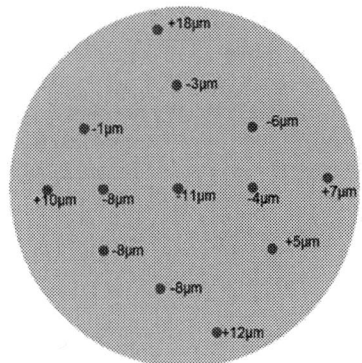

Figure 3. Dispersion cavity depths on a wafer.

It has to be pointed out that a slight curve is observed on cavities (Fig. 4). This must be taken into account in the layout to prevent dies from touching the top.

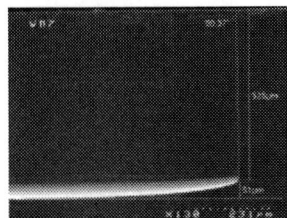

Figure 4. Profil of cavity etched by DRIE in lid wafer.

Interposer wafer manufacturing:

Interposer wafer was also made from 200mm doped silicon wafer. First a dedicated SiO2 layer was deposited and windows were opened to allow contact on Silicon bulk. AlSi/WN/Au metallic layer was deposited and patterned in order to form bonding ring of the cavity and lines which allow connecting dies to the interposer and to the bonding ring. The wafer was designed to characterize the metal properties and electrical connection to the interposer. Contact resistance on Si bulk measurements were performed using Kelvin probe structure through 210x210μm² via on SiO$_2$ layer (Fig. 5). The average contact resistance which includes 144 structures on 10 wafers, was measured at 116 mOhm with a standard deviation of 25 mOhm. An example of distribution in a ¼ wafer is presented in figure 6.

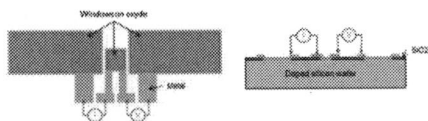

Figure 5. a) Kelvin resistance pattern of the electrical contact measurement on the silicon bulk; b) schematic diagram.

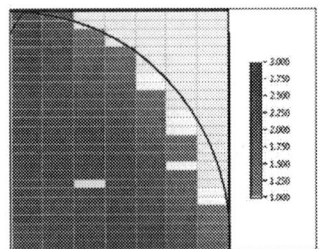

Figure 6. Resistance value distribution of the Kelvin measurements on ¼ wafer.

Gold was etched through photoresist mask in order to open pad on aluminum layer and allow wedge bonding.

An adhesive made of polyimide was used to bond dies on interposer. It was deposited on wafer by spin coating with thickness around 4μm. This material was patterned using photoresist mask and plasma etching. It exhibits thermoplastic behavior and high adhesion to silicon dies after cure. This material was chosen because of its high stability at high temperature. Thermal Gravity Analysis and Differential Scanning Colorimetriy were performed. The results are presented in the following graphs (Fig. 7) and (Fig. 8). The polymer was designed to resist under temperatures up to 450°C. Weight loss is below 0.2% at 450°C for 1hr.

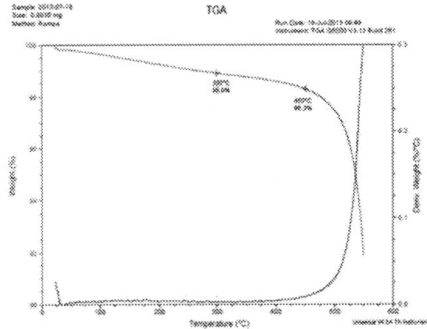

Figure 7. TGA (Thermal Gravity Analysis).

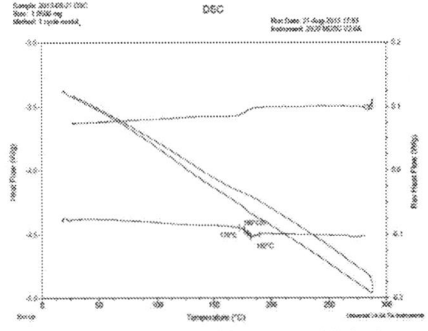

Figure 8. DSC (Differential Scanning Colorimetriy).

ASIC and MEMS pick and place on interposer wafer:

Dies were aligned and attached using a DATACON 220APM+ thermo-compression bonder on adhesive (Fig. 9). Typically, the bonding temperature was 250 °C, the bonding pressure was 15Kg/cm² and the bonding time was 5 seconds. A final collective thermal treatment at 400°C without pressure was necessary to achieve complete bonding on interposer. Shear tests were performed on dies after bonding and show very good adhesion. Indeed, silicon dies were damaged by the shear tool before breaking the adhesion on silicon.

Figure 9. ASIC and accelerometer integration on interposer wafer.

Dies were finally connected each other and to the interposer by wire bonding using wedge aluminum bonding.

Wafer lid bonding:

Lid wafer was bonded to the interposer wafer by eutectic AuSi formation at 400°C. This operation requires specific pretreatment on both wafers to guaranty good bonding. A flash of Ti/Au was deposited on lid wafer and plasma O_2 was performed on wafers just before the bonding. A gas pressure of 5×10^{-3} mbar inside the chamber was maintained and a tool pressure of 30KN was applied on the wafers stack for 30min. The alignment process has been validated on non-doped wafers based on the observation of specific metallic marks on wafers using infrared microscope (Fig.10).

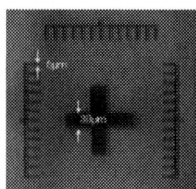

Figure 10. Observation of wafers alignment thanks to metallic marks visible with infrared microscope.

The wafer to wafer alignment was achieved with accuracy on the order of 5μm.

Interposer thinning, metallization, devices labeling and dicing:

A Ti/TiN layer was deposited on the lid side. The interposer was thinned down to 200μm and 1.5 mm edge was removed. Stress release process necessary to improve die strength was performed by dry etching. A Ti/TiN was deposited on interposer side and patterned in order to label individually each device. Finally, the stack wafers was diced.

The first demonstration was done without ASIC and accelerometer. View of dice wafer and device are presented in figures 11.

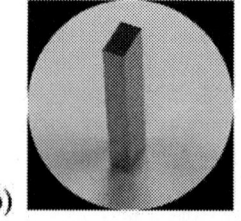

Figure 11. a) bonded wafers after dicing b) Silicon box device.

Hermeticity of the package

Hermeticity is a critical point for medical implantable devices because the life of the electronics parts guarantees the life of the patient. In the case of cardiac stimulation application, the life time target for the devices is 20 years. To preserve electronic from corrosive substances, which constitute the major part of the body, several barriers should be placed. In our case, one of them is the silicon box. In electronic packaging it is generally accepted that the life time limit is reached when the vapor pressure and moisture level inside the cavity induce the formation of water droplet condensation [3]. Indeed, the formation of water dip associated with ionic elements can induce corrosion or short circuit. The dew point is a parameter essential to know. According to Antoine equation approximation (1), the water pressure dew point at 37°C and 100% of humidity rate is 0.062 atm.

$$\log_{10} P = 8.07131 - \frac{1730.63}{233.426+Tb} \qquad (1)$$

Where Tb is the boiling point in degrees Celsius and P the pressure in Torr.

However, as a safety margin and taking into account that some contaminants can promote the condensation before reaching the saturation level of the dew point, it is usual to consider 5/6 of the dew point as an acceptable limit in the cavity [4]. This corresponds to 0.052 atm. So, an estimate of the leak rate leading to this pressure in our package after 20 years was carried out.

The leak rate can be calculated using the following equation, which is described in the literature [5]:

$$\Delta Pt = \Delta Pi. e^{\frac{-Lt}{VP0}} \qquad (2)$$

Where: ΔPt is the partial pressure difference from inside to outside the package at time t, ΔPi is the partial pressure difference from inside to outside the package at the initial time, L is the leak rate, V is the internal volume of the package and P0 is the difference in partial pressure between the inside and outside of the package (P0=1 atm).

The volume of our cavity is ~ 4mm² but it was considered that 90% is filled by dies. The water vapor partial pressure in the human body corresponds to the dew point of 0.062 atm. The water partial pressure inside the package at the initial time is estimated below 0.1 mbar. This was confirmed by gas analysis presented in the following sections. Taking into account these values it is possible to calculate the pressure inside the cavity after 20 years versus water leak rate. Such a plot is shown in figure 12.

978-1-4799-2408-0/14 $31.00 © 2014 IEEE

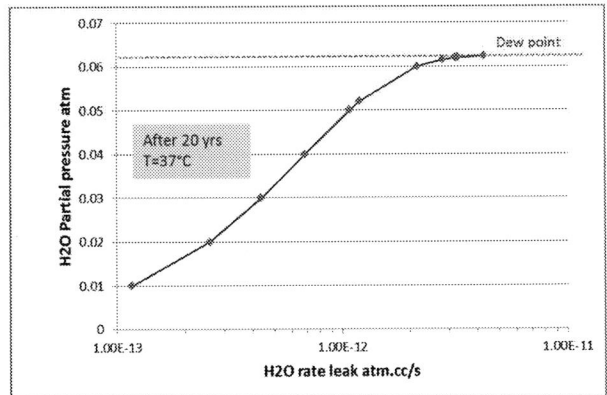

Figure 12. Pressure inside the cavity after 20 years versus water leak rate

Considering this graph, the H_2O leak rate which corresponds to the partial pressure of 0.052 atm in our cavity after 20 years is 1.2×10^{-12} atm cc/s. This value is very small and it is well known that the measurement of such fine leak rate with the addition of very small volume is a technical challenge. Below 10^{-3} cc the limit using standard He fine leak equipment test is approximately 5×10^{-8} atm.cc/s (He).

In our case two different technics were evaluated to test the hermeticity of our package: - Optical leak testing by detecting the deflection of the surface when pressure is applied onto the package; - test with krypton 85 as a gas tracer. Both nondestructive methods are described in literature [5,6]. In order to optimize first our bonding process, devices that were tested were empty silicon package, i.e. without adhesive polyimide, MEMS and ASIC.

Optical leak testing:

The pressure difference between the inside of the package and the bombing chamber was visible by the membrane deflection. The equipment used for this characterization is described in literature [5]. Figure 13 shows the deformation of our 100μm silicon lid for different He pressures in the chamber.

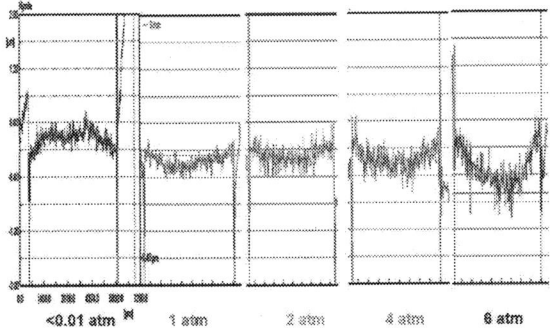

Figure 13. Silicon lid deformation from for different pressures in the chamber.

The impact of the pressure in the chamber is clearly observed. For 6 atm (He), the membrane deflection was measured on 52 devices and the average was 450nm with standard deviation of 100nm. To determine the leak rate three scans of the membrane deflection have to be performed. The first measurement should be performed at the atmospheric pressure as a reference measurement. Then, the chamber should be pressurized with He gas at high pressure and the membrane deflection should be measured for two different times t1 and t2. From the increase of the pressure inside the package, the leak rate can be calculated using this formula:

$$LA = V.P0.\sqrt{\frac{MHe}{MA}} \cdot \frac{1}{\Delta t} \cdot \ln\left(\frac{\Delta PHe,1}{\Delta PHe,2}\right) \qquad (3)$$

Where: L is the equivalent standard leak rate of air in atm cc/s, V is the internal volume of the package, P0 is the atmospheric pressure in atmosphere, MA is the equivalent mass of dry air in g (28.7 g), MHe is the molecular mass of He (4 g). ΔPHe,1 is the partial pressure difference from inside to outside the package at time t1 and ΔPHe,2 is the partial pressure difference from inside to outside the package at time 2.

In our case, it was assumed a linear dependency between membrane bending and pressure difference ΔP so the leak rate can be approximated to:

$$LA \approx V.P0.\sqrt{\frac{MHe}{MA}} \cdot \frac{1}{\Delta t} \cdot \ln\left(\frac{d1}{d2}\right) \qquad (4)$$

Where, d1 and d2 are respectively the membrane bending at t1 and t2.

The membrane deflection was measured on a same device at t0 and 3 days after. The scans are plotted in figure 14.

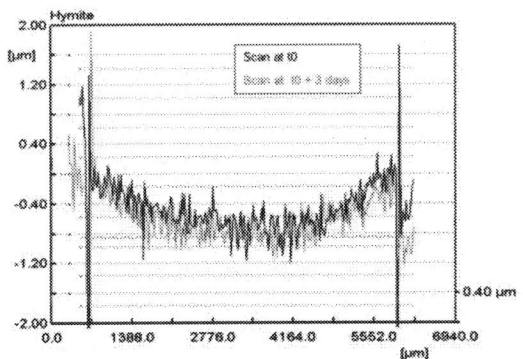

Figure 14. Membrane deflection measured on a device at t0 and 3 days after.

No significant change was observed. Taking into account that the resolution limit of the equipment given by the manufacturer is 20 nm and considering the equation (4), the time required to measure a device with a leak rate of 2.5×10^{-12} atm.cc/s for He (equivalent to 1.2×10^{-12} atm cc/s for H_2O) would be 90 days. This time is too long to be used in an industrial protocol test.

Krypton 85 testing:

Krypton 85 used as gas tracer leak is a testing method which allows to measure fine and gross leak rate with small cavity size. Test devices are placed in a sealed chamber and pressurized with mixed of air and the radioisotope Krypton85 for a specified time which depends on the rejected limit leak rate targeted. The tank is then backfilled with air and devices are removed for screening using X-ray scintillation crystal.

978-1-4799-2408-0/14 $31.00 © 2014 IEEE

Any leak rate measurement is potentially feasible. The reject limit is a function of time and pressure.

For this study, this test was performed by the company Oneida Research Services, Inc. 520 silicon package coming from 3 different wafers were tested. Devices submitted to ORS underwent processes including gross leak and fine leak testing. Devices were pressurized at 100psia (6.89 bar) for 6 days, which corresponds in our case to a reject limit of 5.0×10^{-12} atm cc/sec Kr-85.

Finally, 517 devices from 520 passed the test successfully. Two devices failed gross leak hermeticity testing and one failed fine leak testing. However, the reject limit of 5.0×10^{-12} atm cc/sec Kr-85 corresponds only to 1.1×10^{-11} atm cc/sec H_2O which is not our target. To reach 1.2×10^{-12} atm cc/s H_2O devices should be pressurized for 60 days.

Considering the time required to perform these 2 types of test, it is clear that fine leak hermeticity testing in production is still a challenge.

Residual Gas Analysis:

For the assessment of our device cavity gaseous content, ultra-high vacuum Residual Gas Analysis (RGA) test benches equipped with a mass spectrometer was used. The experiment consists in breaking devices placed under ultra-high vacuum in a first test bench, and analyzing with a mass spectrometer the gaseous species released from the tiny cavity during the opening process and flowing through a calibrated diaphragm. The equipment and procedure are described in literature [7].

The analysis of the time-dependent amplitude decrease of every m/q signal detected allows the calculation of the flowing constants for each m/q ratio and to rebuild the signal of the various gaseous species initially contained inside the cavity at initial breaking time. 6 devices from a same wafer located from center to the periphery and with sealing wall of 100μm and 200μm were tested.

The total average pressure measured inside cavity is 36.8 mbar with standard deviation of 2.2 mbar. The partial pressure for different gas is presented in figure 15.

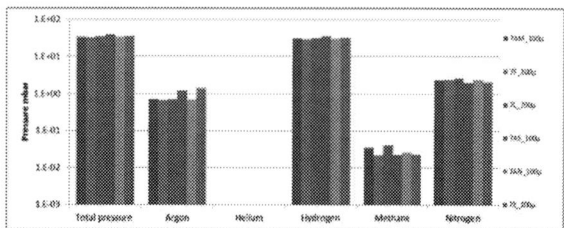

Figure 15. Partial pressures inside the cavity.

It has to be pointed out that pressure reproducibility in our cavity is quite good. No difference between die with sealing wall of 100μm and 200μm was observed. The total pressure is quite higher if is taken into account that the bonding process was performed below 0.1 mbar. Anyhow, the vacuum in the cavities is not necessarily required in our device. Gas detected can have different origin, entrapped in materials during process deposition, surface preparation or bonding procedure.

Bonding resistance in saline solution:

30 devices from a batch were placed in a phosphate buffered saline solution at 67 ° C. Phosphate Buffered Saline (PBS) is a water-based salt solution which ions concentrations match those of the human body. The objective was to appreciate the resistance of the sealing ring without encapsulation in saline water.

First separation of lid and interposer occurred after only 20 days. It was observed that WN present in the bonding ring was dissolved in saline water. This was confirmed by testing different layer deposited on silicon coupon and immersed in PBS. 100nm of WN was dissolved in approximately 4 hours at 67°C (Fig.16).

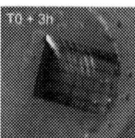

Figure 16. WN after 2 hours, 3 hours and 4 hours of immersion in PBS at 67°C.

Investigations are carrying out to replace WN by another barrier material. However it has to be pointed that a coating of our device will be necessarily implemented in order to prevent short circuit between lid and interposer and to solve biocompatibility issue.

Coating:

Atomic layer deposition (ALD) was chosen for coating because it is very conformal and uniform deposition. Materials such as Al_2O_3, TiO_2 are good candidates because they are known to be biocompatible and can be deposited by ALD. Coating was tested on silicon coupon with WN layer which is very sensitive to PBS. 20 nm of Al_2O_3 was deposited on 100nm of WN. As the ALD deposition is conformal the edges of the coupons are fully covered. However it was found that the Al_2O_3 was not perfectly hermetic. Indeed 100nm of WN was dissolved in approximately 11 hours at 67°C (Fig.17).

Figure 17. WN/Al2O3 before immersion and after 3 hours, 8 hours and 11 hours of immersion in PBS at 67°C.

It is assumed that some isolated defects in the Al_2O_3 layer allow the PBS to infiltrate and to etch the WN. To avoid this issue, multilayer will be investigated in future work.

Conclusions

The development of a new wafer level process flow for the manufacturing of low profile implantable medical device was presented. The packaging approach should enable electronics components, such as an ASIC and a MEMS, to be integrated in a hermetic silicon box sealed by AuSi bonding. The device manufacturing sizing 6.49x1.49x0.85mm³ with an internal cavity of 4mm³ was demonstrated. Taking into account our application, the leak rate which corresponds to reach dew point in our package after 20 years was estimated at 1.2×10^{-12}

atmcc/s (H_2O). Two different technics were evaluated to test the hermeticity of our package, optical leak testing and krypton 85. Both are too time consuming to reach our level of leak rate. However, it was demonstrated that 517 devices from 520 pass the test with rejected limit of 1.1×10^{-11} atm cc/sec (H_2O). This shows the good yield of our process. RGA analysis allows the access to our cavity gas content. The total average pressure measured inside cavity was ~36.8 mbar. Immersion tests in PBS have shown that WN layer present in the bonding ring was very sensitive and must be either replaced or encapsulated. First results using Al_2O_3 encapsulation deposited by ALD are promising but not sufficient. Multilayer will be investigated in future work.

References

1. W. Mokwa, "Medical implants based on microsystems," *Measurement Science and Technolog*, vol. 18, pp. 47–57, 2007

2. M. Op de Beeck, "Design and characterization of a biocompatible packaging concept for implantable electronic devices," *IMAPS conference*, Long Beach, CA Oct, 2011

3. G. Jiang, "Technology Advances and Challenges in Hermetic Packaging for Implantable Medical Devices," *Implantable Neural Prostheses 2, Biological and Medical Physics, Biomedical Engineering*, pp 27-61, 2010,

4. H Greenhouse, *Hermeticity of Electronic Packages*, pp. 54-59.

5. G Elger, "Optical Leak Detection for Wafer Level Hermeticity Testing," *29th International Electronics Manufacturing Technology Symposium*, San Jose, CA, July 14-16, 2004.

6. J.L Murgatroyd, "Leak-Rate Determination Using Krypton 85," Transaction on instrumentation and measurement, vol. Im-21, N° 1, pp 41-48, 1972

7. P-L Charvet, "MEMS packaging reliability assessment: Residual Gas Analysis of gaseous species trapped inside MEMS cavities," *European Symposium on Reliability of Electron Devices, Failure Physics and Analysis*, vol 53, Issues 9–11, pp 1622–1627, 2013.

978-1-4799-2408-0/14 $31.00 © 2014 IEEE

Capping Technologies for Wafer Level MEMS Packaging based on Permanent and Temporary Wafer Bonding

K. Zoschke[1], M. Wilke[1], M. Wegner[1], K. Kaletta[1], C.-A. Manier[1], H. Oppermann[1],
M. Wietstruck[2], B. Tillack[2], M. Kaynak[2], K.-D. Lang[3]

[1]Fraunhofer IZM, Gustav-Meyer-Allee 25, 13355 Berlin
[2]IHP GmbH, Im Technologiepark 25, 15236 Frankfurt (Oder), Germany
[3]Technical University of Berlin, Gustav-Meyer-Allee 25, 13355 Berlin

kai.zoschke@izm.fraunhofer.de

Abstract

This paper describes techniques for the miniaturized, low-cost wafer level chip-scale packaging of MEMS based system in packages (SiPs). The approaches comprise permanent bonding of cap structures using adhesives or solder onto a passive or active silicon wafer which is populated with MEMS components or which is itself a MEMS wafer. The paper addresses different options for manufacturing of lid or cap structures and their subsequent bonding to the partner wafer. Different technologies like bonding of full area cap wafers as well as partial capping approaches based on reconfigured cap structures on a help wafer or cap structures created on a compound wafer are presented. Examples like the selective capping process for RF-MEMS switches are discussed in detail. All processes were performed at 200 mm wafer scale.

Introduction

Further cost reduction, performance increase and miniaturization of electronic systems requires new highly efficient SiP concepts for MEMS components like RF resonators or switches, quartz crystals, bolometers, BAWs etc. Based on the fabrication of cap or lid structures and usage of suitable handling and bonding technologies wafer level packaging is a promising approach to evolve the next package generation for such components. [1, 2, 3]

In the first section of the paper a general overview of the possible wafer level package configurations for different types of MEMS and TSV less cap structures is discussed. The presence or absence of TSVs in the base wafer significantly determines the cap manufacturing and bonding effort. In that context, boundary conditions for TSV based and non-TSV based packaging solutions are generally distinguished here.

The second section of the manuscript focuses on the different wafer level manufacturing possibilities of cap structures. A selected variety of options for cap manufacturing is summarized and discussed here.

Section three presents applications and examples where wafer level capping is done by wafer bonding of full format cap wafers. This is only applicable, if finally TSVs are routing the electrical IOs to the opposite side of the base wafer.

Section four describes solutions how wafer level capping with caps laterally smaller than the corresponding base chips

can be obtained. Two ways are distinguished and described in detail.

The first approach is based on fabrication and singulation of the caps followed by their temporary face up assembly in the desired pattern on a help wafer. In a subsequent wafer to wafer bonding sequence all caps are transferred onto the target wafer. Finally the help wafer is removed from the back side of the bonded caps. This approach of reconfigured wafer bonding is especially used for uniform cap patterns or, if MEMS with own bond frame structure have to be mounted onto their corresponding ASICs. In that case no additional cap is required, since the MEMS can act as their own cap.

The second approach is based on cap structure fabrication on a compound wafer consisting of two temporary bonded wafers. One wafer acts as carrier wafer whereas the other wafer is treated by using processes like thinning, silicon dry etching, deposition and structuring of polymer or metal bonding frames and optional partial pre-dicing to form the cap structures. Thus, the fabrication sequence transforms the original compound wafer into a carrier wafer with singulated, face-up mounted cap structures. A wafer to wafer bonding process is used now to bond all cap structures in parallel onto the target wafer. Finally, the temporary carrier wafer is removed from the backside of the caps. Due to the mask defined fabrication of the caps out of the compound wafer, a fully custom specific selective wafer level capping is possible with irregular areas and locations to be capped on the target wafer.

In order to enable such selective capping approaches high performance and versatile temporary bonding and de-bonding processes are required. The demands in such processes and corresponding adhesive materials are also outlined and discussed in section four.

1. MEMS Packaging Options at Wafer Level

The combination of TSV technologies, silicon bulk micro machining, redistribution, die to wafer assembly and wafer to wafer bonding enables a set of versatile advanced wafer level packaging concepts for MEMS. Thereby packaging is considered as housing of the MEMS, simultaneous connection to their signal processing device and preparation of interconnects to the next system level.

978-1-4799-2408-0/14 $31.00 © 2014 IEEE

As shown in Figure 1, depending on MEMS type and final SiP architecture different kinds of packaging scenarios for MEMS are possible at wafer level.

In row A of Figure 1 non-TSV based packaging solutions for MEMS are shown. In case (A) single components with defined electrical contacts like BAW devices or quartz crystals are mounted in die to wafer mode on the corresponding ASIC wafer. Subsequently, a wafer to wafer bonding process is used to seal all devices with additional cap structures. In case (B) components like silicon resonators which feature own seal frame structures additionally to the electrical contacts are mounted on corresponding ASIC wafers. In that case due to the present seal frame the components act as their own cap and no additional cap structures are required for them. Also here all devices are mounted in parallel using a wafer to wafer bonding process to the base wafer. In case (C) the MEMS components like RF switches or optical sensors are still in wafer form and act as base for the final package. Simple caps are used to seal these components. Also these caps are bonded simultaneously to the base wafer by wafer to wafer bonding. Depending on the requirements of the MEMS, the seal frame structures are either based on adhesive materials for non-hermetic sealing or metal/solder materials for hermetic sealing.

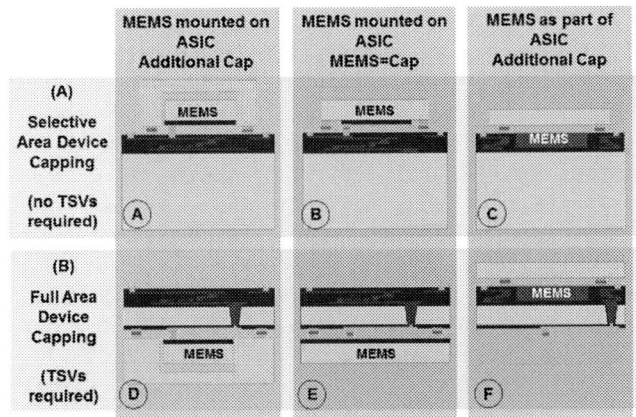

Figure 1: MEMS packaging options by wafer level capping

All scenarios in row A have in common that no TSVs are present either in the cap or base wafer. Due to that, the cap structure needs to be smaller than the base chip, so that electrical contacts outside the cap area can be accessed on the base chip after the caps were bonded. Thus, wafer level fabrication and bonding processes of the cap structures need to allow gaps and spacing between the particular caps. Corresponding scenarios to accommodate such a partial wafer level capping are discussed in section 4 of this manuscript.

In row B of Figure 1 TSV based packaging solutions for the same MEMS types as in row A are shown. Since the electrical signals of the MEMS can be routed though the base substrate to its other side, caps and the corresponding base chips can have the same lateral size. Thus, cap wafers without gaps and spacing between the cap structures can be used which simplifies their fabrication. Application examples for usage of such full area cap wafers are discussed in section 3 of this manuscript.

2. Wafer Level Manufacturing Options of Cap Structures

Based on the usage of standard technologies from wafer level packaging, different cap configurations can be fabricated. Some of the basic possibilities are schematically shown in Figure 2.

Case (A) shows the simplest cap configuration using bond frames directly deposited onto a blank silicon wafer. The frames are either made of adhesive or metal/solder. Adhesive materials are either deposited by dry etching using a photo mask or directly structured, if they are photo sensitive. Typical materials are epoxy and polymer materials like SU8, BCB or polyimide. Metal and solder based frame structures are deposited by semi-additive technology using sputter deposition, lithography, electro-plating followed by resist removal and seed layer differential etching. Typical materials used are Cu, Au, CuSn, AuSn. The height of the bond frames defines the height of the final cavity structure.

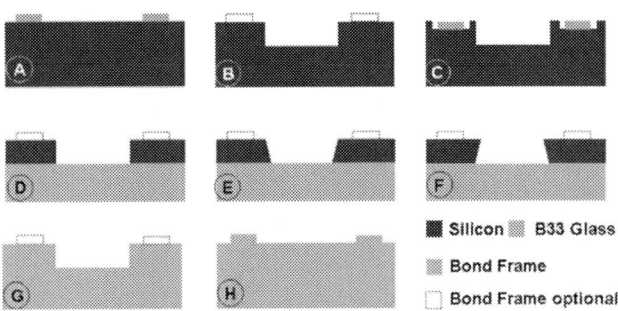

Figure 2: Different configurations of cap structures fabricated at wafer level

If the required cavity depth cannot be purely generated by the height of the bond frames, additional cavities are etched into the silicon surface. Case (B) shows such a structure as an extended version of Case (A). The cavities with vertical side walls are generated by dry etching using a photo resist mask for pattern definition. If an adhesive bond should be formed with the target wafer and the adhesive is already deposited at this partner wafer, additional bond frames on top of the etched silicon frames might be obsolete. It is furthermore possible to transfer adhesive with a stamping process to the frames of the created cap wafer.

Figure 3 shows an example of a processed silicon cap wafer. The frame structures with lateral dimensions of 680 μm x 680 μm represent the original wafer surface. The surrounding and inner area was set 25 μm back by silicon dry etching. The wafer is used as test vehicle to establish a glue transfer print process for later usage with wafers of type (D).

If additional bond frame structures of adhesive or metal are required on top of the cavity frames, they can either be generated before or also after the cavity formation. Boundary conditions as bond frame height and material basically determine the best process order. A generation of the bond frames before cavity formation requires process compatibility of the lithography and silicon dry etching process with the already existing bond frames structures. The bond frames need to be fully covered and sufficiently protected by the photo resist for silicon dry etching as well as with its removal process for example.

If the cavities are formed before the bond frame structures, advanced high topography compatible deposition processes for resists and/or adhesives like spray coating might be required for the definition and deposition of the bond frames.

Figure 4 shows an example of a silicon cap wafer with 200 µm deep silicon etched cavities and 10 µm high bond frames of electro-plated gold. The gold frames were fabricated before the silicon cavities were etched. Such wafers are used for hermetic sealing of MEMS components mounted on corresponding partner wafers.

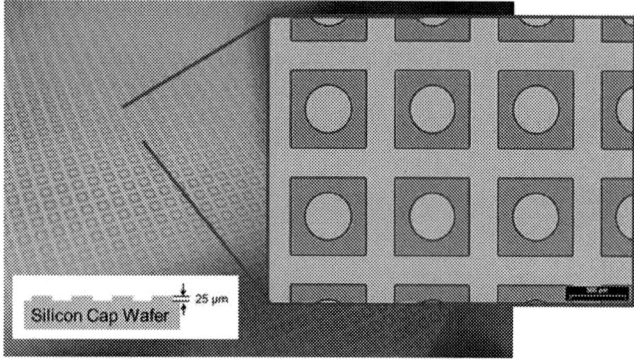

Figure 3: Silicon cap wafer without additional bond frame structure (according Case (B), Figure 2)

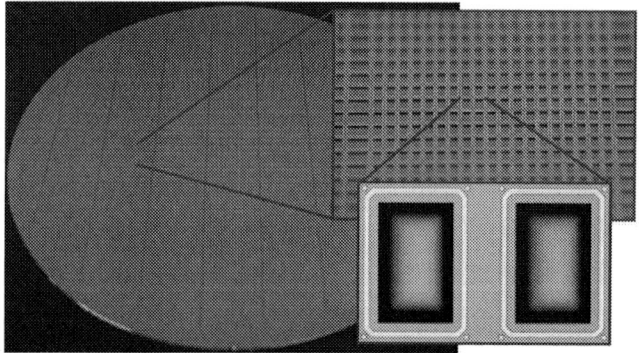

Figure 4: Silicon cap wafer with additional Au bond frames, (according Case (B), Figure 2)

Case (C) in Figure 3 shows an option for the implementation of buried bond frame structures. Such a structure is useful, if no or only a minimized remaining gap is allowed between the bond partners after bonding. In order to create such a structure fist the groves for the buried frames are fabricated. Following, bond frame structures and cavities are generated.

The cases (D-H) describe optical semi or full transparent cap structures which are obtained by using Borofloat33 glass wafers as share or main part of the cap device.

In order to create cap structures of case (D) a silicon and a glass wafer are anodically bonded in the first process step. The height of the final silicon frames can now be adjusted by

thinning of the silicon wafer. Following, processes for bond frame formation as well as cavity formation are performed similar as already described for case (B). The glass represents an etch stop for the silicon dry etching process.

Case (E) is focused on optical semi-transparent cap structures with silicon frames, but defined sloped side wall. The sloped side wall is created by KOH etching of silicon with <100> orientation instead of dry etching, which gives a defined side wall angle of 54.7 °. The conventional process flow to create such a structure would require a silicon wafer with a thickness matching to the final cavity depth which is coated with SiO_2 and Si_3N_4 at both sides. A lithography would define the KOH etch structures. Si_3N_4 etching would be performed followed by resist stripping, SiO_2 etching and KOH etching. Finally, the wafer would be cleaned from Si_3N_4 and SiO_2. The structured silicon wafer would then be anodically bonded to the glass wafer.

A more easy process flow is enabled by using Lithoglas[TM], an evaporated thin glass, as KOH etch mask. In that case, the anodic bond is performed as a first step in the fabrication process. Following, the silicon wafer is thinned to the desired thickness and the KOH etch mask is generated. For that, a photo resist suitable for lift-off processing is coated and structured onto the silicon surface. Now, a thin glass layer is deposited using a low temperature plasma ion assisted evaporation process. [4, 5] The deposited glass is structured by photo resist lift-off resulting in the desired glass masking of the silicon for KOH etching. The anodically bonded glass wafer serves as etch stop for the KOH etching. Finally, the glass masking is removed using a BHF etch step. If required, optional metallic or adhesive bond frame structures can now be deposited on the frames of the silicon cavities by using spray coating for resist or adhesive deposition. Figure 5 shows an example of a cap wafer according to case (E) which was fabricated using the described process. The sloped side wall areas of the KOH etched cavities are directed towards the open side of the cap structure.

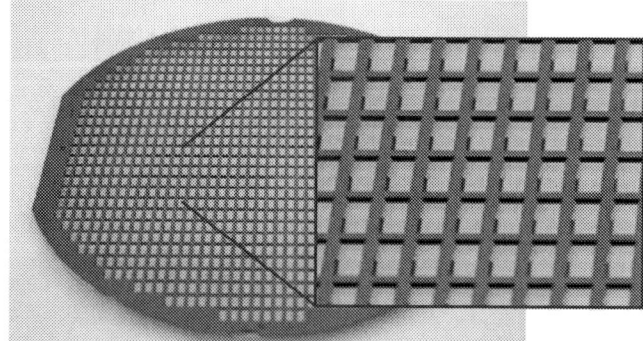

Figure 5: optical semi-transparent cap wafer with KOH etched silicon frames, cavities open away from glass

In some applications it is advantageous that the side wall areas are sloped into the opposite direction and facing towards the glass. According to case (F) also such a cap structure can be fabricated. For that, processing starts with KOH etching on a thick silicon wafer to a depth matching to the final required cavity depth. After removal of the etch masks, the silicon wafer is anodically bonded with the side of the etched cavities onto the glass wafer. Now, wafer thinning processes are applied to remove the bulk silicon. The thinning can either be done until the cavities are opened from their bottom side or stop before the cavities are reached. The second case allows an easy processing of additional metallic or adhesive bond frame structures on the still smooth surface of the wafer. Finally, the cavities are opened by a lithography and subsequent dry etching process. Figure 6 shows an example of a cap structure according to Figure 2 case (F). The photographs are taken from the side with the bonded glass wafer. As can be seen, the sloped side wall areas are facing towards the glass.

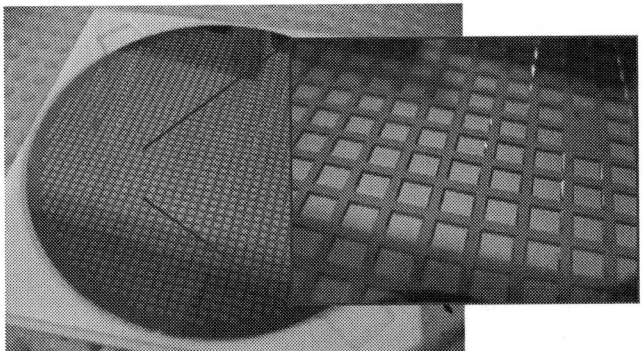

Figure 6: optical semi-transparent cap wafer with KOH etched silicon frames, cavities open towards glass

Cases (G) und (H) represent options of pure glass cap structures including glass base and glass frames.

Case (G) can be manufactured in subtractive technology by glass dry etching using a photoresist as etch mask. However, the depth of the etched cavities is limited due to the low etch rate of the glass material (approx. 1 µm/min).

Case (H) represents a possibility for additive manufacturing of pure glass cap wafers. Also here the Lithoglas[TM] technology is used to deposit glass frame structures with up to 20 µm height onto a glass base wafer using glass evaporation and lift-off processing. Such pure glass cap structures are useful, if UV curable adhesives should be used to bond the glass frames to the device. The adhesive can be directly cured by UV exposure through the glass base and glass frames. One example of a pure glass cap wafer is shown in Figure 7.

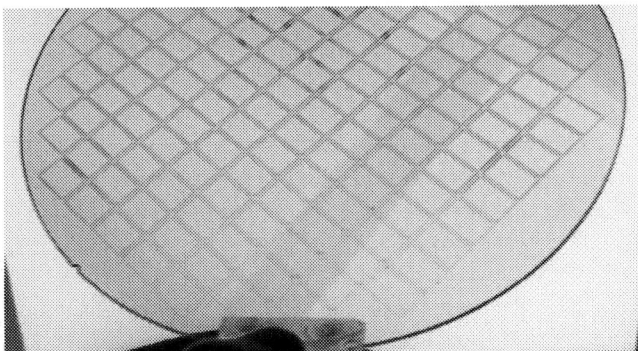

Figure 7: optical transparent cap wafer with glass frames

Cap structures as shown in cases (B, D, E, F, G, H) of Figure 2 can be bonded adhesively. The adhesive can be deposited on top of the cavity frames or on the corresponding partner wafer. If both options for adhesive deposition are not allowed or too risky and costly, adhesive transfer might be an alternative solution. As shown in Figure 8, with that approach glue can be transferred from a help wafer to the topography features of the cap wafer without additional effort for structuring of the adhesive. The help wafer is coated with the corresponding adhesive material and subsequently pressed onto the cap wafer and separated subsequently. Depending on thickness and viscosity of the used adhesives a certain amount can be transferred from the help wafer to the topography features of the cap wafer. Typical spin coating compatible materials like UV curable adhesives, SU8, Cyclotene and also certain polyimide precursors are well suited to work with that approach. [6]

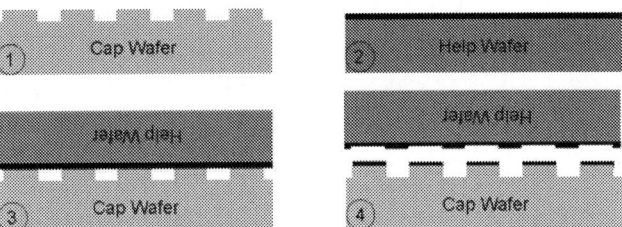

Figure 8: principle process flow of adhesive transfer technology

3. Capping by Full Format Cap Wafer to Wafer Bonding

It was already discussed in the introduction, that caps covering the full device area are only useful, if the electrical connections are feed through the device substrate to its backside. This means, that the TSVs have to be processed into the device substrate either before or after bonding of the cap wafer.

Wafer level image sensor packaging is one example for using full format cap wafers. In the first process step, a glass wafer is bonded onto the active side of an image sensor wafer. The bonding is either done using a transparent glue material which is coated on the entire wafer or a structured adhesive which is not present over optical areas. Figure 9 shows microscopic and schematic cross sectional view of a CMOS image sensor package with bonded glass wafer using full area transparent glue. The bond process of the glass wafer comprised spin coating of the UV curable adhesive onto the CMOS wafer and subsequent application of pressure and UV exposure under vacuum using a wafer to wafer bonder.

After the glass wafer was bonded, the CMOS wafer is backside processed to form TSVs and connect the original front side pads via redistribution with new IO structures on the backside of the CMOS wafer. [7]

A structured adhesive bond is required, if optical sensitive structures as micro lenses are present on the image sensor wafer. A pattered adhesive bond can be obtained by direct coating and structuring of adhesive on the glass wafer or usage of glass wafers with cavity structures in combination with adhesive transfer technologies. Figure 10 shows two examples of structures on cap wafers usable for bonding with image sensor wafers. The left image shows 10 µm high BCB frames, which were directly structured onto the glass wafer. The right image shows the same structures, but implemented as 50 µm high silicon posts on a glass wafer, which are suitable to receive adhesive using an adhesive transfer process. The right structure was fabricated according to the discussed process flow of Figure 2 / case (D) in the previous chapter of this manuscript. Both types of wafers can be bonded to the device wafer by using a low temperature thermo compression wafer to wafer bonding process.

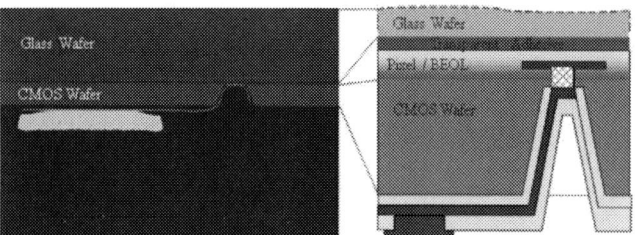

Figure 9: CMOS image sensor package with full area bonded glass wafer using transparent adhesive

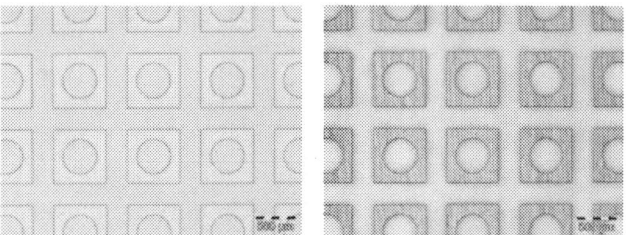

Figure 10: Cap wafers for bonding with image sensor wafers, structured adhesive on glass wafer (left), silicon posts on glass wafer (right)

Further examples for device capping by bonding of full format cap wafers are the MEMS packaging approaches using active or passive silicon interposers with TSVs. Figure 11 shows two corresponding cross sectional views. The images on the left side in Figure 11 show a cross section of a silicon interposer with a bonded cap wafer. Each cavity hosts a quartz crystal, which was assembled onto the interposer before the cap wafer was bonded. The electrical contacts of the quartz crystal are routed by TSVs through the interposer to its back side where the IOs of the package are located. The used cap wafer has a structure according to case (B) in Figure 2. Both wafers are hermetically bonded by soldering of Au and Au+Sn frame structures located on cap and interposer wafer respectively. Each quartz crystal is surrounded by a solder frame structure leading to hermetic sealed cavities. [8]

Figure 11: examples for silicon interposers based packages

The images on the right hand side of Figure 11 show a package structure for BAW filters. The bottom wafer features TSVs and carries the BAW filters. Following, the cap wafer is bonded onto the bottom wafer using an adhesive bonding process. Also this cap wafer has a structure according to case (B) in Figure 2. Due to the severe topography of the cap wafer, a glue transfer process according to Figure 8 is used to easily coat the surface of the cap wafers' silicon frames with adhesive.

4. Partial Wafer Level Capping

This section of the manuscript focuses on approaches which enable an only partial or selective capping of the devices on the base wafer. According to Figure 1 / row (A) such partial capping is required, if cap structure and electrical contacts are located on the same side of the device and no TSVs are used. In the following section two main approaches for selective wafer level capping are presented. Furthermore, a temporary bonding and de-bonding technology is described which is one key enabler for the described scenarios.

4.1. Key Technology – Temporary Bonding / Debonding

The partial capping solutions are based on simultaneous bonding of multiple cap structures to the device wafer using wafer to wafer bonding technologies. Such scenarios require carrier wafers on which the caps are temporary mounted in face up direction at dedicated positions according to the required cap pitch on the target wafer. After the wafer bonding process is done those carrier wafers have to be released from the back side of the cap structures.

One possibility for the establishment of such carrier wafers with face up mounted cap structures is die bonding of pre-processed caps at dedicated positions on the carrier wafer. For that, the carrier wafer is prepared with alignment marks and an adhesive coating.

Direct fabrication of cap structures by processing one side of a temporary bonded wafer sandwich is another possibility to generate carrier wafers with face up mounted cap structures. In that case one wafer of a compound wafer is shaped by certain process steps, so that the cap structures are left behind.

To enable the described approaches a versatile temporary bonding and de-bonding technology is required, which is compatible with the following main process schemes and conditions:

- temporary wafer to wafer bonding
- wafer thinning and thin wafer processing
- temporary die to wafer bonding
- cap transfer bonding using thermo compression like processes at high temperature, high vacuum
- easy and fast release of carrier wafer from backside of bonded cap structures

Due to best matching to the above mentioned requirements, Borofloat33 glass wafers and HD3007 are used as temporary carrier wafers and adhesive material for the required cap formation, fixation and handling purposes. The chosen material combination allows a laser assisted initiation of the release process for the carrier wafers after the cap structures are transfer bonded to the device wafer.

The principle work flow of bonding and de-bonding by using laser release technique is shown Figure 12. Glass wafers are coated with the adhesive. Subsequently wafers or components are mounted by thermo compression processes. The de-bonding is initiated by exposing the adhesive through the glass wafer using a 248 nm (KrF) excimer laser. The adhesive absorbs the laser radiation within thin a layer. The laser energy causes material de-composition and the bond layer is opened. Following, the glass carrier can be detached from the wafer or components. [9]

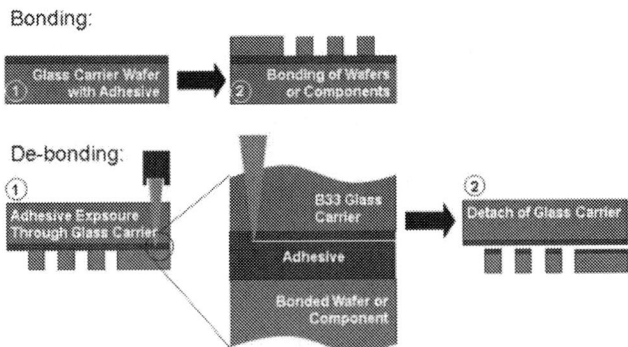

Figure 12: principle work flow of bonding and de-bonding using laser assisted release

4.2. Partial capping using reconfigured help wafer

The general idea of this approach is to reconstitute already singulated MEMS or cap components face-up onto a help wafer which allows their later simultaneous bonding to a target wafer by using wafer to wafer bonding processes.

The schematic process flow in Figure 13 shows two options of reconfigured wafer bonding. The help wafer is equipped with local marks to align the placement of each individual MEMS according to the required reticle pitch on the target wafer (ASIC, interposer...) on which the components will be finally bonded. The components are temporary fixed using an adhesive layer which is coated on the help wafer. Step (1a) and (1b) indicate the possibility of using full area adhesive or structured adhesive to be present only where the dices are placed. The placement of the components is similar to a die bonding process using adapted tools for component pick and place (steps 2a and 2b).

After component placement on the help wafer their positioning pattern is ideally exactly mirrored to the required placement pattern on the target wafer. For the subsequent wafer to wafer bonding the help wafer is flipped and local marks on help and target wafers are used to align both wafers to each other (steps 3a and 3b). By doing the wafer to wafer alignment all components are simultaneously aligned to their required position on the target wafer. This is enabled since the positioning of the global and local marks on the help wafer was done in reference to the required component positions and locations of the global marks on the target wafer. To enable a

sufficient accurate final component alignment high placement and die bonding accuracy are required during reconfiguration.

The following wafer to wafer bonding process is adjusted to the used bond materials to be joined between MEMS components and target wafer. Typically thermo-compression like processes at temperatures up to 300 °C are possible to run. Thus, adhesive bonding but also soldering processes based on Cu/Sn or Au/Sn are possible.

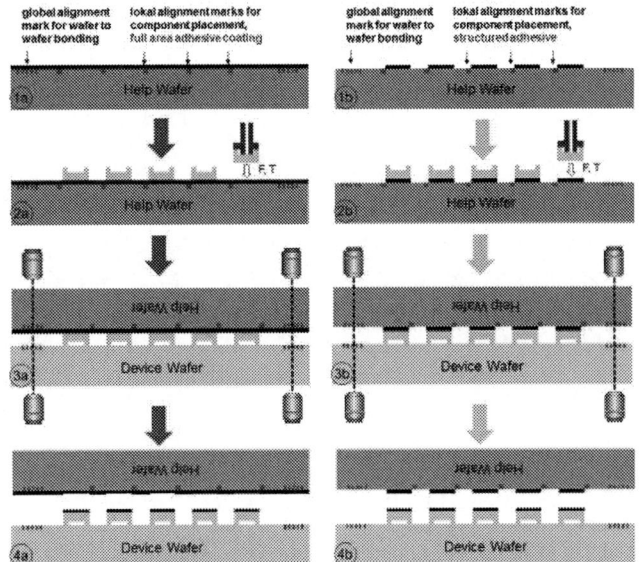

Figure 13: principle work flow of reconfigured wafer bonding using help wafer

After the wafer bonding is finished and the MEMS are bonded to the target wafer, the help wafer has to be removed from their back side. If the adhesive is present on the entire help wafer surface as shown in step (4a), the laser release process needs to be done selectively. This is required to prevent adhesive de-lamination from the entire help wafer. Thus, an aligned laser exposure is required, so that only the areas with placed components are exposed.

If the adhesive was already structured and is only present under a placed component, a full area laser exposure can be done. According to step (4b) adhesive is only present where a component is located. Exposure of the other areas of the help wafer will not cause adhesive delamination.

The pictures in Figure 14 show an example for capping by using reconfigured wafer bonding. The upper pictures show silicon resonators provided from VTT Technical Research Centre of Finland which are reconfigured on a help wafer. The help wafer is a 500 μm thick glass wafer. The required alignment marks were fabricated by subtractive structuring of a sputtered Ti:W layer. As adhesive layer 5 μm thick HD3007 was used. The bottom pictures in Figure 14 show the transfer bonded silicon resonators on the ASIC target wafer after removal of the glass carrier wafer. The 200 μm thick silicon resonators with Pt/Ti/Mo seal rings were bonded with corresponding Au/Sn rings on the ASIC wafer. The maximum process temperature of the wafer to wafer bonding process was in the range of 300 °C.

The principle of reconfigured wafer bonding is well suited for all types of singulated components like pre-tested MEMS but also passive cap structures according to the options shown in Figure 2. Due to the usage of die bonding equipment for placement of the components on the help wafer, a regular pitch of the components is desired. Multiple pitches and various types of cap forms and dimensions are difficult to process with the described approach.

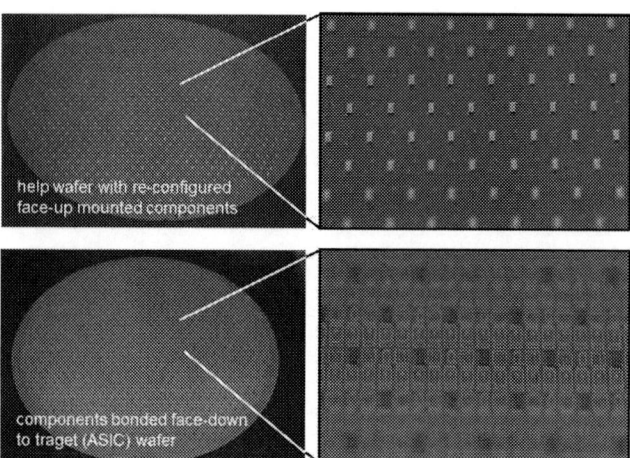

Figure 14: reconfigured components at help wafer (top) and components bonded to device wafer (bottom)

4.3. Partial capping by fabrication of cap structures at compound wafer and transfer bonding

The main idea of this approach is the fabrication of cap structures out of one wafer of a bonded wafer stack and their subsequent transfer to the target wafer. Due to that, caps can be fabricated fully custom specific with arbitrary shapes, dimensions and pitches which is well suited for capping of non-homogeneous device distributions on the target wafer.

Figure 15 shows the principle work flow of full selective wafer level capping. In step (1) a glass help wafer is coated with adhesive material and bonded in step (2) to a second wafer which is silicon. In subsequent process steps the silicon wafer is processed to generate the cap structures. Processes like back grinding, lithography, silicon dry etching, polymer dry etching, sputtering, electroplating etc. are used to form silicon cap structures according to Figure 2 cases (A, B, C). Even optical transparent caps according to Figure 2 / cases (D-H) can be generated. In that case, instead of a silicon wafer a second glass wafer or a silicon/glass compound wafer would have to be bonded to the glass help wafer in process step 1.

The structuring of the caps can include a removal of the material in between the particular cap structures by using an additional lithography and dry etching of the silicon. In that case the adhesive layer acts as etch stop. Additionally, even the adhesive can be removed next to the caps (step 3a). An alternative way to separate the caps from the surrounding material is cutting by mechanical dicing (step 3b). In the next process step the created cap wafer is aligned and bonded to the corresponding target wafer (steps 4a and 4b). The required alignment marks on the cap wafer were created during the cap processing.

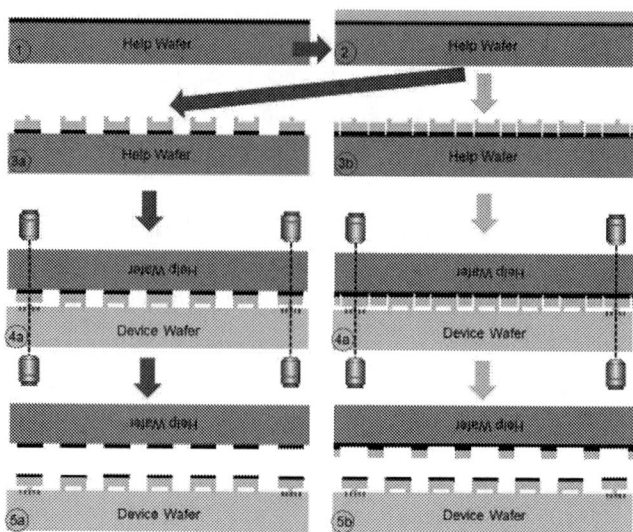

Figure 15: principle work flow of full selective wafer level capping

Figure 16 shows an example of a processed cap wafer. The photographs clearly show the arbitrary form and location of the cap structures on the help wafer. The caps were structured by dry etching into a 50 µm thin silicon wafer which was created by back grinding from full thickness (725µm) after bonding. Each cap has an adhesive bonding frame around the rim, with a height of 5 µm. The frames were structured by lithography and dry etching before the silicon was dry etched. The adhesive layer on the help wafer around the caps was also removed by dry etching. The caps were transferred to the target wafer using a wafer to wafer thermo compression bonding process at 300 °C and pressure of 0.1-0.2 MPa.

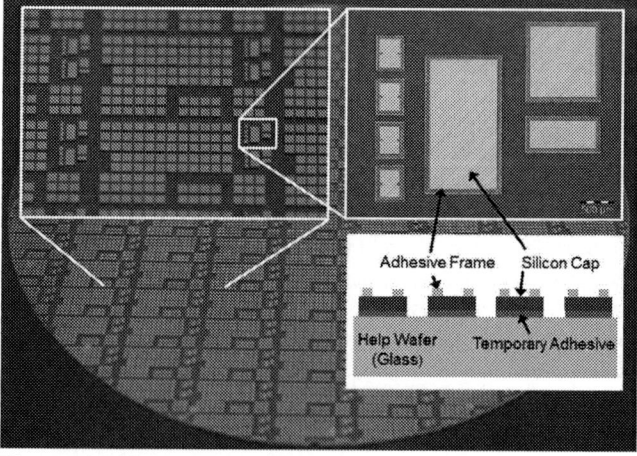

Figure 16: Help wafer with arbitrary shaped cap structures ready for cap transfer bonding

After the caps were bonded to the target wafer, the help wafer needs to be removed from the backside of the caps. According to Figure 15, two different options are possible here. In case the glue is only present underneath the caps, a full area laser exposure of the help wafer can be performed (step 5a). If the intermediate pieces between the caps are still present, then a structured laser release is required which exposes only the cap areas (step 5b). In that case, the intermediate pieces remain on the help wafer during help wafer release.

Figure 17 shows an active wafer after cap transfer bonding and removal of the help wafer. The caps were transferred from a similar help wafer as shown in Figure 16. The arbitrary shaped and 50 µm high caps can be clearly seen in the shown images. The target wafer hosts RF-MEMS switches which are embedded into the metallization levels of the BiCMOS wafer. RF-MEMS switches are very promising for fully integrated solutions for radar and imaging applications. The monolithic integration provides lowest parasitics which makes it very attractive for mm-wave applications. In such systems RF-MEMS switches provide superior RF performance and high linearity for reconfigurable circuits, Tx/Rx-switches and phased-array systems. By electrostatic actuation, the position of the suspended membrane can be controlled changing the capacitance between membrane and signal line. The capacitive switch provides a capacitance ratio of ~1:10 between the membrane and the signal line with a pull-in voltage <40 V. The adhesively bonded caps provide a non-hermetic protection of the MEMS structures.

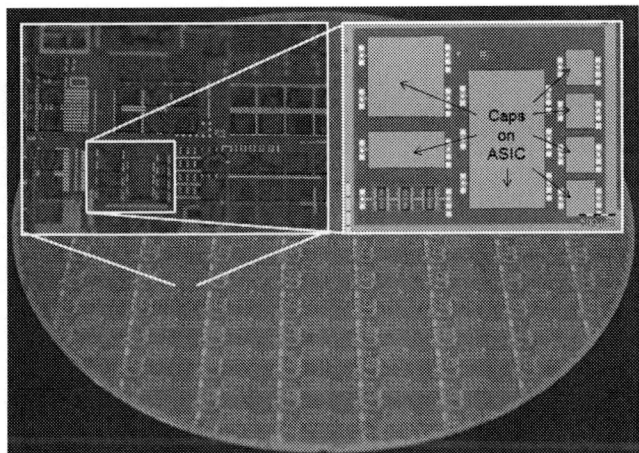

Figure 17: Active wafer with packaged MEMS switches after cap transfer bonding from help wafer

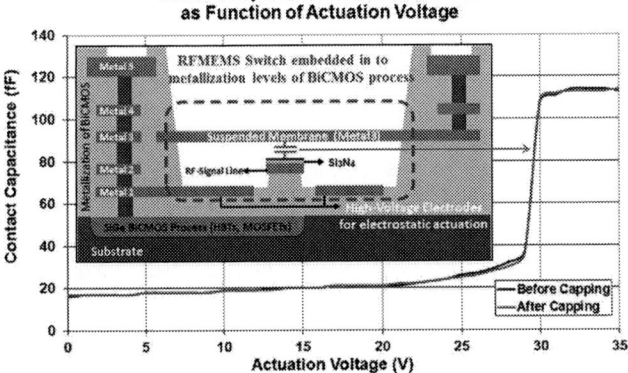

Figure 18: influence of capping process on electrical performance of MEMS switches

Figure 18 shows a plot of the measured contact capacitance as function of the actuation voltage for a typical RF-MEMS switch before and after the capping process was performed. The measurements clearly indicate that no influence of the capping process on the device performance could be observed.

Conclusion

A broad variety of capping technologies for wafer level packaging of MEMS is available. Different types of cap structures with or without optical transparence can be fabricated by combining standard technologies like silicon bulk micro machining, redistribution, wafer bonding and deposition and structuring of polymer or metal layers.

Full format cap wafer to wafer bonding is possible, if the device IOs are feed by TSVs though its backside so that the cap can cover its entire surface. Partial capping is required, if no TSVs are present so that the IOs and cap are located at the same side of the device. Two options to obtain partial capping were presented. In the first option pre-processed caps are reconfigured on a help wafer to enable their subsequent simultaneous transfer bonding to the target wafer. In the second option the help wafer with mounted caps is created by processing one wafer of a bonded wafer stack. Based on that full custom specific arbitrary shaped cap structures can be created and transferred to the target wafer. The technology was demonstrated by capping of RF-MEMS switches which are embedded into the BEOL stack of a BiCMOS Wafer.

References

1. D. Ruffieux et al., "A Versatile Timing Microsystem based on Wafer-level Packaged XTAL/BAW Resonators with sub-µW RTC Mode and Programmable HF Clocks", IEEE Journal of Solid State Circuits, Volume 49, January 2014, pp. 2012-222

2. R. N. Candler et al., "Single wafer encapsulation of MEMS devices," in IEEE Transactions on Advanced Packaging, Vol. 26, Aug. 2003, pp. 227-232

3. M. Kaynak et al., "RF-MEMS Switch Module in a 0.25µm BiCMOS Technology", 12th Topical Meeting on Silicon Monolithic Integrated Circuits in RF Systems, SiRF 2012, 16-18 January, 2012, Santa Clara, California, pp 25-28

4. J. Leib et al., "Wafer-Level Glass-Caps for Advanced Optical Applications", Proc. 61st ECTC Conference, Lake Buena Vista, FL, USA, 2011, pp. 1642-1648

5. K. Zoschke et al., "Evaluation of Micro Structured Glass Layers as Dielectric- and Passivation Material for Wafer Level Integrated Thin Film Capacitors and Resistors", 57th Electronic Components and Technology Conference, May 29 – June 1, 2007, Reno, Nevada USA, pp. 566-573

6. K. Zoschke et al., "CycloteneTM based Low Temperature Wafer to Wafer Bonding for Advanced Wafer Level Packaging Solutions", Wafer Bond 2013, December 5-6, 2013, Stockholm, Sweden, pp.107-108

7. M. Wilke et al., "Prospects and Limits in Wafer-Level-Packaging of Image Sensors," in ECTC, 2011, vol. 49, no. 0, pp. 1901–1907

8. K. Zoschke et al., "Silicon Interposers with TSVs - A basis for Wafer Level 3D System Integration", Smart Systems Integration 2013, 13-14 March 2013, ISBN-978-3-8007-3490-0

9. K. Zoschke et al., "Polyimide based Temporary Wafer Bonding Technology for High Temperature Compliant TSV Backside Processing and Thin Device Handling", 62nd Electronic Components and Technology Conference, May 29 – June 1, 2012, San Diego, California, USA, pp. 1054-1061

The Novel Assembly Method of a Field Deployable Biosensor Unit

P. Xu, F. M. Guo, X. Y. Liu, J. H. Shen, L. Ding, W. Wang, Y. Q. Li, Y. P. Ge, S. H. Zhang, M. J. Wang, H. Z. Zheng
Laboratory of Polar Materials & Devices, School of Information Science Technology, East China Normal University
J. T. Ye, L. Luo
Shanghai Institute of Microsystem and Information Technology, Chinese Academy of Sciences
No.500, Dongchuan Road, Shanghai
fmguo@ee.ecnu.edu.cn, 86-21-54345169

Abstract

In this paper we report the novel assembly method of a biosensor unit based on 64 pixel photodetector array with a proprietary semiconductor quantum dots (QDs) quantum well (QW) hybrid structure. There is significantly lower dark current for optimized high sensitivity detector. The capacitive trans-impedance amplifier (CTIA)-correlated double sampling (CDS) readout circuit can readout weak optical signal which respond by photodetector. To minimize the packaging size of our biosensor unit, a double-sided printed circuit board designed to decrease noise by special package making photodetector array can operate at room temperature and lower temperature for get better performance. The readout circuit is bonding on the backside of substrate through silicon via (TSV) structure, and the photodetector array bonding on the front side of substrate for making the optic signal through to photosensitive window of detector. A highly integrated Cortex-M4 MCU (STM32F407) has build the data acquisition and analysis unit providing Wi-Fi interface to communicate with the PC software for biosensor unit rapid diagnosis infectious disease and more easy carry.

Introduction

With the rapid progress of microelectronics and biological technology, the biological detection technology has rapidly developed and gradually applies in the air quality monitoring, water treatment and food safety field and so on. Other types of biosensors include fluorescence characterization and Raman spectral identification have been used point-of-care disease diagnosis. In the case of infectious disease diagnosis, the signal measured is often extremely weak, but rapid diagnosis is especially important to prevent the diseases from spreading. However, the critical amount of pathogen that can cause illness is far less than the minimal amount required for a reliable diagnosis. In many cases, the concentration difference could be in 4 to 5 orders of magnitude [1-5].

In this paper we show the design and assembly of a field deployable biosensor unit based on a photodetector array fabricated from a proprietary design of low dimensional quantum effects. Due to its high sensitivity, low power consumption, fast response speed, adjustable range of spectral response and peculiar property, the photodetector of quantum dots QDs- quantum well QW hybrid structure has been used in a wide range of applications such as infrared and multicolor detection [6-9]. The key part of biosensor unit, semiconductor QDs detector has been designed carefully to significantly decrease noise such as dark current and thus enable the biosensor to operate at high detection sensitivity [10]. Its responsivity is comparable with commercially avalanche photodiodes (APDs) when operated at no-Geiger mode. However, the advantage of our detector is that it may operate at around -1 volt, while APDs are normally operated at over -100 volts [11, 12]. This puts our detector at a hugely advantageous position in portable applications. In addition, we have designed specially a variable gain readout circuit, a specially package for 64 pixel photodetector array and its signal readout, and a highly integrated Cortex-M4 MCU (STM32F407) to build the data acquisition and analysis unit, providing Wi-Fi interface to communicate with the PC software. It can complete the tasks like data acquisition, digital filtering, spectral display, network communication, human-computer interaction etc.

Photodetector structure and properties

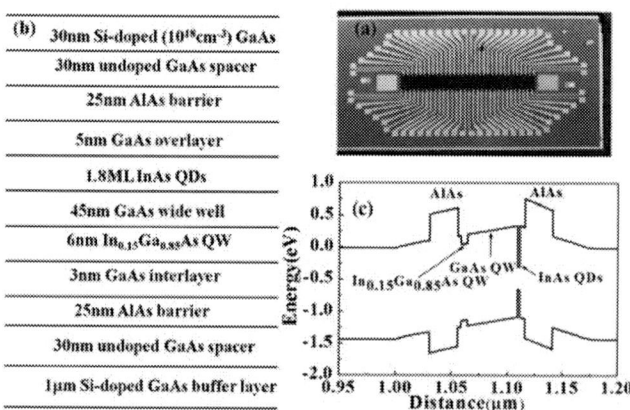

Figure 1. (a) The 64 pixel photodetector array photo, (b) A schematic diagram of the device structure, (c) The band diagram of the device simulated

Detection sensitivity to bio-related optical signals of our sensor mainly depends on a semiconductor photodetector array which special designed. Fig. 1 (a) is the 64 pixel photodetector array photo. The array was fabricated from a novel proprietary semiconductor QDs-QW hybrid structure [6, 8]. The deposition sequence is generally as follows: a 1μm thick Si-doped ($10^{18}cm^{-3}$) GaAs buffer layer at the bottom, an undoped 30nm GaAs spacer, a first 25nm AlAs barrier, a 3nm GaAs interlayer, a 6nm $In_{0.15}Ga_{0.85}As$ QW, a 45nm GaAs wide well, a 1.8ML self-assembled InAs QDs layer with a 5nm GaAs overlayer, a second 25nm AlAs barrier. Another 30nm undoped GaAs spacer and finally a 30nm Si-doped ($10^{18}cm^{-3}$) GaAs as the capping layer. Ohmic contacts were made both on the top and at the bottom of the device. For each photodetector in the array, a rectangular aperture

(50×500) μm^2 was left on the top surface for optical access. A schematic diagram of the device structure is shown in Figure 1 (b).

Figure 1 (c) is the band diagram of the device simulated using Crosslight Apsys software at thermal equilibrium. The double AlAs barriers make it difficult for electrons generated during thermal fluctuations to tunnel through. Whilst photoexcited electrons and holes generated in the wide GaAs quantum well are dissociated quickly, and captured in the narrow $In_{0.15}Ga_{0.85}As$ QW and the InAs QDs respectively. When bias voltage applied, the average electric field in 120nm active region of the device is about 10^5V/cm. Because of the mobilities of electron and hole are 8000 $cm^2/V \cdot s$ and 400 $cm^2/V \cdot s$, their transit times are about 1×10^{-14}s and 1×10^{-13}s respectively, which are on a much shorter time scale than the electron-hole recombination time of about 1×10^{-9}s. Thus the carriers generated in the GaAs wide well shall be swept to the two opposite sides of the GaAs without being recombined [6]. This process changes the device built-in voltage profile and affects tunneling mechanism through the barriers in the GaAs-AlAs-GaAs structure, which leads to an increment of the photocurrent [13, 14].

Fig. 2 shows we measured the photoelectric test platform diagram. A HeNe laser of 633nm wavelength was used for illumination. The laser beam was focused onto the detector array in insert graph, and the focal spot size was 50 µm in diameter. To obtain a better result from the photodetector, the sample was mounted on the cold finger of a Dewar and cooled down by liquid nitrogen and tested at low temperature.

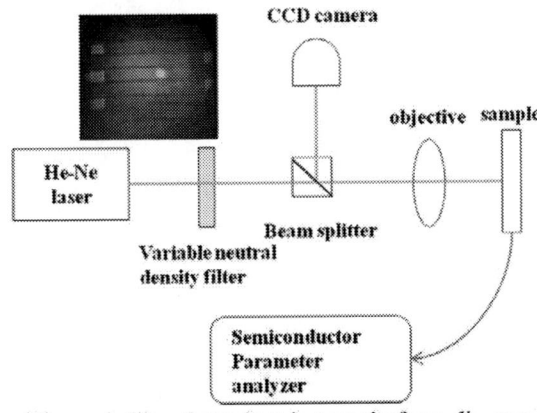

Figure 2. The photoelectric test platform diagram, insert picture is the focal spot onto the detector

Fig. 3 plotted respectively the I-V characteristics for one detector in the array under light illumination power ranging from 1 milliwatt (mW) to 10 pecowatt (pW). The difference between the maximum and minimum illumination light power can reach 10^8 orders of magnitude, the detector maintained good signal-to-noise ratio at the lower end but showed no sign of saturation at the higher end. This demonstrated its wide dynamic range. The measurement was done using Keithley 4200-SCS semiconductor parameter analyzer. The calculated current responsivity at light illumination power of 10pw at 633nm was about 0.76A/W. The inset plot in (b) is the dependence of detector current on the illumination power at a fixed bias voltage of -1.5V. The average response rate is

about 2.62 A/W at 760nm wavelength illumination and average quantum efficiency can reach 427%, which shows distinct quantum mechanism of multiplication.

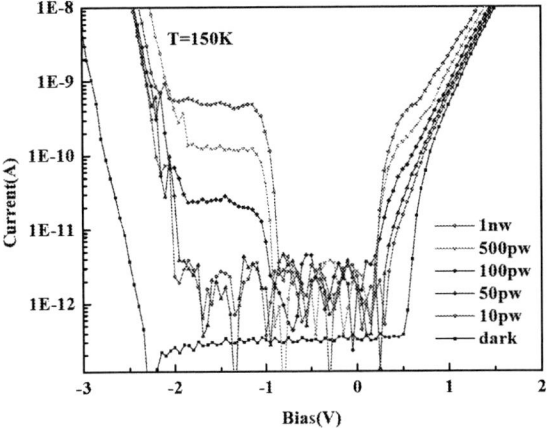

(a) The light illumination from 1 mW to 10 pW

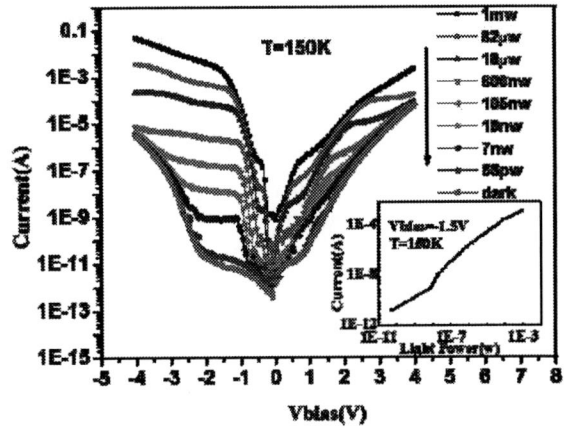

(b) The light illumination from 1 mW to 50 pW

Figure 3. I-V characteristics for one detector in the array

Readout design

To convert the photoelectric current generated in the photodetector array into voltage signals, we designed a readout circuit based on Capacitive Trans-Impedance Amplifier (CTIA) and Correlated Double Sampling (CDS) structures as shown in Fig.4. The readout circuit consists of row select switch matrix, column amplifier, correlated double sampling (CDS) circuit, column select multiplexer and output driver. Features of CTIA structure such as lower noise, more stable voltage offset, and higher injection efficiency [15], it make the circuit a promising candidate to process the weak photocurrent generated in the photodetector array. The purpose of the CDS structure [16] is to decrease the noise induced by the circuit, thus further improving the ability of readout circuit to read out weak signal. To photoelectric response signal of an individual photodetector, the CTIA structure first converted the photoelectric current into voltage signal by charging the corresponding built-in capacitor over a selected integration time period. One branch of the CDS structure was set to obtain the voltage signal at the start of the integration; and another branch at the end of the integration. This voltage pair was then simultaneously sent to an output buffer for subsequent data progressing, including subtraction and AD conversion. For the entire photodetector array, the readout circuit takes photoelectric current signal from the detector array in parallel and outputs the voltage pairs in serial. This was controlled by digital switches and shift registers in the readout circuit and an external clock signal provided in the data processing unit.

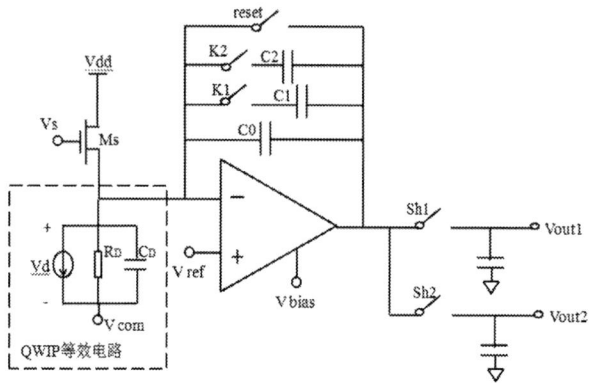

Figure 4. The readout structure

Because specially designed QDs-in-QW detector has high sensitivity and wide dynamic characteristic, the gain adjustable readout circuit structure was particularly designed base on the CTIA structure by two switches K1 and K2 to controlling two capacitors respectively. It may forms three integration capacitors readout mode. When testing the weak light, the smaller capacitor 2pF was employed to obtain higher sensitivity. When testing the nature light, the bigger capacitor can be employed respectively to increasing the dynamic range and difficult readout saturation.

The assembly for photodetector and readout circuit

The layout design, simulation and process of the gain adjustable readout circuit chip has been implemented using CMOS 0.35μm mixed integrated technology step by step. The photodetector array or area array connect with readout circuit chip has direct package and indirect hybrid package mode. Direct package mode need strict solder joint area, chip size matched, and higher technology requirements. We choose more flexible indirect hybrid package mode. The current hybrid packaging is mostly planar package form which detector array is connected with readout circuit chip by wire lead with the some side. To minimize the package size of our biosensor unit, we designed specially a double hybrid packaging substrates to put the readout circuit chip on the substrate backside and the photodetector array on the front side. Making 64 input ends and 16 control ends of the 64 pixel photodetection array through hole processed on PCB substrate connect with readout circuit chip at substrate backside. This configuration enables optic access to photosensitive window of photodetector array.

Considering biosensor unit often work at room temperature, the ultra-thin double-sided printed circuit board (PCB) is chosen to design package lead circuit for low manufacture cost. The PCB substrate is width 22mm and length 34.4 mm, and thickness only 0.4mm. The crosstalk between transmission lines can be optimized to reduce parasitic parameters (capacitance, inductance, resistance) by ADS software. Fig. 5 shows the front side of PCB substrate and the ADS simulation model for transmission lines. There are 28 leads jack divided into both sides for tube base package further. The simulations reveal that increase wire lead width and line spacing could reduce crosstalk. If signal line made in PCB layer, reduced signal and power supply layer, dielectric layer more thickness, used terminating technology and so on, they all can reduce the crosstalk.

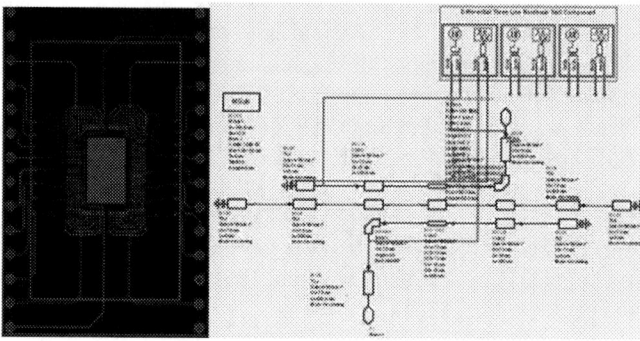

Figure 5. The ADS simulation model for transmission lines on the front side substrate

(a) The readout circuit chip packaged on PCB

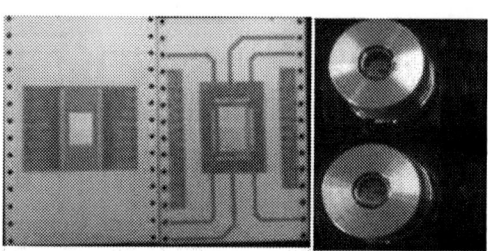

(b) Ceramic substrates and photodetector array packaged in Dewar

Figure 6. The photo for readout circuit chip and photodetector array packaged

In Fig. 6 (a), the photo on the left is a microscopic image about the PCB substrate backside which bonded readout circuit chip. The photo on the right is after welding the readout circuit chip which surface covered with a layer of transparent protective film. The readout circuit chip is fixed between two white supports which making ease of bonding photodetector array on front side substrate. To embody the high sensitivity of the QDs-QW detector at low temperature, the ceramic substrates with the good thermal conductivity has been chosen double-sided design too. Fig. 6 (b) shows the photos of the processed ceramic substrates and packaged 64 pixels photodetector array in Dewar. Now, the white supports in Fig. 6 （a）should be changed ceramic material too.

The advanced package form

We choose still silicon substrate with the some good thermal conductivity and mature large scale integrated circuit technology. Especially, the TSV (Through Silicon Via) is an enabling technology of three-dimensional integration with high density on system in package (SIP).

Toolbox of MATLAB was used to solve partial differential equation (PDE) on the heat conduction problem for getting the simulation results of stress field about dislocation deformation. Fig. 7 is the temperature field and deformation tendency of the TSV in the wet process and the heat conduction vector simulation results from high temperature to low temperature [17]. The vector represents the direction of decreasing temperature. Using MATLAB can precision grid division through the boundary condition of different model structure, solving the heat equation to obtain thermal distribution diagram, further to obtain the thermal expansion deformation trend.

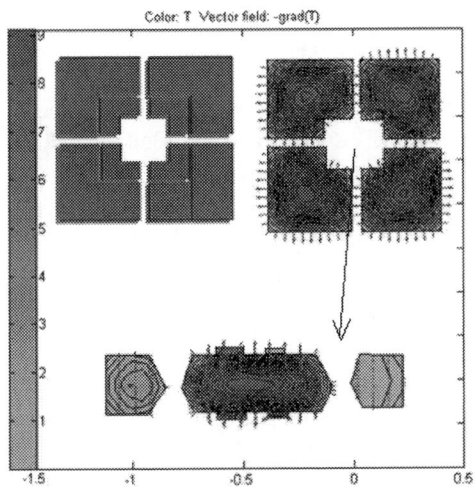

Figure 7. The heat conduction vector simulation from high temperature to low temperature in TSV

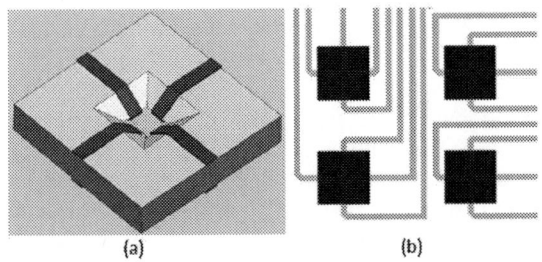

Figure 8. An TSV with four leading wire

We specially designed the double etch and groove process TSV technology. Every TSV process with four leading wire at different directions as shown in Fig. 8. One is three-dimensional graph, and another is plane graph. The lead pattern on silicon substrate has been specially designed. Making the area of silicon printed circuit board with one TSV - four leading wire is near half than the PCB substrate, width 17.08 mm and length 25.08mm. The TSV structure was fabricated with double-sided wet chemical etching process. Although the technology of wet chemical etching process is simple, the size of TSV is proportional to the thickness of silicon. So, via-in-groove technique has been introduced.

Result and development discussion

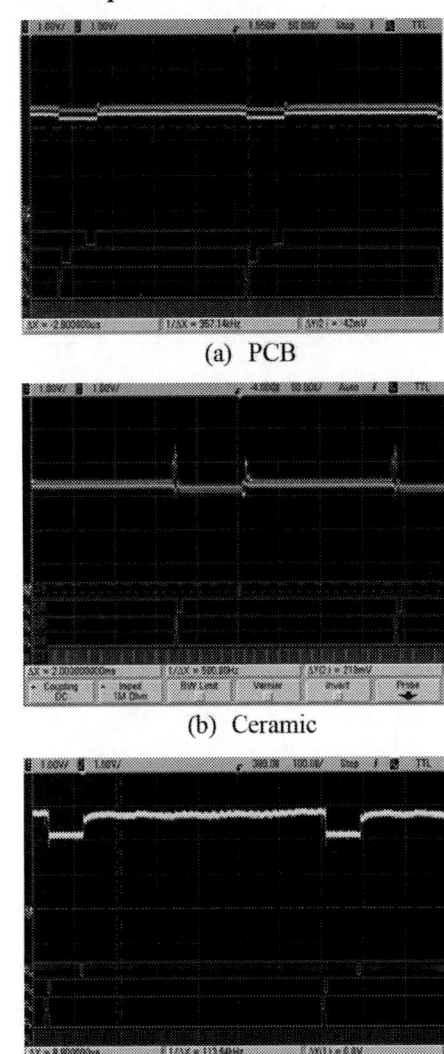

(a) PCB

(b) Ceramic

(c) Silicon substrate with one TSV – four leading wire at low temperature

Figure 9. Three substrates test result

Specially designed gain adjustable readout circuit chip has been bonded respectively on silicon substrate with one TSV – four leading wire, ceramics substrate with laser drilling, and PCB substrate backside. Fig. 9 shows their output test result using Agilent MSO6052A. The ceramic substrate has some crosstalk problems because some process problem, the optimization design of the printed circuit makes the noise and crosstalk of the PCB and silicon substrate very small. Further, the 64 pixels photodetector array was bonded on the substrate front side.

The spectral response of the 64 pixels photodetector array can been measured by response spectrum measurement

978-1-4799-2408-0/14 $31.00 © 2014 IEEE   1215

platform as shown in Fig. 10. To obtain a better result from the photodetector, the experiment was conducted at liquid nitrogen temperature. The photodetector array can read out a good response in the spectral range from 475nm up to over 840nm, with peak value around 650nm as shown in Fig 11. This spectral range can be useful for applications such as fluorescence labeling, monitoring water pollution, the growth of crops and so on. When the photodetector array illuminated from 10pw to 100pw, we can see the response voltage was 4mV at laser light power 10pw and the voltage responsivity reached about 4×10^8V/W. This weak light detection ability of our photodetector array makes advantageous choice for applications such as extremely low concentration of bio-sample or the signal amplitude from the bio-samples is inherently low.

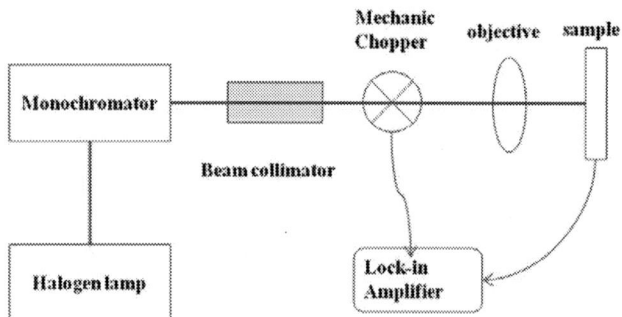

Figure 10. Response spectrum measurement platform diagram

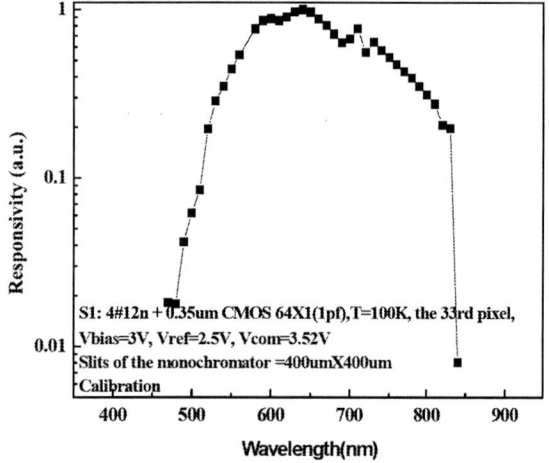

Figure 11. Response spectrum of the QDs- QW photodetector

Having established how the 64 pixels photodetector array responds to weak light illumination and its high sensitivity, we are now ready to assemble the field-deployable biosensor unit and discuss the result. When the laser light is shone on the photodetector array, a change in the photoelectric voltage immediately shows on the LCD screen as shown in Fig. 12. The signal processing unit computes the result and displays the on the LCD screen. The Micro Controller Unit (MCU) and processing unit layered configuration has decreased the size to 9.6cm×6.3cm×3.5cm [18].

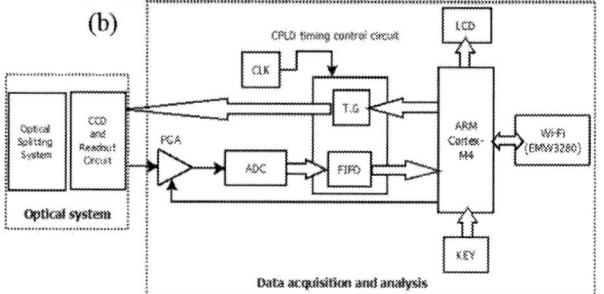

Figure 12. The structure diagram of the data acquisition and analysis unit

The data acquisition and analysis unit can be divided into three parts: CPLD timing control circuit, data acquisition and digital filtering, spectral display and data analysis. We use UCGUI of small space, good portability and high efficiency for interface development. It encapsulates complex drawing functions, provides API to user, and solving the most problems in drawing. Human-computer interaction is realized by the touch screen.

The driver sends the data information to the application when receives the coordinates, then the application will handle the data as needed. The graphical interfaces have four parts: configuring, bar chart, line chart and data chart. By configuring the start and stop value in the interface of configure, we can see corresponding graph in bar chart and line chart, also we can read the data from data chart directly as shown in Fig. 12 (a).

The filtered data is the response spectrum intensity for 64 pixels, so the data without processing which distribution in different wavelengths. The abscissa axis map to the position order for pixel, while the vertical axis stand for the signal amplitude. Adjust the angle of the grating, so that the central wavelength would fall on the central region of the pixels. We use the light whose wavelength is known for calibrating, and then we can get the wavelength-intensity relationship graph.

The miniaturized biosensor unit consists of two parts: the optical system and the data acquisition and analysis unit. The optical system is composed of SMA-entrance connector, collimating mirror, grating and focussing mirror. If applied MOEMS process technology, the optical system would be miniaturized further.

To obtain a better result from the photodetector unit, the biosensor unit should be conducted at liquid nitrogen temperature. The dismountable minidewar would be designed for matching specially a minipump.

Conclusions

A field deployable biosensor unit have been designed and developed by novel assembly for applicating signal detection and analysis. We have analyzed in detail and demonstrated experimentally the sensitivity of the particular QDs –QW hybrid structure photodetector for weak light power density as low as 10pw. The gain adjustable readout circuit was specially designed on the basic of the CTIA structure by two switches K1 and K2 to controlling two capacitors respectively which forms three integration capacitors readout. A double hybrid package mode has been specially designed to put the readout circuit on the substrates backside and the photodetector array on the front side. The 64 input ends and 16 control ends of the 64 pixel photodetection array through hole on PCB substrate are connected with readout circuit chip at backside. We specially designed silicon substrate with the double etch and groove process TSV technology too. Every TSV has processed with four leading wire at different four directions. The photodetector array has read out a good response in the spectral range from 475nm up to over 840nm, with peak value around 650nm. The response voltage for the photodetector array illuminated at laser light power 10pw was 4mV and the voltage responsivity has reached about 4×10^8V/W. By reducing the size of data processing unit, the biosensor unit can be truly portable. In the next step, we plan to apply the sensor in various biological samples and demonstrate its advantage in biological monitor. Furthermore, the photodetector array could also be tailored to the specific spectral windows of various bacteria and pathogens for Raman or fluorescence detection.

Acknowledgments

National Scientific Research Plan (2006CB932802, 2011CB932903) and State Scientific and Technological Commission of Shanghai (No. 078014194, 118014546) supported this work.

References

1. S. Li, Q. Yuan, B. Morshed, C. Ke, J. Wu and H. Jiang, "Dielectrophoretic Response of DNA and Fluorophore in PhysiologicalSolution by Impedimetric Characterization," *Biosensors &Bioelectronics*, Vol. 41, pp. 649–655, 2013.

2. Q. Yuan and J. Wu, "Thermally Biased AC Electrokinetic Pumping Effect for Lab-on-a-chip Based Delivery of Biofluids," *Biomedical Microdevices*, Vol. 15, pp. 125-133, 2013.

3. J. S. M. Peiris, W. C. Yu, C. W. Leung, C. Y. Cheung, W. F. Ng, J. M. Nicholls, T. K. Ng, K. H. Chan, S. T. Lai, W. L. Lim, K. Y. Yuen and Y. Guan, "Re-emergence of fatal human influenza A subtype H5N1 disease," *Lancet*, 2004, 363, (9409), pp.617-619.

4. M. Lian and J. Wu, "Ultra Fast Micropumping by Biased AC Electrokinetics," *Appl. Phys. Lett.*, 94, 064101.

5. L.M. Bellan, D. Wu, and R. S. Langer, "Current trends in nanobiosensor technology," Wiley Interdisciplinary Reviews: *Nanomedicine and Nanobiotechnology*, 2011, 3, (3), pp.229-246.

6. B. Hu, X. Zhou, Y. Tang, H. D. Gan, H. Zhu, G. R. Li and H. Z. Zheng, "Photocurrent response in a double barrier structure with quantum dots-quantum well inserted in central well," *Physica E: Low-dimensional Systems and Nanostructures*, 2006, 33, (2), pp.355-358.

7. GUO Fangmin, LI Ning, LU Wei, et al., "The theory and experiment of very-longwavelength 256×1 GaAs/Al$_x$ Ga$_{1-x}$ As quantum well infrared detector linear arrays," *China Series G-Physics Mechanics and Astronomy* 51 (7) (2008) 1－2.

8. S.B. Bian, Y. Tang, G.R. Li, Y. X. Li, F. H. Yang, H. Z. Zheng and Y. P. Zeng, "Photon-Storage in Optical Memory Cells Based on a Semiconductor Quantum Dot-Quantum Well Hybrid Structure," *Chinese Physics Letters*, 2003, 20, (8), pp.1362-1365.

9. Ning Li, Fangming Guo, W. Lu, Y. Fu, et al., "Detection wavelengths and photocurrents of very long wavelength quantum-well infrared photodetectors", *Infrared Physics & Technology* 47 (2005) 29－36.

10. Lin Ding, Yan-Qui Li, Fang-Min Guo, et al., "Weak light characteristics of potential biosensor unit", *Micro & Nano Letters*, (2013), pp. 1–4.

11. D. Renker, "Properties of avalanche photodiodes for applications in highenergy physics, astrophysics and medical imaging," *Nuclear Instruments and Methods in Physics Research A*, 2002, 486, pp.164-169.

12. A. Dorokhov, A. Glauser, Y. Musienko, C. Regenfus, S. Reucroft and J. Swain, "Study of the Hamamatsu avalanche photodiode at liquid nitrogen temperatures," *Nuclear Instruments and Methods in Physics Research A*, 2003, 504, pp.58-61.

13. D. Landheer, H. C. Liu, M. Buchanan, and R. Stoner, "Tunneling through AlAs barriers: Γ–X transfer current," *Applied Physics Letters*, 1989, 54, pp.1784-1786.

14. S.C. Shen, "Comparison and competition between MCT and QW structure material for use in IR detectors", *Microelectronics Journal* 25 (8) (1994) 713－739.

15. G. Z. Zhan, F.M.Guo, W. Lei, J. Huang, Z. Q. Zhu, and J. H. Chu, "A Specifc Architectures of CMOS Readout for Resonant-cavity-enhanced Devices," *Piers Proceedings, Hangzhou, China*, 2008, pp.912-915.

16. X. Chen, "Theoretical analysis and experimental application of CDS CMOS integrated circuit for uncooled infrared focal plane arrays," *Optik - International Journal for Light and Electron Optics*, 2011, 122, (9), pp.792-795.

17. Xiao Chen, Jiajie Tang, Gaowei Xu, Le Luo, "Process development of a novel wafer level packaging with TSV applied in high-frequency range transmission", [J]. *Microsyst Technol* (2013) 19:483–491.

18. L. Ding, M. J. Wang, Y. Q. Li, X. Y. Liu, J. H. Shen, F. M. Guo, " Weak Light Characteristics of a New Photoelectric Sensor with Potential Biosensor Application", *NEMS2013 Proceedings, Suzhou, China*, 2013, pp. 1261-1264.

SIMEIT-Project: High Precision Inertial Sensor Integration on a Modular 3D-Interposer Platform

Wolfram Steller [1], Christoph Meinecke [2], Knut Gottfried [3], Gregor Woldt [4],
Wolfgang Günther [5], M. Juergen Wolf [1], and K. Dieter Lang [1]

(1) Fraunhofer IZM, ASSID, HDI & WLP, Ringstrasse 12, 01468 Moritzburg, Germany
(2) Technische Universität Chemnitz, ZfM, Reichenhainer Str. 70, 09126 Chemnitz, Germany
(3) Fraunhofer ENAS, Technologie-Campus 3, 09126 Chemnitz, Germany
(4) Microelectronic Packaging Dresden GmbH, Grenzstraße 22, 01109 Dresden, Germany
(5) GEMAC, Zwickauer Straße 227, 09116 Chemnitz
Wolfram.Steller@assid.izm.fraunhofer.de

Abstract

The applications of inertial sensors have a wide variety in terms of accuracy and costs. A new technology approach is joining higher sensor accuracy and lower production costs by using a new Interposer / sensor interconnect technology applied on 300mm wafer diameter without changing the sensor element itself.

The **higher accuracy** is mainly covered by a multiple point program: (1) stress less assembly due interface silicon Interposer to silicon MEMS; (2) better Signal to Noise Ratio (SNR) by polymer redistribution layer on the interposer (due to better wiring geometry and less parasitic capacities / inductivities); (3) reduction of mechanical stress by using flexible bar springs for mechanical decoupling of sensor and Interposer substrate; (4) additional stress reduction by using a polymer layer for mechanical decoupling of metal redistribution layer (RDL) and Interposer substrate.

The **cost efficiency** even in small scale serial production based on: (1) 300mm multi project wafer technology including warehouse ready system packaging; (2) a new MEMS contact technology, which gives technical benefit, smaller dimensions and simplifies the assembly of MEMS and ASIC (which are placed on a 2.5D-Interposer in order to enable a System In Package (SiP) as well as for higher sensor accuracy); (3) the flexible ASIC feature enables the integration of different MEMS with analogue signal output; (4) minor costs for integration of different sensors into the existing package.

The heterogeneous 3D integration is a key enabler and justifies the additional process steps (mainly TSV-processing, thin wafer handling) by implementing the advantages of the polymer RDL. This integration approach results leads to improved mechanical and electrical properties.

This paper will give an overview about the current achievements in the SIMEIT-project, which are predestined to improve the accuracy of different MEMS-applications with analogue signal transfer to the ASIC as well as MEMS-applications with need of stress less integration.

Introduction

The project name **SIMEIT** stands for "Stress less Integration of high precision MEMS- and Electronic components via new Interposer Technology". Background of the project is the application of the **2.5D-Interposer** approach including TSV-technology for the realization of a SiP which is fully integrated for assembly on PCB or substrate. **Main**

driving factor is the separation of process steps which are requiring for high accuracy (needed to provide the improved performance) and of assembly steps on PCB level (with wide assembly tolerances, dependent of the respective tool configuration of the final component user (e.g. at device manufacturers, OEMs, OSATs).

The concept of a 2.5D-Interposer platform includes a vertical wiring by using of Through Silicon Vias (TSV). This feature allows in a future situation of More-than-Moore requirements an **easy vertical extension** from the 2.5D to the 3D stacking approach, e.g. for the integration of additional functionality in the same SiP. For that realization just the integration of additional die stack is required without the need of additional footprint.

The developments in the project were done on a **real application** of an inertial low-g sensor system. The **initial integration concept** was the placement of MEMS, ASIC on a ceramic substrate, the electrical linkage (between MEMS, ASIC and socket) via wire bonds and the package was encapsulated (Fig. 1). In generally the system worked with a good performance but has some obviously limitations in mechanical stress, assembly technology, footprint, and system height.

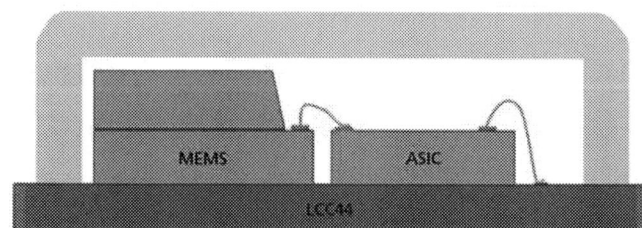

Figure 1. Starting point for the project: initial architecture of the intertial sensor system, using wire bond technology as electrical interconnect; the structured cap wafer of the MEMS covers the sensor element only, the MEMS pads are exposed.

Challenge of the SIMEIT project is to overcome these mentioned limitations by using the 3D-integration approach which will provide some functional improvements as well as a volume compatible assembly technology. Actually the simplest description of the assembly background is the changeover from the wire bond technology to the flip chip technology. This causes explicit changings of the component

978-1-4799-2408-0/14 $31.00 © 2014 IEEE   1218   2014 Electronic Components & Technology Conference

requirements and interfaces (initial state / new requirements) as follow:

(1) limited MEMS height (sensor thickness & hermetic lid)
o   initial state: height of MEMS and hermetic lid of ~700µm
o   new SiP requirement: thin stack including Interposer, micro bumps, MEMS and housing

(2) MEMS contact pad concept
o   initial state: lid covers the MEMS partially and the pads are located on MEMS top surface level (Fig. 1)
o   new flip chip assembly requirement: micro bumps on a plane chip surface

(3) MEMS/ASIC wiring concept
o   initial state: wire bond and encapsulation or integration into hermetically sealed ceramic carrier
o   new requirement: bumped components (including Under Bump Metallization (UBM), wiring via RDL and MEMS rewiring from chip top to back side without increasing of intrinsic stress

(4) MEMS/ASIC performance
o   initial state: MEMS bonded on carrier and connected via wire bond (causes intrinsic stress), final assembly of single components at device manufacturers, OEMs, OSATs
o   new requirement: MEMS mechanically decoupled from substrate, wiring via RDL and copper TSV's on Interposer, final product is a fully integrated SiP

The **assessment of the described 3D-integration approaches** reveals further improvements (additional values) concerning the system performance and cost: (1) improved sensor performance due to the use of RDL (no parasitic capacities between the flexible wires used for the analog sensor signal transmission); (2) less mechanical stress to the MEMS element due to exchange of organic substrate (strong CTE-mismatch) by silicon Interposer and mechanical decoupling via free standing silicon pillars; (3) volume compatible process due replacement of wire bonding by wafer level processing and flip chip assembly; (4) better signal to noise ratio due to reduced wire material (use of small RDL wiring instead of bond wires i.e. less resistance). Furthermore the SIMEIT project allows the adaption of the MEMS-element and the ASIC-circuits. This allows to prepare a system platform, applicable for the described inertial low-g MEMS as well as for different MEMS applications.

System Integration Concept

The initial decision regarding the realization was concerned about the **system integration concept**. One possibility would be the using of the **VIA last approach** applied to the ASIC. The ASIC would be an active 2.5D-Interposer and the MEMS would be assembled on the ASIC top side. No Interposer means fewer components which is good regarding costs. On the other hand main drawbacks are the required matching of ASIC and MEMS footprint (MEMS need to be smaller than ASIC and additional footprint needs for the housing). On the other hand the additional VIA formation into the ASIC requires free CMOS area and the thin wafer processing may diminish the ASIC yield as well.

Despite this inflexibility the approach is very suitable for a high production volume with a tight parameter alignment of components.

The **selected integration concept** based on the preparation of a separate **2.5D-Interposer with the VIA first approach**. The main benefit is the preparation of an **Interposer platform** by integrating of electrical wirings (e.g. TSV's) more than needed in the current application. This allows the adaption of the 2.5D-Interposer to different components just by redesigning of the Interposer top side processing. Less changings means lower costs. The achieved yield remains stable because setup and fine tuning of the critical processes of the 3D-Integration remains untouched (e.g. TSV-formation, thin wafer processing) even in case of a different MEMS application. Furthermore there are no changes necessary on active components itself even in case of any feature improvement (e.g. regarding single processes or integration of additional parts). But the consequence of the selected integration approach is the need of a flip chip assembly compatibility of the sensor chip.

Interposer Platform

The design rules and specifications (for the selected 2.5D-Interposer approach) were defined in the awareness to establish a **stress less component assembly** and an **open Interposer platform** for the focus application of the project (a low-g inertial sensor) as well as for different sensor systems requirements.

The **stress less assembly** is mainly driven by spatial separation of the silicon devices and the PCB (with big differences in the CTE between both). Thereby the silicon Interposer acts as an additional silicon stack between them and provides the same CTE as by the silicon components. Therefore the components and the PCB are decoupled and stress input from PCB is minimized. As well the lower stress (respective the lower CTE) of the silicon Interposer allows a much finer wiring layout as feasible on the PCB.

This starting point of the Interposer definition was the most critical **process interaction between TSV geometry and thin wafer handling** in order to achieve equilibrium of costs, risks and allowing of further technology developments. The material of the TSV fill is fixed to copper regarding the in house applied TSV technology. In general the TSV geometry itself is limited by the aspect ratio of diameter and depth, mainly because the best sidewall coverage & conformity of the barrier/seed liner deposition in standard processes is about 1:10 to 1:15. Therefore in case of 5µm TSV-diameter the TSV-depth is about 50µm, which means the Interposer thickness is 50µm too. Actually it is possible to handle this thin wafers but the mechanical stability of the SiP is weaker. This might limit the sensor performance (under rough environment conditions) and besides the subsequent assembly of a system housing became more complicated regarding the lower mechanically stability.

One the other hand a bigger TSV diameter allows a flexible wafer thickness if desired (as long as the ratio TSV-diameter / wafer thickness is lower than 1:10). But the bottleneck is the much longer electrochemical metal deposition process time, for a complete filled TSV and the bigger mechanical stress due the recrystallization (with volume increase) as well as the different coefficient of thermal expansion (CTE) of silicon and copper [1]. This means increasing TSV-diameters are correlated to increasing stress. These problems may bypass by a TSV metal liner deposition only but the remaining hole is a new obstacle with the need of a TSV plugging process regarding the feasibility of the following spin on processes. Available plugging processes are still not standard, especially regarding thermal budgets or quality requirements of following processes (mainly vacuum ones).

Considering that limitations a **TSV-diameter of 10µm** with a respective **final wafer thickness of 100µm** were selected and realized as **VIA first approach**. As already described this avoids the footing phenomena (TSV side wall under etching during pad opening at TSV-bottom) as well as the process influence of different thermal properties of slight variations in the layer uniformity of the temporary bond adhesive.

Interposer process flow (Fig. 2)

The first main process on the silicon wafer is the **TSV-etch**. It was carried out in an inductive coupled plasma (ICP) source reactor using a modified Bosch process (with <1 second cycles of DRIE and side wall passivation to prepare smooth side walls shape; [2]. Thereafter the TSV-depth are measured (using an interferometric SRM-Camtek sensor) which is required for the later TSV backside reveal. For the TSV **isolation liner deposition** the SA-CVD was used (sub-atmospheric chemical vapor deposition) because the side effect of that process is a certain smoothing of the TSV side walls profiles [3]. The relative high processing temperature of 350°C is uncritical for the silicon.

The setup of the **barrier/seed liner** material was a stack of Titanium nitride, Titanium and Copper, which is sufficient for the needed electrical isolation of the ASIC's digital signal transmission directed (via TSV's & BGA) to the board. The TSV **copper fill process** uses a bath of electrolytes and certain additives to ensure a void free bottom-up fill mechanism which was monitored via 3D µCT X-ray. An annealing step has to cure the copper crystal structure. This recrystallization step is necessary to drive out the plating additives but comes together with a volume increase and causes mechanical stress. Chemical mechanical polishing (CMP) is the next process step needed for the **removal of the TSV-copper overburden**. Afterwards were formed **redistribution layers (RDL)** for the wiring between TSV's and interconnect layer. Instead of the dual damascene approach for RDL-formation the polymer RDL approach was used. This means instead of CMP planarization steps and thin liner isolation polymer layers were structured and applied via spin on coating. This polymer RDL approach is predestinated for limited wiring complexity and much more cost efficiency due omitting of expensive CMP process steps. The first RDL-layer is placed on a 5µm thick structured polymer layer of an **Inter Level Dielectric** (ILD / structured for an electrical connection to the TSV). This layer enables a certain mechanical decoupling between Interposer and the assembled components as well as it provides a better isolation of the

Figure 2. Regulare Interposer process flow (gray – Silicon; yellow/red – Isolation & Barrier/Seed liner; orange – copper; TSV & RDL & pad/copper pillar; light blue – adhesive; blue - solder).

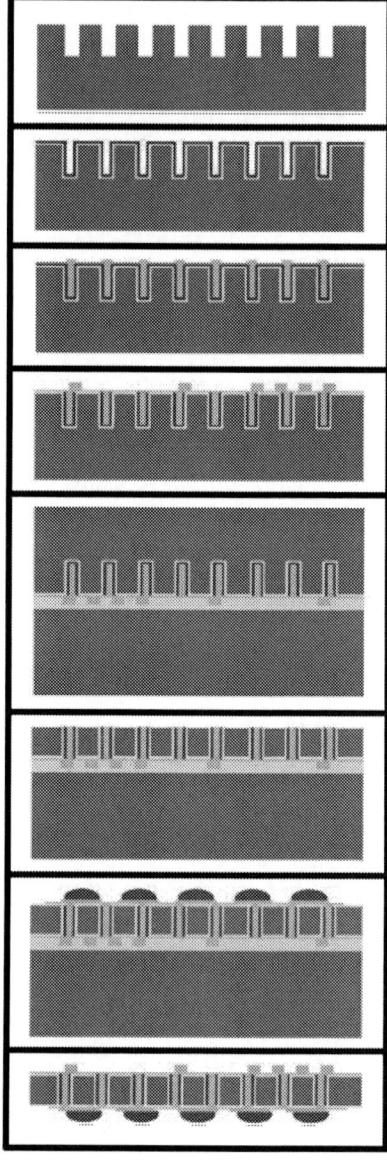

TSV-etch

Isolation and Barrier/Seed liner deposition

TSV copper fill

CMP of copper overburden; RDL and copper pillar (or pad) plating

Temporary wafer bond (device wafer top side down on carrier wafer)

TSV back side reveal by wafer thinning (Grinding & CMP)

Back side RDL and BGA pad (alternatively including solder studs)

Device wafer debond and die singulation

RDL for analog signal transmission between MEMS and ASIC (less parasitic capacitance due increasing the isolation thickness). The 5µm thick RDL-wiring was deposited equivalent to the TSV's filling with Ti/Cu liner as barrier/seed and electrochemical deposition. Final Steps of RDL formation are the barrier/seed liner removal and isolation by using of ILD-layer. The formation of the **interconnection layer** is the next step. Usually solder bumps are placed on the component regarding a reproducible flux wetting of the bumps which is important for the interface reliability. Despite (for the initial demonstrator) an interconnect layer of plated SnAg with Ni/Au cap was prepared on the Interposer because the process development of the current MEMS developments is limited by the under bump metallization. Therefore for the purpose of system evaluation the solder deposition was made partly on the Interposer. Just for the ASIC mounting copper pillar were

978-1-4799-2408-0/14 $31.00 © 2014 IEEE 1220

formed with Ni/Au finish (as passivation). Whereas the copper pillar height is adjustable (regarding the needed component standoff / e.g. for underfill or special assembly). **Temporary bonding** (needed for thin wafer processing) is following the finalization of the top side processing (top side down facing to the adhesive layer). The ZoneBond approach (just the outer wafer perimeter provides strong adhesion, the wafer center is covered by a anti sticky layer) were used because the wafer debonding process (Edge Zone Release and Edge Zone Debond) can be done at room temperature. This is important for a later wafer debonding in presence of solder bumps. The zone bond process itself requires slightly elevated temperatures for reduction of adhesive viscosity in order of an easier embedding of the device wafer topology. Especially the void free and uniform deposition of the adhesive is prerequisite for the uniform TSV back side reveal as well as for the uniform results of the following back side processes (especially for processes with heat removal via chuck or vacuum processes). The initial step of **wafer thinning** is the **grinding**. It stops close before TSV opening (predefined by TSV-depth and wafer TTV measurements) and is followed by the **TSV reveal** via **CMP** process with special slurry which removes copper smears on the silicon wafer as well. The final **electrical TSV back side connecting** is performed simultaneously with the formation of the **backside-RDL and pads**, which are isolated via polymer ILD as well. The separation of the device wafer is realized with the edge zone release and **edge zone debond** (EZR & EZD). Main advantage is the processing at room temperature and the adhesive properties (actually it is a thermoplastic) which enable a clear separation of solder and adhesive without loss of solder material. **Stealth dicing** was used for the Interposer **die singulation**. This results in smoother die edges (less chipping/cracks) and improves the Interposer stability due to less micro defects than with conventional sawing.

MEMS developments

The MEMS (used in this project) is a two axes low-g inertial sensor working as differential capacitor with a respective final size of 3280x5000x400µm. This quite large footprint was selected regarding the physical need of a big seismic mass in order to achieve a high accuracy with high SNR using the AIM-technology.

One of the **main challenges** of the project is the transformation from an already working inertial low-g sensor into an improved MEMS which matches the needs of volume production. The **initial situation of the project** is shown in Fig. 1. The **cap-wafer** bonding is related to the need of a **hermetic sealing** of the sensor element as well as an accessible sensor pad row. Therefore a perforated cap wafer was bonded (using glass frit) on the sensor wafer.

New sensor requirements regarding the MEMS assembly on the 2.5D-Interposer (which is the preferred integration concept) are **micro bumps** and a flip chip compatible die shape. Practically this means the sensor lid has to cover the complete MEMS-bulk, with the consequence of covering the sensor pads too. The **first approach** to prepare related notches in the cap wafer failed. Background was the hermetic wafer bonding using glass frit and the need to form

interconnects structures. Because the glass frit is not allowed to cover the sensor pad areas an air gap remains between bulk and cap wafer. This gap makes the preparation of a consistent metal seed layer via sputtering impossible which is prerequisite for the electrochemical deposition of interconnects. Furthermore the **thinning of the sensor cap** wafer from 300µm thickness to only 50µm is another task in the project in order to enable a SiP as shallow as possible.

An **alternative approach** (regarding the limited access on the cap side of the sensor) is the rewiring through the sensor bulk wafer. The initial idea to use a VIA last approach with copper filled TSV's did not convince. On the one hand the recrystallization of the copper fill is related to stress caused by the volume increase. On the other hand the additional processes as well as the different CTE of copper and silicon will intensify the intrinsic stress [1, 4, 5]. Larger distances and a bigger bulk thickness might be a solution but it is contrary to the project objective which is also directed to minimize the dimensions.

Considering all this restrictions, the variation of the via last process from a copper filled TSV to **free standing silicon pillars** (Fig. 3) allows to **overcome this limitations**. No additional intrinsic stress due materials or layers. Additional the stress absorption of the silicon Interposer will be supported by the integration of the freestanding silicon pillars. They realize the mechanical and electrical connection between MEMS and Interposer and are used as flexible spring bars as well.

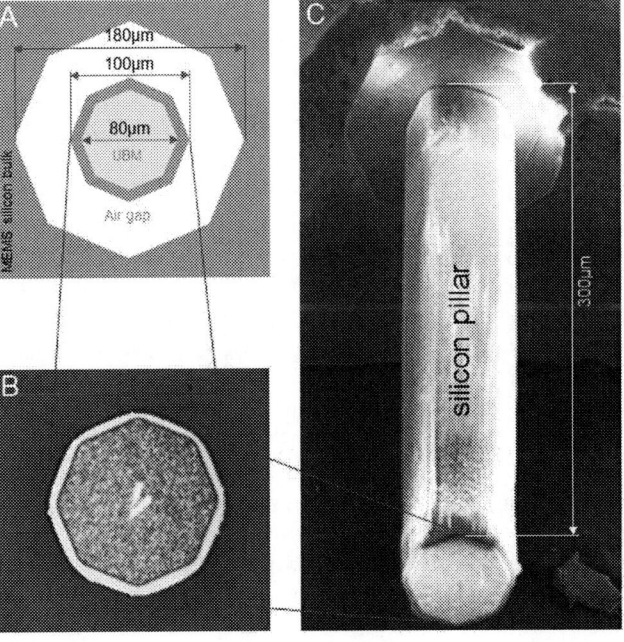

Figure 3. Freestanding flexible silicon pillars of the MEMS. They realize the mechanical and electrical connection between MEMS and Interposer. A: schematic view on si-pillars with UBM (yellow); B: microscope image of si-pillar (with probe marks); C: SEM image of a prearated Si-pillar.

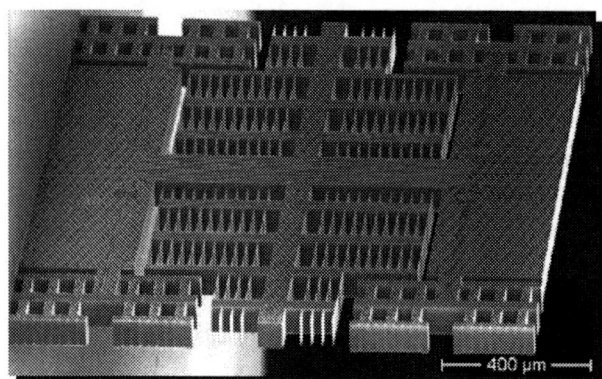

Figure 4. Microstructure of the AIM sensor element (around the AIM element are just about 500μm bulk material used for glass frit (cap wafer bonding) and the freestanding silicon pillars as mechanical & electrical sensor interface)

The **process flow of the sensor realization** (Fig. 5) starts with the preparation of oxide-nitride **isolations** and the sensor aluminum **wirings**. Thereby the isolation has openings towards the silicon for the electrical contact with the later formed freestanding silicon pillars. The **sensor element** is formed as **A**ir gap **I**nsulated **M**icrostructures (AIM) into the silicon bulk material (Fig. 4). The detailed **AIM processing** itself is described in literature [6, 7]. For reliable measurements a hermetic isolation is realized via **cap wafer bonding** with a mesh of glass frit rails around the AIM sensor elements. The cap wafer gets **thinned via grinding** to reduce the system total height. The process is challenging regarding the ambitious final cap thickness. Challenges here are the grinding of the wafer compound consisting of a glass frit bonded cap and sensor wafer down to a cap thickness of 50μm and the vibrations during the grinding process. This can influence the reliability of the AIM sensor respectively the glass frit bond. The **formation of the free standing vertical electrical feedthrough** started with galvanic deposition of the UBM and is followed by the DRIE of the silicon pillars. The etch is landing on the isolation of the aluminum wires on wafer top side which is mechanically strengthen by the glass frit and the sensor lid on top of the wiring. Challenging was the 300μm deep DRIE in straight neighborhood of UBM-topology.

It could be handled with spray coating of a thick resist layer as DRIE mask equally as the later removal of the resist masks. In this context the protection of a **solder bump** on UBM top is a future challenge to face. In the current application the solder is provided by the Interposer but later it can be applied on top of the silicon pillars via using of special galvanic deposition or with special bumping solutions (e.g. Wafer Level Balling as offered by PacTech). Finally the dies become **singulated** for shipment.

Figure 5. Process flow of the integration of silicon pillars into the MEMS (gray – Silicon; light blue – Isolation; yellow – Al-wiring; green – glass frit; red/yellow – UBM; blue - solder).

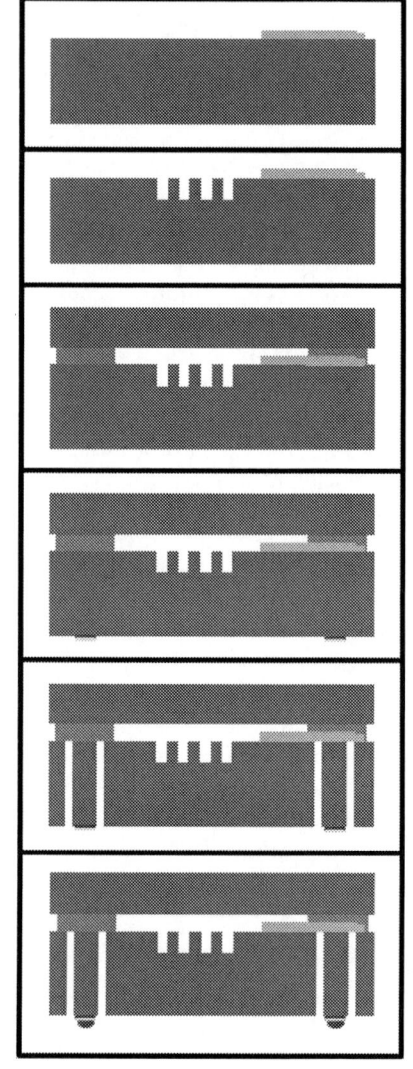

Isolation (Oxide-Nitride) deposition & Al-wiring

AIM formation

Cap wafer bond (using glass frit) and cap wafer thinning

Galvanic deposition of UBM (on position of future silicon pillars)

DRIE of silicon pillars (dummy pillars on the left side)

Solder deposition (use of external bumping solution) & die singulation

Mechanical stress is one of the biggest performance risk for the MEMS measurements. The silicon Interposer absorbs the main stress concerning the different CTE from silicon to the polymer PCB/substrate. **Additional stress compensation** (e.g. influence of different CTE's between copper TSV's and silicon) allows the using of the **freestanding, flexible silicon pillars** (which realize the mechanical and electrical connection between MEMS and Interposer as well. The approach of free standing pillars for all 18 contacts (9 electrical and 9 dummy contacts) were done for an assessment of this promising "mechanical decoupling approach". **Preliminary shear tests** showed that the lowest stability might be located in the glass frit interface but a final assessment of the stability requires the full integration on Interposer including the ASIC. The assessment of the **electrical sensor parameter** requires the same. Preliminary electrical tests performed on the MEMS (regarding vibration and isolation) confirmed the sensor functionality but does not allow assessing the system performance.

978-1-4799-2408-0/14 $31.00 © 2014 IEEE

ASIC developments

The SIMEIT-project deals with an Interposer platform which is capable to **host different MEMS** combined with an improving of their performance. Therefore an ASIC's with suitable features is required too. The capability of the market regarding available ASIC's is quite big but very limited regarding free **scalable ASIC parameters** in one device. The new ASIC (developed in this project) is fulfilling the integration concept in order to have a flexible data treatment and evaluation for different kind of sensors.

The **ASIC developments** were mainly focused on

o space reduction via technology down scaling and circuit under pad technology (CUP)
o variability regarding different kind of sensors and their configurations
o high sensitivity and SNR
o low energy consumption

Basis of the developments was the **resizing of the existing technology** with structure size of 0.6µm. The most economical relation (between increasing technology costs and smaller dimension) was found with the 0.35µm technology. Anyway the resizing was applied only to the digital signal processing, because the analog signal processing requires big components (e.g. transistors) in order to achieve a high SNR. Another size factor was the development of the **circuit under pad technology** (CUP), which are now applicable due the technology change from wire bond to the bumping technology (the mechanical stress of the ultrasonic supported wire bond can damage the circuits under pads).

The saved space was used for the integration of **extended functionalities**: (1) variable control of amplitude and frequency of the sensor element excitation - e.g. it allows the control of inertial or vibration sensors with different hardware configurations, (2) the charge amplifier allows the offset correction and the adjustment of the amplification level regarding to the needs. (3) 12bit analog digital converter, digital FIR signal filter and EEPROM for the algorithm setup of data evaluation, (4) internal oscillator, (5) reference voltage source for voltage setup and an precise conversion of analog to digital signals, (6) PTAT-circuit for correction of sensor temperature dependencies, (7) test and supervision circuits as power-on-reset, serial tests or redundant measurement mode, (8) Serial Peripheral Interface (SPI) for communication to the system.

The entirety of structural setups and available functions allows the **reduction of the ASIC energy consumption**. Hardware related are the smaller 0.35µm technology (applied on the digital circuits) and the integration of usually external components (e.g. oscillators, capacitors). Furthermore flexible settings of ASIC parameter reveal further energy saving potentials, especially at the frequency setup of the signal input frequency and the data processing. If needed high frequencies are available otherwise lower frequencies (related to lower energy consumption) are available too.

The described functionalities were **realized in different ASIC versions** (with circuits for an open loop setup or a close loop setup), in order to match the different application needs regarding measurement method and costs. The **close loop control** integrates additional control loop circuits for the compensation of the sensor element deflection. The needed voltage for electrostatic compensation (max. 40V) is measured instead the capacity change caused by deflection. This method allows a much wider measurement range, provides linear signals, has nearly no temperature dependency but due to the control loop the maximal measurable acceleration is limited to about 10g (dependent from sensor element geometries / weight). The **open loop control** measures the deflection of the sensor element. The missing electrostatic compensation of the sensor element deflection limits the maximal sensor measurement range, causes a certain temperature dependency and nonlinear signals. Otherwise it allows higher input frequencies (e.g. 10-20kHz / for vibration sensors), is less failure affected, and the lower production effort are reducing the costs.

System package

The **package** as interface to the main system has to cover the needs of the maximal stress reduction (by decoupling of PCB and silicon devices with big differences in the CTE's via Interposer), the component integration as volume process as well as to match the requirements for the final assembly at device manufacturers, OEMs or OSAT's. Actually there is the question what price is acceptable for which benefit. The **initial approach** (available to project start) of wire bonded and molded components is working but on cost of reliability (e.g. due to mechanical stress), precision (e.g. due to no mechanical decoupling) and limited durability (e.g. due to hygroscopic properties of polymers). The **alternative initial solution** of a system with wire bonds encapsulated inside of (too expensive) ceramic packages gains the precision in terms of minimized parasitic capacities due to the exclusion of humidity around the analog bond wires. Otherwise the wires are flexible which is inducing parasitic capacities and inductivity (dependent of the strength of system movements).

The **selected approach** of an redistribution layer (RDL) **on silicon Interposer** and **micro bumping** instead of component wire bond provides short, symmetric, fixed, small and isolated wirings for the analog signal transmission between sensor and ASIC. Even the hermetic properties could be improved (in terms of long-term signal stability) by placing silicone gel with low viscosity or a thin metal layer on top of the wiring layer. Advantages of these combinations are the long term minimization of parasitic capacities, inductivities, simultaneously improved SNR and reduction of the Interposer footprint (due to flip chip assembly instead of wire bond). The once **alternative approaches** with similar advantages are the via last integration into the ASIC (which is limited as described in paragraph "system integration concept") or the monolithic integration of MEMS on top of the ASIC [8, 9] and within the CMOS process flow - but it is perfect for very specialized system in volume production only (as well as much more expensive).

The **flip chip assembly approach** includes the partial **definition of the 3D process flow** regarding the suitable integration of the interconnect layer with solder bumps on the top side of the 100μm thick Interposer chip (interconnect of MEMS & ASIC to the Interposer) and additional a BGA with solder studs on the back side (interconnect between Interposer and substrate). In the final demonstrator the solder will be deposited on the components (MEMS & ASIC) and related pads are placed on the Interposer top side. After top side processing the Interposer wafer became temporarily bonded top side down in adhesive and the back side preparation takes place (including solder studs formation). The separation of solder and adhesive is uncritical due to the use of a room temperature debond process.

The **final system buildup** (Fig. 6) is quite short and will be used for the testing and assessment of the sensor system. It starts with the placement, reflow and a low stress underfill of the Interposer with copper studs on a ceramic LCC44 (lid less chip carrier). A reflow process follows after MEMS and ASIC flip chip assembly. The final step is the assembly of a polymer cap over the system connected to the LCC44 to avoid the entry of additional mechanical stress to the 100μm thin Interposer.

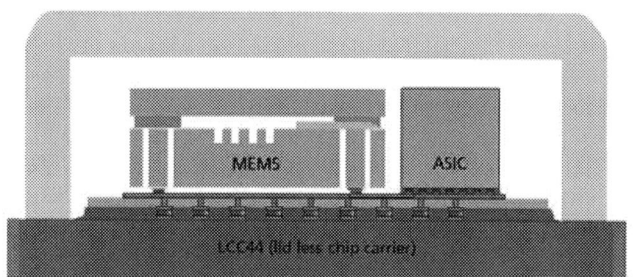

Figure 6. System buildup for assessment of the sensor performance.

An open topic is the **replacement of the LCC-carrier** as external interface by the Interposer itself. Therefore a thicker Interposer is needed (stable enough for direct assembly of the Interposer) but requires respectively larger TSV-diameter too (Fig. 7). The advantages of omitting the LCC-Carrier are costs and the additional hermetic sealing of components becomes available (by soldering of a silicon cap directly on the silicon interposer). In consequence there are principally additional investigation needed regarding the process flow and the system performance related to the complete TSV-fill (e.g. concerning warpage & higher mechanical stress) or related to the partial fill (e.g. concerning material influence of an appropriated hole plugging). Despite, for these improvements are reference results (especially regarding the additional stress input into the Interposer) of a working stress reduced Interposer sensor system required (as the system buildup on LCC-Carrier). Finally this approach will enable the production of the sensor system as wafer level packaging (WLP) process on 300mm wafers which improves the cost-benefit ratio additionally.

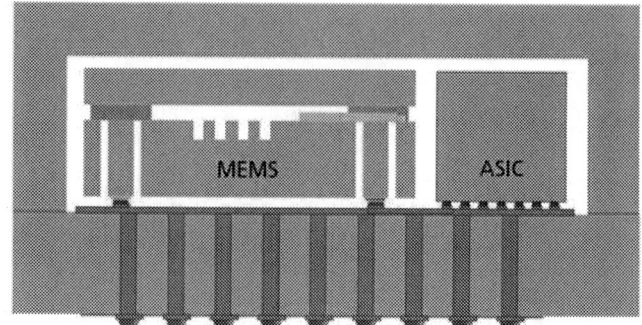

Figure 7. Final sensor package prepared by WLP-Technology

Conclusions

The Interposer approach is good for mechanically separation of materials with different CTE (e.g. organic substrate and silicon devices) and enables a redistribution which improves the sensor performance. **Objective** of the SIMEIT project (currently still in process) is a 3D-integration approach which enables a functional improved and economic small scales serial production. This paper gives an overview of the currently achieved results. The results (presented in this paper) reflect the **technology development progress** of the components (Interposer, MEMS, ASIC, Packaging), which have seen a strong maturation compared to the initial state.

The savings in terms of **cost efficiency** are related to balance the costs (of the additional Interposer) to the market needs regarding sensor systems in small scale volume production. The developed platform allows the usage of different MEMS just by redesigning the top side wiring layer (according the sensor pad positioning) without new costs for the lithographic mask sets of the whole system. Even all the applied process technologies can be used without losing time regarding process setup. This allows an Interposer preprocessing (warehouse / adjustable to different sensors) and reduces the delivery time significantly. Additional the processing of the Interposer in a 300mm wafer pilot line, the future assembly with WLP technologies and the possibility of multi project wafer (assembly of different components) allows lower process costs related to the larger die quantities per wafer.

The achievements regarding the **functional performance** are hard to assess in the moment because a full system buildup of MEMS, ASIC and Interposer is necessary which is currently in progress. Therefore the next step in the project will be the system integration of the components by using standard technologies of volume production and their functional assessment.

Acknowledgments

The authors would like to thank the Europäischer Fonds für regionale Entwicklung (EFRE) and the Sächsische Aufbau Bank (SAB) for the financial funding of the project "Stress less Integration of high precision MEMS- and Electronic components via new Interposer Technology" (SIMEIT / Förderkennzeichen SAB 100097073).

978-1-4799-2408-0/14 $31.00 © 2014 IEEE

References

1. C. S. Selvanayagam, et al., "Nonlinear Thermal Stress/Strain Analyses of Copper Filled TSV (Through Silicon Via) and Their Flip-Chip Microbumps", Electronic Components and Technology Conference, ECTC 2008.

2. Robert Bosch GmbH, U.S: Patent 4,855,017 and 4,784,420.

3. R. Puschmann, et al., "Via last technology for Direct Stacking of Processor and Flash", Electronic Components and Technology Conference, Electronic Components and Technology Conference, ECTC 2012.

4. S. Warnat, et al., "Through silicon via in micro-electromechanical systems", Mater. Res. Soc. Symp. Proc., 2009, pp 65-70.

5. K.-H. Lu, et al., "Thermomechanical reliability challenges for 3D interconnects with through silicon vias", AIP Conference Proceedings 2010, Vol. 1300, No. 1, pp 189f.

6. M. Nowack, et al., "Micro arc welding for electrode gap reduction of high aspect ratio microstructures", Sensors and Actuators A: Physical, Volume 188, December 2012, Pages 495–502.

7. A. Bertz, et al., "A novel high aspect ratio technology for MEMS fabrication using standard silicon wafers", Digest Tech. Papers, Transducers'01, Munich Germany, June 10–14, 2001, pp. 1128–1131.

8. TEXAS Instruments, "Digital micromirror device and its application to projection displays", Journal of Vacuum Science & Technology B: Microelectronics and Nanometer Structures, Volume:12, Issue: 6, 2009.

9. Lin, Y. et al., "A monolithic CMOS MEMS accelerometer with low noise gain tunable interface in 0.18µm CMOS MEMS technology", Sensors 2012, Taipei, Taiwan, 28-31 Oct. 2012.

Mechanical Stress Management for Electrical Chip-Package Interaction (e-CPI)

Wei Zhao, Mark Nakamoto, Vidhya Ramachandran, Riko Radojcic
Qualcomm Technologies, Inc.
5775 Morehouse Drive, San Diego, CA 92121
weizhao@qti.qualcomm.com

Abstract

e-CPI has emerged as a new risk in modern chip design as silicon dies become increasingly thinner and packages become increasingly more complex. e-CPI is different from traditional mechanical reliability related chip-package interaction, as it focuses on package stress impact on electrical circuit performance. A complete e-CPI modeling flow has been demonstrated. Both package FEA models and silicon piezoresistance coefficients are important for evaluating e-CPI risks. A case study of a product level circuit yield loss issue due to e-CPI has been discussed and modeled. The model has successfully reproduced the failure mechanism. Circuit designers, package designers, silicon foundries and package assembly houses should all take part in managing e-CPI risks.

Keywords: mechanical stress, transistor mobility, modeling, chip-package interaction, reliability

Introduction

Mechanical stress chip-package interaction (CPI) is a widely recognized phenomenon that is largely driven by CTE mismatch between silicon dies and package materials. The stress can be as high as a few hundred MPa. Low-κ dielectric layers in silicon die, Pb-free solder balls and Cu pillars are vulnerable to CPI stress due to various failure mechanisms, such as cracking, de-lamination and fatigue. Although CPI is a serious risk for reliability concern, it is well known in the package design world. Many practical measures have been taken by silicon foundries and package assembly houses to detect and mitigate the CPI related reliability risks. [1-6]

On the other hand, mechanical stress is not a new concept for silicon device engineers and circuit designers either. Mechanical stress as high as a few GPa has been intentionally and carefully introduced to local areas in silicon transistors, using high stress liners and/or SiGe strained Si. Through various mechanisms, such high mechanical stress boosts transistor performance significantly. [7]

Although the effects of mechanical stress have been widely recognized among both package designers and circuit designers from different perspectives, the impact of package induced mechanical stress on transistor electrical performance is hardly acknowledged and managed. However, since the package induced mechanical stress can be as high as a few hundred MPa, it is more than enough to disturb electrical performances of critical transistors and fail sensitive circuits due to piezoresistance effects.

Unlike traditional CPI related reliability issues, electrical circuit performance degradation due to e-CPI is difficult to be detected early on. Without modeling work, e-CPI issues can only be detected near the end of the product design cycles when the final packaged silicon dies have been tested. At this point, any attempt to redesign silicon dies and/or packages can be a prohibitive task.

Even though many empirical solutions for various e-CPI issues have already been implemented by circuit designers for stress sensitive circuits, those empirical solutions can be costly as they are often overly pessimistic. As a result, die area has been wasted and circuit performance has been compromised. Those empirical approaches are not sufficient for novel package solutions either, when new stress sources have been introduced, such as 2.5D interposer, and 3D dies stacking packaging technology.

On the other hand, e-CPI modeling can help detect potential issues and provide guidelines to circuit designers and package designers in the early design stage to prevent any potential e-CPI related issue.

In this paper, we will first compare traditional CPI vs. e-CPI modeling, then explain a typical e-CPI modeling flow. Later, a case study of a product level circuit yield loss issue due to e-CPI will be discussed. Finally, the semiconductor industry-wide collaborations are needed to manage e-CPI risks.

Traditional CPI vs. e-CPI Modeling

Similar as the traditional mechanical CPI modeling for reliability issues, the e-CPI modeling relies extensively on finite element analysis (FEA) models. However, there are several major differences between them.

First of all, traditional CPI modeling focused on "weak" areas on a silicon die, such as ultra-low-κ back-end-of-line (BEOL) dielectric layers, or solder joints. On the other hand, e-CPI focused exclusively on the very top surface of silicon dies, where transistors are manufactured.

Second, the most common goal of traditional CPI modeling is to identify the area with the highest absolute stress values, and then optimize packaging processes, structures or material properties to reduce the peak stress values. Thus, the maximum shear or tensile stress vector is usually the key modeling concerns.

On the other hand, the goal of e-CPI modeling is to evaluate the package stress impact on the electrical circuit performance on silicon dies. Electrical circuit performance is a function of transistor mobility. The transistor mobility deviation from its nominal value is in turn a weighted average of component stress. There are three component stresses. They are in-plane component stress that is parallel (S_{xx}) or perpendicular (S_{yy}) to the current flow direction of silicon transistors, and out-of-plane (S_{zz}) component stress. Thus, the relationship between mobility deviation and component stress can be approximated as the equation below,

$$\Delta\mu\,(\%) = S_{xx} \cdot P_x + S_{yy} \cdot P_y + S_{zz} \cdot P_z$$

where $\Delta\mu$ is the mobility deviation, P_x, P_y, P_z are piezoresistance coefficient values in respective directions. [7]

Transistor Corner Models

Silicon process variation is significant in the state-of-the-art technology nodes. Besides SPICE models for nominal transistor performance, corner models have already been widely adopted nowadays. They are usually used to validate circuit functionality and evaluate performance degradation in the presence of large amount of die-to-die, wafer-to-wafer process variations. They are largely intended for handling global systematic process variations, since all NMOS/PMOS will become faster or slower simultaneously in a corner model. As a result, modern circuit designs usually have significant margins to protect them from those global systematic process variations.

Nevertheless, e-CPI effect will produce large systematic variation within a relatively short distance. Some transistors will become faster or slower than other nearby transistors. This cannot be efficiently handled by a global corner model, and further widen model "corners" are also very expensive.

Stress Gradient

In other words, steep mobility deviation in local areas due to e-CPI can still lead to circuit failure even if circuits have been verified with SPICE corner models, especially for sensitive analog circuits, such as digital-to-analog converters (DACs). DAC is a common analog circuit block that has been widely used in modern SoC chips. Inside a DAC core, there are many arrays of large transistors. DAC typically demands nearly perfect matching among these large transistors in its cores. In other words, every transistor in a DAC core should not be any faster or slower than any other transistor in the same DAC core. Even if there is a tiny amount of mobility deviation among transistors in the DAC cores, the DAC performance will degrade. And DAC will even fail when the mobility mismatch exceeds what its self-compensation circuits can tolerate.

Thus, the more specific goal of e-CPI modeling is to identify the area with the highest gradient of mobility deviation, or where the gradient of mobility deviation is higher than certain criteria on silicon die surface. As explained before, the gradient of mobility deviation is a function of the gradient of component stress.

e-CPI Modeling Flow

A complete e-CPI modeling flow include both package FEA model and silicon FEA model as shown in Fig.1. The package FEA model needs package design information, such as package geometries, solder ball locations, package material properties, such as Young's Modulus, CTE, Tg, and process conditions, such as processing temperature, time, pressure etc..

When package FEA simulation has been finished, it will pass displacements near the top active silicon surface of the interested area to silicon FEA models as BCs. Then silicon FEA models convert component stress gradient maps into a mobility deviation map using piezoresistivity equations.

Based on the mobility deviation map, either linear current deviation or saturation current deviation of transistors in the interested region can be estimated.

In the end, circuit designers can determine how much circuit performance has degraded due to e-CPI impact based on current deviation. Then a keep out zone (KOZ) can be defined and sensitive circuits are moved out of the KOZ. This is similar as modeling of the stress impact from through-silicon-vias (TSVs). [3] A KOZ is defined in the end to protect circuits in both cases. The difference is that now stress comes from packages, instead of TSVs in silicon dies.

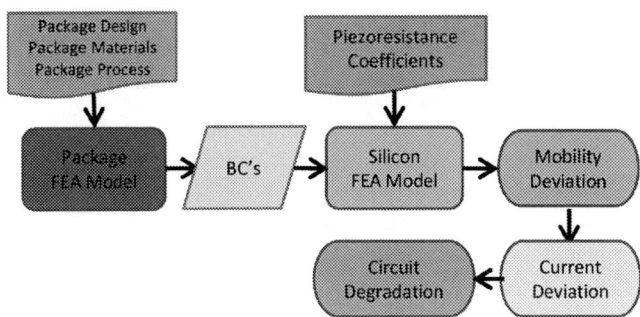

Figure 1. e-CPI modeling flow

Package FEA Model

However, IC design houses usually do not own detailed package process information and material properties needed to build the package FEA models. The knowledge resides with the package assembly houses, but is typically considered as their proprietary IP. One feasible solution is to have the package FEA model built, calibrated and maintained by the package assembly houses, but is freely accessible to IC design houses. IC design house is only responsible for providing package and silicon die geometry information, and the choice of package materials. The package FEA model outputs must also be designed to be compatible to IC houses silicon FEA model. Package assembly house is then responsible for the quality of package FEA models. This has often been accomplished by model calibrations with shadow-moire measurement of package/die warpage at different process steps, such as after reflow, end-of-line, etc.. [10]

Silicon Piezoresistance Coefficients

Package FEA models essentially calculate component stress values at top active silicon surfaces, then those stress values need to be translated into mobility deviation in percentage so that circuit designers can comprehend.

Thus, the silicon piezoresistance coefficients to feed silicon FEA model must be accurate. For many research studies, both the bulk silicon, and long channel MOSFET piezoresistance coefficients have been widely used, especially in many TSV KOZ studies. Their values are summarized in the table 1.

Table 1. Piezoresistance Coefficient Values (%/GPa) [7]

<110> Channel	Type	P_x	P_y	P_z
PMOS	Bulk	-71.8	66.3	1.1
	MOSFET	-71.7	33.8	6.2
NMOS	Bulk	31.6	17.6	-53.4
	MOSFET	35.5	14.5	20.7

In reality, those values can deviate from the reported ones due to different levels and methods of silicon transistor stress engineering. As a result, piezoresistance coefficients values will vary from foundry-to-foundry, process-to-process, and even device flavor-to-flavor. For example, a low-V_{th} core transistor from foundry A is likely to have different values from high-V_{th} IO transistor from foundry B. Eventually, silicon foundries should be responsible for providing those piezoresistance coefficient values to IC design houses, since they own silicon process. But this has yet to become a common practice in the semiconductor industry.

Research institutes and universities, such as IMEC, have already started to measure piezoresistance coefficient values from silicon die samples. The most common approach is using die bending equipment to carefully bend silicon die samples in the direction either in parallel or perpendicular to current flow of transistors, and then measure the response of linear currents or saturations currents of transistors. However, research institutes and universities have difficulty in accessing the state-of-the-art silicon die samples, thus the piezoresistance coefficient values they measured can be obsolete.

IC design houses, on the other hand, can access the latest silicon die sample. So they can always calibrate piezoresistance coefficient values using the same die bending method by their own.

A Case Study of e-CPI

There are many e-CPI stress sources. For example, stress from flip chip bumps can modulate the mobility in the region near bumps. [10] In this paper, an interesting case, package stress from a DRAM memory die on top of a logic die, will be discussed and modeled. Fig.2 and Fig. 3 show the top view and side view of the package structure respectively. This is a standard laminated substrate based package technology. The DRAM memory die has been glued on top of the logic die using die attach materials, and it has been wire bonded to the package substrate. Then both the logic and memory dies have been over molded by industry standard molding compound.

Notice that there are two DACs on the bottom logic die. Both DACs have been well designed and verified in a traditional single die package. This time, package designers wanted to add the DRAM die in the same package as a product enhancement. They tested several different DRAM die placements for the new package. In one of the test cases as shown in Fig. 2 and Fig. 3, the right edge of the DRAM die has been placed right on top of DAC1. When this particular package came back, circuit designers observed significant yield loss due to DAC1 failure, but no failure was observed in DAC0. Historically, this is not a problem as the bottom logic die is usually thick enough so that sensitive circuits such as

DAC have been shielded by the thick bottom die itself from any stress gradient coming from the back side of the die. However, there is a continuous trend, driven by consumer electronics market, of thinning silicon dies so that the form factor of the entire package can be smaller. As in this case, the bottom logic die has finally been thinned so much that it is no longer immune to stress gradient from the top memory die. Detailed analysis reveals that there is a large gradient of linear current shifts in the transistors array in DAC1 as shown in Fig.4. Meanwhile, the DAC0 is quiet. And for other test cases, DRAM die edges have not been placed near DACs, sure enough, the DACs are working as designed. Clearly, DAC1 on the bottom logic die in this particular test case has been disturbed by the DRAM die edge.

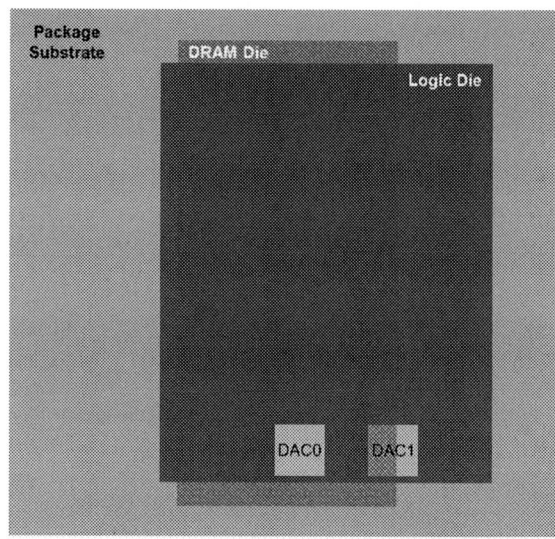

Figure 2. Top view of a DRAM die on top of a logic die

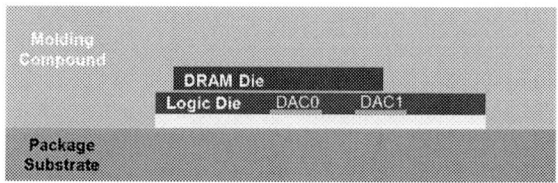

Figure 3. Side view of a DRAM die on top of a logic die

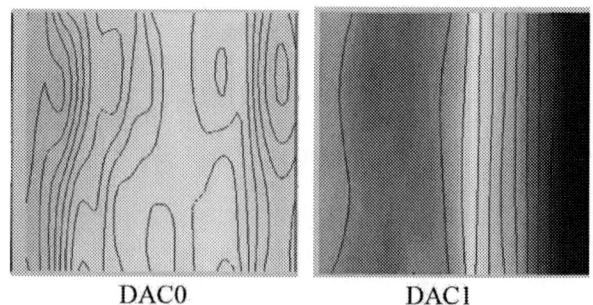

Figure 4. Color contour map of linear currents deviation in DAC transistors arrays

e-CPI Modeling Validation

To verify and evaluate this new e-CPI stress phenomenon, a package FEA model has been built in the Amkor web-based stress simulator [11], since Amkor is the package vendor of this particular case and it has all the detailed knowledge of the packaging process and material properties. The details of this web-based stress simulator and package FEA model have been discussed in [10].

Fig. 5 shows the color contour of the component stress on the bottom logic die from the Amkor web-based package FEA model. The package FEA model confirms that top DRAM die has produced considerable stress gradient on the bottom logic die, as the footprint of the DRAM die can be clearly seen in the stress contour plot. Particularly, there is significant stress gradient near one of the DRAM die edges, while the stress gradient is much smaller in the region that has been completely covered by the DRAM die. To investigate the interested region near DAC1, the package FEA model has been further refined with denser meshes in the region inside the rectangle box as shown in Fig.5. Accordingly, more accurate and steeper stress gradient can be calculated in the interested region.

Figure 5. Color contour map of component stress on the bottom logic die

Then displacements data of this refined region from the package FEA model are feed into the silicon FEA model. The piezoresistance coefficient values for the long channel MOSFETs as reported in [7] have been used for converting component stress to mobility deviation. The final color contour map of the mobility deviation on the bottom die near DAC1 is shown in Fig. 6. The models have successfully reproduced the pattern of the mobility deviation of DAC1 as illustrated in Fig.4. Clearly, either DAC1 needs to be moved away from DRAM die edge, or the DRAM edge needs to be placed so that it is not near either one of the DACs.

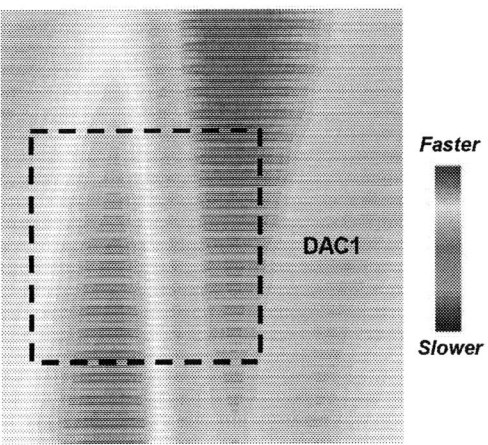

Figure 6. Color contour map of mobility deviation on bottom logic die near DAC1

Managing e-CPI Stress

Both the yield data from the DACs and the FEA models in our case study have showed that e-CPI does pose a new risk to modern circuit design. And the thinner silicon dies become, the more vulnerable they will be to e-CPI package stress. The risk can further increase due to various emerging types of advanced package solutions, such as fan-out technology, 2.5D interposer technology and 3D die stacking technology. As a result, the e-CPI risk should not be ignored and it needs to be carefully monitored and managed for future circuit and package design.

On the other hand, e-CPI stress risk needs to be managed through the collaboration of the entire semiconductor industry, including silicon foundries, IC design houses and package assembly houses. And it cannot be single-handed managed by any party alone for they all lack a complete set of skills, information and capability to detect and solve the e-CPI issues.

For example, there are variety types of packaging solutions even for the same silicon process and circuit design. Silicon foundries have no knowledge nor any control of packages choices of their customers. Silicon foundries thus are not able to tweak transistor SPICE models to include package stress impact since different packages will have different stress consequences. However, they should be ultimately responsible for calibrating piezoresistance coefficients values for all technology nodes, transistor types and flavors. They are also responsible for controlling these values not to deviate from their nominal values.

Package assembly houses now not only need to consider manufacturability and reliability issues of their packages, but should also keep e-CPI stress risks in mind when designing new package structures and choosing/optimizing package materials and process. After all, packaging is the source of e-CPI stress risks. But they also cannot solve the issue alone as they do not have sufficient knowledge on circuit designs and know accurate piezoresistance coefficient values for a given process. In turn, they are responsible for building and calibrating package FEA models for all package types and provide access of their models outputs to IC design houses.

Circuit designers in IC design houses should be more aware of the package designs and their stress implications to circuit performance than before. They not only need to verify their design with transistor process corner models, but also need to take steps to protect their designs from e-CPI package stress gradients. They can either build more robust circuits to immune from local stress gradient, which can be costly. Or they can evaluate the maximum amount of the stress gradient their design can tolerant, and do not place their circuits at locations where stress gradient exceeds the circuit tolerance.

In summary, managing e-CPI risks requires a comprehensive understanding of package structure, package process, package material, silicon piezoresistance coefficient, and circuit design, thus relies on the collaborations of the entire semiconductor supply chain and design community.

Conclusions

In this paper, the e-CPI risk has been identified as a new stress risk outside silicon die due to material properties mismatching between packages and silicon dies. The high stress gradient will change transistor motilities within a short distance, thus degrades circuit performance, or even fail stress sensitive circuits. The e-CPI modeling flow has been discussed. A product level e-CPI induced circuit yield loss has been studied and modeled. The e-CPI models are able to reproduce the failure mechanism. Managing e-CPI risks needs collaborations of the entire semiconductor industry.

References

1. Chip-Package Interaction Understanding, Identification and Evaluation, JEP156, JEDEC Standards, Mar 2009.
2. Stress-Test-Driven Qualification of and Failure Mechanisms Associated with Assembled Solid State Surface-Mount Components, JEP150, JEDEC Standards, May 2005.
3. Jamil Kawa, "TSV Stress Management," Synopsys Inc, 2010.
4. X.H. Liu, T.M. Shaw., M.W. Lane, E.G. Liniger, B.W. Herbst and D.L. Questad, "Chip-Package-Interaction Modeling of Ultra Low-k/Copper Back End of Line," *International Interconnect Technology Conference*, Burlingame, CA, June 4-6, 2007, pp. 13-15.
5. Xuefeng Zhang, "Chip Package Interaction (CPI) and its impact on the Reliability of Flip-Chip Packages," Dissertation, The University of Texas at Austin, Dec., 2009.
6. G. Wang, P. S. Ho, and S. Groothuis, "Chip-packaging interaction: a critical concern for Cu/low k packaging," *Microelectronics Reliability 45*, pp. 1079–1093, Jan. 2005.
7. S. E. Thompson, G. Sun, Y. S. Choi, and T. Nishida, "Uniaxial-process-induced strained-Si: extending the CMOS roadmap," *IEEE Trans. on Electron Devices*, vol. 53, no. 5, pp. 1010–1020, May 2006.
8. M.Nakamoto, R. Radojcic, W. Zhao, V.K. Dasarapu, A.P. Karmarkar, and X. Xiaopeng, "Simulation methodology and flow integration for 3D IC stress management", *Custom Integrated Circuits Confernce (CICC)*, San Jose, CA, Sept. 19-22, 2010.
9. R.Radojcic, M. Nowak, and M. Nakamoto, "TechTuning: stress management for 3D Through-Silicon-Via stacking technologies", *American Institute of Physics, AIP Conference Proceedings, Stress Management For 3D ICs Using Through Silicon Vias*, vol. 1378, pp. 5-20, 2011.
10. K. Dhandapani, M. Nakamoto, W. Zhao, A. Syed, W. Lin, and R. Radojcic, "A methodology for Chip - Package Interaction (CPI) modeling in 3D IC structures," *IMAPS 8th International Conference and Exhibition on Device Packaging (DPC)*, Fountain Hills, AZ, Mar. 6-8, 2012.
11. Amkor Web.Data, https://webdata.amkor.com

Cu Pillar Flip Chip Assembly: Chip Attach Process Failure Mode Study

Shengmin Wen[1], Bora Baloglu[1], Guangfeng Li[2]
Amkor Technology
[1]1900 S. Price Road, Chandler, AZ 85286, USA
[2]Amkor Assembly and Test (Shanghai) Co, Ltd., China
shengmin.wen@amkor.com

Abstract

An experiment is conducted to study failure mechanism during flip chip attach process for Cu Pillar bumped Si device that uses mass reflow assembly technology. A three-leg design of experiment (DOE) is conducted, which includes two UBM sizes, two different Cu pillar height, and with / without polyimide option to collect basic failure information. Finite element software is used to correlate the failure mode and identify the Si – Cu pillar bump – substrate interactions. Based on the experiment and finite element analysis results, a simple shallow beam mechanical model was recommended to be a basic start point of Cu Pillar flip chip assembly technology application. Results are discussed in details.

Introduction

Cu pillar is becoming the bump of choice for most flip chip packages that are used in today's mobile applications. Such packages include those for baseband (BB), application processor (AP), memory, and front end radio frequency multi-chip modules. The driving force for Cu Pillar flip chip is the number of signal input/output (IO) increase which in turn has resulted in bump pitch decrease to less than 150um, with some designs having bump pitch all the way down to around 50um. To make the Cu Pillar assembly technology even more challenging, most such packages typically requires thin package, i.e., less than 0.9mm for chip scale packages (CSP) and around 1.0mm for package-on-package (PoP) that has both the bottom AP and top memory package combined. There is a trend that many such packages are now adopting two layer and thin core substrates with the substrate thickness goes under 200um total thickness in large strip format. Such structure is totally different from the traditional flip chip BGA type packages where thicker substrate and large pitch, large size of bumps are typically used. Theoretical and experimental studies to understand the basics of Cu pillar assembly technology are essential to provide basic guidelines that can be used to design a specific Cu pillar structure.

In comparison to SnAg solder or eutectic solder bump, Cu pillar bump comprises copper pillar and solder cap on top of the pillar, with both the pillar's and the cap's geometry highly customized per application and substrate technologies. For example, to make a total of 80um high Cu pillar bump the bump's structure can be either 50um Cu pillar plus 30um solder cap or 40um Cu pillar plus 40um solder cap, and the Cu pillar cross section can be made rectangular, circular, square, octagon, or any type of polygon shape with adjustable sizes. In addition, these bumps may have different under bump metallurgy (UBM) shape/size, with PI or without PI layer option. Refer to Fig. 1 for various types of Cu Pillar bump shape / sizes. There is a need to understand the mechanics and the failure mode of such highly customized

bump structure so that first time right optimized design and reliable package can be achieved. As a first step, the mass reflow flip chip attach process before underfill application is studied in this paper through a 3-leg DOE. A simple shallow beam mechanical model and finite element model is used to analyze the results.

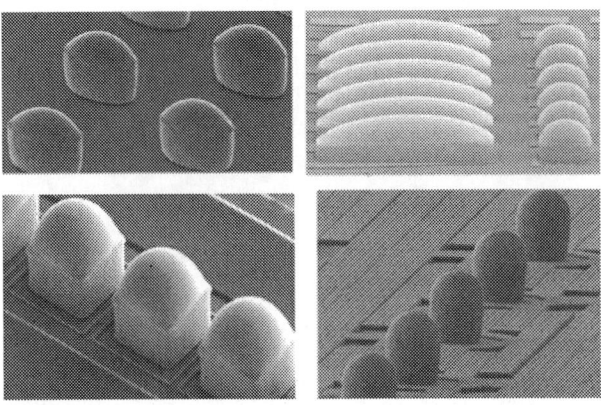

Figure 1. Various type of Cu Pillar bumps. Courtesy of Powertech Technology Inc., and Amkor Technology Inc.

Experimental Setup and Results Discussions

A most commonly used circular type bump structure was selected to do the experiment. Since polyimide (PI) has been in use for flip chip package that uses mass reflow process for a long time, while the recently developed Cu pillar flip chip package that uses thermal compression process does not use PI, this experiment included a PI option to see how much the PI benefits. Cu pillar height is another factor because it will affect the shear strain magnitude induced by the thermal expansion coefficient mismatch during cool down process within the bump and most importantly within the Si die-electric layers. Details of the design are listed in Table 1. The solder cap height is the same for all legs.

Table 1. DOE legs definition

Leg #	Cu Height (um)	UBM (um)	PI (um)
1	40	45	5
2	40	40	NA
3	45	40	NA

The packages are assembled using copper pillar bump-on-lead technology and mass reflow process with capillary underfill. The packages went through regular assembly process steps until the end of line. The packages were then taken to do scanning acoustic microscope (C-SAM) images to check on any crack or delamination. Once any delamination or crack is observed, a physical cross section is performed and secondary scanning electron microscope (SEM) is used to reveal the failure details. In this study, there was no actual ELK structure or device built. Instead, dummy Si that had gone through specially treatment process was used to achieve better failure mode demonstration. Figs 2a and 2b show a representative DOE results. Fig. 2a shows a crack that was initiated in the area that has highest stress intensity factor. Fig 2b shows that underfill material is within the crack in a different failed bump through EDX confirmation – an indication that the crack was formed before the underfill process and right after the flip chip attach process.

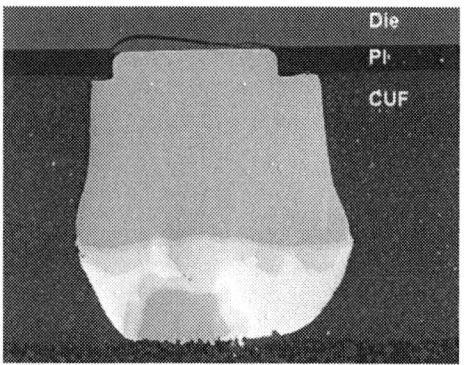

Figure 2a. Crack initiation location is at the die pad/bump contact corner area.

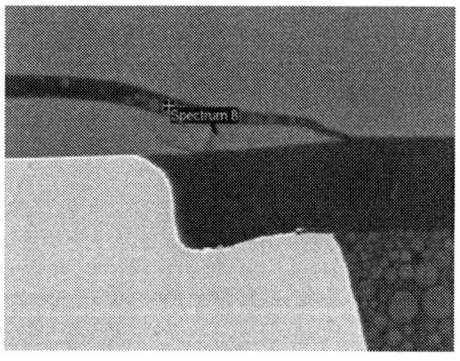

Figure 2b. On a different bump crack, EDX confirmed that the material within the crack was underfill, indication of the fact that crack was formed before underfill process.

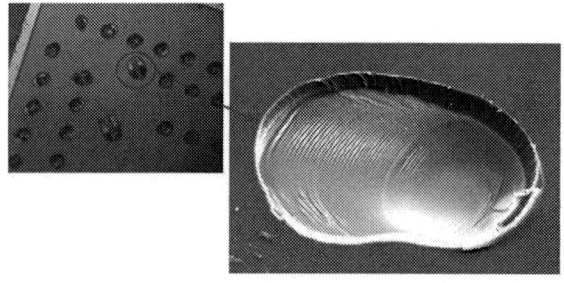

Figure 3. Multiple cracks initiated, and the die falls off from the substrate after flip chip attach process.

There was one extreme case, referring to Fig. 3, where the die just fell off from the substrate, leaving the bumps bonded on the substrate's bumping leads. The fractured surface topology is consistent with the observation from the cross sections (Figs 2a and Figures 2b). These results show that the failure mode during flipchip attach process is most likely type I crack (open crack), that is resulted from normal stress.

Finite Element Analysis

To confirm the stress state and understand the failure mechanism better, a full scale finite element analysis (FEA) structural model that includes ELK, Al pad, and PI is constructed. Refer to Fig. 4 for the FEA model details. Commercial FEA package ANSYS was used to explore the deformation, stress and strain distribution within the ELK layer during the flip chip process. The elements used here is in general 8-node 3D solid element, with viscoelastic material constitutive relationship being used for solders that joins the Cu pillar and the lead. Lead free reflow profile is used, and after cooling down, the residual stresses, the deformation, and the relative stress levels within the ELK layer are show in Figs. 4, 5 and 6.

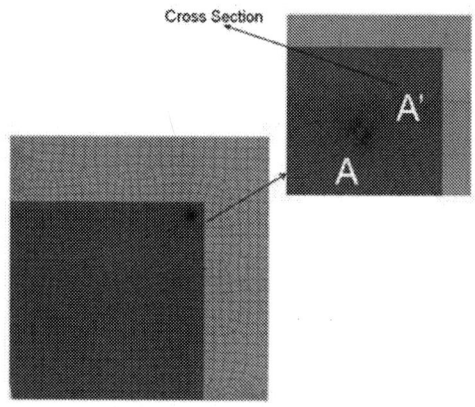

Figure 4. FEA model uses quarter of the package. The stress and deformation are extracted along the cross section plan direction A-A'.

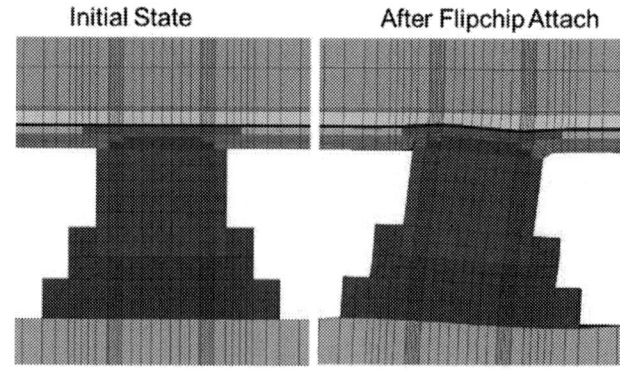

Figure 5. The deformed joint, Al Pad, and ELK layer after flip chip attach process.

Above deformation shows that in the die Al pad and underneath ELK area, the deformation shows clear shear effect, which means the plane does not keep as plane and perpendicular lines do not keep as perpendicular after the deformation. Fig. 6 shows the stress distribution within the ELK layer along across section line A – A' direction. In the figure, all the stresses are normalized to the maximum Z direction normal stress σ_{zz}.

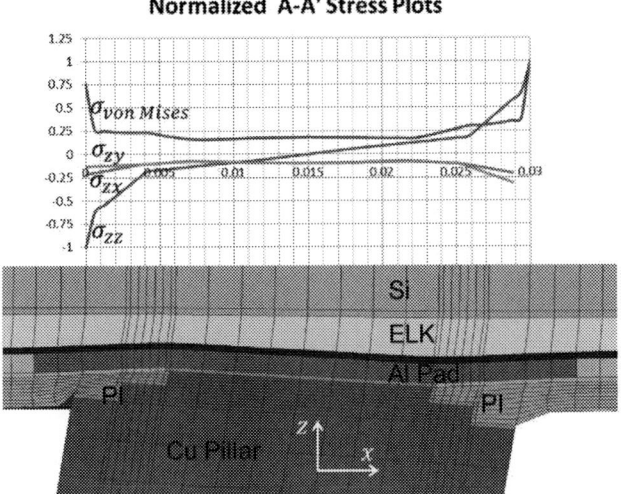

Figure 6. Overlap plot of stress and deformation.

It can be observed that normal stress loading is the major stress applied onto the ELK layer by the Cu pillar. This normal stress loading is center point symmetric along the bump's cross section line with one half being tensile and the other half compressive. Such stress interaction forms a local coupled moment. The magnitude of this moment, however, is related to the overall loading from CTE mismatch between the die and the substrate during the cooling down process after joint's formation. Since no external Z direction traction and no external moment is applied to this structure as a whole, the shear stress and the moment couple is locally in equilibrium condition respectively. It must be noted that exactly how the stress, strain and deformation is distributed within this area is far more complicated than above FEM model results. In this study, what is interested in is not the local distribution but rather how the bump design will affect the flipchip attach process' failure mode. Therefore a resultant force and resultant moment is used for the study per St. Venant's principle, which states that the local distribution of forces can be replaced with a local resultant force when considering stress/strain distributions sufficiently far away from the local area.

Shallow Beam Theory

Per above FEA analysis results and analysis, the die and Cu pillar bump interaction can be describes in Fig. 7. The die acts like a plate in plane stress condition, with one end fixed and the other end supported by the Cu pillar bump. The bump can be treated as a shallow beam, with one end (the joint to the lead or bump pad side) being fixed while the other end being loaded with the shear force from the die. The couple moments M_{BD} and M_{DB} are locally balanced and they are

reactive moments to balance the shear stresses τ_{BD}, τ_{DB}, which has the resultant shear force Q through below equation:

$$M_{DB} + Qh = 0 \tag{1}$$

Assume that the CTE difference between the die and the substrate is $\Delta\alpha$, the temperature difference is ΔT, the stiffness of the die's in-plane equivalent deformation is K_D, and the bump's overall bending stiffness is K_B, the force Q can be derived per below eqution:

$$Q = -\frac{AE\Delta\alpha\Delta T}{1.0 + \frac{K_D}{K_B}} \tag{2}$$

In the equation, A, E are the die's equivalent thermal deformation cross section area and Young's modulus and respectively.

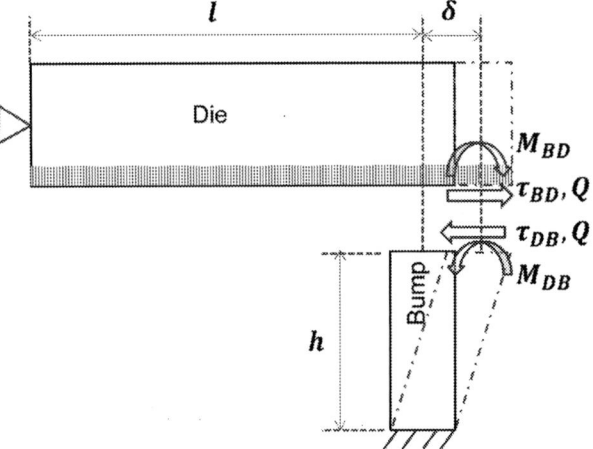

Figure 7. Simplified die / bump interaction diagram

The negative sign of equation (2) shows that the shear force Q is in the opposite direction during the cooling down process of flipchip attach step. This equation reveals an important bump design principle that can be used to optimize the Cu pillar bump design and prevent chip attach failure, and that is: the bump's deformation stiffness K_B is the major factor that will affect the shear force Q if $\Delta\alpha$, ΔT, and K_D is given. Take two extreme cases as examples: for a rigid bump, which has K_B being infinite, the die takes the total force to accommodate zero thermal expansion due to temperature change, while for a totally soft bump, which has K_B being zero, the die takes no force at all – the die has virtually free end movement under thermal expansion / contraction. It is therefore important to manage the bump's bending stiffness by optimizing the bump design to prevent the failure during flip chip attach process.

In general, for Cu pillar bump, the ratio of bump height and bump diameter is around 1.0 with mostly being less than 1.0. The bump can be approximately modeled as shallow Timoshenko beam, and the shear effect has to be considered, as it is clearly showed in the FEA model (Fig. 6) that the end plane is warping. Refer to Fig. 7, the bump's deformation stiffness K_B includes two terms K_{Bb} and K_{Bs}, with K_{Bs} to account for additional shear stiffness consideration, see equation (3a). The first term K_{Bb} is the bending stiffness, and it is straight forward, per equation (3b). The second term K_{Bs} was studied in very detail by J. D. Renton [1]. Various formulas were given per different cross section type, which may be found in today's highly customized Cu pillar shape. They are listed here for convenience as 3(c) ~ 3(e).

$$\frac{1}{K_B} = \frac{1}{K_{Bb}} + \frac{1}{K_{Bs}} \tag{3a}$$

$$\frac{1}{K_{Bb}} = \frac{h^3}{3EI} \tag{3b}$$

For a rectangular bump that has depth of $2a$ and breadth of $2b$, and for $\frac{b}{a} \le 1.0$ being usual the case (a is bump size along the bumping trace lead direction or shear direction),

$$\frac{1}{K_{Bs}} = \frac{1}{GA} \left[1.2 + 0.2 \left(\frac{\mu}{1+\mu} \right)^2 \left(\frac{b}{a} \right)^4 \right] \tag{3c}$$

For a circular bump,

$$\frac{1}{K_{Bs}} = \frac{1}{6GA} \left[7 + \left(\frac{\mu}{1+\mu} \right)^2 \right] \tag{3d}$$

For an elliptic or oval bump with semi-axes a and b, with a being the axe along the trace direction (shear direction),

$$\frac{1}{K_{Bs}} = \frac{1}{6GA} \left[6 + \frac{2(a^2+b^2)}{3a^2+b^2} + \left(\frac{\mu}{1+\mu} \right)^2 \frac{4b^4}{a^2(3a^2+b^2)} \right] \tag{3e}$$

In above formulas, μ, G are Poisson's ratio and shear modulus respectively.

Conclusions

Failure mode during Cu pillar flip chip attachment process was studied with experiment. Based on the failure mode observation, a finite element model is constructed and used to correlate and simulate the failure mode. Theoretically a shallow beam deformation model that accounted for both bending and shear elements was proposed for the design of highly customized Cu pillar bumps.

Acknowledgments

The authors would like to thank Mr. Min Yoo, Mr. William Huang from Amkor Technology Taiwan, Inc., for their generous support of providing experimental materials.

References

1. J. D. Renton, "Generalized Beam Theory Applied to Shear Stiffness," Int J Solids Structures, vol 27, no 15, pp. 1955 – 1967, 1991.

Mechanical and Thermo-mechanical Stress Considerations in Applying 3D ICs to a Design

Jia-Shen Lan, and Mei-Ling Wu*
Department of Mechanical and Electro-mechanical Engineering
National Sun Yat-sen University
70 Lien-Hai Rd. Kaohsiung, Taiwan (R.O.C)
E-mail: d023020004@student.nsysu.edu.tw
E-mail: meiling@mail.nsysu.edu.tw*

Abstract

This paper provides the physics of failure (PoF) analysis methodology in a three-dimensional integrated circuit (3D IC) integration based on mechanical and thermo-mechanical concepts. The majority of research on the 3D IC package has focused on the Coefficient Thermal Expansion (CTE) mismatch and heat junctions. The primary problems of CTE mismatch and heat dissipation cause failures or fatigues in 3D IC integration, and they become critical reliability issues. However, mechanical stress induced by mechanical loading has a significant effect on the strength of a material, causing, for example, interfacial cracking or the failure of through-silicon-vias. The strategy environment, pressure, and application of mechanical loading all lead to failure concerns. Thus, full 3D IC package modeling needs to be developed to achieve a more reliable 3D IC integration. In this paper, we will discuss the different physics of insight between thermo-mechanical and mechanical loading for 3D IC integration.

Introduction

Three-dimensional integrated circuit (3D IC) technology had emerged due to the requirements of high performance, low cost, and small sizes in 3D packaging. The enhancement of device density per volume and the utilization of short vertical interconnections for improved electrical characteristics in packages characterize certain unique advantages. Previously research has focused on thermo-mechanical stress induced by the Coefficient Thermal Expansion (CTE) mismatch between through-silicon-vias (TSV) and the substrate material. Further research has been carried out on thermal stress caused by high junction temperature and temperature concentration phenomenon inside the stacking chips. One critical problem of 3D IC is how to improve the interconnection (TSV and micro bump) reliability. There has been significant effort in TSV fabrication and electrical design, as TSV plays a key role in enabling technology for 3D IC stacking, silicon interposer technology, and advanced wafer level packaging (WLP). This study focuses on the analysis of thermo-mechanical and mechanical with physics of failure (PoF) for 3D IC structure design.

In general, 3D integration consists of 3D IC packaging, 3D IC integration, and 3D Silicon (Si) integration. The difference among them is that TSV uses the latter two interconnections. The 3D-WLP structure in this paper is selected, which is low-cost and small-sized and has high performance and low profile features of the 3D chip stacking package. The 3D-WLP with ten layers of chips stacking is divided into four major parts: TSV, Ajinomoto Built-up Film (ABF), silicon chips, and metal bumps. In this paper, using 3D Finite element analysis (FEA) modeling, we are concerned with the induced thermal stress due to temperature change and the application of mechanical loading in 3D WLP.

In this research, ten layers of thinner stacking chips in the 3D package, as proposed by EOL/ITRI, are covered by the insulation in the form of soft polymer film-ABF, and the FE method (FEM) analysis is subsequently performed to assess the failure of the 3D-WLP package. The 3D-WLP package has been developed by EOL/ITRI [1-8], who introduced the laser drilling process to form a TSV on chip. The authors have subsequently investigated the delamination behavior between copper bumps in the 3D-WLP package. Moreover, they have analyzed the thermo-mechanical phenomena in the 3D-WLP package in order to investigate temperature changes, and have discussed the thermal stress on TSV and copper trace under thermal cycling loading. Aside from the 3D WLP package, the EOL/ITRI has developed simulation models for 4-layer stacked IC package [9] and 3D vertical stacked IC package [10, 11] to improve the reliability of the 3D IC package and enhance the thermal management by employing underfills. Thermo-mechanical reliability is a critical issue for TSV in 3D IC package. Thermal stress on TSV for 3D IC package induced by the mismatch in CTE between the copper TSV and the surrounding silicon chip has been presented elsewhere [12-17]. In this paper, 3D WLP package is used in the analyses, whereby the loading condition (in terms of temperature) changes from 260 °C to 25 °C. Three main points are emphasized and discussed in this paper: (1) Direct and indirect coupled-field analyses are introduced in the 3D WLP package for the aforementioned temperature change, (2) The thermal stress in the TSV and bumps in the 3D WLP package is investigated when the temperature changes from 260 °C to 25 °C, and (3) Design of Experiment (DoE) and Response Surface Method are used to develop a rapid assessment approach model and analyze the optimal region for reducing concentrated stress. The analysis results can provide an effective approach to designing the guidelines for 3D IC packaging or support for EDA design tools.

Methodology

As previously noted, in this work, ten thinner stacking chips in the 3D package proposed by EOL/ITRI are covered by the insulation based on a soft polymer film-ABF, and the FE method (FEM) analysis (by using ANSYS software) is

978-1-4799-2408-0/14 $31.00 © 2014 IEEE

subsequently performed in order to better understand the 3D-WLP package failure mechanism. First, in order to identify the differences between direct coupled-field analysis and indirect coupled-field analysis, four element types (solid 45, solid 5, solid 70, and solid 185) and two methods for setting the thermal boundary conditions are discussed. Second, we analyze the detailed thermo-mechanical stress induced by the CTE mismatch between the copper TSV and the silicon chip under temperature loading. Third, the ABF material has four types of ingredients—GX13, GX92, GX-T31 and GZ41—each characterized by different material properties and mechanisms. Thus, the aim is to identify a new way to isolate the electrical signal effectively and prevent leakage from the TSV. Finally, a new design is proposed, mitigating the thermo-mechanical reliability problems currently present in TSV for the 3D IC package.

Thermal Structural Coupling Analysis

For the temperature change from 260°C to 25°C, the thermal structural coupling analysis in the 3D WLP package is analyzed, as shown in Fig. 1. Development of a correct model for the DoE method necessitates employing two methods of thermal boundary condition and thermal structural analysis.

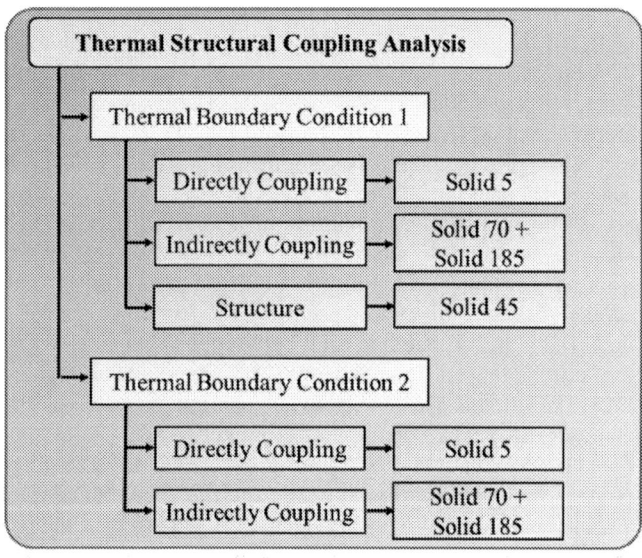

Figure 1: Flowchart of Thermal Structural Coupling Analysis

For the first thermal boundary condition, the 3D WLP package body temperature is set to decrease from 260 °C to 25 °C. On the other hand, for the second thermal boundary condition, the temperature is set at 260 °C in the area around the 3D WLP package, and at 25 °C of the initial temperature. In the first of thermal boundary condition, the effects of the temperature change on the 3D WLP package are evaluated by directly coupling, indirectly coupling, and structure problem. The structure problem is not employed in the second thermal boundary condition because solid 45 does not support the thermal conductivity for heat conduction. The difference between the directly coupled method and the indirectly coupled method is depicted in Fig. 2. The indirect coupling method is a sequential process, whereby thermal and structural output are calculated separately. On the other hand,

the directly coupled method is based on an interaction between the processes used for calculating the thermal and structural problems.

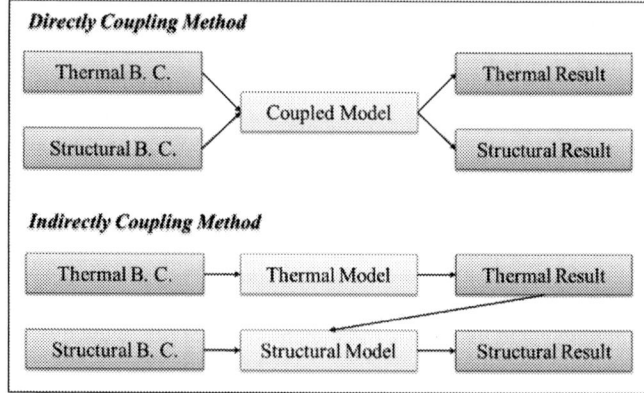

Figure 2: The Process of Directly Coupling Method and Indirectly Coupling Method

Table 1: The Comparison Among of Structure, Directly Coupling Method, and Indirectly Coupling Method in Thermal B. C. 1 and Thermal B. C. 2

	Thermal B. C. 1		Thermal B. C. 2	
	W	σ	W	σ
Structure	0.017352	439.864		
Directly Coupling	0.019198	486.658	0.017354	439.865
Indirectly Coupling	0.019218	507.029	0.017370	458.276

Hint: W is the deformation of 3D WLP package, and σ is the maximum of von-Mises stress in TSV

As shown in Table 1, in the first of the two thermal boundary conditions, the direct coupling method and the indirect coupling method produce significantly different results. This indicates that both methods should use heat transfer for thermal boundary condition in order to yield accurate output. In this research, the structural method applied to the 3D WLP package is presented, as it is deemed most efficient and suitable for obtaining accurate solution of simulation.

Investigation of Thermal Stress in 3D WLP Package

The 3D WLP package is employed to analyze the failure of the TSV and the bump. For this purpose, the ten thinner stacking chips surrounded by TSVs and bumps in the 3D package proposed by EOL/ITRI are covered by the insulation in the form of soft polymer film-ABF. The displacement distribution of the 3D WLP package due to the temperature change from 260 °C to 25 °C is shown in Figure 3. Due to the CTE mismatch among the silicon chip, ABF material, and copper, 3D WLP package is subject to shrinkage, which causes failure of both the TSVs and the bumps. Figure 4 and Figure 5 show the von-Mises stress contour of the 3D WLP package and the ABF material, repsectively. The analysis results indicate that the stress is concentrated on the TSV and near the TSV on the ABF material. As shown in Figure 6, the maximum stress occurs on the first layer at the corner of the TSV.

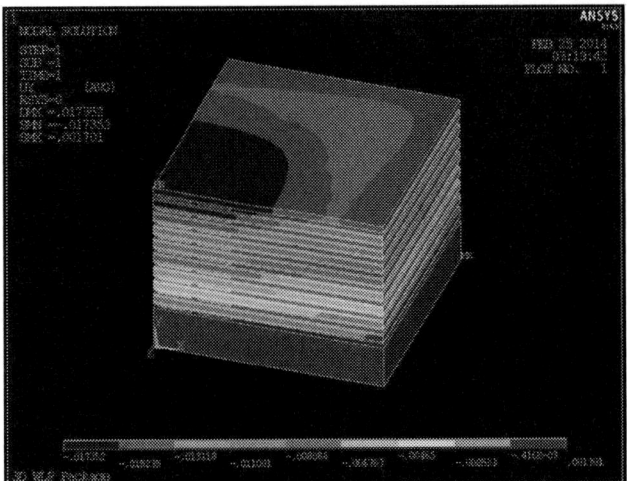

Figure 3: The Deformation of 3D WLP Package under Thermo-mechanical Loading

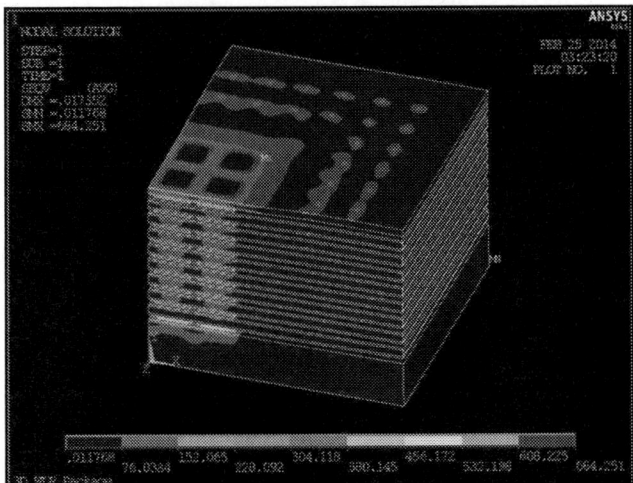

Figure 4: The von-Mises Stress of 3D WLP Package under Thermo-mechanical Loading

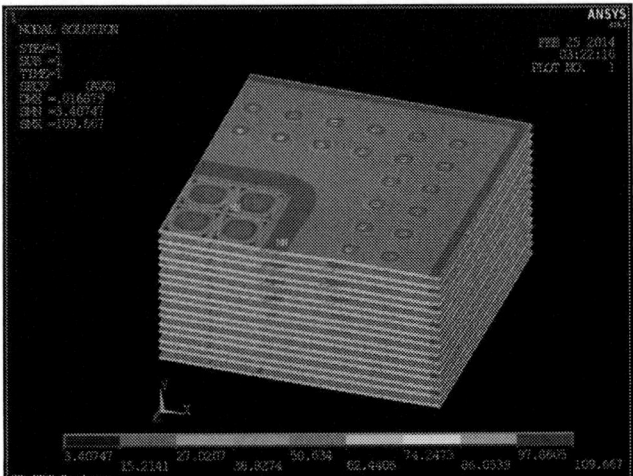

Figure 5: The von-Mises Stress of ABF material in 3D WLP Package under Thermo-mechanical Loading

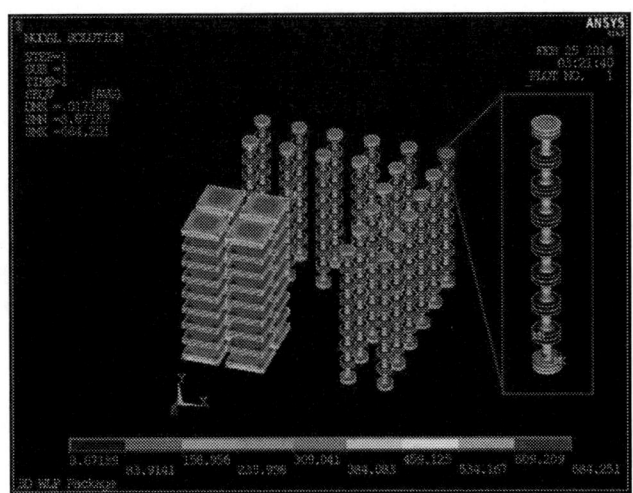

Figure 6: The von-Mises Stress of TSV and Bump in 3D WLP Package under Thermo-mechanical Loading

The stress to which the TSV and the bump are subjected is further investigated by applying the DoE method, as this enables identification of the critical parameters. A DoE with factorial analysis is thus adopted to obtain the important factor by 3D FEM. The critical factors that are calculated and selected by FEM and DoE for 3D WLP package are presented in Table 2. These findings enable us to propose a design optimization of a parametric 3D WLP package for the temperature change from 260 °C to 25 °C.

Table 2: The Critical Factors in 3D WLP Package

Design Factors		TSV Stress	Bump Stress
Material Parameters			
Chip	E		
	ν		
	α	✓	
Copper	E	✓	✓
	ν		
	α	✓	✓
Tin	E		
	ν		
	α		
ABF	E		✓
	ν		✓
	α		✓
Geometry Parameters			
TSV	D		✓
ABF	T_1		
	T_2		✓
Bump	D		✓
	H	✓	
Tin	H		
Chip	L		
	H		✓
Substrate	H		

Hint: E is Young's modulus, ν is Poisson's ratio, α is CTE, D is diameter, T_1 is the thickness of ABF material covered on silicon chip, T_2 is the thickness of ABF material covered on TSV, H is height, and L is the length of silicon chip.

Design of Experiment Analysis for 3D WLP Package

The objective of the optimization is to improve the stress profile of the TSV and the bump by using the aforementioned ABF material to insulate the silicon chip and the TSV in order to prevent the signal loss. FEM software ANSYS is used to predict the stress the TSV and the bump are subjected to. It is coupled with DoE software Design Expert in ANSYS for providing the detail on selecting design factors. The ABF material properties are given in Table 3. For the purpose of benefiting from the FEM capabilities in the design process, reduced models based on Design of Experiment (DoE) and Response Surface Method (RSM) can be constructed. These models allow us to undertake fast evaluations of the response for different design standards. Therefore, three important 3D WLP package characteristics that are known to affect the TSV failure are selected for the three-level analysis of factorial design, listed in Table 4.

Table 3: The Material Property of ABF (Source: Ajinomoto Fine-Techno Co., Inc.)

	GX13	GX92	GX-T31	GZ41
CTE (25-150°C)	46	39	23	20
CTE (150-240°C)	120	117	78	67
Tg	156	153	154	171
Young's modulus (GPa)	4	5	7.5	9

Table 4: Three Levels for each of the Factors

Factor	Name	Unit	Level 1	Level 2	Level 3
A	E_{ABF}	G Pa	4	6.5	9
B	v_{ABF}		0.2	0.25	0.3
C	α_{ABF}	ppm/°C	20	70	120

The 3D response surface for the stress of TSV and bump are generated by the regression model with actual factors as:

(1)

$$Stress\_TSV = 180.424 + 21.032E_{ABF} - 2702.042v_{ABF} - 0.626\alpha_{ABF}$$
$$+ 66.559E_{ABF} \cdot v_{ABF} + 0.628E_{ABF} \cdot \alpha_{ABF}$$
$$+ 26.608v_{ABF} \cdot \alpha_{ABF} - 3.545E_{ABF}^2 + 3919.2v_{ABF}^2$$

(2)

$$Stress\_Bump = 389.082 - 10.253E_{ABF} - 223.384v_{ABF} - 2.005\alpha_{ABF}$$
$$+ 24.767E_{ABF} \cdot v_{ABF} + 0.196E_{ABF} \cdot \alpha_{ABF}$$
$$+ 2.164v_{ABF} \cdot \alpha_{ABF} + 5.329\alpha_{ABF}^2$$

The CTE of ABF material and Young's modulus of ABF material has a profound impact on the stress of TSV and bump. As shown in Fig. 7, the contour of stress distribution are curved, which illustrates the interaction with the CTE of ABF material and Young's Modulus of ABF material. From Fig. 7(a), the optimum region of stress in TSV is obtained with the lower CTE of ABF material. From Fig. 7(b), the optimum region of stress in bump is obtained with the lower

CTE of ABF material and Young's modulus of ABF material. The von-Mises stress in TSV and bump can be predicted by regression model equation.

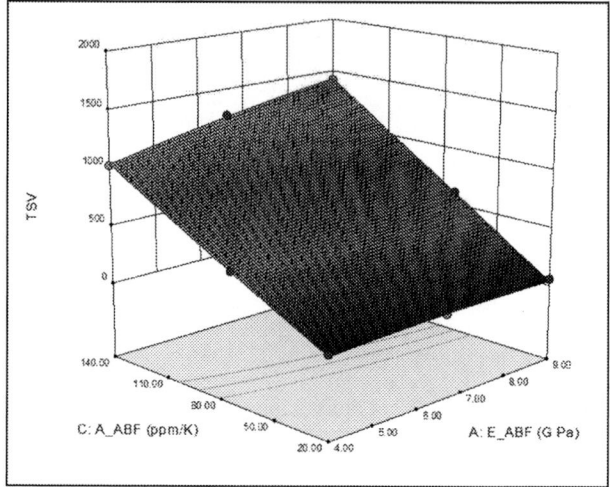

(a) Response Surface Curve of the Stress of TSV

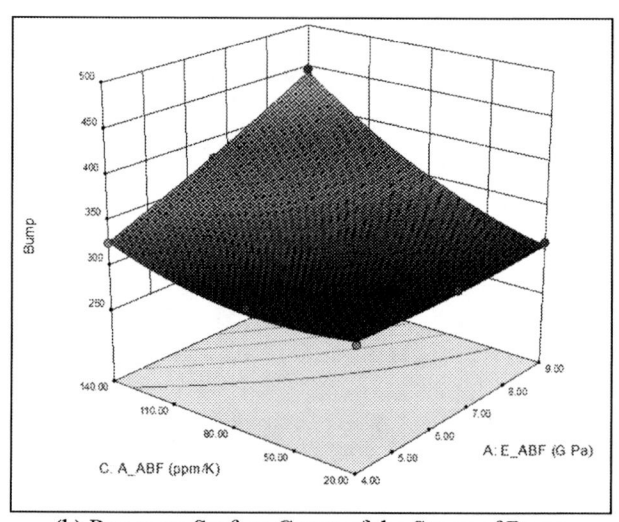

(b) Response Surface Curve of the Stress of Bump

Figure 7: The 3D Response Surface of the Interaction to CTE_{ABF} and E_{ABF}. (a) Response Surface Curve of the Stress of TSV, and (b) Response Surface Curve of the Stress of Bump.

To identify the robust of the regression model, the statistics results needed to be analyzed. The results indicate that, by optimizing the design, the von Mises stress measured at the TSV and the bump is significantly reduced. As shown in Fig. 8, the residuals can measure how many standard deviations each point in the dataset is away from the line fitted by the model. Figure 9 gives the accurate and stable RS-predicted model with actual conditions.

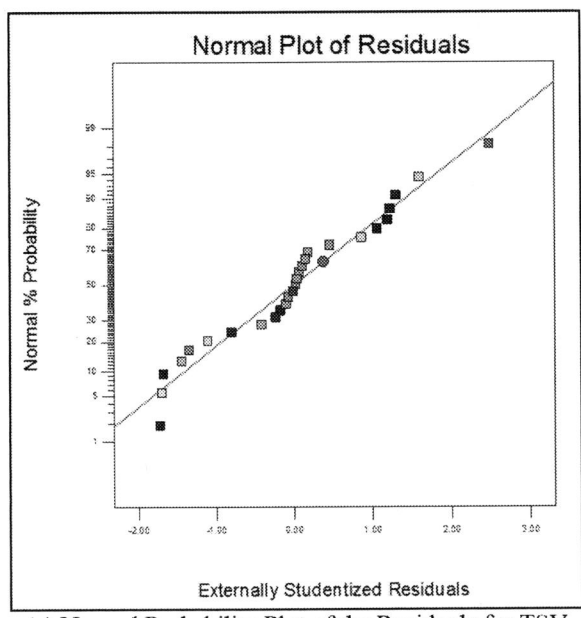

(a) Normal Probability Plot of the Residuals for TSV

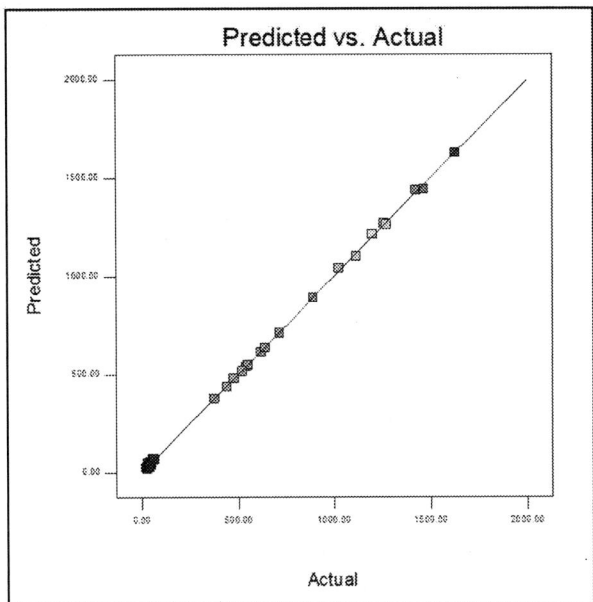

(a) Predicted Vs. Actual for TSV

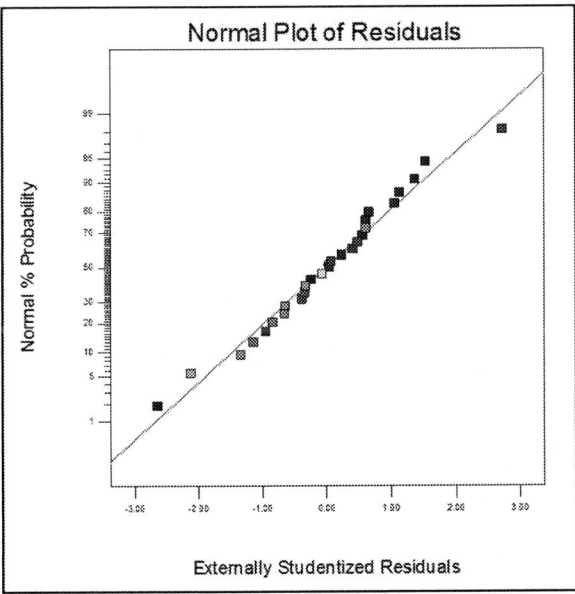

(b) Normal Probability Plot of the Residuals for Bump

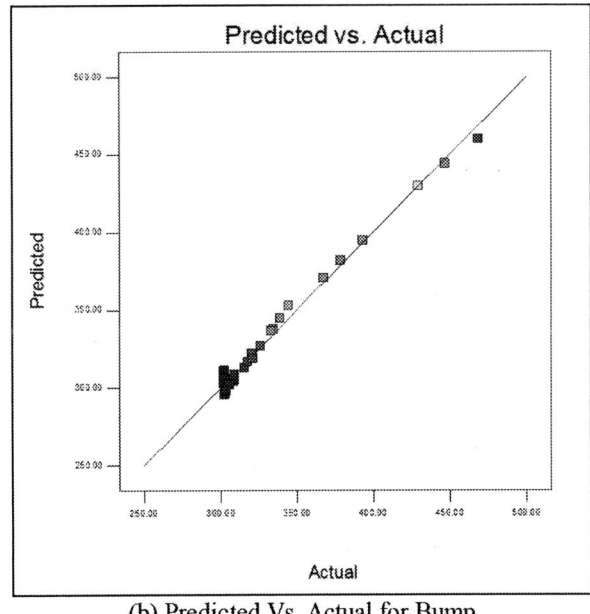

(b) Predicted Vs. Actual for Bump

Figure 8: Normal Probability Plot of the Residuals. (a) the Stress of TSV, (b) the Stress of Bump

Figure 9: Predicted Vs. Actual. (a) the Stress of TSV, (b) the Stress of Bump

Conclusion

As the stress-free temperature is set at 25 °C, a significant temperature difference between 260 °C to 25 °C, to which the package was subjected during the decreased temperature loading induces higher von-Mises stress on the TSV, and micro bumps would result from the mismatch in the coefficients of thermal expansion. Under thermo-mechanical loading, the results show that the von Mises stress is concentrated on the corner of the copper via. This paper aimed to investigate the behavior of the copper via and a micro bump in 3D IC integration when subjected to temperature loading. The results reported here indicate that the new materials can be successfully utilized to improve the strength of the copper via and the micro bump in a 3D IC package. This conclusion is supportable by PoF.

References

1. Wu, C. J., Hsieh, M. C., & Chiang, K. N. (2010). Strength Evaluation of Silicon Die for 3D Chip Stacking Packages Using ABF as Dielectric and Barrier Layer in Through-Silicon Via. *Microelectronic Engineering*. 87 (3). doi: 10.1016/j.mee.2009.08.010.
2. Wu, C. J., Hsieh, M. C., Chiu, C. C., Yew, M. C., & Chiang, K. N. (2011). Interfacial Delamination Investigation between Copper Bumps in 3D Chip Stacking Package by Using the Modified Virtual Crack Closure Technique. *Microelectronic Engineering*. 88 (5). doi: 10.1016/j.mee.2010.05.009.
3. Hsieh, M. C., Yu, C. K., & Lee, W. (2009). Effects of Geometry and Material Properties for Stacked IC Package with Spacer Structure. Proceedings of *the EuroSimE 2009 in Thermal, Mechanical and Multi-Physics simulation and Experiments in Microelectronics and Microsystems*, 26-29 April.
4. Wu, S. T., & Hsieh, M. C. (2009). Design and Simulation Study for Stacked IC Packages with Spacer Structure. Proceedings of *the IMPACT 2009 Microsystems, Packaging, Assembly and Circuits Technology Conference*, 21-23 Oct. Taipei.
5. Kuo, T. H., Su, Y. F., Wu, C. J., & Chiang, K. N. (2011). Stress/stain Assessment and Reliability Prediction of Through Silicon Via and Trace Line Structures of 3D Packaging. Proceedings of *the EuroSimE 2011 in Thermal, Mechanical and Multi-Physics Simulation and Experiments in Microelectronics and Microsystems*, 18-20 April.
6. Hsieh, M. C., & Lee, W. (2008). FEA Modeling and DOE Analysis for Design Optimization of 3D-WLP. Proceedings of *ESTC 2008*, 1-4 Sept.
7. Kuo, T. Y., Chang, S. M., Shih, Y. C., Chiang, C. W., Hsu, C. K., Lee, C. K., ... & Lo, W. C. (2008). Reliability Tests for a Three Dimensional Chip Stacking Structure with Through Silicon Via Connections and Low Cost. Proceedings of *ECTC 2008*, 27-30 May. Lake Buena Vista, FL.
8. Lo, W. C., Chen, Y. H., Ko, C. T., & Kao, M. G. (2009). TSV and 3D Wafer Bonding Technologies for Advanced Stacking System and Application at ITRI. Proceedings of *VLSI Technology*, 16-18 June. Honolulu, HI
9. Hsieh, M. C., & Yu, C. K. (2008). Thermo-mechanical Simulations for 4-layer Stacked IC Packages. Proceedings of *EuroSimE 2008 in Thermal, Mechanical and Multi-Physics Simulation and Experiments in Microelectronics and Micro-Systems*, 20-23 April. Freiburg im Breisgau.
10. Hsieh, M. C., Yu, C. K., & Wu, S. T. (2010). Thermo-mechanical Simulative Study for 3D Vertical Stacked IC Packages with Spacer Structures. Proceedings of *SEMI-THERM 2010 in Semiconductor Thermal Measurement and Management Symposium*, 21-25 Feb. Santa Clara, CA.
11. Yu, C. K., Hsieh, M. C., Liu, C. K., Dai, M. J., & Tain, R. M. (2009). The Numerical Study for the Thermal Characteristics of 3D Vertical Stacked Die Packages. Proceedings of *ASME 2009 InterPACK Conference collocated with the ASME 2009 Summer Heat Transfer Conference and the ASME 2009 3rd International Conference on Energy Sustainability*, 18-20 April. Linz
12. Lau, J. H., & Yue, T. G. (2012). Effects of TSVs (Through-silicon Vias) on Thermal Performances of 3D IC Integration System-in-package (SiP). *Microelectronics Reliability*. 52 (11). doi: 10.1016/j.microrel.2012.04.002.
13. Ryu, S. K., Lu, K. H., Jiang, T., Im, J. H., Huang, R., & Ho, P. S. (2012). Effect of Thermal Stresses on Carrier Mobility and Keep-out Zone around Through-silicon Vias for 3-D Integration. *Device and Materials Reliability, IEEE Transactions on*. 12 (2). doi:10.1115/1.4024169.
14. Savidis, I., Alam, S. M., Jain, A., Pozder, S., Jones, R. E., & Chatterjee, R. (2010). Electrical Modeling and Characterization of Through-silicon Vias (TSVs) for 3-D Integrated circuits. *Microelectronics Journal*. 41 (1). doi: 10.1016/j.mejo.2009.10.006.
15. Liu, X., Chen, Q., Sundaram, V., Tummala, R. R., & Sitaraman, S. K. (2013). Failure Analysis of Through-silicon Vias in Free-standing Wafer under Thermal-shock Test. *Microelectronics Reliability*. 53 (1). doi: 10.1016/j.microrel.2012.06.140.
16. Frank, T., Moreau, S., Chappaz, C., Leduc, P., Arnaud, L., Thuaire, A., ... & Poupon, G. (2013). Reliability of TSV Interconnects: Electromigration, Thermal Cycling, and Impact on above Metal Level Dielectric. *Microelectronics Reliability*. 53 (1). doi: 10.1016/j.microrel.2012.06.021.
17. Cheng, E. J., & Shen, Y. L. (2012). Thermal Expansion Behavior of Through-silicon-via Structures in Three-dimensional Microelectronic Packaging. *Microelectronics Reliability*. 52 (3). doi: 10.1016/j.microrel.2011.11.001.

Modeling Microstructure Effects on Electromigration in Lead-free Solder Joints

Jiamin Ni[1], Yong Liu[2], Jifa Hao[2], Antoinette Maniatty[1] and Barry O'Connell[2]

Rensselaer Polytechnic Institute[1], Troy, NY

Fairchild Semiconductor[2], South Portland, ME

Abstract

This paper studies the microstructure effects on electromigration in lead-free solder joints in wafer level chip scale package (WL-CSP). It is an extension of an earlier isotropic model [1]. The three dimensional finite element model for solder joints is developed and analyzed in ANSYS®. A sub-modeling technique combined with an indirect coupled electrical-thermal-mechanical analysis is utilized to obtain more accurate simulation results in solder bumps. . Four representative microstructures of the solder bumps are modeled and anisotropic elastic, thermal and diffusion property data are used. The results obtained from the fourrepresentative microstructures are compared with each other. The microstructure effects on electromigration are drawn from the plots of the atomic flux divergence (AFD) and the time to failure (TTF) with respect to microstructure parameters.

1. Introduction

The microelectronics industry continues to push for higher performance and downscaling of device dimensions, leading to an increase in current density that must be carried out by metallic interconnects and solder bumps. With increased current density, electromigration induced failure becomes a major reliability concern. Electromigration is a mass diffusion process attributed to momentum transfer from conducting electrons to atoms. This gradually leads to voids forming at the cathode, which causes an increase in resistance and may ultimately result in failure due to loss of connection. Many experiments and simulations have been done to study electromigration in interconnects and solder bumps. Due to environmental concerns, tin-based solders have now replaced lead-based solders in microelectronic devices. SnAgCu is one of the most promising candidates because of its competitive price and good mechanical properties. A typical solder bump in a wafer level chip scale package (WL-CSP) has a complex under bump metallurgy (UBM) on the chip side and a bump pad on the substrate side. The complex geometry of the bumps leads to current crowding. The bumps are under the loads of a non-uniform temperature field and thermal stress, which induces thermal and mechanical forces interacting with the electrical force [2].

The near eutectic Sn-Ag-Cu solder joints are more than 95 atomic percent Sn and comprised of very limited β-Sn grains [3]. The small number of grains in lead-free solder joints is caused by the difficulty in β-Sn nucleation and the associated large undercooling during solidification. β-Sn has a body-centered-tetragonal (BCT) lattice structure, with parameters a=b=583pm and c=318pm. Direction [001] (side c) is almost half the length of the basal plane sides ([100] and [010]) which leads to anisotropic behaviors in mechanical, thermal,

electrical, and diffusion properties. Therefore, the microstructure of Sn grains of the Sn-Ag-Cu solder joints has a significant effect on the electromigration process, and it has been reported that the anisotropy of the Pb-free Sn-based solders can lead to anomalous early failure [4]. However, very limited work has been done to study the microstructural effects on electromigration in lead-free solder joints.. This paper represents a first attempt at addressing the effects of microstructure.

The 3D electromigration model is further developed with consideration of representative microstructures of the lead-free solder joints in WL-CSP. The indirect electrical-thermal-structural coupled field analysis is carried out on ANSYS multi-physics simulation platform [5], and a submodeling technique is utilized to get a more accurate simulation of the critical solder bump regions. Three mechanisms are considered in the calculation of the atomic flux divergence, i.e. the elctromigration, the thermomigration and the stress migration. Four representative microstructures are modeled and analyzed with anisotropic material properties. Isotropic viscoplastic behavior is also modeled with the viscoplastic model describe in Brown et al.[6] The distributions of current, temperature, thermal gradient and hydrostatic stress are compared among those four microstructures. The atomic flux divergence is calculated and element birth/death function in ANSYS is used to show the void locations in the bumps. The effects of grain orientation and grain size on electromigration are investigated with the discussion of the simulation results.

2. Basic migration formulation and algorithm

Three driving forces for atomic migration are considered, namely electromigration, thermal gradients and gradients of mechanical stress. The atomic flux can be expressed as follows [7]:

$$\vec{J}_{Tol} = \vec{J}_{Em} + \vec{J}_{Th} + \vec{J}_{S}$$

$$= \frac{ND}{kT} Z^* e \rho \vec{j} - \frac{ND}{kT} Q^* \frac{\nabla T}{T} + \frac{ND}{kT} \Omega \nabla \sigma_h \quad (1)$$

where N is the atomic concentration; k is Boltzman's constant; e is the electronic charge; T is the absolute temperature; ρ is the temperature dependent electric resistivity tensor; $\vec{j}$ is the current density vector; D is the diffusivity tensor; Z^* is effective charge number; Q^* is heat of transport; Ω is the atomic volume; σ_h is the local hydrostatic stress. The resistivity tensor with respect to the Sn crystal lattice has two independent components, each modeled as linearly varying with temperature $\rho = \rho_0 (1 + \alpha (T - T_0))$, α is the temperature coefficient and ρ_0 is the electrical resistivity measured at temperature T_0. Likewise, there are two

978-1-4799-2408-0/14 $31.00 © 2014 IEEE

2014 Electronic Components & Technology Conference

independent components of the diffusivity tensor, where each may be expressed as $D = D_0 \exp\left(-\dfrac{E_a}{kT}\right)$, D_0 is the pre-exponential factor and E_a is the activation energy.

The time dependent evolution of the local atomic concentration is given by the continuity equation:

$$div(\vec{J}_{Tol}) + \frac{\partial N}{\partial t} = 0 \tag{2}$$

The above relationship allows for the evolution of the atomic concentration to be computed. We assume an element becomes a void when it reaches the criterion that the atomic concentration is 10% of the initial concentration.

The divergence of the total atomic flux may be approximately expressed as:

$$div(\vec{J}_{Tol}) = N \cdot F(\vec{j}, T, \sigma_m, E_a, D_0, E, \cdots) \tag{3}$$

The above equation reveals that the divergence of atomic flux is proportional to the atomic concentration, and to a function F in which different physical parameters are included. In this way, equation (2) is equivalent to

$$NF + \frac{\partial N}{\partial t} = 0 \tag{4}$$

The theoretical evolution of the atomic concentration can be obtained by:

$$N = N_0 e^{-F\Delta t} \tag{5}$$

where N_0 is the initial value of the atomic concentration and Δt is the time to get to an atomic concentration of N.

From equation (5), we can get the expression for Δt as:

$$\Delta t = -\frac{1}{F} \ln\left(\frac{N}{N_0}\right) \tag{6}$$

3. Modeling

The flip-chip CSP package in reference [8] is modeled in the commercial FEA software ANSYS. The whole structure has 36 solder bumps with 500 μm pitch. The exterior 20 solder bumps are assumed to connect with each other in a daisy chain. Due to the symmetry of the structure, a quarter of it is actually modeled. The submodeling technique in ANSYS is utilized to yield more accurate results in the critical solder bump regions. At first, a quarter global structure is modeled and analyzed with a relatively coarse mesh as in Fig. 1. Then a refined submodel of solder bumps with an UBM (Al/Ni/Cu) layer is modeled as in Fig. 2. The thermal electrical coupled simulation is carried out on the submodel to get the current density and temperature field followed by a thermal-elasto-viscoplastic submodel simulation to get the stress distribution. Thetin-based solder is modeled as anisotropic elastic, isotropic viscoplastic in the current formulation. Then atomic flux may be calculated from equation (1) and the distribution

of the atomic flux divergence can be subsequently calculated. Since the highest divergence values correspond to the nucleation locations of voids, the 30 elements with the largest atomic flux divergence will be deleted using the live/death function in ANSYS. And the time by which the element will be deleted can be calculated from equation (6) by setting N/N_0 equal to 10%. Then, the structure is automatically modified, analyzed and calculated again, until the failure condition is reached. Fig. 3 shows the flow chart of the analysis procedure.

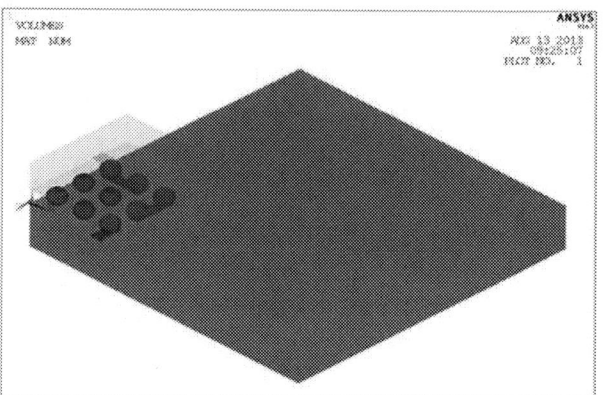

Fig. 1 Quarter global model

Fig. 2(a) Submodel

Fig. 2(b) Front view of the fine meshed solder bump

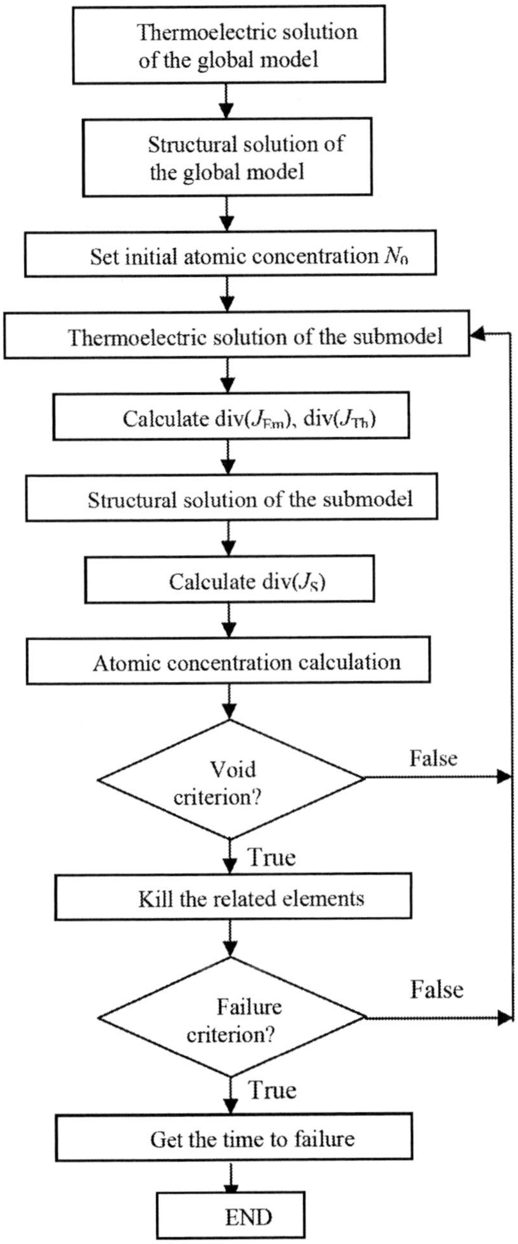

26.75GPa, $G_{yz} = G_{xz} = 2.56$GPa, $\nu_{xy} = 0.473$, $\nu_{xz} = 0.170$ and $\nu_{yz} = 0.208$ [8]. Table 1 lists the viscoplastic ANAND model parameters for SnAgCu. And the anisotropic thermal, electrical and diffusion properties are listed in the Table 2.

Table 1 Brown et al. [6] model parameters for SnAgCu

Description	Symbol	95.5Sn4.0Ag0.5Cu [9]
Pre-exponential factor	A(1/s)	325
Activation energy	Q/R(K)	10561
Stress multiplier	ξ	10
Strain rate sensitivity of stress	m	0.32
Coeff. For deformation resistance saturation value	$\hat{S}$ (MPa)	42.1
Strain rate sensitivity of saturation value	n	0.02
Hardening coefficient	h_0(MPa)	800000
Strain rate sensitivity of hardening coeff.	a	2.57
Initial value of s	s_0(MPa)	20

Table 2 anisotropic thermal, electrical, diffusion properties

	$\perp c$	$// c$
Coefficient of thermal expansion (/°C)	15.8×10^{-6}	28.4×10^{-6} [10]
Electrical resistivity at RT ($\Omega \cdot$m)	9.9×10^{-8}	14.3×10^{-8} [11]
Temperature coefficient of resistivity (/°C)	0.00469	0.00447 [11]
Effective charge number	-16	-10 [12]
Self diffusion coefficient (m^2/s)	0.0021	0.00128 [13]
Activation energy (joules/molecule)	$25.9 \times 6.95 \times 10^{-21}$	$26 \times 6.95 \times 10^{-21}$ [13]

Fig. 3 Flow chart of the analysis

3.1 Material Parameters

The anisotropic material property data of pure β-Sn is used here for the lead-free solder bump. The elastic behavior of β-Sn single crystal can be described by the engineering constants with $E_x = E_y = 76.20$GPa, $E_z = 93.33$GPa, $G_{xy} =$

3.2 Microstructure

Four representative microstructures of the lead-free solders are modeled as in Fig. 4. These microstructures are based on observed microstructures in refs. [14]-[16]. Fig. 4(a) shows a bump of two grains with the grain boundary perpendicular to the substrate. The c direction of the left grain is almost parallel to the current direction, while the c direction of the right grain is perpendicular to that. Fig. 4(b) shows a tri-grain bump and the disorientation angle between the grains is almost 60°. Fig. 4(c) shows a bump of two grains with the grain boundary inclined at 45° to the substrate. Fig. 4(d) shows a twining structure of 60° rotations with 3 dominant cyclic twin orientations cyclically repeating around the nucleus which is known as a Kara's beach ball structure.

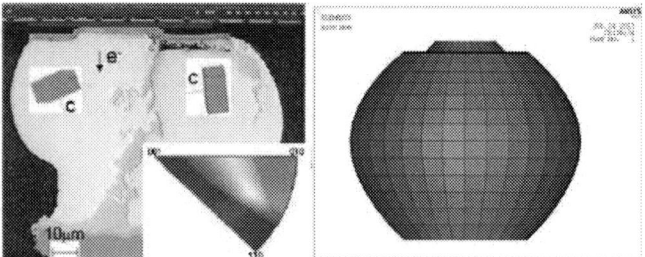

Fig. 4(a) 3D Ansys finite element model of a bi-crystal bump similar to observed structure in [14]. (Delete images from other papers unless you have permission from the publishers to reprint)

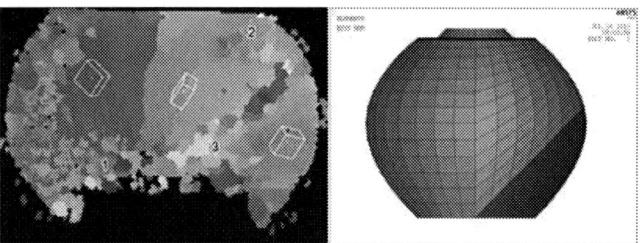

Fig. 4(b) 3D Ansys finite element model of a tri-crystal bump [15].

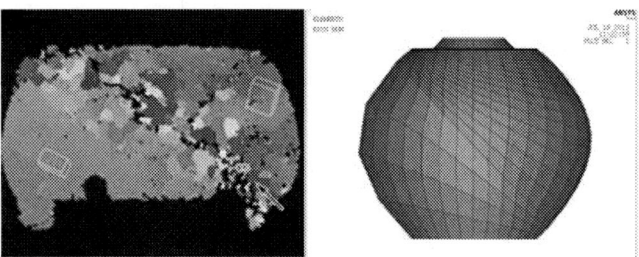

Fig. 4(c) EBSD 3D FE model of a bump with grain boundary inclined at 45° to substrate [15]

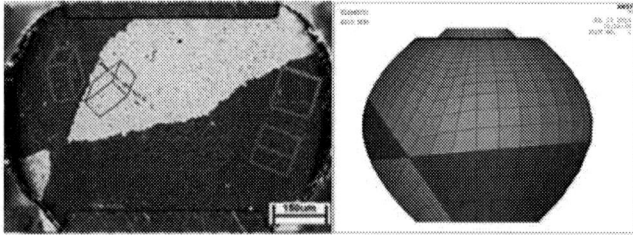

Fig. 4(d) 3D finite element model of a beach-ball bump [16]

4. Simulation results and discussion

Fig. 5 shows the distribution of the temperature of the four representative microstructures. The difference in temperature of the solder bumps is small, about 2 degree Celsius. There is little difference in the temperature distribution for the four microstructures. Fig. 6 shows the current distribution in the four microstructures. There is current crowding where the current enters the bump. And the tri-grain model has the highest current density among those four microstructures. Fig. 7 shows the distribution of thermal gradient of the four microstructures. We can see that the thermal gradient is going in the opposite direction of the current. Fig. 8 shows the

distribution of the hydrostatic stress of the four microstructures. There is a large variation in the hydrostatic stress field among the four microstructures. The maximum hydrostatic stress in the 45° model is more than twice of that in the bi-crystal model.

Fig. 9 shows the dynamic distribution of the atomic flux divergence of the four representative microstructures and the nucleation of voids. We can see that the beach ball model has the minimum nucleation void volume and the bi-grain model has the maximum voids among the four microstructures. Since elements with higher atomic flux divergence will void faster, this distribution gives us information about the nucleation of voids and time to failure.

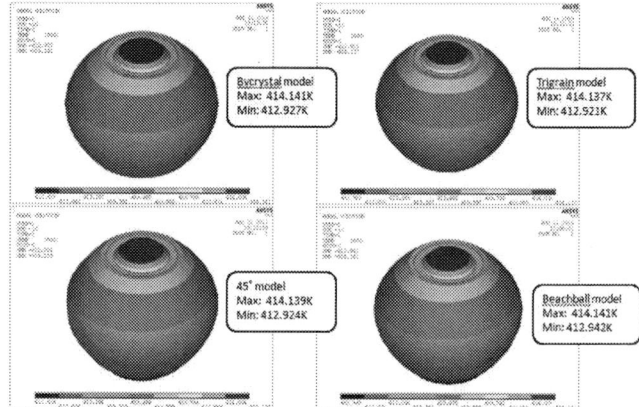

Fig. 5 Temperature distribution of 4 microstructures

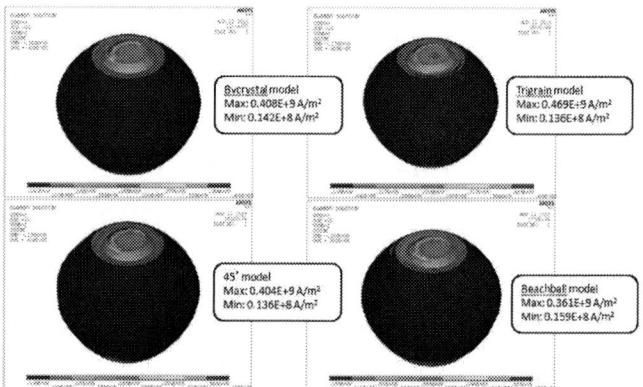

Fig. 6 Current distribution of 4 microstructures

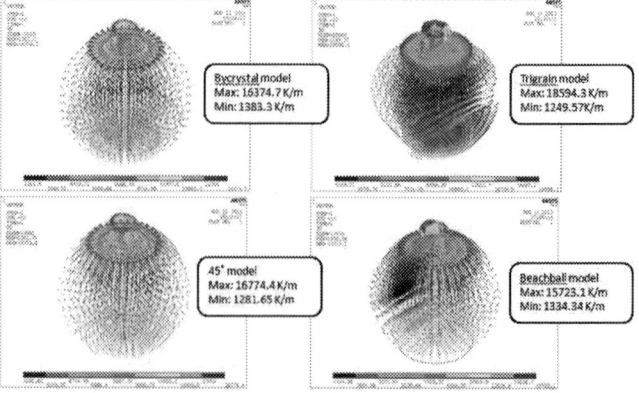

Fig. 7 Thermal Gradient of 4 microstructures

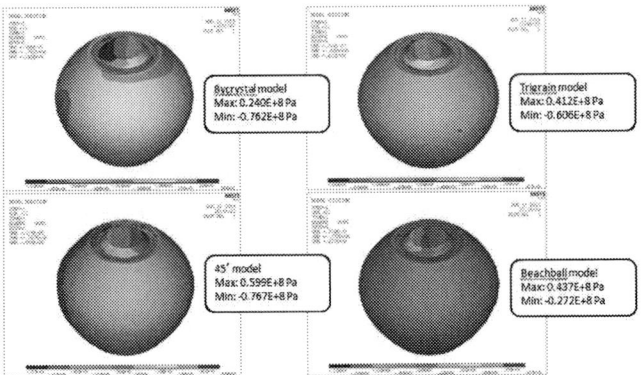

Fig. 8 Hydrostatic stress of 4 microstructures

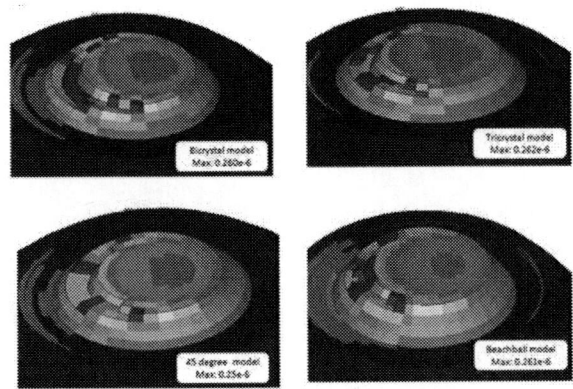

Fig. 9 Atomic flux divergence of 4 microstructures

Table 3 gives the time to failure for the four representative microstructures. The bi-grain model has the minimum time to failure followed by the tri-grain model and the 45°, and the maximum TTF is from the beach-ball model. This has something to do with the orientation of the grains. For the bi-grain model to the beach ball model, the grain boundary layers increase, which might induce the slow mass diffusivity in the current flow direction. Therefore, it results in longer time to failure.

Table 3 TTF for 4 different microstructures

Model	Bi-crystal	Trigrain	45°	Beach-ball
TTF (h)	1137	1296	1370	1471

In order to better understand the effects of the grain orientation on electromigration, we run the single crystal case with different grain orientations. Fig. 10 shows the life of time to failure versus the angle of crystal orientation. It shows when c-direction is aligned with the current flow direction, the TTF is the largest. As the c-direction gradually deflects from the current direction, the TTF decreases until the c-direction becomes perpendicular to the current direction. when the c-direction is perpendicular to the current flow direction, the TTF reaches the minimum, which makes sense because the self diffusivity of tin along a-direction is faster than that along c-direction at the working temperature. One interesting phenomenon is at the angle of c-direction 60 degree, the TTF raises a little longer as compared to the TTF at the c-direction 30 degree. This possibly might be the coupling effect of thermal mechanical stress and the pure

electro-wind induced migration at the 60 degree, which induces a slight smaller mass migration.

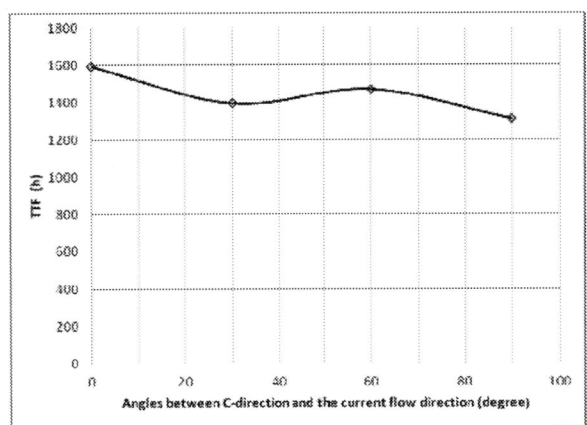

Fig. 10 TTF vs. angles between c-direction and current flow direction in a single crystal

5. Conclusions

The three dimensional in-direct electrical, thermal and structural coupled analysis for the solder bumps in WL-CSP is carried out with consideration of representative microstructures. The distributions of the current, thermal gradient, hydrostatic stress and atomic flux divergence are presented and compared among four representative microstructures of lead-free solder bumps. The effects of grain orientation and grain size on electromigration are studied and discussed. From the curve of TTF vs. crystal orientation angles, we can see that as the c-direction gradually deflects from the current flow direction, the TTF will decrease, which means that the bump is more prone to electromigration.

In the future, Electromigration test to measure the migration of different microstructures of lead-free solder bumps will be conducted for correlation between the simulation results and the experiment data.

Acknowledgments

The authors wish to acknowledge the support from Reliability Department of Fairchild Semiconductor Corp. This work was also supported by the National Science Foundation under Grant No. DMR-1207291.

References

1. Y. Liu, S. Irving, T. Luk, L. Liang, S. Wang, "3D modeling of electromigration combined with thermal-mechanical effect for IC device and package", EuroSime 2007.
2. Y. Liu, "Finite element modeling of electromigration in solder bumps of a package system", EPTC2008 .
3. S. Yang, Y. Tian, C. Wang, T. Huang, "Modeling Thermal Fatigue in Anisotropic Sn-Ag-Cu/Cu Solder Joints", International Conference on Electronic Packaging Technology & High Density Packaging (ICEPT-HDP), 2009.
4. K. N. Subramanian, and J. G. Lee, "Effect of Anisotropy of Tin on Thermomechanical Behavior of Solder Joints",

J. Mater Sci: Mater Electron, Vol. 15, No. 4 (2004), pp.235-240.

5. http://www.ansys.com

6. S. B. Brown, K. H. Kim, and L. Anand, "An internal variable constitutive model for hot working of metals", Int. J. Plasticity, Vol. 5, pp. 95-130, 1989.

7. D. Dalleau, K. Keide-Zaage, "Three-Dimensional Simulation in chip Metallization Structures: a Contribution to Reliability Evaluation", Microelectronics Reliability 41 (2001) 1625-1630.

8. S. Gee, N. Kelkar, J. Huang and K. Tu, "Lead-free and PbSn Bump Electrmigration Testing", Proceedings of IPACK2005, IPACK2005-73417, July 17-22.

9. Q. Wang, et al, "Experimental determination and modification of the Anand model constants for 95.5Sn4.0Ag0.5Cu", Eurosime 2007, London, UK, April, 2007.

10. J. Zhao, P. Su, M. Ding, S. Chopin, and P.S. Ho, "Microstructure-based stress modeling of tin whisker growth," IEEE Transactions on Electronics Packaging Manufacturing, , Vol. 29, No. 4, October 2006.

11. K. J. Puttlitz, and K. A. Stalter, "Handbook of Lead-free solder technology for microelectronic assemblies", pp. 920-926.

12. H. B. Huntington, "Effect of driving forces on atom motion", Thin Solid Films, Vol. 25, No. 2, pp. 265-280, 1975.

13. F. H. Huang and H. B. Huntington, "Diffusion of Sb124, Cd109, Sn113, and Zn65 in tin", Physical Review B, vol. 9, Issue 4, pp. 1479-1488, 1974.

14. M. Lu, D. Shih, P. Lauro, C. Goldsmith, D. W. Henderson, "Effect of Sn grain orientation on electromigration degradation in high Sn-based Ph-free solders", Applied Physics Letter 92, 211909 (2008).

15. J. Xu, "Study on lead-free solder joint reliability based on grain orientation", Acta Metallurgica Sinica, Vol. 48, No. 9, pp. 1042-1048 (2012).

16. S. Park, R. Dhakal, J. Gao, "Three-dimensional finite element analysis of multiple-grained lead-free solder interconnects", Journal of Electronic Materials, Vol. 37, No. 8, 2008

Experimental Demonstration of the Effect of Copper TPVs (Through Package Vias) on Thermal Performance of Glass Interposers

Sangbeom Cho, Yoichiro Sato*, Venky Sundaram, Yogendra Joshi and Rao Tummala
3D Systems Packaging Research Center, Georgia Institute of Technology
813 Ferst Drive N.W., Atlanta GA 30332
*Asahi Glass, Japan
E-mail:scho84@gatech.edu

Abstract

Glass has been proposed to be an ideal material for interposers to address the limits of both organic and silicon, except for its poor thermal conductivity compared to silicon. This paper proposes to use large number of copper through package vias (TPVs) that can be fabricated at low cost to improve this shortcoming. We report on experimental fabrication and evaluation of the effect of copper TPVs on thermal performance of glass interposer structures. Copper via arrays with different via densities (0.62 vias/mm^2 ~ 22 vias/mm^2) were fabricated in glass interposer structures by using low cost laser drilling and electroplating. The thermal performance of such structures was quantified by measuring spatial temperature distribution, including the maximum value around the heat source. This study provides fundamental understanding of heat transfer within thermally-enhanced glass interposers, and offers guidelines for the design of copper via arrays to improve their thermal properties.

Introduction

Glass has been demonstrated as an ideal package material [1], except for its poor thermal conductivity compared to silicon. Its thermal conductivity (~1 W/m·K) is an order of magnitude higher than organics, but two orders of magnitude lower than silicon (~150 W/m·K). However, this limitation of glass interposers can potentially be overcome by implementation of large number of high thermal conductivity (~400 W/mK) copper through package vias (TPVs) [2]. If routing space allows, additional copper TPVs can be utilized as purely thermal vias, which can further improve thermal conduction in glass interposers.

Recent simulations have reported the favorable effect of copper Through-Silicon-Vias (TSVs) on thermal performance of silicon interposers for Flip Chip Ball Grid Array (FCBGA) package [3], network system application [4], and 3D IC integration [5]. Thermal influence of the array density of TSVs, and the proximity to a heat source have been studied through modeling, and validated through experiments by using back-end-of-line (BEOL) structures [6]. Analytical model for thermal via network design has been developed to optimize various parameters including diameter, pitch, plating thickness, and trapped air void level of the filled copper inside the vias [7]. More recently, integrated thermal via planning with 3D IC floor planning algorithm has been developed [8]. However, most of the published literature deals with TSVs in silicon interposer or silicon based 3D IC structures, and is based only on simulations, without experimental validations.

This paper experimentally examines thermal performance of TPVs in glass interposer structures, with different design parameters by using infrared microscopy.

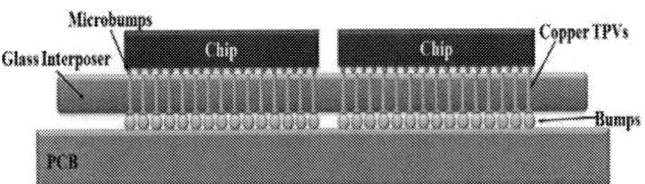

Figure 1. 2.5D glass interposer with copper TPVs

TPV Fabrication and Metallization on Glass

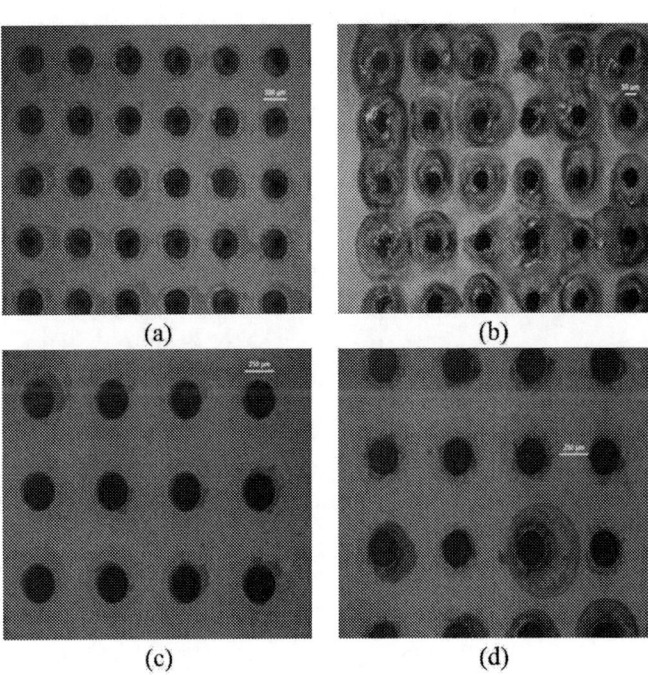

Figure 2. Optical image of via entrance with (a) 100 μm, (c) 260 μm diameter and via exit with (b) 70 μm, (d) 240 μm diameter formed by UV laser.

Various TPV arrays were fabricated on a 114 mm x 114 mm x 100 μm size borosilicate glass panel. The panel was divided into 36 18 mm x 18 mm samples with via entrance diameters of 100 μm, 180 μm and 260 μm, and pitches of 200 μm, 400 μm, 600 μm, and 900 μm. Prior to via formation, 22.5 μm thick polymer was laminated on both sides of the glass, which to protect it from the direct impact of laser during the via drilling. The polymer laminated glass sample was then subjected to UV laser ablation for via formation. Using UV laser resulted in tapered via profile, and Figure 2 shows different diameter of via entrance (100 μm and 260 μm) and exit (70 μm and 240 μm) on polymer laminated glass.

To achieve good metallization on glass samples, the surface of polymer was roughened through micro etch process. Then a copper seed layer was formed, followed by copper electrolytic plating. Via pads were patterned through photolithography process on both sides of the sample. Via pad diameters were designed to be 40 μm larger than via diameters. Via dimensions including plating thickness and pad size for each sample are tabulated in Table 1.

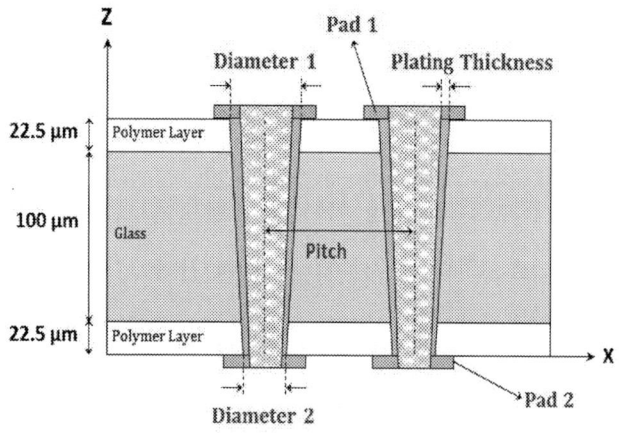

Table 1. Via geometry parameters

Diameters (1 & 2)	Pad (1 & 2)	Pitch	# of vias
1:100 μm, 2: 70 μm	1: 140 μm, 2: 110 μm	200 μm	144
		400 μm	36
		600 μm	16
1: 180 μm, 2: 160 μm	1: 220 μm, 2: 200 μm	400 μm	36
		600 μm	16
		900 μm	4
1: 260 μm, 2: 240 μm	1: 300 μm, 2: 280 μm	600 μm	16
		900 μm	4

Test Setup

A 2.5 mm x 2.5 mm heater was attached to each fabricated glass sample (18 mm by 18 mm) with vias. Via area in the sample was carefully aligned with the heater, so that most of the heat generated could be applied to the via zone as shown in Fig 3. The heater consists of two resistors for heating and four diodes for temperature measurement across the die. Each resistor dissipates a maximum power of 6 W. The heater was wire bonded to a PCB, which was connected to a power supply. After bonding, the heater and wire bonds were sealed with epoxy to protect them from mechanical and electrical

impact. Thermally conductive paste with thermal conductivity of 1.49 W/m·K was applied between heater and sample to reduce contact resistance.

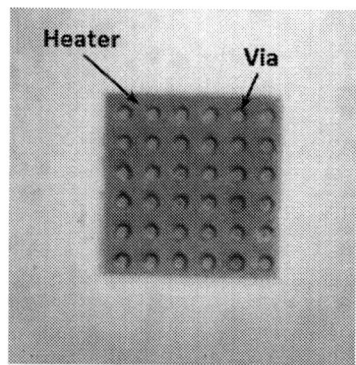

Figure 3. Infrared image of via zone on heater before heating (D: 180 μm, P: 400 μm)

Diodes embedded in the heater chip were used to measure temperature changes. For diode calibration, the chip assembly was placed in a temperature controlled oven. The assembly was connected to a constant current power supply. 1 mA was selected for diode forward current during the measurement, to avoid excessive self-heating, which causes measurement errors. Thermocouples were utilized to check the temperature stability and uniformity around the chip. When the temperature changes measured by thermocouples became less than 0.5 °C over 10 minutes of measurement, the diode forward voltage was recorded. The calibration was carried out between 20°C and 100 °C, over 10 °C intervals.

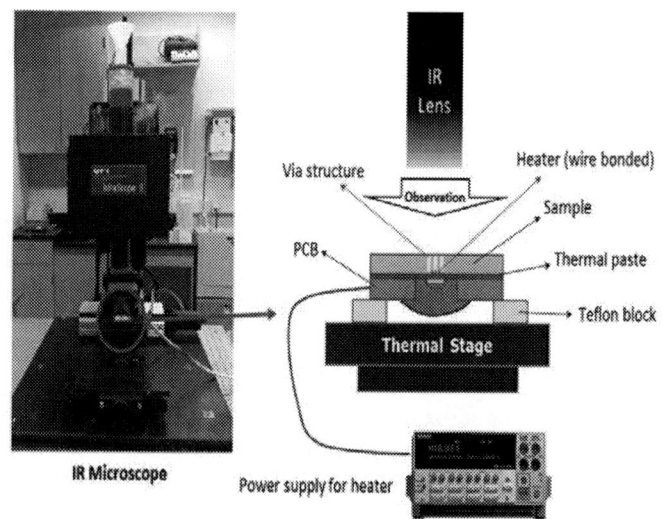

Figure 4. Schematic of test setup

A Quantum Focus Instruments Infrared (IR) microscope, with a spatial resolution of 2.8 μm and a pixel size of 1.6 μm, was used to measure thermal performance of the each sample with different via parameters. While diodes embedded in the chip captured its temperature changes, the infrared microscope detected surface temperature field of the sample.

Figure 4 shows the diagram of the test setup used for the measurement.

Test Procedure

Teflon bocks were placed between the test-chip-sample assembly and thermal stage of the IR microscope to thermally isolate the assembly from the microscope system. The epoxy which protects the heater has a low thermal conductivity (~ 0.3 W/m·K), to minimize heat loss from the back side of the heater to the ambient. The chip heater was turned on after measuring reference radiance from the sample for IR microscope calibration. The IR microscope image was acquired when the assembly reached steady state, as inferred from five thermocouples attached at different locations. Steady state was assumed when the temperature change at each location was within 0.5 °C during the last 10 minutes of observation. The chip temperature was also measured by recording the diode forward voltage. Nine different types of via samples, including a no via sample were measured during the tests.

During the first experiment, average surface temperatures of a sample were measured at heater powers of 130 mW, 180 mW, 197 mW and 216 mW to check the sample thermal resistance, and the repeatability of results. The power range was chosen after measuring the heater temperature without the sample to prevent excessive heat from damaging the epoxy around the heater. For the experiment, a via array sample with 100 μm diameter and 200 pitch μm was used. Figure 5 shows a thermal resistance network representing the test setup. At each power condition, surface temperature profile of the sample, and heater temperature were recorded to calculate overall thermal resistance for the thermal paste and glass sample.

During the second experiment, surface temperatures of samples with different via patterns, and corresponding chip heater temperatures were recorded at 180 mW heater power. Thermal resistance for each sample was calculated for performance comparison.

During the third experiment, temperature changes of a sample which has via arrays connected to Cu spreader (2.5 mm x 2.5 mm x 10 μm) were measured to study the effect of copper TPVs in decreasing the assembly temperature.

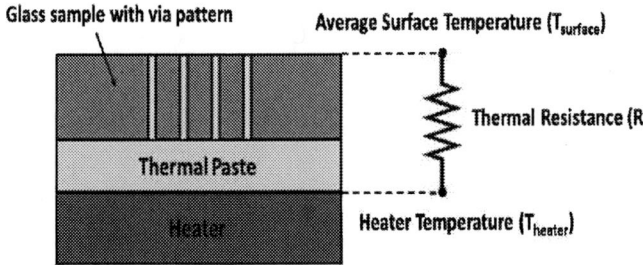

Figure 5. Thermal resistance model for test sample

Experimental Results and Discussion

Figure 6 shows the surface temperature profile of the sample with 100 μm diameter and 200 μm pitch via array, subjected to 4 different heater powers. The color in the temperature image shows that temperature increased as the heater power increased.

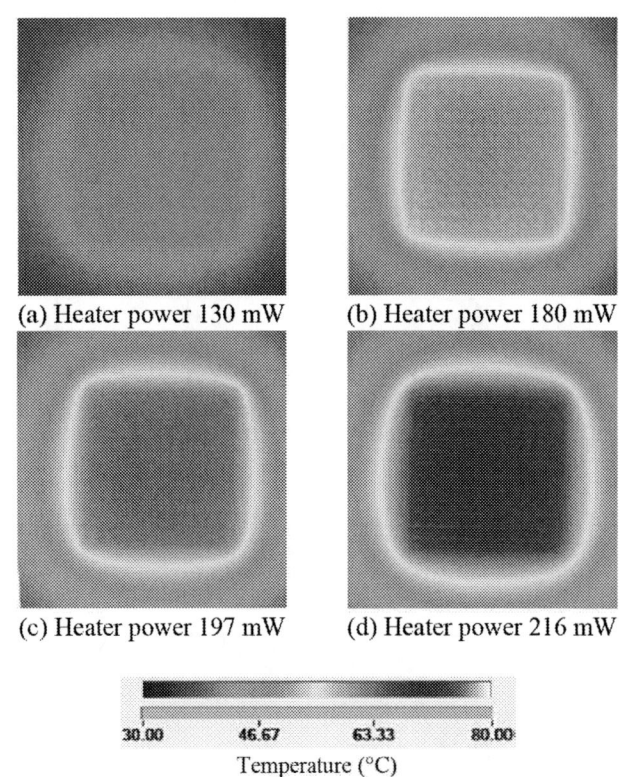

Temperature (°C)

Figure 6. Sample surface temperature with different heater power

Before each measurement, the sample and heater assembly were cooled down to ambient temperature. After 4 steady state temperature measurements, thermal resistance R of each case was calculated by using Eq (1), where T_{heater} is heater temperature measured by diode, $T_{surface}$ is average surface temperature of heated area and Q is heater power used for each measurement (130 mW, 180 mW, 197 mW and 216 mW).

$$R = \frac{T_{heater} - T_{surface}}{Q} \qquad (1)$$

Figure 7 compares thermal resistance values calculated from 4 different heater power conditions. All thermal resistance values are normalized by maximum value (thermal resistance at 130 mW).

Measured thermal resistance decreased as the heater power increased and the maximum difference in thermal resistances appeared between 130 mW (minimum power) and 216 mW (maximum power), which was ~3% of maximum power. This trend is mainly due to the enhanced convective and radiative heat transfer occurring at the surface of a sample

978-1-4799-2408-0/14 $31.00 © 2014 IEEE

at higher temperature, resulting in a slight decrease in the internal thermal resistance. This result confirms that using the thermal resistance concept to compare thermal performance of different via structures will be acceptable, as long as the same heating power condition is applied to each sample.

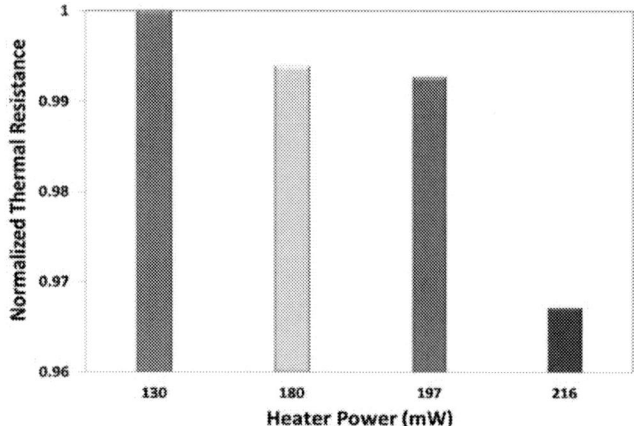

Figure 7. Normalized thermal resistance calculated at different heater power (Via sample diameter: 100 µm, Pitch: 200 µm)

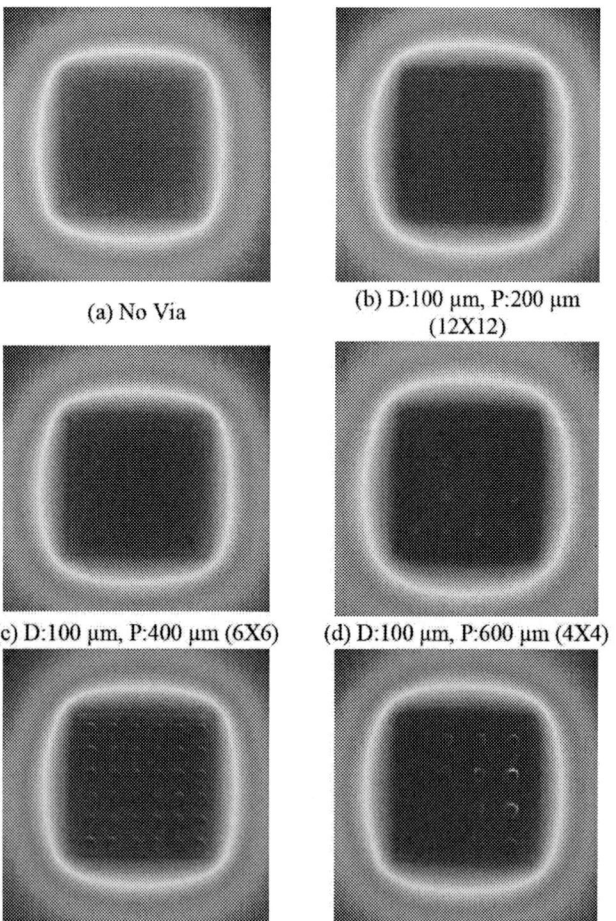

(a) No Via

(b) D:100 µm, P:200 µm (12X12)

(c) D:100 µm, P:400 µm (6X6)

(d) D:100 µm, P:600 µm (4X4)

(e) D:180 µm, P:400 µm (6X6)

(f) D:180 µm, P:600 µm (4X4)

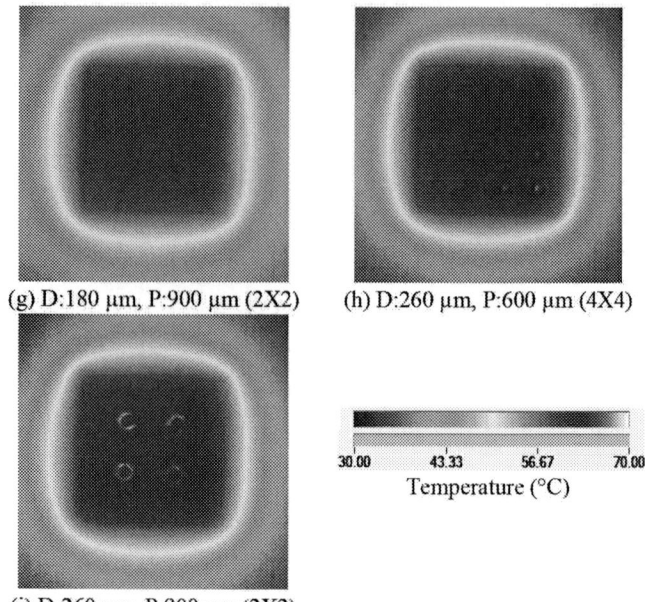

(g) D:180 µm, P:900 µm (2X2)

(h) D:260 µm, P:600 µm (4X4)

(i) D:260 µm, P:900 µm (2X2)

Figure 8. Surface temperature of different via samples

Figure 8 shows surface temperature of various via samples. The heater power was controlled to 179 mW ~ 181 mW to avoid large surface temperature differences between samples. Power generated by heater was not exactly the same for all cases since the heater resistance slightly increased with temperature. Heater temperature measured by the diode showed similar values for all sample cases, ranging from 72 °C to 75 °C. As shown in Figure 8, surface temperature of sample without via structure, (a), was found to be the lowest due to the sample's low effective thermal conductivity. Sample with 100 µm diameter and 600 µm pitch via array showed the highest surface temperature due to the higher heater power applied than any other samples. At each sample's surface, copper via pads exhibited the highest temperature, showing that copper structures were major heat flow paths in the sample structure. Average surface temperature was extracted from the recorded data and thermal resistances of all 9 samples were calculated by using Eq (1) and they are plotted in Figure 9. The resistance values are normalized by maximum value.

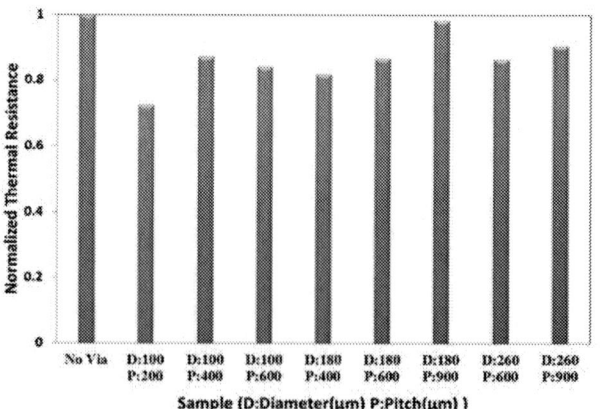

Figure 9. Normalized thermal resistance of the samples

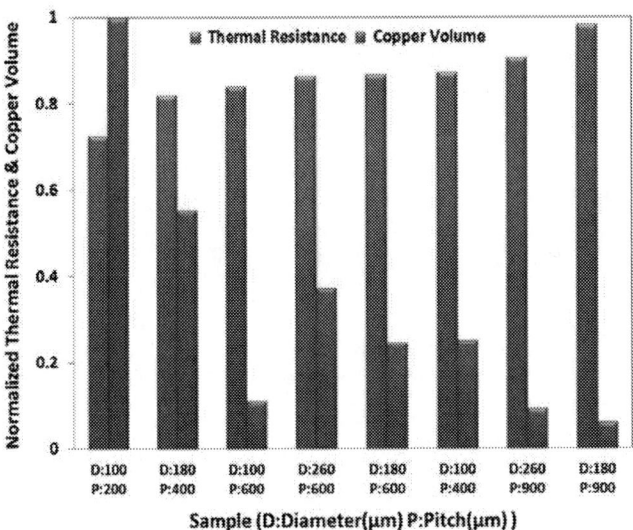

Figure 10. Normalized thermal resistance of samples and copper volume in the samples

As shown in Figure 9, thermal resistance of sample without via structure presented the highest value, while the sample with the highest via density (D:100 μm, P:200 μm) showed about 30% decreased value compared to it. Figure 10 rearranges the thermal resistance data in ascending order. Copper volume (mm^3) in each sample is also normalized by its maximum, and presented together with thermal resistance as the out-of-plane thermal performance is closely related to the amount of the copper used for vias. The thermal performance of various samples can be arranged in accordance with the amount of copper, except for a sample with via diameter of 100 μm and via pitch of 600 μm. Further discussions on the results are presented in the following uncertainty analysis section.

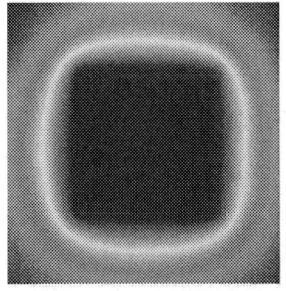

(a) D:100 μm, P:200 μm (12X12)

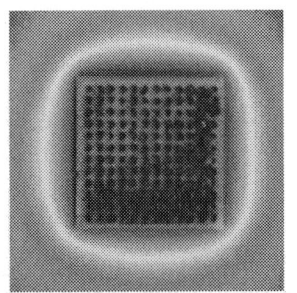

(b) D:100 μm, P:200 μm connected to copper metal layer(12X12)

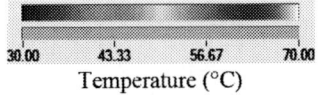

Temperature (°C)

Figure 11. Surface temperature profile of via structure (D: 100 μm, P:200 μm)(a) without metal layer, and (b) with metal layer

A Cu metal layer with 10 μm thickness (2.5 mm x 2.5 mm) was connected to via arrays with 100 μm diameter and 200 μm pitch to observe the effect of copper TPVs on thermal performance, when connected to a metal heat spreading layer. Figure 11 shows the surface temperature of the structure and compares it with the sample with the same condition but without metal layer. As the heat from the heater flowed through vias, and was dissipated in metal layer, average surface temperature as well as heater temperature dropped by ~4 °C and ~5 °C, respectively. Figure 12 compares its thermal resistance with that of the sample without via structure, and the sample without heat spreader.

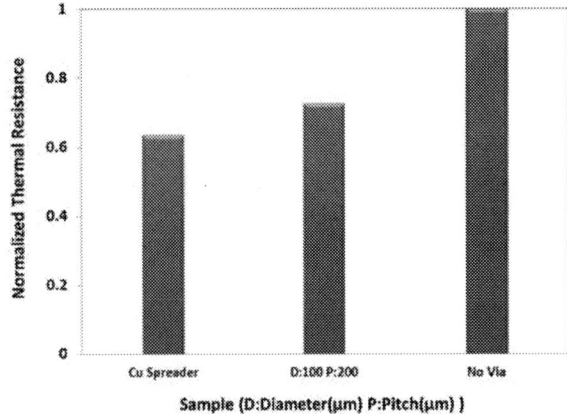

Figure 12. Normalized thermal resistance of 3 different samples (Sample connected to Cu spreader, sample with via diameter of 100 μm and pitch of 200 μm, and sample without via structure)

By connecting vias to the copper spreader structure, thermal resistance was reduced by ~40 % compared to sample without via structure. The resistance can be further reduced when additional cooling is provided on the spreader surface.

Uncertainty Analysis

Uncertainty analysis has been performed on the measurement setup, including DC power supply, heater, diode, and IR microscope. The uncertainty of calculated thermal resistance is determined by Eq (2):

$$U_R = \sqrt{\frac{\partial R}{\partial \Delta T}U_{\Delta T} + \frac{\partial R}{\partial Q}U_Q} \qquad (2)$$

where ΔT is temperature difference, Q is heater power, U_R is total uncertainty for thermal resistance, U_T is uncertainty associated with temperature measurement, and U_Q is uncertainty with power input for heater. The temperature uncertainty is based on the reported sensitivity of 0.1 K of the Infrascope II and the power input uncertainty is based on a separate error propagation analysis as shown in Eq (3).

$$U_Q = \sqrt{\frac{\partial R}{\partial V} U_V + \frac{\partial R}{\partial I} U_I} \qquad (3)$$

where V is voltage, I is current, U_V is uncertainty associated with voltage input and U_I is uncertainty with current reading. The uncertainty in the voltage input is 0.05%+20 mV and the readback accuracy of the current across the heater is 0.15%+4 mA. Based on the voltage input for each test case, total uncertainty for thermal resistance is calculated and presented in Figure 13.

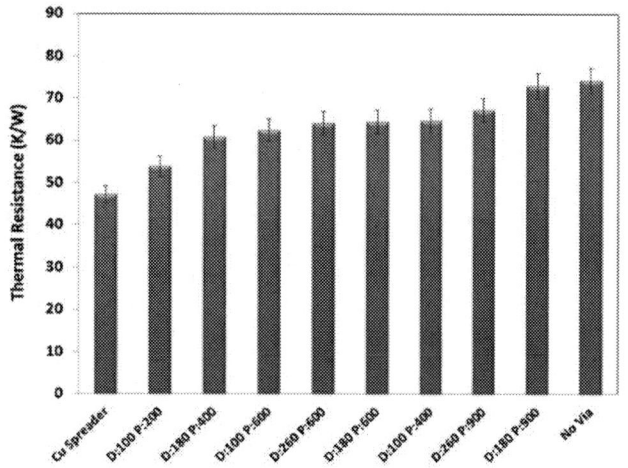

Figure 13. Measured thermal resistances of all samples and their uncertainties

The uncertainties from the test setup make it difficult to clearly rank the thermal performance of via structures among the samples with 400 µm, 600 µm and 900 µm via pitches, as they showed close thermal performance to each other. The uncertainty could affect the result for a sample with via diameter of 100 µm and via pitch of 600 µm, which showed higher thermal performance than was expected. However, it is shown that the implementation of 144 copper vias (D: 100 µm P: 200 µm) increased the thermal performance of glass interposer structure by 20 % - 33 % and by 31 % - 41% with Cu spreader compared to a sample without via structure. Additional measurements are under way to reduce overall measurement uncertainties.

Summary

The effect of copper TPV design parameters including diameter and pitch on the thermal performance of glass interposer structure was measured. The results showed that incorporating internal heat transmission pathways, such as copper TPVs in the glass interposer structures where needed selectively, mitigated the heat conduction problem caused by low thermal conductivity of glass, making glass interposers an attractive option for emerging microsystems packaging applications. Further measurements are needed to characterize thermal performance of via arrays.

Acknowledgements

This research was supported by the Glass and Silicon Interposer Industry Consortium at the Georgia Tech 3D Systems Packaging Research Center.

References

1. Vijay Sukumaran, Qiao Chen, Fuhan Liu, Nitesh Kumbhat, Tapobrata Bandyopadhyay, Hunter Chan, Sungwhan Min, Christian Nopper, Venky Sundaram, and Rao Tummala, "Through-Package-Via Formation and Metallization of Glass Interposer," Proc. IEEE Electronic Components and Technology Conference (ECTC), Las Vegas, NV, June 1-4, 2010.

2. Sangbeom Cho, Yogendra Joshi, Venky Sundaram, Yoichiro Sato, and Rao Tummala, "Comparison of Thermal Performance Between Glass and Silicon Interposers," Proc. IEEE Electronic Components and Technology Conference (ECTC), Las Vegas, NV, May 28-31 , 2013.

3. Yen Yi Germaine Hoe, Tang Gong Yue, Pinjala Damaruganath, "Effect of TSV Interposer on the Thermal Performance of FCBGA Package," Proc. Electronic Packaging Technology Conference, pp. 778-786, 2009.

4. Heng-Chieh Chien, John H. Lau, Yu-Lin Chao, Ming-Ji Dai, Ra-Min Tain, "Thermal Evaluation and Analyses of 3D IC Integration SiP with TSVs for Network System Applycations," Proc. IEEE Electronic Components and Technology Conference (ECTC), San Diego, CA, May 29 - June 1, 2012.

5. John H. Lau, and Tang Gong Yue, "Thermal Management of 3D IC Integration with TSV (Through Silicon Via)," Proc. IEEE Electronic Components and Technology Conference (ECTC), San Diego, CA, May 26 - 29, 2009.

6. H. Oprins, A. Srinivasan, M. Cupak, V.Cherman, C. Torregiani, M. Stucchi, G. Van der Plas, P. Marchal, B. Vandevelde, E. Cheng, "Fine Grain Thermal Modeling and Experimental Validation of 3D-ICs," Microelectronics Journal, 42, pp. 572-578, 2011.

7. R. S. Li, "Optimization of Thermal Via Design Parameters Based on an Analytical Thermal Resistance Model," Thermal and Thermomechanical Phenomena in Electronic Systems, pp. 475-480, May 1998.

8. Zhuoyuan Li, Xianlong Hong, Qiang Zhou, Shan Zeng, Jinian Bian, Hannah Yang, Vijay Pitchumani, Chung-Kuan Cheng, "Integrating Dynamic Thermal Via Planning with 3D Floorplanning Algorithm," Proceedings of the international symposium on Physical design, pp.178-185, 2006.

Failure Mechanism Investigation of Stacked Via Cracking in Organic Chip Carrier

Shidong Li, Yi Pan, Sushumna Iruvanti, David L. Questad, Randall J. Werner
IBM Corporation
2070 Route 52, Hopewell Junction, NY, 12533-6531

Abstract

The increasing demand for high density interconnects leads to the adoption of stacked via technology. By layering multiple vias directly on top of each other, via stacking allows for more compact and flexible routing. However, due to the geometric discontinuity and non-uniform stiffness, stacked vias also present significant reliability challenges.

In the investigation of via stack cracking mechanism in packaging applications, 16 types of stacked and staggered via chain structures were designed and fabricated in an organic chip carrier test vehicle. The experiments were also designed to evaluate other effects such as stacked via location, laminate materials, etc. Comparison of fail counts versus via chain types after 1000 cycles of deep thermal cycling (DTC) revealed that some types of stacked via structures are significantly more robust than others. Strong location dependency of stacked via fail was also observed by comparing the identical stacked via structure in different locations: out of 75 modules, 31 fails were detected in the stacked via chain under the chip center, but none under the chip corner.

This paper focuses on the development of a predictive model with finite element method. Modeling activities were carried out to investigate the effect of via structure, package geometry, laminate material and other form factors on via cracking. The thermal-mechanical modeling methodology will be described in this paper. The discussion of failure mechanism and the correlation of simulations with experimental results will be presented.

Introduction

Organic substrates are widely implemented for silicon packaging due to low cost and electrical performance enhancements. The key technology elements of an organic substrate include build-up layers containing most of the wiring, a core and the surface finish for soldering and adhesion. Volume production and low cost of organic substrates were enabled by breakthroughs in laser drilling technology for micro-via, a metalized connecting channel that provides a layer-to-layer connection [Blackshear 2005].

The increasing demand for computation performance calls for higher chip frequencies, higher bandwidth, and lower latencies that lead to continuous technology scaling. Moreover, continuous scaling and system integration (adding new and more functionality, such as RF communication, sensors, etc.) require higher I/O density both at the wafer level and at the package level. The 2012 ITRS roadmap calls for organic substrates with lines and spaces below 10um for 16nm technology and beyond [ITRS 2012].

Stacked via technology has been used in substrates for wiring efficiency. It allows for small form factors, high I/O densities and high speed[Liu 2003; Lo 2008].

As increased I/O and decreased pitch reduce signal wiring channels, more conductive layers are therefore introduced to increase wiring capacity. Microvias are implemented to provide a conductive path between layers. There are three main types of via-to-via connections: stacked via, staggered via and spiral via. One of the advantages of stacked via is that it allows for greater wiring densities [Liu 2003]. Staggered via can increase routing flexibility. As packaging functionality and I/O density increase, there is a demand to decrease stacked via size, from 60-150um in 2005 to 30-80um in 2014 [Nikolova 2008], and to even smaller sizes in the future as lines and spaces in substrate decrease to10um.

Stacked vias represent a delicate interconnect structure in an organic package that can be adversely affected by both manufacturing processes and field applications. CTE mismatch between the wiring metal, the dielectric material and the silicon chip is responsible for stacked via failures. Material discontinuity within the substrate also contributes to stress concentration. These stresses will eventually decrease the reliability of the organic package. Efforts to address the challenges include development of low CTE chip scale package [Yamada et al. 2013], design and selection of dielectric materials with compatible properties through thermal-mechanical modeling [Nakanish 2007, Banerji 2002, Iannuzzelli 1991] and fabrication of robust stacked-via structures [Sundaram 2004].

The primary factors decreasing stacked via robustness include: (1) high aspect ratio structures which are susceptible to cracking, 2) high stresses due to mismatch of material properties, 3) process related factors such as multiple curing cycles that result in residual stresses and 4) poor via geometry and via to via connection. Sensitivity analysis showed that stacked via reliability increases with increasing copper ductility, which can be achieved by increasing the minimum plating thickness and drilled hole diameter and decreasing plating non-uniformity. The reliability also increases by decreasing the imposed loading or total strain range [Iannuzzelli 1991, Nakanishi 2007]. To achieve reliable stacked via design, the following factors need to be considered: (1) stacked via locations, (2) via connectivity management (i.e., height of stacked, via centered or offset on PTH), (3) via geometry (such as diameter and height) (4) CTE differences of materials used in the substrate, and (5) field conditions [Iannuzzelli 1991, Lo 2008, Nakanishi 2007]. A reliable stacked via formation is dependent upon a combination of design, material set selection and process control of via connectivity.

In this paper, a finite element analysis of stacked via cracking risk and mechanism are presented. Sixteen types of stacked and staggered via structures were designed and fabricated in an organic chip carrier test vehicle. The experiments were also designed to evaluate other effects such as stacked via locations, laminate materials, etc. Comparison of fail counts versus via chain types after 1000 cycles of deep thermal cycling (DTC) reveals that some types of stacked via structures are significantly more robust than others. Strong location dependency of stacked via fail is also observed by comparison of two identical stacked via structures in different locations.

Experimental Setup

A FCPBGA package with a 55mm×55mm body size was chosen as the test vehicle to evaluate the robustness of various stacked via structures. A 5-2-5 structure which consists of five layers of build-up and copper on each side of a copper clad core was used. Core thickness is nominally 400µm, while build-up dielectric layers are in the 33µm range. Build-up copper thicknesses are nominally 15µm except for the core copper which is 20µm.

The silicon chips used were 19mm×19mm×0.785mm. Flip chip bump compositions are lead free Sn-Ag-Cu type. Nominal bump pitch is 185µm.

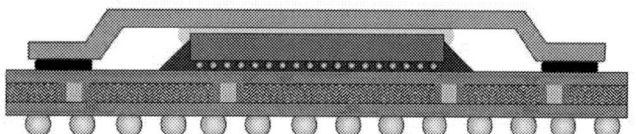

Figure 1 Schematic of Flip-chip Module Cross Section

The silicon dies were joined to the chip carriers using a reflow solder process. After cool down, the chips were underfilled using a capillary type encapsulant to provide a stronger mechanical connection. Monitoring post-underfill die camber versus temperature with digital image correlation (DIC) method indicated the nominal stress-free temperature of the chip-on-laminate assembly to be125°C, which provided a reference for finite element modeling. Thermal dissipation was achieved with a nickel-plated copper lid and a high conductivity thermal interface material, between the die and the lid. The lid was bonded to the carrier with a silicone adhesive. Figure 1 shows the schematic of module cross-section.

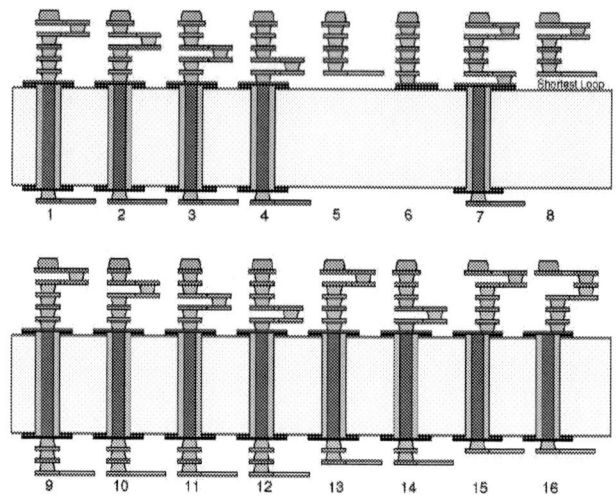

Figure 2 Stacked Via Structures

16 types of stacked and staggered via chain structures were designed and fabricated in the aforementioned test vehicle to investigate the via-cracking mechanism. Figure 2 shows the schematic plots of all the stacked via types (1 to 16) tested. Types1 to 4 represent 5 vias stacked upon resin-filled plated through-hole (RFP) but staggered in FV5, FV4, FV3, and FV2 layers, respectively. Types 5 and 6 are 4-high and 5-high via stacks offset from RFP. Type 7 is staggered at both FV1 and FV5 and it is connected to but not directly stacked on RFP. Type 8 breaks at FV1 and is the shortest loop among all 16 via stack types. Types 9 to 12 have exactly the same front copper structure as Type 1 to 4 respectively. The only difference lies in the back copper side, where Types 9 to 12 have 2 more stacks than their counterparts 1 to 4. Type 13 and 14 are similar to Type 1 and Type 4 but with 1 more stack in the back side of core. Type 15 and Type 16 are located off the center of RFP in order to evaluate the impact of eccentricity on stacked via robustness.

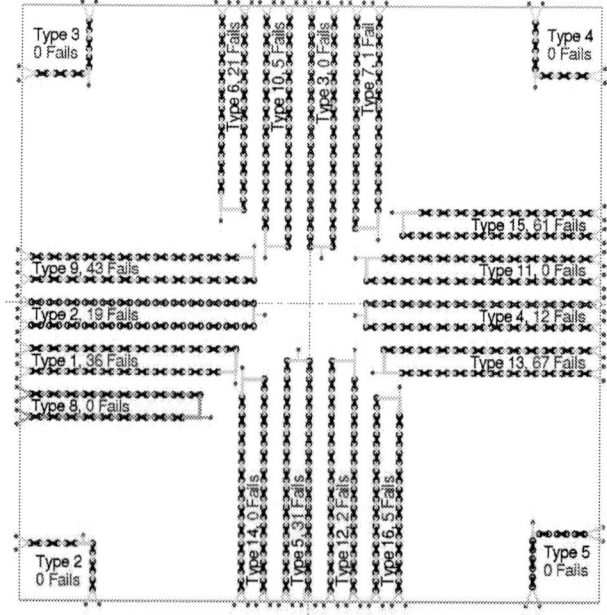

Figure 3 Layout of Test Sites and DTC Fail Counts

978-1-4799-2408-0/14 $31.00 © 2014 IEEE   1254

Figure 3 shows the locations of each via stack type and the fail counts after 1000 deep thermal cycling stress (DTC). Distribution of DTC fail counts indicates that certain via stack structures are more robust than the others. For example, although located at similar regions of die shadow, the fail count for daisy chain of Type 13 was 67 vs. zero for type 8.

Experimental results also revealed that stacked via crack is highly location dependent. In this test vehicle, Types 2, 3, 4 and 5 daisy chains were included in both chip corner and chip center locations. Via stacks in the chip center clearly were associated with higher risk of DTC fail than in the chip corner.

Modeling

The complete simulation of stacked via crack mechanism involves features across a broad length scale. The size of the stacked via is a few tens of microns compared to the 55x55mm body size of the carrier (test vehicle). It is impractical to contain all the desired details within a single model. Models at two levels, global and local, were thus developed to fit the size requirements. In the global model, all the material models were temperature dependent linear elastic. The entire structure was meshed with ABAQUS solid element (mesh size: ~1mm). The initial temperature of the entire flip chip module was set to be 125°C, based on the measurement of 125°C stress-free temperature. All the components were assumed to be stress free and flat at initial condition. The model was then cooled down to -55°C to calculate the most detrimental deformation of module under DTC.

Figure 4 Global Model (A Quarter of the Entire Test Vehicle with Biaxial Symmetric B.C.)

The local model was focused on the via stack of interest. The surrounding region beyond 0.5mm from via center was virtually cut off; the upper boundary was set at the mid-section of silicon die; the lower boundary was at the bottom surface of chip carrier. The entire local model was 1mm ×1mm × 1.2mm in size, as is shown in Figure 5. In order to reflect the geometric details of the desired via stack, the local model was meshed with 10um tetrahedral elements. Kinematic hardening model was used to simulate the plastic deformation of stacked via. Utilizing submodeling technique, the constraint on the cut boundaries of the local model was interpolated from the deformation at the corresponding spot of the global model. Local stress state in the vias was thereby solved.

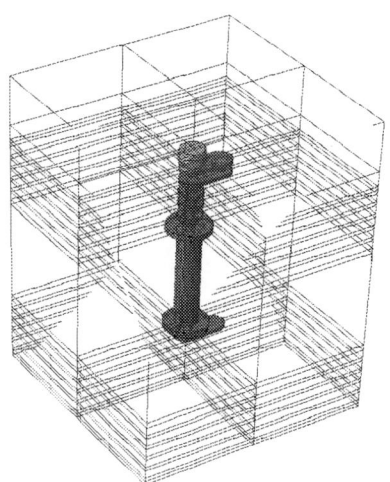

Figure 5 Example of the Local Model

Results and Discussion

Thermal fatigue is believed to be responsible for via crack in DTC. Thermal cycling causes expansion and contraction, hence repeated thermal stresses. If the localized thermal stresses are above a certain threshold, microscopic cracks will initiate and propagate. The stacked via will eventually fracture when the cracks reach a critical size. Plastic strain was utilized as the metric for fatigue damage as both the plastic strain and fatigue damage behave similarly in that they are not triggered until stress reaches a certain level; also, higher stress causes both increased plastic strain, and fatigue damage. In this work, the copper was assumed to be fully annealed. The plastic strain due to one thermal cycle was calculated and compared with the stacked via fail counts detected after DTC, as is shown in Figure 6. The correlation between plastic strain and the corresponding DTC fail counts suggests that plastic strain is a reasonable indicator for the cumulative damage caused by thermal fatigue.

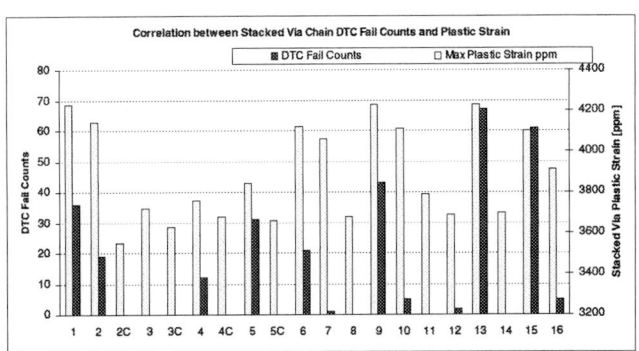

Figure 6 Correlation between Simulation Results and Experimental Data. ("C" is abbreviation for "Corner", i.e. "2C" means Type 2 via stack located at chip corner)

Figure 6 shows that although there is some fluctuation, in general, the DTC stacked via crack probability follows the same trend with the maximum plastic strain calculated. Comparison indicates that stack via crack are observed when the calculated plastic strain is above 3,650ppm, indicating a threshold. It should be noted that this threshold value is empirical and may vary with environmental material, manufacturing process and other factors. However, as long as such factors stay consistent, this is a valid method to evaluate the risk of stacked via crack and to perform design optimization as necessary.

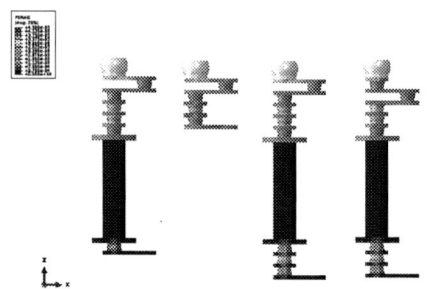

Figure 7 Contour Map of Plastic Strain in Selected Types of Via Stacks. (from left to right: Type 1, Type 8, Type 9 and Type 10)

The number of vias directly stacked upon one another, and the existence of RFP in the loop are both important factors that affect the robustness of via stack. For example, Type 1, 8 and 9 have similar upper structures; Type 1 and Type 9 are directly connected to RFP's, while Type 8 has no RFP in the loop. Difference between Type 1 and 9 lies in that Type 9 has 3 straight stacked via in the back copper layers but Type 1 has only 1. Figure 7 shows that damage accumulates much faster in Type 1 and Type9 than in Type 8. This is because vias directly stacked upon a RFP are less compliant to thermal deformation, and are subjected to higher thermal stress.

The stress state in the upper structures of Type 1 and Type 9 are very similar, despite the difference in structure of the back side via, indicating that the front vias are not influenced by the back vias due to the rigid core layer in between. Comparison of the back vias in Type 1 and Type 9 shows that the greater the number of vias directly stacked, the higher the stress will be. From a mechanical perspective it is recommended to replace high stacked vias with staggered vias.

Across Type 1, 9 and 10, the highest fatigue damage in all cases occurs at the 2nd or 3rd via above the RFP, showing these are weak points that need extra caution in the via design. Referring to Type 10, the plastic strain in the 3-high vias of the front copper side is much higher than that in the 3-high vias of the back copper side, indicating that front vias are more susceptible to DTC failure than the back vias.

Typically higher stress is expected at higher DNP locations. In this case, it seems counter-intuitive that via stacks near the chip center are associated with higher crack risk than those near the chip corner, as observed by both experimental data and simulation results. Simulation shows that center vias and corner vias are subjected to different types of cyclic loads. More specifically, the center vias are subjected to "longitudinal" cyclic load, while the corner vias are subjected to cyclic load in the "transverse" direction. As the organic laminate has much higher in-plane strength than out-of-plane strength, vias are more susceptible to longitudinal cyclic loads than to the transverse cyclic loads, hence more DTC fails were found in the chip center than in the chip corner.

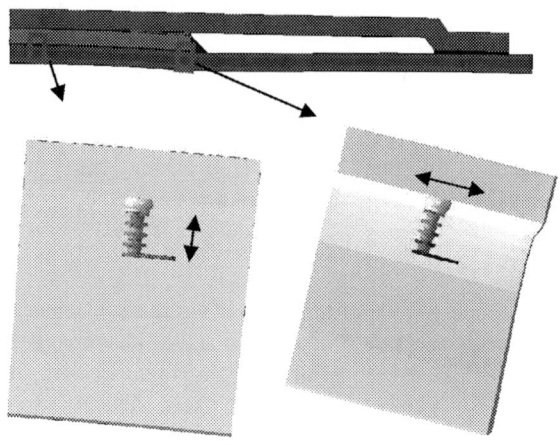

Figure 8 Deformation of Type 5 Stacked Via in Package a) vias stack near chip center; b) vias stack near chip corner

Conclusion

This paper describes the modeling and experimental work conducted in the investigation of the thermal reliability of stacked via in organic chip carriers. A test vehicle with 16 different types of stacked and staggered vias was designed and subjected to deep thermal cycle stress. The experimentally observed via-crack fail counts of each via type were compared with the calculated maximum plastic strain and found to have a general correlation.

Finite element method was employed to investigate the failure mechanism that causes stacked vias to crack. Noticing that plastic deformation and fatigue damage are very similar in behavior, plastic strain was utilized as indicator for thermal fatigue damage by intentionally equating material yield strength to its fatigue strength. This method was validated by a good correlation between simulation results and the observed DTC fail counts. Some of the conclusions from this study include:

Thermal fatigue induced via crack is location dependent. Via stack located near the chip center is associated with higher risk than that near the chip corner. Numerical analysis indicates that the organic laminate's weak out-of-plane strength is the root cause for center via being prone to thermal fatigue.

High stacked vias, especially vias stacked directly upon RFP, increase the via crack risk during DTC. From reliability perspective it is recommended to replace the high stacked vias with staggered vias if possible.

Acknowledgement:

The authors would like to thank Douglas Powell, David Hawken, and Rick Heath for their support and advice to this work.

978-1-4799-2408-0/14 $31.00 © 2014 IEEE

References

1. Blackshear, E.D., et al. "The evolution of build-up package technology and its design challenges". *IBM Journal of Research and Development*, Volume: 49 , Issue: 4.5, 2005, 641 – 661.

2. Ray Iannuzzelli. "Predicting Plated-Through-Hole Reliability in High Temperature Manufacturing Processes". *Electronic Components and Technology Conference (ECTC), IEEE*, 1991, 410-421.

3. Maria Nikolova, et al. "New Generation Solution for Micro Via Metallization and Through Hole Plating", *3rd International Microsystems, Packaging, Assembly & Circuits Technology Conference*,2008, 417-420.

4. Nakanishi, T.; Ohkuma, H. ; Ohira, H., "Research of Stacked VIA's Mechanical Stress Thermal, Mechanical and Multi-Physics Simulation Experiments in Microelectronics and Micro-Systems", *International EuroSime*, 2007, 1-8

5. Filippi, R.G. et al. "Thermal cycle reliability of stacked via structures with copper metallization and an organic low-k dielectric", *IEEE International Reliability Physics Symposium Proceedings*, 2004. 61 – 67.

6. Larry Lo, "Stacked Via Technology For substrate, 3rd International Microsystems", *Packaging, Assembly & Circuits Technology Conference (IMPACT)*, 2008, 164-166.

7. Banerji, S. et al. "The role of stiff base substrates in warpage reduction for future high-density-wiring requirements", *Proceedings of 8th International Symposium on Advanced Packaging Materials*, 2002, 221 – 225.

8. Yamada, T. et al. "Development of a Low CTE chip scale package", *IEEE 63rd Electronic Components and Technology Conference (ECTC)*, 2013, 944-948

9. Sundaram, V. et al. "Next-generation microvia and global wiring technologies for SOP", *IEEE Transactions on Advanced Packaging*, Vol.27, No.2, 2004, 315 – 325.

10. International Technology Roadmap for semiconductors (ITRS) 2012 Update. Assembly and Packaging.

11. Fuhan Liu et al. "A novel technology for stacking microvias on printed wiring board", *Proceedings of 53rd Electronic Components and Technology Conference*, 2003, 1134 – 1139.

A Novel Method to Predict Fluid/Structure Interaction in IC Packaging

Chih-Chung Hsu[1]; Tzu-Chang Wang[2]; Yen-Chi Chen[2]; Yang-Kai Lin[2]

1. Department of Chemical Engineering, National Tsing Hua University
2. CoreTech System (Moldex3D) Co., Ltd.
1. No. 101, Section 2, Kuang-Fu Road, Hsinchu, Taiwan 30013, R.O.C.
2. 8F-2, No. 32, Taiyuan St., Chupei City, Hsinchu County 302, Taiwan
E-Mail: finitevolume@gmail.com, Phone: +49 351 463 33056, Fax: +49 351 463 37035

Abstract

This paper reports a perspective investigation of computational modeling of fluid-structure interaction (FSI) in molded integrated-circuit(IC) packaging. The investigation is carried out through two aspects, respectively on interaction between the fluid and structure in the encapsulation process and appropriate methodology for modeling. We present a novel and integrated method to predict the FSI during the encapsulation process. This method not only provides more accurate melt front and pressure result but also predict precisely the FSI behavior through the dynamic mesh deformation technique simultaneously in accordance with continually deformed geometry (two way FSI). This is different from previous study that only one way considered fixed geometry (one way FSI) during encapsulation. Moreover, the experimental data for single- and stacked-chip were compared with the simulation results for two way FSI implementation to verify flow front advancement. From a real paddle shift case study, the result indicates that the deflection prediction is well predicted and could predict void formation well when it considers two way FSI effect. It is expected that this paper could clarify relevant issues in prediction of FSI in IC packaging and induce more considerations for modeling FSI using two way FSI multiphase flow method.

Introduction

With the tendency of microchip encapsulation technologies continuously moving toward smaller scale and higher density, the existed defects problems during fabrication becomes more and more important. In the packaging of plastic encapsulated microelectronics (PEM), transfer molding is the dominant technique for encapsulation processes, especially for the microchip encapsulation. Moreover, improper selection of processing conditions, complexity of material properties during processing, and lead frame layout or molding design, raise manufacturing challenges. During fabrication of IC packaging, stress-induced problems such as chip deformation and paddle shift are the most common. Therefore, the fundamental understanding of the FSI during molded packaging is important for the engineers and package designers. The fluid/structure interaction (FSI) phenomenon was also extensively reported in the encapsulation[1][2].

Conventionally, these problems can only be solved by means of using "trial-and-error" method in molding design optimization. However, for most FSI problems, analytical solutions to the model equations are impossible to obtain, whereas laboratory experiments are limited in scope; thus to investigate the fundamental physics involved in the complex interaction between fluids and solids, numerical simulations

may be employed[3]. Over the last decade, progress in both hardware and software has made simulation an effective tool for analyzing the complicated physical phenomena inherent in processes of plastic encapsulation of microchips[4][5]. However, these studies only consider one way FSI on fixed geometry during encapsulation which may lead to different predictions under some situations.

In this paper, the novel two way FSI simulation method is adopted to evaluate the design of microchip encapsulation. As an integrated technology to structure analysis through the dynamic mesh deformation technique simultaneously in accordance with continually deformed geometry, it gives a comprehensive FSI solution for microchip encapsulation. After incorporating structure analyses, the results of chip deformation and paddle shift can be obtained. Moreover, the information about the melt front evolution, the position of air trap, the stress tensor distribution, and the transfer pressure on deformed chips and paddle are obtained. The results can provide good guideline for the microchip encapsulation development and the FSI phenomenon for different molding design of microchip encapsulation can be easily obtained painlessly.

Implementation details

Governing Equations:

Theoretically, microchip encapsulation process is a three-dimensional, transient, reactive problem with moving resin front. The non-isothermal resin flow in mold cavity can be mathematically described by the following equations [6]:

$$\frac{\partial \rho}{\partial t} + \nabla \cdot \rho \mathbf{u} = 0 \tag{1}$$

$$\frac{\partial}{\partial t}(\rho \mathbf{u}) + \nabla \cdot (\rho \mathbf{u}\mathbf{u} - \boldsymbol{\sigma}) = \rho \mathbf{g} \tag{2}$$

$$\boldsymbol{\sigma} = -p\mathbf{I} + \eta(\nabla \mathbf{u} + \nabla \mathbf{u}^T) \tag{3}$$

$$\rho C_P \left(\frac{\partial T}{\partial t} + \mathbf{u} \cdot \nabla T \right) = \nabla(k \nabla T) + \Phi \tag{4}$$

where u is the velocity vector, T is the temperature, t is the time, p is the pressure, σ is the total stress tensor, ρ is the fluid density, k is the thermal conductivity, Cp is the specific heat, and Φ is the energy source tem. In this work, the energy source contains two contributions:

$$\Phi = \eta \dot{\gamma} + \dot{\alpha} \Delta H \tag{5}$$

where η is the viscosity, $\dot{\gamma}$ is the magnitude of the rate of deformation tensor, $\dot{\alpha}$ is the conversion rate and ΔH is the exothermic heat of polymerization.

Chemorheology:

The curing reaction of epoxy resins has received much attention using different analyses. In this work, we apply the

978-1-4799-2408-0/14 $31.00 © 2014 IEEE   1258   2014 Electronic Components & Technology Conference

combined model to investigate the curing kinetics of the given EMC because of its ability to accurately predict the experimental data. The combined model can be expressed as follows:

$$\frac{d\alpha}{dt} = \left(k_1 + k_2\alpha^m\right)\left(1-\alpha\right)^n \tag{6}$$

$$k_1 = A_1 \exp\left(-\frac{E_1}{RT}\right) \tag{7}$$

$$k_2 = A_2 \exp\left(-\frac{E_2}{RT}\right) \tag{8}$$

Where α is the conversion of reaction, A_1, A_2, E_1, E_2, m, n are model parameters.

During the curing process, the viscosity of epoxy resins changes with temperature and conversion rate. The Castro-Macosko model is adopted to describe the rheological properties of epoxy resins:

$$\eta(\alpha, T) = \eta_0(T)\left(\frac{\alpha_g}{\alpha_g - \alpha}\right)^{C_1 + C_2\alpha} \tag{9}$$

$$\eta_0(T) = A \exp(E_a/RT) \tag{10}$$

Where A, E_a, C_1, C_2 are model parameters, α_g denotes gelation conversion at which viscosity curve grows up because of the formation of three-dimensional network structure of the epoxy resins.

The FSI phenomenon during encapsulation could be distinguished into two situations: (1) the viscous drag forces on chip or paddle exerted by the resin melt flow causes the shift problem. (2) While the non-uniform loading on chip or paddle system applied by uneven melt flow within cavities results in deformation problem. Both of these two forces should be considered well to balance the flow behavior and avoid the deformation and shift problems. Hence the FSI phenomena can be obtained from the following force balance:

$$\nabla\sigma + F = 0 \tag{11}$$

$$\sigma = \mathbf{C}(\varepsilon - \varepsilon^0 - \alpha_{CLTE}\Delta T) \tag{12}$$

$$\varepsilon = \frac{1}{2}(\nabla\mathbf{U} + \nabla\mathbf{U}^T) \tag{13}$$

where σ is the stress, F is the external loading due to melt pressure and drag forces, $\mathbf{C}$ is a 4th tensor, ε is the strain tensor, α_{CLTE} is CLTE tensor and $\mathbf{U}$ is the displacement vector, respectively. Therefore the force balance can be derived by differential equations of equilibrium.

Numerical Method

The collocated cell-centered FVM (Finite Volume Method)-based 3D numerical approach developed in our previous work is further extended to simulate the mold filling in IC packaging [6]. The numerical method is basically a SIMPLE-like FVM with improved numerical stability. Furthermore, the volume-tracking method based on a deformed mesh framework is incorporated in the flow solver to track the evolutions of melt front during molding.

After filling analysis is done, the force loading on the chip or paddle exerted by this fluid will be used for further coupling region both structure and mesh deformation analysis. While a new solid mesh deformation technique is developed

to disperse the deformation displacement to all cavity meshes. Hence high quality deformed cavity meshes after each deformation are generated for the filling analysis to simulate melt front precisely. Once the deformation mesh impressed by stress is updated, the resin filling analysis will iterate into the next step by using newest deformation mesh. The detailed coupling algorithm is clearly depicted in figure 1.

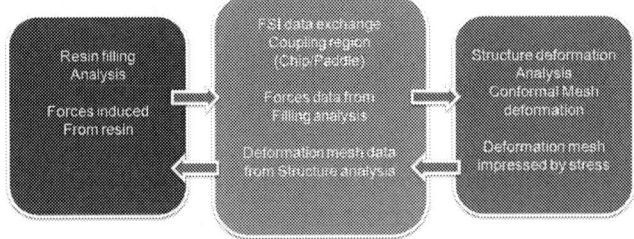

Figure 1. Coupling algorithm for two way FSI simulation

Results and discussions

Validation for two way FSI coupling

The fluid-structural interaction experimental results published in 2012 [7] have been introduced to validate the Two-way FSI modeling results of single- and stacked-chip packages. The scaled-up package for both models is located at the cavity and the dimension is the same as the reference. The geometry of both experimental models has been depicted including cavity, chip, solid ball, and substrate. The models of single- and stacked-chip packages for simulation have been developed in figure 2.

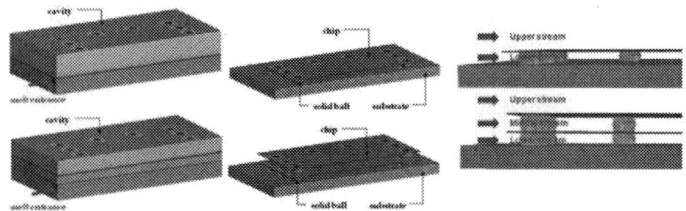

Figure 2. 3D meshed model with flow path (a) single-chip model (b) stacked-chip model

Figure 2 illustrates the fluid flows to the cavity through the entrance at the left lower corner of cavity and separated to two flow streams, upper and lower streams for single-chip model. In stacked-chip model, the fluid separated to upper, middle, and lower streams to flow around the chips. A test fluid is Newtonian fluid with constant viscosity of 4 Pa-s. The mechanical properties of chip, solid ball, and substrate are tabulated in Table I. And the material properties for fluid are tabulated in Table II respectively.

Parameter	chip	Solid ball	substrate
Elastic modulus, E (GPa)	1.571	71.7	68.98
Poisson ratio, ν	0.33	0.33	0.33
Solid density, ρ_s (kg/m^3)	4454	2850	2711

Table 1. Mechanical properties of chip, solid ball and substrate

Parameter	
Density (kg/m³)	937
Heat capacity (J/(kg · K))	1460
Thermal conductivity (W/(m · K))	0.182
Elastic modulus (MPa)	548
Poisson ratio	0.33
Thermal expansion coefficient (1/K)	0.000131
Viscosity(Pa · sec)	4

Table 2. Material properties of fluid

When injection molding process is performed in the system, a non-uniform force loading around the chip results in chip deformation, and a pervasive problem especially in fluid-structure interaction happened. The details will be addressed in the follows. To accurately modeling fluid-structural interaction effect, the filling procedure is divided into 2 time steps. The nodes displacements on the chip are simulated by the previous structure analysis at each time. Simultaneously, the solid meshes of cavity are deformed to maintain the mesh quality as well as possible after each chip deformation as figure 3 shown.

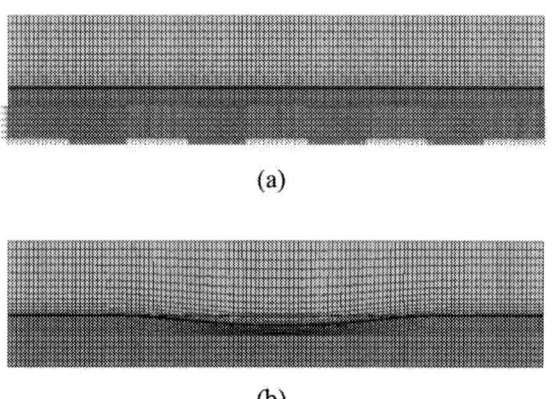

(a)

(b)

Figure 3. Deformation of elements for chips and cavity simultaneously during encapsulation:
(a) at initial time step (b) at the end of filling

In the following case study in this paper, the current two way FSI simulation method will automatically detect the boundary for fixed node areas where no displacement boundary condition is assigned as figure 3(a) shown in blue arrow. And other regions will be free to move as the initial condition when the mesh deformation analysis is launched. To ensure the numerical stability need, we adopt the hexagon or prism as the element type for all the simulation. The tetrahedron element is not recommended for the two way FSI simulation due to its less degree of freedom for node entity. Current study will ensure the mesh quality both the aspect ratio and orthogonality in the mesh deformation analysis.

Comparison of experiment with simulation in flow profiles

The comparison of predicted deflection of chip with the deformation shape is depicted in figure 4. The flow front advancement of modeling nearly corresponds to that of experiment in shape and flow behavior. The deformation of chip by simulation (1.718 mm) agreed with that by experiment (1.72-1.74 mm) at the middle region of chip without supporting by the solid balls. Therefore, the present modeling results by utilizing two way FSI can provide substantial results to predict the flow front and the deformation of chip or paddle.

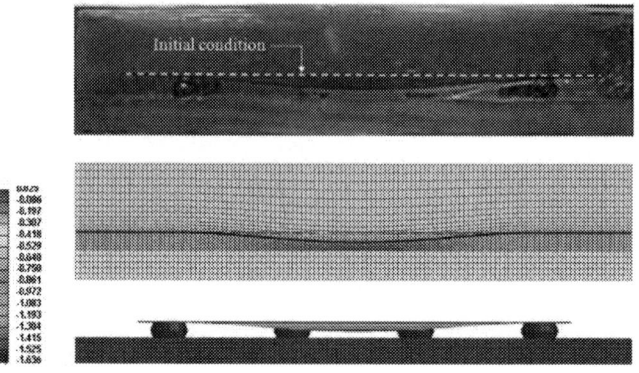

Figure 4. Deformation of chip in experiment[7] and current simulation results

In the injection process, flow profile and deformation of chip at the middle region were presented with various filling time percentage (5 − 90%) in figure 5. At the 5% of filling volume, the fluid enters and reaches to the edge of chip which demonstrates the initial condition without initial deformation in the chip. The fluid continues to flow toward the center of the chip and the contour of tiny deflection of chip develops slightly around the 30% filling volume which nearly agrees with realistic deformed one. In the meanwhile, we found the race-track effect and cross flow occur during the filling stage which results from unbalanced flow because of the discrepancy in the two free spaces at the upper and lower stream region. The fluid flows faster at the upper space region than that at the lower region of chip. At the 45% of fluid filling cavity, the fluid reaches to center region and flows more freely through the large space at the upper region of the chip than those at the lower stream region; meanwhile, this movement causes the small and apparent deformation in the chip.

The obvious deformation of chip occurs at the 60% of chip covered by the fluid. The foregoing race-track effect still results in the unbalanced pressure distributing onto the chip so that it deflects downward at the center region without supporting by solid balls. At the last of filling process, the fluid covers the middle region of chip and forces the chip close to the substrate. The deformation shape by simulation agrees with experimental results. At the end of filling stage, the nearly 90% of fluid fill the cavity as well as attribute to middle region of chip approach to substrate.

978-1-4799-2408-0/14 $31.00 © 2014 IEEE   1260

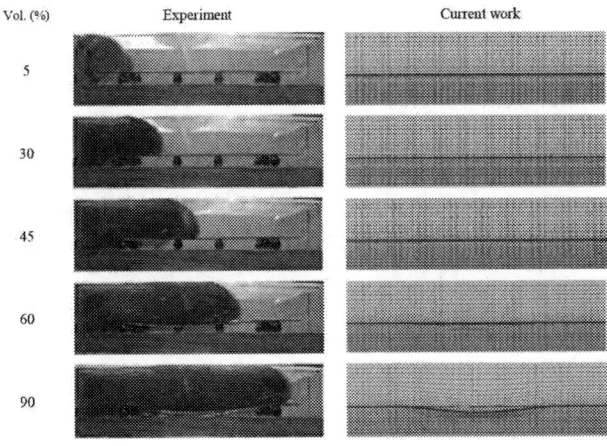

Figure 5. The entire FSI process record and modeling deformation results of chip with various filling volume (%)

Moreover, the comparison of experimental with modeling results of stacked-chip package is tabulated in figure 6. The fluid starts to reach the chip nearly 15% of filling stage and separate to three flow paths, upper, middle, and lower stream with balanced flow behavior at the 20% stage. The fluid does not result in the deformation of both chips. However, the tiny deformation develop at the nearly 50% of filling volume. At the end of filling from 75 to 90%, the race-track effect is not obvious during filling stage and unbalanced pressure does not occur so that the fluid does not exert forces on the chips to develop larger deformation because the gap height at the upper, middle, and lower region is almost the same without larger discrepant distance around 0.3 to 0.4 mm as in figure 2. Therefore, the deflection of both chips is smaller than that at the single-chip mold.

Figure 6. Comparison of flow profiles and deformation chip of experiment with simulation for stacked-chip at various stage

Application for two way FSI coupling

Since the paddle shift is also a flow-induced phenomenon, paddle shift analysis can be realized by current two way FSI framework. The deformation of paddle changes with time can be examined. Side view and top view of our 3d geometry model used for paddle shift analysis are shown in figure 7 respectively, and the predicted results could be validated by experiment results. However, in cases with long slender paddle such as 2 cavity model shown in figure 7, paddle contact with each other leading to short may be a severe problem during encapsulating. The material property for the fluid is tabulated in Table III respectively.

Figure 8 (a) ~ (c) shows the comparison of filling pattern between experiment and simulation results at distinct filling time, and the results show a close agreement. Also, from the simulation result shown in figure 8, we can see that paddle with larger deflection occurs at positions where melt surges into cavity from the entrance and melt merges at the end of filling. However, large deflection of paddle suffered from large lateral force exerted by flow overwhelms the deflection of paddle suffered from pressure drop within cavity. Therefore, paddles used to support chips can be analyzed individually.

Comparison of deformed shape of paddle at end of filling between simulation and experiment results are shown in figure 9, and there is a coincident in position where deformed paddle exposed outside the package between these two results. Side view of deformed shape of paddle shown in figure 10 makes it clearer about the tendency of paddle shift. Therefore, from the results of filling and paddle shift analyses, the feasibility of our novel 2 way method to analyze mold designs for microchip encapsulation can be demonstrated.

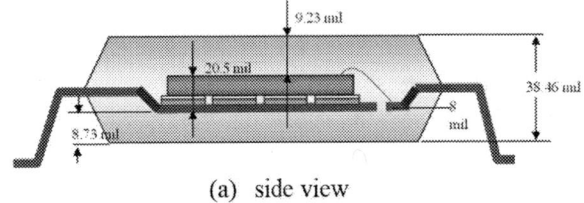

(a) side view

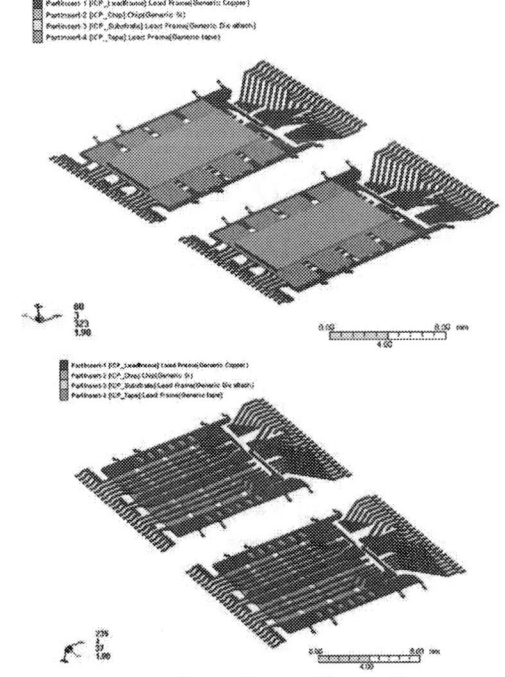

(b) front view and back view

Figure 7. 3D geometry model for paddle shift

Parameter	
A (g/(cm · sec))	1.72
C_1, C_2	5.734, -7.66
α_g	0.5592
E_a (erg/mole)	1213
A_1, A_2 (1/sec)	0.1, 8.117e+6
E_1, E_2 (erg/mole)	2405.5, 1012.8
m, n	0.4671, 1.116

Table 3. Material properties of fluid

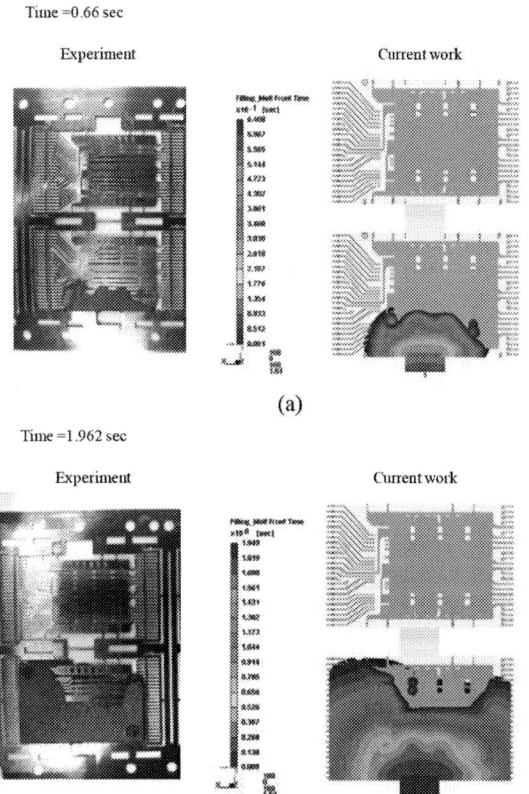

Figure 8. Comparison of filling pattern between experiment and simulation results at distinct filling time

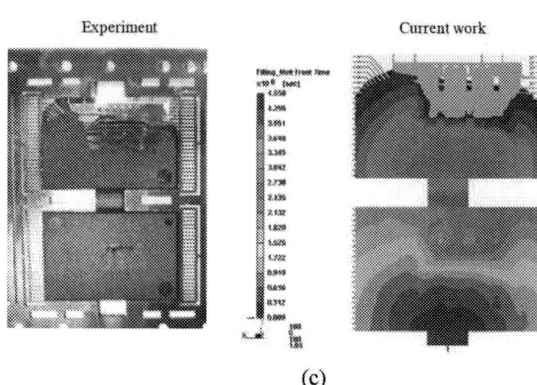

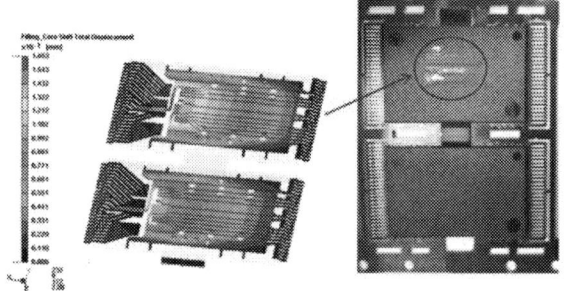

Figure 9. Comparison of deformed shape at end of filling between simulation and experiment results

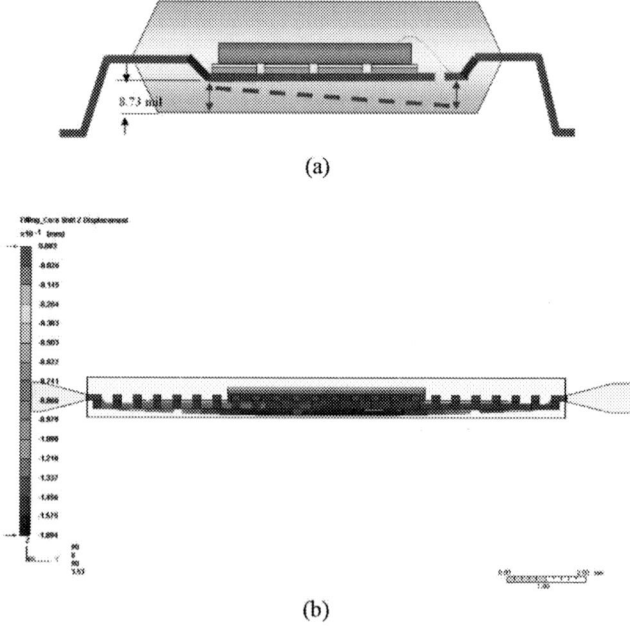

Figure 10. Side view of paddle shift phenomenon:
(a) Illustration of paddle shift within package
(b) Predicted paddle shift result

978-1-4799-2408-0/14 $31.00 © 2014 IEEE

Conclusions

Two-way FSI simulation approach has been implemented for both single- and stacked-chip package models, and paddle analysis using two way FSI framework. Furthermore, a novel approach was introduced to make further investigation for both packages models. Deformation of chips at the single-chip packages can be obviously observed during experiment as well as corresponding to the modeling results using current two way FSI code. The race-track effect results in the unbalanced flow-profiles at the upper and lower stream through the single chip. In addition, for the stacked-chip package, the fluid flows separately to three flow paths, upper, middle, and lower region along the chips. Owing to nearly the same gap height between the chips, the stream trend at the three regions is closer than at the single-chip package. Therefore, the more balanced flow front advancement it is, the less unstable pressure exert to the surface of chips.

Furthermore, by comparing simulation results to experimental data, the prediction of paddle shift can be validated, which shows the two way FSI approach can provide more realistic interaction between fluid and structure so that the deformation of paddle can be observed simultaneously during encapsulation process. This study not only provides a novel and alternative approach to investigate the FSI phenomenon during encapsulation but also demonstrates the feasibility and usefulness of introducing current simulation technology into mold design for microchip encapsulation.

References

[1] D. Ramdan and M. Z. Abdullah, "Fluid/Structure Interaction Investigation in PBGA Packaging," *IEEE Trans. Components, Packag. Manuf. Technol.*, vol. 2, no. 11, pp. 1786–1795, Nov. 2012.

[2] D. Ramdan, Z. M. Abdullah, M. A. Mujeebu, W. K. Loh, C. K. Ooi, and R. C. Ooi, "FSI Simulation of Wire Sweep PBGA Encapsulation Process Considering Rheology Effect," *IEEE Trans. Components, Packag. Manuf. Technol.*, vol. 2, no. 4, pp. 593–603, Apr. 2012.

[3] Hou, "Numerical Methods for Fluid-Structure Interaction - A Review," *Commun. Comput. Phys.*, vol. 12, no. 2, pp. 337–377, 2012.

[4] S.-Y. Teng and S.-J. Hwang, "Simulations and experiments of three-dimensional paddle shift for IC packaging," *Microelectron. Eng.*, vol. 85, no. 1, pp. 115–125, Jan. 2008.

[5] E. E. S. Ong, M. Z. Abdullah, C. Y. Khor, W. K. Loh, C. K. Ooi, and R. Chan, "Fluid–structure interaction analysis on the effect of chip stacking in a 3D integrated circuit package with through-silicon vias during plastic encapsulation," *Microelectron. Eng.*, vol. 113, pp. 40–49, Jan. 2014.

[6] R.-Y. Chang, W.-H. Yang, S.-J. Hwang, and F. Su, "Three-Dimensional Modeling of Mold Filling in Microelectronics Encapsulation Process," *IEEE Trans. Components Packag. Technol.*, vol. 27, no. 1, pp. 200–209, Mar. 2004.

[7] C. Y. Khor, M. Z. Abdullah, and W. C. Leong, "Visualization of Fluid/Structure Interaction in IC Encapsulation," *IEEE Trans. Components, Packag. Manuf. Technol.*, vol. 2, no. 8, pp. 1239–1246, Aug. 2012.

Modeling, Design and Demonstration of Multi-Die Embedded WLAN RF Front-End Module with Ultra-miniaturized and High-performance Passives

Srikrishna Sitaraman, Yuya Suzuki[ς], Christopher White, Vijay Nair *, Telesphor Kamgaing *, Frank Juskey[Ω], Sung Jin Kim,
P. Markondeya Raj, Venky Sundaram, and Rao Tummala.
3D Systems Packaging Research Center, Georgia Institute of Technology, 813 Ferst Dr N.W., Atlanta, GA 30332.
* Intel Corporation, Chandler, AZ 85226, USA. , [Ω] TriQuint Semiconductor, FL, USA.,
[ς]Zeon Corporation, Kawasaki, Kanagawa, Japan
Email: srikrishna@gatech.edu

Abstract

This paper demonstrates, for the first time, a Wireless Local Area Network (WLAN) radio frequency (RF) front end module (FEM), incorporating the smallest, high-performance band-pass filter (BPF) on a 110μm-thin organic substrate with chip-last embedded actives and thin-film passives. The FEM consists of a power amplifier (PA) die, a switch die, and two low-noise amplifier (LNA) dies, integrated with a BPF and a low-pass filter (LPF). Full-wave electromagnetic (EM) simulations are employed to study the signal path loss, EM radiation and coupling. The BPF and LPF have 0.25dB and 0.5dB insertion loss respectively, with in-substrate dimensions of 1mm x 1mm x 0.05mm. The PA die shows a gain of around 10.8 dB at 2.4GHz. The path between the antenna and the amplifiers is also characterized to have a loss of 3dB. The electromagnetic coupling from the PA output to the LNA input and to the PA power supply is simulated using full wave EM solver HFSS and found to be higher than 60dB, indicating very good EM isolation. Each block of the FEM is individually characterized and combined using Agilent ADS to obtain the complete S-parameter performance. Both the transmitter and receiver chains have gain of 9dB.

1. Introduction

The expanding foray of smart mobile systems into different aspects of life has propelled the need for miniaturization leading to higher functional density and improved performance, at an affordable cost. To communicate with other devices, such mobile systems utilize a number of application-specific wireless technologies namely: GPS (navigation), GSM (cellular), WiMAX (mobile internet), WLAN (localized internet), and Bluetooth (short range data transfer). Hence, to achieve the higher functional density required of tomorrow's smart mobile systems, the existing wireless technology modules need to be highly miniaturized and integrated [1].

The RF front end module in a typical WLAN system is illustrated in Figure 1. Miniaturization of RF actives (LNA, PA, Switch) involves developing efficient on-chip designs followed by their realization on RF substrate materials such as gallium arsenide (GaAs) and gallium nitride (GaN). Although realizing the complete RF module on a single silicon die using System-on-Chip (SOC) approach can help miniaturize to a great extent, the performance of such modules is below par owing to different substrate-material requirements for RF actives and passives [2]. Miniaturized WLAN FEMs excluding the passives have been demonstrated on a single

GaAs die [3]. Further, to demonstrate a complete FEM with integrated RF filters, in-process substrate-embedded filters have been developed [4-6]. These modules are mostly based on either low-temperature co-fired ceramics (LTCC) or organic substrates. However, such filters either have high loss (>2dB) or are of fairly large thickness (>200μm) due to the inherent substrate thickness and multi-layer design approaches. Although alternatives such as FOWLP [7-9] have achieved good miniaturization and performance, they are challenged by low substrate yield loss arising from chip-first embedding, inability to embed multiple heterogeneous actives, and thermal dissipation issues with densely packed components.

GT-PRC has been pioneering the System-on-Package concept (SOP) [1, 10] as the strategic basis of system scaling for smart systems. SOP integrates and miniaturizes the entire system in a small package with embedded double-side actives and thin-film passive in ultrathin low-loss packages. Chip-last embedded actives with thin-film passives is one example of the SOP concept, which has the following benefits over SOC and chip-first approaches: 1) no substrate yield loss due to heterogeneous die embedding; 2) intermediate substrate testability to isolate defective units before and after assembly; 3) very low interconnection parasitics, leading to better electrical performance; 4) improved thermal performance enabled by die backside accessibility; and, 5) ability to use different substrate materials to ensure optimal performance of each component. Hence, a SOP-based WLAN RF FEM is demonstrated in this paper.

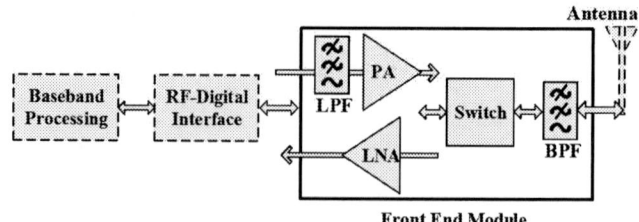

Figure 1. Components of a WLAN sub-system.

Previous work at GT-PRC has demonstrated functional WLAN RF modules on multi-layer organic substrates [11-13] using chip-last SOP. This paper extends the WLAN RF module integration further with multiple actives for LNA, PA and switch dies integrated with miniaturized embedded band-pass and low-pass filters based on a novel design, on a single ultra-thin organic package using ZEONIF™ XL – advanced dielectric material developed by Zeon Corporation.

Section 1 of the paper is the introduction, followed by the module schematic, substrate and component specifications in section 2. Modeling, design and analysis of the transmission lines, band-pass filter and low-pass filters are presented in section 3. In section 4, the module layouts and full-wave EM analyses for the PA module and FEM are detailed. Following this, the fabrication and assembly of the designs are discussed in section 5. In section 6, the characterization results are presented and the paper is summarized in section 7.

2a. Module schematic and specifications

The module schematic is shown in Figure 2. Here, 'LNA' indicates the low-noise amplifier dies; 'PA', the power amplifier die; and 'Switch', the switch die. The RF low-pass filter and band-pass filters are indicated as 'LPF' and 'BPF' respectively. Each LNA block represent one die of the LNA module. The first die in the signal path is for input-matching and the second is the amplifier. The DC power rails are shown for the actives and the ground connections are common to all the components, although they are represented individually on each component. The two thick black arrows represent the direction of RF signal through the components.

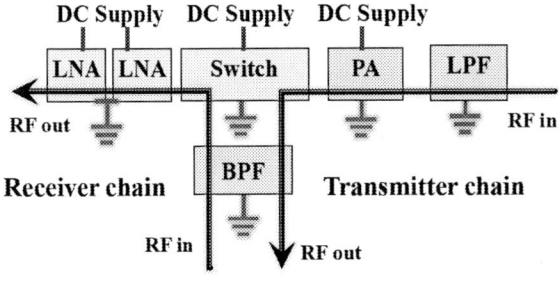

Figure 2. Schematic of WLAN front-end module.

The FEM is designed for 2.4GHz WLAN, and constitutes a receiver chain and a transmitter chain. The signals in the receiver chain originate at the antenna terminal (GSG RF-probe pads), and are routed to the LNA dies, through the BPF and the switch die. The transmitter signal-path comprises the LPF, PA die, switch die, and BPF, terminating at the RF probe pads. The single-pole double-throw (SPDT) switch die directs the signals from the antenna to the receiver section, or from the transmitter section to the antenna. Transmission lines with an impedance of 50Ω are used to provide a reflection-free RF path between the different components.

2b. Substrate stack-up and design rules

The substrate stack-up comprises a XL high dielectric constant (high dk) organic core of thickness 35 microns, from Zeon Corporation. The core has a dielectric permittivity (dk) of 6.2, with a loss tangent (df) of 0.0031. A build-up layer of thickness 50 microns is used. The build-up polymer material is from DuPont (dk=2.9, df=0.008). The stack-up is shown in Figure 3. Two layers of metallization are featured on either side of the core. There is no metallization on the build-up

layer since the build-up is employed mainly to demonstrate a functional embedded RF FEM.

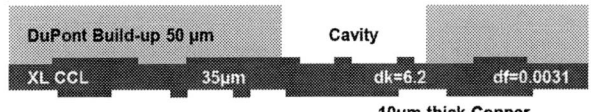

Figure 3. Stack-up for the WLAN front-end module substrate.

The design rules for the above stack-up are summarized in Table I. The copper thickness on both M1 and M2 layers is 10 μm. The smallest feature size and spacing is 30 μm. Through vias are drilled in the core to enable interconnection between the two sides. The vias have a diameter of 50 μm and are conformally plated to realize a thickness of 8 μm.

Table I. Summary of Design Rules.

Parameter/ Property	Target
Chip-Cavity Clearance	100 μm on all sides
Chip Height	100μm – 500μm
Copper thickness	10 μm
Min. Line width-Spacing	30 μm
Via diameter	50μm
Via pitch	100 μm

2c. Schematics of the dies

All the dies were obtained from TriQuint Semiconductor. The LNA comprises of two gallium arsenide (GaAs) dies – one for input matching and the other functioning as the amplifier. The datasheet diagram indicating the pin configuration of the bare die is shown in Figure 4. A bias voltage of 3.3V is required for the LNA. It has a datasheet gain of 17dB with a noise figure of 1.45dB. Two decoupling capacitors are required for ripple filtering on the LNA power supply network.

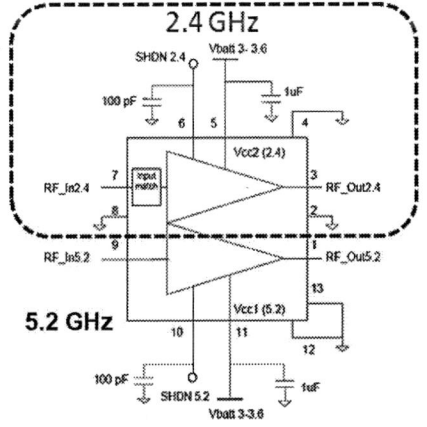

Figure 4. LNA schematic; from TriQuint datasheet [14]

The PA is a single GaAs die with on-chip input and output matching networks. The PA die-pin configuration from its datasheet is depicted in Figure 5. The PA requires a bias voltage of 3.3V, and draws a current of 180mA. The PA is specified to have a gain of 30dB at 2.45GHz. The PA module

978-1-4799-2408-0/14 $31.00 © 2014 IEEE

requires a number of surface-mount capacitors, inductor and resistors, as indicated in the datasheet schematic.

The PA and the LNA dies are all 100 μm thick, with metallized back-side ground. The PA and LNA dies were designed for wire-bonding, although are employed here in a face-down configuration.

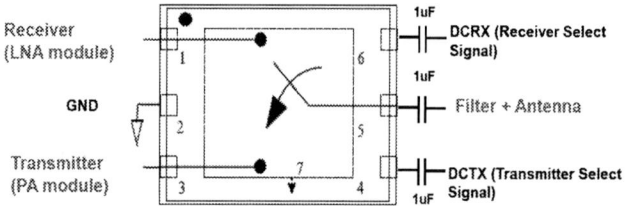

Figure 5. PA schematic; from TriQuint datasheet [15]

The switch comprises a single flip-chip silicon die of thickness 500 μm, designed for Single-pole double throw (SPDT) operation. It is designed for flip-chip interconnections to the package. The schematic of the switch die is shown in Figure 6. The switch has an insertion loss of 0.6dB with more than 25dB isolation. A control voltage of 3V is required on the control pins of the switch to select between the transmitter and the receiver paths.

Figure 6. Switch schematic; from TriQuint datasheet [16]

The LNA, PA and the switch are all dual-band dies, with a 2.4GHz chain and a 5GHz chain. Since the two paths could be operated independent of the other, the 5GHz chain was not considered for the design of this FEM module.

3a. Transmission Line Design

In order to achieve very low transmission loss between different components, it is critical to design a 50-ohm impedance-matched transmission line. The design of the transmission line depends on the substrate stack-up and material properties. Since the substrate used here comprises of two metal layers, a microstrip line (MSL) or a co-planar waveguide (CPW) configuration is possible. Further, the minimum line and spacing of the copper structures that can be realized on this substrate governs whether a MSL or a CPW can be used. Since the MSL radiates lesser than CPW and also

is less susceptible to EM interference, it is a better choice. To achieve 50-ohm impedance using a dielectric material with a dk of 6.2 and ground separation of 35μm, a line width of 32 μm is required. However, to realize such thin lines over reasonable lengths results in lower substrate yield. On the other hand, designing a CPW for 50-ohm impedance requires a trace width of more than 500μm. Hence, combining MSL and CPW, co-planar ground was included on both sides of a microstrip line to achieve the target impedance using a reasonable trace width. Such a conductor-backed CPW configuration allows for a trace width of 60 μm with the co-planar ground spaced at 50 μm. The structure of the designed 50-ohm transmission line is shown in Figure 7.

Figure 7. Structure of the 50-ohm transmission line.

Such a configuration also contains the fields of the transmission line better, and offers higher immunity to EM interference. The simulated S-parameters of this transmission line structure are shown in Figure 8. Even for a line of length 10mm, the insertion loss is around 0.3dB and the return loss more than 30dB. Moreover, having such a design with coplanar ground enables the measurement of RF signals at any point in the module, using GSG probes.

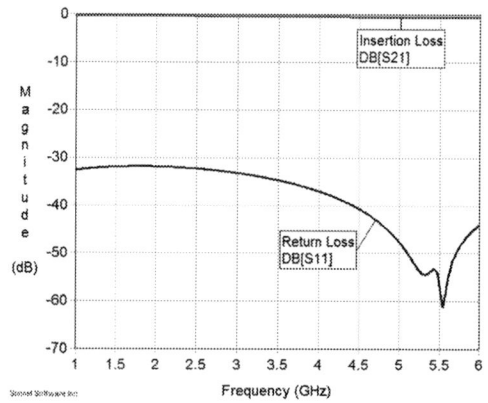

Figure 8. Simulated response of 50-ohm transmission line.

3b. Modeling, design and Layout of Band Pass Filter

A novel structure was employed to realize a miniaturized band-pass filter. The WLAN filter specification included a pass-band centered on 2.4GHz with less than 2dB insertion loss, and more than 15dB return loss. The BPF structure is shown in figure 9. This structure utilizes the ultra-thin substrate by employing vertical coupling and optimizes its occupied area through horizontal coupling as well.

978-1-4799-2408-0/14 $31.00 © 2014 IEEE 1266

Figure 9. 3D view of the BPF structure.

Considering the different coupling in this structure, the circuit level schematic is shown in Figure 10. This BPF is tuned by varying the line-width of the spirals, and the dimensions of the central metal patch. The basic schematic is tuned in Agilent ADS circuit simulator, followed by optimization using full-wave EM simulator- Ansoft HFSS.

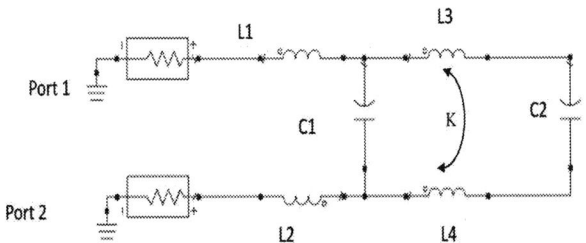

Figure 10. Circuit Schematic of novel BPF structure

The full-wave EM-optimized S-parameter response is shown in Figure 11. As can be observed, the simulated insertion loss at 2.35 GHz is around 0.25dB, with a return loss of more than 30dB. The out-of-band rejection at 4.9 GHz was around 35dB. The BPF is designed for a lower frequency so as to adjust for expected frequency shifts (towards higher frequencies) after fabrication. The dimensions of the BPF are 0.8mm x 0.9mm.

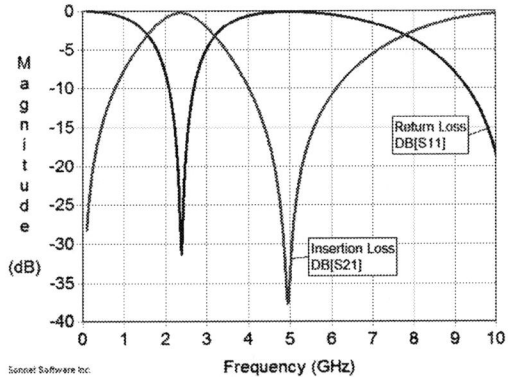

Figure 11. Full-wave 3D-EM simulated BPF response.

3c. Modeling, design and Layout of Low Pass Filter

The schematic employed to design the Low Pass Filter is shown in Figure 12.

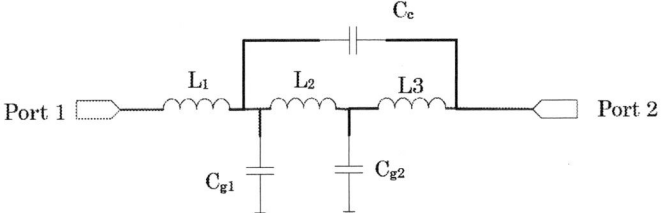

Figure 12. Circuit Schematic of the LPF

The schematic values were tuned in Agilent ADS and optimized using Sonnet- full wave EM solver, to achieve transmission below 3GHz and a rejection near 5.2GHz. The optimized structure of the LPF is shown in Figure 13. This LPF measures 1.1mm x 1mm x 0.05mm.

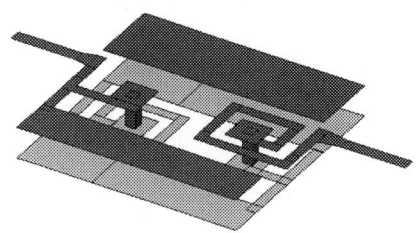

Figure 13. Full-wave EM optimized layout of LPF

The simulated S-parameter of the LPF is shown in Figure 14. The insertion loss at 2.4 GHz is -0.18dB with more than -15dB return loss.

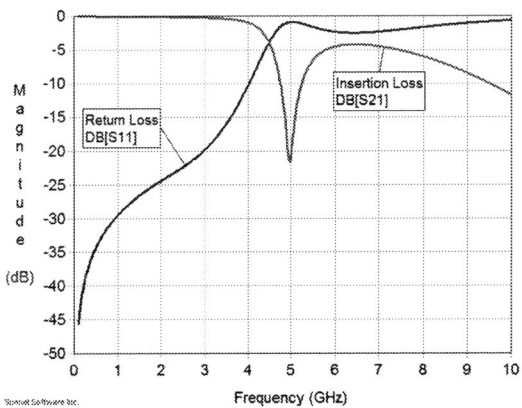

Figure 14. Simulated S-parameters of the LPF.

4a. PA module design

To test the isolated PA performance on the organic substrate, a module with only the PA is designed. The module layout is shown in Figure 15a. The dimensions of this module substrate was 10mm x 11mm x 0.05mm. The die was an additional 100 microns thick, surface assembled using thermo-compression copper-copper bonding. There was no cavity included in this 2 metal-layer design. This module was fabricated, assembled and characterized to obtain the performance of the PA. This layout was employed as the basis for the design of the FEM.

978-1-4799-2408-0/14 $31.00 © 2014 IEEE

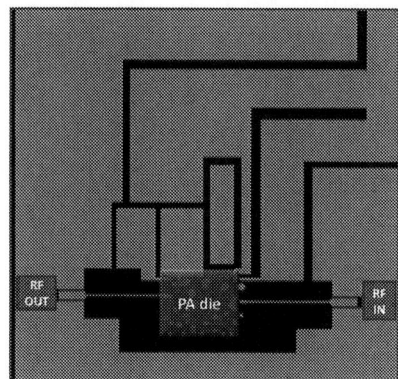

Figure 15a. Layout of PA module

4b. FEM layout and full-wave EM analysis

The Front-End Module layout was performed in Sonnet EM suite. The different elements such as the TL, BPF and LPF were directly integrated with the die-pad designs. Since the LNA and PA dies required back-side wire-bonding to ground, the ground plane near the dies were allotted for this purpose. The various parameters that had to be considered for the module layout were: 1) miniaturization of the area occupied by the module, 2) minimal parasitics between the components, 3) spaced-out placement of the surface-mount components to facilitate SMD assembly, and 4) sufficient clearance for the GSG RF probe pads to facilitate probe landing. The top-view of the two-metal layer FEM layout is shown in Figure 15b. The individual blocks are indicated as well.

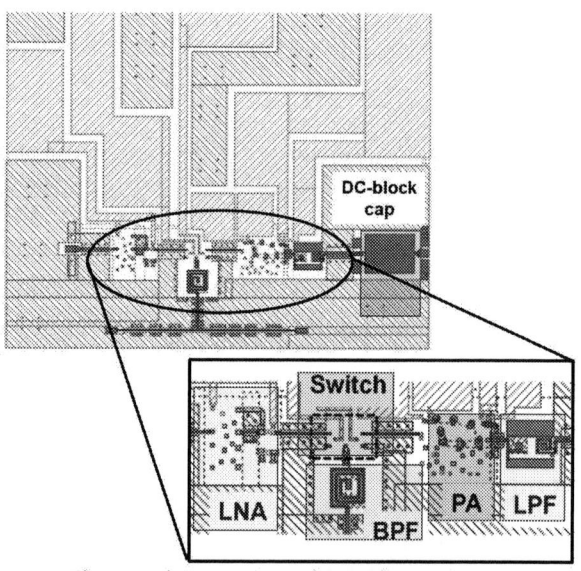

Figure 15b. Top view of FEM layout in Sonnet

At the input of the receiver chain, the BPF itself provides a DC block. However, since a LPF does not block DC currents, a parallel-plate capacitor was designed at the PA input for DC-blocking function. This parallel-plate capacitor was simulated in Sonnet EM suite separately to ensure a cut-off frequency higher than 2.5GHz.

Sufficient ground-vias were included throughout the layout to provide a low-inductance ground path. Simulation of the EM coupling from the PA to LNA, BPF and input transmission lines was studied through full-wave 3D EM analysis of this layout. The EM coupling between the PA output and the LNA input is shown in Figure 16. It can be observed that the highest coupling is -65dB, indicating very good isolation.

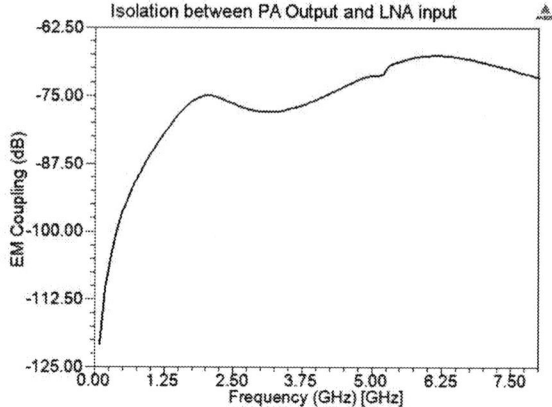

Figure 16. EM isolation between PA and LNA.

The EM coupling between the PA RF output and the PA power supply network is shown in Figure 17. It is observed that even the highest coupling is less than -70dB.

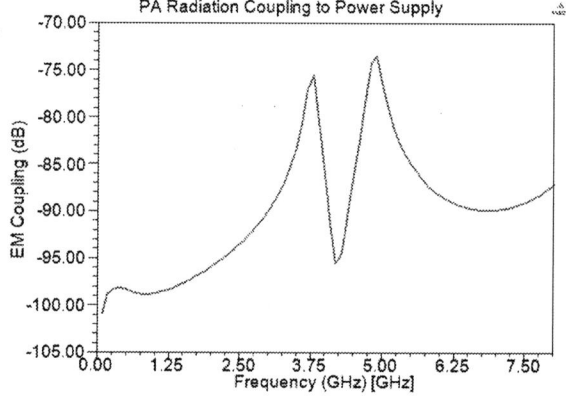

Figure 17. EM isolation from PA output to Power supply.

5. Fabrication and assembly

The first step of the fabrication of the PA module and the FEM design is the through-via drilling in the 35 μm XL laminate clad with 4 μm of copper on both sides. The vias are drilled using LASER ablation. Subsequently, the vias are metallized through electro-less plating of a seed layer of thickness 1μm, followed by electrolytic plating up to a surface copper thickness of 10 μm. A photo-resist is then laminated on both sides of the core and the designed photo-mask is used to pattern the photo-resists. After photo-patterning, the exposed copper is completely etched away to obtain the final pattern on both sides. Then, a layer of immersion gold was applied to the exposed metal regions on the sample, through electro-less deposition. This was the last step for the PA module. For the FEM fabrication, after the immersion gold deposition, the build-up polymer material is laminated using roll-lamination, and the cavities are made.

After fabrication, the dies and surface-mount components were assembled. The LNA and PA dies were bonded using thermo-compression copper-copper bonding. The switch and the surface-mount components were assembled through solder reflow. Finally, the LNA and PA backside was connected to the ground islands on the package, using a thin copper foil. The foil was mechanically and electrically connected to the substrate and die backsides using a conductive silver epoxy paste. Images of the assembled modules are shown in Figure 18.

Figure 18a. Top-view image of assembled PA module.

Figure 18b. Top-view image of assembled FEM.

6. Characterization and analysis

S-parameter characterization was performed using a Vector Network Analyzer. The BPF, LPF and the PA were characterized individually. Further, the signal path from the antenna terminal (GSG pads) through the BPF and Switch was characterized. The measured BPF and LPF responses are shown in Figure 19 and Figure 20 respectively, correlated with simulation. The BPF has a very low insertion loss of 0.25dB at 2.4GHz, with more than 15dB return loss. For the LPF, the insertion loss was 0.5dB, with 10dB return loss, at 2.4GHz.

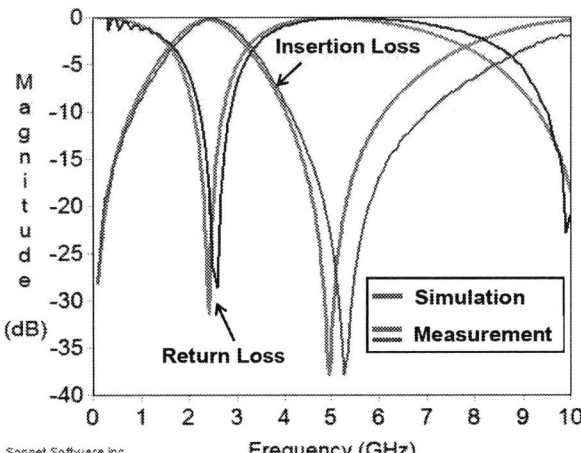

Figure 19. Measured response of BPF

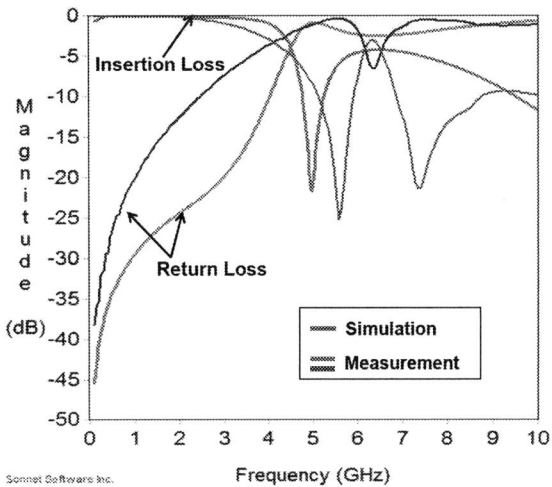

Figure 20. Measured response of LPF

A 3V DC signal is provided to the appropriate select pin of the switch die, to enable either the transmitter or the receiver paths. The measured 2-port S-parameter response for the Switch and BPF including the transmission lines is shown in Figure 21. It can be observed that the loss is around 6dB with return loss of 11dB.

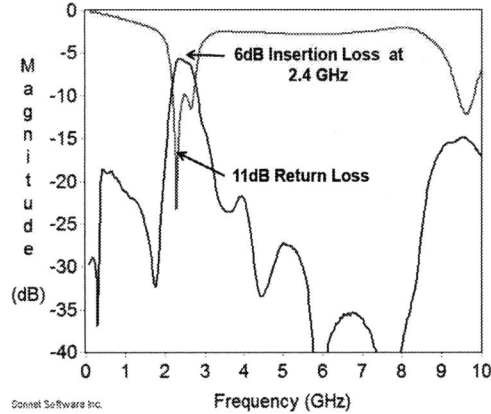

Figure 21. Measured response of Switch+BPF.

978-1-4799-2408-0/14 $31.00 © 2014 IEEE   1269

Previously, the LNA dies were demonstrated on XL substrate and characterized to have a gain of 14dB at 2.4GHz [13]. Further, from the individual PA module fabricated, the PA gain performance is shown in Figure 22. The gain is around 10.8dB with return loss of 20dB. The PA drew a current of 190mA at a supply of 3.3V.

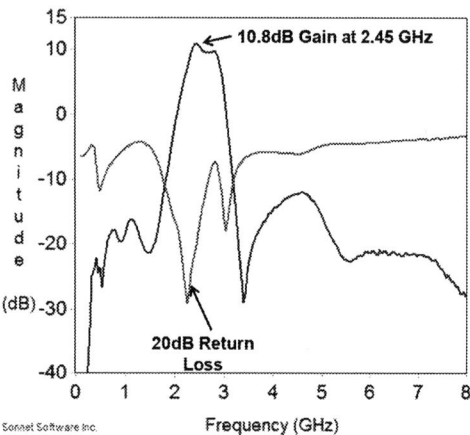

Figure 22. Measured response of PA.

The gain of the receiver chain includes the gain of the LNA and the loss from the switch and BPF. From the measured performances of the BPF, switch, LNA and transmission line sections, the gain of the receiver chain can be estimated to be 9dB, by merging the individually characterized s-parameter models in Agilent ADS, as shown in Figure 23. A similar set-up was used for the transmitter chain.

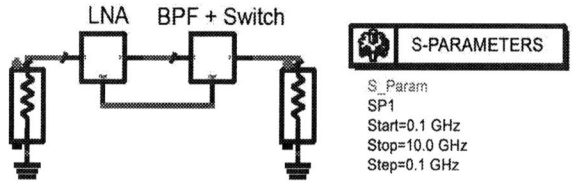

Figure 23. ADS setup to combine the s-parameters.

The merged response is shown in Figure 24. It can be observed that the rejection at 5.2 GHz is more than 30dB., indicating very good rejection of the adjacent band (5.2GHz WiFi) signals.

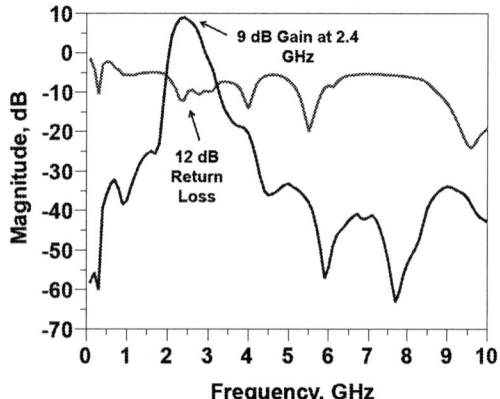

Figure 24. Response of receiver chain, obtained from ADS.

The gain of the transmitter section includes the PA gain and the loss from the switch, LPF and BPF. From the measured s-parameters of the BPF, switch, LNA and transmission line sections, the gain of the receiver chain can be estimated to be 9.6dB by merging the individual measured S-parameter models in Agilent ADS. The merged response is shown in Figure 25.

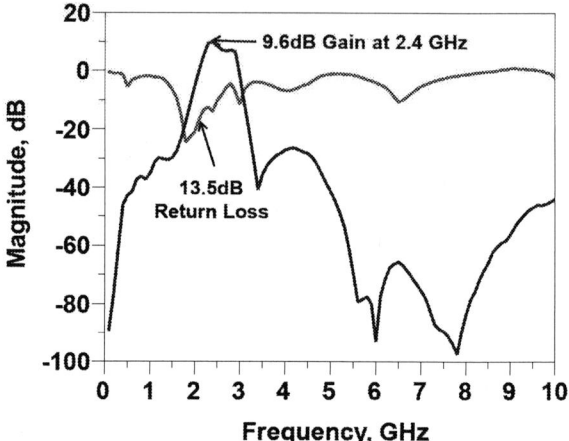

Figure 25. Response of transmitter chain, merged using ADS.

7. Conclusions

This paper presents the integration of chip-last embedded LNA, PA and switch dies, with embedded passives on a single low-loss advanced organic substrate towards a miniaturized and integrated WLAN FEM. The ultra-miniaturized BPF has a footprint of 1mm x 1mm x 0.05mm, and a measured insertion loss of 0.25dB, with return loss greater than 15dB. The LPF has 0.5dB insertion loss, and 10dB return loss. The PA shows a gain of about 11dB, which when merged in ADS with the other blocks in the transmitter chain result in 9.6dB of gain. Similarly, the receiver gain obtained by combining the individual performances of the LNA, switch and passives is 9dB. Thus, this is the first demonstration of a WLAN FEM with the smallest low-loss BPF, on ultra-thin organic substrate.

Acknowledgments

The authors wish to acknowledge Jason bishop for help with fabrication and assembly, Gokul Kumar of PRC and Dr.–Ing. A. Cagri Ulusoy of GEDC for help with RF measurements; Additionally, we would like to thank the industry sponsors of the EMAP consortia program at GT-PRC for their technical guidance.

References

1. Tummala, R.R.; Laskar, J.; , "Gigabit wireless: system-on-a-package technology," Proceedings of the IEEE , vol.92, no.2, Feb. 2004, pp. 376- 387,
2. Kamgaing, T.; Rao, Valluri R., "Passives partitioning for single package single chip SoC on 32nm RFCMOS technology," IEEE MTT-S International 2012, pp.1,3.
3. Vaidya, R.;, et al., "A Miniature Low Current Fully Integrated Front End Module for WLAN 802.11b/g Applications," Compound Semiconductor Integrated

Circuit Symposium, 2007. CSIC 2007. IEEE , vol., no., pp.1-4, 14-17 Oct. 2007

4. Tao Yang; et al., "Super Compact Low-Temperature Co-Fired Ceramic Bandpass Filters Using the Hybrid Resonator," Microwave Theory and Techniques, pp.2896,2907, Nov. 2010.

5. Chien-Hsiang Huang; et al., "Compact bandpass filter using novel transformer-based coupled resonators on integrated passive device glass substrate," Microwave and Optical Technology Letters 2012, pp.257-262.

6. Young-Joon Ko; et al., "A miniaturized LTCC multi-layered front-end module for dual band WLAN (802.11 a/b/g) applications," Microwave Symposium Digest, 2004 IEEE MTT-S International , vol.2, no., pp. 563- 566 Vol.2, 2004

7. Brunnbauer, M. et al., "Embedded Wafer Level Ball Grid Array (eWLB)," Electronic Manufacturing Technology Symposium (IEMT), 2008 33rd IEEE/CPMT International , vol., no., pp.1,6, 4-6 Nov. 2008

8. Durand, C, et al., "High performance RF inductors integrated in advanced Fan-Out wafer level packaging technology," Silicon Monolithic Integrated Circuits in RF Systems (SiRF), 2012 IEEE 12th Topical Meeting on , vol., no., pp.215,218, 16-18 Jan. 2012

9. Kamgaing, T.; Davies-Venn, E.; Radhakrishnan, K.; , "A compact 802.11 a/b/g/n WLAN Front-End Module using passives embedded in a flip-chip BGA organic package substrate," Microwave Symposium Digest, 2009. MTT '09. IEEE MTT-S International , vol., no., pp.213-216, 7-12 June 2009

10. Pinel, S.; Lim, K.; Maeng, M.; Davis, M.F.; Li, R.; Tentzeris, M.; Laskar, J.; , "RF System-on-Package (SOP) Development for compact low cost Wireless Front-end systems," Microwave Conference, 2002. 32nd European , vol., no., pp.1-4, 23-26 Sept. 2002

11. Sridharan, V.; et al., "Ultra-miniaturized WLAN RF receiver with chip-last GaAs embedded active," ECTC, 2011 IEEE 61st , pp.1371,1376, May 31 2011-June 3 2011

12. Kumar, G., et al., "Modeling and design of an ultra-miniaturized WLAN sub-system with chip-last embedded PA and digital dies," ECTC, 2012 IEEE 62nd , pp.1015-1022, May 29 2012-June 1 2012

13. Suzuki, Y., et al., "Low cost system-in-package module using next generation low loss organic material," ECTC, 2012 IEEE 62nd , pp.1412-1417, May 29 2012-June 1 2012.

14. TriQuint TQM3M7001 802.11a/b/g Dual-Band, Low Noise Amplifier Module.

15. TriQuint TQM7M7012 802.11a/b/g Dual-Band, Power Amplifier Module.

16. TriQuint TQS5200 802.11a/b/g SPDT Switch Module.

A Compact 4-Chip Package with 64 Embedded Dual-Polarization Antennas for W-band Phased-Array Transceivers

Xiaoxiong Gu[1], Duixian Liu[1], Christian Baks[1], Alberto Valdes-Garcia[1], Ben Parker[1], MD R Islam[1],
Arun Natarajan[1,2], and Scott K. Reynolds[1]
[1] IBM Thomas J. Watson Research Center, Yorktown Heights, NY, USA
[2] Now with Oregon State University, Corvallis, OR, USA

Abstract

A fully-integrated antenna-in-package (AiP) solution for W-band scalable phased-array systems is demonstrated. We present a fully operational compact W-band transceiver package with 64 dual-polarization antennas embedded in a multilayer organic substrate. This package has 12 metal layers, a size of 16.2 mm × 16.2 mm, and 292 ball-grid-array (BGA) pins with 0.4 mm pitch. Four silicon-germanium (SiGe) transceiver ICs are flip-chip attached to the package. Extensive full-wave electromagnetic simulation and radiation pattern measurements have been performed to optimize the antenna performance in the package environment, with excellent model-to-hardware correlation achieved. Enabled by detailed circuit-package co-design, a half-wavelength spacing, i.e., 1.6 mm at 94 GHz, is maintained between adjacent antenna elements to support array scalability at both the package and board level. Effective isotropic radiated power (EIRP) and radiation patterns are also measured to demonstrate the 64-element spatial power combining.

Introduction

Millimeter-wave (mmWave) packaging technology with embedded antenna is a key enabling factor for the implementation of high data-rate wireless communication and high-resolution radar imaging applications. Significant efforts have been made to develop mmWave phased-array modules with integrated off-chip antennas to take advantage of the large available bandwidths and the short wavelengths at these frequencies. For example, antenna arrays have been developed for 60 GHz (V-band) communication applications using low-temperature co-fired ceramic (LTCC) substrate [1-3] and multilayer organic substrate [4-5]. Embedded wafer level ball grid array (eWLB) technology has also been used to build phased-array modules for 60 GHz industrial radar [6] and 77 GHz (E-band) automotive radar [7] applications.

W-band (75-110 GHz) transceivers implemented in SiGe [8-11] have also recently been reported, including our work demonstrating a fully-integrated 16-element 94 GHz phased-array transceiver [11]. The reported IC integrates complete transmitter and receiver functions along with digital circuits on the same die. The front ends also support two antenna polarizations, which provide advantages to imaging systems in environments with degraded visibility [12-13]. Here, the mmWave package design required to integrate both multiple transceiver ICs and a large number of dual-polarized antennas is a major challenge that must be met to realize the vision of a highly-versatile and scalable W-band phased array system.

In this paper, we present a compact multi-chip package with embedded W-band antennas using an organic chip carrier technology. The package design is suitable for building larger scalable arrays at the board level by tiling packaged ICs adjacent to one another. Aggressive physical dimensions were designed and implemented to integrate all the desired functionality in a form factor small enough to be compatible with half-wavelength (~1.6 mm) antenna spacing. Specifically, 64 dual-polarized W-band patch antennas fed by 4 SiGe-based RFICs have been integrated in a 292-pin 0.4mm-pitch ball grid array. Compared to other W-band modules with integrated antennas, e.g., in [14], this work presents the highest level of system-in-package integration with off-chip antennas so far reported at W-band frequencies. In addition, compared to the W-band wafer-scale phased-array concept with on-chip antennas [9], our system-in-package solution is compliant with industrial assembly constraints to enable volume production.

This paper is organized as follows. First, we describe briefly our mmWave package design and test flow and identify the major technical challenges associated with our approach. Next, we discuss the package concept and design details, including substrate stack-up, package layout and simulation, as well as antenna prototyping and integration. We then discuss the processes of hardware assembly, screening and test. Finally, we present measurement results, including antenna array patterns and EIRP after spatial power combining at the IC- and package-level, respectively.

Overview of mmWave Package Design and Test Flow

Fig. 1 illustrates a typical design and test flow of our mmWave phased-array packages, including the 64-element W-band transceiver package and our previous 60 GHz 16-element transmitter/receiver packages [3-5]. The interactions between key design steps are highlighted in red; these include substrate and material selection, antenna prototyping and package-circuit-antenna co-design.

For the W-band phased-array module, we selected an organic chip carrier technology that supports a large number of metal layers with fine ground rules to implement mmWave antenna structures, signal (IF, baseband, digital, etc.) routing, and power and ground distribution. Dielectric properties of the package substrate were measured and characterized up to 100 GHz. The subsequent antenna prototyping process involved both extensive electromagnetic simulation and hardware validation, i.e., gain and reflection coefficient measurement for various antenna coupons. Another major challenge was co-designing the package iteratively with circuit layout and antenna placement to support the phased-array scalability without compromising radiation performance. Details of these design aspects, as well as system assembly and test flows, are discussed in the following sections.

978-1-4799-2408-0/14 $31.00 © 2014 IEEE

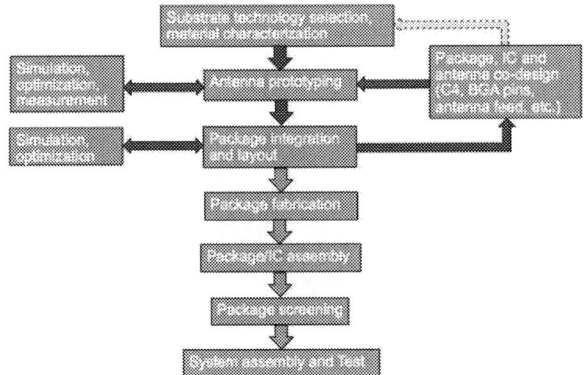

Figure 1. mmWave package design and test flow.

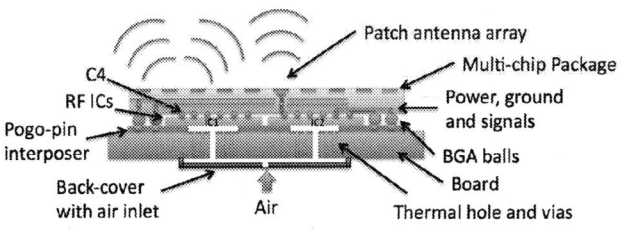

(a)

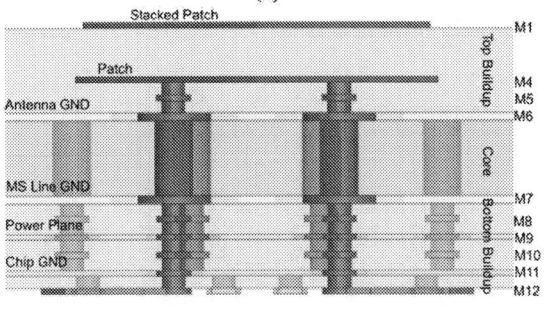

(b)

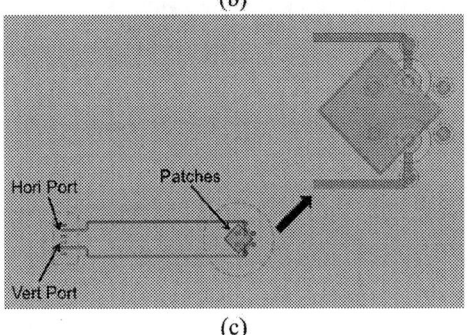

(c)

Figure 2. (a) Illustration of a W-band phased-array MCM module; (b) Package substrate stack-up; and (c) antenna test site with probe launch [15].

Antenna-in-Package Stackup and Design Concept

Fig. 2(a) illustrates a cross-sectional view of the assembled multi-chip module with embedded antennas on a system board. The patch antenna array is on the top of the package. Four SiGe ICs are flip-chip attached to the bottom of the package. The module is mounted to a system board via a pogo-pin-based interposer, which allows air cooling and supports easy removal for module screening.

A detailed zoom-in view of the antenna cross-section is shown in Fig. 2(b). The stackup consists of three portions: the core, the top buildup, and the bottom buildup. The top and bottom substrate buildups have the same dielectric properties and each portion contains five metal layers. The core portion has two metal layers. The total layer count is 12 metal layers (M1-M12), with 6 layers in the top and bottom portions, respectively. Each antenna is formed using a patch on the M4 metal layer; a stacked patch is formed using the M1 layer.

The top view of an antenna test structure for probe-based measurement is also illustrated in Fig. 2(c). It can be seen that the stacked patch antenna has two feeds, since dual polarizations are required in this design. The antenna feed lines go through from the M4 metal layer and the antenna ground plane to the M12 metal layer. Since the antenna feed lines have to travel vertically, coaxial-like sections are used for the vertical portions. In addition to the antenna feed lines, the package also includes additional critical components, such as the power planes on the M9/M10 metal layers and the RFIC ground plane on the M11 metal layer. The antenna and the RFIC chip ground planes are connected together using core and build-up vias. Measurements on individual antenna test structures show ~3dBi peak gain in both polarizations and ~8GHz bandwidth [15].

Iterations of circuit-package-antenna co-design were performed under severe physical dimension constraints to support array scalability at the package- and board- level. Fig. 3(a) shows a top view of the package, including antenna patches, BGA pins and C4 pins. The size of the package containing 4 SiGe-based RFICs and a 292-pin 0.4mm-pitch BGA is 16.2 mm × 16.2 mm × 0.75 mm.

Fig. 3(b) illustrates the overlay of a SiGe die image with a quadrant of the package layout. Two rows and columns of BGA pins provide all signal, power and ground connections to the C4s on north and east sides of the die. Signal and power integrity are taken into account in the IC-package co-design.

For example, high-speed differential signals are routed from the inner BGA row as short microstrip pairs on the M12 layer to avoid via transition, whereas low-speed single-ended signals are routed from the outer BGA row as striplines. Furthermore, two groups of voltage supply pins, as well as ground pins, are placed evenly on the periphery to ensure good power distribution to the chip.

The front-end C4s for the W-band antenna feed are laid out using a 225-µm-pitch GSGSG configuration. In order to minimize the RF antenna feed line length, the locations of these C4s were optimized together with the circuit layout for the front-end, core and digital macros.

100 (10 × 10) patch structures at 1.6 mm spacing (λ/2 at 94 GHz) cover the surface of the package. Out of these, 64 are actual dual-polarized patch antennas and 36 are dummy structures. Therefore, the effective array fill factor is 64%. The dummy structures are placed at randomized locations to minimize the impact of the reduced fill factor on side lobes. Fig. 4 illustrates the proposed W-band scalable phased array concept. Here, the 64-element packages housing four 16-element transceiver ICs can be further tiled at the board level as unit cells. However, only one single package and its array performance are described in this paper.

978-1-4799-2408-0/14 $31.00 © 2014 IEEE   1273

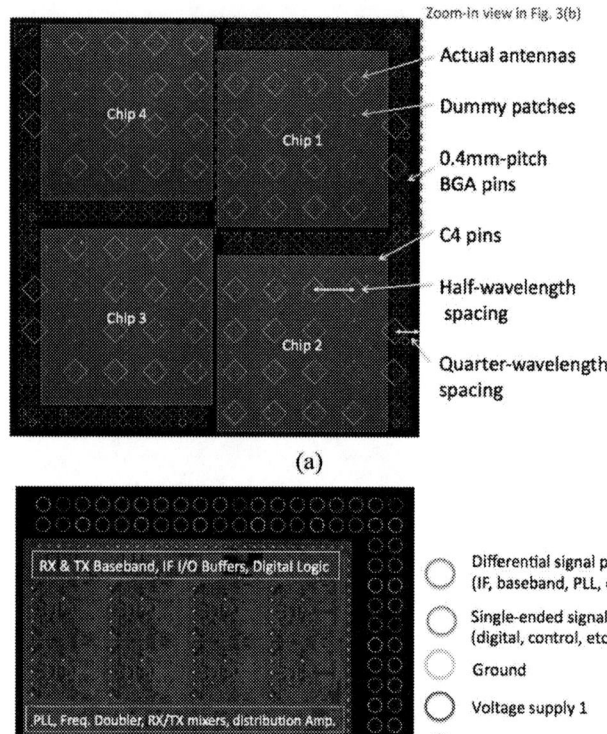

Zoom-in view in Fig. 3(b)

Actual antennas

Dummy patches

0.4mm-pitch BGA pins

C4 pins

Half-wavelength spacing

Quarter-wavelength spacing

(a)

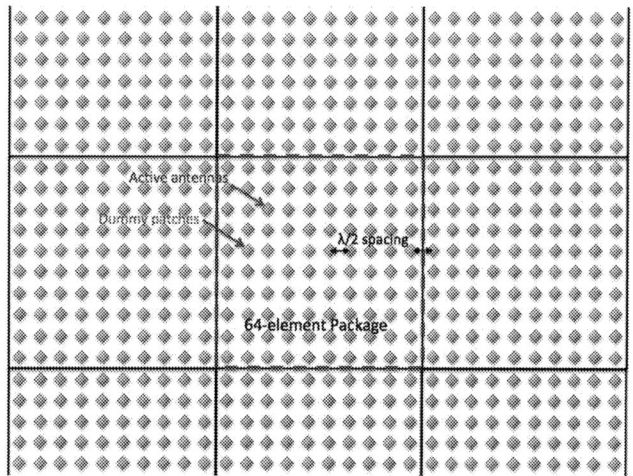

Differential signal pairs (IF, baseband, PLL, etc.)

Single-ended signals (digital, control, etc.)

Ground

Voltage supply 1

Voltage supply 2

(b)

Figure 3. (a) Phased-array package layout concept housing 4 SiGe ICs with 64 dual-polarization antennas; (b) BGA and C4 pin assignment based on circuit-package-antenna co-design.

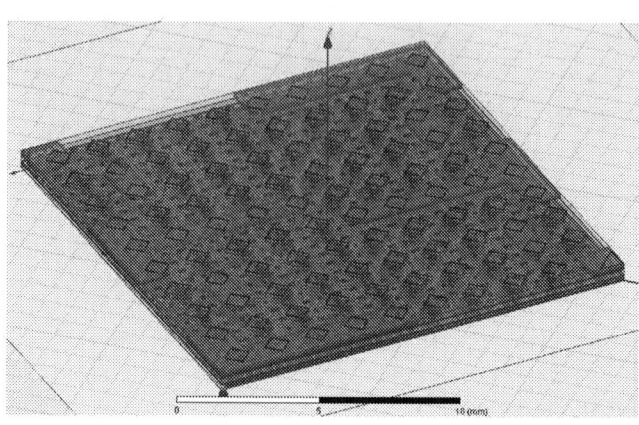

Figure 5. A large-scale 128-port 3-D full-wave HFSS model for the package simulation.

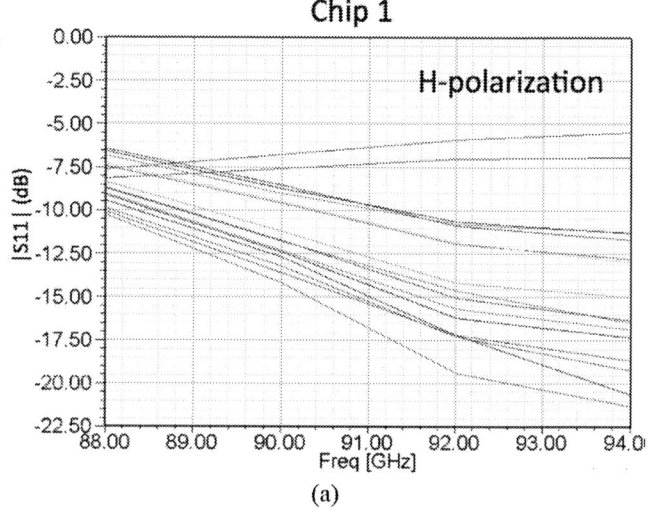

(a)

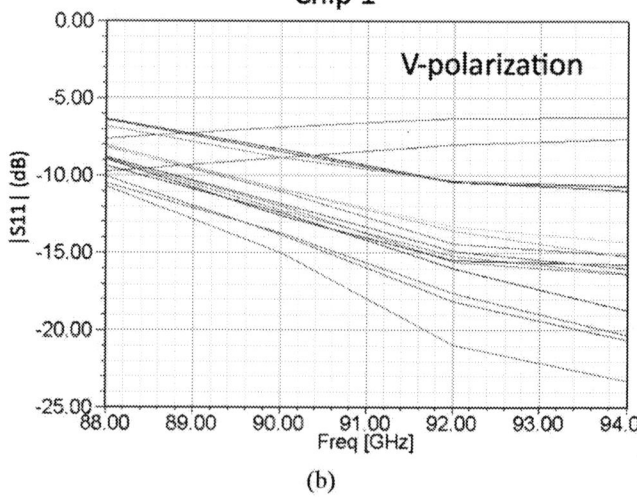

(b)

Figure 6. Simulated reflection coefficients (both polarizations) of the 16 elements associated with one SiGe IC.

Figure 4. A scalable phased array concept on board level based on the implemented 64-element package.

Package Modeling and Simulation

Full-wave 3-D simulation was performed to optimize and verify the package layout. Fig. 5 shows a finite-element-method (FEM)-based HFSS model for the entire package with 128 ports defined for all 64 of the embedded antennas.

Due to the small wavelength at the target frequencies and the large number of antennas in the package, it is computationally expensive to perform the full-wave analysis

978-1-4799-2408-0/14 $31.00 © 2014 IEEE

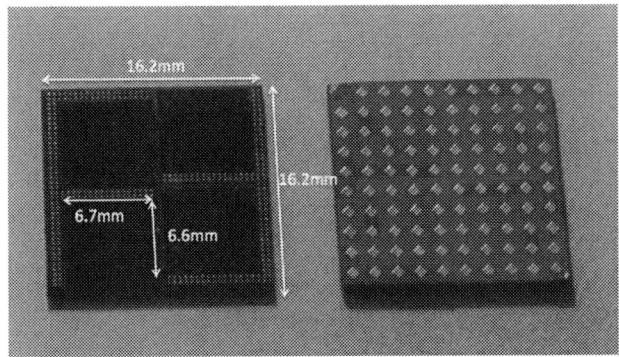

Figure 7. Flip-chip assembled packages with 64 integrated antennas (package and IC dimensions included).

of the package. In particular, the adaptive volume mesh generates ~5.3 million tetrahedra. The simulation was performed using 10 Xeon processors. It took approximately 9 hours to analyze one frequency with 146 GB peak memory usage.

Fig. 6 plots the simulated reflection coefficients of the 16 antennas associated with one of the four chips. Both horizontal and vertical polarizations are shown. Notice that most antennas here show better than 10 dB return loss at 94 GHz. The only two antennas exhibiting relatively higher reflection levels are the ones which are intentionally placed with a λ/2 offset to randomize the antenna array, as shown in Fig. 3. As a trade-off, these two antennas have more complex and longer feed lines to enable proper impedance matching.

All the antennas associated with the other 3 SiGe ICs exhibit similar characteristics. Antenna gain was simulated as well and the results are plotted below along with the relevant measurement to show model-to-hardware correlation.

Package Assembly and Test

The IC package assembly was performed using standard flip-chip attach processes with lead-free solder reflow and underfill. Fig. 7 shows a picture of the fully assembled packages with both front and back side. The antenna patches, four SiGe ICs and BGA solder balls are clearly visible.

Assembled packages were subsequently tested and screened in a socketed evaluation board, as shown in Fig. 8. A high-speed pogo pin test socket with air cooling allows functional verification of assembled packages by monitoring synthesizer locking and voltage/current consumption of each power supply.

Fig. 9 shows a fully assembled package mounted on a test board using a polymer cover which has a center window to allow measurement of the output power and radiation pattern of each antenna element. SMP connectors are populated on the board to provide PLL reference and IF signals to the four ICs. In addition, a daisy-chain configuration is implemented so only one PLL reference input and one IF input are required for antenna pattern and radiated power measurement.

Figure 8. A socket with high-speed pogo pins for package screening and test.

Figure 9. An assembled package mounted to the test board for antenna radiation pattern and power measurement.

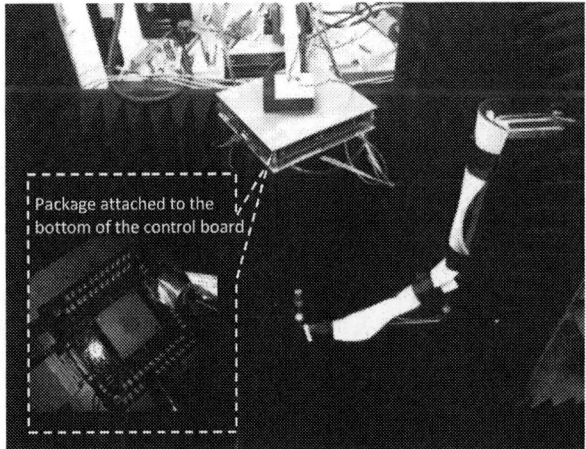

Figure 10. W-band antenna chamber measurement set-up.

Antenna Pattern and Radiated Power Measurement

The radiation pattern of each individual antenna in the package has been measured in both polarizations using a setup shown in Fig. 10. Two motors are used to control the rotation of the package and the receiving horn (i.e., azimuth and elevation angles), respectively.

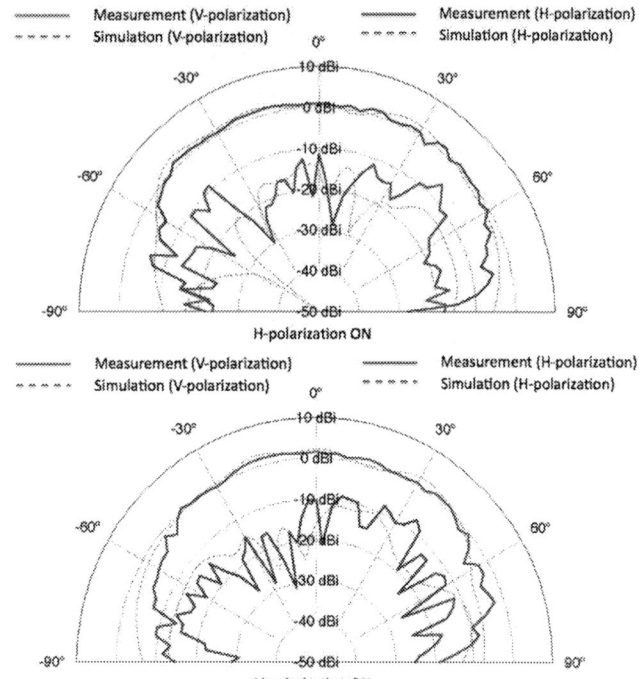

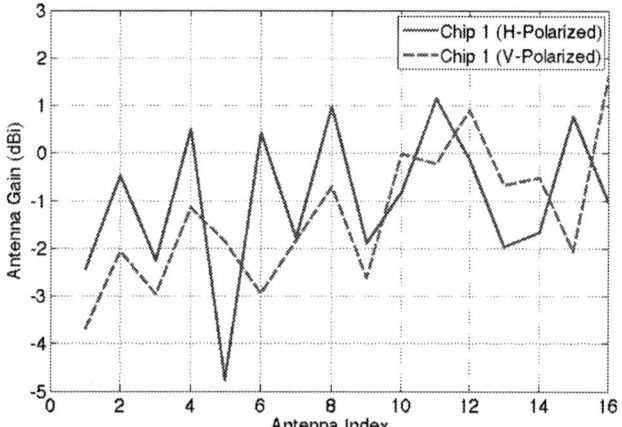

Figure 11. Model-to-hardware correlation of antenna radiation pattern for 1 element in package (1/64).

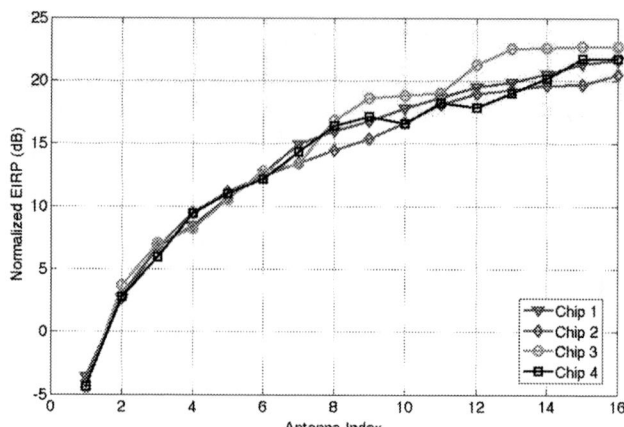

Figure 13. Measured spatial power combining of each IC (16 elements) normalized by the average power.

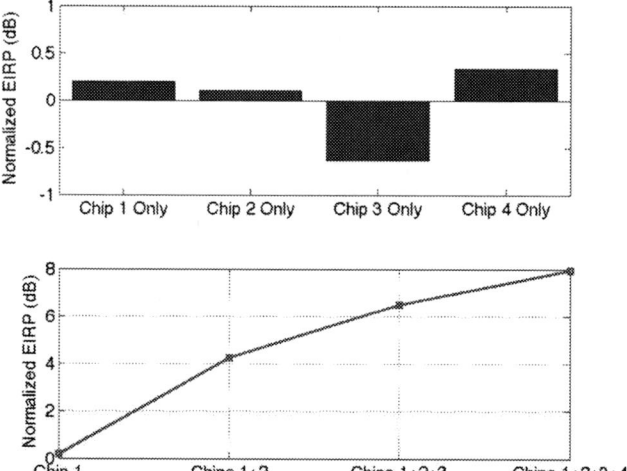

Figure 14. Variation of EIRP of four ICs is within 1dB (top); Spatial power combining of 4 ICs at the package level (bottom). Both plots are normalized by the average EIRP of 4 ICs.

Figure 12. Measured antenna gain variations for one SiGe IC.

Fig. 11 plots an example of the radiation pattern of one antenna element. In this experiment, the array is operated in TX mode, activating one element and one polarization at a time. The measured radiation patterns are in good agreement with simulations and the measured cross-polarization isolation is ~ 15 dB. The measured antenna gain varies between -5 dBi and +2 dBi across elements, as shown in Fig. 12. This variability is comparable to that measured in state-of-the-art packages for V-band applications fabricated in organic substrates [4-5].

EIRP in TX mode was measured across frequency. The measured average EIRP across 64 elements is over -1 dBm from 90 to 94 GHz, with some antenna elements radiating as much as 4 dBm EIRP at 92 GHz. Spatial power combining experiments have been performed with the assembled module in TX mode at 90 GHz. For each of the four ICs, 16 TX

elements are activated sequentially, adjusting their phase for optimum power combining. The resultant sequential increase in total EIRP is recorded. Fig. 13 presents the results of this experiment. In the plot, the results are normalized with respect to the average EIRP of each IC (each sub-set of 16 elements). As can be observed, in the four cases a spatial power combining gain of 20~23 dB is measured, close to the expected ideal value of 24 dB with 16 identical elements. The maximum EIRP for each group of 16 elements is ~20 dBm.

Next, we demonstrate the spatial power combining of all 64 elements in the package. Here, only one PLL reference input and one TX IF input are employed, and these signals are distributed through the integrated daisy-chain circuitry. To minimize temperature variation effects on the measurement, we keep all four ICs on and maintain constant supply voltage and current for this test. In this case, switching the control register to enable or disable the TX output does not change the overall power consumption (~12.3W) on the module. Furthermore, we adjust the IF VGA of every IC to equalize

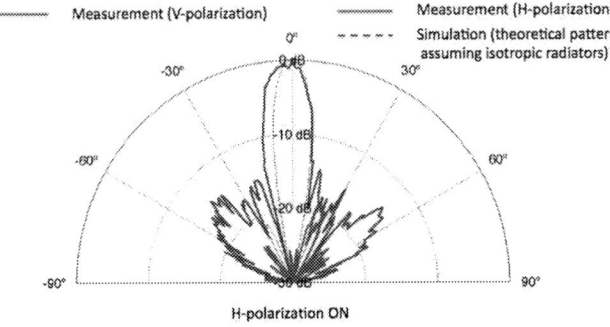

Figure 15. Measured radiation pattern after spatial power combining of all 64 elements.

their EIRP after power combining. Fig. 14(top) shows that the EIRP variation of the four ICs is well controlled, namely, to within 1 dB. Under these conditions, we sequentially adjust the phase difference of the 16 TX elements of each IC. Fig. 14(bottom) shows that a total ~8 dB increase of EIRP is achieved after power combining the four ICs. Here, the expected gain with ideal antennas and 100% fill factor is 12 dB.

Finally, Fig. 15 shows the measured radiation pattern after spatial power combining of all 64 elements. Compared to the single-element pattern in Fig. 11, a much sharper beam is evident. The measured cross-polarization isolation is reduced to ~25 dB. Fig. 15 also includes the comparison with a theoretical radiation pattern, where 64 isotropic radiators are assumed in the simulation. The measured pattern exhibits a similar but slightly wider main lobe, as well as higher side lobe levels. The non-ideal power combining effects seen here can be further lessened by reducing element-to-element variation and increasing the resolution of phase discretization.

Conclusions

A fully-integrated antenna-in-package solution is demonstrated to implement a compact 64-element W-band phased-array transceiver module that supports dual polarization with 4 SiGe ICs. Detailed package design concept and modeling, antenna integration and module-level active measurement results are presented. Circuit-package-antenna co-design challenges for the implementation of a silicon-based scalable mmWave phased array have been tackled for the first time. To the best of the authors' knowledge, this work presents the first demonstration of a W-band scalable phased array based on SiGe ICs and a multi-layer organic package.

Acknowledgments

This work has been partially funded by DARPA Strategic Technology Office (STO) under contract # HR0011-11-C-0136 (Si-Based Phased-Array Tiles for Multifunction RF Sensors, DARPA Order No. 8320/00, Program Code 1P30). The views, opinions, and/or findings contained in this presentation are those of the author/presenter and should not be interpreted as representing the official views or policies, either expressed or implied, of the Defense Advanced Research Projects Agency or the Department of Defense.

References

1. S. Emami, *et al.*, "A 60 GHz CMOS phased-array transceiver pair for multi-Gb/s wireless communications," in IEEE Int. Solid-State Circuits Conf. Tech. Dig., Feb. 2011, pp. 164–166.
2. E. Cohen, *et al.*, "A CMOS Bidirectional 32-Element Phased-Array Transceiver at 60 GHz With LTCC Antenna", IEEE T-MTT vol. 61, no. 3, pp. 1359-1375, March 2013.
3. D. G. Kam, *et al.*, "LTCC packages with embedded phased-array antennas for 60GHz communications", IEEE Microw.Wireless Compon. Lett., vol. 21, no. 3, pp. 142–144, Mar. 2011.
4. D. G. Kam, *et al.*, "Organic Packages with Embedded Phased-Array Antennas for 60-GHz Wireless Chipsets", IEEE T-CPMT, vol. 1, no. 11, pp. 1806–1814, November 2011.
5. X. Gu, *et al.*, "Enhanced Multilayer Organic Packages with Embedded Phased-Array Antennas for 60-GHz Wireless Communications", IEEE ECTC, pp. 1650-1655, May 2013.
6. R. Agethen, *et al.*, "60 GHz industrial radar systems in silicon-germanium technology", in IEEE MTT-S Symp. Dig., 2013.
7. M. Wojnowski, *et al.*, "A 77-GHz SiGe Single-Chip Four-Channel Transceiver Module with Integrated Antennas in Embedded Wafer-Level BGA Package", IEEE ECTC, pp. 1027-1032, 2012.
8. S. Shahramian, *et al.*, "A 70-100 Direct-Conversion Transmitter and Receiver Phased Array Chipset in 0.18µm SiGe BiCMOS Technology", IEEE RFIC Symp. Dig., pp. 123-127, June 2012.
9. W. Shin, *et al.*, "A 108–114 GHz 4x4 Wafer-Scale Phased Array Transmitter With High-Efficiency On-Chip Antennas", IEEE JSSC, vol. 48, no. 9, pp. 2041-2055, September 2013.
10. F. Golcuk, T. Kanar, and G. Rebeiz, "A 90-100-ghz 4X4 sige bicmos polarimetric transmit/receive phased array with simultaneous receive-beams capabilities", IEEE T-MTT, vol. 61, no. 8, pp. 3099-3114, August 2013.
11. A. Valdes-Garcia, et al., "A fully-integrated dual-polarization 16-element W-band phased-array transceiver in SiGe BiCMOS", IEEE RFIC, pp. 1-4, June 2013.
12. K. Sarabandi, and M. Park, "Extraction of Power Line Maps from Millimeter-Wave Polarimetric SAR Images", IEEE T-AP, Vol. 48 (12), pp. 1802-1809, December 2000.
13. L. Schulwitz, A. Mortazawi, "A Compact Dual-Polarized Multibeam Phased-Array Architecture for Millimeter-Wave Radar", IEEE T-MTT, Vol. 53 (11), pp. 3588-3594, March 2012.
14. W. T. Khan, *et al.*, "Packaging a W-Band Integrated Module With an Optimized Flip-Chip Interconnect on an Organic Substrate", IEEE T-MTT, vol. 62, no. 1 pp. 64-72, January 2014.
15. D. Liu, *et al.*, "A Dual Polarized Stacked Patch Antenna for 94 GHz RFIC Package Applications", to appear in IEEE APS, 2014.

978-1-4799-2408-0/14 $31.00 © 2014 IEEE

Active Die Embedded Small Form Factor RF Packages for Ultrabooks and Smartphones

Vijay K. Nair[a], Carlton Hanna[b], Ronald Spreitzer[c], and Johanna Swan[a]

[a]Components Research, Chandler, Arizona

[b]Connectivity Engineering and Manufacturing Services, Santa Clara, California

[c]Design Group, Assembly & Test Technology Development, Chandler Arizona

Email: vijay.k.nair@intel.com

Abstract

This paper reports on the design, fabrication and characterization of active die embedded ultra slim system-in-packages suitable for integrating RF and digital integrated circuits, and discrete components. Portable communication devices such as Ultrabooks, tablets and smart phones require very small form factor radio subsystems. To address this need, several proof of concept (POC) packages with embedded active die have been designed and characterized. A 50% form factor reduction compared to packages currently used in radio system half minicards is achieved by using this packaging approach. Test results show that the radio frequency (RF) performance of this small form factor (SFF) SiP is within the system specification.

1. Introduction

The current wireless industry trend is to maintain separate subsystems for broadband radio vs. local connectivity. Broadband protocols would consist of voice and data over EDGE*, 3G*, LTE* and WiMax*. Local connectivity protocols would consist of WiFi*, Bluetooth, NFC*, WiGig*, and reception only systems such as Global Positioning Systems (GPS) and FM Radio. Maintaining this separation allows the OEM to substitute different broadband devices per carrier. To be successful in these market segments, it is necessary to decrease the form factor of radio subsystems. Special attention should be given to package z-height reduction in addition to x-y form factor reduction, while also integrating an increasing number of wireless protocols.

The integration trend within connectivity device is System on Chip (SOC) with multiple protocols on a single chip. RF ICs and RF passives do not follow Moore's law scaling as digital devices, and therefore can present a challenge with SOC integration. Package level integration offers a cost effective alternative and it is amenable for integrating multi-protocol radio components in one package. Due to a shorter cycle time in package design and fabrication, new wireless system product families can be introduced to market at a faster pace to meet the rapidly changing consumer electronics device form factors. Research and development work in the passive and active die embedded substrate technology have been reported in literature [1-6].

In this paper we report on the design, fabrication and characterization of low profile packages suitable for integrating RF and digital ICs, and passive components. Active and passive components can be embedded in the thin core of the organic substrate to reduce overall z-height of the package. Surface mount components were wire bonded and assembled on the top of the package. Section 2 describes the package architectures and their attributes. Section 3 details various package designs and simulation results. In Section 4 the design of experiment (DOE) results and the test vehicle fabrication is discussed. Section 5 describes the test set up and measured results.

2. System in Package (SiP) Architecture

A schematic of the embedded die package architecture is shown in Figure 1. The package structure consisted of a 250 um substrate core and two build up layers of 30 um each on either side of the core. A cavity was created in the core to embed the die. Plated through holes (PTHs) and microvias were utilized to interconnect embedded die to the BGA and other components on the top of the package. As shown in the figure, the walls of the cavity in the substrate core can be metal plated to provide an RF shield for the die.

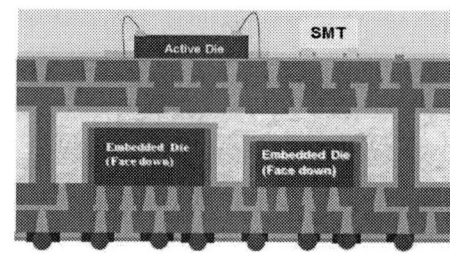

Figure 1. Embedded Die Package Architecture.

This package structure has many salient features. The silicon die can be embedded with the die pad up (face up) or die pad down (face down) orientation. This will help in realizing the optimum architecture to minimize the RF insertion loss and to improve the signal integrity of the interconnect lines. In-situ shielding of the die can be accomplished by surrounding the embedded die with ground metal. This minimizes the pickup of spurious noises by the die and also prevents unwanted signals that might emanate from the die radiating through the package. This metal layer also enhances a thermal dissipation from the die. This package technology enables embedding dies of different thickness in the same substrate core by controlling the depth of the cavity to match the thickness of the dies.

In order to demonstrate active die embedded system in package technology (e-Die SiP), test vehicles were designed utilizing a baseband (BB) chip and the radio frequency IC (RFIC) chip used in commercial products. These silicon chips were originally designed with peripheral I/Os to facilitate wire bond assembly. The wire bond pad sizes varied from 50 to 70 um and the pitch varied from 70 to 100 um. These small bond pads and the tight pitch posed a serious challenge to die embedding. Two approaches were adopted to overcome this challenge. One approach was to create small microvia to connect directly on top of the die pads. Another option was to fabricate redistribution layers (RDLs) to convert the peripheral I/Os on these dies to distributed I/Os. Cu pillars, with a diameter of 100 um and a pitch of 180 um and height 10 um were fabricated to achieve reliable micro via connection to these dies.

Several e-Die SiP package architectures were identified and designed to prove out the optimum package configuration for the SFF multi-radio system. Figure 2 (a) shows the schematic of the two-package configuration and Figure 2 (b) shows equivalent embedded die package implementation. The size of the BB and RFIC packages was 10 mm X 10 mm each. With embedded die package architecture proposed in this paper, the integration of these two ICs into one 10 mm X 10 mm package was enabled, thereby achieving 50% form factor reduction.

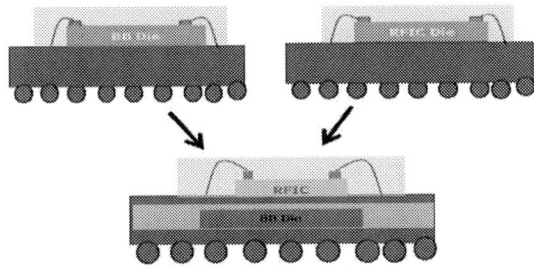

Figure. 2 (a) Schematic of the two-package system and (b) the equivalent embedded die package.

The embodiment shown in this figure 2 (a) is one in which the BB die is embedded in the substrate core and the RFIC is wire bonded on top of the package. Other configurations are also possible. 4 different versions of the embedded die packages analyzed are shown in Table 1. The first design, P1, consisted of embedded RFIC having no RDL layers and a wire bonded BB. The second design, P2, consisted of embedded RFIC with RDL layers and wire bonded BB.

Table 1. Embedded Die Package configurations.

ID	Embedded Die Type	Die size (mm) Package size (mm)	Package Configuration
P1	RFIC -No RDL layer/	5.34 x 6.52 x 0.15 10 x 10	
P2	RFIC with RDL layer	5.34 x 6.52 x 0.15 10 x 10	
P3	BB -No RDL layer	5.81 x 6.35 x 0.15 10 x 10	
P4	BB with RDL layer	5.81 x 6.35 x 0.15 10 x 10	

The third design, P3, consisted of embedded BB having no RDL layers and wire bonded RFIC. The fourth design, P4, consisted of embedded BB with RDL layers and wire bonded RFIC.

3.0 Package Design

Continuous reduction in RF multi-chip package form factors poses unique challenges in package architecture and design. Most notably, the package designer must conceptualize a 3D solution before ever starting the actual design. This is where architecture plays a predominant role. Establishing connectivity and comprehending signal integrity and power delivery are also paramount to a successful package.

The die used in this study contained only wire bond pads which posed design challenges while attempting to interconnect them. The package used in this study was a 10 mm x10 mm Molded-Matrix Array BGA with 221 balls. As discussed in the previous section, both P1 and P3 packages employed silicon dies designed with peripheral I/O pads. Microvias using a 70um capture pad on the bottom buildup layers with 40um drill were utilized to make direct connection to the wire bond pads without a redistribution layer on embedded dies within substrate. The package size was limited by the embedded die cavity size and the ability to escape route between cavity and package edge. In addition, in embedding RFICs, special care had to be taken to ensure that the RF characteristics of the inductors were not changed due to the proximity of ground metal just above the inductor coils. This was accomplished by removing the ground metal in the layer just above the inductors.

The P3 package used an embedded BB die that was very large for the package form factor which increased routing density, making escape routing even more challenging. The mousebites in the corner of the cavity metal along with highlighted alignment fiducials (see figure 3) that were required for accurate embedded die placement reduced the routing area which in turn increased routing complexity.

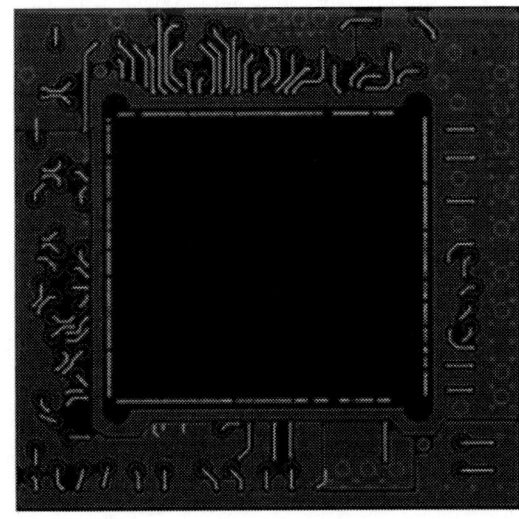

Figure 3. Embedded BB die without RDL in substrate core showing cavity corner mousebites and alignment fiducials.

The P4 package utilized BB die with RDL layers (Fig. 4). The routing challenges faced with this package originated from the increased number of I/Os on the BB die. The fan in RDL pads were the most complex since RDL pads had to be arranged and assigned to ensure the interconnection could be made reliably.

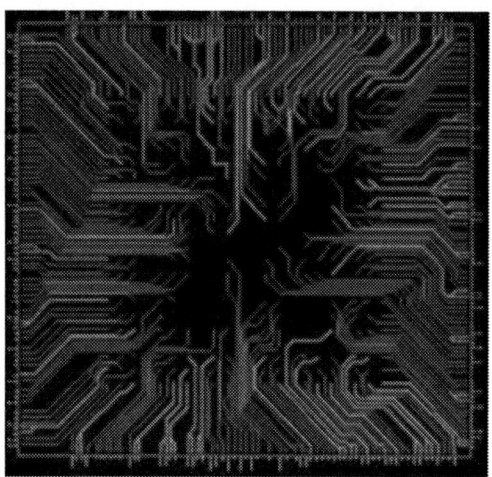

Figure 4. Layout showing fan in from original wire bond pads to RDL pads in BB die.

3.1 Electrical Simulations

RF characteristics of transmission lines were simulated using full wave high frequency structure simulator (HFSS*) software. The line width, spacing and the package routing were optimized by reducing the impedance discontinuities to achieve lowest return loss in the frequency bands of interest. The embedded structure enabled us to reduce the breakout routing length from die bump to regular trace in substrate and shorten the routing length in substrate. The transmit signal lines for the 2.4 GHz (TX2), 3.5 GHz (TX3) and 5.2 GHz (TX5), were simulated. RF return loss characteristics of these lines are shown in Figure 5. It shows better than -10 dB return loss at the WLAN frequency band.

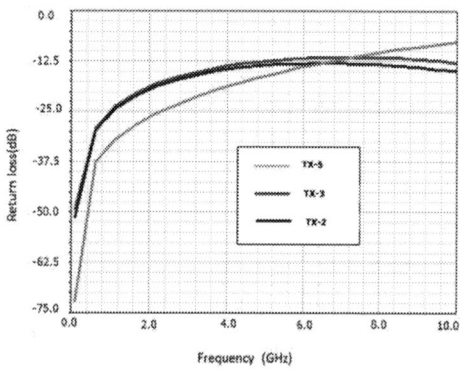

Figure 5. Return loss of the embedded die package.

The return loss characteristics of the differential pairs were also simulated by using HFSS. The trace routings were modified to achieve better than 10 dB of return loss for the differential pair by providing better return current path (reference plane) and reducing the trace length. Figure 6 shows the simulated result. It shows better than -20 dB return loss at the WLAN frequency band.

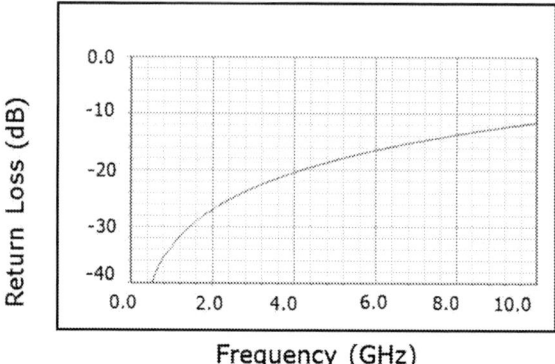

Figure 6. Simulated return loss of a differential pair lines

3.2 Thermal Simulations

Power dissipations of two versions of the embedded die packages were simulated. In one configuration the RFIC die was embedded and the BB was wire bonded on to the top of the package. In the other configuration the BB die was embedded and the RFIC was wire bonded on to the top of the package. Power dissipation characteristics of both packages were simulated under the same power consumption condition of 1x1 Transmit/Receive modes. The power dissipation in each mode is summarized in Table 2.

Table 2 Power dissipation of RFIC and BB/BAC die
(1X1 Transmit Receive Mode)

	RFIC Power (milliwatt)	BB Power (milliwatt)	Total Power (milliwatt)
1x1 RX	345	256	601
1x1 TX	354	206	560

The junction temperatures of the BB die when it was wire bonded on to the surface of the package (Figure 7(a)) were 104.5 °C in receive mode and 101.1 °C in transmit mode. When the same die was embedded (Figure 7(b)) its junction temperatures were 99.7 °C in receive mode and 97.7 °C in

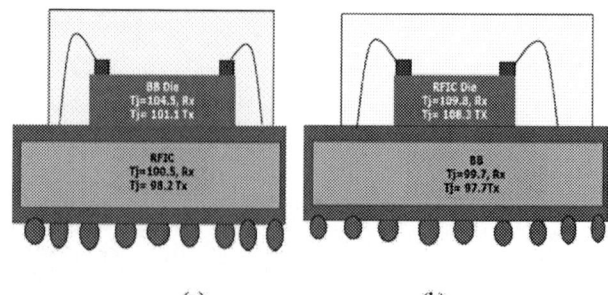

(a)   (b)

Figure 7. Junction temperature of silicon die in 1x1 transmit/Receive mode

978-1-4799-2408-0/14 $31.00 © 2014 IEEE

transmit mode respectively. These results indicate that when the die is embedded in the face down configuration inside the package, its junction temperatures tend to be lower.

Similarly, the junction temperatures of the RFIC die when it was wire bonded on to the surface of the package (Figure 7(b)) were 109.8 °C in receive mode and 108.3 °C in transmit mode. When this RFIC die was embedded (Figure 7 (a)) its junction temperatures were 100.5 °C in receive mode and 98.2 °C in transmit mode respectively. These results also indicate that when the RFIC die is embedded face down inside the package, its junction temperatures tend to be lower. This can be attributed in part to shorter distance of the die pads to the BGA balls achieved by the face down configuration of the embedded die, which gives the lowest thermally resistive path to the board.

It is interesting to note that both RFIC and BB die had higher temperature in the receive mode than in transmit mode. During the receive mode, the devices remain ON for longer periods of time and they process more information than in transmit mode. It should also be pointed out the power amplifier module which is external to this SFF package is not included in this analysis. More rigorous temperature modeling and junction temperature measurements have to be performed to reach a definitive conclusion on the thermal dissipation characteristics of embedded die packages.

4.0 Package Fabrication

Several new process technologies were developed to embed silicon RF die in thin substrate core. Since BB and RFIC dies were originally designed with peripheral pads for wire bond assembly, the die pad size was only 50 um and the pitch was 75um. Making connection directly to the wire bond pads on the die required 30 um diameter microvias. Figure 8 shows the cross section of the fabricated fine pitch microvia of diameter of 30 um with a pitch of 75 um fabricated to make connection to the embedded die packages P1 and P3.

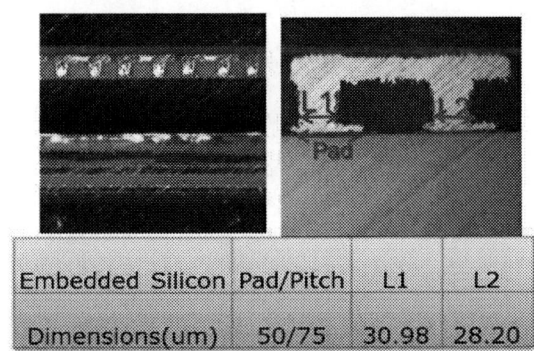

Embedded Silicon	Pad/Pitch	L1	L2
Dimensions(um)	50/75	30.98	28.20

Figure 8. Fine pitch microvia DOE Result.

Another process technology developed for the embedded die fabrication was the precise drilling of the substrate core to create a cavity at a given depth with very tight tolerance. Based on the DOE experiments performed cavity depth of 175 um was selected to embed a 150 um thick silicon die. The variation in die thickness, die attach film thickness, RDL layer thickness and the cavity routing tolerance were considered in calculating the substrate core thickness. Figure 9 (a) shows the

cavity that was created to embed 150 um thick silicon die. Figure 9 (b) shows the profilometer measurement. The cavity depth was measured to be 176.25 um and it is very uniform across the entire the cavity.

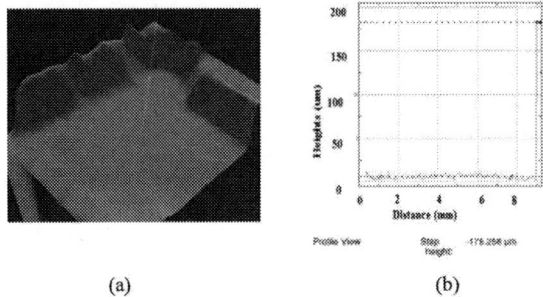

(a)   (b)

Figure 9. Depth profile of cavity drilled to embed BB die (a) X-Ray image (b) profilometer measurement.

Figure 10 shows the cavity edge profile and cavity metal thickness. Slight tapering of the cavity edge was observed in some cases.

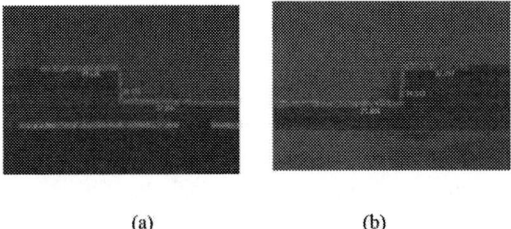

(a)   (b)

Figure 10. Edge profile of cavity to embed BB dies.

Figure 11 (a) shows the top view of four cavities created in the substrate core to enable embedding smaller dies and Figure 11 (b) shows the cross sectional view of the cavity.

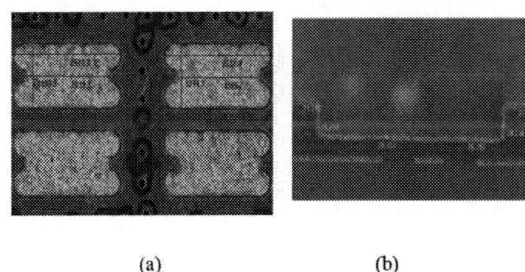

(a)   (b)

Figure 11. Multi cavity substrate core profiles (a) top view of four cavities and (b) cross sectional view of a cavity.

The substrate fabrication consisted of the following steps. (1) The cavity routing to achieve the required depth, followed by PTH drilling (2) the metallization of the cavity walls to provide in situ shielding of the embedded die, (3) patterning of the plated metal, (4) Die pick and place and die attach in the cavity, (5) build up layer lamination of both sides of the substrate core, (6) microvia drilling of the buildup layers to make connection to the die and to other signal lines, (7) microvia plating, (8) second build up layer lamination and microvia drilling, (9) 2nd layer microvia plating, (10) third build up layer lamination, (11) surface layer plating to wire bond the die and to assemble SMT components, (12) solder

978-1-4799-2408-0/14 $31.00 © 2014 IEEE   1281

mask opening for attaching BGA on the bottom side of the package and to assemble SMT components and wire bond die on top side of the package.

After the die attachment in the cavity and lamination of the first build up layer, the package substrate was flipped upside down to achieve the face down configuration of the embedded die. Figure 12 shows the picture of RFIC and BB die embedded in cavity of the substrate core material.

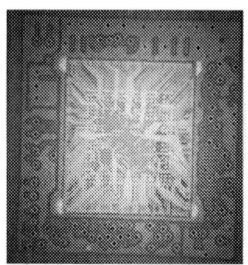 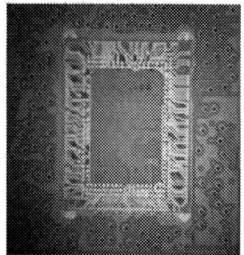

(a)   (b)

Figure 12. Embedded silicon Dies the substrate core (a) BB die and (b) RFIC die.

5. Package Assembly and Test

In order to test the embedded die packages, half minicards were designed and fabricated. They were designed to test individual embedded die packages and to compare its performance with conventional package. Fig. 13 shows the half mini card designed to test the P4 package consisting of embedded BB IC and wire bonded RFIC. All other components on the board are the same as in the WLAN half minicard currently used in products (Figure 14).

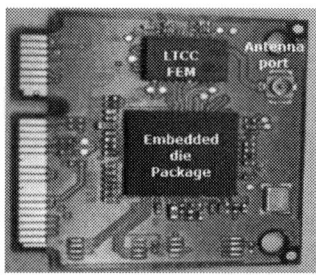 

Figure 13. Half minicard board, (side 1 and 2), used to test embedded die package P4

Figure. 14. Half minicard board, (side 1 and 2), used in current products

Printed circuit board area saving due to the integration of BB and RF ICs in one package using the e-Die SiP technology is evident from the comparison of figures 13 and 14.

Figure 15 shows the test system that was used to test the embedded die packages. The system idle current was measured to be 110 mA. When the BB IC and RFIC were turned on the total current increased to 230 mA, which was within the range of the WLAN radio system specifications. For the five packages tested, the idle current and total system current varied from 90-115 mA and 190-240 mA respectively. RF functionality test was performed on several embedded die packages. The system functionality test was performed in the WLAN frequency band and specifically at channel 6 of the WLAN band operating at 2.438 GHz.

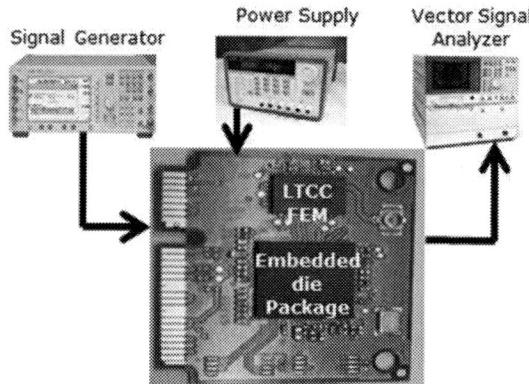

Figure 15. Embedded die package under test in RF system setup.

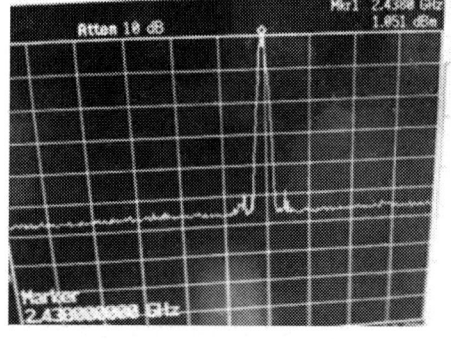

Figure 16. RFIC Transmit signal output from embedded die package, P4. Freq. = 2.438 GHz, Power output = 1.05 dBm.

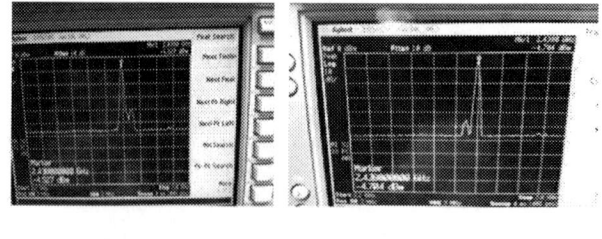

RFIC TX-N   RFIC TX-P

Figure 17. Transmit signal from RFIC differential output ports (a) Tx- Negative and (b) Tx Positive.

978-1-4799-2408-0/14 $31.00 © 2014 IEEE   1282

Measured RFIC Transmit signal strength is shown in Figure 16. RFIC output signal strength was observed to be + 1.0 dBm for the embedded package operating at a total current of 230 mA. The transmit signal strength varied from -5 dBm to +1.0 dBm for the packages that were tested. The WLAN system specification is -5 dBm to +5 dBm. These test results showed that the embedded die packages are RF functional and within specifications.

Figures 17 (a) and (b) show the signal output observed at each of the differential pair output (N and P port) of the RFIC. The spectrum shows small side band signals on both sides. These seem to be the signals bleeding from the adjacent channels. This needs to be further investigated at the next prototyping effort.

6. Summary

The focus of this research effort was to prove the viability of the active die embedded RF packages for wireless applications, to identify the issues and offer solutions to remedy them. The embedded die packages discussed in this paper demonstrated many salient features that are suitable for small form factor packages for wireless communication devices.

Active die embedding in face up or face down fashion was designed and fabricated. In situ shielding of the device was achieved by metalizing the walls of the cavity in which the die is embedded. A process technology capable of embedding dies of different thickness and sizes were demonstrated by creating cavities of varying sizes and depths in the same substrate core. Innovative package design approaches were implemented to embed RF devices having very small peripheral bond pads at very tight pitch with and without redistribution layers. This package design and fabrication also demonstrated techniques to integrate components designed for flip chip, wire bond and SMT technologies in a very small form factor package.

This active die embedded system in package design approach achieved a 50% reduction in package form factor compared to the single die packages used in current systems. Electrical simulation results showed that RF characteristics such as return loss, signal integrity and insertion loss were as good as or better than system specification. Thermal simulation results indicated that when the dies are embedded in the face down configuration inside the core, their junction temperatures tend to be lower.

RF system functionality tests of the embedded SiPs were performed at WLAN channel-6 frequency of 2.438 GHz. The embedded die package achieved power output as high as 1.05 dBm at this frequency band. Five packages tested exhibited the system idle current (i.e. during no transmission) and the total system current (i.e. during transmission) within the range of the WLAN system specification. The RFIC transmit signal strengths of the embedded SiP varied from -5 dBm to +1.0 dBm and were also within the system specification.

The highly integrated RF silicon fabrication technology is advancing rapidly and SOC integration of radio system is in progress. However the integration of RF passives and multi-protocol radio system integration in SOC is very challenging.

Small form factor system in package with embedded die is a promising option for ultra slim portable communication devices such as Ultrabooks, tablet and smartphones.

Acknowledgments

We would like to acknowledge S. Lotz, T. Kamgaing, L. Liu, D. Hackett, D. Bruneau, M. Megahed, D. Russell, J. Huang, M. Strunk, J. Chen, R. Perry, T. Xong and P.Vin Haa for their contributions in package design, substrate fabrication and test.

References

[1] B. K. Appelt, B. Su, D. Lee, U. Yen, M. Hung, "Embedded Component Substrates Moving Forward," 2011 13th Electronic Packaging Technology Conference, pp. 558-561.

[2] T. Braun, K. F. Becker, L.Bottcher, J. Bauer, T. Thomas, M. Koch, R. Kahle, A. Ostmann, R. Aschenbrenner, H. Reichl, M. Brindel, J. F. Haag, U. Scholz, "Large Area Embedding for heterogenous System Integration," 2010 Electronic Components and Technology Conference, pp.550-556.

[3] C.E. Patterson, T.K. Thrivikraman, S/K. Bhattacharya, C.T. Coen, J.D. Cressler and J Papapolymerou, "Multilayer Organic Packaging Technique for a Fully Embedded T/R Module," Proceedings of European Microwave Conference, Oct. 2011, pp. 10-13

[4] T. Kamgaing, R. Vilhauer, V. K. Nair and D. Choudhury, "Embedded RF Passives Technology Using a Combination of Multilayer Organic Package Substrate and Silicon-based Integrated Passive Devices," 2010 Electronic Components and Technology Conference,, pp.1547-1551.

[5] G. Kumar, S. Sitaraman, V. Sridharan, N. Sankaran, F. Liu, N. Kumbhat, V. Nair, T. Kamgaing, F. Juskey, V. Sundaram, R. Tummala, "Modeling and design of an ultra-miniaturized WLAN sub-system with chip-last embedded PA and digital dies," 2012 Electronic Components and Technology Conference, pp.105-1022..

[6] S. Sitaraman, Y. Suzuki, V. Nair, T. Kamgaing, F. Juskey, F. Liu, M. Hashimoto, V. Sundaram, S. Kim, and R. Tummala, "Modeling, Design and Demonstration of Multi-Die Embedded WLAN RF Front-End Module with Ultra-miniaturized and High-performance Passives," to be published in 2014 ECTC conference digest, Orlando, FL, USA.

Intel® is a registered trademark of Intel Corporation or its subsidiaries in the United States and other countries.

*Other brands and names are the property of their respective owner.

Design and Material Contributions to Second-Harmonic Nonlinearities in RF Silicon Integrated Passive Devices

Robert Frye, Robert Melville* and Kai Liu**

RF Design Consulting, LLC
334 B Carlton Avenue
Piscataway, NJ 08854 USA
bob@rfdesignconsulting.com

* Emecon, LLC
1756F Springfield Ave.
New Providence, NJ 07974
bobmelville1@gmail.com

** STATS ChipPAC, Inc.
1711 West Greentree, Suite 117
Tempe, Arizona 85284, USA
kai.liu@statschippac.com

Abstract

Second harmonic distortion in silicon integrated passive devices is sufficiently low to be of no concern in most applications. However, recent changes in frequency allocation for 4G cellular applications have raised concerns for interference with sensitive GPS signals. Previously, we reported a number of measurements on low-pass filters around 750MHz. These measurements suggested that harmonic distortion was mainly attributable to lateral surface fields in the circuit's devices interacting with the substrate depletion layer. If this is the case, the nonlinearities in some cases could be reduced by design changes.

We have performed further experiments and measurements aimed at improving our understanding of these issues and the relative importance of design versus substrate material. We have evaluated low pass filters with design variations aimed at reducing the interaction of the filters with the substrate. The designs were fabricated on substrates of different nominal resistivity, and with and without surface treatments aimed at reducing the so-called "parasitic surface conduction" effect.

In substrate wafers that do not have surface passivation, harmonic distortion is generally high. In this case, we find that minor design variations may significantly reduce the second harmonic power. Furthermore, we observe in these devices that the harmonic power increases under conditions of illumination, suggesting that surface conduction plays a role. Importantly, however, in surface-passivated wafers we find that the harmonic power is significantly reduced and that design variations make less difference. Furthermore, we observe insensitivity to illumination, indicating that the photoconductive effect in these passivated samples is negligible.

We have compared samples fabricated on different substrate resistivities. We find a very strong dependence. Comparison of samples fabricated on silicon substrates with 3500Ωcm resistivity, versus 1000Ωcm, shows roughly an order of magnitude reduction in second harmonic power. For 4G cellular applications in the 700MHz frequency band, substrate resistivity appears to be a critical consideration.

As in other studies, we find that 2nd-harmonic generation is closely correlated with losses in passive components and circuits. We have examined inductor quality factor, as well as insertion losses in transmission lines and filters, and find that higher losses are generally associated with higher harmonic power levels.

Introduction

To increase the available bandwidth for 4G cellular telecommunications, the FCC in 2008 made available new spectrum in the range from 668 to 806MHz (the so-called "700MHz" band). Signal propagation in this band is relatively good at penetrating walls, making it especially desirable for cellular applications in dense, urban areas.

A drawback, however, is that for signals near the middle of this band, second harmonic power falls within the sensitive Global Positioning Satellite (GPS) frequency of 1575MHz, as shown in Fig. 1. To help avoid interference, within the 700MHz the sub-band around 788MHz is not used for cellular communications, and serves as a guard band.

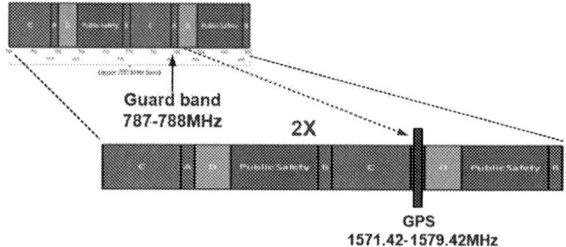

Fig. 1: 700MHz band frequency allocation

This frequency allocation effectively prevents GPS interference from radiated emissions in cellular applications. However, nearly all modern cellular handsets have internal GPS receivers. Even very weak signals *inside* the handset can saturate the front-end of these receivers. Despite the guard band, weak intermodulation tails from active adjacent channels may leak into the guard band. With harmonic upconversion, these small signals may cause interference within the handset because of their physical proximity to the GPS receiver. Consequently, the specifications for second harmonic generation in passive components are especially stringent in this band.

Previous Investigation

Previously [1], we reported on initial experiments aimed at identifying the mechanism that is responsible for these nonlinearities in Silicon Integrated Passive Devices (IPDs). Measured 2nd-harmonc levels in commercial low-pass filters designed for use in the 700MHz band showed little dependence on substrate thickness. Since vertical field strengths are significantly different for different substrate thicknesses, this observed insensitivity led us to suspect that the nonlinear effects giving rise to the harmonics arose mainly through the interac-

978-1-4799-2408-0/14 $31.00 © 2014 IEEE   1284   2014 Electronic Components & Technology Conference

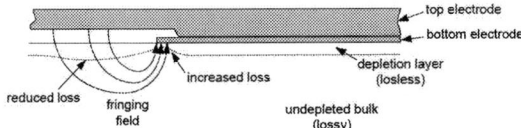

Fig. 2: Proposed nonlinear mechanism

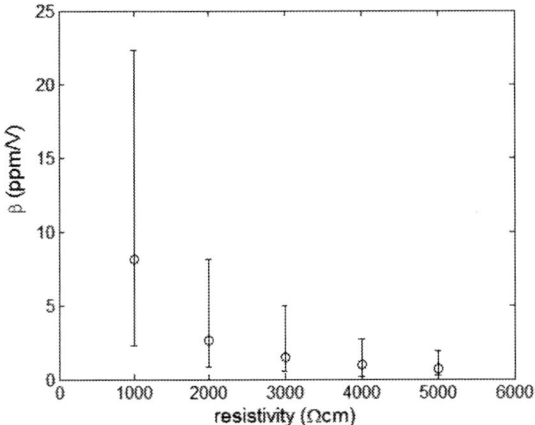

Fig. 3: Estimated first-order voltage coefficient of capacitance for an example capacitor.

tion of lateral electronic fields with the depletion layer in the substrate. This mechanism is illustrated in Figure 2.

The results of electromagnetic simulation indicated that for these types of IPD filters, lateral electric fields near the surface are much stronger than the vertical fields. This is especially true near the edges of parallel plate capacitors, as shown in the cross-section in Fig. 2. These lateral fields can modulate the depth of the substrate depletion layers. This proposed mechanism was used to estimate the first-order voltage coefficient of capacitance for a typical example capacitor as a function of substrate resistivity.

The key thing to note in this result is that the nonlinearity is expected to fall sharply with increased substrate resistivity. Furthermore, these strong interactions arise at the periphery of capacitors, where the top-level metal passes over the edge of the bottom-level metal capacitor electrode. In many cases, minor design changes can be made to minimize the area of these strong interactions without significantly affecting the IPD's overall electrical characteristics.

Substrate Resistivity Dependence

Figure 4 shows measured 2nd-harmonic power levels in commercial low-pass filters that are designed to operate in the 700MHz cellular band. These filters had a packaged substrate thickness of 250μm, and were attached to test substrates using bond-wire interconnection. Two samples of each filter were tested to verify the uniformity of the result.

At typical cellular power levels, around 20 to 24dBm, the filters fabricated on 3.5KΩ-cm substrates showed about 13dB lower levels of 2nd-harmonic power. This is consistent with the estimated resistivity dependence suggested by our previous estimates, shown in Fig. 3. At higher power levels, it can be seen that the slope of the curve for the 1KΩ-cm material begins to roll off, and the relative difference between the two substrates is less. This kind of behavior in the power depend-

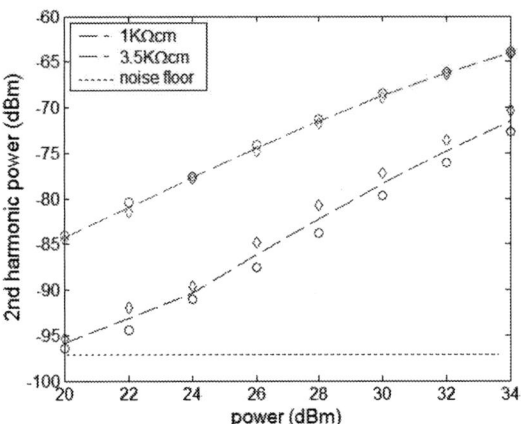

Fig. 4: Measured 2nd harmonic power levels in commercial low-pass filters designed for use in the 700MHz band.

ence is typical of nonlinear mechanisms that saturate (like, for example, MIS diodes).

Surface Passivation Experiments

It is well-known that high resistivity silicon substrates, without special treatment, exhibit parasitic surface conduction that causes increased conductive losses at high frequency [2-5]. The losses can be significantly reduced by the formation of a trap-rich surface layer through various methods of ion-implantation or film deposition. More recently, these same surface treatment methods have also been shown to significantly reduce the levels of 2nd-harmonic generation in coplanar transmission line measurements [6, 7].

We have observed an insensitivity of 2nd-harmonic power generation on the device substrate thickness, leading us to propose the model described above for nonlinear conduction driven mainly intense lateral electric fields near the wafer surface. This is also consistent with the other studies linking 2nd-harmonic generation to the presence of parasitic surface conduction. The results shown in Fig. 4 suggest that there is also a dependence on substrate resistivity, but since the two types filters compared in that study were fabricated in separate lots, the results are not conclusive. It is possible that there is some difference in the condition of their surfaces, arising from variations in their processing.

To help address these questions, we ran further experiments comparing filters with design modifications aimed at reducing the lateral electric field intensity at the substrate surface. Fig. 5 shows maps of the simulated lateral electric field tangential to the substrate surface in these filters. The legend on the right of the figure shows the relative field strength, in dB. In the nominal design, strong electric fields (shown in orange) originate at the edges of capacitor plates connected to the filter input and output terminals. In the modified design, the capacitor shape and placement were rearranged to reduce the strength of these fields by nearly 10dB. In all other respects, the electrical characteristics of the two filters are nearly indistinguishable. Consequently, we may ascribe any observed difference in the 2nd-harmonic power generation to effects originating near the surface.

Fig. 6 shows a comparison of 2nd-harmonic generation in filters fabricated on high resistivity silicon substrates with

978-1-4799-2408-0/14 $31.00 © 2014 IEEE

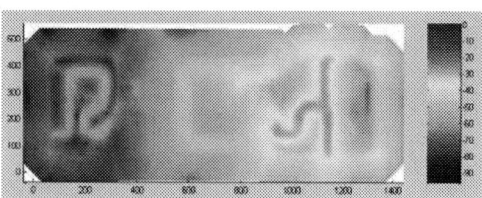

Nominal Design

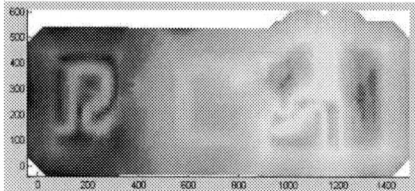

Modified Design

Fig. 5: Lateral electric field intensity map for alternative 700MHz-band low-pass filter designs used in this study

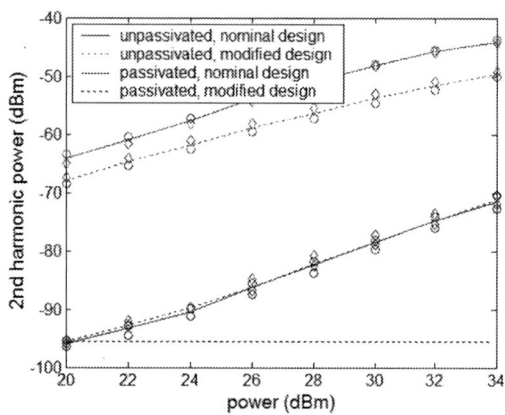

Fig. 6: Comparison of 2nd-harmonic power generation for unpassivated and passivated substrates, showing the effects of design modification.

(shown in blue) and without (shown in red) surface passivation treatment. The results for the nominal design are shown by the solid lines, and those for the modified design are shown by the dotted lines. The substrate wafers in all cases were from a common lot, specified to have resistivity greater than 1000Ω-cm, and the filters were made in a common fabrication run.

In these results, it can be seen that the design modification resulted in reduction of harmonic power levels of 5 to 6dB in the unpassivated samples. This is roughly consistent with the amount of reduction in the lateral field strength, as seen in the plots in Fig. 5. However, in wafers that received surface passivation treatment, there was no appreciable difference in the characteristics of the two designs.

Surface conduction is generally known to contribute to harmonic generation. For the unpassivated samples, which presumably have a conductive surface layer, it is not surprising that the modified design, which interacts less strongly with the surface, would show reduced harmonic power. The surprising result is that the passivated samples show similar behavior for both design variants. This suggests that the observed nonlinearities in these devices originate primarily in the bulk.

In addition, in the unpassivated samples we observed that 2nd-harmonic power levels could be increased by several dB by illumination, whereas the passivated samples were insensitive. These results suggest that surface conduction mechanisms, either inherent or photo-induced, have nonlinear aspects that contribute to 2nd-harmonic generation.

Correlation with other characteristics

More recently, we have performed a variety of measurements on IPD wafers of different types. In addition to examining the issues of nonlinear effects, we have also examined loss mechanisms in a number of different test devices. For these experiments, we fabricated IPDs on three different silicon substrate types, listed in Table 1. Substrate types A and B in these experiments were of very high resistivity silicon, and were available with two different surface treatment methods for passivation. Substrate type C was the standard material used in routine IPD production.

Coplanar transmission lines have been used in several of the past studies cited above to characterize loss and harmonic generation in silicon substrates. The layout of the test device and measured insertion loss for the three wafer types listed above are shown in Fig. 7. These results demonstrate the importance of surface passivation. Note that the insertion loss for lines fabricated on the 1000Ω-cm resistivity wafer (C) was intermediate between the two 4000Ω-cm substrates having different passivation treatments.

The fundamental circuit components in IPD technology are inductors and capacitors. In the useful operating frequency range of IPD circuits, such as filters, diplexers and baluns, the capacitors are relatively high Q devices, and losses are domi-

Table 1: Silicon substrates used for IPD evaluation

Substrate	Description
A	>4000Ω-cm, surface passivation method 1
B	>4000Ω-cm, surface passivation method 2
C	>1000Ω-cm, std surface passivation (control)

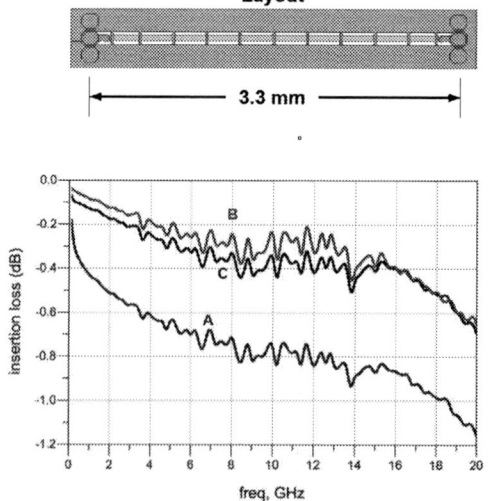

Fig. 7: Coplanar transmission line test device layout and measured insertion loss.

Layout

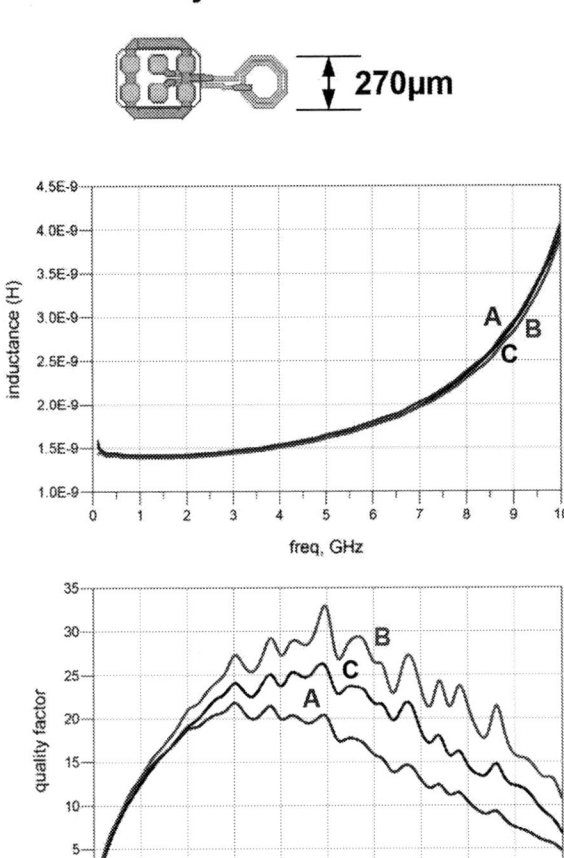

Fig. 8: Planar spiral inductor test device layout and measured characteristics.

Layout

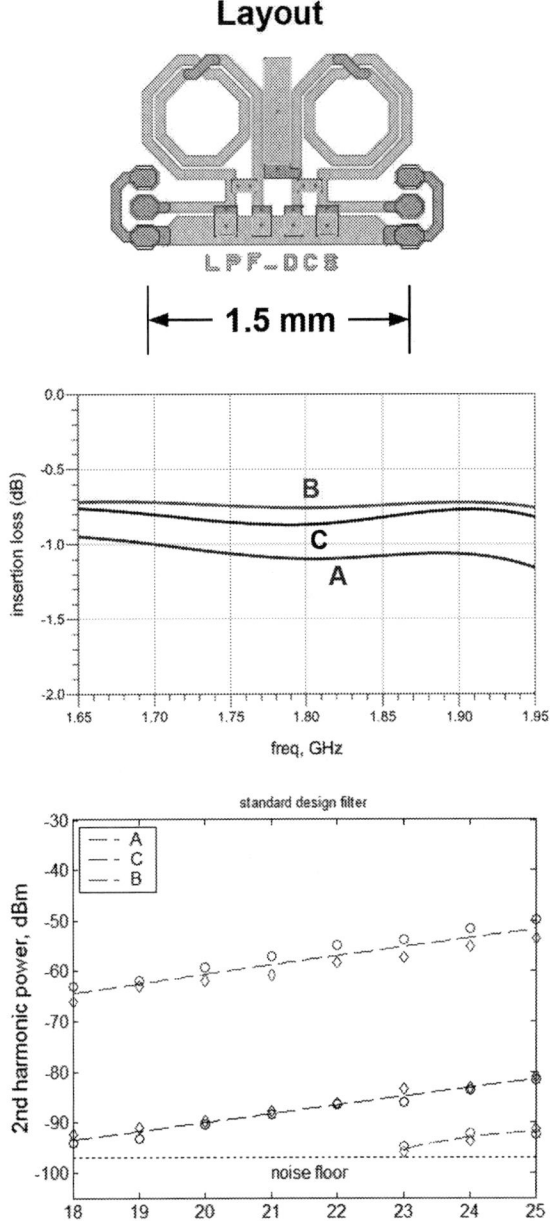

Fig. 9: 5th-order DCS low-pass filter test device layout and measured characteristics.

nated by the inductors. For this reason, inductor quality factor (Q) is often a figure of merit for technology comparisons. Figure 8 shows the layout of the wafer-probe test inductor used in the evaluation, along with the measured inductance and quality factor for the three substrate types. The trend in these devices, as expected, is the same as in the coplanar transmission lines with higher quality factor correlating with lower transmission line losses.

The loss characteristics of IPD circuits show varying degrees of sensitivity to component loss, depending on the type of circuit and design. It can be seen above that substrate loss has negligible effect on the inductance. Consequently, in IPD filters the frequency response is not generally very sensitive to low levels of substrate loss. The main effects are seen in the pass-band insertion loss.

For comparison of IPD loss characteristics, we used a common 5th-order low-pass filter design, shown in Fig. 9. This particular filter is designed for operation in the cellular DCS band from 1710-1910 MHz. The figure shows insertion loss around the center of the pass-band. Here, the same trend as in the above results can be seen.

Figure 9 also shows the results of 2nd-harmonic measurements for these devices. In these measurements, the input

power was at 750MHz and the harmonic was measured at 1500MHz. (This test used the same setup and procedure as the tests for the commercial 700MHz-band filters discussed above.) Note that the harmonic measured results follow the same trend as the loss measurements: Devices fabricated on substrate A show the highest insertion losses, and also show the highest 2nd-harmonic power, and devices fabricated on substrate B show the lowest losses and lowest harmonic power levels.

In an experiment similar to the one described above for the commercial filters, we made design modifications to these DCS filters aimed at reducing the surface lateral electric fields. Figure 10 shows a comparison of the original design and the modification. In the original design, the wide horizontal trace

Layout

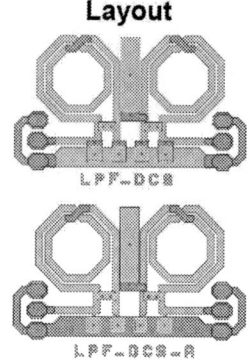

Surface lateral electric field intensity

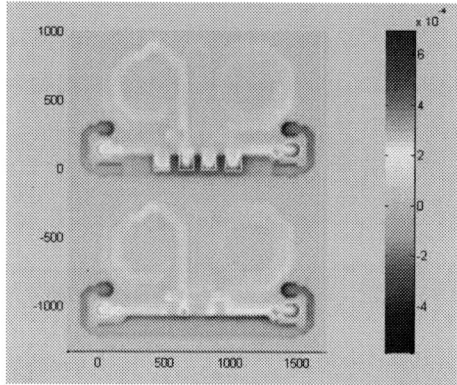

Fig. 8: Layout design modification and simulated surface lateral electric field intensity distributions.

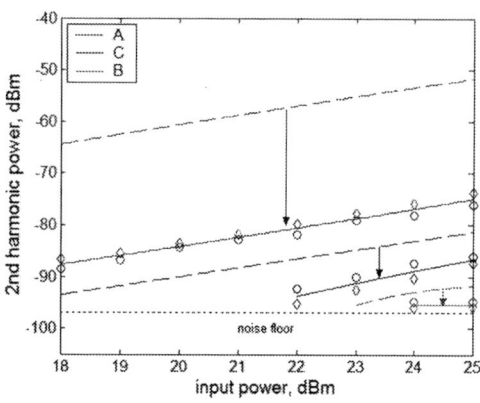

Fig. 9: Measured 2nd-harmonic power levels for the modified DCS low-pass filter design.

across the bottom of the filter forms the ground connection between the input and output pads. This trace is on the top layer of metal, and it also forms the top electrode of several capacitors that are formed beneath this trace, and are consequently exposed to the substrate. In the modified design, the ground trace is made instead using the bottom level metal, and the capacitors lie on top of it. This effectively shields them from interaction with the substrate. As the simulated field intensity distribution maps show, the field intensity is significantly reduced in the modified design. Importantly, these two types of design were located adjacent to each other on the test wafer, so their conditions of bulk resistivity and surface passivation should be nearly identical.

The design modifications shown above did not significantly change the filter frequency response. Figure 11 shows the measured 2nd-harmonic power characteristics for the modified design. For comparison, the dashed lines show the results for the standard design. In this comparison, it can be seen that the modified design resulted in reduced levels of 2nd-harmonic power for all three substrate types.

In the case of substrate B, the measured harmonic levels were very near the noise floor of the experimental setup, so these particular results are less accurate and may be lower than indicated. More remarkably, in the case of substrate A the design modification reduced the 2nd-harmonic power levels by more than 20dB.

As in the earlier experiments using the commercial filters, in these filters we observe photoconductive dependence in the case of substrate A. This suggests that the surface passivation method used in this wafer was not completely effective. Typically, methods that create a trap-rich layer on the surface effectively pin the Fermi level and reduce surface mobility. Both of these factors suppress photoconductivity.

Table 2 summarizes the results of the various loss and harmonic power measurements.

Table 2: Measurements summary

measurement	A	B	C
CPW insertion loss @ 5GHz (dB)	0.64	0.21	0.27
test inductor Q @ 5GHz	20.1	32.6	25.7
DCS filter loss @1.8GHz (dB)	1.10	0.76	0.87
DCS filter 750MHz 2nd-harmonic power @ 24dBM input	-52	-83	-92
modified DCS filter 750MHz 2nd-harmonic power @ 24dBM input	-76	-88	-95

Discussion

The correlation between loss and 2nd-harmonic power generation in silicon IPDs has been noted in past studies, and is generally ascribed to nonlinear effects arising from the semiconductor-insulator interface at the top of the substrate. Very high resistivity substrates tend to behave more like ideal insulators, and consequently these nonlinear mechanisms are less prevalent in them.

In our previous study [1] we proposed a mechanism of depletion layer depth modulation by strong lateral electric fields to explain the behavior that we observed, particularly the lack of dependence on substrate thickness. The additional results described above suggest that the mechanism of nonlinear power generation is more complex.

As in other studies, we see a direct correlation between loss and 2nd-harmonic power. However, we also observe that high substrate resistivity alone is insufficient to suppress harmonic generation. Surface passivation also plays a very important role. We have seen examples, like the result shown in Figure 6, in which there is no discernible improvement from design modifications. By contrast, we have also seen results like those shown in Figure 11, in which design modifications make a significant difference.

These results suggest that both surface losses and bulk losses contribute to 2nd-harmonic power generation in varying degrees depending on the details of design, materials and processing. Side-by-side comparison of design variants provides some useful insight into the relative contributions of surface and bulk nonlinearities.

Acknowledgements

The authors wish to acknowledge the help of Hlaing Ma Phoo Pwint for help with device characterization and Jae Hun Ku and M. Pandi Chelvam for device fabrication.

References

1. R. Frye, K. Liu and R. Melville, "Second-Harmonic Nonlinearities in RF Silicon Integrated Passive Devices," *Proc. 63rd. IEEE Electronic Components and Technology Conference,* May 28-31, 2013, pp. 1667-1674.

2. H. S. Gamble, B. M. Armstrong, S. J. N. Mitchell, Y. Wu, V. F. Fusco, and J. A. C. Stewart, "Low-loss CPW lines on surface stabilized high-resistivity silicon," *IEEE Microwave Guided Wave Letter,* 9, Oct. 1999, pp. 395–397.

3. E. Valetta, J. van Beek, A. den Dekke, N. Pulsford, H. F. F. Jos, L. C. N. de Vreede, L. K. Nanver, and J. N. Burghartz, "Design and characterization of integrated passive elements on high-ohmic silicon," *in MTT-S Dig.,* 2, 2003, pp. 1235–1238.

4. A. B. M. Jansman, J. T. M. van Beek, M. H. W. M. Van Delden, A. L. A. M. Kemmeren, A. D.A. Den. Dekker, and F. P. Widdershoven, "Elimination of accumulation charge effects for high-resistivity silicon substrate," *in Proc. ESSDERC,* 3–6, 2003.

5. B. Rong, J. N. Burghartz, L. K. Nanver, B. Rejaei, and M. van der Zwan, "Surface-Passivated High-Resistivity Silicon Substrates for RFICs," *IEEE Electron Device Letters,* 25, April 2004, pp. 176-178.

6. D. C. Kerr, J. M. Gering, T. G. McKay, M. S. Carroll, C. Roda Neve, and J.–P. Raskin, "Identification of RF Harmonic Distortion on Si Substrates and its Reduction Using a Trap-Rich Layer," *IEEE Topical Meeting on Silicon Monolithic Integrated Circuits in RF Systems,* 2008, pp.151 – 154.

7. Cesar Roda Neve and Jean-Pierre Raskin, "RF Harmonic Distortion of CPW Lines on HR-Si and Trap-Rich HR-Si Substrates," *IEEE Trans. Electron Devices,* 59, April 2012, pp. 924-932.

Integration of Magnetic Materials into Package RF and Power Inductors on Organic Substrates for System in Package (SiP) Applications

Hao Wu[1], Donald S. Gardner[2], Cheng Lv[1], Zhihua Zou[3] and Hongbin Yu[1]

[1]Ira A. Fulton Schools of Engineering, Arizona State University, Tempe, AZ 85287, United States.
[2]Intel Labs, Intel Corp., Santa Clara, CA 95052, United States.
[3] Assembly Materials Characterization Lab, Intel Corp., Chandler, AZ 85226, United States.
E-mail: yuhb@asu.edu

Abstract

In this paper, soft ferromagnetic materials were deposited on organic packaging substrates to explore their potential applications in in-package inductors. Amorphous cobalt alloy, Co-Zr-Ta-B, was chosen due to its high saturation magnetization, low coercivity and small magnetostriction. As-deposited films were characterized by vibrating sample magnetometer (VSM) showing comparable magnetic properties in comparison to the films on quartz and silicon wafers. Stripline and spiral inductors with Co-Zr-Ta-B films were fabricated on package substrates to explore their potential in-package RF and power inductor applications.

Introduction

With the fast development of portable electronics, such as smart phone, ultrabook, as well as high performance computation, integrating discrete passive components in package to miniaturize the overall form factor has been demanded by system in package (SiP) technology.[1, 2] Among passive components, inductors are the most challenging to be integrated due to their large occupied area of current air-core structure and potential electromagnetic interference (EMI) to adjacent electronic circuits and components. Introducing soft ferromagnetic materials can effectively miniaturize in-package inductors and reduce EMI. [3-5] Although extensive efforts have been made to incorporate magnetic materials to on-chip inductors [6-9], less attentions were given to the magnetic materials on package substrates mainly because magnetic properties could be affected by packaging substrate surface, stress and temperature resulting in unfavorable degradation of device performance.[10] In previous work, the authors have demonstrated on-chip inductors, fabricated on quartz substrate with soft ferromagnetic material (Co-Zr-Ta-B), capable of operating at radio frequency (RF) up to 2 GHz.[8, 11] In this paper, we investigated the feasibility of applying such ferromagnetic materials in package inductors for both power and RF applications.

This paper characterized soft amorphous Co-Zr-Ta-B films on standard organic package substrates including ABF and polyimide. Effects of substrate roughness and stress were analyzed and simulated which provide strategies for integrating Co-Zr-Ta-B into package inductors and improving inductors performance.

Film Deposition and Characterization

Amorphous Co-4%Zr-4%Ta-8%B (at. %) films were deposited with a constant DC magnetic field by DC magnetron sputtering on two kinds of organic packaging substrates, i.e. ABF[12] and polyimide substrates. The substrates investigated here are 500 μm thick commercial available polished polyimide substrate, 100 μm thick ABF laminated on glass substrate, along with our standard polished quartz substrate. Boron in Co-Zr-Ta-B film was used to increase the film resistivity for high power density applications as in previous experiments.[5, 8, 11] The unlaminated films were 500 nm thick while the laminated films were ten 50 nm thick layers with a few nanometers of cobalt oxide insulation layers in between to reduce eddy current loss. Vibrating sample magnetometer (VSM) was used to measure the hysteresis loops of CZTB film. Figure 1 presents the results from laminated 50 nm × 10 film with well defined easy and hard axis indicating a uniaxial anisotropy in the films.

Surface roughness effect was first examined by measuring the roughness of bare substrates then after depositing Co-Zr-Ta-B films using an optical profiler with white light interferometer in the vertical scanning mode (Bruker Corp.). It has 1 nanometer resolution in z direction and optical lateral resolution about 0.6 μm. Measurement results are summarized in Table I. It shows that the surface roughness of Co-Zr-Ta-B films deposited on ABF or polyimide substrates are on the order of tens of nanometer, larger than those deposited on Si substrate, which is less than 10 nm. This is due to originally rougher ABF/polyimide substrate compared to polished Si substrate.

Table I Roughness of as-deposited Co-Zr-Ta-B films on different substrates (nm)

	Bare	500 nm Co-Zr-Ta-B	50 nm × 10 Co-Zr-Ta-B
SiO$_2$/Si	--	9.4	8.0
ABF/Glass	42.0	29.6	20.4
Polyimide	34.4	37.3	62.1

Both ABF and polyimide substrates present tens of nanometer roughness, resulting in slightly sheared and broaden hysteresis loops compared to the films deposited on Si substrate. The coercivity of both laminated and nonlaminated films on ABF/glass are all smaller than 1 Oe, see Table II. The values are slightly larger but very similar to those on SiO$_2$/Si substrate, indicating their high quality. Similarly, the anisotropy H$_k$ are also similar to those on SiO$_2$/Si. Overall, the quality and the softness of the Co-Zr-Ta-B films on ABF/glass remain the same as those deposited on SiO$_2$/Si.

However, the coercivity measured from laminated film on polyimide substrate is 1.25 Oe which is larger than that from film on ABF substrate (0.2 Oe). The rougher polyimide surface shown in Table I is the main reason for this increase in coercivity. Moreover, for 500 nm film on polyimide, measured coercivity reaches 75 Oe accompany with the disappear of uniaxial anisotropy, which is mainly due to the poor chamber vacuum during sputtering as a result of outgassing from polyimide substrate. Prebaking of polyimide substrates might be a possible solution for this out gassing issue.

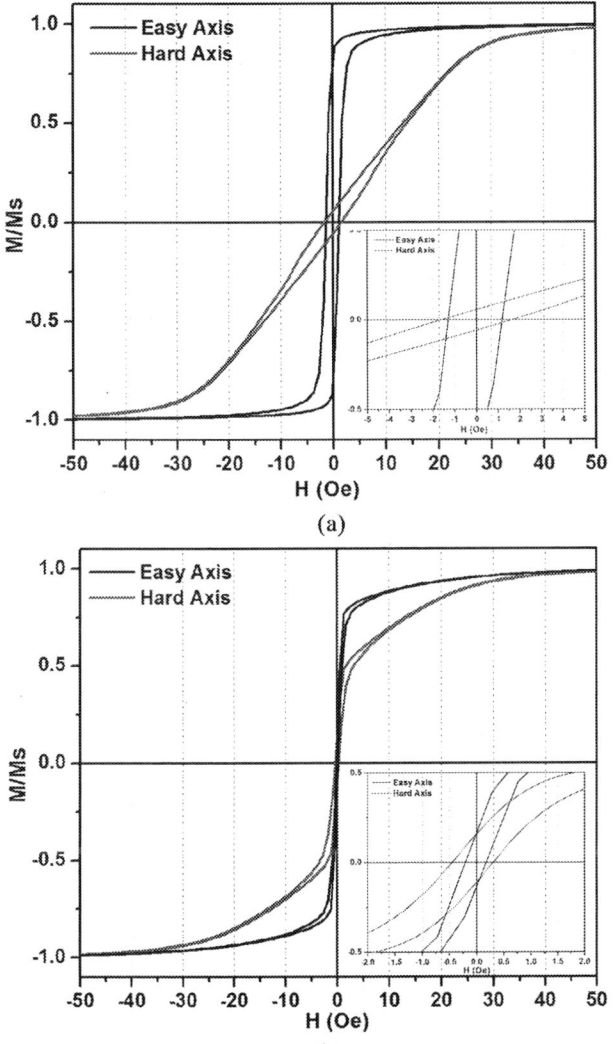

Figure 1. Measured hysteresis loops of laminated Co-Zr-Ta-B films on (a) polyimide substrate; (b) ABF/glass substrate.

Post deposition thermal magnetic annealing is helpful to remove magnetic structure disorder improving device performance. The films were then annealed at 200 °C in N_2 ambient for 2 hours with a DC magnetic field around 1000 Oe along predefined easy axis. Only films on Si substrates showed improvement in coercivity and no obvious improvement can be seen from the films on package substrates. Coercivity and anisotropy field were extracted from the hysteresis loops and listed in Table II. Increased coercivity was observed especially for the films on polyimide

substrates. For laminated films on both substrates, in-plane anisotropy tends to disappear after annealing which is partially due to the change of Co oxide insulation layers. Based on the experiment results, post deposition annealing is not recommended for films on package substrates.

Table II H_c and H_k of Co-Zr-Ta-B films on different substrates (Oe)

		500 nm		50 nm × 10	
		as-deposit	annealed	as-deposit	annealed
H_c	SiO₂/Si	0.1	0.1	0.18	0.05
	ABF/glass	0.4	0.4	0.2	0.5
	Polyimide	75	100	1.25	7
H_k	SiO₂/Si	17.8	19	22.5	21
	ABF/glass	27	24	27	15
	Polyimide	--	--	27	--

Thermal Stress Analysis

Stress in magnetic films has significant consequences for both device fabrication and performance. For device fabrication, large stress can cause deformation of inductor coils and package substrates such as warping, therefore is undesired. From a device performance perspective, stress in the magnetic films can introduce extra anisotropy resulting in magnetic properties deterioration. Before fabricating inductors with magnetic films on package substrates, it is necessary to analyze the stress in the magnetic films.

Commercial software Abaqus was applied to perform the finite element analysis (FEA) for thermal stress analysis of magnetic thin film on organic substrates. Simulated structure is shown in Figure 2. As the geometry is symmetric in two of its dimension, only ¼ of the structure was built in Abaqus to reduce the cost of calculation. The model consists of two parts, on top is a cobalt thin film and at bottom is the polyimide substrate. These two parts are bounded together by the TIE constraint. As shown in Figure 2, the polyimide substrate is 500 μm thick while the cobalt film is 0.5 μm thick. The material constant applied here are $E_{Co} = 209 GPa$, $\gamma_{Co} = 0.31$, $\alpha_{Co} = 1.3e^{-5}$, $E_{polyimide} = 2.2 GPa$, $\gamma_{polyimide} = 0.34$, $\alpha_{polyimide} = 2e^{-5}$ where E, γ, α are Young's modulus, Poisson's ratio and coefficient of thermal expansion, respectively. The applied temperature change is 250 °C. The boundary condition is set to fix the bottom layer of Cobalt film without any displacement during the entire thermal process. Here 24647 3D 20-node quadratic, reduced integration (C3D20R) elements are used to mesh the whole model. The reason why the C3D20R element is chosen is it is high order element, which ensures high accuracy results with relative high efficiency calculation.

Figure 3 shows the contour of the von Mises stress of the cobalt layer. The reason why this equivalent stress, instead of any individual one on a specific direction, is chosen is that it provides the most proper stress information for the ductile material such as metals. It can be seen that the von Mises stress on the edges are much higher than the ones in the middle. To further illustrate the change of the stress, Figure 4 shows the increase of von Mises stress with the increase of the distance from center along the diagonal of the cobalt layer. The value is almost constant for about 90% of the length. In the rest part closed to the edge, the stress increase sharply. This indicates that there is no significant stress induced anisotropy in films which is helpful in keeping the uniaxial anisotropy of the magnetic film with high saturation field.

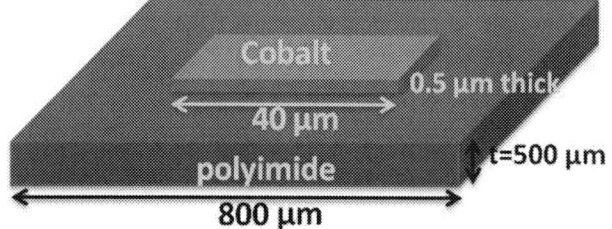

Figure 2. Simulated structure for thermal stress analysis

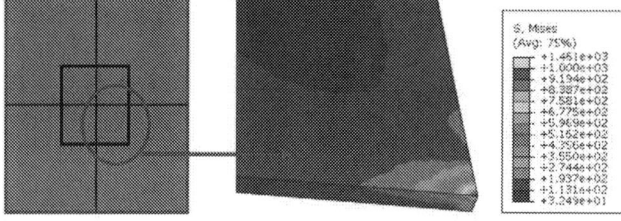

Figure 3. Contour of the thermal strain on cobalt layer.

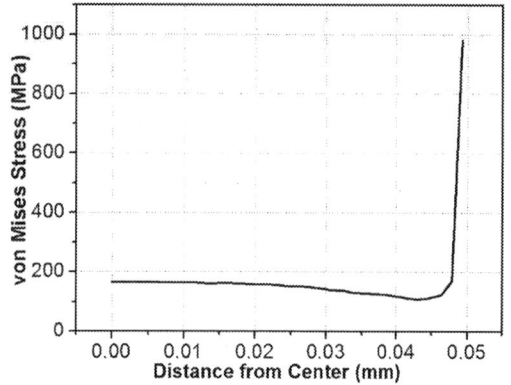

Figure 4. Relationship between the thermal strain and the distance from center along the diagonal of the cobalt layer.

Inductor Fabrication

For demonstration of in-package magnetic thin film inductors the stripline inductors incorporating 500 nm thick Co-Zr-Ta-B film were fabricated using the identical procedure and conditions such as process temperatures onto three different substrates, quartz, polyimide and laminated ABF film on glass, using electron beam lithography (EBL) and magnetron sputtering for pattern definition and metallization, respectively, a standard procedure used in prior experiments.[5] The typical fabricated stripline inductors without magnetic film, with film and with patterned magnetic film on ABF/glass substrate are shown in Fig. 5 (a), along with the schematic of the cross-section of the fabricated inductor and a representative scanning electron microscopy (SEM) image. The length of the stripline inductor is 450 μm. Various magnetic film structures were fabricated to obtain a comprehensive understanding. Spiral inductors with 500 nm Co-Zr-Ta-B films were fabricated on polyimide substrates as well. The spiral inductors are 4-turn rectangular-shaped with outer diameters of 88 μm by 160 μm and an inductance of 1.9 nH without magnetic material. Copper wires are 2 μm thick and 5 μm wide wrapped around by Co-Zr-Ta-B thin films (see Figure 5c). Polyimide was used as insulating layers separating the copper conductor and the magnetic material. Two layers of Co-Zr-Ta-B films were integrated into both spiral and stripe inductors by joining the two layers through magnetic vias to form a continuous magnetic circuit for maximum flux enhancement. The thickness for each layer is 500 nm determined by the skin depth of magnetic material at GHz frequency range.[13] Thicker films will induce larger eddy currents that will deteriorate the quality factor. HP8720D network analyzer and Cascade GS probes were utilized for one-port measurements.

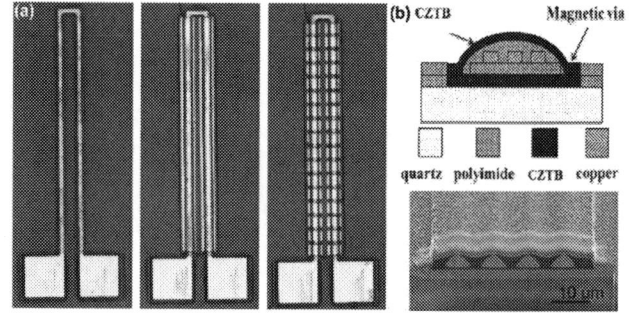

Figure 5. Stripline inductor structure. (a) Pictures of fabricated bare, film and patterned Co-Zr-Ta-B film inductors on ABF films laminated on glass substrate. Length of the stripline inductor is 450 μm. (b) Schematic view of the cross-section of the inductor structure (top) and representative scanning electron microscopy image of the inductor cross-section (bottom).

Inductor Characterization and Discussion

Measurement of inductance and quality factor versus frequency of stripline inductors on three different substrates using non-laminated single layer 500 nm films are compared and shown in Figure 6-8. With 500 nm thick non-laminated film the inductance increases to 2.7x, 1.6x, and 1.9x respectively, on quartz, polyimide and ABF/glass substrates compared to bare (air-core) inductors. All fabricated inductors show very good frequency response, with inductors on quartz substrate only showing near-constant value until sharp drop-off at frequency greater than 1 GHz, while inductance starting to decrease at around 1 GHz and 500 MHz respectively, for inductors on polyimide and ABF/glass. In the meantime, the quality factor of the inductor on polyimide and ABF/glass typically greater than 3, matches with those on quartz substrate. The quality factor can be further improved by suppressing eddy current in the magnetic core. Instead of large long magnetic core shown in the middle image of Figure 5 (a), patterned magnetic core with multiple bars can effectively avoid large eddy current loops leading to an improved quality factor. Such improvement is more obviously

978-1-4799-2408-0/14 $31.00 © 2014 IEEE

observed in inductors on ABF/glass substrates (more than 30% increase in quality factor), as shown in Figure 7.

The slight smaller values in the inductance increase on polyimide and ABF/glass could due to rougher surface and the stress due to different materials involved in the inductor structure. It is possible that repeated temperature cycles during the device fabrication process may have caused the stress in the magnetic film thus degrading the permeability and inductance increase. This effect, however, can be reduced if we introduce an intermediate layer between magnetic materials and packaging substrates. In addition, different substrate materials may lead to different thickness of spin-coated polyimide layer, which is used in the inductor as insulating layer. Such variations could affect magnetic layer structure, therefore the inductance values.

Field annealing was also performed on fabricated inductors on various substrate under these conditions: 200 °C for 2 hours under magnetic field estimated to be between 0.1 to 0.2 T. For inductor fabricated on ABF, this annealing did not degrade the performance of the inductors, with the same inductance value and quality factor. This indicates the robustness of the device, and it is promising to integrate inductor on packaging substrate, even with magnetic core materials integrated.

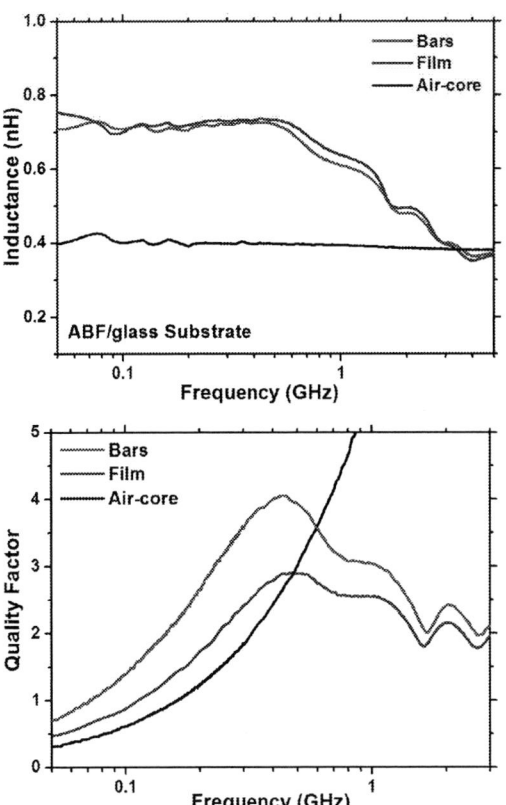

Figure 7. Measured inductance and quality factor from stripeline inductors on ABF/glass substrates.

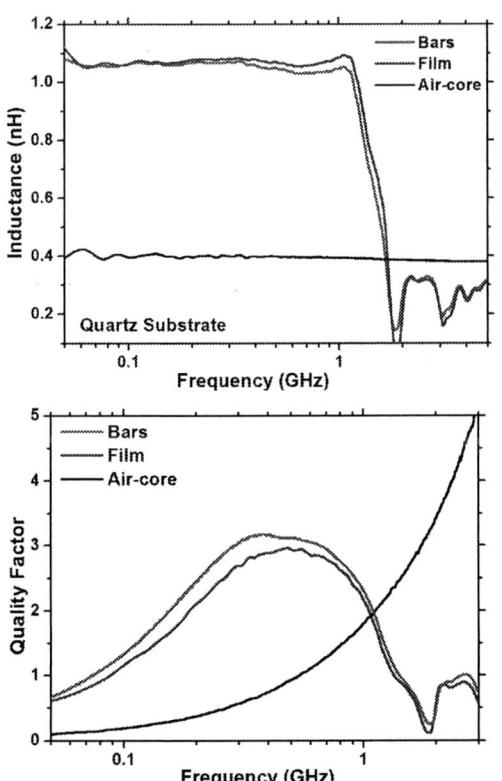

Figure 6. Measured inductance and quality factor from stripeline inductors on quartz substrates.

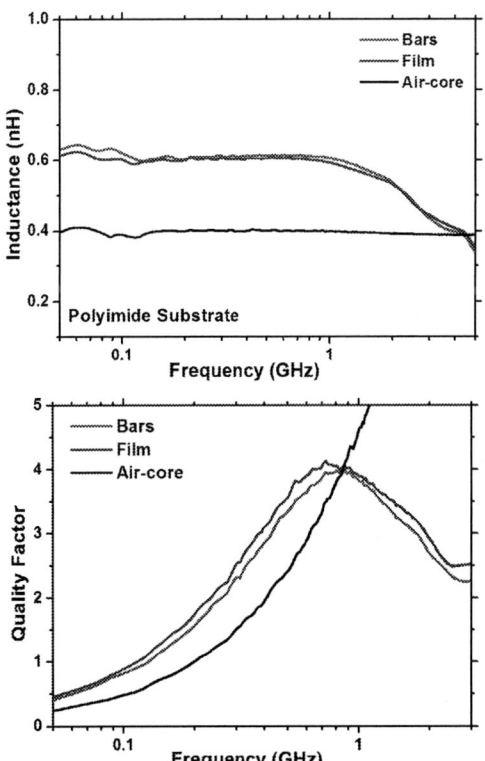

Figure 8. Measured inductance and quality factor from stripeline inductors on polyimide substrates.

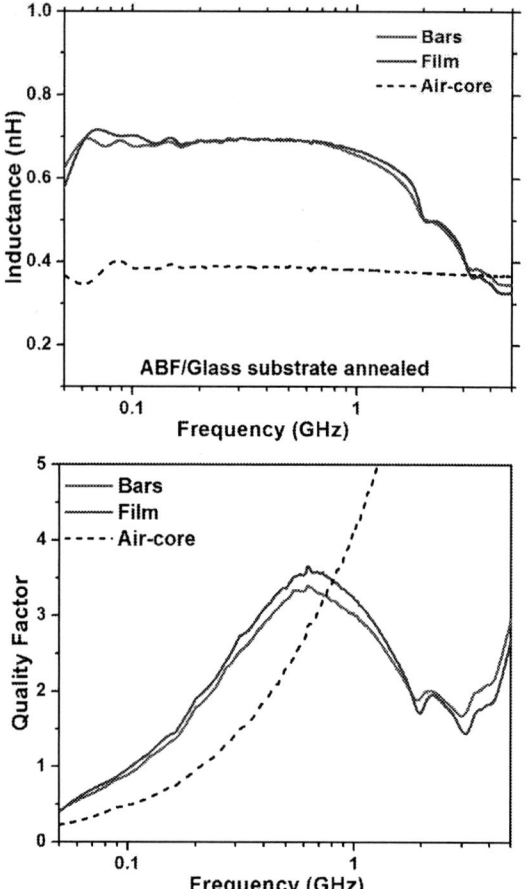

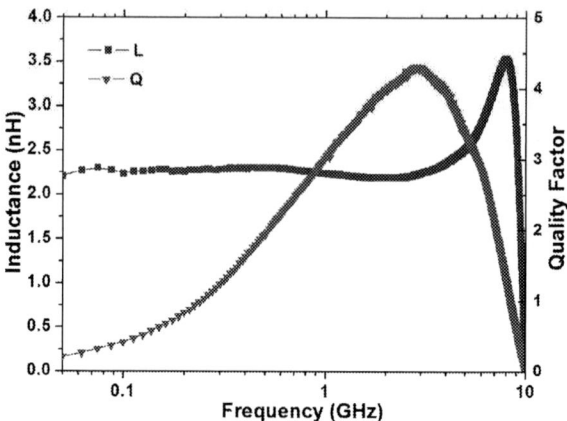

Figure 10. Measured inductance and quality factor from spiral inductors on polyimide substrates.

Figure 9. Measured inductance and quality factor from stripeline inductors on ABF/glass substrates after annealing under magnetic field.

The air-core spiral inductor on the polyimide substrate maintains the theoretical inductance value of 1.9 nH indicating that the process on quartz substrates is fully compatible with organic packaging substrates. For the inductors with magnetic materials, nearly 20% inductance enhancement was observed as shown in Fig. 10. Compared to the stripline inductors, such low inductance increase is mainly because more complicated fabrication process for spiral inductor results in the deterioration of the magnetic films which can be improved by the following ways.

1. Inserting an intermediate layer, such as spin-coated polyimide, between substrate and magnetic layer to reduce the possible thermal stress effect.

2. Reducing the process temperature, which is primarily determined by the curing temperature of polyimide layers, in order to reduce the stress in the magnetic layer.

3. Extending the experience gained from fabricating inductors using CoZrTaB as magnetic layers to the incorporation of NiFe layers, as it is a commonly used material in many forms of inductors, including spiral, solenoid, and toroid structures, and one could therefore further evaluate the performance of such magnetic inductor on packaging substrate.

Conclusions

Ferromagnetic materials have been deposited onto organic package substrates. Magnetic characterization results showed comparable magnetic properties in terms of coercivity and anisotropy field to the films on Si substrates. Inductors with magnetic films fabricated on package substrates presented maximum near two times inductance increase. Such in-package magnetic thin film inductors show promising applications in SiP design.

Acknowledgments

The authors would like to acknowledge funding from Intel Corporation through ASU Connection One and Semiconductor Research Corporation (SRC) Global Research Collaboration program. The authors would also like to thank Dr. Shamala Chickamenahalli for helpful discussions, Shawn Hansen for assistance and Dr. Nicholas D. Rizzo for providing measurement facilities.

References

[1] S. A. Chickamenahalli, H. Braunisch, S. Srinivasan, H. Jiangqi, U. Shrivastava, and B. Sankman, "RF packaging and passives: design, fabrication, measurement, and validation of package embedded inductors," Advanced Packaging, IEEE Transactions on, vol. 28, pp. 665-673, 2005.

[2] M. F. Davis, A. Sutono, Y. Sang-Woong, S. Mandal, N. Bushyager, L. Chang-Ho, et al., "Integrated RF architectures in fully-organic SOP technology," Advanced Packaging, IEEE Transactions on, vol. 25, pp. 136-142, 2002.

[3] C. O. Mathuna, N. N. Wang, S. Kulkarni, and S. Roy, "Review of Integrated Magnetics for Power Supply on Chip (PwrSoC)," Ieee Transactions on Power Electronics, vol. 27, pp. 4799-4816, Nov 2012.

[4] C. R. Sullivan, "Integrating magnetics for on-chip power: Challenges and opportunities," in Custom Integrated Circuits Conference, 2009. CICC '09. IEEE, 2009, pp. 291-298.

[5] H. Wu, D. S. Gardner, W. Xu and H. Yu, "Integrated RF on-chip inductors with patterned Co-Zr-Ta-B Films," IEEE Trans. Magn. 48 (11), 4123-4126 (2012).

[6] D. S. Gardner, G. Schrom, F. Paillet, B. Jamieson, T. Karnik, and S. Borkar, "Review of On-Chip Inductor Structures With Magnetic Films," Ieee Transactions on Magnetics, vol. 45, pp. 4760-4766, Oct 2009.

[7] B. Jamieson, J. Godsell, N. Wang, and S. Roy, "Device Geometry Effects in an Integrated Power Microinductor with a $Ni_{45}Fe_{55}$ Enhancement Layer," IEEE Trans. Magnetics, vol. 49, no. 2, pp. 869–873, Feb. 2013.

[8] H. Wu, S. R. Zhao, D. S. Gardner and H. B. Yu, "Improved high frequency response and quality factor of on-chip ferromagnetic thin film inductors by laminating and patterning Co-Zr-Ta-B Films," IEEE Trans. Magn. 49 (7), 4176-4179 (2013).

[9] P. R. Morrow, C. M. Park, H. W. Koertzen and J. T. DiBene, "Design and Fabrication of On-Chip Coupled Inductors Integrated With Magnetic Material for Voltage Regulators," IEEE Trans. Magn. 47 (6), 1678-1686 (2011).

[10] P. Zou, W. Yu, and J. A. Bain, "Influence of stress and texture on soft magnetic properties of thin films," Ieee Transactions on Magnetics, vol. 38, pp. 3501-3520, Sep 2002.

[11] H. Wu, D. S. Gardner, S. R. Zhao, H. Huang and H. B. Yu, "Control of magnetic flux and eddy currents in magnetic films for on-chip radio frequency (RF) inductors: Role of the magnetic vias," Journal of Applied Physics, in press.

[12] Ajinomoto Fine-Techno Co., Inc.
http://www.aft-website.com/en/electron/abf

[13] W. Xu, H. Wu, D. S. Gardner, S. Sinha, T. Dastagir, B. Bakkaloglu, Y.Cao, and H. B. Yu, "Sub-100 μm scale on-chip inductors with CoZrTa for GHz applications, " J. Appl. Phys. 109 (7) (2011).

Through Silicon Capacitor Co-integrated with TSV as an Efficient 3D Decoupling Capacitor Solution for Power Management on Silicon Interposer

O. Guiller[a], S. Joblot[a], Y. Lamy[b], A. Farcy[a], E. Defay[b], K. Dieng[c]

[a] STMicroelectronics, 850 rue Jean Monnet, 38926 Crolles, France.
[b] CEA, LETI, MINATEC Campus, 17 rue des Martyrs, 38054 Grenoble Cedex 9, France.
[c] IMEP-LAHC, Université de Savoie, 73376 Le Bourget du Lac Cedex, France.

Abstract

First part of this paper discusses decoupling method limitation within the Power Delivery Network of a classical circuit and challenges introduced by 3D integrated circuit in term of power management. Solutions are exposed, such as integration of decoupling capacitor on silicon interposer.

Second part of the paper focuses on the Through Silicon Capacitor (or TSC) as an alternative decoupling solution co-integrated with Through Silicon Vias on silicon interposer. TSC realization is described and architectural benefits of adding a partial copper-filling prior to the Metal-Insulator-Metal stack deposition are discussed.

A distributed analytical model is used to quantify partial filling resistance contribution, pointing out a 6 decade decrease in ESR value of the structure. TSC process and matrix design parameters impact on capacitance density are studied. Finally, electrical performances of TSC modules are evaluated showing a low intrinsic impedance behavior granted by TSC parallel structure.

Introduction

In a search of increased performances and mobility, traditional architectures of microelectronic devices are evolving toward 3D integrated circuit (3DIC), where heterogeneous dies are stacked on top of each other. 3D stacking solutions will take time to reach maturity as they require the evolution of the whole industry ecosystem. An intermediate step in terms of design and process maturity can be found in side by side integration on silicon interposer, usually referred as 2.5D integration.

Silicon interposer allows several heterogeneous dies to be stacked on its surface without thermal mismatch, thus increasing their communication bandwidth thanks to higher interconnection densities offered by the silicon substrate compared to conventional organic substrate. Vertical electrical connections from the front side of the interposer carrying the large scale integration (LSI) dies to its backside are provided by Through Silicon Vias (TSV). More relaxed routing densities are exhibited by the interposer's backside to ensure compatibility with the underlying organic substrate (usually Ball Grid Array) reached through copper pillars or bumps. Interposer then acts as a buffer layer between LSI and the outer world in terms of routing densities. A passivation at BGA level finalizes the packaging of the module which can be reported on the printed circuit board (PCB), as represented in Figure 1.

Interposer realization is not straightforward and several types of interconnects processes are developed in parallel. Those processes differ in architecture, via geometry and interposer thickness. Alongside advantages already discussed,

the addition of the interposer brings complexity to the structure by introducing new elements such as TSVs, µ-bumps, front-side and back-side redistribution layers (RLD) that act as parasitic elements in the Power Delivery Network (PDN) of the system.

In the first part of this paper, challenges of power management in 3DIC and the need of new decoupling solutions will be discussed. In a second part, Through Silicon Capacitor (TSC) is presented as an alternative solution for the integration of decoupling capacitors in a Si-interposer.

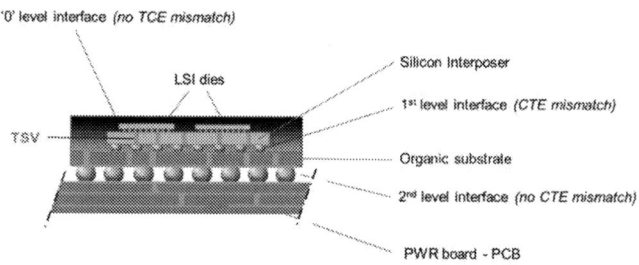

Figure 1. Silicon interposer 2.5D integration

PDN decoupling solutions for 3D integration

PDN's role is to deliver a stable power supply from the voltage regulator module (VRM) to all components in the system. In CMOS circuits, logic die draws current when its transistors are switching, leading to a ripple voltage in the PDN. This effect known as simultaneous switching noise (SSN) is the main source of noise in digital IC. Since the high and low logic states are defined by sensing the voltage (with an acceptance margin α), voltage ripple in the PDN exceeding this margin can lead to logical errors in the core process.

A method widely used by designer to ensure PDN's reliability is the definition of target impedance Z_{TARGET}. The network impedance response must remain under this value over the whole operating frequency range where current transient exists. Z_{TARGET} value is defined by Equ.1 [1], where V_{dd} represents the logic core voltage, α the allowed ripple voltage ratio, I_{max} the maximum current flowing in the circuit and I_{min} the minimum current during idle state. The transient current in the circuit is the difference between I_{max} and I_{min}.

$$Z_{TARGET} = \frac{V_{dd} \cdot \alpha}{I_{max} - I_{min}} \qquad Equ.\ 1$$

Table 1 sums up the industry trend up to 2019. Data for technological node (half-pitch of a flash memory cell), allowable maximum power P_{max} and core voltage V_{dd} have been extracted from the *International Roadmap for Semiconductors* (ITRS) update 2012 [2], for a high performance system with heat sink. Current has been worked out with Ohm's law from P_{max} and V_{dd}. Z_{TARGET} has been evaluated with Equ. 1 assuming 5% ripple voltage α and 50% transient current. The constant lowering of logic core voltage V_{dd} leads to lower PDN impedance requirements.

Year	Node[1] (nm)	P_{max}[1] (W)	V_{dd}[1] (V)	I_{max}[1] (A)	Z_{Target} (mΩ)
2011	22	161	0.90	179	0.50
2013	18	149	0.85	175	0.48
2015	15	143	0.80	179	0.45
2017	13	130	0.75	173	0.43
2019	10.9	133	0.71	187	0.38

Table 1. Industry trend up to 2019 in terms of technological nodes, allowable maximum power P_{max}, core voltage V_{dd}, maximum current I_{max} and target impedance Z_{TARGET}. [1] : ITRS 2012 data for high performance system with heat sink.

In the PDN, decoupling capacitors act as local energy storages providing electrons to the switching transistors, thus lowering ripple voltage in the network by lowering its impedance. Decoupling performances are driven by the capacitor value and its access impedance as seen by the logic, which depends on its position in the PDN.

Most commonly used decoupling capacitors type in classical circuits are Surface Mounted Devices (SMD) capacitors on PCB (C_{PCB}), and on-chip capacitors (C_{FE}) located in the transistor planes of the logic die (front-end). Those different types of decoupling capacitors dominate the PDN at different frequency ranges, C_{PCB} allows the introduction of large capacitance values, but their high access impedance limits their response to lower frequencies (~100MHz [3]). On the other hand C_{FE} exhibits limited capacitance values with very low access impedance allowing the decoupling of higher frequencies (>2GHz [3]).

A major issue in power management comes from chip/package anti-resonance taking place when a parallel LC resonator circuit is formed between on-chip capacitance C_{FE} and package inductance [4]. Figure 2 illustrates a simplified resonator model of a classic circuit including the two previously discussed decoupling capacitor types (C_{PCB} and C_{FE}) and their associated access inductance (L_{PCB} and L_{FE}). f_1 and f_3 are intrinsic LC series circuit resonance frequencies of each capacitor and f_2 is the anti-resonance frequency of LC parallel circuit formed between the on-chip capacitor C_{FE} and the package inductance L_{PCB}. This anti-resonance effect results in a peak in PDN impedance profile than can exceed Z_{TARGET} at intermediate frequencies and leads to logical error in the core process [1].

If chip/package anti-resonance already is a major concern in classical circuit [3] [4] [5], it becomes critical in 3DIC. The complexity induced by stacking dies has several noticeable

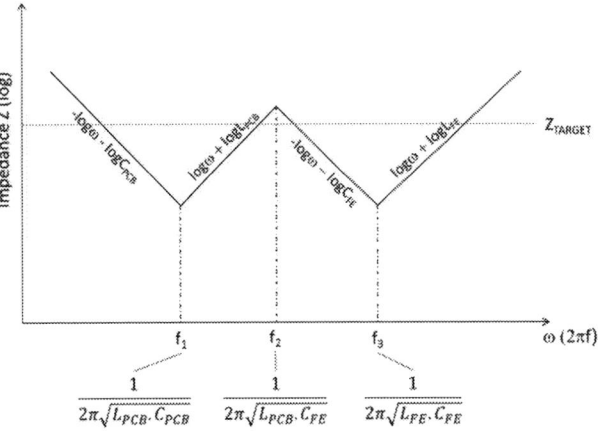

Figure 2. Simplified resonator model including two decoupling capacitor. C_{PCB} mounted on the power board surface and C_{FE} integrated to the logic die.

effects on PDN's quality. If multiple logic dies are integrated on the same platform, the current drawn by transistors during switching is increased resulting in higher SSN [6]. Moreover, new elements such as the silicon interposer bring along parasitic interconnection in the PDN, whose inductance takes part in the chip/package resonance, promoting higher impedance peak at intermediate frequencies [7].

Structural complexity induced by 3DIC requires an increased control over PDN impedance. Solutions to keep PDN impedance under Z_{TARGET} are the following [8]:
- Increase on-chip decoupling C_{FE}.
- Reduce package impedance.
- Increase on-package decoupling.

Increasing on-chip capacitance leads to a prohibitive increase in the size, thus the cost of logic die. A careful PDN design can reduce its inductance up to a certain limit defined by the intrinsic impedance of the interconnections composing it. Adding on-package decoupling capacitor seems to be a good solution to limit anti-resonance at intermediate frequencies.

Solutions to add decoupling capacitance to the package are numerous. Figure 3 represents a logic die/Si-interposer/BGA stack reported on PCB, including several types of decoupling solutions at different floors of the package: beside C_{FE} and C_{PCB} already discussed, C_{BEOL} is situated in the back-end levels of the logic die [9], C_{INT} is situated in silicon interposer and C_{BGA} in the BGA. Two kinds of capacitors are distinguished at each floor: SMD and substrate embedded. Several studies points out the benefits of embedded over discrete capacitors (at the same package floor) due to the removal of parasitic inductance induced by mounting pads, signal/ground traces and signal/ground vias, either on PCB [10], BGA [3] or interposer [6].

Figure 4 shows a simplified PDN RLC electrical model of the structure depicted in Figure 3. Decoupling capacitors connected between signal and power planes bring energy to switching transistors but this power supply is limited by the parasitic impedance of the interconnections in the circuit loop inducing anti-resonance effects. A smart placement of those

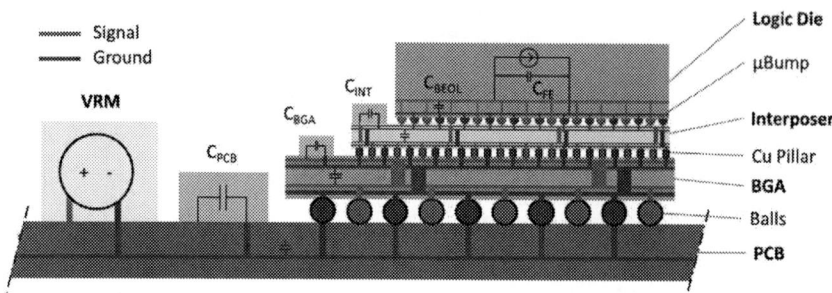

Figure 3. Diagram of a logic die/Si-interposer/BGA stack reported on PCB including several types of decoupling solutions at different floor of the package.

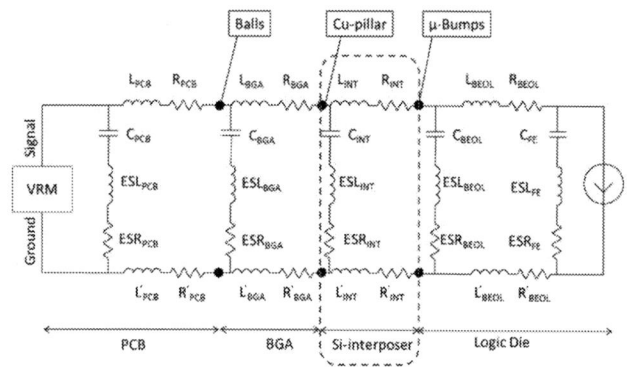

Figure 4. Simplified PDN RLC electrical model of logic die/Si-interposer/BGA stack reported on PCB including decoupling capacitors at each floor of the structure.

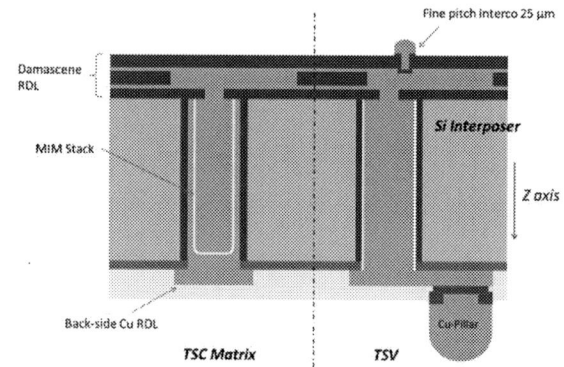

Figure 5. TSC principle in a via-middle Si-interposer integration.

various capacitors in the PDN allows designer to mitigate anti-resonance peaks in order to keep PDN impedance under Z_{TARGET} over the whole operating frequency of the device. Introducing Si-interposer in the structure requires associated decoupling capacitance to be integrated as well. This is why several capacitors types are being investigated for integration on interposer, either planar [11] [6] [12] [7] [13] thin-film or 3D trench [14] [15]. *Heeseok Lee et al.* [12] demonstrated a successful decoupling of intermediate frequencies from ~10MHz to 2GHz by integrating a 40nF (1nF.mm⁻²) planar capacitor on the interposer. *Zhe Li et al.* [7] showed a significant reduction of the anti-resonance peak between C_{FE} and L_{BGA} situated around 100MHz by integrating a planar capacitor (2nF.mm⁻², 140mΩ) on the interposer.

The rest of the paper is dedicated to the Through Silicon Capacitor (TSC), whose purpose is to decouple intermediate frequencies (~100MHz-2GHz) by introducing a high-density and low impedance capacitance to the interposer.

Through Silicon Capacitor

The major feature concerning TSC fabrication lies in its co-integration with Through Silicon vias: both processes share common steps such as Deep Reactive Ion Etching (DRIE), passivation, copper filling and backside contact realization. This co-integration leads to an overall cost reduction of the process.

TSC is a 3D Metal-Insulator-Metal (MIM) capacitor, which originality consists in the fact that it crosses over the

whole thickness of Silicon interposer, as illustrated in Figure 5. Layers constituting the MIM stack are deposited on a matrix of deep holes realized by DRIE, developing the capacitive area along the cavities sides. The upper electrode is connected to interposer front-side, whereas the lower electrode is connected to its backside.

Since each deep hole is contacted individually, the overall TSC matrix then consists of a multitude of individual capacitors connected in parallel. Such a structure allows the overall equivalent series resistance (ESR_{TOT}) and equivalent series inductance (ESL_{TOT}) to be reduced proportionally to the numbers of individual capacitors n in the matrix, while adding up their capacitance following :

$$ESR_{TOT} = \frac{ESR_{TSC}}{n} \qquad Equ.\ 2$$

$$ESL_{TOT} = \frac{ESL_{TSC}}{n} \qquad Equ.\ 3$$

$$C_{TOT} = C_{TSC}.n \qquad Equ.\ 4$$

where ESR_{TSC}, ESL_{TSC} and C_{TSC} stand for the individual contributions of a TSC and n the number of TSCs in the matrix.

In the case of a via-middle interposer type, TSC/TSV co-integration is made easier thanks to the partial copper filling of the TSC. This electrolytic growth of copper is operated in

978-1-4799-2408-0/14 $31.00 © 2014 IEEE

the TSC prior to the MIM stack layer deposition, it consists in a partial bottom-up filling of the cavity. This process is illustrated in Figure 6, where a bottom-up cooper filling has been operated on a 10x100 μm TSV: inhibitors on the side of the TSV and on wafer surface combined with an accelerator at the via bottom favour an evolving copper growth from bottom to top, thus avoiding the formation of voids in the final structure. In the partial-filling case, bottom-up process is stopped after ~10% of completion, leaving the cavity partially filled at the bottom. After MIM stack deposition, TSCs are filled with a standard bottom-up copper growth, a chemical-mechanical-polishing (CMP) planarizes the structure before front-side damascene redirection layer realization.

Ongoing studies are evaluating the addition of a lithography step that allows the etching of the seed layer around the TSVs before the partial-filling growth step in order to cut their electrical access, so partial-filling is localized in the TSC only. The resulting structure shifts the following MIM stack deposition higher in the TSC since its bottom is filled with copper. This disparity between TSC and TSV proves to be useful during backside contact realization: MIM layers deposited at TSVs bottom are suppressed by CMP allowing a direct electrical contact between the two sides of the interposer but those layers are protected by the partial-filling in the TSC.

Although via-middle integration could be done without partial-filling step, backside contact would be delicate. Furthermore TSC ESR would be limited by lower electrode resistivity, as it will be detailed in in the next paragraph.

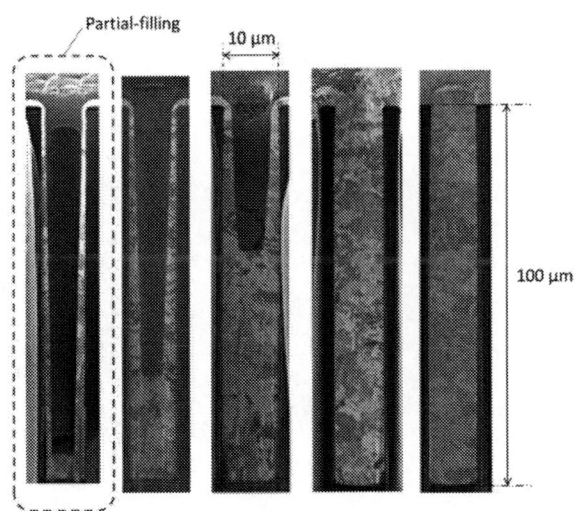

Figure 6. Bottom-up copper filling in 10x100 μm TSV.

Copper partial-filling impact on ESR

Partial filling is a key feature concerning TSC/TSV co-integration, however its contribution is not limited to architectural purposes. As discussed in a previously communicated work [16], Metal Organic Chemical Vapor Deposited titanium nitride (MOCVD TiN) is used to obtain a conformal electrode in deep holes during TSC process. This material shows a major drawback if used for electrical conduction as exposed hereinafter.

During the TiN MOCVD process, NH_3 densification plasma is used in alternation with deposition steps in order to lower its resistivity. Densified TiN resistivity has been measured at 194 μΩ.cm on wafer top against 8000μΩ.cm for material deposition without plasma treatment (PT). If this treatment is viable for planar integration, plasma high directionality limits TiN densification on the side of the TSC, leaving the deposited material with high resistivity. Partial-filling introduction underneath bottom TiN electrode lowers the ESR of the structure due to copper low resistivity.

In order to quantify this effect, a distributed analytical RC model of the TSC has been adapted from *A. Bajolet et. al's* work [17]. This model, originally developed for a 3D MIM structure in the interconnection levels of an LSI die [9], takes into account electrical distribution along the electrodes and is in agreement with both 3D numerical simulations and experiments [17].

Modifications have been brought to the original model to fit TSC cylindrical structure by using radial resistance elements in the calculation, contribution from via bottom and inner copper-filling has been implemented as well. Layer deposition non-uniformity along TSC sides has been modeled by dividing the whole structure into 10 size-equivalent sections, each contribution takes into account material thickness variation assuming a linear decrease along via side (Z axis on Figure 5). Longitudinal electrical transport along z-axis of each electrode is exclusively attributed to copper, except in the case without partial-filling where it is attributed to TiN.

TSC materials properties input for the model relative to ESR calculations are summed up in Table 2, TiN material on TSC side is considered undensified and its deposition conformity has been worked out from SEM cross section measurements in 10x80 μm via.

	Cu	TiN	TiN + PT*
Thickness max (nm)	600	70	70
Resistivity (μΩ.cm)	1.9	8000	194
Side uniformity (%)	100%	44%	44%

Table 2. TSC material input for TSC ESR calculation. *Plasma Treatment

ESR has been calculated for 10 μm TSC-diameter and 80 μm-depth, 200 nm-conformal passivation, 5 μm-copper growth at the bottom of the cavity and 600 nm on its side in the case including partial filling. Without partial-filling, highly resistive bottom TiN electrode dominates the ESR of the whole structure for a total resistance of 1.5 kΩ. Introduction of the copper layer underneath the TiN mitigates this value since it carries out the longitudinal electrical path along the TSC side, resistance value is then lowered to 40 mΩ for a single TSC. Model highlighted a 6 decade ESR decrease of a single TSC with the introduction of a less resistive metal layer beneath TiN bottom electrode, pointing out benefits of adding a copper layer toward the realization of a low impedance device.

978-1-4799-2408-0/14 $31.00 © 2014 IEEE   1299

TSC process and matrix design impact on capacitance

In this section, both TSC and matrices geometrical parameters' impact on capacitance are exposed. Figure 7 illustrates TSC and hexagonal matrix design parameters: width x, length y, via diameter Φ and depth L, inter-via spacing S and pitch P the TSC repetition step defining matrix density.

As stated earlier, various interposer realization processes are studied in parallel, leading to different TSV geometries. Since TSV and TSC are etched during the same process step, differences in TSV process lead to the same difference in TSC geometries, thus capacitance values.

Two ranges of diameter Φ/depth L couples have been evaluated in order to fit either "via-middle" or "via-last" TSV realization processes. TSC aspect ratio (AR) has been fixed to 3 for via-last process and 8 for via-middle process, capacitance density has been calculated using a 20 nm-thick Al_2O_3 dielectric with a relative permittivity ε_r of 8.5 extracted from previous electrical measurements [16], a 200 nm-SiO_2 passivation and 80 nm-thick TiN top and bottom electrode. TSC are arranged in 1x1 mm hexagonal matrices. For comparison purposes, exhibited results do not take partial-filling effect into account and inter-via spacing is set to 10 μm.

Figure 8 illustrates capacitance density disparity in regards to TSC geometrical variations in hexagonal matrices. If TSC architecture allows its co-integration with multiple TSV types, its capacitance density is highly dependent on the TSV process used. Furthermore, according to interposer TSV type, the amount of parallel TSC in the matrix for a given area varies, which leads to disparity in ESL and ESR values as discussed before. Previously published work showed that calculation method is in agreement with capacitance measurements operated on patterned $TiN/Al_2O_3/TiN$ stack deposited on DRIE deep hole matrices of various geometries [16].

If via diameter Φ and depth L depends on TSV process, some independent parameters influence capacitance density such as dielectric material used, matrix type and density.

Figure 9 illustrates capacitance density variation function of pitch for 3 different 10x80 μm TSC matrices of 1mm² with partial filling. In the first two cases, a 20 nm-thick Al_2O_3 dielectric has been used for calculation and TSC are arranged in square (green) or hexagonal (blue) matrix. For the third case (red), a 40 nm-thick Ta_2O_5 dielectric ($\varepsilon_r = 25$) has been used for calculation on a hexagonal matrix.

Dielectric thicknesses have been defined according to electrical reliability of TiN/Dielectric/TiN stack in 3D structure reported in literature, especially their breakdown voltages V_{BD} corresponding to the highest achievable polarization voltage before the creation of irreversible conductive path within the dielectric: 20 nm-thick Al_2O_3 exhibits a $V_{BD} > 11V$ in [9] and 40 nm-thick Ta_2O_5 a $V_{BD} > 15V$ in [18].

Hexagonal matrices exhibit higher capacitance density than square one since this configuration allows denser matrices. Using Ta_2O_5 results in higher capacitance values even with a layer thickness doubled compared to Al_2O_3 due to reliability issues: for a pitch set to 15μm, Al_2O_3 hexagonal matrices exhibit a capacitance density of 39 nF.mm⁻² where

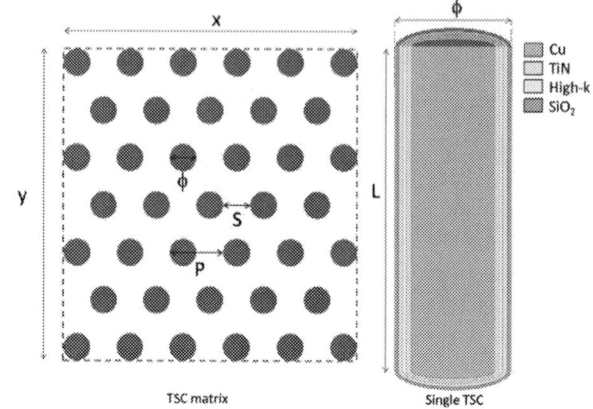

Figure 7. Design parameters of a TSC and hexagonal matrix

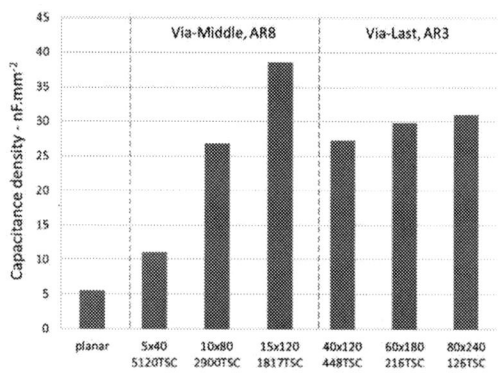

Figure 8. Modeled capacitance densities of various TSC hexagonal matrices geometries using 20nm Al_2O_3 ($\varepsilon_r = 8.5$) dielectric. Hexagonal matrices area is 1 mm² and inter-via spacing is 10 μm.

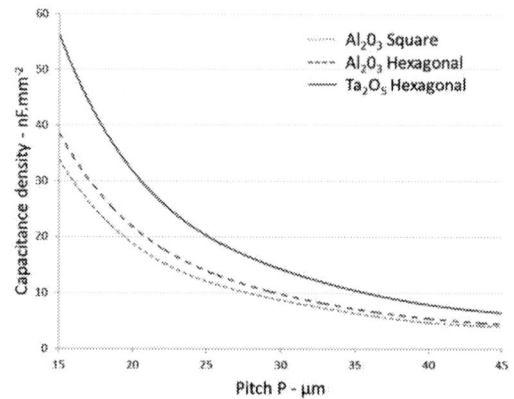

Figure 9. Capacitance density versus matrix pitch for 10x80 μm TSC matrices of 1mm²

Ta_2O_5 increases this value up to 56 nF.mm⁻² in the same configuration.

This capacitance density shows a great improvement compared to classical 2D structures [7] [12] [11], however higher value could be reached using a thinner dielectric layer up to a limit defined by device reliability such as V_{BD} and leakage current. Furthermore multiple dielectric layers structure increases capacitance density as in [19], at the expense of dedicated additional process steps.

Another crucial parameter impacting capacitance values is the matrix density: using aggressive pitches greatly increases the number of TSC in the matrix developing capacitive surface of the device. Reducing matrix repetition pitch from 20 to 15 µm results in a capacitance density increase of 56% in all three scenarios.

High capacitance density devices can be achieved using high permittivity dielectric such as Ta_2O_5 over dense matrices. However such structures may have a significant mechanical impact on the final interposer due to induced stress to the silicon substrate [20] and DRIE process may be limited by TSC proximity within the matrix.

TSC device electrical performances evaluation

Previous section highlighted process and design requirements to achieve high density capacitor, this final paragraph evaluates intrinsic electrical performances of such devices.

Table 3 summarizes major properties of TSC matrices in correlation with their area occupied on the interposer. Number of TSC n_{TSC} has been worked out for a hexagonal matrix of 10x80 µm TSC with a 15 µm repetition pitch. This geometry corresponds to via-middle TSV process, which tends to be widely used in industry's cleanrooms for Si-interposer realization. Device capacitance of structures including partial filling has been calculated for a 40 nm-thick Ta_2O_5 dielectric and Equ. 4, while its ESR has been worked out using the previously discussed model and Equ. 2 assuming ideal interconnections between TSCs. ESL of a single TSC has been evaluated to 10±2 pH, this value has been extracted from electromagnetic simulation of an equivalent TSC structure. ESL of the whole device has been worked out with Equ. 3, assuming ideal interconnections between single TSCs.

Size mm²	nTSC	C (nF)	ESR (µΩ)	ESL (fH)
0.5	2511	27.57	16.42	3.98
0.75	3795	41.67	10.87	2.64
1	5120	56.22	8.05	1.95
2	10246	112.51	4.02	0.98

Table 3. Main intrinsic properties of TSC matrices function of their size.

As expected, device impedance drops with increasing matrix size since more individual TSC are connected in parallel: 2 mm² device with nominal capacitance of 112 nF owns 10246 TSCs, its ESR and ESL have been evaluated to 4 µΩ and 1 fH respectively. Those low values points out benefits offered by the TSC parallel structure compared to other 2D [11] [7] or 3D [21] structures. Intrinsic capacitor module resonant frequency is size independent and has been evaluated to ~15GHz.

Those values must be considered carefully as they are intrinsic to the capacitor module and do not directly reflect decoupling performances since they do not take into account

neither interconnection between single TSCs within the matrix nor interconnection impedance in the logic decoupling loop.

Conclusion

In the first part of this paper, classic decoupling method has been discussed. Its main limitation comes from chip/package anti-resonance taking place when a LC parallel circuit is formed between on-chip capacitance and package inductance resulting in an impedance peak in the Power Delivery Network at intermediate frequencies (~100MHz-2GHz). 3DIC introduces new parasitic elements to the structure, making power management even harder. New decoupling capacitance must be added in 3DIC packages to obtain low impedance PDN, and integration of decoupling capacitor on interposer is a successful way to achieve it.

The second part of the paper introduces the Through Silicon Capacitor as an alternative solution for integration of decoupling capacitance in a silicon interposer. TSC is co-integrated with Through Silicon Vias as they share most fabrication process steps.

TSC realization in a via-middle silicon interposer has been exposed and architectural benefits of adding a partial copper-filling prior to the Metal-Insulator-Metal stack deposition has been discussed. An ESR decrease of 6 decades due to Cu partial filling introduction has been demonstrated through a distributed analytical RC model.

TSC process and matrix design impact on capacitance has been modeled pointing out TSV process dependence. Beside, capacitance density of 56 nF.mm⁻² should be achieved by using high permittivity dielectric such as Ta_2O_5 deposited on dense 10x80 µm deep holes matrices (pitch=15µm).

Finally, electrical performances of such devices have been evaluated for different allocated area on silicon interposer leading to low ESR (µΩ range) and ESL (fH range) values.

TSC flexible architecture proves to be an interesting way of adding the decoupling function silicon interposer crucially needs to achieve low PDN impedance in 3DICs.

Acknowledgments

The authors would like to thank Mathilde Gottardy for the partial-fill cross section (Figure 6) and Monica Larissa Djomeni Weleguela for TiN resistivity measurements.

References

1. L. Smith, S. Sun, P. Boyle, B. Krsnik, "System power distribution network theory and performance with various noise current stimuli including impacts on chip level timing," in *IEEE Custom Integrated Circuits Conference (CICC)*, 2009, pp. 621 - 628.
2. (2012) International Roadmap for Semiconductors. [Online]. http://public.itrs.net
3. P. Muthana, M. Swaminathan, E. Engin, P.M. Raj, R. Tummala, "Mid frequency decoupling using embedded decoupling capacitors," *in IEEE Electrical Performance of Electronic Packaging*, 2005, pp. 271 - 274.
4. M.M. Corbalan et al., "Power and signal integrity challenges in 3D systems," in *IEEE Design Automation Conference* (DAC), 2013, pp. 1 - 4.

5. L. Smith, S. Sun, M. Sarmiento, Z. Li, K. Chandrasekar, "On-Die Capacitance Measurements in the frequency and time domains," in *DesignCON*, 2011.

6. K. Kikuchi et al., "Wideband ultralow power distribution network impedance evaluation of decoupling capacitor embedded interposers for 3-D integrated LSI system," in *IEEE Electronic Components and Technology Conference (ECTC)*, 2013, pp. 1190 - 1196.

7. Zhe Li, Hong Shi, J. Xie, A. Rahman, "Development of an optimized power delivery system for 3D IC integration with TSV silicon interposer," in *IEEE Electronic Components and Technology Conference (ECTC)*, 2012, pp. 678 - 682.

8. D. Amey, K. Dietz, "Application of embedded capacitor technology for high performance semiconductor packaging," in *DesignCon*, 2007.

9. S. Jeannot et al., "Toward next high performances MIM generation: up to 30fF/µm2 with 3D architecture and high-κ materials," in *IEEE Electron Devices Meeting (IEDM)*, 2007, pp. 997 - 1000.

10. H. Kim, B. K. Sun, J. Kim, "Suppression of GHz range power/ground inductive impedance and simultaneous switching noise using embedded film capacitors in multilayer packages and PCBs," *IEEE Microwave and Wireless Components Letters*, vol. 14, no. 2, pp. 71 - 73, Feb. 2004.

11. A. Takano et al., "Development of Si interposer with low inductance decoupling capacitor," in *IEEE Electronic Components and Technology Conference (ECTC)*, 2011, pp. 849 - 854.

12. H. Lee et al., "Power Delivery Network Design for 3D SIP Integrated over Silicon Interposer Platform," *in IEEE Electronic Components and Technology Conference (ECTC)*, 2007, pp. 1193 - 1198.

13. J.M. Yook, J.C. Kim, S.H. Park, J.I. Ryu, J.C. Park, "High density and low-cost silicon interposer using thin-film and organic lamination processes," in *IEEE Electronic Components and Technology Conference (ECTC)*, 2012, pp. 274 - 278.

14. B. Dang et al., "Three-Dimensional Chip Stack With Integrated Decoupling Capacitors and Thru-Si Via Interconnects," *IEEE Electron Device Letters*, vol. 31, no. 12, pp. 1461 - 1463, Dec. 2010.

15. H. Jacquinot and D. Denis, "Characterization, modeling and optimization of 3D embedded trench decoupling capacitors in Si-RF interposer," in *IEEE Electronic Components and Technology Conference (ECTC)*, 2013, pp. 1372 - 1378.

16. O. Guiller et al., "Through Silicon Capacitor co-integrated with TSVs on silicon interposer," *Microelectronic Engineering*, December 2013. ISSN 0167-9317, http://dx.doi.org/10.1016/j.mee.2013.12.017.

17. A. Bajolet et al., "Low-Frequency Series-Resistance Analytical Modeling of Three-Dimensional Metal–Insulator–Metal Capacitors," *IEEE Electron Devices, Transactions on*, vol. 54, no. 4, pp. 742 - 751, apr. 2007.

18. M. Thomas et al., "Impact of TaN/Ta copper barrier on full PEALD TiN/Ta2O5/TiN 3D damascene MIM capacitor performance," in *IEEE International Interconnect Technology Conference*, 2007, pp. 158-160.

19. J.H. Klootwijk et al., "Ultrahigh Capacitance Density for Multiple ALD-Grown MIM Capacitor Stacks in 3-D Silicon," *IEEE Electron Device Letters*, vol. 29, no. 7, pp. 740 - 742, July 2008.

20. Y. S. Chan, H. Y. Li, X. Zhang, "Thermo-Mechanical Design Rules for the Fabrication of TSV Interposers," *IEEE Components, Packaging and Manufacturing Technology, Transactions on*, vol.3, no.4, pp.633 - 640, Apr. 2013.

21. F. Lallemand and F. Voiron, "Silicon interposers with integrated passive devices, an excellent alternativ to discrete components," *in Microelectronics Packaging Conference (EMPC)*, 2013, pp. 1 - 6.

Design of RF and Thermal Pads of CMOS PAs using Copper to Copper Bonding Technology

Lih-Tyng Hwang[1] and An-Yu Kuo[2]

National Sun Yat-Sen University[1] and Cadence Design Systems, Inc.[2]

70 Lianhai Road, Kaohsiung, Taiwan[1], 2655 Seely Avenue, San Jose, CA 95134, USA[2]

FiftyOhm@mail.nsysu.edu.tw[1]

Abstract

Different radios require different levels of power output; for example, GSM demands up to 4 W (36 dBm) for its RF operations. CMOS RFICs are frequently designed up to 2 W (32+ dBm) [1, 2]. Recently, copper to copper bonding technology has been fiercely pursued in C2W and W2W domains [3-5]. However, the technology is equally applicable in other packaging platforms where similar materials are packaged; for example, CMOS PAs (Power Amplifiers) being integrated onto other silicon platforms, such as Si-IPD and Silicon interposer. It is because the high stress occurred at the low profiled copper to copper bonding interface can be avoided (or properly managed) when the thermal expansion coefficients of the two materials are the same or similar. Here we study the RF and thermal pad designs for CMOS PA packaged onto Si-IPD or Si-interposer using copper to copper bonding. The results will be compared to those using flip chip bumps. For RF I/O pad design, G-S I/O configuration is assumed. The frequency band of interest is up to 60 GHz. For thermal design, the PA die sizes vary from 0.7 mm x 0.7 mm to 2 mm x 2 mm. Thermal pad sizes are 50 μm x 50 μm and 75 μm x 75 μm. We assumed two isolated heating sources; each heating source outputs 1 W (total 2W, 33 dBm). Using Cadence's thermal design tools, steady state and transient results were obtained for Cu-Cu bonding. Results for flip chip bump technologies were then extrapolated from the Cu-Cu bonding results. The junction temperature difference between the systems employing these two technologies was significant; using Cu-Cu bonding technique, the junction temperature of PA can be significantly lower (about 50 °C) than 110 °C, the simulated junction temperature of the PA system employing flip chip technology. This represents a huge benefit on the reliability of device using Cu-Cu bonding technology.

Introduction

Power amplifier (PA) is a key component in many wireless communication devices. It usually resides in the front end module that drives the antenna(s). The purpose of PA is to boost the transmitting signals, so they can reach an RF receiver at a distance. Take cellular phones as an example, GSM demands up to 4 W (36 dBm) for its RF operations. CMOS RFICs are frequently designed up to 2 W (32+ dBm) [1, 2]. The actual size of a PA IC is small, varying from 0.5 mm x 0.5 mm to 2 mm x 2 mm. According to [6], the temperature rise and gradient have strong effects on both chip performance and reliability. In [7], temperature-sensitive reliability is discussed. Therefore, it is important for a PA to be packaged in a configuration that dissipates heat efficiently. For example, in a wire bonding package, the entire backside of the PA die is used as heat dissipating area. When flip chip

technology is employed, the heat is removed through the thermal bumps. This heat dissipation technique is not as efficient as those configurations employing wire-bonds, since the total area of the bumps is less than the entire back side of the PA.

Recently, copper to copper bonding technology has been fiercely pursued in C2W and W2W domains [3]. Copper to Copper bonding, compared to flip chip technique, can provide a more efficient heat dissipation mechanism. It is because the thermal resistance in Cu-Cu interconnection scheme is almost negligible, while finite thermal resistance certainly exists in flip chip bumps. We will illustrate this point later in this article.

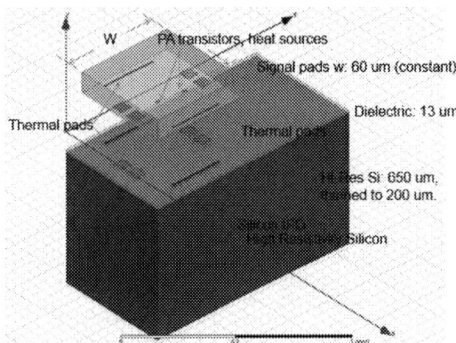

Figure 1. A CMOS PA Cu-Cu (or FC) is bonded to Si IPD, which is thinned to 200 μm when attached to a QFN.

IPD has been a versatile passive (including, R, L, C, filter, balun, antenna, etc..) integration technology. An introduction of Hi Res Silicon IPD (Si HRS Substrate), known as HighQ™ copper on silicon IPD, can be found in [8]. The IPD, which employs high resistivity silicon as the substrate, is ideal for the production of passive devices such as baluns, filters, couplers, and diplexers that are used in portable, wireless and RF applications. STATSChipPAC employs a copper metallization process capable of depositing 8 microns or more of copper on a silicon wafer. This results in higher Q components that reduce loss in the RF signal transmission path, thereby increasing battery life of the wireless system and improving reception. The size of matching circuitry and filters is often reduced by 40% [9]. Recently, IPD technology, termed 3D IPAC, is even applied to integrate combination of passive and active devices [10]. That is, an IPD is actually becoming an Active Device Carrier (ADC).

In this paper, a CMOS PA is attached on hi-resistance silicon IPD; forming a PA IPD, Fig. 1. The PA IPD is then packaged in a QFN, which is in turn assembled on a printed circuit board, Fig. 2. In order to analyze the entire thermal path, an IC-package-board co-simulation approach was adopted. In reality, the PA devices are usually housed in a product, such as a cellular phone, a tablet computer, or a desktop. It makes sense to evaluate the junction temperature of the PA relative to the ambient temperature, for example, 25°C in an environment. To take into consideration various environments, we also included ambient air flow in the simulation. The model and the power output patterns, as the inputs to the model, are introduced and discussed first, followed by results and discussions.

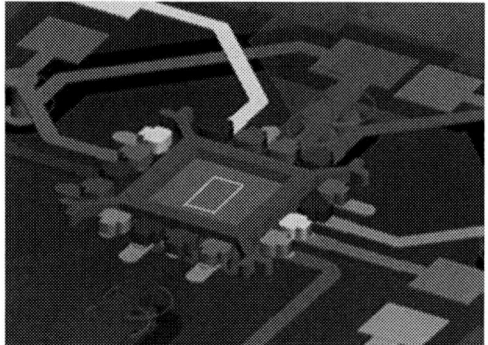

Figure 2. 16-pin QFN package on a 4-layer 55mm x 55mm board; the white border inside the QFN indicates the location of the PA IPD

Model of the Full PA IPD Co-Simulation

Fig. 1 shows the PA IPD model that is used in this article. On the PA die, there are RF signal pads and thermal pads, in addition to the two heat sources. The signal pads are of G-S configuration. The thermal pads are arranged on two sides of the PA die. In this article, only the dimensions of the thermal pads were varied, while the signal and heat source sizes were kept unchanged through the study. The signal pad has a dimension of 60 μm at a side; the heated source has a dimension of 25 μm by 45 μm. The following table, Table 1, shows the simulation space adopted in this study. The IPD size was about twice the size of CMOS PA. The extra area may be used to integrate other passive or active components, as described in the last section [10]

Case #	CMOS Size (mm2)	IPD Size (mm2)	Thermal Pad Size (um2)	# of Pads, Total Area (um2)
1	0.7 x 0.7	0.8 x 1.525	75 x 75	10, 56,250
2	0.7 x 0.7	0.8 x 1.525	50 x 50	14, 35,000
3	1.0 x 1.0	1.1 x 2.125	75 x 75	10, 56,250
4	1.0 x 1.0	1.1 x 2.125	50 x 50	14, 35,000
5	1.5 x 1.5	1.6 x 3.125	75 x 75	10, 56,250
6	1.5 x 1.5	1.6 x 3.125	50 x 50	14, 35,000

Table 1. Cases simulated in SS and Transient

The last column in Table 1 shows the number of the thermal pads and the total area of the pads. This data will be used later in estimating the <u>average</u> <u>additional</u> temperature rise due to the use of flip chip bumps, relative to Cu-Cu bonding technology.

PA IPD was integrated into a 16-pin QFN package, Fig. 2. The white outline indicates the location of the PA IPD. The IPD sits on a 4-layer 55mm x 55mm printed circuit board. In

this simulation, Cadence PowerDC was used. The PowerDC thermal analysis tool is based on 3D finite element method [11]. The ambient temperature was set at 25 °C, the ambient air flow was set at 1 m/s, the CMOS power amplifier power source was set to 1 W per source (i.e.,, 2 Watts, about 33 dBm, in total). The material properties in this simulation are listed in Table 2.

	Conductivity (W/m-K)	Density (Kg/m3)	Heat Capacity (J/Kg-K)
Copper	400	8933	385
FR4	0.3	1900	1200
Mold Compound	0.99	1200	1000
BCB	0.29	96	218
Si-IPD	149	2330	712
CMOS	30	2330	712
Air	0.027	1.77	1005

Table 2. Material properties used in simulation

Profiles of Transient Power of CMOS PA

It is nearly impossible to obtain the true junction temperatures of the CMOS PA through simulation, since the temperatures depend on the actual working conditions of the PA. Rarely a PA is working at 100% duty cycles; instead, it bursts regularly at 1/8 or 1/4 duty cycles, according to GSM communication protocols, Fig. 3, [12]. In Profile 1 (top), 1/8 duty cycle at the full power (1 W on each heating source) was assumed; and in Profile 2 (bottom), 1/4 duty cycle at 72% of the full load was used.

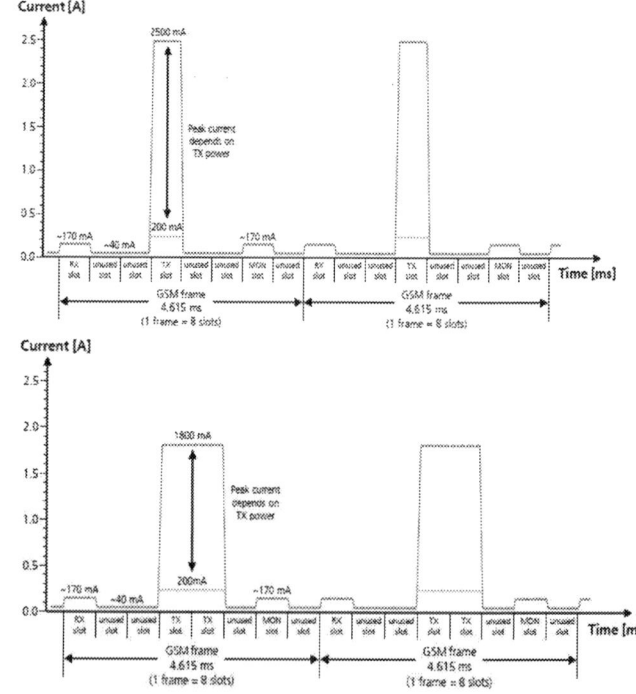

Figure 3. Power vs. Time transients: 1/8 duty cycle with full power (top, Profile 1), 1/4 duty cycle with 72% full power (bottom, Profile 2) [12].

After a period of long talk time (i.e., PA at work), steady state (SS) simulations using <u>average</u> power (1/8 of full power, or 1/4 of 72% power) can predict accurately the temperature rise of the PA device. That is, the SS prediction is accurate only when the PA is operated (the talk time) long enough, so the average power makes sense. Here, we define long talk time as a time span that is greater than the response time of the IC-package-board system.

In the following section, steady state results for system employing Cu-Cu bonding technology are shown first. The results were then extrapolated to obtain the junction temperature for system using flip chip technology. We will emphasize the advantages of Cu-Cu bonding technology over flip chip bump technology in both thermal and RF aspects. At the end, transient simulation was performed to obtain the response time of the IC-package-board system. The significance of the response time is also discussed.

Results and Discussions

Steady state junction temperatures are shown in Table 3. Junction temperature for 1/8 duty cycle (Profile 1), as expected, is lower than those for 1/4 duty cycle (Profile 2), since the total heat accumulated (Power times burst duration) for Profile 2 was greater than that for Profile 1. In all the cases considered, the smaller the CMOS PA die size or the smaller thermal pad size are, the greater the junction temperature becomes. Case #2 has the maximum junction temperature in both Profile 1 and Profile 2. It makes sense, since a smaller thermal pad size will result in a larger thermal spreading resistance (to be explained later). Also, smaller thermal mass makes it hard to transport heat; thus, resulting in less heat to reach the system heat sink (the board surface and the forced air flow).

Case #	Junction Temperature (°C), Full duty	Junction Temperature (°C), 1/8 duty	Junction Temperature (°C), 1/4 duty
1	186.5	45.19	54.07
2	189.9	45.63	54.70
3	167.5	42.81	50.65
4	169.6	43.13	51.10
5	144.9	40.00	46.60
6	146.1	40.13	46.78

Table 3. Results for Cu-Cu bonding, at different duty cycles

Fig. 4 shows temperature plot for Case# 2. The red hot area near the heat sources clearly indicates thermal spreading phenomenon [13]. In thermal spreading, the heat generated from a finite, isolated two dimensional spot has to spread out first, before the heat can be efficiently and fully dissipated through the three dimensional thermal volume under the isolated hot spot, Fig. 5. One immediate consequence of the spreading is an increase in thermal resistance, due to the constricted thermal path. A constricted path usually forms a hot spot.

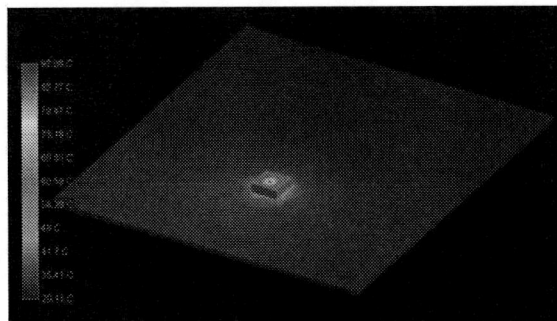

Figure 4. SS Temperature plot for PA, QFN, and Board, Case 2

From Fig. 4 plot, the IPD is seen at an elevated temperature, while the majority of the board remains essentially at the room temperature.

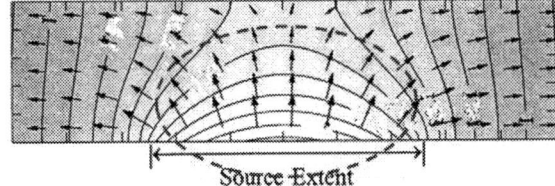

Figure 5. Illustration of thermal spreading phenomenon. Spreading exists between a 2-D isolated heat source and a larger 3D thermal mass.

The junction temperature results presented so far are for Cu-Cu bonding technology. Using a simple cylindrical model, Fig. 6, the effect of flip chip (FC) bumps on the junction temperature can be estimated. In essence, the bumps are shaped as a truncated sphere. The larger volume of a truncated sphere effect (lower temperature rise) in heat transport can be compensated by a larger cross sectional area (square area was used on top and bottom, as opposed to the circular area in a truncated sphere). Here the square area shown in Table 1 was used to estimate the additional temperature rise due to the FC bumps.

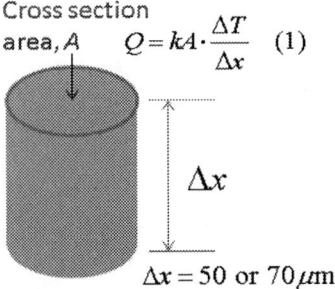

Figure 6. Model and formula used to determine the average temperature rises for flip chip bumps

Depending on the two-dimensional flows, the additional junction temperature rise is not the same for all the thermal pads (or, the FC bumps). Here, we have assumed each thermal pad (and, thus each FC bump) dissipates the same amount of heat (the total heat divided by the number of pads).

After taking into consideration of the flip chip bumps, the junction temperatures (steady state) are shown in Table 4.

978-1-4799-2408-0/14 $31.00 © 2014 IEEE

Here, eutectic solder bumps were used. The thermal conductivity for eutectic solder is 50 W/mK at 25°C). In Case # 2, the junction temperature has risen above 110°C. Elevated temperature is critical (should be avoided), since it related to the reliability of the device [14].

Case #	Junction Temperature (°C), 1/8 duty	Junction Temperature (°C), 1/4 duty
1	94.97	103.85
2	102.77	111.84
3	92.59	100.43
4	100.27	108.24
5	89.78	96.38
6	97.27	103.92

Table 4. Results for flip chip technology, at different duty cycles

The benefits of the Cu-Cu bonding in propagating RF signals were also studied. In Fig. 7, the definition of RF transition for Cu-Cu interconnection is shown. The top die is a CMOS IC based on 0.18 µm technology, and the bottom substrate is hi-resistivity silicon IPD. The Cu-Cu pads are located between the CMOS die and hi-resistivity silicon IPD. As mentioned, the RF signal pads are 60 µm by 60 µm in size.

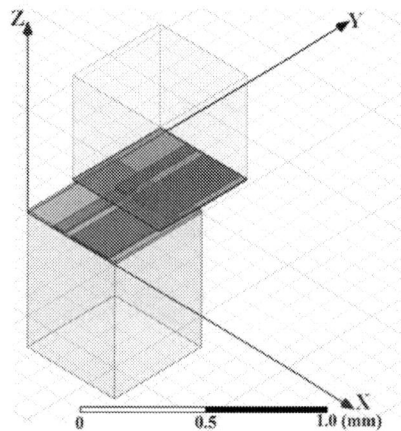

Figure 7. Definition of the RF transition for Cu-Cu interconnection, the scale is 1 mm.

The HFSS tool was employed to obtain the RF characteristics of the transition. To obtain the RF characteristics of the transition for FC bump interconnection, the Cu-Cu pads in Fig. 7 were replaced by FC bumps. In HFSS, realistic truncated sphere structure was used for FC transition signal integrity study.

The characteristics of the RF transitions (Cu-Cu pads and FC bumps) were obtained by propagating a wave front at the both sides of the G-S transmission lines. The s-parameters were then de-embedded to the edges of the RF transition. The return loss (S11) and the insertion loss (S21) are shown in the following, Figs. 8 and 9.

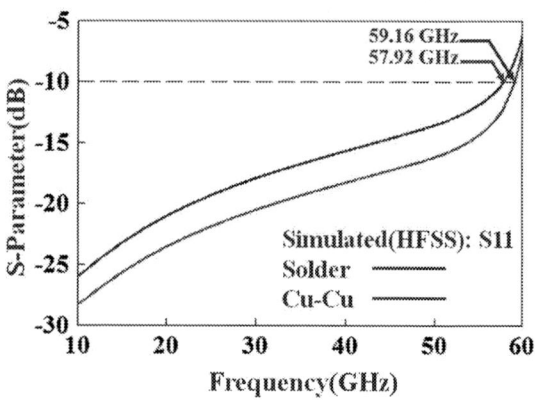

Figure 8. Comparison of RF transitions for Cu-Cu and FC bumps, using G-S configurations, return loss S11. Red line for Cu-Cu, Blue line for FC bump.

Fig. 8 shows that Cu-Cu bonding technology has a better return loss at all frequencies (from low GHz to 60 GHz). The return loss for Cu-Cu bonding is consistently better than that of flip chip technology by 2-3 dB. Use 10 dB rule for return loss, the Cu-Cu bonding is valid to 59 GHz, and the FC bump to 58 GHz.

Fig. 9 shows that the Cu-Cu bonding technology has a better insertion loss at all frequencies (from low GHz to 60 GHz). The low insertion loss translates a saving in battery consumption. Use 1 dB rule for insertion loss, the Cu-Cu bonding is valid to 60 GHz, and the FC bump to 59.5 GHz.

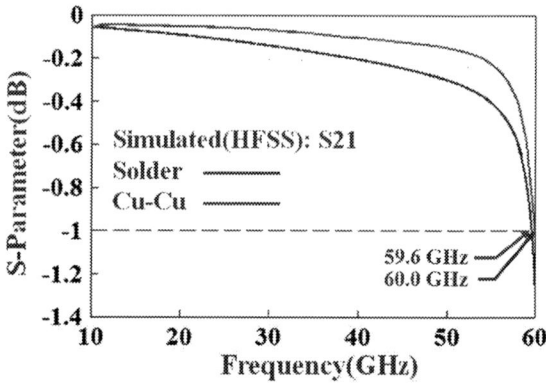

Figure 9. Comparison of RF transitions for Cu-Cu and FC bumps, using G-S-G configurations, insertion loss S21. Red line for Cu-Cu, Blue line for FC bump.

It is important to obtain the response time for the system. The response time indicates the time when the system reaches a steady state. Only when the operating time of the PA is greater than the system response time, should our steady state results be considered valid.

From Fig. 10, the system response times are about 150 seconds. This time span can be easily reached for normal cellular operations; thus, our steady state results are considered valid most of the times. For talk time less than the system response times, the heat generated from the PA may just be absorbed by the system thermal mass.

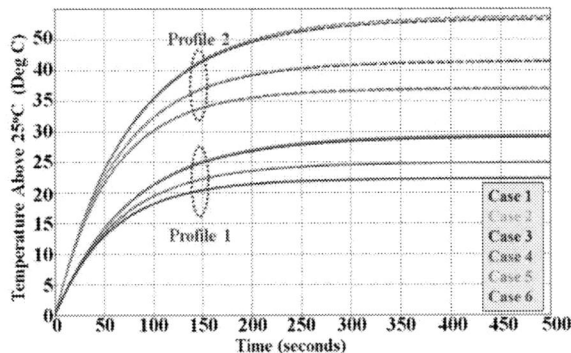

Figure 10. Junction temperature and response time of the IC-package-board system from transient simulations

Conclusions

PA is a key component in many wireless systems. Inside a GSM phone, a PA can output as much as 4 W (36 dBm). Cu-Cu bonding technology has been widely applied in C2W W2W arenas. Here, this bonding technology is applied in an RF package. First, the PA is attached on an IPD; forming a PA IPD. The PA IPD is then integrated onto a QFN, Figs 1 and 2.

Elevated temperatures (should be avoided) are critical since they are related to the reliability of the device [6, 7, 14]. Full IC-package-board co-simulation was performed to obtain the junction temperature of the PA device. In co-simulation, the air flow rate of 1m/s was assumed. Steady state and transient results were obtained. Here, we found from various simulated cases, the system response times were about 150 seconds, which is easily satisfied in daily RF device uses. That is, the SS prediction can be considered accurate most of the time.

Here, we show-case the benefits of Cu-Cu bonding in thermo-mechanical reliability and RF signal propagation. Assuming 2W (33 dBm) for PA power output, Cu2Cu bonding had much lower junction temperature, compared to that using flip chip bump technology, Tables 3 and 4. Using Cu-Cu bonding technique, the junction temperature of PA can be significantly lower (50°C) than 110 °C, the junction temperature of the PA that employs flip chip interconnection technology. Moreover, the temperature difference has significant implications on device reliability [6, 7, 14]. The lower the device junction temperature is, the more reliable the device is.

Please note that the junction temperatures predicted would be higher if the power output is greater than 2W. In realistic cellular phone operation, the air inside the casing is practically stagnant, i.e., v=0, and the junction temperatures can be even higher. In obtaining the junction temperatures for system using flip chip interconnection, we assumed an equal (average) amount of heat passed the bumps. However, in reality, some bumps may carry more heat than others. Efforts to reduce the junction temperature will be explored; for example, flexible thermal pad locations, relative to the heat sources; employing Cu studs, instead of using eutectic bumps, etc..

Cu-Cu bonding has better RF performance, too, over the FC counterpart. Even though the differences in performance indicators measured by the conventional return and insertion loss criteria are not significant; however, the superior characteristic of Cu-Cu bonding in signal transmission is evident throughout the simulation frequency band, i.e., from low to 60 GHz.

Acknowledgments

The first author would like to thank the support of Taiwan's National Science Council through the grant NSC102-2221-E-110-013-MY3. RF simulation work by NSYSU's Ian Feng is also appreciated.

References

1. Jonas Fritzin, Christer Svensson, and Atila Alvandpour, "A +32 dBm 1.85 GHz Class-D outphasing RF PA in 130nm CMOS for WCDMA/LTE," *IEEE ESSCIRC* 2011.

2. Melina Apostolidou, Mark P. van der Heijden, Domine M. W. Leenaerts, Jan Sonsky, Anco Heringa, and Iouri Volokhine, "A 65 nm CMOS 30 dBm Class-E RF power amplifier with 60% PAE and 40% PAE at 16 dB back-off," *IEEE Journal of Solid-State Circuits*, Vol. 44, No. 5, May 2009.

3. Yu-San Chien, et al., "Low temperature (<180°C) wafer-level and chip-level In-to-Cu and Cu-to-Cu bonding for 3D integration," in *Proc. Electronic Components and Technol. Conf. (ECTC)*, 2013, Las Vegas, USA.

4. Ya-Sheng Tang, Yao-Jen Chang, Kuan-Neng Chen, "Wafer-level Cu-Cu bonding technology," Microelectronics Reliability, Vol. 52, No. 2, pp. 312-320, Feb. 2012.

5. Cheng-Ta Ko and Kuan-Neng, Chen, "Low temperature bonding technology for 3D integration," Microelectronics Reliability, Vol. 52, No. 2, pp. 302-311, Feb. 2012.

6. Yi-Kan Cheng, *Electrothermal Simulation and Temperature-Sensitive Reliability Diagnosis for CMOS VLSI Circuits*, Thesis, UIUC, Urbana, Illinois, 1997.

7. Mustafa Acar, *Power Amplifier in CMOS Technology*, Centre for Telematics and Information Technology, CTIT Ph.D. Thesis Series No. 10-187.

8. On-Semi High-Q™ IPD technology, ww.onsemi.com.

9. Silicon based IPD technology from StatsChipPac, http://www.statschippac.com.

10. P. Markondeya Raj, Saumya Gandhi, Srikrishna Sitaraman, Venky Sundaram, and Rao Tummala, "3D IPAC: a new concept in integrated passive and active components," *Chip Scale Review*, November/December 2013.

11. PowerDC simulation tool by Cadence Design System, https://www.cadence.com/products/sigrity/powerdc

12. LEON-G1/G2 series quad-ban GSM/GPRS data & voice modules System Integration Manual, version H, September 10, 2013, www.u-blox.com.

13. Gordon N. Ellison, "Maximum thermal spreading resistance for rectangular source and plates with nonunity aspect ratios," *IEEE Trans. Components and Packaging Technologies*, vol. 26, no. 2, June 2003.

14. *Thermal Considerations of RF Power Amplifier Devices*, Texas Instruments Applications Report, pp.1998.

Wafer IMS (Injection Molded Solder)
- A New Fine Pitch Solder Bumping Technology on Wafers with Solder Alloy Composition Flexibility

Jae-Woong Nah[1*], Jeffrey Gelorme[1], Peter Sorce[1], Paul Lauro[1], Eric Perfecto[2], Mark McLeod[2], Kazushige Toriyama[3], Yasumitsu Orii[3], and Peter Brofman[1]

[1]IBM T. J. Watson Research Center, Yorktown Heights, NY 10598
[2]IBM Systems & Technology Group, 2070 Route 52, Hopewell Junction, NY 12533
[3]IBM Tokyo Research Laboratory, Kawasaki, Japan 212-0032
*E-mail: jnah@us.ibm.com, Phone: 914-945-1875

Takashi Nauchi[4], Akira Takaguchi[5], Kazuya Ishiguro[5], Tomoyasu Yoshikawa[6], Derek Daily[6], and Ryoichi Suzuki[4]

[4]Senju Metal Industry Co., Ltd., Tokyo, Japan 120-8555
[5]Senju System Technology Co., Ltd., Toyama, Japan 939-2708
[6]Senju Comtek Corp., Campbell, CA 95008

Abstract

In this paper, we will describe a new low cost solder bumping technology for use on wafers. The wafer IMS (injection molded solder) process can form fine pitch solder bumps on wafers, while offering greater solder alloy flexibility. This method is also applicable to form uniform solder bump heights when a wafer has different size and shape of I/O pads. The wafer IMS bumping process uses a solder injection head that melts the desired bulk solder alloy composition and then dispenses the molten solder into resist material cavities on wafers within a nitrogen environment. The injected molten solder contacts and wets to the metal pads without flux, thus forming intermetallic compounds at the solder/pad interface. After stripping the resist material, solder bumps exhibit straight side walls and round tops as the solders have solidified inside the cavities of this resist film. This particular geometry is unique and offers a ready-for-substrate bonding condition without an additional reflow step. In the case of using Cu pillars, one resist material is used for both Cu electroplating and molten solder injection. After patterning the resist material, the Cu pillars are electroplated to the desired height, and the remaining cavities of resist material are filled by the injection of molten solder. The final bump height is defined by the thickness of the resist material. Therefore, any non-uniformity of Cu pillar height across a wafer is masked by the final solder bump uniformity. A prototype tool for wafer IMS bumping technology has been developed and solder bumping has successfully been demonstrated with Sn-3.0Ag-0.5Cu solder on 200mm wafers. The test wafer employed interconnects pads of four different diameters and three different shapes. Other solder compositions have also been tried successfully.

Introduction

Since the solder bumps interconnection of flip chips was first developed nearly 40 years ago, various solder bumping methods and solder materials have been developed for C4 (Controlled Collapse Chip Connection, aka Flip Chip Solder Bump) on wafers. Each bumping method has its own advantages for certain applications. For example, the solder paste printing method is used for lower cost and large pitch products with relatively wide solder alloy composition flexibility. An electroplating method is often utilized for fine pitch products with rather limited alloy compositions such as pure Sn or a binary solder. Recently, the micro-ball mounting method has been widely used in manufacturing for fine pitch wafer bumping to enable advantages in both solder alloy composition flexibility and fine pitch, while also minimizing entrapped solder bump voids. Although the micro-ball mounting method may process spheres less than 100 microns in diameter, the throughput decreases and the ball cost becomes relatively higher as the solder ball size decreases further. Additionally, micro-ball mounting is not capable of forming a uniform solder bump height on a wafer having I/O pads/pillars with various size and shapes. Table 1 summarizes some major advantages/disadvantages and the technical limitations of these three solder bumping technologies.

Table 1. Comparison of three solder bumping technologies.

	Solder paste Printing [1]	Micro-ball mounting [2]	Electro-plating [3]
Pros.	- Low cost - Solder alloy flexibility	- Solder alloy flexibility with fine pitch capability - Minimum voids	- Easy to achieve fine pitch
Cons.	-Voids inside bumps - Bump height uniformity	- High cost of perform balls, masks, and processes	- Susceptible to Micro voids - Handling chemical waste
Technical limitation	Limitation in fine pitch bumping	- Question of manufacturing tool availability for bumping 25um dia. balls - Not applicable when one wafer has I/Os with various sizes and shapes	Only pure Sn or binary composition is possible

IBM developed a tool and process to directly inject molten solder for bumping on wafers. The fill head contains a reservoir of molten solder of desired composition and a slot through which the molten solder is injected with an optimized combination of pressure and temperature. Through careful control of the fill environment in low oxygen, the molten solder is injected into the holes in the resist material and it will wet and solidify on the pads without the use of flux or formic acid. After solidification of the solder bump, the resist material is then stripped from the wafer.

While the IMS process is in some ways very simple and similar to the stencil printing method, the advantage of IMS over the solder paste process is that it can make smaller size solder bumps with uniform volume for very fine-pitch applications (below 40 microns pitch) due to the use of pure molten solder without flux. The solder bump volume can be easily controlled for different applications by simply changing the resist material thickness and/or photolithography patterned hole sizes. In addition to the fine pitch application using IMS, the use of precise alloy compositions (ternary, quaternary solder alloys, and specially formulated alloys with doping elements) is possible by simply switching an IMS head having solder ingots of various target compositions inside. In contrast, it is generally difficult to form ternary or quaternary solder compositions using the electroplating method.

Another key advantage of IMS bumping is that IMS makes different solder volumes in a single pass and still maintains good co-planarity when the wafer has different sizes of I/O pads. The technology allows the design flexibility of applying larger volume interconnects for power joints and smaller volume interconnects for signal to meet both performance and reliability requirements.

This paper discusses the development challenges and solutions of wafer IMS technology, describing the process feasibility for 200mm wafer bumping.

Experiments

Figure 1 schematically illustrates an IMS procedure for wafer bumping including the electroplating of Cu pillars [4]. The electroplated Cu pillars are, ideally, substantially uniform in height, but in practice the heights of the pillars may not be uniform from channel to channel and one side of a pillar in a particular channel could be higher than another side. The IMS process can use the same resist material used for electroplating Cu and still compensate for non-uniformity in the Cu pillar height. This occurs as IMS fills up the remaining volume of the channels of the photoresist material after Cu plating.

A detailed description of the molten solder injection tool and processes can be found elsewhere [5, 6]. The solder injection slot of the IMS head is configured of a portion of compliant material and a low friction material [7] such that good wiping characteristics are allowed. In addition this design allows the IMS head to better track surface topography and warpage of wafers that it scans over, therefore no solder material is left on the surface of the resist material.

In this study, several different dry films and solvents had been evaluated for wafer IMS technology. The photoresist material for wafer IMS technology should be stable without burning or delamination during the injection process of molten solder. Also, it has to be strippable after solidification of solders by solvents. This means that the resist material must

include strong thermal resistance at Pb-free solder melting temperatures, but remain only partially cured before the stripping process.

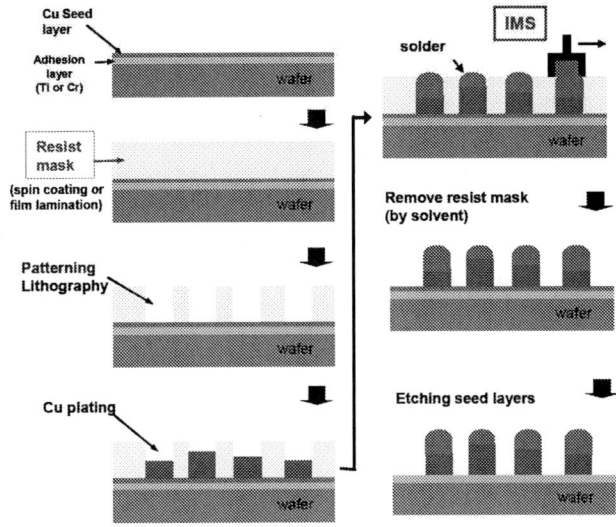

Figure 1. IMS wafer bumping steps of electroplating Cu, solder injection, and stripping of resist material to produce bumps of equal height when electroplated Cu pillars have non-uniform height.

Wafer IMS wafer bumping technology has been demonstrated on a single 200mm wafer having diverse pitches with several different opening sizes and various opening shapes of resist material. Also, wafer IMS were processed on 2, 20, and 40 microns Cu bumps for a different Cu/Solder ratio feasibility demonstration.

The solder material mostly used in this study is SAC 305 (Sn-3.0 wt% Ag-0.5 wt% Cu) due to its commonality. In addition, In-Bi-Sn solders and quaternary SAC solders with Zn were used to demonstrate a low melting temperature capability and to highlight the IMS solder alloy flexibility, respectively. The wafer IMS bumping technology has greater alloy flexibility because the target solder composition comes directly from solder ingots which are then injected from the IMS head.

Results and Discussion

Wafer IMS results on different types of dry films.

Table 2 shows a simple summary of wafer IMS results of SAC 305 solders on four different types of dry films. Depending upon the properties of these dry films, the IMS showed four different results. The first result for the type "A" resist film exhibited delamination during the IMS step due to low heat resistance. In the case of the second type "B", the molten solder smeared at the interface between the dry film and Cu seed layer as the film had low adhesion on Cu seed layer at solder melting temperature. When the dry films have high thermal resistance, there is no damage to them during the

IMS step. However, after IMS, some films are fully cured and can not be stripped by the solvent while the other films can be stripped by the solvent. This result is shown in the type "C" and "D" processes, respectively. Therefore, proper selection of the dry film and solvent is important for the success of IMS. Also, the change of dry film process parameters may alter the thermal resistance and the adhesion of dry films without fully curing them.

Table 2. Wafer IMS results difference based on types of dry films. The type D is acceptable for wafer IMS.

Dry film	IMS – SAC 305	Stripping dry film
Type A	Film delamination	N/A
Type B	Solders smear at film/Cu interface	O
Type C	O	X
Type D	O	O

Figure 2 shows wafer IMS results using type "D" dry film. As shown in Figure 2 (a), the type D dry film did not show delamination, nor solder leaking at the dry film/Cu interface. Also, the IMS head provides good wiping characteristics and there is not any solder residue on the surface of dry film after solder injection. This effective wiping characteristic enables IMS method to be extended to very fine pitch bumping without solder bridging. Also, since this type D dry film is not fully cured during the SAC305 IMS process, the solvent may easily strip the dry film after IMS. Figure 2 (b) shows SEM images of SAC 305 IMS solders on Cu pillar bumps after stripping the dry film. The dry film is stripped clearly without any residue and there is not any solder leaking at the interfaces.

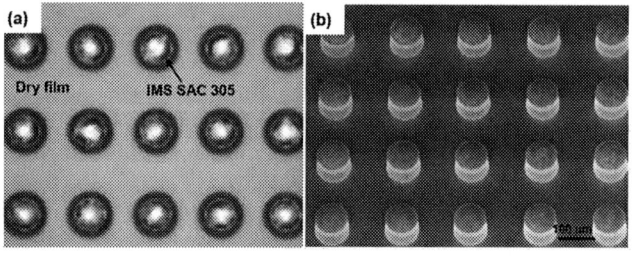

Figure 2. (a) Optical microscope image of top view of dry film type D after SAC 305 solder injection and (b) SEM image after stripping dry film, 30° tilted.

Wafer IMS using SAC 305 on two different Cu thicknesses.

Figure 3 shows, respectively, the wafers with 40 microns thick Cu pillars and 2 microns Cu UBM (Under bump metallurgy) in 60 microns thick dry film; after molten solder injection and solidification; and after the dry film has been stripped. The diameter of holes in the dry film is 80 microns.

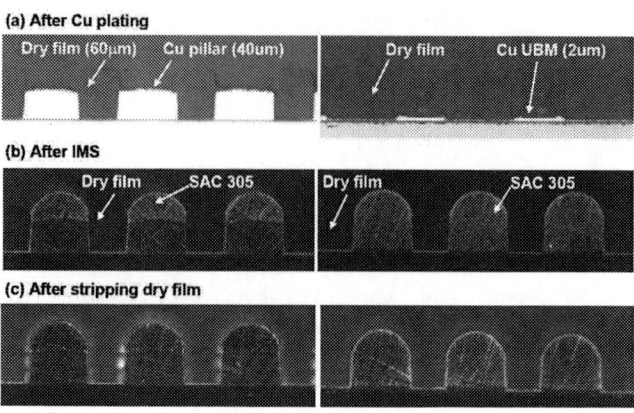

Figure 3. Cross sectional images of (a) 40 μm thick Cu pillars and 2 μm thick Cu UBM, (b) after IMS, and (c) striping of dry film resist material.

As shown in Figure 3, the molten solders were filled into the opening holes of resist material and the IMS bumps on the Cu have columnar shaped side walls with round tops. These column shape side wall solder bumps occurred due to solidification of molten solder inside the resist opening holes. Initially the molten solder was filled up at the top of resist material, but the surface tension of solder material formed the round tops of the bumps during the solidification of solder in a nitrogen environment. Therefore, the IMS method does not need an additional reflow process to form sphere shape bumps. In addition, pure molten solder injection without flux makes void-free solder bumps and the top height of solder bumps is higher than the top of resist material as shown in Figure 3 (b). This particular geometry of column side wall and round top achieved by IMS is unique and this IMS method can be applied to very fine pitch applications. Also, even though the solder volume differs greatly in these two cases, the throughput is identical because a single scan with the IMS head can cover the whole area of 200mm wafer.

Figure 4 shows SEM microscope images after stripping resist material. The resist material was stripped clearly by using the solvent after IMS bumping step. As shown in Figure 4 (b) and (d), the surfaces of Cu seed layer as well as solder and Cu bumps show very clean surface. There is not any leaking of molten solder at the interfaces of neither dry film/Cu seed layer nor dry film/Cu pillar bump. Also, as show in Figure 4(e), Sn-Cu intermetallic compounds were formed at IMS solder/Cu interface which means very strong adhesion between IMS bump and plated Cu. The consumption of Cu forming IMCs was a very minor amount.

978-1-4799-2408-0/14 $31.00 © 2014 IEEE

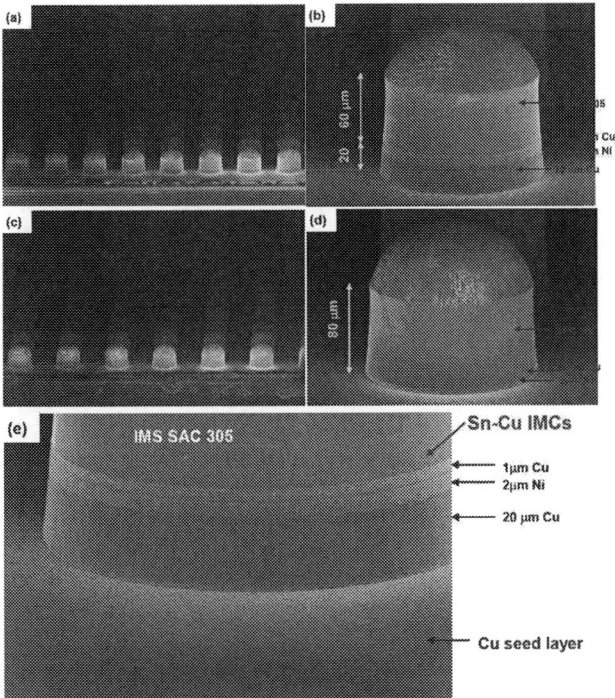

Figure 4. SEM images of IMS bumps on 1Cu/2Ni/20Cu and 1Cu/2Ni plated UBMs after stripping the dry films.

Wafer IMS prototype tool for 200mm wafers

Figure 5 shows a schematic diagram and photo image of wafer IMS prototype tool for 200mm wafers. This tool is composed of four stages, loading/unloading, pre-heating, solder injection, and cooling stages. These four stages enable the handling of four wafers at the same time inside the tool. The index table can rotate wafers form one stage to the other stages step by step. A wafer is loaded into the tool through the 1st stage and it moves to the 2nd stage for pre-heating. At that time, a next wafer is loaded at the 1st stage. The solder injection process is performed at the 3rd stage. At this stage, the IMS head moves from the right side to the left side and the molten solders are filled into the holes of resist material and wet on the pads of the wafer. As explained in experiments, the IMS head is provided with a compliant material which spreads the compressive force between the head and resist material, resulting in good contact between tool head and a wafer, such that gaps that could cause leakage are eliminated. Finally, the wafer is moved to the 4th stage for cooling down, and then the IMS bumped wafer exits through the initial stage. All these four stages are operated in a nitrogen environment and it serves to make the solder "ball-up" during the solidification of the injection molten solder. Thus, the presence of a nitrogen environment has the potential to eliminate the subsequent reflow process typically used in other cases such as electroplating, paste printing, and micro-ball mounting methods. This further reduces the number of process steps and thus cost.

The tool shown in figure 5 is a prototype tool which can handle up to 200mm size wafers. We are working on developing a manufacturing tool for 300mm wafers.

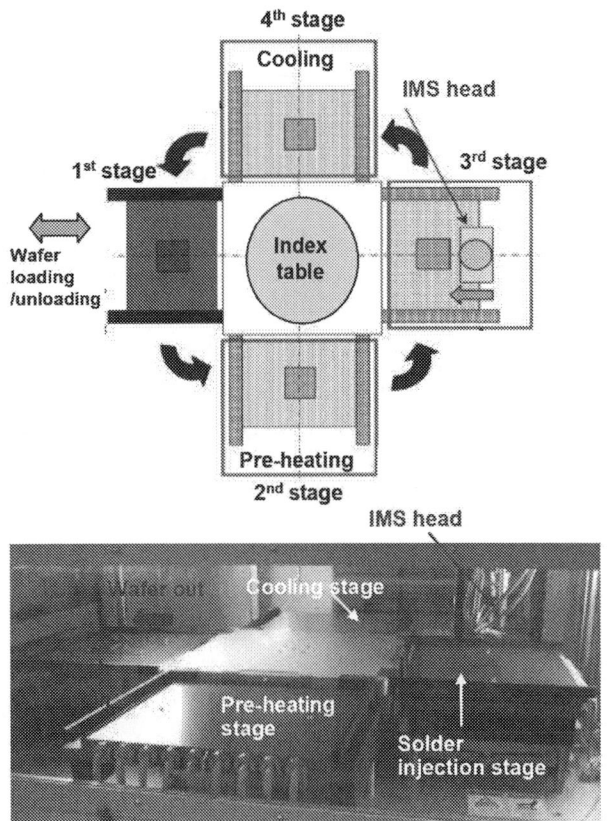

Figure 5. schematic diagram and photo image of wafer IMS prototype tool for 200mm wafers.

Wafer IMS bumping results on 200mm wafers

Solder bumping on 200mm wafers with diverse I/Os designs were accomplished in the wafer IMS prototype tool as shown in Figure 5.

One of the key advantages of wafer IMS technology is that the solder bump volume can be easily controlled for different applications by simply changing the resist material thickness and/or holes size. Also, even though a wafer has different pad opening size and/or different pad shape, uniform height of bumps can be achieved because the final bump height is decided by the resist material thickness.

Figure 6 shows one example of wafer IMS result on a 200mm wafer with several different pad opening sizes and shapes. As shown in Figure 6, wafer IMS can make various solder bump volumes and shapes in various pitches by using a single pass for one wafer and wafer IMS bumps still maintain good bump height uniformity. The solder bump volume/shape follows the volume/shape of holes opening in resist material, but the bump height is determined by the thickness of resist material. Since one wafer uses one thickness of resist material and 100% of the soldering happens without flux while solidifying within the resist material, the solder bump height is uniform independent of hole size/shape.

The wafer IMS process is simple and similar to the stencil printing method, but IMS can achieve higher volume solder bumps with uniform height for a given pitch and can be applied for finer pitch applications because only molten solder is used instead of solder pastes which have 50 volume % of flux.

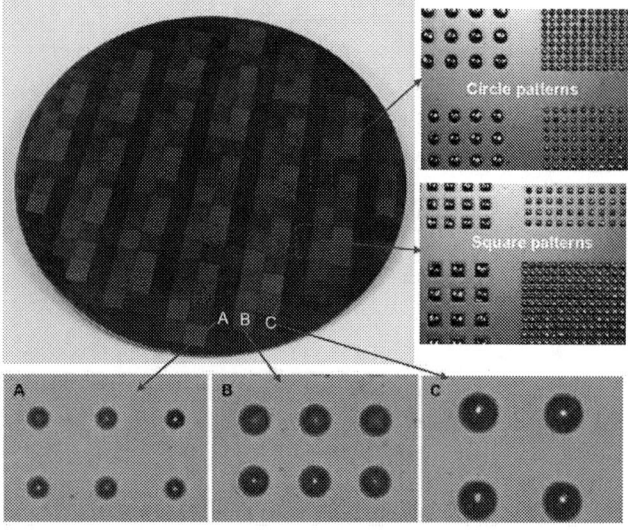

Figure 6. Wafer IMS bumping result on a 200mm wafer with resist material. There are several different pad pitch/opening sizes and shapes in one wafer.

Figure 7 shows one example of making diverse solder bump volume with uniform height in one wafer by using 50 μm thick resist material and varying the hole sizes with different pitch. As shown in Figure 7 (a), there are many different bump sizes/shapes in one wafer, no missing pads nor solder bridging was found; a 100 % solder bumping yield was achieved even though the diameter difference among bumps in a wafer is bigger than 10 times. Figure 7 (b) and (c) show that the resist material has been stripped clearly after IMS bumping and two bumps with different diameters of 50μm and 75μm show same height.

Wafer IMS bumped 200mm wafer after stripping resist material

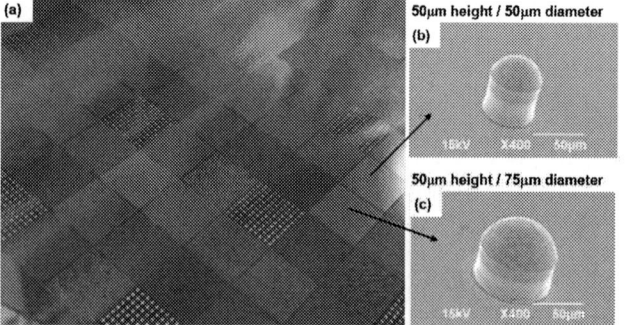

Figure 7. Wafer IMS bumping result on a 200mm wafer after stripping resist material.

Wafer IMS bumping results of various solder compositions

Wafer IMS technology has solder alloy flexibility which includes ternary, quaternary, or higher component alloy. Figure 8 shows IMS bumped In-Bi-Sn solder and SAC with small amount of Zn alloy. A solder vendor prepares the alloy by combining the component metals in the desired weight percentages and manufactures ingots for use in the head solder chamber. Then, changing the solder composition requires only that an operator swap to another head. Of course the process parameters, such as process temperature, needs to be optimized for each different composition of solder materials. Still, this head change in wafer IMS technology is very easy and fast compared to the other solder bumping methods. The wafer IMS eliminates solder waste by injecting only the required solder into the holes of resist material of each wafer.

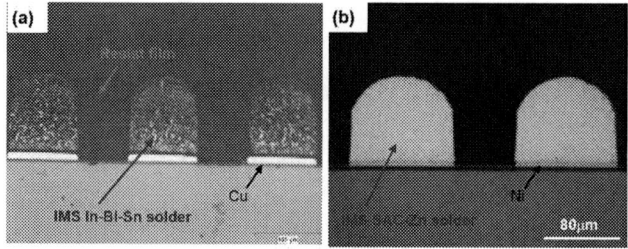

Figure 8. Wafer IMS bumping results of (a) In-Bi-Sn solder and (b) SAC-Zn solder alloys.

IMS technology for solder bumping on organic laminates

In general flip chip technology, the solder bump volume on the chip side is twice as large as those bumps on the organic substrate side. Therefore, the solder bump volume on the laminate side was not that big. IMS technology also can be applied to form bumps on the laminate side by using a similar method and tools which used for wafer bumping [8]. One advantages of laminate IMS is to form very high volume solder bumps on the laminates. For example, laminate IMS method can form full solder volume on the laminate side for making solder joints between a chip and a laminate, thereby eliminating the need for solder bumping on the chip side.

In the case of 3D packaging, the thinned 3D IC wafers usually require temporary bonding to a handler wafer (glass or silicon). If any wafer processing is needed on a 3D wafer after the solder bumping, the temporary bonding adhesive must be thick enough to accommodate these 60-80μm tall bumps. This will significantly increase the materials cost as well as the complexity in cleaning afterwards. Therefore, if the non-bumped pre-stacked 3D chips can be joined on the "full solder volume bumped" laminates as though it is a regular 2D chip, it could have many advantages in terms of cost and process simplicity. Detailed information about the bumping and assembly results of this laminate IMS for 3D packaging are to be presented in another paper [9].

978-1-4799-2408-0/14 $31.00 © 2014 IEEE   1312

Summary

We are presenting a new low cost solder bumping technology for use on wafers, the wafer IMS (injection molded solder) process. Four different types of dry film have been investigated and have demonstrated a successful wafer IMS bumping processes by achieving particular geometry of solder bumps with column side wall and round top. A low friction compliant material is used at the bottom of the IMS tool head and it prevents leakage of molten solder between the IMS head and the mask. The head material selection also allows the IMS head to better track surface topography and warpage of wafers that it scans over.

Wafer IMS technology demonstrates its advantages, including, but not necessarily limited to; very fine pitch bumping, solder alloy composition flexibility and solder volume/shape diversification with uniform height. In addition, all of these benefits may be available at a potentially low cost as a result of the process simplicity. Since only molten pure solder is used instead of solder paste, this method can achieve higher volume solder bumps for a given pitch, and can be used for fine-pitch applications in addition to solder alloy flexibility and void-free bumps. Also, the wafer IMS eliminates solder waste by injecting only the required solder into the holes of resist material of each wafer. We successfully built a prototype tool which can form fine pitch solder bumps on wafers, offering solder alloy flexibility and bump size variation on 200mm wafers.

The wafer IMS bumping technology has many advantages and is not just an incremental improvement to allow finer pitches with alloy composition flexibility, but is a fundamental change in the overall solder related interconnect strategy.

Acknowledgments

The authors would like to thank Dr. Eric Lewandowski for supporting InSnBi solder ingots and Dr. Sung Kang for supporting SAC+Zn solder ingots. We also would like to acknowledge management support from Dr. Ghavam Shahidi and Dr. Tze-Chiang of IBM research.

References

[1] Nah, Jae-Woong et al, "A strudy on coning processes of solder bumps on organic substrates," *IEEE Trans-EPM*, Vol. 26, No. 2 (2003), pp.166-172.

[2] Pang, Mengzhi et al, "Method of providning mixed size solder bumps on a substrate using a solder delivery head", US 7,569,471 B2.

[3] Lee, Ning-Cheng, "Reflow soldering process",Butterworth - Heinmann 2002, p.162.

[4] Nah, Jae-Woong *et al*, "Injection molded solder process for forming solder bumps on substrates", US 8,496,159 B2

[5] Nah, Jae-Woong *et al*, "Mask and mask-less injection molded solder (IMS) technology for fine pitch substrate bumping", Proceeding of IMAPS 2010-43rd International Symposium on Microelectronics, Raleigh, NC, October 31-November 4, 2010, pp.348-354.

[6] Nah, Jae-Woong *et al*, "Injection molded solder – A new fine pitch substrate bumping method", Proceeding of 2009 ECTC meeting, San Diego, CA, May 26-May 29, 2009, pp.61-66.

[7] Nah, Jae-Woong *et al*, "Micro-fluidic injection molded solder (IMS)", US 7,980,446 B2, US 8,136, 714 B2.

[8] Nah, Jae-Woong *et al*, "IMS (Injection Molded Solder) with two resist layers for forming solder bumps on substrates", US 20120318855.

[9] Dang, Bing et al, "Assembly and Packaging of Non-bumped 3D Chip Stacks on Bumped Substrates", Proceeding of 2014 ECTC meeting, Orlando, FL, May 27-May 30, 2014, will be published.

Reliability of Paste Based Transient Liquid Phase Sintered Interconnects

Hannes Greve, S. Ali Moeini, and F. Patrick McCluskey

CALCE/Department of Mechanical Engineering, University of Maryland, College Park, Maryland

CALCE/Department of Mechanical Engineering
Clark School of Engineering
University of Maryland, College Park, MD 20742
mcclupa@umd.edu

Abstract

In this paper we present Transient Liquid Phase Sintering (TLPS) as a technology that shows the potential for reliable interconnection in electronic systems under extreme temperature conditions. Ni-Sn and Ni-Cu-Sn TLPS sinter pastes have been developed that enable the formation of joints with a microstructure that is characterized by pure metallic particles (Ni, Cu) embedded in a matrix of (Ni,Cu)-Sn intermetallics. They can be processed at low temperatures, but possess high melting temperatures upon process completion. No application of vacuum or reducing atmosphere is required during processing. Good wetting capabilities are demonstrated for Cu-, Ni-, and Ag-metallizations and low levels of voiding are achieved. Depending on paste composition, the joints can possess high melting temperatures of above 600°C. Superior drop test reliability of Ni-Sn sinter joints compared to those formed with Sn3.5Ag solder is demonstrated.

Introduction

Electronic systems are increasingly used in harsh environments under elevated thermal conditions. This includes applications with high absolute temperatures as well as those with very large temperature swings. Some examples of these types of applications include deep well drilling, military and aerospace, or automotive products.

In automotive applications the integration of microelectronic systems in the engine compartment, referred to as the under-the-hood environment, increases the thermal load on the system due to heat dissipated by the engine during operation. With the commercial success of electric vehicles (EV) and hybrid electric vehicles (HEV), power electronic systems for energy storage and conversion are now used ubiquitously in automotive applications. With temperatures of the liquid coolant of 75°C and more, to ensure reliable operation of silicon (Si) power devices at temperatures below 150°C, power densities have to be limited, or considerable effort concerning the thermal management systems is required, e.g. a secondary liquid cooling loop. Additional cost as well as increased weight and volume requirements are associated with these efforts.

Another example for the increasing requirements electronic systems have to fulfill is the planned exploration of Venus. On its surface the atmospheric pressure is more than 90 times higher than on earth and the average temperature is 467°C and contains highly corrosive sulfuric acid.

Increasing power densities, the economic urge for miniaturization, and raised application environment temperatures necessitate the design of electronic systems for higher temperatures. Si devices are limited to low voltage operation at elevated temperatures and can generally not be used at temperatures above 175°C due to its material properties as a low band-gap semiconductor. Devices based on the wide band-gap (WBG) semiconductors (viz. silicon carbide (SiC) and gallium nitride (GaN)) have been introduced. They can be operated at higher temperatures, possess higher breakdown voltages, and enable higher switching frequencies with reduced switching losses. Multiple systems operating in temperature ranges up to 200°C and above have been presented. The operation of a DC-DC converter based on SiC JFETs and SBDs was successfully demonstrated at temperatures up to 450°C [1].

Yet the packaging and interconnect technologies applied in these microelectronic systems were developed for conventional Si-based devices with their limited application temperatures. This limits their temperature range for reliable operation to temperatures below 200°C.

- Eutectic Sn37Pb solders have melting temperatures (T_m) of 183°C, too low for high temperature applications. Additionally they show high creep rates at elevated temperatures.

- Eutectic Sn3.5Ag and SAC (Sn-Ag-Cu) alloys possess comparatively low T_ms of 221°C and 217°C respectively. They are frequently used as substitute materials for Sn37Pb solder (phased out by regulatory restrictions) due to their similar T_ms and wetting behavior. Like Pb-based solders they are highly ductile and soften close to their T_m, limiting their fatigue life. Furthermore they have been shown to form considerable IMCs with Cu metallization, leading to the formation of Kirkendall and Champagne voids and reduced joint strength.

- Bi-Ag solder alloys possess liquidus temperatures of 262°C. They exhibit brittle behavior with limited ductility and elongation, which can be slightly mitigated by increased ratios of silver at higher costs. They furthermore have limited wetting capabilities and low thermal conductivities.

- Solders with Pb as the main constituent (such as Pb5.0Sn2.5Ag) have T_ms of approx. 290°C. Their use is restricted by regulations for most applications with some exceptions such as military and as die attach. Their availability has reduced due to the shrinking market and the possibility of them being completely restricted in future version of regulations such as RoHS. They must be processed at comparably high temperatures.

- A few alloys with Au as a main constituent exist, among them Au20Sn, Au12Ge, and Au3.2Si with T_ms of 280°C, 361°C, and 363°C respectively. They form strong joints with good fatigue life, but must be processed at high temperatures. Furthermore the high costs associated with these alloys inhibit their use in most applications.

It can be seen that established technologies suffer significant drawbacks for reliable high temperature operation and alternatives are required. In this paper, we will introduce Ni-Sn and Ni-Cu-Sn sinter pastes as an interconnect technology for the Transient Liquid Phase Sintering (TLPS) of electronic joints. Their microstructures will be analyzed, their application temperature limits determined, and their toughness under drop conditions analyzed and compared to that of conventional interconnect technologies.

Transient Liquid Phase Sintering

Transient Liquid Phase Sintering (TLPS) is a liquid-assisted sintering process for the joining of metallic or metallized surfaces. During processing, the temperature is raised above the melting temperature of a metallic low melting-temperature constituent A, which melts causing the liquid A phase to surround and diffuse into a solid metal B with a high melting-temperature. The formation of intermetallic compounds (IMCs) is initiated. The diffusion and growth of IMCs continues until the liquid phase is completely consumed and all of A transformed to IMCs, which in the following is referred to as process completion. The joint now consists of pure B bridged by A_xB_x IMCs. Adequate amounts of B must be present to achieve full consumption of A and process completion.

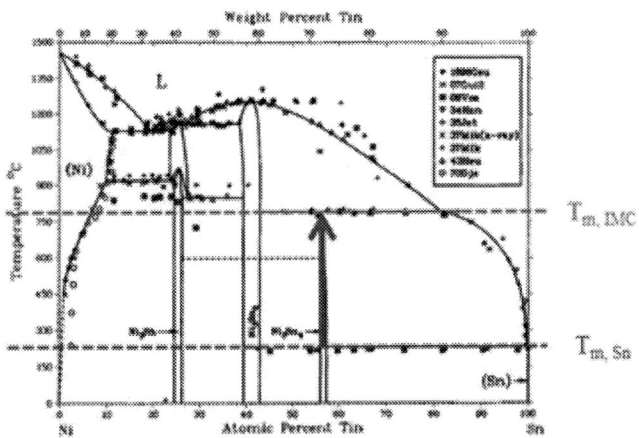

Figure 1. The Ni-Sn phase diagram, compare [2]

TLPS can be used to form high melting temperature joints at low process temperatures. This requires that the IMCs formed during sintering possess a melting temperature $T_{m,IMC}$ significantly higher that of the liquid phase $T_{m,A}$. Two examples of these are the Cu-Sn and Ni-Sn systems that are used in this work. Fig. 1 shows the Ni-Sn phase diagram. Sn

and Ni have melting temperatures of 232°C and 1455°C respectively. If the process temperature is raised above 232°C, the Sn melts. Solid-liquid interdiffusion is initiated, and Ni_3Sn_4 IMCs form and grow. If the percentage of Ni in the Ni-Sn joint exceeds 43 at-%, all Sn can be consumed by the formation of these IMCs. Upon process completion the joint consists solely of Ni and Ni-Sn IMCs with T_ms higher than 798°C. At this point the melting point has shifted by 566°C.

In this work, two types of TLPS sinter joints, based on Ni-Sn and Ni-Cu-Sn sinter pastes, are assessed. Table 1 summarizes the T_ms and the required minimum Ni/Cu percentages of the Ni_3Sn_4 IMC for the Ni-Sn system as well as those of the Cu_6Sn5 and Cu_3Sn IMCs for the Cu-Sn system. Ni_3Sn_4 is the IMC with the highest percentage of Sn in the Ni-Sn system and the one that initially forms during the TLPS process, compare Fig. 1. Cu_6Sn_5 is the first IMC that forms in the Cu-Sn system, and the one with the highest Sn percentage. It possesses a T_m of 416°C, which results in a melting temperature shift of 184°C. With additional annealing time at elevated temperatures, Cu_6Sn_5 IMCs can be transformed to Cu_3Sn IMCs, which possess a T_m of more than 640°C, depending on the Cu percentage.

Lower percentages of high melting temperature constituents facilitate the sintering process, as more liquid phase is present that can fill the voids between the solid surfaces. When comparing the weight percentages required for the formation of the three types of IMCs, it becomes clear that the Ni-Sn system has a distinct advantage in this aspect over the Cu-Sn system. Yet, Ni-Sn IMCs form considerably slower than Cu-Sn IMCs due to the lower diffusivity of Sn and Ni compared to that of Sn and Cu. This might decrease the time required for process completion of Cu_6Sn_5 sinter joints compared to Ni_3Sn sinter joints. The formation of Cu_3Sn IMCs on the other hand will require extensive annealing time, or an increase of process temperature to above 416°C to melt the Cu_6Sn_5 IMCs.

IMC	Ni_3Sn_4		Cu_6Sn_5		Cu_3Sn	
T_m	798°C		416°C		> 640°C	
	Ni	Sn	Cu	Sn	Cu	Sn
at-%	43	57	55	45	75	25
wgt-%	28	72	40	60	62	38

Table 1. Melting temperatures of Ni_3Sn_4, Cu_6Sn_5, and Cu_3Sn IMCs and their associated at-% and weight-% percentages

The majority of work performed on TLPS systems for electronic applications has focused on the joining of two layers of a high melting temperature metallization by consuming a layer of low melting temperature constituent sandwiched between them, see Fig. 2. A summary of the work that has been performed on TLPS is given in the literature [3]. The IMCs grow from the metallized layers of B to the center line of A. The IMC thickness is identical to the joint thickness. The Disadvantage of this approach is that it is limited to thin joint thicknesses of 10µm or less. Fick's second law of diffusion states that the diffusion length, which, in this case, corresponds to the thickness of IMCs, is proportional to the square root of time. Thicker joints therefore require considerable process times or high process temperatures to accelerate diffusion.

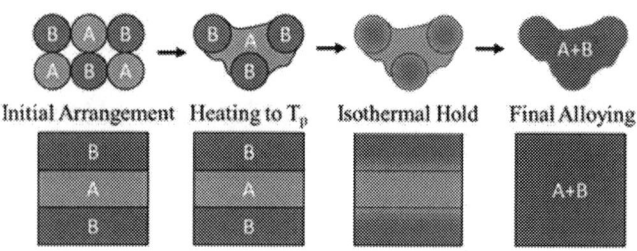

Figure 2. Formation of TLPS joints with a sinter paste (top) and by a layer-based approach (bottom). *A* and *B* are the low- and high-melting temperature constituents respectively

Sinter pastes on the other hand enable the simultaneous and homogeneous formation of IMCs throughout the joint, see Fig. 3. As particles of the high melting temperature constituent are present throughout the joint, IMCs are formed at multiple interfaces. The required individual IMC thicknesses are small compared to the layer-based approach; the joint IMC is formed by multiple IMCs bridging the individual high melting temperature particles. This can reduce the process completion time by orders of magnitude. It can furthermore be reduced, if excess high melting temperature material is present in the sinter paste, and IMCs are only required for forming small bridges between these particles. This is the case for most joints assessed in this work, see below.

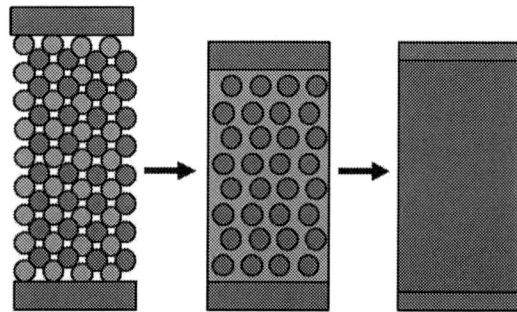

Figure 3. Schematic of a paste based TLPS process: Homogenous and simultaneous IMC formation

Processing of Transient Liquid Phase Sinter Pastes

The procedure for the preparation of sinter pastes is described in detail in [3] and will here be explained briefly for the convenience of the reader:

1. Mixing of high and low melting temperature particles by predefined weight ratios
2. Blending of the dry powder mix into an organic binder material to form sinter paste
3. Stencil printing of sinter paste on substrate.
4. Placing of a die and processing

The sinter pastes designed in [3] were optimized for pressure-less sintering. To reduce joint voiding in this work, a low pressure < 0.5MPa was applied to the die during sintering. Whereas for the pressure-less pastes, Cu-substrates and Cu-

dice were used, in this work, Ni with a side length of 6.35mm as well as power diodes with Ag-metallization and a side length of 7mm were joined to Ni plates. The different metallization types enable an assessment of the wetting behavior of the TLPS process for Ni, Cu, and Ag surfaces.

Ni-Cu-Sn and Ni-Sn sinter joints show process completion after less than 10 and 30 minutes above the melting temperature of Sn, respectively. The joints manufactured for this work were sintered for 30 minutes at 300℃ to ensure process completion for both types of pastes. Fig. 4 shows the sinter process temperature profile. It includes two isothermal steps: The first step is a 5 minute hold at 180℃, below the Sn melting temperature, to activate the organic binders. The second step is the 30 minute hold at 300℃ during which the Sn melts, and the TLPS process is initiated. The sinter pastes consisted of Ni, Cu, and Sn3.5Ag solder spheres.

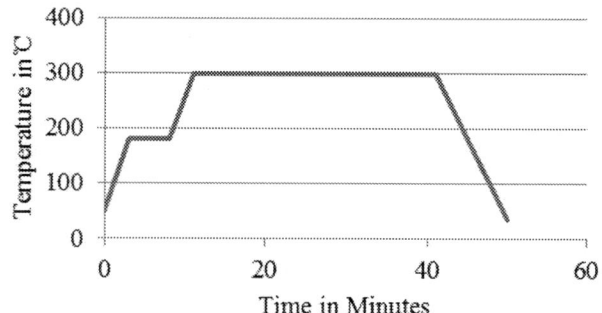

Figure 4. Sinter process temperature profile

Microstructural Analyses

Fig. 5 and Fig.6 show cross-sections of a Ni-Cu-Sn sinter joint at magnifications of x54 and x100 respectively. Very low voiding is present within the joint; only one void of significant size is visible. The sinter paste shows excellent wetting capabilities and has formed continuous IMCs bonding substrate, die, and Cu and Ni particles of the sinter paste.

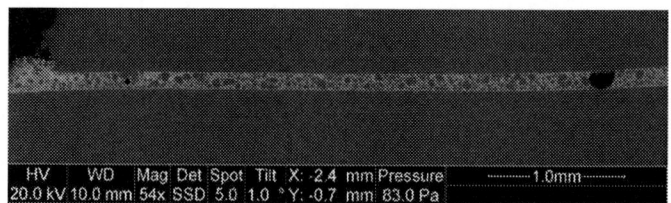

Figure 5. Cross-section of a Ni-Cu-Sn interconnect joining a Ni-substrate and a Ni-die; magnification x54

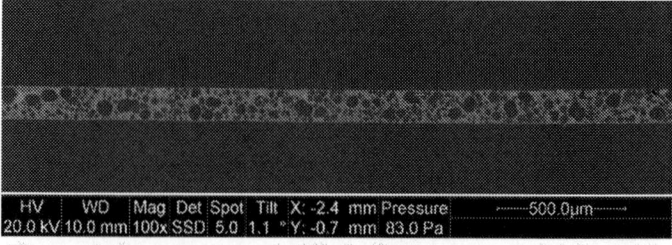

Figure 6. Cross-section of a Ni-Cu-Sn interconnect joining a Ni-substrate and a Ni-die; magnification x100

Fig. 7 shows a detailed cross-section of the Cu-Ni-Sn sinter joint. A complex microstructure has formed. Full wetting of substrate and die surfaces is visible. Residual Ni and Cu particles that have only been partially consumed are present in the joint. Ag_3Sn IMCs are present throughout the joint. They originate from the Sn3.5Ag solder used as low melting temperature constituent. As the Cu-Ni-Sn IMCs grow, the free space for the Ag_3Sn IMCs becomes more constrained, and small particles coalesce.

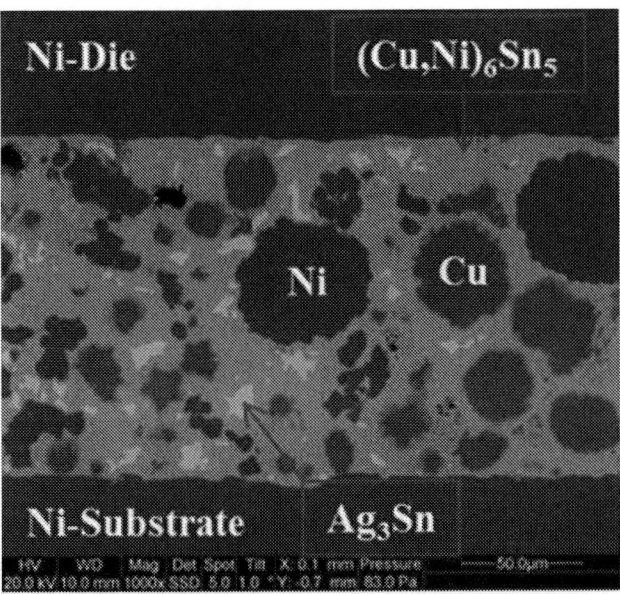

Figure 7. Cross-section of a Ni-Cu-Sn interconnect joining a Ni-substrate and a Ni-die; magnification x1000

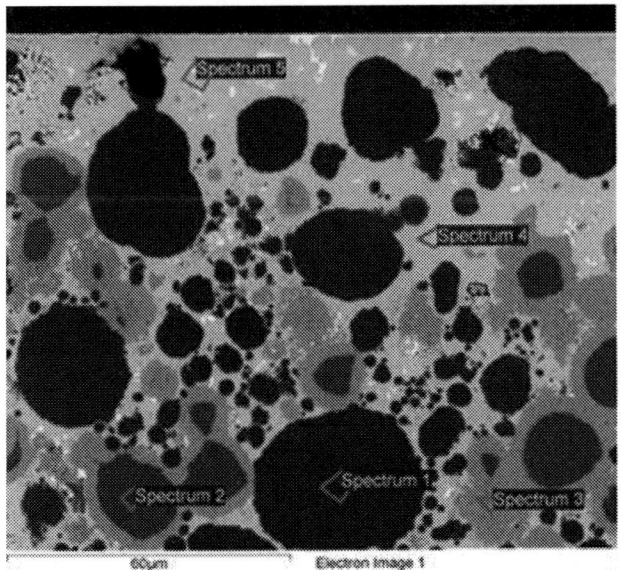

Figure 8. Cross-section of an EDS analysis performed on a Cu-Ni-Sn sinter joint

Fig.8 shows a cross-section of a Cu-Ni-Sn sinter joint that has been used to analyze its material composition. The dark gray spherical particles of spectrum 1 and spectrum 2 were identified as Ni and Cu respectively. Ni-Cu-Sn IMCs with two different stoichiometries can be found: A slightly darker

$(Cu,Ni)_3Sn$ IMCs (Spectrum 3), and a brighter $(Cu,Ni)_6Sn_5$ IMC (Spectras 4 and 5). The percentage of Ni:Cu in these IMCs was found to be between 1:6 and 1:3. The size, distribution, and Ni-to-Cu ratios are expected to change under conditions of significant additional diffusion, e.g. thermal cycling with high dwell temperatures and thermal annealing.

Fig. 9 shows a cross-section of a Ni-Sn sinter interconnect joining a silicon power diode with an Ag-metallization and a Ni-substrate. Ideal wetting of Ni and Ag can be seen. As with the Ni-Cu-Sn joint, low levels of voiding are present throughout the joint. EDS analyses confirmed that two types of IMCs are present. Ag_3Sn IMCs from residual Ag of the Sn3.5Ag solder, and Ni_3Sn_4 IMCs.

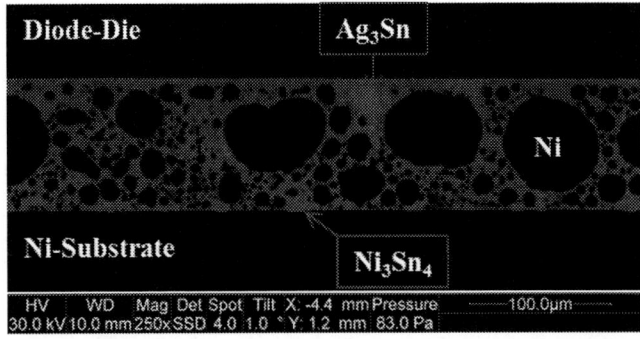

Figure 9. Cross-section of a Ni-Sn interconnect joining a Ni-substrate and a die of a Si power diode; magnification x250

Softening Behavior

To ensure long fatigue life under elevated thermal conditions, interconnect materials need to possess good strength at high temperatures. A high temperature shear test has been developed; details on the setup and shear analysis specimens can be found in [3]. A test procedure for the analysis of the joint softening behavior has been developed. A shear load of 10MPa is applied to the joints. The temperature is increased continuously. The temperature is detected when a drop in shear strength below 10MPa occurs. This temperature is designated the softening temperature.

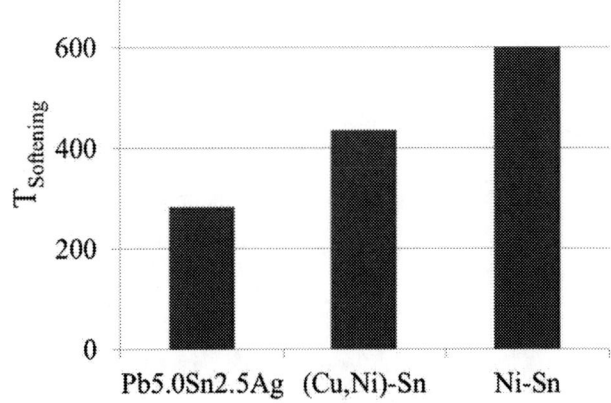

Figure 10. Softening temperature of Pb5.0Sn2.5Ag, Ni-Cu-Sn, and Ni-Sn joints; criteria is a drop of shear strength below 10 MPa

Fig. 10 shows the softening temperature of a Pb5.0Sn2.5Ag high temperature solder, a Ni-Cu-Sn, and a Ni-Sn sinter joint. The softening temperature of Pb5.0Sn2.5Ag was detected as 282°C, which is in good agreement with its solidus temperatures of 287°C. Ni-Cu-Sn joints soften at 435°C, which is close to the melting temperature of Cu_6Sn_5 IMCs, which is 416°C. No reduction of shear strength could be detected for the Ni-Sn sinter joints within the test setup temperature limit of 600°C.

Mechanical Shock Tests

The ability of an interconnect material to survive dynamic loads is a crucial factor for reliability of electronic packages. From prior vibration and shock tests of solder interconnects it is generally known that joint failure occurs at or within the IMCs that form between the solder and the substrate during processing or operation. It has therefore been hypothesized that the IMC layer is the limiting factor for the reliability of interconnects under dynamic loads.

As shown above, IMCs are main constituents of TLPS sinter joints. If the reliability of interconnects is limited by the amount of IMCs, TLPS interconnects would be expected to exhibit limited life time under dynamic drop loads. The formation of intermetallic in these joints thus requires more analysis to qualify interconnect reliability under dynamic loading conditions [4].

Most of the consumer electronics manufacturers focus on the mechanical reliability of component to board connections (solder joints). There are standard methods available to assess useful life of these connections, such as JEDEC Standard: JESD22-B111. However, there is no specific standard method which includes the mechanical reliability evaluation of die attach materials. In most cases, it is expected that the component to board connections fail faster than the die attach material because of the smaller contact area. Increasing the intermetallic content of the die attach materials in transient liquid phase sintering joints requires more investigation into the behavior these joints under mechanical shock loads. In this study, drop shock test is considered as a qualification method to simulate highly accelerated mechanical loads on TLPS sinter joints.

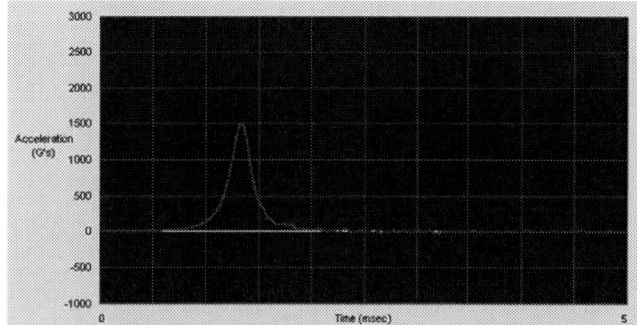

Figure 11. Drop shock test acceleration over time

Due to the lack of a standard method to assess mechanical reliability of die attach materials under drop shock loading, a similar method to Board Level Drop Test Method of Components for Handheld Electronic Products (JEDEC Standard: JESD22-B111) and the qualification method for TLPS adhesive composites is used [5]. For this purpose, a drop tower capable of providing 1,500 G is used. The acceleration pulse duration is 0.5 milliseconds and it follows a half sine curve similar to JEDEC Standard requirements, Fig. 11. To mount the samples on the drop tower an aluminum fixture was designed. Fig. 12 shows the fixture and how it is mounted on the drop tower. Samples are prepared by attaching a metal die on metal strips with Ni-Sn sinter pastes and Sn3.5Ag solder, as shown in Fig. 13. To define the reliability of the samples some exploratory tests were designed.

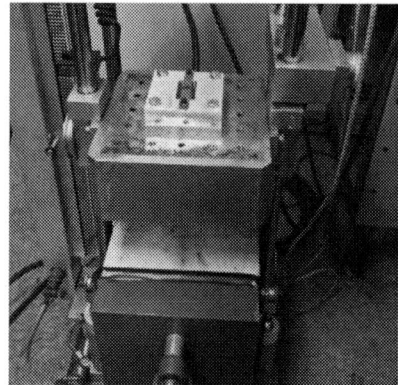

Figure 12. Sample fixture mounted on drop tower

Figure 13. Ni-Sn TLPS drop test sample

The Ni-Sn sinter specimen was tested for a combination of 500 cycles under 1,500 G (i.e. drop from 19 inches) and 10 drops from 62 inches (i.e. at higher G loading). The samples were cross-sectioned after the shock cycles and investigated for initiated cracks. Environmental Scanning Electron Microscope (ESEM) was used to detect the cracks.

Fig. 14 shows the ESEM image of a Ni-Sn TLPS sinter joint after drop cycling. The image shows two cracks initiated at the right corner of the joint at the corner of the die. Both cracks seem to start in the solder material region. The first crack propagated on the edges of intermetallic; however, the second one went through the intermetallic and stopped at a void. Fig. 15 shows small cracks formed near a nickel particle at the die corner. No other cracks were observed in the cross section.

A test sample made of Sn3.5Ag solder was subjected to the same loading conditions and then cross-sectioned analyzed using ESEM to assess joint condition. The solder joint condition is presented in Fig. 16. A crack started at the edge of the die and propagated though the interface of solder material and Ni substrate. In this figure, it is obvious that the solder material crack growth under mechanical drop shock loading is faster compared to that of Ni-Sn sinter pastes; the joint was close to detachment. Figure 7 shows the other corner of Ni

978-1-4799-2408-0/14 $31.00 © 2014 IEEE   1318

die. It is obvious that another crack is initiated at this corner, too, and is propagating through the joint material. It is interesting to note that in both cases the crack initiation location was not the IMC layer, but the solder. It might be that some IMCs are the location of crack propagation, but that sintered joints comprising (Ni,Cu) Sn IMCs with metallic particles, can prohibit the initiation and blunt the propagation of cracks and thereby improve reliability under dynamic loads.

These results indicate that Ni-Sn TLPS sinter joints possess higher reliability under drop loading. However, the exploratory nature of this study necessitates that further study is required to define the reasons behind this behavior.

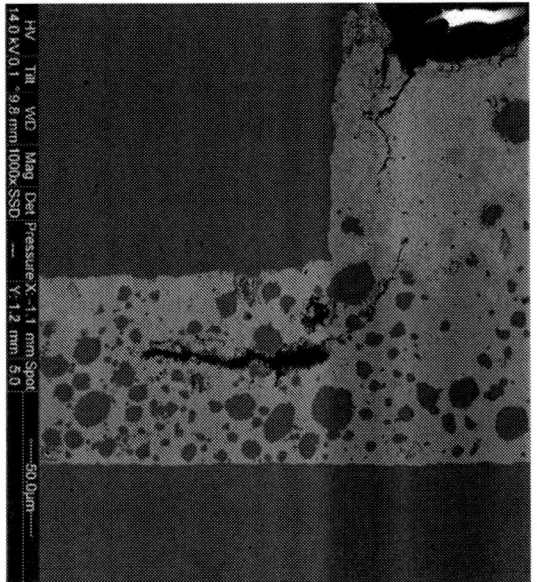

Figure 14. ESEM image of Ni-Sn sinter joint between Ni-die and Ni-stripe) after drop tests; magnification x1000

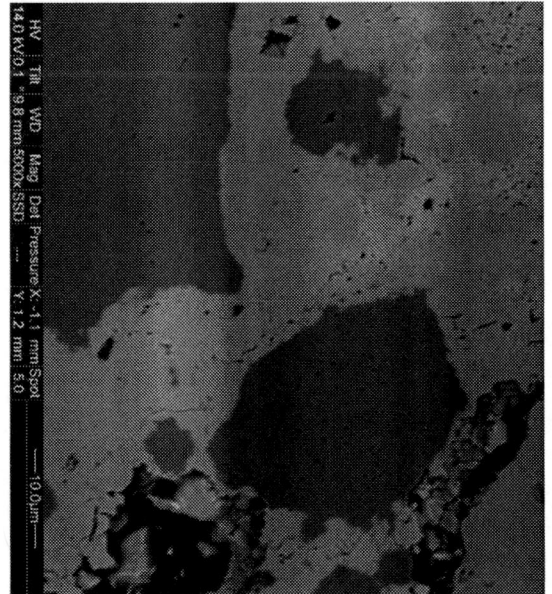

Figure 14. ESEM image of Ni-Sn sinter joint between Ni-die and Ni-stripe) after drop tests; magnification x5000

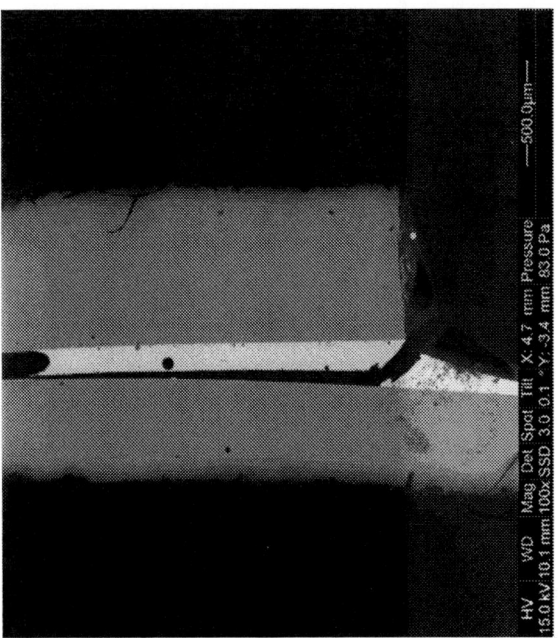

Figure 14. ESEM image of Sn3.5Ag solder joint between Ni-die and Ni-stripe) after drop tests; magnification x1000

Conclusions

In this paper we introduced Ni-Sn and Ni-Cu-Sn Transient Liquid Phase Sintered (TLPS) joints as an interconnect technology for electronic packaging, which shows the potential for reliable operation in environments with extreme thermal or mechanical loads. Sinter pastes were developed that enable to formation of joints at low process temperatures that are capable of high strength operation at elevated temperatures. They show limited voiding and excellent wetting capabilities on different metallizations common in electronic systems, such as Cu, Ni, or Ag. It was demonstrated that interconnects manufactured with these pastes exhibit high strength up to temperatures of 600°C, depending on paste composition. Cyclic drop tests show that Ni-Sn sinter joints show better drop test reliability for high loads than Sn3.5Ag solder joints, which is counterintuitive to the idea that the formation of large IMCs is detrimental to dynamic load fatigue life.

Acknowledgments

The authors would like to thank the members of the Center for Advanced Life Cycle Engineering (CALCE) at the University of Maryland for their support of this work.

References

1. T. Funaki, J.C. Balda, J. Hunghans, A. S. Kashyap, H.A. Mantooth, F. Barlow, T. Kimoto, T. Hikihara, "Power Conversion with SiC Devices at Extremely High Ambient Temperatures", *IEEE Transactions on Power Electronics, vol. 22, no. 4,* 2007
2. P. Nash, A. Nash, „The Ni-Sn (Nickel-Tin) System", *Bulletin of Alloy Phase Diagrams,* 1985
3. Greve, H., McCluskey, F.P., "Transient Liquid Phase Sintered Attach for Power Electronics", *63rd IEEE Electronic Components and Technology Conference (ECTC),* 2013

4. H. K. Kim, H. K. Liou, and K. N. Tu, "Three-dimensional morphology of a very rough interface formed in the soldering reaction between eutectic SnPb and Cu," *Appl. Phys. Lett.*, vol. 66, no. 18, pp. 2337–2339, May 1995.

5. C. Shearer, B. Shearer, G. Matijasevic, and P. Gandhi, "Transient liquid phase sintering composites: Polymer adhesives with metallurgical bonds," J. Electron. Mater., vol. 28, no. 11, pp. 1319–1326, Nov. 1999.

A Lead Free Joining Technology for High Temperature Interconnects Using Transient Liquid Phase Soldering (TLPS)

Christian Ehrhardt[1], Matthias Hutter[2], Hermann Oppermann[2], Klaus-Dieter Lang[1]

[1] Technical University of Berlin
[2] Fraunhofer IZM
Gustav-Meyer-Allee 25, 13355 Berlin, Germany
E-mail: christian.ehrhardt@izm.fraunhofer.de, phone: +49 (0)30 46403-159

Abstract

This Paper reports an emerging lead-free joining technology for high temperature application, which can be used for operating temperatures above 200 °C. It is called: "Transient Liquid Phase Soldering (TLPS)". The TLPS paste used contains a tin-copper powder mixture and is almost completely transformed into Cu_6Sn_5 and Cu_3Sn intermetallic phases after soldering. Due to the reaction between the liquid tin and the copper powder a skeleton of intermetallic phases are formed immediately during soldering and prevents the paste from collapsing so that a lot of voids remain in the solder line. The challenge for this investigation was to understand the mechanism of the skeleton formation, describe them in detail and find possibilities to avoid the skeleton formation. In this paper a new TLPS paste and two processes are described as a means to manufacture an almost void-less joint. Furthermore, a model was developed that describes the TLPS process in detail. The activation of the TLPS joint is crucial and will be described. Temperature cycling results, failure mechanisms and conclusions to increase the lifetime as well as reliability of such TLPS interconnects will be presented in this paper.

Introduction

The requirements for highly conductive thermally stable and reliable joining technologies in the field of power electronics, LED lighting and board interconnection have sharply risen in recent years [1]. It is well known that Sn-based solder alloys like SnAg3.0Cu0.5 (SAC305) are the most common lead-free solders. Lead-free Sn-based solders reach their thermal and mechanical load limit due to the ever increasing power densities and rising life-time requirement. It is known that a longer life time and a higher reliability can be obtained by adding certain elements into the solder alloy [2]. The maximum operating temperature can be increased up to 165 °C [3]. For even higher operating temperatures the use of tin based solders is limited.

Therefore novel interconnection technologies have to be developed to withstand operating temperatures above 165 °C. This requires joints that do not fail due to thermo-mechanical fatigue. Therefore, an emerging lead-free joining technology for high temperature stable interconnects has been developed, referred to as "Transient Liquid Phase Soldering (TLPS)".

The Transient Liquid Phase Soldering is based on a phase transformation using a low melting powder and a high melting powder [4]. During reflow the powder with the low melting point melts, wets and reacts with the high melting powder forming the intermetallic phases. The intermetallic phases have a significantly higher melting point compared to element with the low melting point. They are brittle and have a high

strength. For using Transient Liquid Phase Soldering as an alternative joining technology for high operating temperature applications different requirements should be complied with. In general it is necessary that a material system is used whereas the low melting element in the liquid phase has a certain solubility for the element with the higher melting point. The low melting powder should have a melting point that is in the temperature range of typical solder alloys, which should be below 250 °C. Another requirement is a high diffusion rate of the high melting element into the liquid solder so that the intermetallic phases can quickly form and grow. Other points are the cost of the material, the environmental compatibility and the processability of the elements. The Cu-Sn system meets the specific requirements best, so that this material combination is the preferred one.

In our first studies few years ago it was found that during soldering the liquid tin wets the copper powder and reacts immediately into a skeleton formation of Cu_6Sn_5 and Cu_3Sn intermetallic phases, as shown in Figure 1. Due to the skeleton structure in Figure 1 on the right hand side the TLPS paste cannot collapse so that a lot of voids remain after soldering [6]. This skeleton formation was also observed in the cross section images by Greve et al. who manufactured TLPS joints using a variation of three tin-copper sinter pastes (Cu60Sn, Cu50Sn and Cu40Sn) [5].

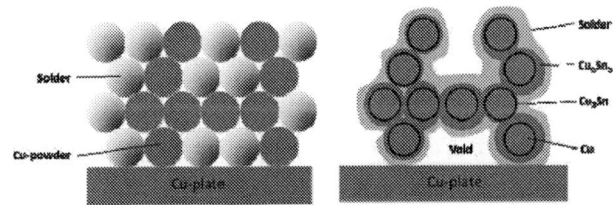

Figure 1: Schematic drawing of the intermetallic phase skeleton formation using a TLPS paste with Sn-based solder powder, Cu-powder and flux

Another option was to fill the voids with an epoxy material. Ormet developed such a paste for joining vertical interconnects in multilayer structures [7]. After soldering a temper step follows to cure the polymeric binder. For high temperature applications such an epoxy filling of the void formation is unsuitable, because the polymer has a substantially lower thermal conductivity compared to metal joints and the epoxy will decompose at high temperatures. In addition, it is also possible that cracks can prematurely initiate due to the much higher CTE of the polymer.

In this paper a new paste and process is described to manufacture an almost void less full metal TLPS joint. The thermal conductivity of such a TLPS joint is higher compared to Greve [5] and Gallagher [7], the re-melting temperature is approximately 180 K higher than the solder temperature and no fatigue of the TLPS layer can be observed as in conventional solders.

Model for a void-less TLPS joint

The skeleton formation can be avoided by delaying the reaction between the liquid solder and the Cu-powder. Figure 2 shows a TLPS modeling description for a void-less and almost completely transformed intermetallic joint.

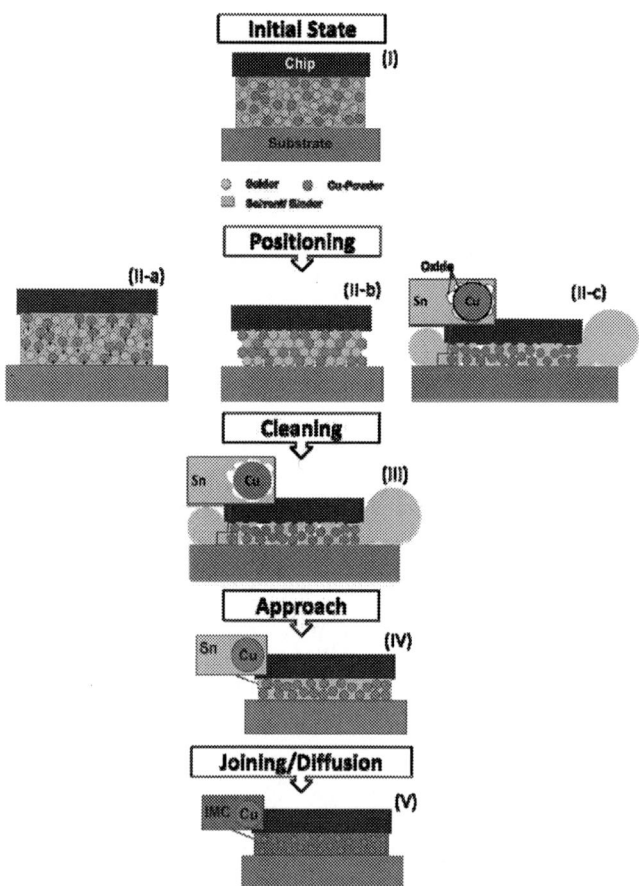

Figure 2: TLPS modeling description for a void-less and almost completely transformed intermetallic joint

The TLPS paste contains a solder material like Sn-based solder powder, a Cu-powder and a solvent without reducing properties. This paste is printed on a substrate and a die is placed on the top (Figure 2, initial state). The next step is called positioning. Positioning means to bring the die and the Cu-powder in the final position, before the reaction between the liquid tin and copper powder starts (Figure 2, positioning). A movement of the Cu powder and the die is not allowed during soldering. Subsequently the cleaning step takes place (Figure 2, cleaning). The existing oxide-layer must be removed so that the liquid tin can wet the copper powder in the next step (Figure 2, approach). In the last step the tin and copper react to form the Cu_6Sn_5 and Cu_3Sn intermetallic phases (Figure 2, joining/ diffusion). This model could be applied as a two-step process, which will be explained in the following.

Experiment

TLPS-paste and 2-step process

The TLPS paste contains a Cu powder, a Sn-based solder powder for example SAC405 solder and a solvent without reducing properties. This means that any kind of flux inside the TLPS paste is not allowed otherwise the skeleton will be formed, as shown in Figure 1. The amount of Cu powder and the diameter of the Cu spheres significantly influence the quality of a TLPS joint. A large Cu surface, which is equivalent to a small copper powder diameter, promotes a formation of intermetallic phases. An agglomeration of Cu-powder could be observed using Cu-diameters less than 5 μm. Copper powder diameters of 8-45 μm showed the best results in the experiments. Larger diameters are not well printable and tend to form stand-offs. The amount of Cu powder depends on the formation of intermetallic phases. Cu_6Sn_5 and Cu_3Sn are the existing intermetallic phases in the temperature range up to 300 °C. Assuming that Cu_6Sn_5 is the most dominated intermetallic phase after soldering, six copper and five tin atoms are needed. The copper powder weight fraction can be determined according to Equation 1. The molar mass of copper $m_A(Cu)$ with 63.546 g/mol and of tin $m_A(Sn)$ with 118.71 g/mol were used.

$$w\%(Cu) = \frac{6 \cdot m_A(Cu)}{6 \cdot m_A(Cu) + 5 \cdot m_A(Sn)} \quad (1)$$

$$w\%(Cu) = \frac{6 \cdot 63{,}546\,g/mol}{6 \cdot 63{,}546\,g/mol + 5 \cdot 118{,}71\,g/mol}$$

$$\underline{w\%(Cu) = 0{,}391}$$

Mathematically, there is a copper powder weight fraction of about 39.1 percent. Due to the mechanical pressure in the first process step (refer to Figure 2, step II-c), the copper powder in the TLPS paste will be compressed so that less than 39.1 w% Cu powder is needed. First tests showed that approximately 30 to 35 w% Cu powder are sufficient. The solvent comprises an alcohol and a thickener to make a printable TLPS paste.

After some investigation, a two-step process was developed, which fits well with the just described model in Figure 2. The TLPS paste is printed on a substrate and a die is placed on top. In the first process step the TLPS paste is compressed applying a certain pressure. The maximum soldering temperature lies above the melting point of the SAC405 solder. During heating the solder gets liquid and due to the pressure applied is partially pressed out of the solder line. Owing to the fact that the solvent has no reducing substances almost no reaction between the liquid Sn and the Cu powder or the Cu surfaces takes place. After the first step the die adheres to the substrate but is only partly connected to the substrate. In Figure 3 the joint gap after the first process step is illustrated.

978-1-4799-2408-0/14 $31.00 © 2014 IEEE

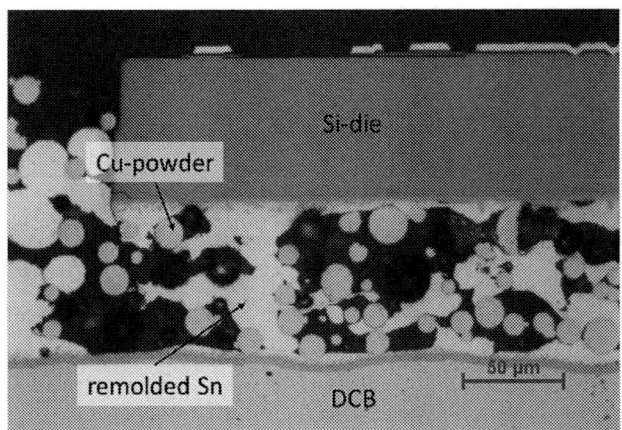

Figure 3: Cross section SEM image of a TLPS interconnection after the first process step

In the second process step the activation of the joint is performed. Activation means that in the porous interconnection gap the surface oxides are reduced. The process time of the activation step depends on the size of the die, the activation gas used (CO, SF6, HCOOH) and the flow rate of the process gas. As soon as all parts of the joint gap are activated a second reflow is performed. The liquid solder that is partly located outside of the solder line after the first process step infiltrates the voids in the joint gap and reacts with the Cu to form the Cu_6Sn_5 and Cu_3Sn intermetallic phases. In Figure 4 a cross section image after the second TLPS step is shown. The Cu plates are connected by a continuous layer of intermetallic phases. There are only a few small existing areas of pure tin, which can be transformed in intermetallic phases in a subsequent heat treatment or during soldering. Therefore it can be concluded that using the Cu/Sn system the process parameters during the soldering process and the geometry of the Cu spheres are the most important factors to achieve a total phase transformation.

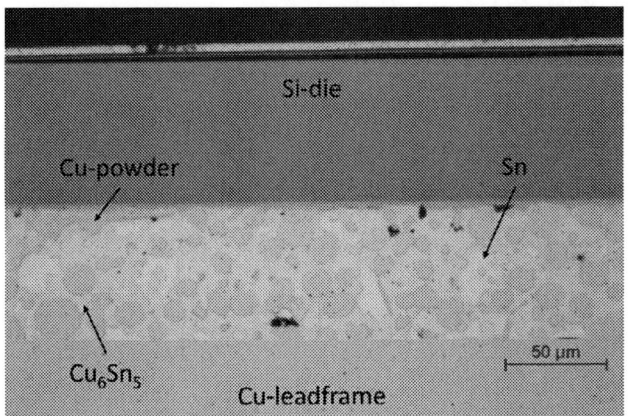

Figure 4: Cross section SEM image of a TLPS interconnection after the second process step

TLPS joint made by lateral infiltration of pure Cu-powder using Sn-based solder preform

A second option how to manufacture a void-less TLPS joint having a re-melting temperature of about 400 °C is based on an infiltration effect of the melted solder [8]. In the

schematic drawing in Figure 5 the concept of the development is shown.

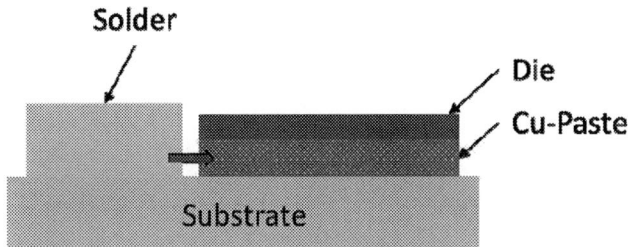

Figure 5: Schematic drawing of the lateral infiltration process using a pure Cu powder and a Sn-based solder preform

The Cu paste is printed on a substrate and a die is placed on top. The solder deposit is located next to the chip. The Cu paste contains pure Cu powder only and a flux-free solvent. After placing the die on the substrate the oxide layers have to be reduced. Subsequently, the solder melts, infiltrates the space between the Cu powder and immediately reacts to form the Cu_6Sn_5 and Cu_3Sn intermetallic phases without forming large voids in the solder line. In Figure 6 a cross section image of a TLPS joint between an IGBT and a DCB with Ni/Au-finish is shown using the lateral infiltration mechanism as illustrated in Figure 5. The TLPS layer predominantly consists of copper powder and Cu_6Sn_5 intermetallic phase which suggests a low creep resistance and a high thermal conductivity compared to a tin based solder joint.

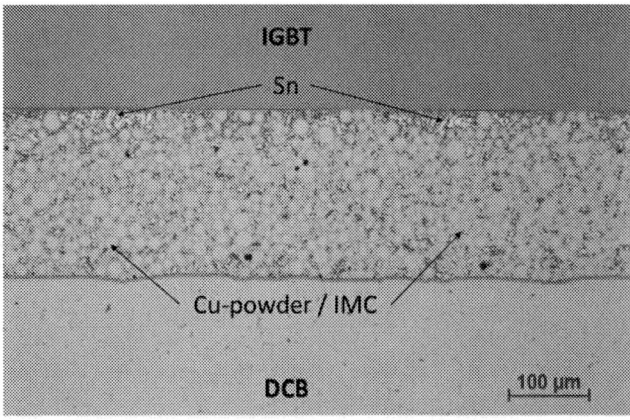

Figure 6: Cross section image of a TLPS joint between an IGBT and a DCB with Ni/Au-finish using lateral infiltration process. The TLPS joint contains a high content of Cu powder

Influences on activation of the Cu powder depots

The process time of the activation step depends on the size of the die, the used activation gas (e.g. CO, SF6, HCOOH) and the flow rate of the process gas. Due to the gas formation of a flux during soldering an activation with flux is not preferred. The lateral infiltration process as illustrated in Figure 5 was used for testing the influences of gas activation. A copper powder paste which contains pure Cu powder and a solvent without reducing properties was printed on a DCB

978-1-4799-2408-0/14 $31.00 © 2014 IEEE   1323

with Ni/Au-finish. The copper spheres have a diameter of 20 μm to 45 μm. The printing area had the size of the IGBT which was 9.73x10.23 mm² and a wet layer thickness of approximately 150 microns. The die was placed on the top of the printing surface. Two eutectic SnAg solder balls were placed next to the chip and Cu-powder printing area as shown in Figure 7.

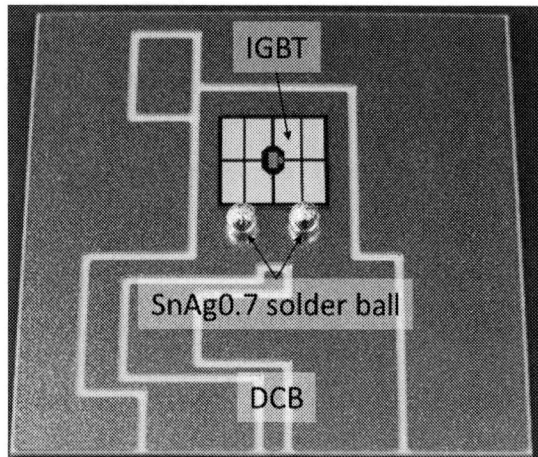

Figure 7: Image of the positioning of the power semiconductor (600 V IGBT) and the SnAg0.7 solder balls on the DCB. The IGBT has been placed on the top of the Cu printing depot

Each solder ball has a volume of about 10 mm³. After placement the sample was put into the reflow-oven. In the first step the sample was activated using an active gas. Afterwards the sample was reflowed at 260 °C with a peak time of about 60 seconds. During soldering the SnAg0.7 solder balls melt, wet the substrate and infiltrate the Cu powder layer due to the capillary force.

The following parameters were varied in order to determine an impact on the activation:

- Substrate metallization: Ni/Au-finish and Cu-finish
- Chip metallization: Ni/Ag-finish and Cu-finish
- Cu powder diameter: type 3 copper powder (20-45 μm) and type 6 copper powder (5-15 μm)
- Cu wet printing layer thickness: 50 μm and 150 μm
- Splitting of the Cu printing layer: standard (9.73x10.23 mm²) and four times (4 x 4,4x4,6 mm²)

The activation time was constant for all experiments. In Figure 8 the X-ray results after lateral infiltration are shown. All light gray areas in the images (Figure 8a-f) represent non infiltrated regions. Figure 8a shows the standard using an average Cu powder size between 20-45 microns, a stencil thickness of 150 microns with a width to length ratio of 9.73 to 10.23 mm², a DCB with Ni/Au-finish and a IGBT with Ni/Ag backside metallization.

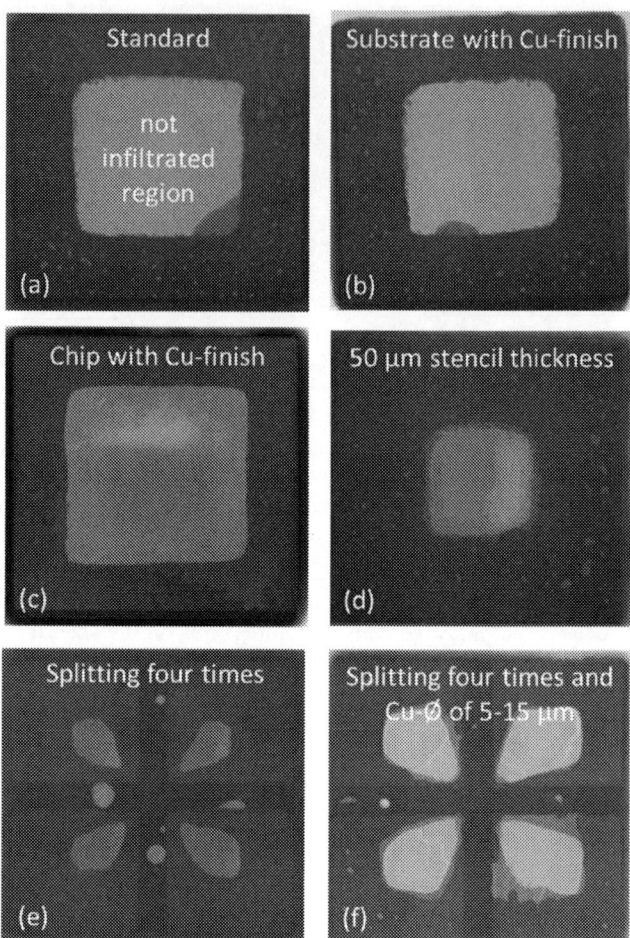

Figure 8: X-ray images after lateral infiltration of pure Cu powder using active gas and SnAg0.7 solder balls depending on (a) the standard, (b) the substrate surface, (c) chip metallization, (d) thickness of Cu printing layer, (e) splitting the printing layer four times and (f) Cu powder size.

Change surface

An influence of the substrate metallization (Figure 8b) and the chip metallization (Figure 8c) could not be observed. The metallization of the substrate was pure Cu (Figure 8b) compared to the standard assembly with a Ni/Au-finish (Figure 8a). The bottom side IGBT metallization was changed from Ni/Ag (Figure 8a) to pure Cu (Figure 8c).

Change stencil thickness

Figure 8d shows the influence of the Cu powder wet stencil thickness of approximately 50 μm compared to the standard with about 150 μm. The infiltration distance by 50 μm stencil thickness has significantly increased.

Change printing mask

Also a splitting of the printing area in four parts showed an improvement (Figure 8e). The non-infiltrated areas are symmetrically formed and the dark gray regions between the copper depots are filled with Sn-based solder.

Change copper powder size

In Figure 8f a finer copper powder size (Cu powder diameter between 5 μm to 15 μm) was used compared to Figure 8e (Cu powder diameter between 20 μm to 45 μm).

The non-infiltrated area is much bigger for the experiment with a Cu powder size between 5 to 15 microns as shown in Figure 8f.

For all experiments of lateral infiltration it was found that the Cu powder surface has a significant influence on the activation rate. With increasing the copper powder surfaces the non-infiltrated area decreases. This was demonstrated in Figure 8a compared to Figure 8d, where a three times higher Cu powder amount by the same Cu powder size were used. The non-infiltrated area in Figure 8d is with 13 percent approximately three times smaller compared to Figure 8a with 35 percent (as shown in Table 1). A comparable effect can be observed in Figure 8e and Figure 8f. The copper powder size in Figure 8f with 5-15 μm is approximately three times smaller compared to Figure 8e with 20-45 μm. A much larger non infiltrated region can be observed in Figure 8f, which is the result of the three times higher copper powder surface compared to Figure 8e.

A further activation optimization could be done by splitting the Cu printing depot. The activation gas can more efficiently flow along the non-printed Cu powder regions which increases the activation rate (compare results from Table 1, Figure 8e with Table 1, Figure 8a).

Table 1: Infiltrated area in percent based on the area ratio of the printed Cu powder layer under the die

Figure	Variation	Non-infiltrated area
8a	Reference	35%
8b	Cu surface on substrate	31%
8c	Cu surface on die	40%
8d	Decrease of stencil thickness	13%
8e	Splitting stencil mask	17%
8f	Splitting stencil mask and increase Cu size	28%

Re-melting temperature measurement

The re-melting temperature measurement is a feasible method to determine if a continuous intermetallic layer between the component and the substrate could be formed during soldering. For the re-melting temperature measurement an approximately 20 mm² TLPS-layer was established between two Cu sheets. The TLPS paste contains about 35 w% of Cu powder, SAC405 solder powder and a solvent without reducing properties. For joining the two-step TLPS process was used. The sample was clamped in a tensile tester and was loaded with a constant force of 15 N as shown in Figure 9. The heating rate was about 6 K/min.

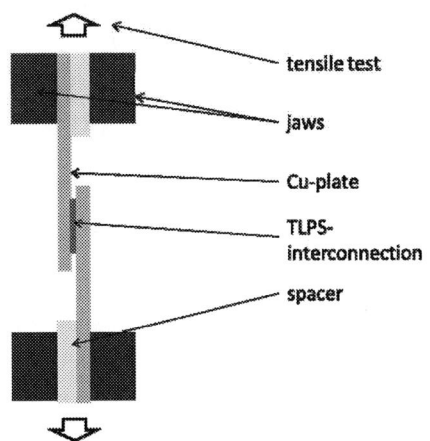

Figure 9: Schematic drawing of the re-melting temperature measurement using a tensile tester with a constant force of 15 N and a heating rate of 6 K/min

Then, the drop in force could be determined as a function of the temperature. In Figure 10 the force-temperature curve is shown. A reduction in force was observed at a temperature of about 391 °C. This suggests that a continuous intermetallic layer of Cu_6Sn_5 has formed between the two copper sheets. The Cu_6Sn_5 intermetallic phase is thermodynamically stable up to 415 °C and then decomposes in Cu_3Sn and liquid tin [9]. Accordingly, a decrease in force should be detected at approximately 415 °C. Because of the temperature-dependent mechanical properties of the Cu_6Sn_5 intermetallic phase a plastic fatigue of the Cu_6Sn_5 phase could be a possible reason for the premature drop-in force. It is also conceivable that at high temperatures (400 °C) a liquid phase may form due to available Sn-residues.

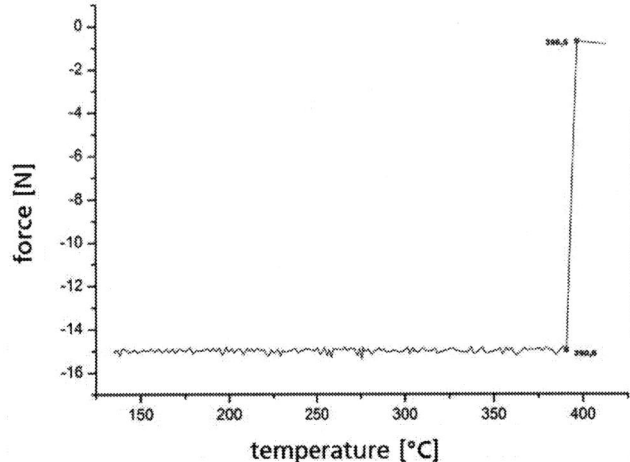

Figure 10: Force-temperature curve of the re-melting temperature measurement using a tensile tester with a constant force of 15 N and a heating rate of 6 K/min

Temperature cycling test and approaches for optimizing the lifetime of TLPS interconnects

In power applications, the most widely used semiconductor material is silicon which has an expansion coefficient of about $2\text{-}3 \cdot 10^{-6}$ K^{-1} [10]. For high temperature application SiC, GaAs and other semiconductors will be used which have a slightly higher CTE of about $4\text{-}7 \cdot 10^{-6}$ K^{-1} [11] [12]. These power semiconductors are usually connected to a DCB and mounted on a heatsink. Due to the different thermal expansion coefficients of the materials used thermomechanical stresses will occur at higher operating temperatures. These stresses can be reduced either by plastic or by elastic deformation of the material. If none of these events happened a crack will form and grow until it comes to a total failure of the module. Sn-based solder joints fail due to creep relaxation of the die interconnect. A TLPS joint has a brittle material behavior so that a plastic deformation of the transformed intermetallic solder joint could not be observed. In the following three failure mechanisms will be presented, which may occur in TLPS joints if the bonding partners have no matched CTEs.

Figure 11 shows a cross section of a TLPS joint between a Si-die and a direct aluminum bonded substrate (DAB) after 100 passive temperature cycles between -55 °C and +125 °C. Due to the thermal-mechanical stress during cycling vertical cracks have formed in the TLPS interconnect. The crack stops after a few micrometers in the aluminum layer of the DAB and propagated almost completely through the silicon. This crack propagation is not comparable to Sn-based solder failures. A characteristic solder fatigue crack starting at the edge of the die attach and propagate inwards was not be observed in TLPS joints.

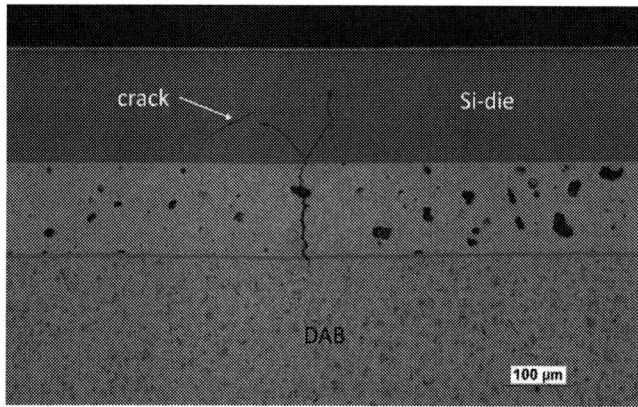

Figure 11: Cross section image of a TLPS joint between a Si-die and a DAB-substrate after 100 passive temperature cycles between -55 °C and +125 °C

Another failure mechanism of a TLPS layer could be detected between a Si-die and a Cu leadframe after 100 passive thermal cycles between -55 °C and +125 °C as shown in Figure 12. The Ti/Ni chip back-side metallization peels off completely. A vertical crack propagation could not be found in this case compared to Figure 11.

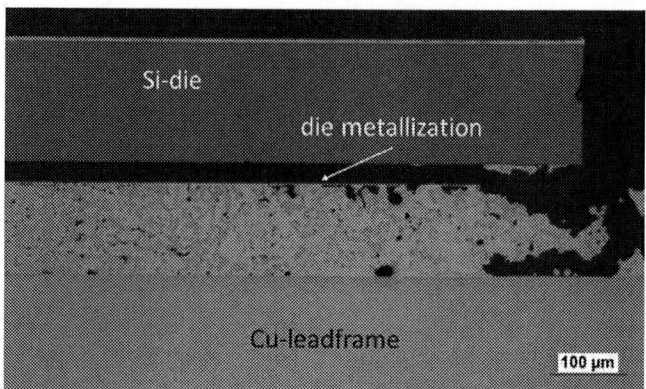

Figure 12: Cross section image of a TLPS joint between a Si-die and a Cu-leadframe after 100 passive TC (-55 °C / +125 °C)

In Figure 13 a third failure mode was observed. A diode with Cu top side metallization was joined on a DCB using a TLPS interconnect in between. The acting forces during cooling shall be responsible for the peeling of the approximately 40-50 µm Cu/Al chip top-side metallization. In the cross section image the still available silicon residues can be clearly seen.

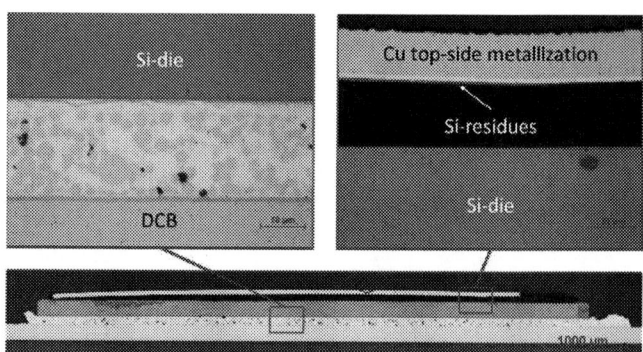

Figure 13: Cross section image of a TLPS joint between a diode with Cu top-side metallization and a DCB in the initial state after cooling

To sum up, besides die attach cracks shown in Figure 11 two other failure mechanisms have been detected, namely delamination between die and interconnect (Figure 12) as well as delamination on die top side (Figure 13). The almost completely transformed solder joint into Cu_6Sn_5 and Cu_3Sn intermetallic phases does not degrade by plastic deformation compared to Sn-based solder joints. The failure mechanism can be expanded on the whole module if a TLPS solder joint is used. In the following some ways how to improve the lifetime of power electronic modules and assemblies with large thermal expansion differences will be presented.

The first option is to use buffer layers. These layers can be applied on the bottom-side of a semiconductor as shown in Figure 14. Such a buffer layer substantially fulfills two tasks. On the one hand the CTE mismatch between the silicon die and the substrate material for instance Cu can be compensate by using a Cu buffer layer on the bottom side of the chip. On the other hand the thermo-mechanical stresses may be reduced by a plastic deformation of the copper buffer layer.

978-1-4799-2408-0/14 $31.00 © 2014 IEEE

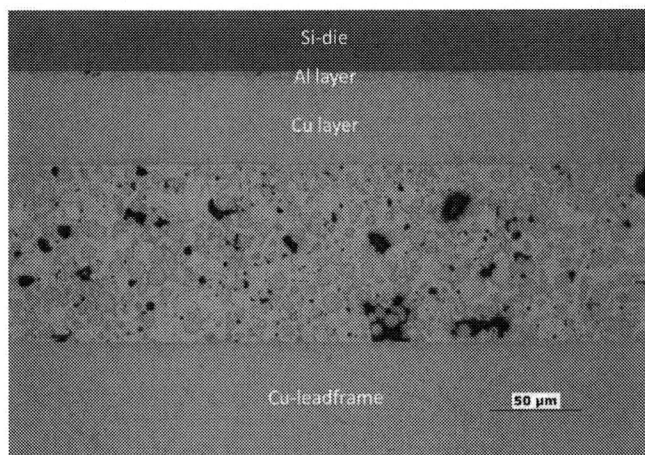

Figure 14: Cross section image of a TLPS interconnection using a Cu buffer layer on the bottom side of the Si-die

A second possibility to increase the lifetime of a power electronic module is to adjust the thermal expansion of the joining partners. For example, the substrate material could be matched to the Si-die by using a substrate alloy like FeNi42 which has approximately the same CTE as the semiconductor material.

Conclusions

A novel joining technology for high operating temperatures of electronic devices has been introduced in this paper. It is called: "Transient Liquid Phase Soldering (TLPS)". A model for a void-less TLPS joint has been presented. This model was transferred to an application-related two-step-process and a new TLPS paste was developed, which contains 35 w% Cu-powder, tin-based solder powder and a solvent without reducing properties. An alternative process using the lateral infiltration effect of a pure Cu powder printing layer by liquid solder have been shown too. For activation a reducing gas was used. It has been shown that the activation for the lateral infiltration process mostly depends on the amount of available copper powder surface. Splitting the Cu printing layer reduce the time for activation, too. The re-melting temperature measurement test has been shown that a continuous intermetallic layer between the component and the substrate could be formed. The TLPS joint is soldered at approximately 250 °C and is thermally stable up to 400 °C. Reliability tests demonstrate that the almost completely transformed TLPS joint into Cu_6Sn_5 and Cu_3Sn does not degrade by plastic deformation. Therefore, the failure mechanism may occur somewhere else in the module if a TLPS solder joint will be used. Finally, approaches for optimizing the lifetime of TLPS interconnects have been presented.

The main advantage of TLPS using copper spheres and solder powder mixed in one paste compared to other high temperature joining technologies like silver sintering or TC/TC Au-bonding is that common stencil printing machines, pick & place equipment and reflow ovens with gas activation can be used. Rough surfaces on substrates of several micrometers can easily be compensated. Because of the low raw material costs and the robust process capability the TLPS is well suitable as a high temperature joining technology.

Acknowledgments

The author would like to thank the German Federal Ministry for Education and Research (BMBF) for funding the project "Hot Power Connection" under contact No. 13N11513 and the project "ProPower" under contact No. 16N11896.

References

[1] M. Wrosch and A. Soriano, "Sintered conductive adhesives for high temperature packaging," *Electron. Components Technol. Conf.*, no. 760, 2010, pp. 973–978.

[2] D. A. Shnawah, S. Binti, M. Said, M. Faizul, M. Sabri, I. A. Badruddin, and F. A. X. Che, "High-Reliability Low-Ag-Content Sn-Ag-Cu Solder Joints for Electronics Applications," vol. 41, no. 9, 2012.

[3] A.-Z. Miric, "New Developments in High-Temperature, High-Performance Lead-Free Solder Alloys," in *SMTA International Conference*, 2010.

[4] X. Qiao and S. . Corbin, "Development of transient liquid phase sintered (TLPS) Sn–Bi solider pastes," *Mater. Sci. Eng. A*, vol. 283, no. 1–2, May 2000, pp. 38–45.

[5] H. Greve, L. Chen, I. Fox, F. P. Mccluskey, G. L. M. Hall, and C. Park, "Transient Liquid Phase Sintered Attach for Power Electronics," *ECTC*, vol. 63, 2013, pp. 435–440.

[6] C. Ehrhardt, M. Hutter, O. Hermann, and K.-D. Lang, "Transient Liquid Phase Soldering for lead-free joining of power electronic modules in high temperature applications," in *High Temperature Electronics Conference (HiTEC)*, 2012, pp. 1–9.

[7] C. Gallagher, G. Matijasevic, P. Gandhi, D. Pommer, S. Sargeant, R. Kumar, and D. Neuburger, "Vertical interconnect in multilayer applications using Ormet conductive composites," *Proc. 3rd Int. Symp. Adv. Packag. Mater. Process. Prop. Interfaces*, 1997, pp. 35–37.

[8] C. Ehrhardt, M. Hutter, H. Oppermann, and K.-D. Lang, "Transient Liquid Phase Soldering – An emerging joining technique for power electronic devices," in *PCIM*, 2013, pp. 1–8.

[9] J. F. Li, P. a. Agyakwa, and C. M. Johnson, "Interfacial reaction in Cu/Sn/Cu system during the transient liquid phase soldering process," *Acta Mater.*, vol. 59, no. 3, 2011, pp. 1198–1211.

[10] Z. Yang, X. He, L. Wang, R. Liu, H. Hu, L. Wang, and X. Qu, "Microstructure and thermal expansion behavior of diamond / SiC /(Si) composites fabricated by reactive vapor infiltration," *J. Eur. Ceram. Soc.*, vol. 34, no. 5, 2014, pp. 1139–1147.

[11] R. Johnson and C. Wang, "Power device packaging technologies for extreme environments," *Electron. Packag.*, vol. 30, no. 3, 2007, pp. 182–193.

[12] J. D. Adam, H. Buhay, M. R. Daniel, M. C. Driver, G. W. Eldridge, M. H. Hanes, and R. L. Messham, "Monolithic integration of an X-band circulator with GaAs MMICs," in *Microwave Symposium Digest*, 1995, pp. 97–98.

Developments of High-Bi Alloys as a High Temperature Pb-Free Solder

Sandeep Mallampati[1], Harry Schoeller[2], Liang Yin[3], David Shaddock[3], and Junghyun Cho[1]

[1]Binghamton University (State University of New York), Mech Eng. & MSE, Binghamton, NY 13902-6000
[2]Universal Instruments Corporation, Conklin, NY 13748
[3]GE Global Research, Niskayuna, NY 12309
jcho@binghamton.edu, (607) 777-2897

Abstract

Predominant high melting point solders for high temperature electronics (operating temperatures from 200 to 250°C) are Pb-based but these are hazardous to environment. In this study, we designed and characterized the 'new' Bi-based solder alloys, 78Bi-14Cu-8Sn and 70Bi-20Sb-10Cu, to understand their performance for high temperature die attach applications. The specific alloy compositions were determined by thermodynamics calculations via the Thermo-Calc software to have a solidus temperature above 271°C. Microstructures of the selected as-cast alloys have shown Cu_3Sn and Cu_2Sb intermetallic compounds (IMCs) spread in Bi or Bi-Sb matrix, respectively, which are essential for maintaining good thermal conduction and high temperature strengths. Die-attach solder joints were replicated using a Cu substrate, a solder preform and a Ti/Ni/Au metallized Si die, to study interfacial reactions and joint microstructure. In this paper, microstructure – property relationships of high-Bi alloys are highlighted to assess their potential as a high temperature Pb-free solder. In particular, we examined the reactivity of these solders with common surface finishes (e.g., Ni/Au, Cu) of the die and the substrate at various reflow conditions. It was shown that strong reaction exists between Bi and Ni (BiNi), between Sb and Ni (NiSb), and between Sb and Cu (Cu_2Sb, Cu_4Sb), but the reaction in the case of 78Bi-14Cu-8Sn was rather weak with Cu (forming Cu_3Sn) due to the limited supply of Sn from the Bi matrix. The die shear testing was also performed on these sandwiched coupons to evaluate their solder performance at various reflow conditions. The die attach with 70Bi-20Sb-10Cu showed as-reflowed die shear strengths over 30 MPa, which were comparable to or better than high-Pb alloys.

Introduction and Background

High Pb solders (90Pb-5Sn, 92.5Pb-5Sn-2.5Ag) are some of the most commonly used high temperature solders for operating temperatures above 200°C [1]. Toxicity of Pb poses serious threat to environment. Although currently exempt from the Restriction of Hazardous Substances (RoHS) Directive, the new regulation (RoHS2 Directive 2011/65/EU) drafted in 2011 will soon ban the Pb-containing solders in broader electronics including high temperature electronics. Given the urgency in need of such affordable Pb-free solders in high temperature electronics including SiC power modules and downhole drilling electronics, a "new" class of solder alloys, with high Bi content ($\geq$ 70 wt. %), were designed to evaluate their potential as high temperature Pb-free solder alloys. Pb based solder alloys have been the preferred interconnect materials due to their superior mechanical performance [2,3]. Prospective replacement alloys are required to match their mechanical and thermal performance.

Pure Bi has a higher melting point than common eutectic Sn-Pb or Sn-based solder alloys, thereby making it be a feasible alternative. Poor electrical and thermal conductivity and limited plasticity are, however, of concern. Low ductility is due to its rhombohedral crystal structure. Alloying elements such as Cu, Sn, Zn, and Sb are needed in Bi to negate those shortcomings by creating innovative alloy compositions and microstructures. Cu helps enhance thermal and electrical conductivity. Sn facilitates interfacial reactions as it forms CuSn type intermetallic phases (Cu_3Sn, Cu_6Sn_5) with Cu substrate and NiSn type intermetallic phases (Ni_3Sn_4) with Ni metallized surface. Sb has the same crystal structure as Bi and very similar electronegativity to Bi, so forms a complete solid solution with Bi. In our previous study, Sn and In form solid solution with Pb and contribute to its high temperature strength by solid solution strengthening [2]. Sb also has high reactivity with Cu and Ni.

In this study, the specific alloy compositions were determined by thermodynamics calculations via the Thermo-Calc software to have a solidus line above 271°C, which is the melting point of Bi. Among them, two alloy compositions (78Bi-14Cu-8Sn, 70Bi-20Sb-10Cu) were initially selected for detailed studies of solderability, microstructure features, and mechanical reliability [4]. This paper represents for an initial study of the alloying behavior and associated microstructure developments that can potentially tailor the heat dissipation capacity and plasticity of these relatively "unknown" but high-impact potential Bi alloy solders. Depending on the performance, this new alloy development will provide one possible option for high-temperature Pb-free solders without major changes to existing manufacturing processes.

Calculation of Phase Diagram

The initial alloy composition of high-Bi solders for use at 200°C or above was determined with the CALPHAD (**Cal**culation of **Pha**se **D**iagrams) method using the Thermo-Calc software. This software package deals with Gibbs free energy minimization that allows various thermodynamic and phase diagram calculations. In this study, we employed the most recent solder database, TCSLD1, which includes more elements (Ag, Al, Au, Bi, Co, Cr, Cu, Ge, In, Ni, Pb, Pd, Pt, Sb, Si, Sn, and Zn) than the other two old solder databases: NPL Solder solutions database (NSLD2) and NIST Solder solutions database (USLD1).

Various phases present in the Bi-Cu-Sn alloy can be seen in the vertical section of the ternary phase diagram, as shown in Figure 1. The 'cd' line in the diagram represents for the solidus line, where melting starts. The selected alloy composition at room temperature consists of Bi matrix, Cu_3Sn and Cu rich phases. Figure 2 shows the vertical section of the ternary phase diagram of a Bi-Sb-Cu alloy. Because of complete solid solution between Bi and Sb, the melting temperature of the alloy increases with increasing Sb concentration. The solidus line of this alloy becomes 294°C with 20% Sb (Bi-20Sb-10Cu), which is higher than that of common high-Pb alloys (Pb-5Sn-2.5Ag for 287˚C).

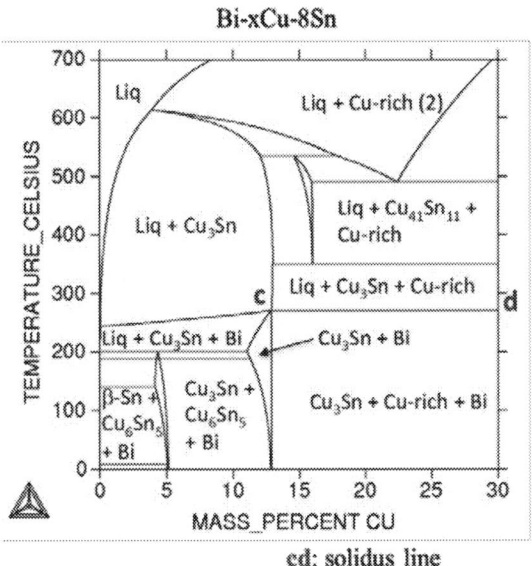

Figure 1. Calculated vertical section of a Bi-Cu-Sn ternary phase diagram at 8 wt. % Sn with varying Cu concentration.

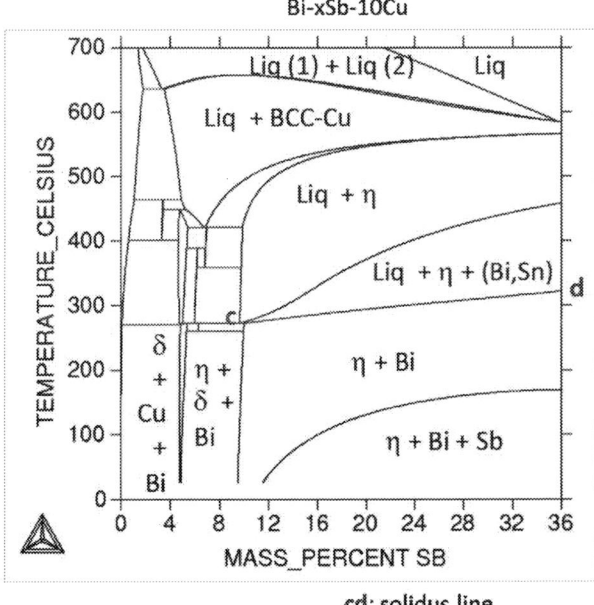

Figure 2. Calculated vertical section of Bi-Sb-Cu ternary phase diagram at 10 wt. % Cu, with varying Sb concentration.

Based upon these phase diagrams, a partial melting is expected for both alloys since some of primary IMCs will not be melted at reflow temperatures (around 350˚C). The thermodynamic calculation predicts the relative amount of liquid (i.e., molten Bi) to be ~80% at 350˚C.

High Bi Alloy: Ingots, Preforms, and Die Attach

The two alloys (78Bi-14Cu-8Sn, 70Bi-20Sb-10Cu) examined in this study were acquired as ingots with 3.5"L x 0.75"W x 0.313"D from Indium Corporation, USA. For observing microstructures of as-received solder alloys, these ingots were cut into pieces that were subsequently polished mechanically using silicon carbide papers of various grit sizes, then with diamond slurry and finally with colloidal silica solution.

The die attach assembly joint was made using a solder preform between the substrate and the die. High conductive Cu was used as a substrate material and a bare Si with backside metallization (BSM) sputtered with Ti (70 nm) / Ni (300 to 2000 nm) / Au (70 nm) was used as a die. The dimensions of the die were 1/8'' x 1/8''.

The weight of the alloy sample used to make a solder preform was determined using the dimensions of the die and bond line thickness (BLT) of the joint desired. A BLT close to 100 µm was aimed in this work. These alloy samples were then heated to 375˚C on a hotplate and pressed between glass slides. The thickness of the preforms was controlled by placing shims between the glass slides. This process was performed inside a glove box having nitrogen environment with moisture content < 10 ppm and O_2 < 0.1 ppm.

The substrate, preform and die stacked in order from bottom to top were then subjected to a reflow process to make a joint. This reflow process is carried out on a hotplate inside the glove box, or in a reflow oven with N_2 flow. A small amount of pressure was manually applied on the top of the die to ensure good flow of the molten alloy. The reflow condition that will be specified throughout this paper refers to the temperature measured on the surface of the substrate, using a thermocouple, and the length of time the sample stack was placed on the hotplate surface. Once the reflow was completed, the die attach sample is immediately taken off from the hotplate and allowed to cool at ambient temperature.

Characterization of Bulk Solders of High-Bi Alloys

i) Microstructure

Bulk alloy microstructure analysis was the first test done on the alloys received from the solder supplier. Figure 3 shows the as-received microstructure of 78Bi-14Cu-8Sn ingot. This alloy has almost pure Bi matrix with IMCs that are as large as ~100 µm. SEM clearly shows the morphologies of these IMCs spread in Bi matrix (Fig. 3 bottom). The IMCs were indeed confirmed by EDS as Cu_3Sn.

Figure 4 shows the as-received microstructure of 70Bi-20Sb-10Cu ingot. The polarized optical microscope (OM) image shows the grain orientation contrast in the matrix, along with IMCs in the form of dendrites. The elongated needle shapes in the matrix phase may be attributed to mechanical twin formation. The two-phase regions are also visible at higher magnification in SEM using a backscattered electron (BSE) imaging mode, in which lighter contrast regions represent for Bi-rich regions (with lower Sb-content) and darker contrast regions represent for Bi grains with higher Sb-content. In the bulk ingot microstructure, the Sb-rich regions were not clearly observed. Even though true equilibrium predicts the matrix consisting of a Bi-rich phase and a Sb-rich phase, the latter may not be easy to obtain due to the limitation of diffusion.

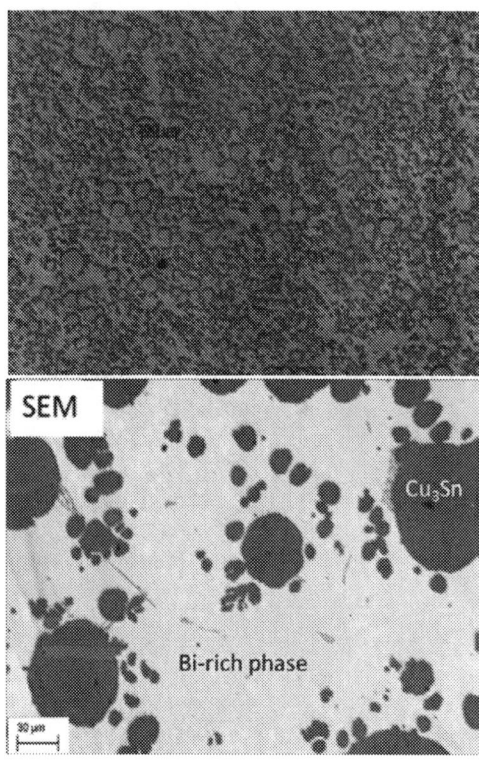

Figure 3. Optical microscope image (top) and SEM image (bottom) of the as received 78Bi-14Cu-8Sn alloy.

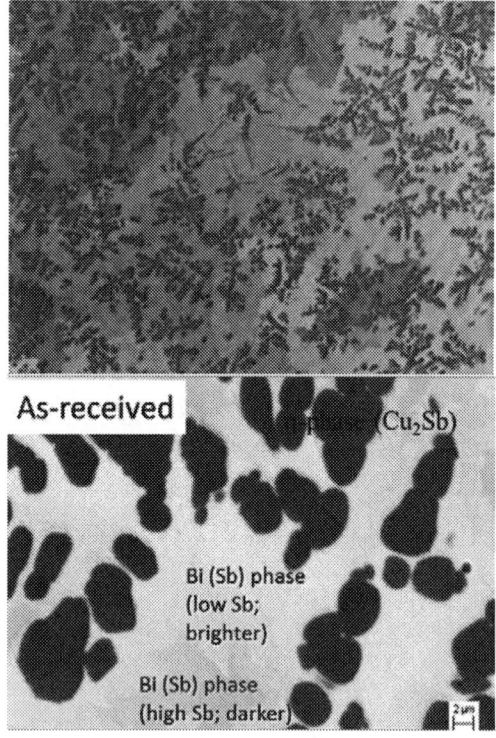

Figure 4. Optical microscope image (top) and SEM image (bottom) of the as received 70Bi-20Sb-10Cu alloy.

ii) Melting Temperatures

The onset of melting was observed at 270.5°C and 280.5°C for 78Bi-14Cu-8Sn and 70Bi-20Sb-10Cu, respectively, in a DSC measurement (Fig. 5). For the 78Bi-14Cu-8Sn alloy, melting at 270.5°C is in a good agreement with the thermodynamic predictions of the solidus line of the phase diagram. Unlike the sharper valley for this alloy, the 70Bi-20Sb-10Cu alloy displayed the broader valley that can be attributed to the presence of the two phase mixture (or variation in the concentration of Sb) within the matrix. From the DSC curves the alloy started melting at 280.5°C although the equilibrium thermodynamic predicts the melting at 294°C. Hence, the Bi matrix may remain in partially molten state at the reflow temperatures (350°C).

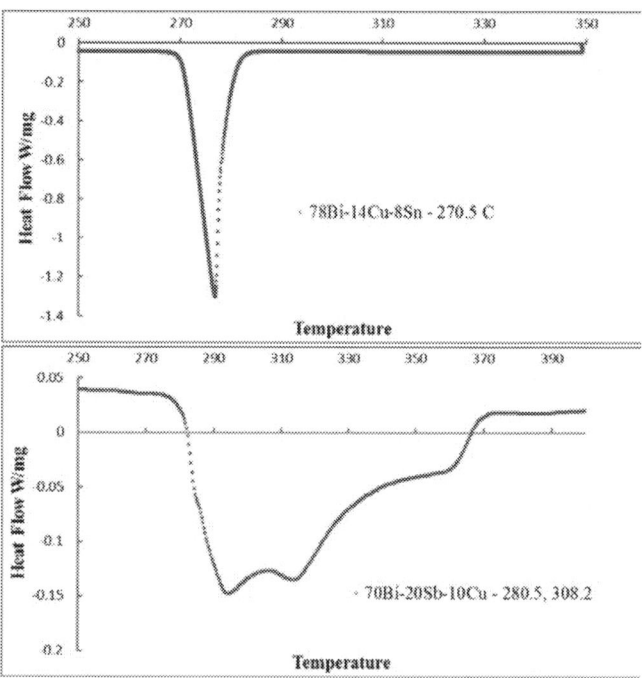

Figure 5. Melting transition regions in DSC for both alloys

iii) Nanomechanical Behavior

To understand the mechanical behavior of the constituent phases, elastic modulus and nanohardness of the Bi matrix and the IMCs were measured using a nanoindenter (TriboIndenter from Hysitron, Inc., Minneapolis, MN, USA). This system measures the indentation response during the sequence of loading and unloading when the sharp tip penetrates the surface of the material (i.e., indentation load vs. displacement). For this work, we used a Berkovich tip. Table 1 shows the nanoindentation results of Bi alloys, compared to those in high-Pb alloys [5].

Elastic modulus of Bi phase in the 78Bi-14Cu-8Sn alloy had a value close to 39.9 GPa but in 70Bi-20Sb-10Cu, it was close to 60 GPa. The higher value for the matrix of the latter alloy is because of solid solution strengthening effect from Bi-Sb. Reported value of elastic modulus of Bi is ~34 GPa [6].

978-1-4799-2408-0/14 $31.00 © 2014 IEEE

Table 1. Indentation hardness and elastic modulus of Bi [5].

Solders		Nanoindentation		Tensile Testing	
		Elastic Modulus (GPa)	Nano-hardness (MPa)	Elastic Modulus (GPa)	Yield Strength (0.02% offset) (MPa)
Pb-3Sn	matrix	28 ± 1.4	172 ± 9.50	18.3	10
	β-Sn	36 ± 2.4	252 ± 32.0		
Pb-3Sb	Matrix	25 ± 1.5	162 ± 17.7	18.1	12.3
	β-Sn	46 ± 4.2	398 ± 32.9		
Pb-3In	Matrix/band	23 ± 1.0	222 ± 32.9	16.1	11.8
Pb-5In-2.5Ag	On Crater	23 ± 1.7	273 ± 24.5		
	On Band	24 ± 2.0	238 ± 31.5	22.4	11.8
	On IMC (Ag$_9$In$_4$)	39.5	1620		
78Bi-14Cu-8Sn	On IMC (Cu$_3$Sn)	141 ± 31	4566 ± 301	NA	NA
	Matrix	39.9 ± 3.2	226 ± 7		
70Bi-20Sb-10Cu	On IMC (Cu$_2$Sb)	101 ± 4.3	2870 ± 90	NA	NA
	Matrix 1	60 ± 11	480 ± 30		
	Matrix 2	59 ± 11.4	850 ± 77	NA	NA

Nanohardness values of Bi-Sb solid solution were also higher than pure Bi. Both IMC phases showed quite high values compared to the matrix phase, so they will work as a reinforcing phase as in composite materials. In particular, Bi-Sb alloy shows two representing values for the matrix, for which high values (matrix 2) is believed to be Sb-rich (or higher Sb content) phase as Sb is harder than Bi. From the AFM built in the nanoindenter, it was not feasible to differentiate these two phase regions.

iv) Thermal Diffusivity Measurements

Thermal diffusivities of Bi alloys were measured by Anter Flashline 2000. For this, a specimen of certain predetermined thickness was subjected to a radiant energy pulse and a thermogram of the temperature rise on the other face of the specimen was recorded. Thermal diffusivity was then calculated from the half rise time (half of the time taken for the temperature to reach maximum on the rear face) and the thickness of the specimen.

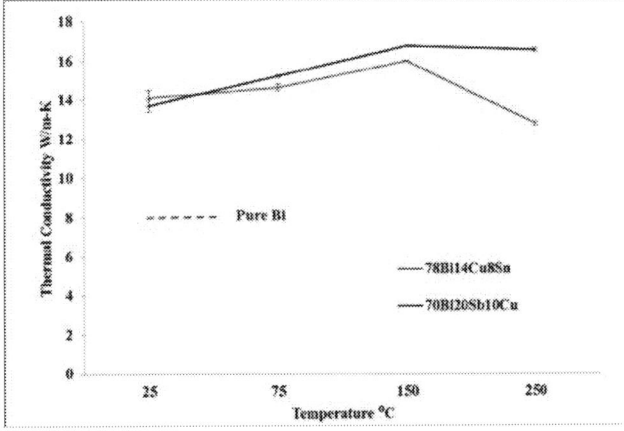

Figure 6. Thermal conductivity vs. temperature for High – Bi alloys, compared to that of pure Bi.

Alloying of Bi with the elements that have higher thermal conductivities has resulted in the effective increase in the conductivity of these alloys, as shown in Fig. 6. Thermal conductivity of pure Bi is also compared in this plot (~ 8 W/m·K) [7].

Solder Reactions with Cu and Ni

Interfacial reactions with Cu (from Cu substrate) and Ni (from Ti/Ni/Au backside metallization of Si die) were studied for both of the aforementioned alloys. Figure 7 shows the interfacial reactions of 78Bi-14Cu-8Sn with BSM of the die and the Cu substrate. Unlike its bulk microstructure, the primary Cu$_3$Sn IMCs are smaller, which indicates the dissolution during the reflow process. BiNi-type IMC was observed on the die side and Cu$_3$Sn intermetallic was formed on the substrate side. The Ni layer on the die side seemed to be completely consumed, which indicated that the reaction between Ni and molten Bi was quite active during the reflow process. On the other hand, the reaction with Cu was minimal forming a very thin Cu$_3$Sn layer. This is due to the amount of available Sn in the Bi matrix is very low, which therefore should be provided from the dissolution of Cu$_3$Sn followed by Sn diffusion that can be a rate-limiting process during the reflow.

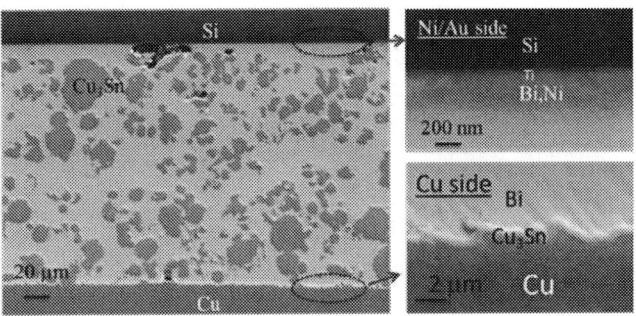

Figure 7. Reactions: 78Bi-14Cu-8Sn sandwiched between Si (w/ Ti/300 nm Ni/Au) and Cu (Reflow: 350˚C – 120Sec)

To better understand the reaction of the solder with Ni, electroless nickel, electroless palladium, immersion gold (ENEPIG) on a polyimide substrate was used for the same reflow profile for 78Bi-14Cu-8Sn. This substrate had a 2-3 μm thick Ni layer. It can be seen in Figure 8 that upon a reflow of the alloy, the IMC layer formation was more apparent. The entire Ni layer was again consumed but the original Ni layer was replaced with BiNi, suggesting very aggressive reaction kinetics between Ni and Bi.

Earlier studies have reported that BiNi is very brittle and can be transformed from Bi$_3$Ni via a solid-state reaction [8, 9]. In the case of ENEPIG substrates, residual Cu-rich phase and IMC-free zone were observed within the bulk solder close to the substrate. Presence of Cu-rich was indeed predicted by thermodynamics calculations as shown in Fig. 1. It suggests that primary Cu$_3$Sn be decomposed during the reflow process to provide Sn for Ni layer and Cu for Bi matrix.

978-1-4799-2408-0/14 $31.00 © 2014 IEEE

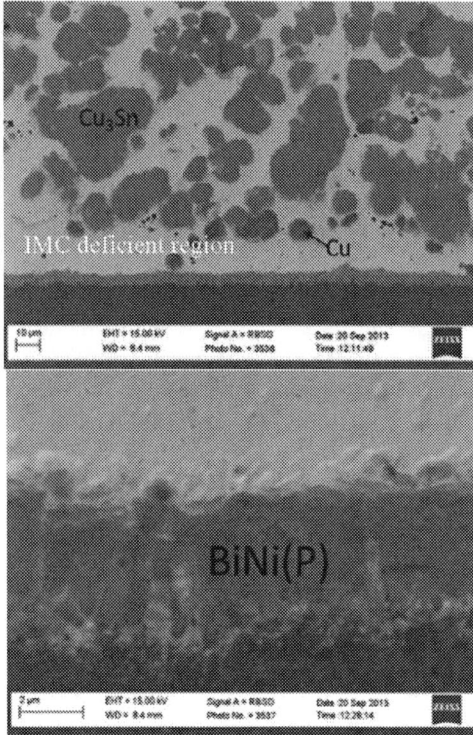

Figure 8. Reaction: 78Bi-14Cu-8Sn reflowed on ENEPIG (Reflow: 350˚C – 120Sec)

Reflow was performed for the 70Bi-20Sb-10Cu preform with various thicknesses (300 nm to 2,000 nm) of Ni layer in the BSM stack and the Cu substrate. Figure 9 shows the cross-sectional microstructure and interfacial reaction of 70Bi-20Sb-10Cu sandwiched between the die (70-nm Ti / 1000-nm Ni / 70-nm Au) and the Cu substrate.

Reflow was performed on the hotplate at 350˚C for 120 sec including the ramp time from room temperature. The matrix region on the die side was devoid of the Cu_2Sb IMCs, similar to the ENEPIG substrate cases with 78Bi-14Cu-8Sn. The possible origin of this is again attributed to the decomposition of primary Cu_2Sb IMCs that supply Sb for the interfacial reactions with Ni. Indeed, NiSb interfacial IMCs were observed, which made ~ 600 nm of the Ni barrier layer remained. There are platelet structures grown out of the NiSb layer that was not clearly identified due to Si contamination from the polishing media.

A thick Cu_2Sb interfacial IMC layer was observed on the substrate side. This indicates a plenty of Sb available in the Bi matrix to react with Cu surface. The spalling of Cu_2Sb layer was also observed, where Cu_4Sb forms between Cu_2Sb and Cu. The thickness of Cu_2Sb was over 4 ⱱm. As explained later, this side of interface is in general stronger than the die side interface.

Die Shear Strengths

This test was performed on Dage 4000 plus Bondtester from Nordson Dage, which has a 100 kg load cell. A shear height of 30 μm above their respective BLTs was used for all the samples. A shear speed of 500 μm/sec was used. The shear force measured when the joint failed was divided by the area of the die face to calculate the shear strength of the joint.

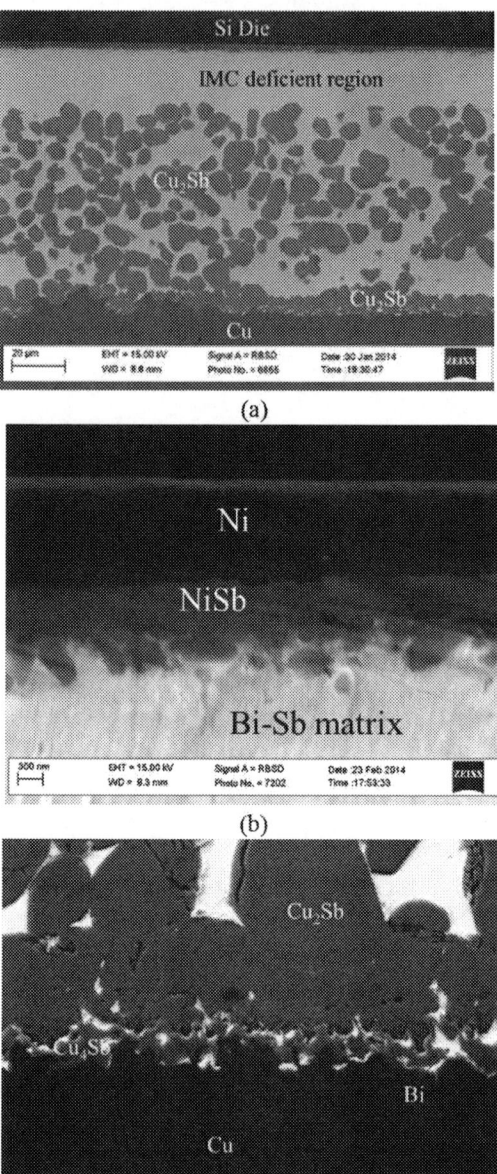

(a)

(b)

(c)

Figure 9. (a) Reactions: 70Bi-20Sb-10Cu sandwiched between Si (w/ Ti/1000 nm Ni/Au) and Cu (Reflow: 350˚C – 120Sec) (b) die side IMC layer (c) substrate side IMCs

Figure 10 shows the fracture surface of Si die/70Bi-20Sb-10Cu/Cu substrate (the arrows indicate the viewing direction). As seen in the picture, the failure (indicated by a dotted line) has not occurred entirely within the bulk solder; instead it also has occurred in the die stack that exposed solder and Ni. A possible reason for the failure of the die side interface was the presence of the regions, where Ni was completely consumed or NiSb was formed, has weakened the interface. The failure in the die stack, however, suggested that the solder was stronger than the strength measured during the failure.

Figure 10 Die attach failure surface attached to Cu side (Reflow: 350°C – 120Sec) Si (w/ Ti/1000 nm Ni/Au) – (70Bi-20Sb-10Cu) – Cu

The failure for the case of 78Bi-14Cu-8Sn, on the other hand, occurs on the substrate side, where the Cu_3Sn IMC formation was very sluggish because of no immediate supply of Sn for the Cu substrate. Since molten Bi can accommodate more Sn than in solid Bi, the primary IMCs can be dissolved during the reflow process to provide Sn for the reaction with Cu substrate. But the process involving the dissolution and the Sn diffusion can be quite slow; resulting in very few spots of reaction on Cu substrate as shown in Figure 11 (the arrows indicate the viewing direction).

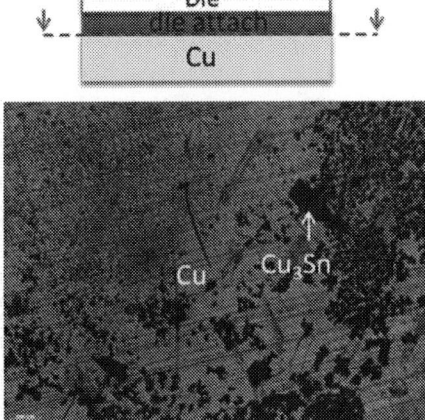

Figure 11. Cu surface after failure (Reflow: 350°C – 120Sec) for Si (w/ Ti/300 nm Ni/Au) – (78Bi-14Cu-8Sn) – Cu

Figure 12 shows the die shear strength comparison of 70Bi-20Sb-10Cu alloys, along with the strengths of some of the commercially available solders for similar applications. The shear strength values of this alloy under RF2 showed a wide variation due to a change in the dominant failure mode among the samples, which occurred either through bulk solder or along the interface between Ni and NiSb. When bulk solder failure was dominant, the strength was close to ~50 MPa, exceeding those of the Pb based alloy (PB-A that refers to 92.5Pb-5Sn-2.5Ag).

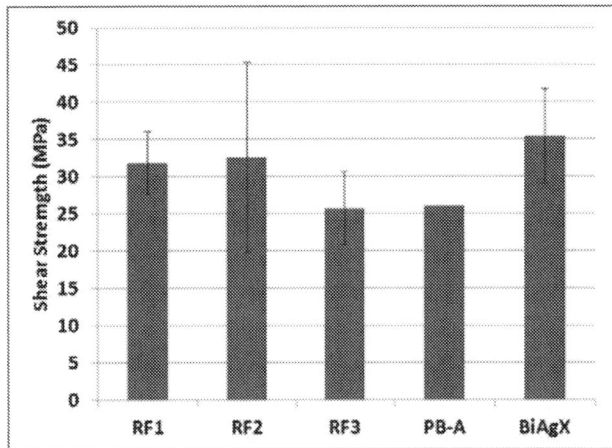

Figure 12. (70Bi-20Sb-10Cu) shear strength comparison (Reflow 1 (RF1): 370°C – 120 Sec), (Reflow 2 (RF2): 350°C – 120 Sec) and (Reflow 3 (RF3): 330°C – 180 Sec). The data for BiAgX are from Ref. [10].

Die attach samples using all the three reflow temperatures for 70Bi-20Sb-10Cu have failed at the die side (through bulk solder or along the NiSb-Ni interface). While the error bars of these strength values are overlapped, high temperature reflow conditions (350°C, 370°C) showed higher average strengths. The microstructures of the joints for these three cases are under further investigation to understand their respective interfacial reactions and their influence on the shear strength.

Conclusions

Bi based solders were designed for high temperature Pb-free solders (operating temperature over 200°C) using computer simulation. The alloy ingots were cast, from which the solder preforms were fabricated. Various phases predicted in the Bi matrix were verified by microstructure observation: Cu_3Sn IMCs in Bi matrix for 78Bi-14Cu-8Sn and Cu_2Sb IMCs in clustered Bi-Sb matrix for 70Bi-20Sb-10Cu. These IMCs will have an opportunity to prevent grain growth and maintain the strength at high temperature operations. Thermal conductivities of both alloys increased compared to that of pure Bi. The wetting of these solders on Ni and Cu surfaces was promising. In particular, a strong reaction exists between Bi and Ni (BiNi), Sb and Ni (NiSb), Sb and Cu (Cu_2Sb, Cu_4Sb), but the reaction with Cu in the case of 78Bi-14Cu-8Sn was rather weak due to the limited supply of Sn from the Bi matrix. On the Ni side, there was strong reaction with molten Bi, so alloy compositions and reflow conditions needed to be optimized to prevent complete consumption of the Ni layer. The die shear testing was also performed on the sandwiched coupons to evaluate the soldering performance of these two solder compositions at various reflow conditions. 70Bi-20Sb-10Cu has performed well in the shear strength test (over 30 MPa) at room temperature, compared to the commercial high-Pb solders.

Acknowledgments

This work was sponsored by Integrated Electronics Engineering Center (IEEC) of Binghamton University (State University of New York). In particular, user facilities from Analytical and Diagnostics Laboratory (ADL) at Binghamton University were used for some of materials characterization work presented here. Helpful discussions with Drs. Arun Gowda and Kaustubh Nagarkar (from GE Global Research) and Dr. Martin Anselm (from Universal Instruments Corporation) are greatly acknowledged.

References

1. R. Tummala and E. J. Rymaszewski, Microelectronics Packaging Handbook, Van Nostrand Reinhold, New York, 1989, pp. 374-375.
2. H. Schoeller, S. Bansal, A. Knobloch, D. Shaddock, J Cho. "Microstructure evolution and the constitutive relations of high-temperature solders," J Electron Mater, vol. 38, no. 6, pp. 802–809, Mar 2009.
3. H. Schoeller, S. Bansal, A. Knobloch, D. Shaddock, J Cho. "Effect of alloying elements on the creep behavior of high Pb-based solders". Mater. Sci. Eng, A, vol. 528, pp. 1063–1070, 2011.
4. J Cho, S. Mallampati, and H. Schoeller, " Bismuth Alloys for a Pb-Free Solder," unpublished work (2014).
5. S. Maganty, H. Schoeller, and J.Cho, "Microstrucutres of high temperature solders and their relationship to mechanical behavior," unpublished work (2013).
6. F. Cardarelli, Materials Handbook: A Concise Desktop Reference, 2nd Ed. Springer-Verlag, London, 2008.
7. D. R. Lide, CRC Handbook of Chemistry and Physics, 88th Ed., p.12-200, CRC Press, 2007.
8. M. S. Lee, C. Chen, and C. R. Kao, "Formation and Absence of Intermetallic Compounds during Solid-State Reactions in the Ni-Bi System," Chem. Mater., vol. 11, no. 2, p. 292 (1999).
9. J. Song, H. Chuang and Z. Wu "Interfacial reactions between Bi-Ag High – Temperature solders and metallic substrates," Journal of Electronic Materials, vol. 35, no. 5, pp. 346–351, Jan. 2006.
10. H. Zhang and N.-C. Lee, "High reliability high melting mixed lead-free BiAgX solder paste system," pp. 1-7 in Proceedings of the Electronic Manufacturing Technology Symposium (IEMT), 2012 35th IEEE/CPMT International. 2012), 2012.

The Quantum Theory of Solid-state Atomic Bonding

Chin C. Lee and Lianxi Cheng*
Electrical Engineering and Computer Science
Materials and Manufacturing Technology
University of California, Irvine, CA, 92697-2660, USA
E-mail: lianxic1@uci.*edu*

Abstract Solid-state bonding refers to the bonding of solid material A and solid material B. Numerous experimental data have shown this possibility. Neither principle nor theory at the atomic level has been reported. How is solid-state bonding possible? Fundamentally, it is possible only if materials A and B can be brought within atomic distance. Over the years, we have proposed the principle: "As A atoms and B atoms are brought within atomic distance so that they can see each other, they will bond provided that they are willing to share the outer electrons." This is qualitative statement. In this research, we took it one step further and established a quantitative bonding theory. It has been proved that Cu, Ag, and Au atoms do share outer electrons to form molecules: Cu_2, Ag_2, Au_2, CuAg, AgAu, and CuAu. The binding energy, equilibrium distance, and vibrational frequency of the molecule have been measured. They are used to fit the Morse potential energy (E) vs. atomic separation S_{atm} curve. In our model, A atoms and B atoms on the bonding interface share electrons like molecules A:B, where ":" designates 2 shared electrons. The interface is emulated as 2-D array of A:B molecules. The A molecules connect to metal A represented by conventional model of ion core submerged in an electron sea. Same is true for molecules B. The breaking strength of the bonding interface is obtained by multiplying the binding force and the number of atoms per area. The Young's modulus can also be calculated. For Cu:Cu bonding, the Young's module is 261GPa. The experimental value is 110Gpa. The strength of Cu-Ag bonding interface depends on S_{atm} between Cu and Ag atoms. The maximum strength is 25.5Gpa at S_{atm} =0.283nm. The strength decreases to 2.55 GPa at 0.481nm, 255 MPa at 0.635nm, and 25.5MPa at 0.788nm. In bonding experiments, S_{atm} is determined by bonding conditions and the surface conditions. The bonding theory allows us to estimate how close the interface atoms have to be to achieve adequate bonding strength.

1. Introduction

In bonding a material to a solid substrate, the material can be in vapor state, liquid state, or solid state. Solid-state bonding refers to the bonding action for which the material to be bonded is in solid-state. Examples of vapor-phase bonding are sputtering and chemical vapor deposition techniques. Examples of liquid-state bonding are soldering and welding processes. A popular process for achieving solid-state bonding is thermo-compression. The actual bonding action is solid-state bonding. The fundamental mechanism and principle of solid-state bonding were seldom reported.

Over the past decade, we have proposed the solid-state atomic bonding principle as follows: "As A atoms and B atoms are brought within atomic distance so that they can see each other, they will bond provided that they are willing to share the outer electrons. Diffusion of A and B atoms alone does not guarantee bonding if A and B atoms do not want to share electrons. 'A' and 'B' atoms may share electrons to form A_xB_y compound. The A_xB_y compound may or may not bond with A or B atoms, depending on whether they are willing to share electrons." Our experimental bonding results have indicated that this bonding principle is correct [1-3]. At present, the ability to share electrons is determined by the experimental data. The broader questions are: "how do we determine whether A atoms and B atoms would share electrons and how closely they need to be brought together to achieve bonding?" We searched the literature and could not find any publications in this subject.

In this research, we establish a bonding model based on quantum mechanics. We begin with diatomic metallic molecules reported by others. It was experimentally discovered that diatomic molecules such as Cu_2, Ag_2, and Au_2 do exist in vapor state as determined by mass spectrometer study [4]. The dissociation energy of these molecules was measured by spectroscopic techniques. The energy versus atomic distance curves are obtained by fitting the experimental data to the Morse potential [5]. In our model, a two-dimensional array of molecules such as A:B is set up on the bonding interface. "A" atoms of A:B molecules in the array are connected to the metallic material A. "B" atoms of A:B molecules are connected to the metallic material B. Using this model, important properties of the structure of material A bonded to material B can be analyzed and obtained, including the breaking strength versus atomic separation and the Young's modulus.

In section 2, we introduce the basic wave function approach to the hydrogen-like molecules. By solving the *Schrödinger* equation, we can get the ground state energy as a function of the atomic separation, which is the essential part of the two-atom molecule system. In section 3, bonding model of diatomic molecules based on Morse energy is presented. In section 4, we build the solid-state atomic bonding model using experimental data reported on the binding energy, the equilibrium distance, and the vibrational frequency. Solid-state bonding strength and Young's modulus are then calculated. In section 5, experimental bonding model and data are presented and discussed.

2. Schrödinger equation for hydrogen-like molecules

The wave function is the quantum approach to studying the bonding of metallic molecules such as copper molecules. The ground state electronic configuration of a copper molecule is described by $d_A^{10} d_B^{10} \sigma^2$ [6]. The d-subshells are completely filled. To form a molecule, the atoms must share the outer σ electrons. The wave function of copper atoms

978-1-4799-2408-0/14 $31.00 © 2014 IEEE   1335   2014 Electronic Components & Technology Conference

does not exist in an analytical expression. However, numerical wave functions of the ground state of copper and other transition metallic atoms can be obtained using the Herman-Skillman program which was written based on the Hartree-Fock method [7]. The Hartree-Fock method is an approximation of the wave function and the energy of a quantum many-body system in a stationary state.

Similar to hydrogen molecules, the eigenstates of a diatomic metallic molecule are either singlet or triplet spin states. The normalized wave function is given by,

$$\psi_s(r_1, r_2) = \frac{\psi_A(r_1)\psi_B(r_2) + \psi_B(r_1)\psi_A(r_2)}{\sqrt{2(1+S^2)}} \quad (1)$$

$$\psi_t(r_1, r_2) = \frac{\psi_A(r_1)\psi_B(r_2) - \psi_B(r_1)\psi_A(r_2)}{\sqrt{2(1-S^2)}} \quad (2)$$

where $\psi_s(r_1, r_2)$ is the wave function for singlet state, $\psi_t(r_1, r_2)$ is the wave function for the triplet state, $\psi_A(r)$ is the normalized ground state centered on proton A, $\psi_B(r)$ is the ground states energy centered on proton B, and S is the overlap integral of the two wave functions.

The Hamiltonian operator of the singlet state and the triplet state can be written as:

$$H = -\frac{\nabla_1^2}{2m} - \frac{\nabla_2^2}{2m} - \frac{e^2}{|r_1 - R_A|} - \frac{e^2}{|r_2 - R_B|} + \frac{e^2}{|R_A - R_B|} + \frac{e^2}{|r_1 - r_2|} \quad (3)$$

where R_A and R_B are the position of the two protons, respectively, and r_1 and r_2 are the position of the $4s$ electrons.

The energy of a state is the expectation value of the Hamiltonian and can be calculated using the Heilter-London method [8], which results in,

$$E_\pm(R) = \frac{<AB|H|AB> \pm <BA|H|AB>}{1 \pm S^2} \quad (4)$$

$$<AB|H|AB> = \int d^3r d^3r' \psi_A(r)\psi_B(r') H \psi_A(r)\psi_B(r') \quad (5)$$

$$<BA|H|AB> = \int d^3r d^3r' \psi_B(r)\psi_A(r') H \psi_A(r)\psi_B(r') \quad (6)$$

The singlet state is a low energy election state yielding an energy function, $E_+(R)$, that has a minimum value. $E_-(R)$ is the energy of the triplet state that has higher energy. It rises monotonically with decreasing R. Thus, the molecule binds in the singlet state but not in the triplet state.

In this wave function approach, it is assumed that the core of the atom neither changes the Coulomb potential distribution nor interacts with other cores. However, this assumption is not true for a copper molecule which is actually much more complicated than a hydrogen molecule. Many theorists have made great effort to develop the first principle of calculating the potential energy of transition metal molecules. For example, effective core potentials and configuration interaction were used to calculate the bond length, vibrational frequency, and dissociation energy of Cu_2 [9]. Later on, theorists made more accurate calculations using the density functional theory [10].

3. Morse energy model of two atoms in a diatomic molecule and the breaking force

The Morse potential, as exhibited in Fig. 1, is a popular model for the potential energy of a diatomic molecule [5]. It presents a better approximation for the vibrational structure of a molecule than the quantum harmonic oscillator since the electronic potential energy curve is not purely quadratic. Most importantly, it contains the information of dissociated energy, which is related to atomic bonding. It is widely used to study other interactions, such as the interaction between an atom and a surface. It is also applied to fitting experimental spectroscopic data to obtain the vibrational frequency and binding energy, and to further plot the potential energy versus atomic separation curve of a molecule. It is determined by three parameters in the equation below,

$$V(r) = D_e(e^{-2\alpha(r-r_e)} - 2e^{-\alpha(r-r_e)}) \quad (7)$$

where r is the distance between the atoms, r_e is the equilibrium bond distance, and D_e is the dissociation energy, α is a parameter that relates to the force constant. D_e is often referred to as the well depth of the well seen in the energy versus distance curve. Fig. 1 exhibits the shape of the Morse potential energy curve showing the equilibrium bond distance r_e and the well depth D_e.

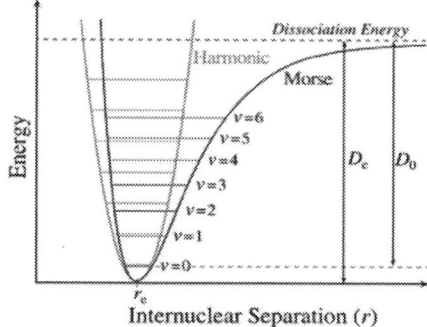

Figure 1. Morse potential versus bond distance of a diatomic molecule.

To further simplify the Morse potential energy curve, we can expand $V(r)$ around $r = r_e$ into Taylor series and keep up to the second-order term as follows:

$$V(r) = V(r_e) + \frac{\partial V(r)}{\partial r}\bigg|_{r=r_e}(r - r_e) + \frac{1}{2}\frac{\partial^2 V(r)}{\partial r^2}\bigg|_{r=r_e}(r - r_e)^2 \quad (8)$$

The coefficient of the second order is defined as the spring constant. Plugging equation (7) into equation (8) leads to

$$k_e = \frac{\partial^2 V(r)}{\partial r^2}\bigg|_{r=r_e} = 2\alpha^2 D_e \quad (9)$$

$$\alpha = \sqrt{\frac{k_e}{2D_e}} \quad (10)$$

The Morse potential expresses well the potential energy and bonding energy. To relate the energy to the spring constant and vibrational frequency, we use the harmonic oscillator approximation around the equilibrium distance. Its potential energy curve is also displayed in Fig. 1. The vibrational (oscillating) frequency is then calculated from the harmonic oscillator model as,

$$v = \frac{1}{2\pi}\sqrt{\frac{k_e}{\mu}} = \frac{\alpha}{2\pi}\sqrt{\frac{2D_e}{\mu}} \qquad (11)$$

where μ is the effective mass of the molecule with nuclei 1 and 2, which is defined as:

$$\mu = \frac{m_1 m_2}{m_1 + m_2} \qquad (12)$$

Experimentally measured vibrational frequency is defined as,

$$\omega_e = \frac{v}{c} = \frac{1}{c}\frac{\alpha}{2\pi}\sqrt{\frac{2D_e}{\mu}} \qquad (13)$$

where c is the speed of light. Please be aware that, here, ω_e is the vibrational frequency used in spectroscopic measurements with unit of cm^{-1}. It is the wave-number (1/wavelength) of the electromagnetic wave that resonates with the harmonic oscillator. It is different from the ordinary definition of angular frequency.

Table 1. Experimental data of D_e, r_e and ω_e. α is calculated based on ω_e.

Molecules	Experimental results			Calculate
	D_e / eV	$r_e / Angstrom$	ω_e / cm^{-1}	α
Cu₂	1.98[5]	2.22[5]	266.4[5]	1.453
Ag₂	1.65[5]	2.48[5]	192.4[5]	1.497
Au₂	2.29[5]	2.47[5]	190.9[5]	1.704
Cu-Ag	1.76[5]	2.37[5]	231.8[5]	1.504
Cu-Au	2.41[5]	2.33[5]	250.0[5]	1.519
Ag-Au	2.06[11]	2.64*[12]	176.7[5]	1.230

* The equilibrium distance for Ag-Au molecules is obtained from theoretical calculation since no experiment datum is available.

Table 1 gives all the experimental data of transition metal molecules. Taking Cu₂ molecules as an example, $\omega_e = 266$ cm^{-1}, the actual oscillating frequency is $v = \omega_e \times c = 8$ TeraHz. Based on the experimental data and the equations we derived above, the curves of energy vs. atomic separation of molecules are plotted as in Fig. 2.

To find the strength of two materials bonded together using solid state technique, we first need to find the breaking force of two atoms in a molecule. Near the equilibrium bond distance, the molecule can be treated as a harmonic oscillator.

$$F = \frac{dE}{dr} \qquad (14)$$

Based on Fig. 2, we calculated the restoring force versus atomic separation of six different molecules, as shown in Fig. 3. It is seen that the breaking force depends on the atomic separation. The minimum force required to break these two atoms apart is the maximum point of the $F - r$ curve. Taking Cu_2 as an example, the maximum point is at (1.439 eV/$\mathring{A}$, 2.697 $\mathring{A}$). Thus, the breaking force is 1.439 eV/$\mathring{A}$, or $2.30 \times 10^{-9} N$.

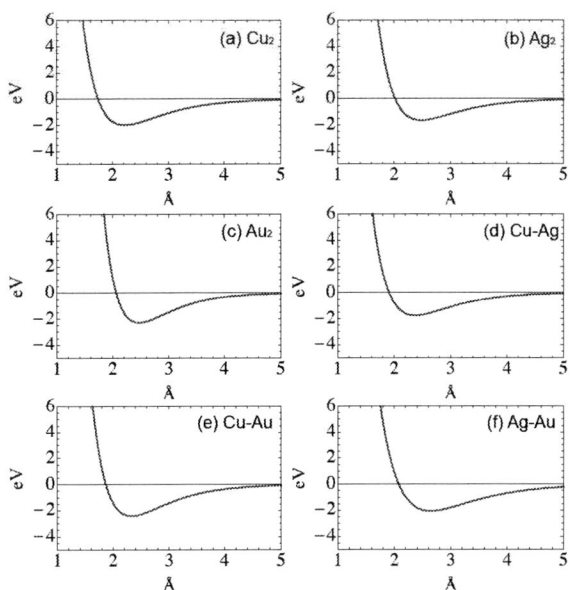

Figure 2. Energy vs. atomic separation.

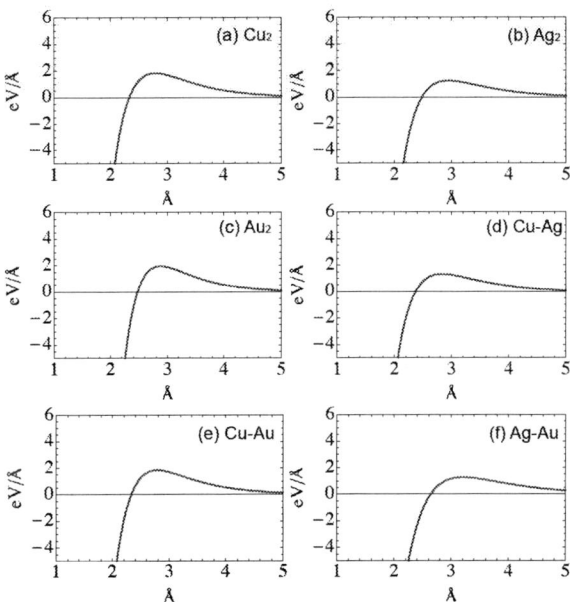

Figure 3. Restoring force between two atoms versus atomic separation.

4. Quantum bonding model, bonding strength, and Young's modulus

Using the diatomic metallic molecules presented above, a two-dimensional array of molecules such as A:B is constructed on the bonding interface, as exhibited in Fig. 4. "A" atoms of A:B molecules in the array are connected to the metallic material A. "B" atoms of A:B molecules are connected to the metallic material B. In Fig. 4, metals A and B are presented as the traditional free electron model consisting of ion core submerged in an electron sea [13].

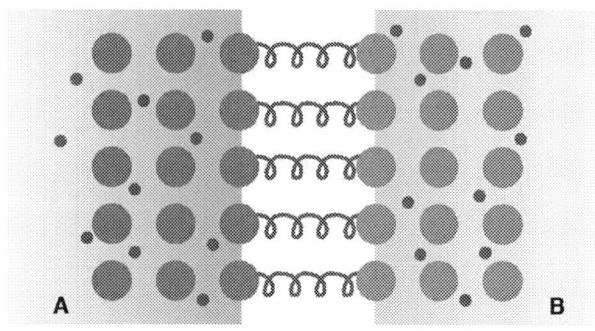

Figure 4. Our Quantum Solid-State Bonding Model Showing the Atomic Bonding Interface.

The breaking force between atoms A and B of the A:B molecule was discussed in Section 3.2. To find the breaking strength of the bonding interface represented by the 2-dimensional array of A:B molecules, we need to estimate the number of bonding molecules per interface area. Metals are seldom single crystal. But to make this model manageable, we assume that the bonding interface is on one of the principal planes of a cubic lattice, i.e., (100), (110) or (111). This should be good enough to illustrate the essence of the model. The density of atoms per surface area on these planes for Cu, Ag, and Au are shown in Table II. The calculated breaking strengths of the bonding interfaces formed by combinations of Cu, Ag, and Au on (100) plane are presented in Table 2. The strength values fall within a relatively narrow range: 23.8 to 37.6 GPa, as shown in Table 3. The strength of Cu:Cu, Ag:Ag, and Au:Au bonding interfaces is 27.8, 23.8, and 37.6GPa, respectively. The measured ultimate tensile strength of Cu, Ag, and Au is 220, 170, and 100M Pa, respectively. The calculated strength is 126 to 376x of the measured ultimate tensile strength. In our model, it is assumed that the interfacial molecules form a perfect 2-dimensional array on a plane. Real metals are far from perfection. There are numerous dislocations, sliding planes, and grain boundaries that can weaken its structure. Thus, it is probably too early to tell that the calculated strength values are far from the truth.

Table 2. Atomic density per area on different planes of Cu, Ag, and Au crystals in fcc structure.

plane	Atomic pattern	Cu/ m^2	Ag/ m^2	Au/ m^2
		$r = 128\,pm$	$r = 144\,pm$	$r = 144\,pm$
(111)		1.762×10^{19}	1.392×10^{19}	1.392×10^{19}
(100)		1.526×10^{19}	1.206×10^{19}	1.206×10^{19}
(110)		1.082×10^{19}	8.50×10^{18}	8.50×10^{18}

For hetero-structures such as Cu-Ag, there is a little mismatch on the number of atoms on the interface because of difference in lattice constants. This means that not all outer

electrons of Cu atoms can find outer electrons of Ag atoms to form binding pairs. A small percentage of outer electrons will be left unbound. In that case, the breaking strength is determined by the bound pairs.

Table 3. Breaking strength of the bonding interfaces formed by combinations of Cu, Ag, and Au.

Breaking strength /GPa	Cu	Ag	Au
Cu	27.8	25.5	35.3
Ag	25.5	23.8	24.4
Au	35.3	24.4	37.6

From the spring constant of the springs that emulate interface molecules exhibited in Fig. 4, we can calculate the Young's Modulus. The spring constant is given by,

$$k = \frac{\partial^2 E}{\partial r^2} \tag{15}$$

The calculated curves of the spring constant versus atomic separation of Cu_2, Ag_2 and Au_2 are displayed in Fig. 5. Since the molecules are not a harmonic oscillator, the spring constant is a function of displacement. In other words, the spring is nonlinear. Near the equilibrium atomic separation, the spring can approximated as a harmonic oscillator. As we stretch the 2-dimensional array of springs in Fig. 4 in the direction perpendicular to the interface, we assume that the lattice will only stretch in that one direction as portrayed in Fig. 6. The separation of atoms parallel to the interface is kept the same while the separation of atoms in the perpendicular direction becomes larger. Based on our quantum bonding model, we try to figure out how the change of atomic separation affects the breaking force.

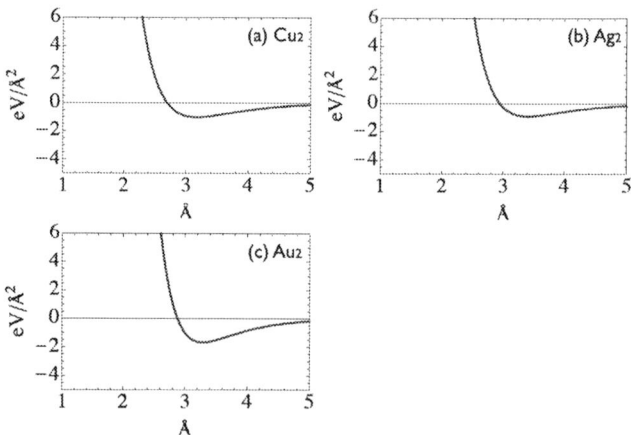

Figure 5. Spring constant between two atoms versus atomic separation of molecules Cu_2, Ag_2 and Au_2 from (a) to (c).

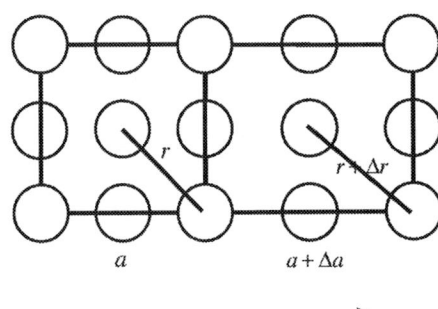

Figure 6. FCC lattice being stretched in one direction.

For fcc lattice, it can be shown that,

$$r = \frac{\sqrt{2}a}{4}, \quad \Delta r = \frac{\Delta a}{2\sqrt{2}} \quad (16)$$

where a is the lattice constant and r is the separation between two closest atoms. Hooke's Law gives,

$$f = k \times \Delta r \quad (17)$$

Multiplying both sides by the total number of atomic bonding pairs, N, we get,

$$fN = (k \times \Delta r)N = kN \frac{a}{2\sqrt{2}} \frac{\Delta a}{a} \quad (18)$$

The Young's modulus is the slope of the stress versus strain curve. We obtain the expression of Young's modulus as,

$$E = kN \frac{a}{2\sqrt{2}} \quad (19)$$

The calculated values and experimental values are compared in table 4. Calculated values are higher than experimental values by a few times. Again, the model assumes perfect crystal while real metals posses numerous defects, dislocations, sliding planes which make them less stiff.

Table 4. Young's Modulus of Cu, Ag and Au solids.

Young's Modulus /GPa	Cu	Ag	Au
Calculation	261	206	370
Experiment	120	83	79

5. The Experimental bonding model and bonding strength data

In a perfect situation, the atomic separation of the interface molecules should be the equilibrium atomic distance, r_e, as shown in Table 1. For Cu-Ag, it is 0.237nm. When materials A and B are brought into atomic distance during a solid-state bonding process, the atomic distance that can be achieved on the bonding interface would depend largely on surface conditions and cleanness. Fig. 7 depicts an experimental bonding model. In reality, the binding pairs on the interface are far from idea. As shown in Fig. 3, the restoring force of an interface molecule depends on the atomic separation. For Cu-Ag interface molecules, the maximum restoring force is $2.12 \times 10^{-9} N$, which is obtained at an atomic separation of $S_{atm} = 0.283$nm. During bonding, if the Cu and Ag atoms in the *CuAg* molecules are held at atomic separation larger than

the separation at peak restoring force, the force needed to pull the atoms apart will be smaller than the maximum restoring force. It will be the restoring force at the specific atomic separation shown in Fig. 3. Based on this concept, the breaking strength of Cu-Ag bonding interface is calculated versus interface atomic separation and exhibited in Fig. 8. It is seen that the maximum strength is 25.5GPa at S_{atm} =0.283nm. The strength decreases to 2.55GPa at 0.481nm, 255MPa at 0.635nm, 25.5MPa at 0.788nm, and 2.55MPa at 0.942nm. We can see that interfacial bonding conditions must be near perfect to achieve adequate bonding strength.

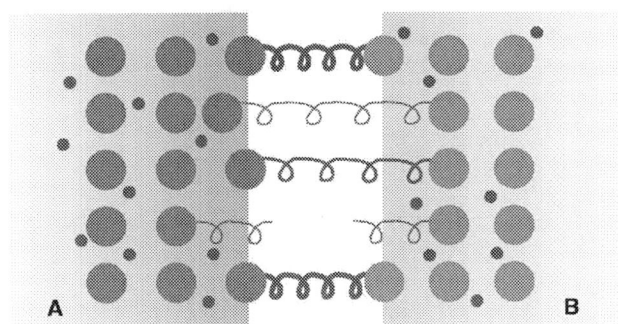

Figure 7. Experimental quantum bonding model showing imperfect bonding interface.

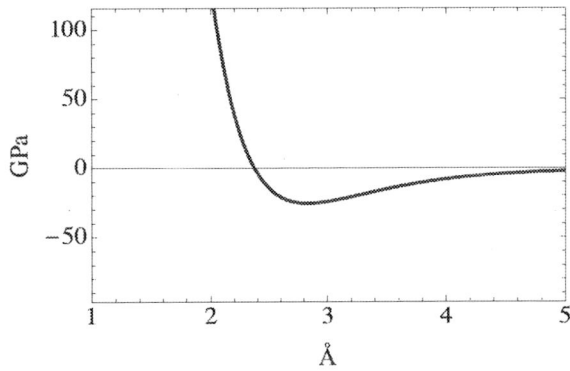

Figure 8. Breaking strength of Cu-Ag bonding calculated using our quantum model.

Among the many solid-state bonding processes that we developed, three are cited here [2-4]. The bonding results seem to indicate that the solid-state bonding principle that we established is correct. Experimental results of Ag-Cu bonding will be cited and discussed here because shear strength values and fracture surfaces are available [4]. Fabrication processes are briefly reviewed. 99.9% 12mmx12mmx0.8mm Cu substrates were first electroplated with 50μm thick Ag. They were annealed at 400°C for 5 hours to grow the Ag grains from 30nm to several μm, thus making the Ag layer more ductile. The 8mmx10mm Cu chip to be bonded was slightly polished to remove oxides and placed over the Cu substrate. They were held together with 1,000psi pressure and heated to 300°C in a vacuum chamber pumped to 0.1 torr. The bonding time at 300°C is 3 minutes which is limited by the equipment. In principle, bonding action should occur in seconds rather than in minutes. It is worth mentioning that the pressure applied here is at least an order of magnitude smaller than what used in industrial thermo-compression processes. Fig. 9 shows the cross section SEM images. The entire Cu-Ag bonding interface on the cross section is seen to be perfect without voids, gaps, or cracks.

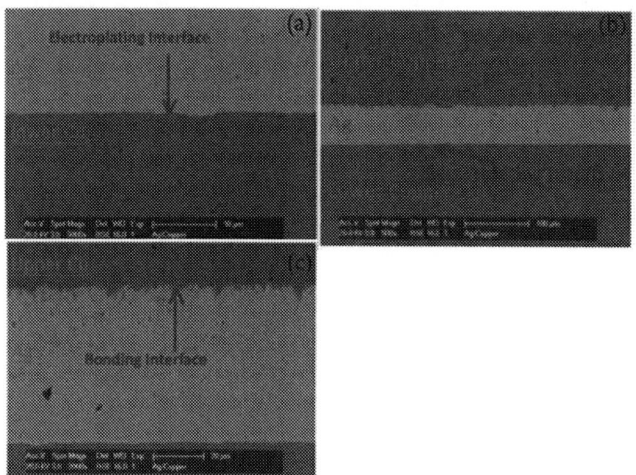

Figure 9. Cross-section SEM images of the sample bonded at 300°C with 1000 psi static pressure (a) Ag/Cu region, (b) Cu/Ag/Cu region, and (c) Cu/Ag region.

To evaluate the bonding strength, tensile test was not used because the tensile tester cannot grip on the sample. Instead, shear tester was used. Of the 5 samples tested, the breaking force ranges from 45 to 90kg, far exceeding 5kg shear breaking force required by MIL-STD-883G standard method 2019.7 in semiconductor die attachment. Sample 5 has 90kg breaking force, corresponding to shear strength 10.5MPa. This is only 0.4×10^{-3} of the 25.5GPa tensile strength predicted by the quantum bonding model presented in last section. If the quantum bonding model can serve as guidance, it would mean that there is a lot of room to improve in increasing the bonding strength. One thing to investigate is to observe if Cu and Ag atoms on the interface are brought within atomic distance and form bonding pairs. The cross section sample for SEM examination cannot provide this information because the bonding interface cannot be atomic sharp due to smear of Ag and Cu in the polishing process. So, we turn to an indirect observation, looking at the fracture surface. Fig. 10 is the optical microscopy image of the Cu chip that was sheared off sample 5. A vast majority of the fracture surface seen is Cu. This indicates that most of the Cu surface was not bonded to the Ag because the Cu surface did not conform to the Ag surface within atomic scale. Many small patches of Ag in sizes from 5 to 25µm remained bonded on the Cu chip. These represent the actual bonded regions. An estimate shows that the total bonded area is only 8.3%. If the entire interface were bonded, the shear strength could be increased to 127 MPa.

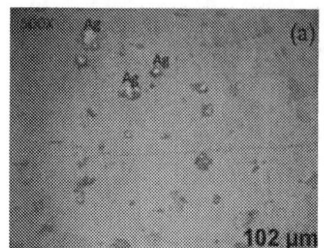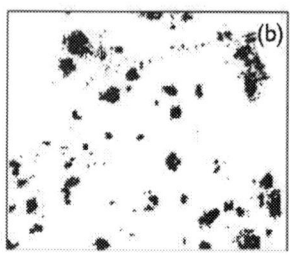

Figure 10. Cu chip that was sheared off sample no. 5: (a) optical microscopy image of a typical region, (b) Bonding patches for calculating the percentage of bonding on the Ag-Cu bonding interface.

One way to increase the percentage of bonding is to increase the pressure. But any pressure higher than 1,000psi is not acceptable in our applications. We are looking for other ways to improve the bonding strength without increasing the pressure or the temperature.

Summary

We established this bonding theory based on quantum mechanics of diatomic molecules. Metal atoms such as Cu, Ag, and Au do form diatomic molecules in vapor state: Cu_2, Ag_2, Au_2, CuAg, AgAu, and CuAu. The well depth, equilibrium separation, and vibrational frequency had been measured and reported. We took three parameters and construct the Morse energy vs. separation curves for all these molecules. The bonding model consists of 2-D array of molecules such as A:B to emulate the bonding interface. Molecules A connect to metal A that is represented by the conventional model of ion core submerged in an electron sea. Same is true for molecules B. We then calculated the bonding strength as a function of atomic separation on the interface. In bonding experiments, the atomic separation on the interface is not the equilibrium value, but rather is controlled by bonding and surface conditions. The theory allows us to calculate the bonding strength at specific atomic separation. It tells us how close the interface atoms have to be brought together to achieve any bonding. It provides a guideline on how strong a bonded structure can be. Furthermore, a model was developed to calculate the Young's modulus of these meals.

Acknowledgments

This research was supported by II-VI Foundation.

References

1. Pin J. Wang, Jong S. Kim, and Chin C. Lee, "Direct silver to copper bonding process." *ASME J. Electronic Packaging*, 130, pp. 45001-1 to -4, Dec. 2008.
2. Chu Hsuan Sha, Pin J. Wang, Wen P. Lin, and Chin C. Lee, "Solid state bonding of silver foils to metallized alumina substrates at 260°C." *ASME J. Electronic Packaging*, 133, pp. 041007-1 to 3, Dec. 2011.
3. Yi-Ling Chen and Chin C.Lee, "Strength of solid-state silver bonding between copper." *Proc. IEEE Electronic Components and Technology Conference*. pp. 1773-1776, Las Vegas, Nevada, May 28-31, 2013.

4. Jean Drowart and Richard E. Honig, "Mass spectrometric study of copper, silver, and gold." *The Journal of Chemical Physics*, vol. 25, pp. 581-582, 1956.

5. Michael D. Morse, "Clusters of transition-mental atoms." *Chemical Reviews*, vol. 86, pp. 1049-1109, 1986.

6. Gregory A. Bishea, Ninette Marak, and Michael D. Morse, "Spectroscopic studies of jet-cooled CuAg," *The Journal of Chemical Physics*, vol. 95, 5618, 1991.

7. Herman-Skillman Program http://hermes.phys.uwm.edu/projects/elecstruct/hermsk/HS.TOC.html

8. Baym Gordom, *Lectures on Quantum Mechanics*, Westview Press, New york, 1990, pp. 480-481.

9. Michel Pelissier, "Bonding between transition metal atoms. Ab initio effective potential calculations of Cu", *The Journal of Chemical Physics*, vol. 75, 775, 1981.

10. E Y. Zarechnaya, Natalia V. Skorodumova, SI Simak, Boïje Johansson, and Eyvas I. Isaev, "Theoretical study of linear monoatomic nanowires, dimer and bulk of Cu, Ag, Au, Ni, Pd and Pt," *Computational materials science*, vol. 43, pp. 522-530, 2008.

11. Gregory A. Bishea, Jacqueline C. Pinegar, and Michael D. Morse, "The ground state and excited d-hole states of CuAu," *The Journal of Chemical Physics*, vol. 95, 5630, 1991.

12. Richard B. Ross and Walter C. Ermler, "Ab initio calculations including relativistic effects for diatomic silver, diatomic gold, silver-gold (Ag-Au), silver hydride (AgH), and gold hydride (AuH)," *The Journal of Physical Chemistry*, vol. 89, 5202-5206, 1985.

13. Neil W. Ashcroft and N. David Mermin, *Solid State Physics,* Cengage Learning, New York, 1976, pp. 32-40.

Effective Method to Disperse and Incorporate Carbon Nanotubes in Electroless Ni-P Deposits

Sha Xu [1], Yan Cheong Chan [1*], Xiaoxin Zhu [2], Hua Lu [2], Chris Bailey [2]

[1] Department of Electronic Engineering, City University of Hong Kong,
83 Tat Chee Avenue, Kowloon Tong, Hong Kong, China
[2] School of Computing and Mathematical Sciences, University of Greenwich,
30 Park Row, London SEI0 9LS, UK
* Contact e-mail: eeycchan@cityu.edu.hk

Abstract

The ever-increasing demand for higher I/O counts on chip requires the finer pitches to improve the performance, cost effectiveness and higher yield. The conventional under bump metallization (UBM) technology may not guarantee the required performance due to higher current density and diffusivity nowadays. In order to overcome these challenges, a UBM with higher strength and resistance to diffusion, better thermal and microstructural properties are required. Recently, nano reinforced Ni-P alloy has been identified as potential approach. Especially, the addition of Carbon nanotube (CNT) was expected to provide strengthening to the resulting UBM layers. CNTs have exceptional and attractive thermal, electrical, and mechanical properties, and are believed to be ideal material for fabricating composites. However, due to the large surface volume ratio, CNTs have strong tendency to agglomerate, which results in clusters. CNT clusters may lead to void formation and further cracking. Therefore, it is of great importance to figure out methods to disperse and incorporate CNTs into Ni-P UBM. In order to disperse carbon nano-tubes in electroless Ni-P coatings, two approaches were employed. Chemical modification, which includes acid oxidation treatment and surfactant treatment were conducted to disperse CNT clusters. Meanwhile, magnetic stirring and ultrasonic agitation were also employed to prepare the CNT/Ni-P composite coatings with CNTs homogeneously embedded. SEM and TEM were used to observe the dispersion after adopting the aforementioned treatment as well as the surface morphology of the deposited layers. It was verified that the dispersion can be significantly improved after proper treatment process, and surface of the CNT/Ni-P was quite smooth ; CNTs were equably dispersed throughout the matrix. In addition, the interfacial bonding between CNTs and Ni-P coatings was good and firm. In these cases , an understanding of the effects of the acid oxidation, surfactant dispersant, magnetic stirring and ultrasonic agitation on CNT dispersion in both solution and deposited layer were ascertained.

Introduction

In the modern microelectronic industry, interconnections play an important role to provide electrical, mechanical and thermal function during assembly process. As fine pitch and tiny size are required by high-density packaging, the conventional interconnection technology cannot meet the new standard even under the same service condition [1]. Thus, the reliability of miniaturized solder interconnections are crucial issues. Moreover, with the implementation of Waste Electrical and Electronic Equipment (WEEE) and Restriction of the Use of Hazardous Substances in Electrical and Electronic Equipment (RoHS Directive), many new interconnections reliability issues arise along with the application of lead-free solders, due to their high reaction temperature and high dissolution rate. Because the lead-free alloy material normally have higher melting temperature and longer peak temperature time (time above melting point), it may not only expose the IC components to the challenge of overheating, but also lead to rapid and excessive growth of brittle and porous intermetallic compounds (IMC).

A great deal of research on improving the reliability of the lead-free solders have been carried out, and one of the most effective and low-cost approach is to incorporate small amount of reinforcing particles into the solder alloy [2]. These foreign particles normally act as reinforcement phase. Shen et al. studied eutectic Sn_9Zn solder alloys with nano ZrO_2 particles and significantly improved their hardness as well as shear strength [3]. A.K. Gain et al. studied the microstructure, thermal property and hardness of $Sn_{3.5}Ag_{0.25}Cu$ solders containing 1 wt.% TiO_2 nanoparticles, and showed that after the addition of nano TiO_2, the solder alloys had finer microstructure and higher hardness than plain $Sn_{3.5}Ag_{0.25}Cu$ solders [4]. These previous research are very effective in enhancing the mechanical properties of joints, but most of them focused on the improvement of solder material, while too little attention has been paid to under bump metallization (UBM), that will lead to many failure modes. In this study, a novel method of incorporating carbon nanotubes into UBM was proposed to improve mechanical integrity of solder interconnections.

Nowadays, electroless Nickel (phosphorus) coating is the most widely used diffusion barrier film which act as UBM on the BGA Flip Chip. Nickel (phosphorus) plating has great advantages of uniform surface, low cost, but the mechanical strength and stability of the Ni-P UBM still has potential to fail under high reaction temperature [5]. Therefore, electroless composite deposition was introduced as a promising approach to improve the properties of Ni-P UBM by incorporating reinforcement particles. The conventional electroless composite deposition are conducted by adding micro-sized SiC, BN, Si_3N_4, Al_2O_3, graphite, diamond, Polytetrafluoroethylene (PTFE), etc. to plating bath [6]. However, micro-sized composite coatings cannot meet the requirements of modern electronic industry, and nano-scaled reinforcing particles come in age owing to their superior properties. Among all the nano-sized candidates, carbon nanotubes (CNTs) have received great attention due to their high Young's modulus, superb chemical stability [8][9][10]. Moreover, CNTs are low mass density material, which can be

978-1-4799-2408-0/14 $31.00 © 2014 IEEE

loaded into almost any host matrix of polymer, metal and ceramic, without increasing original weight. Zarebidaki et al. reported that CNTs can improve both corrosion resistance and micro hardness of pristine electroless Ni-P layer [7]. And similar results have been proved by many other researchers. However, there are little work studied the deposition mechanism of CNTs, and the dispersion is not satisfactory, since the homogeneous CNT distribution in Ni-P UBM is the key factor for superior mechanical property. In this study, acid oxidation, surfactant dispersion, magnetic stirring and ultrasonic agitation have been introduced to improve the dispersion, and the effects of each additive or assisted method have been investigated. Besides, the morphology of Ni-P/CNT UBM have been examined.

Experiment Procedures

Materials

Commercial Multiwall CNTs (grown by chemical vapor deposition with purity $\geq$ 95%, diameter of about 60 nm and length of 1-2 μm) were provided by ShenZhen Nano Tech. Port Co., Ltd (China). The surfactants polyvinylpyrrolidone (PVP,PVP10-500G,Sigma-Aldrich) and The surfactants-sodium dodecyl benzene sulfonate (SDBS, 90%, TianJin FuChen Chemical reagents Co., Ltd) were used as-received to fully de-bundle the CNTs. Cu plates (6 mm thick, 99.98 wt.%) were used as substrate for composite plating.

Pre-treatment and Surfactants

Liquid-phase chemical oxidative pre-treatment is employed in this experiment to remove impurities and. Firstly, the CNTs were purified with HCl (37 wt. %) under ultrasonic bath (BRANSON B5210E, 47 kHz) for 0.5 hour at room temperature, then the purified CNTs were refluxed in concentrated HNO_3 (67 wt. %) at various temperature(25°C, 50°C, 80°C) for 10 hours. After the above treatments, the CNTs were dispersed in water with a concentration of 1.0 wt.%. After mixing CNTs into the solvent, all the samples were treated with centrifuge to remove large tangles of CNTs. Finally, the CNTs were filtered, washed and dried at 80°C in incubator. Each of the above steps were followed by de-ionized water washing. After the acid treatment, CNTs were treated with different surfactants with different ratio.

The Cu substrates were polished to micrometer surface finish and etched with HNO_3 (30 vol.%) for 30 s to eliminate surface scratches and metallic oxide impurities. Then the Cu substrate was immersed in commercial Ru activation solution for surface activation.

Electroless plating

The electroless Ni-P/CNT plating bath was prepared by dispersing CNTs into the commercial Nickel (phosphorus) plating solution (from MacDermid). Before plating process, the surfactant PVP/SDBS was added into the CNT suspension solution. Then the solution was magnetic stirred at 600 rpm for 10 hours to de-bundle the CNTs clusters by further eliminating the van der Waals force. the CNT/Ni-P plating bath was ultrasonic processed for another 3 hours to avoid agglomeration, and the ultrasonic agitation continued in the entire process. This alkaline Ni-P bath contains nickel sulphate ($NiSO_4.6H_2O$) as nickel sources, sodium hypophosphite($NaPO_2H_2$) as a reducing agent along with sodium citrate($Na_3C_6H_5O_7$) as the complexion and buffering agent.

Microstructure and Morphology

The surface morphology of all coating layers was observed by atomic force microscope (AFM, Park Systems). The microstructure and surface morphology observation were carried out by scanning electron microscope (SEM, JEOL, JSM-6340F). The CNTs were observed by Transmission Electron Microscopy (SEM, JEOL, JEM-2010).

Results and Discussion

Dispersion Effects of Acid Oxidation

The as-received CNTs were shown in Fig.1. As can be seen from the SEM image, the raw CNTs have strong tendency to agglomerate. The agglomeration can be attributed to two main reasons. The first one is owing to the existence of impurities. There are always amorphous carbon, fullerene, graphite particles and even metal catalysts adhering to the CNTs no matter what the synthesis method is. These impurities act as nucleation center that attracted CNTs by van der Waals force. The second reason is that CNTs have high surface energy and large curvature. Just like the headset wires often get twisted, the CNTs are easily to tangle because of its large length-diameter ratio.

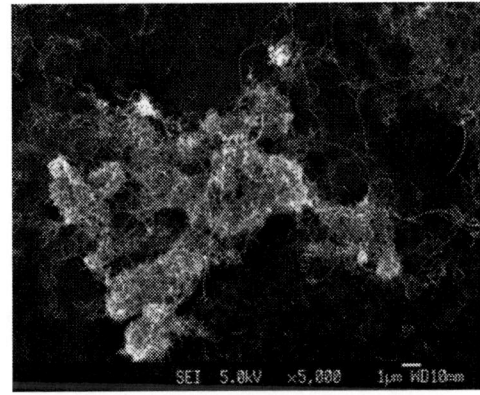

Figure 1. SEM image of raw CNTs

These agglomerations have disastrous impact on the quality of deposited layer. Because the large agglomerated clusters will result in voids, which will transform into the starting point of crack propagation. Besides, gliding effect will appear between individual CNT, which will degrade the mechanical strength further.

The image of HCl treated and HNO_3 treated CNTs are shown in Fig. 2. Compared to HNO_3, HCl is not strong oxidative acid, which cannot fully reacted with carbon byproducts. However, two-step acid treatment is more effective than using HNO_3 only. Generally, HCl can react with transition elements, which often act as catalysts during CNTs synthesis; while strong oxidative HNO_3 can melt away other impurities. Fig. 2(a) demonstrates that the tangling phenomenon still exist after HCl treatment, and the morphology of each CNTs are quite different. Thin and short CNTs have tendency to tangle with thick and long CNT, which forms clusters finally. Fig. 2 (b) shows the CNTs after

HNO₃ treatment for 10 hours. The long and thick CNTs were cut into smaller and thinner fragments, and the morphology of CNTs become more uniform. Meanwhile, the CNTs clusters become loose, and it suggests that disintegration or de-bundling reaction may happen during HNO₃ treatment.

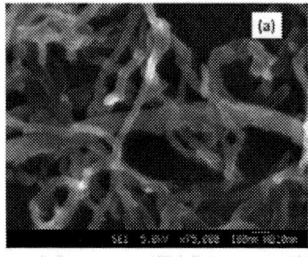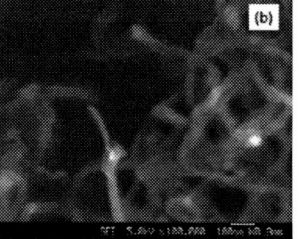

Figure 2. SEM image of (a)HCl-treated CNTs (b)HNO₃-treated CNTs

Because the carbon atoms that form CNTs are fixed in a hexagonal network, each carbon atom is covalent bonded to three other carbon atoms, which are more chemically stable than amorphous carbon and graphite particles. If the reaction time and processing condition is carefully and accurately controlled, only carbon impurities and residues will be consumed.

Even though, acid treatment can eliminate unwanted impurities, the total yield of CNTs are relatively low after two-step treatment, and 25%~30% yield was very common after acid treatment. In this experiment, the total yield was measured for different HNO₃ treatment temperature (25°C, 50°C, 80°C) for 10 hours, while the HCl treatment remain the same. The weight percentage of residual CNTs are calculated by dividing the weight of the dried acid-treated CNTs on the filter paper that was used to filter the deionized water-diluted mixture after each acid treatment by the original weight of CNTs before treatment. As plotted in Fig. 3, the total yield of CNTs after two-step acid treatments decrease with the treatment temperature. The decrease rate is much higher when temperature is below 50°C, while the yield rate is much milder when above 50°C.

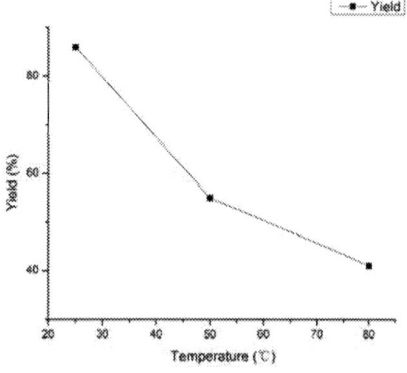

Figure 3. The total yield of CNTs after two-step acid treatment

Acid treatment not only eliminated away impurities and carbon byproducts, but also change the morphology of CNTs. It is because the crude CNTs are capped on both ends by half-spherical fullerene fragments, which is not in hexagonal structure, so the fullerene will be peeled off after the acid reaction. Then the acid filled in the hollow space of the CNTs, and reacted with its inner walls. After the inner wall become thinner, the pristine CNTs will be etched into bamboo type, as shown in Fig.4. Actually, this artificially introduced defects are quite beneficial to the deposition quality. It is reported by G Yamamoto et al, these defects which are in nano-scale have interlocking effect with the metal matrix [11], the Nickel atoms can easily fill up the nan defects, and form firm bonds with CNTs. These acid-introduced nan defects can decrease the poor CNT–matrix connectivity and severe phase segregation.

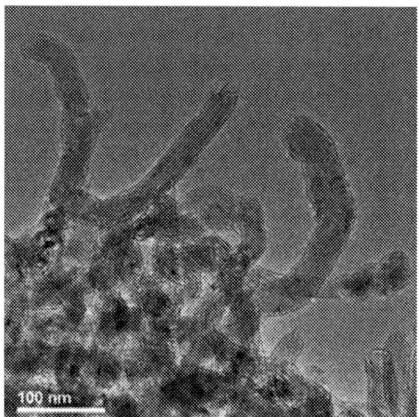

Figure 4 TEM image of acid-treated CNTs with nan defects in bamboo structure

Dispersion Effects of Surfactant

Even though acid treatment is very effective in disintegrate and de-bundle the CNT cluster, it is not enough for high quality Ni-P deposition. Because, there is still residual van der Waals force that attracting individual CNT, even after the impurity particles have been removed by acid cleaning. Therefore, surfactant dispersant is necessary to distribute CNTs homogeneously, and form stable dispersion solution, because surfactant can accelerate the breakup of the agglomerated CNTs. It is well know that van der Waals force made the CNTs aggregate into clusters, so adding surfactant is an effective approach to de-bundle them. In this study, two different surfactants were employed to obtain stable CNT solution. Both of PVP and SDBS are noncovalent modification, which do not destroy the structure of CNTs. The difference is that SDBS are randomly adsorbed on CNTs, which is shown in Fig. 5(d), while PVP are molecular chain type that surrounded and adhered to the outer walls of CNTs, which is shown in Fig. 5(e). Both of the surfactant can enhance the wettability of CNTs, and help to obtain stable dispersion solution.

Fig. 5(a) shows the SEM image of pristine CNTs without adding any surfactant. It is clear that agglomerates and closely packed clusters are very common in the image. After the acid cleaning, metallic and amorphous carbon impurities has been melted away, thus no great agglomerations exist. However, CNTs still tangled with each other, and the concentration is too high for electroless plating bath. Compare Fig. 5(b), which shows the CNTs treated by SDBS with the CNTs without surfactant, the CNT clusters become significantly loosen, while the concentration is much lower than pristine CNT solution. PVP shows the most superb disperse ability, which is confirmed by Fig. 5(c), even better dispersion and lower concentration was achieved after adding PVP to CNT solution. It is an effective method to separate CNTs through surfactant treatment.

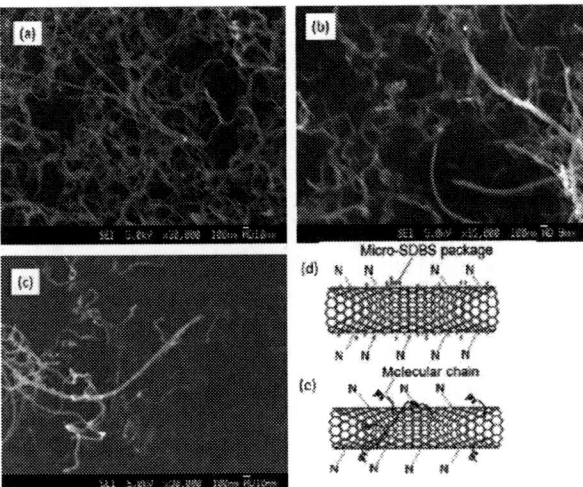

Figure 5 SEM images of (a)pristine CNTs (b)CNTs treated by SDBS (c)CNTs treated by PVP (d)schematic picture of SDBS adsorption (e)schematic picture of PVP molecular chain

In order to intensify the disperse ability and keep the CNT solution more stable, mixed surfactant treatment is employed and examined in this study. PVP and SDBS were mix with the ratio of 1:1 before adding to the CNT solution, after 30 minutes sonication, one drop of solution was scrutinized under TEM examination. In Fig. 6(a), the CNTs were treated without surfactant, and they still remain as aggregates. While the CNTs treated with mixed PVP/SDBS can be fully exfoliated as single CNT. It indicates that, almost CNTs have been peeled off, and the dispersion is achieved.

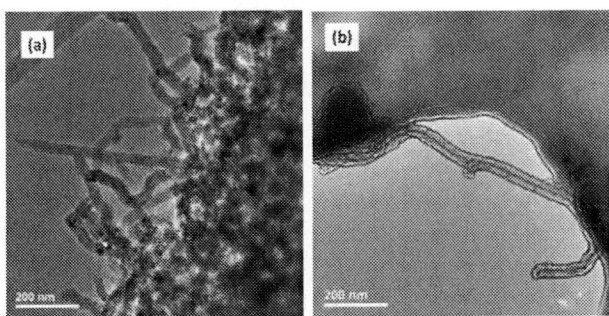

Figure 6 TEM images of (a) pristine CNTs (b)CNTs treated with mixed PVP-SDBS

Dispersion Effects of Magnetic Stirring

Generally, acid treatment and surfactant are chemical dispersion methods. In order to get higher dispersion level, mechanical methods should necessarily be applied before plating process. Ball milling is a convenient and cost effective approach to disintegrate CNTs clusters by exfoliating. However, ball milling will damage the natural structure of

CNTs. The outer walls will become irregular after ball milling, while the high aspect ratio of CNTs will be destroyed. Therefore, magnetic stirring is a more gentle method to remove aggregated clusters, without changing the microstructure of CNTs. In this study, magnetic stirring has been conducted for 10 hours before deposition, and it has direct influence on the deposition quality.

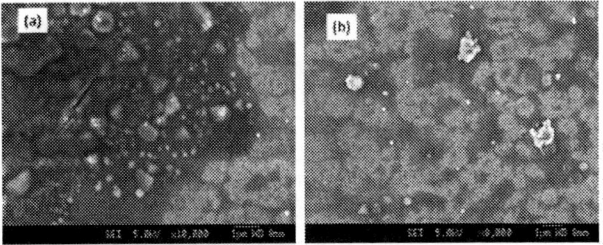

Figure 7 SEM image of deposited CNT/Ni-P layer (a) without magnetic treatment (b) with magnetic treatment

As depicted in Fig.7(a), the surface morphology of deposited CNT/Ni-P layer without magnetic stirring treatment is poor. The surface is covered with irregular particles, and a crack is also very obvious on the surface. Besides, part of the surface is rich in CNTs, which are the white nano dots in the left of Fig.7(a); while CNTs are absent from other part of the surface. The surface is not smooth because of these uneven bumpy nodules. It is because metal atoms can hardly cover CNT agglomerations continuously, they can only cover CNT agglomerations as isolated particles, which made the surface unsmooth and uncontinuous.

In Fig. 7(b) shows the surface morphology of deposited CNT/Ni-P layer with magnetic stirring treatment. The surface was compact and exhibits an even nodular shape with a typical cauliflower-like morphology. There was no irregular shaped particles on the surface and the CNTs were evenly distributed and embedded in the coating matrix. However, there are several bumpy nodules, which are small CNT clusters covered by metal atoms. As can be seen, there are more than one CNTs exist inside of the bumpy nodules, and the size of the nodules are normally 1 μm.

In order to further examine the surface quality of deposited CNT/Ni-P layer with and without magnetic stirring, AFM is employed to measure the surface smoothness and continuity. In Fig. 8 (a), The topographic image of CNT/Ni-P coating without magnetic stirring showed a variation in height, the root mean square roughness of the coating was 72.4 nm, with a peak to peak roughness of 287.8 nm. The surface cross-section analysis was marked with red line in the image, which showed distribution of hills and valleys. Besides, the bumpy area occupies almost half of the surface. On the other hand, the root mean square roughness of CNT/Ni-P coating with magnetic stirring was 21 nm, and the peak to peak roughness was 94 nm. The surface cross-section exhibits a more uniform and finer distribution of hills and valleys as compared to coatings without magnetic stirring. The size of bumpy nodules were around 1 μm, which is in accordance with Fig. 7(b). The decreased roughness indicates that magnetic stirring treatment is effective in forming uniform, continuous and compact surface.

978-1-4799-2408-0/14 $31.00 © 2014 IEEE

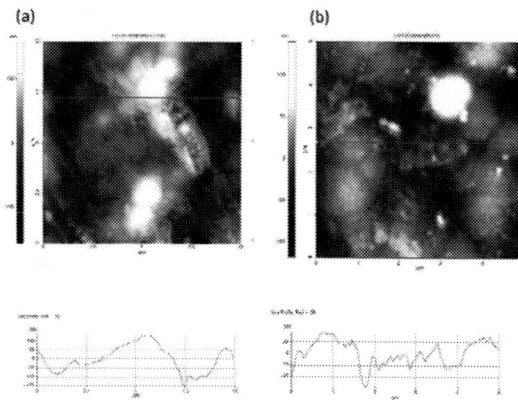

Figure 8 AFM topological image of CNT/Ni-P deposited layer (a)without magnetic stirring (b) with magnetic stirring

CNT incorporation with ultrasonic agitation

Ultrasonic agitation is another mechanical dispersion method, which is continuing in the entire plating process. Fig. 9(a) is a SEM micrograph of CNT/Ni-P composite coatings without ultrasonic agitation. The surface is compact and continuous, shown a typical cauliflower-like morphology, with a few CNTs dispersed in it. However, even though the CNTs have been fully de-bundled, a bumpy nodule still exist on the surface of the coating layer. The concentration density is not satisfactory, only a few CNTs can be found, which is not enough for desired reinforcing effect. Besides, the CNTs are not deeply embedded in the metal matrix, instead the CNTs are only parallel connected with the deposited layer, and the connection between CNTs and coatings is relatively weak.

Fig. 9(b) is a SEM micrograph CNT/Ni-P composite coatings without ultrasonic agitation. Similarly, the surface is compact and continuous, shown a typical cauliflower-like morphology. Several CNTs can be found in the image, and the CNTs are not presented in agglomeration morphology, it was deeply contacted with the metal matrix and no voids existed around the CNT. This proved that metal atoms can continuously deposit on individual CNTs, which is beneficial to form void-free surface. The concentration density is higher than the deposited layer without ultrasonic agitation, which is beneficial to high mechanical strength and corrosion resistance. It can also be observed that the diameter of CNTs have slightly increased, because some of the CNTs were wrapped and covered by Ni particles during the plating process. This is very advantageous in forming bonds between CNTs and Ni matrix.

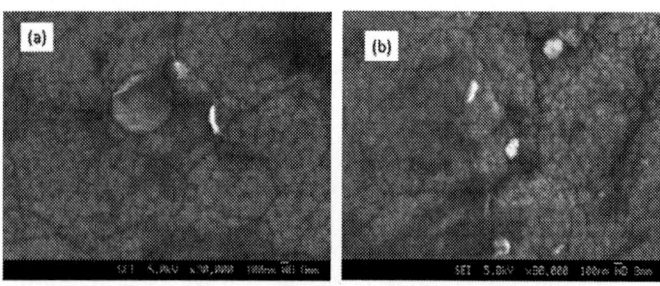

Figure 9 SEM image of the deposited CNT/Ni-P coatings (a) without ultrasonic agitation (b) without ultrasonic agitation

The formation of stable CNT solution is shown in Fig.10. After the CNTs are coated with surfactant, space or gap will be formed, while the surfactant will propagate along the gap with the assistance of ultrasonic agitation. Finally, individual CNT will be fully exfoliated from the clusters. It proves that, ultrasonic agitation is effective in assisting the propagation of surfactant, and then exfoliate single CNT [12].

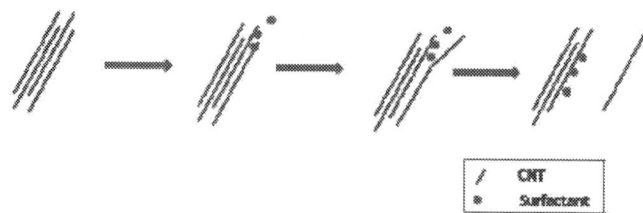

Figure 10 schematic illustration of CNT exfoliation with the assistance of surfactant

Studying the composite deposition mechanism, it can be assumed that the plating of CNTs and Ni-P includes four stages: (1) Homogeneous CNT suspension solution is obtained; (2) The CNTs moved towards the substrate by diffusion force; (3) Weak adsorption between CNTs and substrate; (4) CNTs covered by Ni particles, and embedded in Ni matrix gradually. The composite plating process with ultrasonic agitation was illustrated in Fig. 11. (a) Firstly, the CNTs moved towards the substrate by diffusing. (b) Secondly, some of the CNTs arrived at the substrate surface and some of them are weakly adsorbed. (c) The CNTs which are weakly adsorbed were shaken and driven away. (d) The Ni particles began to deposited on the surface of the substrate and covered the CNTs simultaneously. Finally, all the CNTs were embedded in the Ni matrix, and a strong bonds between CNTs and deposited layers were formed. It can be concluded that, ultrasonic agitation is beneficial to plating process by reducing agglomerates, which made the deposited layer more smooth and the distribution is more uniform.

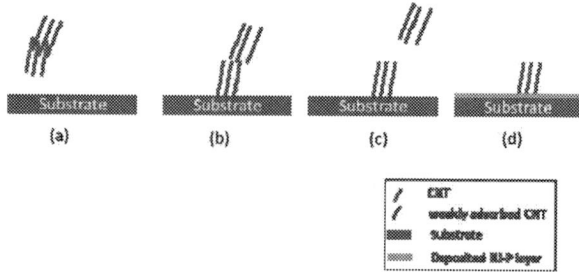

Figure 11 Schematic illustration of how ultrasonic agitation prevent CNT agglomerations

The electrolessly deposited CNT/Ni-P composite coatings were shown in Fig. 12. It can be observed that a uniform and continuous coatings were formed and the surface is uniformly covered. In Fig. 12(a), the CNTs, which are white dots in the image, were very uniformly distributed in the electrolessly deposited coating, no obvious CNT rich zone and CNT poor zone existed. In Fig.12(b) All CNTs were well bonded and embedded in the metal matrix. The CNTs were not in agglomeration morphology, and no voids exist beside CNT. The deposition quality is smooth and compact.

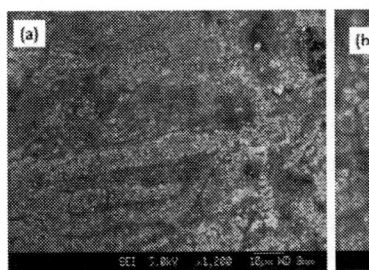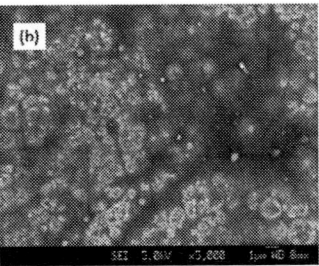

Figure 12 SEM image of CNT/Ni-P deposited coatings (a) low magnification (b) high magnification

Factors affecting plating process

Composite CNT/Ni-P plating is more subtle than plain Ni-P plating, several factors will affect the quality of deposited layers.

(1) The higher pH level will accelerate the depositing speed. However if the pH level is too high, the plating solution is easy to decompose, and the deposited layer will be porous and incompact.

(2) The magnetic stirring speed also affects the plating quality. If the magnetic stirring speed is less than 200 rpm and the stirring time is less than 3 hours, the CNTs cannot disperse in the solution. If the striking speed is higher than 800 rpm, the plating solution is unstable and easy to decompose.

(3) The CNT concentration in the plating bath has direct effect on the CNT content in the composite coatings. After a certain "saturation value", the CNTs are difficult to disperse in the plating bath, which will result in agglomerations. Meanwhile, if the CNT concentration is too high, the CNTs will become active nucleation site itself, and the Ni particles tend to deposit on CNTs rather than Cu substrate.

(4) The plating temperature is key parameter. The deposit rate is very low under 80°C, because both diffusion and activity is very limited under that temperature. If the temperature is higher than 95°C, the plating bath will become very unstable and easy to decompose. So The optimum plating temperature should be adjusted.

Conclusions

In order to disperse carbon nanotubes in electroless Ni-P coatings, different approaches were employed in this study. Acid oxidation is a covalent chemical modification method which can cut long and thick CTNs into short fragments as well as introduce nan defects on the outer walls of CNTs and result in good interlocking effect with metal matrix. Appropriate amount of surfactant addition can significantly improve the dispersion wrapping individual CNT with molecular chains. Besides, surfactant PVP has better dispersion ability than surfactant SDBS when applying the same amount. conducted to disperse CNT clusters. Magnetic stirring and ultrasonic agitation are non-destructive dispersion method. Magnetic stirring can help de-bundle large clusters while ultrasonic agitation can accelerate the breakup of CNT agglomerations and let CNTs homogeneously distributed on the deposited coatings. It was verified that the dispersion can be significantly improved after proper treatment process, and surface of the coating layer is quite smooth ; CNTs are equably dispersed throughout the matrix. In addition , the interfacial bonding between CNTs and Ni-P alloy is good. In these cases, an understanding of the effects of the acid oxidation, surfactant dispersant, magnetic stirring and ultrasonic agitation on the CNT dispersion in both solution and deposited layer were ascertained.

Acknowledgments

The authors would like to acknowledge the financial support provided by the Research Grants Council, Hong Kong, Ref. No. 9041636 (A study of nanostructured electronic interconnects-preparation, characterization and integration), City University of Hong Kong Research project: 7002848 (A study of functionalized CNT/grapheme reinforced composite electronic interconnects: preparation, characterization and integration for green nanoelectronic applications".

References

1. Y. C. Chan, "Failure mechanisms of solder interconnect under current stressing in advanced electronic packages", Prog. Mater. Sci., vol. 55, issue 5, pp. 428-475, Jul. 2010.

2. J. Shen, "Research advances in nano-composite solders", Microelectron. Reliab., vol. 49, issue 3, pp. 223–234, Mar. 2009.

3. J. Shen, "Effects of ZrO2 nanoparticles on the mechanical properties of Sn–Zn solder joints on Au/Ni/Cu pads", J. Alloy Compd., vol. 477, issue 1-2, pp. 552-559, May 2009

4. A. K. Gain, "Microstructure, thermal analysis and hardness of a Sn-Ag-Cu-1wt% nano-TiO2 composite solder on flexible ball grid array substrates", Microelectron. Reliab., vol. 51, issue 5, pp. 975–984 , May 2011.

5. K. H. Krishnan , "An overall aspect of electroless Ni-P depositions—A review article", Metall. Mater. Trans. A, vol. 37, issue 6 , pp. 1917-1926, Jun. 2006.

6. R. C. Agarwala, "Electroless alloy/composite coatings: A review", Sadhana , vol. 28, issue 3-4, pp. 475-493, Jun. 2003

7. M. Alishahi, "The effect of carbon nanotubes on the corrosion and tribological behavior of electroless Ni–P–CNT composite coating", Appl. Surf. Sci., vol. 258, issue 7, pp.2439-2446, Jan. 2012.

8. S. Iijima, "Helical microtubules of graphitic carbon", Nature, vol. 354, pp. 56-58, Nov. 1991.

9. T. W. Ebbesen, "Electrical conductivity of individual carbon nanotubes", Nature, vol. 382, pp. 54-56 Jul. 1996.

10. P. Kim, "Thermal Transport Measurements of Individual Multiwalled Nanotubes", Phys. Rev. Lett. vol. 87, issue 21, pp.5502, Oct. 2001.

11. G Yamamoto, "A novel structure for carbon nanotube reinforced alumina composites with improved mechanical properties", Nanotechnology, vol. 19, issue 31, no.31, pp.5708, Jun. 2008

12. Linda Vaisman, "The role of surfactants in dispersion of carbon nanotubes", Adv. in Colloid Interfac. Sci., vol. 128–130, pp. 37–46, Dec. 2006.

Electroless Ni-W-P Alloy as a Barrier Layer between Zn-based High Temperature Solders and Cu Substrates

Li Liu[1], Longzao Zhou[2], and Changqing Liu[1]

[1]Wolfson School of Mechanical and Manufacturing Engineering, Loughborough University, Loughborough, UK

[2]Huazhong University of Science and Technology, Wuhan, China

E-mail: l.liu2@lboro.ac.uk, Phone: +44-1509-227639

Abstract

A need for a drop-in lead-free high temperature solder to replace high lead solders has been stringently desired for the die attachment in SiC power devices. Currently, the gold-based solders (Au-20Sn, Au-12Ge, and Au-3Si eutectic solders) and nano-silver paste are widely applied for their excellent high-temperature properties. But due to high price and high-demand of a processing facility, their applications are hindered. Zn-based solders, which exhibit numerous advantages such as low cost, high melting point, high thermal and electrical conductivity, show their potential uses to replace high lead solders. However, the oxidation and significantly reactions of Zn-based solders on copper substrates lead to excessive Cu-Zn intermetallic compounds (IMCs) at elevated temperature and further makes the joints brittle and easy to fracture. In this work, in order to effectively prohibit the growth speed of Cu-Zn IMCs under harsh environment, a ternary Ni-W-P coating (6-7 wt% of P and 18-19 wt% of W) on copper substrate was developed for its good thermal stability. The interfacial reactions at Cu/Ni-W-P/Zn-5Al interface and Cu/Zn-5Al interface were studied and compared after soldering at 450°C for different time. It was found that Cu-Zn IMCs grew rapidly after soldering for 30 minutes and cracks were observed in IMC layer of Cu/Zn-5Al interface. However, no voids and cracks were found at the Ni-W-P/Zn-5Al interface with the same thermal treatment. Moreover, the thickness of γ-Ni_5Zn_{21} and Al_3Ni_2 formed in Ni-W-P/Zn-5Al solder joints were quite thin and stable during soldering for different time, which proves the barrier-effect of Ni-W-P layer. These results indicate that Ni-W-P metallization is a promising under bump metallization (UBM) to improve reliability in long term aging and multiple reflow.

Introduction

Lead-rich solders, such as Pb95-Sn and Pb90-Sn with melting points of 310°C and 305°C, respectively, have been widely used in high power electronics for good wettability, high ductility and low cost[1-2]. However, recent regulations such as RoHS (Restriction of the Use of Certain Hazardous Substances in Electrical and Electronic Equipment) and WEEE (Waste Electrical and Electronic Equipment) directives prohibit the use of lead (Pb) due to its toxicity. Continuous effects have been made in developing a promising drop-in lead-free solder such as gold-rich solders and nano-silver particles to replace lead-rich solders recently. Gold-rich solders (Au-20Sn, Au-12Ge and Au-3Si eutectic solders), particularly Au-20Sn solder, can substitute high lead solders for the advantages of excellent thermal and electrical conductivity, good creep resistance and fluxless bonding

process [3-4]. However, high cost and brittle Au-Sn intermetallic compounds (IMCs) prevent its widespread application [5-6]. Regarding nano-silver paste, the advantages are excellent electrical and thermal conductivity, high tensile strength and low process temperature, but high pressure during sintering process limits the wide use of nano-silver particles technology [7-9].

Zn-based solders show potential uses to replace high lead solders, which exhibit numerous advantages such as low cost, high melting point, high thermal and electrical conductivity [10-12]. However, Zn-based solders oxidize easily, and also react actively with copper substrate, forming excessive Cu-Zn IMCs at elevated temperature to make the joints brittle and easy to fracture [13]. To solve this issue, Ni deposit can be utilized between the Zn-based solder and copper substrate to prevent the Cu-Zn IMCs forming [14]. Because the crystallization temperature of ternary Ni-W-P (450°C) is much higher than that of Ni-P (350°C), for which the Ni-W-P can keep an amorphous structure below 450°C, possessing a better heat-resistivity compared to binary Ni-P deposition [15]. Therefore, Ni-W-P layer is quite adaptive in high temperature electronics.

In this work, a ternary electroless Ni-W-P coating on copper substrate was developed for soldering of Cu/Zn-5Al, which can effectively prohibit the growth speed of Cu-Zn IMCs under harsh environment. The interfacial reactions and IMC morphologies at interface of Cu/Zn-5Al joint and interface of Cu/Ni-W-P/Zn-5Al joint were investigated. The growth and characteristics of IMCs at Cu/Ni-W-P/Zn-5Al interface were compared with that at the Cu/Zn-5Al interface.

Sample Preparation

In this work, the substrates used for Ni-W-P electroless plating were copper sheets (1 mm thickness, 99.9% purity). The copper sheet was cut into 20×20 mm square and then ultrasonically cleaned with acetone for 5 minutes. Before chemical-plating, the copper was rinsed into 50% nitric acid for 15 seconds to remove the oxidation of the copper substrate and then washed by deionized water for 10 seconds. Afterwards, these copper substrates were immersed into electroless plating bath, and an aluminum wire was subsequently attached to the copper surface to initiate the deposition and then took out after a thin Ni-W-P layer formed on the copper. The plating solution was an alkaline bath, making up with deionized water and sodium tungstate for tungsten source, nickel sulphate for nickel source, sodium hypophosphite for reducing agent and sodium acetate for buffering agent. The pH of the bath was adjusted to be 8.2 by 20% sulfuric acid and 25% ammonium hydroxide during plating process. The plating temperature was $88 \pm 2°C$ and the plating time was 90 minutes.

Zn-5Al casting ingots (Zn: 95 wt%, melting point: 382°C) bought from the Brock Metal Company Limited were used as a Zn-based high-temperature solder. Zn-5Al sheets of 1 mm thickness were sliced from ingots and then polished with 1 μm alumina powder to remove oxidation on the surface. After that, the solder sheets were diced into pieces of 0.15 g and rinsed in absolute ethanol to prevent further oxidation prior to soldering. Finally, these preformed solder pieces was plated on the surface of copper substrates and Ni-W-P coated copper substrates which were covered with flux and heated on the hotplate (IKA C-MAG HP4) at 450°C for 1-30 minutes in the air. This will enable the investigation on the formation of Cu-Zn IMCs on copper and Ni-W-P coating layer.

After soldering, the specimens were hot mounted, grinded and polished down to 1 μm finish to reveal the cross section of specimens. The thickness of Ni-W-P and IMCs layers were measured which can be associated with barrier properties of Ni-W-P deposit, and the interfacial reactions and IMC morphologies at interface of Cu/Zn-5Al joint and interface of Cu/Ni-W-P/Zn-5Al joint were then investigated. Under the same condition, the growth and characteristics of IMCs at Cu/Zn-5Al interface were compared with that at the of Cu/Ni-W-P/Zn-5Al interface.

The surface morphologies of Ni-W-P layer were observed by Scanning Electron Microscopy (SEM, Cambridge Stereoscan 360 SEM). The thickness of Ni-W-P interlayer and different IMCs was measured by optical microscopies (Leica DMI 5000M) and ImagePro Plus software. Finally, the interfacial microstructure of Ni-W-P/Zn-5Al solders joints after soldering was observed and the composition of each type of IMCs at the interfaces was analyzed as well by Energy dispersive X-ray (EDX) incorporated with the SEM to understand the microstructural evolution at the interface with Ni-W-P deposit as a barrier layer.

Results and Discussion

As-deposit Ni-W-P metallization

The composition of the deposited Ni-W-P layer was measured at 6-7 wt% of phosphors and 18-19 wt% of tungsten by EDX. In Fig. 1(a), the surface of Ni-W-P deposit was pore-free with different sizes of micron-sized nodular characteristics. The micron-sized nodular structure (spectrum A) was analyzed to be as similar as the flat deposit (spectrum B) with Electron Probe X-ray Microanalysis (EPMA) in table 1. The reason why the micron-sized nodular exists is because the aggregation of Ni-W-P ternary compound. An anionic surfactant, 0.3 mg/L sodium lauryl sulphate (SLS) can be used as a pin-hole reliever in Ni-W-P plating to achieve a largely compact Ni-W-P deposit [16]. The thickness of deposited Ni-W-P interlayer was measured to be approximately 10 μm in the Fig. 1(b). It can be observed that the coating was adhered to the copper substrate and no pores were found at the interface. Overall, the coating was crack-free, compact and adherent, showing a good quality as a barrier layer.

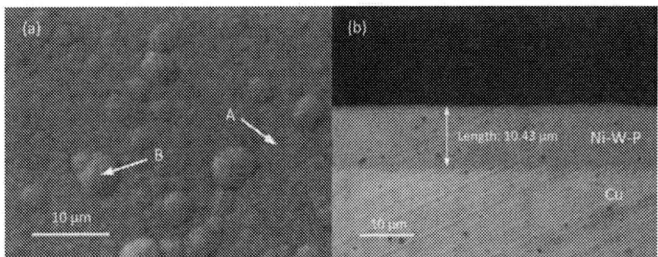

Fig.1 As-deposit Ni-W-P layer: (a) surface morphology, (b) cross-sectional micrograph

Table 1 EPMA analyses of the spectrums in Fig.1 (a). (wt%)

Spectrum	Ni	W	P
A	74.29	18.80	6.91
B	74.83	18.39	6.78

The interfacial reactions and IMC morphologies of Cu/Zn-5Al interface

During the joining process, Zn-5Al solder was heated up and melt at 450 °C for different time to investigate the interfacial reactions and IMC morphologies at Cu/Zn-5Al interface. According to Cu-Zn phase diagram (Fig. 2), three IMCs, β′ -CuZn, γ -Cu$_5$Zn$_8$ and ε -CuZn$_4$ can form at the interface in order between copper substrate and Zn-5Al solder. The sequence of IMCs formation could be explained as follow: At first, zinc atoms in liquid Zn-5Al alloy react with copper substrate directly and lead to formation of ε-CuZn$_4$, which hinders the direct reaction between zinc atoms and copper substrate. Then, with continues growth of ε -CuZn$_4$, zinc atoms in ε -CuZn$_4$ diffuse into copper and generate γ-Cu$_5$Zn$_8$ in the middle layer between ε-CuZn$_4$ and copper substrate. Finally, small amount of zinc atoms in γ-Cu$_5$Zn$_8$ diffuse continuously into copper to form β′-CuZn.

The cross-sectional micrograph of as-soldered Cu/Zn-5Al interface was demonstrated in Fig. 3. Combined with EDX analyses, it can be observed that ε-CuZn$_4$ formed in dendritic shape near Zn-5Al solder. In the meanwhile, γ-Cu$_5$Zn$_8$ clearly formed as a leveled layer between ε -CuZn$_4$ and copper substrate. This was also verified by EPMA that ε-CuZn$_4$ and γ-Cu$_5$Zn$_8$ formed in the order as discussed above in Cu-Zn phase diagram (Fig. 2). For the reason that the melting time is quite short (1 second after Zn-5Al solder melting), there was barely any β′-CuZn can be seen at the as-soldered Cu/Zn-5Al interface. The thickness of ε -CuZn$_4$ and γ -Cu$_5$Zn$_8$ were measured at 2.57 μm and 6.69 μm, respectively, as shown in Fig. 3.

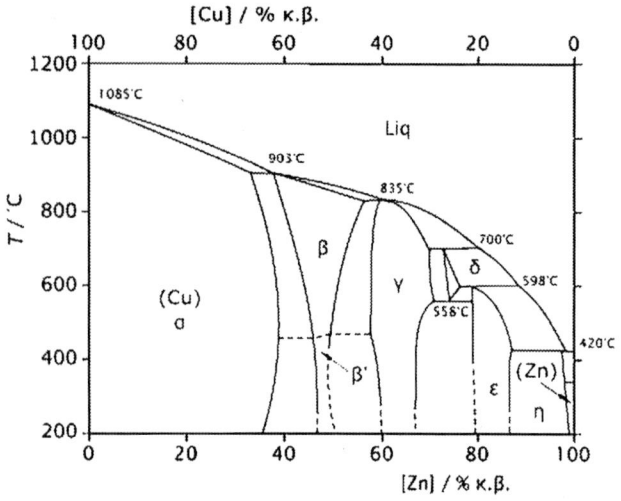

Fig.2 Phase diagrams of Cu-Zn binary system [17].

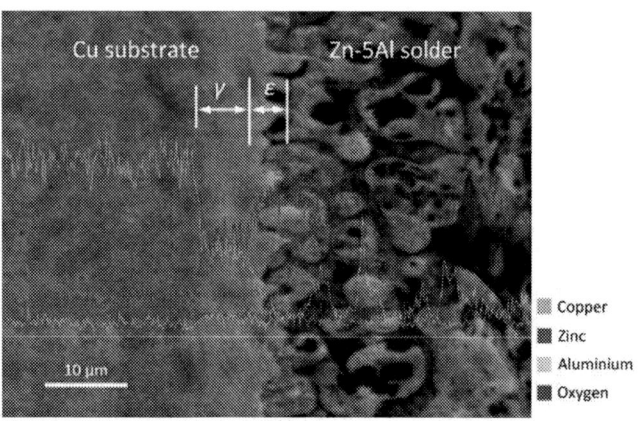

Fig.3 Back-scattered SEM image showing Cu-Zn IMCs formed at the as reflowed Cu/Zn-5Al interface with line-scanned EDX results

The interfacial reactions and IMC morphologies of Cu/Ni-W-P/Zn-5Al interface

When Zn-5Al solders melt on Ni-W-P layer, the reactions would be more complicated than the reactions on copper substrate. According to the related Ni-Zn and Ni-Al phase diagrams [18-19] (Fig. 4), β_1-NiZn, γ-Ni$_5$Zn$_{21}$ and δ-NiZn$_8$ in the Ni-Zn system, and Al$_3$Ni, Al$_3$Ni$_2$, Al$_3$Ni$_5$ and AlNi$_3$ in the Ni-Al system are stable at the temperature studied herein. Fig. 5 shows the microstructure at the as-soldered Ni-W-P/Zn-5Al interface formed at 450°C and no voids or cracks can be found at this interface. Two IMC layers formed between Zn-5Al solder and Ni-W-P interlayer. The line-scanned EDX results show that zinc, aluminum, copper atoms were effectively blocked by Ni-W-P ternary layer. Therefore, zinc and copper could not react with each other directly to form brittle Cu-Zn IMCs. According to the related research and EDX analyses [20-22], only γ-Ni$_5$Zn$_{21}$ and Al$_3$Ni$_2$ can be observed at Ni-W-P/Zn-5Al interface. However, the layers of two IMCs at as-soldered Ni-W-P/Zn-5Al interface were too thin to be confirmed with EPMA. The thickness of the black and the grey IMCs in the joint were measured at 0.48 µm and 0.97 µm, respectively. Afterwards, an additional EPMA was carried out in the sample heated at 450°C for 30 minutes to

confirm the IMCs beside Zn-5Al solder in next section.

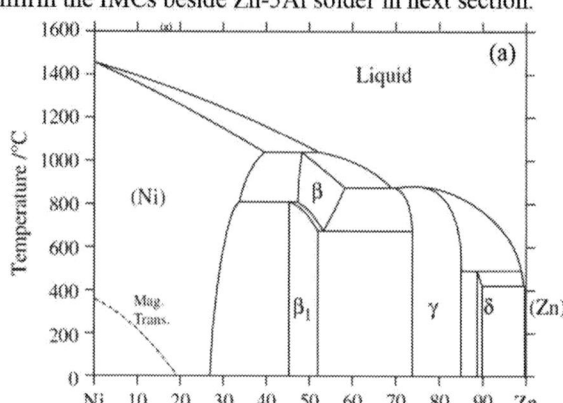

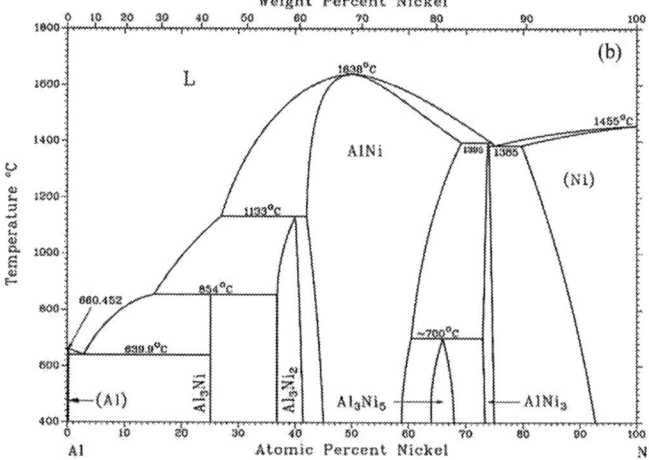

Fig.4. Phase diagrams (a) Ni-Zn system; (b) Ni-Al system

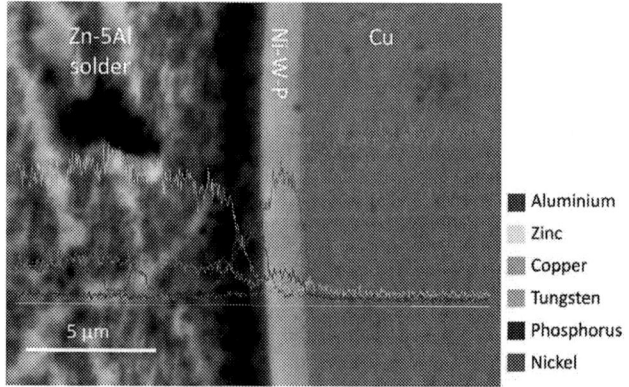

Fig.5. Back-scattered SEM image showing the IMCs formed in the as reflowed Ni-W-P/Zn-5Al interface with line-scanned EDX results

The barrier capability of Ni-W-P layer on copper after long-term soldering

The growth mechanisms of Cu-Zn IMCs during soldering process and thermal aging below melting temperature were studied that they were both controlled by the volume diffusion of zinc and copper atoms. The Ni-P compound in Ni-W-P deposit started to crystallize until 450°C, which is 100°C higher than the crystallizing point of Ni-P compound in bare Ni-P plating, thus indicating the superiority to electroless Ni-P plating layer for high temperature electronics [15]. In addition, as liquid-solid reaction is more intense than solid-solid diffusion during thermal aging, the barrier property of Ni-W-P layer under harsh environment was investigated. In this study, Zn-5Al solders were soldered on Ni-W-P coating at 450 °C for 1-30 minutes in comparison with that on pure copper in the equal condition.

Fig. 6 illustrates a severe crack across the entire cross section of a failed sample without Ni-W-P metallization. It can be observed that zinc atoms dissolved into copper substrate to form excessive Cu-Zn IMCs and 46 μm thick (value d shown in Fig. 6) copper was consumed after 30 minutes soldering. As the rapidly growth of Cu-Zn IMCs, Cu/Zn-5Al solder joints became quite brittle and fractured along the IMCs. It is also noteworthy that cracks started forming at the edge of the interface and then extended through γ-Cu_5Zn_8 across the interface of the sample. This could be explained as follow: when the solder joint formed, the specimens were heated up to 450°C and then cool down to room temperature; due to the higher thermal conductivity of the copper substrates compared to Zn-5Al solder and IMCs, a temperature difference will cause internal stresses in the joints; then the Coefficient of Thermal Expansion (CTE) difference between the copper, ε-$CuZn_4$, γ-Cu_5Zn_8 and β′-CuZn phases will result in the internal residue stress in the structure, leading to the formation of cracks. According to Fig. 6, the cracks formed through the whole γ-Cu_5Zn_8 as such the γ-Cu_5Zn_8 should be the weakest phase at this interface which reduces the reliability of the connection between Zn-5Al solder and copper substrate. Therefore, it can be firmly concluded that the brittleness of γ-Cu_5Zn_8 and its CTE mismatch among the three Cu-Zn IMCs has led to the cracks in the solder joints.

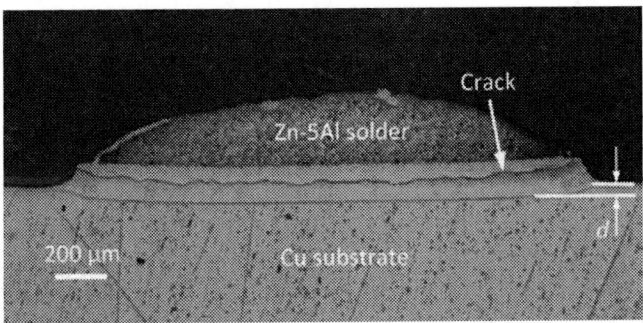

Fig.6 Cross section of the Cu/Zn-5Al solder joint with cracks.

The Cu-Zn IMCs were characterized and defined respectively by EPMA as shown in table 2. According to Fig.3 and Fig. 7, after a long term of soldering at 450 °C, Cu-Zn IMCs formed regularly and grew up rapidly, where the ε-$CuZn_4$ changed its shape from dendrite (Fig. 3) to scalloped crystal (Fig. 7), with β′-CuZn can be observed near copper substrate. After 30 minutes soldering, the thickness of ε-$CuZn_4$, γ-Cu_5Zn_8 and β′-CuZn were measured, and they were approximately 33.62 μm, 114.81 μm and 2.14 μm, respectively. Therefore, the growth speed of these IMCs could be calculated, in which the growth speed of γ-Cu_5Zn_8 was the fastest among these three IMCs, followed by the ε-$CuZn_4$ and β′-CuZn.

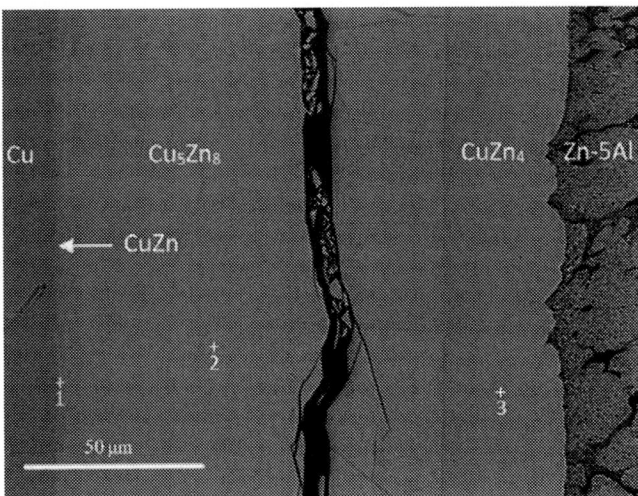

Fig.7 Back-scattered SEM micrograph showing Cu-Zn IMCs formed at the Cu/Zn-5Al solder interface after 30 minutes soldering at 450°C.

Table 2 EPMA analyses of the spectrums in Fig. 7. (wt%)

Spectrum	Cu	Zn	Al	Phase
1	38.68	59.48	1.84	β′-CuZn
2	36.59	61.54	1.87	γ-Cu_5Zn_8
3	74.29	18.80	6.91	ε-$CuZn_4$

The investigations on the Ni-W-P/Zn-5Al solder joints after soldering at 450°C for 30 minutes were conducted in contrast with the Cu/Zn-5Al solder joints to investigate the barrier property of Ni-W-P. The IMC morphologies at Ni-W-P/Zn-5Al interface after 30 minutes-soldering were almost as the same as that at the as-soldered Ni-W-P/Zn-5Al interface discussed above. Two thin IMC layers formed between solder and Ni-W-P layer, the grey layer near Ni-W-P deposit was confirmed to be γ-Ni_5Zn_{21} and the black layer near Zn-5Al solder was Al_3Ni_2 according to the EPMA results (Table 3), which proved the previous hypothesis in the as-soldered Ni-W-P/Zn-5Al joints. It should be noted that some elements such as copper, tungsten, phosphorus and oxygen were not listed in table 3 for their limited contents (less than 15 wt% in total) and non-reactions with Ni-W-P interlayer. From Fig. 8, the thickness of γ-Ni_5Zn_{21} and Al_3Ni_2 were measured at 1.30 μm and 1.91 μm, which were 0.48 μm and 0.97 μm in the as-soldered specimen. It is also noteworthy that the growth speed of two IMCs and the consumption speed of Ni-W-P layer during soldering were extremely slow, particularly compared with Cu-Zn IMCs (e.g. around 150 μm thickness) at the Cu/Zn-5Al interface in the same condition. Even with high magnification, no voids and cracks can be observed at Ni-W-P/Zn-5Al interface in Fig. 8, so that the Ni-W-P ternary layer as a barrier layer to resist the solder reactions with copper substrate is proved valuable for the improvement of reliability of Zn-5Al high temperature soldering at elevated temperature.

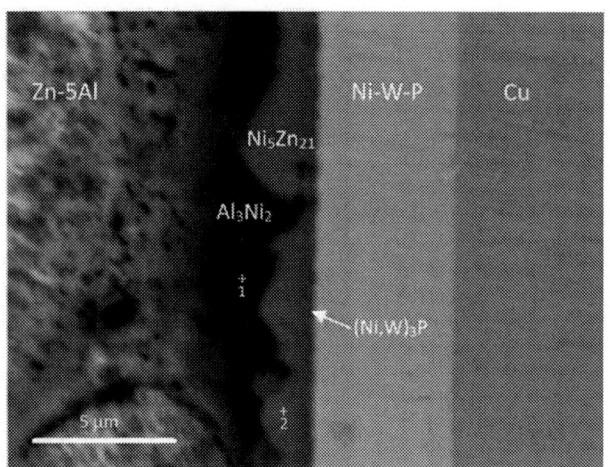

Fig.8 Back-scattered SEM cross-sectional micrograph showing the IMCs at the Ni-W-P/Zn-5Al solder interface after 30 minutes soldering at 450°C.

Table 3 EPMA analyses of the spectrums in Fig. 8 (wt%)

Spectrum	Zn	Ni	Al	Phase
1	17.37	42.08	26.83	Al_3Ni_2
2	71.38	15.42	8.69	Ni_5Zn_{21}

Moreover, a dark layer pointed by the arrow in Fig. 8 has been observed between γ-Ni_5Zn_{21} and unreacted Ni-W-P layer, which were found to likely be $(Ni,W)_3P$, as reported elsewhere [15-16][23]. This $(Ni,W)_3P$ layer will not crystallize but can prevent the consumption of Ni-W-P layer, which further contributes to the barrier-effect of Ni-W-P coating. However, as the characteristics and mechanism of $(Ni,W)_3P$ amorphous layer has not been fully understood and its limited size (less than 0.5 µm), some future in-depth investigation is needed to further elaborate the details in terms of the mechanism of $(Ni, W)_3P$ at the interface.

Conclusions

In this work, an electroless ternary Ni-W-P alloy was developed to be utilized as an UBM. EDX compositional analysis illustrated that tungsten and phosphorus content in Ni-W-P coating were 18-19 wt% and 6-7 wt%, respectively. At Cu/Zn-5Al interface, three Cu-Zn IMCs, ε-$CuZn_4$, γ-Cu_5Zn_8 and β'-CuZn formed from Zn-5Al solder to copper substrate in order. After heating up to 450°C for 1-30 minutes, ε-$CuZn_4$, γ-Cu_5Zn_8 rapidly grew into thick layers and γ-Cu_5Zn_8 showed brittleness or high CTE mismatch in Cu/Zn-5Al solder joints, which has caused cracks during soldering. In the meanwhile, Ni-W-P coating exhibited excellent barrier-layer effects to prevent copper from reacting with zinc directly and has demonstrated its excellent barrier property enabling a high reliability with no cracks and no voids formed at 450°C even after a long term of soldering. Two thin IMC layers, γ-Ni_5Zn_{21} and Al_3Ni_2, were found from Ni-W-P layer to solder in order, and the thickness of each IMC was quite stable even after long-term soldering. In addition, a dark amorphous layer was observed between γ-Ni_5Zn_{21} and unreacted Ni-W-P layer which is assumed to be $(Ni,W)_3P$,

acting as a diffusion barrier, which should be the fundamental cause of the high barrier capability of Ni-W-P ternary layer for Zn-based high-temperature solders.

Acknowledgments

The authors would like to acknowledge the 7th European Community Framework Programme for financial support through a Marie Curie International Research Staff Exchange Scheme (IRSES) Project entitled "Micro-Multi-Material Manufacture to Enable Multifunctional Miniaturised Devices (M6)" (Grant No. PIRSES-GA-2010-269113). Additional support from China-European Union technology cooperation programme (Grant No. 1110) is also acknowledged. The authors wish to thank Dr. Trevor Pearson and Dr. Keming Chen from Macdermid Corporate for providing helpful advice on electroless plating. Deep gratitude is expressed to Mr. Shaun Fowler for his kind support in EPMA analyses.

References

1. H. , 2006, "Sn Concentration on the reactive wetting of high-Pb solder on Cu substrate," *Journal of Material Chemisty Physics*, Vol. 99, pp. 202-205, 2005.
2. V. Chidambaram, "High-Temperature Lead-free Solder Alternatives," *Journal of Microelectronic Engineering*, Vol. 88, pp. 1926-1931, 2011.
3. A. Hartnett, "Process and Reliability Advantages of AuSn eutectic die attach," *in Proc. 42nd IMAPS*, Finland, Nov. 2009, pp. 281-287.
4. D. G. Ivey, "Microstructural Characterization of Au/Sn Solder for Packaging in Optoelectronic Applications," *Journal of Micron*, Vol. 29 (4), pp. 281-287, 1998.
5. A. Torleif, "Au-Sn SLID Bonding – Properties and Possibilities," *Journal of Metallurgical and Materials Transactions B*, Vol. 43, pp. 397-405, 2011.
6. R. R. Chromik, "Mechanical Properties of Intermetallic Compounds in the Au-Sn System," *Journal of Materials Research Society*, Vol. 20 (8), pp. 2161-2172, 2005.
7. T. Wang, "Low-Temperature Sintering with Nano-Silver Paste in Die-Attached Interconnection," *Journal of Electronic Materials*, Vol. 36, No. 10, pp. 1333-1340, 2007.
8. K. Suganuma, "Low-temperature Low-pressure Die Attach with Hybrid Silver Particle Paste," *Journal of Microelectronics Reliability*, Vol. 52, No. 2, pp. 375-380, 2012.
9. G. Bai, "Processing and Characterization of Nanosilver Pastes for Die-attachment SiC Devices," *IEEE Transactions on Electronics Packaging Manufacturing*, Vol. 30, No. 4, pp. 241-245, 2007.
10. T. Shimizu, "Zn-Al-Mg-Ga Alloys as Pb-Free Solder for Die-Attaching Use," *Journal of Electronic Materials*, Vol. 28, No. 11, pp. 1172-1175, 1999.
11. M. Rettenmayr, "Zn-Al Based Alloys as Pb-Free Solders for Die Attach," *Journal of Electronic Materials*, Vol. 31, No. 4, pp. 278-285, 2002.
12. Y. Takaku, "A review of Pb-free high temperature solders for power semiconductor devices: Bi-based

composite solder and Zn-Al solder," *Journal of ASTM International*, Vol. 8, No. 1, pp. 1-18, 2011.

13. Y. Takaku, "Interfacial Reaction between Cu Substrate and Zn-Al Based High Temperature Pb-Free Solders", *Journal of Electronic Materials*, Vol. 37, No. 3, pp. 314-323, 2008.

14. J. Li, "Advanced Materials for Drop in Solution to Pb in High Temp Solders: The Next Gerneration of Zinc Based Solder alloy", *in proc. 63rd Electronic Components & Technology Conference(ECTC)*, Las Vegas, 28th - 31st May, 2013, pp.1628-1633.

15. Y. Yang, "Electroless Ni-W-P Alloy As A More Enduring And Reliable Soldering Metallization", *in proc. 13th Electronics Packaging Technology Coference(EPTC)*, Singapore, Dec. 7-9, 2011, pp. 44-48.

16. K. Chen, "Electroless Ni-W-P alloys as barrier coating s for liquid solder interconnects", *in proc. 1st Electronics System integration Technology Conference(ESTC)*, Germany, Sep. 5-7, 2006, pp. 421-427.

17. M. Kowalski, "Thermodynamic reevaluation of the Cu-Zn system", *Journal of Phase Equilibria*, Vol. 14, No. 3, pp. 432-438, 1993.

18. P. Nash, "Phase Diagrams of Binary Nickel Alloys", *ASM International, Materials Park, OH*, pp. 382-390, 1991.

19. P. Nash, "Phase Diagrams of Binery Nickel Alloy", *ASM International, Materials Park, OH*, pp. 3-11, 1991.

20. W. Zhu, "Interfacial Reactions Between Sn-Zn Alloys and Ni Substrates", *Journal of Electronic Materials*, Vol. 39, No. 2, pp. 209-214, 2010

21. Y. Takaku, "Interfacial Reaction between Zn-Al-Based High-Temperature Solders and Ni Substrate", *Journal of Electronic Materials*, Vol. 38, No. 1, pp. 54-60, 2009.

22. C. Wang, "Study of the Effects of Zn content on the Interfacial Reactions Between Sn-Zn solders and Ni Substrates at 250°C", *Journal of Electronic Materials*, Vol. 39, No. 11, pp. 2375-2381, 2010.

23. D. Jang, "Tungsten alloying of the Ni(P) films and the reliability of Sn-3.5Ag/NiWP solder joints", *Journal of Materials Research Society*, Vol. 26, No. 7, pp. 889-895, 2011.

Fabrication and Reliability Evaluation of A Novel Package-on-package (PoP) Structure Based on Organic Substrate

Xiaofeng Sun[1,2], Lixi Wan[2], Yuan Lu[1,2]
1- National Center for Advanced Packaging (NCAP China)
Building D1, China Sensor Network International Innovation Park, 100 Linghu Boulevard, Wuxi City, Jiangsu, PR China
2-Institute of Microelectronics, Chinese Academy of Sciences
No. 3, BeiTuCheng West Road, Chaoyang district, Beijing, PR China
Email: sunxiaofeng@ime.ac.cn

Abstract

The package-on-package(PoP) stacking technique with the advantages of printed-circuit-board(PCB) space saving, flexible combination and assembly has become a kind of mainstream solution for the high density package. However, the conventional PoP technology is limited with constraint aspect ratios of I/O density between the top and base packages under conventional solder ball interconnection approach. In this paper, a multilayer PoP structure was designed and fabricated. This design has an organic substrate with a cavity which allowed the package using fine pitch interconnection solder balls and consequently improved the interconnection I/O density between the neighboring packages and reduced the thickness. Also four reliability tests were applied to three-layer PoP samples to evaluate the reliability of this design.

Keywords: PoP, organic substrate, cavity, reliability test

1 Introduction

Microelectronic package technology has evolved rapidly over the past two decades to keep pace with the demand of emerging business market segments driven by advanced digital handheld device applications such as smart phones, tablets, personal digital assistant(PDA) and mobile internet device(MID). But the traditional packaging technologies in the manufacture process cannot meet the great demands. The package-on-package(PoP) stacking technology with advantages of printed-circuit-board(PCB) space saving, flexible combination and assembly has become a kind of feasible solution for high density package[1]-[3]. PoP has become an attractive solution for high density package used in various products such as mobile phone, PDA and digital camera. However, the conventional PoP technology is limited with constraint aspect ratios of I/O density between neighboring layers under the traditional solder ball interconnection approach as the inter-package connection system, which is shown in Fig.1 [4].

Figure 1. Conventional PoP with solder ball interconnecting the neighboring layer

In order to improve the I/O density and reduce the size and thickness of PoP package, great effort was devoted and several novel PoP structures were designed and studied which are shown in Fig.2 [5]-[9]. These novel structures can improve the I/O density, but they all need complicated processes or extra assembly steps and consequently lead to higher cost.

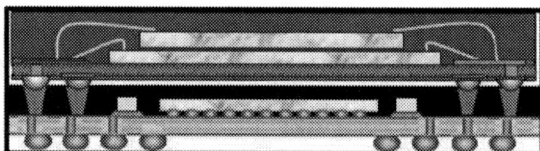

(a) Cross-section view of Amkor's PoP

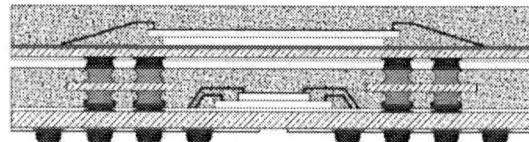

(b) Cross-section view of ASTRI's PoP

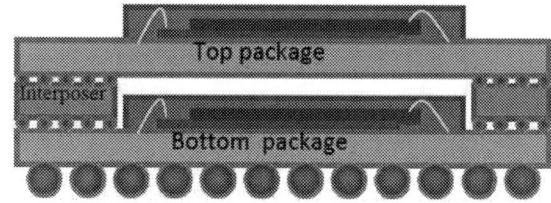

(c) Cross-section view of Endicott's PoP
Figure 2. Improved PoP structures

In this study, a new PoP structure was designed and fabricated and its reliability was evaluated through several reliability tests. This design has an organic substrate with a cavity and chips can be assembled on both sides of the substrate, which is shown in Fig. 3. Because of the cavity, this design allowed the package using fine pitch interconnect solder balls and consequently improved the interconnection I/O density between the neighboring packages and reduced the thickness.

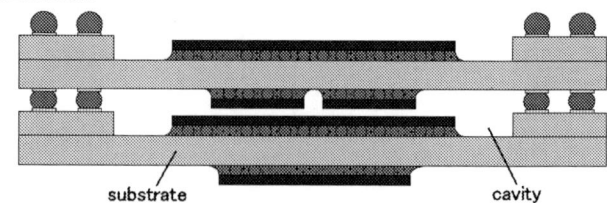

Figure 3. Schematic of the new PoP structure

2 Design and Fabrication

Using the design mentioned above, three-layer and four-layer PoP samples were fabricated. For the reason of simplifying the design and fabrication process, every layer of the PoP package is the same which has one chip A assembled in the cavity and four chip B on the other side of the substrate. Fig.4 displays the schematic of four-layer PoP structure.

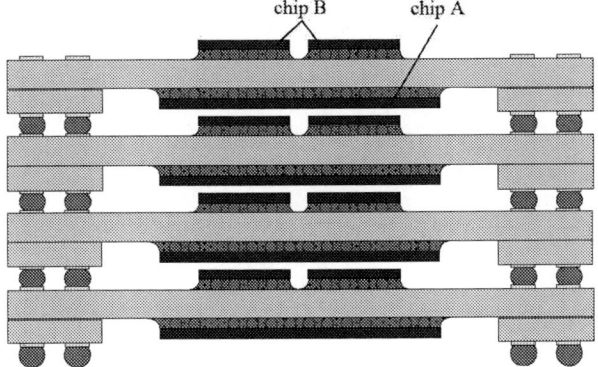

Figure 4. Schematic of four-layer PoP

The size of chip A and chip B is 5.95mm × 4.35mm × 0.2mm and 1.95mm × 1.95mm × 0.2mm respectively. The diameter of BGA balls is 250um. Daisy chain circuits were fabricate on both chip A and chip B for electrical test. Fig.5 shows the layout and photograph of chip A and chip B.

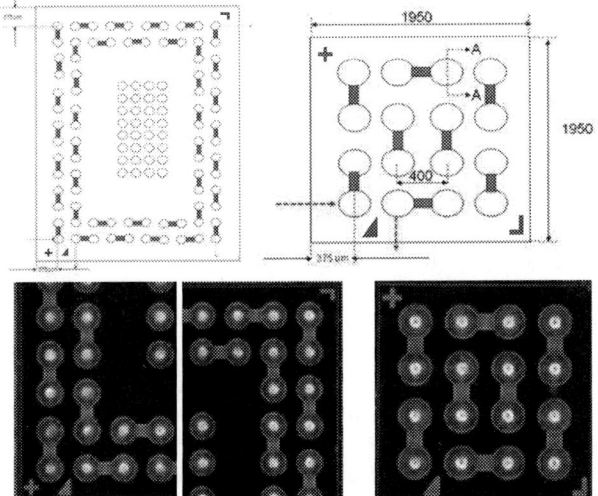

Figure 5. Layout and photograph of chip A and chip B

Fig.6 shows the schematic of the organic substrate and its parameters shows in table 1. The size of the substrate is 12.4mm × 12.4mm × 1mm and the deep of the cavity is 0.5mm. The material of the substrate is FR-4. Fig.7 is the photograph of the substrate which also shows the distribution of the one chip A and the four chip B. Every chip can be tested individually through the pads on both sides of the substrate which is also shown in Fig. 7.

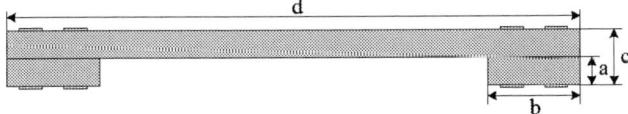

Figure 6. Schematic of the organic substrate

Table 1. Parameters of the organic substrate

a	b	c	d
0.5mm	2mm	1mm	12.4mm

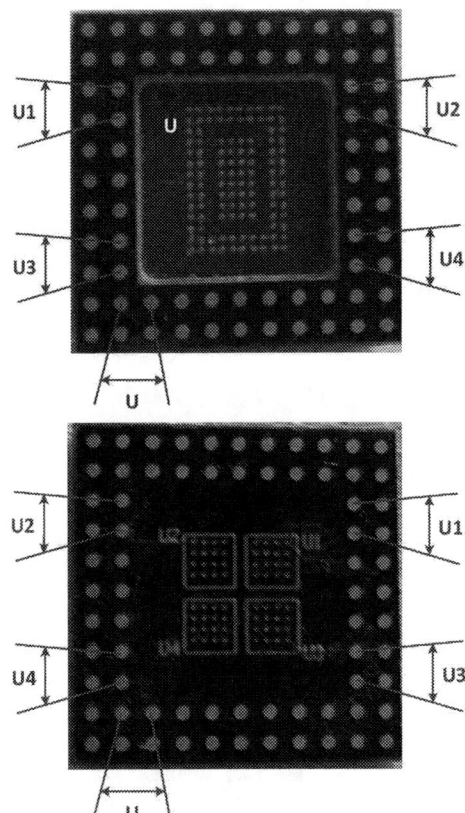

Figure 7. Photograph of the organic substrate

The fabrication process of multi-layer PoP packages is shown in Fig. 8 and the processes can be described as:

(a) Organic substrate prepared
(b) Chip A assembled
(c) Chip A underfilled
(d) Four chip B assembled
(e) Chip B underfilled
(f) Interconnection solder ball printed
(g) Two-layer PoP package stacked
(h) Three-layer PoP package stacked
(i) Four-layer PoP package stacked

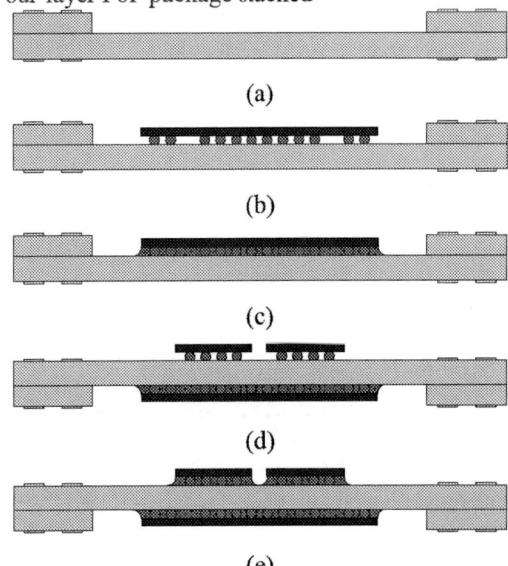

978-1-4799-2408-0/14 $31.00 © 2014 IEEE   1355

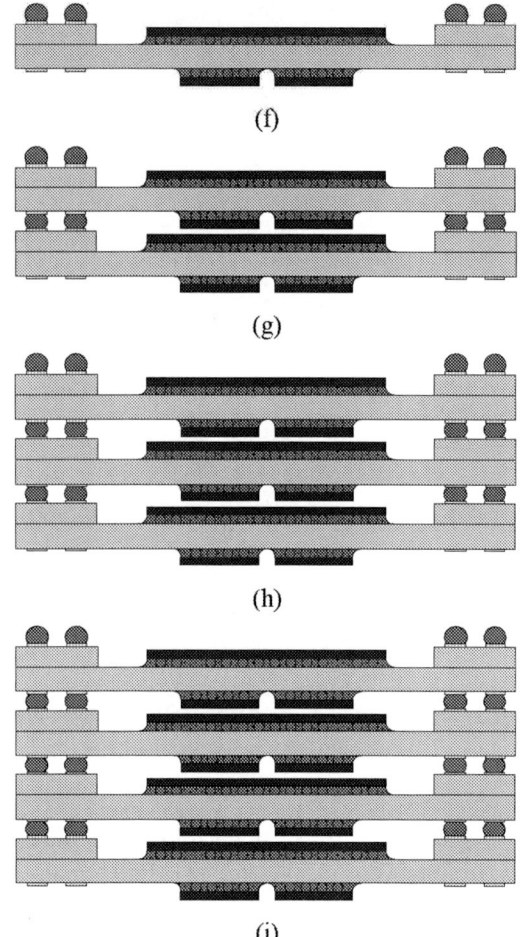

(f)

(g)

(h)

(i)

Figure 8. Fabrication process of multilayer PoP package

Fig.9 shows the photograph of three-layer and four-layer PoP samples.

Figure 9. Photograph of PoP samples

3 Reliability test

Four reliability tests were applied to three-layer PoP samples to evaluate their reliability. As shown in Fig.10, because every layer of the PoP is exactly the same in the design, chips on the same location of every layer are connected in parallel. When not all of them are failed during the reliability test, the electrical measurement cannot give the correct result. So the PoP samples for reliability tests were fabricated as the design of Fig.11, which only has chips on outer surface of the PoP samples. Take chip A as example, it can be measured through pad A1 and B1 or pad A2 and B2 respectively. The conducting state can be tested from A1 to B1, A2 to B2 and A1 to A2 to recognize whether chip A or interconnection solder balls or both are failed.

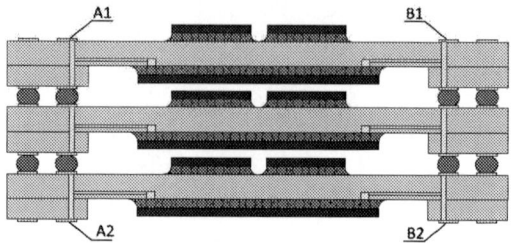

Figure10. Chips on the same location of every connected in parallel

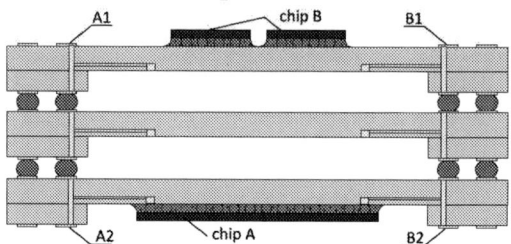

Figure11. Schematic of PoP samples for reliability tests

The resistance of daisy chain circuit can be measured through test pad. As to chip A, the resistances measured from A1 to B1 and from A2 to B2 are represented by R1 and R2 respectively and the resistance from A1 to A2, R, can be calculated as：

$$R = (R1-R2)/2 \qquad (1)$$

The interconnection solder balls and their interconnection status with pad will be changed during the reliability test and consequently lead to the variation of resistance R. In this study the reliability of three-layer PoP samples are evaluated through calculating the variation of resistance R of different chips before, during and after the reliability tests.

(1) Highly accelerated temperature and humidity stress test (HAST)

This test is designed according to JESD 22-A110D. In this test there are 10 samples which are numbered from 1 to 10. A precondition test is applied to the samples before the HAST and the experiment process and test conditions are shown in table 2.

Table 2. HAST experiment process and test conditions

Step	Details
1	Measurement
2	Precondition: 125°C /24h→85°C /85%/168h→reflow 3 times
3	Measurement
4	HAST: 130°C /85%/96h
5	Measurement

The resistance R was calculated three times during the test from the measurement of chip U, U1, U2, U3 and U4 of the 10 samples respectively which are shown in Fig.12. It can be seen from Fig.12 that most of the calculated resistance R changed little after precondition process and increased obviously after the HAST process. The resistance R increased about 10mΩ after HAST test. The solder joints changed as the temperature and humidity increased which can lead to the increase of the resistance R. Fig. 13 shows the cross-sections of interconnection solder balls and BGA balls and there is no obvious cracking.

978-1-4799-2408-0/14 $31.00 © 2014 IEEE   1356

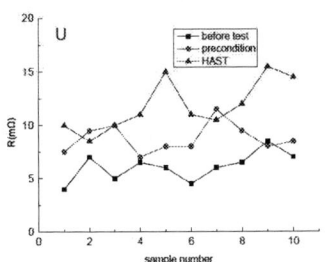

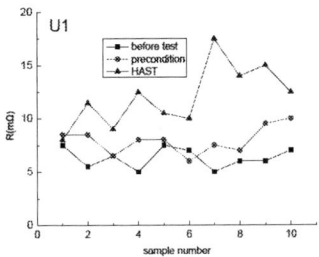

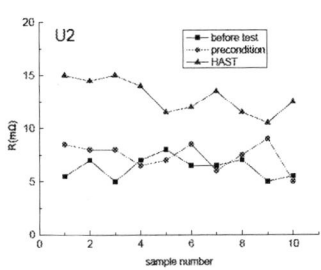

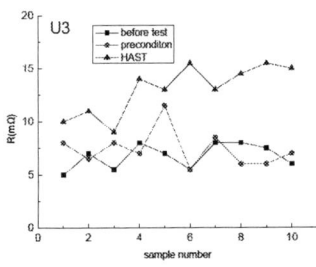

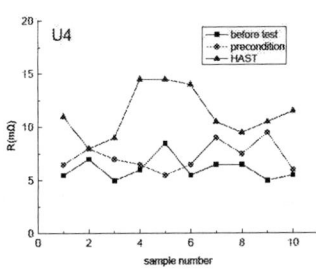

Figure 12. The calculated resistance R of HAST

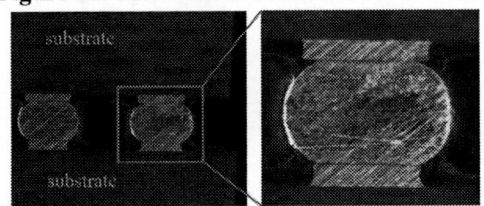

(a)Cross-section view of interconnection solder balls

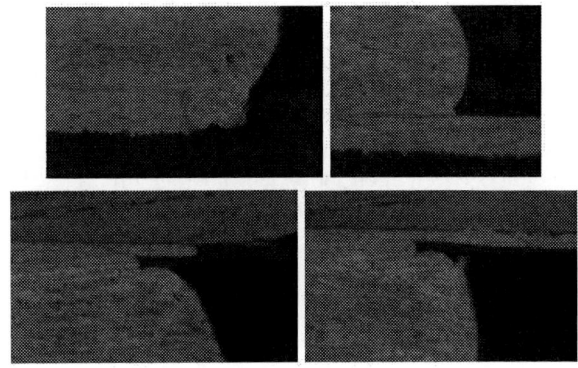

(b) Cross-section view of BGA balls

Figure 13. Solder balls cross-sections after HAST

(2)Temperature cycling(TC)

This test is designed according to JESD 22-A104 and also has 10 samples which are numbered from 11 to 20. Before the temperature cycling test there is a same precondition process as the HAST and the experiment process is shown in table 3. The temperature changed from -40°C to -125°C and the soak time at each extreme temperature is 15minutes. The ramp rate is 16.5°C /minute and the temperature curve is shown in Fig. 14.

Table 3. TC experiment process

Step	Details
1	Measurement
2	Precondition: 125°C /24h→85°C /85%/168h→ reflow 3 times
3	Measurement
4	TC: Temperature range: -40°C～125°C soak time at each extreme temperature: 15minutes ramp rate: 16.5℃/minute 500 cycles
5	Measurement
6	TC: Temperature range: -40°C～125°C soak time at each extreme temperature: 15minutes ramp rate: 16.5°C /minute 250 cycles
7	Measurement

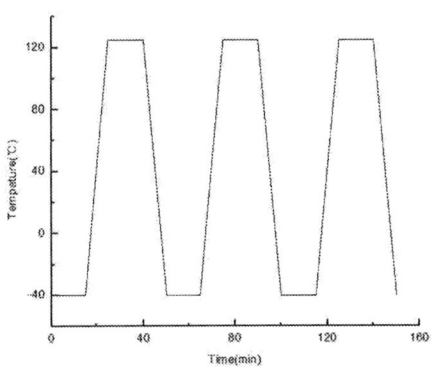

Figure 14. Temperature curve of TC

978-1-4799-2408-0/14 $31.00 © 2014 IEEE

The resistance R was calculated four times in the same way which is shown in Fig.15. As shown in Fig.15, the resistance R changed little after the precondition process and increased obviously after 500 cycles. The resistance R increased about 8 to 10mΩ after 500 cycles and there is little variation between after 500cycles and after 750cycles. Fig. 16 displays the cross-sections of interconnection solder balls and BGA balls.

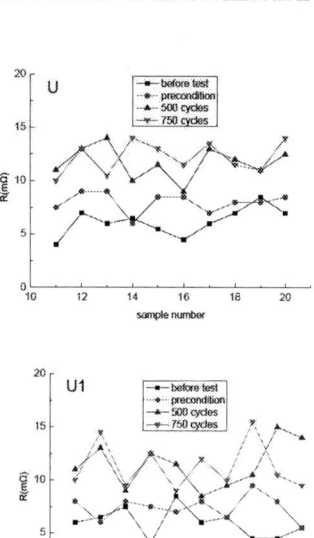

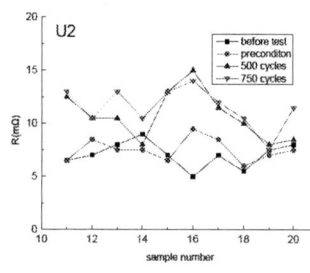

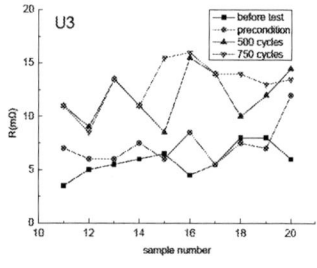

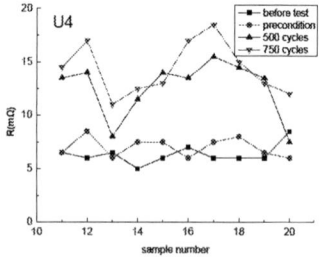

Figure 15. The calculated resistance R of TC

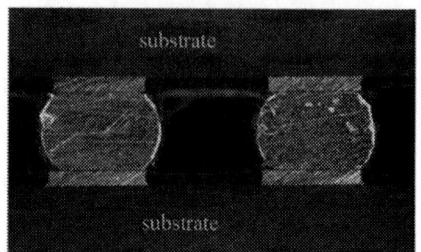

(a) Cross-section view of interconnection solder ball after TC

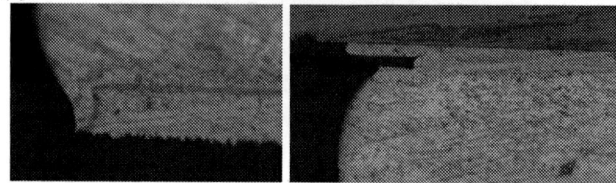

(b) Cross-section view of BGA balls

Figure 16. Solder balls cross-section view after TC

(3)High temperature storage(HTS)

High temperature storage test is designed according to JESD 22-A103D and the 10 samples are numbered from 21 to 30. The HTS experiment process and test conditions are shown in table 4. The 10 samples were placed in a temperature chamber of 150°C for 750hours.

Table 4. HTS experiment process

Step	Details
1	Measurement
2	HTS: 150°C /500 hours
3	Measurement
4	HTS: 150°C /250 hours
5	Measurement

The samples were measured three times and the calculated resistance R is shown in Fig.17. From Fig.17 it can be seen that the resistance R increased obviously after 500hours' storage. Compared to be stored for 500hours, most of the resistance R stored for 750 hours varied not so obvious.

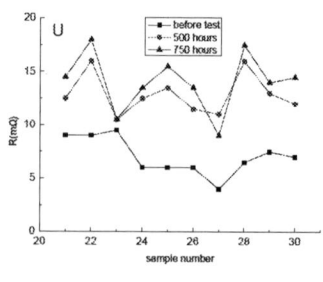

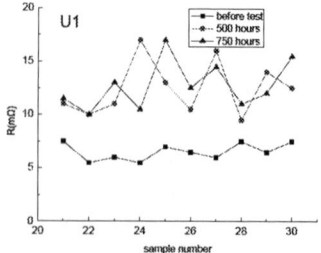

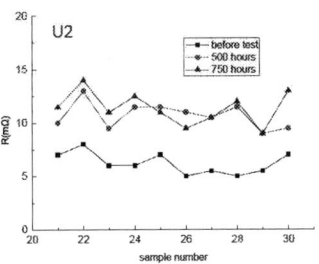

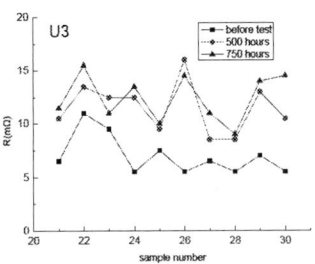

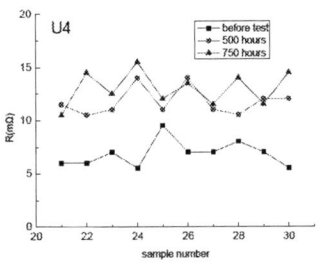

Figure 17. The calculated resistance R of HTS

Fig. 18 displays the cross-section view of interconnection solder balls and BGA balls.

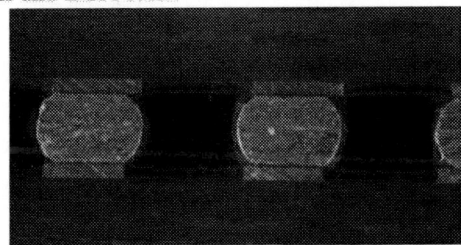

(a) Cross-section view of interconnection solder balls

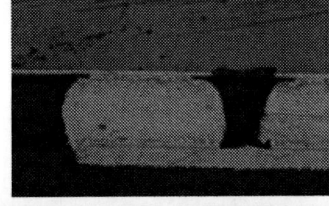

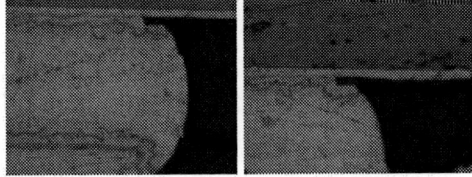

(b) Cross-section of BGA balls

Figure 18. Solder balls cross-section view after HTS

(4)Board level drop test

The board level drop test was designed according to JESD 22-B111 and two PoP samples were tested. Fig. 19 displays the standard test board and the sample was assembled on location U8 which is the most prone to failure during the drop test. According to the JESD 22-B111 and the drop test equipment abilities, the drop height is 130mm and the peak acceleration is 1500G. The pulse duration is 0.5ms. During the test, the current of the sample was monitored all the time.

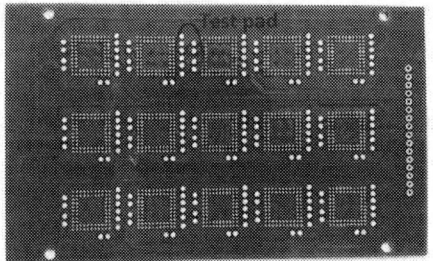

Figure 19. Drop test board

The first sample was assembled to the test board through solder balls without underfilled and it flaked away from the test board after dropped 378 times which is shown in Fig.20. But the electrical test of the first sample showed that it is not failure. In order to avoid this situation, the second sample was assembled with underfilled and it didn't flaked away from the test board after dropped 500 times and the monitoring system showed that it is still not failure.

Figure 20. The sample flaked away from the test board

Conclusions

In this work a new PoP structure was designed and studied. This structure has an organic substrate with a cavity and chips can be assembled on both sides. This cavity allowed the PoP package using fine pitch interconnection solder balls and consequently improved the I/O density between the neighboring packages. The size of the substrate is 12.4mm × 12.4mm × 1mm and the cavity is 0.5mm deep. There are two kinds of chips which sizes are 5.95mm × 4.35mm × 0.2mm and 1.95mm × 1.95mm × 0.2mm respectively. In every layer five chips were assembled through flip-chip technology.

In order to evaluated the reliability of the PoP package, HAST, TC, HTS and board level drop test were applied to three-layer PoP samples.

Acknowledgments

The research funding from National Science and Technology Major Project numbered 2013ZX02501 is greatly acknowledged. The authors would like to thank Professor Fei

Xiao and Jia Xi both from Fudan University for their help in board level drop test.

References

1. David Geiger, Dongkai Shangguan, Samuel Tam, Dan Rooney, "Package Stacking in SMT for 3D PCB Assembly", 28th International Symposium on Electronics Manufacturing Technology, 2003, pp. 261-264.

2. Moody Dreiza, Akito Yoshida, Jonathan Micksch and Lee Smith, "Stacked Package-on-Package Design Guidelines", Chip Scale Review, July 2005. 3. Joanna Kristine Wildhart, Moody Dreiza, "Challenges for high density PoP (package on package) utilizing SoP (solder on pad)", Global SMT & Packaging, April 2008.

3. Joanna Kristine Wildhart, Moody Dreiza, "Challenges for high density PoP (package on package) utilizing SoP (solder on pad)", Global SMT & Packaging, April 2008.

4. Bok Eng Cheah, Jackson Kong, Shanggar Periaman, Kooi Chi Ooi, "A Novel Inter-Package Connection for Advanced Package-on-Package Enabling", Electronic Components and Technology Conference, 2011:589-594

5. Jinseong Kim, Kiwook Lee, Dongjoo Park, Taekyung Hwang, Kwangho Kim, Daebyoung Kang et al., "Application of Through Mold Via (TMV) as PoP Base Package". Electronic Components and Technology Conference, 2008:1089-1092.

6. Akito Yoshida, Jun Taniguchi, Katsumasa Murata, Morihiro Kada, Yusuke Yamamoto et al., " A Study on Package Stacking Process for Package-on-Package (PoP)", Electronic Components and Technology Conference, 2006:825-830.

7. Tae-Kyung Hwang, Dong-Joo Park, Jin-Seong Kim, Jin-Young Kim, Jae-Dong Kim, Choon-Heung Lee, "Board Level Reliability Assessments of Thru-Mold Via Package on Package (TMVTMPoP)", International Conference on Electronic Packaging Technology & High Density Packaging, 2009:1124-1129.

8. Peng SUN, Vincent LEUNG, Debbie YANG, Robin LOU, Daniel SHI ,Tom CHUNG, "Development of a New Package-on-Package (PoP) Structure for Next-Generation Portable Electronics", Electronic Components and Technology Conference, 2010:1957-1963.

9. Rabindra N. Das, Frank D. Egitto, Barry Bonitz, Mark D. Poliks, Voya R. Markovich, "Package-Interposer-Package (PIP): A Breakthrough Package-on-Package (PoP)Technology for High End Electronics", Electronic Components and Technology Conference, 2011:619-624.

Strip Grinding Introduction for Thin PoP

[*]Jinseong Kim, [*]Yesul Ahn, [*]Gyuwan Han, [*]Byoungwoo Cho, [*] Dongjoo Park, [*] Juhoon Yoon, [*]Choonheung Lee,
[**]Lou Nicholls, [**]Shengmin Wen

[*]Research and Development Center, Amkor Technology Korea Inc.,
151, Dongil-ro, Sungsu-dong, Sundong-gu, Seoul 133-706, Korea
[**]Amkor Technology Inc., 1900 South Price Road, Chandler, AZ 85286
[*]E-mail: jskim@amkor.co.kr
[*]Telephone: +82-2-460-5042
Fax: +82-505-460-5968

Abstract

Due to rapid growth in the mobile industry, Package-on-Package (PoP) has been widely adopted for 3D integration of logic and memory devices within mobile handsets, and other portable multimedia products, etc. Typical PoP configuration includes a logic function in the bottom package while memory dice are assembled into the top package. The TMV solution is widely adopted to reduce package warpage, to achieve a fine pitch PoP, and to stabilize stacking performance. Another big benefit of the TMV is to generate a thin structure by exposing the back side of the die using a film assist mold system. However, the trend of mobile devices is going thinner and thinner, and there is a limitation to achieve the thin PoP structure. The most difficult barrier to generate the thin PoP structure is the warpage control. In this paper, a strip grinding process is introduced as a solution to generate thinner PoP structures in overmolded packages. Applying the strip grinding process to exposed die, reducing mold clearance, and exploring double sided mold structures to reduce die/package height were also investigated.

Introduction

Package-on-Package stacking has become a preferred method for 3D integration of logic devices utilizing high-performance memory in mobile products like smart phone handsets and tablets. The trend of mobile devices is toward thinner platforms. To this end, PoP enables stacking both within the package as well as between packages. The total stack-up height is the most critical factor to PoP, and many activities are being performed to reduce the stack-up height.

Another critical factor is the package warpage. Package stacking is the key process for PoP and the most important factor influencing package stacking is package warpage. In general, warpage control is more difficult when the package is thinner. The construction of thin, planar packages can be difficult (especially as size increases) because of process capability limitations. For example, in an overmolded structure, there is a limit as to how much the mold cap thickness can be reduced since material properties such as viscosity and its effect on flow rate and coverage can influence quality (e.g., void creation in the mold compound). Another example of a process limitation is the chip attach process. If a very thin chip is handled, die cracking or chipping can more easily occur. Thus, there is a minimum required chip thickness to maintain the structural integrity of the device.

Currently, exposed die structures are being widely used in thin PoP solutions through film assist molding. Figure 1 shows the standard TMV® and exposed die type TMV® structures.

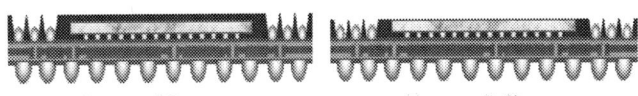

Overmold   Exposed die

Figure 1. PoP base package structures.

Under the exposed die structure, the mold cap thickness is the same as the die height. Normally, the mold cap thickness is thinner than the overmolded cap. In the thin mold cap case, it's not easy to control the package warpage. The warpage can be controlled with a thicker substrate, but this increases the package thickness. The other issue concerns mold flash on the back side of the die. In case of film assist molding, we need to control the mold and die thickness very tightly or mold flash can happen when the die thickness is thinner than the mold cap. Figure 2 shows the mold flash on the die backside.

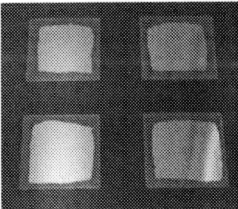

Figure 2. Mold flash on die back side of four dice.

Another issue is thin die handling. To reduce the package height with a thin mold cap, die thickness should be thin; however, as die area increases, thin die are more susceptible to cracking or chipping during die pick and place.

To resolve the above issues and generate thin PoP structures, a strip grinding process was developed. The concept of strip grinding is to grind the mold compound and die together. The advantage of strip grinding is to use normal die thickness and mold cap thickness, thus reducing the risk of thin die handling and narrow mold clearance. Mold flash is eliminated through the grinding methodology. By applying a strip grinding process, we can easily generate a very thin die and mold cap. In addition, the process can be applied to other structures, similarly reducing warpage. Double sided mold structures and two layer mold structures can be considered. In this paper, the strip grinding process and system is applied to various package structures. Finally, in order to generate the thin PoP, package structure and warpage simulation results are presented.

Strip Grinding System

The function of the strip grinding system is to remove the mold compound and thin the silicon (Si) die down to the target thickness without incurring any damage to the package. The system consists of a loader, grinding and cleaning zone, vision inspection zone and unloader. Grinding is performed by a blade wheel and after that, the cleaning process is performed. This is followed by a DI water rinse.   Vision inspection is performed to check for package damage and then finally, the molded strip is loaded into the magazine. In this grinding machine, there is a measuring system to check package thickness, and to guarantee this measuring system, accuracy and repeatability checks are performed. For accuracy check, 7 different thickness block gauges are used and the results show maximum 1.8um tolerance. For the repeatability check, the measuring system reads 10 times repeatedly on the same point of the molded strip, and the results show maximum 4um variation. Figure 3 shows the process sequence of the grinding.

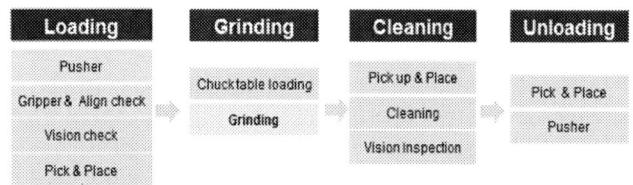

Figure 3. Strip grinding process sequence.

The strip grinding process is performed by contacting the blade and package which may cause excessive static electricity on the package surface. To check for ESD after grinding, a contact voltmeter is used. Acceptable voltage levels must measure below 10 volts.

Strip Grinding Test

In order to check the actual system performance, a 74mm x 240mm, 160um thickness substrate consisting of three mold blocks was used (25 units per block). Solder bumped die measuring 10 x 10mm with 150um thickness was used, and a 330um mold cap thickness was applied. Ten substrate strips were used for thickness measured after grinding and 10 points were measured in each strip. Figure 4 shows the thickness distribution. The target thickness after grinding was 250um for the mold cap. To achieve this target thickness, 80um mold compound and 50um die were ground off. The result of thickness after grinding showed a maximum 259um, minimum 249um, and average 256um. In general, thickness was well controlled within a +/- 10um range.

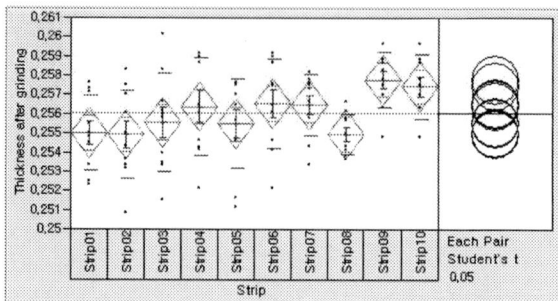

Figure 4. Thickness distribution after grinding.

Die surface roughness after grinding was measured at 627nm as shown in Figure 5.

Figure 5. Die surface roughness.

Application 1 - Exposed Die Package

PoP trends are going thinner and the exposed die solution is being widely applied through the use of film assist mold, but it's difficult to control the mold flash on the die because of the relative impact of the mold cap and die tolerances to the total structure. When the mold cap is thicker or the die is thinner than the target thickness, mold flash readily happens. Therefore, very tight control of the mold cap and die thickness is required in order to prevent the mold flash. It's difficult to control because of the limitations in process capability for the wafer grinding process and mold process. Therefore, more precise tools and processing is necessary but this tends to increase cost.

To resolve the mold flash issue, strip grinding can be recommended as a solution, because strip grinding can complete the die and mold grinding together. This process enables a mold flash free package without tight control for die and mold cap thickness resulting in a wider process window.

This solution can be applied to the multichip package where it is difficult to control the stacked thickness because of the added variation inherent with multiple die. Each die will have its own thickness tolerance, and the thickness should be controlled for all dice and mold cap to prevent mold flash. Strip grinding enables flash-free mold even when multidie levels are different, because the grinding process is independent of the die height level. Another application is utilized for exposed heat sinks, where strip grinding can expose the heat sink on the die without mold flash while keeping the heat sink flush with the mold and die surface. This provides for optimal external heat spreader attachment.

Figure 6 shows the mold surface after grinding and there is no mold flash on the die back side and no damage including die crack.

Figure 6. Exposed die with flash-free mold by grinding.

Application 2 - Thin Die Package

One of the methods to reduce the PoP thickness is to reduce the die thickness then generate the thinner PoP with a thinner mold cap. However, there is a limitation in process capability regarding thin die handling, and this is more critical in case of large die. Normally, the logic die consists of an SoC of large size and it's difficult to perform a chip attach process with large, thin die due to silicon cracking or chipping. Another issue is chip attach yield during reflow because a thinner die induces more severe warpage. With the strip grinding process, it doesn't require thin die handling. The chip attach process is done with thick die followed by the mold process. This normal process flow is then followed by strip grinding. This grinding generates the thin die as an exposed die structure. An overmold structure is also possible by applying an additional mold process. During the additional mold process, either the same mold compound as the first mold compound or a different mold compound can be used. In this evaluation, 30um die thickness was targeted and different mold compound was used during the second mold process. After grinding down to a die thickness of 30um, there was no damage related to die and bump. Figure 7 shows the thin die structure as exposed die and overmold structures.

a) Exposed die structure

b) Overmold structure

Figure 7. Thin die structure.

Application 3 - Narrow Mold Clearance

Another method to reduce the package thickness is to generate a narrower mold clearance between die top to mold top. In order to avoid the mold void, a proper mold clearance is required, and this clearance is more critical above a large die. By applying the strip grinding process, it's possible to follow the normal process flow such as chip attach and mold process with a proper gap, and then grind down to the target clearance. In this evaluation, a 30um mold clearance was targeted and maximum 24um filler cut mold compound was used. The die was 11 x 11mm with 130um thickness. After grinding, there were no mold related defects including mold void.

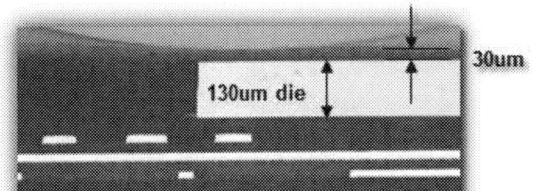

Figure 8. Narrow mold clearance.

Application 4 - Double Sided Molded Package

By applying a grinding process, double side molded structures are possible. Standard format is having a molded structure on the die side of the substrate only. However, if the bottom side mold is feasible, then this structure will help make a balanced structure on top and bottom which tends to improve the warpage performance. So far, bottom side mold is difficult, because the BGA ball is mounted on the bottom area, and the bottom mold interferes with bottom ball exposure and, therefore, its ability to solder to the motherboard. For the double sided mold process flow, chip attach on the top side and BGA ball attach on the bottom side need to be done first. This is followed by double side mold using mold chase which has a cavity on the top and bottom chase. For the bottom chase, the cavity should be deeper than the BGA ball height in order to avoid ball damage during mold. The bottom mold is ground until the bottom ball is exposed. After grinding, the bottom area seems like a metal pad. To make a BGA standoff for motherboard interconnection, a second ball attach needs to be performed. Finally, a double sided molded structure can be generated. Figure 9 shows the bottom area molded structure. As described above, the ball attach was performed twice to generate a proper BGA standoff.

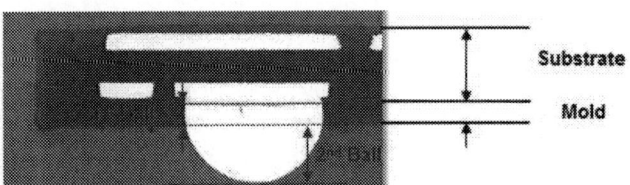

Figure 9. Bottom side molded structure created with a ball attach process that is performed twice.

Thin PoP Package

PoP thickness is trending thinner and thinner and the most critical issue is the warpage control with the thin package structure. PoP requires SMT processing to stack the memory package on top of the logic package. SMT stacking performance is directly affected by package planarity. That's why it's more important to control the warpage for PoP structures. Currently, the thinnest structure is the exposed die, but the mobile industry requires an even thinner structure and it's difficult to generate a thinner structure with existing methods and materials. To achieve the acceptable warpage performance for the thin PoP structure, a new approach with a new structure and material set is likely needed. Table 1 shows the warpage performance comparing two structures. The package size is 15 x 15mm with a die size of 11 x 11mm and 100um thickness utilizing copper pillar bumping. Overmold and exposed die structures were considered for warpage performance comparison.

Substrate thickness	Die thickness	Mold thickness	Mold type	25°C warpage	260°C warpage
300um	100um	250um	Overmold	45um	-55um
210um	100um	100um	Exposed	169um	-214um

Table 1. Warpage performance comparison per structure.

The overmolded structure is the existing structure and showed stable warpage performance, but to make thin PoP, the mold cap and substrate thickness were reduced and the warpage performance rapidly decreased. This result shows how difficult it is to control the warpage with the thin PoP structure.

Warpage simulation was performed prior to actual sample evaluation for various package options as introduced above.

Warpage Simulation

As shown in Table 1, in order to meet the warpage specification for thin PoP, a new structure is required. To predict the warpage for thin PoP, warpage simulation was performed and various structures were considered, including double side molded structures, two layer molded structures, etc. The applied test vehicle was a 15 x 15mm package outline with 11 x 11mm copper pillar bumped die. Considered variables were substrate thickness, mold cap thickness, die thickness and molding method. A 4-layer substrate with a 60um core was applied to this simulation and total substrate thickness was 210um. For mold cap thickness, 50um and 70um were considered, and die thicknesses of 30um, 50um and 70um were considered. For package structure, overmold with a 300um thickness substrate was selected as a reference, which was the existing structure. The exposed die structure, thin die with overmold structure, thin die with exposed die structure, and double sided mold structure were considered. Table 2 shows the considered package structures for simulation and Table 3 shows the DOE matrix.

Table 2. Package structure comparison.

	Package type	Substrate thickness	Die thickness	Mold thicknes
Leg 1	Overmold	300	100	250
Leg 2	Exposed die	210	100	100
Leg 3	Thin die+overmold	210	30	70
Leg 4	Thin die+overmold	210	50	70
Leg 5	Thin die+exposed die	210	70	70
Leg 6	Thin die+double sided mold	210	50	70 (btm 30)

Table 3. DOE matrix for warpage simulation.

Figure 10 shows the simulation result. Regarding package type, the overmolded structure showed the best warpage performance and Leg 1, which was the reference leg, showed the best warpage performance with thick substrate and mold cap. With the exposed die structure, the simulation didn't show significant warpage difference by mold cap thickness as shown in Leg 2 vs Leg 5. But even under the same mold cap thickness, die thickness significantly affected the warpage. Comparing Leg 3 to Leg 5, the same mold cap was applied with 70um, and different die thickness was applied. The result showed that thinner die displayed better warpage performance for 25°C and 260°C as shown in Leg 3 which used a 30um thick die. This may be caused by mold compound volume which is an important factor to the warpage since thinner die means a higher mold compound volume ratio that may help to improve the warpage performance. To build a more balanced structure, a double sided mold structure was considered in Leg 6. For the bottom side mold, a 30um mold was applied while the top side mold was 70um. This leg showed the best warpage performance with the thin substrate application, and this warpage performance was almost close to the current standard structure in Leg 1. Based on the result, it showed the possibility to control the warpage within the specification even with the thin PoP structure.

978-1-4799-2408-0/14 $31.00 © 2014 IEEE

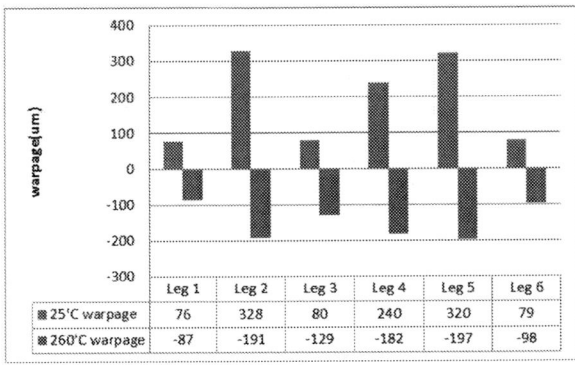

Figure 10. Warpage simulation per structure.

	Leg 1	Leg 2	Leg 3	Leg 4	Leg 5	Leg 6
25°C warpage	76	328	80	240	320	79
260°C warpage	-87	-191	-129	-182	-197	-98

Discussion

Various strip grinding applications such as mold flash free exposed die, narrow mold clearance, thin die, two layer mold and double sided mold have been demonstrated. These technologies were combined to generate thin PoP structures with warpage simulation performed to verify results. At this time, warpage simulation was focused on the package structure only.

Future work will focus on simulation for material optimization especially for mold compounds. Different mold compounds will be considered for two layer mold structures and double sided mold structures. Based on the simulation study, selected structures and materials will be applied to the actual sample where empirical data will verify the thin PoP generation.

Conclusions

According to the simulation study, the overmolded structure is expected to perform better than the exposed die structure in terms of warpage. A higher ratio of mold compound volume to die volume is another key factor to improve the warpage performance. In order to generate the thin PoP, it is prudent to combine the substrate and mold cap and utilize a concurrent thinning procedure. Strip grinding is a viable solution to generate the structure.

References

1. JinSeong Kim, et al., "Fine pitch PoP introduction," IEEE CPMT Symposium Japan 2013.
2. JinSeong Kim, et al, "Application of Through Mold Via (TMV®) as PoP Base Package," The 58th Electronic Components and Technology Conference, Lake Buena Vista, Florida, May 2008.
3. Zwenger, C., et al, "Next Generation Package-on-Package (PoP) Platform with Through Mold Via (TMV®) Interconnection Technology," IMAPS Device Packaging Conference, Scottsdale, Arizona, March 2009.
4. Zwenger, C., et al, "Next Generation Package-on-Package (PoP) Platform with Through Mold Via (TMV®) Interconnection Technology," IMAPS Device Packaging Conference, Scottsdale, Arizona, March 2009.
5. Yoshida, A., et al, "Design and Stacking of an Extremely Thin Chip-Scale Package," Electronic Components and Technology Conference, 2003.
6. Dreiza, M., et al, "High Density PoP (Package-on-Package) and Package Stacking Development," The 57th Electronic Components and Technology Conference, Reno, Nevada, May 2007.
7. Smith, Lee, "Driven by Smartphones, Package-on-Package Adoption and Technology Are Ready to Soar" By Lee Smith, Amkor Technology, Chip Scale Review Magazine, July 2008.
8. Yoshida, A., et al, "A Study on Package Stacking Process for Package-on-Package (PoP)," ECTC 2006.

978-1-4799-2408-0/14 $31.00 © 2014 IEEE 1365

Cost and Performance Effective Silicon Interposer and Vertical Interconnect for 3D ASIC and Memory Integration

[1]Li Li, [2]Mitsutoshi Higashi, [2]Akihito Takano, [1]Jie Xue and [2]Gary Ikari
[1]Cisco Systems, Inc.
[2]Shinko Electric Industries Co., Ltd.

Abstract

To enable three-dimensional (3D) ASIC and memory integration, large-size silicon interposer is a critical technology [1]. Currently most silicon interposers are manufactured by wafer foundries and are limited in size by the wafer lithographic processing. In this study, manufacturing of cost- and performance-effective, large-size silicon interposers are investigated. The existing supply chain and infrastructure of high-performance flip-chip packaging substrates is leveraged.

A 3D System-in-Package (SiP) is designed and manufactured that includes a large-size silicon interposer with Through-Silicon-Vias (TSV) and Cu wiring layers on both sides of the silicon interposer. To develop the assembly process for the 3D SiP, thermal deformation of each constituent component is analyzed using metrological techniques such as Digital Image Correlation (DIC). With the assembly process developed, an ASIC die is attached on top of the silicon interposer while two smaller memory dice are attached to the bottom of the silicon interposer with micro-bump interconnection. The 3D IC stack is then assembled on an organic package substrate with two Vertical Interconnect Spacers (VISs) and "regular" solder bumps so the bottom dice can be accommodated without any interference to the planar organic substrate. The completed 3D SiP module is finally assembled on a test board using a lead-free surface mount process. Thermo-mechanical reliability of the 3D SiP assembly is studied using temperature cycling testing. Both the thermal deformation analysis and stress testing results are used to gain insights into the 3D IC technology and to enable ASIC and memory integration for next generation high-performance network systems.

Introduction

The innovation of 3D IC integration has been successfully applied to address the gap seen between the slowdown of Moore's law scaling and ever increasing system integration requirements. Leveraging 3D IC integration, a concept of 3D ASIC and memory integration was proposed and a feasibility study was conducted [1]. 3D ASIC and memory integration is an extension of the Flip Chip and Memory Package (FCAMP) that was developed about a decade ago [1]. It has the advantages of reducing power consumption, improving the bandwidth between the ASIC and memory and modularizing system hardware designs. Critical components for enabling 3D ASIC and memory integration include large size silicon interposer, 3D stacked memory and micro-bump or micro-pillar interconnects.

The introduction of 3D IC integration also brought new requirements and sometimes disruptions to the existing microelectronics manufacturing supply chain. One example is the manufacturing and supply of silicon interposers (Si-IP). Currently there are three commonly used interposer manufacturing process flows. The first can be referred to as the "Foundry Process" flow. The silicon interposer is fabricated completely by the wafer foundry and sometimes it even includes packaging the 3D IC sub-assembly to the package substrate [2].

The second can be referred to as the "Middle End Of Line (MEOL) Process" flow. MEOL is often used by the Outsourced Assembly and Test (OSAT) suppliers for processing TSV wafers with active circuits. It is usually started at the wafer foundry for via generation and filling, and fabrication of front side metal wiring layers. Then the full thickness wafers are delivered to the OSAT for further downstream processes that include wafer thinning, TSV reveal, backside metal layer generation, passivation and bumping. The third process can be called as the "Substrate Process" flow. In this case, the silicon interposer is fabricated and supplied by the traditional packaging substrate suppliers. Figure 1 shows schematically a comparison of the three aforementioned silicon interposer manufacturing and supply chain flows.

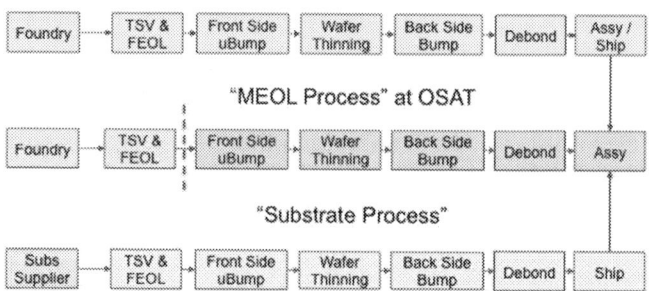

Figure 1. A comparison of Si-IP manufacturing and supply flows.

For both the Foundry and MEOL processes, TSVs are generated using the Deep Reactive Ion Etch (DRIE) process [1]. The front-side interconnects or wiring layers are made with Cu damascene techniques. For the backside interconnection, MEOL usually uses Redistribution Layer (RDL) process [3]. For true 3D wafers with active circuits and TSVs, MEOL may be preferred but for passive interposers, an alternative and cost effective way may exist. This alternative, "Substrate Process" flow is the focus of this study.

Interposer Manufacturing

In this study, manufacturing of cost- and performance-effective, large-size silicon interposer is investigated. The existing supply chain and infrastructure of high-performance flip-chip packaging substrates is leveraged. There are several advantages in this approach. One is minimal disruption to the existing supply chain. The silicon interposer is considered as a packaging material rather than another piece of silicon chip. Secondly, large size silicon interposers can be manufactured with a line width and line spacing in the range of a few micrometers. This type of silicon interposer (Si-IP) is often referred to as coarse pitch silicon interposer in order to distinguish itself to the ones made by wafer foundries. Table 1 shows a comparison between the coarse pitch silicon interposer to the fine pitch silicon interposer.

Table 1. Comparison of coarse pitch and fine pitch silicon interposers.

Features	Coarse pitch Si-IP	Fine pitch Si-IP
Cu Wiring (Dielectric)	SAP (Polyimide)	Damascene (Oxide)
uBump Material	Cu/Ni/SnAg etc.	Cu/Ni/SnAg etc.
uBump size / pitch (min)	30 / 50um	20 / 40um
FS wiring line / space / thickness (min)	3 / 3 / 3um	0.5 / 0.5 / 1.0um
RDL via size	20um	1.0um
TSV size / pitch /depth	60 / 150 / 200um	10 / 50 / 100um
BS wiring line / space / thickness (min)	10 / 10 / 3um	10 / 10 / 1.0um
Back side pad or bump size	Ni/Au 100 / 150um	Ni/Au 100 / 150um

For coarse pitch silicon interposers, the Semi Additive Process (SAP) is used to fabricate Cu wiring on either side of the interposer. The SAP method has fewer process steps and uses conventional equipment that is also used for fine pitch printed wiring board fabrication [4]. On the other hand, fine pitch silicon interposer relies on the damascene technique for Cu wiring fabrication that requires both Chemical-Mechanical Polishing (CMP) and dry etching processes. Because it involves fewer process steps and uses conventional equipment for fabrication, coarse pitch silicon interposers will be less costly compared to fine pitch silicon interposers [4].

Figure 2 shows the top and bottom views of a 28 mm x 28 mm silicon interposer fabricated with the SAP method. Major fabrication steps used are shown schematically in Figure 3.

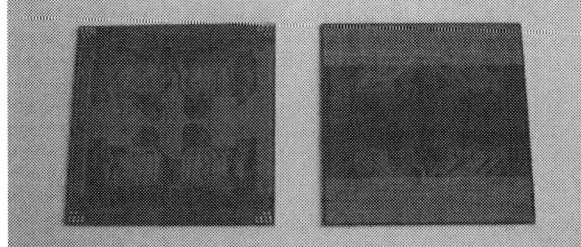

Figure 2. Top and Bottom views of a 28 x 28 mm silicon interposer.

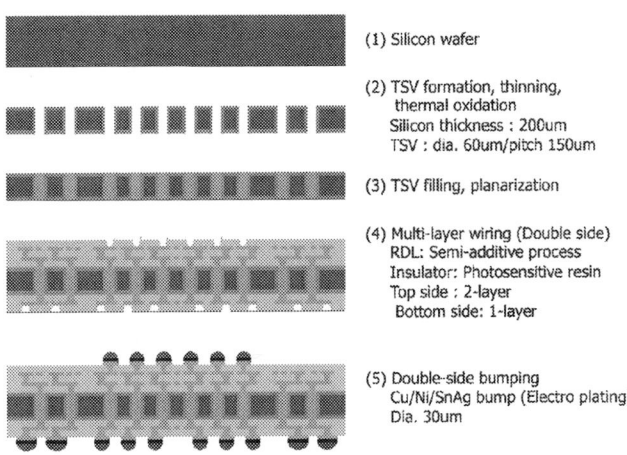

(1) Silicon wafer

(2) TSV formation, thinning, thermal oxidation
Silicon thickness : 200um
TSV : dia. 60um/pitch 150um

(3) TSV filling, planarization

(4) Multi-layer wiring (Double side)
RDL: Semi-additive process
Insulator: Photosensitive resin
Top side : 2-layer
Bottom side: 1-layer

(5) Double-side bumping
Cu/Ni/SnAg bump (Electro plating)
Dia. 30um

Figure 3. Major fabrication steps used to fabricate the coarse pitch silicon interposer.

Figures 4 (a) and 4 (b) show the micro-bumps formed and two layers of Cu wiring on the front side (FS) of the interposer. Figures 4 (c) and 4 (d) show the micro-bumps, as well as the "regular" bumps formed along with one layer of wiring on the back side (BS) of the interposer.

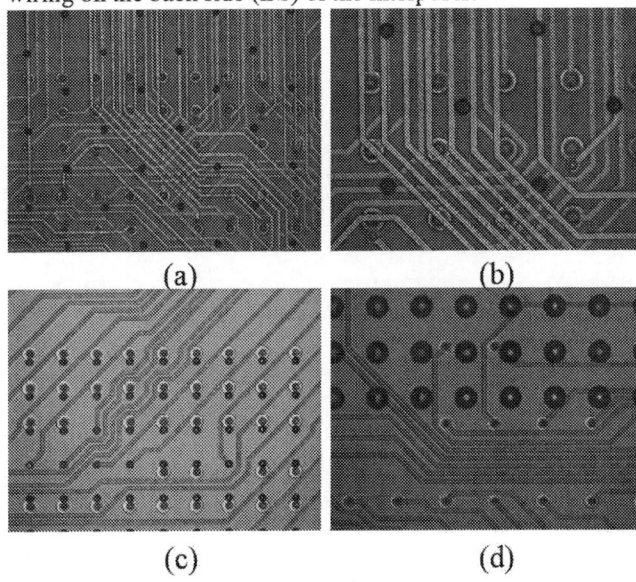

(a)   (b)

(c)   (d)

Figure 4. Top view and bottom view of the 28 mm x 28 mm silicon interposer fabricated with the SAP method and the "Substrate Process" flow.

Figures 5(a) and 5(b) show the cross-sectional views of the 28 mm x 28 mm silicon interposer.

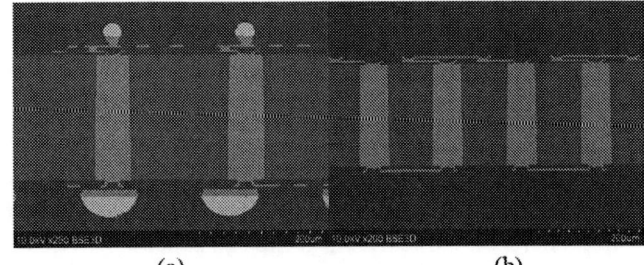

(a)   (b)

Figure 5. Cross-sectional views of the silicon interposer manufactured.

3D SiP Assembly

A 3D System-in-Package (SiP) is designed and manufactured that includes a large-size silicon interposer with Through-Silicon-Vias (TSV) and Cu wiring layers on both sides of the silicon interposer.

For electrical characterization, manufacturing and reliability evaluation, a 20 mm x 20 mm x 0.3 mm ASIC die is attached on top of the silicon interposer while two smaller memory dice, each 10 mm x 10 mm x 0.3 mm, are attached to the bottom of the silicon interposer with the micro-bump interconnection and a uniquely developed double-sided Chip to Chip (C2C) joining process. The large size silicon interposer manufactured is measured at 28 mm x 28 mm x 0.200 mm overcoming the size limitation often seen due to the reticle size used in wafer lithographic processing.

The 3D IC stack is then assembled on an organic package substrate with a couple of unique Vertical Interconnect Spacers (VIS) and solder bumps so the bottom dice can be accommodated without any interference to the planar organic substrate. Communication between the top ASIC die and the bottom memory dice is made through the TSVs and the wiring layers of the silicon interposer. Figure 6 shows a schematic of the 3D SiP design.

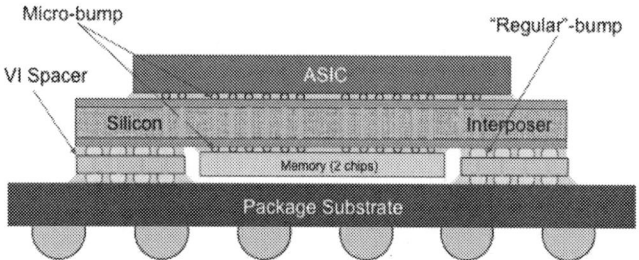

Figure 6. A schematic view of the 3D SiP.

In order to optimize the chip joining and packaging assembly processes, the DIC method was used to analyze the thermo-mechanical behavior of each constituent component of the 3D SiP. Figure 7 shows the warpage measurement results for the top die at room temperature and 260°C. As expected, the warpage of the silicon die was small and did not change with temperatures.

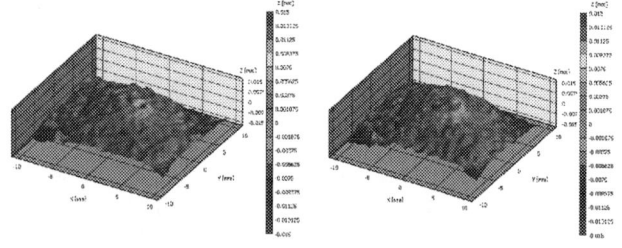

Figure 7. Warpages measured for the top die at room temperature (left) and at 260°C (right).

Figure 8 shows the warpage measurement results for the silicon interposer at room temperature and at 260°C. It can be seen the warpage of the silicon interposer was relatively small and was further reduced when the temperature is raised up to 260°C.

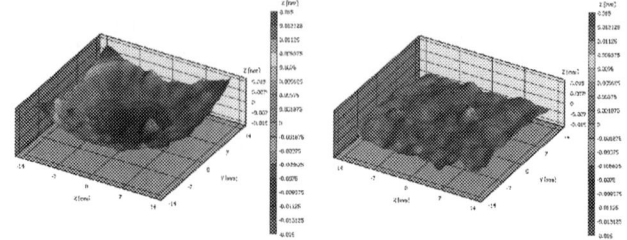

Figure 8. Warpages measured for the 28 mm x 28 mm silicon interposer at room temperature (left) and at 260°C (right).

The DIC method was also used to measure warpages of the package substrate. Figure 9 shows the measurement results for the package substrate.

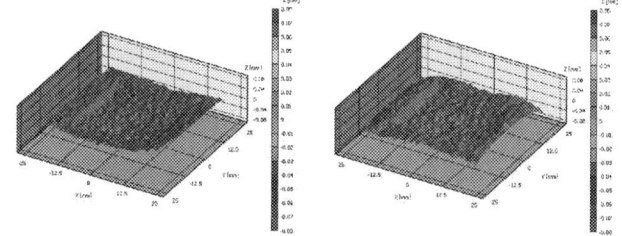

Figure 9. Warpages measured for the package substrate at room temperature (left) and at 260°C (right).

Based on the thermal deformation analysis for each components of the 3D SiP, an assembly process flow was developed. Since the warpages of the silicon dice and silicon interposer were small and did not change with temperature, the dice were assembled to the two sides of the interposer to form the 3D IC stack first. Underfill encapsulation was used to protect the joints made of micro-pillars and micro-bumps. A top view and a bottom view of the 3D IC stack are shown in Figures 10 (a) and 10 (b), respectively.

(a)   (b)

Figure 10. Top and bottom views of the 3D IC stack.

The warpage measurement of the finished 3D IC stack is shown in Figure 11. The warpage was relatively small at about 15 micron. This was expected and this confirmation raised the confidence on making the final 3D SiP module. In the next step of the assembly process, two Vertical Interconnect Spacers (VISs) were attached to the package substrate to form the substrate subassembly using the "regular" solder bump interconnection. Figure 12 shows a top view of the package substrate with the VIS attached.

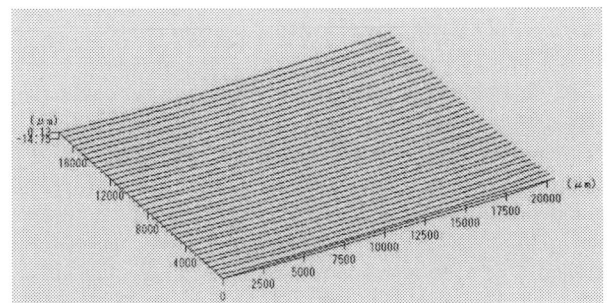

Figure 11. Warpage measured for the 3D IC stack at room temperature.

Figure 12. A top view of the VIS – package substrate subassembly.

The warpage for the VIS – package substrate subassembly was measured from the bottom BGA side and the result measured at room temperature is shown in Figure 13.

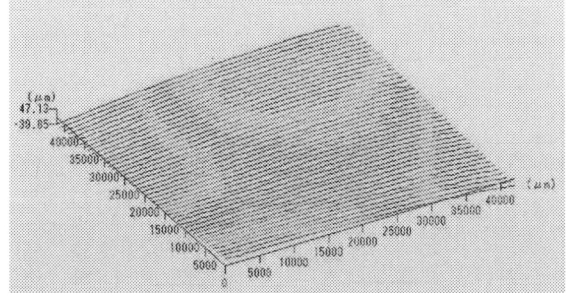

Figure 13. Warpage measured for the VIC – package substrate subassembly at room temperature.

Figure 14. A top view and a bottom view of the 3D SiP assembled.

In the final assembly process step, the 3D IC stack was attached to the VIS – package substrate subassembly using the "regular" solder bump interconnection. The "regular" solder bumps were then encapsulated using the underfill material. A top view and a bottom view of the finished 3D SiP are shown in Figure 14.

Warpage of the 3D SiP was also measured at room temperature. The result shown in Figure 15 indicates the warpage of 3D SiP module could be over 200 um at room temperature; this can potentially impose challenges for mounting the module to the Printed Circuit Board (PCB).

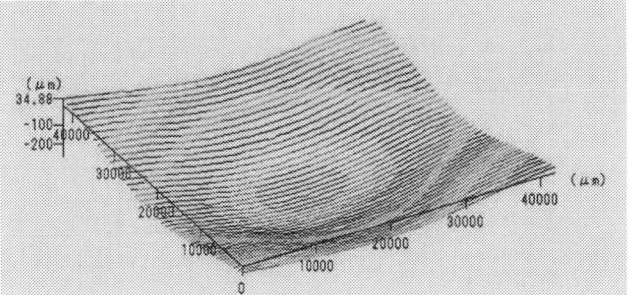

Figure 15. Warpage measured for the finished 3D SiP module at room temperature.

The structure of 3D SiP module was analyzed using both Optical Microscopy (OM) and SEM. Pictures of the cross-sections for the silicon dice, interposer, VIS, micro-bumps are shown in Figures 16 and Figure 17, respectively.

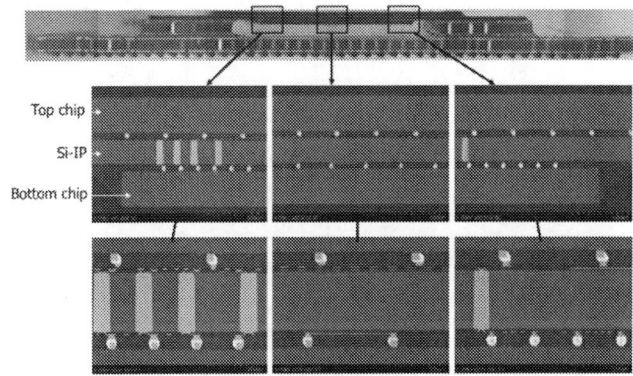

Figure 16. Optical microscopic and SEM cross-sectional views of the finished 3D SiP module.

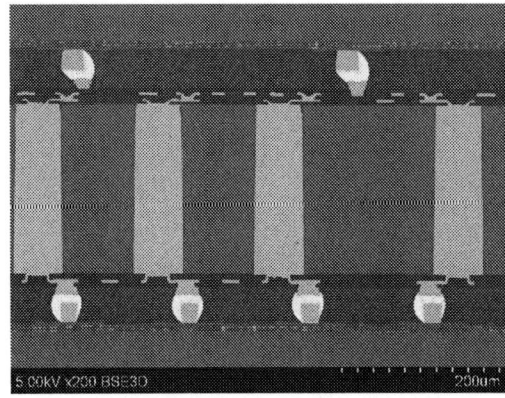

Figure 17. A cross-sectional view showing the top and bottom dice, silicon interposer with TSVs and micro-bumps.

978-1-4799-2408-0/14 $31.00 © 2014 IEEE

Post assembly analysis and characterization also included continuity measurements for all the daisy chains and test features designed. A summary is given in Table 2 for all the daisy chains designed for the 3D SiP module.

Table 2. Summary of daisy chains included in the 3D SiP module.

Type	Route	Net Count (per Module)
1	BGA – VIS – Si-IP – Top Chip	26
2	BGA – VIS – Si-IP – Top Chip – Bottom Chip 1	10
3	BGA – VIS – Si-IP – Top Chip – Bottom Chip 2	10
4	BGA – VIS – Si-IP – Bottom Chip 1	10
5	BGA – VIS – Si-IP – Bottom Chip 2	10
6	BGA – VIS – Si-IP – (TSV and Via Chains)	20
	Total	86

Board Level Assembly and Reliability Evaluation

To evaluate electrical performance and reliability of the assembled 3D SiP, a test board was designed and fabricated. The test board has a total of 6 layers and is 2.36 mm (93mil) thick. A high Tg printed circuit board material was used. The surface finish of the board was Cu OSP.

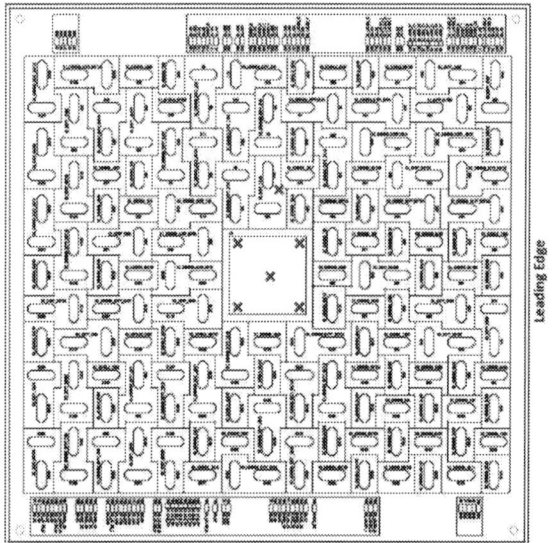

"X" = Thermocouple Locations shown for TC Assembly

Figure 18. The red cross marks indicate the thermo-couple locations on the test board.

The completed 3D SiP module has a 45mm x 45 mm body size, 1848 I/O, and its BGA pitch is 1.0 mm. The modules were assembled to the test board using a standard production SMT line with a Pb-free, no-clean assembly process. A total of 20 boards were built. Prior to the build extensive set-up trials were conducted to optimize the reflow conditions and the solder paste printing process. For the topside of the board a 0.125 mm thick laser cut stainless steel stencil was used. A Pb-free solder paste was selected for printing solder paste to the front side of the test board.

Six thermocouples were used for monitoring solder joint temperatures to develop the optimal reflow profile. The thermocouples were attached by drilling through the PCB and by placing the metal contacts onto the PCB pads. After solder paste deposition, a spare component was placed onto the setup board. Figure 18 shows the locations of the thermocouples attached. For the test board assembly build, all pads were inspected for proper solder volumes prior to component placement and reflow.

Figure 19 shows a top view of a completed 3D SiP – test board assembly. A combination of continuity measurement and X-ray inspection was employed to detect opens and shorts and to assess the BGA solder joints quality and assembly yield. Figure 20 shows a group of four X-ray pictures taken for the four corners of the BGA for the 3D SiP – test board assembly. No solder ball bridging or open joint was detected.

Figure 19 A top view of the 3D SiP – test board assembly.

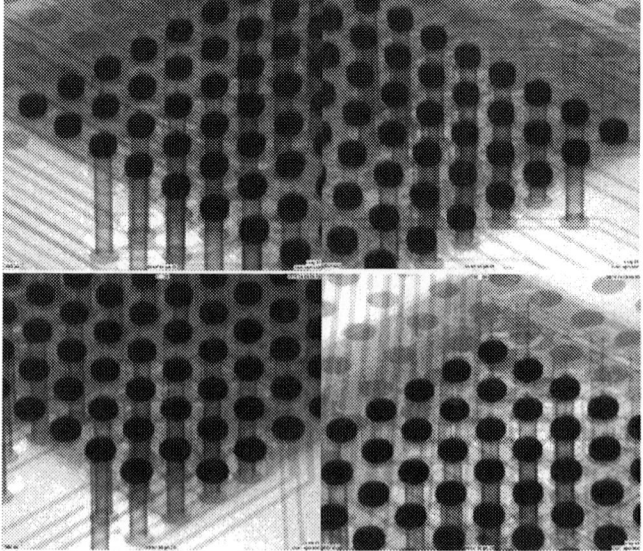

Figure 20. X-ray inspection of BGA solder joints.

Thermo-mechanical performance and reliability of the 3D IC stack was studied using environmental stress testing. Both the modeling and stress testing results are used to gain insights into the 3D IC technology and to enable ASIC and memory integration for next generation high-performance network systems.

Accelerated board-level temperature cycling test (0 to 100°C with 10 minute ramps and dwell times at 10 minutes) was used to assess long term reliability of the 3D SiP modules selected for the study. Daisy chains were resistance-monitored continuously using a data logger. Cycles to failure for each component were recorded. A total of 3000 cycles were completed for the temperature cycling test. Except for a few very early fails on some chains in the sub 400 cycle range, all the other daisy chains monitored passed 3000 cycles with no failures. Failure analysis on these early fails is planned. It was suspected that potential manufacturing issues leading to isolated poor quality solder joints were a major contributor.

Summary and Conclusions

Leveraging the existing supply chain and infrastructure of high-performance flip-chip packaging substrates, a 3D System-in-Package (SiP) is designed and manufactured that includes a large-size silicon interposer with Through-Silicon-Vias (TSV) and Cu wiring layers on both sides of the silicon interposer.

- The large size silicon interposer is manufactured and supplied with the SAP method and the "Substrate Process" flow so it can be cost- and performance-effective.

- Thermal deformation of each constituent component of the 3D SiP was analyzed using real-time, metrological techniques such as Digital Image Correlation (DIC) to optimize material selection, assembly process and reliability.

- A series of OM and SEM analyses were conducted to demonstrate that the 3D SiP module can be built with micro-bump interconnects, double-sided C2C joining, VIS and the final packaging assembly processes developed.

- The completed 3D SiP module was assembled on a test board using a lead-free surface mount process. Thermo-mechanical performance and reliability of the 3D SiP assembly was studied using accelerated temperature cycling testing.

- Both the thermal deformation analysis and stress testing results are used to gain insights into the 3D IC technology and to enable ASIC and memory integration for next generation high performance network systems.

Ongoing activities include electrical performance analysis of the 3D SiP and comparison with electrical simulation using field solver. Results of these studies will be published in a future paper.

Acknowledgments

The authors would like to thank their colleagues for all the discussions and support.

References

1. Li, L., S. Peng, J. Xue, et al, "Addressing Bandwidth Challenges in Next Generation High Performance Network Systems with 3D IC Integration," Proceedings of the 62nd Electronic Components and Technology Conference (ECTC), San Diego, CA, May 2012, pp. 1040-1046.

2. CoWoS Service, www.tsmc.com

3. Yoon, S., D. Na, et al, "2.5D/3D TSV Processes Development and Assembly/Packaging Technology," Proceedings of the 11th Electronics Packaging Technology Conference (EPTC), Singapore, December 2011, pp. 336-340.

4. Sunohara, M., A. Shiraishi, Y. Taguchi, K. Murayama, M. Higashi, and M. Shimizu, "Development of Silicon Module with TSVs and Global Wiring (L/S=0.8/0.8um)", Proceedings of the 59th Electronic Components and Technology Conference (ECTC), San Diego CA, USA, May 2009, pp. 25-31.

Assembly and Packaging of Non-bumped 3D Chip Stacks on Bumped Substrates

Bing Dang, Joana Maria, Qianwen Chen, Jae-Woong Nah, Paul Andry, Cornelia Tsang, Katsuyuki Sakuma, Christy Tyberg, Raphael Robertazzi, Michael Scheuermann, Michael Gaynes, and John Knickerbocker

IBM T.J. Watson Research Center

1101 Kitchawan Road, Rte 134, Yorktown Heights, NY 10598

dangbing@us.ibm.com, +1 914-945-1568

Abstract

In this paper, a novel assembly and packaging approach is proposed for 3D/2.5D chip stacks based on bumped substrates. The thinned chips are stacked using thermal compression bonding with "flat" metallization to reduce assembly complexity associated with conventional controlled-collapse-chip-connection (C4) solder bumps. Meanwhile, the laminate substrates are bumped with C4s using injected molten solder (IMS) processes. The pre-stacked chips are then assembled and packaged on the bumped laminates successfully.

Introduction

While 3D/2.5D ICs continue to move into more commercial applications, many challenges still remain in their integration, assembly and packaging. For instance, handling technologies for thinned TSV wafers are still evolving [1-3]. These technologies must provide temporary support of the fragile thin wafers during processing and testing. Typically, conventional C4 solder bumps are fabricated at wafer level for joining and connecting IC chips to a substrate in flip chip assembly processes. Today, solder bumps or copper pillars up to 60~80µm in height and 120~150µm in pitch are commonly used in mainstream flip chip organic packages. Special handling issues are then raised when such a conventional bumping technology is applied to 3D/2.5D integration. For example, polymeric adhesive is usually used to temporarily bond a 3D/2.5D wafer to a handler wafer (glass or silicon) prior to its thinning. If the solder bumps are already present before the temporary wafer bonding, the adhesive must be thick enough to accommodate these 60~80µm tall bumps. This will significantly increase the materials cost as well as the complexity in adhesive removal post handler wafer debonding. A temporary wafer handling solution with thin adhesive coating for flat pads may offer more than 4X material cost reduction per wafer compared to that with thick adhesive for the conventional tall bumps or pillars.

In addition, stacking and assembly of thin chips on a base chip with the presence of large C4 bumps is a non-trivial challenge. First of all, the C4 solder bumps create a non-planar surface, so they can not be directly chucked with vacuum in a flip chip bonder. Second, the C4 bumps at the very bottom of a chip stack may melt and collapse when top chips are bonded unless a complex solder temperature hierarchy is used or C4 bumps are coated with pre-applied underfill. Therefore, the assembly and packaging processes of 3D/2.5 chip stacks need to be carefully investigated.

In this work, we explored several process options for test vehicles in various die sizes from 6mm x 6mm to 21mm x 21mm.

Then we proposed an approach based on the bumped substrates and the non-bumped 3D chip stacks for a large chip size. We characterized the thin chip warpage under various conditions such as free-standing, joined on laminate first, stacked first, etc. Using Injection Molded Solder (IMS) technology, we added Pb-free solder C4 bumps on the packaging substrate side instead of the IC wafer side [4]. Meanwhile, we replaced the conventional solder bumping process on the 3D IC wafers with thin metallization. Such a process modification enables a simpler handling and assembly process of the 3D chip stacks because of the "flat" bottom surface of a chip stack. We demonstrated a fluxless bonding process with ~8s of bonding time for each layer of thin chip. Such a bonding process is suitable for both chip-on-chip (COC), chip-on-wafer (COW), and multi-chip bonding with high throughput and high-yield. Then the pre-stacked chips can be treated and joined to the "bumped" substrates similarly to a conventional 2D chip.

"Chip Stack first" vs. "Substrate first"

It is well-known that wafer and die warpage may be present when a silicon wafer is thinned and it is exacerbated with any unbalanced stress from through-silicon-via (TSV) and on-chip metal wiring and dielectrics [5]. There are basically two main options to assemble and package thin chips: build up the chip stacks directly on a substrate (namely "substrate first" approach), or to form the stack first and then join the chip stacks on the substrate last, (namely "chip stack first approach"). Figure 1 shows the schematic description of the "substrate-first" and "chip-stack first" approaches. In both cases, the greatest challenge is to accommodate the warpage of the thinned 3D/2.5D chips as well as the warpage of the organic substrates.

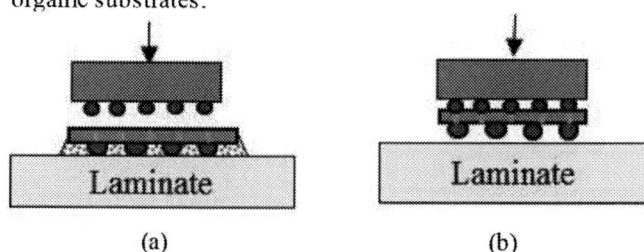

(a)   (b)

Figure 1. Chip stacking processes options (a) by stacking up sequentially onto a substrate (b) by stacking chips first and then join chip stack onto a substrate

For individual TSV chips, the overall warpage is proportional to their x-y dimensions for a given chip thickness. For a very small chip size, either process can accommodate the relatively small warpage and non-planarity during chip stack assembly. As shown in Figure 2, we successfully demonstrated chip stack assembly using both "substrate first" and "chip stack first" approaches for small chip sizes of 6 x 6mm and 6.7 x 12.6mm with a thin TSV base chip in the thickness of ~70um.

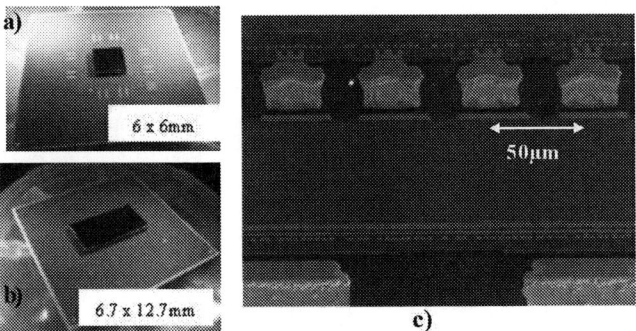

Figure 2. A chip stack formed by thermal compression bonding using: a) substrate first approach, b) chip stack first approach, c) cross-section SEM picture of b)

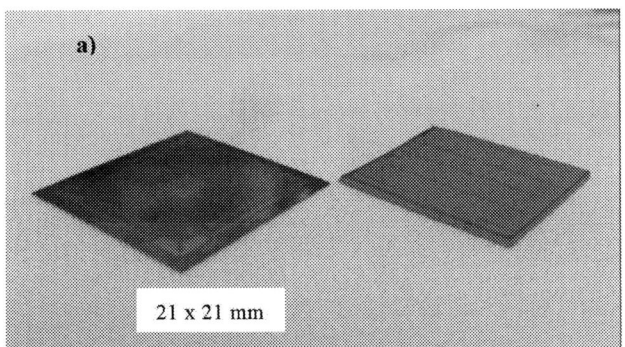

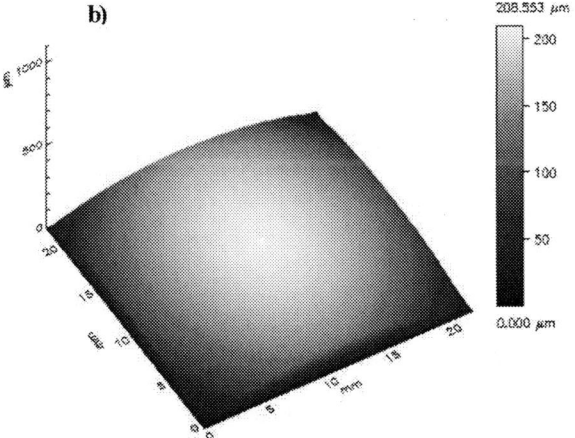

Figure 3. (a) Comparison of a free-standing thinned TSV chip and a regular CMOS chip at the size of 20mm x 20mm (b) topographic measurement of the thinned 3D TSV chip

However, chip stacking and assembly became much more challenging when the chip size and the warpage increased. Figure 3 shows images of a large size (~20mm x 20mm) free-standing thinned TSV chip (~70µm thick) along with a regular full thickness CMOS chip (~ 760µm). It is noticeable that a thinned TSV chip can be very warped and fragile to handle,

while a full thickness chip is relatively rigid and flat. The optical topographic measurement shows warpage of ~ 200µm across the thin chip area.

In addition, organic laminate substrates usually have some intrinsic warpage that is temperature dependent. All these factors lead to significant challenges for assembling large size thinned TSV chips onto a laminate substrate. Nevertheless, using a flat bonding tool interface, it is still possible to join a large size thinned TSV chip onto a laminate through thermal compression bonding, as shown in Figure 4a. However, after the chip joining on the laminate, a large warpage of ~ 206µm can still be observed, as shown in Figure 4b. The warpage leads to significant C4 solder interconnect deformation across the chip.

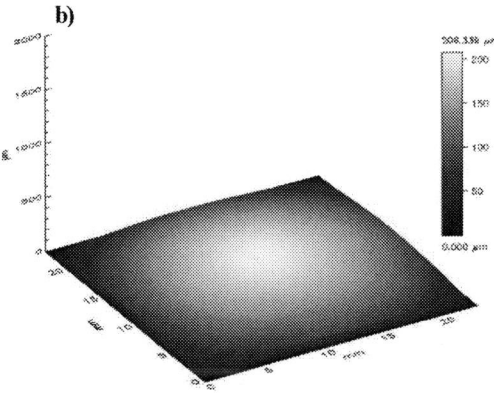

Figure 4. a) Image of a thinned TSV chip joined directly onto a laminate substrate b) topographic measurement across the chip area

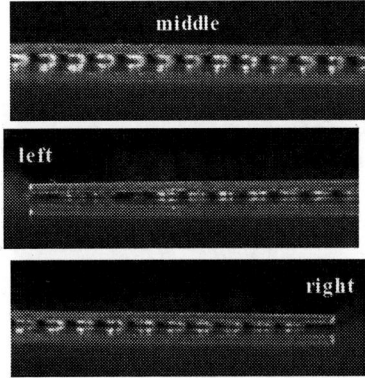

Figure 5. Side view of conventional C4 interconnects (200µm pitch) after bonding a free-standing thin chip directly onto a laminate substrate (middle, left, right)

978-1-4799-2408-0/14 $31.00 © 2014 IEEE   1373

Figure 5 shows the side view of the C4 interconnects at various locations of a chip edge. The C4 interconnects near the middle of the chip edge (top image) maintained a relatively normal "truncated" shape, while the C4 interconnects at the ends of the chip edge (middle and bottom images) are seriously compressed into "pancake" shape due to the temperature dependent laminate warpage during chip joining.

The warpage of a large thinned TSV chip joined on a laminate indicates a significant obstacle for subsequent stacking of any additional chips. Micro-bumps of height typically around 20-30μm and pitch less than 50μm [6,7] are often used in typical 3D chip stacks for maximum interconnect density. Such a small micro-bump height is obviously not sufficient to accommodate the warpage of the thinned TSV chip on laminate. Therefore, in this work, we propose using a "chip stack" first approach to improve the micro-bump bonding yield and efficiency. Unlike conventional flip chip assembly, the bottom chip (or base chip) of the chip stack is deposited with "flat" metallization instead of C4 solder bumps. The flat bottom surface allows the use of a vacuum chuck in any flip chip bonder without any special fixture. Multi-layer chip stacks with micro-bump interconnects can then be built on this base chip.

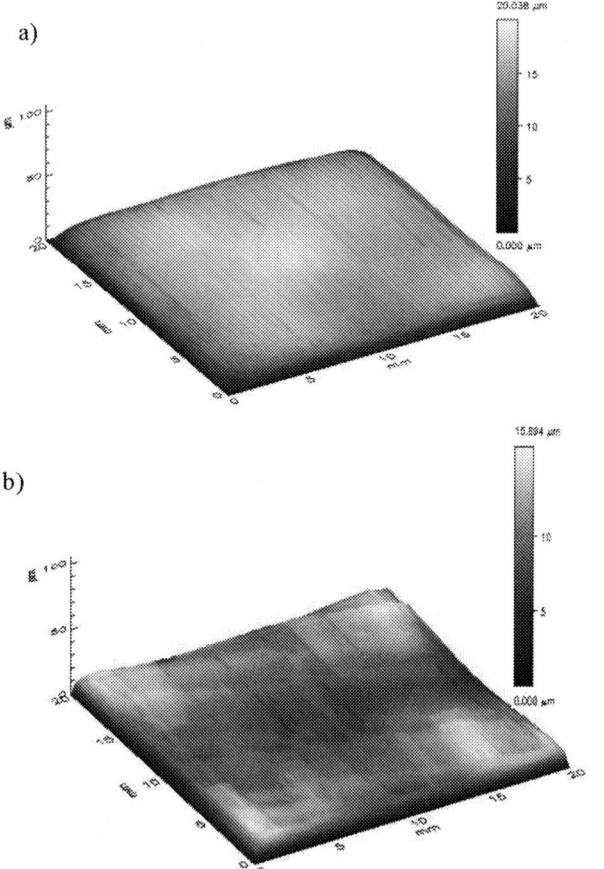

Figure 6. Topographic warpage measurement on a base TSV chip with (a) 1-layer chip stacked (b) 4-layer chip stacked.

Figure 6 shows the topographic measurement results for a chip stack of 1 layer and a chip stack of 4 layers bonded onto a thinned base chip. In fact, the warpage across the chip area decreases when the total chip stack thickness increases with the number of layers in the stack. In these large sizes as well as multi-layer chip stack form factors, it is advantageous to build the chip stack separately in order to achieve the best assembly yield with the very small micro-bumps.

Figure 7 plots the warpage data across the chip area diagonally for a free-standing thin base chip, 1-layer chip stack, 2-layer chip stack, as well as a 4-layer chip stack on a base chip. A free-standing base TSV chip shows its intrinsic center-to-corner warpage of more than 200μm as mentioned previously. With 1-layer thin chip stacked on a base chip, the overall warpage is largely reduced to ~ 20μm because of the added stiffness. However, its warpage distribution still shows a similar downward direction. The overall warpage is further reduced to ~ 16μm if 2-layer thin chips are stacked on a base chip. With 4-layer chip stack, the overall warpage distribution became more complex because of the increased stiffness as well as the non-planarity of the inter-chip interconnects. There is no longer a clear warpage direction. Nevertheless, the much reduced overall warpage is advantageous for final assembly on a laminate substrate. In fact, such a pre-stacked multi-layer chip stack may be simply treated as a regular 2D chip with conventional flip-chip reflow processes.

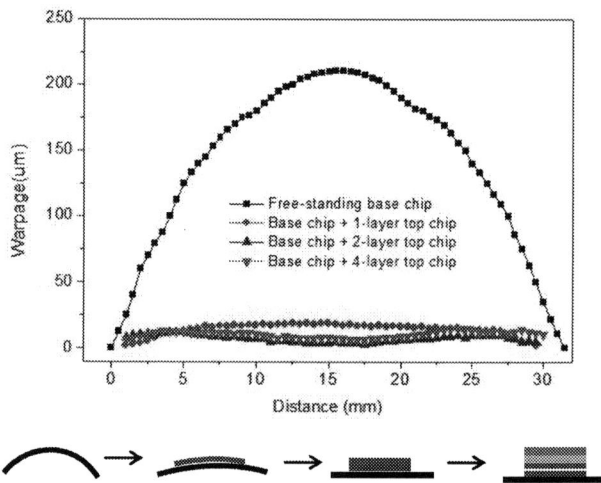

Figure 7. Plot of warpage measurement on a base TSV chip without chip stack, 1-layer chip stack, 2-layer chip stack, and 4-layer chip stack, respectively.

IMS bumping processes of organic laminates

As discussed above, we proposed the use of 3D chip stacks without the bottom layer bumped with conventional C4 solder bumps. This means that the laminate substrate side needs to provide the solder bump interconnection for the final assembly. In this work, we studied the bumping processes of organic laminate substrates to enable the non-bumped 3D chip stacks [4]. Figure 8 schematically illustrates a method for forming large volume solder bumps on an organic substrate. A dry film of photoresist is laminated to the substrate above the solder resist layer and holes are formed in the dry film by photolithography as shown in step (b) and (c), respectively. The desired solder volume above the solder resist is defined by the thickness of the dry film and the opening size of the hole. A solder fill head moves across the substrate during the IMS process shown in step (d). Molten flux-free solder is injected and filled into the openings of a dry film resist and solder resist, and then wet on the contact pads of the substrate. The IMS process and solder solidification are conducted in a nitrogen environment. Step (d) shows the solder bumps following solidification, the bumps having rounded top ends that extend above the top surface of the dry film resist. The dry film resist is removed following solder solidification to produce the structure shown in step (e).

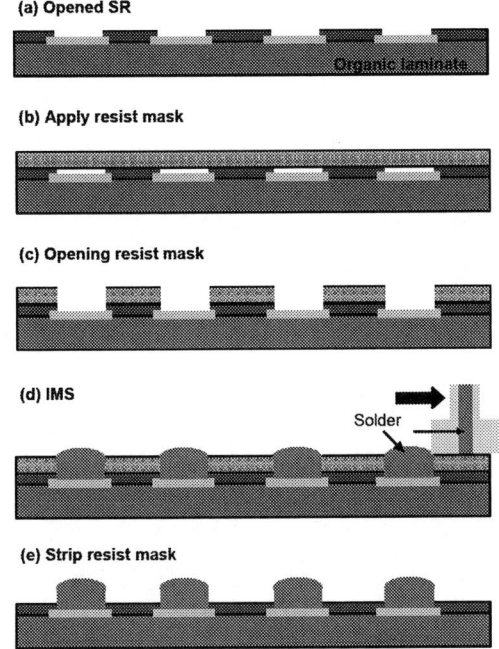

Figure 8. Laminate IMS bumping process flow

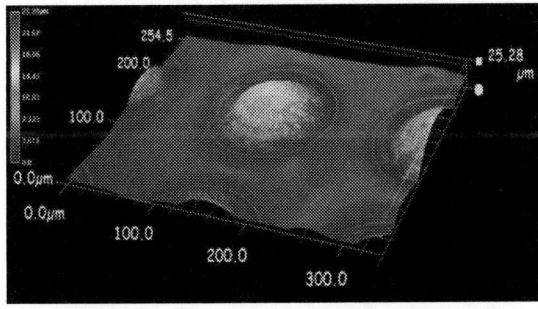

a) IMS bumps on dry film photoresist

b) IMS bumps after stripping

Figure 9. Three-dimensional measurement of IMS bumps on the organic substrate.

Figure 9 shows the three dimensional measurement images of IMS bumps on the 50μm thick dry film and after stripping dry film. Figure 9 (a) demonstrates that the IMS bump height is about 25μm above the dry film and Figure 9 (b) shows that the 75μm height IMS bump is formed above the solder resist of the substrate after stripping 50μm thick dry film resist. The IMS process leads to 100% fill of the dry film and solder resist opening with pure solder since molten solder itself is injected without flux. Such a laminate IMS bumping technology has shown that the solder bump volume can be easily controlled for different applications by simply changing the dry film resist thickness and/or holes size. Also, this technology can form high volume solder bumps at fine-pitch application.

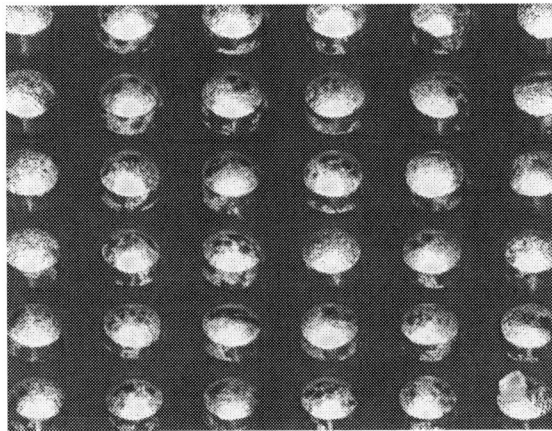

Figure 10. Side view images of IMS bumps on the laminate. The IMS bumps have column shape sidewalls and round tops.

Chip stack assembly on laminate substrates

In this work, two types of test vehicles (TV) were used for the assembly demonstration of non-bumped 3D chip stacks. As shown in the schematic drawing in Figure. 11a, TV1 contains 2 layers of chips. The top chip is a conventional CMOS chip and the base chip is a thinned TSV chip with 32nm technology node. As shown in Figure 11b, TV2 contains 4 layers of thinned 3D test chips on a slightly larger base chip to form a chip stack.

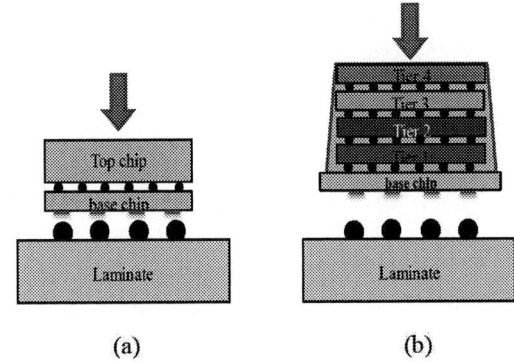

Figure 11. Schematic drawing of the chip stack structures for TV1 (a) and TV2 (b)

In TV1, the base chip has a similar dimension of the top chip. Therefore, there is no sufficient space for underfill dispensing. The micro-bumps between top and base chips provide the mechanical connection during final assembly. After the chip stack is joined on a laminate substrate, a underfill is used to fill both chip-to-chip gap and the chip-to-laminate gap to complete the final packaging.

In TV2, the base chip is larger than the top tiers of the chips, which allows underfilling prior to the final assembly on a laminate. Since the total thickness of the entire chip stack is still relatively thin (~300μm) and the mechanical strength is not very high, it is helpful to have the underfill between the 4 layers of thin chips to enhance the rigidity. Nevertheless, in both cases of TV1 and TV2, the chip stacks hold their mechanical integrity well during the final assembly and the packaging processes.

Figure 12 and Figure 13 show the cross-section view of interconnects in the 2-layer chip stacks and 4-layer chip stacks after the final assembly. Both micro-bumps within the chip stacks and the C4 bumps were well joined. All the gaps including were successfully underfilled in order to protect the interconnects and maintain mechanical integrity. Small amount of voids are visible near the solder-to-laminate interfaces because the in-coming laminate substrates were pre-soldered using conventional solder paste technology. In fact, the IMS bumping technology can eliminate such an issue if in-coming laminate substrates are not pre-soldered.

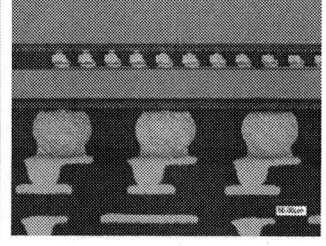

Figure 12. Cross-section of a 2-layer chip stack (TV1)

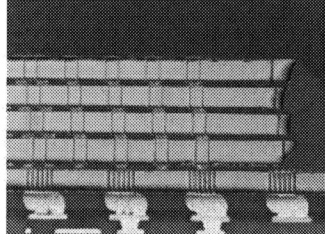

Figure 13. Cross-section of a 4-layer chip stack (TV2)

Testing summary

As described above, several test vehicles in different form factors have been studied using "chip-stack first" or "substrate first" methods. The test vehicles were designed for various testing purposes including power/noise, I/O macro, as well as chip-package interaction. As summarized in Table 1, relatively small chip stacks including TV1a (6mm x 6mm) and TV1b (6.7mm x 12.7mm) were assembled in both "substrate first" and "chip stack first" approaches and passed the electrical testing successfully.

However, the larger test vehicles including TV1c (19mm x 19mm) and TV2 (20mm x 20mm) only were successfully assembled using the "chip stack first" approach. The laminate "substrate first" was not suitable for these large chip sizes because of the large warpage.

Table 1. Description of assembly and test results of various test vehicles

Test vehicle	Form factor /stack structure	Testing purpose	"Substrate first"	"Chip stack first"
TV1_a	6mm x 6mm (a thick chip on a thin base chip)	I/O characterization	Pass	Pass
TV1_b	6.7mm x 12.7mm (a thick chip on a thin base chip)	Power /Noise	Pass	Pass
TV1_c	19mm x 19 mm (a thick chip on a thin base chip)	Chip-package interaction	Failed	Demonstrated, in testing
TV2	20mm x 20 mm (4 layers of thin chips on a thin base chip)	Thermal resistance between chips	Failed	Demonstrated, in testing

Future work

Based on the assembly and packaging approaches developed in this work, we are exploring chip-on-wafer integration for the formation of 3D/2.5D chip stacks. Future work will include a study on yield, reliability, and throughput.

Conclusions

Various chip stack assembly processes have been studied. A "chip stack first" approach is more promising for large-size chip stack and multi-layer chip stack assembly. In addition, we successfully demonstrated the use of "bumped substrates" to simplify the chip stacking processes for the first time.

Acknowledgments

The authors would like to thank Robert Polastre, Evan Colgan, Gerald Advocate, Glenn Pomerantz, CSS staff for their help or discussion. Thanks are extended to Ghavam Shahidi and Tze-Chiang Chen for their management support.

References

1. S. Pargfrieder, et al., "Temporary Bonding and DeBonding Enabling TSV Formation and 3D Integration for Ultra-thin Wafers," in proc. *10th Electronics Packaging Technology Conference, (EPTC)*, pp.1301-1305, 2008.
2. A. Phommahaxay, et al., "Ultrathin wafer handling in 3D Stacked IC manufacturing combining a novel ZoneBOND™ temporary bonding process with room temperature peel debonding," in proc. *2011 IEEE International 3D Systems Integration Conference (3DIC)*, pp.1-4, 2012.
3. A. Jourdain, et al., "Integration and manufacturing aspects of moving from WaferBOND HT-10.10 to ZoneBOND material in temporary wafer bonding and debonding for 3D applications," in proc. *63rd IEEE Electronic Components and Technology Conference (ECTC)*, pp.113-117, 2013.

4. Jae-Woong Nah and Mark McLoed, US 20120318855, IMS (Injection Molded Solder) with two resist layers for forming solder bumps on substrates.

5. D. W. Kim, et al., "Development of 3D through silicon stack (TSS) assembly for wide IO memory to logic devices integration," in proc. *63rd IEEE Electronic Components and Technology Conference (ECTC)*, pp.77-80, 2013.

6. J. Maria, et al.,"3D Chip stacking with 50 μm pitch lead-free micro-c4 interconnections," *61st IEEE Electronic Components and Technology Conference (ECTC)*, pp.268-273, 2011

7. Jing-Ye Juang, et al."Effect of metal finishing fabricated by electro and Electroless plating process on reliability performance of 30μm-pitch solder micro bump interconnection," *63rd IEEE Electronic Components and Technology Conference (ECTC)*, pp. 28-31, 2013

The Miniaturization of a Micro-Ball Endoscope by SiP Approach

Xunxun Zhu[1], Jian Cai[1, 2,*], Yu Chen[1], Yingke Gu[1], Xiang Xie[1], Qian Wang[1], Zhihua Wang[1, 2], Xiaofeng Sun[3], Lixi Wan[3]

[1]Insitute of Microelectronics, Tsinghua University
[2]Tsinghua National Laboratory for Information Science and Technology
Tsinghua University, Haidian District, Beijing, 100084, P.R.China
[3]Institute of Microelectronics of Chinese Academy Sciences
3 Beitucheng West Road, Chaoyang District, Beijing, 100029, P.R. China
*E-mail: jamescai@tsinghua.edu.cn

Abstract

The use of capsule endoscope greatly contributes to the painless and hygiene diagnoses of the small intestine, which cannot be reached easily by traditional upper endoscopy or colonoscopy. However, most of current commercial capsule endoscopes only carry one or two cameras, which may lead to missing some interesting spots due to the limited field of view. A micro-ball endoscope with six CMOS image sensors mounted on a cubic carrier could reduce the miss rate. Some initial prototypes were built with PCBs and connectors. The minimum size of initial prototypes was 22mm×22mm×22mm, which limited the practical use in human medicine. Thus, it is essential to minimize the prototype cube size. In the present work, a SiP (System in Package) approach was proposed to promote the miniaturization of the micro-ball endoscope. Utilization of rigid-flex substrate of special configuration, which is bendable other than PCBs and connectors, achieved seamless connection between adjacent boards and higher package density by employing three-dimension (3-D) space. The technology of multi die stack was also applied to lower the footprint. With this new SiP design, the size of complicated micro-ball endoscope, which involves ASIC (Application Specific Integrated Circuits) chips, image sensors, Flash chips, and wireless transceiver module, was successfully reduced to 16mm×16mm ×16mm

Keywords: System in Package, Miniaturization, Rigid-Flex substrate, Micro-ball endoscope

1. Introduction

The development of electronics products, particularly consumer electronics, is a huge driving force to the innovation of IC packaging technology. To meet the demands of small outline and multifunctional integration from market, SiP (system-in-package) has become one of the most promising technologies for the miniaturization of electronic products in recent years. According to ITRS (The International Technology Roadmap for Semiconductors), SiP is defined as a combination of multiple active electronic components of different functionality, assembled in a single unit that provides multiple functions associated with a system or sub-system. A SiP may optionally contain passives, MEMS, optical components and other packages and devices [1]. Besides significant size and weight reductions, SiP provides other obvious advantages such as lower power consumption and noise, higher bandwidth, and improved performance, etc. compared with normal traditional packaging technology.

The invention of medical endoscopes, such as gastroscope and colonoscope, is valuable as it has made diagnose of the internal side of some section of digestive tract possible.

However, the current traditional endoscopes still have many drawbacks mostly due to their structure and configuration. One of the biggest concerns is their limited inspecting area, which makes it not easy to reach some small organs. For example, it's difficult to conduct intestine examination using traditional inspecting devices or methods. Uncomfortable examination experience and potential in-hospital cross-infection for the patients are the drawbacks as well.

The updated technology capsule endoscope, also called wireless capsule endoscope, greatly changes this situation. The capsule endoscope is a disposable, small, swallowable, wireless, miniature camera which allows getting a direct visualization of the gastrointestinal mucosa [2]. The capsule endoscope can go through the digestive tract and capture images of the inside after swallowed. Thus it can provide, for the first time, painless optical imaging of the whole small intestine [3]. Nowadays, there are several companies throughout the world focused on developing capsule endoscope. Given Imaging, a world famous Israeli medical technology company, has issued several series of PillCam capsule endoscope. The PillCam has been used in many hospitals and clinics for practical visualization and detection. Another well-known pioneering solutions provider for capsule endoscope is Olympus, whose products are available in many places throughout the world. Invention and application of these patient friendly endoscopes facilitate painless inspection of the gastrointestinal tract.

However, most of current commercial capsules are only installed with one or two cameras, which may result in missing some interesting spots [3, 4] and leading to insufficient image data and inaccurate diagnoses. Thus, a micro-ball endoscope, with a full-view image capture system, has been proposed as a better alternative medical examination methodology [5, 6]. Despite having similar workflow, the micro-ball endoscope significantly reduces the miss rate by using six CMOS image sensors compared with the current capsule endoscope. A complete micro-ball endoscope contains at least several modules of system control and coordination, image capturing and storing, power supply and management, motion control, wireless transmitting, and antennas, all of which are integrated on a cubic carrier and packaged in a transparent ball shell made of biocompatible material. Meanwhile, the subsystem integrated on each board is capable of capturing and saving images under master control. Validated cubic system prototype samples [6] have been realized based on PCBs and connectors. Six pieces of PCB boards form the cubic package carrier and more than 160 components in forms of bare dies, packaged chips, other active devices, and passives are all included in the cubic system. Details of the main components mentioned above are listed in the Table 1.

Table 1. Detailed information of the main components

Chip Name	Size (mm)	Package	Amount	Pad Distribution
Master	3.8×3.7×0.30	Die	1	Peripheral
Power	1.5×1.5×0.30	Die	1	Peripheral
Slave	3.8×3.7×0.30	Die	6	Peripheral
CMOS image sensor	4.2×4.0×0.82	CSP	6	BGA
Flash	12.9×9.0×0.30	Die	6	Single row
Accelero-meter	3.0×3.0×1.0	LGA	1	LGA
Magneto-meter	2.0×2.0×0.85	DFN	1	DFN
Wireless transceiver	7.0×7.0×0.80	QFN	1	QFN
Camera cap	Ø=13.00	—	—	—

In this system, the master and slave dies form the control module. Images are captured by CMOS image sensors and then stored in flash memories. The energy supply of different voltage volumes is distributed and managed by the power chip. An attitude sensing unit including magnetometer and accelerometer is also embedded in the micro-ball. Finally, the image data captured can be transmitted by wireless transceiver with antennas [6].

The ideal size of a micro-ball endoscope is considered to be easily swallowed by human. The prototype size of the first version is around 78mm×78mm×78mm, which is too big to swallow. After adjustment and optimization, size of the second version was reduced to 22mm×22mm×22mm. The comparison of two versions of prototype samples is shown in Fig.1. However, size of the second version of prototype sample is still not feasible. The miniaturization mainly utilizing traditional packaging method and SMT（Surface Mount Technology） has reached its limit given the system components sizes and numbers.

Figure 1. Validated prototype sample system cube based on PCBs and connectors.

The oversize of this micro-ball endoscope limits its practical use in medicine. So, further miniaturization of the system cube is essential. Considering the limit of traditional packaging technology and the significant advantages of SiP as mentioned above, a SiP approach using rigid-flex substrate is presented to reduce the size in the present work.

2. Package Solution

Based on the existing system architecture and electrical constitution [6], this study mainly focuses on the optimization of the package solution and substrate design. Rigid-flex substrate was selected as the packaging substrate and multi die stack technology was applied to fully use the vertical space and reduce planar routing area on substrate.

A. Rigid-flex substrate

Rigid-flex is a multilayer printed circuit board with both rigid and flex electric interconnecting layers laminated together [7]. The rigid substrate often uses BT or FR4 as the base material and components can be packaged on the rigid section as normal. The flex section is usually fabricated from polyester or polyimide (PI), which allows high interconnecting density. The composition of rigid and flexible substrate in one single circuit board brings some novel advantages over traditional rigid substrate or flexible substrate. First of all, this technology allows designers free from connectors, wires, cable ribbons or other hardware when trying to connect multiple substrates. The flex substrate allows high interconnecting density and thus flexible inner layers could connect a number of boards as one piece and provide a continuous electrical path between boards. In addition, flex substrate is relatively soft, which allows the rigid-flex substrate bent or folded at the flexible section to utilize the third dimension space. Thereby, it offers an ideal package solution for applications with a tight form factor. Besides, significant reduction in package weight can be achieved with rigid-flex substrate due to the thinness and lightness of flexible layers. Demonstrated superior mechanical reliability, good electrical performance, tighter form factor, and lower weight have made rigid-flex substrate enjoy great popularity in high end applications ranging from military and aerospace electronics to implantable medical devices [8].

As described in the introduction part, the prototype system of the micro-ball endoscope is cubic. High density electrical interconnections exist between components packaged on different boards. Thus a suitable system packaging carrier needs to be both strong and soft enough, and to meet the form factor requirements at the same time. In addition, considering the special and complex application environment in human body of the micro-ball endoscope, a packaging substrate with high connection reliability is expected. Compared with normal packaging technology based on PCBs, rigid-flex substrate is a better solution to the high demands of the micro-ball endoscope by providing more reliable 3-D interconnection packaging and higher system miniaturization.

B. Multi die stack

In applications that need high integration, planar components distribution cannot meet the demands. It is well known that multi die stack is an effective way to utilize 3-D space and minimize X-Y package dimensions in high density package field [9], in which dies can be stacked on the Z axis direction. Traditional stacking methods are often classified into two types: pyramidal and cantilever shape. For dies with different sizes, pyramidal structure is used to place smaller die on the bigger one. Cantilever stacking usually applies to dies of similar sizes to expose the bonding pads. However, if the sizes contrast is huge, neither of the stacking structures mentioned above is proper. In this situation, just as shown in Figure2 (a), long bonding wires are required to connect the upper die with the package carrier, which may lead to subsidence and cause bad signal transmission. Other direct defects also include bigger carrier and higher package cost.

Table1 shows that the discrepancy in the size of some dies is large. For example, the flash dies are much bigger than the master, slave, and power dies. If these dies were stacked in the normal pyramidal manner, it would be certain to bring the defects mentioned earlier. To improve this situation, an inverted pyramidal die stack structure is presented in this work. Firstly, a die with smaller size was attached and wire bonded to the substrate. Then spacers were attached beside the die. And finally, the bigger die was placed on the spacers. When interconnecting the upper bigger die to the substrate, the length of bonding wires will not be too big. Figure2 (b) illustrates the top and cross section views of the inverted pyramidal die stack structure. In this system, dies such as master, slave, and power are smaller in size compared with flash die, so they can be stacked beneath the flash die in the inverted pyramidal structure.

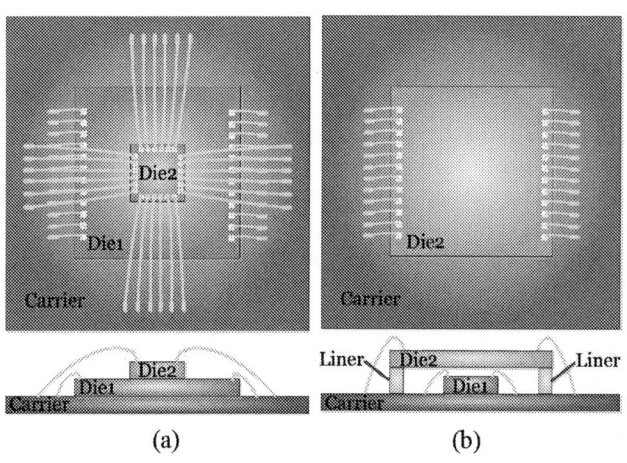

Figure 2. Comparison between (a) traditional pyramidal structure and (b) inverted pyramidal structure.

Figure 2 compares the traditional pyramidal die stack structure and the inverted one applied in this paper. Die1 and Die2 stand for the stacked dies on the package carrier. From the top and cross section views of two different die stack structures, it's clear that the new stack method reduces the length of wire loops efficiently and lowers footprint. Furthermore, risks of signal integrity and unreliability are decreased to some extent because of shorter bonding wires.

C. Substrate outline and layout

As introduced earlier, multilayer double-side rigid-flex substrate was applied in this work due to its advantages to

minimize the size of the cubic system. Components were placed on both surfaces of the substrate. In addition, special substrate configuration was designed. Herein, the rigid-flex substrate consists of seven rigid boards and several flexible sections. Among the seven rigid sections, six pieces are of the same size and the other one is slightly smaller. When bending the substrate after components packaging into a cube, the smaller rigid board could be folded inside. The special configuration didn't only affect the final cubic form, but also expanded space available for packaging. Figure3 shows the substrate outline designed in this paper. Boards numbered 1, 2, 3, 4, 5, 6 are the same sized and bigger than board numbered 7.

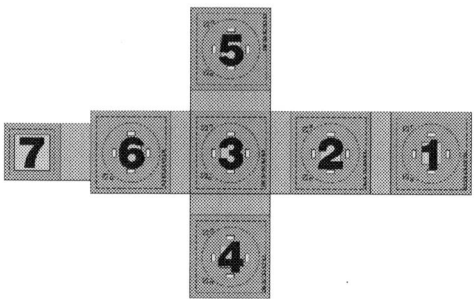

Figure 3. Substrate outline and rough components distribution.

The micro-ball endoscope system involves different types of electrical components, such as bare dies, packaged chips, passives, and other actives. The compatibility of practical packaging and assembly process should be considered to arrange the components properly. Thus, bare dies and other packaged devices are better to be placed separately on different sides of the substrate. Also, components interconnected or belonged to the same module ought to be placed nearby. Facilitation for routing is another concern. Distribution following rules above reduces the length of interconnecting trace and decreases the possibility of signal integrity problems caused by long transmission line. Meanwhile, according to the system demands, the subsystems on all six boards could obtain and store images respectively. To meet the requirements and balance components distribution, necessary components for these functions including one CMOS image sensor, four LEDs, a flash chip, and a slave chip were placed on the substrate first. Then others were placed based on functions and practical substrate space capability.

Trials of layout were conducted on the rigid-flex substrate. Considering fabrication error, the minimum distance from board edges to the nearest components was set around 1mm. The final minimum sizes of rigid boards were 16mm×16mm (6 pieces) and 12mm×12mm (1 piece). To avoid collision between elements attached on two adjacent faces when forming a cube, safe lengths of flex substrate were set according to the components' heights and the minimum radius of the flexible materials.

3. Substrate Design and Electrical Simulation

In contrast to simple single die packaging substrate design, the SiP approach for the complicatedly interconnected micro-ball endoscope brought in complex substrate routing. In this work, circuits schematic and nets list were created for substrate design and error checking. As the package solution describes, a rigid-flex substrate shown in Fig. 3 should be finished. The rigid sections were concerned with six layers: signal 1, ground, signal 2, signal 3, power, and signal 4; and the flexible sections only concern two intermediate layers: signal 1 and signal 2. All bare dies were packaged with inverted pyramidal die stack method introduced in section 2. Bonding fingers for two levels of bare dies were placed near them. It's necessary to reserve enough space for dummy spacers between the bonding fingers of flash dies and the first level dies. Packaged chips or SMDs were preferred to be placed on the different side of the substrate from bare dies. The ground layer was kept undivided to provide full current loop plane. On the power plane, areas representing different voltage volumes were divided for diverse inputs. Through vias were applied where traces need to change layer. Before routing, some detailed design rules were raised according to the general assessment of the system package solution and mature packaging process. The details about the substrate design rules were illustrated in Table 2.

Table 2. Substrate design rules

Rigid substrate size(mm×mm)	16×16 （6 pieces） 12×12 （1 piece）
Bonding wire(μm)	20
Bonding finger(μm × μm)	90×300
Finger opening(μm × μm)	100×400
Min trace width/space(μm)	50/75
Diameter of via/via pad(μm)	100/250

The whole substrate design was worked out with Cadence SPB. Figure 4 shows the routing of surface layer signal1 and signal 4 as an example.

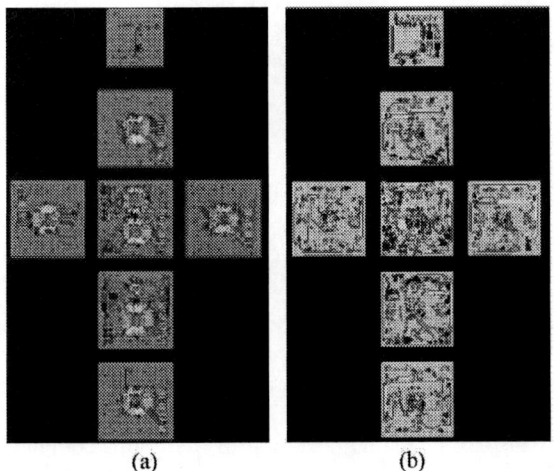

(a)   (b)

Figure 4. Routing of (a) signal 1 layer and (b) signal 4 layer.

The signal integrity problems are known to happen in high interconnecting density electronic systems, especially when the system is working at high frequency. To decrease signal integrity problems, electrical simulations including S parameters and time domain wave conformity were conducted on some key signal traces on the substrate design. A correct clock signal is important for one electronic system. So the traces of the key clock signals were selected to perform the simulation. S parameters computation results with sweep frequency ranging from 0 to 3 GHz are displayed in Fig. 5 (a). Figure 5(b) shows the partial enlarged details of (a).

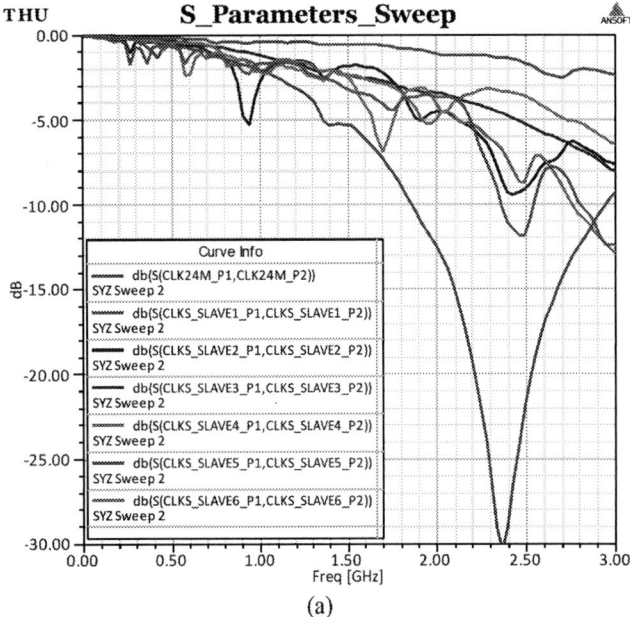

(a)

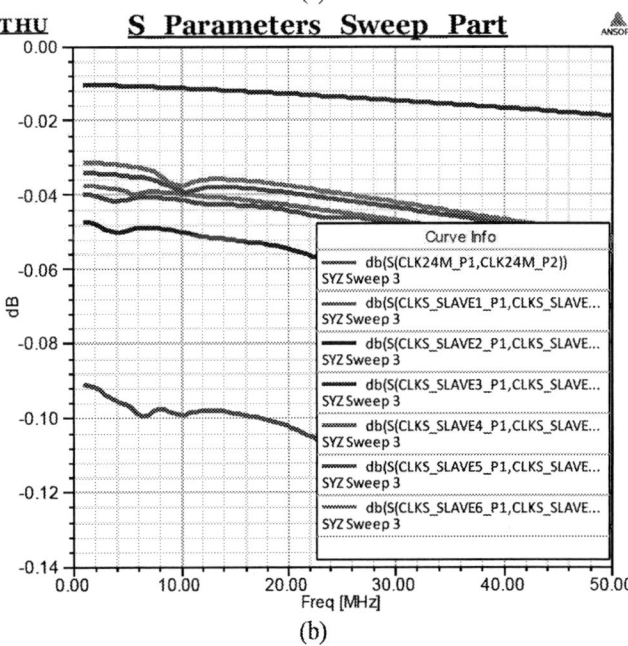

(b)

Figure 5. S parameters sweep of key clock signal traces.

It's clear that attenuation along the clock traces increases at high frequency as expected. However, the standard clock frequency, also the highest working frequency of the micro-ball endoscope was 24 MHz. So, in the frequency range the system concerns, the attenuation is negligibly small. The time domain wave conformity shown in Fig. 6 verifies the small attenuation at low frequency. Electrical simulation above verified the reasonability of substrate routing to some extent.

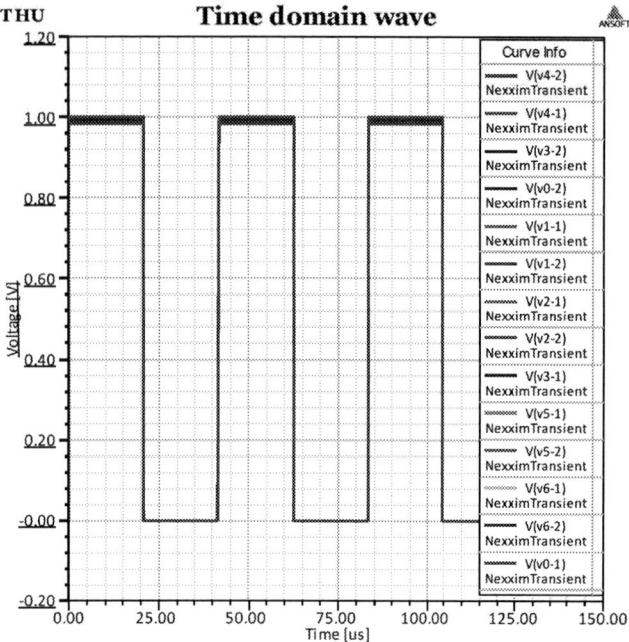

Figure 6. Time domain wave of key clock signals

4. Fabrication and Practical Experiment

The rigid substrate chose material BT and PI was the material of the flex substrate. Total thickness of the substrate was required to be less than 0.5mm. Soldermask was filled through vias. For the rigid substrate, the treatment decided to be NiPdAu to satisfy the demands of wire bonding.

The whole micro-ball endoscope system contains bare dies, packaged chips, passives, and some other actives. So a compatible packaging process was proposed for the prototype realization: SMT, die attach and wire bonding on one side of the substrate, and then SMT on the other side. The whole packaging process can be described by the flow chart shown in Fig. 7.

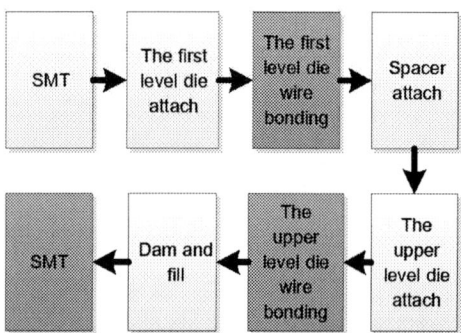

Figure 7. The whole packaging process.

More specifically, the SMDs were welded on one side of the substrate and then the first level bare dies were attached and wired. The next step is to attach the dummy spacers. Care was taken to keep clear of the first level dies' bonding wires

and any bonding finger of the upper dies when attaching the spacers. The second level dies were then attached onto the spacers and wire bonding was conducted for them. Technique of dam and fill was used to protect the delicate bare dies and gold wires. Afterwards, the rest SMDs were mounted onto the other side of the substrate. The whole package process was completed so far. Figure 8 shows one finished package with the image-capturing surface being displayed. When making the samples, process involving only wire bonding and SMT on second side, which are labeled with blue boxes in Fig. 7, were finished by machines. Other processes marked with white boxes in Fig. 7 were done by hand.

Figure 8. Package on the substrate.

The package was eventually built as shown in Fig. 8. The size of the biggest rigid board is 16mm×16mm and the maximum length of flex films is 5mm. Compared with the initial cubic system based on PCBs (21mm×21mm×21mm), current design successfully reduces the volume by more than half. The comparison between the new cubic packaged system and the initial one based on PCBs is shown in Fig. 9.

Figure 9. Comparison between new cubic packaged system and the initial one based on PCBs.

Experimental test was conducted on prepared testing vehicles to check the function and performance. As illustrated in Fig. 10, the testing vehicles include a power supply module numbered 1, an external wireless transceiver numbered 3, and a tiny display screen numbered 4. In addition, a computer was used to compose the man-machine interface for choosing testing instructions with the tiny display screen. Through antennas, the endoscope system and the external wireless transceiver can communicate with each other, such as sending indexes and captured images.

978-1-4799-2408-0/14 $31.00 © 2014 IEEE

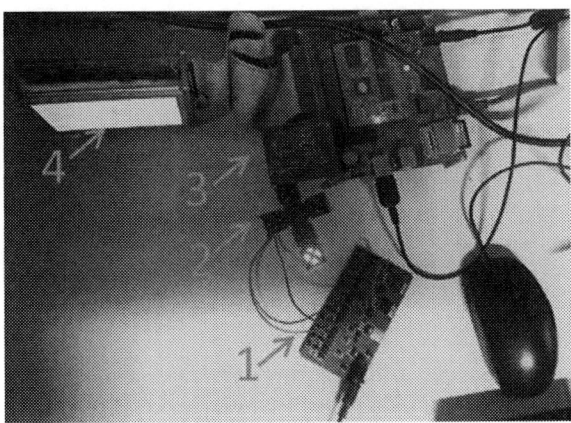

Figure 10. Experimental test scene.

Figure 11 shows some images obtained from the test and their references. The objects included the school badge of Tsinghua University and the institute logo of the Department of Microelectronics and Nanoelectronics, Tsinghua University.

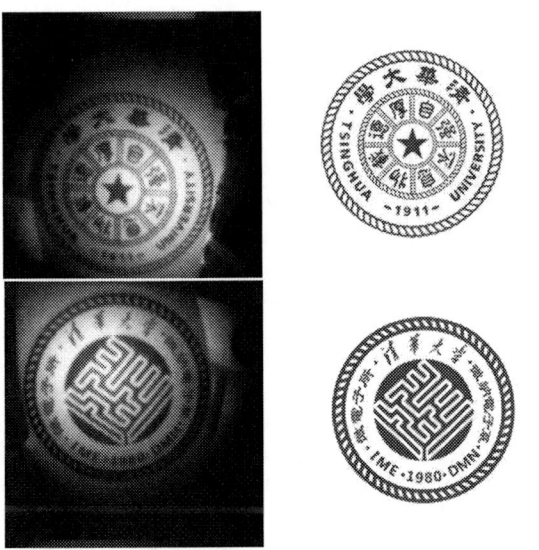

Figure 11. Images obtained from the test and their references.

The above results indicate that the new micro-ball endoscope prototype packaged by SiP approach works very well. The captured images are clear and can present key features. The successful experimental test validated the feasibility of the SiP approach based on rigid-flex substrate. An ideal micro-ball endoscope system with proper size for practical internal application can be achieved using the SiP approach with further optimization in the system constitution.

Conclusions

For the complex micro-ball endoscope system, the miniaturization based on PCBs and traditional packaging technologies has reached its limit. In the present work, a SiP approach using rigid-flex substrate was proposed and tested. An inverted pyramidal die stack structure, which was applicable to dies with great difference in size, was introduced to lower footprint. A double-sided substrate concerned six layers was designed and manufactured. The results of electrical simulation manifested the reasonability of the substrate layout and routing. Since the micro-ball endoscope system comprises diverse types of electrical component, a compatible packaging process flow was worked out. The final size of the packaged cubic system prototype was around 16mm×16mm×16mm, which is around 56% volume smaller than the one based on PCBs. Practical experimental test was conducted on the testing vehicles and clear images were obtained. In summary, miniaturization of the micro-ball endoscope was realized by the SiP approach. Meanwhile, SiP showed its strength and potential in a complex electronic system.

Acknowledgement

The research funding from National Science and Technology Major Project numbered 2009ZX02038-003 is greatly acknowledged.

The authors would like to thank Dr. Xin Gu and Jing Jiang both from Shennan Circuit Ltd. for their advice in substrate design and manufacturing. The help in experimental test work from Zheng Yin and Shan Chen, both of whom are from Institute of Microelectronics, Tsinghua University, is appreciated as well.

References

1. http://www.itrs.net/papers.html.
2. Nakamura T.Terano A. Capsule Endoscopy: Past, Present, and Future. J Gastroenterol, vol 43, pp.93–99, 2008.
3. G. Iddan, G. Meron.A. Glukhovsky, et al. "Wireless capsule endoscopy". Nature, vol 405, pp. 417-418, May 25, 2000.
4. T Sun, X Xie, G Li, et al. An Asymmetric Resonant Coupling Wireless Power Transmission Link for Micro-Ball Endoscopy. Annual International Conference of the IEEE EMBS, pp.6531-6534, Aug.31-Sept.4 2010.
5. Y. Gu, X. Xie, Z. Wang, et al. A new Globularity Capsule Endoscopy System with Multi-Camera.IEEE Biomedical Circuits and System Conference, pp.289-292, Nov. 2009.
6. Y. Gu, X. Xie, G Li, et al. Design of Micro-Ball Endoscopy System. IEEE Biomedical Circuits and System Conference, pp.208-211, Nov. 2012.
7. http://www.mst.com/dyconex/index.html.
8. James Keating. Transition of MCM-C applications to MCM-L using rigid flex substrates. Microelectronics Reliability, vol 39, pp. 1399-1406, 1999.
9. Kenneth M Brown. System In Package "The Rebirth of SIP". IEEE Custom Integrated Circuits Conference, pp. 681-686, 2004.

Design and Demonstration of Paper-Thin and Low-Warpage Single and 3D Organic Packages with Chip-Last Embedding Technology for Smart Mobile Applications

Sung Jin Kim, Zihan Wu, *Makoto Kobayashi, Fuhan Liu, Vanessa Smet, P. Markondeya Raj,
Venky Sundaram, and Rao Tummala
3D Systems Packaging Research Center, Georgia Institute of Technology,
813 Ferst Drive, NW, Atlanta, Georgia 30332-0560 U.S.A.
*NAMICS Corporation,
3993 Nigorikawa, Kita-ku, Niigata City, 950-3131 Japan

Abstract

This paper presents innovations and advances in demonstrating paper-thin organic packages with low warpage. These advances include: 1) Reduction in over-all substrate warpage, 2) Ultra-low stand off interconnection height, 3) Ultra-thin (30 μm core thickness) and ultra-low CTE (1–5 ppm/°C) organic substrates, 4) Assembly of large die onto the ultra-thin substrate, and 5) Assembly to form advanced package-on-package using conventional batch-type SMT reflow processes. Design, fabrication, assembly and characterization of test-vehicles demonstrating all these innovations are described in this paper. Comprehensive warpage modeling is performed by taking into account all substrate fabrication and assembly steps. The measured warpage data after package fabrication and assembly was used to refine and validate the models. Optimized geometry and material parameters are designed from the validated models.

I. Introduction

Portable and smart mobile products continue to demand miniaturized semiconductor packages at lower cost to enable higher functionality at system level. Recent advances in organic packages to achieve these demands have primarily focused on embedding actives and passives into substrates and stacking the independently tested packages. In spite of these advances, there is no integrated package technology reported to meet the emerging needs for packaging of ultra-thin, large, and multiple dies with low-warpage and low-cost using the existing package infrastructures. Current single and PoP organic packages [1] are limited both in overall thickness reduction and in electrical performance [2]. Chip-first embedding [3] and fan-out wafer level packaging [4] can attain thin and comparatively smaller package dimensions, but face fundamental barriers in large-die packaging, Known-Good Die (KGD), thermal dissipation and die commitment issues. Figure 1 shows typical wafer level fan-out (WLFO) package and its typical thickness is in the range from 300 to 600 microns. In the WLFO package structure, the conflict of CTE mismatch between die, substrate, and epoxy molding compound, makes warpage as a leading concern for further thickness reduction.

Georgia Tech PRC (Package Research Center) proposed and demonstrated paper-thin organic packages with low-warpage and low-cost. It also addressed warpage issues in spite of the cavity that is used resulting in an unbalanced mechanical structure. Section II describes these aforementioned challenges and the proposed chip-last

embedding in paper-thin organic package. Sections III shows FEM simulations to address the warpage challenge resulting in optimum material and package designs. Section IV describes the fabrication details of ultra-thin organic substrates. The overall results are analyzed in Section V. This paper is summarized in section VI.

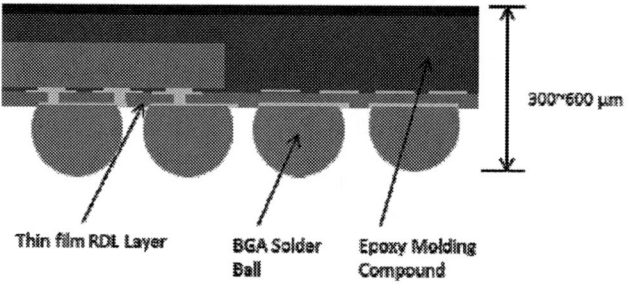

Figure 1. Structure of Wafer Level Fan out Package

II. Paper-Thin Organic Package Structure

A typical plastic laminate package with EMC (Epoxy Molding Compound) has warpage challenges due to intrinsic CTE mismatch between the EMC, substrate, die-attach adhesive and chip. It may appear that the elimination of EMC element in thin package structure helps in controlling the overall package warpage performance. In reality, relative contribution of the die, die-attach material and laminate substrate in the thin package, must be considered. Ultra-low CTE laminate material is designed to minimize the CTE mismatch with the die. Other factors such as volume of the die-attach material, BNUF (B-Stage Non-flow Underfill) material, cavity structure must be considered. Low-temperature exposure of substrates before die-attach is also very critical in controlling the warpage. Low-temperature (180°C) Cu interconnection and assembly, previously-reported, is applied in this research for chip-level interconnections to minimized the warpage [5]. The paper-thin package structure, taking into account all these factors is illustrated in Figure 2. The thickness of laminate substrate is 95 μm comprising of 30μm ultra-low CTE LαZ laminate core (provided by Sumitomo Bakelite), 50 μm cavity, and 15μm solder mask layer.

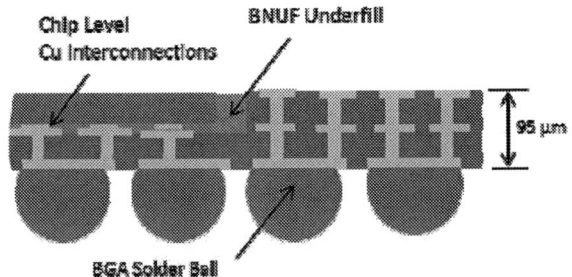

Figure 2. Structure of Sub-100μm Chip-Last Embedding Package

III. Simulations

FEM-based simulations (Finite element model) were used to analyze CTE-induced package warpage in order to optimize the package materials and designs in this study. The key material properties used, both in the fabrication and analysis are listed in Table 1.

Table1: Major Material Properties

Material	Young's modulus (GPa)	Poisson's ratio	CTE (ppm/K)
Laminate Core	21	0.3	2
Prepreg	19	0.3	13
RXP-4	1.88	0.3	45
Underfill	3	0.34	55

Simplified half-symmetrical FEM models were built based on the package structure shown in Figure 3. In this figure, height of chip is little bit higher than package substrate. In chip-last embedding packages such as this study, there are two major CTE-related concerns when core thickness is dramatically reduced to paper thin thickness (sub-100 μm): 1) Cavity, and 2) Solder mask.

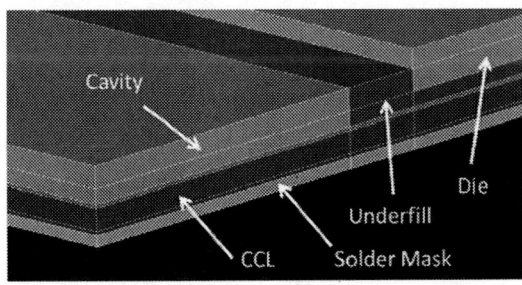

Figure 3. FEM Simulation Overview

Cavity Structure Simulations

Bare substrate warpage with cavity is due to two factors, CTE-induced, and asymmetrical structure of substrate. Warpage analysis of bare substrate was carried out to compare the EOL (End of Line) package warpage by varying cavity thicknesses from 30μm to 50μm of low CTE prepreg material sets. Optimum cavity thickness and structure were obtained from the warpage analysis. Cavity-manufacturing options included C-stage laminate core and with RXP-4, as

shown in the substrate cross-section schematic in Figure 4. In these simulation studies, 7mm x 7mm die with a thickness of 55 μm and a stand-off height of 20 μm after first-level interconnections were used. The FEM simulation result for each step is listed in Figure 5, with warpage results summarized in the figure caption. Simulation results show that as the cavity thickness is increased, it resulted in significantly reduced the warpage of the substrate before die assembly, and all the substrates warped to form a concave shape.

Figure 4. 50 μm cavity manufacturing option with 20 μm RXP-4 and 30 μm LαZ

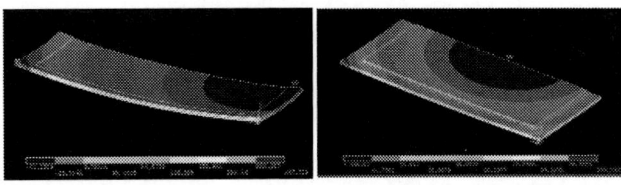

(a) 30μm cavity: 319μm before and 109μm after die-attachment

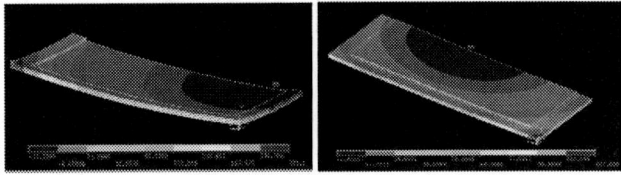

(b) 40μm cavity: 258μm before and 119μm after die-attachment

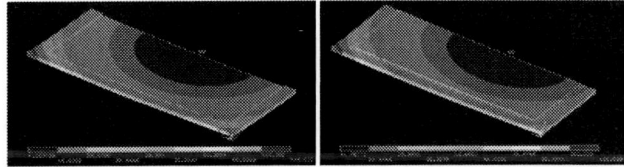

(c) 50μm cavity: 147μm before and 131μm after die-attachment

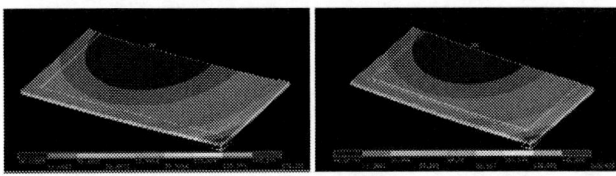

(d) 50μm with 2 material stack-up cavity : 182μm before and 171μm after die-attachment

Figure 5. FEM simulation results with different cavity thicknesses

Solder Mask Effectiveness Simulations

In thin-core laminate manufacturing processes, solder-mask process control is a well-known challenge in the industry. This is expected with paper-thin packages as well. To quantify and minimize the overall substrate and EOL package warpage behavior, warpage contribution factor from solder mask area with different solder mask design approaches were studied. The basic concept for this solder mask design approach is depicted in Figure 6. In this figure, the black-colored area represents the actual solder mask coverage in the substrate. Minimum solder mask layer contribution to warpage is expected by implementing a 'Ring cover' concept as illustrated in Figure 6 (b). If there are any traces on the bottom layer, 'Ring cover' concept cannot be applied because this exposes the underneath Cu traces. Therefore, 30% coverage was assumed for simulating the 'Ring cover' solder mask design.

As seen from the simulation results in Figure 7 (c), we could notice that, unlike with the conventional symmetrical organic package substrates, more solder mask area can help reduce undesired warpage in asymmetrical cavity structures in ultra-thin substrates.

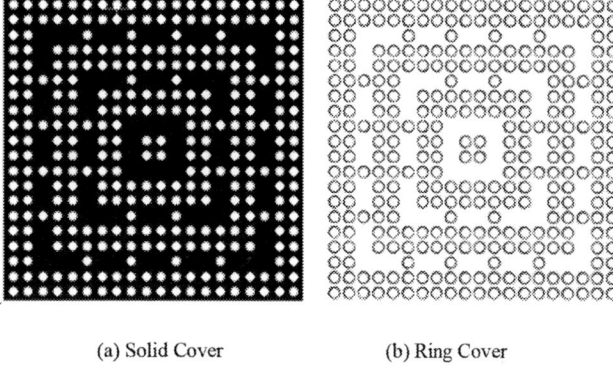

(a) Solid Cover   (b) Ring Cover

Figure 6. Two different solder mask design approaches for paper-thin package application

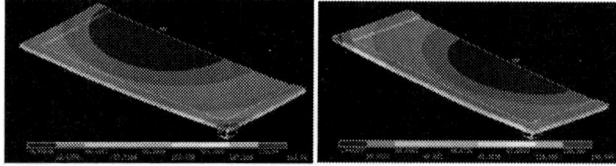

(a) 193 μm without solder mask (b) 130 μm with 70% solder mask coverage

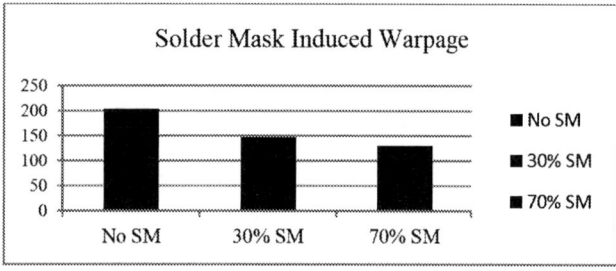

(c) Overall summary for the solder mask coverage

Figure 7. Bare substrate warpage with 30 μm LαZ Core and 50 μm prepreg cavity stack-up

Cu Trace Area Effectiveness Simulations

With the asymmetric substrate layouts, different factors should be considered to alleviate the warpage in the 'unbalanced' structures. The original test vehicle design in this study only had daisy chain connections on the top layer, with around 30% top-side covered with Cu traces, while the bottom side of the traces and solder ball metal pads have 80% coverage. By applying balanced-Cu trace coverage area, that can be easily implemented in a functional package design, the overall substrate warpage was reduced down to 70 μm. This simulation result is shown in Figure 8.

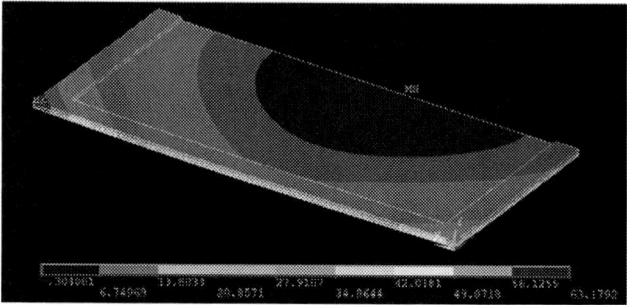

Figure 8. The FEM simulation result with balanced Cu trace area 70 μm for bare substrate

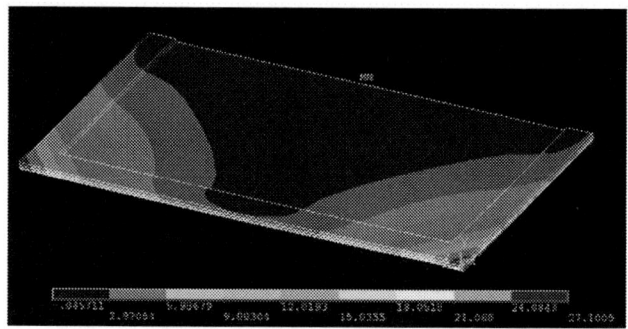

(a) 28μm before die assembly

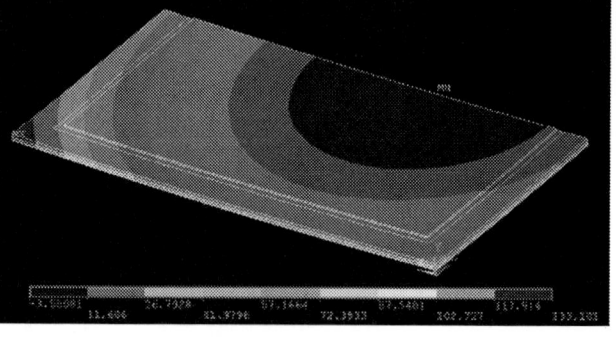

(b) 132μm after die assembly

Figure 9. Warpage Behavior for 30 μm LαZ Core and 50 μm 2ppm/K CTE cavity stack-up with 70% solder mask coverage

Simulation Summary

By combining all the predictions in the aforementioned simulations, the optimum design and material factors for the package are obtained: 1) 50 μm cavity thickness with 2ppm/K CTE stack-up, 2) 70% solder mask coverage area, 3) balanced Cu traces for top and bottom layers. The package simulation results for this optimized stack are shown in Figure 9.

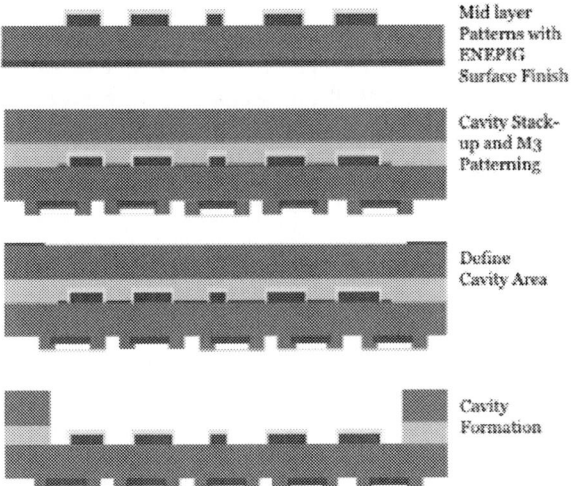

Figure 10. Paper-Thin Substrate Fabrication Process Flow

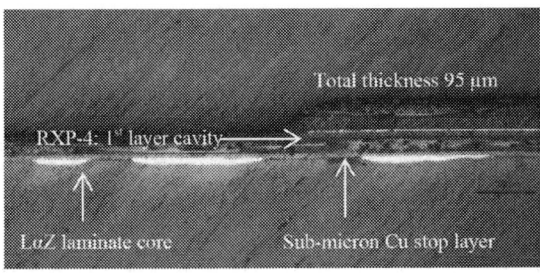

(a) Cross section view

(b) Topside over view

Figure 11. Overview of Fabricated Paper-Thin Organic Substrate

IV. Paper-Thin Organic Substrate Fabrication

Substrate fabrication process used for the paper-thin package is compatible with standard package fabrication infrastructure. A brief process flow is shown in Figure 10 and its major steps are: a) Bare substrate core with laser ablated through vias, b) Electroless plated 0.5 µm thin copper layer, c) Copper electroplating to fill the vias, d) Subtractive pattern etching to thin down the copper thickness and pattern both pad layer and routing layer, e) Cavity layer lamination, f) Cavity formation, g) ENEPIG surface finish, and h) solder mask layer formation.

Based on simulation results, a 50 µm cavity showed minimal warpage and ideal cavity material combination were found to be 'low CTE laminate plus prepreg combination'. But considering the relative higher manufacturing cost of prepreg etching process, we also added one more process option, 30 µm ultra-low CTE LαZ laminate and 20 µm RXP-4. The RXP-4 material is a pure resin material without any glass-cloth, used as an adhesion layer with easy etching process. Submicron thick eletroless Cu seed layer was plated on the cavity area before cavity stack-up to protect fine-line patterns during cavity formation process. Figure 11 (a) shows the cross section view of fabricated paper-thin substrate for this study. Figure 11 (b) shows the overall strip overview for 'laminate plus prepreg combination type' with overall warpage at minimal level, considering that the measured thickness for the whole substrate is only 95 µm.

V. Test Vehicle Demonstrations

With 40 µm thick ultra-low CTE laminate, two types of 50 µm cavity structures were fabricated in this study: 1) 30 µm ultra-low CTE LαZ laminate plus 20 µm RXP-4 as an adhesion layer, 2) 30 µm ultra-low CTE LαZ laminate plus 25 µm LαZ prepreg as an adhesion layer. For the first-level interconnection, the assembly processes used involved the GT-PRC's unique low temperature copper interconnection and assembly process [6]. This low-temperature process is the biggest contributor to overall EOL package warpage reduction. The low warpage characteristic of this process makes it suitable for conventional batch-type SMT reflow to other substrates such as package-on-package stacking. Base on the simulation result in this study, we could notice that, unlike with the conventional symmetrical organic package substrate, more solder mask area can help to reduce undesired warpage in asymmetrical cavity structures fabricated with, ultra-thin substrates of ultra-low CTE laminate and prepreg. And if we use pure resin RXP-4 as an adhesion layer, we can foresee warpage behavior of this substrate will be similar to normal symmetrical organic substrates. In this case, solder mask will play negative role in overall warpage characteristics. The overall warpage measurements were performed by using profile measurement instrument. The scanning direction of measurement probe started from the center of package as shown in Figure 12 (a). There are total 4 different test vehicle sets were measured. 1) Bare substrate with solder mask, 2) Bare substrate without solder mask, 3) EOL package with solder mask, and 4) EOL package without solder mask. Different from the cavity stack-up options in the simulation, the low CTE cavity was replaced by this two-layer stack-up configuration, resulting in the different result.

978-1-4799-2408-0/14 $31.00 © 2014 IEEE

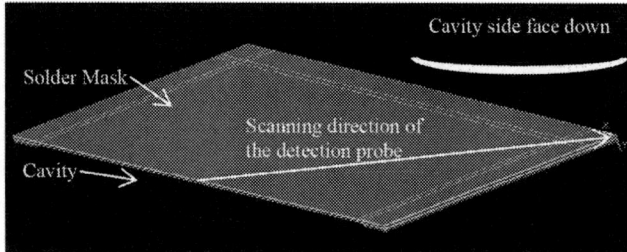

(a) Scanning direction of measurement probe

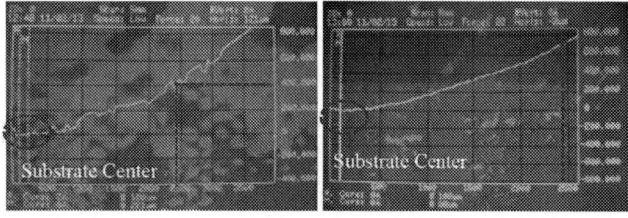

(b) Bare substrate warpage measurement with and without solder mask

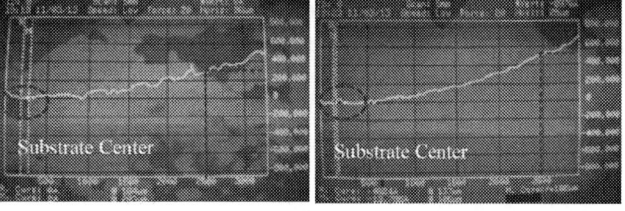

(c) EOL Package warpage measurement with and without solder mask

Figure 12. LαZ laminate plus RXP-4 Combination Cavity Type Package Warpage Measurement

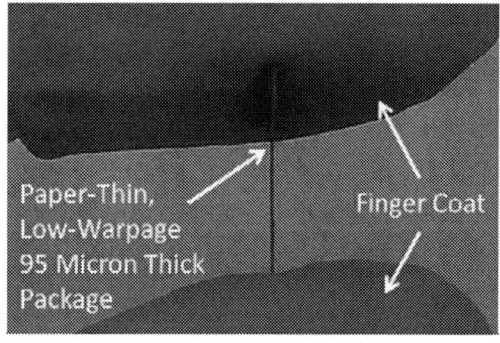

Figure 13. Fabricated low warpage paper-thin package with chip-last embedding

The ultra-low CTE laminate which serves as the second layer of the cavity, also functions to compensate the unbalanced caused by CTE mismatch between ultra-low CTE laminate core and high CTE RXP-4. Also a large percentage of Cu coverage at the bottom side compared with the interconnection side, also contributed to the substrate warpage towards the cavity side. After the die attachment, the warpage of the package was released, as demonstrated in Figure 12 (c), since large area of low CTE silicon die dominates the warpage behavior for the entire package. It was also noticed that the package with solder mask showed more

warpage. So we can tell that ring type solder mask design which is proposed in this study will greatly beneficial to minimize the overall warpage. Actual super flat 95 μm paper-thin package is shown in Figure 13.

VI. Conclusions

This paper reported the design and demonstration of ultra-thin organic package with low warpage. To achieve paper-thin (less than 100 μm thickness including die) package, there are five major key innovations and their technical detailes were described: 1) Reduction of over-all substrate and EOL package warpage and its validation by test vehicle verifications as well as FEM modeling, 2) Low-temperature Cu interconnections and assembly with ultra-low off-chip interconnection height, 3) Fabrication of ultra-thin (40 μm thickness) with, ultra-low CTE (1–5 ppm/°C) organic substrates, 4) Assembly of 49 mm² die onto the ultra-thin substrate, and 5) Assembly process to form advanced package-on-package using conventional batch-type SMT reflow processes. Characterization of 30 μm ultra-low CTE LαZ laminate plus 25 μm LαZ prepreg cavity type EOL package will be updated in future study.

References

[1] Dreiza, M., Yoshida, A., Ishibashi, K. Maeda, T, "High Density PoP (Package-on-Package) and Package Stacking Development", Electronic Components and Technology Conference (ECTC), 2007

[2] Sung Jin Kim, Honrao, C, Raj P.M, Sundaram, V, Tummala, "Ultra-thin and ultra-high I/O density package-on-package (3D Thin PoP) for high bandwidth of smart systems." pp. 406-11, 2013 IEEE 63rd Electronic Components and Technology Conference (ECTC). Las Vegas, NV, USA

[3] A.Kriechbaum, H.Stahr, M.Biribauer, N.Haslebner, M.Morianz, "ECP® – Embedded Component Packaging Technology", Semicon Europe, 2011

[4] Palm P., Tuominen R., Kivikero A., "Integrated Module Board (IMB); an advanced manufacturing technology for embedding active components inside organic substrate," Electronic Components and Technology Conference, 2004. Proceedings, 54th, pp. 1227- 1231, June 2004

[5] K., Nitesh; C., Abhishek; M., Gaurav; Raj, P. M.; S., Venky; T., Rao. IEEE Transactions on Components, Packaging & Manufacturing Technology. Sep2012, Vol. 2 Issue 9, p1434-1441

[6] Khan, S.A.; Choudhury, A.; Kumbhat, N.; Pulugurtha, M.R.; Sundaram, V.; Meyer-Berg, G.; Tummala, R., "Multichip Embedding Technology Using Fine-Pitch Cu–Cu Interconnections," Components, Packaging and Manufacturing Technology, IEEE Transactions on , vol.3, no.2, pp.197,204, Feb. 2013

Manufacturing Readiness of BVA Technology for Ultra-High Bandwidth Package-on-Package

Rajesh Katkar, Rey Co and Wael Zohni
Invensas Corporation
3025 Orchard Parkway, San Jose, CA 95134
wzohni@invensas.com

Abstract

Bond Via Array (BVA) technology has been developed to enable more than 1000 vertical connections between memory and processor components in a standard outline Package-on-Package (PoP) configuration. This higher density interconnect more than doubles current PoP capability and thereby addresses next generation wide IO mobile device demands for increased bandwidth [1]-[3]. In this paper, we discuss BVA manufacturing process details and associated demonstration test vehicle design. This prototype BVA PoP demonstrates 1020 vertical interconnects at 0.24mm pitch within an industry standard 14 x14mm package footprint. Information regarding test vehicle design optimization, critical assembly process steps, package reliability, and testing will be discussed along with the overall high volume manufacturing (HVM) readiness of BVA.

Introduction

Package-on-Package has become the most popular way to assemble the processor and memory subunit in today's smartphones and tablets. Existing standard outline PoP solutions are able to provide only around 300 interconnects between the processor and the memory package. This arrangement limits memory data interconnects to 64 lines, for a corresponding bandwidth of up to 25.6 GB/s at 1600 MHz DDR signal speeds. With a trend towards System-on-Chip (SoC) mobile processors with multi-core CPU, memory bandwidth requirements are sharply increasing. To meet these emerging requirements, a wide IO memory standard has been proposed that provides up to 512 memory data interconnects. This standard provides about 4 times current bandwidth (>100 GB/s) even at lower 800 MHz DDR signal speeds. For memory devices to offer 512 data lines, about 1000 interconnects are required in order to include the accompanying address, control, power and ground signals required for operation. Existing PoP solutions are not capable of offering >1000 interconnects as they cannot achieve the required pitch values of ≤0.24mm within the standard 14mm x 14mm package footprint. Although Through-Silicon-Via (TSV) technology is an ideal high-bandwidth solution, it is still being developed, is expensive and not expected to be widely available for the next few years.

A new high-density PoP interconnect technology called "Bond Via Array" (BVA™) is developed to provide high-bandwidth interconnect capability today. A BVA test vehicle demonstrating 1020 PoP interconnects at 0.24mm pitch is assembled within the conventional 14mm x 14mm outline package specification. The fine pitch vertical interconnects are achieved by utilizing well established wire bond equipment and processes.

In this paper, equipment and process development engineering results related to HVM readiness of BVA technology are presented along with brief details of the design optimization, critical assembly process steps, reliability performance and testing. The core processes developed for BVA include free standing vertical wire-bond formation and fine-pitch package-on-package stacking. In both cases, manufacturing studies were conducted in cooperation with 3rd party equipment vendors in order to validate approaches for HVM. This paper also provides details from these HVM studies.

1020 IO Bond Via Array PoP

BVA is developed as a cost effective advanced PoP solution that is able to provide a very high density of ultra-fine pitch high aspect ratio interconnects between the bottom processor and the top memory package. As depicted in Figure 1, five rows of free-standing ultra-fine pitch wire-bonds are formed along the periphery of the processor package as terminals to connect to the topside memory package.

Figure 1. *Top and angled view of free-standing array of wire bonds at the periphery of the bottom processor package.*

Figure 2. *Bottom view of 240µm pitch memory package*

A 432 IO BVA prototype was built and its details along with the reliability data have been published [2], [3]. In this first prototype build the array of free standing wire bonds were achieved by forming the stitch bond on the logic substrate. The stitch bond process introduces an offset between the bond pad and the wire tail position due to the angular motion required for this type of bonding. In the updated 1020 IO test vehicle, the wire bonding process has been modified to utilize a conventional EFO ball bond at the substrate. The vertical motion typical of ball bonding correspondingly results in reduced offset in the resulting vertical interconnect.

BVA makes achieving ultra-fine pitch interconnections that are out of reach of existing PoP technologies possible without the need for new assembly equipment or materials. This advantage is realized by fully leveraging the well-established wire bonding assembly infrastructure. Wire bond pitches below 100µm are easily achievable today on polymer substrates and hence BVA technology provides a solution that would be viable for the next few generations of finer pitch PoP interconnects.

Design Optimization

BVA prototype design optimization was performed with comprehensive finite element analysis (FEA) modeling of package structure and construction materials. FEA simulations are instrumental to establishing acceptable package warpage characteristics critical for successful assembly of thin PoP package. Stepwise FEA simulations were created to represent each assembly process step including flip chip, underfill, molding and solder ball attach. Accurate material property datasets as a function of temperature were also generated (e.g. modulus Vs temperature curves) to a more accurate representation of the prototype. The materials and structural design were selected to achieve package warpage <u>within a maximum of</u> 100µm, and minimize the differential warpage between the memory and the logic packages within a stack height <u>below</u> 1.3mm for

the standard 14mm x 14mm package footprint. The simulation methodology was validated with a 432 I/O prototype having total package thickness of 1.65mm prior to optimizing the 1020 IO package at ≤ 1.4mm [4].

Over-mold compound was identified as one of the critical materials affecting package warpage and so a wide range of options were selected for analysis [4]. The effect of die thickness and mold gap thickness on overall package warpage was also examined. It was noted that both the processor and memory packages warp in the same direction with temperature increase. However, as the mold gap and the die thickness are decreased, the room temperature warpage increases; making it extremely challenging to manage package warpage as the package thickness is reduced.

Several prototype memory packages with a total silicon thickness varying from 100, 200 and 250µm (all with 100µm mold cap) were built for experimental verification. The experimental results indicated a significant reduction in warpage by increasing the die thickness from 100 to 200µm. However, further increases of the die thickness to 250µm brought only nominal improvement in the warpage at both reflow and room temperatures. Hence, a total die thickness of 200µm with 100µm mold cap was selected for the topside memory package. The resulting total height of this memory component of the 1020 IO BVA PoP prototype is 510µm [4].

The FEA simulations were then performed for the optimized logic package to select the mold compund along with with the substrate core material and core thickness. Two different core materials were analyzed including standard Sub-A and LCTE substrate with a Low Coefficient of Thermal Expansion. Approximately 23 different combinations were analyzed as shown in Table 1. It was found that increasing the substrate core thickness decreases the overall package warpage. Moreover, as the mold gap thickness above the die is reduced, the warpage at both room and reflow temperatures decreases. In order to minimize package warpage as well as profile height, a core thickness of 60µm was selected.

Option	Mold Compound	Mold Gap (um)	Die Thickness (um)	Mold Cap (um)	Core Material	Core Thickness (um)	Warpage at 25C	Warpage at 217C
1	A	400	125	575	Sub-A	40	133	-26
2	A	261	125	436	Sub-A	40	114	-23
3	A	160	125	335	Sub-A	40	89	-17
4	G	160	125	335	Sub-A	40	119	-85
5	G	100	125	275	Sub-A	40	95	-84
6	G	50	125	225	Sub-A	40	74	-81
7	G	50	125	225	Sub-A	80	68	-66
8	G	50	125	225	Sub-A	120	62	-54
9	G	50	125	225	LCTE	40	83	-50
10	G	20	125	195	Sub-A	40	73	-78
11	G	50	125	225	LCTE	120	72	28
12	H	100	125	275	Sub-A	60	93	-74
13	I	100	125	275	Sub-A	60	94	N.A.
14	G	100	125	275	Sub-A	60	91	-77
15	G	120	125	295	Sub-A	60	98	-77
16	H	100	100	250	Sub-A	60	107	-77
17	H	150	100	300	Sub-A	60	132	-86
18	H	100	150	250	Sub-A	60	91	-86
19	H	80	150	280	Sub-A	60	83	-85
20	H	100	100	250	LCTE	60	118	-47
21	H	150	100	300	LCTE	60	140	-49
22	H	80	150	280	LCTE	60	98	-55
23	H	100	150	300	LCTE	60	90	-55

Table 1. *Structural iterations within the FEA design of experiment matrix for the 1020IO logic package*

The most promising design consideration turned out to be option 23 which provided a warpage of less than 100μm. This profile meets the original criteria of < 1.3mm total stack height and individual warpage ≤ 100μm at both room and reflow temperatures. The differential warpage between the memory and the logic was then analyzed to expose any potential assembly yield issues. For option 23, the warpage in the memory and logic at room temperature and reflow temperature is shown in Table 2 for comparison.

Package	25°C (μm)	217 °C (μm)
Memory	90	-37
Logic (Sub-A)	83	-85
Logic (LCTE)	90	-55

Table 2: *FEA predicted warpage values for the 1020 I/O logic (2 core options) and overlying memory package.*

As indicated above, the memory and logic display similar warpage characteristics. The differential warpage between the memory and the logic with a low CTE substrate is 0μm and 18μm at 25°C and 217°C, respectively. The warpage difference between the memory and the logic with a standard substrate is 7μm and 48μm, respectively.

Test Vehicle

An optimized 1020 IO daisy chain test vehicle design was finalized based on the FEA modeling results for a standard 14mm x 14mm package footprint. Figure 3 below provides a cross section view drawing of the test vehicle while Table 3 lists associated package dimensions.

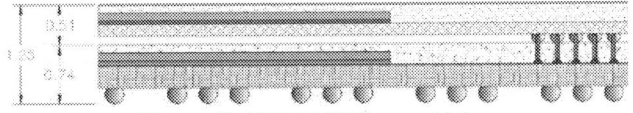

Figure 3. *1020 IO BVA test vehicle*

As indicated in table 3, there are a total of 5 rows of bond via interconnects at 240μm pitch distributed about the periphery of a 200μm thick 7.5mm x 7.5mm logic flip-chip die. These bond-vias provide a total of 1020 connections between the bottom logic and top memory package. The topside of the logic substrate corresponding has flip-chip pads at the central region of the component while providing wire-bond pads at the periphery for planting of the vertical bonds. The base of the topside memory substrate provides a solder pad arrangement matching the logic package's peripheral bond via array.

Package foot print	14mm x 14mm
Package Thickness	1.25mm
Logic Package Thickness	0.74mm
Memory Package Thickness	0.51mm
Logic BGA Pitch	0.4mm x 0.4mm
Number of Logic IOs	916
PoP IO Pitch	0.24mm
Number of PoP IO rows	5
Number of PoP IOs	1020

Table 3. *Summary of 1020 IO BVA PoP Package*

BVA Manufacturing Process

Figure 4 briefly describes the major process steps required to assemble a BVA PoP package.

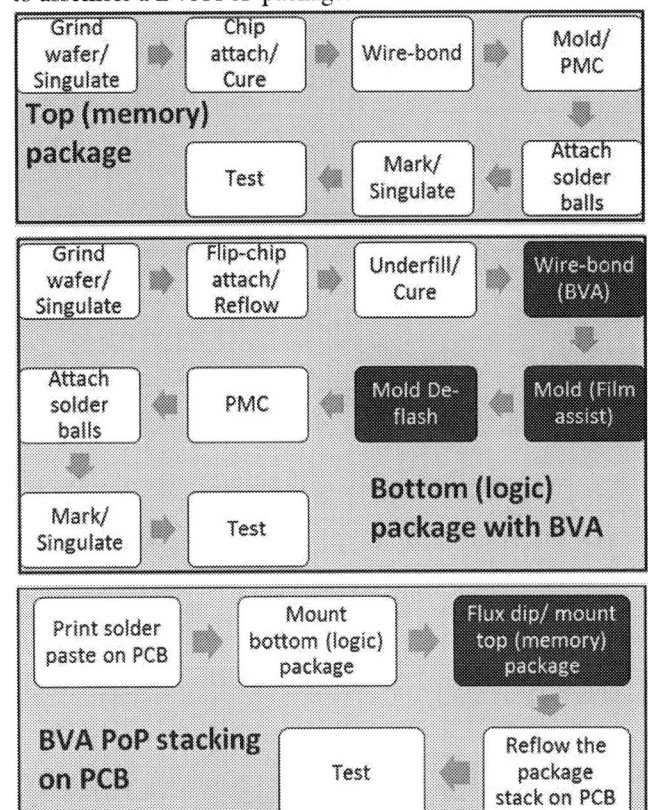

Figure 4. *Process flow for BVA PoP components and stacking.*

As indicated, the BVA process introduces minimal changes to the existing PoP assembly process. Key process conditions for BVA include:

- Forming the free standing array of vertical wire bonds on the logic substrate while maintaining the positional accuracy of the wire tails
- Encapsulating the logic package while making the wire tails protrude above the encapsulation mold compound
- Post-mold cleaning of the wire tails to facilitate uniform solder wetting around the wire and the package stacking.

The logic die is first flip chip bonded to the substrate with a standard assembly process. The array of free standing wire bonds are then formed at the periphery of the logic substrate surrounding the flip-chip die representing a processor (per Figure 5).

Figure 5. *Peripheral array of free standing wire bonds surrounding a central flip-chip logic die on substrate.*

The vertical wire bond process was developed in cooperation with Kulicke and Soffa (K&S) utilizing the ICONN wire bond system and 50μm diameter Pd coated Cu wires. This equipment was used to verify High Volume Manufacturing (HVM) feasibility of this approach to forming high-density vertical connections for PoP. Extensive run trials in quantity of 500K consecutive wire bonds at <40ms cycle-time were successful in verifying production capability.

Achieving a high degree of XYZ positional accuracy with the BVA wire tails is essential for alignment and stacking of the top memory package. The wire tail position was measured and verified to be within specifications using automated optical monitoring systems by both OGP's SmartScope and View Micro-Metrology's Benchmark 450 tool. Measurements indicated XY positional accuracy of +/-10μm around the wire bond pad. The wire bonds were optimized to be nominally 440μm tall with a height variation of no more than +/-20μm about this mean.

After completing flip-chip attach and BVA bonding, the processor package is then molded using a standard film assist transfer molding process. Film assist is commonly used to accomplish exposed device backside molding. In the case of BVA, when the mold tool is clamped over the partially assembled logic package, the tips of the bond vias pierce through the film on the top mold chase. As a result, the film acts as a barrier to mold flow only for these wire-tip regions of bond vias. The mold cavity is then filled via transfer mold process and the mold is cured. As the mold tool is opened, the film is pulled away from the package to reveal the aforementioned wire-tip regions exposed at uniform height over the mold cap line (as illustrated in Figure 6). The post-molding wire tail position was measured again and was found to be within the required tolerance specification for stacking.

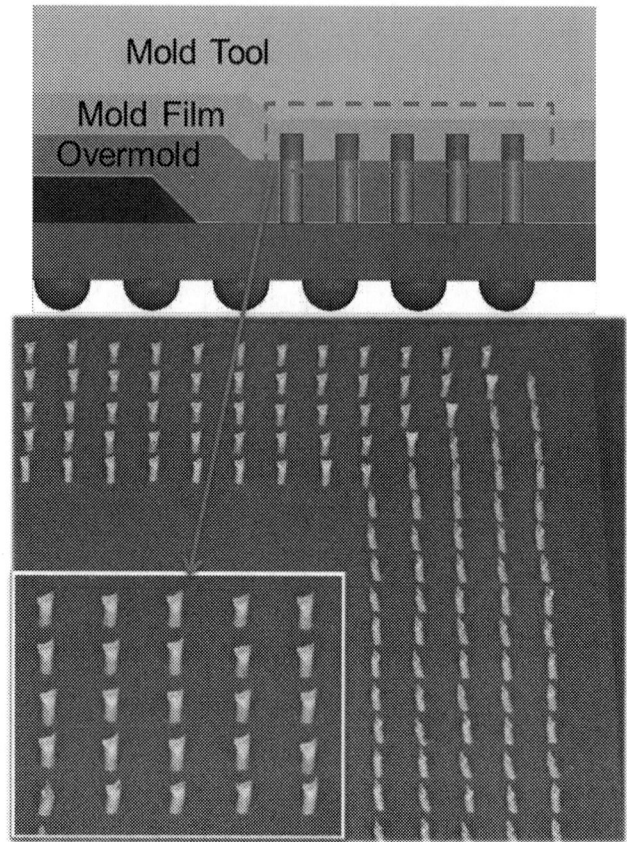

Figure 6. *Top: cross-section drawing depicting wire-tip penetration into film material during molding process. Bottom: scanning electron microscope (SEM) photo of resulting exposed wire-tip formations over mold cap line from actual prototype.*

The BVA wire protrusions above the mold of the bottom processor package are designed to form lands for the base solder balls of the topside memory package. Stacking is performed in similar fashion to current industry practices for PoP using a standard reflow operation.

Although the film assist mold process enables exposure of wire-tip regions over the mold cap line, there is a measure of mold flash that occurs during processing. Such mold residue may accumulate along wire-tip surfaces to compromise the wettable surface intended for use in reflow of the adjoining memory BGA package to the BVA. To avoid such issues at package stacking, any mold bleed is removed after curing. Removal can be accomplished using various conventional mechanical and chemical methods including wet blast, chemical wet etch as well as plasma cleaning. Inspection and process results after CF_4/O_2 and Ar plasma based de-flash procedure demonstrates the effectiveness of a readily available process to eliminate mold residue from wire tips, as shown in Figure 7.

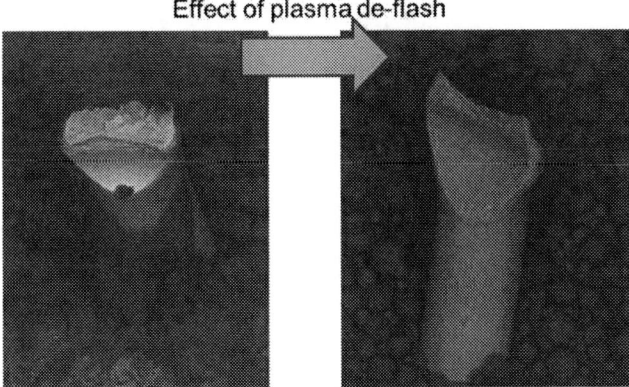

Figure 7. *Plasma de-flash process was developed to remove the mold bleed on wire tails*

After mold deflash, the memory package is then stacked on top of the bottom logic using a conventional SMT-style PoP assembly process. A high yielding and re-workable soldering process was developed to obtain uniform and consistent joints. Solder paste is first printed on PCB board, followed by pick and place of the logic package. The solder paste dipped memory is then stacked on top of the logic package and the entire assembly reflowed. Package stack SMT of the BVA prototype yielded uniform and consistent joints at fine pitch of 0.24mm as shown in Figure 8 below.

Figure 8. *Top: optical scope side-view of BVA to memory joints. Bottom: SEM micrograph of the cross sectioned BVA PoP package. SMT mounting was demonstrated at Universal Instruments Corporation.*

Testing this ultra-fine pitch logic component as well as the overall PoP stack introduces some challenges as the interconnect pitch exceeds current mainstream PoP design limits. Objectives for engineering demonstration of BVA include development and demonstration of a low-cost test socket hardware solution that accommodates 240µm pitch while remaining fully compatible with existing automated test equipment. A prototype bi-level socket assembly has been fabricated and is currently under evaluation (Figure 9).

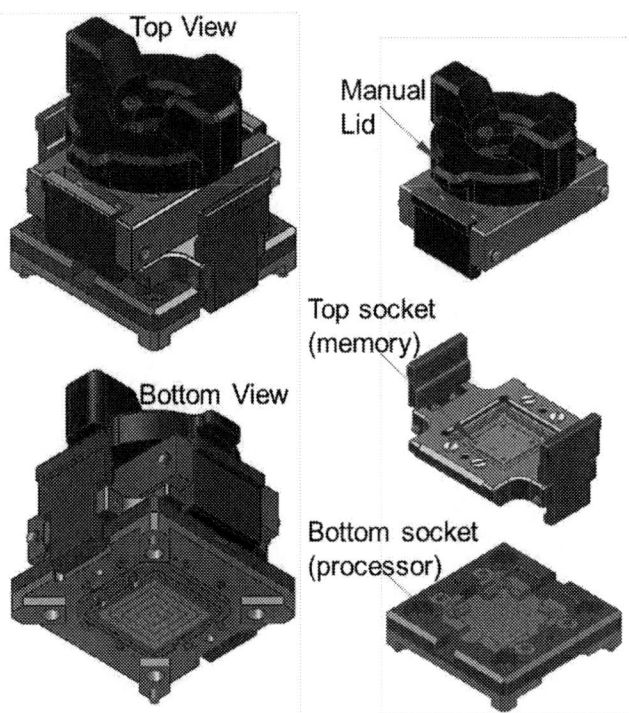

Figure 9. *Test socket design (Courtesy of Leeno and SemiQual)*

The top socket of this hardware holds the 0.24mm 1020BGA memory package. The top socket also connects the BGA of the memory device to the top of the exposed BVA wire tips of the underlying logic device. The bottom (base) socket is mounted to a test board which is interfaces with a programmable open/short tester. The shape and size of the wire tips, their orientation, and relative position was measured with respect to the BGA grid reference outline for individual units. The positional tolerance was measured to fall within specifications and the alignment capability of the socket.

Reliability Testing

The packages were subjected to a complete set of reliability tests including JEDEC level 3 moisture sensitivity, board level temperature cycling, high temperature storage and drop testing. The test details are summarized in Table 4 below:

Test	Standard	Test Condition	Sample Size
Moisture Sensitivity Level 3	IPC/JEDEC J-STD-020C	125°C for 24hrs; 30°C / 60%RH for 192 hrs, 3X Pb-free reflow	22
Temperature Cycling (Board Level)	JESD22-A104D Condition G	-40°C to 125°C, 1000 cycles	45
High Temperature Storage	JESD22-A103D Condition B	150°C, 1000 hrs	22
Drop Testing	JESD22-B111	>30drops, 1500G, 0.5mS of 1/2 sine pulse	20

Table 4. *Reliability test and results of 1020 IO BVA PoP*

At the time of writing, a total of 900 cycles board level temperature cycling testing were completed without any failure. The packages have also passed rest of the reliability tests including MSL3, high temperature Storage and drop testing.

Figure 10 shows the cross section of parts that completed 1000hrs of high temperature storage at 150°C. It was reported earlier [1] that bare Cu wire is completely converted into Cu_6Sn_5 intermetallic compound (IMC) when embedded into solder from the top memory package. Such interconnects are not reliable and hence Pd coated Cu wires were instead used. Still, even with Pd coating, tip formation of wire bonds may have also resulted in exposed underlying copper material as a result of bonder mechanical forces. Therefore evaluating the high temperature testing performance again for Pd-coated Cu wire was critical, particularly for BVA interconnects. As seen in Figure 10, although some degree of intermetallic mold compound (IMC) is formed around the wire tails within the memory solder ball, the Pd nonetheless provides an effective diffusion barrier thus creating a very reliable interconnect in this regard.

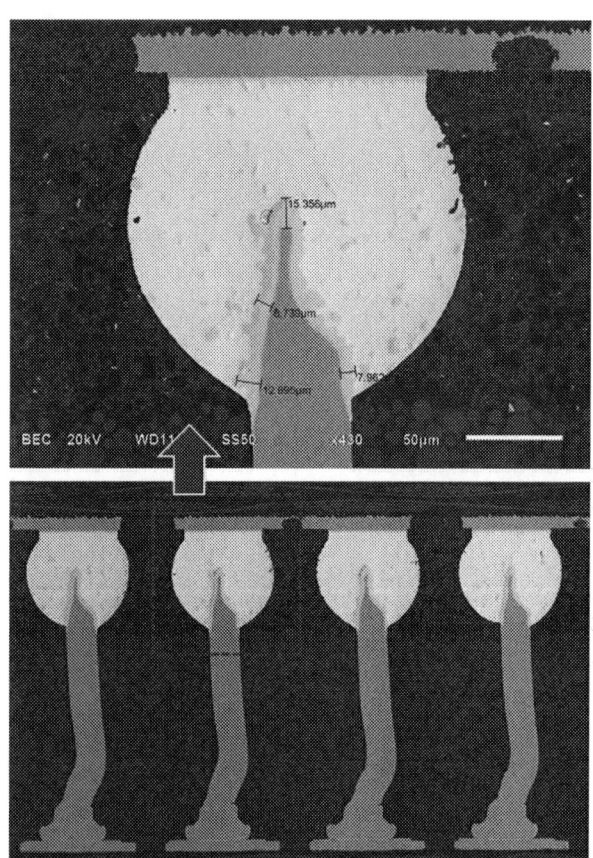

Figure 10. *Cross sectional view of BVA PoP interconnect after 1000hrs of high temperature storage.*

For accurate drop testing performance of the BVA interconnects, only the BGA portion of the bottom processor package was underfilled on the test board while BVA interconnects between the bottom logic and top memory package were left without any form of encapsulation. By underfilling the logic BGA, the drop failures were forced to move to the BVA interconnects. BVA interconnects displayed notably robust reliability performance in extended drop testing. The first failure was reported after 181 drops while multiple units survived 500 drops at which time the test was terminated. Figure 11 shows the Weibull plot for the drop test failures.

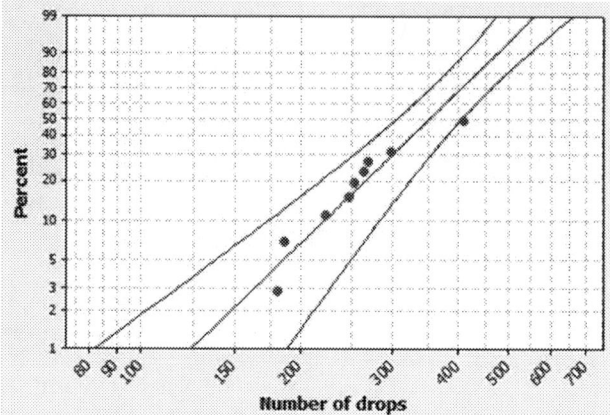

Figure 11. *Weibull plot for drop test failures*

Mean Time to Failure (MTTF) was calculated to be approximately 354 drops which is significantly improved over existing PoP technologies. All the failures occurred at the corner BVA interconnects of the failed packages as expected. Extensive failure analysis found crack failures in corner BVA interconnects, while no BGA failures were observed on the underfilled bottom processor package. Figure 12 shows cross sections of the drop test failures. Figure 12a shows a part failing after 270 drops while Figure 12b shows the corner BVA bump of the part which survived 500 drops.

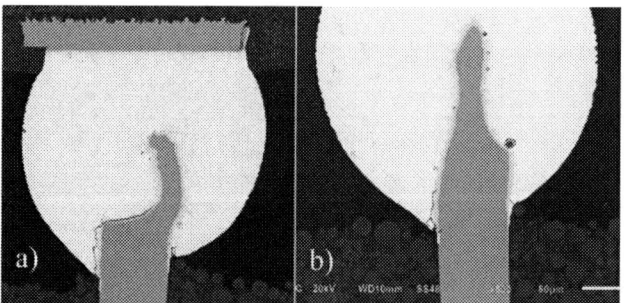

Figure 12. *Cross-section SEM of drop test samples: a) 270 drop fail, b) 500 drop pass*

It is evident from Figure 12 that the crack propagation is affected by the shape of the wire tail, amount of IMC in the vicinity, and mechanical orientation within the memory solder ball. The crack initiated in the corner BVA bumps and propagated vertically along the Cu wire. Failure occurred as the crack moved across the thinner and weaker (IMC dominant) areas of the wire tail. This elongated the crack length required to cause a failure, thereby improving the interconnect reliability.

Conclusions

BVA PoP has been developed as a wire bond based interconnect technology that enables very high density vertical interconnects at low cost by utilizing existing infrastructure and assembly practices. BVA PoP with 1020 IO logic interface was successfully assembled at 0.24mm pitch and passed all the reliability tests. High volume manufacturing feasibility studies covering various critical process steps were completed and verified with automated assembly and test equipment suppliers.

References

1. P. Damberg, et al, "Fine Pitch copper PoP for Mobile Applications," *Proceedings of IEEE Electronic Components and Technology Conference (ECTC),* San Diego, CA, June 2012

2. I Mohammed, et al, "Package-on-Package with Very Fine Pitch Interconnects for High Bandwidth" *Proceedings of IEEE Electronic Components and Technology Conference* (ECTC), Orlando, FL, May 2013

3. R. Katkar, et al, "Ultra-Fine Pitch Package on Package Solution for High Banwidth Mobile Applications", *9th International Conference and Exhibition on* Device Packaging, Scottsdale, AZ, March 2013.

4. L. Mirkarimi, et al, "A Non-Through Silicon Via 1000+ IO Package On Package Solution for Wide IO Applications", *Proceedings of 46th International Symposium on Microelectronics* (IMAPS), Orlando, FL, October 2013.

Improvement of Substrate and Package Warpage by Copper Plating Process Optimization

Omar Bchir[1], Houssam Jomaa[1], Chin Kwan Kim[1], Layal Rouhana[1], Kuiwon Kang[2], Milind Shah[1], Steve Bezuk[1]

Qualcomm Technologies, Inc.

[1] 5775 Morehouse Drive, San Diego, CA 92121 USA

[2] POBA Gangnam Tower, 119, Nonhyeon-dong, Gangnam-gu, Seoul, KO 135-820 (Korea)

obchir@qti.qualcomm.com, 858-658-2813

Abstract

High substrate warpage can lead to unacceptable yield loss during chip attach in assembly, and cause high yield fallout during package mount on the circuit board. For the first time, through this work, the electrolytic copper (Cu) plating process in substrate manufacturing was shown to contribute significantly to package warpage. For a 14x14mm package, reducing the Cu plating rate (within the manufacturing operating window) resulted in 21% package warpage reduction, while a change in Cu plating solution provided an additional 6% reduction (total 27% reduction). Hence the Cu plating process and solution must be carefully scrutinized to minimize package warpage, specifically for thin packages (<1mm) where Cu stresses become a large contributing factor.

Introduction

As the industry moves to thinner packages, control of package warpage is an increasing area of concern. The substrate is a large factor in determining overall package warpage performance. High substrate warpage can lead to unacceptable yield loss during chip attach in assembly, and cause high yield fallout during package mount on the circuit board.

Substrate warpage improvements are typically approached through modification of dielectric material properties (such as CTE, Tg, modulus), layer thicknesses (core, prepreg, solder resist and Cu thickness), and Cu areal density per layer. An overlooked factor in the warpage improvement effort is the impact of the electrolytic Cu plating process. Electroplated Cu thin films tend to have porous grain boundaries, wherein grain boundary volume is strongly dependent on electroplating conditions and subsequent thermal processing [1]. During thermal processing, Cu grains grow and merge, eliminating grain boundaries; Cu deposits with larger grain boundary volume shrink more. The residual stress in the initial deposit, coupled with shrinkage during subsequent thermal processing, strongly impacts the warpage response of the substrate and package. This is compounded by the inherent front-to-back Cu density imbalance which is typical in substrate design.

Test Device and Processing Description

The first step in the analysis was to check the magnitude and mode of stress in the deposited Cu. This was done using deposit stress test strips and a Deposit Stress Analyzer system from Specialty Testing & Development Co. (York, PA, USA). Test strips were plated in a bench-top electrolytic plating cell, using various chemistries and plating current density conditions of interest. After plating, the thickness of the deposited Cu, as well as the increments of spread between the test strip legs, were measured. These values, along with the strip calibration constant and direction of spread, were used to determine deposit stress magnitude and mode (compressive vs. tensile).

Next, substrates were manufactured for use in warpage testing. Substrates were built in panel format, using a typical electrolytic copper plating configuration (Figure 1). The inner layers on the front and back of the panel were plated simultaneously in an initial step. The outer layers on the front and back of the panel were also plated simultaneously in a subsequent step.

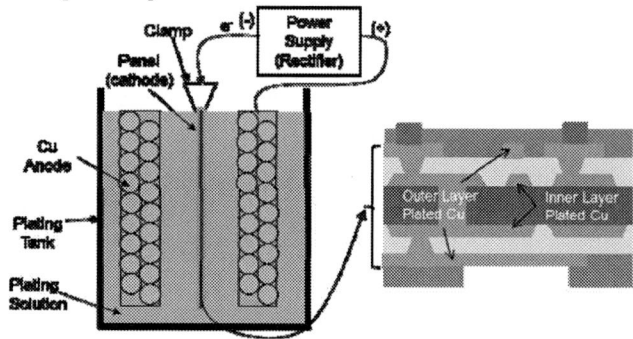

Figure 1. Typical setup for substrate electrolytic Cu plating process used in this evaluation.

Two different substrate designs, both with 14x14mm body size, were used for this evaluation. The first design (DES1) used a 4-layer substrate with 18um Cu thickness per metal layer, a 100μm core with CTE <12ppm/°C, and was manufactured by a first substrate supplier. The second design (DES2) used a 4L substrate with 18um Cu thickness per metal layer, a 100 μm core with CTE <5ppm/°C, and was manufactured by a second substrate supplier. The substrate material sets and layer thicknesses were fixed for each design; the only variables explored in this evaluation were the electrolytic Cu plating solution and plating current density in amperes/dm² (ASD). A total of three different electrolytic Cu plating solutions were tested: solutions A and B (on DES1) and solutions B and C (on DES2). Several different ASD skews were also tested on each substrate design.

The substrates were then assembled into overmolded PoP (package on package) bottom packages. The assembly material set and thickness was fixed for each design. These packages were then used to conduct warpage evaluations.

Unit-level high temperature package warpage data was collected using an Akrometrix Shadow Moiré tool and 100 lines per inch (LPI) grating. HT warpage data was collected using a peak temperature of 260°C.

978-1-4799-2408-0/14 $31.00 © 2014 IEEE

2014 Electronic Components & Technology Conference

Cu Deposit Stress Results

Deposit stress analysis results generally indicated compressive stress in the as-plated Cu, with stress transitioning to tensile mode after either long sit times at room temperature or post-plating bake. Figure 2 shows example photos of plated test strips before and after baking. The change in stress state after baking or long sit times is due to the metastable nature of the deposited Cu grains, which tend to coarsen into larger grains with time to eliminate grain boundaries and reduce overall interfacial energy of the system. This phenomenon of a compressive deposit relaxing to a tensile state has been reported previously for electroless and electroplated Cu films [2,3].

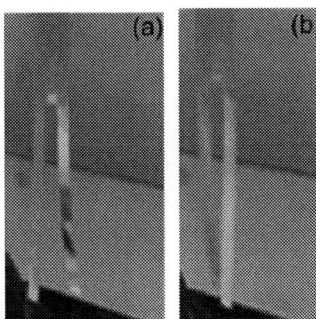

Figure 2. Example Cu deposit stress strips measured (a) after plating (compressive mode) and (b) after baking (tensile mode) at 200°C for 2 hours.

For a given plating current density, the electrolytic Cu plating solution impacted the magnitude and evolution of stress in the plated deposit. Figure 3 shows the impact of successive thermal treatments on the evolution of Cu stress plated by solution A vs. solution B. The as-plated Cu stress is compressive for both chemistries, but the magnitude is higher for solution B than for solution A. After aging for 6 hours at room temperature (RT), the deposit from solution A has already relaxed to tensile mode, while the deposit from solution B has relaxed to a lesser degree, and is still in compressive mode. After baking, deposits from both solution A and B are in tensile mode; their respective stress levels remain constant after an additional 10 days aging at RT. Between 10 and 30 days aging at RT, the stress drifts upward for samples from both plating solutions, but is more significant for solution A (67% increase) than B (15%). This continued relaxation after 30 days is indicative of the degree of residual stress in the as-plated Cu, and suggests that the stress state in the Cu layers of the substrate continues to evolve well after manufacturing is completed. The higher degree of relaxation for Cu from solution A suggests that deposits from this solution have more residual stress than those from solution B, and therefore have lower metastability (i.e. more prone to grain coarsening). Note that the surface of the plated Cu is fully exposed in the deposit stress coupon, allowing the Cu to diffuse/shrink freely to minimize the overall energy of the system. The Cu layers in the substrate are constrained by neighboring dielectric layers, however, which would inhibit the rate and degree of change from compressive toward tensile mode. The final stress state in the substrate's Cu would therefore be more strongly influenced by the as-deposited stress state.

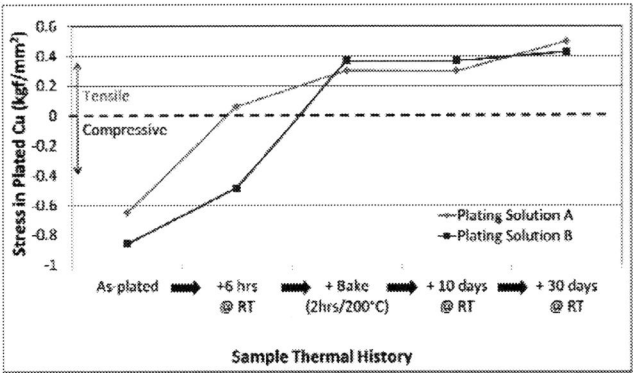

Figure 3. Impact of plating solution and thermal history on stress in plated Cu at a fixed current density.

For a given plating solution, the plating current density also impacted the magnitude and mode of stress in the plated deposit. Fundamentally, increasing the plating current density leads to reducing the size of the deposited grains [1], wherein the smaller grains are less thermodynamically stable [4]. Smaller grains have more grain boundaries per unit volume of deposited Cu, and higher plating current density should therefore give rise to higher compressive stress.

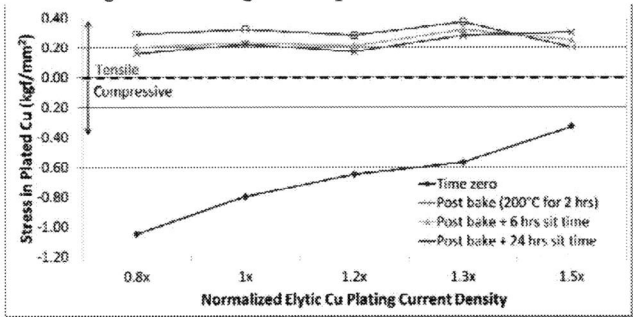

Figure 4. Impact of electrolytic Cu plating current density on stress in the deposit for as-plated and post bake / aging conditions.

Interestingly, however, the time zero data in Figure 4 shows that increasing the current density value results in a shift from more compressive to less compressive stress in the as-plated deposit. This trend was confirmed for multiple vendor chemistries (not shown), as was the shift to tensile mode after baking for all current densities. For the time zero deposit, the reduction in compressive stress with increasing current density is counter intuitive, and is believed to be due to compressive and tensile regions coexisting in the deposit, coupled with limitations of the deposit stress analyzer method. This method provides "net" deposit stress results, which can vary depending on the ratio of compressive to tensile stresses in the plated deposit. A reduction in net compressive stress could indicate that compressive stress is indeed lower, or that the ratio of the deposit under tensile stress has increased. Our hypothesis is that the Cu grains nucleating at the deposit surface are under compressive stress, and grain coarsening and relaxation into tensile mode occurs in part of the bulk layer as plating continues. Figure 5 shows a pictorial explanation. At low plating current density, the initially deposited Cu grains would be relatively large and stable. As plating continues, these grains would continue to nucleate at the deposit surface, while a relatively small percentage of grains in the underlying bulk Cu would be coarsening and relaxing into tensile mode. At high plating current density, the initial deposited grain size is small, increasing the compressive nature of the surface deposit, and increasing the driving force for grain coarsening and relaxation in the bulk of the Cu. Similarly, a transition from small grains at the as-plated Cu surface to larger grains in the bulk layer was reported previously [5].

Relative Plating Current Density	Initial Elytic Cu Grain Nucleation	Elytic Cu Layer After Plating
Low		Compressive region
Mid		Compressive region / Tensile region
High		Compressive region / Tensile region

Figure 5. Schematics depicting impact of plating current density on initial grain size and final grain size/stress state after full layer plating.

Package Warpage Results

Results for DES1 (Figure 6) show the impact of plating solution and current density magnitude on package warpage. For solution A, reduction of plating current density on inner and outer layers (leg 1 vs. 2) provides an 11μm /19μm mean/max warpage reduction. For solution B, reduction of plating current density on inner and outer layers (leg 3 vs. 4) provides a 6μm reduction in both mean/max warpage. At higher plating current density, a change in plating chemistry from solution A to B (leg 1 vs. 3) provides a 12 μm /26 μm mean/max warpage reduction. At lower plating current density, a change in plating chemistry from solution A to B (leg 2 vs. 4) provides a 7 μm /13 μm mean/max warpage reduction. Overall, a change in plating chemistry from solution A to B, coupled with a reduction in plating current density (leg 1 vs. 4), provides a very substantial 18 μm /32 μm mean/max warpage reduction. Plating solution B with lower plating current density on the inner/outer layers was demonstrated to have the best warpage performance on DES1. While plating solution and current density strongly affected warpage magnitude for DES1, they had no impact to warpage shape change (i.e. smiling vs. crying) with temperature, as per the signed warpage graph in Figure 6.

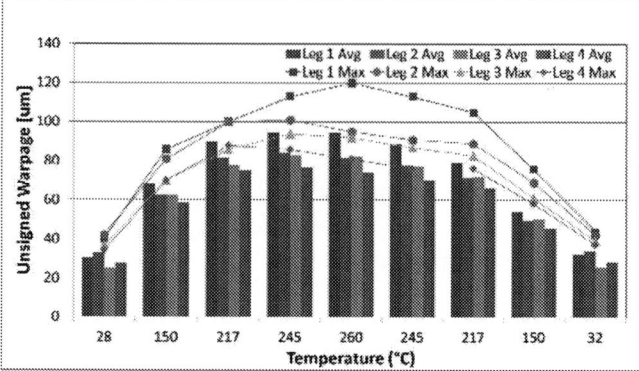

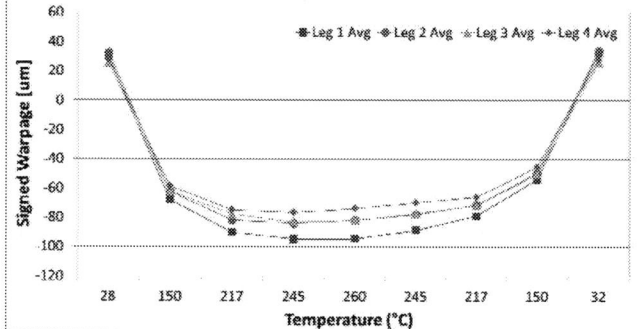

Leg	Cu Plating Solution	Normalized Plating Current Density		Avg Warpage Above Liquidus (um)	Max Warpage Above Liquidus (um)
		Inner Layers	Outer layers		
1	A	1.2x	1.6x	95	120
2	A	1x	1x	84	101
3	B	1.2x	1.6x	83	94
4	B	1x	1x	77	88

Figure 6. DES1 high temperature warpage results (unsigned and signed) for Cu plating solution and ASD skew builds.

Results for DES2 (Figure 7) show the impact of plating solution, current density magnitude, and current density balancing (inner vs. outer layers) on package warpage. At lower plating current density, a change in plating chemistry from solution C to B (leg 5 vs. 8) provides a 14 μm /9 μm mean/max warpage reduction. For solution B, using a higher current density on the inner vs. outer layers (leg 6 vs. 7) provides a 7 μm /8 μm mean/max warpage reduction. As with DES1, solution B with lower plating current density on the inner/outer layers was demonstrated to have the best warpage performance on DES2. Moreover, the data indicates the importance of balancing plating current density on inner vs. outer layers. Use of current density on the outer layers which is equal to or lower than that on the inner layers was better for warpage.

While plating solution and current density strongly affected warpage magnitude for DES2, they had no impact to warpage shape change (i.e. smiling vs. crying) with temperature, as per the signed warpage graph in Figure 7.

978-1-4799-2408-0/14 $31.00 © 2014 IEEE

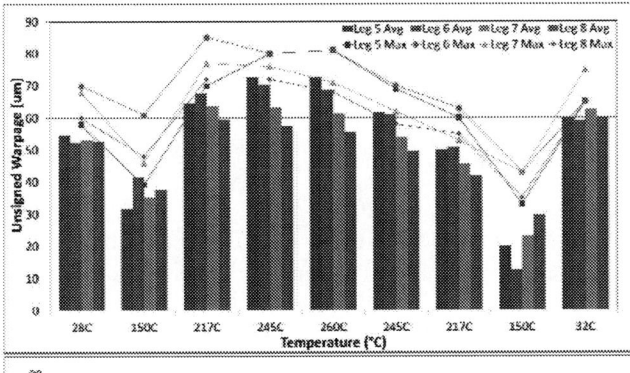

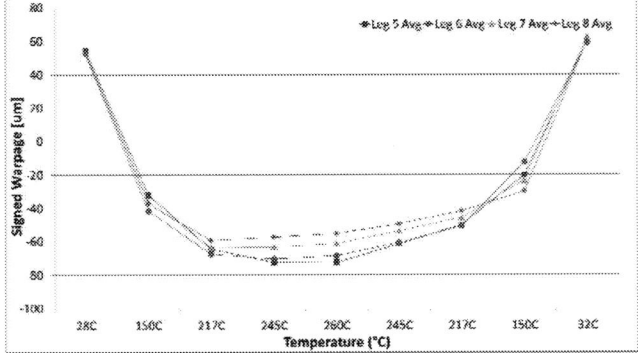

Leg	Cu Plating Solution	Normalized Plating Current Density		Avg Warpage Above Liquidus (um)	Max Warpage Above Liquidus (um)
		Inner Layers	Outer layers		
5	C	1y	1y	73	81
6	B	1y	1.2y	71	85
7	B	1.2y	1y	64	77
8	B	1y	1y	59	72

Figure 7. DES2 high temperature warpage results (unsigned and signed) for Cu plating solution and ASD skew builds.

Conclusions

Results of the analysis show that choice of substrate electrolytic Cu plating solution has significant impact on the magnitude of package warpage. The influence of Cu plating solution on warpage is related to the resulting grain size distribution and stress state deposited from a given chemistry. Additives such as levelers and brighteners are used in the plating solution to control deposition rate across features on the panel. Additives are intended to, in part, foster hydrogen evolution, change the electrode potential at the plating surface, and brighten the deposit surface [6]. These same additives (or fragments thereof) can be co-deposited as impurities into the Cu layer, and have been shown to strongly impact residual stress and grain coarsening behavior of the Cu deposit [7]-[9]. The additive makeup and chemistry in a given plating solution can therefore influence the deposited grain size, the amount of impurities and residual stress in the plated Cu, the rate/extent of grain coarsening, and the overall package warpage. This is evident in Figure 3, which shows a difference in the initial deposit stress and rate of stress change for solution A vs. B.

Based on warpage data from DES1 and DES2, Cu plating solution B provides lower package warpage. Linking Cu stress and warpage results together for plating solution B (Figures 3, 6 and 7), we believe that higher compressive stress (and higher metastability) in the as-plated Cu, coupled with

slower change to tensile mode, provides lower package warpage.

Results from this study also show that electrolytic Cu plating current density has significant impact on the magnitude of package warpage. For both DES1 and DES2, reducing the plating current density for a given plating solution led to substantial reduction in package warpage (Figures 6 and 7). As mentioned previously, an increase in the plating current density causes a reduction in the deposited grain size [1], hence a reduction in current density would lead to larger deposited grains. Larger grains would mean reduced grain boundary volume, less "shrink" in the Cu layer as it relaxes to tensile mode, and potentially lower residual stress in the Cu. While reducing current density improves package warpage, it negatively affects plating process throughput in high volume manufacturing. Since current density is related to throughput and package warpage, a balance must be maintained to ensure good warpage and good plating productivity. Figure 8 depicts the target operating range for current density to balance warpage performance against throughput, and the associated as-deposited Cu stress.

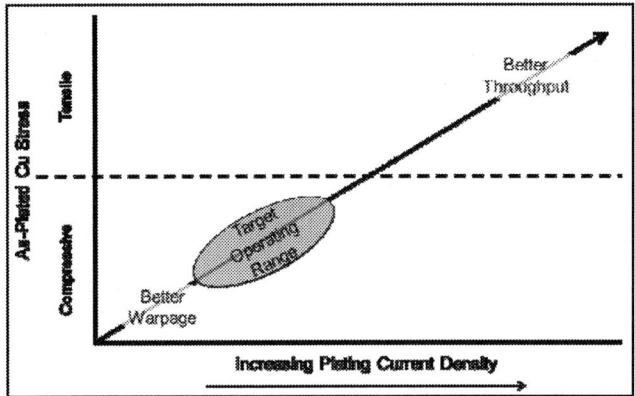

Figure 8. Proposed operating range for Cu plating current density to control package warpage.

Apart from the plating current density magnitude, balancing of the current density used to plate the inner vs. outer layers of the substrate was also important for warpage. Results from DES2 (legs 6 vs. 7) show that use of current density on the outer layers which is equal to or lower than that on the inner layers was better for warpage. Substrate designs typically have an inherent front/back Cu density imbalance, which when coupled with higher current density on the outer layers, appears to exacerbate warpage issues. If we assume that the center of the core layer is the substrate's neutral axis (Figure 9), we can consider the moment caused by residual stress in each Cu layer based on the following relation:

$$M = F * d$$

where M is the moment, F is the force and d is the distance from the neutral axis to the Cu layer of interest.

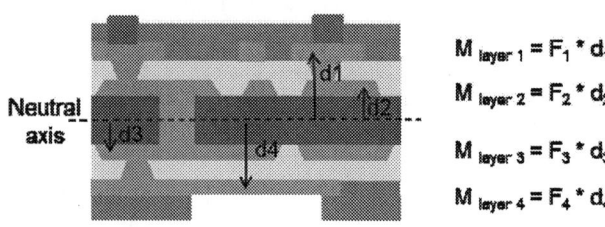

$$M_{layer\,1} = F_1 * d_1$$
$$M_{layer\,2} = F_2 * d_2$$
$$M_{layer\,3} = F_3 * d_3$$
$$M_{layer\,4} = F_4 * d_4$$

Figure 9. Schematic showing moment equation as applied to Cu layers in the substrate.

With equivalent residual stress in all Cu layers ($F_1=F_2=F_3=F_4$), the outer layers by definition have a larger moment than the inner layers, due to their greater distance (d1,d4) from the neutral axis. Higher residual stress in the outer layers (F1,F4), coupled with larger distance to neutral axis (d1,d4), would cause more curling/warpage. Higher residual stress on the inner layers (F2,F3) has less impact to warpage, as the closer proximity to the neutral axis (d2,d3) counterbalances the force in the moment equation. By this logic, and based on empirical results, we believe that use of a plating current density on the outer layers which is less than or equal to the inner layers is best for warpage.

One typical approach to control package warpage is to bake the substrate before assembly to eliminate residual stresses. Prior testing on DES1 showed this to have no impact on package warpage. While baking the substrate may reduce residual stress in the Cu layers (through Cu shrink), this stress would be transferred into neighboring dielectric layers, and is therefore not eliminated from the substrate. Hence baking a substrate which already has a Cu areal density imbalance (by design) is not expected to provide a benefit for package warpage.

From initial warpage modeling work, the magnitude of the net deposit stress values reported above were considered to be small, having little impact on simulated warpage behavior of the final substrate. The empirical warpage results, however, conclusively show that the residual stress state in the Cu is important for package warpage. The sources of this disconnect between modeling and empirical data is unclear at this time, and will be explored in future work.

Based on the complete analysis, the following variables in the substrate Cu plating process proved to be critical for package warpage control:

1. Cu plating solution chemistry
2. Magnitude of Cu plating current density
3. Cu plating current density balancing on inner vs. outer layers

For warpage-sensitive packages, the above variables should be carefully considered and evaluated to determine the best operating window for a given package/substrate design.

Acknowledgments

The authors would like to thank Nancy Bailey at Qualcomm for her work to standardize our warpage measurement methodology, and our substrate and assembly partners for executing the builds in this study. We would also like to thank Qualcomm's management team for their support of this work.

References

1. N. Saito, N. Murata, K. Tamakawa, K. Suzuki and H. Miura, "Evaluation of the Crystallinity of Grain Boundaries of Electronic Copper Thin Films for Highly Reliable Interconnections," in *Proc. IEEE Electronic Components and Technol. Conf. (ECTC)*, San Diego, CA, US, May 29-June 1, 2012.
2. R. Brüning, B. Muir, E. McCalla, É. Lempereur, F. Brüning, J. Etzkorn, "Strain in electroless copper films monitored by X-ray diffraction during and after deposition and its dependence on bath chemistry," Thin Solid Films **519** (2011), pp. 4377-4383.
3. H. Lee. S. Wong, S. Lopatin "Correlation of stress and texture evolution during self- and thermal annealing of electroplated Cu films", Journal of Applied Physics **93-7** (2003), pp. 3796-3803
4. C.C. Hu, C.M. Wu, "Effects of deposition modes on the microstructure of copper deposits from an acidic sulfate bath," Surface and Coatings Technology **176** (2003), pp. 75-83.
5. H. Miura, K. Suzuki, and K. Tamakawa, "Fluctuation Mechanism of Mechanical Properties of Electroplated-Copper Thin Films Used for Three Dimensional Electronic Modules, " in *Proc. International Conference on Thermal, Mechanical and Multi-Physics Simulation Experiments in Microelectronics and Micro-Systems*, London, England, April 16-18,2007.
6. T.C. Franklin, "Review: Some Mechanisms of Action of Additives in Electrodeposition Processes," Surface and Coatings Technology **30** (1987), pp. 415-428.
7. M. Stangl, V. Dittel, J. Acker, V. Hoffmann, W. Gruner, S. Strehle, K. Wetzig, "Investigation of organic impurities adsorbed on and incorporated into electroplated copper layers," Applied Surface Science **252** (2005), pp. 158–161.
8. T.G. Woo, I.S. Park, K.W. Seol, "The Effect of Additives and Current Density on Mechanical Properties of Cathode metal for Secondary Battery," Electronic Materials Letters Vol.9, No. 4 (2013), pp. 535-539.
9. S. H. Brongersma, E. Richard, I. Vervoort, and K. Maex, "A Grain Size Limitation Inherent to Electroplated Copper Films," in *Proc. of the IEEE 2000 International Interconnect Technology Conference*, Burlingame, CA, US, June 5-7, 2000.

978-1-4799-2408-0/14 $31.00 © 2014 IEEE

Coreless Substrate with Asymmetric Design to Improve Package Warpage

Wei Lin, Bora Baloglu, Ken Stratton

Amkor Technology

1900 S Price Rd, Chandler, AZ 85286

wei.lin@amkor.com

Abstract

Coreless substrates have been used in more and more advanced package designs for their benefits in electrical performance and reduction in thickness. However, coreless substrate causes severe package warpage due to the lack of a rigid and low CTE core. In this paper, both experimental measured warpage data and model simulation data are presented and illustrate that asymmetric designs in substrate thickness direction are capable of improving package warpage when compared to the traditional symmetric design. A few asymmetric design options are proposed, including Cu layer thickness asymmetric design, dielectric layer thickness asymmetric design and dielectric material property asymmetric design. These design options are then studied in depth by simulation to understand their mechanism and quantify their effectiveness for warpage improvement. From the results, it is found that the dielectric material property asymmetric design is the most effective option to improve package warpage, especially when using a lower CTE dielectric in the bottom layers of the substrate and a high CTE dielectric in top layers. Cu layer thickness asymmetric design is another effective way for warpage reduction. The bottom Cu layers should be thinner than the top Cu layers. It is also found that the dielectric layer thickness asymmetric design is only effective for high layer count substrate. It is not effective for low layer count substrate. In this approach, the bottom dielectric layers should be thicker than the top dielectric layers. Furthermore, the results show the asymmetric substrate designs are usually more effective for warpage improvement at high temperature than at room temperature. They are also more effective for a high layer count substrate than a low layer count substrate.

Introduction

Coreless substrates have been used in more and more advanced package designs[3][5][6]. They eliminate signal routings through a thick core and provide superior electrical performance demanded by the next generation package[4][6]. At the same time, they make the substrate thinner, which is very important for today's mobile device applications[3][6][7]. However, compared to a cored substrate, a coreless substrate is far more mechanically challenged for package design and assembly[1][8]. The core in a substrate usually has much lower CTE than the build-up dielectric material and the Cu metal layers. Without a core, a coreless substrate has much higher effective CTE than a cored substrate. Furthermore, core material usually has much higher stiffness (modulus) than buildup dielectric material. The core provides the mechanical rigidity of the entire substrate. Without a core, a coreless substrate is far less

rigid and prone to deform. Due to the higher CTE and less rigidity, packages using coreless substrate usually have severe warpage issues. The increased package warpage may lead to failures during assembly processes. As a result, warpage control becomes even more critical for any package design using coreless substrate[2][3][5].

Traditionally in substrate design, the layer stack-up structure, layer thickness, and the materials used in each layer are usually symmetric in thickness direction to keep the substrate balanced and flat. In this paper, a few new asymmetric layer design approaches for coreless substrate are studied to effectively improve the warpage performance, as shown in Figure 1. The new asymmetric designs use different dielectric materials and thickness for the build-up layers; for example, high CTE ABF dielectric in the top layers and a much stiffer and lower CTE glass cloth reinforced dielectric in the bottom layer. While increasing the rigidity of the entire substrate, the asymmetric design also creates CTE mismatch within the substrate layers so that the substrate offers the ability to counteract the package warpage. In addition, the copper signal layers in the substrate can also be designed to have different thickness in top and bottom layers to further enhance the ability to reduce package warpage.

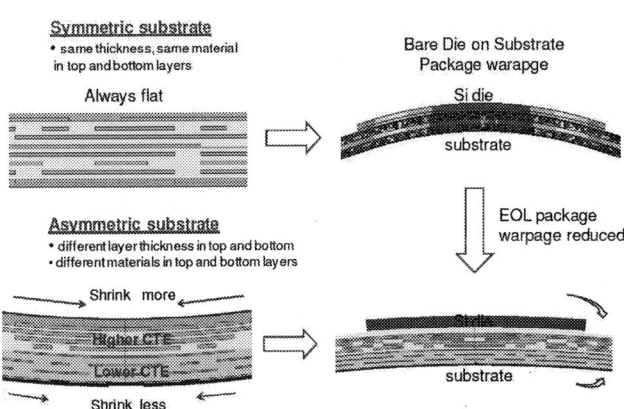

Figure 1. Warpage improvement using asymmetric substrate design

In this paper, a 23x23mm package with a 0.15mm thick 4-layer coreless substrate is used as a test vehicle for experimental study. Shadow moiré warpage data for both the conventional symmetric design and the new asymmetric design are presented and compared to illustrate the warpage improvement of the new design over the conventional design. Finite element warpage modeling is also performed to correlate with the measured data. Since asymmetric design can be implemented in quite a few different ways such as asymmetric in Cu layer thickness, dielectric layer thickness, dielectric layer materials, solder mask layer thickness and materials, etc., warpage simulation evaluations are further conducted in depth to systematically understand the mechanism and compare the effectiveness of the various asymmetric design options mentioned above. The purpose is to identify the critical parameters and provide design guides for effective asymmetric substrate design.

Experimental Study

Figure 2 is the package used as the test vehicle for experimental study. It is a 23x23mm molded package. The coreless substrate is a 4 layer design with a total thickness of 0.159mm. Both the conventional symmetric design and the new asymmetric design are implemented for this package, as shown in Figure 3. In the asymmetric design, the bottom dielectric layer uses a glass cloth reinforced dielectric material instead of the ABF material which is used in the top two dielectric layers.

Figure 2. 23x23mm package

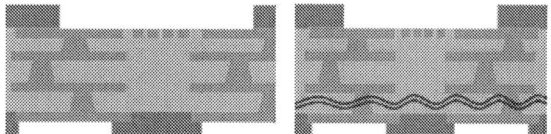

Figure 3. Symmetric vs asymmetric substrate design

Table 1 shows the material properties comparison between the ABF and the glass reinforced dielectric material. The glass reinforced dielectric has much lower CTE than the ABF. Meanwhile, its modulus is much higher than the ABF.

Table 1. Material properties comparison

	ABF	glass reinforced dielectric
CTE1 (ppm)	23	16
CTE2 (ppm)	78	9
Modulus (Gpa)	7.5	16.7

Figure 4 shows the shadow moiré measured warpage data of the package at the end of line (EOL). Warpage at different temperatures from 25C to 260C are shown in the plot. Both the warpage data for the symmetric design and asymmetric design are plotted together for the comparison. The measured data show that the asymmetric substrate design does improve the package warpage, especially in the high temperature range. This is because in high temperature the CTE

difference between the glass reinforced dielectric and the ABF is much more significant.

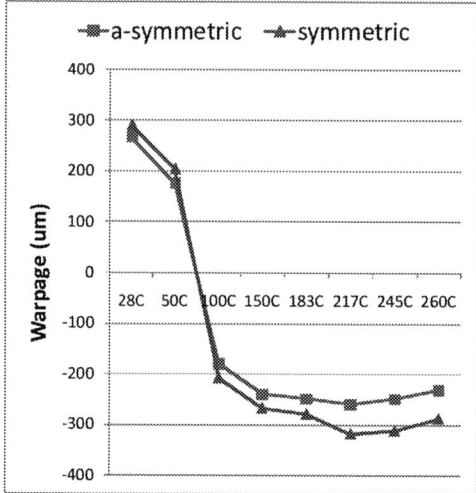

Figure 4. Shadow moiré warpage data comparison between symmetric and asymmetric substrate designs.

Finite Element Warpage Model

Finite element warpage model is built to correlate to the above experimental results. Traditionally, the substrate is usually modeled as a uniform composite material. However, to capture the asymmetric substrate design effect on warpage, the substrate cannot be modeled as a composite material anymore. Each individual layer in the substrate has to be modeled to represent their differences. Figure 5 shows the warpage model with detail layered substrate.

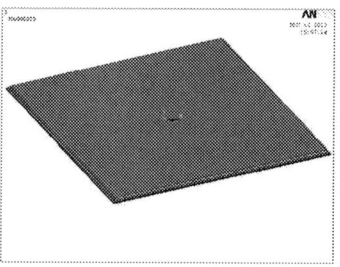

(a) Package model

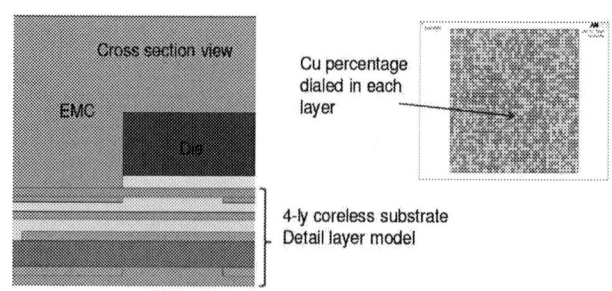

(b) Substrate model

Figure 5. Finite element warpage model

Figure 6 shows the warpage simulation results for the test vehicle presented in the last section. The simulation data indicates the asymmetric design has lower warpage than the symmetric design, which shows the same trend as the test data.

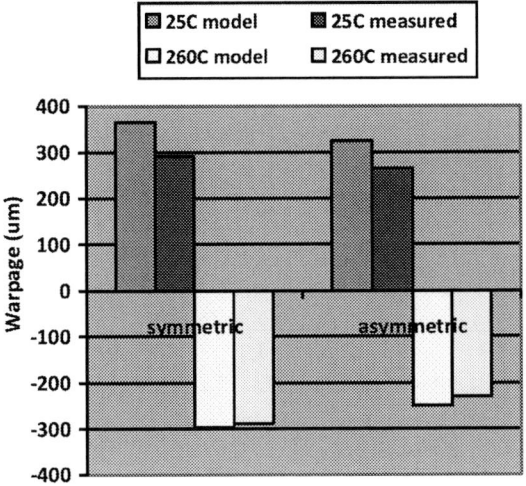

Figure 6. Warpage correlation between modeling and test data.

Simulation Study

The above experimental and simulation data both show that the asymmetric substrate is capable of improving package warpage. As we know, asymmetric design can be implemented in quite a few different ways, such as varying Cu layer thickness, varying dielectric layer thickness and materials, varying solder mask layer thickness and materials, etc. Among these options, which one has the most significant impact? Furthermore, for applications with high layer count and low layer count coreless substrates, does the asymmetric design have the same level of impact?

To better understand and quantify the impacts of different design parameters in an asymmetric substrate for warpage improvement, extensive simulations are carried out to evaluate all these options. To eliminate other effects and just focus on the substrate, a simple bare die flip chip underfilled package is selected for this simulation study. This bare die package is assumed to be 15x15mm body size with die size of 11x11mm. For the substrate, an 8 layer design (0.335mm total thickness) and a 4 layer design (0.175mm total thickness) are evaluated separately to see any different impacts of asymmetric design between high layer count and low layer count substrate. The models are shown in Figure 7. A symmetric substrate design is used as a reference for the other asymmetric design options to compare to. For the reference symmetric design, all Cu layer thickness is 15um and all dielectric layer thickness is 25um. A prepreg material with 6ppm CTE value is used as the dielectric.

Effect of Copper Layer Thickness Asymmetric

The first approach for an asymmetric design is to use thinner Cu metal in the bottom layers and thicker Cu metal in the top layers. Table 2 lists the simulation legs conducted

to understand the effect of Cu layer thickness asymmetric design for the 8 layer substrate, and figure 8 shows the warpage simulation results for these legs.

(a) Bare die package model (quarter model)

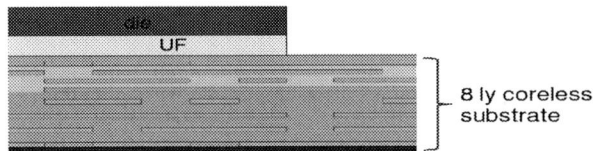

(b) 8 layer substrate design

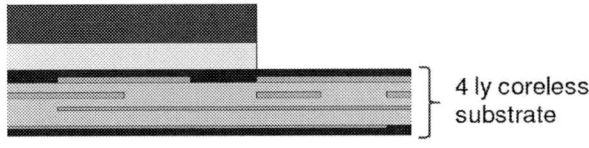

(c) 4 layer substrate design

Figure 7. Bare die package for simulation study

Table 2. Simulation legs for Cu layer thickness effect

Leg	ly1-4 Cu thk	ly5-8 Cu thk	Dielectric thk	Total thk
0 (symmetric)	15 um	15 um	25 um	335 um
1	15 um	8 um	25 um	307 um
2	22 um	8 um	25 um	335 um
3	15 um	8 um	29um	335 um

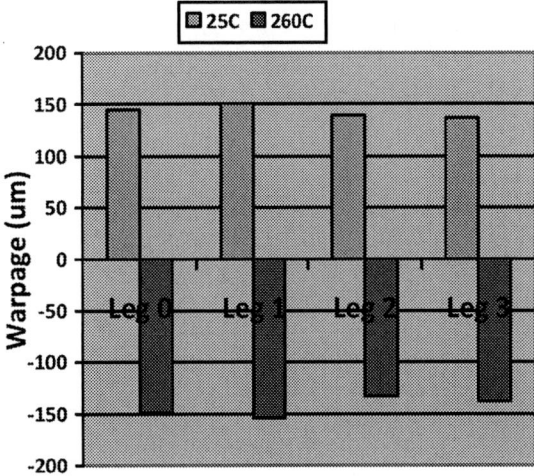

Figure 8. Cu layer thickness asymmetric effect

978-1-4799-2408-0/14 $31.00 © 2014 IEEE 1403

Leg 0 is the reference symmetric design. Leg 1, 2, 3 are three different cases for Cu thickness asymmetric. There are some interesting findings from the above results:

(1) When comparing Leg 1 with Leg 0, the results show that it does not improve warpage if only the thickness of the bottom 4 Cu layers is reduced from 15um to 8um while the top 4 Cu layers remain the same thickness at 15um. Actually the warpage is even worse. This is because the total thickness of the substrate is also reduced from 335um to 307um. The reduced total thickness makes the substrate less rigid, which is a negative impact to warpage and cancels the benefit of asymmetric design.

(2) Comparing Leg 2 and Leg 3 to Leg 0, the results show that, to improve warpage by Cu layer thickness asymmetric design, the thickness of the bottom 4 Cu layers should be reduced, while at the same time, the thickness of the top 4 Cu layers or the dielectric layers should be increased to keep the total substrate thickness the same. Increasing the thickness of the top 4 Cu layers (Leg 2) is slightly more significant than increasing the dielectric layer thickness (Leg 3).

Effect of Dielectric Layer Thickness Asymmetric

The second approach to create an asymmetric design is to use different dielectric layer thicknesses in the top and bottom layers. Table 3 lists the simulation legs conducted to understand the effect of dielectric layer thickness asymmetric design for the 8 layer substrate, and Figure 9 shows the warpage simulation results for these legs.

Table 3. Simulation legs for dielectric layer thickness effect

Leg	ly1-3 Dielec. thk	Ly4-7 Dielec. thk	Total thk
0 (symmetric)	25 um	25 um	335 um
1	35 um	15 um	335 um
2	15 um	35 um	335 um

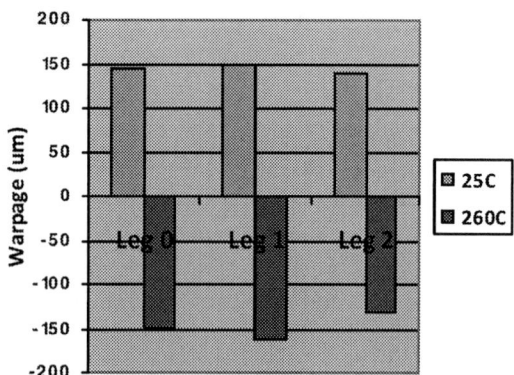

Figure 9. Dielectric layer thickness asymmetric effect

The results show only Leg 2 improves the warpage, while Leg 1 actually makes the warpage even worse. This means that, to improve warpage by dielectric layer thickness asymmetric, the bottom dielectric layers must be thicker than the top dielectric layers while keeping the total substrate thickness the same. This is because using thicker dielectric in the bottom layers makes the effective Cu density lower in the bottom portion of the substrate. Therefore, the effective CTE of the bottom portion is also lower than that of the top portion of the substrate. This causes the substrate to warp towards the smiling face direction which counteracts the package warpage.

The dielectric asymmetric design has a similar level of significant impact on warpage improvement as the Cu metal layer asymmetric design. This can be found when comparing Leg 2 in Figure 8 and Leg 2 in Figure 9.

Effect of Dielectric Layer Material Asymmetric

The third approach for asymmetric design is to use different dielectric materials in the top and bottom dielectric layers. In this approach, the mechanism is to use lower CTE dielectric in the bottom layers and higher CTE dielectric in the top layers to generate CTE mismatch within the substrate. However, the combination of high and low CTE dielectric may also change the effective CTE of the whole substrate, which is another factor that also affects the package warpage. Therefore it is important to understand how to select the best combination. Table 4 lists the simulation legs conducted to understand the effect of dielectric layer material asymmetric design, and Figure 10 shows the warpage simulation results for these legs.

Table 4. Simulation legs for dielectric layer material effect

Leg	ly1-3 Dielec. CTE/modulus	Ly4-7 Dielec. CTE/modulus
0 (symmetric)	6ppm, 18Gpa	6ppm, 18Gpa
1	6ppm, 18Gpa	2ppm, 18Gpa
2	20ppm, 7Gpa	6ppm. 18Gpa
3	20ppm, 7Gpa	2ppm, 18Gpa
4	2ppm, 18Gpa	2ppm, 18Gpa

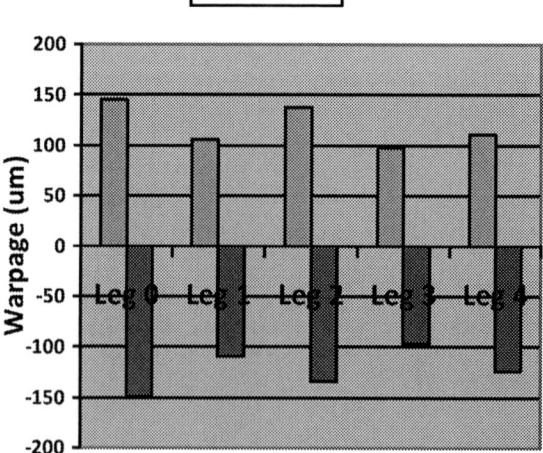

Figure 10. Dielectric layer material asymmetric effect

There are a few important findings from the above results:

(1) When comparing Leg 1 with Leg 0, as expected, changing the bottom 4 dielectric layers to a lower CTE material improves the package warpage significantly.

(2) When comparing Leg 2 with Leg 0, changing the top 3 dielectric layers to a higher CTE material would still improve the warpage, even though the overall effective CTE of the whole substrate is higher. This means the asymmetric effect in the substrate has more impact than the overall effective CTE effect. On the other hand, when comparing Leg 2 with Leg 1, both legs are asymmetric design, but it is obvious that Leg 1 is much better than Leg 2 because the overall effective CTE of Leg 1 is lower than Leg 2.

(3) The best warpage improvement is from Leg 3, which is a combination of the lowest CTE (2 ppm) prepreg dielectric in the bottom layers and the highest CTE (20 ppm) ABF in the top layers, and thus has the biggest CTE difference between the bottom and the top dielectric layers. It means using two dielectric materials with the biggest delta CTE difference in the top and bottom layers has the most significant effect for warpage improvement.

(4) Leg 4 is also a symmetric design but uses the same lowest CTE (2 ppm) dielectric material in all layers. However, comparing Leg 4 with either Leg 1 or Leg 3, its warpage is higher. This means, a symmetric design with lower CTE dielectric in all layers is not better than an asymmetric design with higher CTE dielectric in the top layers for warpage improvement. Again, this indicates that the asymmetric effect is more significant than the overall effective CTE effect.

Effect of Substrate Layer Count

In the previous sections, various asymmetric design options and their impacts on warpage improvement have been discussed based on an 8-layer substrate. To understand whether the substrate layer count has any impact on the trends and the level of significance for those various asymmetric designs, a 4-layer substrate is also simulated and the results are compared to those of the 8-layer's. Table 5 lists the 5 different asymmetric design options which are used to compare the 4-layer vs 8-layer substrates. Figure 11 shows the results. In Figure 11, the level of significance is presented as percentage warpage improvement over their corresponding reference symmetric designs for 4-layer and 8-layer respectively.

Table 5. Simulation legs for 4-layer and 8-layer comparison

Leg	Top/low Cu thk	Top/low Dielec thk	Top/low Dielec CTE
1	22/8 um	25/25 um	6 / 6 ppm
2	15/15 um	15/35 um	6 / 6 ppm
3	15/15um	25/25 um	6 / 2 ppm
4	15/15um	25/25 um	20 / 6 ppm
5	15/15um	25/25 um	20 / 2 ppm

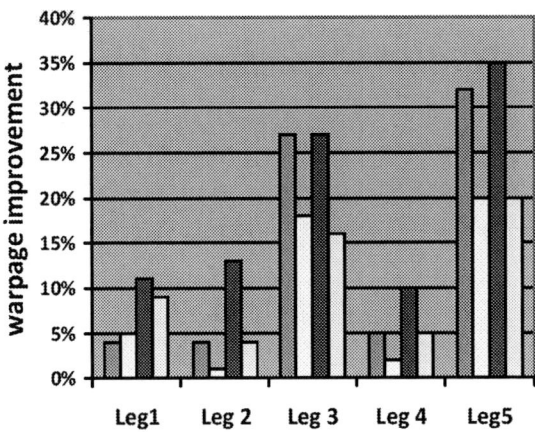

Figure 11. 4-layer vs 8-layer comparison for various asymmetric design options

A few important conclusions can be found from results in Figure 11:

(1) For most of the legs, 8 layer substrate has more significant warpage improvement than 4 layer substrate. This means, in general, asymmetric designs are more effective for a high layer count substrate than a low layer count substrate. This is because high layer count substrate has more layers to create more substantial CTE difference between the top and the bottom portions of the substrate.

(2) For 8 layer substrate, the dielectric layer thickness asymmetric design (Leg 2) is as effective as the Cu layer thickness asymmetric design (Leg 1). However, for 4 layer substrate, the dielectric thickness asymmetric design is much less effective than the Cu thickness asymmetric design.

(3) Asymmetric design is usually more effective for warpage improvement at high temperature than at room temperature.

(4) In terms of significance on warpage improvement, dielectric material asymmetric design, especially using a lower CTE dielectric in the bottom layers (Leg 3 and Leg 5), has the most significant impact, followed by Cu thickness asymmetric design (Leg 1).

Summary

Coreless substrate causes severe package warpage due to the lack of a rigid and low CTE core in the substrate. Both experimental measured warpage data and model simulation data show that asymmetric designs in substrate are capable of improving package warpage compared to the traditional symmetric design. A few asymmetric designs are proposed and studied in depth by simulation to understand the mechanism and the effectiveness of various asymmetric design options. From the results, it is concluded that:

(1) The most effective asymmetric substrate design option to improve package warpage is the dielectric material property asymmetric design, especially when using a lower CTE dielectric in the bottom layers of the substrate and high CTE dielectric in top layers to have the delta CTE difference as large as possible. Asymmetric effect is more significant than the overall effective CTE effect.

(2) Cu layer thickness asymmetric design is another effective way to reduce package warpage. The thickness of the bottom Cu layers should be thinner than the top Cu layers. For this design option to be effective the reduced thickness in the bottom Cu layers should be added to the top Cu layers or the dielectric layers to keep the total substrate thickness remain the same.

(3) Dielectric layer thickness asymmetric design is as effective as Cu thickness asymmetric design for high layer count substrate. However, it is not effective for low layer count substrate. In this approach, the bottom dielectric layers should be thicker than the top dielectric layers.

(4) In general, asymmetric substrate designs are more effective for warpage improvement at high temperature than at room temperature.

(5) Asymmetric designs are more effective for a high layer count substrate than a low layer count substrate. This is true for all types of asymmetric design options.

(6) Due to the other design considerations and processing issues, there are limitations on each option of the asymmetric designs in actual implementation. Therefore, an optimized combination is more likely. Further studies are needed to understand the interaction and effectiveness of such a combination.

Acknowledgments

The authors would like to thank the production and the test teams for their great support. The authors would also like to thank Shengmin Wen and John McCormick for their helpful inputs and discussions.

References

1. Kim, Jinho, et al, "Warpage Issues and Assembly Challenges Using Coreless Package Substrate", *IPC APEX EXPO Proceedings,* 2012.
2. Baloglu, Bora, Lin, Wei, et al, "Warpage Characterization and Improvements for IC Packages with Coreless Substrate," *IMAPS Symposium,* 2013
3. Sun, Y., et al, "Development of Ultra-thin Low Warpage Coreless Substrate," *ECTC,* 2013
4. Kim, GW, et al, "Evaluation and Verification of Enhanced Electrical Performance of Advanced Coreless Flip-chip BGA Package with Warpage Measurement Data," *ECTC,* 2012.
5. Kurashina, M., et al, "Low Warpage Coreless Substrate for Large-size LSI Packages," *ECTC,* 2012
6. Nickerson, R., et al, "Application of Coreless Substrate to Package on Package Architectures," *ECTC,* 2012
7. Lin, Wei, "Warpage Challenges for Next Generation Package-on-Package(PoP)," *IMAPS Think Thin Workshop,* 2011
8. Fujimoto, D., et al, "New Fine Line Fabrication Technology on Glass-Cloth Prepreg without Insulating Films for PKG Substrate," *ECTC,* 2011.

Ultra Low CTE (1.8 ppm/°C) Core Material for Next Generation Thin CSP

Tomohiko Kotake, Hikari Murai, Shin Takanezawa, Masato Miyatake,
Masaaki Takekoshi, Masahisa Ose[*]
Tsukuba Research Laboratory, Hitachi Chemical Co., Ltd.,
[*]Printed Wiring Board Materials Business Unit, Hitachi Chemical Co., Ltd.
1919 Morisoejima Chikusei-shi, Ibaraki, 308-0861, Japan
Phone: +81-296-23-6252, Fax: +81-296-25-5505
E-mail: t-kotake@hitachi-chem.co.jp,

Abstract

Along with the advancement in miniaturizing of mobile devices, typified by smart phones and tablet PCs, the semiconductor PKG substrate installed in these devices is demanded to be thinner and higher in density. As one of the most innovative solutions, the PoP (package on package) technology, which has the three-dimensional construction, has been expanding rapidly in recent years.

However, the thinner PKG such as PoP tends to warp at the assembly process and cause the decrease in the connection reliability. Therefore ultra low CTE (coefficient of thermal expansion) core materials have been needed as a key solution for the reduction of the warpage for PoP.

Recently, we have developed new ultra low CTE core material named E-770G for next generation thin CSP, applying new resin systems, featuring low shrinkage and low residual stress. In particular, E-770G has achieved ultra low CTE of 1.8 ppm/°C which leads to significant reduction of the warpage. Furthermore, it has low dissipation factor at high frequencies (Df: 0.005 at 1 GHz). So it's also applicable to high speed PKG applications.

Confirming the warpage property, we evaluated the warpage behavior of the bottom PKG before/after assembly process. E-770G showed the much lower warpage than the conventional ultra low CTE core material.

Introduction

As the demand for the portable handheld products and devices, such as smart phones and tablet PCs, still continues to increase, higher density packaging technologies have been required to reduce the assembly area of substrate. From this point of view, three-dimensional packaging is a key technology to minimize the total size of products and devices.

PoP technology, which enables us to achieve three-dimensional packaging, has been accelerated in adoption particularly for smaller systems. However, because of its thinner construction in PoP, warpage of the substrate at soldering process, which may result in poor connection reliability, has been one of the key challenges to overcome. In many cases, the main factor of the warpage is the mismatch of CTE between the substrate and the silicon chip, and therefore, ultra low CTE core materials are required for thinner PKG applications.

We have already developed ultra low CTE core material named E-705G (CTE value:2.8 ppm/°C) for this application, and it is now in commercial mass production. However, the demand for ultra low CTE core materials is still growing for thinner PKG applications, and the CTE needs to be lower that of E-705G in these days.

In this paper, we introduce new ultra low CTE core material named E-770G for next generation thin CSP, and confirmed the relationship between the warpage and material properties (CTE and shrinkage).

Warpage in Assembly Process by Simulation

The basic idea of the warpage behavior of PKG substrates is understood by the CTE of the substrate and the silicon chip as follows. In the assembly process, the PKG substrate shows a concave shape at reflow temperature, since the thermal expansion of the substrate is larger than that of the chip. On the contrary, when the substrate is cooled down to the room temperature, it shows a convex shape due to the shrinkage of the substrate.

In order to investigate the relationship between the warpage and material properties (CTE and modulus), mathematical simulations were done. The PKG model and the substrate construction are shown in Figures 1 and 2. The thicknesses of core material and prepreg for the simulation are 100 μm and 40 μm, respectively. The simulations of the warpage behavior of PKG substrates were done by a three-dimensional elastic analysis method at the temperatures of 25 °C and 260 °C.

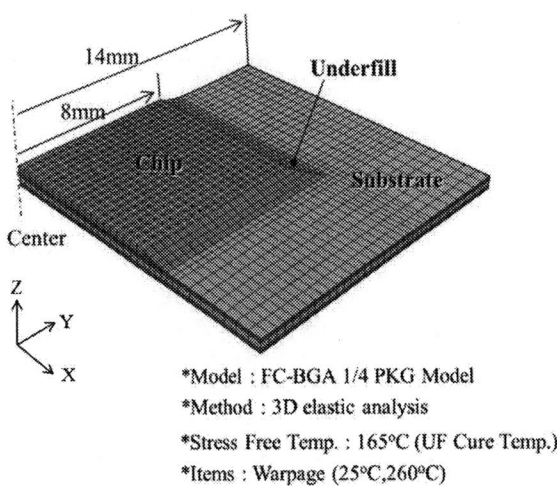

*Model : FC-BGA 1/4 PKG Model
*Method : 3D elastic analysis
*Stress Free Temp. : 165°C (UF Cure Temp.)
*Items : Warpage (25°C,260°C)

Figure 1. PKG model for simulation

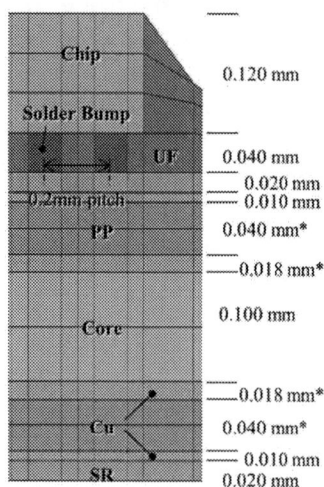

Figure 2. Construction of substrate for simulation

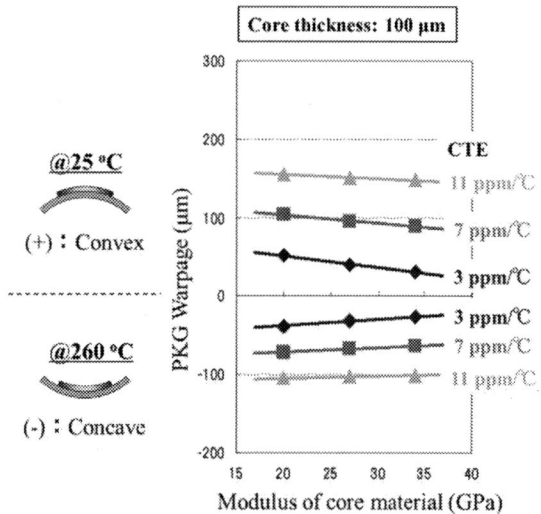

Figure 3. Simulation results of warpage

The simulation results are shown in Figure 3. From the results, the low CTE core material was confirmed to decrease the warpage at both temperatures. We have also confirmed that the contribution of the low CTE core material to reduce the warpage is more effective than that of the high modulus core material.

Molecular Design of New Resin System

As the simulation result demonstrates, lower CTE core materials can reduce the warpage of thinner PKG substrate effectively. We have already developed two types of technical concepts for the further ultra low CTE (type A and type B) [1].

In the technical concept of type A, the resin system has a planer stack structure of aromatic ring and the strong intermolecular force between the stack can bring ultra low CTE and low shrinkage. The resin system of type A has been adopted for E-705G. On the other hand, we adopted another resin system for type B, introducing soft segments in the

system to realize the low elastic modulus. By applying the soft segments, the elastic modulus of the resin itself becomes much lower than that of type A. Therefore, the resin can follow approximately consistent with the thermal behavior of glass fabric when heated and cooled. Consequently, the material can show the ultra low CTE which is similar to that of glass fabric itself. Moreover, the low elastic modulus of resin also shows the low residual stress.

Recently, we have developed new resin system for ultra low CTE core material (E-770G), featuring low shrinkage and low residual stress. The polymer of E-770G is formed by our original resin having hard segments and soft segments as illustrated in Figure 4. Hard segments of the new resin system lead to the ultra low CTE and low shrinkage. And soft segments of the new resin system lead to the ultra low CTE and low residual stress. Moreover, the resin system contains higher amount of inorganic filler to attain ultra low CTE and high modulus by applying the FICS (filler interphase control system) technology [2-4].

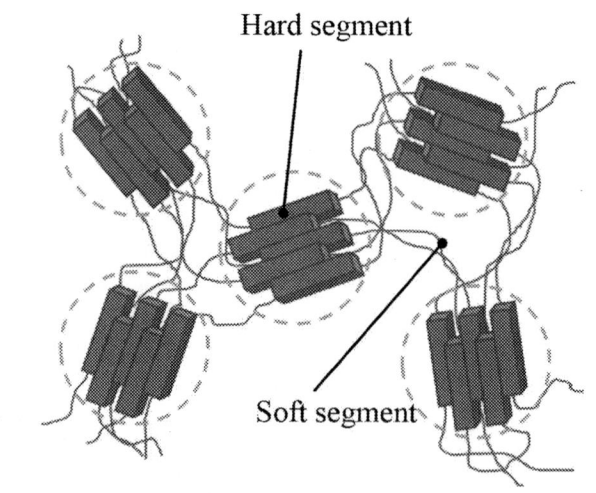

Figure 4. Model of molecular design for new resin system

General Properties of New Ultra Low CTE Core Materials

The general properties of our newly developed core material (E-770G) are shown in Table 1. As shown, we also evaluated HD (high density) S-glass in addition to E-glass and S-glass.

From the table, E-770G had ultra low CTE, high modulus, and low shrinkage. The CTE of E-770G (4.9 ppm/°C with E-glass, 2.3 ppm/°C with S-glass, 1.8 ppm/°C with HD S-glass) was lower than that of E-705G. The flexural modulus (33 GPa with E-glass, 36 GPa with S-glass, 38GPa with HD S-glass) at room temperature was similar to that of E-705G. The resin shrinkage of E-770G (0.08 %) was considerably lower than that of E-705G (0.23 %). Moreover, E-770G had excellent dielectric properties of low dielectric constant and low dissipation factor, at high frequencies. Especially, the low dissipation factor (Df: 0.005 at 1 GHz) of E-770G can be applicable to high speed PKG applications.

Table 1. General properties of ultra low CTE materials

Item	Condition	Unit	E-705G			E-770G		
Glass cloth	-	-	E-glass	S-glass	HD S-glass	E-glass	S-glass	HD S-glass
Tg	TMA	°C	260	260	260	270	270	270
CTE	a1 (X)	ppm/°C	5.9	3.3	2.8	4.9	2.3	1.8
Flexural modulus	A	GPa	33	36	38	33	36	38
Solder heat resistance	288 °C, Float	s	>300	>300	>300	>300	>300	>300
Dk	1 GHz	-	4.3	4.1	4.1	4.2	4.0	4.0
	10 GHz		4.2	4.0	4.0	4.0	3.8	3.8
Df	1 GHz	-	0.008	0.008	0.008	0.005	0.005	0.005
	10 GHz		0.010	0.010	0.010	0.007	0.007	0.007
Resin shrinkage	TMA	%	0.23	0.23	0.23	0.08	0.08	0.08

Influence of Resin Shrinkage

Generally, in cooling process after the reflow, the PKG substrate shows a convex shape due to the shrinkage of the substrate which largely depends on the core shrinkage.

So, the contribution of the resin shrinkage to the core shrinkage was examined following figure 5.

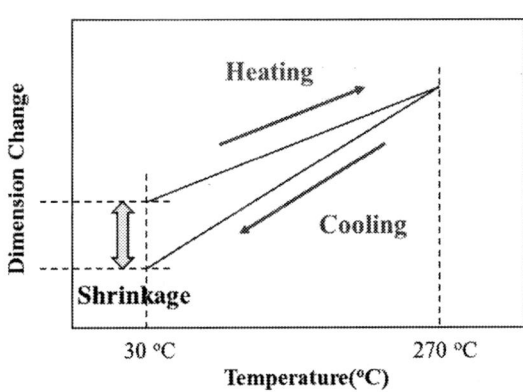

Figure 5. TMA measurement of the shrinkage of resin and core

Figure 6 shows the relationship between the shrinkage of resin and core. As obviously shown, the decreasing of resin shrinkage effectively lowers the core shrinkage.

Figure 7 shows the CTE in cooling process when varying core shrinkage. In this evaluation, core materials that have the CTE of 1.8 ppm/°C in heating process were used. As shown in the figure, the decreasing of core shrinkage brings the ultra low CTE of the substrate in the cooling process which was confirmed to reduce the warpage.

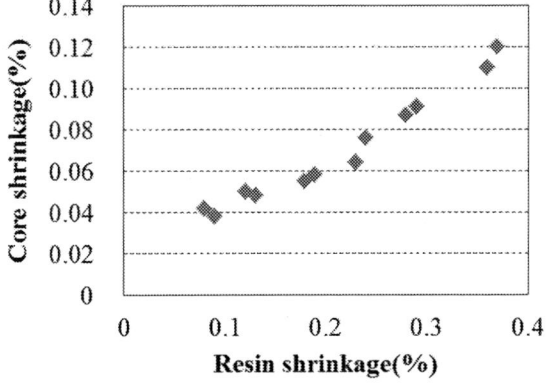

Figure 6. Relationship between the shrinkage of resin and core

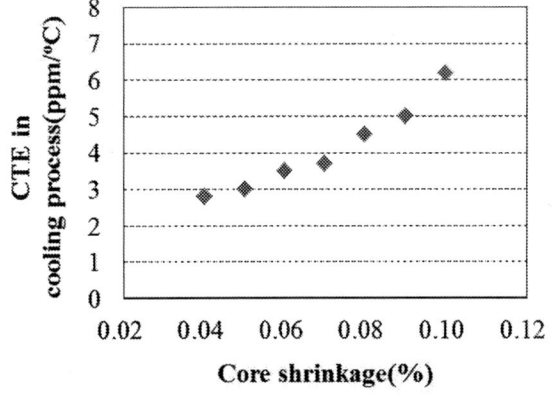

Figure 7. Relationship between the CTE in cooling process and core shrinkage

Warpage Property of Thinner PKG Substrate

The warpage property of thinner PKG substrate with simple design using E-705G and E-770G with HD S-glass was examined. Figure 8 shows the overview and cross-section of the PKG substrate (4-layer PKG construction). In this evaluation, the prepreg was used as build-up layers. After the patterning, the chip was assembled and underfilled. The sizes of the PKG substrate and the chip were 14x14x0.2 mm and 7.3x7.3x0.15 mm, respectively. The thickness of the core material and the prepreg were 200 μm and 30 μm, respectively. The warpage at the initial state of substrate, after underfilling and after dicing were measured by Shadow-Moire system. The overview of the measurement by Shadow-Moire system is shown in Figure 9.

The results of the warpage measurement at the solder ball area of PKG substrates at 25 °C and 260 °C are shown in Figure 10. As shown, the warpages of PKG substrates using core of E-770G were much smaller than those of E-705G at both temperatures. Especially, PKG substrate using E-770G material for both core and prepreg, which has much lower CTE and lower shrinkage than E-705G material, showed the lowest warpage. The warpage of E-770G material was approximately 60 % of that of E-705G material. That means the CTE and shrinkage of material will affect the warpage more strongly. We think these results explain the simulation results of the warapge and the evaluation of shrinkage well.

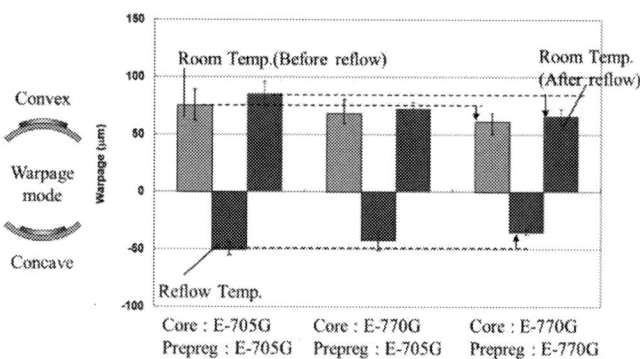

Figure 10. Measurement results of warpage of PKG substrate by Shadow-Moire system

Conclusions

We have developed new ultra low CTE core material (E-770G) and examined the warpage property of the material.

(1) The ultra low CTE and low shrinkage was achieved by applying our original resin system having hard segments and soft segments .

(2) The new ultra low CTE core material showed much lower warpage than the conventional ultra low CTE core material.

(3) From the results of the simulation and the evaluation of shrinkage and measurement of the warpage, it was demonstrated that ultra low CTE and low shrinkage lead to smaller warpage in thinner PKG substrate.

Note: The contents of this paper are based on the results of experiments and do not represent a guarantee of the values for each property.

Reference

1. M. Miyatake, "Newly Developed Ultra Low CTE Materials for thin core PKG", ECTC, 2012
2. N. Takano, "Low CTE & High Elastic Modulus Substrate for Mounting Semiconductor", The 3rd IEMT/IMC Symposium, 1999
3. N. Takano, "Filler Interface Control System for PWB Substrate", European PCB Convention, 1999.
4. K. Ikeda "New Halogen-free materials for PWB, HDI and advanced package substrate" IPC, 2000

*Package size : 14x14 μm
*Chip size : 7.3x7.3 μm
*Chip thickness : 150 μm
*Underfill thickness : 60 μm
*Solder resist : 20 μm
*Core thickness : 200 μm
*Prepreg thickness : 40 μm

Figure 8. Overview and cross-section of the PKG structure for warpage evaluation

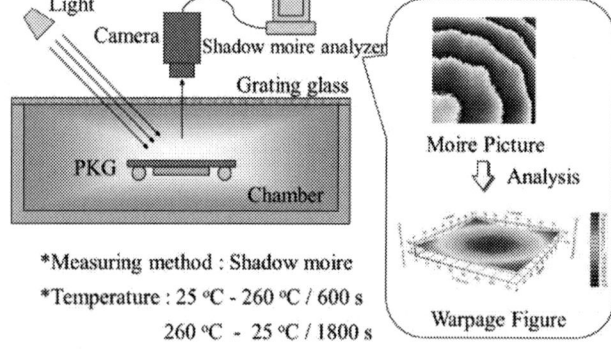

*Measuring method : Shadow moire
*Temperature : 25 ℃ - 260 ℃ / 600 s
260 ℃ - 25 ℃ / 1800 s

Figure 9. Overview of warpage evaluation by Shadow-Moire system

A Novel Redistribution Layer Tailored by Nanotwinned Copper Decreases Warpage in Wafer Level Packaging

Heng Li, Wenguo Ning, Chunsheng Zhu, Gaowei Xu, Le Luo

State Key Lab Transducer Technol, Shanghai Inst Microsyst & Informat Technol, CAS

No. 865 Changning Road, Shanghai , 200050, China

leluo@mail.sim.ac.cn

Abstract

As a common problem in wafer lever packaging(WLP), wafer warpage caused by heat process should be carefully controlled in case of product inaccuracy or yield loss, and redistribution layer (RDL), as a key structure of WLP, is one of the major concerns that causes warpage. In this paper, a novel RDL tailored by pulsed electrodeposited nanotwinned copper (nt-Cu) was introduced into WLP. It was found that grains grew larger and nt-Cu became rare in our novel RDL when it underwent 300°C annealing. Compared with traditional RDL consisting of normal electroplated copper, the novel RDL revealed quite different warpage characteristics when heating to 300°C for the first time, which was probably due to thermal stability and high yield stress of nt-Cu. Namely, twin lamina growth rather than grain growth during annealing helps nt-Cu avoid the sharp decrease of yield stress. It's very promising to take the advantage of nt-Cu to reduce wafer warpage.

I. Introduction

In wafer level packaging(WLP), warpage is an annoying problem that may lead to overall yield declining and cost rising because large warpage may cause a lot of problems during a variety of process such as photolithography, wafer bonding, chips stacking[1, 2], etc. This problem could be thornier when larger and thinner wafers are adopted for further cost down. It is crucial to control wafer warpage.

Redistribution layer (RDL) is a fundamental structure in packaging (the other one is solder bump) that relocates the IO pads of an integrated circuit. Because RDL is usually consists of large amount of metals with complex pattern, it is easy to accumulate stress and as a result, RDL becomes one of the major concerns that causes wafer warpage in WLP. For the same reason, it effects the wafer warpage in a complex way by various parameters such as line geometrical shape, angle of corner, area ratio, etc.

Generally, the stress in RDL is composed by intrinsic stress (micro-stress) and thermal stress (macro-stress) [3, 4], both of which causes warpage. The former is generated due to lattice mismatch between film and substrate, lattice defect such as dislocation, and nonuniform deformation between grains during film growth. The latter is generated by coefficient of thermal expansion (CTE) mismatch when undergoing temperature changes.

In many cases, intrinsic stress in RDL is much smaller than thermal stress when undergoing high temperature cycles, and a large amount of intrinsic stress can be released by recovery, recrystallization or grain growth when the temperature is high enough(i.e., higher than $0.4T_m$ where Tm refers to melting point of the metal) and holding time is long enough. This process is driven by thermal stress (strain energy) and largely alters materials properties such as Young's Module and yield stress, and causes elastoplastic deformation.

The evolution of mechanical parameters during heating is not entirely symmetrical with that during cooling, which results that compressive strain during heating could not be entirely counteracted by tensile strain during cooling. This is the reason of warpage accumulation during temperature cycling.

Supposing if the metal in RDL is superior thermal stable under a certain temperature (critical temperature), then its mechanical properties almost remain the same during temperature cycling, and very little plastic deformation occurs. As a result, the warpage caused by thermal cycle could be largely reduced.

Nanotwinned copper (nt-Cu), with grains tailored by high density of twin boundaries (TB), is a very suitable candidate material that meet the requirement mentioned above, because it simultaneously demonstrates superior thermal stability, ultrahigh strength and high electrical conductivity. These properties are certainly desirable in microelectronics but usually they are mutually exclusive in traditional metals [5, 6]. Varies of methods can be taken to get nt-Cu such as sputtering, electroplating by directed current (DC) or pulse reverse current, etc. [7-9].

Our work aims at introducing nt-Cu into RDL and promote its integrated performance. Compared with traditional RDL, it demonstrates that nt-Cu RDL has the capability to decrease warpage in 3D wafer level packaging (WLP).

II. Experimental

Firstly, nt-Cu RDL was fabricated, followed by microstructure analyzing. Then, wafer warpage during thermal cycling was measured in situ.

The key process of RDL fabrication is as follows. About 2000 Å Silicon dioxide was firstly deposited on the wafer, followed by sputtering 500Å TiW/2000Å Cu as a seed layer with (111) preferred orientation by DC magnetron sputtering. To reveal metal warpage properties more apparently, polyimide was omitted. Afterwards, 20um photosensitive polyimide was coated and trenches were etched as a mask for electro deposition. At last, 20um thick and 400um long copper line array were deposited by pulse electroplating (PED) on the wafer, and the width of line array was range from 10um to 58um.

The pulsed electrodeposition was conducted on Pulse Reverse Power Supply from Advanced interconnected Technology Ltd. High density of nanotwinned copper lines was achieved under the peak current density of 140 ASD (A/dm2) with 5ms on-time and 99ms off-time. The composition of the electrolyte is 6 L of Liquid Copper Sulfate, 2.4 L of concentrated Sulfuric Acid, 15L DI water, and additives such as brighter. The electrolyte was kept at room temperature and mechanically stirred.

To evaluate nt-Cu warpage properties, two more samples were prepared for comparison. All the samples were same processed except electrodeposition parameters, as shown in the following table. J_p and J_a represents peak current density and average current density respectively, and duty cycle refers to $T_{on}/(T_{on}+T_{off})$. Sample A and B were DC electroplated and sample C and B were under the same average current density, as shown in Table 1.

Table 1. Three samples and their electroplating parameters.

Sample	Electroplate	J_p (asd)	Duty cycle	J_a (asd)
A	DC plate	0.7	1	0.7
B	DC plate	7	1	7
C	PED	140	5/(99+5)	7

Microstructure analyses were conducted before and after annealing. Focused ion beam and scanning electron microscopy dual system (FIB/SEM) gave a view of RDL cross section, illustrating nanoscaled twins clearly. Dark field TEM images and HRTEM images with a clear twin electronic diffraction pattern demonstrated explicit coherent twin boundaries with nanoscaled thickness.

Warpage of three groups of samples were measured in situ during RT-300°C temperature cycling by Multi-beam Optical Sensor (MOS) system from k-Space associate Inc. At last, the stress distribution was analyzed.

III. Results And Discussion

(A) In situ warpage measurement

Traditionally, a high temperature annealing process is taken to enlarge copper grains in RDL in order to decrease electrical resistance, so it is important to analyze wafer warpage during thermal cycles.

Twelve laser points (3 rows times 4 column)was projected on the center of the wafer all the time and the distance change of reflected spot on CCD was measured to calculated warpage. Therefore, the warpage just represent the center area of the wafer, as shown in figure 1.

During the first time of heating, the curve of sample A and B could approximately be divided by 4 stages. (1) Linear rising of elastic deformation. The range for A was 110°C and B was 120°C. (2)Nonlinear rising to the maximum. Warpage of sample A reached climax at 140°C and B at 160°C, and large elastoplastic deformation occurred during this period. (3) Sharp declining. Vigorously evolution happed in grains and yield stress decreased sharply. This range for A is 280°C and B is 210°C. (4)Stable stage. Grains equilibrated under new temperature without large new warpage accumulated.

However, for sample C in Fig. 1(c), no sharp declining stage (3th stage) could be found. It meant no vigorously grain evolution happed and no yield sharply stress decreased. This difference showed the thermal stability of nt-Cu and could be used to decrease warpage caused by thermal cycle. It can assume that if the nt-Cu is more stable, then the warpage can be decreased further.

By microstructure observation, it was found that grains grew large after 300°C thermal cycle, and no twins could be found again in sample C. Perhaps it was twin lamina growth rather grain growth that avoided sharp decrease in figure4(c).

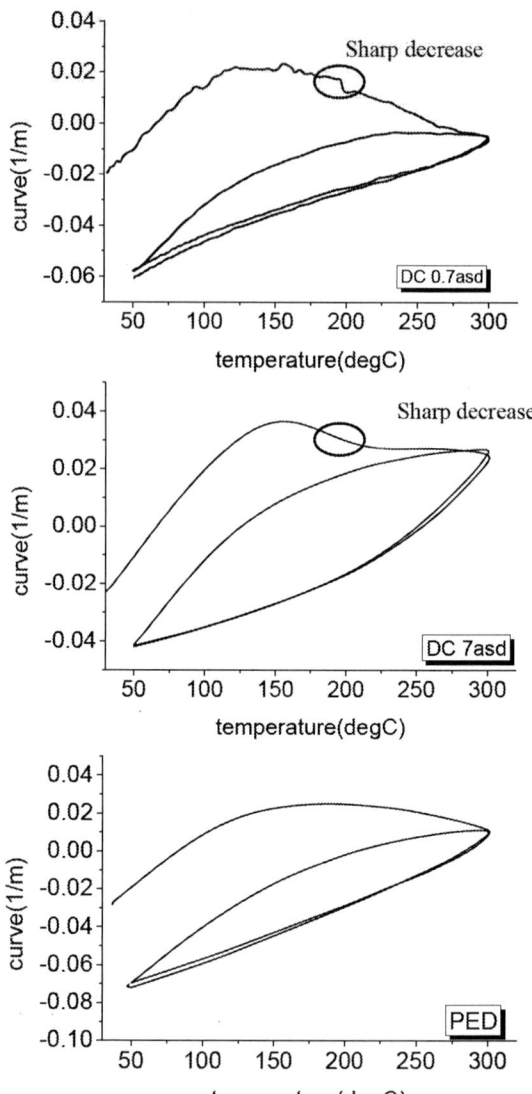

Figure 1. Three samples warpage during thermal cycle. Fig (a) (b) and (c) represented sample A (0.7asd DC), B (7asd DC) and C (7asd PED) successively. A sharp decrease stage (in the circle of the figure) could be found in A and B but not C. All profiles were smoothed by box filter and B-spline.

In the following cycles, a circle was repeatedly appeared in all samples and no obvious difference can be found, suggesting all grains were stable and twin lamina may disappeared gradually during high temperature.

(B) Nt-Cu observation

SEM/FIB dual beam system is a very useful tool in microelectronics industry, such as nanomanipulation like welding nanowires, drilling holes, and take ion/electric images. Ion image of FIB has good contrast to grain orientation that can be very timesaving and convenient for materials analyses.

In the present work, FIB was used to make a cross section of RDL, and take ion image to detect nanoscale twins. In sample C, a high density of TBs was observed before annealing, as showed in figure 2, while in sample A and B, few TBs was found before and after anneal, and the FIB images without TBs are omitted here. To enhance the contrast of

978-1-4799-2408-0/14 $31.00 © 2014 IEEE

different grain orientation, a shallow ion etch was conducted by FIB with 80pA etching ion current for 45seconds. Ion images were taken with 10pA ion current and 300μs dwelling time, and the magnification was ×12000.

Figure 2. FIB image of sample C before annealing, showing the thickness of paralleled twin lamination was from several to hundreds of nanometers. (a) Top view with 52°tilt and (b) cross section view of RDL.

In Fig. 2(a), it was found high density of TBs on the surface of RDL lines, with its orientation distributed randomly, and in fig. 1(b), the size of lower grains was very small while the upper grains were much larger. Similarly, there were scarce TBs within lower grains but high density of TBs within the upper grains. This phenomenon was also seen by Tao-Chi Liu, etc.[8]

Grains at the bottom of the line were very small because grain size of sputtered seedlayer was usually very small, which offered lots of nucleation center during electroplating. In the subsequent deposition, film growth followed the volmer-weber (VW) mode, leading to growth of grain size and increment of inner stress or distortion energy within a grain. While the stress was high enough to drive atoms to form TBs but not enough to form grain boundaries (GBs), TBs would appear. This also meant larger grains with higher inner stress could lead to higher density of TBs.

To take an inner observation and further analyses of TBs, High-resolution TEM (HRTEM) images were taken, as showed

in figure 3. A 59.4nm thick TEM sample was prepared by FIB. Figure 3 (a) was a TEM dark field image of the GBs and TBs of the nt-Cu, and figure 3(b) was a HRTEM image of nanoscale twins, and the inset is an electron diffraction pattern.

Though many defects were found in the HRTEM image, it was still easy to distinguish TBs. The electron diffraction pattern gave a convinced demonstration of the TBs. Combined with these images, it could tell that there were high density nanoscale twins in our RDL.

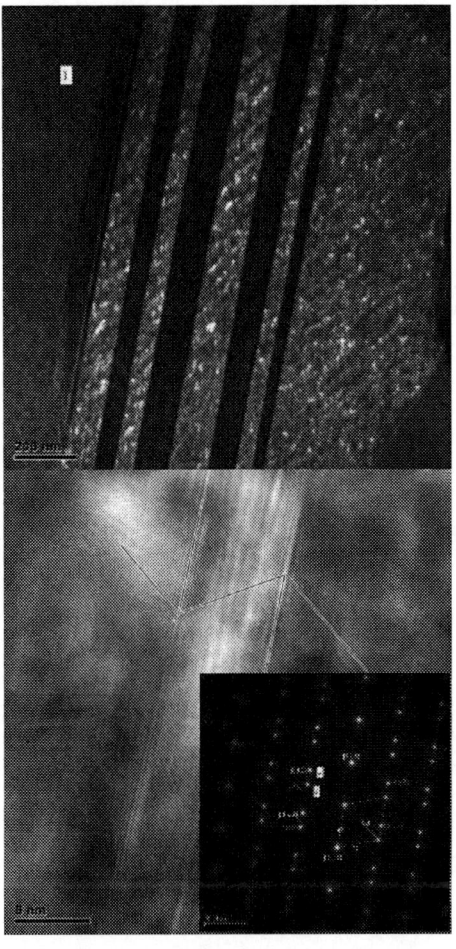

Figure 3. TEM image of sample C. (a) A TEM dark field image of the GBs and TBs of the nt-Cu. (b) HRTEM image of nanoscale coherent TBs with an electron diffraction pattern at lower right corner.

(C) Grain evolution during thermal cycle

During thermal cycle, grain evolution occurred and microstructures changed. SEM image of sample B and C was displayed in figure 4, as shown below.

SEM image of Sample A was similar with C, so its SEM images were omitted.

Sample B was amorphous copper electroplated at 7asd by DC. Amorphous copper was formed under too large current density, which increased the nucleation rate sharply and there was not enough time for grains growing up. After thermal cycle, the surface was smoothed.

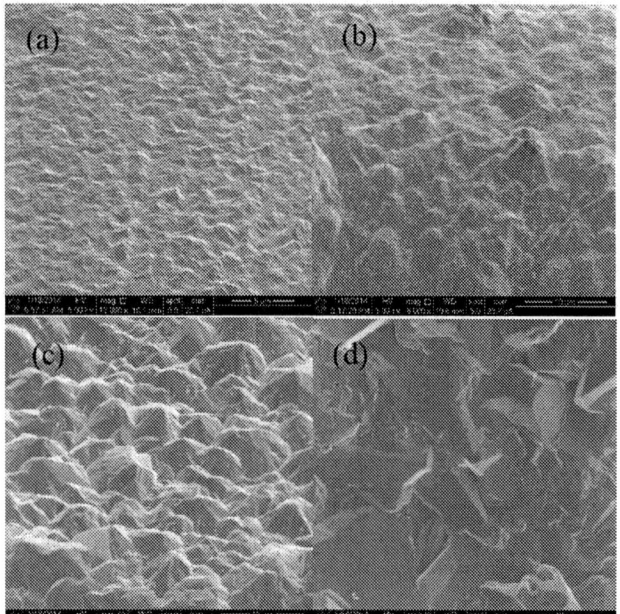

Figure 4. SEM image of copper grains. Sample A were similar with C and its SEM images were omitted here. Fig (a) and (b) were SEM image of sample B before and after 300°C thermal cycle respectively, and Fig (c) and (d) were sample C before and after 300°C thermal cycle respectively.

Sample C was PED nt-Cu. After thermal cycle, grains apparently grew large, which caused plastic deformation. The average grain diameter grew from 6.27um before anneal to 12.77um. At the same time, twins were almost disappeared, as revealed by FIB image. From Hall-Petch principle (yield stress decreases with the grain size enlarging), it could be estimated that the yield stress decrease, which was also demonstrated in figure 1(c). For the same reason, it could be concluded that much more vigorous grain evolution was occurred in sample A and B during the sharp warpage decline stage, as figure 1 showed. This phenomenon demonstrated that nt-Cu was relatively thermal stable.

In fact, energy of CTBs (coherent TBs) and ITBs (incoherent TBs) are about one-tenth and half of GBs energy respectively[9], and hence TBs have a drastically reduced driving force for the coarsening of twins. It could also be explained by atoms movement. During heat, TBs atoms just need a short distance to get to the new equilibrium state.

Nanotwins were not only thermally stable by themselves, but their existence may also improve the thermal stability of fine grains where GBs migration is limited by the low propagation rate of intercepting CTB [10-12].

(D) Stress distribution analyze

Experiments show that wafer curvature changes nonlinearly with temperature increases/decreases because of inelastic deformation occurred at high temperature due to evolution of grain structures or plastic deformation. Inelastic deformation during high temperature (e.g. the annealing process) leads to residual stress when cooling down to room temperature (RT), and residual stress leads to warpage of wafer, which could be found by Stoney Formula indicated.

$$\sigma_f = \frac{E_s h_s^2}{6 h_f (1 - \nu_s)} (\kappa - \kappa_0) \qquad (1)$$

Where $\sigma, E, h, \nu, \kappa$ refers to stress, Young's Modulus, film thickness, poisson ratio, and warpage, and subtitle s and f refers to substrate and film (i.e., the RDL).

As mentioned above, RDL stress σ_f was formed by two parts, the intrinsic stress σ_{int} and thermal stress σ_{th}, and thermal stress was generated by CTE mismatch.

$$\sigma_f = \sigma_{int} + \sigma_{th} \qquad (2)$$

$$\sigma_{th} = E_{ef} \int_{T_0}^{T} (\alpha_s - \alpha_f) d_T = \frac{(\alpha_s - \alpha_f)(T - T_0) E_f}{1 - \nu_f} \qquad (3)$$

Where E_{ef} refers to effective Young's Modulus and α refers to CTE. During temperature cycling, all parameters can be considered as constant except Young's Modulus of the metal, which will change with temperatures and deformation. When combined formula 1-3 and omitted σ_{int}, a formula could be derived to estimate accumulated warpage:

$$\kappa = \kappa_0 + \frac{6 h_f (1 - \nu_s)(\alpha_s - \alpha_f)}{E_s h_s^2 (1 - \nu_f)} \int_{T_0}^{T} E_f d_T \qquad (4)$$

The analyses above suggests that if E_f is stable during thermal cycle, new warpage can be limited when temperature return to T0. The Nonlinear and asymmetric relationship between E_f and temperature would cause large inelastic deformation of traditional RDL and warpage in wafer level packaging. However, traditional strengthen method such as grain refining and precipitation strengthening is useless for it either sacrifices conductivity or plasticity. Moreover, because of more stable thermal stability, nt-Cu could limited wafer warpage during thermal cycle.

IV. Future work

As for Young's Modulus of nt-Cu, it is highly correlated with twin lamina thickness and nt-Cu with critical twin space displays the best mechanical properties[13], so it's very meaningful to control twin lamina thickness.

Nt-Cu can be formed by PED with a large range of pulse electronic density, but in this experiment, best nt-Cu were made by PED with 140asd, which is a relatively high current. Kinds of additives will be adopted in the future to decrease current density when depositing nt-Cu[13].

V. Conclusions

High density nt-Cu RDL was achieved by PED in WLP, which was the first ever successful attempt. In situ wafer warpage measurement during thermal cycle revealed that during temperature rising, a general sharp warpage decrease stage was omitted in nt-Cu samples. Twin lamina growth rather than grain growth at high temperature helps nt-Cu avoid the sharp decrease of yield stress. Nt-Cu properties is quite acceptable such as larger elastic range, higher Young's modulus, higher yield stress and better thermal stability. In fact, nt-Cu is a very promising material candidate for future microelectronics.

Acknowledgments

The work is supported by National Science and Technology Major Project (NO.2009ZX02025, NO. 2011ZX02602).

References

1. Tu, K.N., *Reliability challenges in 3D IC packaging technology.* Microelectronics Reliability, 2011. **51**(3): p. 517-523.

2. van Driel, W.D., et al., *Prediction and verification of process induced warpage of electronic packages.* Microelectronics Reliability, 2003. **43**(5): p. 765-774.

3. Vinci, R.P., E.M. Zielinski, and J.C. Bravman, *Thermal Strain and Stress in Copper Thin-Films.* Thin Solid Films, 1995. **262**(1-2): p. 142-153.

4. Keller, R.M., S.P. Baker, and E. Arzt, *Quantitative analysis of strengthening mechanisms in thin Cu films: Effects of film thickness, grain size, and passivation.* Journal of Materials Research, 1998. **13**(5): p. 1307-1317.

5. Lu, L., et al., *Ultrahigh strength and high electrical conductivity in copper.* Science, 2004. **304**(5669): p. 422-6.

6. Jang, D., et al., *Deformation mechanisms in nanotwinned metal nanopillars.* Nat Nanotechnol, 2012. **7**(9): p. 594-601.

7. Xie, Q., et al., *The effect of sputtered W-based carbide diffusion barriers on the thermal stability and void formation in copper thin films.* Microelectronic Engineering, 2010. **87**(12): p. 2535-2539.

8. Liu, T.-C., et al., *Fabrication and Characterization of (111)-Oriented and Nanotwinned Cu by Dc Electrodeposition.* Crystal Growth & Design, 2012. **12**(10): p. 5012-5016.

9. Lu, L., et al., *Revealing the maximum strength in nanotwinned copper.* Science, 2009. **323**(5914): p. 607-10.

10. Zhang, X., et al., *Thermal stability of sputter-deposited 330 austenitic stainless-steel thin films with nanoscale growth twins.* Applied Physics Letters, 2005. **87**(23).

11. Saldana, C., et al., *Stabilizing nanostructured materials by coherent nanotwins and their grain boundary triple junction drag.* Applied Physics Letters, 2009. **94**(2).

12. Zhang, X. and A. Misra, *Superior thermal stability of coherent twin boundaries in nanotwinned metals.* Scripta Materialia, 2012. **66**(11): p. 860-865.

13. Zhu, L., et al., *Modeling grain size dependent optimal twin spacing for achieving ultimate high strength and related high ductility in nanotwinned metals.* Acta Materialia, 2011. **59**(14): p. 5544-5557.

Demonstration of 3-5 μm RDL Line Lithography on Panel-based Glass Interposers

Hao Lu, Yutaka Takagi+, Yuya Suzuki, Brett Sawyer, Robin Taylor^, Venky Sundaram, Rao Tummala

3D Systems Packaging Research Center

813 Ferst Drive NW, Georgia Institute of Technology, Atlanta, GA, USA

+ NGK Spark Plug Co., Ltd. Japan

^Atotech GmbH, Berlin, Germany

Abstract

Interposer technology is becoming important to interconnect ultra-high performance ICs with ultra-high density I/Os. Silicon interposers fabricated by back-end of line (BEOL) wafer processes address these wiring density requirements, but are limited by their high cost and by their high electrical losses. Organic interposers have limitations too. Their limitations are due to their poor dimensional stability, which require larger capture pads, which limit the I/O density Glass has been proposed by Georgia Tech [1-4] as a superior interposer material to address the limitations of both silicon and organic interposers in recent years. This paper describes the first demonstration of low cost and double sided glass interposer with 3-5 μm line lithography to form multilayer redistribution layers (RDL) to achieve 20 micron bump pitch, ready for chip-level copper interconnections. Unlike prior work using wafer based RDL processes or thin film wiring applied to organic cores, this research applies low cost laminate-like processes, scalable to large panels for lowest cost. To achieve this, semi-additive plating (SAP) process, combined with high resolution dry film photoresists, was optimized to fabricate fine-pitch copper traces with 3-5 μm lines on thin polymer films, laminated on thin glass. Such an ultra-high I/O density of interconnections form the backbone of 2.5D interposers, interconnecting multiple chips. Such ultra-small pitch copper traces can reduce the number of wiring layers required, thus reducing the cost. Compared to sub-micron Cu traces on Si interposers, these 3-5 μm lines are lower in resistance.

I. Introduction

The current approaches to high-density 2.5D and 3D interposers are either based on incrementally extending organic substrate technologies, or based on silicon carriers utilizing back-end of the line (BEOL) tools and processes. Traditional organic substrates are limited by coarse pitch wiring due to their poor dimensional and thermal stabilities and by high surface roughness and non-planarity. Recent advances in low coefficient of thermal expansion (CTE) organic substrates have pushed the line width down to 8 μm [5]. To achieve much higher wiring densities resulting in much smaller pitch, silicon interposers have been developed. Silicon, unlike organic substrates, has excellent dimensional and thermal stabilities as well as matched-CTE to the silicon chip, in addition to providing ultra-smooth surfaces to form sub-micron wiring structures. Xilinx used a 65 nm back-end of line (BEOL) processes to form the first 2.5D silicon interposer technology with 40 μm interconnection bump pitch [6,7]. However, this technology, requiring BEOL wafer processes on small 300 mm wafers, make Si interposers too expensive for most applications. Thin film process technology using liquid dielectrics, sputtered metals, and chemical-mechanical polish (CMP) processes have also been applied to form fine Cu wiring traces on silicon interposers with 0.8μm wiring lines using Semi-Additive Plating (SAP) method [8]. However, such fabricated Cu wiring with very low copper thickness, limited by wafer processing, raises line resistance concerns.

Georgia Tech Packaging Research Center (PRC) and its industrial partners have pioneered large, panel-based glass, as a lower cost and higher performance alternative, to overcome the challenges with organic and silicon interposers. Figure 1 shows the schematic of the proposed 2.5D glass interposer.. Glass has dimensional and thermal stabilities very similar to silicon as well surface smoothness to form ultra-small wiring traces, but superior in electrical performance than silicon due to its low loss and ultra-high resistivity, and in cost, due to very large area (8X) processability [1-3]. The insertion loss of TPVs in glass is lower than in traditional silicon interposers. Modeling, design and fabrication of ultra-thin glass interposers with fine pitch through-package-vias (TPVs) have been successfully demonstrated [1-4]. However, ultra-fine pitch and low cost RDL layers on glass interposers have not been demonstrated. This paper presents the first demonstration of 3-5 μm RDL wiring structures on glass interposers. The key materials, processes and tool advances to form low-cost RDL wiring layers with 3-5 μm lines, leading to a first demonstration of a 2.5D glass interposer are described.

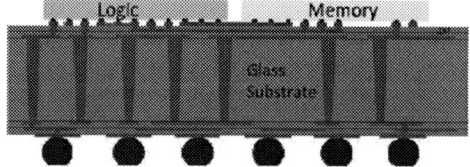

Figure 1. Schematic of 2.5D glass interposer

Semi-Additive Plating (SAP) process was advanced to demonstrate 3-5 μm RDL structures. The challenging steps in SAP processes are photo lithographic process and copper seed layer etching step. These challenges limit the minimum feature size of RDL structures. In lithographic processes, the pattern-resolution is determined by the lithography tool, resolution of photoresist, adhesion of photoresist to the substrate, and the surface roughness of the substrate. These requirements drive advances in large-area and high-resolution lithographic tools, photoresists, and new dielectric polymers with low surface roughness. The seed layer etching step determines the copper RDL structure integrity. Isotropic wet etching processes attack both seed copper layer and

978-1-4799-2408-0/14 $31.00 © 2014 IEEE

electroplated copper at the same speed. In reality, the copper seed layer between two adjacent copper traces is more difficult to etch than in open area, resulting in either under-etch or over-etch.

The rest of this paper is organized as follows. In Section II, the glass substrate preparation is discussed. Section III demonstrates a low-cost, high-resolution lithography process. Fine-line RDL fabrication was demonstrated through SAP using a novel seed-layer etch technology in Section IV. The key findings are summarized in Section V.

II. Glass Substrate Handling for RDL Layer Fabrication

Thin glass substrates are brittle and can easily be broken during the fabrication processes, compared to organic substrates. An innovative approach to handling of glass during processing by using polymer buildup layers on glass was developed by Georgia Tech PRC [1]. The laminated polymer material acts as a buffer layer between the metal layer and glass core, mitigating the CTE mismatch issues between copper and glass. Besides, this polymer layer also has good adhesion strength to both glass core and copper metal layer, with appropriate surface treatments. ZEONIF™ ZS-100 (ZIF) polymer from ZEON Corporation is applied as the polymer material [9]. The glass substrate is covered by ZIF polymer, which is compatible with existing electroless copper seed plating, and yet has a smoother surface than traditional build-up polymers such as epoxies, which enables improved lithographic resolution. The glass core was cleaned by a custom surface treatment process to ensure sufficient adhesion between polymer and glass. If this important step is skipped, the laminated ZIF polymer may delaminate during the following wet processes and in reliability testing. The lamination process was performed in a vacuum-laminator, followed by a hot-press process to ensure surface planarity of the polymer film. After lamination, the ZIF polymer was thermally cured in an oven at 180C for one hour. The cured ZIF polymer was then treated by permanganate desmear process to enhance the copper-to-polymer adhesion, ensuring the reliability of fine RDL structures on the ZIF polymer layer. SAP process requires a copper seed layer to enable electrolytic plating. In the current approach, the copper seed layer was formed by electroless plating, which is a fast and low-cost wet process, scalable to large panels and double side processing.

III. High-Resolution Lithography

The high resolution lithography process is an important factor that determines the limits of the SAP method for fine line RDL fabrication. The resolution is determined by the lithography tool, the photoresist, adhesion of photoresist to the substrate, and the roughness of the substrate. Compared to traditional liquid photoresists, dry film resists (DFR) have lower cost and large panel scalability for double side processing. A high resolution dry film resist from Hitachi Chemical [10] with a thickness of 15 μm was used in the current study. In addition, an adhesion promoter treatment [11] called Novalink™ from Atotech was applied to increase the adhesion of the photoresist to the smooth copper seed layer surface. With the appropriate lithography machine, high

resolution DFR, and proper surface treatment, the resolution of lithography process for fine-line patterns can be ensured.

A glass photomask consisting of comb structures, escape-routing structures, and long coplanar waveguide (CPW) lines to form the RDL on glass interposers, was designed. Two different lithography processes were tested in this study as shown by the process flows in Figure 2. The difference between the two processes was in the surface treatment before DFR lamination. In process A, the sample was chemically-treated in a tank containing sulfuric acid to remove any oxidized copper on the surface. In process B, Atotech's adhesion promoter treatment (Novalink) was applied. This surface treatment forms a chemically active adhesion promoter layer for bonding enhancement of the photoresist to the copper surface. The developed DFR structures are, therefore, less likely to delaminate, and thus increase the yield.

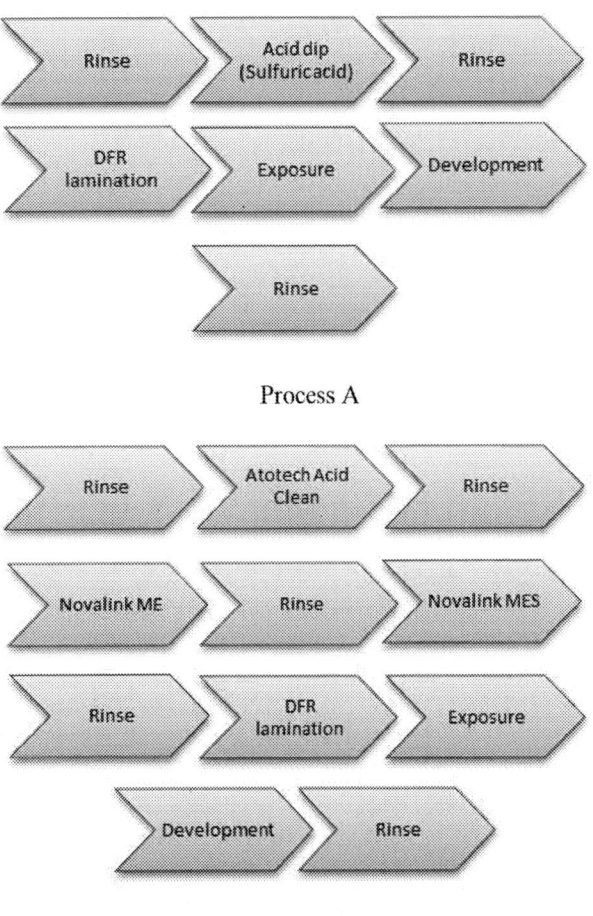

Figure 2. Comparison of two lithography processes

A mask with cross-patterns containing different line-widths was utilized to test the yield of the lithography processes A and B. The mask consisted of 36 patterns with the same line width to assess the process yield and uniformity. A Tamarack 152R high resolution contact mode lithographic tool with a collimated UV light source was used for imaging. A good DFR pattern on the copper seed layer, and a typical failed pattern where the DFR structure delaminated and collapsed, is shown in Figure 3. The comparison of lithography yield is shown in Figure 4. For 5 µm line width, the Novalink adhesion promotion treatment increased the yield from 16.7% to 66.7%. The adhesion of DFR on the copper surface is an essential factor to obtain high yield in fine-pattern lithography. When the line width becomes larger, the adhesion strength of DFR becomes less important in the lithography yield, as indicated by similar yields for 7 µm lines with and without surface treatment.

To demonstrate the processability of fine line RDL on glass interposers, a photo mask with fine line and space comb structures, escape routing patterns, and CPW lines was designed and the adhesion promoter treatment was applied for high resolution lithography process. Some of the mask patterns are shown in Figure 5, and the SEM images of DFR structures are shown in Figure 6.

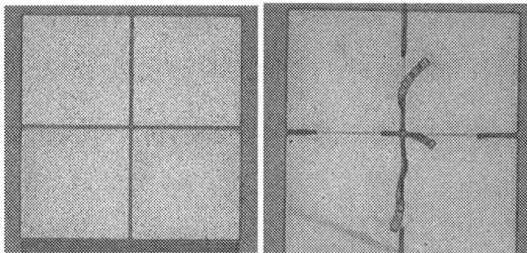

Figure 3. Left: 5 µm width DFR cross pattern. Right: delaminated DFR cross pattern

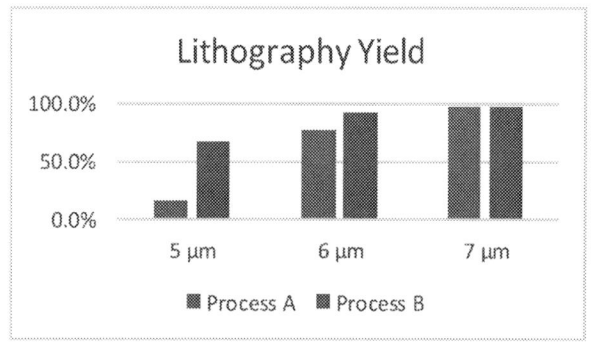

Figure 4. Lithography yield of cross patterns

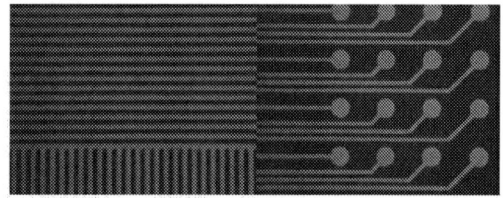

Figure 5. Left: 5/5 µm line/space patterns; Right: escape routing of 50 µm pitch bump and 4/4 µm line/space

To further increase the lithographic resolution, Ushio's new advanced projection lithography tool Ushio UX-44101 [12] was set up at Georgia Tech PRC. A resolution of 3 µm was achieved using this tool with the same DFR and Novalink treatment described above.

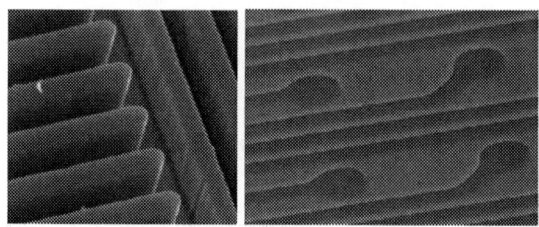

Figure 6. Left: 6/6 µm line/space DFR patterns with 15 µm thickness; Right: 5/5 µm line/space escape routing with 50 µm bump pitch

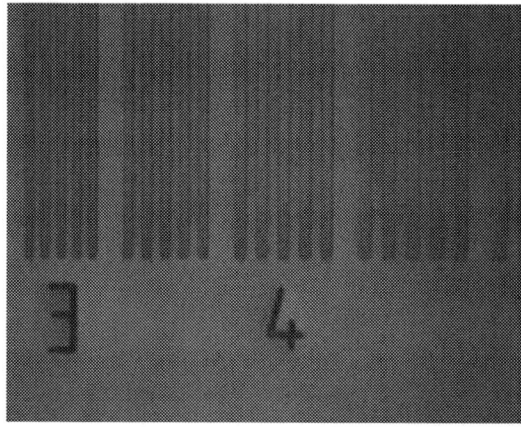

Figure 7. DFR patterns down to 3 µm line and space exposed by Ushio's lithographic tool.

IV. Fine Line Metallization on Glass

Following the seed layer deposition and lithography process steps, the copper structures on glass interposer were formed by electrolytic pattern plating. To improve the plating quality and yield, the sample was subjected to a short oxygen plasma etching process to remove any DFR residue in the open areas. The copper structure thickness is controlled by plating time and current, while the thickness uniformity is determined by the additives in the plating chemistry. With an optimized ratio of inorganic and organic additives, the plated copper structures are smooth, bright, and have good thickness uniformity. After electrolytic copper plating, the DFR was removed by a stripper solution.

The copper seed layer was then chemically etched to obtain the designed RDL patterns. This final step is critical and determines the processability and the yield of the SAP method for fine line RDL fabrication. The general wet etching process is isotropic, and the electroplated copper lines are also etched during the removal of the copper seed layer. Ideally, an anisotropic process is needed where the etching speed of the electroless plated seed layer is significantly higher than that of the electroplated copper. For ultra-fine lines and spaces below 10 µm, the seed layer etching process has further challenges in the ability of the etching solution to penetrate the narrow and high aspect ratio trenches between the copper traces. As a result, the copper traces are etched faster than the seed layer in the gaps between the traces. This causes over etching of the copper traces and undercut, leading to loss of line width and damage to ultra-fine traces. Figure 8 shows 6 μm/6 μm line/space test comb-structures, and Figure 9 shows the 6 μm/6 μm line/space escape-routing structures. Figure 10 shows the CPW transmission line on a glass interposer with 7.84 μm signal line width and 6.37 μm gap. The results suggest that with the current isotropic seed-layer etching process, the RDL traces become narrower and the line cross-sectional shape is not rectangular.

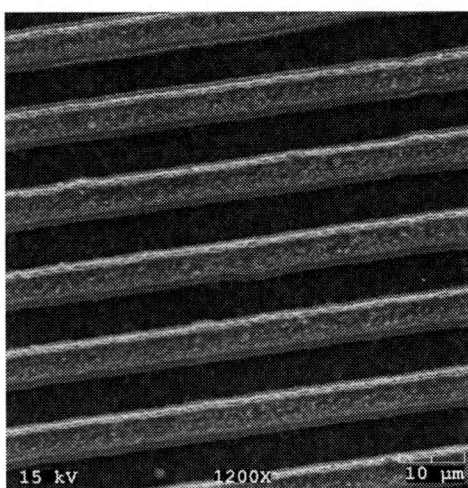

Figure 8. SEM image of 6/6 μm line/space copper traces obtained by isotropic seed layer etching

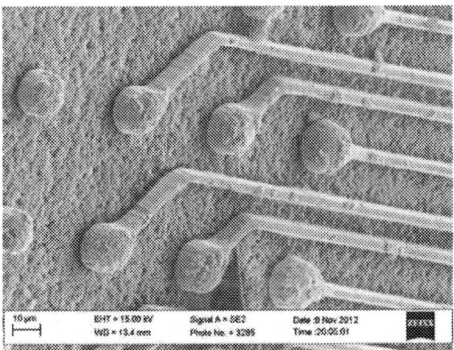

Figure 9. SEM image of 6/6 μm line/space escape routing obtained by isotropic seed layer etching

Two novel anisotropic etching methods were used to address this challenge, SPS spray etching and differential etching, provided by Atotech. A comparison of these two etching methods with the isotropic etching described earlier is shown in Table 1. The differential etch method has higher etch-selectivity than the SPS spray etch method, and is preferred for fine line RDL fabrication. The SEM image and cross-section of 4/4 μm line/space comb structures obtained by differential etch is shown in Figure 11. No undercut is observed after seed layer etching, and the line profile is sharp. However, the line width is slightly smaller than the space, indicating a slight over etch. This can be optimized by controlling the etching time and pressure.

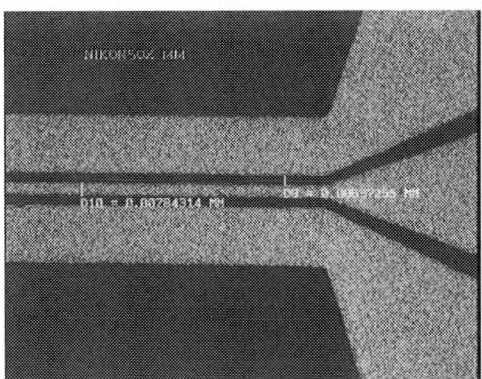

Figure 10. CPW transmission line on glass interposer with 7.84 μm signal line width and 6.37 μm gap

Table 1. Comparison of different etching methods.

Etching method	Copper etch -speed	Etch-speed control
Isotropic: Same etch speed	Independent of copper type or quality	Independent of the direction (vertical=lateral)
SPS Spray Etch: Etch -speed varies	Independent of copper type or quality	Dependent on the direction (vertical>lateral)
Differential Etch: Etch-speed varies	Dependent on copper quality or type	Dependent on the direction (vertical>lateral)

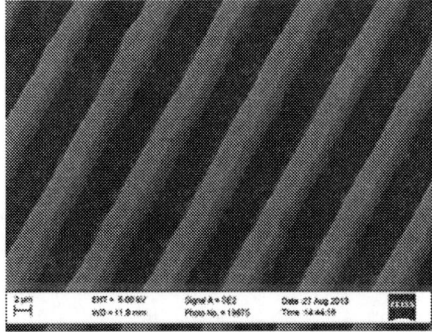

(a)

(b)

Figure 11. (a) SEM image of 4/4 μm line/space copper traces; (b) Cross-sectional image of 4/4 μm line/space copper traces

A third anisotropic etch process from Solid State Equipment Corporation (SSEC) with end point detection built into a spray etch tool was also evaluated. This unique system uses high resolution optical inspection technology to monitor the sample during etching and compares it to the pre-defined endpoint. The etching process stops when the sample reaches the end point. The endpoint reference setting is important and determines the final etch process parameters. The copper comb structure with 5/5 µm line/space fabricated with the SSEC etching tool is shown in Figure 12. Due to the precise control during seed layer etching, over-etch was minimized, and the line width was equal to the line space.

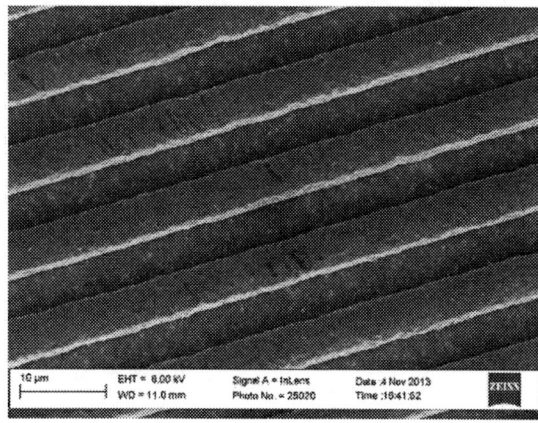

Figure 12. SEM image of 5/5 µm line/space copper traces

V.   Summary and Conclusion

This paper reports the first demonstration of 3-5 µm RDL wiring structures on glass interposers. High-resolution lithography process with up to 3 µm pattern-definition on DFR was carried out. Different copper seed-layer etching technologies were demonstrated to improve the line-definition and minimize the undercut. In addition, this paper, describes tools, materials, and process advances in SAP methods to fabricate RDL on glass with up to 3-5 µm structures. The reliability testing of this RDL on glass is ongoing.

Acknowledgments

The authors wish to acknowledge the industry sponsors (Atotech, SSEC, Hitachi Chemical and Ushio) for supporting this task and PRC staff (Chris White and Jason Bishop) for help with fabrication.

REFRENCES

1. Vijay Sukumaran, Qiao Chen, Fuhan Liu et al., "Through-package-via formation and metallization of glass interposers," in *Proceedings of 61st Electronic Components and Technology Conference (ECTC)*, 2011, pp.557-pp.563.
2. Vijay Sukumaran, et al., "Low-Cost Thin Glass Interposers as a Superior Alternative to Silicon and Organic Interposers for Packaging of 3-D ICs," *IEEE Transactions on Components, Packaging and Manufacturing Technology*, Vol. 2, No. 9, 2012, pp.1426-pp.1433.
3. Vivek Sridharan, Sunghwan Min, Venky Sundaram et al., "Designing and fabrication of bandpass filters in glass interposer with through-package-vias (TPV)," in

Proceedings of 60th Electronic Components and Technology Conference (ECTC), 2010, pp.530-pp.535.
4. Vijay Sukumaran, Tapobrata Bandyopadhyay, Qiao Chen et al., "Design, fabrication and characterization of low-cost glass interposers with fine-pitch through-package-vias," in *Proceedings of 61st Electronic Components and Technology Conference (ECTC)*, 2011, pp.583-pp.588.
5. Katsura Hayashi, et al., "Advanced surface laminar circuits using newly developed resins," *IEEE Transactions on Components, Packaging and Manufacturing Technology*, Vol. 1, No. 12, 2011, pp.1908-pp.1915.
6. Kirk Saban, "Xilinx stacked silicon interconnect technology delivers breakthrough FPGA capacity, bandwidth, and power efficiency," in WP380 (v1.2), December 2012.
7. Bahareh Banijamali, et al., "Outstanding and innovative reliability study of 3D TSV interposer and fine pitch solder micro-bumps," in *Proceedings of 62nd Electronic Components and Technology Conference (ECTC)*, 2012, pp.309-pp.314
8. Masahiro Sunohara, et al., "Development of silicon module with TSVs and global wiring (L/S=0.8/0.8µm)," in *Proceedings of 59th Electronic Components and Technology Conference (ECTC)*, 2009, pp.25-pp.31.
9. Zeon Corp. (2013) "ZEONIF Insulation Materials for Printed Circuit Board," http://www.zeon.com.jp/business_e/enterprise/imagelec/zeonif.html
10. Hitachi Chemical Co., Ltd. (2013). "RY Series for PKG Board," http://www.hitachi-chem.co.jp/english/products/pm/017.html
11. Atotech Inc. (2013) "Products," http://www.atotech.com/products.html
12. USHIO Inc. (2013) Lighting Edge Technologies, http://www.ushio.co.jp/en/index.html

Characterization of Thin Polymer Films with the Focus on Lateral Stress and Mechanical Properties and their Relevance to Microelectronics

Markus Woehrmann[1], Thorsten Fischer[2], Hans Walter[2], Michael Toepper[2]; Klaus-Dieter Lang[1]

[1]Technische Universität Berlin, Berlin, Germany
[2]Fraunhofer Research Institution for Reliability and Microintegration IZM, Berlin, Germany
Markus.Woehrmann@tu-berlin.de

Abstract

Thin film polymers play an essential role in system integration. The mechanical properties of the polymers are crucial for 3-D-Integration and advanced WLP because with the thinning of the silicon wafers, i.e. chips to less than 150 µm, the influence of the polymer layers gets an increasing impact on the mechanical stability of the electronic device. Next generation polymers have entered the market which are tailored to reach the further optimized mechanical property parameter set. This paper will give a guideline for the choice of the optimal polymer based on the demands of the application in relation to the material properties.

The main material properties for high reliability are the Young's modulus, tensile strength, elongation at break and coefficient of thermal expansion. Aside of the material properties of the polymer the interaction of the polymer layer with the substrate is important. The material mismatch causes for example warpage and material cracking; the main impact factor being the residual stress in the layers in relation to the fracture toughness of the material and the interface. The warpage is an issue for the processing and the assembly process.

The focus of our investigation is on properties of polymers on silicon substrates. The development of stresses in the polymer layers is measured and analyzed for different polymers (BCB, PI, PBO). The residual stress in a thin polymer film is measured by the warpage of the substrate in relation to different temperatures depending on the application. The estimation of stress-temperature behavior allows to develop processing concepts for a stress reduction being essential for 3-D integration. The generated stress drives cracking which leads to the effect that the impact of the forces should be taken into account for the quantification of the fracture toughness. The relation between the stress as the driving force and the fracture toughness are further discussed in details. A comprehensive study of the mechanical polymer properties is essential for high reliable devices.

Introduction

Polymer thin films play an important role as dielectric for WLP. In contrast to the inorganic layers like silicon dioxide the material is mostly photostructurable and the layer thickness in the range from 1 up to 20 µm is easy to produce. A mayor advantage is the planarization effect which is important for multilayer build-ups. The deposition of inorganic materials is mainly pure conformal due to CVD or PVD processes.

The polymer layers have in general a lateral stress which is based on interaction of the polymer layer and its underlying substrate. This interaction leads to stress in the substrate and polymer layer. For the reliability and analyzing aspects the stress in the polymer layer is the more important value because it is typical 10 up to 100 times higher.

The stress in the polymer layers are known but in the most cases not marked as a critical value for the reliability or handling because on the one hand the value is far away from the limits of the material toughness and on the other hand the handling ability of the tools could compensate the stress caused deformation of the substrate which is relative stiff with a typical thickness in the range of 500 up to 800 µm. Related to the changes in the substrate dimensions and behavior the stress in the system becomes more impact.

Different methods are used to decrease the stress or to face the effects of the stress like bending. The possibility of the following methods is discussed in relation to production and reliability issues:

- **Stress Management**
- **Cure Management / Cure Method**
- **Polymer tailoring (Next Generation Polymers)**
- **Temporary Bonding**

The concepts of the methods are based on different functional principles and have also different goals. To estimate the constraints of the methods it is necessary to take into account the addressed issue and the origin of stress generation in thin film polymer layers. For this an advanced characterization of the polymer layers is necessary.

Definition of the stress in thin polymer layers

The stress in a thin film on a substrate is a combination of compression and bending stress. The stress in the thin polymer layer is much higher than in the substrate which is based on the layer thickness ratio of the layers and the different material properties. Also related to the thickness ration the stress in the polymer layer could be postulated as nearly constant over the whole thickness. The stress in the polymer layer is the main indicator for the reliability and analyzing aspects.

The stress arises during the deposition of the polymer film on the substrate. The stress is mainly influenced by the deposition process and the material mismatch between polymer and substrate. The stress in the polymer can be estimated by the following formulation:

(1)

$$\sigma_f = E'_f(\alpha_f - \alpha_{si})\Delta T$$

where E'_f is the biaxial modulus of the polymer. α_f and α_{si} are the CTE of the polymer and the substrate. ΔT is the temperature difference of the current temperature related to the temperature of the stress-free state which is specific to the material and the process. The result of (1) is named *thermal stress*, because of the CTE and temperature dependency. The CTE value of thin film polymers is typical in the range of 30 up to 80 ppm/K [1], which leads to lateral stress value for polymers on silicon substrate in the range from 28 to 45 MPa at room temperature. The stress will be increased for temperatures below room temperature.

There are also other reasons for stress generation in the material. Aside of the thermal stress, the stress in the polymer could also generated by shrinkage effects during the polymerization process (the cure) where this is often described as intrinsic stress. Intrinsic stress effects based on shrinking have for the most thin film polymer materials only a minority effect, because the shrinkage are often done during the rubber state, where no stress could arise.

Origins of the stress in polymer layers

Stress decreasing concepts are based among other things on the understanding of its development during the process. Nearly all thin film polymers are delivered by the chemical industry as a pre-polymer dissolved in an organic solvent. The pre-polymer is applied by a spin coating process on to the substrate. After coating the solvent is evaporated by a post bake at typical temperatures around 100°C. In case of photosensitive formulations a structuring process is followed. Final step is the curing where the polymerization process forms a tough and insoluble solid state. The curing process is the important step where the stress is generated.

The Figure 1 shows the stress development in a polymer layer during a cure process. When heating starts, the material is in rubber state (gel phase) where nearly no stress could be applied based on the low modulus [2].

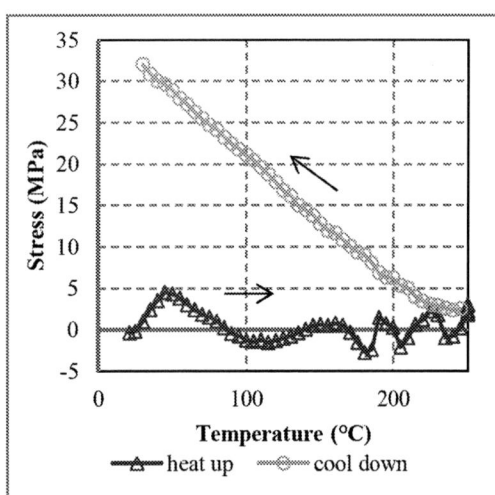

Figure 1: The stress in the material during a cure

The polymer changes from a solid state into the rubber state if the ambient temperature is higher than the glass transition temperature T_g. The pre-polymer which has typical a weak molecule crosslinking has also a low T_g. But with the increased degree of polymerization also the T_g increase during the cure. The polymer layer becomes a solid state when the T_g reach the ambient temperature which is also called as vitrification. The polymer is no longer a liquid but a fully formed solid material with a fairly high thermal expansion

coefficient. The vitrification point fixes the stress state in the polymer in relation to the substrate and it can be set as the stress free point. During the cool-down phase the thermal stress increases by the linear relation of the current temperature to the temperature where the stress-free state is frozen (see (1)). This point is not in general the soak temperature of the cure process, like it is shown in Figure 1. As a kind of cure management the process could be tailored to freeze the stress free state at lower temperature.

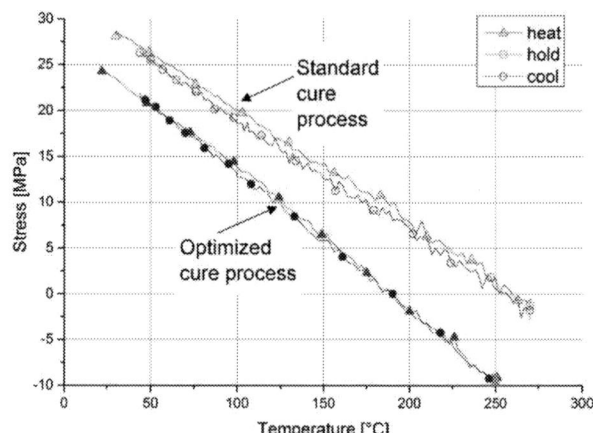

Figure 2: Stress decreasing by cure process adoption of BCB

The stress over temperature of a thermoset is shown for a standard cure and an optimized cure in Figure 2. The lowering of the set point for the zero stress state leads to shifting of the stress curve with the result of a 10% decreased stress value at room temperature. The effect of the cure management depends on the cure reaction properties and an economical cure process time. For this polymer system a decreasing of around 30% was demonstrated [2]. To estimate the optimized cure profile a kinetic model is necessary. The cure management is only usable for polymers where no reentry into the rubber state during the cure occurs, like it is typical for thermosets.

For thermoplastics like the polyimides the cure is typical done above the T_g of the polymerized polymer (see Figure 3). For this case the stress free point are coincident with the glass transition point. This mean as long as the soak temperature is set above the T_g any cure process changes will show no effect for the stress behavior.

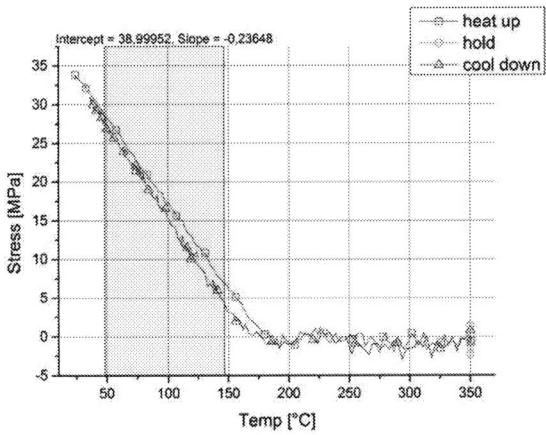

Figure 3: Stress behavior of a PI with a low Tg

The stress could also influenced by the change of the cure method. This could be done for example by a microwave [3] or e-beam. In the case of microwave the energy for the polymerization is induced by molecule vibration which allows to lower the cure temperature. The decreased cure temperature has the potential to decrease also the vitrification point. This shifting of the stress-free state down to lower temperatures influence the stress behavior, like it is done by cure management. Alternative cure methods need in comparison to conventional convection oven more sophisticated equipment. The applicability for any kind of sample needs to be verified. In addition there is the risk of damaging the devices underneath by the alternative contribution of energy.

Stress tailoring by polymer engineering

Instead of tailoring the process, the stress behavior could be influenced by changes of the polymer chemistry or adding additives. Based on the increased demands for polymer layers there are ongoing activities by the material supplier (like HD Microsystems, Dow Chemicals, FujiFilm, Asahi Kasei etc.) to bring the next generation polymers into the market. Prime focus is on enhanced mechanical properties and low temperature cure.

In relation to the stress behavior the polymers are tailored with the focus of an in-plane low CTE value. The stress behavior of two optimized photosensitive polymers (XP-603 and XP-604), which are under development from HD Microsystems (Hitachi Chemical DuPont MicroSystems), are shown in Figure 4.

The stress behavior is compared with a conventional PI. The stress at room temperature is 30% lower for XP-604 and the XP-603 shows a decrease of even around 50%.

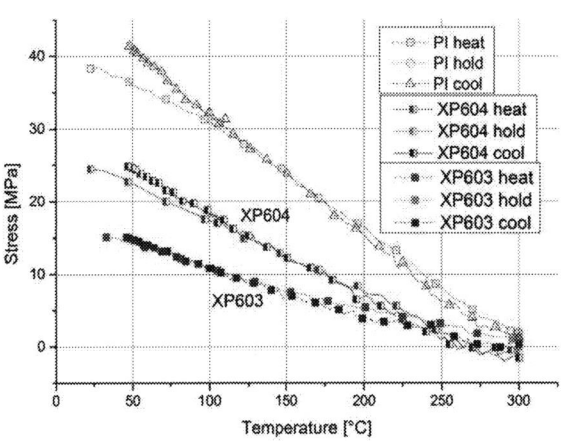

Figure 4: New PI formulations from HDM

The measurements indicate that the CTE value is below 30 ppm/K for XP-604 and below 20 ppm/K for the XP-603 material. Nearly all commercial photostructurable polymers have a higher CTE Value in the range from 42 up to 80 ppm/K [1]. Only a few applications demand a zero stress build-up, so there is a trade-of where the CTE is lowered and the material has already a high resolution by photostructuring. These next generation polymers are already in the evaluation phase and the commercialization will be soon. The reliability tests for the new HD Microsystems polymers are ongoing for RDL build-ups at the moment.

There are already polymers which are well fitted to the silicon substrate, like the PI-2611 from HD Microsystems. The PI-2611 has a CTE of 3 ppm/K, but these polymers are in general not photosensitive. The application of non photostructured polymers will become more attractive since companies like SUSS bring new tools into the market which allow a laser ablation of the polymer with a high throughput. Without the high demands of the photostructuring process to the polymer properties, the materials could be better tailored in relation to the electrical and mechanical demands.

Impact of the stress in polymer for the devices

The booming mobile device markets with the trend to higher integration demands more and more thin substrates. Because of this the main issue in relation of the stress is the warpage of the thinned substrate. A typical limit of the warpage of a 200 mm wafer is set to 200 µm [4]. A large bow leads to issues with the automatic handlers and could also negatively influence the processes on the substrate. Figure 5 shows an analytical calculation of the substrate bending for a 200 mm substrate where an unstructured RDL polymer layer of around 10 µm is deposited on top. The bow is calculated for a stress of around 30 MPa in the polymer.

978-1-4799-2408-0/14 $31.00 © 2014 IEEE   1423

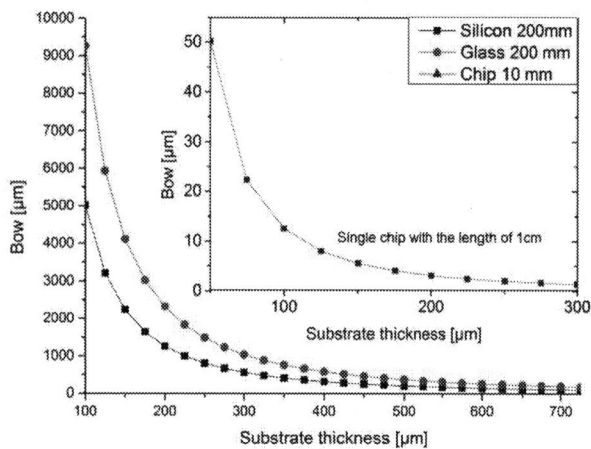

Figure 5: Analytical calculation of the substrate bow

Silicon interposers and thinned chips have typically a thickness below 150 µm. Related to the weak stiffness and the bow issues, the handling problems are common solved by applying a temporary bonded carrier. A thick silicon or glass substrate carrier with a thickness of 500 up to 725 µm is glued on the back side of sample. After processing the carrier will be released.

Aside silicon based interposers the glass becomes more attractive as interposer material related to better high frequency properties. Glass as substrate for interposer applications is much more influenced by the polymer layer due the lower Young's modulus of the glass instead of the silicon. Meanwhile the handling of thin wafers has been solved by temporary bonding but the stress could also be critical for the singulated chip (Figure 5). Bending of the chip leads to issues for the assembly process [5] [4]. Figure 6 shows an open interconnect due to chip bending at reflow temperature.

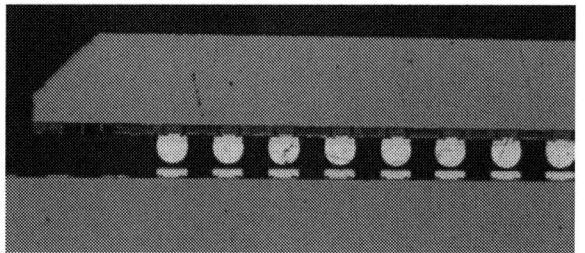

Figure 6: Cross section of hybrid pixel module with open interconnections [6]

The warpage could be decreased by a stress management using an additional layer. This is applied to balance the bending moment of the whole layer stack which is the sum of the layer stress in the layers and the distance to the neutral bending phase [7] [8]. To compensate the polymer layer, which has a tensile load, a layer with compressive stress could applied on top of the stack or a layer with tensile stress is applied on the back side. The stress management technic has the disadvantage, that the additional layers induce more strain energy into the whole stack. The strain energy could be roughly estimated by the following formulation:

(2)

$$E_{strain\ sample} = \sum_i^n \frac{\sigma_i^2 h_i}{E'_i}$$

where σ_i, h_i and E_i are the stress, the thickness and the modulus of the i-th layer. The strain energy could drive cohesive or adhesive failures if it overwhelms the energy release rate ERR, which is the necessary energy to generate two new surfaces and crack tip related energy dissipations:

(3)

$$E_{strain} \geq ERR$$

This sets the limits of the usability of the stress management, because to high strain energy values will drive cracks and delaminations in the layer stack. The strain energy depends on the stress in the film, which means for the application that the crack risk increases, if the samples will apply on low temperatures. But it also depends on the film thickness with the result, that it could be estimated a limit for layer thickness where a critical cracking or delamination will occurs, like it is reported for a glass interposer from KOIZUMI at al. [9] or for a silicon interposer from ARKALGUD at al. [4].

The stress in a polymer film in relation to the cohesive failure is also an important reliability factor for WLCSP build-ups. Figure 7 shows a solder interconnect where a crack in the polymer layer occurs after thermal cycling.

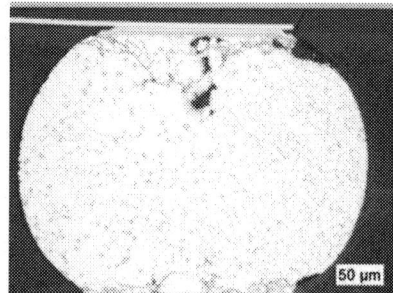

Figure 7: Bump connection with a crack failure in the RDL

The crack in the polymer is located at the corner of the bump interconnects. The CTE-mismatch between the PCB and the silicon introduce stress during thermal cycling where the stress singularities are located at the solder bump edges. The polymer is highly stressed at this point, which is visualized by simulation. The stress deviation is shown in Figure 8, where red areas marked a stress concentration which is in correlation with other publications [10].

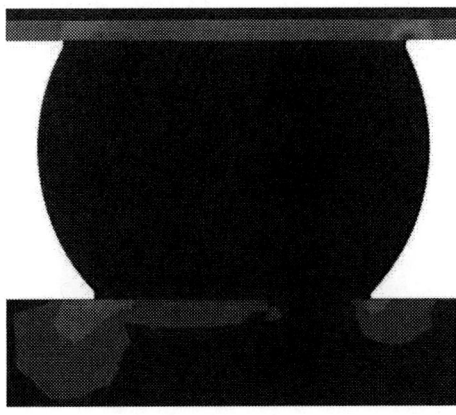

Figure 8: Stress distribution for a Bump connection

A crack in the polymer would arise if the limit of the tensile strength is reached. The stress tolerance of a polymer film is decreased by the lateral thermal stress, which arise from the coating. For most of the PI's this is not critical because the lateral stress from coating is in the range from 15 to 20% of the tensile strength [1]. But for materials like BCB or Epoxies this value could be around 40 to 50% of the tensile strength. This limits the toughness of the layer and also the application range. Figure 9 shows a substrate with a polymer coating which was tested for a deep temperature application. The sample was cooled down to -150°C, where the thermal stress exceeds the tensile strength and leads to cohesive failures in the polymer coating.

The role of the polymer layer as a stress layer becomes more and more importance for covered brittle ULK (ultra low k) layer. where cohesive crack is critical because the stress is applied to the brittle ULK. FUJII et al. outlined that the next generation polymers for stress buffer application should have a higher Modulus [10]. This leads in relation to (1) to a higher thermal stress in the layer during the curing process.

Figure 9: Cohesive failure of the polymer coating

Stress behavior analyzing

The characterization of the lateral mechanical stress in the polymer layer was done with the wafer bow measurement tool FLX 2320 from the company Toho. An unstructured polymer layer was coated on silicon substrates in the range of 5 to 10 μm. The stress is estimated by the warpage change of the sample due the coating of the polymer layer. The stress calculation is based on the Stoney-formulation:

(4)

$$\sigma_f = \frac{E'_s t^2_s}{6 \; t_f}\left(\frac{1}{R_2} - \frac{1}{R_1}\right)$$

where E'_s is the biaxial modulus of the substrate. t_s and t_f are the thickness of the substrate and the polymer layer. R_1 are the initial radius of the substrate and R_2 are the radius which is affected by the mismatch of the substrate and the layer. Due to the curve calculation by subtraction any influences by pre-deformations of the substrate can be avoided. Only 80% of the wafer length is measured to exclude inhomogeneous warpage at the wafer edge.

The tool uses a contactless laser based measurement for the warpage estimation. The sample can be placed in a temperature chamber which allows the simulation of a convection oven cure. In contrast to a stress measurement only at room temperature the stress measurement over the temperature allows a better analyzing and understanding of the material behavior and material property characterization. The slope of the stress curve allows the estimation of the CTE and Young's modulus relation of the polymer layer due the formulation (1).

Stress prediction and simulation

Stress in microelectronic devices and their negative effects like warpage and adhesive/cohesive failure are often neglected during the design. The described methods are typical applied by a try and error process, if an issue occurs. The FEA could assist during the design to localize in an early development phase any stress issues. In common FEA tools, like ANSYS, the setting of definition of a discrete stress value as a parameter in a model is limited in the accessibility because special elements are needed. Most of the elements could only handle the stress as an output value. The stress is produced as an output parameter of the simulation by the definition of the thermal parameters of the materials. There are at least three main polymer parameter needed, which are the young's modulus, the CTE value and the temperature of the stress-free state (vitrification point). The young's modulus, the CTE are essential parts of the datasheet but the temperature of zero stress is also depended by the process parameters and often not described in a datasheet. To get a realistic model behavior the parameters should related to measured stress versus temperature plot of the polymer, like Figure 2. The stress in a polymer layer is simulated by a quarter wafer model (Figure 10) to verify the procedure for applying lateral stress into a model.

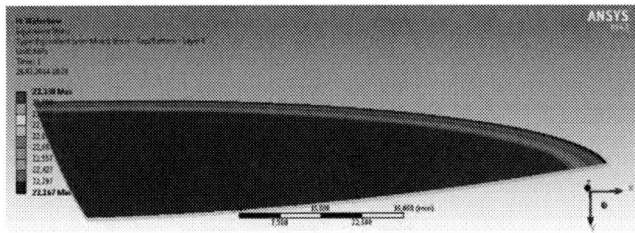

Figure 10: Simulation of the stress behavior of a polymer layer on a silicon substrate

The stress is homogeneous over the whole wafer, which is in accordance to the Stoney formulation. There are only stress changes at the wafer edge. Related to the high aspect ratio for a thin layer stack on a 200 mm wafer the special shell elements are used. Such a model could be used to examine the usability of stress and warpage compensation methods. Also the reliability investigation of the layer build-up could be done. The described procedure for applying lateral stress could be easy implement into more complexes interconnect models where the lateral stress is normally neglected.

Also the risk of stress driven delaminations could be investigated in FEA by using special elements like cohesive zone modelling. The adhesion related parameters for this special element need to be estimated by special thin film adhesion tests [11].

Conclusions

With the growing business of thinned chips and increased 3-D integration the demands of the polymer dielectric layers also increased. This study presents an advanced polymer material characterization and analysis which set the basics for optimized processes. The stress in the polymer layers is outlined as a key material parameter for the manufacturability and for the reliability of the system. The temperature depended stress measurement of polymer layers allows a fast extraction of the material parameters like young's modulus, CTE and vitrification point. The extracted parameters set the basics for a more realistic thermo mechanical simulations of the polymer behavior.

This study gave a compressive overview about the complex stress topic in thin film polymers. The understanding of the stress origin and their development set the basics for methods which should avoid the stress or the stress related issues. A FEA supported investigation could avoid expensive and time consuming try and error process, because it allows to examine stress related issues during an early state of system design.

One of the mayor issues is the warpage of the thinned substrates. The stress management is commonly used. Only a weak knowledge about the stress states in the system is necessary and it could be applied to any finished system build-up by adding additional layers. The stress management has the disadvantage that the warpage adaption fits normally only for a discrete temperature range which could be an issue for high temperature processes like soldering. Also the risk of delamination and cracking is increased by the higher strain energy.

A good process and material understanding is necessary for the cure management because of this it will be mainly applied for special application where sensitive devices are used and special polymers demanded.

The usage of the temporary bonding is reasonable in combination with the stress management where the bending issues is avoided as long as the bending moment is neutralized by an additional layer or process.

The polymer tailoring seems to have the brightest future because the stress as the origin of the issues is influenced. It has in comparison to the cure management a higher stress decreasing potential which was demonstrated by the measurement of the two next generation polymers from HD Microsystems.

Acknowledgments

The authors want to emphasize the contributions from the colleagues at TU Berlin and Fraunhofer IZM who are not mentioned here by name. HD Microsystems is gratefully acknowledged for the cooperation and material support.

Financial support from the Deutsche Forschungsgemeinschaft (International Research Training Group GRK 1215, "Materials and Concepts for Advanced Interconnects") is gratefully acknowledged.

References

[1] M. Töpper, T. Fischer, T. Baumgartner und H. Reichl, „A Comparison of Thin Film Polymers for Wafer Level Packaging," in *ECTC*, 2010.

[2] M. Woehrmann und M. Toepper, „Polymerization of Thin Film Polymers," in *New Polymers for Special Applications*, Intech, 2012, pp. 113-138.

[3] R. L. Hubbard und B.-S. Lee, „Reduced Stress and Improved 2.5/3DIC Process Compatibility with Stable Polyimide Dielectrics," in *IWLPC*, 2013.

[4] S. Arkalgud, A. Cao, E. Chau, G. Gao, H. Katske, L. Wang, B.-S. Lee, H. Shaba, H. Shen, Z. Sun und C. Uzoh, „Reaching A Low Cost, Manufacturable Interposer," in *3D ASIP*, 2013.

[5] I. Limansyah, M. J. Wolf, K. Zoschke, R. Wieland, M. Klein, H. Oppermann, L. Nebrich, A. Heinig, A. Pechlaner, H. Reichl und W. Weber, „3D Image Sensor SiP with TSV Silicon Interposer," in *ECTC*, 2009.

[6] T. Fritzsch, K. Zoschke, M. Woehrmann, M. Rothermund, O. Ehrmann, H. Oppermann, K.-D. Lang und F. Hügging, „Flip Chip Assembly of Thinned Chips for Hybrid Pixel Detector Applications," in *Proceedings of 15th International Workshop on Radiation Imaging Detectors (IWORID)*, Paris, France, 2013.

[7] M. J. Yim, R. Strode, J. Brand, R. Adimula, J. J. Zhang und C. Yoo, „Ultra Thin POP Top Package using Compression Mold: Its Warpage Control," in *ECTC*, 2011.

[8] K. Crook, M. Carruthers, D. Archard, S. Burgess und K. Buchanan, „Dielectric Stack Engineering for Via-Reveal Passivation," in *ECTC*, Las Vegas, 2013.

[9] N. Koizumi, „Basic Study of Packaging Strucuture using Glass Material," in *GIT*, Atlanta, 2013.

[10] A. Fujii, „Photosensitive Polyimide Buffer Coat Material for FC-BGA," in *The Symposium on Polymers for Microelectronics*, 2012.

[11] M. Woehrmann, M. Toepper und K.-D. Lang, „Strain energy driven adhesion test for adherence characterization of thin polymer films for microelectronic applications," in *Thermosets 2013*, Berlin, 2013.

Thin Polymer Dry-Film Dielectric Material and a Process for 10 um Interlayer Vias in High Density Organic and Glass Interposers

Yuya Suzuki[α,β], Yutaka Takagi[γ], Venky Sundaram[β], Rao Tummala[β]

α: Zeon Corporation, Research and Development Center, 1-2-1, Yako, Kawasaki-ku,
Kawasaki-shi, Kanagawa, 210-9507, Japan

β: 3D Systems Packaging Research Center, Georgia Institute of Technology, 813 Ferst Drive, Atlanta, GA 30332, USA

γ: NGK Spark Plug Co., LTD., 2808 Iwasaki, Komaki-shi, Aichi, 485-8510, Japan

ysuzuki3@mail.gatech.edu

Abstract

This paper describes the first demonstration of 10 μm diameter interlayer vias in low- moisture uptake and low surface- roughness dry film polymer dielectric for multi-layered re-distribution layer (RDL) structures to achieve 50 μm bump pitch in high density organic and glass interposers. A new series of polymer dry films, ZS-100, at 10 μm thickness were deposited on thin and low CTE organic or glass cores using double- sided vacuum lamination processes. The ultra-small vias were fabricated by 248nm KrF excimer laser drilling, followed by electroless and electrolytic copper plating. Fully-filled via structures were successfully fabricated without any chemical-mechanical polishing. The processes demonstrated in this paper achieve much finer bump pitch than current organic packages, and can be scaled to large panels leading to lower cost than previous work in fine pitch Si interposers using back-end of line (BEOL) wafer processes.

Keywords: excimer laser, micro-via, fine-pitch via, thin dielectric film

Introduction

Increasing demand for miniaturized, high-performance and highly functional electronic systems requires high density I/O interposers and packages which require multi-layer wiring redistribution layers with fine pitch wiring with ultra-small interlayer micro-vias. Such high density packages can be fabricated on silicon IC using silicon dioxide or nitride dielectric layers on silicon wafers with sub-micron lithographic lines and vias by damascene or dual-damascene processes. [1-2]. Photo-sensitive dielectric materials, on the other hand, are developed for wafer level packaging to achieve below 10 μm vias [3]. Although extremely small bump pitch is achieved with Si interposers using small 300 mm wafers, they are expensive and low in electrical performance. On the other hand, organic package technology utilizes dry-film build-up polymer dielectric materials to form multi-layer wiring layers on large panels yielding 8X more packages than from wafers. Therefore, organic packaging technology can provide a low cost approach, with a typical panel size of 500mm x 500mm [4]. One of the biggest limiting factors in this organic, panel-based approach, however, is the formation of small micro-vias that can be stacked from layer to layer with minimum via-misregistration. The multilayer RDL structure with these stacked micro-vias is essential to achieve ultra-small bump I/O pitch. Current line and micro-via dimensions in the industry are about 12-15 μm for wiring width and 40-50 μm for micro-via diameter, using CO_2 lasers. Long wavelength (10.2 and 10.6 μm) and large- beam spot size of CO_2 laser are limiting factors in reducing the via diameter. Recent advances in semi-additive processing and in Nd-YAG lasers led to the demonstration of 9 μm line and 25 μm micro-vias [5-6]. To achieve even smaller dimensions, new advances have been proposed and reported such as trench filling processes and thin-film processes, adopted from wafer level packaging. Atotech and Amkor have demonstrated Via[2] processes down to 10 μm line features with trench filling processes by excimer laser and copper plating processes [7-8]. Fujikura reported one of the most advanced trench circuit technology to create 2 μm line width and 10 μm micro-via diameter [9]. However, this demonstration is limited to single layer vias and there was no demonstration of a multilayer wiring structure leading to fine-bump pitch micro-via structures. Furthermore, trench filling processes are similar to, damascene processes, requiring expensive chemical-mechanical polishing (CMP) after copper plating, thus increasing the process cost. Shinko recently demonstrated 2 μm lines and 10 μm diameter vias using liquid photosensitive thin film process [10], but this process is limited in cost reduction by wafer tools and single- sided processing.

This paper presents the first demonstration of an advanced polymer dielectric material with an advanced process to form 10 μm diameter micro-vias in 10 μm thick polymer build-up dry-films. Dry-film build-up materials enable panel size and double-sided processes to maximize the throughput. The first section of the paper describes the thin polymer dry film along with its properties. The second section discusses ultra-small via formation by KrF excimer laser process. The third section explains the method for via metallization by electroless and electrolytic plating processes. Excimer laser drilling can be carried out with mask projection processes, therefore multiple vias can be prepared simultaneously. There are some reports using excimer lasers to make vias in dielectric layers, [11-12] however, they applied the laser to thicker polymer materials and as a result, via openings were relatively large (30- 50 μm). In this paper, 10 μm thin polymer dry-film was used for interposer applications to achieve ultra-small micro-vias. This technology enables panel-based processing to form the basis of 50 μm bump pitch RDL wiring.

978-1-4799-2408-0/14 $31.00 © 2014 IEEE

Low Loss Thin Polymer Dry Film Dielectric

The dielectric material used is a dry-film polymer material, ZEONIF™ ZS-100 (ZS-100). The ZS-100 is a build-up polymer film that is compatible with semi-additive processes (SAP), currently used in organic substrate panel manufacturing. The advantages of this material are many that include low moisture absorption, low electrical loss (Df), very low surface roughness (Ra<100 nm) and strong chemical bond to electroless plated copper seed layers (figure 1). This extremely smooth surface is beneficial to forming fine wiring and lowering transmission loss.

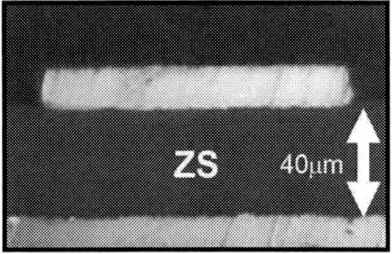

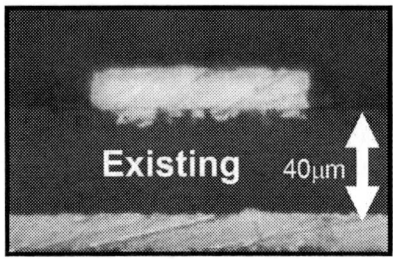

Figure 1. Difference in surface roughness in ZS-100 (top) and existing material (bottom)

Due to the ultra-smooth surface, 4 μm line and space structure by SAP was successfully fabricated on ZS-100 [13]. Figure 2 shows the copper lines and spaces structures fabricated on ZS-100 by SAP.

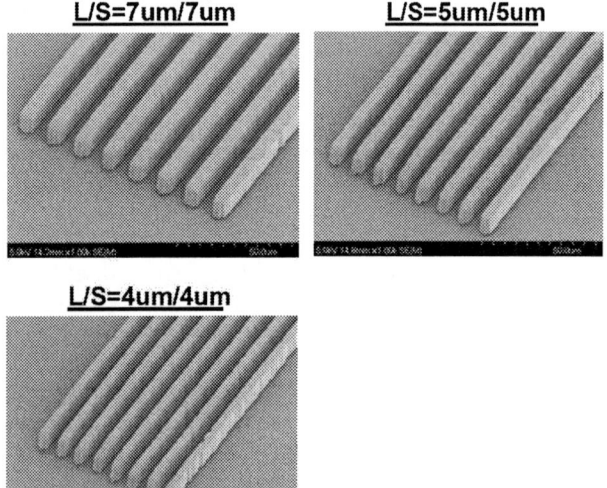

Figure 2. Fine pitch lines and spaces wiring structures by SAP

The ZS-100 dielectric is available in a wide range of thicknesses from 10 to 40 μm. The dielectric thickness required to achieve 50 ohm impedance for transmission lines of 10 μm width and 10 μm height in microstrip configuration was calculated to be 6.2 μm by equation 1 [14]. Figure 3 shows the thickness of the dielectric under transmission line to meet 50 ohm impedance matching.

$$w = \frac{7.463h}{\exp\left(\dfrac{Z_0\sqrt{0.475\varepsilon_r + 0.67}}{60}\right)} - 1.25t \quad \text{(Eq. 1)}$$

w: width of copper trace
t: height of copper trace
h: thickness of dielectric layer
Z0: impedance (normally 50 ohm)
εr: permittivity of the dielectric

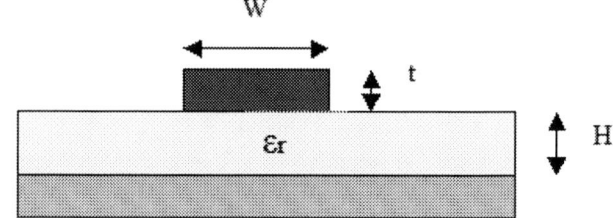

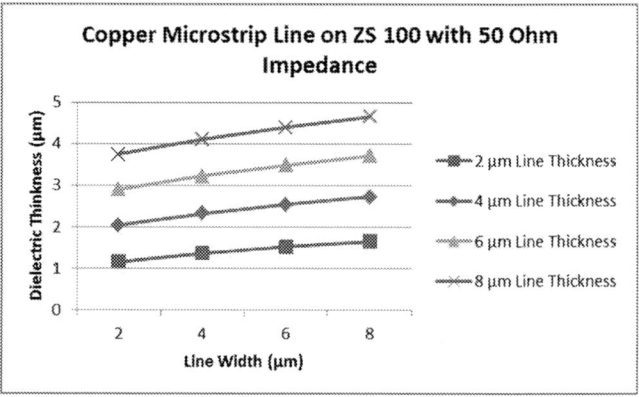

Figure 3. Cross section of microstrip line used for equation 1 (top), calculated dielectric thickness for 50 ohm impedance matching

This calculation indicates the thin dielectric layer is required in order to minimize electrical loss from impedance mis-matching in transmission line.

The 10 μm thick ZS-100 film was deposited on both sides of the copper clad laminate FR-4 core by vacuum lamination at 100 °C. The panel was then hot pressed for a short time using low pressure to planarize the surface of the build-up material. After the lamination, the sample was thermally cured in an oven, ramping up from room temperature to 180 °C, and kept at 180 °C for 30 min. The ZS-100 dry film was also laminated on 100 μm thickness low CTE glass core using vacuum lamination, without any damage to the glass and organic panel as shown in Figure 4.

Figure 4. Pictures of ZS-100 laminated on glass (top) and organic core

Properties of the fully cured ZS-100 polymer dielectric are summarized in Table 1.

Table 1. General properties of ZS-100

Properties	Unit	ZS-100
Dk (10GHz)	-	3.0
Df (10GHz)	-	0.006
Tg (TMA)	deg. C	162
CTE	ppm/deg.C	25
Copper surface roughness	nm	< 100
Peel strength	N/cm	> 7
Modulus	GPa	7
Water absorption	wt%	0.2

Ultra-Small Via Formation by Laser Ablation

Current methods of CO_2 and UV lasers cannot shrink the via diameter to below 10 μm. Although liquid photosensitive dielectrics have been shown to achieve sub-10 μm vias, the choice of dielectrics is limited and most of these materials require sputter deposited barrier and seed layers, thus increasing the process cost and limiting the scalability to large panels. Excimer laser ablation was selected as the front-up process to form ultra-small vias in ZS-100. The main advantages of excimer lasers are 1) high absorption by polymer to generate chemical interactions for efficient drilling process 2) minimal thermal damage enabling clean vias, and 3) the availability of projection tools for high through-put and large panel scaling [15]. It is important to match the wavelength of the laser source and the peak absorption of the polymer in order to minimize thermal effects and damage. The ZS-100 polymer has a characteristic absorption around 250 nm wavelength. The light absorbance at 250 nm is four time of the one at 355nm (2.0 at 250 nm and 0.5 at 355 nm). This means absorption efficiency of KrF excimer laser is several orders of magnitude higher than Nd-YAG laser 3rd harmonics. Excimer laser with shorter wavelength such as ArF (193 nm) or F_2 (150 nm) cannot be used for this application because copper has light absorption at their wavelength and bottom pad would be damaged during the laser process. Hence, KrF excimer laser with 248 nm wavelength was selected as a highly efficient laser for via drilling in RDL. When a laser beam is irradiated on a sample, attenuation of light intensity is described by equation 2, where α is absorbance of the material at the wavelength and z is penetration depth.

$$I(z) = I_0 \exp(-\alpha z) \quad \text{(Eq. 2)}$$

Absorbed energy per unit mass Em is calculated by equation 3. (t_p: laser pulse width, ρ: density of mass)

$$Em(z) = I(z)t_p\alpha / \rho \quad \text{(Eq. 3)}$$

This energy should be larger than the threshold energy E_{th}, in order to obtain ablation of the material. Threshold light intensity I_{th} is calculated by $I_{th}, = \rho E_{th} /\alpha t_p$

Ablation ratio per one pulse $L_a \sim z$ can be assessed by equation 4.

$$La = \frac{1}{\alpha}\ln\left(\frac{I}{I_{th}}\right) \quad \text{(Eq. 4)}$$

To ensure the via formation by laser drilling, it is required to use the laser system with the fluence higher than the threshold. On the other hand, too much fluence has also negative effect on the via; fabricated via holes would be larger than designed and distorted from round shape. In this study, a 248 nm KrF excimer laser system (Coherent LPX210i, figure 5) with 1 mm x 1 mm beam size at 200 mJ fluence was used for micro-via making. A 1" x 1" quartz mask with sputtered aluminum opening of 10 x 10 array (opening size of 100 μm at 500 μm pitch) was used for the projection. The mask image was demagnified by a factor of 10x, therefore 10 μm via array structures could be formed.

Figure 5. Picture of the excimer laser system Optec /Coherent LPX210i

Figure 6-a shows top view of the drilled vias. The shape of the vias is round due to the non-thermal process of excimer laser. Cross section of the formed laser via is shown in figure 6-b. (Copper seed layer was deposited on the build-up layer to make the interface visible) No significant damage on the bottom copper pad was observed after laser drilling and almost no taper was confirmed. Number of the laser shot is important input since it defines the throughput of the process. Obtained via openings vs. number of shots are summarized in table 2. Here, time of the operation was calculated from the frequency of the laser 100 Hz. Opening of the via after 10 shots was not sufficient to ablate 10 μm polymer layer to expose the bottom copper layer. Irradiation of 20 or more shots was able to penetrate the polymer and create via opening.

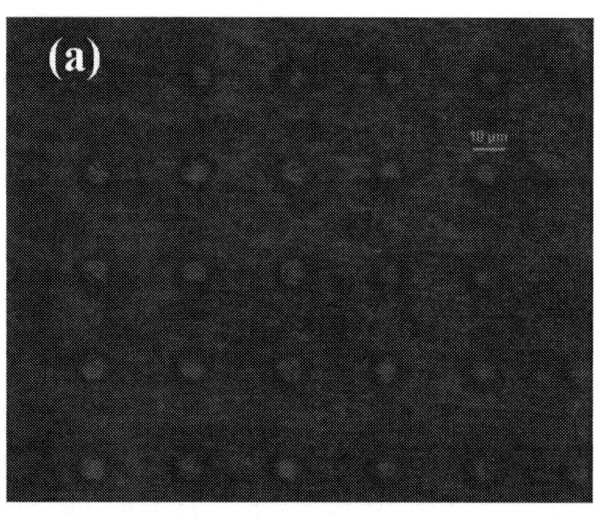

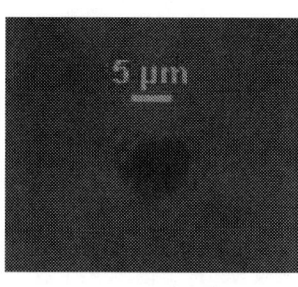

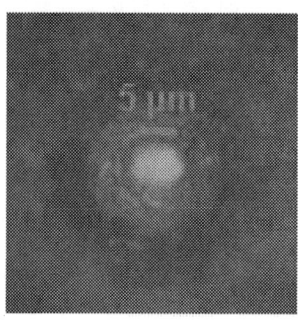

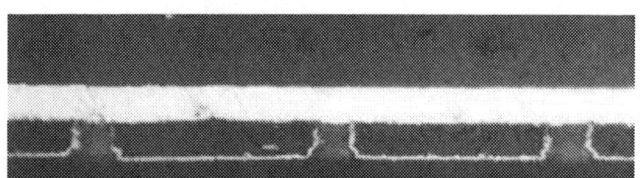

Figure 7. Drilled via before (top) and after (bottom) desmear cleaning

Figure 6. Via array drilled by excimer laser (a: top view, b: cross-sectional view)

Table 2 Number of the shot and top view of the vias created by laser irradiation

Shots	10	20	30	40	50	100
Time (sec)	0.1	0.2	0.3	0.4	0.5	1.0
Picture						
Open/Not	No	Yes	Yes	Yes	Yes	Yes

After via hole drilling, desmearing process was applied to clean the via surface. After 15 min of desmear process in the solution of permanganate and sodium hydroxide, via edge was cleaned (figure 7). No severe side wall erosion in the drilled vias was observed during the desmear process.

Via Metallization

Metallization of the micro-via was carried out by electroless copper plating followed by electrolytic copper plating. After desmear of the drilled sample, Pd catalyzed electroless plating process was applied and 0.5 µm thick copper seed layer was formed. After 40 min of the electrolytic plating of the sample, fully-filled vias were successfully fabricated (figure 8). Thickness of the copper deposited during the plating on top was 8 µm, which is thinner than the general thickness of the dry-film photo-resist (15-20 µm) used for SAP process. Furthermore, there are no dimples on top of the vias and high surface planarity was achieved. Therefore, this technology is beneficial to compose the combination of fine-pitch via and wiring in one step by SAP, without polishing process. Copper filling by electrolytic plating process was successfully performed due to relatively small aspect ratio (1:1) of the vias.

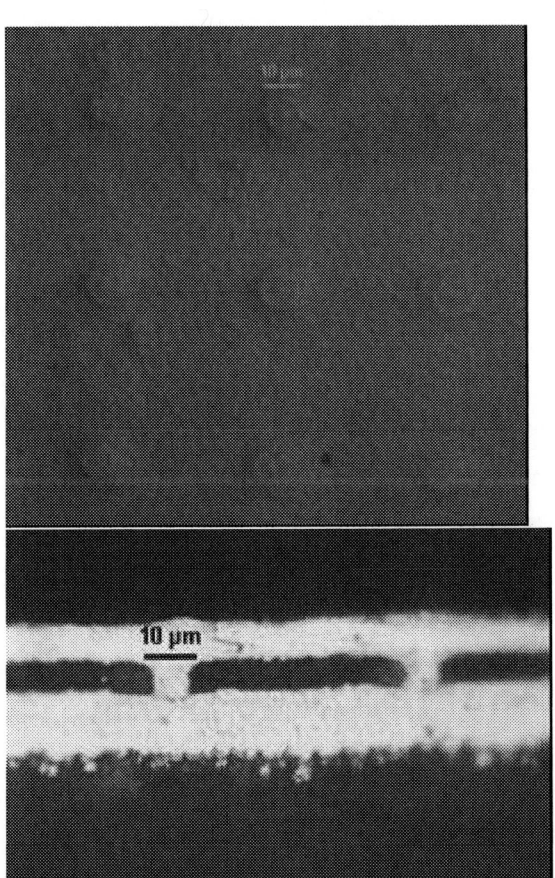

Figure 8. Cross-sectional and top view of the fully filling plated vias

Conclusions

In summary, ultra-small 10 µm micro-vias at 50 µm bump pitch in ultra-thin dry-film dielectrics was successfully demonstrated, for the first time. Thin dry-film dielectric

material was laminated onto low CTE cores. Multiple ultra-small micro-vias were formed by projection excimer laser ablation processes, followed by copper plating to fabricate fully-filled interlayer micro-vias. The processes reported in this paper extend current organic packaging to 50µm bump pitch using traditional lamination processes.

Acknowledgments

The authors would like to thank to Richard Shafer, Georgia tech for the support with excimer laser process.

References

1. P. J. Tzeng, John H. Lau, C. Zhan, Y. Hsin, P. Chang, Y. Chang,J. Chen, S. C. Chen, C. Wu, C. Lee, H. Chang, C. Chien, C. Lin, T. Ku, M. J. Kao, M. Li, J. Cline, K. Saito, M. Ji "Process Integration of 3D Si Interposer with Double-Sided Active Chip Attachments ", in Proc. Electronic Components and Technology Conference (ECTC), Las Vegas, NV, May 28-31, 2013, pp 86-93.

2. H.Y. Li, H. M. Chua, F. X. Che, A. D. Trigg, K. H. Teo, S. Gao, "Redistribution layer (RDL) process development and improvement for 3D interposer", in Proc. Electronics Packaging Technology Conference (EPTC) IEEE, Singapore, Dec. 7-9, 2011, pp 341-344

3. A. Tanimoto, K. Abe, S. Nobe, H. Matsutani, "Development of Low Temperature Curable Positive Tone Photosensitive Dielectric Material" in Proc. CPMT Symposium Japan, 2012 2nd IEEE, Kyoto, Japan, Dec. 10-12, 2012, pp 1-3.

4. E.D. Blackshear et al. "The evolution of build-up package technology and its design challenges", IBM Journal of Research and Development, vol. 49, 2005, pp 641-661

5. F. Liu, V. Sundaram, B. Wiedenman, R. R. Tummala, "Advances in High Density Interconnect Substrate and Printed Wiring Board Technology" in Proc. IEEE. 6th International Conference on Electronic Packaging Technology, Aug. 30-Sep. 2, 2005, pp. 307-313

6. K. Yamanaka, K. Kobayashi, K. Hayashi, M. Fukui, "Materials, Processes, and Performance of High-Wiring Density Buildup Substrate with Ultralow-Coefficient of Thermal Expansion", IEEE Transactions on components and packaging technologies, vol. 33, 2010, pp 453-461

7. D. Baron, "Via2 - Laser Embedded Conductor Technology", in Proc. Microsystems, Packaging, Assembly & Circuits Technology Conference (IMPACT), Taipei, Taiwan, Oct. 22-24 2008. pp. 106 - 109

8. Atotech GmbH, "Via² Technology - Copper Trench Filling for Ultra Fine Lines", http://www.atotech.com/products/electronics/panel-pattern-plating/horizontal-systems/via2-technology.html

9. T. Hondo, Y. Nitta, K. Nakamura, H. Hirano, M. Saruta, T. Inoue, O. Nakao, "Ultra-fine Trench Circuit on Polymer Film", in Proc. Electronic Components and Technology Conference (ECTC), Las Vegas, NV, May 28-31, 2013, pp 136-139

10. N. Shimizu, "Development of Organic Multi Chip Package for high performance application" International

Microelectronics Assembly and Packaging Society (IMAPS), Orlando, FL, Sep. 30- Oct. 3, 2013

11. H. Zheng, E. Gan, G. C. Lim "Investigation of laser via formation technology for the manufacturing of high density substrates", Optics and Lasers in Engineering, vol. 36, 2001, pp 355-371

12. D. J. Chuang, S. K. Bhattacharya, J. Papapolymerou, "Low Loss Multilayer Transitions using Via Technology on LCP from DC to 40 GHz", in Proc. Electronic Components and Technology Conference (ECTC), San Diego, CA, May 26-29, 2009, pp 2025-2029

13. Zeon Corporation, "ZEONIF™: Insulation Materials for Printed Circuit Board", http://www.zeon.co.jp/business_e/enterprise/imagelec/zeonif.html

14. B.C. Wadell, Transmission Line Design Handbook, 1991, Artech House Publishers

15. J. Meijer, "Laser beam machining (LBM), state of the art and new opportunities" Journal of Materials Processing Technology, vol. 149, 2004, pp. 2-17.

Pad Crater Detection Using Acoustic Waveform Analysis

W. Carter Ralph[1], Elizabeth E. Benedetto[2], Aileen M. Allen[3], Keith Newman[3]
[1]Southern Research Institute, Birmingham, Alabama
[2,3]Hewlett-Packard, Houston, Texas and Palo Alto, California
[1]ralph@southernresearch.org

Abstract

Initial studies have shown that acoustic emission detection may prove an effective technique for pad crater monitoring. Implementation concerns include pad crater locating accuracy and the ability to discriminate between different types of events. This study found that the direction of wave propagation in the circuit board laminate has a significant effect on the event locating calculation due to orientation-dependent acoustic velocity, and that adjustment of the acoustic velocity based on the orientation between event location and transducers can improve locating accuracy to approximately 1 mm. In addition, the use of cumulative acoustic energy is demonstrated as a simple and effective metric for determining the onset of pad cratering. An alteration to the present test method is proposed that would improve the precision, decrease the throughput time, and decrease the cost of the test.

Introduction

Detection of the precise moment of interconnect damage initiation during mechanical loading of BGA components has been difficult to achieve using industry standard methods such as strain gages, electrical continuity chains, and destructive failure analysis. These processes are typically imprecise and time-consuming, and are seldom feasible on printed circuit assemblies (PCAs). Initial assessments of acoustic emission detection indicate that the technique is able to identify circuit board pad craters, and has the potential to improve the precision, speed, and flexibility of interconnect crack-initiation testing.

The IPC/JEDEC-9707 test method establishes a test approach to determine a safe bending level [1,2]. Engineering test units are loaded to failure in a spherical bending mode while instrumented with strain gages and continuity chains in order to estimate the likely range of damage initiation. Additional units are then tested to one or more strain levels that are estimated to be below the level of damage, and are then inspected for interconnect damage using traditional failure analysis techniques. This is a resource-intensive approach that is only as precise as the number of strain levels that are tested.

Acoustic emission detection is a well-established technology, but has not found much application in the microelectronics industry. The sudden release of energy in brittle fracture events results in transient elastic waves that propagate outward from the crack site. Multiple transducers can be used to detect the waves and calculate the source location. This technique has been used for detecting cracks in aircraft components during flight, welds during formation, and cracks in pressure vessels.

A study by Cisco in 2011 demonstrated that acoustic emission detection has potential application for crack detection in microelectronic assemblies [3]. In that study, test units were loaded in four point bend and monitored by two transducers for acoustic events and locations (in one dimension), and found a sudden increase in acoustic events which were located near the edge of the component. A study by the authors of this paper used four transducers (to allow for two-dimensional event location) on test units made from two different laminates loaded in spherical bend [4]. That study found a similar sudden increase in acoustic activity located at the component corners which appeared to correspond to the onset of pad cratering. A follow-on study by the authors of the Cisco study found similar results [5].

The present study performed spherical bend tests to identify the onset of damage acoustically, validated that result with bend tests above and below the acoustic onset, and improved the accuracy of acoustic event locating. The bending tests were performed in accordance to the test standard, using the spherical bending mode and bending rates comparable to production line in-circuit tests. The locating accuracy tests were performed with pencil lead break (PLB) tests, and determined that more appropriate acoustic velocity values, based on the direction of wave propagation through the circuit board laminate, can significantly improve locating accuracy.

The bend test, PLB test, and failure analysis methods are described, and acoustic data analysis methods are presented. A method for determining the damage threshold using cumulative acoustic energy is presented and validated. Acoustic velocity measurements using PLBs are demonstrated, and the event locating results for different velocities are quantified. A method for applying the acoustic detection techniques to the standard test method is proposed, and the benefits are discussed. Finally, further applications and improvements to the test methods are identified.

Test Methods

Several tests were performed in this study: spherical bend, pencil lead breaks, cross section, and dye penetrant inspection. The spherical bend test is a flexure test, similar to three point bend, which is described in the IPC/JEDEC-9707 test standard and is designed to induce a bending mode that is considered to be worst-case for printed circuit assemblies. Pencil lead break is a common method for inducing an acoustic wave. Cross-section and dye penetrant inspection are failure analysis techniques for determining the type and location of cracks.

The specimens were PBGA packages reflowed onto a test PCA as shown in Figure 1. The packages were 35x35 mm and incorporated organic substrates with fully-encapsulated die.

978-1-4799-2408-0/14 $31.00 © 2014 IEEE

The packages were mounted to the printed circuit assemblies with 1 mm pitch SAC305 solder balls, SAC305 solder paste and circuit board metal-defined pads. The circuit boards were 160x160 mm, 1.6 mm thick, 8 layers, with three different FR-4 laminate materials. The laminates were fabricated with alternating 0/90 glass cloth orientations. Daisy-chain networks and strain gage lead soldering points were designed into the PCAs and specimens at each corner. Strain gages were mounted to the PCA, offset 5 mm diagonally from the corners of the package in accordance with IPC/JEDEC-9704 [6], and as shown in Figure 1.

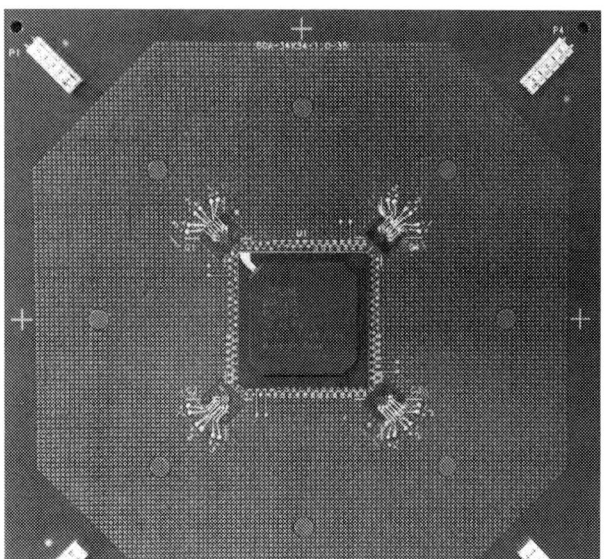

Figure 1. Test specimen.

The spherical bend tests used the fixture shown in Figure 2. The plate was mounted on a uniaxial test frame and the support pins were installed on a 55mm radius. The reaction pin was mounted to the upper side of the test frame and centered on the plate. A test specimen was centered on the fixture and aligned so that support pins were at the corners and edges of the package. Angle brackets were installed using 0.2 mm spacers at the edges of the PCA to ensure repeatable specimen alignment without binding against the brackets. Four acoustic transducers were mounted with accelerometer wax to the PCA at a radius of 38 mm from the center of the BGA, and were positioned every ninety degrees either at the edges or at the corners of the package. Load, displacement, and strain gage data were collected at a sampling rate appropriate for the rate of the test, in addition to the acoustic data. A typical test setup is shown in Figure 3. A screw-driven United FM-20AE test frame was used for the quasi-static tests, and a servo-hydraulic MTS 810 was used for the transient tests. The PCAs were loaded in displacement control at approximately 1 mm/minute for the quasi-static tests and approximately 5mm/second for the transient tests, producing strain rates of approximately 700 με/minute and 5,000 με/second, respectively. The quasi-static test rate is low enough to allow the test to be stopped manually when acoustic events were detected, while the transient test rate is aligned with the test standard.

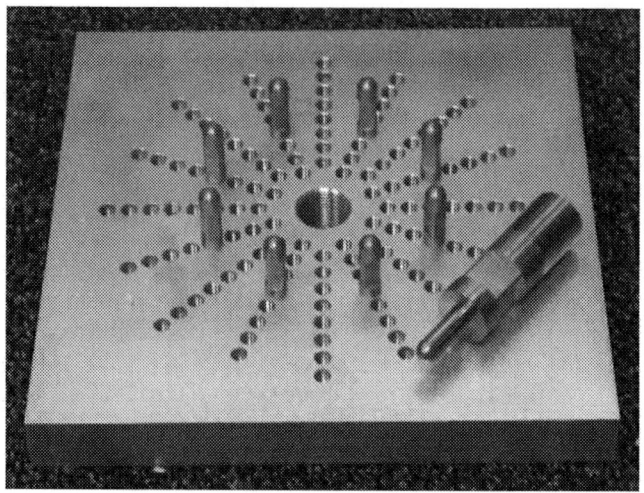

Figure 2. Spherical bend test fixture.

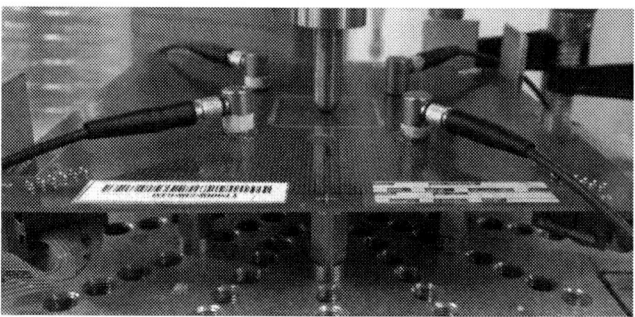

Figure 3. Test setup

PLB tests were performed to establish the sonic velocities of the test specimens, to cause acoustic events at known locations, and to verify that the transducers were functioning correctly. PLBs were performed before each bend test at twelve locations: each of the four corner-most plated through holes at the corners of the BGA, each of the four corners of the PCA, and at each of the four edge centers of the PCA. Lead breaks were performed with a 0.3 mm diameter 2H lead that was 2.4 mm long.

Acoustic data were measured with a DigitalWave FM-1 system using four B1025 transducers, which are approximately 9mm in diameter. This system records a pre-set amount of waveform data whenever the signal exceeds a threshold. The sampling rate was 20 MHz, and either 4096 or 8192 acoustic waveform data points were collected post-trigger, while 1,024 points were collected pre-trigger. The overall pre-amp was set at 12, 18, or 24 dB; the recorded signal was amplified 12 db; and the trigger was amplified 15 dB.

Both cross section and dye penetrant inspection were used to check for cracks post-test. Cross-section was performed on a limited number of test PCAs in order to verify the failure mode. Dye penetrant inspection was performed on the majority of PCAs to determine the extent of damage. Dykem red dye penetrant was applied in a vacuum and dried in an oven before package removal, and both the PCA and package sides of the BGA were inspected under an optical stereo-microscope at a magnification of no less than 25X.

Acoustic Data Reduction Methods

Two acoustic metrics were used in this study: event location and measured event energy. If the locations of three or more acoustic transducers are known, the time delay between arrival of an event at each transducer can be used to calculate the location of the event origin in two dimensions. For each waveform recorded at a transducer, the area under the wave can be integrated to calculate a relative measure of energy. The methods for calculating these data are described in this section.

Precise locating of events involves knowing the transducer locations, the wave propagation velocity through the PCA, and the arrival time of the wave front at each transducer. Transducer locations were measured and marked on each PCA before the test, and a circle slightly larger than the transducer was drawn at each location. This allowed the transducers to be placed on the PCAs during setup on the test fixture and quickly aligned to the correct location.

The wave velocity was determined by measuring the arrival times of a PLB wave. A single PLB waveform is shown for four transducers in Figure 4. This waveform is for a PLB event in the upper right-hand corner of the test PCA, where the transducers are located at the corners of the package. The wave crosses the upper right transducer first, the upper left and lower right transducers simultaneously second, and the lower left transducer last. The distance between the upper right and lower left is divided by the measured time delay between the arrival times at those transducers to determine the velocity.

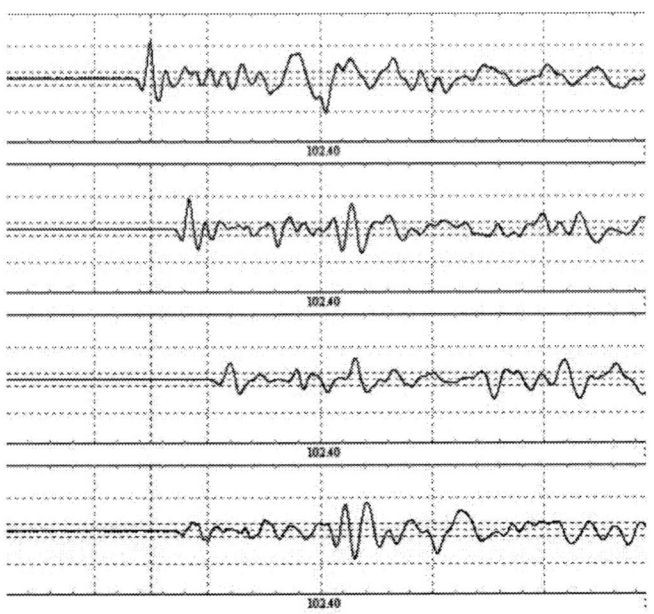

Figure 4. PLB waveform arrival times.

Arrival time is when the transducer signal crosses a threshold value, as shown in Figure 5. In this figure the black line is the signal voltage and the red dotted lines are the positive and negative threshold. Arrival time was determined automatically during the test using DigitalWave's SmartThreshold routine, which estimates the noise floor for each event and sets the threshold. After the test, the event locations were refined by manually setting the threshold value to the lowest level possible above the pre-event noise, as shown in the figure. The DigitalWave system can resolve arrival time to 0.01 microsecond.

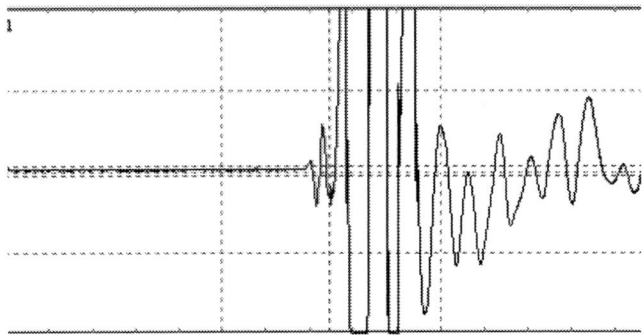

Figure 5. Arrival time determined by threshold crossing.

Energy was calculated by the DigitalWave system. The energy value measured at each transducer depends upon a number of factors. First is the amount of energy released by the event in the form of an acoustic wave. As the wave propagates through the PCA some of its energy is dissipated through reflections and losses. Upon arrival, the efficiency of conversion to electrical signal is determined by each transducer. The system then amplifies the signal by an amount determined by the user; however, too much amplification will saturate the signal and crop the peak of the wave. The amplified signal is recorded for a pre-set period of time. A waveform that is slightly saturated and truncated is shown in Figure 6. Since the energy is the area beneath the curve, the saturation and truncation will artificially decrease the calculated energy.

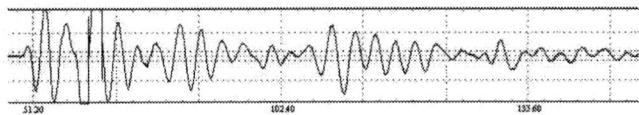

Figure 6. A waveform that is slightly saturated and truncated.

Energy is measured in units of $V^2 \cdot \mu s$. In this study, the energy at the transducer of first arrival is used as the measured energy. Due to all the factors listed above, the energy was treated as a relative value in this study rather than estimating the original energy of the source event. As will be shown, relative energy can be an effective metric for determining the onset of interconnect damage.

Damage Threshold Results

In all tests, acoustic activity followed a consistent trend with three stages: 1) many low energy events in the early stage of the test, 2) an intermediate stage with relatively few, low energy events, and 3) a final stage of a sudden increase in the number and energy of events, as shown in Figure 7. In this figure, the total detected acoustic energy is summed at each event and plotted as a function of test time. This final stage typically began with a high energy event that was 2-3 orders of magnitude more energetic than the events in the first two stages, and included a mixture of low, intermediate, and high energy events. These three stages have been interpreted as follows: 1) settling of the PCA, 2) energy storage, and 3) energy release through interconnect cracking.

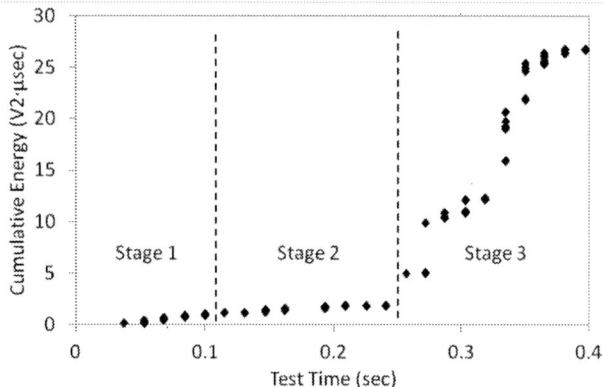

Figure 7. Typical cumulative energy plot showing three stages of acoustic activity.

Several rounds of validation were performed in an effort to determine the extent of damage based on acoustic activity. In the prior published study [3], several PCAs were bent to different levels and inspected with dye penetrant. These PCAs were bent to four different strain levels of 400, 500, 600, and 700 με. Dye penetrant inspection showed no dye stain for the three lower strain levels, and stain in only one BGA corner of one PCA for the 700 με tests. The PCA with the dye stain was the only test that showed a spike in acoustic activity, which was located in the same quadrant as the dye stain. Figure 8 shows the cumulative energy as a function of strain for the PCA with dye stain.

In the present study, two rounds of validation testing were performed. In the first round, three PCAs made from the second laminate material were bent at the low rate of 700 με/minute until two high energy events were identified. All three PCAs showed between 4 and 29 partial and full pad craters in each corner. The cumulative energy is plotted in Figure 9. Even though the bending rate was low, significant damage had occurred by the time two high energy events could be identified. The test was then stopped, and the PCA was unloaded. It is interesting to note that the number of events is less than the number of damaged interconnects, suggesting that single acoustic events, especially high energy events, may represent pad cratering at multiple interconnects. A high energy waveform from one of these tests is shown in Figure 10. If this waveform does represent multiple pad craters, the fractures would have to occur essentially simultaneously, since the entire high energy waveform lasts only about 200 microseconds before damping out.

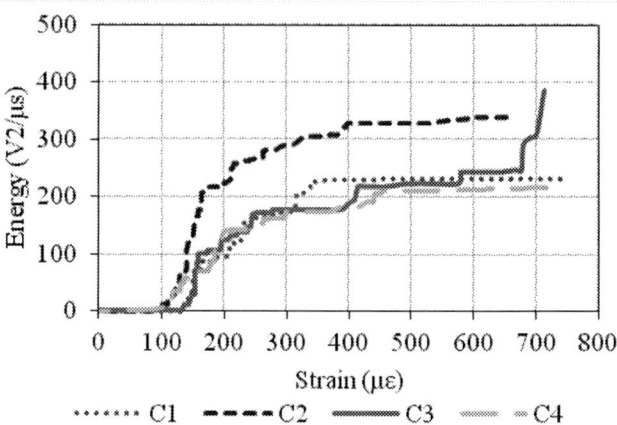

Figure 8. 700 με cumulative energy plot for the damaged PCA from the first study [3].

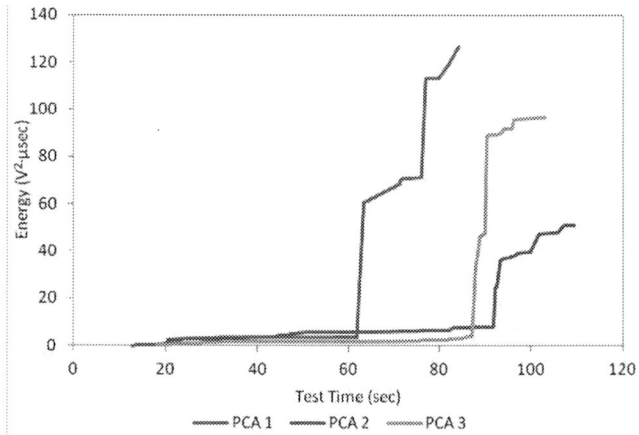

Figure 9. Cumulative energy for third validation test showing large amounts of dye stain.

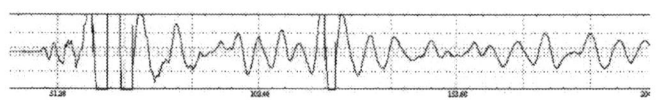

Figure 10. High energy waveform.

The fourth validation test was performed on twelve PCAs made from the third laminate material at a higher rate of approximately 5,000 με/second. In addition to validating the damage as a function of cumulative acoustic energy, the goals of this test were to demonstrate that acoustic measurement is feasible at IPC/JEDEC-9707 strain rates and that strain rate does not significantly change the results. The first six PCAs were used to set up the test and to determine the approximate displacement and strain level at which the third stage (energy release through interconnect cracking) occurs. The last six PCAs were bent to lower displacements estimated to be below the third stage threshold for half of the PCAs and above the threshold for the other half. The first three PCAs were bent to 0.89, 1.02, and 1.14 mm and did not produce any high energy

978-1-4799-2408-0/14 $31.00 © 2014 IEEE   1436

events. The last three PCAs were bent to 1.27, 1.40, and 1.27 mm and all produced high energy events. The cumulative energy curves for these six tests are plotted in Figure 11, and show a small spike in energy for the fourth test and larger spikes for the fifth and sixth tests. Dye penetrant inspection for these PCAs detected no dye stain for the first three PCAs, partial dye stain on two interconnects for the fourth PCA, dye stain on 53 interconnects for the fifth PCA, and dye stain on five interconnects for the sixth PCA. The board-side dye stain from the fourth PCA is shown in Figure 12, and a board-side cross section is shown in Figure 13. All failure analysis revealed board-side pad crater failures as the only failure mode. The calculated locations of the events did not always correspond well to the locations of dye stain. This could be explained by multiple pad crater cracks occurring simultaneously.

These tests support the conclusion that the sudden spike in the number and energy of acoustic events corresponds to pad cratering. The spike can be used in destructive tests to identify the onset of damage in the BGA.

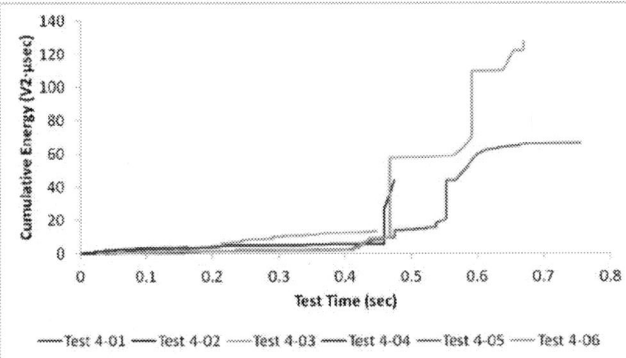

Figure 11. Cumulative energy for the high rate validation test.

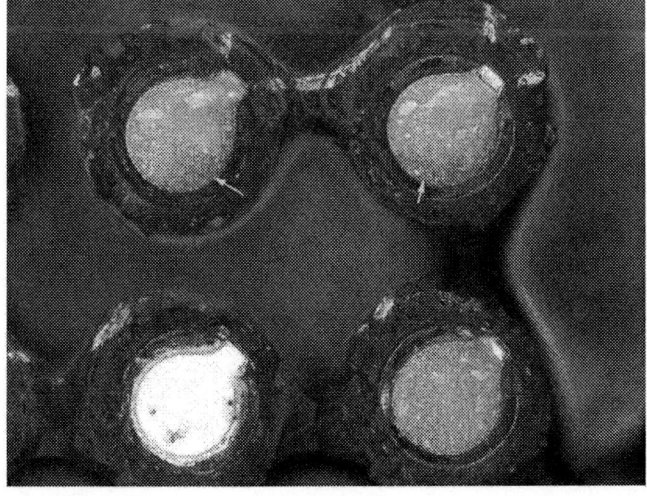

Figure 12. Dye stain on the fourth PCA from the final validation test.

Figure 13. Cross section showing pad cratering.

Event Locating Results

Pencil lead breaks (PLB) were used to determine the locating accuracy since the locations of the lead breaks were known. In the prior study, the PLB velocity in the on-axis fiber (0/90) direction was used. When PLB locations at the four BGA corners were determined on three PCAs from the same laminate material, the average location error was 3mm. Considering that the transducers are 9mm in diameter and that the spatial resolution of the DigitalWave system is 1mm, this accuracy is reasonable for typical acoustic emission applications. In the case of a 1 mm pitch ball grid array, this accuracy represents approximately 27 interconnects, which provides spatial accuracy sufficient to determine the corner from which the event originated. The event locations are shown in Figure 14, where the package shadow is shown with black lines, the transducer locations are shown with black circles, the actual PLB locations are marked with red diamonds, and the calculated locations are marked with blue diamonds.

The event locations in Figure 14 are all skewed outward from the BGA center, which indicates that the acoustic velocity is too high. Since the transducers are all positioned either approximately 45 or 15 degrees from the package corners, it was hypothesized that the acoustic velocity could be lower in the off-axis direction. This could be due to sound travelling faster in the direction of the fiber bundles than at a diagonal to the fiber directions. PLBs at the corners of these PCAs were used to determine the off-axis velocity, which was consistently 11% lower than the on-axis velocity. When the PLB locations were recalculated with the off-axis velocities, the average error was reduced to 1 mm, and the standard deviation reduced from 1.6 to 0.9 mm. The new event locations are plotted in Figure 15, and show that the events are now scattered evenly about the PLB locations.

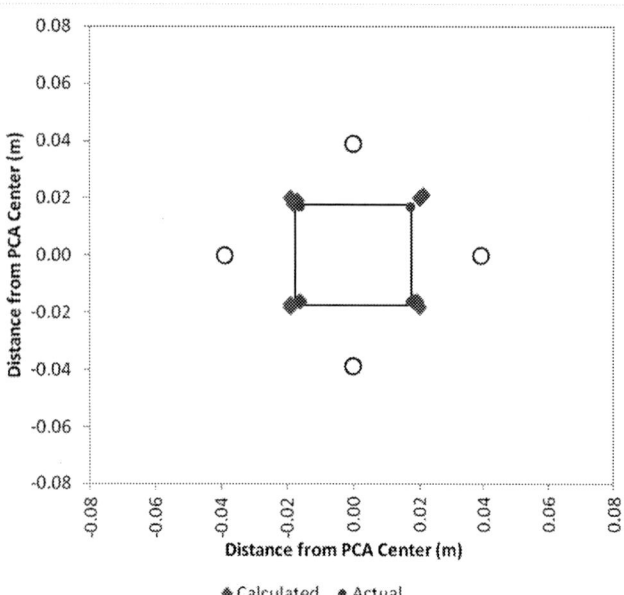

Figure 14. PLB locations using on-axis fiber velocity.

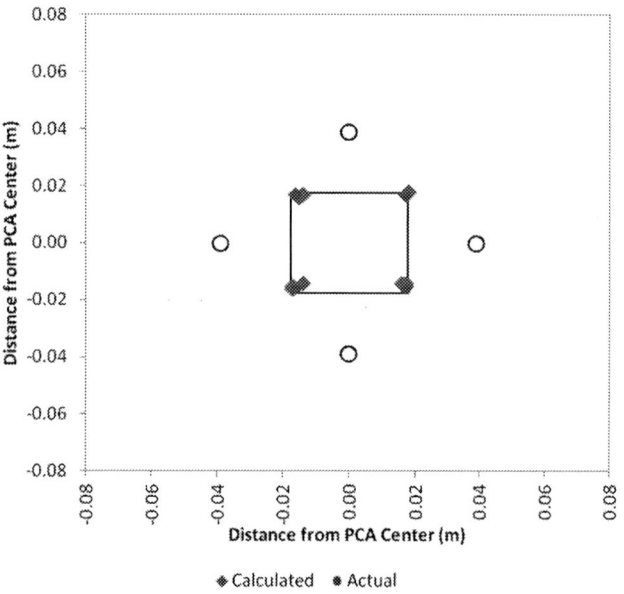

Figure 15. PLB locations using off-axis velocity.

Three additional acoustic transducer arrangements were evaluated in order to determine whether this would improve locating precision. The four arrangements are shown in Figure 16. The first arrangement, listed as B in Figure 16, used the same 38 mm radius as before, but placed the transducers at the BGA corners, rather than edges. The benefit to this arrangement is that events at any corner will arrive first at a single transducer near the same corner, rather than nearly simultaneously at two transducers placed at the nearby edges. In terms of direction, the transducers that are nearest and farthest from each corner are 45 degrees to the fiber orientation, and the two intermediate transducers are about 15 degrees to the fiber orientation, giving no advantage over the original arrangement. The other two arrangements

used a 71 mm radius (the maximum for the board design) with the transducers at the edges and corners of the PCA— arrangements C and D, respectively. In arrangement C, the transducers are located approximately 10 and 20 degrees off-axis from the PLB locations, so the on-axis velocity was used. In arrangement D, the transducers are approximately 45 and 25 degrees off-axis, so the off-axis velocity was used. The disadvantage to these last two arrangements is that the transducers are farther away from the PLBs, so the acoustic wave attenuates more and the extensional and transverse wave components separate further (due to their different velocities), making analysis more complicated. Twelve PLBs were performed on the same PCA for each arrangement. There was no significant difference in the average location error between the four arrangements. Arrangement B was chosen for the tests in this study since it was not much different than Arrangement A, but offered the data analysis advantages mentioned above.

A locating algorithm that incorporated an orthotropic velocity model should further improve the locating accuracy, especially if the calculating resolution was improved to less than 1 mm. This is complicated by the fact that velocity is likely to vary throughout the laminate, especially in the BGA region, due to differences in local elastic modulus. Since arrival times can be measured by this system to 0.01 microsecond, and the acoustic velocity in these materials is on the order of 3,000 m/s, locations should be able to be resolved to less than 0.1 mm (0.01 microsecond * 3,000 m/s = 0.03 mm).

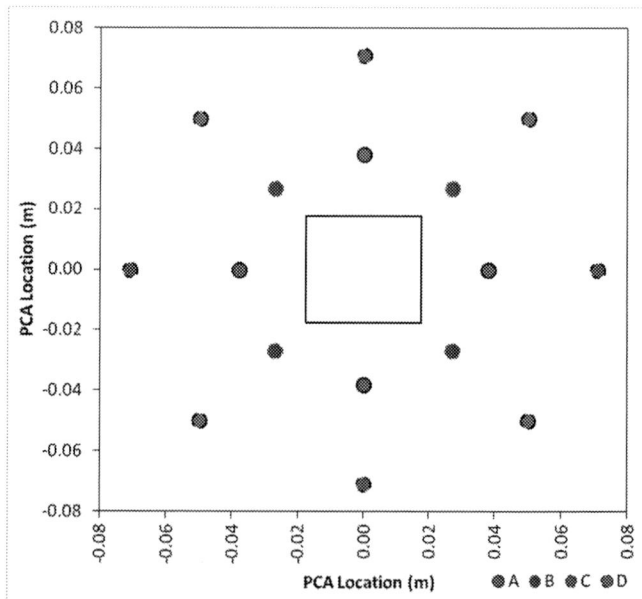

Figure 16. Four transducer arrangements that were evaluated.

Application to the Standard Test Method

Acoustic emission detection appears to be immediately applicable to the IPC/JEDEC-9707 test method and could potentially improve the precision of the test, decrease the time to complete the testing, and decrease the cost. This could be done by adding acoustic transducers to the test setup while keeping all other aspects of the setup unchanged. The test method calls for a small number of PCAs to be tested to obvious interconnect failure, and then a statistically significant number of PCAs to be tested to one or more (typically two) lower levels in order to determine the level at which damage occurs using failure analysis. Instead, a single set of a statistically significant number of PCAs could be tested to obvious failure, and acoustic emission could be used to determine strain at which the increase in acoustic activity occurs. For validation, a smaller number of PCAs could be bent to the safe strain level and destructively inspected for damage.

This method could increase the precision of the test by providing the strain at which damage initiates, rather than only a pass/fail judgment of particular strain levels. In addition, the strain level can be pinpointed at the BGA corner in which damage initiates, rather than grouping the strain values at all four corners. If the strain state at damage initiation is not significantly different between the corners, then multiple corners, or even all four corners could be analyzed separately, providing as much as four times the number of strain values per PCA and reducing the required number of test units and tests accordingly. An example of this is shown in Figure 17, in which cumulative energy for a single test is plotted versus strain and grouped by corner. The test would provide a distribution of damage strain values, which could be used to quantify the safe strain level in a number of ways: the minimum damage strain, the mean damage strain minus a particular number of standard deviations, the mean damage strain minus a pre-determined safety-factor, etc. As an example, five acoustic threshold events were identified from PCAs made from the same laminate material and tested using the methods described above. The strain value was extracted from the strain gage located 5 mm from the corresponding package corner at the same point in time as the acoustic threshold event. This strain distribution is shown in Figure 18.

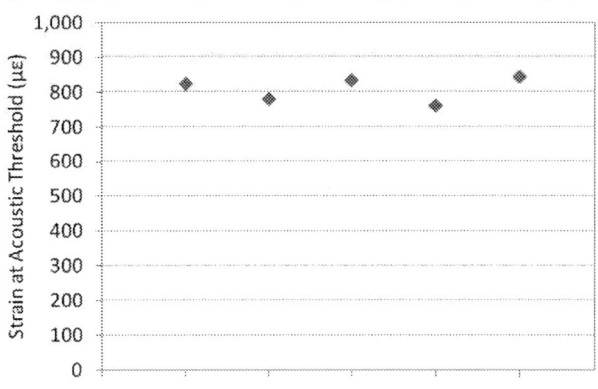

Figure 18. Damage strain distribution for PCAs made from the same laminate material.

This testing approach could also increase the speed of testing. The overall number of tests is likely to be decreased, which decreases instrumentation time, test time, and data reduction time. Probably most significant is the reduction of failure analysis time, since failure analysis is a labor-intensive step. In addition, preliminary data would be available soon after the bending test, rather than having to wait for failure analysis to be completed.

The decrease in the number of tests could also translate to a cost decrease. The savings would be due to a reduction in labor hours spent on strain gage instrumentation, bend testing, and failure analysis.

Conclusions

Acoustic emission detection appears to be a promising technique for detecting pad cratering, and potentially multiple forms of interconnect damage events. While acoustic activity can be detected throughout PCA bending, the increase in acoustic event energy and frequency allows interconnect damage to be discriminated from the background noise. An improvement in event locating accuracy to 1 mm has been demonstrated by using the off-axis acoustic velocity, which is more appropriate for most transducer arrangements.

The addition of acoustic monitoring techniques may improve the IPC/JEDEC-9707 test method in terms of precision, through-put time, and labor cost. It is also possible that acoustic monitoring could be used on production printed circuit assemblies during manufacturing and testing to check for damage. If this is feasible, it may be able to significantly reduce the need for strain-based monitoring.

Further refinements to the acoustic emission testing techniques are recommended. Event waveform frequency distribution could be used to further discriminate between different kinds of events, such as settling of the laminate, pad cratering, and intermetallic fracture. Cold ball pull tests could be used to relate acoustic energy to mechanical energy, measured event energy to actual event energy, and test event locating to pencil lead break locating accuracy. An improved algorithm for orthogonal velocity properties is likely to further improve locating accuracy and precision.

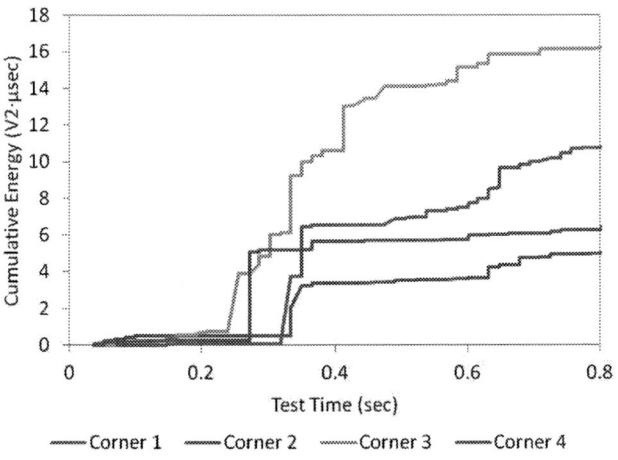

Figure 17. Cumulative acoustic energy grouped by BGA corner.

Acknowledgments

The authors would like to George Raiser at Medtronic for his generous assistance with the cold ball pull testing, which will be used for future development.

Pradeep Lall and his team at Auburn University's CAVE3 laboratory provided micro-CT scanning failure analysis assistance.

Mark Nickerson, Mark Gelber, and Gregory Daspit at Southern Research Institute assisted significantly in the test development and data analysis.

References

1. IPC/JEDEC-9707, "Spherical Bend Test Method for Characterization of Board Level Interconnects," September, 2011.

2. R. Reinosa, "Characterizing Mechanical Performance of Board Level Interconnects for In-Circuit Test," International Test Conference (ITC), Austin, TX, November 2-4, 2010.

3. A. Bansal, "A New Approach for Early Detection of PCB Pad Cratering Failures." IPC/APEX Conference, Las Vegas, NV. 2011.

4. W. C. Ralph, "Acoustic Emission Detection of BGA Components in Spherical Bend," in Proc. IEEE Electronic Components and Technol. Conf. (ECTC), Las Vegas, NV, May 28-31, 2013, pp. 208-213.

5. A. Bansal, "Investigation of Pad Cratering in Large Flip-Chip BGA using Acoustic Emission." IPC/APEX Conference, San Diego, CA, April 2012.

6. IPC/JEDEC-0704A, "Printed Circuit Assembly Strain Gage Test Guideline," January, 2012.

High Acceleration Board Level Reliability Drop Test Using Dual Mass Shock Amplifier

Andy Zhang
Texas Instruments, Inc.
13020 TI Blvd, MS 3636, Dallas, TX, 75243, USA
azhang@ti.com

Abstract

The advance of mobile electronics applications has been demanding higher drop/shock reliability performance under more severe drop/shock impacts. This in turn is requiring the board level reliability (BLR) drop test to be able to reach higher peak acceleration, particularly higher than 5000 G. One method to reach this high acceleration is to use an apparatus called Dual Mass Shock Amplifier (DMSA) attached onto a conventional drop tester. The conventional drop tester typically generates a primary impact in the range of 1000 ~ 2000 G, and the DMSA, which hangs above the drop table of the drop tester and continues to fall down immediately following the primary impact, is capable to generate a secondary impact with the drop table with magnitude 2~5x of the primary impact in terms of peak acceleration. In this study, the mechanics of drop test and DMSA was introduced and modeled with Newton physics. Two DMSAs with various different features were custom-designed, fabricated, assembled, and tested. Various aspects of the DMSA, including materials, layout, weight, gap height, holding mechanism, were evaluated and addressed. The effects of drop height, weight of DMSA, and gap height on the change of velocity and peak acceleration of secondary impact were studied. The strain on BLR drop test board under high acceleration impact was measured and discussed. Consistency of the shock amplification characterized by C_{pk} was studied and method to improve the consistency was discussed. It is demonstrated through this study that using DMSA can be a simple, economical, and consistent method to achieve high acceleration for BLR drop test.

Introduction

Touch screen smart phones and tablet computers have become so popular these years and have changed people's life in a significant way. Engineers have managed to put more and more functionalities into reduced footprints of these handheld mobile devices. Reduction in space, especially in the out of plane z axis can have a strong influence on product reliability, particularly the board level reliability (BLR). The small tolerances used between materials in a z-stack, i.e., between a printed circuit board (PCB) and a semi rigid plane such as a metal frame, may not be enough to prevent contact during a drop event [1]. Studies have shown that impact of a flexible structure and stiff frame can generate very high accelerations in the range of 10^4~10^5 G [2]. There are several methods for obtaining high acceleration in reliability testing. These include rail gun, air gun, centrifuge, and pneumatic shock [3]. The rail gun and air gun methods are very expensive and time consuming. Drop test is a very commonly used method for BLR thanks to its low cost and simplicity. However, due to the technical difficulties of reaching high acceleration using drop test, industry BLR drop test standards typically limit acceleration levels to 3,000 G and less.

In recent years, an apparatus called Dual Mass Shock Amplifier (DMSA) has been given more and more attention for high acceleration drop test [4~7]. Using DMSA to generate high acceleration for the purpose of mechanical test has been reported as early as since 1988 [8]. Schematic of a typical DMSA attached to a conventional drop test machine is showed in Figure 1. It consists of a base, 2 or 4 guide rods, a table that moves along the guide rods freely, springs or bungee cords that hold the table above the base with a certain gap height, and some accessories. A special flat sheet material called programmer is inserted into the gap between DMSA table and base in order to generate desired pulse duration. The DMSA base is rigidly attached to the drop table of a conventional drop test machine. To perform the high acceleration drop test, the drop table, DMSA base and DMSA table, will free fall all together from a certain drop height. The drop table and DMSA base will then collide with drop base in an impact called primary impact, and then bounce upward. At the same time the DMSA table will continue moving downward and collide with the upward-moving DMSA base in an impact called secondary impact, and then bounce upward with a large change of velocity which generates high acceleration. The device under test (DUT) is attached to the DMSA table using some sort of fixture.

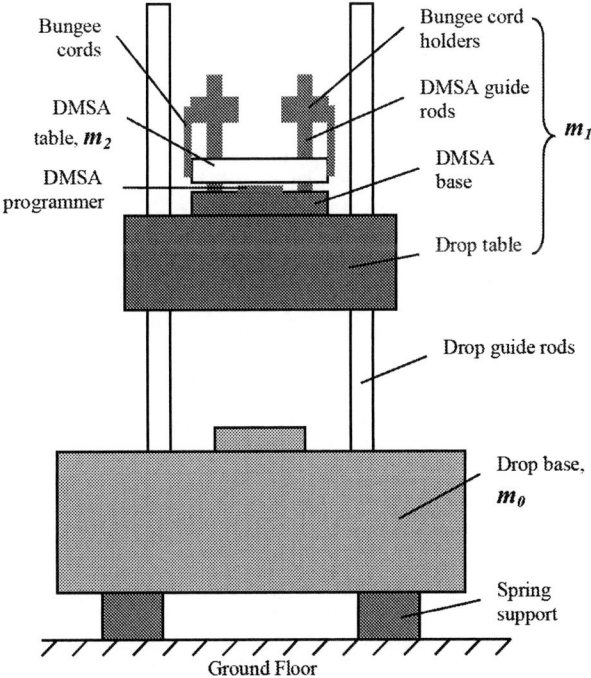

Figure 1. The schematic of a typical drop test machine and DMSA.

Analytical models of the DMSA have been reported. Rodgers *et al* [2] proposed a model for determining the dynamics of multiple pair-wise collisions in a chain, and demonstrated that a shock amplifier with mass ratios based on a power law relationship is optimal and most energy-efficient. Douglas [4] in a thesis provided models ranging from simple analytic closed-form rigid-body mechanics to detailed nonlinear dynamic finite element analysis. The effects of different equipment design parameters (table mass, spring stiffness, table clearance) were investigated through parametric modeling [4].

In this paper, the process of dual mass shock amplification will be analyzed and modeled using simple rigid-body mechanics. Analytical equations for estimating the change of velocity and peak accelerations will be established. How the change of velocity and peak accelerations are affected theoretically by various parameters will be discussed. Guidelines for designing DMSA will be provided based on the model. In the experiment section, two DMSAs with different features were custom-designed, fabricated, assembled, and tested. Various aspects of the DMSA, including materials, weight, drop height, gap height, holding mechanism, were evaluated by experiments and commented. The strain on a BLR drop test board under high acceleration impacts was measured and analyzed. Consistency of the shock amplification characterized by C_{pk} was studied. Methods that aim to improve the repeatability and consistency of high acceleration drop test using DMSA will be introduced and discussed.

Analytical Model of Shock Amplification

In conventional drop test, the drop table, which the DUT is attached to, falls freely and collides with a stationary base, generating a modest change of velocity on the drop table. The key to produce high acceleration in dual mass shock amplification is to allow a light rigid body which the DUT is attached to (i.e., the DMSA table) to move and collide with a heavy rigid body moving at opposite direction (i.e., the drop table), generating a large change of velocity on the light DMSA table. To make this possible, the DMSA table will collide with the drop table when the drop table is bouncing upward after it collides with the drop base. The process of dual mass shock amplification will be analyzed and modeled in the remaining of this section. Table 1 lists the nomenclatures to be used in this paper.

Table 1. Nomenclatures used in the DMSA modeling.

m_0	Mass of the drop base
m_1	Mass of the sum of drop table, DMSA base, and all other accessories rigidly attached to DMSA base.
m_2	Mass of the DMSA table
h	Drop height
k	Spring constant of the bungee cords
d	Gap height between DMSA table and DMSA base
g or G	Gravity
v_i	Velocity right before primary impact for body i, where subscript 0 stands for drop base, 1 for drop table, and 2 for DMSA table
v_i'	Velocity right after primary impact for body i
v_i''	Velocity right after secondary impact for body i
Δv_1	Change of velocity of drop table during primary impact
Δv_2	Change of velocity of DMSA table during secondary impact
A_1	Peak acceleration on drop table in primary impact
A_2	Peak acceleration on DMSA table in secondary impact
t_1	Pulse duration of primary impact
t_2	Pulse duration of secondary impact

The initial condition is that the drop base is stationary right before primary impact, i.e.,

$$v_0 = 0 \qquad (1)$$

The velocity of drop table and DMSA table right before primary impact can be obtained using free fall assumption:

$$v_1 = v_2 = -\sqrt{2gh} \qquad (2)$$

where the negative sign denotes the downward direction of the velocity.

During the primary impact, the velocity of drop table and drop base change, governed by conservation of momentum

$$m_0 v_0 + m_1 v_1 = m_0 v_0' + m_1 v_1' \qquad (3)$$

and conservation of kinetic energy,

$$\frac{1}{2} m_0 v_0^2 + \frac{1}{2} m_1 v_1^2 = \frac{1}{2} m_0 v_0'^2 + \frac{1}{2} m_1 v_1'^2 \qquad (4)$$

assuming no other forms of energy are generated.

By solving Eq. (1) ~ (4), we can derive the velocity of drop table right after primary impact as

$$v_1' = \frac{m_0 - m_1}{m_0 + m_1} \sqrt{2gh} \qquad (5)$$

The change of velocity of drop table during primary impact can be described by

$$\Delta v_1 = v_1' - v_1 = \frac{2m_0}{m_0 + m_1} \sqrt{2gh} \qquad (6)$$

Assuming the pulse shape is perfect half-sine shape, the peak acceleration of drop table in primary impact can be expressed by

$$A_1 = \frac{\pi \Delta v_1}{2t_1} = \frac{\pi m_0 \sqrt{2gh}}{(m_0 + m_1)t_1} \qquad (7)$$

Due to the very short duration (~0.5 ms) of the primary impact, it is reasonable to assume that the velocity of DMSA table does not change during the primary impact, i.e.,

$$v_2' = v_2 = -\sqrt{2gh} \qquad (8)$$

Due to the very short duration (<2 ms) between primary and secondary impact, it is again reasonable to assume that for the DMSA table the velocity right after primary impact is the same as the velocity right before secondary impact. Similarly, by the conservation of momentum and kinetic energy, we have:

$$m_1 v_1' + m_2 v_2' = m_1 v_1'' + m_2 v_2'' \qquad (9)$$

$$\frac{1}{2}m_1{v_1'}^2 + \frac{1}{2}m_2{v_2'}^2 = \frac{1}{2}m_1{v_1''}^2 + \frac{1}{2}m_2{v_2''}^2 \qquad (10)$$

By solving Eq. (5), (8), (9) and (10), we can derive the velocity of DMSA table right after secondary impact as

$$v_2'' = \frac{3m_0m_1 - m_1^2 - m_0m_2 - m_1m_2}{(m_0+m_1)(m_1+m_2)}\sqrt{2gh} \qquad (11)$$

The change of velocity of DMSA table during secondary impact can be written by

$$\Delta v_2 = v_2'' - v_2' = \frac{4m_0m_1}{(m_0+m_1)(m_1+m_2)}\sqrt{2gh} \qquad (12)$$

Again assuming the pulse shape is perfect half-sine shape, the peak acceleration of DMSA table in secondary impact can be described by

$$A_2 = \frac{\pi \Delta v_2}{2t_2} = \frac{2\pi m_0m_1\sqrt{2gh}}{(m_0+m_1)(m_1+m_2)t_2} \qquad (13)$$

To further simplify Eq. (13), when $m_0 \gg m_1$ and $m_1 \gg m_2$, which is true for the drop machine and DMSA used in this study, the peak acceleration of DMSA table (A_2) can be estimated roughly using the following equation

$$A_2 \approx \frac{2\pi\sqrt{2gh}}{t_2} \qquad (14)$$

From the stand point of DMSA design, for a given drop test machine, a given drop height, and a specified pulse duration (t_2), we can achieve maximum peak acceleration (A_2) when the following condition is met:

$$m_1 = \sqrt{m_0m_2} \qquad (15)$$

Eq. (15) is in agreement with the model derived by Rogers et al [2]. By carefully designing the mass of DMSA base and table to meet Eq. (15), we can achieve maximum peak acceleration for a given drop machine, a given drop height, and a specified pulse duration, or in other word, we can achieve maximum shock amplification efficiency (i.e., lowest needed drop height to achieve a given peak acceleration and pulse duration). On the other hand, in a system where m_0 and m_1 are fixed and unchangeable, it is preferable to have m_2 as small as possible in order to generate high acceleration, according to Eq. (13).

Design of DMSA

Two custom designed DMSAs were fabricated, assembled, used and assessed in this study. DMSA1 was equipped with a table made with high-strength aluminum alloy, and DMSA2 was with a table made with high-strength magnesium alloy. Both DMSAs were equipped with bases and accessories made with general-purpose aluminum alloy. Outline dimensions for both tables were the same. The base of DMSA1 was slightly thicker than the one of DMSA2. A picture of DMSA2 (Mg alloy table) is showed in Figure 2. The DMSA table is designed to be a "porous" and light-weight structure which is believed to able to produce higher peak acceleration than that produced by a heavier one, for given and unchangeable m_1 and m_0, according the Eq. (13). The top surface of the DMSA table is wide enough to allow attaching a PCB test vehicle of up to 100x100 mm. The DMSA table is hung above the DMSA base by two bungee cords hooked up to a number of thumb screws on both sides of the DMSA table. The other end of each bungee cord is attached to a holder, whose position can be adjusted in vertical direction. The DMSA table can be adjusted to be perfectly horizontal and parallel to the surface of the DMSA base with desired gap height by adjusting each holder's position independently. The DMSA table runs along two case hardened and precision ground shafts with self-aligning ball bearings in between. There is a polymer tube inserted into each shaft in the section above the DMSA table, serving as a shock buffer. Polymer programmer with desired pulse duration characteristics will be inserted in between the DMSA table and base prior to use (not showed in Figure 2).

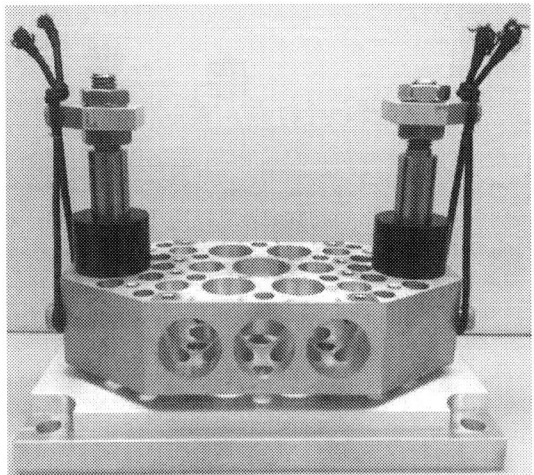

Figure 2. A custom designed DMSA made with Mg alloy

A commercially available free-fall shock/drop machine (Model # SD-10, L.A.B Equipment Inc.) with highest drop height of 0.7 m was used. The DMSA base was designed to fit the mounting holes on the drop table. The guide rods of the drop machine were cleaned and lubricated according the owner's manual provided by the manufacturer. The base of the drop machine is supported by four springs underneath and is movable during primary impact. No programmer was used between the drop table and drop base. The strike surface of primary impact was rigid plane on both impacting bodies; therefore loss of kinetic energy during primary impact was minimized. There is a rebounce brake which stops the drop table when the drop table bounces up to approximately 20 cm, and therefore prevents a second impact between the drop table and base.

The weights of parts of the drop machine and the two DMSAs were listed in Table 2.

Table 2. The weights of drop machine and DMSA parts.

Item		Weight (kg)
Drop base		228
Drop table		20.4
DMSA1	Base and accessories	2.38
	Table (Aluminum)	1.07
DMSA2	Base and accessories	1.94
	Table (Magnesium)	0.70

Table 3. Comparison of the change of velocity of drop table and DMSA table (Δv_1 and Δv_2, respectively) and peak acceleration of DMSA table (A_2) as predicted by model and measured in experiments.

Drop Parameters		Model Predictions			Experiment Measurements			Ratio of Experiment to Model for		
h (m)	t_2 (ms)	Δv_1 (m/s)	Δv_2 (m/s)	A_2 (G)	Δv_1 (m/s)	Δv_2 (m/s)	A_2 (G)	Δv_1	Δv_2	A_2
0.3	0.231	4.42	8.57	5944	4.28	5.18	4361	0.97	0.60	0.73
0.4	0.193	5.10	9.89	8215	5.19	6.64	6787	1.02	0.67	0.83
0.5	0.133	5.70	11.06	13328	5.65	7.20	9736	0.99	0.65	0.73
0.6	0.124	6.25	12.11	15660	6.12	8.21	13056	0.98	0.68	0.83
0.7	0.120	6.75	13.09	17478	6.54	8.83	13665	0.97	0.68	0.78

Methodology of Measurements

Acceleration of the drop table was measured by an IEPE type single-direction accelerometer (Model # 8704B5000, Kistler AG) with sensitivity of approximately 1 mV/g, stud-mounted to the DMSA base which was rigidly attached to the drop table. Acceleration of the DMSA table was measured by a shear ICP® type single-direction accelerometer with sensitivity of approximately 0.1 mV/g (Model # 350D02, PCB Piezotronics Inc.), stud-mounted to the top surface of DMSA table. Both accelerometers were connected to National Instruments NI 9234 acceleration data acquisition module using excitation current of 2 mA with sampling rate of 51.2 kS/s. Pulse duration was measured as "10% to 10%", i.e., the duration between the instant the acceleration rises to 10% of the peak level and the instant the acceleration drops to 10% of the peak level for the first time after reaching the peak. Change of velocity was calculated as the area under the acceleration-time curve within the pulse duration.

The ability of a process to produce output within specification limits can be characterized by using a parameter commonly referred as C_{pk}, which is calculated by

$$C_{pk} = \min\left(\frac{USL - \mu}{3\sigma}, \frac{\mu - LSL}{3\sigma} \right) \quad (16)$$

where USL and LSL are upper and lower specification limits, respectively, and μ and σ are the estimate mean and standard deviation of the process, respectively. In this study, USL and LSL of DMSA peak acceleration or change of velocity for a given drop height are unknown. However, given that tolerance of +/-10% is generally accepted by industry drop/shock test standards, it is reasonable to assume that USL and LSL are equal to 100%+/-10% of μ, and therefore C_{pk} in this study was calculated by

$$C_{pk} = \frac{\mu}{30\sigma} \quad (17)$$

Drop Height

Drop test using the magnesium alloy DMSA at drop height (h) from 0.3 to 0.7 m was conducted. An inelastic programmer with pulse duration (t_2) variable in the range of 0.23 to 0.12 ms depending on the peak acceleration was used in between the DMSA table and base. 15 drops were conducted at each drop height. The measured data of the change of velocity of drop table and DMSA table (Δv_1 and Δv_2, respectively) and peak acceleration of DMSA table (A_2) was averaged (at each drop height) and listed in Table 3, and compared with the data predicted by the model. For the Δv_1, the model prediction was in very good agreement with the experimental measurements, which suggests that the assumptions of no drop friction and no kinetic energy loss during primary impact are very close to reality. On the other hand, the Δv_2 and A_2 measured from experiments were much lower than those predicted by the model. The fact that the ratio of experimental measurement to model was much lower than 1 for Δv_2 (and therefore A_2, which is linear to Δv_2 for a given pulse shape) may be mainly due to the assumption of no kinetic energy loss during secondary impact, which was far from reality when inelastic programmer was used. In addition, the neglect of the effects of gravity and the pulling of the bungee cords during and in between primary and secondary impacts also contributed to the discrepancy between model and experiment for both Δv_2 and A_2. Further comparison found that the ratio of experiment to model for A_2 was slightly higher than that for Δv_2, which may be explained by the possibility that the pulse of secondary impact was not half-sine shape but "slimmer", leading to higher peak acceleration if change of velocity was unchanged. Even though the absolute numbers predicted by the model were far lower than experimental measurement, the ratios of experiment to model for Δv_2 and A_2 remained at relatively constant numbers, with 0.60 to 0.68 for Δv_2 and 0.73 to 0.83 for A_2, respectively. This implies that by simply applying coefficients of 0.66 and 0.78 to the model-predicted data of Δv_2 and A_2, respectively, the real change of velocity and peak acceleration obtained from the magnesium DMSA drop test system can be estimated quite accurately. The same coefficients can be applied to the model-predicted data when the aluminum DMSA drop test system is used.

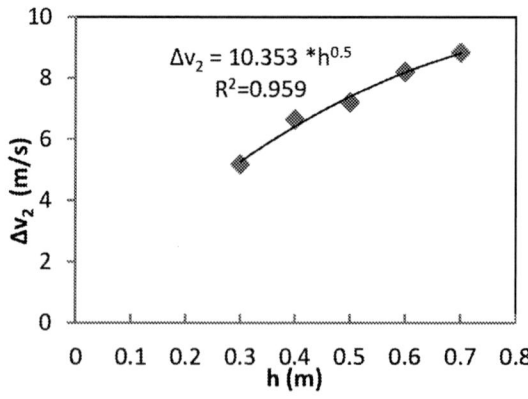

Figure 3. Change of velocity of DMSA table during secondary impact (Δv_2) measured from experiments with various drop heights (h).

Eq. (12) suggests that for a given DMSA/drop machine system, the change of velocity of DMSA table during secondary impact (Δv_2) is linearly proportional to the square root of drop height (h). To verify this, the Mg DMSA was used and tested at various drop heights with 15 drops at each height. The measured Δv_2 with h in the range of 0.3~0.7 m was plotted using polynomial regression with order of 0.5, as showed in Figure 3. The high R^2 value (0.959) indicates that the linear relationship between Δv_2 and square root of h as predicted by the model in Eq. (12) is verified by the experiment result.

Weight and Material of DMSA Table

As predicted by the model, the difference between the weight of the Aluminum DMSA table and that of the Magnesium one (1.07 kg vs. 0.70 kg) results in only approximately 2% difference in the change of velocity and peak acceleration. Experiments using these two DMSA tables have confirmed that this small difference in change of velocity and peak acceleration is true. There is no indication that the difference in material characteristics between aluminum and magnesium alloys, such as hardness and strength, would lead to noticeable difference in the pulse shape.

Pulse Shape

Eq. (13) in the analytical model for calculating the peak acceleration was based on the assumption of half-sine pulse shape in secondary impact. In reality, experiment results indicated that this assumption is valid only when the DMSA is well balanced during the impacts. The pulse shape would be distorted from half-sine shape if there are imperfections including imbalanced DMSA weight distribution, non-symmetrical placement of accelerometer or test vehicle on DMSA table, high stiffness of the monitoring cables, different friction characteristics of the two bearings in DMSA, different tensions on the two bungee cords, etc. Efforts shall be made carefully to minimize all these imperfections before a near half-sine pulse shape can be obtained. In the situation where the pulse is not close to half-sine shape, using Eq. (13) to estimate peak acceleration would be incorrect. A coefficient for adjusting the effect of pulse shape on peak acceleration shall be added into Eq. (13). On the other hand, change of velocity instead of peak acceleration can be used to characterize the pulse in secondary impact, due to the fact that change of velocity is not sensitive to pulse shape or pulse duration.

A representative waveform of the acceleration of DMSA table during secondary impact, obtained with magnesium DMSA at drop height of 0.6 m, is showed in Figure 4(a). Momentarilly after reaching the peak of 12000 G, the acceleration continued to ossocilate with numerous small peaks with amplitudes typically within 30% of the main peak, until approximately 4 ms later when it stablized around zero. Figure 4(b) showed a magnified view of the same waveform at the main peak, where a half-sine pulse shape is confirmed.

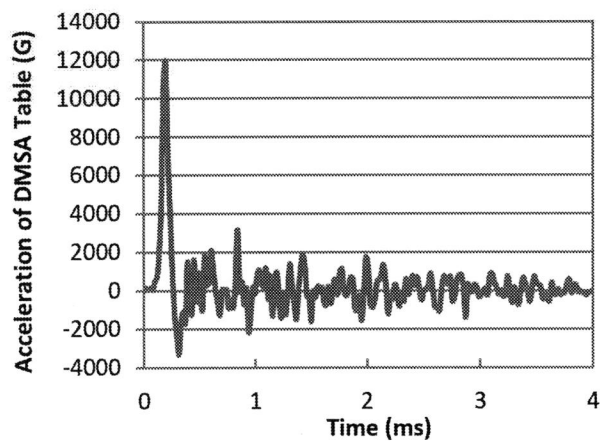

Figure 4(a). A representative waveform of the acceleration on DMSA table during secondary impact.

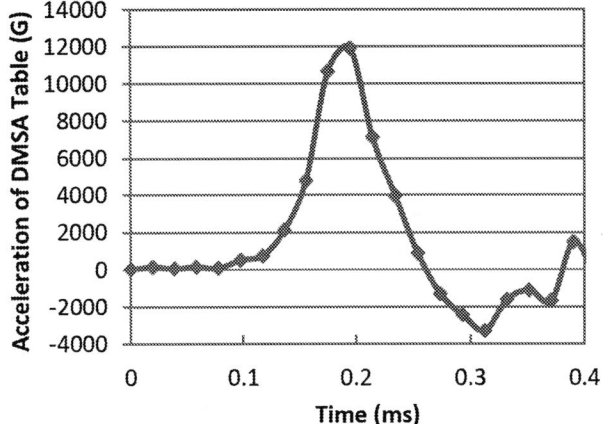

Figure 4(b). A magnified view of Figure 4(a) showing the half-sine shape of the pulse during secondary impact.

Effects of Gap Height and Bungee Cords

The analytical model presented in previous section is based on an assumption that the period between primary and secondary impacts (called transitional period) is short enough so that the change of velocity for drop table and DMSA table in this period is negligible. If this change of velocity during the transitional period has to be considered, the effect of spring constant of bungee cord and gap height would have to be examined carefully. During the transitional period, the drop table moves upward and decreases its velocity due to gravity, while the DMSA table continues to move downward under two opposite-direction forces from gravity and from bungee cords, until the two bodies collide in the secondary impact. A numerical model that describes the kinetics in this period was established. The weights of drop machine and magnesium DMSA used in the experiment were applied in the model. Velocity right before secondary impact of drop table ($v_1{}'$) and DMSA table ($v_2{}'$), and peak acceleration of DMSA table (A_2) at various combinations of bungee cord spring constant (k) and gap height (d), were adjusted accordingly by the model. For a typical drop height of 0.7 m and pulse duration of 0.2 ms, the non-adjusted and adjusted values were listed in Table 4(a) and 4(b), respectively.

Table 4. Non-adjusted (a) and adjusted (b) velocities right before secondary impact of drop table (v_1') and DMSA table (v_2'), and peak acceleration of DMSA table (A_2), at various combinations of bungee cord spring constant (k) and gap height (d).

(a) Without adjustment for transitional period				
k (N/m)	d (mm)	v_1' (m/s)	v_2' (m/s)	A_2 (G)
N/A	N/A	3.0430	-3.7041	10486.85
(b) With adjustment for transitional period				
k (N/m)	d (mm)	v_1' (m/s)	v_2' (m/s)	A_2 (G)
100	6	3.0354	-3.7105	10484.96
250	6	3.0354	-3.7086	10482.12
1000	6	3.0354	-3.6995	10467.93
5000	6	3.0353	-3.6508	10392.24
250	2	3.0412	-3.7056	10486.42
250	6	3.0354	-3.7086	10482.12
250	10	3.0295	-3.7083	10472.54
250	20	3.0150	-3.6925	10425.47

Comparing Table 4(a) and 4(b), it can be concluded that the differences in velocity and peak acceleration between the non-adjusted and adjusted were so small that it is accurate enough to make the assumption that there is no change of velocity during the transitional period, for an analytical model aiming to provide DMSA design guideline and estimate of output acceleration. On the other hand, the variations in Table 4(b) suggests that a higher spring constant (i.e., stiffer bunger cord), or a larger gap height, with any other parameters remaining unchanged, would decrease the peak acceleration of DMSA table slightly. Therefore, it is favorable to use bungee cord with smaller spring constant and/or smaller gap height in the DMSA design whenever permitted. In this study, spring constant of 250 N/m and gap height of 6 mm were used as a baseline.

DMSA Table Holding Mechanism

Two mechanisms for holding the DMSA table and bungee cords were designed, used, and assessed. In Figure 5(a), the DMSA table is held by only one thumb screw on each side, and while in Figure 5(b), it is held by two thumb screws. In practice we found that the DMSA table with the dual thumb screw mechanism is more likely to maintain parallel to the base during the drop movement, and is more likely to generate consistent values of peak acceleration and pulse duration, and maintain the half-sine pulse shape. C_{pk} of the change of velocity (Δv_2) for the two mechanisms at various drop heights were showed and compared in Figure 6. The dual thumb screw mechanism showed consistently higher C_{pk} value than the single thumb screw mechanism did, which indicates that the dual mechanism would be preferable to the single mechanism in the consideration of repeatability. In addition, the screws are less likely to fracture over time in the dual thumb screw mechanism than in the single one.

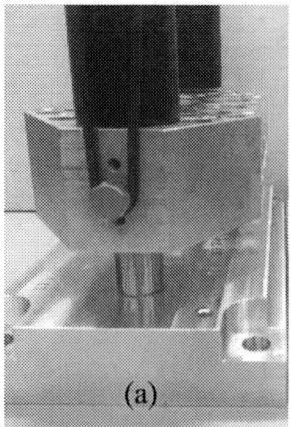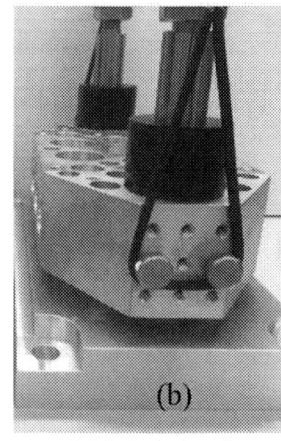

Figure 5. (a) Single thumb screw, and (b) dual thumb screw mechanisms for holding the DMSA table above the base.

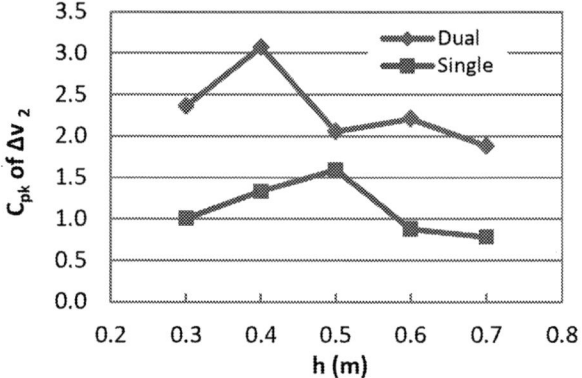

Figure 6. C_{pk} of change of velocity of DMSA table during secondary impact (Δv_2) at various drop heights (h) for single and dual thumb screw mechanism.

Strain of PCB under High Acceleration Drop Test

Strain response of PCB under high acceleration drop test using DMSA was studied by experiments. A 77x77 mm square 1 mm thick 8 layers FR4 PCB assembled with four 12x12 mm 547 pin 0.4 mm pitch plastic BGA devices, as showed in Figure 7, was used. According to the symmetry of the PCB layout, there are two types of BGA corners, one called "inner corner" which is the BGA corner closer to the PCB center, the other called "outer corner" which is the BGA corner farther away from PCB corner. Triaxial 0-45-90° strain gage rosettes (Model # C2A-06-062WW-350, Vishay Micromeasurements Group) were mounted right underneath the inner and outer corner solder balls, on the back side of the PCB. The strains measured from these two strain gage rosettes are, theoretically, equal to the strains on the corner PCB pads, where strains are believed to be the maximum among all the BGA balls. The PCB was attached to a light and rigid aluminum fixture with device-side facing down, using 4 point support in accordance with JEDEC JESD22B111 standard. The fixture was then attached to the DMSA table for drop testing. The strain gages were connected to National Instruments NI 9237 strain data acquisition module with sampling rate of 51.2 kS/s. Principal strains ($\varepsilon_{1,2}$) were calculated from the strains obtained in three directions, using the following equation:

$$\varepsilon_{1,2} = \frac{\varepsilon_0 + \varepsilon_{90}}{2} \pm \sqrt{\frac{(\varepsilon_0 - \varepsilon_{45})^2 + (\varepsilon_{90} - \varepsilon_{45})^2}{2}} \qquad (18)$$

where ε_0, ε_{45}, and ε_{90} are the individual strains in 0, 45, 90 degree, respectively.

Drop test of this PCB using the magnesium DMSA with drop height of 0.7 m was repeated 10 times. The measured average change of velocity, peak acceleration and pulse duration were 8.034 m/s, 10672 G and 0.139 ms, respectively. The average peak principal strains for inner and outer corners were -3610 and -6150 microstrain, respectively. The negative strain value denotes that the strain was compressive, due to that the PCB side mounted with strain gages was facing up during drop test. The strains of inner and outer corners from one representative drop were showed in Figure 8. The measured fundamental mode PCB vibration frequency in this drop condition is approximately 430 Hz. This result indicates that by using DMSA high peak strain on PCB can be generated in drop test.

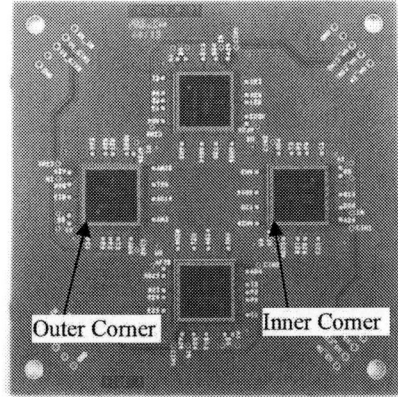

Figure 7. Photograph of the 77x77 mm square PCB used in drop test of 10000 G peak acceleration.

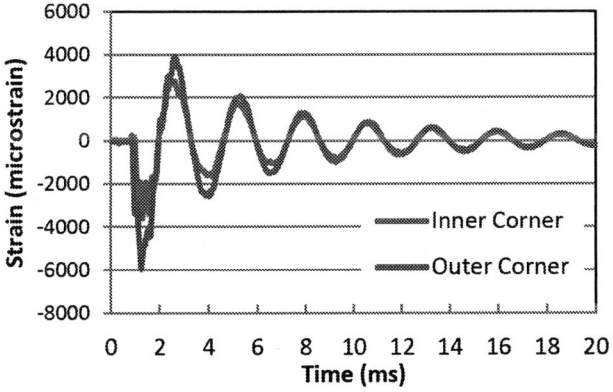

Figure 8. Representative strain response of the 77x77 mm square board under 10000 G 0.14 ms drop test with DMSA.

Conclusions

A simple analytical model for dual mask shock amplification was created and presented in this paper. Even though the numbers for change of velocity (Δv_2) and peak acceleration (A_2) in secondary impact predicted by the model were far higher than those measured from experiments due to the model's assumption of no kinetic loss, the model does propose a relation between change of velocity (Δv_2) and masses of the three bodies in a DMSA/drop machine system,

which can be employed as a guideline for designing high efficiency DMSA. The model also provided a relation between change of velocity (Δv_2) and drop height (h), which can be useful when adjusting the drop height to reach the specified change of velocity (Δv_2) and peak acceleration (A_2) for a given DMSA/drop machine system. For shock pulse with half-sine shape, the peak acceleration can be calculated from the model-predicted change of velocity once the pulse duration is determined. However, for shock pulses with irregular shapes, calculating peak acceleration from change of velocity and pulse duration would be risky; in this situation, change of velocity instead of peak acceleration shall be used to characterize a shock pulse.

The model predicts that the two custom-built DMSA tables with weights of 1 and 0.7 kg respectively due to different material density would not generate significant difference in the shock output, which was also verified by experiment. The model also predicts that even though lower gap height and smaller bungee cord spring constant would result in higher peak acceleration, the effect is very small compared to the effect of drop height and is almost negligible.

The mechanism for holding the DMSA table in a DMSA exhibits a large impact on the repeatability of drop output and durability of the DMSA. The dual thumb screw mechanism is superior to the single thumb screw mechanism in this regard. When the former mechanism is used on the custom-built DMSA, C_{pk} of change of velocity can be well above 1.5 when tolerance of +/-10% is required.

In conclusion, this paper provided a practical example that shows how a carefully designed DMSA can be employed together with a conventional drop machine to generate very high acceleration in the range of 5000~15000 G for the application of BLR drop test, in a simple, economical, and repeatable way. Preliminary result suggested that strain of up to 6000 microstrain can be obtained on a BLR drop test PCB at the drop condition of 10000 G peak acceleration and 0.14 ms pulse duration.

Acknowledgments

The author would like to thank Donnie Laughery of Texas Instruments for conducting the drop test in this study. The layout of the PCB presented in this study was created by JEDEC JESD22B111 task group. The PCBs were procured and provided generously by Hewlett-Packard Company.

References

1. S. Douglas, J. Meng, J. Akman, I. Yildiz, M. Al-Bassyiouni, A. Dasgupta, "The effect of secondary impacts on PWB-level drop tests at high impact accelerations," in *Proc. 12th International Conference on Thermal, Mechanical and Multi-Physics Simulation and Experiments in Microelectronics and Microsystems (EuroSimE)*, April 18-20, 2011, vol., no., pp.1/6-6/6.

2. B. Rodgers, S. Goyal, G. Kelly, and M. Sheehy, "The Dynamics of Multiple Pair-Wise Collisions in a Chain for Designing Optimal Shock Amplifiers," *Shock and Vibration*, vol. 16, no. 1, pp. 99-116, 2009.

3. S. Zhang, "Survey on High-G Testing Methodology", http://www.empf.org/empfasis/june05/g0605.htm.

4. S. T. Douglas, "High Accelerations Produced Through Secondary Impact and Its Effect on Reliability of Printed Wiring Assemblies," *Master Thesis*, University of Maryland, College Park, 2010.

5. P. Lall, K. Patel, R. Lowe, M. Strickland, D. Geist, R. Montgomery, "High-G Shock Reliability of Ceramic Area-Array Packages," in *Proc. 13th IEEE Intersociety Conference on Thermal and Thermomechanical Phenomena in Electronic Systems (ITherm)*, May 30-June 1, 2012, pp.1028-1036,

6. P. Lall, K. Patel, R. Lowe, M. Strickland, J. Blanche, D. Geist, R. Montgomery, "Modeling and Reliability Characterization of Area-Array Electronics Subjected to High-G Mechanical Shock Up to 50,000G," in *Proc. IEEE Electronic Components and Technology Conference (ECTC)*, May 29-June 1, 2012, pp.1194-1204.

7. J. Karppinen, J. Li, J. Pakarinen, T.T. Mattila, M. Paulasto-Kröckel, "Shock impact reliability characterization of a handheld product in accelerated tests and use environment", *Microelectronics Reliability*, Vol. 52, no. 1, Jan. 2012, pp. 190-198.

8. J.W. Berglund, "A Modified Dual Mass Amplifier for High-Strain-Rate Testing of SIFCON in Uniaxial Compression," *Experimental Mechanic*, vol. 28, no. 3, pp. 281-287, Sep. 1998.

978-1-4799-2408-0/14 $31.00 © 2014 IEEE   1448

Non-Destructive Crack and Defect Detection in SAC Solder Interconnects Using Cross-Sectioning and X-Ray Micro-CT

Pradeep Lall, Shantanu Deshpande,
Junchao Wei, Jeff Suhling
Auburn University
NSF-CAVE3 Electronics Research Center
Department of Mechanical Engineering
Auburn, AL 36849
Tele: (334)844-3424
E-mail: lall@auburn.edu

Abstract

The industry is going through a transition in material sets for second level interconnects including adoption of leadfree solders. High-rel systems may often have a mix of components with different solder alloys in the printed circuit assemblies including both leaded and leadfree solders because some original leaded components may only be available in leadfree configurations. In this paper, the potential of x-ray micro-computed tomography (μCT) to fulfill the need for non-destructive three dimensional imaging of electronic assemblies has been evaluated. A number of defect seeded assemblies have been used for detection of a number of common solder interconnect defects and failure modes. In addition, the ability of the x-ray μCT for examination of complete products has been examined. Three-dimensional rendered versions of the board assemblies have been constructed for visualization of the defects and failure modes. Void sizes have been measured using Volume Graphics reconstruction and Matlab modules. In each case, the assemblies have been cross-sectioned after imaging by x-ray μCT to ascertain the morphology of the defect or failure mode using optical imaging. Results indicate that x-ray μCT is capable of providing high resolution imaging of the common defect types and failure modes in electronic assemblies and has potential for risk mitigation in sustainment of long-life high-rel systems.

Introduction

Ball-grid array packaging is widely used package format in fine-pitch electronics. Several of these components are placed in ever increasing proximity of the product housing, often exposing them to damage due to shock, vibration or thermal fatigue in normal operation. Further, the transition of electronics to leadfree solders poses an additional set of challenges. The high-rel community has product lifetimes much longer than consumer grade electronics. It is common to have products with a mix of SnPb and leadfree components primarily necessitated by the lack of availability of the SnPb formulation from manufacturers. Continued reliable operation and life extension of the long-life high-rel products requires the uses of non-destructive tools and techniques which will allow defect detection and failure mode isolation in complete assemblies. BGA format packages use solder balls for mechanically and electrical attachment to PCB. These solder balls are located at the bottom of the package and are attached to PCB using surface mount technique and are often not available for direction visual inspection without cross-sectioning the package. When subjected to thermal cycling, the mismatch of coefficient of thermal expansion between the PCB, and the package causes crack initiation and propagation in the solder interconnects. Further, the transition to new leadfree solders has brought a number of defect types which may have been eliminated because of the maturity of the SnPb solder technology. Examples include the head-in-pillow defect or non-reflowed solder joints because of the use of mixed technology containing solders with different compositions and different liquidus temperatures. There is need for tools and techniques which will allow non-destructive inspection of the solder joints in packages, board assemblies and complete products. A variety of non-destructive techniques have been used for delamination and crack-initiation reliability studies of electronics including 2D x-ray imaging and scanning mode acoustic microscopy [Liu 2003, Tohmyoh 2003, Sato 2005]. Acoustic microscopy provided 2-dimensional information of the interface based on differences in the acoustic impedance of the interface. 2D x-ray imaging provides 2-dimensional image of the complete object in the x-ray field based on differences in the density and material thickness. Neither of the techniques allow reconstruction of the complete object under test to evaluate the three-dimensional topology of the object.

In this paper, x-ray microscopic computed tomography (μCT) system has been used to examine a number of defect types and failure modes in electronic assemblies. Information extraction in the x-ray machine depends on the material density and thickness dependent attenuation of the x-rays passing through the specimen. The attenuated x-rays passing through the specimen are detected by the imaging device. X-ray μCT has the ability to provide sub-micron level resolution measurements of the object through three-dimensional re-construction of the imaged object. Further, slice-by-slice inspection of the μCT image stack enables non-destructive inspections and location identification of package-defect and the damage accrual during operation of the assembly. Previously researchers have implemented industrial 2D x-ray inspection techniques to locate defects in automobile engines [Izumi 1993], two-dimensional detection of defects in solder joints [Sumimoto 2002], two-dimensional inspection of electronic package wirebonds [Maur 2003; Kovacs 2005] successfully. Automated solder inspection technique has been developed using Artificial Neural Network trained algorithm alongside 2D X-ray imaging [Teramoto 2007]. Real time X-ray radiography was coupled with reconstruction algorithms and 3D models of packages were reconstructed to find defects in solder joints of discrete components [Oppermann 2009, Tsuritani 2011]. The inspection of ball-grid arrays defects and failure modes in complete board assemblies and assembled products with three-dimensional reconstruction and measurement of void size reported in this paper is new. In current methodology, micro-computed tomography technique is used for inspection of BGA's subjected to thermal cycling and different defects are located and reported by using high resolution 3D reconstructions of the packages. Packages with defect were cross-sectioned and inspected by using metallographic microscope, and optical images and µCT images are compared. Special reconstruction code was developed using MATLAB to render located defects in 3D and in different color combinations.

Microscopic Computed Tomography (µCT)

Industrial Micro-Computed Tomography (µCT) has found applications in defect detection and failure analysis of electronics components. Using the same basic principles as used in medical computed tomography techniques, µCT offers very high resolution, magnification and data processing [Izumi 1993]. Parts tested in the industrial µCT usually have no adverse effect of prolonged x-ray exposure, which enables user to perform quality scans and take thousands of projections of the sample. These images have been combined together and converted into 3D rendered volume providing internal details about package architecture and the ability to inspect the volume on a slice-by-slice basis at a user specified angle to detect defects. Quality of x-ray imaging and reconstructed volume depends on lot of parameters, for e.g. acceleration voltage of electron beam, filament current, mono-chromaticity, distance between sample-x-ray tube-detector and spot size [Tsuritani 2007, 2011]. For analysis purpose, x-rays are made pass through the specimen as the sample is rotated in the x-ray field. When x-rays penetrate the sample, some amount of energy is absorbed, i.e. x-rays are attenuated. This attenuation depends on density, thickness, wavelength of x-ray and atomic number of the material medium through which X-rays are passing. This attenuation is then detected by detector and represented on computer screen in the form of different grayscale values. In order to make 3D reconstruction, object is rotated in the x-ray field with the help of chuck. The resulting projections are then stitched together using sophisticated reconstruction algorithms to form 3D rendered version of the object. The YXLON µCT Cougar system located at CAVE facility in Auburn University has been used for the examination of samples in this study. The YXLON µCT Cougar Machine is a state-of-art machine with a detail detectability of less than 1 µm and a measurement resolution of 0.1 µm [Y.Cougar Technical Data 2011]. Figure 1 shows the interior of µCT system. A server grade workstation with Volume Graphics provides 3D reconstruction capability. A Step-up transformer provides the necessary acceleration voltage in the range of 70KV to 160KV to accelerate and direct electron beam. Tungsten filament has been used to produce electron beam. X-rays are generated from electron beam in X-ray tube. Vacuum is maintained in the tube by rotary vacuum pump.

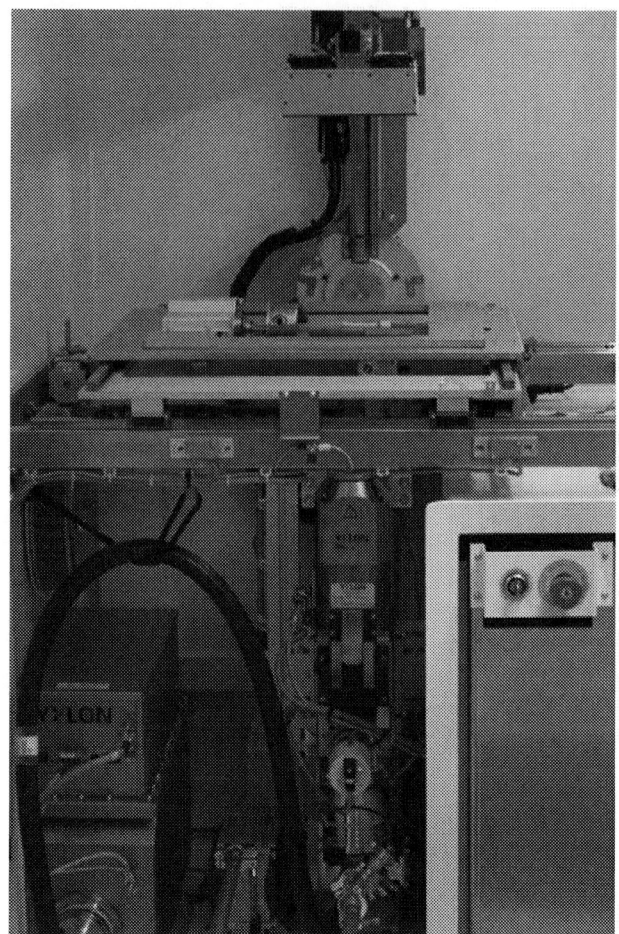

Figure 1: YXLON µCT Cougar System

Coherent and mono-chromatic X-rays are generated in X-ray tube. Magnetic focus lenses are provided to have sharp focus and to avoid blurry images. Because of these characteristics, very small defects can be easily identified readily. X-ray tube can move in the "z" direction, i.e. vertically up and down to achieve desired geometric magnification. Three jaw chuck is provided to hold samples, and it is mounted on holding tray. Chuck can rotate in clockwise and counterclockwise direction by 360°, and holding tray can move in "x" and "y" direction. The machine is equipped with a digital flat panel detector (1004 x 1004 pixels), which can rotate either clockwise, or anticlockwise by 75°. Voltage and current can be varied to change the focal spot size, and achieve higher resolution. Volume Graphics Software has been used for the 3D reconstruction process. The software takes the still images of the object acquired as the object is rotated in the x-ray field and uses the data to reconstruct the three dimensional version of the object under examination. High resolution measurements of the geometric dimensions can be made non-destructively.

µCT Applications

In this section, electronic assemblies have been examined for various defects and commonly encountered failure modes to understand the ability of the µCT technique for identification of defects. Figure 3 shows the 3D reconstruction of a plastic ball grid array package in which the traces, solder interconnects, traces, and the metallization under the chip are

978-1-4799-2408-0/14 $31.00 © 2014 IEEE   1450

clearly visible. The package is a PBGA324 ball-grid array package with package parameters shown in Table 1.

Table 1: Attributes of PBGA 324 Test Vehicle.

I/O Count	324
I/O Pitch	1mm
Body Size	19mm
Ball Matrix	18x18
Ball Alignment	Full Array
Package Type	PBGA
Board Finish	ImAg
Substrate Pad Type	SMD
Solder Ball Material	Sn3Ag0.5Cu
Substrate Pad Dia.	0.45mm
Ball Dia.	0.4mm
Mold Cap thickness	1.17mm

Figure 2: Physical Picture of the Test Vehicle

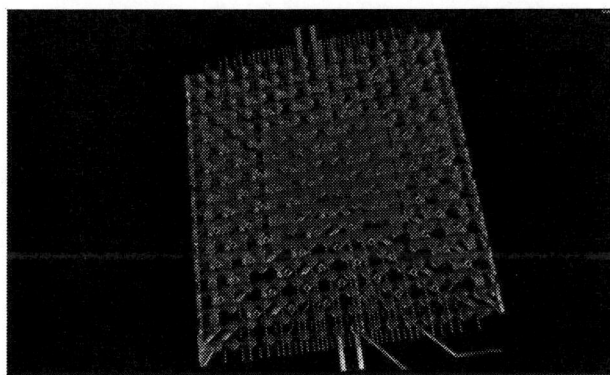

Figure 3: 3D Reconstructed BGA Package of the Test Vehicle shown in Figure 2

Fatigue Crack Detection

Fatigue crack nucleation and propagation in solder joints is one of the major reliability issues. The crack is formed because of excessive stress formation in solder ball region caused by thermal mismatch [Lau 1995]. Size of the crack is usually from one to few microns. To capture these micro-cracks, high quality scan is required. BGA 324 packages, that had failed during thermal cycling test were selected for crack detection. Packages were scanned at high magnification and its architecture was reconstructed using volume graphics software. The reconstructed volume was then analyzed slice by slice from all three directions, x, y and z. Once the crack was located, parts were polished till that plane and inspected under metallographic microscope to compare the morphology

of the imaging in μCT versus that observed from cross-sectioning and mechanical polishing. Figure 4a shows thermal fatigue cracking in the solder ball and Figure 4b shows crack found at the same location in slice-by-slice inspection of reconstructed volume. Figure 5 also shows microscopic and μCT reconstructed crack images for a different package fatigue crack. Planar cracks in general will have much clearer definition in the μCT slices then non-planar cracks. In order to verify the non-planarity of the crack, the 3D volume was re-constructed.

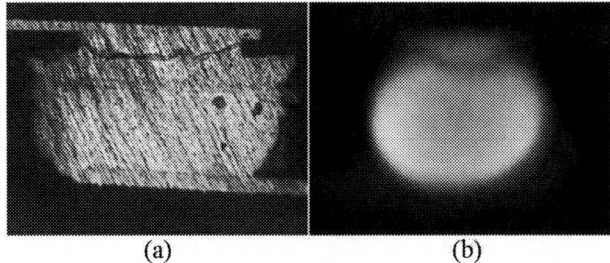

(a)   (b)

Figure 4 – Crack detection (a) Metallographic Microscopic image (b) Slice-by-Slice inspection of μCT images.

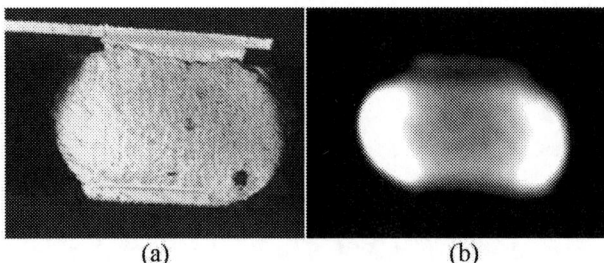

(a)   (b)

Figure 5: Crack detection (a) Metallographic Microscopic image (b) Slice-by-Slice inspection of μCT images.

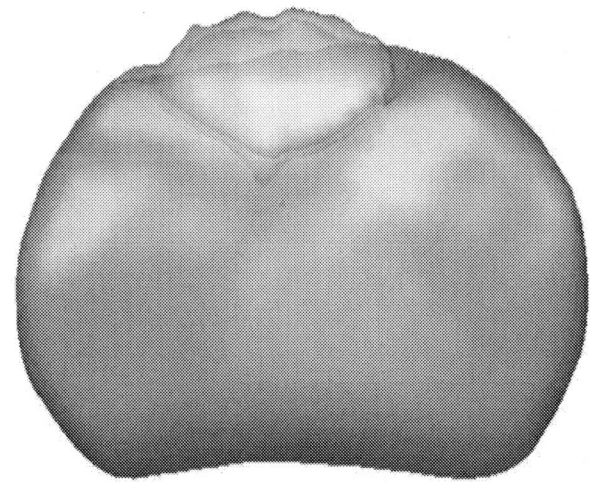

Figure 6: 3D Rendered crack.

MATLAB Code was developed to render selected area as a three-dimensional object using output information from the μCT. Code is developed in such way that density of solder will have particular color and defects in solder ball which have low density is represented in different color. Figure 6 shows 3D volume rendered by using MATLAB code. The non-planar definition of the crack can be easily seen on top side of solder ball. Figure 6 shows that the crack is non-planar in the crack plane dipping in the middle of the solder ball.

Void Detection

Voiding is often cited in electronics manufacturing literature as a major contributor to the reduction in fatigue life of the solder joint. Manufacturing processes are often optimized to minimize the amount of voiding in the solder interconnects. Further, detection of voids in electronics assembly is not only a part of failure analysis but also part of ongoing process quality control. Voiding is governed by many factors, like type of solder paste, board finish, reflow pattern, and substrate type. Previous researchers have used 2D x-ray techniques for void detection [Moore 2002; Sumimoto 2002]. For detailed BGA inspection, usually oblique view is preferred [Teramoto 2007; Oppermann 2009]. Main disadvantage of oblique view is that though it provides attention to details, inspecting each solder ball in this view is tedious task, and is time consuming. However, if there are more than one BGA devices on either side of PCB, then oblique view approach cannot be implemented.

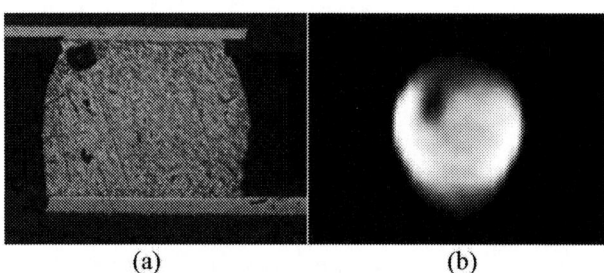

(a)   (b)

Figure 7 - Void detection (a) Metallographic Microscopic image (b) Slice-by-Slice inspection of reconstructed volume.

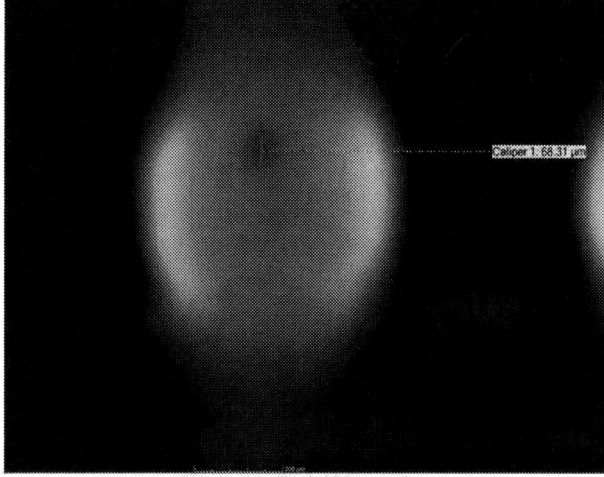

Figure 8: Void Detection

Under these circumstances, void detection from 3D reconstructed volume is suitable option. Image processing tools and filters help to get a better view of micro-cracks. Figure 7 shows voids found while performing slice-by-slice inspection on the BGA324 board assembly. Pass-fail type automated inspection system can be implemented for 3D rendered volume which calculate percentage void area and evaluate quality of manufactured BGA. Figure 7a shows void located using optical microscopy of the cross-sectioned

sample. Figure 7b shows same void in the slice-by-slice inspection of the μCT image stack. Figure 8 shows another void found during slice-by-slice inspection of package. Void diameter has been measured using software toolbox. Solder ball with void was then rendered using MATLAB code, as shown in Figure 9. Voids in solder ball contain air and thus have lower density than solder material. Low density part is represented by dark green color in Figure 10. It can be seen that definition of the void is noticeably better in the 3D rendered image shown in Figure 9, compared to the single slice of the μCT.

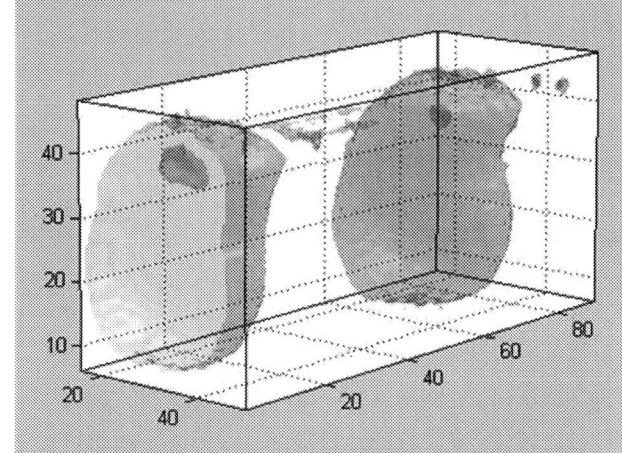

Figure 9: 3D Rendering of Voids – Zoomed-in view of a Pair of Solder Balls.

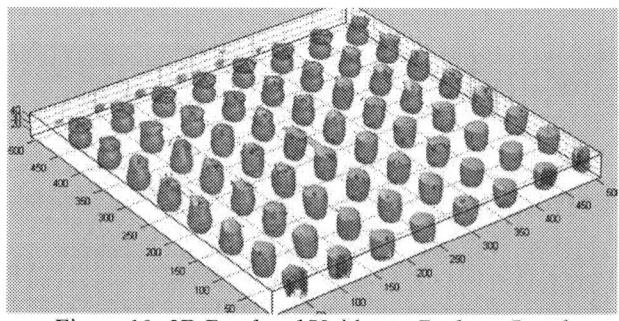

Figure 10: 3D Rendered Voids at a Package Level

The reconstructed solder ball opacity can be varied to see internal details. In Figure 9 and Figure 10, the solder balls have been intentionally kept transparent to allow easier identification of the voids in solder ball at a package level as shown in Figure 10. Voids in solder ball were successfully found with the help of Volume Graphics and MATLAB code.

Head-in-Pillow Defect

Head-in-pillow (HiP) defect also known as a ball in socket defect involves solder paste which wets the pad but does not completely wet the ball (Figure 11a). HiP defect morphology is more common leadfree solder components. The joint provides limited electrical connectivity but lack mechanical integrity. Solder joint with head-in-pillow defects may fail under low magnitudes of mechanical stress or thermal stress. The components may pass electrical test but may prematurely fail in the field. Components often under warpage during reflow process causing some of the balls to lift up and oxidize.

978-1-4799-2408-0/14 $31.00 © 2014 IEEE

In the subsequent stages of temperature exposure during the reflow process, the component may flatten again making contact with the solder paste. However, there may not be enough flux activity left to break down the oxide layer producing a head-in-pillow defect [Seeling 2008; Scalzo 2009]. Previously, oblique view 2D x-ray imaging was used to find these defects by [Arazna 2010]. In this paper, 3D rendered volume has been used to identify HiP defects in BGA 324 packages. Figure 11a,b and Figure 12a,b shows the head and pillow defect detected by visual inspection after polishing, and morphology of the same defect location prior to cross-sectioning using µCT slice-by-slice analysis.

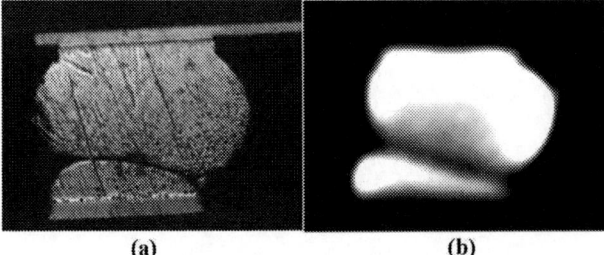

(a)   (b)

Figure 11 – Head and Pillow detection (a) Metallographic Microscopic image (b) µCT reconstructed image.

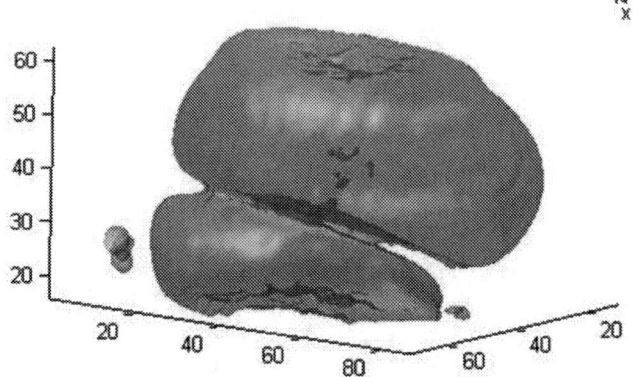

(a)   (b)

Figure 12 - Head and Pillow detection (a) Metallographic Microscopic image (b) µCT reconstructed image.

Figure 13 shows 3D head and pillow defect rendered using MATLAB code using Volume Graphics Data. We were able to find this defect non-destructively by using Volume Graphics and MATLAB code, and it is in good agreement with result obtained from optical microscopic imaging.

Figure 13 – 3D µCT Rendering of Head and Pillow Defect

Improper Reflow of Solder Joints

Non-optimal reflow profile may produce unreflowed solder joints and solder paste. In such cases, the solder paste may not reflow completely producing solder joints which retain their original shape prior to reflow. The defect may be caused by a number of reasons including the higher thermal mass of the assembly causing the assembly to lag the set points of the reflow profile or because of improper reflow of assemblies containing a mix of components with different solder compositions. The defect could also occur in common copper core assemblies with high thermal mass of the printed circuit board in presence of improper reflow profile or faster than needed belt speed. Figure 14b shows same defect in a complete package identified using the slice-by-slice examination in the µCT. the defect morphology has been verified through cross-sectioning of the identified defect location. Figure 14a shows a non-reflowed solder joint in which the solder joint and the solder paste have not merged during the reflow process. Figure 15 and Figure 16 shows additional similar defects, which were located by non-destructive method, and then verified by actually polishing the same part till the plane of the defect. Figure 17 and Figure 18 shows 3D reconstructed solder balls using MATLAB code based on Volume Graphics data. Crack at the bottom side of the solder ball is clearly visible, and it matches well with the cracks found by using Volume graphics software, and also by performing actual cross-sectioning.

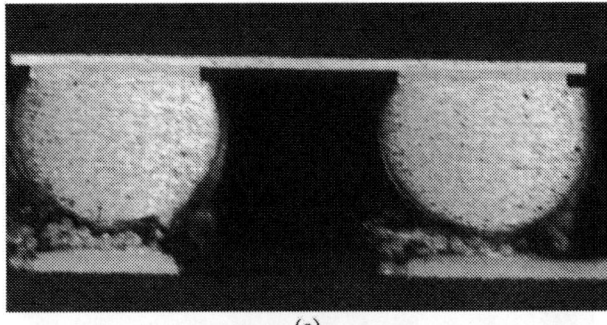

(a)

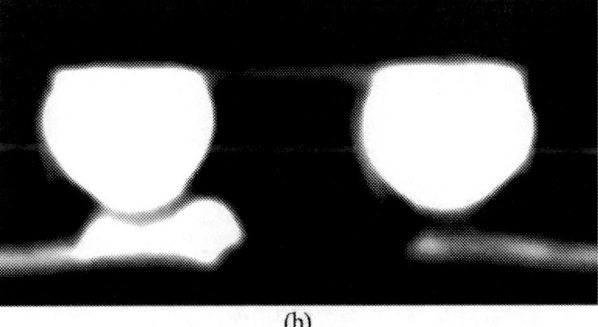

(b)

Figure 14: (a) Metallographic Microscopic image (b) Slice-by-Slice inspection of µCT images.

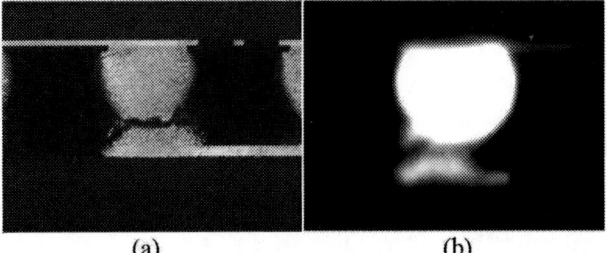

(a)   (b)

Figure 15: (a) Metallographic Microscopic image (b) Slice-by-Slice inspection of µCT images

Figure 16 – Slice-by-slice inspection of μCT Reconstructed volume showing detached solder bump and pad.

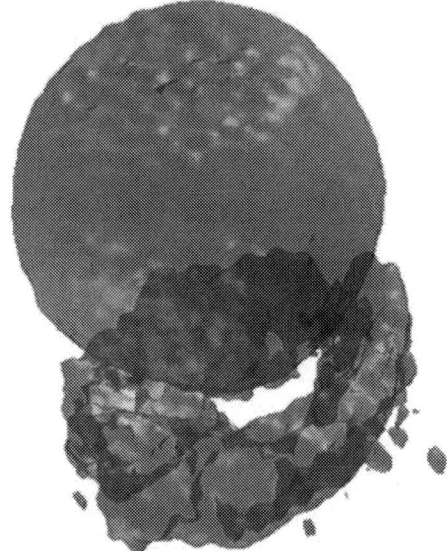

Figure 17: 3D Rendered solder ball lift.

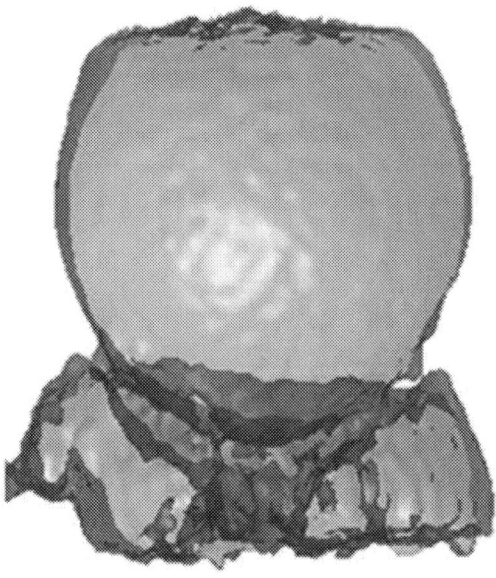

Figure 18: 3D Rendered solder ball lift.

μCT Inspection of an Assembled Product

Failed flash drive was selected for study to understand the ability of the μCT in identification of defects in products. In the process of usage the flash drive was accidently dropped and failed. The μCT was used to root cause the failure. μCT quality scan of flash drive was performed in two different steps to get better magnification. In first step, μCT scan of connecting pin was done, and in next step μCT scan of circuitry that's on PCB was performed. Before scan, capacitor that was on circuit was detached. μCT scan of working flash

drive was also performed at the same settings to compare two datasets. A flash-drive complete with the plastic housing and USB connector was examined in μCT.

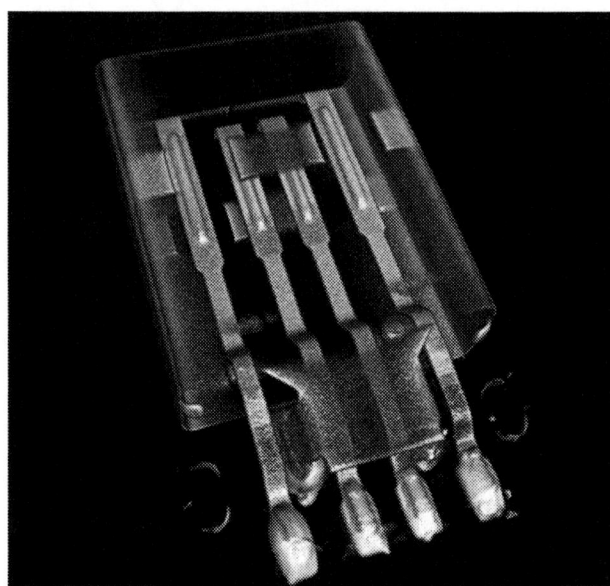

Figure 19: 3D-Rendered Connectors of Good Flash Drive

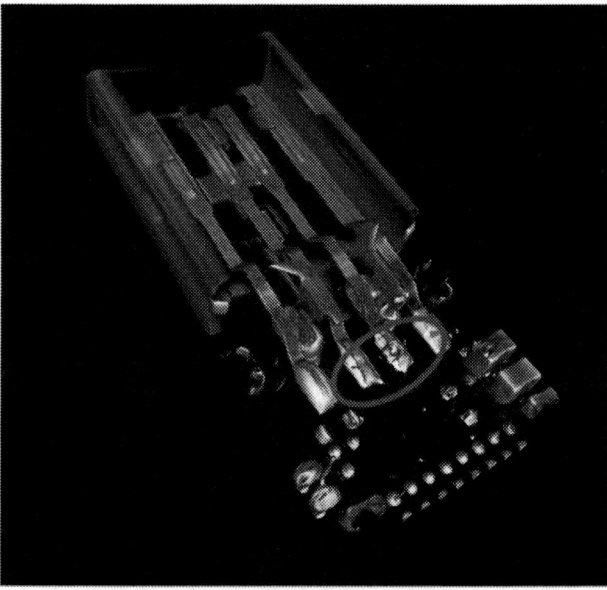

Figure 20: 3D-Rendered Connectors of Failed Flash Drive

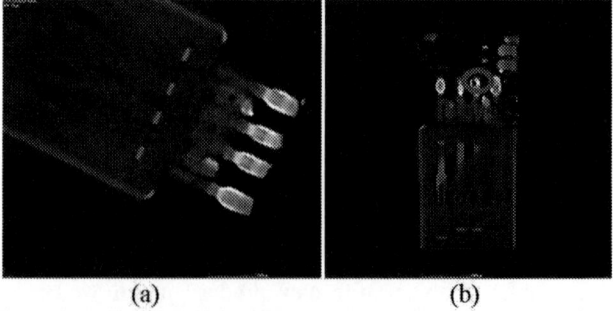

Figure 21: Slice-by-Slice inspection of volume. (a) Good flash drive (b) Failed flash drive

978-1-4799-2408-0/14 $31.00 © 2014 IEEE   1454

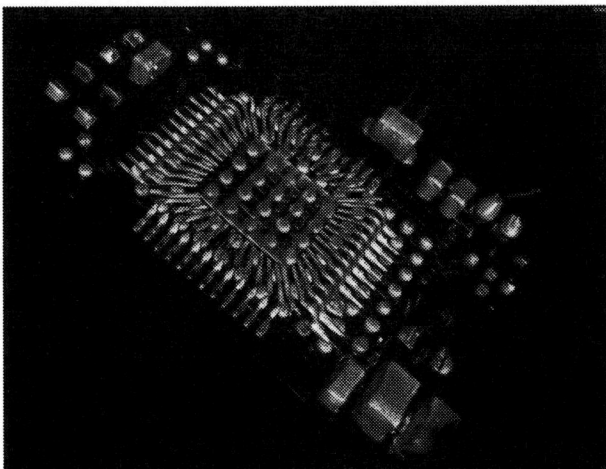

Figure 22 – Architecture on PCB.

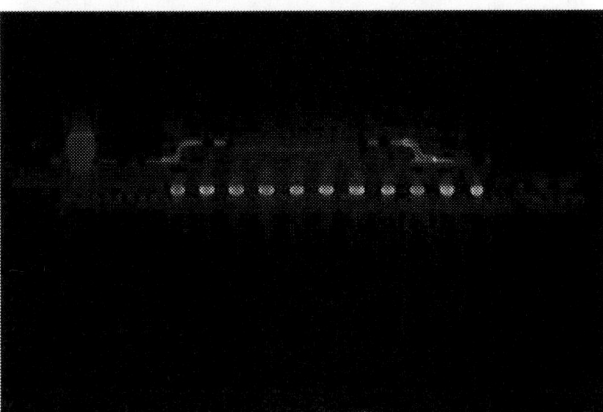

Figure 23 – Slice-by-Slice inspection of 3D volume.

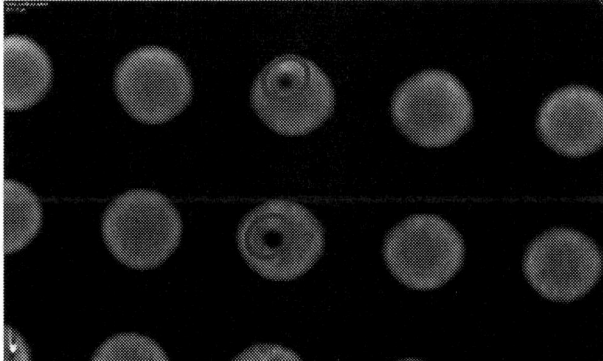

Figure 24 – Voids located in Solder Ball during Slice by Slice inspection of reconstructed volume.

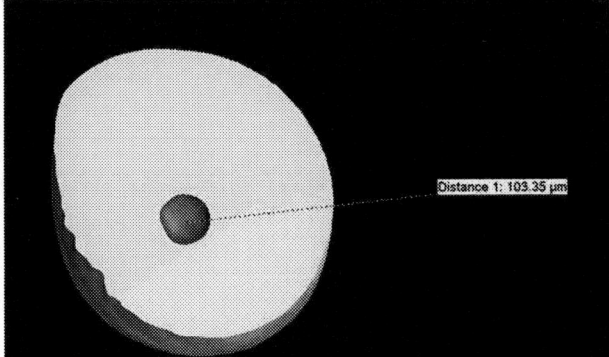

Figure 25 – 3D clipped volume with void measurement.

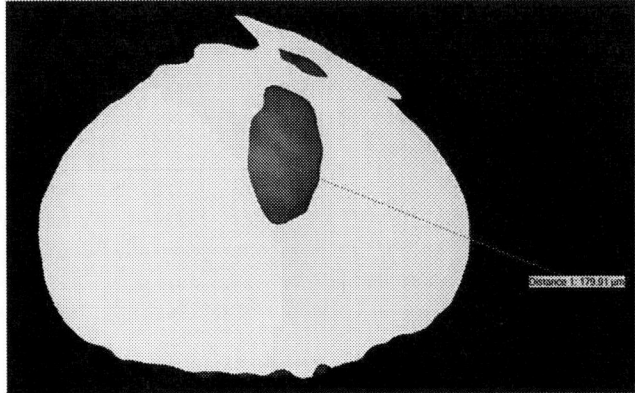

Figure 26 - 3D clipped volume with void measurement

Figure 19 and Figure 20 show 3D μCT reconstructions of connecting pins of good flash drive and failed one. Figure 20 shows discontinuity in one of the connecting leads, marked with red circle. One of the leads showed minor cracking. This might be due to rough handling or drops of the flash drive. For working flash drive, leads were continuous. Figure 21a shows top sectional view of those volumes. In Figure 21b the solder cracking can be easily seen (marked by red circle). Figure 22 shows internal architecture of flash drive seen captured in the μCT through the plastic housing of the flash drive. The flash drive has packages on either side of PCB. A QFP-style lead frame package can be seen on one side of the PCB. A plastic BGA package can be seen on the other side of the PCB. Figure 23 shows slice-by slice inspection of assembly from side view. No discontinuity in the leads of lead frame package was found. However, while inspecting BGA, multiple voids were found. Figure 24 shows 2D sectional top view of solder balls with voids. Figure 25 and Figure 26 shows 3D rendered voids with their measurement. The void diameter has been measured using virtual calipers in Volume Graphics.

Summary and Conclusions

A new non-destructive approach based on microscopic computed tomography has been used for evaluation of solder-joints. The approach involves using mono-chromatic finely focused x-rays for 2D imaging of an object rotating in the x-ray field and uses sophisticated reconstruction algorithms to convert 2D images into 3D rendered volume. Thousands of X-ray projections are recorded during rotational scan of specimen, and these images were combined together to form 3D reconstruction of specimen. Reconstruction was done by using Volume Graphics software package and also by using developed MATLAB code. The potential of the technique in detecting various defects including cracks, voiding, head and pillow defect was studied. In addition to single package, board assemblies, μCT scan was done on failed flash drive complete in the housing. In each case, once the packages or product assemblies had been subjected to μCT scan, the area of interest was cross-sectioned to optically verify the location and type of defect. Results of μCT were in good agreement with optical microscopic imaging.

978-1-4799-2408-0/14 $31.00 © 2014 IEEE

Acknowledgments

The research presented in this paper has been supported by NSF-FRS-1127913 and members of NSF-CAVE3 Electronics Research Center.

References

Arazna, A., Koziol, G., Steplewski, W., Lipiec, K, Head on Pillow Defects in BGA's Solder Joints, Conference Proceedings of 3rd Electronic System-Integration Technology Conference, 2010

Dössel, O., Bildgebende Verfahren in der Medizin, Springer (Berlin, 2000).

Izumi S, Kamata S, Satoh K, Miyani H, "High Energy X-ray Computed Tomography for Industrial Applications", IEEE Transactions On Nuclear Science, Vol. 40, No. 2, April 1993

Kovacs, R., X-Ray Inspection of Micro-wire Bonds, Proceedings of IEEE 28th International Spring Seminar on Electronics Technology, pp 448-451, 2005

Lau J, Ball Grid Array Technology, McGraw-Hill Inc Publications, 1995.

Liu, S., and Ume, I. C., Digital Signal Processing in a Novel Flip Chip Solder Joint Defects Inspection System, Transactions of ASME Journal of Electronic Packaging, Vol. 125, No. 1, pp. 39-43, 2003.

Luebbehuesen J, Roth, H., Neubrand, T., Brunke, O., X-ray nanoCT of interconnections in IC packages: Visualizing of internal 3D-Structures with Submicrometer Resolution, European Microelectronics and Packaging Conference, 2009. EMPC, pp. 1-4, 2009.

Maur F., X-Ray Inspection for Electronic Packaging Latest Developments, Proceedings of IEEE ICEPT Conference, 2003.

Moore T., Vanderstraeten D., Forssell M., Three Dimensional X-Ray Laminography as a Tool for Detecting and Characterization of BGA Package Defects, IEEE Transactions on Components and Packaging Technologies, Vol. 25, No. 2, pp. 224-229, 2002.

Motalab M., Cai Z., Suhling J., Zhang J., Evans J., Bozack M, Lall P., "Improved Predictions of Lead Free Solder Joint Reliability that Include Aging Effects", Proceedings of 62nd Electronic Components and technology Conference (ECTC), 2012, pp 512-531, 2012.

Oppermann M., Zerna T, Wolter K.J., "X-ray Computed Tomography on Miniaturized Solder Joints for Nano Packaging", 11th IEEE Electronics Packaging Technology Conference, pp 70-76, 2009

Sato Y., Miura H., "Non-Destructive Inspection Method for Detecting Open Failure in Flip Chip Structures," Conference Proceedings of InterPACK Conference, pp 1587-1592, 2005.

Scalzo, M., "Addressing the Challenge of Head-in-Pillow Defects in Electronics Assembly", Indium Corporation Technical Library, 2009

Seeling, K. "HIP Defects in BGAs", Circuits Assembly, Vol. 16, No. 12 (2008), pp. 28- 32 2008

Sumimoto T., Maruyama T. "Detection of Defects at Solder Joints by Using X-Ray Imaging", Proceedings of IEEE ICIT Conference, Bangkok, THAILAND, pp 238-241 2002.

Teramoto A, Murakosh T., etc. "Automated Solder Inspection Method by Means of X-ray Oblique Computed Tomography", Proceedings of IEEE ICIP Conference, pp 433-436, 2007

Tohmyoh, H., and Saka, M., A High-resolution Dry contact Acoustic Imaging of the Solder joints for Ball Grid Array Assembly, Proceedings of ASME InterPACK Conference, 2003.

Tsuritani H., Sayama T., Uesugi K., Takayanagi T., Mori T., "Nondestructive Evaluation of Thermal Phase Growth in solder Ball Microjoints by Synchrotron radiation X-ray Microtomography", Transactions of ASME Journal of Electronics Packaging, Vol 129, Issue 4, pp 434-440, 2007.

Tsuritani, H., Okamoto Y., Uesugi, K., Sayama, T., Takayanagi, T., and Mori, T. Three Dimensional And Nondestructive Evaluation of Thermal Fatigue Crack Propagation Process in Complex-Shaped Solder Joints by Synchrotron Radiation X-Ray Micro-Tomography, Proceedings of ASME InterPACK Conference, July 6-11, 2011.

High Resolution and Fast Throughput-time X-ray Computed Tomography for Semiconductor Packaging Applications

Yan Li [1], Mario Pacheco [1], Deepak Goyal[1], John W. Elmer[2], Holly D. Barth[2], Dula Parkinson[3]

[1]Assembly Test and Technology Development Failure Analysis Labs, Intel Corporation, 5000 W. Chandler BLVD. Chandler, AZ 85226
[2] Materials Engineering Division, Lawrence Livermore National Laboratory, 7000 East Avenue Mail Stop L-342 Livermore, CA 94551-0808
[3]Advanced Light Source, Lawrence Berkeley National Laboratory, 1 Cyclotron Road, Berkeley, CA 94720
Contact email: mario.pacheco@intel.com

Abstract

The recent applications of 3D X-ray computed tomography (CT) in microelectronic packages, including non-destructive failure analysis, defect monitoring in solder joints and Cu vias, and progressive reliability study of solder voids, electron migration induced void nucleation in solder joints, and void evolution in Cu vias are reviewed. The high resolution and non-destructive 3D X-ray CT data has proven to be highly valuable in package assembly process development, quality control and reliability risk assessment; however, the field of view of current lab-scale 3D X-ray CT technology is limited to about 1-2mm^2 localized area at micron level resolution, due to its low brightness and non-parallel X-ray beam resulting in long data acquisition time. Synchrotron X-ray sources, on the other hand, can provide large area collimated beams with high brightness, which allows imaging within 3-20 minutes an entire 3D package, including Si, underfill, multiple levels of solder joints, and dielectric layers, Cu vias as well as through holes in multiple substrates. The limitation of current 3D X-ray CT techniques as well as directions for next generation 3D X-ray CT techniques provided by the synchrotron X-ray study of 3D packages are discussed in this paper.

Introduction

Semiconductor industry requirement for higher levels of integration, lower costs, and a growing need for a complete system configuration has resulted in a continued drive towards 3D electronic package solutions with higher circuit density, smaller size and lower Z height. Due to the increased complexity and reduced interconnect dimensions in semiconductor packages, non-destructive and high resolution analytical techniques are highly desirable for semiconductor package failure analysis, assembly technology development, quality control, and reliability risk assessment.

Lab-scale 3D X-ray computed tomography (CT) technologies have been successfully proven to be capable of detecting defects in package components with resolution of micron range in intact samples. Even though good progress has been made in terms of resolution, the through put time (TPT) still remains long requiring several hours of data acquisition time per 1-2 mm^2 region of interest, which gravely limits the application of 3D X-ray CT in semiconductor packages as the size of the packages is usually in the range of 100-3600 mm^2.

Applications in microelectronic packages using current lab-scale 3D X-ray CT, including non-destructive failure analysis, solder joint or Cu via defect monitoring, and progressive reliability study are reviewed. The limitations of the current X-ray CT techniques are discussed.

Recent studies of 3D packages using synchrotron x-ray CT that provide directions for future improvement are presented. High brightness synchrotron x-ray sources have large area collimated beams that provide high resolution x-ray imaging over much larger areas than conventional sources. Results indicate that an entire 16x16 mm System In Package (SIP) can be imaged within 3 to 20 minutes.

Lab-Scale X-ray CT Operation Principle

A wide variety of industrial x-ray CT systems for semiconductor applications have been proposed and developed, with defect detection capabilities ranging from one to a few microns, depending on the specific configuration that is used. The basic setup of x-ray CT systems is shown in Fig. 1(a) and is formed with an x-ray source that radiates the object, which is mounted on a rotating stage that provides a discrete angular displacement, and a detection sub-system that captures the 2D x-ray images at each angle. All images are then mathematically processed to reconstruct the object in 3D [1].

Since the processed data contain the volumetric information of the sample under test, the analyst can manipulate it to display virtual cross-sections or slice views at any given location of the three dimensional data set. This provides the capability of removing interfering features in the field of view that may have been masking out the signal produced by the defect in single 2D x-ray images; the outstanding property of x-ray CT as analytical method comes from the fact that the subtle information of the defect contained in each 2D image becomes important when it is added to signal contributions at the specific xyz location in the whole set of images that were collected at equally spaced angles. Fig. 2 shows an example of typical virtual cross-sections, slice or planar, and orthogonal views of a capacitor; notice how the virtual cross sections in Fig. 2(a) and Fig. 2(b) provides details of shape and xyz localization of solder voids, while the virtual plane view Fig. 2(c) reveals voids that were hiding underneath the highly x-ray absorbing capacitor. The orthogonal view, Fig. 2(d), helps the analyst to visualize the

978-1-4799-2408-0/14 $31.00 © 2014 IEEE

xyz location of voids. Fig.2(e) shows a 3D rendering of the capacitor, which is a capability that can be used to visualize the external surface in case the capacitor in not optically accessible, like inside a heat sink or molding compound [1].

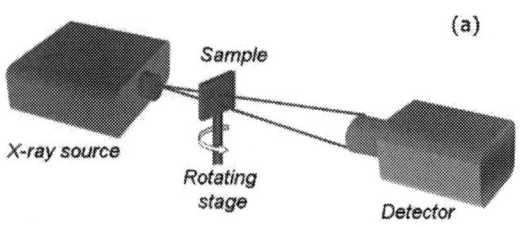

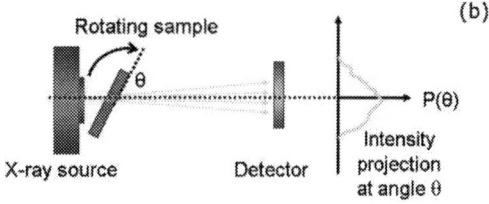

Figure 1. (a) Basic schematic of the x-ray CT setup consisting of x-ray source, rotating stage and detector. (b) The projections at each angle of the intensity of collected 2D x-ray images are superimposed and mathematically processed to generate the three dimensional image.

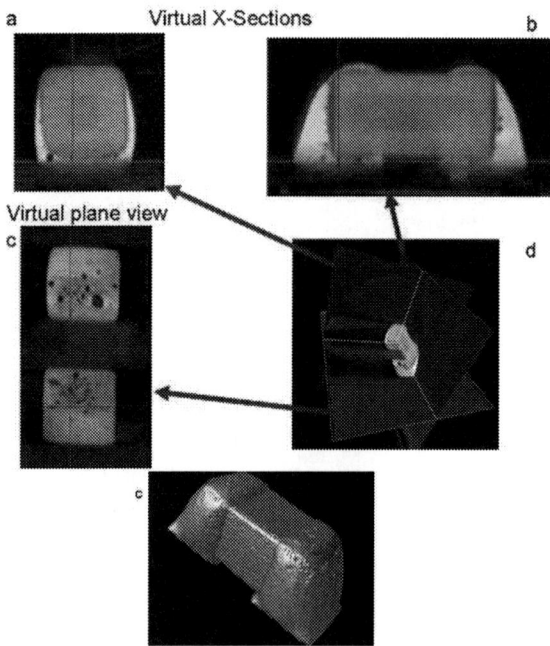

Figure 2. Typical example of x-ray CT data display of a capacitor that was mounted on a package and not optically accessible because of metallic heat sink: (a) and (b) are virtual cross sections, (c) is a virtual planar or slice view, (d) is an orthogonal rendering showing the location of the planes, and (e) the 3D rendering.

The Applications of Current 3D X-ray CT Technology in semiconductor packages

As 3D X-ray CT techniques could provide high resolution 3D information non-destructively, they have been integrated into routine failure analysis flow for semiconductor packages, especially for 3D packages where multiple failures could exist in one unit because of the increased complexity [2]. The combination of having nondestructive fault isolation techniques, such as Time-Domain Reflectometry (TDR), Electro Optical Terahertz Pulse Reflectometry (EOTPR), Scanning SQUID microscopy (SSM), to isolate the open or short locations and 3D X-ray CT to image the failures has been proven to be very effective in 3D packages [2]. Fig. 3 illustrates a Cu trace crack found around the location identified by EOTPR in a 3D package. Fig. 4 demonstrates a leakage failure caused by very subtle solder extrusion between solder joints.

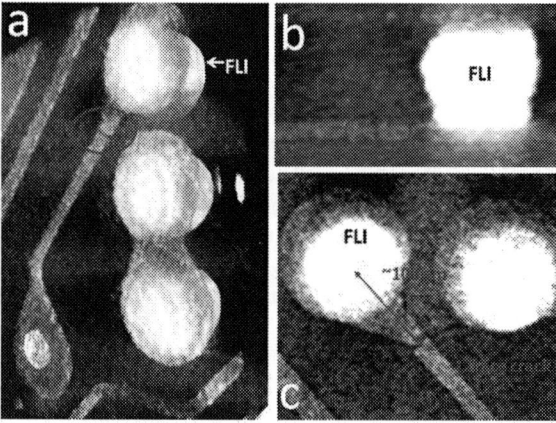

Figure 3. (a) 3D view, (b) virtual X-sectional view, and (c) virtual planar view of the 3D X-ray CT showing trace crack in a 3D package.

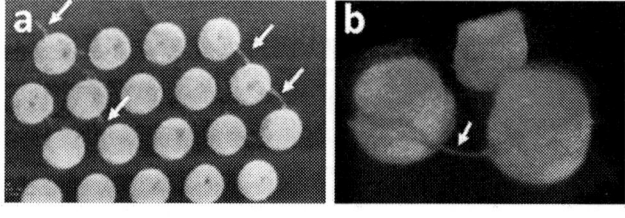

Figure 4. (a) Virtual planar view, (b) 3D view of solder extrusion induced leakage failure in a 3D package.

3D X-ray CT techniques have found application in the semiconductor package assembly process for monitoring defects in solder joints and substrate vias, such as non-contact open, non-wet, and voids in solder joints, and voids or cracking in Cu vias. Fig. 5 (a) shows a virtual X-sectional 3D X-ray image of partial non wet solder joints connecting between the die and the substrate. Compared with good solder joints shown in Fig. 5 (b), the partially non wet solder joints have a pear shape and show small gaps between the die bumps and the bulk solder.

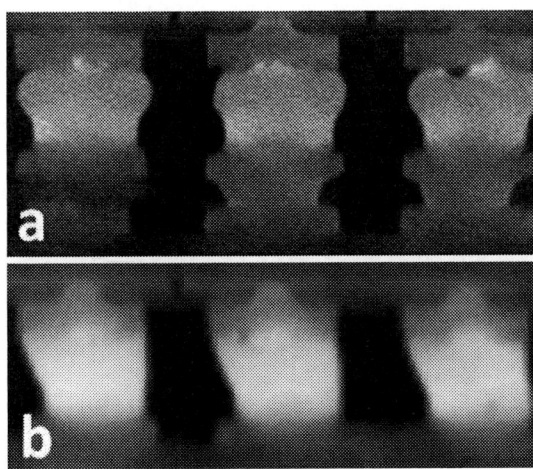

Figure 5. (a), (b) virtual X-sectional view of solder joints having partial non wet and good solder joints, respectively.

Eliminating or reducing the amount of voids in solder is becoming critical for 3D packages as the diameter of solder joint interconnects are driven to be much smaller than those in conventional semiconductor packages and the first level solder joints usually need to go through more reflows during the assembly process. The traditional 2D X-ray technology cannot capture voids smaller than 5μm in diameter in best case scenario, and cannot differentiate overlapping voids either, thus monitoring solder voids using 3D X-ray CT is very important for developing the soldering process for 3D packaging. Fig. 6 illustrates the 3D view of solder joints having multiple solder voids, the void size and location could be obtained from the 3D X-ray CT images, which provides valuable information for further process improvement.

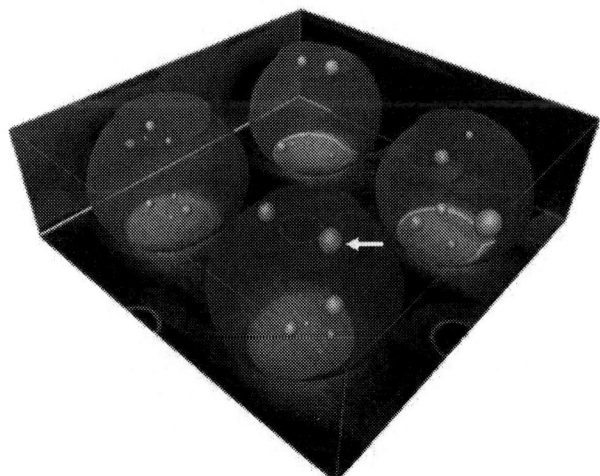

Figure 6. 3D view of solder joints having multiple solder voids.

Miniaturization and portability of consumer electronics has driven the next generation substrate technology to have higher circuit density, smaller size and lower Z height. Due to the increased complexity, defects like substrate Cu via voids and cracks are commonly seen during the process development

stage. As the defects usually are smaller than 5μm, monitoring the defects with 3D X-ray CT is usually applied in the substrate assembly process. Fig.7 demonstrates that 3D X-ray CT distinctly captures voids and cracks in small substrate Cu vias.

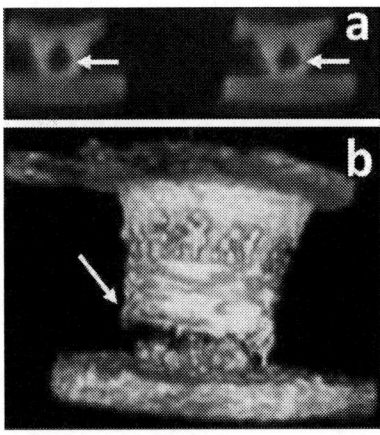

Figure 7. (a) Virtual X-sectional view of voids in substrate Cu vias. (b) 3D view of cracks in Cu vias.

3D X-ray CT can provide high resolution information of semiconductor packages non-destructively, which enables the progressive study of package components during reliability stress, such as reflow, temperature cycling, and consistent current flow at elevated temperatures. Solder void growth kinetics during multiple reflows has been studied using the 3D X-ray CT [3]. As indicated in Fig. 8, the growth of same solder voids is very obvious during the additional reflow post assembly.

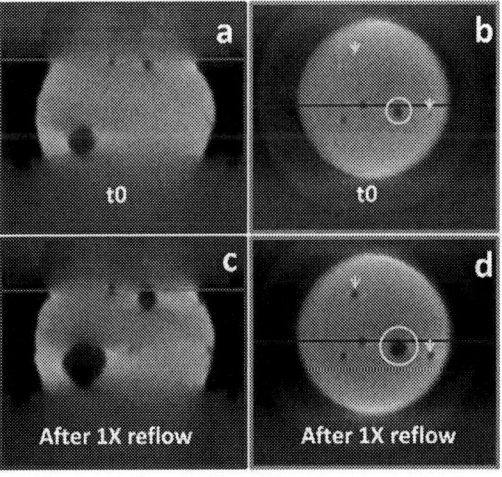

Figure 8. (a), (b), virtual x-sectional view and virtual lapping view of a solder joint before reflow, respectively. (c), (d), the same virtual X-sectional and virtual lapping view of the solder joint post reflow, respectively showing the growth of solder voids during additional reflow post assembly.

Void nucleation and growth in solder joints under current flow at elevated temperatures is one of the focused areas

978-1-4799-2408-0/14 $31.00 © 2014 IEEE

during the development of semiconductor packages having different design and form factors. Progressive 3D X-ray CT study can provide the location and size of the voids at time zero and at intermediate read outs of the electrical test. As shown in Fig. 9 (a)-(d), voids in the solder joints start to nucleate and grow at the cathode during the electrical test due to the electron migration. The void growth kinetics can be obtained from the 3D X-ray CT images, which could be used for effective reliability risk assessment.

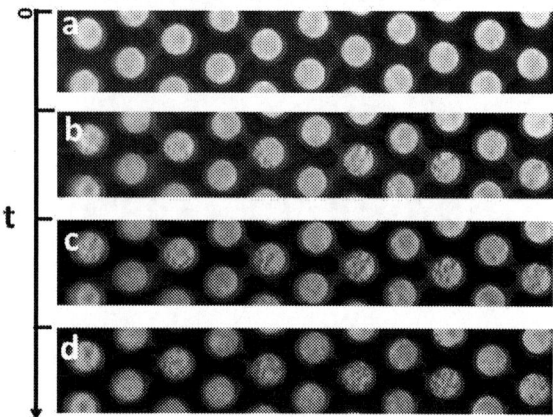

Figure 9. (a), (b), (c), (d) virtual lapping view of the same solder joints under consistent current flow with elevated temperature at time zero, and intermediate read outs, respectively.

Voids could form during the electrolytic Cu filling process of Cu vias with relatively large aspect ratio. 3D X-ray CT has been used to characterize the voids in Cu vias before and during the electrical test where consistent high current is applied to the Cu vias at elevated temperature to capture the nature of the void evolution [4]. As shown in the X-ray CT virtual cross sectional image taken at time zero (Fig. 10 (a)), there is an inverted keyhole shaped void in the Cu via between the Bottom Layer (BL) and Top Layer (TL) on either sides of the substrate core. Fig. 10 (b) shows the same virtual x-sectional view of the Cu via (shown in Fig. 10(a)) post 1000 hours of 3A current flow at 150°C, while Fig. 10 (c) and 10 (d) show another virtual x-sectional view and virtual lapping view of the same Cu via post the test, respectively. The 3D X-ray CT data clearly indicates that the void migrates and agglomerates as a large void at the end of the Cu via in the TL layer of the substrate under applied high current, which is the anode of the Cu via. The void evolution kinetics has been studied based on the progressive 3D X-ray CT data [4].

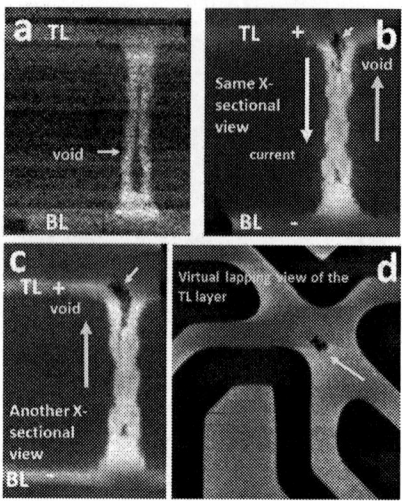

Figure 10. (a) 3D X-ray CT virtual X-sectional image of a Cu via at time zero. (b) The same virtual X-sectional view of the same Cu via after 1000 hours of 3A current flow at 150°C. (c) Another X-sectional view of the same Cu via post the electrical test. (d) Virtual planar view of the same Cu via after 1000 hour of electrical test.

The limitations of current 3D X-ray CT techniques

The current 3D X-ray CT techniques can provide high resolution 3D information inside semiconductor packages non-destructively and have been applied in the field of failure analysis, process control, and kinetic study of electronic packages; however, the technique do have some limitations which make them less desirable as compared to destructive analysis methods.

First of all, the flux and brightness of the X-ray sources are low, which results in relatively long exposure or image capture time, thus longer TPT for high resolution imaging. To get 3D X-ray images with reasonable TPT, the analyst has to trade-off field of view (FOV) and image quality to compensate the long exposure time. Long TPT is usually the show stopper when either multiple FOVs in a single sample or a single FOV in mulitple samples are needed. In these cases, alternative destructive analytical methods provide faster time to information.

On the other hand, X-ray beam with energy higher than 100 kV is usually used in the semiconductor packge imaging to cope with highly absorbing Cu or solder components. This sacrifices the resolution due to the beam spot size blooming at higher energy. Additionally, high energy X-ray imaging reduces the phase contrast, and also makes the organic packaging materials invisible. These factors limit the use of 3D X-ray CT in the detection of defects in packaging materials other than metals with high Z numbers, for instance, Si cracking, and voids or delaminations in underfill, molding compound, solder resist, and other dielectric materials in electronic packages.

Because of these limitations, synchroton 3D X-ray CT studies of 3D packages have been initiated to image an entire 3D SIP package with synchrotron radiation microtomography as a starting point to explore the application possibilities of 3D X-ray CT with a desired X-ray source [5].

3D Synchrotron Radiation CT Operation

Synchrotron Radiation CT, illustrated in Fig. 11, operates in much the same way as the lab-scale CT except that the source of x-rays comes from a synchrotron radiation source rather than a lab-scale x-ray tube source. The principle differences between a synchrotron radiation beam and a lab-scale beam are that the synchrotron radiation has an extremely high flux, has the capability of being monochromatic, has a large area footprint that can be optically manipulated to produce different resolutions, and has a long focal length so that the x-rays are highly parallel with low divergence [6]. These factors combine into a term called beam brightness or brilliance (photon flux divided by divergence, beam size, and energy band width), which may be 10 orders of magnitude higher than lab-scale sources [6]. The key advantages of synchrotrons for CT are that TPT can be significantly reduced due to the high flux, larger regions of interest can be examined due to the large footprint of the beam, the image quality can be preserved over large part dimensions due to the low divergence of the beam, and the possibility of using relatively low energy level enables phase contrast imaging. Here, a synchrotron beam is used to image an entire multi-level SIP package with synchrotron radiation microtomography as a point of reference and comparison to conventional 3D CT methods.

The synchrotron work was performed at the Lawrence Berkeley National Laboratory Advanced Light Source (ALS) beamline 8.3.2. This beam line is dedicated to microtomography with X-rays being produced from a superconducting bend magnet to produce a high flux x-ray beam from a storage ring with energy of 1.9 GeV and current of 500 mA. Details about this beam line can be in the literature [7]. In these experiments a polychromatic beam is used to maximize the x-ray flux on the sample that is mounted on a rotating stage in front of the imaging optics. The X-rays that pass through the sample impinge on a LuAg single crystal scintillator that fluoresces to produce visible light. The visible light image is viewed using different magnification lenses (1, 2,5, and 10x) by one of two different cameras referred to as camera A (PCO Edge CCD imaging camera with 2560 x 2160 pixels), and camera B (PCO 4000 CCD imaging camera with 4008 x 2672 pixels). Different combinations of cameras and lenses produce different resolutions and fields of view (FOV) as summarized in Table 1. Note that the FOV decreases with increasing resolution, and that the highest resolution with this setup is ~ 0.65 μm.

3D CT synchrotron imaging of a SIP package was performed by rotating the sample over 180° in angular increments of 0.175°, and at each angular step an image is recorded. The rapid imaging time of the synchrotron allowed an entire scan to be completed in 3 to 20 min as summarized in Table 1 for the different optical arrangements. Each scan resulted in 1024 images, which are reconstructed using a

filtered back projection method at ALS using the Octopus software [8]. Once the reconstruction is completed, visualization and analysis of the 3D image is performed using the commercial software package Avizo.

Source	Resolution Option	Camera/ Lens Mag.	Pixel Size (mm)	FOV Width (mm)	FOV Height (mm)	Imaging Time (min)
Synchrotron ALS BL 8.3.2	low	A/1X	8.7	36.0	6	3
	low	B/1X	6.5	16.6	6	3
	med	A/2X	4.5	18.0	6	20
	med	B/2X	3.3	8.3	6	3
	high	B/5X	1.3	3.3	2.8	5
	high	B/10X	0.65	1.7	1.4	11
Lab-based source	high	-	1.5-2	1.5-2	1.5-2	180-240

Table 1. Summary of resolutions, field of view (FOV) and imaging time for different cameras and lens. The highlighted rows are used in this investigation in order to image an entire 16x16 mm chip in as little as 3 min.

Synchrotron X-ray CT imaging of a SIP package

The microelectronic package to be imaged is a 3D SIP component measuring 16x16 mm and approximately 3 mm thick. The package is placed in the beam line with a horizontal orientation, as indicated in Fig. 11, which allows the entire package to be imaged in one scan. The SIP package (Fig. 11) consists of a FPGA silicon die attached to a FPGA substrate with first level interconnect (FLI) solder joints of approximately 100μm in diameter; mid-level interconnect (MLI) solder joints of approximately 350μm connecting the FPGA substrate to the SIP substrate; and second level interconnect (SLI) solder balls of approximately 650μm on the back side of the SIP substrate.

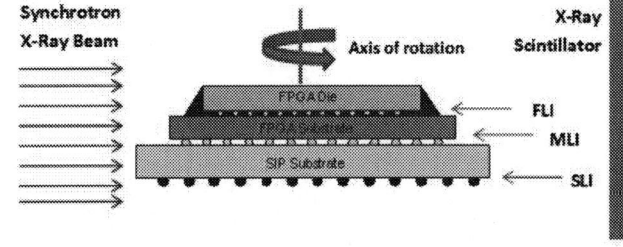

Figure 11. Schematic of a SIP package that is used in the synchrotron CT study. The 16x16 mm package is placed horizontal relative to the x-ray beam, allowing the entire package to be imaged at once.

Fig. 12(a) shows one 3D CT view of the entire package that is imaged with an 8.7 μm resolution in 3 min of scan time. In this image, the plated through holes (PTHs), copper vias, and some of the substrate structures are clearly seen. Note that this is just one image of a complete 3D CT reconstruction of the package that can be rotated and flipped, and zoomed in to inspect different regions of interest (ROIs) in the package. Fig. 12(b) shows the same reconstruction, with three orthogonal sectioning planes in place that can be moved independently to inspect specific interconnect and package details. Here, the FLI and MLI solder connections can be seen in the vertical planes, along with PTHs in both substrates.

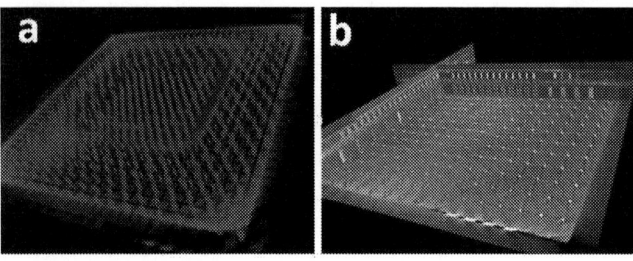

Figure 12. Entire FPGA SIP package imaged with 8.7 µm resolution and a scan time of 3 min. a) 3D reconstructed image, and b) orthogonal sections through the reconstructed image showing all levels of interconnects.

Fig. 13 shows an image from reconstruction with the contrast adjusted to show all three levels of the SIP chip solder joint interconnections at 8.7 µm resolution. This reconstruction is performed in one 3 minute scan, and demonstrates the ability of synchrotrons to penetrate through multiple layers of interconnects quickly and at high resolution with the ability to inspect multi-layer and multi-scale features.

Figure 13. Entire FPGA SIP package imaged with 8.7 µm resolution and a scan time of 3 minutes, highlighting the FLI, MLI, and SLI solder ball interconnects.

Each of the above images can be zoomed in to provide details about small regions of interest from any portion of the entire SIP package. This is demonstrated in Fig. 14, which shows a zoomed in portion of the SIP package reconstruction, and highlights the major features that can be observed. In this figure the entire SIP package is scanned with a resolution of 4.5 µm in 20 minutes. The major interconnect features are highlighted, where the image clearly shows all levels of interconnects. The synchrotron white light beam covers a wide range of energies from 4-100 kV, allowing different materials to be imaged. In this case, the Si die and underfill are both visible, along with the metal components of the substrates and interconnects.

The above examples highlight some of the first 3D CT images produced of an entire multi-level microelectronic

package and demonstrate that high resolution images can be taken rapidly and over large fields of view.

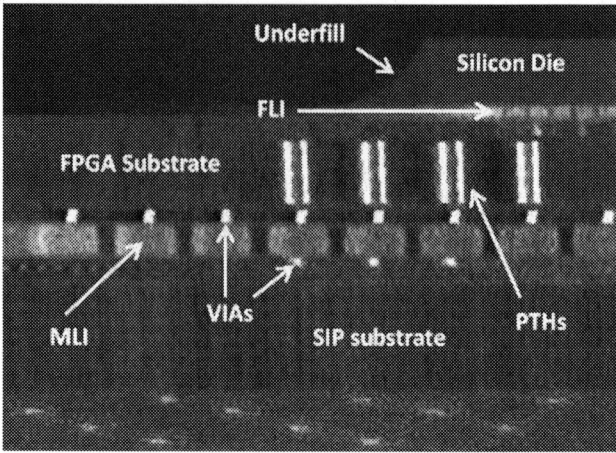

Figure 14. Zoomed in region of the entire FPGA SIP package imaged with 4.5 µm resolution and a scan time of 20 min. The silicon die, underfill, both substrates, and all levels of interconnects can be observed.

Desired features for X-ray CT Systems in Microelectronics

The synchrotron X-ray CT study of the SIP package reveals the possibility of expanding the applications for 3D X-ray CT in microelectronics. As the whole package could be imaged with high resolution within less than one hour, all the defects in various packaging materials can be inspected at the same time. Thanks to the extremely short data collection time in synchrotron based CT, the in-situ study of package components in real packages under reliability stress, such as reflow, temperature cycling, and mechanical shear or tensile stress, could be performed, which will be providing highly valuable information for further process improvement or save time in reliability tests.

The synchrotron X-ray CT study also provides some suggestions to the next generation lab-scale 3D X-ray CT systems. A suitable x-ray CT system for microelectronics needs to meet the following highlights: (a) It should have a way to compensate for magnification and resolution deterioration due to large sample sizes, so there is no need to trim the samples to achieve micron level resolution, (b) the x-ray source needs to provide high brightness x-ray beam with an energy spectrum ranging from 40 kV to 150kV to deal with highly absorbing copper or solder packaging components, as well as being able to imaging dielectric materials in packages, and (c) the data acquisition TPT should be less than 10 minutes per FOV to compete with alternative destructive analysis methods.

Conclusions

Current 3D X-ray CT technology have been widely used in microelectronic package failure analysis, package assembly process control, substrate via quality control, and progressive reliability study of solder voids, solder joint electromigration, and Cu via void evolution. The high resolution 3D images of defects are obtained non-destructively, thus can provide valuable information for failure analysis as well as assembly process development. However, the TPT of current lab-scale 3D X-ray CT technology is quite long and not practical when applications involve a large number of samples, as typically required in assembly process optimization. On the other hand, the relatively low brightness of lab-scale sources doesn't allow using its lower keV energy that is needed for phase contrast defect highlighting in the presence of copper or solder surrounding features, which disables many other applications for X-ray CT in microelectronic packages.

To explore the possible applications of X-ray CT techniques in semiconductor packages with desired X-ray sources, synchrotron 3D X-ray CT studies of 3D SIP packages have been performed. The results indicate that an entire 3D package with 16x16mm in size can be imaged within 3-20 minutes having about 4-8 μm resolution. Details of the package within each layer, including Si die, underfill, multiple levels of solder joints, dielectric materials, Cu vias, and through holes in multiple substrates, can be obtained at the same time.

The results obtained by using synchrotron 3D X-ray CT suggest a path forward for new applications in microelectronics, and also provides directions for next generation lab-scale X-ray CT technology. The ability to compensate for magnification and resolution deterioration due to large sample sizes, an x-ray source having wider energy spectrum with high brightness, and faster data acquisition TPT are of the upmost importance factors for semiconductor applications and highly needed in the next generation 3D X-ray CT technology.

Acknowledgments

The LLNL portion of this work was performed under the auspices of the U.S. Department of Energy by Lawrence Livermore National Laboratory under Contract DE-AC52-07NA27344.

The Advanced Light Source is supported by the Director, Office of Science, Office of Basic Energy Sciences, of the U.S. Department of Energy under Contract No. DE-AC02-05CH11231.

The authors would like to thank Pilin Liu, Liang Hu, William Hammond, and Carlos Orduno from Intel Corporation for some of the data collection and helpful discussions.

References

1. M. Pacheco, and D. Goyal, "Detection and Characterization of Defects in Microelectronic Packages and Boards by Means of High-Resolution X-Ray Computed Tomography (CT)". *Proc. IEEE Electronic Components and Technology Conference (ECTC)*, 2011, pp. 1263-1268.
2. Y. Li, Y. Cai, M. Pacheco, R. C. Dias, and D. Goyal, "Non-destructive failure analysis of 3D electronic packages using both Electro Optical Terahertz Pulse Reflectometry and 3D X-ray Computed Tomography".

Proceedings from the 38th International Symposium for Testing and Failure Analysis (ISTFA, ASM International); 2012, pp. 95-99.
3. Y. Li, J. S. Moore, B. Pathangey, R. C. Dias, and D. Goyal, , "Lead-Free Solder Joint Void Evolution During Multiple Subsequent High-Temperature Reflows", *IEEE Transactions on Device and Materials Reliability*, Vol. 12, No. 2 (2012), pp. 494-500.
4. Y. Li, L. Xu, P. Liu, B. Pathangey, M. Pacheco, R. C. Dias, and D. Goyal, "Void migration in Cu vias under current flow detected by 3D X-ray computed tomography" *Presented at the 143rd TMS Annual meetings and exhibition; San Diego, CA,* 2014.
5. J. Elmer, Y. Li, H. Barth, D. Parkinson, M. Pacheco, and D. Goyal, "Synchrotron radiation microtomography study of 3D microelectronic packages" *Presented at the 143rd TMS Annual meetings and exhibition; San Diego, CA,* 2014.
6. H. Winick, *Synchrotron Radiation Sources: A Primer,* World Scientific Publishing, 1994.
7. J.H. Kinney and M.C. Nichols, "X-ray tomographic microscopy (XTM) using synchrotron radiation," *Annu. Rev. Mater. Sci,* 22, pp. 121-152, 1992.
8. M. Dierick, B. Masschaele, L. Van Hoorebeke,"Octopus, a fast user-friendly tomographic reconstruction package developed in LabView (R)," *Meas. Sci. and Technol.,* 15, pp. 1366–1370, 2004.

In-Situ Measurements of the Relative Thermal Resistance: Highly Sensitive Method to Detect Crack Propagation in Solder Joints

Gordon Elger* – Technische Hochschule Ingolstadt, Germany
Shri Vishnu Kandaswamy – Technische Hochschule Ingolstadt, Germany
Maarten von Kouwen –Philips Technology GmbH, Germany
Robert Derix – Philips Technology GmbH, Germany
Fosca Conti – University of Padova, Italy
* Corresponding author: gordon.elger@thi.de

Abstract

The crack propagation in solder joints is detected by measuring the relative thermal resistance using transient thermal analysis. The method can be applied on high power devices which have to dissipate a significant amount of thermal load through the solder joint. The thermal load in the device is switched and the forward voltage of the junction is measured in time resolved modality. As published earlier, to obtain the relative thermal resistance the dissipated power of the device and the proportional factor (k-factor) between temperature and forward voltage are not required when the time resolved forward voltage curves are normalized [1].

Aim of the research was to investigate the concept and the sensitivity of the measuring method using high power ceramic LED packages. For prove of sensitivity, two batches of high power LEDs were soldered on PCBs, i.e. aluminum insulated metal boards (Al-IMS): one batch with SAC305 and the other one with a SAC+ (Innolot) solder. The test modules were exposed to typical temperature cycles of -40°C / +125°C as required for automotive applications. With standard test methods differences in reliability, i.e. thermo mechanical fatigue, of the two solders were very difficult to detect for the module design under these test conditions [2].
First, we investigated the relative thermal resistance after defined number of cycles at room temperature. To define a realistic failure criterion for calculation of the cumulative failure probability, we have set-up a finite element model and simulated the transient temperature curves. The increase of measured relative thermal resistance is critically compared and calibrated with transient finite element simulations. The Weibull curves were determined based on the failure criteria describing a 70% cracked solder area..

Then, after resolving a significant difference in the increase of thermal resistance of the two solders, an In-Situ test system was set up and measurements were performed, i.e. a set of samples was measured in the temperature chamber under hot and cold condition.

Introduction

Many electronic systems, i.e. high power electronic devices, have to operate at high temperature and are exposed to temperature cycles with large temperature difference, e.g. automotive application, and have to withstand significant cyclic thermo-mechanical stress. One typical failure mode is the cracking of solder joints. A typical example are LED lighting modules, i.e. ceramic LED packages on PCBs. Typically, lifetimes between 25.000h and 50.000h (even up to 100.000h) are required for LED modules. Due to thermal management and cost, aluminum insulated metal boards (Al-IMS) are often preferred. The LED industry has to develop appropriate test methods to detect the cracking of the solder joints. The first and simplest test is the so called "light-on-test" of the LED [3]. As long as the LED "lights on", the solder contact is defined as consistent. However, this test is rather rough and not adequate to detect solder cracks in their early state. A crucial problem is related to the physical dimensions and configurations: the LEDs have quite large solder contacts for thermal management. The solder contacts start to crack from the corner of a package and the crack grows in direction to the center because the stress is highest in the corner where the effective length for the thermo-mechanical mismatch is highest. In the center of the package the solder often holds the package on the board. As long as the package is mechanically hold in place, an electrical contact can be formed by pressure, even if there is significant solder-crack. The phenomenon implies that a LED can pass the light-on-test even with severe cracks in the solder contact.

Another possibility to measure the quality of the solder joint is the destructive shear test [4]. The shear force of the device is reduced when a crack grows through the solder joint. However, besides the fact that the shear test is a destructive method, the shear strength of the solder material itself degrades with temperature driven intermetallic phase growth.

At Philips Technology GmbH, an electrical resistance test method was introduced for solder joint analysis of LED packages based on the electrical resistance measurements described in the IPC-785 and IPC-9701 [5,6].

For high power devices, in which a significant amount of heat has to be conducted through the solder joints to the printed circuit board, changes in the thermal path can be detected by the change of the thermal resistance. The calculation of the thermal resistance for a transient thermal measurement requires measuring of the real thermal power step (P_{th}). In addition, a linear factor k (called k-factor), between forward voltage (V_f) and temperature (T), has to be measured for each LED [7-9]. Until now the approach to measure the thermal resistance to detect crack formation in the solder joint is not applied in reliability tests of SMD packages because the measurements and data evaluation are considered as time consuming and expert know how is required. However, the measuring effort can be significantly reduced because during reliability testing solely the evaluation of the changes in the transient temperature curves are required but not the absolute thermal resistance. Due to cracks in the solder joint, the thermal resistance of the joint degrades and a cracked joint can be identified by measuring the change of thermal resistance compared to the initial situation.

In the first part of the paper we describe the application of transient thermal analysis on the solder joint integrity without measuring k-factor and real thermal power step.

In the second part the sensitivity of the method is demonstrated by investigating two batches of ceramic LEDs packages soldered with two different solders (SAC305 and Innolot: Sn91,175/Ag3,5/Cu0,7/Ni0,125/Sb1,5/Bi3) on Aluminum Insulated Metal Substrates (Al-IMS). The increase of relative resistance is measured.

In the third part a failure criteria is derived by transient thermal finite element (FE) simulation. Based on the failure criteria, the culminated failure function is calculated and a Weibull diagrams are plotted for the two solders.

Finally, first In-Situ measurements are presented and potentials and future applications of the method are described and discussed.

Instrumentation: Test Vehicles and Thermal Equipment

Ceramic LED Packages (AlN, 200 lm at 0.7A drive current and V_f=3.1V) were vacuum soldered on Al-IMS boards with hard dielectric layer (see Fig. 1).

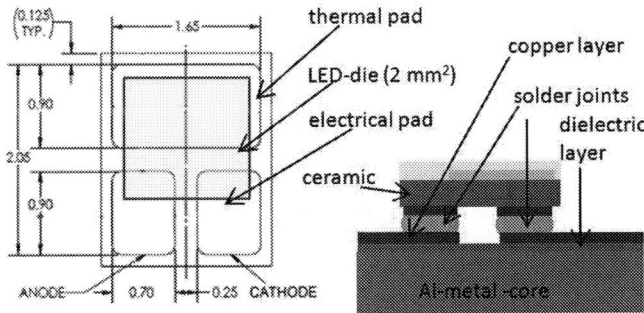

Fig. 1 Ceramic LED package and Al-IMS board

To eliminate the statistical occurrence of voids vacuum soldering was used. Almost void free solder joints were obtained. Two test batches were assembled: one with standard SAC305 (Sn96.5/Ag3/Cu0.5) and the other with Innolot (Sn91,175/Ag3,5/Cu0,7/Ni0,125/Sb1,5/Bi3). Every test group contained 40 LED.

Temperature cycling is performed by placing the LEDs in a TST (Temperature Shock Test) chamber with a temperature setting -40°C / +125°C, dwell time in the hot and cold

conditions 30 minutes respectively along with a transfer time of 10 seconds. One cycle is approximately 1 hour long.

The transient thermal measurements are made by using T3Ster equipment from Mentor Graphics (former MICRED). The samples were taken out after defined cycle intervals (cycle number: 0, 100, 214, 399, 640, 802, 1009) and measured. The device under test is mounted to the temperature controlled heat sink of the T3Ster. The heat sink is maintained at a temperature of 25°C. Initially, a junction to case thermal equilibrium is ensured by applying a drive current (I_{drive}) for short duration. Then the I_D is switched off and a small sense current (I_{sense}) is applied. When the system transfers to its new thermal equilibrium, the cooling down of the junction is detected. A heating current of 0.7A was applied for 40s and the $V_f(t)$ was measured for 40s using a sense current of 20mA..

In a second step, In-Situ measurements were performed. Test procedure and equipment are described in the next sections.

Theoretical approach: Relative Thermal Resistance

The transient thermal method, i.e. the measurement of junction temperature and thermal resistance of LED package [10, 11], is a well established method for thermal characterization, . An overview can be found in JESD51-14 [12]. The thermal response of a system like an LED package on a printed circuit board is measured in time resolved modality after switching a heat load. The temperature T(t) is detected at a significant location time resolved while the system transfers into its new thermal equilibrium. For example, initially a constant heat flux (I_{drive}) is applied until the thermal equilibrium in the LED package is reached. The thermal equilibrium is then changed by switching from I_{drive} to a small sense current I_{sense}, i.e. the heat flux is switched almost off. The forward voltage V_f is measured by the small current I_{sens} (see Fig. 2). The temperature is measured by V_f which depends linearly from the temperature in a limited temperature range. The proportional factor is the so called k-factor [7-9].

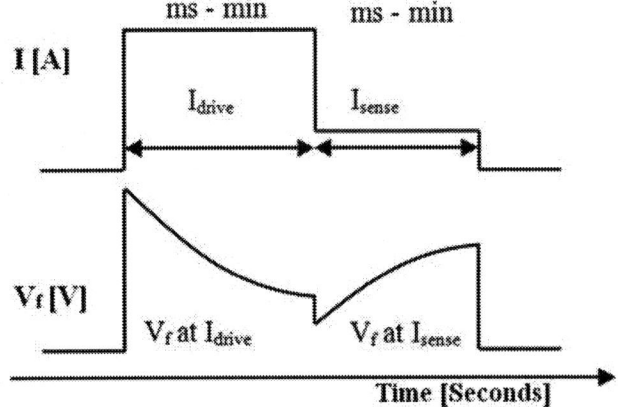

Fig. 2 Transient thermal measurement: V_f is measured time resolved after switching the drive current.

978-1-4799-2408-0/14 $31.00 © 2014 IEEE

The transient thermal measurement method works by reconstructing the heat flow path: in our case, from the LED-die to the heat sink. If there is a crack in the LED assembly, the heat flow path would change and the phenomenon can be detected by the change of the thermal response. Due to the fact that in microelectronic packages the thermal masses increase from die over heat spreader, the thermal resistances of the different package levels can be usually separated in the time domain.

The calculation of the thermal resistance for a transient thermal measurement requires the measurement of the real thermal power step P_{th} and of the k-factor. P_{th} and the k-factor can change during the reliability test. Therefore, the measuring effort is large because P_{th} and the k-factor have to be measured for every measurement interval which is not applicable for effective In-Situ or In-line measurements. However, this measuring effort can be eliminated when evaluating only the relative changes in the time-derived transient temperature curves. Due to cracks in the solder joint, the thermal resistance of the joint degrades and by measuring the change of thermal resistance compared to the initial situation, cracks in the joint can be identified. By adopting a suitable data, post-processing P_{th} and k-factor are eliminated [1,13].

The thermal network is physically described by a Cauer RC-network. However, the Cauer network can be transformed into a Foster network for which the transfer function can be analytically calculated [10,12]. For n-discrete RC nodes of a Foster network, the temperature response is:

$$T(t) = -\frac{V_f(t)}{k} = P_{th} \sum_{i=1}^{N} R_{th_i}(1 - e^{-\frac{t}{R_i C_i}}) \qquad (1)$$

Important is to note that P_{th} and the k-factor are solely linear factors. The temperature response function $Z_{th}(t)$ is obtained by dividing $T(t)$ through P_{th}. The next steps are to substitute $z=\ln(t)$ (logarithmic time) and to calculate the derivative of $Z_{th}(z)$ which we call $b(z)$:

$$b(z) = -\frac{1}{k \cdot P_{th}} \frac{d}{dz} V_f(z) \qquad (2)$$

As mentioned earlier, the linear factors P_{th} and k can change during temperature cycling. Therefore we change from the linear to logarithmic representation:

$$B(z) = \ln\left(-\frac{1}{k \cdot P_{th}} \frac{d}{dz} V_f(z)\right) = \ln\left(\frac{1}{k \cdot P_{th}}\right) + \ln\left(-\frac{d}{dz} V_f(z)\right) \qquad (3)$$

In the logarithmic representation, the linear factors become an offset and therefore its effect on the transient curves can be eliminated by normalization. This can be done by dividing every time-derived curve by its value at a selected normalization time window or moving the curves on top of

each other for the defined normalization window by a simple fit algorithm.

Experimental Results: Relative thermal resistance

The time dependent signals for an LED measured after different cycle numbers are depicted in Fig. 3.

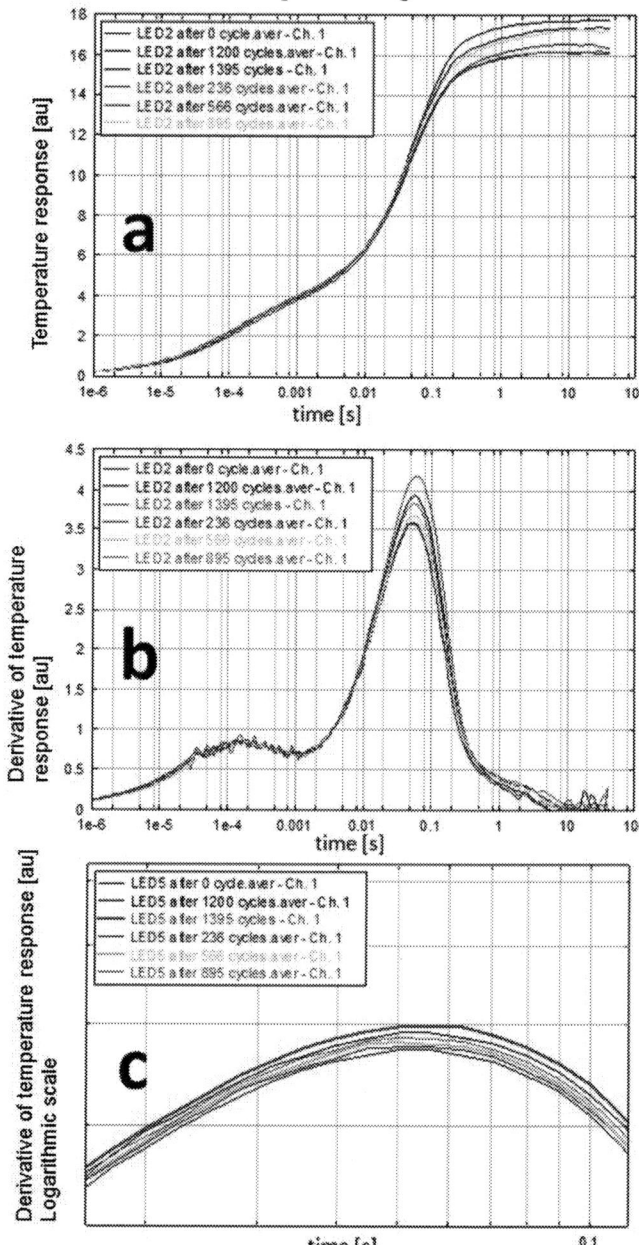

Fig. 3 Time dependent signals with z=ln(t): (a) T(t), (b) b (z), (cB(z)

An increase of the peak around 70ms is observed. In the following this peak is called "solder joint peak". This increase of the peak can be correlated to an increase of the thermal resistance of the solder joint and will be discussed in detail in the section regarding finite element simulations. A time interval between 0.1 ms and 1 ms was selected for normalization. During this time interval, the heat flows from the junction to the ceramic, i.e. the temperature gradient between junction and ceramic decays. As long as the heat flows in the LED die itself don't change by failure in the die (e.g. delamination or cracks), the result of the fit is pure residual free noise and the normalization is fine. As soon as a significant residual is obtained, a potential die failure occurred and the normalization can't be done automatically. Closer inspection is required to verify the potential LED die failure. In Fig. 4 the die and phosphor of the LED were damaged by purpose (red curve) with the bond tool during assembly. A significant difference is observed in the time range of 10^{-4} s. From the plot of Fig. 4, it can be seen that, in the time range between 0.1ms and 1ms, the LED (red curve) which is damaged shows a different behavior. In contrast, the influence of the solder joint is located in the time range of 70ms. Since the resolution of transient thermal method is very high (0.1ms) it is possible to detect also some LED-die damages.

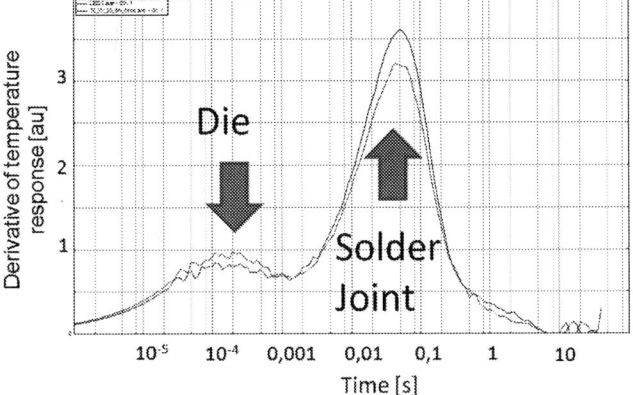

Fig. 4 Die failure: In the 10^{-4}s time region the thermal resistance of the die itself is located. Blue curve: LED without die failure. Red curve: Die and phosphor were by purpose damaged by the bond tool.

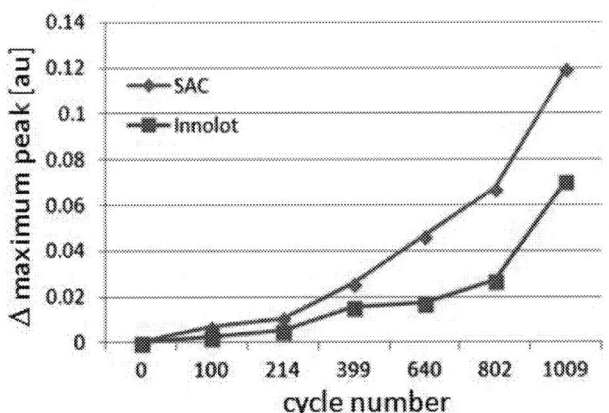

Fig. 5 Average increase of peak maximum (between 60ms and 89ms)

In Fig. 5 the average increase of the solder joint peak is plotted in dependence of the cycle number. A significant difference can be observed for SAC305 and Innolot. A comparison between SAC and Innolot is discussed in the next section in relation to the sensitivity.

Finite Element Simulation and Calibration

To correlate the increase of the solder joint peak during cycling to a failure criteria transient FE simulation were performed, i.e. transient temperature curves as experimentally measured. A FE model was set-up within Solid Works using the integrated FloEFD solver. Only heat conduction was enabled because the heat, which is dissipated by radiation and convection, is small (below 5%) and not relevant for investigating the change of the transient signals by the solder interconnect. Thermal resistance and heat capacity of the materials were taken from literature. As thermal boundary condition a heat flux condition of 1.5W were applied and a heat transfer condition at the bottom of the Al-IMS. The model size was about 800 million cells so that a transient simulation process is approximately 1h long (Fig. 6). The complex die structure was simplified to limit the model size resulting in a systematic error of the model for the time domain under 1ms when the transient signal is determined by the heat flow through the die.

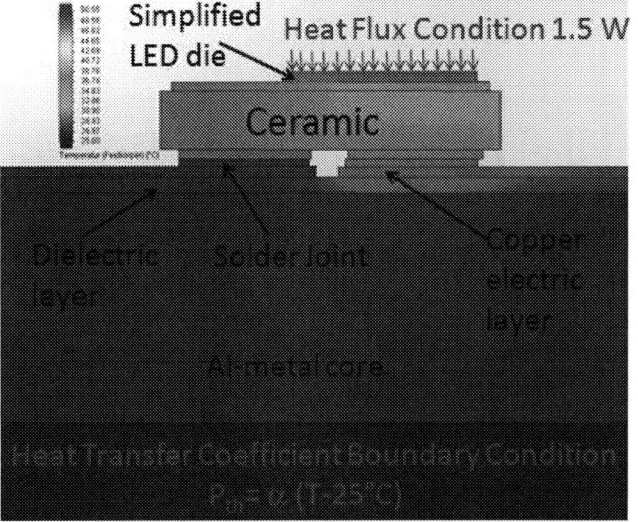

Fig. 6 Finite Element Model

After achieving the matching of the experimental and the simulated dime derived curve, a variation of the thermal conductivity of the solder joint was performed to model good and bad solder joints, i.e. the thermal conductivity of the SAC solder was varied from the nominal value of 56W/mK to 5.6W/mK. The increase of thermal resistance is calculated by the FE simulation and also by simply using the Fourier law:

$$P_{th} = \frac{\lambda \cdot A}{d} \Delta T \quad \rightarrow \quad R_{th} = \frac{\Delta T}{P_{th}} = \frac{d}{\lambda \cdot A} \quad (4)$$

Simulation and calculation fit very well because the AlN ceramic acts as a good heat spreader and the heat flows almost homogeneously through the area of the solder pads. Post-processing the heat flow through the solder joints in the simulation proves that: one obtains an almost homogeneous heat flow through the solder pad area at the bottom of the AlN ceramic. In the solder joint itself, because of its small

978-1-4799-2408-0/14 $31.00 © 2014 IEEE

50-80μm thickness, almost no heat spreading occurs. Reducing the thermal conductivity from 56 W/mK to 5.6 W/mK represents a 90% cracked solder joint whereby the cracks are homogeneous distributed. Indeed, this is a strong simplification because the cracks will be definitely not homogeneous distributed but grow from the outer package areas to the inner ones. However, this first simplified analysis is already able to reproduce the experimental effect of the increase of the solder joint peak during temperature cycling (Fig. 8). Presently cracks are simulated in the solder joints to obtain the influence of the crack location on the time derived curve. Results will be presented in a follow-up paper.

The R_{th} increase is correlated to the increase of the maximum of the peak for the simulated data. In Fig. 9 the R_{th} increase is plotted in dependence of the maximum of the peak. Based on this simulation we define as failure criteria an increase of the maximum of the solder joint peak of 0.05 (no units necessary because we work with normalized curves). The value 0.05 represents a "homogeneous cracked" area of around 70%.

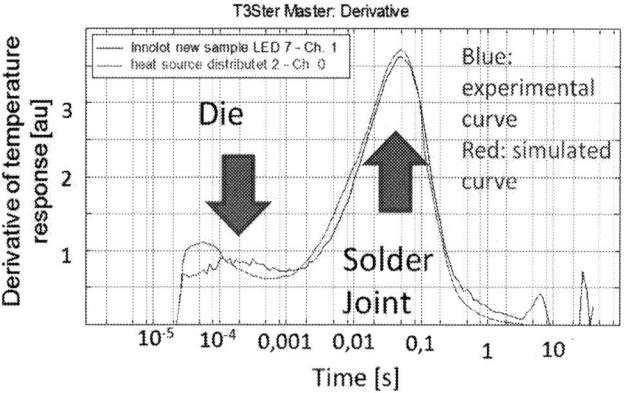

Fig. 7 Simulated (red) and experimental (blue) curve. Due to the simplification of the die, the curves do not match at die position. The solder joint peak is well simulated.

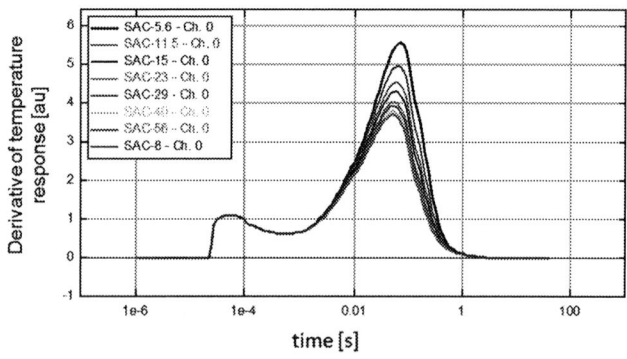

Fig. 8 Simulation of cracked solder joint by reducing the thermal conductivity of the SAC solder from nominal 56W/mK stepwise to 5.6W/mK (numbers in insert indicate the thermal conductivity in W/mK)

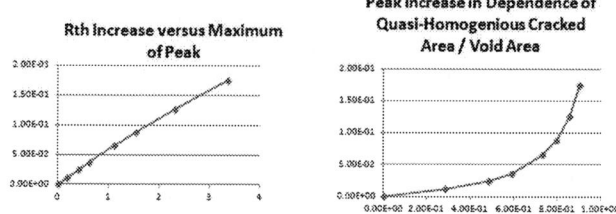

Fig. 9 Right: Rth increase in dependence of increase of maximum of solder peak at 70ms. Left: Peak increase of "Homogeneous Cracked area".

Correlation of R_{th} and Solder joint degradation

Cross-section analysis was made to experimentally verify the solder joint degradation and its effect on the increase of thermal resistance. It was found that LEDs with a stronger increase of solder joint peak maximum had larger cracks and LEDs with a smaller increase of the peak maximum revealed less cracks. The left part of Fig. 10 shows the time derived normalized curves of two LEDs which presented significant but different increase of the solder joint peak maximum during temperature cycling. The crack formation is visible in the cross section depicted on the right part of Fig. 10. LED 3 is characterized by the larger increase of the maximum of the peak and shows very large cracks in the thermal and electrical pads. It presents a high thermal degradation. LED2, which has a relatively less thermal degradation, shows less cracks.

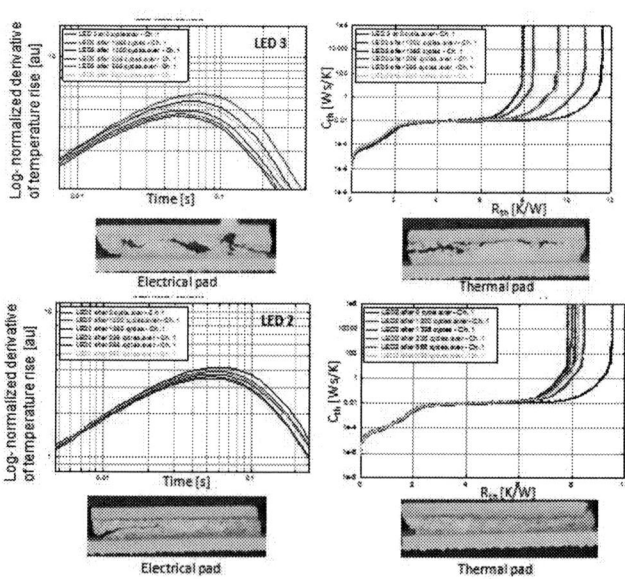

Fig. 10 Increase of Peak at 70ms and crack formation

A potential second failure mode, delamination of the dielectric layer, was not found. The peak high but also the peak position depends on the dielectric layer, too. We are presently running simulation to distinguish between crack in the solder joint and delamination/degradation of the dielectric layer.

Sensitivity: Comparison SAC305 and Innolot

SAC305 and Innolot are among the most widely used interconnect materials for surface mount technology. It is well known that the Innolot has a higher creep resistance under high temperature condition due to solid solution hardening of the Sn matrix by the dopants. However, for the test condition (TST, -40 / +125°C, 30 min dwell time) it was so far not resolved for the LED application under investigation in this paper. Based on the failure criteria: Increase of peak hight by 0.05 (70% of cracked area) the cumulative failure function is calculated. In Fig. 11 the Weibull plots for SAC and Innolot are depicted with the obtained Weibull parameters. The lifetime obtained with Innolot is significantly higher, i.e. T=1360 for Innolot compared to 740 using SAC. The next task is to correlate the standard failure criteria when using light on test or electrical resistance measurements with those one defined by the thermal measurements.

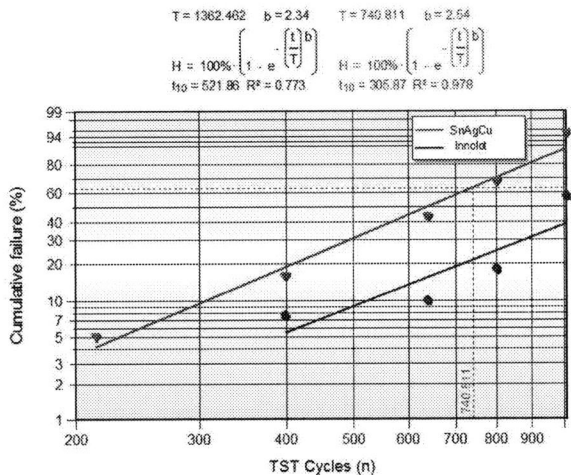

Fig. 11 Weibull Plot – SAC and Innolot

In Situ Measurements

The first transient thermal In-situ measurements were performed and are reported in this paper. An additional test batch of 10 samples was placed into the temperature cycle chamber. The samples were electrically connected to measure them during cycling. Because a multiplexer was so far not available, the samples were measured during cycling only every morning switching manually between the samples. For measuring, a home build test equipment described in [1] was used. Looking on the transient temperature curves (see Fig. 3) it can be concluded that already after 2s the temperature gradient between ceramic and Al-IMS is almost fully decayed. Therefore, it is sufficient to reduce heating and sensing time to 2s each for the In-Situ tests. This conclusion is important, because the samples are placed without any heat sink in the temperature cycling chamber. Within 2s heating, he heating up of the Al-IMS board can be neglected. The increase of the maximum peak under cold and hot condition is depicted in Fig. 12. Interesting is the shift of the peak between hot and cold condition. The shift can be explained by the temperature dependence of the heat capacity of the

AlN ceramic. At higher temperature the heat capacity increases roughly by a factor of 2.

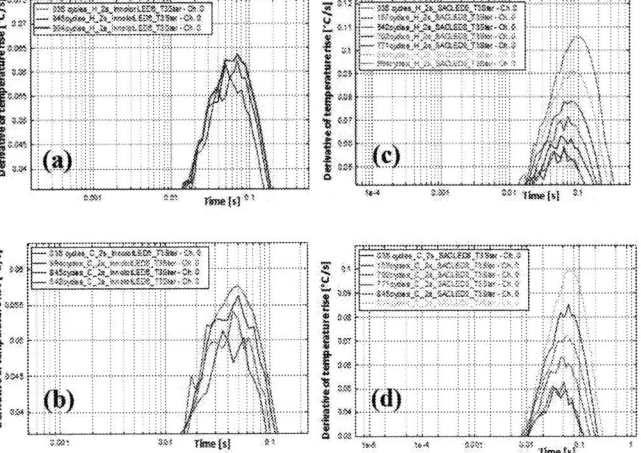

Fig. 12 Insitu measurements of SAC and Innolot (a) Innolot measured at +125 °C (b) Innolot measured at cold condition, -40 °C (c) SAC measured at +125 °C (d) SAC measured at -40 °C

Conclusions

The sensitivity of the transient thermal measurements is demonstrated by the comparison between SAC305 and Innolot. The transient thermal measurement of the relative thermal resistance for package integrity has a high potential to be used for In-Situ or In-Line measurements because measurement of the thermal load and the k factor is not required. They also can be applied for investigation of interconnect materials. Crack propagation can be studied by using an appropriate test chip with heat sources, e.g. LEDs or other small heat producing semiconductors, on different positions. By that the location of cracks and their propagation can be resolved.

References

1. G. Elger, R. Lauterbach, K. Dankwart, C. Zilkens, „*Inline thermal transient testing of high power LED modules for solder joint quality control*", in Proc. Electronic Components and Technology Conference (ECTC), 2011 IEEE 61st, pp. 1649 – 1656.

2. G. Elger, M. Hutter, S. Rauschenbach, H. Willwohl, „*Performance of improved SAC solders under high thermo-mechanical stress condition*", 13th Electronic Circuits World Convention (ECWC13), Nürnberg 7th-9th May 2014, Proceedings

3. R. Raut, R. Bhatkal, W. Bent, B. Singh, S. Chegudi, R. Pandher, J. Kolbe and S. Misra "*Assembly Interconnect Reliability in Solid State Lighting Applications – Part I*", Pan Pacific Symposium 2011, Proceedings

4. M. Zitzlsperger, J. Reill, "*Next Generation SMT LED for Automotive Signaling and Forward Lighting Functions*", ISAL 2011, Proceedings

5. "*Guidelines for Accelerated Reliability Testing of Surface Mount Solder Attachment*", IPC-SM-785,1992 and "Performance Test Methods and Qualification

Requirements for Surface Mount Solder Attachments", IPC-9701

6. J. Pan and J. Silk, "*A Study of Solder Joint Failure Criteria*", in Proc. 44th International Symposium on Microelectronics, Long Beach, CA, October 9-13, 2011

7. E. F., Schubert, "Light Emitting Diodes", Cambridge University Press, 2006, ISBN-13 978-0-521-86538-8

8. H. Diekera, C. Miesnera, D. Püttjera and B. Bachla, "Comparison of different LED Packages", in "Manufacturing LEDs for Lighting and Displays", edited by Thomas P. Pearsall, Proc. of SPIE, Vol. 6797, 67970I, (2007)

9. K.P. Streubel, H.W. Yao, E.F. Schubert, "Light-Emitting Diodes: Research, Manufacturing, and Applications" X Proc. of SPIE, Vol. 6134, 613405, (2006)

10. M. Rencz, V. Székely, "Measuring Partial Thermal Resistances in a Heat-Flow Path", IEEE Transactions on Components and Packaging Technologies, Vol. 25, No. 4 (2002), pp. 547-553

11. D. Schweitzer, "The Junction-To-Case Thermal Resistance: A Boundary Condition Dependent Thermal Metric", 26th IEEE SEMI-THERM Symposium, 2010

12. JESD51-14, "Transient Dual Interface Test Methodfor the Measurement of the Thermal Resistance Junction to Case of Semiconductor Devices with Heat Flow Trough a Single Path"

13. G. Elger, S. V. Kandaswamy, R. Derix and J. Wilde, "*Transient Thermal Analysis as a Test Method for the Reliability Investigation of High Power LEDs during Temperature Cycle Tests* "Journal of Microelectronics and Electronic Packaging", First Issue 2014, in press

Reliability Testing of Wire Bonds Using Pad Resistance with van der Pauw Method

Michael Mayer*, Samuel Kim

University of Waterloo, 200 University Ave. W. Waterloo, ON N2L3G1, Canada
*email: mmayer@uwaterloo.ca

Abstract

In microelectronic wire bonding, a reliable electrical connection is of utmost importance. With the advent of advanced bonding wire materials such as Cu, Pd coated Cu, and Ag alloys, there are substantial ongoing efforts in process development and reliability characterization with these materials. To measure bond quality and especially bond reliability is limited with respect to sample size and number of different conditions that can be covered with available resources. Methods that apply non-destructive bond quality measurements during thermal aging could be useful to increase characterization throughput while keeping the need for resources under control.

A non-destructive pad resistance method based on the well known van der Pauw method is introduced that indicates bond aging more conveniently than previously reported bond resistance methods. With four connection lines designed at each corner of a standard square bonding pad, the pad resistance was monitored at various stages during reliability testing by high temperature storage at 250 °C of a Au ball bond ≈50 μm in diameter and ≈15 μm high on a standard Al pad. The pad resistance increased during aging up to 30.8 h due to intermetallics (IMCs) formation, then dropped ≈5 % during 14 h down to 50.1 mΩ, and then rose again to reach a plateau above 55 mΩ after 76 h that remain unchanged for the remainder of the aging which lasted 150 h.

To explain the unexpected resistance drop a number of finite element (FE) models was developed to simulate various IMC distributions in the bond zone and their effect on the pad resistance. It was found that the resistance drop can be explained by the occurrence of Al-rich IMCs at the periphery of the bond in later stages of bond aging. Compared to the earlier Au-rich IMCs, Al-rich IMCs have lower resistivity. Therefore, the time of the pad resistance drop can indicate the beginning of a late stage of bond deterioration of the Au/Al system and effectively serve as a non-destructive reliability indicator.

Introduction

Wire Bonding is a robust interconnection method for high quality microelectronics [1]. Recently, the use of novel materials for bonding wires has reduced the cost of wire bonding significantly [2-9]. For example, coated bonding wires are being used in large quantity for their improved bondability. In particular, palladium coated copper wire (Pd/Cu) has been successfully introduced in mass production. Not only Pd/Cu, but also palladium coated silver wire, gold coated copper wire, insulated gold and copper wires, and other coating/wire combinations have been tried out and reported [2-14]. When studying several new coating/wire combinations, the need for accelerated testing methods is significant. Bond resistance [15-17] and residual stress [18] measurements can be done non-destructively. When bond aging is accelerated with high temperature storage (HTS)

[19], non-destructive measurements can reduce the required sample size in experiments to optimize bond reliability.

In this work, the feasibility of a pad resistance method for the non-destructive determination of bond quality during a reliability test is studied. The reliability test used here is accelerated aging at high temperature.

Measurement Concept for Pad Resistance

The method by van der Pauw [20] is extremely popular for determining resistances of thin films. The basic concept is illustrated in Fig. 1. The pad voltage V is measured with a nanovolt meter. Together with the known current level, 1 mA, the pad resistance, R_{pad}, can be determined using the van der Pauw equation valid for a square structure with contacts at the corners [20]:

$$R_{Pad} = \frac{\pi}{\ln 2} \cdot \frac{V}{I} \qquad (1)$$

When a ball is bonded to the pad and aged, the value of R_{pad} is expected to go through a succession of changes as outlined in Figs. 2 (a) to (d). In (b) R_{pad} drops as part of the current goes

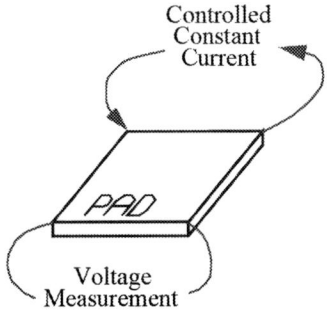

Fig. 1   Illustration of Van der Pauw method used for quadratic pad resistance measurement. Pad has four contacts, one in each corner.

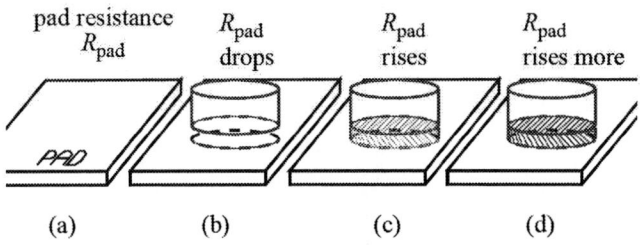

Fig. 2   Overview of evolution of pad resistance before/after ball bonding and bond aging. (a) Pad only. (b) ball bond on pad, pure metals only. (c) Au rich and (d) Al rich intermetallics below ball bond.

978-1-4799-2408-0/14 $31.00 © 2014 IEEE

through the bonded ball which corresponds to a resistance in parallel to the pad. Upon aging of the ball bond at an elevated temperature, Au/Al intermetallics form with higher resistivities than those of pure gold and aluminum, as illustrated in Figs. 2 (c) and (d), causing the value of R_{pad} to rise. Thus, R_{pad} is proposed as a non-destructive indicator of bond aging.

Testchip

A testchip is developed in a conventional commercial CMOS technology with two typical Al metal layers and a standard test pad design that has 75 µm square opening. It has four electrical connections, one in each pad corner, allowing for a con,uration of the van der Pauw type [20], as shown in Fig. 3. In this con,uration, a constant current I of 1 mA is applied through two neighboring pad corners while a pad voltage V is sensed across the two opposite pad corners.

The value of R_{pad} before ball bonding is expected to be

$$R_{Pad} \text{ (before bonding)} = \frac{1}{\frac{1}{R_1} + \frac{1}{R_2}} \qquad (2)$$

where R_1 and R_2 are the sheet resistances of the metal 1 and metal 2 layers of the CMOS process used, respectively. The effect of the diffusion barriers in this CMOS technology on pad resistance is assumed negligible. The two metal layers are placed on top of each other, effectively being in parallel for any current flowing in and out of the pad via two adjacent pad corners.

Test Bond and Thermal Aging of Bond

The ball bond was made on the test pad with 25 µm diameter Au wire and typical bonding parameters, resulting in a bonded ball of approximately 50 µm in diameter and 15 µm in height, as shown in Fig. 4 (a) .

The aging was achieved using an integrated microheater [21-23]. The microheater is a silicon resistor implanted below the testpad with a nominal resistance of 75 Ω at room temperature [22, 23]. As it needs to be powered by voltages above 10 V to achieve high local temperatures, it interferes with the low voltage measurement required for the van der Pauw method. This matter needs to be addressed by intermittent cooling.

The value of R_{pad} is therefore measured with the heater switched off for 15 min, allowing for reliable low level measurements between heating intervals that last 45 min. I.e., there is one measurement of pad resistance per hour. An image of an example bond after aging this kind of intermittent heating/cooling is shown in Fig. 4 (b).

Experimental Results

Before wire bonding, the pad voltage V measured with the van der Pauw method was 4.5 µV, translating into a pad resistance value R_{pad} of 20.4 mΩ at room temperature, using equation (1). Assuming the two metal layers have the same square resistances, their value can be determined by rearranging equation (2), resulting in $R_1 = R_2 = 2 \cdot R_{Pad} = 40.8$ mΩ. This value lies within the range specified by the CMOS technology used.

Bonding a gold ball on the aluminum testpad using a standard thermosonic bonding process caused the value of R_{pad} to drop to 18.6 mΩ as the bonded ball presents an additional resistor in parallel to the pad, effectively lowering the total resistance. A drop far below 10 mΩ was expected for a perfect bond. The difference to the lower ideal value indicates the partial nature of the testbond due to remaining interfacial gaps, debris, and its partial aging during any waiting time on the hot process area on the wire bonder before the sample was removed and brought to room temperature, leading to first intermetallics formed at the interface.

In the accelerated aging, the bond was heated by the microheater to 250 °C for 200 h, resulting in massive formation of intermetallics consuming all pad metal below the ball bond and also a part of the pad metal at the periphery next to the ball bond. Due to aging the value of R increased continuously for 30 h up to 55 mΩ, dropped for more than ten hours down to below 50 mΩ, and again increased for about 50 h up to a final value of 51 mΩ that remained constant until 200 h total aging time. This result in shown in Fig. 5.

The drop of R observed between 30 and 40 h aging at 250 °C is remarkable and counterintuitive and has not been reported before to the best knowledge of the authors. Therefore, a numerical study was carried out to understand this drop.

Electrical Finite Element Model for Bond Resistance

Finite element models were developed using COMSOL Multiphysics® software (COMSOL, Inc., Burlington, MA, USA). The components of an example model with dimensions, bottom view, and subcomponent definition are shown in Figs. 6 (a), (b), and (c), respectively. However, the pad thickness is not given yet. As described in a later paragraph, it is adjusted to match numerical with experimental results. The mesh is shown in Fig. 6 (d). Up to 38049 mesh elements and 59080 degrees of freedom were used in the models.

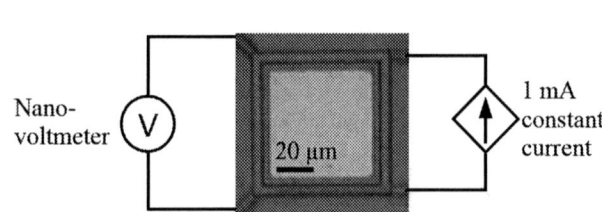

Fig. 3   Micrograph of testpad with four corner contact lines and schematic circuit.

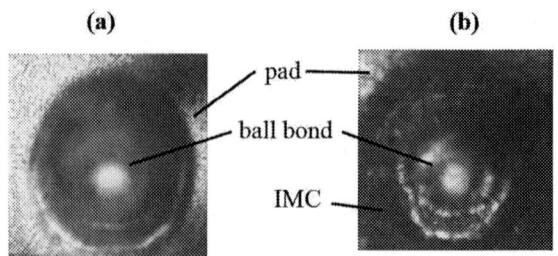

Fig. 4   Micrograph of Au ball bond on t test pad before (a) and after (b) 24 h of 250 °C aging in air.

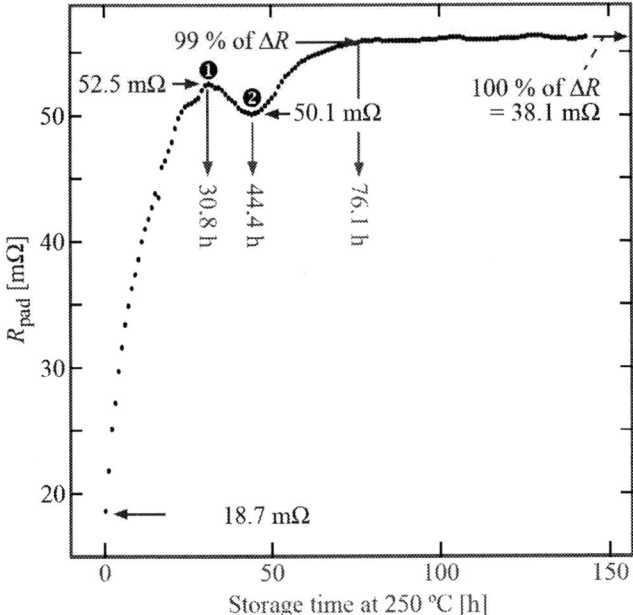

Fig. 5 Example measurement of pad resistance at room temperature after various aging times at 250 °C. Local maximum and minimum indicated with ❶ and ❷, respectively.

To inject current into the model, 1 μm wide blocks are placed at the corners of the pad providing small areas α and β, as shown in Fig. 6 (a), where boundary conditions for ground and 1 mA are applied, respectively. All other surface areas are insolating, and internal interfaces are continuous with respect to the flow of current. The pad voltage V is the potential difference between two isolated surfaces from the corners γ and δ, as shown in Fig. 6 (a).

The current density of an example result is visualized in Fig. 7 using a streamline plot offered by Comsol. The two pad corners used for current supply are where the streamlines converge. The bonded ball attracts the streamlines as it provides the path of least resistance.

A total of six finite models were constructed in a way to follow the known evolution of bond interface at high temperature between an Au ball and an Al pad, as described in [24]. The six models are:

(0) pad only,

(A) pad with ball bond unaged,

(B)-(E) aged ball bond with various Al-Au intermetallics below the bond and in its peripheral region, partly and completely consuming the pad area close to the bond, as expected for short, medium, and prolonged periods of aging, and

(F) bond aged to near destruction with voids simulated between ball and pad with an interlayer [subcomponent 1 in Fig. 6 (c)] that has a technical conductivity of 775 kS/m, a value adjusted to produce qualitative agreement with the experiment.

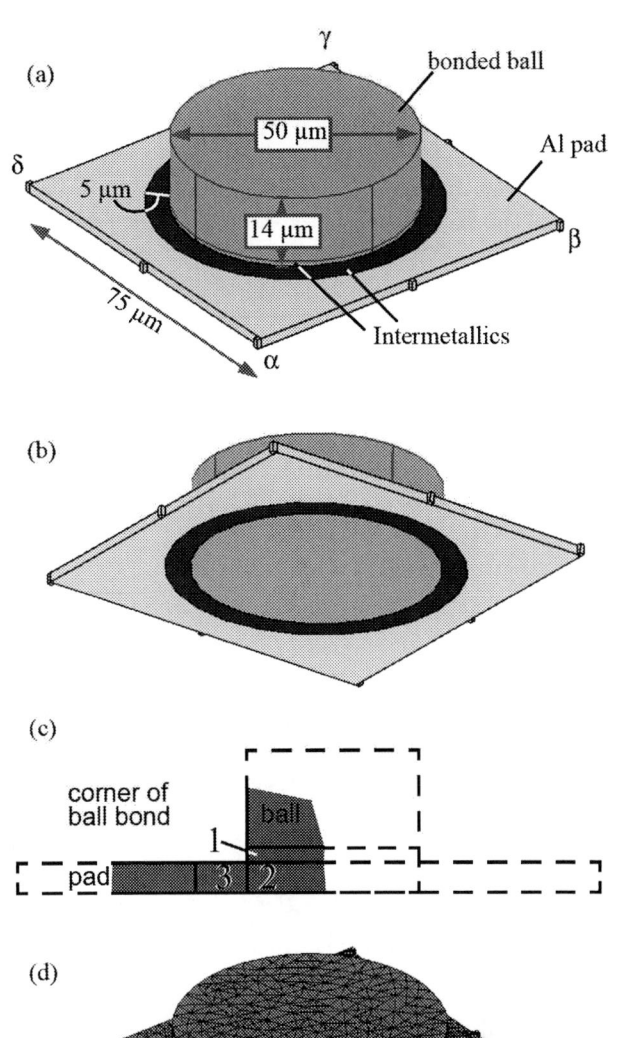

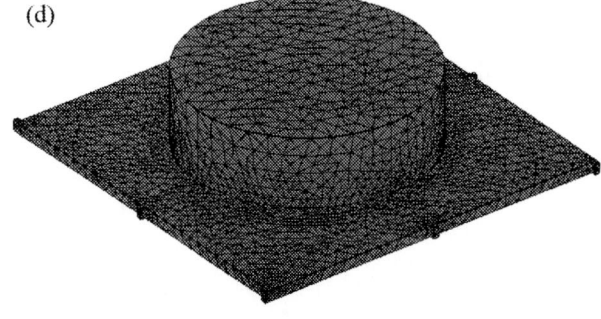

Fig. 6 Overview of main model used in FE study. (a) Components. (b) Bottom view. (c) Corner detail defining components 1-3, not to scale. (d) Mesh.

Table 1 gives an overview of the models together with the definition of subcomponents 1-2 for bonded ball bottom layer, pad in bond zone under ball, and pad ring peripheral to bond, respectively. The list of all materials used and their conductivity values are given in Table 2. The values for pad resistance resulting from the simulations are given in Fig. 8. They vary between 3.1 mΩ and 24.3 mΩ.

Discussion

The numerical result for model 0, R_{pad} = 20.4 mΩ, is brought to complete agreement with the experimental value by adjusting the pad thickness to 1.46 μm which is very similar to the experimental value. This thickness value was then used for the remaining models.

Models A and B are in qualitative agreement with the experiments, but their values are substantially lower, between 3 mΩ and 5 mΩ compared to the experimental value of 18.6 mΩ. The reason for this difference is that the real bond has already partially been aged on the bonder process zone which can be long enough to consume a big part of the Al pad underneath the bond by IMC formation. Moreover, a real bond is a partial bond at best in contrast to the ideal bond in the models.

The results for models C and D reflect the increase in resistance observed experimentally but to a lower degree. A reason for this difference might be voiding at the interface which is not included in the models.

In model E, the Al rich IMCs are taken into account. Al-rich IMCs have lower specific resistivity than Au-rich IMCs [16]. Therefore, the Al-rich IMCs are responsible for producing a drop in resistance compared to models A-D. This numerical drop is in good agreement with the experimentally observed

drop, leading us to conclude that it indicates the point in time when there is a lot of Al rich IMC formation, an event usually observed in late stages of aging.

In model F, the conductivity of subcomponent 1 is adjusted to a value that results in an increase of pad resistance of approximately 5 Ω which is the same as that observed experimentally, leading to the conclusion that voiding in that subcomponent can be the reason for the observed increase.

The models are very effective in visualizing the flow of current and clarify the importance of the bonded ball in providing a preferred path for current parallel to the pad itself. Compared to the experiment, the simulated R_{pad} values are low and this outlines the limitation of the model, i.e. the discrepancy between

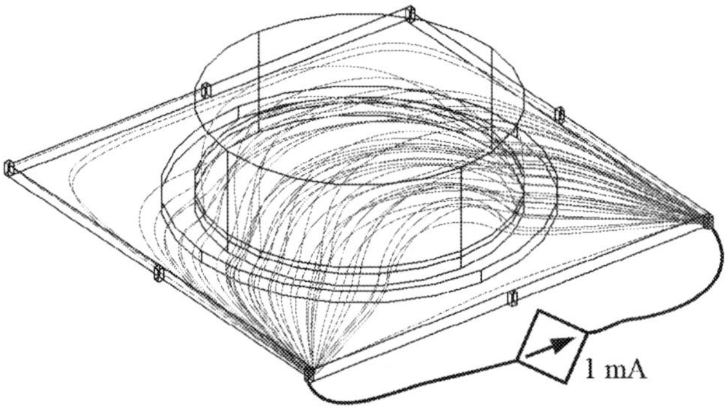

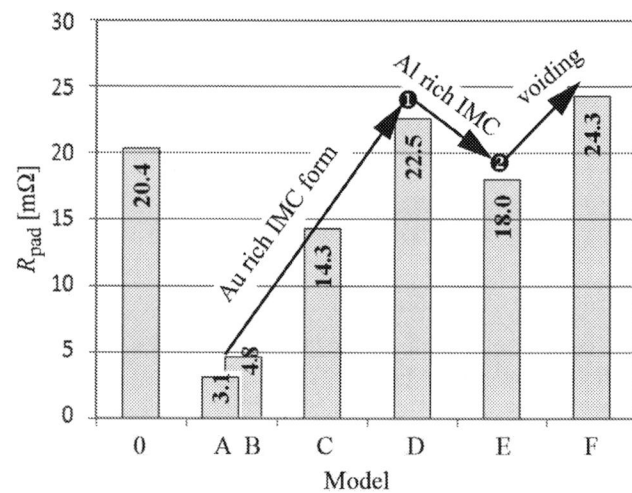

Fig. 7 Example result: streamline plot of current density. Path of least resistance is through bonded ball.

Fig. 8 Simulated values for pad resistance for models 0 and A-F. Symbols ❶ and ❷ as in Fig. 5.

Table 1: Model Definitions

sub-component #s 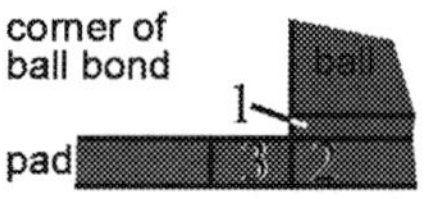	Model					
	A	B	C	D	E	F
1	Au	Au-IMC 1	Au-IMC 1	Au-IMC 1	Au-IMC 1	air (voids)
2	Al	Al	Au-IMC 1	Au-IMC 1	Au-IMC 1	Au-IMC 1
3	Al	Al	Al	Au-IMC 2	Al-IMC	Al-IMC

Table 2: Conductivity of Model Materials

	Au	Al	Au-IMC 1	Au-IMC 2	Al-IMC
Conductivity [S/m]	4.517×10^7 (bonding wire)	3.333×10^7 (Al-1%Si pad metal)	0.2667×10^7 (Au_4Al)	0.7634×10^7 (Au_2Al)	1.266×10^7 ($AuAl_2$)

ideal connection and realistic connection with its bond imperfections (cracks, unbonded areas, intermetallics).

Conclusions

We demonstrated the feasibility of the van der Pauw method for pad resistance measurement to monitor bond quality during high temperature reliability testing. The method can detect specific intermetallics formation mechanism based merely on the non-destructive measurement of pad resistance. If multiplied and used on several testchips, this method will allow the non-destructive measurement of many bond samples continuously during high temperature aging. Compared to previously reported bond resistance methods with double-bonds, the van der Pauw based method can be used with a single bond only, greatly improving the ease of use.

Acknowledgements

This work is supported in part by Microbonds Inc. (Markham, Canada), MK Electron Co. Ltd. (Yongin, S. Korea), the Initiative for Automotive Manufacturing and Innovation (IAMI), Ontario, Canada, and the Natural Sciences and Engineering Research Council of Canada (NSERC).

References

1. G. G. Harman, Wire Bonding in Microelectronics, Third Edition, McGraw-Hill, 2010, USA

2. Clauberg, Horst, Bob Chylak, Nelson Wong, Johnny Yeung, and Eugen Milke. "Wire bonding with Pd-coated copper wire." In CPMT Symposium Japan, 2010 IEEE, pp. 1-4. IEEE, 2010

3. Kaimori, Shingo, Tsuyoshi Nonaka, and Akira Mizoguchi. "The development of Cu bonding wire with oxidation-resistant metal coating." Advanced Packaging, IEEE Transactions on 29, no. 2: 227-231, 2006

4. Uno, Tomohiro, Shinichi Terashima, and Takashi Yamada. "Surface-enhanced copper bonding wire for LSI." In Electronic Components and Technology Conference, 2009. ECTC 2009. 59th, pp. 1486-1495. IEEE, 2009.

5. Kai, Liao Jun, Liang Yi Hung, Li Wei Wu, Men Yeh Chiang, Don Son Jiang, C. M. Huang, and Yu Po Wang. "Silver alloy wire bonding." In Electronic Components and Technology Conference (ECTC), IEEE 62nd, pp. 1163-1168. IEEE, 2012.

6. Cho, Jong Soo, Yong Jin Park, Jeong Tak Moon, Eun Kyu Her, and Kyu Hwan Oh. "Au-Ag based alloy wire for semiconductor package." U.S. Patent 8,022,541, issued September 20, 2011.

7. Guo, Rui, Tao Hang, Dali Mao, Ming Li, Kaiyou Qian, Zhong Lv, and Hope Chiu. "Behavior of intermetallics formation and evolution in Ag–8Au–3Pd alloy wire bonds." Journal of Alloys and Compounds 588: 622-627, 2014

8. Tanna, Suresh, Jairus L. Pisigan, W. H. Song, Christopher Halmo, John Persic, and Michael Mayer. "Low cost Pd coated Ag bonding wire for high quality FAB in air." In Electronic Components and Technology Conference (ECTC), 2012 IEEE 62nd, pp. 1103-1109. IEEE, 2012.

9. Persic, John, Jairus L. Pisigan, Suresh Tanna, Yong Guo, W. H. Song, and Michael Mayer. "Low-cost Palladium coating process and its effect on free-air-ball softness and second bond strength of Cu bonding wires." In Electronic Components and Technology Conference (ECTC), IEEE 62nd, pp. 1169-1173. IEEE, 2012.

10. Cheng, C. H., H. L. Hsiao, S. I. Chu, Y. Y. Shieh, C. Y. Sun, and C. Peng. "Low cost silver alloy wire bonding with excellent reliability performance." In Electronic Components and Technology Conference (ECTC), 2013 IEEE 63rd, pp. 1569-1573. IEEE, 2013.

11. Moazenzadeh, A., N. Spengler, R. Lausecker, A. Rezvani, M. Mayer, J. G. Korvink, and U. Wallrabe. "Wire bonded 3D coils render air core microtransformers competitive." Journal of Micromechanics and Microengineering 23, no. 11: 114020, 2013

12. Fischer, Andreas C., Jan G. Korvink, N. Roxhed, G. Stemme, U. Wallrabe, and Frank Niklaus. "Unconventional applications of wire bonding create opportunities for microsystem integration." Journal of Micromechanics and Microengineering 23, no. 8: 083001, 2013

13. Nan, Chunyan, Michael Mayer, Norman Zhou, and John Persic. "Golden bump for 20micron diameter wire bond enhancement at reduced process temperature." Microelectronic Engineering 88, no. 9: 3024-3029, 2011

14. Lee, Jaesik, Michael Mayer, Norman Zhou, and John Persic. "Microelectronic wire bonding with insulated Au wire: Effects of process parameters on insulation removal and crescent bonding." Materials transactions 49, no. 10: 2347-2353, 2008

15. Blish, R. C., Parobek, L., "Wire Bond Integrity Test Chip", Proc. Reliability Physics Symposium, pp. 142-147, 1983

16. Maiocco, L., Smyers, D., Munroe, P. R., Baker, I., "Correlation Between Electrical Resistance and Microstructure in Gold Wirebonds on Aluminum Films", IEEE Trans. on Components, Hybrids and Manufacturing Technology, Vol. 13, No. 3, pp. 592–595, 1990

17. Murcko, R. M., Susko, R. A., Lauffer, J. M., "Resistance Drift in Aluminum to Gold Ultrasonic Wire Bonds", IEEE Trans. Comp. Hybrid Manuf. Technol., vol 14, no. 4, pp. 843–847, 1991

18. Mayer, M. "Non-Destructive Monitoring of Au Ball Bond Stress During High-Temperature Aging". Proc. ECTC, 2008.

19. Rossi, Carole, Pierre Temple-Boyer, and Daniel Estève. "Realization and performance of thin $SiO2/Si3Nx$ membrane for microheater applications" Sensors and Actuators A: Physical 64, no. 3: 241-245, 1998

20. van der Pauw, L. J., "A Method of Measuring Specific Resistivity and Hall Effect of Discs of Arbitrary Shapes", Philips Res. Repts. 13, 1-9, 1958

21. Mayer, M., McCracken, M., Persic, J., "Development of Accelerated Method for Thermal Cycling in Electronic Packaging Application", ASME J. Electronic Packaging, Vol. 135, pp. 021007-1 to 021007-6, 2013.

22. Mayer, M., Kim, S., Persic, J., Moon, J. T., "Simplifying Reliability Testing of Wire Bonds Using On-Chip Heater and Pad Resistance Method", ASME J. Electronic Packaging (accepted), 2014.

23. Kim, S., "Novel Methods in Ball Bond Reliability Using In-Situ Sensing and On-Chip Microheaters", thesis, University of Waterloo (http://hdl.handle.net/10012/7217), 2013

24. McCracken, M.J., Koda, Y., Hyoung Joon Kim, Mayer, M., Persic, J., June Sub Hwang, Jeong-Tak Moon, "Explaining Nondestructive Bond Stress Data From High-Temperature Testing of Au-Al Wire Bonds," Components, Packaging and Manufacturing Technology, IEEE Transactions on , vol.3, no.12, pp.2029,2036, Dec. 2013

Colour Shift in Remote Phosphor Based LED Products

M. Yazdan Mehr[1,2*], W.D. van Driel[2,3], G.Q. Zhang[2]

1 Materials innovation institute (M2i), Delft, The Netherlands
2 Delft University of Technology, EEMCS Faculty, Delft, The Netherlands
3 Philips Lighting, Eindhoven, The Netherlands
m.yazdanmehr@m2i.nl

Abstract

In this paper, the thermal stability and the colour shifting of remote phosphor plates, made from Bisphenol-A polycarbonate (BPA-PC), are studied. In this study, the remote-phosphor and lens of BPA-PC samples of 3 mm thickness were thermally aged at temperature range 100 to 140 ºC. Results show that thermal ageing leads to a significant decrease in the luminous flux and chromatic properties of plates. Lumen depreciation up to 30% reduction is extrapolated to temperatures lower than 100 °C. It is shown that the lifetime, defined as 30% lumen depreciation at 40 °C, is around 35 khrs. It is also shown that by increasing the temperature, the reaction rate becomes faster, inferring that lumen depreciation takes place at shorter time. Results also illustrate the colour shifting of white light towards yellow region.

Keywords: LED, Reliability, Remote phosphr,

Introduction

Phosphor down conversion efficiency is one of the common methods to generate white light [1-3]. In the commercial white LEDs phosphor is mixed with the encapsulant [2-4]. Nowadays, several remote phosphor ideas have been proposed [4-6], mostly based onputing phosphor away from the chip. Remote phosphor plates are widely used in LED-based products to produce white light by converting blue light using yellow phosphor. The application of these white light LEDs is considered as an alternative for conventional lighting products, which significantly contribute to the worldwide emission of CO2 and the global warming [7]. The use and development of high efficient LED light source would certainly decrease the energy consumption, with subsequent economic and environmental benefits [8]. Over the last decades, GaN-based light-emitting diodes have been shown to be a good candidate for the high efficiency light sources for general applications [7-9]. Although LEDs are more reliable than conventional light sources, several reports [10-13] have shown that package and phosphor layer of white LEDs can degrade, resulting in the reduction in the light efficiency. The main reason for phosphor damage is the generated heat by LED chip during operation.

Moisture or humidity is another stress for degradation and yellowing of lens/remote phosphor plates. Chan et al. [10] Studied the effect of humidity on LED package and shown that temperature and humidity stresses lead to light output decay, discoloration of the encapsulating material, and the formation of bubble in the package. On the other hand, there are also some results which show that moisture does not have any significant effect on the discoloration of remote phosphors [13]. Yanagisawa et al. [13] also found that yellowing is not significantly affected by a high humidity test environment.

To characterize and express "color", some terms are used frequently. The Commission Internationale de l'Eclairage (CIE) system is the most common method to describe the composition of any colour in terms of three primaries [14]. Artificial "colours", denoted by X, Y, Z, also called tristimulus values, can be added to produce real spectral colors. By a piece of mathematic legerdemain, it is necessary only to quote the quantity of two of the reference stimuli to define a colour, since the three quantities (x, y, z) are made always to sum to 1 [14]. The x, y, z, i.e. the ratios of X, Y, Z of the light to the sum of the three tristimulus values, are the so-called chromaticity coordinates [14]. (x, y) is usually used to represent the colour.

To obtain the reasonably equidistant chromaticity scales that are better than the CIE 1931 diagram, the CIE 1976 uniform chromaticity scale (UCS) diagram which is also called (u', v') . The (u', v') coordinates are related to the (x, y) coordinates by the following equations:

$$U' = \frac{4x}{-2x \times 12y \times 3} \tag{1a}$$

$$V' = \frac{9y}{-2x \times 12y \times 3} \tag{1b}$$

Based on Eq. (1), Δu'v', which defines the colour shifting, at any two positions (0 and 1) can be calculated using the following formula,

$$\Delta u'v' = \sqrt{(u_1' - u_0') + (v_1' - v_0')} \tag{2}$$

The yellowing index (YI) is calculated according to ASTM D1925 [14] with the following equation:

$$YI = \frac{100(1.28\,X\,CIE - 1.6\,Z\,CIE]}{Y\,CIE} \tag{3}$$

Materials and methods

Two types of 3 mm-thick plates which are the lens plate and remote phosphor in which the phosphor is laminated on the same lens are used in this study. The Correlated Colour Temperature (CCT) of remote phosphor is 4000 K, which shows that it produces the warm white light. The samples were aged under high temperature stresses at 100, 120, and 140 ºC for 3000 h. Testing temperatures for accelerated lumen depreciation test is determined in such a way that the temperature does not go above the glass transition temperature of the plastics. Optical properties of thermally-aged plates, i.e. Luminous flux depreciation, were studied at room temperature, using an integrated sphere.

Reliability model for the life time assessment is based on an exponential luminous decay equation, where the time-to-failure can be calculated as [15]

$$\phi(t) = \beta \exp(-at) \qquad (4),$$

where $\Phi(t)$ represents the lumen output, α is the rate of reaction or depreciation rate parameter, t is time and β is a pre-factor. When lumen output, Φ, is equal to 70%, t is time-to-failure [15]. The rate of reaction, α, is related to the activation energy of the reaction and to the ageing temperature as follows [15,16]

$$a = A \exp(\frac{-E_a}{KT}) \qquad (5)$$

where A is a pre-exponential factor, Ea is the activation energy (ev) of the degradation reaction, K is the gas constant, and T is the absolute temperature (K).

Results

Stress at high temperature levels can induce thermal ageing and therefore a strong optical power dropping and reduction of light output. Figure 1 illustrates the spectral power distribution (SPD) of both lens and remote phosphor plates for the case of thermal ageing at 140 °C. It is obvious that there is a reduction in blue light transmission (450 nm) and also the phosphor conversion yellow light (600 nm). One can see that the reduction in the blue peak has almost the same trend in both lens and remote phosphor.

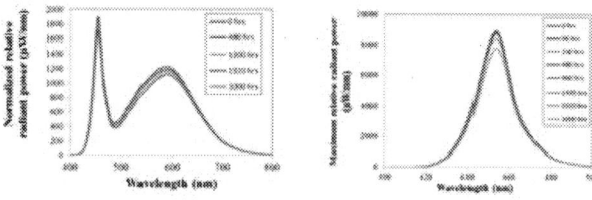

Figure 1: Spectral power distribution (SPD) of a) Remote Phosphor, and b) Lens, aged at 140 ºC

After the measuring the colour values of aged remote phosphor, the phosphor layer is mechanically removed from the remote phosphor and the discolouration of plate is measured. Surprisingly the lens is browner than the aged lens without phosphor.

The normalized blue light intensity over the stress time for both remote phosphor and lens is shown in Figure 2. One can clearly see that the reduction of light output in the plate with phosphor is higher than that in plates without phosphor.

Obviously the transparency of lens with phosphor is reduced more than that in aged lens without phosphor.

Figure 3 illustrates more details of the effects of thermal-stress on the performance of remote phosphor and lens. This figure shows the evolution of the normalized flux intensity and the degradation rate of the phosphor plates and lens plate. Clearly, the degradation rate shows a significant dependence on the stress temperature level in both plates, the higher the

ageing temperature, the higher the lumen depreciation and the degradation kinetics. It is also noticeable that the extent of degradation in remote phosphor plates is more than that of lens plates.

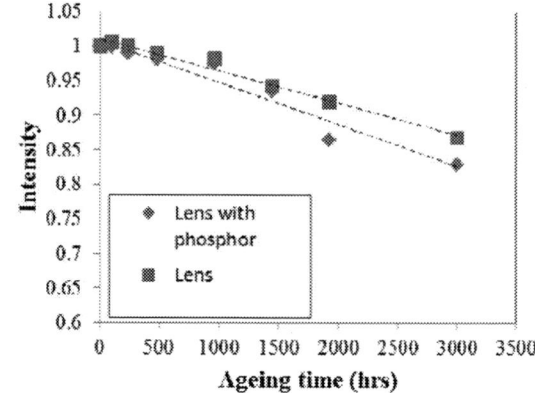

Figure 2: Normalized light intensity of aged lens with and without phosphor

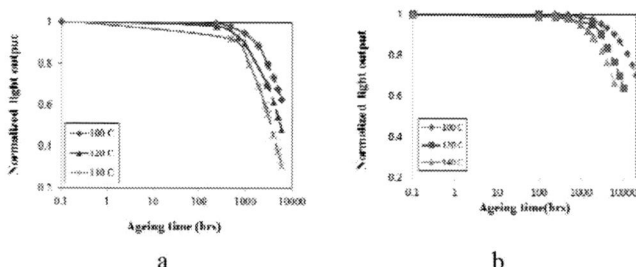

a   b

Figure 3: Normalized flux of a) remote phosphor plates, and b) Lens at 100 up to 140 °C

The activation energy of the degradation reaction in LEDs depends on the materials and the working conditions. Figure 4 illustrates the ln (α) vs 1/KT (calculated from equation 5) for both remote phosphor and the substrate. The slope is multiplied by the negative of the gas constant to obtain the activation energy, Ea, in the eV. Activation energy for remote phosphor and lens is 0.333 and 0.39eV respectively. One can see that activation energy of degradation for remote phosphor plates is slightly less than that in lens plates, showing that the decay of phosphor conversion efficiency is also have a noticeable influence on the lumen depreciation.

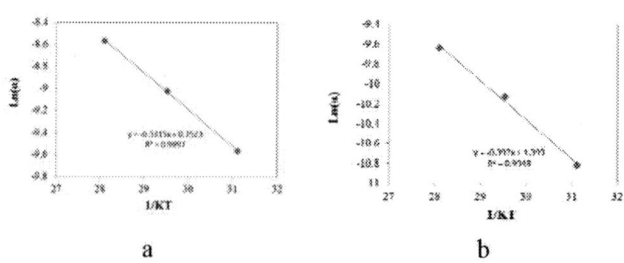

a   b

Figure 4: Activation energy of a) remote phosphor plates, and b) Lens

According to the alliance for solid state illumination system and technology (ASISST) standard, lifetime of LEDs is defined as time to reach 70% of its initial lumen output [10]. The experiments at 100 °C were performed up to 20% reduction in light output in the case of remote phosphor and extrapolated the lumen output up to 30% for aged lens plates at 100 °C. The lumen output is extrapolated to higher depreciation by the model that is explained in our previous paper [17]. Table 1 illustrates the calculated values for the reaction rate (α) for each temperature for remote phosphor plates and lens.

Table 1: Reaction rate a for remote phosphor plates A and lens at temperature up to140 °C

Temp (C)	Remote Phosphor	Lens
40	1.03E-05	1.97E-06
60	2.15E-05	4.73E-06
80	4.12E-05	1.03E-05
100	7.38E-05	2.0 E-05
120	1.24E-04	4.0 E-05
140	2.0E-04	6.5E-05

Obviously, by increasing the temperature the reaction rate becomes faster, inferring that shorter time is needed to reach the same level of lumen depreciation. Also is shown that activation energy of remote phosphor is less than that in lens substrate.

Thermal-ageing test also have some important effects on the CCT. The variation of CCT during high temperature stress test is shown in Figure 5. It is seen that CCT decreases by increasing the thermal ageing time. One can also notice that the higher the ageing temperature, the higher the degradation kinetics. The reduction in CCT follows the same kinetics as the luminous flux decay and can therefore be ascribed to the thermally activated degradation mechanism discussed above. The reduction in Colour Temperature suggests that the degradation of the remote phosphor plates has consequences not only on the light extraction efficiency but also on the colour of the emitted light.

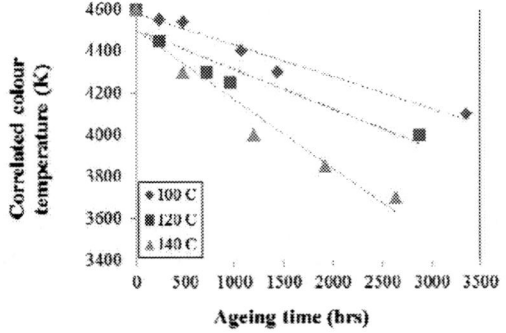

Figure 5: Correlated colour temperature (CCT) variation during high thermal-stress tests

The calculated color shifting of aged remote phosphor with time is shown in Figure 6. As is illustrated in this figure, the Duv increases by increasing ageing time.

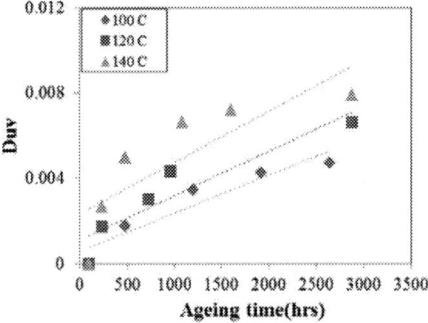

Figure 6: Colour shifting of remote white light

Figure 7 depicts the ratio of the intensity of the yellow to the blue peak with ageing time, for the sample aged at 140 °C. The increasing of ratio of yellow peak to blue peaks clarified that the light turns towards yellow.

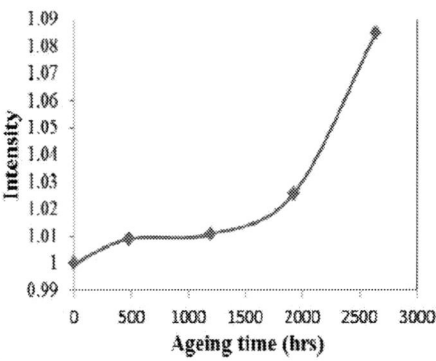

Figure 7: Variation of the ratio of the intensity of the yellow to the blue peak measured on a white LED submitted to stress at 140 °C

Effect of Humidity

The evolution of YI of BPA-PC plates at 140 °C as a function of thermal ageing time for both normal humidity and 85% RH is shown in Figure 8. As it is seen, there are no differences between the normal humidity condition and 85% RH.

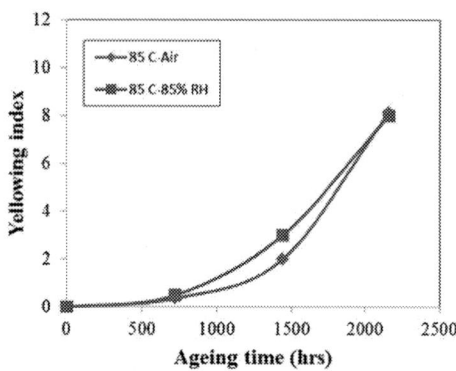

Figure 8: Variation in yellowing index (YI) of lens, aged at 140 °C for different thermal ageing times (in hrs)

A more quantitative description of the effects of humidity and moisture on the performance of remote phosphor A and B is given in Figure 9. This Figure illustrates the evolution of the normalized flux intensity and therefore the degradation kinetics of the phosphor plates. It is clearly seen that the humidity does not have a big influence on the lumen output.

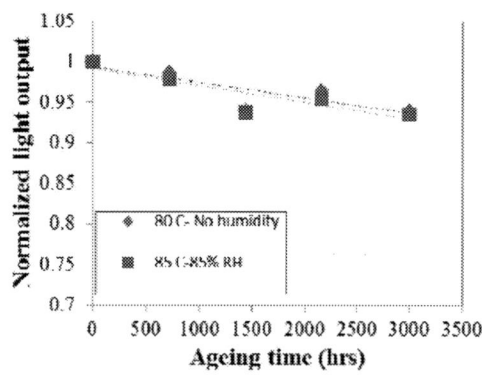

Figure 9: Normalized flux of remote phosphor plates at 80 °C in Air and in 85% RH

In Figure 10 the variation of CCT during high temperature stress test is shown for both remote phosphor plates A and B. It is obvious that humidity does not have big influence on CCT.

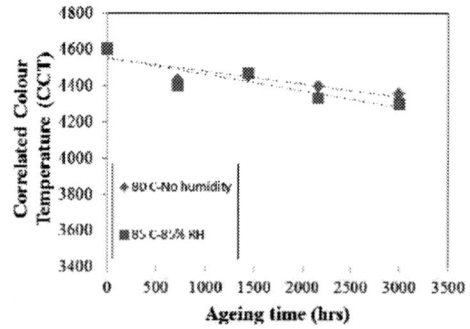

Figure 10: Correlated colour temperature (CCT) during humidity tests for remote phosphor

Life time prediction

The real working temperature of LDEs is much lower than the applied temperatures for the accelerated tests [15]. Therefore, the kinetics of lumen depreciation to 30% of its initial value by using exponential luminous decay model and Arrhenius equation should be extrapolated to temperatures lower than 100 °C. This can be done using Equation 4 by equating ϕ to 0.7, knowing that α can be obtained from Equation 5 The values of α, calculated for 40, 60 and 80 °C, are given and shown in Table 1, as it is seen that the higher the temperature the faster the lumen depreciation is.

Figure 11 illustrates time-to-failure (70% lumen decay) of both remote phosphor and lens, calculated at different temperatures. It is seen that lens has a longer life time compared to remote phosphor; i.e. at 40 °C the light output from lens A reduces to 70% of its initial value after 100khrs, while for remote phosphor time-to-failure is 35 khrs.

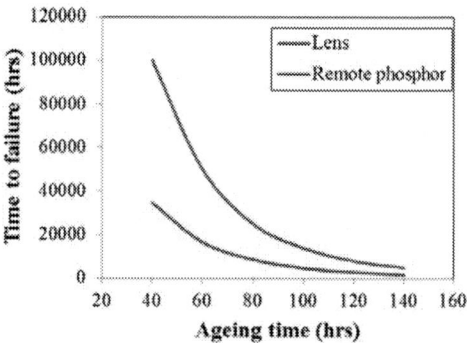

Figure 11: Time-to-failure (70% lumen decay) of remote phosphor at different temperatures

Discussion

Temperature is a very significant controlling parameter in LED reliability. High temperature levels can damage the optical properties of the package and of the material used for the encapsulation [1-5]. This can result in a significant reduction in the luminous flux, emitted by the devices. Spectral power distribution (SPD) method is used to study the effect of high temperature stress test on the optical degradation of remote phosphor. The aim was to investigate the effect of temperature on the lumen depreciation of LED-based products and on their CCTs. It is shown that the degradation mechanism is thermally-activated. As it is published already in our previous paper [13] and also shown in this paper the main reason of the decreasing intensity of blue light is the yellowing and discoloration of the lens. Comparing the results of lens and remote phosphor plates, one can see that the activation energy of the yellowing reaction in commercial lens plates (substrate plates) are slightly higher than that of remote phosphor plates, inferring that the activation energy for remote phosphor plates has contributions both from the worsening of substrate plates and the reduction in the phosphor conversion efficiency. In another words, presence of phosphor would accelerate the degradation kinetics or slightly decrease the activation energy.

It is clearly seen that the lower the depreciation rate, the better the performance a remote phosphor could have. The results also show that there is a direct relation between the temperature and kinetics of degradation. It is also shown that decreasing the transsimity of PC plates together with the reduction in phosphor efficiency limits the reliability of remote phosphor light sources and there is a colour shift towards yellow.

The real working temperature of LEDs depends on the working conditions (whether it is outdoor or indoor application). The LED temperature in a sunny day in a hot summer at currents as high as 700 mA can reach 90 °C [3,18]. However in this paper the mentioned temperature as an example (40 ºC) is more an average over the whole year. So, lumen depreciation up to 30% reduction is extrapolated to temperatures lower than 100 °C. It is shown that the lifetime, defined as 30% lumen depreciation at 40 °C, is around 35 khrs, for remote phosphor and 100 khrs for substrate.

Also it is shown that moisture diffusion and humidity does not have any significant effect on the lumen depreciation, color shifting, and accelerating of ageing.

Conclusions

BPA-PC lens plates and remote phosphor plates are exposed to temperature in the range of 100 to 140 °C. Exponential luminous decay model and Arrhenius equation are used to predict the lumen depreciation the lifetime of plastic lens in LED lamps in real service conditions. The photometric properties of thermally-aged plates, monitored during the stress thermal ageing tests, showed a significant change both in the correlated colour temperature (CCT) and in the chromaticity coordinates (CIE x,y). The decrease in the luminous flux is strongly correlated to the deterioration of the chromatic properties of the phosphor plates. Increasing the exposure time is associated with the discolouration, decrease in the relative radiant power value, and increase in the yellowing index (YI) of PC plastic lens. By increasing the temperature, the reaction rate becomes faster, meaning that lumen depreciation takes place at shorter time. The reaction rate follows the Arrhenius acceleration law.

The lifetime of the plastic lens and remote phosphor plates, defined as 30% lumen depreciation at 40 °C, is around 100khrs and 35khrs, respectively.

Acknowledgments

This research was carried out under project number M71.9.10380 in the framework of the Research Program of the Materials innovation institute M2i (www.m2i.nl). The authors would like to thank M2i for funding this project. Authors would also like to acknowledge "TNO innovation for life" company for SPD measurements.

References

1. D. F. Downey, Ion Implantation Technology, Prentice-Hall, New York, 1993, pp. 65–67. [book reference example]
2. Y. Wasserman, "Integrated single-wafer RP solutions for 0.25-micron technologies," IEEE Trans. CPMT-A, vol. 17, no. 3, pp. 346–351, Mar. 1995

3. M. Meneghini, M. Dal Lago, N. Trivellin, G. Meneghesso, E. Zanon, Thermally Activated Degradation of Remote Phosphors for Application in LED Lighting, IEEE transaction on device and materials reliability,vol.13, march 2013
4. W. K. Shu, "PBGA wire bonding development," in Proc. IEEE Electronic Components and Technol. Conf. (ECTC), Orlando, FL, May 28–31, 1996, pp. 219–225. [conference paper example]
5. Q.Y. Zhang, K. Pita, W. Ye, W.X. Que, Chem. Phys. Lett. 351 (2002) 163.
6. Q.Y. Zhang, K. Pita, S. Buddhudu, C.H. Kam, J. Phys. D 35 (2002) 3085.
7. W.R. Stevens, Building Physics: Lighting, Pergamon Press, London, 1969.
8. J.E. Kaufman, J.F. Christensen, Lighting Handbook, Waverly Press, Maryland, 1972.
9. Accelerated life test of high power white light emitting diodes based on package failure mechanisms
10. S.I. Chan , W.S. Hong, K.T. Kim, Y.G. Yoon, J.H. Han, J.S. Jang, Microelectronics Reliability 51 (2011) 1806–1809
11. Ugo Lafont, Henk van Zeijl, Sybrand van der Zwaag, Increasing the reliability of solid state lighting systems via self-healing approaches: A review, Microelectronics Reliability xxx (2011) xxx–xxx
12. E. Nogueira, M. Vjzquez, N. Nooez, Evaluation of AlGaInP LEDs reliability based on accelerated tests, Microelectronics Reliability 49 (2009) 1240–1243
13. Yanagisawa T, Kojima T. Long-term accelerated current operation of white light-emitting diodes. J Lumin 2005;114:39–42[5] Long-term accelerated current operation of white light-emitting diodes, Journal of Luminescence 114 (2005) 39–42
14. American Society for Testing and Materials. Test method for yellowness index of plastics. Annual book of standards, 8.01, ASTM D1925-70. Philadelphia: ASTM, 1970
15. S. Koh, C. Yuan, B. Sun, B. Li, X. Fan, G.Q. Zhang, Indoor SSL product level accelerated lifetime test, eurosim Confere.
16. Illuminating Engineering Society, TM-21-11 Projecting Long Term Lumen Maintenance of LED Light Sources, 2012
17. Yazdan Mehr M, van Driel W.D, Jansen K.M.B, Deeben P, Zhang G.Q. Lifetime Assessment of Plastics Lenses used in LED-based Products. Journal Microelectronics Reliability 2013, in press
18. IES Approved Method for Measuring Luminous Flux and Color Maintenance of Remote Phosphor Devices' LM86

Multifunctional System Integration in Flexible Substrates

K. Bock, E. Yacoub-George, W. Hell, A. Drost, H. Wolf,
D. Bollmann, C. Landesberger, G. Klink, H. Gieser, C. Kutter
Hansastraße 27d, D-80686 Munich, Germany
Email: karlheinz.bock@emft.fraunhofer.de

Abstract

In this paper we present a technology developed for reliable electrical interconnection on film substrates and between vertically stacked film layers. Applying through-hole via technologies for 3D foil stacks enables multi-functionality and RF performance combined with open form-factor and very cost-efficient manufacturing of conformable electronic modules. The manufacture of fine line metal patterns (line /space geometries below 20μm) on film substrates is performed by cost-effective roll-to-roll technology. Furthermore procedures and technologies for handling and lamination of film based sub-modules have been developed. Also the manufacture and handling of ultra-thin and flexible integrated circuits has been combined with placement of SMD type passive and with integrated printed passive components in the same technology.

Introduction

Modern system technology requires multi-functionality and beneath the performance requirements often open form-factor and very cost-efficient solutions are required, therefore integration in organic substrates and foil substrates gain interest in the last years. Selective structuring processes are developed to fulfil the requirement for high resolution i.e. integrated organic electronics in a foil as well as lithography of metal layers to fulfil the requirements of high performance power or data busses in the foil substrates [1]. In the meanwhile it is possible to hetero-integrate full systems in a foil with organic integrated circuits and PV, display, passive components, RF devices, wave-guides, flexible battery, thin silicon IC, as well as printed sensors and actuators [1, 2, 4, 6]. The combination of several of these components will be the key to product application in near future.

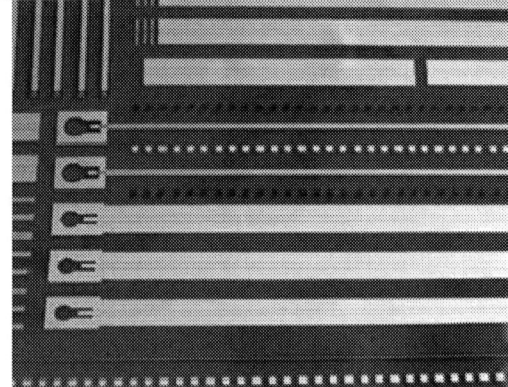

Figure 1. Interconnects and RF wave guides on PI film up to 60GHz

With components manufactured by optimized modular process technology applying organic substrates, devices and component technologies heterogeneous integration of functions from different technology environments will become possible and a significant advancement in functionality, flexibility and profitability will be reached. System integration technology requires multi-functionality and in many cases energy autarkic systems, very cost-efficient or open form factor solutions. Integration in plastic or foil substrates by a flex-to-flex integration concept shows the potentially free form factor which allows placing of film based systems on curved surfaces or in housings of very low thickness.

Interconnects on film substrates

A technology has been developed for reliable electrical interconnection over the film substrate and between vertically stacked film layers including through-hole via technologies for film stacks [2]. The EMFT reel to reel technology applies a pattern plating process for the metallization. It starts with sputtering an adhesion layer of few nm Cr followed by Cu with a thickness of 500 nm. A reel-to-reel lithography using a dry resist defines the metal pattern. A standard acid Cu bath is used for the electroplating process step. Applying this technology it is possible to manufacture fine line metal patterns (line /space geometries below 20μm) on film substrate, preferably to be done by cost-effective roll-to-roll technology and with the same technology including the via process discussed later to integrate standard interconnects RF wave guides in PI and LCP foil technologies showing very promising performance of 40GHz at - 3db [6]. The coplanar wave guides fabricated show extremely low roughness of the metal interface to the substrate is in the range of 2 μm for commercially available technologies and thus far higher than for the newly developed technology showing a roughness in the range of 150 nm (Fig. 2). For frequencies higher than 1 GHz the skin depth is lower than 2 μm. Thus, for commercially available technology the roughness will significantly influence the RF behavior. In the range up to 60 GHz the skin depth decreases to approximately 260 nm which is still above the roughness by our integrated foil wave guide technology.

Potentially even much higher frequencies are possible to be handled in flexible systems. At present up to 170GHz is possible with polymer co-planar wave guides developed for on-top integration to silicon interposer for GaAs IC integration and interconnect [7].
Such technology, in-spite of being developed for silicon interposer substrate at present, by principle could be transferred to a flex film substrate too.

Interconnection holes (vias) through film substrate

For a two layer interconnect metallization via holes have to be fabricated in the 50 μm thick polyimide substrate. A diode pumped solid state Nd:YVO Laser with a wavelength of 355 nm, 20 ns pulse length and an average power of 5 W is

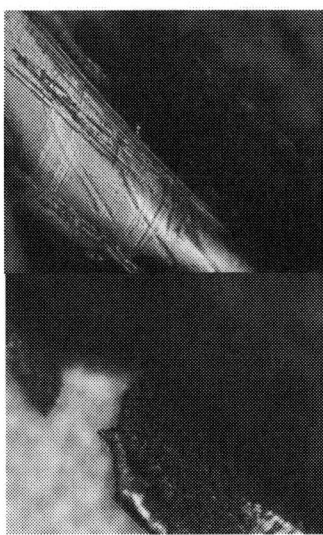

Figure 2. RF wave guide fabricated with extremely low roughness (Cu/Cr/PI) metal interface 150nm (left) comparison to a commercial (Cu/CuO/PI) metal interface of typical 2μm roughness (right)

used to ablate the polyimide material to fabricate a blind via. Via processing and the subsequent metallization process is rather critical, since it can strongly impact yield of the whole system. For a reliable interconnection process it has to beensured that the via hole is cleared from any substrate material that prevent contact to the second metallization, but on the other hand laser power has to be tuned to a range where it stops efficiently at the metallization without destroying the free standing copper landing pad on the bottom side. The power density of the laser beam has been adjusted accordingly, but to take into account process variations it turned out that a metallization thickness of at least two μm is needed for using the available laser equipment.

By defined defocusing of the laser beam the surface of copper remains smooth and a defined side wall geometry can be obtained. Via holes with a side wall slope of 60° and smooth edges on top have been fabricated, which eases sputtering and patterning of the second metallization layer.

However it has been observed that debris from the laser ablation leads to many bad contacts (Fig. 3 lower right). For this the laser ablation process has been combined with plasma cleaning with a combination of O2 and CF4 process atmosphere. All the processing steps have been done in equipment operating roll-to-roll. For characterization of the processing resistance of multiple vias has been measured with a Kelvin configuration.

As result a good interconnection resistance with an average value of 623 μOhm and an acceptable standard deviation of ±66 μOhm (1 σ) is achieved (Fig. 4).

Thin chip integration and "film package"

Silicon based integrated circuits (IC) offer a huge variety of electronic functionalities at rather low cost. In order to prepare very thin and also flexible ICs device wafers need to be thinned to a final thickness in the range of 10 to 30 μm. Technologies for handling and electrical interconnection of ultra-thin chips already have been demonstrated. Thin dies

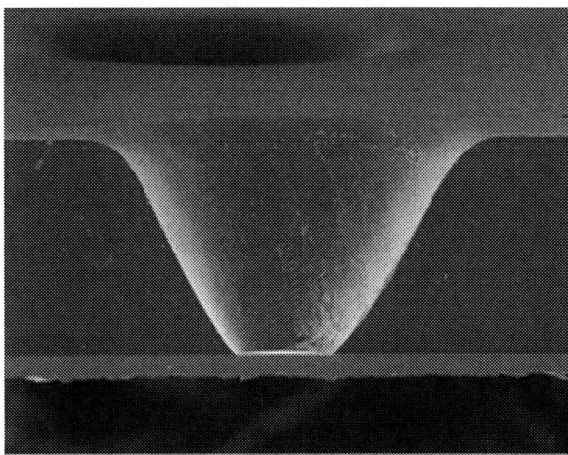

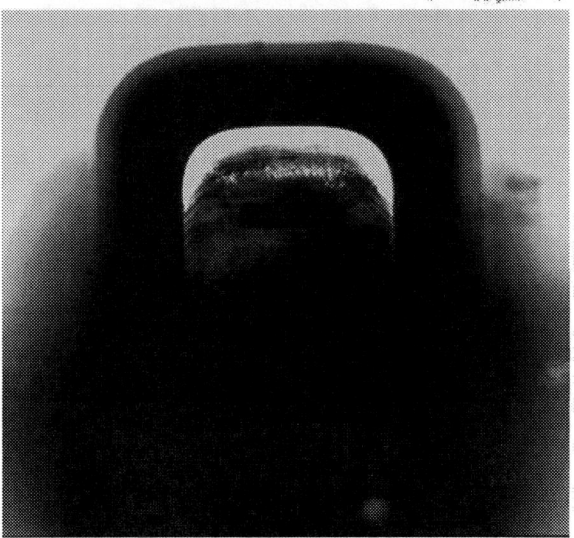

Figure 3. (top) Laser drilled circular via hole with smooth

sidewalls and opened rear side metallization (30μm scale only for the top picture. (middle) squared vias before cleaning and (lower) after plasma cleaning.

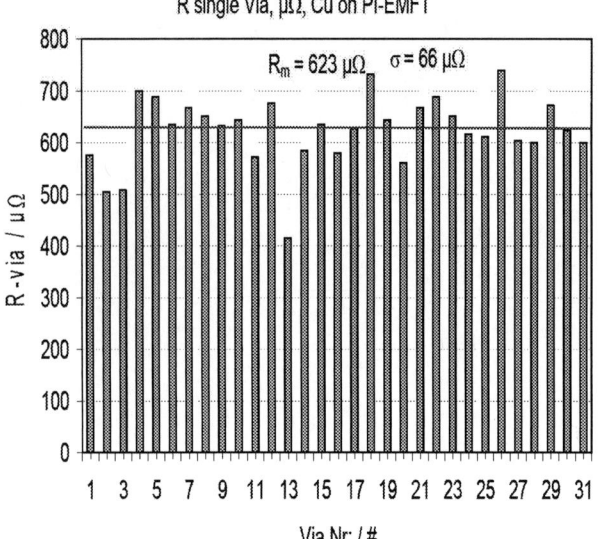

Figure 4. Measurement of the via resistance of testmodul with 31 vias, mean resistance Rm=623 μΩ and standard deviation of ±66 μOhm (1 σ)

can be embedded in a substrate either by first preparing cavities and then placing the dies at the recessed bottom area or by first placing the dies and then apply additional process steps for overall planarization by spin-coating or lamination of polymeric layers. Electrical interconnection can be done either by flip-chip bonding using anisotropic adhesives or solder attach. Recent publications from various research groups show really interesting and promising results in the development of thin chip film packages [8,9]. Chip alignment and assembly in foil has been developed towards a chip in foil package solution enabling also multi-chip modules in foil. Assembly of dies can be done with conventional pick&place and with newly developed self-assembly processes with assembly accuracies of less than 1μm and typically of less than 3μm with different sizes and topographies.

Our thin IC package concept is based on the following technical targets: Very thin and fragile IC devices should be located along the central line of a film based multilayer structure, any topographic structures (e. g. solder balls, bumps) should be avoided or completely planarized in order to minimize mechanical forces acting onto thin IC devices, processing temperatures should be limited to below 200 °C, all processes should be compatible with roll-to-roll manufacture and the possibility to integrate further active or passive devices in the assembly flow.

These targets could be achieved by a cavity concept for the location of electronic components and a direct thin film metallization and lithography technology for electrical interconnection. Fig 5 illustrates the design of the targeted foil package with fan-out contacts; the related basic process steps are shown in the following (Fig. 6). An outstanding advantage of the face-up die mounting and wiring concept is given by the fact that all electrical interconnects of components and the fan-out wiring layer are realized in just one process step. We propose the name "one step interconnect" for this metallization scheme.

A first focus is on the embedding process and the key enabling process steps. Here, the cavity concept in conjunction with the continuous movement of the web substrate and a blade coating system for the application of the polymer for planarization enables a rather simple process technology.

Figure 5. Concept for thin chip foil package.

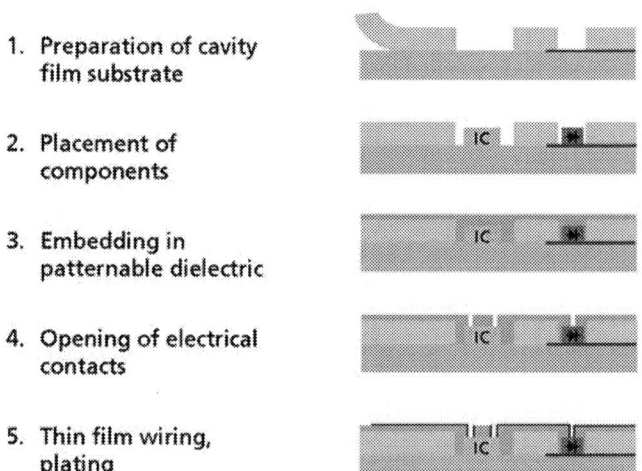

Figure 6. Schematic of the targeted technology platform for embedding of thin components in flexible film substrates

A second key process step focuses on the opening of via holes above the contact pads of IC devices and further components. Two possibilities can be considered for this step: local laser ablation or the use of photo-patterned dielectric materials. The latter choice would allow for parallel processing, clean surfaces and minimum impact to the metal pads of ICs.

Specific requirements for all chip integration technologies on polymer films originate from non-uniform substrate shrinkage or expansion. Consequently, lithographic patterning needs to be performed with respect to local alignment marks given by the individual IC devices on the web. This can be achieved either by a roll-to-roll stepper lithography concept or by laser direct writing in combination with an automated pattern recognition system.

In principle, the latter technique is already available today from several laser patterning equipment companies. An extension to roll-to-roll machinery can be expected in the near future.

Figure 7. Microscopic top view of an ultra-thin μ-controller IC (thickness < 30 μm) accurately placed in a 40 μm deep cavity on polyimide film substrate array of cavities (top) single cavity (below).

Die placement and self-alignment

Placement of IC and further components can be done in different ways. Metal patterns on the base film or lithographically defined fiducials in the cavity layer (for instance the corners of the cavities themselves) allow for accurate die alignment (Fig. 7). An even more promising approach for highly precise positioning of chips would use self-alignment techniques. Such concept and technique has already been shown [3]. It is based on selective wetting behavior of low viscosity liquids on metal areas in contrast to surrounding polymeric materials after fluorine plasma activation [10]. Accordingly, self-alignment of dies will require a lithographically defined metal pad at the bottom area of cavities which show the same size as the IC device. This can be realized perfectly by using plasma dicing for separation of very thin silicon devices which is in fact also the best solution for the preparation of mechanically robust and flexible ultra-thin dies [11].

By such concept we found self-alignment accuracy for 50 μm thin dies in the range of just 1 μm [5]. This will perfectly support a highly parallel and cost effective roll-to-roll based

manufacture technique for die placement on flexible film substrates.

3D Foil integration and lamination

A new process sequence "pick & laminate" has been developed for vertically (3D) stacking of film based sub-modules. This technique allows stacking of multi-layer foil modules of different size and without restrictions on the target position based on adhesive tapes. An alignment accuracy of 50 μm could be confirmed repeatedly. Electrical interconnects have been realized by jetting of silver filled adhesives into the laser drilled through film vias which typically show diameters of 500 μm. The resistance of vias is within 20 – 40 Ω per daisy chain corresponding to 36 via holes. Performed reliability tests of double layer film assemblies with electrically active through film vias showed very promising results. A first decrease of the electrical performance has been observed at a bending radius of 5 mm and the daisy-chains of vias changed their resistance by roughly 5 % only.

The development of reliable technologies for 3D-integration of foil based wiring layers and foil components is a main task to bring flexible electronic devices into high volume applications. 3D-integration or 3D-foil assembly means the building of the 3D-structure of a multilayer system-in-foil. In principle, the development of 3D-foil assembly technologies has to deal with three questions:

- alignment,
- mechanical connection,
- electrical interconnection

This technology for aligned foil-to-foil lamination was developed within the EU project Interflex and is called pick&laminate. In combination with a foil-to-foil via fill technology for electrical interconnection it enables the fabrication of high functional, multilayer systems-in-foil that are still flexible. The pick&laminate approach picks up the basic principle of pick&place for die attach but uses instead of a rigid placement tool a framed carrier foil (foil chuck) as handling tool for the foil sheet to be laminated.

The concept bases on a flexible, transparent foil chuck as carrier for a wiring foil or foil component, an x, y, φ-device for foil alignment and a video camera to control the alignment process. The actual lamination step is done after the alignment procedure by a roller device that is placed on top of the foil chuck. The concept for the electrical 3D-interconnection of two foil layers bases on foil-to-foil vias and a via fill process.

This presumes that the vias are drilled through the top foil before lamination. After lamination they are filled with a conductive material e. g. by dispensing, jetting or screen printing and cured by a temperature treatment.

The pictures in figure 9 demonstrate how the pick&laminate approach is used for foil-to-foil lamination. They are taken from the fabrication process for the Interflex demonstrator and show the two most important steps in the pick&laminate working sequence: the camera controlled alignment and the roller based lamination.

The evaluation of the alignment accuracy of the pick&laminate approach was done experimentally in the following way: Two pieces of wiring foils with identical copper pattern were congruently laminated together. To determine the misalignment in x- and y-direction the shift of the top and the bottom foil was measured at 6 different positions on the foil laminate by a microscope with reticule.

In figure 9 the mean values and the standard deviations for misalignment in x- and y-direction over six measurement positions are shown for seven laminated foil samples.Figure9 illustrates that the extent of misalignment (mean values μ) and also the extent of misalignment over the measurement positions within one sample (standard deviations σ) differs from sample to sample. These fluctuations in misalignment values can be explained by the fact that the pick&laminate machine is currently a manually operated tool and the experience and skill of the user have a significant effect on the quality of the foil laminate. Nevertheless, the experiments demonstrate that the alignment accuracy of the pick&laminate method is satisfactory. The misalignment over all values is 28 ± 16 μm in x-direction and 32 ± 20 μm in y-direction.

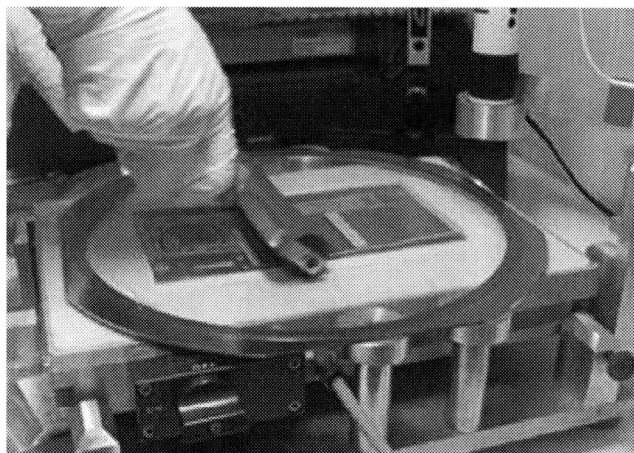

Figure 8. Two important steps in the working sequence of the pick&laminate approach: alignment step (upper) and roller lamination (lower).

Conclusions

In conclusion, it is pointed out that the newly developed pick&laminate technology is a valuable tool for 3D-foil-to-foil integration. In the meanwhile the lamination concept is transferred into a automatic pilot line equipment. The Interflex project proved that even complex foil systems containing several foil layers as well as integrated active and passive components can be assembled and interconnected reliably with this technology.

The status of film hetero-integration technologies has been introduced and linked with wave guide manufacturing in film substrates, silicon RF interposer and handling of thin semiconductor substrates. Full system heterointegration for products is in reach for foil substrates merging classical board and RF technologies with organic integrated circuits and PV, display, passive components, flexible battery, thin silicon IC, as well as printed sensors and actuators.

It should be mentioned that such foil technologies can also be applied for rigid modules to reduce the topography or to improve the heat sink and for flexible and open form factor applications. The combination of several of these components is demonstrated applying RF wave guides in film substrates enabling autarkic multi-functional wireless systems for product application, but also for new semiconductor handling and packaging technology. This enables the heterointegration of different technologies like compound semiconductor with silicon technologies and MEMS and shows the potential to improve form factor, performance (like cooling, etc.) and cost.

Figure 9. Mean value and standard deviation for

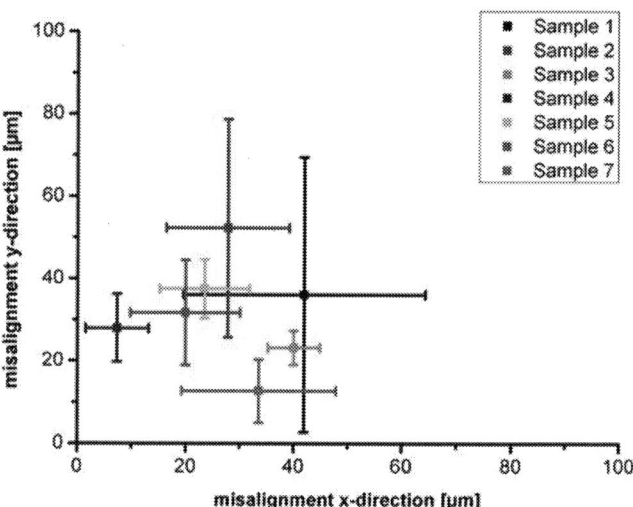

misalignment in x and y-direction are shown over the 6 measurement positions for all samples.

Acknowledgments

The authors would like to thank the Interflex project partners (Bosch, STMicroelectronics, CEA-Liten, Henkel and Infotech Automation) for providing components and materials, and the European Commission for funding the Interflex project.

The authors acknowledge the cooperation projects with Sony Corporation Labs. in Stuttgart Germany on the silicon transmission line technology and the company Rosenberger in Germany on the RF wave guide technology in film substrates as well as Dieter Hemmetzberger, Axel Wille, Uli Schaber , Martin König, S. Scherbaum at EMFT for their technological, experimental and manufacturing support.

References

1. Bock, K.; Polymer Electronics Systems – Polytronic, Proceedings of the IEEE, Volume 93, Issue 8, Aug. 2005 Page(s):1400 – 1406, Digital Object Identifier 10.1109/JPROC.2005.851513

2. E. Yacoub-George, A. Ohlander, L. Meixner, D. Bollmann, R. Faul, C. Landesberger, K. Bock; Large area multilayer foil assembly for flexible electronic systems, Smart Systems Integration 2011; Dresden, Germany, 22-23 March 2011.

3. Bock, Karlheinz; Scherbaum, Sabine; Yacoub-George, Erwin; Landesberger, Christof, Selective one-step plasma patterning process for fluidic self-assembly of silicon chips, Electronic Components and technology Conference ECTC, Lake Buena Vista, Florida, USA, May 2008

4. K. Bock, Modular Solid State Technologies for a Multi-functional System Integration, CS Mantech Conference, Indian Wells (Palm Springs), California, May 16- 19, 2011, paper 10a.4, published in the proceedings and n the web-site.

5. Mitsuru Hiroshima, Kiyoshi Arita, Hiroshi Haji, Bernhard Oberhofer, Christhof Landesberger, Sabine Scherbaum, Josef Weber, Karlheinz Bock, "A robustness study on self-alignment of thin-Si dies using surface tension", MES 2012 in Osaka Japan.

6. Karlheinz Bock, Erwin Yacoub-George, Henry Wolf, Christof Landesberger, Gerhard Klink, Horst Gieser, Heterointegration technologies for high frequency modules based on film substrates", CS Mantech 2013, NewOrleans, May 13- 17, 2013, paper 10b.4 published in the proceedings and on the website.

7. Eray Topak, Joo-Young Choi, Thomas Merkle, Stefan Koch, Shin Saito, Christof Landesberger, Robert Faul and Karlheinz Bock; „Broadband Interconnect Design for Silicon-Based System-in-Package Applications up to 170 GHz; to be published, European Microwave Week 2013, 6.-11. October 2013.

8. Ranjan Rajoo, Lim Y.Y, Chong S.C , Myo Paing, Fernandez Daniel Moses, Justin See Toh Wai Hong, Vasarla Nagendra Sekhar, Soon Wee Ho, Serene Thew, Justin Wee Kim Soon, Xiaowu Zhang: Embedding of 15um Thin Chip and Passives in Thin Flexible Substrate; 12th Electronics Packaging Technology Conference, 2010.

9. S. Priybadini, T. Sterken, M. Op de Beeck, J. Vanfleteren; Photo-definable polyimide-based Flat UTCP technology for 3D-stacking application; Smart Systems Integration (SSI) Conference, Amsterdam, NL, 13-14 March 2013.

10. Christof Landesberger, Sabine Scherbaum, Josef Weber, Karlheinz Bock, Mitsuru Hiroshima; Bernhard Oberhofer, Plasma dicing enables high accuracy self-alignment of thin silicon dies for 3D-device-integration, Electronics System Integration Technology Conference ESTC; September 13-16, 2012; Amsterdam, Netherlands.,

11. S. Takyu, T. Kurosawa, N. Shimizu, S. Harada, "Novel Wafer Dicing and Chip Thinning Technologies Realizing High Chip Strength", MRS Proceedings MRS, Volume 970, 2006.

Preparation of a Micro Rubidium Vapor Cell and Its Integration in a Chip-Scale Atomic Magnetometer

Yu Ji[1], Jintang Shang[*1], Youpeng Chen[1], Ching-Ping Wong[2]

1.Key Laboratory of MEMS of Ministry of Education, Southeast University
Sipailou 2, Nanjing, Jiangsu, China, Email: jshang@seu.edu.cn, Tel: +86 13913869603

2. College of Engineering, The Chinese University of Hong Kong

Abstract

Micro Rubidium vapor cell has very important research values in many fields including atomic magnetometer[1-2], atomic clock[3] ,atomic gyroscope[4] and so on. In this study, a new method to fabricate the micro Rubidium vapor cell has been investigated and uniform wafer-level micro Rubidium vapor cells, which could be used in atomic devices, have been prepared successfully by the proposed process. First of all, the fabrication process is introduced. Then the prepared uniform wafer-level micro Rubidium vapor cell has been characterized by Energy Dispersive Spectroscopy (EDS) and Ultraviolet spectrophotometer. And the result demonstrates the existing of Rubidium and shows that the SNR (Signal to Noise Ratio) of the micro Rubidium vapor cell is high , which suits for atomic devices as well as other applications. At last, its integration in a Chip-Scale atomic magnetometer is presented.

Introduction

Miniaturized atomic magnetometers[1-2][4] characterized by a smaller size and drastically reduced power consumption compared to large scale atomic magnetometer and SQUIDs (superconducting quantum interface devices) exhibit an increasing interest mainly for applications in many fields, such as geophysical mapping, underground deposit detection, navigation, and physiological mapping [1-2]. Micro-fabricated alkali vapor cell is the key part of miniaturized atomic magnetometers. The micro Rubidium vapor cell[10], the heart of an atomic magnetometer, consists of a sealed cavity contains alkali metal and light windows. The key challenge during the fabrication of vapor cells is placing alkali metal into a hermetically sealed cavity due to the volatile character of alkali metals and the reactivity of alkali metals with oxygen. As a result, all handling of the alkali metals has to be done under high vacuum conditions , which complicates the fabrication of the atomic vapor cells.

There are several fabrication approaches to fabricate a micro Rubidium vapor cell which can be classified into four different groups: a) direct injection method [5]; b) chemical reaction of barium azide and rubidium chloride [6]; c) alkali metal azide deposited by vacuum thermal evaporation followed by UV decomposition [7]; d) electrolytic decomposition of alkali metal enriched glass [8]. Though each approach above has been proved successfully to fabricate a micro Rubidium vapor cell and they can be potentially scaled-up to wafer-level filling, each approach has its own drawback. For example, the methods of a) and d) are complicated in the fabrication process and may be cost

prohibitive for commercialization; the disadvantage of b) is that unreacted barium tends to form different forms of nitride with the released nitrogen ,causing pressure fluctuation inside the cell , which affects the stability of the atomic vapor cell [5]; and the drawback of c) is that the azide has to be heated above its melting point, favoring uncontrolled decomposition and explosion[9] ; and it is difficult to find Rubidium enriched glass ,which makes the method of d) difficult to achieve.

To address these challenges, we describe a rubidium batch-filling technique based on anodic bonding and thermal decomposition of rubidium azide. Rubidium azide is a solid at room temperature and is stable in the air , and the melting point of the Rubidium azide is about 275°C according to our experiment, what's more, we have proven that rubidium azide decomposes into pure rubidium and buffer gas nitrogen under the condition of high temperature about 310°C. In our approach, rubidium azide can be handled under normal atmospheric conditions and high purity rubidium will exist after the second bonding. The micro Rubidium vapor cell is characterized by EDS (Energy Dispersive Spectroscopy) and ultraviolet spectrophotometer. Results are demonstrated that wafer-level micro Rubidium vapor cell is fabricated successfully and it is perfect in SNR (Signal to Noise Ratio). As a result, the fabrication process is easily scalable to wafer-level filling.

Fabrication Process

The fabrication process is illustrated in Fig.1.

The micro Rubidium vapor cell has been carefully designed, which consists of two cavities of which the smaller one is used as a thermal decomposition chamber and the larger one is taken as a working chamber. This design preventing the residual rubidium azide from entering into working chamber improves the purity of rubidium in the working chamber.

First of all, a double-sided polished <100> oriented p-type silicon wafer with a thickness of 600μm was cleaned before wet etching. Then the 4-inch silicon wafer was photo-lithographically patterned in a clean room and wet etched in TMAH to obtain cavities and micro-channel through the wafer.

Secondly, the etched wafer was anodic bonded onto a 4-inch Pyrex7740 glass at the temperature of 400°C under vacuum creating what we refer to as a "perform"[12] ,as was shown in Fig. 1 (a).

Thirdly, rubidium azide was dissolved into at least one solvent such as DI water. Then a dispenser, which has the ability to control the solution accurately and is widely used in

978-1-4799-2408-0/14 $31.00 © 2014 IEEE   1488   2014 Electronic Components & Technology Conference

industry, was used to transfer the solution into the smaller cavities, as was shown in Fig.2 (b). After that, the hot plate was heated in stages and at a rate of about 5°C/min and held at 150°C for about two hours to evaporate the solvent for forming solid powder. And in this way, the quantity of rubidium azide could be precise controlled by this method. Uniform micro Rubidium vapor cells could be achieved on a wafer level simultaneously.

After azide deposition , the "perform" was placed on the chuck of the bonding machine EVG501. And then bonding machine was heated in stages and at a rate of about 2°C/min and held at 200°C[7] ,the "perform" was baked at the temperature of 200°C for several hours to ensure there is no water residual. In the mean time, the chamber of the anodic bonding machine was evacuated to 10^{-4} mbar which took about 1 hour. Upon reaching the pressure of 10^{-4} mbar , the pump was switched off, a Pyrex wafer was then anodic bonded to the silicon side of the "perform"[13] at a periodic voltage which had been proven to have higher bonding strength in our previous research, as was shown in Fig. 1 (c).

Finally ,after the second bonding, the wafer-level rubidium vapor cells had been successfully fabricated , and then the wafer stack was diced into individual cells, each micro Rubidium vapor was successfully fabricated and each one could be used to integrate in Chip-Scale atomic devices, as was shown in Fig.1.(d).

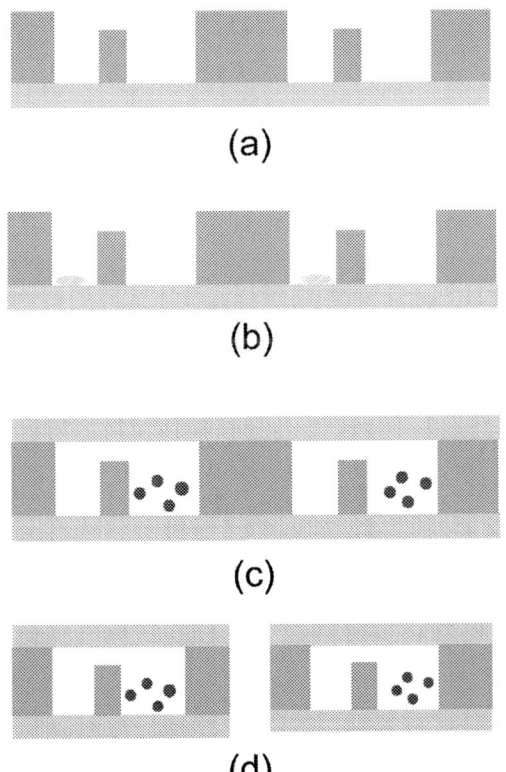

Fig.1 Schematic process of wafer level micro-fabricated Rubidium vapor cells

To confirm there was rubidium metal existing in the vapor cell, we could observe the micro Rubidium vapor cell

under a microscope. And then it was characterized by EDS, using the Zeiss Ultra Plus scanning electron microscope[11], as shown in Fig.2. In this case, we used EDS to analyze the elementary composition of substance resided on the internal surface of the second bonding glass.

And to confirm the quality of the micro Rubidium vapor cell is good enough and could be used in atomic devices , the micro Rubidium vapor cell was characterized by Ultraviolet spectrophotometer.

Results and Discussions

The wafer-level Rubidium vapor cells were successfully fabricated, an array of micro Rubidium vapor cells was diced into single one with the external volume of $4 \times 5 \times 1 mm^3$. The results show that the quality of the micro Rubidium vapor cell is perfect and it could be integrated in atomic devices, such as Chip-Scale atomic magnetometer.

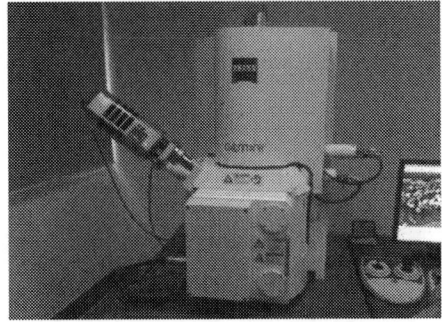

Fig.2. Ultra Plus scanning electron microscope

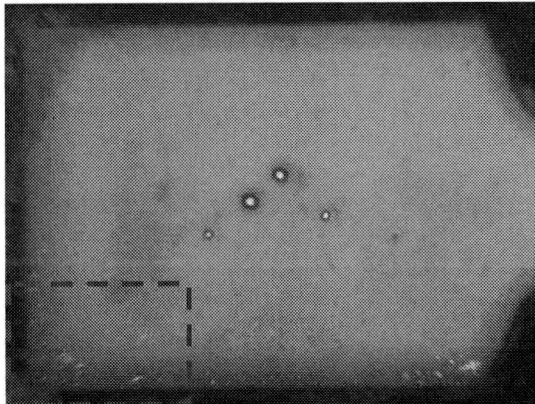

Fig.3 (a) metallic droplets on the inner cell window

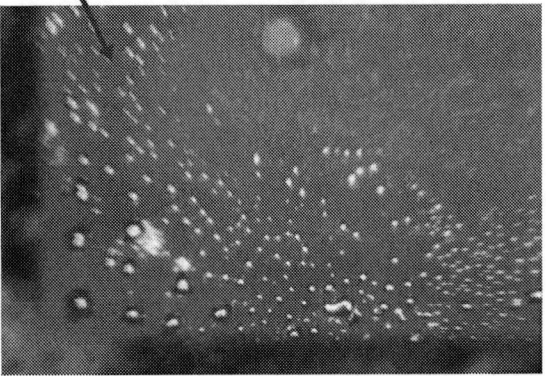

Fig.3 (b) enlarged metallic droplets on the corner of the cell

When the micro Rubidium vapor cell was observed under a microscope, we could find that metal produced by the thermal decomposition is visible , like gold-colored metallic droplets on the inner cell window, and we guessed it was rubidium metal. It was shown in the Fig3(a) that there are many rubidium particles on the inner cell window, and the Fig3(b) enlarged the corner of the Fig3(a).

To ensure the observed gold-colored metallic droplet was rubidium metal, the micro vapor was characterized by EDS. We should break the second bonding glass before EDS test, and we found that the gold-colored metallic droplets come into oxidation rapidly. Then a glass fragment was characterized by EDS. The SEM image of the glass fragment was shown in Fig.4.

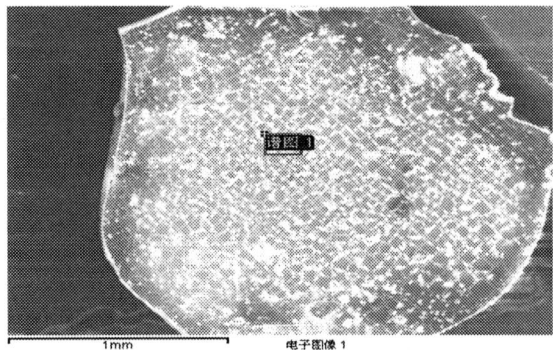

Fig.4 The SEM image of a glass fragment

The result of EDS was shown in Table 1 and Fig.5. As we all known, Rubidium is a very active substance. So when we broke the micro vapor cell, Rubidium was in reaction with oxygen immediately. From the result of table1 and Fig.5, we can suppose that a part of rubidium existed on the internal surface of the glass initially, and then the oxide was residue on the internal surface of the glass. So this result proved that the gold-colored metal was Rubidium.

TABLE 1. ELEMENTARY COMPOSITION OF SUBSTANCE RESIDED ON THE INTERNAL SURFACE OF THE GLASS FRAGMENT.

Element	Weight percent	Atomic percent
Rb	3.53	0.80
Si	34.55	23.93
O	61.91	75.27

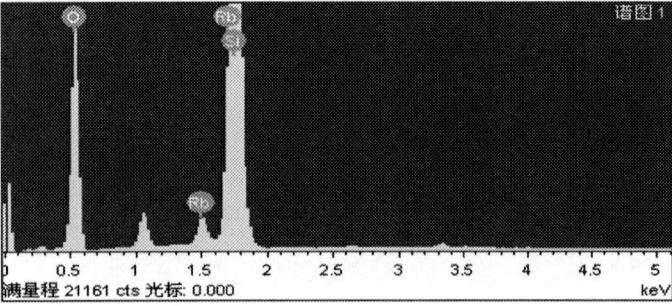

Fig.5 Spectrogram of the surface of the glass fragment

As we know, the quality of the micro Rubidium vapor cell has a great effect on the performance of the atomic devices such as the stability and the sensitivity. The quality of the micro Rubidium vapor cell was confirmed by optical spectroscopy and we used a UV-Vis spectrophotometer. In order to increase the density of the rubidium ,the vapor cell should be heated to about 80-120°C. And Fig 6 shows an absorption spectra of our micro Rubidium vapor cell where the horizontal axis corresponds to the wavelength of the laser exiting the rubidium atoms, and the vertical axis corresponds to the transmission intensity of the laser. From the figure ,we have known that the spectrum consists of two peaks at 780nm and 795nm, which demonstrated that pure rubidium were prepared successfully. What's more ,the sharp peaks show that the SNR (Signal to Noise Ratio) of the micro Rubidium vapor cell is perfect , which suits for atomic devices as well as many other applications.

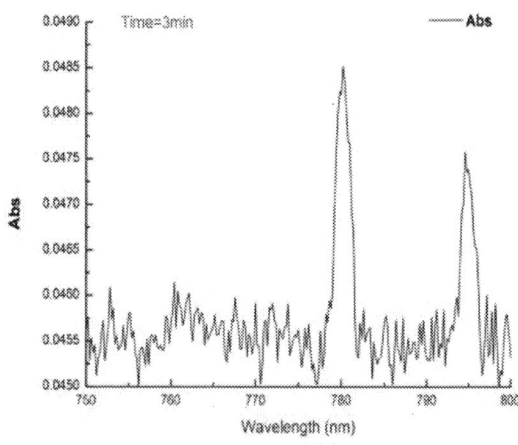

Fig.6 An absorption spectra of the micro Rubidium vapor cell

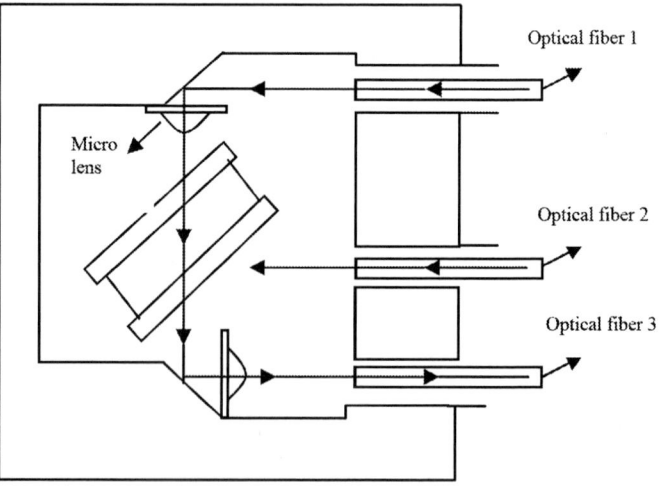

Fig.7 A schematic diagram of the Rubidium vapor cell's integration in Chip-Scale atomic magnetometer

At last, we introduce one method to integrate the micro Rubidium vapor cell in a Chip-Scale atomic magnetometer[1-2]. As is shown in Fig.7, a first circularly polarized beam of laser light at 780nm passes through the micro Rubidium vapor cell to pump the alkali vapor and this light emits from the optical fiber 2; a second linear polarized beam of laser light at 795nm emitting from optical fiber 1, takes the action of probing the magnetic field, through optical fiber 3,the light is detected.

In this integration, micro lens is used to parallel the light beam, and we have studied on the fabrication of low cost wafer-level micro lens arrays[14], which is beneficial to the optical integration of the Chip-scale atomic magnetometer.

Conclusion

In this paper, uniform wafer-level micro Rubidium vapor cells have been prepared successfully by the proposed process, and we will explore its integration in Chip-scale atomic magnetometer further, which comes to the following conclusion.

1. Uniform wafer-level micro Rubidium vapor cells are fabricated by a new process successfully.

2. The characterization results demonstrate the existing of Rubidium and show that the SNR (Signal to Noise Ratio) of the micro Rubidium vapor cell is high, which is suitable for integration in atomic devices

3. The proposed method is a low cost and simple wafer level process which suits for volume production of Chip-Scale atomic devices, such as Chip-Scale atomic clock ,Chip-scale atomic magnetometer and Chip-scale atomic gyroscope.

4. The micro Rubidium vapor cell's integration in Chip-scale atomic magnetometer is also presented and will be demonstrated in future.

Acknowledgment

This work is supported by the National Science Foundation of China (No. 51275091, No.50775038), the National High-Tech R&D Program of China (863 Program, No. 2009AA04Z306). The research funding from National S&T Major Project with contract No. 2009ZX02038 is also greatly acknowledged.

References

1. I.K. Kominis, T.W. Kornack, J.C.Allred, and M.V.Romalis, Nature (London) 422, 596 (2003).

2. H. B. Dang, A. C. Maloof, and M.V. Romalis, Appl. Phys.Lett. 97, 151110 (2010).

3. S. knappe et al., A microfabricated atomic clock, APPLIED PHYSICS LETTERS, 2004, pp.1460-1462.

4. John Kitching, Svenja Knappe, and Elizabeth A. Donley IEEE SENSORS JOURNAL, VOL. 11, NO. 9, SEPTEMBER 2011

5. S. knappe et al., Atomic vapor cells for chip-scale atomic clocks with improved long-term frequency stability, Optics Letters, 2005,pp.2351-2353.

6. Li-Anne Liew et al., Microfabricated alkali atom vapor cells, APPLIED PHYSICS LETTERS, 2004, pp.2694-2696.

7. LA Liew et al., Wafer-level filling of microfabricated atomic vapor cells based on thin-film deposition and photolysis of cesium azide, APPLIED PHYSICS LETTERS, 90(2007), 114106

8. F Gong et al., electrolytic fabrication of atomic clock cells, International Frequency Control Symposium and Exposition, IEEE, 2006, pp.711-714.

9. US patent NO US 2012/0301631A1

10. M. Hasegawa et al., Microfabrication of cesium vapor cells with buffer gas for MEMS atomic clocks, Sensors and Actuators A: Physical, 2011, pp.594-601.

11. Youpeng Chen, Jintang Shang et al., "Microfabricated Low Cost Wafer-Level Spherical Alkali Atom Vapor Cells for Chip-scale Atomic Clock by a Chemical Foaming Process （CFP）," in Proceedings of 14th Electronic Packaging Technology & High Density Packaging (ICEPT-HDP), Dalian, China, Aug. 2013.

12. Adriana Cozma Lapadatu and Kari Schjlberg – Henriksen   Anodic Bonding

13. J wei et al., Low temperature wafer anodic bonding, J. Micromech. Microeng. 13(2003), 217-222.

14. Shunjin Qin, JintangShang, Li Zhang et al, "Fabrication of Low Cost Wafer-Level Micro-Lens Arrays with Spacers using Glass Molds by Combining a Chemical Foaming Process (CFP) and a Hot Forming Process (HFP)," in Proceedings of 62nd Electronic Components and Technology Conference, San Diego, USA, May 2012, pp. 213-217.

Nanowires-based High-density Capacitors and Thinfilm Power Sources in Ultrathin 3D Glass Modules

Saumya Gandhi, Liyi Li, Ho-Yee Hui, Parthasarathi Chakraborti, Himani Sharma,
P. Markondeya Raj, C.P. Wong and Rao Tummala
3D Systems Packaging Research Center
Georgia Institute of Technology,
Atlanta, GA 30332-0560

Contact: sgandhi30@gatech.edu, Phone: 404 580 8698

Abstract:

This paper explores silicon nanowire technology for ultrathin high-density capacitors, supercapacitors and batteries. Development of such thin power components on glass or silicon will allow integration with other passive components as well as actives such as decoupling capacitors close to locic Ics to form 3D integrated passive and active devices (3D IPACs) that could then be surface-assembled onto glass packages leading to ultrathin self-powered modules or subsystems. Thinfilm integration of power components and passives on ultrathin glass 3D IPD substrates also leads to highly-efficient power distribution for miniaturized and high-performance electronic systems.

The first part of the paper presents an analytical model to highlight the benefits of nanowire electrode-based high-density capacitors in capacitance density and operating frequency. The second part of the paper describes nanowire synthesis and a novel fabrication process for nanowire-electrode capacitors, and their characterization. Results indicate that nanowires enable a major breakthrough in thinfilm capacitors with ultrahigh volumetric capacitance densities of about 100 μF /mm^3, 10X higher than all current capacitor technologies including trench, MLCC and tantalum capacitors.

I. INTRODUCTION

The primary drivers for electronic systems have been higher bandwidth and lower power, both enabled by the integration of ultra-thin active components in ultra-thin packages with higher interconnect density [1]. While major advances are being made in active devices and the interconnect densities with which to interconnect them with novel interposer and off-chip interconnection technologies, other system components for power storage, power conversion and power-supply with low impedance still remain as major fundamental bottlenecks. [2]. These components include batteries, storage components such as inductors and capacitors in VRM (voltage regulator modules) or DC-DC convertors, and decoupling capacitors.

Power components are discretely manufactured as individual components and surface-mounted on the board, far away from the active devices they serve. Their low volumetric densities and manufacturing limitations with today's microscale materials also results in thicker components. High-density passives such as Ta and Al capacitors are usually above 300 microns in thickness. MLCCs are being made thinner but are still discretely manufactured and assembled as SMDs. This leads to increased parasitics and reduced system performance. In the recent past, different methods to achieve high capacitance density and reduced component thickness have been explored. Etched Al foils help achieve the high capacitance densities in reduced component size, but suffer from lower volumetric density, frequency roll-off, reliability issues and integration challenges. Si trench capacitors have been explored, but have met limited success because of the very high costs. Integrating tantalum electrodes on silicon or glass creates process-compatibility challenges that still need to be solved. This paper explores silicon nanowires as an alternative electrode system to address these limitations. Silicon nanowires can achieve much higher volumetric density because of their nanoscale dimensions and much higher aspect ratio. Their process-compatibility with large glass panels also helps in substantially scaling down the cost. Silicon also forms a natural oxide with the highest electric field strength, which enables the thinnest dielectrics with the highest breakdown voltage. Hence, Si nanowires offer several advantages over the aforementioned technologies as they provide for higher densities, better reliability and manufacturability at lower-cost.

Silicon, in nanowire form, is a strategic nanomaterial for energy storage too. Silicon is an excellent host for lithium intercalation. By deposition of conformal thin lithium electrolytes and Li-Si anodes, high-energy density thinfilm batteries can be monolithically grown on glass substrates. The open structure of silicon nanowires allows easy expansion and contraction with lithium intercalation, thereby addressing the major fatigue concern with traditional silicon-based battery cathodes. Because of the low deposition temperature, nanowire batteries can be easily scaled and integrated using existing technologies, making them very attractive to both the silicon and packaging industry. Nanowires, therefore, offer unique benefits and can be used as novel building blocks for ultra-miniaturized passive components such as integrated passive devices (IPDs) with higher device integration [3,4,5,6,7].

Glass forms an ideal candidate for IPD substrate for thin components because of its ultra-thinness, low electrical loss, high resistivity, low-cost based ultrathin and large-panel processability. Recent innovations in through-glass vias will further improve the performance, reduce the final size and module cost. This paper advances the traditional IPD technology in two ways, 1.) Bringing heterogeneous component technologies to integrate batteries, storage capacitors and decoupling capacitors as double-side thinfilms that are compatible with through-vias,, as schematically envisioned in Fig. 1a and 1b, 2.) Enhancing individual component characteristics, such as capacitance density using high aspect ratio nanowire electrodes. These 3D IPDs also enable double-side integration of actives and passives on an ultra-thin glass substrate to form a 3D Integrated Passive and Active Component (3D IPAC) module.

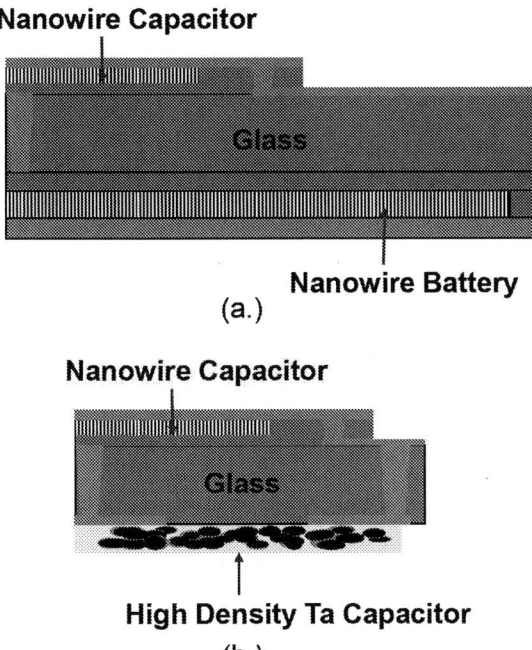

(a.)

(b.)

Fig. 1: Different configurations of Si-nanowire capacitors as 3D IPDs (a.) with Nanowire batteries to form self-sustained sub-system and (b.) with high density Ta capacitor to form a heterogeneous VRM+decoupling solution.

This paper primarily focuses on exploring and demonstrating silicon nanowires for power components, using capacitor as an example, which can be then extended to batteries and other storage components such as supercapacitors. The paper is organized as follows: Section II presents an analytical model to determine the voltage levels required to achieve maximum capacitance based on intrinsic doping concentration in silicon. Section III describes the fabrication process for the Si nanowire capacitor. Section IV presents the characterization and Section V summarizes the key findings.

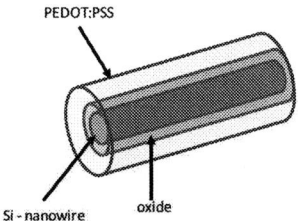

Fig. 2: Schematic representation of an individual nanowire capacitor.

II. Analytical Model

Silicon nanowire capacitors are essentially metal-oxide-semiconductor type capacitors. Fig. 2 shows a simple schematic of the structure. The capacitance of a MOS structure depends on the voltage bias applied to the metal (gate). The capacitor can operate in three different states based on the voltage applied, as is seen in Fig. 3. The states can be classified as (i) Accumulation: Surface accumulation of carriers that are same as the majority carriers in the bulk, (ii) Depletion: No carriers present on the surface with only a space charge or depletion region is present and (iii) Inversion: Opposite charge carriers to those present on the body accumulate on the surface. Two voltages can be used to effectively separate these three different regions. These are (a) Flatband Voltage: It helps separate the accumulation region from the depletion region and (b) Threshold voltage: Helps separate the depletion region from the inversion region. This paper focuses only on the accumulation region for a MOS capacitor. Based on the calculated capacitance and resistance, the RC time constant for such capacitors can be estimated.

1. Nanowire Length approximation:
When Au is used as a catalyst, the nanowire length can be estimated using Equation 1, where L is the length of the nanowire, r is the radius at the base, a^3 is the atomic volume of Au and theta is the average Au coverage on the sidewalls. Theta has been estimated to be 1 – 1.5 monolayers [8]. Table 1 summarizes the approximate length of nanowires that can be grown based of the Au catalyst size.

$$L = \frac{r^2}{2a\theta} \qquad (1)$$

2. Nanowire Resistance:
Given a certain conductivity of the silicon, the resistance of the nanowires can be estimated. The following equations allow to calculate the resistance based on the doping concentration.

$$R = \frac{\Delta l}{\sigma S} \qquad (2)$$

Where R is the resistance in ohms, Δl is the length of the nanowire in meters, σ is the conductivity in Siemens/m and S is the cross-sectional area of the nanowire. The conductivity of an intrinsic semiconductor is defined as

$$\sigma = n_i(\mu_e + \mu_h)q \qquad (3)$$

where n_i is the intrinsic carrier concentration, q is the charge of the electron (1.6×10^{-19} C), and μ_e and μ_h are the mobilities of the electrons and holes respectively. For an extrinsically doped semiconductor, either the electron or hole concentration will dominate and hence the other can be ignored while calculating the conductivity. It is well known that for an intrinsic semiconductor there is no net charge. This remains same even for extrinsically doped semiconductors. This means that even after doping, the total number of positive charges equals the total number of negative charges. This is derived from the Mass-Action law, which states that

$$np = n_i{}^2 \qquad (4)$$

For extrinsically doped n-type semiconductor:
$$\sigma = n\mu_e q \qquad (5)$$

For extrinsically doped p-type semiconductor:
$$\sigma = n\mu_h q \qquad (6)$$

where n in the above equations can be replaced with N_D or N_A based on the doping type.

Table 1: Analytical modeling data showing capacitance and cut-off frequency as a function of Au catalyst size

Si – type	Au catalyst size (nm)	Length of wire (m)	Resistance (Mohms)	Capacitance (fF)	RC time constant (nS)	Cut-off frequency (MHz)
5 ohm-cm	30	1.731	30	2.2	66	2.4e6
	50	4.8	31	9.4	0.3×10^3	0.53
	100	19.23	31	70	2.1×10^5	75.7×10^{-3}
10 ohm-cm	30	1.731	60	2.2	132	1.2
	50	4.8	62	9.4	0.6×10^3	0.26
	100	19.23	62	70	4.2×10^3	37.5×10^{-3}

The volumetric capacitance density was calculated based on the capacitance from Table 1. The total number of nanowires per unit area was estimated based on the nanowire diameter and minimum distance between two wires. The density was estimated to be $100 - 150 \ \mu F/mm^3$. This value is 10X higher than current capacitor technologies, and 2X higher than what is projected as the maximum achievable density with existing capacitor technologies.

3. Flatband voltage and Cut-off frequency:

This phenomenon takes place when the voltage applied is greater than the flatband voltage (positive or negative depends on substrate type). The flatband voltage is the voltage at which no charge is present on the capacitor electrodes leading to no electric field across electrodes. The flatband voltage depends on the doping concentration of Si. It is also highly dependent on any residual charge that may be present at the interfaces and across the electrodes. Hence, when a voltage greater than the flatband (large negative voltage for a p-type substrate) voltage is applied, it causes the holes to be attracted to the interface causing accumulation. The opposite is true for an n-type substrate [9]

$$V_{FB} = \phi_m - \phi_s \qquad (7)$$

$$\phi_s = \chi - \frac{E_g}{2q} - V_t Ln\left(\frac{N_a}{n_i}\right) \qquad (8)$$

$$\phi_s = \chi - \frac{E_g}{2q} + V_t Ln\left(\frac{N_d}{n_i}\right) \qquad (9)$$

$$V_t = \frac{KT}{q} \qquad (10)$$

A p-type substrate is considered in this analysis; hence the threshold voltage is defined as

$$V_T = V_{FB} + 2\phi_F + \frac{\sqrt{4\varepsilon_s N_a \phi_F}}{C_{ox}} \qquad (11)$$

$$C_{ox} = \frac{\varepsilon_o \varepsilon_r A}{d} \qquad (12)$$

where Φ_m is the work-function of the metal gate, Φ_s is the work-function of Si, χ is the electron affinity, Φ_F is the bulk potential of Si, ε_s is the relative permittivity of Si, C_{ox} is the oxide capacitance (under accumulation), K is the Boltzmann's constant, T is the temperature in K and q is the charge of an electron in C. It is important to calculate the threshold voltage and the flatband voltage, as this will allow capacitor operation in accumulation, resulting in maximum capacitance [10, 11].

Based on the above equations, the capacitance and resistance is calculated and the RC time constant can be estimated as

$$\tau = RC \qquad (13)$$

And the cut-off frequency is estimated as

$$\tau = \frac{1}{2\pi f_c} \qquad (14)$$

$$f_c = \frac{1}{2\pi\tau} \qquad (15)$$

Fig. 3: Representative MOS capacitance. Graph shows voltage-dependent capacitance.

Table 2: Analytical flatband and threshold voltages

	Copper	N⁺ Poly	P⁺ Poly
V_{FB} (V)	-0.36	-0.87	0.23
V_T (V)	0.37	-0.14	0.96

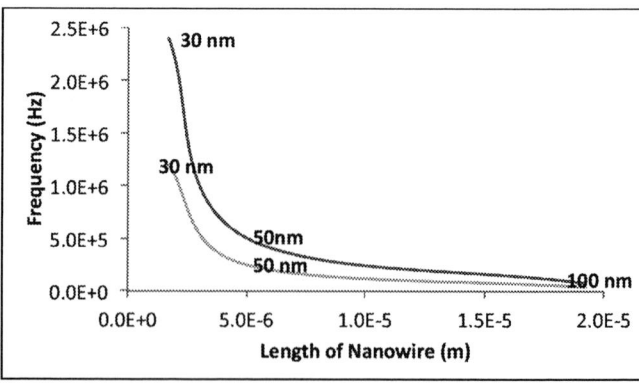

Fig. 4: Change in cut-off frequency with increasing nanowire length. It is important to note that the length of the nanowire changes based on the size of the Au catalyst.

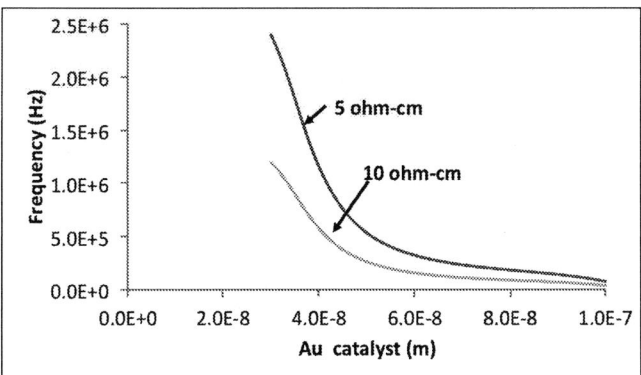

Fig. 5: Change in cut-off frequency with respect to Au catalyst size.

Cut-off frequency, the frequency above which the capacitor performance degrades, is an important factor in power supply design. Using Equations (13-15), Fig. 4 and Fig. 5 show the cut-off frequency with changing wire length and catalyst size respectively. The cut-off frequency is inversely proportional to increasing R and C. Hence, with increasing wire length, both resistance and capacitance increase but cut-off frequency decreases.

III. Fabrication

Capacitors were fabricated with Si-nanowires as the bottom electrode, thin-oxide as the dielectric and PEDOT:PSS (poly(4,5 ethylene dioxy thiophene) – poly(styrene sulphonate)) as the top electrode. The nanowires were fabricated using two different techniques: 1. CVD growth and 2. Etching. The oxide was formed using a novel thermal oxidation technique and finally the top electrode was formed dispensing PEDOT:PSS nanosuspension. Next, each of the steps in the fabrication process is explained in detail. A schematic of the entire fabrication process is seen in Fig. 6.

1. Nanowire Fabrication:
The bottom electrode of the capacitor was formed by the nanowires. Nanowire fabrication was carried out using two different methods.

a. CVD Growth:
The experiments to grow Si nanowires were carried out using a CVD furnace. The nanowires were grown on a Si(111) substrate. The substrate was cleaned to remove any natural oxide using 10% hydrofluoric acid. Gold catalyst was then deposited using a colloidal gold suspension. This allows for the Au to be distributed on the substrate. Gold particles with various diameters were used as the gold catalyst. It was seen that the 50 nm particles consistently produced nanowires. The substrate with gold catalyst was placed in the CVD chamber. The temperature was maintained between 410 – 430 °C. A mixture of silane gas (15%) and hydrogen gas (85%) (carrier) was used. The process was run for 10 minutes. The nanowires grown from these experiments are completely random in orientation. More control over the process parameters can yield extremely directional wires. For capacitors, the random orientation of the wires can be an advantage as it provides considerably more surface area than vertical wires and increases the overall capacitance density. Hence, there was no further effort to improve the orientation of the nanowires.

b. Etching:
An alternative approach based on etching process was also utilized to form the Si nanowires. The etching process uses Au as the catalyst, which is then patterned into nanoscale islands using e-beam nanolithography. It was seen that the gold was more stable compared to colloidal gold catalyst used for the CVD process. It did not move and diffuse into other gold droplets. The etching solutions consisted of a mixture of hydrofluoric acid and hydrogen peroxide. The Si was etched only in locations were Au catalyst was present, leaving the Si around it unaffected. This allowed for extremely vertical Si nanowires. This process requires more control than the CVD process as any change in the etching mixture could result in unwanted etching and pit formation on the Si surface. This is highly undesirable when trying to achieve high aspect ratio structures.

2. Oxidation:
The next step in the capacitor fabrication process after nanowire fabrication is the formation of the dielectric. Silicon oxide has the highest electric breakdown strength because of its large bandgap, making it withstand sufficient voltages even when thinned down to 10s of nm. It is also important to note that the deposition of high-k dielectrics such as barium strontium titanate in high-aspect ratio structures such as these nanowires would be extremely challenging. Hence, silicon oxide was the dielectric of choice for these capacitors using a novel technique for low-temperature (<500 C) oxidation.

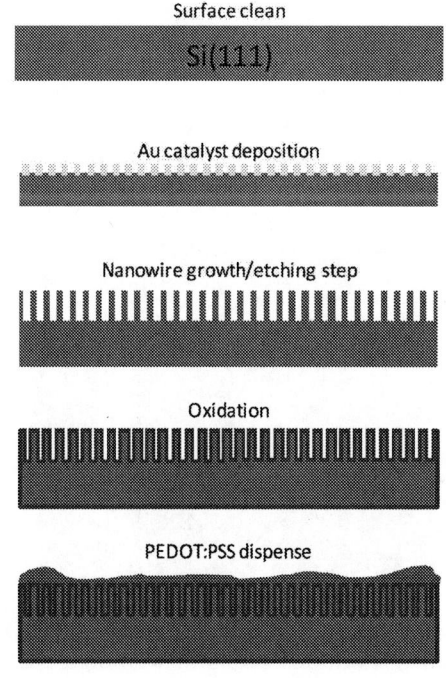

Fig. 6: Fabrication process for capacitor formation starting from surface preparation to top-electrode formation.

3. Top Electrode using PEDOT:PSS

Conducting polymer (PEDOT:PSS – poly(4,5 ethylene dioxy thiophene) – poly(styrene sulphonate)) was used as the cathode for the silicon-nanowire systems. PEDOT:PSS is the most widely used conducting polymer in the capacitor industry. PEDOT:PSS has a high conductivity (~600 S/cm), which results in extremely low ESR compared to other cathodes such as MnO2 and is self-healing in nature. Self-healing in PEDOT:PSS leads to burning of the cathode next to a defect site in dielectric, thus, preventing a short between the cathode and the anode. This also helps in minimizing the leakage current.

IV. Characterization

The morphology of the nanowires was studied using scanning electron microscopy (SEM). The oxide was studied using SEM and X-ray photon spectroscopy (XPS). The capacitor was tested using a liquid electrolyte as the cathode. Each step is briefly discussed below.

a. Morphology of Si-Nanowires

The Si-nanowire samples were characterized using scanning electron microscopy (SEM) to study the morphology. It was seen that nanowires that were grown using the VLS (Vapor-Liquid-Solid) technique in the CVD chamber were randomly oriented. This is desirable as the resultant porous structure allows for extremely high surface area that can lead to very high capacitance density. SEM image of the randomly-oriented nanowires is shown in Fig. 7. The length of the nanowires was measured to be around 3 – 3.5 micron, with a Au catalyst size of 50 nm. The analytical modeling shows that the length should be around 5 micron. The limited growth is attributed to the small processing times used (< 10 minutes). Nanowires formed using the etching processes were also characterized, as shown in Fig. 8. It was seen that the wires in this case were completely vertical. The length of all the wires was almost the same, indicating minimal standard deviation. In case of the etching process, the length of the wires was highly dependent on the etching times.

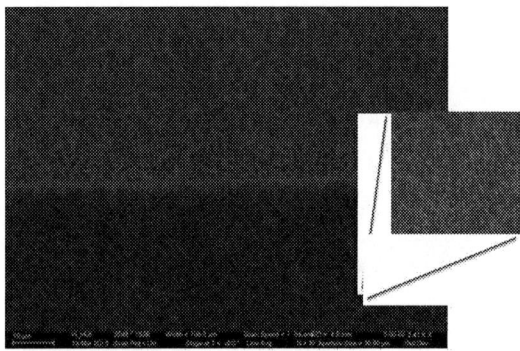

Fig 7. Morphology of Si-nanowires using FE-SEM. Image shows randomly orientated nanowires grown using CVD.

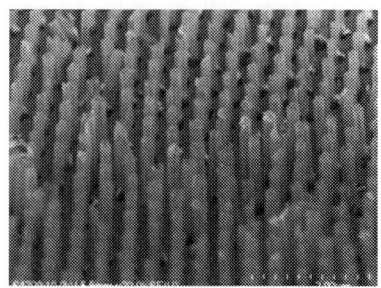

Fig 8. Morphology of Si-nanowires using FE-SEM. Image shows extremely vertical nanowires formed using an etching process.

b. Capacitor Characterization:

The fabricated devices were characterized for capacitance and leakage currents. For measuring the ideal capacitance density of the device utilizing all the electrode area, liquid electrolyte testing with a set-up shown in Fig. 9 was used. The silicon nanowire capacitor was used as the anode whereas sulfuric acid with a concentration of 0.5 M was used as the electrolyte, in addition to working as a cathode. The presence of an electrolyte leads to additional capacitance due to the formation of the electrochemical double layer at the oxide-electrolyte interface. A schematic of the interfaces involved in the capacitor system is presented in Fig. 9 to illustrate the capacitance contribution from the silicon oxide and the electrochemical double layer. As seen from the figure, the capacitance contribution from the double layer and the oxide are in series. With such series capacitors, the smaller capacitor dominates the capacitance. The net capacitance for this capacitor series is therefore determined by the contribution from the nanowire because the double-layer capacitance is much higher. The capacitance is calculated as:

$$C = \frac{A\varepsilon\varepsilon_r}{t}$$

(16)

where A = Surface area of the electrode, ε = Permittivity in vacuum, ε_r = Relative permittivity of the dielectric, and t = thickness of the dielectric.

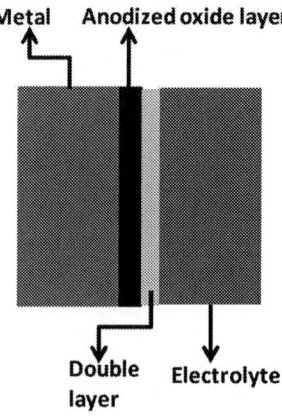

Fig. 9 Interface involving two dielectrics in the wet-system *(metal oxide and electrochemical double layer)* [12].

Assuming the surface area (A) of the double-layer and oxide capacitors to be the same, ε_{r1} and t_1 = permittivity and thickness of the oxide layer, ε_{r2} and t_2= permittivity and thickness of the double layer, C1 = capacitance from the oxide layer, C2 = capacitance from the double layer;

$$\varepsilon_{r1} \ll \varepsilon_{r2}; \quad t_2 \ll t_1 => C_1 \ll C_2$$

Here, since C1 and C2 are in series,

$$C_{eff} = \frac{(C_1 \times C_2)}{(C_1 + C_2)} \cong C_1 \qquad (17)$$

Hence, the capacitance measurement obtained using this set-up can be approximated as the capacitance from the silicon nanowire electrode. Capacitors with solid cathodes based on PEDOT-PSS were characterized by directly probing the cathode and anode surfaces.

The measurement results showed a maximum capacitance density of 20 $\mu F/cm^2$ with liquid-electrolyte testing. A leakage current of 1 $\mu A/\mu F$ was recorded. When compared to planar silicon oxide capacitors with 50 nm oxide films, this approximates to a 40 X enhancement in surface area in spite of being only 2 microns thin. The nanowire electrodes result in a high volumetric density compared to other thin capacitor approaches. As seen in Fig. s 4 and 5, it is estimated that the capacitors will be limited in frequency performance to lower kHz. This is attributed to the extremely high resistance of the nanowires with test silicon wafers (1-10 ohm cm resistivity). With low-resistivity doped silicon, the conductivity can be further enhanced to improve the operation frequency of these capacitors.

Compared to alternative substrate-compatible high surface area electrode techniques such as trench capacitors or nanoparticle electrodes, the etched nanoelectrode process is much simpler and allows easy scale-up at low cost. The results therefore represent a significant breakthrough and advance in nanocapacitor technologies.

V. Summary

A novel process to achieve ultrahigh-density capacitors was demonstrated using Si nanowire electrodes. An analytical model was developed to predict the length of the nanowire, the capacitance density and frequency-stability that can be achieved. The flatband voltage and the threshold voltage were also calculated. This helps to determine the required voltage for the MOS device to operate in accumulation mode for maximum capacitance density.

Nanowires were formed using two distinct processes: 1). VLS growth technique using CVD and 2) Wet-etching process. SEM was used to characterize the wires. The VLS growth technique produced randomly-oriented wires that are desirable as they enable very high capacitance densities. The etching process, on the other hand, provided vertical nanowires. Thin oxides were grown using a novel thermal oxidation process. Finally, top electrodes were formed using PEDOT:PSS (conducting polymer). A capacitance density of 20 $\mu F/cm^2$ was measured for a 2-micron film, indicating a volumetric capacitance density of 100 $\mu F /mm^3$, 10X higher than other current capacitor technologies. The new,

nanowire-based approach can also be extended to supercapacitors and batteries to enable completely self-powered modules.

REFERENCES:

[1] G. Kumar, *et al.*, "Power delivery network analysis of 3D double-side glass interposers for high bandwidth applications," in *Electronic Components and Technology Conference (ECTC), 2013 IEEE 63rd*, 2013, pp. 1100-1108.

[2] S. Gandhi, *et al.*, "A new approach to power integrity with thinfilm capacitors in 3D IPAC functional module," in *Electronic Components and Technology Conference (ECTC), 2013 IEEE 63rd*, 2013, pp. 1197-1203.

[3] C. K. Chan, *et al.*, "High-performance lithium battery anodes using silicon nanowires," *Nature nanotechnology*, vol. 3, pp. 31-35, 2008.

[4] K. Peng, *et al.*, "Silicon nanowires for rechargeable lithium-ion battery anodes," *Applied Physics Letters*, vol. 93, p. 033105, 2008.

[5] A. I. Hochbaum, *et al.*, "Enhanced thermoelectric performance of rough silicon nanowires," *Nature*, vol. 451, pp. 163-167, 2008.

[6] B. Tian, *et al.*, "Coaxial silicon nanowires as solar cells and nanoelectronic power sources," *Nature*, vol. 449, pp. 885-889, 2007.

[7] S. Hofmann, *et al.*, "Gold catalyzed growth of silicon nanowires by plasma enhanced chemical vapor deposition," *Journal of Applied Physics*, vol. 94, pp. 6005-6012, 2003.

[8] J. Hannon, *et al.*, "The influence of the surface migration of gold on the growth of silicon nanowires," *Nature*, vol. 440, pp. 69-71, 2006.

[9] R. V. Jones (2002, January). Electronic Devices and Circuits. Harvard University. [Online]. Available: http://people.seas.harvard.edu/~jones/es154/lectures/lecture_4/pdfs/6.152J.FT01.MOSCap01.pdf

[10] S. M. Sze, and K .K. Ng. Physics of semiconductor devices. Hoboken, NJ:John Wiley & Sons, 2006.

[11] B. V. Zeghbroeck,(2004, December). Principles of semiconductor devices. Colorado University. [Online]. Available: http://ecee.colorado.edu/~bart/book/book/chapter6/ch6_3.htm

[12] P. Chakraborti, *et al.*, "High-density capacitors with conformal high-k dielectrics on etched-metal foils," in *Electronic Components and Technology Conference (ECTC), 2012 IEEE 62nd*, 2012, pp. 1640-1643.

Development of a High Density Glass Interposer Based on Wafer Level Packaging Technologies

Michael Töpper[1], Markus Wöhrman[2], Lars Brusberg[1], Nils Jürgensen[1], Ivan Ndip[1], Klaus-Dieter Lang[2]

[1]Fraunhofer Research Institution for Reliability and Microintegration IZM, Berlin, Germany

[2]Technische Universität Berlin, Berlin, Germany

Michael.toepper@izm.fraunhofer.de

Abstract

Currently glass is mainly used as unstructured wafers or panels with the highest market share in glass capping applications. Higher functionality in glass is driven by the applications in RF and Photonics. Since the technologies of via interconnects in Si and glass are completely different, it is challenging to perform a direct and fair comparison. Mainly laser technology and electrical discharge are used for forming the vias into the glass. Slightly modified thin film technologies already in mass production in WLP can be used to fill the vias with a copper metallization. Conformal metallization and full via plating are options. High yield and excellent reliability have been achieved. Generally, due to the lossy nature of silicon and complex polarization mechanism that occurs at the Si-SiO2, TSVs may suffer from severe signal integrity and EMI problems such as huge insertion loss, delay and cross-talk, depending on the Si-resistivity considered. Therefore, regarding the dielectric material, TGVs have significant advantages over TSV, especially when either LRS or MRS is used. In summary TGVs show excellent RF characteristics over TSVs. This has been proven for a test design up to 40 GHz.

Introduction

The importance of glass for the microelectronic industry is strongly growing. Currently glass is mainly used as unstructured wafers or panels. It will migrate to higher functionality using electrical through hole interconnects (TGV, Through Glass Vias) and other additional structures. One major aspect for the material and equipment suppliers is the much higher revenue which can be achieved. Blank 300 mm wafers are sold for around 30 USD per wafer. This can go up to 100 USD if cavities or channels are structured into the glass reaching a very high value of 400 USD if TGVs are added [1]. The overall glass substrate market size is proposed by Yole to grow 24% CAGR.

Currently the main application is glass capping for MEMS and MOEMS applications. Starting in 2015 a major part will be going to glass carriers with and without TGV. Therefore the formation of the glass vias and the filling of these through holes with a conductive material are of tremendous importance. Most of the application area is covered by the so-called interposer technology. Interposer are substrates with high density electrical wiring which differs therefore from PWB (printed wiring boards). The origin of this technology is in the MCM business coming up 20+ years ago for military application. Basic function is a high speed interconnection between FC-bonded ICs. Basic materials have been Silicon, ceramic, and laminates for MCMs (called MCM-D, MCM-C and MCM-L). In the early days through holes technology was only an option for multilayer ceramics. This has changed for

the laminate technology being now the major IC substrate for high pin-count IC like microprocessors. Further miniaturization and higher performance in the front-end technology is further pushing interposer technology. Therefore silicon and glass are now being prime candidates for interposers due to excellent planarity for highest routing density and matched CTE to the ICs. This paper will focus on TGV (through glass vias) being a very important process step for a successful implementation of glass as an interposer technology with a focus on the RF performance

A Comparative Analysis at RF/Microwave Frequencies of Through Glass Vias and Through Silicon Vias

In recent years, enormous research effort has been done by academia and industry to understand and optimize the performance of Through Silicon Vias (TSVs) and Through Glass Vias (TGVs) at RF/microwave frequencies. Especially in the frequency range above 1 GHz the material of the interposer is very important for high performance. Since wave propagation along TSVs and TGVs occurs predominantly in the dielectric between the interconnects, these dielectrics play a significant role in the RF properties of the interconnects.

Since glass is not a semiconductor, and it is not as lossy as silicon, it is expected that the impact of the dielectric materials of glass on TGVs will not be as severe as the impact of silicon on TSVs. As a result of the fact the alkaline content of glass materials are different, it is expected that TGVs will have changes in the RF performances when different glass materials are used. If the geometrical dimensions of the TGVs are not optimized considering different glass materials, then severe signal integrity problems may occur. The via configurations in glass substrates have excellent signal transmission characteristics up to 100 GHz. Simulation has shown that approximately less than 1.5% of signal power propagating through the TGVs is lost for frequencies up to 10 GHz. Less than 3% is lost up to 60 GHz and less than 5% is lost up to 100 GHz [2].

It can be concluded that when the same conductor is used for fabricating TGVs in different glass substrates (having the same relative permeabilities), and the geometrical configuration of the TGVs is the same, then there is practically no difference in the resistances and inductances of the TGVs. The main difference is in the capacitance and the insertion loss of the TGVs. The higher the ε_r value of the glass substrate, the higher is its capacitance and higher its insertion loss. The higher the $\tan \delta$ value of the glass is, the higher its insertion loss will be. For high-speed signal transmission, it is recommended that glasses with smaller ε_r and $\tan \delta$ should be used, whenever possible [2]. Example of glass properties of the ENA 1 Glass from AGC are given in Figure 1:

Properties			ENA1
Thermal	Thermal Expansion Coefficient	ppm/°C	3.8
	Thermal Conductivity	W/mK	1.0
	Softening Point	°C	950
	Strain Point	°C	720
Optical	Refractive Index		1.52
Mechanical	Density	g/cm³	2.51
	Young's Modulus	GPa	77
	Poisson's Ratio		0.24
Electrical	Bulk Resistivity	log(Ω·cm)	13.6
	Dielectric Constant	at 1MHz	5.8
		at 10GHz	5.4
	Loss Tangent	at 1Mhz	0.002
		at 10GHz	0.005

Figure 1: Glass Material Properties of ENA Glass from AGC [11]

The impact of the radius and pitch of TGVs on their electrical performance, has been investigated in a previous ECTC publication [3].

A test layout has been designed to measure the high frequency performance of the TGV together with via chains to measure the yield of the through substrate metallization (Figure 2).

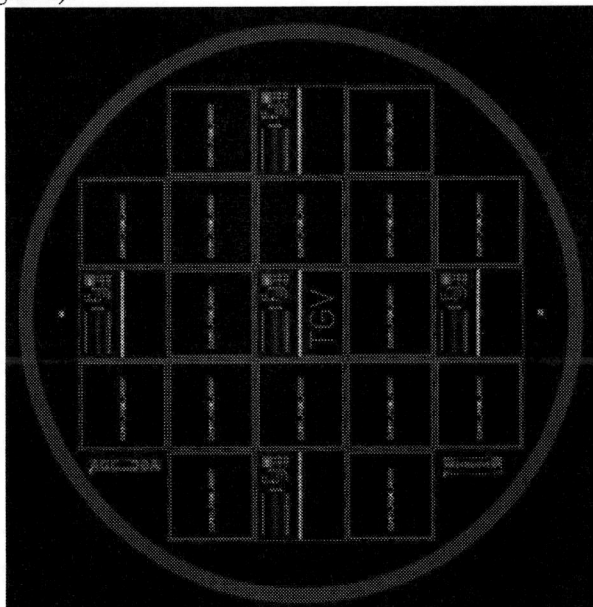

Figure 2: Test layout for TGV

Different glass via drilling technologies and development status:

Different laser options

The major driver for Si with TSVs was the existing via Si etching technology based on the BOSCH-process. This process has been successfully transferred to the VIA etching process for Si. Such a DRIE process is not yet working for glass due to the much slower etch rate of SiO_2 compared to Si.

The BOSCH process has its origins in the MEMS industry for the fabrication of different kind of sensors and was picked up by the packaging industry [4].

Laser via drilling and the resulting quality are very dependent on the glass type and laser system characteristics. If the glass is not transparent for the incident laser beam, the beam generates energy direct on the glass surface (surface absorption). If the glass is transparent for the laser beam, the laser has to be focused on top or bottom side of the substrate. The laser beam induces heat which changes the material by melting and evaporation. Due to mechanical stress in the glass substrate material cracks out. Vias can be drilled by moving the laser beam focus from bottom to top side. By overcoming a certain energy threshold level of the material a direct phase change from solid to vapor aggregate state occurs, in other words a direct material sublimation. The threshold energy is defined by the fluence (incident energy per irradiated area) of the laser beam. The ablation process can be divided into three main ablation mechanisms: Cold ablation, hot ablation and indirect ablation through thermic or pressure shocks. Hot ablation is mostly encouraged by continues wave or quasi continues wave lasers. The glass melts through the laser energy and evaporates. Recast layers and a large heat affected zone around the vias are formed during this process. Cold ablation needs short (ns) or ultra-short (ps/fs) laser beam pulse length and not a match of laser wavelength and glass absorption spectrum. In summary the ablation result is very dependent on laser wavelength, laser energy, laser beam pulse length, threshold fluence and glass type.

A laser system for glass via drilling has to be configured in a way that laser source and beam shaping optics support the targeted ablation process. Vias can be drilled from incident side. If the glass is transparent for the laser wavelength also starting the drilling from bottom side is feasible for improved material evacuation e.g. using vacuum cleaners.

The resulting via shape and dimension are very dependent on the laser process parameters as well as the position of the focus point during the drilling process.

UV excimer laser ablation is based on the production of plasma, and evaporation as a result of the laser radiation. The necessary threshold fluence depends on the glass- and wavelength-specific absorption characteristic.

Solid State Lasers can be either q-switched diode pumped solid state lasers (DPSSL) or lamp pumped lasers. Ablation occurs through multi-photon absorption or other non-linear effects. At low wavelengths and with specific glasses, direct absorption and thus glass evaporation might occur.

Short or ultra-short pulses (ps, fs) are necessary to avoid volume absorption and the resulting large heat affected zone. Because of the transparency of the glass in the working wavelength of Solid State Lasers, both bottom-up and top-down drilling are possible.

The drilling speed is largely defined through the laser fluence, pulse duration and repetition rate. The threshold for each individual glass has to be exceeded in order to achieve the cold ablation. The improvement of short-puls lasers make it more and more interesting for via drilling, because of excellent drilling results and improved drilling speed.

Carbon dioxide lasers offer low investment and maintenance costs. For the ablation of glass only the most common wavelength of 10600nm has been evaluated.

For cw and quasi-cw, the TGV process suffers from the large heat affected zone and the accompanying effects like tension, recast layers and cracks in the glass. Because of the induced mechanical stress a post-annealing around Tg is necessary for improving reliability.

So far the CO_2-Laser drilling is the fastest drilling technology for single shot processes compared with Excimer and Solid-State Lasers. The research goal is to have a crack-free drilling in combination with post thermal annealing. Another improvement could be a short-pulsed CO_2 laser.

Discharge for Via Formation

The electrical discharge process has been developed by the Asahi Glass Corporation (AGC). Mainly it consists of two steps that focused and controlled electrical discharging created locally molten region of glass, and finally it induced dielectric breakdown together with internal high pressure by Joule heat and ejection of glass. The process is applicable for a lot of types of glass such as fused silica, soda-lime glass, alkali-free glass, alkali-containing glass [5].

The process is maskless and it allows a high throughput by applying multi-head systems. Via formation time is between 200 to 500 ms depending on the glass thickness. It is possible to apply large size glass substrate. A aspect ratio of 10 is reported which allow relative small vias for thick substrates. An example is given in Figure 3.

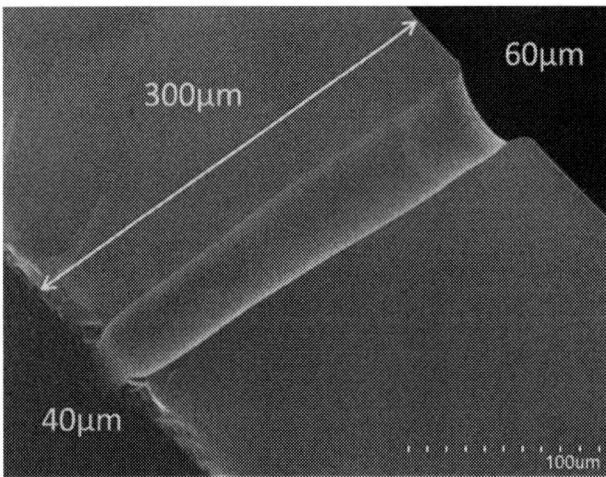

Figure 3: Example of a TGV processed by electrical discharge

Other Technologies for Via Formation

Other technologies like mechanical drilling (micro ultrasonic machining) or abrasive drilling are limited in the resolution (above 200 μm) and are not an option for TGVs.

Wet etching is mostly used in production for applications like MEMS caps, glass spacers for wafer-level optics or fluidic channels.

Photo-structurable glass (PSG) and glass ceramic (PSGC) like Foturan have been manufactured by Schott Glass Corp and distributed by Invenios. Similar glasses are available from Corning (Fotoform), from Hoya (PEG-3) and Life Bioscience Inc./ 3D Glass Solutions (APEX). The PSG is exposed using mask aligner or laser followed by a high temperature treatment (600°C). The UV-light induces crystallization in the exposed area which have a much higher etch rate in HF-solutions. Based on the etching process the surface and the via sidewall are relative rough. The Foturan has a relatively high CTE in the range of 8 to 10 ppm/K. The costs for the substrate material are in general much higher than normal glass substrates and are still not an option for TGVs.

Glass hole filling technologies

The different technologies could be divided into two main parts. One category is the sequential opening and filling of the glass via which are done by different technics and the other part are the hermetic sealed TGV creation in more or less one step. A hermetic sealed TGV technology using metal as a conductive material for the wiring through the glass has been published by Schott/NEC named HERMES [6] The approach from Schott is an additive technology: Glass is melted over W-plugs or FeNi-plugs which eliminate all drilling and filling processes. The metal vias are made of tungsten that can be 100 μm in diameter with a spacing of 250 μm. Finer pitch is under development. A process of PlanOptik is using highly doped silicon inside borosilicate glass which is also done in an additive technology. Through Glass Vias with doped silicon inside are in application for MEMS production from Murata/ VTI. Tecnisco provides also glass substrates with tungsten- or doped silicon-plugs. All of these additive technologies are focusing on MEMS application which has only a few I/O's.

Paste filling

Paste printing of conductive lines is typical used on ceramic substrates, silicon solar cell or printed circuit boards. The substrate is covered by a stencil where the paste is applied by blade. The paste is fired after printing.

Through glass via filling by conductive paste filling has been demonstrated for electronic applications by AGC & nMode and further research at universities [7]. The through silicon via filling by conductive paste is in production for silicon solar cells, where the printing process was improved. The typical via for solar cell application has a diameter of 100 μm and aspect ratio of 2.5. Conductive paste contains metal powder as the conductive ingredient in an organic binder or plasticizer and inorganic additives, normally low melting glass and reactive oxides are used. The inorganic additives are used for matching the CTE to the substrate and creating a bonding contact to the substrate.

Filling by electrical copper deposition

The copper electroplating is a standard process for copper filling of through silicon vias (TSV). Optimized electrolytes are used which allow a bottom fill up of blind vias. The technology is adopted from the well-known RDL formation processing: Copper is electroplated into photoresist openings followed by seed layer etching and polymer passivation. The TSV manufacturing process has to be cost effective and CMOS compatible and comprise a void-free metal filling step. On one hand, the via filling depth is limited to a specific

978-1-4799-2408-0/14 $31.00 © 2014 IEEE   1500

aspect ratio depending on the filling process, on the other hand, process filling time is shorter for smaller vias. This set the demand for small diameter vias and thin substrates for the silicon interposer applications aside of a high interconnect density. The plating time for a substrate with 20 μm diameter vias with a depth of 100 μm is around 100 min-200 min.

To achieve a defect-free via fill, the copper deposition rate at the via bottom has to be higher than at its top and aperture area. Primary and tertiary current distribution do counteract this "super-conformal" or "bottom-up" growth, on the other hand suitable electrolyte additive chemistries and/or pulse reverse plating current profiles do promote a void-less fill.

There are a couple of bottom-up filling electrolytes commercial availability from companies like Dow Chemical, Atotech, BASF, DuPont and Enthone.

For the metallization by electroplating a seed layer of copper or other conductive material must be plated or coated onto the side walls. This is usually done with modified sputtering processes which used ionized plasma. A combination of thin adhesion layer based on titan or titan/tungsten and a thin copper layer are mostly used. This mature technology produce reliable interconnects. The sputter tools are relative costly. Therefore alternative methods that may offer improved cost and environmental performance are emerging to challenge the existing practice. Based on the experience from the printed circuit process electroless copper has been demonstrated for seed layer deposition [8]. None of the chemistries present in the electroless copper line are particularly expensive (with the possible exception of the palladium-based catalyst), but the typical production line is space consuming (17 or more tanks, depending on rinse configurations) and may have 8 or more process cells.

The bottom-up copper electroplating process from one side is adopted on the silicon via creation which is done from front side by Bosch-process. After the filling the silicon substrate is thinned from the backside till the copper vias are open. This process flow is not applicable for glass substrate because the thinning of glass is not practicable. In addition there are through substrate vias in glass instead of blind vias in silicon substrates. The through glass vias set the potential of alternative electroplating process.

It is possible to realize a filling-up process by contacting the substrate from the backside. The process flow is shown in Figure 4. First step is a Cu seed layer creation only on the surface of one side. The seed layer will not cover the whole side wall of the via. The copper layer is enforced by electroplating until the via is closed near the surface. The wafer is now flipped to the backside in the plating tool. In addition the contact is done from the backside. The deposition of Cu by electroplating leads now to a filling-up effect. There is a low risk of trapping voids in contrast to typical TSV electroplating process and higher aspect ratios are possible. After deposition a CMP is done to remove the overburden. This process reduce the effort for the seed layer sputtering because no side wall coating in the via is necessary. Furthermore a simple electrolyte with the potential of a higher deposition rate could be used, which lead to a cost effective filling of vias with larger diameters. The absence of a seed layer in the via sidewall could result in adhesion issues which might be overcome by mechanical dominated adhesion by

rough surface or trapezoidal via shape. Hoya seems to use a process like this for glass via filling with copper.

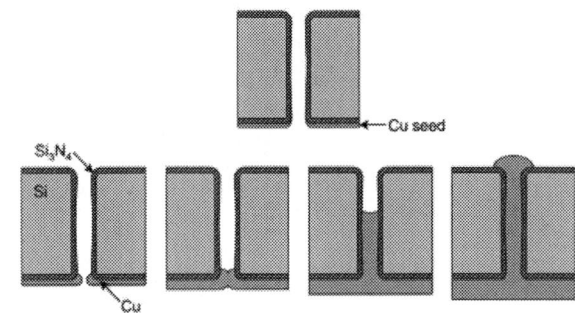

Figure 4: Electroplating of through vias by backside contacting [9]

A through via has the potential of a double side plating process, which is usual done for PWB. The process flow is shown in Figure 5. The starting process step is a through via which is conformal coated by a seed layer. After this the substrate is plated simultaneously from both sides. The electrolyte contains additives which suppress the deposition on the top and rear side of the substrate. The copper growing is stronger in the via which leads to a via closing followed by a bottom-up filling from both sides of the via. The double side processing allows a high aspect ratio and a high deposition speed.

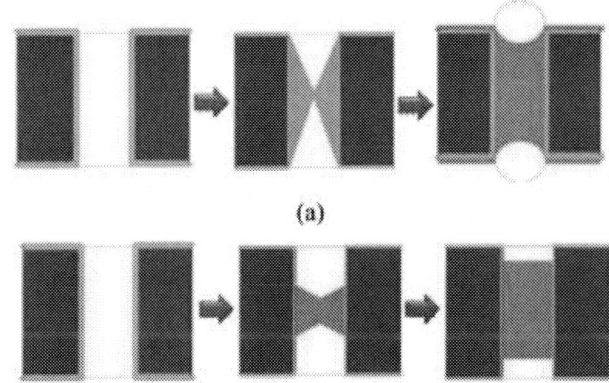

Figure 5: Process flow of a through via double side electroplating [10]

The demands for the via filling process/ technology depends strongly on the shape of the via. There are two main options: Blind holes or through hole which is opened on both sides.

The established silicon based interposer technology use only blind holes. There are a couple of reasons for that: The technology is based on well-established processes like the Bosch process, sputtering and galvanic copper deposition which are advanced or reconfigured. Especially the galvanic process was strongly improved to realize a bottom-up filling of the hole with high aspect ratios in substrate. These special electrolytes have often a relative low deposition rate. The creation of the redistribution layer on the substrate surface is realized in a sequential process for the blind via condition. This allows an easier handling process because the opposite side is covered during the process. Also standard equipment in

978-1-4799-2408-0/14 $31.00 © 2014 IEEE   1501

production works in general only with single side processing. After the filling process from one side the silicon substrate is thinned by grinding from the backside which also contains the opening of the vias from the back side. The single side filling results in relative high production costs and further limitations for the substrate thickness and via diameters if standard TSV-Processes are used.

The filling of through holes could be attractive for glass based substrate because of alternative via forming technologies. Through holes are well applied for PCB's the plating being done from both sides. This allows a higher aspect ratio and reduces the processing time. The double side processing set higher handling demands for following process steps. The filling of through holes has a much higher potential for lowering the costs for the through substrate vias.

Example of TGV filled by a conformal metallization (left) and full via filling (right) both by electroplating are shown in Figure 6.

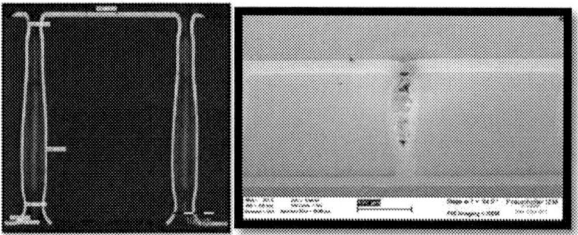

Figure 6: Cross-cuts of TGVs filled by a conformal metallization (left) and full via filling (right) both by electroplating (glass wafer from AGC)

The conformal filling has been achieved by sputtering an adhesion and seed layer of Ti:W and Cu followed by an electroplating process. A different electrolyte has been used for the full via plating. A slight optimization has to be done to avoid any kind of voids because the via shape is different to TSV the electrolyte has been developed for.

The yield and the reliability has been measured by via chains (Figure 7 and Figure 8):

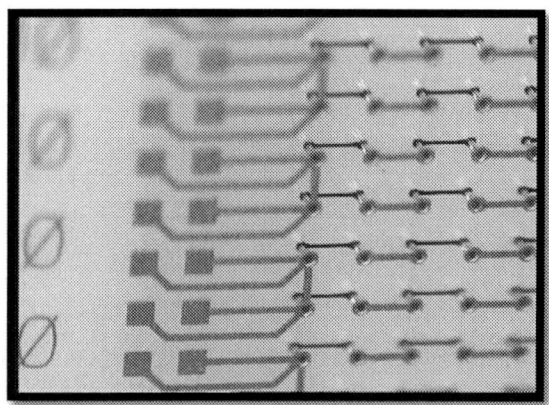

Figure 7: Via chains of TGVs

A 3 step cycling (-55°C ; room temperature ; 125°C) has been performed for the reliability test. The measurement of daisy-chain is based on 100 vias. Only 4 of 250 measured lines have an open circuit. 3 daisy-chains are measured without significant changes, which consist of 5000 vias. The reliability tests are ongoing.

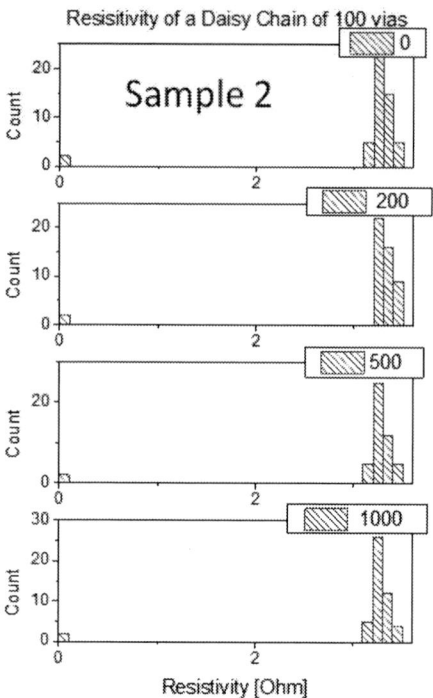

Figure 8: Reliability of conformal metallized TGVs (up to 1000 cycles AATC)

The electrical resistivity of single vias has been measured using 4-point Kelvin-type structures. Values between 14.5 and 19 mOhm per vias have been measured for the conformal metallization and 4.5 mOhm per via for the full plated vias.

Glass Interposer for RF

The same wafers from AGC metallized by a conformal Cu plating process has been used for a demonstrating the RF performance. The GSG-pads on one side of the glass wafers are shown in Figure 9 and the measurement of S11 and S12 up to 40 GHz on Figure 10 and Figure 11:

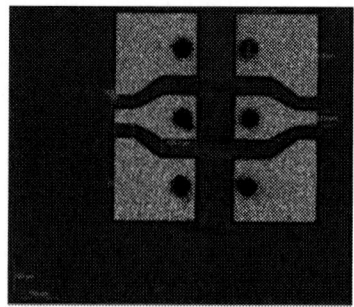

Figure 9: GSG pads on the glass wafer with TGVs

The results proofs the excellent performance of TGVs.

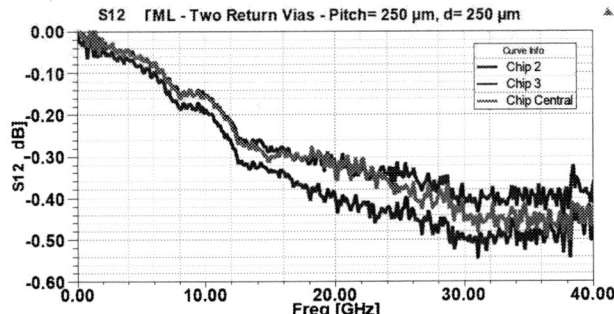

Figure 10: Measurement of S12 till 40 GHz (AGC ENA1 wafers conformal metallized with Cu structured Cu metallization on both sides)

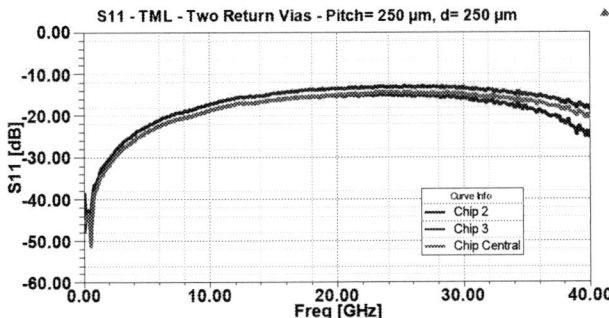

Figure 11: Measurement of S11 till 40 GHz (AGC ENA1 wafers conformal metallized with Cu, structured Cu metallization on both sides)

Summary and Conclusion: Comparison of TGVs and TSVs

Since the technologies of both via interconnects are completely different, it is challenging to perform a direct and fair comparison. Generally, due to the lossy nature of silicon and complex polarization mechanism that occurs at the Si-SiO₂, TSVs may suffer from severe signal integrity and EMI problems such as huge insertion loss, delay and cross-talk, depending on the Si-resistivity considered. Therefore, regarding the dielectric material, TGVs have significant advantages over TSV, especially when either LRS or MRS is used. 3D full-wave electromagnetic field simulations using HFSS has been published [12]. As can be seen in Figure 12 TGVs show excellent RF characteristics compared to TSVs, considering all three silicon resistivities. Glass wafers ENA 1 with TGV done by electrical discharge have been used for the evaluation till 40 GHz. The metallization process was based on processes widely used in WLP: Sputtering TiW/Cu and electroplating Cu. Major difference is in the optimized electrolyte for the through substrate metallization process. In addition both sides of the wafer have to be processed. 1000 cycles AATC have been based without any change of the electrical resistance.

Acknowledgments

The company AGC is greatly acknowledged for supporting this project

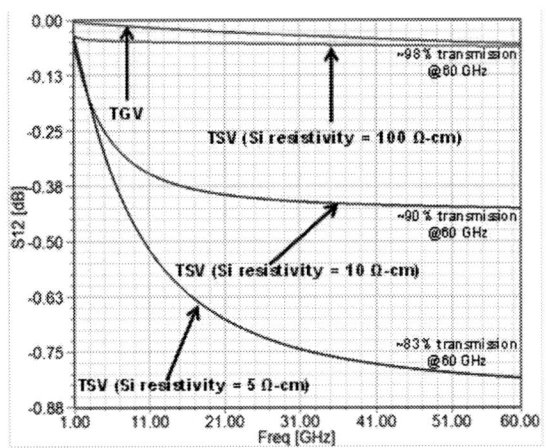

Figure 12: Comparison of insertion losses of GS configurations of TSVs and TGVs

References

[1] Yole, "Glass Substrates for Semiconductor Manufacturing," 2013

[2] I. Ndip, M. Töpper, K. Löbicke, A. Öz, S. Guttowski, H. Reichl and K.-D. Lang, "Characterization of Interconnects and RF Components on Glass Interposers," in *In Proc. 45th International Symposium on Microelectronis (IMAPS 2012)*, San Diego, CA, USA, 2012

[3] M. Töpper, I. Ndip, R. Erxleben, L. Brusberg, N. Nissen, H. Schröder, H. Yamamoto, G. Todt and H. Reichl, "3-D Thin Film Interposer Based on TGV (Through Glass Vias): An Alternative to Si-Interposer," ECTC 2010

[4] M. Wilke, M. Töpper, H. Quoc Huynh, K.-D. Lang „Process Modeling of dry etching for the 3D-Integration with tapered TSVs" Procedings ECTC 2012

[5] S. Takahashi, K. Horiuchi, K. Tatsukoshi, M. Ono, N. Imajo and T. Mobeley, "Development of Through Glass Via (TGV) Formation Technology Using Electrical Discharging for 2.5/3D Integrated Packaging," AGC & nMode, 2013

[6] M. Töpper, H. Schröder, L. Brusberg, H. Yamamoto, G. Todt, H. Reichl "3-D Thin film Interposer based on TGV (Through Glass Vias): An Alternative to Si-Interposer" Proceedings ECTC 2010, Las Vegas 2010

[7] S. Takahashi, K. Horiuchi, K. Tatsukoshi, M. Ono, N. Imajo and T. Mobeley, "Development of Through Glass Via (TGV) Formation Technology Using Electrical Discharging for 2.5/3D Integrated Packaging," AGC & nMode, 2013

[8] S. Bamberg, M. Merschy, T. Bernhard, F. Bruening, R. Taylor „Challenges of Adhesion Promotion for the Metallization of Glass Interposer" Int. Symp. IMAPS 2013, USA

[9] J. H. Wu, J. Scholvin and J. A. d. Alamo, "A Through-Wafer Interconnect in Silicon for RFICs," ECTC, 2004

[10] J.-J. Yan, Y.-T. Lin and W.-P. Dow, "A Novel Cu Plating Formula for Filling Through Holes," in 224th ECS Meeting, 2013

[11] N. Imajyo "Path to the Future of Glass Interposer" GIT 2013, Nov. 2013, Atlanta, USA

[12] I. Ndip and M. Töpper, "High-Frequency Modeling and Optimization of Interconnects in Electronic Packaging," in *Professional Development Course (PDC) ECTC 2013*, Las Vegas; NV; USA, 2013

978-1-4799-2408-0/14 $31.00 © 2014 IEEE

Novel Sealing Technology for Organic EL Display and Lighting by Means of Modified Surface Activated Bonding Method

Takashi Matsumae, Masahisa Fujino, Tadatomo Suga
The Department of Precision Engineering, The School of Engineering, The University of Tokyo.
Hongo 7-3-1, Bunkyo-ku, 113-8656 Tokyo, Japan.
Mail: matsumae.takashi@su.t.u-tokyo.ac.jp Phone: +81-3-5841-6495

Abstract

For a sealing of organic electro luminescence displays and lightings, a room temperature bonding of polymer films without organic adhesives is required. This paper describes a new bonding technique called modified surface activated bonding (mSAB) method using nano-adhesion layer that meets the requirement. Polymer films such as PEN, PET and PI can be bonded by this method. Especially PEN and PET films are so strongly bonded that the bond interface has tolerance for bending and loci of fracture after a peeling test are located in the polymer bulk. The organic electro luminescence lighting bonded by the proposed method clears environmental tests such as high temperature and high humidity tests.

Introduction

Organic electro-luminescent display (OELD) and lighting have attracted great interest because of their advantages such as thin profile, lightweight, and low driving voltage. However moisture and oxygen that may permeate the OLED cause detachment between the organic luminescent layer and the cathode electrode, cracking of the organic materials, and oxidation of the electrodes. Therefore, the commercial OELDs are made of glass substrate and the substrates are sealed by glass frit bonding. However increasing the size of the OELD, the process of glass frit bonding will be hardly adopted because of the brittleness of the glass sealing. The other potential applications of OELD are flexible displays and lightings, which are fabricated on polymer substrates. For those applications, polymer films with high gas barrier characteristics such as polyethylene naphthalate (PEN) have been developed. However there is no suitable adhesive for good sealing agent against permeation of water and oxygen from the atmosphere. Although a thermal compression bonding method realizes bonding without organic adhesives, polymer films and components of organic electronic are easily damaged at annealing process. Therefore, a new bonding technique of polymer films at room temperature without organic adhesives is required for the application of organic electronics.

Surface activated bonding (SAB) method is a promising candidate for the room temperature bonding without organic adhesive[1]-[3]. In the SAB method, contamination and native oxide layer on the material surface are removed by Ar ion beam bombardment in an ultra-high vacuum (UHV) pressure. The activated surface becomes reactive to form chemical bond even at room temperature. It is reported that various materials can be bonded by SAB method, such as various metals[4], [5] and semiconductor materials[6], [7] at room temperature.

Recently, a new modified SAB (mSAB) method was developed for bonding ionic materials that have been hardly bonded by the conventional SAB method. This is because the surface of ionic materials becomes inhomogeneous polarized after the surface cleaning. Ionic materials such as SiO_2 and SiN can be bonded by the mSAB method by means of Fe nano-adhesion layer and Si sputter deposited intermediate layers on the bond surface[8], [9]. The Fe atoms make chemical bond to atoms at material surfaces. Moreover, the Fe atoms diffuse into the Si layers. These chemical bonding and diffusion enhance the bonding of ionic materials. This method realizes room temperature bonding with an inorganic bond interface that is suitable for the sealing of water and oxygen proof.

However, it is still challenging to bond polymer films by means of SAB and mSAB. This is because the process of Ar ion beam surface cleaning damages polymer surfaces. Therefore, a further modification of the mSAB process is required for the bonding of polymer films.

The purpose of this study is to develop the mSAB method for the room temperature bonding of polymer films without organic adhesives. Furthermore, the gas barrier property of the interface bonded by the proposed method is evaluated by environmental tests.

Bonding Procedure

Figure 1 shows an overview of the process flow of the mSAB for polymer films. This process consists of three steps; deposition of intermediates layers by ion beam sputtering, post-surface activation and bonding at room temperature. In bond procedure of the previous study about SiO_2/SiO_2 bonding [9], material surfaces are cleaned by Ar ion beam irradiation to remove contamination, and Fe nano-adhesion layer is deposited. However, this step is skipped to avoid surface damage of polymer films by Ar ion beam irradiation.

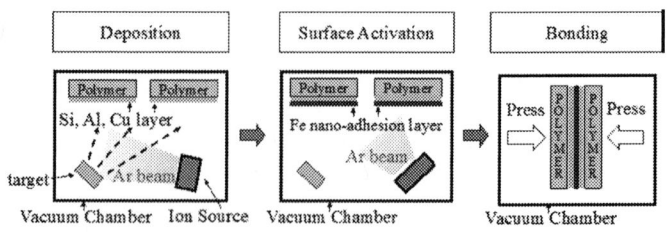

Figure 1. Bonding process of the proposed method. The deposited layer protects polymer surfaces from the damage caused by Ar ion beam.

At the step of deposition of the intermediate layer, intermediate layers are formed on polymer surfaces by means of ion beam sputtering using a sputtering target in the process chamber. The deposited layers are around 10 nm thickness. In this study, materials used for the intermediate layer can be changed to Si, Al or Cu. Next, in the post-surface activation step, the surface of deposited layers are cleaned by Ar-ion beam irradiation, and simultaneously Fe nano-adhesion layer formed. Fe atoms are sputtered from a wall of the ion source that is made from stainless steel by accelerated Ar ion beam. Finally, activated surfaces are brought into contact and pressed to each other. All processes are conducted at room temperature and in ultra high vacuum. Because the polymer surface isn't damaged by the direct ion beam bombardment in this method, it is expected that polymer films can be bonded by this method.

Samples

In this study, bonding experiments of polyethylene naphthalate (PEN), polyethylene terephthalate (PET), polyimide (PI) and glass wafer are conducted. PEN film used in this study is Q65F and PET film is KEL86W. The thickness of these films are 125 µm. The type of PI used in this study is Kapton-100H. Kapton is trademark of DuPont film and Type H is the standard type of Kapton film. The thickness of PI film is 25 µm. The glass wafers used in this study were non-alkali glass from Asahi-glass Co. and the thickness is 500 µm.

Bonding of PEN film

A void-free bond interface is desirable for sealing of organic electronics. Thus, bonding conditions were optimized to bond polymer films without voids. We optimized several parameters at the surface cleaning, deposition, surface activation and bonding load. Table 1 shows conditions we tried and PEN films are bonded without voids at the condition 1. Figure 2 (a) is the picture of the bonded PEN/PEN at the condition 1. There are no visible void at the bonded area and bond interface is enough strong to have tolerance for bending as shown in Figure 2 (b). This result indicates that the proposed method can be applicable to sealing of flexible devices. Following paragraphs describe discussions for each bond parameter.

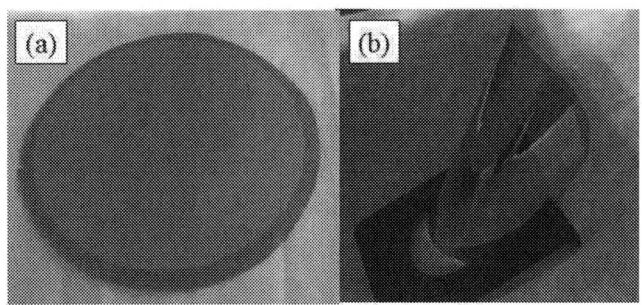

Figure 2. (a) Bonded sample of PEN films using Si deposited layer and Fe nano-adhesion layer. (b) The sample is bendable.

The process of the surface cleaning before the deposition of intermediate layer is required for the bonding of SiO_2 wafers to remove surface contaminations. This process is called pre-surface activation. However, this process causes the surface damage for polymer films by direct Ar ion beam irradiation. This damage is supposed to make an adhesion between the deposited layer and the polymer surface weak. This is why bonding failed at the condition 2. Therefore the bonding without the pre-surface activation process is preferable.

In the bond procedure of the proposed method, the sputter-deposited intermediate layers are formed in order to prevent the damage of polymer surface from Ar ion beam irradiation. The bonding experiments were conducted in several deposition conditions shown in the condition 3-5 in Table 1. The results of bonding experiments in this conditions show that the preferable energy of ion beam during sputter-deposition is 1.2 kV and the suitable process time is 90 sec. The bonding failed at the process of too short or excessive deposition of intermediate layer. The reason for this is the excessive deposition causes rough surface, which is hard to bond by room temperature bonding. The surface roughness profile using Atomic Force Microcopy (AFM) shows that the surface at condition 5 with excessive deposition (RMS ~2.17 nm) becomes rough compared with the surface prepared at optimized condition (RMS ~1.97 nm). This is because deposited Si atoms starts nucleation and Si grains grows. Therefore, moderate deposition is required.

Table 1. Experimental conditions of the bonding of PEN films. PEN films are bonded without voids at the condition 1.

Condition No.	Surface cleaning before deposition	Ion source voltage during Deposition [kV]	Deposition time [sec]	Ion source voltage during Surface Activation [kV]	Surface Activation time [sec]	Bond load [MPa]	Bond time [min]	Pretreatment (80°C, 1 hour)	Void-free bonding
1	**without**	**1.2**	**90**	**1.0**	**30**	**1**	**3**	**with**	**Achieved**
2	**with**	1.2	90	1.0	30	1	3	with	Failed
3	without	**1.5**	90	1.0	30	1	3	with	Failed
4	without	1.2	**45**	1.0	30	1	3	with	Failed
5	without	1.2	**135**	1.0	30	1	3	with	Failed
6	without	1.2	90	**1.2**	30	1	3	with	Failed
7	without	1.2	90	1.0	**without**	1	3	with	Failed
8	without	1.2	90	1.0	**60**	1	3	with	Failed
9	without	1.2	90	1.0	30	**0.7**	3	with	Failed
10	without	1.2	90	1.0	30	1	**1**	with	Failed
11	without	1.2	90	1.0	30	1	3	**without**	Failed

978-1-4799-2408-0/14 $31.00 © 2014 IEEE

Post-surface activation, which is the activation process of the bond interfaces after the deposition of intermediate layer, is required for void-free bonding. Without this process, bonding fails as shown at the condition 7 in Table 1. However, the polymer bonding also fails at the excessive surface activation like condition 6, 8. This is because the excessive Ar ion beam etching makes the bond interface rough. The RMS of the surface roughness prepared at optimized conditions is 1.97 nm. However, the RMS of that prepared at condition 8 with excessive etching increases 2.13 nm due to surface damage by strong Ar surface treatment. Therefore, modulate irradiation at the Post-Surface Activation is suitable for the void-free polymer bonding.

Strong bonding load and long bonding time are preferable to enhance the connection between bond interfaces and reduce voids at the interface. The bond load over 1 MPa and the time over 3 min are required for the void-free bonding of polymer films.

Furthermore, a volatile nature or other moisture absorb on the polymer surface and inside of the film. These absorbates on polymer films deteriorate the adhesion between polymer surface and the deposited layer. Moreover, gases form these absorbates prevent high-vacuum condition. Thus degassing treatment for removing the absorbates is necessary to achieve strong adhesion strength. Before introduce load-lock chamber, PEN films are annealed at 80°C, which is below the grass transition point of PEN (120°C), in 1 hour for degasification. Void-free bonding of PEN films is realized with this pretreatment.

Bonding of PET, PI and Glass

This process is applied to the bonding of PET, PI and the bonding of polymer and glass wafer.

PET films are successfully bonded without void by the proposed method. Figure 3 (a) is the picture of PET/PET bonded using Si deposited intermediate layer. PEN films are also bonded without voids using Al and Cu intermediate layers.

By contrast, PI films are bonded but there are unbonded areas as shown in Figure 3 (b). This figure shows the PI/PI sample that is bonded using the Al intermediate layer and the right side of the sample isn't bonded. Furthermore, the PI films are hardly bonded using the Si deposited intermediate layer.

The bonding of dissimilar polymer films is successfully performed by the proposed method. The pair of PET/PEN is bonded without void and PET/PI and PEN/PI are bonded at almost area as shown in Figure 3 (c). This picture shows the bonded sample of PEN/PI using Al layer. There are small voids at the underside of Figure 3 (c).

Moreover, the proposed method is successfully applied to a bonding of polymer film to glass substrate. The bond procedure is optimized for the void-free bonding of polymer/glass. The pre-surface activation process is necessary for the surface of glass wafer to remove contaminations and deposit the Fe nano-adhesion layer. However, the pre-surface activation process causes the surface damage for polymer films. Therefore, glass wafer is treated at the step of the pre-surface activation. Contrary to this, polymer films are brought into the process chamber after the pre-surface activation step and not treated by this process to avoid the surface damage. All process after the pre-surface activation is the same as the bonding process for polymer/polymer.

Figure 3 (d) shows the ponded sample of PEN/Glass using the Si deposited layers. PEN/Glass, PET/Glass and PI/Glass are bonded without void. The bonded sample using Si deposited layer is almost transparent but little colored due to the deposited layers. By contrast, the bond interface of samples using Al or Cu layers become like a mirror.

Taken together, PEN, PET, PI and Glass are bonded by the proposed method. Especially PEN and PET films can be bonded without void. However, PI films are supposed to be hardly bonded compared with PET and PEN. The results are summarized in Table 2.

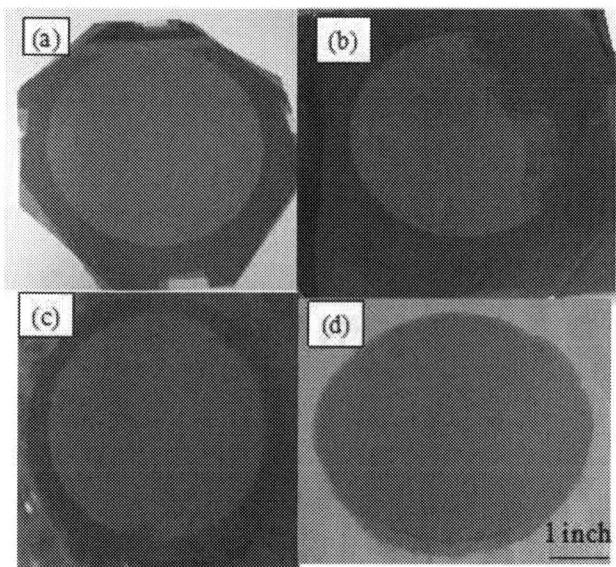

Figure 3. Pictures of bonded samples.
(a) PET/PET using Si layer. (b) PI/PI using Al layer.
(c) PEN/PI using Al layer. (d) PEN/Glass using Si layer.

Table 2. Summary of bonding experiments. PEN, PET films and Glass are successfully bonded.

	PEN	PET	PI	Glass
PEN	⊚ void-free	⊚ void-free	○ small void	⊚ void-free
PET	–	⊚ void-free	○ small void	⊚ void-free
PI	–	–	◖ unbonded area	⊚ void-free

978-1-4799-2408-0/14 $31.00 © 2014 IEEE

Peeling test and Fracture analysis

To evaluate strength of the bond interface, adhesion strength of bonded polymers were investigated. Adhesion strength of the bonded polymer films is measured by T-peel test. T-peel test is used to investigate adhesion strength of flexible adherents. The bonded films were cut into 25 mm width rectangle and the specimens were peeled at an angle of 90°. T-peel test are performed at crosshead speeds of 10 mm/min and at room temperature.

The results of T-peel test are summarized in Table 3. Since adhesion strength of PET/PET is so strong that the samples are fractured from PET bulk rather than the intermediate layer, adhesion strength of PET/PET is immeasurable. The first feature is that there is little difference between the samples bonded using Al and Si deposited layer. The reason for this is supposed that the thickness of the deposited layers is very thin (~ 20 nm).

Another feature is that adhesion of PI/PI is very weak compared with other bonded films.

Table 3. Summary of the adhesion strengths of the samples bonded by the proposed method.

Sample	using Al layer [N/cm]	using Si layer [N/cm]
PEN/PEN	2.8	2.7
PET/PET	Immeasurable	Immeasurable
PI/PI	0.68	0.63
PET/PI	1.45	Not tried
PEN/PI	0.93	Not tried

Bond interface of PET/PET is too strong.

In order to determine loci of fracture after the peeling test, fracture surfaces are investigated using. Figure 4 shows the fracture surfaces after the peeling test.

Figure 4 (a) is the fracture surface of PEN/PEN using Si deposited layer. This sample is relatively strongly bonded. The fracture surface is very rough. The height difference of the surface roughness is in the range of 5-15 μm. During the peeling test, fracture from the PEN bulk is observed.

Contrary, Figure 4 (b) is the fracture surface of PI/PI using Al deposited intermediate layer. The result of T-peel test indicates that adhesion strength of the bonded PI films is relatively low. The fracture surface of this sample is plain and patterns on the surface transfer to other side.

To identify loci of failure, atomic ratios of the fracture surfaces are analyzed using X-ray photoelectron spectroscopy (XPS). Table 4 shows atomic ratios of the selected areas in Figure 4.

From the atomic ratio of the area 1 in Figure 4, C and O are mainly exposed and the ratio of Si from deposited layer is low. Detail curve fitting calculation reveals that the C and O are derived from PEN molecule. The area 2 is the opposite side of the area 1 and the atomic ratio of the area 2 is almost same as that of the area 1. These results indicate that the fracture of this sample is mainly located in PEN bulk.

Conversely, the fracture surface of PI/PI using Al intermediate layer is plain. The area 3 looks covered with metal and this area mainly consists of Al from deposited layer. The area 4 is the opposite side of the area 3 and this area looks yellow which is the color of PI film. From the atomic ratio, N is detected at the area 4. Since the N atoms are derived from PI bulk, PI bulk is exposed at the area 4. These results indicate that fracture is located between PI and Al layer.

The fracture paths are summarized in Figure 5. Summarizing these results, the weakly bonded sample is fractured between polymer and the deposited layer whereas the strongly bonded sample is fractured from the polymer bulk. Therefore, strong adhesion between polymer and deposited layer is important for the strong bonding.

Table 4. Atomic ration of selected areas in Figure 4.

Data	C [%]	O [%]	Al [%]	Si [%]	N [%]	Fe [%]
(1)	69.7	18.7	–	11.6	–	–
(2)	74.1	20.2	–	5.7	–	–
(3)	16.5	34.9	44.6	–	4.0	–
(4)	56.6	22.9	6.4	–	14.1	–

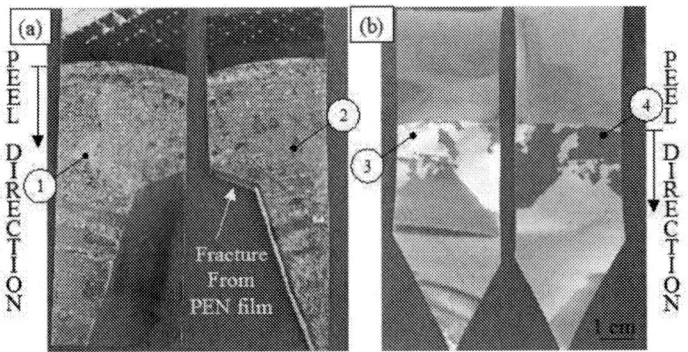

Figure 4. Fracture surfaces after T-peel test.
(a) Fracture surface of PEN/PEN (b) Fracture surface of PI/PI

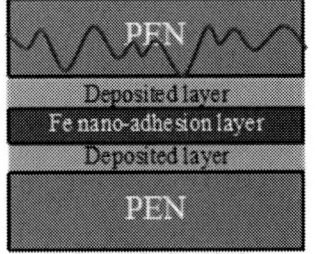 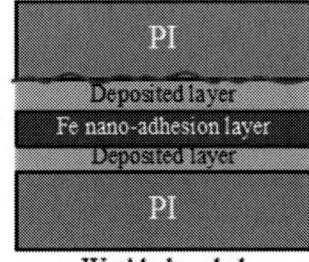

Figure 5. Summary of fracture path. Weakly bonded sample is fractured between polymer and the deposited layer. The strong adhesion between polymer and deposited layer is important for the strong bonding.

Sealing performance

The method was applied to sealing a model OELD and thermal cycle tests were conducted. A sealing frame was formed by PI surrounding the OELD element area, covered by Si layer with Fe and bonded at room temperature by the proposed method. The reliability test was carried out at a temperature 85°C in humidity of 85% and it run more than 300 hours, which satisfies the requirement of the practical implementation.

Conclusions

Polymer films such as PEN, PET and PI are bonded by the proposed method at room temperature with inorganic bond interface. Especially PEN and PET films are so strongly bonded that bond interfaces have tolerance for bending and loci of fracture after peeling test are located in polymer bulk. The result of the reliability test indicates that the organic electro luminescence lighting bonded has high gas barrier property.

Acknowledgments

This work was supported by Grant-in-Aid for Scientific Research (A)(22360025).

References

1. T. Suga, K. Miyazawa, and Y. Yamagata, "Direct bonding of ceramics and metals by means of a surface activation method in ultrahigh vacuum," *MRS Int. Meet. Adv. Mater. Mater. Res. Soc.*, vol. 8, pp. 257–262, 1988.
2. T. Suga, Y. Takahashi, H. Takagi, B. Gibbesch, and G. Elssner, "Structure of Al-Al and Al-Si3N4 interfaces bonded at room temperature by means of the surface activation method," *Acta Metall. Mater.*, vol. 40, pp. 133–137, 1992.
3. T. Suga, T. Fujiwaka, and G. Sasaki, "Surface activated bonding and its application on micro-bonding at room temperature," *in Proc. 9th Euro. Hybrid Microelectron. Conf.*, pp. 314–321, 1993.
4. T. Akatsu, N. Hosoda, and T. Suga, "Atomic structure of Al / Al interface formed by surface activated bonding," *J. Mater. Sci.*, vol. 4, pp. 4133–4139, 1999.
5. A. Shigetou, T. Itoh, M. Matsuo, N. Hayasaka, K. Okumura, and T. Suga, "Bumpless Interconnect Through Ultrafine Cu Electrodes by Means of Surface-Activated Bonding (SAB) Method," *IEEE Trans. Adv. Packag.*, vol. 29, no. 2, pp. 218–226, May 2006.
6. H. Takagi, K. Kikuchi, R. Maeda, T. R. Chung, and T. Suga, "Surface activated bonding of silicon wafers at room temperature," *Appl. Phys. Lett.*, vol. 68, no. 16, p. 2222, 1996.
7. T. Chung, L. Yang, N. Hosoda, H. Takagi, and T. Suga, "Wafer direct bonding of compound semiconductors and silicon at room temperature by the surface activated bonding method," *Appl. Surf. Sci.*, vol. 18, pp. 808–812, 1997.
8. R. Kondou, C. Wang, and T. Suga, "Room-temperature Si-SiN wafer bonding by nano-adhesion layer method," *in Proc. 60th Electron. Components Technol. Conf.*, pp. 357–362, 2010.
9. R. Kondou and T. Suga, "Si nanoadhesion layer for enhanced SiO2–SiN wafer bonding," *Scr. Mater.*, vol. 65, no. 4, pp. 320–322, Aug. 2011.
10. T. Matsumae, M. Fujino, and T. Suga, "Direct Bonding of PEN at Room Temperature by Means of Surface Activated Bonding method using Nano-adhesion Layer," *in Proc. IEEE CPMT Symp. Japan 2013*, vol. 16, p. 2, 2013.
11. Y. De Puydt, P. Bertrand, and P. Lutgent, "Study of the Al / PET Interface in Relation with Adhesion," *Surface and interface analysis*, vol. 12, pp. 486–490, 1988.
12. J. Silvain, A. Arzur, M. Alnot, J. Ehrhardt, and P. Lutgen, "XPS study of Al/polyethylene terephtalate interface," *Surf. Sci.*, pp. 787–793, 1991.
13. A. Calderone, R. Lazzaroni, J. L. Brddas, Q. Toan, and J. J. Pireaux, "A joint theoretical and experimental study of the aluminium / polyethylene terephthalate interface," *Synth. Met.*, vol. 67, no. 1–3, pp. 97–101, 1994.

Solder Joint Inspection with Induction Thermography

Johannes Bohm, Klaus-Juergen Wolter, Henning Heuer
Technische Universitaet Dresden, IAVT
Helmholtzstr. 10, D-01069 Dresden, Germany
johannes.bohm@tu-dresden.de, Tel. +49 351 463 32079

Abstract

Defect solder joints on SMD packages can lead to the immediately fail of an electronic device or to instable operation. Apart from electrical conductive tests of the device, there is no appropriate non-destructive evaluation (NDE) method to test the functionality of such solder joints. The paper will present an alternative NDE method based on inductively excited thermography. QFP samples are prepared and test measurements are performed. Parameters like excitation frequencies and possible coil orientations are investigated. The measurements demonstrate that the presented method can be used to test the functionality of solderings. Application limits (e.g. required measurement time) are discussed. This new NDE method is capable of being used as an add-on in electronic manufacturing and failure analysis.

Introduction

Defect solder joints on SMD packages result from multiple reasons like gull wing bending (lifted leads), bad wettability, solder printing, solder paste or stress-induced cracks. If loose contacts are not identified during manufacturing, they can cause instable operation or the complete fail of the device in the field. Lifted leads are hard to detect by conventional optical inspection. Angle view optical inspection cannot be used if the sight is constrained by high neighboring components like big capacitors [1]. Electrical conductive function tests usually require a device specific adapter and design for testability.

The presented NDE approach is a kind of heat flow thermography. It has already been demonstrated for the inspection of electronic connections like PCB vias and band solderings on solar cells [2]. Induction thermography does not need any board-specific adapter. Although the used planar coils have a small working distance the method can be implemented contactless. Heat is induced locally by a coil into the package lead frame and diffuses through the solder joints to the connection pads and further to connected conductor structures on the carrier board. Solder cracks or lifted leads constrain the heat transport and reduce the observed temperature on the board surface near the solder joints compared to defect-free connections. By recording the temperature distribution it is possible to verify the electrical function of the solder connection. The coil design ensures that the semiconductor inside the package gets not damaged by induced overvoltage.

Recently the method was also applied on wire bondings of IGBT modules [3] to detect bond cracks but there the whole test sample was excited by a very strong electromagnetic field which might destroy semiconductors depending on the orientation within the field.

Samples and Preparation

To simulate solder defects induced by bad wetting or bad solder printing, two QFPs of different sizes were mounted on a double sided PCB by selective manual soldering. The PCB has been taken from a real application und thus the connections vary from near large area supply planes over long line signal wire and vias to unconnected solder pads. Fig. 1 shows the top view.

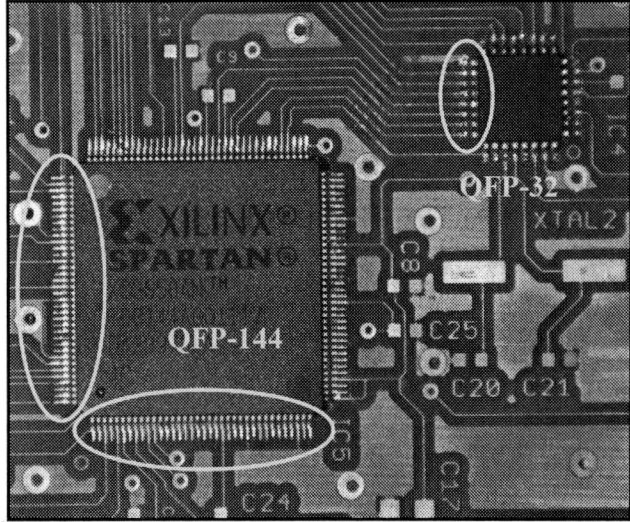

Figure 1. Overview of prepared sample with wetting defects: PCB with two QFPs, inspected edges marked

The first inspected component is a FTDI FT232BL in a QFP-32 package with an 800 µm lead pitch. The second IC tested is a Xilinx XC3S50AN packaged as a QFP-144 with a 500 µm lead pitch. For both packages the lead thickness varies from 90 to 200 µm and the package thickness is about 1.4 mm.

Three connection types were implemented (see Fig. 2 and Fig. 3). The two defect types have a higher electrical and thermal resistance depending on the lead bending.

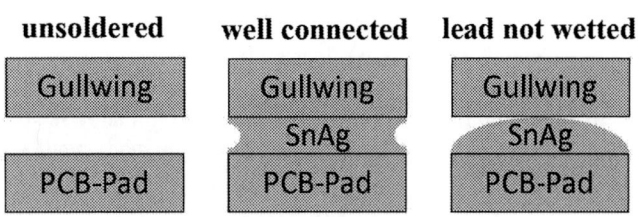

Figure 2. Repeating soldering scheme
(for QFP-144 two leads for each type)

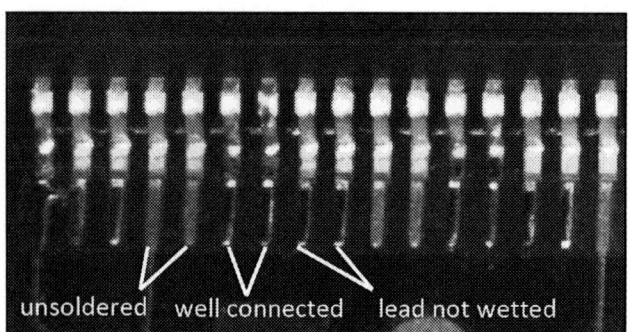

Figure 3. Prepared solderings on QFP-144

At a third investigated QFP the lifted lead defects were generated by bending of a complete lead row which enlarges the pad-pad-distance and results in defects during the reflow soldering (see Fig. 4). The used dummy QFP has a lead pitch of 400 µm and it features an internal daisy chain (every other lead is connected to its neighbor). The board-level connection between the solder pads was cut before the thermal measurements as this connection would allow heavy heat exchange between the pads which reduces the lateral resolution.

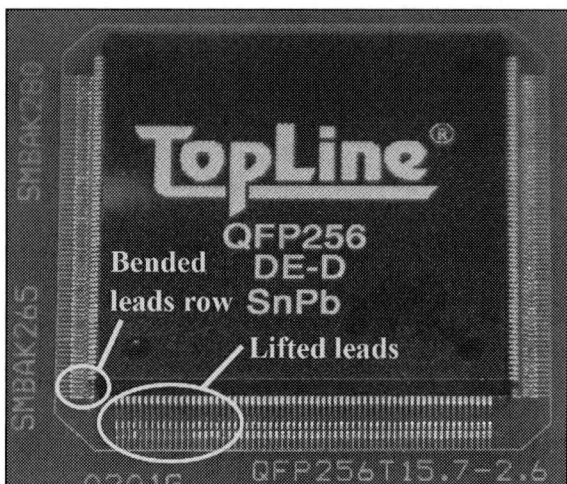

Figure 4. Prepared lifted leads on QFP-256

The resistance measurement of the contacts is difficult as there is only little contact area at the leads. Thus only an imprecise two-wire measurement was performed. The resistance of the two defect connection types depends heavily on the probe pressure.

Measurement Setup

The coil used for this work is a planar line array of circular spirals (see Fig. 5). It has already been presented in [2]. Neighboring spirals have opposing current orientations. This limits the induced voltage while covering a large heated area. The cooling is convective. The amplitude modulated coil current is driven through a resonant circuit. The induction frequency was set to 1.37 MHz which leads to a skin depth in copper about 57 µm which is below the lead frame thickness.

It is advantageous to heat the sample from the package side instead of exciting the connected wires and pads on the PCB because the mold thickness can be larger than the lead pitch and thus it would not be possible to distinguish neighboring leads in the thermal image. In the presence of neighboring components the package top is better accessible by a coil than the PCB surface.

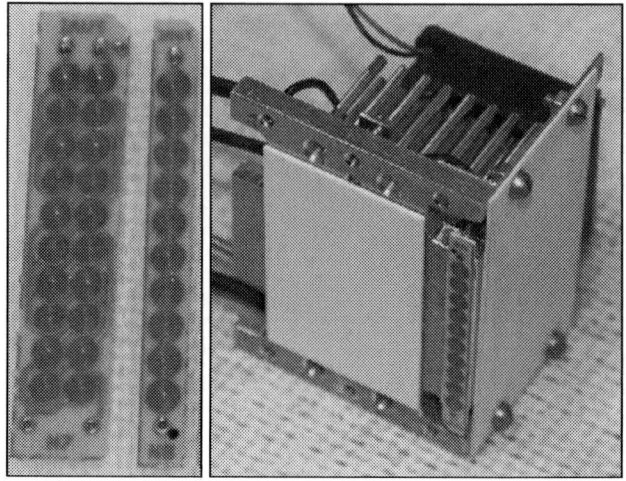

Figure 5. Used coil arrays and mounted coil head [2]

The heat is induced into the lead frame. In addition, there is an unavoidable heat transfer from the warm coil through the mold compound. Two sheets of paper are placed between the coil and the sample to limit this unwanted warming. Fig. 6 shows the measurement setup. The coil edge lays superimposable with the mold edge of the package. A cooled infrared camera (Flir Silver 660M) observes the visible area next to the coil focusing at the PCB surface.

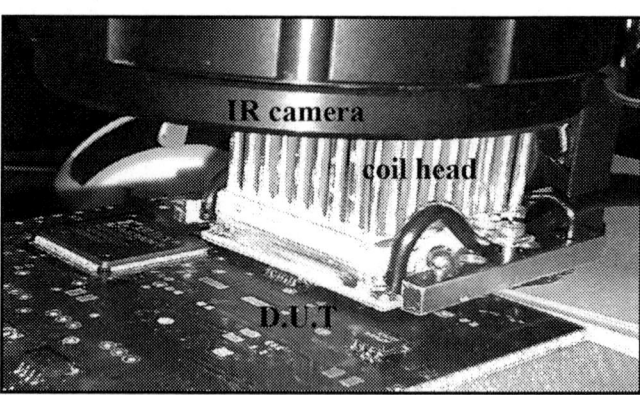

Figure 6. Measurement setup

Experimental Results on Samples with Wetting Defects

Fig. 7 and Fig. 8 show the results for the QFP-32. The four intact low-resistance contacts heat the connected wires on the PCB and their surrounding area whereas the other four ones contacts stay cool. The surrounding of the large wire (No. 17) is heated less than for thin signal wires because of the higher thermal capacitance. A lock-in frequency f_{lock} of 0.2 Hz is suitable for the large pitch and the generated temperature amplitude allows use of uncooled infrared cameras but the lateral resolution is better at higher

978-1-4799-2408-0/14 $31.00 © 2014 IEEE   1510

frequencies. As expected the phase image is more robust und gives analyzable information even for low amplitudes.

Fig. 8 also includes a comparison with thermal conductive-only heating by the same coil. At the lock-in frequency of 1 Hz the inductive heating fraction is larger than the conductive one. The inspection principle also works without induction but the mold thickness limits the amplitude especially for high lock-in frequencies.

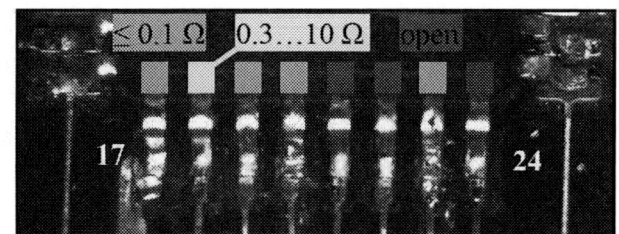

Figure 7. Resistance measurement at QFP-32 leads 17 to 24, value of lead No. 18 depends on probe pressure

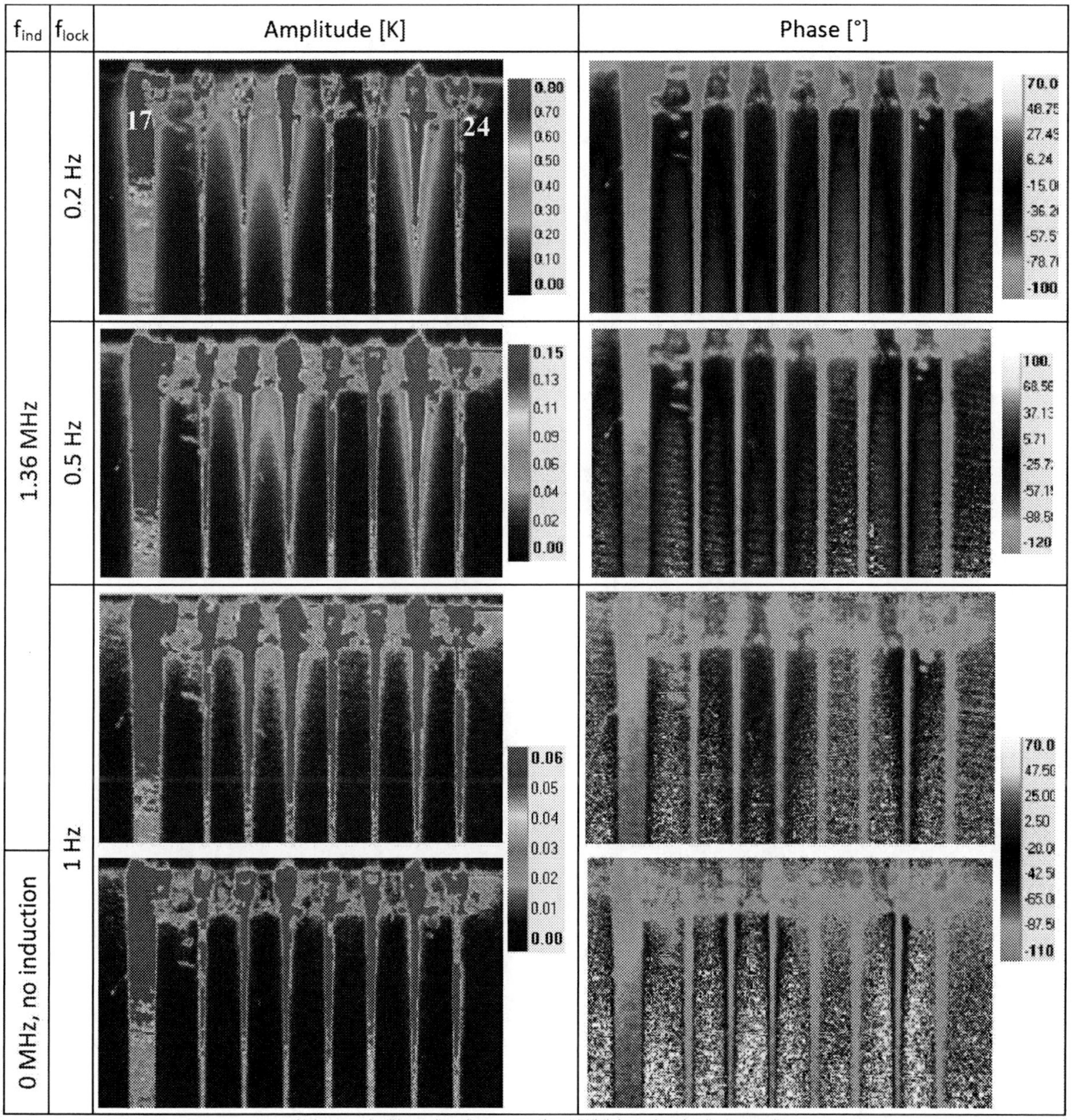

Figure 7. Result pictures for QFP-32, leads 17 to 24

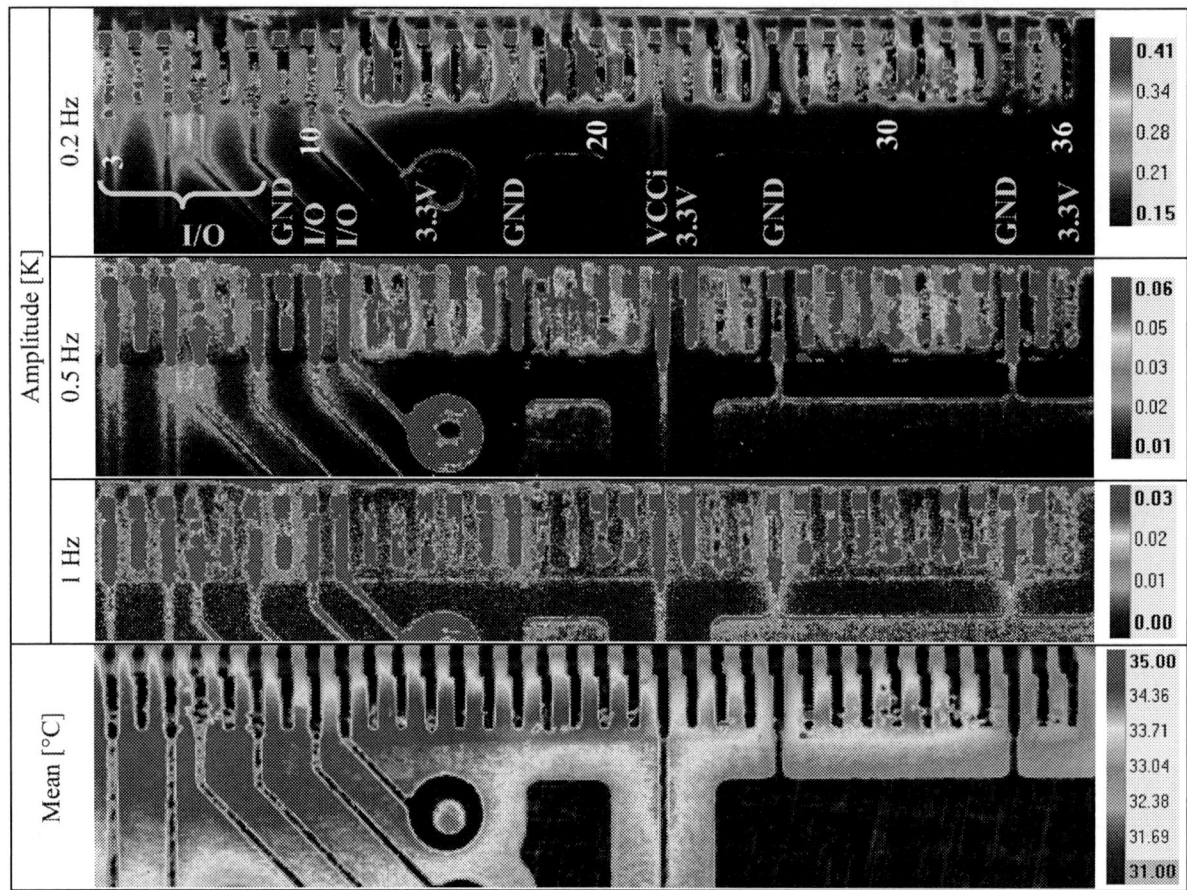

Figure 9. Amplitude images for QFP-144 with defects, leads 3 to 36, single-row coil, f_{ind} 1.37 MHz

Fig. 9 shows the amplitude images for the QFP-144 at multiple lock-in frequencies. As the samples were not blackend, the solder covered leads and pads reflect the coil heat worse than uncovered ones due to a lower emissivity than blank copper and the round solder shape. Thus solder areas appear dark in the amplitude image and the three prepared soldering types can be distinguished (for the selected temperature ranges only at f_{lock} 0.2 Hz well visible).

The temperature at the board next to the leads depends mainly on the PCB layout (see Fig. 9 top and Fig. 10). Ground 'GND' pads have the lowest amplitude due to the high thermal capacity of the connected large copper areas on the PCB, whereas unconnected pads are heated most. At 0.2 Hz neighboring pads and wires interfere as the thermal diffusion length $\mu_{th,FR4}$ (570 µm) is higher than the isolation distance (200 µm). But at 1 Hz the heat transport through the mold cap to the leads is to low for a senseful evaluation. Weak eddy currents develop through the multiple ground loops leading to slightly heated ground pads. This effect depends on the lateral coil position as the opposing coil orientations in the array compensate each other.

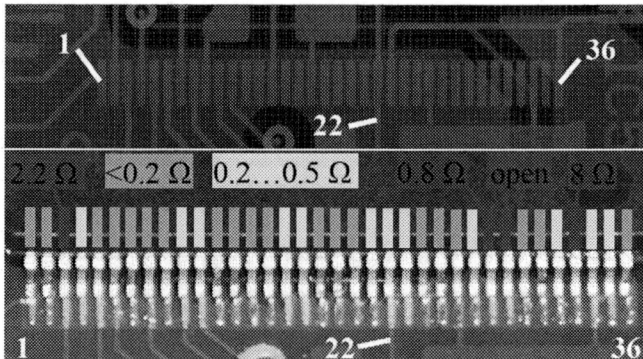

Figure 10. PCB layout and joint resistance at QFP-144, leads 1 to 36

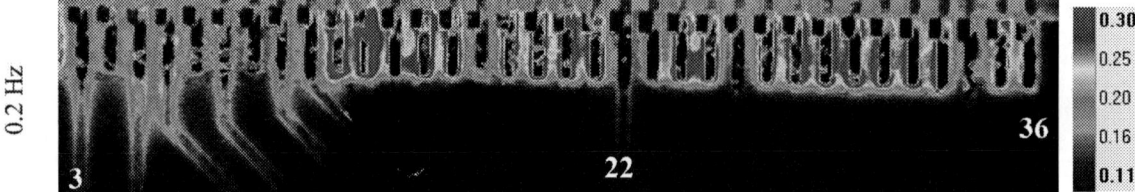

Figure 11. Reference measurement of correctly soldered QFP-144, amplitude image, leads 1 to 36, f_{ind} 1.38 MHz

The electrical resistance of the lead-to-pad connection has only minor influence (Fig. 10). Around the electrically open lead No. 29 the amplitude is slightly lower than around its neighbors. Within a large group of neighboring unconnected pads the correctly soldered ones are heated more than the two defect types. The reference measurement of a correctly soldered package in Fig. 11 shows a more uniform amplitude distribution for the unconnected pads. The resistance of loose wire-connected leads (e.g. No. 3) is too low to generate a measureable difference to correctly soldered ones.

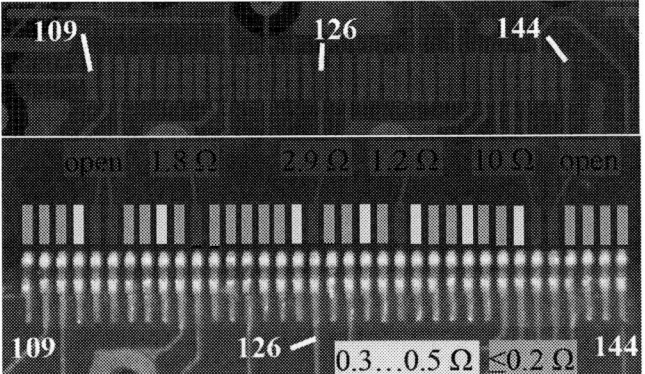

Figure 12. PCB layout and joint resistance at QFP-144, leads 109 to 144

The solder stop edge is visible as a horizontal border. The lateral coil position and the PCB connections determine the phase distribution between the solder pads more than the

connection resistance. Thus the pase image gives no additional information and is not shown.

Another edge of the QFP-144 has generally stronger defects and even three open contacts for low probe pressure (see Fig. 12). These three defects and one high-resistance contact No. 139 can be found in the amplitude images of Fig. 13 at 0.2 and 0.5 Hz as low-amplitude surrounded pads. In addition, the two used signal leads No. 139 and 140 can be detected by viewing at the PCB wires as seen at the QFP-32. These defects stick out particularly because they occur in two pairs. The better conducting loose contacts like leads No. 119, 126 and 131 cannot be found in the thermal images.

Again the phase image gives no additional information about the connection resistances because the PCB wires exchange heat due to the small isolation distance. The generated amplitudes are too low for the needed higher lock-in frequencies above of at least 1 Hz (see Tab. 1).

Tab 1. Thermal and electrical penetration depths

	eddy current penetration depth d_{Skin} in µm			thermal penetration depth μ_{th} in mm			
f_{ind} / f_{lock}	1	1.37	5	0.2	0.5	1	2
	MHz			Hz			
Cu	67	57	30	13.7	8.6	6.11	4.32
SnAg3.5	176	150	79	8.45	5.34	3.78	2.67
FR4 lateral	$\approx \infty$			0.57	0.36	0.25	0.18
FR4 axial				0.44	0.28	0.20	0.14

Figure 13. Amplitude images for QFP-144 with defects, leads 109 to 144, two-row coil, f_{ind} 1 MHz

For both inspected packages the signal-to-noise-ratio between solder pads is worse than around connected wires because the flux residues near pads bring disturbing reflections and a non-uniform emissivity. Thus it is easier to evaluate the temperature of connected PCB wires. The inspected connections of the QFP-32 are connected to long uncovered PCB wires which have a large isolation distance and therefore these solderings can easily be tested.

The larger pitch of the QFP-32 leads to less thermal interference between neighboring leads and wires on the PCB but it also results in higher temperature amplitudes because it the lead frame is different. The X-ray images in Fig. 14 show that the molded leads of the QFP-32 are comparably short and wide. The induction of eddy currents and the heat transfer through the mold are more effective for wider leads. In addition, the thermal resistance is lower and the capacity is almost equal compared to the QFP-144. Large packages like the inspected QFP-144 with long molded leads should be excited by a coil that covers a large fraction of the molded lead area.

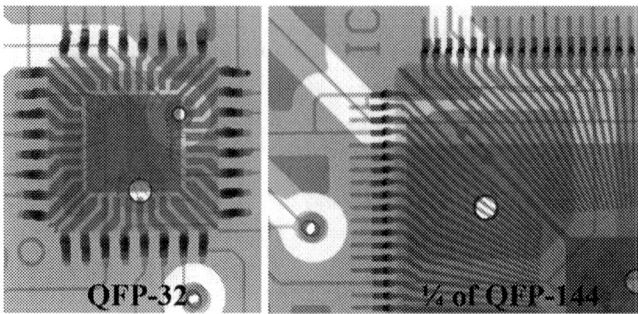

Figure 14. X-ray images of test samples

Experimental Results on Sample with Lifted Lead Defects

Fig. 15 shows the results of the lifted leads sample. The principle behavior is the same as described above but the amplitudes are lower due to the higher mold thickness and the small lead frame wire width (visible in Fig. 16). Thus only a quite low lock-in frequency could be used and the lateral resolution is accordingly bad.

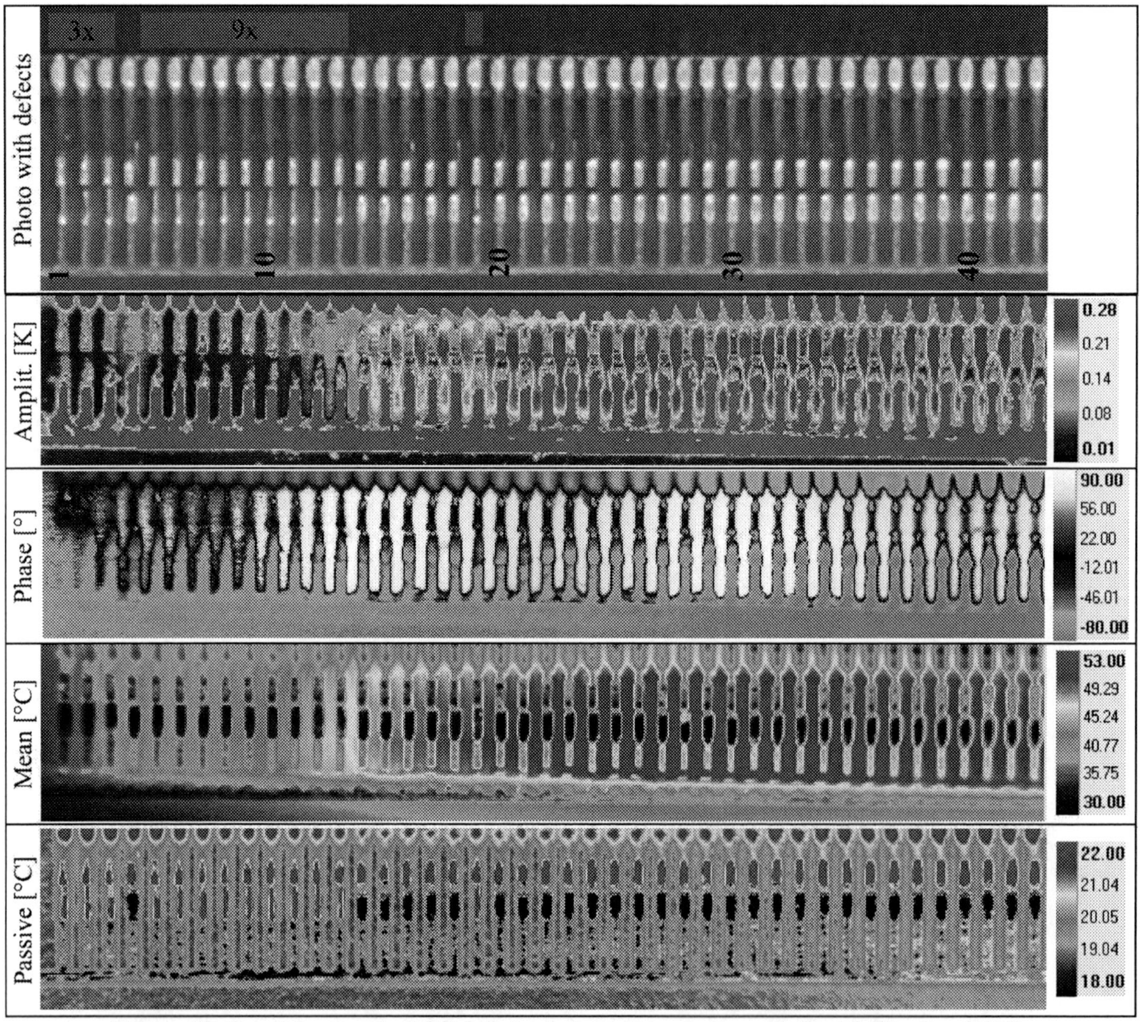

Figure 15. Result images for QFP-256, leads 1 to 43, two-row coil, f_{ind} 1 MHz, f_{lock} 0.2 Hz

Only the two groups of neighboring defects can be detected by viewing at the PCB surface. The single defect lead No. 19 is inconspiciuos in Fig. 15 as its pad is heated by the neighboring contacts. Again, the phase image gives no additional information.

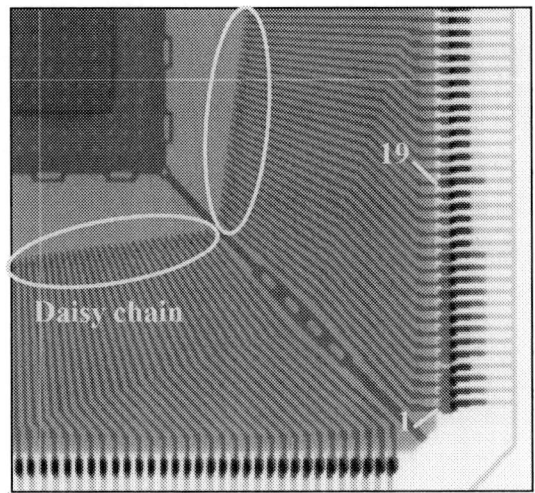

Figure 16. X-ray image ¼ of the QFP-256

Proposed Improvements

Each presented measurement took about 30 seconds plus a tuning duration of about 60 seconds which limits the inline-applicability. For the QFP-32 the acquisition time can be reduced to fewer lock-in periods because the signal-to-noise-ratio is high enough. But for industrial use with uncooled infrared cameras the shown amplitudes are barely detectable at f_{lock} 0.2 Hz. Thus, longer acquisition times are helpful. Inductive pulse excitation seems to require a very strong electromagnetic field strength which increases the risk of overvoltage damages to the IC. A stamp-like coil head in the package size could excite all four package sides at once to inspect the whole QFP in a single measurement. Such a small coil head would also be insensitive to high neighboring packages. A more uniform coil shape is useful to evaluate the phase images.

Conclusions

Solderings on QFP of two different sizes were investigated using induction thermography. Completely open solder connections can be detected. A connection with a resistance dimension of $10\,\Omega$ can be found when it is connected to a visible PCB wire. But loose connections with smaller electrical and thermal resistance could not be found. Solderings on pads without uncovered PCB wire connections are hard to test because of flux residues around the pads. The temperature distribution depends heavily on the PCB layout because copper wires strongly dissipate the heat from the solder connections. In general a reference measurement of a defect-free sample is useful. It was shown that the developed NDE principle works. The measurements are repeatable.

Outlook – Area Solderings

The presented NDE method may be further developed to be used as an add-on to automatic optical inspection in SMD

inspection systems. It has also the potential to test area solderings, e.g. on power electronics. Fig. 17 demonstrates this for thyristor dies soldered on an Al_2O_3 substrate. The die edge length is about 10 mm. One "good" sample was mounted on a full area thick film pad with full area solder printing whereas the other "bad" sample was mounted on a smaller and split pad covering less than 50% of the die area. As there are no wire bonds, the planar coil was placed directly on the die top similar to the QFP measurements above.

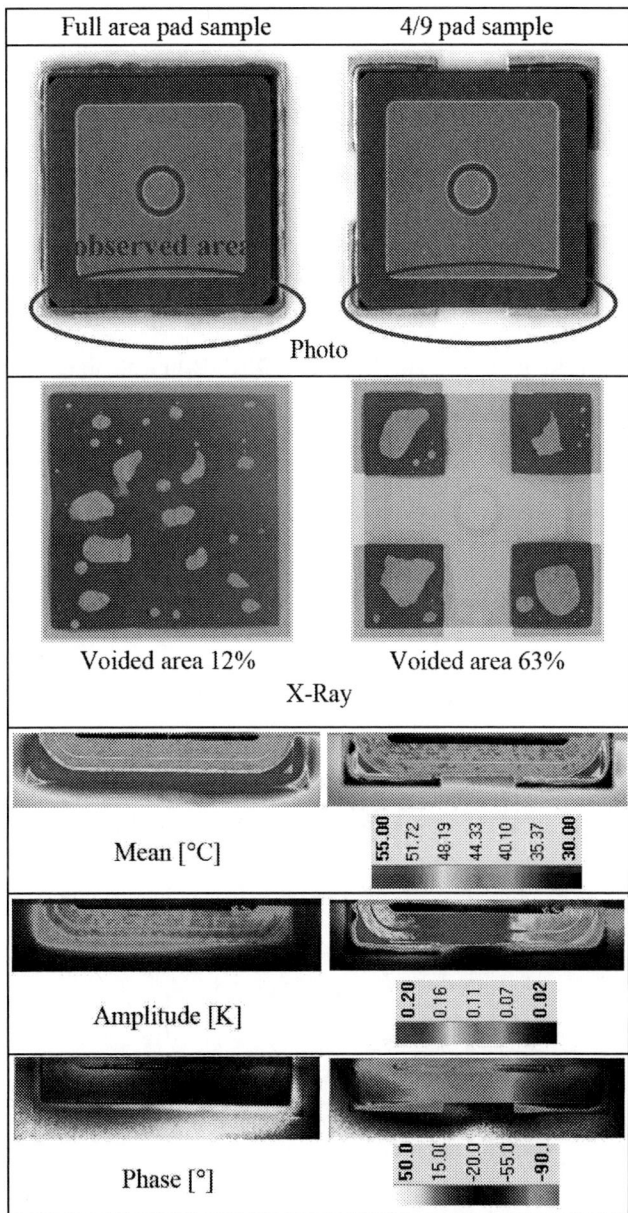

Figure 17. Test of area solderings, f_{ind} 1 MHz, f_{lock} 2 Hz

The eddy currents are induced into the die and its top metallization but at the chosen induction frequency of 1 MHz the electromagnetic field partly reaches the solder layer. Voids reduce the induced power and thus the mean temperature next to the solder pad in Fig. 17 is lower for the bad sample. The thermal capacity of the solder volume and the substrate keeps the amplitude of the die temperature oscillation overall lower and later for the good sample. For the shown extremely bad sample even a lateral localization of the defect is possible in the amplitude and phase image. Further investigations are needed to quantify the detectability of the void content and to estimate applicability for realistic defects and sintered bonds.

Acknowledgments

Thanks to Mr. M. Trusch for preparing test samples and to Goepel GmbH Jena for providing lifted lead sample.

References

1. Kokott, J.: "Safe detection of hidden faults - Layout independent fault detection by AOI systems with high-end angled-view inspection", Goepel Electronic GmbH Jena, 2009, http://www.goepel.com/nc/en/articles/ce/5263/action/download/e/safe-detection-of-hidden-faults.html

2. Bohm, J; Wolter, K.-J.: "Anregungstechnik für die Induktiv Angeregte Thermografie in der Aufbau- und Verbindungstechnik" *DGZfP Jahrestagung,* Dresden, May 6-8, *2013,* http://jt2013.dgzfp.de/Portals/jt2013/BB/poster40.pdf

3. Li, K.; Tian, G.; Cheng, L.; Yin, A.; Cao, W.; Crichton, S.: "State Detection of Bond Wires in IGBT Modules using Eddy Current Pulsed Thermography" *IEEE Transactions on Power Electronics,* vol.PP, no.99

AUTHOR INDEX

A

Aasmundtveit, Knut	139, 498
Abe, Hidenori	1523
Agarwal, Rahul	590
Agrawal, Akash	2014
Agrawal, Jai P.	1838
Ahn, Billy	2027
Ahn, Byungki	1616
Ahn, Yesul	1361
Aizawa, Mitsuhiro	1166
Akaike, Masakate	1143
Akin, Meriem Ben-Salah	1595
Alatorre, Roseann	215
Alcira, Cecille	47
Allen, Aileen M.	1433
Allen, Craig	878
Almeida, Rodrigo	935
Althoff, Simon	1549
Alvanos, Tyson	452
Alvarado, Rey	100, 925, 1173
Amara, Karima	1198
Aminov, Parvizso	1838
Amirkhany, Amir	560
Anderson, R.	1641
Andreassen, Erik	139
Andry, P.	576, 883, 1372, 1880
Angyal, Matthew	647
Anselm, Martin	119
Anzai, Nobuhiro	829
Aoki, Issei	1776
Aoyagi, Masahiro	62, 1915, 1926, 2003
Araga, Shinji	2194
Arai, Yoshiyuki	913
Arakere, Guruprasad	395
Arfaei, Babak	425, 655, 1769
Arik, Mehmet	209
Arkalgud, Sitaram	862
Armutlulu, Andac	1098
Arnold, Kim	894
Arvin, Charles	1940

Aryasomayajula, Lavanya 348
Asahi, Noboru 913
Asai, Kosuke 186
Asai, Osamu 1885
Asai, Satoshi 1028
Asano, Tanemasa 2303
Aschenbrenner, R. 940
Atsumi, Koichiro 2227
Audoin, M. 1714
Auerswald, E. 1134
Augustine, Anne E. 528
Aung, Kyaw Oo 596
Awatsuji, Yasuhiro 186

B

Baba, Mikio 763
Baba, Shunji 68
Bader, V. 940
Bae, Byeong-Hyun 1635
Bae, Byung-Hyun 1735
Bae, Hyun-Cheol 1154
Bae, In-Tae 1562
Bae, J.-C. 304
Bae, Jangyong 973
Baek, Hyunho 554
Bailey, Chris 1342
Bailey, Susan 894
Baillin, X. 1714
Bajwa, A.A. 1707, 2181
Bakir, Muhannad S. 13
Baks, Christian 1016, 1272, 1955
Ball, Sabrina 2035
Baloglu, Bora 1231, 1401
Balucani, M. 194, 1571, 1842, 2137
Bandarenka, H. 194
Bao, Andy 47
Barros, Isabel 935
Barth, Holly D. 1457
Barwicz, Tymon 179
Bauer, J. 940
Bchir, Omar 1396
Bea, JiChel 636
Becker, K.-F. 940
Beer, Gottfried 1183
Belardini, A. 194
Beleran, John D. 490

Benedetti, A. 194
Benedetto, Elizabeth 425, 1433
Benjamin, Shuki 1021
Bennett, N.S. 1517
Berg, J. 1071
Berger, Daniel 647
Berger, François 478
Berghuvud, A. 2220
Bernardi, D. 1571, 1842
Berry, C.J. 1965
Beyene, Wendemagegnehu 730
Beyer, G. 33, 309, 572, 850, 894
Beyne, E. 26, 33, 309, 572, 613, 850, 894
Bezuk, Steve 47, 925, 1173, 1396
Bian, Yuan 1902
Bieler, Thomas R. 697
Björnängen, T. 2220
Blair, Justin 609
Böck, J. 956
Bock, K. 1482
Bohm, Johannes 1509
Bollmann, D. 1482
Bondarenko, V. 194
Borgesen, P. 371
Bösch, Wolfgang 1183
Böttcher, M. 625
Bowen, Terry 1054
Boyer, Nicolas 179
Boyne, Dan 1539
Bozack, Michael J. 379
Bozack, Mike 242
Brady, D. 74
Brand, Sebastian 850
Brandt, Lutz 279, 1652
Braun, T. 940
Bravin, Julian 888
Breach, Christopher 1803
Brofman, Peter 255, 1308
Brusberg, Lars 1033, 1498, 1759
Budd, R. 883
Burggraf, Jürgen 888

C

Caffey, Kevin 1173
Caggiano, Michael 2159
Cai, Jian 1378

Calvert, Jeffrey 342
Campos, José 935
Cangellaris, Andreas C. 717, 2094
Cao, Andrew 862, 1810
Cao, Liqiang 1116, 1932, 2008, 2131
Cao, Zhihua 464
Caplet, S. 1714
Capuz, G. 572
Cardoso, Paulo 935
Castagné, Laetitia 1198
Castro, José 935
Cate, S. 74, 952
Cha, Seungyong 354
Chada, Arun Reddy 2081
Chahal, Premjeet 775, 2144, 2168
Chainer, Timothy 1955
Chakraborti, Parthasarathi 541, 1492
Chan, M. 952
Chan, Yan Cheong 1342
Chang, Chih-Wei 512
Chang, David 947
Chang, Gee-Kung 1054
Chang, Hong-Da 947
Chang, Hsiang-Hung 1891
Chang, Hung-Hsein 868, 1628
Chang, Jenny 1562
Chang, Nistec 81
Chang, Pai-Cheng 290
Chang, Po-Chih 1853
Chang, Shih-Chieh 316
Chang, Xin 2200
Chang, Yiu-Hsiang 1853
Chao, Chun-Chieh 868
Chao, Shu-Han 1908
Chao, Yu-Lin 290
Chau, Ellis 1810
Chen, Agassi 1750
Chen, Cheng-Fu 963, 2020
Chen, Chien-Chou 1853
Chen, Chih 1908
Chen, Chunwei 26
Chen, Dong 2299
Chen, Eason 81
Chen, Erh-Hao 1853
Chen, Gang 1080
Chen, George 470

Chen, Guang **566, 748**

Chen, Hsien-Wen **1**

Chen, Jing.. **1902, 2099**

Chen, Jui-Chin **1853**

Chen, K.M. **297**

Chen, Kim-Hui **2299**

Chen, Kuan-Neng **512**

Chen, Kuo-Hua **512**

Chen, M.T. **572**

Chen, Qi.. **1992**

Chen, Qianwen **1372**

Chen, Quan **1992**

Chen, Scott **1562**

Chen, Shang-Chun **1853**

Chen, Shi-Ching **868**

Chen, Shih-Ching **1628**

Chen, Stephen.. **1**

Chen, Su-Mei **1908**

Chen, Tai-Yu **2069**

Chen, Xu.. **717, 1080**

Chen, Yan-Heng **1750**

Chen, Yan-Siang **1584**

Chen, Yen-Chi **1258**

Chen, Yi.. **1584**

Chen, Yi-Ling **1529**

Chen, Yin-Fa.. **419**

Chen, Youpeng **1488**

Chen, Yu **1378**

Chen, Yu Hua **360, 1590**

Chen, Zhaoqing **723, 2104**

Cheng, Hung-Hsiang **2112**

Cheng, Lianxi... **1335**

Cheng, M.D. **572**

Cheng, Ren-Shin **290, 1908**

Cheng, Xiaojin **1926**

Cheng, Yu-Mei **290**

Chéramy, S. **906**

Cherman, V. **309, 1517**

Chia, Pierre **1122**

Chiang, Tzu-Hsing **419**

Chien, Chun-Hsien **290, 1891**

Chien, F.L. **56**

Chien, Heng-Chieh **290, 963**

Chin, Wai Foo **1790**

Chiou, Jin-Chern **512**

Chiu, Chi-Tsung **512, 2112**

Chiu, Steve 1, 947
Chiu, Ying-Ta 419
Chlieh, Outmane Lemtiri 2293
Cho, Byoungwoo 1361
Cho, Jounghyun 541
Cho, Junghyun 1328
Cho, Sangbeom 1247
Choi, Alan 925
Choi, BaekKyu 2075
Choi, Eun-Kuk 1735
Choi, Hoi Wai 919
Choi, Hyeseon 1004
Choi, Joonyoung 836
Choi, Kwang-Seong 1154
Choi, Won Kyung 596
Choi, Yongwon 1128, 1661, 1755, 2241
Choi, Youjoung 836
Choi, Young Won 1765
Choki, Koji 1148
Chong, Chan Kai 490
Choo, Joung-Hoon 1735
Chou, Bruce C. 1054
Chou, Lei-Chun 512
Chua, S.L. 324
Chuang, Ching-Te 512
Chuang, Shih-Keng 1944
Chung, Ming-Feng 2112
Chung, YoungSuk 582, 1816
Chylak, Bob 1523
Ciarniello, Danilo 1571
Clauberg, Horst 1523
Co, Rey 215, 1389
Cochet, Philippe 20, 523
Collado, Ana 796
Collins, Sian 878
Conti, Fosca 1464
Cook, Jeffery 684
Cordes, Steven A. 1782
Cotts, Eric 655, 690, 1769
Coutier, C. 1714
Cowley, A. 1517
Coyle, Richard 425, 655
Cremaldi, Joseph 255
Crescenzi, Rocco 1571
Croes, Kristof 613
Crouthamel, D. 74

Cui, Tong	100
Cunningham, G.	8
Czurratis, Peter	850

D

Daerhan, Daerhan	759
Dai, Ming-Ji	290
Daily, Derek	1308
Daily, R.	165, 309, 572
Dal Molin, Renzo	1198
Dalal, Mitul	145
Dan, Qiang	1010, 1690, 2245
Dang, B.	576, 883, 1372, 1880
Dang, J.	74
Danilewsky, A.N.	1517
Darveaux, Robert	703
Daubenspeck, T.	1949
Davis, Roy I.	1996
Davis, Taryn	236
De Messemaeker, J.	33, 613
De Vos, J.	309
De Wolf, I.	309, 613, 850, 1517
Defay, E.	1296
Dehag, G.	1714
Dej, Sebastian	590
Demir, Kaya	1098
Demosthenous, Andreas	2213
Derix, Robert	1464
Deshpande, Shantanu	242, 1449, 1973
Desmaris, V.	1071
Detalle, Mikael	33
D'hiver, Philippe	1198
Dieng, K.	1296
Dinan, Thomas	862
Ding, L.	1212
Ding, Wen	2236
Djuric, Tatjana	850
Doany, Fuad E.	1016
Dobashi, Masahiro	763
Dobritz, Stephan	873
Donaldson, Nick	2213
Dong, Jianwei	342
Dong, Mingzhi	1192
Dornala, Kalyan	85
Dragoi, Viorel	888
Dressler, C.	1714

Drewniak, James L. 2081
Drost, A. 1482
Du, Ellen 748
Duffy, Daniel 1803
Durfee, Loren 236
Dzarnoski, John 157

E

Eastep, Brian 279
Eggen, Trym 498
Ehrhardt, Christian 1321
Eichstadt, David 1803
Eisenstadt, William R. 554
Elger, Gordon 1464
Ellis, Charles D. 1086
Elmer, John W. 1457
Elsherbini, Adel A. 1610
Emerich, S. 74, 952
Endo, Yoshinori 1523
Engelmann, Sebastian 179
Eom, Yong-Sung 1154
Esfahani, Zahra Kolahdouz 1556
Eto, Michiyuki 2003

F

Fan, Chia-Wen 1908
Fan, Chuanhong 1885
Fan, Jun 2081
Fan, Xuejun 967
Fang, Runiu 641, 1902, 1986, 2099
Fanget, S. 1714
Farahmand, Farid 2159
Farcy, A. 1296
Farrugia, Mark-Luke 411
Fay, Owen 2014
Feng, June 748
Feng, Wei 1915
Fiedler, C. 625
Fischer, Thorsten 1421
Fischer, U.H.P. 1862
Flack, Warren 26
Fortier, Paul 179
Franiatte, R. 906
Frank, Torrey W. 1610
Frye, Robert 1284, 2027
Fu, Hailuo 1652

Fu, Huan-Chun 290, 1891
Fu, Shancan 1080
Fu, Xianzhu 464
Fu, Xingming 170, 1189
Fu, Yifeng 459
Fujii, Atsushi 829
Fujimaru, Koichi 913
Fujimoto, Koji 1853
Fujino, Masahisa 1504
Fujita, Mitsuru 829
Fukuoka, Yoshitaka 1673
Fukushima, T. 304, 636, 856, 1110, 1148
Fukuzono, Kenji 68
Furuya, Ryosuke 1885

G

Gaherty, Lee 279
Gandhi, Saumya 541, 1492
Gao, Guilian 862, 1810
Garant, John 452
Gardner, Donald S. 1290
Garnier, A. 906
Gaynes, Michael 255, 1372, 1955
Ge, Y.P. 1212
Ge, Yun 684
Gebara, Edward 2293
Gelorme, Jeffrey 1308
Georgakopoulos, Stavros V. 759
Georgiadis, Apostolos 796
Ghaffari, Roozbeh 145
Gharaibeh, Mohammad 119
Giagka, Vasiliki 2213
Gieser, H. 1482
Gilham, David 504
Gill, Harpreet 354
Gissila, T. 2220
Glodde, M. 883
Goldsmith, Charles 1940
Golick, L. 952
Gonzalez, M. 33, 309, 1517
Goodwin, S. 8
Gottfried, Knut 1218
Goumans, Leon 411
Goyal, Deepak 1457
Grafe, Juergen 873
Grams, A. 625

Graves-Abe, Troy 647
Greco, F. 1714
Gregory, C. 8
Greve, Hannes 1314
Grosinger, Jasmin 1183
Grymyr, Ole Johannes 139
Gu, Ping 464
Gu, Sam 609
Gu, Xiaoxiong 548, 1272
Gu, Yingke 1378
Guan, Yong 1902
Guerin, Luc 647
Guerrero, Alice 894
Guevara, Gabe 215
Guidotti, Daniel 1932
Guiller, O. 1296
Gundurao, Anil 354
Günther, Wolfgang 1218
Guo, F.M. 1212
Guo, Maofeng 2236
Guo, W. 309
Guo, Xing 1010, 2283
Gurrum, Siva P. 821
Guthrie, William 647
Guven, Ibrahim 2035

H

Hagelauer, A. 956
Hahm, Yeon-Chang 730
Hale, Cassandra 236
Halonen, Eerik 151
Halvorsen, Per Steinar 139
Hamasha, S. 371
Hamilton, Michael C. 441, 1086
Han, C.J. 1641
Han, Gyuwan 1361
Han, KiJin 2124
Han, Kyu 782
Han, Michael 47
Han, Minghui 560
Han, SeungChul 582, 1816
Han, Sungwon 973
Hanna, Carlton 1278
Hao, Jifa 1241
Harada, Takeshi 1857
Harel, Stephane 179

Hartner, W.	956
Hasharoni, Kobi	1021
Hashiguchi, H.	856, 1110
Hashimoto, H.	304, 636
Hashimoto, Masakazu	1955
Hashimoto, Tomoaki	596
Hasnine, Mohammad	379
Haupt, M.	1862
Hau-Riege, Christine	1173
He, Hongwen	1116
He, Huanyu	548
He, Yi	2008, 2131
Heinrich, T.	108
Hell, W.	1482
Hemsel, Tobias	1549
Henriques, Vitor	935
Herbold, Christian	203
Herbst, Christian	1033
Hernandez, George A.	1086
Heuer, Henning	1509
Heumann, Wolfgang	1595
Higashi, Mitsutoshi	1166, 1366
Higgins, Leo	1539
Hill, Michael J.	528
Hilton, A.	8
Hiner, Dave	590
Hirai, Hiroyuki	1729
Ho, Heidi S.Y.	1965
Ho, Lung-Hua	316
Ho, Paul S.	1122
Hoff, Lars	139
Hoffrogge, Peter	850
Hoivik, Nils	139
Hölck, O.	625, 1134
Höll, S.	1862
Holmes, Pat	47
Holweg, Gerald	1183
Hong, J.B.	1533
Hong, Shengping	1622
Honrao, Chinmay	1160
Horibe, A.	803
Hoshino, Manabu	1955
Hoshiyama, M.	803
Hou, Fei	1690, 2048
Hou, Fengze	2008
Howell, Keith	425

Hsiao, Zhi-Cheng 290
Hsieh, Robert 26
Hsieh, Wan-Lin 1908
Hsu, Chao-Kai 1891
Hsu, Chih-Chung 1258
Hsu, H.S. 947
Hsu, Yung-Yu 145
Hu, Dyi-Chung 360, 1590
Hu, Hao 759
Hu, Je-Ping 1944
Hu, Y.H. 572
Hu, Yating 1189
Huang, Chen-Yu 1
Huang, Cui 1722
Huang, Hsiao-Chun 868, 947
Huang, Joseph 1533
Huang, Kuo-Hsin 620
Huang, Louie 419
Huang, Po-Tsang 512
Huang, Rui 1122
Huang, Shin-Yi 1908
Huang, Timothy 2266
Huang, Ting-Chia 1160
Huang, Yaping 2236
Huang, Yu-Wei 290, 1891, 1908
Huffman, A. 8, 20
Hui, Ho-Yee 1492
Hummler, Klaus 1648
Hunegnaw, Sara 1652
Hung, Chih-Pin 2112
Hung, Yin-Po 360
Huppert, Gil 145
Hutter, Matthias 1321
Hutzler, Aaron 1700
Huylenbroeck, S.V. 572
Hwang, Lih-Tyng 1303
Hwang, Seung Min 74
Hwang, Wei 512
Hwang, Yuchul 973

I

Iijima, Yu 452
Ikari, Gary 1366
Im, Jay 1122
Imai, H. 899
Imajyo, Nobuhiko 2271

Imanari, Masaaki 342
Imenes, Kristin 139, 498
Imura, Fumito 1915
Indyk, Richard 452
Inomata, Katsumi 1796
Iruvanti, Sushumna 236, 1253
Ishida, Hiroya 404
Ishigure, Takaaki 1042
Ishiguro, Kazuya 1308
Ishino, H. 1179
Islam, Md. R. 1272
Islam, Nokibul 50, 836
Ito, Yuka 1148
Itoh, Toshihiro 1857
Ivankovic, A. 309, 1517
Iwanabe, Keiichiro 2303
Iwata, Yoshiharu 2227
Iyer, Subramanian 647
Izzi, M. 194, 1842

J

Jang, BoRa 582, 1816
Jang, Myong-Gi 447
Jao, Pitfee 789
Jeong, Min-Su 1735
Jeong, Yonghyuk 836
Jeong, Youchul 753
Ji, Jason 1750
Ji, Liang 1116
Ji, Yu 1488, 2299
Jiang, D.S. 1533
Jiang, Feng 1740
Jiang, Hanqing 470
Jiang, Tengfei 1122
Jimbo, Toshihiko 1955
Jin, Howard (Hwa II) 1790
Jin, Yufeng 641, 1902, 1986, 2099
Jin, Zhenrong 535
Jing, Xiangmeng 1116, 1740
Jinka, Oblesh 1782
Jo, Chanmin 2075
Joblot, S. 1296
Johansson, A. 1071
Johansson, Susie 157
John, P. 625
Johnson, Christopher 1577

Johnson, Mark 2014
Johnson, Michael 1965
Jomaa, Houssam 1396
Jonah, Olutola 759
Joshi, Gaurang 119
Joshi, Yogendra 1247
Jourdain, Anne 894
Jouve, A. 906
Juang, Jing-Ye 1908
Jung, Dae Young 1562
Jung, Seong-Yoon 2255
Jürgensen, Nils 1498
Juskey, Frank 1264

K

Kabir, M.S. 1071
Kahle, R. 940
Kaletta, K. 1204
Kamgaing, Telesphor 1264, 1610
Kamlapurkar, Swetha 179
Kanagavel, Senthil 1790
Kandaswamy, Shri Vishnu 1464
Kang, Dongchul 1523
Kang, Kuiwon 1396
Kang, Sung K. 1782
Kang, SungGeun 590
Kang, Un Byung 1128, 1755
Kannan, Sukeshwar 590
Kao, C.R. 2277
Kao, Ming-Jer 290, 1853
Kao, Nicholas 1750
Karikalan, Sam 2119
Karlicek, Jr., Robert F. 2207
Karsli, Kivanc 209
Kashyap, Anirudh 279
Kata, Keiichirou 596
Katkar, Rajesh 1389
Kato, Fumiki 62, 1926
Katoh, Y. 1179
Kaur, Amanpreet 2144, 2168
Kaushik, B.K. 1091, 1921
Kawamoto, S. 803
Kawanami, Satoshi 186
Kawano, Kenji 2003
Kaynak, M. 1204
Ke, Chang-Bo 2249

Keech, John...20
Keezer, David..2260, 2287
Kenyon, Gareth...26
Keser, Beth...100, 925, 1173
Khan, Wasif T..2293
Khim, JooHyun..582
Kholostov, K...194, 1571, 1842
Ki, Wing-Hung..919
Kida, Tsuyoshi...596
Kijkanjanapaiboon, Kasemsak..967
Kikuchi, Katsuya..62, 1915, 1926
Kim, Cheolbok...1103
Kim, Cheolgyu...1004
Kim, Chimyung...1616
Kim, Chin Kwan...1396
Kim, Choong-Un...133, 697
Kim, Chunho...263
Kim, DoHyeong..582, 590, 1816
Kim, Dong Wook..609
Kim, Dongsu..41
Kim, Ga Won...354
Kim, Gwang..50, 2027
Kim, Hakgu..1616
Kim, Henry..2119
Kim, Hui Joong..712
Kim, HyunTai..2027
Kim, Hyup Jong..1103
Kim, Il...1661
Kim, Jaemin..753
Kim, Jinseong...1361
Kim, JooHyun..1816
Kim, Joungho...541
Kim, Jun Chul...41
Kim, June-Bum...1635
Kim, Kwonil..279
Kim, KyungOe..50
Kim, Min Sung...1004
Kim, Namhoon..2119
Kim, Nam-Seog...1122, 1635
Kim, Samuel...1471
Kim, Seung-Ho...841, 2241
Kim, Seung-Hyun..1635
Kim, Sung Jin...1264, 1384
Kim, Sungjoo..2075
Kim, Tae Wan..1060
Kim, Taek-Soo...1004

Kim, Taewan	271, 1616
Kim, YongSang	1816
Kim, Yoosun	841, 2241
Kim, Young Soon	1128, 1661, 1755, 1765
Kim, Young Suk	1853
Kim, Younghoon	354
Kimura, Kazushi	230
Kimura, Michitaka	596
King, A.	601
Kino, H.	856, 1110
Kintaka, Kenji	186
Kitada, Hideki	1853
Klink, G.	1482
Klyshko, A.	194, 1571
Knickerbocker, J.	576, 647, 883, 1372, 1880
Ko, Cheng-Ta	1891
Ko, Wen-Ching	1944
Kobayashi, Makoto	284, 365, 484, 742, 1160, 1384, 1959
Kobayashi, Yuta	913
Kodama, Shoichi	1853
Kodani, K.	1077
Kodemura, Junji	1955
Kohara, S.	647, 803
Koide, Masateru	68
Koiwa, Sumio	1673
Kono, Tsutomu	2042
Kotake, Tomohiko	1407
Koyama, Toshinori	348
Koyanagi, M.	304, 636, 856, 1110, 1148
Kraft, Silke	1700
Krieger, William E.R.	983
Krüger, Michael	114
Ku, Tzu-Kun	1853
Kubo, A.	899
Kuchta, Daniel M.	1016
Kumar, Gokul	541
Kumar, Santosh	712
Kumar, Vobulapuram Ramesh	1091
Kumari, Archana	1921
Kumpulainen, Tero	1685
Kunimoto, Yuji	348
Kuo, An-Yu	1303
Kuo, C.L.	297
Kuo, Chih-Ming	316
Kuo, H.J.	572
Kuo, Kuei Hsiao (Frank)	56

Kuramochi, Satoru 1673
Kurihara, Takashi 348
Kurokawa, Tsuyoshi 1776
Kutlu, Zafer 348
Kutter, C. 1482
Kwark, Young 1955
Kwatra, Abhishek 983

L

La Manna, A. 33, 309
Lachner, R. 956
Laflamme, Simon 179
Lagae, Liesbet 165
Lai, Chiming 2299
Lai, J.Y. 1
Lai, Yi-Che 947, 1750
Lall, Pradeep 85, 242, 379, 666, 990, 1449, 1973, 2053
LaManna, A. 572
Lambert, William J. 528
Lamy, Y. 1296
Lan, Albert 81, 1533
Lan, Jia-Shen 1235
Landesberger, C. 1482
Lang, K.-D. 114, 625, 873, 940, 1033, 1204, 1218, 1321, 1421, 1498, 1759
Langlois, Richard 236, 647
Lannon, J. 8, 1641
Larson, Lyndon 236
Lau, John 56, 290, 1908
Lau, Kei May 919
Lauro, Paul 1308
LaVoie, Annique 236
Law, Edward 518
Lazerand, Thierry 1577
Le, Fuliang 919
Le, Taoran 769
Lee, Bong-Sub 862, 1810
Lee, Byeong Rok 1635, 1735
Lee, Chang-Chi 2112
Lee, Chin C. 1335, 1529
Lee, Ching-Kuan 290, 1891
Lee, ChoonHeung 582, 1361, 1816
Lee, Chu-Chung (Stephen) 1539
Lee, Chung-Fa 2069
Lee, DongHoon 1816
Lee, DongHun 582
Lee, Fred C. 504

Lee, Haksun	1154
Lee, Heng	815
Lee, Inho	342
Lee, Jae Hong	712
Lee, Jason	56
Lee, K.W.	304, 856
Lee, Kangwook	636, 1110, 1148
Lee, Kenny	47
Lee, KiWook	590
Lee, Minkyu	1616
Lee, Ning-Cheng	655
Lee, Rick	56
Lee, S.W. Ricky	919
Lee, Sang Hoon	271, 1060
Lee, Sangmin	2075
Lee, Seungbae	354
Lee, Seungyong	1616
Lee, Shih-Wei	512
Lee, Taeik	1004
Lee, Tae-Kyu	133, 697
Lee, Yil-Hak	342
Lee, YongTaek	2027
Lee, Young Woo	712
Lee, Yuan-Chang	290
Lee, Yuh-Lang	1584
Levine, Lyle E.	1648
Lewandowski, Eric	255
Leyrer, Benjamin	203
Li, Bin	2249
Li, C.	1641
Li, Guangfeng	1231
Li, Heng	1411
Li, Jun	2131
Li, K.H.	324
Li, Lang	1876
Li, Leo Hongyu	2119
Li, Li	1122, 1366
Li, Liyi	631, 1492, 1745, 1848
Li, Long	1690
Li, Menglu	609
Li, Qiang	504
Li, Shidong	1253
Li, Xin	1080
Li, Y.Q.	1212
Li, Yan	1457
Li, Yuan	338

Li, Yuefa	170
Li, Zhe	338
Li, Zhuo	1745, 1828
Liang, Hanshuang	470
Liang, Jiang-Long	1584
Liao, Anmou	2131
Liao, Jash	1750
Liao, Li-Ling	290
Liao, Mark	81
Liao, Sue-Chen	1853
Libsch, Frank	1016
Lii, M.J.	572
Liimatta, Toni	151
Lin, C.F.	297
Lin, Cha-Hsin	1853
Lin, Chun-Tang	947
Lin, Chun-Te	1891
Lin, Chun-Ting	1590
Lin, Edward	748
Lin, Frank M.-S.	620
Lin, Kung-An	316
Lin, Kwang-Lung	419, 1584
Lin, M.J.	297
Lin, Puru	1590
Lin, Wei	647, 1401
Lin, Y.C.	297
Lin, Y.J.	952
Lin, Yang-Kai	1258
Lin, Yaojian	931
Lin, Yi-Fan	1944
Lin, Yu-Min	1891, 1908
Lin, Ziyin	447, 769, 1828, 2231
Lindner, Paul	888
Ling, Feng	2094
Liou, Chang-Ho	1944
Liu, C.S.	572
Liu, Changqing	1348
Liu, Chaojun	1010, 1189
Liu, Cheng-Hsiang	868, 1628
Liu, Chengxun	165
Liu, Duixian	1272
Liu, Fengman	2008, 2131
Liu, Fuhan	1384
Liu, Hsichang	647
Liu, Hung-Wen	1750
Liu, Johan	459

Liu, Kai	1284, 2027
Liu, Kenny	947
Liu, Li	1348
Liu, Sheng	170, 1010, 1189, 1690, 1876, 2048, 2245, 2283
Liu, X.Y.	1212
Liu, Y.	1880
Liu, Yan	447, 2231
Liu, Yingxia	609
Liu, Yi-Wei	1750
Liu, Yong	808, 1241, 1601, 2063
Liu, Yumin	808, 1601
Liu, Zhiming	1652
Lo, Wei-Chung	290, 360, 1891
Lofrano, M.	309
Loh, Chooi Ian	748
Longgood, Stuart	425
Longworth, Hai	236
Lopez-Montesinos, Pedro	342
Lu, Chang-Lun	868, 947, 1628
Lu, Guo-Quan	1080
Lu, Hao	742, 1416
Lu, Hsin-Chia	1944
Lu, Hua	1342
Lu, James J.-Q.	2207
Lu, Jian	1857
Lu, Jian-Qiang	548
Lu, Minhua	690, 1940
Lu, PingHung	26
Lu, Terren	1
Lu, Wengao	2099
Lu, Yongqiang	878
Lu, Yuan	1354, 2008
Lu, Yung Jean (Rachel)	1891
Luckenbach, Thomas	1606
Lueck, M.	8, 20
Luesebrink, Helge	1021
Luo, L.	815, 1212, 1411
Luo, Xiaobing	170, 1189, 1876, 1992
Luo, Yihua	242
Luo, Yu	1782
Luo, Zhang	1690, 2048, 2245
Lv, Cheng	470, 1290
Lwo, Ben-Je	620

M

Ma, Shenglin 1902
Ma, Teng 470
Mackenzie, Kenneth D. 1577
Madenci, Erdogan 973
Maeda, Nobuhide 1853
Maeda, Ryutaro 1857
Maekawa, Katsuhiro 2194
Magaya, Tafadzwa 1652
Maikowske, Stefan 203
Majeed, Bivragh 165
Majumder, M.K. 1921
Maldonado, Alberto 2159
Mallampati, Sandeep 1328
Malta, D. 8
Maman, Avi 1021
Manhas, S.K. 1921
Maniatty, Antoinette 1241
Manier, C.-A. 1204
Mäntysalo, Matti 151, 1685
Mao, Cindy 56
Marcoux, Phil 1071
Maria, Joana 1372
Mariappan, Murugesan 636
Martin, Bill 2124
Martinez, Linnell 1577
Marzano, Frank Silvio 2137
Mason, Maurice 1782
Mathewson, Alan 1064
Matsubara, Takahiro 1028
Matsumae, Takashi 1504
Matsumoto, Rie 1870
Matsushita, Kiyoto 404
Matsuura, Ikkei 1955
Matthias, Thorsten 888
Mauer, Laura B. 878
Mavinkurve, Amar 411
Mawer, Andrew 1539
Mayer, Michael 1471
McCann, Scott R. 2189
McCarson, Brian 2035
McCleary, Roger 523
McCluskey, F. Patrick 1314
Mclaughlin, Kevin 878
McLeod, Mark 1308

McMullen, Tom 126

McNally, P.J. 1517

Mehr, M. Yazdan 1477

Mehta, Gaurav 490

Mei, Yunhui 1080

Meindl, Manfred 1183

Meinecke, Christoph 1218

Melville, Robert 1284

Mendoza, Nestor 1523

Menezes, Gary 1959

Mesh, Michael 1021

Miao, Min 641, 1902, 1986, 2099

Middendorf, Andreas 114

Milanes II, Ninoy 490

Miller, A. 33

Miller, Allen 1033

Miller, Andy 26, 894

Milton, Basil 1523

Min, Max (Sungwan) 354

Minami, Shigenobu 1729

Mino, Masayuki 2042

Mirza, Kazi 990

Misra, E. 1949

Mita, Mamoru 2194

Mitachi, Seiko 230

Miura, Hideo 1885

Miura, Testunosuke 186

Miyamoto, Yoshinori 913

Miyatake, Masato 1407

Miyazaki, Tomokazu 165

Mizuno, Hikaru 1796

Mizusawa, Ryuma 1870

Mizutani, Daisuke 68

Moeini, S. Ali 1314

Mok, Philip K.T. 919

Möller, Eike 1707

Moon, Jeong Tak 712

Moon, Kyoung-Sik 447, 1828, 2231

Moon, Seongjae 2075

Morales, Jorge Mario Herrera 478

Morey, Briana 145

Mori, H. 803

Morikawa, Y. 846, 1697

Morris, Jeffrey 221

Mu, Mingkai 504

Mukai, Kenichiroh 279

Mullen, Don 730
Murai, Hikari 1407
Murayama, Kei 1166
Murayama, T. 846, 1697
Murtuza, Masood 242, 1973
Murugesan, M. 304, 1148
Mustafa, Muhannad 666
Mutnury, Bhyrav 2081
Mutuku, Francis 425, 655, 1769

N

Na, Duk Ju 596
Na, Nanju 2151
Nad, Suddhasattwa 684
Nagao, S. 1077, 1179
Nah, Jae-Woong 1308, 1372
Nair, Vijay 782, 1264, 1278
Nakagawa, Hiroshi 1915
Nakamoto, Mark 1226
Nakamura, Shinya 1523
Nakamuta, Y. 1697
Nakazuru, Kazumi 1028
Nam, Kiyoung 1616
Natarajan, Arun 1272
Nauchi, Takashi 1308
Ndip, Ivan 1498
Neitz, Marcel 1759
Nemoto, Shunsuke 1915
Nenzi, Paolo 1571, 2137
Neo, Chong-Wei 518
Neuhaus, Jan 1549
Newman, Keith 1433
Nguyen, Anh Tuan Thai 139
Nguyen, Hoa 470
Nguyen, Hoang-Vu 498
Nguyen, Luu 242, 1973
Ni, Chih-Hsien 316
Ni, Jiamin 1241
Nicholls, Lou 1361
Nicolas, S. 1714
Niittynen, Juha 151
Niizeki, Shoichi 913
Nimura, Masatsugu 913
Ning, Wenguo 815, 1411
Niotaki, Kyriaki 796
Nishio, Kenzo 186

Nishizono, Shinji 763
Niwa, Hiroyuki 913
Nolmans, P. 33
Nonaka, Toshihisa 913
Noriki, A. 856
Numata, Hidetoshi 179, 1870
Nuss, M. 625

O

Obeng, Yaw 1648
Ochiai, Toshihiko 596
Ochoa, Juan S. 717
O'Connell, Barry 1241
Oh, Dan 566, 748, 1896
Oh, Tac-Keun 1635
O'Halloran, G.M. 411
Ohba, Takayuki 1853
Oi, Kiyoshi 348
Okada, Hironao 1857
Okamoto, Kazuya 2227
Okamoto, Keishiro 68
Okamoto, Kenji 1796
Okoro, Chukwudi 1648
Onishi, M. 304
Ooshima, Nobuhiro 1729
Oppermann, H. 1204, 1321
Orii, Y. 803, 1308
Osaka, Masahiko 1523
Ose, Masahisa 1407
Osenbach, J. 74, 952
Oster, Sasha N. 1610
Ostrowicki, Gregory T. 821
Otaka, S. 348, 899
Oterkus, Erkan 973
Oterkus, Selda 973
O'Toole, Eoin 935
Ou, Jack 2159
Ouyang, Eric 836
Oya, S. 899

P

Pacheco, Mario 1457
Pachler, Walther 1183
Paek, JongSik 590
Pagliucci, Roberto 1571
Paik, Kyung-Wook 271, 841, 1060, 1128, 1661, 1755, 1765, 2241, 2255

Pakbaz, Faraydon 535
Palma, Fabrizio 2137
Pan, Jie 2163
Pan, Po-Chih 2112
Pan, Shiji 2088
Pan, Yi 332, 1253
Pang, Cheng 2163
Papakyrikos, Cole 145
Papapolymerou, John 2293
Parat, Guy 1198
Park, Dongjoo 1361
Park, John 1667
Park, JungSoo 582, 1816
Park, Kyoung Youl 775
Park, S.W 1077, 1179
Park, Susan 2027
Park, Yongsun 1616
Park, Young-Bae 1635, 1735
Parker, Ben 1272
Parker, Richard 425
Parkinson, Dilworth Y. 609
Parkinson, Dula 1457
Parks, Gregory 690
Patnaik, Amalendu 1091
Paul, Jens 590
Pavlidis, Spyridon 2293
Pays-Volard, David 1577
Pedreira, Olalla Varela 613
Pei, Min 395, 684
Peng, Jr-Wei 1584
Peng, Shih-Liang 1
Perfecto, Eric 647, 690, 1308, 1606, 1940
Pesika, Noshir 255
Peterson, Andrew F. 2271
Peterson, K. 1641
Pham, Nam 535
Philipsen, Harold 613, 850
Phommahaxay, Alain 850, 894
Pinho, Nelson 935
Pitarresi, James 119
Pitwon, Richard 1033
Plante, David 236
Pleyer, Wolfgang 1021
Polastre, R. 883, 1880
Pollard, Scott 20
Prange, Jonathan 342

Pressel, Klaus 1183
Prewitz, T. 625
Privett, Mark 894
Prorok, Barton C. 379
Pucha, Raghuram 1098
Pufall, R. 1134

Q

Qasaimeh, A. 371
Qian, Jiangyuan 2088
Qian, Richard 2063
Qin, Hong-Bo 2249
Qin, Ivy 1523
Qin, Y. 342, 2181
Qin, Zheng 2163
Qiu, Delong 1932
Qiu, Jinghao 1690
Qu, Shichun 808
Quay, R. 2181
Queisser, Marco 1759
Questad, D. 332, 1253, 1949

R

Radhakrishnan, Kaladhar 528
Radojcic, Riko 1226
Raghavan, Sathyanarayanan 983
Rahimi, Arian 736, 789
Raj, Milan 145
Raj, P. Markondeya 284, 484, 541, 782, 1160, 1264, 1384, 1492, 1959, 2266
Rajendran, H. 1134
Rajoo, Ranjan 490
Ralph, W. Carter 1433
Ramachandran, Koushik 647
Ramachandran, Vidhya 1226
Ramm, P. 1134
Ranjan, Manish 26
Rao, Valluri R. 1610
Rastjagaev, Eugen 1707
Rastogi, Rajiv 2035
Ratel, David 478
Razdan, Sandeep 1054
Razeeb, Kafil M. 1064
Razzaq, A. 324
Rebibis, K.J. 572
Rebibis, Kenneth 894
Reiner, R. 2181

Ren, Xiaoli 2163
Reynolds, Scott K. 1272
Rimolo-Donadio, Renato 1955
Rissing, Lutz 1595
Roa, Fernando 1656
Robertazzi, Raphael 1372
Roberts, Jordan C. 666
Röder, Julia 1033
Rodriguez, Daniel 2151
Rogoff, Rich 523
Romankiw, Lubomyr T. 1782
Rong, Aosheng 2094
Rongen, René 411
Rosenthal, Christopher 452
Rouhana, Layal 1396
Roy, Rajiv 523
Ruhmer, Klaus 20, 523
Rylyakov, Alexander V. 1016
Ryu, Woonghwan 354, 2075

S

Sabuncuoglu, Deniz 165
Saeidi, Nooshin 2213
Saephan, Chio 2159
Sakai, Taiji 742
Sakalaukus, Peter 2053
Sakamoto, S. 1077
Sakuishi, T. 846, 1697
Sakuma, Katsuyuki 647, 1372
Sakurai, Tomohiko 1796
Sakuyama, Seiki 68
Saleem, A.M. 1071
Sammakia, Bahgat 2175
Samoilov, Arkadii 1679
Sandhu, Javed 518
Santagata, Fabio 1192
Sasaki, Hideki 763
Satio, Yukihiro 2227
Sato, Osamu 452
Sato, Ryohei 2227
Sato, Y. 304
Sato, Yasuhiro 1729
Sato, Yoichiro 365, 1247, 2271
Sato, Yutaka 636
Sawyer, Brett 742, 1416
Scheuermann, Michael 1372

Schletz, Andreas	1700
Schmitz, D.	371
Schmitz, Stefan	114
Schneider, Marc	203
Schoeller, Harry	1328
Schow, Clint L.	1016
Schröder, Henning	1033, 1759
Schultz, Mark	1016
Schutt-Ainé, José E.	717
Schwarz, Mark	100, 925
Secker, Dave	730
Seki, Toshitake	365
Seler, E.	956
Senior, David E.	789, 1103
Seo, Hochul	1616
Seo, SeongMin	582, 1816
Seo, Sun Kyung	1128, 1755
Seo, YoungChul	582
Serenelli, L.	1842
Sextro, Walter	1549
Shaba, Hala	215
Shaddock, David	1328
Shafiee, S.	1071
Shah, Milind	1396
Shan, Lei	1955
Shang, Jintang	1488, 1833, 2299
Shang, Wenya	2163
Shangguan, Dongkai	1116, 1740, 2008, 2131, 2163
Sharifi, Reza	2119
Sharma, Himani	782, 1492, 2266
Shen, Hong	862
Shen, J.H.	1212
Shen, Wen-Wei	1891
Shi, Shawn	263
Shibuya, Hiroki	763
Shigemoto, Takumi	2227
Shih, W.L.	2277
Shim, Yujeong	1896
Shimizu, Kozo	68
Shimizu, Noriyoshi	348
Shimizu, Tadashi	763
Shin, Jaemin	2075
Shin, Jiwon	841, 1128, 1661, 1755, 1765
Shiraishi, Takashi	1048
Shirangi, M.H.	108
Shirazi, S.	371

Shlepnev, Yuriy	730
Shorey, Aric	20, 1103
Shoukrun, Aki	1606
Shuto, Takanori	2303
Sibilia, C.	194
Sikka, Kamal K.	332
Sillanpää, Hannu	151
Simon, Gilles	478, 1198
Simons, V.	309
Sin, Johnny K.O.	919
Sinha, Pankaj	2035
Sitaraman, Srikrishna	1264
Sitaraman, Suresh K.	983, 2189
Skidmore, G.	1641
Skordas, Spyridon	647
Slabbekoorn, John	26
Sleeper, Scott	263
Smet, Vanessa	284, 365, 484, 742, 1054, 1160, 1384, 1959
Smith, Dan	590
Snyder, S.	601
Soler, Vilmarie	1606
Soltani, Ayat	1685
Soma, Kazutomo	1042
Son, Ho-Young	1122, 1635
Son, Yong	590
Sone, H.	803
Song, Chongshen	1116
Sorce, Peter	1308
Souchon, F.	1714
Souriau, Jean-Charles	478, 1198
Spreitzer, Ronald	1278
Sprenger, Daniel J.	1996
Steffek, Roland	1021
Steller, Wolfram	1218
Sten-Nilsen, Bjørnar	498
Stepanov, Stanislav	1021
Stewart, Aaron	119
Stömmer, Christian	1021
Stratton, Ken	1401
Strothmann, Tom	931
Struyf, Herbert	850
Su, Meiying	1116
Su, Peng	1122
Su, Quang	119
Su, Yipeng	504
Suaoke, Kuniaki	1870, 647

Suga, Tadatomo 1143, 1504
Sugahara, T. 1077, 1179
Suganuma, Daisuke 1042
Suganuma, K. 1077, 1179
Sugase, Naoki 452
Sugawara, Mariko 1048
Sugawara, Yohei 1110
Sugiura, K. 1179
Suh, Min-Suk 1122, 1635
Suhling, Jeff 379, 666, 990, 1449
Suk, Kyung-Lim 1661
Sun, Jibin 1054
Sun, Minggang 641, 1986
Sun, Rong 464
Sun, Xiaofeng 1354, 1378
Sun, Xin 641, 1902, 1986, 2099
Sun, Yangyang 47
Sun, Zhuowen 862
Sundaram, Venkatesh 1098, 2189
Sundaram, Venky 284, 365, 541, 742, 1054, 1247, 1264,
.... 1384, 1416, 1427, 1959, 2266, 2271
Sung, Baegin 753
Suthau, Eike 173
Suthiwongshunthorn, Nathapong 490
Suu, K. 846, 1697
Suzuki, A. 1697
Suzuki, Hiroko 763
Suzuki, Ken 1885
Suzuki, Kousuke 1673
Suzuki, Naoya 2042
Suzuki, Ryoichi 1308
Suzuki, Takuji 1729
Suzuki, Yuya 742, 1264, 1416, 1427
Swaminathan, Madhavan 782, 2124, 2260, 2287
Swan, Johanna 1278
Sweatman, Keith 425, 655
Syed, Ahmer 100, 1173
Sze, Henry 518

T

Taddei, John 878
Tai, Rui-Feng 947
Tain, Ra-Min 290, 360
Taira, Yoichi 179, 1870
Takagi, Yutaka 365, 742, 1416, 1427
Takaguchi, Akira 1308

Takahashi, Hiroshi 2042
Takahashi, Naoki 1028
Takahashi, Shintaro 2271
Takanezawa, Shin 1407
Takano, Akihito 1366
Takasugi, Hiroshi 1776
Takegami, Toshifumi 913
Takekoshi, Masaaki 1407
Takenobu, Shotaro 179
Takizawa, Hideo 452
Tamura, K. 899
Tan, C.S. 324
Tan, Keith 518
Tanaka, Kazuhiro 1048
Tanaka, Masato 348
Tanaka, T. 304, 636, 856, 1110, 1148
Taner, Ozgur 967
Tang, Rui 470
Tang, Yin 236
Tanikawa, Seiya 1110
Tao, Jing 1064
Taylor, Robin 1416
Telikepalli, Satyanarayana 2260, 2287
Temple, D. 8, 1641
Tentzeris, M.M. 759, 769
Terajima, Katsushi 763
Teraki, Shin 1776
Thangaraju, Sara 590
Thomas, T. 940
Thomas, Windsor 1103
Thompson, J. 601
Tian, Shurong 1016
Tillack, B. 1204
Tjulkins, Fjodors 139
Tkachenko, Anton 2207
Toepper, Michael 1421
Tokarski, Adam 1700
Tong, Ho-Ming 512
Tong, Jialing 1098, 2271
Töpper, Michael 1498
Toriyama, Kazushige 1308
Toukhy, Medhat 26
Trampert, Stefan 114
Tran, TuAnh 1539
Tsai, Jensen 1533
Tsai, Jyun-Ling 868, 1628

Tsai, Wen-Li 290
Tsang, C. 576, 883, 1372
Tsang, Leung 2200
Tsebo, Simo G. 108
Tseng, Stephen 1
Tsukuda, Tatsuaki 763
Tsunoda, Masatoshi 1028
Tsuruta, K. 1179
Tu, Chia-Jung 316
Tu, K.N. 609
Tuan, Chia-Chi 447, 2231
Tucci, M. 194, 1842
Tummala, Rao 284, 365, 484, 541, 742, 782, 1054, 1098, 1160, 1247,
.... 1264, 1384, 1416, 1427, 1492, 1959, 2189, 2266, 2271
Tung, Bui Thanh 62, 1926
Tunga, K. 1949
Tyberg, Christy 1372
Tyler, P. 601

U

Uhrmann, Thomas 888
Ulusoy, A. Cagri 2293
Unnikrishnan, R. 108
Ura, Shogo 186
Uwataki, R. 1179
Uzoh, Cyprian 862

V

Valdes-Garcia, Alberto 1272
Van Acker, Lut 165
Van der Donck, Tom 613
Van der Plas, G. 309
Van Driel, W.D. 1477
van Zeijl, Henk 1556
Vandevelde, B. 33, 1517
Vanstreels, K. 309
Vardakas, John 796
Varlamava, Volha 2137
Velenis, D. 572
Verbinnen, Greet 894
Vick, E. 8, 1641
Vlasov, Yurii 179
Vogel, D. 1134
Voges, S. 940
Voitsekhivska, Tetiana 173
von Kouwen, Maarten 1464
Vujosevic, Milena 395, 684

W

Wagner, Rebecca 236
Wahrmund, Wieland 873
Walls, Lloyd 535
Walter, H. 625, 1421
Waltereit, P. 2181
Walters, E. 601
Wan, Lixi 1354, 1378, 1932
Wang, Chen-Chao 2112
Wang, Daijiao 1965
Wang, Guanjiang 641, 1986, 2099
Wang, Hanguo 221
Wang, Howard 2175
Wang, Huijuan 2163
Wang, Liang 215, 862
Wang, M.J. 1212
Wang, Nan 459
Wang, Qian 1378
Wang, Qibin 1740
Wang, Qidong 1932
Wang, Shiqiang 1086
Wang, Shuling 1932
Wang, T. 309, 572
Wang, Tao 284, 484
Wang, Tzu-Chang 1258
Wang, W. 1212, 2048, 2245
Wang, Xianyan 145
Wang, Xugang 1932
Wang, Yiman 1876
Wang, Z. 108
Wang, Zheyao 1722
Wang, Zhihua 1378
Wassick, T. 1940, 1949
Watanabe, Manabu 68
Watanabe, Naoya 62, 1915, 1926, 2003
Watanabe, Shoji 348
Weatherspoon, M.R. 601
Webb, Bucknell 576
Weber, Daniel 1033
Wee, K.H. 324
Wegner, M. 1204
Wei, Frank 452
Wei, Jia 1192
Wei, Junchao 1449, 2053
Wei, Pinghung 145

Weigel, R.	956
Wen, Ming	1010, 2048, 2245
Wen, Shengmin	1231, 1361
Wentlent, L.	371
Werner, Randall J.	1253
Westerman, Russell	1577
Whalley, Simon	1033
White, Christopher	1264
Wietstruck, M.	1204
Wilde, Jürgen	1707, 2181
Wilke, M.	1204
Wimplinger, Markus	888
Win, Si T.	2151
Winstel, Kevin	647
Wittler, O.	625
Woertink, Julia	342
Wöhrmann, Markus	1421, 1498
Wojewoda, Leigh	528
Wojnowski, M.	956
Woldt, Gregor	1218
Wolf, H.	1482
Wolf, M.J.	625, 873, 1218
Wolter, Klaus-Juergen	173, 1509
Wong, C.P.	447, 464, 631, 769, 1488, 1492, 1745, 1828, 1833, 1848, 2231, 2299
Wong, C.S.	1517
Wong, K.	952
Woo, Min	126
Wood, James	1606
Wright, S.L.	1880
Wu, C.Y.	297
Wu, Chenglin	1122
Wu, Chuan-Yu	316
Wu, Chung-Hsi	512
Wu, Dong	1722
Wu, Fei-Jain	316
Wu, Hao	470, 1290
Wu, Li Wei	1533
Wu, Mei-Ling	1235
Wu, Peng	2131
Wu, Shang-Lin	512
Wu, Sheng-Tsai	290, 963
Wu, Wei-Hsin	316
Wu, Xiaomeng	2008, 2131
Wu, Xin	1010, 1189, 2048, 2245, 2283
Wu, Yuan-Yun	1529

Wu, Yung Shen ... 316
Wu, Zhongming ... 126
Wu, Zihan ... 1384
Wunderle, B. ... 1134

X

Xia, Deyang ... 2236
Xie, Dongji ... 85, 126
Xie, John ... 338
Xie, Weidong ... 697
Xie, Xiang ... 1378
Xiong, Wei ... 560
Xu, Cheng ... 1116
Xu, Chunlin ... 2283
Xu, Gaowei ... 815, 1411
Xu, Hui ... 1523
Xu, Huili ... 133
Xu, Jiafeng ... 1833
Xu, P. ... 1086, 1212
Xu, Ruqing ... 1648
Xu, Sha ... 1342
Xu, Steven ... 100
Xu, Yichao ... 641, 1986
Xu, Yong ... 170, 1010, 1189, 2048, 2245, 2283
Xue, Jie ... 1366
Xue, Kai ... 1116, 1740

Y

Yacoub-George, E. ... 1482
Yagisawa, Takatoshi ... 1048
Yamada, Hiroshi ... 1729
Yamaguchi, Mitsugu ... 2194
Yamamoto, Tsuyoshi ... 68
Yamasaki, Kazuhiko ... 2194
Yan, Xiaolong ... 1622
Yang, Hyung Suk ... 13
Yang, Melinda (Ling) ... 354
Yang, Ming-Hsien ... 1
Yang, Ping-Feng ... 419, 2112
Yang, T.L. ... 2277
Yang, Wenhua ... 1143
Yang, Xianbo ... 2144, 2168
Yau, YouWen ... 1173
Yazdani, Farhang ... 1667
Ye, J.T. ... 1212
Ye, Jiaotuo ... 815

Ye, Lilei 459
Ye, Tiachun 1932
Yeap, Geoffrey 47
Yeh, C.T. 297
Yeung, Tak-Sang 518
Yin, L. 371, 1328
Yoneta, Kiyoto 2227
Yong, Andy Chang Bum 596
Yoo, Sehoon 447
Yook, Jong-Min 41
Yoon, Jong-Sun 1735
Yoon, Juhoon 1361
Yoon, S.W. 596, 931, 952
Yoon, Yong-Kyu 736, 789, 1103
Yoshikawa, Tomoyasu 1308
Yoshioka, T. 899
You, Eileen 354
You, Se-Ho 354
Youssef, Ramey 878
Yu, Daquan 1116, 1740, 1932, 2163
Yu, Doug C.H. 572
Yu, H. 324, 470, 1290
Yu, J.J. 2277
Yu, Man 2245
Yuan, Cadmus 1192
Yue, C. Patrick 919
Yue, Wu 2249
Yuen, Matthew M.F. 464, 1822
Yun, Feng 2236

Z

Zandén, Carl 459
Zhan, Chau-Jie 290, 1891, 1908
Zhang, Andy 85, 1441
Zhang, Chaoqi 13
Zhang, David C. 2260
Zhang, Di 85
Zhang, G.Q. 1477, 1556
Zhang, Gaugping 464
Zhang, Guoqi 1192
Zhang, Haipeng 1054
Zhang, Jinshen 170
Zhang, Kai 464
Zhang, Li 2299
Zhang, Mingchuan 2163
Zhang, Minyan 2236

Zhang, Pengtu 459

Zhang, Ron 215

Zhang, S.H. 1212

Zhang, Shengzhi 1010, 1690, 2283

Zhang, Tingting 2175

Zhang, Tonglong 1622

Zhang, Weihan 2236

Zhang, Weihong 1622

Zhang, Wenli 504

Zhang, Wenqi 1116

Zhang, Xia 2008

Zhang, Xin-Ping 2249

Zhang, Xuefeng 47

Zhang, Ye 2236

Zhao, Lily 47

Zhao, Wei 1226

Zhao, Xueying 1745

Zhao, Yukun 2236

Zhao, Zhili 1876

Zheng, H.Z. 1212

Zheng, Huai 1876

Zheng, Min 2236

Zhong, Jie 1086

Zhong, Xiao 1902

Zhong, Yinsheng 1822

Zhou, Bite 684

Zhou, Eric 221

Zhou, Feng 1622

Zhou, Longzao 1348

Zhou, Min-Bo 2249

Zhou, Tiao 1679

Zhu, Chunsheng 815, 1411

Zhu, Xiaoxin 1342

Zhu, Xunxun 1378

Zhu, Yunhui 641, 1902, 1986, 2099

Zilch, Christian 1183

Zischler, Sigrid 1700

Zitz, Jeffrey 332, 647

Zohni, Wael 1389

Zoschke, K. 1204

Zou, Simin 441

Zou, Yu 2299

Zou, Zhihua 1290

Zschenderlein, U. 1134

2014 IEEE 64th Electronic Components and Technology Conference

(ECTC 2014)

Lake Buena Vista, Florida, USA
27-30 May 2014

Pages 1517-2307

IEEE Catalog Number: CFP14ECT-POD
ISBN: 978-1-4799-2408-0

Copyright © 2014 by the Institute of Electrical and Electronic Engineers, Inc
All Rights Reserved

Copyright and Reprint Permissions: Abstracting is permitted with credit to the source. Libraries are permitted to photocopy beyond the limit of U.S. copyright law for private use of patrons those articles in this volume that carry a code at the bottom of the first page, provided the per-copy fee indicated in the code is paid through Copyright Clearance Center, 222 Rosewood Drive, Danvers, MA 01923.

For other copying, reprint or republication permission, write to IEEE Copyrights Manager, IEEE Service Center, 445 Hoes Lane, Piscataway, NJ 08854. All rights reserved.

***This publication is a representation of what appears in the IEEE Digital Libraries. Some format issues inherent in the e-media version may also appear in this print version.**

IEEE Catalog Number: CFP14ECT-POD
ISBN 13:   978-1-4799-2408-0

Additional Copies of This Publication Are Available From:

Curran Associates, Inc
57 Morehouse Lane
Red Hook, NY 12571 USA
Phone: (845) 758-0400
Fax: (845) 758-2633
E-mail: curran@proceedings.com
Web: www.proceedings.com

TABLE OF CONTENTS

1: Interposer Technologies
Chairs: Subhash L. Shinde, *Sandia National Laboratory*
John Knickerbocker, *IBM Corporation*

Integration Study of Die Strength and Various Bumping Volume and Reliability Performance on 2.5D Silicon Interposer Assembly 1
Shih-Liang Peng, *Siliconware Precision Industries Co., Ltd.*; Chen-Yu Huang, *Siliconware Precision Industries Co., Ltd.*; Ming-Hsien Yang, *Siliconware Precision Industries Co., Ltd.*; Stephen Tseng, *Siliconware Precision Industries Co., Ltd.*; J.Y. Lai, *Siliconware Precision Industries Co., Ltd.*; Terren Lu, *Siliconware Precision Industries Co., Ltd.*; Hsien-Wen Chen, *Siliconware Precision Industries Co., Ltd.*; Steve Chiu, *Siliconware Precision Industries Co., Ltd.*; Stephen Chen, *Siliconware Precision Industries Co., Ltd.*

Process Integration, Improvements, and Testing of Si Interposers for Embedded Computing Applications 8
S. Goodwin, *RTI International*; J. Lannon, Jr., *RTI International*; A. Hilton, *RTI International*; A. Huffman, *RTI International*; M. Lueck, *RTI International*; E. Vick, *RTI International*; G. Cunningham, *RTI International*; D. Malta, *RTI International*; C. Gregory, *RTI International*; D. Temple, *RTI International*

Mechanically Flexible Interconnects with Highly Scalable Pitch and Large Stand-off Height for Silicon Interposer Tile and Bridge Interconnection 13
Chaoqi Zhang, *Georgia Institute of Technology*; Hyung Suk Yang, *Georgia Institute of Technology*; Muhannad S. Bakir, *Georgia Institute of Technology*

Advancements in Fabrication of Glass Interposers 20
Aric Shorey, *Corning Incorporated*; Philippe Cochet, *Rudolph Technologies*; Alan Huffman, *RTI International*; John Keech, *Corning Incorporated*; Matt Lueck, *RTI International*; Scott Pollard, *Corning Incorporated*; Klaus Ruhmer, *Rudolph Technologies*

Large Area Interposer Lithography 26
Warren Flack, *Ultratech, Inc.*; Robert Hsieh, *Ultratech, Inc.*; Gareth Kenyon, *Ultratech, Inc.*; Manish Ranjan, *Ultratech, Inc.*; John Slabbekoorn, *IMEC*; Andy Miller, *IMEC*; Eric Beyne, *IMEC*; Medhat Toukhy, *AZ Electronics Materials USA Corporation*; PingHung Lu, *AZ Electronics Materials USA Corporation*; Chunwei Chen, *AZ Electronics Materials USA Corporation*

Minimizing Interposer Warpage by Process Control and Design Optimization 33
Mikael Detalle, *IMEC*; B. Vandevelde, *IMEC*; P. Nolmans, *IMEC*; J. De Messemaeker, *IMEC*; M. Gonzalez, *IMEC*; A. Miller, *IMEC*; A. La Manna, *IMEC*; G. Beyer, *IMEC*; E. Beyne, *IMEC*

High Performance IPDs (Integrated Passive Devices) and TGV (Through Glass Via) Interposer Technology Using the Photosensitive Glass 41
Jong-Min Yook, *Korea Electronics Technology Institute*; Dongsu Kim, *Korea Electronics Technology Institute*; Jun Chul Kim, *Korea Electronics Technology Institute*

2: Advances in Copper Pillar & Solder Based Flip Chip Technologies
Chairs: Tom Gregorich, *Micron*
Bernd Ebersberger, *Intel Corporation*

Challenges and Opportunities of Chip Package Interaction with Fine Pitch Cu Pillar for 28nm 47
Andy Bao, *Qualcomm, Inc.*; Lily Zhao, *Qualcomm, Inc.*; Yangyang Sun, *Qualcomm, Inc.*; Michael Han, *Qualcomm, Inc.*; Geoffrey Yeap, *Qualcomm, Inc.*; Steve Bezuk, *Qualcomm, Inc.*; Pat Holmes, *Qualcomm, Inc.*; Cecille Alcira, *Qualcomm, Inc.*; Xuefeng Zhang, *Qualcomm, Inc.*; Kenny Lee, *Qualcomm, Inc.*

Electromigration for Advanced Cu Interconnect and the Challenges with Reduced Pitch Bumps 50
Nokibul Islam, *STATS ChipPAC, Inc.*; Gwang Kim, *STATS ChipPAC, Inc.*;
KyungOe Kim, *STATS ChipPAC, Inc.*

Electromigration Performance of Cu Pillar Bump for Flip Chip Packaging with Bump on Trace by Using Thermal Compression Bonding 56
Kuei Hsiao (Frank) Kuo, *Siliconware Precision Industries Co., Ltd.*; Jason Lee, *Siliconware Precision Industries Co., Ltd.*; F.L. Chien, *Siliconware Precision Industries Co., Ltd.*; Rick Lee, *Siliconware Precision Industries Co., Ltd.*; Cindy Mao, *Siliconware Precision Industries Co., Ltd.*; John Lau, *ITRI*

Flip-Chip Bonding Alignment Accuracy Enhancement Using Self-Aligned Interconnection Elements to Realize Low-Temperature Construction of Ultrafine-Pitch Copper Bump Interconnections 62
Bui Thanh Tung, *Nanoelectronics Research Institute; Institute for Photonics-Electronics Convergence System Technology*; Naoya Watanabe, *Nanoelectronics Research Institute*; Fumiki Kato, *Nanoelectronics Research Institute*; Katsuya Kikuchi, *Nanoelectronics Research Institute*; Masahiro Aoyagi, *Nanoelectronics Research Institute; Institute for Photonics-Electronics Convergence System Technology*

Development of Second-Level Connection Method for Large-Size CPU Package 68
Shunji Baba, *Fujitsu Advanced Technologies, Ltd.*; Masateru Koide, *Fujitsu Advanced Technologies, Ltd.*; Manabu Watanabe, *Fujitsu Advanced Technologies, Ltd.*; Kenji Fukuzono, *Fujitsu Advanced Technologies, Ltd.*; Tsuyoshi Yamamoto, *Fujitsu Advanced Technologies, Ltd.*; Seiki Sakuyama, *Fujitsu Laboratories, Ltd.*; Kozo Shimizu, *Fujitsu Laboratories, Ltd.*; Keishiro Okamoto, *Fujitsu Laboratories, Ltd.*; Daisuke Mizutani, *Fujitsu Laboratories, Ltd.*

Development of Fine Pitch Area Array Cu Pillar/Lead Free Solder Bumps for Large 28nm Die in Large Organic Flip Chip Packages 74
John Osenbach, *LSI Corporation*; Sue Emerich, *LSI Corporation*; S. Cate, *LSI Corporation*; D. Brady, *Amkor Technology, Inc.*; Seung Min Hwang, *Amkor Technology, Inc.*; J. Dang, *Kyocera America Inc.*; D. Crouthamel, *LSI Corporation*

ELK Delaminate Improvement Methodology on Cu Pillar Interconnect BOP Structure 81
Nistec Chang, *Siliconware Precision Industries Co., Ltd.*; Albert Lan, *Siliconware Precision Industries Co., Ltd.*; Mark Liao, *Siliconware Precision Industries Co., Ltd.*; Eason Chen, *Siliconware Precision Industries Co., Ltd.*

3: Dynamic Mechanical Characterization
Chairs: Darvin R. Edwards, *Edwards Enterprises*
Tim Chaudhry, *Broadcom Corporation*

Transient Dynamics Model and 3D-DIC Analysis of New-Candidate for JEDEC JESD22-B111 Test Board 85
Pradeep Lall, *Auburn University*; Kalyan Dornala, *Auburn University*; Di Zhang, *Auburn University*; Dongji Xie, *Nvidia Corporation*; Andy Zhang, *Texas Instruments, Inc.*

Interconnect Reliability Prediction for Wafer Level Packages (WLP) for Temperature Cycle and Drop Load Conditions 100
Tong Cui, *Qualcomm Technologies, Inc.*; Ahmer Syed, *Qualcomm Technologies, Inc.*; Beth Keser, *Qualcomm Technologies, Inc.*; Rey Alvarado, *Qualcomm Technologies, Inc.*; Steven Xu, *Qualcomm Technologies, Inc.*; Mark Schwarz, *Qualcomm Technologies, Inc.*

A Novel Drop Test Methodology for Highly Stressed Interconnects in Automotive Electronic Control Units 108

M.H. Shirangi, *Robert Bosch GmbH*; Simo G. Tsebo, *Robert Bosch GmbH*; Z. Wang, *Bosch Automotive Products (Suzhou) Co., Ltd.*; R. Unnikrishnan, *RWTH Aachen University*; T. Heinrich, *Robert Bosch GmbH*

Early-State Crack Detection Method for Heel-Cracks in Wire Bond Interconnects 114

Michael Krüger, *Technical University Berlin*; Stefan Trampert, *Fraunhofer IZM*; Andreas Middendorf, *Technical University Berlin*; Stefan Schmitz, *Fraunhofer IZM*; Klaus-Dieter Lang, *Technical University Berlin*

Accelerated Vibration Reliability Testing of Electronic Assemblies Using Sine Dwell with Resonance Tracking 119

Quang Su, *Binghamton University*; James Pitarresi, *Binghamton University*; Mohammad Gharaibeh, *Binghamton University*; Aaron Stewart, *Binghamton University*; Gaurang Joshi, *Binghamton University*; Martin Anselm, *Universal Instruments Corporation*

Crack Monitoring and Life Modeling Technique Towards High Thermal Cyclic and Mechanical Reliability of fcBGA Solder Joint 126

Dongji Xie, *Nvidia Corporation*; Zhongming Wu, *Nvidia Corporation*; Min Woo, *Nvidia Corporation*; Tom McMullen, *Nvidia Corporation*

Fatigue Properties of Lead-Free Solder Joints in Electronic Packaging Assembly Investigated by Isothermal Cyclic Shear Fatigue 133

Huili Xu, *University of Texas, Arlington; Intel Corporation*; Tae-Kyu Lee, *Cisco Systems, Inc.*; Choong-Un Kim, *University of Texas, Arlington*

4: Bio & Flexible Electronics

Chairs: Joana Maria, *IBM Corporation*
C.S. Premachandran, *GLOBALFOUNDRIES*

MEMS-Based Implantable Heart Monitoring System with Integrated Pacing Function 139

Fjodors Tjulkins, *Buskerud and Vestfold University College*; Anh Tuan Thai Nguyen, *Buskerud and Vestfold University College*; Erik Andreassen, *Buskerud and Vestfold University College; SINTEF Materials and Chemistry*; Nils Hoivik, *Buskerud and Vestfold University College*; Knut Aasmundtveit, *Buskerud and Vestfold University College*; Lars Hoff, *Buskerud and Vestfold University College*; Ole Johannes Grymyr, *Oslo University Hospital Intervention Centre*; Per Steinar Halvorsen, *Oslo University Hospital Intervention Centre*; Kristin Imenes, *Buskerud and Vestfold University College*

Archipelago Platform for Skin-Mounted Wearable and Stretchable Electronics 145

Yung-Yu Hsu, *MC10, Inc.*; Cole Papakyrikos, *MC10, Inc.*; Milan Raj, *MC10, Inc.*; Mitul Dalal, *MC10, Inc.*; Pinghung Wei, *MC10, Inc.*; Xianyan Wang, *MC10, Inc.*; Gil Huppert, *MC10, Inc.*; Briana Morey, *MC10, Inc.*; Roozbeh Ghaffari, *MC10, Inc.*

Inkjet Printing in Manufacturing of Stretchable Interconnects 151

Toni Liimatta, *Tampere University of Technology*; Eerik Halonen, *Tampere University of Technology*; Hannu Sillanpää, *Tampere University of Technology*; Juha Niittynen, *Tampere University of Technology*; Matti Mäntysalo, *Tampere University of Technology*

Ultra Small Hearing Aid Electronic Packaging Enabled by Chip-in-Flex 157

John Dzarnoski, *Starkey Hearing Technologies*; Susie Johansson, *Starkey Hearing Technologies*

Fabrication of Silicon Based Microfluidics Device for Cell Sorting Application 165

Bivragh Majeed, *IMEC*; Chengxun Liu, *IMEC*; Lut Van Acker, *IMEC*; Robert Daily, *IMEC*; Tomokazu Miyazaki, *JSR Micro NV*; Deniz Sabuncuoglu, *IMEC*; Liesbet Lagae, *IMEC*

A Novel 3D Neural Probe with Integrated Channel and Its Package 170

Xingming Fu, *Wuhan University*; Yong Xu, *Wayne State University*; Yuefa Li, *Wayne State University*; Jinshen Zhang, *Wayne State University*; Xiaobing Luo, *Huazhong University of Science & Technology*; Sheng Liu, *Wuhan University*

CMOS Multiplexer for Portable Biosensing System with Integrated Microfluidic Interface 173

Tetiana Voitsekhivska, *Technical University, Dresden*; Eike Suthau, *Technical University, Dresden*; Klaus-Juergen Wolter, *Technical University, Dresden*

5: Silicon Photonics & LEDs

Chairs: Fuad Doany, *IBM Corporation*

Stefan Weiss, *II-VI Laser Enterprise GmbH*

Assembly of Mechanically Compliant Interfaces between Optical Fibers and Nanophotonic Chips 179

Tymon Barwicz, *IBM Corporation*; Yoichi Taira, *IBM Corporation*; Hidetoshi Numata, *IBM Corporation*; Nicolas Boyer, *IBM Corporation*; Stephane Harel, *IBM Corporation*; Swetha Kamlapurkar, *IBM Corporation*; Shotaro Takenobu, *Asahi Glass Corporation*; Simon Laflamme, *IBM Corporation*; Sebastian Engelmann, *IBM Corporation*; Yurii Vlasov, *IBM Corporation*; Paul Fortier, *IBM Corporation*

Proposal of Integrated-Optic Wavelength-Selective Modulator Based on Coupling-Efficiency Control of Distributed Bragg Reflector in Straight Waveguide 186

Shogo Ura, *Kyoto Institute of Technology*; Testunosuke Miura, *Kyoto Institute of Technology*; Satoshi Kawanami, *Kyoto Institute of Technology*; Kenji Kintaka, *National Institute of Advanced Industrial Science and Technology*; Kosuke Asai, *Kyoto Institute of Technology*; Kenzo Nishio, *Kyoto Institute of Technology*; Yasuhiro Awatsuji, *Kyoto Institute of Technology*

Porous Silicon Technology, a Breakthrough for Silicon Photonics: From Packaging to Monolithic Integration 194

M. Balucani, *Sapienza University of Rome*; A. Klyshko, *Sapienza University of Rome*; K. Kholostov, *Sapienza University of Rome*; A. Benedetti, *Sapienza University of Rome*; A. Belardini, *Sapienza University of Rome*; C. Sibilia, *Sapienza University of Rome*; M. Izzi, *Enea Casaccia Research Centre Rome*; M. Tucci, *Enea Casaccia Research Centre Rome*; H. Bandarenka, *Belarusian State University of Informatics and Radioelectronics*; V. Bondarenko, *Belarusian State University of Informatics and Radioelectronics*

High Power Density LED Modules with Silver Sintering Die Attach on Aluminum Nitride Substrates 203

Marc Schneider, *Karlsruhe Institute of Technology*; Benjamin Leyrer, *Karlsruhe Institute of Technology*; Christian Herbold, *Karlsruhe Institute of Technology*; Stefan Maikowske, *Karlsruhe Institute of Technology*

Effect of Optical Design on the Thermal Management for the Smart TV LED Backlight Systems 209

Kivanc Karsli, *Vestel AS*; Mehmet Arik, *Ozyegin University*

Wafer Level LED Packaging with Optimal Light Output and Thermal Dissipation for High-Brightness Lighting 215

Liang Wang, *Invensas Corporation*; Gabe Guevara, *Invensas Corporation*; Hala Shaba, *Invensas Corporation*; Roseann Alatorre, *Invensas Corporation*; Rey Co, *Invensas Corporation*; Ron Zhang, *Invensas Corporation*

High Power Laser Packaging Challenges and Standardization 221

Eric Zhou, *LDX Optronics, Inc.*; Jeffrey Morris, *LDX Optronics, Inc.*; Hanguo Wang, *University of California, Los Angeles*

6: Adhesives, Underfills, and Thermal Interface Materials
Chairs: Don Frye, *ATMI*
C. Robert Kao, *National Taiwan University*

Novel Highly Moisture Resistant Optical Adhesives and Their High Power Resistivity 230
Seiko Mitachi, *Tokyo University of Technology*; Kazushi Kimura, *Yokohama Rubber Co. Ltd.*

Engineered Thermal Interface Material 236
Lyndon Larson, *Dow Corning Corporation*; Yin Tang, *Dow Corning Corporation*; Loren Durfee, *Dow Corning Corporation*; Cassandra Hale, *Dow Corning Corporation*; David Plante, *Dow Corning Corporation*; Sushumna Iruvanti, *IBM Corporation*; Rebecca Wagner, *IBM Corporation*; Taryn Davis, *IBM Corporation*; Hai Longworth, *IBM Corporation*; Annique LaVoie, *IBM Corporation*; Richard Langlois, *IBM Corporation*

Degradation Mechanisms in Electronic Mold Compounds Subjected to High Temperature in Neighborhood of 200°C 242
Pradeep Lall, *Auburn University*; Shantanu Deshpande, *Auburn University*; Yihua Luo, *Auburn University*; Mike Bozack, *Auburn University*; Luu Nguyen, *Texas Instruments*; Masood Murtuza, *Texas Instruments*

Time, Temperature, and Mechanical Fatigue Dependence on Underfill Adhesion 255
Joseph Cremaldi, *Tulane University*; Michael Gaynes, *IBM Corporation*; Peter Brofman, *IBM Corporation*; Noshir Pesika, *Tulane University*; Eric Lewandowski, *IBM Corporation*

Study on Isotropic Electrically Conductive Adhesive for Medical Device Applications 263
Shawn Shi, *Medtronic, Inc.*; Scott Sleeper, *Medtronic, Inc.*; Chunho Kim, *Medtronic, Inc.*

Effect of Aligned Nanofiber in Nanofiber Solder Anisotropic Conductive Films (ACFs) on the Solder Ball Movement for Flex-on-Flex (FOF) Assembly 271
Tae-Wan Kim, *Korea Advanced Institute of Science and Technology (KAIST)*; Sang-Hoon Lee, *Korea Advanced Institute of Science and Technology (KAIST)*; Kyung-Wook Paik, *Korea Advanced Institute of Science and Technology (KAIST)*

Adhesive Enabling Technology for Directly Plating Metal on Molding Resin 279
Kwonil Kim, *Atotech USA, Inc.*; Kenichiroh Mukai, *Atotech USA, Inc.*; Brian Eastep, *Atotech USA, Inc.*; Lee Gaherty, *Atotech USA, Inc.*; Anirudh Kashyap, *Atotech USA, Inc.*; Lutz Brandt, *Atotech USA, Inc.*

7: Interposers & 3D Integration
Chairs: Katsuyuki Sakuma, *IBM Corporation*
Lou Nicholls, *Amkor Technology, Inc*

Modeling, Design, and Demonstration of Low-Temperature Cu Interconnections to Ultra-Thin Glass Interposers at 20 µm Pitch 284
Tao Wang, *Georgia Institute of Technology*; Vanessa Smet, *Georgia Institute of Technology*; Makoto Kobayashi, *Namics Corporation*; Venky Sundaram, *Georgia Institute of Technology*; P. Mardkondeya Raj, *Georgia Institute of Technology*; Rao Tummala, *Georgia Institute of Technology*

Low-Cost TSH (Through-Silicon Hole) Interposers for 3D IC Integration 290

John H. Lau, *Industrial Technology Research Institute (ITRI)*; Ching-Kuan Lee, *Industrial Technology Research Institute (ITRI)*; Chau-Jie Zhan, *Industrial Technology Research Institute (ITRI)*; Sheng-Tsai Wu, *Industrial Technology Research Institute (ITRI)*; Yu-Lin Chao, *Industrial Technology Research Institute (ITRI)*; Ming-Ji Dai, *Industrial Technology Research Institute (ITRI)*; Ra-Min Tain, *Industrial Technology Research Institute (ITRI)*; Heng-Chieh Chien, *Industrial Technology Research Institute (ITRI)*; Chun-Hsien Chien, *Industrial Technology Research Institute (ITRI)*; Ren-Shin Cheng, *Industrial Technology Research Institute (ITRI)*; Yu-Wei Huang, *Industrial Technology Research Institute (ITRI)*; Yuan-Chang Lee, *Industrial Technology Research Institute (ITRI)*; Zhi-Cheng Hsiao, *Industrial Technology Research Institute (ITRI)*; Wen-Li Tsai, *Industrial Technology Research Institute (ITRI)*; Pai-Cheng Chang, *Industrial Technology Research Institute (ITRI)*; Huan-Chun Fu, *Industrial Technology Research Institute (ITRI)*; Yu-Mei Cheng, *Industrial Technology Research Institute (ITRI)*; Li-Ling Liao, *Industrial Technology Research Institute (ITRI)*; Wei-Chung Lo, Industrial Technology Research Institute (ITRI); Ming-Jer Kao, *Industrial Technology Research Institute (ITRI)*

Cu Pattern Density Impacts on 2.5D TSI Warpage Using Experimental and FEM Analysis 297

C.T. Yeh, *United Microelectronics Corporation*; C.Y. Wu, *United Microelectronics Corporation*; C.F. Lin, *United Microelectronics Corporation*; K.M. Chen, *United Microelectronics Corporation*; M.J. Lin, *United Microelectronics Corporation*; Y.C. Lin, *United Microelectronics Corporation*; C.L. Kuo, *United Microelectronics Corporation*

A Resilient 3-D Stacked Multicore Processor Fabricated Using Die-Level 3-D Integration and Backside TSV Technologies 304

K.W. Lee, *Tohoku University*; H. Hashimoto, *Tohoku University*; M. Onishi, *Tohoku University*; Y. Sato, *Tohoku University*; M. Murugesan, *Tohoku University*; J.-C. Bae, *Tohoku University*; T. Fukushima, *Tohoku University*; T. Tanaka, *Tohoku University*; M. Koyanagi, *Tohoku University*

3D Stacking Induced Mechanical Stress Effects 309

V. Cherman, *IMEC*; G. Van der Plas, *IMEC*; J. De Vos, *IMEC*; A. Ivankovic, *IMEC; KU Leuven*; M. Lofrano, *IMEC*; V. Simons, *IMEC*; M. Gonzalez, *IMEC*; K. Vanstreels, *IMEC*; T. Wang, *IMEC*; R. Daily, *IMEC*; W. Guo, *IMEC*; G. Beyer, *IMEC*; A. La Manna, *IMEC*; I. De Wolf, *IMEC*; E. Beyne, *IMEC*

Six-Die Stacking: Three-Dimensional Interconnects Using Au and Pillar Bumps 316

Fei-Jain Wu, *Chipbond Technology Corporation*; Lung-Hua Ho, *Chipbond Technology Corporation*; Chih-Ming Kuo, *Chipbond Technology Corporation*; Chia-Jung Tu, *Chipbond Technology Corporation*; Chih-Hsien Ni, *Chipbond Technology Corporation*; Shih-Chieh Chang, *Chipbond Technology Corporation*; Chuan-Yu Wu, *Chipbond Technology Corporation*; Kung-An Lin, *Chipbond Technology Corporation*; Wei-Hsin Wu, *Chipbond Technology Corporation*; Yung Shen Wu, *Chipbond Technology Corporation*

TSV-Less 3D Stacking of MEMS and CMOS via Low Temperature Al-Au Direct Bonding with Simultaneous Formation of Hermetic Seal 324

S.L. Chua, *Nanyang Technological University*; A. Razzaq, *Nanyang Technological University*; K.H. Wee, *DSO National Laboratory*; K.H. Li, *Nanyang Technological University*; H. Yu, *Nanyang Technological University*; C.S. Tan, *Nanyang Technological University*

8: Flip Chip Packaging & Advanced Substrate
Chairs: Young-Gon Kim, *IDT*
Omar Bchir, *Qualcomm, Inc.*

Chip Package Interaction: An Experiment Study on White Bump Mitigation Using Flat Laminates 332

Yi Pan, *IBM Corporation*; Jeffrey A. Zitz, *IBM Corporation*; David L. Questad, *IBM Corporation*; Kamal K. Sikka, *IBM Corporation*

Design and Package Technology Development of Face-to-Face Die Stacking as a Low Cost Alternative for 3D IC Integration 338

Zhe Li, *Altera Corporation*; Yuan Li, *Altera Corporation*; John Xie, *Altera Corporation*

From C4 to Micro-Bump: Adapting Lead Free Solder Electroplating Processes to Next-Gen Advanced Packaging Applications 342

Julia Woertink, *Dow Electronic Materials*; Yi Qin, *Dow Electronic Materials*; Jonathan Prange, *Dow Electronic Materials*; Pedro Lopez-Montesinos, *Dow Electronic Materials*; Inho Lee, *Dow Electronic Materials*; Yil-Hak Lee, *Dow Electronic Materials*; Masaaki Imanari, *Dow Electronic Materials*; Jianwei Dong, *Dow Electronic Materials*; Jeffrey Calvert, *Dow Electronic Materials*

Development of New 2.5D Package with Novel Integrated Organic Interposer Substrate with Ultra-Fine Wiring and High Density Bumps 348

Kiyoshi Oi, *Shinko Electric Industries Company, Ltd.*; Satoshi Otake, *Shinko Electric Industries Company, Ltd.*; Noriyoshi Shimizu, *Shinko Electric Industries Company, Ltd.*; Shoji Watanabe, *Shinko Electric Industries Company, Ltd.*; Yuji Kunimoto, *Shinko Electric Industries Company, Ltd.*; Takashi Kurihara, *Shinko Electric Industries Company, Ltd.*; Toshinori Koyama, *Shinko Electric Industries Company, Ltd.*; Masato Tanaka, *Shinko Electric Industries Company, Ltd.*; Lavanya Aryasomayajula, *GLOBALFOUNDRIES, Inc.*; Zafer Kutlu, *GLOBALFOUNDRIES, Inc.*

Package Embedded Decoupling Capacitor Impact on Core Power Delivery Network for ARM SoC Application 354

Ga Won Kim, *Samsung Semiconductor Inc.*; Max (Sungwan) Min, *Samsung Semiconductor Inc.*; Melinda (Ling) Yang, *Samsung Semiconductor Inc.*; Anil Gundurao, *Samsung Semiconductor Inc.*; Eileen You, *Samsung Semiconductor Inc.*; Harpreet Gill, *Samsung Semiconductor Inc.*; Seungyong Cha, *Samsung Electronics Corporation*; Younghoon Kim, *Samsung Electronics Corporation*; Se-Ho You, *Samsung Electronics Corporation*; Seungbae Lee, *Samsung Electronics Corporation*; Woonghwan Ryu, *Samsung Electronics Corporation*

Embed Glass Interposer to Substrate for High Density Interconnection 360

Dyi-Chung Hu, *Unimicron Technology Corporation*; Yin-Po Hung, *Unimicron Technology Corporation*; Yu-Hua Chen, *Unimicron Technology Corporation*; Ra-Min Tain, *Unimicron Technology Corporation*; Wei-Chung Lo, *Industrial Technology Research Institute (ITRI)*

First Demonstration of a Surface Mountable, Ultra-Thin Glass BGA Package for Smart Mobile Logic Devices 365

Venky Sundaram, *Georgia Institute of Technology*; Yoichiro Sato, *Asahi Glass Company*; Toshitake Seki, *NGK Spark Plug Co., Ltd.*; Yutaka Takagi, *NGK Spark Plug Co., Ltd.*; Vanessa Smet, *Georgia Institute of Technology*; Makoto Kobayashi, *Namics Corporation*; Rao Tummala, *Georgia Institute of Technology*

9: Interconnect Reliability

Chairs: Tz-Cheng Chiu, *National Cheng Kung University*
Vikas Gupta, *Texas Instruments*

Towards a Quantitative Mechanistic Understanding of the Thermal Cycling of SnAgCu Solder Joints 371

D. Schmitz, *Binghamton University*; S. Shirazi, *Binghamton University*; L. Wentlent, *Binghamton University*; S. Hamasha, *Binghamton University*; L. Yin, *GE Global Research*; A. Qasaimeh, *Tennessee Tech University*; P. Borgesen, *Binghamton University*

Exploration of Aging Induced Evolution of Solder Joints Using Nanoindentation and Microdiffraction 379

Mohammad Hasnine, *Auburn University*; Jeffrey C. Suhling, *Auburn University*; Barton C. Prorok, *Auburn University*; Michael J. Bozack, *Auburn University*; Pradeep Lall, *Auburn University*

Accessing Adhesive Induced Risk for BGAs in Temperature Cycling 395

Guruprasad Arakere, *Intel Corporation*; Milena Vujosevic, *Intel Corporation*; Min Pei, *Intel Corporation*

Characteristics of Ceramic BGA Using Polymer Core Solder Balls 404

Hiroya Ishida, *Sekisui Chemical Co., Ltd.*; Kiyoto Matsushita, *Sekisui Chemical Co., Ltd.*

Lifetime Prediction of Cu-Al Wire Bonded Contacts for Different Mould Compounds 411

René Rongen, *NXP Semiconductors*; G.M. O'Halloran, *NXP Semiconductors*; Amar Mavinkurve, *NXP Semiconductors*; Leon Goumans, *NXP Semiconductors*; Mark-Luke Farrugia, *NXP Semiconductors*

The Corrosion Performance of Cu Alloy Wire Bond on Al Pad in Molding Compounds of Various Chlorine Contents under Biased-HAST 419

Ying-Ta Chiu, *ASE Group*; Tzu-Hsing Chiang, *ASE Group*; Yin-Fa Chen, *ASE Group*; Ping-Feng Yang, *ASE Group*; Louie Huang, *ASE Group*; Kwang-Lung Lin, *National Cheng Kung University*

The Effect of Nickel Microalloying on Thermal Fatigue Reliability and Microstructure of SAC105 and SAC205 Solders 425

Richard Coyle, *Alcatel-Lucent*; Richard Parker, *iNEMI*; Babak Arfaei, *Universal Instruments*; Francis Mutuku, *Binghamton University*; Keith Sweatman, *Nihon Superior Co., Ltd.*; Keith Howell, *Nihon Superior Co., Ltd.*; Stuart Longgood, *Delphi*; Elizabeth Benedetto, *Hewlett Packard Company*

10: Novel Materials & Processes
Chairs: Ivan Shubin, *Oracle*
Bing Dang, *IBM Corporation*

Flexible Non-Volatile Cu/CuxO/Ag ReRAM Memory Devices Fabricated Using Ink-Jet Printing Technology 441

Simin Zou, *Auburn University*; Michael C. Hamilton, *Auburn University*

Ultra-High Refractive Index LED Encapsulant 447

Chia-Chi Tuan, *Georgia Institute of Technology*; Ziyin Lin, *Georgia Institute of Technology*; Yan Liu, *Georgia Institute of Technology*; Kyoung-Sik Moon, *Georgia Institute of Technology*; Sehoon Yoo, *Korea Institute of Industrial Technology*; Myong-Gi Jang, *EI Lighting Co. Ltd.*; Ching-Ping Wong, *Georgia Institute of Technology*; *Chinese University of Hong Kong*

A Novel Methodology for Wafer-Specific Feed-Forward Management of Backside Silicon Removal by Wafer Grinding for Optimized Through Silicon Via Reveal 452

Tyson Alvanos, *Disco Hi Tec America, Inc.*; John Garant, *IBM Corporation*; Yu Iijima, *Disco Hi Tec America, Inc.*; Richard Indyk, *IBM Corporation*; Christopher Rosenthal, *Lasertec USA, Inc.*; Osamu Sato, *Lasertec Corporation*; Naoki Sugase, *Lasertec Corporation*; Hideo Takizawa, *Lasertec Corporation*; Frank Wei, *Disco Hi Tec America, Inc.*

Thermal Characterization of Power Devices Using Graphene-Based Film 459

Pengtu Zhang, *Chalmers University of Technology*; *East China University of Science and Technology*; Nan Wang, *Chalmers University of Technology*; Carl Zandén, *Chalmers University of Technology*; Lilei Ye, *Smart High Tech AB*; Yifeng Fu, *Smart High Tech AB*; Johan Liu, *Chalmers University of Technology*

High Performance Phase Change Thermal Interface Materials Based on Porous Graphitic Carbon Spheres-Paraffin Wax Composite 464

Zhihua Cao, *Shenzhen Institutes of Advanced Technology, Chinese Academy of Sciences*; *University of Science and Technology of China*; Kai Zhang, *Hong Kong University of Science and Technology*; Gaugping Zhang, *Shenzhen Institutes of Advanced Technology, Chinese Academy of Sciences*; Matthew M.F. Yuen, *Hong Kong University of Science and Technology*; Ping Gu, *University of Science and Technology of China*; Xianzhu Fu, *Shenzhen Institutes of Advanced Technology, Chinese Academy of Sciences*; Rong Sun, *Shenzhen Institutes of Advanced Technology, Chinese Academy of Sciences*; C.P. Wong, *Chinese University of Hong Kong*

High Sensitivity In-Plane Strain Measurement Using a Laser Scanning Technique 470

Hanshuang Liang, *Arizona State University*; Teng Ma, *Arizona State University*; Cheng Lv, *Arizona State University*; Hoa Nguyen, *Arizona State University*; George Chen, *Arizona State University*; Hao Wu, *Arizona State University*; Rui Tang, *Arizona State University*; Hanqing Jiang, *Arizona State University*; Hongbin Yu, *Arizona State University*

Biophysicochemical Evaluation of Passivation Layers for the Packaging of Silicon Microsystems in Medical Devices 478

Jorge Mario Herrera Morales, *CEA-LETI*; Jean-Charles Souriau, *CEA-LETI*; David Ratel, *CEA-LETI*; François Berger, *CEA-LETI*; Gilles Simon, *CEA-LETI*

11: Innovative Packaging Technologies
Chairs: Paul Tiner, *Texas Instruments*
Shichun Qu, *Fairchild Semiconductor*

A New Era in Manufacturable, Low-Temperature and Ultra-Fine Pitch Cu Interconnections and Assembly without Solders 484

Vanessa Smet, *Georgia Institute of Technology*; Makoto Kobayashi, *Namics Corporation*; Tao Wang, *Georgia Institute of Technology*; Pulugurtha Markondeya Raj, *Georgia Institute of Technology*; Rao Tummala, *Georgia Institute of Technology*

Enabling Fine Pitch Cu & Ag Alloy Wire Bond Assessment for 28nm Ultra Low-k Structure 490

John D. Beleran, *United Test and Assembly Center, Ltd.*; Ninoy Milanes II, *United Test and Assembly Center, Ltd.*; Gaurav Mehta, *United Test and Assembly Center, Ltd.*; Nathapong Suthiwongshunthorn, *United Test and Assembly Center, Ltd.*; Ranjan Rajoo, *GLOBALFOUNDRIES, Inc.*; Chan Kai Chong, *GLOBALFOUNDRIES, Inc.*

Assembly of Multiple Chips on Flexible Substrate Using Anisotropic Conductive Film for Medical Imaging Applications 498

Hoang-Vu Nguyen, *Buskerud and Vestfold University College*; Trym Eggen, *GE Vingmed Ultrasound AS*; Bjørnar Sten-Nilsen, *GE Vingmed Ultrasound AS*; Kristin Imenes, *Buskerud and Vestfold University College*; Knut E. Aasmundtveit, *Buskerud and Vestfold University College*

High Frequency High Current Point of Load Modules with Integrated Planar Inductors 504

Wenli Zhang, *Virginia Polytechnic Institute and State University*; Yipeng Su, *Virginia Polytechnic Institute and State University*; David Gilham, *Virginia Polytechnic Institute and State University*; Mingkai Mu, *Virginia Polytechnic Institute and State University*; Qiang Li, *Virginia Polytechnic Institute and State University*; Fred C. Lee, *Virginia Polytechnic Institute and State University*

Integrated Microprobe Array and CMOS MEMS by TSV Technology for Bio-Signal Recording Application 512

Lei-Chun Chou, *National Chiao Tung University*; Shih-Wei Lee, *National Chiao Tung University*; Po-Tsang Huang, *National Chiao Tung University*; Chih-Wei Chang, *University of California, Los Angeles*; Shang-Lin Wu, *National Chiao Tung University*; Jin-Chern Chiou, *National Chiao Tung University; China Medical University*; Ching-Te Chuang, *National Chiao Tung University*; Wei Hwang, *National Chiao Tung University; Advanced Semiconductor Engineering, Inc.*; Chung-Hsi Wu, *Advanced Semiconductor Engineering, Inc.*; Kuo-Hua Chen, *Advanced Semiconductor Engineering, Inc.*; Chi-Tsung Chiu, *Advanced Semiconductor Engineering, Inc.*; Ho-Ming Tong, *Advanced Semiconductor Engineering, Inc.*; Kuan-Neng Chen, *National Chiao Tung University*

Material Characterization of a Novel Lead-Free Solder Material - SACQ 518

Tak-Sang Yeung, *Broadcom Corporation*; Henry Sze, *Broadcom Corporation*; Keith Tan, *Broadcom Corporation*; Javed Sandhu, *Broadcom Corporation*; Chong-Wei Neo, *Broadcom Corporation*; Edward Law, *Broadcom Corporation*

Lithography Challenges for 2.5D Interposer Manufacturing 523

Klaus Ruhmer, *Rudolph Technologies, Inc.*; Philippe Cochet, *Rudolph Technologies, Inc.*; Roger McCleary, *Rudolph Technologies, Inc.*; Rich Rogoff, *Rudolph Technologies, Inc.*; Rajiv Roy, *Rudolph Technologies, Inc.*

12: Power Integrity & Passive Component Modeling
   Chairs: Wendem Beyene, *Rambus Inc.*
   Daniel de Araujo, *Nimbic, Inc.*

Package Embedded Inductors for Integrated Voltage Regulators 528
   William J. Lambert, *Intel Corporation*; Michael J. Hill, *Intel Corporation*; Kaladhar Radhakrishnan, *Intel Corporation*; Leigh Wojewoda, *Intel Corporation*; Anne E. Augustine, *Intel Corporation*

Power Supply Filter for PLL Circuit in Digital Systems 535
   Nam Pham, *IBM Corporation*; Faraydon Pakbaz, *IBM Corporation*; Zhenrong Jin, *IBM Corporation*; Lloyd Walls, *IBM Corporation*

Coaxial Through-Package-Vias (TPVs) for Enhancing Power Integrity in 3D Double-Side Glass Interposers 541
   Gokul Kumar, *Georgia Institute of Technology*; P. Markondeya Raj, *Georgia Institute of Technology*; Jounghyun Cho, *Korea Advanced Institute of Science and Technology (KAIST)*; Saumya Gandhi, *Georgia Institute of Technology*; Parthasarathi Chakraborti, *Georgia Institute of Technology*; Venky Sundaram, *Georgia Institute of Technology*; Joungho Kim, *Korea Advanced Institute of Science and Technology (KAIST)*; Rao Tummala, *Georgia Institute of Technology*

Modeling of Switching Noise and Coupling in Multiple Chips of 3D TSV-Based Systems 548
   Huanyu He, *Rensselaer Polytechnic Institute*; Xiaoxiong Gu, *IBM Corporation*; Jian-Qiang Lu, *Rensselaer Polytechnic Institute*

Characterization of On-Die Power Supply Noise in FCBGA (Flip-Chip Ball Grid Array) Packages 554
   Hyunho Baek, *University of Florida*; William R. Eisenstadt, *University of Florida*

An Enhanced Power Integrity Analysis Flow Based on the Interdependence between Simultaneous Switching Output Noise and Static IR Drop 560
   Minghui Han, *Samsung Display*; Amir Amirkhany, *Samsung Display*; Wei Xiong, *Samsung Display*

Improving the Target Impedance Method for PCB Decoupling of Core Power 566
   Guang Chen, *Altera Corporation*; Dan Oh, *Altera Corporation*

13: 3D Process Integration & Die Stacking
   Chairs: Rozalia Beica, *Yole Developpement*
   Jianwei Dong, *Dow Electronic Materials*

Process Development to Enable 3D IC Multi-Tier Die Bond for 20μm Pitch and Beyond 572
   Y.H. Hu, *TSMC*; C.S. Liu, *TSMC*; M.T. Chen, *TSMC*; M.D. Cheng, *TSMC*; H.J. Kuo, *TSMC*; M.J. Lii, *TSMC*; A. LaManna, *IMEC*; K.J. Rebibis, *IMEC*; T. Wang, *IMEC*; S.V. Huylenbroeck, *IMEC*; R. Daily, *IMEC*; G. Capuz, *IMEC*; D. Velenis, *IMEC*; G. Beyer, *IMEC*; E. Beyne, *IMEC*; Doug C.H. Yu, *TSMC*

Factors in the Selection of Temporary Wafer Handlers for 3D/2.5D Integration 576
   Bing Dang, *IBM Corporation*; Bucknell Webb, *IBM Corporation*; Cornelia Tsang, *IBM Corporation*; Paul Andry, *IBM Corporation*; John Knickerbocker, *IBM Corporation*

Optimization and Challenges on TSV MEOL Integration 582
   DoHyeong Kim, *Amkor Technology Korea, Inc.*; DongHun Lee, *Amkor Technology Korea, Inc.*; YoungChul Seo, *Amkor Technology Korea, Inc.*; JungSoo Park, *Amkor Technology Korea, Inc.*; SeungChul Han, *Amkor Technology Korea, Inc.*; BoRa Jang, *Amkor Technology Korea, Inc.*; JooHyun Khim, *Amkor Technology Korea, Inc.*; YoungSuk Chung, *Amkor Technology Korea, Inc.*; SeongMin Seo, *Amkor Technology Korea, Inc.*; ChoonHeung Lee, *Amkor Technology Korea, Inc.*

TSV Integration on 20nm Logic Si: 3D Assembly and Reliability Results 590

Rahul Agarwal, *GLOBALFOUNDRIES, Inc.*; Dave Hiner, *Amkor Technology, Inc.*; Sukeshwar Kannan, *GLOBALFOUNDRIES, Inc.*; KiWook Lee, *Amkor Technology, Inc.*; DoHyeong Kim, *Amkor Technology, Inc.*; JongSik Paek, *Amkor Technology, Inc.*; SungGeun Kang, *Amkor Technology, Inc.*; Yong Son, *Amkor Technology, Inc.*; Sebastian Dej, *GLOBALFOUNDRIES, Inc.*; Dan Smith, *GLOBALFOUNDRIES, Inc.*; Sara Thangaraju, *GLOBALFOUNDRIES, Inc.*; Jens Paul, *GLOBALFOUNDRIES, Inc.*

TSV MEOL (Mid End of Line) and Packaging Technology of Mobile 3D-IC Stacking 596

Duk Ju Na, *STATS ChipPAC, Ltd.*; Kyaw Oo Aung, *STATS ChipPAC, Ltd.*; Won Kyung Choi, *STATS ChipPAC, Ltd.*; Tsuyoshi Kida, *Renesas Electronics Company*; Toshihiko Ochiai, *Renesas Electronics Company*; Tomoaki Hashimoto, *Renesas Electronics Company*; Michitaka Kimura, *Renesas Electronics Company*; Keiichirou Kata, *Renesas Electronics Company*; Seung Wook Yoon, *STATS ChipPAC, Ltd.*; Andy Chang Bum Yong, *STATS ChipPAC, Ltd.*

Thermally Enhanced 3 Dimensional Integrated Circuit (TE3DIC) Packaging 601

S. Snyder, *Harris Corporation GCSD*; J. Thompson, *Harris Corporation GCSD*; A. King, *Harris Corporation GCSD*; E. Walters, *Harris Corporation GCSD*; P. Tyler, *Harris Corporation GCSD*; M.R. Weatherspoon, *Harris Corporation GCSD*

Filler Trap and Solder Extrusion in 3D IC Thermo-Compression Bonded Microbumps 609

Yingxia Liu, *University of California, Los Angeles*; Menglu Li, *University of California, Los Angeles*; Dong Wook Kim, *Qualcomm, Inc.*; Sam Gu, *Qualcomm, Inc.*; Dilworth Y. Parkinson, *Lawrence Berkeley National Laboratory*; Justin Blair, *Lawrence Berkeley National Laboratory*; K.N. Tu, *University of California, Los Angeles*

14: TSV Fabrication & Its Reliability Impact

Chairs: Li Li, *Cisco Systems, Inc.*
Wei-Chung Lo, *ITRI*

Correlation between Cu Microstructure and TSV Cu Pumping 613

Joke De Messemaeker, *IMEC*; Olalla Varela Pedreira, *IMEC*; Harold Philipsen, *IMEC*; Eric Beyne, *IMEC*; Ingrid De Wolf, *IMEC*; Tom Van der Donck, *KU Leuven*; Kristof Croes, *IMEC*

TSV Reliability Model under Various Stress Tests 620

Ben-Je Lwo, *National Defense University*; Frank M.-S. Lin, *National Defense University*; Kuo-Hsin Huang, *National Defense University*

Development of Process and Design Criteria for Stress Management in Through Silicon Vias 625

O. Hölck, *Fraunhofer IZM*; M. Nuss, *Fraunhofer IZM*; A. Grams, *Fraunhofer IZM*; T. Prewitz, *Fraunhofer IZM*; P. John, *Fraunhofer IZM*; C. Fiedler, *Fraunhofer IZM*; M. Böttcher, *Fraunhofer IZM*; H. Walter, *Fraunhofer IZM*; M.J. Wolf, *Fraunhofer IZM*; O. Wittler, *Fraunhofer IZM*; K.-D. Lang, *Technical University Berlin*

High-Speed Wet Etching of Through Silicon Vias (TSVs) in Micro- and Nanoscale 631

Liyi Li, *Georgia Institute of Technology*; Ching-Ping Wong, *Georgia Institute of Technology*; *Chinese University of Hong Kong*

Replacing the PECVD-SiO$_2$ in the Through-Silicon Via of High-Density 3D LSIs with Highly Scalable Low Cost Organic Liner: Merits and Demerits 636

Murugesan Mariappan, *NICHe, Tohoku University*; Takafumi Fukushima, *NICHe, Tohoku University*; JiChel Beatrix, *NICHe, Tohoku University*; Hiroyuki Hashimoto, *NICHe, Tohoku University*; Yutaka Sato, *NICHe, Tohoku University*; Kangwook Lee, *NICHe, Tohoku University*; Tetsu Tanaka, *NICHe, Tohoku University*; Mitsumasa Koyanagi, *NICHe, Tohoku University*

Investigation of a TSV-RDL In-line Fault-Diagnosis System and Test Methodology for Wafer-level Commercial Production 641

Runiu Fang, *Peking University*; Min Miao, *Beijing Information Science and Technology University*; Xin Sun, *Peking University*; Yunhui Zhu, *Peking University*; Guanjiang Wang, *Peking University Shenzhen Graduate School*; Yichao Xu, *Peking University Shenzhen Graduate School*; Minggang Sun, *Beijing Information Science and Technology University*; Yufeng Jin, *Peking University*

Bonding Technologies for Chip Level and Wafer Level 3D Integration 647

Katsuyuki Sakuma, *IBM Corporation*; Spyridon Skordas, *IBM Corporation*; Jeffrey Zitz, *IBM Corporation*; Eric Perfecto, *IBM Corporation*; William Guthrie, *IBM Corporation*; Luc Guerin, *IBM Corporation*; Richard Langlois, *IBM Corporation*; Hsichang Liu, *IBM Corporation*; Koushik Ramachandran, *IBM Corporation*; Wei Lin, *IBM Corporation*; Kevin Winstel, *IBM Corporation*; Sayuri Kohara, *IBM Corporation*; Kuniaki Sueoka, *IBM Corporation*; Matthew Angyal, *IBM Corporation*; Troy Graves-Abe, *IBM Corporation*; Daniel Berger, *IBM Corporation*; John Knickerbocker, *IBM Corporation*; Subramanian Iyer, *IBM Corporation*

15: Solder Joint Reliability

Chairs: Keith Newman, *Hewlett-Packard Company*
Toni Mattila, *Aalto University*

Dependence of Solder Joint Reliability on Solder Volume, Composition and Printed Circuit Board Surface Finish 655

Babak Arfaei, *Universal Instruments Corporation*; Francis Mutuku, *Binghamton University*; Keith Sweatman, *Nihon-Superior*; Ning-Cheng Lee, *Indium Corporation*; Eric Cotts, *Binghamton University*; Richard Coyle, *Alcatel-Lucent*

The Effects of Aging on the Fatigue Life of Lead Free Solders 666

Muhannad Mustafa, *Auburn University*; Jordan C. Roberts, *Auburn University*; Jeffrey C. Suhling, *Auburn University*; Pradeep Lall, *Auburn University*

Solder Joint Height Impact on Temperature Cycle Reliability of BGA Components with Thermal Enabling Load 684

Yun Ge, *Intel Corporation*; Jeffery Cook, *Intel Corporation*; Min Pei, *Intel Corporation*; Milena Vujosevic, *Intel Corporation*; Bite Zhou, *Intel Corporation*; Suddhasattwa Nad, *Intel Corporation*

Controlling the Sn Grain Morphology of SnAg C4 Solder Bumps 690

Gregory Parks, *Binghamton University*; Minhua Lu, *IBM Corporation*; Eric Perfecto, *IBM Corporation*; Eric Cotts, *Binghamton University*

The Impact of Microstructure Evolution, Localized Recrystallization and Board Thickness on Sn-Ag-Cu Interconnect Board Level Shock Performance 697

Tae-Kyu Lee, *Cisco Systems, Inc.*; Weidong Xie, *Cisco Systems, Inc.*; Thomas R. Bieler, *Michigan State University*; Choong-Un Kim, *University of Texas, Arlington*

Thermal Cycle Fatigue Life Prediction for Flip Chip Solder Joints 703

Robert Darveaux, *Skyworks Solutions, Inc.*

High Thermo-Mechanical Fatigue and Drop Impact Resistant Ni-Bi Doped Lead Free Solder 712

Jae Hong Lee, *MK Electron, Ltd.*; Santosh Kumar, *MK Electron, Ltd.*; Hui Joong Kim, *MK Electron, Ltd.*; Young Woo Lee, *MK Electron, Ltd.*; Jeong Tak Moon, *MK Electron, Ltd.*

16: Advances in Signal Integrity & High-Speed System Design

Chairs: Xiaoxiong (Kevin) Gu, *IBM Corporation*
Kemal Aygun, *Intel Corporation*

Optimal Relaxation of I/O Electrical Requirements under Packaging Uncertainty by Stochastic Methods 717

Xu Chen, *University of Illinois, Urbana-Champaign*; Juan S. Ochoa, *University of Illinois, Urbana-Champaign*; José E. Schutt-Ainé, *University of Illinois, Urbana-Champaign*; Andreas C. Cangellaris, *University of Illinois, Urbana-Champaign*

An Accurate and Convenient Lumped/Discrete Port De-Embedding Method for the 3D Integration and Packaging Full-Wave Modeling by Splitting and Absorbing the Error-Cancelling Network 723

Zhaoqing Chen, *IBM Corporation*

Design, Modeling, and Characterization of Passive Channels for Data Rates of 50 Gbps and Beyond 730

Wendemagegnehu Beyene, *Rambus, Inc.*; Yeon-Chang Hahm, *Rambus, Inc.*; Dave Secker, *Rambus, Inc.*; Don Mullen, *Rambus, Inc.*; Yuriy Shlepnev, *Simberian Inc.*

Low Loss Conductors for CMOS and Through Glass/Silicon Via (TGV/TSV) Structures Using Eddy Current Cancelling Superlattice Structure 736

Arian Rahimi, *University of Florida*; Yong-Kyu Yoon, *University of Florida*

Modeling, Design, Fabrication and Characterization of First Large 2.5D Glass Interposer as a Superior Alternative to Silicon and Organic Interposers at 50 Micron Bump Pitch 742

Brett Sawyer, *Georgia Institute of Technology*; Hao Lu, *Georgia Institute of Technology*; Yuya Suzuki, *Zeon Corporation*; Yutaka Takagi, *NGK Spark Plug Co. Ltd.*; Makoto Kobayashi, *Namics Corporation*; Vanessa Smet, *Georgia Institute of Technology*; Taiji Sakai, *Fujitsu Laboratories Ltd.*; Venky Sundaram, *Georgia Institute of Technology*; Rao Tummala, *Georgia Institute of Technology*

Coupling Impact of Single Ended Signals to LVDS Interface 748

June Feng, *Altera Corporation*; Chooi Ian Loh, *Altera Corporation*; Edward Lin, *Altera Corporation*; Ellen Du, *Altera Corporation*; Guang Chen, *Altera Corporation*; Dan Oh, *Altera Corporation*

Analysis on Interference between Multi-Giga Bit Display Serial Link and RF Components in Smart Mobile Device 753

Youchul Jeong, *Silicon Image Inc.*; Jaemin Kim, *Silicon Image Inc.*; Baegin Sung, *Silicon Image Inc.*

17: Emerging Wireless Technologies & Design

Chairs: Amit P. Agrawal, *Cisco Systems, Inc.*
Lih-Tyng Hwang, *National Sun Yat-Sen University*

Novel Highly-Efficient and Misalignment Insensitive Wireless Power Transfer Systems Utilizing Strongly Coupled Magnetic Resonance Principles 759

Daerhan Daerhan, *Florida International University*; Olutola Jonah, *Florida International University*; Hao Hu, *Florida International University*; Stavros V. Georgakopoulos, *Florida International University*; Manos M. Tentzeris, *Georgia Institute of Technology*

A Wireless Charging and Near-field Communication Combination Module for Mobile Applications 763

Hiroki Shibuya, *Renesas Electronics Corporation*; Tatsuaki Tsukuda, *Renesas Electronics Corporation*; Hiroko Suzuki, *Renesas Electronics Corporation*; Tadashi Shimizu, *Renesas Electronics Corporation*; Masahiro Dobashi, *Renesas Electronics Corporation*; Shinji Nishizono, *Renesas Electronics Corporation*; Mikio Baba, *Renesas Electronics Corporation*; Hideki Sasaki, *Renesas Electronics Corporation*; Katsushi Terajima, *Renesas Electronics Corporation*

Enhanced-Performance Wireless Conformal "Smart Skins" Utilizing Inkjet-Printed Carbon-Nanostructures 769

Taoran Le, *Georgia Institute of Technology*; Ziyin Lin, *Georgia Institute of Technology*; C.P. Wong, *Georgia Institute of Technology*; M.M. Tentzeris, *Georgia Institute of Technology*

Novel THz Imaging Array Using High Resistivity Metasurfaces 775

Kyoung Youl Park, *Michigan State University*; Premjeet Chahal, *Michigan State University*

Magneto-Dielectric Characterization and Antenna Design 782

Kyu Han, *Georgia Institute of Technology*; Madhavan Swaminathan, *Georgia Institute of Technology*; P. Markondeya Raj, *Georgia Institute of Technology*; Himani Sharma, *Georgia Institute of Technology*; Rao Tummala, *Georgia Institute of Technology*; Vijay Nair, *Intel Corporation*

Flexible Liquid Crystal Polymer Based Complementary Split Ring Resonator Loaded Quarter Mode Substrate Integrated Waveguide Filters for Compact and Wearable Broadband RF Applications 789

David E. Senior, *Universidad Tecnológica de Bolívar; University of Florida*; Arian Rahimi, *University of Florida*; Pitfee Jao, *University of Florida*; Yong-Kyu Yoon, *University of Florida*

A Dual-Band Power Amplifier Based on Composite Right/Left-Handed Matching Networks 796

Kyriaki Niotaki, *Centre Tecnologic de Telecomunicacions de Catalunya*; Ana Collado, *Centre Tecnologic de Telecomunicacions de Catalunya*; Apostolos Georgiadis, *Centre Tecnologic de Telecomunicacions de Catalunya*; John Vardakas, *Iquadrat S. L.*

18: WLCSP, Flip Chip, and PoP

Chairs: Valerie Oberson, *IBM Corporation*
Sa Huang, *Medtronic Corporation*

Wafer-Level Non Conductive Films for Exascale Servers 803

A. Horibe, *IBM Corporation*; S. Kohara, *IBM Corporation*; H. Mori, *IBM Corporation*; Y. Orii, *IBM Corporation*; S. Kawamoto, *Namics Corporation*; H. Sone, *Namics Corporation*; M. Hoshiyama, *Namics Corporation*

Bump Geometric Deviation on the Reliability of BOR WLCSP 808

Yumin Liu, *Fairchild Semiconductor Corporation*; Yong Liu, *Fairchild Semiconductor Corporation*; Shichun Qu, *Fairchild Semiconductor Corporation*

Experimental Identification of Warpage Origination During the Wafer Level Packaging Process 815

Chunsheng Zhu, *Chinese Academy of Sciences*; Wenguo Ning, *Chinese Academy of Sciences*; Heng Lee, *Chinese Academy of Sciences*; Jiaotuo Ye, *Chinese Academy of Sciences*; Gaowei Xu, *Chinese Academy of Sciences*; Le Luo, *Chinese Academy of Sciences*

A Stress-Based Effective Film Technique for Wafer Warpage Prediction of Arbitrarily Patterned Films 821

Gregory T. Ostrowicki, *Texas Instruments, Inc.*; Siva P. Gurrum, *Texas Instruments, Inc.*

Drop Test and TCT Reliability of Buffer Coating Material for WLCSP 829

Nobuhiro Anzai, *Asahi Kasei E-Materials Corporation*; Mitsuru Fujita, *Asahi Kasei E-Materials Corporation*; Atsushi Fujii, *Asahi Kasei E-Materials Corporation*

Optimization of Compression Bonding Processing Temperature for Fine Pitch Cu-Column Flip Chip Devices 836

Yonghyuk Jeong, *STATS ChipPAC, Inc.*; Joonyoung Choi, *STATS ChipPAC, Inc.*; Youjoung Choi, *STATS ChipPAC, Inc.*; Nokibul Islam, *STATS ChipPAC, Inc.*; Eric Ouyang, *STATS ChipPAC, Inc.*

Reliability Improvement Methods of Solder Anisotropic Conductive Film (ACF) Joints Using Morphology Control of Solder ACF Joints 841

Yoo-Sun Kim, *Korea Advanced Institute of Science and Technology (KAIST)*; Seung-Ho Kim, *Korea Advanced Institute of Science and Technology (KAIST)*; Jiwon Shin, *Korea Advanced Institute of Science and Technology (KAIST)*; Kyung-Wook Paik, *Korea Advanced Institute of Science and Technology (KAIST)*

19: Progress in 3D Integration

Chairs: Shawn Shi, *Medtronic Corporation*
Mark Gerber, *Texas Instruments*

Development of the Technology to Control the Spatial Distribution of Plasma Using Double ICP Coil 846

T. Sakuishi, *ULVAC, Inc.; NMEMS Technology Research Organization*; T. Murayama, *ULVAC, Inc.; NMEMS Technology Research Organization*; Y. Morikawa, *ULVAC, Inc.; NMEMS Technology Research Organization*; K. Suu, *ULVAC, Inc.; NMEMS Technology Research Organization*

Defect Detection in Through Silicon Vias by GHz Scanning Acoustic Microscopy: Key Ultrasonic Characteristics 850

Alain Phommahaxay, *IMEC*; Ingrid De Wolf, *IMEC; KU Leuven*; Tatjana Djuric, *PVA TePla Analytical Systems GmbH*; Peter Hoffrogge, *PVA TePla Analytical Systems GmbH*; Sebastian Brand, *Fraunhofer IWM*; Peter Czurratis, *PVA TePla Analytical Systems GmbH*; Harold Philipsen, *IMEC*; Gerald Beyer, *IMEC*; Herbert Struyf, *IMEC*; Eric Beyne, *IMEC*

Temporary Spin-on Glass Bonding Technologies for Via-Last/Backside-Via 3D Integration Using Multichip Self-Assembly 856

H. Hashiguchi, *Tohoku University*; T. Fukushima, *Tohoku University*; A. Noriki, *Tohoku University*; H. Kino, *Tohoku University*; K.-W. Lee, *Tohoku University*; T. Tanaka, *Tohoku University*; M. Koyanagi, *Tohoku University*

TSV Module Optimization for High Performance Silicon Interposer 862

Andrew Cao, *Invensas Corporation*; Thomas Dinan, *Invensas Corporation*; Zhuowen Sun, *Invensas Corporation*; Guilian Gao, *Invensas Corporation*; Cyprian Uzoh, *Invensas Corporation*; Bong-Sub Lee, *Invensas Corporation*; Liang Wang, *Invensas Corporation*; Hong Shen, *Invensas Corporation*; Sitaram Arkalgud, *Invensas Corporation*

Study of TSV Thinning Wafer Strength Enhancement for 3DIC Package 868

Jyun-Ling Tsai, *Siliconware Precision Industries Co., Ltd.*; Chun-Chieh Chao, *Siliconware Precision Industries Co., Ltd.*; Hsiao-Chun Huang, *Siliconware Precision Industries Co., Ltd.*; Cheng-Hsiang Liu, *Siliconware Precision Industries Co., Ltd.*; Hung-Hsein Chang, *Siliconware Precision Industries Co., Ltd.*; Chang-Lun Lu, *Siliconware Precision Industries Co., Ltd.*; Shi-Ching Chen, *Siliconware Precision Industries Co., Ltd.*

Challenges in 3D Die Stacking 873

Juergen Grafe, *Fraunhofer IZM*; Wieland Wahrmund, *Fraunhofer IZM*; Stephan Dobritz, *Fraunhofer IZM*; Juergen Wolf, *Fraunhofer IZM*; Klaus-Dieter Lang, *Fraunhofer IZM*

Wet Silicon Etch Process for TSV Reveal 878

Laura B. Mauer, *Solid State Equipment, LLC*; John Taddei, *Solid State Equipment, LLC*; Ramey Youssef, *Solid State Equipment, LLC*; Yongqiang Lu, *SACHEM, Inc.*; Sian Collins, *SACHEM, Inc.*; Kevin Mclaughlin, *SACHEM, Inc.*; Craig Allen, *SACHEM, Inc.*

20: 3D Materials & Processing

Chairs: Myung Jin Yim, *Intel Corporation*
Daniel D. Lu, *Henkel Corporation*

Advanced Wafer Bonding and Laser Debonding 883

P. Andry, *IBM Corporation*; R. Budd, *IBM Corporation*; R. Polastre, *IBM Corporation*; C. Tsang, *IBM Corporation*; B. Dang, *IBM Corporation*; J. Knickerbocker, *IBM Corporation*; M. Glodde, *IBM Corporation*

Versatile Thin Wafer Stacking Technology for Monolithic Integration of Temporary Bonded Thin Wafers 888

Thomas Uhrmann, *EV Group*; Jürgen Burggraf, *EV Group*; Julian Bravin, *EV Group*; Viorel Dragoi, *EV Group*; Markus Wimplinger, *EV Group*; Thorsten Matthias, *EV Group*; Paul Lindner, *EV Group*

Temporary Bonding for High-Topography Applications: Spin-on Material versus Dry Film 894

Anne Jourdain, *IMEC*; Alain Phommahaxay, *IMEC*; Greet Verbinnen, *IMEC*; Alice Guerrero, *Brewer Science, Inc.*; Susan Bailey, *Brewer Science, Inc.*; Mark Privett, *Brewer Science, Inc.*; Kim Arnold, *Brewer Science, Inc.*; Andy Miller, *IMEC*; Kenneth Rebibis, *IMEC*; Gerald Beyer, *IMEC*; Eric Beyne, *IMEC*

Development of New Concept Thermoplastic Temporary Adhesive for 3D-IC Integration 899

A. Kubo, *Tokyo Ohka Kogyo Co., Ltd.*; K. Tamura, *Tokyo Ohka Kogyo Co., Ltd.*; H. Imai, *Tokyo Ohka Kogyo Co., Ltd.*; T. Yoshioka, *Tokyo Ohka Kogyo Co., Ltd.*; S. Oya, *Tokyo Ohka Kogyo Co., Ltd.*; S. Otaka, *Tokyo Ohka Kogyo Co., Ltd.*

Underfilling Techniques Comparison in 3D CtW Stacking Approach 906

A. Garnier, *CEA-LETI*; A. Jouve, *CEA-LETI*; R. Franiatte, *CEA-LETI*; S. Chéramy, *CEA-LETI*

High Throughput Thermal Compression NCF Bonding 913

Toshihisa Nonaka, *Toray Industries, Inc.*; Yuta Kobayashi, *Toray Industries, Inc.*; Noboru Asahi, *Toray Industries, Inc.*; Shoichi Niizeki, *Toray Industries, Inc.*; Koichi Fujimaru, *Toray Industries, Inc.*; Yoshiyuki Arai, *Toray Engineering Co., Ltd.*; Toshifumi Takegami, *Toray Engineering Co., Ltd.*; Yoshinori Miyamoto, *Toray Engineering Co., Ltd.*; Masatsugu Nimura, *Toray Engineering Co., Ltd.*; Hiroyuki Niwa, *Toray International America Inc.*

Through Silicon Underfill Dispensing for 3D Die/Interposer Stacking 919

Fuliang Le, *Hong Kong University of Science and Technology*; S.W. Ricky Lee, *Hong Kong University of Science and Technology*; Kei May Lau, *Hong Kong University of Science and Technology*; C. Patrick Yue, *Hong Kong University of Science and Technology*; Johnny K.O. Sin, *Hong Kong University of Science and Technology*; Philip K.T. Mok, *Hong Kong University of Science and Technology*; Wing-Hung Ki, *Hong Kong University of Science and Technology*; Hoi Wai Choi, *University of Hong Kong*

21: Wafer-Level & Fan-Out Packages

Chairs: Christopher Bower, *X-Celeprint Ltd.*
E. Jan Vardaman, *TechSearch International, Inc.*

Board Level Reliability and Surface Mount Assembly of 0.35mm and 0.3mm Pitch Wafer Level Packages 925

Beth Keser, *Qualcomm Technologies, Inc.*; Rey Alvarado, *Qualcomm Technologies, Inc.*; Alan Choi, *Qualcomm Technologies, Inc.*; Mark Schwarz, *Qualcomm Technologies, Inc.*; Steve Bezuk, *Qualcomm Technologies, Inc.*

Encapsulated Wafer Level Package Technology (eWLCS) 931

Tom Strothmann, *STATS ChipPAC, Inc.*; Seung Wook Yoon, *STATS ChipPAC, Ltd.*; Yaojian Lin, *STATS ChipPAC, Ltd.*

Enabling of Fan-Out WLP for More Demanding Applications by Introduction of Enhanced Dielectric Material for Higher Reliability 935

Rodrigo Almeida, *Namium, S.A.*; Isabel Barros, *Namium, S.A.*; José Campos, *Namium, S.A.*; Paulo Cardoso, *Namium, S.A.*; José Castro, *Namium, S.A.*; Vitor Henriques, *Namium, S.A.*; Eoin O'Toole, *Namium, S.A.*; Nelson Pinho, *Namium, S.A.*

24" x 18" Fan-Out Panel Level Packaging 940

T. Braun, *Fraunhofer IZM*; K.-F. Becker, *Fraunhofer IZM*; S. Voges, *Technical University Berlin*; J. Bauer, *Fraunhofer IZM*; R. Kahle, *Technical University Berlin*; V. Bader, *Fraunhofer IZM*; T. Thomas, *Technical University Berlin*; R. Aschenbrenner, *Fraunhofer IZM*; K.-D. Lang, *Technical University Berlin*

Development and Characterization of New Generation Panel Fan-Out (P-FO) Packaging Technology 947

Hong-Da Chang, *Siliconware Precision Industries Co., Ltd.*; David Chang, *Siliconware Precision Industries Co., Ltd.*; Kenny Liu, *Siliconware Precision Industries Co., Ltd.*; H.S. Hsu, *Siliconware Precision Industries Co., Ltd.*; Rui-Feng Tai, *Siliconware Precision Industries Co., Ltd.*; Hsiao-Chun Huang, *Siliconware Precision Industries Co., Ltd.*; Yi-Che Lai, *Siliconware Precision Industries Co., Ltd.*; Chang-Lun Lu, *Siliconware Precision Industries Co., Ltd.*; Chun-Tang Lin, *Siliconware Precision Industries Co., Ltd.*; Steve Chiu, *Siliconware Precision Industries Co., Ltd.*

Development of Exposed Die Large Body to Die Size Ratio Wafer Level Package Technology 952

J. Osenbach, *LSI Corporation*; S. Emerich, *LSI Corporation*; L. Golick, *LSI Corporation*; S. Cate, *LSI Corporation*; M. Chan, *STATS ChipPAC, Ltd.*; S.W. Yoon, *STATS ChipPAC, Ltd.*; Y.J. Lin, *STATS ChipPAC, Ltd.*; K. Wong, *STATS ChipPAC, Inc.*

3D Rectangular Waveguide Integrated in Embedded Wafer Level Ball Grid Array (eWLB) Package 956

E. Seler, *Friedrich-Alexander University Erlangen-Nuremberg*; M. Wojnowski, *Infineon Technologies AG*; W. Hartner, *Infineon Technologies AG*; J. Böck, *Infineon Technologies AG*; R. Lachner, *Infineon Technologies AG*; R. Weigel, *University of Erlangen-Nuremberg*; A. Hagelauer, *Friedrich-Alexander University Erlangen-Nuremberg*

22: System-Level Thermal & Mechanical Models I
Chairs: Yong Liu, *Fairchild Semiconductor Corporation*
Sandeep Sane, *Intel Corporation*

Interplay and Influence of Thermomechanical Stress in Copper-Filled TSV Interposers 963

Sheng-Tsai Wu, *Industrial Technology Research Institute (ITRI)*; Cheng-Fu Chen, *University of Alaska, Fairbanks*; Heng-Chieh Chien, *Industrial Technology Research Institute (ITRI)*

Does Current Crowding Induce Vacancy Concentration Singularity in Electromigration? 967

Ozgur Taner, *Lamar University*; Kasemsak Kijkanjanapaiboon, *Lamar University*; Xuejun Fan, *Lamar University*

Hygro-Thermo-Mechanical Analysis and Failure Prediction in Electronic Packages by Using Peridynamics 973

Selda Oterkus, *University of Arizona*; Erdogan Madenci, *University of Arizona*; Erkan Oterkus, *University of Strathclyde*; Yuchul Hwang, *Samsung Electronics Company, Ltd.*; Jangyong Bae, *Samsung Electronics Company, Ltd.*; Sungwon Han, *Samsung Electronics Company, Ltd.*

Cohesive Zone Experiments for Copper/Mold Compound Delamination 983

William E.R. Krieger, *Georgia Institute of Technology*; Sathyanarayanan Raghavan, *Georgia Institute of Technology*; Abhishek Kwatra, *Georgia Institute of Technology*; Suresh K. Sitaraman, *Georgia Institute of Technology*

Damage Pre-Cursor Based Life Prediction of the Effect of Mean Temperature of Thermal Cycle on the SnAgCu Solder Joint Reliability 990

Pradeep Lall, *Auburn University*; Kazi Mirza, *Auburn University*; Jeff Suhling, *Auburn University*

Methodology Development of Warpage Analysis of Polymer Based Packaging Substrate 1004

Cheolgyu Kim, *Korea Advanced Institute of Science and Technology (KAIST)*; Taeik Lee, *Korea Advanced Institute of Science and Technology (KAIST)*; Hyeseon Choi, *Korea Advanced Institute of Science and Technology (KAIST)*; Min Sung Kim, *Samsung Electro-Mechanics*; Taek-Soo Kim, *Korea Advanced Institute of Science and Technology (KAIST)*

Simulations for the Impact of Warpage on the Accuracy of Attitude and Heading Reference System 1010

Shengzhi Zhang, *Huazhong University of Science & Technology; Wuhan National Laboratory for Optoelectronics*; Qiang Dan, *Huazhong University of Science & Technology; Wuhan National Laboratory for Optoelectronics*; Chaojun Liu, *Huazhong University of Science & Technology; Wuhan National Laboratory for Optoelectronics*; Yong Xu, *Wayne State University*; Xin Wu, *Wayne State University*; Sheng Liu, *Wuhan University*; Xing Guo, *Huazhong University of Science & Technology; Wuhan National Laboratory for Optoelectronics*; Ming Wen, *Huazhong University of Science & Technology; Wuhan National Laboratory for Optoelectronics*

23: Optical Interconnects
Chairs: Hiren Thacker, *Oracle*
Ping Zhou, *LDX Optronics, Inc.*

Multicore Fiber 4 TX + 4 RX Optical Transceiver Based on Holey SiGe IC 1016

Fuad E. Doany, *IBM Corporation*; Daniel M. Kuchta, *IBM Corporation*; Alexander V. Rylyakov, *IBM Corporation*; Christian Baks, *IBM Corporation*; Shurong Tian, *IBM Corporation*; Mark Schultz, *IBM Corporation*; Frank Libsch, *IBM Corporation*; Clint L. Schow, *IBM Corporation*

336-Channel Electro-Optical Interconnect: Underfill Process Improvement, Fiber Bundle and Reliability Results 1021

Shuki Benjamin, *Compass-EOS*; Kobi Hasharoni, *Compass-EOS*; Avi Maman, *Compass-EOS*; Stanislav Stepanov, *Compass-EOS*; Michael Mesh, *Compass-EOS*; Helge Luesebrink, *PVA TePla AG*; Roland Steffek, *PVA TePla AG*; Wolfgang Pleyer, *PVA TePla AG*; Christian Stömmer, *PVA TePla AG*

Development of Optical Multi-Channel Connector for Rigid Waveguide – Fiber Optical Interconnection 1028

Kazumi Nakazuru, *Kyocera Corporation*; Satoshi Asai, *Kyocera Corporation*; Masatoshi Tsunoda, *Kyocera Corporation*; Naoki Takahashi, *Kyocera Corporation*; Takahiro Matsubara, *Kyocera Corporation*

Electro-Optical Backplane Demonstrator with Gradient-Index Multimode Glass Waveguides for Board-to-Board Interconnection 1033

Lars Brusberg, *Fraunhofer Institute IZM*; Henning Schröder, *Fraunhofer Institute IZM*; Richard Pitwon, *Xyratex Technology Ltd.*; Simon Whalley, *ILFA Feinstleitertechnik GmbH*; Allen Miller, *Xyratex Technology Ltd.*; Christian Herbst, *Technical University of Berlin*; Julia Röder, *Fraunhofer Institute IZM*; Daniel Weber, *Fraunhofer Institute IZM*; Klaus-Dieter Lang, *Technical University of Berlin*

Three-Dimensional High-Density Channel Integration of Polymer Optical Waveguide Using the Mosquito Method 1042

Takaaki Ishigure, *Keio University*; Daisuke Suganuma, *Keio University*; Kazutomo Soma, *Keio University*

Novel Trace Design for High Data-Rate, Multi-Channel Optical Transceiver Assembled Using Flip-Chip Bonding 1048

Takatoshi Yagisawa, *Fujitsu Laboratories, Ltd.*; Takashi Shiraishi, *Fujitsu Laboratories, Ltd.*; Mariko Sugawara, *Fujitsu Laboratories, Ltd.*; Kazuhiro Tanaka, *Fujitsu Laboratories, Ltd.*

Modeling, Design, and Demonstration of Ultra-Miniaturized and High Efficiency 3D Glass Photonics Modules 1054

Bruce C. Chou, *Georgia Institute of Technology*; Sandeep Razdan, *TE Connectivity*; Haipeng Zhang, *TE Connectivity*; Jibin Sun, *TE Connectivity*; Terry Bowen, *TE Connectivity*; Vanessa Smet, *Georgia Institute of Technology*; Gee-Kung Chang, *Georgia Institute of Technology*; Venky Sundaram, *Georgia Institute of Technology*; Rao Tummala, *Georgia Institute of Technology*

24: Innovative Interconnections

Chairs: James E. Morris, *Portland State University*
Nathan Lower, *Rockwell Collins, Inc.*

A Study on Nanofiber Anisotropic Conductive Films (ACFs) for Fine Pitch Chip-on-Glass (COG) Interconnections 1060

Sang Hoon Lee, *Korea Advanced Institute of Science and Technology (KAIST)*; Tae Wan Kim, *Korea Advanced Institute of Science and Technology (KAIST)*; Kyung-Wook Paik, *Korea Advanced Institute of Science and Technology (KAIST)*

Study of Fine Pitch Micro-Interconnections Formed by Low Temperature Bonded Copper Nanowires Based Anisotropic Conductive Film 1064

Jing Tao, *University College Cork*; Alan Mathewson, *University College Cork*; Kafil M. Razeeb, *University College Cork*

Carbon Nanofibers (CNF) for Enhanced Solder-Based Nano-Scale Integration and On-Chip Interconnect Solutions 1071

V. Desmaris, *Smoltek AB*; A.M. Saleem, *Smoltek AB*; S. Shafiee, *Smoltek AB*; J. Berg, *Smoltek AB*; M.S. Kabir, *Smoltek AB*; A. Johansson, *Smoltek AB*; Phil Marcoux, *PPM Associates*

Pressure-Less Plasma Sintering of Cu Paste for SiC Die-Attach of High-Temperature Power Device Manufacturing 1077

S. Nagao, *Osaka University*; K. Kodani, *Nissin, Inc.*; S. Sakamoto, *Osaka University*; S.-W. Park, *Osaka University*; T. Sugahara, *Osaka University*; K. Suganuma, *Osaka University*

Bonding 1200 V, 150 A IGBT Chips (13.5 mm x 13.5 mm) with DBC Substrate by Pressureless Sintering Nanosilver Paste for Power Electronic Packaging .... N/A

Shancan Fu, *Tianjin University*; Yunhui Mei, *Tianjin University*; Guo-Quan Lu, *Tianjin University, Virginia Tech*; Xin Li, *Tianjin University*; Gang Chen, *Tianjin University*; Xu Chen, *Tianjin University*

Flip Chip Based on Compliant Double Helix Interconnect for High Frequency Applications 1086

Pingye Xu, *Auburn University*; George A. Hernandez, *Auburn University*; Shiqiang Wang, *Auburn University*; Jie Zhong, *Auburn University*; Charles D. Ellis, *Auburn University*; Michael C. Hamilton, *Auburn University*

Modeling of Crosstalk Effects in Coupled MLGNR Interconnects Based on FDTD Method 1091

Vobulapuram Ramesh Kumar, *Indian Institute of Technology Roorkee*; Brajesh Kumar Kaushik, *Indian Institute of Technology Roorkee*; Amalendu Patnaik, *Indian Institute of Technology Roorkee*

25: Recent Advances in 3D Package Reliability

Chairs: Deepak Goyal, *Intel Corporation*
Jeffrey Suhling, *Auburn University*

First Demonstration of Reliable Copper-Plated 30μm Diameter Through-Package-Vias in Ultra-Thin Bare Glass Interposers 1098

Kaya Demir, *Georgia Institute of Technology*; Andac Armutlulu, *Georgia Institute of Technology*; Jialing Tong, *Georgia Institute of Technology*; Raghuram Pucha, *Georgia Institute of Technology*; Venkatesh Sundaram, *Georgia Institute of Technology*; Rao Tummala, *Georgia Institute of Technology*

Through-Glass Interposer Integrated High Quality RF Components 1103

Cheolbok Kim, *University of Florida*; David E. Senior, *University of Florida; Universidad Tecnológica de Bolívar*; Aric Shorey, *Corning, Inc.*; Hyup Jong Kim, *University of Florida*; Windsor Thomas, *Corning, Inc.*; Yong-Kyu Yoon, *University of Florida*

Minimization of Keep-Out Zone (KOZ) in 3D IC by Local Bending Stress Suppression with Low Temperature Curing Adhesive 1110

Hisashi Kino, *Tohoku University*; Hideto Hashiguchi, *Tohoku University*; Yohei Sugawara, *Tohoku University*; Seiya Tanikawa, *Tohoku University*; Takafumi Fukushima, *Tohoku University*; Kangwook Lee, *Tohoku University*; Mitsumasa Koyanagi, *Tohoku University*; Tetsu Tanaka, *Tohoku University*

Effect of Thermal Annealing on TSV Cu Protrusion and Local Stress 1116

Xiangmeng Jing, *National Center for Advanced Packaging; Chinese Academy of Sciences*; Hongwen He, *National Center for Advanced Packaging; Chinese Academy of Sciences*; Liang Ji, *National Center for Advanced Packaging*; Cheng Xu, *National Center for Advanced Packaging*; Kai Xue, *National Center for Advanced Packaging*; Meiying Su, *National Center for Advanced Packaging; Chinese Academy of Sciences*; Chongshen Song, *National Center for Advanced Packaging; Chinese Academy of Sciences*; Daquan Yu, *National Center for Advanced Packaging; Chinese Academy of Sciences*; Liqiang Cao, *National Center for Advanced Packaging; Chinese Academy of Sciences*; Wenqi Zhang, *National Center for Advanced Packaging*; Dongkai Shangguan, *National Center for Advanced Packaging; Chinese Academy of Sciences*

Effect of High Temperature Storage on the Stress and Reliability of 3D Stacked Chips 1122

Tengfei Jiang, *University of Texas, Austin*; Chenglin Wu, *University of Texas, Austin*; Peng Su, *Cisco Systems, Inc.*; Pierre Chia, *Cisco Systems, Inc.*; Li Li, *Cisco Systems, Inc.*; Ho-Young Son, *SK Hynix, Inc.*; Min-Suk Suh, *SK Hynix, Inc.*; Nam-Seog Kim, *SK Hynix, Inc.*; Jay Im, *University of Texas, Austin*; Rui Huang, *University of Texas, Austin*; Paul S. Ho, *University of Texas, Austin*

A Novel Fine Pitch TSV Interconnection Method Using NCF with Zn Nano-Particles 1128

Ji-Won Shin, *Korea Advanced Institute of Science and Technology (KAIST)*; Yong-Won Choi, *Korea Advanced Institute of Science and Technology (KAIST)*; Young Soon Kim, *Korea Advanced Institute of Science and Technology (KAIST)*; Un Byung Kang, *Samsung Electronics Company, Ltd.*; Sun Kyung Seo, *Samsung Electronics Company, Ltd.*; Kyung-Wook Paik, *Korea Advanced Institute of Science and Technology (KAIST)*

Residual Stress Investigations at TSVs in 3D Micro Structures by HR-XRD, Raman Spectroscopy and fibDAC 1134

U. Zschenderlein, *Technical University Chemnitz*; D. Vogel, *Fraunhofer ENAS*; E. Auerswald, *Fraunhofer ENAS*; O. Hölck, *Technical University Chemnitz*; H. Rajendran, *Technical University Chemnitz*; P. Ramm, *Fraunhofer EMFT*; R. Pufall, *Infineon Technologies*; B. Wunderle, *Technical University Chemnitz*; *Fraunhofer ENAS*

26: 3D Microbumps

Chairs: Kathy Cook, *Ziptronix*
Lei Shan, *IBM Corporation*

Formic Acid Treatment with Pt Catalyst for Cu Direct and Hybrid Bonding at Low Temperature 1143

Tadatomo Suga, *University of Tokyo*; Masakate Akaike, *University of Tokyo*; Wenhua Yang, *University of Tokyo*

Direct Multichip-to-Wafer 3D Integration Technology Using Flip-Chip Self-Assembly of NCF-Covered Known Good Dies 1148

Yuka Ito, *Tohoku University; Sumitomo Bakelite Co., Ltd.*; Mariappan Murugesan, *Tohoku University*; Takafumi Fukushima, *Tohoku University*; Kang-Wook Lee, *Tohoku University*; Koji Choki, *Sumitomo Bakelite Co., Ltd.*; Tetsu Tanaka, *Tohoku University*; Mitsumasa Koyanagi, *Tohoku University*

Maskless Screen Printing Technology for 20μm-Pitch, 52InSn Solder Interconnections in Display Applications 1154

Kwang-Seong Choi, *ETRI*; Haksun Lee, *ETRI*; Hyun-Cheol Bae, *ETRI*; Yong-Sung Eom, *ETRI*

Accelerated SLID Bonding Using Thin Multi-Layer Copper-Solder Stack for Fine-Pitch Interconnections 1160

Chinmay Honrao, *Georgia Institute of Technology*; Ting-Chia Huang, *Georgia Institute of Technology*; Makoto Kobayashi, *Namics Corporation*; Vanessa Smet, *Georgia Institute of Technology*; P. Markondeya Raj, *Georgia Institute of Technology*; Rao Tummala, *Georgia Institute of Technology*

Study of Electro-Migration Resistivity of Micro Bump Using SnBi Solder 1166

Kei Murayama, *Shinko Electric Industries Company, Ltd.*; Mitsuhiro Aizawa, *Shinko Electric Industries Company, Ltd.*; Mitsutoshi Higashi, *Shinko Electric Industries Company, Ltd.*

The Impact of Different Under Bump Metallurgies and Redistribution Layers on the Electromigration of Solder Balls for Wafer-Level Packaging 1173

Christine Hau-Riege, *Qualcomm Technologies, Inc.*; Beth Keser, *Qualcomm Technologies, Inc.*; Rey Alvarado, *Qualcomm Technologies, Inc.*; Ahmer Syed, *Qualcomm Technologies, Inc.*; YouWen Yau, *Qualcomm Technologies, Inc.*; Steve Bezuk, *Qualcomm Technologies, Inc.*; Kevin Caffey, *Qualcomm Technologies, Inc.*

Low-Pressure Sintering Bonding with Cu and CuO Flake Paste for Power Devices 1179

S.W. Park, *Osaka University*; R. Uwataki, *Osaka University*; S. Nagao, *Osaka University*; T. Sugahara, *Osaka University*; Y. Katoh, *Denso Corporation*; H. Ishino, *Denso Corporation*; K. Sugiura, *Denso Corporation*; K. Tsuruta, *Denso Corporation*; K. Suganuma, *Osaka University*

27: Sensors & MEMS Technologies

Chairs: Joseph W. Soucy, *Draper Laboratory*
Daniel Baldwin, *Engent, Inc.*

A Novel 3D Packaging Concept for RF Powered Sensor Grains 1183

Walther Pachler, *Graz University of Technology*; Klaus Pressel, *Infineon Technologies AG*; Jasmin Grosinger, *Graz University of Technology*; Gottfried Beer, *Infineon Technologies AG*; Wolfgang Bösch, *Graz University of Technology*; Gerald Holweg, *Infineon Technologies AG*; Christian Zilch, *Magna Diagnostics GmbH*; Manfred Meindl, *Danube Mobile Communications Engineering GmbH & Co. KG*

A Novel Sound Sensor and Its Package Used in Lung Sound Diagnosis 1189
Xingming Fu, *Wuhan University*; Chaojun Liu, *Wuhan University*; Yong Xu, *Wuhan University; Wayne State University*; Yating Hu, *Wayne State University*; Xiaobing Luo, *Huazhong University of Science & Technology*; Xin Wu, *Wayne State University*; Sheng Liu, *Wuhan University*

Novel System-in-Package Design and Packaging Solution for Solid State Lighting Systems 1192
Mingzhi Dong, *Delft University of Technology; State Key Laboratory of Solid State Lighting*; Fabio Santagata, *Delft University of Technology; State Key Laboratory of Solid State Lighting*; Jia Wei, *Delft University of Technology; State Key Laboratory of Solid State Lighting*; Cadmus Yuan, *Chinese Academy of Sciences; State Key Laboratory of Solid State Lighting*; Guoqi Zhang, *Chinese Academy of Sciences; Delft University of Technology*

Implantable Device Including a MEMS Accelerometer and an ASIC Chip Encapsulated in a Hermetic Silicon Box for Measurement of Cardiac Physiological Parameter 1198
Jean-Charles Souriau, *CEA-LETI*; Laetitia Castagné, *CEA-LETI*; Guy Parat, *CEA-LETI*; Gilles Simon, *CEA-LETI*; Karima Amara, *Sorin CRM SAS*; Philippe D'hiver, *Sorin CRM SAS*; Renzo Dal Molin, *Sorin CRM SAS*

Capping Technologies for Wafer Level MEMS Packaging Based on Permanent and Temporary Wafer Bonding 1204
K. Zoschke, *Fraunhofer IZM*; M. Wilke, *Fraunhofer IZM*; M. Wegner, *Fraunhofer IZM*; K. Kaletta, *Fraunhofer IZM*; C.-A. Manier, *Fraunhofer IZM*; H. Oppermann, *Fraunhofer IZM*; M. Wietstruck, *IHP GmbH*; B. Tillack, *IHP GmbH*; M. Kaynak, *IHP GmbH*; K.-D. Lang, *Technical University Berlin*

The Novel Assembly Method of a Field Deployable Biosensor Unit 1212
P. Xu, *East China Normal University*; F.M. Guo, *East China Normal University*; X.Y. Liu, *East China Normal University*; J.H. Shen, *East China Normal University*; L. Ding, *East China Normal University*; W. Wang, *East China Normal University*; Y.Q. Li, *East China Normal University*; Y.P. Ge, *East China Normal University*; S.H. Zhang, *East China Normal University*; M.J. Wang, *East China Normal University*; H.Z. Zheng, *East China Normal University*; J.T. Ye, *Chinese Academy of Sciences*; L.; Luo Chinese Academy of Sciences

SIMEIT-Project: High Precision Inertial Sensor Integration on a Modular 3D-Interposer Platform 1218
Wolfram Steller, *Fraunhofer IZM*; Christoph Meinecke, *Technical University Chemnitz*; Knut Gottfried, *Fraunhofer ENAS*; Gregor Woldt, *Microelectronic Packaging Dresden GmbH*; Wolfgang Günther, *GEMAC*; M. Juergen Wolf, *Fraunhofer IZM*; K. Dieter Lang, *Fraunhofer IZM*

28: System-Level Thermal & Mechanical Models II
Chairs: Pradeep Lall, *Auburn University*
Xuejun Fan, *Lamar University*

Mechanical Stress Management for Electrical Chip-Package Interaction (e-CPI) 1226
Wei Zhao, *Qualcomm Technologies, Inc.*; Mark Nakamoto, *Qualcomm Technologies, Inc.*; Vidhya Ramachandran, *Qualcomm Technologies, Inc.*; Riko Radojcic, *Qualcomm Technologies, Inc.*

Cu Pillar Flip Chip Assembly: Chip Attach Process Failure Mode Study 1231
Shengmin Wen, *Amkor Technology*; Bora Baloglu, *Amkor Technology*; Guangfeng Li, *Amkor Assembly and Test (Shanghai) Co., Ltd.*

Mechanical and Thermo-Mechanical Stress Considerations in Applying 3D ICs to a Design 1235
Jia-Shen Lan, *National Sun Yat-Sen University*; Mei-Ling Wu, *National Sun Yat-Sen University*

Modeling Microstructure Effects on Electromigration in Lead-Free Solder Joints 1241
Jiamin Ni, *Rensselaer Polytechnic Institute*; Yong Liu, *Fairchild Semiconductor*; Jifa Hao, *Fairchild Semiconductor*; Antoinette Maniatty, *Rensselaer Polytechnic Institute*; Barry O'Connell, *Fairchild Semiconductor*

Experimental Demonstration of the Effect of Copper TPVs (Through Package Vias) on Thermal Performance of Glass Interposers 1247

Sangbeom Cho, *Georgia Institute of Technology*; Yoichiro Sato, *Asahi Glass*; Venky Sundaram, *Georgia Institute of Technology*; Yogendra Joshi, *Georgia Institute of Technology*; Rao Tummala, *Georgia Institute of Technology*

Failure Mechanism Investigation of Stacked Via Cracking in Organic Chip Carrier 1253

Shidong Li, *IBM Corporation*; Yi Pan, *IBM Corporation*; Sushumna Iruvanti, *IBM Corporation*; David L. Questad, *IBM Corporation*; Randall J. Werner, *IBM Corporation*

A Novel Method to Predict Fluid/Structure Interaction in IC Packaging 1258

Chih-Chung Hsu, *National Tsing Hua University*; Tzu-Chang Wang, *CoreTech System (Moldex3D) Co., Ltd.*; Yen-Chi Chen, *CoreTech System (Moldex3D) Co., Ltd.*; Yang-Kai Lin, *CoreTech System (Moldex3D) Co., Ltd.*

29: Integrated RF & Power Modules

Chairs: Rockwell Hsu, *Cisco Systems, Inc.*
P. Markondeya Raj, *Georgia Institute of Technology*

Modeling, Design and Demonstration of Multi-Die Embedded WLAN RF Front-End Module with Ultra-Miniaturized and High-Performance Passives 1264

Srikrishna Sitaraman, *Georgia Institute of Technology*; Yuya Suzuki, *Zeon Corporation*; Christopher White, *Georgia Institute of Technology*; Vijay Nair, *Intel Corporation*; Telesphor Kamgaing, *Intel Corporation*; Frank Juskey, *TriQuint Semiconductor*; Sung Jin Kim, *Georgia Institute of Technology*; P. Markondeya Raj, *Georgia Institute of Technology*; Venky Sundaram, *Georgia Institute of Technology*; Rao Tummala, *Georgia Institute of Technology*

A Compact 4-Chip Package with 64 Embedded Dual-Polarization Antennas for W-Band Phased-Array Transceivers 1272

Xiaoxiong Gu, *IBM Corporation*; Duixian Liu, *IBM Corporation*; Christian Baks, *IBM Corporation*; Alberto Valdes-Garcia, *IBM Corporation*; Ben Parker, *IBM Corporation*; Md. R. Islam, *IBM Corporation*; Arun Natarajan, *IBM Corporation; Oregon State University*; Scott K. Reynolds, *IBM Corporation*

Active Die Embedded Small Form Factor RF Packages for Ultrabooks and Smartphones 1278

Vijay K. Nair, *Intel Corporation*; Carlton Hanna, *Intel Corporation*; Ronald Spreitzer, *Intel Corporation*; Johanna Swan, *Intel Corporation*

Design and Material Contributions to Second-Harmonic Nonlinearities in RF Silicon Integrated Passive Devices 1284

Robert Frye, *RF Design Consulting, LLC*; Robert Melville, *Emecon, LLC*; Kai Liu, *STATS ChipPAC, Inc.*

Integration of Magnetic Materials into Package RF and Power Inductors on Organic Substrates for System in Package (SiP) Applications 1290

Hao Wu, *Arizona State University*; Donald S. Gardner, *Intel Corporation*; Cheng Lv, *Arizona State University*; Zhihua Zou, *Intel Corporation*; Hongbin Yu, *Arizona State University*

Through Silicon Capacitor Co-Integrated with TSV as an Efficient 3D Decoupling Capacitor Solution for Power Management on Silicon Interposer 1296

O. Guiller, *STMicroelectronics*; S. Joblot, *STMicroelectronics*; Y. Lamy, *CEA-LETI*; A. Farcy, *STMicroelectronics*; E. Defay, *CEA-LETI*; K. Dieng, *Université de Savoie*

Design of RF and Thermal Pads of CMOS PAs Using Copper to Copper Bonding Technology 1303

Lih-Tyng Hwang, *National Sun Yat-Sen University*; An-Yu Kuo, *Cadence Design Systems, Inc.*

30: Solders & Bonding

Chairs: Mikel Miller, *Draper Laboratory*
Grace Yi Li, *Intel Corporation*

Wafer IMS (Injection Molded Solder) – A New Fine Pitch Solder Bumping Technology on Wafers with Solder Alloy Composition Flexibility 1308

Jae-Woong Nah, *IBM Corporation*; Jeffrey Gelorme, *IBM Corporation*; Peter Sorce, *IBM Corporation*; Paul Lauro, *IBM Corporation*; Eric Perfecto, *IBM Corporation*; Mark McLeod, *IBM Corporation*; Kazushige Toriyama, *IBM Corporation*; Yasumitsu Orii, *IBM Corporation*; Peter Brofman, *IBM Corporation*; Takashi Nauchi, *Senju Metal Industry Co., Ltd.*; Akira Takaguchi, *Senju System Technology Co., Ltd.*; Kazuya Ishiguro, *Senju System Technology Co., Ltd.*; Tomoyasu Yoshikawa, *Senju Comtek Corporation*; Derek Daily, *Senju Comtek Corporation*; Ryoichi Suzuki, *Senju Metal Industry Co., Ltd.*

Reliability of Paste Based Transient Liquid Phase Sintered Interconnects 1314

Hannes Greve, *University of Maryland*; S. Ali Moeini, *University of Maryland*; F. Patrick McCluskey, *University of Maryland*

A Lead Free Joining Technology for High Temperature Interconnects Using Transient Liquid Phase Soldering (TLPS) 1321

Christian Ehrhardt, *Technical University Berlin*; Matthias Hutter, *Fraunhofer IZM*; Hermann Oppermann, *Fraunhofer IZM*; Klaus-Dieter Lang, *Technical University Berlin*

Developments of High-Bi Alloys as a High Temperature Pb-Free Solder 1328

Sandeep Mallampati, *Binghamton University*; Harry Schoeller, *Universal Instruments Corporation*; Liang Yin, *GE Global Research*; David Shaddock, *GE Global Research*; Junghyun Cho, *Binghamton University*

The Quantum Theory of Solid-State Atomic Bonding 1335

Chin C. Lee, *University of California, Irvine*; Lianxi Cheng, *University of California, Irvine*

Effective Method to Disperse and Incorporate Carbon Nanotubes in Electroless Ni-P Deposits 1342

Sha Xu, *City University of Hong Kong*; Yan Cheong Chan, *City University of Hong Kong*; Xiaoxin Zhu, *University of Greenwich*; Hua Lu, *University of Greenwich*; Chris Bailey, *University of Greenwich*

Electroless Ni-W-P Alloy as a Barrier Layer between Zn-Based High Temperature Solders and Cu Substrates 1348

Li Liu, *Loughborough University*; Longzao Zhou, *Huazhong University of Science & Technology*; Changqing Liu, *Loughborough University*

31: PoP, SiP, and Die Stacking

Chairs: Raj N. Master, *Microsoft Corporation*
Deborah Patterson, *Amkor Technology, Inc.*

Fabrication and Reliability Evaluation of a Novel Package-on-Package (PoP) Structure Based on Organic Substrate 1354

Xiaofeng Sun, *National Center for Advanced Packaging; Chinese Academy of Sciences*; Lixi Wan, *Chinese Academy of Sciences*; Yuan Lu, *National Center for Advanced Packaging; Chinese Academy of Sciences*

Strip Grinding Introduction for Thin PoP 1361

Jinseong Kim, *Amkor Technology Korea, Inc.*; Yesul Ahn, *Amkor Technology Korea, Inc.*; Gyuwan Han, *Amkor Technology Korea, Inc.*; Byoungwoo Cho, *Amkor Technology Korea, Inc.*; Dongjoo Park, *Amkor Technology Korea, Inc.*; Juhoon Yoon, *Amkor Technology Korea, Inc.*; Choonheung Lee, *Amkor Technology Korea, Inc.*; Lou Nicholls, *Amkor Technology Inc.*; Shengmin Wen, *Amkor Technology Inc.*

Cost and Performance Effective Silicon Interposer and Vertical Interconnect for 3D ASIC and Memory Integration 1366

Li Li, *Cisco Systems, Inc.*; Mitsutoshi Higashi, *Shinko Electric Industries Company, Ltd.*; Akihito Takano, *Shinko Electric Industries Company, Ltd.*; Jie Xue, *Cisco Systems, Inc.*; Gary Ikari, *Shinko Electric Industries Company, Ltd.*

Assembly and Packaging of Non-Bumped 3D Chip Stacks on Bumped Substrates 1372

Bing Dang, *IBM Corporation*; Joana Maria, *IBM Corporation*; Qianwen Chen, *IBM Corporation*; Jae-Woong Nah, *IBM Corporation*; Paul Andry, *IBM Corporation*; Cornelia Tsang, *IBM Corporation*; Katsuyuki Sakuma, *IBM Corporation*; Christy Tyberg, *IBM Corporation*; Raphael Robertazzi, *IBM Corporation*; Michael Scheuermann, *IBM Corporation*; Michael Gaynes, *IBM Corporation*; John Knickerbocker, *IBM Corporation*

The Miniaturization of a Micro-Ball Endoscope by SiP Approach 1378

Xunxun Zhu, *Tsinghua University*; Jian Cai, *Tsinghua University*; Yu Chen, *Tsinghua University*; Yingke Gu, *Tsinghua University*; Xiang Xie, *Tsinghua University*; Qian Wang, *Tsinghua University*; Zhihua Wang, *Tsinghua University*; Xiaofeng Sun, *Chinese Academy of Sciences*; Lixi Wan, *Chinese Academy of Sciences*

Design and Demonstration of Paper-Thin and Low-Warpage Single and 3D Organic Packages with Chip-Last Embedding Technology for Smart Mobile Applications 1384

Sung Jin Kim, *Georgia Institute of Technology*; Zihan Wu, *Georgia Institute of Technology*; Makoto Kobayashi, *Namics Corporation*; Fuhan Liu, *Georgia Institute of Technology*; Vanessa Smet, *Georgia Institute of Technology*; P. Markondeya Raj, *Georgia Institute of Technology*; Venky Sundaram, *Georgia Institute of Technology*; Rao Tummala, *Georgia Institute of Technology*

Manufacturing Readiness of BVA Technology for Ultra-High Bandwidth Package-on-Package 1389

Rajesh Katkar, *Invensas Corporation*; Rey Co, *Invensas Corporation*; Wael Zohni, *Invensas Corporation*

32: Substrates
Chairs: Yu-Hua Chen, *Unimicron*
Dong Wook Kim, *Qualcomm, Inc.*

Improvement of Substrate and Package Warpage by Copper Plating Process Optimization 1396

Omar Bchir, *Qualcomm Technologies, Inc.*; Houssam Jomaa, *Qualcomm Technologies, Inc.*; Chin Kwan Kim, *Qualcomm Technologies, Inc.*; Layal Rouhana, *Qualcomm Technologies, Inc.*; Kuiwon Kang, *Qualcomm Technologies, Inc.*; Milind Shah, *Qualcomm Technologies, Inc.*; Steve Bezuk, *Qualcomm Technologies, Inc.*

Coreless Substrate with Asymmetric Design to Improve Package Warpage 1401

Wei Lin, *Amkor Technology*; Bora Baloglu, *Amkor Technology*; Ken Stratton, *Amkor Technology*

Ultra Low CTE (1.8 ppm/°C) Core Material for Next Generation Thin CSP 1407

Tomohiko Kotake, *Hitachi Chemical Co., Ltd.*; Hikari Murai, *Hitachi Chemical Co., Ltd.*; Shin Takanezawa, *Hitachi Chemical Co., Ltd.*; Masato Miyatake, *Hitachi Chemical Co., Ltd.*; Masaaki Takekoshi, *Hitachi Chemical Co., Ltd.*; Masahisa Ose, *Hitachi Chemical Co., Ltd.*

A Novel Redistribution Layer Tailored by Nanotwinned Copper Decreases Warpage in Wafer Level Packaging 1411

Heng Li, *Shanghai Institute of Microsystem and Information Technology, Chinese Academy of Sciences*; Wenguo Ning, *Shanghai Institute of Microsystem and Information Technology, Chinese Academy of Sciences*; Chunsheng Zhu, *Shanghai Institute of Microsystem and Information Technology, Chinese Academy of Sciences*; Gaowei Xu, *Shanghai Institute of Microsystem and Information Technology, Chinese Academy of Sciences*; Le Luo, *Shanghai Institute of Microsystem and Information Technology, Chinese Academy of Sciences*

Demonstration of 3–5 µm RDL Line Lithography on Panel-Based Glass Interposers 1416

Hao Lu, *Georgia Institute of Technology*; Yutaka Takagi, *NGK Spark Plug Co., Ltd.*; Yuya Suzuki, *Georgia Institute of Technology*; Brett Sawyer, *Georgia Institute of Technology*; Robin Taylor, *Atotech GmbH*; Venky Sundaram, *Georgia Institute of Technology*; Rao Tummala, *Georgia Institute of Technology*

Characterization of Thin Polymer Films with the Focus on Lateral Stress and Mechanical Properties and Their Relevance to Microelectronics 1421

Markus Woehrmann, *Technical University Berlin*; Thorsten Fischer, *Fraunhofer IZM*; Hans Walter, *Fraunhofer IZM*; Michael Toepper, *Fraunhofer IZM*; Klaus-Dieter Lang, *Technical University Berlin*

Thin Polymer Dry-Film Dielectric Material and a Process for 10 μm Interlayer Vias in High Density Organic and Glass Interposers 1427

Yuya Suzuki, *Zeon Corporation; Georgia Institute of Technology*; Yutaka Takagi, *NGK Spark Plug Co., Ltd.*; Venky Sundaram, *Georgia Institute of Technology*; Rao Tummala, *Georgia Institute of Technology*

33: Novel Test Methods

Chairs: Lakshmi N. Ramanathan, *Microsoft Corporation*
Sridhar Canumalla, *Microsoft Corporation*

Pad Crater Detection Using Acoustic Waveform Analysis 1433

W. Carter Ralph, *Southern Research Institute*; Elizabeth E. Benedetto, *Hewlett Packard*; Aileen M. Allen, *Hewlett Packard*; Keith Newman, *Hewlett Packard*

High Acceleration Board Level Reliability Drop Test Using Dual Mass Shock Amplifier 1441

Andy Zhang, *Texas Instruments, Inc.*

Non-Destructive Crack and Defect Detection in SAC Solder Interconnects Using Cross-Sectioning and X-Ray Micro-CT Using Cross-Sectioning and X-Ray Micro-CT 1449

Pradeep Lall, *Auburn University*; Shantanu Deshpande, *Auburn University*; Junchao Wei, *Auburn University*; Jeff Suhling, *Auburn University*

High Resolution and Fast Throughput-Time X-Ray Computed Tomography for Semiconductor Packaging Applications 1457

Yan Li, *Intel Corporation*; Mario Pacheco, *Intel Corporation*; Deepak Goyal, *Intel Corporation*; John W. Elmer, *Lawrence Livermore National Laboratory*; Holly D. Barth, *Lawrence Livermore National Laboratory*; Dula Parkinson, *Lawrence Berkeley National Laboratory*

In-Situ Measurements of the Relative Thermal Resistance: Highly Sensitive Method to Detect Crack Propagation in Solder Joints 1464

Gordon Elger, *Technische Hochschule Ingolstadt*; Shri Vishnu Kandaswamy, *Technische Hochschule Ingolstadt*; Maarten von Kouwen, *Philips Technology GmbH*; Robert Derix, *Philips Technology GmbH*; Fosca Conti, *University of Padova*

Reliability Testing of Wire Bonds Using Pad Resistance with van der Pauw Method 1471

Michael Mayer, *University of Waterloo*; Samuel Kim, *University of Waterloo*

Colour Shift in Remote Phosphor Based LED Products 1477

M. Yazdan Mehr, *Materials Innovation Institute; Delft University of Technology*; W.D. Van Driel, *Philips Lighting; Delft University of Technology*; G.Q. Zhang, *Delft University of Technology*

34: Novel Packaging

Chairs: Vasudeva P. Atluri, *Renavitas Technologies*
Jai Agrawal, *Purdue University*

Multifunctional System Integration in Flexible Substrates 1482

K. Bock, *Fraunhofer EMFT*; E. Yacoub-George, *Fraunhofer EMFT*; W. Hell, *Fraunhofer EMFT*; A. Drost, *Fraunhofer EMFT*; H. Wolf, *Fraunhofer EMFT*; D. Bollmann, *Fraunhofer EMFT*; C. Landesberger, *Fraunhofer EMFT*; G. Klink, *Fraunhofer EMFT*; H. Gieser, *Fraunhofer EMFT*; C. Kutter, *Fraunhofer EMFT*

Preparation of a Micro Rubidium Vapor Cell and Its Integration in a Chip-Scale Atomic Magnetometer 1488

Yu Ji, *Southeast University*; Jintang Shang, *Southeast University*; Youpeng Chen, *Southeast University*; Ching-Ping Wong, *Chinese University of Hong Kong*

Nanowires-Based High-Density Capacitors and Thinfilm Power Sources in Ultra-Thin 3D Glass Modules 1492

Saumya Gandhi, *Georgia Institute of Technology*; Liyi Li, *Georgia Institute of Technology*; Ho-Yee Hui, *Georgia Institute of Technology*; Parthasarathi Chakraborti, *Georgia Institute of Technology*; Himani Sharma, *Georgia Institute of Technology*; P. Markondeya Raj, *Georgia Institute of Technology*; C.P. Wong, *Georgia Institute of Technology*; Rao Tummala, *Georgia Institute of Technology*

Development of a High Density Glass Interposer Based on Wafer Level Packaging Technologies 1498

Michael Töpper, *Fraunhofer IZM*; Markus Wöhrmann, *Technical University Berlin*; Lars Brusberg, *Fraunhofer IZM*; Nils Jürgensen, *Fraunhofer IZM*; Ivan Ndip, *Fraunhofer IZM*; Klaus-Dieter Lang, *Technical University Berlin*

Novel Sealing Technology for Organic EL Display and Lighting by Means of Modified Surface Activated Bonding Method 1504

Takashi Matsumae, *University of Tokyo*; Masahisa Fujino, *University of Tokyo*; Tadatomo Suga, *University of Tokyo*

Solder Joint Inspection with Induction Thermography 1509

Johannes Bohm, *Technical University Dresden*; Klaus-Juergen Wolter, *Technical University Dresden*; Henning Heuer, *Technical University Dresden*

Development of B-Spline X-Ray Diffraction Imaging Techniques for Die Warpage and Stress Monitoring inside Fully Encapsulated Packaged Chips 1517

C.S. Wong, *Dublin City University*; A. Ivankovic, *IMEC; KU Leuven*; A. Cowley, *Dublin City University*; N.S. Bennett, *Dublin City University*; A.N. Danilewsky, *Albert-Ludwigs-Universität*; M. Gonzalez, *IMEC*; V. Cherman, *IMEC*; B. Vandevelde, *IMEC*; I. De Wolf, *IMEC; KU Leuven*; P.J. McNally, *Dublin City University*

35: Innovations in Wirebond Technology

Chairs: William Chen, *Advanced Semiconductor Engineering, Inc.*
Gilles Poupon, *CEA-LETI*

Process Optimization and Reliability Study for Cu Wire Bonding Advanced Nodes 1523

Ivy Qin, *Kulicke and Soffa Industries, Inc.*; Hui Xu, *Kulicke and Soffa Industries, Inc.*; Basil Milton, *Kulicke and Soffa Industries, Inc.*; Nestor Mendoza, *Kulicke and Soffa Industries, Inc.*; Horst Clauberg, *Kulicke and Soffa Industries, Inc.*; Bob Chylak, *Kulicke and Soffa Industries, Inc.*; Hidenori Abe, *Hitachi Chemical Co., Ltd.*; Dongchul Kang, *Hitachi Chemical Co., Ltd.*; Yoshinori Endo, *Hitachi Chemical Co., Ltd.*; Masahiko Osaka, *Hitachi Chemical Co., Ltd.*; Shinya Nakamura, *Hitachi Chemical Co., Ltd.*

Silver-Assisted Copper Wire Bonding Using Solid-State Processes 1529

Yi-Ling Chen, *University of California, Irvine*; Yuan-Yun Wu, *University of California, Irvine*; Chin C. Lee, *University of California, Irvine*

Ag Alloy Wire Characteristic and Benefits 1533

Jensen Tsai, *Siliconware Precision Industries Co., Ltd.*; Albert Lan, *Siliconware Precision Industries Co., Ltd.*; D.S. Jiang, *Siliconware Precision Industries Co., Ltd.*; Li Wei Wu, *Siliconware Precision Industries Co., Ltd.*; Joseph Huang, *Siliconware Precision Industries Co., Ltd.*; J.B. Hong, *Siliconware Precision Industries Co., Ltd.*

Copper versus Palladium Coated Copper Wire Process and Reliability Differences 1539

Chu-Chung (Stephen) Lee, *Freescale Semiconductor, Inc.*; TuAnh Tran, *Freescale Semiconductor, Inc.*; Dan Boyne, *Freescale Semiconductor, Inc.*; Leo Higgins, *Freescale Semiconductor, Inc.*; Andrew Mawer, *Freescale Semiconductor, Inc.*

Improving the Bond Quality of Copper Wire Bonds Using a Friction Model Approach 1549

Simon Althoff, *University of Paderborn*; Jan Neuhaus, *University of Paderborn*; Tobias Hemsel, *University of Paderborn*; Walter Sextro, *University of Paderborn*

High Aspect Ratio Lithography for Litho-Defined Wire Bonding 1556
Zahra Kolahdouz Esfahani, *Delft University of Technology*; Henk van Zeijl, *Delft University of Technology*;
G.Q. Zhang, *Delft University of Technology*

**Comprehensive Intermetallic Compound Phase Analysis and Its Thermal Evolution at Cu
Wirebond Interface** 1562
In-Tae Bae, *Binghamton University*; Dae Young Jung, *Binghamton University*; Jenny Chang, *Advanced
Semiconductor Engineering, Inc.*; Scott Chen, *Advanced Semiconductor Engineering, Inc.*

36: Recent Advancement in Manufacturing Technology
Chairs: Paul Houston, *Engent*
Hirofumi Nakajima, *Consultant*

High Uniformity and High Speed Copper Pillar Plating Technique 1571
Konstantin Kholostov, *Sapienza University of Rome*; Aliaksei Klyshko, *Sapienza University of Rome*;
Danilo Ciarniello, *Rise Technology S.r.l.*; Paolo Nenzi, *Rise Technology S.r.l.*; Roberto Pagliucci, *Rise
Technology S.r.l.*; Rocco Crescenzi, *Sapienza University of Rome*; Dario Bernardi, *2BG*; Marco Balucani,
Sapienza University of Rome, Rise Technology S.r.l.

Plasma-Based Die Singulation Processing Technology 1577
Kenneth D. Mackenzie, *Plasma-Therm LLC*; David Pays-Volard, *Plasma-Therm LLC*; Linnell Martinez,
Plasma-Therm LLC; Christopher Johnson, *Plasma-Therm LLC*; Thierry Lazerand, *Plasma-Therm LLC*;
Russell Westerman, *Plasma-Therm LLC*

**Removed Organic Solderability Preservative (OSP) by Ar/O2 Microwave Plasma to Improve
Solder Joint in Thermal Compression Flip Chip Bonding** 1584
Jr-Wei Peng, *ASE Group*; Yan-Siang Chen, *ASE Group*; Yi Chen, *ASE Group*; Jiang-Long Liang,
National Cheng Kung University; Kwang-Lung Lin, *National Cheng Kung University*; Yuh-Lang Lee,
National Cheng Kung University

A PoP Structure to Support I/O over 2000 1590
Dyi-Chung Hu, *Unimicron Technology Corporation*; Puru Lin, *Unimicron Technology Corporation*; Yu
Hua Chen, *Unimicron Technology Corporation*; Chun-Ting Lin, *Unimicron Technology Corporation*

Enabling Eutectic Soldering of 3D Opto-Electronics onto Low Tg Flexible Interposers 1595
Meriem Ben-Salah Akin, *Leibniz University of Hanover*; Lutz Rissing, *Leibniz University of Hanover*;
Wolfgang Heumann, *Leibniz University of Hanover*

Parameter Optimization in Assembly Manufacturing Process for a Power Module 1601
Yumin Liu, *Fairchild Semiconductor Corporation*; Yong Liu, *Fairchild Semiconductor Corporation*

Automated Inspection and Metrology for 2.5D and 3D/TSV Process Assurance 1606
James Wood, *IBM Corporation*; Vilmarie Soler, *IBM Corporation*; Eric Perfecto, *IBM Corporation*;
Thomas Luckenbach, *Camtek USA*; Aki Shoukrun, *Camtek Ltd.*

37: Interactive Presentations 1
Chairs: Mark Poliks, *i3 Electronics, Inc.*
Ibrahim Guven, *University of Arizona*

Investigation of a Photodefinable Glass Substrate for Millimeter-Wave Radios on Package 1610
Telesphor Kamgaing, *Intel Corporation*; Adel A. Elsherbini, *Intel Corporation*; Torrey W. Frank, *Intel
Corporation*; Sasha N. Oster, *Intel Corporation*; Valluri R. Rao, *Intel Corporation*

Design and Fabrication of Low-Pressure Piezoresistive MEMS Sensor for Fuel Cell Electric Vehicles 1616

Minkyu Lee, *Hyundai Motor Company*; Kiyoung Nam, *Hyundai Motor Company*; Seungyong Lee, *Hyundai Motor Company*; Hakgu Kim, *Hyundai Motor Company*; Chimyung Kim, *Hyundai Motor Company*; Yongsun Park, *Hyundai Motor Company*; Byungki Ahn, *Hyundai Motor Company*; Taewan Kim, *Sejong Industrial Company, Ltd.*; Hochul Seo, *Sejong Industrial Company, Ltd.*

Demonstration of TCNCP Flip Chip Reliability with 30μm Pitch Cu Bump and Substrate with Thin Ni and Thick Au Surface Finish 1622

Weihong Zhang, *Nantong Fujitsu Microelectronics Co., Ltd.*; Shengping Hong, *Nantong Fujitsu Microelectronics Co., Ltd.*; Xiaolong Yan, *Nantong Fujitsu Microelectronics Co., Ltd.*; Feng Zhou, *Nantong Fujitsu Microelectronics Co., Ltd.*; Tonglong Zhang, *Nantong Fujitsu Microelectronics Co., Ltd.*

Integrated Process Characterization and Fabrication Challenges for 2.5D IC Packaging Utilizing Silicon Interposer with Backside Via Reveal Process 1628

Cheng-Hsiang Liu, *Siliconware Precision Industries Co., Ltd.*; Jyun-Ling Tsai, *Siliconware Precision Industries Co., Ltd.*; Hung-Hsien Chang, *Siliconware Precision Industries Co., Ltd.*; Chang-Lun Lu, *Siliconware Precision Industries Co., Ltd.*; Shih-Ching Chen, *Siliconware Precision Industries Co., Ltd.*

Structure Effects on the Electrical Reliability of Fine-Pitch Cu Micro-Bumps for 3D Integration 1635

Byeong-Rok Lee, *Andong National University*; June-Bum Kim, *Andong National University*; Seung-Hyun Kim, *Andong National University*; Byeong-Hyun Bae, *Andong National University*; Ho-Young Son, *SK Hynix Inc.*; Tac-Keun Oh, *SK Hynix Inc.*; Min-Suk Suh, *SK Hynix Inc.*; Nam-Seog Kim, *SK Hynix Inc.*; Young-Bae Park, *Andong National University*

Demonstration of Low Cost TSV Fabrication in Thick Silicon Wafers 1641

E. Vick, *RTI International*; D.S. Temple, *RTI International*; R. Anderson, *RTI International*; J. Lannon, *RTI International*; C. Li, *DRS RSTA, Inc.*; K. Peterson, *DRS RSTA, Inc.*; G. Skidmore, *DRS RSTA, Inc.*; C.J. Han, *DRS RSTA, Inc.*

X-Ray Micro-Beam Diffraction Measurement of the Effect of Thermal Cycling on Stress in Cu TSV: A Comparative Study 1648

Chukwudi Okoro, *NIST*; Lyle E. Levine, *NIST*; Ruqing Xu, *Argonne National Laboratory*; Klaus Hummler, *SEMATECH*; Yaw Obeng, *NIST*

Adhesive Enabling Technology for Directly Plating Copper onto Glass/Ceramic Substrates 1652

Hailuo Fu, *Atotech USA Inc.*; Sara Hunegnaw, *Atotech USA Inc.*; Zhiming Liu, *Atotech USA Inc.*; Lutz Brandt, *Atotech USA Inc.*; Tafadzwa Magaya, *Atotech USA Inc.*

Very Thin POP and SIP Packaging Approaches to Achieve Functionality Integration Prior to TSV Implementation 1656

Fernando Roa, *Amkor Technology, Inc.*

A Study on the Fine Pitch Chip Interconnection Using Cu/SnAg Bumps and B-Stage Non-Conductive Films (NCFs) for 3D-TSV Vertical Interconnection 1661

Yongwon Choi, *Korea Advanced Institute of Science and Technology (KAIST)*; Jiwon Shin, *Korea Advanced Institute of Science and Technology (KAIST)*; Young Soon Kim, *Korea Advanced Institute of Science and Technology (KAIST)*; Kyung-Lim Suk, *Korea Advanced Institute of Science and Technology (KAIST)*; Il Kim, *Korea Advanced Institute of Science and Technology (KAIST)*; Kyung-Wook Paik, *Korea Advanced Institute of Science and Technology (KAIST)*

Pathfinding Methodology for Optimal Design and Integration of 2.5D/3D Interconnects 1667

Farhang Yazdani, *BroadPak Corporation*; John Park, *Mentor Graphics Corporation*

Cost Effective Interposer for Advanced Electronic Packages 1673

Satoru Kuramochi, *Dai Nippon Printing Co., Ltd.*; Sumio Koiwa, *Dai Nippon Printing Co., Ltd.*; Kousuke Suzuki, *Dai Nippon Printing Co., Ltd.*; Yoshitaka Fukuoka, *WEISTI*

Thermal Management for Wafer Level Packaging (WLP) 1679

Tiao Zhou, *Maxim Integrated*; Arkadii Samoilov, *Maxim Integrated*

Inkjet Printed Nano-Particle Cu Process for Fabrication of Re-Distribution Layers on Silicon Wafer 1685
    Ayat Soltani, *Tampere University of Technology*; Tero Kumpulainen, *Tampere University of Technology*; Matti Mäntysalo, *Tampere University of Technology*

Design of Multi-Sensor for Safety Monitoring of Heavy Machinery 1690
    Long Li, *Huazhong University of Science & Technology*; Fei Hou, *Dongfeng Automobile Electronics Co., Ltd.*; Jinghao Qiu, *Nanjing University of Aeronautics and Astronautics*; Zhang Luo, *Huazhong University of Science & Technology*; Shengzhi Zhang, *Huazhong University of Science & Technology*; Qiang Dan, *Huazhong University of Science & Technology*; Sheng Liu, *Wuhan University*

Novel TSV Process Technologies for 2.5D/3D Packaging 1697
    Y. Morikawa, *ULVAC, Inc.; NMEMS Technology Research Organization*; T. Murayama, *ULVAC, Inc.; NMEMS Technology Research Organization*; T. Sakuishi, *ULVAC, Inc.; NMEMS Technology Research Organization*; A. Suzuki, *ULVAC, Inc.*; Y. Nakamuta, *ULVAC, Inc.*; K. Suu, *ULVAC, Inc.; NMEMS Technology Research Organization*

Increasing the Lifetime of Electronic Packaging by Higher Temperatures: Solders vs. Silver Sintering 1700
    Aaron Hutzler, *Fraunhofer IISB*; Adam Tokarski, *Fraunhofer IISB*; Silke Kraft, *Fraunhofer IISB*; Sigrid Zischler, *Fraunhofer IISB*; Andreas Schletz, *Fraunhofer IISB*

Comparison of New Die-Attachment Technologies for Power Electronic Assemblies 1707
    Eike Möller, *University of Freiburg*; Adeel Ahmad Bajwa, *University of Freiburg*; Eugen Rastjagaev, *Infineon Technologies AG*; Jürgen Wilde, *University of Freiburg*

High Vacuum Wafer Level Packaging for High-Value MEMS Applications 1714
    S. Nicolas, *CEA-LETI*; F. Greco, *CEA-LETI*; S. Caplet, *CEA-LETI*; C. Coutier, *CEA-LETI*; C. Dressler, *CEA-LETI*; M. Audoin, *CEA-LETI*; X. Baillin, *CEA-LETI*; G. Dehag, *CEA-LETI*; F. Souchon, *CEA-LETI*; S. Fanget, *CEA-LETI*

Thermal and Electrical Tests of Air-Gap TSV 1722
    Cui Huang, *Tsinghua University*; Dong Wu, *Tsinghua University*; Zheyao Wang, *Tsinghua University*

Heterogeneous System Integration Pseudo-SoC Technology for Smart-Health-Care Intelligent Life Monitor Engine & Eco-System (SILMEE) 1729
    Hiroshi Yamada, *Toshiba Corporation*; Yasuhiro Sato, *Toshiba Corporation*; Nobuhiro Ooshima, *Toshiba Corporation*; Hiroyuki Hirai, *Toshiba Corporation*; Takuji Suzuki, *Toshiba Corporation*; Shigenobu Minami, *Toshiba Corporation*

Effects of Various Environmental Conditions on the Electrical Properties and Interfacial Reliability of Printed Ag/Polyimide System 1735
    Byung-Hyun Bae, *Andong National University*; Min-Su Jeong, *Andong National University*; Byeong Rok Lee, *Andong National University*; Joung-Hoon Choo, *HICEL*; Eun-Kuk Choi, *HICEL*; Jong-Sun Yoon, *HICEL*; Young-Bae Park, *Andong National University*

Wafer Level Warpage Characterization for Backside Manufacturing Processes of TSV Interposers 1740
    Feng Jiang, *National Center for Advanced Packaging*; Qibin Wang, *National Center for Advanced Packaging; Chinese Academy of Sciences*; Kai Xue, *National Center for Advanced Packaging; Chinese Academy of Sciences*; Xiangmeng Jing, *National Center for Advanced Packaging; Chinese Academy of Sciences*; Daquan Yu, *National Center for Advanced Packaging; Chinese Academy of Sciences*; Dongkai Shangguan, *National Center for Advanced Packaging; Chinese Academy of Sciences*

Stretchable and Transparent Silicone/Zinc Oxide Nanocomposite for Advanced LED Packaging 1745
    Xueying Zhao, *Georgia Institute of Technology*; Liyi Li, *Georgia Institute of Technology*; Zhuo Li, *Georgia Institute of Technology*; Ching-Ping Wong, *Georgia Institute of Technology; Chinese University of Hong Kong*

Warpage Characterization of Panel Fan-Out (P-FO) Package 1750

Hung-Wen Liu, *Siliconware Precision Industries Co., Ltd.*; Yi-Wei Liu, *Siliconware Precision Industries Co., Ltd.*; Jason Ji, *Siliconware Precision Industries Co., Ltd.*; Jash Liao, *Siliconware Precision Industries Co., Ltd.*; Agassi Chen, *Siliconware Precision Industries Co., Ltd.*; Yan-Heng Chen, *Siliconware Precision Industries Co., Ltd.*; Nicholas Kao, *Siliconware Precision Industries Co., Ltd.*; Yi-Che Lai, *Siliconware Precision Industries Co., Ltd.*

38: Interactive Presentations 2

Chairs: Mark Eblen, *Kyocera America, Inc.*
Michael Mayer, *University of Waterloo*

A Novel Double Layer NCF for Highly Reliable Micro-Bump Interconnection 1755

Ji-Won Shin, *Korea Advanced Institute of Science and Technology (KAIST)*; Yong-Won Choi, *Korea Advanced Institute of Science and Technology (KAIST)*; Young Soon Kim, *Korea Advanced Institute of Science and Technology (KAIST)*; Un Byung Kang, *Samsung Electronics Company, Ltd.*; Sun Kyung Seo, *Samsung Electronics Company, Ltd.*; Kyung-Wook Paik, *Korea Advanced Institute of Science and Technology (KAIST)*

CO2-Laser Drilling of TGVs for Glass Interposer Applications 1759

Lars Brusberg, *Fraunhofer IZM*; Marco Queisser, *Technical University Berlin*; Marcel Neitz, *Technical University Berlin*; Henning Schröder, *Fraunhofer IZM*; Klaus-Dieter Lang, *Technical University Berlin*

Effects of Pad Surface Finish on Interfacial Reliabilities of Cu-Pillar/Sn-Ag Bumps of 2.5D TSV-Interposer on PCB Applications 1765

Youngsoon Kim, *Samsung Electro-Mechanics Company, Ltd.*; Ji-Won Shin, *Korea Advanced Institute of Science and Technology (KAIST)*; Young Won Choi, *Korea Advanced Institute of Science and Technology (KAIST)*; Kyung-Wook Paik, *Korea Advanced Institute of Science and Technology (KAIST)*

Effect of Variation in the Reflow Profile on the Microstructure of Near Eutectic SnAgCu Alloys 1769

Francis Mutuku, *Binghamton University*; Babak Arfaei, *Binghamton University; Universal Instruments Corporation*; Eric J. Cotts, *Binghamton University*

Development of the Thin Film with High Thermal Conductivity for Power Devices 1776

Hiroshi Takasugi, *Namics Corporation*; Shin Teraki, *Namics Corporation*; Tsuyoshi Kurokawa, *Namics Corporation*; Issei Aoki, *Namics Corporation*

Development of Electroless Nickel-Iron Plating Process for Microelectronic Applications 1782

Yu Luo, *IBM Corporation*; Sung K. Kang, *IBM Corporation*; Oblesh Jinka, *IBM Corporation*; Maurice Mason, *IBM Corporation*; Steven A. Cordes, *IBM Corporation*; Lubomyr T. Romankiw, *IBM Corporation*

Novel Conductive Paste Using Hybrid Silver Sintering Technology for High Reliability Power Semiconductor Packaging 1790

Howard (Hwa Il) Jin, *Alpha Advanced Materials*; Senthil Kanagavel, *Alpha Advanced Materials*; Wai Foo Chin, *Alpha Advanced Materials*

Novel Low Temperature Curable Photo-Sensitive Insulator 1796

Kenji Okamoto, *JSR Corporation*; Hikaru Mizuno, *JSR Corporation*; Tomohiko Sakurai, *JSR Corporation*; Katsumi Inomata, *JSR Corporation*

3D and 2.5D Packaging Assembly with Highly Silica Filled One Step Chip Attach Materials for Both Thermal Compression Bonding and Mass Reflow Processes 1803

Christopher Breach, *Kester Inc.*; Daniel Duffy, *Kester Inc.*; David Eichstadt, *Kester Inc.*

Process Compatibility of Conventional and Low-Temperature Curable Organic Insulation Materials for 2.5D and 3D IC Packaging – A User's Perspective 1810

Guilian Gao, *Invensas Corporation*; Bong-Sub Lee, *Invensas Corporation*; Andrew Cao, *Invensas Corporation*; Ellis Chau, *Invensas Corporation*

Optimization of CMP Process for TSV Reveal in Consideration of Critical Defect 1816

DongHoon Lee, *Amkor Technology Korea, Inc.; Sungkyunkwan University*; DoHyeong Kim, *Amkor Technology Korea, Inc.*; SeungChul Han, *Amkor Technology Korea, Inc.*; JooHyun Kim, *Amkor Technology Korea, Inc.*; JungSoo Park, *Amkor Technology Korea, Inc.*; BoRa Jang, *Amkor Technology Korea, Inc.*; YoungSuk Chung, *Amkor Technology Korea, Inc.*; SeongMin Seo, *Amkor Technology Korea, Inc.*; YongSang Kim, *Sungkyunkwan University*; ChoonHeung Lee, *Amkor Technology Korea, Inc.*

High Throughput Roller Type Nano-Pattern Transfer Technique on Both Rigid Flexible Substrates and Mold Deformation Analysis under Atmospheric Imprint Environment 1822

Yinsheng Zhong, *Hong Kong University of Science and Technology*; Matthew M.F. Yuen, *Hong Kong University of Science and Technology*

Capacitive Deionization of Water Coolant Using Hybrid Carbon Electrodes for High Power Electronic Applications 1828

Ziyin Lin, *Georgia Institute of Technology*; Zhuo Li, *Georgia Institute of Technology*; Kyoung-Sik Moon, *Georgia Institute of Technology*; Ching-Ping Wong, *Georgia Institute of Technology; Chinese University of Hong Kong*

A Microfluidic Chip Integrated with a Sono-Transducer Using Combined Resonance between Oscillations of Hemispherical Micro Glass Shell and Enclosed Microfluid .... N/A

Jiafeng Xu, *Southeast University*; Jintang Shang, *Southeast University*; Ching-Ping Wong, *Chinese University of Hong Kong*

RF Energy Harvesting 1838

Parvizso Aminov, *Purdue University*; Jai P. Agrawal, *Purdue University*

Localized Metal Plating on Aluminum Back Side PV Cells 1842

M. Balucani, *Sapienza University of Rome; Rise Technology S.r.l.*; K. Kholostov, *Sapienza University of Rome*; L. Serenelli, *ENEA Casaccia Research Centre*; M. Izzi, *ENEA Casaccia Research Centre*; D. Bernardi, *2BG S.r.l.*; M. Tucci, *ENEA Casaccia Research Centre*

Wet Etching of Deep Trenches on Silicon with Three-Dimensional (3D) Controllability 1848

Liyi Li, *Georgia Institute of Technology*; Ching-Ping Wong, *Georgia Institute of Technology; Chinese University of Hong Kong*

An Innovative Bumpless Stacking with Through Silicon Via for 3D Wafer-On-Wafer (WOW) Integration 1853

Sue-Chen Liao, *Industrial Technology Research Institute (ITRI)*; Erh-Hao Chen, *Industrial Technology Research Institute (ITRI)*; Chien-Chou Chen, *Industrial Technology Research Institute (ITRI)*; Shang-Chun Chen, *Industrial Technology Research Institute (ITRI)*; Jui-Chin Chen, *Industrial Technology Research Institute (ITRI)*; Po-Chih Chang, *Industrial Technology Research Institute (ITRI)*; Yiu-Hsiang Chang, *Industrial Technology Research Institute (ITRI)*; Cha-Hsin Lin, *Industrial Technology Research Institute (ITRI)*; Tzu-Kun Ku, *Industrial Technology Research Institute (ITRI)*; Ming-Jer Kao, *Industrial Technology Research Institute (ITRI)*; Young Suk Kim, *Tokyo Institute of Technology*; Nobuhide Maeda, *Tokyo Institute of Technology*; Shoichi Kodama, *Tokyo Institute of Technology*; Hideki Kitada, *Tokyo Institute of Technology*; Koji Fujimoto, *Tokyo Institute of Technology*; Takayuki Ohba, *Tokyo Institute of Technology*

3D Integration and Assembly of Wireless Sensor Nodes for 'Green' Sensor Networks 1857

Jian Lu, *National Institute of AIST; NMEMS Technology Research Organization*; Hironao Okada, *National Institute of AIST; NMEMS Technology Research Organization*; Toshihiro Itoh, *National Institute of AIST; NMEMS Technology Research Organization*; Takeshi Harada, *NMEMS Technology Research Organization*; Ryutaro Maeda, *National Institute of AIST; NMEMS Technology Research Organization*

New Demultiplexer Component for Optical Polymer Fiber Communication Systems 1862

S. Höll, *Harz University of Applied Sciences*; M. Haupt, *Harz University of Applied Sciences*; U.H.P. Fischer, *Harz University of Applied Sciences*

Nanofiller Based Spin-on Materials for Negligible Reflection of Silicon Photonic External Coupling 1870

Yoichi Taira, *IBM Corporation*; Ryuma Mizusawa, *Tokyo Ohka Kogyo Co., Ltd.*; Rie Matsumoto, *Tokyo Ohka Kogyo Co., Ltd.*; Kuniaki Suaoke, *IBM Corporation*; Hidetoshi Numata, *IBM Corporation*

Effect of Patterned Substrate on Light Extraction Efficiency of Chip-on-Board Packaging LEDs 1876

Huai Zheng, *Huazhong University of Science & Technology*; Zhili Zhao, *Huazhong University of Science & Technology*; Yiman Wang, *Huazhong University of Science & Technology*; Lang Li, *Huazhong University of Science & Technology*; Sheng Liu, *Huazhong University of Science & Technology*; Xiaobing Luo, *Huazhong University of Science & Technology*

39: Interactive Presentations 3

Chairs: Patrick Thompson, *Texas Instruments, Inc.*
Rao Bonda, *Amkor Technology, Inc.*

Transferrable Fine Pitch Probe Technology 1880

Y. Liu, *IBM Corporation*; S.L. Wright, *IBM Corporation*; B. Dang, *IBM Corporation*; P. Andry, *IBM Corporation*; R. Polastre, *IBM Corporation*; J. Knickerbocker, *IBM Corporation*

Improvement of the Crystallinity of Electroplated Copper Thin Films for Highly Reliable 3D Interconnections 1885

Chuanhong Fan, *Tohoku University*; Osamu Asai, *Tohoku University*; Ryosuke Furuya, *Tohoku University*; Ken Suzuki, *Tohoku University*; Hideo Miura, *Tohoku University*

Process, Assembly and Electromigration Characteristics of Glass Interposer for 3D Integration 1891

Chun-Hsien Chien, *Industrial Technology Research Institute (ITRI)*; Ching-Kuan Lee, *Industrial Technology Research Institute (ITRI)*; Chun-Te Lin, *Industrial Technology Research Institute (ITRI)*; Yu-Min Lin, *Industrial Technology Research Institute (ITRI)*; Chau-Jie Zhan, *Industrial Technology Research Institute (ITRI)*; Hsiang-Hung Chang, *Industrial Technology Research Institute (ITRI)*; Chao-Kai Hsu, *Industrial Technology Research Institute (ITRI)*; Huan-Chun Fu, *Industrial Technology Research Institute (ITRI)*; Wen-Wei Shen, *Industrial Technology Research Institute (ITRI)*; Yu-Wei Huang, *Industrial Technology Research Institute (ITRI)*; Cheng-Ta Ko, *Industrial Technology Research Institute (ITRI)*; Wei-Chung Lo, *Industrial Technology Research Institute (ITRI)*; Yung Jean (Rachel) Lu, *Corning Inc.*

Improved PCB Via Pattern to Reduce Crosstalk at Package BGA Region for High Speed Serial Interface 1896

Yujeong Shim, *Altera Corporation*; Dan Oh, *Altera Corporation*

A Wafer Level Through-Stack-Via Integration Process with One-Time Bottom-up Copper Filling 1902

Yunhui Zhu, *Peking University*; Shenglin Ma, *Xiamen University, Peking University*; Xin Sun, *Peking University*; Runiu Fang, *Peking University*; Xiao Zhong, *Peking University*; Yuan Bian, *Peking University*; Yong Guan, *Peking University*; Jing Chen, *Peking University*; Min Miao, *Peking University, Beijing Information Science and Technology University*; Yufeng Jin, *Peking University*

Effect of Joint Shape Controlled by Thermocompression Bonding on the Reliability Performance of 60μm-Pitch Solder Micro Bump Interconnections 1908

Yu-Wei Huang, *Industrial Technology Research Institute (ITRI)*; Chau-Jie Zhan, *Industrial Technology Research Institute (ITRI)*; Jing-Ye Juang, *Industrial Technology Research Institute (ITRI)*; Yu-Min Lin, *Industrial Technology Research Institute (ITRI)*; Shin-Yi Huang, *Industrial Technology Research Institute (ITRI)*; Su-Mei Chen, *Industrial Technology Research Institute (ITRI)*; Chia-Wen Fan, *Industrial Technology Research Institute (ITRI)*; Ren-Shin Cheng, *Industrial Technology Research Institute (ITRI)*; Shu-Han Chao, *National Chiao Tung University*; Wan-Lin Hsieh, *National Chiao Tung University*; Chih Chen, *National Chiao Tung University*; John H. Lau, *Industrial Technology Research Institute (ITRI)*

Development of Micro Bump Joints Fabrication Process Using Cone Shape Au Bumps for 3D LSI Chip Stacking 1915

Fumito Imura, *National Institute of AIST*; Naoya Watanabe, *National Institute of AIST*; Shunsuke Nemoto, *National Institute of AIST*; Wei Feng, *National Institute of AIST*; Katsuya Kikuchi, *National Institute of AIST*; Hiroshi Nakagawa, *National Institute of AIST*; Masahiro Aoyagi, *National Institute of AIST*

Effect of Polymer Liners in CNT Based Through Silicon Vias 1921

Archana Kumari, *Indian Institute of Technology Roorkee*; M.K. Majumder, *Indian Institute of Technology Roorkee*; B.K. Kaushik, *Indian Institute of Technology Roorkee*; S.K. Manhas, *Indian Institute of Technology Roorkee*

Investigation of Low-Temperature Deposition High-Uniformity Coverage Parylene-HT as a Dielectric Layer for 3D Interconnection 1926

Bui Thanh Tung, *National Institute of AIST*; Xiaojin Cheng, *National Institute of AIST; Loughborough University*; Naoya Watanabe, *National Institute of AIST*; Fumiki Kato, *National Institute of AIST*; Katsuya Kikuchi, *National Institute of AIST*; Masahiro Aoyagi, *National Institute of AIST*

Arrays of Millimeter-Wave Silicon Waveguides for Interchip Communication on Glass Interposer 1932

Qidong Wang, *Chinese Academy of Sciences; National Center for Advanced Packaging*; Daniel Guidotti, *Chinese Academy of Sciences; National Center for Advanced Packaging*; Liqiang Cao, *Chinese Academy of Sciences; National Center for Advanced Packaging*; Delong Qiu, *National Center for Advanced Packaging*; Daquan Yu, *Chinese Academy of Sciences; National Center for Advanced Packaging*; Shuling Wang, *Chinese Academy of Sciences; National Center for Advanced Packaging*; Xugang Wang, *Chinese Academy of Sciences; National Center for Advanced Packaging*; Tiachun Ye, *Chinese Academy of Sciences; National Center for Advanced Packaging*; Lixi Wan, *National Center for Advanced Packaging*

Effect of Ag and Cu Content in Sn Based Pb-Free Solder on Electromigration 1940

Minhua Lu, *IBM Corporation*; Charles Goldsmith, *IBM Corporation*; Thomas Wassick, *IBM Corporation*; Eric Perfecto, *IBM Corporation*; Charles Arvin, *IBM Corporation*

Low Loss Transmission Lines on Flexible COP Substrate by Standard Lamination Process 1944

Chang-Ho Liou, *Industrial Technology Research Institute (ITRI)*; Hsin-Chia Lu, *National Taiwan University*; Yi-Fan Lin, *National Taiwan University*; Shih-Keng Chuang, *National Taiwan University*; Wen-Ching Ko, *Industrial Technology Research Institute (ITRI)*; Je-Ping Hu, *Industrial Technology Research Institute (ITRI)*

FBEOL No-Aluminum Pad Integration in Pb-Free C4 Products for Environmental, Cost and Reliability Benefits 1949

E. Misra, *IBM Corporation*; T. Daubenspeck, *IBM Corporation*; T. Wassick, *IBM Corporation*; K. Tunga, *IBM Corporation*; D. Questad, *IBM Corporation*

Preparing 25Gbps Electrical I/O for Exascale Computing Systems 1955

Lei Shan, *IBM Corporation*; Young Kwark, *IBM Corporation*; Renato Rimolo-Donadio, *IBM Corporation*; Christian Baks, *IBM Corporation*; Michael Gaynes, *IBM Corporation*; Timothy Chainer, *IBM Corporation*; Manabu Hoshino, *Zeon Corporation*; Masakazu Hashimoto, *Zeon Corporation*; Toshihiko Jimbo, *Zeon Corporation*; Junji Kodemura, *Zeon Corporation*; Ikkei Matsuura, *Zeon Corporation*

Large Low-CTE Glass Package-to-PCB Interconnections with Solder Strain-Relief Using Polymer Collars 1959

Gary Menezes, *Georgia Institute of Technology*; Vanessa Smet, *Georgia Institute of Technology*; Makoto Kobayashi, *Namics Corporation*; Venky Sundaram, *Georgia Institute of Technology*; Pulugurtha Markondeya Raj, *Georgia Institute of Technology*; Rao Tummala, *Georgia Institute of Technology*

The Study of Bare-Die FCBGA Die Damage in Response to Applied Mechanical Stress During Heat Sink Assembly 1965

Heidi S.Y. Ho, *Broadcom Corporation*; Daijiao Wang, *Broadcom Corporation*; Michael Johnson, *Amkor Technology, Inc.*; C.J. Berry, *Amkor Technology, Inc.*

Prognostication of Copper-Aluminum Wirebond Reliability under High Temperature Storage and Temperature-Humidity 1973

Pradeep Lall, *Auburn University*; Shantanu Deshpande, *Auburn University*; Luu Nguyen, *Texas Instruments*; Masood Murtuza, *Texas Instruments*

Low-Frequency Testing of Through Silicon Vias for Defect Diagnosis in Three-Dimensional Integration Circuit Stacking Technology 1986

Yichao Xu, *Peking University*; Min Miao, *Beijing Information Science & Technology University; Peking University*; Runiu Fang, *Peking University*; Xin Sun, *Peking University*; Yunhui Zhu, *Peking University*; Minggang Sun, *Beijing Information Science & Technology University*; Guanjiang Wang, *Peking University*; Yufeng Jin, *Peking University*

Fast Estimation of LED's Accelerated Lifetime by Online Test Method 1992

Qi Chen, *Huazhong University of Science & Technology*; Quan Chen, *Huazhong University of Science & Technology*; Xiaobing Luo, *Huazhong University of Science & Technology*

Methodology and Apparatus for Rapid Power Cycle Accumulation and In-Situ Incipient Failure Monitoring for Power Electronic Modules 1996

Roy I. Davis, *Fairchild Semiconductor Corporation*; Daniel J. Sprenger, *Fairchild Semiconductor Corporation*

Fine-Pitch Probing on TSVs and Microbumps Using a Chip Prober Having a Transparent Membrane Probe Card 2003

Naoya Watanabe, *National Institute of AIST*; Michiyuki Eto, *STK Technology Co., Ltd.*; Kenji Kawano, *STK Technology Co., Ltd.*; Masahiro Aoyagi, *National Institute of AIST*

40: Interactive Presentations 4

Chairs: Nam Pham, *IBM Corporation*
Rabindra N. Das, *MIT Lincoln Labs*

Thermal Management of 3D RF PoP Based on Ceramic Substrate 2008

Fengze Hou, *National Center for Advanced Packaging; Chinese Academy of Sciences*; Fengman Liu, *National Center for Advanced Packaging; Chinese Academy of Sciences*; Yi He, *Chinese Academy of Sciences*; Xiaomeng Wu, *Chinese Academy of Sciences*; Xia Zhang, *National Center for Advanced Packaging; Chinese Academy of Sciences*; Liqiang Cao, *National Center for Advanced Packaging; Chinese Academy of Sciences*; Yuan Lu, *National Center for Advanced Packaging; Chinese Academy of Sciences*; Dongkai Shangguan, *National Center for Advanced Packaging; Chinese Academy of Sciences*

Bump Pattern Optimization and Stress Comparison Study for DCA Packages 2014

Akash Agrawal, *Micron Technology Inc.*; Owen Fay, *Micron Technology Inc.*; Mark Johnson, *Micron Technology Inc.*

Characterization of In-Plane Stress in TSV Array – A Unit Model Approach 2020

Cheng-Fu Chen, *University of Alaska, Fairbanks*

Electrical-Thermal Characterization of Wires in Packages 2027

Kai Liu, *STATS ChipPAC, Inc.*; Robert Frye, *STATS ChipPAC, Inc.*; HyunTai Kim, *STATS ChipPAC, Inc.*; YongTaek Lee, *STATS ChipPAC, Inc.*; Gwang Kim, *STATS ChipPAC, Inc.*; Susan Park, *STATS ChipPAC, Inc.*; Billy Ahn, *STATS ChipPAC, Inc.*

Computational Investigation of Failure in Anodized Aluminum 2035

Sabrina Ball, *University of Arizona*; Ibrahim Guven, *University of Arizona*; Pankaj Sinha, *Intel Corporation*; Rajiv Rastogi, *Intel Corporation*; Brian McCarson, *Intel Corporation*

Study on Prediction about Residual Position of Void Generated by Resin Flow 2042

Masayuki Mino, *Hitachi, Ltd.*; Naoya Suzuki, *Hitachi Chemical Co., Ltd.*; Hiroshi Takahashi, *Hitachi Chemical Co., Ltd.*; Tsutomu Kono, *Hitachi, Ltd.*

Modeling and Analysis of Temperature Effect on MEMS Gyroscope 2048

Ming Wen, *Huazhong University of Science & Technology*; Weihui Wang, *Huazhong University of Science & Technology*; Zhang Luo, *Huazhong University of Science & Technology*; Yong Xu, *Wuhan University; Wayne State University*; Xin Wu, *Wayne State University*; Fei Hou, *Dongfeng Automobile Electronics Co., Ltd.*; Sheng Liu, *Wuhan University*

Life Prediction and Classification of Failure Modes in Solid State Luminaires Using Bayesian Probabilistic Models 2053

Pradeep Lall, *Auburn University*; Junchao Wei, *Auburn University*; Peter Sakalaukus, *Auburn University*

Modeling for Reliability of Ultra Thin Chips in a System in Package 2063

Richard Qian, *Fairchild Semiconductor Corporation*; Yong Liu, *Fairchild Semiconductor Corporation*

Development of Effective Thermal Characterization on Handheld Devices by Matrix Method 2069

Tai-Yu Chen, *MediaTek Inc.*; Chung-Fa Lee, *MediaTek Inc.*

Comprehensive Design Optimization for 2.133 Gbps LPDDR3 Extension for Mobile Platform System 2075

Chanmin Jo, *Samsung Electronics*; Jaemin Shin, *Samsung Electronics*; BaekKyu Choi, *Samsung Electronics*; Sangmin Lee, *Samsung Electronics*; Seongjae Moon, *Samsung Electronics*; Sungjoo Kim, *Samsung Electronics*; Woong Hwan Ryu, *Samsung Electronics*

Estimation of Mode Conversion and Crosstalk Impact from a Single-Ended Aggressor to a Differential Victim Using Statistical BER Analysis 2081

Arun Reddy Chada, *Missouri S&T EMC Laboratory*; Jun Fan, *Missouri S&T EMC Laboratory*; James L. Drewniak, *Missouri S&T EMC Laboratory*; Bhyrav Mutnury, *Dell, Inc.*

Power Distribution Network Worst-Case Power Noise and an Efficient Estimation Method 2088

Jiangyuan Qian, *Broadcom Corporation*; Shiji Pan, *University of California, Irvine*

Fast Calculation of Electromagnetic Interference by Through-Silicon Vias 2094

Aosheng Rong, *University of Illinois, Urbana-Champaign*; Andreas C. Cangellaris, *University of Illinois, Urbana-Champaign*; Feng Ling, *Nanjing University of Science and Technology*

Electrical Simulation and Analysis of Si Interposer for 3D IC Integration 2099

Xin Sun, *Peking University*; Min Miao, *Beijing Information Science & Technology University*; Yunhui Zhu, *Peking University*; Runiu Fang, *Peking University*; Guanjiang Wang, *Peking University*; Wengao Lu, *Peking University*; Jing Chen, *Peking University*; Yufeng Jin, *Peking University*

A SPICE Model of Multi-Mode Optical Fiber in Mid-Channel Link for Package System SI Transient Simulations 2104

Zhaoqing Chen, *IBM Corporation*

Next Generation Package-on-Package Solution to Support Wide IO and High Bandwidth Interface 2112

Hung-Hsiang Cheng, *Advanced Semiconductor Engineering, Inc.*; Chang-Chi Lee, *Advanced Semiconductor Engineering, Inc.*; Ming-Feng Chung, *Advanced Semiconductor Engineering, Inc.*; Po-Chih Pan, *Advanced Semiconductor Engineering, Inc.*; Ping-Feng Yang, *Advanced Semiconductor Engineering, Inc.*; Chi-Tsung Chiu, *Advanced Semiconductor Engineering, Inc.*; Chih-Pin Hung, *Advanced Semiconductor Engineering, Inc.*; Chen-Chao Wang, *Advanced Semiconductor Engineering, Inc.*

Package-Level Electromagnetic Interference Analysis 2119

Namhoon Kim, *Broadcom Corporation*; Leo Hongyu Li, *Broadcom Corporation*; Sam Karikalan, *Broadcom Corporation*; Reza Sharifi, *Broadcom Corporation*; Henry Kim, *Broadcom Corporation*

A Path Finding Based SI Design Methodology for 3D Integration 2124

Bill Martin, *E-System Design*; KiJin Han, *UNIST*; Madhavan Swaminathan, *Georgia Institute of Technology*

Design and Implementation of a 700-2600 MHz RF SiP for Micro Base Station .... N/A

Yi He, *National Center for Advanced Packaging; Chinese Academy of Sciences*; Fengman Liu, *National Center for Advanced Packaging; Chinese Academy of Sciences*; Anmou Liao, *National Center for Advanced Packaging; Chinese Academy of Sciences*; Jun Li, *National Center for Advanced Packaging; Chinese Academy of Sciences*; Xiaomeng Wu, *National Center for Advanced Packaging*; Peng Wu, *National Center for Advanced Packaging; Chinese Academy of Sciences*; Liqiang Cao, *National Center for Advanced Packaging; Chinese Academy of Sciences*; Dongkai Shangguan, *National Center for Advanced Packaging*

Dielectric Lens Optimization for Conical Helix THz Antennas 2137

Paolo Nenzi, *ENEA Frascati Research Center*; Volha Varlamava, *Sapienza University of Rome*; Frank Silvio Marzano, *Sapienza University of Rome*; Fabrizio Palma, *Sapienza University of Rome*; Marco Balucani, *Sapienza University of Rome*

Embedded Diodes for Microwave and Millimeter Wave Circuits 2144

Xianbo Yang, *Michigan State University*; Amanpreet Kaur, *Michigan State University*; Premjeet Chahal, *Michigan State University*

PCIe Gen3 Link Design and Tuning in Server Systems with End Devices from Multiple IP Suppliers 2151

Si T. Win, *IBM Corporation*; Daniel Rodriguez, *IBM Corporation*; Nanju Na, *IBM Corporation*

A Low-Cost PCB Fabrication Process 2159

Jack Ou, *Sonoma State University*; Alberto Maldonado, *Sonoma State University*; Chio Saephan, *Sonoma State University*; Farid Farahmand, *Sonoma State University*; Michael Caggiano, *Rutgers University*

Novel Band-Pass Filters on Thin Glass Substrate with Through Glass Vias (TGVs) .... N/A

Cheng Pang, *National Center for Advanced Packaging; Chinese Academy of Sciences*; Wenya Shang, *National Center for Advanced Packaging; Chinese Academy of Sciences*; Mingchuan Zhang, *Chinese Academy of Sciences*; Zheng Qin, *National Center for Advanced Packaging; Chinese Academy of Sciences*; Huijuan Wang, *National Center for Advanced Packaging; Chinese Academy of Sciences*; Xiaoli Ren, *National Center for Advanced Packaging*; Jie Pan, *Chinese Academy of Sciences*; Daquan Yu, *National Center for Advanced Packaging; Chinese Academy of Sciences*; Dongkai Shangguan, *National Center for Advanced Packaging; Chinese Academy of Sciences*

Study of Microwave Circuits Based on Metal-Insulator-Metal (MIM) Diodes on Flex Substrates 2168

Amanpreet Kaur, *Michigan State University*; Xianbo Yang, *Michigan State University*; Premjeet Chahal, *Michigan State University*

41: Student Interactive Presentations
Chairs: Mark Poliks, *i3 Electronics, Inc.*
Ibrahim Guven, *University of Arizona*

Nanocomposite Pastes for Thermal and Mechanical Bonding 2175

Tingting Zhang, *Binghamton University*; Bahgat Sammakia, *Binghamton University*; Howard Wang, *Binghamton University*

Assembly and Packaging Technologies for High-Temperature and High-Power GaN HEMTs 2181

A.A. Bajwa, *University of Freiburg*; Y. Qin, *University of Freiburg*; J. Wilde, *University of Freiburg*; R. Reiner, *Fraunhofer Institute IAF*; P. Waltereit, *Fraunhofer Institute IAF*; R. Quay, *Fraunhofer Institute IAF*

Flip-Chip on Glass (FCOG) Package for Low Warpage 2189

Scott R. McCann, *Georgia Institute of Technology*; Venkatesh Sundaram, *Georgia Institute of Technology*; Rao R. Tummala, *Georgia Institute of Technology*; Suresh K. Sitaraman, *Georgia Institute of Technology*

Laser-Based Conductive Film Forming with Gold Nanoparticles for Electrical Contacts 2194

Mitsugu Yamaguchi, *Ibaraki University*; Shinji Araga, *Ibaraki Giken Ltd.*; Mamoru Mita, *M&M Research Laboratory*; Kazuhiko Yamasaki, *Ibaraki University*; Katsuhiro Maekawa, *Ibaraki University*

Analysis of Modes Effect on Signal/Power Integrity in Finite Cavity for Chip and Die Level Packaging Based on a Hybrid Full Wave Method 2200

Xin Chang, *University of Washington*; Leung Tsang, *University of Washington*

Directed Self-Assembly of Mesoscopic Dies Using Magnetic Force and Shape Recognition 2207

Anton Tkachenko, *Rensselaer Polytechnic Institute*; Robert F. Karlicek, Jr., *Rensselaer Polytechnic Institute*; James J.-Q. Lu, *Rensselaer Polytechnic Institute*

Controlled Silicon IC Thinning on Individual Die Level for Active Implant Integration Using a Purely Mechanical Process 2213

Vasiliki Giagka, *University College London*; Nooshin Saeidi, *University College London*; Andreas Demosthenous, *University College London*; Nick Donaldson, *University College London*

Connectors and Vibrations – Damages in Different Electrical Environments .... N/A

A. Berghuvud, *Blekinge Institute of Technology*; T. Björnängen, *Blekinge Institute of Technology*; T. Gissila, *Blekinge Institute of Technology*

Study of Extreme Low Temperature and Load Solid-Phase Sn-Ag System Bonding Mechanism for 3D ICs 2227

Kiyoto Yoneta, *Osaka University*; Ryohei Sato, *Osaka University*; Yoshiharu Iwata, *Osaka University*; Koichiro Atsumi, *Osaka University*; Kazuya Okamoto, *Osaka University*; Yukihiro Satio, *Osaka University*; Takumi Shigemoto, *Osaka University*

Self-Patterning, Pre-Applied Underfilling Technology for Stack-Die Packaging 2231

Chia-Chi Tuan, *Georgia Institute of Technology*; Ziyin Lin, *Georgia Institute of Technology*; Yan Liu, *Georgia Institute of Technology*; Kyoung-Sik Moon, *Georgia Institute of Technology*; Ching-Ping Wong, *Georgia Institute of Technology*; *Chinese University of Hong Kong*

Study of High CRI White Light-Emitting Diode Devices with Multi-Chromatic Phosphor 2236

Min Zheng, *Xi'an Jiaotong University*; Wen Ding, *Xi'an Jiaotong University*; Feng Yun, *Xi'an Jiaotong University*; Deyang Xia, *Xi'an Jiaotong University*; Yaping Huang, *Xi'an Jiaotong University*; Yukun Zhao, *Xi'an Jiaotong University*; Weihan Zhang, *Xi'an Jiaotong University*; Minyan Zhang, *Xi'an Jiaotong University*; Maofeng Guo, *Xi'an Jiaotong University*; Ye Zhang, *Xi'an Jiaotong University*

The Effects of Self-Fluxing Additives in Solder Anisotropic Conductive Films (ACFs) on Solder Wettability and Joint Reliability of Flex-on-Board (FOB) Assemblies 2241

Seung-Ho Kim, *Korea Advanced Institute of Science and Technology (KAIST)*; Yongwon Choi, *Korea Advanced Institute of Science and Technology (KAIST)*; Yoosun Kim, *Korea Advanced Institute of Science and Technology (KAIST)*; Kyung-Wook Paik, *Korea Advanced Institute of Science and Technology (KAIST)*

Modeling and Analysis of Frequency Shift of MEMS Gyroscope Subjected to Temperature Change 2245

Weihui Wang, *Huazhong University of Science & Technology*; Sheng Liu, *Wuhan University*; Zhang Luo, *Huazhong University of Science & Technology*; Ming Wen, *Huazhong University of Science & Technology*; Qiang Dan, *Huazhong University of Science & Technology*; Man Yu, *Huazhong University of Science & Technology*; Yong Xu, *Wuhan University, Wayne State University*; Xin Wu, *Wayne State University*

Interaction Effect between Electromigration and Microstructure Evolution in Cu/Sn-58Bi/Cu Solder Interconnect 2249

Hong-Bo Qin, *South China University of Technology*; Bin Li, *Southern Methodist University*; Wu Yue, *South China University of Technology*; Chang-Bo Ke, *South China University of Technology*; Min-Bo Zhou, *South China University of Technology*; Xin-Ping Zhang, *South China University of Technology*

Effects of Alignment of Graphene Flakes on Water Permeability of Graphene-Epoxy Composite Film 2255

Seong-Yoon Jung, *Korea Advanced Institute of Science and Technology (KAIST)*; Kyung-Wook Paik, *Korea Advanced Institute of Science and Technology (KAIST)*

Characterization of Alternate Power Distribution Methods for 3D Integration 2260

David C. Zhang, *Georgia Institute of Technology*; Madhavan Swaminathan, *Georgia Institute of Technology*; David Keezer, *Georgia Institute of Technology*; Satyanarayana Telikepalli, *Georgia Institute of Technology*

Adhesion and Reliability of Direct Cu Metallization of Through-Package Vias in Glass Interposers 2266

Timothy Huang, *Georgia Institute of Technology*; Venky Sundaram, *Georgia Institute of Technology*; P. Markondeya Raj, *Georgia Institute of Technology*; Himani Sharma, *Georgia Institute of Technology*; Rao Tummala, *Georgia Institute of Technology*

High-Frequency Characterization of Through-Package Vias Formed by Focused Electrical-Discharge in Thin Glass Interposers 2271

Jialing Tong, *Georgia Institute of Technology*; Yoichiro Sato, *Asahi Glass Company*; Shintaro Takahashi, *Asahi Glass Company*; Nobuhiko Imajyo, *Asahi Glass Company*; Andrew F. Peterson, *Georgia Institute of Technology*; Venky Sundaram, *Georgia Institute of Technology*; Rao Tummala, *Georgia Institute of Technology*

Interfacial Reactions between Cu and Sn, Sn-Ag, Sn-Bi, Sn-Zn Solder under Space Confinement for 3D IC Micro Joint Applications 2277

T.L. Yang, *National Taiwan University*; W.L. Shih, *National Taiwan University*; J.J. Yu, *National Taiwan University*; C.R. Kao, *National Taiwan University*

Simulation and Optimization of a Micro Flow Sensor 2283

Xing Guo, *Huazhong University of Science & Technology*; Chunlin Xu, *Huazhong University of Science & Technology*; Shengzhi Zhang, *Huazhong University of Science & Technology*; Yong Xu, *Wayne State University*; Xin Wu, *Wayne State University*; Sheng Liu, *Wuhan University*

Minimizing Coupling of Power Supply Noise between Digital and RF Circuit Blocks in Mixed Signal Systems 2287

Satyanarayana Telikepalli, *Georgia Institute of Technology*; Madhavan Swaminathan, *Georgia Institute of Technology*; David Keezer, *Georgia Institute of Technology*

A Feasibility Study of Flip-Chip Packaged Gallium Nitride HEMTs on Organic Substrates for Wideband RF Amplifier Applications 2293

Spyridon Pavlidis, *Georgia Institute of Technology*; A. Cagri Ulusoy, *Georgia Institute of Technology*; Wasif T. Khan, *Georgia Institute of Technology*; Outmane Lemtiri Chlieh, *Georgia Institute of Technology*; Edward Gebara, *I2R Nanowave Inc.*; John Papapolymerou, *Georgia Institute of Technology*

A Novel Molding Process for Wafer Level LED Packaging Using Uniform Micro Glass Bubble Arrays 2299

Yu Zou, *Southeast University*; Jintang Shang, *Southeast University*; Yu Ji, *Southeast University*; Li Zhang, *Jiangyin Changdian Advanced Packaging Co. Ltd.*; Chiming Lai, *Jiangyin Changdian Advanced Packaging Co. Ltd.*; Dong Chen, *Jiangyin Changdian Advanced Packaging Co. Ltd.*; Kim-Hui Chen, *Jiangyin Changdian Advanced Packaging Co. Ltd.*; Ching-Ping Wong, *Chinese University of Hong Kong*

Analysis of Room-Temperature Bonded Compliant Bump with Ultrasonic Bonding 2303

Keiichiro Iwanabe, *Kyushu University*; Takanori Shuto, *Kyushu University*; Tanemasa Asano, *Kyushu University*

Development of B-spline X-ray Diffraction Imaging Techniques for Die Warpage and Stress Monitoring Inside Fully Encapsulated Packaged Chips

[1]C.S. Wong, [2,3]A. Ivankovic, [1]A. Cowley, [1]N.S. Bennett, [4]A.N. Danilewsky, [2]M. Gonzalez, [2]V. Cherman, [2]B. Vandevelde, [2,3]I. De Wolf and [1]P.J. McNally

[1]School of Electronic Engineering, Dublin City University
Dublin 9, Ireland
mcnallyp@eeng.dcu.ie

[2]imec
Leuven, Belgium

[3]Faculty of Engineering, KU Leuven
Leuven, Belgium

[4]Kristallographie, Institut für Geowissenschaften, Albert-Ludwigs-Universität
Freiburg, Germany

Abstract

Advanced packaging is a key "More than Moore" (MtM) enabling technology [1]. In all of these advanced packaging processes the semiconductor die are becoming much thinner (e.g. 25-50 µm thick) and many packages include multiply stacked silicon die. This leads to very thin packages where there is a trade-off between the thickness of constituent package layers and their rigidity, thus leading to reliability problems. Currently there are no compelling metrologies that can non-destructively measure the stress and/or warpage of the semiconductor die inside these packaged chips. Furthermore, since the thermal processing of these packages leads to the generation of thermal/mechanical stresses a new metrology, which is capable of real-time, or near real-time, monitoring of the generation or amelioration of these stresses during the thermal processing, would be a major advantage.

In this study, we report on recent advances in the development of a new technique, which we describe as B-Spline X-Ray Diffraction Imaging (B-XRDI), which produces a reconstruction of strain field and/or lattice misorientation data from x-ray diffraction data/images of the *in situ* semiconductor die inside a test wirebonded encapsulated BGA package. High-speed digital x-ray topography images are captured at a synchrotron source (ANKA, Germany and Diamond, UK) in times as short as 8 seconds for a full 8 mm x 8 mm semiconductor die inside the fully encapsulated packages. Using a laboratory-based source (Jordan Valley Bede D1 High Resolution X-Ray Diffractometer) and applying the B-Spline technique, maps are also produced of the entire silicon die, which reveal warpage via measurements of x-ray rocking curve full-widths-at-half-maximum (FWHM) as a function of position across the encapsulated packages. These maps are also correlated with warpage measurements performed by mechanical and interferometric profilometry and finite element modelling (FEM).

The B-spline XRDI technique

For laboratory-based x-ray systems the technique stems from early work by Toda *et al.* [2] and in this study we have extended the technique for the study of a complete semiconductor die using a triple-axis Jordan Valley Bede D1 X-ray diffractometer with a copper ($\lambda= 1.5405$ Å) radiation source operated at 45 kV and 40 mA. We generate maps of the complete silicon die, which reveal the lattice misorientation/warpage of the (1 1 0) crystallographic planes in the Si die via mapping of the 004 symmetric rocking curve (RC) full-widths-at-half-maximum (FWHMs) as a function of the position across fully encapsulated packages. The incident X-ray beam dimensions were set to 500 µm x 2000 µm as a reasonable trade-off between acceptable spatial resolution across the die and the time required to acquire the data. Each RC was recorded by rotating the specimen (ω-axis) through a suitable angular range about the Bragg angle of the Si 004 reflection, with the detector (2θ axis) fixed at twice the Bragg angle. The presence of warpage creates a curvature of the Si crystal planes, which leads to a range of angular positions on the distorted Si die in which the Bragg diffraction conditions are satisfied, *i.e.* a broadening of the rocking curve. A spatially resolved RC map can be produced by integrating a series of RCs collected at different positions across the Si die of the package (*e.g.* along the x-direction) [3]. The variation of the RC peak position ($\omega-\omega_0$) across the Si die is a signature of warpage-induced tilt. For example, if one observes a region, say on the left-hand side of the mapped Si die where the diffraction angle occurs at larger diffraction angles than at the central region of the Si die (ω_0), whereas it decreases below ω_0 for the reflections from the right-hand side of the Si die, this is clear evidence that the Si die is warped in a convex shape. In order to map the precise locations of the highly distorted region of the Si die, we convert the RC maps into a FWHM line scan by extracting the FWHM of each individual RC of the RC map using the peak analysis software (Quick Graph) provided with the JV Bede D1 tool. A series of FWHM line scans were repeated at 800 µm steps across the Si, controlled using an automatic motorized x–y sample stage. A full map of RC FWHMs, which reveals the warpage across the whole Si die of the packaged chip, was reconstructed by modifying the 3-dimensional surface modelling technique outlined in reference [4]. The FWHM line scans were imported into Solidworks™ and these formed a series of 3D curves in which the x and y coordinates correspond to the on-chip location where each RC was recorded, and z represents the FWHM extracted from each RC recorded at each x–y position. (See Fig. 1).

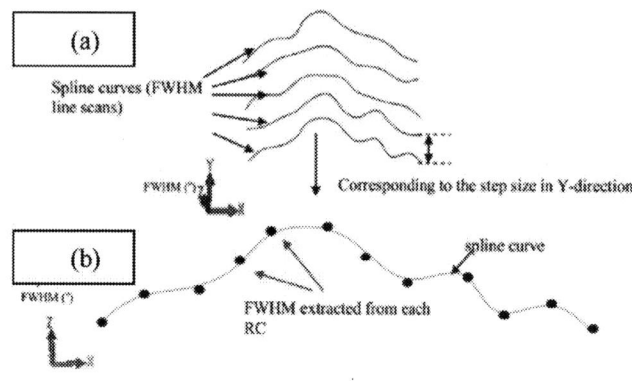

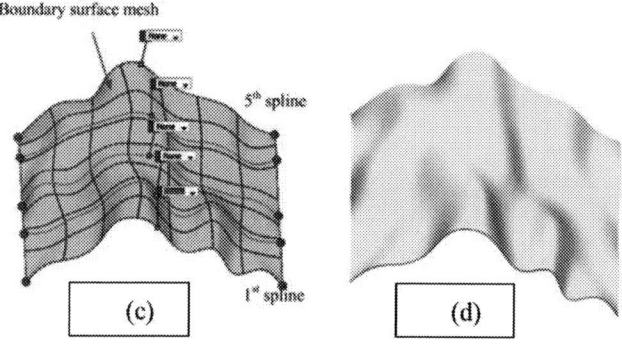

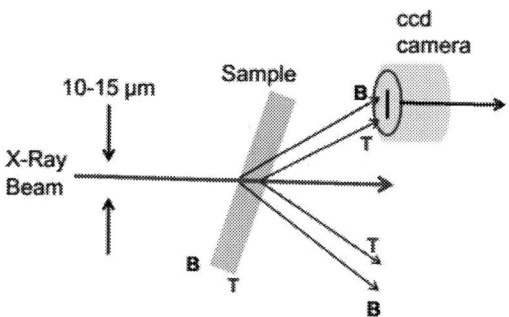

Figure 2. Experimental set-up for synchrotron-based XRDI in section transmission mode.

Figure 1. (a) A series of recorded FWHM line scans are imported into Solidworks™; (b) a spline curve formed automatically by Solidworks™ using the imported FWHM data; (c) boundary surface feature used to connect the spline curves; (d) a solid surface B-spline model of lab-based RC data demonstrating the lattice misorientations of the (110) plane in a Si die, which was fully encapsulated.

For the laboratory-based measurements Rocking Curve (RC) FWHM B-Spline mapping measurements the Si die measurements were performed at die rotation angles of $\phi = 0°$ and $90°$ in order to investigate the lattice misorientations of the two orthogonal (110) planes. For the silicon die used in this study, typically 8 mm x 8 mm, a full mapping experiment takes approximately 4 hours.

However, more rapid measurements are acquired using synchrotron x-ray sources, where the time required to implement an acceptable die map is reduced from hours to seconds. X-Ray Diffraction Imaging (XRDI), also known as x-ray topography, is a rapid non-contact, non-destructive characterization tool able to track an individual die/wafer from virgin substrate through to completed circuit, or

packaged chip. It can provide a comprehensive mapping of the effects of each fabrication level including wafer warpage, crystalline misorientation, process-induced strains, precipitates, inclusions, dislocations, etc. [5-7].

These experiments were performed at the ANKA Synchrotron Source, Karlsruhe, Germany (TOPO-TOMO beamline) and the Diamond Light Source, Oxfordshire, UK (Beamline B16). Details of the experimental setup are outlined in [8,9]. Briefly, for these experiments a series of conventional x-ray diffraction images (topographs) were recorded in section transmission (ST) geometry. The beam was collimated into a narrow ribbon by a slit approx. 15 µm in height.

Radiation passes through the back side of the sample, and reflections are recorded on a CCD camera system [8,9]. The narrow slit size is used to avoid heating effects on the samples. The digital topographs are recorded using a PCO.4000 high resolution 14 bit cooled CCD camera, with magnification optics (pixel size of 2.5 µm) or on a Photonic Science X-ray MiniFDI, 11mm dia, 1392 x 1040 pixels (6.5µm optical pixel size). The sample to camera distance, L, is set to 80-200 mm, depending on beamline configuration and the CCD camera is positioned to record the 220 reflection from the Laue diffraction pattern of topographs.

The sample is mounted on a high precision X-Y stage, which enables the user to step across the sample and thus obtain a series of section topographs across the packaged chip. ST topographs are taken at, typically, 0.5-1.0 mm steps across the chip. In previous experiments we have recorded hundreds of individual ST topographs across an individual chip, with steps of the order of 15 µm, but the reconstruction can be time-consuming, and tends to yield not much more information when examining global wafer warpage and curvature issues for encapsulated die. Typical recording times for the packages studied here in the order of 5 to 10 seconds.

Test case – wirebonded BGA package

The package used to evaluate the efficacy of the B-XRDI techniques was a wirebonded BGA package comprising of imec's thermal/thermo-mechanical stress dedicated test chip

978-1-4799-2408-0/14 $31.00 © 2014 IEEE

[10-12]. This test chip is used at imec for stress related studies covering 3D stacking and 3D packaging impacts, while in this current scenario it acted as a provider of a realistic packaging environment rather than focusing on the exploitation of the chip's advanced features.

The geometry of the package and its constituents is shown in a cross section schematic in Figure 3 along with the corresponding values in *Table 1*.

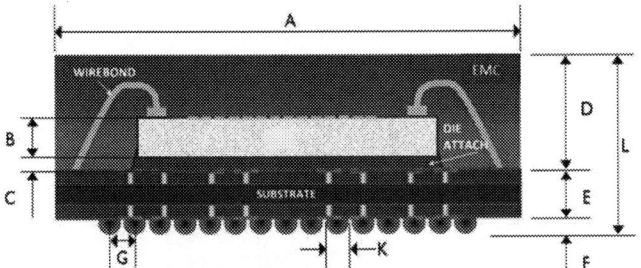

Figure 3. Package cross section schematic.

Table 1. Package dimensions

Feature	Code	Dimensions
Package Type	-	LFBGA
Package Size	A	14 x 14 mm
MC Thickness	D	0.700 mm
Chip Thickness	B	300 µm
Die Attach BLT	C	30 µm
Number of Wirebond Pads	-	12
Substrate Thickness	E	0.240 mm
Substrate Layer	-	2
BGA I/O	-	676
Solder Ball Size	K	0.300 mm
Solder Ball Pitch	G	0.500 mm
Solder Ball Height (collapsed)	F	0.130 mm
Total Package Height	L	1.160 mm

In total, six epoxy mold compounds (EMCs) were under consideration for packaging, all of which went through a particular selection procedure. The purpose of this preselection was multiple:

- Obtain an EMC with sufficient stress impact on Si
- Gather precise material properties for inputs to finite element models
- Calibrate finite element models later expanded to package level.

A straightforward assembly test was employed where the EMCs were placed on top of a laminate material, as shown in Figure 4.

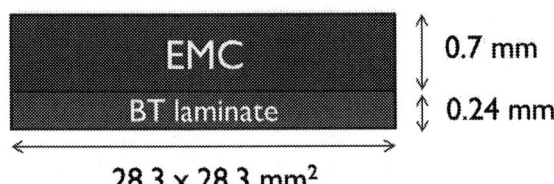

Figure 4. EMC on laminate test structure.

This structure was submitted to heating from room temperature to 250 °C while monitoring the warpage on the EMC side. The warpages of all 6 EMCs are presented in Figure 5. For clarity, only data for the cool-down are shown.

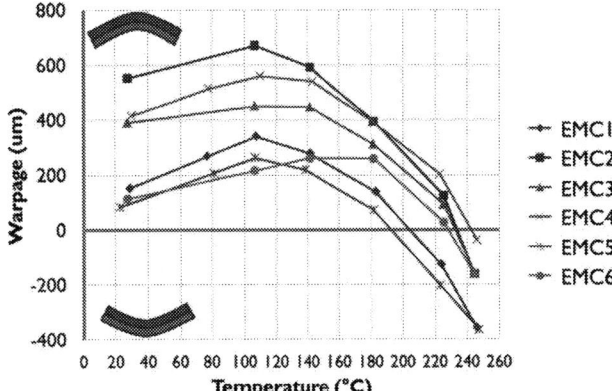

Figure 5. EMC warpage test.

The EMCs follow a similar shape with increasing warpage until the glass transition temperature (T_g) and a decline in warpage, or even the development of opposite warpage, appears above T_g. The distinctively different warpage levels obtained through this test allowed one to rank the EMCs according to the intensity of their impact.

Values from datasheets often do not describe the material behaviour precisely enough which is why these curves, obtained from a realistic setting, were used for material properties extraction. The glass transition temperature was extracted from the inflection point of the curves while the initial Young's modulus and coefficient of thermal expansion (CTE) values were taken from the datasheet and adapted to fit the curve above and below T_g via a finite element model (FEM). This procedure can be viewed as fine tuning of the datasheet given EMC properties but it also provides a strong confidence in the finite element model of the EMC to be used in the overall package model. Figure 6 presents the fitting procedure of the EMC finite element model and the obtained warpages. The initial model with inputs directly from datasheets is at great variance with respect to the actual warpage measurements. The adapted models used the T_g from the measurements and fine tuned the Young's modulus and CTE to follow the actual warpage.

978-1-4799-2408-0/14 $31.00 © 2014 IEEE   1519

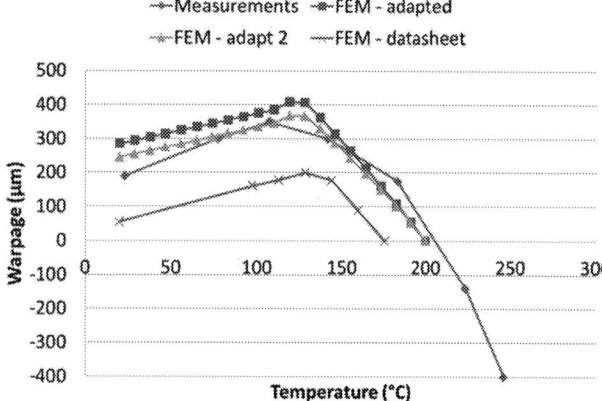

Figure 6. Extracting EMC properties with finite element models.

These EMC models were subsequently used in the wirebonded BGA package simulations. Apart from efforts to obtain realistic EMC material properties, package thicknesses were also measured on real samples. Figure 7 shows measured thickness values for the package cross section while *Table* 2 denotes the differences in nominal and measured values. If disregarded, these differences could eventually contribute to offsets in warpage and stress results in the simulations.

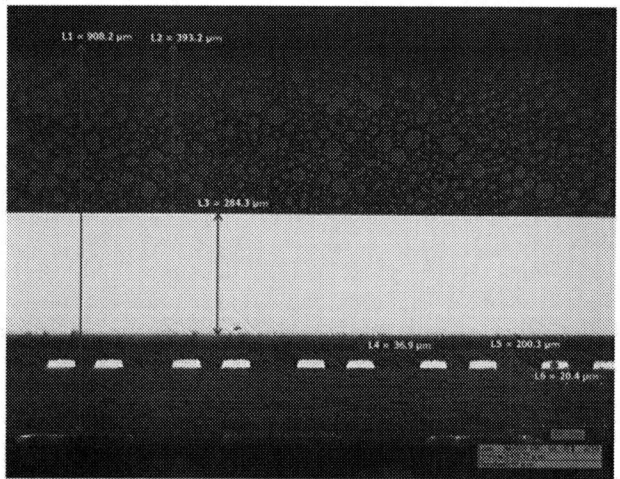

Figure 7. Thickness measurements on package SEM cross section.

Table 2. Variations in nominal and measured package thicknesses

Package thicknesses	Nominal thickness [μm]	Measured thickness [μm]
Die	300	278-284
Glue	30	36-47
Laminate	240	200-213
EMC	700	713-717

The EMC with highest produced warpage, labelled EMC2 in Figure 5, was chosen for these studies. A full package finite element model was built which included the fine-tuned ovemold properties and measured package thicknesses. Figure 8 presents a 1/8 slice of the simulated package at room temperature indicating concave package and inner die bending.

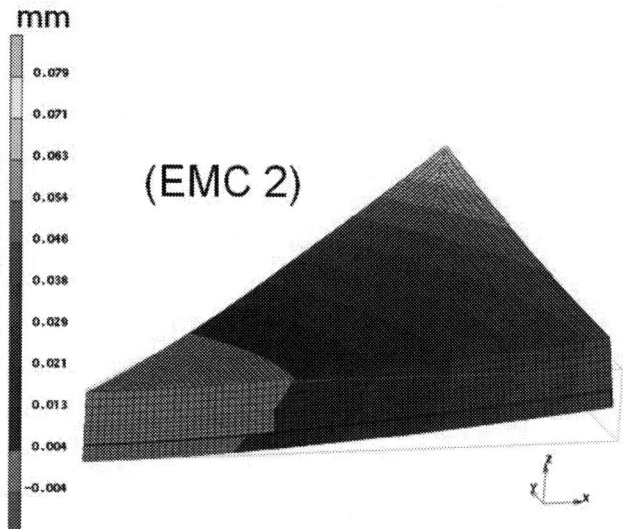

Figure 8. FEM based vertical displacement of package for EMC2.

The EMC2 packages were submitted to analysis with two profilometry techniques: optical and mechanical. The Veeco Wyko NT3300 interferometric profiler and DEKTAK XT Stylus mechanical profilometer measured vertical displacements on the outer package EMC side on a diagonal through the centre of the chip. Figure 9 displays the displacements obtained from both profilers on the EMC2 package.

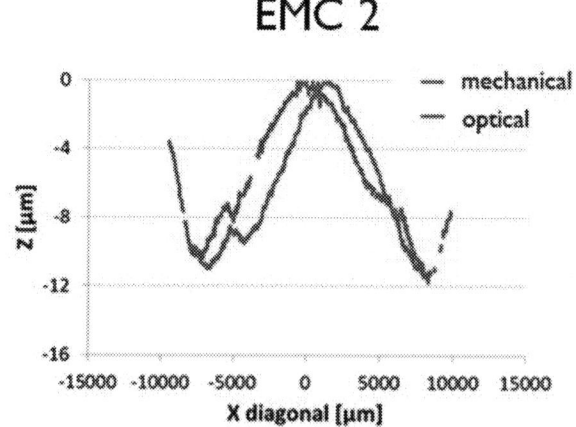

Figure 9. Profilometry based vertical displacement of the EMC2 package.

When compared to the FEM the results, at first glance, may seem peculiar. While FEM suggests the presence of concave warpage of both die and package, the two profilometry techniques reveal a convex warpage of the

978-1-4799-2408-0/14 $31.00 © 2014 IEEE

package when measured from the top side of the package. It should be noted that the FEM represents a thermo-mechanical approach, dependent principally on CTE differences and predicts concave warpage for both the die and package. However, it is much more likely that the mold compound interacts with the die in a more complex manner. This is captured in the profilometry data, where a concave shape is exhibited at the edges of the package in addition to convex warpage towards the central die area. One can visualize this as a mold compound ''bubble'' above the die area, a feature that is commonly observed and is not related to package cooling but, rather, to mold cure shrinkage at the dispensing/processing temperature (180 °C). Figure 10 is a schematic of this hypothesis.

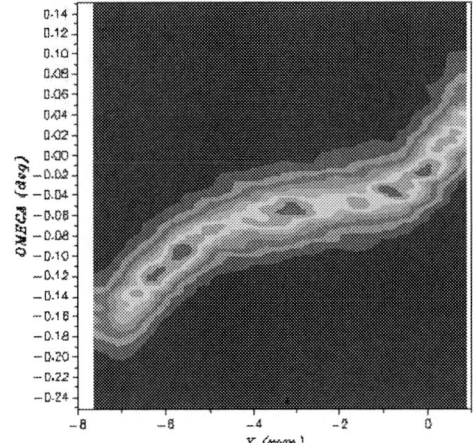

Figure 11. Corresponding RC map indicating overall concave nature of Si die warpage for (110) lattice planes

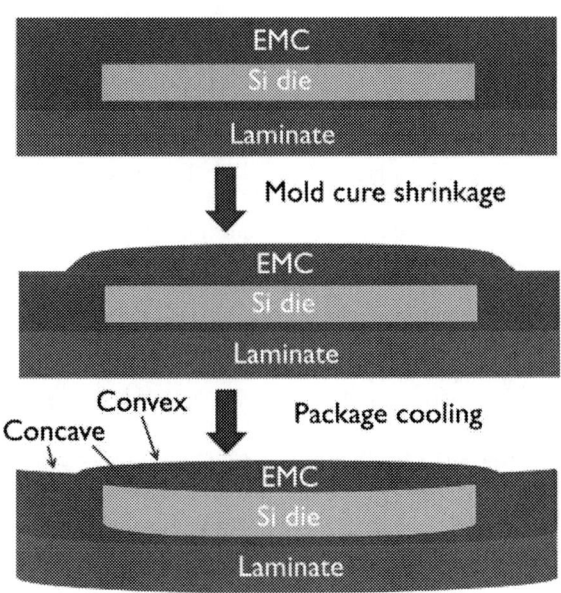

Figure 10. Assumed package deformation process

Initially, mold cure shrinkage causes the mold compound ''bubble'' on top of the die and then cooling of the package occurs, which causes concave overall package warpage, which leads to the die bending in a concave shape as well.

It is at this point that the B-XRDI measurements play a vital role, as they now afford the unique opportunity to see inside the package nondestructively, and address the apparent conflict between "conventional" experimental data and the simulations. As described earlier, non-destructive RC maps were produced to reveal the die warpage. The RC map for (110) lattice planes presented in Figures 11 and 12 confirms the generally concave nature of the Si die warpage, and this aligns to the finite element modelling.

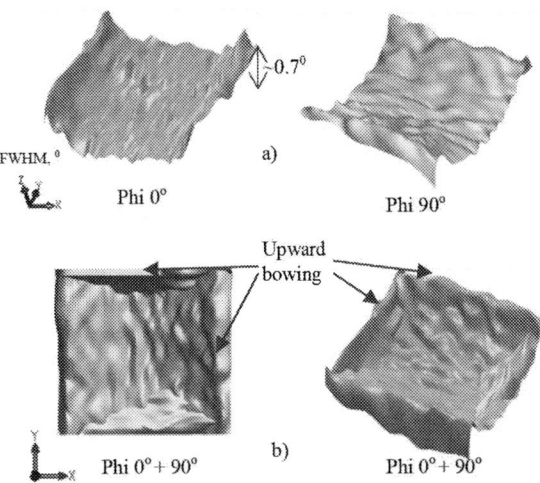

Figure 12. a) Corresponding $\phi = 0°$ and 90° RC warpage maps indicating overall concave nature of Si die warpage for (110) lattice planes. b) Combination of $\phi = 0°$ and 90° RC warpage maps from different viewpoints. The regions indicated by red arrows in the plots correspond to more upward bowing of the Si semiconductor die (in this case towards the die edges).

In order to confirm this further the EMC2 sample was examined using synchrotron-based B-XRDI and the results, presented in Figure 13, reveal *in situ* the shape and nature of the warpage of the (220) lattice planes in the Si die (8mm x 8 mm). Again the synchrotron-based B-XRDI confirms the general concave shape of the encapsulated Si die, with the edges of the die showing mostly upward bowling (max. magnitude of $0.5° \approx 8.6$mrad; comparable to the $\sim 0.7° \approx 12$mrad using the lab-based RC mapping). However the behaviour is complex with one die edge (lower left hand side in Figure 13) appearing to have relaxed somewhat. Further investigation of this is ongoing.

978-1-4799-2408-0/14 $31.00 © 2014 IEEE   1521

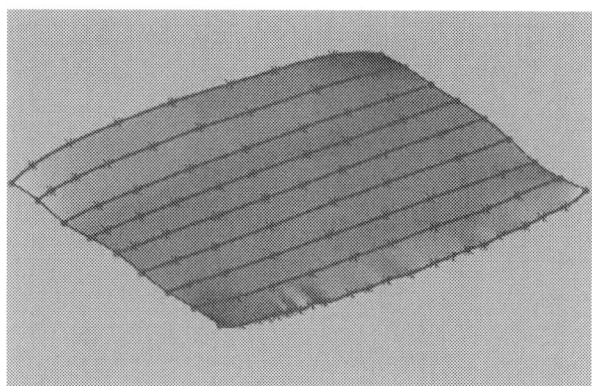

Figure 13. Synchrotron B-XRDI map reveling the *in situ* shape and nature of warpage of (220) lattice planes in the Si die

Conclusions

In conclusion, using a test case BGA package, we confirmed that B-XRDI, both laboratory-based and synchrotron-based, provides a powerful new technique for the non-destructive analysis of Si die warpage in packaged chips. The induced *in situ* warpage in the die displays a more complex behaviour then can be inferred from profilometry-based package warpage measurements. This is most likely due to the fact that the B-XRDI techniques are sensitive to warpage in the die only, and that differing external and internal warpages, particularly those due to overmold post-processing, will not be captured by cooling based finite element models, regardless of how carefully one infers the material properties. Furthermore, although profilometry techniques precisely and directly measure the warpage on the EMC side, they do not measure the *in situ* warpage of the active semiconductor die within the encapsulated package. B-XRDI was able to complement the more conventional techniques and together with them provides a clearer understanding of die and package behaviour.

Acknowledgments

CSW, AC & PMN acknowledge the support of Science Foundation Ireland's Strategic Research Cluster Programme ("Precision" 08/SRC/I1411), the Irish Higher Education Authority INSPIRE programme funded by the Irish Government's Programme for Research in Third Level Institutions, Cycle 5, National Development Plan 2007-2013 and the EU/ERDF/ESF. The research leading to these results has received funding from the European Community's Seventh Framework Programme (FP7/2007-2013) CALYPSO under grant agreement n° 312284. The measurements could not be done without help and input from imec's team and partners.

References

1. www.itrs.net/LINKS/2010ITRS/IRC-ITRS-MtM-v2%203.pdf
2. A. Toda, N. Ikarashi, Jap. J. Appl. Phys. 49 (2010). 04DB03-1.
3. J.Z. Domagala, A. Czyzak, Z.R. Zytkiewicz, Appl. Phys. Lett. 90 (2007) 241904.
4. J. Stopford, A. Henry, D. Manessis, N. Bennett, K. Horan, D. Allen, J. Wittge, L. Boettcher, A. Cowley, P.J. McNally, in: Proceedings of the 18th European Microelectronics Packaging Conference, 2011, pp. 1–8 and arXiv:1204.1466v1 (2012).
5. A. Authier, *Dynamical Theory of X-Ray Diffraction* (International Union of Crystallography Monographs on Crystallography, Oxford: Oxford University Press, UK, 2001, rev. 2003.
6. D.K. Bowen and B.K. Tanner, *X-ray Metrology in Semiconductor Manufacturing*, (CRC Taylor and Francis, Boca Raton, USA), 2006.
7. T. Tuomi, K. Naukkarinen and P. Rabe, *phys. stat. sol. (a)* 25, 93 (1974).
8. A.N. Danilewsky, A. Rack, J. Wittge, T. Weitkamp, R. Simon, H. Riesemeier and T. Baumbach, "White beam synchrotron topography using a high resolution digital X-ray imaging detector", Nuc. Instrum. Meth. Phys. Res. B, 266, pp.2035-2040, (2008).
9. A. Rack, F. Garcia-Moreno, T. Baumbach and J. Banhart, Synchrotron Radiat. 16, 432–434 (2009).
10. V. Cherman *et al.*, "3D Stacking Induced Mechanical Stress Effects", in *Proc. IEEE Electronic Components and Technology Conference (ECTC)*, Orlando, FL, May 28–31, 2014, in press
11. M. Gonzalez *et al.*, "Mechanical Stability of Cu/low-k BEOL Interconnects", *International Reliability Physics Symposium (IRPS)*, Waikoloa, HI, June 1-5, 2014, in press
12. K. Vanstreels *et al.*, "Advanced Experimental Back-End-Of-Line (BEOL) Stability Test: measurements and simulations", *Materials for Advanced Metallization Conference (MAM)*, Chemnitz, Germany, March 2-5, 2014, in press

Process Optimization and Reliability Study for Cu Wire Bonding Advanced Nodes

Ivy Qin, Hui Xu, Basil Milton, Nestor Mendoza, Horst Clauberg, Bob Chylak
Kulicke and Soffa Industries, Inc. Fort Washington, PA, US

Hidenori Abe, Dongchul Kang, Yoshinori Endo, Masahiko Osaka, Shinya Nakamura
Hitachi Chemical Co., Ltd,

Abstract

Cu wire bonding has taken over Au wire bonding due to its cost savings and other performance advantages such as higher mechanical strength for complex looping and better electrical performance. Currently, Cu wire bonding of 28nm node devices has been realized in high volume production [1]. To meet the challenges of device reliability for these advanced node devices, we need to understand how to achieve the optimal wire bonding results for the best condition for reliability tests.

Previous reliability study by the authors identified one of the key factors in achieving good wire bonding reliability outcome is the wire type [2]. Pd coated Cu wire (PdCu) wire showed much better performance than Bare Cu wire. In this paper, traditional bare Cu wire, AuPd coated Cu wire and Pd doped Cu wire are compared. Biased HAST testing is one of the most challenging reliability tests for Cu wire bonded packages. Biased HAST test was performed for up to 336 hours. A medium grade molding compound was used. Coated AuPdCu wire shows the best performing wire in our study. The Pd doped Cu wire showed much better reliability than traditional bare Cu wire, indicating that the Pd doped Cu wire is a lower cost alternative to Pd coated Cu wire.

In advanced node devices such as 28nm and 20nm, the pitch is further reduced to as low as 45um with the bond pad opening of around 40um. Two different wire diameters (15um and 18um) and 6 different ball diameters are tested to understand the correlation between bonded ball diameter, ball to pad contact area and reliability. ProCu5 process was used and all cells achieved over 90% IMC. Bonded ball diameters of 34um and larger passed the reliability tests for both 96 hours and 168 hours. The two smallest bonded ball diameters (30 and 32um) saw low level of failures at 96 hours. This confirms that contact area is an important factor in reliability testing. A 34um bonded ball is suitable for 45um pitch application, which means that reliability can be achieved with a medium grade molding compound for a 45um pitch application. Finer than 45um pitch devices may need higher grade molding compound to ensure reliability.

Another objective of this paper is to understand the critical wire bonding responses for good reliability outcome. The wire bonding responses studied here include contact area between ball and pad, intermetallic coverage percentage (IMC%), and Al remain thickness. We found that higher values for contact area and IMC% result in increased reliability. An IMC% value of 85% is generally recommended to ensure an acceptable reliability outcome. The Al remain % is dependent on the application and the process being used. With an optimal process, Al remain around 50% of the initial Al thickness shows good reliability outcome. A less optimal process tends to require less Al remain % to pass reliability.

Introduction

The main bonding wire types are Au, Cu and Ag alloy wires for ball bonding and Al wire for wedge bonding. Cu wire including Pd-coated Cu wire (PCC) and bare Cu wire have grown in popularity in recent years due to the need for cost reduction and breakthrough in Cu wire bonding and packaging capability [3, 4]. The percentage of Cu wire (including PdCu and bare Cu wire) will continue to grow over time and will become the dominant wire. Figure 1 shows the wire type trend over time (courtesy of Heraeus).

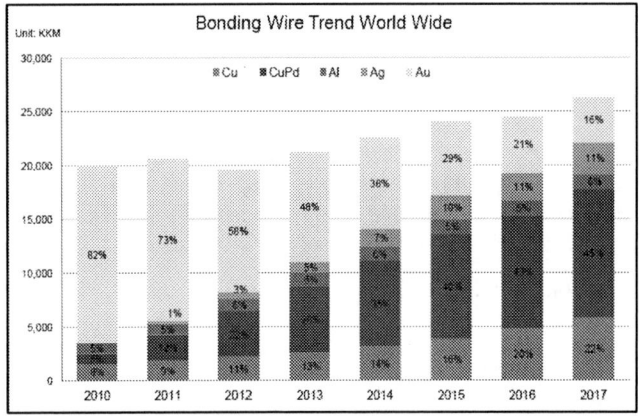

Figure 1. Bonding wire trend over time. (Source: Heraeus Materials).

The reliability and yield for Cu wire bonding have been improved due to a few critical technology breakthroughs in material, equipment and process. In material, the introduction of Pd coated and AuPd coated Cu wire have resulted in improvements in reliability [2, 5, 6]. Molding compounds with low Cl ions and controlled PH levels exhibited better reliability [7, 8]. The refinement of the cover gas kit, and bonding force and ultrasonic control, in the latest Cu specific wire bonders have produced larger process windows, higher throughput and higher yields for Cu wire bonding production [3, 4]. Furthermore, highly optimized Cu processes were developed and implemented. An example of a highly optimized Cu process is the new ProCu5 process on K&S's most recent Cu wire bonder ProCu Plus™. With these new equipment and process improvements, throughput is shown to improve up to 40% compared to previous Cu processes. Pad damage such as pad crack can be more easily eliminated, and a process window and ease of optimization is similar to Au wire bonding [3, 9].

Wire Type Study

In a previous study [2], the two most influential factors for reliability were identified as the wire type and ultrasonic current levels. PdCu wire has shown better reliability performance compared to bare Cu wire. Higher ultrasonic levels showed better performance than lower ultrasonic level for bare Cu wire, but the Pd coated Cu wire shows less sensitivity to USG current level [2]. Pd elements in PCC wire help to prevent Cu rich intermetallic compound corrosion because reactivity of the Cu/Al/Pd IMC to Cl ion was much less than that of Cu rich IMC [2, 8]. This paper further studies the influence of wire type and USG current level on reliability outcome. The device used in this study is a test chip mounted on a SOP alloy 42 leadframe. The bond pad Al thickness is 1.4 um and the metallization is Al-1%Si-0.5%Cu. Biased HAST was carried out and the test conditions were 130ºC, 85% RH with ±5V.

A 2 variable, 3 level, full factorial DOE was run. Three wire types were tested. Wire #1 is an AuPd coated Cu wire (AuPdCu wire). AuPdCu wire is said to further improve the bonding reliability of Pd coated Cu wire [10]. A thin layer of Au (~ 5 nm) is put on the outside of the Pd layer and forms an Au-Pd mixture layer. The addition of Au helps improve the Pd distribution on the FAB surface and therefore has the benefit of better reliability [10]. Wire #2 is a more traditional bare Cu wire. Wire #3 is a Pd-doped Cu wire that is shown to improve reliability [11]. The bonding USG current level is controlled by the Bond Strength Adjust (BSA) parameter in ProCu5 process. BSA of 75%, 85%, and 95% are used in the experiment. The DOE settings and wire bonding results are given in (Table 1). The process used in this study is a fine pitch process with around 36um bonded ball diameter that is suitable for 45 ~ 50 um bond pad pitch applications.

Table 1. DOE setup and wire bonding results. Wire1 is AuPdCu wire, Wire2 is bare Cu wire, Wire3 is Pd-doped Cu wire. BSA Low is 75%, Med is 85%, High is 95%.

Cell	Wire Type	BSA [%]	Ball Dia. [um]	Y-Splash [um]	Shear [g]	Cont. Dia. [um]	IMC [%]	Al remain [%]
1	1	Low	36.6	39.0	14.7	31.9	88.1	53.8
2	1	Med	36.7	40.7	16.2	31.5	94.1	48.1
3	1	High	36.4	41.6	16.5	31.6	96.1	44.8
4	2	Low	36.6	38.8	14.1	32.7	73.8	54.7
5	2	Med	36.9	41.0	16.7	32.7	84.1	48.1
6	2	High	36.9	42.1	18.1	32.6	89.8	41.5
7	3	Low	36.7	38.6	14.5	31.4	83.2	50.9
8	3	Med	36.5	40.1	16.5	31.3	91.9	46.2
9	3	High	36.8	41.4	17.1	31.3	96.0	38.2

Wire bonding responses are given in Table 1. IMC images and measurements are obtained after the Cu ball is etched away. To get reliable optical IMC measurements, the sample is baked for 4 hours at 175°C. IMC images are shown in Figure 2 and cross-section images are shown in Figure 3. In the IMC images, the dark grey color indicates IMC area, while the white color is the Non-IMC area which remains Al or Al oxide after bonding. The green line presents the contact line between bonded ball and Al pad. The red line encloses the Non-IMC area. The dark area outside the green line indicates the amount of Al material being pushed outside the ball edge, which is called Al splash or Al squeeze. Al splash can also be seen in Figure 3. A few observations can be made

from Figures 2 and 3. The IMC area increases with BSA (Ultrasonic current), as does the Al splash. The effect of ultrasonic energy is to scrub the thin Al oxide layer off of the bond pad to create clean metal to metal contact, which is essential to form a desirable intermetallic connection. A negative effect of ultrasonic energy is pad damage such as Al splash and pad crack. Figure 2 shows that higher BSA levels create more Al splash. Some differences between the wire types are also observed. The ball roundness and ball symmetry look the best with AuPdCu wire showing rounder contact area in Figure 2 and more side to side symmetry in Figure 3. The interface flatness of all bonded samples is good as shown in Figure 3. The Al splash is minimal, with the average of 3-4 um for the Med USG setting. In most cases, Al splash with ProCu5 process is at least 50% less than with a traditional process. With a traditional process, 8-10um Al splash is expected for 1.4um Al pad.

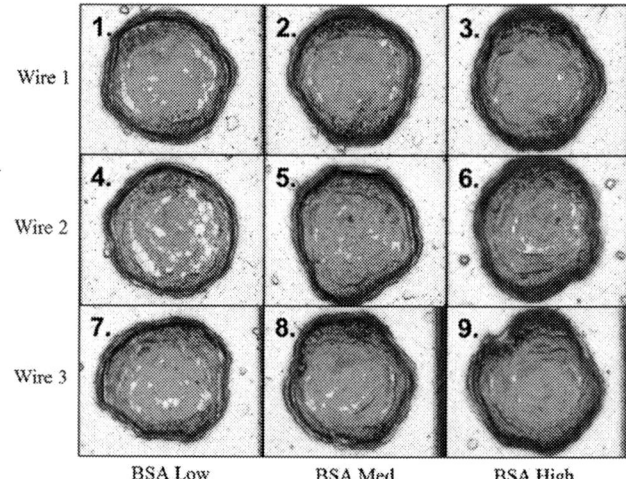

Figure 2. Inter-metallic coverage for cells 1-9.

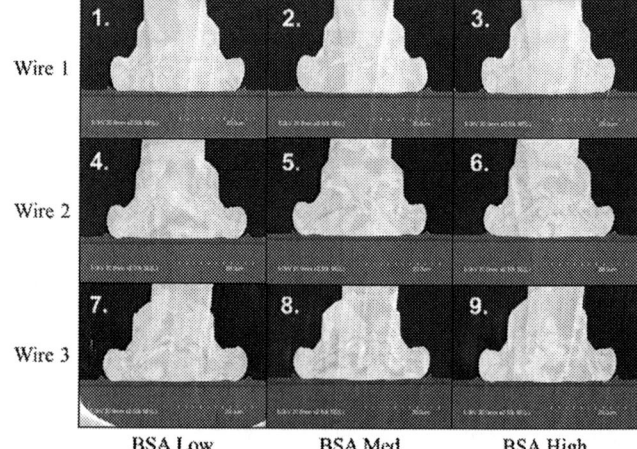

Figure 3. Cross-section images for cells 1-9.

A few key wire bonding responses are examined. These key responses include ball diameter, Y splash, shear, contact diameter, IMC percentage and Al remain percentage. Contact area is the area enclosed by the green line shown in Figure 2. IMC area shows as the dark grey color as in Figure 2. The ratio between the IMC area and contact area is IMC%.

978-1-4799-2408-0/14 $31.00 © 2014 IEEE

Contact diameter is computed as the diameter of a circle with area equal to the contact area. The reason we converted the areas to diameters is for easy comparison to the bonded ball diameter. A higher IMC% and a larger contact area generally equates to a stronger and bigger bond, which theoretically should be better for reliability. Next we verified the theory with our reliability results. The reliability results for the wire type and BSA level test is given in Table 2. JEDEC standard uses 96 hours as the passing criteria while some critical devices require a longer time. We collected HAST reliability data until 336 hours to obtain more data. Any cell failing at or before 96 hours is shown in red. Subsequent failures are shown in yellow. Correlation between input variables and the wire bonding response, and reliability failure, is analyzed with JMP statistical software. The Prediction Profiler plot is shown in Figure 4.

Table 2. Reliability failure over time for different wire types and different USG levels.

Cell	Wire Type	BSA [%]	24Hr	48Hr	96Hr	168Hr	336Hr
1	1	Low	0%	0%	0%	0%	9%
2	1	Med	0%	0%	0%	0%	5%
3	1	High	0%	0%	0%	0%	13%
4	2	Low	0%	3%	13%	28%	28%
5	2	Med	0%	0%	3%	12%	20%
6	2	High	0%	0%	3%	12%	38%
7	3	Low	0%	8%	11%	23%	30%
8	3	Med	0%	0%	0%	13%	28%
9	3	High	0%	0%	0%	16%	25%

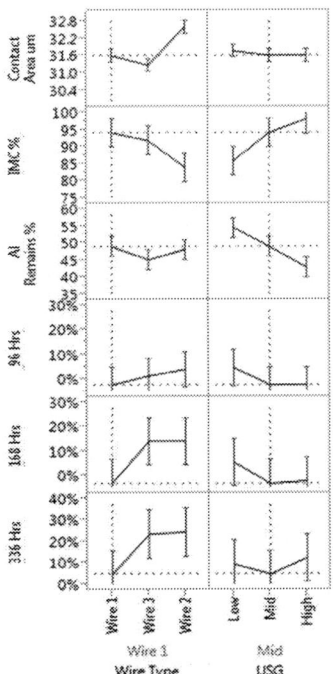

Figure 4. JMP Profiler plot showing the relationship between wire type, USG (BSA) level and the wire bonding responses and reliability failure rate. We re-order the wire type from best performing to worst performing wire, so it goes Wire1, Wire3 and Wire2.

The above results confirmed that wire type has a big influence on reliability. AuPdCu wire has the best reliability performance with all BSA settings passed 96 and 168 hours. Bare Cu wire showed the worst reliability results with all cells failed at 96 hours. Pd doped Cu wire showed very encouraging results. Two out of three cells passed the reliability tests at 96 hours. This indicates that with optimized wire bonding parameters, Pd-doped Cu wire can achieve good reliability for fine pitch applications. The addition of the Pd element shows improvement for reliability as expected. The reason for the improvement has been explained previously in this paper as well as in many previous publications.

Wire type also influences wire bonding responses such as IMC%. AuPdCu wire has the highest IMC%, followed by the Pd-doped wire. The reason for the better IMC% with AuPdCu could be that the AuPd layer provides oxidation protection, as the resulting free air ball is less oxidized and easier to bond to Al pad.

Ultrasonic current level also influences reliability. The lowest USG settings had failures at 48 hours for both bare Cu and Pd doped Cu wire. However, there is no significant difference between the mid and high settings of the ultrasonic level. In fact, the high USG level may have performed slightly worse than the mid USG level. This shows that once a certain level of intermetallic coverage is reached, excessive ultrasonic scrubbing does not further improve the reliability. In Cu wire bonding, excessive ultrasonic energy is often harmful for causing pad damage. Reliability for AuPdCu is not sensitive to the ultrasonic level in the range tested. This means the process window for the ultrasonic level is larger with the AuPdCu wire.

Regarding wire bonding response vs. reliability outcome, IMC% shows good correlation to reliability. For PdCu wire and Pd doped Cu wire, the 5 cells with an IMC% over 85% all passed 96 hours reliability. Average IMC% of Cell 7 is lower than 85% and it failed at 96 hours. Average IMC% of 85% is generally accepted as a target spec by the industry. From our test results, it is a good spec to follow for both coated wire (PdCu and AuPdCu) and Pd doped Cu wire. For bare Cu wire, better molding compound may be needed for passing HAST reliability [2].

Al remain% is defined as the ratio of Al thickness after wire bonding to unbonded Al bond pad thickness. Al remain% is sometimes used as a wire bonding criteria to gauge reliability outcome, for example, there is some specification calling for acceptable Al remain% between 10% - 40%. Too much Al remain is an indicator for an under-bonded ball. It could result in NSOP, ball lift, low IMC and higher HAST open failure. Too little Al remain is an indicator for pad damage, which include bigger Al splash, pad peeling, and pad crack. Higher levels of Al splash may cause bond short. Too low Al remain can also cause higher reliability failures due to the damage to the Al layer. An acceptable range for Al remain% depends on the process being used. The Al remain% for highly optimized processes such as K&S's ProCu process is normally higher than with traditional processes. Less optimal processes need to over-scrub the Al pad to achieve an acceptable level of bonding and IMC%. A reasonable Al remain% target for these less than optimal processes may be 10%-50%. A highly optimized process such as ProCu5 can achieve good IMC coverage, and strong bonds, without removing an excessive amount of the Al pad. This is achieved by applying the optimal combination of normal force, table scrub and ultrasonic energy during different stages of bonding. A process like ProCu5 can achieve stronger bonding (higher IMC%, larger contact area) with less pad damage (lower Al Splash, no pad crack, and higher Al remain%). With the ProCu5 process used in this test, the cells that passed the 96 hours reliability test have Al remain% from 38% to 54%. The cells that passed the 168 hours reliability test have Al remain% from 45% to 54%. For this application, a good Al remain percent target could be 35%-55% for a ProCu5 process. For a traditional process, a lower Al remain% is typically needed to produce reliable bond.

It is hard to draw a relationship between contact area and reliability based on this test since the contact areas in this DOE are very similar to each other. In the next session, the contact area influence to the reliability outcome will be examined.

Finer Pitch Wire Bonding Process and Reliability Study

A failure mechanism of HAST is corrosion of the intermetallic compound from the edge inward. Corrosion first appears to be a gap at the edge of the bond and the gap then progresses toward the center. Once the gap propagates all the way to the center, an open is detected and the bond fails. The interface condition is illustrated in Figure 5.

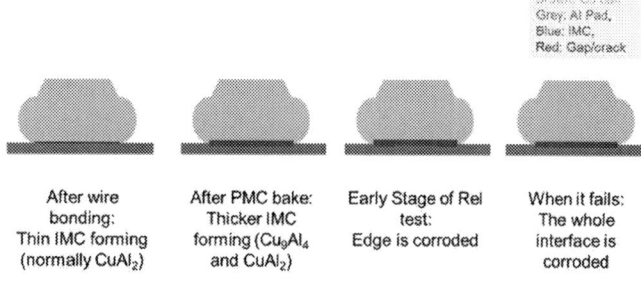

Figure 5: Interface evolution after wire bonding until reliability failure.

Since the attack of the bonded interface continues over time, the bigger the interface, the more reliable the bond is expected to be. Knowing the relationship between interface contact area vs reliability is important to access fine pitch capability of the Cu bonding process. To understand this relationship, we tested different ball diameters with different contact areas. To cover a large range of contact areas, 0.6 mil and 0.7 mil AuPdCu wire are used in the next experiment. Target Ball Diameters are set to 28, 30, and 32um for 0.6mil (15um) diameter wire. Target Ball Diameters are set to 34, 36 and 38um for 0.7mil (18um) diameter wire. The ball diameters tested here cover the normal ball diameter range for advanced nodes such as 28nm and 20nm. For example, a 34 um bonded ball with 0.7 mil wire is suitable for 45um pitch/40um bond pad opening devices.

The processes for different ball diameters were easily established with the ProCu5 process. Target Ball Diameter (TBD) is the main input parameter in K&S's response based ProCu processes. The calculation of bonding parameters including bond force and ultrasonic current level are based on

the user desired TBD setting. The Bond Strength Adjust (BSA) parameter is adjusted slightly to achieve similar IMC% for each cell. IMC% is shown to influence the reliability outcome. The wire bonding results are shown in Table 3. As in the previous test, the wire bonding results include actual bonded ball diameter, Y Splash, shear, contact diameter, IMC% and Al remain%. The actual bonded ball diameter ranges from 29.6um to 40.8um which is slightly larger than the TBD. The IMC% for each cell is above 90% in this test. Al remain% is around 50% with minimal of 49% and maximal of 54%. The IMC images are shown in Figure 6 and the cross section images are shown in Figure 7.

Table 3. Test settings and wire bonding results for different bonded ball diameter process.

Cell	Wire Dia. [um]	TBD [um]	Ball Dia. [um]	Y-Splash. [um]	Shear [g]	Cont. Dia. [um]	IMC [%]	Al remain [%]
10	15	28	29.6	34.8	11.1	24.1	96.2	51.8
11	15	30	31.5	36.3	12.8	26.9	94.6	52.9
12	15	32	33.8	38.1	14.1	28.5	93.6	53.6
13	18	34	37.1	41.2	16.5	30.7	95.0	51.8
14	18	36	38.9	41.7	17.3	33.0	91.2	53.6
15	18	38	40.8	44.1	19.2	35.9	91.6	49.1

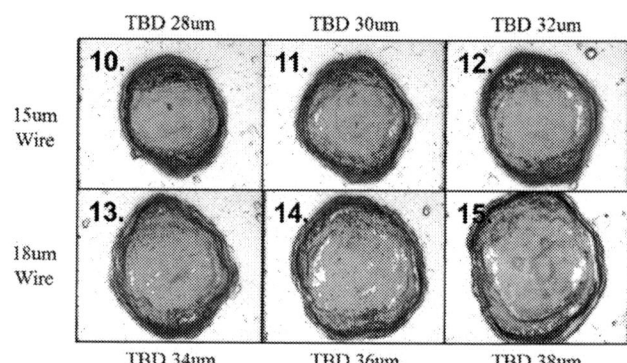

Figure 6 . IMC image for cells 10-15 with varying bonded ball size and contact area.

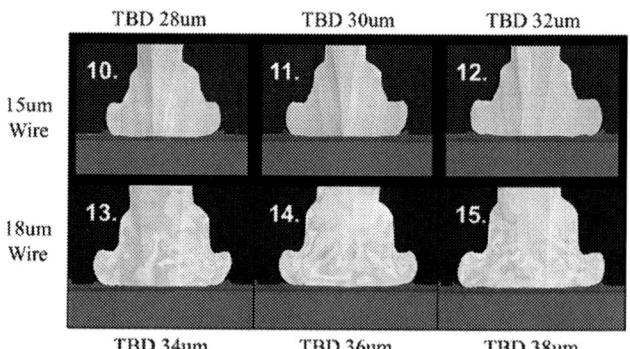

Figure 7. Cross-section images for cells 10-15.

Reliability results confirmed that larger ball diameters with larger contact areas and larger IMC areas have better reliability performance. TBD settings of 32um and above all passed 96 hours and 168 hours. The two lowest diameter ball sizes showed a low level of failure (8%) at 96 hours. Based on our past experience, a better molding compound can significantly improve the reliability outcome [2]. Reliability results are given in Table 4 and a variability chart is shown in Figure 8. In our future test, higher grade molding compound will be used to test with failed cells here.

Table 4. Reliability results

Cell	Wire Dia. [um]	TBD [um]	24Hr	96Hr	168Hr	336Hr
10	15	28	0%	8%	16%	34%
11	15	30	0%	8%	9%	34%
12	15	32	0%	0%	0%	9%
13	18	34	0%	0%	0%	0%
14	18	36	0%	0%	0%	3%
15	18	38	0%	0%	0%	3%

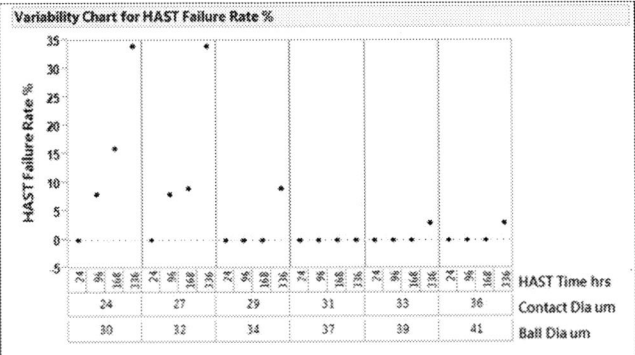

Figure 8. JMP variability chart for HAST failure rate. Reliability failure rate over time vs. contact diameter and actual ball diameter.

Failure Analysis with SEM and EDX

To understand the corrosion over time and the failure mechanisms, samples prior to HAST testing, and at different HAST time intervals, are analyzed by SEM and EDX. Cell 2 for AuPd-Cu wire, cell 5 for bare Cu wire, and cell 8 for Pd-doped Cu wire are used in this study. Besides the wire type, all other conditions are identical.

Our previous HAST reliability study [2] showed corrosion of the intermetallic compound (IMC) during HAST. Corrosion reaction of IMC forms aluminum oxide, and then a gap when Cu and Al oxide separates due to their poor bad connectivity. The Cl content from the molding compound is an important factor influencing the bond reliability, since it accelerates the corrosion reaction. All failures in this HAST test are on bonds with positive-voltage, while all bonds with negative-voltage passed 336 hours HAST. Cl^- is a negative ion, and it moves towards a positive-voltage bond, and therefore accelerates the corrosion at positive-voltage bond.

~100-nm-thick IMC (Cu_9Al_4 near Cu ball and $CuAl_2$ next to Al pad) is present prior to HAST (after molding and post molding cure), as shown in Figure 9. Corrosion reaction during HAST starts with attacking Cu_9Al_4, and Al oxide appears as a product. A gap is formed between Cu and Al oxide when Cu_9Al_4 is corroded. For bare Cu wire, such gap propagates throughout the whole bond interface within 96 hours HAST, so the bond fails (Figure 9b).

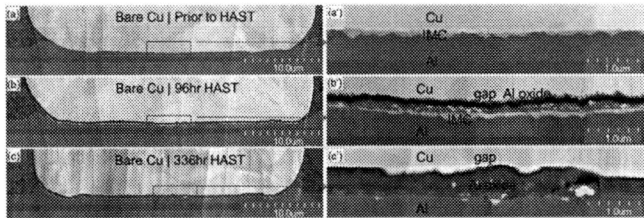

Figure 9. SEM showing corrosion at a bare Cu wire bond and Al pad interface during HAST: (a, a') prior to HAST, ~100nm-thick IMC between Cu ball and Al pad; (b, b') after 96 hours, entire interface is corroded and a gap progresses into the bond center; (c, c') after 336 hours, all IMC was replaced by a gap and aluminum oxide.

AuPdCu wire has shown much better HAST reliability than bare Cu, due to the appearance of Pd and Au at the bond interface (Figures 10 and 11). It is known that Cu rich IMC including Cu_9Al_4 is corrosion sensitive. In AuPdCu wire bonds, (Pd, Au, Cu)-Al IMC, but not Cu_9Al_4, is formed, and therefore improves corrosion resistance. As evidenced in Figure 11b, there is only little corrosion at the edge of the AuPdCu wire bond after 96 hr HAST, while the majority of the bond area is not corroded. After 336 hours, the corrosion is obvious, but the gap is not throughout the whole bond interface, so the bond has not yet failed.

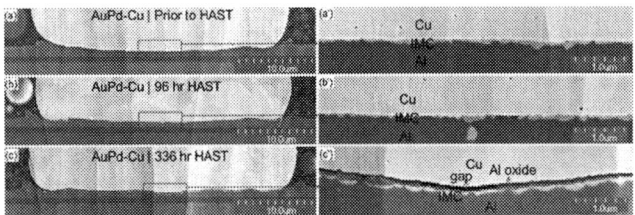

Figure 10. SEM showing corrosion at an AuPd-coated Cu wire bond and Al pad interface during HAST: (a, a') prior to HAST, ~100nm-thick IMC; (b, b') after 96 hr, corrosion is at the edge of bond only; (c, c') after 336 hours, a gap is not throughout the whole bond interface.

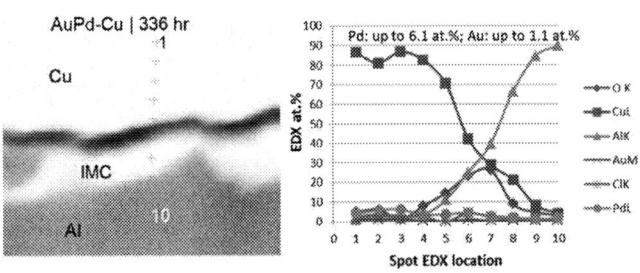

Figure 11. EDX showing up to 6.1 at% Pd and 1.1 at.% Au on an AuPd-Cu bond.

The reliability of Pd-doped Cu wire is not as good as AuPd Cu wire, but is better than bare Cu wire. Pd-doped Cu wire bonds also have Pd at the bond interface, and formation of (Pd,Cu)-Al IMC improves corrosion resistance. Since corrosion reaction consumes IMC, the amount of remaining IMC is an indicator of corrosion grade. As shown in Figures 9, 10 and 12, IMC is not fully consumed in Pd-doped Cu wire and AuPd Cu wire after 336 hours, but it is totally corroded in the bare Cu wire. This indicates that Pd-doped Cu wire has better corrosion resistance than bare Cu wire. Our previous TEM study [12] reported that, for Pd-coated wire, the Pd coating melts and dissolves into the ball during FAB formation. So the Pd distribution in the as-bonded bond is similar between Pd-coated wire and Pd-doped wire. This suggests, from reliability point of view, Pd-doped wire is a good alternative wire to Pd-coated wire, and the manufacturing cost for Pd-doped wire is lower.

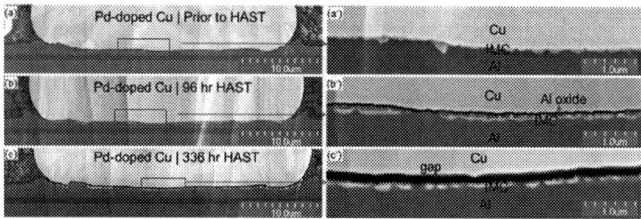

Figure 12. SEM showing corrosion at a Pd-doped Cu wire bond and Al pad interface during HAST: (a, a') prior to HAST, ~100nm-thick IMC; (b, b') after 96 hours, some corrosion; (c, c') after 336 hours, a gap progresses into the bond center; IMC has not fully corroded.

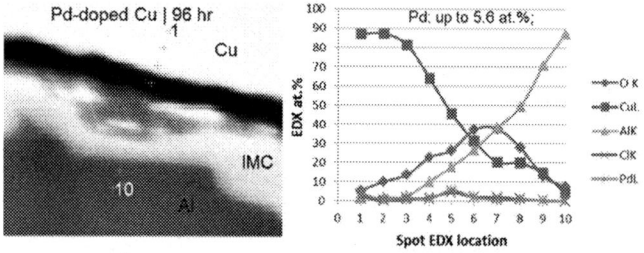

Figure 13. EDX showing up to 5.6 at% Pd on a Pd-doped Cu bond.

Conclusion

For HAST reliability test, we have shown again that the wire type played an important role. The addition of Au and Pd elements improve interface reliability due to more stable intermetallic compound formation. In our test, AuPd coated Cu wire and Pd doped bare Cu wire both showed good reliability for fine pitch, advanced node wire bonding. We found the important wire bonding responses are contact area and IMC%. Higher contact area and higher IMC% are good for reliability. IMC% of 85% is a good spec target. Bonded ball of 34um with contact diameter of 29um can pass 96 and 168 hours reliability with medium grade molding compound. As the pitch and ball size continue to go down, higher grade molding compound may be needed to pass bias HAST reliability test. The outcome in this paper shows that reliable wire bonding on advanced nodes devices can be achieved with 0.6mil and 0.7 mil Cu wire for 45um and 40 um pitch devices using optimized wire bonding processes and adequate molding compound.

Acknowledgments

We would like to thank Heraeus for providing information and wires. We appreciate Son Truong Nguyen, Jim Suttie and Hung Ly from K&S for their assistance with the metrology.

References

1. Bernd K Appelt, Andy Tseng, Shoji Uegaki, and Louie Huang, "Cu wire bonding knows no limit – 28nm is qualified", ICSJ, December 2012

2. Ivy Qin et. al, "Molded Reliability Study for Different Cu Wire Bonding Configurations", *Proc 63nd Electronic Components and Technology Conference (ECTC)*, June 2013, pp 1587-1594.

3. Ivy Qin, Bob Chylak, and Nelson Wong "Wire Bonding Technology for 20 and 28nm Nodes", *2013 Electronics Packaging Symposium,* Binghamton, NY, USA, Oct. 2013.

4. B. Chylak, H. Clauberg, J. Foley, I.Qin and B. Milton, "Copper wire bonding: R&D to high volume manufacturing," *45th International Symposium on Microelectronics IMAPS*, San Diego, Sept. 2012.

5. T. Uno et al., "Surface-enhanced copper bonding wire for LSI," *Proc. 59th Electronic Components and Technology Conference (ECTC),* San Diego, CA, USA, 2009, pp. 1486-1495.

6. T. Uno et al., "Improving humidity bond reliability of copper bonding wires," *Proc 60th Elctronic Components and Technology Conference (ECTC),* Las Vegas, NV, USA, June 2010, pp. 1725-1732.

7. H. Abe et al, "Cu wire package reliability and molding compounds", *Proc 12th IC Packaging Technology Expo*, Tokyo, Japan, Jan. 2011.

8. H. Abe et al., "Cu wire and Pd-Cu wire package reliability and molding compounds," *Proc 62nd Electronic Components and Technology Conference (ECTC)*, San Diego, CA, USA, May 2012, pp. 1117-1123

9. I. Singh et. al, "Pd-coated Cu Wire Bonding Reliability Requirement for Device Design, Process Optimization and Testing," *45th International Symposium on Microelectronics IMAPS*, San Diego, Sept. 2012.

10. M. Takada et. al, "Palladium cladded copper ball bonding wire" , US Patent, NDN- 223-9517-8655-5

11. S. Murali, Johnny Yeung and Roman Perez, "Alloyed Copper Bonding Wire with Homogeneous Microstructure", 35th International Electronic Manufacturing Technology Conference, 2012.

12. H. Xu, I. Qin, H. Clauberg, B. Chylak, V. L. Acoff, "Behavior of palladium and its impact on intermetallic growth in palladium-coated Cu wire bonding," Acta Materialia, 2013, Vol. 61(1), pp. 79–88.

Silver-assisted Copper Wire Bonding Using Solid-State Processes

Yi-Ling Chen*, Yuan-Yun Wu and Chin C. Lee
Electrical Engineering and Computer Science
Materials and Manufacturing Technology
University of California, Irvine CA 92697-2660

Abstract

The copper (Cu) wire has become a new alternative of wires in advanced packaging technology. In this research, we selected large Cu wires to develop a new bonding process based on solid-state process. Upon thorough investigation, we chose silver (Ag) as the bonding medium between Cu wires and silicon (Si) chips, for its ductility and superior physical properties. The end of Cu wires is cut and polished at an angle to form a flat bonding surface. This surface is plated with 50μm thick Ag, following annealing to make it more ductile to facilitate solid-state bonding. For demonstration, Cu substrates and Si chips were used. Other chips such as SiC can be used too. The bonding was performed at 300°C with 1,000psi (6.9MPa) pressure in 0.1 torr vacuum for a few minutes. This pressure is less than 1/10 of what used in industrial thermo-compression processes. The breaking force of resulting wire-bonds was evaluated by pull test. For 1mm Cu wire-bonds on Si chips, the average breaking force is 17kg. The fracture surfaces were examined and analyzed in details. The limit on the operating temperature is the eutectic temperature of the Ag-Cu system, 780°C. Our target operating temperature is 350°C for 1,000 hours, still being verified. At this temperature, no encapsulating material or mounding compound is available on the market to reinforce the wire-bonds, except lead-containing glass. Most likely, the wire-bond has to stand as it is. Accordingly, high breaking force is an essential requirement.

Key Words: copper wire, copper chip, silicon chip, silver, wire bonding, solid-state bonding, electronic packing.

Introduction

In the microelectronic packaging industry, the most common interconnect technology is wire bonding due to its flexibility, low cost and high yield rate [1], [3]. There are three basic wire bonding techniques: thermo-compression bonding, thermo-sonic bonding, and ultrasonic bonding [4]. Thermo-compression bonding utilizes heat and force for the wedge method only. Thermo-sonic bonding utilizes heat and ultrasonic power for the ball/wedge methods. Ultrasonic bonding utilizes ultrasonic power at room temperature and for the wedge method only. Wire materials include aluminum (Al), gold (Au) and Cu. Ultrasonic wedge bonding is primarily used with Al wire [5]. Au wire is used extensively for these basic wire bonding techniques because of its excellent inherent properties until gold price rises sharply. A vast amount of development activities have been invested in Cu wire bonding technology over the past several years. Besides the considerably low cost, Cu wire has superior electrical and thermal conductivity, less intermetallic growths than Au wire on Al metallization surface, which prevents Kirkendall voidings, and great reliability of the bonding structure at elevated temperature [6], [9]. With these advantages, Cu wire is chosen as a suitable alternative to gold wire but also does pose a new challenge.

With the transition from Au wire to Cu wire, the main disadvantage is its high tendency to form a Cu oxides layer on its surface to result in bonding failure. Using the forming gas which is composed of nitrogen and hydrogen to provide an inert environment is one option. Electroplating an oxidation-resistant layer on bare Cu wire to prevent Cu oxides is another one [10], [11]. In this project, electroplating an Ag layer on Cu wire as an anti-oxidation layer was chosen. For the Ag coated Cu wire (Ag/Cu wire), the forming gas is not necessary through the bonding process. Apart from this, Ag has the highest electrical conductivity and thermal conductivity among all metals. Ag is also greatly ductile to accommodate the large CTE mismatch between Si and Cu. The most important of all, the Ag/Cu wire is expected to sustain a high operating temperature for high power devices.

Using the atomic solid-state bonding process we have successfully developed a novel method to bond Ag/Cu wire to Si chip. The bonding process is performed at 300°C with 1000 psi (6.9 MPa) pressure without molten phase. The cross-sectional images of scanning electron microscopy (SEM) display the nearly perfect quality of Ag joints. The fracture modes were also observed and analyzed. The breaking force of the wire-bond is evaluated through destructive in-plane pull test. The pulling direction is on the bonding interface. For 1 mm Cu wire, the highest in-plane breaking force is 18 Kg on the Si chip. The high breaking force can eliminate a concern from industries that is the quality of the wire-bond interconnection used in microchip.

Experimental Designs and Procedures

Corning MACOR machinable glass ceramic was chosen as the winding core due to its excellent physical properties and heat tolerance. Each MACOR rod is 0.5 inch in diameter and 3 inches in length. It is drilled 2 holes to tie up the Cu wire and other 2 holes to fix the device in position. Fig. 1 displays the engineering sketch of the MACOR rod. Cutting grooves in the surface of the MACOR rod was introduced in order to have the same pitch for each Cu wire. After the fabrication process of the MACOR rod, it is ultrasonically cleaned in an acetone bath for cleaning purpose. Using 40 mils Cu wires wind the rod. The Cu wire is cut in half diameter from the center to obtain a flat surface which is slightly polished to remove the Cu oxides and contamination before the electroplating process, as shown in Fig. 2. With this condition, the bonding area of each sample is 0.0422 cm². That the thickness of pure Ag layer is 50µm is coated on the Cu wire by electroplating process. After the electroplating process, a 50µm Ag layer on each Cu wire is shown in Fig 3.

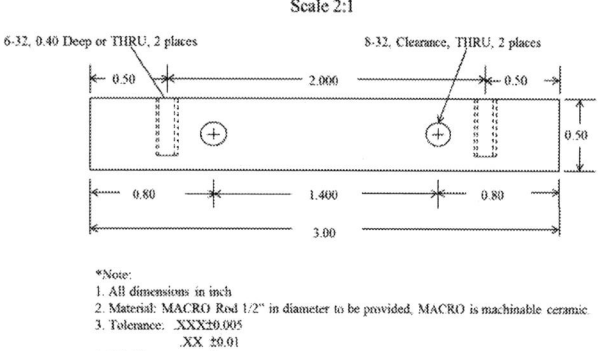

Figure 1 The engineering sketch of the MACOR rod includes diameter, length, and screw sizes.

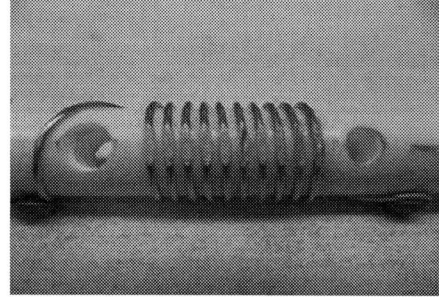

Figure 2 The Cu wire is cut in half diameter from the center to obtain a flat surface which is slightly polished to remove the Cu oxides and contamination.

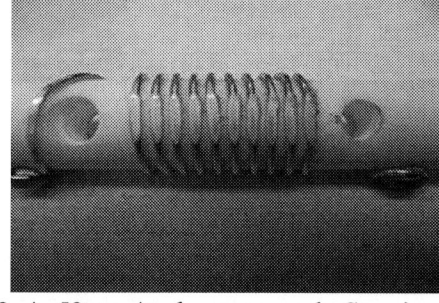

Figure 3 A 50µm Ag layer on each Cu wire, after the electroplating process.

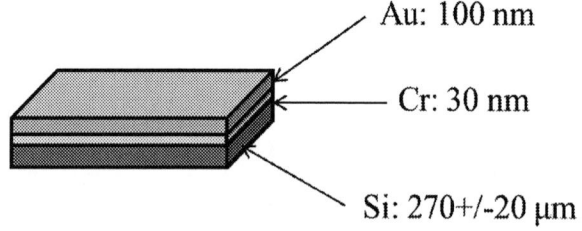

Figure 4 The layer structure of Si chip.

Before the bonding process, the 50 µm Ag onto Cu wire is annealed at 350°C for 5 hours. The Si chips are selected to bond to Ag. Si wafers are deposited 30 nm chromium (Cr) and 100 nm Au by e-beam evaporation system at 3×10^{-6} Torr vacuum. The layer structure of Si chip is presented in Fig. 4. The Cr layer is the adhesion layer and the Au layer protects Cr from oxidation [12]. The assembly of the solid-state bonding is to set samples on the heating graphite platform under 0.1 Torr to restrain oxidations [13]. The Ag layer is used as a bonding medium to bond to Si chips, by using a solid-state bonding process with a static pressure of 1000 psi and heated to peak temperature of 300°C with a dwell time of 3 minutes. The bonding pressure is controlled by the spring-loaded mechanism. The vacuum chamber cools down naturally to room temperature after the bonding process is completed. Some bonded samples are mounted in epoxy-resin and then cut in halves to investigate the cross sectional structure of the Ag joints. SEM is adopted to study the quality of Ag joints, the interface morphologies of Ag joints on Si chips and Cu wires, and the microstructure of fracture modes. The in-plane pull test is used to evaluate the strength of bonded samples.

Experimental results and discussion

Experiment of annealing the Ag layer onto Cu wire at 350°C for 5 hours was designed to increase its grain size for enhancing ductility [14]. The SEM image of annealing the Ag layer onto Cu wire at 350°C for 5 hours is shown in Fig. 5. The solid state bonding process is conducted at 300°C with a static pressure of 1000 psi (6.9 MPa). It is worth noting that this pressure is 1/10 to 1/50 of what used in industrial thermo-compression bonding processes. The bonding temperature is much lower than the melting temperatures of materials in the system, including Cu, Ag, Au, Cr and Si, and there is no molten phase involved during the bonding process accordingly; hence, it is termed solid-state bonding process. The solid-state bonding process forms permanent bonds between the Ag joint and Si chips interface by bringing their mating surfaces in intimate contact. In other words, they may have a chance to share their electrons when the spacing between their mating surfaces is brought within atomic distance [15].

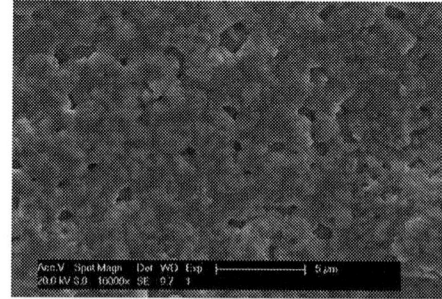

Figure 5 The SEM image of annealing the Ag layer onto Cu wire at 350°C for 5 hours.

We first examined the quality of the Ag joint. From Fig. 6, the cross-sectional SEM images of a typical bonded sample display that the electroplating interface is very smooth and sharp; moreover, no voids and gaps are observed at the bonding interface, which means that the quality of the Ag joint is nearly perfect.

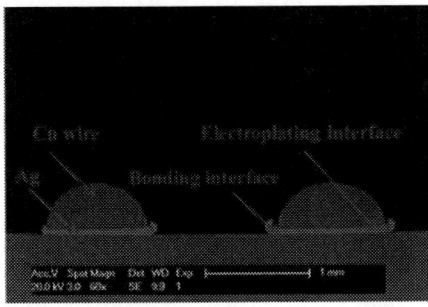

Figure 6 Cross-sectional SEM images of a typical bonded sample. The bonding process was performed at 300°C with a static pressure of 1000 psi.

Table I The breaking forces and modes of bonded samples are measured by in-plane. Each substrate is Si chip.

Sample	Breaking force (kg)	Shear strength (MPa)	Fracture modes
1	15	35	mixed
2	17	39	mixed
3	18	41	mixed

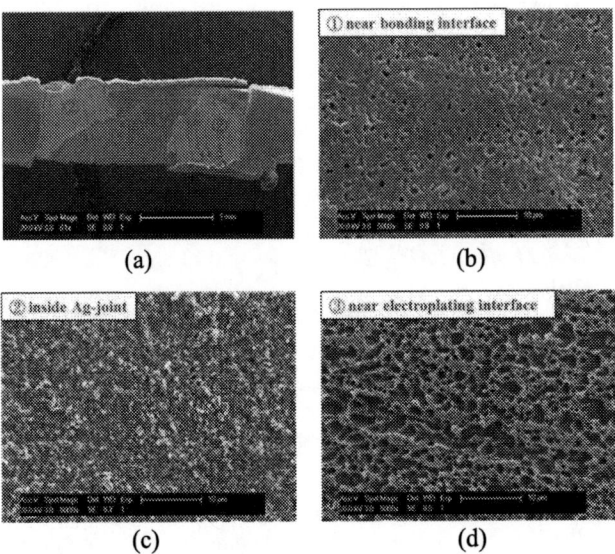

Figure 7 The SEM images of sample 3 on the Cu wire side after the in-plane pull test (a) The whole bonding area, (b) The fracture interface near the bonding interface, (c) The fracture interface inside the Ag joints, and (d) The fracture interface near the electroplating interface.

The in-plane pull test is performed on bonded samples to evaluate the bonding strengths. The pulling direction of in-plane pull test is on the bonding interface. The breaking forces of bonding Cu wire to Si chip, ranging from 15 to 18 Kg, are listed in Table I. Si wafers are deposited 30 nm Cr and 100 nm

Au by e-beam evaporation. The bonding area, 0.0422 cm², is obtained by computing the area of an ellipse. Thus, the shear strengths of bonding Cu wire to Si chip are achievable.

There are three fracture types of bonding Cu wire to Si chip on the Cu wire side after the in-plane pull test. It contains the fracture surfaces near the bonding interface (①), inside the Ag joints (②), and near the electroplating interface (③). For example, the three fracture types are illustrated in Fig. 7. First, most of the Ag joints are flat near the bonding interface to demonstrate that the Ag joints were deformed well under the influence of applied loads during the solid-state bonding process, as shown in Fig. 7 (b). Second, some of the fractures are broken inside the Ag joints shown in Fig. 7 (c). It displays that the bond between Ag and Si chip can be stronger than Ag joint itself. Finally, a quite thin Ag layer near the electroplating interface is shown in Fig. 7 (d). A large amount of Ag atoms was discovered in this region. Apparently, a major part of this fracture type is still inside the Ag joints because it does not break completely along the bonding interface. Fig. 8 shows the high magnification SEM image of sample 3 which includes three fracture types.

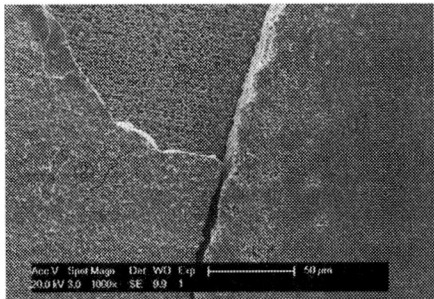

Figure 8 The SEM image of sample 3 includes three fracture types.

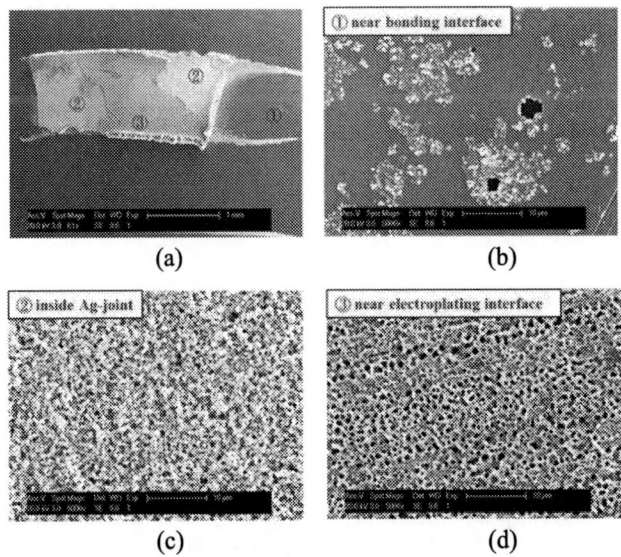

Figure 9 The SEM images of sample 3 on the Si chip side after the in-plane pull test (a) The whole bonding area, (b) The fracture interface near the bonding interface, (c) The fracture interface inside the Ag joints, and (d) The fracture interface near the electroplating interface.

978-1-4799-2408-0/14 $31.00 © 2014 IEEE

We next observe the fracture types on the Si chip side after the in-plane pull test. It, of course, has the three fracture types similar to the Cu wire side, as shown in Fig. 9. Note that there are some holes on the Si chip because the wire-bond between the Ag joints and Si chip is quite strong to result in Si chip broken during the pull test, as shown in Fig. 9 (b). Also, the appearance of fracture surface near bonding interface looks like Ag residues on the Si substrate to infer that this region is partially bonded. Those results could be a part of reason to understand why bonded Cu wire to Si chip after in-plane pull tests have three fracture types.

Conclusions

This paper presents a new concept of electroplating Ag on the Cu wire to bond to Si chips, by adopting the solid-state bonding process. It requires only a single stage operation at low processing temperature and pressure without any molten phase involved. In addition, the Ag joint has many exceptional attributes, including the high operating temperature, a protective barrier against the oxidation on Cu, the highest electrical and thermal conductivity among metals, and is capable of managing large CTE mismatch between Cu and Si to avoid the stress crack. According to the experimental results above, no voids and gaps are observed at the bonding interface which means the quality of Ag joint is excellent. The destructive bond pull tests demonstrate that the bonding strength is quite strong. Furthermore, it eliminates a concern from industries which is the strength of the bonding interface and the Ag joint. In this study, the high breaking force allows designers to use the bond-wires as the terminal wires for interconnect. Our target operating temperature for devices is 350°C. So far, we have not found any encapsulating material that can take this temperature except lead-containing glass. Most likely, the wire-bond has to stand as it is without enforcement by encapsulant. Accordingly, high breaking force is an essential objective. In principle, we should be able to make the wire-bond stronger than the Cu wire so that Cu wire breaks first prior to the wire-bond.

Acknowledgments

This research was supported by the II-VI Foundation.

References

1. Z. W. Zhong, "Overview of wire bonding using copper wire or insulated wire," *Microelectronics Reliability,* vol. 51, pp. 4-12, Jan 2011.
2. G. G. Harman, "Wire bonding in microelectronics," third ed., New York: McGraw-Hill, pp. 33-38, 2010.
3. S. C. Kim and Y. H. Kim, "Flip chip bonding with anisotropic conductive film (ACF) and nonconductive adhesive (NCA)," *Current Applied Physics,* vol. 13, pp. S14-S25, Jul 2013.
4. M. Mayer, O. Paul, D. Bolliger, and H. Baltes, "Integrated temperature microsensors for characterization and optimization of thermosonic ball bonding process," *IEEE Transactions on Components and Packaging Technologies,* vol. 23, pp. 393-398, Jun 2000.
5. A. Shah, H. Gaul, M. Schneider-Ramelow, H. Reichl, M. Mayer, and Y. Zhou, "Ultrasonic friction power during Al wire wedge-wedge bonding," *Journal of Applied Physics,* vol. 106, pp. 013503 1-8, Jul 2009.
6. S. Murali, N. Srikanth, and C. J. Vath, "An analysis of intermetallics formation of gold and copper ball bonding on thermal aging," *Materials Research Bulletin,* vol. 38, pp. 637-646, Mar 2003.
7. P. Liu, L. Tong, J. Wang, L. Shi, and H. Tang, "Challenges and developments of copper wire bonding technology," *Microelectronics Reliability,* vol. 52, pp. 1092-1098, Jun 2012.
8. P. Ratchev, S. Stoukatch, and B. Swinnen, "Mechanical reliability of Au and Cu wire bonds to Al, Ni/Au and Ni/Pd/Au capped Cu bond pads," *Microelectronics Reliability,* vol. 46, pp. 1315-1325, Aug 2006.
9. B. K. Appelt, A. Tseng, C. H. Chen and Y. S. Lai, "Fine pitch copper wire bonding in high volume production," *Microelectronics Reliability,* vol. 51, pp. 13-20, Jan 2011.
10. I. Qin, X. Hui, H. Clauberg, R. Cathcart, V. L. Acoff, and B. Chylak, *et al.*, "Wire bonding of Cu and Pd coated Cu wire: Bondability, reliability, and IMC formation," in *Electronic Components and Technology Conference (ECTC), 2011 IEEE 61st*, pp. 1489-1495, May 2011.
11. H. Clauberg, B. Chylak, N. Wong, J. Yeung, and E. Milke, "Wire bonding with Pd-coated copper wire," in *CPMT Symposium Japan, 2010 IEEE*, pp. 1-4, Aug 2010.
12. C. C. Lee, D. T. Wang, and W. S. Choi, "Design and construction of a compact vacuum furnace for scientific research," *Review of Scientific Instruments,* vol. 77, pp. 125104-1-125104-5, Dec 2006.
13. W. P. Lin, C. H. Sha, and C. C. Lee, "40-μm Cu/Au flip-chip joints made by 200 degrees C solid-state bonding process," *IEEE Transactions on Components Packaging and Manufacturing Technology,* vol. 3, pp. 126-132, Jan 2013.
14. W. P. Lin, C. H. Sha, P. J. Wang, and C. C. Lee, "Microstructures of silver films plated on different substrates and annealed at different conditions," *2011 IEEE 61st Electronic Components and Technology Conference (ECTC),* pp. 1782-1786, May 2011.
15. C. H. Sha and C. C. Lee, "Low-Temperature Solid-State Silver Bonding of Silicon Chips to Alumina Substrates," *IEEE Transactions on Components Packaging and Manufacturing Technology,* vol. 1, pp. 1983-1987, Dec 2011.

Ag Alloy Wire Characteristic and Benefits

Jensen Tsai, Albert Lan, D. S. Jiang, Li Wei Wu, Joseph Huang, J. B. Hong
Siliconware Precision Industries Co., Ltd. No. 153, Sec. 3, Chung-Shan Rd. Tantzu
Taichung 427, Taiwan, R.O.C.
Email: yctsai @spil.com.tw

Abstract

Gold wire has been high volume production in IC packaging industry. With soaring price of gold in recent years and IC packaging search for cost reduction, copper wire offers 2nd alternative for wire bonding type assembly. But copper wire has drawbacks in control issues such as pad crack, aluminum splash, cratering and low throughput, Cu wire need a more complex multi-processing program to solve above problems, even if copper raw material cost is low, that cause the process costs are increased

These limitations of copper wire are related to pad thickness and structure. Silver alloy wire can emerge as 3rd alternative as the cost can compete with copper and the properties of silver are nearly identical to gold while in the die to die package which benefits will be more apparent

Many Study had performed for Silver alloy wire (88% & 95%), included the general workability, wire pull, ball shear, Al splash, pad-to-pad bonding. To get good workability result, Silver alloy wire properties have been studied and done the DOE to determine the best material properties windows. The key properties included the element composition and elongation, hardness, etc.

IMC is another key point for the reliability performance, with the DOE of best wire bonding parameters, the IMC structure was analyzed with different condition of lead-frame based, substrate based. Combined with process flow and compound material types, IMC behavior has been observed. Electron Migration study had been performed as well to check if any side effect for Silver alloy wire bonding. Experiment results showed Silver Alloy wire bonding has better performance than copper. Silver alloy wire can be the mainstream to replace gold and copper wire, especially in pad-to-pad wire bonding type.

Introduction

Gold wire has been used for IC assembly for over 20 years. With the rocketing gold price, currently still keep above 1200USD per ounce (Fig 1), which caused the exorbitant wire bonding packages. About 5-6 years ago, Cu wire showed up to replace gold wire due to its low cost feature, after the hard working for properties control, bonding parameters optimized, Cu wire is currently the mainstream for wire bonding product and in very high volume production. About 55%-60% wire bonded products have been chosen Cu wire as bonding solution, the remaining 40%, majorly still keep as gold wire.

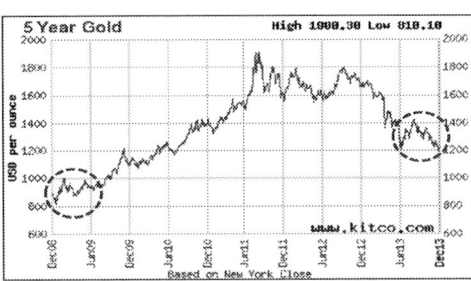

Fig.1 Gold Price Trend

Whoever handled Cu wire bonding, understood that the Cu wire is much critical than gold wire, no matter in raw material storage, forming gas control, Cu ball formation, bonding IMC formation, wire bonder machine alarm rate, reliability test situation, etc, but due to significant cost reduction percentage, around 20% to 25% compared to gold wire used, very critical methodology is required for Cu wire handling and verification and confirmation run. Although the raw material cost is lower, but the engineering effort and resources for Cu wire is much huge, after several year production experiences, people start thinking to survey a wire material, which the cost could compete against copper wire, but the workability could be as easy as gold wire.

Silver alloy wire was developed based on the motives above, after several years trial run, good composition was proven to replace gold wire and copper wire, 88% Silver, the other major elements are Pd and gold. With some high volume production for 88% Ag alloy wire, further study to reduce gold content percentage as cost reduction also implemented. It is around 95% Ag, called 1N Ag alloy wire, followed 88% and has been verified to perform well in workability and reliability, with the high throughput, 1N Ag wire bonding cost could be a little bit cheaper than copper wire, depends on wire counts, pad-to-pad bonding situation, could reduce the cost by 5% at most.

After several years cooking, Ag alloy wire overcame the cost situation and been verified relentlessly for workability and reliability, six quarters ago, Ag alloy wire started high volume production, 1st product application is used on TSOP package, DRAM application. After TSOP experience, Ag alloy wire then used for regular lead-frame products and substrate based products, mainly for handheld application, like baseband, RF. With same verification procedure, Ag alloy no problem to support these logic product and run into production as well start from the beginning of year 2012.

2014 Electronic Components & Technology Conference

Ag alloy wire roadmap

The bond pad conditions are dominated by pad pitch, pad opening, Al thickness, pad structure, currently 0.7mil (17.8um) and 0.8mil (20.3um) wire diameter are most commonly used for production, Ag wire can handle bond pad opening as small as 40um, the same level as Cu wire. Regarding to the composition, due to pure Ag wire has too much difference in diffusion rate with Al element, therefore, Au and Pd elements were adopted into the silver, make it as alloy. As mentioned, 88% and 1N (95%) Ag alloy wire have been verified and in high volume production, with the lower cost situation, 1N Ag alloy wire should be the mainstream for widely use.

Wire Properties Comparison

Fig 2. is the table for wire properties comparison, with 1N Ag wire, Pd coating Cu wire, 2N gold wire. The key point is the hardness for free air ball (FAB), apparently gold wire is the softest 41~43, which is good for bonding process with smaller force to prevent bond pad crack, the hardness of Cu wire with Pd coating is 60~70, which is the hardest, and this is reason to optimized Cu wire parameter to have IMC formation and prevent bond pad crack simultaneously. The hardness of Ag alloy wire is 58~62, softer than Cu wire, but still harder than gold, but due to the softness, Ag alloy wire handling is easier than Cu wire. With the high volume production experience of Cu wire, there is definitely hassle-free for Ag alloy wire usage in the bonding process.

From the data sheet, people are concerned about the high resistance of Ag alloy wire, it is about 2 times higher Cu wire. But after the overall resistance comparison, actually, overall resistance of Ag alloy wire is only 10%~20% higher than Cu wire, because the highest resistance came from IMC, not wire itself. For Cu wire IMC, like Cu_9Al_4, Cu_3Al_2, the resistance is around 13~15, but for Ag alloy wire IMC, like Ag_3Al, Ag_2Al, the resistance is only about 10. After real product samples built and judged by final test program. The samples built by Ag alloy wire could pass the final test program with normal yield.

The other key property, like elongation, it is vital pad-to-pad bonding, Fig 3. shows the situation with pad-to-pad bonding, while making the pad-to-pad bonding, a Cu ball or Ag ball need to be bonded first and stitched off, if the elongation is too small, like 5%~6%, then the pad-to-pad wire is very easy to have snake type due to unreleased stress. The recommended elongation is around 10%~11%.

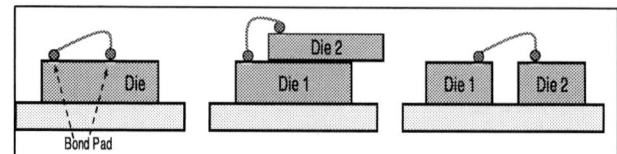

Fig.3 Die to Die bonding

Workability Result

Fig4. expresses a comparison of a non-via bond pad structure using PdCu wire and Ag alloy wire, with this weak pad structure, bond pad crack was observed by using of Cu wire after cratering test, on the contrary, no any pad crack was found by using Ag alloy wire. Not just only bond pad structure, but also Al pad thickness, with IMC difficult to form for Cu wire, minimum bond pad Al thickness 0.8um is required, but for Ag alloy wire, it can easily deal with bond pad Al thickness down to 0.5um, same performance as gold wire. Not only non-via pad structure, others like partial-via, partial metal structure, these are fragile for Cu wire bonding, are liable to have bond pad crack, those bond pad can handled by Ag alloy wire and gold wire.

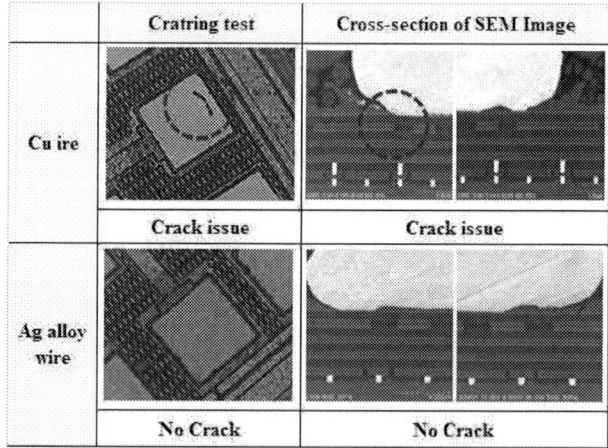

Fig.4 Results of crating test and SEM cross-section image

Normally, the usage of Cu wire required sophisticated wire bonder, like K&S Inconn, Pro Cu, these bonders allow three steps parameter output, which fortify Cu wire bonding and IMC formation. But high-end bonders are no longer a must for Ag alloy bonding, this means the regular bonder can handle Ag alloy wire, just need a forming gas kit. Moreover,

Wire Type		2N Au		AuPdCu		Ag_95%	
Wire Diameter (mil)		0.7	0.8	0.7	0.8	0.7	0.8
Composition (%)	Cu (core)	-		>99.99%		-	
	Ag	-		-		95%	
	Pd	0.95%		1.3~2.3%		4.5	
	Au	99%		<0.02%		0.5	
Density (g/cm³)		19.2		9.03		10.6	
BL(g)		4.5~8.5	6~10	4.5~8.5	6~11	2~7	3~8
EL(%)		2~7	2~7	11~17	11~17	8~12	8~12
Hardness(Hv)	FAB	45.6	46.8	75.1	74.3	51.8	50.9
	Wire	45.6	55.7	53.5	54.8	63.7	63.5
Resistivity (×10⁻⁸Ω.m)		2.9		1.98		3.5	
Fusing Current (mA)		410	460	470	530	390	430

Fig.2 Wire Properties Comparison

due to Cu wire required complicated verification methodology, like optimized parameters, confirmation run, IMC check after baking, remaining Al thickness check, ...etc, the machine and product setting takes much longer than gold wire and Ag alloy wire.

Ag alloy wire is also a boon to throughput, especially in pad-to-pad bonding mode, while forming pad-to-pad bonding, Ag alloy wire throughput is 10%-20% more than Cu wire, and less alarm rate, with this raised throughput, the overall cost will be cut down by using of Ag alloy wire. Fig5. tells the workability benefit comparison table.

Items \ Wire type		Au wire	Cu wire	Ag alloy wire
Forming gas kits		No need	Need	Need
Machine setting time		Short	Long	Middle
Through put Ranking	Normal Bond	NO.1	NO.3	NO.2
	Pad to Pad	NO.1	NO.3	NO.2
Pad structure (Pad crack concern)		No limitation	Can not apply to Nov-VIA, Partial	No limitation

Fig.5 Workability Comparison

Fig6. and Fig7. are the wire pull and ball shear data for gold, Cu and Ag wires, the data reveals Ag alloy wire is comparable with Cu wire and good Cpk, > 1.67 with same criteria, Cu wire and Ag alloy wire are both better than gold wire.

Wire Type	Criteria	Sample Size	Max	Min	Avg
Au wire		2pcs/all wire	5.5	3.5	4.530
Cu wire	> 2.7g	2pcs/all wire	12.1	7	8.714
Ag alloy wire		2pcs/all wire	10.8	5.7	8.064

Fig.6 Wire pull data of Ag alloy wire

Wire Type	Criteria	Sample Size	Max	Min	Avg
Au wire		2pcs/all ball	35.30	19.50	28.343
Cu wire	> 8.5g	2pcs/all ball	40.10	27.70	32.997
Ag alloy wire		2pcs/all ball	39.57	25.23	32.689

Fig.7 Ball shear data of Ag alloy wire

Fig.8 shows the SEM photo of wire bond 1st bond and 2nd bond. The balls are all well-formed. Al splash turns up on Cu wire whereas it is not observed on Ag alloy wire.

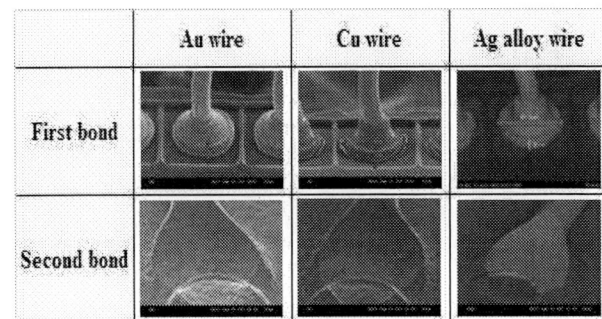

Fig.8 Ball shape and Al pad splash comparison

Fig 9. is the comparison for IMC formation, with same criteria > 80%, all of three wires got the same performance. As a matter of fact, Cu wire IMC is not easy to observe, normally required baking a few hours with high temperature, and observed on the bond pad side. This is why Cu bonding takes longer on machine setting and IMC check..

Wire type	IMC		
	Criteria	Sample Size	Result
Au wire	> 80%	0/2	90% 91%
Cu wire	> 80%	0/2	92% 90%
Ag alloy wire	> 80%	0/2	97% 93%

Fig.9 IMC coverage comparison

Fig 10. is the cratering test, no crack was found for all three wire bond types.

Wire type	Cratering Test		
	Criteria	Sample Size	Result
Au wire	No Crack	0/3	No Crack No Crack
Cu wire	No Crack	0/3	No Crack No Crack
Ag alloy wire	No Crack	0/3	No Crack No Crack

Fig.10 Cratering test

Fig 11. is the IMC verification by using of cross-section and SEM, it shows gold wire IMC thickness is 6.3um, Ag alloy wire is 3.56um, but PdCu wire only 0.7um. The result tells that IMC can be control easily by using gold and Ag alloy wire, IMC control for PdCu wire is really challenging for Cu wire.

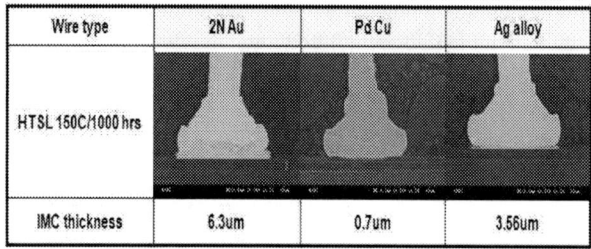

Fig.11 Wire Material IMC Comparison

Wire type	2N Au	Pd Cu	Ag alloy
HTSL 150C/1000 hrs			
IMC thickness	6.3um	0.7um	3.56um

Reliability Result

When it comes to new material, reliability is always a concern. In some papers, 90%Au/10%Ag wire is employed while it failed the reliability test. At this moment, the 90%Au/10%Ag is not cost effective either. Our goal is trying to find out a wire whose cost is similar to Pd coating Cu wire, therefore the candidates are 88% Ag, 92% Ag, 95% Ag, other elements are dominating also, majorly Pd and Au. Owing to alloy material properties are vendor-dependent, thus we should test and qualify vendor by vendor. Currently, there are five vendors offering Ag alloy wire, the recipe of each vendor is different, and need to be verified by workability and reliability separately.

One vendor is chosen as target and The 1st application on DRAM using TSOP package, after the reliability test, the result are all passed, especially the PCT and bias-HAST, PCT didn't see failure and passed 700hours, bias-HAST didn't see failure and passed 400hours. Fig 12. is the reliability test result table. At each read-point, cross-section was performed to examine the IMC situation. The ball shape is satisfying at each read point.

Test item	Read point	Sample Size SAT	Sample Size F/T	Cross-Section
Moisture Sensitivity Test MSL3 (30℃/60%RH/192hrs)	0 hour	0/180	0/180	
	Precondition	0/180	0/180	
Pressure Cooker Test (PCT)	>168 hours	NA	0/45	
Thermal Shock Test (TST)	500cycles	NA	0/45	
Temperature Cycling Test (TCT)	1000cycles	NA	0/45	
Bias High Acc. Stress Test (bias-HAST)	>192 hours	NA	0/45	
High Temperature Storage Life Test (HTSL) (Without precondition)	1000 hours	NA	0/45	

Fig.12 Ag Alloy Wire Reliability Performance

Based on Cu wire experience, bias-HAST is the most unattainable. After the DRAM's success, the next target is to have logic product which pass bias-HAST test, target package types included lead-frame based QFN and substrate based BGA. Real die is used to build the samples and put onto the bias-HAST board, and it passed 168 hours, the 1st failure occurred in 300 hours or so, which is much better than Cu wire bonding. Fig13 Shows the bias-HAST table.

- Lead Frame Package

PKG	Wire Type	Wire Dia (mil)	B-HAST(hrs) 96	B-HAST(hrs) 168	Bias Voltage (V)	Current (A)	BOM
PKG-1 (MQFN)	95% Ag	0.8	F/T Passed	F/T Passed	VDD33 : 3.6V VDD12 : 1.32V	0.15A 0.02A	Ag L/F
PKG-2 (MQFN)	95% Ag	0.9	O/S Passed	O/S Passed	PS 1.2V,3.3V	0.01A, 0.05A	Ag L/F

- Substrate Package

PKG	Wire Type	Wire Dia (mil)	B-HAST(hrs) 96	B-HAST(hrs) 168	Bias Voltage (V)	Current (A)	BOM
PKG-1 (BGA)	95% Ag	0.7	O/S Passed	O/S Passed	PS1 3.3V PS2 1.8V PS3 1.5V PS4 0.9V	0.04A 0.01A 0.01A 0.09A	Au Finger
PKG-2 (BGA)	95% Ag	0.7	F/T Passed	F/T Passed	VDD12 1.2V VDD25 2.5V VDD33 3.3V	0.05A 0.03A 0.02A	Au Finger

Fig 13. Bias-HAST Performance using Ag wire

De-cap Capability

Decapsulation is one of the key capability for failure analysis, following with the same method of Cu wire, laser removed the most compound first, then chemical etching to reveal the die and wires. Fig 14. is real case after de-cap, the performance is as good as Cu wire. Molding compound used in the package is the key for de-cap, due to Ag alloy wire properties are different from Cu wire, in the chemical etching process, wire protection is quiet important. Two molding compounds from Sumitomo are chosen as baseline for Ag alloy wire, one for lead-frame base package, the other for substrate base package. The de-cap successful rate can guarantee more than 95%.

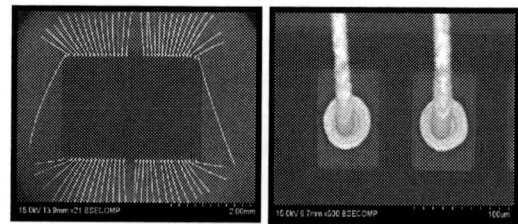

Fig.14 De-Cap Capability

IMC formation Comparison

Compared the IMC formation with the 4N gold wire and 2N gold wire, it could explain why the Ag alloy wire reliability performance is good. A lot of paper already addresses that 4N gold wire and the kirkendall void behavior,

978-1-4799-2408-0/14 $31.00 © 2014 IEEE

the 2N gold wire IMC, Al pad diffusion rate is slower due to the Pd rich layer formation between the wire and IMC, therefore 2N gold wire behavior and performance is better than 4N gold wire. With the Ag alloy wire, the IMC behavior is similar to 2N gold wire, just different IMC and look like Pd rich layer to reduce the diffusion and no kirkendall void, therefore the reliability test performance is the best, refer Fig 15.

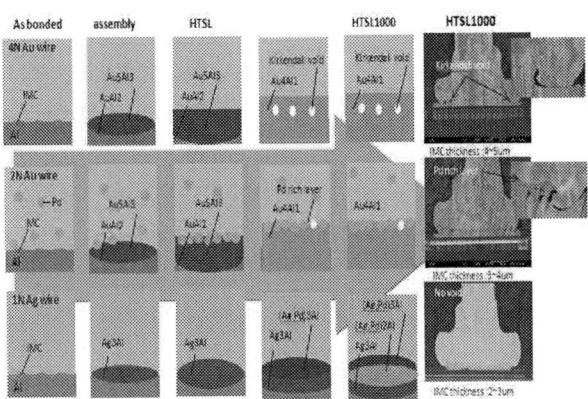

Fig.15 IMC Growth Behavior for Au/Ag Wire

Ag Alloy wires Screen DOE

Fig 16. is wire properties for each difference content of Ag wire, the major doping element of Ag alloy wires are Au & Pd. High doping will increased electric resistivity, 95%Ag wire have lower resistivity then 88% and 92%. All legs are passed reliability test as Fig 17.

Wire type		88Ag	92Ag	95Ag
Composition (%)	Ag	87.3	92.3	95.5
	Au	8.8	4.3	-
	Pd	3.9	3.4	4.5
Melting point (℃)		1020	985	985
Mechanical property	EL (%)	8~12	8~15	8~12
	BL(gf)	2~7	4~8	2~7
	Elastic	87	80	90
Electric resistivity (μΩcm)		5	4	3.5

Fig. 16. Ag alloy wire property comparison

PKG	Wire Type	W/B alarm rate (PPM)	uHAST / PCT (hrs)		TCT(cycle)		HTSL(hrs)	
			96	168	500	1000	500	1000
BGA-1	88Ag	37	0/45	0/45	0/45	0/45	0/45	0/45
	95Ag	44	0/45	0/45	0/45	0/45	0/45	0/45
BGA-2	88Ag	25	0/45	0/45	0/45	0/45	0/45	0/45
	95Ag	53	0/45	0/45	0/45	0/45	0/45	0/45
QNF-1	88Ag	42	0/45	0/45	0/45	0/45	0/45	0/45
	95Ag	32	0/45	0/45	0/45	0/45	0/45	0/45
QFN-2	88Ag	35	0/45	0/45	0/45	0/45	0/45	0/45
	95Ag	42	0/45	0/45	0/45	0/45	0/45	0/45

Fig. 17. Ag alloy wires R/A result

Fig. 18 is for Au & Pd Doping Impact in Ag-Al IMC formation, only one Ag3Al IMC type was found for 99% Ag wire after HTSL 1000hrs, but while doping more Pd and Au elements, there are three kinds of IMC phases at the bonding interface after HTSL 1000hrs@150C, IMC type: (Ag,Au,Pd)3Al, (Ag,Au,Pd)2Al & Ag3Al

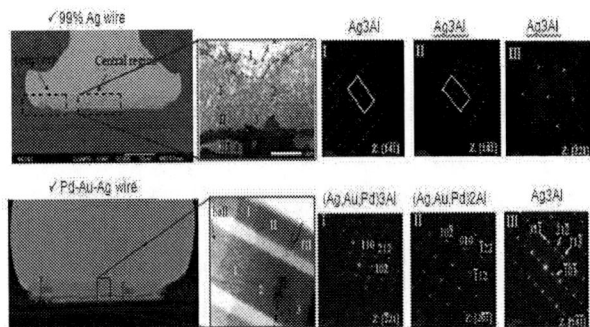

Fig. 18. Ag TEM-EDS Analysis Result

Based on uHAST 168 hrs results, it was found that major factor is Cl to avoid dark layer at IMC, minor factor is to eliminate S element during assembly process. The content of Cl content is recommended less than 20ppm, same as Cu wire requirement.

Fig. 19&20. Ag-Al dark layer in uHAST 168 hrs

All of Ag alloy legs can pass O/S and < 20% dark layer, when Cl ion < 20ppm.

Regarding to Electro migration (EM), Black's equation is a mathematical model for the mean time to failure (MTTF) of a semiconductor circuit due to electro migration, it is a phenomenon of molecular re-arrangement (movement) in the solid phase caused by an electromagnetic field, Fig 21 MTTF of Ag is bigger than Sn & Al. we measure the real materials, there is no electronic migration for 3 kinds of wires after 1200hrs 0.3A/150C. Even post bias HAST 400 hrs, no any Ag migration occurrence as the following SEM images, Fig 22.

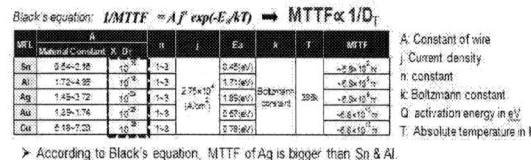

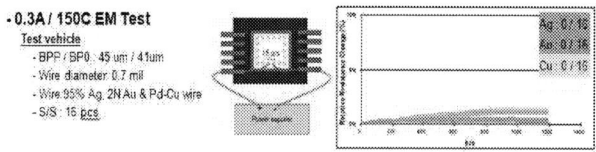

Fig. 21. MTTF for Electro migration

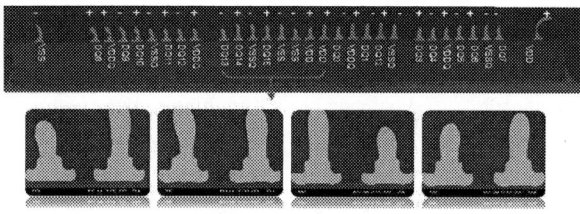

Fig. 22. Electro migration test condition: Bias HAST 400 hrs (3.6V, 130C/85%/400hrs)

Base on above study, 95% Ag wire have lower resistance then 88% and 92%Ag and Ag alloy wire did not has EM issue.

Benefits of Ag alloy wire

Since gold wire and Cu wire have been high volume production for years, why we need Ag alloy wire? Below are the summaries of benefits to use Ag alloy wire:

1. Outstanding bias-HAST performance：Compared to Cu wire, Ag alloy wire passes longer life time with real die and apple-to-apple comparison. For some critical application, Ag alloy wire is a good choice.

2. Bond pad structure is no longer a limitation：With less force which makes a lower risk of bond pad crack. No matter what kind of pad structure and foundry source, Ag alloy can handle it without problem.

3. Longer shelf life：Ag alloy wire could store with a longer shelf time than Cu wire, which is good for process control

4. 95% Ag wire is lower resistance then 88% and 92%Ag and Ag alloy wire did not have any EM issue.

Conclusions

As of today, not only gold wire and Cu wire could be considered, Ag alloy wire could be an option for wire bond products, with the benefits of Ag alloy wire, we do believe in some day, Ag alloy wire can be as 3rd alternative for gold wire and Copper wire.

References

1. Hai Liu, Qi Chen, Zhenqing Zhao, Qian Wang, Jianfeng Zeng, Jonghyun Chae, Jaisung Lee, Samsung Semiconductor China R&D CO., Ltd, Reliability of Au-Ag Alloy Wire Bonding

2. Hao-Wen Hsueh, Fei-Yi Hung, Truan-Sheng Lui a, Li-Hui Chen, Microelectronics Reliability,

3. C.H Cheng, H.L Hsiao, S.I Chu, Y.Y Shieh, C.Y Sun, C. Peng, Elite Semiconductor Memory Technology Inc., Low Cost Silver Alloy Wire Bonding with Excellent Reliability Performance

4. Kyung A. Y. et al., "Reliability Study of Low Cost, Alternative Ag Bonding Wire with Various Bond Pad, Materials", 11th Electronic Packaging Technology, conference, 2009, pp851-857.

5. Hyoung J. K. et al, "Effects of Pd Addition, Jong S. C. et al., "Reliability Study of Low Cost, Alternative Ag Bonding Wire with Various Bond Pad, Materials", 60th Electronic Components and Technology, Conference, 2010, pp1541-1546.

6. A.Kamijo and H.Igarashi, "Silver Wire Ball bonding and Ball/Pad interface characterics," Proceedings of the 35th, Electronic Component Conference, 1985, pp.91-97.

Copper versus Palladium Coated Copper Wire Process and Reliability Differences

Chu-Chung (Stephen) Lee[1], TuAnh Tran[2], Dan Boyne[3], Leo Higgins[4], Andrew Mawer[5]

Freescale Semiconductor Inc

6501 William Cannon Drive West, Austin, Texas 78735 USA

1: stephen.chuchung.lee@freescale.com; 2: Tu.Anh.Tran@freescale.com; 3:Dan.Boyne@freescale.com;
4:Leo.Higgins@freescale.com; 5:Andrew.Mawer@freescale.com

Abstract

Fine pitch copper wire bonding presents challenges to both first and second bond processes. Corrosion of the first bond copper-aluminum (Cu-Al) intermetallic compound bond interface layer can be induced by mobile chlorine in the epoxy mold compound, and the second bond process window can be narrower than with gold wire. Palladium-coated copper wire is believed to overcome these two problems, however, the actual benefits and challenges of the wire need to be considered by the semiconductor industry. The price of palladium-coated copper wire is 2.5 - 3 times higher than bare copper wire. The mechanical properties of palladium-coated copper wire increase the risk of damaging bond pad structures if not bonded correctly. The addition of a thin palladium layer can increase electrical resistivity, which can be a concern for high frequencies and smaller diameter wire applications. Others promote palladium-coated copper wire with reports of improved biased highly accelerated stress test (HAST) results versus bare copper wire. As a consequence, debate over the choice between palladium-coated and bare copper wire is common. Some semiconductor suppliers and original equipment manufacturers (OEMs) incorrectly believe that palladium-coated copper wire is a panacea for all historical concerns with the use of bare copper wire.

A study has been conducted to assess advantages and disadvantages of bare and palladium-coated copper. This paper shows bare copper wire can provide the same level of chlorine-induced corrosion resistance as palladium-coated copper wire if the copper-aluminum intermetallic bond is properly formed and the mold compound is correctly formulated. The basis for the belief that palladium-coated copper wire provides better resistance to chlorine-induced corrosion is explained and the electrical performance difference between palladium-coated and bare copper wires is discussed. High temperature (175°C) storage life testing, up to 7000 hours, was conducted with both wire types to determine the end-of-life failure mechanism. Bond interface cracking, initiating in the bond periphery, was observed. The time dependence of copper-aluminum intermetallic phase transformation for both wire types will be presented. It is shown that choice criteria for each wire type can be defined by product field application requirements and not by perceived advantages. The reported work shows bare copper has equal performance to palladium-coated copper wire under Automotive Electronic Council (AEC) reliability grade 1 in specified package types when the bonding process, substrate / lead frame design and mold compound have been correctly optimized [1].

Introduction

In the past five years, fine gauge ($\leq$ 35µm diameter) copper (Cu) wire has been rapidly replacing fine gold (Au) wire in consumer, commercial, and industrial products, and has recently begun moving into automotive electronics [2-4]. The first wave of Cu wire products used bare, uncoated Cu wire, and soon thereafter palladium-coated Cu (PdCu) wire was introduced to the semiconductor assembly market. This first wave of fine Cu wire components were largely used in consumer products that had to pass lower level reliability testing, normally not requiring voltage-humidity tests (biased HAST, THB), extended temperature cycle (TC) tests, or extended high temperature storage life (HTSL) tests [2,5]. Generally, field returns for consumer electronics is less common than for products with higher reliability requirements, consequently problems with Cu wire bonding were not commonly identified. As Cu wire bond moved into commercial / industrial products with more demanding application environments and reliability stress tests, reports of reliability problems began to appear. Device failures were often attributed to open circuits at the device input-output (IO) pad ("first bonds"), and the most common failure modes were damage in the interlayer dielectric (ILD) stack, and corrosion at the Cu-aluminum (Al) bond interface [2-4].

First bond ILD stack damage occurs when the stress of the wire bonding process causes cracking or delamination of an ILD layer, or results in a high stress at some point in the ILD stack that causes crack formation over time in the use environment. The cracks often result in breaks in conductors in a layer under the ball bond resulting in an open circuit, or less commonly, extrusion of conductor metal through a crack or delamination region, resulting in a short circuit. Cu is harder than Au, and since Cu work hardens much more than Au, the risk of damaging wire bond pad structures is higher with Cu wire bonding. Cu wire bonding requires higher force and ultrasonic energy than Au wire bonding, further increasing the risk of pad damage without properly optimized bonding [4,6,7]. Corrosion of the Cu-Al bond interface was found to occur at the Cu-Al intermetallic (IMC) bond interface. This IMC bond interface has been shown to be comprised of one or more layers of Cu-Al IMC phases, and the most common failures have been shown to be corrosion of the Cu-rich IMC phase in contact with the bottom of the Cu ball bond [2-4,8-9]. Elimination of the ILD pad damage was often accomplished by a combination of a modification of the ball bonding process to reduce mechanical stress on the bond pad and ILD stack and a change in the capillary design, but without careful optimization, this stress reduction could reduce the quality of the IMC layer, increasing risk of corrosion failures.

Palladium-coated Cu (PdCu) wire was introduced into the market at about the time that reports of reliability issues with fine pitch Cu wire bonded parts were causing industry concerns [2,8-13]. Normally, bare Cu wire and the core of PdCu wire are 99.99 wt. % Cu. The Pd coating increases oxidation resistance of the Cu wire, but the PdCu ball bonds are somewhat harder than the ball bonds of the bare Cu wire.

PdCu wire was marketed as the means to resolve corrosion issues, and improve second bonds, and found rapid market acceptance despite a typical 2.5 – 3 times increase in cost versus bare Cu wire. The benefit Pd provides requires the Pd to be uniformly distributed on the surface layer of the ball bond, particularly on the bottom of the ball where contact with the Al pad occurs. Reported data [14] has shown that EFO current and time has a significant effect on the distribution of Pd on the Cu ball, and has also shown that the Pd coating can produce a wider range in the shape of the free air ball formed during the EFO process. The presence of voids in the Cu ball near the bond interface has also been attributed to the presence of the Pd. It has been reported that PdCu wire provides longer life in autoclave, unbiased HAST and biased HAST testing when compared to bare Cu wire [9-12,15].

Copper work hardens more than gold, and this has been identified as a material property that affects the strength of the Cu second bond as measured by wire pull testing at the second bond ('stitch pull'), requiring careful optimization of the Cu second bond. It has been reported that the Pd coating on Cu wire allows formation of good adhesion between the wire and the second bond pad with a lower stress process, resulting in stitch pull strength that may exceed the pull strength of a non-optimized bare Cu wire second bond process [6,16].

Investigations into Cu and PdCu wire bonding advantages and disadvantages are ongoing and new insights into the interaction of fundamental material behavior and bonding process parameters continue to emerge. This paper is intended to provide information from lead frame and organic substrate studies with both bare Cu wire and PdCu wire, and to discuss successful high reliability semiconductor packaging results, and problems overcome, for both wire types in both package platforms.

Experimental Approach:

Samples used for this paper were either 55 or 90 nm wafer technology with Freescale® (Freescale Semiconductor Inc.) BOA (Bond Over Active) structures under the bonding regions which have been described in details in previous papers [2, 4].

Optimizing Palladium Distribution on Bonded Balls

In order to reap the reliability advantage of Pd coating Cu wire on the first ball formation, care must be taken to ensure uniform Palladium distribution on the Free-Air-Ball (FAB). The uniform Palladium distribution on the FAB, specifically on the tip of the FAB, helps protect the more Cu-rich intermetallic compound from being attacked by the mobile halogen ions present in the molding compound under a biased condition.

The Palladium distribution on the FAB was studied by modulating the wire type and the Electro Flame-off (EFO) current. Three 20 µm diameter PdCu wires, A, B and C, from two suppliers were selected for this study. FAB was formed at different EFO current settings ranging from a nominal value (1x mA), to multiples of this value, 2x, 3x, and 4x. EFO time was adjusted accordingly in order to keep the FAB size constant. FAB's were optically inspected for Pd coverage. Figure 1 presents representative optical images of

the FAB's for three PdCu wires at different EFO current settings. Rough surfaces on FABs were observed on all three wires at all EFO current settings. Some level of Cu oxidation, indicated by the copper color at the center of the FAB, was observed on PdCu Wire A at all EFO current settings. PdCu wire B clearly had better Pd coverage, indicated by the gray color of the FAB, from EFO currents ranging from 1x to 3x mA. At 4x mA, Cu exposure was observed. PdCu wire C had slightly less surface roughness than wire B but had a narrower EFO current range with no Cu exposure, from 1x to 2x mA. Figure 2 shows SEM micrographs of the FAB's for the three wire types at 2x mA EFO current. The Scanning Electron Microscope (SEM) images confirmed the higher surface roughness on wire types A and B. Although the surface roughness did not affect the Pd coverage, it could be minimized by not using the Contact Angle parameter in the wire bond recipe.

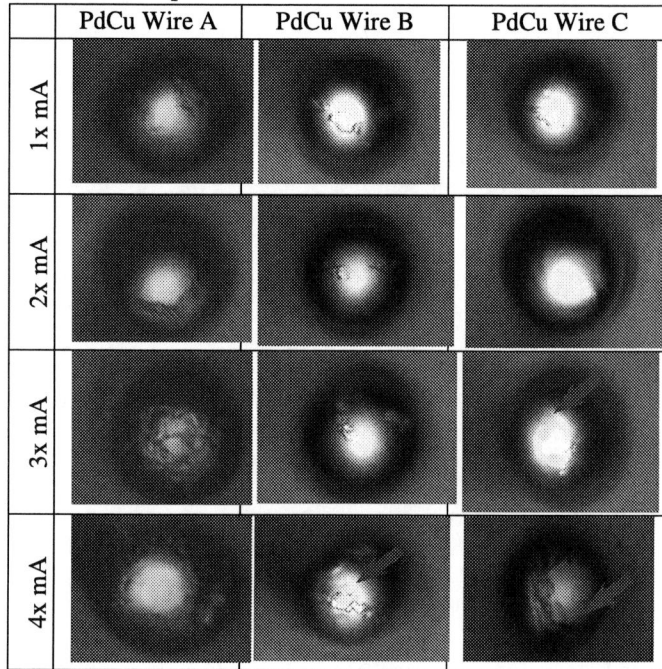

Figure 1. FAB optical pictures at EFO current of 1x, 2x, 3x, and 4x mA.

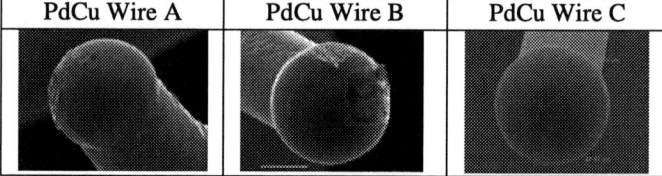

Figure 2. SEM micrographs of FAB's at 2x EFO current

While examining the extent of Pd coverage on the FAB assists wire bond engineers in defining the proper EFO current range, this is not sufficient. The real goal should be achievement of uniform Pd coverage at the foot of the bonded ball. Consequently, the three wire types were wire bonded on Al bond pads. The ball bonds were then removed by etching away the Al pad under the bonds and inspecting for Pd coverage at the bottom of the balls interfacing with the Al pad. Based on the FAB Pd coverage assessment for all three wire types, the EFO current range was narrowed down to 1.67x – 2.33x mA. Figure 3 presents optical images of the foot of the etched bonded balls at EFO current of 1.67x, 2x, 2.27x, 2.33x mA for the three wire types. The Cu colored region indicates exposed Cu whereas the silver colored region indicates Pd coverage. PdCu wire A had full Pd coverage underneath the bonded ball at 1.67x mA, but showed Cu exposure at higher EFO currents. PdCu wires B and C had good Pd coverage at both 1.67x and 2x mA. At higher EFO current, both wires exhibited Cu exposure. Furthermore, closer examination of Pd coverage on PdCu wires B and C at 1.67x and 2x mA revealed a thicker Pd coverage on wire B.

This study illustrated that different Pd coverage can be achieved by changing the PdCu wire type and by optimizing the EFO current setting. In general, lower EFO current setting would produce more uniform Pd coverage on FAB and bonded balls, and consequently higher resistance against chlorine induced corrosion. It is also noteworthy that the EFO current setting for PdCu wire tends to be lower than that of bare Cu or of Au wire.

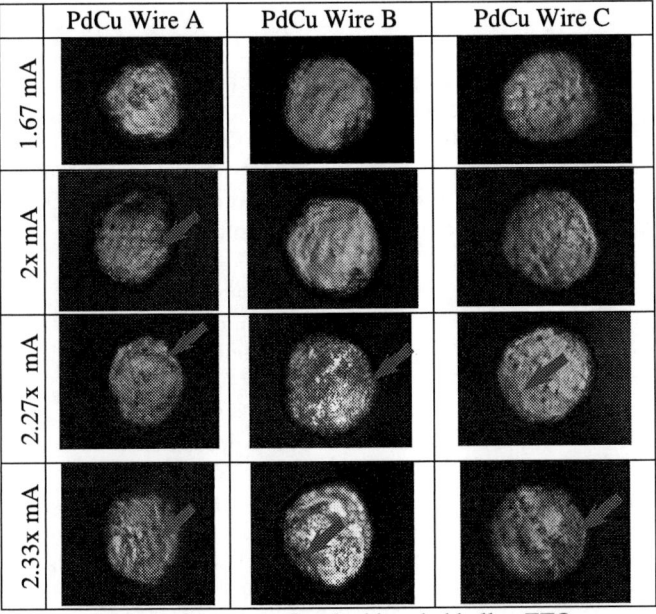

	PdCu Wire A	PdCu Wire B	PdCu Wire C
1.67 mA			
2x mA			
2.27x mA			
2.33x mA			

Figure 3. Optical images of etched bonded ball at EFO currents of 1.67x, 2x, 2.27x, and 2.33x mA.

The effect of FAB Pd coverage on biased HAST performance can be significant. Figure 4 shows an example of one part wire bonded with PdCu wire failing AEC grade 1 biased HAST test (96 hrs at 130°C and 85% RH) when using a conventional EFO setting concept, i.e. high EFO current. Focused Ion Beam (FIB) / SEM images taken from the failed ball bond clearly showed a Cu-Al IMC corrosion layer under the ball bond (Figure 4). Using the same bill of materials (BOM) but with an optimized EFO setting (i.e. low EFO current), a good Pd coverage on the FAB was achieved as proven in Figure 5 optical inspection image. Moreover, Electron Probe Micro-Analysis (EPMA) images taken from bonded balls clearly proved the uniformity of the Pd layer around the bonded ball (Figure 5). Parts with this EFO setting are capable of passing 2x AEC grade 1 biased HAST test (192 hrs at 130°C/85% RH) without showing any sign of Cu-Al IMC corrosion. As shown in Figure 6, FIB/SEM image of an optimized bonded ball at 192 hour HAST test revealed no sign

of IMC corrosion. PdCu wire cannot perform well without a substantial amount of work to obtain an optimized EFO setting ensuring Pd uniformity around FAB's and bonded balls. The requirement of EFO process optimization is less critical for bare Cu than for PdCu wire.

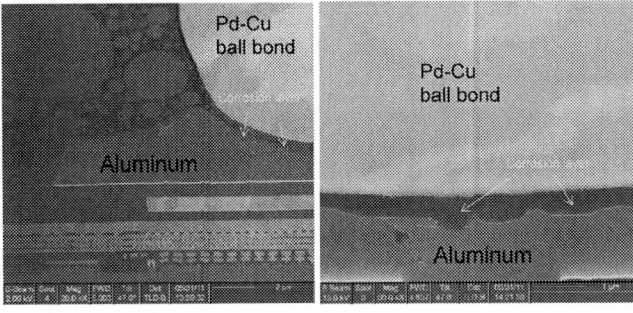

Figure 4. An example of Cu-Al IMC corrosion from PdCu ball bond found in PBGA package in AEC Grade 1 biased HAST (130°C/85%RH) due to non-optimized EFO setting (high EFO current).

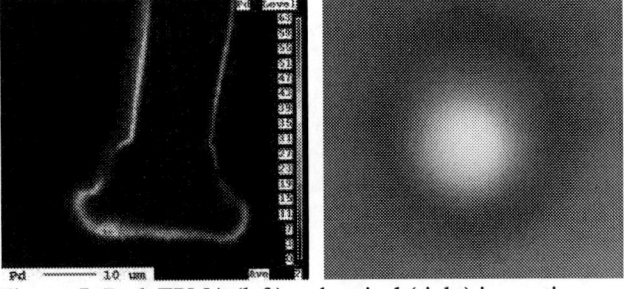

Figure 5: Both EPMA (left) and optical (right) inspection showed uniform Pd distribution around bonded ball and FAB with optimized EFO setting (low EFO current).

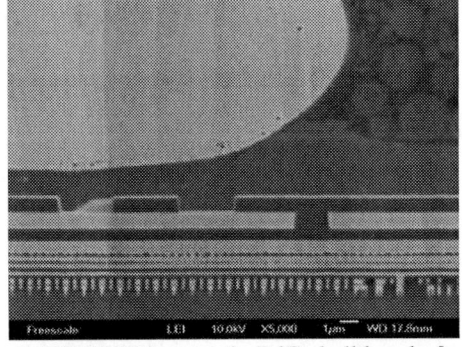

Figure 6: FIB/SEM image of a PdCu ball bond of a part passing 2x AEC grade 1 biased HAST. No Cu-Al IMC corrosion was found.

Cu vs. PdCu Wires in Biased HAST Performance

PdCu wire and mold compound suppliers have proposed that PdCu wire enables parts to survive longer in biased HAST condition than bare Cu wire [10, 13]. That is, the use of PdCu wire can create a better corrosion resistant Cu-Al IMC to chlorine which comes from the epoxy resin [2,17]. Freescale has performed extensive studies on the effects of pH and chlorine content in the mold compound for Cu wire parts in biased HAST performance [2,4,17]. Our work has bare Cu with proper wire bonding parameter optimization to achieve

Freescale proprietary requirements for Cu ball bonds, along with Freescale proprietary formulation of epoxy mold compound, can meet the required biased HAST performance (i.e. AEC grade 1 and grade 0) without the need of using PdCu wire. In the previous section of this paper, poor EFO setting on PdCu wire can lead to biased HAST failure, indicating that the use of PdCu wire may not provide a guaranty of a better biased HAST performance.

A Freescale product qualification provided a good example of a product using bare Cu wires to meet AEC grade 1 biased HAST condition. Automotive AEC grade 1 product is required to be biased at 12 volts in biased HAST test. As shown in Figure 7, when bare Cu wire and a mold compound proven to pass qualification under 6 volt biased HAST conditions were used, biased HAST failure showing typical Cu-Al IMC corrosion under the ball bond leading to open failure were found.

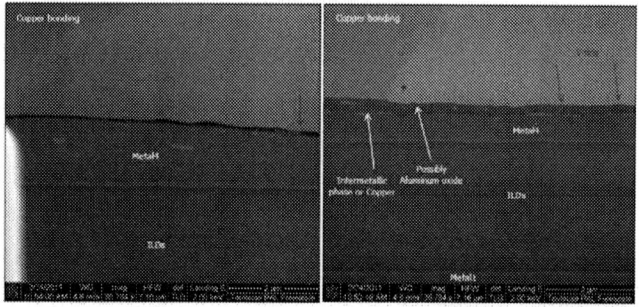

Figure 7: Automotive AEC Grade 1 tier product with biased voltage of 12 Volts failing biased HAST test due to the formation of Cu-Al IMC corrosion layer.

A wire bond DOE was performed to ensure Freescale proprietary acceptability criteria for bare Cu wire bonds were met. The molding compound was also modified meeting Freescale proprietary acceptance criteria for Cu wire. Once these two criteria were met, parts passed 192 hr biased HAST (12 volts, 130°C/85 RH %) without failures (3 lots of 77 units each) using bare Cu wire. It is worth noting that parts using PdCu wires from the DOE with a poor wire bond recipe (i.e. failing Freescale proprietary ball bond requirements) failed 12 volt biased HAST even with the new modified molding compound. Furthermore, Freescale was able to use bare Cu wire with a different molding compound that was reformulated for Cu wire compatibility to demonstrate passing biased HAST tests with biasing voltage of 65 volts. It is concluded that bare Cu wire is capable of passing biased HAST tests assuming that both bonding parameters and mold compound are fully optimized per Freescale proprietary criteria. Thus, the belief that PdCu wire must be used to pass biased HAST qualification requirements may not be correct. Freescale data demonstrated that both wire bond recipe and mold compound selection are the key factors in determining success for a device high biased voltage HAST condition (e.g. 65 Volts), and independent of wire selection (i.e. bare Cu vs. PdCu wire).

Copper Voiding in PdCu Wire

In addition to assessing chlorine induced corrosion resistance of PdCu wire, it is important to ensure good reliability in high temperature bake tests. In order to comply with the requirements of some automotive customers, after high temperature baking packages must pass not only electrical test but also wire pull and ball shear tests after package decapsulation. 20 μm diameter PdCu wire type B described above was wire bonded on Al pads using four EFO current settings of 1.33x, 1.67x, 2.67x, and 3x mA with EFO time adjusted accordingly to achieve the same FAB and squashed ball diameters. One control cell was included with 20μm diameter bare Cu wire bonded using EFO current of 3x mA. Packages were assembled and submitted to high temperature bake (HTB) at 150°C for 1,008 and 2,016 hours, at 175°C for 504 and 1,008 hours, and at 200°C for 288 hours. 2,016 hour bake at 150°C and 1,008 hour bake at 175°C are equivalent per Automotive Electronics Council AEC-Q100 Grade 0 specification. 288 hour bake at 200°C is a condition Freescale uses for a more highly accelerated bake test and is 'equivalent' to AEC grade 0 assuming vacancy diffusion control mechanism for IMC length growth.

All packages passed electrical test at all read-points. Some packages were then decapsulated for wire pull and ball shear tests. Figures 8(a), 8(b) and 8(c) present the wire pull failure mode distributions of the five splits after 2,016 hour bake at 150°C, 1,008 hour bake at 175°C and 288 hour bake at 200°C, respectively. Two primary wire pull failure modes were observed, wire span break (more preferable) and ball lift (i.e. separation of the Cu ball from the intermetallic phases or from Al pad).

For PdCu wire, no ball lifts were found at 1,008 and 2,016 hour HTB-150°C. No ball lifts were found at 504 hour HTB-175°C. However, ball lifts were found at both 1,008 hour HTB-175°C and 288 hour HTB-200°C. This data suggests that for the wire pull ball lift failure mode, the acceleration factor between 150 and 175°C is approximately 4 to 1. Interestingly, the rate of ball lifts decreased with the increasing EFO current setting. Figure 9 presents representative cross sections of the ball bonds of PdCu wire at 3x mA EFO current after 2,016 hours HTB-150°C. Similarly, Figure 10 shows similar cross sections after 2,016 hours HTB-175°C. Copper voiding was found near the ball bond surface in cross-section views when PdCu ball bonds, formed with EFO currents from 1.67x to 3x mA, were baked for 2,016 hours at 175°C. Copper voiding was not found after baking PdCu ball bonds for 1,008 hours at a lower temperature of 150°C. Furthermore, copper voiding was not detected with bare Cu after HTB at any duration or temperature in our evaluation.

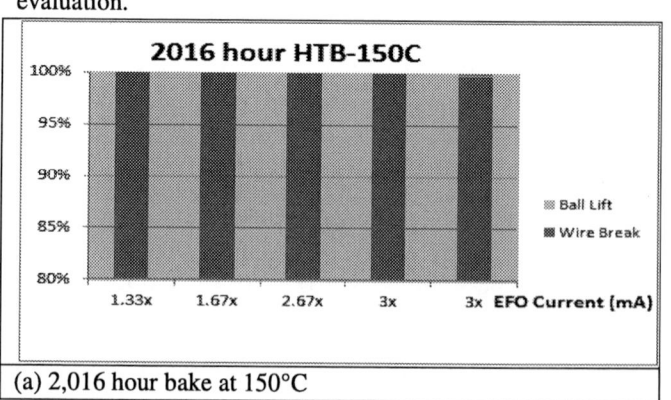

(a) 2,016 hour bake at 150°C

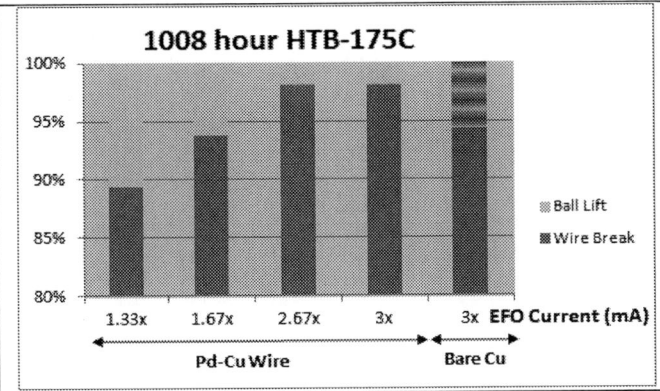

(b) 1,008 hour bake at 175°C

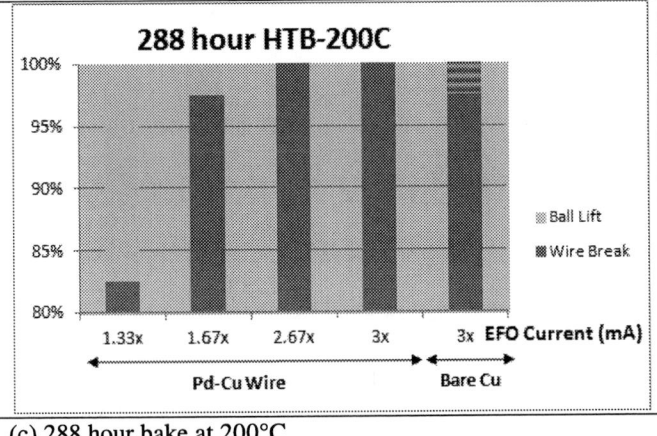

(c) 288 hour bake at 200°C

Figure 8. Wire pull failure mode distribution for PdCu wire at different EFO currents (1.33x, 1.67x, 2.67x and 3x) and bare Cu wire (Control, 3x mA).

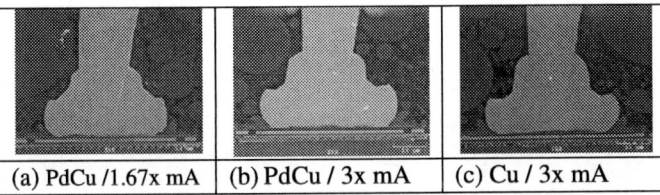

(a) PdCu /1.67x mA | (b) PdCu / 3x mA | (c) Cu / 3x mA

Figure 9. Ball bond cross sections at 2,016 hour HTB-150°C.

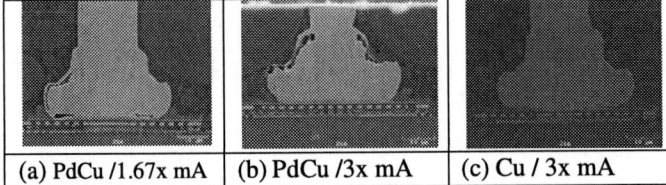

(a) PdCu /1.67x mA | (b) PdCu /3x mA | (c) Cu / 3x mA

Figure 10. Ball bond cross sections at 2,016 hour HTB-175°C.

Figure 11(a) presents a higher magnification of Figure 10(a) focusing on the two Cu voiding regions of the PdCu ball baked at 2,016 hour HTB-175°C. Copper voiding was found at two locations, one close to the capillary chamfer squeezed region on top of the squashed ball, and one at the ball edge interfacing with the intermetallic compound. It is worth noting that the voiding was not found in the Cu-Al intermetallic compound, but was found in the bulk Cu ball above the Cu-Al intermetallic compound. Based on the data

collected, it appears that the Cu voiding above the intermetallic compound reduced in frequency when the EFO current setting increased. Figure 11(b) focused on one Cu voiding area near the chamfer squeezed region on top of the ball bond. Figures 11(c), 11(d) and 11(e) display the EDX spectra taken in the Cu voiding region (location 1), and in the Cu bulk area (locations 2 and 3), respectively. In the Cu voiding area, only Cu and O was detected. In the bulk Cu region, Palladium was detected in location 2 but not location 3.

It is well known that the Pd coverage around the FAB and squashed ball is not uniform or even contiguous. It is therefore possible that in the thinned down Pd region, the bulk Cu is exposed. With large corrosion potential between Pd and Cu, and Cu being at the lower potential, galvanic corrosion would occur preferentially to the more active element (Cu) in a conducting electrolyte. When the dimension of the active area (called the anode, in this case copper) is substantially smaller than the dimension of the noble material (called the cathode, in this case Pd), the Cu corrosion occurs even faster. This might explain the high incidence of Cu voiding around the chamfer squeezed region on top of the ball bond due to high friction experienced by the top of the ball bond under the chamfer region of the capillary. The chamfer squeeze effect is also transferred to the ball edge interfacing the aluminum pad. As presented earlier, increasing the EFO current increases the Cu area exposure at the foot of the bonded ball. The larger Cu area effectively helped reduce the corrosion rate for the bulk Cu above the intermetallic compound. In fact, by increasing EFO current, the bulk Cu voiding at the intermetallic interface reduced, thus lowering the rate of wire pull ball lifts (referring to Figure 8).

The fundamental driving force of this galvanic corrosion in high temperature bake condition for PdCu wire is still being investigated. Although increasing EFO current helped lower the ball lift rate after HTB-175°C, lower EFO current is still preferred because of better Pd coverage and better intermetallic corrosion resistance in biased condition. Further wire bond and mold compound re-optimization helped minimize both intermetallic corrosion and bulk copper corrosion when lower EFO current was used. The specific details of this process optimization are beyond the scope of this publication.

This type of galvanic corrosion should not happen with bare Cu wire if the galvanic cell (i.e. PdCu) theory is correct. In order to prove that, the bare Cu wire parts were baked for extended hours followed by cross section and SEM investigation. Figure 12 shows a cross section of bare copper ball bond from part baked at 175°C for 13,000 hrs and no sign of Cu voiding inside the copper ball bond was observed.

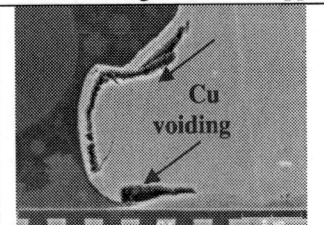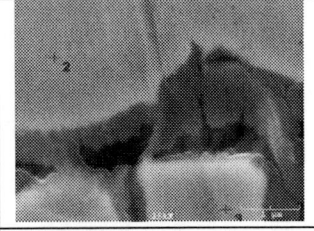

(a) Higher magnification of Cu voiding. | (b) EDX locations of Cu voiding.

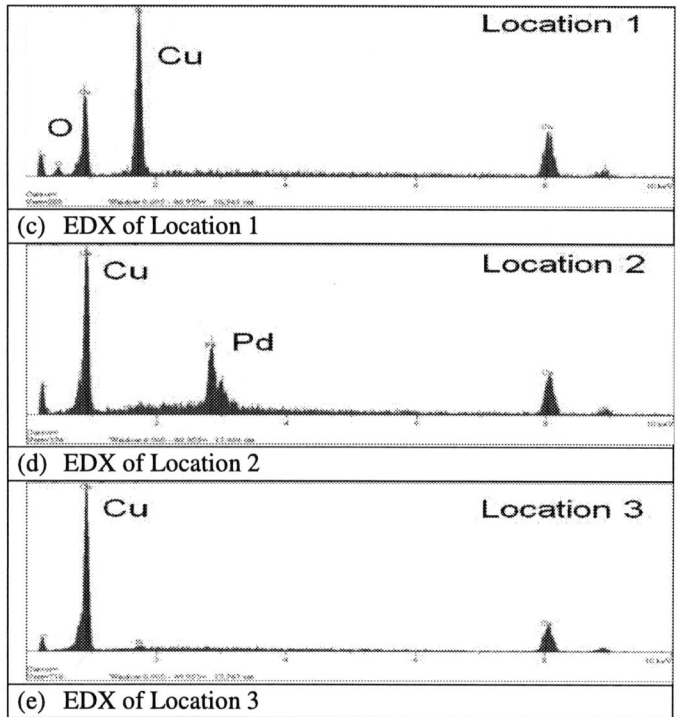

(c) EDX of Location 1

(d) EDX of Location 2

(e) EDX of Location 3

Figure 11. FIB/SEM image and EDX analysis of Cu voiding.

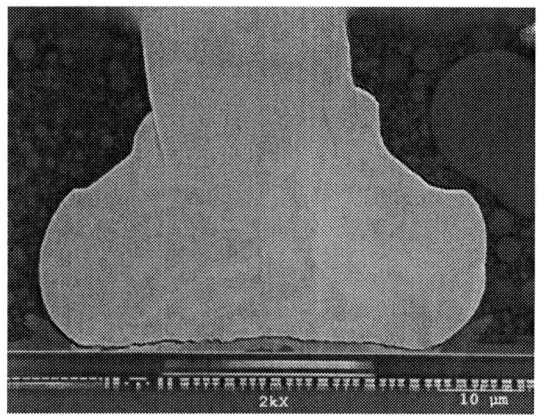

Figure 12. Cross section SEM picture on bare Cu ball bond that has been baked for 13,098 hours at 175°C. Note that there is no Cu voiding around ball bond surface. Cracking at the Cu – IMC interface will be discussed later in this paper.

Although bare Cu did not have bulk Cu voiding, Figure 8 shows that bare Cu had ball lifts after wire pull after HTB-150°C, HTB-175°C and HTB-200°C. In fact, bare Cu wire experienced higher ball lift rate than the PdCu cell running at the same 3x mA EFO current. Figure 13 shows a higher magnification view of the Cu-Al intermetallic compound on the ball edge of PdCu and bare Cu wires after an accelerated 288 hour HTB-200°C. A small crack initiating from the ball edge between the Cu-Al intermetallic compound and the Cu ball bonds was found on the bare Cu sample. This interfacial crack was not observed, or was much smaller on PdCu balls. The IMC thickness was much lower on PdCu ball bonds than on bare Cu ball bonds. Unlike the PdCu balls that experienced ball lifts due to bulk Cu voiding, bare Cu balls had ball lifts due to the interfacial crack causing the Cu ball bond to lift from the intermetallic compound during wire pull testing.

These two types of failure modes are considered cosmetic and did not lead to electrical failures during stress testing.

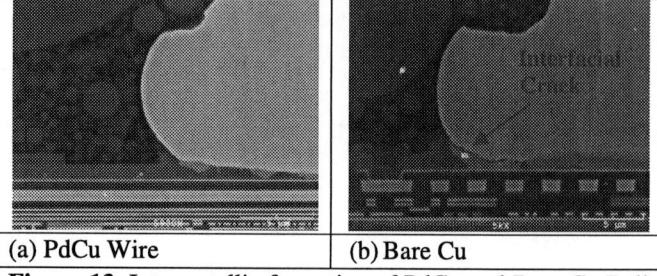

(a) PdCu Wire (b) Bare Cu

Figure 13. Intermetallic formation of PdCu and Bare Cu Ball Bonds at 288 hour HTB-200°C. PdCu does not show any edge ball bond crack while bare copper does.

Extended High Temperature Bake Study

In order to further assess long-term reliability concern of the interfacial crack formed between Cu-Al IMC and Cu ball bond, both PdCu and bare Cu parts were baked at 150 and 175°C. Table 1 summarizes high temperature bake results for 23 µm bare Cu wire devices.

Table 1: Summary of extended high temp bake results using 23 µm bare Cu wire. E-test result is reported as number of failure per total testing sample, e.g. 11/40 means 11 failures out of total of 40 tested units.

Temp (C)	Hours	E-Test (# failure)	Sample Group	Interfacial Crack	Note
85	12,720	0/120	A	No data*	
125	12,888	0/120	B	No data*	
150	1,008	0/40	C	None	
150	2,016	0/40	C	Length: < 20% BBD Location: Edge only	Fig 16 &17
150	3,024	0/40	C	Length: < 20% BBD Location: Edge only	Fig 18
150	4,032	11/40	C	No data*	
150	5,040	67/495	D	No data*	
150	10,080	92/179	E	Length: 100% BBD	
175	1,008	0/45	F	Length: < 5% BBD (estimated from 504 hr data) Location: Edge only	Fig 15
175	2,016	0/45	F	Length: < 20% BBD Location: Edge only	Fig 16
175	8,000	11/45	F	Length: 100% BBD Based on 5000 hr (175C) data	Fig 19
175	13,098	33/45**	F	Length: 100% BBD Based on 5000 hr/175C data	Fig 19
175	14,098	45/45**	F	Length: 100% BBD Based on 5000 hr/175C data	Fig 19

* Cross-section analysis on ball bonds was not performed.
** Failure units were not taken out of chamber and continued going for the next read point.

Units passed HTB at electrical test at 150°C for 3,024 hours, or 175°C for 2,016 hours. The earliest failure was seen at the 150°C 4,032 hours test point. Some parts passed testing after 13,098 hours at 175°C. The point of failure is far beyond the most demanding AEC grade 0 requirements. Therefore, it was determined the peripheral interfacial crack is not a long-term reliability concern.

Moreover, a comparison study was made between the Cu-Al and Au-Al systems. Table 2 summarizes electrical results of the same device using Au and Cu wires baked at 150℃ for different durations. Au wire parts showed failures at 2,016 hours due to the formation of Kirkendall voids. Cu wire parts consistently passed HTB 150℃ 2,016 hours [2]. Bare Cu wire bonding greatly out-performed Au wire bonding in this extended testing, clearly demonstrating that Cu wire is much more reliable than Au wire in high temperature environments required in some automotive applications.

Table 2. A comparison between Cu and Au wire parts (same device with same mold compound) after high temperature bake tests (150°C). Reporting number of sample failure in the below table, e.g. 4/30 means 4 failures out of 30 units in the test).

HTB	Size	504hrs	1008hrs	1512hrs	2016hrs
Cu Lot 1	80	0/80	0/80	0/80	0/80
	80	0/80	0/80	0/80	0/80
Cu Lot 2	80	0/80	0/80	0/80	0/80
	80	0/80	0/80	0/80	0/80
Cu Lot 3	80	0/80	0/80	0/80	0/80
	80	0/80	0/80	0/80	0/80
Au Lot 1	30	0/30	0/30	0/30	4/30
	30	0/30	0/30	0/30	4/30
Au Lot 2	30	0/30	0/30	0/30	1/30
	30	0/30	0/30	0/30	0/30
Au Lot 3	30	0/30	0/29	0/29	0/29
	30	0/29	0/29	0/29	1/29

The AEC Q100 standard specifies the equivalency of HTB at 175°C/504 hours and HTB at 150°C/1,008 hours, by allowing the use of either condition for reliability testing. The peripheral interfacial cracks shown in Figure 13 were found to initiate from the edge of the ball bond much earlier when baked at 175°C than at 150°C for the AEC-specified exposure times. These data indicate that the testing conditions may not be equivalent for this failure mode, with crack initiation being accelerated much more quickly with the 175°C bake than with the longer term bake at 150°C.

Figure 14 shows a PdCu wire part baked at 150°C for 1,008 hrs with no sign of interfacial cracking at the edge of the ball bond. Figure 15 displays a PdCu wire part baked at 175°C for 504 hrs where a visible interfacial crack was observed at the edge of the ball bond. It is proposed that the cracking difference is due to the development of different Cu-Al IMC phases and IMC phase layer thicknesses at 1,008 hour bake at 150℃ and 504 hour bake at 175℃. This suggests that this interfacial cracking is related to Cu-Al IMC phase transformations and associated volume changes at the interface. Previous work has reported a similar phenomenon and concluded the same crack mechanism occurred [4].

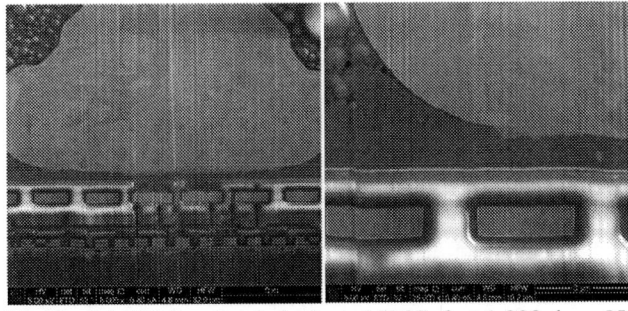

Figure 14. PdCu part baked at 150℃ for 1,008 hrs. No interfacial crack was observed at the edge of ball bond.

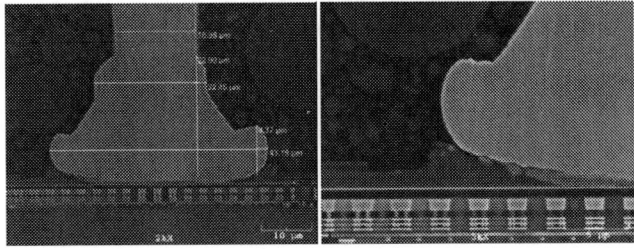

Figure 15. PdCu wire bond after HTB 175°C/504 hrs. A minute peripheral interfacial crack was observed.

As noted earlier, with the same HTB time, the PdCu ball bond has a smaller interfacial crack than does bare copper ball bond. Figure 16 shows a comparsion between bare Cu and PdCu ball bonds baked at 175°C for 2,016 hrs where the PdCu ball bond has a characteristically smaller interfacial crack than does the bare Cu ball bond. It is believed that the Pd coating around the ball bond as seen in Figure 5 reduced the rate of IMC formation. This may have changed the Cu-Al IMC phase transformation rate the delaying formation of interfacial cracks believed to be due to volume change associated with IMC phase transformation during high temperautre bake.

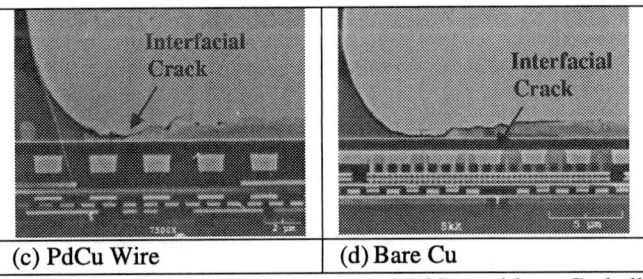

(c) PdCu Wire	(d) Bare Cu

Figure 16. Intermetallic formation of PdCu and bare Cu ball bonds with HTB 175°C 2,016 hours, and cracking at the edge of ball bonds.

Figure 17 shows a bare copper ball bond baked at 150°C for 2,016 hours, where interfacial cracking was observed, but cracking is so small that there is no reliabillity concern. Figure 18 indicates that after 3,024 hours at 150°C the ball bond should pass electrical tests because the crack is very small and is limited to the bond periphery.

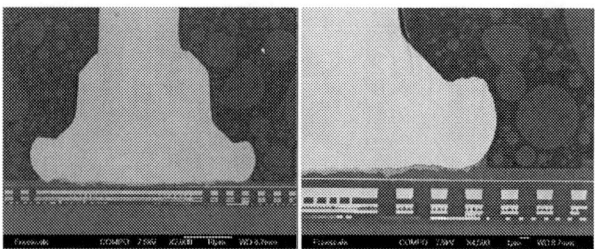

Figure 17. Bare Cu part baked at 150℃ for 2,016 hrs. Interfacial crack was observed at the ball bond edge.

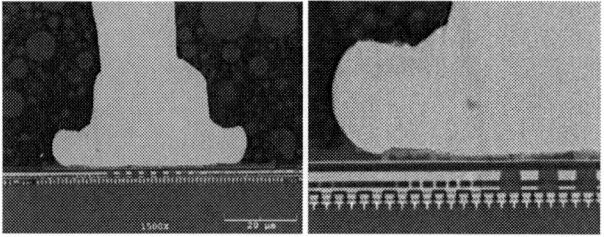

Figure 18. Bare Cu ball bond baked at 150°C for 3,024 hrs. Very small interfacial cracking, initiating from the ball bond edge, was observed. No electrical failures occurred.

Figure 19 shows a bare Cu ball bond baked at 175°C for 5,068 hours. A crack line across the entire bond interface was found. This type of crack formation is likely to explain the first failures after HTB 150°C 4,032 hrs as shown in Table 1. Figure 20 is cross sectional bare copper ball bond baked for 13,098 hours and a crack line was found across the entire ball bond as well.

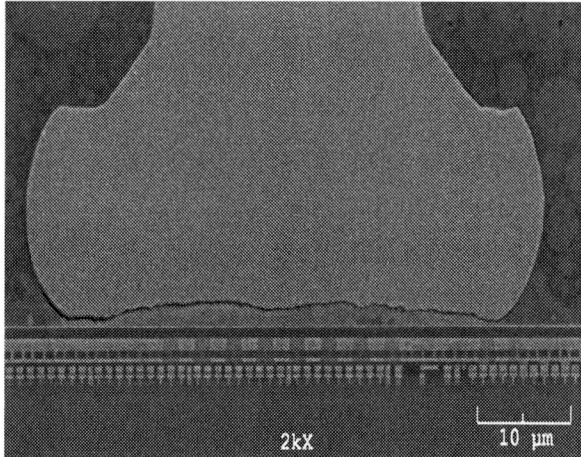

Figure 19: Bare copper ball bond baked at 175C for 5068 hrs. The interfacial crack propagated the ball bond.

Based on observations, it is believed that the high temperature bake failure for Cu-Al system is due to the formation of this type of interfacial crack. This is not the same as with the Au-Al system, where failure is due to the formation of Kirkendall voids. This interfacial crack line in the Cu-Al system is believed not to be the Kirkendall voiding based on two facts:

- The crack initiated from the ball bond edge, and propagated smoothly toward the center of the ball bond and did not form randomly along the interface. If these cracks were due to Kirkendall voids, it would have been

formed as a diffusion controlled process and the crack would not be expected to be so systematically formed along the interface between Cu and Cu-Al IMC.

- The boundary line between Cu and Cu-Al IMC remains the same from time zero to after high temperature bake exposure time. Figure 20 clearly shows that the interfacial line between Cu and Cu-Al IMC does not change. It indicates the crack line is not diffusion control, i.e. not Kirkendall voids.

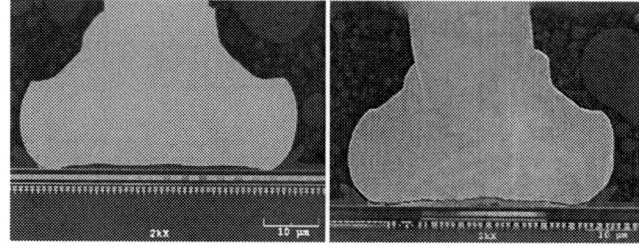

Figure 20. A comparison between a time zero unit (left) and HTB 175°C/13,098 hr unit (right). The interfacial line between Cu and Cu-Al IMC remains unchanged throughout the entire 13,098 hrs at 175°C.

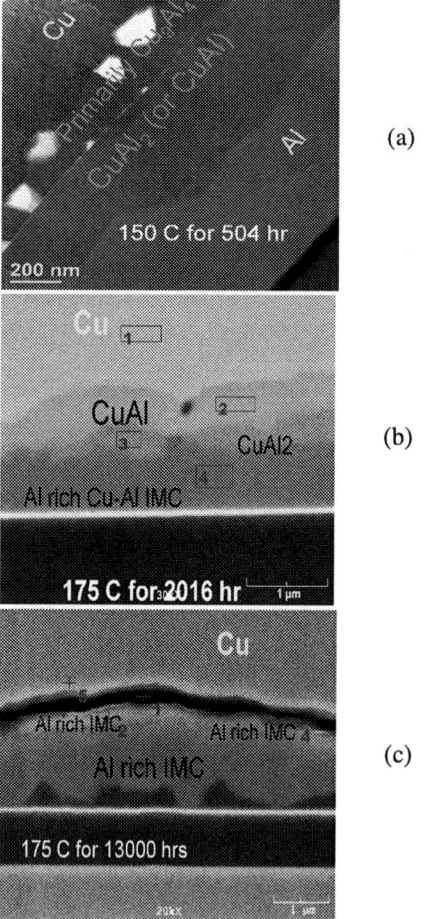

Figure 21. Cu-Al IMC composition has been estimated: (a) 150°C for 504 hr and the Cu-Al IMC composition is confirmed by TEM [2]; (b) 175°C for 2,016 hr; (c) 175°C for 13,000 hrs. The Cu-Al IMC composition is only estimated by SEM/EDX and is semi-quantitative due to the limitation of EDX.

IMC Phase Transformation Study

Figure 21 shows Cu-Al IMC compositions at various baking temperatures and times. The IMC is bi-layer Cu_9Al_4/CuAl after baking at 150°C for 504 hours, as seen in the Transmission Electron Microscope (TEM) image. As the baking time increases, the majority of IMC transforms to the Al rich phase based on the semi-quantitative nature of the EDX spectra from such small phase regions. The study of Cu-Al IMC compositions aims to calculate the theoretical volumetric change from Cu rich phase to Al rich phase in order to prove the mechanism of forming the interfacial crack line. As discussed above, this crack line should not be Kirkendall void mechanism controlled. Since a comprehensive TEM study has not been performed, the accurate IMC composition cannot be provided solely based on EDX spectra.

Electrical Performance

Although not identical, the electrical characteristics of Cu wire and PdCu are similar. Published resistivity values for typical Cu and PdCu wire are both 1.9 μΩ·cm [2,18], or ~10% higher than the 1.7 μΩ·cm value of pure, bulk copper [19]. The resistivity of Pd is more than six times that of Cu [19]. However, the Pd coating is very thin on PdCu wire: approximately 80nm, or 0.5 wt. % of even the finest-pitch, 0.6-mil diameter wire (15 μm). In this extreme example, the Pd coating constitutes only 0.01% of the wire's cross-sectional area, and therefore has a negligible impact on the DC resistance.

At sufficiently high frequencies, however, the electromagnetic *skin effect* causes conduction to occur predominantly near a wire's surface. This effect is common to all good conductors (including Cu, Au, and Pd). At a *skin depth*, δ, beneath the surface, the current density is attenuated by a factor of $1/e$ (0.368). The material property, δ also depends on the frequency of the electrical signal, f. At 100 GHz, for example, Cu has a skin depth of only 210 nm [20]. That is, most of the current flows in the outer 200-300 nm of wire. The 80 nm Pd coating constitutes a significant fraction of this depth. One therefore expects poorer electrical performance from PdCu relative to bare Cu at frequencies above 100 GHz.

Fortunately, most products with bond wires operate well below 100 GHz. The skin depth varies as $1/\sqrt{f}$, and increases to 2.1 μm as the frequency decreases to 1 GHz [20]. This skin depth is more than 20 times the Pd thickness. Below 1 GHz, therefore, the Pd coating can be ignored for electrical resistance. Likewise, electrical inductance and capacitance are not affected by the Pd coating. Both of these parameters depend on the distribution of charge and current in the wires, which are not affected by the Pd coating at frequencies below 1 GHz. The Pd coating therefore has no direct effect on electrical performance below 1 GHz.

An indirect effect comes into play, however, when manufacturers choose smaller wire diameters for PdCu because of its greater hardness relative to bare Cu wire. During wire bonding, PdCu's hardness may require a smaller wire diameter to avoid damage to the bond pad or underlying circuitry. Simulations reveal small changes (<5%) in inductance and capacitance when reducing wire diameter.

However, the increase in resistance can be significant; 20 μm PdCu has 26% greater resistance than 23 μm Cu wire, for the same wire length. In such cases, PdCu can result in poorer electrical performance relative to Cu wire due to the smaller wire diameter chosen for mechanical properties.

High Volume Manufacturing Concern

PdCu does provide a longer factory floor life than bare Cu. Bare Cu wire floor life is less than a week while PdCu wire floor life can be as long as 30 days based on several vendors' recommendations. However, Freescale has conducted a series of studies on floor life for PdCu wire and concluded that the 2nd bond defect rate increases as the actual floor life PdCu wire approaches the floor life limit. It is therefore better to have a shorter floor life for PdCu wire than the floor life recommended by wire vendors.

Second Bond Concern

PdCu wire can produce higher 2nd bond peel strength more easily with less process optimization work than that required for bare Cu due to PdCu wire's higher mechanical strength. In general, PdCu wire has a lower risk of 'fish tail' defect (a high volume manufacturing defect) than bare Cu wire [15,21,22]. However, a reliable 2nd bond can be readily achieved and maintained in high volume manufacturing using bare Cu wire if wire bonding parameters are fully optimized with a proper capillary.

Summary and Conclusion

- Multiple comparative studies have shown bare Cu wire can provide the same level of chlorine-induced corrosion resistance as palladium-coated copper wire if the copper-aluminum intermetallic bond is properly formed and the mold compound is correctly formulated.

- High temperature storage life testing (150 and 175°C), up to 14,098 hours, conducted with both wire types determined that the end-of-life failure mechanism is not Kirkendall voiding controlled. An interfacial crack initiated at the bond periphery was observed and is believed to be due to volumetric change when Cu-Al IMC phases transform from Cu rich phase(s) (Cu_9Al_4) to Al rich phases (CuAl or $CuAl_2$).

- Interfacial crack line propagation and IMC growth are slower with Pd-coated Cu wire than with bare Cu wire under extended high temperature storage life test. The presence of Pd under the ball bond slows down Cu-Al IMC formation even though Pd element may not be easily detected by the FIB cross section/SEM/EDX technique. PdCu wire can be more reliable at the extended high temperature storage life test than bare Cu wire.

- The boundary line between the Cu-Al IMC and the Cu ball remains the same and does not move as the IMC layer grows and transforms. All IMC growth is downward (i.e. into the Al bond pad). This is opposite to the Au-Al system where the IMC can grow toward the Au ball bond.

- Cu voiding issue observed at the chamfer squeezed region on top of the squashed ball bond and at the interface to the IMC is galvanized corrosion controlled

978-1-4799-2408-0/14 $31.00 © 2014 IEEE   1547

when using PdCu wire and should not happen to bare Cu ball bonds. However, this type of Cu voiding should be considered as cosmetic and will not be a long term reliability concern.

- IC component qualifications with bare Cu wire have passed 3 times AEC Grade 1 high temperature storage life test (150°C for > 3,000 hours). A portion of sample lots pass 175°C for 13,098 hrs electrically.

- The small increase in electrical resistivity of PdCu wire is not expected to significantly impact electrical performance for high frequency applications.

- PdCu wire holds advantage over bare copper for higher 2nd bond peel strength. When having difficulty of achieving higher 2nd bond peel strength, PdCu wire would be the choice.

Future Work

A Weibull distribution chart based on the data in Table 1 should be generated to calculate the acceleration factor. Cu-Al IMC composition for extended stress test (Figure 21) needs to be confirmed by TEM in order to have a better understanding of time dependence of the Cu-Al IMC system for high temperature application. The mechanism for Cu – IMC crack initiation and progression needs further study to determine the effects of external stress, IMC phase transformation stress and volume changes, interfacial oxidation, and other possible factors.

Acknowledgments

The authors would like to thank the following people for their support in their respective areas of expertise: Y.K. Au for his wire bonding process expertise; Mike Ascerno and Toan Trinh for providing package reliability stressing and electrical test support; Anne Anderson, Alvin Youngblood and Roy Arldt for performing package cross-sectioning, SEM, FIB and EDX analyses; Les Postlethwait and Arthur Green for package de-processing and performing ball shear and wire pull tests. Thanks also to Dr. Sam Subramanian for performing the TEM.

References

1. http://www.aecouncil.com/AECDocuments.html
2. Lee, C., Higgins, L., "Challenges of Cu Wire Bonding on Low-k/Cu Wafers with BOA Structures", *Proc. 60th Electronic Components & Technology Conference 2010*, pp. 342 – 349.
3. Hiew, P. F., *et al.*, "Development and Qualification of Copper Wire Bond Process for Automotive Applications", *Proc. 14th Int'l Conference on Electronic Materials and Packaging 2012*.
4. Tran Tu.Anh., et al "Copper wire bonding on Low-k/ copper wafers with bond over active structures for automotive customers ", Proc. ECTC 2011, pp.1508-1515.
5. Hang C.J, et al "Growth behavior of Cu/Al intermetallic compound and cracks in copper ball bonds during isothermal aging", Microelectronics Rel 48 (2008), pp. 416-424.
6. Bing An, *et al*, "Improvement of the Second Bond Strength in Copper Wire Bonding on Pre-Plated Leadframe", *2011 International Conference on Electronic Packaging Technology & High Density Packaging*, pp. 391 – 394.
7. Yow Kai Yun, Eu Poh Leng, "Cu Wire Bond Reliability Improvement Through Focused Heat Treatment after Bonding", *33rd IEEE/CPMT Int'l Electronic Manufacturing Technology Symposium (IEMT) 2008*.
8. Boettcher, T., *et al*, "On the Intermetallic Corrosion of Cu-Al wire bonds", *12th Electronics Packaging & Technology Conference 2010*, pp. 585 – 590.
9. Tang, L. J., et al, "Investigation of Palladium Distribution on the Free Air Ball of Pd-coated Cu Wire", 12th Electronics Packaging Technology Conference 2010, pp. 777 – 782.
10. Uno, T., Yamada, T., "Improving Humidity Bond Reliability of Copper Bonding Wires", *Proc. 2010 Electronic Components and Technology Conference*, pp. 1725 – 1732.
11. Lim, A. B. Y., *et al.*, "Palladium-coated and Bare Copper Wire Study for Ultra-Fine Pitch Wire Bonding", Electro-Chemical Society (ECS) Transactions, 52 (1) 717-730 (2013).
12. Carson, F., *et al.*, "Study of Pd Mixing During PdCu Wire Ball Formation and Impact on Wire Bond Quality", Proc. 62nd Electronic Components & Technology Conference 2012, pp. 1939 – 1944.
13. Abe, H., et al, "Cu wire and PdCu wire package reliability and molding compounds", Proc. 62nd ETCT2012 , pp.1117-1123
14. Pequegnat, A., *et al*., "Effect of EFO parameters on Cu FAB hardness and work hardening in thermosonic wire bonding", *J Mater Sci: Mater Electron*, DOI 10.1007/s10854-008-9841-8, 2008.
15. Yauw, O., *et al*, "Wire Bonding Optimization with Fine Copper Wire for Volume Production", *12th Electronics Packaging Technology Conference 2010*, pp. 467 – 472.
16. Kaimori, S., *et al*, "Development of Hybrid Bonding Wire", *SEI Technical Review*, Number 63, pp. 14 – 18, Dec. 2006.
17. Mathew V., et al, "Copper wirebond compatibility with organic and inorganic ion present in mold compounds", 46th IMAPS 2013, TA35, pp.89 - 93.
18. T. Uno, K. Kimura, and T. Yamada, "Surface-Enhanced Copper Bonding Wire for LSI and Its Bond Reliability Under Humid Environment," in *Proceeding of Microelectronics and Packaging Conference*, 2009.
19. D.R. Lide, *Handbook of Chemistry and Physics*, 71st ed. Boca Raton, FL: CRC Press, 1990, pp. 12-23 – 12-24.
20. P. Lorrain and D. Corson, *Electromagnetic Fields and Waves*, 2nd ed. New York: W.H. Freeman and Co, 1970, p. 476-478.
21. Siliconnfareast.com, "Wirebonding Process", http: http://www.siliconfareast.com/wirebond.htm
22. MEPTEC website (Microelectronics Packaging and Test Engineering Council), Tanaka report: by Bud Crockett "Introduction of Ag Alloy Bonding Wire", MEPTEC Luncheon Presentation, 4/10/2013. http://www.meptec.org/Resources/TANAKA_Meptec%202013_Final.pdf

Improving the Bond Quality of Copper Wire Bonds Using a Friction Model Approach

Simon Althoff, Jan Neuhaus, Tobias Hemsel, Walter Sextro
University of Paderborn
Pohlweg 47-49, 33098 Paderborn, Germany
E-mail: simon.althoff@upb.de
Phone: +49(521)601809; Fax: +49(521)601803

Abstract

In order to increase mechanical strength, heat dissipation and ampacity and to decrease failure through fatigue fracture, wedge copper wire bonding is being introduced as a standard interconnection method for mass production [1].

To achieve the same process stability when using copper wire instead of aluminum wire a profound understanding of the bonding process is needed. Due to the higher hardness of copper compared to aluminum wire it is more difficult to approach the surfaces of wire and substrate to a level where van der Waals forces are able to arise between atoms. Also, enough friction energy referred to the total contact area has to be generated to activate the surfaces.

Therefore, a friction model is used to simulate the joining process. This model calculates the resulting energy of partial areas in the contact surface and provides information about the adhesion process of each area. The focus here is on the arising of micro joints in the contact area depending on the location in the contact and time. To validate the model, different touchdown forces are used to vary the initial contact areas of wire and substrate. Additionally, a piezoelectric tri-axial force sensor is built up to identify the known phases of pre-deforming, cleaning, adhering and diffusing for the real bonding process to map with the model.

Test substrates as DBC and copper plate are used to show the different formations of a wedge bond connection due to hardness and reaction propensity. The experiments were done by using 500 µm copper wire and a standard V-groove tool.

Model conception for wire bonding

Current model concepts divide the bonding process into four phases: (I) Touchdown/pre-deformation [2], (II) Sticking friction (III) Slipping friction/surface activation and (IV) diffusion phase [3,4].

Before the tool starts vibrating, the wire is pressed with a defined force and velocity on the substrate. Hereby, the wire is pressed into the V-groove of the tool and an initial contact area between wire and substrate is created. The touchdown velocity has an influence on the size of this contact area. A high touchdown velocity (e.g. 25mm/s) leads to an overshoot of normal force (up to 30%) and therefore to a higher pre-deformation of the wire. Also the hardness of the wire and substrate after the touchdown depends on the touchdown force and velocity. This hardness change is described in [5]. Both contact bodies increase their hardness in and next to the contact area by up to 40% for typical touchdown forces.

After powering the sonotrode, the tool tip starts to vibrate and excites the wire. Depending on the bonding parameter (here: normal bonding force and ultrasonic voltage amplitude) the short sticking friction phase starts and is followed by the slipping friction phase. Due to this, the surfaces grind their oxide layers and start to strain the contact near parts of the bodies. This results in an activation of surfaces.

While activating the contact bodies, the ultrasonic softening effect begins to soften the materials and reduces the yield points. This effect is responsible for the fact that after smoothing the asperities caused by sliding, the residual ripples of both bodies penetrate into each other. If the contact partners are approached sufficiently and enough activation energy per contact area is obtained, atomic bonds can be built up. These micro welds start to fix the wire while it is still plastically deformed [6]. A data driven model to describe the ultrasonic softening effect was built up in [7] and its dependency of bonding parameters was investigated.

After a specific bonding time, the wire deformation saturates and residual stresses in the remained stationary contact area are resolved and result in a higher connection quality.

Friction Model Approach

To understand the ultrasonic joining process, a detailed model of the described processes carried out is needed. Based on the need of relative motion between wire and substrate, the developed model describes the frictional behavior between the two. The core of the model is a discretization of the contact area between wire and substrate into Point Contact Elements [8]. These elements can be used to calculate the frictional behavior of partial areas. Figure 1 shows Point Contact Elements representing partial areas of the contact.

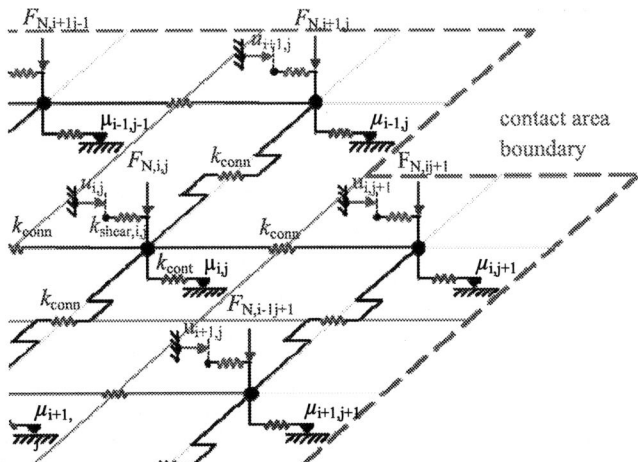

Figure 1. Friction model consisting of single Point Contact Elements in the elastic contact layer between wire and substrate

In the first phase the wire deformation caused by the touchdown has to be determined. This is realized by calculating the wire deformation using Finite Element Method (FEM). This calculation delivers the contact area and its normal pressure distribution with reference to material properties and the used geometries (e.g. the tool geometry). This area is used for the first discretization and determines the initial amount of point contact elements.

In the following phases the wire deformation increases due to the ultrasonic softening effect. The contact areas and normal pressure distributions for different heights must also be calculated using FEM. As a consequence of the time variant wire deformation, the normal forces of the elements have to be updated corresponding to the wire deformation velocity. The amount and position of Point Contact Elements has to be updated as well. This data will be used by the single Point Contact Elements to calculate the occurring friction energy.

One massless Point Contact Element consists of three different types of springs. The stiffnesses are displayed for element i,j in figure 1. The movement of the system is realized by applying the excitation u of the tool tip to the upper spring which represents the shear stiffness (k_{shear}) of the wire. The spring stiffness depends on the wire material used but changes with the change in wire height. This can be achieved by coupling the shear stiffness to the deformation of the wire. With decreasing wire height, the stiffness increases.

The contact stiffness spring (k_{cont}) represents the elasticity of the thin contact layer between the wire and the substrate. This stiffness must be determined for each material combination by experiments. This is done by deflecting the wire while measuring the frictional force. The contact stiffness can be identified using the hysteresis loop and the known material properties.

Additionally, all point contact elements are connected to each other which prevent the elements from moving independently and showing correct elastic behavior. This behavior is realized by adding the connection spring stiffness (k_{conn}) between neighbor elements. Otherwise arising micro joints would not have any influence on the movement of the neighbor elements. The connection stiffness is dependent on the discretization level and the material properties. A higher or lower level and therefore a larger or smaller number of connection springs must always represent the same behavior.

To identify if point contact elements are sticking or slipping, the sticking force of each element has to be calculated. This is done by using coulomb friction law. The normal force for each Point Contact Element ($F_{N,i,j}$) is given by FEM. To simulate a binding between wire and substrate, the friction coefficient of each Point Contact Element is dependent on the frictional energy that each element accumulates. Elements not sliding will not generate friction energy and therefore will not adhere to the substrate. Detailed information about the model is described in [9].

Test aims and test description

Using this model helps to understand the formation of wire connections. This leads to purpuseful bonding parameter adjustments producing reliable copper bond connections. In addition, helpful geometry changes of the bond tool are imaginable for special applications.

The aims of the tests are to validate the described bonding model and to understand the influence of different bonding parameters. Only source bonds are investigated here. For all tests, 500 µm copper wire (Heraeus –PowerCu) and a Hesse Mechatronics BJ 939 wire bonder with ribbon/heavy wire bondhead is used.

For all tests, simple bonding parameters are chosen to identify the dependencies easily (see table 1).

In particular, the touchdown force and therefore the size of the initial contact area are the research topic of this investigations. To demonstrate the effect of pre-deformation of the wire, bond connections are produced using touchdown forces starting at 5 N up to 45 N with an increment of 5 N. The touchdown velocity is limited to 0.01 m/s to avoid the previously mentioned force overshoots. The bond qualities are referred to the initial contact area. Bond quality is defined here as adhesiveness between wire and substrate and can be evaluated through shear tests.

Furthermore, the bond process is interrupted after specific points in time to track the formation of the heavy wire bond connection. This gives information about the influence of the pre-deformation on the bonding process when the effect is particularly strong. Shear forces are measured if the source bond is not lifted off and the print of the contact area or the shear area are inspected. To cover a wide range of touchdown forces, 5, 25 and 45 N are chosen for this experiment.

To validate the bond model also for different substrates, DBC (250 µm copper each side, 250 µm Al_2O_3, 57 HV copper) and normal copper plate (1000 µm thickness, 91 HV) are used. The activation energy to develop a bond connection on a DBC is lower compared to copper plate.

Touch-down speed	Touch-down force	Bonding normal force	Ultrasonic voltage amplitude (ramp)	Bonding time
0.01 m/s	5-45 N	30 N	50 V (20 ms)	300 ms

Table 1. Bonding parameters

Another difference between both materials is that the

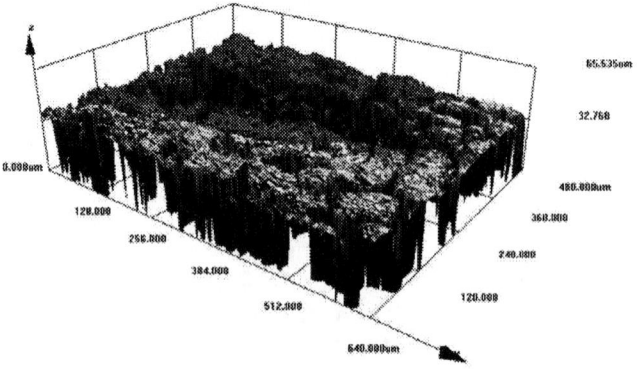

Figure 2. Depth of indentation of 500 µm PowerCu wire into DBC material using a touchdown force of 45 N

Copper plate deforms significantly less compared to the soft copper layers of the DBC and so the wire is plastically more deformed using harder substrate. The plastic deformation after touchdown using 45 N touchdown force is shown in figure 2. The depth here is about 30 μm, whereas the impression on copper plate is within the surface mean roughness (> 5 μm).

The same bonding parameters shown in table 1 are used for DBC substrate and copper plate. Based on the strong influence of the substrate to the results, the experiments are divided by their usage.

Formation of bond connections using DBC substrate

To identify the subjection between initial contact area and bond quality on DBC substrate, shear tests are done with different touchdown forces after a bonding duration of 300 ms. The results are shown in figure 3. No influence or trend can be identified from these data. Therefore, either other effects are covering the influence or no dependency exists.

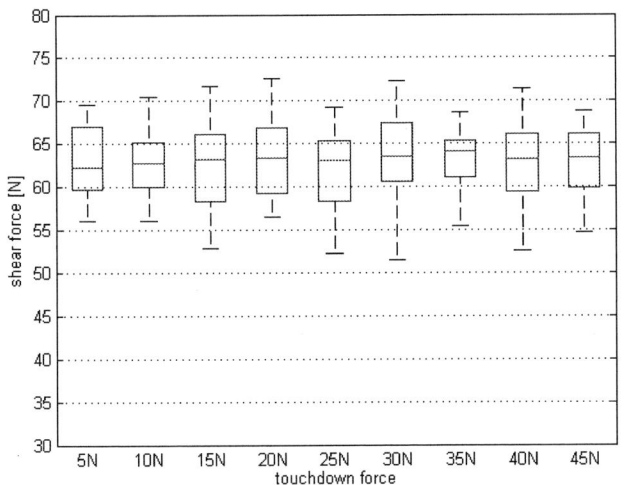

Figure 3. Shear test results of DBC bonds with different touchdown forces (each 50 source bonds tested)

As described before, the development of a bond connection can be shown chronologically by interrupting the bond process after specific points in time. Figure 4 shows the climb of shear values by time setting 5 N touchdown force. Before 75 ms, no source bond is adhering to the DBC. In the first 200 ms, the step size is 25 ms. For good measure, the bonding time was extended to 400 ms to cover a large parameter space.

Using 5 N touchdown force and the standard bonding parameters, only one source bond (out of 50 samples) remains sticking on the substrate after 75 ms and following loop shaping. The yield losses for all touchdown forces using a DBC substrate can be seen in table 2. After 100 ms, no losses can be detected. The source bonds lifted off are not included in the evaluation of the shear values for all measurements.

After 300 ms, the increase of shear values flattens and most probably the diffusion phase begins. At this point, the internal stresses are reduced and the micro structure transformation is continuing.

DBC substrate		Touchdown force [N]		
		5	25	45
Time [ms]	50	100%	100%	100%
	75	94%	14%	6%
	100	4%	0%	0%

Table 2. Yield loss dependent on touchdown force and time for DBC substrate

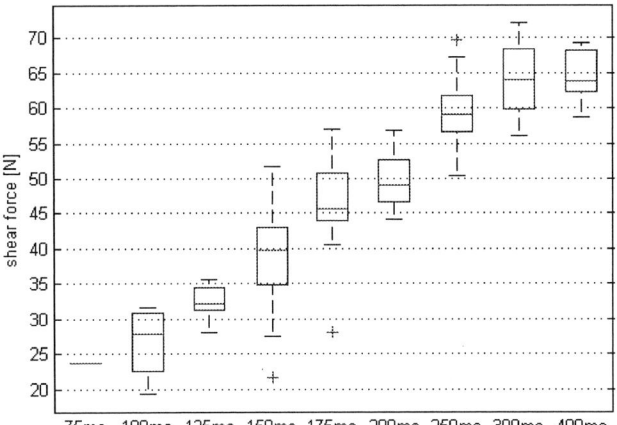

Figure 4. Development of shear strength using 500cN touchdown force and a DBC substrate

Additional information is provided by the time-dependent formation of the bond areas corresponding to the initial plasticization shown in figure 5.

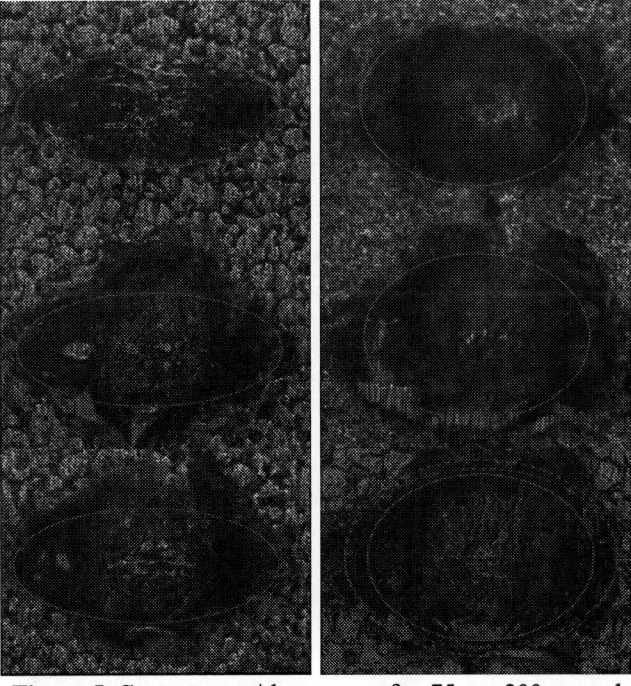

Figure 5. Contact areas/shear areas after 75 ms, 200 ms and 400 ms using 5 N (left) and 45 N (right) touchdown forces on DBC substrate

The connection/contact prints after 75 ms are clearly different. The connections using 5 N touchdown forces form a long drawn elliptical contact which does not show a deep dent in the DBC and is small compared to the 45 N touchdown area. Each initial contact area is displayed in its chronological sequence to show the growth of the connection. The final contact area is marked with dashed edging. In the lower right image both representative final areas are compared. The size, the shape and the location of micro welds can hardly be distinguished in the end.

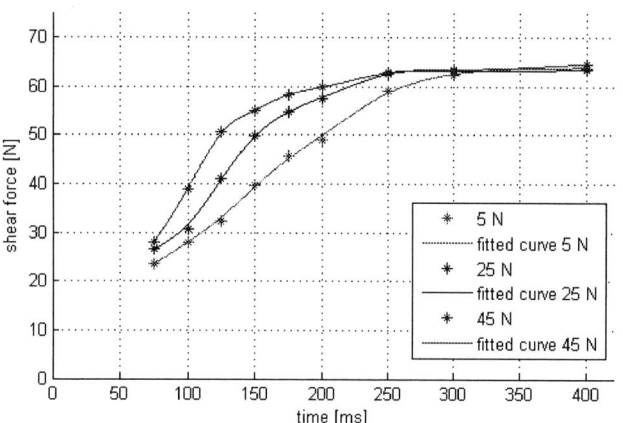

Figure 6. Mean shear values over time using different touchdown forces and DBC substrate

In figure 6, the time dependent mean shear force for all three touchdown forces is displayed. It is noticeable that the higher the touchdown force, the faster the connection formation develops. After 300 ms, the shear forces are on the same level and do not increase with additional bonding time. This is also the case in the chronological contact area development discussed before, where the final contact areas are nearly the same. This, of course, results in similar shear forces. Obviously, higher pre-deformation reduces the activation energy, the bond connection can develop faster and additionally partial contact areas coming into contact, based on the wire deformation, can become welded.

Formation of bond connections using copper plate substrate

In contrast to DBC bond experiments with different touchdown forces, a trend in maximum shear forces for source bonds on copper plate with a bonding time of 300 ms can be identified. Figure 7 shows decreasing shear values for increasing touchdown forces. This may be caused by the fact that the initial contact area for low touchdown forces is smaller and the friction energy is allocated to a smaller contact area. Harder substrates are suspected to require high activation energy to become activated. To generate high friction energy per area using the standard bond parameter, the initial contact area should be as small as possible.

\ Another important fact is that apart from the higher hardness of the substrate, the copper wire and the substrate become even stiffer caused by the high deformation degrees of their bodies. This additionally increases the required energy for developing micro joints.

The chronological development of the shear force values for copper plate substrate and low touchdown force in figure 8 combined with mean shear value development for all touchdown forces in figure 9 confirms the previous

consideration. First of all, the slope of shear force increase is low and the saturation starts at around 400 ms. The duration while micro joints arise is in proportion to other substrates long. This is based on the material properties of the rolled and drawn copper plate and the high deformation degree of the wire.

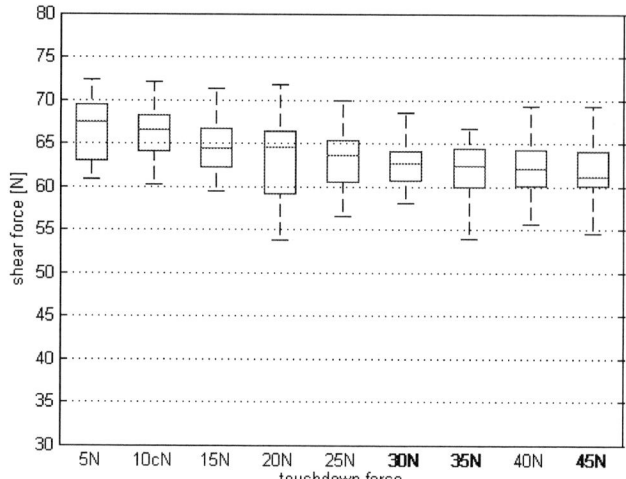

Figure 7. Shear test results of copper plate bonds with different touchdown forces (each 50 source bonds tested)

The source bonds which are less pre-deformed (5 N touchdown force) adhere first to the copper plate (after 75 ms, figure 8) with the smallest amount of lift offs. The yield losses are listed in table 3. The higher pre-deformed bonds need additional time to adhere to the substrate and then show higher yield losses compared to less deformed source bonds.

The curves of the achieved shear forces over time in figure 9 show the early adhering of the bonds with 5 N and 25 N touchdown forces. In contrast to the curves using DBC substrate, the bonds adhering later cannot reach the high bond quality of the early adhering bonds. This implies that the mechanically connected parts of contact areas for low touchdown forces are higher whereas the contact area is smaller.

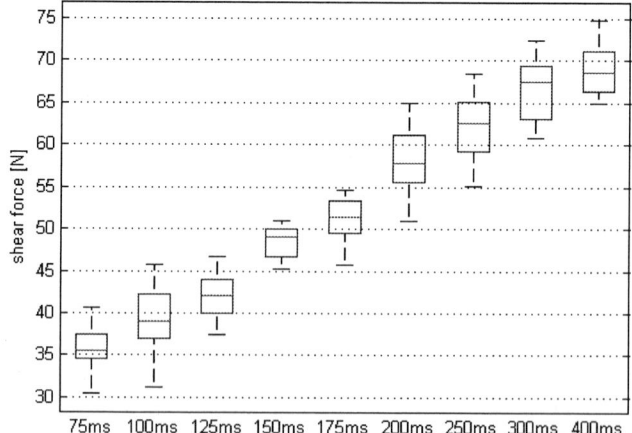

Figure 8. Connection development over time using 5 N touchdown force and copper plate

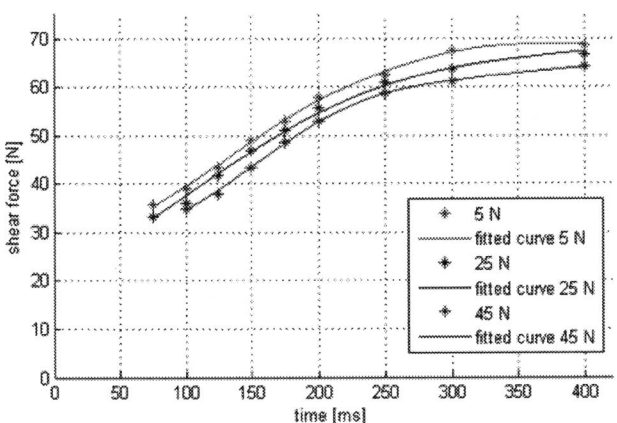

Figure 9. Mean shear values over time using different touchdown forces and copper plate substrate

Copper plate	Touchdown force [N]		
	5	25	45
Time [ms] 50	100%	100%	100%
75	16%	42%	100%
100	0%	0%	10%

Table 3. Yield loss dependent on touchdown force and time for copper plate

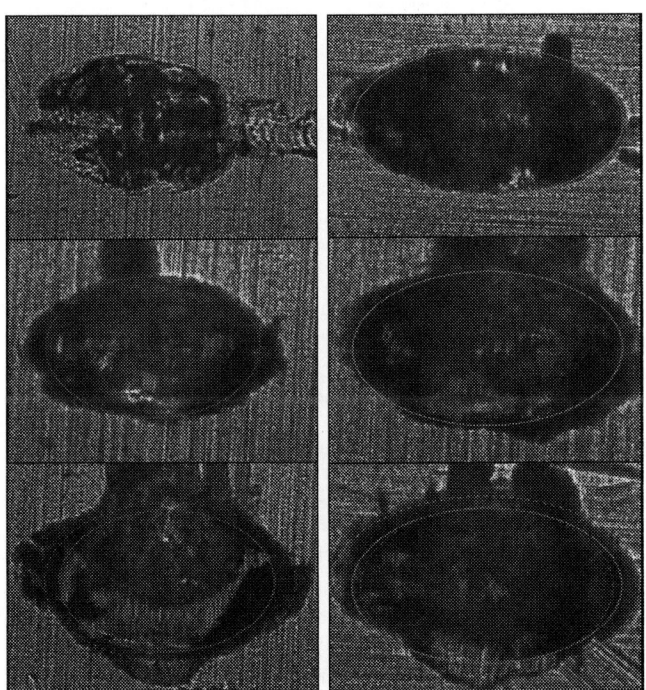

Figure 10. Contact areas/shear areas after 75 ms, 200 ms and 400 ms using 5 N (left) and 45 N (right) touchdown forces on copper plate substrate

The chronological contact areas/shear areas in figure 10 of the low and high pre-deformed contact area substantiate this expectation. The initial contact areas vary explicitly. The lower right image shows the initial contact area for 45 N touchdown force in the continuous blue edging. The dashed edgings are again the final contact area after 400 ms. It is striking that the final contact area of low pre-deformed bonds

after 400 ms is as small as the contact area of high pre-deformed bond connections after 75 ms. The quotient of the mechanically connected area to the total contact area is much higher for low pre-deformed bonds.

Validation of the model concept by using a tri-axial force sensor

To validate the wedge bonding model and the conclusions drawn in this paper, a piezoelectric tri-axial force sensor is built up. This accesses the opportunity to get detailed information about the friction process by measuring its forces. To collect all information, the normal force and the tangential force in oscillation direction and orthogonal to it is measured. The principle construction concept is shown in figure 11.

Figure 11. Construction concept for tri-axial force sensor

Each of the three piezoelectric ceramics (size: 16 by 16 mm) is conducted from top and bottom with copper electrodes. These electrodes are joined to the ceramic coating with special glue. This glue is conductive and shows a very high elasticity module after being baked out. Different glue layer thicknesses between electrode and ceramic are compensated by using stiff polymer layers. These layers can be milled off to get an even surface without tilt angle. Finally, all three conducted sensors are glued together to get the tri-axial sensor. The piezo ceramics, polymer layers and electrodes have a hole in the center to install a vacuum suction. Hereby, the substrate with the same dimensions as the sensor area can be clamped to the sensor.

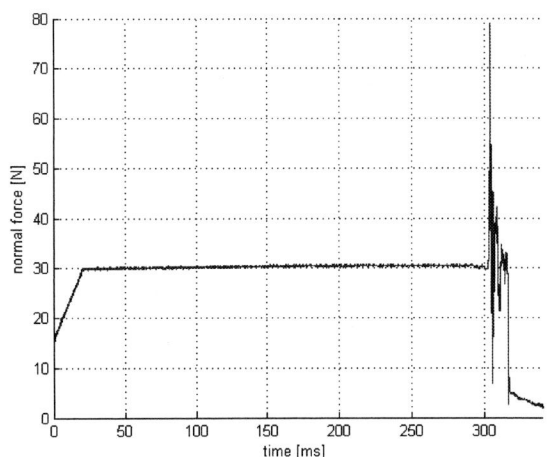

Figure 12. normal force measured while bonding with standard bonding parameters and a touchdown force of 15 N

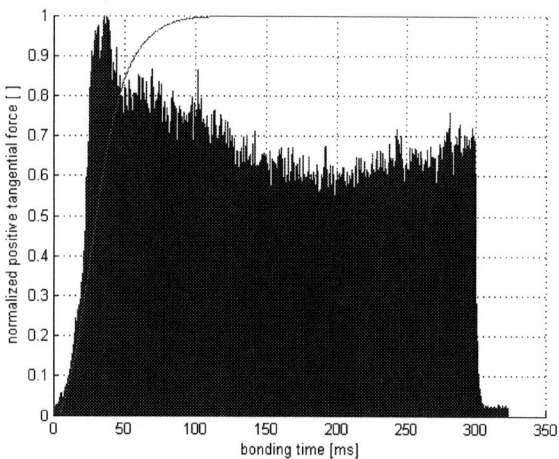

Figure 13. Positive tangential force signal of the ceramic polarized in oscillation direction and the simulated normalized force development (red line)

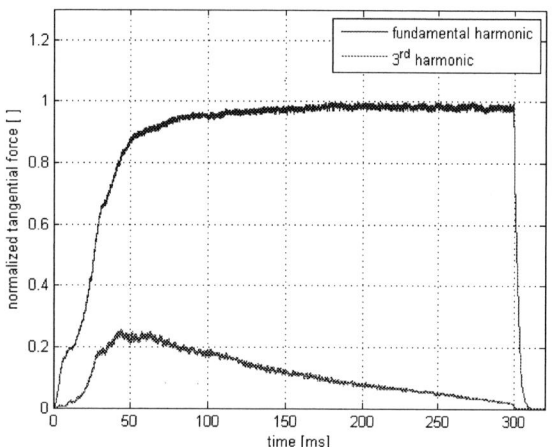

Figure 14. Normalized force signal separated into amplitudes of the existing frequencies

Each sensor is linked to a charge amplifier which transforms the charge produced by the bond process into voltage. These voltages are analyzed by their amplitudes and after transforming the signals into the frequency domain, the spectrum is analyzed. This procedure gives information about the phase the bonding process is located.

The normal force sensor installed at the bottom of the sensor stack delivers the signal in figure 12 for touchdown forces of 15 N and the standard bonding parameters listed in table 1. The distinctive force peak after 300 ms is the force applied to the sensor while cutting the wire. The complete signal is filtered to wipe out the high frequency signal noise. The sensor signals follow the normal forces set in the bonding machine very precisely.

The piezoelectric ceramic located in the middle of the sensor is measuring the forces perpendicular to the oscillation direction. As expected, no considerable signals can be measured here.

The upper ceramic measures the force in oscillation direction and therefore provides most information. Figure 13

shows the positive non-filtered amplitude of the force signal. The negative amplitude is nearly congruent. The force amplitude starts rising in the first 100 ms up to 80% of its maximum value and in additional 100 ms the saturation is reached. Splitting the signal into the amplitude of the basic harmonic and its higher harmonics leads to figure 14. Here, it can be seen that the amplitudes of higher harmonics are relatively large in the first 100-150 ms and disappear with time if the total force signal saturates. Higher harmonics exist in case of slipping friction [10]. Thus, it can be concluded that the wire and the substrate have clear relative motion to each other in the first half of the bonding process. The additional contact area enlargement, caused by the ongoing wire deformation may induce micro slipping, but this does not increase the welded area.

Conclusions

In this paper, two investigations are carried out which, in the beginning, appear independent but together validate the friction model approach.

Using two different copper substrates and varying the touchdown force leads to the knowledge, that depending on the material hardness, a proper touchdown force has to be selected. Especially in case of hard surfaces, smaller initial contact areas should be selected which means lower touchdown forces should be set. Smaller contact areas in combination to the same bonding normal force result in a pronounced leveling of the asperities. The smaller the remaining roughness is, the easier is the approaching of the contact bodies. This approaching is necessary to develop adhesion between bodies.

In case of soft substrate material (compared to the wire), a high touchdown force results also in a marked plastic deformation of the substrate and herewith a clear approach is done. Using small touchdown forces as shown in the example using the DBC and 5 N touchdown force, the bodies do not approach as much as necessary to develop micro joints (high yield loss after 75ms). In this case, the asperities of the surface roughness are touching the wire and the real area of contact is low.

This case can be simulated by using a discrete model for calculating the contact of rough surfaces based on [11]. This model is independent from the friction model presented here and focuses on a more microscopic scale. No tangential movement of grinding is considered here. This model explicitly points out that the approaching of two bodies is dependent on their material behavior and the applied normal force. In the example shown in figure 15, the wire is assumed to be smooth and the roughness of the copper plate is characterized by $R_z = 2\mu m$. The lower line represents the roughness of the substrate and the blue area approaching represents the wire. Using the wire material properties, 5N touchdown force and the size of the contact area results in the penetration seen in the upper figure. The wire is just touching the asperities of the substrate. The actual area of contact is very low. At the time the ultrasonic softening starts in the real bonding process, the bodies can approach much more and the real area of contact increases. Now, the surfaces can more extensively react in case of activation.

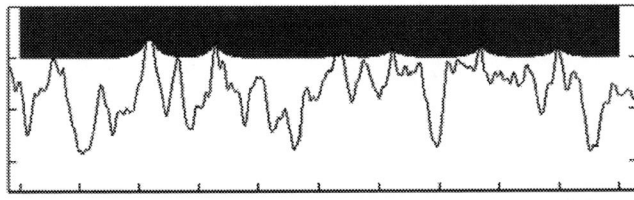

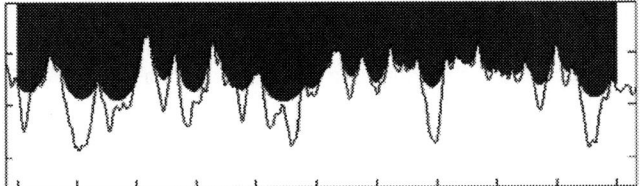

Figure 15. Penetration of roughness without and with ultrasonic softening effect

To sum up, in the first phase of the bonding process high plasticization progresses the approach of the contact partners by far. In case of little approach after the touchdown the wire has to be softened first in the second phase to get to the same level of approach. The experiments also show that a high pre-deformation is not expedient in case the surfaces needing high activation energy and therefore a high friction energy per area has to be created. The touchdown force has to be selected depending on the contact partner showing higher hardness.

The friction model approach can map this by adapting the function between friction energy and adhering level to the touchdown force. Also the variation of the initial contact area can be simulated easily by using FE calculations. These calculations also provide the normal force distribution which is needed to parameterize the Point Contact Elements which result from discretizing the contact areas. The increase of adhesion in the second and third phase can be calculated using this procedure.

The experiments using the tri-axial force sensor are important to identify the processes in the second and third phase. The sensor offers the possibility to divide time of sticking, slipping friction and complete adhering of the contact. In addition, it can be seen that the majority of the resulting shear force is developed in the time while slipping friction in the contact can be measured and the wire has its highest slope in wire deformation.

In the last phase of the bond process, the diffusion phase, additional bonding time is used to generate micro structural changes in and next to the welding area. Based on the measurements to identify the shear forces depending on the bonding time, it can be concluded that the additional 100 ms after 300 ms of previous bonding result in just a slight increase in the shear force. In case there is no micro slipping in the contact, the model cannot consider this small increase which is based on microstructural changes in and next to the contact area. This small increase in bond quality measured through shear force may have bigger impact on the quality regarding fatigue strength [12]. This has to be investigated and in this case the diffusion properties have to be a function of time.

References

1. D. Siepe, "The Future of Wire Bonding is? Wire Bonding!" in Proc. 6th International Conference on Integrated Power Electronics Systems (CIPS), Nurnberg, Germany, 2010, March 16-18, Nuremberg (Germany)

2. H. Gaul, „Berechnung der Verbindungsqualität beim Ultraschall Wedge/Wedge-Bonden",Ph.D. dissertation, Dept. Electric Eng. , Technische Universität Berlin, Berlin, Germany 2009.

3. F. Osterwald, "Verbindungsbildung beim Ultraschall-Drahtbonden/Einfluß der Schwingungsparameter und Modellvorstellungen", Ph.D.dissertation, Dept. Electric. Eng., Technische Univ. Berlin, Berlin, Germany,1999.

4. W. Budweiser, "Untersuchung des Thermosonic-Ballbondverfahrens", Ph.D. dissertation, Dept. Electric. Eng., Technische Universität Berlin,Berlin, Germany, 1992.

5. F. Eacock, "Mikrostrukturuntersuchungen an Al-und Cu-Bonddrähten", diploma thesis, Dept. Mech. Eng., University of Paderborn, Paderborn, Germany, 2013

6. Seppänen, Henri, "Real time contact resistance measurement to determine when microwelds start to form during ultrasonic wire bonding." Microelectronic Engineerin,g vol.104, pp. 114-119, 2013.

7. A. Unger, "Data-driven Modeling of the Ultrasonic Softening Effekt for Robust Copper Wire Bonding", in Proc. 6th International Conference on Integrated Power Electronics Systems (CIPS), ETG-Fachbericht, Nuremberg (Germany), February 25-27,2014.

8. W. Sextro, Dynamical contact problems with friction: models, methods, experiments and applications. Springer, 2007, pp. 136-139.

9. S. Althoff, , "A friction based approach for modeling wire bonding", in Proc. 46th International Symposium on Microelectronics, Orlando, Fl, October 1-3, 2013

10. Shah, A, "Ultrasonic friction power during thermosonic Au and Cu ball bonding," J. Phys. D: Appl. Phys, Vol. 43, No. 32, 2010, 325301.

11. J. Neuhaus, "A Discrete 2D Model for Dynamical Contact of Rough Surfaces", in Proc. 3rd International Conference on Computational Contact Mechanics (ICCCM),Lecce (Italy), July 10-12, 2013

12. J. Goehre, "Influence of bonding parameters on the reliability of heavy wire bonds on power semiconductors", in Proc. 5th International Conference on Power Electronics Systems (CIPS), ETG-Fachbericht, Nuremberg (Germany), March 6-8, 2012, pp. 279-290

High Aspect Ratio Lithography for Litho-Defined Wire Bonding

Zahra Kolahdouz Esfahani, Henk van Zeijl, and G. Q. Zhang

Delft Institute of Microsystems and Nanoelectronics (DIMES), Delft University of Technology,

Feldmannweg 17, 2628 CT Delft, The Netherlands

Abstract:

To overcome some challenges and reliability issues of wire bonding, a study of a new interconnect technique is done .High aspect ratio interconnect called "litho defined wire bonding" can be a promising method to resolve these problems in many of the applications. Using contact aligner and resist multilayer spray-coating over high steps up to 150 µm, we are able to achieve AL interconnect lines with different parameters. Further, using multi step imaging on a ASM PAS 5500/80 projection aligner a resolution down to 2 µm on the bottom of a 150 µm deep cavity was attained. With simulation based on aerial image calculation, different multi step imaging schemes are studied to get higher resolution over topographies.

Introduction:

As a mature technology wire bonding is still the dominant interconnection technique in spite of the rapid developments in chip and wafer level packaging technology because of its high reliability. The process of connecting metal bond pads to substrate leads can be done by different techniques. Despite of the kind of wire bonding, the bond forming process is performed at a relatively high temperature which helps the welding process [1], [2]. However, the increased process temperature of wire bonding may leads to oxidation of the bonding pad, and mechanical issues like stress into the Si active regions. Besides there are various challenges in wire bonding for multi-domain packaging like temperature limitations, deep access capability and limited aspect ratio and bonding on sensitive devices and over cavity and cantilever leads [3]–[6].

Different challenges and reliability issues with conventional interconnect technologies especially with wire bonding, introduce an increasing needs for developing new I/O interconnect techniques. Two of the major drives are: 1) the fragile nature of low-k dielectrics and their relatively poor adhesion to the surrounding materials that makes it critical to minimize the mechanical stresses on the chip. 2) Wire bonding difficulties to obtain fine interconnections with high density.

One of the novel technique recently used for some limited areas is "Lithographic defined wire bonding," which includes the deposition of metal film and then patterning a "wire-like" trace through high aspect ratio lithography. The proposed interconnect method may be deployed where wire bonding, flip-chip (face-down) and TSV interconnections are difficult to carry out. Figure 1 represent schematically an integrated system using this technique.

This technique promises some potential such as:
- Better interconnection reliability
  - Low temperature process compare to typical wire bonding
  - Less stress introduced into the Si active region and mechanical issues
- Higher interconnect density (applicable for RF application)
- H.A.R interconnect with TSV performance
- On chip passive component
- Appropriate for LED SiP and 3D integration applications

Recently some different groups proposed this approach [7]–[12]. In [7] by Joung and Allen, a Cu-electroplating-bonded S-shaped flexible interconnect system was used for chip-to-board interconnection. It was shown that as-fabricated electroplated interconnection was able to withstand the harsh environment and endured the high operating temperature of 153°C. However, in [7], the electrical connection is achieved by an additional structure called polymer core post that is coated with the metal layers

Figure 1: Schematic of system with litho-defined wire bonds.

Figure 2: Using SU8, 20 µm wide lines, 20 µm apart, 60 µm DRIE etched channel [11].

978-1-4799-2408-0/14 $31.00 © 2014 IEEE   2014 Electronic Components & Technology Conference

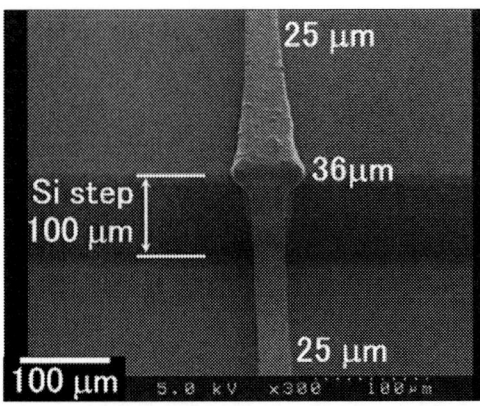

Figure 3: Cu high-step-coverage lateral interconnections over a 100 µm thick chip [10].

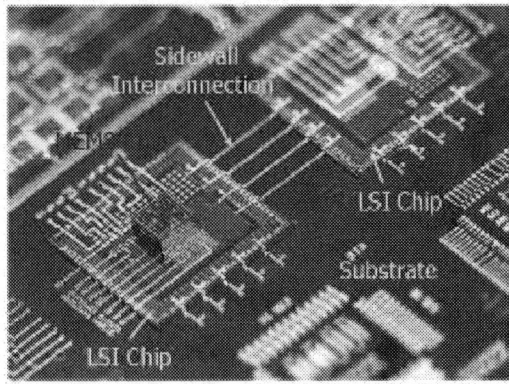

Figure 4: Cu sidewall interconnection crossing-over the chip with step height of 100 µm [12].

of Ti/Cu/Ti, where Ti is titanium. Metallization along few hundred microns high Si sidewall (such as MEMS structure, TSV, etc) has been studied previously. However, in most cases it was carried out on the tapered Si sidewall [8], [9]. In another study by M. Murugesan et al, they fabricated a high-step-coverage Cu-lateral interconnection vertically over 100 µm thick LSI die via Cu electroplating [10]. Figure 2-4 shows some examples from the mentioned works.

However, these works just cover wiring over maximum 100 µm and interconnect density was not that much improved. One of the main challenges in such works is lithography step. It can be said, "If you can litho interconnect, you can make it."

Methodology:

There are many challenges in different phases to pattern a high aspect ratio wire structure over a step height of one to several hundred microns:

1- Resist Apply: having a uniform resist coverage over high topography surfaces.
2- Exposure: defocusing problem and different leveling requirements. The wire lines are getting bigger while going down from the top to the bottom of steps.

3- Development: different resist thickness over a topography needs different time to react in development solution.

(I) Contact Exposure

First sets of experiments were done with contact aligner. It should be considered that in high topography Fresnel diffraction of UV light cause the loss of resolution. For deeper cavities, gap between mask and surface increases and so the effect is more severe [13]. The design was done for different wiring parameters such as wire width and density on a 1.4 µm Aluminum sputtered layer in KOH cavities with different depths from 25 µm up to 150 µm.

To have a good coverage over the topography instead of conventional spin coating, resist apply was done with spray coating. The spray coating experiments were performed on an EVG 101 spray coater on 4 inch wafers. A diluted negative AZNlof 2070 resist solution is used in this experiment and it is necessary in order to get a proper droplet size of resist distribution. The schematic of spray coating setup is shown in Figure 5. An ultrasonic atomizer is used to produce an aerosol of small resist droplets. Pressure tunable air current which carries the aerosol injected the droplets onto the sample surface through Photoresist spray nozzle. Meanwhile the wafer holder chuck is slowly rotating while the spray moving arm scans across from one side to the other side of the wafer with changing speed. Spraying direction is perpendicular to the wafer surface with tunable distance between spray nozzle and the chuck. Moving arm speed variation is applied as the wafer surface that needs to be covered reduces when the nozzle reaches near center of the wafers [8], [14].

Spin rate, spray nozzle scanning speed and profile, spray pressure and resist dispense volume are relevant parameters and were adjusted to optimize the deposition of thick resist layer on topographic wafers. To increase the resist coherency some annealing step can be added during thick coating

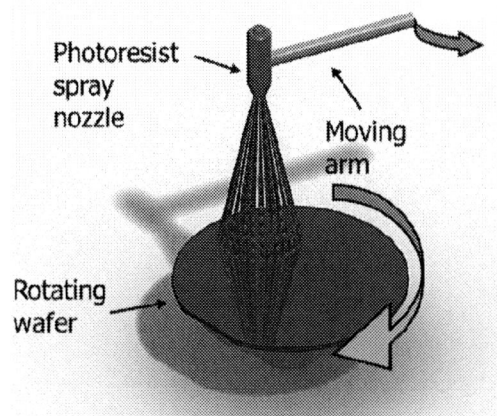

Figure 5: Spray coating setup, an alternative process for high topography.

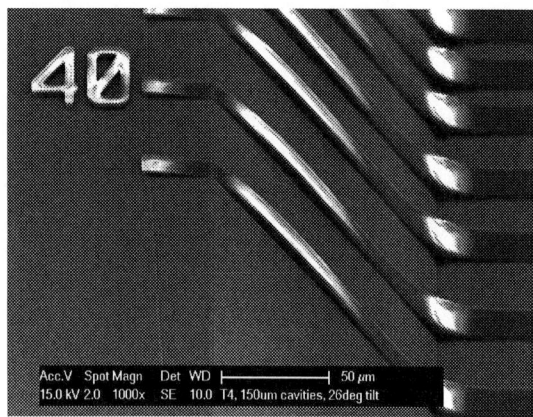

Figure 6: Resist coverage over 150 μm cavity coated with spray coating.

(a)

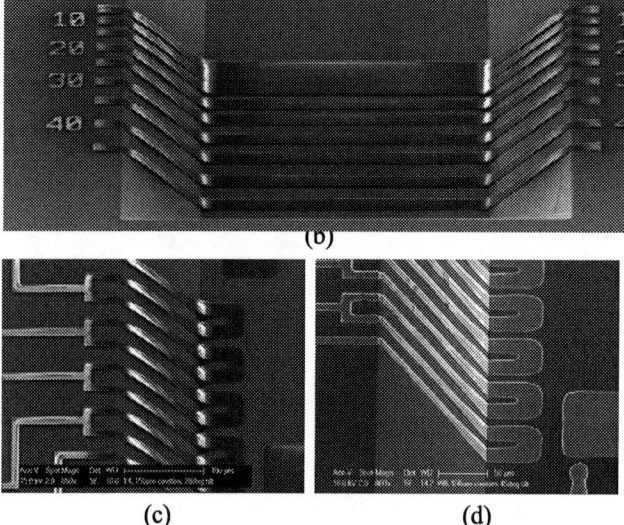

(b)

(c)   (d)

Figure 7: (a) Different structures of 10 μm Lines over 150 μm cavity, (b) Structure with various distance from 10 to 40 μm. It can be observed that the lines with 20 μm pitch are merged, (c) 10 μm wire trace on resist film and the same feature after etching in Aluminum layer.

processes.

For less reflection from sidewalls of the cavities, negative resist was coated in 2 steps of 4 layer spraying. Although there was a considerable thickness difference in top and bottom corners of cavities (2 μm), the good coverage over top corner resolves the problem of pattern splitting (Figure 6).

For wire pitches less than 20 μm, the lines are merged together. For pitches bigger than 20 μm, lines with minimum width of 10 μm were successfully developed and etched. Some of the results are shown in Figure 7(a-c).

Because of low resist thickness in the upper corner of cavities, in Aluminum etching step dry etching does not work selective enough and may cause in line splitting while wet etching process is selective enough. Figure 8 shows the different result of two processes.

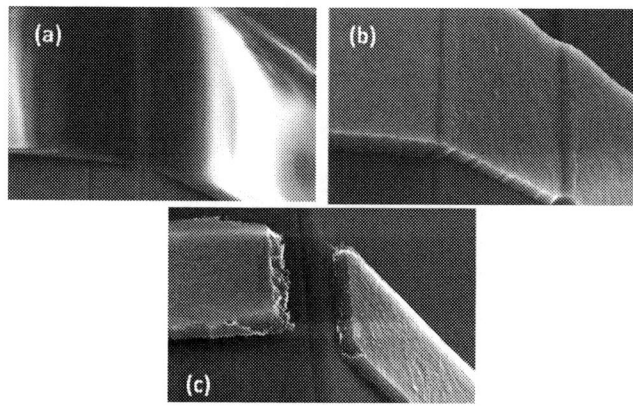

Figure 8: Line over upper corner of the cavity (a) resist pattern (b) Al line after wet etching (c) Al line after dry etching.

(II) Projection:

Although contact aligners are more cost effective compared with wafer steppers, nowadays, wafersteppers are being used for advanced WLP [15]. For non-planar RDL with high resolution, multi-step imaging on a waferstepper [16] is perhaps a future option enable high resolution imaging on high topography substrates.

Multi-step imaging (MSI) functionality, developed by ASML for MEMS industry, allows increasing the global focus offset range up to ±200 μm from the standard specification of ± 30 μm. Additionally, a local focus offset of ±5 μm can be added to each die to the global focus offset to adapt to the process induced thickness non-uniformity of the high topographic wafers [12].

978-1-4799-2408-0/14 $31.00 © 2014 IEEE

In comparison to contact aligner, MSI is more flexible for imaging over high topographies. The exposure dose with MSI can be controlled at different focus levels. It has the potential to adapt the dose in the aerial image at different focus levels to better match local resist thickness and compensate thickness variation of spray coated resist. While the constant dose of exposure in contact aligner causes problem in developing patterns for different resist thickness. For example considering thick resist at the bottom corner of cavity, the exposure dose and developing time should be increased which may cause over exposure and loosing features for thinner part of resist layer.

To make features just at the bottom of cavities, the average depth of the cavities is considered in the global focus offset. The focal plane is thus at the same level as the bottom of the cavities. To keep the same resolution all over the wafer, several issues need to be taken into account: 1) non-uniformity and roughness of KOH etched cavities due to chemical nature of process, 2) spray-coated resist is not uniform all over the wafer and also in different place of the topographies. These types of issues can be addressed in the local focus offset.

For second sets of experiment the same resist coating process was used. Using multi step imaging on an ASM PAS 5500/80 projection aligner (NA = 0.48), we realized a resolution down to 2 μm on the bottom of a 150 μm deep cavity (see Figure 9). However for sidewall imaging processing of wires, further research is needed.

Simulation and Discussion:

Aerial image calculation can help to do exploration and get a better vision of defocus effect over the topography. It is also useful in optimization of multi-step imaging.

Calculation of aerial image for projection exposure was done based on Fraunhofer diffraction equations[17]. Light intensity over 150 μm cavity in different depth from top to the bottom is calculated and 3D color map image is plotted. These calculation can be done for masks including lines with different line parameters.

Figure 10(a,b) show the simulation result for interconnect

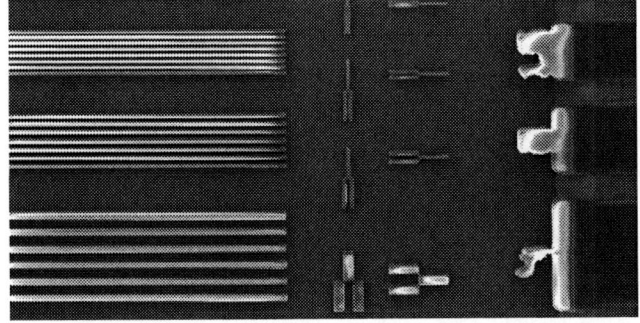

Figure 9: Different line structures at bottom of 150 μm cavity using multi step imaging with a wafer stepper.

lines of 10 μm width and two different pitches of 20 and 30 μm (selected based on previous experiment parameters). The focal plane is the same as the wafer level at the top of the cavities. It is seen that as going deeper to the cavity bottom and so more defocus issues, the lines are getting wider and pattern sharpness is getting worse. For 30 μm pitch the features are reasonably splitting at the bottom while for 20 μm pitch the lines are going too close to each other. An important point to be considered in this case is the significant difference of resist thickness over the cavities. Pattern transfer to thick resist layer at the bottom corner of cavities is more challenging as it was also observed in the experimental results.

In the next step for getting better resolution and pattern sharpness, we applied multiple imaging at different focal planes. In Figure 10(c) for the same pitch of 20 μm imaging was done in double steps with focus at the top and bottom corners of cavity. The light intensity is the total light intensity summed over two exposures. In this case the patterns are formed in good quality at 2 focal planes since in each step of imaging defocusing problem of the other step is compensated. In addition pattern sharpness at sidewall is improving comparing to Figure 10(b).

Getting better pattern quality and resolution at cavity sidewalls, we can add more imaging steps with focal planes at different level of cavity sidewalls. In Figure 10(d), one more imaging is done at mid-level of sidewall. In this case the line patterns are sharply split over the topography.

However for different line structures the optimized imaging steps can be calculated. For different application, there is a trade-off between imaging step numbers (i.e. proportional to the lithography time) and pattern sharpness. Take into account that for normalized resist thickness threshold value of normalized light intensity is practically 0.4. For more accurate simulation we can add resist behavior model versus different thickness over the depth. There is also some space for mask engineering for less imaging steps.

Conclusion:

In this work "Lithographic defined wire bonding" is introduced as a new approach to overcome limitations and challenges of conventional wire bonding for special applications. It includes metal film deposition and patterning the wire trace through high aspect ratio lithography. Different challenges and process details were discussed.

Two sets of experiment was done to make various line structures over KOH cavities of different depth up to 150 μm. Multilayer spray-coating was used to have a good resist coverage over deep cavities. Exposure was done with contact aligner and multi-step imaging on a ASM PAS 5500/80 projection aligner. Aluminum interconnect lines with different parameters were achieved after etching process. For sidewall imaging of wires using MSI on stepper, further research is needed. In the end simulation studies were done

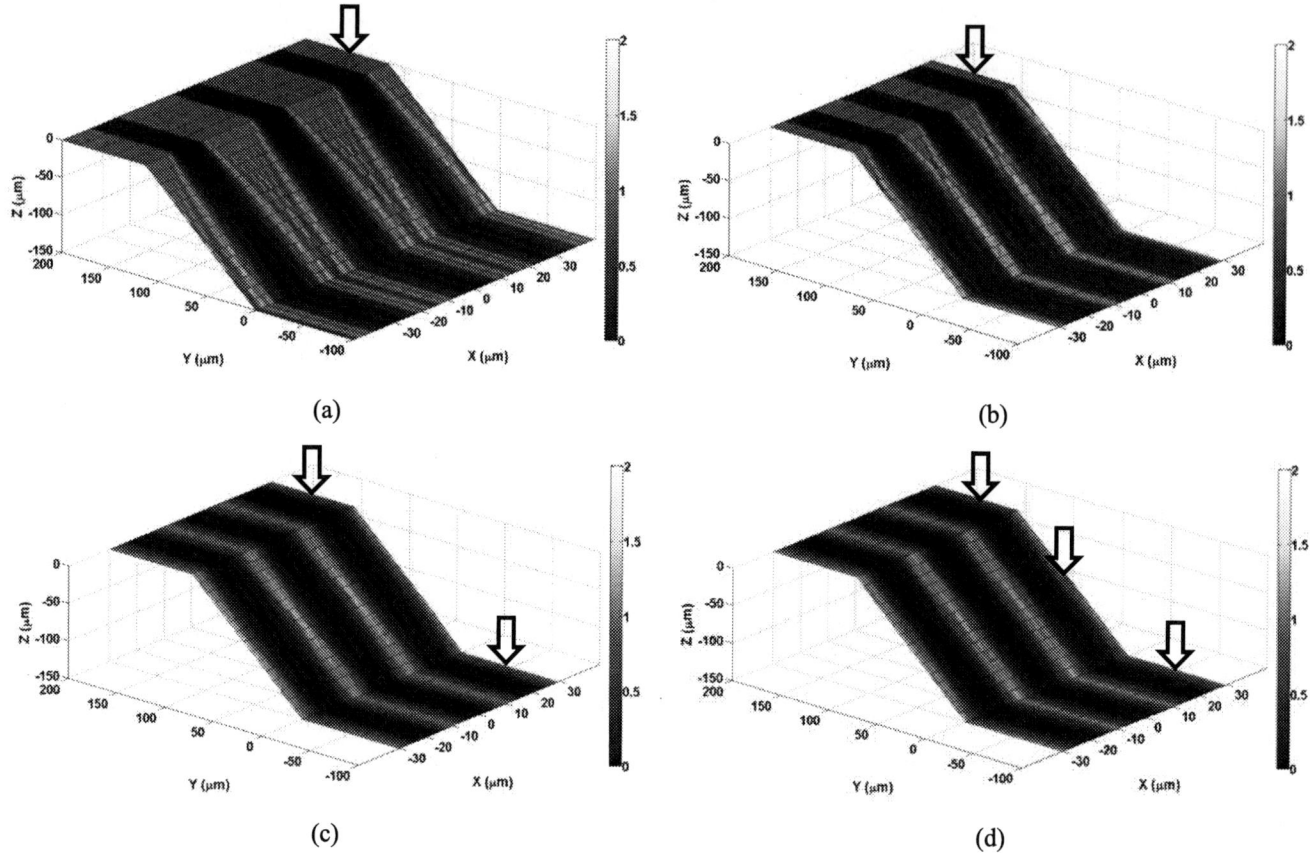

Figure 10: Light intensity simulated plots over 150 μm cavity for mask of lines of 10 μm width for, (a) 30 μm pitch with one step focus at top, (b) 20 μm pitch with one step focus at top, (c) same structure with double focus steps at top and bottom, (d) with triple focus steps at top and bottom and mid-level of sidewall. The arrows point to the focal plane positions.

using aerial image calculation for optimization of MSI module in the further steps. This method can be applied for heterogeneous integration area where a reliable and flexible interconnection method is critical.

Acknowledgements:

The authors would like to acknowledge the whole DIMES-ICP group for technical support and Massoud Tohidian for his fruitful help in simulation coding. Special thanks are due to   at ASML Company for his support over using the MSI.

References:

[1] G. Harman, *Wire bonding in microelectronics*, 3rd ed. McGraw-Hill electronics packaging and interconnection series, 2010, p. 446.

[2] C. D. Breach, "What is the future of bonding wire? Will copper entirely replace gold?," *Gold Bulletin*, vol. 43, no. 3. pp. 150–168, 2010.

[3] P. Liu, L. Tong, J. Wang, L. Shi, and H. Tang, "Challenges and developments of copper wire bonding technology," *Microelectron. Reliab.*, Dec. 2011.

[4] J. Pan and P. Fraud, "Wire Bonding Challenges in Optoelectronics Packaging," *Proc. 1st SME Annu. ...*, no. 313, 2004.

[5] F. Iacopi, S. . Brongersma, B. Vandevelde, M. O'Toole, D. Degryse, Y. Travaly, and K. Maex, "Challenges for structural stability of ultra-low-k-based interconnects," *Microelectron. Eng.*, vol. 75, no. 1, pp. 54–62, Jul. 2004.

[6] J. C. J. Chen, D. Degryse, P. Ratchev, and I. De Wolf, "Mechanical issues of Cu-to-Cu wire bonding," *IEEE Trans. Components Packag. Technol.*, vol. 27, no. 3, 2004.

[7] Y. Joung and M. Allen, "A micromachined chip-to-board interconnect system using electroplating bonding technology," *Adv. Packag. IEEE Trans. ...*, vol. 31, no. 2, pp. 357–366, 2008.

[8] N. Pham, M. Bulcke, and P. Moor, "Spray coating of photoresist for realizing through-wafer interconnects," *2006 8th Electron. Packag. Technol. Conf.*, pp. 831–836, 2006.

[9] C.-H. Ji, F. Herrault, and M. G. Allen, "A metallic buried interconnect process for through-wafer interconnection," *J. Micromechanics Microengineering*, vol. 18, no. 8, p. 085016, Aug. 2008.

[10] M. Murugesan, T. Fukushima, K. Kiyoyama, J. C. Bea, T. Tanaka, and M. Koyanagi, "High step coverage Cu lateral interconnections over 100 μm thick chips on a polymer substrate—an alternative method to wire

978-1-4799-2408-0/14 $31.00 © 2014 IEEE

bonding," *J. Micromechanics Microengineering*, vol. 22, no. 8, p. 085033, Aug. 2012.

[11] T. M. Verhaar, J. Wei, and P. M. Sarro, "Pattern transfer on a vertical cavity sidewall using SU8," *J. Micromechanics Microengineering*, vol. 19, no. 7, p. 074018, Jul. 2009.

[12] K.-W. Lee and M. Koyanagi, "Novel interconnection technology for heterogeneous integration of MEMS–LSI multi-chip module," *Microsyst. Technol.*, vol. 16, no. 3, pp. 441–447, Oct. 2009.

[13] N. P. Pham, E. Boellaard, W. Wien, L. D. M. van den Brekel, J. N. Burghartz, and P. M. Sarro, "Metal patterning on high topography surface for 3D RF devices fabrication," *Sensors Actuators A Phys.*, vol. 115, no. 2–3, pp. 557–562, Sep. 2004.

[14] N. Pham and E. Boellaard, "Photoresist coating methods for the integration of novel 3-D RF microstructures," *J. Microelectromechanical Syst.*, vol. 13, no. 3, pp. 491–499, 2004.

[15] W. Flack, R. Hsieh, and G. Kenyon, "Lithography technique to reduce the alignment errors from die placement in fan-out wafer level packaging applications," in *Electronic Components and Technology Conference*, 2011, vol. 1, pp. 65–70.

[16] P. Maury, J.-M. Quemper, S. Pocas, D. Van Vliet, N. Noordam, P. Ten Berge, and K. Best, "Sub-micron imaging on high-topography wafers using spray coating and projection lithography," *Microelectron. Eng.*, vol. 87, no. 5–8, pp. 904–906, May 2010.

[17] J. D. Plummer, M. D. Deal, and P. B. Griffin, *Silicon VLSI Technology: Fundamentals, Practice and Modeling*. Prentice Hall, 2000, p. 817.

Comprehensive Intermetallic Compound Phase Analysis and Its Thermal Evolution at Cu Wirebond Interface

In-Tae Bae[1,*], Dae Young Jung[1], Jenny Chang[2] and Scott Chen[2]

[1]Small Scale Systems Integration and Packaging Center, State University of New York at Binghamton
Binghamton, NY 13902, USA

[2]Advanced Semiconductor Engineering (ASE), Inc.
1255 E Arques Ave, Sunnyvale, CA 94085, USA

[*]E-mail: itbae@binghamton.edu, phone: +1-607-777-2243

Abstract

Intermetallic compound (IMC) formation in Cu wire bond is of great technological importance since it underpins the bonding characteristics between Cu wire ball and Al pad, which are directly associated with long-term reliability. Thus, extensive works have recently been performed to understand IMC formation at the interface between Cu wire ball and Al pad metallization. Most of the previous works implemented qualitative scanning electron microscopy (SEM)-based analysis techniques such as energy dispersive x-ray spectroscopy (EDS) and electron probe micro-analyzer (EPMA) to identify IMC phases. However, to understand IMC growth and phase evolution in depth, transmission electron microscopy (TEM)/nano-beam electron diffraction (NBED) study is necessary. We have recently performed TEM/NBED investigation on 20 μm (=0.8 mil) Cu wire bond interface and unambiguously verified the presence of metastable θ'-CuAl$_2$ IMC phase as the majority IMC phase. In this study, further comprehensive IMC phase analyses were performed for side-by-side comparative study between 18 μm (=0.7 mil) Pd-coated Cu wire and 18 μm Pd-coated Cu wire that was followed by Au flash coating using SEM and TEM/NBED in conjunction with structure factor (SF) calculation. For SEM examination, cross-section of Cu wirebond samples was prepared by a combination of mechanical polishing and Ar ion cross-section polisher. TEM sample was prepared using dual-beam focused ion beam technique. Cross-sectional SEM result shows that in both of the wirebonds, discontinuous IMC patches were formed at the Cu/Al interface immediately after packaging, and they grow toward Al pad after HTS test at 150 °C for 1000 h. TEM/NBED result combined with SF calculation revealed that in both of the wirebonds, the majority of IMC phase was metastable θ'-CuAl$_2$ phase while Cu$_9$Al$_4$ phase was also found occasionally between metastable θ'-CuAl$_2$ phase and Cu wire in as-packaged wirebonds. After HTS test, both of the phases grow in size. SEM, and TEM/NBED results suggest that there is no Au coating effects in terms of IMC morphology, phase formation and their thermal evolution. Based on NBED and EDS results, possible Cu$_9$Al$_4$ growth mechanism will be discussed.

1. Introduction

Wire bonding is one of the most widely used interconnect techniques to provide electrical paths for power and signal distribution between integrated circuit chips and external circuit boards. While Au wire bonding has been the workhorse for the industry since 1957, rapid and volatile increase of Au price in recent years has triggered industry conversion from Au to Cu wire bonding material in fine pitch wire bonding in high volume production [1]. Cu wire bonding is considered to provide benefits not only for cost saving but also for enhancing electrical and thermal performances [2]. There have been several technological concerns for implementing this technology for high volume manufacturing. They include the need of forming gas during free air ball formation [3-6], Al splash and dielectric damage risk underneath Al pad [7,8], long term reliability concern under humid environments [8,9], and slow Cu-Al intermetallic compound (IMC) formation [3,10-13]. The formation of Cu-Al intermetallic compound is of great interest since it is directly related to the wire bonding materials and bonding process development. Extensive studies have been made to investigate IMC growth behavior at the interface between Cu wire and Al pad metallization. Conceptually IMC growth sites are initiated over discrete positions at the Cu-Al interface. During annealing the IMC spreads across the entire Cu-Al interface as the IMC phase evolve into Cu-rich IMC with Cu as the dominant material present [14]. An early annealing study at a temperature ranging between 150 and 200 °C showed that while Cu$_9$Al$_4$ is the first and major IMC phase CuAl$_2$ showed up with an increased annealing time using micro-beam x-ray diffraction technique (MBXRD) [10]. The presence of CuAl$_2$ phase was not conclusive due to overlapping and splits of Bragg peaks. In a subsequent annealing study with annealing at 250 °C, similar were reported; Cu$_9$Al$_4$ and CuAl$_2$ are the major IMC phases [15]. Recent in-depth studies suggested that Cu-Al IMC phase evolves from CuAl$_2$ and CuAl phases to Cu$_9$Al$_4$ phase from as-bonded state to an annealing temperature of 350 °C [14,16]. The IMC growth kinetics has been well addressed in [14]. Since most of these previous works implemented qualitative techniques such as MBXRD [10,15], fast Fourier transform [16], and scanning electron microscopy (SEM)-based EDS and/or electron probe micro-analyzer (EPMA) for identification of IMC phases [14,17], we have recently made use of TEM/NBED technique in combination with structure factor (SF) calculation for quantitative IMC phase analysis. As a result, metastable θ'-CuAl$_2$ phase was unambiguously verified as the majority IMC phase at Cu/Al bond interface [18,19].

It is known that Pd coating on Cu wire can provide significant benefits compared with bare Cu wire in terms of extended shelf-life of the wire and long-term wirebond reliability [20-21, 23]. However, high hardness of Pd coating causes more damage to wire drawing facility that leads to higher manufacturing costs. There has been an effort to resolve this

978-1-4799-2408-0/14 $31.00 © 2014 IEEE

issue with Au flash coating on Pd-coated Cu wire in a hope that Au can play a roles as a lubricant during wire drawing process.

In this work, a comprehensively comparative study of IMC formation and its subsequent thermal evolution is examined for Cu/Al interfaces made by (1) Pd-coated Cu wire and (2) Pd-coated Cu wire that was followed by Au flash coating, using TEM/ NBED and EDS in combination with SF calculation.

2. Experimental

Cu ball bonds were made using 18 μm (=0.7 mil) Pd-coated Cu wire (hereafter, denoted as "*Pd*-Cu wire") and 18 μm Pd-coated Cu wire followed by Au flash coating (hereafter, denoted as "(*Au*)*Pd*-Cu wire") [22]. Ball bonds were made on typical Al pads on SiO_2 insulation layer on a silicon die mounted on a lead frame. Molding compound was cured at 175 °C for 4 hours. Some samples were further annealed at 150 °C for 1000 h for HTS test.

In order to reveal IMC phase existing at the interface between Cu wire and Al pad, the top surface of the package was gradually polished until it barely showed the Cu ball. Then, the package was carefully polished from one side until the polished cross-section reaches ~25 μm away from the center of the Cu wire. Final polish for SEM observation was performed using an ion cross-section polisher with 4 kV Ar^+ ion beam. The structure of the cross-sections was examined using Zeiss Supra 55 field emission SEM. Cross-sectional TEM samples were directly lifted-out from the cross-section polished samples using FEI Nova 600 dual-beam focused ion beam (FIB). The energy of Ga ion beam was gradually decreased from 20 to 5 kV to minimize ion beam induced damages. These samples were examined using 200 kV operating JEOL JEM-2100F field emission TEM equipped with Scanning TEM, EDS and electron energy loss spectroscopy (EELS) in Analytical and Diagnostics Laboratory at State University of New York at Binghamton.

3. Results and discussion

3.1. Free air ball (FAB) and bare wire morphology for *Pd*- and (*Au*)*Pd*-Cu wires

Figure 1 shows a SEM secondary electron image with EDS elemental mapping results for *Pd*-Cu wire. An oxidation-free

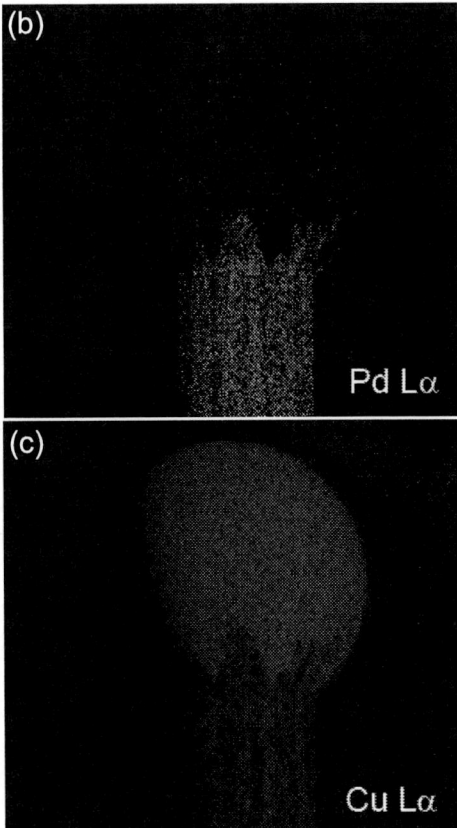

Figure 1. SEM secondary image (a) with EDS elemental mappings for Cu (b), and Pd (c) for *Pd*-Cu wire.

free air ball (FAB) is formed with ~32μm diameter as shown in Fig. 1(a). Note that while Pd coating fades away as it runs from bare wire area across neck area, and completely disappears inside FAB (see Fig. 1(b)), Cu signal becomes stronger near the center of FAB (see Fig. 1(c)). This indicates that Pd coating completely dissolves into FAB forming process except the neck area, so that the concentration of Pd in FAB is below detectability of EDS.

The same examination was performed for (*Au*)*Pd*-Cu wire as shown in Figure 2. While a ~30μm diameter oxidation-free FAB is confirmed (see Figs. 2(a) and 2(b)) with Pd coating at bare wire area as shown in Fig. 2(c), Au signal was not

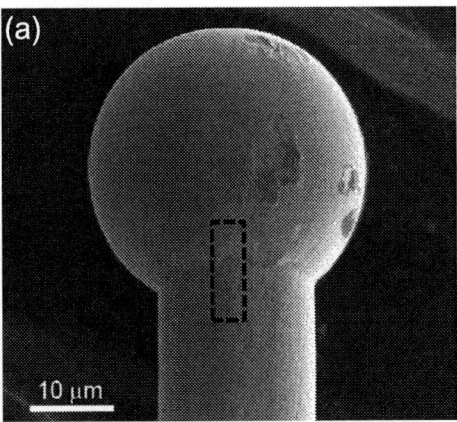

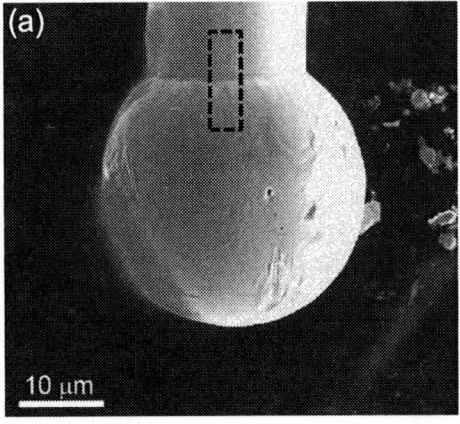

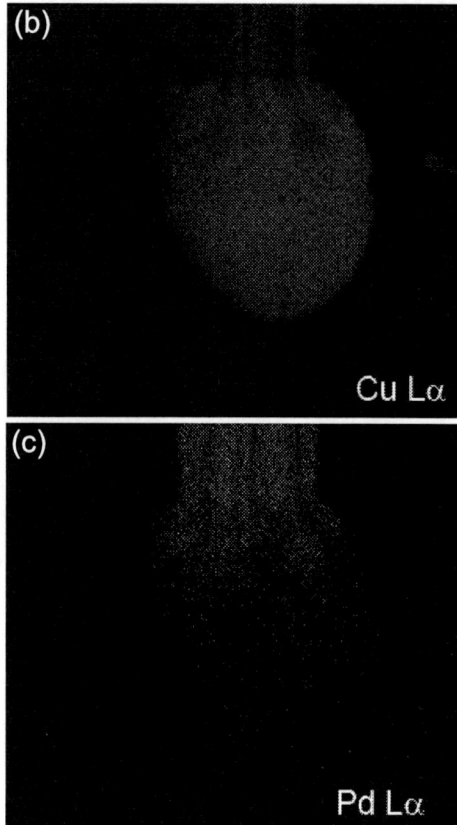

Figure 2. SEM secondary image (a) with EDS elemental mappings for Cu (b), and Pd (c) for (*Au*)*Pd*-Cu wire.

detected by EDS. This indicates that the amount of Au incorporated as flash coating is so small that it is beyond EDS detectability. In Figure 3 are shown vertical cross sectional SEM images of the squared area marked by dotted lines in Figs. 1(a) and 2(a), respectively. In both of Figs. 3(a) and 3(b), Pd coating layers of ~80 nm are clearly visible from bare wire areas. Those coating layers disappear as it runs from neck area through FAB. Meanwhile, Au flash coating was not confirmed in Fig. 3(b) as expected from EDS result, suggesting that the amount of Au flash coating is beyond spacial resolution of SEM (~4 nm). These are in good agreement with the Pd distributions found in Figs 1(b) and

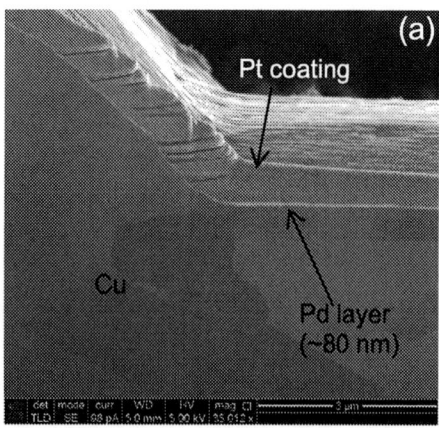

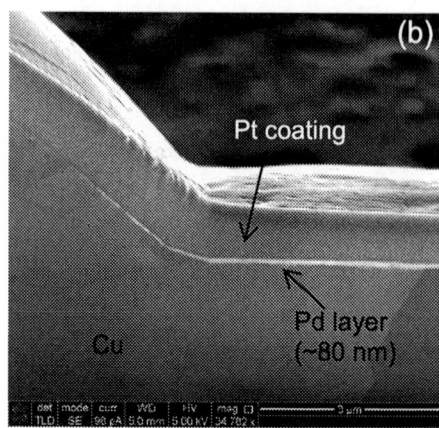

Figure 3. Cross sectional SEM images for *Pd*- (a) and (*Au*)*Pd*-Cu wires.

2(b).

By using ~80 nm of Pd coating on 18 μm-diameter Cu wire that is revealed above together with ~32 μm-diameter FAB, Pd concentration in FAB is calculated to be ~1.8 vol. %. Note that the Pd coating was not completely dissolved in FAB near neck area (see Fig. 1(b) and 2(b)). Thus, the real Pd concentration in FAB is speculated to be lower than 1.8 vol. %. This highly low Pd concentration is considered to be associated with no Pd signal detected in the most of the FAB areas for both of *Pd*- and (*Au*)/*Pd*- coated Cu wires.

3.2. Cross sectional SEM examination of wirebonds made with *Pd*- and (*Au*)*Pd*-Cu wires before HTS

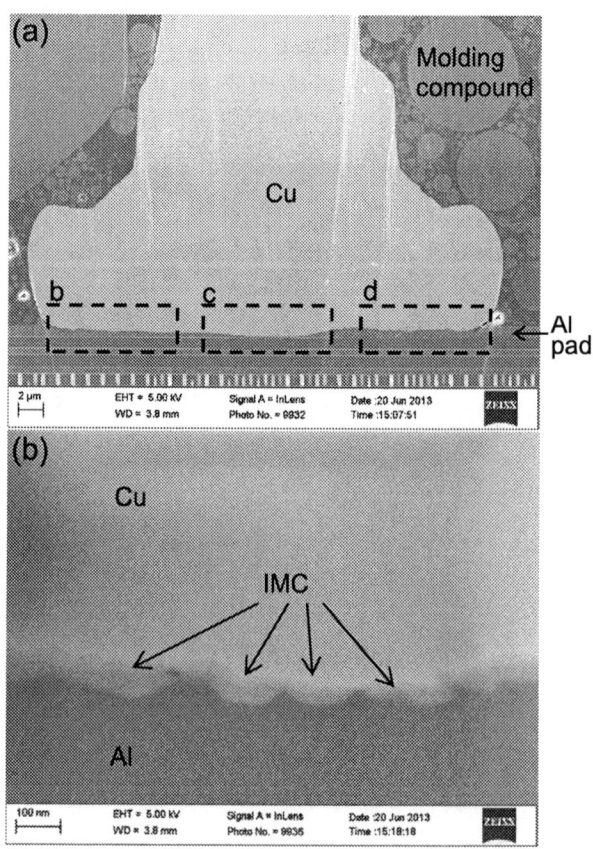

978-1-4799-2408-0/14 $31.00 © 2014 IEEE

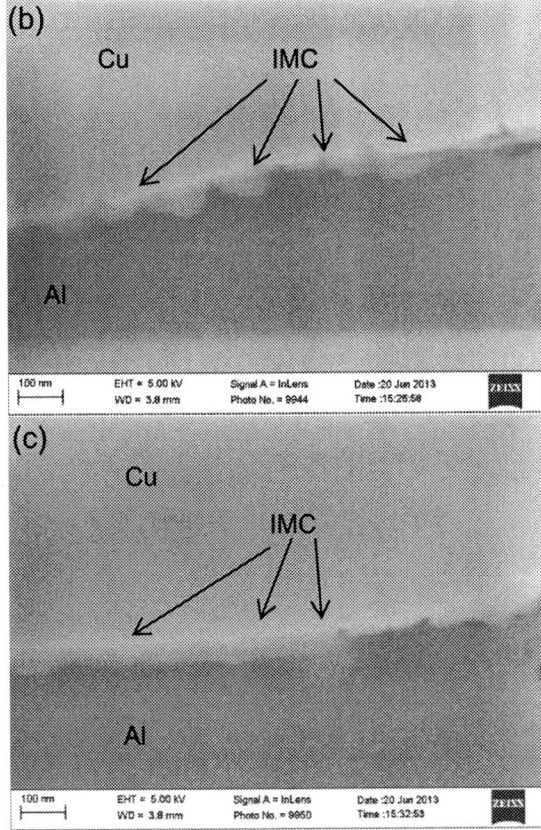

Figure 4. Cross sectional SEM images for *Pd*-Cu wire before HTS test, showing an overall view (a) with Cu/Al interface areas (b), (c) and (d).

Figure 4(a) is a typical cross sectional SEM image that shows an entire wirebond interface made with *Pd*-Cu wire before HTS. Note that bonding between *Pd*-Cu ball and Al is well established with no significant Al splash or cracks/voids. In order to acquire more detailed information, the areas denoted as b,c, and d are enlarged to show in Figs. 4(b), 4(c) and 4(d), respectively. In all three images, IMCs can be clearly observed between Cu ball and Al pad with their thickness varying from 40 to 80 nm. In Figure 5(a) is shown a typical cross sectional SEM image of an entire wirebond made with (*Au*)*Pd*-Cu wire before HTS. Well-established bond can be readily confirmed with no significant Al splash or

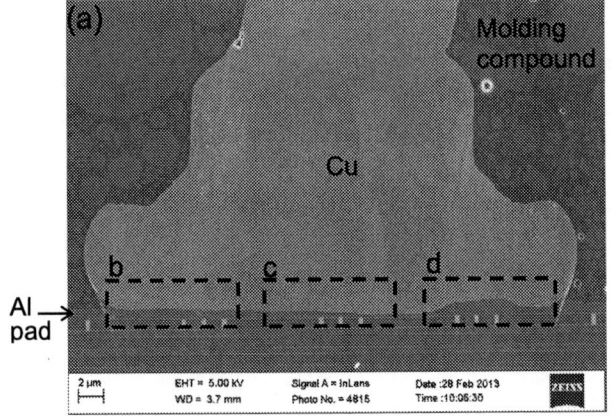

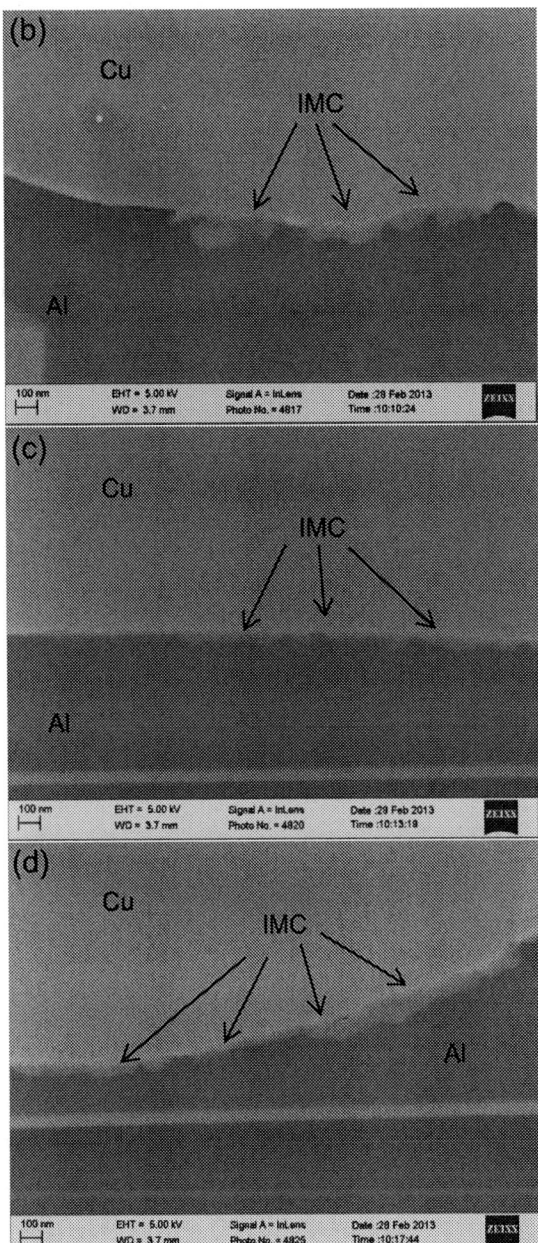

Figure 5. Cross sectional SEM images for (*Au*)*Pd*-Cu wire before HTS test, showig an overall view (a) with Cu/Al interface areas (b), (c) and (d).

cracks/voids. Enlarged images obtained from the areas denoted as b, c, and d are displayed in Figs. 5(b), 5(c) and 5(d), respectively. They clearly show IMC formation between Cu ball and Al pad with IMC thickness ranging from 30 to 80 nm. Based on Figs 4 and 5, it is found that there is no significant differences in wierbond interface structures between with *Pd*- and (*Au*)*Pd*-Cu wire in terms of IMC morphology and thickness.

3.3. Cross sectional SEM examination of wirebonds made with *Pd*- and (*Au*)*Pd*-Cu wires after HTS at 150 °C for 1000 h

978-1-4799-2408-0/14 $31.00 © 2014 IEEE

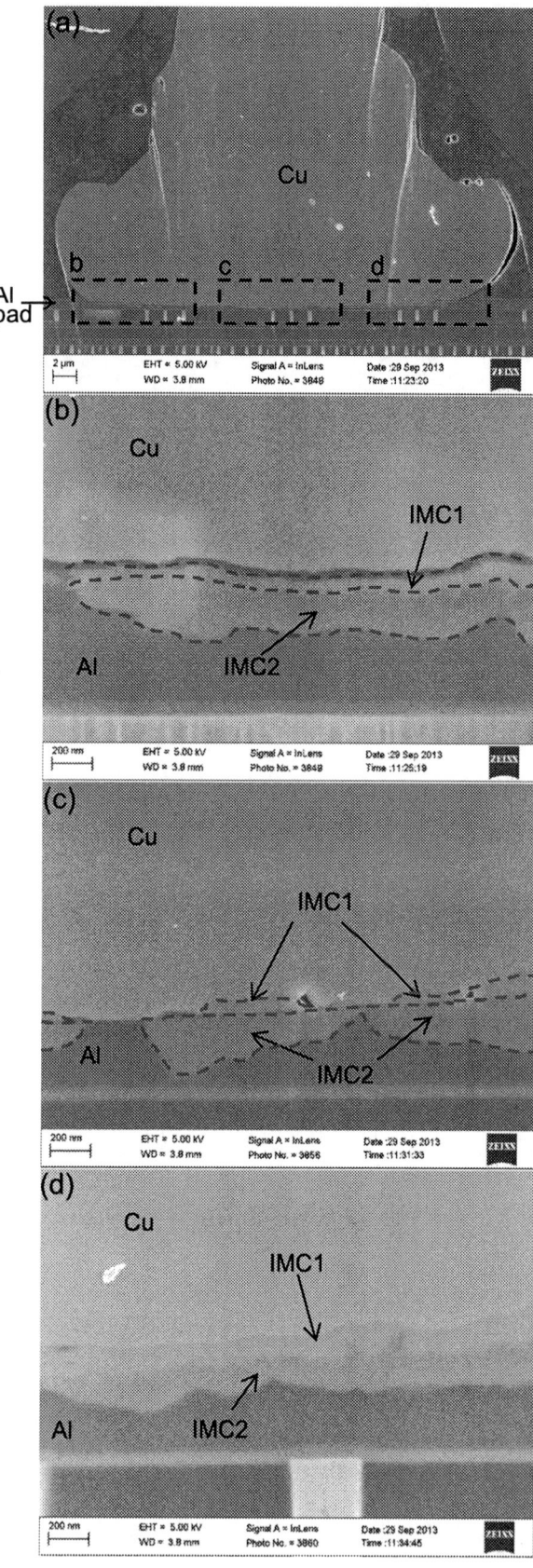

In Figure 6(a) is shown a typical cross sectional SEM image of an entire wirebond made with *Pd*-Cu wire after HTS. It can be readily noticed that IMC at Cu/Al interface grows significantly in size up to ~450 nm as a result of thermal annealing. For more detailed observation, Cu/Al interface areas marked by b, c, and d are enlarged in Figs. 6(b)~(d). Unlike '*before HTS*' sample, two distinctive IMC layers (denoted as IMC1 and IMC2) can be clearly observed particularly in Figs 6(b), 6(c) and 6(d). This indicates that an additional IMC and/or different IMCs have formed as result of HTS test. Figure 7 is a typical cross sectional SEM image of an entire wirebond made with (*Au*)*Pd*-Cu wire after HTS. As a result of thermal annealing, IMC at Cu/Al interface grows significantly in size up to ~400 nm.

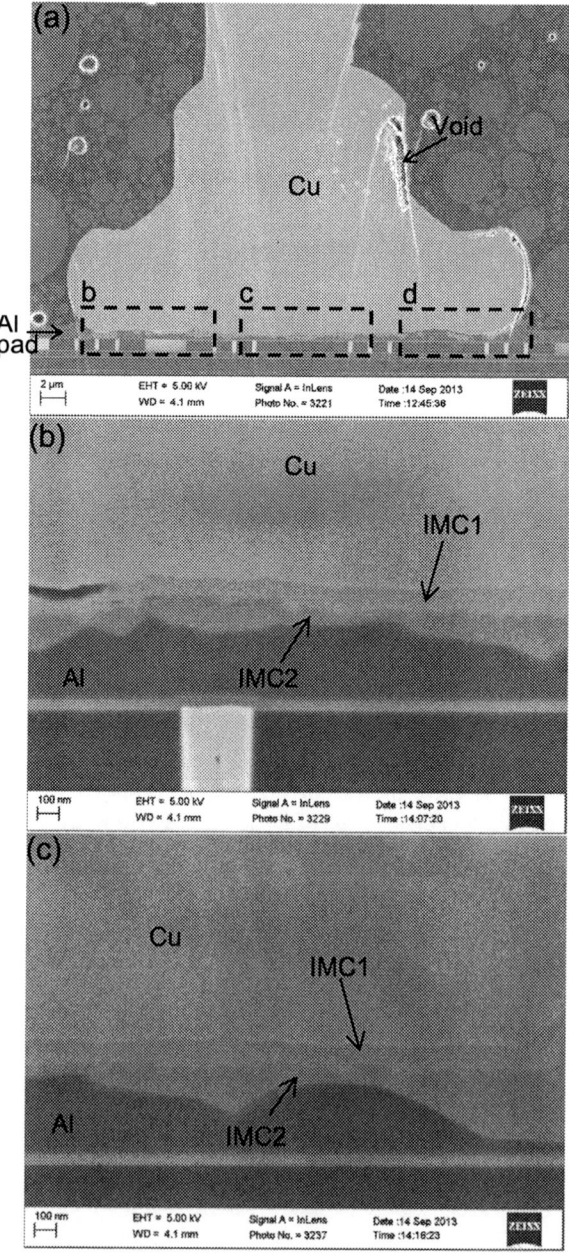

Figure 6. Cross sectional SEM images for *Pd*-Cu wire after HTS test at 150 °C for 1000 h, which show an overall view (a) and Cu/Al interface areas (b), (c) and (d).

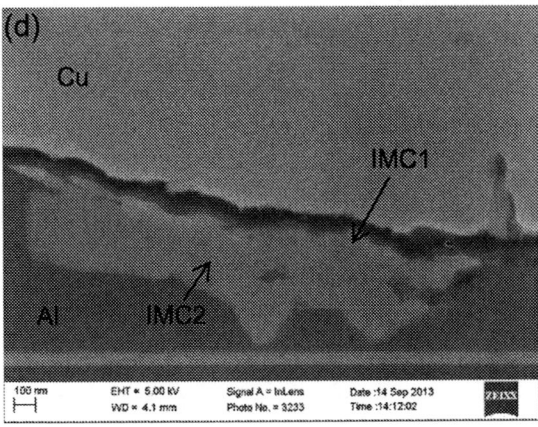

Figure 7. Cross sectional SEM images for (*Au*)*Pd*-Cu wire after HTS test at 150 °C for 1000 h, which show an overall view (a) and Cu/Al interface areas (b), (c) and (d).

In order to acquire more detailed information, enlarged images obtained from the areas denoted as b, c, and d are displayed in Figs. 7(b), 7(c) and 7(d), respectively. Similarly to *Pd*-Cu wire, two distinctive IMC layers (denoted as IMC1 and IMC2) can be clearly observed especially in Figs. 7(b) and 7(c). While relatively brighter contrast of IMC1 than that of IMC2 in Figs.6 and 7 implies that IMC1 might consist of more Cu and less Al, precise IMC phase analysis requires TEM work which follows in the next session.

3.4. Cross sectional TEM examination of wirebonds made with *Pd*- and (*Au*)*Pd*-Cu wires before HTS

In order to understand the phases and crystalline structures of the IMC found before HTS as shown in Figs. 4 and 5, cross sectional TEM examination was performed as shown in Figures 8 and 9. A cross sectional bright-field (BF) TEM image obtained from *Pd*-Cu wirebond (see Fig. 8(a)) clearly shows an IMC layer of ~45 nm (denoted as IMC2) with irregularly-shaped growth front toward Al pad. This corresponds to the IMC found in Fig. 4. It is worth noting that there is another thin IMC layer of ~20 nm (denoted as IMC1) with smooth interface between IMC2 and Cu. Since FIB prepared TEM sample is expected to have rather smooth and homogeneous thickness distribution across its entire electron transparent area, the distinctively darker contrast of IMC1

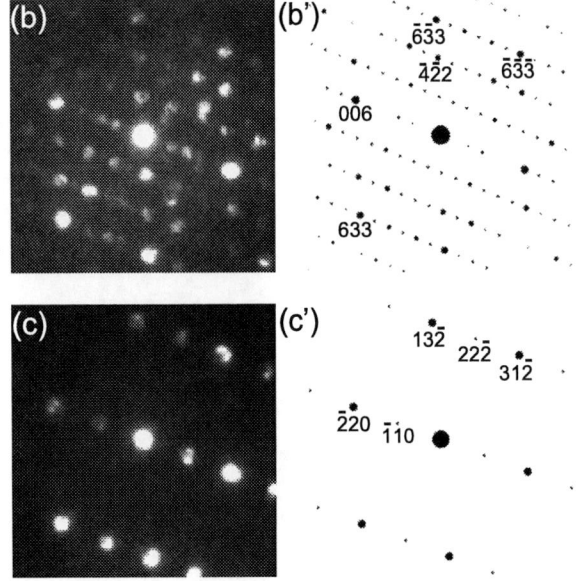

Figure 8. Cross sectional BF TEM image (a) with NBED patterns from IMC1 (b) and IMC 2 (c) areas for *Pd*-Cu wire before HTS test. (b) and (c) correspond to [$\bar{1}$20] of Cu₉Al₄ and [112] of θ'-CuAl₂ as confirmed by SF calculation results shown in (b') and (c'), respectively.

indicates that the chemical composition of IMC1 layer may be different from IMC2. In order to elucidate the phases for IMC1 and IMC2, NBED patterns were acquired from the encircled areas in Fig. 8(a), and show in Figs. 8(b) and 8(c). Bragg reflections in the NBED patterns are carefully calibrated to obtain precise lattice spacing values using undistorted single crystalline sapphire material (a = 0.4753 nm, c =1.298 nm) as a reference. The structure factor, F_{hkl}, where *hkl* represents a specific Bragg reflection, was calculated for all of the known Cu-Al IMC phases, such as θ-CuAl₂ (tetragonal, space group: I4/mcm, a = 0.5949 nm, c = 0.4821 nm), CuAl₂ (orthorhombic, space group: F/mmm, a = 0.848 nm, b = 0.859 nm, c = 0.496 nm), θ'-CuAl₂ (tetragonal, space group: I-4m2, a = 0.404 nm, b = 0.404 nm, c = 0.580 nm), CuAl (monoclinic, space group: C2/m, a = 0.9822 nm, b = 0.4061 nm, c = 0.6807 nm), Cu₉Al₄ (cubic, space group: P-43m, a = 0.8704 nm), and Cu₁.₇Al (hexagonal, space group: P63/mmc, a = 0.4146 nm, c = 0.5063 nm). The calculation of the electron diffraction patterns was based on the kinematical approximation:

$$F_{hkl} = \sum_n f_n \exp[2\pi i(hx_n + ky_n + lz_n)],$$

where f_n is the atomic scattering factor for atom *n* at fractional coordinates (x_n, y_n, z_n). Among the calculation results for various zone axes of the Cu-Al IMC phases listed above, the net patterns of [112] from θ'-CuAl₂ and [$\bar{1}$20] from Cu₉Al₄ are shown in Figs. 8(b') and 8(c'). They match well with Figs. 8(b) and 8(c), respectively. This indicates that there exist two IMC phases between Cu and Al pad:(1) Cu-rich Cu₉Al₄, IMC1, located adjacent to Cu ball and (2) Al-rich θ'-CuAl₂, IMC2, located adjacent to Al pad. In order to verify the NBED and SF calculation results, nano-beam EDS was also performed to the same encircled areas and revealed the Cu:Al

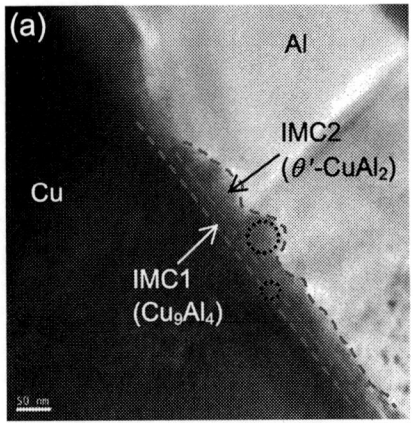

978-1-4799-2408-0/14 $31.00 © 2014 IEEE

ratios are 38:62 and 67:33, which correspond to the compositions of θ'-CuAl$_2$ and Cu$_9$Al$_4$, respectively. Both of NBED and nano-beam EDS results unambiguously confirm the formation of two different IMC layers, i.e. Cu$_9$Al$_4$ and θ'-CuAl$_2$, between Cu ball and Al pad. In Fig. 9(a) is shown a cross sectional BF TEM image of (Au)Pd-Cu wirebond.

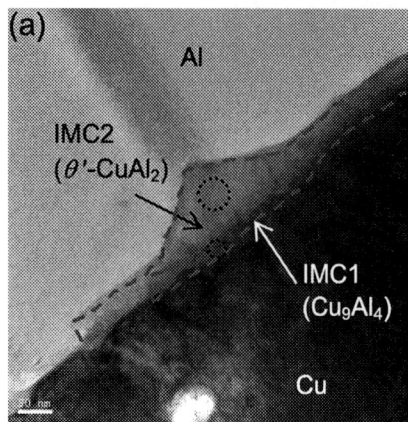

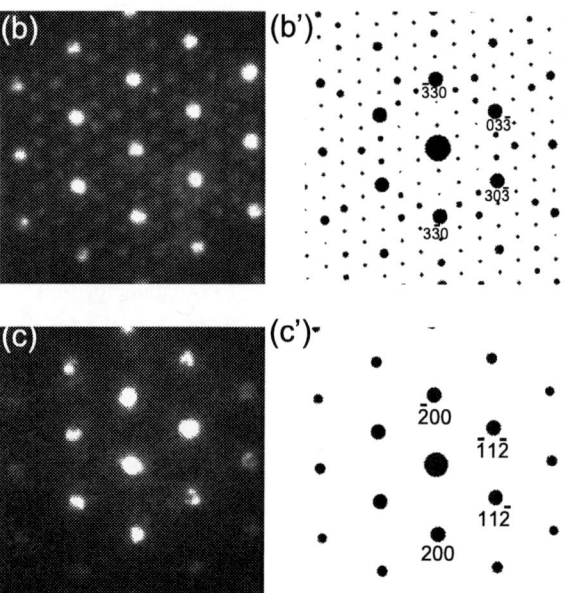

Figure 9. Cross sectional BF TEM image (a) with NBED patterns from IMC1 (b) and IMC 2 (c) areas for (Au)Pd-Cu wire before HTS test. (b) and (c) correspond to [111] of Cu$_9$Al$_4$ and [021] of θ'-CuAl$_2$ as confirmed by SF calculation results shown in (b') and (c'), respectively.

Similarly to Pd-Cu wirebond, two IMC layers are observed between Cu ball and Al pad, i.e., (1) ~20 nm thick IMC1 with smooth growth front located adjacent to Cu ball, (2) ~95 nm thick IMC2 with irregularly shaped growth front located adjacent to Al pad. NBED technique was also employed to the encircled areas and their patterns are shown in Figs. 9(b) and 9(c). SF calculation was performed to obtain the corresponding phases and their crystallographic orientation in the same manner described above. The calculation result indicates that Figs. 9(b) and 9(c) corresponding to [021] of

θ'-CuAl$_2$ and [111] of Cu$_9$Al$_4$, respectively. From Fig. 8 through Fig. 9, two important informations were found: (1) there exist two different IMC phases of Cu$_9$Al$_4$ and θ'-CuAl$_2$ between Cu ball and Al pad; (2) No significant morphological or IMC phase difference between Pd- and (Au)Pd-Cu wirebonds.

Our previous studies on 20μm (=0.8mil) Pd coated Cu wirebond have revealed that IMC phase found at Cu/Al interface was metastable θ'-CuAl$_2$ [18,19]. The possible reasons for the formation of metastable phase were discussed based on two factors: (1) highly short bonding time that can be comparable to quenching effect and (2) epitaxial relationship between Cu ball and θ'-CuAl$_2$ that allow for θ'-CuAl$_2$ to grow on Cu ball by minimizing lattice mismatch [19]. θ'-CuAl$_2$ IMC phase found in the current study is in good agreement with this previous work [19]. In addition, the current study is the first to report the formation of equilibrium Cu$_9$Al$_4$ phase in as-packaged sample, i.e., before HTS test. Cu$_9$Al$_4$ phase formed as a layer with <~20 nm thickness between θ'-CuAl$_2$ and Cu ball is evident from the Figs. 8(a) and 9(a). It is worth noting that Cu$_9$Al$_4$ phase is so thin that it can be confirmed by TEM only. This highly thin nature of Cu$_9$Al$_4$ phase is considered one of the reasons that Cu$_9$Al$_4$ phase has not been reported in previous studies that used SEM-based techniques. Another interpretation of Cu$_9$Al$_4$ formation before HTS test may be associated with Pd concentration in Cu ball. It has been reported that Pd needs to be expelled from θ'-CuAl$_2$ phase toward Cu ball as θ'-CuAl$_2$ phase grows at Cu/Al interface [14,19]. It is speculated that as θ'-CuAl$_2$ phase grows further, more Pd will accumulate at the Cu side of Cu/IMC interface. This leads to prevent further Cu diffusion from Cu ball to θ'-CuAl$_2$ phase, which will eventually reduce θ'-CuAl$_2$ phase growth rate. Recent studies revealed that Pd-enrichment on the Cu side of Cu/IMC interface [23]. One of previous studies revealed that the presence of Pd reduces Cu/Al IMC growth rate [24]. Since Pd concentration in both of Pd- and (Au)Pd-Cu wirebonds used in the present study is so low that it is beyond EDS detectability, it can be speculated that IMC growth in the current study is not as much disturbed as it was in the previous studies. As a result, Cu$_9$Al$_4$ phase which is usually formed at later stages, i.e., after HTS test, of IMC formation can already appear in as-packaged sample in the current work.

3.5. Cross sectional TEM examination of wirebonds made with Pd- and (Au)Pd-Cu wires after HTS at 150 °C for 1000 h

In order to acquire more detailed understanding about the changes in Cu/Al interface structure after HTS, cross sectional TEM examination was performed as shown in Figures 10 and 11. In Fig. 10(a) is shown a BF TEM image obtained from Pd-Cu wirebond. It clearly shows the two IMC layers (that correspond to those found in Fig. 6) between Al pad and Cu ball. NBED technique was employed at the areas encircled to identify there corresponding phases. Figs. 10(b) and 10(c) show their corresponding NBED patterns. SF calculation was performed in the same manner that was described above. SF calculation result confirm that Figs. 10(b) and 10(c) match with [111] net pattern of Cu$_9$Al$_4$ and the [021] net pattern of

978-1-4799-2408-0/14 $31.00 © 2014 IEEE   1568

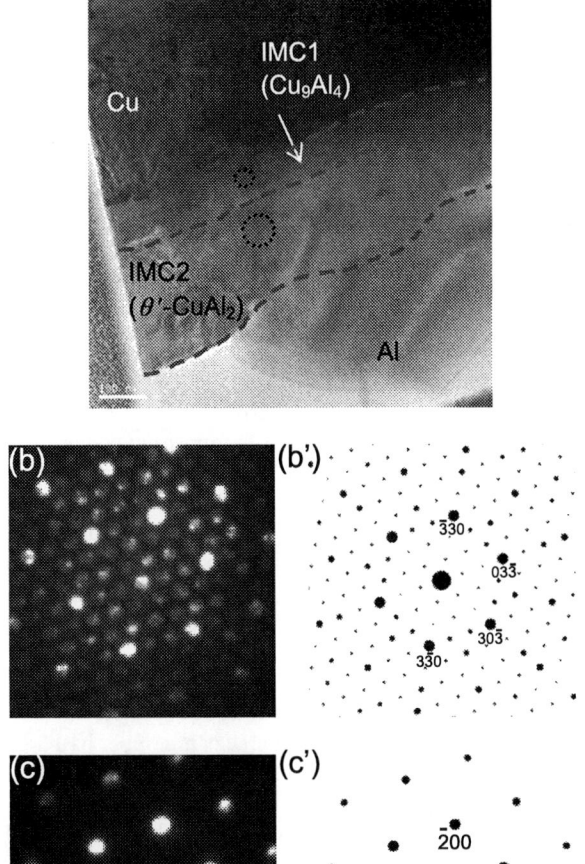

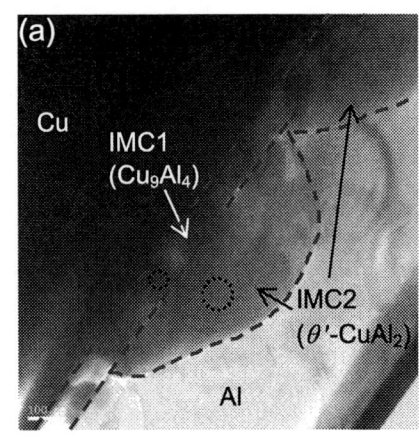

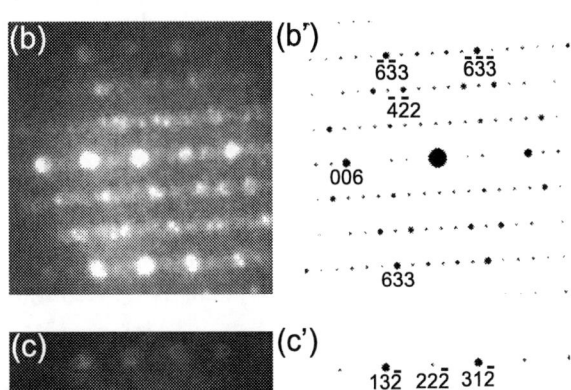

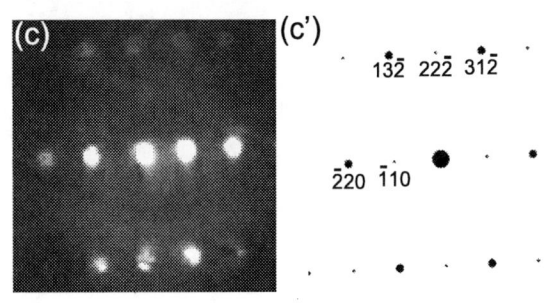

Figure 10. Cross sectional BF TEM image (a) with NBED patterns from IMC1 (b) and IMC 2 (c) areas for *Pd*-Cu wire after HTS test at 150 °C for 1000 h. (b) and (c) correspond to [111] of Cu_9Al_4 and [021] of θ'-$CuAl_2$ as confirmed by SF calculation results shown in (b') and (c'), respectively.

Figure 11. Cross sectional BF TEM image (a) with NBED patterns from IMC1 (b) and IMC 2 (c) areas for (*Au*)*Pd*-Cu wire after HTS test at 150 °C for 1000 h. (b) and (c) correspond to [$\bar{1}$20] of Cu_9Al_4 and [112] of θ'-$CuAl_2$ as confirmed by SF calculation results shown in (b') and (c'), respectively.

θ'-$CuAl_2$, respectively, as shown in Figs. 10(b') and 10(c'). Meanwhile, a BF TEM image from (*Au*)*Pd*-Cu wirebond is shown in Fig. 11(a). Similarly to Fig. 10(a), two distinct IMC layer are observed. NBED patterns are acquired from the encircled areas to show in Figs. 11(b) and 11(c). SF calculation results acquired following the same manner reveal that Figs. 11(b) and 11(c) correspond to [112] of θ'-$CuAl_2$ and [$\bar{1}$20] Cu_9Al_4, respectively. Thus, it becomes clear that after HTS test, θ'-$CuAl_2$ and Cu_9Al_4 grow in size for both *Pd*-or (*Au*)*Pd*-Cu wirebonds. No new Cu-Al IMC phase was formed in either *Pd*- or (*Au*)*Pd*-Cu wirebond after HTS test. Also, no significant difference was found between *Pd*- and (*Au*)*Pd*-Cu wirebonds after HTS test, indicating the effect of of Au flash coating on Cu-Al IMC formation is negligible.

4. Conclusions

In summary, a comparative study on IMC formations at

wirebond interfaces made of *Pd*-Cu and (*Au*)*Pd*-Cu wires have been performed using transmission electron microscopy techniques.

(1) While cross sectional SEM images revealed ~80 nm thick Pd coating layer from both of *Pd*-Cu and (*Au*)*Pd*-Cu wires, Au flash coating layer turned out too thin to be observed by SEM technique; EDS spectrum indicated the presence of Au coating layer.

(2) Before HTS test at 150 °C for 1000 h, i.e. as-packaged state, while irregularly shaped θ'-$CuAl_2$ with its thickness varying from ~40 to ~80 nm was found as a majority IMC, Cu_9Al_4 of highly thin layer (< 20 nm) was also found occasionally between metastable θ'-$CuAl_2$ phase and Cu ball. After HTS test, both of the phases grow in size.

(3) SEM, and TEM/NBED results suggest that there is no difference of IMC formation between *Pd*- and

(*Au*)*Pd*-Cu wirebonds before and after HTS test in terms of its morphology and types of crystallographic phases formed. It is considered that Au flash coating on Cu wire is so small that it hardly makes any effect on IMC formation.

Acknowledgments

This research was funded in part by Small Scale Systems Integration and Packaging Center (S^3IP) at State University of New York at Binghamton. S^3IP is a New York State Center of Excellence and receives funding from the New York State Office of Science, Technology and Innovation (NYSTAR), the Empire State Development Corporation, and a consortium of industrial members.

References

1. B. K. Appelt, A. Tseng, C.-H. Chen and Y.-S. Lai, "Fine Pitch copper wire bonding in high volume production", *Microelectronics Reliability*, Vol. 51, No. 1 (2011), pp.13-20.

2. C. D. Breach and F. W. Wulff, "A brief review of selected aspects of the materials science of ball bonding", *Microelectronics Reliability*, Vol. 50, No. 1 (2010), pp.1-20.

3. S. L. Khoury, D. J. Burkhard, D. P. Galloway and T. A. Scharr, "A comparison of copper and gold wire bonding on integrated circuit devices", *IEEE Transactions on Components, Hybrids, and Manufacturing Technology*, Vol. 13, No 4 (1990), pp.673-681.

4. L. T. Nguyen, D. McDonald, A. R. Danker and P. Ng, "Optimization of copper wire bonding on Al-Cu metallization", *IEEE Transactions on Components, Packaging, and Manufacturing Technology-Part A*, Vol. 18, No.2 (1995), pp.423-429.

5. H. M. Ho, J. Tan, Y. C. Tan, B. H. Toh and P. Xavier, "Modelling energy transfer to copper wire for bonding in an inert environment", *Proc. 7th Electronics Packaging Technology Conference*, Singapore, 2005, pp.292-297.

6. C. J. Hang, W. H. Song, I. Lum, M. Mayer, Y. Zhou, C. Q. Wang, J. T. Moon and J. Persic, "Effect of electronic flame off parameters on copper bonding wire: Free-air ball deformability, heat affected zone length, heat affected zone breaking force", *Microelectronic Engineering*, Vol. 86, No. 10 (2009), pp.2094-2103.

7. A. Shah, M. Mayer, Y. Zhou, S. J. Hong and J. T. Moon, "In situ ultrasonic force signals during low-temperature thermosonic copper wire bonding", *Microelectronic Engineering*, Vol. 85, No. 9 (2008), pp.1851-1857.

8. T. Uno, S. Terashima and T. Yamada, "Surface-enhanced copper bonding wire for LSI", *Proc. 59th Electronic Components and Technology Conference, San Diego, CA, USA*, 2009, pp.1486-1495.

9. S. Kaimori, T. Nonaka and A. Mizoguchi, "The development of Cu bonding wire with oxidation- resistant metal coating", *IEEE Transactions on Advanced Packaging*, Vol. 29, No2 (2006), pp.227-231.

10. H.-J. Kim, J. Y. Lee, K.-W. Paik, K.-W. Koh, J. Won, S. Choe, J. Lee, J.-T. Moon and Y.-J. Park, "Effects of Cu/Al intermetallic compound (IMC) on copper wire and aluminium pad bondability", *IEEE Transactions on Components and Packaging Technologies*, Vol. 26, No. 2 (2003), pp.367-374.

11. F. W. Wulff, C. D. Breach, D. Stephan, Saraswati and K. J. Dittmer, "Characterisation of intermetallic growth in copper and gold ball bonds on aluminium metallization", *Proc. 6th Electronics Packaging Technology Conference, Singapore*, 2004, pp.348-353.

12. M. Drozdov, G. Gur, Z. Atzmon and W. D. Kaplan, "Detailed investigation of ultrasonic Al-Cu wire-bonds: II. Mocrostructural evolution during annealing", *Journal of Materials Science*, Vol. 43, No. 18 (2008), pp.6038-6048.

13. H. Xu, C. Liu, V. V. Silberschmidt and Z. Chen, "Growth of intermetallic compounds in thermosonic copper wire bonding on aluminium metallization", *Journal of Electronic Materials*, Vol. 39, No. 1 (2010), pp.124-131.

14. Y. H. Lu, Y. W. Wang, B. K. Appelt, Y. S. Lai, and C. R. Kao, "Growth of CuAl intermetallic compounds in Cu and Cu(Pd) wire bonding", *Proc. 61th Electronic Components and Technology Conference, Lake Buena Vista, Fl, USA*, 2011, pp.1481-1488.

15. C. J. Hang, C. Q. Wang, M. Mayer, Y. H. Tan, Y. Zhou, and H. H. Wang, "Growth behaviour of Cu/Al intermetallic compounds and cracks in copper ball bonds during isothermal aging", *Microelectronics Reliability*, Vol. 48, No. 3 (2008), pp416-424.

16. H. Xu, C. Liu, V. V. Silberschmidt, S. S. Pramana, T. J. White and Z. Chen, " A re-examination of the mechanism of thermosonic copper ball bonding on aluminium metallization pads", *Scripta Materialia*, Vol. 61, No. 2 (2009), pp.165-168.

17. C.-F. Yu, C.-M. Chan, L.-C. Chan and K.-C. Hsieh, "Cu wire bond microstructure analysis and failure mechanism", *Microelectronics Reliability*, Vol. 51, No. 1 (2011), pp119-124.

18. I.-T. Bae, D. Y. Jung, Y. Du, "Electron microscopy study on intermetallic compound formation in Cu-Al bond interface", *Proc. 62nd Electronic Components and Technology Conference,San Diego, Ca, USA*, 2012, pp.1146-1152.

19. I.-T. Bae, D. Y. Jung, W. T. Chen, and Y. Du, "Intermetallic compound formation at Cu-Al wire bond interface", *Journal of Applied Physics*, Vol. 112, No 12 (2012), pp.123501.

20. T. Uno, "Enhancing bondability with coated copper bonding wire", *Microelectronics Reliability*, Vol. 51, 2011, pp.88-96.

21. T. Uno, "Bond reliability under humid environment for coated copper wire and bare copper wire", *Microelectronics Reliability*, Vol. 51, 2011, pp.148-156.

22. I. Qin, H. Xu, H. Clauberg, R. Cathcart, V. L. Acoff, B. Chylak and C. Huynh, "Wire bonding of Cu and Pd coated Cu wire: bondability, reliability, and IMC formation", *Proc. 61th Electronic Components and Technology Conference, Lake Buena Vista, Fl, USA*, 2011, pp.1489-1495.

23. T. Uno and T. Yamada, "Improving humidity bond reliability of copper bonding wires" *Proc. 60th Electronic Components and Technology Conference, Las Vegas, Nv, USA*, 2010, pp.1725-1732.

24. H. Xu, I. Qin, H. Clauberg, B. Chylak, and V. L. Acoff, "Behavior of palladium and its impact on intermetallic growth in palladium-coated Cu wire bonding", *Acta Materialia* Vol. 61, (2013), pp. 79-88.

978-1-4799-2408-0/14 $31.00 © 2014 IEEE

High Uniformity and High Speed Copper Pillar Plating Technique

Konstantin Kholostov[1,¶], Aliaksei Klyshko[1], Danilo Ciarniello[2], Paolo Nenzi[2], Roberto Pagliucci[2],
Rocco Crescenzi[1], Dario Bernardi[3] and Marco Balucani[1,2]

[1] Department of Information Engineering, Electronics and Telecommunications, University of Rome "La Sapienza",
Via Eudossiana 18, 00184 Rome (RM), Italy
[2] Rise Technology S.r.l.,
Lungomare Paolo Toscanelli 170, 00121 Rome (RM), Italy
[3] 2BG,
Via Monte Bianco 18, 35018 San Martino di Lupari (PD), Italy
[¶] E-mail: kholostov@diet.uniroma1.it

Abstract

In this work we report the application of the selective wet processing technique based on dynamic liquid meniscus for copper pillar bumps (CPB) plating. The industrial plating of copper for CPB process is typically carried out at 2 μm/min. A much higher copper deposition rate is necessary to improve throughput for this process. To achieve higher deposition rates of copper the hydrodynamic issue that is natural for all conventional plating baths processes must be solved. A number of solutions is proposed towards realization of high speed and high throughput CPB plating process. Uniformity of copper pillar over a 6-inches silicon wafer is presented and the morphology and shapes of pillars are investigated by scanning electron microscopy (SEM). Copper pillar height and dimension are investigated within different topology over the wafer showing the robustness of the process for the thickness uniformity. Preliminary investigation of the CPB plating shows the uniformity of better than 2 % within 6" silicon wafer.

Introduction

Since copper pillar bumps (CPB) were introduced in 2006 by Intel, foundries and semiconductor manufacturers are evaluating this new technology. Currently, in semiconductor industry there is no effective countermeasure against plating induced poor uniformity of CPB electroplating that offers the uniformity over a wafer not better than 5 % [1]. To ensure accurate joints, good reliability, high yields, and flexible layouts the uniformity must be better than 1 % along the full wafer. Currently such uniformity is achievable only by introducing a chemical mechanical polishing (CMP) step after the plating process. The CMP process obviously introduces an increase in the total cost of ownership (TCO) of the copper pillar process making it less appealing to the global packaging market.

Nowadays, the highest deposition rate for copper pillar, to our knowledge, is 5 μm/min [2, 3]. Considering that current technology requires approximately 100 μm thick CPB it means very low throughput (i.e. 2 wafer/hour). Further improvements according to laws of the semiconductor industry demand reduced form factors, making CPB smaller and taller. The appropriate technology doesn't exist to fulfill the needs of the industry. In order to increase throughput the only solution is to improve the speed of plating. Several attempts in the optimization of plating baths were already done [2, 3] such as to use higher solubility copper salt (i.e. methane sulfonic) and better performance-enhancing organic

additives. In any case, the plating bath optimization didn't revolutionize the industry.

On the other hand, it is well known that the material deposition rate is governed by the mass transport. It was shown that copper deposition rate can be intensified up to 50 μm/s using jet plating technique [4]. Unfortunately, fine dimensions of nozzles and deposits obtained thereby as well as certain liquid management issues didn't allow this technique to be commercialized. In this work, we utilize the described effect in the new high speed plating technique based on the dynamic liquid drop (DLD) and the dynamic liquid meniscus (DLM) [5, 6] that allows significantly increase the flow speed and, thus, electroplating rates. Using the DLD/DLM technique we also achieve copper pillar plating with accuracy high enough to don't need the further step of CMP.

Preliminary investigation of CPB plating shows the uniformity of better than 2 % resulting for the first time the possibility that a plating process could be sufficient for copper pillar technology avoiding the expensive polishing technique and a suitable TCO process for packaging market.

High speed plating

In electroplating, the deposition rate is defined by current density through the Faraday's law of electrolysis:

$$\frac{d_i}{t} = \frac{M_i}{e z_i N_A \rho_i} jS$$

where d_i is the thickness of a metal, t is the plating time, M_i is the molecular weight of a metal, z_i is the valence of a metal, e is the elementary charge, N_A is the Avogadro constant, ρ_i is the density of a metal, j is the applied current density, and S is the surface area of the cathode.

Higher deposition rates are achievable only through increase of the current density. The current density is limited by the mass transport and the diffusion layer at the interface between the cathode and the bulk solution. Due to the depletion of metallic ions at the cathode area the current density cannot be increased indefinitely. Diffusion issues result in increased hydrogen diffusion that leads to burnt and irregular deposits.

The maximum deposition rate is obtained when all the cathode current is utilized for reduction of metallic ions (Cu^{2+} ions in this case). In practice, deposition rate is limited by side reactions, the reduction of hydrogen in this case, which participating in the cathode current negatively influences the plating efficiency. It was determined experimentally that the plating efficiency is a function of current density and the solution flow rate in the current range where hydrogen evolution occurs at a high rate. The study in [7] showed that the plating efficiency increases with increasing of the flow rate of the solution. It may be deduced that hydrogen evolution tends to participate less in the cathode current with increase of the solution flow rate making the transport issues through the diffusion layer negligible. In this work, we will proof that significantly intensified mass transport during the plating process increases the limiting current density, thus, allowing to achieve higher plating rates.

The mass transport is theoretically described by the Nernst-Planck equation. The equation describes the flux of ionic species under the influence of an ionic concentration gradient, a velocity of the fluid and an electric field [8, 9]:

$$\vec{j}_i = -D_i \nabla c_i + \vec{u}_i c_i - \frac{D_i z_i e}{k_B T} c_i \nabla \varphi$$

where $\vec{j}_i$ is the diffusive species flux density, D_i is the diffusivity of species in the solvent, c_i is the concentration of species, $\vec{u}_i$ is the molar average velocity, z_i is the valence of ionic species, $\nabla \varphi$ is an electric field. The molar average velocity is given by the vector sum of the velocity of the fluid $\vec{u}$ and the electrophoretic velocity of species with respect to the fluid:

$$\vec{u}_i = \vec{u} - \mu_{EP,i} \nabla \varphi$$

where $\vec{u}$ is the velocity of the fluid, $\mu_{EP,i}$ is the electrophoretic mobility of species.

Figure 1 shows the calculated diffusive flux density as a function of the strength of the electric field between cathode and anode, the solution flow speed and the concentration of copper ions during the plating process using DLD/DLM technique. It is seen that at low solution flow speeds (0 and 0.01 m/s) the dependence has clear relation to the electric field and the concentration of copper ions. With increase of the electrolyte flow speed, the influence of the electric field on the mass transport resulting in the electrophoretic mobility of ions reduces. Since that moment the mass transport is defined mainly by the electrolyte flow rate, i.e. the velocity of the fluid. It is clearly seen that the mass transfer of copper ions can be increased three orders of magnitude intensifying the electrolyte flow speed.

Dynamic liquid drop/meniscus technology

The record of copper deposition rate is 50 μm/s that was obtained using jet plating technique at 150 A/cm² current density [4]. Due to the high flow speed of the liquid the main issues for such equipment are to resist erosion and to limit and contain the splashes of the liquid. Sub 0.10 μm copper interconnects already use a jet plating technique but both wafer and anode jet assemblies are immersed in the electrolyte to remove splashes [10]. Nowadays, jet plating in air is mainly used in reel-to-reel plating equipment for the printed circuit board (PCB) industry with a maximum deposition rate of 16 μm/min.

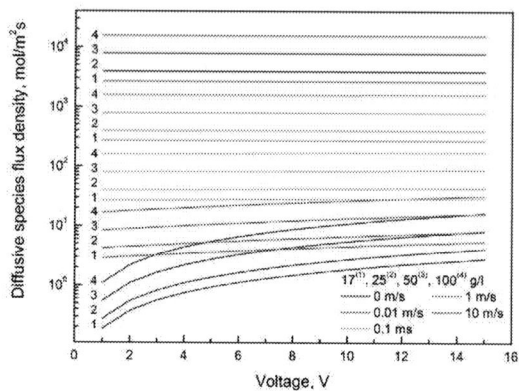

Figure 1. Calculated diffusive flux density as a function of the electric field, the solution flow speed, and the concentration of copper ions during the plating in DLD/DLM configuration.

Localized electrochemical deposition (LECD) was already introduced [11] where a conducting micro-electrode is used to fabricate high aspect ratio metal structures. LECD is performed by placing an electrode tip, which has micrometer dimensions, near a substrate in an electrolyte and applying a potential difference between them. Due to the highly localized electric field in the region between the micro-electrode and the substrate, confined deposition is produced. In any case, there is a severe spreading at columns bases, which limits the resolution of the LECD technique [12]. A liquid static meniscus obtained by a micropipette in close proximity to the substrate solves the spreading problem of the LECD technique. Wire dimensions down to 100 nm where obtained [13]. A static liquid drop is formed at the end of the micropipette, and, as the liquid drop gets in contact with the substrate surface, a static liquid meniscus is formed spreading on the surface depending on the wetting angle. As the dimensions get bigger, gravity influences and, the micropipette, with all the liquid inside, must be fully closed; or as the liquid touches the substrate surface, the liquid spreads out until a new equilibrium is reached between surface tension and gravity. Static meniscus imposes tight control on the distance between micropipette and substrate due to finite quantity of liquid in the meniscus. Increasing such distance breaks the formed meniscus. Furthermore, with a static meniscus is impossible to have mixing of the solution and this will limit seriously the deposition speed and decrease the morphology characteristics of the deposit.

Figure 2 presents 2D view of a DLD. The system is composed, in a principle implementation, by an internal jetting channel where a liquid flow is forced, and an external recalling channel where liquid under a depression (e.g. obtained by a vacuum pump) is recalled back into the system. The input channel, that could be of any shape (e.g. circular, rectangular, etc.), confined by rigid wall (i.e. solid material), pumps a constant liquid flow with the velocity that depends on the input nozzle dimension. Due to a lower pressure in the surrounding of the output channel, an airflow (gas) sustains

978-1-4799-2408-0/14 $31.00 © 2014 IEEE

liquid (figure 2: black arrows in the red liquid) forming a DLD. The confined drop is dynamic due to continuous refreshment of liquid. A schematic view of a 3D DLD for a rectangular input structure is also presented at figure 2.

and, as an example, start a plating process. As finished, a substrate moves away in a lateral plane with speed up to tens of m/s to get positioned in a new processing site.

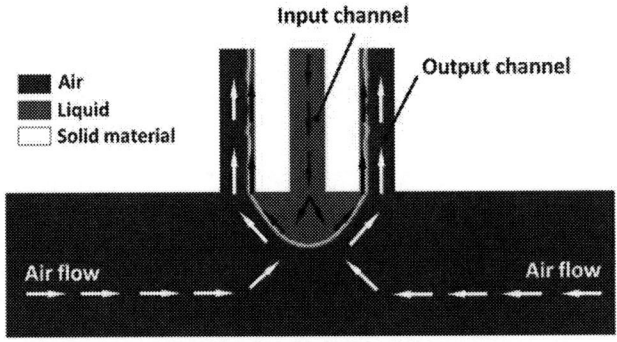

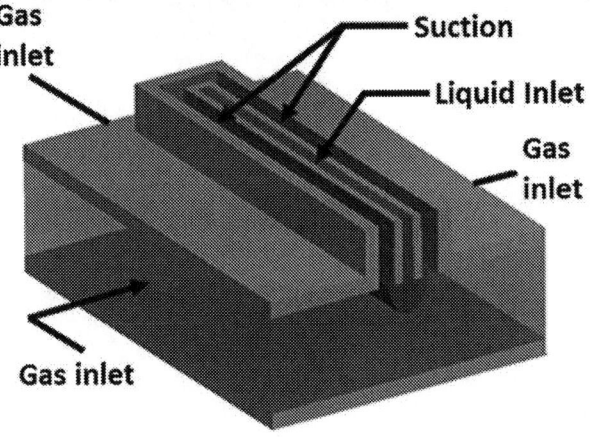

Figure 2. 2D and 3D schemes of the dynamic liquid drop (DLD).

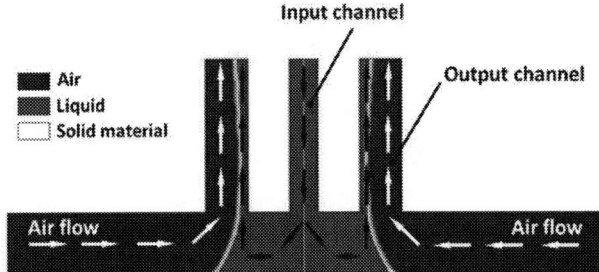

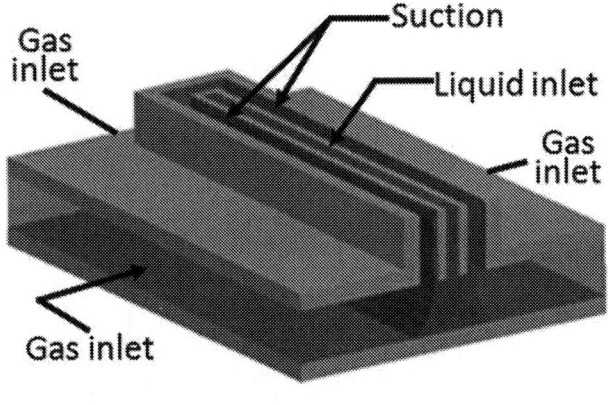

Figure 3. 2D (left) and 3D schemes of the dynamic liquid meniscus (DLM).

As a substrate gets in contact with a DLD, a DLM is formed as shown in figure 3. The contact angle depends on the wettability of the surface. Once DLM is formed and its parameters (i.e. fluid velocity and pressure) are kept constant, lateral movement of the nozzle or substrate is not causing DLM shape to change. In figure 3 the situation when the rectangular DLD touches a substrate is shown. In the same figure, output channels and gas inlet are shown.

In order to avoid liquid losses, a DLD formation has been obtained by, first, applying a pressure drop at the output channel and, second, switching on the pumping system of the liquid input inlet. As the DLD is formed the substrate can be brought in contact by moving the nozzle or a substrate or by a combination of them. In this way it is possible to perform two ways of processing:

1) Continuous process in which the nozzle is fixed and a substrates moves continuously under the nozzle head. Such technique has been tested on substrates kept by vacuum on a belt conveyer with speed up to 50 mm/s without losses of liquid.

2) Stop and go process in which a substrate or the nozzle move up/down in the z-direction, which is orthogonal to the plane of a substrate, to get in contact and form the DLM, stop

Figure 4 illustrates geometrical parameters of the plating head that we were using throughout the present work. Main parameters are the width of the output channel O, the width of walls W, the width of the input channel I, the distance of a substrate H, and the height of the DLD H_d. Using this geometrical parameters properties of the liquid flow were calculated. Table 1 contains configuration geometry of the nozzle and parameters of liquid flowing through it (where ΔP is depression, Q is the flow of a liquid, and V is the velocity of the flow). It is seen that theoretically is possible to obtain very intensive liquid flow with up to 10 m/s speed of liquid. Practically, several obstacles appear. A simplified calculation model is not taking into account the volume of the entire pipelines of the system that actually provides the movement of a liquid. Thus, the calculated depression is a real value only for a certain plating head. Secondly, the momentum conservation between liquid and gas flows should be respected in order to form a stable DLD. Considering the mass ratio between a liquid (e.g. the water) and a gas (usually the air) is approximately 3 orders of magnitude, it has to be compensated by velocities of flows. That, in practice, may not be possible due to very high resulting velocity of a gas flow.

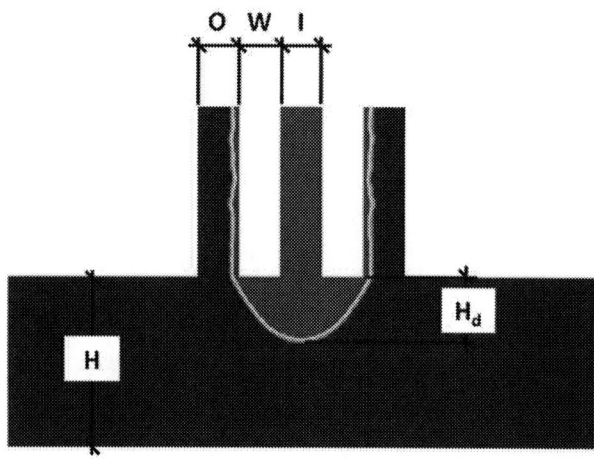

Figure 4. 2D scheme of the DLD, where O is the width of the output channel, W is the width of walls, I is the width of the input channel, H is the distance of a substrate, H_d is the height of the DLD.

Table 1. Geometry of the nozzle of the plating head and calculated parameters of the liquid flowing through it.

Nozzle Dimensions			
O, μm	W, μm	I, μm	H_d, μm
400	300	400	500
Calculated parameters of a liquid flow			
ΔP, kPa	Q, l/min	V, m/s	
1.150	0.072	0.25	
5.250	0.72	2.50	
10.0	1.44	5.0	
20.0	2.88	10.0	

In this work, CPB plating was occurred using the plating head which design is illustrated at figure 5. The plating head was realized by Rapid Prototyping Stereolithography.

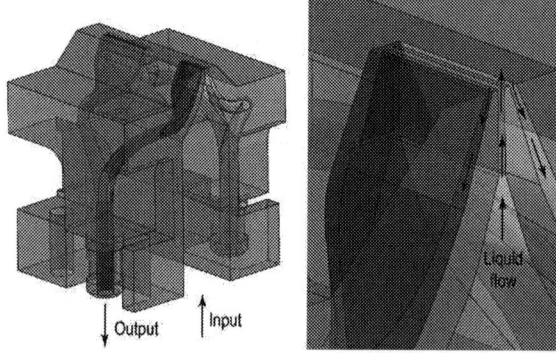

Figure 5. Design of the plating head used in this work.

A plating head has to be connected to the system that provides movement of liquid and gas flows. Inlet channel is connected directly to a tank that contains an electrolyte. In order to start a flow the additional pressure ΔP_I has to be applied. In order to confine a DLD a gas has to start flow immediately. That is reached applying a depression in the outlet channel, allowing continuous gas flow that confines a DLD and suction of an electrolyte. Table 2 summarizes working parameters obtained in this work. It is seen that 1.56 m/s liquid flow is reached using the DLD/DLM configuration.

Table 2. Parameters of the liquid flow obtained experimentally.

Inlet ΔP_I, kPa	Suction ΔP_S, kPa	Q, l/min	V, m/s
30	-50	0.20	0.69
40	-50	0.30	1.06
60	-50	0.35	1.22
80	-50	0.40	1.39
100	-50	0.45	1.56

In this work, we report the continuous mode for CPB plating using the DLD/DLM configuration. A certain area (a chip area) within a wafer (a c-Si wafer 150 mm in diameter, for instance) is processed in continuous mode. Geometrical dimensions of the DLM are as reported in Table 1. The contact area between the DLM and a substrate is designed to be less than the area that have to be plated. In order to plate the entire area, a substrate has to move within one direction forward and backward. This allows to eliminate plating non-uniformity caused by fluid dynamics within the DLM, thus, within a chip area of process and among the entire wafer.

CPB plating using DLM/DLD

In this work, CPB plating was performed through 30 μm thick photoresist mask. A photoresist layer was spinned on a 150 mm c-Si wafer coated with the 300 nm rf-sputtered silver layer, which served as an under bump metallization. The Enthone® Cuprostar LP-1 acid copper plating electrolyte was used in this work. Concentration of copper ions was 17 g/l. CPB plating occurred in the regime when the voltage between anode and cathode was kept constant. Since it's not necessary that a photoresist mask has a uniform distribution of openings, the plating area under a DLM can vary. That issue can be solved in two ways. One solution can be to adjust current according to a certain position. On the other hand, we found that due to fine dimensions of nozzles and channels where passes an electrolyte their electrical resistivity dominates in the electrochemical circuit. It makes a voltage drop on the cathode to be negligible, thus, allows to use voltage control for the plating process, which is able to adjust automatically the current.

The Alfa-Step 300 profilometer and the Carl Zeiss AURIGA CrossBeam FIB-SEM Workstation were used to investigate CPB obtained in this work. Different copper pillar dimensions were investigated within different topology over the wafer showing the robustness of the process for the thickness uniformity.

Figure 6 shows an array of copper pillars obtained using DLD/DLM plating process. Measurements showed height distribution of copper pillars in the range 27.8–28.3 μm and

978-1-4799-2408-0/14 $31.00 © 2014 IEEE

the thickness uniformity is 1.77 % within both a single chip and the entire wafer. Each copper pillar has pin-like shape obtained due to over-plating effect through the 30 μm thick photoresist that had vertical walls for 25 μm and a cone shape for the last 5 μm. The image of a single copper pillar is presented in figure 7, where is evident the over-plated cap due to photoresist non uniformity. The plating current density was approximately 3 A/cm^2 resulting in 1 μm/s deposition rate. It is seen that plating resulted in smooth and dense deposit quality.

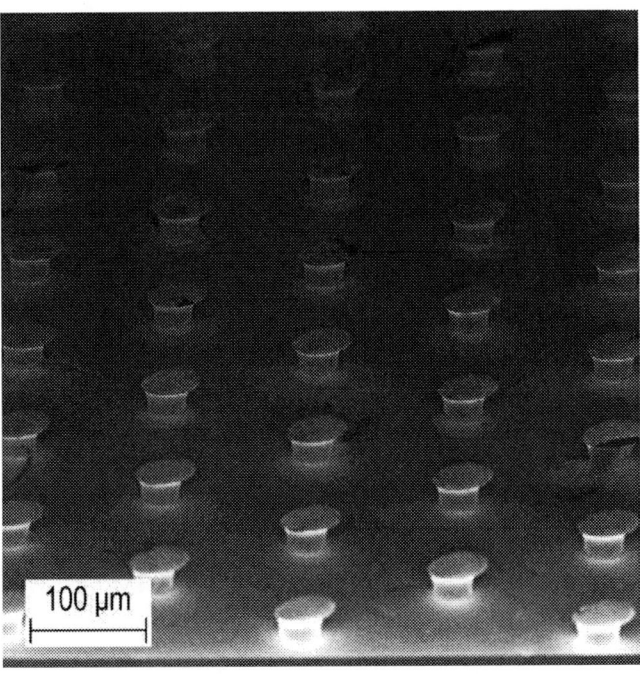

Figure 6. A birds-eye view of an array of copper pillars.

Figure 7. SEM image of a single 28 μm thick copper pillar.

In this work, there are two contributors in non-uniformity of CPB. The first one was obtained due to over-plating effect on the photoresist mask. The second was inherited from satin quality deposit with 0.2–0.8 μm roughness that is provided by the Enthone® Cuprostar LP-1 copper plating electrolyte. In order to further improve the uniformity of the process copper plating electrolytes giving bright deposits, i.e. roughness in the range of 20nm will be used in next processes as well as 100 μm thick dry film photoresist masks.

Conclusions

In this work the mechanism of high speed plating was discussed in terms of hydrodynamics of the fluid. It was hypothesized that significantly intensified mass transport during the plating process increases the limiting current density, thus, allows to achieve higher plating rates. Calculations of diffusive flux density as a function of the electric field, the solution flow speed, and the concentration of copper ions during the plating in DLD/DLM configuration according to Nernst-Plank equation was provided. It was shown that the mass transfer of copper ions can be increased three orders of magnitude intensifying the solution flow speed. CPB plating using the DLD/DLM technique that allows to implement high speed plating of copper was discussed. The thickness uniformity of ~2 % within the wafer at 1 μm/s deposition rate was achieved.

Acknowledgments

The assistance of Dr. Bruno Zaccaria from Enthone®, for the preparation and providing the copper plating solutions is gratefully acknowledged.

References

1. H.-J. Hsu et al., "A novel high coplanarity lead free copper pillar bump fabrication process," in *Proc. IEEE Interconnect Technology Conference (IITC)*, Sapporo, Hokkaido, June 1–3, 2009, pp. 169–170
2. W. Koh and B. Lin, "Process development for high-current electrochemical deposition of copper pillar bumps,", in *Proc. IEEE Electronic Components and Technol. Conf. (ECTC)*, San Diego, CA, May 29–June 1, 2012, pp. 630–635.
3. Y. Zhang et al., "A high speed Cu pillar bump plating process," in *Proc. Microsystems, Packaging, Assembly & Circuits Technology Conference (IMPACT)*, Taipei, October 22–24, 2008, pp. 28–31.
4. R. J. von Gutfeld and D. R. Vigliotti, "High-speed electroplating of copper using the laser-jet technique," *Appl. Phys. Lett.*, vol. 46, no. 10, pp. 1003–1005, May 1985.
5. M. Balucani et al., "New selective wet processing," in *Proc. IEEE Electronic Components and Technol. Conf. (ECTC)*, Las Vegas, NV, May 28–31, 2013, pp. 247–254.
6. M. Balucani et al., "New selective processing technique for solar cells," *Energy Procedia*, vol. 43, pp. 54–65.
7. L. J. J. Janssen, "High-rate electrochemical copper deposition on bars," *J. Appl. Electrochem.*, vol. 18, iss. 3, pp. 339–346, May 1988.

8. R.F. Probstein, *Physicochemical Hydrodynamics. An Introduction (2nd edn)*, Wiley, New York, 1994, pp. 41–45.

9. B. J. Kirby, *Micro- and Nanoscale Fluid Mechanics. Transport in Microfluidic Devices*, Cambridge University Press, New York, 2010, pp. 250–256.

10. M. De Vogelaere et al., "High-speed plating for electronic applications," *Electrochim. Acta*, vol. 47, iss. 1–2, pp. 109–116, Sept. 2001.

11. I. W. Hunter et al., "Three dimensional microfabrication by localized electrodeposition and etching," U.S. Patent 5 641 391, June 24, 1997.

12. R. A. Said, "Microfabrication by localized electrochemical deposition: experimental investigation and theoretical modelling," *Nanotechnology*, vol. 14, pp. 523–531, Mar. 2003.

13. J. Hu and M.-F. Yu, "Meniscus-confined three-dimensional electrodepostion for direct writing of wire bonds," *Science*, vol. 329, no. 5989, pp. 313-316, Jul. 2010.

Plasma-based Die Singulation Processing Technology

Kenneth D. Mackenzie, David Pays-Volard, Linnell Martinez, Christopher Johnson, Thierry Lazerand, Russell Westerman
Plasma-Therm LLC
10050 16th Street North, St. Petersburg, FL 33716, USA
Tel: (727) 577-4999
Email: ken.mackenzie@plasmatherm.com

Abstract

In support of improved productivity for the semiconductor and optoelectronics manufacturing industry, new work is presented on the development of a plasma-based die singulation process technique for thin Si wafers. It is shown that the technique leads to a significant gain in productivity through reduced process times and increased available good die per wafer. Some additional related key benefits are increased yield, die strength, and the potential of non-orthogonal die.

1. Introduction

The semiconductor and optoelectronics industry is driven to lower manufacturing costs while still maintaining high performance devices with good yields and high reliability. At the wafer level, lower manufacturing costs translate into increasing the number of good active die chips that can be extracted from each wafer and made available for final assembly and packaging. To increase die per wafer, it is important to consider utilization of all real estate on the wafer. A non-negligible percentage of the wafer area is reserved for the scribe lines or "streets" used in the die singulation process. Presently, the typical widths of these streets are about 50 to 100 µm and dictated by current die singulation techniques. Any reduction in street widths will free up more real estate to add extra die.

In this paper, we report on a new and efficient plasma-based die singulation technique developed for dicing silicon wafers that achieves a significant reduction in street widths while improving some aspects of the die integrity. These benefits will be demonstrated for reduced street widths of 15µm.

Plasma-based die singulation technology complements the industry direction toward increased implementation of thin (~100µm) and ultra-thin (≤50µm) wafers. By 2017, it is expected that nearly 75% of all wafers shipped in the world will be thin wafers and that their thickness will be at or below 200µm [1]. This inevitable trend is essentially driven by package size reduction, especially for portable electronics. The final die thickness is entirely dominated by the wafer thickness. The contribution to die height from the device circuitry on a die is a small percentage and perhaps only on the order of about 20µm i.e. about 3% of a full thickness Si wafer.

With this motivation for miniaturization with more densely packed devices on a die, thinned Si is critical to improve heat dissipation and device performance [2]. This is not only important in applications involving 3D packaging, and also for GaN-on-Si LEDs, CMOS image sensors, RF, and power devices, but also for ON resistance reduction of MOSFET and breakdown voltage, ON state, and switching energy loss reduction for IGBT [1].

As will be discussed in later sections of the paper, the new singulation technology overcomes some of the significant manufacturing challenges encountered in processing thin wafers by conventional wafer dicing technologies. A review of conventional die singulation methods is given in the next section. In the remaining sections, we present the plasma-based die singulation technology and discuss the performance and capabilities, including some added benefits gained by adoption of the technique.

2. Die Singulation - Overview

The dicing or singulation process is used extensively in the semiconductor industry to separate each die after their fabrication on wafer substrates and prior to package encapsulation and installation on printed circuit boards. Conventional wafer dicing technology is based on two principal methods, mechanical sawing and laser dicing [3]. In the first method, a high-speed rotating saw is used to separate or singulate each die after their fabrication on wafer substrates. Productivity is compromised by both the serial nature of the process and the necessary reduction in linear saw cutting speed to prevent wafer breakage and layer delamination, particularly for substrates with thickness of 100µm or less. A compounding issue with saws is the induced cracking at die edges and resulting decrease in die strength.

Mechanical saw dicing is relatively low cost and used in about 90% percent of applications on thick or full thickness wafers. Dicing using laser ablation has been attempted for a few years to overcome process limitations associated with sawing particularly for thin wafers. A focused laser beam potentially allows die singulation using smaller streets resulting in an increase of the active silicon area on the wafers or more die per wafer. Laser dicing is gaining acceptance for applications with fragile substrates commonly found in compound semiconductor devices such as sapphire, LiNbO$_3$, GaAs, and SiC. In an effort to increase yields, often, a two-step combination of laser ablation and mechanical sawing is used, at the expense of a significant drop in productivity. Furthermore, the heat generated by the laser beam induces thermal damage to the lateral region around the area of contact. This can lead to an unwanted potential for interlayer mixing to occur, and multi-layer delamination by thermal mismatch. This forces the designers to maintain enough real estate in the streets and to build in "seal rings" to protect the die edges from thermal damage induced by laser and crack propagation from the saws. This becomes even a more predominant concern with ultra-thin (~50µm) wafers, especially with respect to die strength. A variant of laser dicing is called stealth dicing in which with the aid of a longer wavelength laser, the beam is focused deeper into the silicon to initiate singulation and to avoid contamination or damage to the wafer surface. With small die on thin wafers, an extra break step after stealth laser dicing is sometimes required.

Due to the chemical nature of the process used in the plasma-based singulation technique, no mechanical force or vibration is applied to the wafer thus eliminating wafer breakage, layer delamination, and lateral damage or chipping of the die. There is also no associated thermal-induced damage with the plasma-based technique.

3. Plasma-based Die Singulation Technology

This technology is based on the highly successfully time-multiplexed deep reactive ion etching technique originally developed in the 1990's for MEMS devices at Robert Bosch GmbH [4]. Applications for this technique have greatly expanded into many new areas and now is recognized as the enabling related technology for TSV (Through Silicon Via) die interconnects in 3D packaging [5]. The technique, commonly referred to as the "Bosch", DRIE, or DSE™ process, is capable of achieving very deep and narrow trenches in silicon with high selectivity to common mask materials such as photoresist, polyimides, and silicon dioxide.

A newly designed production system for plasma dicing of wafers on tape frames has been developed at Plasma-Therm LLC based on an adaptation of the "Bosch" technique. The system, a model MDS™ 100 micro die singulator etch machine is fully automated with a dual cassette tape frame loading station and robotic handler. Following the same procedures as for conventional die singulation, prior to processing the wafers are mounted on tape frames. Wafers sizes up to 200mm can be accommodated on each tape frame. To maintain downstream compatibility with existing packaging work flows, the form factor of the tape frames used follows industry standards. Both from a logistics and a cost perspective, this is extremely important when introducing a new technology into an existing manufacturing process line.

The etch reactor module on the system is equipped with an inductively coupled plasma (ICP) source and RF-biased lower electrode. During processing the wafer and tape frame are clamped and cooled. Process gases SF_6 and C_4F_8 are used in the time-multiplex alternating etching/deposition cycles to etch the silicon in the streets and fully singulate all the die all the way down to the tape.

The system has been specially engineered to handle the wafer mounted on a dicing frame and avoid degradation of the tape during plasma etching. Process termination, when complete wafer dicing has occurred, is achieved by an optical endpoint technique.

Process development has been done on a wide variety of wafer sizes (25 to 200mm) of differing thicknesses from 100 to 700µm, mask materials, die size, and street widths on the MDS 100 machine. The experimental and model calculation results presented in the next sections are based on this work.

4. Plasma Die Singulation – Performance & Capabilities

4.1 Good Die Per Wafer

As discussed earlier, with the restriction on narrowing street width lifted, it is possible to increase the die density on the wafer. Figure 1 shows the calculated percentage gain in good die per wafer when the street width is reduced from a conventional width of 90µm to 15µm for a range of common die sizes from 2mm x 2mm down to 125µm x 125µm. The calculation is for a 200mm wafer with a 3mm edge exclusion. From Figure 1, the gain in available good die per wafer possible by plasma-based singulation is about 8% for the large 2mm x 2mm die and up to about 140% for small 125µm x 125µm die.

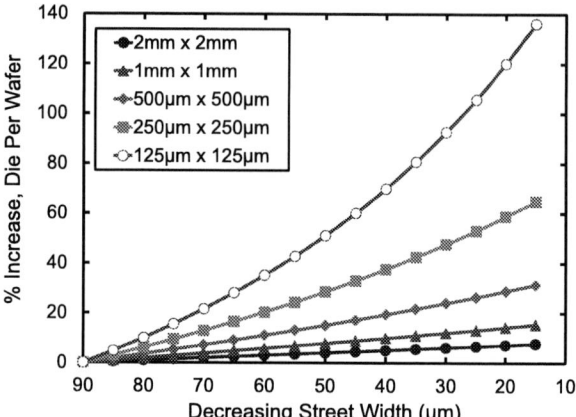

Figure 1. Calculated percentage increase in available complete good die through reduction of street widths from 90 to 15µm for indicated different square die dimensions. Wafer size is 200mm with 3mm edge exclusion.

It is instructive to translate the percentage gain in good die per wafer into numerical values. This can be estimated from the following equation [6].

$$ DPW = \frac{\pi}{4}\left(\frac{D}{d+w}\right)^2 - \frac{\pi}{\sqrt{2}}\left(\frac{D}{d+w}\right) \qquad (1) $$

where DPW is the good die per wafer, D is the useable diameter of the wafer, d is the width of the square die, and w is the width of the streets surrounding the die. It is common to include a few test die on every wafer. This will reduce the number of available good die estimated from equation (1).

Table I shows an example from this calculation comparing number of good die available, for the same series of die sizes from Figure 1, when the street width on a 200mm wafer is reduced from 90 to 15µm.

Table I
Comparison of good die on 200mm wafer for 90 & 15µm streets for various die sizes (3mm edge exclusion).

Die Size	90µm Streets	15µm Streets
2mm x 2mm	6,560	7,070
1mm x 1mm	24,500	28,300
500µm x 500µm	84,200	111,000
250µm x 250µm	254,000	419,000
125µm x 125µm	637,000	1,510,000

As summarized in Table I, the number of additional die that can be achieved to increase manufacturing productivity is significant.

It should be noted that all calculations described above use a minimum street width of 15μm. With the plasma-based singulation technology, narrower street widths down to 5μm or less are possible. Though, downstream process equipment such as pick and place defines the minimum practical street width at currently between about 10 to 15μm before tape expansion.

4.2 Increased Die Active Area

The advantage of narrowing the street widths is not limited to increasing the number of good die per wafer. The additional real estate gained can be used to increase the existing die size. This increase in active area can be beneficial to device performance. As an example, for power devices, the increased active area leads to a highly desirable reduction in ON resistance.

Figures 2 and 3 show a comparative example to illustrate gain in die size between identical die when the street width is reduced from 90 to 15μm. In both cases, plasma-based singulation was used.

Figure 2. Plasma-singulated die separated by 90μm streets.

Figure 3. Plasma-singulated die separated by 15μm streets.

4.3 Silicon Etch Rate

Throughput in semiconductor manufacturing is often expressed in wafers per hour. In the case of plasma-based processing, this usually depends directly on the deposition or etch rate. In this regard, the DRIE process with the capability of high etch rates, controllable feature profiles at high aspect ratios is well suited for high throughput applications. The DRIE Si etch rate is dependent primarily on two parameters, the feature aspect ratio and the exposed Si load. As illustrated in Figure 4 by the compiled literature values for Si etch rate *versus* feature aspect ratio, the etch rate increases significantly as the aspect ratio decreases [7]. This is beneficial for die singulation especially for thin die surrounded by narrow streets. Etch rates on the order of 20μm/min are not uncommon. Typical process times to fully singulate 1mm x 1mm die separated by 15μm streets on a 100μm thick 200mm wafer is about 5 minutes.

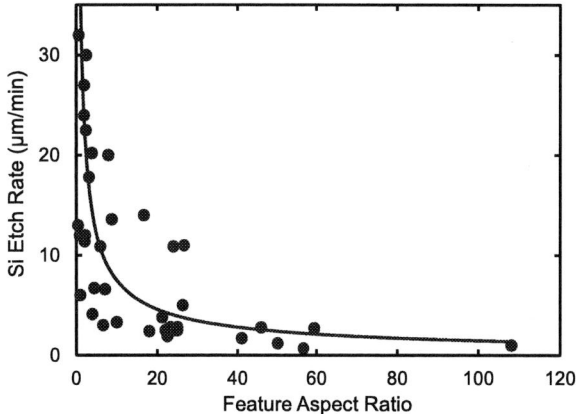

Figure 4. Measured dependence of Si etch rate on feature aspect ratio. Fitted solid line emphasizes monotonic increase in etch rate with decreasing aspect ratio. (Data compiled from published literature for anisotropic features at various Si loads).

As mentioned above, the etch rate is not only influenced by the feature aspect ratio, it is also affected by the amount of exposed Si. Etch rate is inversely proportional to the exposed Si load [8], [9]. Therefore, the higher the percentage of exposed Si then the lower the Si etch rate. As summarized by the calculated results in Figure 5, the ability to narrow the street width greatly reduces the exposed Si load resulting in higher etch rate.

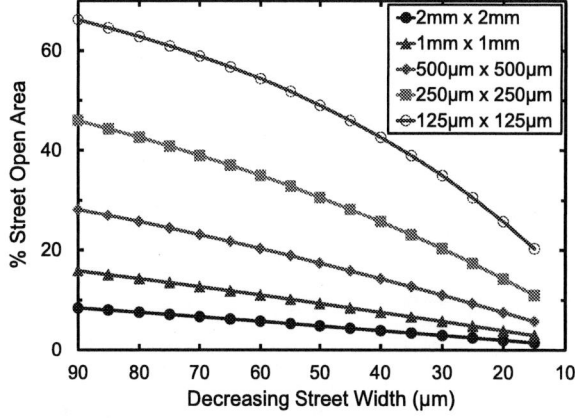

Figure 5. Calculated reduction in exposed Si with decrease in street width for die sizes indicated.

For the reasons discussed above, the DRIE process technology is ideally suited as an efficient technique for wafer dicing and die singulation applications.

4.4 Effective Cut Speed

The plasma-based process operates quite differently to conventional die singulation techniques. Diametrically opposite to the serial nature of saw or laser dicing where each street needs to be cut one after another, the plasma dicing acts as a parallel process eliminating the material in all streets at same time across the entire wafer.

To assess the plasma-based singulation performance with that of existing techniques, it is instructive to calculate an "effective" linear cut speed derived from the actual plasma etch die singulation process times. Figure 6 shows the effective linear cut speed for 1mm x 1mm die on a 200mm Si wafer with streets of 15µm and wafer thicknesses from 300 down to 50µm. For thick wafers, the effective cut speed is comparable with current technology. However, for thin wafers the effective cut speed is considerably higher and increasing rapidly as the wafer thickness decreases. This is opposite to the case of conventional dicing saw which has to be slowed down on thinner wafers solely to minimize cracking, chipping and wafer breakage.

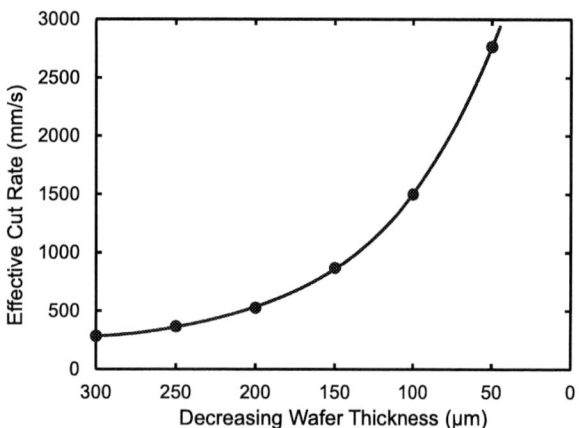

Figure 6. Effective cut speed for plasma singulation method *versus* wafer thickness.

4.5 Productivity: Throughput

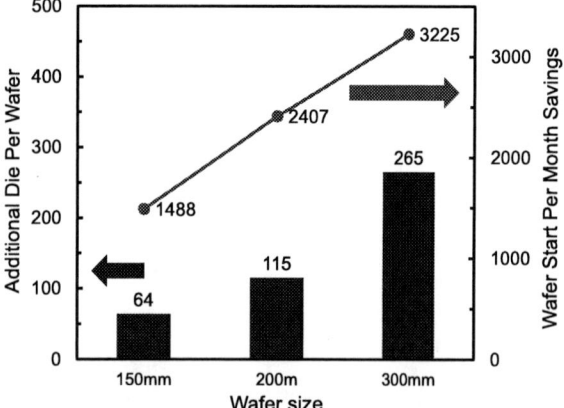

Figure 7. Plasma singulation system throughput showing gain in additional 3mm x 3mm die per wafer and savings in monthly wafer starts.

Figure 7 presents one example of the throughput benefit of the plasma-based singulation process. The calculation is done based on predicted process times for large 3mm x 3mm die on 100µm thick wafers for narrow streets for wafer sizes of 150, 200, and extending calculation to 300mm wafers. As the results indicate even for such large die on a 150mm wafer, the gain in additional die and wafer start per month savings are significant.

4.6 Die Fracture Strength

Fracture due to the brittle nature of silicon occurs via crack propagation along the surface or from within the material itself [10]. Crack propagation leading to breakage can occur at any time and without warning. It is therefore important to minimize the probability of crack formation and its propagation. This is extremely important for thinned Si wafers which are mechanically extremely fragile and can have high propensity to fracture.

To achieve a high die fracture strength, it is therefore critical during the singulation process on these thinned wafers that no cracking or chipping occurs. Figure 8 summarizes the measured relative die fracture strengths for singulated 1mm x 1mm die on 120µm thick Si by mechanical saw, laser, and plasma-based dicing. The die fracture measurements are done by the four-point bending method.

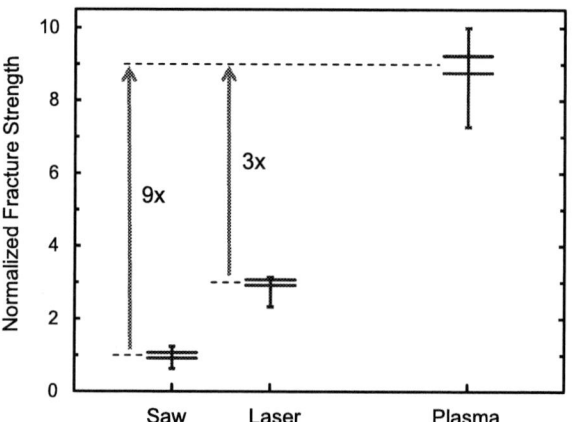

Figure 8. Die strength comparison between saw, laser, and plasma dicing.

As the results indicate, the plasma-based technique has a factor 9 higher die fracture strength compared to conventional mechanical saw, and about a factor 3 higher than laser-based dicing. This asserts a key benefit of the plasma-based technique compared to the existing methods; allowing further die thinning compared to the incumbent dicing methods. Porter and Berfield also report improved die strength with the plasma approach compared to traditional mechanical saw [11].

A weakness of orthogonal die is the localized high concentration of stresses in the small areas at the tips of the die corners. This can lead to fracture. Using the plasma-based singulation method, it is possible to round the corners of the die and thereby minimize the stress. This capability addresses packaging issues especially for very large and ultra-thin die.

Figure 9 shows an example of die with rounded corners achieved by the plasma-based singulation process.

978-1-4799-2408-0/14 $31.00 © 2014 IEEE

Figure 9. SEM image of die with rounded corners.

4.7 Die Edge Quality

Figure 10 compares the die edge quality achieved by the die singulation techniques of mechanical saw, laser, and plasma dicing. As indicated by review of the optical images, in contrast to the saw and laser methods, the die edge quality of the plasma dicing approach is totally free of any defects, debris or cracks. In addition, the final die size is predetermined by the resolution of the lithography step which results in extremely well controlled dimensions.

Technique	Frontside	Backside
Saw		
Laser		
Plasma		

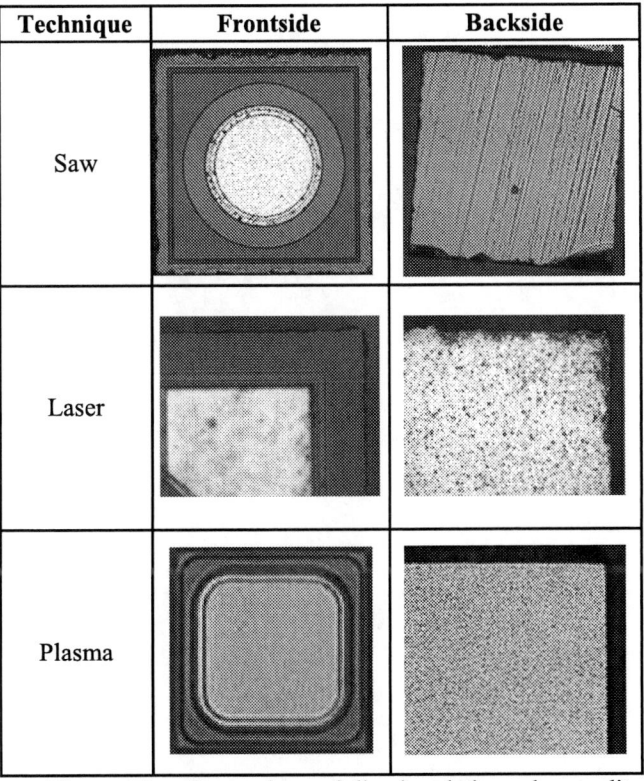

Figure 10. Comparison of die singulation edge quality between saw, laser, and plasma dicing.

4.8 Bond Pad Exposure

In plasma etching of semiconductors and related materials, fluorine and also chlorine based chemistries are used. In order to prevent possible corrosion to any exposed materials or formation of unwanted interfacial layers by these highly reactive elements and their associated species generated in the plasma, it is extremely important to clean the surface of the exposed material on the diced wafers prior to removal from the process vacuum chamber.

In plasma-based die singulation, there may be instances where aluminum bond pads are exposed directly to the fluorine-based Si etch chemistry. Fluorine reacts with aluminum to form a thin and electrically resistive aluminum fluoride layer on top of the bond pad. If present, this AlF layer needs to be removed [12].

As shown by the Auger electron spectroscopy depth profile data presented in Figures 11 and 12 that through implementation of a plasma clean sequence immediately following the die singulation process, the AlF layer can be eliminated leaving a clean aluminum bond pad ready for the next stage in the processing, wire bonding. Wire bond pull strength tests are covered in the next section.

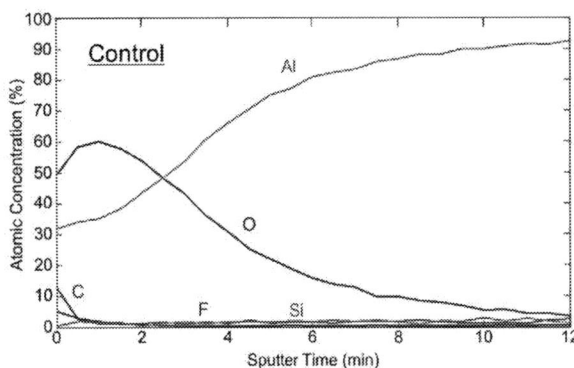

Figure 11. Auger depth profile data on unprocessed control sample for Al bond pad. (Sputter rate of ~10Å/min of equivalent SiO_2.)

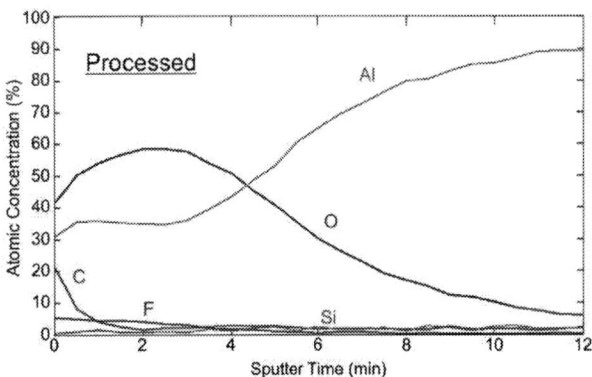

Figure 12. Auger depth profile data for Al bond following plasma-singulation and post etch clean. Note comparable low F levels with control sample (Figure 11).

4.9 Wire Bond Pull Strength Tests

To evaluate the efficacy of the post etch clean process described above on the aluminum bond surface, wire bond pull strength measurements were made under multiple bonding and post plasma die singulation cleaning procedures. In Figure 13, the results from these tests are summarized. This destructive test is to determine the force at which the wire bond fails. For conditions investigated, the wire pull strength exceeds the minimum requirements necessary for successful die bonding. There is no statistical difference between these tests results. The results are similar to that achieved by conventional mechanical saw. In Figure 13, the saw singulation data is marked "STD" and other splits are plasma singulation on the graph.

978-1-4799-2408-0/14 $31.00 © 2014 IEEE   1581

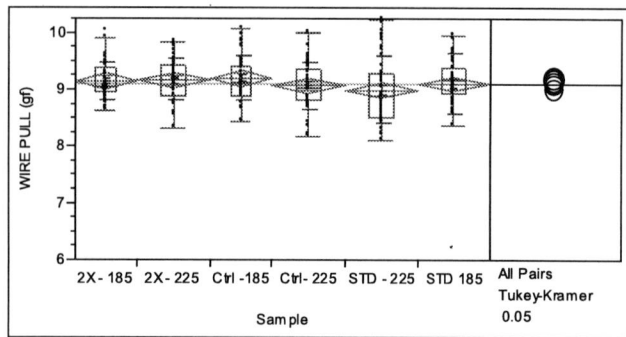

Figure 13. Measured wire bond pull strength on aluminum pads on die measured under multiple bonding and cleaning conditions after completion of singulation process.

4.10 Non-Orthogonal Die Capability

The majority of die singulated by conventional techniques are either square or rectangular in format and orthogonally aligned on the wafer. With plasma-based singulation, any type of die shape is essentially possible e.g. hexagons and round die. This opens up new possibilities for creative designers for many different device types in electronic, optical, and optoelectronic applications, including the repositioning of large die on the same wafer to optimize the active silicon utilization.

Figure 14 shows a patterned 100mm wafer with circular die mounted on tape frame prior to plasma die singulation. Figure 15 shows the extracted circular optical devices with a through wafer central hole from this wafer after application of plasma-based singulation.

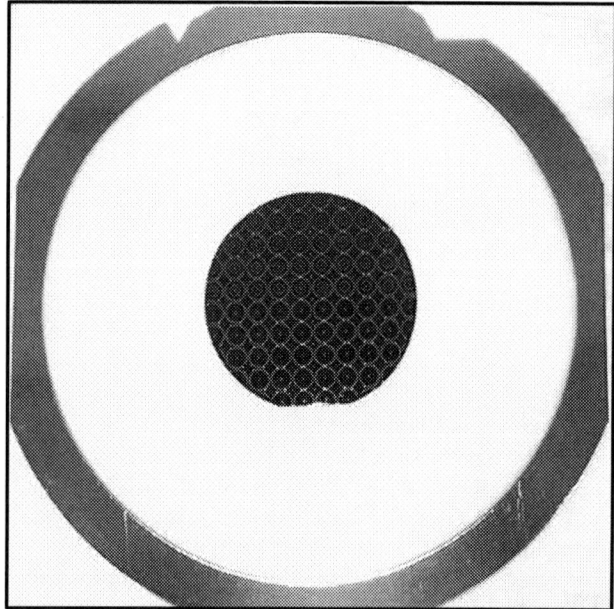

Figure 14. 100mm patterned wafer on dicing tape prior to plasma etching.

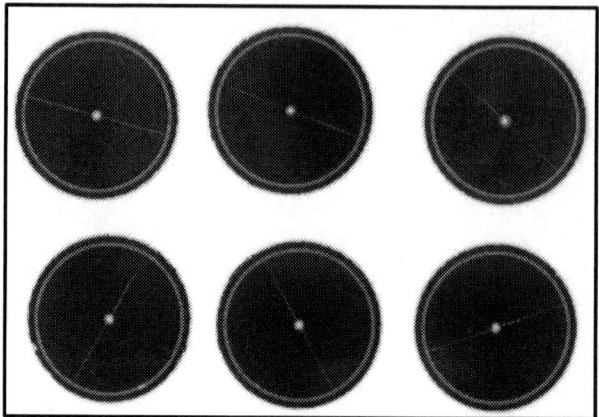

Figure 15. Fully plasma singulated devices with through wafer holes in Figure 14.

Figures 16 and 17 show more examples of non-orthogonal die including multi-shaped die on the same wafer produced by the plasma-based singulation technique.

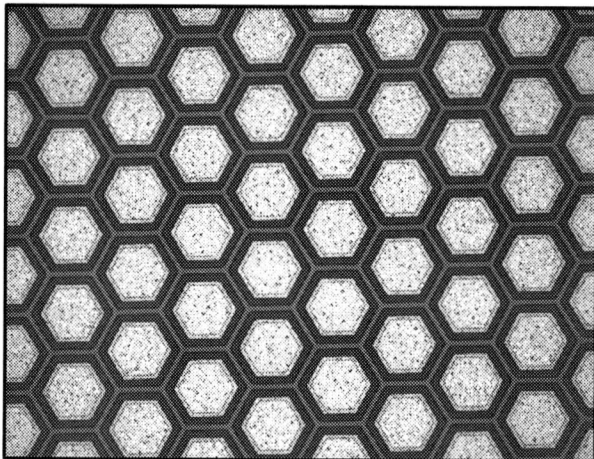

Figure 16. Hexagonal die produced by plasma-based singulation.

Figure 17. Plasma-singulated multi-shaped die on same wafer.

4.11 Device Performance & Finished Product

Multiple types of devices singulated using the plasma-based technique have been qualified. Completed electrical testing on both product and PCM test structures show no significant shift in electrical performance.

Figure 18 shows an example of a packaged wire-bonded plasma-singulated die taken from a product wafer.

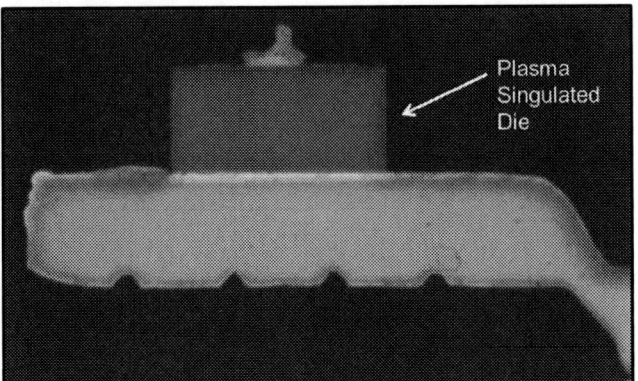

Figure 18. Packaged wire-bonded plasma-singulated die from product wafer.

5. Conclusions

In summary, DRIE-based plasma dicing on tape is an extremely promising technology that is expected to help in manufacture of semiconductor and optoelectronic devices by productivity improvements, by the possibility of higher performance devices, and by the creation of new and novel devices. A very important aspect of this technology is the ability to singulate wafers on conventional dicing tape frames. This makes easier implementation and integration into existing process work flows. The key features of this technology demonstrated relative to conventional die singulation technologies are summarized below.

1. *Higher Throughput*
   - More good die per wafer
   - Reduced process times

2. *Higher Yield*
   - High die fracture strength
   - Low mechanical stress
   - No shift in electrical performance

3. *New Device Design Potential*
   - Exclusivity of orthogonal die lifted

Acknowledgments

The authors gratefully acknowledge Gordy Grivna, Jason Doub, and the team at ON Semiconductor for permission to use the die photographs and die analysis data. Additionally, the authors sincerely appreciate Roger Forrest at Opto Diode Corporation, an ITW company for the use of the round die images shown in the paper.

Finally, the authors wish to recognize Michael Moore, John Nolan, Rich Gauldin, Joe Barraco, Mike Teixeira, and Jason Plumhoff at Plasma-Therm LLC for their continued technical support throughout the course of this work.

References

1. E. Mounier, A. Pizzagalli, and M. Rosina, *"Thin Wafers & Temporary Bonding - Equipment & Material Market"* Yole Development S.A, France, Oct 2012.

2. R. W. Keyes, "Physical limits of silicon transistors and circuits," Rep. Prog. Phys. vol. 68, pp. 2701-2746, 2005.

3. W.-S. Lei, A. Kumar, and R. Yalamanchili, "Die singulation technologies for advanced packaging: A critical review," J. Vac. Sci. Technol. B vol. 30, pp. 040801-1/27, 2012.

4. F. Laermer and A. Schilp, "Method of anisotropically etching silicon" U.S. Patent No. 5501893 (1996).

5. J. H. Lau, *Through-Silicon Vias for 3D Integration*, McGraw Hill, New York, 2013.

6. J. L. Hennessy and D. A. Patterson, *Computer Architecture: A Quantitative Approach*, Morgan Kaufmann, San Francisco, 2007.

7. R. Westerman, L. Martinez, D. Pays-Volard, K. Mackenzie, and T. Lazerand, "Deep silicon etching: current capabilities & future directions," to be published in *Proc. SPIE Photonics West* (2014).

8. C. J. Mogab, "The loading effect in plasma etching," J. Electrochem. Soc. vol. 124, pp. 1262-1268, 1977.

9. J. Karttunen, J. Kiihamaki, and S. Franssila, "Loading effects in deep silicon etching," *Proc. SPIE* vol. 4174, pp. 90-97, 2000.

10. H. M. Rosenberg, *The Solid State*, Clarendon Press, Oxford, 1975.

11. D. A. Porter and T. A. Berfield, "Die separation and rupture strength for deep reactive ion etch silicon wafers," J. Micromech. Microeng. vol. 23, pp. 085020-1/8, 2013.

12. K.-H. Ernst, D. Grman, R. Hauert, and E. Hollander, "Fluorine-induced corrosion of aluminium microchip bond pads: an XPS and AES analysis," Surf. Interface Anal. vol. 21, pp. 691-696, 1994.

Removed Organic Solderability Preservative (OSPs) by Ar/O2 Microwave Plasma to Improve Solder Joint in Thermal Compression Flip Chip Bonding

Jr-Wei Peng[1*], Yan-Siang Chen[1], Yi Chen[1], Jiang-Long Liang[2], Kwang-Lung Lin[2], Yuh-Lang Lee[3]

[1]ASE Group, No. 26, Chin 3rd Rd., Nantze Export Processing Zone, Kaohsiung 81170, Taiwan

[2]Department of Materials Science and Engineering, National Cheng Kung University, No.1, University Road, Tainan 70101, Taiwan

[3]Department of Chemical Engineering, National Cheng Kung University, No.1, University Road, Tainan 70101, Taiwan

*jrweipeng@gmail.com

Abstract

Organic soderability preservative (OSP) is the cheapest metal surface finish to apply for corrosion inhibition of bare copper. In conventional flip chip process, the various OSPs could be dissolved by rosin pre-fluxes before the chip placement to enable solder bonding. The thermal compression flip chip bonding with non-conductive paste (TCNCP) process is one type of no-flow underfill process. In a TCNCP process, the OSP layer must dissolve in self-fluxing agent in NCP, but the process does not have adequate time for allowing complete dissolution to occur. In order to assist in OSP removal, this work attempted to Ar/O$_2$ microwave plasma to remove the OSP layer on Cu trace of a substrate. The SEM-FIB investigation revealed that almost all of OSP layer could be removed with plasma, while only few OSPs remained in the relatively deep concavity on the surface of Cu trace. The analysis results by SEM-EDX showed that the carbon content of the treated plasma residue decreased from 44.69% to 11.5%. The Raman spectrum also showed that the bands of vibration mode in benzimidaozle almost disappeared, indicating the successful removal of the OSP layer by Ar/O$_2$ microwave plasma. The cross-sectional SEM image of the bump on trace produced with the TCNCP process indicated that thin intermetallic compound (IMC) layer formed at the solder/Cu trace interface when incorporated with Ar/O$_2$ microwave plasma pre-treatment. The SEM-DEX composition analysis delineated that the IMC layer consists of Cu$_3$Sn and Cu$_6$Sn$_5$. These results indicate the successful removal of OSP.

Introduction

Organic solderability preservative (OSP) has been widely used for protecting bare Cu on a printed circuit boards (PCBs) and Ball Grid Array (BGA) substrate. The OSP finish offers effective corrosion protection during the thermal excursion of assembly process and lowers processing cost. The organic compounds of OSP are based on aza-aromatic bicyclic molecules, such as benzotriazole and benzimidazole, which will form organometallic compounds with Cu and cover the surface of bare Cu. The thickness of it is usually 100 to 500 nm [1].

It is generally known that the OSP layer would be dissolved in inorganic acid, e.g. hydrochloric acid (HCl), sulfuric acid (H$_2$SO$_4$), or organic acid, e.g. abietic acid, formic acid. The most common flux adopted in conventional flip chip underfill process is organic acid. The acid provides flux appropriate level of hydrogen ion for not only preventing solder and Cu from oxidizing, but also removing the OSP film.

Thermal compression bonding (TCB) with nonconductive paste (TCNCP) process is a no-flow underfill process. In a TCNCP process sequence, the NCP is dispensed on to a preheated substrate. The chip is then picked, aligned to bond on substrate, and then held at elevated temperature under bonding force for a period of bonding time for solder joint formation. Afterwards, the assembly will be post cured. This process eliminates the flux dispensing, dipping and cleaning steps. It also avoids the capillary flow of NCP. In order to inhibit the formation and growth of metal oxide during the TCB step, the NCP must contain self-fluxing function. The 0.5~4% of self-fluxing agent was generally incorporated with NCP [2]. Although the self-fluxing function could also remove the OSP layer, the fast TCB process, typically couple seconds, does not allow enough time for dissolving the OSP completely. Part of the reason of the inadequate removal of OSP is due to the behavior that the NCP would be squeezed out from the top side of Cu trace on substrate. Consequently, the TCNCP process could end up with poor wettability and solder cold joint with OSP finished substrates.

Plasma is a non-equilibrium system composed of various kinds of neutral and charged particles. Plasma has been widely used for surface cleaning, for instance, oxygen ECR (electron cyclotron resonance) plasma was used to remove organic contaminants on silicon wafer [3]. Plasma also can be used to decomposed the structure of organic compounds, for example, poly(ethylene terephalate) (PET) could be aged and degraded in an oxygen plasma [4]. In this work, for improving solder joint in a TCNCP process, the OSP finished substrate was treated by argon-oxygen microwave plasma. The cross sectional image of the OSP residue was investigated with Scanning Electron Microscope-Focused Ion Beam (SEM-FIB) and Raman Spectroscopy to reveal the effectiveness of the OSP removal. The interfacial interaction between Cu trace and the Cu pillar solder bump was investigated to further confirm the effectiveness of OSP removal.

Experimental

Figure 1 presents the bond on trace (BOT) schematic of the TCB process. The NCP was not shown herewith. The interconnection between Si chip and OSP-finished substrate was achieved by directly bonding the Cu pillar-solder cap on the Cu trace. The solder cap adopted in this study composed of Sn-Ag solder. The Cu trace was covered with OSP.

The OSP-finished substrates were treated by microwave (2.45GHz) plasma. The plasma was operated at a power of 800Wunder pressure of 0.5 mbar. The mass flow rates of the carrier gas Ar and active gas O_2 are, respectively, 2000 sccm and 240 sccm into vacuum chamber. The plasma treatment time was 10 minutes. The cross section of the OSP layer was prepared with focused ion beam (FIB) for investigation of the thickness variation with scanning electron microscope (SEM). The OSP film and residue were analyzed with SEM equipped with energy dispersive X-ray (SEM-EDX) and Raman spectroscopy. The Raman spectroscopy analysis was performed with excitation laser with wavelength 532 nm.

The commercial NCP applied contains self-fluxing agent of which the composition was not released. The chips were bonded on pristine OSP-finished substrate and substrate of plasma treated, respectively. The bonding was conducted at bonding time of 5 sec. The formation of intermetallic compounds (IMCs) at the interface between solder and Cu trace were investigated by SEM-EDX after a TCNCP process.

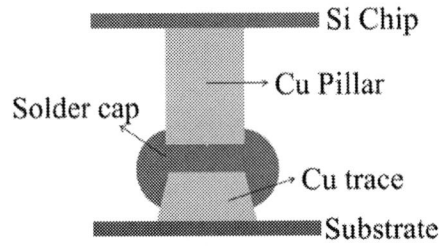

Figure 1. Schematic diagram of Cu pillar bond on Cu trace produced with TCB

Results and Discussion

Analysis on the removal of OSP layer by Ar/O$_2$ microwave plasma

The width of the Cu trace is slightly less than 25 μm as shown in Figure 2(a). The coating of OSP was preceded by acidic cleaning of the Cu trace, produced on the substrate, in an aqueous acid solution. Accordingly, the surface of the Cu trace usually exhibits certain level of roughness. The tilted angle FIB-SEM image of the Cu trace in Figure 2(a) shows that the surface of Cu trace consists of convex morphology throughout the surface. The OSP layer in the concave, hill foot, and the plain areas is generally thicker than that on the top and hill of the convex structures of the Cu trace. This is anticipated in view of the fluid flow behavior of OSP during the coating process. The thickness of the pristine OSP layer is in the order of 50~100 nm (on convex and hill) to in order of 300~450 nm (on plain and concave areas) (Figures 2(b) and 2(c)).

In a series of investigation, the OSP finished substrate was treated with various plasma conditions. Figure 3 presents the residue OSP layer left behind with the plasma powder of 800 W and the Ar-O$_2$ flow rates specified in the Experimental of the study. It is evident from Figure 3(a) that the OSP layer was almost completely removed by the plasmas treatment except in the areas of the hill foot, Figure 3(b), of the surface convex

structure of the Cu trace. The tiny residues remained are of the thickness of 130~150 nm, yet in short cut of less than 100 nm. The activity of plasmas gas is in general degraded by the shape of the object. The convex structure of the Cu trace thus blocked the preceding of the plasma ions and thus reduced the efficacy of plasma treatment on the hill foot area.

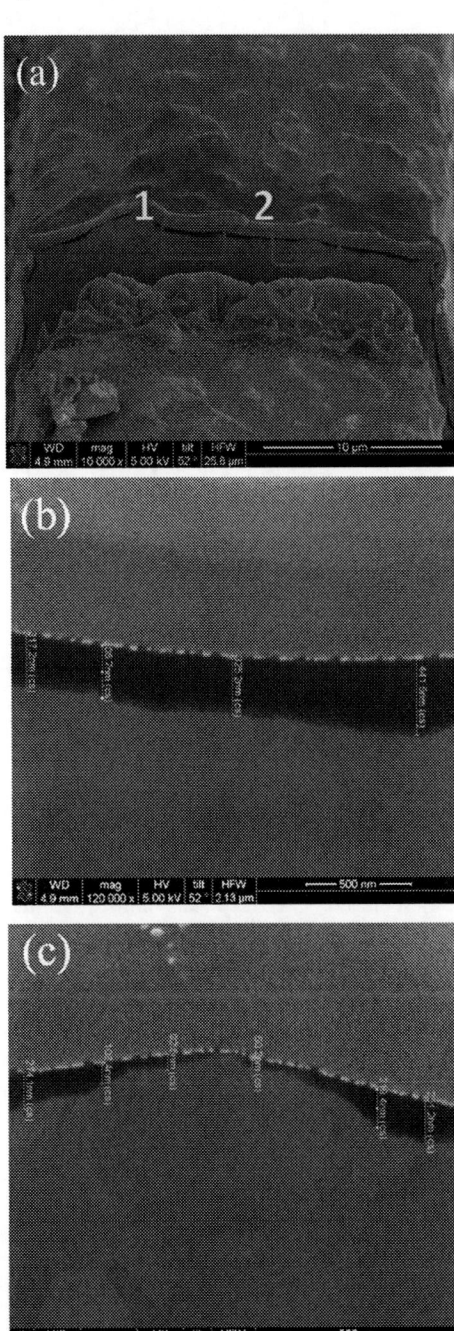

Figure 2. The cross-section FIB-SEM images of (a) pristine-OSP layer on top of the substrate Cu trace, (b) enlargement of of the rectangle region 1 in (a), (c) enlargement of the rectangle region 2 in (a).

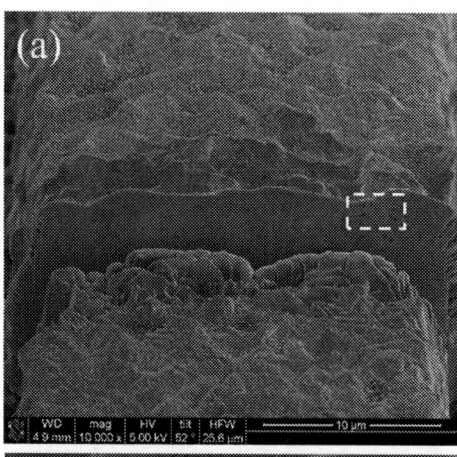

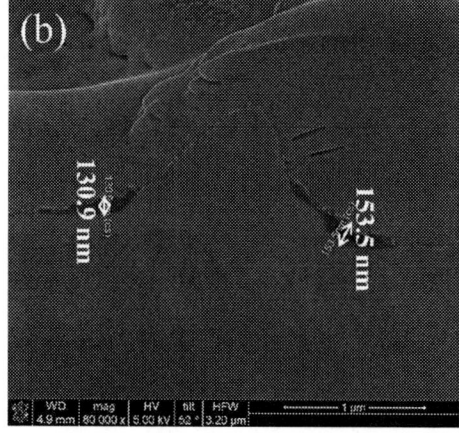

Figure 3. (a)The FIB-SEM image of the cross section of the OSP finished Cu trace after Ar-oxygen plasma treatment, (b)the enlargement of the dotted rectangle area in (a).

The surface appearance of the pristine OSP layer, Figure 4(a), shows that the Cu trace was embedded in the OSP. However, the Cu surface were exposed as seen in the specimen after 10 minutes of plasma treatment, Figure 4(b). The OSP layer became broken network after the plasma treatment. The convex Cu trace morphology also became quite prominent after the plasma treatment. The comparison of Figures 4(a) and 4(b) evidences the decomposition and removal of the OSP by plasma.

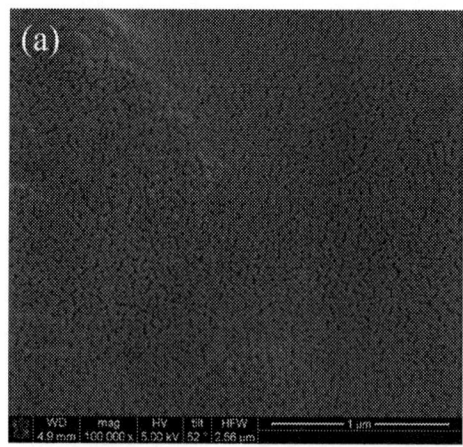

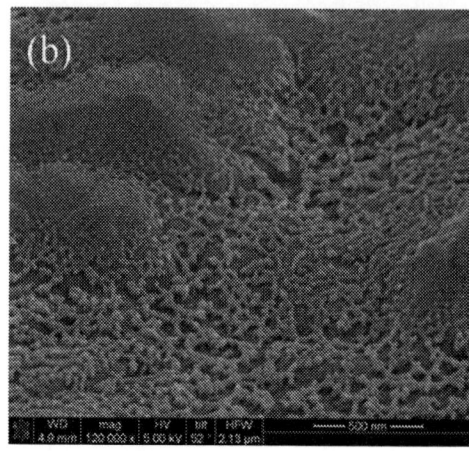

Figure 4 The SEM appearance of (a) the pristine OSP-Cu and (b) OSP-Cu after 10 minutes of plasma treatment.

In an attempt to understand the occurrence during the plasma treatment, this study investigated the OSP layer with FTIR and Raman spectroscopy. It was generally realized that the OSP may consist of benzimidazole as the major constituent. The investigation was conducted on the OSP coated Cu trace and the OSP solution. The results of the FTIR investigation, Figure 5, indicates the possible existence of O-H (3300 cm^{-1}), N-containing functional group (1580 cm^{-1}) and C=C (890 cm^{-1} and 607 cm^{-1}). The results of Raman analysis, Figure 6, indicates that the chemicals are composed of O-H (3062 cm^{-1}), C-H (2917 cm^{-1}), N-H (1538 cm^{-1}), C=C (1278 cm^{-1}) and possibly C-O (1037 cm^{-1}) and C-N (847 cm^{-1})。These bonding are in correspondence with the chemical structure of benzimidazole. The present study was not trying to identify the detail composition and chemical structure of the OSP layer. Rather, it was tried to extract the key structure and to provide understanding to the mechanism or occurrence of the plasma interaction with OSP.

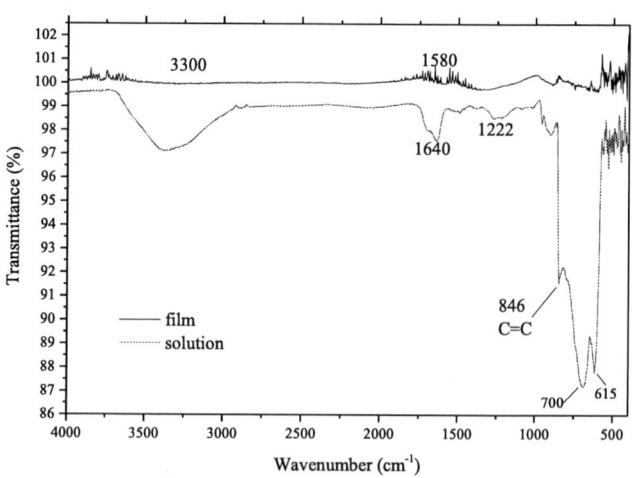

Figure 5 The FTIR analysis of the OSP.

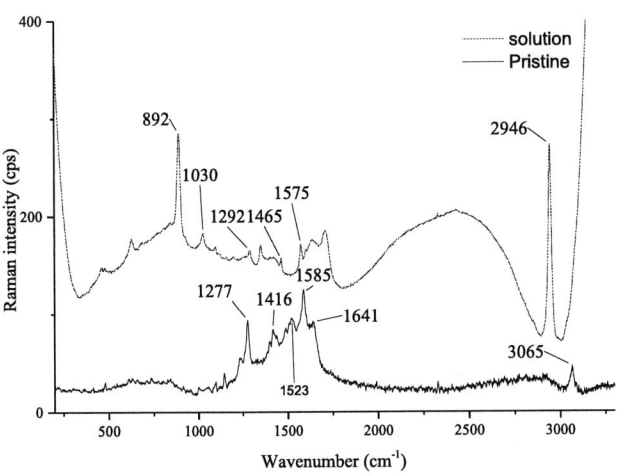

Figure 6 The Raman spectrum of the OSP.

The lone pair electrons of the bonds identified in the above analysis, as for benzimidazole, are easily to bond with transition metal to form metal complexes. This is part of the reasons why the benzimidazole is popularly used as corrosion inhibitor. It has been indicated [5] that the ring structure of the imidazole could be opened with radio frequency (RF) plasma treatment. An earlier study [6, 7] has pointed out that the benzene could be decomposed by plasma to produce carbon depositions or polycyclic aromatic. It was also indicated that benzene could be oxidized to CO_2, CO, and H_2O by oxygen plasma treatment [8]. In view of these, it is reasonable to achieve the decomposition of benzimidazole with oxygen microwave plasma, regardless of bonding with transition metal. Figure 3 shows that OSP layer almost removed by plasma except that partial OSPs in the cavities. Accordingly, the plasma can effectively remove the OSP deposit on Cu trace as delineated by the result in Figure 3. Further analysis on the interaction are described as followings.

Table 1. The elemental composition of the surface of Cu trace before and after plasma treatment.

Substrate	Atomic %		
	Cu	C	O
Pristine	46.43	44.69	7.85
After plasma treatment	78.91	11.5	9.59

Table 1 is the result of SEM-EDX analysis of the elemental composition of the surface of Cu trace prior to and after plasma treatment. The results evidence that the atomic percent of carbon was drastically reduced from 44.69% to 11.5% after treating the substrate by oxygen microwave plasma, while the Cu content increased form 46.43% to 78.91%. It is also noticed that the atomic percent of oxygen only slightly increased. Oxygen might be sourced cupric oxide or cuprous oxide on the surface of Cu trace [9]. It is generally known that the major elemental composition of benzimidazole is carbon. Thus the results of elemental analysis delineate that

the massive amounts of OSP were decomposed and removed by the plasma treatment. The remaining carbon is believed to be from the OSPs which remained in the concavity on the surface of Cu trace.

The Raman spectra of the pristine OSP is further presented in Figure 7(a) along with the plasma treated OSP, Figure 7(b). The previous study [10] indicated that the bands of the C-H bending of the bridgehead carbon on the imidazole were in the region of 950 to 1050 cm^{-1}. In this region, the shift in the position of bands depended on the oxidation state of the copper. The present study, however, did not detect the bands in this region because the intensity of these bands was too weak. It was also indicated [10] that the bands in the 1250 to 1350 cm^{-1} region are related to the nature of the copper-benzimidazole complex. The two bands at 1236 cm^{-1} and 1277 cm^{-1}, are assigned to an imidazole ring stretching mode and an in-plane bending mode of the ring in the copper-benzimidazole, respectively. Three bands at 1422, 1522, and 1588 cm^{-1} were found in Figure. The 1422 and 1588 cm^{-1} bands had been assigned to a C=C stretching in the benzene ring and an aza-aromatic bicyclic ring stretching, respectively [11]. The 1522 cm^{-1} band was assigned to an asymmetric in-plane imidazole ring stretching which might enhanced by that benzimidazole adsorbed on Cu trace [12]. The smaller shoulder band at 1642 cm^{-1} was assigned to C=N stretching in benzimidazole, and the 3065 cm^{-1} was the assignment of C-H stretching in benzene ring [11]. A comparison between the spectra of Figure 7(a) and 7(b) shows that the plasma treatment in the present work seems not result in detectable bond change. In other words, no change in chemistry of the OSP for the remaining OSP residue.

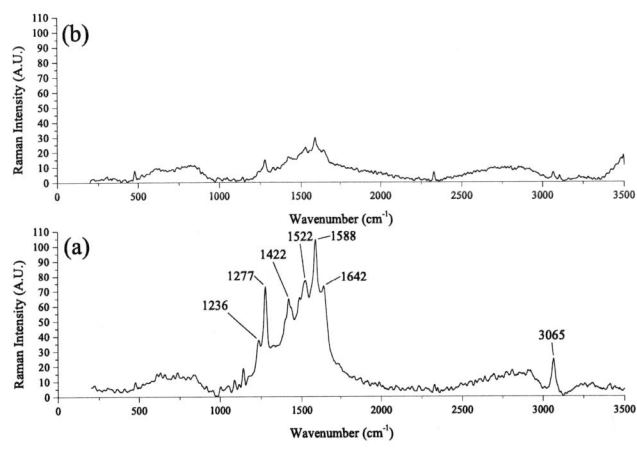

Figure 7 Raman spectra of OSP layer coated on Cu trace both (a) before and (b) after oxygen microwave plasma treatment

The aforementioned result of SEM-EDX analysis and the observation of FIB-SEM indicated that the OSP layer could be removed by Ar/O_2 microwave plasma treatment. The Raman spectra of OSP coated Cu trace also showed the same result. In figure 7(b), the intensity of these stronger bands in the 1200~2000 cm^{-1} region drastically decreased. Some of weaker bands out of this region almost disappeared. Therefore, the analysis by Raman spectrum further prove that OSP layer can be decomposed and removed by Ar/O$_2$ microwave plasma.

The solder joint formed in the TCNCP process

It is well known that the soldering process involves not only the dissolution of base metal in the molten solder but also the formation of IMC between the base metals and solder components. For instance, intermetallic compounds Cu$_3$Sn and Cu$_6$Sn$_5$ usually formed between tin solder and Cu [13, 14]. Accordingly, the formation of these IMCs provides the indication of the joint formation for the TCB process. It was a general practice to remove OSP on the OSP-Cu with rosin flux or non-clean flux embedded in the solder paste [13-15]. The flux also assists in removing surface oxide and thus enhance wetting during the reflow step.

The cross sectional images of the solder joints produced on the pristine OSP-Cu and plasma treated OSP-Cu were investigated for clarifying the efficacy of the plasma treatment on TCB. The SEM images shown in Figure 8 are for the joint produced on pristine OSP-Cu trace. The images of Figure 8(a) and 8(b) clearly show the existence of crack between the solder and Cu trace. The images of higher magnification, Figures 8(a-1) and 8(b-1) further indicate that there is no IMC formation on the Cu trace, the lower Cu structure. It is of interest to notice from Figure 8(b-1) that the profile of the solder right above the crack follows exactly the surface profile of the Cu trace. In other words, the solder front forms a replica of the Cu trace profile. This appearance indicates that the molten solder was compressed to touch with the Cu trace during the TCB. Nevertheless, the compression was not able to induce any interaction between molten solder and Cu trace. This is apparently due the existence of the un-removed OSP which forms a barrier to the reaction between solder and Cu trace and thus a replica of the Cu trace surface profile was formed after solder solidification.

On the other hand, a good solder joint was formed between solder and the sides of the Cu trace as seen in Figures 8(a-1) and 8(b-1). The SEM-EDX analysis results suggested that the IMC formed are Cu$_6$Sn$_5$ on the top layer and Cu$_3$Sn on the inner layer in right contact with Cu trace. It is believed that the flux within the NCP may flow down along the side along with the down flow behavior of molten solder that assists in the removal of OSP to enable solder–Cu interaction. Nevertheless, the flux in the NCP was not given adequate time and mechanical action on the top surface of the Cu trace to remove the OSP. Therefore, the solder was not able to form joint on the top side of Cu trace without Ar/O$_2$ plasma treatment.

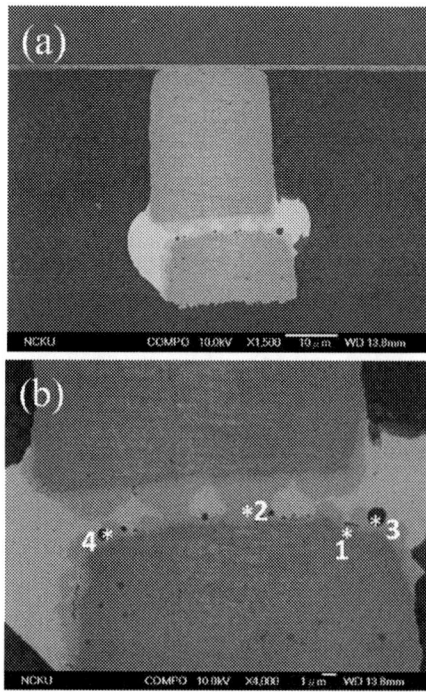

Figure 9. (a) a cross-section SEM images of Cu pillar bump bonding on Cu trace which has been treated by Ar/O$_2$ microwave plasma, (b) enlarged image of (a).

The aforementioned results have verified that OSP layer can be removed by the Ar-O$_2$ microwave plasma. There is very few OSP remained in the concavities on the surface of Cu trace. It is believed that the OSP residue on the top side of Cu trace can be easily dissolved by the self-fluxing agent in the NCP within the short time of TCB process.

Figure 9 (a) show that the solder joint was formed on a plasma treated OSP-Cu trace. The image of higher magnification, Figure 9(b) clearly shows the formation of IMCs between solder and the lower Cu-trace. It is of interest to indicate that there are two types of IMC formed on the Cu trace. A continuous uniform layer of IMC lies right on top of the Cu trace along the entire Cu trace front. Another thick scallop morphology of IMC layer was formed between the continuous layer and the solder bump. The EDX analysis results indicate that the uniform continuous layer (point 1) is Cu$_3$Sn and the scallop IMC (point 2) is Cu$_6$Sn$_5$. The process

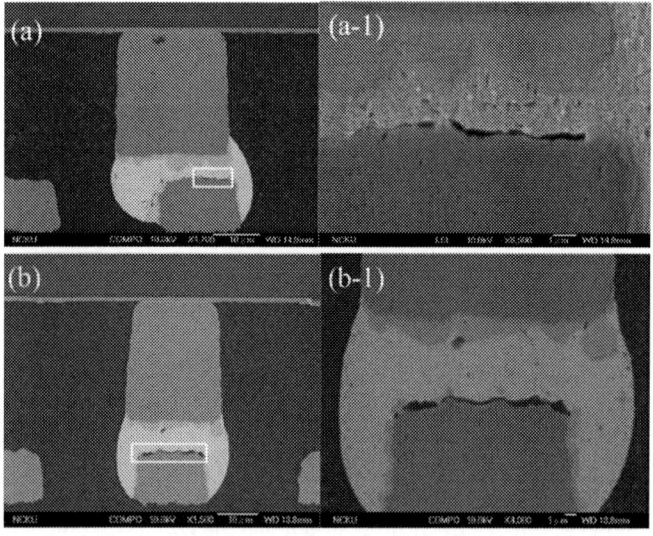

Figure 8. (a) and (b) are the cross-section SEM images of Cu pillar solder bump bond on Cu trace produced on pristine substrate. (a-1) and (b-1) are the enlarged images of the the rectangle regions of (a) and (b), respectively.

978-1-4799-2408-0/14 $31.00 © 2014 IEEE   1588

may left a few filler particle, (points 3 and 4) at the interface. This is because of the character of the NCP. The formation of IMC clearly evidences the successful formation of the Cu pillar-solder bond on OSP-Cu trace with the assistance of plasma treatment of the OSP-Cu substrate.

Conclusions

It has been demonstrated that the OSP layer on top of the Cu trace can be successfully removed with appropriate plasma treatment. The removal of the OSP prior to TCB enable the formation of Cu pillar solder bump bonding on the OSP-Cu trace. The bonding interaction form Cu_3Sn, and the Cu_6Sn_5 on top of the Cu trace. The NCP was not satisfactorily removed by the self-fluxing function of the NCP during the TCB process, which inhibit the interaction between solder and Cu trace under the TCB conditions. The plasma treatment of the OSP will decompose the OSP chemistry and remove the coating that expose the Cu surface for solder joining. With the assistance of appropriate $Ar-O_2$ plasma treatment, TCNCP can be a successful process for manufacturing the Cu-pillar bond on Cu trace.

Acknowledgments

The authors from National Cheng Kung University acknowledge the financial support of this project from ASE Group.

References

1. I. Artaki, U. Ray, H. Gordon, and R. Opila, "Corrosion protection of copper using organic solderability preservatives," *Circuit World*, vol. 19, no. 3, pp. 40–45, 1993.

2. S. H. Shi and C. P. Wong, "Study of the fluxing agent effects on the properties of no-flow underfill materials for flip-chip applications," *IEEE Trans. Components Packag. Technol.*, vol. 22, no. 2, pp. 141–151, Jun. 1999.

3. C. Lee, H. W. Kim, and S. Kim, "Organic contaminants removal by oxygen ECR plasma," *Appl. Surf. Sci.*, vol. 253, no. 7, pp. 3658–3663, Jan. 2007.

4. J. Friedrich, I. Loeschcke, H. Frommelt, H.-D. Reiner, H. Zimmermann, and P. Lutgen, "Ageing and degradation of poly(ethylene terephthalate) in an oxygen plasma," *Polym. Degrad. Stab.*, vol. 31, no. 1, pp. 97–114, Jan. 1991.

5. J. Stille, R. Sung, and J. Kooi, "The Reaction of Benzene in a Radiofrequency Glow Discharge," *J. Org. Chem.*, vol. 1065, no. 6, pp. 3116–3119, 1965.

6. S.-I. Shih, T.-C. Lin, and M. Shih, "Decomposition of benzene in the RF plasma environment. Part I. Formation of gaseous products and carbon depositions.," *J. Hazard. Mater.*, vol. 116, no. 3, pp. 239–48, Dec. 2004.

7. S.-I. Shih, T.-C. Lin, and M. Shih, "Decomposition of benzene in the RF plasma environment. Part II. Formation of polycyclic aromatic hydrocarbons.," *J. Hazard. Mater.*, vol. 117, no. 2–3, pp. 149–59, Jan. 2005.

8. M. Cal and M. Schluep, "Destruction of benzene with non - thermal plasma in dielectric barrier discharge reactors," *Environ. Prog.*, vol. 20, no. 3, pp. 151–156, 2001.

9. C. K. Chung, Y. J. Chen, C. C. Li, and C. R. Kao, "The critical oxide thickness for Pb-free reflow soldering on Cu substrate," *Thin Solid Films*, vol. 520, no. 16, pp. 5346–5352, Jun. 2012.

10. M. L. Lewis, L. Ledung, and K. T. Carron, "Surface structure determination of thin films of benzimidazole on copper using surface enhanced Raman spectroscopy," *Langmuir*, vol. 9, no. 1, pp. 186–191, Jan. 1993.

11. M.-S. Kim, M.-K. Kim, C.-J. Lee, Y.-M. Jung, and M.-S. Lee, "Surface-enhanced Raman Spectroscopy of Benzimidazolic Fungicides: Benzimidazole and Thiabendazole," *Bull. Korean Chem. Soc.*, vol. 30, no. 12, pp. 2930–2934, Dec. 2009.

12. B. H. Loo, Y. Tse, K. Parsons, C. Adelman, a. El-Hage, and Y. G. Lee, "Surface-enhanced Raman spectroscopy of imidazole adsorbed on electrode and colloidal surfaces of Cu, Ag, and Au," *J. Raman Spectrosc.*, vol. 37, no. 1–3, pp. 299–304, Jan. 2006.

13. G.-T. Lim, B.-J. Kim, K. Lee, J. Kim, Y.-C. Joo, and Y.-B. Park, "Temperature Effect on Intermetallic Compound Growth Kinetics of Cu Pillar/Sn Bumps," *J. Electron. Mater.*, vol. 38, no. 11, pp. 2228–2233, Aug. 2009.

14. C.-T. Lin, C.-S. Hsi, M.-C. Wang, T.-C. Chang, and M.-K. Liang, "Interfacial microstructures and solder joint strengths of the Sn–8Zn–3Bi and Sn-9Zn–1Al Pb–free solder pastes on OSP finished printed circuit boards," *J. Alloys Compd.*, vol. 459, no. 1–2, pp. 225–231, Jul. 2008.

15. J.-M. Koo, B. Q. Vu, Y.-N. Kim, J.-B. Lee, J.-W. Kim, D.-U. Kim, J.-H. Moon, and S.-B. Jung, "Mechanical and Electrical Properties of Cu/Sn-3.5Ag/Cu Ball Grid Array (BGA) Solder Joints after Multiple Reflows," *J. Electron. Mater.*, vol. 37, no. 1, pp. 118–124, Oct. 2007.

A PoP Structure to Support I/O over 2000

Dyi-Chung Hu, Puru Lin, Yu Hua Chen, Chun-Ting Lin
NBD, Unimicron Technology Corp.
No.290, Chung-Lun Village, Hsin-Feng, Hsinchu, Taiwan (304)
E-mail: dchu@unimicron.com

Abstract

The rapid growth of smartphones and tablets in mobile market demands the packaging technology to be in a thinner profile with small form factor and reduce power consumption. The PoP structure is widely used in the package of smart phones to connect memory and Application Processor. Even if TSV is the preferred structure for connecting memory and AP, the high cost of TSV process prohibit TSV for wide presence in the smart phone applications.

Recently, I/Os of memory are required to increase from a few hundreds to more than 1000. For wide I/O2, I/Os more than 2000 may be required. We have demonstrated PoP connections over 1000 with pitch of 200 μm in last year's ECTC conference with Unimicron HCP (High Copper Pillar) technology [1]. In preparation to meet future memory requirement and to evaluate the extendibility of Unimicron proposed HCP structure, a test vehicle has been built.

The test vehicle consists of a PoP structure with a fine pitch of 100 μm that can support 2472 I/Os in 6 rows around the peripheral area of the die. The package size is in a 12x12 mm² CSP format. The target copper pillar height is 80 μm. The copper pillars with a designed diameter of 70 μm and space of 30 μm.

The uniformity of copper pillar height is required to assure the reliable interconnections between the top memory die and the bottom AP. For a designed copper pillar height of 80 μm, tolerance of +/- 15 μm is required. In this paper, a method to achieve the copper height uniformity of +/- 15 μm with 100 μm PoP pitch connection will be demonstrated.

Introduction

Portable electronics, such as smartphone and tablet, have been exclusively gain wide popularity in the consumer market. The key features of those smart devices are providing thinner profile [2], faster data transfer rate and lower power consumption at high bandwidth [3]. In stacked memory application, 3.2~6.4 GB/s bandwidth of LPDDR2 DRAM can't provide faster access to the Internet and fluid viewing of high resolution video. Currently, LPDDR3 is emerging in the market and LPDDR4 is in sample base. But in the future a bandwidth of 51.2 GB/s is desirable. New memory format such as wide I/O2 is emerging. Many mobile device manufacturers adopted Package on Package (PoP) technology as the solution to achieve small form factor and high performance requirements.

The PoP is a packaging technology to allow vertically assembling discrete memory packages and logic into a single module. Usually, the memory package is stacked up on top of the application processor. This structure could save motherboard space and offer good signal integrity by reducing the interconnection distance between memory and processor chips.

The advanced memory devices need more than 1000 I/Os to connect to the logic die. However, the traditional solder ball interconnected PoP (Figure 1) cannot achieve the target I/O due to the ball pitch is limited to the range of 0.5 to 0.4 mm. Several PoP solutions were proposed and developed to meet the requirements mentioned above [4], [5].

In the paper published in ECTC 2013 [1], we demonstrated that HCP (High Copper Pillar) with I/O over 1000 can be processed on laminated substrate. In this paper, we would like to demonstrate that I/Os more than 2000 can be achieved by the extension of HCP technology. We designed a test vehicle which has I/O more than 2000. This would give memory designers more freedom to achieve high bandwidth with low power.

Since the numbers of copper pillars in this paper are more than 2000, we name it SHCP (Super HCP). There are several criteria for SHCP, first, it can support I/Os more than 2000. The second is that the SHCP height should be as uniform as possible. In this paper we use +/- 15% as a target. The third one is not related to technology but to cost. We would like to plate the pillars of 80 μm height within one hour.

A 2-layer bottom package with 100 μm pitch of copper pillars and top package with solder cap were prepared. There are 2472 couples of connections in a single PoP unit. The process to fabricate the top and bottom package are described in later part of this paper. The results of package on package assembly of SHCP are also be demonstrated.

Figure 1. Traditional PoP structure

SHCP-PoP Structure

In this study, we will present a top package with Sn plated solder mounted on bottom package with SHCP structure. The SHCP-PoP structure is illustrated in Figure 2.

The bottom part of PoP package is a 2-layer substrate with 0.2 mm thick core material (CTE is around 10 ppm) which the total thickness of the substrate is 0.28 mm. If the surface finish of ENIG or ENEPIG is used, tin solder is easily spread over the whole copper pillar. This may lead to the solder bridging between pillars. Hence OSP (Organic Solderability Preservatives) was selected as the surface finishing method to

2014 Electronic Components & Technology Conference

978-1-4799-2408-0/14 $31.00 © 2014 IEEE   1590

protect SHCP. The design of copper pillars diameter is 70 μm, and the space between copper pillars is 30 μm.

The top package of PoP is another 2-layer substrate with 0.2 mm core material (CTE is around 10 ppm) and the total thickness of the substrate is 0.28 mm. There are 30 μm height copper pillars with 20 μm thick tin electroplating on the backside of top package (Figure 2).

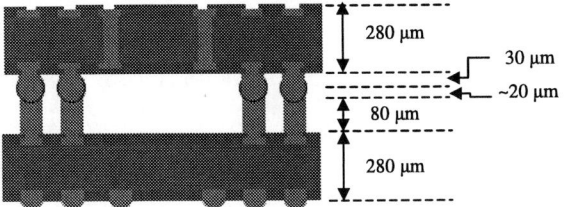

Figure 2. Dimensions of SHCP-PoP structure used in this paper.

Table 1. SHCP-PoP Structure Dimensions

	Top Package	Bottom Package
Body Size (mm)	12*12*0.35	12*12*0.37
Pillar Diameter (μm)	70	70
Pillar Pitch (μm)	100	100
Pillar Height (μm)	30	80
SR opening (μm)	40	40
Number of Pillars	2472	2472
Solder height (μm)	40	NA

Electrical Test Pattern Design

In order to achieve > 2000 I/O counts design, a package size of 12 mm x 12 mm was selected. The PoP interconnection contains 2472 I/Os in 6 rows with pitch of 100 μm. The test vehicle contains 2472 couples of connections. Since there are too many I/O counts in a single unit, for the ease of failure analysis, the test vehicle is divided into 64 zones where there are 6, 8 and 9 rows (each row contains 6 pillars) in different locations (Figure 3). The designed resistances of 9, 8, 6 rows of daisy chain are estimated to be 150 mΩ, 135 mΩ and 100 mΩ, respectively.

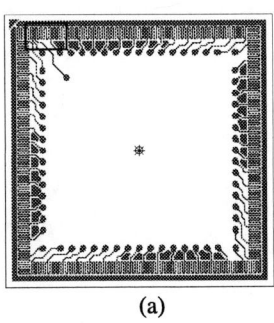

(a) (b)

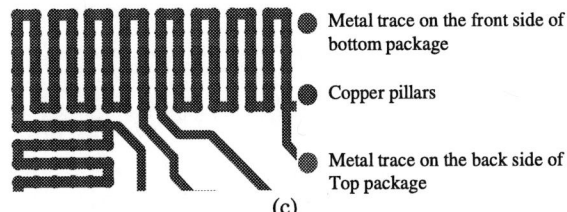

Figure 3. Test vehicle design layout. (a) The overview of test vehicle in a single unit (b) 64 Zones of test vehicle (c) Magnified design of the upper left corner of pattern (a).

Process Flow of SHCP-PoP Structure

The fabrication of SHCP-PoP package follows procedures as shown in Figure 4(a) to Figure 4(c). Figure 4(a) and 4(b) show the process flows of top and bottom packages respectively. Figure 4(c) shows the assembly process flow of the new SHCP-PoP structure.

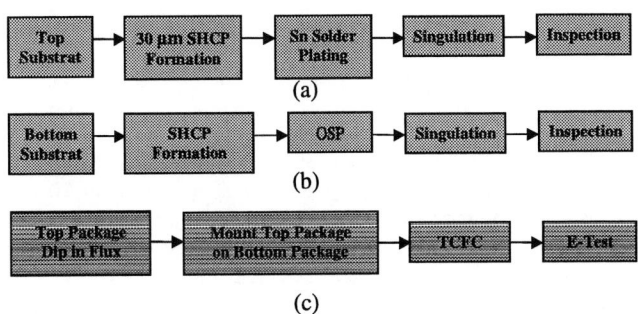

Figure 4. (a) Process flow of top package. (b) Process flow of bottom package. (c) Assembly flow of top & bottom packages.

SHCP Formation on the Bottom Package

Figure 5 shows the process flow of manufacturing SHCP on the bottom package. First, a 2-layer substrate is fabricated by the conventional carrier manufacture process. The pad size is about 60 μm in diameter, and the pitch between the pads is 100 μm. Later a solder mask is laminated and patterned on top of the copper pad. Follow by Electroless Cu on top of the solder mask to provide seed layer for copper pillar plating. A dry film of 120 μm thickness is used to form copper pillar. Laser direct imaging with dose energy in the range of 150 to 300mJ/ cm^2 is used to define the pattern of copper pillars. After exposure, the DF was developed in a solution contains 1 wt% Na_2CO_3 by spraying. Next is the copper plating process. The current density of 7 to 10 ASD is used in this study to achieve the goal of plating time around one hour. DF and seed layer is striped after copper pillar plating. The surface finish of OSP is applied to protect copper pillars. Figure 5 summarizes the process flow of manufacturing the SHCP structure on the carrier substrate.

978-1-4799-2408-0/14 $31.00 © 2014 IEEE

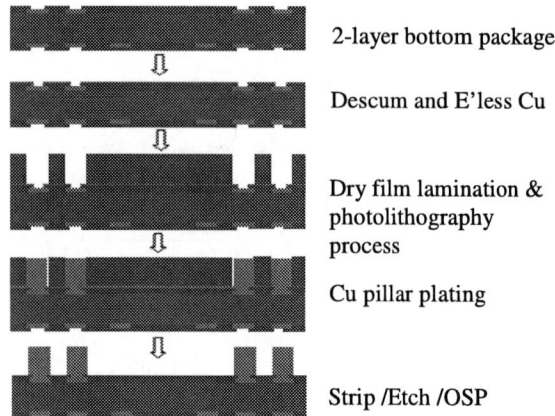

Figure 5. Process flow of SHCP formation on the bottom package

2-layer bottom package

Descum and E'less Cu

Dry film lamination & photolithography process

Cu pillar plating

Strip /Etch /OSP

Solder Cap Formation on the Top Package

Similar to the SHCP formation, the process flows of the top package are almost the same as bottom package shown in Figure 5, except for tin solder plating and surface finish. There are the same numbers of 2472 copper pillars formed on the top package with the copper pillar height about 30 μm. Before striping of the DF, tin solder with thickness of 20 μm is electroplated on top of the shorter copper pillars. The process flow of the top package is listed in Figure 6.

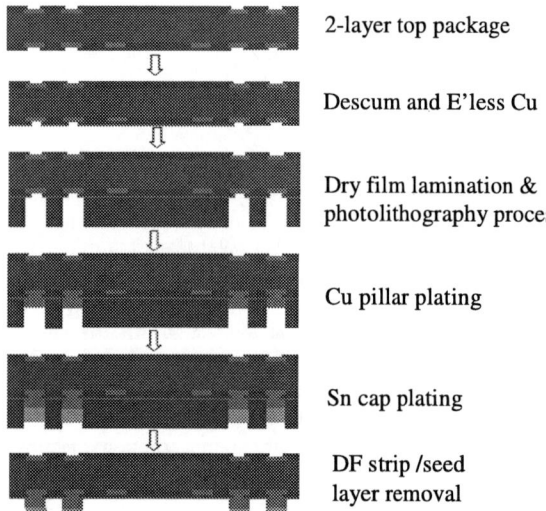

2-layer top package

Descum and E'less Cu

Dry film lamination & photolithography process

Cu pillar plating

Sn cap plating

DF strip /seed layer removal

Figure 6. Process flow of solder cap formation on the bottom package

Bottom Package SHCP Process Results

Figure 7 shows the results of SHCP bottom package. The pitch of copper pillars is 100 μm and the diameter of the copper pillar is 70 μm and the spacing between two pillars is 30 μm. Figure 7 (a) shows the overview of SHCP substrate and the detailed magnification of the copper pillars.

Figure 7 (b) shows the cross-section view of SHCP. The height of SHCP is about 103 μm. This particular sample is

intentionally made to have copper pillars around 100 μm. The diameters of top-side and bottom-side SHCP are 62μm and 80μm respectively. The diameter difference between top and bottom SHCP is due to the over developed condition which could avoid scums remained on top of Cu pad. The over developed conditions can be minimized when SHCP enter mass production phase. The copper trace thickness is about 20 μm. The total thickness of SR is 30~35 μm and the thickness of SR on copper pad is 15 μm.

The SHCP height uniformity is critical in the PoP package assembly. If the height difference between pillars is more than 20 μm, it may cause the cold joint.

Figure 8 shows the height variations of SHCP structure which is measured by 3D Laser Scanning Microscope. The measurement sampling position of full panel, block and single substrate is shown in Figure 9. The sampling size includes three different blocks in one panel and 5 substrates in one block. The within substrate uniformity of the copper pillar are sampled from 8 positions in one substrate, 6 points per position, which come up to of 48 measurements per substrate. The measurement results of SHCP height uniformity within single substrates, blocks and full panel are listed in Figure 10.

The measurements of SHCP pillar height are summarized in Figure 10. And the pillar height range distributions are summarized in Figure 11. The ranges of copper pillar heights per substrate are within +/- 10 um in this study. However, the averaged copper pillar heights various from 90 to 75 μm depend on the substrate location inside the panel.

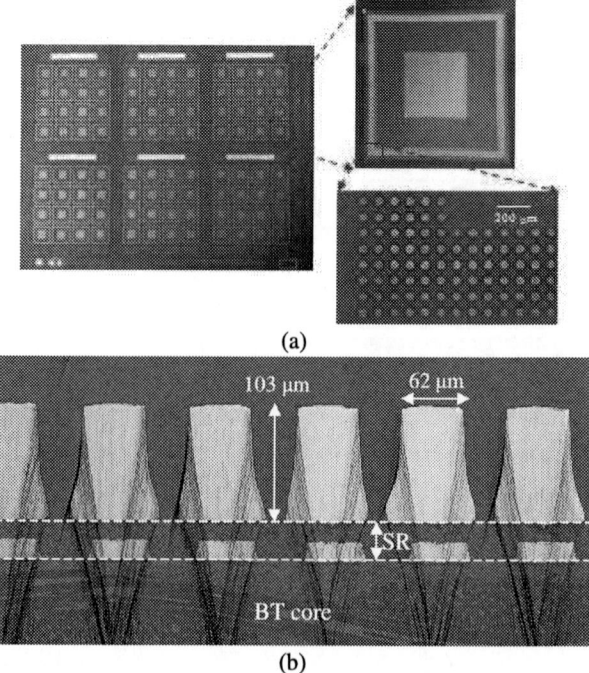

(a)

(b)

Figure 7. The results of SHCP formed on bottom package. (a) Pictures of SHCP on bottom package (b) Cross-section view of SHCP

978-1-4799-2408-0/14 $31.00 © 2014 IEEE 1592

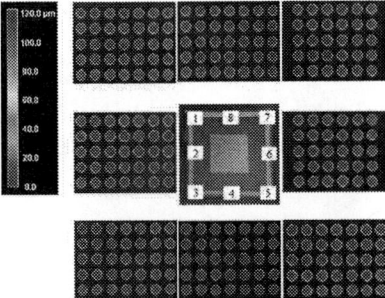

Figure 8. 3D image of SHCP structure

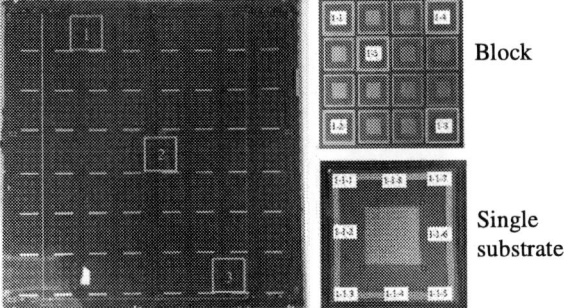

Block

Single substrate

Full Panel
Figure 9. Sampling position of SHCP pillar height measurements.

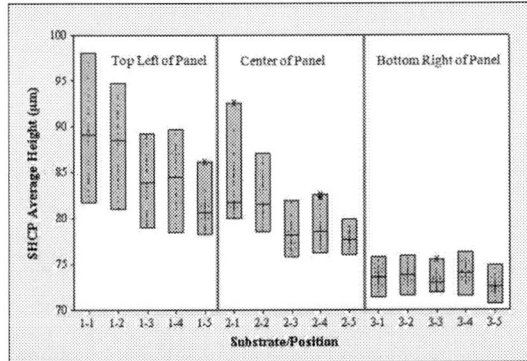

Figure 10. Measurement results of average copper pillar height at different panel locations.

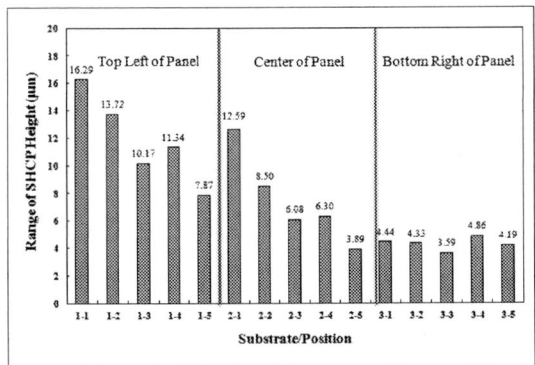

Figure 11. Measurement of the range of copper pillar height on different panel locations.

SHCP-PoP Assembly Results

The SHCP-PoP assembly is made by Thermal Compressing Flip Chip bonding (TCFC) method with the pressure around 1 kg/cm². The temperatures of top and bottom plate are set to 280℃ and 210℃, respectively. Before assembly, the SHCP is coated with flux which could help to remove metal oxide at the joint interface and improve solder wetting ability. After assembly, X-ray images of PoP structure are shown in Figure 12. It can be seen that the alignment between top and bottom package is very good. Figure 13 shows the micro section of the joined SHCP-PoP package. It shows good alignment of the top and bottom packages. Due to the nature of TCFC, solder was pushed out during the assembly process.

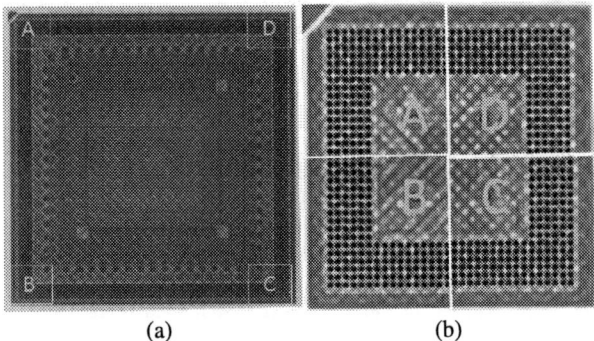

(a)   (b)

Figure 12. X-ray image of assembled SHCP-PoP package (a) Top-view of assembled SHCP-PoP (b) Magnification pictures of the four corners from (a)

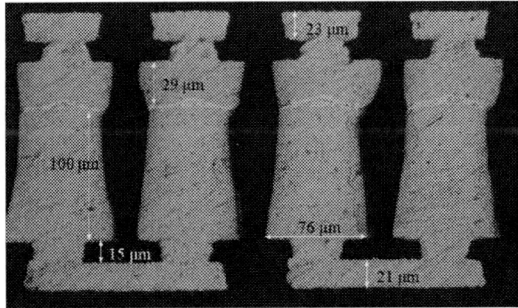

Figure 13. Cross-section view of assembled SHCP-PoP. It shows good top/bottom package alignment.

Electrical Test Results

Figure 14 shows the mapping results of the open-short testing. The mapping was measured by the 2-wire method (Keithley 2400 Sourcemeter) on one assembled SHCP-PoP package. Circuit open in zone 37 was observed. The shorting was also found since the Sn solder was pushed out during the TCFC process. The gaps between copper pillars were too narrow and not optimized in this study and thus increases the risk of short between pillars.

978-1-4799-2408-0/14 $31.00 © 2014 IEEE   1593

Figure 14. Open-short testing results of one assembled SHCP-PoP package. Red indicates open.

Components and Technol. Conf. (ECTC), San Diego, CA , May 29–June 1, 2012, pp. 1361–1367.

Conclusions

In this paper, we demonstrated a fine pitch SHCP structure, which could provide 2472 I/Os with 100 μm PoP pitch in a 12mm x 12mm package. The pillar height of 80 μm can be achieved around one hour plating time. The feasibility of SHCP-PoP assembly has also been demonstrated by TCFC method.

However, there are still improvements needed in two major areas for the realization of SHCP-SPOP technology. First, the copper pillar height uniformity inside the panel. The range of copper pillar height per substrate is within 20 μm in this study. However, the averaged copper pillars height various depend on the substrate location inside the panel. This needs to be further improved by the modification of plating equipment and plating chemicals. The second improvement needed is the development of a robust SHCP-PoP assembly technology. Molding of copper pillar prior to the SHCP assemble could help avoid shorting between pillars. Various innovative assembly methods can be further developed with the help of the assembly industry.

Acknowledgments

The authors wish to thank Unimicron's New Business Division team for their stimulated discussions and the strong support from the management of this project. Also thanks to ITRI's Packaging Technology Division team for their support.

Reference

1. Dyi-Chung Hu, Tsung-Si Wang, Chun-Ting Lin, "A PoP Structure to Support I/O over 1000," in *Proc. IEEE Electronic Components and Technol. Conf. (ECTC)*, Las Vegas, NV, May 28–31, 2013, pp. 412-416.
2. Eslampour H., et al, "Comparison of Advanced PoP Package Configurations," in *Proc. IEEE Electronic Components and Technol. Conf. (ECTC)*, Las Vegas, NV, Jun 1–4, 2010, pp. 1946-1950.
3. Matt Nowak, "3D Packaging Magazine", No. 10, pp.2, Nov, 2010.
4. Curtis Zwenger, et al, "Surface Mount Assembly and Board Level Reliability for High Density PoP Utilizing Through Mold Via Interconnect Technology", SMTA International Conference, Orlando, Florida, August 2008.
5. Philip Damberg, et al, Invensas Inc., "Fine Pitch Copper PoP for Mobile Applications," in *Proc. IEEE Electronic*

Enabling Eutectic Soldering of 3D Opto-Electronics onto Low Tg Flexible Polymers

Meriem Ben-Salah Akin, Lutz Rissing, Wolfgang Heumann
Institute of Micro-production Technology, Leibniz University of Hanover
An der Universitaet 2, 30823 Garbsen, Germany
bensalahakin@impt.uni-hannover.de

Abstract

We present a cost-efficient and reproducible technique for assembling 3D components to mechanically bendable low glass transition temperature (Tg) polymeric interposers. First, we propose localized soldering using a focused hot air gun. Second, we use eutectic solders, which permit reflow at temperatures close to the Tg of the interposer. Then, we adopt differential heating and cooling in order to enable rapid heat dissipation through the interposer during soldering. Furthermore, we apply a metal jig as a load to maintain contact between the interposer and the 3D component during soldering. Also, the jig serves as a shield to protect the interposer surface outside the soldering zone from thermal loading, and more importantly as a passive heat sink. We showcase our approach by manufacturing a test vehicle on a polyethylene terephthalate (PET) interposer having a Tg of 71 deg. C. We solder a 30 mA DC current SMD LED by means of a eutectic compound of 48 w. % Sn and 52 w. % In at 118 deg. C.

Related Work

As the world is moving towards the internet of everything, smart structures, e.g. mechanically bendable opto-electronic sensors, will be mounted everywhere to serve night and day. That being the case, and in order to solve the bottleneck associated with the manufacturing cost of the next generation of smart structures, ultra low-cost organic materials like lightweight polymers at the interposer or component level are to be employed. Currently, lightweight inexpensive polymers like polyethylene terephthalate (PET), polycarbonate (PC), polymethyl methacrylate (PMMA) and polyvinyl chloride (PVC) are dedicatedly employed in numerous applications such as RFID tagging [2], medical implants [5] and electronic displays [1]. Thereupon, polymeric substrates offering a wide range of optical, thermal and mechanical properties are researched and manufactured [8]. Moreover, the use of the aforementioned polymers as a platform for the deposition of electronic interconnects is extensively investigated ([3], [4], [6], [9], [16]).

Nonetheless, practices of surface mounting of three-dimensional (3D) components to the deposited interconnects remain limited due to the restricted thermal budgets of the substrate. Clearly, inexpensive polymers like PET, PC, PMMA and PVC have glass transition temperatures (Tg) below 150 deg. C. Low rigidity and thermal sensitivity of these aforementioned polymers at classically low temperatures constitute the brainteaser. In order to ensure an effective roll-to-roll mass production, warpage of the polymer interposer that is due to coefficient of thermal expansion (CTE) mismatches of the assembly and thermo-mechanical properties of the interposer need to be monitored rigorously.

While low heat thermo-compressive ([10], [12]) or ultraviolet adhesive bonding [13] comply with the thermal budget of the substrate, limitations on the shear strength and electrical resistivity of the adhesive joint persist. Hence, various non-ceasing efforts on adapting conventional surface mount technologies such as thermal soldering to inexpensive polymers are expended. For instance, recent investigations on sintering Ag-based solder pastes onto PC substrates using a continuous wave laser did lead to promising, but not to fully satisfactory, results [7]. Besides, in [11], a conductive paste based on carbon nanotube composites was successfully bonded to heat stabilized PET at 150 deg. C. Likewise, terminals of field effect transistors were effectively soldered to heat stabilized PET using a silver paste [14]. Furthermore, SnAgCu solder pixels were successfully applied by means of laser induced forward transfer approach, and reflowed onto PET substrates [15].

In order to cope with restricted thermal budgets for soldering purposes, a number of approaches exist. First, eutectic compounds, which conventionally melt at lower temperatures than the individual components that constitute the compound itself, are used as soldering mediums [17]. Further, localized soldering reduces thermal loading at the system level and focuses the application of heat exclusively to the region that is to be soldered [18]. In addition, since temperature sensitive materials exhibit an exponential increase in thermal expansion when thermally loaded, differential heating and cooling is proven to alleviate the effect of mismatches in thermal coefficients on the soldered product [19].

Based on the previously mentioned techniques, we propose a procedure for soldering 3D opto-electronics onto low Tg polymers. In particular, we preserve available conventional processing tools inducing minimal modifications and special handling. Additionally, we enable a horizontally integrated supply chain that uses distinct suppliers and manufacturers for raw materials and components.

Test Vehicle and Materials Characterization

We showcase our approach by manufacturing a test vehicle on a commercially acquired PET interposer. The interposer has the chemical composition -[-C10-H8-O4-]-, a material density of 1,32 g/cm^3, a VICAT softening point of 71 deg. C., and a melting temperature Tm of 255 deg. C. Moreover, the interposer has a thickness of 200 μm.

Furthermore, the ultimate tensile strength of the interposer is 45 N/mm², and the tensile impact strength is 250 kJ/m².

Copper (Cu) electronic interconnects of a thickness of 200 nm are sputter deposited on the PET interposer. Furthermore, 2 μm of Tin (Sn) are applied onto the region of the copper interconnects to-be-soldered through evaporation. Specifically, structuring of interconnects and soldering pads is achieved using a shadow mask. Particularly, the shadow mask is made of steel, structured means laser and 100μm thick.

In order to improve the adhesion of the electronic interconnects to the PET interposer by increasing the surface roughness and enhancing the purity of the surface, the interposer surface is subject to oxygen plasma treatment. Consequently, the roughness of the interposer is characterized by the arithmetic mean Ra=24.64 nm, the quadratic mean Rq=34.81 nm and the maximum roughness Ra,max = 36.03 nm. However, no significant change in

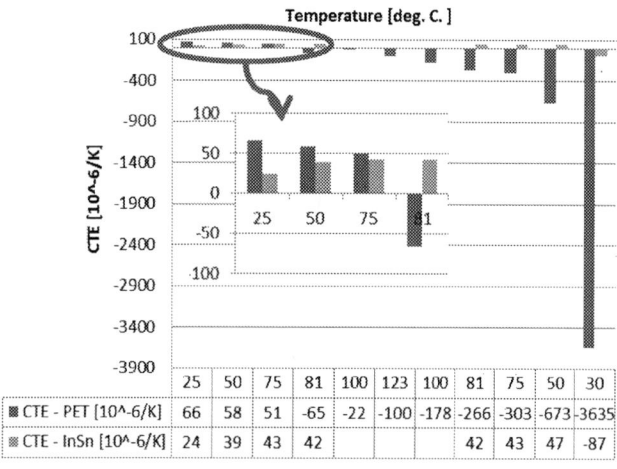

Temperature [deg. C.]	25	50	75	81	100	123	100	81	75	50	30
▨ CTE - PET [10^-6/K]	66	58	51	-65	-22	-100	-178	-266	-303	-673	-3635
▨ CTE - InSn [10^-6/K]	24	39	43	42				42	43	47	-87

Figure 2: Coefficients of thermal expansions of PET and InSn measured during a heating and cooling cycle.

thickness is recorded. It is to be noted that the plasma treatment intentionally does not induce a lasting chemical activation of the surface. Empirically obtained, the chains of the PET film are observed to reversibly decay back to equilibrium once thermally loaded under 130 deg. C. for not more than a couple of seconds, which we refer to later as relaxation time.

A commercial top light emitting diode (LED) is soldered to the electronic interconnects. Notably, the LED is based on InGaN and packaged as a surface mount device in a plastic housing of the size 5.0x5.0x1.6 mm³. Pre-tinned copper legs connect the LED to the outside of the package. Besides, the LED can sustain up to 260 deg. C. in reflow soldering temperature for a duration of 4 seconds, and up to 80 deg. C. during ongoing operation. A maximum of 30 mA forward direct current operates the LED. Markedly, the test vehicle does not comprise a current-limiting serial resistor connected to the LED. Therefore, forward direct currents of the order of 0.5 mA are not to be surpassed.

A eutectic compound of 48 w. % Sn and 52 w. % Indium (In), later referred to as InSn, which is commercially available as a 50 μm thick ribbon, is employed for soldering the legs of the LED to the interconnects. In particular, the liquidus temperature of the solder lies at around 118 deg. C. Moreover, the eutectic solder is characterized by a tensile strength of 11 N/mm² and shear strength of 11.2 MPa. Since soldering is conducted under ambient conditions, the use of soldering grease (DIN EN 23451-1 F-SW21 according to German Institute for Standardization) becomes necessary in order to break the oxide layers at interconnects and solder.

So far, the mentioned characteristic figures are obtained from the manufacturer or the distributor except the surface roughness data, which was obtained in house. Moreover, using an optical dilatometer, coefficients of thermal expansions of the PET film and the InSn ribbon were measured (Figure 2). As a matter of fact, the glass transition temperature of the PET film was verified to be less than 80 deg. C. In addition, the InSn material maintains a fairly constant coefficient of thermal expansion during the heating and cooling cycle. Furthermore, the mismatch of coefficients of thermal expansions between InSn and PET is alleviated when heated above 50 deg. C. and under the Tg of PET.

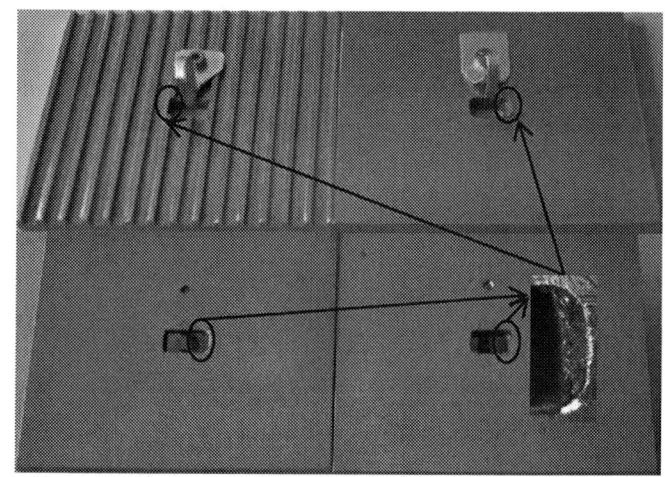

Figure 1: Metal jigs with and without grooves, hook/contact spring, and detail of flow chamber

Additionally, the PET film deforms exceedingly, and irreversibly, once completely subject to temperatures above 80 deg. C. Further, it is to be noted that large negative coefficients of thermal expansion are an indication for a curling of the free extremities of the sample, which is detected to be a shrinking of the material by the optical dilatometer.

Setup of Eutectic Soldering

Typically used for rework purposes, a focused hot air gun with a cylindrical vent with a diameter of 0.6 mm is used as a localized heat source for soldering purposes. The temperature of the hot air arriving at the region to-be-soldered depends on the distance of the opening of the hot air gun to the solder;

and the angle with which the hot air collides. For instance, hot air emanating from the gun at a temperature of 250 deg. C. arrives at a temperature of 140 deg. C. after traveling a vertical distance of 1.5 centimeters. During soldering, the amount of hot air is measured and adjusted by means of a precision flow meter.

Next, a metal jig made of stainless steel with an aperture for the LED is used. In particular, the jig serves as a shield to protect the interposer surface outside the soldering zone from thermal loading. In addition, the jig functions like an iron load to maintain planarity of the polymeric interposer. As importantly, the jig acts as a passive heat sink for accelerating heat dissipation by the polymer film. In this regard, two variants of the metal jig are investigated: (a) a homogeneous board with an aperture for LED, and (b) a homogeneous board with an aperture and pin fin enlarging the surface area of the jig (Figure 1). Moreover, both jigs dispose of a small flow chamber located at the corner of the LED aperture. Hence, hot air emanating from the air gun is trapped into the flow chamber and localized. In order to maintain contact between interconnects and LED during soldering, and in addition to avoiding chuck vacuum, a slender plate in the shape of a hook serves as a contact spring. Specifically, a light spring load becomes active once the hook is pushed down to touch the center of the upper surface of the LED. Consequently, contact between the LED and the polymer film is ensured during solder drop formation, solder wetting and thermal shrinkage. Last but not least, geometrical dimensions of jigs (60 mm x 35 mm x 4 mm), pins (1.25 mm x 2.5 mm), distance between pins (3.3 mm), and aperture (5.5 mm x 5.5 mm) were adopted.

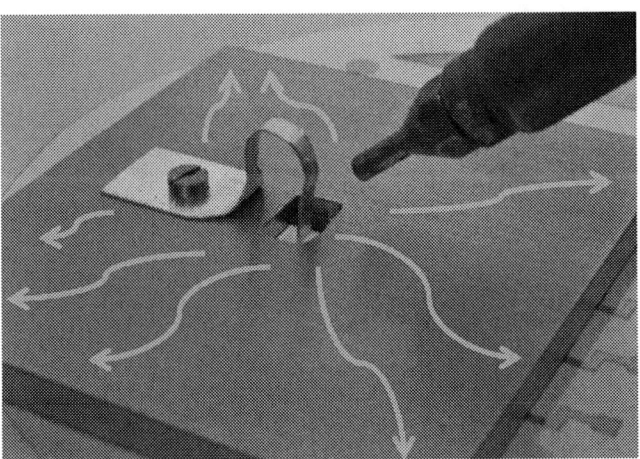

Figure 4: Soldering setup (Note: metallic platform is not depicted) and example of heat flow pattern

Differential heating and cooling are enabled through the use of a metallic platform, which is made of stainless steel and 7 mm thick, under the polymeric film. According to the coefficients of thermal expansion discussed in the previous section, and while the hot air is required to meet the melting temperature of the InSn eutectic compound, 118 (+20) deg. C., the configuration of the soldering setup is designed to facilitate the polymeric substrate to maintain low thermal expansion. During soldering, the temperature of the polymeric film around the soldering zone is verified using a digital quick-response thermocouple.

The soldering setup (Figure 4) is mounted under ambient conditions. The surrounding temperature is maintained at constant 21 deg. C. Relative humidity is kept within the range of 40-70%.

Process of Eutectic Soldering

Using the setup described above, the LED is eutectically soldered according to the following procedure (Figure 3):

1. InSn ribbon is folded multiple times in order to attain the required amount of solder.
2. InSn is immersed in soldering grease.
3. InSn is positioned onto the Sn pads. Soldering grease works as a temporary mechanical fixation of InSn to the Sn pads.
4. LED is positioned on the polymer film and fixed prior to soldering by instant bonding of the bottom face of the LED to the polymer film. Conventional cyanoacrylate adhesive is used.
5. Soldering jig is applied on the test vehicle with LED located in the aperture of the jig.
6. Contact spring/hook is applied on the LED.
7. Hot air vent is focused at soldering region at an angle of 45 degrees.
8. Hot air flow is set to required level.
9. Hot air is released in 2 second pulses, which is less than the relaxation time of the PET.
10. Hot air is released again once temperature of polymer is reset to 21 deg. C, i.e. heat is dissipated from soldering zone through jig and metallic platform. Temperature is verified using the thermocouple.
11. Repeat steps 9 and 10 until solder drop formation is complete.

Figure 3: Process order (from left to right: (1) Cu interconnects and Sn pads, (2) deposition of folded InSn ribbon immersed in soldering grease, (3) instant bonding of LED, (4) LED in operation after soldering)

Results and Discussion

Besides from successful operation of the LED under direct current, optical inspection of the soldering zone discloses successful wetting and formation of intermetallic phases between interconnects, solder and pads (Figure 6). Soldered regions remain connected after thermal shrinkage, which is an indication for acceptable stresses within the assembly.

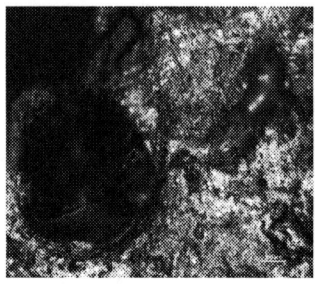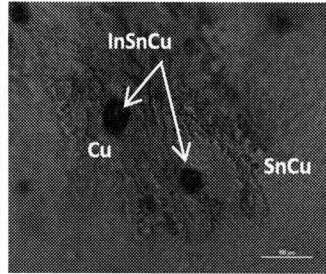

Figure 6: Top view of solder wetting on Sn pads (left), Bottom view through transparent polymer of formation of inter-metallic phases between Sn, Cu and InSn (right)

Due to low thickness (200 nm) of Cu, microscopic inspection of interconnects before and after soldering enable the investigation of material deformation inside and outside of soldering zone. Apart from handling aftereffects, e.g. scratches through tweezers, the Cu layer remains intact after soldering (Figure 8). In particular, formation of micro-cracks does not occur.

Measurements of temperature during soldering reveal that the temperature gradient decays rapidly over the soldering region constraining the soldering zone to minimal CTE mismatches. Therefore, the soldered product exhibits minimal warpages. Moreover, the use of soldering jig is proven to be beneficial by shrinking the heat affected zone. Average diameters of heat affected zones for variations of usage of soldering jig are listed in Table 1.

Figure 8: Cu interconnects before (left) and after (right) soldering (Note: different regions are depicted)

The chemical reaction between soldering grease and PET is investigated by subjecting the PET-soldering grease assembly to thermal testing at 60 deg. C., in particular below glass transition of PET, for about 2 hours. Despite ultrasonic cleaning in an ethanol bath, residues of the soldering grease remain on the surface of the PET film (Figure 5), which is a clear indication for a thermo-chemical reaction between soldering grease and PET. In this regard, either compatible soldering grease should be determined or soldering grease should be completely eliminated from the assembly. In the latter case, special care needs to be taken in order to avoid oxidation of interconnects and solder.

Even though cyanoacrylate adhesive is used for temporary mechanical fixation and is not needed any longer after soldering is complete, it is significant to consider the chemical degradation of the adhesive due to thermal loading during soldering. Here, deterioration of the instant adhesive is not observed due to low soldering temperature and short thermal loading. In terms of mechanical bendability, the flexibility of the system is prohibited at the contact region between LED and interposer.

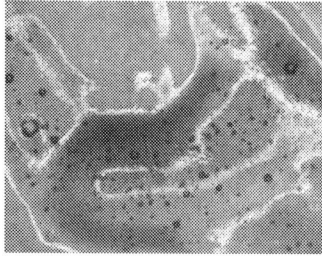

Figure 5: Grease drop on PET before thermal testing (left), Residues of soldering grease on bare PET after thermal testing and cleaning (right)

The amount of hot air required for soldering depends on the soldering setup and product, in particular materials and geometry. For the test vehicle presented in this work, a hot air flow of 1792 sccm is determined to be optimal. On one hand, smaller amounts of hot air are observed to be dissipated quickly, and so solder drop formation does not occur. On the other hand, larger amounts of hot air take longer to be dissipated, and the polymer relaxation time is consequently exceeded.

The use of hot air induces side effects. First, hot air can arrive at regions outside of soldering zone, and cause the formation of sand dunes of material (Figure 7). Second, particles of solder and soldering grease can be blown outside of soldering region by hot air, which may cause short circuits. By maintaining perfect contact between the soldering jig and the polymer film, these artifacts can be easily avoided.

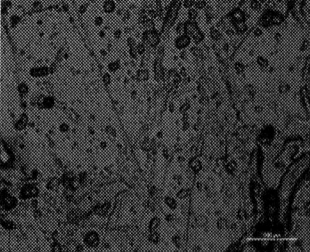

Figure 7: Sand dune effect (left), Blown solder and grease particles (right)

978-1-4799-2408-0/14 $31.00 © 2014 IEEE   1598

In terms of manufacturing costs, while InSn is more expensive than classical solders such as SnCu, the solder cost is substantially compensated by the inexpensive interposer. Moreover, the investment of soldering setup, i.e. the soldering jig and metallic plate, can be counterbalanced by roll-to-roll mass production. It is to be noted that the size of the aperture is dependent on the component and combination of components on the board as well.

	Without jig	Jig	Jig with fin
Diameter	8 mm	5 mm	2 mm

Table 1: Diameter of heat affected zone for various usage of jig at 118 (+20) deg. C. and after 10 seconds

Conclusion

We developed an inexpensive and reproducible assembly approach for 3D components onto low Tg polymers based on four principles:

(1) Localized application of heat,
(2) Low temperature eutectic soldering,
(3) Differential heating and cooling,
(4) Ironing during heating and thermal shrinkage.

We successfully demonstrated a test vehicle based on PET, which has a Tg of 71 deg. C. Notably, solder wetting and formation of inter-metallic phases prevailed. Moreover, the heat affected zone was minimized, such that interposer warpage was limited.

Acknowledgments

The authors gratefully acknowledge the financial support by Deutsche Forschungsgesellschaft (DFG) within the Collaborative Research Center "Transregio 123-Planar Optronic Systems."

Moreover, the authors would like to acknowledge Prof. Elke Pichler from the Institute of Energy Research and Physical Technologies at the Technical University of Clausthal for coordinating the measurements of the coefficients of thermal expansion.

References

1. Jeong In Han, 2003, "Flexible Display; Low Temperature Processes for Plastic LCDs", Transactions on Electrical and Electronic Materials, vol. 4, pp. 10-14.
2. S. Cichos, J. Haberland, H. Reichl, "Performance Analysis of Polymer based Antenna-Coils for RFID", Proc. IEEE Polytronic, 2002, pp. 120-124.
3. K. Koski, J. Hölsä, P. Juliet, Z.H. Wang, R. Aimo, K. Pischow, 1999, „Characterization of aluminum oxide thin films deposited on polycarbonate substrates by reactive magnetron sputtering", Materials Science and Engineering: B, vol. 65, issue 2, pp. 94-105.
4. F. Bodino, G. Baud, M. Benmalek, J.P. Besse, H.M. Dunlop, M. Jacquet, 1993, "Alumina Coating on

polyethylene terephtalate", 1994, Thin Solid Films, vol. 241, pp. 21-24.
5. Manal M. Shalabi, Johannes G.C. Wolke, Vincnet M. J.I. Cuijers, John A. Jansen, 2007, "Evaluation of bone reponse to titanium-coated polymethyl methacrylate resin (PMMA) implants by X-ray tomography", Journal of Materials Science: Materials in Medicine, vol. 18, pp. 2033-2039.
6. Ana Maria Oliveira Brett, Frank Michael Matysik, M. Teresa Vieira, 1997, "Thin-film gold electrodes produced by magnetron sputtering. Voltammetic characteristics and application in batch injection analysis with amperometric detection", Electroanalysis, Vol. 9, pp. 209-212.
7. Jussi Putaala, Maciej Sobocinski, Saara Ruotsalainen, Jari Juuti, Petri Laakso, Heli Jantunen, 2014, "Characterization of laser-sintered thick film paste on polycarbonate substrates", Optics and Lasers in Engineering, vol. 56, pp. 19-27.
8. W. A. MacDonald, M. K. Looney, D. MacKerron, E. Eveson, R. Adam, K. Hashimoto, K. Rakos, 2012, "Latest advances in substrates for flexible electronics", Journal of the Society for Information Display, vol. 15, pp. 1075-1083.
9. Frederik C. Krebs, Mikkel Jorgensen, Kion Norrman, Ole Hagemann, Jan Alstrup, Torben D. Nielsen, Jan Fyenbo, Kaj Larsen, Jette Kristensen, 2009, „A complete process for production of flexible large area polymer solar cells entriely using screen printing – First public demonstration", Solar Energy Materials and Solar Cells, vol. 93, pp. 422-441.
10. Seung-Ho Kim, Kiwon Lee, Kyung-Wook Paik, 2010, "High speed touch screen panels (TSPs) assembly using anisotropic conductive adhesives (ACAs) vertical ultrasonic bonding", 60th proceedings of IEEE Elecronic Components and Technology Conference (ECTC), pp. 1964-1967
11. J.H. Choi, J.H. Park, J.S. Moon, J.W. Nam, J.B. Yoo, C.Y. Park, J.H. Park, C.G. Lee, D.H. Choe, 2006, "Fabrication of carbon nanotube emitter on the flexible substrate", Diamond and Related Materials, vol. 15, pp. 44-48
12. G. Connell, R. L. D. Zenner, J. A. Gerber, 1997, „Conductive adhesive flip-chip bonding for bumped and unbumped die", Proceedings of 47th IEEE Electronic components and Technology Conference, pp. 274-278.
13. Lin Hui, Yu Junsheng, Wang Nana, Huang Chunhua, 2008, "Flexible organic light-emitting diodes with improved performance by insertion of an UV-sensitive layer", Journal of Vacuum Science and Technology B: Microelectronics and Nanometer Structures, vol. 26, pp. 1379-1381.
14. B. Nalini, D. Nirmal, J. Cyril Robinson Azriah, 2013, "Fabrication and characteristics of flexible thin film depletion mode field effec transistor (FET) using high k dielectric nano zirconia", International Journal of

978-1-4799-2408-0/14 $31.00 © 2014 IEEE

Emerging Trends in Engineering and Development, vol. 2, pp. 295-299.

15. K.S. Kaur, J. Missine, B. Vandecasteele, G. Van Steenberge, S.M. Perinchery, E.C.P. Smits, R. Mandamparabil, 2013, "Laser-induced forward transfer-assisted flip-chip bonding of optoelectronic components", IEEE Conference on Lasers and Elecro-Optics- International Quantum Electronics Conference, Abstracts.

16. Yinxiang Lu, Suhua Jiang, Yongming Huang, 2010, "Ultrasonic-assisted electroless deposition of Ag on PET fabric with low silver content for EMI shielding", Surface and Coatings Technology, vol. 204, pp. 2829-2833

17. J. Glazer, 1994, "Metallurgy of low temperature Pb-free solders for electronic assembly", International Materials Reviews, vol. 40, pp. 65-93

18. Y.T. Cheng, L. Lin, K. Najafi, 1999, "Localized bonding with PSG or indium solder as intermediate layer", 12th IEEE International Conference on Micro Electro Mechanical Systems, pp. 285-289.

19. K. Sakuma, E. Blackshear, K. Tunga, Lian Chenzhou, Li Shidong, M. Interrante, O. Mantilla, Jae-Woong Nah, 2013, "Flip chip assembly method employing differential heating/cooling for large dies with coreless substrates", IEEE 63rd Electronic Components and Technology Conference, pp. 667-673.

20. Jongman Kim, Harry Schoeller, Junghyun Cho, Seungbae Park, 2007, "Effect of oxidation on indium solderability", Journal of Electronic Materials, vol. 27, pp. 483-489.

Parameter Optimization in Assembly Manufacturing Process for a Power Module

Yumin Liu and Yong Liu
Fairchild Semiconductor Corp.
82 Running Hill Rd, South Portland, ME 04074
Tel: (207) 761-3155, Fax: (207) 761-6339, Email: yong.liu@fairhildsemi.com

Abstract

In this paper, the trimming process in manufacture of a power module is investigated by using the commercial FE code Ansys/LS-dyna®. The setup of the trimming process consists of an automated system of cutting die, stripper, punch, and etc. The design and interworking of these tools are quite critical. On one hand, when the gap between the punch and stripper/cutting die is too big, the dambar may not be fully cut as expected. On the other hand, when the gap is too small, the punch might be broken after some cycles due to flash contamination on the tools. If the stripper clamp is too close to the package body, high stress might be generated in the package edge or even in the silicon chips to induce package crack or die crack. The DoE legs with regard to the gap between the punch & stripper and the distance between the stripper clamp & the package body are conducted by the FE dynamic simulations. The simulation results provide good guidance for the assembly process parameter setting. .

Introduction

Power modules are now commonly used in both converter and inverter circuits for power supply and motor drive applications in industrial products and consumer appliances. The structure of the power modules usually contains the IGBTs/Mosfets, IC dice, and passive components [1], [2]. For such system in package, generally more than 20 I/O pins are required, with pins for power I/Os at one side, and pins for IC I/Os at the other side [3], [4], [5]. The assembly manufacture processes become more challenging in order to manufacture robust products and achieve high yields [6]. In this paper, the trimming process in manufacture of a power module is investigated by using the commercial FE code Ansys/LS-dyna®, which is powerful for explicit dynamic analysis.

The trimming process manages cutting the dambars that short the leads together. The machine used in this process consists of an automated system of cutting die and upper die set. The upper die set includes the stripper and punch. During the dambar trimming process, the molded package strip is loaded on the cutting die, then the upper die set moves down. While the stripper firmly holds the leads, the punch cuts the dambars immediately. The stripper and punch are usually driven by hydraulic system, and a spring is applied between the stripper load adjuster and the top of the stripper set to control the clamping force. The design and interworking of these tools are quite critical. On one hand, when the gap between the punch and stripper/cutting die is too big, the dambar may not be fully cut as expected. On the other hand, when the gap is too small, the punch might be broken after some cycles. If the stripper clamp is too close to the edge of the package body, high stress might be generated in the package edge or even in the silicon chips to induce package crack or die crack. Therefore, these parameters should be set properly with optimized values in the trimming process.

In this paper, the trimming process of a lead frame based power module with ceramic substrate is simulated, with the detailed drawing of lead frame, package, and tools in the FE model. The eroding technique of Ansys/LS-dyna® is adopted to simulate the dambar cutting process. The eroding surface to surface contact is defined between the punch and the dambars, and failure strain criterion is defined for the lead frame material. With this technique, the elements of dambars will be removed when the calculated strain reaches the failure strain, and finally all the dambars are punched off. The DoE legs with regard to the gap between the punch & stripper and the distance between the stripper clamp & the package body are conducted through the FE dynamic simulations.

FEA Model Setup

The trimming/forming process is an End of Line (EOL) process in semiconductor package assembly. The equipments for this process are usually automatic, and the main tool includes the cutting die, the stripper and punch. During the trimming process, the molded strip is loaded on the cutting die, and the stripper moves down and holds the leads, then the punch cuts the dambars through the hydraulic system. In order to simulate the trimming process, some assumptions and simplifications are made. A schematic drawing of the trimming process is illustrated in Fig. 1, in which the gap between the punch and stripper, and the gap between the punch and cutting die are highlighted with red circles. The punch load driven by the hydraulic system is assumed as a constant velocity.

The trimming process of a lead frame based power module with ceramic substrate is simulated, with the detailed drawing of lead frame, package, and tools with real dimensions in the FE model, as shown in Fig. 2. The cutting of the dambars at the side of IC pins is studied. In order to save simulation time, some simplifications are made to the FE model, and the simplified model is shown in Fig. 3. The cutting die is neglected, accordingly, the lead frame bottom area which is supported by the cutting die, is assumed to be fixed in the Z direction. Only a certain length of the stripper and punch in the Z direction are considered, with equivalent densities by keeping the total masses same as the real case. When cutting the dambars at the side of IC pins, very small stresses may transfer across the package to the other side, and the stresses of lead root area and mold compound edge area at the side of IC pins are critical. Therefore, only half molded strip is considered with symmetrical boundary conditions applied. It is supposed that the impact on the IC chips is small due to long distance to the package edge, so the IC chips are also neglected in the simplified model.

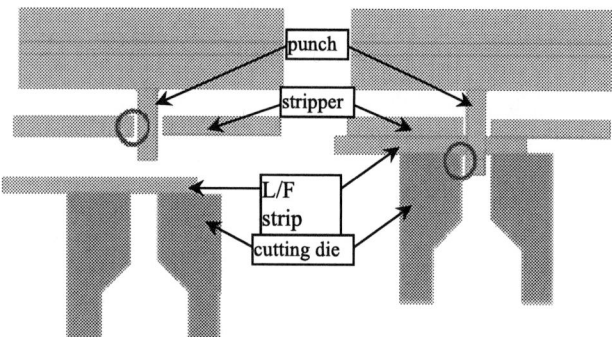

Figure.1 Schematic of trimming process.

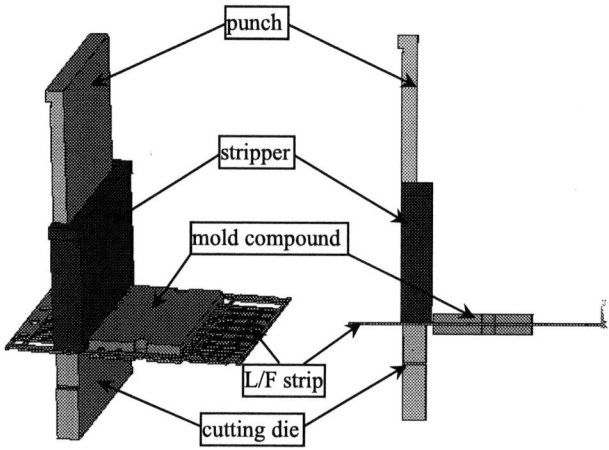

Figure.2 FEA model for trimming process.

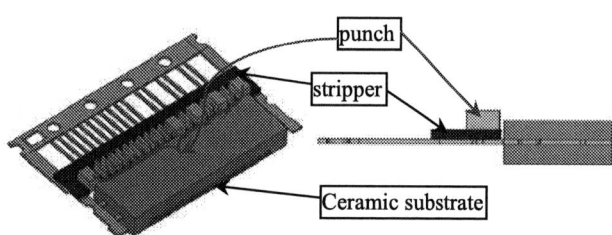

Figure.3 Simplified FEA model for trimming process.

The commercial FE code Ansys/LS-dyna® is used for the dynamic simulations. The eroding surface to surface contact pairs are defined between the punch and the dambars, and failure strain is defined for the lead frame material. The normal surface to surface contact pairs are defined between the stripper and the lead frame strip. The lead frame bottom area which is supported by the cutting die, is assumed to be fixed in the Z direction. Both the stripper and punch are defined as rigid material. A constant velocity of 40 mm/s in the negative Z direction is applied to the punch, with zero acceleration during the trimming process. The material property data of the power module package and tools are

listed in Table 1. The lead frame is defined as elastic linearly plastic material with 5% failure strain, and all the other parts are defined as elastic materials.

Table 1 Material property data

	Ceramic adhesive	ceramic	EMC	Lead frame	punch
Density (Kg/mm^3)	2.96e-8	3.78e-8	1.970e-8	8.94e-8	14e-8
Young's Modulus (KPa)	19.75e6	340e6	19.566e6	1.18e6	5.28e6
Poisson's ratio	0.3	0.22	0.3	0.345	0.29
Yielding stress (KPa)	--	--	--	3.45e5	--
Tangent modulus (KPa)	--	--	--	1.0e6	--
Failure strain	--	--	--	0.05	--

Impact of the Gap between Punch and Stripper

The design and interworking of the punch, stripper and cutting die in trimming process are quite critical. On one hand, when the gap between the punch and stripper/cutting die is too big, the dambar may not be fully cut as expected, and some lead burrs may be left at the root area of the dambars. On the other hand, when the gap is too small, the punch might be broken after some cycles due to flash contamination on the tools. In the assembly line, it has been proved that a 25 um gap between the punch and cutting die has very good performance for the current power module during the trimming process. So the gap between the punch and cutting die is fixed at 25 um in the simulations. Two cases regarding the gap between the punch and the stripper are studied. One is the ideal case with zero gap, the other with a 10 um gap, as listed in Table 2.

Table 2 DoE legs of gap between punch and stripper

	Gap btw punch and stripper (mm)	Gap btw punch and cutting die (mm)
Model 1	0	0.025
Model 2	0.01	0.025

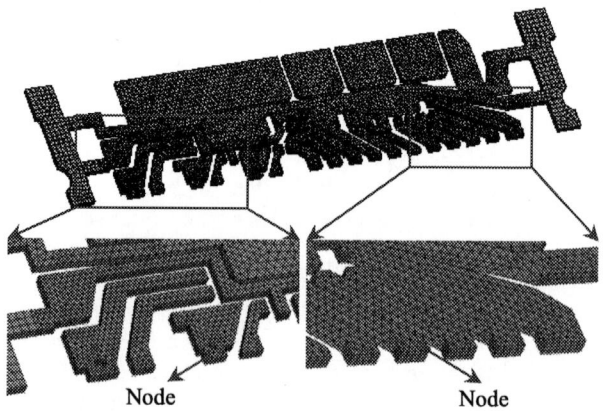

Figure.4 Locations with max von Mises stress.

During the trimming process, when the punch touches the dambars, the stresses are transferred along the dambars to the lead frame and package body. When the calculated strain of the elements of dambars reaches the failure strain of the lead frame material, the elements are removed according to the eroding technique of Ansys/LS-dyna®. The stresses generated at the lead root area and mold compound edge area fluctuates with respect to time during the trimming process. From the simulation results, the max von Mises stress at the lead root area occurs at different locations for the 2 models (Node A and Node B as shown in Fig. 4). Therefore, the time history of the von Mises stress at both locations is outputted. For model 1, the von Mises stress of lead root area at Node B reaches the max value of 77.0 MPa at 4.96 ms, as shown in Fig. 5. For model 2, the max von Mises stress of lead root area is 94.0 MPa, occurring at 4.95 ms, locating at Node A, as shown in Fig. 6.

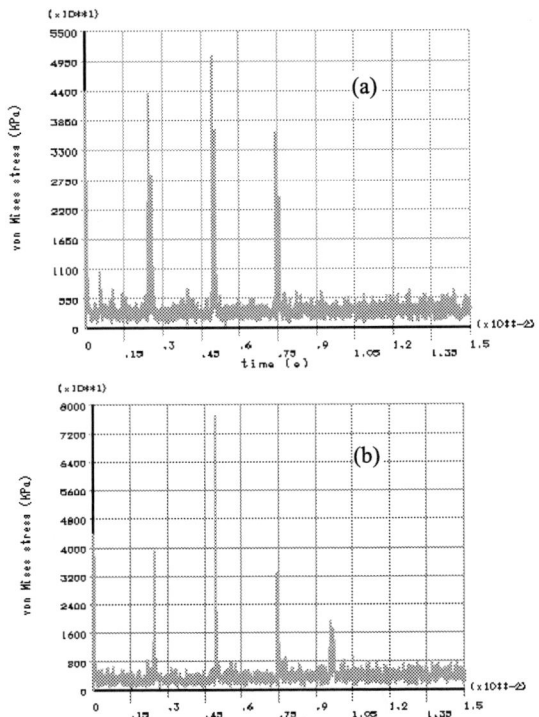

Figure.5 Time history of von Mises stress of lead root area for model 1: (a) at Node A (b) at Node B.

For both models, the max tensile stress of mold compound occurs at the same location Node C, a corner point at the interface of lead frame and the mold compound, as shown in Fig. 7. The time history of the 1st principal stress of the mold compound at Node C is illustrated in Fig. 8. For model 1, the 1st principal stress of the mold compound reaches its maximum value of 15.6 MPa at 4.95 ms during the trimming process (see Fig. 8a). For model 2, the max 1st principal stress of the mold compound is 13.2 MPa at 4.96 ms (see Fig 8b).

The summary of the maximum stresses of lead root area and mold compound edge area for both models are listed in Table 3. It can be seen that the maximum stresses occur at almost same time (4.95 ms vs 4.96 ms). Model 1 with 10 um gap between the punch and stripper, produces larger stresses at the lead root area and mold compound edge area,

comparing with model 2 with zero gap for the ideal case. This indicates that the ideal case with zero gap between the punch and stripper may have the smaller impact on the package body from the viewpoint of induced stresses at the package body edge area. In reality, the punch tends to be broken after some cycles if the gap is too small. Therefore, this should be balanced off when designing the punch and stripper system of the trimming tools for each package product.

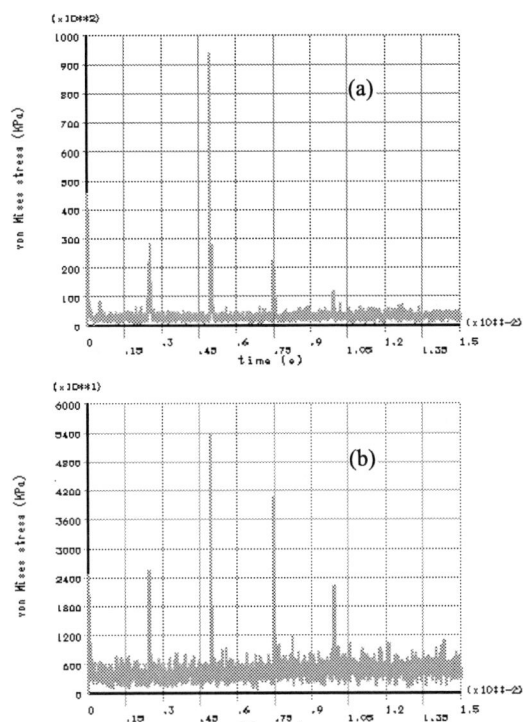

Figure.6 Time history of von Mises stress of lead root area for model 2: (a) at Node A (b) at Node B.

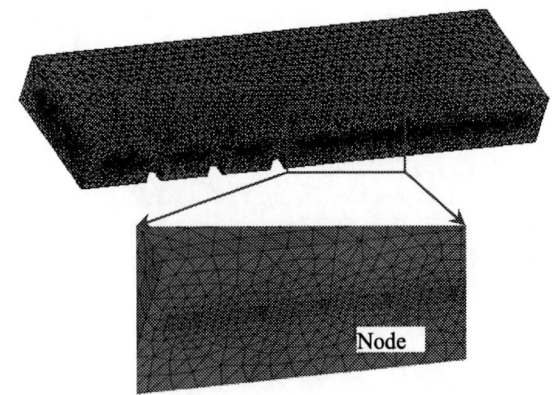

Figure.7 Location with max S1.

Table 3 Summary of max stresses for model 1 & 2

	Gap btw punch and stripper(mm)	von Mises stress at L/F root (Mpa) (node A)	von Mises stress at L/F root (Mpa) (node B)	Max S1 at the EMC edge (Mpa) (node C)
Model 1	0.01	94 @ 4.95 ms	53.7 @ 4.95 ms	15.6 @ 4.95 ms
Model 2	0	50.4 @ 4.97 ms	77 @ 4.96 ms	13.2 @ 4.96 ms

978-1-4799-2408-0/14 $31.00 © 2014 IEEE

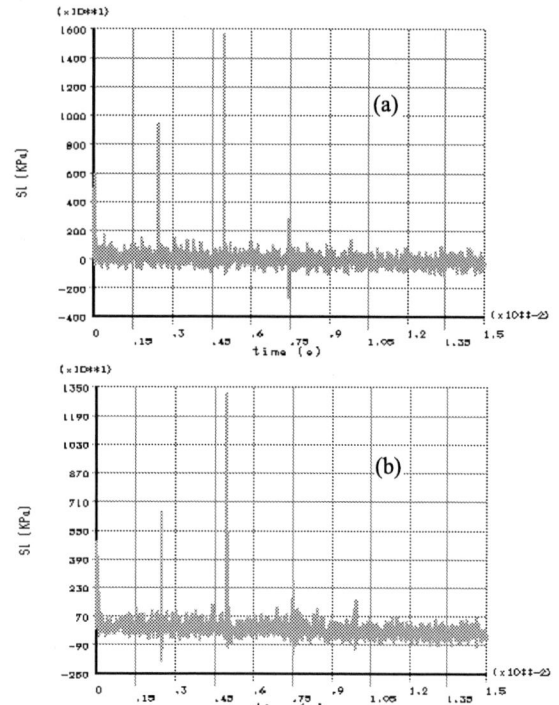

Figure.8 Time history of S1for mold compound at Node C: (a) model 1 (b) model 2.

Impact of Distance between Stripper and Package Edge

The distance between the stripper and package edge (see Fig. 9) impacts the stripper clamping area of the lead frame, which may have some influence on the stress at the package edge area during the trimming process. For model 1 & 2 in the previous section, this distance is fixed at 0.45 mm. In this section, two more cases with larger distance values are studied as listed in Table 4 for model 3 & 4. The gap between the punch and stripper is fixed at 10 um, and gap between the punch and cutting die is fixed at 25 um.

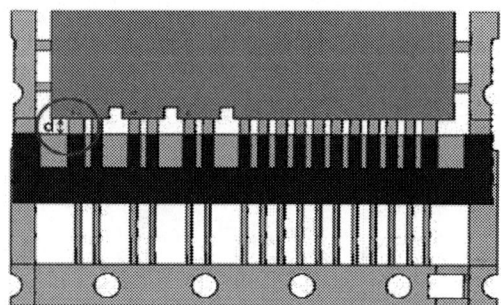

Figure.9 FE model with the distance between the stripper and pkg edge.

Table 4 DoE legs regarding distance btw stripper and pkg

	Distance btw stripper and pkg edge (mm)	Gap btw punch and stripper (mm)	Gap btw punch and cutting die (mm)
Model 1	0.45	0.01	0.025
Model 3	1.1	0.01	0.025
Model 4	1.75	0.01	0.025

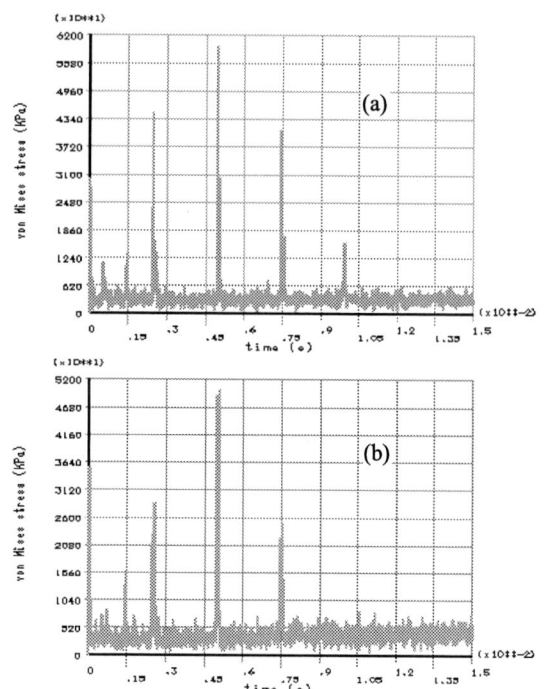

Figure.10 Time history of von Mises stress of lead root area at Node A: (a) model 3 (b) model 4.

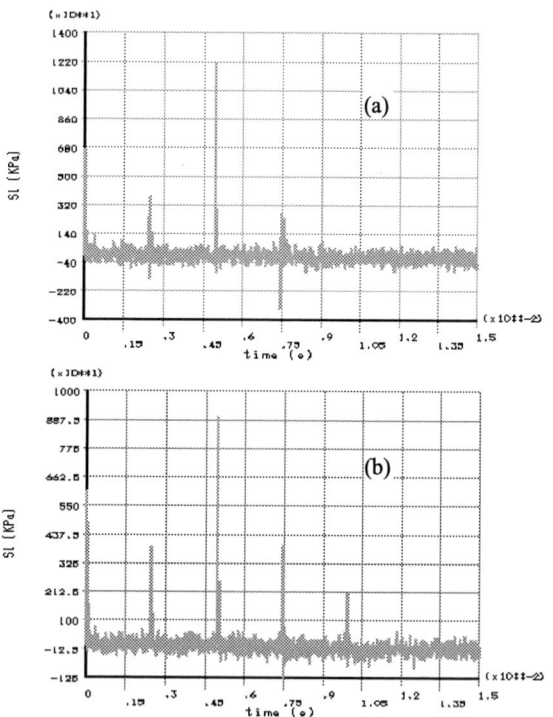

Figure.11 Time history of S1for mold compound at Node C: (a) model 3 (b) model 4.

The simulation results for the impact of distance between the stripper and package edge are shown in Figs. 10 & 11. Figure 10 shows the time history of von Mises stress of the lead root area at Node A for models 3 & 4. It can be seen that the von Mises stress reaches the max value at a little later time for model 4 (at 5.04 ms), comparing with model 1 & 3 (at 4.95 & 4.96 ms). Figure 11 shows the time history of the 1st principal stress of the mold compound edge area at Node C for models 3 & 4. Model 3 induces relatively larger tensile stress in the mold compound edge area during the trimming process, comparing with model 4, but the max stresses occur at the same time (4.95 ms) for both models.

The max stresses for model 1, 3 & 4 with different distances between the stripper and package edge, are summarized in Table 5. With the increase of the distance between the stripper and package edge, both the stresses at the lead root area and mold compound edge area decrease during the trimming process. This indicates that the stripper clamp has bigger impact on the package edge area when it moves closer to the package body. On the other hand, if the stripper clamp is too far from the package body, the package body may have some small bouncing during the trimming process. This should be avoided in the real assembly process.

Table 5 Summary of max stresses for model 1, 3 & 4

	Distance btw stripper and pkg edge (mm)	Gap btw punch and stripper (mm)	von Mises stress at L/F root (Mpa) (node A)	Max S1 at the EMC edge (Mpa) (node C)
Model 1	0.45	0.01	94.0 @ 4.95 ms	15.6 @ 4.95 ms
Model 3	1.1	0.01	58.8 @ 4.96 ms	12.1 @ 4.95 ms
Model 4	1.75	0.01	50.3 @ 5.04 ms	9.0 @ 4.95 ms

Discussion & Conclusions

In this paper, the dambar trimming process of a power module is simulated in Ansys/LS-dyna®. The impact of the gap between the punch & stripper and the distance between the stripper & the package edge is investigated. From the FE simulation results, it can be concluded:

(1) Smaller gap between the punch and stripper may have benefit to the package body from the viewpoints of stresses induced at the package edge area. But too small gap may damage the tools due to flash contamination after some cycles.

(2) The stripper clamp has significant impact on the package edge area when it moves closer to the package body. If the stripper clamp is too far from the package body, it may induce some bouncing issue in the assembly.

The simulation results can give the guidance for the manufacture trimming process, although some assumptions and simplifications are made during the modeling. In real assembly process, the situation is more complicated with multiple factors interacting and influencing. Therefore, these parameters should be balanced off in order to manufacture more robust products and achieve higher yields.

Acknowledgments

The authors wish to thank the support from Package Development Group and Package Assembly line (Suzhou site), Fairchild Semiconductor Corp.

References

1. Y. Liu, and D. Kinzer, "Challenges of Power Electronic Packaging and Modeling," in *Proc. Thermal, Mechanical and Multi-Physics simulation and Experiments in Microelectronics and Microsystems (EuroSimE)*, 2011, pp. 1-9.

2. Y. Liu, S. Irving, T. Luk, D. Kinzer, "Trends of power electronic packaging and modeling", In *proc. Electronics Packaging Technology Conference*, 2008, pp. 1-11.

3. Y. Liu, Y.M. Liu, Z.F. Yuan, T. Chen, K.H. Lee and S. Belani, "Warpage analysis and improvement for a power module," in *Proc. IEEE Electronic Components and Technol. Conf. (ECTC)*, 2013, pp. 475-480.

4. Y. Liu, D. Desbiens, S. Irving, T. Luk, C. Lolar, Y.M. Liu and Q.X. Qian, "Systematic evaluation of die thinning application in a power SIPs by simulation," in *Proc. IEEE Electronic Components and Technol. Conf. (ECTC)*, 2006, pp. 981-989.

5. R. Qian, Y.M. Liu, Y., Liu, S. Martin, and O.S., Jeon, "Thermal mechanical modeling and assessment for a novel power system module with vertical input capacitor," in *Proc. Thermal, Mechanical and Multi-Physics simulation and Experiments in Microelectronics and Microsystems (EuroSimE)*, 2011, pp. 1-8.

6. Y. Liu, *Power electronic packaging*, Springer, 2012

Automated Inspection and Metrology for 2.5D and 3D/TSV Process Assurance

James Wood, Vilmarie Soler, Eric Perfecto
IBM, 2070 Rt 52, Hopewell Jct, NY 12533
woodj@us.ibm.com

Thomas Luckenbach
CamtekUSA, 2000 Wyatt Dr., Santa Clara, CA 95054

Aki Shoukrun
Camtek Ltd., Ramat Gavriel Ind. Zone, Migdal Haemak, 23150, Israel

Abstract

The microelectronics industry continues to witness an ever-increasing movement towards heterogeneous integration leveraging the advantages of 2.5D and 3D/TSV in advanced semiconductor packaging and integration. New processes and physical structures are being developed to incorporate these advantages into fully functional manufacturable, reliable and cost effective devices. 3D wafer processing will rely heavily on advanced inspection and metrology capability in order to achieve the affordability and reliability necessary for commercial success.

Automatic inspection is playing a key role in reducing the cycles of learning needed to develop these new processes and structures. Inspection strategies have been implemented which provide the required monitoring and process control. This paper will cover various inspection and metrology examples used in the monitoring of wafer finishing sector through the use of available Automated Optical Inspection (AOI) equipment. Focus is placed on those processes required to make the interconnect structures, as opposed those needed to create the internal TSVs themselves. Photolithography control, TSV reveal, redistribution, and metal feature dimensions, bump metrology, and defect detection, will be topics discussed in this paper.

Lateral (in plane) dimensions were measured using a CCD camera, while Triangulation and Chromatic Confocal scanning where used for height measurement. Of course, in order to maximize yield and provide high reliability, defect detection at various steps in the overall process flow is critical. The AOI played a critical role in accelerating the end to end learning during wafer finishing fabrication.

Introduction

The introduction of 2.5D and 3D/TSV technology, which will be referred to as 3D throughout the remainder of this discussion, continues to accelerate. Processes and structures which have not previously been incorporated into mainstream semiconductor devices are necessary for the various 3D implementation schemes that are arising. The Through Silicon Vias (TSV) which are the fundamental enabling features in 3D are the subject of much study and discussion, but the accompanying physical connections are also of extreme importance.

While the TSV structures are the key element by which chip stacking is achieved, the full wafer finishing structures needed to make the interconnections functional and reliable include much more. 3D wafer finishing commonly requires radical transformation of the wafer itself due to the requirements of wafer bonding and thinning used to make the contact of the TSVs possible. This adds to the complexity of performing and controlling the processing steps necessary to complete these interconnecting structures. For the sake of clarity, the terminology "wafer finishing" (WF) will refer to the processing needed to transform a "thick" wafer with buried TSVs, into a thinned wafer with complete interconnection structures on both sides.

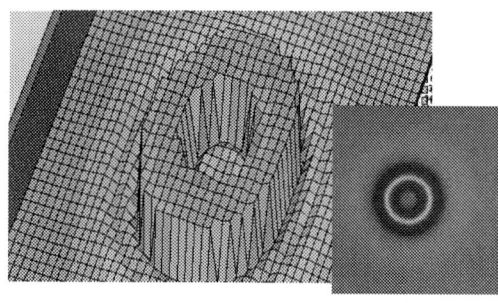

Figure 1. Chromatic Confocal height of exposed TSV.

Recently, IBM has been conducting extensive work on developing viable manufacturing schemes to bring true 3D integration about. [2,3,4] To develop this emerging technology, a "Via Middle" integration scheme where the TSVs are fabricated during the Back End of Line (BEOL) processes, has been pursued. Automated optical inspection and metrology have been incorporated into the processing flow to assure compliance to required product specifications, and provide on-going process control. An inspection and measurement strategy tailored to the process flow, as the process transitions into and through the development phase, and into full manufacturing, is critical. Competing potential process flows were compared during the process development and refinement. Data rich feedback was provided enabled informed process flow decisions. The entire WF process flow was monitored during initial process flow development.

As part of WF, the wafer is bonded to a handler with the use of an adhesive. Buried TSVs are then exposed to allow for interconnecting structures. This is achieved by thinning of the 300mm Si wafer to reveal the Cu TSV. This via height has been measured using both available sensors as described in the equipment description section below. CMP is employed to level the Cu TSV against the Si surface. [1,5,6] A key goal of

978-1-4799-2408-0/14 $31.00 © 2014 IEEE   1606   2014 Electronic Components & Technology Conference

the TSV reveal process was to achieve uniform height at the Si surface.

Advancement of the micro-bump technology has accelerated with 3D adoption, in particular Si to Si interconnects. Micro-bump technology, in which bump dimensions and spacing can be reduced to 25 microns and below, allows a large increase in the number of possible I/O connections over conventional C4 (controlled collapse chip connection) bumping schemes. These small bump interconnects have been difficult to implement when devices are bonded to traditional laminate substrates due to well known CPI (chip to package interaction) issues on very large die, and the high bow of the thinned wafers and the laminates.

With 3D architecture, the bottom die is bonded to the laminate substrate and will use traditional bump dimensions. Any die stacked above the bottom die would be bonded to another Si IC, in which case, micro-bumps are a viable option. Although the major stumbling blocks which have long thwarted the reduction in bump dimensions are eliminated with 3D technology, other challenges such as efficient metrology will remain. Assurance of the specification for 100% of all bumps for very large devices with large numbers of bumps per die is a challenge, especially given the criticality of tight coplanarity specifications. In this paper we review metrology and defect data obtained in measurements and inspections of high bump count 300mm wafers. In the following sections we will very briefly discuss these topics.

Both Face to Face and Face to Back stacking schemes have been developed. In the case of Face to Face, the device sides of each wafer are bonded together with TSVs through one of the devices leading to bumped pads on the opposite side. In Face to Back, each device has through vias, allowing the staking of several die. In this case, it is necessary to create terminal connections on the device side of the wafer, followed by wafer thinning operations to expose the buried ends of the TSVs and create connection structures on the side opposite of the devices. Some of the structures which are required are pads for bump attach, redistribution lines (RDL), thermal or fill patterns, pads connecting directly to TSVs, and pads with solder bumps. These structures are often created using top surface photolithography methods with plate-up and seed metal subetch. In order to assure correct electrical characteristics and to accommodate reliability requirements, tight dimensional and overlay process control is a requirement.

Control charts were created using dimensional measurement data collected with Automated Optical Inspection equipment. Following the photo resist develop step, an inspect/measure operation was incorporated in which a "critical dimension" is measured and at the same time, the entire resist pattern is inspected for potential defects. Variable factors such as photo resist thickness and exposure dose were optimized to center the dimension size. Maintaining tight control of these variables minimizes deviation from the target values. Additionally, overlay alignment of the current pattern to underlying features can be determined.

Various sampling schemes can be determined once baseline data has been collected to establish process capability. Features and structures created during "wafer finishing" are top surface structures that are ideal for characterization in an automated fashion. Careful utilization of these process monitoring operations in the photolithography sector will assure high yields and high reliability. Automated measurement allows data from many sites across the wafer surface to be collected quickly, in order that any dimensional variation across the wafer surface can easily be determined and corrected. For example a > 525mm2 interposer can have more than 20 K connection pads.

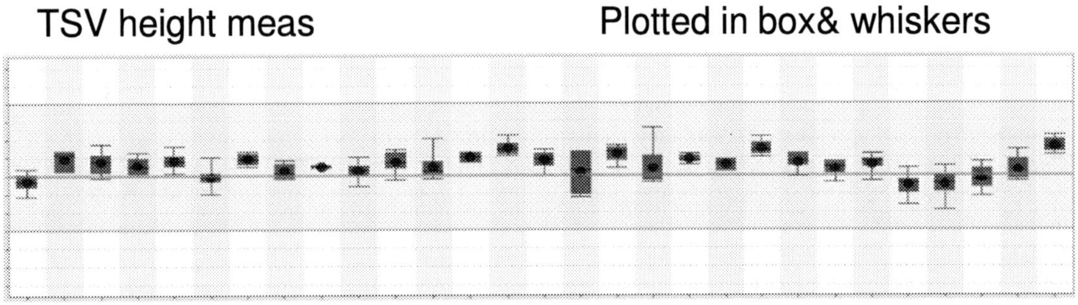

Even with careful storage, photo masks occasionally can also be subject to collecting contamination which can result in errors in the photo resist pattern. These errors can largely be eliminated by creating "dummy" patterns on blank wafers and inspecting these for defects. An effective strategy of inspecting monitor wafers and sample inspection of produce wafers can be put in place and tuned to produce high yield with minimal inspection activity. Inspection of the photo patterns can be increased or decreased depending on the results, and on the results of inspections of the completed structures. Early in process development higher rates of surface inspections may be required to identify failure mechanisms and access remedial actions.

Equipment

All measurement and inspection operations described herein are contained in a single unit. The equipment is comprised of double load ports, a pre-alignment station, a two end effector robot, and stage. Sensors include a high resolution CCD camera with multiple objective lenses, a two sensor triangulation system and a chromatic confocal sensor. The CCD camera is used for defect detection inspection, and lateral dimension metrology. The triangulation is used for

Figure 2. Sample TSV expose height plotted per wafer.

measuring bump and exposed via heights, and the confocal sensor measures individual feature heights such as pad or line thickness.

Figure 3. A) CCD camera objective, B) Chromatic Confocal sensor, C) Triangulation illumination source.

Process Control - Photolithography

Process control in the photolithography sectors is a key example of the use of Automated Optical Inspection (AOI) to monitor critical process steps. Metrology data from automatically measured features was uploaded to established databases from which control charts are automatically derived. Process control chars were readily implemented and give information needed to center the process on the desired dimensions, and reduce variability within wafer and between wafers.

In the same scan which generates the wafer metrology data, the photolithography image is inspected and any detected defects are reported. The photo resist pattern largely determines the lateral dimensions, alignment and continuity of the pads and wiring lines making up the interconnecting structures (figure 4). If errors are determined in the photolithography it is possible to rework and correct the issues leading to the out of spec condition. Dimensional control of the resist image is critical since it directly affects the plated thickness and co-planarity.

Process Control - Metallization

The plated metal features which are defined by the photolithography processes were also checked for within specification lateral dimensions and thicknesses. Again, the CCD scan which provides the lateral dimension measurements also runs a defect scan to locate any potential pattern errors.

As successive layers are inspected, defect map overlay was useful to determine the earliest step in the process where the anomalies were encountered. Plated metal thickness was monitored with sample height measurement of features using the chromatic confocal sensor. Both wafer to wafer variation and cross wafer uniformity was characterized for over via connector pads and redistribution lines for example.

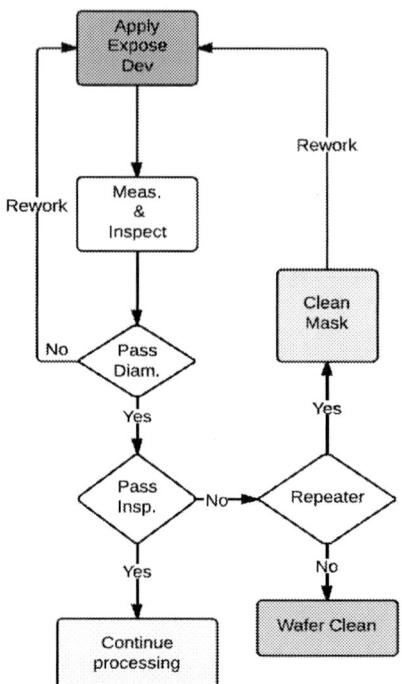

Figure 4. Photolithography sector process and measure/inspect flow diagram.

TSV Measure and Inspect

A key challenge in the wafer finishing process is thinning the wafer to the point where the TSV buried ends, can be uniformly and controllably revealed. There are several important dimensional factors such as via depth, wafer thickness and variation, carrier thickness and variation, and bonding adhesive thickness and variation, which need to be understood and controlled.

This is an area of much interest in the 3D arena, but is not discussed further in this paper. However, once the thinning and via reveal operations have been completed, accurate measurement of the exposed vias is possible and provides feedback on the success of the reveal. Each of the height measurement sensors have been utilized to provide this feedback measurement. For a Go, No-Go measurement of each wafer during process development, sample vias in strategic locations have been taken in order to quickly determine a complete reveal. Additionally, using the triangulation sensor, the entire wafer surface can be scanned giving reveal heights for all of the TSVs. This full wafer scan (figure 5) gives the critical feedback to the several upstream processes.

978-1-4799-2408-0/14 $31.00 © 2014 IEEE

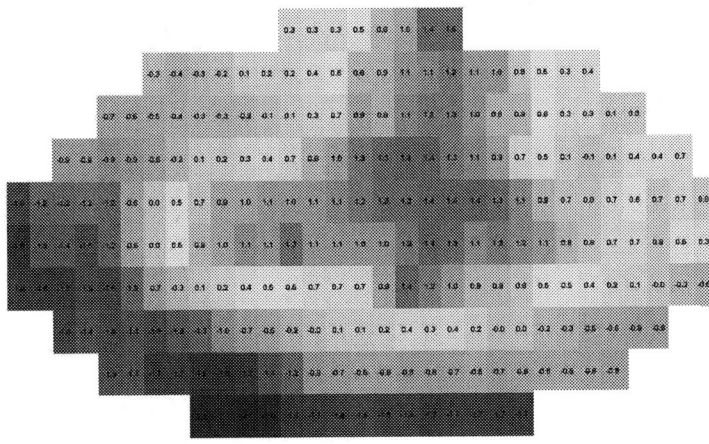

Figure 5. Exposed TSV height variability across wafer.

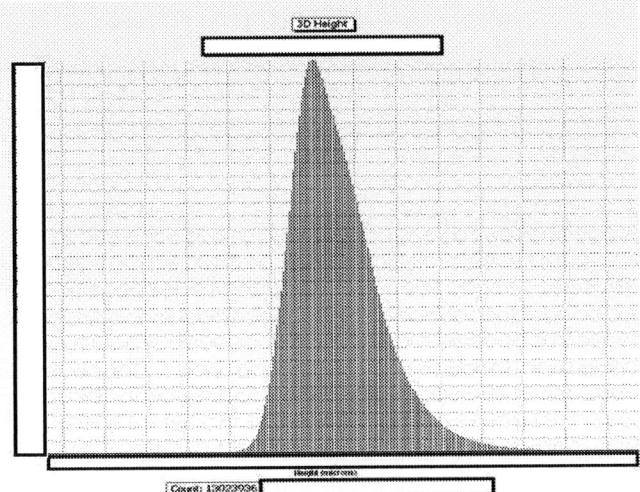

Figure 6. Histogram of height measurements of all micro-bumps on one 300mm wafer. No. of bumps vs. bump height, 13M bumps measured.

C4 Wafer Bumping

The final phase in the wafer finishing process flow is plating C4 bumps. The process here is not different from the standard bumping process with the exception that this is a thinned wafer bonded to a carrier substrate. A photolithography step defines the features where a metal pad and solder metallurgy will be plated. The photo developed wafer is inspected and the pad opening diameter is measured for product specification assurance. Once the C4 metallurgy has been deposited, a reflow operation creates the finished C4 bumps. The final measure and inspect operation is required to assure that each of the bumps meets specification for height, diameter, and volume (figure 6). Additionally, in order to assure proper downstream joining processes, die coplanarity is calculated for each die. The bump diameters are measured during the CCD scan, and a second scan with the triangulation sensor records the height of each bumps.

As mentioned in the introduction, die to be stacked onto the bottom die may be terminated with micro-bump interconnects. These bumps have also been inspected and measured with the same system but using higher resolution objectives mounted on rotating turrets. Bump size is reduced so that the number of bumps, or I/O connections, can be much larger than traditional bumps would allow. Measurements were taken for diameter and height for die designed with over 200,000 bumps each.

Coplanarity must be controlled to a tight specification of only a few microns in order to produce reliable bonding with such small solder bumps. In order to achieve good coplanarity values for each die, each bump must be accurately measured for height. In addition to obtaining coplanarity values, all diameters are recorded and defects such as missing or bridging bumps and misshapen bumps are discovered. This information becomes critical as die are stacked and a defective connection would lead to a lost stack.

Conclusions

Incorporation of a comprehensive metrology strategy is crucial to the success of 3D chip stacking process flows. Key elements of the manufacturing scheme which are intimately tied to this strategy include process control, engineering specification assurance, process yield, and reliability assurance. Metrology and inspection of C4 bumps, especially micro-bumps, are integral components of a successful 3D stacking program.

References

1. *Halder, S. et al.,"Metrology and Inspection for Process Control During Bonding and Thinning of Stacked Wafers for Manufacturing 3D SIC's," IEEE 61st ECTC, Lake Buena Vista, FL, USA, May 31, 2011*

2. *Sakuma, K., et. al.,"Bonding Technologies for Chip Level and Wafer Level 3D integration," IEEE 64th ECTC*

3. *Farooq, M.G., et. al., "3D Copper TSV Integration, Testing and Reliability," 2011 IEEE IEDM*

4. *Knickerbocker, J. et al., "Development of next generation systems-on-package (SOP) technology based on silicon carriers with fine pitch interconnections," IBM Journal of Research and Development, vol. 49, no.5, 2005.*

5. *Halder, S. et al., "In-line Metrology and Inspection for process control during 3D stacking of IC's," IEEE International 3D Systems Integration Conference (3DIC), Osaka, Japan, Jan. 31-Feb.2, 2012.*

6. *Arkalgud, S.R. "Leading edge 3D technology for high volume manufacturing," 2009 Symposium on VLSI Technology, Kyoto, Japan, 16-18 June, 2009.*

Investigation of a Photodefinable Glass Substrate for Millimeter-Wave Radios on Package

Telesphor Kamgaing, Adel A. Elsherbini, **Torrey W. Frank, Sasha N. Oster, Valluri R. Rao
Components Research, **Assembly and Test Technology Development, Intel Corporation
5000 W. Chandler Blvd, Chandler, AZ 85226, USA
telesphor.kamgaing@intel.com, adel.a.elsherbini@intel.com, torrey.w.frank@intel.com,
sasha.n.oster@intel.com, valluri.r.rao@intel.com

Abstract

Photodefinable glass has been investigated for applications at mm-wave frequencies. Test structures including through-glass vias, transmission lines and microstrip patch antennas have been fabricated and fully characterized up to 67 GHz. Good correlation was obtained between simulation and measurements. The antenna exhibited return loss better than 10 dB in the frequency band of interest. Via transition losses were less than 0.15 dB/via and the transmission line insertion loss was ~ 3.3 dB/cm at 60 GHz.

I. Introduction

The ISM frequency band around 60 GHz presents a huge opportunity for short range wireless communication with datarate in the order of several Gigabits per second (Gbps). Those data rates are key enablers for applications such as wireless docking, wireless sync-and-go, rapid video downloading, and the display/streaming of uncompressed high definition video [1-3]. Radio circuits such as amplifiers have the disadvantage that the output power of typical CMOS transistors drops significantly with increasing frequency; hence, it's necessary to combine several amplifiers in a phased array for applications at those frequencies. The CMOS phased array is supported by antennas that are implemented on a package substrate. Taking into account that the power amplifiers of the CMOS circuits already have low gains, it's critical that the package exhibit low routing losses and deliver antennas with high efficiency.

Over the last decade, significant research has been carried out on different package material for mm-wave applications. Similar to WiFi, most early adopters selected low temperature co-fired ceramic (LTCC) as package substrates for mm-wave phased arrays [4], [5]. In addition to the cost, the co-firing process does not yield high tolerance for the package stackup, which means most test structures such as antennas have to be overdesigned to compensate for potential process variations. In [6], an alumina substrate was used, but that substrate only had one level of interconnect that enabled flip-chip die assembly without on-package antenna integration. Several organic substrate materials have been investigated including liquid crystal polymers (LCP) [7] and BT laminate [8], [9]. While LCP usually presents excellent electrical characteristics, e.g. low dielectric constant and low loss tangent, its fabrication has been limited to only a few dielectric layers mostly due to its coefficient of thermal expansion (CTE), which in a multilayer stackup may lead to via failures during processing or SMT assembly.

Many dielectric materials that are more suitable for mainstream printed circuit board (PCB) manufacturing still suffer from performance degradation at mm-wave frequencies. This can be generally attributed to their high dielectric loss tangent and the high surface roughness realized with typical fabrication processes. Dielectric constants of glass are strongly dependent of the glass structure and can vary anywhere from 4.9 to 12. The best loss tangent reported for amorphous glass is about 0.004 at 10 GHz. In addition, amorphous glass can have incoming surface roughness below 10 nm, which is strongly desired for low loss transmission lines operating at mm-wave frequencies. Despite these advantages, most manufacturing techniques used to create vias in amorphous glass, as required by multilayer packages, utilize some kind of mechanical drilling or laser ablation. This typically results in the risk of fractures and cracks and hence limits both the via size and via density.

In this work, we investigate photodefinable glass as a potential substrate for standalone and phased array antennas operating at mm-wave frequencies. Unlike amorphous glass, photodefinable glass [10] is a special type of glass in which microstructures can be formed without the use of any conventional drilling or micromachining process. When exposed to ultra-violet (UV) radiation and baked at a certain temperature and duration, the photodefinable glass transforms into its crystalline-phase and becomes very reactive to hydrofluoric acid. By using this or similar techniques, through glass holes with diameters as small as 5 um and pitch of 6 um can be created with relatively high yield. A test vehicle with various test structure has been fabricated and fully characterized.

Thin die with TSV

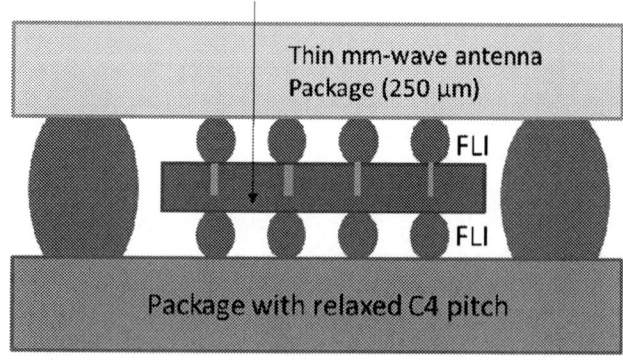

Figure 1. 3D mm-wave phased array antenna architecture using glass as mm-wave antenna module

2014 Electronic Components & Technology Conference

The rest of the paper is organized as follows. Section II discusses the mm-wave phase array architecture and substrate requirements. The package stackup and fabrication results are discussed in section III. Session IV discusses the test vehicle overview. The transmission lines and antenna characterization characterizations are presented in section V and VI, respectively.

II. Millimeter-Wave Phased Array Module Architecture and Substrate Requirements

Figure 1 shows an illustrative cross-sectional view of the proposed phased array antenna module. It consists of a millimeter-wave radio die that is sandwiched between two organic packages. The bottom package is mainly dedicated to the power supply and low frequency control signals. The upper package is a dedicated millimeter wave substrate that carries both the antenna elements and the high frequency feeding lines. The connection between the antennas and the actual circuits is provided by through silicon vias (TSVs). One specific advantage of this architecture is the decongestion of the first level interconnect (FLI) by using the TSVs.

The lower substrate can be lossy and have very relaxed C4 pitch. The upper package on the backside of the silicon die requires extremely fine pitch and CTE compatibility with the silicon. In this case, using a photodefinable glass with a CTE comparable to that of silicon is more appropriate. In addition, the low surface roughness of glass is important in reducing the losses due to conductor surface roughness and can also contribute to improving the antenna efficiency.

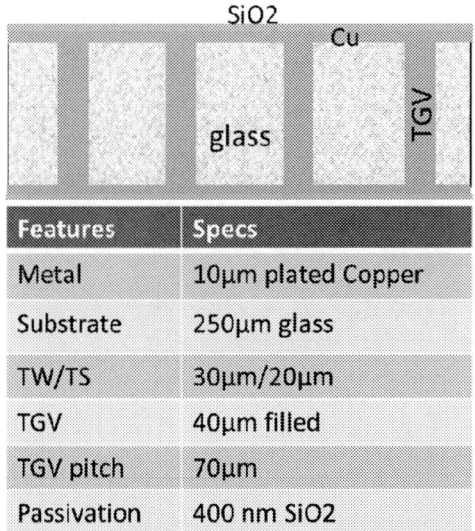

Features	Specs
Metal	10μm plated Copper
Substrate	250μm glass
TW/TS	30μm/20μm
TGV	40μm filled
TGV pitch	70μm
Passivation	400 nm SiO2

Figure 2. Stackup of photodefinable glass substrate

III. Substrate Stackup and Fabrication Results

a- Package Stackup

For the purpose of this study, a photodefinable glass was selected. Unlike amorphous glass, the photodefinable glass can return to its crystalline phase after exposure to UV light, which enables the creation of through glass holes in a typical semiconductor fabrication environment. Figure 2 shows the stackup of the glass package substrate along with the physical dimensions. It consists of a 250 um thick photodefinable glass, 10 um copper was used as metallization on both sides of the dielectric substrate. As passivation, a thin layer of silicon dioxide was used. Gold plating was used as surface finish for the exposed pads. The copper thickness was driven by both the electrical RF and DC performance requirement. The TGV pitch is driven by both the copper plating thickness and the need for creating die attach pads directly on top of the TGVs.

b- Package Fabrication Results

The photodefinable glass process was investigated by fabricating several test structures implemented on a research test vehicle. Figure 2 shows 3D X-ray examples of plated TGVs and a millimeter-wave microstrip patch antenna. The dimensions of the vias were all within the expected specs. Good alignment and registration tolerances were obtained between the vias and the vias pads.

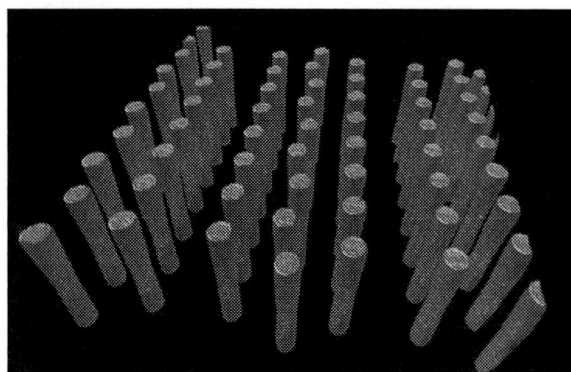

(a) Plated through glass via

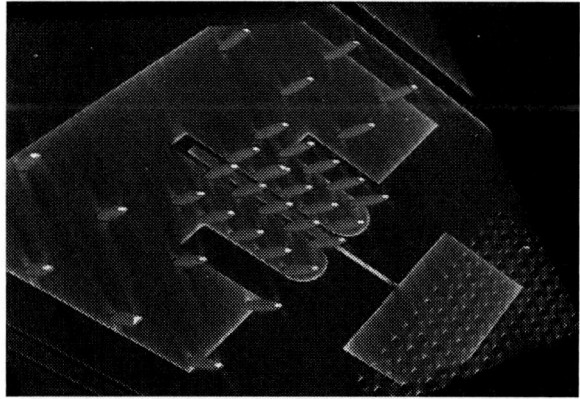

(b) Microstrip patch antenna

Figure 3. 3D view of fabricated (a) through glass vias and (b) microstrip patch antenna.

IV. Test Vehicle Overview

In order to validate the electrical properties of the glass substrate, a 15 mm x 15 mm test package was designed and included coplanar waveguides (CPW), grounded coplanar

978-1-4799-2408-0/14 $31.00 © 2014 IEEE

waveguides (GCPW), via transitions and microstrip patch antennas. Figure 4 shows 3D models of the transmission lines and transition test structures. The transition test structure consists of two back-to-back TGVs that are connected on the bottom side by a 100 um long 50 Ohm GCPW. Several ground TGVs are implemented on both sides of the signal TGV and serve as signal return path. Each transmission line was implemented in two different lengths and two different configurations. In the first configuration, both the signal line and the test pads are on the same side of the substrate. In the second configuration, the test pads are on the opposite side of the substrate and are connected to the signal trace using TGVs. The comparison between the two configurations enables a direct evaluation of TGV losses in multilayer glass-based packages.

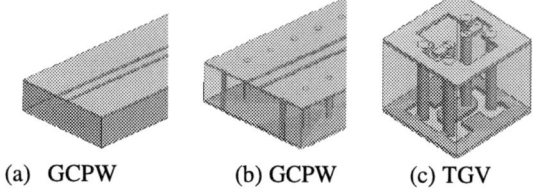

(a) GCPW   (b) GCPW   (c) TGV

Figure 4. 3D models of evaluated transmission lines and transition test structures

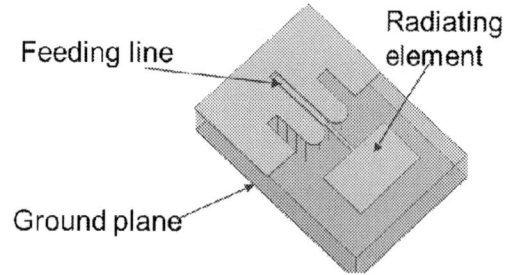

Figure 5. 3D model of microstrip patch antenna on photodefinable glass

A basic microstrip patch antenna that comprises a ground plane on the back side of the substrate and a radiating element (patch) on the top side was used to evaluate the antenna integration on glass at millimeter-wave frequencies as shown in Figure 5. The radiating element was fed using a planar transmission line structure that also acts as impedance matching network between the 50 Ohm test point and the patch. In one case, the transmission line feed was designed to fit a ground-signal ground on-wafer probe. In the other case, it was designed to match a micro edge-mount connector designed for operation up to 70 GHz. The antenna was repeated several times on the test package while varying either the feed line topology or the ground plane structure below the antenna. Solid and patterned ground planes were designed to both reduce the surface current and enhance the gain of the microstrip patch antenna.

Top views of the layout and fabricated packages are shown in Figure 6.

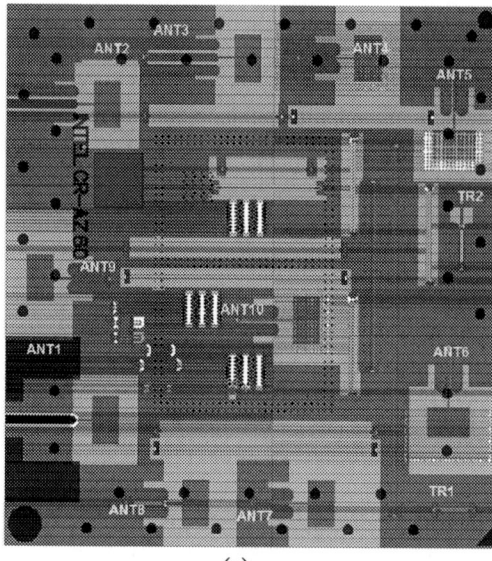

(a)

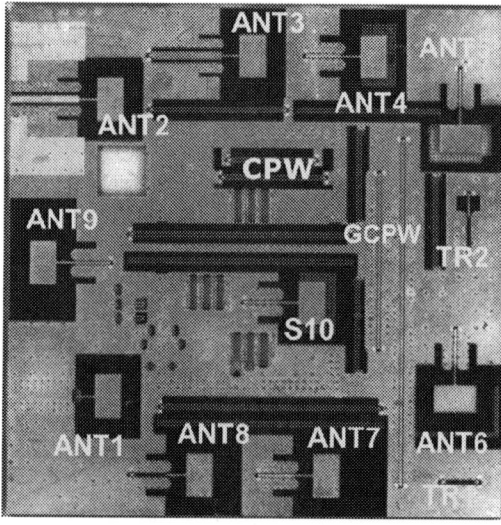

(b)

Figure 6. Top views of (a) layout and (b) fabricated test package on photodefinable glass

V. Transmission Lines and Material Characterization

a- Transmission Line Characterization

The transmission lines and via transitions were measured on a semi-automated probe station using 150um pitch GSG probes and a 70 GHz performance network analyzer (PNA). SOLT calibration was used. Figure 7 shows the modeled and measured electrical performance of two back-back vias separated by a 100 um long 50-Ohm transmission line on the backside. At 60 GHz, the total insertion loss is 0.35 dB, which translate to ~ 0.15 dB per via transition. In Figure 8, we show the measured electrical performance of both the coplanar waveguide (CPW) and the grounded coplanar waveguide (GCPW). At those frequencies, the GCPW is less susceptible to package resonances and exhibits a much cleaner signal.

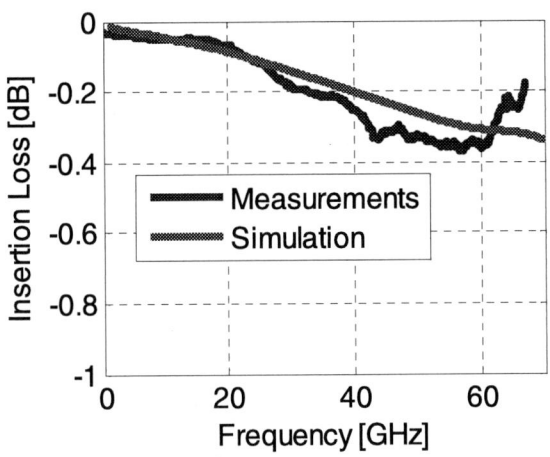

Figure 7. TGV electrical performance

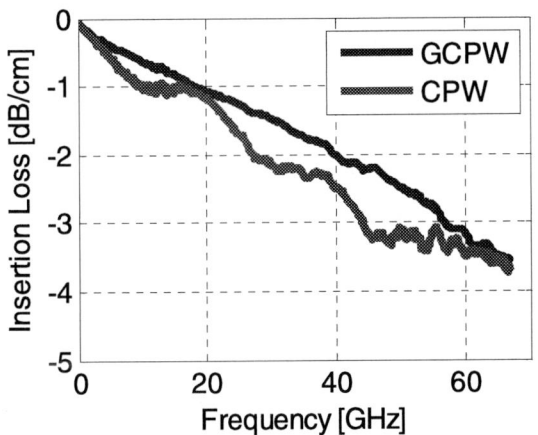

Figure 8. Measured transmission line performance

b- Dielectric Property Extraction

Using the fabricated transmission lines, both the dielectric constant and loss tangent of the photodefinable glass were extracted following the transmission line inversion method described in [11]. At 60 GHz, the effective dielectric constant was 6.4, whereas the loss tangent was 0.022. Those effective values also include any surface roughness at the dielectric/metal interface, which was estimated to be less than 50 nm in our fabrication process. Figure 9 shows the extracted dielectric characteristics extracted from the transmission lines along with measured dielectric properties of an unprocessed photodefinable glass slab using a low frequency resonant cavity. The discrepancy between the dielectric constant is less

than 5%, whereas a very good correlation is obtained for the loss tangent up to 10 GHz.

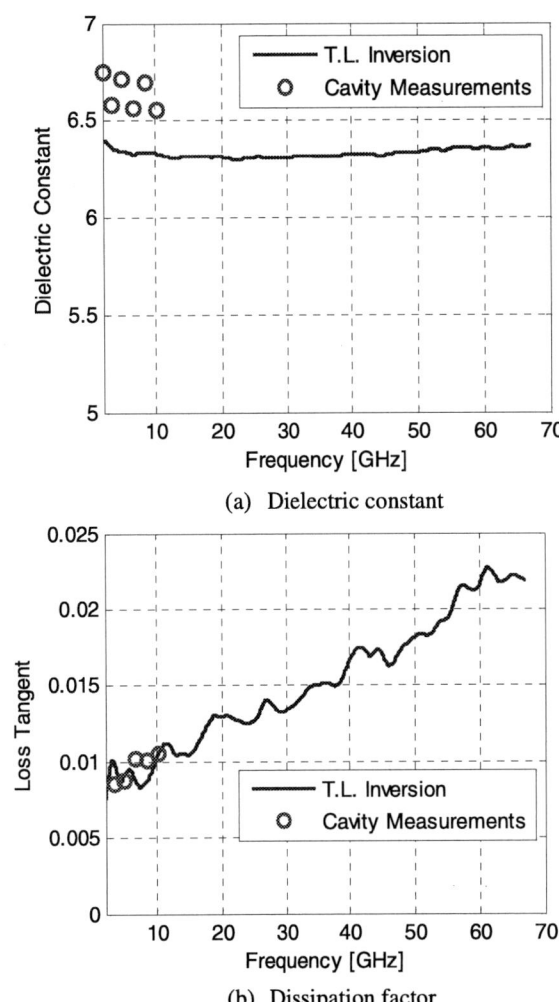

(a) Dielectric constant

(b) Dissipation factor

Figure 9. Extracted material properties of the photodefinable glass using inverse transmission line and resonant cavity methods

VI. Microstrip Patch Antenna Characterization

The first level characterization of the microstrip patch antenna consists of evaluating the input return loss using S-parameters. Those measurements were performed in a probe station environment using the GSG probes. Figure 10 shows the measurement vs. simulation data for our reference design. The antenna exhibits a fractional bandwidth of about 3.6% around 60 GHz.

Systematic measurements were performed for the remaining antennas on the package. A comparison between ANT3, ANT7, and ANT9 shows that the feeding line has no major impact on the antenna matching characteristics. The higher bandwidth obtained for ANT9 is mainly due to the removal of the coplanar ground plane on one side for the

978-1-4799-2408-0/14 $31.00 © 2014 IEEE   1613

patch. The patterning of the ground plane had significant to moderate impact on the electrical performance of the different patch antennas Antenna ANT5 did not resonate at all, whereas antennas ANT3, ANT6 and ANT10 showed a slight improvement in the fractional bandwidth. Table 1 shows a summary of key antenna characteristics obtained from the return loss measurement.

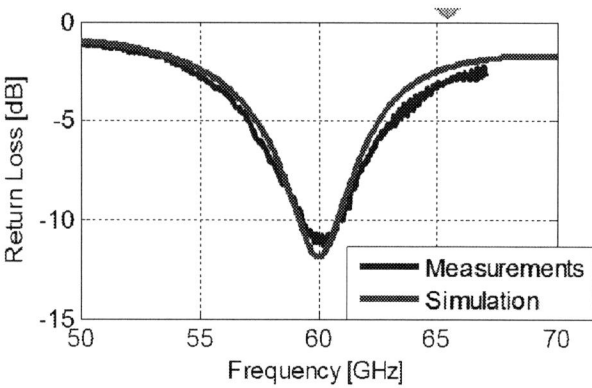

Figure 10. Simulated vs. measured return loss of microstrip patch antenna on glass

Device	Center Frequency f₀ [GHz]	Absolute Bandwidth [GHz]	Fractional Bandwidth [%]
ANT9	59.7	2.43	4.07
ANT7	60.0	1.93	3.22
ANT6	59.3	2.34	3.95
ANT4	59.9	2.26	3.77
ANT3	60.3	2.18	3.61
ANT10	59.5	1.84	3.1

Table 1. Measured performance of microstrip patch antennas with various ground patterns and feeding lines

The second level of measurement consists of evaluating the antenna efficiency and radiation pattern. In this case, the antenna is fed by an edge mount micro-connector designed for V-band applications as shown on the inset of Figure 11a. The measurements are then performed in an anechoic chamber. Figure 11a shows the normalized radiated power vs. frequency. The distortion seen in the measured radiation pattern for Figure 11b can be attributed to additional reflection due to the presence of the other structures on the package.

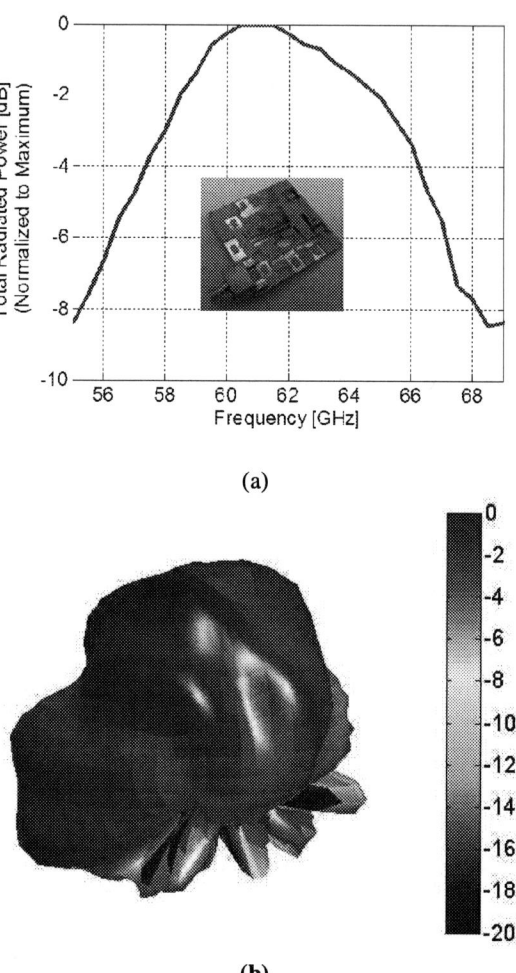

Figure 11. Measured (a) radiated power and (b) normalized radiation pattern (in dB) of fabricated microstrip patch antenna on photodefinable glass (PDG).

VII. Conclusions

In this paper we investigated photodefinable glass for applications in millimeter wave radio on package. Using a 2-layer package substrate stackup, transmission lines, via transitions and microstrip patch antennas were designed, fabricated and fully characterized up to 67 GHz. The transmission lines exhibit losses of 3.3dB/cm at 60 GHz, which is mostly due to the increased dielectric loss tangent with frequency. The through glass vias' (TGV) losses were less than 0.15dB at the same frequency. Great modeling to simulation correlation was obtained for the implemented microstrip patch antennas. The analysis of different ground patterns showed only minimum impact on the antenna performance. The efficiency and radiation pattern of the antennas were further characterized using a spherical near field anechoic chamber.

Acknowledgments

The authors would like to acknowledge Mario Pacheco, Jeffrey Lee, Stefanie Lotz, Chuan Hu, Henning Braunisch, Aleksandar Aleksov and Shawna Liff for their technical contributions. Acknowledgments also go to Intel management for supporting the project as well as all Intel suppliers for their contributions in the material selection and package substrate prototyping.

References

1. C. J. Hensen, "WiGig: Multi-gigabit wireless communication in the 60 GHz band," IEEE Wireless Communication, Dec. 2011, pp. 6-7.

2. Y. Katayama, C. Haymes, D. Nakano, T. Beukema, B. Floyd, S. Reynolds, U. Pfeiffer, B. Gaucher, K.Schleupen, "2-Gbps uncompressed HDTV transmission over 60-GHz SiGe radio," 4th Annual IEEE Consumer Communications and Networking Conference, CCNC 2007, Pages 12-16.

3. N. Saito et al., " A fully integrated 60-GHz CMOS transceiver chipset based on WiGig/IEEE 802.11ad with built-in self calibration for mobile usage," IEEE Journal of Solid-State Circuits, Vol. 48, NO. 12, pp. 3146-3159, December 2013.

4. A. Balankutty, S. Pellerano, T. Kamgaing, K. Tantwai and Y. Palaskas, "A 12-Element 60GHz CMOS Phased Array Transmitter on LTCC Package with Integrated Antennas," 2011 IEEE Asian Solid State Circuits Conference (A-SSCC), pp. 273 – 276, 14-16 Nov. 2011

5. D. G. Kam, D. Liu, A. Natarajan, S. Reynolds, H. Chen, and B. A. Floyd, "LTCC packages with embedded phased-array antennas for 60 GHz communications," IEEE Microw.Wireless Compon. Lett., vol. 21, no. 3, pp. 142–144, Mar. 2011.

6. E. Cohen, M. Ruberto, M. Cohen, O. Degani, S. Ravid and D. Ritter," A CMOS Bidirectional 32-Element Phased-Array Transceiver at 60 GHz With LTCC Antenna," IEEE Transactions on Microwave Theory & Techniques. Mar2013, Vol. 61 Issue 3, p1359-1375.

7. D. J. Chung, A. L. Amadjikpè, and J. Papapolymerou," Multilayer Integration of Low-Cost 60-GHz Front-End Transceiver on Organic LCP," IEEE Antennas and Wireless Propagation Letters, Vol. 10, 2011, pp. 1329-1332.

8. P. Talebbeydokhti and M. A. Megahed, "Low cost BT organic material for wireless 60 GHz application," 2013 IEEE 63rd Electronic Components and Technology Conference (ECTC), pp. 1634-1639.

9. Kam, D.G., Liu, D., Natarajan, A., Reynolds, S.K., Floyd, B.A., "Organic packages with embedded phased-array antennas for 60-GHz wireless chipsets," (2011) IEEE Transactions on Components, Packaging and Manufacturing Technology Vol. 1, Issue 11, pp. 1806 – 1814, November 2011

10. K. H. Tantawi, E. Waddel, J. D. Williams, "Structural and composition analysis of Apex™ and Foturan™ photodefinable glasses," Journal of Materials Science, August 2013, Volume 48, Issue 15, pp 5316-5323.

11. H. Braunisch and D.-H. Han, "Broadband characterization of package dielectrics," in Proc. IEEE Electronic Components and Technology Conf. (ECTC), New Orleans, LA, May 27-30, 2003, pp. 1258-1263.

Design and Fabrication of Low-Pressure Piezoresistive MEMS Sensor for Fuel Cell Electric Vehicles

Minkyu Lee[1], Kiyoung Nam[1], Sengyong Lee[1], Hakgu Kim[1], Chimyung Kim[1], Yongsun Park[1], Byungki Ahn[1],
Taewan Kim[2], Hochul Seo[2]
Hyundai Motor Company[1], Sejong Industrial co.ltd[2]
104, Mabuk-Dong, Giheung-Gu, Yongin-Si, Gyeonggi-Do, Korea 446-912[1]
1029, Yeongdeok-Dong, Giheung-Gu, Yongin-Si, Gyeonggi-Do, Korea[2]
mk2lee@hyundai.com, 82-31-899-3254

Abstract

This research focuses on the design and fabrication of a MEMS-based pressure sensor to measure the internal pressure of a thermal management system for fuel cell electric vehicles. The principle of measurement of the pressure is converting the piezoresistor value on a diaphragm deformed by a pressure difference into an output voltage. The diaphragm should operate normally for pressure sensing even harsh environments such as those in automobiles. However, the pressure sensor outputs abnormal values with an offset voltage due to physical and chemical factors during driving. One of the reasons is the polymer reaction originating from the coolant of the thermal management system between the polymer and sensor diaphragm. In this paper, the reason for abnormal operation is analyzed by measurement with a microscope, computational fluid dynamics, finite element method, and reproducible tests. A solution is suggested and a revised pressure sensor is proposed.

Introduction

Generally, piezoresistive pressure sensors are produced by MEMS-based methods at low cost through mass production and used to measure air or oil pressure in the automotive industry. In fuel cell electric vehicles, the sensor is required to monitor coolant pressure inside the thermal management system. The purpose of monitoring is to check the status of the stack which generates electric power for the vehicle, in terms of leakage among unit cells. The stack structure consists of unit cells which are isolated to provide air in the cathode channel, hydrogen in the anode, and the coolant in coolant channels. The sensing element can be exposed to coolant pressure with the driving force from the pump, which causes the sensor to operate abnormally because of physical and chemical effects by the coolant. Optimization of the packaging process is being researched for detecting the pressure of lubricant in automobile engines [1]. The effect on the offset of pressure sensors with silicon oil by has been demonstrated by FEM and experimental tests based on a TO base to minimize the zero offset. High-resolution MEMS pressure sensors were analyzed and characterized by Haiyang Yang et al. [2] in order to understand the mechanical responses of the sensor according to applied pressure through FEM and experiments. The coolant and sensor diaphragm reaction should be considered in fuel cell electric vehicles using coolant which has anti-freezing and electrical isolation characteristics. During coolant drying, the coolant components act on the pressure sensor diaphragm, which has been verified by experiments [3]-[5].

This current study handles abnormal operation by failure analysis based on operational principles of the sensor, and a solution is proposed.

Manufacturing and Fabrication Process

The manufacturing process of pressure sensor largely is divided into 3 steps. First, a sensing element is attached to a PCB by die bonding. In order to avoid thermal mismatch, a ceramic material is selected for the PCB. This is because the ceramic substrate is generally robust in terms of hardness. Epoxy adhesive is applied to silicon and the PCB [6]. Second, wire bonding is performed between the sensing element and PCB pad to connect the electronic path. Lastly, curling is done to mechanically mount the pin connector and pressure sensor sub-assembly, which includes the PCB and sensing element.

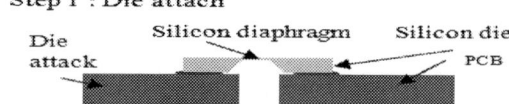

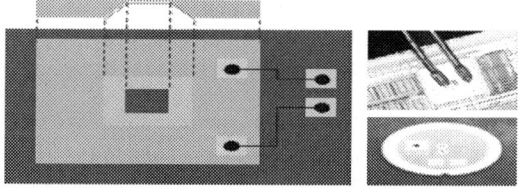

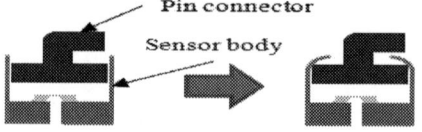

Figure 1. Manufacturing and Fabrication Process.

Failure Analysis

The performance and life cycle of developed product basically is validated by performing reliability tests such as thermal shock cycle, thermal humidity cycle, and vibration tests. Unfortunately, a sensor offset value is found with excessive range during vehicle operation. The offset value means that the sensor signal is out of range, which is ±60mV. This study tried to clear the error caused in the failure analysis process, as shown in Fig. 2.

978-1-4799-2408-0/14 $31.00 © 2014 IEEE

For analyzing pressure sensor failure, structural analysis is performed by finite element method to check the sensor damage by excessive stress loading.

Fig. 3 shows the result of structural analysis using the CATIA FEM tool. The maximum stress is about 364Mpa at maximum stress near the corner of the sensor trench, and the stress is enough to have a safety margin of about 20 times the yield stress. This means that the sensor offset does not originate from the structural damage and deformation.

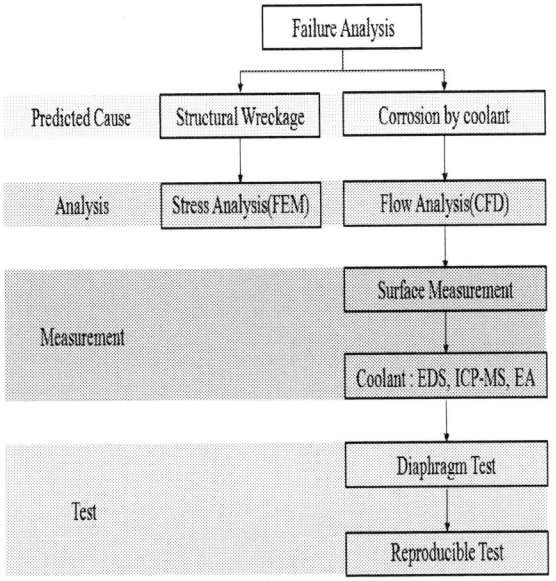

Figure 2. Failure Analysis Flow Chart.

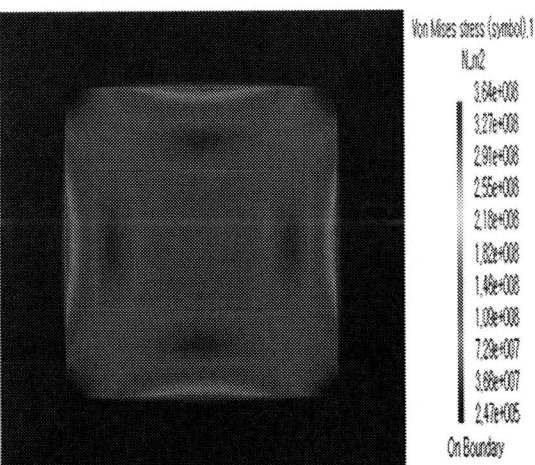

Figure 3. Structural Analysis Results(FEM).

One of the predicted reasons for the offset is corrosion of the sensor surface by coolant. The possibility of coolant contact on the sensor surface should be considered. This is because the sensor port diameter to introduce coolant is as small as 1.5 mm. Computational Fluid Dynamics was performed by the volume of fraction (VOF) method, which is capable of evaluating the interface position between fluids with commercial FLUENT software. Fig. 4 presents the analysis results for the interface between the air and the coolant. The results show that not only is the possibility of

coolant contact high, but the TMS is also vacuumed before filling with coolant. Coolant eventually contacts the sensor surface. The blue section is in the figure is the coolant portion, and the red one is air. The analysis area is the diluted gray section of the left side in Fig. 4.

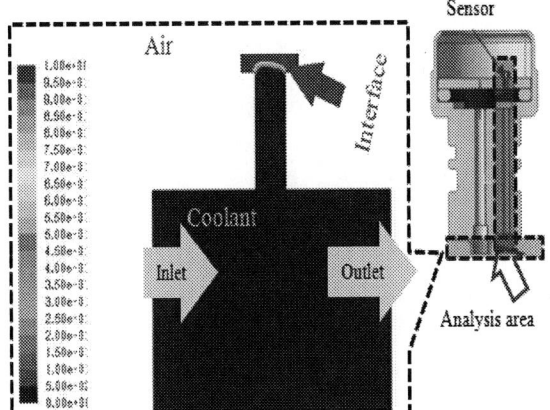

Figure 4. Internal Flow near Sensor(CFD VOF).

The CFD results show the possibility of contact with coolant by the sensor surface as theoretical evidence. The substance on the sensor surface should be identified to research the relation between the substance and diaphragm through measuring with a microscope. Fig. 5 shows a stain on the sensor surface. The measurement of the sensor surface was simply performed by a taking picture using a microscope. It looks like there is debris from TMS circulating in the coolant. But, the ingredients of the stain have to be identified for the causes of failure either corrosion by chemical reaction or physical accumulation.

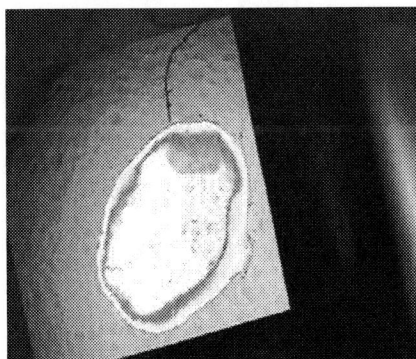

Figure 5. Sensor Surface with stain obtained by microscope.

In order to do so, two things are considered. One is to analyze the stain ingredients by SEM (Scanning Electron Microscope) and EDS (Energy Dispersive Spectrometer), and the other is to analyze the coolant components by EDS, ICP-MS (Inductively Coupled Plasma Mass Spectrometer), and EA (Elemental Analyzer).

First of all, the sensor with stain was measured by SEM, as shown in Fig. 6. The diaphragm is thinner than the substrate. It was predicted that the diaphragm would be easily deformed by external disturbances. The stain ingredients were

978-1-4799-2408-0/14 $31.00 © 2014 IEEE

measured by EDS, as shown in Fig. 7. In the fiure7, spots 1 and 2 are the measuring positions. The stain includes copper, zinc, and silicon.

Second, the ingredients of the coolant are analyzed with EDS, ICP-MS, and EA. The reason why several measurement methods were used is that the coolant is not a solid but a mixture with glycol family and water in a type of liquid, and the coolant is a type of polymer.

Figure 6. Cross-Section View of Sensor Diaphragm (SEM).

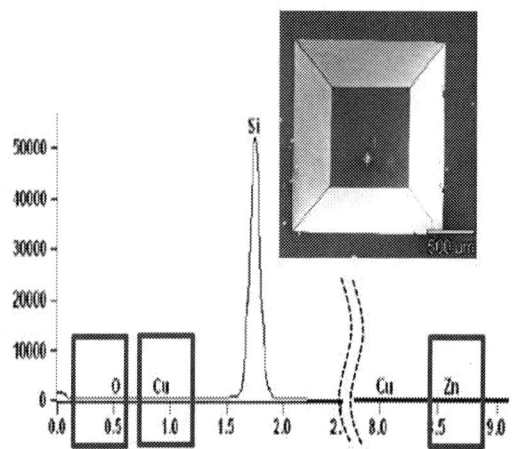

Figure 7. EDS Results for Sensor Surface.

Table 1 presents the coolant ingredients. The quantities of ingredients are intentionally omitted for company confidentiality.

Table 1. Ingredients of Coolant According to measurement Method

	ICP	ICP-MS	EA
Silicon	O	-	-
Copper	-	O	-
Zinc	-	O	-
Hydrogen	-	-	O

Based on the two ingredient analysis results, it is estimated that the coolant contacted the sensor, and that there

is no corrosive phenomenon. The coolant contact with the sensor surface was verified through theoretical and experimental methods.

As mentioned earlier, the malfunction of the sensor is estimated by coolant contact, and the cause of malfunction should be clarified for simple coolant contact on the sensor. This phenomenon was reported by previous studies, in that polymer (coolant) remained as a type of film on the substrate after drying over time stuck to the substrate [7], [8]. If the polymer is dried by evaporating while remaining as a film, the solvent of the polymer results in deformation. Then this deformation generates bending force on the substrate by removing water through drying. The reason for this still remains controversial [9]-[17]. Fig. 8 briefly explains this phenomenon.

After putting the polymer on the substrate and then drying, the water and polymer media generate capillary force. The capillary force bends the substrate under horizontal shear stress. The capillary forces depend on particle characteristics, as shown in Fig. 8.

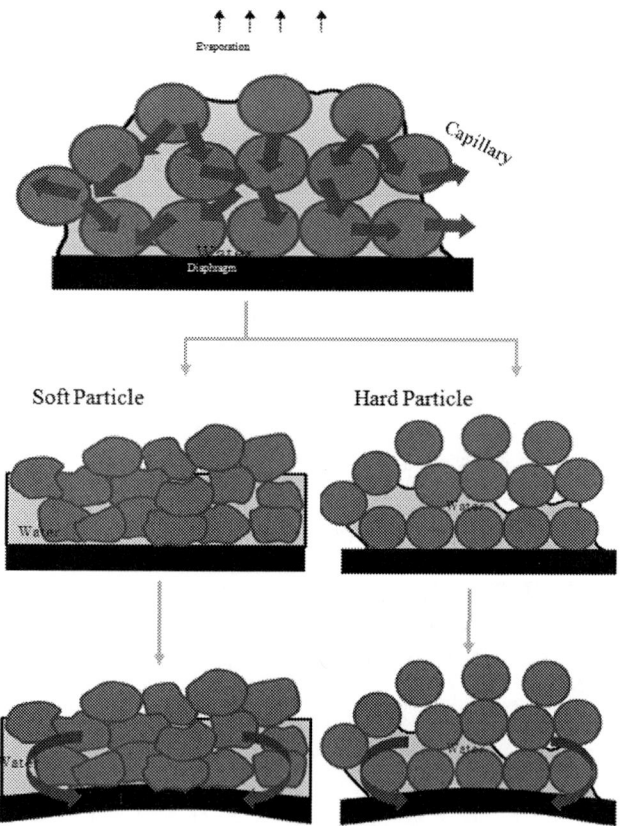

Figure 8. Stress Loading Mechanism by Polymer. (Singh and Tirumkudulu, 2007 [16])

In order to experimentally verify this phenomenon, one method that can be used is the cantilever method, which measures stress by an output signal obtained using a laser beam to recognize the substrate curvature during polymer drying [18]-[20]. Fig. 9 indicates how to measure the stress by the cantilever method.

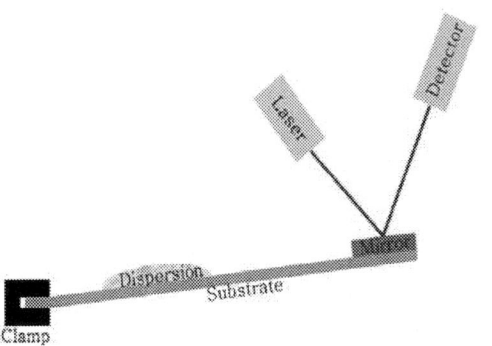

Figure 9. Experimental Equipment for Cantilever Method.

The average equivalent stress is calculated with poly vinyl acetate (PVAc), bronze (Goodfellow), a cantilever (28 mm length, 100 μm thickness, 5 mm width), a helium-neon laser beam, a mirror (5 mm X 10 mm, silica), and Quadrant detector (JQ 5OP 's) [21].

The result shows that about 0.3MP is applied on the substrate after 1 hour of drying. The pressure sensor was investigated in regard to whether or not the force is able to deform the pressure sensor.

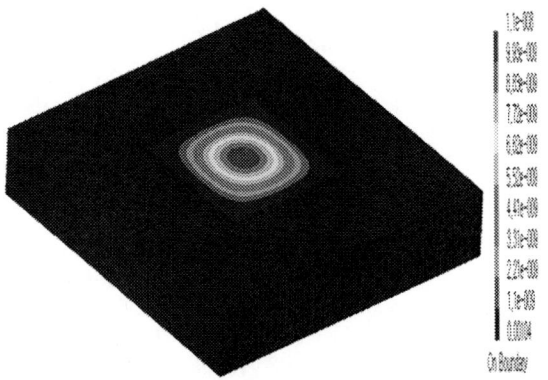

Figure 10. Displacement Loaded with 1 Bar Gauge.

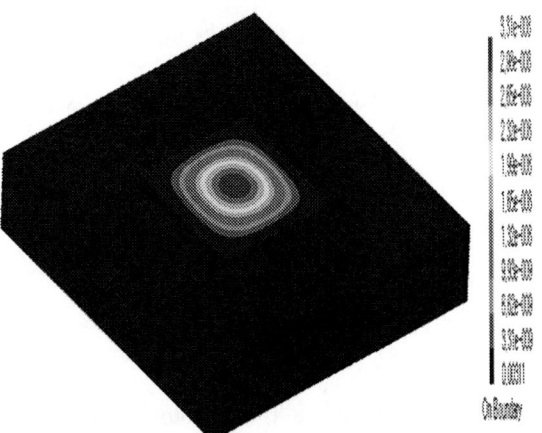

Figure 11. Displacement Loaded with Additional Pressure by Polymer.

As can be seen in Fig. 10 and Fig. 11, when applying the force to the pressure sensor, the displacement was

investigated. In the case of the 1 bar gauge pressure load, the displacement was 11μm, but in the stress loaded by the polymer, the displacement was 33μm. With this additional stress, a sensor offset will be applied by the polymer coolant.

Reproducible Experiment

As stated earlier, the sensor diaphragm thickness is about 20μm for targeting 1 bar gauge pressure measurement. This thin diaphragm is vulnerable to debris. In order to obtain stable operational characteristics, a buffer material is needed to reduce the external disturbance. The buffer material disperses and absorbs the focused stress by the coolant drying, and the structure should be able to transfer the TMS internal pressure to the sensor surface. Silicone gel was selected to act a buffer material (Shin-Etsu SIFEL 8370). To reproduce the abnormal drying operation, the samples were prepared as shown in Fig. 13. First, coolant is dropped onto the sensor surface for each group of samples. Then, the samples are exposed to a highly drying environment at 120°C for 1hour. After drying, the sensor is measured with debris without pure liquid, as shown in Fig. 14. The test times were about 60, and there was no offset limit in the case of Group C.

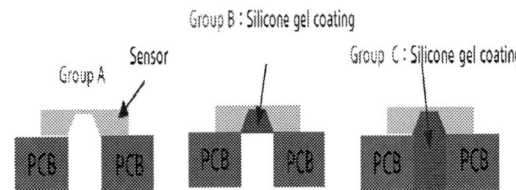

Figure 12. Sensor Sample Configuration with Silicone Gel.

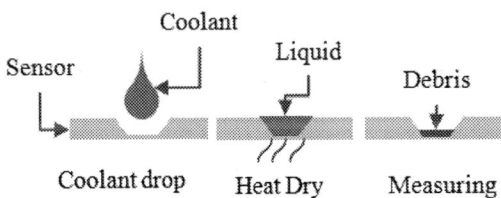

Figure 13. Drying Cycle for reproducible Phenomenon.

Fig. 14 presents the test results performed for 3 groups with respect to gel application on the sensor surface. Group A (Samples #1, 2) did not receive gel on the surface, and group B(Samples #3, 4) had gel applied only to the sensor surface, while group C (Samples #5, 6, 7) had gel applied to the PCB. Fig. 14 shows the offset values with respect to voltage. The normal offset range is from 0.44 to 0.56V. Group A and Group B samples were out of scope. In Fig. 14, the red solid line represents the offset limit, and the offset values are dealt with using an absolute function for easy identification.

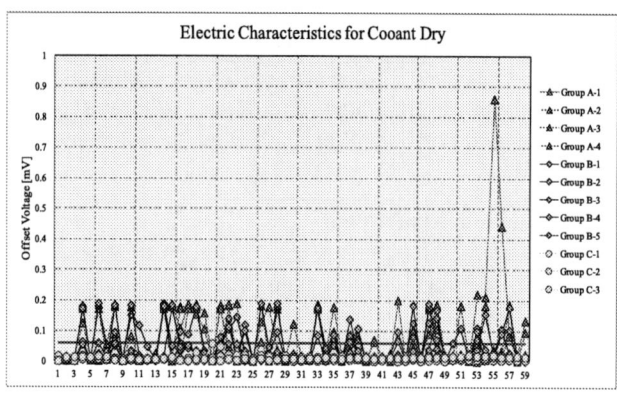

Figure 14. Verification Test Results According to Sample.

Group C has a relatively small offset compared with group A and group B, and the feasibility was verified for silicone gel application on the sensor surface to obtain stable sensor operation. At the beginning of the experiments, the number of samples was 5 for each group, but the sample number was decreased because of carelessness.

Even though the previous dry cycle test was verified, additional durability tests are required to ensure that the sensor operation is normal under vehicle driving. High/low temperature environment and thermal shock tests were therefore performed. The offset limit considered to be ±80mV after durability test. The results are shown in Figs. 15 to 17.

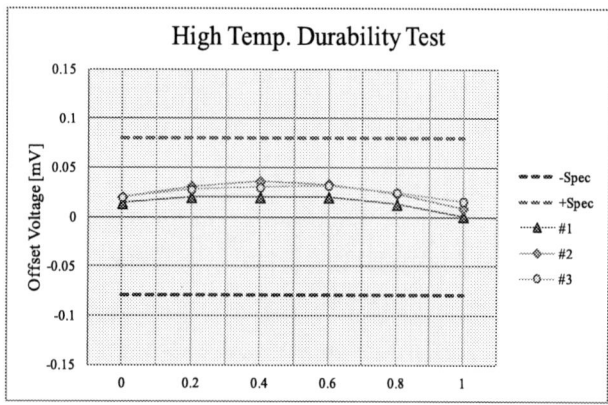

Figure 15. High Temp. Durability Test Result.

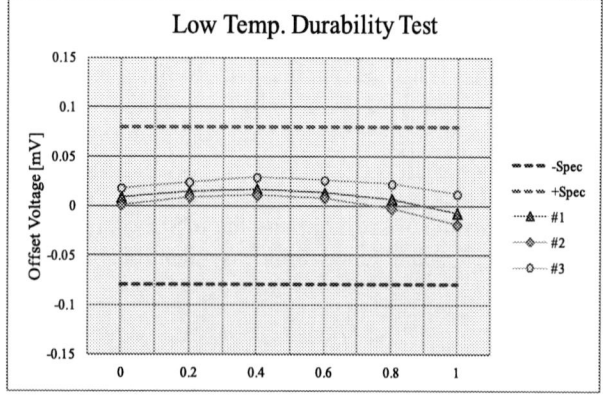

Figure 16. Low Temp. Durability Test Result

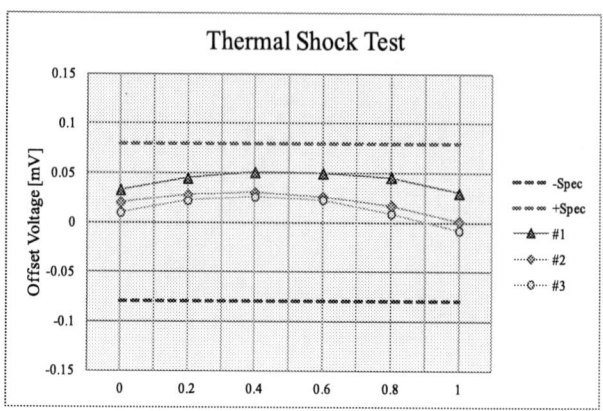

Figure 17. Thermal Shock Test Result

Conclusions

In this study, a physically and chemically robust pressure sensor was designed and fabricated utilizing a MEMS-based element. The effect of coolant on the sensor surface was analyzed regarding structural and chemical aspects, and the main reason was determined to be stress exerted on the surface by polymer drying. To overcome this malfunctioning, a buffer material was applied to the sensor surface and evaluated for feasibility through testing.

Acknowledgments

This work was supported by the New & Renewable Energy Core Technology Program of the Korea Institute of Energy Technology Evaluation and Planning (KETEP) granted resource from the Ministry of Trade, Industry & Energy, Republic of Korea (No. 20113010030050)

References

1. Zongyang Zhang, Chojun Liu, Zhimin Wan, Gang Cao, Yun Lu, Bin Song, Sheng Liu, "Optimization of Packaging Process Piezoresistive Engine Oil Pressure Sensor", International Conference on Eletronic Packaging Technology & High Density Packaging, pp. 1362-1365, 2010.

2. Yang, Haiyang, and Cosme Furlong. "Analysis and Characterization of high-resolution MEMS pressure sensor." Proceedings of the 11th International Congress and Exposition. 2008.

3. K. B. Singh and M. S. Tirumkudulu, "Measurement of critical thickness for cracking in colloidal films", Physical Review Letters, 98, 218302 2007).

4. E. R. Dufresne, D. J. Stark, N. A. Greenblatt, J. X. Cheng, J. W. Hutchinson, L. Mahadevan, D. A. Weitz, "Dynamics of fracture in drying suspensions", Langmuir 22, 7144 (2006).

5. For a review see Butt, H.-J.; Raiteri, R. In Surface Characterization: Principles, Techniques and Applications; Milling, Ed.

6. Dr. Christian Wohlgemuth, Dr. Anton Leidl, Dr. Gregor Feiertag, Dipl.-ing. Wolfgang Pahl, "Miniaturised Ceramic Package for Piezoresistive Pressure Sensors", Sensor+Test Conference, Proceedings II, P1.14(2009)

7. bharat Bhushan, "Adhesion and sticktion: Mechanism, measurement techniques, and methods for reduction", J. Vac. Sci. Technol. B 21(6), 1071-1023, 2003.

8. Niels Tas, Tonny Sonnenberg, Henri Jansen, Rob Legtenberg and Miko Elwenspoek, "Stiction in surface micromachining", J. Micromech. Microeng. 6 385-397, 1996

9. W.B. Russel, N. Wu, W. Man, "A generalized Hertzian model for the deformation and cracking of colloidal packings saturated with liquid", Langmuir 24(5) 1721 2008)

10. W. Man, W. B. Russel, "Direct measurements of critical stresses and cracking in thin films of colloid dispersions", Physical Review Letters 100(19) 198302 (2008)

11. M.S. Tirumkudulu, W.B. Russel, "Cracking in drying latex films", Lnagmuir 21 4938 (2005)

12. A.M. Koenig, T.G. Weerakkody, J.L. Keddie, D. Johannsmann, "Heterogeneous drying of colloidal polymer films: Dependence on added salt", Langmuir 24 7580 (2008).

13. A. Turshatov, J. Adams, D. Johannsmann, "Interparticle contact in drying polymer dispersions probed by time resolved fluorescence", Macromolecules 41 5365 (2008).

14. V.R. Gundabala, C-H Lei, K. Ouzineb, O. Dupont, J.L. Keddie, A.F. Routh "Lateral Surface Nonuniformities in Drying Latex Films", AIChE J 54 3092 (2008).

15. L. Xu, S. Davies, A.B. Schofield, D.A. Weitz, "Dynamics of drying in 3D porous media", Phys. Rev. Lett. 101, 094502 (2008).

16. K. B. Singh and M. S. Tirumkudulu, "Measurement of critical thickness for cracking in colloidal films", Physical Review Letters, 98, 218302 2007).

17. E. R. Dufresne, D. J. Stark, N. A. Greenblatt, J. X. Cheng, J. W. Hutchinson, L. Mahadevan, D. A. Weitz, "Dynamics of fracture in drying suspensions", Langmuir 22, 7144 (2006).

18. For a review see Butt, H.-J.; Raiteri, R. In Surface Characterization: Principles, Techniques and Applications; Milling, Ed.

19. Cahn, J. W.; Hannemann, R. E. Surf. Sci. 1964, 1, 387.

20. Finn, M. C.; Gatos, H. C. Surf. Sci. 1964, 1, 361.

21. Christian Petersen, Carsten Heldmann, and Diethelm Johannsmann, "Internal Stresses during Film Formation of Polymer Latices", Langmuir 15(22) 7745-7751, (1999)

Demonstration of TCNCP Flip Chip Reliability with 30μm Pitch Cu Bump and Substrate with Thin Ni and Thick Au Surface Finish

Weihong Zhang, Shengping Hong, Xiaolong Yan, Feng Zhou, Hailun Lu, Haijun Shen, Tonglong Zhang
Nantong Fujitsu Microelectronics Co., Ltd.
#228 Chongchuan Road, Nantong, Jiangsu, China 226006
zhang.wh@fujitsu-nt.com

Abstract

Recently, TCNCP (Thermal Compression with Non-Conductive Paste) flip chip technology has been adopted for various devices because of high demand for large bump density. Fine pitch micro Cu bumps are usually connected to trace on the substrate (Bump on Trace, BOT) due to space limitation on substrate. In this study, TCNCP technology was used to attach die onto substrate trace with surface finish of thin Ni (0.1um) and thick Au (0.4um). The micro Cu bump used in the device under test has a dome shaped Sn-based solder cap of 13um in height. The ratio of Au layer plated on the substrate trace to Sn cap on the Cu bump is about 10 wt%, which is far beyond threshold value of 3 wt% for Au embrittlement. However, it was proved that the micro joint obtained with TCNCP can pass TCT 1000 cycles, HTS 1000hours and uHAST 192hrs without failure. SEM/EDX study on HTS samples (0hr, 500hr, 1000hr) showed that the joints had a large amount of IMC formed mainly consisting of $(Cu_xAu_{1-x})_6Sn_5$ phase after reflow and independent Cu_3Sn phase formed near to die bump at a later stage during HTS test. Although $AuSn_4$ was present in 0hr sample, it disappeared later. No Ni was detected. Phase segregation appeared near to die bump when HTS time went up to 1000hrs. No crack and void were found at that point.

Introduction

The massive adoption of handheld terminals with more sophisticated functions accelerates the demand for smaller and lighter ICs. The shrinking size of ICs with advanced wafer nodes drives development of highly dense and fine pitch bumps for flip chip application. When bump pitch is lower than 80um, usually thermo-compression bond technology is used to ensure reasonable yields. In this paper, copper pillars of 30um staggered pitch were investigated using TCB technology.

Many relevant papers were published in IC packaging have been published [1-6]. In one paper published in ECTC 2013[1], Juang et al from ITRI investigated the intermetallic for micro Cu pillars with Ni limiting layer on the Cu pillars and solder cap connecting to ENIG or ENEPIG. It was observed that interconnect of micro Cu pillars with ENEPIG is less reliable during TC test. The poor performance is due to crack along the interface between $(Pd, Ni)Sn_4$ and Ni_3Sn_4 caused by large CTE mismatch between the two IMC phases.

In another study, ENEPIG was found to have more Cu consumption, crack and void defects as compared with ENIG and Sn plating surface finish after multi reflow, HTS and thermal cycling [2]. In this case, crack was observed near the interface of Cu, UBM, and $(Cu,Ni)_6Sn_5$ induced by Kirkendall voids gathering. ENEPIG without Ni layer has better u-joint performance than with Ni layer. Ni_3Sn_4 was found hard and brittle as compared with Cu_6Sn_5 and Cu_3Sn and spalling occurred easily on Ni_3Sn_4 phase [4, 5]. Y.S. Park reported that the formed IMC composition of SAC solder on Cu/Ni pad finish changes with solder ball size. As solder ball size decreased and aging time increased, Ni-Sn IMCs changed from Cu-rich phases to Ni-rich phases [6]. Therefore, though in similar joint structure, such as, Cu/Ni-Sn-ENEPIG, IMC compositions could be different for different ball sizes. For micro bumps, thick Ni layer can easily lead to Ni-rich phases rather than Cu-rich phases. Therefore, we adopted thin Ni (0.1um) and thick Au (0.4um) as surface finish on substrate trace and no Ni deposited on Cu pillars for TCNCP with bump pitch as fine as 30um staggered and 60um in line. From manufacturing point of view, the thick Ni layer is hard to realize for fine bump pitches due to Ni plating limitation. Fine bump pitch requires fine line and space on substrate normally smaller than 20um L/S. This makes some of commonly used surface finishes such as thick nickel (3um to 5um) followed by Au layer difficult to apply.

Packaging Structure and Process Development

In our study, die size was about 6mm x 6mm. Bump pitch was 30um staggered and 60um in line with bump diameter and height of 28um and 30um respectively. The height of solder cap on top of Cu column is 13um. No Ni layer was deposited on Cu pillars. Also, on substrate pad, Ni/Au was electro-plated with Ni thickness of 0.1um and Au thickness of 0.4um. Solder was Sn-based lead free solder. Its schematic structure is shown in Fig. 1.

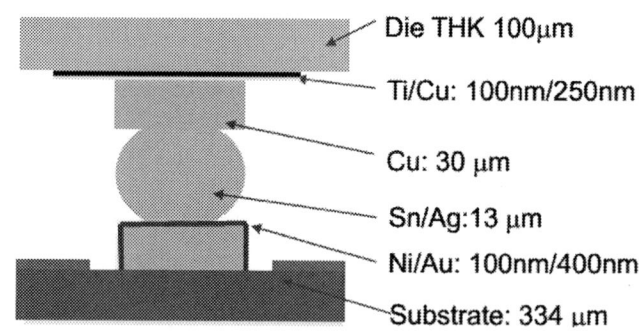

Figure 1. The structure of micro bump joint.

TCNCP process was used to firmly join die on substrate. Equipment with +/-3um bump accuracy was used with bump parallelism kept below 6um during TCB bonding. Development of the process is given in detail as below.

Figure 2. TCB process flow.

TCB (Thermo-Compression Bonding) process is a die bonding technique, also known as diffusion bonding or pressure bonding. Two metals form a joint when force and heat is simultaneously applied. It includes dispensing NCP (Non-Conductive Paste) onto substrate and then flipchip bonding with IC die displayed in Fig. 2. In the TCB process, a low CTE non-conductive paste (NCP) is pre-dispensed onto the substrate chip attach pads. A copper pillar (with reflowable solder cap) bumped die is then picked and aligned to the substrate pads. Heat and mechanical force is applied to the die surface to form the solder joint (typical time 5 secs). Heat (~250。 C peak) is applied to melt the solder caps on the Cu pillar bumps, and mechanical force is applied to push the silica fillers in NCP away from the joint area.

Good control of dispensing volume and pattern are very important: too much NCP could cause die crack since flipchip bonding head could be coated with NCP. 100% coverage of NCP underneath IC is important for reliability. We observed that package reliability performance was compromised by less than 100% NCP coverage due to insufficient NCP volume dispensed onto substrate before flipchip bonding. NCP dispensing pattern could affect NCP coverage and void. In our study, three types of dispensing patterns were evaluated to determine the best pattern. It was observed that pattern "米" could help achieve uniform spreading of NCP without trapped voids. With the optimized pattern, dispensing volume was also optimized and process window defined to ensure good NCP coverage and no NCP creeping. The optimized dispense weight in this study was less than 10mg.

For the TCB process where the IC was attached onto substrate with interconnects formed between them, a few key factors were studied such as parallelism between IC and substrate, bonding temperature and bonding force. First, the parallelism between IC and substrate during bonding is very important to ensure proper connection for all bumps. It also affects NCP coverage and creeping. It is recommended that the parallelism between IC and substrate is within 6um. Temperature profile is another key factor for TCB process. The NCP pre-applied onto substrate has various reactions during TCB process. It has to be prepared for its good flow property by heating up the substrate to a proper temperature; Then during bonding, when die presses the NCP on substrate to spread it over and cover all the die area, temperature on die bond head shall ensure NCP material could flow smoothly without voids within a few seconds. At last, when temperature is about the solder's liquidus point, the NCP should stop spreading completely to ensure molten solder is not pushed away from Cu pillars and Ni/Au traces leading to open connections. Also, by the time temperature is about solder liquidus, the oxidation on solder cap will need to be completely removed by acidity of NCP. A proper pressure needs to be applied when solder is melted to ensure good solder joint quality. The bonding force profile needs to ensure the smooth and voidless flow of NCP before solder melting. Cooling rate affects UPH and usually a fast cooling after solder wetting substrate pad is desirable. Therefore, material selection is also very important.

Reliability tests were performed on the package structure: Preconditioning (MSL3), TCT (Thermal Cycling Test), HTS (High Temperature Storage test) and uHAST (un-biased highly accelerated storage test). Their conditions are given in Table 1. The joints passed reliability tests (TCT 1000 cycles, HTS 1000hours and uHAST 192hrs) without failure.

Table 1. Reliability Test Conditions

Test	Reference	Condition/Duration
MSL 3	JEDEC 22-A103	30℃/60%RH, 192 hrs
TCT	JEDEC 22-A104-C	500/1000 cycles,-55~125℃
uHAST	JEDEC 22-A118	130℃/85%RH,33.5 PSI,96 hrs, 192hr
HTS	JDEC-22-A110	150C, 1000hrs

Microstructures of the HTS samples were analyzed using SEM (Scanning Electron Microscope) / EDS (Energy Dispersive X-ray spectroscope) to achieve good understanding on intermetallic compound (IMC) development in the micro bump joints.

Results and Discussions

Joint microstructures, composition analysis and element mapping of the HTS samples (0hr, 500hr, 1000hr) are given in Fig. 3 ~ 6. #1, #2 and #3 is location close to die bump, middle of the joint, close to substrate respectively. Fig. 7 shows area mapping of each whole joint for the three scenarios.

We can see that at 0 hr before HTS test, the joint already had a large amount of $(Cu_xAu_{1-x})_6Sn_5$ phase with residual Sn as indicated by black arrows (Fig 3 top). EDS analysis shows that the IMC formed in the area near to die side has Cu-rich Cu_3Sn + (Cu) though it is quite thin, and Cu_3Sn + $(Cu_xAu_{1-x})_6Sn_5$ is seen at the other side close to substrate. Line mapping on 0hr sample shows that its Cu distribution across the white line has almost linearly increasing slope at sides due to element inter-diffusion and a platform in the middle indicating a stable IMC, $(Cu_xAu_{1-x})_6Sn_5$ phase, formed between them. However, the phase in the middle is not quite uniform, Au-Sn phase was formed indicated by black arrow in the middle in Fig. 3 bottom. It can be seen that Au and Sn lines both fluctuate around this location. No Ni presence was found in the joint because only a quite thin Ni layer was deposited on substrate Cu trace.

In this study, Au layer was 0.4um thick taking about 10 % of the Sn-solder cap in weight, which is 3 times Au embrittlement threshold value of 3%. To check embrittlement effect, traditional flip chip process with underfill was used for comparison purpose.

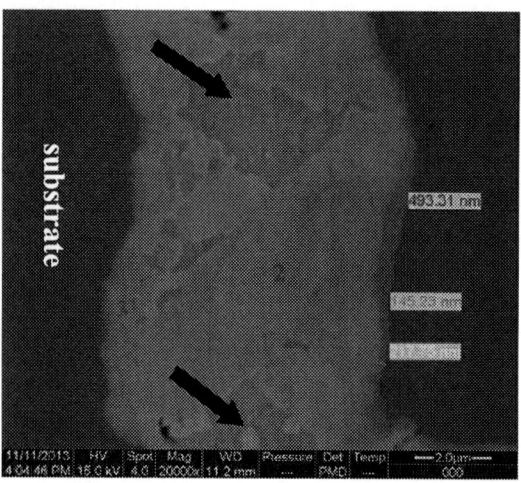

EDS analysis (at%)

	1#	2#	3#
Au	0.54	16.46	8.86
Sn	9.57	45.05	36.29
Cu	89.89	38.49	54.84
Structure	Cu_3Sn+ (Cu)	$(Cu_xAu_{1-x})_6Sn_5$	Cu_3Sn+ $(Cu_xAu_{1-x})_6Sn_5$

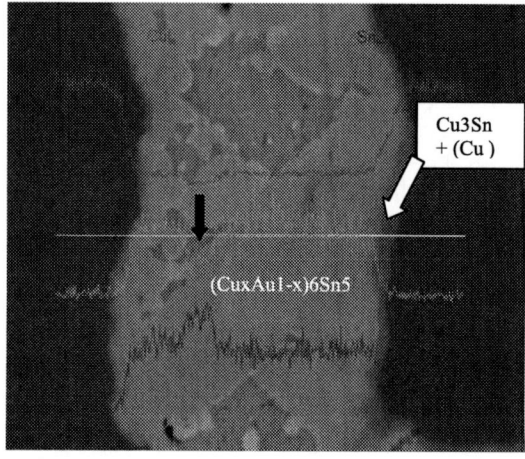

Figure 3. The microstructure (top), EDS analysis and line mapping (bottom) of 0 hr HTS sample.

It turned out that most bumps were disconnected using normal flip chip process where die was attached on substrate and then went to reflow followed by underfilling. Fig.6 is microstructure of a cracked joint along with EDS analysis result showing weight percentage of each element. The AuSn4 phase was identified near the fracture boundary (point 4) among the phase of $(Cu_xAu_{1-x})_6Sn_5$ (point 2) as shown in Fig. 4. Point 1 has large Cu wt% as high as 82% in weight ratio of Cu/(Cu+Sn). Point 3 is located on organic filling material. Point 5 (45%) has higher Cu wt% than the middle (point 2, 35%). The Cu wt% data at different locations of the cracked joint highly correspond to the EDS analysis result on the 0hr HTS sample using TCNCP (see Table 2 later). We can see that Cu_3Sn was already formed next to die side and substrate side, but in metastable status. At die side, Cu-rich metastable Cu_3Sn was formed with Cu element coming from

the Cu bump on die, while at substrate side Cu in Cu_3Sn came out of $(Cu_xAu_{1-x})_6Sn_5$ in the middle, i.e., from Cu bump on die but not from copper trace on substrate. It showed that presence of Ni deposited on substrate trace likely retarded Cu diffusion from the copper trace underneath and also suppressed Cu_3Sn growth. This is why only a thin and non-uniform dark layer containing Cu_3Sn was seen at die side with thickness up to 500nm.

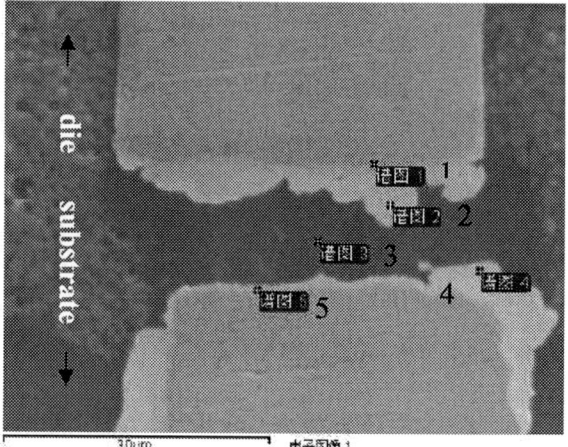

谱图	在状态	C	N	O	Cu	Sn	Au	总的
谱图 1	是	7.72			75.05	17.24		100.00
谱图 2	是	7.62	-1.92		26.73	48.47	19.10	100.00
谱图 3	是	71.75		28.25				100.00
谱图 4	是	5.69	-1.61			69.23	26.69	100.00
谱图 5	是	9.65			35.29	41.98	13.08	100.00
最大		71.75	-1.61	28.25	75.05	69.23	26.69	
最小		5.69	-1.92	28.25	26.73	17.24	13.08	

Figure 4. EDS analysis on a cracked joint obtained using normal flip chip process instead of TCNCP.

From this experiment, Au embrittlement due to presence of large amount of Au was exposed and the support function of NCP material during reflow process was fully demonstrated. How to appropriately choose NCP material appeared very important in this case to ensure good alignment between solder cap and substrate trace with low stress applied on solder joints. Its viscosity is important. Also, its glass transition temperature (Tg) and thermal expansion coefficient (CTE) are very important to ensure a good support when needed. It also suggests that the Au layer could be somehow thick in this case. A thinner layer of Au could help to yield better joints even without NCP support.

After HTS 500 hrs, the dark layer at die side became much thicker (up to 1800nm) as shown in Fig. 5 top and EDS analysis detected more stable Cu_3Sn phase formed there. Both Cu and Sn lines become further smooth in the middle and flatten out at sides of the joint (Fig. 5 bottom). Cu content near the die side is still higher than near the substrate side but

978-1-4799-2408-0/14 $31.00 © 2014 IEEE

getting close due to continued diffusion of Cu from die direction. Au mapping line appears relatively spiky with a sloped line at sides, but it looks a bit smoother than 0 hr anyway. It reflects dynamic diffusion activity of Au element across the joint up to HTS 500hr.

disappeared completely from the joint and Au completely diffused back from the interface with die bump to form more $(Cu_xAu_{1-x})_6Sn_5$ in the middle. However, where did Cu come from since Cu diffusion across joint had slowed down near to die bump side?

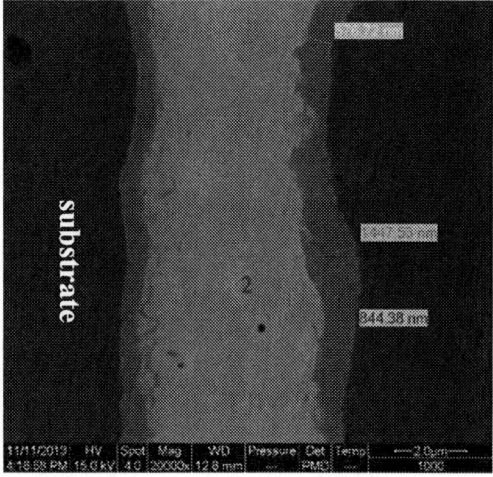

EDS analysis (at%)

	1#	2#	3#
Au	1.20	13.79	9.50
Sn	21.00	42.08	29.61
Cu	77.80	44.13	60.89
Structure	Cu_3Sn	$(Cu_xAu_{1-x})_6Sn_5$	$Cu_3Sn +$ $(Cu_xAu_{1-x})_6Sn_5$

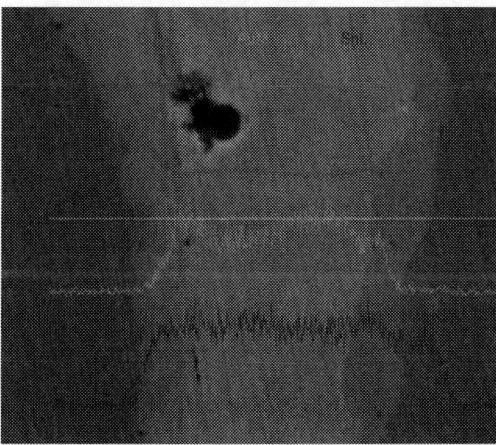

Figure 5. The microstructure (top), EDS analysis and line mapping (bottom) of 500hr HTS sample.

After HTS 1000hrs, all elements distributed in plateau and lowland with a clear cut line along the boundary at die side, which exhibits element saturation and phase segregation at this location (Fig. 6). It seems that all elements slowed down diffusing here. The dark layer of pure Cu_3Sn phase became further uniform and continuous without further increase in maximum thickness but a decrease instead. This indicates diffusion in the direction parallel to die was more active than that perpendicular to die at that location. It could be likely because Cu gradient along the direction perpendicular to die became relatively low up to that point. Au-Sn phase

EDS analysis (at %)

	1#	2#	3#
Au	0.02	20.17	0.79
Sn	27.42	33.12	10.06
Cu	72.56	46.71	89.15
Structure	Cu_3Sn	$(Cu_xAu_{1-x})_6Sn_5$	$Cu_3Sn +$ (Cu)

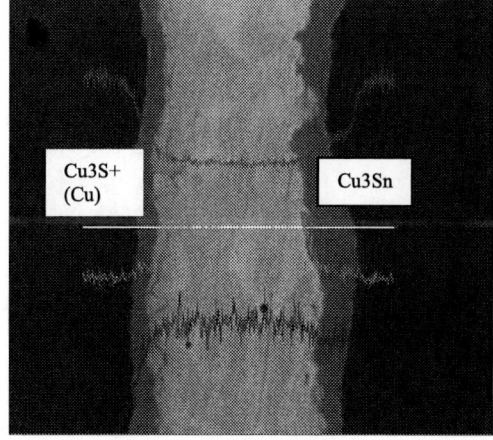

Figure 6. The microstructure (top), EDS analysis and line mapping (bottom) of 1000hr HTS sample.

Interestingly, at substrate side, diffusion activity appears quite intense, which is reflected by sloped lines of all elements (Fig. 6 bottom). A continuous and thick dark layer is seen at the side near to substrate as well. The EDS analysis result (Fig. 6 middle) tells that metal stable Cu–rich Cu_3Sn was formed similar to the situation near to die side at 0hr. We can see a sudden phase change at substrate side during period between 500 hr and 1000 hr in HTS test: the transformation from the previous metastable phase, $(Cu_xAu_{1-x})_6Sn_5 + Cu_3Sn$, to another metastable phase, $Cu_3Sn + (Cu)$. This jump was likely due to sudden sufficient Cu supply coming from Cu trace on substrate when Ni barrier layer completely disappeared before HTS 1000 hr. Although Ni layer was not detected due to its extremely low thickness, its presence was still displayed by the serial change of Cu-contained phases which is related to Ni barrier function. At initial HTS period, Cu diffusion was going across the joint from die to substrate due to presence of Ni barrier; at a later stage after 500hrs, Cu atoms moved from substrate to die due to absence of Ni barrier.

Fig. 7 is area mapping result of HTS samples (0hr, 500hr, 1000hr). During reflow, Au was dissolved into the solder joint and diffused over it anywhere except for some locations indicated by arrows in Fig. 7a. Those parts without Au coverage were residual Sn solder. With further development at high temperature in HTS test, the solder was fully converted and the Au distribution became more and more uniform across the joint. Also, some voids seen in the joint microstructure after HTS 500hr indicated by arrows in Fig. 7b were gone after HTS 1000hr (Fig. 7c). This is highly likely attributed to sufficient Cu supply from substrate side after 500hr. The voids in the joint due to Cu depletion before HTS 500hr were refilled by Cu atoms coming from the substrate after HTS 500hr.

Based on those obtained from the HTS samples, it can be predicted that joint crack could initiate in the area near to die side if HTS time further increased due to its earlier element saturation and phase segregation than the substrate side.

Overall, phase development of the micro bump joint in this study was summarized in Table 2 and Fig. 8 as well as Cu wt%. The Cu wt% is Cu/ (Cu + Sn) x100% in weight.

After reflow, most of the micro bump joint was converted into IMC due to its small joint. Au was dissolved into solder joint during reflow to form relatively stable $(Cu_xAu_{1-x})_6Sn_5$ with Cu_6Sn_5 between die and substrate indicated by point 2. Its phase change was slight through the HTS test up to 1000hrs. Its Cu wt% moved from a bit left side to a bit right side with reference to Cu_6Sn_5 illustrated in Fig. 8 (see #2).

Near to die Cu bump side, Cu dissolution was fast and Cu-rich phase, $Cu_3Sn + (Cu)$, formed at 0 hr, and then it continued moving steadily towards Cu_3Sn and settled down at Cu_3Sn before HTS 500hr and firmly stay there before HTS 1000hr illustrated in Fig. 8 (see #1).

Table 2. Review on Cu wt% and phase change in the solder joints with change of HTS time.

HTS Time	0hr	500hr	1000hr
#1 close to die	83% Cu3Sn + (Cu)	66% Cu3Sn	59% Cu3Sn
#2 middle	32% (CuxAu1-x)6Sn5	36% (CuxAu1-x)6Sn5	43% (CuxAu1-x)6Sn5
#3 close to substrate	45% Cu3Sn + (CuxAu1-x)6Sn5	53% Cu3Sn + (CuxAu1-x)6Sn5	82% Cu3Sn + (Cu)

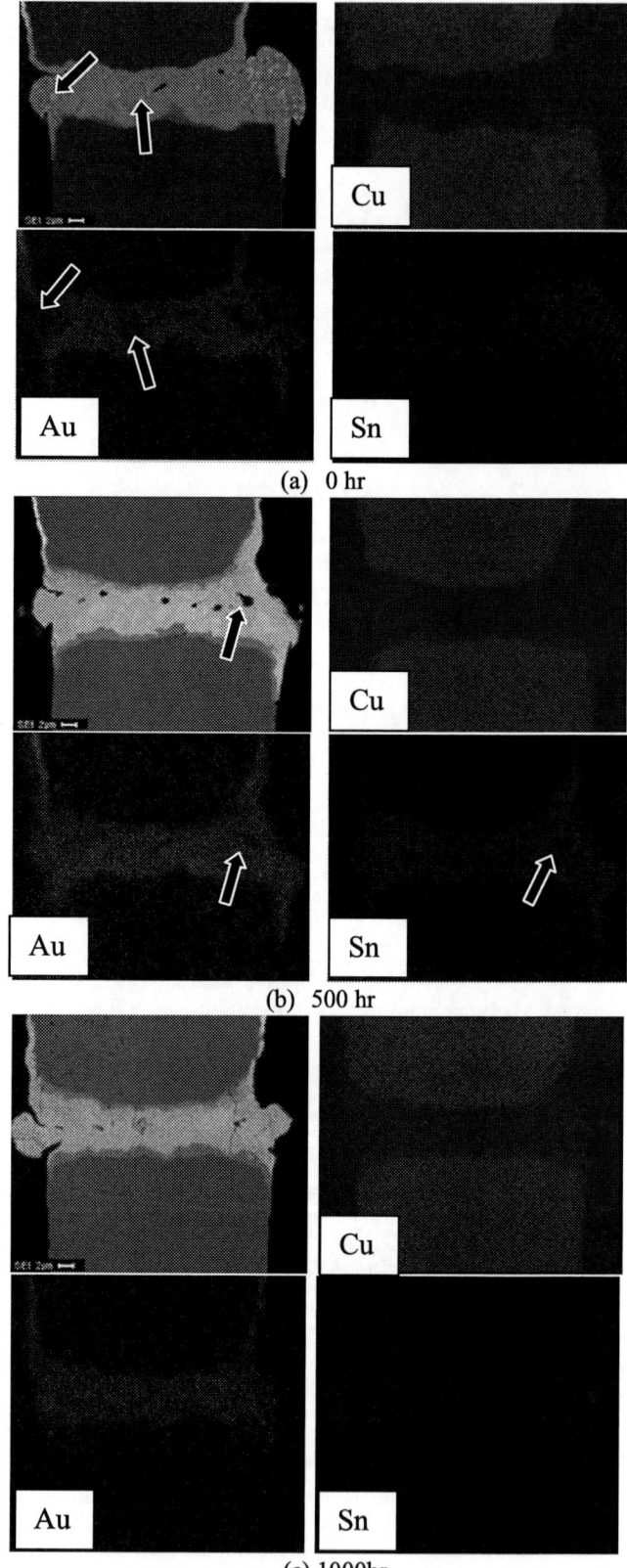

(a) 0 hr

(b) 500 hr

(c) 1000hr

Figure 7. The element area mapping across whole joint of HTS samples (a) 0 hr; (b) 500 hr; (c) 1000hr.

Due to Ni presence on substrate Cu trace, Cu3Sn formation was delayed near to substrate. It appeared always unstable through the HTS process with largest Cu wt%

change jumping from far left to far right referring to Cu3Sn position, which is illustrated in Fig. 8 (see #3). Its future change could follow history of point 1 (#1).

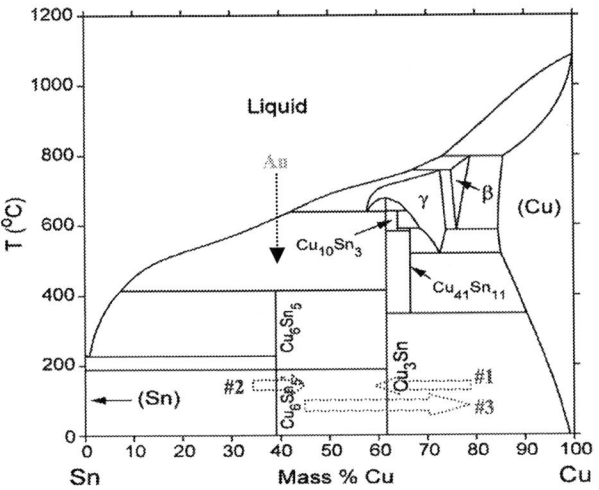

Figure 8. Phase diagram of Cu-Sn.

Conclusions

TCNCP process was used in packaging die with bump pitch of 30um staggered and 60 um in line. The micro bumps had size of 28um in diameter and height of 30um. The joint structure was Cu pillar- Sn solder- thick Au/thinNi surface finish on Cu trace in this study. Excellent performance of the joint structure formed using TCNCP process was fully demonstrated. It passed all reliability tests (TCT 1000 cycles, HTS 1000hours and uHAST 192hrs) without failure.

The micro Cu bump used in the device under test has a Sn cap of 13um in height. The weight ratio of Au layer plated on the substrate trace to Sn cap on the Cu bump is about 10 %, which is far beyond threshold value (3 wt%) for Au embrittlement. However, with NCP support during TCB process, good joints were formed firmly without bump crack..

Compositional analysis on the joints before (0 hr) and after HTS test (500 hr, 1000hr/150。 C) was performed using SEM/EDX. It showed that the joint mainly had stable $(Cu_xAu_{1-x})_6Sn_5$. At 0 hr, most of the joint was converted to $(Cu_xAu_{1-x})_6Sn_5$ mingled with some $AuSn_4$ phase and non-converted solder (residual Sn). With further development during HTS test, the joint microstructure appeared denser after HTS 1000hr than after HTS 500hr. Stable Cu3Sn was seen at die bump side after HTS 500hr and phase segregation was seen at die bump side after HTS 1000hr. No Ni was detected in this analysis.

From this study, we found that Cu supply in Cu-contained IMC formation came from Cu bumps on die at initial stage and then from Cu trace on substrate at later stage. Though Ni could not be detected, its presence was displayed by IMC formation near to substrate side. Its presence retarded Cu3Sn formation at this location. Cu-rich IMC found here after HTS 1000hr reflected its absence after long time aging. And also, the phenomenon that voided microstructure after HTS 500hr appeared dense after HTS 1000hr can be explained by Cu-refill from Cu trace on substrate.

Acknowledgments

We would like to express our sincere gratitude towards Institute of Microelectronics at Tsinghua University for their help in SEM/EDS work. Also, we would like to extend our deep appreciation to the FC team and FA team of NFME for their great support.

References

1. Jinf-Ye Juang, et al, "Effect of Metal Finishing Fabricated by Electro and Electro-less Plating Process on Reliability Performance of 30um Pitch Solder Micro Bump Interconnection", in *Proc. IEEE Electornic Components and Technol. Conf. (ECTC)*, pp. 653-659, 2013

2. Mu-Hsuan et al, "Thermal Cycling Effect on Intermetallic Formation with Various Surface Finish of Micro Bump Interconnect for 3D Package", in *Proc. IEEE Electornic Components and Technol. Conf. (ECTC)*, pp. 2163-2167, 2013

3. S.L.Wright et al, "Characterization of Microbump C4 Interconncects for Si-Carrier SOP Applications" in *Proc. IEEE Electornic Components and Technol. Conf. (ECTC)*, pp. 633-640, 2006

4. Ping-Feng Yang et al, "Mechanical Properties of Cu6Sn5, Cu3Sn and Ni3Sn4 Intermetallic Compounds Measured by Nanoindentation", Electronic Packaging Technology, International Conference on ICEPT, 2007 (IEEE)

5. Per-Erik Tegehall, "Review of the Impact of Intermetallic Layers on the Brittleness of Tin-Lead and Lead-free Solder Joints", IVF Project Report 06/07

6. Y. S. Park et al, "Effect of Fine Solder Ball Diameters on Intermetallic Growth of Sn-Ag-Cu Solder at Cu and Ni Pad Finish Interfaces during Thermal Aging", in IEEE Electronic Components and Technology Conf. (ECTC), PP. 1870-1877, 2011.

Integrated Process Characterization and Fabrication Challenges for 2.5D IC Packaging Utilizing Silicon Interposer with Backside Via Reveal Process

Cheng-Hsiang Liu, Jyun-Ling Tsai, Hung-Hsien Chang, Chang-Lun Lu, Shih-Ching Chen

Siliconware Precision Industries Co., Ltd. (SPIL)

No.153, Sec.3, Chung Shan Rd., Tantzu Dist., Taichung 42756, Taiwan, R.O.C.

seanliu@spil.com.tw +886-4-2534-1525 ext. 4615

Abstract

Conventional IC packaging requires device chips or dice to be packaged at the same level in a way we generally imagined, while newly developed and thriving 3D IC packaging utilizes skyscraper concept to stack numerous types of device chips with different functions occupying the exact same or similar footprint. This approach not only reduces overall package dimension and thickness, but also improves electronic interconnection performance, as well as provides other advantages like lower power dissipation and greater signal bandwidth. Nevertheless, because of the fact that there were tremendous amounts of money and integration efforts spent on fabrication process development, another variant of 3D IC packaging had started to emerge and rapidly flourished in recent years, and it's been referred to as a hybrid between 2D IC and 3D IC packaging, or more specifically, 2.5D IC packaging. Compared with 3D IC, 2.5D IC possesses the benefits of easier process configuration and lower production cost while maintains similar electrical performance as well as reduces certain obstacles where 3D IC packaging is prone to generate. In this paper, our current development of 2.5D IC packaging was demonstrated and displayed, followed by further elaboration of detailed process flow, including device wafer and interposer wafer fabrication in bumping part of process, intermediate die sawing process, and final die level assembly part of process. During the development stage, there were many challenges we had encountered, such as thin wafer handling, carrier bonding and debonding uniformity, warpage alleviation, material stress control, film delamination, as well as other lesser issues. We have offered and proposed certain approaches to particular challenges and the explanation of these challenges as well as proposed solutions was addressed in this paper to properly demonstrate the progression of our 2.5D IC packaging.

Introduction

Semiconductor industry has grown and prospered rapidly in the last few decades. As consumer electronics have quickly produced and bloomed in recent years, the demands from customers or end-users became much crucial and prominent in the advancement of technology and the aspect of marketing. Cellular phones, for example, have inevitably followed this trend and played a transitional role from luxurious apparatus to inexpensive necessities in less than 15 years. Easy-to-carry and compact-size cellular phones have been transformed and evolved to smartphones with larger display, thinner body, and additional integrated functions. In order to achieve this kind of demands transition, innovative packaging technology and process techniques have emerged and greatly evolved lately.

Sometime around the beginning of the 1990s, people noticed about the forthcoming appearance of devices known as Multi-Chip Modules (MCMs), where multiple ICs were packaged onto a combined substrate, operating as a single component [1]. This concept is similar to System-in-Package (SiP), and the main difference is that while MCM indicates numerous dice in a single component, SiP usually implies numerous dice in a single package, whether it is mounted on substrate or not. There are a variety of materials which can be utilized as package substrate, ranging from silicon to thin laminates, and all these materials may presented in ceramic, metallic, or plastic packages. As semiconductor technology development experienced transitional change from industry-driven to customer-driven, device dimension became a much crucial consideration ever than before. Therefore, the concept of stacked IC was originated and started to shine for the past few years. In 2004, Intel had developed a 3D type of their renowned Pentium CPU, where the IC chip was manufactured with two dice using face-to-face stacking [2]. Backside Through-Silicon-Via (TSV) technology was used for I/O portion and power supply. This new 3D type of CPU had attributed to about 15% performance improvement and power saving due to the elimination of pipeline stages and the reduction of wiring compared to the original 2D type.

3D IC packaging provides the benefits of smaller footprint, shorter interconnection, larger bandwidth for data transfer, greater circuit security, lower power consumption, and so forth. Nevertheless, there are also many challenges currently existed, including design complexity, electrical signal integrity, heat dissipation, heterogeneous die integration, and fabrication difficulties resulted from TSV implementation. Figure 1 shows the TSV chip wafer forecast on all 3D platform, and it is observed that logic 3D IC and 3D stacked DRAM will continue to dominate overall wafer amount towards 2017, with around 56% increase of Compound Annual Growth Rate (CAGR), predicted by Yolé Developpment in middle 2012.

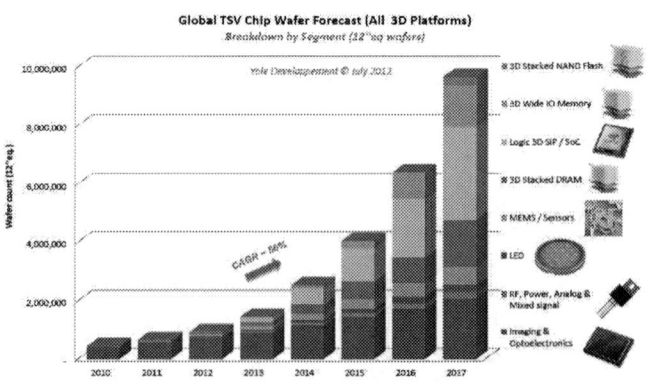

Figure 1. Global TSV chip wafer forecast on 3D platforms

Over the past few years, it was noted that the evolutionary transition from 2D IC to 3D IC integration was inevitable and steadily progressing. Since 3D IC packaging indeed possessed numerous obstacles during development, wafer foundries and OSATs had started to focus on a simpler approach known as interposer-based IC packaging, or simply called 2.5D IC. This term has already been well established in packaging field. Originally meant to be an interim solution for 3D IC, this kind of evolutionary silicon interposer technology solved many of the 2D IC problems and did not tend to generate many 3D IC new problems. This technology utilized an interposer made of silicon, glass, or ceramics to make interconnection between bottom substrate and top device die. Thus, instead of die-to-die stacking, single die or several dice were placed side by side and all stacked on interposer first and the die-interposer assembly was then aligned and stacked on substrate afterwards. Microbumps on device wafers were fabricated and connected to micropads on interposer wafers. By creating one or more layers of Redistribution Layers (RDL), electrical signals can be transferred from device side through interposer to package side, as shown in Figure 2. Recently, several companies and institutes have announced and declared about their progress and breakthrough on 2.5D IC and 3D IC packaging, including the four-DRAM stacked Hybrid Memory Cube (HMC) on an organic substrate [3], the integrated SiP which utilized organic interposer [4], and so forth. These announcements have made professionals and specialists around the globe pour huge amounts of resources in the field of interposer development, including fabrication method optimization, material selection, process limitations, and whatnot.

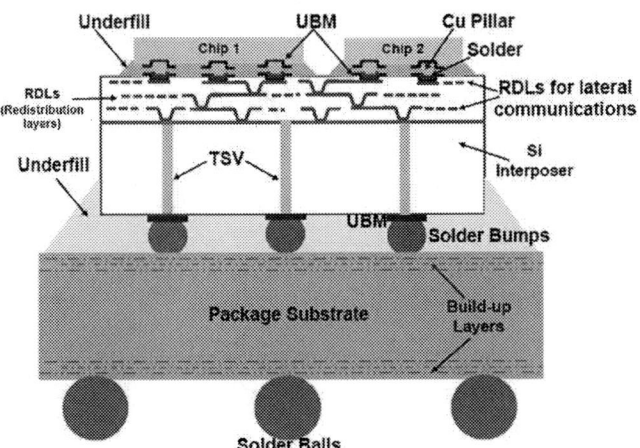

Figure 2. Typical structure of 2.5D IC package utilizing interposer (not to scale)

Process Flow

Before elaborating the process challenges we encountered, process flow should be described in advance for much clearer comprehension. Our overall process flow hierarchy is shown in Figure 3. After device wafers and interposer wafers were obtained, the first step was to fabricate microbumps for device wafers and micropads for interposer wafers to create surface interconnection. In addition to microbumps and micropads fabrication, backside via reveal process as well as Controlled

Collapse Chip Connection (C4) process for interposer wafers were performed. After that, both device wafers and interposer wafers were sawed into separated dice for further die bonding process in order to combine dice into an assembly. This die-interposer assembly was then bonded again with substrate and then undergone assembly part of process before final package was finished, as shown in Figure 3. Since the bumping part of process for interposer wafers involved many fabrication steps, more process details will be described later.

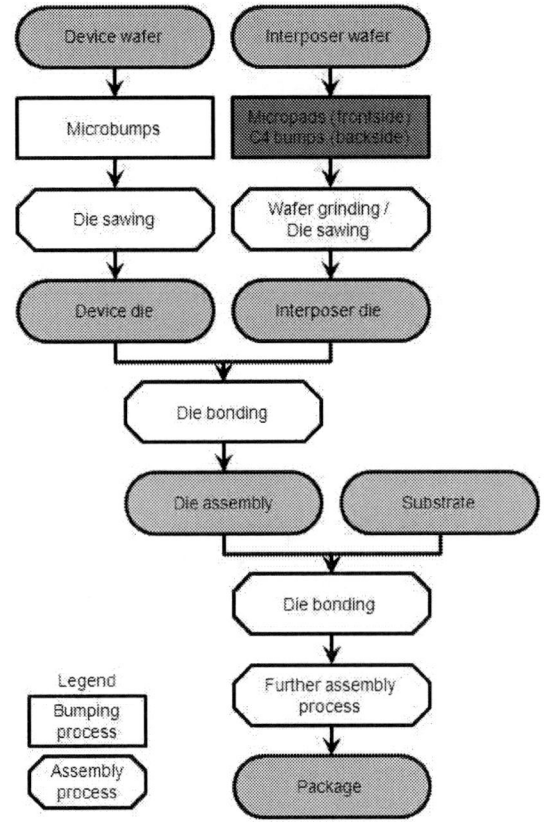

Figure 3. 2.5D IC package process flow hierarchy. (Orange ellipse represents semi-finished product at different stages, while white rectangle and octagon represent the bumping part of process and the assembly part of process respectively)

The bumping part of process mentioned above, or the blue rectangle marked in Figure 3, contains many fabrication steps. Detailed process flow is shown in Figure 4. The first part was the frontside process, and the primary purpose was to fabricate micropads in order to achieve the interconnection with microbumps on device wafers. The process was initiated by sputtering a thin layer of metal as seed layer, followed by lithography for pattern definition and electroplating for metal shaping. Wet etching and PR strip were then utilized to remove undesired parts for micropads forming. This stage of process is shown in Figure 4(a) ~ 4(d). The next part was backside via reveal process to expose through via metal for further interconnection with C4 bumps. Temporary carrier bonding was essential to prevent surface damage and provide protection for previously fabricated frontside micropads when backside process was underway. Glass wafers were employed

978-1-4799-2408-0/14 $31.00 © 2014 IEEE

as carrier wafers for much easier inspection of possible defects during fabrication steps. Mechanical wafer grinding was then implemented to further reduce interposer wafer thickness while the next Deep Reactive Ion Etching (DRIE) process was utilized for silicon etching specifically. This stage of process is shown in Figure 4(e) ~ 4(g). After DRIE process, TSVs were then exposed and protective isolation films were then deposited on wafer surface to fully cover silicon and TSVs. The composition and thickness of the isolation films were pretty crucial to device die strength and warpage performance, and these parameters also had huge impact on operation and handling of latter assembly part of process. This challenge and our proposed approach will be addressed later. After isolation films deposition, Chemical Mechanical Polishing (CMP) was introduced to further flatten and polish the wafer surface, revealing only TSVs without exposing silicon surface. This stage of process is shown in Figure 4(h) ~ 4(i). Backside via reveal process was thus concluded, and revealed TSVs were prepared for further C4 bumps connection.

Under Bump Metallurgy (UBM) and bottom isolation films. This passivation layer also contributed to overall die strength because of its reinforcement of adhesion. The following process consisted of UBM sputter, lithography, metal and solder electroplating, wet etching, and PR strip in series, as shown in Figure 4(j) ~ 4(l). There were combinations of several metal layers and solder contained in C4 bumps to form pillar structure after final reflow process. This reflow process was meant to reshape and harden the solder part of pillar structure. Finally, carrier debonding was utilized to remove temporary carrier, as shown in Figure 4(n), and consequent interposer wafer was available for assembly part of process.

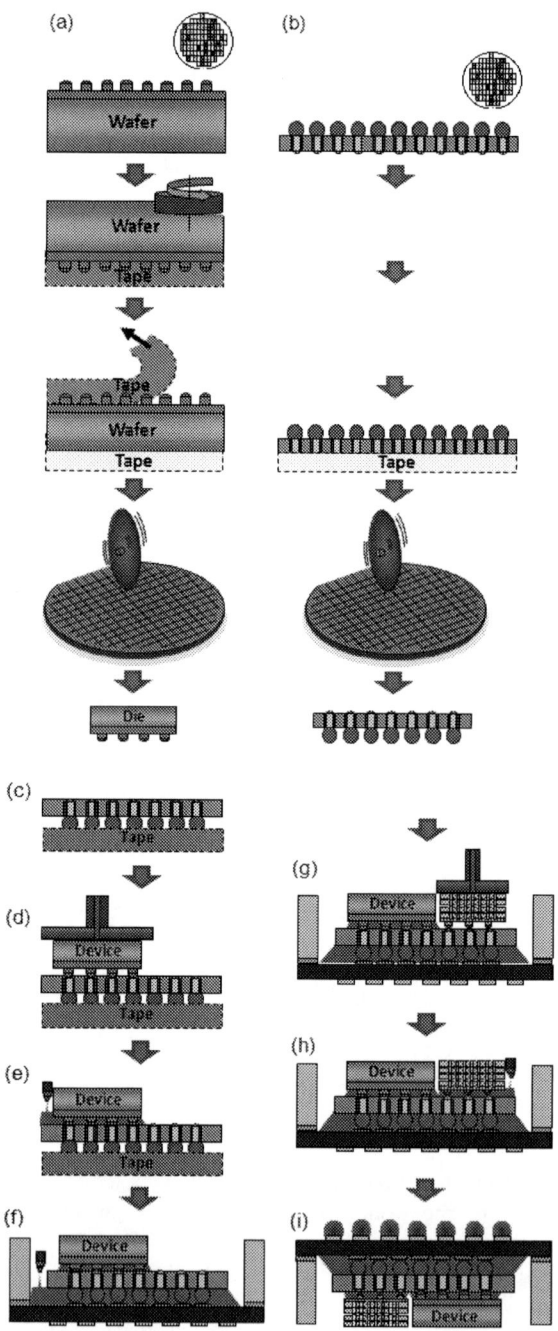

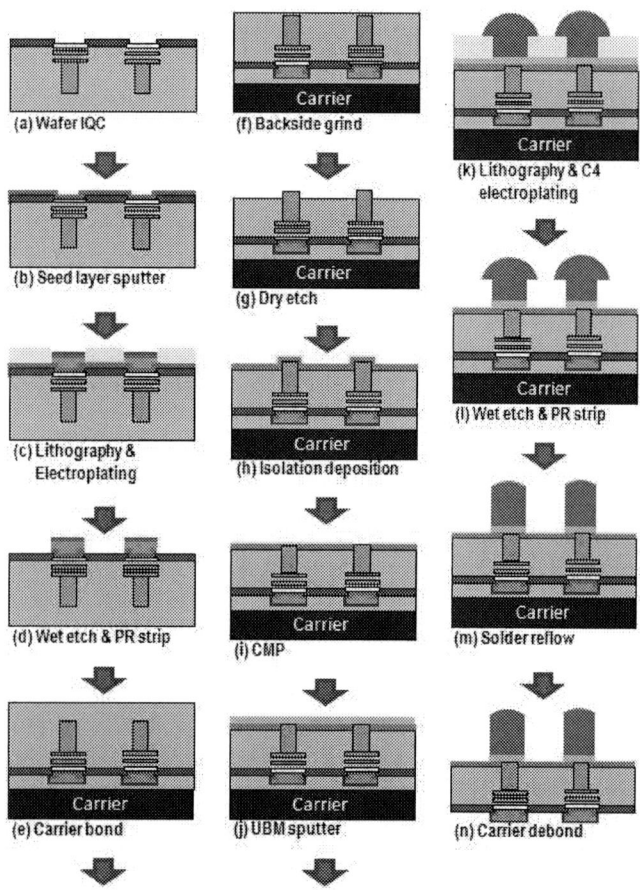

Figure 4. Bumping part of process flow for interposer wafers (not to scale)

After backside via reveal process was completed, backside C4 bumps were then fabricated by series of process somewhat similar as frontside micropads process mentioned before. Prior to seed layer sputter, passivation layer was coated and patterned to not only act as a protection to TSVs, but also act as an intermediate layer to increase adhesion between upper

Figure 5. Assembly part of process flow from interposer and device wafer to die package (not to scale)

978-1-4799-2408-0/14 $31.00 © 2014 IEEE

In the beginning of assembly part of process, interposer wafers and device wafers were both diced into single dice for further die bonding process. Nevertheless, device wafers may require additional grinding if required, as shown in Figure 5(a). After both interposer wafers and device wafers were diced into single dice, interposer die was first bonded with device die to form the die assembly using underfill glue injection to simultaneously encapsulate microbumps on device die and micropads on device die, and the die assembly was then bonded on substrate afterwards, using the same method above, as shown in Figure 5(c) ~ 5(f). Another similar die bonding process was implemented again for the placement of another device die or memory IC on interposer die, based on applications and eventual product requirements, as shown in Figure 5(g) ~ 5(h). Eventually, this multi-die assembly was completed after solder balls were directly placed or screen printed on substrate to accomplish the whole 2.5D IC package with TSV interposer.

Warpage Alleviation

As mentioned before, there were numerous challenges we have encountered during development, and one of the primary challenges was thin wafer warpage control. Since warpage increase is a natural behavior for film stacking structure under various times of thermal treatments within fabrication process, especially for thin wafer, it is pretty significant to search for an approach to alleviate warpage in order to lessen potential difficulties originated in assembly part of process, especially for die bonding process where void or minor vacancy may be existed because of uneven die surface caused by warpage, resulting in poor adhesion between interposer die and device die. In this paper, the terms of warpage for interposer die was defined as "positive" where interposer die was curled up, and "negative" where interposer die was curled down, under the condition that C4 bumps were placed downward, as shown in Figure 6.

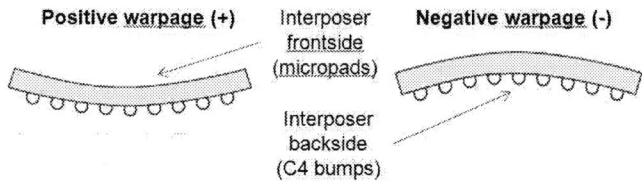

Figure 6. Warpage terms and definitions used in the paper to describe different directions of curl

Die level warpage was measured and acquired by using Shadow Moiré technique. The Moiré concept was initiated in early days, however, its employment in warpage measurement was not proposed and developed until around 1990s and the first warpage measurement system was manufactured and commercialized in 1998. The primary mechanism of Shadow Moire is to utilize the projection of light passing through a Ronchi grating, which casts shadows on the surface of test samples [5]. By using camera to capture the interference patterns created by grating itself and shadows of the grating, the patterns are then converted to digital data via the usage of software. Combined with temperature difference caused by

exterior heater, Shadow Moiré warpage measurement can be implemented to provide us with information and clues to predict and estimate warpage variations during reflow process or under thermal treatment process, and also further interpret possible failure mechanism or latent influences.

Our setup was to measure interposer die warpage under the temperature ranging from room temperature (25°C) to high temperature (250°C) in order to verify warpage variation and characterize the mutual relationship between temperature and warpage. In our research, it was observed that the composition of isolation films we used to cover silicon and TSVs surface were pretty much related to die warpage. The original material we took into consideration to cover wafer surface after CMP process was silicon dioxide (SiO_2), which was commonly used in electrical signal isolation. However, silicon nitride (SiN_x) was proven to be a much satisfactory material for isolation, encapsulation, and as a capping layer, and it also possessed advantages such as highly resistant to molecular diffusion, impenetrable to moisture and ionic contaminants.

The magnitude of SiN_x stress can be further adjusted by the addition of Helium (He) as well as Radio Frequency (RF) power density. Since SiN_x was prepared from standard gas mixtures of silane (SiH_4), ammonia (NH_3), and nitrogen (N_2), if concentration of He in the above gas mixtures became higher, SiN_x stress may incline to much compressive behavior instead of much tensile behavior. Larger RF power density also contributed to much compressive stress, with apparent difference observed at higher He concentration, as shown in Figure 7. It is shown that three different magnitude of RF power have contributed to SiN_x stress variation and the larger the RF power, the lower the stress at same He concentration. Although SiN_x possessed several advantages, it did generate undesired side effects. As shown in Figure 4(e), temporary carrier bonding was required during backside C4 bumps fabrication to further protect previously fabricated frontside micropads. Bonding glue coating was utilized during carrier bonding process and would be removed during final carrier debonding process. Since SiN_x stress was greatly related to warpage performance, different stress behavior thus caused different directions of warpage, resulting in distinct behaviors of bonding glue.

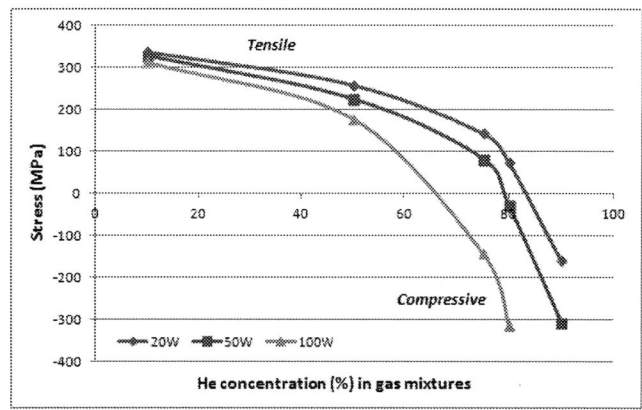

Figure 7. Stress variation of SiN_x by different Helium (He) concentration in gas mixtures at different RF power density

In the beginning, our interposer wafers exhibited negative warpage with non-altered SiNx deposited as isolation films. By means of He concentration adjustment, stress-altered SiNx, which was denoted as A-SiNx, was chosen and deposited, and interposer wafers had curled slightly upward, resulting in smaller absolute value of warpage. Based on interposer wafer design and layout, various magnitude of warpage alleviation were demonstrated. For interposer wafer which still performed large negative warpage even after A-SiNx deposition, isolation films with much highly stress-altered SiNx, which was denoted as B-SiNx, was required to further enhance the influences of warpage alleviation. As shown in Figure 8(a), slightly negative warpage was what we have expected for the behavior of interposer wafers. However, if the effects of B-SiNx were excessively great, interposer wafers might curl upward too much, giving rise to possible positive warpage. This kind of positive warpage might cause the spill of intermediate bonding glue, which was used in carrier bonding process, as shown in Figure 8(b). The spill, or referred to as bleeding, of bonding glue jeopardized process operability, contaminated fabrication equipments, and also gave rise to other handling difficulties. Enlarged top view of glue bleeding is shown in Figure 8(c), where bonding glue was spilled and remained at wafer rim. Nevertheless, the lack of bonding glue resulted in possible wafer crack during final debonding process. Hence, it was pretty essential to select desirable SiNx isolation films and optimized film thickness in order to maintain and further control warpage in acceptable range. It was our anticipation that the desirable warpage of interposer die after wafer dicing, was in the range of -20 μm ~ 0 μm.

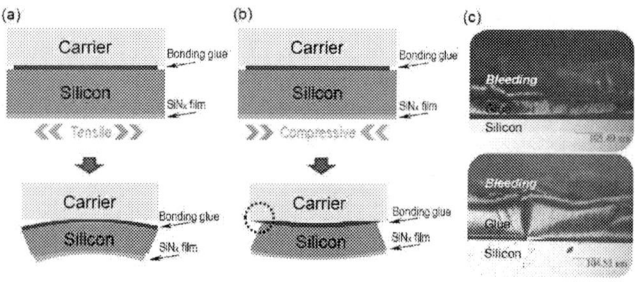

Figure 8. Effects of different behaviors of SiNx stress on interposer wafers warpage. (a) tensile; (b) compressive; (c) enlarged top view of dotted circle showing glue bleeding

Film Thickness Optimization

After selecting SiNx as the desirable material of isolation films, thickness optimization was another task which should be tackled with. We have proposed combinations of several materials and thickness of isolation films, including A-SiNx, B-SiNx, and another type of SiNx which combined with an extra buffer layer. The main purpose of including this extra buffer layer in isolation films was to increase further adhesion between SiNx and bottom silicon surface, which will also be mentioned later. As shown in Table 1 and Figure 9, three different types of isolation films materials and thickness were evaluated. Condition A utilized only A-SiNx, while condition B and condition C both utilized bi-layer structure of buffer

layer and B-SiNx together. It can be observed and compared that condition A had the greatest warpage among all these conditions, while condition B and condition C had much lower warpage, which was 0.21x and 0.35x of A-SiNx warpage at room temperature, respectively. Warpage at high temperature are also shown in this table.

Conditions		Buffer layer thickness ratio	SiNx thickness ratio	Normalized die warpage @ 25°C	Normalized die warpage @ 250°C
A	A-SiNx only	None	1	-1x	-3.44x
B	B-SiNx combined (thick)	1	1	-0.21x	-2.95x
C	B-SiNx combined (thin)	3	2	-0.35x	-2.84x

Table 1. Room temperature (25°C) and high temperature (250°C) warpage results of three different types of isolation films materials and thickness

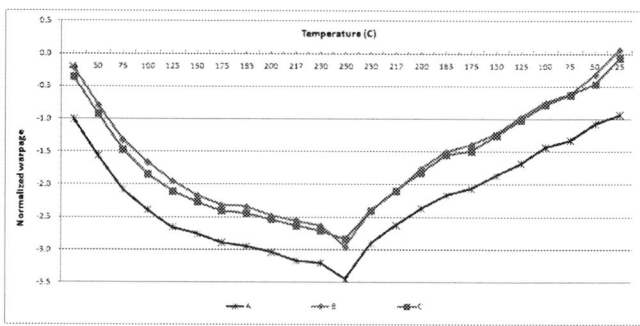

Figure 9. Shadow Moiré warpage results (in average value) measured at various temperatures of three different types of isolation films materials and thickness

As shown in Table 1, it can be compared that there was only SiNx thickness difference between condition B and condition C, and condition B have contributed to warpage rate for about 40% decrease comparing to warpage of condition C. These warpage data were collected from wafer edge die to wafer center die, and then average warpage was calculated and listed in Table 1 and plotted in Figure 9. Actual warpage value was transformed to normalized warpage and displayed in ratio for better comprehension of the effect of warpage alleviation. Since bi-layer condition demonstrated much better warpage performance than A-SiNx only condition, hence, A-SiNx only condition was omitted for further regional warpage analysis. The correlation between B-SiNx regional thickness and their corresponding warpage is summarized in Figure 10. Regional warpage including top, bottom, center, left, and right areas, were all gathered and matched with respective local isolation films thickness for comparison. It is apparent that the warpage was in a way related to B-SiNx thickness. From regional warpage data analysis, with every 1000 Å of thickness increase of B-SiNx, warpage can be alleviated about 0.35 in normalized value, making warpage direction curl towards positive direction. This alleviation was highly related to device layouts and structure parameters, where warpage database should be created for separate devices in order for much precise estimation. However, B-SiNx thickness should be kept under 4000 Å to prevent the possibility of reaching positive warpage, leading to bonding glue spill as mentioned.

978-1-4799-2408-0/14 $31.00 © 2014 IEEE

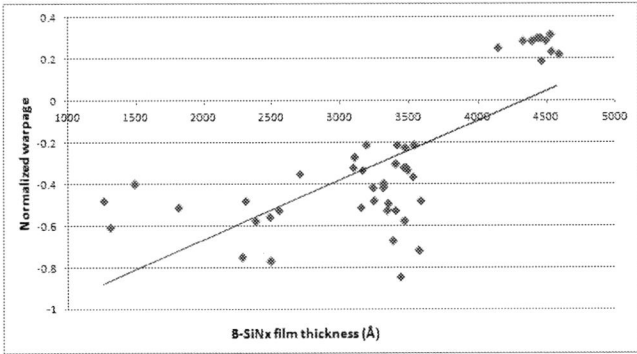

Figure 10. The correlation between warpage and thickness of B-SiNx at different thickness conditions

Delamination Prevention

Delamination usually happened when film to film adhesion was poor in the beginning such as material mismatch or surface roughness difference. Nevertheless, there were still certain conditions where delamination did not happen until post process stage, including the effect of thermal treatment or material residue stress existence. Initially, no passivation layer was planned on isolation films, and the most two frequently happened delamination interface occurred were either between the surface of UBM and isolation films or between the surface of isolation films and silicon, as shown in Figure 11(a) and 11(b) respectively. Since the isolation films were flattened by CMP process to expose TSVs, there were conditions where partial isolation films were over-polished using non-optimized CMP parameters, which resulted in no isolation films covered on certain areas, as shown in Figure 11(c). In such case, delamination happened between the interface of UBM and silicon instead.

There were several different approaches for the prevention of delamination, including material adjustment such as metal composition optimization or the addition of extra materials, like an extra intermedium between isolation films and silicon or between UBM and isolation films, acting as buffer layers to enhance adhesion between layers. The first buffer layer was implemented during warpage tuning stage, not only for SiNx warpage control, as mentioned previously, but also for the prevention of delamination stated herein and for the adhesion increase between SiNx and silicon surface. Besides the first buffer layer, a second buffer layer was also added to strengthen the adhesion between UBM and isolation films. After material property analysis and experimental trial runs, Polybenzoxazole (PBO), one kind of passivation material which has been extensively investigated was selected because of its excellence and merits in dielectric, electrical, thermal, and mechanical properties, such as low dielectric constant, low stress, high modulus, chemical resistance, high electrical resistivity, and thermal stability and low coefficient of thermal expansion [6]. In many cases, PBO was preferred over polyimide (PI), as it absorbed less water, and can be cured at lower temperature and still retained its excellent mechanical characteristics. In addition, PBO has been shown to be a better gap filling material and has better planarizing characteristics.

Figure 11. Three kinds of delamination happened between different interfaces. (a) UBM/isolation; (b) isolation/silicon; (c) UBM/silicon

Three kinds of indexes were utilized to further validate the results of implementing PBO as second buffer layer, including peeling test, shear test, and von Mises test, as shown in Figure 12. The von Mises test was utilized to predict materials yielding under any loading conditions from results of uniaxial tensile tests [7]. Several layouts were designed and put into simulation first to evaluate the testing results. Besides original none passivation type, two kinds of PBO layouts were also demonstrated and the difference between these two types lay in its opening size. The larger the PBO opening size, the relative smaller area the UBM was directly adhered with isolation films. As shown in Table 2, it can be concluded that this layer of passivation addition indeed contributed to stress decrease, whether in peeling, shear, or von Mises test. As for design B and design C in Table 2, it is observed that design with larger PBO opening showed slightly lower peeling stress and von Mises stress than the one with smaller PBO opening, suggesting that UBM delamination may have lesser possibility to occur. In addition, there were still numerous kinds of PBO layouts with opening size difference, and they were utilized to compare stress reduction capability and to examine other side effects which might be generated.

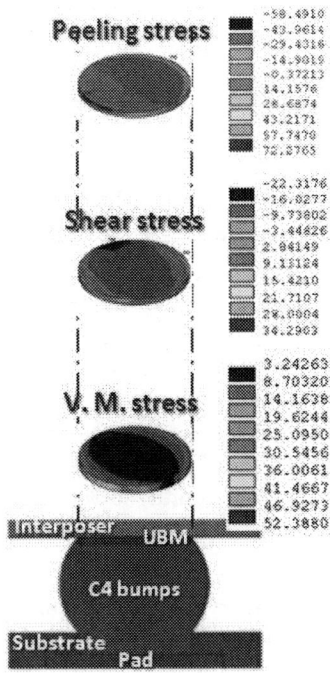

Figure 12. Three kinds of UBM stress simulation contours for device with none passivation design

Items \ Design		A	B	C
PBO parameters	Opening (µm)	NA	70	30
	Width (µm)	NA	20	40
	Over UBM width (µm)	NA	10	10
UBM stress	Peeling (kg/mm²)	72	16	21
	Shear (kg/mm²)	34	11	6
	von Mises (kg/mm²)	52	20	23

Table 2. Three kinds of UBM stress simulation results

Conclusion

In this paper, we proposed certain solutions to numerous challenges we have encountered during development of 2.5D IC packaging. Our primary focus was concentrated on three main challenges, warpage alleviation, isolation film selection and thickness adjustment, and delamination prevention. For warpage alleviation, several isolation films were utilized to compare their wapage alleviation extent, including stress-altered SiN_x only, and two bi-layer structures of buffer layer and highly stress-altered SiN_x with different thickness. As a result, highly stress-altered SiN_x was proven to be the most desirable isolation films because of its warpage alleviation and film adhesion improvement, with the inclusion of buffer layer. After specified SiN_x was chosen as isolation films material, thickness adjustment was then under evaluation and has been finely tuned to further decrease warpage to near-zero value while still maintained sufficient die strength. A certain range of applicable SiN_x thickness was thus determined and utilized. As for delamination prevention, PBO was selected as the desirable passivation layer to act as the buffer layer between interface of UBM and isolation films. Different PBO layout designs were taken into consideration, where actual opening dimension had obvious impact on UBM stress results, and combinations of PBO opening dimension and stress have been exhibited. All of these improvements indeed contributed to final package yield, and were able to eliminate some undesired failures or defects we confronted during development. Several minor and lesser defects and process issues are still required to tackle with and more efforts should be made in order for the preparation of future mass production stage in the near future.

Acknowledgments

All bumping part of process and assembly part of process were developed and realized in SPIL internally with a few supports from foreign material and equipment vendors on certain experimental confirmation runs and consultation. Moreover, the authors would like to appreciate internal RD teams about their fully support and hard work from all process module team members in order to accomplish this interposer type 2.5D IC package technology development.

References

1. Charles A. Harper *et al.*, *Electronic Packaging and Interconnection Handbook 4th Edition*, McGraw-Hill, New York, 2005, pp. 8.83-8.85.
2. B. Black, D. Nelson, C. Webb, and N. Samra, "3D Processing Technology and Its Impact on iA32 Microprocessors," in *Proc. on International Conference on Computer Design (ICCD)*, San Jose, CA, Oct. 11-13, 2004, pp. 316-318.
3. GLOBE NEWSWIRE, Boise, Idaho, 2013, "Micron's Hybrid Memory Cube Earns High Praise in Next-Generation Sercomputer," https://globenewswire.com/news-release/2013/11/07/587823/10056650/en/Micron-s-Hybrid-Memory-Cube-Earns-High-Praise-in-Next-Generation-Supercomputer.html
4. Rabindra N. Das *et al.*, "3D Integration of System-in-Package (SiP) Using Organic Interposer: Toward SiP-Interposer-SiP for High-End Electronics," *The 46th International Symposium on Microelectronics (IMAPS)*, Orlando, FL, Sep. 30 - Oct. 3, 2013, pp.531-537.
5. Bongtae Han *et al.*, *Handbook of Experimental Solid Mechanics*, Springer Science+Business Media, New York, 2008, pp. 611-614.
6. J. Yota et al., "Variable Frequency Microwave and Convection Furnace Curing of Polybenzoxazole Buffer Layer for GaAs HBT Technology," *IEEE Transactions on Semiconductor Manufacturing*, vol. 20, issue 3, pp. 323-332, Aug. 2007.
7. S. M. A. Kazimi *et al.*, *Solid Mechanics 1st Revised Edition*, Tata McGraw-Hill, New York, 2006, pp. 120-122.

Structure Effects on the Electrical Reliability of Fine-Pitch Cu Micro-Bumps for 3D Integration

Byeong-Rok Lee[1], June-Bum Kim[1], Seung-hyun Kim[1], Byeong-Hyun Bae[1], Ho-Young Son[2], Tac-Keun Oh[2], Min-Suk Suh[2], Nam-Seog Kim[2] and Young-Bae Park[1*]

1 School of Materials Science and Engineering, Andong National University, Andong, 760-749, Korea

2 SK Hynix Inc., Icheon, 467-701, Korea

e-mail : ybpark@andong.ac.kr

Abstract

Recently, flip chip solder bumps have been replaced by the fine-pitch solder micro-bump due to the miniaturization of electronic devices and the high performance requirement, and so on. Because of the fast decreasing size of the micro-bump and increasing power consumption needs in logic through-Si via applications, the significance of electromigration among major reliability issues have been increased. There are several important electrical current-induced reliability issues such as current crowding, polarity effect and thermomigration. And the excessive intermetallic compound (IMC) growth and Kirkendall voiding or micro voiding in micro-bump can degrade the mechanical reliability as well as electrical reliability. Therefore, the understanding of fundamental IMC growth mechanism is essential. This study systematically investigated the effects of bump structures such as solder height and UBM structure on the IMC growth kinetics and electromigration performance of Cu micro-bump. Quantitative analyses on the IMC growth kinetics during in-situ electromigration test were performed in a scanning electron microscope chamber under current stressing conditions with current density of 1.5×10^5 A/cm^2 at 150°C. Under high temperature and electric current stressing, the IMCs growth is accelerated by electron wind force. And the IMC phase transition time became shorter because IMC growth rates increased. In Cu/Sn-Ag and Cu/Ni/Sn-Ag system, the effect of current crowding and Joule heating was negligible in fully IMC-transformed micro-bump. Finally, microvoid formation mechanisms during IMC growth in the Cu/Ni/thin Sn system were discussed in detail.

1. Introduction

Three dimensional chip stacking technology has received recently much increasing attentions because of the increased demands for high performance, small form factor and multi-functions in electronics devices. Interconnections with high I/Os and fine pitch were needed in other to achieve the requirement of multi-function for next generation electronics [1-6]. Among the interconnection technologies, through-silicon-via (TSV) provides especially the shortest length and the highest density which lead to significantly reduced signal delay and power consumption [7]. TSV integrated structure is stacked by using micro-bumps and TSV, moreover, the diameter of micro-bumps will be about 10μm to 20μm, which is one order of magnitude smaller than that of a flip-chip solder joint [7]. With the reduction of micro-bump size, the current density in micro-bump will be increased at least by a factor of 100 under the same power demand. Furthermore, when Si chips are stacked together, joule heating would be

serious and the actual operation temperature of 3D IC packaging will be significantly increased [7]. As a result, the newly 3D IC technology is expected to confront electrical reliability problems associated with electromigration, thermomigration. Electromigration is the enhanced diffusion of atoms in the current direction [8]. The concentrated joule heating near the electron-entering positions from interconnect to micro-bump due to excessively high local current density during the accelerated electromigration test can be significant [9]. In addition, the excessive IMC growth in micro-bump can degrade the mechanical and electrical reliability. Therefore, it is essential to understand the fundamental mechanisms of IMC growth kinetics [10].

In this paper, we have systematically investigated the effects of bump structures such as solder height and UBM structure on the IMC growth kinetics and electromigration performance of Cu micro-bump.

2. Experimental procedures

Figure 1 shows a schematic of the cross-sectioned bump structure to investigate the effect of microstructure of Cu micro-bump structure under annealing and current stressing conditions. Figure 1(a) is Cu pillar /Sn-Ag structure. Cu and Sn-Ag layer were electroplated with thickness 10 and 6μm respectively. Cu trace is sputtered on top Si-chip. The micro-bump diameter and height were 20μm, 16μm, respectively. The method of bonding is reflow process, and peak temperature is 260 °C [11].

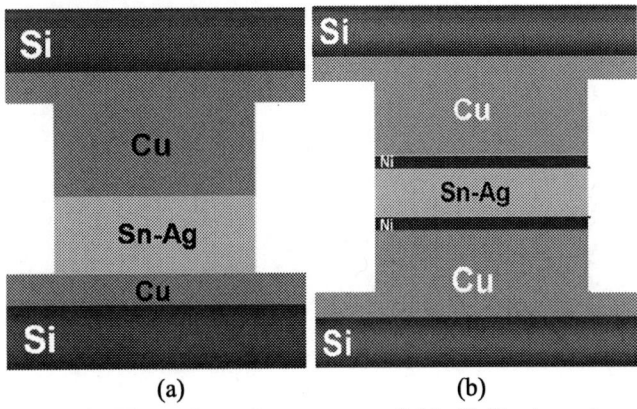

Figure 1. The schematic structure of (a) Cu/Sn-Ag micro-bump [10], (b) Cu/Ni/Sn-Ag micro-bump

Figure 1(b) is Cu pillar/ Ni/ Sn-Ag structure. Cu, Ni and Sn-Ag layer are formed by electroplating. The bonding process was thermo-compression bonding. The IMC growth kinetics were investigated by in-situ technology [7,12]; that is, the cross-sectioned Cu micro-bump samples were stressed electrically in an SEM chamber at 150°C with current density of 1.5×10^5 A/cm^2 conditions [11]. The transformation of interfacial IMC phases in the micro-bumps was analyzed using SEM in backscattered electron (BSE) mode and using energy-dispersive x-ray spectroscopy (EDS) [11]. Ni-Sn IMC thickness was quantified using an image analyzer and defined as the area of the IMC divided by the interface length [11].

3. Results and Discussion

Figure 2 shows cross-sectional BSE image of the as-bonded Cu micro-bump. Figure 2(a) is Cu/Sn-Ag micro-bump and Fig. 2(b) is Cu/Ni/Sn-Ag micro-bump.

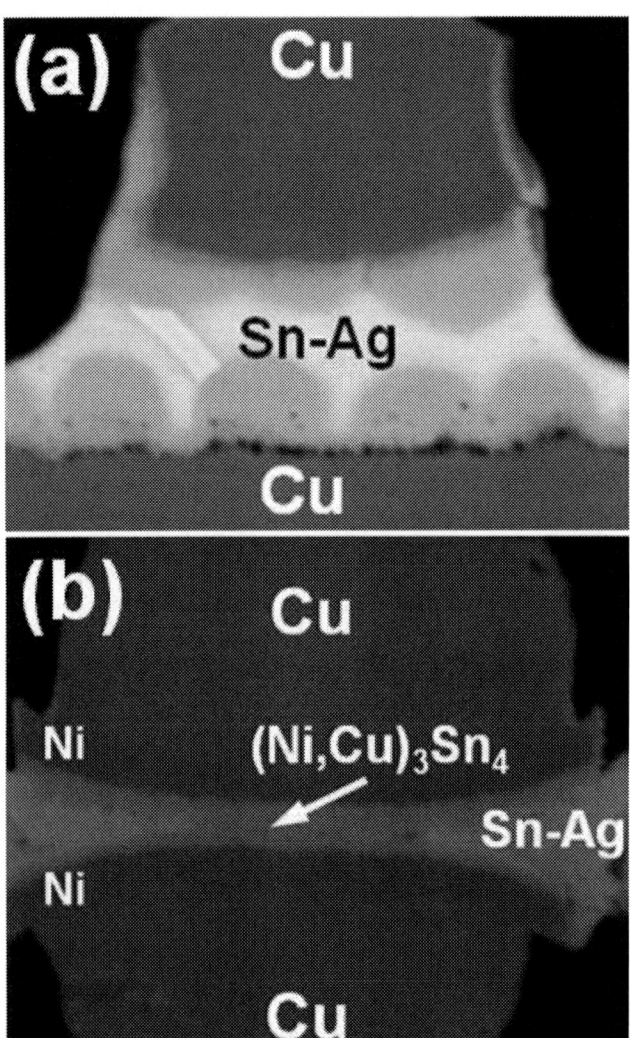

Figure 2. Cross-section BSE image of as-bonded (a) Cu/Sn-Ag/Cu [10], (b) Cu/Ni/Sn-Ag/Ni/Cu micro-bump.

The primary difference between the two samples is the existence of Ni layer. While Fig. 2(a) shows the bonding between Cu and solder, Fig. 2(b) shows the existence of Ni layer between the solder and Cu. Therefore, these structures are observed using the in-situ SEM during current stressing.

Figure 3 showed the enlarged BSE image of the cross-sectioned Cu/Sn-Ag micro-bump interface before and after current stressing at 1.5×10^5 A/cm^2, 150°C for 150h [10]. Most of Sn was transformed into the Cu_6Sn_5 phase (Figure 3b). As the test time increases, only a few regions of Sn were not consumed and isolated between the transformed regions (Figure 3b) [13]. In the Sn limited system, formation of Cu_3Sn phase at the Cu/Cu_6Sn_5 interface result from the continued diffusion of Cu atoms. [7,12]. Rapid growth of Cu_3Sn phase and decrease of Cu_6Sn_5 phase thickness occur, because IMC phase evolution depends on the relative amounts of the reactants [14].

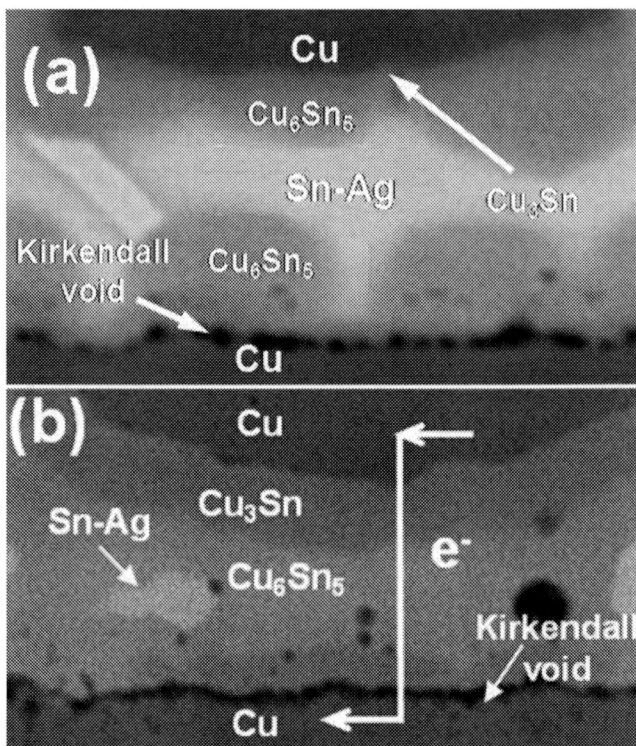

Figure 3. Enlarged BSE image of the cross-sectioned Cu/Sn-Ag micro-bump interface (a) as-bonded, (b) electron downstream bump at 1.5×10^5 A/cm^2, 150°C for 150h [10].

In Fig. 4, SEM image of BSE mode and EDS results show that $(Ni,Cu)_3Sn_4$ phase is formed at the interface between Ni and Sn-Ag. Cu/Sn-Ag micro-bump and Cu/Ni/Sn-Ag micro-bump are different from the interfacial reaction. In the case of Cu/Sn-Ag micro-bump, the phases of Cu_6Sn_5 or Cu_3Sn are formed due to rapid reaction between Cu and solder. However, in the case of Cu/Ni/Sn-Ag micro-bump, thin layer of $(Ni,Cu)_3Sn_4$ phase was confirmed due to Ni layer as a diffusion barrier layer. Ni_3Sn_4 phase gradually grows as the experiment progresses (Figure 4a). It is occurred to connect between top side of $(Ni,Cu)_3Sn_4$ phase and bottom side of $(Ni,Cu)_3Sn_4$ phase for growth. The microvoid can be occurred at this connected location (Figure 4b). The resistances of the test samples were measured by in-situ to investigate the situation of the micro-bump degradation [15]. Figure 5 is the curves of the resistance increase ratio as a function of electromigration testing time for 150°C at 1.5×10^5 A/cm^2. This shows that the initial resistance can be increased rapidly.

978-1-4799-2408-0/14 $31.00 © 2014 IEEE   1636

Over time, the resistance increase ratio is gradually decreased. This phenomenon is similar to resistance change of general micro-bumps.

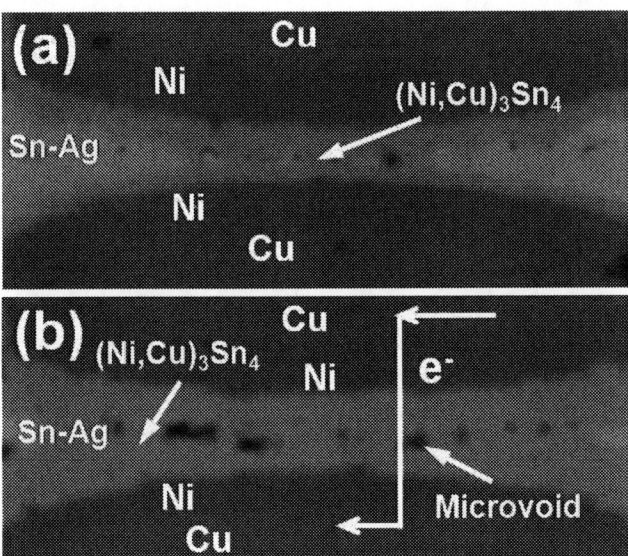

Figure 4. Enlarged BSE image of the cross-sectioned Cu/Ni/Sn-Ag micro-bump interface (a) as-bonded, (b) electron down-stream bump at 1.5×10^5 A/cm², 150℃ for 600h.

The diameter of the flip-chip solder bump is larger than that of the micro-bump so the current density through the flip-chip solder bump is less than the micro-bump [10]. Thus, the high current density of micro-bumps results in faster IMC formation and growth and leads to the rapid resistance change of the micro-bump [10].

Figure 6 shows the variations of $(Ni,Cu)_3Sn_4$ phase thickness as a function of current stressing time for 150℃ at 1.5×10^5 A/cm². As mentioned earlier, the $(Ni,Cu)_3Sn_4$ phase grows rapidly initially due to initial fast IMC phase changes, and then saturated in the late stage due to the restrained growth of IMC phase by limited Sn amount.

Table 1. The resistivity of the materials of micro-bump [16].

Materials	Resistivity ($\mu\Omega$-cm)
Sn	11
Ni	6.99
Ni_3Sn_4	28.5

The behavior of curve in Fig.5 is similar to that of curve in Fig.6. Table 1 show that the resistivity of Sn is $11\mu\Omega$-cm, the resistivity of Ni is $6.99\mu\Omega$-cm and the resistivity of Ni_3Sn_4 phase is $28.5\mu\Omega$-cm [16]. The resistivity of Ni_3Sn_4 phase is two times of that of Sn and four times of that of Ni. Therefore the resistance of micro-bump is increased by the transformation from Sn and Ni to Ni_3Sn_4 phase. Thus, the

initial resistance of micro-bump is abruptly increased due to early growth of $(Ni,Cu)_3Sn_4$ phase in Cu/Ni/Sn-Ag structure.

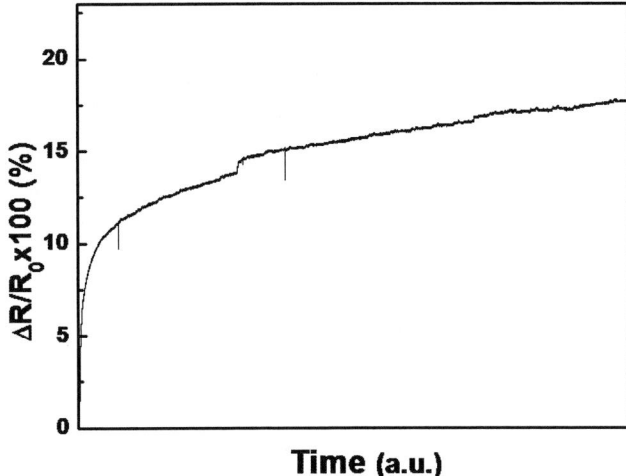

Figure 5. Relative resistance increase ratio as a function of current stressing time at 150℃, 1.5×10^5 A/cm².

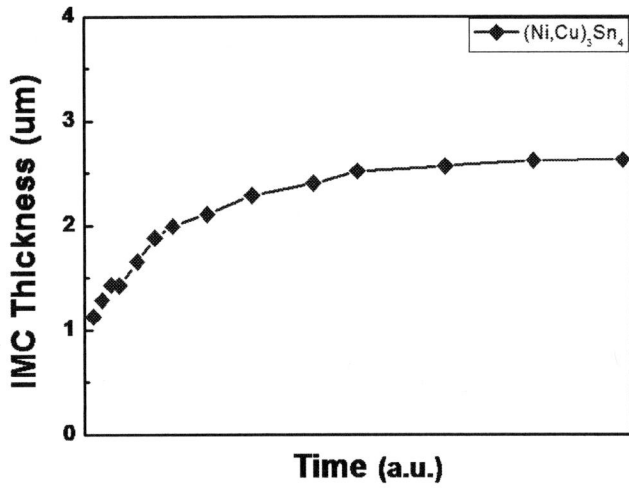

Figure 6. $(Ni,Cu)_3Sn_4$ IMC thickness as a function of current stressing time at 150℃, 1.5×10^5 A/cm².

Figure 7(a), (b) show cross-sectional SEM images of Cu/Sn-Ag micro-bump at 150℃, 1.5×10^5 A/cm². Figure 7(a) shows that most of Sn transformed into the Cu_6Sn_5 phase, and only a few regions of Sn were not consumed and trapped between the transformed regions before electromigration test [13]. And voids were observed at the interface between Cu and solder as shown in Fig. 7(a). After current stressing for 150 h, in the Sn-limited system, Cu_3Sn phase grows and Cu_6Sn_5 phase decreases as shown in Fig. 7(b) [10]. Remained solder phase was completely consumed because the atomic rearrangement to form the Cu_6Sn_5 phase of the regular structure occurs at the Cu/solder interface [17]. Kirkendall void is formed along Cu_3Sn/Cu interface due to the difference in intrinsic diffusivities of two diffusing species. And then Kirkendall void is combined.

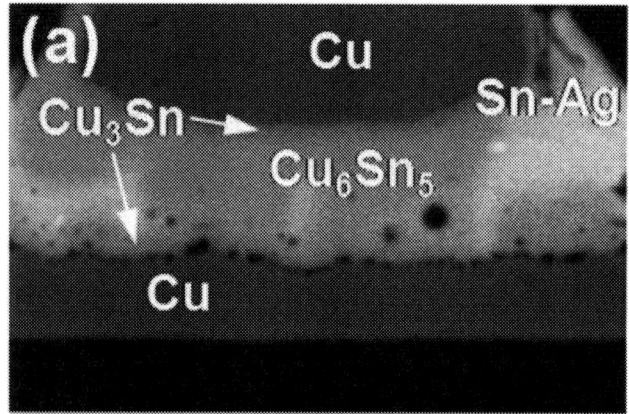

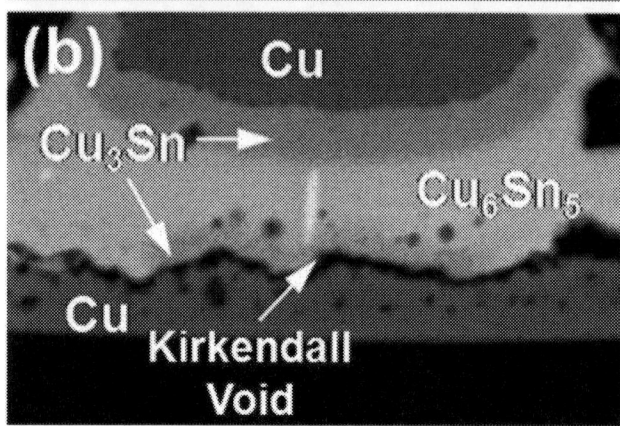

Figure 7. Enlarged SEM images of the cross-sectioned Cu/Sn-Ag micro-bump at 150℃, $1.5 \times 10^5 A/cm^2$: (a) as-bonded, (b) 150h [10].

Figure 8 (a) ~ (c) show the Cu/Ni/Sn-Ag structures at 150 ℃ for 0h, 90h and 500h, respectively. During current stressing, both top and bottom sides of $(Ni,Cu)_3Sn_4$ phase grow into solder, as shown in Fig. 8(a). And $(Ni,Cu)_3Sn_4$ grains growing from the opposite directions started to influence each other at certain locations [18]. After current stressing time for 90h, as showed in Fig. 8(b), $(Ni,Cu)_3Sn_4$ phase become thicker and grows in both directions. Sn is isolated between $(Ni,Cu)_3Sn_4$ phases. When the current stressing time for 120h, as showed in Fig. 8(c), all Sn is transformed into $(Ni,Cu)_3Sn_4$. One key observation of this study was the existence of many voids into the $(Ni,Cu)_3Sn_4$ phase [18]. These voids appeared consistently in all sample. The existence of these voids occur serious concerns of electrical and mechanical reliability [18].

Table 2. Molar volumes of the materials of micro-bump [18, 19].

Material	molar volumes (cm^3/mol)
Sn (bulk)	16.26
Cu	7.12
Ni	6.59
Cu_3Sn	27.30
Cu_6Sn_5	118.01
Ni_3Sn_4	75.25

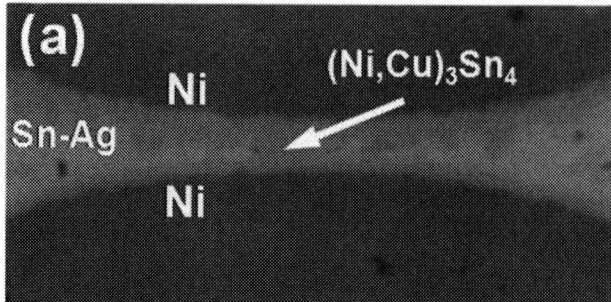

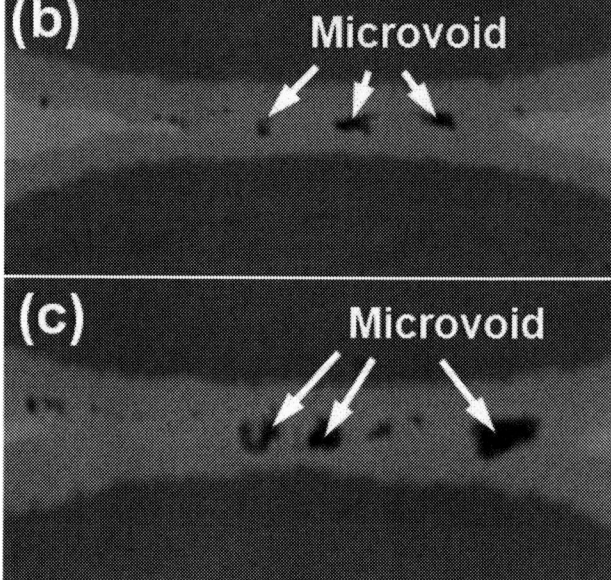

Figure 8. Enlarged SEM images of the cross-sectioned Cu/Ni/Sn-Ag micro-bump at 150℃, $1.5 \times 10^5 A/cm^2$: (a) as-bonded , (b) 90h, and (c) 500h.

In order to understand mechanism of void formation when Ni_3Sn_4 phase is formed by reacting interface between Sn and Ni, the volume change should be noted. The molar volumes for Ni, Sn and Ni_3Sn_4 are 6.59, 16.26 and 75.25 $cm^3 \, mol^{-1}$, respectively as shown in Table 2 [18, 19]. The volume change of Ni_3Sn_4 phase is calculated to be $\{75.25 - [3(6.59) + 4(16.26)]\}/[3(6.59) + 4(16.26)] = -0.113$, or -11.3% [18]. This value is greater than volume change value of Cu_6Sn_5 or Cu_3Sn phase as shown in Table 3 [18].

Table 3. Volume change value rate according to the reaction [18].

Reactions	Volume Change
$6 \, Cu + 5 \, Sn \rightarrow 1 \, Cu_6Sn_5$	-5.0%
$9 \, Cu + 1 \, Cu_6Sn_5 \rightarrow 5 \, Cu_3Sn$	-4.3%
$3 \, Ni + 4 \, Sn \rightarrow 1 \, Ni_3Sn_4$	-11.3%

From these results, the mechanism of void formation can be expected [18]. When the $(Ni,Cu)_3Sn_4$ phase is connected on each other, Sn was isolated between $(Ni,Cu)_3Sn_4$ phases. It is difficult that the stress on volume change is emitted through different means. Because the plastic deformation of

$(Ni,Cu)_3Sn_4$ phase is very difficult. Thus, this stress is dissipated by diffusing Sn atoms and microvoids are formed.

4. Conclusions

The IMC growth behaviors in fine-pitch Cu/Sn-Ag and Cu/Ni/Sn-Ag micro-bump for TSV integration were systematically investigated using the in-situ SEM during annealing and current stressing. Due to Sn-limited characteristics of micro-bump, IMC growth behaviors of Cu/Sn-Ag and Cu/Ni/Sn-Ag micro-bumps were different each other. Thicknesses of Cu_6Sn_5 and Cu_3Sn phases rapidly grew in Cu/Sn-Ag micro-bump. And Kirkendall voids were formed along the interface between Cu_3Sn and Cu due to the difference in intrinsic diffusivities of two diffusing species. In the other hand, IMC of $(Ni,Cu)_3Sn_4$ phase was formed in Cu/Ni/Sn-Ag micro-bump due to existence of Ni layer as a diffusion barrier effect, which led to relatively slow growth of $(Ni,Cu)_3Sn_4$ phase. Electrical resistance of Cu/Ni/Sn-Ag micro-bump continuously increased due to the continuous growth of $(Ni,Cu)_3Sn_4$ phase with relatively high resistivity compared to Sn phase. Finally, isolated Sn phases between $(Ni,Cu)_3Sn_4$ phases were replaced with microvoids due to large volume change when Sn is transformed into $(Ni,Cu)_3Sn_4$ phase by reacting between Sn and Ni. Therefore, symmetric UBM structure with Ni barrier layer seemed to be effect to improve electrical reliability of micro-bump even though micro voiding issues between IMC phases inside micro bump should be solved in future works.

Acknowledgments

This study was supported by SK Hynix Inc. and Nano. Material Technology Development Program(Green Nano Technology Development Program) through the National Research Foundation of Korea(NRF) funded by the Ministry of Education, Science and Technology (2011-0019986).

References

1. A. Yu, J. H. Lau, S. W. Ho, A. Kumar, D.Q. Yu, M. C. Jong, V. Kripesh, D. Pinjala, D.-L. Kwong, "Study of 15µm Pitch Solder Microbumps for 3D IC Integration," in Proc, IEEE Electronic Components and Technol. Conf. (ECTC), San Diego, California , May 26 - 29, 2009, pp. 6-10.

2. A. Yu, J. H. Lau, S. W. Ho, A. Kumar, H. W. Yin, J. M. Ching, V. Kripesh, D. Pinjala, S. Chen, C.-F. Chan, C.-C. Chao, C.-H. Chiu, C.-M. Huang and C. Chen, "Three Dimensional Interconnects with High Aspect Ratio TSVs and Fine Pitch Solder Microbumps," in Proc, IEEE Electronic Components and Technol. Conf. (ECTC), San Diego, California , May 26 - 29, 2009, pp.350- 354.

3. A. Yu, A. Kumar, S. W. Ho, W. Y. Hnin, John H. Lau, C. H. Khong, P. S. Lim, X. W. Zhang, D. Q. Yu, N. Su, B. R. Chew, M. C. Jong, T. C. Tan, V. Kripesh, C. Lee, J. P. Huang. J. Chiang, S. Chen, C.- H Chiu, C.-Y Chan; C.-H. Chang, C.-M. Huang and C.-H. Hsiao, "Development of Fine Pitch Solder Microbumps for 3D Chip Stacking," in Proc, IEEE Electronic Components and Technol. Conf. (ECTC), Orlando, Florida, May 27 - 30, 2008, pp. 387-392.

4. K. Sakuma, P.S. Andry, B. Dang, J. Maria, C.K. Tsang, C. Patel, S.L. Wright, B. Webb, E. Sprogis, S.K. Kang, R. Polastre, R. Horton, and J.U. Knickerbocker, "3D Chip Stacking Technology with Low-Volume Lead-Free Interconnections," " in Proc, IEEE Electronic Components and Technol. Conf. (ECTC), Reno, Nevada, May 29 - June 1, 2007, pp. 627-632.

5. M. Umemoto, Tanida, K., Nemoto, Y., Hoshino, M., Kojima, K., Shirai Y. and Takahashi, K., "High-Performance Vertical Interconnection for High-Density 3D Chip Stacking Package," in Proc, IEEE Electronic Components and Technol. Conf. (ECTC), Las Vegas, Nevada, June 1-4, 2004, pp. 66-623.

6. P. S. Andry, C. K. Tsang, B. C. Webb, E. J. Sprogis, S. L. Wright, B. Dang and D. G. Manzer, "Fabrication and characterization of robust through-silicon vias for silicon-carrier applications," IBM J. R. D, vol. 52, no. 6, Nov. 2008, pp. 571-581.

7. F. Y. Ouyang, H. Hsu, Y. P. Su and T. C. Chang, "Electromigration induced failure on lead-free micro bumps in three-demensional integration circuits packaging," journal of applied physics, vol. 122, pp. 023505, Jul. 2012.

8. Y. D. Lu, X. Q. He, Y. F. En, X. Wang, Z. Q. Zhuang, "Polarity effect of electromigration on intermetallic compound formation in SnPb solder joints," Acta Materialia, vol. 57, no. 8, pp. 2560–2566, May 2009.

9. M. Lu, S. L. Wright, G. M. Vicker, S. M. Sri- Jayantha, "Effect of Joule Heating on Electromigration Reliability of Pb-free Interconnect," in Proc, IEEE Electronic Components and Technol. Conf. (ECTC), San Diego,CA, May 29, 2013, pp. 590-596.

10. Y. B. Park, S. H. Kim, J. J. Park, J. B. Kim, H. Y. Son, K. W. Han, J. S. Oh, N. S. Kim and S. H. Yoo, "Current Density Effects on the Electrical Reliability of Ultra Fine-Pitch Micro-Bump for TSV integration," in Proc, IEEE Electronic Components and Technol. Conf. (ECTC), Las Vegas, Nevada, May 28-31, 2013, pp. 1988-1993

11. M. H. Jeong, J. W. Kim, B. H. Kwak, Y. B. Park, "Effects of annealing and current stressing on the intermetallic compounds growth kinetics of Cu/thin Sn/Cu bump," Microelectron Eng. vol. 89, pp.50-54, Jan. 2012.

12. G.T. Lim, B.J. Kim, K.W. Lee, J.D. Kim, Y.C. Joo, and Y.B. Park, "Temperature Effect on Intermetallic Compound Growth Kinetics of Cu Pillar/Sn Bumps," J. Electron. Mater. vol. 38, no. 11, pp. 2228-2233, Nov. 2009.

13. Lee, J. Park, S.J. Jeon, K.W. Kwon, and H.J. Lee, "A Study on the Bonding Process of Cu Bump/Sn/Cu Bump Bonding Structure for 3D Packaging Applications," J. Electrochem. Soc, vol. 157, no. 4, pp. H420-H424, Feb. 2010.

14. K. .N. Tu, J.W. Mayer, and L.C. Feldman, Electronic Thin Film Science, Macmillan, New York, 1992.

15. T. H. Lin, R. D. Wang, M. F. Chen, C. C. Chiu, S. Y. Chen, T. C. Yeh, Larry C. Lin, S. Y. Hou, J. C. Lin, K. H. Chen, S. P Jeng, and Douglas C. H. Yu, "Electromigration Study of Micro Bumps at Si/Si Interface in 3DIC Package for 28nm Technology and Beyond," Proc 61th Electronic Components and Technology Conf. (ECTC), Lake Buena Vista, FL, May. 2011, pp.346-350.

16. D.R. Frear, S.N. Burchett, H.S. Morgan, and J.H. Lau, eds., The Mechanics of Solder Alloy Interconnects, Van Nostrand Reinhold, New York, 1994, pp. 60.

17. B.H. Kwak, M.H. Jeong, and Y.B. Park, "Effects of Temperature and Current Stressing on the Intermetallic Compounds Growth Characteristics of Cu Pillar/Sn–3.5Ag Microbump," Jap. J. Appl. Phys, vol. 51, no. 5, pp. 2264-2270, May. 2012.

18. H.Y. Chuang, J.J. Yu, M.S. Kuo, H.M. Tong and C.R. Kao, "Elimination of voids in reactions between Ni and Sn: A novel effect of silver," Scripta Mater., vol. 66, no. 3 pp. 171–174, Feb. 2012.

19. J.Y. Song a, Jin Yu, T.Y. Lee, "Effects of reactive diffusion on stress evolution in Cu–Sn films," Scripta Materialia, vol. 51, no. 2, pp.167–170, Jul. 2004.

Demonstration of Low Cost TSV Fabrication in Thick Silicon Wafers

E. Vick[1], D. S. Temple[1], R. Anderson[1], J. Lannon[1], C. Li[2], K. Peterson[2], G. Skidmore[2] and C.J. Han[2]

[1] RTI International

Research Triangle Park, NC, USA

[2] DRS RSTA, Inc.

Dallas, TX, USA

Abstract

Low cost wafer-level chip-scale vacuum packaging (WLCSVP) imposes unique constraints on potential implementation of through-silicon vias (TSVs). A WLCSVP requires a relatively thick substrate to prevent mechanical failure. Two approaches for integrating TSVs in thick silicon wafers have been successfully demonstrated. Both approaches enable TSV formation from the backside of a device wafer and are compatible with the requirements of subsequent packaging operations. We achieved low contact resistance between TSVs and frontside Ti/Cu and Al metallization, while demonstrating high isolation resistance and high TSV yield.

1. Introduction

Three-dimensional (3D) through-silicon via (TSV) technology has been converging toward several common implementations. Most of these include small diameter TSVs (≤ 10 μm). However, since even the most advanced etch and deposition processes used to form and fill TSVs can become problematic at high aspect ratios (>10:1) [1], [2], the corresponding TSV depth and, subsequently, the final silicon (Si) substrate thickness must be between 10 and 100 μm. Furthermore, the process technologies which enable higher aspect ratio TSV implementations, such as metal-organic chemical vapor deposition (MOCVD) or atomic layer deposition (ALD), are typically expensive and/or slow, thereby increasing cost and cycle time. Consequently, TSV applications requiring a relatively thick substrate (on the order of 400 μm) and low cost of implementation pose a unique challenge.

Integrating TSVs with a wafer-level chip-scale vacuum package (WLCSVP) represents such a challenge. The package, created at the wafer level by bonding a cap wafer to a device wafer, requires a minimum substrate thickness to maintain mechanical integrity. The minimum thickness corresponds to the thickness at which a mechanical fracture of the silicon layer could occur due to the pressure difference between the interior of the wafer-level vacuum enclosure and the ambient atmosphere. For a typical package pressure of ~0.1 Torr, the minimum Si thickness must be on the order of 400 μm. Consequently, integrating TSVs with WLCSVP can pose significant processing and cost challenges.

While TSV integration is not absolutely required for a wafer-level vacuum package, TSVs formed either in device wafers or cap wafers enable several important improvements. Compared to wire-bond interconnections, TSVs offer the potential for higher density input/output (I/O) configurations as well as a significant reduction of the size and weight of the package [3-6]. The combination of both can translate into more functionality and a smaller overall form factor.

In order to address the requirements of the WLCSVP application, two TSV fabrication processes have been developed and successfully demonstrated. Both approaches involve the formation of TSVs after completion of the device wafer, but before the capping process. In both cases, the TSVs were designed to be integrated outside of the vacuum cavity footprint. The two approaches will be described in detail, and the electrical results of multiple test vehicle lots will be discussed.

2. TSV Fabrication Processes

2.1 Standard Low Aspect Ratio TSV Approach

The first approach involves forming low aspect ratio (AR) TSVs in thick (400 μm) Si from the backside of the IC wafer. This TSV approach is relatively mature [1-4], conceptually straight-forward, and inherently compatible with the WLVP requirements discussed above. As shown in the general schematic in Figure 1, TSVs are etched from the backside of the device wafer, terminating (or landing) on the backside of the lowest level of the frontside metallization. The etching of TSVs is followed by the deposition of the TSV insulator, which is then selectively removed from the TSV landing pad. Next, a Cu layer is deposited. This Cu layer forms the TSV interconnect, and is subsequently patterned to form the first level of the backside redistribution layer. Finally, solder bump pads are formed, completing the backside processing. This approach is compatible with implementations where a limited number of interconnects are required between the vacuum-packaged devices and other integrated circuit (IC) die.

Figure 1. Low aspect ratio (AR) vias-last schematic.

2.2 Low-Aspect-Ratio Hybrid TSV Approach

The second approach involves forming low AR TSVs in thin (<100 μm) Si. Reducing the Si substrate thickness enables a lower AR to be achieved with smaller diameter TSVs. Consequently, a higher TSV density can be implemented relative to the previously described approach. Since a stand-alone device wafer with a thickness smaller than 100 μm would not survive the vacuum packaging, a novel variation of this approach was conceived and is described below.

First, the device wafer is thinned to a thickness of less than 25 μm, and low AR Cu TSVs are fabricated from the backside of the wafer to land on the frontside metal. Due to the low AR enabled by the use of thin Si, conventional planar IC processing techniques can be utilized. A thick substrate is then created by bonding a Si interposer with TSVs to the thinned device wafer. The structure is illustrated in Figure 2a for the case of ICs fabricated on silicon-on-insulator (SOI) substrates. After bonding, the carrier wafer used to temporarily support the thinned IC wafer is removed. The TSV process is completed by metalizing the TSVs of the interposer. Figure 2b shows a schematic of the completed TSV after Cu metallization.

This hybrid approach eliminates the need for the

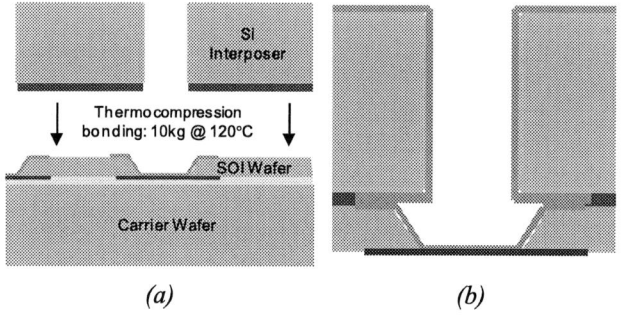

(a)   *(b)*

Figure 2. (a) Si interposer bonding to thinned SOI wafer with Cu contacts to frontside metal, (b) Cu TSVs formed through thick Si interposer.

deposition of dielectric and barrier layers in high aspect ratio vias. In the work described in this paper, we used MOCVD Cu for the TSV metallization. While MOCVD Cu represents a higher cost fabrication step, it could be used simply as a thin sidewall plating base for electrochemical deposition (ECD) of thicker Cu since the contact to the frontside metal has been previously formed. Such an implementation would minimize the required thickness of the MOCVD Cu layer.

3. Test Vehicle Description
3.1 Standard Low AR TSV Approach

Two passive test vehicle lots were designed and completed to exercise the standard low AR vias-last approach described in Section 2.1. The wafer size for both lots was 150 mm, and the bulk Si substrate thickness was 400 μm. To obtain an AR of 1:1, a nominal TSV diameter of 400 μm was selected.

The first test vehicle lot (Test Lot 1) utilized standard bulk Si test wafers with a single frontside metal layer. Two frontside metals were tested: (1) Ti/Cu and (2) Ti/Al-0.5%Cu.

For simplicity, Ti/Al-0.5%Cu metallization will be referred to as Al in this paper. The Ti/Cu frontside metal was selected as a control, since it has been well characterized for similar 3D TSV interposer applications [3]. The Al split was included to replicate standard foundry metallization. For this first lot, the TSV passivation layer was an organic dielectric, and the TSV metal was MOCVD Cu. Figure 3 shows representative scanning electron microscope (SEM) cross-section images of the TSVs formed in the first test vehicle lot.

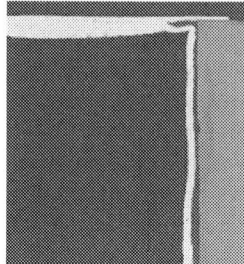

Figure 3. TSV profile for the first low AR test vehicle lot.

For the second test vehicle lot (Test Lot 2), the bulk Si wafers were replaced with SOI substrates designed to replicate a specific IC design and technology. The surrogate IC wafer was a 150 mm SOI wafer with a 2 μm thick device Si layer and a 1 μm thick buried oxide (BOX) layer. The alternating SiO_2/Si layers of the SOI substrate make TSV contact formation and fill more challenging. In the second test lot, the organic TSV passivation layer was replaced with an inorganic PECVD SiO_2 layer to provide a greater thermal budget for downstream processing. PECVD oxide is also a low cost solution and the deposition tools are standard in most microfabrication facilities.

The TSV metallization process used in the second lot was Cu ECD on a standard sputtered Ti/Cu seed layer. Electroplated Cu is a more common and cheaper method of Cu metallization compared to MOCVD Cu. Also, this low cost ECD Cu process facilitates thicker metal layers, so the Cu thickness could be increased to 10 μm. Figure 4 shows two representative cross-sectional SEM images of a sample from the second test vehicle lot.

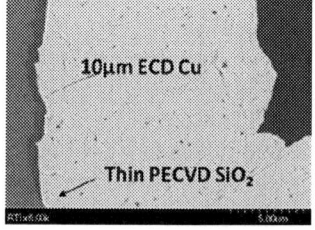

Figure 4. TSV profile for the second low AR test vehicle lot.

Both test vehicle lots incorporated test structures to evaluate TSV electrical performance. The wafer and die layout of the first test vehicle is shown in Figure 5. The test die were comprised of standard test structures that we had

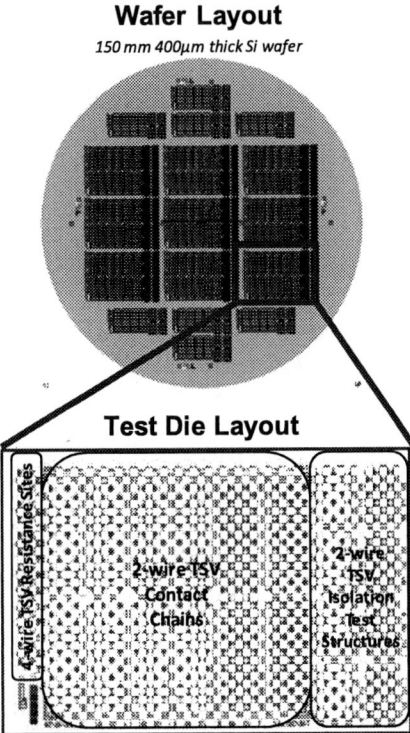

Figure 5. Wafer/die layout for the first test vehicle lot.

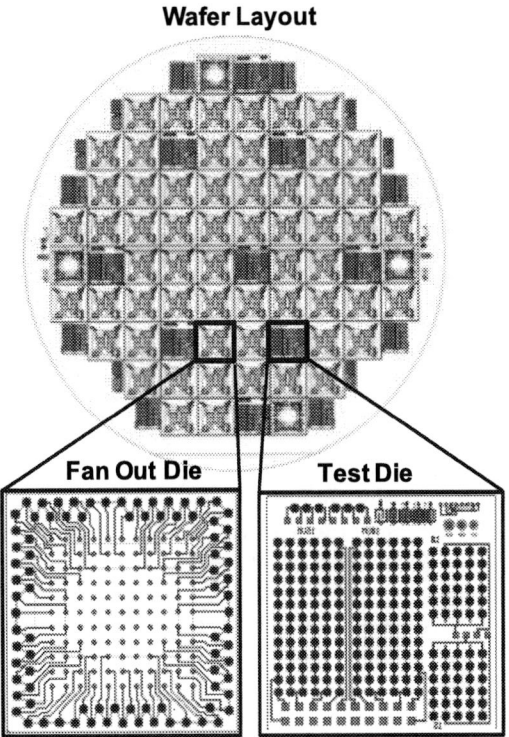

Figure 6. Wafer/die layout for the second test vehicle lot.

used to develop and characterize similar high AR vias-last technologies [3]. Each test die contained 4-wire TSV contact resistance sites, 2-wire contact chains, and 2-wire TSV isolation structures. These structures enable measurement of TSV contact resistance, TSV yield, and TSV isolation resistance, respectively.

The second test vehicle layout, shown in Figure 6, contained test die similar to the ones used in the first lot, as well as fan-out die designed to resemble a typical application. The fan-out die enabled further testing of the design rules developed in the first lot. The TSVs in the fan-out design were staggered to ensure that these large aspect ratio TSVs did not create a susceptible cleave path on the die edge.

3.2 Low AR Hybrid TSV Approach

A TSV test vehicle lot was designed to demonstrate the hybrid approach illustrated schematically in Figure 2. Again, we used passive 150 mm SOI wafers with a 2 μm device Si layer and a 1 μm BOX layer as surrogates for device wafers. Al and Ti/Cu frontside metal layers were patterned to emulate the foundry frontside metals, with the Ti/Cu frontside metal used as a control. The demonstration lot was completed in 2 phases. In the first phase, the bulk Si of the SOI wafer was thinned to approximately 25 μm. Next, the TSV etch mask for the deep reactive ion etch (DRIE) process was created by patterning a photoresist layer on the backside of the thinned Si. The TSV formation consisted of 4 DRIE process steps: (1) DRIE of 25 μm thick Si, (2) DRIE of 1 μm thick BOX, (3) DRIE of 2 μm thick device Si, and (4) DRIE of 1 μm thick frontside SiO_2. These DRIE steps created low AR TSVs which landed on the backside of the frontside metal. Figure 7 shows a representative result of the TSV formation process.

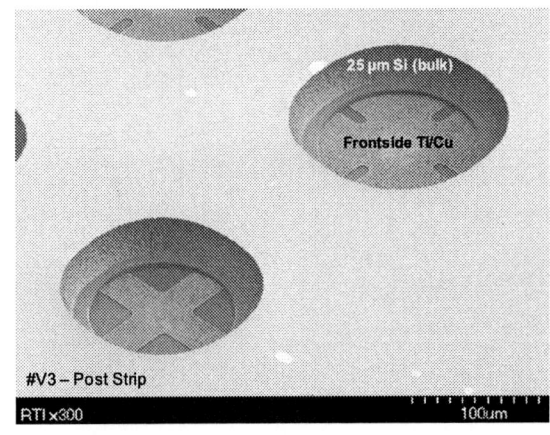

Figure 7. Nominal 90 μm (bottom left) and 100 μm (top right) TSVs landing on frontside metallization.

Following the TSV etch, low AR electrical connections were made to the frontside metal layers by Cu metallization, as shown in Figure 8a. The nominal TSV diameter was 100 μm, but could easily be scaled to < 50 μm [7], [8].

In the next phase of the process flow, a 400 μm thick Si interposer with TSVs was fabricated and bonded to the thinned die. The TSV diameters were nominally 100 μm (easily scalable to < 50 μm) and the TSV insulator was 0.5 μm thermal SiO_2. Microchem SU8 was used as the bond adhesive and spray-coated on the Si interposer. Aligned die-level thermo-compression bonding was performed as shown in Figure 2a. Cu connections were formed, as shown schematically in Figure 2b, and routed on the backside of the

978-1-4799-2408-0/14 $31.00 © 2014 IEEE

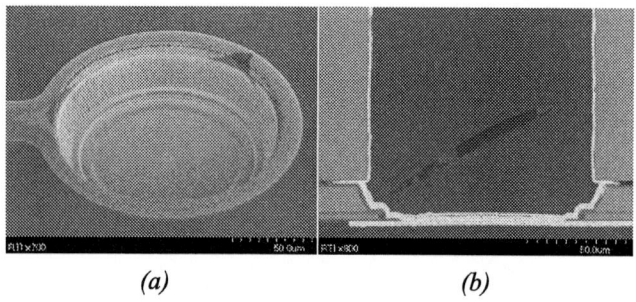

(a)   *(b)*

Figure 8. (a) Backside TSV contact to the frontside metal of a thinned SOI wafer, (b) Cu TSV in Si interposer bonded to a thinned SOI wafer.

interposer. A cross-sectional SEM of the completed TSV test vehicle is shown in Figure 8b.

The wafer layout and test die configuration for the hybrid approach were similar to that shown in Figure 5. The test die contained test structures similar to those of the standard low AR test vehicle lots. The test die was smaller due to the smaller TSV diameter and pitch; as a result the wafers could be populated with more die

4. Test Vehicle Results and Discussion

4.1 Standard Low AR TSV Process

The first test vehicle lot was completed and tested. Figure 9 illustrates the nomenclature for TSV-related resistances used in the electrical characterization of the test vehicle. Using the individual test structures shown in Figure 5, 4-wire

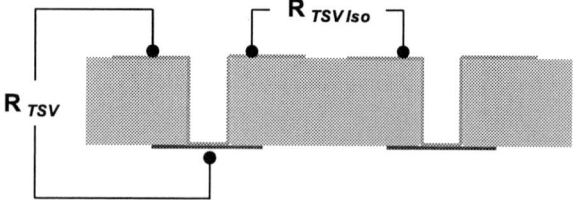

Figure 9. Nomenclature used in the electrical characterization of the test vehicle.

TSV contact resistance structures were measured for 9 die on 2 wafers. Of the 2 wafers, one incorporated the Ti/Cu frontside metal and the other the Al frontside metal. The 4-wire contact resistance, R_{TSV}, measurements are shown in Figure 10. All measured R_{TSV} values were below 5 mΩ. There were minor differences between TSVs landing on Al and Ti/Cu pads. This result demonstrates that low contact resistance TSV interconnects can be made to Al frontside metals in spite of the propensity of Al to oxidize in process plasmas.

Next, the TSV isolation structures were tested. For all sites (9 sites/wafer) of both frontside metal splits (2 wafers), $R_{TSV\ Iso}$ exceeded 1 GΩ, demonstrating that the process of removing the insulator from the landing pad did not affect the quality of the sidewall passivation. Next, the 2-wire TSV chains were also tested on the same 2 wafers. Six TSV chains consisting of 48 TSVs in series were tested for 5 die per wafer, for a total of 60 contact chain structures. Of these 60

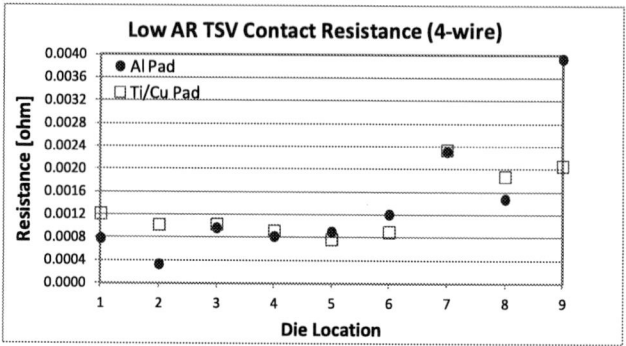

Figure 10. TSV contact resistance for the first low AR test lot.

structures, 59 showed resistance values within 5.5 Ω of the theoretical resistance values based on the nominal geometry of the interconnects. These 59 chains were considered operational, translating into a 98.3% chain yield. Using the chain length and a statistical model assuming a random distribution of defects, we calculated the probability of a TSV being functional to be equal to 99.96%.

Figure 11 compares values of R_{TSV} for TSVs with Al landing pads for both test lots. The average R_{TSV} for Test Lot 2 was 0.6 mΩ compared to 1.4 mΩ for Test Lot 1. This difference is due to the use of a thicker Cu layer for the TSV metallization in Test Lot 2. However, $R_{TSV\ Iso}$ was approximately 6 MΩ for Test Lot 2, which was substantially lower than the isolation resistances measured on the first test lot. As can be seen from cross-sectional SEM images of the TSV profiles for Test Lot 1 and Test Lot 2 (Figures 3 and 4), the lower isolation resistance for Test Lot 2 was likely a result of a lower than desired PECVD oxide thickness near the bottom of the TSV. This issue can be addressed by increasing the oxide thickness, or by utilizing a SiO$_2$ deposition method which results in higher comformality. For example, a TEOS sub-atmospheric ozone chemical vapor deposition process has significantly better conformality than that of a standard PECVD process. Of the 18 TSV contact chains tested, all were operational, translating into a TSV yield of 100%.

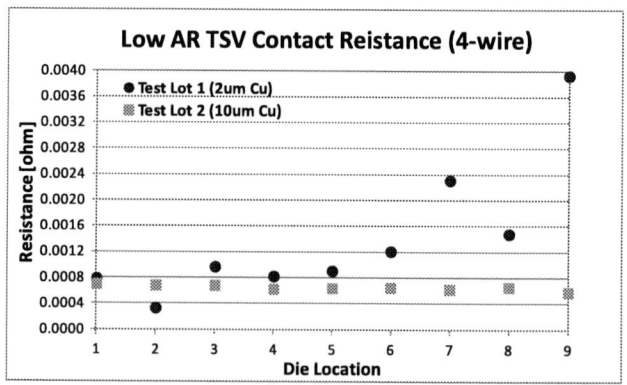

Figure 11. TSV contact resistance comparison for standard low AR test lots.

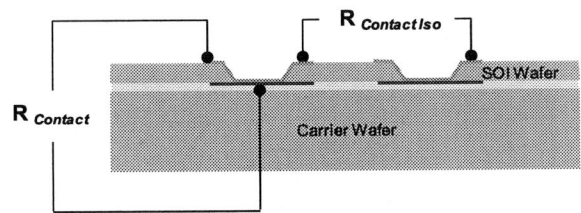

Figure 12. TSV resistance nomenclature for the hybrid TSV contacts to the frontside metal.

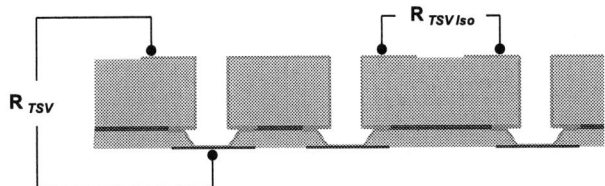

Figure 13. TSV resistance nomenclature for the completed hybrid TSV test vehicle.

4.2 Low AR Hybrid TSV Approach

Electrical characterization of the hybrid TSV test vehicle lot was performed after formation of the low AR Cu connections to the frontside metal, but before Si interposer bonding/interconnection. The nomenclatures for the resistances measured at this point are shown in Figure 12. The contact resistance of the Cu interconnections to the backside of the frontside metal, $R_{Contact}$, was 10 mΩ for both Ti/Cu trials and one of the Al trials. Isolation resistance was greater than 1 GΩ for all trials. For trials with Si thickness below 20 µm, 100% yield was achieved on contact chains up

to 504 TSVs in length. Table 1 shows a summary of the electrical test results.

Die were selected from the 2 wafers with 100% yield (#M3 and V1), and bonded at the die-level to Si interposers. After deposition and patterning of the TSV metal, the die were electrically tested. The TSV resistance nomenclature for these measurements is shown in Figure 13. Using the 4-wire test structures, TSV resistances were measured and the results summarized in Table 2. As shown in Table 2, low R_{TSV} was achieved for both Ti/Cu and Al front-side metals. R_{TSV} less than 10 mΩ was measured for die with Ti/Cu frontside metal (wafer #V1), and $R_{TSV} \approx 1$ Ω was achieved for die with Al

Trial Description				TSV Contact to Front Metal		
Wafer #	Metal	SOI	Si Thk	$R_{Contact}$ (Ω)	$R_{Contact\,Iso}$ (Ω)	Average Yield
M3	Al	No	16.9	0.609	2.7E+11	100%
V1	Ti/Cu	Yes	13.0	0.001	7.7E+11	100%
V3	Ti/Cu	Yes	26.5	0.001	5.2E+12	26%
V7	Al	Yes	25.8	2.908	4.1E+11	44%
V8	Al	Yes	26.7	0.0064*	4.0E+11	24%

* Measurement from 1 die due to low yield

Table 1. Electrical characterization results for Cu contacts to frontside metal in thin SOI wafers.

Device Wafer							R_{TSV}	$R_{TSV\,Iso}$ (@ 3.3 V)	Yield	
Stack	Wafer - Die	Type	Metal	Si Thick	Dicing Config	Average $R_{Contact}$	Ohm	Ohm	Average - 84 Link Chains	Longest Chain
2	V1 - [U]	SOI	Ti/Cu	13.0	Centered	0.0007	0.010	1.2E+09	50.0%	504
3	V1 - [Q]	SOI	Ti/Cu	13.0	Centered	0.0007	0.015	7.1E+10	44.0%	504
4	V1 - [D]	SOI	Ti/Cu	13.0	Centered	0.0007	0.012	1.1E+08	0.0%	22
12	V1 - [H]	SOI	Ti/Cu	13.0	Centered	0.0007	0.011	6.8E+09	33.3%	504
13	V1 - [R]	SOI	Ti/Cu	13.0	Centered	0.0007	0.010	4.0E+04	70.0%	378
14	V1 - [Y]	SOI	Ti/Cu	13.0	Centered	0.0007	0.103	3.1E+09	20.0%	210
6	M3 - [C]	Std	Al	16.9	Std	0.609	0.805	5.4E+07	93.3%	504
8	M3 - [E]	Std	Al	16.9	Std	0.609	1.032	7.2E+09	95.0%	504
9	M3 - [F]	Std	Al	16.9	Std	0.609	1.116	< 100	0.0%	42
16	M3 - [J]	Std	Al	16.9	Std	0.609	1.065	4.6E+09	80.0%	504

Table 2. Electrical characterization of bonded low AR TSV test vehicle.

978-1-4799-2408-0/14 $31.00 © 2014 IEEE

frontside metal (wafer #M3). The higher TSV contact resistance for Al is largely due to the higher $R_{Contact}$ (0.6 Ω) of wafer #M3 (Table 1). Based on the lower contact resistance observed on wafer #V8, it is likely that TSV contact resistances for Al frontside metallization can be similar to those of the Ti/Cu trials (< 20 mΩ).

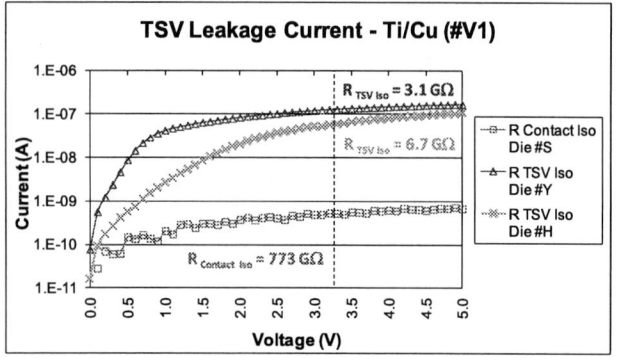

Figure 14. TSV leakage current for low AR hybrid test vehicle with frontside Ti/Cu metallization.

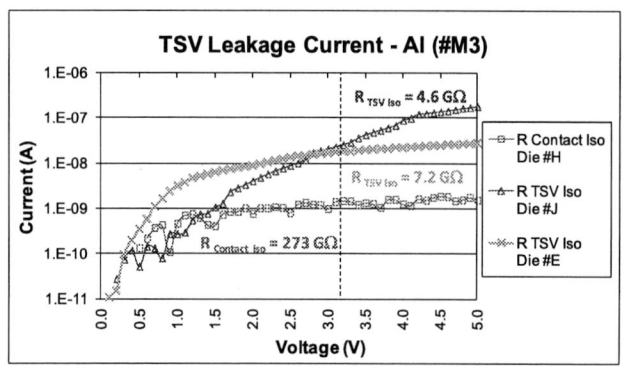

Figure 15. TSV leakage current for low AR hybrid test vehicle with frontside Al metallization.

TSV leakage current values are plotted as a function of voltage for Ti/Cu and Al metallization trials in Figures 14 and 15, respectively. For both frontside metals, leakage current increased after bonding and TSV metallization. While higher, the TSV leakage current is acceptable, as TSV isolation resistance at 3.3V was still above 3 GΩ. TSV isolation resistance for all trials is shown in Table 2 (labeled "$R_{TSV\,Iso}$"). While the majority of the isolation resistances exceeded 1 GΩ, cross-sectional SEM analysis was performed to determine the source of the higher leakage current (relative to $R_{Contact\,Iso}$ values). Figure 16 shows the result of the cross-sectional SEM analysis. In forming the TSV contact from the backside of the thinned Si, the etch profile was slightly reentrant. As a result, a high conformality parylene gap fill process was used to attempt to smooth the DRIE etch profile. However, the subsequent PECVD oxide deposition bread-loafed slightly at the top, leaving an area of poor oxide coverage just below the interface. The PECVD oxide coverage in this area was marginal and did not protect the

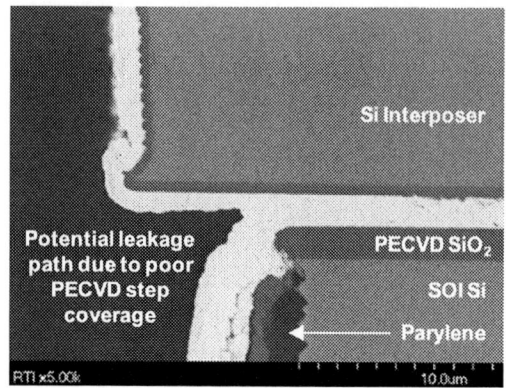

Figure 16. Cross-sectional SEM of Si interposer/SOI interface for the low AR bonded test vehicle.

parylene from the post-bond O_2 plasma surface treatments. Consequently, the parylene was slightly eroded, creating a leakage path to the SOI Si. This issue can be easily addressed by improving the DRIE etch profile and/or improving the conformality of the insulation layer.

Finally, 2-wire TSV contact chains were tested at multiple sites on each bonded die. Comparing measured resistances to an appropriate model, the TSV chain yield was calculated [3]. The longest TSV chain was composed of 504 TSVs, with taps at 84 TSV intervals. The average yield of the 84-link TSV chains is shown in Table 2 for each bonded die, along with the longest working chain. Based on the same TSV yield calculation performed for the standard low AR test lots, greater than 99% TSV yield was achieved for 6 of the 10 trials. For 2 trials (#6 and #8), TSV yield exceeding 99.9% was achieved

Based on the results shown in Table 2, yields are generally higher for the Al metallization split (wafer #M3). This was not attributed to the frontside metal, but to the dicing configuration of the SOI wafer, indicated in the column labeled "Dicing Config" in Table 2. As shown in Figure 17a, wafer #V1 was diced such that the individual TSV test structures were at the center of the die. Alternatively, wafer #M3 was diced in the standard (Std) configuration shown in Figure 17b. It was observed during the development of the bonding process that bonding results were very pattern dependent. Based on trials with glass substrates, higher and more uniform pattern density generally resulted in better SU8 bonding results. This likely explains the strong correlation between the dicing configuration and TSV yield.

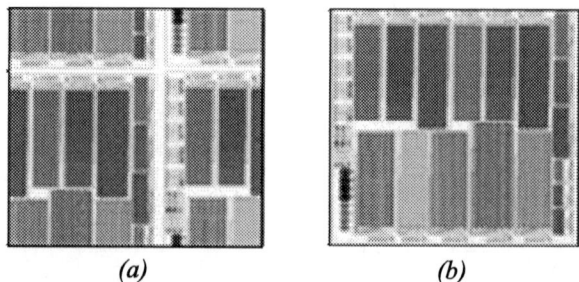

Figure 17. (a) Centered die configuration after dicing (b) Standard (Std) die configuration after dicing

978-1-4799-2408-0/14 $31.00 © 2014 IEEE   1646

Successful demonstration of this novel hybrid approach for TSV integration has some notable features beyond those already described. Since the high AR interposer formation is de-coupled from the process of forming contacts to the ROIC foundry metal, the interposer TSV fabrication can use high-throughput, low cost insulation/barrier processing techniques, like furnace oxidation/CVD. Also, since the only limitation on the interposer is the planarity of the bonding surface, SiO_2 (or most any other material) wafers or die can be used. As noted above, the bonding for this test lot was performed at the die-level. The process is scalable to wafer-level bonding, which is necessary for the WLCSVP application. However, die-level bonding has some inherent advantages which could be useful for other applications. Namely, die-level bonding enables bonding of known-good die (KGD). This can substantially reduce the cost and improve the process yield when working with low-yield wafers, such as those from leading edge technologies and applications. Also, die-level bonding can enable rapid turn-around demonstrations for emerging and niche applications.

Conclusions

Two low cost approaches to TSV integration with ROIC wafers intended for WLCSVP applications have been successfully demonstrated. The first low cost TSV approach involves fabrication of standard low aspect ratio TSVs from the backside of a ROIC wafer. The ROIC wafer thickness and TSV diameter were both 400 µm in order to yield an AR of 1:1. Two TSV test vehicle lots were successfully completed using this process approach. Optimum process conditions were identified and resulted in a TSV resistance of less than 5 mΩ and isolation resistance greater than 500 MΩ. Via yield exceeding 98% was achieved on the first lot, while 100% via yield was demonstrated on the second lot. Additionally, the second lot showed the feasibility of using standard planar IC processing techniques for TSV integration.

The second approach involved thinning the bulk Si of the ROIC to less than 25 µm, and forming low AR Cu TSV contacts to the backside of the foundry metal. A Si interposer was then bonded to the thinned ROIC for mechanical support, and Cu TSV interconnections were formed. Cu TSV interconnects were successfully integrated with TSV resistance (R_{TSV}) less than 10 mΩ for the Ti/Cu frontside metal split and approximately 1Ω for the Al frontside metal split. For both frontside metals, TSV isolation resistance greater than 500 MΩ was achieved. Finally, greater than 99% TSV yield was achieved for 6 of the 10 trials, and greater than 99.9% TSV yield was achieved on 2 parts.

In addition to compatibility with the WLCSVP application, both low AR process approaches offer lower cost implementations relative to other vias-last approaches. The low aspect ratio approaches enable the use of conventional planar wafer-level processing, thereby offering the benefits of traditional IC processing – high throughput, low cost, and high yields. Further, elimination of any frontside ROIC processing reduces cost, and saves precious ROIC design/routing space. Finally, the hybrid approach for TSV integration offers unique flexibility by de-coupling the interposer TSV fabrication from the process of forming TSV connections to the frontside ROIC foundry metal.

Acknowledgments

This work was supported in part by DARPA. The views expressed are those of the author and do not reflect the official policy or position of the Department of Defense or the U.S. Government.

The authors wish to thank staff members of the RTI microfabrication lab for support of wafer processing, and the RTI analytical lab for sample preparation and characterization. We would also like to acknowledge Entrepix for their support for the temporary bonding and thinning steps.

References

1. D. Malta, C. Gregory, D. Temple, C. Wang, T. Richardson, and Y. Zhang, "Optimization of Chemistry and Process Parameters for Void-Free Copper Electroplating of High Aspect Ratio Through-Silicon Vias for 3D Integration", *Proc. of 59th Electronic Components and Technology Conference*, San Diego, CA, May 2009, pp. 1301-1306.
2. R. Beica, C. Sharbono, T. Ritzdorf, "Through Silicon Via Copper Electrodeposition for 3D Integration," *Proc. of 58th Electronic Components and Technology Conference*, Lake Buena Vista, FL, May 2008, pp.577-583.
3. E. Vick, E, S. Goodwin, D. Temple, "Electrical Demonstration of TSV Interconnects and Multilevel Metallization for 3D Si Interposer Applications," *Proc. of IMAPS 43rd International Symposium on Microelectronics*, Research Triangle Park, NC, Oct 31-Nov 4, 2010
4. E. Vick, S. Goodwin, G. Cunnigham, D. S. Temple, "Vias-last process technology for thick 2.5D Si interposers," *3D Systems Integration Conference (3DIC), 2011 IEEE International* , vol., no., pp.1,4, Jan. 31 2012-Feb.2 2012 doi: 10.1109/3DIC.2012.6262990
5. J. M. Lannon, J. A. Hilton, A. Huffman, M. R. Lueck, E. P. Vick, S. H. Goodwin, G. B. Cunningham, D. M. Malta, C. W. Gregory, & D. S. Temple (2012), "Process Integration and Testing of TSV Si Interposers for 3D Integration Applications," *Proc. 62th IEEE Electronic Component and Technology Conference, ECTC 2012,* pp. 268–273. doi: 10.1109/ECTC.2012.6248839
6. H. Yoshikawa et al., "Chip-scale camera module (CSSM) using through-silicon via (TSV)", *Digest of 2009 IEEE International Solid State Circuits Conference*, p. 476, 2009.
7. D. Temple, C. A. Bower, D. Malta, J. E. Robinson, P. R. Coffman, M. R. Skokan and T. B. Welch, "High Density 3-D Integration Technology for Massively Parallel Signal Processing in Advanced Infrared Focal Plane Array Sensors", *IEDM Digest*, p. 1-4, Dec 2006.
8. D. Temple, C.A. Bower, D. Malta, J.E. Robinson, P.R. Coffman, M.R. Skokan, and T.B. Welch, "3-D Integration Technology for High Performance Detector Arrays", *Mater. Res. Soc. Symp. Proc. 970*, p. 115-124, 2006.

X-Ray Micro-Beam Diffraction Measurement of the Effect of Thermal Cycling on Stress in Cu TSV: A Comparative Study

Chukwudi Okoro[a], Lyle E. Levine[b], Ruqing Xu[c], Klaus Hummler[d], Yaw Obeng[a]

[a]Semiconductor and Dimensional Metrology Division,
[b]Material Science and Engineering Division,
National Institute of Standards and Technology (NIST), Gaithersburg, MD 20899, USA.
Phone: 301-975-2040, Email: chukwudi.okoro@nist.gov
[c]Advanced Photon Source, Argonne National Laboratory, Argonne, Illinois 60439-4800, USA.
[d]SEMATECH, 257 Fuller Road, Albany, NY 12203, USA

Abstract

Microelectronic devices are subjected to constantly varying temperature conditions during their operational lifetime, which can lead to their failure. In this study, we examined the impact of thermal cycling on the evolution of stresses in Cu TSVs using synchrotron-based X-ray micro-diffraction. Two test conditions were analyzed: as-received and 1000 cycled samples. The principal and shear stresses in the 1000 cycled sample were five times greater than in the as-received sample. This was attributed to the increased strain hardening upon thermal cycling. The variation in stresses with thermal cycling is a clear indication that the impact of Cu TSV proximity on front-end-of-line (FEOL) device performance will fluctuate throughout the lifetime of the 3D stacked dies, and thus should be accounted for during FEOL keep-out-zone design rule development.

INTRODUCTION

Stress-related reliability challenges are known to be one of the main causes of failure in electronic devices. This failure type arises due to the mismatch in the mechanical properties of the constituent materials. One of the causes of stress-related failures is the continuous fluctuation of temperature in microelectronic devices during their in-service usage [1]. This leads to thermal fatigue occurrence, and subsequently to failure.

The emergence of three-dimensional stacked integrated circuits further increases the risk of thermal fatigue failures. This is because the electrical connection between the stacked dies are achieved using mainly Cu through-silicon vias (TSVs), which are fabricated through the active silicon. The large dissimilarity in the material properties of the Cu TSV and the surrounding Si can result in huge stresses.

In order to understand and mitigate thermal fatigue concerns in Cu TSV interconnects, many studies have been reported that have assessed the thermal cycling effect on their reliability performance [2], [3], [4], [5]. These reported studies were focused on understanding how thermal cycling impacts the electrical characteristics of Cu TSVs, as well as the use of focused ion beam (FIB) and scanning electron microscopy (SEM) failure analysis tools to identify damage growth and propagation. However, no reported study has been able to experimentally relate the underlying root cause of the changes in the Cu TSV, being the buildup of stresses, with the witnessed number of thermal cycles.

Thus, it is critical to quantify the stresses in the Cu TSV. Last year at ECTC 2013, we reported a novel method, using synchrotron-based X-ray micro-diffraction to determine the depth-dependent, full stress tensor in Cu TSVs [7]. Using this developed technique, we determined the effect of thermal cycling on the evolution of stresses in Cu TSV in this present study.

SAMPLE AND EXPERIMENT DESCRIPTION

Sample

For this study, a SEMATECH built two-level stacked die bonded together with Benzocyclobutene (BCB) was used [6]. The top die contained a daisy chain of 60 TSVs, with a pitch of 16 μm. Each TSV has a diameter and depth of about 5 μm and 50 μm, respectively. During the processing stage of the Cu TSVs, the wafers were subjected to a maximum annealing treatment of 400 °C for about 2 min. The samples were stored at room temperature for more than nine months before being used for this experiment. For this work, two dies from the same Cu TSV wafer were studied; one without any thermal cycling, called the as-received sample, while the other sample was subjected to 1000 thermal cycles. Each thermal cycle was completed in about 6 min, and involved reversal heating and cooling between minimum and maximum temperatures of 30 ºC and 150 ºC, respectively. Only one Cu TSV was measured for each test condition. Figure 1 is a cross-sectional image of the stacked die used in this study; measurements were performed on the top die that contained the TSVs.

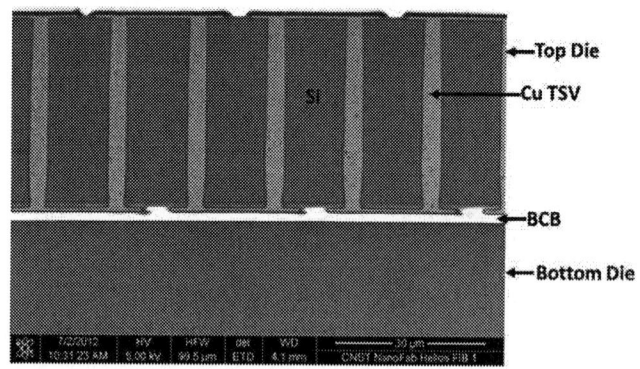

Figure 1: FIB-SEM cross-sectional image of the stacked die used in this study

978-1-4799-2408-0/14 $31.00 © 2014 IEEE

X-Ray Micro-Diffraction Experiment

This synchrotron-based experiment was performed using the X-ray micro-diffraction instrument on sector 34-ID E at the Advanced Photon Source (APS), Argonne National Laboratory. In this work, three area detectors surrounded the intersection of the horizontal X-ray beam with the sample. The central detector was perpendicular to the vertical direction, while the other two detectors were inclined at 45°, each at the opposite sides of the central detector, thus maximizing the angular range of the detected diffracted beams. This arrangement minimizes measurement uncertainties for the extracted full strain and stress tensors from the Cu TSV. Depth resolution along the beam direction was made possible by the use of a platinum wire based depth profiler. A total of 11 measurements were performed along the length of the Cu TSV, at a translation step interval of 5 μm. The beam size was about 0.5 μm by 0.5 μm, with a depth resolution of about 1 μm. Both monochromatic and polychromatic X-rays were used to scan the Cu TSV.

From this X-ray micro-diffraction experiment, we obtain depth and detector dependent data such as the matching reflection indices, peak intensity coordinates and their uncertainties as well as their lattice parameter and uncertainties. This information was used to obtain the full strain and stress tensors in the Cu TSV. Euler angles describing the orientations of the grains in the laboratory frame coordinate system were also determined. A detailed description of the measurement procedure used in this study was reported elsewhere in [7], [8], [9].

Even though all six component stresses were obtained, only the principal stresses are reported in this study.

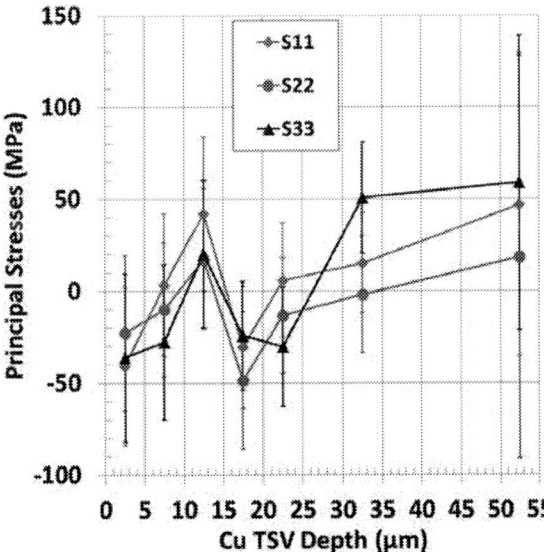

Figure 2: The experimentally-determined principal stresses for the as-received Cu TSV sample with respect to its depth. The one standard deviation uncertainties result from uncertainties in the diffracted X-ray peak positions on the area detectors and uncertainties in the lattice spacing determination.

RESULTS

Figure 2 shows the depth-dependent principal stresses in the as-received sample. The orientations of the principal directions are the orthogonal basis vectors of the Cu unit cell and do not reflect the sample geometry. The principal stresses are all overlapping within the measurement uncertainties, consistent with a hydrostatic stress state (all uncertainties reported in this paper are one standard deviation). The top region of the Cu TSV is compressive, with an average value of about -30 MPa. A maximum hydrostatic stress of about 41 MPa was measured at the bottom of the Cu TSV. The principal stresses fluctuated along the entire Cu TSV depth, having an average absolute value of ≈ 27 MPa.

Figure 3 presents the results for the 1000 cycled sample. Similar to the as-received sample, the principal stresses were overlapping and varied along the length of the Cu TSV. With the exception of the stresses at 17.5 μm depth, the absolute minimum stresses were observed at the two ends of the Cu TSV, having absolute average values of 30 MPa and 17 MPa, at the top and the bottom of the Cu TSV, respectively. A maximum hydrostatic stress of 276 MPa was measured at the center of the Cu TSV (22.5 μm depth). By averaging the absolute hydrostatic stress along the entire Cu TSV, a value of about 123 MPa was obtained for the 1000 cycled Cu TSV. The principal stresses in the 1000 cycled Cu TSV were mainly tensile.

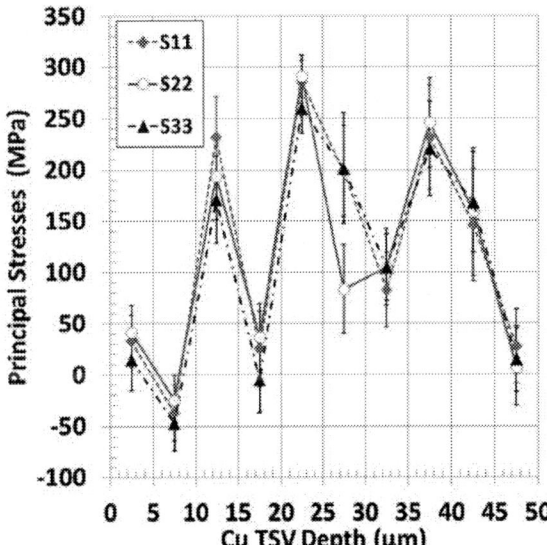

Figure 3: The experimentally-determined principal stresses for the 1000 thermally cycled Cu TSV sample with respect to its depth. The one standard deviation uncertainties result from uncertainties in the diffracted X-ray peak positions on the area detectors and uncertainties in the lattice spacing determination.

DISCUSSION

From Figure 2 and Figure 3, it is observed that, independent of the thermal cycling condition, the principal stresses were approximately equal. Additionally, the minimum stresses were observed to occur at the top region of the Cu TSV. For the 1000 cycled sample (Figure 3), the minimum stresses occur at both ends of the Cu TSV, thus

978-1-4799-2408-0/14 $31.00 © 2014 IEEE

displaying an approximately "bow-like" stress profile. This occurs because the ends of the Cu TSV have less constraint than the inner regions, thereby leading to less stress at the ends. This measured "bow-like" profile of the principal stresses is in agreement with FEM studies reported by the authors in [3].

Both the as-received (Figure 2) and the 1000 cycled (Figure 3) show variation in the values of the principal stresses along the depth of the Cu TSVs. This observed fluctuation is attributed to microstructural inhomogeneity in the Cu TSV, arising from its polycrystalline nature.

The measured principal stresses in the as-received Cu TSV (Figure 2) were found to be very low, having an average hydrostatic stress of $\approx$ 27 MPa along the entire Cu TSV. The low stress values were attributed to stress relaxation. Since the sample was measured more than nine months after fabrication, the stresses in the Cu TSV are expected to relax even at room temperature as a result of self-annealing [10], [11]. Self-annealing which occurs due to microstructural evolution, results in the decrease in stresses with time under room temperature storage conditions [10]. In the author's earlier reported work based on nano-indentation measurements, it was found that self-annealing results in an apparent change in the elastic modulus of Cu TSVs, which was attributed to the variation in the residual stresses in the Cu TSV [11].

On the other hand, for the 1000 cycled sample, the principal stresses (Figure 3) were observed to be high, with a maximum hydrostatic stress of about 276 MPa in the central region and an average hydrostatic stress of about 123 MPa along the entire length of the Cu TSV. This means that stresses in the 1000 cycled Cu TSV are five times greater than that of the as-received Cu TSV (Figure 2). This increase in the stresses after 1000 thermal cycles was attributed to strain hardening during cyclic loading. It is expected that thermal cycling will initially lead to the increase in stress in the Cu TSV due to the increase in dislocation density and their entanglement [10], [13]. This leads to the strain hardening of the structure; however, continued thermal cycling will lead to the growth and the coalescing of defects such as micro-cracks or micro-voids. These defects are known to aid the stress relief in metals, and when they attain a critical size, they will result in a decrease in the stresses in the Cu TSV [10], [13]. This decrease in stress is expected to continuously increase as the cracks and voids in the structure increase with thermal cycling.

As a further evidence, the authors in a prior study performed failure analyses to determine the impact of thermal cycling on crack and void growth on the dies from the same wafer as those used in this present study [3]. It was found that the measured crack length at the Cu TSV sidewall remains statistically unchanged even after 2000 thermal cycles. This is in agreement with our proposition that the measured increase in the stresses in the Cu TSVs after 1000 cycles is due to strain hardening, as no substantial growth in cracks was observed.

This observed variation in the stresses in the Cu TSV with thermal cycling is anticipated to have an impact on the in-service performance of neighboring front-end-of-line (FEOL) devices such as transistors [14]. This is because 3D stacked dies will be subjected to continuously fluctuating cyclic temperatures during their lifetime. This thermal cycling condition will result in stress fluctuations in the Cu TSVs, as such, leading to variation in the performance of integrated FEOLs over their lifetime. While today's design rules are based on the stresses in the TSV in the as-received state, there is an eminent need to account for the stress evolution in Cu TSVs over their projected lifetime.

CONCLUSIONS

In this study, a synchrotron-based X-ray measurement was used to study the impact of thermal cycling on the stress evolution in Cu TSVs. The full stress tensor in Cu TSVs was determined with respect to position along its length. Two conditions were studied; as-received (no thermal cycling) and 1000 cycled Cu TSVs.

For both test conditions, the principal stresses along the depth of the Cu TSVs were equal, within the experimental uncertainties, demonstrating that there was a large hydrostatic stress component. The minimum stresses were found to occur at the top region of the Cu TSV, due to the limited constraint in this region. For the 1000 cycled sample, the profile of the principal stresses was found to exhibit an approximately "bow-like" stress profile, although with substantial fluctuations.

The results show that the 1000 cycled sample had an average hydrostatic stress of about 123 MPa, which is approximately five times greater than the as-received sample. This large disparity in the stresses is attributed to strain hardening, caused by dislocation multiplication and entanglement. The measured low principal stress (27 MPa) in the as-received sample was attributed to self-annealing due to its storage at room temperature for nine months.

ACKNOWLEDGMENTS

The XOR/UNI facilities on Sector 34 at the APS are supported by the U.S Department of Energy (DOE), Office of Science, Office of Basic Energy Sciences, under Contract No. DE-AC02-06CH11357.

REFERENCES

[1] Min Pei, Ru Han, Daeil Kwon, Alan Lucero, Vasu Vasudevan, Robert Kwasnick, Praveen S Polasam, "Define Electrical Packing Temperature Cycling Requirement with Field Measured User Behavior Data," 63th Proc., Electronics Components and Technology Conference (ECTC), May 2013, Las Vegas, NV, pp. 159 – 165.

[2] C. Okoro et al., "Accelerated Stress Test Assessment of Through-Silicon Via using RF Signals," IEEE Transaction on Electron Devices. Vol. 60, No.6, 2013, pp 2015-2021

[3] C. Okoro, F. Golshany, J. W. Lau, K. Hummler, Y. S. Obeng, "A Detailed Failure Analysis Examination of the Effect of Thermal Cycling on Cu TSV Reliability". IEEE Transactions on Electron Devices, Vol. 61, No. 1, 2014, pp. 15 - 22.

[4] M. G. Farooq et al., "3D Copper TSV Integration, Testing and Reliability," International Electron Devices Meeting (IEDM), Dec. 2011, Washington DC, USA, pp. 7.1.1 – 7.1.4

[5] Xi Liu, Qiao Chen, Venkatesh Sundaram, Rao R. Tummala, Suresh K. Sitaraman, "Failure Analysis of Through-Silicon Vias in Free-Standing Wafer under Thermal-Shock Test," Microelectronics Reliability, Vol. 53, 2013, pp. 70 -78.

[6] Certain commercial equipment, instruments, or materials are identified in this paper to specify experimental or theoretical procedures. Such identification does not imply recommendation by NIST nor the authors, nor does it imply that the equipment or materials are necessarily the best available for the intended purpose.

[7] C. Okoro, L. E. Levine, J. Z. Tischler, R. Xu, W. Liu, O. Kirillov, K. Hummler, Y. S. Obeng, "X-Ray Micro-Beam Diffraction Determination of Full Stress Tensors in Cu TSVs", 63th Proc., Electronics Components and Technology Conference (ECTC), May 2013, Las Vegas, NV, pp. 648 – 652.

[8] Chukwudi Okoro, Lyle E. Levine, Ruqing Xu, Klaus Hummler, Yaw Obeng, "Non-destructive Measurement of the Residual Stresses in Copper Through-Silicon Vias using Synchrotron-Based Micro-beam X-ray Diffraction," IEEE Transactions on Electron Devices. Submitted

[9] Lyle E. Levine, Chukwudi Okoro, Ruqing Xu, "Full Elastic Strain Tensor Measurements from Individual Dislocation Cells in Deformed Copper." In preparation

[10] S. Lagrange, S. H. Brongersma, M. Judelewicz, I. Vorvoort, E. Richard, R. Palmans, K. Maex, "Self-annealing characterization of electroplated copper films," Microelectronic Engineering, Vol. 50, 2000, pp. 449 – 457.

[11] C. Okoro, K. Vanstreels, R. Labie, O. Lühn, B. Vandevelde, B. Verlinden, D. Vandepitte, "Influence of annealing conditions on the mechanical and microstructural behavior of electroplated Cu-TSV," J Micromech Microengineering, Vol. 20, 2010, pp. 045032.

[12] J. Koike, S. Utsunomiya, Y. Shimoyama, K. Maruyama, H. Oikawa, "Thermal Cycling Fatigue and Deformation Mechanism in Aluminum Alloy Thin Films on Silicon," Journal of Materials Research, Vol. 13, No. 11, 1998, pp. 3256 – 3264.

[13] R. Schwaiger, O. Kraft, "High Cycle fatigue of thin Silver Films Investigated by Dynamic Microbeam Deflection," Scripta Materialia, Vol. 41, No. 8, 1999, pp. 823 – 829

[14] A. Mercha et al., "Comprehensive Analysis of the Impact of Single and Arrays of Through Silicon Vias induced Stress on High-k / Metal Gate CMOS Performances," International Electron Devices Meeting (IEDM), Dec. 2010, San Francisco, CA, USA, pp. 2.2.1 -2.2.4.

Adhesive Enabling Technology for Directly Plating Copper onto Glass/Ceramic Substrates

Hailuo Fu, Sara Hunegnaw, Zhiming Liu, Lutz Brandt, Tafadzwa Magaya
Atotech USA Inc.
369 Inverness Parkway Suite 350, Englewood, CO 80112

Abstract

This study showcases that metal oxide adhesion promoters (MOAP) can function as strong adhesive layers between plated Cu and glass/ceramic. In this new approach a 10-200 nm thick metal oxide layer is deposited by a modified sol gel process followed by sintering. This enables reliable electroless and electrolytic metallization of glass or ceramic substrates.

With the new approach, Cu can be plated on a variety of glass types. Substrate roughness appears to have only limited impact. The new approach also can be extended to ceramics such as Al_2O_3 and $BaTiO_3$. Cu film of over 50 µm thickness can be deposited without delamination.

Adhesion of 15 µm thick Cu layers as measured by 90° peel strength tests can achieve well above 5 N/cm. The plated layer stands up well to solder reflow shock (260°C) and HAST without significant loss of adhesion.

Good coverage of the MOAP layer and excellent copper adhesion inside the via holes of patterned substrates have been also demonstrated. There is no indication of blockages of holes with diameters >20 µm by the process.

Introduction

An economical and reliable copper metallization of non-conductive inorganic substrates, such as glass and ceramic, is becoming one of the major targets in packaging technologies [1]. Conventionally, adhesion on these substrates is achieved either by sputtering a thin metallic adhesive (Ti) and copper seed layer [2], or by mechanical roughening [3] followed by wet chemical plating.

However, sputtering requires a vacuum system which has low throughput and high capital investment cost. Moreover, it is difficult to obtain a noble metal layer with strong adhesion directly on glass/ceramic. Mechanical anchoring requires strong roughening of the substrate surface which negatively impacts the functionality of the metallized surface. Also the roughening process is difficult to control due to the variable glass compositions. Moreover, the improvements in adhesion that come from such roughening treatments on glass substrate are often insufficient.

In contrast to sputtering, electroless plating uses a simple chemical reaction and is known to metallize insulating substrates [4]. This method is suited for mass production processes and is well established for fabrication of printed circuit boards.

We have developed a new electroless plating process in which a metal layer with strong adhesion can be obtained on various insulating substrates. A thin metal oxide layer between the substrate and the copper metal layer was used as a novel adhesive. It creates the required adhesion for the subsequent electroless plated copper, which subsequently can be plated electrolytically either by a conformal process or filled by a through-hole filling process. The adhesion is thought to be based both on nano-scale mechanical anchoring and chemical interaction.

MOAP Coating and Copper Plating Procedure

A soluble metal salt was used as the metal oxide precursor. It was dissolved in a solvent and a sol-gel stabilizer was added to the solution. The mixture was stirred vigorously and the obtained sol-gel was clear, uniform and stable for at least several months.

Glass and ceramic substrates from different manufacturers were used as to demonstrate the versatility of the approach. Properties of the "as-received" substrates are shown in Table I. Before coating, substrates were cleaned by an alkaline cleaner and ultrasonic treatment (optional). Uniform MOAP deposition was accomplished by dip coating, followed by baking to form the metal oxide film.

The MOAP coated substrates were then plated in the sequence as follows:
(1) Electroless Copper (Atotech) with thickness of 0.3-0.5µm
(2) Annealing
(3) Electrolytic Copper (Atotech) with variable thickness
(4) Annealing

The annealing process after copper plating is standard in the IC Substrate industry. It releases the stress in the plated deposition by recrystallizing the copper structure.

Table I. Properties of "as-received" substrates

Substrate	CTE (ppm/°C)	Tg °C	Roughness (nm)	Manufacturer
Glass - Type 1	--	low	200	A
Glass - Type 2	3-4	high	<10	B
Glass - Type 3	3-4	high	<10	C
Glass - Type 4	3-4	high	10-1000	C
Ceramic - Al_2O_3	7-8	--	460	E
Ceramic - $BaTiO_3$	6-12*	--	160	F

*CTE value depends on temperature.

Results

1. MOAP Characterization and Control

Being able to adjust the roughness and thickness of MOAP coating is crucial in order to meet requirements such as roughness, adhesion, or translucency for different applications. The experiment was done on glass substrate with surface roughness less than 10nm.

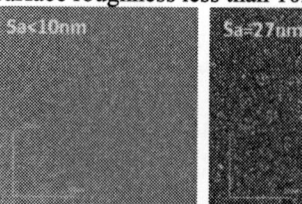

Figure 1: MOAP roughness control

With the concentration of the metal oxide precursor being constant, variation of post-treatment allowed the control of the resulting average surface roughness (Sa) from transparent coatings (Sa<10nm) to translucent coatings (Sa>10nm) as shown in Figure 1.

Figure 2 shows control of the MOAP thickness by adjusting the dip coating parameters, including concentration of the process solution, pull-out speed, and number of applied coatings. By using a low concentration, lower pull-out speed and few coatings, thickness of the MOAP is minimized.

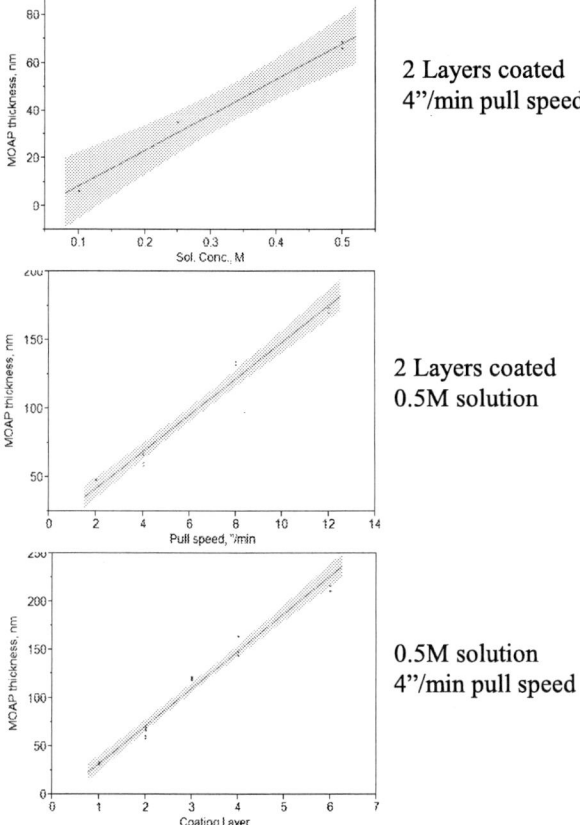

2 Layers coated
4"/min pull speed

2 Layers coated
0.5M solution

0.5M solution
4"/min pull speed

Figure 2: MOAP thickness as function of dip process parameters.

2. Peel Strength dependence on MOAP layer thickness

The MOAP film can be applied in a wide thickness range. In general, there is a reduction of peel strength with lower MOAP thickness as shown in Figure 3. However, at 30nm of MOAP values are still above 6 N/cm. Thinner MOAP layer of 5 – 20 nm can still yields around 2-5 N/cm of peel strength, which appears to be adequate to withstand subsequent processing. Thin adhesive layers are especially attractive in ultra fine-line patterning. While the MOAP layer itself is non-conductive, it can be fully removed in an acidic etch step.

If needed, the thickness of the adhesive layer can be increased to up to 200 nm for exceptionally good peel strength in the range of 10 - 13 N/cm on very smooth glass substrate of around 2nm roughness. This peel strength is maintained even after 5 cycles of reflow stress test at 260°C as discussed below.

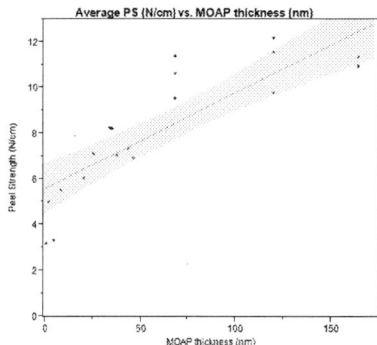

Figure 3: Peel strength vs. MOAP layer thickness

3. Good adhesion independent of glass roughness

Glass substrates of widely varying roughness were tested to assess the effect of surface roughness on adhesion with and without the MOAP surface treatment as shown in Figure 4.

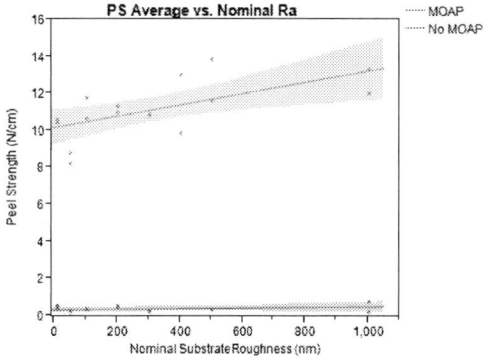

Figure 4: Peel strength vs. surface roughness for MOAP treated and untreated glass samples.

Samples with a surface roughness (Sa) ranging from less than 10 nm to up to 1000 nm showed uniform coating and copper deposition without any blistering. The conformal adhesive layer did not add significantly to the roughness of the substrate. An average peel strength of 11 N/cm was obtained on samples treated with MOAP. Samples that were not treated with MOAP showed low to no adhesion (0.0 -0.5 N/cm). No significant impact was seen with respect to substrate roughness on peel strength.

No blisters or delamination occurred during processing of the highly roughened glass substrates. This indicates the process is amenable to a variety of substrate morphologies without compromise to performance. Feasibility has also been demonstrated on ceramic materials which tend to be much rougher than glass.

4. Good adhesion achieved on different glass types

The MOAP process was shown to yield good peel strength results for most types of glass with different CTE, roughness index and composition. Amorphous glass with high glass transition temperature provides the most flexible working profile for the MOAP process. Samples with low glass transition temperatures limit the process and achieve comparatively moderate peel strengths as shown in Figure 5. Uniform plating free of delamination was achievable on all glass samples thus far. Results for several glass samples are summarized in Table II.

Table II. Adhesion results on various supplier glass types

Glass Supplier	Special Property	Roughness (nm)	MOAP (nm)	Acid Cu (µm)	Peel Strength (N/cm)
1		3	95	15	9.2
2		3	90	15	8.7
2	Thick Electrolytic Cu	3	92	40	14.5
3		3	160	15	5.5
4	Low glass Tg °C	188	90	15	4.3
4		2	120	15	6.5
5	High glass CTE	10	160	15	10.5
5	High glass Roughness	400	160	15	14.0
5	Low glass CTE	2	160	30	8.0
5	Low glass CTE	2	70	15	11.4

Table III. Peel strength vs. Cu thickness.

Copper Overburden µm	Initial Peel Strength N/cm	HAST Peel Strength N/cm
15	6.3	6.3
30	8.0	8.5
40	14.0	14.5

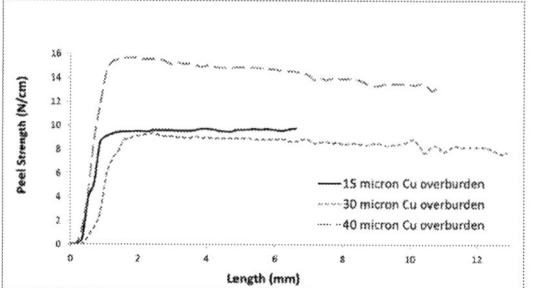

Figure 7: Peel profiles after HAST vs. copper thickness. Slight drop caused by copper thickness variation

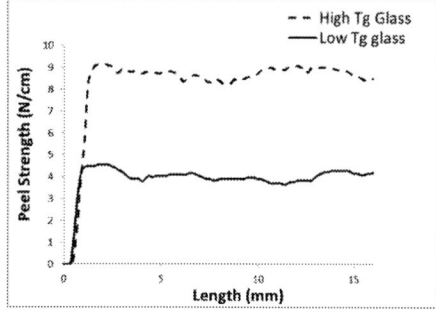

Figure 5: Example of Tg effect on achievable peel strength

5. Good adhesion after accelerated thermal and humidity tests

Samples plated with MOAP test maintain exceptionally high peel strength even after being subjected to two types of environmental stress tests common in the electronics industry: the highly accelerated stress test (HAST, 96hr, 85%RH, 130°C) test and solder reflow test (5x260°C). As shown in Figure 6, almost no decline observed in the average peel strength before and after the environmental tests.

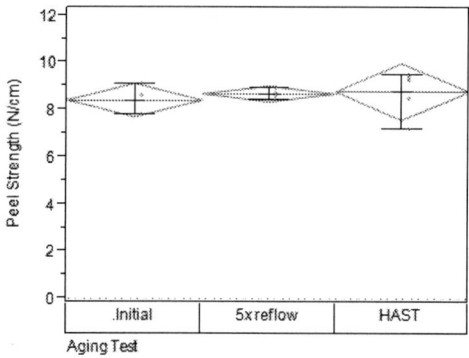

Figure 6: Minimal impact on peel strength b accelerated aging tests

Maintaining high peel strength and good mechanical integrity of glass are essential to ensure the success of subsequent filling steps without delamination of the relatively thick copper overburden. Some applications also require plating thick Cu with very good adhesion and reliability. This has been tested with a Cu build-up of up to 40 µm electrolytic copper exhibiting very good peel strength. Customer samples were subjected to 96hr, 85%RH, 130°C HAST and showed no degradation of peel strength, as shown in Table III and Figure 7 below.

6. Good adhesion on Al_2O_3 and $BaTiO_3$ ceramic

Al_2O_3 and $BaTiO_3$ substrates with a surface roughness (Sa) of ~460nm and ~160nm were subjected to MOAP treatment in order to assess adhesion of plated metal to ceramic materials. Figure 8 shows a blister free Al_2O_3 ceramic plated with electroless copper (0.3-0.5 µm) and electrolytic copper (15µm).

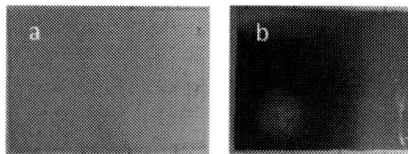

Figure 8: Blister free copper plating on Al_2O_3 ceramic. (a) Electroless copper (b) 15 µm electrolytic copper.

A peel strength of ~8N/cm was obtained on Al_2O_3 samples treated with MOAP. After 5x reflow (260°C) the 15 µm copper layer showed no delamination or blistering. Similarly, a peel strength of ~4N/cm was obtained on $BaTiO_3$ samples treated with MOAP. In contrast, untreated samples show no adhesion to plated metal. The peel strength profiles are shown in Figure 9.

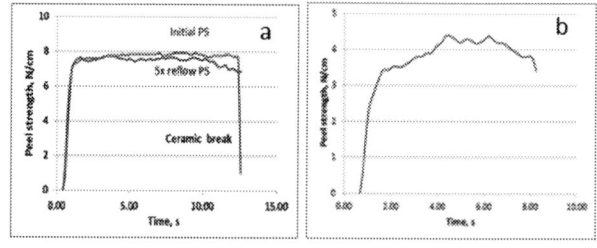

Figure 9: Peel strength profiles of 15µm copper on ceramics. (a) Al_2O_3, (b) $BaTiO_3$.

7. Via structure plating

Since many applications (such as interposers) require reliable metallization not only on the surface of the glass but also in the vias, glass substrates with various hole diameters were investigated. Special emphasis was given to the prevention of through-glass-via (TGV) blockage during the MOAP and electroless copper deposition. The TGV's were free of blockages and of blistering down to a diameter of 20μm. Glass thickness in the last instance was 200μm as shown in Figure 10. Very uniform coatings of the via walls are achievable as shown in Figure 11.

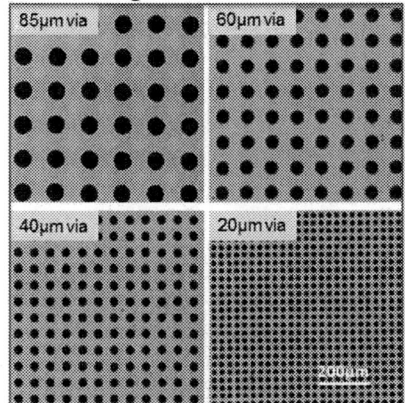

Figure 10: TGVs after MOAP and electroless Cu plating.

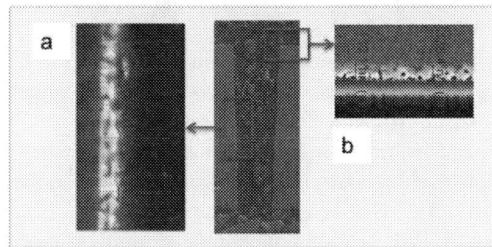

Figure 11: Good throwing power and coverage of MOAP in (a) middle of TGVs (b) on surface of wafers.

Other glass interposer designs require blind micro vias (BMVs) for subsequent metallization and filling. A variety of BMV aspect ratios have been successfully plated after applying a MOAP seed layer and electroless copper, as shown in Figure 10.

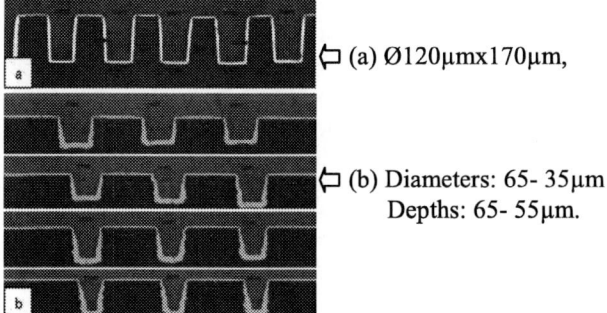

(a) Ø120μmx170μm,

(b) Diameters: 65- 35μm
Depths: 65- 55μm.

Figure 10: BMVs with various dimensions after MOAP and electroless plating.

Based on our current setup, wafer sizes up to 8" can be processed as shown in Figure 12. Neither blisters nor blocking in TGVs was observed.

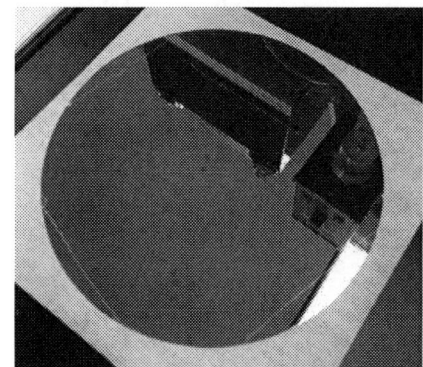

Figure 12: MOAP coating and electroless Cu plating for 8" wafer with Ø 85 μm TGVs

Conclusions

A novel glass and ceramic metallization process targeting excellent adhesion was introduced. Cross-sectioning of TGV and BMV structured substrates showed uniform coverage and Cu deposition without blistering and delamination. Thickness and morphology of the adhesive layer (MOAP) can be controlled such as to make the process compatible with both fine-line patterning and optical applications.

The MOAP process is compatible with a range of glass and ceramic substrates varying in composition, CTE and other physical properties. A somewhat reduced peel strength was observed for samples with exceptionally low glass transition temperatures. Pre-treatment such as roughening is not necessary in order to achieve good adhesion. Thus metallized glass and Al_2O_3 substrates withstand HAST and reflow stress tests without delamination or significant loss in adhesion.

References

1. V. Sukumaran et al., Through-package-via formation and metallization of glass interposers, *Electronic components and technology conference (ECTC) 2010, 60th, pp. 557-563*

2. Han C. Shih et al. Trench gap-filling copper by ion beam sputter deposition. *Materials Chemistry and Physics 97, 2006*

3. D. E. Packham, The Mechanical Theory of Adhesion. *Handbook of Adhesive Technology, 2nd Edition 2003*

4. M. Paunovic, Plating, 55, 1161 (1968)

Very Thin POP and SIP Packaging Approaches To Achieve Functionality Integration Prior To TSV Implementation

Fernando Roa, PhD
Amkor Technology, Inc.
1900 s Price Rd, Chandler, AZ
Fernando.roa@amkor.com

Abstract

Discrete function integration is achieved at the package level with a variety of packaging solutions, mainly package on package. In this paper we review mainstream package on package technology offerings and compare them against improved versions of these packages where advances in materials properties and processing techniques are leveraged to meet increasingly tighter overall package thickness requirements while maintaining or improving the reliability, yield and cost effectiveness metrics set by the existing technologies.

Different avenues to achieve very thin POP and system-in-a-package constructions are explored in detail, showing the implications in terms of material properties, new or improved processing steps and design trade-offs. Similarly, we show examples where these new techniques have been used successfully to achieve the stricter thickness targets and to prove that these technologies are ready to be applied and leveraged immediately for functionality integration without requiring extended development times or completely new processing methodologies.

Package on Package: Background and Challenges

Package-on-package (PoP, Fig1) applications have seen a year over year growth rate that has exceeded any other package family over the last few years. This trend, mainly driven by the commercial popularity of wireless handsets, smart phones in particular, stemmed from the technology's ability to offer a package shortcut to the previous chip integration path (SOC); one which accelerated design and development cycles considerably while being very cost effective and providing great supply chain flexibility.

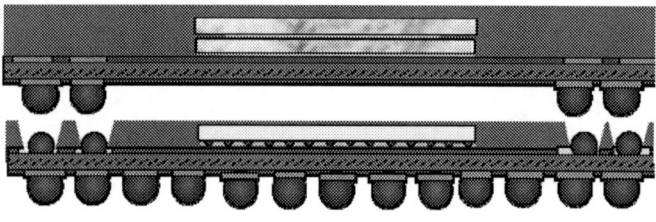

Figure 1. Typical Packag on Package configuration, showing both top and bottom packages separately.

PoP continues to be quite relevant and cost effective packaging solution for application processors in the Smartphone application area; although clearly alternatives are available, these somehow have missed the mark on critical areas, which has allowed PoP to remain widely leveraged in this space. Although not exactly trivial in design, construction or final assembly, PoP manufactures have cleverly dealt with many previous obstacles, insuring the package remained viable for each successive device generation. It has been extended several generations DRAM and most likely will continue to be demanded for the foreseeable future even if other 3D packaging techniques like TSV become more mainstream and cost competitive to combine apps processors with memory for phone applications. [1,2]

In general, the industry is striving for a much thinner, flatter, thermally cool, easily workable PoP. Each industry participant is working from different angles to deal with these issues. For example, memory vendors are working on thinner die, thin molding and small ball sizes for reduced thickness footprint [3]. OEMs are looking at ways to streamline the process for PoP, from the inception of multicomponents to the combination of these packages in different flows that streamline the operation for reduced lead times which help to minimize the need for building large inventories. Nowadays with cellphone life measured in months rather than years, it is important to control the inventory of these parts to minimize the exposure to fluctuations on customer reception, commodity pricing, etc.

Yet challenges to the continuous relevance and applicability of this packaging technology are routinely evidenced with each new generation of devices. Some of these challenges might yet to become hard limitations which make pundits claim the need for other packaging alternatives. One such challenge which package assembles face on a daily basis is how to make the package thinner. Being typically the tallest package on the phone's board, the PoP package is continuously scrutinized for ways to reduce its profile. 2.5 and 3D integration aim to address this issue by interconnecting the die directly through silicon vias. [5]

Another typical limitation is that PoP packages offer limited space for memory IO interconnection slots because in its construction the memory are constrained to the periphery of the bottom package (see Fig 1) while the center is occupied by the silicon die. This fact is quickly becoming a more relevant barrier as new DDR interfaces are expected to increase the memory count beyond any possible accommodation on the periphery.

Although with the prevalence of tablets and wide format phones, it would seem there isn't a necessity to shrink the size of the package, there are other appliances where the board area occupied by the package is becoming a hot commodity due to other chips being added or to further isolate chips from one another and prevent signal degradation from EM radiation or yet still because alternative encasing is used, thus increasing pressure to reduce the typical sizes of PoP packages.

Still another challenge comes in the way of expected cost reduction and progressive margin erosion. Efforts have to be made in all areas of the package, design, manufacture, supply chain, materials to insure the package is still the most cost effective possible.

Challenging on their own, these different and, sometimes, conflicting mandates result on upsetting delicate balances on multiple interconnected factors, thus compounding complexities inherent to the way this package is designed and constructed. For example, the mandated reduction of thickness carries with it the necessity for the package to maintain the coplanarity/warpage specs required to insure high stacking yields. But warpage is a multifaceted challenge; careful and deliberate choice of highly engineered materials, process conditions and balanced designs are at the heart of achieving a solution that meets the strict requirements for warpage to insure high stacking yields. Thus, driving thickness reduction blinded to these previous considerations is likely to result in severe room temp warpage and low stacking yields.

Likewise, driving increases on IO count to enable faster transfer rates between top and bottom packages could result in 'hidden' costs: impedance increase, reduced signal integrity, higher costs due to more advanced design rules for substrates and advanced processing techniques and so on.

And finally driving cost reduction while more aggressive requirements for all these other stretch factors are mandated, lies in odds with the unexpressed interest by most stakeholders to extend the technology as much as possible by utilizing and leveraging deployed assets with the aim of remain as competitive with alternate solutions. Still if thinner but costlier packages result in lower total ownership costs for the end users, there is enough justification to drive these new requirements.

Indeed, at Amkor, we are following many of these same avenues to streamline and improve our PoP offerings and will show that in the next section. However, we want to draw attention to alternate package configurations which although tried in the past, are becoming more and more prevalent as potential solutions to the cost dilemma presented by TSV and 3D integration.

Thinning the existing PoP

Perhaps the most common request from customer is for a thinner PoP package. Typical constructions today start at about 1.2mm for a bottom package using a 4 layer substrate. Inherent to this value is obviously the selection of DRAM package (for simplicity, will be referenced as PoPt in the ensuing discussion), substrate, solder ball, mold cap, etc. Given the tight requirements for coplanarity and warpage, any changes on any of these variables is thought to have a significant effect on the final package of the bottom package (PoPb) with the intent to target thinner appliances.

Table 1 presents a typical PoP construction for regular materials based on the JEDEC standard. This uses a 4L

substrate, typical mold with through mold vias (TMV) for the PoP interconnection [2].

Package Element	Ref.	Unit	Thickness
0.25mm ball, 0.4 pitch)	A1	mm	0.180
4L laminate	A2	mm	0.300
Mold Cap	A3	mm	0.350
Total Bottom Pkg height		mm	0.830
package stand-off	B1	mm	0.400
Top package (mold+sub)	B2+B3	mm	0.320
Overall PoP Stack Height		mm	1.200

Table 1. Typical POP package construction

Figure 2 shows a graphical sketch/representation for this construction.

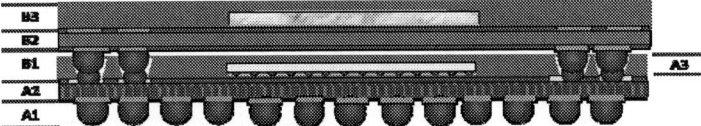

Figure 2. Typical PoP package construction

In the proceeding discussion it is assumed that the top package has a fixed thickness and no attempts are made to achieve further thickness reductions on that package. So the discussion will be relevant only for the bottom PoP package.

When investigating what elements are easily modifiable, Roggerman and collaborators have shown elsewhere [4] that an easy way to reduce the thickness is to use a smaller solder ball. Considerations in this case are related to the reliability of the solder ball connection in regards to the 2nd level interconnection (with the cellphone board). Clearly reducing the ball thickness is easily achievable with minimal change to design, assembly processing or package construction. A typical 0.4mm ball pitch package might typically use 0.25mm solder ball; so going to 0.20mm solder ball size would reduce the package thickness by about 60um. Table 2 shows the calculation.

After reaping the easy thickness reduction possible with solder ball size selection, additional thickness reduction can only be achieved by tackling the more challenging package components in the structure, for example, rethinking the substrate composition and structure. As discussed in the previous section, the substrate forms the backbone for the package, being greatly responsible for stability and flatness. Material selection has to account for these needs in addition to any desired thickness reduction.

Package Element	Ref.	Unit	Thickness
0.20mm ball, 0.4 pitch)	A1	mm	0.120
4L laminate	A2	mm	0.300
Mold Cap	A3	mm	0.350
Total Bottom Pkg height		mm	0.770
package stand-off	B1	mm	0.400
Top package (mold+sub)	B2+B3	mm	0.320
Overall PoP Stack Height		mm	1.140

Table 2. POP package thickness with 0.2mm solder ball

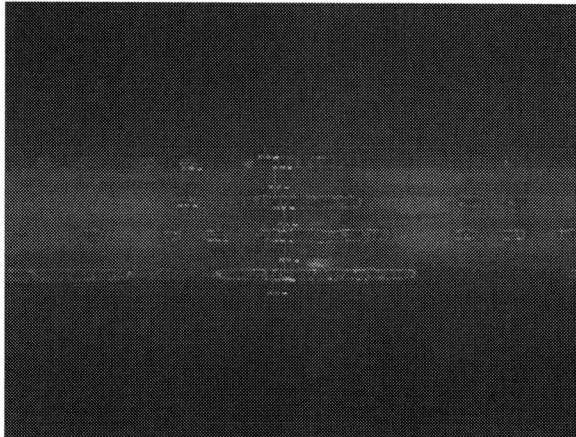

Figure 3. Cross section of typical 4layer coreless PoP substrate

Given that core thickness represents the bulk of the thickness on most 2L and 4L designs, one can see opportunities with coreless substrates which have become more available. Yet, issues with these constructions aren't trivial: for obvious reasons, coreless substrates lack stiffness and dimensional stability at higher temperatures which runs counter to the tight coplanarity/warpage requirements mandated for stacking applications, specially for lower layer count substrates. Figure 3 shows a typical x-section for a 4 layer coreless substrate used for simple PoP:

For more complex devices, requiring buildup substrates, coreless substrates are more successful since they don't require large capture pads typical of cored designs. If they are used in combination with finer bonding methodologies (for example, Cu pillar) layer reduction is possible, therefore not only achieving the desired thickness reduction but also offering some immediate cost savings.

If the coreless path is not a viable option, then the alternative is to use thinner core and build-up layers. Core layers as thin as 30um and build-up layers (both pre-preg and ABF) as thin as 15um have been successfully tried at Amkor. As with the coreless case, thinning the layers always result in a slight loss of stiffness, which typically ends causing warpage for the final package. To alleviate or even mitigate these issues, new films are in development with super low coefficients of thermal expansion (CTE), some as low as 1ppm. Still, a balancing act of thin layers and materials has to

be carefully studied to avoid creating stress concentration leading to other problems for the package. Table 3 shows a comparison of advanced core material properties for different suppliers

Beyond reductions in thickness for the substrate and solder ball, one can target the mold cap itself. Encapsulation of the package occurs by using a filler-reinforced epoxy molding compound (EMC), which is injected at high pressure and temperature into an enclosed cavity where the packages lie (mold chase). Tight process tolerances are required to mitigate formation of voids or incomplete molding for very thin mold cap packages, due to concerns with mold flow obstructions; these tolerances dictate a min clearance from the die backside to the top of the mold cavity (typically no less than 100um). Older POP package offerings dealt with these issues by simply removing the cap altogether, in what was called bare die PoP. Yet, that approach would fail to produce sufficiently flatter packages, due the tendency of the unmolded areas to warp severely; as mentioned before, this issue was successfully addressed by the introduction of the TMV process. For this reason, we believe any proposed mold cap thickness reduction has to consider a molding over the entire package area.

In our experience, a most innovative way to solve this problem is through the use of the film assisted molding process. This molding process differs from the standard molding process in that the backside of the die is covered or protected by a special film during the mold injection phase which blocks the flow of mold compound over the die. Yet the sides of the die remain unblocked and allow for mold material to flow unobstructed. The end result is a package where the backside of the die is exposed but in every other respect is like a typical molded csp package. Clear benefit of this mold process is that the thickness of the mold cap is now defined by the overall die stand-off, or the distance from the die top side to substrate.

Table 3. Core material type and properties for very rigid PoP construction.

Property	MGC	Hitachi	Sumitomo	Panasonic	Doosan
Core type	HL832NSI (LC)	E770GLH	LAZ4785T HJ ver.B	R-1515C	DS7409 HGB (Z)
CTE (<Tg)	1.5	1.8	2	1.5~2.5	1.8
Tg (DMA)	300	-	290	275	270~290
Modulus (RT, GPa)	40	38	36	30~35	31~34

Theoretically, with this process, mold cap thickness can be tailored to be very thin by combining thin die and small collapsed bumps. In reality, other factors have to be considered in selection of these attributes: for example, handling thin die can become problematic due to warpage issues or due to fragility of the wafers. Also, small bumps can become a challenge to control due the potential for opens, again due to substrate or die warpage. Also controlling collapsed height is no trivial matter as solder behavior during reflow can cause some variability in standoff height even for small die. In our experience, judicious design study accounting for comprehensive tolerance sensitivity together with sensible equipment and material choices as well as process control are important factors to secure a stable, healthy process outcome.

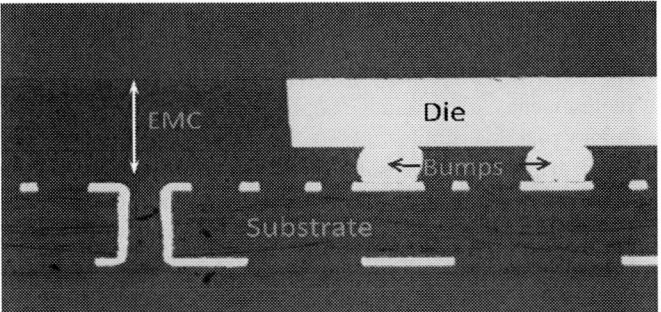

Figure 4. A typical exposed die crosssection.

It has to be mention that an indirect benefit from the use of the exposed mold die process is the enhanced heat dissipation observed for these packages. It not only reduces the resistance associated with thick mold caps but also allows for direct heat to escape directly from the die backside. In this way, the thermal efficiency of the exposed die package is greater than for the molded case.

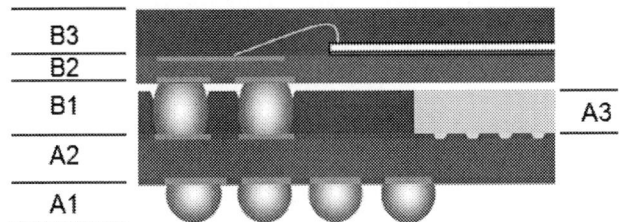

Package Element	Ref.	Unit	Thickness
0.25mm ball, 0.4 pitch)	A1	mm	0.180
4L laminate	A2	mm	0.260
Mold Cap	A3	mm	0.190
Total Bottom Pkg height		mm	0.630
package stand-off	B1	mm	0.220
Top package (mold+sub)	B2+B3	mm	0.320
Overall PoP Stack Height		mm	0.980

Table 4. Exposed die POP package thickness analysis with 0.25mm solder ball

For the exposed die scenario, Table 4 summarizes the potential final PoP thickness. In this calculation, we assumed the use of 0.25mm ball, a 4 layer laminate with typical construction at 0.26m. The big change however, is observed in the mold cap thickness. As mentioned before, the die standoff is controlling now this value and for this particular calculation we assumed the total thickness plus bump stand-off to be 190um. With these values, we observed a package

whose total nominal thickness falls under 1mm; a total gain of more than 200um over the standard PoP construction reviewed in the introduction a summarized in Table 1. It should be noted that 190um is a current production mold cap thickness, however much thinner constructions have been made possible by using thinner die (<100um) in combination with lower stand-off die bump metallurgies (CuP). Figure 5 shows such a construction is possible, which could result in about 50um of additional thickness reduction.

Figure 5. Exposed die POP package with Cu pillar

SIP package for functionality integration

The preceding discussion was focused on Package on package technology, as it remains the most prevalent package solution to achieve functionality integration due its relative flexibility and widespread adoption. There is however, a great deal of package offerings which also aim to achieve integration without the assistance of 2.5D or 3D technologies. The gamut of possibilities is too varied to be covered in this paper and the interested reader is encouraged to consult other package review sources. Yet, to complement the discussion, and to be able to compare the relative thickness reductions achieved by the means described above, one additional package which is going to be covered here.

System in a package (SIP), consists of integrating multiple silicon chips into one single package. This packaging offering has been given other various names in addition to SIP: hybrid, flipstack, multistack, etc. and it has been used quite extensively for memory packaging where the chips are all of the same size, functionality and interconnection methodology. In this particular discussion, however, the interest is in integrating chips of different function, with the added requirement to achieve the lowest package profile possible. This is typically achieved by using mixed interconnection technologies: flip chip for the bottom chip(s) and wirebonding for any stacked chip thereafter. Figure 6 shows a thin SIP construction which is in production currently.

In making thin SIP packages, mostly the same principles discussed for PoP thinning are open with the obvious advantage that all integration is conducted by one operator . This allows for a greater degree of control over the materials and processes, enabling very thin packages. Table 5 shows a typical thin SIP package, with two layers of integration (one FC interconnection level and one WB interconnection).

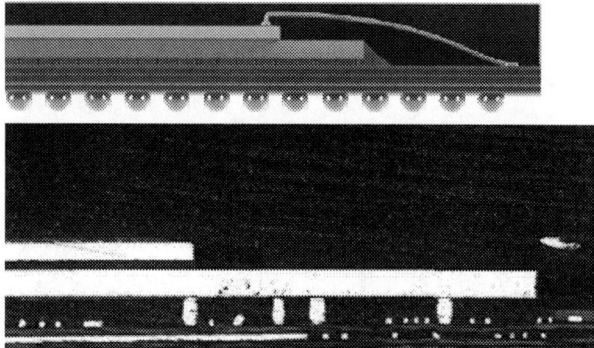

Figure 6. Different views of a SIP package showing 3 die in a package

Package element	Item	Unit	Thickness
D2 Thickness	A	mm	0.080
D1 Thickness	B	mm	0.100
FC interconnection	C	mm	0.050
Mold Cap	D	mm	0.450
Substrate	E	mm	0.180
Solder Ball	F	mm	0.120
Package Thickness	J	mm	0.750

Table 4. Very thin SIP package thickness analysis

As shown in Table 4, very thin packages can indeed result from SIP package integration as compared to PoP. In this particular case, flip chip die thickness of 100um combined with wirebond die package thickness of 80um and copper pillar flip chip bonding allows for the use of a thin mold cap (0.45mm). Moreover a quite thin 4 layer substrate can be used here more easily than in the POP application because the stacking itself will help to control the warpage of the package and warpage does not need to be as tight since there is no need to stack individual packages which was the limitation for

POP. Finally, a 200um ball size complement the package solution, which overall is around 200um thinner than the thinner PoP achieved; and there is still room for further thinning as some other offerings are now below 600um overall package thickness.

Conclusions

Extending PoP package technology continues to be a priority for assembly houses as a way to achieve 3D function integration for important applications while TSV implementation work its way through the many challenges it still faces. Continous PoP package structure optimization continues at a fast pace in order to deal with new tighter requirements in z-height, coplanarity, warpage and cost, among other constraints.

It has also been shown that innovative materials in combination with existing but refined assembly processes can be used in innovative ways to create packaging offerings for both package on package and system in package in order to tailor solutions for highly demanding signal count and integrity applications which are very competitive in terms of pricing and flexibility. This offers the additional benefit of extending the useful life of deployed assets for assembly.

Detailed construction analysis was presented for a few examples to demonstrate the principles required to achieve thinner solutions as well as the development challenges the industry is still facing to extend the life of current high volume packaging solutions to meet the requirements of mechanical robustness, yield, long term reliability, product performance and cost. With the further benefit of enabling the use of current assembly technology, the cost benefits achieved reduction is immediate.

Acknowledgments

The author acknowledges the support of the factory, sales and business unit teams for their support in the material used for this paper.

References

1. A. Yoshida, J. Taniguchi, et Al. "A study on Package Stacking Process for Packag-on-Package (PoP)", Electronic Components and Technology Conference, 2006. Proceedings. 56th, San Diego, CA, 10.1109/ECTC.2006.1645753

2. L. Smith, C. Zwenger, "Next Generation Package-on-Package (PoP) Platform with Through-Mold Via (TMVr) Interconnection Technology, " IMAPS Device Packaging Conference, March 10-12, 2009.

3. S. Deo, "Thin PoP for Mobile Devices", in Proc IMAPS 2013, Orlando, FL, Oct 1-3, 2013.

4. B. Roggerman, R. Kumar, M. Schwarz " Reducing Package Thickess to Accommodate Next Generation Smarthphone Designs" in Proc. IMAPS 2013, Proc IMAPS 2013, Orlando, FL, Oct 1-3, 2013.

5. S.Q. Gu, D.W.Kim, V. Ramachandran, et Al, "Wide IO Memory on Logic 3D Integration using Through Silicon Stacking," Advancing Microelectronics, Vol. 40 No. 3, May/June 2013, Pg 16-19,

A Study on the Fine Pitch Chip Interconnection Using Cu/SnAg Bumps and B-stage Non-conductive Films (NCFs) for 3D-TSV Vertical Interconnection

Yongwon Choi, Jiwon Shin, Young Soon Kim, Kyung-lim Suk, Il Kim, and Kyung-Wook Paik*
Dept. of materials science and engineering, KAIST,
373-1, Gu-seong dong, Yu-seong gu, Daejeon, 305-701, Korea,
E-mail: kwpaik@kaist.ac.kr *

Abstract

The increasing demand for high performance integrated circuit devices has been leading the development of 3-D stacking technologies. One of the state-of-art 3-D stacking methods is the through silicon via (TSV) interconnection which may facilitate very high density memories or ASIC modules. Industrial mass production and academic research to use the through silicon via(TSV) in the 3D interconnection has brought the matured technology for forming the TSV in the chip. However, former method using flux and underfill for interconnection between chips have several drawbacks such as flux residues or voids trap along the bonding interface causing reliability issues. As one of the solutions, chip interconnection using Cu/SnAg bump and non-conductive film has been gaining a lot of interest as the one of the promising ways for 3D TSV interconnection.

In this paper, a study is made for the relationship between the viscosity of pre applied non-conductive film and loading force to predict the gap change. The existing theories are adapted to predict the gap change of a real chip and a substrate during bonding with using simplified model. A gap changes from real bonding of dies were matched to check the validity of prediction. As a summary, 3D-TSV vertical interconnection using Cu/SnAg bump and wafer-level NCFs was theoretically and experimentally investigated. Through the theoretical investigation, bondings were explained using the rheological properties of NCFs, chip size, and bonding parameters. And the real chip bonding was matched to the prediction from the theory. Therefore, chip bonding using Cu/SnAg bump and NCFs could be the promising solution for the fine pitch TSV interconnection.

Introduction

3D packaging technology by through silicon via (TSV) has been considered as one of the promising methods to enhance the performance and packaging efficiency. Nowadays, TSV technology has been adapted to some applications such as memories, camera modules, and so on. Therefore, vertical interconnection methods for TSV chips are becoming an important issue. Through previous reports, hybrid bump with Cu post and SnAg cap using pre-applied non-conductive films (NCFs) provides advantages of easier processing, fine pitch handling capability, and lower cost. [4]

There have been few studies reported to analyze the effects of NCF properties on the Cu/SnAg hybrid joint interconnection and the effects of the chip sizes on the bonding conditions. The main factors for the good electrical interconnection between the Cu/SnAg bumps and Cu pads are the bonding forces, loading time, and temperatures applied to NCFs laminated between chips and substrates. Figure 1 showed the schematic diagram of the typical processes of the TSV chip bonding. The bonding force and loading time cause the NCF polymer resin to be flowed out toward the edge of chip at certain temperature when the viscosity of NCFs becomes the lowest, and at the same time, the initial thickness of NCF decreases until the bumps physically contact to the metal pads on a substrate. Figure 1 showed the good and not-good shapes of the deformed bumps. If the bonding forces are too high or loading time is too long, SnAg cap would be deformed too much, and squeezed Sn may cause the SnAg wetting on the side wall of Cu bumps or even the electrical shortage between neighboring bumps. The temperature not only make NCF resin flow easily but also cures NCF material resulting in mechanical holding of chips and substrates. Therefore, it is necessary to find out the major bonding process factors and the governing equation of bonding condition for any types of NCFs and chip sizes to be bonded.

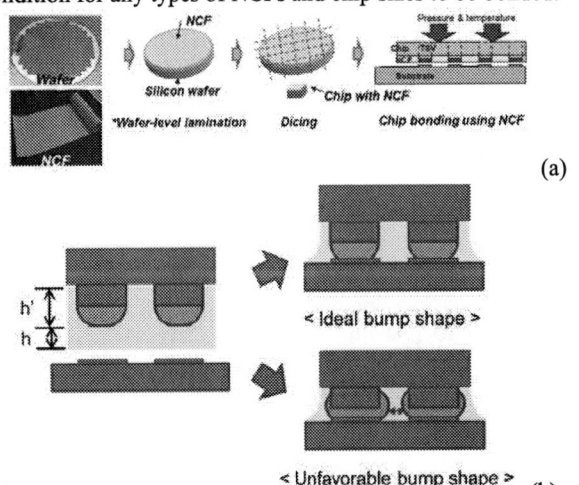

(a)

(b)

Figure 1. The typical NCF bonding processes and bump shapes after the bonding (a) NCF bonding processes and (b) two kinds of bump shapes after the bonding

NCFs are B-stage polymer films which can both flow easily under pressure at low temperature and be rapidly cured at higher temperature. The flow of NCFs resin between chips and substrates before the start of curing reaction is very similar as the squeeze of fluids between two parallel disks. Therefore, TSV chip bonding using NCFs can be explained using the existing flow formula with some adjustments. In the

2014 Electronic Components & Technology Conference

Scott's paper, viscoelastic fluids went through the isothermal time-dependent radial flow between two disks after a constant force has been applied normally to the disks.[1] By considering a power law fluid with the mass balance, momentum equation, and rheological characteristics, the force acting on the upper plate can be obtained by the instantaneous rate of descending the upper plate [2,3].

In this paper, the effects of NCF viscosity, bonding force, loading time, and the chip size on the mechanism of TSV chip bonding using pre-applied NCFs will be investigated using the Scott equation. Also, by using the acquired relationship, an ideal NCF thickness changes during the whole bonding processes will be predicted and compared with the real NCF thickness changes during the bonding processes. Furthermore, the bonding of real test vehicles with the array of Cu/SnAg bumps will be matched to the simulated results to verify the correctness of prediction.

Experiments

Measurement of NCF thickness changes with Si dummy dies

Two kinds of NCFs with different minimum viscosities and the same curing behaviors were prepared. NCF 1 had the minimum viscosity of 20,000 Pa·s and NCF 2 had the minimum viscosity of 500 Pa·s, and the thicknesses of each NCFs were 58 and 78 μm, respectively. A Si dummy wafer was cut into square dies with 7 x 7 mm² and 10 x 10 mm² sizes for upper and lower dies. NCFs were laminated on the upper die followed by the thermo-compression bonding. The heating rate was 5 °C/s until reached to 250 °C followed by 30 seconds of maintaining at 250 °C, and then cooled to room temperature. For every two seconds during the bonding, the gap changes between two dies were measured. All bonded dies were cross-sectioned to observe the NCF thickness changes.

Measurement of NCF thickness changes with Cu/SnAg bumping chip

In addition, two kinds of real test vehicles with sizes of 5 x 5 and 10 x 10 mm² were designed and fabricated. Cu post and SnAg cap were electroplated on silicon wafers with 60 and 110 μm pitch, respectively. The heights of Cu posts were 10 and 50 μm, and 10 and 20 μm for the height of SnAg cap, respectively, as shown in the Figure 2.

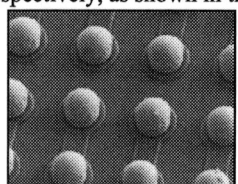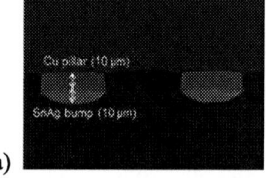

(a)

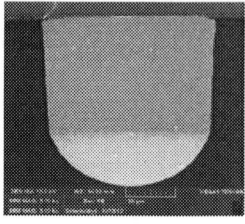

(b)

Figure 2. SEM images of the test vehicle with Cu post and SnAg cap bump structure (a) 5 x 5 mm² (b) 10 x 10 mm²

Rheological properties measurement

The rheological properties of two NCFs with 1 mm thickness and 25 x 25 mm² size were measured using a rheometer. Viscosity changes according to the temperature and strain rate were measured with temperature increased from 40 to 250 °C. And the strain rates increased from 0.0001 to 10/s at 40 °C.

Results & discussion

Rheological calculation

Consider the squeezing flow where initially the fluid completely fills the gap between the two disks of radius R, and the initial disks separation is h_0. A constant force F is applied vertically to the upper disk. A power law fluid is one in which the shear stress τ depends on the shear rate γ in the liquid via a power law, which is a typical behavior of polymer. It is defined as the equation 1.

$$\tau = \eta_0 (\dot{\gamma})^n \quad (1)$$

$$\eta = \frac{\tau}{\dot{\gamma}} = \eta_0 (\dot{\gamma})^{n-1}$$

$$\ln \eta = \ln \eta_0 + (n - 1) \ln(\dot{\gamma})$$

From the Scott equation [1], the required time for reducing the gap from h to h_0 by force F is expressed by the equation 2.

$$\frac{dh}{dt} = -\frac{n}{2n+1} \left(\frac{(n+3)F h^{2n+1}}{2\pi \eta_0 R^{n+3}} \right)^{\frac{1}{n}}$$

$$t = \frac{2n+1}{n+1} \left(\frac{2\pi \eta_0 R^{n+3}}{F(n+3) h_0^{n+1}} \right)^{\frac{1}{n}} \left(\left(\frac{h_0}{h} \right)^{\frac{n+1}{n}} - 1 \right) \quad (2)$$

For the ramping up profile, temperature will vary linearly with time like the equation 3.

$$T = T_1 + kt \quad (3)$$

And, reasonable approximation for the NCFs is that the viscosity depends exponentially on temperature before the curing starts as shown in the equation 4.

$$\eta = \eta_0 e^{-b(T-T_1)} (\dot{\gamma})^{n-1} \quad (4)$$

By substituting the equations 3 and 4 into 2, the equation 5 can be obtained.

$$t = \frac{n}{bk} \ln\left(1 + \frac{bk}{n}\xi\right) \quad (5),$$

$$\xi = \frac{2n+1}{n+1}\left(\frac{2\pi\eta_0 R^{n+3}}{F(n+3)h_0^{n+1}}\right)^{\frac{1}{n}}\left(\left(\frac{h_0}{h}\right)^{\frac{n+1}{n}} - 1\right)$$

To obtain the thickness of NCF after applying force F during certain time, the equation 5 can be rewritten into the equation 6.

$$h = \frac{h_0}{\left\{\frac{n+1}{2n+1}\xi\left(\frac{2\pi\eta_0 R^{n+3}}{F(n+3)h_0^{n+1}}\right)^{\frac{1}{n}} + 1\right\}^{\frac{n}{n+1}}} \quad (6),$$

$$\xi = \frac{n}{bk}\left(e^{\frac{bkt}{n}} - 1\right)$$

"η_0" and "n" can be obtained from the result of viscosity variation with strain rate. "b" can be obtained from the result of viscosity variation with temperature.

Measurement & prediction of NCF thickness changes with Si dummy dies

Figure 3 shows the thickness changes of two NCFs during the bonding process. From the initial values of 58 and 78 μm, NCF thickness decreased to around 25 and 10 μm. The decreased volumes are to be same as the squeezed amount of NCFs out of the Si dies. Until 9 seconds of loading, NCF thickness was gradually reduced, and then followed by an abrupt decrease.

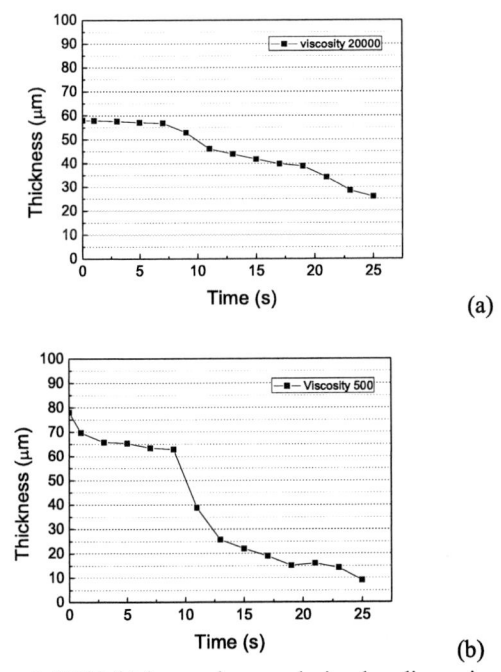

(a)

(b)

Figure 3. NCF thickness changes during bonding using (a) NCF 1 and (b) NCF 2

Figure 4 and 5 shows the viscosity variation according to the strain rate after taking natural log values of viscosity and strain rate of two NCFs, and the viscosity variation at various temperatures of two NCFs. From the natural log of the equation 1,

$$\ln \eta = \ln \eta_0 + (n-1)\ln(\dot{\gamma})$$

the slope of the curve represents "n-1" and the intercept to the $\ln \eta$ represents the natural log of η_0. From the natural log of the equation 4,

$$\ln \eta = \{\ln \eta_0 + bT_1 - (n-1)\ln(\dot{\gamma})\} - bT$$

the slope means –b. The values of η_0, n, and b for each NCF are listed in the Table 1.

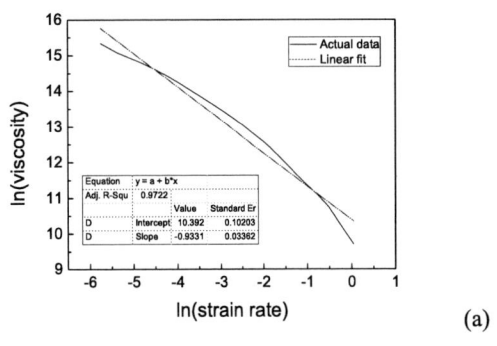

(a)

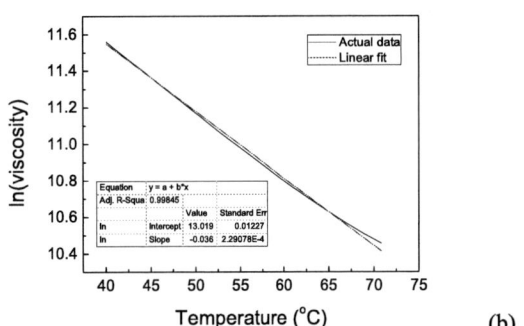

(b)

Figure 4. (a) Linear fitted graphs of the ln(viscosity) versus the ln(strain rate) and (b) versus temperature of NCF 1

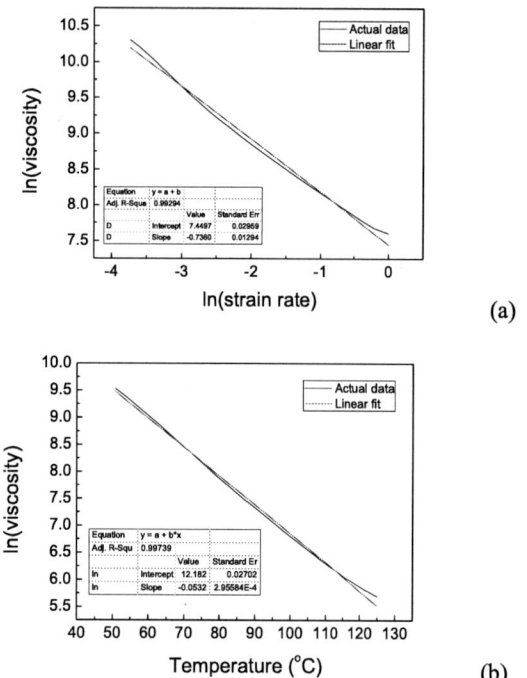

(a)

(b)

Figure 5. (a) Linear fitted graphs of the ln(viscosity) versus the ln(strain rate) and (b) versus temperature of NCF 2

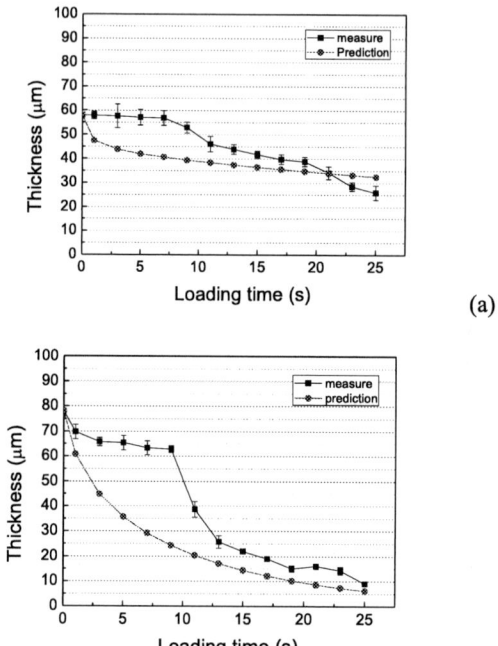

(a)

(b)

Figure 6. Measured and predicted thickness changes of NCFs as a function of loading times of (a) NCF1 and (b) NCF2

Table 1. η_0, n, and b values for NCF 1 and 2 from the equation 1 and 4

	η_0	n	b
NCF 1	32597.797 38	0.0669	2.4084 x10^{-3}
NCF 2	11336.694	0.264	0.0127 148

Then, the predicted thickness changes of NCFs during the bonding are obtained for every two seconds by applying those η_0, n, and b values to the equation 6. Figure 6 shows measured and predicted NCF thicknesses during the force were loaded for two NCFs. Predicted NCF thickness at later loading times well corresponds to the measured value, however, the predicted NCF thickness at early loading times shows much higher values than the measured data.

It can be explained by two possible reasons. One is the tolerance of the heating rate of the bonding profile. The targeted heating rate was 5 °C/s and the real heating rate was measured as 4.7 °C/s. However, the early heating rate was only 1.5 °C/s which was much slower presumably due to the slow initial heating of the bonding machine.

The other reason is presumably due to the thixotropy of the polymer. Thixotropy is a property of non-Newtonian fluids which is time-dependent change in viscosity. Due to the thixotropy, the actual viscosity of NCF right after the loading of force would be much higher than the measured value by a rheometer. Therefore, the viscosity variations of two NCFs according to the loading times were measured under the fixed strain rate as shown in Figure 7. By taking the initial values of viscosity at different strain rate including from Figure 7, the results are re-plotted as ln(viscosity) versus ln(strain rate) as shown in the Figure 8 with the linear fitting. Therefore, new n, η_0, and b from Figure 8 were obtained as shown in the Table 2.

	η_0	n	b
NCF 1	125366.9104	0.351	0.012 636
NCF 2	113210.0249	0.611	0.032 5

New results with considering the heating rate and thixotropy were shown in the Figure 9. The prediction values were well matched to the real measurement of thickness changes of two NCFs.

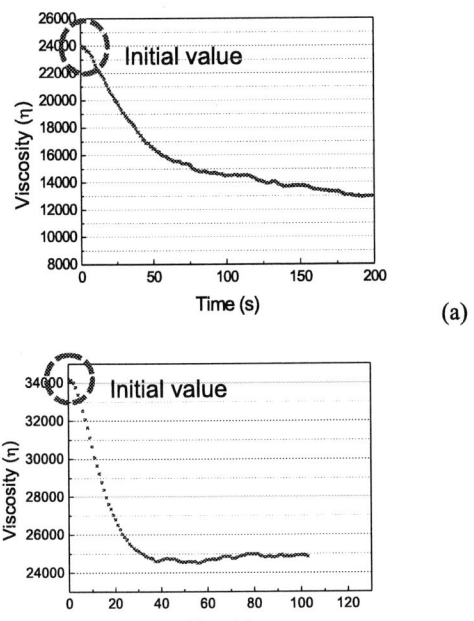

Figure 7. Viscosity decrease of (a) NCF 1 and (b) NCF2 during the measurement by thixotropy effect

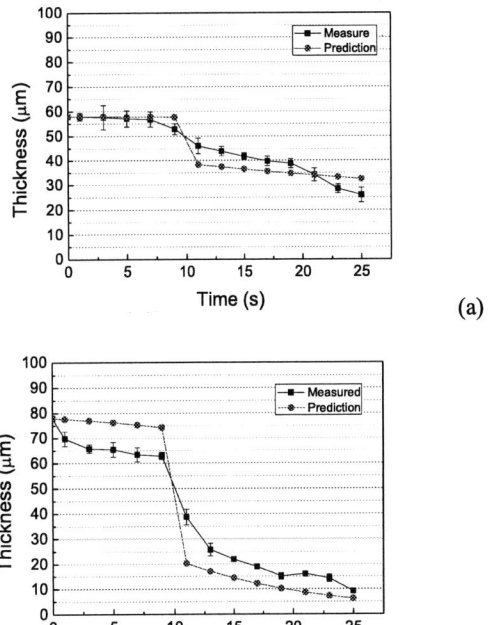

Figure 9. Measured and predicted thickness changes during loading times of (a) NCF 1 and (b) NCF2 by considering the thixotropy and heating rate effects

For NCF 1 and 2, the calculated loading times are listed in the Table 3. Figure 10 showed the cross-sectioned images of bonded samples using NCF 1 and 2 with the calculated loading times. Due to the difference in the rheological properties of two NCFs, calculated loading time was different. However, as shown in the Figure 10, the results were almost the same. The gap thickness was same for both NCFs. The difference in the shape of wetted Sn is presumably due to the viscosity difference of NCFs. When the loading time was 5 seconds longer than the calculation, the gap was too narrow that most of Sn squeezed and wetted on the wrong side of bumps. For NCF 2, the near bumps were even electrically bridged. Chips with the size of 5 x 5 mm^2 were bonded using NCF1 to see the validity of formula on the chip size difference. Figure 14 showed the cross-sectioned bump images of bonded chips. With the calculated loading time, the bumps ideally interconnected with the least bump

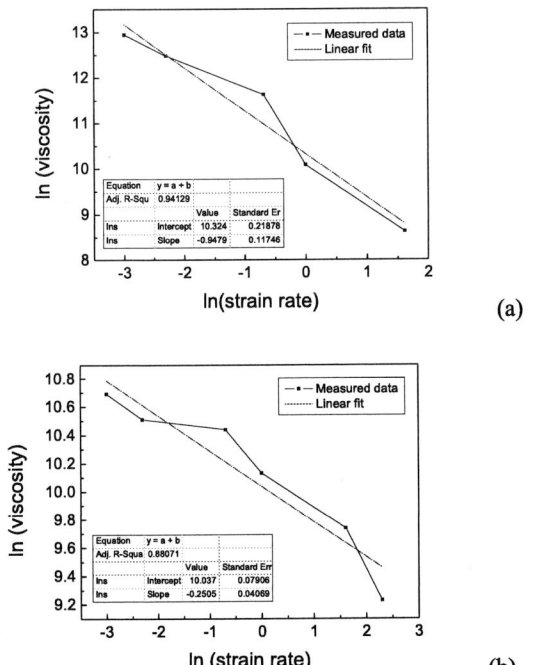

Figure 8. Linear fitted graphs of ln(viscosity) at various ln(strain rate) of (a) NCF 1 and (b) NCF 2

Table 2. New η_0, n, and b values considering the thixotropy of NCFs and heating rate changes during bonding process

978-1-4799-2408-0/14 $31.00 © 2014 IEEE

deformation. However, excess loading time caused the squeezed bump shape.

Table 3. Calculated loading time for the bonding using NCF 1 and 2

NCF	Initial thickness (μm)	Calculated loading time (s)	Loading force (N)
NCF1	80	15	80
NCF2	80	10	

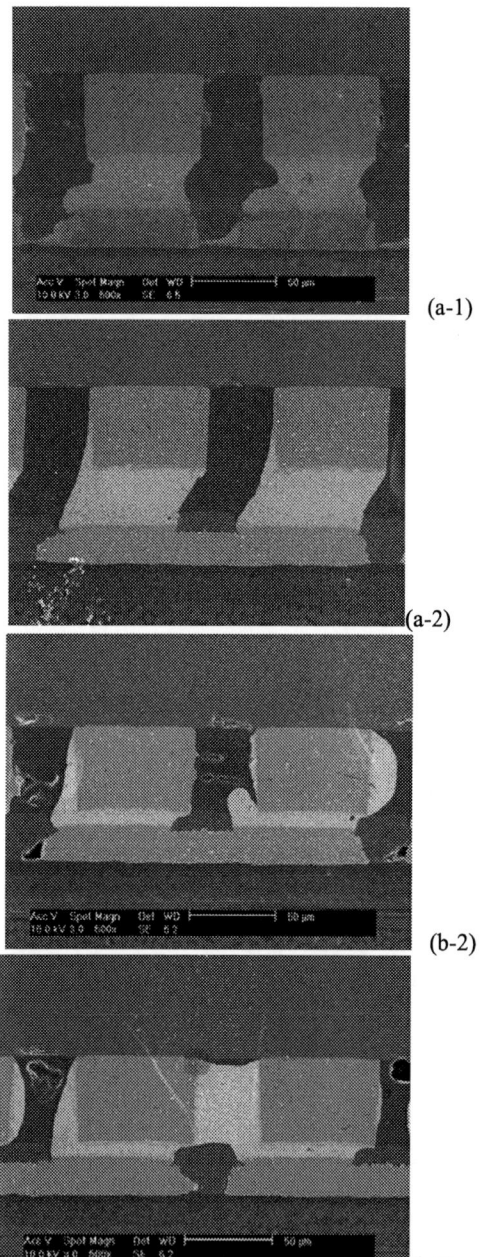

Figure 10. Cross-sectioned SEM images of bonded test vehicles using NCF 1 and 2 (a-1) NCF 1 during 15 seconds (a-2) NCF 2 during 10 seconds (b-1) NCF 1 during 20 seconds (b-2) NCF 2 during 15 seconds

Conclusion

The theoretical prediction using the Scott equation on the process condition for the TSV chip bonding using pre-applied NCFs was investigated. As the formula, the NCF thickness changes was governed by the bonding force, loading time, chip size, and the constant from the rheological properties of NCFs. The formula was verified by bonding dummy Si dies. Measured and predicted NCF thickness changes were compared under the same bonding conditions. With considering the thixotropy of NCFs, the prediction on the NCF thickness changes along the loading times well matched with the actual measurement. For the real chip bonding, the formula also well explained the NCF thickness changes even the presence of bumps structure. Chips with two different sizes were bonded by following the formula. Bumps were electrically interconnected with least physical deformation with the predicted loading time by the formula. Also, in-situ resistance measurement was correlated to the behavior of bump interconnection.

In this paper, the validity of 3D TSV chip interconnection using pre applied NCF was investigated using both of theoretical and experimental analysis. With the Scott equation, flow of NCF resin can be precisely predicted. This can give the optimized bump interconnection no matter what NCFs and chip sizes are used.

Reference

[1] J.R. Scott, Trans. Inst. Rubber Ind., vol. 8, pp. 481-493, 1932.

[2] Philip J. Leider and R. Byron Bird, Ind. Eng. Chem., Fundam., Vol. 13, No.4, pp. 336-341, 1974

[3] Samjid H Mannan, David C Whalley, Yinka O Ogunjimi, David J. Wiliams, "Modelling of the initial stages of the anisotropic adhesive joint assembly process" in Proceedings of Electronic Manufacturing Technology Symposium, 1995, pp.142-145.

[4] L. K. Teh et al., "Some characteristics of anisotropic conductive and non-conductive adhesive flip chip on flex interconnections", Thin Solid Films, vol. 462-463, pp. 446-453, 2004.

Pathfinding Methodology for Optimal Design and Integration of 2.5D/3D Interconnects

Farhang Yazdani[1], John Park[2]

[1]BroadPak Corp., 1735 N 1ST ST STE 301A, San Jose, CA 95112, USA

E-mail: farhang.yazdani@broadpak.com, Tel: +1-408-922-9006

[2]Mentor Graphics Corp., 1811 Pike Road, Longmont, CO 80501, USA

E-mail: john_park@mentor.com, Tel: +1-720-494-1057

Abstract

Given the cost per transistor is to grow below 22nm node, SOC partitioning is emerging as a viable solution compared to a monolithic solution. Ubiquitous 2.5D/3D heterogeneous integration is evolving as an eminent approach to achieve lower cost, higher bandwidth, smaller footprint and lower power. System partitioning schemes and heterogeneous integration mechanisms directly impacts performance, cost and time to market. Partitioning a single die into multiple partitions for further heterogeneous integration requires an ultra-dense connectivity.

One of the key challenges in designing a low cost and high performance 2.5D/3D package is the system-level ultra-dense connectivity exploration and pathfinding. Planning an I/O ring structure and defining I/O buffer cell placement for multiple logic partitions at the early stages of 2.5D/3D product development is not trivial. Moreover, end-to-end optimization of inter-partition's I/O cell placement, Cu pillar bump matrix and package BGA interfaced with numerous components on the PCB with fixed ball patterns is a daunting task.

In this communication, we present a pathfinding methodology for the design and optimization of 2.5D/3D interconnects. We demonstrate our methodology in a 2.5D/3D design where inter-partition's I/O buffer cells placement, Cu pillar bump matrix and package BGA are optimized in a hierarchical fashion with respect to the Wide I/O memory's fixed bump pattern interfaced with the PCB level components placement and connectivity. We further demonstrate the cross domain flexibility and robustness of our methodology by performing an end-to-end pathfinding on ultra-dense 2.5D/3D silicon interposer and single SOC device interfaced with fixed components on a PCB.

Keywords

Silicon Interposer, 2.5D/3D, heterogeneous, integration, Interconnect, Packaging, PCB, System Design, Pathfinding, Methodology, Optimization, Wide I/O, Partition, Logic, Memory, I/O buffer cell, Flip Chip, die, System in Package, SIP, BGA, MCM, 3D IC, Device, Substrate.

1. Introduction

For the first time ever the semiconductor industry has experienced an increase in wafer price, eliminating the traditional process shrink benefits beyond 22nm node [1-2-3]. This increase is part due to increase in equipment cost, extensive process R&D and with lithography contributing to nearly 50% of the overall wafer cost [4-5-6]. Cost constrains combined with complex devices has forced designers to seek alternative architectures. This has prompted the chip designers to move from traditional 2D monolithic System on

Chip (SOC) type structure to 2.5D/3D heterogeneous type integration. The benefits of 2.5D/3D integrations are notably, increase in performance, shorter interconnect length, higher speed, lower delays, reduced power consumption, smaller form factor, reduced weight and volume. More so, 2.5D/3D integration results in reduced costs by allowing integration of mixed node dies from high volume processes and higher reuse of IP cores [7-8-9-10].

However, these benefits comes at a price. Planning and defining 2.5D/3D devices at the early stages of product development and ultimately architecting a 2.5D/3D device requires an advanced pathfinding and optimization methodology. Such methodology directly impacts the performance, cost and time to market. Planning, defining, managing and optimizing off chip ultra-dense connectivity resulting from 2.5D/3D integration requires a novel and unique pathfinding and optimization methodology which is limit the use of conventional package/PCB design tools and methodologies practiced worldwide.

Conventional IC package design tools in practice today are mainly used for 2D package design. Considering the design aspect only, most EDA tools segregate the package and PCB design flow and lack system level pathfinding capability to address the ultra-dense connectivity resulting from 2.5D/3D integration. A 2.5D/3D pathfinding and integration methodology must be capable of performing an early feasibility and evaluation of ultra-dense device connectivity, routability, risks, performance and cost in the context of multiple system configuration (Fig. 1).

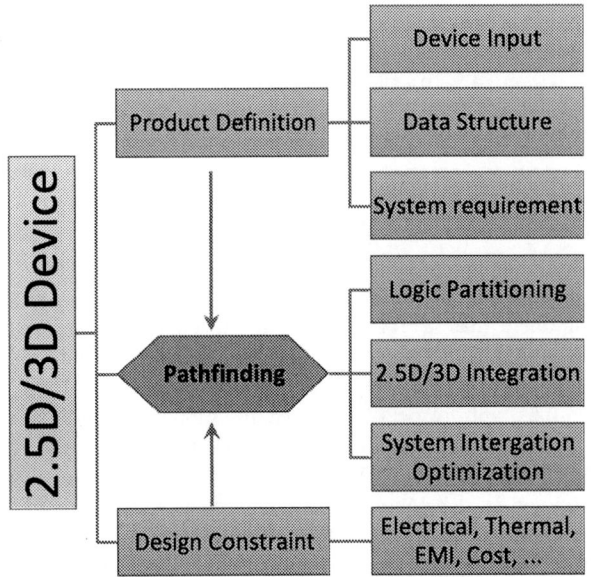

Figure 1: Pathfinding in 2.5D/3D system design.

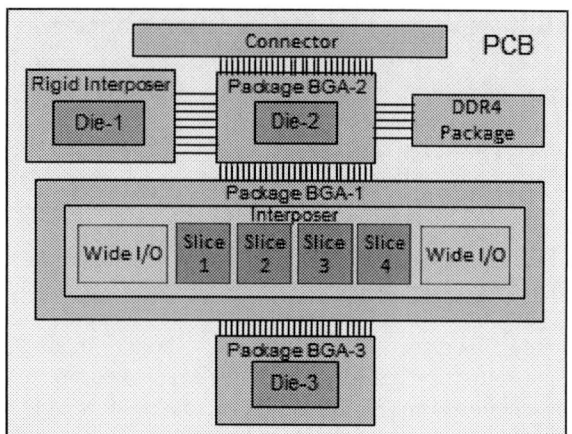

Figure 2: A 2.5D/3D package interfaced with a flip chip package, a rigid interposer package, a DDR4 Package, a connector and a wirebonded package.

In this manuscript, we present an end-to-end pathfinding and optimization methodology for the design and optimization of a 2.5D/3D system interconnect with capability for simultaneous exploration of various package types, various stacking technologies, various assembly technologies and various PCB technologies considering performance criteria such as signal/power integrity, timing, temperature, etc.

We demonstrate our methodology by integrating Wide I/O memory [11] and multi-partition logics on a 2.5D/3D package interfaced with a monolithic SOC flip chip die realized on a multi-layer organic build-up substrate. The flip chip SOC package is subsequently interfaced with a rigid silicon interposer package, a DDR4 package, a connector and a wirdebonded BGA laminate package on the PCB (Fig. 2). Finally, we will perform an end-to-end pathfinding and optimization on the overall system. In particular, the robustness of our methodology strive for early evaluation of, routability, risks, cost, performance and product development.

2. Methodology

Normally, multi-layer build-up substrate contributes to more than 40% of the total package cost, thus, a poorly designed substrate can easily exceed the cost of the die it contains; this problem is amplified considering a 2.5D/3D silicon interposer based package. In system design, often one or more packages comes with a fixed ball pattern, this is normally due to customer requirement/specification, system constraint, legacy product, or an off the shelf component.

Integrating multiple components where one or more component has a fixed ball pattern creates a need for pathfinding and co-design methodology. The ultimate goal is to design a 2.5D/3D device so that its bump pattern is optimized with respect to the system components with fixed ball patterns (off the Shelf packages), more precisely, the 2.5D/3D partitions I/O buffer cell placement must be optimized in the context of system components with fixed ball patterns. This means that components with fixed ball pattern drives the system connectivity (Fig. 3). In this demonstration, components with fixed ball pattern/netlist

are, Wide I/O memory, DDR4 and Connector. We will show how I/O buffer cells placement within the device can be optimized to match the components with fixed pattern mounted on both interposer and PCB.

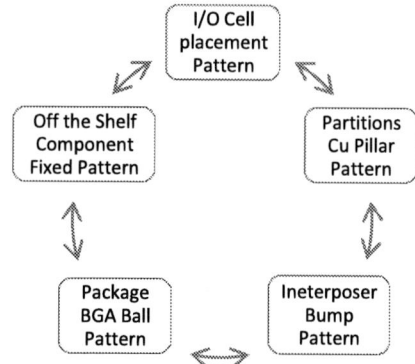

Figure 3: 2.5D/3D pathfinding flow. Components with fixed ball pattern drives the system connectivity.

Xpedition Path Finder from Mentor Graphics was selected to demonstrate our methodology due to its ability to provide a unique TCL based customizable command programming capability, interactive I/O cell placement editor, interactive rule editor for constraint based net assignment, automated symbol generator, interface to VHDL and Verilog languages as well as the ability to interface with other EDA Physical Design Tools.

2.1 Wide I/O Memory

Wide I/O memory is a low power device, targeted mainly for mobile/tablet market. It comes with fixed Cu pillar pattern, pads are located at the center of the die and defined by JEDEC standards (JESD229) [12]. Wide I/O-1 has large bus interface (512 bits), four independent asynchronous channels, pillar pitch of 50um horizontal and 40um vertical, with total of ~1200 pillars.

2.2 Pathfinding & Optimization

Pathfinding and optimization sequence for this study is illustrated in Figure 4. Considering system development is driven from the wide I/O memory perspective, we begin the journey with placement of wide I/O memory dies and logic partitions as described in Figure 4. Slices 1 and 4 are optimized with respect to wide I/O memory die (Fig. 5) followed by optimizing slices 2 and 3 with respect to slices 1 and 4. Subsequently, BGA-1 package and interposer are optimized with respect to logic slices (Fig. 6, Fig. 7).

Next, BGA-2 Package ball pattern is optimized with respect to BGA-1 Package, connector and DDR-4 package (Fig. 8). Die-2 bump pattern is then optimized with respect to BGA-2 package ball-out (Fig. 9). Subsequently, rigid interposer ball pattern is optimized with respect to BGA-2 package ball-out (Fig. 10) followed by optimizing the Die-1 bump pattern with respect to rigid interposer ball-out (Fig. 11). Next, we will optimize the I/O buffer cell placement with respect to Die-1 bump pattern (Fig. 12). At this point all components situated above the BAG-1 package depicted in Figure 2 should be fully optimized with respect to BGA-1 package (Fig. 13).

978-1-4799-2408-0/14 $31.00 © 2014 IEEE   1668

Optimize/Unravel logic slices with respect to Wide I/O memory fixed Cu pillar pattern

Optimize/Unravel Silicon Interposer bump pattern with respect to logic slices Cu pillar pattern

Optimize/Unravel Package BGA-1 ball pattern with respect to interposer bump pattern

Optimize/Unravel BGA-2 ball pattern with respect to BGA-1, DDR4 ball pattern and connector pin pattern

Optimize/Unravel Die-2 bump pattern with respect to BGA-2 ball pattern

Optimize/Unravel rigid interposer, Die-1 bump pattern and I/O ring cell placement with respect to BGA-2 ball pattern

Optimize/Unravel Die-3 bump pattern and BGA-3 ball pattern with respect to BGA-1 ball pattern

Figure 4: Pathfinding and optimization sequence for integrating components depicted in Figure 2.

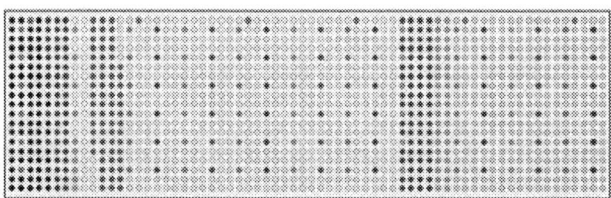

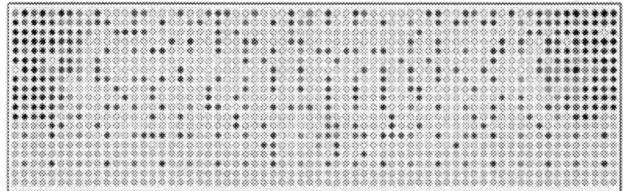

Figure 5: Logic partition interfaced to Wide IO memory. (Top) before optimization, (Bottom) after optimization.

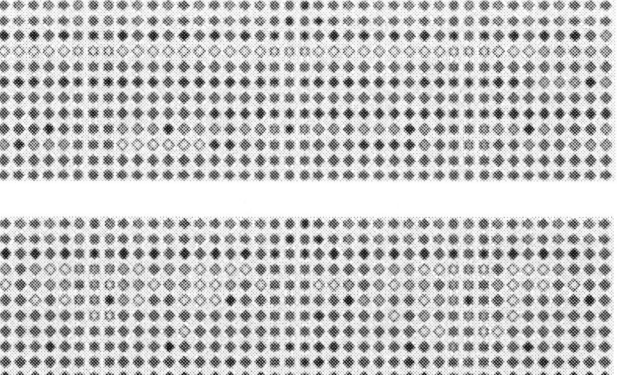

Figure 6: Portion of silicon interposer depicting, (Top) before optimization, (Bottom) after optimization.

Finally, BGA-3 package ball pattern is optimized with respect to BGA-1 package ball-out (Fig.14) followed by optimizing the Die-3 bonding pads with respect to BGA-3 package ball pattern (Fig. 15, Fig. 16).

To enforce signal/power integrity across the system, Xpedition Path Finder's Rule Editor was used to create and enforce constraint for placement and shielding of critical high speed signals.

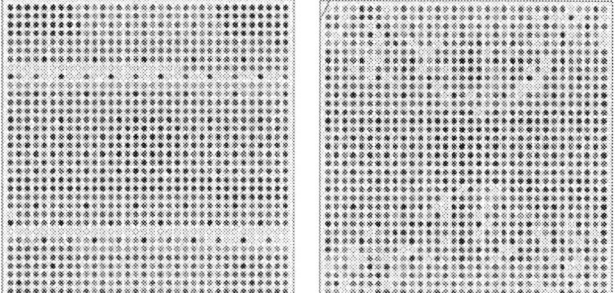

Figure 7: BGA-1 package ball-out. (Left) before optimization, (Right) after optimization.

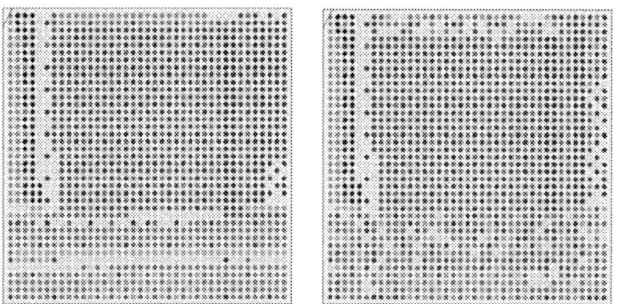

Figure 8: BGA-2 package ball-out. (Left) before optimization, (Right) after optimization.

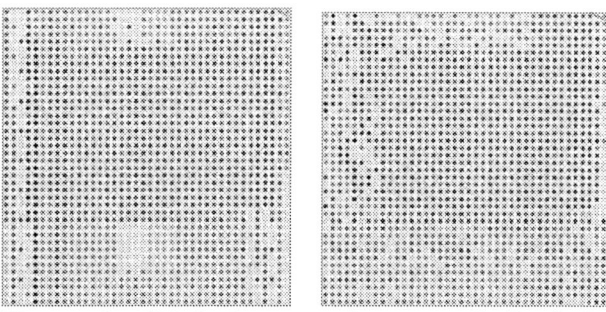

Figure 9: Die-2 bump assignment. (Left) before optimization, (Right) after optimization.

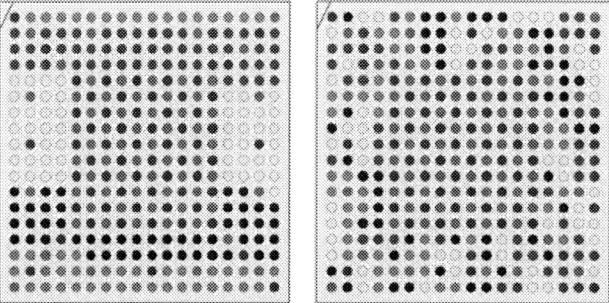

Figure 10: Rigid Interposer package ball-out. (Left) before optimization, (Right) after optimization.

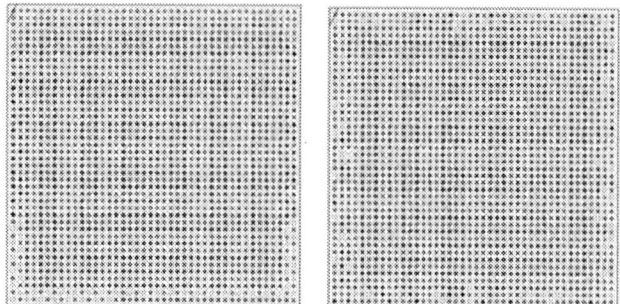

Figure 11: Die-1 bump pattern optimization. (Left) before optimization, (Right) after optimization.

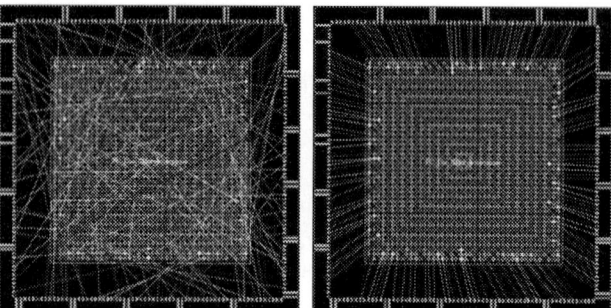

Figure 12: Optimizing Die-1 I/O buffer cell placement with respect to bump. (Left) before optimization, (Right) after optimization.

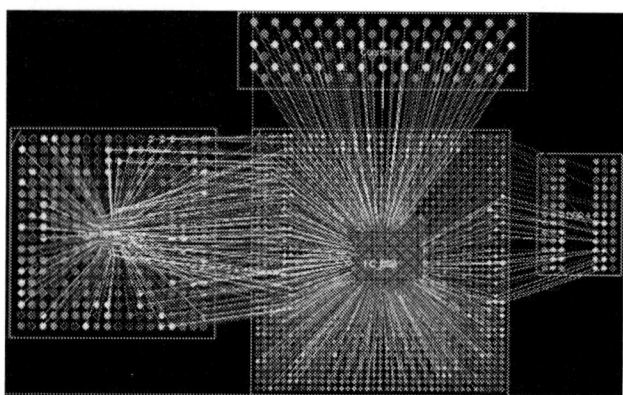

Figure 13: Rigid silicon interposer, Die-1, Die-1 I/O buffer cells, connector, DDR4 package, Die-2 and BGA-2 package fully optimized with respect to BGA-1 package.

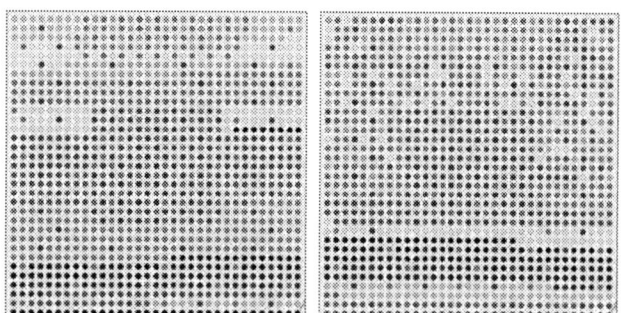

Figure 14: BGA-3 package ball pattern optimized with respect to BGA-1 Package. (Left) before optimization, (Right) after optimization.

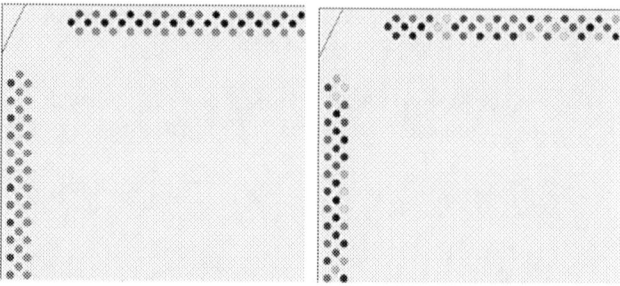

Figure 15: Die-3 bonding pads are optimized. (Left) before optimization, (Right) after optimization.

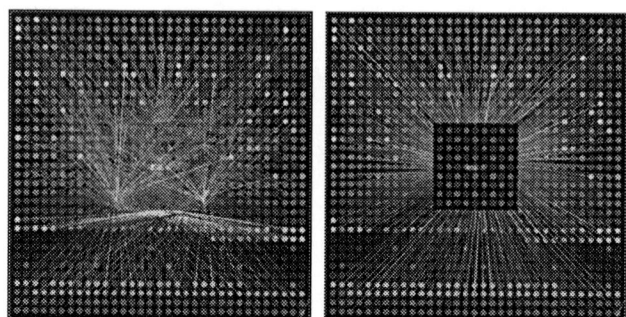

Figure 16: Die-3 optimized with respect to BGA-3 package. (Left) before optimization, (Right) after optimization.

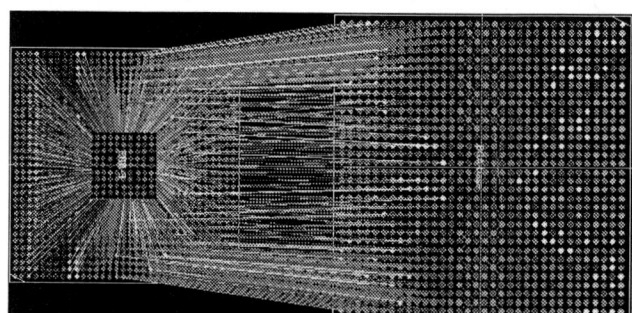

Figure 17: Die-3 and BGA-3 package are optimized with respect to BGA-1 package.

At this point the system should be fully optimized as illustrated in Figure 18. A full system optimization in this case includes optimization across the devices and components. Figure 19 shows the fully optimized high speed Ser/Des differential pairs across the devices and components. Shielding of these pairs were enforced by applying shielding constraints during placement.

3. Conclusion

In this study we demonstrated a methodology for pathfinding and design optimization of a 2.5D/3D package interfaced with a monolithic SOC flip chip die, a rigid silicon interposer package, a DDR4 package, a connector and a wirebonded BGA package.

This methodology can be used to optimize wide variety of devices and systems, furthermore, it can be used to perform trade-off analysis and product feasibility analysis. Such pathfinding and optimization should result in lower cost and higher performance devices and systems.

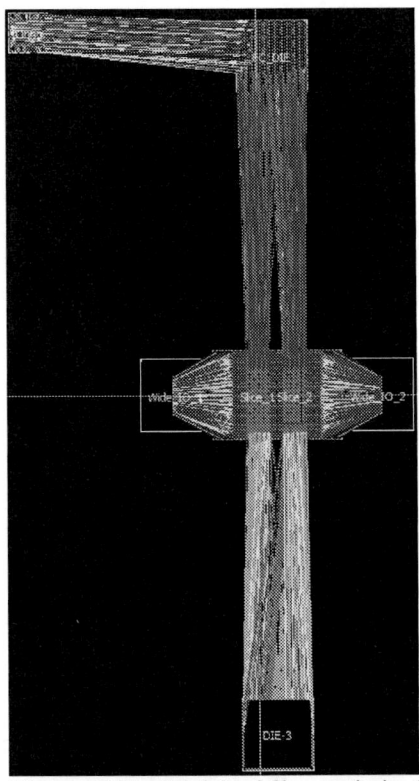 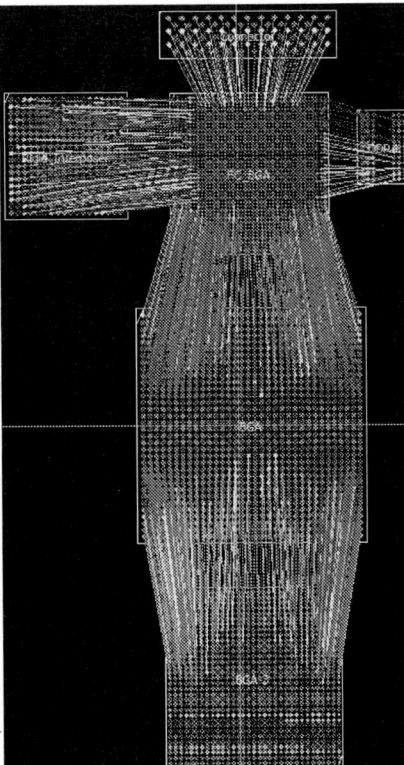 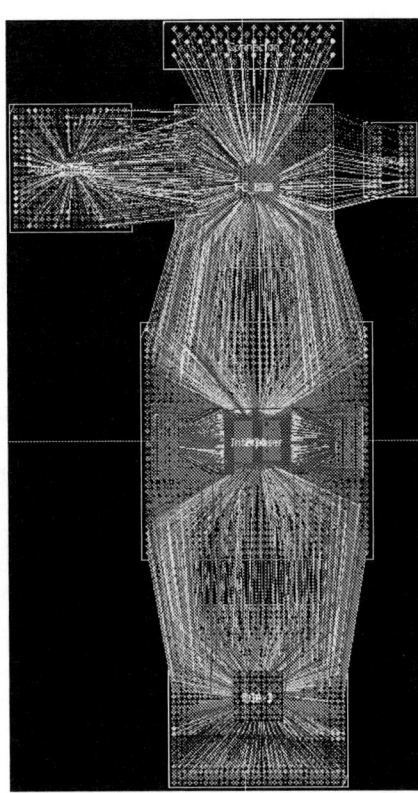

Figure 18: (Right) fully optimized system, (middle) optimized components ball pattern across the system, (Left) all devices optimized across the system.

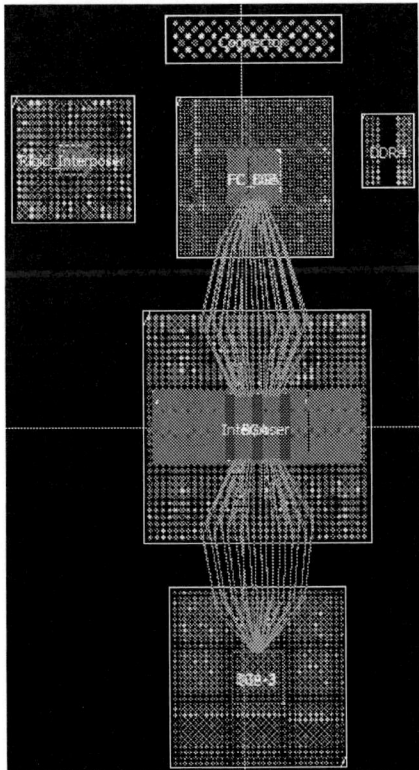

Figure 19: High speed Ser/Des differential pairs are fully optimized across devices and components.

References

1. Solid State Technology, " *High cost per wafer, long design cycles may delay 20nm and beyond*," http://electroiq.com/petes-posts/2014/01/22/high-cost-per-wafer-long-design-cycles-may-delay-20nm-and-beyond/.

2. EE Times, "*Silicon wafer prices increase again*," http://www.eetimes.com/document.asp?doc_id=1161241

3. EET Asia, " *Polysilicon costs forces wafer makers to increase price*," http://www.eetasia.com/ART_8800689375_480200_NT_8c571f50.HTM.

4. Kuan H. Lu, Suk-Kyu Ryu, Qiu Zhao, Xuefeng Zhang, Jay Im, Rui Huang, and Paul S. Ho, "*Thermal Stress Induced Delamination of Through Silicon Vias in 3-D Interconnects*", Proceedings of the 60h IEEE Electronic Components and Technology Conference, pp.40-45, las vegas, Nevada, June1-4, 2010.

5. J. Cho et al, "*Modeling and Analysis of Through-Silicon Via (TSV) Noise Coupling and Suppression Using a Guard Ring,*" accepted for publication at IEEE Trans. Advanced Packaging, 2011.

6. J. U. Knickerbocker et al, "*Three-Dimensional Silicon Integration,*"IBM J. Res. & Dev., Vol. 52, No. 6 (November 2008), pp. 553-569.

7. E. Vick, S. Goodwin, and D. Temple, "*Electrical Demonstration of TSV Interconnects and Multilevel Metallization for 3D Si Interposer Applications,*" IMAPS 2010- 43rd International Symposium on Microelectronics, Research Triangle Park, NC, Nov. 2010, pp. 7-14.

8. Thadesar, P.A.; Dembla, A.; Brown, D.; Bakir, M.S., *"Novel through-silicon via technologies for 3D system integration,"* Interconnect Technology Conference (IITC), 2013 IEEE International, Page(s): 1 - 3.

9. Heegon Kim; Jonghyun Cho; Kim, J.J.; Jung, D.H.; Sumin Choi; Joungho Kim; Junho Lee; Kunwoo Park, " *Eye-diagram simulation and analysis of a high-speed TSV-based channel,"* 3D Systems Integration Conference (3DIC), 2013 IEEE International, Page(s): 1 - 7.

10. Sunohara M, Sakaguchi H, Takano A, Arai R, Murayama K, Higashi M, *"Studies on Electrical Performance and Thermal Stress of a Silicon Interposer with TSVs,"* ECTC, Conference Proceedings, 2010.

11. Karim, M.A.; Franzon, P.D.; Kumar, A., *"Power comparison of 2D, 3D and 2.5D interconnect solutions and power optimization of interposer interconnects,"* Proceedings of the 63rd IEEE Electronic Components and Technology Conference, pp.860-866, 2013.

12. JEDEC, Global Standards for the Microelectronics Indudstry, http://www.jedec.org/category/technology-focus-area/mobile-memory-lpddr2-3-wide-io-memory-mcp.

Cost Effective Interposer for Advanced Electronic Packages

Satoru Kuramochi[1], Sumio Koiwa[1], Kousuke Suzuki[1], Yoshitaka Fukuoka[2]

[1]Dai Nippon Printing Co., Ltd., 250-1 Wakashiba, Kashiwa-shi, Chiba-ken, Japan

[2] Weisti(Worldwide Electronic Integrated Substrate Technology Inc)

E-mail: Kuramochi-S2@mail.dnp.co.jp

Tel: +81-4-7134-2108, Fax: +81-4-7134-1265

Abstract

As electronic product becomes smaller and lighter with an increasing number of function ↵ the demand for high density and high integration becomes stronger.! Interposers for system in package will became more and more important for advanced electronic systems. Silicon interposers with through silicon vias (TSV) and back end of line (BEOL) wirring offer compelling benefits for 2.5D and 3D system integration ;! however, they are limited by high cost and high electrical loss. On the other hand, glass has many properties that make it an ideal substrate for interposer substrates such as; ultra high resistivity, adjustable thermal expansion (CTE) and manufacturability with large panel size. Furthermore, glass via formation capabilities have dramatically improved over the past several months. Fully populated wafers with >100,000 through holes (50μm diameter) are fabricated today with 300μm thick glass. This paper presents the demonstration of glass interposers with fine pitch through glass vias(TGV), ! with 6um RDL lithography. TGVs of 50μm in diameter and 200μm in pitch were formed successfully on 300μm thick alkali-free glass by Focused Electrical Discharging Method (FEDM). The TGVs were filled with solid Copper (Cu) using a void-free electroplating of optimized periodic pulse reverse(PPR) process and chemical mechanical polishing (CMP) as well as the TSVs. Highly insulating TGV with double side polymer insulation resulted in TGV with an insertion loss of less than 0.23dB at 20GHz. Excellent through via reliability was demonstrated, due to double side thick polymer insulator that buffers the stress created by CTE mismatch between glass, copper vias and copper traces, and TGV at 200μm pitch passed 1000 thermal cycles from -40Cdeg to 85Cdeg.

Introduction

A silicon interposer with through silicon via (TSV) is one of the key technologies in microelectronic packaging, system integration and energy saving. This concept has specific advantages in terms on heterogeneous integration of multiple devices such as logics, memories, sensors and so on. The increasing demand for product miniaturization, high package density, high performance and heterogeneous integration of different functional chips has lead to the development of 3D packaging technologies. 3D packaging technology is emerging technology for many SiP designs because of the advantages of smaller size, weight reduction, lower parasite, higher silicon efficiency, lower noise, smaller delay and lower power consumption due to its shorter length connection and reduced wiring density [1]. Wire bonding is one of the conventional interconnect technology in 3D packaging because of its cost effectiveness and flexible interconnect [2]. Recently, silicon interposer with through silicon via (TSV)

interconnects is one of the most promising technologies for heterogeneous stacking of multi functional chips in 3D SiP packaging due to its shorter length connection compared to the conventional wire bonding interconnects. Identical coefficient of thermal expansion (CTE) between silicon interposer and silicon based device chip also increase the reliability of packaging. In other hand, flip chip packaging with the conventional organic buildup package substrate is facing a bottleneck in fine pitch wiring as the interconnect density is continuing to shrink, and the fabrication cost of finer pitch organic substrate is increasing significantly. For the solution of these needs, TSV interposer has been emerged as a key technology for high wiring density interconnection, low CTE mismatch between the Cu/low-k die and the organic buildup substrate, and improved electrical performance of interconnect delay due to its shorter length interconnects from the device chip to the interposer substrate [3].

In recent years, SiP stacks using silicon carrier based substrate is focused on due to its routing capabilities and improved reliability [4-8]. Silicon interposers with through silicon vias (TSVs) are developed to meet this demand, but face two major barriers; high electrical loss, and cost [9].

Table1. Comparison of Si interposer and Glass interposer

Technology	Si interposer	Glass interposer
Material	Si	Glass
Thickness	60-200 μ m	100-300 μ m
Hole	DRIE	FEDM
Metal	Cu	Cu
RDL	Cu Damacine	Cu-Polyimide Build-up
Dielectric	SiO2	Polyimide
Size	300mm wafer	200mm wafer – 730x900mm Panel
Chip Number	47(30mm□)	667(30mm□)

On the other hand, glass has many properties that make it an ideal substrate for interposer substrates such as; ultra high resistivity, adjustable thermal expansion (CTE) and manufacturability with large panel size. Furthermore, glass via formation capabilities have dramatically improved over

2014 Electronic Components & Technology Conference

the past several months [10]. Fully populated wafers with >100,000 through holes (50μm diameter) are fabricated today with 300μm thick glass.

Table1 shows comparison of Si interposer and Glass interposer. Conventional Si interposer fabricated using 300mm wafer BEOL. On the contrary Glass interposer extend to fabricate G4.5(730mmx900mm) large panel line. This paper presents the demonstration of glass interposers with fine pitch through glass vias(TGV),with 6um RDL lithography. TGVs of 50μm in diameter and 200μm in pitch were formed successfully on 300μm thick alkali-free glass by Focused Electrical Discharging Method (FEDM). The TGVs were filled with solid Copper (Cu) using a void-free electroplating of optimized periodic pulse reverse(PPR) process and chemical mechanical polishing (CMP) as well as the TSVs. Void free TGV interconnects were successfully formed. In order to check the connectivity of TSV, the X-ray inspection was done and no significant failures were observed. Highly insulating TGV with double side polymer insulation resulted in TGV with an insertion loss of less than 0.23dB at 20GHz. Excellent through via reliability was demonstrated, due to double side thick polymer insulator that buffers the stress created by CTE mismatch between glass, copper vias and copper traces, and TGV at 200μm pitch passed 1000 thermal cycles from -40Cdeg to 85Cdeg. The fabrication process of TGV based glass interposer was successfully demonstrated in terms of cost effective production.

Experimental result

A glass interposer with TGV interconnects was fabricated on 200 mm glass wafer. The process flow was shown in Figure 1. First of all, TGVs of 50μm in diameter and 200μm in pitch were formed on 300μm thick alkali-free glass by Focused Electrical Discharging Method (FEDM). Ti/Cu seed layer deposition. Then the via was filled up with Cu by electroplating of PPR and Cu overburden from the wafer surface was removed by Cu chemical mechanical polishing (CMP). Thick dielectric polymer layer of 8 μm was coated on the wafer as RDL passivation film for metallization using photosensitive polyimide followed by low temperature cure of 200 degree Celsius. PI was shrink to 4 μm by cure process followed by Cu seed layer deposition. Redistribution lines were patterned with positive photo resist of 10 μm in thickness. Cu RDL line of 3 μm thickness was deposited by Cu electroplating followed by photo resist and Cu seed layer removal

High aspect ratio TGV interconnect fabrication

The Cu filled TGV interconnects were fabricated on 200 mm glass wafer using via first approach for interposer of 300 μm in thickness. The diameter of the TGV was 50 μm and the aspect ratio of the TGV was 6. The via formation process was optimized to achieve smooth surface and no defect was observed on the via sidewall. The Cu electroplating process was optimized for the high aspect ratio via.

Via formation

TGVs of 50μm in diameter and 200μm in pitch were formed on 300μm thick alkali-free glass by Asahi Glass corporation with Focused Electrical Discharging Method (FEDM). Mainly it consists of two steps of that focused and controlled electrical discharging created locally molten regions of glass, and finally it induced dielectric breakdown together with internal high pressure by Joule heat and ejection of glass[11].

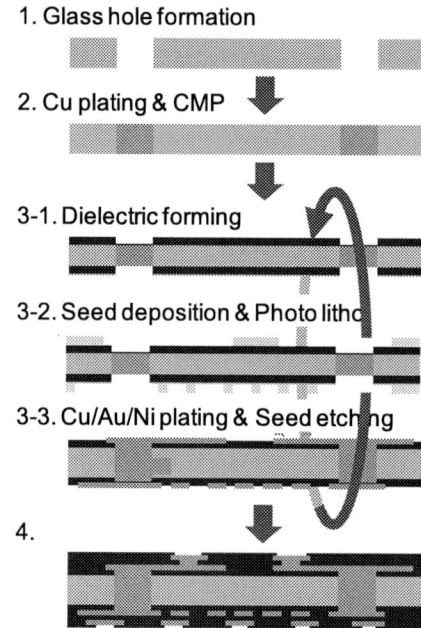

Figure 1. Process flow.

Figure 2 shows the result of TGV formation for 300μm thick glass. In the case of TGV for 300μm thick glass, the top diameter is approximately 60μm and the bottom diameter is approximately 40μm. TGV side wall were smoothed by fine polishing because the process using electrical discharging made glass locally heated and melted by high temperature.

Figure 2. Cross sectional view of via formation FEDM.

Via filling

Uses of conductive interconnect material such as polysilicon, tungsten, or copper, 3D-IC technology enables z-axis interconnection between die stacks. Many factors can influence the via filling performance. Non-optimized filling will lead to the voids occurring inside the via. Such void defects will impact current density, chip performance and the life time of the device. Therefore, it is very important to understand the quality of the Cu filling within TGV substrate [12]. The key requirement for void free via filling was the hydrophilic surface of the barrier and Cu seed layer [13]. After depositing the seed layer, plasma ashing was done from the both sides of the wafer to improve the hydrophilic property of the surface of Cu seed layer. Figure 3 shows the result of Cu film thickness distribution by electroplating with 200 mm glass wafer. Our electro chemical deposition (ECD) system was optimized for 300 µm deep via and void free Cu via filling in 200 mm wafer level was successfully achieved by electroplating of PPR.

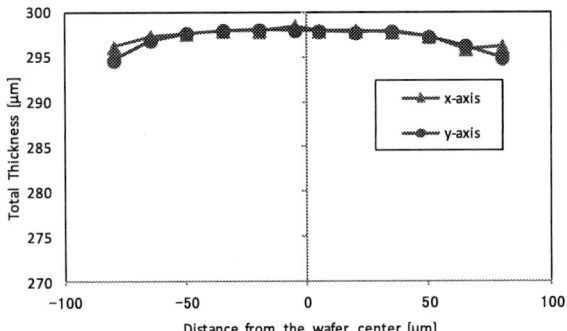

Figure 4. Distribution of dishing.

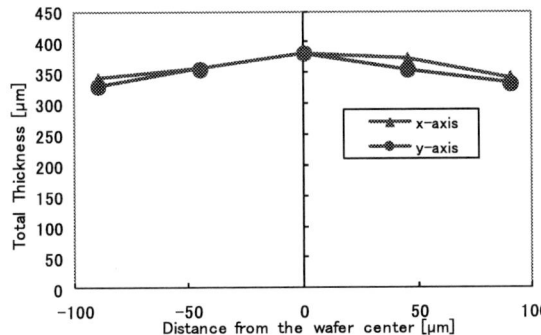

Figure 3. Distribution of the Cu overburden thickness

The Cu overburden thickness from the wafer surface after Cu via filling was measured (Figure 3). The measured TGVs were on horizontal and vertical axis through the wafer center. The Cu overburden thickness was less than 100 µm and the thickness of the wafer center was larger than that of the wafer edge.

Cu CMP

Thick Cu overburden layer of 50-100 µm was formed on the glass wafer surface due to its long Cu electroplating to fill the deep and high aspect ratio vias with void free interconnects and it caused large stress in the wafer which lead wafer warpage. The highly stressed Cu overburden was removed by Cu CMP process. The Cu overburden was removed by Cu CMP, but the dishing occurred as shown in Figure 4. The dishing amount of the Cu via were less than 6µm in wafer.

Figure 5 shows optical image of TGVs filled with Cu. In the case of TGV for 300µm thick glass, the top diameter is approximately 60µm and the bottom diameter is approximately 40µm. Figure 6 shows X-ray image of Cu filling via. The X-ray inspection shows that void free TGV interconnects were successfully formed.

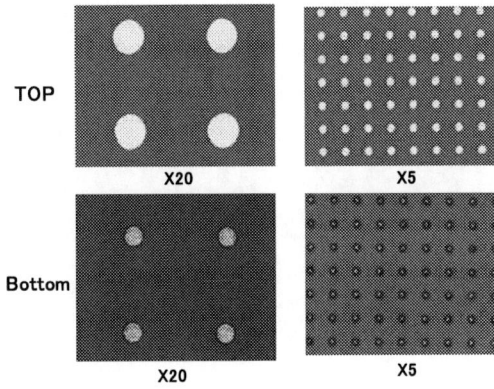

Figure 5. Fabricated TGV

Figure 6. Fabricated TSV and X-ray inspection.

Redistribution layer

The application of glass interposer to 2.5D integration of multiple chips requires ultra-fine line and space. RDL fine pitch copper traces can reduce the number of wiring layers required to interconnect area array micro-bumps at fine pitch. A semi-additive process consisting of copper seed layer deposition, photolithography, and electrolytic copper pattern

978-1-4799-2408-0/14 $31.00 © 2014 IEEE   1675

plating used for RDL. As the copper conductor width is reduced below 5µm, signal loss increase significantly[14]. Thick dielectric polymer layer of 8 µm was coated on the glass wafer as RDL passivation film for metallization using photosensitive polyimide followed by low temperature cure of 200 degree Celsius. Polyimide layer was shrink to 4 µm by cure process followed by Cu seed layer deposition. Redistribution lines were patterned with positive photo resist of 10 µm in thickness. Cu RDL line of 3 µm thickness was deposited by Cu electroplating followed by photo resist and Cu seed layer removal. Figure 7 shows overview of TGV with RDL line. Figure 8 shows the electroplated Cu lines with 6 µm line width.

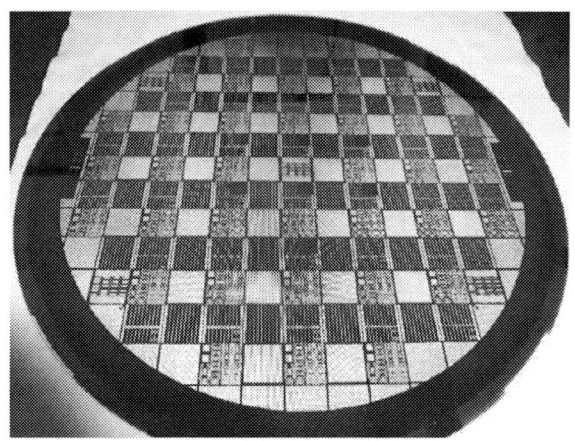

Figure 7. Fabricated RDL line.

Figure 8. Fabricated 6/6 µm line and space RDL lines.

Thermal Cycle Reliability Testing

The TGV array daisy chain structures consisting of double side metal layers were design to assess thermo-mechanical reliability of TGVs in glass. The TGV array daisy chain had 100 TGVs. The TGVs had 60µm via diameter on top and 40µm via diameter on bottom, with a center –to-center pitch 200µm. The TGVs were connected with 50µm width double side metal layer. The top view of the TGV array daisy chain sample is Figure 9.

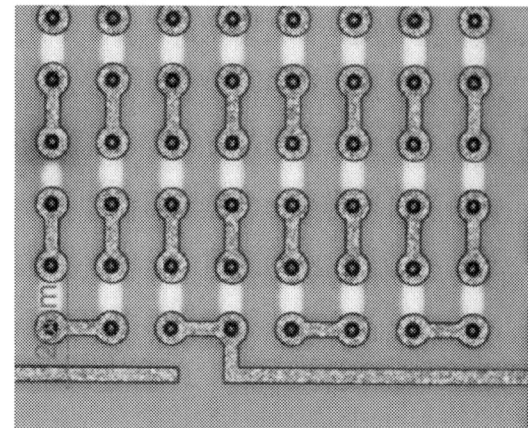

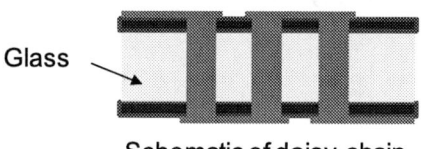

Schematic of daisy-chain

Figure 9. Top view of the TGV array daisy chain.

The electric characteristics of fabricated TGV array daisy chain were evaluated before thermal cycling test. Figure 10 shows the via resistance distribution with 26 arrays measured by four-terminal method. The calculated resistance of the TGV array daisy chain is 2.61ohm. The graph shows that via resistances were normally distributed around the calculated resistance and tangible electric characteristics were achieved.

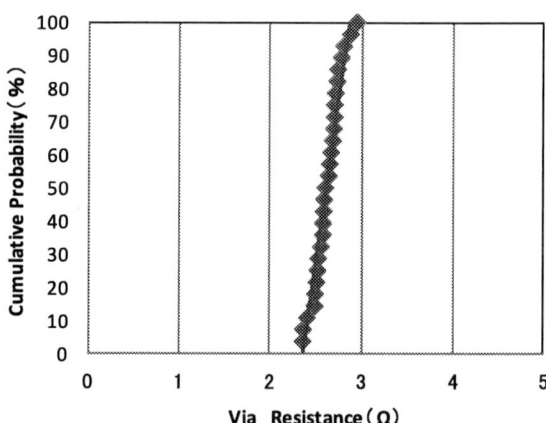

Figure 10. Distribution of the resistance of the TGV array.

The test vehicles were subjected to thermal cycles between -40Cdeg to 85Cdeg with a dwell-time of 30 minutes at each temperature (MIL standard 833). The sample was taken out at 100, 200, 500, 1000cycles, and the daisy chain resistance was measured to monitor TGV failures. Figure 11 shows resistance change s for 10 daisy chains consisting of 100 TGVs each, during thermal cycling up to 1000 cycles. Figure 11 shows no significant resistance changes were observed during the test. Excellent through via reliability was demonstrated, due to double side thick polymer insulator that buffers the stress created by CTE mismatch between glass, copper vias and copper traces, and TGV at 200μm pitch passed 1000 thermal cycles from -40Cdeg to 85Cdeg.

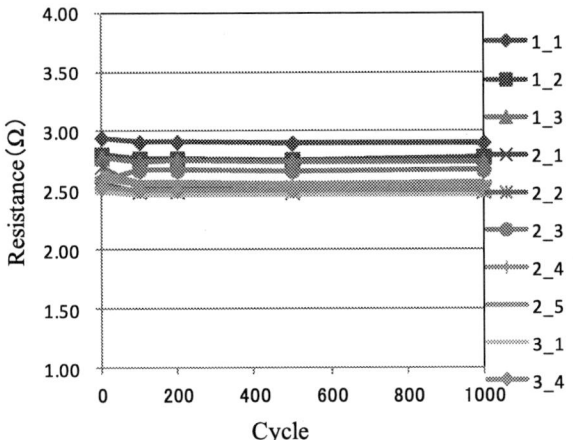

Figure 11. Resistance Changing during Thermal Cycling

High frequency electrical characterization of glass interposer

High frequency electrical characteristics of glass interposer were evaluated. Double side metal layer test vehicle containing co-planer waveguides (CPW) with TGV transmission were designed. Figure 12 shows double side metal layer co-planer waveguide test vehicle for high frequency electrical characterization.

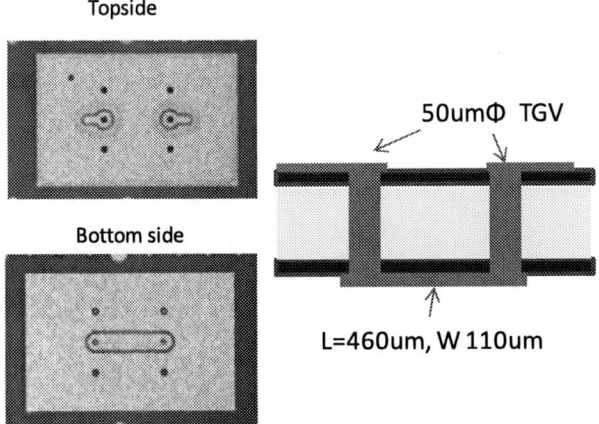

Figure 12. Co-planer waveguide test vehicle

Electronic performances were measured by a network analyzer (Agilent Technologies Inc., N5230A) with a GSG probe. The probe pitch is 250μm. The frequency was swept from 100MHz to 20GHz, and S11 (reflection) and S21 (transmission) were extracted.

Electrical measurement were performed after SOLT calibration and the CPW transmission with TGV transitions were characterized up to 20GHz. Figure 13 shows measurement result of transmission(S21). Figure 14 shows measurement result of reflection(S11). The network analyzer S-parameter measurements indicate that good result obtained a very low loss in transmission. Highly insulating TGV with double side polymer insulation resulted in TGV with an insertion loss of less than -0.23dB at 20GHz.

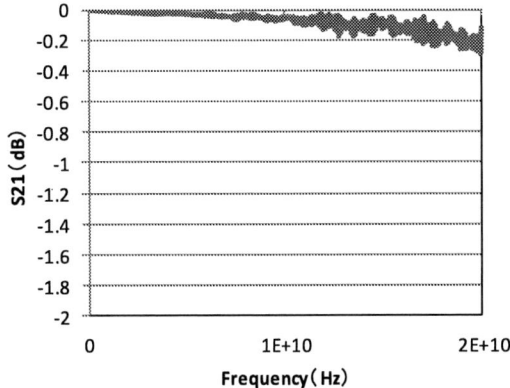

Figure 13. Measurement result of transmission(S21)

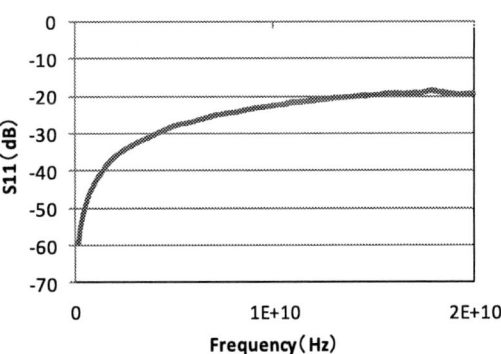

Figure 14. Measurement result of reflection(S11)

Conclusions

A glass interposer with TGV interconnects was demonstrated with fine pitch through glass vias(TGV),with 6um RDL lithography. TGVs of 50μm in diameter and 200μm in pitch were formed successfully on 300μm thick alkali-free glass by Focused Electrical Discharging Method (FEDM). The TGVs were filled with solid

978-1-4799-2408-0/14 $31.00 © 2014 IEEE

Copper (Cu) using a void-free electroplating of optimized periodic pulse reverse(PPR) process and chemical mechanical polishing (CMP). Excellent through via reliability was demonstrated, due to double side thick polymer insulator that buffers the stress created by CTE mismatch between glass, copper vias and copper traces, and TGV at 200µm pitch passed 1000 thermal cycles from -40Cdeg to 85Cdeg. Highly insulating TGV with double side polymer insulation resulted in TGV with an insertion loss of less than 0.23dB at 20GHz. The fabrication process of TGV based glass interposer was successfully demonstrated in terms of cost effective production.

Acknowledgments

The authors would like to thank Asahi glass corporation for supplied glass substrate with TGVs.

References

1. K. Zoschke et al., "TSV based Silicon Interposer Technology for Wafer Level Fabrication of 3D SiP modules," Proceedings 61st Electronic Components and Technology Conference, 2011, pp. 836-843.

2. Masamichi Ishihara et al., "A Dual Face Package Using a Post with Wire Component: Novel Structure for PoP, Wafer Level CSP and Compact Image Sensor Packages," Proceedings 58th Electronic Components and Technology Conference, 2008, pp. 1093-1098.

3. Xiao Wu Zhang et al., "Development of Through Silicon Via (TSV) Interposer Technology for Large Die (21x21 mm) Fine Pitch Cu/low-K FCBGA Package," Proceedings 59th Electronic Components and Technology Conference, 2009, pp. 305-312.

4. Vempati Srinivasa Rao et al., "TSV Interposer Fabrication for 3D IC Packaging," Proceedings 11th Electronics Packaging Technology Conference, 2009, pp. 431-437.

5. Kouichi Kumagai et al., "A Silicon Interposer BGA package with Cu-Filled TSV and Multi-Layer Cu-Plating Interconnect," Proceedings 58th Electronic Components and Technology Conference, 2008, pp. 571-576.

6. Masahiro Sunohara, Takayuki Tokunaga, Takashi Kurihara and Mitsutoshi Higashi, "Silicon Interposer with TSVs (Through Silicon Vias) and Fine Multilayer Wiring," Proceedings 58th Electronic Components and Technology Conference, 2008, pp. 847-852.

7. Seung Wook Yoon et al., "Reliability Studies of a Through Via Silicon Stacked Module for 3D Microsystem Packaging," Proceedings 56th Electronic Components and Technology Conference, 2006, pp. 1449-1453.

8. David Henry et al., "Development and Characterization of High Electrical Performances TSV for 3D Applications," Proceedings 11th Electronics Packaging Technology Conference, 2009, pp. 528-535.

9. Venky Sundaram et al., "Low Cost High Performance, and High Reliability 2.5D Silicon Interposer," Proceedings 63rd Electronic Components and Technology Conference, 2013, pp. 342-347

10. John Keech et al., "Fabrication of 3D-IC Interposers," Proceedings 63rd Electronic Components and Technology Conference, 2013, pp. 1829-1833

11. Shintaro Takahashi et al., "Development of Through Glass Via (TGV) Formation Technology Using Electrical Discharging for 2.5/3D Integrated Packaging," Proceedings 63rd Electronic Components and Technology Conference, 2013, pp. 348-352

12. John Keech et al., "Fabrication of 3D-IC Interposers," Proceedings 63rd Electronic Components and Technology Conference, 2013, pp. 1829-1833

13. Seiichi Yoshimi et al., "Development of 300mm TSV Interposer with Redistribution Layers on Both Sides using MEMS Process," Proceedings 63rd Electronic Components and Technology Conference, 2013, pp. 2168-2172

14. Venky Sundaram et al., "Low Cost High Performance, and High Reliability 2.5D Silicon Interposer," Proceedings 63rd Electronic Components and Technology Conference, 2013, pp. 342-347.

Thermal Management for Wafer Level Packaging (WLP)

Tiao Zhou[1], and Arkadii Samoilov[2]

Maxim Integrated

[1]Dallas, TX, USA

[2]San Jose, CA, USA

Abstract

In this study, the thermal performance of wafer level packaging (WLP) in still air environments is characterized with thermal measurements. Thermal test dice with built-in heaters and temperature sensors are used. Effects of WLP size, WLP design, power dissipation (Pd) area on die, and heat spreaders are investigated. Temperature sensors at different die locations are used to map the die temperature. WLP and board resistance contributions to the overall thermal resistance are also assessed.

It is found that WLP package resistance is only a small portion of total junction to ambient thermal resistance. The heat spreading capability of the PCB significantly affects the overall thermal resistance. WLP design details do not make a significant difference since heat spreading in WLP is carried out in Si. Small WLP has higher thermal resistance. Small Pd area results in higher thermal resistance. Furthermore, localized heating causes "hot spots". Heat spreaders can enhance the thermal performance.

1. Introduction

Wafer level packaging (WLP) now plays an important role in portable electronics. WLP is preferred in many applications mainly due to its small form factor, better signal integrity, and low package cost. The trend of more functionality and higher density results in higher power dissipation per unit die area. Thermal management of WLP is more challenging than conventional plastic packages since there is no additional packaging material to spread the heat. It is important to understand the heat dissipation limit and thermal enhancement options for WLP, so this package can be used in applications where the heat generated on die is safely removed.

WLP thermal studies found in the literature have been limited to numerical modeling. Gualandris and Villa [1] studied the thermal performance of fan out wafer level packaging (FO-WLP). It is found that FO-WLP has lower junction to ambient thermal resistance (θ_{JA}) than flip chip BGA due to the lower vertical thermal resistance. There was a visible thermal resistance increase with reduced power dissipation (Pd) area. Thermal vias on PCB reduced thermal resistance by 15%. Lau and Yue [2] investigated thermal resistance of 3D IC integration with TSV (through silicon via) with finite element modeling. It was found that the θ_{JA} increases with reduced Pd area on die. Localized heat sources caused hot spots. Spacing out the localized heat sources helps to reduce the maximum junction temperature. Empirical equations and guidelines for the equivalent thermal conductivity of chips and thermal resistance in different design scenarios were provided for practical engineering convenience.

Thermal experimental characterization for WLP is not available in the literature. Thermal measurement data are useful to establish the baseline of WLP thermal performance of different configurations. They can be used to validate future modeling work. This study focuses on thermal measurement of WLP in JEDEC standard [3-8] and application like configurations using thermal test vehicles.

In the following sections, test vehicles and test setup are described first. Test data obtained in different design and environment scenarios are presented next, followed by discussions and conclusions.

2. Test vehicle and test setup

2.1 Test dice

Thermal test dice are used for this study. The design of the test dice follows JEDEC guideline [6]. Two die sizes 2.4x2.4 mm and 7.2x7.2 mm are considered. The test dice are processed into 0.4 mm pitch 6x6 and 18x18 array WLP for these two die sizes, respectively. Multiple heater and temperature sensitive parameters (TSP) or temperature sensors are implemented onto the test dice. Figure 1 shows the locations of heater and temperature sensors for the 2.4x2.4 mm test die. It is seen that the area heater covers almost the entire surface. To simplify the discussion, this area heater will be referred to as 2.4x2.4 mm distributed heater. Localized heaters and temperature sensors have sizes of 25x25 μm. They are placed at die center A and die corner B. When a localized heater is powered up, the Pd area is only 0.01% of the total die area. A large temperature gradient is expected. The heater and sensor locations for the 7.2x7.2 mm test die are depicted in Figure 2. As is seen, there are nine 2.4x2.4 mm distributed heaters. They can be powered up individually to achieve different Pd areas. Four temperature sensors are used for this WLP test die. One is at center of the corner heater (A), one is at the corner of the corner heater (B), one at the die center (C), and one at the die corner (D). With these four sensors, the temperature map of 7.2x7.2 mm WLP can be achieved. Figure 3 shows three Pd area options considered for 7.2x7.2 mm WLP, 2.4x2.4 mm, 4.8x4.8 mm, and 7.2x7.2 mm. For these options, the Pd areas are 11%, 44% and 100% of the entire test die area, respectively. Thermal measurements for all these Pd area options are performed.

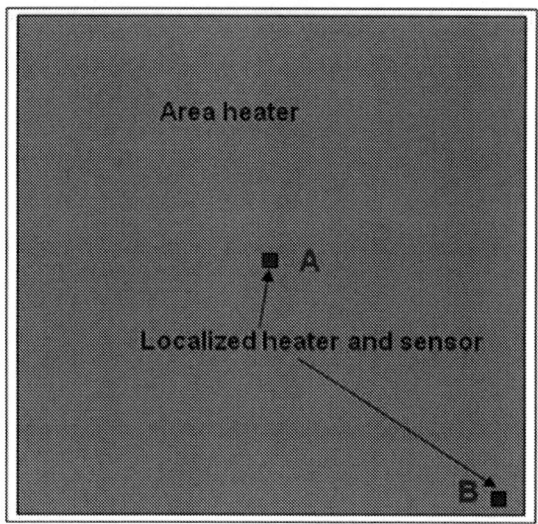

Figure 1. Heater and sensor locations for 6x6 array 2.4x2.4 mm WLP thermal test dice A, B, and C.

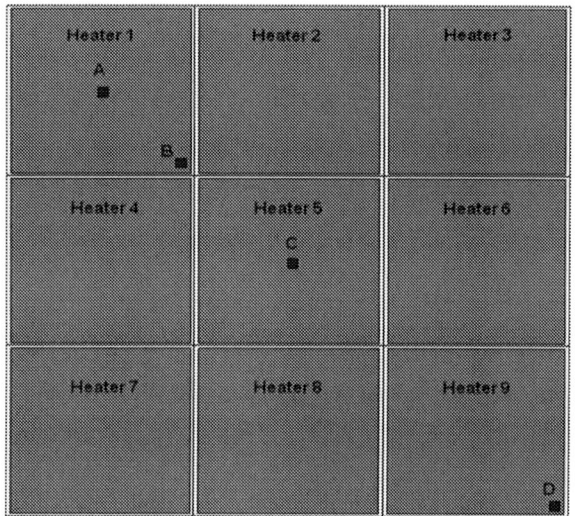

Figure 2. Heater and sensor locations for 18x18 array 7.2x7.2 mm pitch WLP thermal test die D.

In order to understand the effect of WLP design and die design, different WLP via sizes (baseline and four times the baseline) and redistribution layer (RDL) widths (baseline and 10 times the baseline) are considered. The WLP parameters of the four test vehicles are listed in table 1.

Table 1. WLP thermal test die design parameters.

Test vehicle	A	B	C	D
WLP array	6x6	6x6	6x6	18x18
WLP size (mm)	2.4x2.4	2.4x2.4	2.4x2.4	7.2x7.2
WLP via area	1x	1x	4x	1x
WLP RDL width	1x	10x	10x	1x
Temperature sensor locations	Figure 1			Figure 2

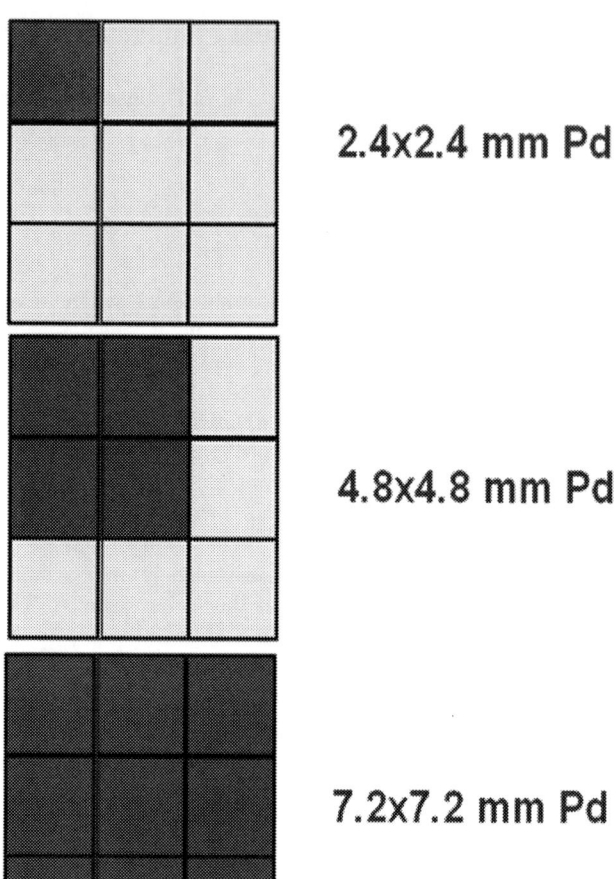

2.4x2.4 mm Pd

4.8x4.8 mm Pd

7.2x7.2 mm Pd

Figure 3. Illustration of power dissipation (Pd) areas that are achieved by powering up selected heaters (marked red).

2.2 Test board

Test boards are designed according to JESD51-9 [9]. The layout of the test board for the 2.4x2.4 mm WLP is depicted by Figure 4. The board size is 101.5x114.5 mm (4x4.5"). There are four metal layers. Inner layers are solid planes with 1 Oz Cu, and the outer layers have 1.5 Oz Cu. Trace width is 200 μm. The total board thickness is 1.6 mm. Standard FR4 material is used. A thermocouple is soldered to an exposed metal pad next to the package footprint for board temperature measurement.

2.3 Test environment

JEDEC still air junction to ambient resistance test [5] is employed in this experimental study. The test environment is a 1 ft^3 box with low thermal conductivity sidewalls. A 40 gage type T thermocouple (TC) is soldered to PCB metal next to each WLP for board temperature measurement. A 36 gage type K TC is used to monitor ambient temperature.

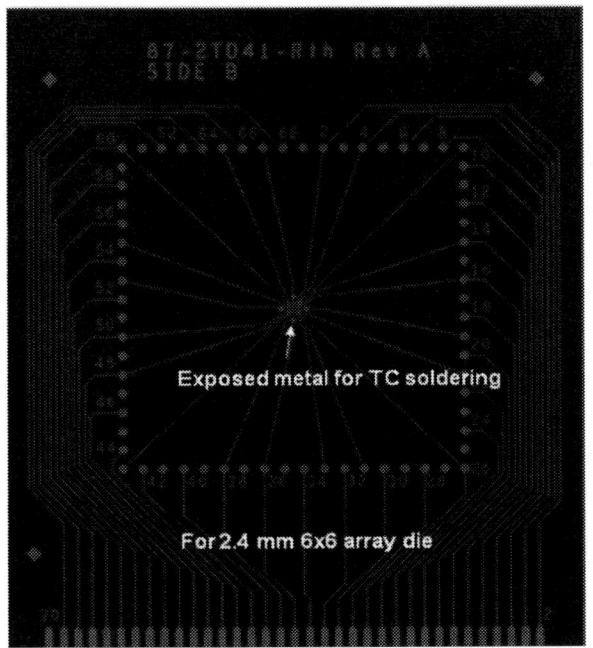

Figure 4. Layout of thermal test board for 6x6 array 2.4x2.4 mm WLP.

2.4 Temperature sensor calibration

The temperature sensors built into the thermal test dice were calibrated for a temperature range of 20 – 120°C using a liquid bath. The calibration curves for nine sensors from 3 different WLP thermal test dice are shown in Figure 5. It is seen from this figure that sensitivity of these temperature sensors is very consistent.

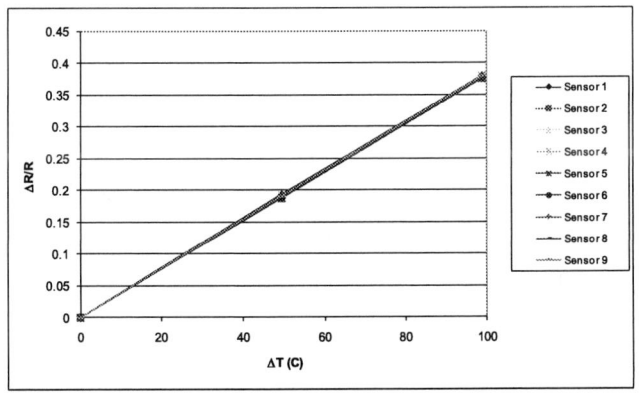

Figure 5. Built in temperature sensor calibration curves.

3. Test Data

Thermal tests are performed for various configurations to study the effects of Pd, die size, WLP design, and heat spreader (HS) applications. Results for different Pd levels are reported first.

3.1 Effect of Pd level

Thermal resistance of the 7.2x7.2 mm WLP at different power dissipation levels is plotted in Figure 6. It is seen that thermal resistance is not sensitive to Pd. To simplify the subsequent discussions, thermal resistance at only one Pd level

is presented. The Pd level is chosen such that the maxim junction temperature is approximately 100°C.

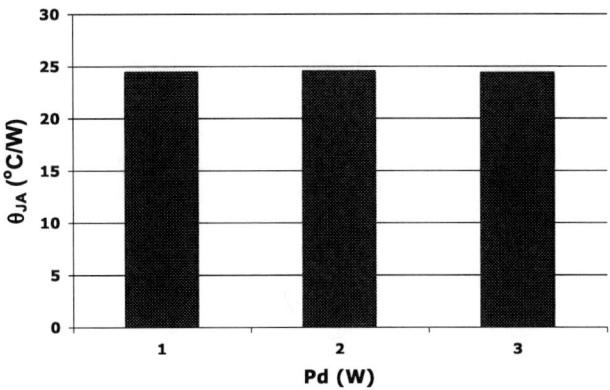

Figure 6. Junction to ambient thermal resistance θ_{JA} for 7.2x7.2 mm WLP.

3.2 Effect of WLP designs

WLP test dice A, B, and C have different WLP via size and RDL widths (Table 1). The measured θ_{JA} values for these three options are plotted in Figure 7. It is seen that θ_{JA} is not sensitive to WLP via size and RDL width. This is probably because thermal resistance of via and RDL is only trivial portion of θ_{JA}. Therefore, WLP thermal performance is not sensitive to WLP design for a given WLP size. It is not necessary to increase the WLP via size and RDL width to improve WLP thermal performance.

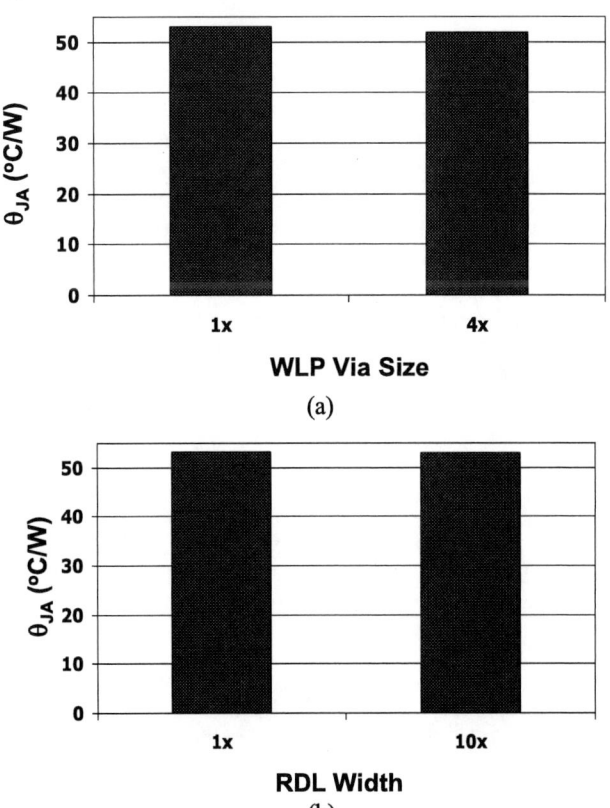

Figure 7. Measured thermal resistance of 2.4x2.4 mm WLP with different WLP via sizes (a) and RDL widths (b).

3.3 Effect of WLP size

Thermal resistance of 2.4x2.4 mm WLP and 7.2x7.2 mm WLP are plotted in Figure 7. Here uniformly distributed Pd is considered. It is seen that the thermal resistance is a strong function of WLP size. It is lower for larger size WLP. This observation is inline with the conclusions for conventional packages.

The θ_{JA} for WLP size of 2.4 to 7.2 mm is in the range of 24 – 54 $^{\circ}$C/W. For 70 $^{\circ}$C maximum allowable junction temperature rise, a WLP within this size range may dissipate 1.3 – 2.9 W, depending on the die size.

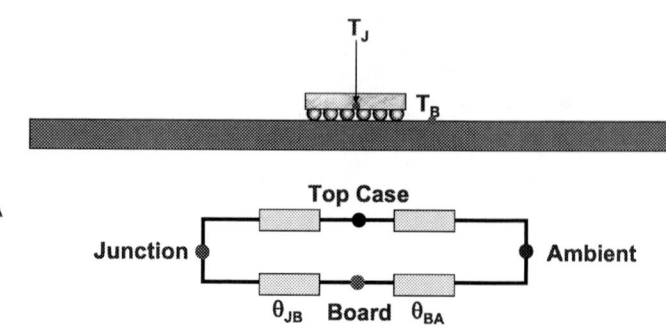

Figure 9. Thermal resistor model and primary thermal path.

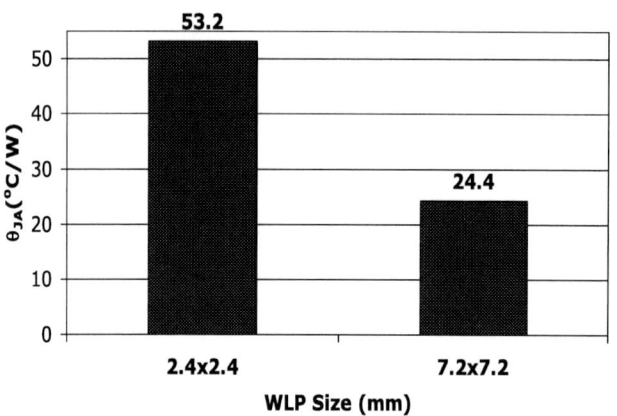

Figure 8. Thermal resistance for 2.4x2.4 and 7.2x7.2 mm WLP. Uniformly distributed Pd is considered.

3.4 Thermal resistance contributors

It is of interest to understand the contributions of the WLP and the PCB in the junction-to-ambient thermal resistance.

In absence of heat sink or heat spreader, the primary thermal path is Junction => WLP solder balls => PCB => ambient. Based on previous work [9], more than 90% of heat is dissipated through the primary thermal path. The secondary thermal path is Junction => WLP top => ambient. This is illustrated by Figure 9. Here the secondary thermal path can be ignored since minimum heat is dissipated through this path. And overall thermal resistance can be viewed as combination of WLP thermal resistance θ_{JB} and board thermal resistance θ_{BA}.

$$\theta_{JA} \simeq \theta_{JB} + \theta_{BA}$$

where θ_{JB} and θ_{BA} are the junction to board and board to ambient thermal resistance, respectively. They can be calculated by

$$\theta_{JB} \simeq (T_J - T_B)/Pd$$
$$\theta_{BA} \simeq (T_B - T_A)/Pd$$

where T_J, T_B, and T_A are junction, board, and ambient temperatures. Pd is the power dissipation. θ_{JA}, θ_{JB}, and θ_{BA} for 2.4x2.4 and 7.2x7.2 mm WLP are plotted in Figure 10.

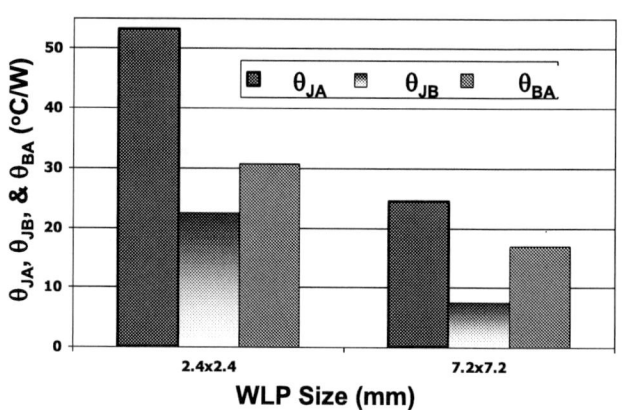

Figure 10. Thermal resistance components for 2.4x2.4 and 7.2x7.2 mm WLP.

It is seen that as WLP size increases, both WLP and board resistance θ_{JB}, and θ_{BA} drop. WLP thermal resistance is lower for a larger WLP due to better lateral heat spreading inside the Si. The lower board resistance θ_{BA} for a larger WLP is due to a larger power/heat input area on the PCB, which in turn reduces the temperature gradient at the WLP footprint.

For both die sizes, WLP thermal resistance θ_{JB} is only small portion of the total resistance θ_{JA}. The majority of θ_{JA} is from board resistance θ_{BA}. WLP thermal resistance is as low as it can be for a given WLP size. And little improvement can be realized at the WLP level. Effective thermal enhancement should be done at board and system level.

3.5 Effect of Pd area

In previous sections, uniform dissipation across the entire WLP die was considered. In this case the die center temperature is the highest and it is considered as the junction temperature. In real applications, Pd on die is often non-uniform. Two investigations are conducted to examine the effect of non-uniform Pd. The first investigation studies the distributed Pd on the 7.2x7.2 mm test die. The second investigation examines localized heat sources using the 2.4x2.4 mm WLP as the test vehicle. Results of these two investigations are presented next.

978-1-4799-2408-0/14 $31.00 © 2014 IEEE   1682

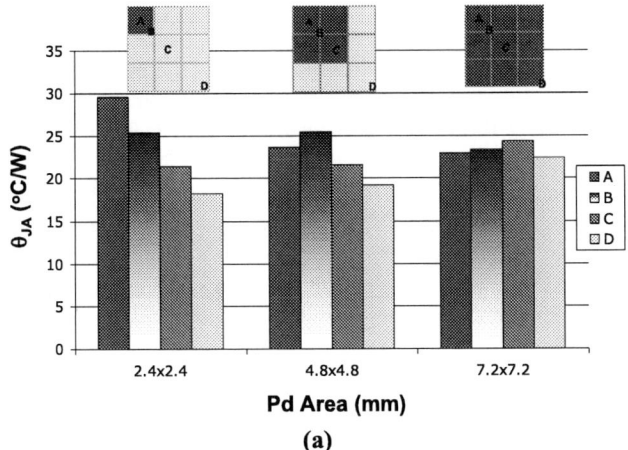

(a)

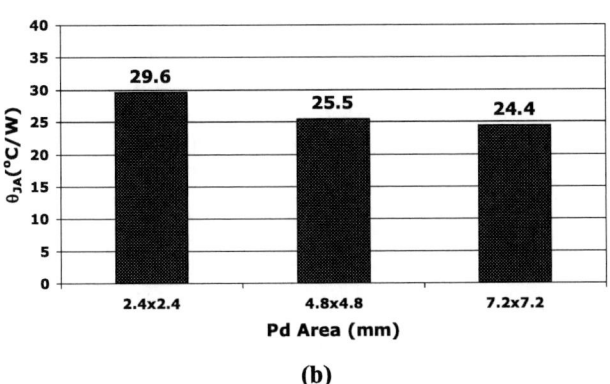

(b)

Figure 11. Thermal resistance of 7.2x7.2 mm WLP due to different Pd areas (marked red). (a) θ_{JA} at observations A, B, C, and D. (b) θ_{JA} at maximum junction temperature.

Selected area heaters on a 7.2x7.2 mm WLP test die are powered up to achieve pd area 1/9, 4/9, and 9/9 of the entire WLP die size. Die/junction temperatures at observations A, B, C, and D are measured. Thermal resistance θ_{JA} based on these observation temperatures are calculated and plotted in Figure 11 (a). It is seen that the maximum temperature is always at the center of the Pd area. The die temperature decreases as the distance from the observation point to Pd area center increases. In the case of 4.8x4.8 mm Pd area, die temperature at C is lower than A. This is because the lateral heat spreading from die center is better than die corner. Thermal resistance calculated based on maximum die temperature is plotted in Figure 11 (b). It is seen that smaller Pd area results in higher thermal resistance. However the difference among these three cases is within 20%. Evidently, as long as the Pd area is at the same order of magnitude as the WLP die size, thermal resistance does not increase significantly to a different order of magnitude.

Localized Pd is studied by using 25x25 μm heaters on the 2.4x2.4 mm WLP test die. Temperatures at die center A and corner B are measured with the temperature sensors. Figure 12 shows the comparison between uniformly distributed Pd and localized Pd. It is seen that when there is uniformly distributed Pd on the entire die surface, temperature gradient across the die is trivial. When the Pd is localized however, a very prominent hot spot is seen at the location of heat source. The maximum θ_{JA} is increased by 20 – 36 times when the Pd area

is changed from uniform heating to localized heating. When the localized heat source is at die corner, the worst case hot spot is seen and it is close to 2000 °C/W. If the maximum allowable junction temperature rise is 70°C, only 35 mW power can be safely dissipated. However, if the heat source is split to two that are far apart (case d), the θ_{JA} is reduced to approximately 800 °C/W, and 85 mW heat can be dissipated with the same allowable junction temperature rise.

Therefore, localize heat sources generate significant hot spots. Spacing out the localize heat sources instead of lumping them together can significantly reduce the prominence of the hot spots.

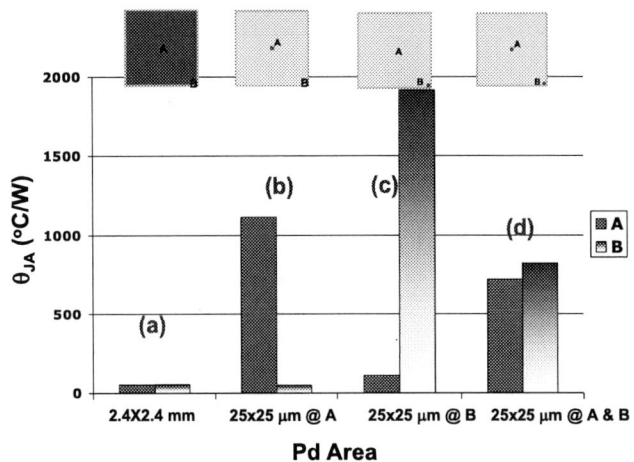

Figure 12. Thermal resistance due to different Pd area for 2.4x2.4 mm WLP. Pd areas are marked red. (a) uniformly distributed Pd, (b) 25 um localized Pd at die center, and (c) 25 um localized Pd at die corner.

3.6 Effect of Heat Spreaders (HS)

Heat sinks are often used for high power applications. For portable electronics however, available space is limited to accommodate the heat sinks. Only thin heat spreaders (HS) are practical for thermal enhancement. In this study, flexible Cu HS sheets with pressure sensitive adhesive are considered. Three HS options are considered and are listed in Figure 13. In option A, the flat HS is attached on top of the WLP. For option B, the same HS is bent upwards to a "U" shape. For option C, a smaller HS is attached to WLP top case.

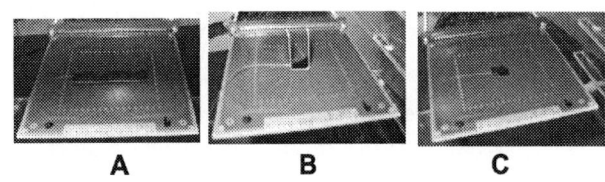

A   B   C

HS Option	Heat spreader size (mm)	Heat spreader shape
A	12.7x50.8	flat
B	12.7x50.8	U-shape
C	12.7x12.7	flat

Figure 13. HS options.

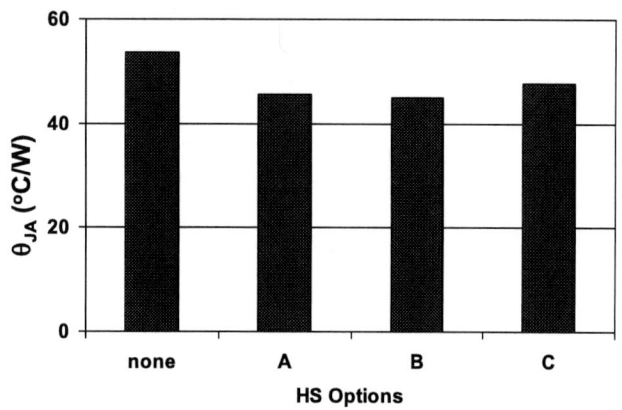

Figure 14. Thermal performance comparison between no heat spreader and with heat spreader.

The measured thermal resistance θ_{JA} for these options are plotted in Figure 14. Compared to no HS case, θ_{JA} is reduced by 15%, 16%, and 11% for options A, B, and C, respectively. U shaped HS does not seem to have significant advantage. Larger HS gives greater improvement.

Conclusions

Thermal measurements of 0.4 mm pitch WLP are conducted using thermal test vehicles under JEDEC still air environment. Effects of WLP size, WLP design, Pd area, and HS are studied. The following conclusions are made:

1. θ_{JA} is higher for smaller WLP than larger WLP. Under JEDEC still air condition, θ_{JA} is between 24 and 55 °C/W for WLP size ranging from 2.4 to 7.2 mm.

2. θ_{JA} is not sensitive to WLP design. It is not necessary to increase WLP via size and RDL width for thermal enhancement alone.

3. Maximum junction temperature is a function of Pd area. Smaller Pd area results in higher θ_{JA}. As long as the Pd area is at the same order of magnitude as the die size, the θ_{JA} value stays in the same order of magnitude.

4. Localized Pd generates hot spot where the temperature is much higher than die average temperature. θ_{JA} can be 36 times higher than uniformly distributed Pd. The worst hot spot is created when the localized Pd is at die corner. Distributing the localized heat sources to multiple locations can significantly reduce the hot spot peak temperature.

5. WLP thermal resistance is only a small portion of the overall resistance θ_{JA}. Effective thermal enhancement may be done at board and system level.

6. Heat spreader mounted on WLP top case can reduce θ_{JA} by 15%.

References

1. Gualandris, D., and Villa, C. M., Wafer Level Packaging Fan out Thermal Managemement: Is Smaller Always Hotter?, Proc. Of Microelectronics and Packaging Conf., June 2009.

2. Lau, J. H., and Yue, T. G, Thermal Management of 3D IC Integration with TSV (through silicon via), pp 635 – 640. Proc. of ECTC 2009.

3. JESD51, Methodology for the Thermal Measurement of Component Packages (Single Component Device), Dec-1995

4. JESD51- 1, Intergrated Circuit Thermal Measurement Method – Electrical Test Method (Single Semiconductor Device), Dec-1995.

5. JESD51-2A, Integrated Circuits Thermal Test Method Environmental Conditions - Natural Convection (Still Air), JANUARY 2008

6. JESD51-4, Thermal Test Chip Guideline (Wire Bond Type Chip), FEBRUARY 1997.

7. JESD51-9, Test Boards for Area Array Surface Mount Package Thermal Measurements, JULY 2000.

8. JESD51-12, Guidelines for Reporting and Using Electronic Package Thermal Information, MAY 2005

9. Tiao Zhou, Board and System Level Effects on Plastic Package Thermal Performance, Proc. of ECTC 1996.

Inkjet Printed Nano-particle Cu Process for Fabrication of Re-distribution Layers on Silicon Wafer

Ayat Soltani[1], Tero Kumpulainen[2], and Matti Mäntysalo[1]
Tampere University of Technology, [1]Department of Electronics and Communications Engineering,
[2] Mechanical Engineering and Industrial Systems
Korkeakoulunkatu 3, P.O. BOX 692, FI-33101 Tampere, Finland
ayat.soltani@tut.fi

Abstract

As of late, attention has been paid towards usage of copper nano-particle inks instead of silver or gold nano-particle inks in the field of printed electronics, the main reason being its good conductivity with respect to its bulk material price. However, there are inherent challenges with using copper inks. One of the most crucial challenges is the reactive nature of copper nano-particles, which easily forms a non-conductive oxide layer. To combat this, the process of sintering the printed copper inks should be performed either in vacuum or an inert environment, which is difficult and costly. Laser sintering provides digital high energy-density alternative for oven sintering, which is fast enough so that the oxide layer does not have enough time to form. In this work, we have studied the sintering of inkjet printed copper ink on silicon substrate using continuous-wave 808nm diode laser. The output power and the scanning velocity of the laser as the main sintering parameters were varied in order to study their effect on the electrical resistance of the samples. A sheet resistance of about 300 mΩ/$\square$ was measured. The tests were conducted in room conditions and the sintered patterns were analyzed using optical microscope and scanning electron microscope (SEM).

Introduction

The focus of printed and organic electronics has been mainly in roll- to-roll mass production technologies but nowadays, inkjet printing technology has also been gaining a strong foothold in printed electronics field. The agility and the relative low cost of the inkjet printing technology have guaranteed such a widespread interest [1, 2]. The ability of inkjet printers to use conductive inks has been explored since the 1980`s [3,4].

Various kinds of conductive inks are available, but most contain a suspension of nano-particles in some form of liquid (e.g. water, toluene, ethylene etc.). Other than this, inks also may contain other components such as dispersants, adhesion promoters, surfactants and other additives that give specific properties. The most commonly used conductive materials are gold [5], silver [6] and copper [7]. The conductivity is achieved after the sintering process in which the additives such as dispersants are removed, usually by heat. The heat will then enable diffusion across the surface boundaries of the metal nano-particles while fusing them together. This fusion will create the conductive path across the metal. Silver inks are widely used in printed electronics on the account of silver`s low resistivity. The sintering process for silver inks is rather simple and can be carried out simply in an oven [8] or alternatively, more complicated methods [9]. These days, due to being less costly relative to the conductivity copper has become more popular and its possibility to replace silver is

been investigated. However since the inherent oxide layer formed on copper at high temperatures negatively affects its resistivity, the sintering process becomes very challenging. It is not possible to simply heat the samples in the oven or hot plate to achieve adequate resistivity.

In order to prevent the oxide layer from forming, the sintering of copper should be carried out either in an inert environment; or at high speed so that oxide layer does not have time to form. Fast sintering methods have been explored such as pulsed light [10, 11], electrical sintering [12], and microwave [13]. This work focuses on laser sintering. In laser sintering, the laser beam is used to scan over the printed area. High energy-density enables the fusion of the nano-particles. Processing speed is controllable by the velocity of the laser beam movement, beam area, and scanning sequence [14]. In addition, digital driven process gives the flexibility to anneal the specific areas on the substrate. Laser sintering of various nano-particle inks on different substrates has already been conducted successfully, such as silver on polyimide substrate [15], gold on glass substrate [16] and copper ink on polyimide substrate [17] and glass substrate [18].

Copper post processing using flash sintering has been investigated by Kim et al [19]. However, not much work has been directed toward laser sintering of nano-particle copper ink on top of silicon substrates. One investigation of copper sintering on top of silicon substrate has been conducted by Ko et al. [20], in which a thin layer of dielectric was spin coated on the substrate prior to ink jetting. We investigated the similar method by printing a thin dielectric layer on top of the silicon wafer before printing copper ink and results confirm that the sintering process of copper ink improves. The reason for this is that the dielectric layer acts as a thermal barrier layer and prevents heat conductance through the silicon. In this work, however, our goal was to achieve adequate sintering without any additional insulating layers except the oxide layer on the silicon substrate. In this paper, laser sintering of copper nano-particle ink on top of silicon substrate is carried out and analyzed. The tests were conducted in room temperature and ambient environment using the 808nm continuous-wave diode laser.

Materials and samples

The ink used in this research was supplied by Intrinsiq Materials, Inc.; model number CI-002. The ink contains coated copper nanoparticles which are dispersed in a solvent. The function of the coating is to prevent agglomeration and ensure long shelf life of the ink. According to the data sheet provided by the manufacturer the particle size for the ink was ~50 nm with the solid content of 25 wt.%. The substrate was 6 inch diameter silicon wafer with 380 μm thickness and 1 μm oxide layer.

978-1-4799-2408-0/14 $31.00 © 2014 IEEE

The printing setup was an inkjet printer, iTi XY MDS2.0 with a drop on demand printhead; Dimatix Spectra SE-128 comprised of 128 nozzles, which results in approx. 30 picoliter droplets. The printing speed was 100 mm/s and printing voltage, 80 V. Prior to printing, surface of the wafers were cleaned using isopropanol.

Four-point measurement patterns were used for the test structure. Shape and dimensions of the four-point pattern is presented in Figure 1. Since the drop diameter on silicon wafers were around 40 μm at 50 °C, these structures were printed using 900 dpi resolutions. The printing parameters are listed on Table 1. Due to the low volume content of the ink, patterns were printed four times.

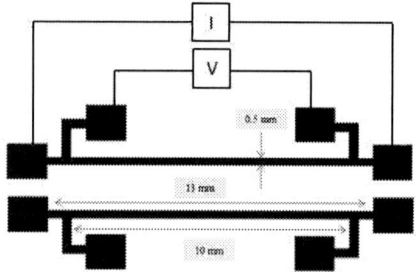

Figure 1. Schematics of the four-point measurement pattern.

Table 1. Print parameters.

Parameter	Value
Ink	CI-002, Intrinsiq Cu
Particle size	50 nm
Solid content	25 wt.%
Printing speed	100 mm/s
Printing distance	1.0 mm
Printing plate temperature	50 °C
Printhead	Dimatix Spectra SE-128
Printing voltage	80 V
Printhead temperature	55 °C
Drop volume	30 pl
Drop diameter on substrate	40 μm
Print resolution	900 dpi

Laser Equipment

An 808-nm continuous wave semiconductor laser HLU35C10x2-808-CD by Lissotchenko Mikro-optik (Limo) was used for the sintering process. The output of the laser is adjustable up to 35W with the spectral beam width of 2.5 nm and the peak wavelength of 808.6 nm. A spherical lens with a focal length of 30 mm was used; this reduced the laser spot for the sintering to 1.5 mm x 0.3 mm. To achieve optimum sintering, the longer edge of the laser spot was placed parallel to the sintering direction.

The laser setup and its schematic are presented in Figure 2 a) and b). The laser diode is attached to a cooling unit over the focusing lens system and the whole setup is placed inside of a static frame. For scanning the laser across the substrate, laser diode was placed over a motorized XY table. This table gives the ability to move the substrate in X and Y directions under the laser at a speeds up to 300 mm/s. So in essence, scanning was done by moving the substrate, not the laser itself.

In order to make sure the entire surfaces of the printed patterns were scanned by the laser, a stepwise movement was used which is shown in Figure 2 c). Each step was set to 0.2 mm, which is smaller than the height of the laser spot. Using this scanning method guaranteed that the entire surface was scanned equally.

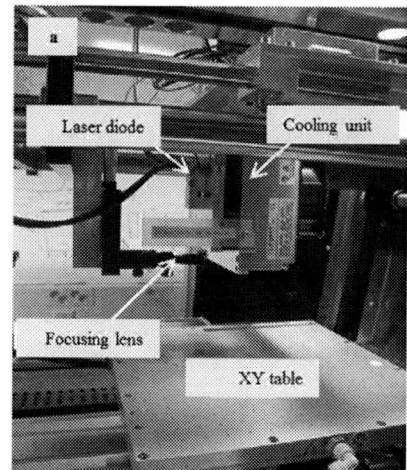

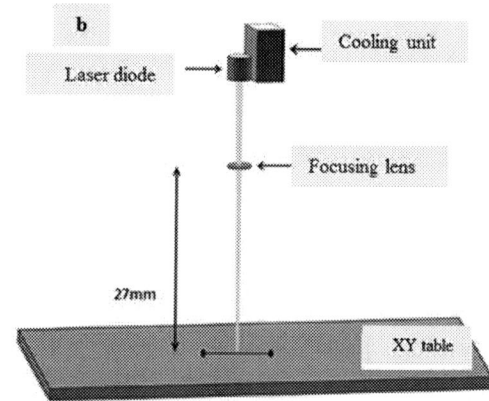

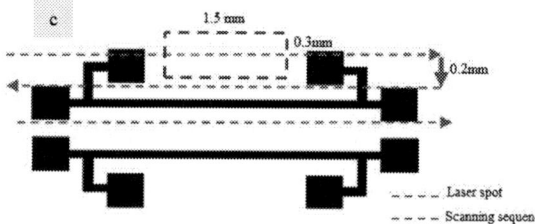

Figure 2. a) The picture of the laser setup, b) the schematics of the laser setup, and c) the scanning sequence and laser spot size.

Structural analysis and electrical measurement

The structures were analyzed using both an Olympus BX60M optical microscope and a Zeiss ULTRA-55 scanning electrode microscope.

For sheet resistance measurement, a Keithley 2400 multi-meter attached to a probe station was used with needles placed on the pads at the ends of the printed patters. This setup is presented in Figure 3 a) and b).

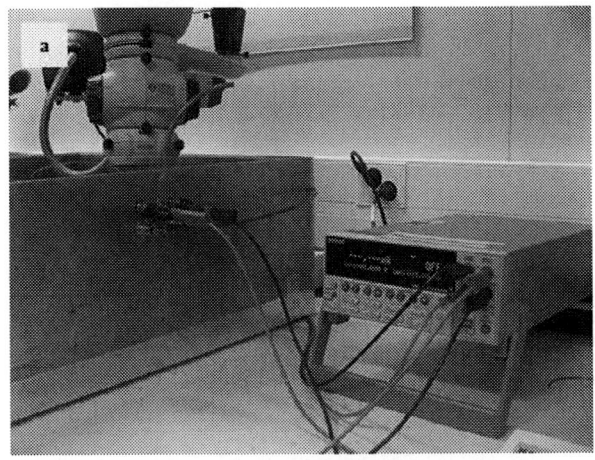

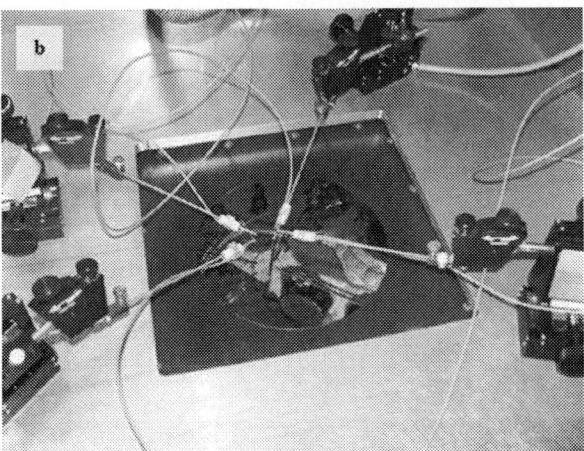

Figure 3. a) Keithley 2400 multi-meter and the probe station b) four point measurement setup.

Results

The initial tests with laser output power less than 35 W showed no resistivity. Therefore, the maximum power of 35 W was used for the experiments and only the scanning velocity was varied. No conductivity was achieved with scanning speed from 1mm/s to 4 mm/s and the annealed patterns showed signs of burning. Microscope images showed that the center of the laser spot has the highest amount of energy and by scanning at low speeds there is enough time for some of the particles to actually burn instead of diffusing together. Hence there are burned parts visible separately from the rest of the sections. A sample of a burned pattern is presented in Figure 4

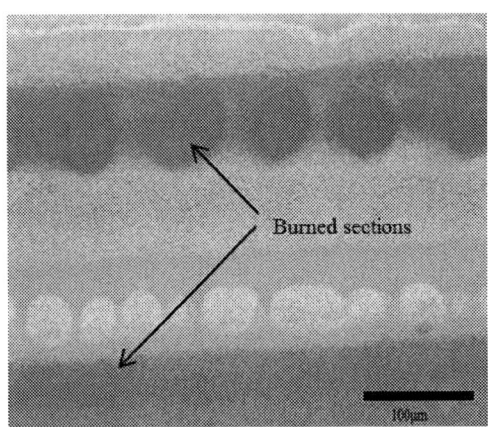

Figure 4. Optical picture demonstrating burn signs on pattern sintered with scanning velocity below 5 mm/s.

In order to prevent the samples from overheating and eventually burning, higher scanning velocities were used. The velocity was increased from 5 mm/s to 300 mm/s in 20 increments and after each run the resistivity was measured (unfortunately 300 mm/s was the limit achievable by our laser setup). Figure 5 presents the sheet resistance versus scanning velocity. While the sheet resistance shows a distribution between 250 mΩ/□ and 470 mΩ/□, an increasing trend can be seen in the values when the fitting line is drawn.

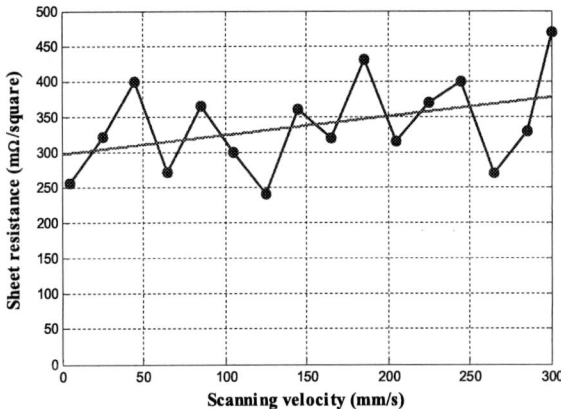

Figure 5. Sheet resistance versus velocity for 35 W constant laser power output.

The variance in the measured values or the apparent increasing trend for that matter; cannot be attributed to the change in the sintering velocity with certainty. Based on our previous experience this is likely related to the normal process variations in printing which is inevitable [21]; thus rarely two printed patterns have the same quality despite using same printing parameters.

In order to get an idea about the structural changes during the sintering process, optical and SEM pictures were taken from a test sample before and after sintering, which are presented in Figure 6 and 7, respectively.

Before sintering is commenced, the printed structure looks uniform under the optical microscope with brownish color as seen in Figure 6 a). However, looking at the SEM picture of the same structure as shown in Figure 6 b), it appears to be porous with the particles stacked on top of each other. This is more clearly visible throughout the structure as shown in Figure 6 c). Overall the density of the structure is relatively small. Since there is no continuous path along the copper particles, no conductivity is measurable.

Upon comparing Figure 6 and Figure 7, marked differences between the sintered and non-sintered structures are discernable. From the optical picture presented in Figure 7 a), it is obvious that the structure has changed. First indicator is that the structure has changed color from brown to somewhat yellowish. Looking at a higher magnification SEM images from the same area in Figures 7 b) and 7 c), the differences to Figure 6 b) and c) are even more pronounced. In Figures 7 b) and 7 c) the particles have fused together quite well and have formed a coherent structure without any individual particles visible. More importantly, numerous pores available in the non-sintered structure are absent, and overall, the structure has become much denser. There are some visible empty spaces, as seen in Figure 7 c), that have separated sintered particles from each other. This is probably due to the remaining liquids such as dispersants that were stuck in the lower layers of the printed patterns. These remaining liquids were evaporated rather quickly under the extreme heat produced by the laser, forming empty spaces in the sintered structure in the process. These voids decrease the resistivity of the sintered structure.

As a final step in the experiment, measurements were made after repeating laser annealing of the samples. This was done in order to find out if scanning an already sintered pattern will have any effect on the resistivity. For this a pattern which was initially sintered using 35W output power and 10mm/s velocity was scanned three more times using the same annealing parameters. After each scan run, resistivity was measured and compared to the previous values. No noticeable change in the measured resistivity was observed in the consequent runs.

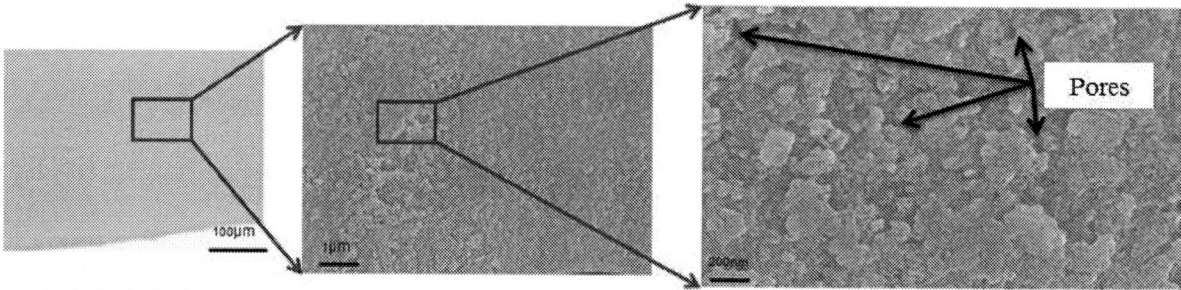

Figure 6. a) Optical picture of the non-sintered specimen, b) and c) Higher magnification SEM picture of non-sintered **specimen.**

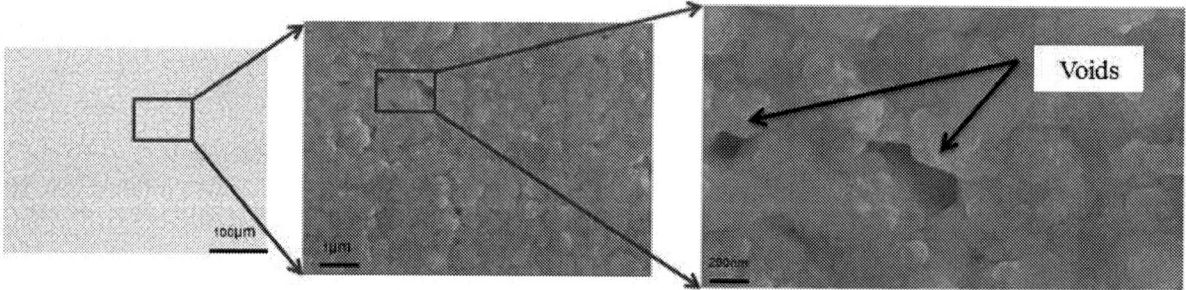

Figure 7. a) Optical picture of the sintered specimen, b) and c) Higher magnification SEM picture of the sintered specimen.

Summary

In this work, copper nano-particle ink was printed on silicon wafer with 1 µm oxide layer. An 808-nm continuous-wave diode laser was used to sinter the printed patterns. The sintering was conducted in room temperature and ambient environment. The wafers were cleaned with isopropanol prior to printing without any additional processes. Different scanning speeds and laser optical powers were used in conjunction with each other. Based on the initial results it was apparent that anything less than 35W of laser power was insufficient for the sintering process. Considering this 35W optical power was used with scanning speeds up to 300 mm/s. Overall an average sheet resistance of 300 mΩ/□ was measured. The measured sheet resistance for scanning speeds from 5 mm/s to 300 mm/s showed a variance between 250 mΩ/□ and 470 mΩ/□ with an increasing trend. The results also showed that repeating the number of scanning times on the already sintered pattern did not improve the resistivity.

Acknowledgments

This work is supported by ENIAC-JU Project Prominent grant No. 324189. M. Mäntysalo is sponsored by Academy of Finland grant No. 251882.

References

1. L. Gasman, "A Roadmap for Printable Electronics," *in IMI's 4th Annual Printable Electronics and Displays Conference & Trade Fair*, Las vegas, USA, 2005.

2. M. Mäntysalo and P. Mansikkamäki, "Inkjet-deposited Interconnections for Electronic Packaging," *in NIP 23 23rd International Conference on Digital Printing Technologies, and Digital Fabrication*, AK, USA, Sept. 16th-21st, 2007, pp. 813-817.

3. C. Gerard and M. P. Desmulliez, "Inkjet Printing of Conductive Materials a Review," *Circuit World*, vol. 38, no. 4, pp. 193-213, 2012.

4. S. M. Bidoki, D. M. Lewis, M. Clark, A. Vakorov, P. A. Millner and D. McGorman, "Ink-jet Fabrication of Electronic Components," *Journal of Micromechanics and Microengineering*, vol. 17, no. 5, 2007.

5. G. C. Jensen, C. E. Krause, G. A. Sotzing and J. F. Rusling, "Inkjet-printed Gold Nanoparticle Electrochemical Arrays on Plastic. Application to Immunodetection of a Cancer Biomarker Protein," *Phys Chem Chem Phys*, vol. 13, no. 11, p. 4888-4894, 2011.

6. J. F. Salmerón, F. Molina-Lopez, D. Briand, J. J. Ruan, A. Rivadeneyra, M. A. Carvajal, L. F. Capitán-Vallvey, N. F. de Rooij and A. J. Palma , "Properties and Printability of Inkjet and Screen-Printed Silver Patterns for RFID Antennas," *Journal of Electronic Materials*, vol. 43, no. 2, pp. 604-617 , 2014.

7. J. S. Kang, H. S. Kim, J. Ryu, H. T. Hahn, S. Jang and J. W. Joung , "Inkjet Printed Electronics Using Copper Nanoparticle Ink," *Journal of Materials Science: Materials in Electronics*, vol. 21, no. 11, pp. 1213-1220, 2010.

8. E. Halonen, T. Viiru, K. Ostman, A. Cabezas and M. Mäntysalo, "Oven Sintering Process Optimization for Inkjet-Printed Ag Nanoparticle Ink," *IEEE Transactions on Components, Packaging and Manufacturing Technology*, vol. 3, no. 2, pp. 350-356, 2013.

9. J. Niittynen, R. Abbel, M. Mäntysalo, J. Perelaer , U. S. Schubert and D. Lupo , "Alternative Sintering Methods Compared to Conventional Thermal Sintering for Inkjet Printed Silver Nanoparticle Ink," *Thin Solid Films*, article in press, 2014. http://dx.doi.org/10.1016/j.tsf.2014.02.001

10. R. Jongeun, K. Hak-Sung and H. H. Thomas, "Reactive Sintering of Copper Nanoparticles Using Intense Pulsed Light for Printed Electronics," *Journal of Electronic Materials*, vol. 23, no. 48, pp. 42-50, 2011.

11. H.-J. Hwang, W.-H. Chung and . H.-S. Kim, "In Situ Monitoring of Flash-Light Sintering of Copper Nanoparticle Ink for Printed Electronics," *Nanotechnology*, vol. 23, no. 48, 2012.

12. Mark.L, A. Mikko, M. Tomi, A. Ari, O. Kimmo, S. Mika and S. Heikki, "Electrical Sintering of Nanoparticle Structures," *Nanotechnology*, vol. 19, no. 17, 2008.

13. J. Perelaer, B.-J. de Gans and U. Schubert, "Ink-jet Printing and Microwave Sintering of Conductive Silver Tracks," *Advanced Materials*, vol. 18, no. 16, pp. 2101-2104, 2006.

14. T. Kumpulainen, J. Pekkanen, . J. Valkama, J. Laakso, . R. Tuokko and M. Mäntysalo, "Low Temperature Nanoparticle Sintering with Continuous Wave and Pulse Lasers," *Optics and Laser Technology*, vol. 43, no. 3, pp. 570-576, 2011.

15. P. Peng, A. Hu and Y. Zhou, "Laser Sintering of Silver Nanoparticle Thin Films: Microstructure and Optical Properties," *Applied Physics A*, vol. 108, no. 3, pp. 685-691, 2012.

16. J. Chung, S. Ko, N. R. Bieri, C. P. Grigoropoulos and D. Poulikakos, "Conductor Microstructures by Laser Curing of Printed Gold Nanoparticle Ink," *Applied Physics Letters*, vol. 84, pp. 801-803, 2004.

17. E. Halonen, E. Heinonen and M. Mäntysalo, "The Effect of Laser Sintering Process Parameters on Cu Nanoparticle Ink in Room Conditions," *Optics and Photonics Journal*, vol. 3, no. 4A, pp. 40-44, 2013.

18. M. Zenou, O. Ermak, A. Saar and Z. Kotler, "Laser Sintering of Copper Nanoparticles," *Journal of Physics D*: Applied Physics, vol. 47, no. 2, 2014.

19. H.-S. Kim, S. R. Dhage, D.-E. Shim and H. H.Thomas, "Intense Pulsed Light Sintering of Copper Nanoink for Printed Electronics," *Applied Physics A*, vol. 97, no. 4, pp. 791-798, 2009.

20. S. H. Ko, H. Pan, C. P. Grigoropoulos, C. K. Luscombe, J. M. J. Fréchet and D. Poulikakos, "Air Stable High Resolution Organic Transistors By Selective Laser Sintering of Ink-jet Printed Metal Nanoparticles," *Applied Physics Letters*, vol. 90, 2007.

21. M. Mäntysalo, V. Pekkanen, K. Kaija, J. Niittynen, S. Koskinen, E. Halonen, P. Mansikkamaki and O. Hameenoja, "Capability of Inkjet Technology in Electronics Manufacturing," *in Electronic Components and Technology Conference*, San Diego, CA, US, 26th- 29th May, 2009, pp. 1330-1336.

Design of Multi-sensor for Safety Monitoring of Heavy Machinery

Long Li[1], Fei Hou[4], Jinghao Qiu[3], Zhang Luo[1], Shengzhi Zhang[1], Qiang Dan[1], Sheng Liu[2]*

1. Huazhong University of Science & Technology, Wuhan, 430074, China
2. Cross-disciplinary Institute of Engineering Sciences, School of Power and Mechanical Engineering, Wuhan University, Wuhan, Hubei, 430072, China
3. Nanjing University of Aeronautics and Astronautics, Nanjing, China
4. Dongfeng Automobile Electronics Co., Ltd
* Corresponding Author: Sheng Liu, Fax: 86-27-87557074, Email: victor_liu63@126.com

Abstract

In this paper, a multi-sensor system in package (SiP) is designed for the safety monitoring of the heavy machinery. The multi-sensor is made up of several parts, including strain gauge array, acceleration sensor, temperature sensor, and a pair of ultrasonic detector and actuator. Strain gauge array in a certain pattern is used to monitor the strain and stress level at some critical positions of the heavy machinery. At the same time, the acceleration sensor is used to examine the status of the vibration and monitors possible events such as impact or shock, providing the needed data for loading monitoring and liability, while the ultrasonic detector/actuator is adapted to detect the crack formed within the parts of the heavy machinery for those easy-to-fail parts. All parts of the system are integrated on a special flexible substrate with micro vias, through which the signal transmission and electrical interconnection among all devices are achieved. Besides, the total system is fixed at the heavy machinery through the flexible substrate, so that a real-time safety monitoring for the important parts of the heavy machinery can be accomplished, even when the heavy machinery is at work. The system is powered by two methods. For the lower power devices like the strain gauges, some materials in the system with the property of piezoelectricity can supply the energy for them, while for the larger ones such as the ultrasonic detectors, wire is needed. Apart from those listed above, wireless transmission devices are also included in the system so that the signal can be sent and received. The system prototype has been made and it is expected that the module will be integrated into the internet of things, a service network good for maintaining and servicing of engineering product systems in the field.

Introduction

With the advancement of the science and technology, many tools, devices and machineries have been invented and created by human to help or replace human to do some work which is usually dull, difficult or dangerous, for instance, the computer rescues human from the tedious numerical calculation. Compared with human, they have the advantage of improved productivity, high efficiency and long-time working persistency. Likewise, thanks to these features, the heavy machinery is employed pervasively and plays an important role in various engineering projects.

However, some defects, such as cracks, wear and tear, may occur on or within the steel of the heavy machinery owing to the long-term utilization and the neglect to proper maintenance and examination periodically. There is no doubt that they pose a potential but large threat on all members joined in the engineering project. Consequently, a big concern has been imposed on them and the safety monitoring to the heavy machinery is necessary in those engineering projects.

Nowadays, a variety of sensors and actuators aimed at monitoring and controlling some critical factors in the machinery have been created. Thus, some important information about the status of the heavy machinery can be obtained in real time so that people are able to learn more things related to the machinery. Nonetheless, the sensors usually focus on a relatively important factor in the eyes of people, ignoring more information associated with the defects. In addition, various sensors are distributed randomly, lacking the integration of the device and the sharing of the information.

The focus of the work in this paper is to design a multi-sensor package for the safety monitoring of the heavy machinery. The multi-sensor package is composed of strain gauge array, acceleration sensor, temperature sensor, and a pair of ultrasonic detector and actuator. By wide employment of the multi-sensor and high integration of the system, on one hand, the full-scale and real-time safety monitoring for the heavy machinery can be achieved. On the other hand, the system with low profile can be easily located on the heavy machinery. In this paper, we take the boom of the concrete pump truck as a concrete example of the heavy machinery.

Finite Element Modeling for Silicon Strip

Strain gauges have been used pervasively across all engineering fields to detect the mechanical deformation and stress of a solid object for many years. Fabricated by a patterned metal foil fixed between two thin polymer sheets, the strain gauge matures gradually from a uniaxial form adopted to measure the uniaxial tension and compression to a strain gauge array used when the status of stress in the solid object is complicated and uncertain. Due to the intrinsic property, however, the gauge factor of the metallic foil is fractional typically ranging from 2 to 5, inducing limits to the metal strain gauge to some degree. At the same time, as the piezoresistive effects of the semiconductor are discovered, semiconductor gauges are preferred over metal foils because of its great gauge factor, for example, the gauge factor of p-type (110) single crystalline silicon can be as high as 200 [1]. But the semiconductor gauge is just used for the small deformation on the flat plane of the stiff object owing to its natural stiffness and brittleness. If the surface of the target object is curvilinear, the gauge will be too stiff to completely conform to the surface [2]. And if the target object is deforming too much, semiconductor will be easily broken. Aiming to this shortcoming, a proposal that ultrathin sheets single crystalline silicon is integrated on a polymer substrate to detect the strain has been proposed. A 3D schematic of a thin silicon strip supported by polymer substrate is illustrated in Figure 1 below.

Recent work reveals that ultrathin sheets single crystalline silicon, silicon nanomembrances, can survive severe bending or stretching when supported by polymer substrates [3]. The Young' modulus and the thickness of the substrate, the length and the thickness of the semiconductor gauge are critical factors for utilization. Therefore a finite element modeling is conducted to help us find a set of feasible factors.

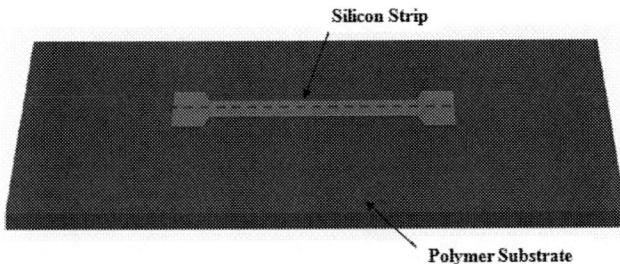

Figure 1. 3D Schematic of A Thin Silicon Strip on Polymer Substrate

In the modeling, geometries are built to resemble the actual conditions. Plane element is chosen to conduct the meshing and the material properties of the silicon and polymer are used for the modeling. Considering the material properties, silicon is simulated as linear elastic material and polymer is modeled as Neo-Hookean material, respectively. Effective materials properties concerning Young' modulus and Poisson's ratio are listed in Table 1. Because of symmetry, only the right half of the system is modeled and the symmetric boundary conditions are imposed. Perfect bonding is assumed between silicon and the substrate. The sample is subject to uniaxial tension which is applied horizontally to the free edge of the substrate as the main load. Here we designate 10% length of substrate as tensile strain and always make substrate length 1.5L (L is the length of silicon strip) in order to decrease the variables. The FEM profile is shown in Figure 2.

Table 1. Material properties used in FEM

Name	Material	Young' modulus	Poisson's ratio
Silicon	Silicon	165.5GPa	0.25
Substrate	Polyimide	2.5GPa	0.34

The ratio $\varepsilon_{Si}/\varepsilon_{app}$ represents the ratio between the strain in both X and Y direction on silicon strip and the applied strain on polymer substrate. In Figure 3, the ratio affected by substrate thickness, silicon length and silicon thickness is mainly discussed. In general, we hope that the displacement in X direction on silicon have a relatively large value because it means that a large fraction of the applied strain has been transferred to silicon [2]. It is also hoped that the displacement in Y direction on silicon is small lest bending failure occurs. Therefore, according to the Figure 3, a relatively thick

substrate (0.5mm) and thin silicon strip ($1\mu m$) are good. But to the length of silicon strip, a trade-off is needed, for example, 0.5mm. The FEM contour plots when substrate thickness, silicon length and silicon thickness are 0.5mm, 0.5mm and $1\mu m$ respectively are shown in Figure 4.

Figure 2. 2D Plane Model Used in FEM

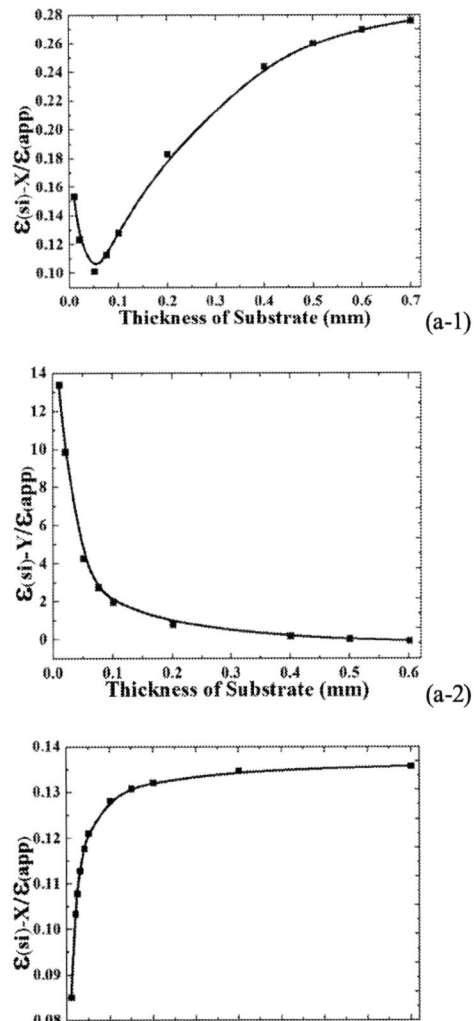

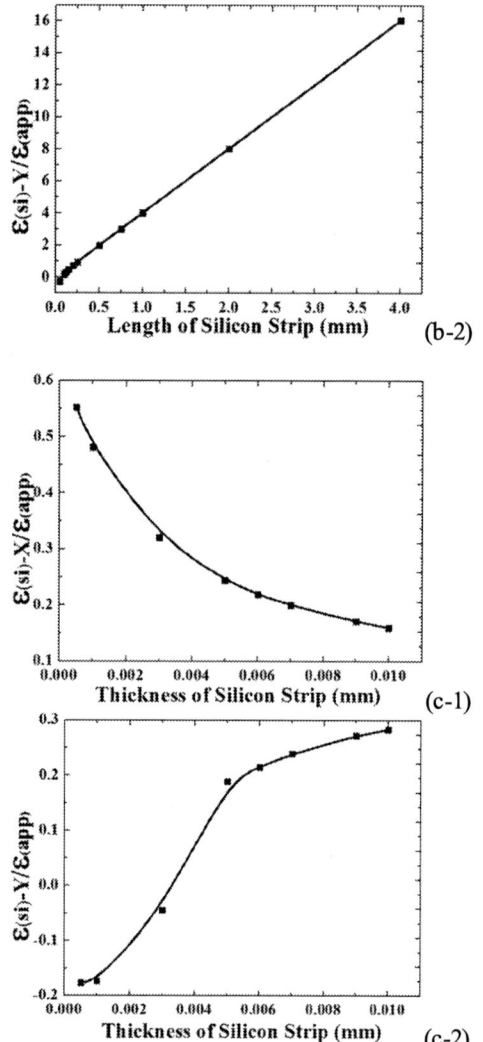

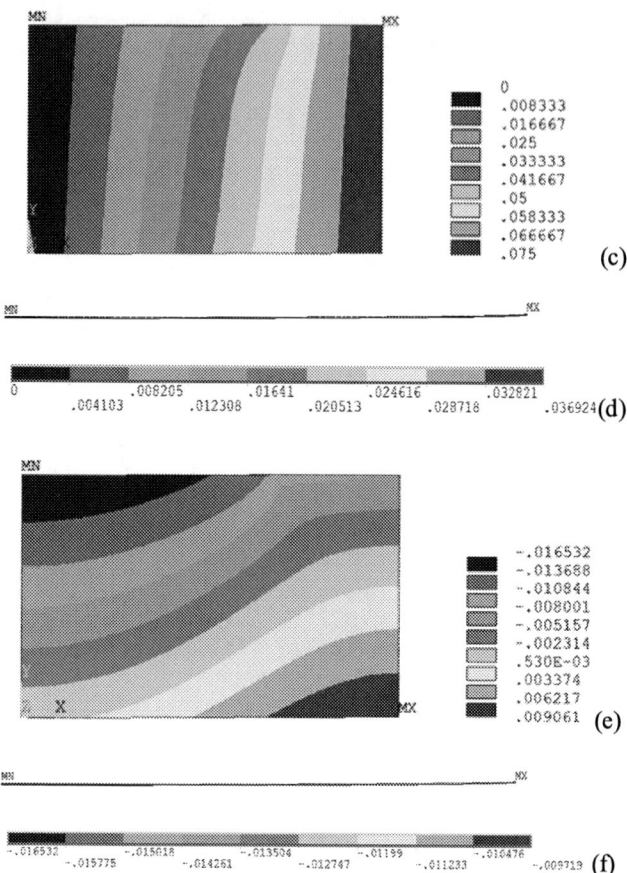

Figure 4. FEM Contour Plots (a) Displacement Vector Sum on Whole Model (b) Displacement Vector Sum on Silicon (c) X-Component of Displacement on Whole Model (d) X-Component of Displacement on Silicon (e) Y-Component of Displacement on Whole Model (f) Y-Component of Displacement on Silicon

Figure 3. Relation between $\dfrac{\varepsilon_{Si}}{\varepsilon_{app}}$ and Substrate Thickness, Silicon Length and Thickness

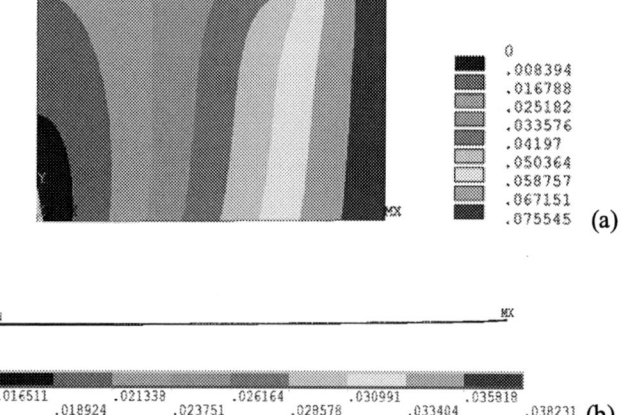

The maximum displacement on the whole model is 0.382mm and that in X direction on silicon is 0.369mm, which accounts for 96.6% of the whole displacement. Besides, the displacement in Y direction on silicon is small so that fracture cannot occur. It should be mentioned that displacement in X direction also should keep moderate lest the elongation of the silicon strip is too large.

Configuration of the Strain Gauge Array

In general, the status of stress throughout the whole structure of the heavy machinery is complicated and uncertain. Through finite element modeling and practical experience, however, the monitoring point and the corresponding status of stress on the point where the stress is usually significant or defects often occur can be achieved, to some degree. For instance, for the boom system of the concrete pump truck, the stress within the pin mounted between every two conjoint components is prone to be arbitrary and the stress at the end of the biggest boom is inclined to lie parallel to the axial direction of the boom. Therefore according to the feature of the stress distribution, a strain gauge array also named strain gauge rosette arranged by several silicon strips in a certain configuration is required.

978-1-4799-2408-0/14 $31.00 © 2014 IEEE

It should be mentioned that silicon is an anisotropic material whose property, for example the piezoresistivity, depends strongly on the orientation of the wafer plane in which it is fabricated. Through calculation, the resistance change of an in-plane sensor fabricated on (100) silicon depends on four components of stress ($\sigma_{11}, \sigma_{22}, \sigma_{33}, \sigma_{12}$) and the orientation of the sensor, while the resistance change of an in-plane sensor fabricated on (111) silicon depends on all six components of stress and the orientation of the sensor [4]. At the same time, the piezoresistive coefficients of the n-type and p-type silicon are also different. The full potential of multi-element sensor rosettes to measure up to six stress components can be achieved by using dual-polarity sensing elements fabricated with both n-type and p-type silicon [4].

Compared to sensors fabricated by (100) silicon, properly designed strain gauge made by (111) silicon has more advantages. Figure 5 shows an optimized eight-element array made by (111) silicon. Dual-polarity elements incorporating n-type and p-type are employed and the configuration with the angle of 45° between every two strain gauges is designed. For an arbitrary state of stress, the six stress components in terms of the measured resistance changes are presented in (2) [4]. Here $B_1 = (\pi_{11} + \pi_{12} + \pi_{44})/2$, $B_2 = (\pi_{11} + 5\pi_{12} - \pi_{44})/6$, $B_3 = (\pi_{11} + 2\pi_{12} - \pi_{44})/3$.

$$[\pi_{\alpha\beta}] = \begin{bmatrix} \pi_{11} & \pi_{12} & \pi_{12} & 0 & 0 & 0 \\ \pi_{12} & \pi_{11} & \pi_{12} & 0 & 0 & 0 \\ \pi_{12} & \pi_{12} & \pi_{11} & 0 & 0 & 0 \\ 0 & 0 & 0 & \pi_{44} & 0 & 0 \\ 0 & 0 & 0 & 0 & \pi_{44} & 0 \\ 0 & 0 & 0 & 0 & 0 & \pi_{44} \end{bmatrix} \quad (1)$$

is the piezoresistive coefficient matrix, and α is the temperature coefficient of resistance and only the first order temperature terms have been retained, which can be obtained using thermal cycling calibration experiments. From the expressions in (2), it is explicit that the three shear stresses $\sigma_{12}, \sigma_{13}, \sigma_{23}$ are irrelevant to the temperature T. So is the difference between the normal stress σ_{11} and σ_{22}.

$$\sigma_{11} = \frac{(B_3^p - B_2^p)[\frac{\Delta R_1}{R_1} - \frac{\Delta R_3}{R_3}] - (B_3^n - B_2^n)[\frac{\Delta R_5}{R_5} - \frac{\Delta R_7}{R_7}]}{2[(B_2^p - B_1^p)B_3^n + (B_1^p - B_3^p)B_2^n + (B_3^p - B_2^p)B_1^n]}$$
$$+ \frac{B_3^p[\frac{\Delta R_1}{R_1} + \frac{\Delta R_3}{R_3} - 2\alpha^n T] - B_3^n[\frac{\Delta R_5}{R_5} + \frac{\Delta R_7}{R_7} - 2\alpha^n T]}{2[(B_1^n + B_2^n)B_3^p - (B_1^p + B_2^p)B_3^n]}$$

$$\sigma_{22} = -\frac{(B_3^p - B_2^p)[\frac{\Delta R_1}{R_1} - \frac{\Delta R_3}{R_3}] - (B_3^n - B_2^n)[\frac{\Delta R_5}{R_5} - \frac{\Delta R_7}{R_7}]}{2[(B_2^p - B_1^p)B_3^n + (B_1^p - B_3^p)B_2^n + (B_3^p - B_2^p)B_1^n]}$$
$$+ \frac{B_3^p[\frac{\Delta R_1}{R_1} + \frac{\Delta R_3}{R_3} - 2\alpha^n T] - B_3^n[\frac{\Delta R_5}{R_5} + \frac{\Delta R_7}{R_7} - 2\alpha^n T]}{2[(B_1^n + B_2^n)B_3^p - (B_1^p + B_2^p)B_3^n]}$$

$$\sigma_{33} = \frac{-(B_1^p + B_2^p)[\frac{\Delta R_1}{R_1} + \frac{\Delta R_3}{R_3} - 2\alpha^n T] + (B_1^n + B_2^n)[\frac{\Delta R_5}{R_5} + \frac{\Delta R_7}{R_7} - 2\alpha^p T]}{2[(B_1^n + B_2^n)B_3^p - (B_1^p + B_2^p)B_3^n]}$$

$$\sigma_{13} = \frac{\sqrt{2}}{8}[\frac{(B_2^p - B_1^p)[\frac{\Delta R_4}{R_4} - \frac{\Delta R_2}{R_2}] - (B_2^n - B_1^n)[\frac{\Delta R_8}{R_8} - \frac{\Delta R_6}{R_6}]}{(B_2^p - B_1^p)B_3^n + (B_1^p - B_3^p)B_2^n + (B_3^p - B_2^p)B_1^n}]$$

$$\sigma_{23} = \frac{\sqrt{2}}{8}[\frac{-(B_2^p - B_1^p)[\frac{\Delta R_1}{R_1} - \frac{\Delta R_3}{R_3}] + (B_2^n - B_1^n)[\frac{\Delta R_5}{R_5} - \frac{\Delta R_7}{R_7}]}{(B_2^p - B_1^p)B_3^n + (B_1^p - B_3^p)B_2^n + (B_3^p - B_2^p)B_1^n}]$$

$$\sigma_{12} = \frac{-(B_3^p - B_2^p)[\frac{\Delta R_4}{R_4} - \frac{\Delta R_2}{R_2}] + (B_3^n - B_2^n)[\frac{\Delta R_8}{R_8} - \frac{\Delta R_6}{R_6}]}{2[(B_2^p - B_1^p)B_3^n + (B_1^p - B_3^p)B_2^n + (B_3^p - B_2^p)B_1^n]} \quad (2)$$

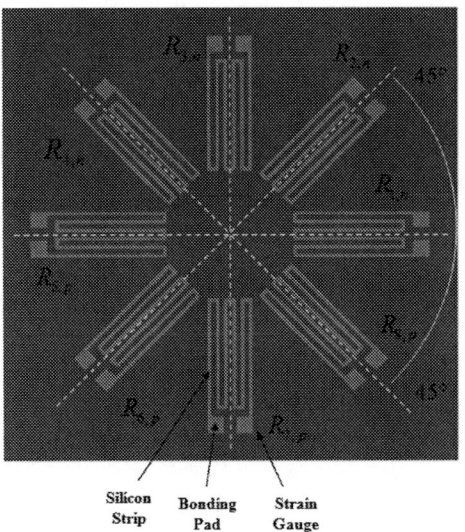

Figure 5. Optimized Eight-Element Array Made by (111) Silicon

Temperature Compensated Measurements

It is well known that silicon resistors have relatively large temperature coefficients, which imposes a great limit to the utilization of silicon as sensing element for strain measurement. There are several factors resulting in temperature errors, for example, actual temperature measurement and uncertainty in resistor measurements.

The best method for eliminating the influence of the temperature on strain measurement is temperature compensation on itself by sum and difference among several terms. In contrast to (100) silicon, optimized strain array on (111) silicon is capable of measuring four temperature compensated stress components, namely, the three shear stresses and the difference between the normal stress σ_{11} and σ_{22}, just as it is shown in expression (2). For the rest of the stress measurement, the temperature sensor is needed.

Acceleration Sensor

Acceleration sensor has been used pervasively in many fields for measurement such as acceleration, vibration and angles of inclination. Accelerometers all share a basic structure consisting of an inertial mass suspended by a spring, but differ in the sensing of the position of the inertial mass when it moves under the influence of the applied acceleration.

The common method is capacitive, while another uses the property of piezoelectricity. For the sake of the high accuracy, the suspended inertial mass, as the core component of acceleration sensor, is usually encapsulated in vacuum in order to minimize the motion resistance affected by the air. Given this point, the packaged acceleration sensor is preferred to the bare chip for our utilization. Besides, it is easy to image that the positions where to integrate the acceleration sensor include the free end of the whole boom system and the pin between the conjoint boom.

Ultrasonic Detector/Actuator

The ability to locate damages or sources in structures like plates is one of the attractions of structural health monitoring systems based on Lamb waves [5].The basic principle of the ultrasonic detector and actuator is delineated in Figure 6 below. In general, the detector and actuator stay on stand-by with low power consumption shown in Figure 6(a). Once a crack occurs somewhere in the steel, an accompanying ultrasonic signal will emerge and be received by the detector mounted on the steel, just like the picture exhibited in Figure 6(b). Then, the ultrasonic actuator will begin to work and send the ultrasonic wave into the steel. The ultrasonic signal will interact with the crack and then be received by the detector, which is presented in Figure 6(c). In this way, the position of the crack can be achieved.

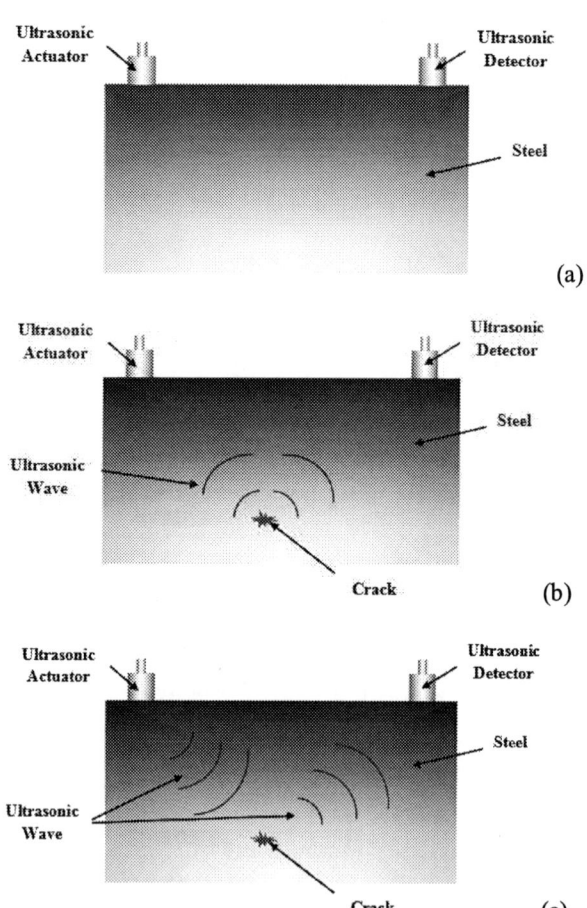

(a)

(b)

(c)

Figure 6. Basic Principle of Ultrasonic Detector and Actuator

The basic structure of the ultrasonic detector and actuator is exhibited in Figure 7 below. They share a similar configuration because of the principle of piezoelectric effect and reciprocal piezoelectric effect. That is, when ultrasonic wave is recognized by the piezoelectric membrane, electricity will be generated, and now it serves as the ultrasonic detector, while when electricity is imposed to the piezoelectric membrane, ultrasonic wave will be produced, and now it will be the ultrasonic actuator.

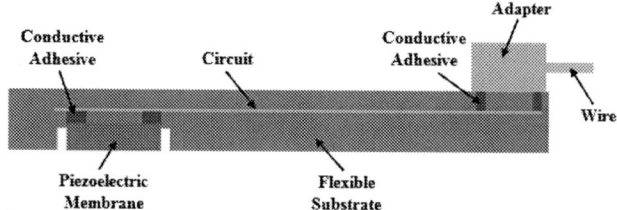

Figure 7. Basic Structure of Ultrasonic Detector and Actuator

Numerous experiments have been conducted to verify the application ability in practice, just shown in Figure 8. In the experiment, damage identification, localization and imaging are all included. The basic procedure is outlined below: first, considering the high attenuation of Lamb wave, the whole monitoring zone is divided into many quadrangular sub-zones with the distance between two PZTs as 120 and 220mm in the horizontal and vertical directions, respectively. Then, a modulated five-cycle sine burst actuating signal with frequency of 110 kHz is imposed on actuator PZT for examination. It is followed to compare the current signals with the health signals recorded from the undamaged structure, shown in Figure 9 and 10, to quantify how the signals have changed by DI [6]. Here DI is expressed as:

$$DI = 1 - \sqrt{\frac{\left[\int_{t1}^{t2} H^{**}(t) D^{**}(t) dt \right]^2}{\left[\int_{t1}^{t2} H^{**2}(t) dt \int_{t1}^{t2} D^{**2}(t) dt \right]}}$$

(3)

Here t_1 and t_2 represent the starting and ending time of the direct arrival wave, respectively. H^{**} and D^{**} are calculated below

$$H^* = \frac{H}{\max(abs(H))}, H^{**} = \Re(shanWT(H^*))$$

$$D^* = \frac{D}{\max(abs(H))}, D^{**} = \Re(shanWT(D^*))$$

(4)

Here max(abs(H)) is the maximum amplitude of the health signal, $\Re(shanWT)$ represents the real part of complex Shannon wavelet transform to extract the narrowband signal from the signals [7].

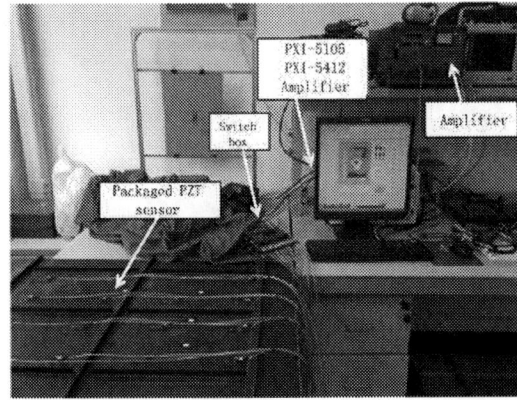

Figure 8. Experiment for Ultrasonic Detection

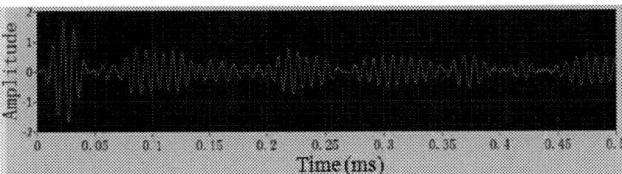

Figure 9. Health Signal Recorded from the Undamaged Structure

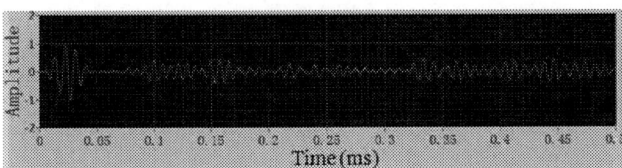

Figure 10. Monitoring Signal Recorded from the Damaged Structure

Damage localization and imaging are explored by delay-and-sum multi-damage imaging algorithm, which includes the calculation for the delayed time in (5), the wave envelops of the scattering signals through Shannon wavelet transform in(6) and the pixel value of the imaging in(7). The point with the pixel peak value of the imaging result is often regarded as the damage location [8-10]. And the result of the positioning and imaging is presented in Figure 11 below.

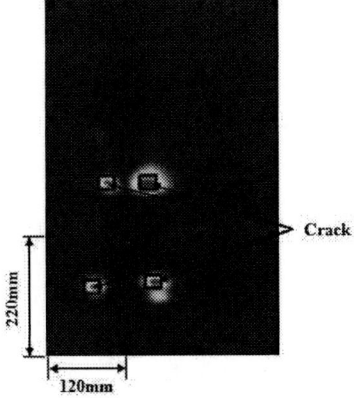

Figure 11. Result of the Positioning and Imaging

$$t_{ij}(x,y) = t_{off} + \frac{\sqrt{(x_i - x)^2 + (y_i - y)^2} + \sqrt{(x_j - x)^2 + (y_j - y)^2}}{c} \quad (5)$$

Here (x_i, y_i) and (x_j, y_j) are the coordinate of the actuator and the sensor, respectively. t_{off} is the time offset corresponding to the actuation and c is the average velocity of the Lamb wave in the structure.

$$E_{ij} = abs(shanWT(H^* - D^*)) \quad (6)$$

Here abs(shanWT) denotes the modulus value obtained from the complex Shannon wavelet transform.

$$E(x,y) = \sum_{i=1}^{3} \sum_{j=i+1}^{4} E_{ij}(t_{ij}(x,y)) \quad (7)$$

Here the pixel value is expressed by E(x,y).

Package

System in package (SiP) is a trend of electronic package for multiple functions [11]. It enables the continued increase in functional density with small profile, low cost and high performance at the same time. Given kinds of sensors and actuators listed above, a multi-sensor system in package (SiP) based on TSV is proposed in the design.

Figure 12 shows the configuration of the multi-sensor SiP. According to the testing principle, ultrasonic detector/actuator and strain gauge should be in contact with the surface of the steel structure directly. Therefore they are located at the bottom of the package. Because strain gauge could not be influenced by other mechanical deformations when it works, it is isolated from the main package component and covered by polymer which protects the strain gauge from the harmful environment factors. The electrical interconnection between the strain gauge array and the substrate is achieved by the circuit embedded into flexible polymer bands, while that between the ultrasonic detector/actuator and the substrate is obtained through TSV. The interconnection among the devices above the substrate is also completed by TSV formed in polymer structure. The signals obtained by the strain gauge, ultrasonic components and acceleration sensor is processed by ASIC simply, for example amplification, and then they are sent to the data processing center for further processing, display and alarm by the wireless RF.

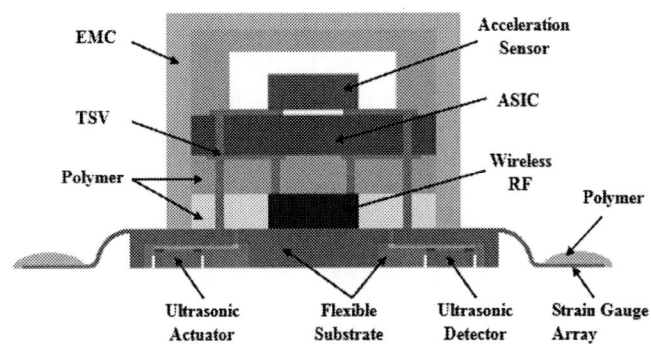

Figure 12. Configuration of the Multi-Sensor SiP

978-1-4799-2408-0/14 $31.00 © 2014 IEEE   1695

Energy Harvesting

Many novel and promising methods for energy supply to devices which are usually in small size with limited stored energy have been proposed and realized, for example the property of piezoelectricity and thermoelectricity. In our design, the devices with small power, for instance the strain gauge array, can be powered by this way, also known as energy harvesting, while for the ultrasonic components with large power, electric wire is needed. The basic circuit of the system is presented in Figure 13.

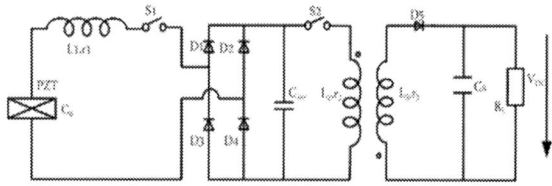

Figure 13. Basic Circuit of Energy Harvest System

Given the vibration when the machinery works, it is easily obtained that the piezoelectricity takes precedence to be the generator for electrical energy in the design, just like it shown in Figure 13. At the same time, the converted energy in intermittent and spontaneous spurts is ineffective and impractical for utilization, so a more-practical approach is to store the energy in energy storage for example capacitors and batteries. By using various passive devices, the damping in the circuit can be reduced efficiently.

Summary

In this paper, a multi-sensor system in package (SiP) is preliminarily designed for the safety monitoring of heavy machinery.In the system, strain gauge array, acceleration sensor and ultrasonic components are used to examine the level of strain and stress on the surface of the steel, the vibration and the crack within the steel, respectively. Temperature sensor is also instrumental in the system for the necessary compensation and measurement. Apart from the monitoring module, wireless RF as the signal transmission device and the advanced method for energy supply are also included. What's more, a package (SiP) based on TSV and flexible substrate provides optimal integration and interconnection for all components. Numerous simulations and experiments have been conducted and the program is still under way. It is expected that the module will be integrated into the internet of things for more widely utilization.

Acknowledgments

The support of High-tech Program (863) with a contract number of 2013AA041105 to SANY, HUST, and NUAA and MEMS-volume Manufacture Technology of National Basic Research Project (973) with a contract number of 2011CB309504 of Ministry of Science and Technology of PR China are highly appreciated.

References

1. Kanda,Y. "Piezoresistance effect of silicon" . Sens. Actuators A Phys.1991.
2. Shixuan Yang, Nanshu Lu. "Gauge factor and stretchability of silicon-on-polymer strain gauges". *Sensors* 2013.
3. Rogers,J.A, Lagally,M.G, Nuzzo,R.G. "Synthesis, assembly and applications of semiconductor nanomembranes". *Nature* 2011.
4. Jeffrey C.Suhling, Richard C.Jaeger. "Silicon piezoresistive stress sensors and their application in electronic packaging". *IEEE SENSORS JOURNAL*,vol.1,no.1,June 2001.
5. Jinling Zhao, Jinhao Qiu, Hongli ji, Ning Hu. "Four vectors of Lamb waves in composites: Semi analysis and numerical simulation".*Journal of Intelligent Material Systems and Structures*.2013.
6. Qiu L, Yuan SF. "One development of a multi-channel PZT array scanning system and its evaluating application on UAV wing box". *Sensor Actuat A*: Phys 2009.
7. Qiu L, Yuan SF, Zhang XY, et al. "A time reversal focusing based impact imaging algorithm and its evaluation on complex composite structures". *Smart Mater Struct* 2011.
8. Wang CH, Rose JT and Chang FK. "A synthetic time-reversal imaging algorithm for structural health monitoring". *Smart Mater Struct* 2004.
9. Michaels JE. "Detection, localization and characterization of damage in plates with an in situ array of spatially distributed ultrasonic sensors". *Smart Mater Struct* 2008.
10. Cai J, Shi LH, Yuan SF, et al. "High spatial resolution imaging for structural health monitoring based on virtual time reversal". *Smart Mater Struct* 2011.
11. Sheng Liu, Yong Liu. *Modeling and Simulation for Microelectronic Packaging Assembly.* John Wiley&Sons, Ltd, 2011.

Novel TSV Process Technologies for 2.5D / 3D Packaging

Y.Morikawa[1,2], T. Murayama[1,2], T. Sakuishi[1,2], A. Suzuki[1], Y. Nakamuta[1] and K.Suu[1,2]
[1]ULVAC Inc., [2]NMEMS technology research organization
1220-1 Suyama, Susono, Shizuoka, 410-1231 Japan
Tel:+81-55-998-1592 , E-mail:yasuhiro_morikawa@ulvac.com

Abstract

"2.5D silicon interposers" and "Hetero 3D stacked" technology for high-performance LSI are gathering the most attention from now on. These technologies can solve interconnection problems using TSV (Through Silicon Via) to electrically connect stacked each function devises. 2.5D and hetero-3D Si integration has great advantages over conventional 2D devices such as high packaging density, small wire length, high-speed operation, low power consumption, and high feasibility for parallel processing. But, the radical problem about the long-term reliability of TSV production is not still solved. In particular, the management of barrier metal film deposition on the smooth surface is most important technology for Cu diffuse protection [1]. On the other hand, TSV isolation liner materials with high step coverage and lower temperature deposition on the smooth surface for high frequency devices will be necessary in the future. "Scallop-free" etching process has developed for TSV fabrication [2]. As a result, the smooth-sidewall had proved shorten PVD process time [3]. At first, it investigated a cost correlation of taper-shape etching and Cu-ECP (electro-chemical plating) in this paper. And then, a polyurea film using a vapor deposition polymerization technology (which is Ulvac's FPF/PV large panel technology) tried introduction as isolation liner for next-generation high frequency device. And, it performed the film formation to a TSV pattern.

Introduction

In order to realize the manufacturing and cost benefits of via last technology for the 3D stacked integration, creation of TSV spanning all layers of fully formed chips must be realized. This is particularly new etching method challenging for TSV devices. The etcher system is using NLD (magnetic Neutral Loop Discharge) plasma that is a kind of high

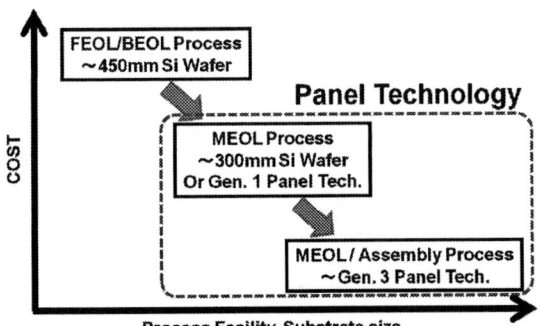

Figure 1. Cost reduction model of packaging.

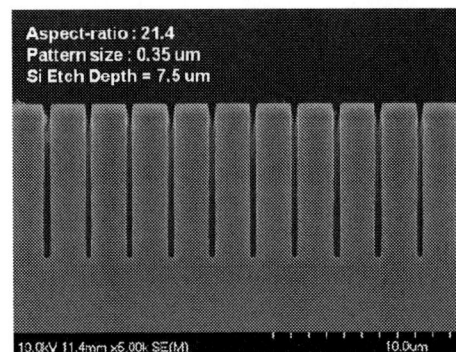

Figure 2. "Scallop-Free" etching processes

electron density ICP (Inductive Coupling Plasma). This plasma etch system can be used for oxide, Si and polymer, for very uniform etching, respectively [2, 3, 4]. In this paper, study on direct (non Bosch) Si etch method to get a smooth sidewalls with no scalloping and a good control of the etched profile using new ICP plasma source. And, it's also trying high selectivity to photo-resist mask for TSV etching to provide the lowest cost processes. Lower cost in the Cu TSV fabrication is most important issue for widespread adoption to various consumers' devices. The issue of the leakage current, which could be caused by sidewall roughness, has been studied. [5, 6] Conventional ICP and CCP plasma are difficult to optimize smooth sidewall etching and high selectivity to photo-resist for TSV fabrication.

Results and Discussions

Deep Silicon Tapered Via Etch Process using by novel ICP plasma for 200 / 300 mm wafer

The need of the method to lower TSV production cost includes the innovation of substrate facilities (fig. 1). The FEOL and BEOL facilities are very expensive for MEOL (middle end of line; like TSV packaging process). TSV fabrication should be prepared by the MEOL method. And it should begin preparations to apply a panel technology to TSV process of manufacture for the cost reduction. "Scallop-free" etching process has developed for TSV fabrication (fig. 2). And, the smooth-sidewall had proved shorten PVD process time. A novel ICP plasma for Si etching is high electron density (10^{12} / cm^3), and very uniform on the Si wafer shown in fig. 3. This plasma characteristic of ICP source above 5Pa process operated with dual rf antenna coils for TSV and bottom-SiO2 etching. Improved plasma characteristics such as higher plasma density and very uniform and high aspect

2014 Electronic Components & Technology Conference

ratio anisotropy TSV etching process were realized in 300mm

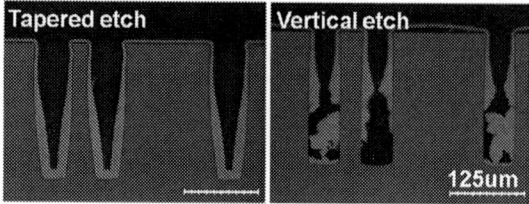

Figure 3. Schematic view of the TSV etcher apparatus.

wafer. Which plasma source is kind of planar type ICP. 13.56MHz or 2MHz of rf for dual antenna coils and low frequency rf bias can operate in independently. Mechanism of Si etching is mainly fluorine radical reaction. High fluorine radical density plasma is need to get high etch rate. On the other hand, management of radical diffusion from around rf antenna is important for very uniform process of high aspect ratio TSV as well. Center gas injection on the rf window is induced instead of side gas injection to avoid of the rf electric field effect. Therefore, when Si was etched using dual rf antenna coil with SF6 / O2 / additional gases mixture injection from center of rf window, the high etch rate and selectivity of Si over the photo-resist and very uniform process were observed. And, the Si wafers used in this work are 200/300 mm in diameter with TSV holes patterned. The diameter of TSV is 10 to 50 um with photo-resist / SSS or GSS substrate structure. When the pressure increased about 20 Pa, the Si etch-rate is increased. On the other hand, Taper-angle can get positive at high-pressure process in 300mm wafer (fig. 4), and when another parameter changed, taper-angle going negative profile shown. These results demonstrate that taper angle is able to controllable by scallop-free etching process. And, which method found that it was better continuous growth of Cu-ECP than the vertical profile (fig. 5).

TSV fabrication for high frequency devises

The surface roughness affects the signal in a high frequency as show in Table 1. Therefore, it is important to form the TSV surface extremely smoothly to reduce a

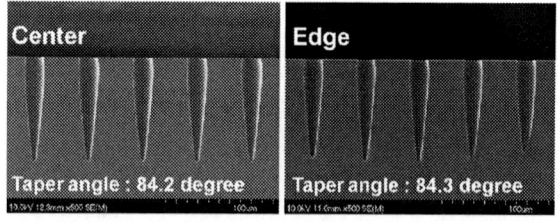

Figure 4. Tapered control with scallop-free etching in 300mm wafer.

conductor loss. As definitions, α is conductor loss, R(f) is

Figure 5. Etched profiles and Cu ECP.

resistivity of conductor skin effect, ε is dielectric constant. Generally, α is proportional to $R(f) * \sqrt{\varepsilon}$.

As for the deep etching technology of the silicon, the cycled etching method (Bosch method) of repetitious deposit and etching is generally. But, the sidewall roughness is accompanied. It can control this roughness weather extremely small by cycle fast method [7]. However, even the existence of small scallops has an effect on signal transmission speed. Therefore, the high Si etching rate is indispensable because general Si interposer is large-size which diameter is 10 - 30um, and aspect ratio is around ten. It is hard to avoid the surface-roughness value of below 100nm by cycled method. The side-wall roughness of the 100nm levels leads to enough conductor losses when it considers skin-effect from Table 1.

Next, the wiring-delay by the dielectric constants of the liner film is important problem. SiO2 film is adopted generally in TSVs, but is fatal to the high frequency device of the GHz band. Therefore, kind of polymer film is introduced as a low dielectric constant for a high frequency band. A novel vapor deposition polymerization system as for TSV and an example of the film deposition on the TSV pattern as shown in Fig.6. A film is polyurea. The dielectric constant is about 3. There are three advantages about that polymer film. First, the film deposition temperature is below 100 degrees. The low temperature film deposition can reduce damage to the temporary adhesion layer. Second, the film stress is quite small. And finally, re-smoothness process form rough-surface (Fig.7).

And then, novel TSV fabrication flow about Via-last TSV for high frequency applications is achieved shown in fig. 8. First, scallop-free etching, and next, resist removal, and polyurea and barrier metal / seed Cu deposit on the TSV, and also Cu ECP deposition. All dry processes are blow 150 degree.

Table 1: Frequency and Skin Depth.

Frequency (GHz)		0.1	1	10	30
Skin Depth (um)	Ag	6.44	2.04	0.64	0.37
	Cu	6.61	2.09	0.66	0.38
	Pt	7.86	2.49	0.79	0.45
	Al	7.96	2.52	0.8	0.46

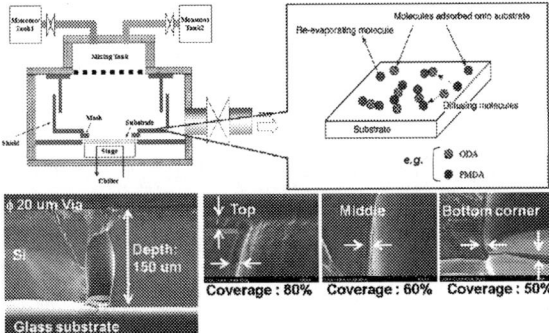

Figure 6. Characteristic of polyurea and deposition on the TSV pattern.

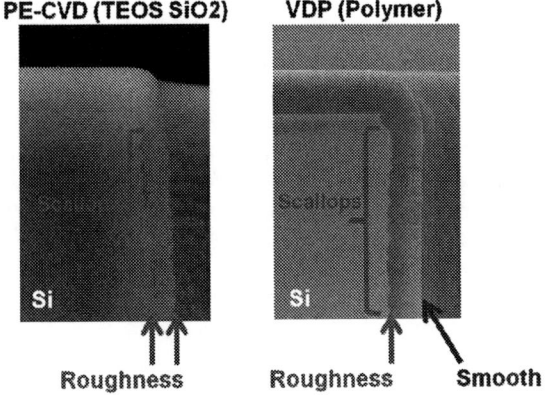

Figure 7. Comparison between PE-CVD and Polymer

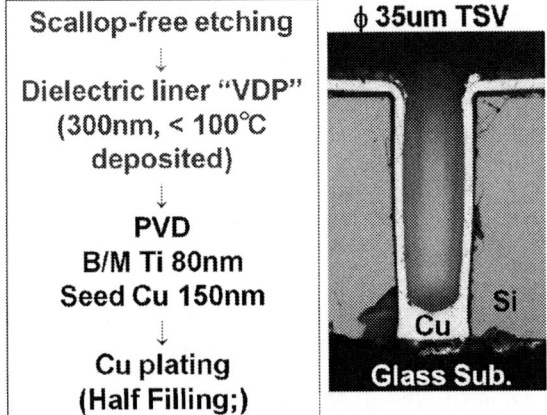

Figure 8. TSV Solutions for high-frequency 2.5D / 3.0D - IC

Conclusion

The taper angle control etching technology with scallop-free has provided for lower cost TSV fabrication which is the epoch-making and practical new technology without cycled-etch method in 300mm and larger substrate. The direct etch is environment conscious process. This is because it does not use fluorocarbon gases in direct Si etching, which main gases are SF6 and O2. The scallop-free with taper angle etching processes have brought about low cost of deposition processes for PE-CVD, PVD and Cu-ECP.

As a result, a novel TSV fabrication flow for 2.5 / 3D and for high frequency applications can provide. "Scallop-Free" & "Tapered-Shape", and the vapor deposition polymerization "polyurea" dielectric liner film are key factor.

Acknowledgments

This work of Si etch process was partly supported by NEDO (New Energy and Industrial Technology Development Organization of Japan).

References

[1] K. W. Lee et al., "High Reliable and Fine Size of 5-um Diameter Backside Cu Through-Silicon Via (TSV) for High Reliability and High-End 3-D LSIs" *IEEE 3D IC 2011*, Osaka, Japan, 2012.

[2] Y. Morikawa, et al., "Very Uniform and High Rate TSV Etching Process in Advanced NLD Plasma" *Int. Symp. on AVS 57th*, Albuquerque, New Mexico, 2010.

[3] Y. Morikawa, et al., "A Novel Scallop Free TSV Etching Method In Magnetic Neutral Loop Discharge Plasma" *Proc. IEEE Electronic Components and Technol. Conf. (ECTC)*, San Diego, CA, May 29 – June 1, 2012, pp. 794–795.

[4] Y. Morikawa et al., "Scallop Free TSV Etching Method For 3-D LSI Integration" *Int. Symp. on AVS 58th*, Nashville, Tennessee, 2011.

[5] H. Kitada, et al., "Surface microroughness-induced leakage current in through-silicon via interconnects" ; *AMC 2011*, San Diego, CA, 2011.

[6] T. Nakamura et al., "Comparative Study of Side-Wall Roughness Effects on Leakage Currents in Through-Silicon Via Interconnects" *IEEE 3D IC 2011*, Osaka, Japan, 2012.

Increasing the Lifetime of Electronic Packaging by Higher Temperatures: Solders vs. Silver Sintering

Aaron Hutzler, Adam Tokarski, Silke Kraft, Sigrid Zischler, Andreas Schletz
Fraunhofer Institute for Integrated Systems and Device Technology IISB
Landgrabenstr. 94, 90443 Nuremberg, Germany
aaron.hutzler@iisb.fraunhofer.de

Abstract

Increasing the temperature in power electronic applications usually causes a decreasing lifetime and reliability. This study shows that packaging materials and technologies like silver sintering or gold germanium solders can easily deal with temperatures above 150°C. Furthermore the power cycling capability at increased temperatures can be much better than at room temperature.

Active power cycling tests with **240 devices** offered more cycles to failure at 120°C cooling temperature than at 40°C. The three tested sample groups consisted of silicon carbide diodes which were soldered (gold germanium/ tin lead) or silver sinter to copper-ceramic-substrates (DBCs).

The reason behind this effect is the decreasing of Young's modulus, yield strength and ultimate strain over temperature. The materials are getting much more ductile and robust against load cycling at higher temperatures. The three mentioned material properties were measured by nano-indentation and tensile tests up to 200°C.

In summary, packaging materials and their properties should be adopted to the intended application and its requirements, starting with a temperature-dependent analysis.

Introduction

In power electronics, increasing the application temperature has a negative influence on the lifetime of power modules like dc-dc converters or inverters. According to popular empirical models, the load cycling capability of these modules will *always decrease* by raising the coolant temperature [1,2].

With high melting materials and bonding techniques like silver sintering or copper-heavy wire bonding the portfolio of packaging technologies in power electronics changed during the last years [3,4]. New combinations of packaging materials caused new effects and interactions at lifetime tests, especially their response to temperature [5]. Active power cycling runs can result in a higher number of cycles to failure at increased temperatures compared to room temperature test [5]. This study extends previous investigations with further material combinations and explains why the best working conditions for new materials are at raised junction temperatures.

Good examples for high temperature applications are hybrid electric vehicles, which use one and the same coolant system for both the power electronic devices and the combustion engine. This increases the temperature conditions for the packaging from 70°C to 120°C [6]. In that environment ordinary soldered silicon semiconductors are reaching their limits due to a maximum junction temperature of 150°C. That is why silicon carbide diodes were used in this study.

For higher temperatures, degradation of large area die-attach joints were identified as the lifetime-limiting failure mechanism [7,8]. Therefore, this research concentrates on the die attach layer. The first part describes the lifetime tests (active power cycling) with silicon-carbide diodes and three different die attach materials (silver sintered, AuGe12 solder and PbSn5 solder). Additionally, temperature-dependent nanoindention measurements were performed with die attach materials up to 200°C to explain this effect.

Materials and Methods

The Fraunhofer IISB power cycling test bench (PCT3) consists of a power supply, a computer for data logging and a measurement acquisition unit for each device under test. A chamber for 20 samples and a tempering unit are also included into the rack. The advantage of this system is the online measurement of electrical data for each device under test at every instant. The captured values per sample are listed in the following table 1. Current, voltage and temperature are measured continuously during the test. Power losses and thermal resistance, as well as thermal impedance are calculated after each cycle.

Table 1: Captured data per sample during active power cycling test.

Name	Abbreviation	Unit
Current (heating phase)	I_{Heat}	Amps (A)
Voltage (heating)	U_{Heat}	Volts (V)
Voltage (cooling)	U_{Cool}	Volts (V)
Temperature (heating)	T_{Heat}	Kelvin (K)
Temperature (cooling)	T_{Cool}	Kelvin (K)
Temperature swing	ΔT	Kelvin (K)
Power losses	P	Watt (W)
Thermal resistance	R_{th}	K/W
Thermal impedance	Z_{th}	K/W

The devices are mounted on a heat sink and current conduction was realized via spring contacts. A sense line inside the springs connects the device with the data logging hardware. The junction temperature is measured indirectly by voltage gauging at a small current (e.g. 30 mA). This method uses the nearly linear relationship between temperature and forward voltage of semiconductor diodes [9]. Measurement current and the determination of the junction temperature are provided by the data acquisition unit and the computer. A schematic setting is shown in Figure 1.

978-1-4799-2408-0/14 $31.00 © 2014 IEEE

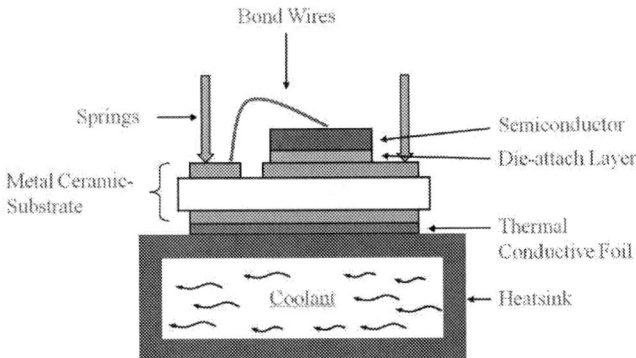

Figure 1. Schematic power cycling test setting.

For the power electronic assemblies, 120 silicon carbide diodes were soldered with Pb95Sn5 and Au88Ge12 (60 samples per solder) on a DBC-substrates with Al_2O_3-ceramic. The test run with silver sintered samples also consisted of 120 devices. The devices under test had no molding compound on top and no baseplate on their backside. A sample with gold germanium solder after the power cycling test is shown in Figure 2.

Figure 2. Power cycling test sample with gold germanium solder after the test.

Table 2 provides detailed data on assembly materials and manufacturers for the power cycling test. The Young's modulus and the yield strength of the solder preforms and the sinter layers were analyzed by nanoindentation at room temperature, 100°C and 200°C. Tensile tests of the solder preforms and silver ribbons at the same temperatures should evaluate the data of indentation method.

Table 2: Used materials for power cycling test, nanoindentation and tensile tests.

Type	Manufacturer	Product name
Ag (100 μm)	Agosi	Silver Ribbon
Pb95Sn5 (100 μm)	Pfarr	Solder Preform
Au88Ge12 (100 μm)	Pfarr	Solder Preform
Sinter Paste	Heraeus	LTS043
Bond Wires	Heraeus	Al-H11 (125 μm)
Semiconductor	Infineon	SiC-1111-DD
Substrate	Curamik	Al_2O_3-DCB

The nano-indentation system G200 by Agilent uses a Bercovich indenter tip (nano-scale), which penetrates into the material of interest. The tip and the work-holder of the system can be heated up to 500°C by a laser. The result of the measurement is a load-displacement diagram from which the Young's modulus and yield strength can be calculated [10,11,12]. Three curves of gold germanium preforms at different temperatures is shown in the following figure.

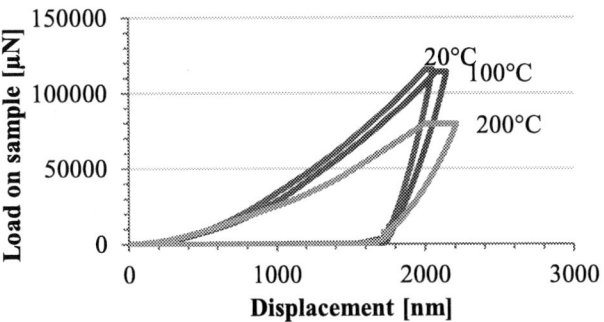

Figure 3. Nanoindentation measurement curve of a AuGe12 solder at different temperatures.

The Young's modulus E can be calculated by the gradient of the unloading curve (stiffness of the material). In the first step the *reduced elastic modulus E_r* has to be determined via the following formula [10].

$$S = \frac{dP}{dh} = \frac{2}{\sqrt{\pi}} E_r \sqrt{A_{projected}} \qquad (1)$$

The projected area A of the indenter tip on the surface of the sample is shown in Figure 4.

$$\frac{1}{E_r} = \frac{(1-v^2)}{E} + \frac{(1-v_i^2)}{E_i} \qquad (2)$$

E_r consists of the sample- and indenter-tip-modulus (E and E_i) as shown in Formula 2. The Poisson's ratio v for the silver material can be assumed as 0.3 [11].

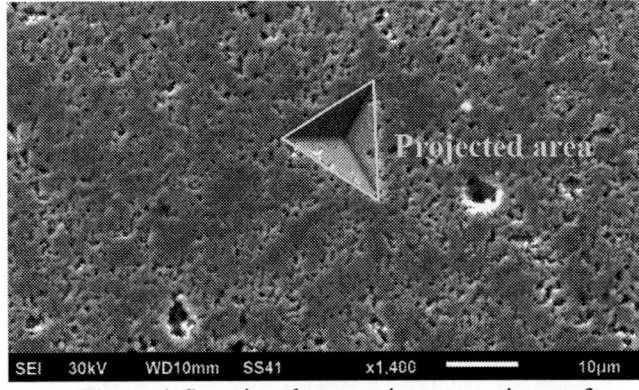

Figure 4. Scanning electron microscopy picture of a nanoindentation footstep.

According to the procedure of [12], the yield strength can be extracted of a nanoindentation curve when prior knowledge of the hardness or the Young's modulus of the tested material is given. The system of equations in [12] was solved with *python*. For the sinter layer, nanoindenation is a good method to analyze material properties due to the small layer thicknesses and hard to manufacture bulk material.

To get the yield strength, ultimate strain and Young's modulus of metals, tensile tests are a common method [13]. For these tests a material testing machine *LRXPlus* by *LLOYD Instruments* combined with a custom-made oven is available at Fraunhofer IISB. Temperature-dependent measurements can be taken up to 300°C. From the stress-strain curve of a tensile test, the Young's modulus, the ultimate strain and the yield strength of the tested material can be extracted [13].

Experimental

The soldered samples (60 with Pb95Sn5 and 60 with Au88Ge12) were assembled in an inert-gas reflow oven with preforms. Formic acid was used instead of a flux. The other devices were silver sintered with 10 MPa pressure at 255°C for 120s.

Active power cycling tests were performed with the PCT3 test system. The temperature of the heat sink, i.e. the minimum temperature was set to 25°C, 40°C, 80°C and 120°C, respectively. The temperature swing ΔT was always 130 K. To achieve this swing, the heating current was individually set for each run and then kept constant during the test. To achieve thermal steady state conditions, cycle times t_{on} and t_{off} were set to 15 seconds each. For the end-of-life-criteria, the common method of a 20% increase of ΔT, the thermal resistance, or the heating voltage (whatever comes first) was used [9]. The test matrix is also shown in Table 3.

All devices failed due to solder/ sinter-layer degradation. Bond wire liftoffs were prevented by bonding smaller wires (125 µm diameter) instead of the commonly used 375 µm or thicker wires [14]. The samples were analyzed after the test with cross sections and light microscopy to verify the failure mechanisms.

For temperature-dependent material characterization of die-attach layers, nanoindenation and tensile tests were performed. By combining both methods, it was possible to measure the Young's modulus, ultimate strain and yield strength of silver ribbons, gold germanium and tin-lead preforms as well as sinter layers at 20°C, 100°C and 200°C. For every test material and test temperature 15 measurements were taken and the mean was calculated.

Scanning electron microscopy and optical microscopy were used to analyze the indentation method and measurement effects like sink-in or pile-up [10].

Results of the Power Cycling Tests and Statistical Analysis

All tested devices showed a higher number of cycles to failure at increased coolant temperatures. The statistical analysis confirmed a significant difference between all test runs and consisted of the following steps:

First, the distribution of each test result (20 samples per run) was identified via *maximum likelihood method* and

Anderson-Darling value [15]. For all test data the Weibull distribution fitted well and was applied for further steps.

Then the Weibull scale and shape parameters of each run were used for a Bayes analyses to find significant differences between the shape parameters, which described the number of cycles where 63,2% of the devices failed [15]. The Bayes analyses showed highly significant differences between the cycles to failure of all runs. Table 3 shows the median life of the test matrix.

Table 3: Results of the active power cycling tests.

T_{min}	Die Attach	Current	Median Life
25	$Au_{88}Ge_{12}$	18	121,500
40	$Au_{88}Ge_{12}$	18	120,100
120	$Au_{88}Ge_{12}$	15	500,000[1]
25	$Pb_{95}Sn_5$	18	2,500
40	$Pb_{95}Sn_5$	18	3,300
120	$Pb_{95}Sn_5$	15	19,600
25	Silver-Sintering	18	45,000
40	Silver-Sintering	18	47,500
80	Silver-Sintering	16	74,900
120	Silver-Sintering	15	85,900

For the statistical analysis the failure mechanisms of the tested devices was very important. If different causes (e.g. bond wire lift off, semiconductor damage or solder degradation) would have been observed during the test, the analysis had to be divided into the different mechanisms. Therefore, the power cycling test data (Figure 5) gives a first hint about for what reason the devices failed.

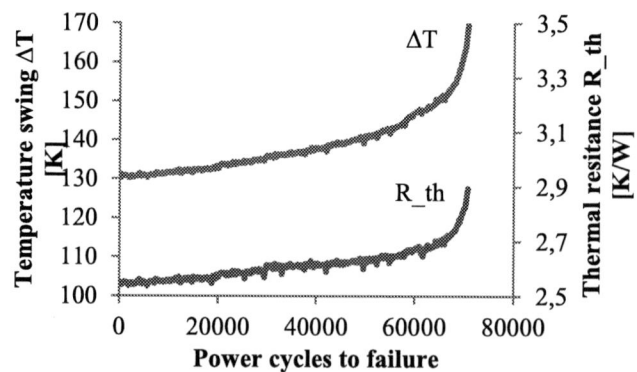

Figure 5. Active power cycling signal with increasing temperature swing ΔT and thermal resistance R_th.

If the thermal resistance and temperature swing increased synchronal, the bottom connection (e.g. solder layer) of the devices often degraded. Figure 5 shows a power cycling signal with this relationship. A damage of the semiconductor

---

[1] No failure after 500,000 power cycles

can easily be detected during the power cycling test due to a change of its measured electrical values (Table 1) during each cycle.

All devices showed the behavior of die-attach degradation like in Figure 5. To prove this first hint, samples of all test runs were analyzed via cross sections and light microscopy. The following Figure 6 shows a cross section of the sinter layer power cycled at 120°C minimum temperature.

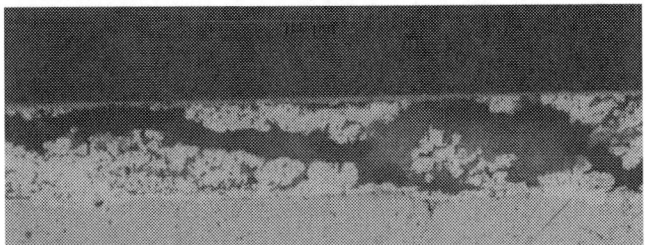

Figure 6. Cross section of a sinter layer after active power cycling at 120°C coolant temperature.

Cracks in the silver sinter die-attach grow from pore to pore right through the layer. The interface between the silver and the copper of the substrate was still in a good condition after the test.

Finally, the bond wires were analyzed for lift offs. The cross sections showed a slight damage of the wires, however, 50 % of the interface area was still intact (Figure 7).

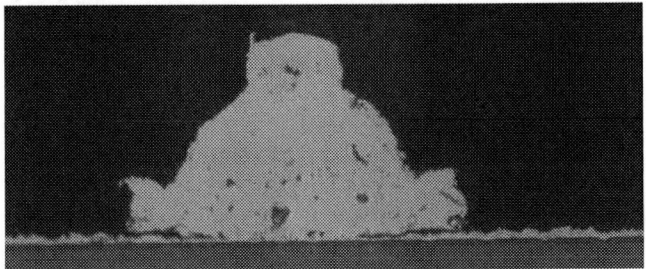

Figure 7. Cross section of a bond wire after active power cycling at 120°C coolant temperature.

Results of the Nanoindention and Tensile Tests

The Young's modulus and the yield strength of silver, gold germanium and tin-lead were measured by nanoindentation and tensile tests. The results of both methods were in accordance with each other, if the different velocities were taken into account. Due to its small thickness (40 µm), the silver sinter layer could only be analyzed by the nano-tip. The indentation measurement curves showed a large variation due to the porosity of the silver sintered layers. Therefore 15 indents were performed on one sample and the average was calculated. At the silver ribbons and solder preforms the standard deviation of the material was much lower and matched to the results of the tensile tests.

In the following Figure 8, the Young's modulus of the mentioned die attach materials from room temperature to 200°C is depicted. The data were taken from nanoindentation, but the results of the tensile tests were quite similar.

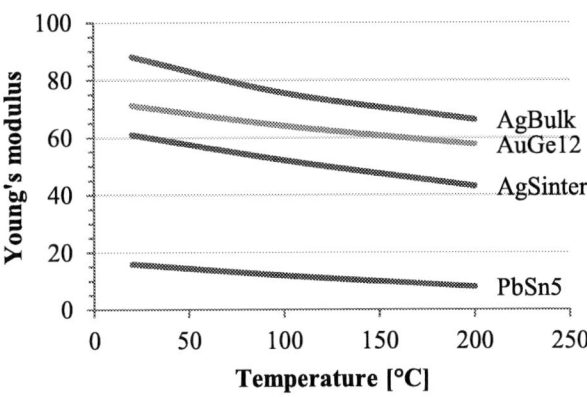

Figure 8. Young's modulus of silver, gold germanium and tin-lead measured by nanoindentation (mean values).

The elastic modulus of all materials decreased over temperature, which is a common behavior of metals [13]. Considering analytical models for stress calculation in substrate materials, a decreased Young's modulus leads to lower stresses inside the sandwiched material (e.g. die attach) [17]. If all other material properties are neglected, a higher minimum temperature reduces the stress at thermo-mechanical load and therefore increases the lifetime of assemblies with different material combinations.

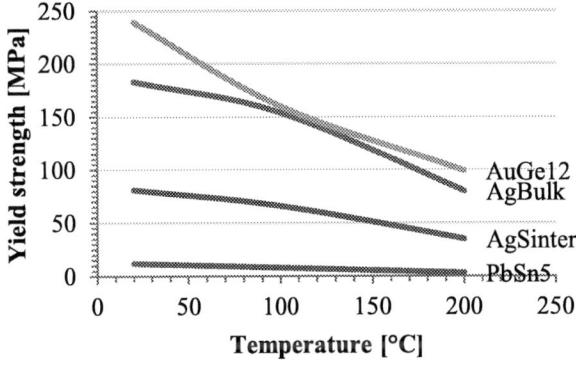

Figure 9. Yield strength of silver bulk material, sintered silver, gold germanium and tin-lead measured by nanoindentation (mean values).

The elastic modulus is only one of many material properties which influence the thermo-mechanical capability of an electronic packaging. A more important factor for die attaches is the ductility of the material. To characterize the ductility, the yield strength, i.e. the beginning of plastic deformation, is an appropriate parameter. Its response to temperature is shown in Figure 9.

Above all, silver gets more ductile between 100°C and 150°C. Including the ultimate strain (Figure 10), silver can withstand much more plastic deformation at temperatures above 100°C than at room temperature. This was already shown by [16].

978-1-4799-2408-0/14 $31.00 © 2014 IEEE   1703

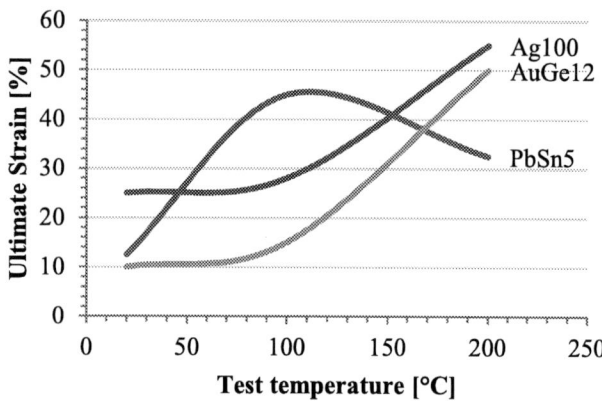

Figure 10. Ultimate strain of silver, gold germanium and tin-lead measured by tensile tests (mean values).

Gold germanium solders showed the same behavior as silver, but due to its lower thermal expansion (see Table 4) and its higher yield strength over temperature (Figure 9), AuGe12 had the best power cycling performance of all tested materials.

Table 4: Material properties of tin lead and gold germanium solder, as well as silver bulk material and sintered silver [3,18,19].

Die Attach	Melting Point [°C]	Thermal Conductivity [W/m/°C]	Coefficient of Thermal Expansion
PbSn5	305-315	35	29
AuGe12	356	44	13
Silver	961	429	19
Silver Sinter	961	*[2]	19[2]

Creep effects are very common at homologous temperatures (see Figure 11) above 0.4 [13]. For high melting materials like silver or gold germanium it was already shown that these effects play only a minor role [20]. The creep behavior of material can also be seen from nanoindentation curves like a characteristic "fingerprint" [21]. For silver material no changes in the curves were detected.

The tin lead solders showed creep mechanisms already at room temperature. By increasing the test temperature these effects increased. Taking the power cycling results and ultimate strain into account, the optimum working point for that material was around 120°C, where the ultimate strain has its maximum peak and creep effects did not yet dominate the main failure cause. By further raising the temperature, the influence of creeping increased so much that ultimate strain lowered again.

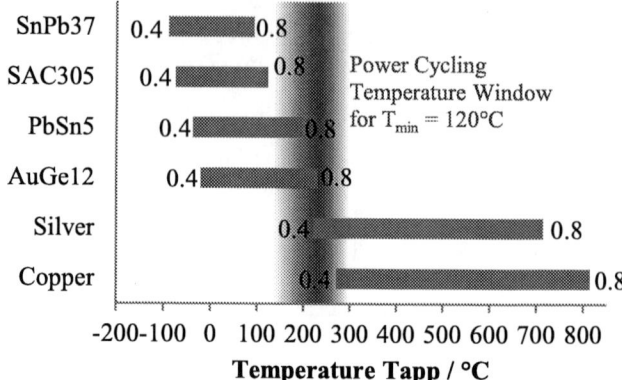

Figure 11. Homologous temperature (application temperature/ melting point) for different packaging materials.

The ultimate strain could only be obtained via tensile tests and the silver sinter die attach analysis could only be realized by nanoindentation.

Further Discussion of Test Results

Active power cycling tests were performed at different coolant temperatures, but with the same temperature swing. To achieve a ΔT of 130 K for every run, the current had to be set individually for each cycling condition. At higher temperatures less current was needed due to increasing power losses of semiconductor devices. For applications such as hybrid electric cars, the maximum current of a power module or rather its price determines the material of choice. The temperature swing is more like an intermediate operand for a better understanding of material effects behind the power cycling test.

It has to be mentioned that the *absolute number* of cycles to failure is *only* valid for the shown assembly. The samples were simplified and the focus of this study was on the die attach material. No molding compound, no housing and no base plate was used and the current was applied via springs. Especially the molding would change the thermal conditions of the tested devices significantly. A thermal conductive foil was used to avoid paste pump-outs. But this also changed the resistance from chip to coolant and therefore changed the number of cycles compared to an application-oriented test.

Another point for discussion is the active load on the assembly. For automotive applications passive temperature cycling is also an important test for lifetime and reliability of power modules. The found relationship between cooling temperature und lifetime is only valid for *active power cycling* tests. Passive loads might cause very different effects.

The solders and sinter materials were analyzed by nanoindentation and tensile tests to explain the experimental results. A combination of both methods can lead to a pool of temperature-dependent material properties for a better understanding of the effects behind the test results. The *absolute values* of the material parameters strongly depend on the test velocity, i.e. tensile test and indentation speed. In this study a low speed was applied to get a creep-similar response of the tested materials.

In addition to that, the material characterization was performed with bulk materials. Their properties might differ

---

[2] Depends on further material parameters (e.g. porosity) [18,19]

significantly from die attaches after the assembly process. For further investigations more application-oriented test dummies have to be used. This can be achieved by indenting into cross sections or manufacturing of sintered preforms for tensile tests.

A major problem of nanoindentation is its handling of inhomogeneous materials. Pores or different phases like inter-metallic compounds can lead to different measurement results. Therefore, further analysis via scanning electron microscopy is important for failure detection. This fact makes the nanoindentation method very time-consuming and expensive.

Temperature-dependent parameters like the Young's modulus, the yield strength and the ultimate strength are needed for simulation models (e.g. finite element analysis). One of the future goals is to create an application-oriented database for power electronic materials.

Conclusion

Active power cycling tests were performed with 240 silicon carbide devices. The samples were soldered with Au88Ge12, PbSn5 or silver sintered to DBC substrates. The temperature swing at all test runs was kept constant at 130 K, but the coolant temperature was altered from 25°C to 120°C. The results showed more cycles to failure at 120°C than at 25°C.

The gold germanium die attach achieved the highest number of cycles to failure at 120°C. With 500,000 power cycles AuGe12 showed a nearly 8 times better performance than the silver sintering technique with 85,000 cycles. Due to its lower CTE and high temperature material properties this effect can be explained. By raising the coolant temperature the die attach layers became more ductile and more capable against thermo-mechanical stress.

The tin lead solder also showed an increased cycle number at 120°C. The reached 20,000 were only a quarter of the silver sintering result. The much lower yield strength caused the lower lifetime. The ultimate strain of PbSn5 has its peak at nearly 120°C. Therefore, the best power cycling capability was reached at that temperature.

To explain the test results, nanoindentation measurements and tensile tests were performed from 25°C to 200°C with all die attach materials. The Young's modulus and the yield strength showed a decrease over temperature, while the ultimate strength of the materials increased. This effect can be very useful for high temperature applications like hybrid electric vehicles or aero-space devices. If the homologous temperatures are taken into account, every material might have its optimum operation temperature, where the material properties are best suited.

Acknowledgments

The authors also would like to thank the German *Federal Ministry of Education and Research* (BMBF), the *Bavarian State Ministry of Economic Affairs and Media, Energy and Technology* and the *European Union* for the financial support of this work.

References

1. R. Bayerer, et. al, "Model for Power Cycling lifetime of IGBT Modules– various factors influencing lifetime", *Integrated Power Electronics Systems (CIPS)*, 2010, pp. 37-42

2. U. Scheuermann, R. Schmidt, "A New Lifetime Model for Advanced Power Modules with Sintered Chips and Optimized Al Wire Bonds", *PCIM Europe Proceedings*, 2013, pp. 1-6

3. M. Knörr, A. Schletz, "Power Semiconductor Joining through Sintering of Silver Nanoparticles: Evaluation of Influence of Parameters Time, Temperature and Pressure on Density, Strength and Reliability", *Integrated Power Electronics Systems (CIPS)*, 2010, pp. 1-6

4. S. Haumann, "Novel Bonding and Joining Technology for Power Electronics", *Applied Power Electronics Conference and Exposition (APEC)*, 2013 Twenty-Eighth Annual IEEE, pp. 622 – 626

5. A. Hutzler, A. Tokarski, A. Schletz, „Extending the lifetime of power electronic assemblies by increased cooling temperatures", *Microelectronics Reliability Vol. 53, Issue 9-11*, 2013, pp. 1774-1777

6. M. März, et. al., "Power Electronics System Integration for Electric and Hybrid Vehicles", *Integrated Power Electronics Systems (CIPS)*, Nürnberg – Germany, 2010, pp. 1-10

7. R. Bayerer, R. John, "High temperature power electronics IGBT modules for electrical and hybrid vehicles", *IMAPS High temperature electronics network*, 2009, pp. 199 – 204

8. L. Feller, S. Hartmann, D. Schneider, "Lifetime analysis of solder joints in high power IGBT modules for increasing the reliability for operation at 150°C", *Microelectronics reliability Vol. 48*, pp. 1161 – 1166

9. J. Lutz, H. Schlangenotto, U. Scheuermann, R. de Donker, „Semiconductor Power Devices: Physics, Characteristics, Reliability", Springer, 2011

10. A. C. Fischer-Cripps, "Nanoindentation", Springer New York, 2011

11. S. Kraft, S. Zischler, N. Tham, A. Schletz, "Properties of a novel silver sintering die attach material for high temperature high lifetime applications", *Sensor + Test Conference Proceedings*, Nürnberg - Germany, 2013, pp. 1-6

12. O. Casals, J. Alcalá:, "The duality in mechanical property extraction from Vickers and Berkovich instrumented indentation experiments", *Acta Materialia Vol. 53*, 2005, pp. 3545-3561

13. N. E. Dowling, "Mechanical Behavior of Materials", Pearson Edingburgh, 2013

14. A. Hutzler, A. Wright, A. Schletz, "Increasing the lifetime of power modules by smaller bond wires", *IMAPS Workshop Wirebonding*, San Jose USA, 2014

15. K. S. Stephens, "Reliability Data Analysis With Excel and Minitab", Amer Society for Quality, 2011

16. Kewei Xiao et.al., "Creep Behavior of Sintered Nano-Silver Paste", *CPES Conference, Blacksburg*, 2010, pp. 1-5.

17. E. Suhir, "Analysis of interfacial thermal stresses in a trimaterial assembly", *Journal of Applied Physics, Volume 89, Number 7*, 2001, pp. 3685-3694

18. J. G. Bai, Z. Z. Zang, J. N. Calata, G. Q. Lu, "Characterization of Low-Temperature Sintered Nanoscale Silver Paste for Attaching Semiconductor Devices", *High Density Microsystem Design and Packaging and Component Failure Analysis*, Shanghai, 2005, pp. 1-5

19. A. A. Wereszczak et. al., "Properties of Bulk Sintered Silver As a Function of Porosity", Oak Ride, 2012

20. A. Drevin-Bazin, F. Lacroix, J.-F. Barbot, "SiC Die Attach for High-Temperature Applications", *Journal of Electronic Materials, Vol. 43, No 3.*, 2014, pp. 695-701

21. M. L. Oyen, R. F. Cook, "A practical guide for analysis of nanoindentation data", *Journal of mechanical behavior of biomedical materials, Vol. 2, 2009*, pp. 396-407

Comparison of New Die-Attachment Technologies for Power Electronic Assemblies

Eike Möller[1], Adeel Ahmad Bajwa[1], Eugen Rastjagaev[2], Jürgen Wilde[1]

[1]Laboratory for Assembly and Packaging Technology, Department for Microsystems Engineering (IMTEK),
University of Freiburg, [2]Infineon Technologies AG, Austria
Georges-Köhler-Allee 103, 79110 Freiburg, Germany
eike.moeller@imtek.uni-freiburg.de

Abstract

Recently, different die-attachment technologies for power modules such as electrically conductive gluing, silver sintering and transient liquid phase bonding have been investigated as alternatives to soldering. In this paper a comparison of these three die-attach technologies is given. For each an overview of the manufacturing process, as well as key performance parameters like thermal resistance and shear strength are presented. In addition the influence of temperature shock cycles on the modules is investigated. Overall, this work showed that all three die-attachment techniques can be alternatives to replace soldering in specific assembly types of power electronic devices.

Introduction and Motivation

Due to the increasing application of power electronic systems as converters, commutators or half-bridges, the demand for more efficient assembly techniques for such systems is rising steadily. Die-attachment technologies have two major challenges:

- First the bonding temperature should be low, because of the induced stress in the chip.
- Second the reliability of the modules should be as high as possible.

Different techniques, such as conductive adhesive bonding, silver sintering or transient liquid phase bonding show a high potential for the use in die-attachments [1, 2, 3, 4, 5]. A main target of research is to increase the maximum achievable power of the chip and to minimize the thermal and thermo-mechanical impact on the assembly. For instance, the thermal resistance of the die-attachment material should be minimal in order to dissipate the heat out of the system [6, 7]. Furthermore a robust production process and low investments are required.

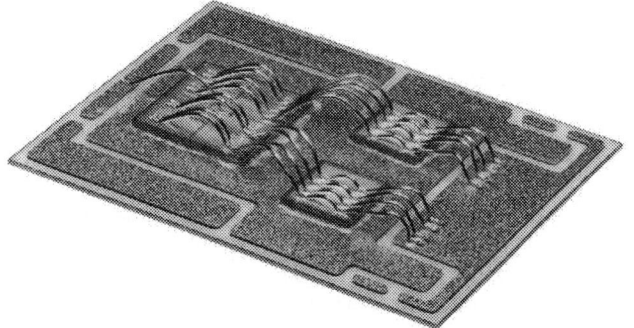

Figure 1: Typical power module with an IGBT and two diodes connected as a half-bridge circuit

Various applications of automobiles, aerospace and deep oil and gas well logging require the electronic circuits to operate at high power and under harsh high temperature environment i.e. 250 °C to 350 °C. Silicon based electronic circuits are unable to operate above 250 °C. Hence, SiC and GaN based electronic devices are regarded as suitable for high power and high temperature applications and operating temperatures of up to 500 °C are mentioned. The choice of die-attach materials, which can endure harsh environmental conditions, is still very limited. In our laboratory, silver sintering and transient liquid phase bonding (TLP) are developed as the die-attachment techniques for high temperature packaging. Both die-attachments can survive temperature up to 500 °C. In addition, their process temperatures are in the range from 210 °C to 240 °C are lower than those of conventional high temperature die-attach solders such as AuSn, AuGe and PbSn, with process temperatures of 300 °C to 380 °C. This in turn implies that thermal and thermo-mechanical stresses coming from the CTE mismatch might be reduced. Moreover, as both techniques are based on silver instead of gold, this reduces the materials costs and yields superior electrical and thermal properties.

Experimental Methods

In this work three die-attachment techniques were investigated with respect to relevant properties for assembling of power modules. In table 1 we present the most important requirements and the corresponding measurement methods for assembled power modules.

Table 1: Overview of relevant properties for assembled power devices and their measurement methods

Requirement	Measurement method
low process temperatures	temperature monitoring
high bonding stability	shear test
high thermal conductivity	thermal resistance
defined layer thickness	metallurgical cross section
high reliability	temperature shock test

First an overview of the different manufacturing processes is given. After that investigations of bonding stability and thermal properties are presented. Furthermore cross sectional views of the different layers were shown and finally reliability aspects were considered.

In order to compare the results all investigations were made with DBC (direct bonded copper) substrates. The DBC consists of a layer combination of 300 µm Cu / 630 µm Al₂O₃ / 300 µm Cu. Furthermore the substrates were electroplated with additional Ni/Au layers.

Test chips were used during the optimization of the different processes. Silicon wafers were cut to various chip sizes of 4 mm², 7.84 mm² and 20.25 mm². Some test chips had a backside metallization of Ti/Ni/Ag. The thickness of Ti, Ni and Ag were 50 nm, 100 nm and 200 nm respectively. After process optimization, thermal resistance measurements of assemblies were made with silicon based power-diodes. The chip size was 4.5 x 4.5 mm² and they were provided by Diotec-Semiconductor, Germany.

Electrical conductive adhesive

The assembly with electrical conductive adhesive (ECA) is an efficient process, which mainly requires a pick & place machine and a curing oven. The advantage of ECAs is a low curing temperature, between 120 °C and 180 °C, while lead-free soldering is performed at 240 °C. Because of the lower process temperature, the induced thermal stress in the assembly is much smaller. Several investigations gave rise to the assumption that electrical conductive adhesive can be an opportunity for solder replacement in the usage of power electronic devices [1, 3].

Investigations with the adhesives were made with a standard ECA and additionally a development product. Both ECAs were made available from Heraeus, Germany. The adhesives were filled with silver flakes with mass contents of approximately 80 %.

The conductive adhesive was deposited on the DBC substrate with a pressure dispensing setup. Afterwards the chip was placed with a vacuum pick & place tool. The adhesives were hardened in an oven for 45 min at 150 °C in a special ramped program. Via wire bonding with a 300 μm Al wire (Heraeus AlX) the electrical connection was achieved. Figure 5 presents an adhesive assembled diode on the DBC substrate.

Silver sintering

The sintering might become the state-of-the-art die-attachment method for high temperature electronic devices. In our laboratory, this method is successfully employed for high power high temperature GaN high electron mobility transistors (HEMT). A novel micro-scaled silver sintering paste LTS 275-3P2, provided by Heraeus GmbH was used. The sintering paste is either dispensed or screen printed on the of substrate. The dies are then placed on the paste with a pick & place tool. In the first drying step, the assembly is placed inside the oven and heated at a rate of 2.5 °C/min up to 160 °C. It is then kept at this temperature for 30 min. In the next sintering step the assembly is heated at a rate of 10 °C/min up to 240 °C, which is monitored with a thermocouple. At 240 °C, the pressure i.e. 10 MPa is applied to the assembly for 5 min. Thereafter the pressure is removed and the assembly is cooled at a rate of 5 °C/min to room temperature. The sinter silver joint is expected to have similar electrical and thermal properties as bulk silver. It has a very high melting point of 961 °C.

Transient liquid foil bonding (TLP)

An alternative die-attachment technique for high power high temperature devices is transient liquid phase bonding. It is based on the diffusion of a liquid bonding material into a metal which is in a solid state. The bonding metals such as tin and indium etc. have lower melting points such as 232 °C and 157 °C respectively. Upon heating above the melting point and with the help of external pressure, the inter-layer metal melts and diffuses into the base metal such as silver or gold, which have high melting points of 961 °C and 1064 °C, respectively. The whole process is ideally performed under vacuum to minimize the reactions with oxygen. Diffusion and isothermal solidification result in the formation of highly temperature stable intermetallic phases that have melting points, which usually are 200 °C to 400 °C higher than the process temperatures. An optional annealing step can be performed to homogenize the joints. A detailed process scheme is given in [4].

A novel foil based TLP bonding process is developed in our laboratory, where TLP interlayer material is used as a pre-form made of a multilayer foil which is produced using electroplating. The weight ratio between the base metal silver and the low melting metal tin or indium is decided based on the binary phase diagrams Ag-Sn and Ag-In. The inter-layer metal of the sandwich structure faces both the substrate and chip backside. Both also exhibit metal finishes like silver or gold, to form the interface connection. In this work, a silver-tin based binary system is investigated for high temperature stable joints. This resulted in the formation of high temperature stable i.e. 500 °C intermetallic phases. The interface was void free and physical material properties are expected to be close to silver after a bond homogenization [4].

Figure 2: IXYS RF transistor assembled on AlN DBC using Sn-Ag-Sn based TLP bonding

Bonding stability

A first investigation was performed to determine the mechanical robustness of the bonded systems. For that purpose shear tests of all investigated die-attachment technologies were performed. Standard silicon wafers were cut in chips with area sizes of 4 mm² and 7.84 mm² were bonded on DBC-substrates with Ni/Au metallization.

For power electronics the mechanical robustness in harsh environments is high importance. For that reason the modules were exposed to passive temperature shocks from -40 °C to 150 °C with a dwelling time of 30 min at each temperature. Afterwards the shear force was measured after 300 cycles and a comparison with the untested specimens was made. In Figure 3 the shear stress data before and after temperature shock tests are presented.

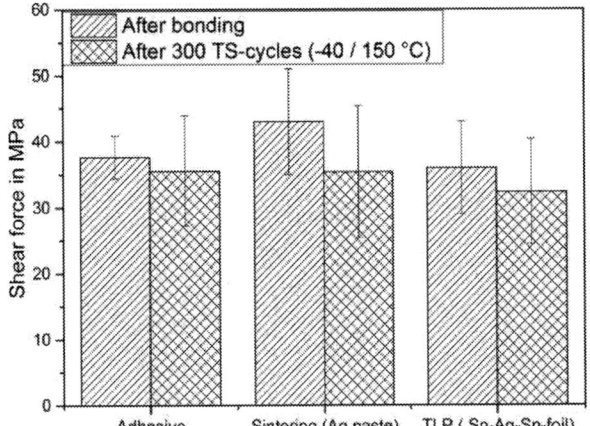

Figure 3: Shear forces of the investigated materials on DBC substrates before and after temperature shock cycling. Chip size 2 x 2 mm²

The highest average shear forces could be achieved with the silver sintering 43 MPa. The shear stress of the TLP foil was found to be 36 MPa, which is also in the range of the adhesive i.e. 38 MPa. After 300 temperature cycles the shear force decreases. Furthermore it was found out that also the DBC substrate degrades, under the temperature loads, due to the CTE-mismatch of the AlN and the Cu.

Thermal analysis

For power electronics heat dissipation from the chip is a real challenge. It must be ensured that the heat dissipation from electrical devices is sufficient to keep the temperature below the critical value of the chip. This requires that the thermal conductivity of the die-attachment material is as high as possible. For investigations of the thermal conductivity the thermal resistance of an assembled device was measured. The thermal resistance of a material is comparable to the electrical resistance. It indicates the heat flow rate $\dot{Q}$ through a component for a measured temperature difference. The heat flow $\dot{Q}$ corresponds to the dissipated power P due to ohmic or switching losses of the device.

The thermal resistance is defined by Equation 1, with ΔT as the temperature difference between the top surface of the diode and the bottom side of the DBC device and P the dissipated power in the active area of the chip.

$$R_{th} = \frac{\Delta T}{P} = \frac{T_1 - T_2}{P} \qquad (1)$$

Here, T_1 is the junction temperature, T_2 is the temperature of the DCB at the bottom.

Typically many suppliers of adhesives or sinter pastes provide data of the bulk thermal conductivity which have not been measured on die bonds. These bulk data are often significantly higher than values generated on assemblies. In our investigation, we measured the thermal resistance using assembled modules, to get representative data. In figure 4 the used measurement setup for thermal resistance measurements is presented. The system consists of a DBC-substrate with a Si-diode assembled on the Ni/Au-metallization layer of the DBC. To measure the temperature difference two temperatures have to be determined. Temperature T_1 is measured with a Pt-1000 temperature sensor, which is glued on top of the diode with a thermal conductive adhesive. T_2 is measured on the bottom copper side of the DBC substrate with a constantan wire, which has a sensitivity of 39.5 µV/K against Cu. The substrate is cooled directly with water.

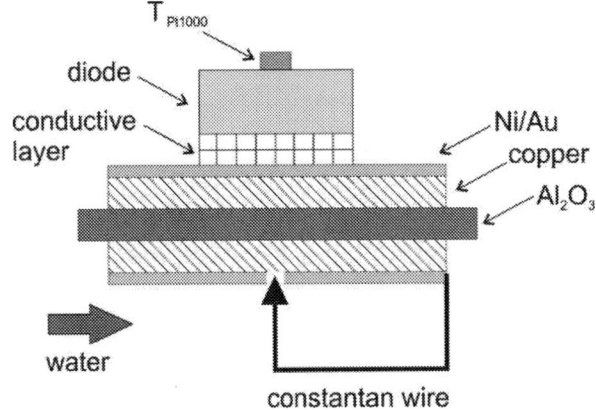

Figure 4: Principle for thermal resistance measurements with the combination of substrate, adhesive and diode

Figure 5: DBC substrate with an assembled diode using electrical conductive adhesive

In this study the thermal resistance of every investigated die-attachment technique was measured with the same substrate, metallization, diode and settings. The thermal resistance of the modules was measured at currents of 2 A, 6 A, 10 A, 14 A and 18 A. All measurements have been done with a 4.5 x 4.5 mm² Si-diode with Ni/Au metallization provided by Diotec Semiconductor, Germany. For comparison the thermal resistance results at 18 A are shown in figure 6.

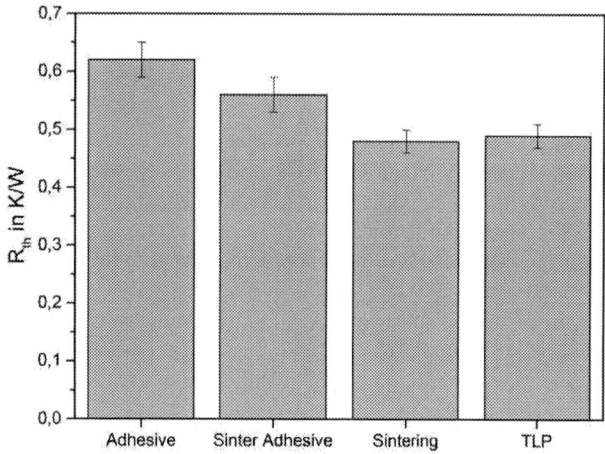

Figure 6: Comparison between measured thermal resistances of investigated die-attachment techniques at a current of 18 A with diodes 4.5 x 4.5 mm² on DBC.

The highest thermal resistance of 0.631 K/W was measured with an adhesively bonded device. A lower thermal resistance can be achieved with a sinter adhesive, which we also investigated at our lab. With this development product it is possible to reduce the thermal resistance to 0.561 K/W at a current of 18 A. Transient liquid phase bonding and sintering technology were nearly similar in performance with 0.495 K/W and 0.481 K/W, respectively.

Furthermore we investigated the influence of the layer thickness of the used material on the thermal resistance and then calculated the thermal conductivity of the material. With Equation 2 the thermal resistance of a block-shaped layer can be calculated by the layer thickness d, the area A and the thermal conductivity λ.

$$R_{th} = \frac{d}{\lambda \cdot A} \qquad (2)$$

The total thermal resistance of the die bond is defined by the resistance which is proportional to the layer thickness and additionally the contact resistance at the interfaces. It can be represented by Equation 3:

$$R_{th} = R_{th,contact} + R_{th,bulk} = R_{th,contact} + \frac{d}{\lambda \cdot A} \quad (3)$$

The layer thickness was determined by cross-sectional specimens for each die bond material. In figures 7-8 images of such die-attachments with ECA and TLP are shown. The values of the mean thickness for each layer are presented in Table 2.

Table 2: Measured thicknesses of the die-attachment materials after curing

Material	Mean value of layer thickness (µm)
Adhesive	25
Sinter material	20
TLP	28

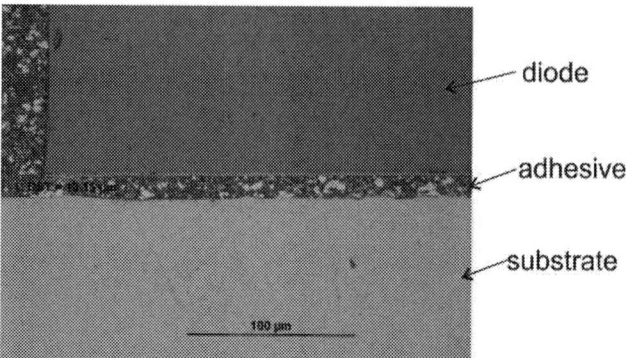

Figure 7: Cross section through the DBC substrate with adhesively bonded diode on top

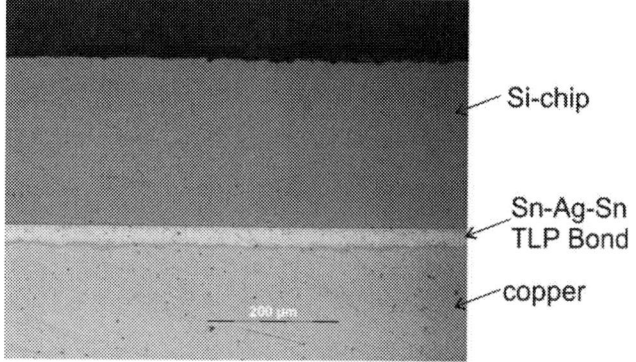

Figure 8: Cross section of the TLP-bonded Si-Chip on DBC substrate

With the measured layer thicknesses it is possible to estimate the thermal conductivity of each material. For that we choose the sinter-material as a reference material, because it has the highest thermal conductivity. Based on previous work, λ = 200 W/mK [8] can be regarded as a reasonable guess. With this value and a mean layer thickness of 30 µm we can calculate the ideal thermal resistance for sintered material with equation 2. In a next step we added this value to the difference of the two measured thermal resistances. Afterwards the total thermal conductivity could be calculated with Equations 4 & 5.

$$\frac{R_{th,sin}}{R_{th,sin}+\Delta R} = \frac{\lambda_x}{\lambda_{sin}} \qquad (4)$$

$$\lambda_x = \frac{R_{th,sin} \cdot \lambda_{sin}}{R_{th,sin}+\Delta R} \qquad (5)$$

$$\frac{1}{\lambda_x} = \frac{1}{\lambda_{sin}} + \frac{\Delta R \cdot A}{d} \qquad (6)$$

It must be noted, that this calculation is only valid for the assumption of the ideal thermal conductivity of the sinter material and an assumed layer thickness of 30 µm. Because λ_{sin} is a literature data we calculated the conductivities also for a lower value of λ = 100 W/mK to present the influence of the thermal conductivity of the reference material. Thus a comparison of the different thermal conductivities is presented in Table 3.

Table 3: Calculated and given thermal conductivities of different die-attach materials. * shows own measured values, ** is based on literature data.

Material	Thermal conductivity λ (W/mK)	Thermal conductivity λ (W/mK)
Adhesive 1	9.4* (> 5 data sheet)	8,98*
Adhesive 2 (sinter adhesive)	17*	15,61*
Sinter material	200**	100**
TLP	69*	51,38*

The sintered silver layers have superior electrical and thermal properties. The effective thermal conductivity of the TLP bonding and the electrically conductive adhesive was estimated based on equation 5 on the basis of measured values of R_{th}. TLP bonding also has low thermal resistance as compared to the conductive adhesives. Low thermal resistance is the prime requirement for the packaging prospects of the power devices.

For the standard adhesive the thermal resistance depending on the layer thickness was determined. The measurements have been done on the standard DBC copper substrate and with a diode of 5.4*5.4 mm², so this is the thermal resistance of a assembled device. A nearly linear correlation was found out. It is obvious that a thinner layer of material induces a lower thermal resistance.

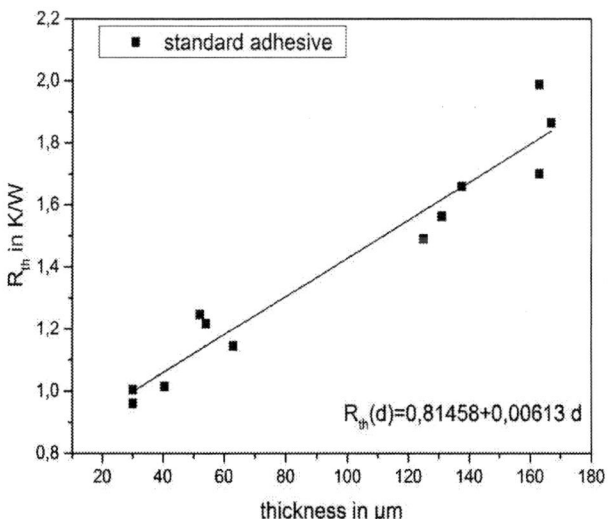

Figure 9: Change of the thermal resistance R_{th} of adhesive assemblies as a function of layer thickness with diodes of 5.4x5.4 mm² on DBC substrate [1]

From the slope of the thermal resistance it is possible to calculate the bulk thermal conductivity of the material. The value is $\lambda = 6.5$ W/mK, for this adhesive the value corresponds to the data which is given in the data-sheet ($\lambda > 5$ W/mK). It must be noted, that it is not the effective thermal conductivity as it neglects the contact resistance as mentioned in equation 3.

Reliability aspects

Long term reliability after passive and active cycling (power or temperature-shock) is a standard criterion to compare different die-attachment technologies. For a first evaluation of the behavior of large power semiconductor under temperature shock cycles, IGBT and diodes were assembled onto copper substrates. Because of their large area and low thickness these devices could not be shear tested.

For that reason the crack length before and after temperature shock cycles was measured for adhesively bonded and silver sintered modules [1, 9, 10]. The crack length was determined via metallographic cross sections on the diagonal direction.

Figure 10: Cutting line for crack length determination on an adhesively bonded IGBT on copper substrate

It was found out that the cracks starts at the edge of the device and propagates towards the center. In figure 11 a crack in an adhesively bonded 12x12 mm² IGBT on a copper substrate after 500 TS cycles at -40/175 °C is presented. For an adhesively bonded IGBT a crack propagation rate of 3.2 µm per cycle at a temperature shock profile of -40/175 °C [1].

Figure 11: Crack propagation under an IGBT adhesively bonded von 300 µm copper, after 500 TS -40/175 °C [1]

For silver sintered devices similar results were observed. Again due to CTE – mismatch of the materials cracks appeared up at the edges of the chip and propagated through the sinter material and the silicon.

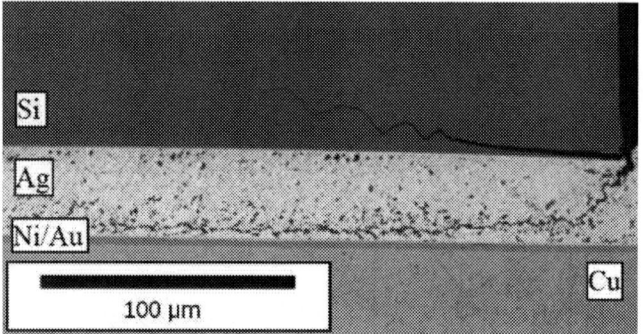

Figure 12: Crack propagation in silver sintered 5.4x3.4 mm² silicon chips after 250 TS -40/200 °C [10]

Crack propagation in solder materials have been investigated in [11]. In this work a 15 x 15 mm² silicon chip was soldered with a (Pb92.5Sn5Ag2.5) solder on a Al_2O_3 - DBC-substrate. 2500 cycles of temperature shock tests have been performed and after every 500 cycles the crack length was measured with an ultrasonic microscope. From these data the crack propagation rate was calculated to 2.7 µm per cycle.

An overview of the different materials, chips and results of the investigated crack propagation is presented in table 4 [1, 4, 9, 10, 11].

Table 4: Overview on investigations on crack propagation for investigated die-attach materials

Material	ECA Adhesive	Sintered silver paste	PbSnAg-Solder
Substrate	Cu	Ni/Au on Cu	DBC
Chip area (mm²)	12x12	5.4x3.4	15x15
TS profile (°C)	-40/175	-40/200	55/155
Crack propag. (µm/cycl.)	3.2 @ 35 µm	1.2 @ 40 µm 2 nm @ 20 µm	2.7 @ 60 µm
Source	[4]	[9, 10]	[11]

It was observed that a smaller bond layer thickness leads to a higher crack propagation rate, due to increase of local stress and strains near to the chip edges, where the cracks started. Furthermore the crack propagation is affected by the CTE-mismatch of the materials, so the crack propagation is also high. For DCB substrates both the crack propagation in the ceramic layer and the CTE-mismatch of the Al_2O_3 and copper material has to be considered, which leads to an enhanced crack propagation rate.

Conclusions

In this paper we have presented a comparison of three different new die-attachment technologies for power electronic devices. The aim of our work was to investigate the capability of electrical conductive adhesive, silver sintering and transient-liquid-phase bonding for application as an assembling technology for power electronic modules.

A first step was to determine the bonding stability of the investigated materials. It was found out that all technologies have a high mechanical robustness, which is comparable to the robustness of a standard solder alloy. Even after 300 temperature shock cycles only little decrease of the shear force was detectable. Only degradation of the DBC-substrate was determined, which induces lower shear forces.

The second requirement to the new die-attachment technologies was the low thermal resistance and consequently a high effective thermal conductivity. The investigations showed that the materials lead to a low thermal resistance, which is nearly in the same range as a standard solder material. Measurements provide the lowest thermal resistance with 0.481 K/W for sinter material and highest value with 0.631 K/W for adhesively bonded modules on DBC substrates. Out of the thermal resistances we calculated the thermal conductivity of the materials.

Also there is strong layer thickness dependency on the thermal resistance, which we investigated for the adhesive. Higher layer thickness leads to an increasing of the thermal resistance. For sintering and transient liquid phase bonding similar dependencies can be expected.
Furthermore a first investigation on crack propagation for adhesive and silver sintering in assembled power modules after temperature shock cycling was presented. The results showed values which were similar to standard solder alloys.

Also the comparison of the manufacturing processes showed the high potential for future applications in power electronic systems. From the investigated technologies, especially conductive adhesive bonding is a very simple process, which will not need special machines. Silver-sintering is presently developed towards a pressure-free method. Nevertheless, like TLP it still needs research work. Table 5 presents a final overview of the investigated materials and their properties for application for assembling of power electronic devices.

Table 5: Comparison of investigated die-attachment technologies and their properties for assembling power electronic devices

	Adhesive	Silver sintering	TLP
Process complexity	++	o	o
Low bond temp.	++	+	+
Bond stability	+	+	+
Thermal conduct.	o/+	++	+
High power appl.	o/+	++	++
High temp. appl.	o	++	++

Summarizing, we are investigating different die-attach materials for power electronic devices at our lab. We are able to handle new die-attachment technologies for standard silicon power electronic devices, new state of the art GaN-modules as well as for high-power and high temperature devices. We think all investigated die-attachment technologies showed a high potential for power applications, although there are still some aspects for further research. For instance chips with larger area have to be assembled and investigated with respect to their thermal mechanical stress during high temperatures. Furthermore active and passive long-time reliability has to be investigated for all material systems, which is our next topic of investigation at our lab.

References

1. J. Ocklenburg, E. Rastjagaev, E. Möller, J. Wilde, *Is conductive adhesive bonding suited for the die-attachment of power devices?* 8th International Conference on Integrated Power Electronics Systems CIPS, Feburary 25-27, 2014.

2. A.A. Bajwa, Y.Y. Qin, R. Zeiser, J. Wilde, *Foil based transient liquid phase bonding as a die-attachment method for high temperature devices*, CIPS, Feburary 25-27, 2014.

3. J. Ocklenburg, E. Rastjagaev, J. Wilde, *Investigation of modern electrically conductive adhesives for die-attachment in power electronics application*, in Proc. IEEE Electronic Components and Technol. Conf. (ECTC), Las Vegas, May 28–31, 2013, pp. 2189-2195.

4. A.A. Bajwa, R. Reiner, Y.Qin, R. Quay, O.Ambacher, J.Wilde, *Assembly and packaging technologies for higth temperature and high power GaN HEMTs*, IEEE ECTC 2014, Orlando, May 27-30, 2014.

5. A.A. Bajwa, R. Zeiser, J. Wilde, *Process optimization and characterization of a novel micro-scaled silver sintering paste as a die-attach material for high temperature high power semiconductor devices*, 36th International Spring Seminar on Electronics Technology (ISSE), Alba Lulia, May 8-12, 2013, pp. 53-58.

6. K. Guth, N. Oelscher, L. Böwer, R. Speckels, G. Strotmann, N. Heuck, S. Krasel, A. Ciliox, *New assembly and interconnect technologies for power modules*, CIPS, March 6-8, 2012.

7. J. Wilde, W. Staiger, *Integration of liquid cooling thermal und thermomechanical design for the lifetime prediction of electrical power modules*, Electronic Packaging SP—1345, Feb.1998.

8. J. G. Bai, Z. Z. Zhang, J. N. Calata, G. Q. Lu, *Characterization of low temperature sintered nano-scaled silver paste for attaching semiconductor devices*, Conference on High Density Microsystem Design and Packaging and Component Failure Analysis, Shanghai, June 27-29, 2005, pp.1-5.

9. T. Herboth, M. Guenther, A. Fix, J. Wilde, *Failure mechanisms of sintered silver interconnections for power electronic applications*, in Proc. IEEE Electronic Components and Technol. Conf. (ECTC), Las Vegas, May 28–31, 2013, pp. 1621-1627.

10. T. Herboth, C. Früh, M. Günther, J. Wilde, *Assessment of thermo-mechanical stresses in low temperature joining technology*, 13th EuroSimE Conference, April 16-18, 2012, Cascais.

11. M. Thoben, *Zuverlässigkeit von großflächigen Verbindungen der Leistungselektronik*, Fortschritt-Berichte VDI Reihe 9 Nr.363, Dissertation 2002.

High Vacuum Wafer Level Packaging for High-value MEMS Applications

S. Nicolas, F. Greco, S. Caplet, C. Coutier, C. Dressler, M. Audoin, X. Baillin, G.Dehag, F. Souchon, S. Fanget
CEA / LETI, MINATEC
17, rue des Martyrs – 38054 Grenoble Cedex 9, France
stephane.nicolas2 @cea.fr, +33 (0) 4 38 78 92 01

Abstract

In this paper, our latest developments on high vacuum packaging technology at the wafer level are presented. The main objective of our works has been to demonstrate the feasibility of hermetic high vacuum wafer level packaging of MEMS (vacuum target is 10^{-3} to 10^{-4} mbar) using either anodic or eutectic bonding (AuSi).

In the frame of this work, a test vehicle based on a MEMS resonator is used to characterize the vacuum inside the packaging by measuring the Quality factor (Q factor) of the resonator.

The resonator is fabricated using 200 mm BSOI wafers. For capping, either 200 mm glass wafer or 200 mm silicon wafer are used depending on the bonding technology chosen. A deep cavity is created in the cap wafer. A getter is deposited inside the cavity to maintain a high vacuum level after bonding. The wafers are then bonded together using anodic or eutectic bonding and electrical interconnections (TSV) are finally carried out.

First, the calibration curve i.e., Q Factor vs residual pressure has been achieved using our resonator with a non-hermetic package and a specific vacuum chamber that allowed us to measure the Q factor of the resonator at different residuals pressures (from 10^{-1} to 10^{-5} mbar).

Then, anodic and eutectic bonding processes have been developed and optimized in order to reach vacuum level as low as 10^{-3} mbar after interconnection. In the case of anodic bonding, it turned out that the main issue that has to be addressed was the permeation of noble gas through the glass wafer during interconnection process. For eutectic bonding, outgassing turned out to be the more significant parameter to control. Finally, residual pressure down to 10^{-4} mbar has been achieved.

Introduction

The packaging of MEMS at the wafer level has been a field of interest for many years now and numerous solutions are available [1].

In this area, MEMS devices dedicated to consumer market generally required hermetic package with "medium" vacuum level (up to 10^{-1} mbar). On the other hand, for high-value applications, with components such as high precision gyroscope or µbolometer, very high vacuum level (below 10^{-3} mbar) is mandatory. Such high vacuum level remains difficult to achieve and to control due to phenomenon like leak, outgassing, permeation and moreover, it generally requires a thin film getter.

In the past years, results with vacuum close to 10^{-3} mbar have already been published with Chip Scale Packaging (CSP) technologies, for µbolometers application as an example [2].

But, to the best of our knowledge, very little works have been published about such high vacuum level obtained with Wafer Level Packaging (WLP) technologies.

From a technological point of view, a wide range of wafer bonding solutions that leads to hermetic package are available: among them, anodic [3-7], glass frit [8-10], metallic [11-15] and direct bonding [16] are the most reported. But, from our best knowledge, none of them has demonstrated a vacuum level below 10^{-3} mbar.

Our previous reported works on anodic bonding [17] showed that a vacuum level of 10^{-2} mbar could be obtained after TSV integration. At that time, our conclusion was that outgassing phenomenon has to be understood to explain the rise of the residual pressure during the TSV process.

This paper addresses the development of a high vacuum wafer level packaging technology that targeted vacuum level below 10^{-3} mbar. For this purpose, the same test vehicle described earlier [17] has been used and combined with an optimized anodic bonding solution but also with an alternative eutectic bonding solution (AuSi). The final structure after packaging with eutectic bonding is reported on figure 1.

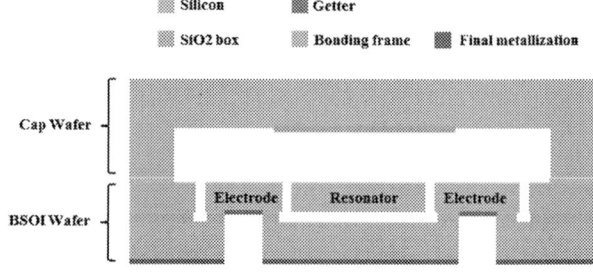

Figure 1. Final structure after WLP – Eutectic option

It has to be noticed that, due to significant outgassing during bonding step, a getter layer inside the cavity is required if the residual pressure target is below 10^{-1} mbar. That is the reason why a getter layer (about 1 µm thick) is deposited on the inner face of the cap wafer [18].

Device fabrication

Our test vehicle is a resonator fabricated from a 200mm BSOI wafer (Fig 2).

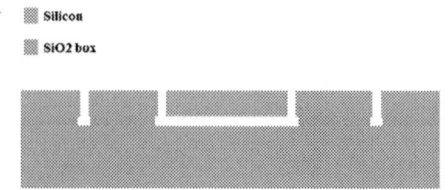

Figure 2. Resonator structure – BSOI wafer

978-1-4799-2408-0/14 $31.00 © 2014 IEEE

The fabrication process has been already described previously [17]. The top silicon layer thickness is 60 µm.

This resonator has sense and drive electrodes required for subsequent electrical characterizations of the Q Factor.

Cap wafer fabrication

For anodic bonding, 200 mm glass wafers have been used and fabrication steps have been described previously [17]. The final glass cap structure is reported on figure 3.

Figure 3. Cap wafer structure for anodic bonding

For eutectic bonding, starting from a 200 mm silicon wafer, process steps are described on figure 4.

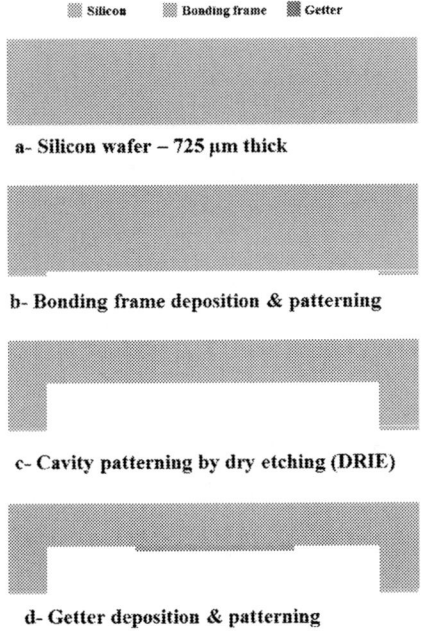

Figure 4. Cap wafer fabrication steps for eutectic bonding

After metal deposition (CrAu) and patterning (Fig 3-b), a deep cavity is etched by DRIE (Fig 3-c). Cavity depth is ranging from 50 µm to 500 µm. Getter layer is then deposited at the bottom of the cavity (Fig 3-d).

WLP and via for electrical interconnection

After resonator and cap wafer fabrication, the wafers are aligned and bonded together. A specific recipe has been used with outgassing steps at different temperatures below 300°C in order to avoid any activation of the getter. The getter is activated during the bonding step at 400°C either for anodic and eutectic bonding. Step duration at 400°C is typically around 30 min.

Once the assembly has been done, a specific process is performed for electrical interconnection (Figure 5). It has to be noticed that the same process has been used for anodic and eutectic wafers assembly.

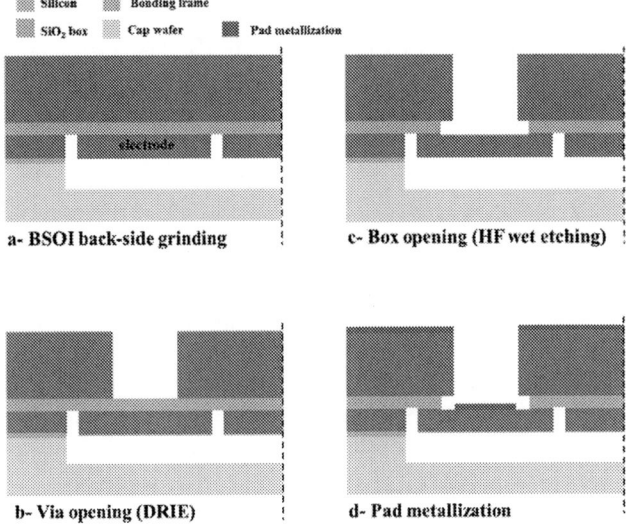

Figure 5. Process description for interconnection

After BSOI wafer back-side grinding (down to 150 µm - Fig 5-a), via are opened by DRIE (Fig 5-b) and the SiO₂ box is wet etched by HF (Fig 5-c). Pad metallization is then carried out using evaporation tool (Fig 5-d).

Q factor measurement

As it has been already reported [16], the Q factor can be evaluated using optical microscopy in the case of anodic bonding. For eutectic bonding, Infrared (IR) microscopy has been used in order to be able to see through the silicon wafer. The idea is to record the resonator's amplitude variation with time after a brief mechanical shock. The higher the vacuum is, the longer the resonator vibration last.

Both optical and IR methods are very useful to check the vacuum level inside the cavity after bonding and during interconnection process steps. However, it has to be reminded that those methods are not very accurate (+/- 20 %) and are only used for an estimation of the vacuum inside the cavity.

For more precise measurements, electrical tests are performed after pads metallization. The set-up used for electrical test has been also described elsewhere [17].

Q factor vs Residual pressure – Experimental results

The well-known theoretical model showed that the Q factor is inversely proportional to the residual pressure inside the cavity. But the model also exhibits a strong dependence of the Q factor with the resonator's design. Thus, the theoretical curve Q factor vs residual pressure could be obtained but some approximations have to be done.

In order to be as close as possible from our final MEMS structure, an experimental set-up has been carried out to measure the Q factor in function of the residual pressure.

Using functional anodic bonded devices, a feedthrough has been etched by Focused Ion Beam (FIB) through the remaining BSOI silicon layer in order to have the same pressure inside and outside the cavity. Devices are then put inside a vacuum chamber provided with a probe card that allowed us to measure the Q factor.

The pressure inside the chamber has been tuned from 10^{-1} to 10^{-5} mbar and Q factor has been measured. It has to be noticed that Q factor measurement is performed after a waiting time of 30 min in order to be sure that the pressure inside the cavity is the same as those in the vacuum chamber. This work has been carried out on 4 devices and results are reported on figure 6.

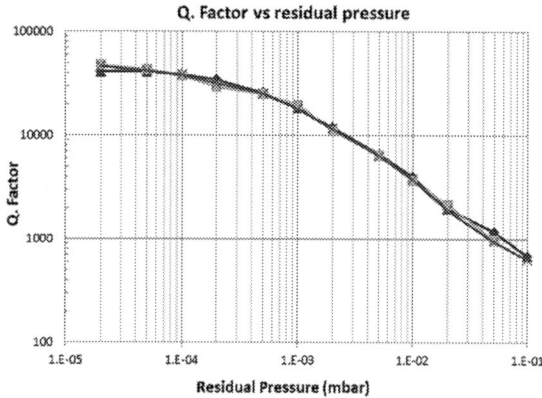

Residual Pressure (mbar)	Q factor (mean value)
1.E-01	600
1.E-02	3700
1.E-03	18 000
1.E-04	38 000
2.E-05	45 000

Figure 6. Q Factor vs Residual pressure – Experimental results on 4 functional devices

Some comments have to be done regarding these results:

- Good repeatability among the different devices
- Resonator sensitivity range = 10^{-1} to 10^{-4} mbar (below 10^{-4} mbar, the Q factor variation is not significant enough)
- Results applicable only for "air" equivalent residual pressure

Thereafter, these curves will be used to convert the measured Q factor into residual pressure.

Experimental results – Anodic bonding

Our previous works [17] has pointed out that Q factor dropped during interconnection process steps (Fig 7) and more specifically during via etching by DRIE.

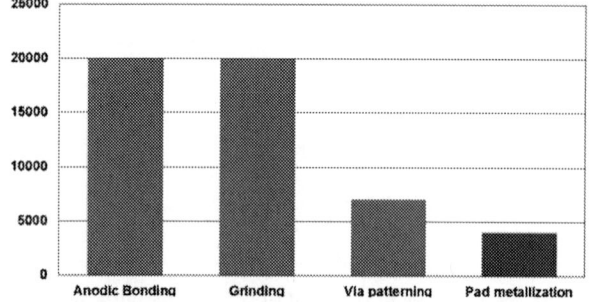

Figure 7. Q Factor evolution (optical method) during interconnection process steps – Anodic Bonding

Regarding these results, it has been first suggested that some outgassing phenomenon occurred within the cavity due to either thermal or mechanical stresses.

However, taken into account materials properties and more particularly glass properties, an alternative mechanism has been suggested to explain the raise of the pressure during via etching: our assumption is that Helium permeation through glass wafer take place during DRIE step. It is well-known in the literature that glass materials are permeable to noble gas [19]. Yet, it turned out that in the DRIE tool, Helium is used to cool down the wafer during the etching step.

In order to validate this assumption, a wafer stack has been anodically bonded and Q factor has been evaluated (Q > 10.000 – measured by optical microscopy). The wafer stack has been then subjected to a pressure of 0.5 bar of Helium during 30 min. It turned out that after this treatment, Q factor dropped dramatically (Q < 500). This result showed that Helium permeation through the glass wafer should be mainly responsible for the raise of the residual pressure inside the cavity.

At this stage, it has to be taken into account that under a residual pressure of Helium, the relationship between Q Factor and residual pressure presented on figure 6 is no more valid because it was obtained under "Air" equivalent residual pressure.

For a resonator operating under air equivalent residual pressure, Q factor (Q_{air}) can be evaluated by [5]

$$Q_{air} = \frac{1}{2.\zeta_{air}} \qquad (1)$$

where ζ_{air} is the coefficient of damping force under air equivalent residual pressure. Then, the Q factor can be expressed as follow

$$Q_{air} = \frac{\omega}{S.P.\sqrt{\dfrac{M_{air}}{2\pi.RT}}} \qquad (2)$$

with

$$\lambda_{air} = S.P.\sqrt{\frac{M_{air}}{2\pi.RT}} \qquad (3)$$

where ω is the natural frequency of the system, λ_{air} is the damping factor under air equivalent residual pressure, S is the resonator surface, P is the residual pressure, M_{air} is the molecular mass of air, R = 8.31 kg m^2/s^2/K is the universal molar gas constant and T is the absolute temperature.

Then, the relationship between Q_{air} in Air equivalent and Q_{He} in Helium equivalent residual pressure is

$$\frac{Q_{air}}{Q_{He}} = \frac{\lambda_{He}}{\lambda_{air}} = \sqrt{\frac{M_{He}}{M_{air}}} = \frac{1}{2.64} \qquad (4)$$

Using the relationship obtained in (4) and experimental curves reported in figure 6, it is then possible to extract the Q factor evolution under Helium residual pressure (Figure 8).

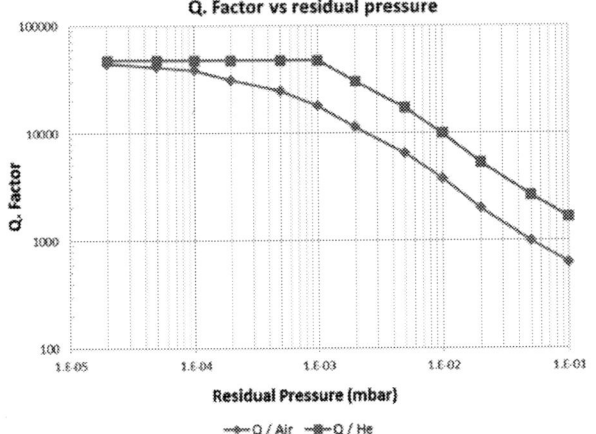

Figure 8. Q Factor vs Air and Helium equivalent residual pressure

It has to be noticed that under Helium residual pressure, the operating range of our resonator is going from 10^{-1} mbar to 10^{-3} mbar, i.e. one decade less than under air residual pressure.

Using the experimental results reported on figure 7 and 8, it is now possible to monitor the evolution of the residual pressure during interconnection steps (Figure 9).

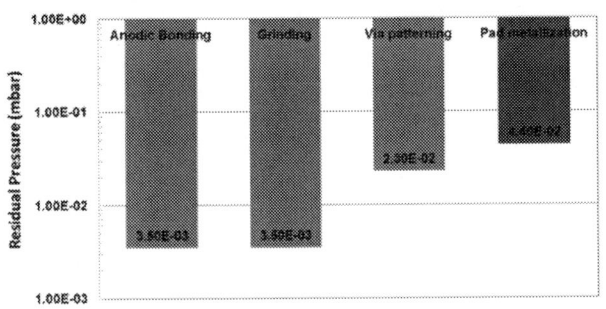

Figure 9. Helium residual pressure evolution during interconnection process steps – Anodic Bonding

It can be concluded that, due to Helium permeation through the glass wafer during DRIE, residual pressure raise from 3.10^{-3} mbar after bonding to almost 5.10^{-2} mbar after pads metallization.

In order to prevent any Helium permeation during the DRIE step and preserve low residual pressure inside the cavity, it has been suggested to add a protective layer on the backside of the glass wafer. Best candidate turned out to be a metallic layer. This "shield" layer has been deposited after anodic bonding by sputtering (TiCu / 0.2 µm - 1 µm). This wafer has been subjected to the same treatment previously described (0.5 bar of Helium for 30 min). Q factor measurement by optical microscopy showed no major degradation on this wafer which suggested the good quality of the protective layer.

The wafer stack has been then processed as usual up to metallization step. Electrical tests have been performed prior dicing and results are reported on figure 10 for several dies located from the edge (die # 02) to the center (die # 17) of the wafer.

Q factor higher than 20 000 have been obtained for most of measured dies. This appears to confirm our previous key assumption concerning Helium permeation. Nevertheless, in order to definitely validate the protective layer efficiency, the wafer stack has been diced. Electrical tests have been performed again just after dicing and 6 months later. Results are reported on figure 11.

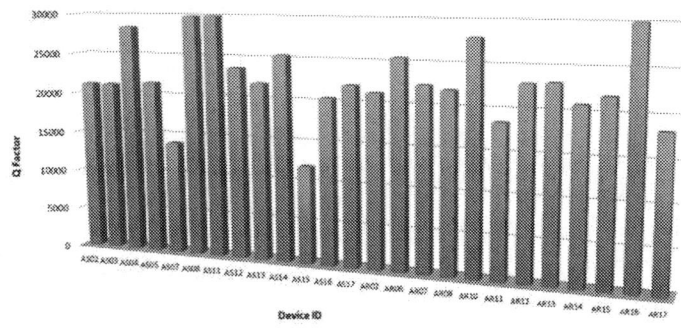

Figure 10. Q factor measurement after pad metallization – Anodic bonding – TiCu protective layer

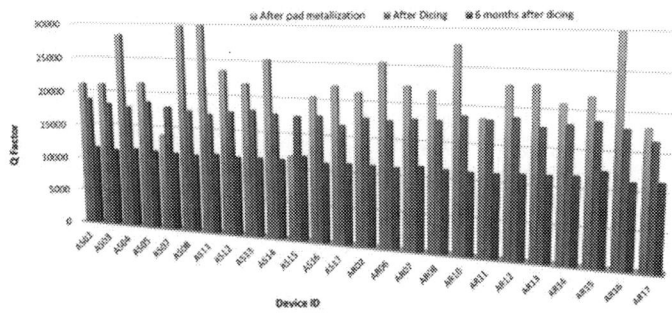

Figure 11. Q factor measurement after pad metallization / after dicing / 6 months after dicing – Anodic bonding – TiCu protective layer

In view of these results, it clearly appears that some outgassing took place during dicing since a substantial drop of the Q factor is observed. Mean Q factor is around 11 000 six months after dicing whereas it was higher than 20 000 prior dicing. But at this point, it is difficult to explain why the residual pressure increased inside the cavity.

In order to help us to understand this issue, Residual Gas Analysis (RGA) [20-21] have been performed on several dies among those previously measured. It turned out that the main residual gas inside the cavity is Helium. Results are reported on figure 12. Residual pressure measured either by electrical test and RGA are reported.

die ID	µRGA	Electrical test	
	P_{He} (mbar)	Q	P (mbar)
AS05	2.16E-02	11305	7.83E-03
AS11	1.86E-02	11470	7.68E-03
AR06	9.93E-03	11338	7.80E-03
AR10	1.57E-02	11342	7.80E-03

Figure 12. µRGA results – 6 months after dicing – Anodic bonding – TiCu protective layer

RGA measurements exhibited a residual pressure of Helium of 10^{-2} mbar (no other gas has been detected) which is rather close to the residual pressure measured by electrical test. Helium still seems to be the main issue.

Mechanisms that take place are not clearly understood for the time being. However, some points are under investigation:

- Protective layer efficiency potential failure
- Helium content in the glass material by RGA

Concerning this last point, our assumption is based on the fact that maybe, a significant amount of Helium is already originally trapped within the glass material and could be released during mechanical or thermal stress induced by the dicing step.

Experimental results – Eutectic bonding

For the first eutectic assembly, the same bonding recipe than for anodic has been used except that, obviously, no voltage has been applied during bonding. It means that outgassing steps remained the same, bonding and getter activation are occurred at 400°C.

After bonding step, Q factor has been estimated par IR microscopy. It turned out that Q factor was roughly between 15 000 to 20 000 corresponding to a residual pressure of 10^{-3} mbar and even lower.

Process steps for interconnection have been done (see Figure 5) and electrical tests has been performed at the wafer level. Results are reported on Figure 13 for more than 700 dies coming from the first wafer assembly.

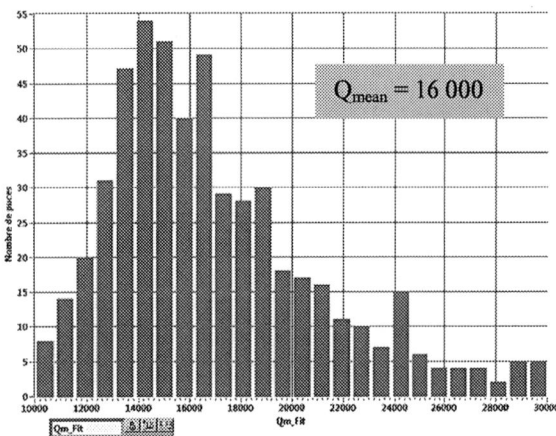

Figure 13. Q factor measurement & distribution on the 1st eutectic wafer assembly

The mean Q factor is around 16 000 corresponding to a residual pressure of 10^{-3} mbar with a large and non-perfect Gaussian distribution. It has also to be noticed that a significant number of dies have a Q factor higher than 20 000 (~15 %). These dies, for which residual pressure is below 10^{-3} mbar, are generally located at the edge of the wafer. It could be explained, in our opinion, by the fact that outgassing steps might not be sufficiently optimized. Then, some outgassing still occurred during the bonding step and residual gas cannot be fully pumped by the getter material. Dies that are located at the edge are less impacted than dies located at the center of

the wafer because residual gas can be evacuated more easily in the vacuum chamber of the bonding tool in the first case (note that vacuum level in the bonding chamber is 10^{-5} mbar).

Nevertheless, these first results with a eutectic assembly are very promising and the target of residual pressure below 10^{-3} mbar is already achieved but with a low yield.

As for anodic bonding, this first eutectic assembly has been diced and electrical tests performed just after dicing and 2 months later. The monitoring of Q factor evolution for 10 dies is reported on figure 14.

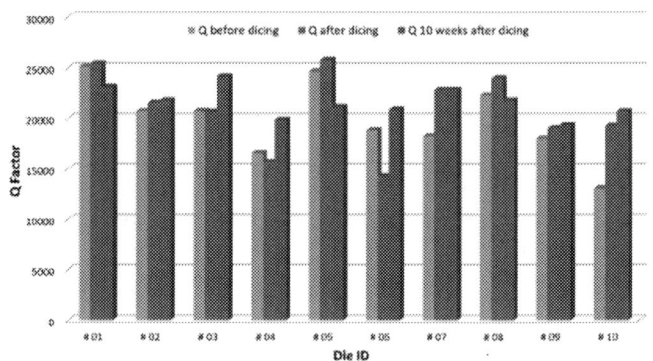

Figure 14. Q factor measurement before / after dicing for 10 dies coming from the 1st eutectic assembly

No significant degradation of the Q factor is observed even 10 weeks after dicing, with originals Q factor values ranging from 15 000 to 25 000. So, it seems that no outgassing phenomenon occurs during and after dicing with the eutectic solution in contrast to the results obtained with anodic bonding.

Taking all these results into account, a second run of 3 wafer assemblies has been processed. Several changes have been made on the bonding recipe mainly by increasing the time of all outgassing steps. After eutectic bonding, Q factor has been evaluated by IR microscopy and a mean Q factors from 20 000 to 25 000 have been found on all eutectic assemblies.

Once interconnection has been achieved, electrical tests could be performed. Q factor distribution results are reported for the 3 assemblies on figure 15.

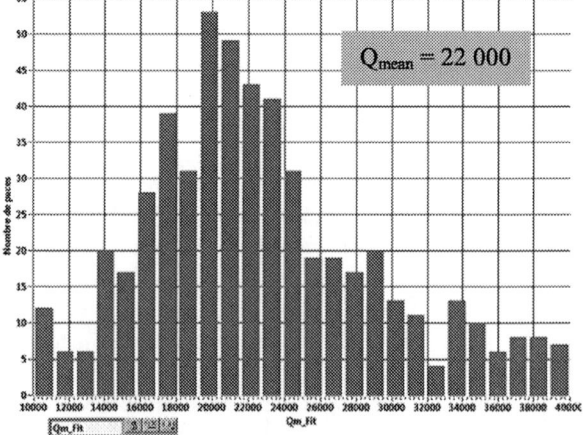

Fig 15-a: wafer #01

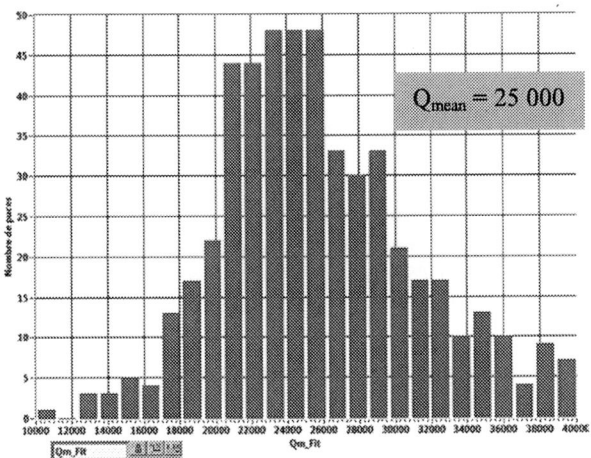

Fig 15-b: wafer #02

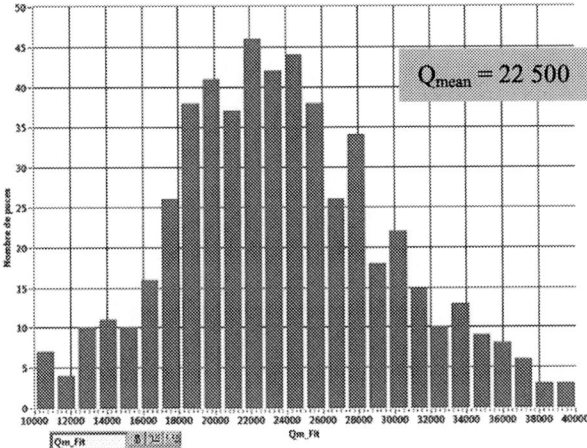

Fig 15-c: wafer #03

Figure 15. Q factor measurement & distribution on the 2nd run of eutectic assemblies

The electrical tests have confirmed what has been previously seen using IR microscopy characterizations: Q factor higher than 20 000 related to residual pressure below 10^{-3} mbar have been obtained and this time, for more than 80 % of the dies that covered the whole wafer surface.

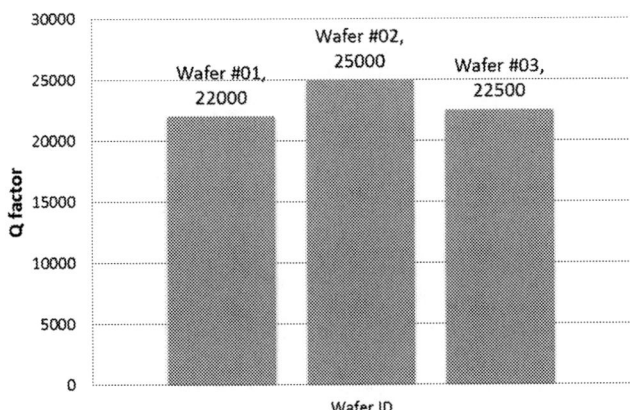

Fig 16-a: 2nd run eutectic – mean Q factor

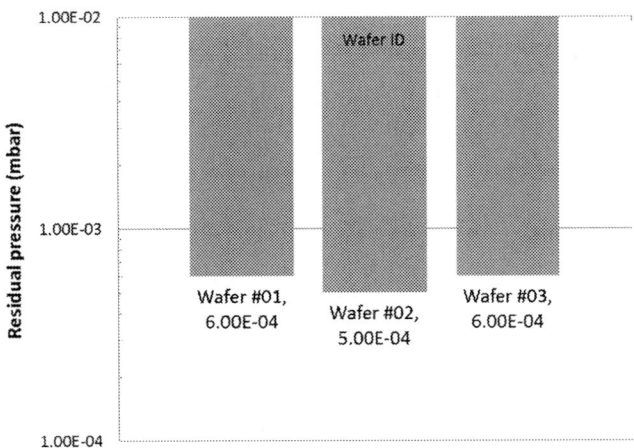

Fig 16-b: 2nd run eutectic – mean residual pressure

Figure 16. Mean Q factor & residual pressure - 2nd run of eutectic assemblies

Mean Q factor and residual pressure values are reported on figure 16. Through this 2nd run, a mean residual pressure of 5.10^{-4} mbar has been achieved which is well within our original target.

That being said, some improvement still have to be done in order to obtain more homogenous results whatever the die location on the wafer. Indeed, dies located at the center of the wafer (20 % of the whole dies) still exhibited smaller Q factor values despite outgassing steps optimization.

Considering that outgassing effects are mainly responsible for the non-homogeneous Q factor value on a whole wafer, we came up with a new optimization proposal: by making the assumption that the getter material is not fully saturated after bonding, meaning that it still could absorbed more gas, it has been proposed to increase the activation time during the bonding step. Thus, activation time has been raised from several tens of minutes to up than 1 hour.

Furthermore, different getter sizes have been implemented with covering surface representing 5 to 50 % of the original cavity surface.

This has been tested on a new eutectic assembly and IR microscopy characterization showed Q factor ranging from 10 000 to 40 000 (depending on the getter size) related to a residual pressure as low as 10^{-4} mbar.

Interconnection process has been applied on this last eutectic assembly and electrical tests have been performed.

Mean Q factor and mean residual pressure are reported on figure 17 for four different getter sizes: G4 is the largest size (~ 50 % of the cavity surface), G3 = 1/2 G4, G2 = 1/4 G4 and G1 = 1/8 G4.

As expected, mean Q factor increased with getter size and a mean residual pressure of 10^{-4} mbar has been obtained with larger getter surfaces (G3 & G4).

For smaller getter sizes (G1 & G2), it seems that the getter material is fully saturated.

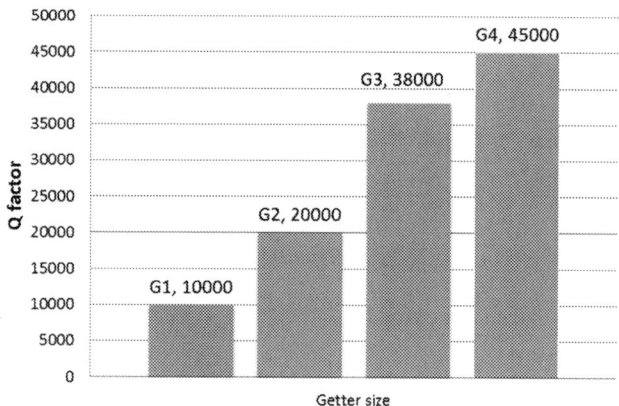

Fig 17-a: mean Q factor vs getter sizes

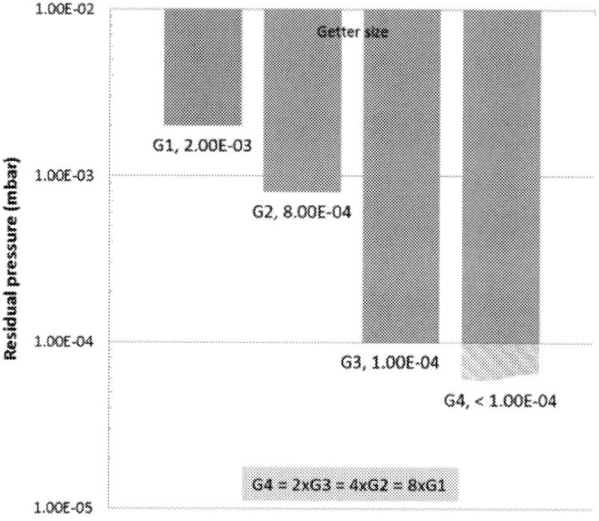

Fig 17-b: mean Residual pressure vs getter sizes

Figure 17. Mean Q factor & residual pressure – optimized eutectic bonding & getter activation process

It has to be noticed that for G4, it is not possible to have an accurate estimation of the residual pressure since measured residual pressure are below the sensitivity range of our resonator (10^{-1} to 10^{-4} mbar).

Finally, no significant Q factor difference from the edge to the center of the measured wafer has been observed. This seems to validate our previous assumption saying that getter activation step time should be increased. However, in order to have a better understanding of gettering mechanisms, it is planned to use a characterization tool specifically developed at CEA-LETI call RGA-TDS (TDS: Thermal Desorption Spectrometry). This should allow us to measure the pumping capacity of the getter for different residual gas composition at different temperature and time [19-20].

Conclusions

A high vacuum wafer level packaging technology has been developed for high-value MEMS applications. The targeted residual pressure range in the MEMS cavity was 10^{-3} to 10^{-4} mbar.

Using anodic bonding, a residual pressure of 10^{-2} mbar has been demonstrated. The target of 10^{-3} mbar hasn't been achieved due to Helium permeation issues. However, once this problem will be solved, it has been showed that the 10^{-3} mbar target could be reached.

With the eutectic bonding solution, residual pressure below 10^{-3} mbar has been demonstrated with a yield higher than 80%. It has even been demonstrated that residual pressure down to 10^{-4} mbar could be obtained which is, to our best knowledge, a high-level achievement in the vacuum wafer level packaging area.

In order to optimize the reliability of these proposed vacuum packaging technologies, it is planned to focus our following works on getter absorption / desorption mechanisms in order to increase the gettering performances.

Acknowledgments

The authors would like to thank Pierre Nicolas and Pierre-Louis Charvet for their works and expertise in RGA.

We would like also to thank David Bouchu for the FIB works done on our functional devices.

References

1. M. Esashi, "Wafer level packaging of MEMS", Journal of Micromechanics and Microengineering, 18 (May (7)) (2008), p. 073001
2. Garcia-Blanco, S., Topart, P., Desroches, Y., Caron, J.S., Williamson, F., Alain, C. & Jerominek, H. 2008, "Low-temperature vacuum hermetic wafer-level package for uncooled microbolometer FPAs", Proceedings of SPIE - The International Society for Optical Engineering.
3. Langa, S., Drabe, C., Kunath, C., Dreyhaupt, A. & Schenk, H. 2013, "Wafer level vacuum packaging of scanning micro-mirrors using glass-frit and anodic bonding methods", Proceedings of SPIE - The International Society for Optical Engineering
4. Zhang, J., Jiang, W., Wang, X., Zhou, J. & Yang, H. 2012, "Design and fabrication of high performance wafer-level vacuum packaging based on glass-silicon-glass bonding techniques", Journal of Micromechanics and Microengineering, vol. 22, no. 12.
5. Zhao, Q.C., Yang, Z.C., Guo, Z.Y., Ding, H.T., Li, M. & Yan, G.Z. 2011, "Wafer-level vacuum packaging with lateral interconnections and vertical feedthroughs for microelectromechanical system gyroscopes", Journal of Micro/ Nanolithography, MEMS, and MOEMS, vol. 10, no. 1.
6. Tachibana, H., Kawano, K., Ueda, H. & Noge, H. 2009, "Vacuum wafer level packaged two-dimensional optical scanner by anodic bonding", Proceedings of the IEEE International Conference on Micro Electro Mechanical Systems (MEMS), pp. 959.
7. S.H. Choa, "Reliability of MEMS packaging : vacuum maintenance and packaging induced stress», Microsyst. Technol., 11, pp.1187–1196 (2005)
8. Sun, X., Xu, D., Xiong, B., Wu, G., & Wang, Y. (2013). A wide measurement pressure range CMOS-MEMS based integrated thermopile vacuum gauge with an XeF2 dry-

etching process. Sensors and Actuators, A: Physical, 201, 428-433.

9. Wu, G., Xu, D., Sun, X., Xiong, B., & Wang, Y. (2013). Wafer-level vacuum packaging for microsystems using glass frit bonding. IEEE Transactions on Components, Packaging and Manufacturing Technology, 3(10), 1640-1646.

10. Wu, G., Xu, D., Xiong, B., Jing, E. & Wang, Y. 2012, "Wafer level vacuum packaged resonator with in-situ Au-Al eutectic Re-Distribution layer", Proceedings of IEEE Sensors.

11. Zoschke, K., Manier, C.-., Wilke, M., Jurgensen, N., Oppermann, H., Ruffieux, D., Dekker, J., Heikkinen, H., Piazza, S.D., Allegato, G. & Lang, K.-. 2013, "Hermetic wafer level packaging of MEMS components using through silicon via and wafer to wafer bonding technologies", Proceedings - Electronic Components and Technology Conference, pp. 1500.

12. Forsberg, F., Roxhed, N., Fischer, A.C., Samel, B., Ericsson, P., Hoivik, N., Lapadatu, A., Bring, M., Kittilsland, G., Stemme, G. & Niklaus, F. 2013, "Very large scale heterogeneous integration (VLSHI) and wafer-level vacuum packaging for infrared bolometer focal plane arrays", Infrared Physics and Technology, vol. 60, pp. 251-259.

13. Yu, A., Premachandran, C.S., Nagarajan, R., Kyoung, C.W., Trang, L.Q., Kumar, R., Lim, L.S., Han, J.H., Jie, Y.G. & Damaruganath, P. 2010, "Design, process integration and characterization of wafer level vacuum packaging for MEMS resonator", Proceedings - Electronic Components and Technology Conference, pp. 1669.

14. Mitchell, J.S.; Najafi, K., "A detailed study of yield and reliability for vacuum packages fabricated in a wafer-level Au-Si eutectic bonding process," Solid-State Sensors, Actuators and Microsystems Conference, 2009. TRANSDUCERS 2009. International , vol., no., pp.841,844, 21-25 June 2009

15. Merz, P.; Reimer, K.; Weiss, M.; Schwarzelbach, O.; Schroder, C.; Giambastiani, A.; Rocchi, A.; Heller, M., "Combined MEMS inertial sensors for IMU applications," Micro Electro Mechanical Systems (MEMS), 2010 IEEE 23rd International Conference on , vol., no., pp.488,491, 24-28 Jan. 2010

16. Langa, S., Utsumi, J., Ludewig, T. & Drabe, C. 2013, "Room temperature bonding for vacuum applications: Climatic and long time tests", Microsystem Technologies, vol. 19, no. 5, pp. 681-687.

17. Nicolas, S.; Caplet, S.; Greco, F.; Audoin, M.; Baillin, X.; Fanget, S.; "3D MEMS high vacuum wafer level packaging," Electronic Components and Technology Conference (ECTC), 2012 IEEE 62nd, pp.370-376, May 29 2012-June 1 2012

18. Lionel Tenchine et al, "NEG thin films for under controlled atmosphere ...Actuators A: Physical, Volume 172, Issue 1, December 2011, Pages 233-239

19. Francis J. Norton "Permeation of gases through solids" J. Appl. Phys. 28, 34. 1957.

20. P. Nicolas, PL. Charvet, D. Bloch, B. Savornin. "Mass spectrometry residual gas analysis: a key tool to optimize the packaging of MEMS", International Vacuum Congress, Paris 2013.

21. P-L. Charvet, P. Nicolas, D. Bloch, B. Savornin. "MEMS packaging reliability assessment: Residual Gas Analysis of gaseous species trapped inside MEMS cavities". ESREF 2013.

Thermal and Electrical Tests of Air-Gap TSV

Cui Huang, Dong Wu and Zheyao Wang[*]
The Institute of Microelectronics, Tsinghua University, Beijing, China
[*]Corresponding author: z.wang@tsinghua.edu.cn

Abstract

Using air-gaps instead of conventional silicon dioxide as the insulations of TSV has the potential to improve the electrical performance and solve some thermal reliability issues of TSV. Problems regarding the air-gap TSV process are studied and solutions for electrical parameters and reliability are investigated. Thanks to large difference of dielectric constant between air and silicon dioxide, the capacitance of air-gap TSV is considerably reduced to 48 fF, and the leakage current is quiet low (1.22 pA at 20 V). Finite element analysis (FEA) shows the stresses in the substrate and insulations are reduced significantly by air-gaps, evaluating the temperature stresses and deformation on the air-gap liner. Because of the air-gap, the Cu plugs are free from shear constraint and stress interaction from the sidewalls. The thermal shock test (TST) shows satisfactory barrier integrity and good stability of air-gap liner, since sudden changes of temperature stresses didn't deteriorate the insulation ability of air-gaps. Thermal variation underlines an impact of temperature on the electrical parameters of air-gap TSV, where the C-V and I-V characteristics are measured from 25 °C to 100 °C with 25 °C increment. A typical hysteresis occurred, which is considered to be attributed to the interface states and traps in the insulator adjacent to the interface. Those simulation and electrical tests demonstrate good thermo-mechanical stability and intrinsic dielectric property, fulfilling the requirement of applications.

1. Introduction

There-dimensional (3-D) integration technology based on through-silicon-via (TSV) strives to shorter vertical interconnect, lower latency and heterogeneous integration as a promising alternative to continuous scaling down for high-performance system [1]–[3].

For mainstream direct contact in 3-D integration applications, TSV interconnect confronts serious challenges on the electrical properties, the thermal stresses and the resulting reliability located at interfaces of TSV and adjacent metallization levels. Due to the large mismatch in the coefficients of thermal expansion (CTE) of Cu plugs (17.5 ppm), silicon substrates (2.5 ppm), and silicon dioxide (SiO₂) liners (0.5 ppm), large stresses are caused in Cu plugs, SiO₂ liners, and silicon substrates along the lateral and vertical directions when experiencing high temperature during fabrication and operation. With thermo-mechanical simulations, copper bump interfacial delamination [4], copper pumping [5], [6], or cracks in silicon substrates and silicon oxide [7] may be caused during fabrication process and thermal cycling of TSV structures [8]. Since the presence of inherent TSV sidewall scalloping roughness would impact forming a conformal liner, and even difficulties of conformal thin barrier and seed layer deposition, they induce stress concentration in SiO₂ liners and barriers, and could lead to potential reliability issues such as failure of insulators, current leakage, copper diffusion, and electro-migration [9]–[11].

To alleviate thermal stresses, a stress buffer layer have been proposed to release the thermal stresses, and TSVs with polymer insulators [12]-[15] have been developed. Polymers are effective in reducing thermal stresses due to the low elastic modulus [16], [17], and achieve low capacitance of TSV as the low relative dielectric constants [13]. To further reduce the parasitic capacitance and increase the mechanical stability, TSVs with air-gap insulators has been proposed recently [18]–[20]. Compare with a conventional silicon dioxide and various low-κ polymer insulators, the air-gap insulator provides a lower capacitance and reduce damages induced by plasma processing and the coefficient of thermal expansion (CTE) mismatch between the copper and the substrate. The use of air-gap integration scheme is also beneficial for multiple technology nodes without too many infrastructure investments and fabrication development.

Air-gap TSVs have been realized based on all-in-one sacrificial material removing technologies using a spin-on heat-depolymerizable polymer as deep trench sacrificial layer [18]. However, the scheme still lack experimental techniques applied to evaluate the thermal stability and structure integrity. Due to the unique structure of air-gap TSVs, air-gap TSV interconnect may confront serious challenges. Collapse over air gaps and bridging the separate layers may probably happen during thermal processes. The minimal solid residue after decomposition can also be a potential reliability concern. As the effect of free-space mobile charges is sensitive to the temperature, temperature dependent electrical characteristics of TSV interconnections should be recognized as a serious reliability issue for air-gap structures.

In this paper, concerns of numerical simulation, thermal shock, and thermal variation TSV on adjacent air-gap, are addressed to evaluate the thermal stability, leakage current and parasitic capacitance of air-gap TSVs. The stresses in the substrate and insulations under heating are simulated by finite element analysis (FEA) to evaluate the temperature stresses and deformation on the air-gap insulator. The fabricated air-gap TSV samples have been tested under thermal-shock test from -65 °C to +150 °C to study the insulation ability and thermo-mechanical stability. To study the temperature dependence of the air-gap TSV, capacitance-voltage hysteresis characteristics are measured at the different temperatures. The present work focuses extensively on air gap's impact on the the thermal-mechanical performance for TSV interconnection.

2 Configuration and Fabrication of Air-gap TSV

Considering that reliability is determined by TSV configuration, materials, and fabrication process, a brief introduction to the fabrication process are presented for completeness.

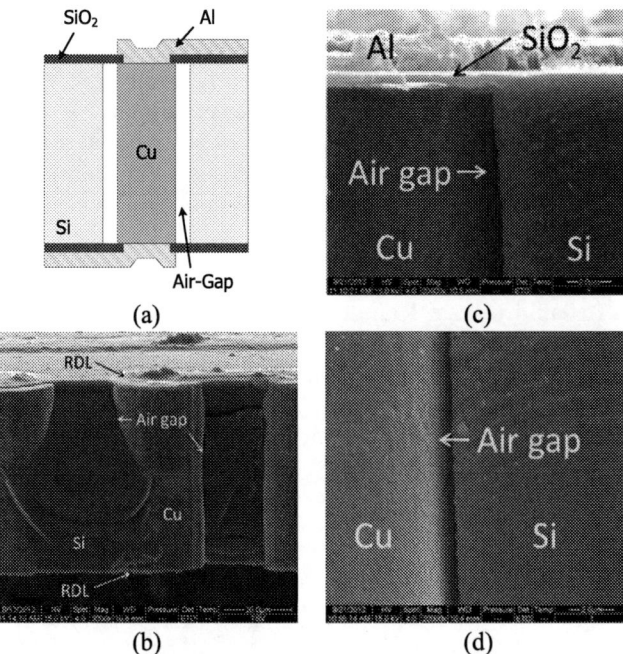

(a)   (c)

(b)   (d)

Figure 1. Conceptual drawing and fabrication results of an air-gap TSVs. (a) Configuration of the air-gap TSV. (b) Cross section SEM photo of a complete air-gap TSV, (c) Detail of the right-top corner of the air-gap, (d) Detail of the middle of the air-gap.

2.1 Configuration and Fabrication Design

Figure 1(a) shows the schematic cross-section view of an air-gap TSV. The air-gap TSV consists of a Cu plug suspended in a through-via and a thin and annular air-gap surrounding the Cu plug, and the Cu plug is supported at the both ends by SiO_2 ILD and the Al redistribution layers (RDLs). This type of TSV demonstrates an ultimate "porous" inter-metal dielectric between the copper interconnect and the silicon substrate. This structure of fabrication involves the polymer coating on the sidewall of deep vias, the polymer-releasing products diffusing through the encapsulating layer.

Air gaps have been fabricated using a number of different techniques, utilizing physical, chemical and thermal fabrication techniques. A straightforward approach of air gap fabrication is CVD-based dielectrics deposition process with poor gap filling properties, but the gap formation depends on the DRIE process, which is associated with gap width and depth [21]. Considering manufactured high-aspect-ratio air-gap liner is narrow and deep, shaped trench etching and via misalignment are major concerns. Another promising fabrication method is via all-in-one sacrificial material removing after the upper metal damascene wiring establishment [22], which prevents misalignment issue. There are however several approaches to release sacrificial materials as well, such as RIE etching, wet etching, photodecomposition, and thermal decomposition. For better control and releasing of sacrificial materials in narrow and deep trench, thermal decomposition has been chosen.

2.2 Sacrificial Material Selection

In order to fabricate an air gap around TSV, a promising all-in-one sacrificial material removing method is desired after metal damascene wiring establishment in intra / inter-layer dielectrics (ILDs). For better control and releasing of sacrificial materials in narrow and deep trench, thermal decomposition has been chosen. Based on the results of polymers filling condition and the comparison of materials decomposition analysis, the sacrificial materials used for the micromachining process was poly(propylene) carbonate (PPC) from Sigma-Aldrich. Solutions consisted of PPC powder dissolved in volatile anisole solvent. Through experimental evaluation, the concentration of PPC solution is optimized to 10% (in weight ratio), allows for performing good condition of cladding around the sidewall of the TSV, such as smooth inner-surface and measurable thickness [23].

2.3 Fabrication Process and Samples Preparation

A brief overview of the fabrication process is presented as follows. For ease of fabrication, the target TSV is with 25 μm of diameter, an aspect ratio of 1 : 2 and a 1-μm-thickness air gap. Even though TSV with higher aspect ratio is often required, our TSV structure is sufficient to study the electric characteristics and reliabilities in the air-gap liner.

The air-gap TSVs are fabricated on 4-inch p-type silicon wafers with a resistivity of 10-20 Ωcm. The blind vias is etched using bosch-process DRIE through thick photoresist mask. Then, with an optimal concentration and volume, PPC solution (in anisole solvent) is dispensed on the patterned substrate. Under vacuum environment, PPC coating on TSV sidewalls is formed after solvent volatilization. To achieve conformal sidewall cladding, a compensated solvent spin-coating is employed with a low rotation speed, thinning the redundant solidified PPC coating. Subsequently, TiW barrier and Cu seed are sputtered followed by Cu electroplating. Next, the chemical mechanical polishing (CMP) removes overburdens of Cu and PPC on substrate surface. Afterwards, a thin layer of SiO_2 is deposited using a PECVD process as a passivation and isolation layer, followed by deposition and patterning of aluminum electrical interconnects. The entire wafer is then temporary stacked with a glass carrier to enhance its mechanical strength. Next, the wafer is thinned down on its backside by mechanical polishing. A similar fabrication process to form SiO_2 layer and Al interconnect is performed on the backside. Finally, the sacrificial polymer PPC decomposes into volatile products upon heating to ~250 °C in vacuum environment. The decomposition products diffuse through the encapsulating silicon dioxide and leave behind a hollow cavity with minimal solid residue.

Figure 1 (b) to (d) shows the cross section of a completed air-gap TSV structure. As seen, the TSVs have no separation or delamination or cracking, and the air-gaps are clean from visible debris. Residue-free of PPC is highly desired to achieve clean and low dielectric constant air-gap liners, while forming the air gaps without destroying the TSV structures. The SEM images clearly demonstrate a firm structure of air-gapped TSV, and no bridging between copper and substrate is spotted.

3 Electrical Performance

High frequency capacitance and leakage current are fundamental electrical parameters of TSVs. Measure of TSV capacitance between the air-gap TSV and the substrate provides information on the quality of vertical electrical connection. TSV essentially forms a typical cylindrical MIS (Metal-Insulation-Semiconductor) capacitor to the substrate. With the designed TSV dimensions, TSV capacitance is expected to be in the order of 45 fF in accumulation region. The C-V characteristics of the air-gap TSV is measured with HP4284A at 1 MHz. The capacitance is measured by sweeping the voltage from -20 V to 20 V at the Cu pillar, with an increasing step of 0.5 V. From the measured values in Figure 2(a), the air-gap TSV is exhibiting the accumulation capacitance of ~48 fF, the minimum depletion capacitance of ~15 fF at 9.5 V, and the capacitance density of ~1.22 nF/cm^2. The TSV C-V curves change from accumulation to depletion and then to inversion region as the bias voltage increases, demonstrating a typical p-Si behavior.

It should be noted that at 0 V bias voltage, the capacitance is in accumulation region, which implies that fixed charges are stored in the air-gap. This is similar to the SiO$_2$ liner TSVs, in which the capacitance at 0 V is either in accumulation region [24] or in inversion region [25]. The accumulation region and the inversion region correspond to positive and negative fixed charges, respectively.

The leakage current through the liner to the substrate is an essential parameter for 3-D IC design. The TSV leakage current values are measured with Keithley 4200 semiconductor characterization system at room temperature, sweeping the voltage from 0 V to 20 V at the probe pad. As shown in Figure 2(b), the measured current-voltage (I-V) plot of a single air-gap TSV structure demonstrates a I$_{TSV}$ ~1.22 pA at electric fields of 200 kV/cm (20 V bias voltage). The largest leakage current is similar to that of 100 nm thick SiO$_2$ cladding [26]. That indicates a good insulation capability of air-gap liner between the copper pillar and the silicon substrate.

4 Reliability Tests, Results and Discussions

Concerning that the mechanical strength of air-gap structure may be poor, thermal and electrical tests of air-gap TSVs should be performed. The reliability of the air-gap TSVs are evaluated in terms of numerical simulation, thermal shock test (TST) or thermal cycling, and thermal variation.

4.1 Numerical Simulation

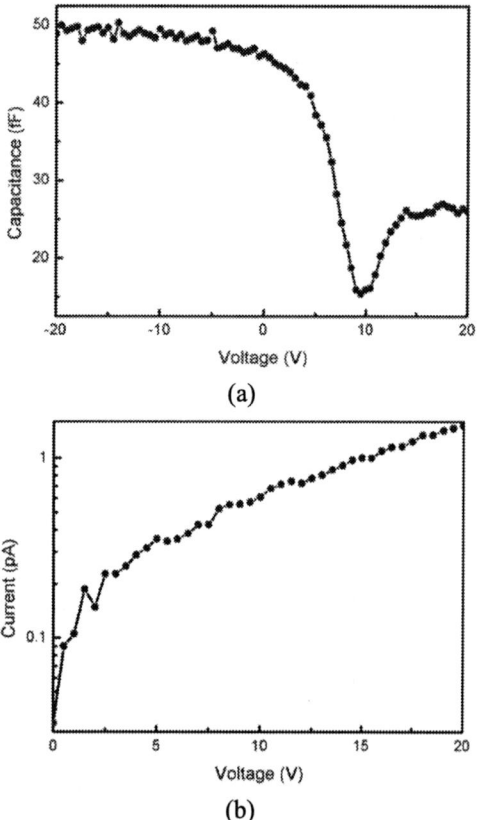

Figure 2. Measured electrical performance of an air-gap TSV. (a) C-V characteristics; (b) I-V characteristics.

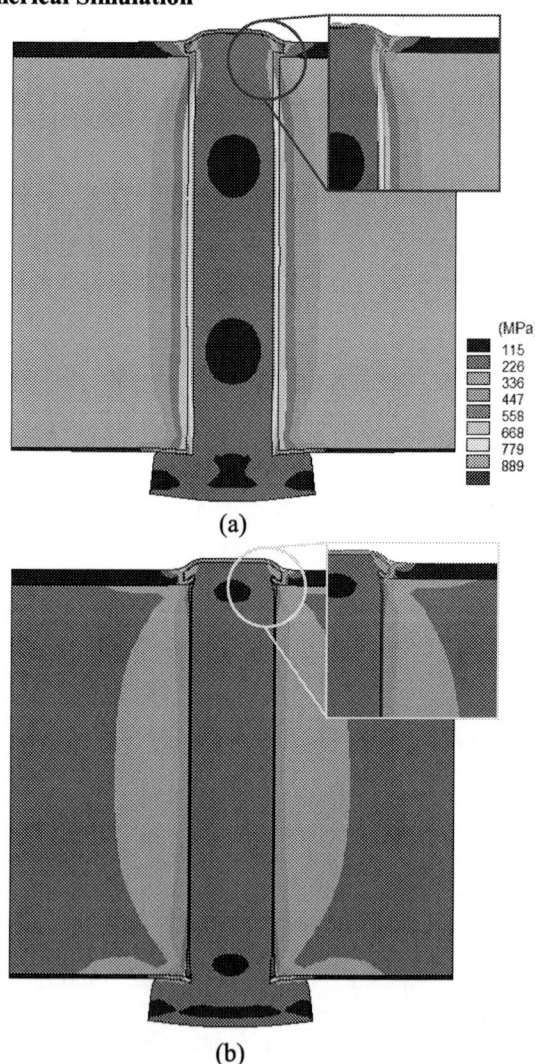

Figure 3. Comparison of thermal stress distribution and structural displacement in the TSVs between SiO$_2$ liner and air-gap liner of 1 μm thickness. (a) SiO$_2$ insulator; (b) air-gap insulator.

In 3-D integration, the thermal stress field in the air-gap TSVs has to be taken into account as this could trigger reliability failure. For the fine and delicate TSV structures, thermo-mechanical modeling of the extrusion in annealing stage is carried out. Considering thermal treatments may happen up to 673 K (400 °C), the heating process is simulated at 673 K.

Figure 3 shows the thermal stress distribution and structural displacement in the TSVs between SiO_2 liner and air-gap liner of 1 μm thickness. The stress coming from adjacent oxide layers in SiO_2 liner TSVs is suppressed in the air-gap TSVs with uniform sizes. The maximum Von Mises stress in the substrate with air-gap TSVs (~380 MPa) is about half that of the TSV using a SiO_2 liner (~700 MPa). This is because the radial stress along the via and silicon interface is compressive, when heating, which also contributes to the driving force for extrusion. However, compared with SiO_2 liner TSVs, the stress at the top and bottom regions of vias is slightly intensified in the air-gap TSVs, where the maximum Von Mises stress happens. It suggests that the stress in the substrate reduces significantly using circular air gap, offering a free space for thermal expansion in radial direction, which may reduce wafer warpage.

From the figure, the maximum displacement occurs at the top and bottom of the TSV. Such displacement, large extrusion at TSV center, is anticipated since the CTE of copper is larger than that of other materials. Due to air gaps, the displacement of the interface between copper and silicon is discontinuous. It is also noted that the interact between the via and the silicon has little influence on the axial displacement of TSVs.

4.2 Thermal Shock Test

To evaluate the temperature impact on the air-gap TSVs and study the thermo-mechanical solvability of air-gap TSVs under external thermal excursions, thermal shock test (TST) is performed and the capacitances and the leakage currents are measured after TST. The TST consists of 3 cycles per hours in a range of -65 °C to +150 °C according to JEDEC standards (JESD22-A106B). The thermal shock is repeated 30 cycles and 90 cycles. The capacitance and leakage current of the air-gap TSVs are measured before and after the thermal shock treatment. For comparison, the results are plotted in Figure 4.

It can be seen that after 30-cycle and 90-cycle thermal shock, the overall capacitances in the accumulation region are ~47 fF and ~46 fF, respectively, nearly equal to the accumulation capacitance before thermal shock, ~48fF. The shift of C-V curves may arise from the drift of free-space mobile charges, induced by injection and drift of Cu atoms into the air-gap low-κ dielectrics. The C-V characteristics before and after TST show independent capacitance variations, and indicate satisfactory barrier integrity and good air-gap stability.

Besides, it can be found that the leakage current decreases sharply after 30-cycle thermal shock, however changes little after 90-cycles. This indicates a stable leakage current can be obtained after tens of thermal cycles. The reduction in leakage current could be attributed to the high temperature treatment,

which further reduces the few residue of PPC decomposition. This could also be the reason for flat and stable inversion capacitance after TST treatment.

The slight shifts in the overall capacitance and leakage current before and after TST indicate that thermal shock does not deteriorate the insulation ability of air-gaps, and the air-gap TSV structure is thermal stable.

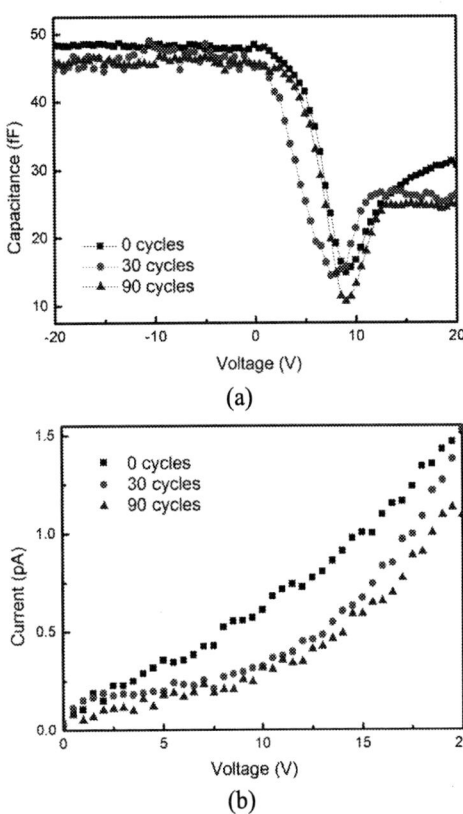

Figure 4. Measured electrical performance of air-gap TSV after 30 cycles and 90 cycles TST, and comparison with original curve. (a) C-V characteristics; (b) I-V characteristics.

4.3 Thermal Variation

To evaluate the performance of air-gap TSVs at different temperatures, the C-V and I-V characteristics are measured using a semiconductor measurement system Agilent B1500A and a temperature control module Espec ETC-200L. The air-gap TSV capacitance sensitivity is observed at 1 MHz within a wide range of temperature varying from 25 °C to 100 °C, and the measurements are taken every 25 °C increment. And the leakage current characteristics are are measured from 25 °C to 125 °C with 25 °C increment.

Figure 5 shows the nature of the C-V hysteresis characteristics for four different temperatures (25 °C, 50 °C, 75 °C and 100 °C). It can be seen that the TSV capacitance changes with temperature in a wide range, and the main influences of temperature on C-V curves are the hysteresis that increases with temperature and the shift in the flat-band voltage, as well as the increase in the minimum capacitance. Those implied a significant electron charging effect.

978-1-4799-2408-0/14 $31.00 © 2014 IEEE   1725

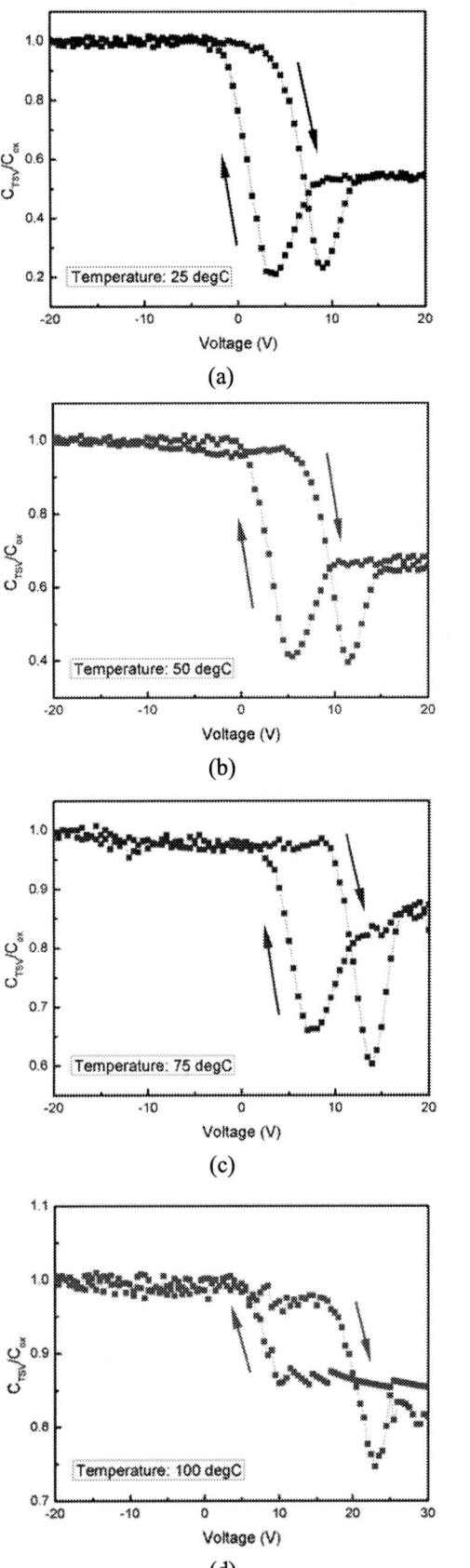

It is obvious that the principal effect of increasing temperature on the C-V curves of the air-gap TSV is the increase in the minimum capacitance and the shift in the flat-band voltage. Additionally, all the C-V curves clearly show that there is a clockwise C-V hysteresis between ascending voltages and descending voltages, and the hysteresis increases with temperature. The depletion capacitance and the inversion capacitance increase as the temperature increases. It is considered to be attributed to the interface states and traps in the insulator adjacent to the interface, which give rise to a charge interchange across the interface. By increasing the intrinsic carrier concentration, the increasing temperature shifts the shape of the C-V.

A typical hysteresis is considered to be attributed to the interface states and traps in the insulator adjacent to the interface, which are induced by the residue from PPC pyrolysis and a thin Teflon like (nCF_2) polymer layer left by DRIE etching. With temperature increase, more traps and charges are activated and injected to the air-gap interfaces and the hysteresis increases with temperature. Besides, the extrinsic charges, mainly located at interface between the air-gap and the semiconductor, will dominate the flat-band voltage shift.

The temperature-dependent capacitance of the air-gap TSVs are compared with the literature reported TSVs using silicon oxide liners to evaluate the contribution of the air-gap liners. The C-V hysteresis characteristics of a TSV with 5 μm diameter, 20 μm height, and 100-nm thickness silicon dioxide liner, are similar in both directions below 100 °C [27], which indicates negligible impact of mobile ions. Besides, the effect becomes prominent at 150 °C. However, the impact of mobile ions in air-gap TSV is obvious at low temperatures, and it becomes more significant as the temperature rising. For capacitance variation at varying temperatures, the C-V curves of TSVs using silicon oxide liners are typical high frequency type. On the contrary, the C-V curves of air-gap TSVs transit from high frequency type to low frequency type as the temperature increases.

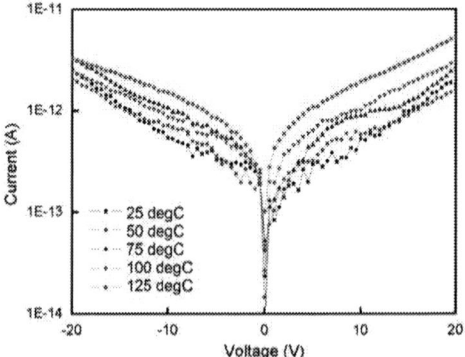

Figure 6. Measured I-V characteristics of air-gap TSV in different temperatures (ramping up).

Figure 6 is a plot of the leakage current versus voltage characteristics at different temperatures. The figure demonstrated that the leakage current of air-gap TSV increases with the temperature at 20 V bias voltage from 1.6

Figure 5. Measured C-V characteristics of air-gap TSV in different temperatures (ramping up). (a) 25 °C, (b) 50 °C, (c) 75 °C, and (d) 100 °C.

pA at 25 °C to 3.1 pA at 125 °C. The relative changes by temperature increasing is close to 100%. However, the absolute leakage current at 125 °C and 20 V bias voltage is still quite low. These results indicate that the isolation property of air gap is preserved, maintaining low leakage current (<3.5 pA) even at 125 °C.

Conclusions

TSVs with low capacitance density have been demonstrated by employing a sacrificial layer technology to fabricate air-gap liners, which the annular polymer PPC cladding that surrounds the copper plugs is removed and air-gaps are formed. Thanks to the relatively large thickness and the extremely low dielectric constant, the capacitance densities of TSVs are reduced by more than an order of magnitude by replacing silicon dioxide liners with air-gap liners. FEA simulation shows the thermal stresses in the substrate and insulations are reduced significantly by air-gaps, evaluating the temperature stresses and deformation on the air-gap liner. Rigorous thermal shock tests between -65°C and +150°C have been performed. The results without significant changes in capacitance and leakage current show the satisfactory electrical performance and thermo-mechanical stability. Thermal variation in a wide range of temperature impacts the capacitance of air-gap TSVs, which indicates that the impact of mobile ions in air-gap TSV is obvious at low temperatures, and it becomes more significant as the temperature rising, but the isolation capability of the air-gap are preserved, fulfilling the requirement of applications. After all, those simulation and electrical tests demonstrate good thermo-mechanical stability and intrinsic dielectric property for air-gap TSV.

Acknowledgments

This work was supported in part by 973 Program under Grant 2011CBA00603 and Beijing Municipal Science and Technology Project under Grant D13110100290000.

References

1. J.-Q. Lu, "3-D hyperintegration and packaging technologies for micro-nano systems," *Proc. IEEE*, vol. 97, no. 1, pp. 18–30, Jan. 2009.

2. R.S. Patti, "Three-dimensional integrated circuits and the future of system-on-chip designs," *Proc. IEEE*, vol. 94, no. 6, pp. 1214–1224, June 2006.

3. J. U. Knickerbocker, P. S. Andry, B. Dang, R. R. Horton, M. J. Interrante, C. S. Patel, et al., "Three-dimensional silicon integration," *IBM J. Res. Dev.*, vol. 52, no. 6, pp. 553–569, Nov. 2008.

4. S.-K. Ryu, K.-H. Lu, X. Zhang, J.-H. Im, P. S. Ho, and R. Huang, "Impact of Near-Surface Thermal Stresses on Interfacial Reliability of TSVs for 3-D Interconnects," *IEEE Trans. Dev. Mat. Reliab.*, vol. 11, no. 1, pp. 35–43, Mar. 2011.

5. I. De Wolf, K. Croes, O. V. Pedreira, R. Labie, A. Redolfi, M. Van De Peer, et al., "Cu pumping in TSVs: Effect of pre-CMP thermal budget," *Microelect. Reliab.*, vol. 51, no. 9, pp. 1856–1859, Sep. 2011.

6. M. Jung, X. Liu, S. K. Sitaraman, D. Z. Pan, and K. L. Sung, "Full-chip TSV interfacial crack analysis and optimization for 3D IC," in *IEEE/ACM Intl. Conf.*

Computer-Aided Design (ICCAD), San Jose, CA, Nov. 7–10, 2011, pp. 563–570.

7. L.-C. Shen, C.-W. Chien, H.-C. Cheng, and C.-T. Lin, "Development of 3-D chip stacking technology using a clamped TSV interconnection," *Microelect. Reliab.*, vol. 50, no.4, pp. 489–497, Apr. 2010.

8. A. P. Karmarker, X. Xu, and V. Moroz, "Performance and reliability analysis of 3D-integration structures employing TSV," in *IEEE Intl. Reliab. Phys. Symp.*, Montreal, QC, Apr. 26–30, 2009, pp. 682–687.

9. N. Ranganathan, D. Y. Lee, Y. Liu, G.-Q. Lo, K. Prasad, and K. L. Pey, "Influence of bosch etch process on electrical isolation of TSV structures," *IEEE Trans. Comp. Packag. Manuf. Tech.*, vol. 1, no. 10, pp. 1497–1507, Oct. 2011.

10. Y. Yang, R. Labie, F. Ling, C. Zhao, A. Radisic, J. Van Olmen, et al., "Processing assessment and adhesion evaluation of copper TSVs for three-dimensional stacked-integrated circuit (3D-SIC) architectures," *Microelect. Reliab.*, vol. 50, no. 9, pp. 1636–1640, Sep. 2010.

11. J. Bea, K. Lee, T. Fukushima, T. Tanaka, and M. Koyanagi, "Evaluation of Cu diffusion from Cu TSV in 3-D LSI by transient capacitance measurement," *IEEE Electron. Dev. Lett.*, vol. 32, no. 7, pp. 940–942, July 2011.

12. M. Bouchoucha, P. Chausse, D. Henry, and N. Sillon, "Process solutions and polymer materials for 3D-WLP through silicon via filling," in *Proc. IEEE Electron Comp. Tech. Conf. (ECTC)*, Las Vegas, NV, June 1–4, 2010, pp. 1696–1698.

13. Y. Civale, D. S. Tezcan, H.G.G. Philipsen, F.F.C. Duval, P. Jaenen, Y. Travaly, et al., "3-D wafer-level packaging die stacking using spin-on-dielectric polymer liner through-silicon vias," *IEEE Trans. Comp. Pack. Manuf. Technol.*, vol. 1, no. 6, pp. 833–840, June 2011.

14. C. Huang, Q. Chen, D. Wu, and Z. Wang, "High aspect ratio and low capacitance TSVs with polymer insulation layers," *Microelect. Eng.*, vol. 104, pp. 12–17, Apr. 2013.

15. Q. Chen, C. Huang, Z. Tan, and Z. Wang, "Low capacitance TSVs with uniform BCB insulation layers," *IEEE Trans. Comp. Pack. Manuf. Technol.*, vol. 3, no. 5, pp. 724–731, May 2013.

16. K. H. Lu, X. Zhang, S. Ryu, J. Im, R. Huang, and P.S. Ho, "Thermo-mechanical reliability of 3-D ICs containing TSVs," in *Proc. IEEE ECTC*, San Diego, CA, May 26–29, 2009. pp. 630–634.

17. M. Jung, J. Mitra, D. Z. Pan, and S. K. Lim, "TSV stress-aware full-chip mechanical reliability analysis and optimization for 3-D IC," *IEEE Trans. Computer Aided Design Integrated Circuits Syst.*, vol. 31, no. 8, pp. 1194–1207, Aug. 2012.

18. C. Huang, Q. Chen, and Z. Wang, "Air-gap TSVs," *IEEE Electron. Dev. Letts.*, vol. 34, no. 3, pp. 441–443, Mar. 2013.

19. M. Sunohara, H. Sakaguchi, A. Takano, R. Arai, K. Murayama, and M. Higashi, "Studies on electrical performance and thermal stress of a silicon interposer with

TSVs," in *Proc. IEEE ECTC*, Las Vegas, NV, June 1–4, 2010, pp. 1088–1093.

20. R. Sharma, E. Uzunlar, V. Kumar, R. Saha, X. Yeow, R. Bashirullah, et al., "Design and fabrication of low-loss horizontal and vertical interconnect links using air-clad transmission lines and TSVs", in *Proc. IEEE ECTC*, San Diego, CA, May 29–June 1, 2012, pp. 2005–2012.

21. L. G. Gosset, F. Gaillard, D. Bouchu, R. Gras, J. de Pontcharra, S. Orain, et al., "Multi-level Cu interconnects integration and characterization with air gap as ultra-low K material formed using a hybrid sacrificial oxide / polymer stack," in *Proc. IEEE Intl. Interconnect Tech. Conf. (IITC)*, Burlingame, CA, June 4–6, 2007, pp. 58–60.

22. N. Nakamura, N. Matsunaga, T. Kaminatsui, K. Watanabe, and H. Shibata, "Cost-effective air-gap interconnects by all-in-one post-removing process," in *Proc. IEEE IITC*, Burlingame, CA, June 1–4, 2008, pp. 193–195.

23. C. Huang, Q. Chen, and Z. Wang, "Polymer liner formation in high aspect ratio TSVs for 3D integration", *IEEE Trans. Comp. Packag. Manuf. Technol.*, vol. 3, no. 7, pp. 1107–1113, July 2013.

24. L. Zhang, H. Y. Li, S. Gao, and C. S. Tan, "Achieving stable TSV capacitance with oxide fixed charge," *IEEE Electron Dev. Lett.*, vol.32, no.5, pp. 668–670, May 2011.

25. G. Katti, M. Stucchi, D. Velenis, S. Thangaraju, K. De Meyer, W. Dehaene, et al., "Technology assessment of TSV by using C–V and C–t measurements," *IEEE Electron Dev. Lett.*, vol.32, no.7, pp. 946–948, July 2011.

26. G. Van der Plas, P. Limaye, I. Loi, A. Mercha, H. Oprins, C. Torregiani, et al., "Design issues and considerations for low-cost 3-D TSV IC technology," *IEEE J. Solid-State Circuits*, vol. 46, no.1, pp. 293–307, Jan. 2011.

27. G. Katti, A. Mercha, M. Stucchi, Z. Tokei, D. Velenis, J. Van Olmen, et al., "Temperature dependent electrical characteristics of TSV interconnections," in *Proc. IEEE IITC*, Burlingame, CA, June 6–9, 2010, pp. 1–3.

Heterogeneous System Integration Pseudo-SoC Technology for Smart-health-care Intelligent Life Monitor Engine & Eco-system (Silmee)

Hiroshi Yamada[1], Yasuhiro Sato[2], Nobuhiro Ooshima[2], Hiroyuki Hirai[3], Takuji Suzuki[4], Shigenobu Minami[4]

[1] Corporate R&D Center, Electron Devices Laboratory, Toshiba Corporation
[2] Toshiba Hokuto Electronics Corporation
[3] Toshiba Design & Manufacturing Service Corporation
[4] Healthcare Business Development Division, Toshiba Corporation

1, Komukai Toshiba-cho, Saiwai-ku, Kawasaki 212-8582, Japan
E-mail: hiroshi.yamada@toshiba.co.jp

Abstract

Heterogeneous devices integration technologies constitute one of most promising driving forces of system integration as they enhance the system performance of electronics products. Currently, many technologies for heterogeneous devices integration have been reported [1]-[2].

The heterogeneous devices integration technologies that have been reported are mainly implemented for SoC (System on Chip) by applying the advantages of process compatibility between the devices. However, it has been impossible to integrate them in the case that the processes are incompatible. Also, many integration technologies applying SiP (System in Package) technology with the interposer substrate have been reported. However, SiP technology, it has been impossible to achieve high integration density comparable to that of monolithic integrated SoC because the interposer substrate occupies a large area in SiP. Accordingly, development of an advanced heterogeneous devices integration technology is required.

In the previous work, pseudo-SoC realized by a wafer-level heterogeneous devices system integration technology, which incorporates MEMS and CMOS-LSI, has been studied to verify the validity [3]-[6]. In the work, the flexible pseudo-SoC for mobile terminal application was also verified [5]. Furthermore, in the previous work, high-density pseudo-SoC of AFE (Analog Front End) for Smart-health-care Intelligent Life Monitor Engine & Eco-system (Silmee™) application was developed [7]-[8]. In the application, 2 kinds of pseudo-SoC of AFEs which are for ECG (Electrocardiogram) and for Pulse were developed. However, developed pseudo-SoC of AFE has not sufficient density because the 2 kinds of pseudo-SoC of AFE occupy a large area in the Silmee engine.

This paper describes the pseudo-SoC which integrated AFE circuit of ECG and Pulse in one microchip for Silmee application that incorporates several types of heterogeneous devices.

Heterogeneous System Integration Pseudo-SoC Technology

Figure 1 shows a schematic view of a pseudo-SoC in the case of incorporating MEMS and CMOS-LSI. The pseudo-SoC is set up to realize one microchip with heterogeneous devices made by using individual processes for epoxy resin, insulating layer and redistribution layer, respectively [3]-[6].

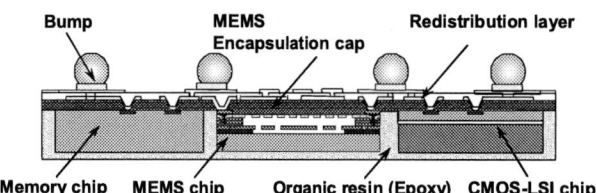

Figure 1. Cross-sectional diagram of pseudo-SoC.

Figure 2 shows a schematic diagram of the wafer-level heterogeneous devices system integration technology for the pseudo-SoC. The Known Good Dies (KGD) are embedded in the organic (Epoxy) resin to realize the high-yield KGD reconfiguration wafer [3]-[6].

As the redistribution layers are formed by semiconductor post-process without interposer substrate, the pseudo-SoC enables integration density identical to that of SoC. Also, as heterogeneous devices are proximally redistributed with short distance, the pseudo-SoC enables signal transmission speed and EMI noise identical to that of SoC. Furthermore, as commercially available heterogeneous devices and peripheral passive components can be used for the individual circuit blocks, the pseudo-SoC offers the advantage of time-to-market and the integration yield identical to that of SiP and at low cost.

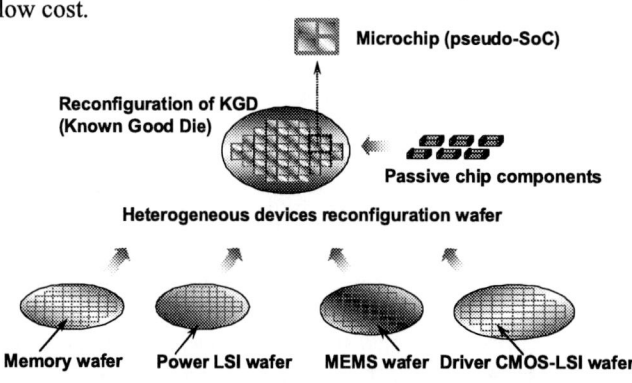

Figure 2. Schematic view of heterogeneous system integration pseudo-SoC technology.

Figure 3 shows a fabrication process of pseudo-SoC [3]-[6]. The process consists of three main steps: heterogeneous device reconfiguration step, redistribution layer formation step and pseudo-SOC wafer thinning step. In the heterogeneous device reconfiguration step, the KGD devices are allocated on an adhesive layer of a support substrate. The heterogeneous devices are encapsulated with the epoxy resin by applying the vacuum printing process. In the redistribution layer step, an insulating layer (polyimide film) is formed on the reconfiguration wafer as a planar layer to the step height between the resin and the chips. Subsequently, the thin-film metal layer is formed by the semiconductor wafer process to achieve the redistribution layer. In the pseudo-SoC wafer thinning process, the wafer is grinded in order to achieve thin pseudo-SoC.

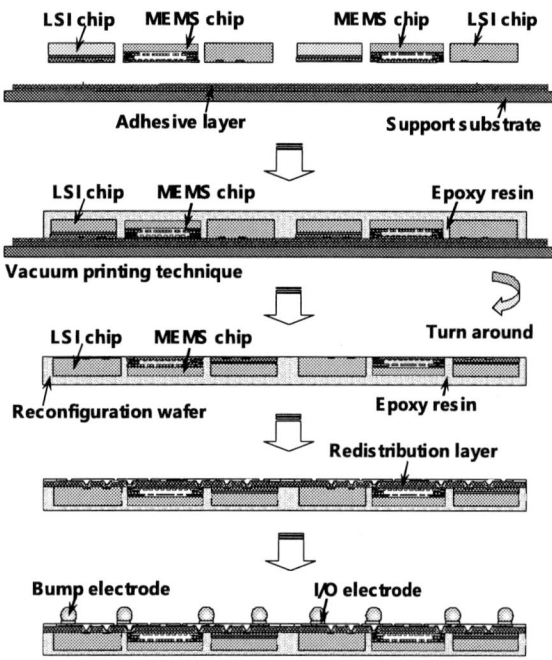

Figure 3. Fabrication process of pseudo-SoC.

Smart-health-care Intelligent Life Monitor Engine & Eco-system (Silmee)

Figure 4 shows the Silmee (Smart-health-care Intelligent Life Monitor Engine & Eco-system) which is the intelligent wearable vital signs sensors for smart-health-care applications. The Silmee simultaneously senses information on key vital signs which are ECG, Pulse, body surface temperature and movements. The Silmee delivers the data to smart-phones and tablet PCs with wireless communication technology to storage the data in the main host computer [7]-[8].

Figure 5 shows Silmee Egg (prototype model). Figure 6 shows the block diagram of Silmee platform. As shown in the figures, Silmee have smart-health-care intelligent life monitor engine, sensing unit for ECG monitoring, sensing unit for Pulse monitoring, temperature sensor for dermatherm, rechargeable button battery. The ECG is monitored by difference of voltage at the 2 point of skin on the body, and the Pulse is monitored by reflective photoplethysmographic sensor system, which detects blood flow change, with LED (Light Emitting Diode) and PD (Photo Diode). Each analog

signal is converted to digital signal by AFE circuit [6]-[7]. Furthermore, the body surface (skin) temperature is monitored by dermatherm system and movement of body is monitored by triaxial acceleration sensor. All data are controlled by the MCU (Micro Control Unit) and send to smart-phones and tablet PCs with Bluetooth™ technology.

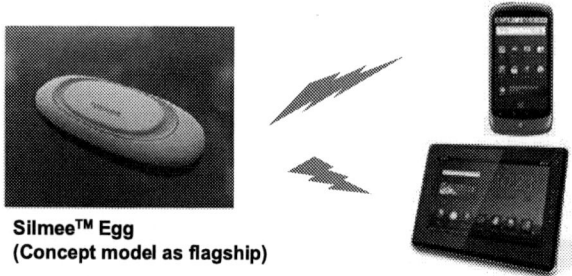

Silmee™ Egg (Concept model as flagship)

Notice: Silmee™ prototype terminal in this demonstration has not applied to medical regulations in Japan and is only for demonstration purpose not for medical use. Medical regulations in each country should be taken care to develop medical products using Silmee™ engine.

Figure 4. Smart-health-care Intelligent Life Monitor Engine & Eco-system (Silmee).

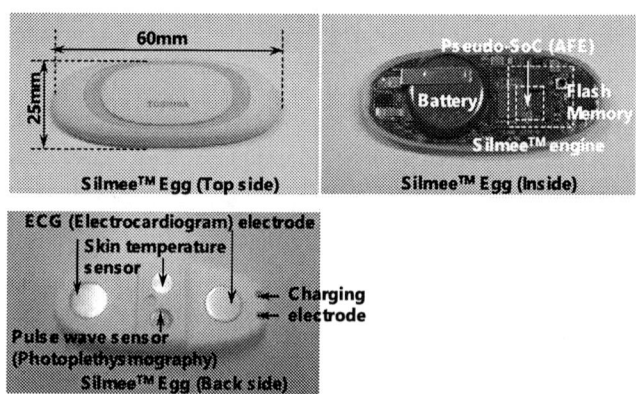

Figure 5. Silmee Egg (prototype model).

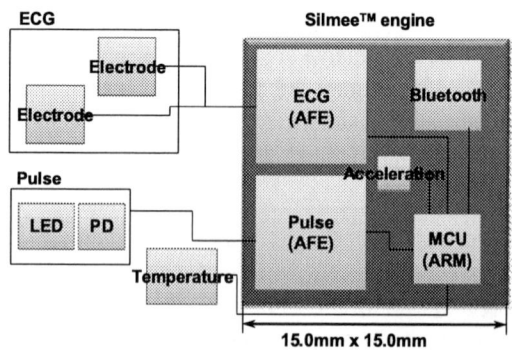

Figure 6. Block diagram of Silmee platform.

978-1-4799-2408-0/14 $31.00 © 2014 IEEE

Pseudo-SoC AFE

As the Silmee has small size of only 60mm × 25mm × 10mm, the AFE circuit has been required to shrink the size because the AFE circuits incorporates a lot of heterogeneous devices and occupies a large area in the Silmee engine.

In the previous work, 2 kinds of pseudo-SoC of AFEs which are for ECG and for Pulse were developed [9]. However, it has been required to realize more high-integrated pseudo-SoC of AFE because the 2 kinds of pseudo-SoC of AFE occupy a large area in the Silmee engine.

Figure 7 shows design of pseudo-SoC of AFE which has AFE circuits for ECG and Pulse in one microchip of 6.3mm × 7.3mm in size. The pseudo-SoC incorporates 3.3V CMOS dual power supply type low current consumption and low noise type of operational amplifier (10 chips), silicon gate C2MOS type high-speed, low-voltage drive quad bilateral switch (2 chip), discrete chip resistor (36 chips), and discrete chip capacitors (25 chips). The pseudo-SoC has 3 multi-layers (Signal/GND/Signal) as redistribution layers.

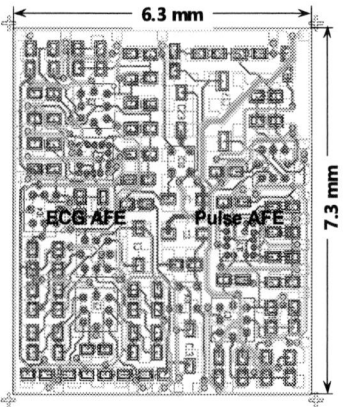

Figure7. Design for pseudo-SoC of AFE.

Figure 8 shows the reconfiguration wafer incorporated 36 pseudo-SoC of AFE (ECG and Pulse) in the 75mmø diameter epoxy resin. The pseudo-SoC has 3 multi-redistribution layers (Al=1.0μm/Ti=0.1μm).

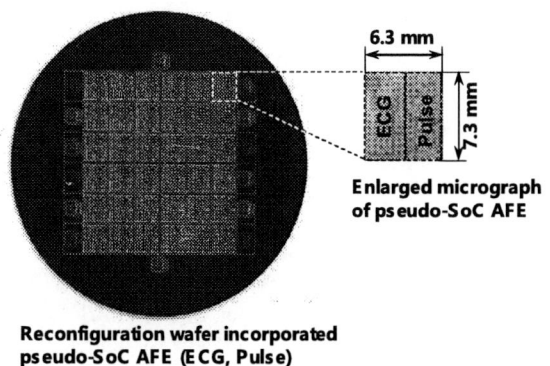

Figure 8. Reconfiguration wafer incorporated pseudo-SoC of AFE (ECG, Pulse).

Figure 9 shows the cross-sectional micrographs of the pseudo-SoC of AFE in the circumference of the embedded devices. The left micrographs show the portions of LSI chips and right micrographs show the portions of the discrete passive chip components. Also, the upper micrographs show the cross-sectional view of each device, and the lower micrographs show the enlarged portion of contact pads. As shown in the micrographs, reliable contact was confirmed between contact pads and redistribution layers. Meanwhile, stepwise shape of redistribution layers around the contact pads is due to multi-layers of polyimide which achieves sufficient polyimide film thickness as insulation of redistribution layers.

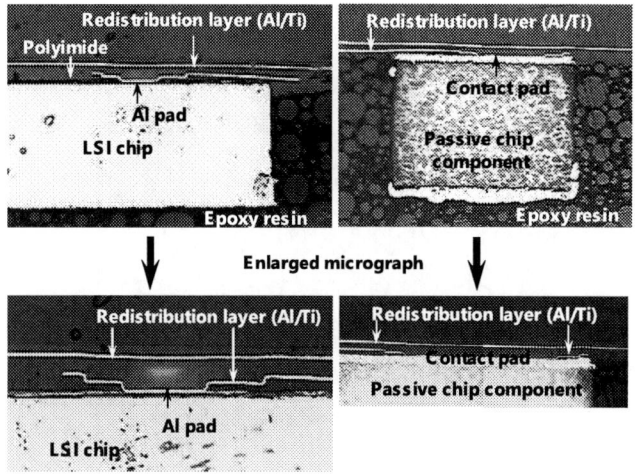

Figure 9. Cross-sectional micrograph of pseudo-SoC of AFE (circumference of the embedded devices).

Figure 10 shows the cross-sectional micrographs of the pseudo-SoC of AFE in the circumference of redistribution layers. The upper micrograph shows the peripheral portion of the reconfiguration wafer and the lower micrograph shows the center portion of the reconfiguration wafer.

As shown in the micrographs, excellent conditions of the redistribution layers were formed on the wafer. Furthermore, the stepwise shape of redistribution layers on the pads is due to multi-layers of polyimide to achieve sufficient polyimide thickness as same as in the case of figure 9.

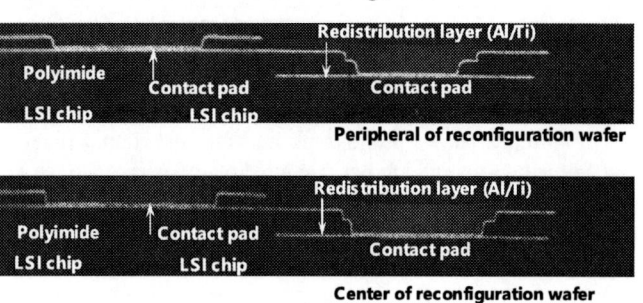

Figure 10. Cross-sectional micrograph of pseudo-SoC of AFE (circumference of redistribution layers).

Figure 11 shows the enlarged micrograph of the pseudo-SoC which has ECG AFE and Pulse AFE circuit on Silmee engine (15.0mm × 15.0mm × 5.2mm). Table 1 summarizes the specifications of Silmee engine.

The Silmee engine has pseudo-SoC of AFE (ECG, Pulse) which size is 6.3mm × 7.3mm × 1.0mm, MCU (ARM Cortex™-M3 core) for the sensor control and signal processing, Bluetooth (4.0 Dual mode: WLP) for wireless communication, digital triaxial acceleration sensor (LGA package) for sensing body movement, and peripheral discrete chip components for power control unit circuit.

The pseudo-SoC has redistribution layers of Al/Ti thin-film metals (Al=1.0μm/Ti=0.1μm) and low cure temperature polyimide (ε=3.4). All of heterogeneous devices for AFE circuits are embedded in the high-content silica filler epoxy resin.

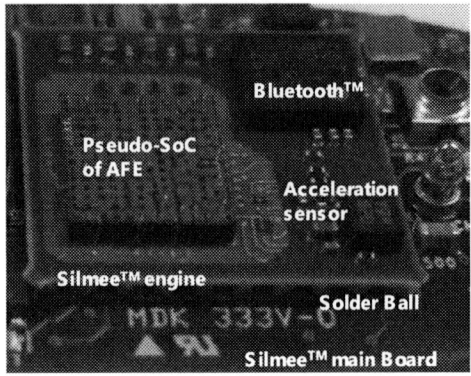

Figure 11. Enlarged micrograph of pseudo-SoC of AFE on Silmee engine.

Table 1. Specifications of Silmee engine.

Smart-health-care Intelligent Life Monitor Engine & Eco-system: Silmee™ Egg		
Silmee™ Engine	15.0 mm x 15.0 mm x 5.2 mm Pseudo-SoC of AFE (ECG, Pulse) integration	
Integrated devices	MCU chip (ARM Cortex™-M3 core)	
	Bluetooth™ chip (4.0 Dual mode)	
	Pseudo-SoC of AFE (6.3mm x 7.3mm x 1.0mm)	
	Digital triaxial acceleration sensor (LGA package)	
	Chip resistor / Chip capacitor	
Pseudo-SoC of AFE (ECG, Pulse)		
Redistribution layer (RDL)	Thickness = Al(1 μm)/Ti(0.1 μm)	
	Minimum line/space =15 μm/15 μm	
Insulating layer for RDL	Low cure temperature polyimide (ε=3.4)	
Core material	High-content silica filler epoxy resin	

Figure 12 shows photograph and cross-sectional diagram of Silmee engine which has pseudo-SoC of AFE. Silmee has size of 25mm × 60mm × 12mm and has weight of 10g.

Silmee engine has the pseudo-SoC of AFE, triaxial acceleration sensor and Bluetooth on the main surface of the substrate. The MCU is integrated on the backside surface of the substrate.

Figure 13 shows monitoring wave-form results of ECG and Pulse by Silmee Egg. Because the monitoring wave-forms show good results as the intelligent wearable vital signs

sensors, it is confirmed that the pseudo-SoC has appropriate performance as AFE circuit.

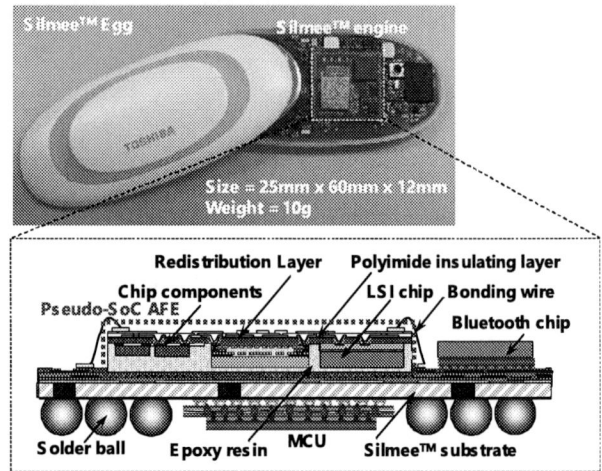

Figure 12. Photograph and cross-sectional diagram of Silmee engine.

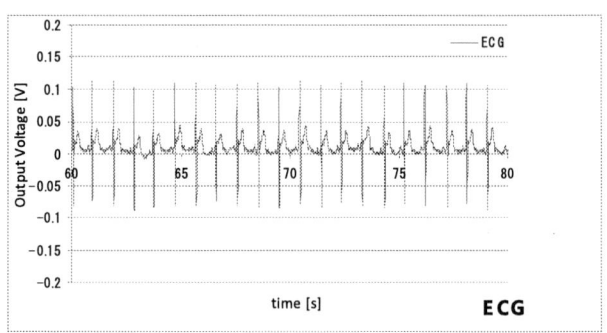

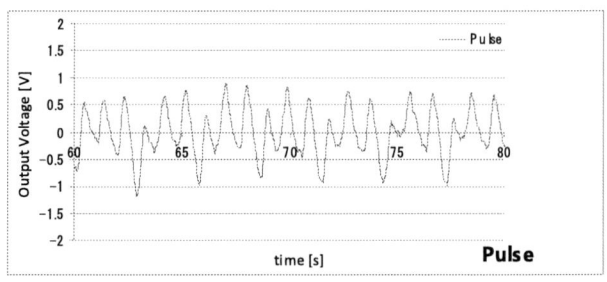

Figure 13. Monitoring waveform results of ECG and Pulse with pseudo-SoC of AFE.

Reliability Test for Pseudo-SoC of AFE

Figure 14 shows reliability test diagram of pseudo-SoC of AFE. The test was performed by applying pseudo-SoC of AFE-TEG (Test Element Group) which has 4 kinds of test circuits for insulation resistance test, diode test, daisy chain resistance test, and migration test. The upper figures show the design and micrograph of pseudo-SoC of AFE-TEG. The lower figure shows the circuit block of pseudo-SoC of AFE-TEG. The AFE-TEG incorporates LSI chips and chip resistors, and has 3 redistribution layers (L1, L2, L3).

The insulation resistance test was carried out for the measurements of dielectric resistance among L1, L2 and L3. The diode test was carried out for the measurements of

Signal-GND voltage in the LSI chip. The daisy chain resistance test was carried out for the measurements of total resistance of chip resistors which have serial interconnections in the pseudo-SoC. The Migration test was carried out for the measurements of dielectric resistance among L1, L2 and L3.

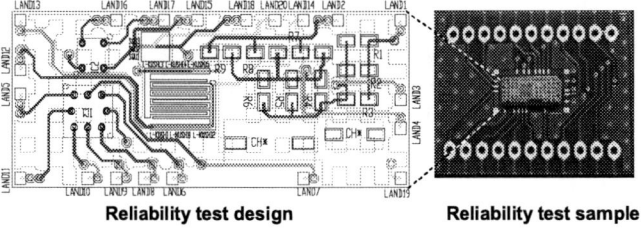

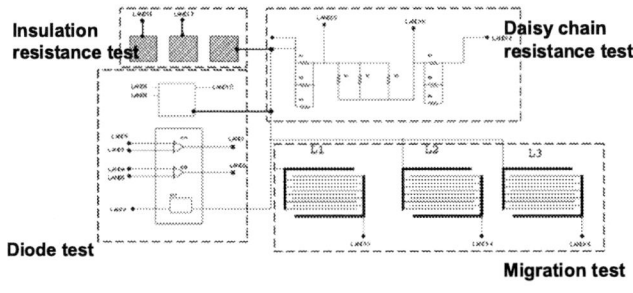

Figure 14. Reliability test diagram for pseudo-SoC AFE.

Table 2 shows reliability test conditions of pseudo-SoC of AFE. The reliability tests are performed under the conditions of high-temperature storage test (125°C, 1000 hours), thermal shock test (-55°C/125°C, 500 cycle), high-temperature and high-humidity biased test (85°C/85% RH, 5.5V, 1000 hours), High-temperature and high-humidity storage test (60°C/90% RH, 1000 hours) and Low-temperature storage test (-40°C, 1000 hours) respectively.

Each reliability test failure criteria were defined ±10% for initial value. As shown in the table, all reliability tests were carried out for 15 samples and passed perfectly.

Table 2. Reliability test conditions.

Test	Conditions	Number of Samples	Time
High-temperature storage test	Temperature: 125°C	15/15 pass	1000 H
Termal shock test	Temperature: -55°C / 125°C	15/15 pass	500 cycles
High-temperature and high-humidity biased test	Condition: 85 °C / 85% RH Voltage: 5.5V	15/15 pass	1000 H
High-temperature and high-humidity storage test	Condition: 60°C / 90% RH	15/15 pass	1000 H
Low-temperature strage test	Temperature: -40°C	15/15 pass	1000 H

Figure 15 shows reliability test result for high-temperature storage test. The test result shows the measurement of daisy chain resistance. As shown in the figure, initial value of resistance was 187.5kΩ and the percentages of change were -0.2%~4.9% (within ±10%). The increase of the daisy chain resistance is considered due to initial failure of contact between passive chip components and redistribution layers.

Figure 16 shows reliability test result for thermal shock test. The test result shows the measurement of daisy chain resistance. As shown in the figure, initial value of resistance was 187.5kΩ and the percentages of change were -0%~8.3% (within ±10%). The increase of the daisy chain resistance is considered due to initial failure of contact between passive chip components and redistribution layers.

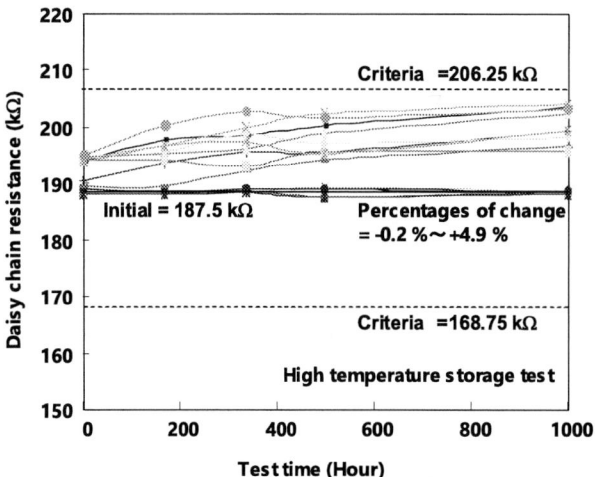

Figure 15. Reliability test result for high temperature storage test.

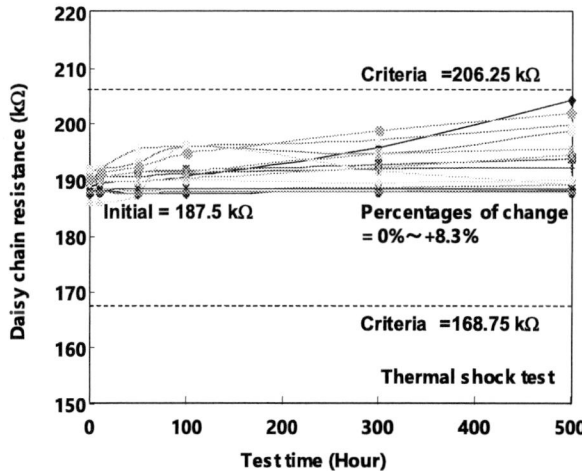

Figure 16. Reliability test result for thermal shock test.

Integration Density of Heterogeneous Devices

Figure 17 shows comparison of integration density of pseudo-SoC of AFE. The left micrograph shows Silmee engine which has AFE (ECG, Pulse) circuit blocks integrated as SiP by using COB (Chip on Board) technology. The AFE circuit of ECG occupies 12.0mm × 6.0mm and that of Pulse occupies 10.0mm × 5.0mm as footprint area.

The right micrograph of upper portion shows Silm e engine which has pseudo-SoC of AFE (ECG) and pseudo-SoC of AFE (Pulse), respectively. The pseudo-SoC of ECG-AFE occupies 8.2mm × 4.2mm and that of Pulse-AFE occupies 7.0mm × 3.0mm as footprint area.

On the other hand, the right micrograph of lower portion shows the Silmee engine which has one microchip AFE (ECG, Pulse) integrated by pseudo-SoC. The pseudo-SoC of AFE occupies 6.3mm × 7.3mm as footprint area. Compared with each Silmee engine, it is found that the AFE circuit enabled as far as 38% shrinkage of ship size by pseudo-SoC technology.

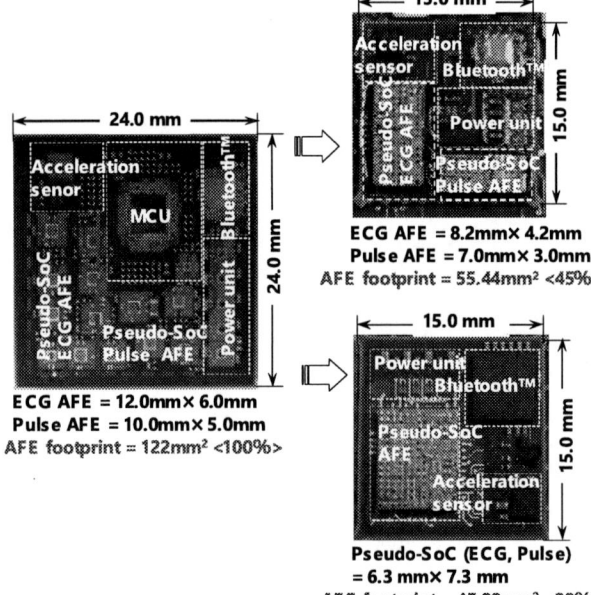

Figure 17. Comparison of integration density of pseudo-SoC of AFE.

Conclusions

The high-density integrated one microchip AFE circuit for the Smart health care Intelligent Life Monitor Engine & Eco-system (Silmee) was developed by applying heterogeneous system integration pseudo-SoC technology.

The developed pseudo-SoC realizes one microchip, which integrates AFE circuit for ECG and Pulse, within 6.3mm × 7.3mm × 1.0mm in size by applying 3 multi-redistribution layers (Signal/GND/Signal).

It is confirmed that the developed pseudo-SoC of AFE shows appropriate performance on Silmee engine and has sufficient reliability. Furthermore, it is confirmed that the developed pseudo-SoC achieved 38% shrinkage of system size as compared to the conventional SiP technology.

References

1. R.R. Tummala Editors, "Fundamentals of Microsystems Packaging," McGraw-Hill, 2001.

2. JIEP Technical Committee Editors, "Prospects on Electronics Packaging Technology," Journal of Japan Institute of Electronics Packaging (JIEP), Vol.17, No.1, pp.1-62, 2014.

3. H.Yamada et al., "MEMS-LSI Heterogeneous Device Integration Technology for System-on-Chip Applications," presented at 5th International Nanotechnology Conference on Communication and Cooperation (INC5), UCLA, Los Angles, California, May 18-21, 2009.

4. H. Yamada, et al., "Wafer-Level Heterogeneous Technology Integration for Flexible Pseudo-SoC," Digest of Technical Papers of IEEE ISSCC 2010, San Francisco, California, February 7-11, 2010, pp.146-147.

5. H. Yamada, et al., "A wafer-level system integration technology for flexible pseudo-SOC incorporates MEMS-CMOS heterogeneous devices," Proceedings of IEEE CPMT Symposium Japan 2010, Tokyo, August 24-26, 2010, pp.11-14.

6. H. Yamada, "A Wafer-level System Integration Technology Incorporates Heterogeneous Devices," Proceedings of 45th International Symposium on Microelectronics, San Diego, California, September 11-13, 2012, pp.793-800.

7. T. Suzuki et al., "Wear able Wireless Vital Monitoring Technology for Smart Health Care," Proceedings of 7th International Symposium on Medical Information and Communication Technology (ISMICT), March 6-8, Tokyo, Japan, 2013, pp.1-4.

8. T. Suzuki et al, "Wearable Intelligent Vital Signs Senor and High-density Packaging Technology with Pseudo-SoC," Proceedings of 15th IC Packaging (ICP) Technology Expo Technical Conference, January 15-17, Tokyo, Japan, 2014, pp.1-18.

9. H. Yamada, "A wafer-level system integration (Pseudo-SoC) technology for intelligent smart senor module applications," Proceedings of 15th Printed Wiring Boards Expo Technical Conference, January 15-17, Tokyo, Japan, 2014, pp.133-172.

Effects of Various Environmental Conditions on the Electrical Properties and Interfacial Reliability of Printed Ag / Polyimide System

Byung-Hyun Bae[1], Min-Su Jeong[1], Byeong Rok Lee[1], Joung-Hoon Choo[2] and
Eun-Kuk Choi[2], Jong-Sun Yoon[2], Young-Bae Park[1]*
[1] School of Materials Science and Engineering, Andong National University, Andong, 760-749, Korea
[2] Technical Research Center, HICEL, Pyeongtaek, 451-830, Korea
e-mail : ybpark@andong.ac.kr

Abstract

Effects of 200 °C annealing and 85 °C/85% relative humidity (R.H.) temperature/humidity environments on the electrical resistivity and interfacial adhesion energy between screen-printed Ag film and polyimide substrate were systematically investigated. Measured peel strength was 22.2 gf/mm before environmental treatment, and then decreased to 0.1 and 0.5 gf/mm for annealing and temperature/humidity treatment during 500 hours, respectively. Cu oxide formation at Cu/Ag interface and the degradation of polyimide itself seem to be responsible for adhesion decrease during annealing and temperature/humidity treatments, respectively. Screen printed Ag film shows the initial sharp decrease in resistivity during annealing due to effective binder removal effect and partially sintering effect.

1. Introduction

The polyimide has many advantages such as flexibility, high temperature resistance, high mechanical strength, good chemical stability, and a low dielectric constant. Therefore, it is applied to flexible printed circuit boards (FPCB) for flexible electrodes.

The conventional lithography technology is a general method of printing for desired patterns on the semiconductor and FPCB, however, is expensive, complex, and time-consuming procedures [1-3]. Here, screen printing technology is one of the alternative processes for simple manufacturing, reduced cost, and silent, less waste and environmentally friendly. In this process, Ag paste flows through a screen mesh after a squeegee stroke. The Ag paste can be printed locally on the selective substrate area. The hygroscopic property leads to reliability problems in printed circuits, such as a reduced conductivity, lower adhesion, and lower resistance to corrosion [4-9]. Some studies reported that functional groups, such as carboxyl and amide bonding may affect the interfacial bonding of a metal to polyimide treated with potassium hydroxide [9].

Previous reports were studied in interface adhesion on metal film/polyimide systems with various environmental conditions [10-14]. After the temperature/humidity treatment, the adhesion strength of the printed Ag/polyimide systems were weaken by formation of the metal oxide at the inkjet-printed Ag/polyimide interface. The major reason of oxidation resulted from the moisture at the inkjet-printed Ag/polyimide interface and polyimide degradation with temperature/humidity treatment [10-14]. In addition, the effect of annealing conditions on the interfacial adhesion energy between electross-plated Ni and polyimide film were systemically analyzed. The adhesion energy after annealing process seemed to be controlled by carbonyl oxygen bonding

in near the cohesive failure region [15]. According to Jung group, they revealed that the electrical property and micro structure on ink jet printing Ag with annealing [16]. However, systematic studies on the effect of humidity on the interfacial adhesion of screen printed Ag /polyimide systems have been little reported.

In this study, the effect of annealing and temperature/humidity treatment conditions on the interfacial adhesion between screen printed Ag and polyimide was quantitatively measured by 180° peel test method. Resistivity of screen printed Ag/polyimide were measured by 4-point probe during annealing at 200 °C. The peeled surfaces of the metal and polyimide substrate were analyzed with scanning electron microscopy (SEM), energy dispersive spectroscopy (EDS) and X-ray photoemission spectroscopy (XPS).

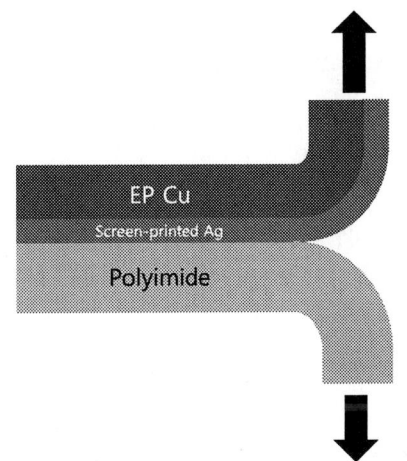

Figure 1. Schematic diagram of 180° peel test.

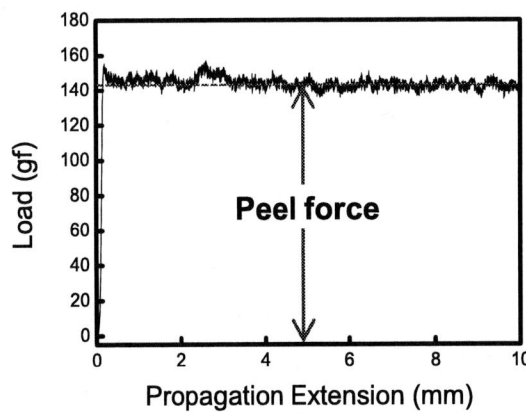

Figure 2. Typical load-displacement curve during steady-state peeling state.

978-1-4799-2408-0/14 $31.00 © 2014 IEEE

2014 Electronic Components & Technology Conference

2. Experimental procedures

In order to understand the effect of annealing and temperature/humidity between screen-printed Ag film and polyimide substrate, the sample structure shown in Fig. 1 was fabricated to evaluate the interfacial adhesion of 180° peel test. The substrate of specimen is PMDA-ODA polyimide (Kapton-H, Dupont). Ag paste was composited 0.3~9.0 μm size of Ag powder, epoxy binder, mixing of the various material and solvent. To perform peel test and 4-point probe, Ag paste was printed in 5 mm × 50 mm and 40 mm × 40 mm in width and length on polyimide. Curing treatment was performed for 30 minutes at 150 °C in oven in order to evaporated solvent. In order to grip the metal film in peel test instrument (Lloyd Instruments, model LRX plus), an electroplated Cu film around 24 μm- thick was deposited on curried Ag paste which is performed in $CuSO_4 \cdot 5H_2O$ solutions with a mixture of brightener at ambient temperature. Specimens were comprised of electroplated Cu (24 μm), screen printed Ag (1 μm) and polyimide (25 μm). Some samples were treated at 85 °C/85% relative humidity for 500hours. Other samples were treated at 200 °C for annealing for 500hours. Films used for peel test at a cross-head speed of 2 mm/min. The peel strength can be calculated as follows.

$$\text{Peel strength(gf/mm)} = \frac{\text{Peel Force (gf)}}{\text{Width (mm)}} \quad (1)$$

Peel strength is defined by dividing the sample width into the average value of several measurements of peel load as shown in Fig. 2. From the graph, load-displacement curve shows steady-state behavior after initial load increase. It has been known that peel strength includes substantial amount of plastic deformation of peeled metal films and polyimide during delaminating film to bend, it has been proposed for the interfacial adhesion energy from measured peel strength by evaluating the permanent deformation energy of peel metal films and polyimide from energy-balance relationship and pure bending assumption during steady-state peeling which can be found in detail elsewhere.

In addition, to measure peel strength, the maximum curvature (K_B) at peeled crack tip as shown in Fig. 1. The stress-strain relationship of peeled film should be determined to obtain the interfacial adhesion energy from peel test [17, 18]. SEM was used to observe the microstructure peeled surface of metal film and polyimide. EDS and XPS were performed on the peeled metal surface and peeled polyimide surface to understand failure locus.

3. Results and Discussion

Figure 3 shows the measured peel strength values of screen printed Ag /polyimide system, which values were 22.2 gf/mm without treatment, and 0.1gf/mm and 0.5 gf/mm with annealing at 200 °C and temperature/humidity treatment, respectively. Peel strength was decreased after annealing and temperature/humidity treatment for 500 hours. Peeled surfaces of metal film and polyimide after peel test are analyzed by SEM, EDS, and XPS, in order to investigate the change of failure locus with annealing and temperature/humidity treatments. Figure 4 represents SEM images of peeled metal and polyimide surface without treatments. The peeled metal

and polyimide surface were roughed, as is shown in Fig. 4. Figure 5 and 6 show SEM images of peeled metal and polyimide surface after temperature/humidity treatment and annealing, respectively. The peeled metal and polyimide surface after environmental treatments were smoother than no -treatments. EDS analyses were performed at peeled surface of metal film and polyimide to investigate the change of failure locus according to annealing and temperature/humidity treatments. Table 1 shows EDS atomic concentrations with annealing and temperature/humidity treatments. Without treatment, C and Ag atomic concentrations were 12.3 and 71.9 at.% on the peeled metal surface, and 58.3 and 10.0 at.% on the peeled polyimide surface, respectively.

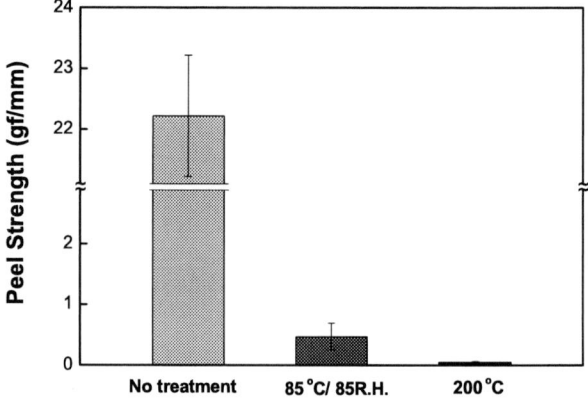

Figure 3. Peel strength of screen-printed Ag /polyimide system before and after annealing and temperature/humidity treatments for 500 h.

EDS and XPS analyses results in without treatments, the value of atomic concentration of Ag is larger than that of polyimide at metal film and the significant amount of C atoms is detected at polyimide. It seems that the interface failure was occurred between screen printed Ag and polyimide by the results.

C atomic concentrations were 31.0 at.% on the peeled metal surface, and 70.9 at.% on the peeled polyimide after 500 hours temperature/humidity treatment, respectively. The amount of Ag peak is hardly detected at metal film. At metal film and polyimide, the significant amount of C is detected by EDS and XPS analyses. The cohesive failure was occurred in polyimide. The carbonyl bonding was broken in polyimide caused by the temperature/humidity treatment.

Humidity may cause the formation of weak Coulomb interactions or hydrogen bonds near the cohesive failure region in screen printing Ag and polyimide system [15, 19]. C, O, Cu and Ag atomic concentrations were 0 and 31.9 at.% on the peeled metal surface, and 35.1 and 3.8 at.% on the peeled polyimide after 500 hours annealing, respectively.

The significant amount of Cu is detected at metal film. Also, the significant amount of C is detected at polyimide by EDS and XPS analyses. It seems that the interface failure was occurred between electroplated Cu and screen printed Ag by the results. Resistivity of screen printed Ag/polyimide system was measured by 4-point probe so as to understand the correlation between microstructure and electrical properties.

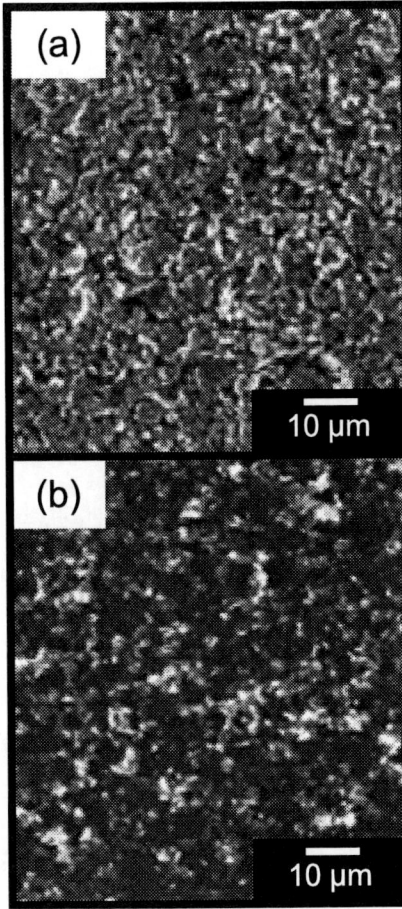

Figure 4. SEM images of peeled (a)metal surface and (b)polyimide surface without treatment.

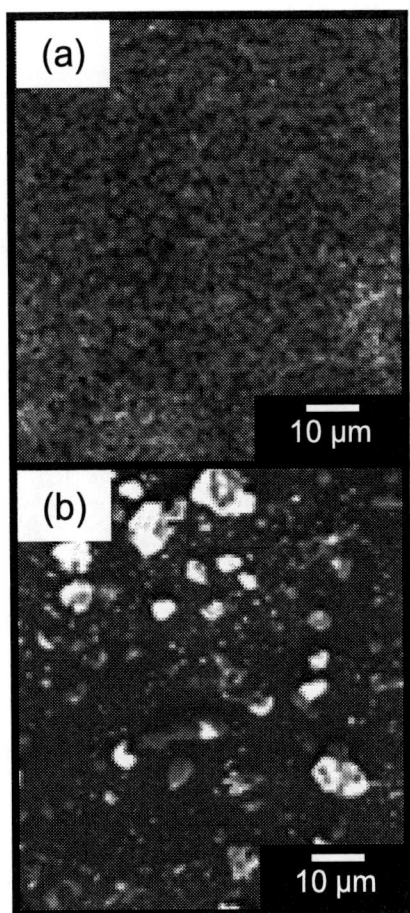

Figure 5. SEM images of peeled (a)metal surface and (b)polyimide surface with temperature/humidity treatment for 500 hours.

The resistivity of screen printed Ag/polyimide system was 9.1 μΩ·cm without treatment and 0.1 μΩ·cm with annealing for 500 hours, respectively. Figure 7 represents SEM images of screen printing Ag surface before and after annealing. Figure 7(b) shows SEM image of printed Ag surface after annealing, it seems to that Ag particles were sintered by high temperature. Table 2 shows carbon concentration by EDS.

C atomic concentrations were decreased after annealing at screen printed Ag surface by EDS and XPS analyses. The Ag particles are coated by the binder. Thus, no electrical conduction is made at all. However, under annealing at the low temperature, several binders begin to be removed locally.

Table 2. EDS atomic concentrations of the screen printed Ag surface before and after annealing at 200 °C.

Treatment condition	Atomic percentage (at.%)		
	C	O	Ag
No Tretment	15.6	0.6	83.8
500 hours @ 200 °C	2.1	5.0	92.9

Table 1. EDS atomic concentrations of peeled surfaces before and after environmental treatments.

Treatment condition	Peeled surface	Atomic percentage (at.%)			
		C	O	Cu	Ag
No Treatment	metal	12.3	2.1	13.7	71.9
	polyimide	58.3	30.7	1.0	10.0
500 hours @ 85□/85% R.H.	metal	31.0	9.3	16.2	43.5
	polyimide	70.9	28.7	0.3	0.1
500 hours @ 200 □	metal	-	1.7	31.9	66.4
	polyimide	35.1	1.2	3.8	59.9

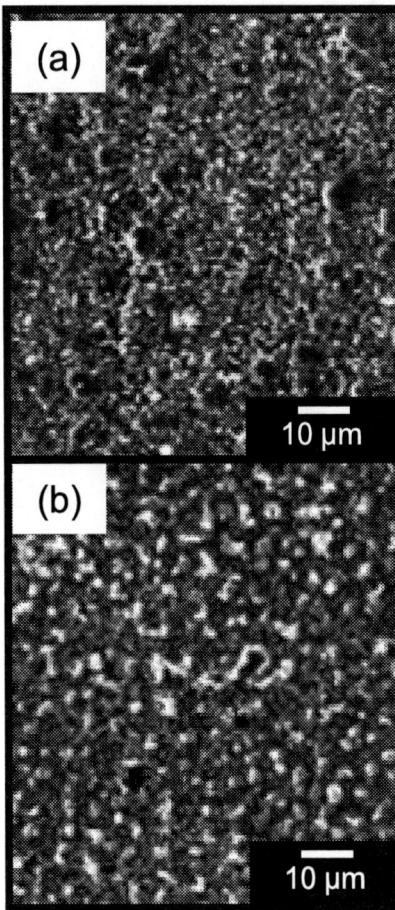

Figure 6. SEM images of peeled (a)metal surface and (b)polyimide surface with annealing for 500 hours.

Figure 7. SEM images of screen printed Ag surfaces (a) before and (b) after annealing for 500 hours.

Therefore, the electrical resistivity can be drastically decreased due to a little area of contact between Ag particles [20-22]. In this results carbon atomic concentration was decreased with annealing condition. It is expected that the resistivity is decreased due to removed binder in screen printing Ag layer during annealing as shown in Fig. 8.

4. Conclusions

Effects of 200 °C annealing and 85 °C/85% R.H. temperature/ humidity environments on the electrical resistivity and interfacial adhesion energy between screen-printed Ag film and polyimide substrate were systematically investigated. Measured peel strength dramatically decreased during both annealing and temperature/humidity treatment for 500 hours. The decreased adhesion strength during annealing seems to be closely related to Cu oxide formation at Cu/Ag interface, while degradation of the polyimide by both the activation agents and the water seem to be responsible for adhesion decrease during temperature/humidity treatments. Screen printed Ag film shows the initial sharp decrease in resistivity during annealing due to effective binder removal effect and partially sintering effect.

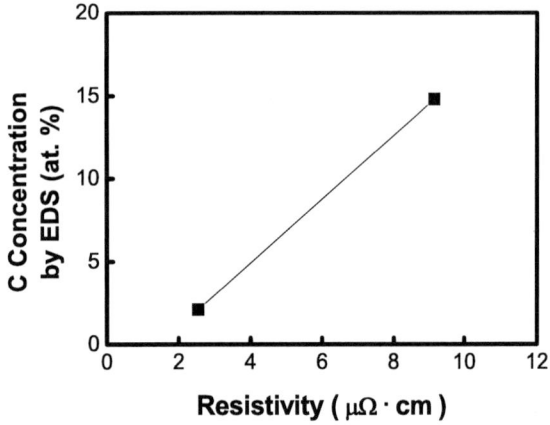

Figure 8. Correlation between resistivity and C concentration by EDS before and after annealing at 200 °C.

978-1-4799-2408-0/14 $31.00 © 2014 IEEE

Acknowledgments

This work was supported by the Global Leading Technology Program (10042421) of the Office of Strategic R&D Planning(OSP) funded by the Ministry of Knowledge Economy, Republic of Korea, and by the Technology Innovation Program Industrial Strategic Technology Development Program (10035430), Development of Reliable Fine-pitch Metallization Technologies.

References

1. B. I. Noh, J. W. Yoon, J. H. Choi and S. B Jung, "Effect of Cr Thickness on Adhesion Strength of Cu/Cr/Polyimide Flexible Copper Clad Laminate Fabricated by Roll-to-Roll Process," J. Electron. Mater. vol. 38, pp. 46-53, Jan. 2009.

2. J. Jang, T. Earmme, "Interfacial study of polyimide/copper system using silane-modified polyvinylimidazoles as adhesion promoters," Polymer, vol. 42, pp. 2871-2876(6), Mar. 2001.

3. K. W. Lee and A. Viehbeck, "Wet-process surface modification of dielectric polymers: Adhesion enhancement and metallization," IBM J. Res. Dev. vol. 38, Issue 4, pp. 457-474, July. 1994.

4. S. Lee, S. Park, and H. K. Lee, "Improvement of Adhesion between Copper Layer and Polyimide Films Modified with Alkaline Potassium Permanganate and/or Alkali Surface Treatments," Macromol. Symp. vol. 249-250, Issue 4, pp. 583-590, Mar. 2007.

5. W. Yu and T. M. Ko, "Surface characterizations of potassium-hydroxide-modified Upilex-S® polyimide at an elevated temperature," Eur. Polym. J. vol. 37, Issue 9, pp. 1791-1799, Sep. 2001.

6. S. M. Ho, T. H. Wang, H. L. Chen, K. M. Chen, S. M. Lian, and A. Hung, "Metallization of polyimide film by wet process," J. Appl. Polym. vol. 51, Issue 8, pp. 1373-1380, 22. Feb. 1994.

7. S. Ikeda, H. Yanagimoto, K. Akamatsu and H. Nawafune, "Copper/Polyimide Heterojunctions: Controlling Interfacial Structures Through an Additive-Based, All-Wet Chemical Process Using Ion-Doped Precursors," Adv. Funct. Mater. vol. 17, pp.889, 2007.

8. Z. Wang, A. Furuya, K. Yasuda, H. Ikeda, T. Baba, M. Hagiwara, S. Toki, S. Shingubara, H. Kubota and T. Ohmi, "Adhesion improvement of electroless copper to a polyimide film substrate by combining surface microroughening and imide ring cleavage," J. Adhes. Sci. Technol. vol. 16, Issue 8, pp. 1027-1040, Apr. 2012.

9. R. Faddoul, N. R. Bruas, A. Blayo, "Formulation and screen printing of water based conductive flake silver pastesonto green ceramic tapes for electronic applications," Mater. Sci. Eng. B, vol. 177, Issue 13, pp. 1053-1066, Aug. 2012.

10. E. C. Ahn, J. Yu, I. S. Park and W. J. Lee, "Adhesion improvement of electroless copper to a polyimide film substrate by combining surface microroughening and imide ring cleavage, " J. Adhes. Sci. Technol. vol. 10, Issue 12, pp.1343-1357, Apr. 2012.

11. A. C. Callegari, H. M. Clearfield, B. K. Furman, T. G. Graham, D. Neugroschl and S. Purushothaman, "Adhesion durability of tantalum BPDA-PDA polyimide interfaces," J. Vac. Sci. Technol. vol. 12, no. 1, pp. 185, 1994.

12. D. D. Denton, M. C. Buncick and H. Pranjoto, "Effects of process history and aging on the properties of polyimide films," J. Mater. Res. vol. 6, Issue 12, pp. 2747-2754, Jan. 2011.

13. D. C. Hu and H. C. Chen, "Humidity effect on polyimide film adhesion," J. Mater. Res. vol. 27, Issue 19, pp. 5262-5268, Oct. 1992.

14. S. C. Park and Y. B. Park, "Effect of Temperature/Humidity Treatment Conditionson Interfacial Adhesion Energy between Inkjet-Printed Ag and Polyimide," Jpn. J. Appl. Phys. vol. 48, 08HL02, Aug. 2009.

15. S. C. Park, K. J. Min, K. H. Lee, Y. S. Jeong, and Y. B. Park, "Effects of Temperature and Humidity Treatment Conditions on the Interfacial Adhesion Energy between the Electroless-Plated Ni and Polyimide," Jpn. J. Appl. Phys. Vol. 49, 08JK01, Ang. 2010.

16. J. K. Jung, S. H. Choi, I. Kim, H. C. Jung, J. Joung and Y. C. Joo, "Characteristics of microstructure and electrical resistivity of inkjet-printed nanoparticle silver films annealed under ambient air," Philos. Mag. vol. 88, no. 3, pp. 339-359, Apr. 2008.

17. J. Y. Song and Jin Yu, "Analysis of the T-peel strength in a Cu/Cr/Polyimide system," Acta Meter. vol. 50, pp. 3985- 3994, Sep. 2002.

18. S. C. Park, S. H. Cho, H. C. Jung, T.W. Joung, and Y. B. Park, "Effect of Temperature/Humidity Treatment Conditions on the Interfacial Adhesion Energu of Inkjet Printed Ag Film on Polyimide," J. Kor. Inst., Met. & Mater. vol. 45, no. 9, pp. 520-526, 2007.

19. K. S. Kim and N. Aravas, "Elastoplastic analysis of the peel test," Int. J. Solid Struct. vol. 24, Issue 4, pp. 417-435, 1988.

20. J. M. Cheon, J. H. Lee, Y. S. Song and J. R. Kim, "Synthesis of Ag nanoparticles using an electrolysis method and application to inkjet printing," Colloids and Surfaces A: Physicochem. Eng. vol. 389, pp. 175-179, Sep. 2011.

21. A. Kosmala, R. Wright, Q. Zhang and P. Kirby, "Synthesis of silver nano particles and fabrication of aqueous Ag inks for inkjet printing," Mater. Chem. Phys. vol. 129, pp. 1075-1080, 2011.

22. I. H. Lee, S. H. Kim, J. H. Yun, I. K. Park and T. S. Kim, "Interfacial toughening of solution processed Ag nanoparticle thin films by organic residuals," Nanotechnology, vol.23, pp. 8, Nov. 2012.

Wafer Level Warpage Characterization for Backside Manufacturing Processes of TSV Interposers

Feng Jiang[1,a], Qibin Wang[1,2], Kai Xue[1,2], Xiangmeng Jing[1,2], Daquan Yu[1,2,b], Dongkai Shangguan[1,2]

[1]National Center for Advanced Packaging, New district, Wuxi, 214135, PR China
[2]Institute of Microelectronics Chinese Academy of Sciences, Beijing, 100029, PR China
[a]Email: fengjiang@ncap-cn.com
[b]Email: yudaquan@ime.ac.cn

Abstract

TSV (through-silicon-via) has been regarded as a key technology for 2.5D and 3D electronic packaging. However, the manufacturing of the through silicon interposer (TSI) is very challenging and costly. The minimization of the warpage of the TSV interposer wafer is crucial for successful subsequent processing，for example, thin wafer handling, backside via revealing and copper pillar bumping. In this paper, warpage was tested before and after most of the backside process steps. Wafer level warpage modeling methodology has been developed by finite element analysis (FEA) using equivalent material model. The warpage was simulated and analyzed by considering different process factors.

1. Introduction

Through silicon via (TSV) has been a hot topic for both academia and industry since it provides solutions for 2.5D and 3D stacked IC [1-2]. 2.5D integration using TSV interposer has been developed intensively in recent years for both front-end foundries and back-end OSAT. The manufacturing of TSV interposer is very complex. As shown in Fig. 1, typically, the process consists of TSV formation, RDL, bumping, thin wafer handling, backside via reveal and debonding.

There are several challenges associated with TSV wafer processing, such as void-free TSV filling with Cu, elimination of Cu protrusion, wafer warpage control, wafer thinning process, bonding/debonding process, backside via reveal process. [3-4]. One of the most challenging tasks in the TSV process is to reduce wafer warpage as wafer flatness is important in subsequent processing. There are many factors contributing to wafer warpage, such as differences in the coefficient of thermal expansion (CTE) among different materials, variations of film thickness and pattern density of vias, RDL, bumps. Stress induced by wafer warpage is one of the root causes leading to process and device failure, such as delamination, cracking, and decrease in device performance [5-6]. Therefore, it is crucial to minimize TSV wafer warpage through processes optimization.

This paper focuses on wafer level warpage characterization during the TSV via reveal process. Currently the TSV front-side wafer processes are mature, but the backside wafer processes are not clear yet. In present study, the processes of backside TSV interposer wafer were introduced as followings. After completion of the front-side wafer processes the interposer wafer is temporarily bonded, face down, onto a

carrier wafer. Here ZoneBOND technology was chose for thin wafer handling, which was proposed by Brewer Science and has been employed and further developed by bonding tool suppliers [7]. The back side of the temporarily bonded wafer is then ground down to a plane where there is 10 to 15um distance away the tips of the buried TSV. After the grinding process, the silicon is dry etched to "reveal" the vias that are still encapsulated and passivated by the using a dielectric via liner stack. After that CMP is generally used to expose the copper. This is followed by backside RDL and under bump metallurgy and bump formation.

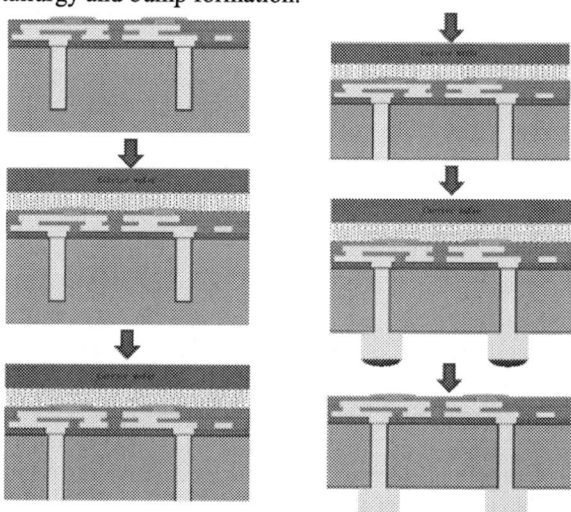

Fig. 1 Typical process flow of TSV interposer

Wafer warpage is measured after backside silicon CMP and debonding. In addition finite element analysis (FEA) modeling has been developed to verify and understand those factors that most contribute to wafer warpage.

2. The Warpage after Wafer Thinning and CMP

After via formation, typically using a through silicon interposer（TSI）approach, finished interposer wafers are temporarily bonded, face-down to the silicon carriers. ZoneBOND technology is an open materials platform because it enables room temperature debonding process by edge zone release which is independent of adhesive and carrier type. In this paper, it was chose for thin wafer handling from wafer thinning, Si CMP, Si recess etching, backside passivation and backside bump formation.

Most of the bulk Si is removed in a backgrind process. About 5 to 10 um of silicon remains over the TSVs after the backgrinding. The TTV of the remaining silicon thickness

(RST) could not be modified by grinding process. Therefore RST was influenced by variations in TSV depth and bonding performance.

Fig 2 shows the wafer thickness before and after CMP process. The data shows that backgrinding TTV is 3um and after CMP the TTV dropped to within 1um. Silicon CMP after Grind is used to achieve a uniform thickness profile and also a smooth surface. Fig. 3 shows the surface roughness after thinning and after silicon CMP. Before CMP roughness is easily observed. After CMP process, the roughness is decreases to 7Å.

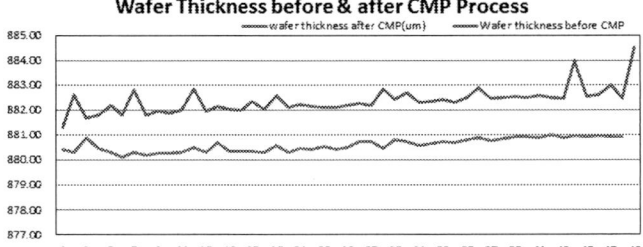

Fig. 2. Wafer thickness before & after CMP process

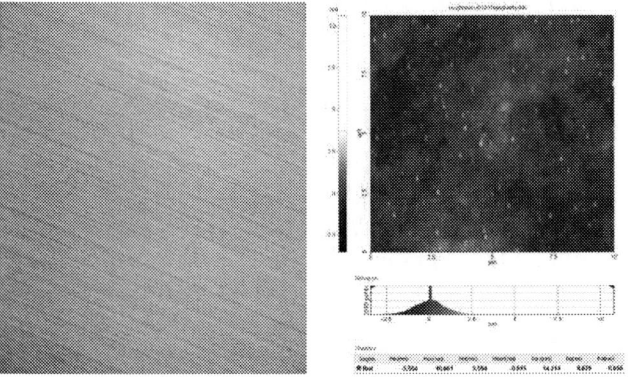

Fig. 3. Surface roughness before & after backside CMP

To accurately describe the warpage, TSI were tested by kSA MOS Thermal-Scan, by which MOS (Multi-beam Optical Sensor) stress measurement technology was used. In the test machine a single laser is used to generate a two-dimensional laser array. In this way, it can measure more than one thousand points on one wafer, and at the same time it can keep 1um spatial scan resolution over the entire wafer. Through the use of sophisticated image processing and data analysis algorithms, it provides unprecedented warpage images directly and simultaneously in the room temperature. For the test theory it only can work on a smooth surface without any topography larger than 1um.

In order to simulate the stress induced on the TSI wafer by the post thinning backside CMP process, a FEA modeling method was established by simulating the warpage of the TSI wafer.

The optimization criterion is to reduce the warpage of the TSI wafer. As shown in Fig. 4, without the CMP process, the warpage of the TSI wafer is 227.6um, while the actual warpage tested on the TSI wafer is about 139.5um as shown in Fig. 5. At the same time the FEA result shows the warpage characterization of TSI wafer is tensile stress from the topside surface of the TSI wafer.

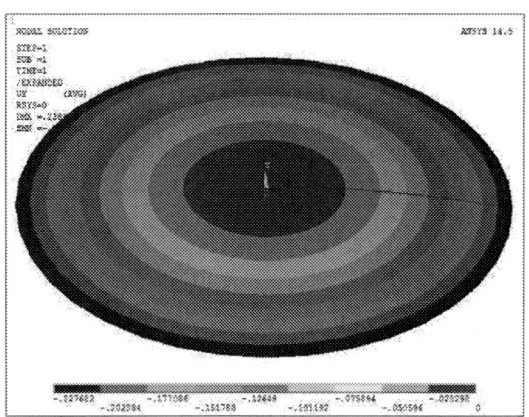

Fig. 4. Simulated warpage without CMP process

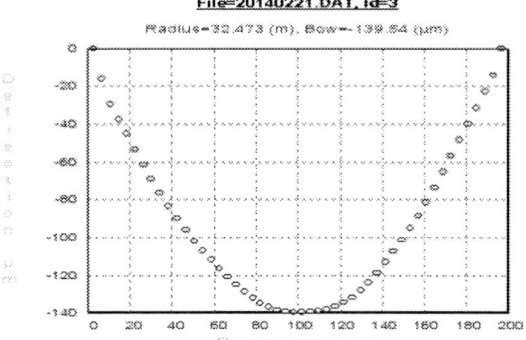

Fig. 5. Measured warpage without CMP process

As shown in Fig. 6, with CMP process, the simulated warpage of the TSI wafer is 66.5um, while the measured warpage post CMP process is about 61.25um as shown in Fig. 7.

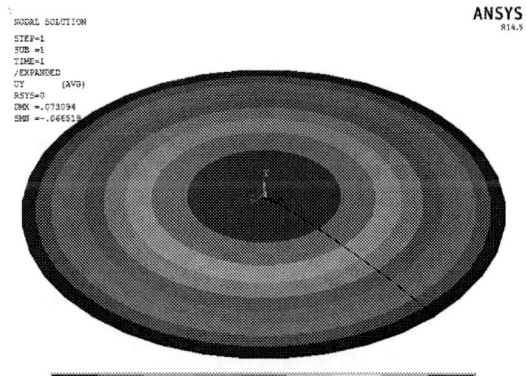

Fig. 6. Simulate warpage post CMP

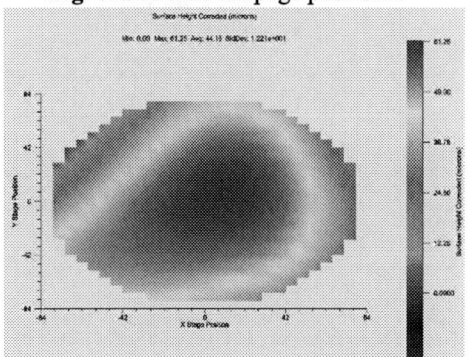

Fig. 7. Measured warpage after CMP

978-1-4799-2408-0/14 $31.00 © 2014 IEEE

3. The Warpage of the rest Processes (Dry etch, Passivation, Revealed Via and Bumping and Debonding)

A major challenge of the via-reveal process is to control the exposed via height, because incoming wafers for via reveal can have significant variations in TSV depth, carrier wafer thickness, adhesive thickness, and silicon thickness after the grind and CMP processes. The silicon plasma etch step is critical to control the height uniformity of the exposed copper vias. Fig. 8 is an image of the TSV wafer after DRIE. High etch selectivity to the TSV liner dielectric is required to prevent oxide liner damage. The exposed height of vias was measured from the surface of the silicon to the top of the via. As shown in Fig. 9, the height was smaller than 5um.

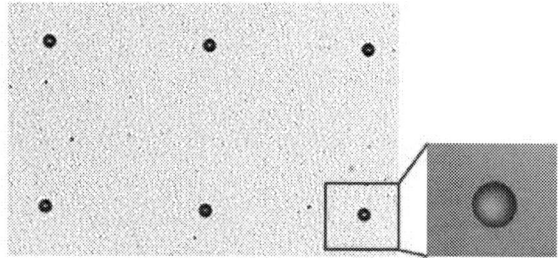

Fig. 8. The TSV image after DRIE

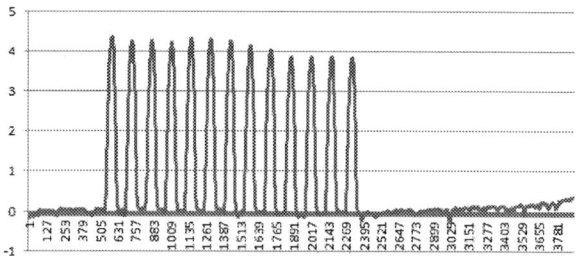

Fig. 9. Exposed height of revealed via

After the copper via is revealed, the passivation process was carried out to protect the TSV electric insulation. In addition, stresses induced by the backside grinding and CMP processes must be carefully compensated to prevent wafer Warping. So a polymeric layer need to coat on the backside surface and shrink during a curing step to produce intrinsic compressive stress in the active TSI wafer.

A polymer layer was spray-coated on the backside surface followed by low temperature (below 150 °C) curing. The passivation layer has a thickness under 8um after curing, as shown in Fig. 10.

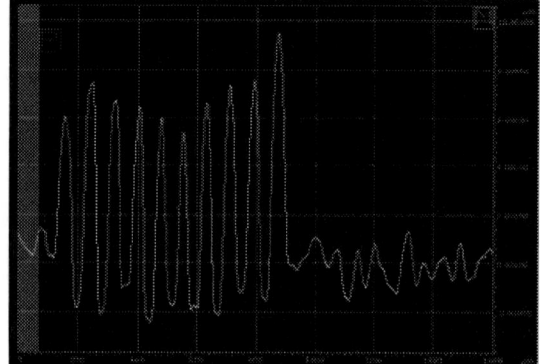

Fig. 10. The passivation layer after curing

After passivation, CMP was used to planarize the TSV protrusion and expose the TSV Cu to enable backside RDL. Fig. 11 illustrates the TSI wafer after passivation and CMP planarization.

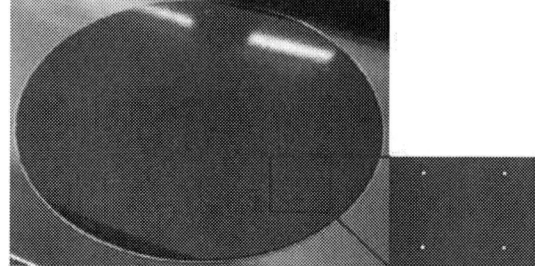

Fig. 11 The TSI wafer after passivation and CMP planarization.

The opening size at the bottom of TSV is slightly larger than the target value, e.g., 20um in diameter as shown in Fig. 12. The dielectric CMP process planarizes the resultant pillars and exposes copper TSVs.

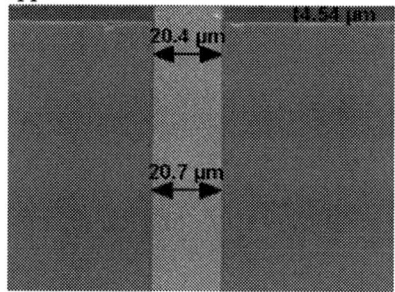

Fig. 12 the SEM image of the TSV after CMP process

After TSVs were revealed from the TSI wafer backside, the bumping process was carried out for the interconnection. Copper pillar bumps were formed by plating and then reflow after tin cap plating. After the reflow process, the pillar bumps formed the shape as shown in Fig. 13.

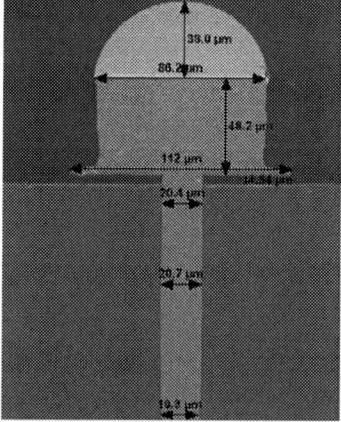

Fig. 13. The pillar bumps after the electroplating process

After Cu pillar electroplating the thin TSI wafer is debonded from the carrier wafer and transferred to a dicing tape on a frame. Debonding was carried out at room temperature which is favorable for low temperature thermal budget TSV wafer processing. Edge bevel trimming with width 1mm on bonded TSV wafer was done to allow the solvent to flow into the strong adhesion wafer edge zone. The special solvent was used to dissolve the adhesive to allow mechanical release and carrier separation. Cleaning of the TSI wafer is required to remove the residue, and the carry wafer after debonding without any edge chipping or cracking.

The warpage of the TSI wafer (with the ZoneBOND process or with the thermal releasing process after the bumping process) is shown in Fig. 14. Without the pulling force of the wafer dicing tape, the TSI wafer is more warping. A ruler indicates that the warpage of TSI wafer was reduced dramatically while the ZoneBOND process, from more than 15.0mm down to less than 8.0mm. But the warpage of the TSI wafer by the FEA simulation was 2.7mm as shown in Fig. 15. The result may be affected by the parameters of the materials, such as the PI passivation material and the copper material, and so on. At the same time the FEA result shows the warpage characterization of TSI wafer is tensile stress from the topside surface of the TSI wafer.

Fig. 14 The warpage of the TSI wafer with (up) or without (down) the CMP process after backgrinding process

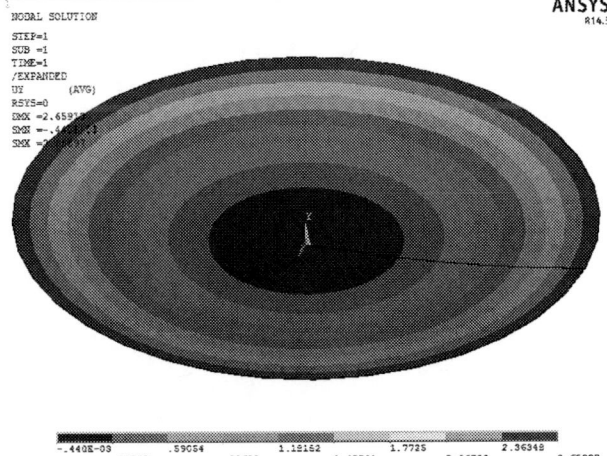

Fig. 15 The warpage of the TSI wafer by the FEA simulation

After the debonding process, the TSI wafer was ready for dicing and stacking on the wafer sawing tape while the carrier wafer can be recycled for a new temporary bonding. Fig. 16 illustrates the TSI wafer after debonding and cleaning processes.

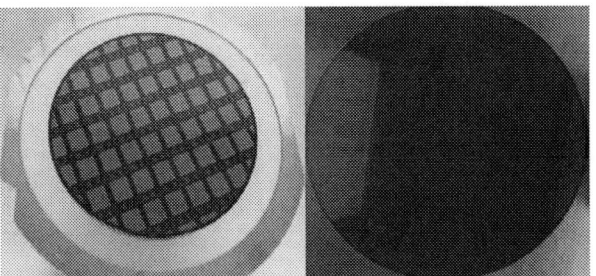

Fig. 16. The TSI wafer and the carry wafer after debonding and cleaning process

Conclusions

We have successfully demonstrated the backside processes of TSI wafers, such as backside grinding, silicon CMP, dry etch, passivation, PI CMP, bumping and debonding. The wafer level warpage was characterized during the processes and FEA simulation was also conducted. Some key conclusions are summarized as follows:

(1) The CMP process after backgrinding process is important for reducing the wafer level warpage.

(2) Base on the FEA and measured data, the wafer stress changed from tensile stress to compression stress after backside passivation.

(3) The room temperature debonding process is better than the thermal releasing process in thin wafer handling for reducing the wafer level warpage.

Acknowledgments

The authors acknowledge the support from the National Science and Technology Major Project under contract No. 2011ZX02709-2.

References:

[1] X.F. Pang, T.T. Chua, H.Y. Li, E.B. Liao, W.S. Lee, and F.X. Che, "Characterization and Management of Wafer Stress for Various Pattern Densities in 3D Integration Technology," Proc. 60th Electronic Components and Technology Conf., June 1-4, 2010, pp. 1866–1869.

[2] Bo Kai Huang, et al., "Integration Challenges of TSV Backside Via Reveal Process，" ECTC, 2013. pp. 915-917

[3] F.X. Che, W.N. Putra, A. Heryan to, A. Trigg, X. Zhang, C.L. Gan, "Study on Cu Protrusion of Through-Silicon Via (TSV)," IEEE Transactions on Components, Packaging and Manufacturing Technology. Vol. 3, No. 5, May 2013, pp. 732–739.

[4] F.X. Che, H.Y. Li, Xiaowu Zhang, S. Gao, and K.H. Teo., "Development of Wafer Level Warpage and Stress Modeling Methodology and Its Application in Process Optimization for TSV (Through-Silicon Via) Wafers," IEEE Transactions on Components, Packaging and Manufacturing Technology. Vol. 2, No. 6, June 2012, pp. 944–955.

[5] Y. Kim, S.-K. Kang, and S. E. Kim, "Study of thinned Si wafer warpage in 3-D stacked wafers," Microelectron. Rel., vol. 50, no. 12, pp. 1988–1993, 2010.

[6] R. P. S. Thakur, N. Chhabra, and A. Ditali, "Effects of wafer bow and warpage on the integrity of thin gate oxide,"Appl. Phys. Lett., vol. 64,no. 25, pp. 3428–3430, 1994.

[7] http://electroiq.com/blog/2011/10/brewer-science-evgcommercialize-temporary-wafer-bonding-with-zoning-laws/

Stretchable and Transparent Silicone/Zinc Oxide Nanocomposite for Advanced LED Packaging

Xueying Zhao[1], Liyi Li[1], Zhuo Li[1], Ching-Ping Wong[1, 2, *]

[1]School of Materials Science and Engineering, Georgia Institute of Technology
771 Ferst Drive, Atlanta, GA 30332
[2]Department of Electronic Engineering, The Chinese University of Hong Kong, Hong Kong
*Corresponding Author: (404)894-8391, cp.wong@mse.gatech.edu, cpwong@cuhk.edu.hk

Abstract

For current light-emitting diode (LED) packaging technology, one of the key challenges is light extraction, due to the difference in index of refraction between LED chip and air. Silicone nanocomposites have been widely researched for applications in LED encapsulant to reduce the difference in refractive index. Silicone is desirable for LED encapsulant because of its optical transparency and photothermal resistance. However, less attention has been paid to the elastic properties of silicone which would enable a stretchable LED encapsulant. Hence the objective of this study is to examine the stretch ability of silicone/zinc oxide (ZnO) nanocomposites for LED packaging. Wurtzite ZnO nanoparticles were prepared in colloids and subjected to silane treatment. Effects of both *ex situ* and *in situ* silane treatment on the final mechanical and optical properties of the silicone/ZnO nanocomposites were examined. Silicone/ZnO nanocomposites exhibit significantly more compliant stress-strain behavior than silicone control; silicone/silane-treated ZnO nanocomposites, in particular, exhibit more serrated stress-strain curves. Silicone/silane-treated ZnO nanocomposites exhibit higher transmittance than silicone/unmodified ZnO nanocomposites, indicating improved dispersion of the nanoparticles. The silicone/5% silane-treated ZnO nanocomposite using *in situ* method is able to deform over a range of up to 160%; its film (~40 microns thick) exhibits transmittance >70% throughout the visible range.

1. Introduction

In high power light-emitting diode (LED) packaging, light extraction is the primary focus of LED encapsulation and has a direct impact on device efficiency [1]. Low light extraction efficiency results from difference in refractive index (RI) between LED chip and air which narrows the light escape cone [2].

To enhance light extraction efficiency, extensive research has been done to reduce the difference in index of refraction by incorporating high-RI nanoparticles (NPs) into optically clear polymer encapsulant. Epoxy resins [2] and silicone [3] have been widely used in recent studies to prepare the polymer matrix of LED encapsulant; nano zinc oxide (ZnO) [4], titania [5], zirconia [6], *etc.* have been used as fillers of the polymer encapsulant to further increase refractive index.

Perhaps due to the high transparency and photothermal resistance of silicone which are the highlighted desired properties of LED encapsulant, less attention has been paid to the elastic properties of silicone which would enable a stretchable LED encapsulant. Silicone is a transparent elastomer that consists of Si-O bonded backbones which form a three-dimensional network after being cured through interchain covalent bonding, shown in Figure 1. This unique network structure enables the stretch ability of silicone.

Figure 1 Three-dimensional network structure of cured silicone

ZnO is a non-toxic and cost-effective semiconductor with high refractive index of 2.0 [7]. Incorporation of nano ZnO is expected to increase the refractive index of the silicone-based encapsulant which is limited to 1.45-1.55 [4]. Sun *et al.* reported synthesis of colloidal ZnO nanoparticles by hydrolyzing zinc acetate dihydrate in basic methanol solution, followed by a precipitation-redispersion washing procedure to purify the nanoparticles [8]. This route was adopted in the study to prepare the nanofiller of the silicone-based encapsulant.

Incorporating dried nanoparticles into the polymer matrix can lead to aggregation of the particles and hence impair transparency of the nanocomposites. Solution mixing is a direct dispersion method [3] to reduce particle agglomeration in the polymer matrix. However, this approach requires the particles to be uniformly dispersed in a solvent that is miscible with the polymer matrix [9]. The ZnO nanoparticles precipitated from the basic methanol solution are surrounded by hydroxyl groups which are incompatible with the silicone matrix. Hence phase separation between polymer and nanofiller can still occur after solvent extraction. Silane treatment can modify the surface of the ZnO nanoparticles by reacting with the hydroxyl groups and coating the nanoparticles with alkyl groups instead; the nanoparticles are therefore stabilized. In this study, the silane-treated ZnO nanoparticles are dispersed in toluene which is miscible with silicone. Since toluene is highly volatile, it is readily removed during solvent evaporation, making it easy to process.

2. Procedure

The complete fabrication process of silicone/ZnO nanocomposite film is shown in Figure 2. The chemicals were purchased from VWR International and directly used without further processing.

2014 Electronic Components & Technology Conference

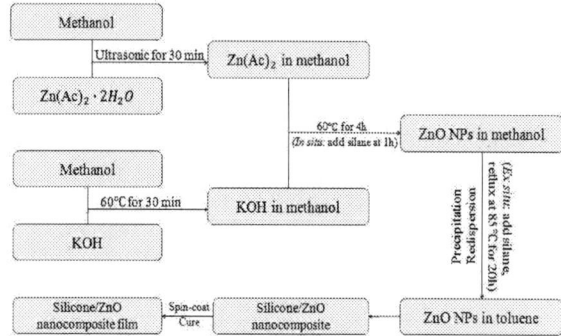

Figure 2 Fabrication process of silicone/ZnO nanocomposite film

2.1 Synthesis of ZnO Nanoparticles

7 mmol zinc acetate dihydrate ($Zn(Ac)_2 \cdot 2H_2O$) in 50 mL methanol solution was placed in Branson 3510 ultrasonic cleaner for 30 min; 14 mmol potassium hydroxide (KOH) in 150 mL methanol solution was oil-bathed at 60°C with constant magnetic stirring for 30 min. Both solutions were then mixed and oil-bathed at 60°C with constant magnetic stirring at 500 rpm for 4 hours.

2.2 Purification of ZnO Nanoparticles

The suspension of nanoparticles was concentrated by means of rotary evaporation at 55°C under vacuum. Isopropyl alcohol and hexane were added at a ratio of 1:5 to the concentrated suspension. Subsequently, the mixture was left to sit in the fridge for at least 10 hours and the supernatant was removed. After the above process was repeated three times, the concentrated suspension of nanoparticles was centrifuged at 5000 rpm for 15 min and redispersed in toluene.

2.3 Silane treatment by *ex situ* and *in situ* methods

ZnO nanoparticles were stabilized through surface modification by coating them with (3-chloropropyl)triethoxy silane. In the *ex situ* method, ZnO nanoparticles were first prepared in methanol and concentrated via rotary evaporation. The silane solution was then added in 1 weight percent (wt.%), followed by refluxing at 85°C for 20 hours with constant magnetic stirring.

In the *in situ* method, the preparation of ZnO nanoparticles was basically the same, except that the silane solution was added to the reaction mixture ($Zn(Ac)_2$ and KOH in methanol) when the mixture had been baked for 1 hour. The silane-treated ZnO nanoparticles were then centrifuged and redispersed in toluene.

2.4 Fabrication of Silicone/ZnO Nanocomposite Films

The toluene dispersion of ZnO nanoparticles was added to Sylgard 184 silicone elastomer (silicone 184) base at 5 wt.%. Toluene solvent was evaporated at room temperature and atmospheric pressure with constant magnetic stirring. After the curing agent was added at a ratio of 1:10 to the base, the nanocomposite was spin-coated onto a glass substrate and crosslinked through thermal curing at 150°C for 10 minutes.

2.5 Characterization

Uniaxial tensile tests were performed using Instron Microtester 5548 at room temperature. Transmittance measurements were carried out using a Shimadzu UV-2450 spectrophotometer. The crystallinity of ZnO nanoparticles was characterized using X-ray diffraction (XRD). Thickness of the nanocomposite films was measured using Heidenhain ND 281 B display unit.

3. Results and Analysis

Figure 3 shows absorbance spectra of the reaction mixture at 0h and 1h. The clear maxima between 300 and 350 nm is due to the electronic transition from top of the valence band to bottom of the conduction band. The absorption peak within this range demonstrates an obvious blue shift compared with bulk ZnO (373 nm), suggesting that the particles are in the quantum regime [10]. Pesika et al. observed that the shape of the absorption spectrum for a suspension of quantum particles is determined by the particle size distribution [11]. Hence the sharp absorption edge and prominent peak position shown in the absorbance spectrum of the reaction mixture at 1h reflect the narrow size distribution of ZnO nanoparticles.

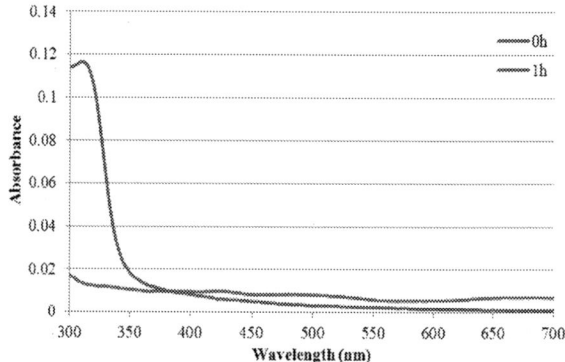

Figure 3 Absorbance spectra of the reaction mixture at 0h and 1h

Figure 4 demonstrates the XRD profile of the as prepared ZnO nanoparticles. The diffraction peaks are attributed to the presence of hexagonal wurtzite crystallites.

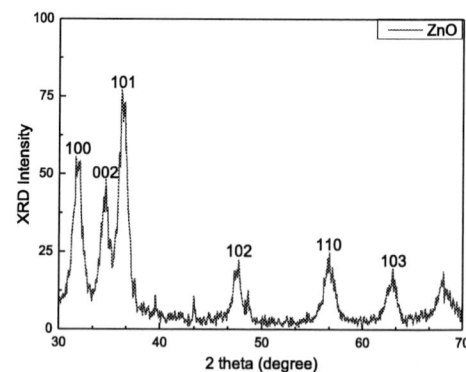

Figure 4 XRD profile of the as prepared ZnO nanoparticles

978-1-4799-2408-0/14 $31.00 © 2014 IEEE

Figure 5 shows stress-strain curves of silicone 184 control and silicone 184/5% unmodified ZnO nanocomposite before failure. Silicone 184 control exhibits nonlinear stress-strain behavior which is characteristic of an elastomer, while silicone 184/5% unmodified ZnO nanocomposite exhibits linear stress-strain behavior and lower stiffness. Hertz suggested that addition of fillers to an elastomer automatically puts the network under strain [12]. Qi and Boyce also pointed out that a pre-stretched material exhibits a significantly more compliant response than that of the virgin material [13].

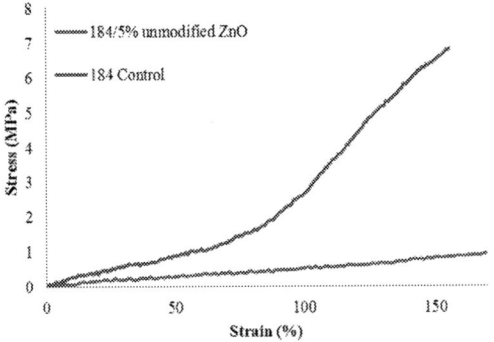

Figure 5 Stress-strain curves of silicone 184 control and silicone 184/5% unmodified ZnO nanocomposite before failure

Figure 6 shows stress-strain curves of silicone 184 control and silicone 184/5% silane-treated ZnO nanocomposites before failure. Silicone 184/5% silane-treated ZnO nanocomposites exhibit more serrated stress-strain behavior than silicone 184 control, indicating inhomogeneous deformation.

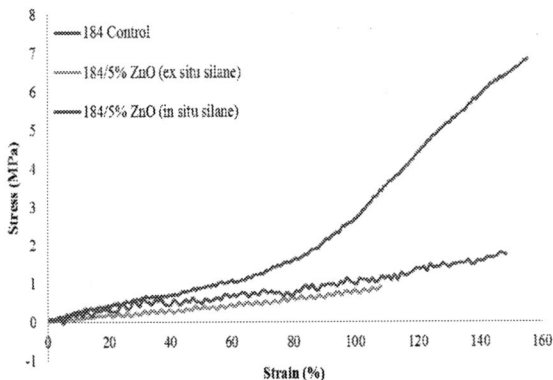

Figure 6 Stress-strain curves of silicone 184 control and silicone 184/5% silane-treated ZnO nanocomposites before failure

Figure 7 shows the breaking strains of silicone 184 control and silicone 184/5% ZnO nanocomposites. Silicone 184/5% silane-treated ZnO nanocomposite using *ex situ* method exhibits lowest breaking strain of 107%. Silicone 184 control and the other two silicone 184/5% ZnO nanocomposites can deform over a wider range of up to 160%.

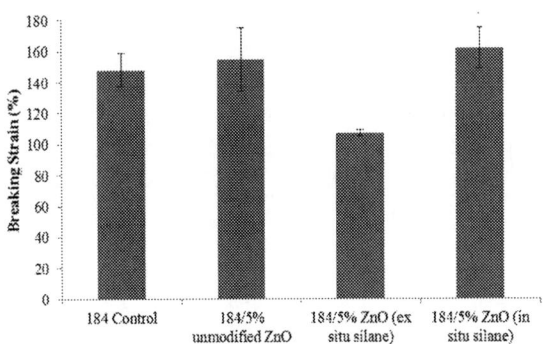

Figure 7 Breaking strains of silicone 184 control and silicone 184/5% ZnO nanocomposites

Figure 8 shows the maximum stresses of silicone 184 control and silicone 184/5% ZnO nanocomposites. Silicone 184 control is able to achieve maximum stress of 6.5 MPa. Silicone 184/5% ZnO nanocomposites, however, break at much lower stress.

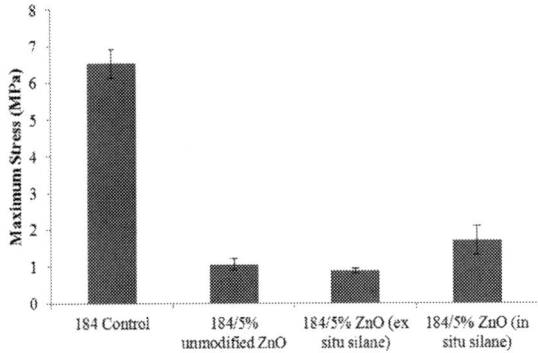

Figure 8 Maximum stresses of silicone 184 control and silicone 184/5% ZnO nanocomposites

Figure 9 shows the elastic moduli of silicone 184/5% ZnO nanocomposites. Table 1 displays the maximum stress to elastic modulus ratio (σ_{max}/E) of silicone 184/5% ZnO nanocomposites.

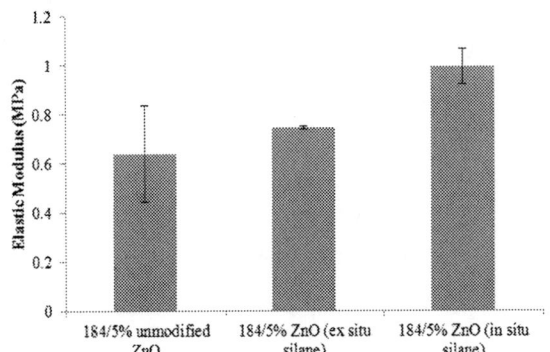

Figure 9 Elastic moduli of silicone 184/5% ZnO nanocomposites

Table 1 Maximum stress to elastic modulus ratio of silicone 184/5% ZnO nanocomposites

Nanocomposite	σ_{max}/E
Silicone 184/5% unmodified ZnO	1.644
Silicone 184/5% silane-treated ZnO (*ex situ*)	1.172
Silicone 184/5% silane-treated ZnO (*in situ*)	1.707

Figure 10 displays the transmittance spectra of silicone 184 control and silicone 184/5% ZnO nanocomposite films in the visible range. The silicone/silane-treated ZnO nanocomposite films exhibit higher transmittance in the visible range than the silicone/unmodified ZnO nanocomposite film, suggesting less particle agglomeration. Silane treatment reduces the polarity of ZnO nanoparticles by replacing the surrounding hydroxyl groups with alkyl groups. The reduced polarity makes the particles split more easily with agitation, thereby breaking down agglomerates [14].

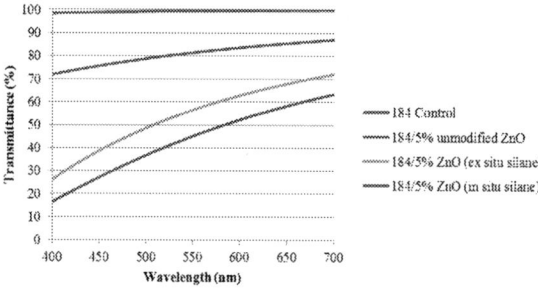

Figure 10 Transmittance spectra of silicone 184 control and silicone 184/5% ZnO nanocomposites

Figure 11 shows the extinction coefficient of silicone 184 control and silicone 184/5% ZnO nanocomposites in the visible range calculated from Beer-Lambert's law (Eq.1), where T is transmittance; k is the extinction coefficient; d is the thickness of the film; and λ is the incident wavelength.

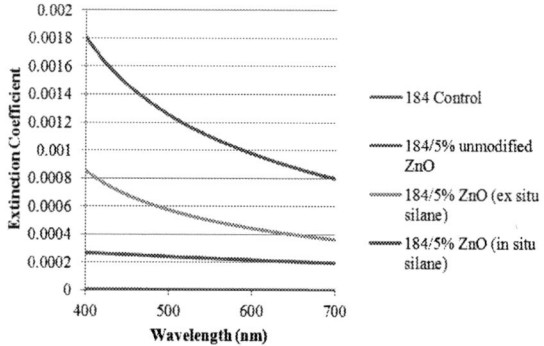

Figure 11 Extinction coefficient of silicone 184 control and silicone 184/5% ZnO nanocomposites in the visible range calculated from Beer-Lambert's law

$$T = \exp(-4\pi kd/\lambda) \qquad (Eq.1)$$

Figure 12 shows extrapolated transmittance of silicone 184 control and silicone 184/5% ZnO nanocomposites as a function of film thickness at a wavelength of 589 nm using Beer-Lambert's law (Eq.1). Transmittance of silicone 184/5% ZnO nanocomposite films at thickness between 30 and 100 microns can thus be predicted from this figure.

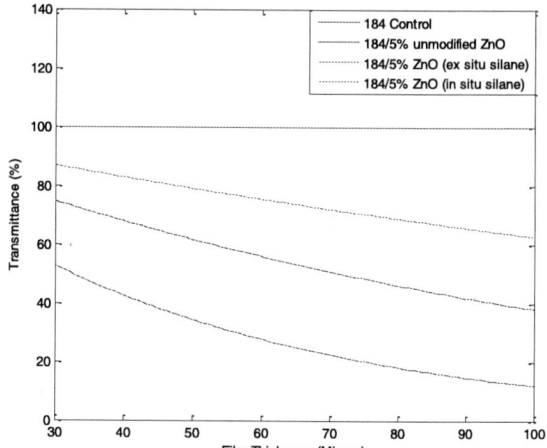

Figure 12 Extrapolated transmittance of silicone 184 control and silicone 184/5% ZnO nanocomposites as a function of film thickness at a wavelength of 589 nm

Conclusions

Stretchable and transparent silicone/ZnO nanocomposites were fabricated as LED encapsulant. Dispersion of the nanoparticles was improved through silane treatment, as evidenced by higher transmittance in the visible range compared with that of the silicone/unmodified ZnO nanocomposites. The silicone/5% silane-treated ZnO nanocomposite using *in situ* method exhibits elastic modulus of ~1 MPa and breaking strain of up to 160%. The reported results demonstrate that the silicone/ZnO nanocomposite may serve as a promising component for future stretchable electronics.

Acknowledgments

It is a great pleasure to acknowledge help from Dr. Malavika Shetty of Georgia Institute of Technology in her guidance on writing. Authors gratefully acknowledge financial support from the Korea Institute for Advancement of Technology.

References

1. R. F. Karlicek, "High power LED packaging," 2005 Conference on Lasers & Electro-Optics (CLEO), Vols 1-3, pp. 337-339, 2005.
2. Y. Liu, Z. Y. Lin, X. Y. Zhao, S. Yoo, K. S. Moon, and C. P. Wong, "ZnO Quantum Dots-filled Encapsulant for LED Packaging," 2012 IEEE 62nd Electronic Components and Technology Conference (ECTC), pp. 2140-2144, May 2012.

3. Y. Yang, Y. Q. Li, S. Y. Fu, and H. M. Xiao, "Transparent and light-emitting epoxy nanocomposites containing ZnO quantum dots as encapsulating materials for solid state lighting," Journal of Physical Chemistry C, vol. 112, pp. 10553-10558, Jul 17 2008.

4. Y. Yang, W. N. Li, Y. S. Luo, H. M. Xiao, S. Y. Fu, and Y. W. Mai, "Novel ultraviolet-opaque, visible-transparent and light-emitting ZnO-QD/silicone composites with tunable luminescence colors," Polymer, vol. 51, pp. 2755-2762, May 28 2010.

5. P. Tao, Y. Li, A. Rungta, A. Viswanath, J. N. Gao, B. C. Benicewicz, et al., "TiO2 nanocomposites with high refractive index and transparency," Journal of Materials Chemistry, vol. 21, pp. 18623-18629, 2011.

6. P. T. Chung, C. T. Yang, S. H. Wang, C. W. Chen, A. S. T. Chiang, and C. Y. Liu, "ZrO2/epoxy nanocomposite for LED encapsulation," Materials Chemistry and Physics, vol. 136, pp. 868-876, Oct 15 2012.

7. Y. Zhang, X. Wang, Y. X. Liu, S. Y. Song, and D. Liu, "Highly transparent bulk PMMA/ZnO nanocomposites with bright visible luminescence and efficient UV-shielding capability," Journal of Materials Chemistry, vol. 22, pp. 11971-11977, 2012.

8. D. Z. Sun, M. H. Wong, L. Y. Sun, Y. T. Li, N. Miyatake, and H. J. Sue, "Purification and stabilization of colloidal ZnO nanoparticles in methanol," Journal of Sol-Gel Science and Technology, vol. 43, pp. 237-243, Aug 2007.

9. D. Sun, W. N. Everett, M. Wong, H. J. Sue, and N. Miyatake, "Tuning of the Dispersion of Ligand-Free ZnO Quantum Dots in Polymer Matrices with Exfoliated Nanoplatelets," Macromolecules, vol. 42, pp. 1665-1671, Mar 10 2009.

10. E. A. Meulenkamp, "Synthesis and growth of ZnO nanoparticles", J. Phys. Chem. B, vol. 102, No. 29, pp. 5566-5572, Jul. 1998.

11. N. S. Pesika, K. J. Stebe, P. C. Searson, "Determination of the Particle Size Distribution of Quantum Nanocrystals from Absorbance Spectra," Adv. Mater., vol. 15, no. 15, pp. 1289-1291, Aug. 2003.

12. D. L. Hertz. Dec 1991, An analysis of rubber under strain from an engineering perspective. *Elastomerics*.

13. H. J. Qi and M. C. Boyce, "Constitutive model for stretch-induced softening of the stress-stretch behavior of elastomeric materials," *Journal of the Mechanics and Physics of Solids,* vol. 52, pp. 2187-2205, Oct 2004.[1]

14. E. K. R. K. Gupta, K. J. Kim, Polymer Nanocomposites Handbook. Boca Raton, FL: CRC Press, 2010.

Warpage Characterization of Panel Fan-out (P-FO) Package

Hung-Wen Liu*, Yi-Wei Liu, Jason Ji, Jash Liao, Agassi Chen, Yan-Heng Chen, Nicholas Kao, and Yi-Che Lai
Siliconware Precision Industries Co., Ltd. (SPIL)
No. 153, Sec.3, Chung Shan Rd., Tantzu Dist., Taichung 42756, Taiwan, R.O.C.
*e-mail: hwliu@spil.com.tw, Tel: +886-4-2534-1525 ext.7895/4653

Abstract

Panel Fan-out (P-FO) packaging technology is known as a new generation FO technology because of high throughout and low cost superiority comparing to first-generation Fan-out Wafer Level Package/Packaging (FO-WLP). However, the process induced warpage is one critical issue needed to be solved. In this paper, the P-FO packages were assembled on a carrier in size of 370×470 mm like a panel, and the focus is how to reduce the warpage of FO panel in several major processes, such as lamination, photolithography and ball placement (BP) process through numerical simulation and experimental study. The FO panel encapsulated by lamination film has a large warpage of 9 mm after lamination film post-curing and 4 mm after full reconstruction process completion respectively, which is primarily attributed to the Coefficient of Thermal Expansion (CTE) mismatch of constituent materials. An effort on material selection for CTE mismatch purpose was applied to reduce the panel warpage to the extent of 3 mm and 1 mm, after lamination film post-curing and full reconstruction process respectively. Besides, the CTE effect of dielectric layer and metal trace on panel warpage before ball placing is found to be more than 1 mm. An approach of reducing carrier size is further proposed to control the warpage within 1 mm for BP process.

Introduction

Miniaturization and higher functionality are the major development trend of portable electronic devices. Therefore, the chip has evolved into size miniaturization and higher functionality with increased Input/Outputs (I/Os) density by the semiconductor node technology evolution. This is the driving force for the development of Wafer Level Package/Packaging (WLP) which has improved thermal performance, electrical performance and higher interconnection density [1].

Conventional Fan-in (FI) concept WLP is designed to fan-in I/Os within chip area by Redistribution Layer (RDL) technology. However, with the dramatic increases of I/Os and pad pitch reduction inside a chip, Fan-in WLP (FI-WLP) is encountering routability issue restricted in chip area and causes the significant I/O pitch mismatch between FI-WLP and Printed Circuit Board/ Printed Wire Board (PCB/PWB), as shown in figure 1 [1]. Therefore, FO-WLP was addressed to overcome routability problem serving adequate I/O pitches between FO-WLP and PCB, and provide higher interconnection density. It has been developed [2], [3] and massed production on 8"/12" wafer carrier as the first-generation FO-WLP.

FO-WLP is one category of Embedded Wafer-Level-Packages which is based on a reconstructed molded wafer infrastructure and the carrier is 8"/12" round wafer; the other homogeny package is based on a PCB infrastructure called Embedded Die Package based on the rectangular carrier, as shown in figure 2. First-generation FO-WLP and Embedded Die Package is a high volume reality and two infrastructures are now clearly settled and proven in high volume mass production in each of their application space. However, this situation will totally change in the future with "Next generation" derivatives of the technologies that are currently under development for future System in Package (SiP) and Package on Package (PoP) [4]. Therefore, with the maturation of FO-WLP packaging technology on 8"/12" round wafer, a novel FO packaging technology on rectangular platform called Panel Fan-out (P-FO) Package/Packaging was considered to be a high throughput and low cost solution due to its large carrier area and high carrier area usage ratio comparing to Embedded Die Package. However, as the carrier size of P-FO package increases, in-process warpage control becomes very critical to bring the success of final fabrication due to warpage is even much worse than 8"/12" wafer.

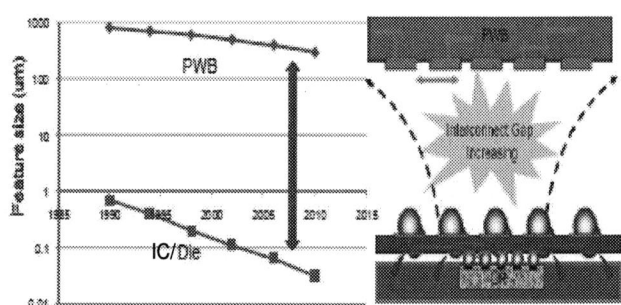

Figure 1. Schematic of I/O pitch mismatch or "interconnection gap" between IC Package and PCB/PWB [1].

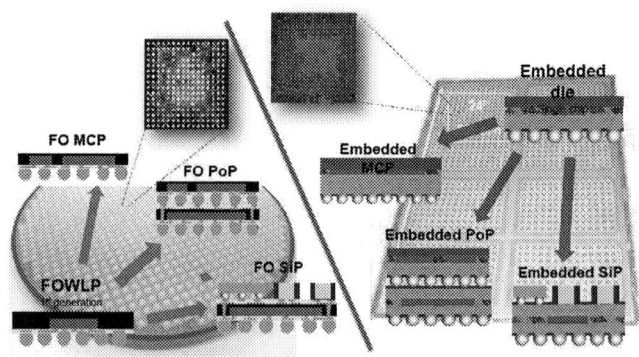

Figure 2. Main categories of Embedded Wafer-Level-Packages [4].

Panel Level Fan-out Packaging Approach

Different infrastructure concept with Embedded Die Packaging technology that supporting by rectangular PCB, P-FO packaging technology using TFT-LCD 2.5 generation (2.5G) 370 × 470 mm material as carrier provides larger platform for FO processing. P-FO package combines with several hetero-technologies such as PCB, semiconductor back-end, semiconductor WLP and TFT-LCD large area processing know-how to construct the new generation FO Package/Packaging. The carrier area of P-FO package is around 5.5 times to 8" and 2.5 times to 12" wafer area respectively and it's capable for higher throughput of FO packages, as shown in figure 3. Besides, this P-FO packaging process is capable of producing over 5.5 times and 2.5 times FO packages than 8" and 12" FO-WLP due to its high carrier area usage ratio. The carrier area usage ratio of 8"/12" FO-WLP is less than 85% due to the influence of edge incomplete package. On the contrary, the carrier area usage ratio of P-FO can achieve 95% due to rectangular chips bonding on rectangular carrier, as shown in figure 4.

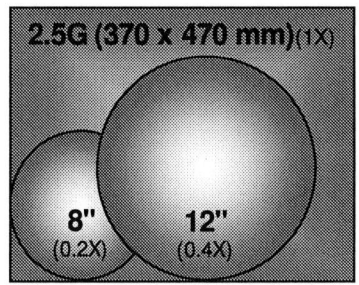

Figure 3. Carrier area comparison between 2.5G (370 × 470 mm) panel carrier and 8"/12" wafer carrier.

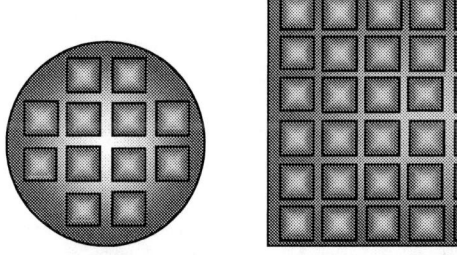

Figure 4. Carrier area usage ratio comparison between FOWLP and P-FO package.

Process Flow

Major processes of P-FO package are similar to FO-WLP which consists of reconstruction, RDL and semiconductor back-end processes. Reconstruction process is combined with semiconductor front-end assembly processes and PCB lamination process. First of all, carrier 1 is bonded with adhesive, and then re-layout the chip with Al pad face-down bonding on the carrier to enlarge chip-to-chip space for package fan-out area forming, as shown in figure 5(a), (b). The following lamination process shown in figure 5(c) is the application of dry film lamination in PCB industry. Comparing to wafer level molding of FO-WLP technology, chip encapsulated by lamination with sheet and film type

material is more suitable for large and rectangular area encapsulation due to corner filling capability. Before RDL process, carrier 2 is bonded as a supporting layer for FO panel handling and carrier 1 is de-bonded for chip Al pad exposed, as shown in figure 5(d). RDL process combines with semiconductor WLP technology and TFT-LCD large area processing know-how to re-layout IO positions within and without chip area, as shown in figure 5(e), (f). Finally, packages is marked by laser marking process and sawed to unit package by singulation process, as shown in figure 5(g).

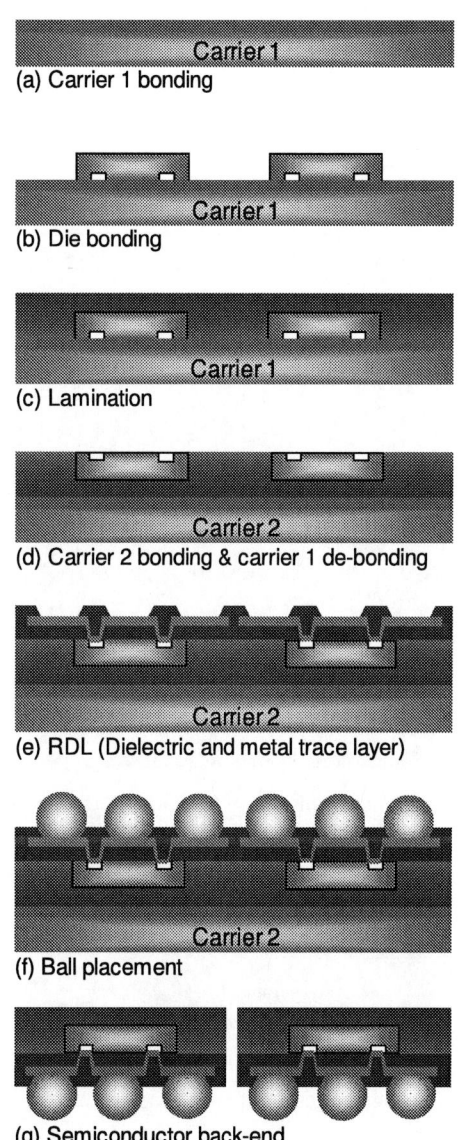

Figure 5. Major process flows of P-FO packaging technology. (a) Carrier1 bonding; (b) Die bonding; (c) Lamination; (d) Carrier2 bonding & carrier1 de-bonding; (e) RDL (Dielectric and metal trace layer); (f) Ball placement; (g) Semiconductor back-end process (Package marking and singulation).

Test vehicle

The test vehicle is a 6 × 6 mm chip encapsulated inside a 9 × 9 mm package with 356 I/Os and one-layer RDL, and totally 1862 pcs gross packages are assembled on a 370 × 470 mm carrier which is 5.8 times to 8" and 2.6 times to 12" wafer respectively, as shown in table 1. The layout of test vehicle is designed in a full matrix of 49 columns by 38 rows on the 370 × 470 mm panel carrier, as shown in figure 6. Besides, the P-FO package structure and RDL layout are shown in figure 7 and figure 8, respectively.

Table 1. Basic information of P-FO package test vehicle.

Basic information		
Package dimension		9 x 9 x 0.475 mm
Chip dimension		6 x 6 x 0.15 mm
RDL		1 layer
Ball size		0.25 mm
Ball pitch		0.4 mm
I/O		356
Gross package	8" (200mm)	318 pcs
	12" (300mm)	723 pcs
	2.5G (370*470mm)	1862 pcs (Full matrix: 49 x 38)

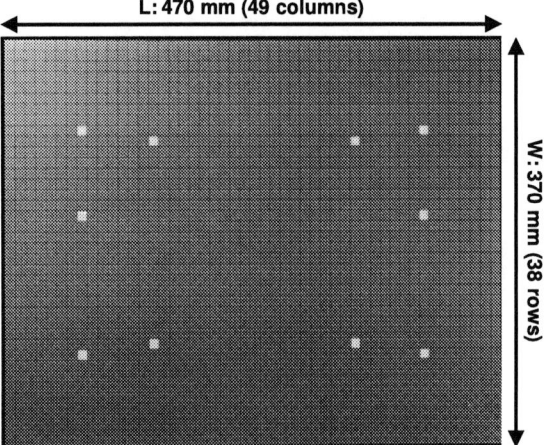

Figure 6. Full matrix layout of P-FO package on 370 × 470 mm carrier, the gross package is 1862 pcs.

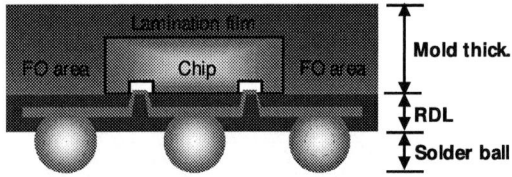

Figure 7. Package structure of P-FO package.

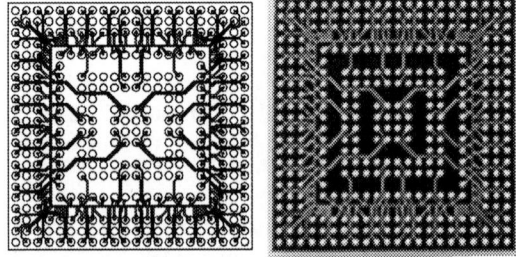

Figure 8. RDL layout of P-FO package (Left: Mask drawing, Right: Test vehicle).

Process Challenges from Panel Warpage

In this paper, the CTE mismatch in package constituent materials is considered as the major parameter to the panel warpage which significantly affect handling and processing ability. For 370 × 470 mm P-FO package, the carrier area is around 5.5 times and 2.5 times to 8" and 12" wafer respectively, as shown in figure 3. As a matter of course, process-induced warpage for the P-FO packages on the 370 × 470 mm carrier is more critical than that fabricated on 8"/12" wafers, especially for lamination and RDL process.

I. Warpage Control for Reconstruction Process

In this study, the CTE of target lamination film is larger than silicon chip and supporting carrier, the residual deformation of the laminated panel after post-curing and full reconstruction process are represented as shown in figure 9 and Figure 10. For P-FO packages fabricated with low CTE carrier 1, the warpage measured after lamination film post-curing was more than 9 mm (Leg 2). Therefore, how to control the warpage caused by CTE effect of lamination film is the main issue for P-FO package fabrication in reconstruction process.

Based on CTE balance concept between supporting carriers and lamination film, two approaches were studied by simulation and experiment to control process warpage, as shown in table 2 and figure 11. First approach was to adjust the CTE of lamination film; the other was to adjust the CTE of carrier. In this study, two types of lamination films and three types of carriers were studied to balance the CTE difference for warpage control. Based on the warpage simulation and experiment data after lamination process, high CTE carrier was more suitable for warpage control comparing to low CTE carrier for either low or high CTE lamination film then the warpage could be controlled within 3 mm after lamination film post curing. Furthermore, the FO panel warpage must be well controlled for photolithography process. Therefore, a complex carrier 2 (low CTE carrier with CTE balance layer on the backside) was applied to enable better warpage result, which was practically less than 1 mm, for either low or high CTE lamination film.

Figure 9. Smiling warpage behavior after lamination film post curing and after full reconstruction process.

Figure 10. Side view of P-FO package panel after lamination film post curing.

Table 2. Warpage simulation and experiment results of P-FO package fabrication during reconstruction process. (Carrier type: L: Low CTE carrier, H: High CTE carrier, C: Complex carrier combined by low CTE carrier and CTE balanced layer; Lamination film: A: Low CTE lamination film, B: High CTE lamination film).

Leg	Process (Structure)	Carrier 1	Lamination film	Carrier 2	Warpage (mm) Sim.	Exp.
#1	Lamination (Carrier 1)	L	A		3.5	5 ~ 5.8
#2		L	B		8	9 ~ 10.1
#3		H	A		-1.9	-2.3 ~ -3.5
#4	Lamination	H	B		2.2	2.3 ~ 2.9
#5			A	L		4.1~4.9
#6	Carrier 2		A	H	-2.3	-1.3 ~ -2.4
#7			B	H	2.1	2 ~2.6
#8	Carrier1 bond/		A	C	-0.4	-0.3 ~ -0.8
#9	carrier2 de-bond		B	C	-0.2	-0.1 ~ -0.7
Warpage driection	Smiling: " + ", Crying: " - "					

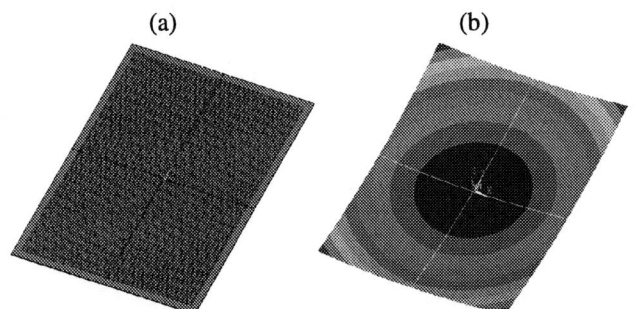

(a)   (b)

Figure 11. (a) Simulation model with full matrix package (b) P-FO warpage contour in lamination and full reconstruction process.

II. Warpage Control for RDL Process

With the previously study, it was found the warpage of P-FO package increases after photolithography and Cu plating process due to the CTE effect of dielectric and metal trace layer. Therefore, the warpage behavior needs to be understood and controlled well before ball placement process. The panel warpage of P-FO package before ball placement is larger than 1 mm and it would be the processing limitation for ball placement. As shown in figure 12, the idea to reduce the panel size to strip panel like is demonstrated to control the warpage within 1 mm for ball placement and to have a good ball placing yield.

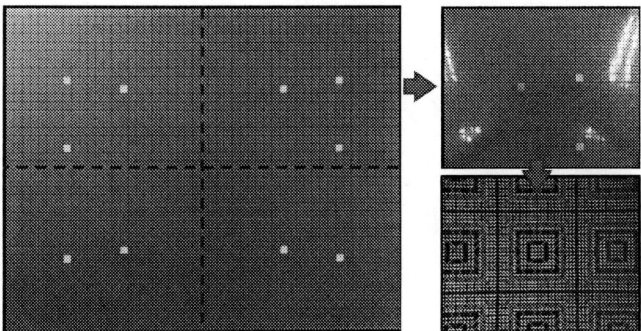

Figure 12. Schematic diagram of strip panel for ball placement process.

III. Temperature-dependent Warpage of Unit Package

In regard to the warpage variation with temperature, the unit P-FO packages laminated with high CTE lamination film are picked from five different positions in a panel and their warpage are measured by Shadow Moiré system. As shown in Figure 13, there is no significant warpage difference between these packages from carrier center and corners. Besides, the warpage variation of P-FO package performs within 20um and it shows lower material property mismatch between die, lamination film and RDL material.

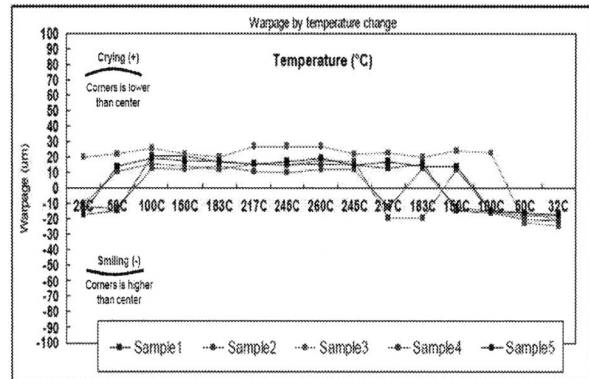

Figure 13. Temperature-dependent warpage measurement of unit P-FO package by Shadow Moiré system.

Figure 14. Schematic diagram of unit P-FO package.

Conclusions

The Panel FO packages with single-layer RDL were successfully developed and practiced on TFT-LCD 2.5 generation (2.5G) 370 × 470 mm carrier for the demand of higher throughput and lower cost. The panel warpage of reconstruction process can be well controlled by CTE balance approaches through the correlation of carrier and lamination film which with different CTE and carrier structure. Besides, the warpage before ball placing process can be well controlled by strip panel. Finally, the warpage variation of unit P-FO package shows lower material properties mismatch between die, lamination film and RDL material.

Future Outlook

In this study, process warpage of P-FO package with single-layer RDL is well controlled by this approach. However, with the dramatic increases of I/Os of a chip, IC package with multi-layer RDL must be the developing trend in the near future. Furthermore, based on the driving force of lower cost and higher throughput of IC packaging, larger carrier size than 370 × 470 mm carrier must be the developing trend. Therefore, this process warpage control approach can be a good reference and it can be applied for the development of P-FO package with multi-layer RDL and larger carrier size than 370×470 mm carrier.

Acknowledgments

All the warpage control technologies were developed and realized by SPIL R&D engineers. I would like to deeply appreciate all related project development integration engineers, process technology development engineers and simulation engineers for their dedication and contribution on the development of P-FO package. Besides, I also deeply appreciate technical supports of Industrial Technology Research Institute (ITRI) and suppliers that provided materials and equipments support.

References

1. S. W. Ho, et al., "Double Side Redistribution LayerProcess on Embedded Wafer Level Package for Package on Package (PoP) Applications", in *Proc. IEEE Electronic Components and Technol. Conf. (ECTC)*, Las Vegas, Nevada, USA, 01 Jun - 04 Jun, 2010, pp. 383.

2. H.S Hsu, David Chang, Kenny Liu, Nicholas Kao, Mark Liao, Steve Chiu, "Innovative Fan-Out Wafer Level Package using Lamination Process and Adhered Si Wafer on the Backside," in *Proc. IEEE Electronic Components and Technol. Conf. (ECTC)*, San Diego, CA, USA, 29 May - 01 Jun, 2012, pp. 1384-1387.

3. Seung Wook Yoon, Yaojian Lin and Pandi C. Marimuthu, "Development and Characterization of 300mm Large Panel eWLB (embedded Wafer Level BGA)," IMPAS-European Microelectronics and Packaging Conference (EMPC-2011), Brighton, UK, 12 Sep - 15 Sep, 2011.

4. Jerome Baron, Lionel Cadix, "FOWLP & Embedded die Packages," Yole Développement (2012).

A Novel Double Layer NCF for Highly Reliable Micro-Bump Interconnection

Ji-Won Shin[1], Yong-Won Choi[1], Young Soon Kim[1], Un Byung Kang[2], Sun Kyung Seo[2], and Kyung-Wook Paik[1]

1) Dept. of Materials Science and Engineering
Korea Advanced Institute of Science and Technology (KAIST)
2) Package Development team
Semiconductor R&D Center
Semiconductor Business
Samsung Electronics Co.,LTD

Abstract

40 μm pitch Cu-pillar/Sn-Ag bump to Ni pad thermo-compression bonding was performed using non-conductive films (NCFs) with different curing agents. The joint morphology of Cu-pillar/Sn-Ag bump/Ni pad was dependent on the viscosity and curing speed of NCFs. Imidazole NCF with high viscosity and fast curing speed showed no solder wetting on both Cu-pillar and Ni pad. However, anhydride NCF with low viscosity and slow curing speed showed both solder wetting on Cu-pillar and Ni pad. Double layer NCF consisting of imidazole NCF as top layer and anhydride NCF as bottom layer was designed and optimized to eliminate solder wetting on Cu-pillar. Joints bonded using double layer NCF showed ideal joint shape with no solder wetting on Cu pillar and good wetting on pad. Thermal aging at 150˚C was performed and IMC growth rate of Cu_6Sn_5 and Cu_3Sn was slower for double layer NCF compared to single layer NCF due to reduced reacting area for Cu and Sn.

Introduction

As demands for smaller and faster electronic products increased rapidly, 3D-packaging became inevitable in electro-packaging industry. Among various 3D packaging materials, 3D chip-stacking using through silicon via (TSV) technology becomes one of the best candidates to make further reduction in size and improve electrical performance [1]. Due to endless effort of many research organizations and industries, state-of-art TSV forming and Copper via filling can be successfully developed. And Cu pillar/Sn-Ag bump structure for vertical interconnection has been widely used for 3D TSV chip stacking [2-8]. In order to perform highly reliable 40 um or less pitch TSV interconnection, bonding process using non-conductive film (NCF) which is pre-applied type of underfill is required [9,10]. Figure 1 shows the advantages of NCF process compared to conventional process. NCF process has advantage of process minimization and cost reduction since it is wafer-level process and NCF-bonding process can perform all the fluxing, underfill, and bonding process at once. Additionally, underfill void problem due to limitation of capillary flow and flux residue can be solved by using NCF-bonding process. However, since NCF bonding process is performed using thermo-compression method, it causes deformation of solder bump, and the solder wets on side of Cu-pillar during the bonding process. The side wetting of solder causes even faster growth of IMC, and hence reliability of the joint is decreased. In order to solve such problem, the idea of double layer NCF is studied. Double layer basically consists of two layers: top layer and bottom layer. When laminated on the chip with Cu pillar/Sn-Ag bump, the top layer covers Cu-pillar and the bottom layer covers Sn-Ag bump. Top layer has high viscosity and fast curing speed, and it blocks solder wetting over Cu-pillar. The bottom layer has relatively lower viscosity and slower curing speed so that joint between solder and pad could be well achieved without resin or silica trapping. In this study, and process optimization of double-layer NCF is performed and reliabilities of joints using conventional NCF and double layer NCF are compared and analyzed.

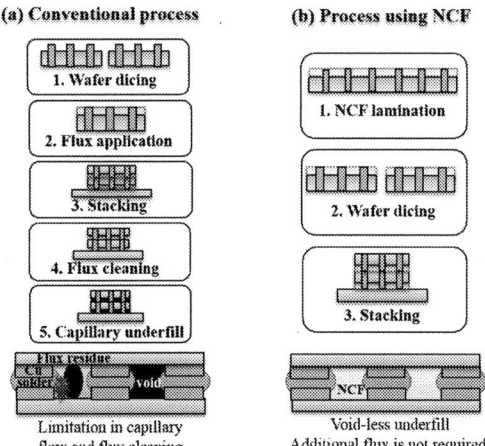

Figure 1. Process comparison of conventional process and process using NCF

Experiments

Test vehicle

As a test vehicle for the experiment, a top chip with dimensions of 8 mm x 8 mm x 0.2 mm and a TSV-formed chip with dimension of 12 mm x 12 mm x 0.06 mm were used. The chip with TSV was already bonded on the PCB substrate with a conventional SMT process, and the top chip was bonded on the chip with TSV using NCFs. The top chip consists of Cu pillar/Sn-Ag bumps with pitch of 40 μm and height of 20 μm and the chip with TSV has Cu pads with pitch of 40 μm and height of 5 μm as shown on Figure 2.

2014 Electronic Components & Technology Conference

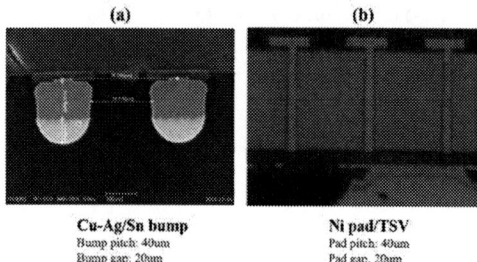

Cu-Ag/Sn bump
Bump pitch: 40um
Bump gap: 20um

Ni pad/TSV
Pad pitch: 40um
Pad gap: 20um

Figure 2. Images and specification of test vehicle

Materials for NCF

NCF is formulated combining mainly three components: resin, curing agent, and additives. In the resin part, epoxy resin, phenoxy resin, and multifunctional resin are mixed to achieve appropriate film properties. The curing agent is the portion that cares most of curing speed and viscosity change. Three types of latent catalysts of imidazole, dicyandiamide (DICY), and anhydride were used in this experiment. Heat flow of NCFs depending on different curing agents are shown on figure 4. Additives such as silica filler and rubber were added to control modulus, CTE, and mainly viscosity. Specification of NCFs with different curing speed and viscosity are listed on Table 1.

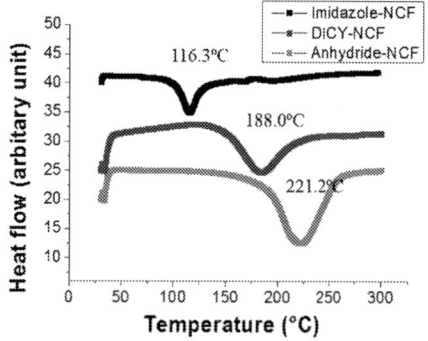

Figure 3. Heat flow curves of fabricated NCFs

Table 1. Specification of fabricated NCFs

NCF type	Curing speed		Min. Viscosity	
Imidazole NCF	Fast	Peak T 116.3°C	High	3500 Pas
DICY NCF	Moderate	Peak T 188.0°C	Moderate	1200 Pas
Anhydride NCF	Slow	Peak T 221.2°C	Low	500 Pas

Lamination & Bonding condition

Lamination of NCFs on chip was performed at 80°C using the vacuum laminator. The laminated chips are bonded using thermo-compression bonding method with heating rate of 6°C/s and peak temperature of 250°C/s for 10 seconds as shown in figure 3.

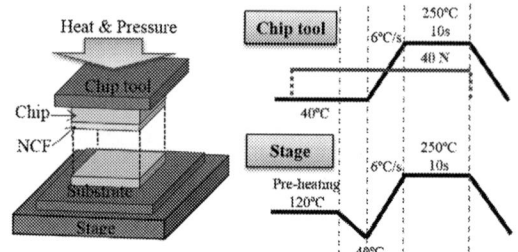

Figure 4. Bonding profile using thermo-compression method

Results and discussion

Analysis on joint morphologies

The test vehicles were bonded using NCF process. NCFs materials of imidazole NCF, DICY NCF, and anhydride NCFs were used. The cross-sectional images of bonded joints using different NCFs are shown on figure 5. Morphologies of joints bonded using different NCFs were analyzed in terms of Sn-Ag bump deformation and Sn-Ag wetting on Ni pad. Joint of Imidazole NCF showed no deformation of Sn-Ag bump shape and no solder wetting on Ni pad. This is due to high viscosity and fast curing nature of imidazole NCF. DICY NCF showed huge solder deformation and bad solder wettability on Ni pad. Lower viscosity as well as slower curing speed of DICY NCF compared to imidazole NCF resulted high deformation of solder; however, since there is no fluxing ingredient to remove solder oxide in DICY NCF, bad solder wettability on Ni pad is achieved. In case of anhydride NCF, huge solder deformation and good solder wettability on Ni pad are achieved due to slowest curing speed, lowest viscosity, and fluxing nature of anhydride NCF.

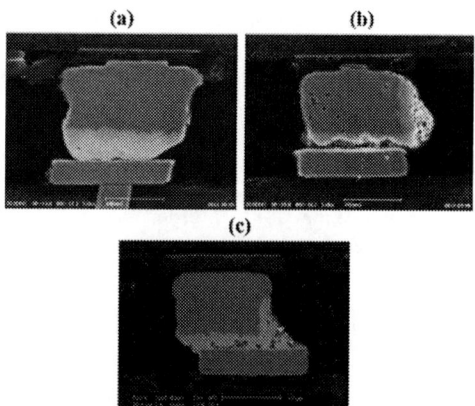

Figure 5. Cross-sectional images of joints bonded using (a) Imidazole (b) DICY (c) Anhydride NCFs

Double layer NCF

Large deformation of solder and subsequent solder wetting on Cu pillar increase reaction between Sn and Cu, and hence faster growth of IMCs are achieved. In order to solve this problem, NCFs with properties that minimize solder deformation and guarantee good wettability on Ni pad are required. Double layer NCF consisting of two layers with different properties is a solution for this problem. Figure 6 shows schematic design and structural specification of double layer NCF. Top layer covers up to the height of Cu-pillar and bottom layer covers Sn-Ag bump. The top layer has the purpose of not wetting solder on Cu-pillar by having fast curing and high viscosity. The bottom layer has the purpose of achieving good wettability of solder on the pad by having slow curing speed and low viscosity. With these two layer combined, no solder wetting on Cu-pillar and good solder wetting on the pad can be achieved resulting ideal joint morphology for fine pitch interconnection. As materials for top layer, highly viscous and fast curing imidazole NCF was used and anhydride NCF with low viscosity and slow curing speed was used as bottom layer.

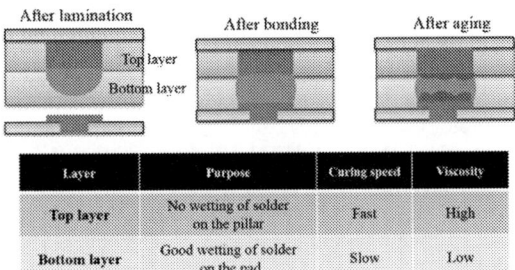

Layer	Purpose	Curing speed	Viscosity
Top layer	No wetting of solder on the pillar	Fast	High
Bottom layer	Good wetting of solder on the pad	Slow	Low

Figure 6. Design and structural specification of double layer NCF

Thickness optimization

In order to achieve optimal joint shape, thickness of each layer of double layer NCF was optimized by controlling the thickness of each layers. With the total thickness of 24 µm, each layer was changed from 6 µm to 18 µm and four double layer samples were fabricated with different thickness of each layer. Table 2 shows thickness of top and bottom layer NCF of double layer NCFs. With the fabricated double layer NCF, the bonding was performed and cross-sectional images of the joints are shown in figure 7. Sample #1 consisting of 8 µm of top layer and 16 µm of bottom layer showed solder wetting on Cu pillar, indicating that top layer was too small to perform its role to prevent wetting on Cu-pillar. However, in case of sample #2 with 12 µm of thickness for both layers, ideal joint morphology with no solder wetting on Cu-pillar and good wetting on the pad was observed. Both of the layers successfully performed their roles in this condition. Sample #3 and #4 with thicker top layer than that of sample #2 showed no solder wetting on Cu-pillar, however bad solder wetting on pad was observed. This indicates that too much thick top layer even affects solder deformation near the pad, and cause bad wettability of solder on pad. Therefore, with thickness of 12 µm for both layers, the ideal joint shape was achieved.

Table 2. Specification of double layer NCFs with different thickness

Layer/Leg#	#1	#2	#3	#4
Imidazole NCF (µm)	8	12	16	18
Anhydride NCF (µm)	16	12	8	6

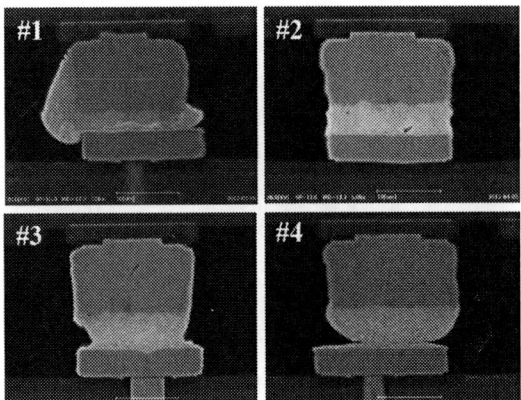

Figure 7. Cross-sectional images of joints bonded using double layer NCFs with different thickness

Analysis on IMC growth

Thermal aging at 150°C was performed with double layer NCF sample #2 and IMC growth was analyzed and compared with that of single layer anhydride NCF. Figure 8 shows cross-sectional image of thermally aged joints bonded using single layer and double layer NCF and ratio of grown Cu_6Sn_5 IMCs is organized on figure 9. At as bonded state, generally a large portion of unreacted Sn remains within the Sn-Ag bump. Since the joints bonded using single layer NCF has larger reacting area for Sn and Cu due to solder wetting on Cu-pillar, the ratio of IMCs was higher in single layer NCF than double layer NCF. At 50 hours of aging, more Cu_6Sn_5 IMCs was grown due to thermally induced diffusion between Sn and Cu. The average area ratio of IMCs was as high as 93.7% for single layer NCF whereas double layer NCF showed relatively lower value of 72.3%. At 250 hours of aging all the pure Sn has reacted with Cu and turned into Cu_6Sn_5 and secondary IMCs of Cu_3Sn were formed near Cu-pillar. However, since single layer NCF has larger reacting area for Cu and Cu_6Sn_5, higher quantity of Cu_3Sn IMCs were formed. Compared to single layer NCF, double layer NCF showed slower growth rate for Cu_6Sn_5 and Cu_3Sn IMCs by reducing solder wetting on Cu pillar.

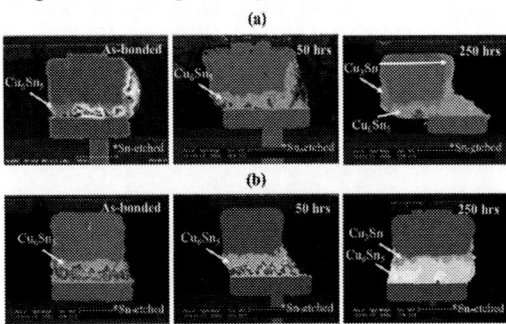

Figure 8. Cross-sectional images of joint using (a) single layer anhydride NCF and (b) double layer NCF after thermal aging

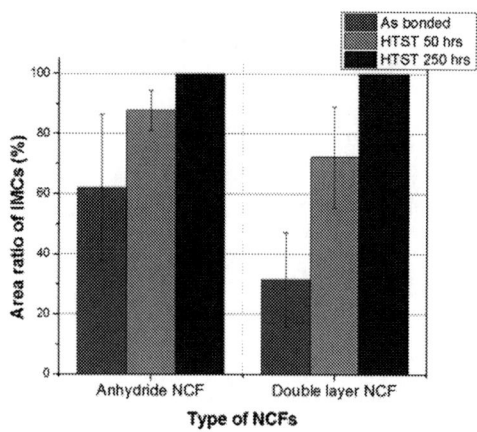

Figure 9. Area ratio of Cu_6Sn_5 IMCs

Conclusion

Cu pillar/Sn-Ag bump to Ni pad bonding was performed using NCF and thermo-compression method. The joint morphology was dependent on the viscosity and curing speed of NCFs. Imidazole NCF with high viscosity and fast curing speed showed no solder wetting on both Cu-pillar and Ni pad. However, anhydride NCF with low viscosity and slow curing speed showed both solder wetting on Cu-pillar and Ni pad. Double layer NCF consisting of imidazole NCF as top layer and anhydride NCF as bottom layer was designed and optimized to eliminate solder wetting on Cu-pillar. Joints bonded using double layer NCF showed ideal joint shape with no solder wetting on Cu pillar and good wetting on pad. IMC growth rate of Cu_6Sn_5 and Cu_3Sn was slower for double layer NCF compared to single layer NCF due to reduced reacting area for Cu and Sn.

References

[1] M.J. Wang, C.Y. Hung, C.L. Kao, P.N. Lee, C.H. Chen, C.P. Hung, H.M. Tong, TSV technology for 2.5D IC solution, Proceedings - Electronic Components and Technology Conference, (2012) 284-288.

[2] H.Y. Son, G.J. Jung, B.J. Park, K.W. Paik, A study on the thermal reliability of Cu/SnAg double-bump flip-chip assemblies on organic substrates, Journal of Electronic Materials, 37 (2008) 1832-1842.

[3] R. Dunne, Y. Takahashi, K. Mawatari, M. Matsuura, T. Bonifield, P. Steinmann, D. Stepniak, Development of a stacked WCSP package platform using TSV (through silicon via) technology, Proceedings - Electronic Components and Technology Conference, (2012) 1062-1067.

[4] J. Hwang, J. Kim, W. Kwon, U. Kang, T. Cho, S. Kang, Fine pitch chip interconnection technology for 3D integration, Proceedings - Electronic Components and Technology Conference, (2010) 1399-1403.

[5] K. Takahashi, M. Umemoto, N. Tanaka, K. Tanida, Y. Nemoto, Y. Tomita, M. Tago, M. Bonkohara, Ultra-high-density interconnection technology of three-dimensional packaging, Microelectronics Reliability, 43 (2003) 1267-1279.

[6] B. Ebersberger, C. Lee, Cu pillar bumps as a lead-free drop-in replacement for solder-bumped, flip-chip interconnects, Proceedings - Electronic Components and Technology Conference, (2008) 59-66.

[7] A. Yu, J.H. Lau, S.W. Ho, A. Kumar, W.Y. Hnin, W.S. Lee, M.C. Jong, V.N. Sekhar, V. Kripesh, D. Pinjala, S. Chen, C.F. Chan, C.C. Chao, C.H. Chiu, C.M. Huang, C. Chen, Fabrication of high aspect ratio TSV and assembly with fine-pitch low-cost solder microbump for Si interposer technology with high-density interconnects, IEEE Transactions on Components, Packaging and Manufacturing Technology, 1 (2011) 1336-1344.

[8] H. Gan, S.L. Wright, R. Polastre, L.P. Buchwalter, R. Horton, P.S. Andry, C. Patel, C. Tsang, J. Knickerbocker, E. Sprogis, A. Pavlova, S.K. Kang, K.W. Lee, Pb-free micro-joints (50 μm pitch) for the next generation micro-systems: The fabrication, assembly and characterization, Proceedings - Electronic Components and Technology Conference, 2006 (2006) 1210-1215.

[9] C. Feger, N. LaBianca, M. Gaynes, S. Steen, Z. Liu, R. Peddi, M. Francis, The over-bump applied resin wafer-level underfill process: Process, material and reliability, Proceedings - Electronic Components and Technology Conference, (2009) 1502-1505.

[10] K. Honda, T. Enomoto, A. Nagai, N. Takano, NCF for wafer lamination process in higher density electronic packages, Proceedings - Electronic Components and Technology Conference, (2010) 1853-1860.

CO$_2$-Laser Drilling of TGVs for Glass Interposer Applications

Lars Brusberg[1], Marco Queisser[2], Marcel Neitz[2], Henning Schröder[1], Klaus-Dieter Lang[2]

[1] Fraunhofer Institute for Reliability and Microintegration, Gustav-Meyer-Allee 25, 13355 Berlin, Germany,
[2] Technical University of Berlin, Gustav-Meyer-Allee 25, 13355 Berlin, Germany,
lars.brusberg@izm.fraunhofer.de

Abstract

Glass as a substrate material for interposer application has many benefits compared to conventional packaging materials like silicon, ceramic or polymer based laminates because of its excellent dielectric and transparent properties. Furthermore, the integration potential of glass is superior because of the dimensional stability under thermal load and the coefficient of thermal expansion (CTE) matching to that of silicon ICs. A small pitch size of conductor traces, small scale through-vias and high alignment accuracy are the key requirements that will be achieved from glass based packaging. Also the transparency of glass has benefits for photonic packaging. Glass substrates are available in wafer and large scale panel formats.

Very fast CO$_2$-laser drilling of holes and thermal post-treatments for reducing mechanical stress are very promising for fast processing and high reliability. Holes with a diameter smaller 100 µm in different glasses with thicknesses between 145 and 500 µm have been achieved by CO$_2$-laser drilling. The holes have been metallized by sputtering a seed layer and galvanic copper platting. The CO$_2$-laser drilling in combination with copper metallization has high potential for through glass via forming in glass substrates for interposer applications.

Introduction

Currently, decreasing pitch size of conductor traces, small scale through-vias and high alignment accuracy are the key requirements for high-density integrated packages. The integration potential of organic laminates (e.g. FR4) is limited because of dimensional instability under thermal load. The alignment of interconnects between different layers is challenging where oversized patterns have to compensate the process tolerances. As a result the pitch size for devices assembled on FR4 is limited. Alternatively, the 3D-System-in-Package (SiP) based on silicon interposer platform is a very active area of ongoing research. Due to the availability of wafer level processing, silicon substrates can be processed with the same pitch size and accuracy as the highly integrated circuit (IC) components that will be assembled. Furthermore, the CTE matches perfectly if the substrate and IC's are made of the same material. The resulting stress on the solder join is reduced. However, a drawback is the semiconductor property of silicon which requires conductive through-vias and wiring to be isolated from the semiconductor bulk material. For optical applications the silicon substrates are only transparent for wavelengths larger 1107 nm, limiting the applicability to the infrared wavelength range.

A glass based packaging technology overcomes the limitations mentioned above. The benefits of glass are the excellent optical properties and its high potential in the field of integrated optics. Beyond that, glass has dimensional stability under thermal load, alignment benefits as a result of transparency, compatibility to wafer level processing, coefficient of thermal expansion matching to silicon and good dielectric properties. Thin glass in wafer form is available at thickness of 30 µm or even below. Wafer level processes can be adapted from CMOS processing while new processes for through glass vias (TGV) have to be explored in more detail.

The high potential of the glass based packaging was already demonstrated for high-speed data transmission applications. A thin film glass interposer made of commercial TGV wafers with tungsten plugs was already fabricated as shown in Figure 1 [1].

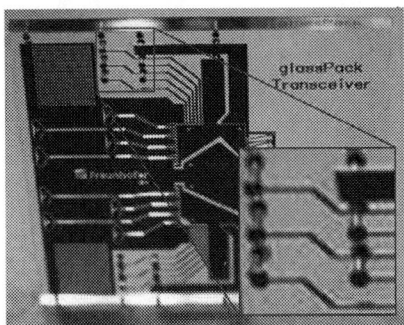

Figure 1. Glass interposer with TGV for electro-optical transceiver application [1].

A different approach by opening the glass by a CO$_2$-laser and metallization afterwards is the focus of our ongoing research. The CO$_2$-laser is well known in the industry as a laser for metal machining. Because of its wavelength and relatively long pulses the drilling and processing in general is based on thermal ablation. Consequently, when drilling glass, the laser has a thermal impact on the glass causing mechanical stress in the drilling zone which can result in crack formation during the cooling or any time later. As a result the reliability of CO$_2$-laser drilled glass substrates can be reduced. In this paper we have explored very fast CO$_2$-laser drilling of holes with diameters smaller than 100 µm in combination with thermal post-treatments for reducing mechanical stress and increasing reliability. The drilling has been proven in different glass types and thicknesses between 145 and 500 µm. Also the metallization of the holes by sputtering a seed layer and galvanic copper metallization is presented in detail. The paper will show the high potential of CO$_2$-laser drilling and copper metallization for through glass via forming in thin glass substrates for interposer applications.

Table 1. A selection of glass brands suitable for glass based packaging offered by SCHOTT, AGC and CORNING

Company	SCHOTT					AGC	CORNING				
Brand	**Lithosil**	**B33**	**B 270**	**D 263T eco**	**AF 32eco**	**EN-A1**	**Pyrex**	**0211**	**Eagle XG slim**	**Gorilla**	**Jade**
Type	fused-silica	boro-silicate	crown-glass	boro-silicate	boro-silicate	boro-silicate	boro-silicate	boro-silicate	boro-silicate	boro-silicate	boro-silicate
Process	micro-float		up-draw	down-draw		n.a.	micro-float	down-draw	down-draw	down-draw	down-draw
Min. Process Thickness	700 µm	700 µm	800 µm	30 µm	100 µm	n.a.	700 µm	50 µm	<400 µm	500 µm	n.a.
Format	Panel and wafer						Panel and wafer				
Alkaline Content	alkali-free	4 wt%	17 wt%	13 wt%	alkali-free	alkali-free	4 wt%	13 wt%	alkali-free	n.a.	alkali-free
CTE	0.5 ppm	3.3 ppm	9.4 ppm	7.2 ppm	3.2 ppm	3.3 ppm	3.3 ppm	7.4 ppm	3.2 ppm	9.1 ppm	3.8 ppm
tanδ (1MHz)	14·10-4	37·10-4	n.a.	61·10-4	28·10-4	n.a.	50·10-4	46·10-4	30·10-4	100·10-4	20·10-4
ε_r (1 MHz)	3.8	4.6	7.0	6.7	5.1	5.8	4.1	6.7	5.3	7.3	6.0

Through glass via drilling technologies

A selection of commercially glass brands suitable for glass based packaging offered by SCHOTT, ASG and CORNING are summarized in Table 1. B33, AF32eco, EN-A1, Pyrex, Eagle XG slim, Jade has a CTE matches to silicon ICs which is beneficial to high reliability of solder joints. The integration of optical functions inside the substrate using the ion exchange technology requires an adequate alkaline content provided by glasses like D263Teco, B270, Gorilla, B33, Pyrex, 0211. Furthermore glass has excellent dielectric properties as well as ceramic materials that are often used for high-frequency applications (Al_2O_3: $\varepsilon_r = 9.6$, tanδ = 5·10^{-4}). But ceramic suffers from high surface roughness which increases high-frequency losses. In contrast, glass has excellent surface quality with R_a below 1 nm which makes it attractive for high-frequency applications. In summary, the right glass has to be selected dependent on the process and application. The introduced glass brands are just a selection of available glasses and further glass compositions will be developed depending on the market needs.

Key requirements for glass interposers are drilling and metallization of TGVs for electrical interconnection through the substrate. Besides commercial available TGV wafers [2], TGVs can be produced using drilling technologies like ultrasonic drilling, sandblasting, wet etching, dry etching or laser drilling in combination with filling the holes with conductive materials. Limitations are long structuring times, disadvantageous via shapes or aspect ratios, e.g. holes with large diameters (>100 µm) in glasses having a thickness of 500 µm or below. Laser drilling is very flexible regarding layout and variation of pitch and diameter on the same substrate.

Talking about laser drilling a few options come to mind, namely cold ablation, hot ablation and indirect ablation through thermic or pressure shocks. Hot ablation is mostly encouraged by continuous wave or quasi continues wave lasers. The glass melts through the laser energy and evaporates. Recast layers and a large heat-affected zone around the holes are formed during this process. Cold ablation needs short (ns) or ultra-short (ps, fs) laser beam pulse lengths and no match of laser wavelength and glass absorption spectrum. In summary the ablation result is very dependent

upon laser wavelength, laser energy, laser beam pulse length, threshold fluence and glass type [3].

Thus, feasibility of via drilling and quality are very dependent on the glass and laser system characteristics. Depended on the process and hole specifications the best laser process has to be selected. If the glass is not transparent for the incident laser beam, the beam generates energy direct on the glass surface (surface absorption). If the glass is transparent for the laser beam, the laser has to be focused on top or bottom side of the substrate. The laser beam induces heat locally changing the material by melting and evaporation. Due to mechanical stress in the glass substrate material cracks out. Holes can be drilled by moving the laser beam focus from bottom to top side. By overcoming a certain energy threshold level of the material a direct phase change from solid to gaseous state occurs, in other words a direct material sublimation. The threshold energy is defined by the fluence (incident energy per irradiated area) of the laser beam. An overview of the different lasers related to the wavelength characteristic, resonator material, and pulse length of a selected glass (B33) is depicted in Figure 2.

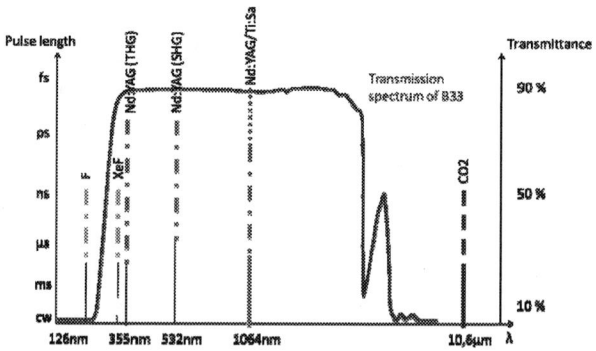

Figure 2. Transmission spectrum of B33 and selection of drilling lasers [4].

Each laser has its own very specific characteristics. For instance Excimer lasers are always pulsed and provide wavelengths from VUV to UV, respectively λ=157 nm Fluorine (F2) laser to λ=351 nm Xenon-Difluoride (XeF) laser. UV Excimer laser ablation is based on the production of plasma, and evaporation as a result of the laser radiation. The needed threshold fluence depends on the glass- and wavelength-specific absorption characteristic [5]. For wavelengths above 300 nm, non-linear effects like multi-photon absorption can occur, depending on the glass characteristics. Due to the high variety of wavelengths, energy and beam sizes, multiple setups are realized in lab-based environments or commercially available. Glass via drilling using an Excimer laser can be done by a sequential process or by using a mask for parallel drilling [6, 7, 8, 9, 10, 11, 12]. Tapered vias with diameters smaller 100 µm have been realized as shown in Figure 3. Drilling speed per via is reported with 5 s in 100 µm thin glass [6].

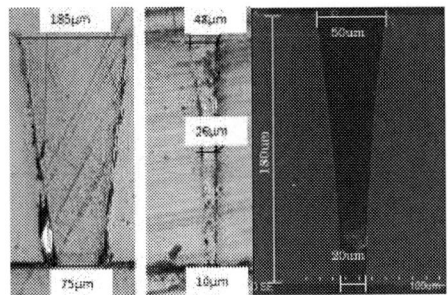

Figure 3. Vias in 500 µm D263Teco (left, middle) [11] and 180µm EN-A1glass (right) [13] drilled with 192 nm Excimer laser.

Another group of lasers as shown in Figure 2 with working wavelengths in the transparent spectral range of the glass are solid state lasers which can be driven in continuous wave (cw) or pulse mode, by diode- or flash-lamp-pumping. As result very short pulse lengths (e.g. pico-seconds, femto-seconds) are achievable. Because of the transparency of the glass, both drilling methods, bottom-up and top-down, are possible with solid state lasers. The drilling speed is largely defined through the laser fluence, pulse duration and repetition rate. The threshold for each individual glass has to be exceeded in order to achieve the cold ablation. The improvement of short-pulse lasers make them more and more interesting for via drilling, because of excellent drilling results and improved drilling speed. Drilling vias in 500 µm glass in times below 1s/via is reported [15]. These lasers are producing very narrow vias and metallization will be challenging. However the via diameter can be adjusted as shown in Figure 4 [11] for better aspect ratio. The drilling of holes with a cylindrical shape can be done by a laser system with ns-pulse duration and trepanning optics with an ablation starting from the bottom side of the glass as shown in Figure 4 [14]. In that case the shape can be accurately controlled by the optics but hole diameter is limited (e.g. 200 µm).

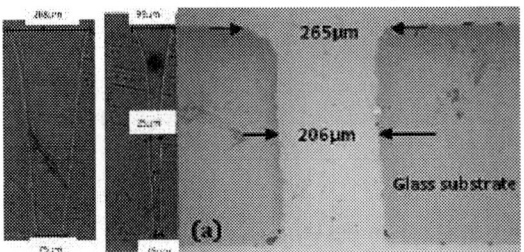

Figure 4. Vias in 500 µm D263Teco glass drilled with 775 nm-fs-pulse laser (left) [11]. Also in 500 µm D263Teco glass but drilled with 532nm ns-pulse-laser (right) [14].

Compared to other laser processes, Carbon dioxide (CO_2) lasers, traditionally at λ=10600 nm (Figure 2), which are available both cw and pulsed need only small initial equipment investments. Because of its wavelength and relatively long pulses, the drilling is based on thermal ablation. Consequently the laser has a thermal impact on the glass that generates mechanical stress in the drilling zone which can result in crack formation during the cooling or any time later.

CO_2-laser drilling of through glass vias

The aim of the investigations was to produce through glass vias in 500, 300 and 145 µm SCHOTT D263Teco thin glass and 300 µm SCHOTT B33 with the CO_2-laser. A major challenge here lies in coping and minimizing thermally induced stresses due to the hot ablation by the IR radiation used. These stresses can reach critical values and lead to cracking and fracturing. Decisive here are the extent and the stress state value of the heat-affected zone around the hole. Furthermore, the objective was to minimize geometric parameters such as hole diameter, conicity and pitch. All this was done by the appropriate choice of relevant process parameters, as exemplified focus diameter and position, as well as laser pulse frequency, duration and number.

The constructed and used laboratory setup is shown in Figure 5. It essentially consists of a CO_2-laser with beam guiding and forming elements, a motorized 2-axis translation stage, a corresponding computer control and cameras for visualization.

As a result of the experiments and investigations for CO_2-laser drilling no general optimal process parameters could be determined because the change in individual parameters has partly opposing effects. For example the optimization leads towards low conicity and smooth, cylindrical walls but also to an increased heat-affected zone. A broadened heat-affected zone in turn increases the minimum reliable pitch. The same applies to the optimization

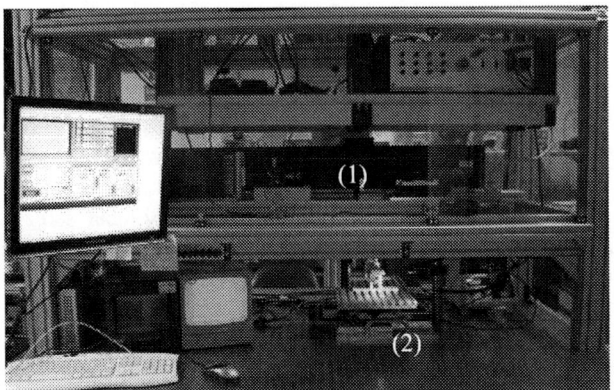

Figure 5. Laboratory setup with CO_2-laser (1) and 2-axis translation stage (2).

978-1-4799-2408-0/14 $31.00 © 2014 IEEE   1761

in the direction of process time. A wide heat-affected zone has a positive impact, because the peak of the mechanical stress and therefore the risk of cracking are reduced by the wider, flatter stress gradient. In consequence the ideal parameter combinations must or can be determined depending on the requirements and layout.

A thermal post-treatment of the drilled substrate removes the stress completely. The entire substrate was placed into an oven for thermal relaxation of the induced stress. Experiments with different treatments using peak temperatures between 300°C and 557°C (annealing point of SCHOTT D263Teco) and peak times between 30 minutes and 24 hours were performed. For instance, curing at a temperature of 529°C for 30 minutes removes the stress completely.

Subsequently, the results for CO_2-laser drilling of through glass vias in 500, 300 and 145 µm SCHOTT D263Teco respectively SCHOTT B33 thin glass are presented in detail.

CO_2-laser drilling of 500 µm thin glass

The results for an optimization run leading to holes with small diameter, high straightness (low conicity) and a narrow heat-affected zone in 500 µm thin glass is shown in Figure 6. The laser process time of such a single TGV is about 0.25 seconds. In addition, holes with higher conicity and/ or larger, even a little smaller, hole diameters are feasible.

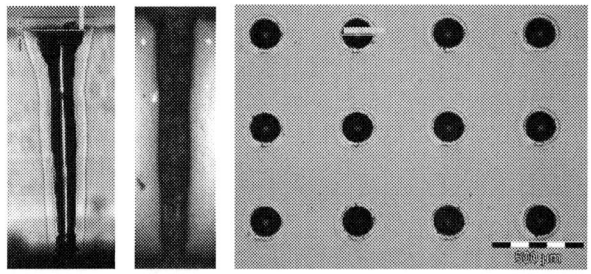

Figure 6. Through glass vias in 500 µm SCHOTT D263Teco thin glass. Micrograph of the hole profile, $d_{in} \approx 120$ µm, $d_{out} \approx 50$ µm (left). Cross section of the hole profile, $d_{in} \approx 80$ µm, $d_{out} \approx 50$ µm (middle). Top view of a TGV array, pitch = 500 µm (right).

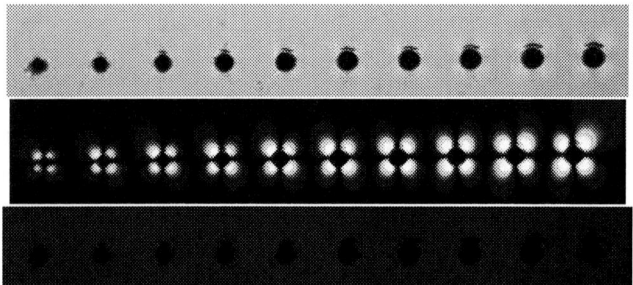

Figure 7. Series of experiments for TGV structuring with visualization of thermally induced stresses. Top view micrograph of a row of holes (top). Photoelastic micrograph, stresses as light areas around the hole visible (middle). Photoelastic micrograph after annealing, no stresses are present (bottom).

Thermally induced stresses are created around the TGV in the glass substrate because of the hot laser ablation. However,

this can be completely eliminated by a subsequent annealing step, and thereby be excluded later defects. In Figure 7, these stresses before and after annealing are photoelastically visualized before and after annealing.

CO_2-laser drilling of 300 µm thin glass

Due to the lower glass thickness the hole profile and conicity could be slightly improved in 300µm SCHOTT D263Teco, see Figure 8.

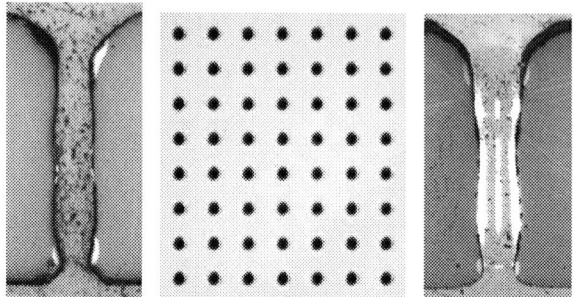

Figure 8. Through glass vias in 300 µm SCHOTT D263Teco and B33 thin glass. Cross section of a hole profile, $d_{mean} \approx 45$ µm, (left) and top view of an array, pitch = 500 µm, in D263Teco (middle). Cross section of a hole profile, $d_{mean} \approx 55$ µm, in Borofloat 33 (right).

In addition, experiments with 300 µm SCHOTT B33 were carried out. With the same process parameters comparable hole geometries resulted with a process time of 0.13 s. However, in this case the thermally induced stresses in the B33 substrate were much less pronounced and extended as shown in Figure 9. This can be attributed to the lower CTE of B33. With this type of glass, or generally glasses with a small CTE, a defect-free, reliable machining with a smaller pitch is possible. This also applies, for example, to fused silica glass.

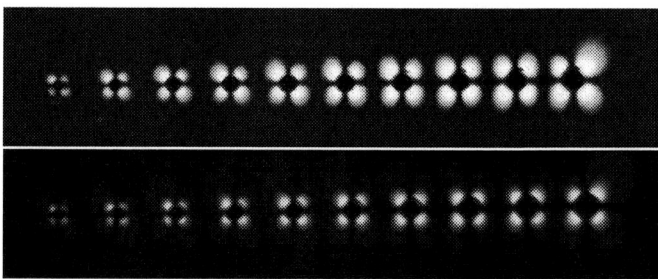

Figure 9. Photoelastic micrographs to compare the thermally induced stresses between SCHOTT D263Teco (top) and SCHOTT B33 (bottom) after identical processing.

All laser drilling experiments were consistently performed with a pitch of 500 µm. With the optimized process parameters no more pitch-related cracks occurred, only selectively to a few holes, possibly reasoned by local defects. The problem of thermally induced stresses as described above applies to all generated holes independent of the glass thickness. To allow a better statement on the reliability of the holes, numerous test layouts, each with 3136 holes having a pitch of 500 µm, have been processed and examined as shown in Figure 10. For this, the cracked holes were counted before and after a thermal annealing step. The overall process time for one wafer including wafer movement and drilling was 42 min and 42 s while the sole drilling time of all holes is only 5 min and 14 s. Across all experimental wafers the reliability and accuracy was constant at about 99.5%, in the best case at 99.93%. This is basically a very high, satisfactory yield. However, no 100% accuracy can be guaranteed so far.

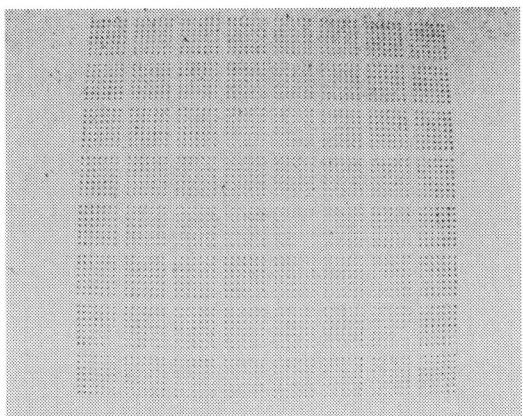

Figure 10. Matrix of 3136 holes in 300μm D263Teco thin glass.

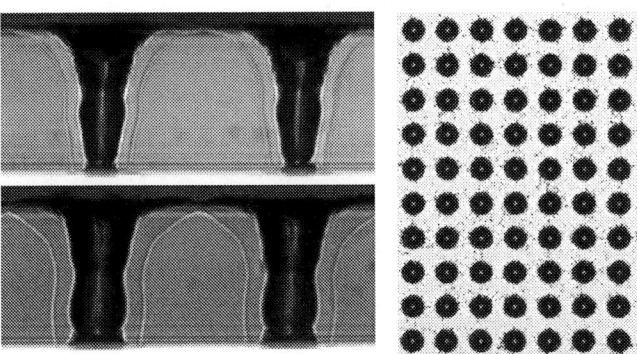

Figure 11. Through glass vias in 145 μm SCHOTT D263Teco thin glass. Micrograph of the hole profile, $d_{in} \approx 55 \, \mu m$, $d_{out} \approx 25 \, \mu m$ (top left) and $d_{in} \approx 60 \, \mu m$, $d_{out} \approx 40 \, \mu m$ (bottom left). Top view of a TGV array with pitch = 150 μm (right).

CO$_2$-laser drilling of 145 μm thin glass

Finally, the experiments were continued on 145μm thin glass. The decreasing glass thickness is here, as expected, accompanied by a reduction of the hole geometry. As well the laser process time of such a TGV is lowered to 0.11 seconds. The minimum pitch (without/ before annealing) could be reduced in this case to 150 μm. The result is shown in the Figure 11.

Metallization

The metallization of the through-via wall is done after CO$_2$-laser drilling and then interconnected on the surface by a semi-additive process. Compared to conventional metallization processes on substrate surfaces, the metallization of the holes is more demanding due to the high aspect ratio which has a negative impact on the creation of a homogeneous surface. High speed signal lines with frequency above 10 GHz require very lower transmission loss also for TGVs. An earlier work showed that a small hole diameter of

100 μm or less and slightly tapered or better cylindrical holes characterizes low transmission loss [16]. The drilling can be optimized in that direction. Of course a challenge is the metallization for high aspect ratio. Thinner substrates like the 145 μm thin glass reduce the aspect ratio. For further processing it has to be studied the complete filling in combination with chemical mechanical polishing (CMP) processing.

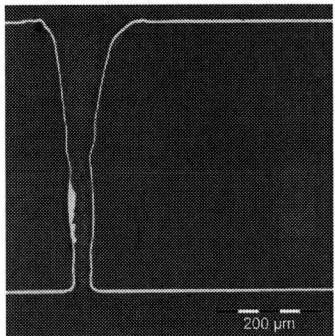

Figure 12. Cross section of a metallized TGV.

Conclusions

The feasibility of CO$_2$-laser drilling of glass was successfully demonstrated in a thickness range of 145 to 500 μm. The challenges of induced thermo-mechanical stress could be overcome by applying a thermal post-treatment. So far an absolute crack-free drilling of the whole TGV array could not be achieved. In the best case a wafer of totally 3136 drilled holes has only two damaged holes. Also first plating results show the high potential of CO$_2$-laser drilling for TGV interposer application.

Acknowledgments

Some of the research leading to these results has received funding from German Ministry of Education and Research (BMBF). The authors would like to thank all the colleagues who have supported this work. Special thanks go to Michael Töpper, Markus Wöhrmann, Robert Rund and Norbert Arndt-Staufenbiel.

References

1. Brusberg, L., Schröder, H., Erxleben, R., Ndip, I., Töpper, M., Nissen, N., Reichl, H., "Glass Carrier Based Packaging Approach Demonstrated on a Parallel Optoelectronic Transceiver Module for PCB Assembling", Proc. 60th Electronic Components and Technology Conference, Las Vegas, NY, USA, 2010.
2. SCHOTT Online http://www.schott.com/epackaging/english/auto/others/hermes.html
3. Hans-Peter Wunde; Time-Bandwith Products, "Präzise Glasbearbeitung mit dem Pikosekundenlaser," in *Laserbearbeitung von Glaswerkstoffen*, Hannover, 2011.
4. PGO Online http://www.pgoonline.com/de/katalog/kurven/boro_kurve.html
5. B. Braren and R. Srinivasan, "Controlled etching of silicate glasses by pulsed ultraviolet laser radiation,"

Journal of Vacuum Science & Technology B: Microelectronics and Nanometer Structures, Vol. 6, No. 2, 1988.

6. D. Bhatt, D. A. H. K. Williams and P. P. Conway, "Process Optimisation and Characterization of Excimer Laser Drilling of Microvias in Glass", *EPTC 2007.*

7. P. M. Mendes, A. Polyakov, M. Bartek, J. N. Burghartz and J. H. Correia, "Integrated Chip-Size Antenna for Wireless Microsystems: Fabrication and Design Considerations," *Sensor and Actors,* 2006.

8. V. Sukumaran and et.al., "Through-Package-Via Formation and Metallization of Glass Interposers," *ECTC,* 2010.

9. H. Exner, B. Keiper, U. Löschner and T. Kuntze, "Drilling of Glass by Excimer Laser Mask Projection Technique," *Journal of Laser Applications,* 2000.

10. F. Hähnel, S. Weißmantel, G.Reiße, "Mikrostrukturierung von Quarzglas und Kalziumflourid mittels Fluorlaser," Laserinstitut der Hochschule Mittweida, Mittweida, 2010.

11. L. Brusberg, H. Schröder, M. Töpper and H. Reichl, "Photonic System-in-Package Technologies Using Thin Glass Substrates", *EPTC 2009.*

12. C. Buerhop, R. Weissmann, "Ablation of silicate glasses by laser irradiation: modelliong and experimental results," *Applied Surface Sciences,* Vol. 54, p. 6, 1992.

13. T. Shintaro, T. Kentaro, K. Kenji , "TGV Technology for Glass Interposer," *Proceedings ESTC,* 2012.

14. A. Lemke; Laser und Medizin-Technologie Berlin, "Mikrobohren und –schneiden mit einer rotierenden Laserbearbeitungsoptik," in *15. Seminar für Laser in der Elektronikproduktion & Feinwerktechnik (LEF)*, Fürth, 2012.

15. B. Keiper, R. Ebert and H. Exner, "Mikrobohren von PYREX mittels fs-Laser," Laserinstitut der Hochschule Mittweida, Mittweida, 2012.

16. Töpper, M.; Ndip, I.; Erxleben, R.; Brusberg, L.; Nissen, N.; Schroder, H.; Yamamoto, H.; Todt, G.; Reichl, H., "3-D Thin film interposer based on TGV (Through Glass Vias): An alternative to Si-interposer," *Proceedings of 60th Electronic Components and Technology Conference (ECTC), 2010*, pp. 66-73, 1-4 June 2010.

Effects of Pad Surface Finish on Interfacial Reliabilities of Cu-Pillar/Sn-Ag Bumps of 2.5D TSV-Enterposer on PCB Applications

Youngsoon Kim[1], Ji-won Shin[2], Young won Choi[2], Kyung-Wook Paik[2]

1) ACI development team, Samsung Electro-Mechanics Co.,LTD
2) Dept. of Materials Science and Engineering
Korea Advanced Institute of Science and Technology (KAIST)
phone: +82-42-350-3375, fax:+82-42-869-8624
e-mail: ysoon@kaist.ac.kr

Abstract

As technology advances, electronic device requires package with higher I/O, faster speed, and higher density. 3D-packaging is one of the solutions for such these needs. The key technology which make 3D-packaging more powerful is the through silicon via (TSV) technology. It is the method of forming vertical electrical path through Si chip, and it is one of the most promising packaging structures for next generation. 2.5D TSV interposer on BGA substrates is one of the most commonly used structure using TSV technology, and advanced packaging companies such as Samsung and Qualcomm are currently working on 2.5D TSV interposer on BGA package structure. However, conventional flip chip process using solder balls is not suitable for 2.5D TSV-interposer to BGA substrate interconnection, since the aspect ratio of solder ball cannot withstand fine-pitch of TSV interconnection. In order to solve this problem, interconnection structure should be replaced to Cu pillar/Sn-Ag bumps on Cu pad of BGA substrates.

However, two major problems occurs using Cu-pillar/Sn-Ag bump structure in 2.5D TSV interposer on BGA substrates: undefill voids problem due to fine pitch and narrow gap and bad interfacial reliability due to decreased solder volume. To solve underfill voids problem, non-conductive film(NCF) bonding process, a pre-applied type underfill film is used. This allows fluxing, bonding, and underfilling at single bonding process eliminating voids issue in the underfill. The second major problem, bad interfacial reliabilities of Cu-pillar/Sn-Ag on Cu pad joint, is often induced by Cu pad oxidation. And this problem can be solved using pad surface finishes such as OSP, ENIG, and ENEPIG. By use of the NCF bonding process and the surface finishes on Cu pad, highly reliable joints as well as voidless underfill can be obtained. In this research, effects of Cu pad surface finishes on interfacial reliabilities of Cu-pillar/Sn-Ag bump on Cu pad of BGA substrates using NCF bonding process are investigated using three types of surface finish (OSP, ENIG, ENEPIG).

All the bonding in this experiments was performed using NCF bonding process with lamination temperature of 80℃ and bonding temperature of 250 °C. As test vehicle, top chip with Cu-pillar/Sn-Ag bump and BGA board with three different types of surface finish (OSP, ENIG, ENEPIG) were used. At OSP surface finish, two commonly known Cu_6Sn_5 and Cu_3Sn IMCs were formed at Sn-Ag bump/Cu pad interface. In the case of ENIG and ENEPIG surface finish, chunky type IMC phase of $(Cu, Ni)_6Sn_5$ and needle type IMC phase of $(Ni,Cu)_3Sn_4$ were formed respectively. Bonded test vehicles were tested under thermal cycle test up to 1000 cycles, and the tendency in contact resistance reliability was observed: OSP>ENEPIG>ENIG. Formation of different IMC phases depending on pad finishes are presumably the main reasons for the failures. Additionally, further analysis on the reliabilities of joints will be achieved by high temperature storage test (HTST) at 150℃ for 1000 hours.

1. Introduction

Requirement of high technology electronic device such as high I/O, speed, and density are increasing rapidly. To satisfy these needs, packaging technology became more important. TSV (Through Silicon Via technology) 3D packaging is one of the solutions to those requests. To investigate the reliability of Cu-pillar/Sn-Ag with BGA substrate which is SAC solder bump, especially, effects of PCB bump pad surface finish was performed[1]

When performing bonding, top Si chip and PCB solder bump with three different types of surface finish was bonded through TC (Thermal compression) bonding using NCF (Non Conductive Film). Bump joint shapes and IMCs (Inter metallic compound) compositions were different when comparing to conventional reflow flip chip bonding.

Three different surface finish effects to interfacial reliability were evaluated through TCT (Thermal Cycle Test, -55 ~ 125C up to 2000cycle) which showed district results of PCB Cu bump pad surface finish. Even the best case (OSP surface finish) did not endure TCT 2000cycle but was stable up to 1500cycle. ENEPIG surface finish was also stable up to TCT 1000cycle but after TCT 1500cycle, the contact resistance increased. On the other hand, ENIG circuit opened after 750cycles of TCT. Normally, mobile electro devices require TCT 1000 cycle, Leg 1 and Leg 3 can be used for PCB surface finish for 2.5D TSV interposer on PCB applications.

Surface finish on PCB solder bump affects the interconnection reliability due to different IMC at bump joint. OSP surface finish showed 2 types of IMCs, Cu_6Sn_5 and Cu_3Sn, In case of Ni layer plated surface finish such as ENIG and ENEPIG, Leg 2 & 3 showed chunky type IMC phase of $(Cu, Ni)_6Sn_5$ and needle type IMC phase of $(Ni,Cu)_3Sn_4$.

In order to understand the IMC at the joint and evaluate high temperature stability, HTST (High Temperature Storage Test, 150C 1000hr) was performed. Circuit open modes were different depending on the surface finish of PCB bump pad.

2. Experimental

2.1. Material preparation & equipment

The top Si chip (1 X 1cm) bump contains Cu-pillar (50μm)/Sn-Ag(20 μm), and the PCB has a solder bump with a height of 15μm after coining process. Composition of PCB solder bump is SAC305 which is formed on three different types (OSP, ENIG, ENEPIG) of surface finishes on the Cu bump pad. In this experiment, top Si chip and PCB pitch is 130μm. Figure 1 shows the top chip structure where Cu pillar and Sn/Ag cap were electroplated on Al pad. The final structure was then reflowed. Coined solder bump of PCB formulated on Cu bump pad using SAC 305 paste. Figure 2 is PCB solder bump top and cross view.

Table 1 shows the surface finish for each PCB substrate layer composition and thickness.

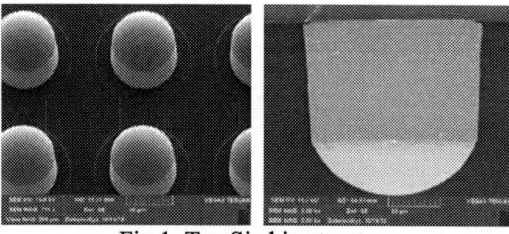

Fig 1. Top Si chip structure

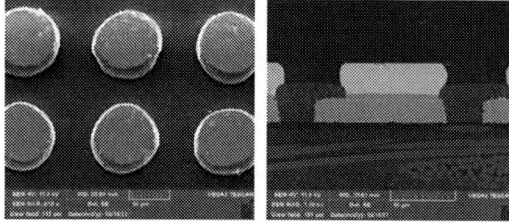

Fig 2. Solder bump of PCB substrate

	Surface finish	Au(μm)	Pd(μm)	Ni(μm)
Leg 1	OSP	-	-	-
Leg 2	ENIG	0.58	-	4.04
Leg 3	ENEPIG	0.08	0.06	7.26

Table 1.　Information of PCB surface finish

2.2 Experimental process

In this experiment, all bonding is done by thermal compression type bonding using NCF after lamination at 80C. The NCFs are formulated so that it contains a flux function, preventing void formation during NCF curing and easily laminated shown in figure 3.

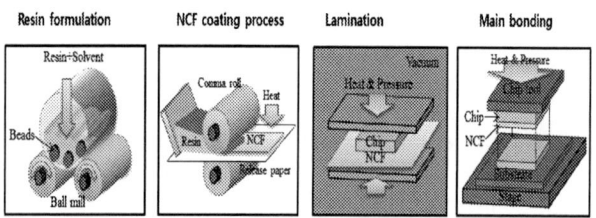

Fig 3. Process of TC bonding (NCF formulation, Lamination, and Main bonding)

Also, the bonding conditions were optimized to consider bump joint shape and CTE mismatch between the top Si chip and the PCB substrate.

The Interfacial reliabilities of interconnection between Cu-pillar/Sn-Ag and three different types of surface finished PCB bump which can be applied for 2.5D TSV interposer were evaluated by TCT and HTST.

After the testing was completed at each step, the electrical data was confirmed through 4points contact resistance measurement. Also, open failure mode and IMC composition were investigated by using SEM and EDX after cross-section.

3. Results and discussion

For high reliability of Cu pillar/Sn-Ag interconnected with PCB solder bump structure, TCT and HTST were performed. After TCT 1000cycle which is the JEDEC guide line JEDEC for mobile electronic devices, Leg 1 & leg 3 contact resistances were stable up to TCT 1000cycle but Leg 2 circuits are opened after TCT 750cycle.(Figure 4)

Distinct result of reliability under TCT condition, all test legs evaluated TCT up to 2000cycle. Only Leg 1(OSP surface finish) was stable up to TCT 1500cycle but after 2000cycle, all test interconnection were opened.

Leg 1 shows the conventional IMC structure, Cu_6Sn_5 and Cu_3Sn. Crack propagation of OSP proceeds at the interface between Cu_6Sn_5 IMC and Sn at center of joint during TCT due to thermal stress.(Figure 5(a)) In case of Ni layer plating test legs (ENIG and ENEPIG surface finish), Ni component diffused into the solder bump and formed a brittle Ni_3Sn_4 IMC which is located at the PCB solder bump pad interface and split into Sn at the center of joint.(Figure 5(b & c))[2]

Circuit of Leg 2(ENIG surface finish) and Leg 3(ENEPIG surface finish) was opened easily compared to Leg 1 due to Ni_3Sn_4 IMC in Sn which can be the crack propagation path during TCT.

Point of interconnection shape, crack should be progress around interface between Cu-pillar and Sn component when applied thermal stress by TCT because Solder bump are wetting on Cu pillar wall during TC bonding so bump joint gap decreased which make Cu pillar located at same level SR surface. Movement of Brittle Cu pillar during TCT make crack at interface between IMC and ductile solder at center of joint.

Among evaluation legs, Ni layer plating legs (ENIG and ENEPIG surface finish) showed a clear contact resistance difference after TCT where Au is plated on Ni layer due to Ni oxidation prevention. Au plating method of leg 2(ENIG surface finish) used displacement plating that makes Ni layer damage and thicker Ni-P layer which gives bad effect to the interconnection reliability because of its brittle property.[3,4,5]

On the other hand, Leg 3 used reduction plating method to plate Au layer on Ni layer. To prevent Ni-P layer increase at interface between Ni layer and SAC solder layer, Pd layer is plated on Ni layer as diffusion barrier about P component. Due to Pd layer of Leg 3(ENEPIG surface finish), PCB solder bump is less brittle leg 2(ENIG surface finish), contact resistance after TCT showed definite results.

For evaluated joint reliability at high temperature, HTST at 150C 1000hr was performed for all test legs which bonds Si chip with three different surface finish type PCB bump pads. All bonding structure showed Kirkendal void at the interface between Cu pillar and Sn-Ag cap plating. When focusing on PCB bump pad, they showed different IMC and crack

propagation

OSP surface finish of PCB solder bump showed 2 types of IMCs. IMC compositions were Cu6Sn5 and Cu3Sn when investigating through EDS(Figure 6(a)). The cross section image which was observed by SEM, Voids were observed at the center of the bump joint because Sn components that were consumed to make IMCs reacted with Cu at Cu pillar and Cu PCB solder bump pad.

After HTST 1000hr, Cu pillar of Leg 2(ENIG surface finish) divided with solder bump totally and all Sn components changed to brittle IMC layer(Figure 6(b)). Even for Leg 3(ENEPIG surface finish), Sn component was observed at the center of the bump joint. However, when Pd layer is present, the Sn component at center of bump joint does not change the IMC since it acts as a barrier layer.(Figure 6(c))

When comparing the mechanical properties between Sn and IMC component, Sn component is more ductile which improves the interfacial reliability of Cu-pillar/Sn-Ag interconnection with PCB solder bump.

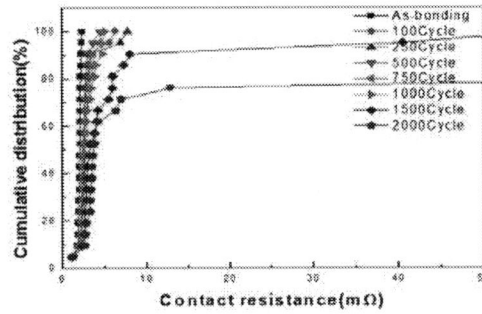

(c) Contact resistance results after TCT 2000cylce of Leg 3(ENEPIG surface finish)

Fig 4. TCT results up to 2000cycle by PCB solder bump pad surface

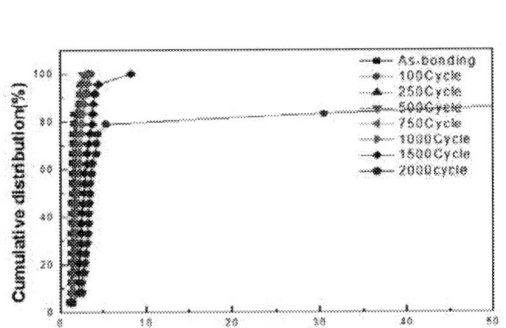

(a) Contact resistance results after TCT 2000cylce of Leg 1(OSP surface finish)

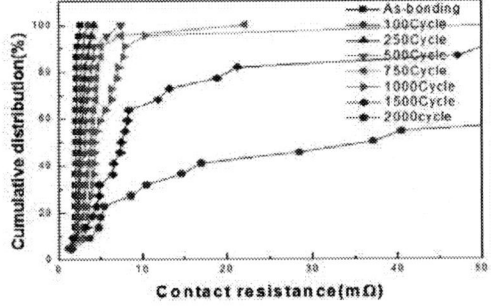

(b) Contact resistance results after TCT 2000cylce of Leg 2(ENIG surface finish)

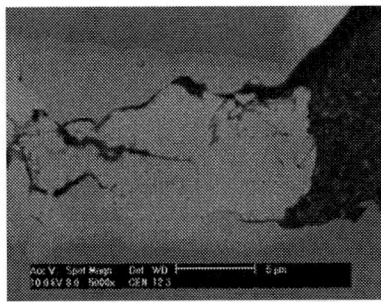

(a) TCT failure mode results of Leg 1(OSP surface finish)

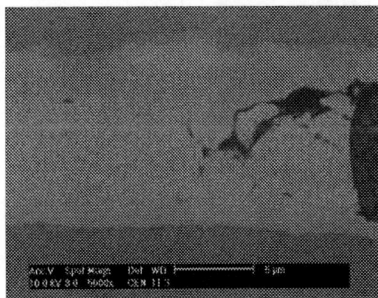

(b) TCT failure mode results of Leg 2(ENIG surface finish)

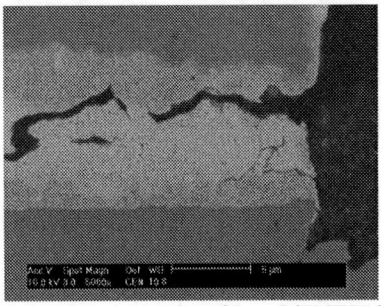

(c) TCT failure mode results of Leg 3(ENEPIG surface finish)

978-1-4799-2408-0/14 $31.00 © 2014 IEEE   1767

Fig 5. TCT failure mode results up to 2000cycle by PCB solder bump pad surface finish

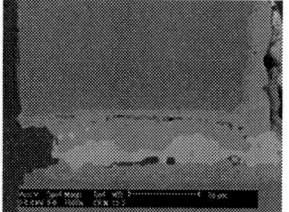

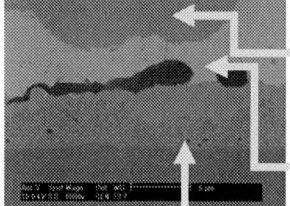

Element	Wt %	At %
SnL	55.43	39.97
CuK	44.57	60.03

Element	Wt %	At %
SnL	96.96	94.46
CuK	03.04	05.54

Element	Wt %	At %
SnL	55.58	40.11
CuK	44.42	59.89

(a) HTST results of Leg 1(OSP surface finish)

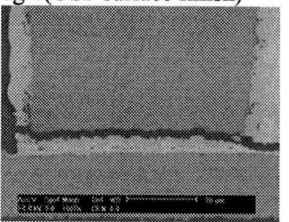

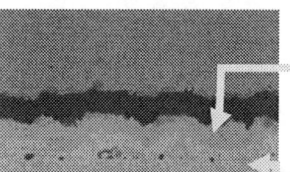

Element	Wt %	At %
SnL	45.00	30.37
NiK	02.74	03.74
CuK	52.26	65.89

Element	Wt %	At %
SnL	54.23	38.65
	03.70	05.34
	42.07	56.01

Element	Wt %	At %
SnL	34.09	21.68
CuK	65.91	78.32

surface finish)

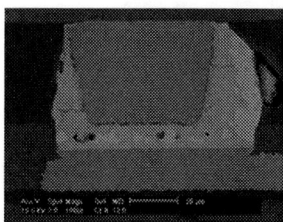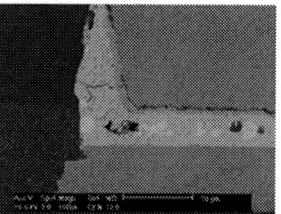

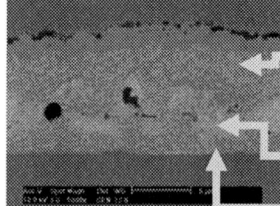

Element	Wt %	At %
SnL	56.70	41.05
NiK	03.49	05.11
CuK	39.82	53.85

Element	Wt %	At %
SnL	92.42	86.71
CuK	07.58	13.29

Element	Wt %	At %
SnL	57.72	41.69
NiK	11.33	16.54
CuK	30.95	41.77

(c) HTST results of Leg 3(ENEPIG surface finish)

Fig 6. HTST results by different surface finish of PCB

4. Conclusion

In this experiment, surface finish effect of PCB Cu bump pads were evaluated for interfacial reliability between Cu-pillar/Sn-Ag bump of top Si chip and PCB solder bump which can be used as 2.5D TSV interposer by using TCT and HTST.

Cu-pillar/Sn-Ag top Si chip bump were successfully bonded by using TC bonding and NCF after lamination on three different PCB solder bump.

Reliability tests (TCT and HTST) showed district effects of PCB surface finish on Cu bump pads. OSP surface finish was suitable for 2.5D TSV interposer where no electrical open failure was observed up to TCT 1500 cycles.

In case of Ni layer plated surface finish such as ENIG and ENEPIG, P content was a key factor to the joint reliability, where it allows brittle fracture to occur. Lastly, for ENIG and ENEPIG surface finish showed different TCT results due to the Ni-P layer.

5. References

[1] ZHANG Hao, CAI Jian,. WANG Qian, WANG Tao, WANG shuidi, "Development of a BGA package Based on Si interposer with Through Silicon Via" TSINGHUA SCIENCE AND TECHNOLOGY ISSN 1007-0214 11/15 pp408-413 Volume 16, Number 4, August 2011

[2] Sung K, Kang., D. Y. Shih, K.Fogel, P. Lauro, Myung-Jin Yim., " Interfacial Reaction Studies on Lead(Pb)-Free Solder Alloys" IEEE TRANSACTIONS ON ELECTRONICS PACKAGING MANUFACTURING, VOL. 25, NO. 3, JULY 2002

[3] Y.C.SOHN, JIN YU, S.K. KANG, D.Y. SHIH, and W.K. CHOI. " Effects of Phosphorus Content on the Reaction of Electroless Ni-P with Sn and Cry Crystallization of Ni-P" Journal of ELECTRONIC MATERIALS, Vol. 33, No. 7, 2004

[4] Doosoo Kim, James Jungho Pak, "Micro void growth in NiSnP layer between (Cu,Ni)6Sn5 intermetallic compound and Ni3P by reflow terperature and multiple reflow" J Mater Sci : Mater electron(2010) 21:1337-1345

[5] Men Shian Li, Jen Tsung Luo, Wei Cheng Lin, Yun Min Yang " IMC investigation of fracture surface between solder joints and different surface finish processes" Microsystems Packaging Assembly and Circuits Technology Conference(IMOACT) 2010 5th International

978-1-4799-2408-0/14 $31.00 © 2014 IEEE

Effect of Variation in the Reflow Profile on the Microstructure of Near Eutectic SnAgCu Alloys

Francis Mutuku[1], Babak Arfaei[1,2], Ph. D. and Eric J Cotts[1], Ph.D

[1]Physics Department and Materials Science Program, Binghamton University, Binghamton, NY

[2]Universal Instruments Corporation, Conklin, NY

fmutuku1@binghamton.edu

ABSTRACT

The effects of changes in thermal history (changes in the reflow profile) on solder joint microstructure are examined. The effects of variations in the reflow temperature, and in the cooling rate from the melt, on the microstructure of near eutectic lead free SnAgCu solder joints were investigated. Changes in precipitate or Sn grain morphologies have previously been correlated with changes in the failure rates in these Pb free alloys. Thus correlations were sought between changes in reflow parameters and in both precipitate and Sn grain morphologies. Precise reflows were conducted in a differential scanning calorimeter and it was found that changes in microstructure were correlated with large changes in the reflow temperature, and that the precipitate microstructure was a strong function of the cooling rate. This effect was similar in magnitude to that observed in relatively large changes (1 to 3wt%) in the Ag concentration of these near eutectic SnAgCu alloys.

INTRODUCTION

The microstructure of a Pb free solder joint affects its fatigue life [1-6]. The fatigue life of a near eutectic SnAgCu solder joint is a function of thermal history and applied stress, and also initial solder joint microstructure. Both the Sn grain morphology and the precipitate size and number are important [1-18]. Therefore, the control and prediction of the performance and reliability of these materials would be facilitated by an understanding of the factors which affect their initial microstructure. The present study focuses on how variation of reflow parameters affects the initial microstructure of near-eutectic SnAgCu solder.

Near eutectic compositions (Ag and Cu concentrations near 3.5 and 0.9 weight percent respectively) of SnAgCu reveal β-Sn dendrites surrounded by a network of Ag_3Sn and Cu_6Sn_5. Large differences in the number and size of the secondary Ag_3Sn and Cu_6Sn_5 precipitates found in the pseudo eutectic phase between dendrites in the SnAgCu solder can exist from sample to sample and within given SnAgCu samples. Large primary Ag_3Sn and Cu_6Sn_5 precipitates, more than one thousand times bigger than the inter-dendritic precipitates, often form during solidification of near eutectic SnAgCu solder alloys (such large precipitates are less likely in alloys with lower Ag compositions, such as SAC105). Previous work has shown that in near eutectic SnAgCu solder joints generally only three dominant Sn grain orientations are present, with large Sn grain sizes observed in larger (500 micron diameter) solder joints [1]. Some such solder joints can be characterized as displaying essentially one single Sn grain, or three dominant Sn grain orientations. For tri-grained samples, the mis-orientation of neighboring Sn grains corresponded to a sixty degree rotation around the [010] axis, consistent with six-fold cyclic twinning at the initial Sn nucleation point [1]. Smaller SnAgCu solder joints reveal an interlaced Sn grain morphology, where still only three Sn grain orientations are observed, with the same mis-orientation as found in larger samples, but with many more grain boundaries. This interlaced Sn grain morphology has been correlated with lower solidification temperatures in SnAgCu solder joints [21].

Previous investigations [22] have revealed a dependence of the mechanical properties of near eutectic SnAgCu solder joints upon their microstructure, including precipitate size and number, Sn grain number and orientation, and on the nature of the intermetallic compounds (IMC) at a solder/substrate interface [1]. The Cu_6Sn_5 and Ag_3Sn precipitates which form in near eutectic SnAgCu strengthen the Sn matrix material. A large number of uniformly distributed, small precipitates are most effective. [3-6]. There have been some previous examinations of the precipitate size distribution and its variation as a function of cooling rate, and the effect of this distribution on the strength of bulk solder [7]. A different study found that the size and population of Ag_3Sn are dependent on the cooling rate and on the amount of undercooling of the Sn during reflow [17].

Sn, the major component in SnAgCu solder, displays large anisotropies in its mechanical properties. Thus the mechanical response of a large Sn grained, near eutectic, Sn-Ag-Cu solder joint depends upon Sn grain morphology[18]. For instance, the orientation of single Sn grain solder joints affects their performance [18]. There is some indication of superior mechanical performance of solder joints being associated with interlaced Sn grain morphologies.

Previous studies have shown that changes in the thermal history of a Pb free solder joint affects solder joint microstructure. For instance, large (order of magnitude) variations of the cooling rate from the melt were observed to change the number density of precipitates. Because such variations in thermal history affect solder joint microstructure and properties, a concerted attempt is made to control the thermal history of solder joints during manufacture (Fig. 1); a specific reflow profile is imposed with specific heating and cooling rates, peak temperature and reflow time (or time above liquidus). Never-the-less, some variation occurs from joint to joint in a package, and from process to process. In general, ramp rates are between 1 ^{o}C/s and 3°C/s, though rates as low as 0.5^{o}C/s may occur; cooling rates are similar. Peak temperatures for near eutectic SnAgCu are generally near 240°C, with reflow times less than 100s. The present study further investigates the nature of such effects. Variations in the reflow temperature and cooling rate (see Fig. 1) are imposed upon samples, and correlated with changes in microstructure.

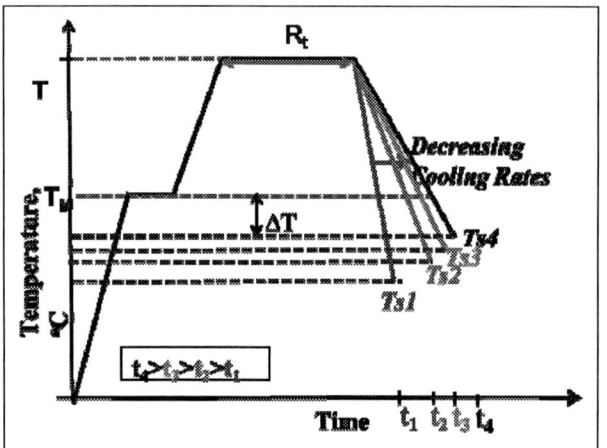

Fig. 1 A plot of temperature versus time reflecting a typical reflow profile in the processing of solder joints. The melting temperature, T_M, the reflow temperature T, and the solidification temperature T_s are represented. Various cooling rates are portrayed; Ts1, Ts2, Ts3 and Ts4.

Experimental

The goal of this project is to better understand the solidification microstructure of SnAgCu solder joints. The nature of the Sn grain morphology and precipitate microstructure were examined as a function of carefully controlled reflow parameters. Precise reflows were conducted in a differential scanning calorimeter. Samples were cross sectioned and examined by means of optical and electron microscopy (including electron backscattered diffraction, EBSD), so as to characterize Sn grain size and orientation, as well as mean precipitate size and number.

5mg free standing samples were cut from ingots commercially supplied by the IBM Corporation, and then reflowed in the differential scanning calorimeter. The samples were reflowed with a peak temperature of 245°C, held at this peak temperature for 600s, and then cooled from the melt at rates ranging from 0.05°C/s to 2.0°C/s. After reflow, the samples were mounted in epoxy for cross-sectioning; ground using silicon carbide paper with different grain sizes and polished to 0.02μm colloidal silica. The carefully cross sectioned solder balls were examined by optical microscopy in both bright and cross-polarized fields. The bright field images were used to examine the network of β-Sn dendrites, the eutectic regions and the primary precipitates in the β-Sn matrix while the cross polarized images were used to examine the Sn grain morphology in each solder ball. After optical microscopy, the samples were imaged by a high resolution Zeiss 55 VP Scanning Electron Microscope equipped with Energy Dispersive Spectrometer (EDS) for precipitate morphology and chemical analysis. Images were taken at a magnification of x4k and at sizes of 76μmx56μm.

Quantitative microstructural analysis was performed using Imagej software and the results were reported in a log-log plot of precipitate spacing and the precipitates number

density for the various cooling rates. The inter-particle spacing, λ between any two neighbor precipitates in an array of precipitates can be calculated if we can locate the co-ordinates of their centers of mass which coincides with the centroid of the particle of uniform density.

Results

The microstructure of Sn-based Pb free solder depends strongly on its precise composition. High purity Sn samples of intermediate sizes (50 to 500 micron diameter) generally display only one Sn grain, with some low angle grain boundaries, and of course, no precipitates. The addition of small percentages of Cu and Ag dramatically changes the microstructure. As seen in Fig. 2(a), an optical micrograph with crossed polarizers of a polished SAC105 sample, three distinct Sn grain orientations are present. Furthermore, Sn dendrites are clearly decorated by prolific distributions of small (approximately one to two micron) Ag_3Sn and Cu_6Sn_5 precipitates in pseudo eutectic phases.

The microstructures of free standing SnAgCu solder samples of a range of compositions were examined. Optical micrographs of cross sections of samples ranging in composition from SAC105 to SAC807 are displayed in Fig.2 (the samples were all reflowed at a temperature of 245°C for 600s and cooled from the melt at a rate of 1°C/s). These optical micrographs (Fig.2) of polished cross sections all reveal a plethora of Sn dendrites decorated by regions of pseudo eutectic which contain Sn and small precipitates of Ag_3Sn and Cu_6Sn_5 ranging in sizes of magnitude one micron.

In solder samples with higher Ag concentrations (e.g SAC807, Fig. 2(i)), much larger precipitates were also apparent, with lengths of several 100 microns. These large precipitates are orders of magnitude longer than the small precipitates observed in the pseudo eutectic regions found between Sn dendrites. These large precipitates were generally found to have compositions close to Ag_3Sn. Judging by their relatively large size, many times larger than Sn dendrite arm sizes, it is concluded that they formed initially in the melt, i.e. before the solidification of Sn. In the series of samples portrayed in Fig. 2, some such large precipitates are observed in alloys with Ag concentrations as low as 3.7wt% (cf. Fig. 2(e)). For samples with low Ag concentrations, i.e. for Ag concentrations below 3wt% (e.g Figs. 2(a) through Fig.2(c)), such large Ag_3Sn precipitates were generally not observed.

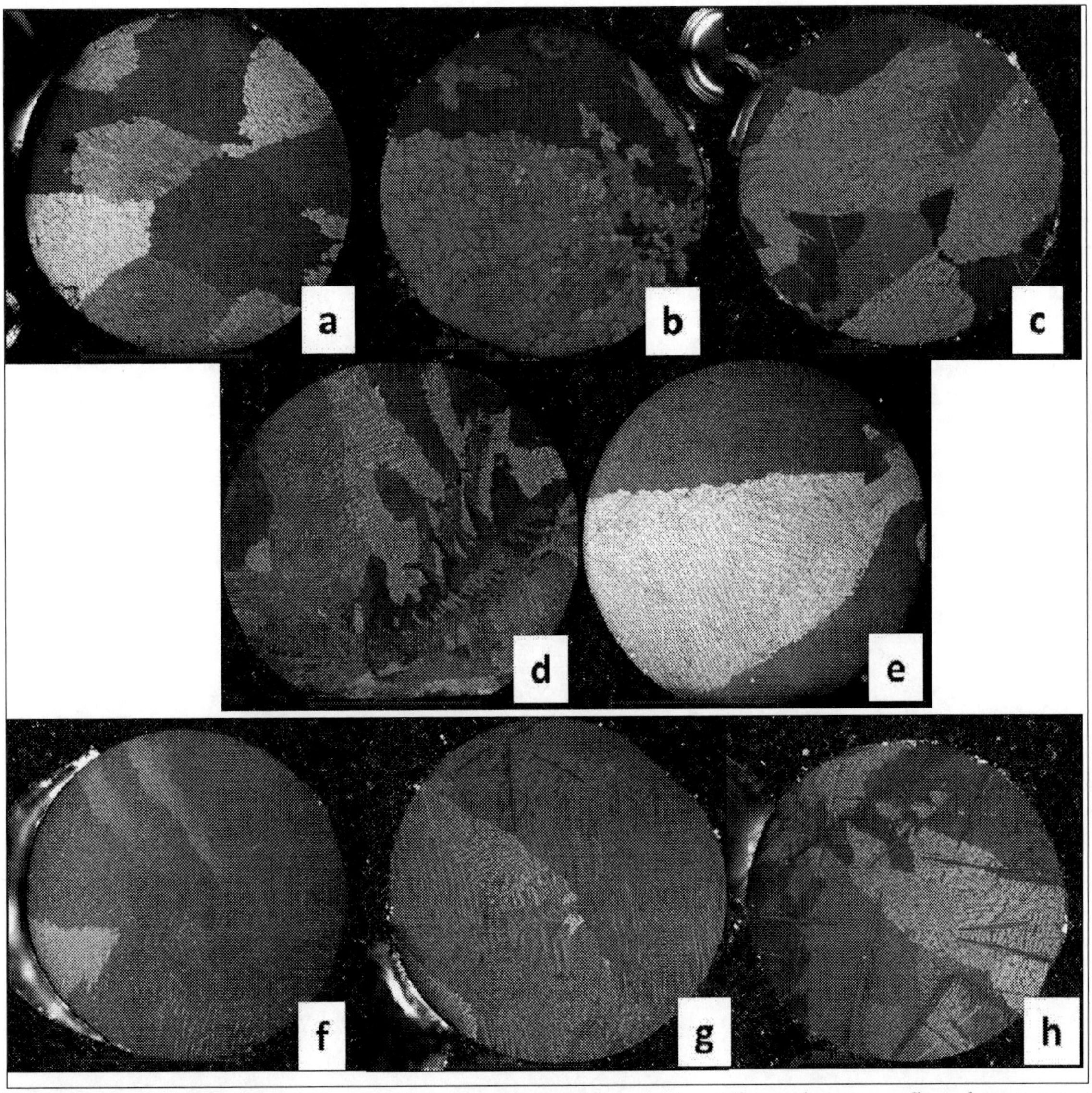

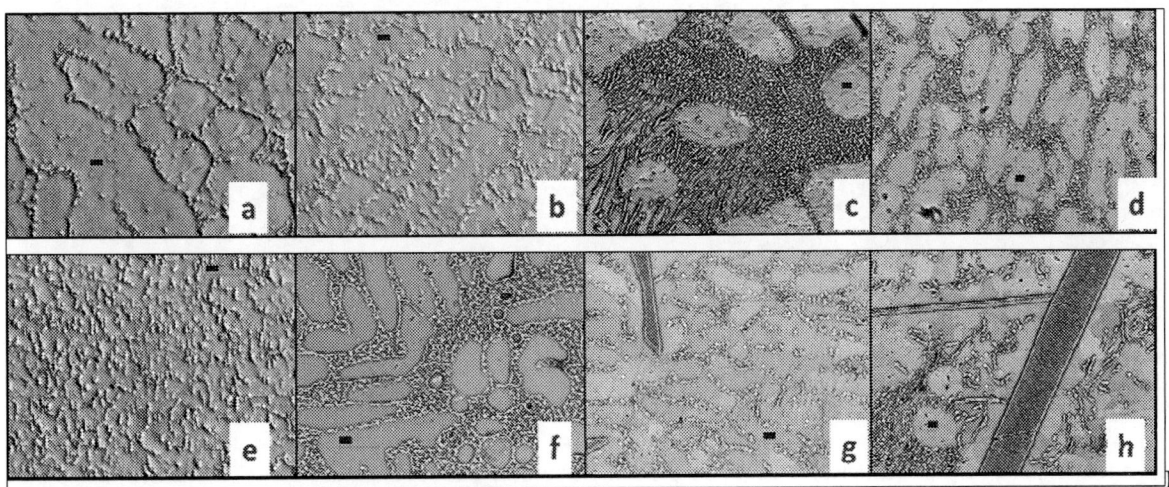

Fig.2 Cross polarized micrographs of 5mg free standing solder samples. All samples were reflowed at a temperature of 245°C for 600s and cooled from the melt at a rate of 1°C/s. Sample compositions were: (a) SAC105 (b) SAC205 (c) SAC207 (d) SAC287 (e) SAC305 (f) SAC327 (g) SAC387 and (h) SAC807.

Fig.3 SEM micrographs of 5mg samples: (a) SAC105 (b) SAC205 (c) SAC207 (d) SAC287 (e) SAC305(f) SAC327 (g) SAC387 and (h) SAC807. All samples were reflowed at a temperature of 245°C for 600s and cooled from the melt at a rate of 0.7°C/s.

978-1-4799-2408-0/14 $31.00 © 2014 IEEE   1771

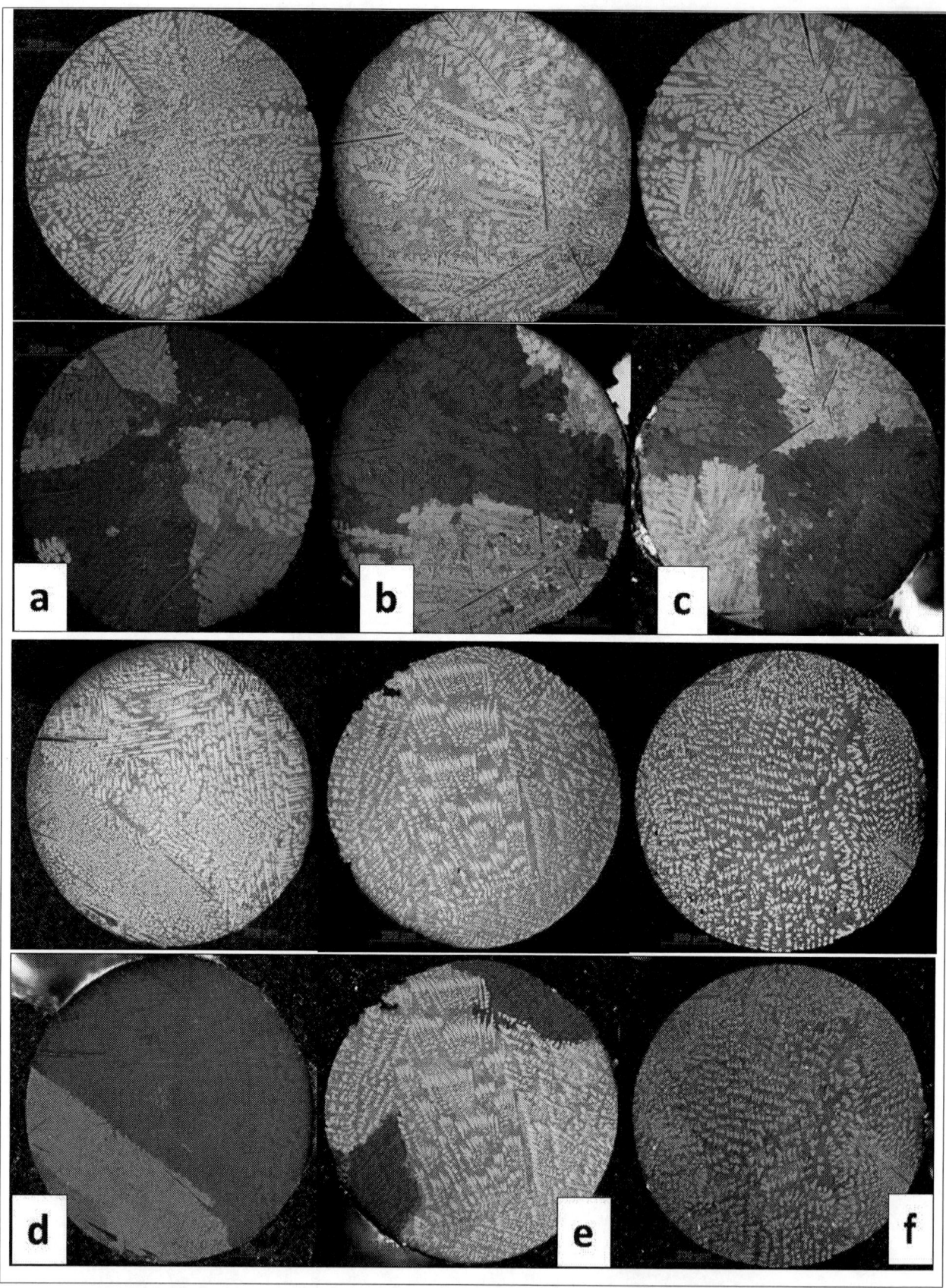

Fig.4 – Optical micrographs of cross sections of SAC387 solder samples produced at cooling rates of (a) 0.1°C/s (b) 0.2°C/s (c) 0.4°C/s (d) 0.8°C/s (e) 1.2°C/s and (f) 1.5°C/s both in bright and cross polarized fields. All samples were reflowed at a temperature of 245°C for 600s.

It is apparent that the number density of interdendritic, small Ag_3Sn and Cu_6Sn_5 precipitates increases with increasing Ag and Cu concentration (Fig.2), up to a concentration of approximately 3.7wt% Ag. This is also reflected in a series of Scanning Electron Microscopy micrographs for the same samples of Fig.2 (presented in Fig.3). For solder alloys with higher Ag concentrations (e.g SAC387, Fig.3(g) and SAC807, Fig.3(h)), large primary Ag_3Sn precipitates are observed, along with a lower number density of smaller precipitates. Presumably, the large primary Ag_3Sn precipitates form early in the melt in such number and breadth that super saturation of the melt was avoided during cooling.

Changing reflow parameters can result in distinctly different solder microstructures. Optical micrographs of free standing SAC387 solder samples cooled from the melt at rates ranging from 0.1°C/s to 1.5°C/s are displayed in Fig. 4. The optical micrographs with crossed polarizers reveal that generally all of these samples displayed multiple Sn grain orientations. Examination of the bright field optical micrographs show that the small precipitates in these samples increased in number density as the cooling rate increased. This is more apparent in scanning electron microscopy micrographs, for example those presented in Fig. 5 for six different cooling rates, ranging from 0.05°C/s to 1.5°C/s , for these SAC387 samples.

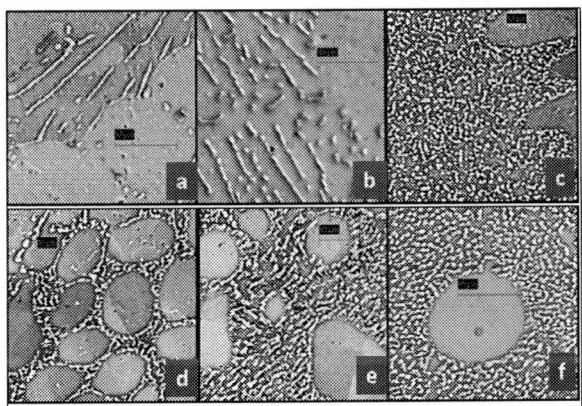

Fig.-5 Scanning electron micrographs of SAC387 solder samples for cooling rates (a) 0.05°C/s (b) 0.2°C/s (c) 0.4°C/s (d) 0.8°C/s (e) 1.2°C/s (f) 1.5°C/s. All samples were reflowed at a temperature of 245°C for 600s.

Scanning electron microcopy micrographs of a series of SAC387 free standing solder samples which were cooled at different rates from the melt were portrayed in Fig.5. Distinctively different precipitate morphologies are seen in these samples; it is apparent that the interdendritic spacing decreases as the cooling rate increases. A more quantitative characterization of this trend is found through analysis of the number density of these precipitates. The Scanning Electron micrographs of Fig.5 were analyzed, as well as those in micrographs from other regions of the samples. Analysis of the number density of the precipitates using standard image analysis techniques revealed a monotonic dependence on cooling rate, as reflected in the plot of number density versus cooling rate of Fig. 6.

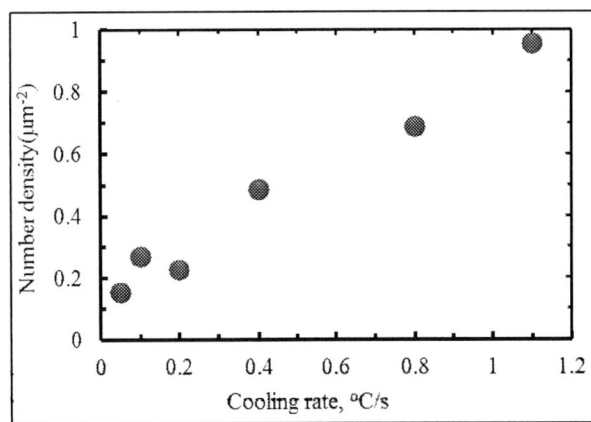

Fig. 6 –A plot of the measured number density of precipitates versus the cooling rate in free standing 5mg samples of SAC387 reflowed at peak temperature 245°C for 600s.

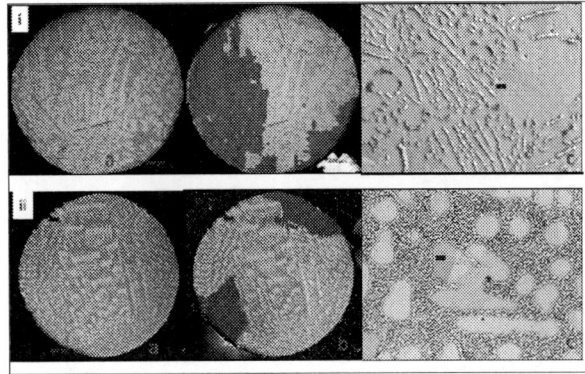

Fig. 7 The effect of cooling rate on the microstructure of two samples of SAC387 with same solidification temperature. The samples were cooled at different rates: (i) 0.2°C/s and (ii) 1.2°C/s (a) Bright field (b) Cross Polarized optical micrographs and (c) scanning electron microscopy micrographs.

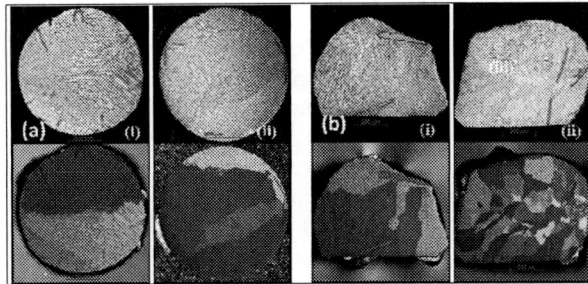

Fig.8 Optical micrographs of cross sections of SAC397 solder samples produced at one of two different cooling rates: (a) 1.5°C/s and (b) 0.02°C/s. Two different reflow temperatures were used: (i) 245°C and (ii) 330°C. Optical micrographs in both bright and cross polarized fields are presented for each sample condition (bright field micrographs are above micrographs with crossed polarizers for each condition).

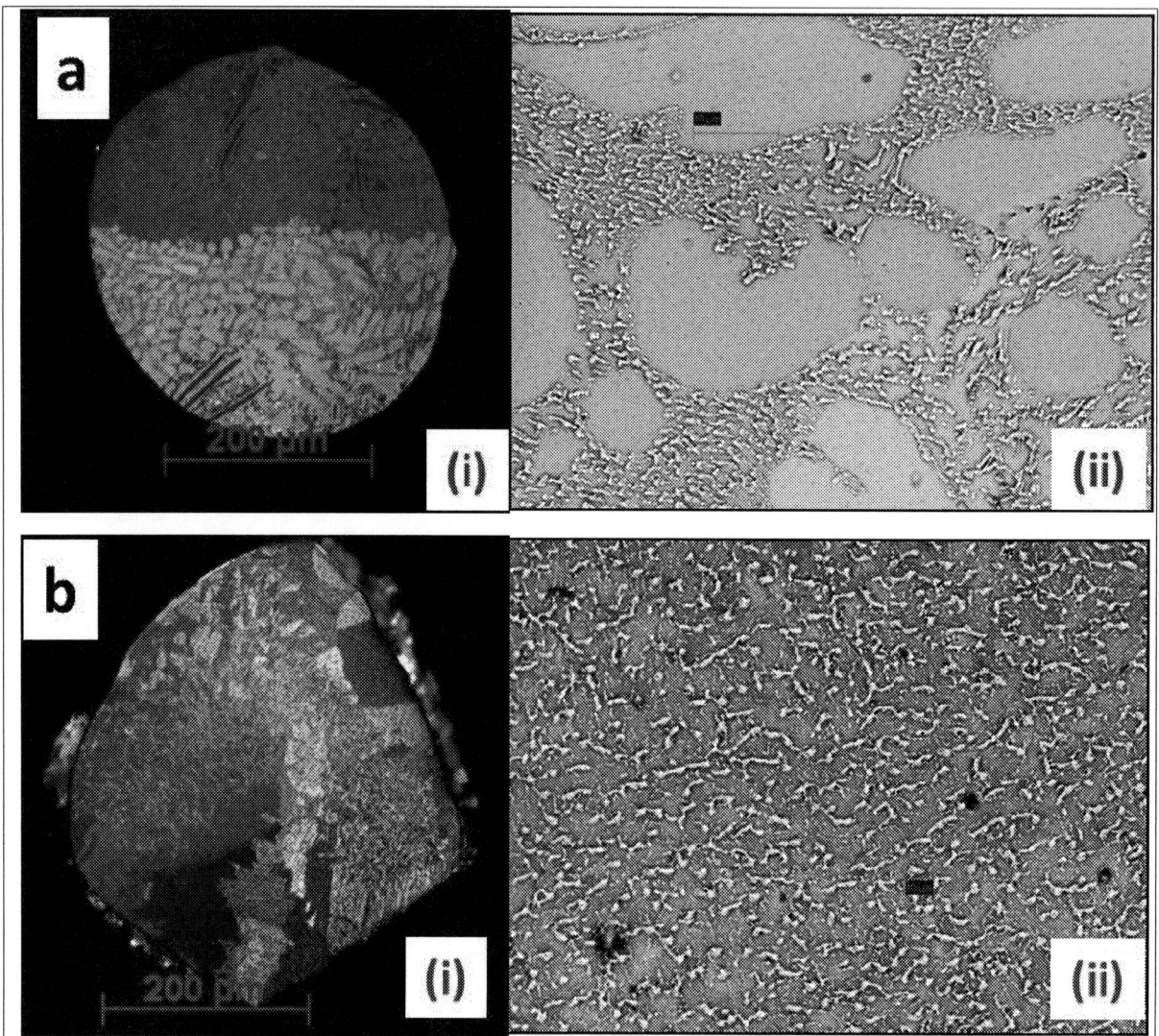

Fig.9 -Micrographs of free standing SAC397 solder samples which were reflowed at different temperatures: (a) 245°C (b) 330°C. (i) Optical micrographs with crossed polarizers and (ii) Scanning electron micrographs.

Further illustration of the effect of cooling rate on the microstructure of SAC387 solder is provided in Fig. 7, where the cooling rate was varied by a factor of six from 0.2°C/s to 1.2°C/s. Although these samples have the same solidification temperature, their microstructures are different. For the more slowly cooled sample, large Ag$_3$Sn precipitates are observed (Fig.7 (i)) which are absent in the bright field, optical micrograph for the cross section of a similar sample cooled at a rate of 1.2°C/s (Fig.7 (b). Corresponding examples of scanning electron microscopy micrographs for the same two samples are also included in Fig. 7. Analysis of these and other SEM micrographs for these samples indicate that the concentration of Ag in solution was significantly less in the slow cooled samples, as compared to the faster cooled sample, before Sn solidification.

The microstructure of near eutectic SnAgCu Pb free solder varied somewhat with changes in the reflow temperature of samples before solidification (cf. Fig. 1). The microstructure was examined for SAC397 samples reflowed with one of three distinctively different reflow temperatures:

245°C or 330°C (Fig. 8). The optical micrographs of Fig. 8 are for either a cooling rate from the melt of 1.5°C/s. (Fig. 8(a) or 0.02°C/s (Fig. 8(b)). For the higher cooling rate samples, some primary precipitates are observed, but these are smaller and less prevalent than in the case of the samples reflowed with slower cooling rates Fig. 8(b).

The effect of reflow temperature on the number density of small precipitates in near eutectic SnAgCu solder is illustrated in Fig. 9, which includes scanning electron microscopy micrographs. Two disparate reflow temperatures were selected, 245°C and 330°C. Some differences in precipitate size and number are observed. For the higher reflow temperature, a higher density of precipitates is evident in scanning electron microscopy micrographs. Further study is planned to further examine such a possible correlation.

The observation that cooling rate affects the distribution of precipitates in the SnAgCu solder microstructure is consistent with previous observations [2, 4-8]. In this study a clear monotonic dependence of the number density of precipitates on cooling rate was observed (Fig. 6). These precipitates, primarily Ag₃Sn, where found to be fairly evenly distributed in the interdentritic spaces. Variations in reflow temperature also appeared to influence the microstructure of these near eutectic SnAgCu solder joints. Further study should further examine the effects of reflow temperature and cooling rate on solidification temperature and solder joint microstructure.

Conclusion

The microstructures of SnAgCu solder alloys were observed to depend upon details of the reflow profile. Precise reflows were conducted in a differential scanning calorimeter and it was found that the precipitate microstructure was a strong function of the cooling rate. The number density of precipitates in the pseudo eutectic region increased monotonically with cooling rate. This effect was similar in magnitude to that observed in relatively large changes (1 to 3wt%) in the Ag concentration of these near eutectic SnAgCu alloys. Some changes in microstructure were correlated with large changes in the reflow temperature, as well.

Acknowledgments

The Authors acknowledge Universal Instruments Co. for support. The help of Mr. James Woods on metallography of selected samples is greatly appreciated. We acknowledge Dr. Larry Lehman from S3IP lab in Binghamton University for important discussions. This work is partially supported by U.S. Department of Defense, SERDP program.

References

1. Lehman, L. P., Xing, Y., Bieler, T. R., & Cotts, E. J. (2010). Cyclic twin nucleation in tin-based solder alloys. Acta Materialia, 58(10), 3546-3556.
2. Korhonen, T., Turpeinen, P., Lehman, L et al (2004). Mechanical properties of near-eutectic Sn–Ag–Cu alloy over a wide range of temperatures and strain rates. Journal of Electronic Materials 33 (12), 1581–1588.
3. Y Kariya, T Hosoi, S.Terashima, M. Tanaka, and M Otsuka, J. Electron. Mater. 33,321 (2004).
4. Kim, K.S., Huh, S.H. and Suganuma, K. et al. (2003), "Effects of intermetallic compounds on properties of Sn-Ag-Cu lead-free soldered joints", J. of Alloys and Compounds, Vol. 352, p. 226.
5. Ochoa, F., Deng, X., & Chawla, N. (2004). Effects of cooling rate on creep behavior of a Sn-3.5 Ag alloy. Journal of electronic materials, 33(12), 1596-1607.
6. Kerr, M., & Chawla, N. (2004). Creep deformation behavior of Sn–3.5 Ag solder/Cu couple at small length scales. Acta materialia, 52(15), 4527-4535.
7. Ochoa, F., Williams, J. J., & Chawla, N. (2003). Effects of cooling rate on the microstructure and tensile behavior of a Sn-3.5 wt.% Ag solder. Journal of Electronic Materials, 32(12), 1414-1420.
8. Kang, S.K., Lauro, P.A., Shih, D-Y., Henderson, D.W. and Puttlitz, K.J. (2005), "Microstructure and mechanical properties of lead-free solders and solder joints used in microelectronic applications", IBM J. Res. Dev., Vol. 49, p. 607.
9. Snugovsky, L., Snugovsky, P., Perovic, D.D. and Rutter, J.W. (2005), "Effect of cooling rate on microstructure of

Ag-Cu-Sn solder alloys", Mater. Sci. Tech., Vol. 21, p. 61.
10. Wiese, S. and Wolter, K. (2004), "Microstructure and creep behaviour of eutectic SnAg and SnAgCu solders",Microelectronics Reliability, Vol. 44, p. 1923.
11. Mueller, M.; Wiese, S.; Roellig, M.; Wolter, K.-J.; "Effect of Composition and Cooling Rate on the Microstructure of SnAgCu-Solder Joints," Electronic Components and Technology Conference, 2007. ECTC '07. Proceedings. 57th , vol., no., pp.1579-1588, May 29 2007-June 1 2007 doi: 10.1109/ECTC.2007.374006
12. Plumbridge, W., Gragg, C., Peters, S., 2001. The creep of lead free solders at elevated temperatures. Journal of Electronic Materials 30 (9), 1178–1183.
13. M. Reid, J. Punch, M. Collins, C. Ryan, (2008). Effect of Ag content on the microstructure of Sn-Ag-Cu based solder alloys, Soldering & Surface Mount Technology, Vol. 20 Iss: 4 pp. 3 – 8
14. Sung K. Kang, Won Kyoung Choi, Da-Yuan Shih, Donald W. Henderson,Timothy Gosselin, Amit Sarkhel, Charles Goldsmith, and Karl J. Puttlitz,(2003) "Ag3Sn plate formation in the solidification of near ternary eutectic SnAgCu," JOM]
15. B. Arfaei ,T. Tashtoush, N. Kim, L. Wentlent, E. Cotts, P. Borgesen " Dependence of SnAgCu Solder Joint Properties on Solder Microstructure".
16. Liang, J., Dariavach, N., & Shangguan, D. (2007).Solidification Condition Effects on Microstructures and Creep Resistance of Sn-3.8 Ag-0.7 Cu Lead-Free Solder. Metallurgical and Materials Transactions A, 38(7), 1530-1538.
17. S. Wiese., K.J. Wolter (2004). Microstructure and creep of eutectic SnAg and SnAgCu solders. Microelectronics reliability, Volume 44, 1923-1931. DOI: 10.1006/j.microrel.2004.04.016
18. Sidhu, R. S., Deng, X., & Chawla, N. (2008). Microstructure Characterization and Creep Behavior of Pb-Free Sn-Rich Solder Alloys: Part II. Creep Behavior of Bulk Solder and Solder/Copper Joints. Metallurgical and Materials Transactions A, 39(2), 349-362.
19. Kang, S. K., Shih, D. Y., Donald, N. Y., Henderson, W., Gosselin, T., Sarkhel, A., ... & Choi, W. K. (2003). Ag3Sn plate formation in the solidification of near-ternary eutectic Sn-Ag-Cu. JOM, 55(6), 61-65.
20. Arfaei, B., & Cotts, E. (2009). Correlations between the microstructure and fatigue life of near-eutectic Sn-Ag-Cu Pb-free solders. Journal of electronic materials, 38(12), 2617-2627.
21. Arfaei, B., Wentlent, L., Joshi, S., Anselm, M., & Borgesen, P. (2012, November). Controlling the Superior Reliability of Lead Free Assemblies with Short Standoff Height Through Design and Materials Selection. In ASME 2012 International Mechanical Engineering Congress and Exposition (pp. 467-473). American Society of Mechanical Engineers.
22. Mutuku.F., Arfaei, B., Eric J.Cotts. (2013, May) 'Effect of Variation in the Reflow Profile on the Microstructure of Near Eutectic SnAgCu Alloys' In Proc.International Conference on Soldering and Reliability(SMTAi), Fort Worth, TX, May14-17,2013.

Development of the Thin Film with High Thermal Conductivity for Power Devices

Hiroshi Takasugi, Shin Teraki, Tsuyoshi Kurokawa, and Issei Aoki
NAMICS Corporation
3993 Nigorikawa, Kita-ku, Niigata City, Niigata Prefecture
TEL: +81-25-258-5577, E-mail: takasugi@namics.co.jp

Abstract

In recent years, heat generation and heat density of electronic devices have been increasing due to highly integrated and high working frequency of semiconductors.

Power devices have a cooling system to release its generated heat efficiently. For cooling system low thermal resistance material is required in order to conduct generated heat from semiconductor to the heat sink.

This paper reports a very low thermal resistive film material developed for thin, high thermal conductive, good compliance to the surface to reduce thermal contact resistance and high resistance for dielectric-breakdown material for power devices.

High thermal conductive films have been promoted in the market. However they have limitations of 1) thermal contact resistance due to low compliance to the surface caused by high modulus as a result of high loading volume of inorganic fillers, and 2) thickness to ensure the dielectric-breakdown resistance. For these challenges we have applied a low modulus resin to satisfy both high filler loading and compliance to the surface. We also have tried optimizing particle size and loading volume of inorganic filler. As a result of development we achieved to have a film material which can suppress corona discharge, has high dielectric-breakdown resistance even thickness is less than 100 µm, and has more than 5 W/m·K of thermal conductivity.

To confirm the heat dissipation performance of this material we performed to measure thermal resistance which has to be a key factor for the thermal design of power devices. The thermal resistance was 0.86 K/W when 50 µm compare to 1.45 K/W with conventional 125 µm film in the market. It demonstrates that this material has very low thermal resistance.

It is expected that using this film material may reduce the joint temperature of semiconductor and also improve thermo-mechanical reliability of the power device by a good stress relaxation performance given by low modulus.

Introduction

Fig.1 represents a typical power device structure [1]. In this structure the heat from a chip dissipates through an insulation layer. Power devices work as a switch using a semiconductor and control the power by being switched on and off. Smaller packages, which are a recent trend, have been increasing the amount of heat generation of the devices. Thus improving the heat dissipation from the devices has become a critical challenge; not to exceed their operating temperature limits.

Insulation is another important function for the power devices. Poor insulation may cause operation failures or break peripheral devices such as a motor. In this study, we aimed to develop a material that ensures the dielectric-breakdown resistance and provides a low thermal resistance, a high thermal conductivity, and high adhesion to decrease contact resistance.

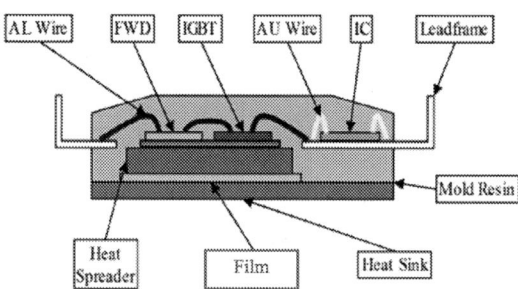

Figure 1. A typical power device

1. Development concepts

The rate of heat transfer through a substance is proportional to the area of a surface perpendicular to the heat flow and also the temperature gradient along a path of the heat flow. One-dimensional, steady-state heat transfer is expressed by Fourier equation (Eq.) [2] :

$$Q = \lambda \times A \times \Delta T / L \quad \text{(Eq. 1)}$$

where,

λ = thermal conductivity (W/m·K)

Q = heat transfer (W)

A = surface (heat transfer) area

L = thickness (distance)

ΔT = temperature difference

Thermal conductivity (λ) is a material's ability to conduct heat, which is an intrinsic property of substances. This property does not depend on the material size or shape.

There is another intrinsic thermal property of substances; thermal resistance (R) expressed by Eq. 2 below.

$$R = \Delta T / Q \quad \text{(Eq. 2)}$$

This can determine how a material resists the heat flow. The relation between λ and R (Eq. 3) can be expressed by substituting Eq. 2 into Eq. 1.

$$R = L / (\lambda \times A) \quad \text{(Eq. 3)}$$

Eq. 3 represents the thermal resistance is directly proportional to the thickness of a substance. In other words, changes in the thermal resistance of a heat-transfer device are proportional to the distance between a heat-generating component and a heat-dissipating component. The thermal resistance increases as the heat-transfer distance increases. The shorter the distance, the lower the thermal resistance becomes.

2014 Electronic Components & Technology Conference

Fig.2 shows the thermal resistance of different thicknesses, calculated by thermal conductivity. The differences in the thermal conductivity decrease with thinner thickness. For instance, the following will result in the same thermal resistance.
- thermal conductivity of 10 W/ m·K, thickness of 500 μm
- thermal conductivity of 3 W/ m·K, thickness of 150 μm

To reduce thermal resistance, which is desired for power devices, thermal conductivity needs to be improved. For films, reducing film thickness is the most effective way to reduce the thermal resistance.

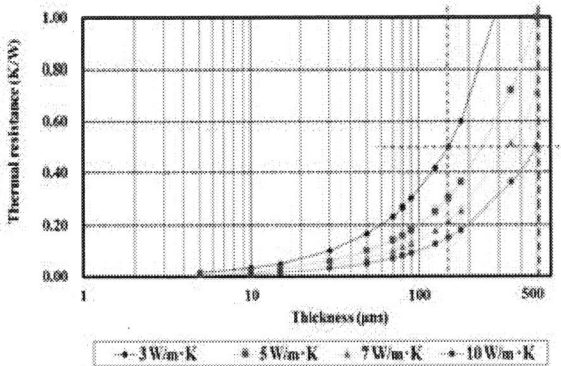

Figure 2. Relation between film thickness and thermal resistance for different thermal conductivities

2. Simulating heat dissipation of a substrate

To confirm the relation between thermal conductivity and film thickness, we simulated heat dissipation of a substrate. The substrate used has a three-layer structure, consisting of an aluminum base, a film, and a cupper foil. We heated the substrate from the bottom to observe how the heat would transfer through the cupper and the aluminum up to the top (Fig. 3). A surface temperature of the cupper foil was measured when its heat source was 150 degrees Celsius (°C). If its junction temperature is low, it means the heat dissipation is high.

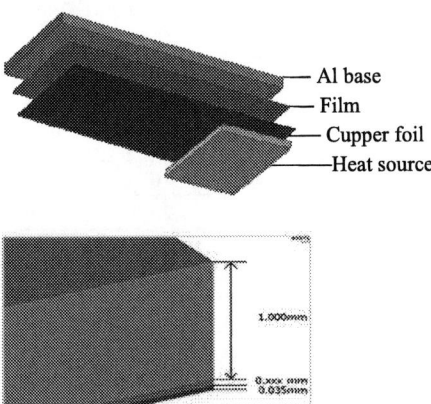

Figure 3. Substrate model

Fig. 4 shows the results simulating different film thicknesses (500 μm, and 80 μm) and thermal conductivities

(10 W/ m·K, 7 W/ m·K, and 5 W/ m·K). Although C and B has the same thermal conductivity, C shows a lower cupper-surface temperature; a higher heat-dissipation. This is because C has a thinner film thickness than B. The comparison between A, B, and C suggests that reducing film thickness is more effective in improving heat dissipation than increasing thermal conductivity.

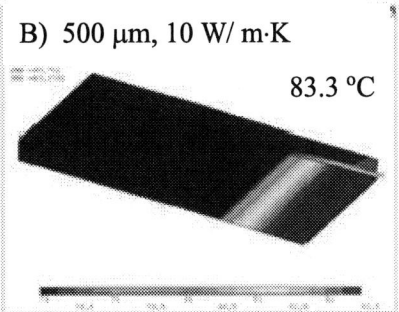

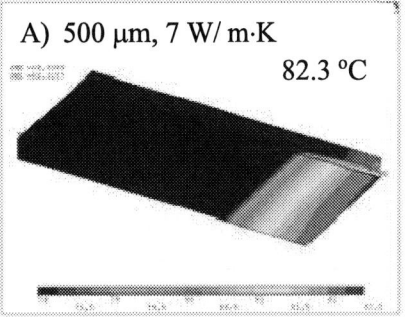

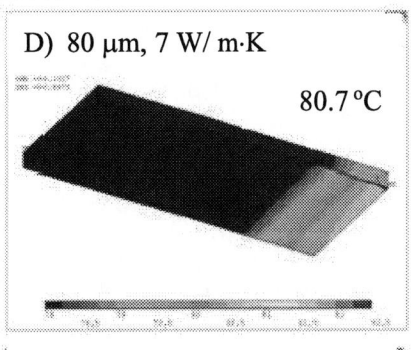

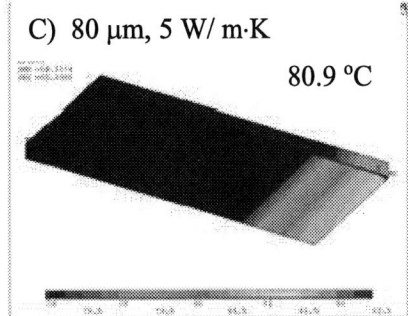

Figure 4. Simulation results

We targeted 100 μm or less for the film thickness and aimed to achieve a higher heat-dissipation than that with a thickness of 500 μm and a thermal conductivity of 10W/ m·K.

3. Designing the materials

We focused on obtaining a high thermal conductivity and reducing film thickness.

3.1. Obtaining a high thermal conductivity

We explored several approaches to increasing the thermal conductivity.

3.1.1. Determining a suitable filler loading volume by Bruggeman's equation

To increase the thermal conductivity of an organic substrate, the thermal conductivities of fillers and resins used and their compounding rate are needed to be considered. These can be expressed by Bruggeman's equation (Eq. 4) [3].

$$1 - \frac{\Phi}{100} = \frac{\lambda c - \lambda f}{\lambda r - \lambda f} \left(\frac{\lambda r}{\lambda c} \right)^{1/3} \quad \text{(Eq. 4)}$$

where,

Φ: volume fraction of filler

λf: thermal conductivity of filler (W/m·K)

λr: thermal conductivity of resin

λc: thermal conductivity of compound

The preconditions are:
- true-spherical filler
- isotropic thermal conductivity
- thermally-stable interface

Fig. 5 shows the relation between the filler volume fraction (Φ) and the compound thermal conductivity (λc), increasing the filler thermal conductivity (λf) from 20 to 40 and 180 and the resin thermal conductivity (λr) from 0.22 to 0.31. Increasing λf affected and enhanced the thermal conductivity of the filler itself in the area with a higher filler volume fraction, leading to an increase in the compound thermal conductivity. Increasing λr enhanced the thermal conductivity of the substrate itself and this also led to an improvement in the compound thermal conductivity even in the area with a relatively-low filler volume fraction. Increasing these two can raise the compound thermal conductivity. Combining them effectively can yield even a higher thermal conductivity.

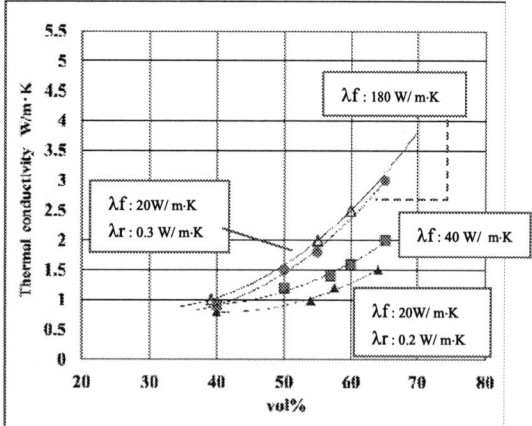

Figure 5. Relation between filler loading volume and thermal conductivity

3.1.2 Obtaining close packing of the fillers

One of the techniques to raise the filler loading volume is to build a close-packed structure. The closed-packed model given by Horsfield is a well-known theory. Given that small particles are packed between primary spheres in the hexagonal close-packed structure, this model determines the diameter ratio and the ratio of the number of particles so as to achieve the closest-packed structure. For this study, we combined the particles at a ratio of 1 large particle to 0.15 small particle (Fig.6 b), using three different particle sizes. Table 1 compares the thermal conductivity between one particle size and three particle sizes. This confirms that packing small particles closely so that the gaps can be filled raises the thermal conductivity.

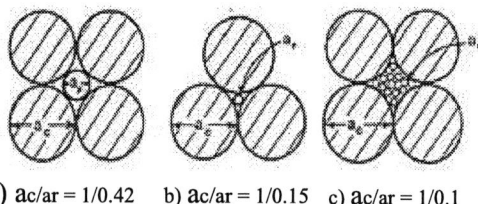

a) ac/ar = 1/0.42   b) ac/ar = 1/0.15   c) ac/ar = 1/0.1

Figure 6. Packing of fine spheres in a planar interstice among coarse particles

Table 1. Comparison of thermal conductivity between different size combinations

Filler size	1.0 r = 100 vol%	6.5 r/1.0 r/0.15 r = 67/23/10
Thermal Conductivity	1.8 W/ m·K	2.1 W/ m·K

Note: "r" represents a filler diameter.

Next, we changed the ratio of the three-particle-size combination to determine the thermal conductivity. The ratio of 60 to 20 to 20 showed the highest thermal conductivity (Table 2).

Table 2. Thermal conductivities at different combination ratios

	6.5 r	1.0 r	0.15 r	λ: m·K
1	80	10	10	2.1
2	70	10	20	2.2
3	70	20	10	2.0
4	67	23	10	2.1
5	60	20	20	2.3
6	60	30	10	2.0
7	50	30	20	2.2

3.1.3 Reducing the stress

We designed a base resin to be flexible because stress relaxation is a key to high reliability in heat cycles.

3.2 Reducing the film thickness

For thinner films, dielectric-breakdown resistance will critically be needed. Without insulation, an undesired leak of the voltage to power an insulated gate bipolar transistor (IGBT) or leak current due to fast switching may occur.

In the measurement of a film dielectric-breakdown voltage, there are two possible causes of the dielectric breakdown. The one is the voids inside a film. The other is an air layer between a film surface and the electrodes to be measured, which is created by protrusions of fillers on the film surface. The air layer originally has a low dielectric-breakdown voltage. Electric charge increases in the air layer with a low dielectric constant and it increases even more as the air layer is more sharply angled. The air layer will cause partial discharge when a voltage is applied (Fig. 7). At this stage of the partial discharge no breakage will occur between electrodes yet. If it lasts, however, active oxygen or high-energy charged particles will be generated. These will degrade a film, resulting in the dielectric breakdown.

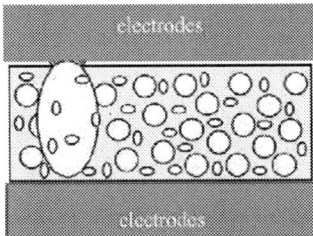

Model A: Filler particles exceed the film thickness

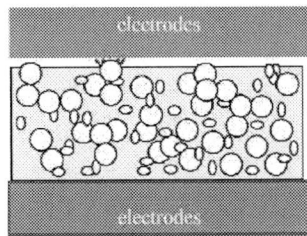

Model B: Filler dispersion is not uniform

Figure 7. Models possibly leading to dielectric breakdown

Table 3 compares the dielectric-breakdown voltage between Model A and Model B. Either case resulted in a very low dielectric-breakdown voltage. To prevent the dielectric breakdown, film surfaces need to be flat and smooth.

Table 3. Dielectric-breakdown voltage using Model A and B

	Model A	Model B
Dielectric-breakdown voltage	3 kV/mm	3 kV/mm

3.2.1 Controlling the maximum particle size

We adopted the fillers that does not exceed the film thickness; that create sufficient room in the particle size distribution.

3.2.2 Distributing fillers uniformly through a dispersion process

To prevent fillers from aggregating, fillers are mixed at high speed and a powerful shearing machine is used during the dispersion process. This has led us to produce the films with a flat and smooth surface (Fig. 8).

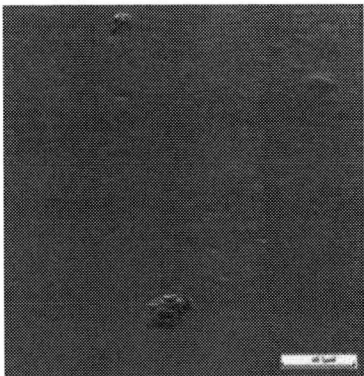

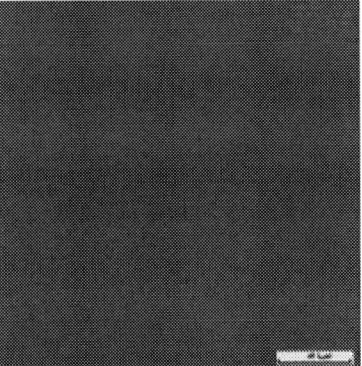

Figure 8. Film surface images

4. Making desirable film samples

Our next step was to optimize film formability and curing properties by changing the compounding ratio. We developed two samples with different thermal conductivities. Table 4 shows their properties. They have a thickness of 100 μm or less, sufficient flexibility, and high performance in thermal conductivity and insulation.

Table 4. Film sample properties

		Sample A	Sample B
Thermal conductivity	W/ m·K	3	5
Cupper Peel strength	N/cm	10.0	6.3
Tensile strength	MPa	18	15
Elongation	%	20	11
Tensile modulus	MPa	350	2000
Storage modulus	GPa	6	12
Volume resistance	Ωcm	6.0×10^{11}	5.0×10^{11}

5. Measuring the dielectric-breakdown voltage

The dielectric-breakdown voltage was measured using only films (Sample A and B) with different thicknesses. In accordance with Japanese Industrial Standards (JIS), a film was placed between electrodes in Fluorinert (Fig. 9). A voltage was applied and raised in stages to determine the dielectric-breakdown voltage. Fig. 10 shows the measurement result of Sample B. The dielectric-breakdown voltage rises with the film thickness. This result confirms that our film samples have a high dielectric-breakdown voltage, 65 kV/mm or greater.

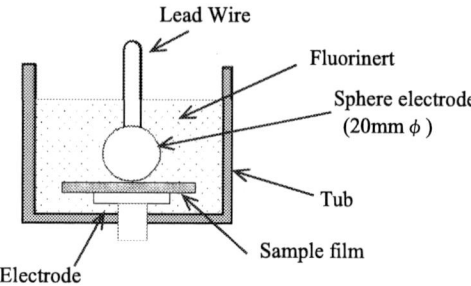

Figure 9. How to measure the dielectric-breakdown voltage

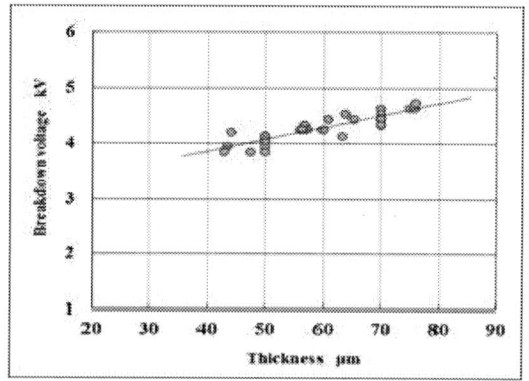

Figure 10. Results of the dielectric-breakdown voltage

6. Evaluating a thermal property

Although thermal conductivity is considered as a representative thermal property, thermal resistance will be the one when a film is applied in an actual module. This is because heat dissipation varies depending on film thickness and adhesion.

We measured the thermal resistance through thermal transient analysis. Analyzing heat changes transiently can obtain highly-accurate structure functions (thermal resistance vs. thermal capacity). Additionally, the thermal resistance can be analyzed individually for components of a device.

The test piece used in this evaluation consisted of three layers: an aluminum base, a film, and a cupper foil (Fig. 11). We observed the thermal history; how the heat from a chip would dissipate through the substrate.

This analysis provides two types of graphs: integral structure function (Fig. 12) and differential structure function

(Fig. 13). *a*) shown in the graphs represents the thermal resistance of the film. Changes in the gradient mean different components.

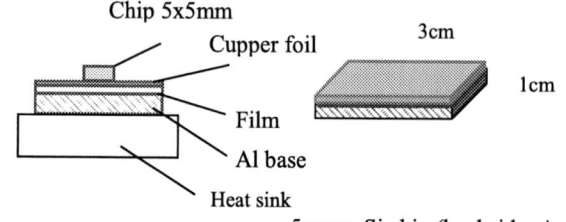

5mm□_Si chip (backside: Au)
Measurement capacity: 2W

Figure 11. Test piece for thermal transient analysis

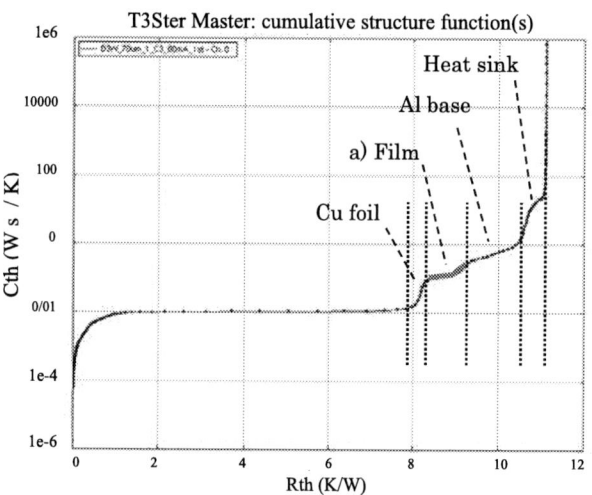

Figure 12. Integral structure function

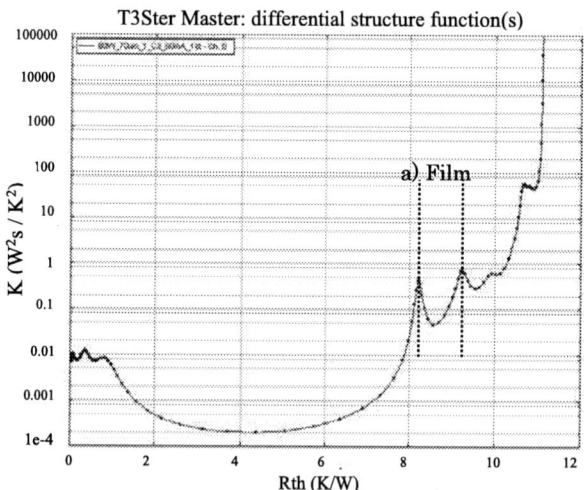

Figure 13. Differential structure function

Measurement method: JEDEC JESD51-1 static test.
Measurement tool: T3Ster, Mentor Graphics-made

Through this thermal transient analysis, the thermal resistance was compared between different film thicknesses and thermal conductivities (Table 5). This result demonstrates that reducing the film thickness can lead to a low thermal resistance and raise the heat dissipation.

Table 5. Analysis results of the thermal resistance

Item	unit	sample A		sample B		Ref.
Themal conductivity	W/m·K	3	3	5	5	7
Thickness	μm	70	90	52	90	125
Thermal resistance	K/W	1.010	1.372	0.858	1.064	1.451

Conclusion

We aimed to develop a thin film with a high thermal conductivity. Through this development, we were able to provide the films with not merely the high thermal conductivity, but also a low thermal resistance and a high dielectric-breakdown voltage. This will definitely expand the applicability of our films for use in power devices.

The power device industries are expected to grow rapidly. More solutions for heat or more requirements for smaller devices with a high current capability will be upcoming challenges. Thus we will be further working on developing a better insulating film; higher thermal conductive and thinner films.

Acknowledgments

The authors would like to thank their colleagues for their technical assistance and valuable insights.

References

1. Hozoji, H."Encapsulation Materials for Power devices", *Journal of Japan Institute of Electronics Packaging*, vol. 15, No. 5, pp. 374-378, 2012.
2. Ono, Tosiaki. "Silicon-Hirogaru ouyoubunya To Gijyutu-doko" [Silicon-Expansion of Application Field and Technical Trend].Tokyo, 2003:*Kagaku Kougyo Nipposha*. pp.127-128.
3. D. A. G. Bruggeman, Ann.Phys., No.24, 1935, p.636.
4. Takahashi, Kazuo,"Kyujyo Filler No Miryoku Wo Ikasu!"[Use the Charm of Spherical Filler!].Tokyo, 1994. :*Kogyo Zairyo*.42(15),pp112-116.
5. Randall. M,"Particle Packing characteristics ", pp. 190-191, 1989
6. "IGBT Kara ECU Made Arayuru Netusokutei Kano Na Souchi T3Ster" [All heart measurement is possible by the device of T3Ster for IGBT to ECU].
   http://www.mentorg.co.jp/products/mechanical/techpubs/t3ster-thermal-characterization/index.html (Accessed Nov. 2012)

Development of Electroless Nickel-Iron Plating Process for Microelectronic Applications

Yu Luo, Sung K. Kang, Oblesh Jinka, Maurice Mason, Steven A. Cordes, Lubomyr T. Romankiw
IBM T.J. Watson Research Center,
1101 Kitchawan Road, Yorktown Heights, NY 10598, USA
E-mail: luoyu@us.ibm.com

Abstract

Electroless nickel-phosphor (Ni-P) plating has been extensively used as a diffusion barrier layer in printed circuit board applications. However, the interfacial reactions between electroless Ni-P and lead (Pb)-free Sn solders during high temperature reflow are known to dissolve Ni-P excessively into Sn leaving behind a brittle Ni_3P compound at the interface. This often causes various reliability issues in the solder joints. To mitigate this problem, an application of NiFe alloy is proposed with the concentration of iron (Fe) greater than 20% in the Ni barrier. The NiFe alloy produces a more robust diffusion barrier in the microelectronic applications. Development of a stable electroless NiFe plating process is therefore essential for its successful implementation.

In this work, we developed an electroless NiFe alloy plating process with a high deposition rate and superior bath stability. The deposition rate of three micron per hour was achieved and five weeks of bath aging did not deteriorate the bath performance. Stable iron content around 30% in NiFe deposit was achieved in this timeframe. Factors that impact deposition rate and alloy composition were identified. The valence of the iron ions in the electroless NiFe bath was found to have a significant impact on the bath stability. The interfacial reaction between Pb-free SnAgCu solder and electroless NiFe(P) is discussed and compared with the electroless Ni-P process. Energy dispersed X-ray spectroscopy (EDS) and scanning electron microscopy (SEM) were utilized to characterize electroless NiFe deposits and to analyze the interfacial reactions between Pb-free solder and electroless Ni-P or NiFe after multiple reflows.

Introduction

As Pb-free solders are being widely implemented in microelectronic applications, several critical reliability issues related with Sn-rich solder joints have been identified, such as interfacial reactions, drop impact resistance, tin whiskers, thermal fatigue, electromigration, chip-package interaction, and others [1-4].

Among them, the interfacial reactions in Sn-rich solder joints have been extensively investigated because they are very aggressive compared to those in eutectic Sn-Pb joints, mainly due to the higher Sn content (>95 vs 62%) and the higher reflow temperature used in Pb-free soldering [5-6].

During the solder reflow, two basic reactions occur at the soldering interfaces; dissolution of interface metallization into a molten solder, and concomitant intermetallic formation at the interfaces. Both reactions have serious impacts on the integrity and reliability of solder joints, and should be controlled for reliable solder joints. The intermetallic phases continue to grow in the solid state during a high temperature aging, often accompanied with interfacial void formation

when a disparity of diffusing species across the interface would exist [3].

To control the interfacial reactions in Pb-free solders, and thereby to improve their reliabilities, a thin Ni layer has been commonly deposited on top of Cu metallization [5-7], because Ni has a much lower solubility than Cu in Sn. The solubility of Cu is approximately seven times higher than Ni in Sn at a reflow temperature of 260°C, and is about five times higher at the typical aging temperature of 150°C [8, 9].

For a chip-side metallization, electrolytic or sputtered Ni can be used as a part of UBM (under bump metallization), while electroless Ni(P) with immersion Au (ENIG) is commonly applied on the bond pads in a printed circuit board (PCB). ENIG is produced by a relatively low cost process with many process/performance advantages, such as selective deposition without a ground connection, uniform deposition thickness, excellent solderability, good diffusion barrier, corrosion resistance, and others [10].

However, several reliability concerns have also been reported with Ni(P) or ENIG when they react with pure Sn or Sn-rich solders, such as excessive interfacial reactions, IMC spalling, "black pad" defect, and others [11-13]. In addition, a severe consumption of Ni layer in Sn-rich solders has been reported during multiple reflows and high current electromigration tests [4].

Several UBM structures (with diffusion barrier layers, such as Ni, Co, Fe, Ti, Ta, V, W, Zr, NiFe, NiCoFe, CoWP, etc) have been proposed for Pb-free, flip-chip structures [14-16]. Among them, NiFe alloys were found to be a most effective reaction barrier layer for Sn-rich solders [14, 17-19]. A superior electromigration resistance has also been found with NiFe metallization [20].

To fully utilize various advantages reported with NiFe alloys as a reaction barrier layer for Pb-free solder applications, development of a low-cost, electroless NiFe process is desirable, especially for PCB applications. A very few investigations have been reported in this area so far [21-23].

In the present work, several formulations of electroless NiFe alloys previously reported are evaluated in search for stable bath chemistry with a practical deposition rate to be applied for PCB applications. The factors affecting alloy composition, deposition rate, and bath stability are identified and thereby optimized. To demonstrate the robust interfacial reactions in Sn-rich solders, solder reflow experiments are conducted with electroless NiFe deposits on Cu pads in PCBs in comparison with a conventional electroless Ni(P) deposit.

Experimental

For the present experiments, oxygen-free copper foils (OFHC-102), 0.010" thick, (from Kamis Inc, Mahopac Falls,

NY) are used and are diced to small blanks of 1"x1" each. Initial experiments are performed using the formulation for NiFe(P) reported in the literature [22-23], where sodium hypophosphite is used as a reducing agent. The present experimentations are performed using the formulation for NiFe(B), provided in table 1 [21].

	Raw Material	Conc. (g/L)	Molar Conc. M
Nickel source	$NiSO_4. 6H_2O$	9.13	0.035
Iron Source	$Fe(NH_4)(SO_4)_2.12H_2O$	5.79	0.012
Complexing Agent	$NaKC_4H_4O_6.4H_2O$	33.86	0.12
	Ammonia	3.48	0.06
Reducing Agent	Dimethylamine Borane (DMAB)	3.01	0.05

Table 1. Formulation of the electroless NiFe(B) plating solution [21].

The solution from Table 1 is prepared by dissolving all chemicals, except the reducing agent, in water. The reducing agent is added just before performing the actual plating to minimize deterioration of bath stability. Copper blanks are degreased using acetone, ethyl alcohol, and isopropyl alcohol, followed by DI water rinse and are subsequently immersed in 10% hydrochloric acid for duration of 30 seconds at room temperature. Then, the blanks are activated using 200ppm palladium chloride and 10% hydrochloric acid mixture for duration of 30-45 seconds at room temperature. Activated blanks are immersed into the plating solution for durations of 30 and 60 minutes respectively at pH of 11. Plating solution is loaded with extra copper blanks to saturate the bath with hydrogen bubbles for more efficient reduction of nickel and iron metals. NiFe alloys are deposited at different temperatures (35, 50 and 65°C) with and without agitation and at varying concentrations of the reducing agent. A double-jacketed beaker connected to a water-heated circulator was used to maintain the bath temperature between 35 to 65°C. Agitation was provided with a magnetic stir bar (at 0-500 RPM) in a 150 mL beaker. The effect of deposition rates with bath aging is performed for a period of 1-29 days. The deposition rates are calculated using the difference in weight measurements.

After establishing the plating process using copper foil samples, the process is applied to the PCB substrates. The FR-4 laminated PCB substrates with copper finish are degreased using acetone, ethyl alcohol, and isopropyl alcohol followed by immersion in 10% HCL for duration of 30 seconds at room temperature. The surface of Cu pads is activated using 55ppm palladium sulfate in 10% sulfuric acid solution for duration of 30-45 seconds at 30°C. The same plating apparatus is used to plate PCB substrates and blank copper samples are immersed to load the bath with hydrogen bubbles before PCB substrates are immersed. Activated PCB samples are immersed into the plating solution for 30 and 60 minutes respectively at pH of 11 with no agitation. The process is performed at different operating temperatures (50 and 65°C) and different concentration of the reducing agent to achieve defect-free deposit and to optimize the plating rate and alloy composition of NiFe(B) plated onto the PCB substrates.

NiFe deposit was characterized with Scanning Electron Microscope (SEM), Energy Dispersive X-ray Spectroscopy (EDS), and Transmission Electron Microscope (TEM). SEM and TEM were used to examine the morphology of NiFe and EDS was used to measure alloy composition.

Solder reflow test was performed on copper blanks with NiFe deposits from the present NiFe(B) process and a commercially available electroless Ni(P) bath. Different thickness (1μm and 2μm) of Ni(P) and NiFe(B) were plated and were reflowed for the durations of 2, 10 and 30 min, respectively. One solder ball of SAC 305 alloy(5 mm in diameter) was placed on each copper foil substrate plated with either NiFe(B) or Ni(P) UBM layer, and reflowed subsequently at 250°C under a nitrogen environment with rosin flux. Reflow was performed on a hot plate enclosed in a quartz chamber that was filled with an inert gas. After reflow, cross-sectional samples were prepared with standard epoxy-mounting and polishing method, and then were examined with SEM/EDS to understand the interfacial reactions.

Results and Discussion

Electroless NiFe Plating Processes

In this paper, the electroless NiFe(B) plating solution reported by Romankiw, et al was benchmarked against the baseline NiFe(P) process reported previously [22, 23].

Formulation	Baseline NiFe (P)	Present NiFe (B)
Source of Iron	$Fe(NH_4)_2(SO4)_2.6 H_2O$ (0. 15M)	$Fe(NH_4)(SO_4)_2.12 H_2O$ (0.012M)
Source of Nickel	$NiSO_4.6H_2O$ (0.05M)	$NiSO_4.6H_2O$ (0.035M)
Ni/Fe Molar Ratio	1:3	3:1
Complexing agent	Sodium Citrate (0.3M)	Potassium Sodium Tartrate (0.12M)
Reducing Agent	Sodium hypophosphite (0.2M)	Dimethylamine borane (DMAB) (0.02 – 0.05M)
pH	11	11
Temp, °C	65	35-65

Table 2. Comparison of the baseline and the present processes

There are four major differences between these two processes in terms of bath formulation; a) the source of Fe, b) the Fe/Ni ratio in the solution, c) the complexing agent for Fe, and d) the reducing agent. In the electroless NiFe(B) plating solution, the source of Fe is ferric ammonium sulfate, Fe/Ni

978-1-4799-2408-0/14 $31.00 © 2014 IEEE

ratio in the solution is kept at 1:3, the complexing agent for Fe is potassium sodium tartrate, and the reducing agent is DMAB. In the baseline NiFe(P) solution, the source of Fe is ferrous ammonium sulfate, Fe/Ni ratio in the solution is 3:1, the complexing agent for Fe is sodium citrate, and the reducing agent is sodium hypophosphite, as listed in Table 2. The highest Fe composition in the NiFe deposit achieved from these two solutions is comparable, which is around 40%. The pH values of both solutions were adjusted to 11 before electroless plating.

Morphology and elemental spectrum of the NiFe deposit plated with electroless NiFe(B) solution are shown in Figure 1. Individual grain size of NiFe film is in the range of 20 to 60 nm. The alloy composition of Fe achieved with the present NiFe(B) electroless solution is as high as 42 % by weight.

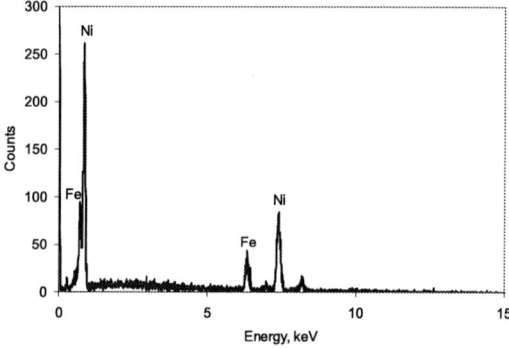

Figure 1. Morphology and elemental spectrum of NiFe(B)

In general, bath composition, solution pH, temperature, agitation, and bath life have significant impact on deposit characteristics and deposition rate of the electroless Ni(P) deposition process,. For metal alloy deposition, these factors may also have strong impact on alloy composition. In this study, the effects of reducing agent concentration, temperature, agitation, and bath life on deposit morphology, alloy composition, and deposition rate are examined.

Due to the brittle Ni$_3$P film formation during the solder reflow process, Ni(P) as a under-bump metallurgy (UBM) layer for Pb-free solder bumps has a reliability issue. Ni(B) is an alternative finish to Ni(P) that would alleviate this issue, however, it still has extensive intermetallic compound (IMC) growth. Addition of Fe in the Ni barrier layer was reported to be effective in reducing the amount of IMC formation [17-20].

In this study, DMAB is used as a reducing agent in electroless NiFe(B) bath to produce NiFe(B) deposit. DMAB concentration investigated varies from 0.02 M to 0.05 M. As shown in Figure 2, the deposition rate of NiFe(B) increases significantly with increasing DMAB concentration, especially at an elevated temperature. This is due to an increased reducing power of DMAB, which increases hydrogen concentration in the solution and facilitates the formation of nickel hydride, the actual reducing species to reduce Ni^{2+} ion to nickel metal. The deposition rate of NiFe(B) increases from 0.1μm/hr to 0.5 μm/hr at 35°C, and increases from 2.5 μm/hr to 5.5 μm/hr at 65°C.

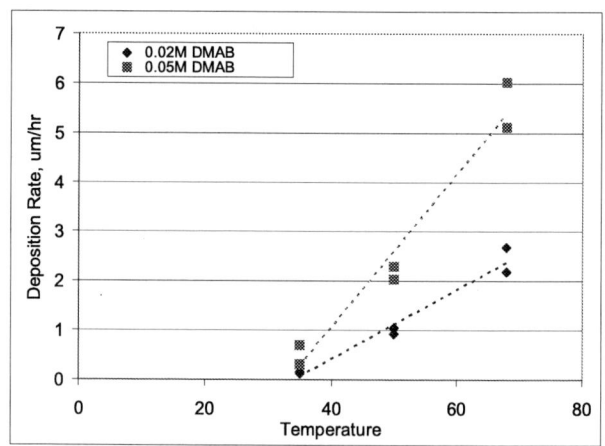

Figure 2. Deposition rate of NiFe(B) process as a function of operating temperature. (Blue diamond: DMAB=0.02M; Magenta square: DMAB=0.05M)

Figure 3 shows the effect of temperature and DMAB concentration on NiFe(B) alloy composition. Fe content in the NiFe(B) deposit also increases with increasing amount of DMAB, while Fe content generally reduces as the temperature increases.

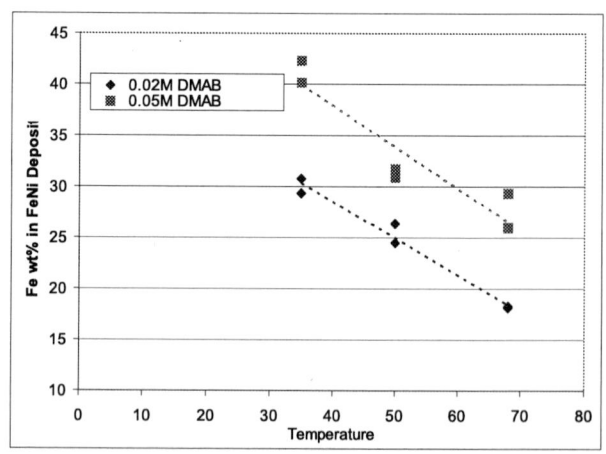

Figure 3. Fe content in NiFe(B) deposit as a function of operating temperature. (Blue diamond: DMAB=0.02M; Magenta square: DMAB=0.05M)

Figure 4 shows the effect of agitation on the deposition rate of electroless NiFe(B) film. As the rotating rate of a magnetic stir bar in a 100 mL plating solution goes up, the deposition rate of NiFe(B) decreases significantly. During electroless NiFe plating, DMAB decomposes water to form H_2 in the presence of a catalytic surface. With agitation, H_2 gas escapes from the solution quickly. The hydrogen overpotential on Ni decreases as the H_2 bubbles are not allowed to develop fully. Hence, not enough reducing power is generated as Ni is not sufficiently saturated with H_2 to allow the formation of NiH, the actual species that provides the reducing power. For most of our work reported in this paper, we did not apply agitation unless specified.

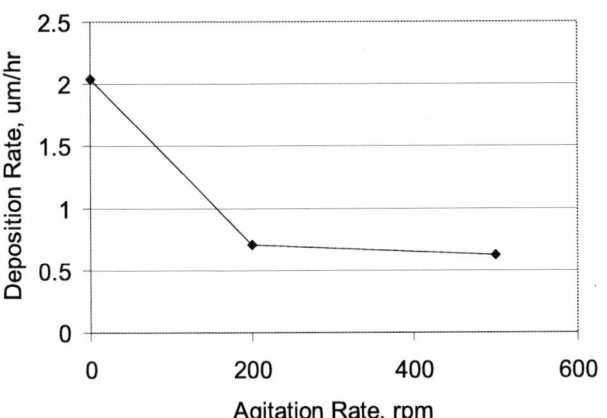

Figure 4. Deposition rate as a function of agitation

The present process was benchmarked against the baseline process in terms of bath performance over its bath life. At day 1, we made a bulk amount of plating solution and kept it in a sealed container. A certain amount of solution was taken at a pre-defined interval to plate NiFe deposit, alloy composition and deposit thickness of NiFe films were examined and reported below.

Stability of the electroless NiFe plating solution improved significantly by using ferric salt as the source of Fe. With the baseline NiFe(P) process, yellowish precipitates were observed at the bottom of the container right after plating. During idling, there were also yellow/brownish precipitates formed in a couple of days, the color of the solution changed from dark green to yellowish/green color. This is understandable as ferrous ion is thermodynamically unstable in aqueous solution and it tends to oxidize to ferric form. Oxidation of ferrous ion with oxygen dissolved in the bath caused formation of ferric hydroxide at a high pH, which is the yellow/brownish precipitate observed after plating and during bath idling.

Figure 5 exhibits the morphology of NiFe deposit as a function of bath aging. For the baseline process, NiFe deposit was a crystalline structure at day 1 and day 3, and became amorphous after day 7. For the present NiFe(B) process, NiFe deposit maintained the crystalline structure from day 1 to day 29.

The baseline process, NiFe(P)

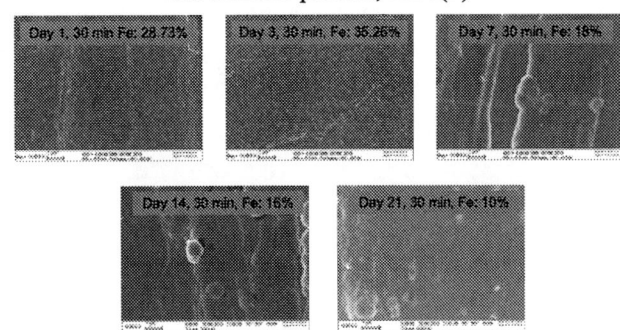

The present process, NiFe(B)

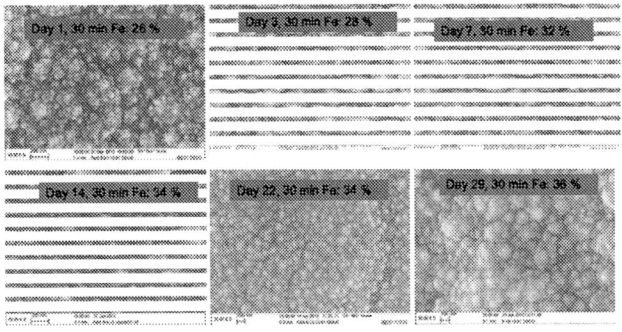

Figure 5. Morphology of NiFe vs. bath aging

Figure 6 shows the alloy composition varying as a function of bath aging. For the baseline process, as the bath aged, Fe content decreased significantly from 40% to 15% in a week and flattened out afterwards. For the present process, Fe content increased slightly from 26% to 35% in 4 weeks. This correlates well with the morphology of NiFe as shown in Figure 5, the deposit becoming amorphous when Fe content was lower than 25%. This indicates that the alloy composition has significant impact on the crystal structure.

The baseline NiFe(P) process

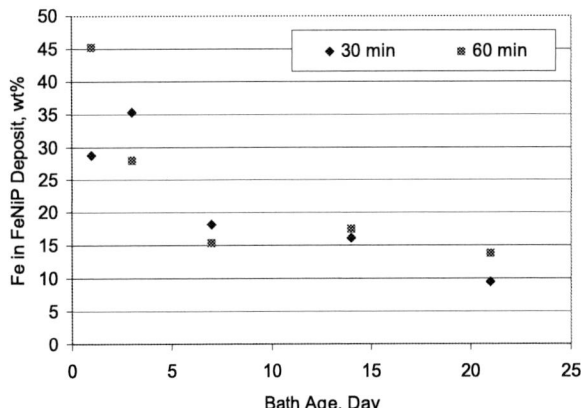

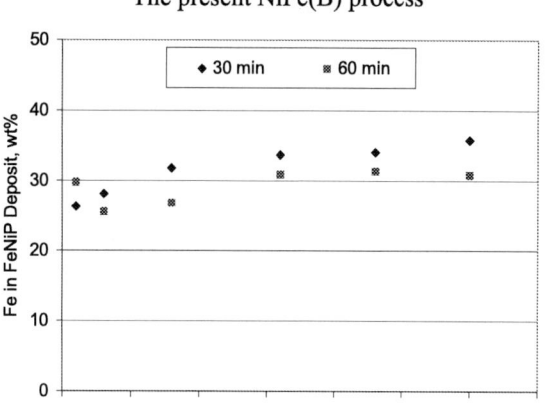

Figure 6. Fe composition in NiFe vs. bath aging

For the baseline process, Fe(II) and citrate form a relatively weak complex after initial bath makeup (logK=4.4), hence, initial free ferrous ion concentration in the bath is high, which results in a high Fe content in the NiFe(P) deposit (up to 45%). After the bath was aged for a few days, a significant change in alloy composition was observed. This is due to instability of ferrous ion in the aqueous solution. As bath ages, ferrous ion is oxidized to ferric ion by dissolved oxygen in the bath, which in turn is either complexed with citrate or precipitates out as $Fe(OH)_3$. As shown in Table 3, the stability constant of ferric-citrate complex is quite high (logK = 11.5), so the concentration of free Fe^{3+} ions in the solution is rather low, which explains the abrupt drop in Fe content in the first few days of bath aging. Afterwards, Fe^{2+}/Fe^{3+} pair reaches an equilibrium and the Fe content in the NiFe(P) deposit remains constant. For the present process, tartrate is used as a complexing agent for ferric ions. Tartrate partially complexes with ferric ions and forms a relatively weak complex, hence, the free Fe^{3+} in the plating solution is relatively high, which explained why the Fe content in the present process is higher. The gradual increases in Fe content over bath aging is due to the ferric/ferrous equilibrium reaction. As a small amount of ferric ions converts to ferrous over time, ferrous ions bond to tartrate less strongly than ferric, so free ferrous ions increase in the plating bath and results in a slightly higher Fe content in the NiFe(B) deposit. Citrate may not replace tartrate in the present bath due to the extremely strong complex formed between citrate and ferric ions, and the free ferric ions in the bath would be too low to yield the desired alloy composition.

Metal ion \ Ligand	Citrate	Tartrate
Fe^{2+}	4.4 (0.1M)	2.2 (0.1M), 1.4 (1M)
Fe^{3+}	11.5 (0.1M)	6.5 (0.1M), 5.7 (1M)
Ni^{2+}	5.4 (0.1M), 5.1 (1M)	2.06 (1M)

Table 3. Stability Constants of Fe(II), Fe(III), and Ni(II) with Citrate and Tartrate (logK, 25°C)

The deposition rate of the baseline NiFe(P) processes as a function of bath aging is shown in Figure 7. The deposition rate of the baseline process is in the range of 1-1.5 μm/hr in the first two weeks and drops to 0.7 μm/hr in the third week, (Fig.7a) The deposition rate of the present NiFe(B) process decreases from 4.5 μm/hr to 2.6 μm/hr after two weeks, and remains at this value afterwards (Fig.7b). The higher deposition rate of the present process is believed to be due to the reducing agent DMAB. DMAB is a stronger reducing agent than sodium hypophosphite. The higher deposition rate of the present process also provides a favorable manufacturing condition suitable for the PCB applications.

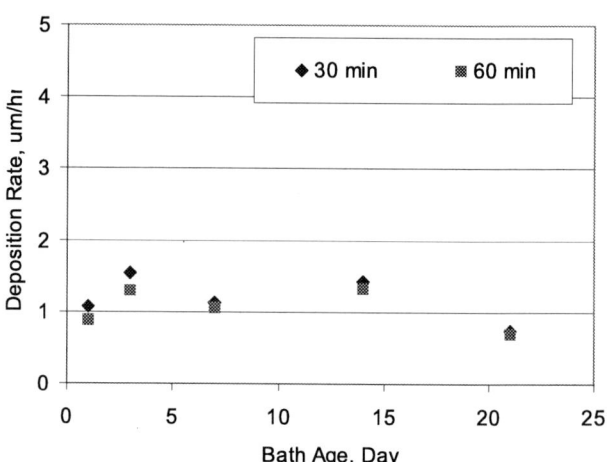

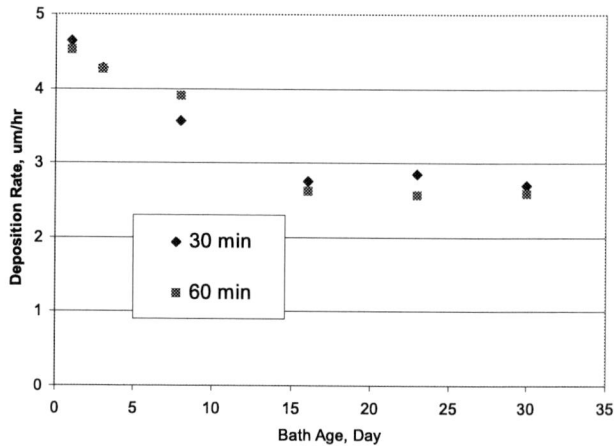

Figure 7. Deposition rate vs. bath aging (Fig. 7a. the baseline process; Fig. 7b. the present process)

Figure 8 shows the microstructure of NiFe(B) deposit examined with a high resolution TEM. The deposit shows columnar grains for the higher Fe content (36%), while no significant grain boundary is observed for a lower Fe content (18% wt).

0.02M DMAB, 18% Fe

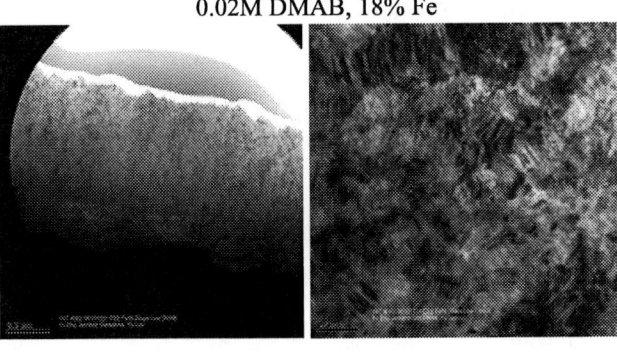

0.05M DMAB, 36% Fe

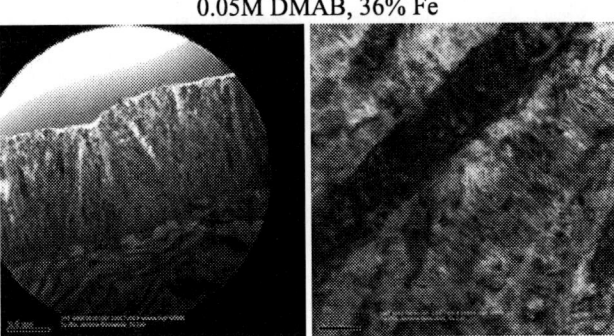

Figure 8. TEM Pictures of NiFe(B) Samples

Run #	UBM Finish	UBM Thk, μm	Reflow Time, min	Wetting Angle, °
1	NiFe(B)	1	2	33
2	NiFe(B)	1	10	29
3	NiFe(B)	1	30	23
4	NiFe(B)	2	2	29
5	NiFe(B)	2	10	27
6	NiFe(B)	2	30	18
7	Ni(P)	1	2	49
8	Ni(P)	1	10	42
9	Ni(P)	1	30	43
10	Ni(P)	2	2	44
11	Ni(P)	2	10	42
12	Ni(P)	2	30	40

Table 4. Matrix of Reflow Test

Interfacial Reactions with Pb-free Solders

Reflow study was carried out with both NiFe(B) and Ni(P) deposited Cu blanks not to investigate the interfacial reactions with Pb-free solder, SAC305. The matrix of reflow study is shown in Table 4, where the thickness of UBM films are 1 μm and 2 μm, and the durations of the reflow time are 2, 10, and 30 min, respectively. After reflow, the cross-sectional samples were prepared by the standard epoxy-mounting and polishing method to examine the interfaces of UBM and lead-free solder.

The contact angle between the reflowed solder ball was measured with these cross-sectional samples (Table 4). The contact angle between Ni(P) and reflowed SAC305 solder balls is in the range of 40 to 49, while it is significantly reduced with NiFe(B) samples, especially for an extended amount of time. This observation is in agreement with the previous work [22].

The Interfaces of UBM and lead-free solder where various IMCs formed during the reflow process was characterized with SEM/EDS for morphology, elemental composition, and thickness of IMC. IMC spalling from the NiFe(B) interface was observed (Figure 9a), while no IMC spalling was observed with Ni(P) samples (Figure 9b). NiCuSn containing IMC was floating above the interface. We believe IMC spalling is due to the relatively high stress of NiFe(B) deposit. Further experiments will be carried out to address this issue. Figure 9 shows elemental maps of IMC formed in Run 6 (Figure 9a) and Run 12 (Figure 9b).

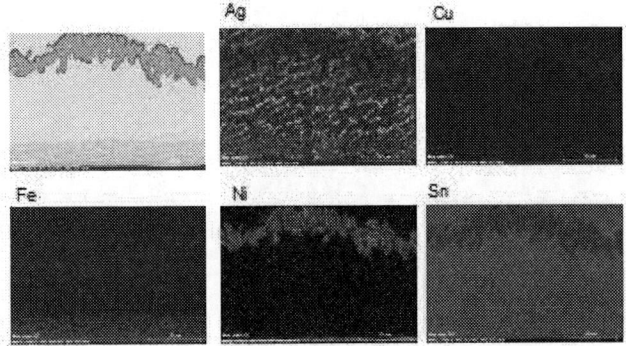

Figure 9a. IMC formed at SAC305 and NiFe(B) interface after 30 min reflow.

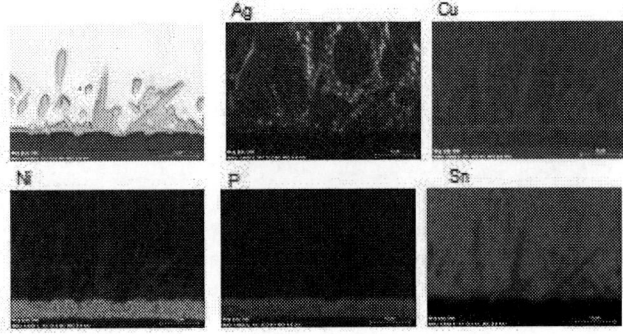

Figure 9b. IMC formed at SAC305 and Ni(P) interface after 30 min reflow

Application of Electroless NiFe Plating to PCB

To demonstrate a practical application of the present NiFe(B) process, the plating evaluation was performed with a FR-4 PCB containing a rectangular array of BGA pads with OSP/Cu finish. A small section of the PCB was plated in the NiFe(B) bath at 65°C for a duration of 30 and 60 min. As shown in Figure 10a, for the initial plating trial, severe bridging between pads as well as skip plating were observed when the bath was not optimized. It was found that bridging was caused by the combination of incomplete rinsing after catalyzing, high operating temperature of electroless NiFe(B) bath, and high concentration of reducing agent in the bath. In the subsequent plating, two approaches were taken to eliminate pads bridging: reduce the amount of reducing agent or lower the temperature. Both approaches were very effective to resolve the bridging issue. Figure 10b shows uniform plating of a NiFe(B) deposit without bridging or skip-plating. Slight difference in alloy composition was observed with the two approaches. The NiFe(B) deposit plated at 65°C with a lower amount of reducing agent has 33% wt Fe, while the one plated at 50°C with the same amount of reducing agent has 40% wt Fe in the deposit.

Without bath optimization (Fig. 10a)

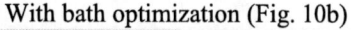

With bath optimization (Fig. 10b)

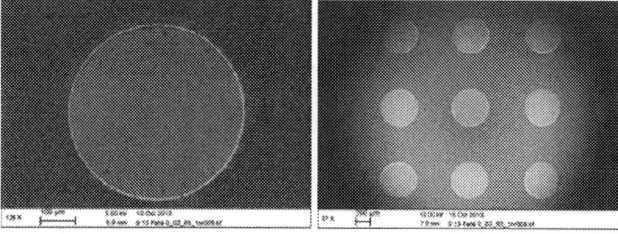

Figure 10. Morphology of NiFe(B) Deposit on BGA Pads

Conclusions

In this work, we developed an electroless NiFe alloy plating process with a high deposition rate and superior bath stability. Ferric salt as the source of iron in the electroless NiFe(B) bath improved the bath stability significantly. Temperature, agitation, and concentration of the reducing agent had a significant impact on deposition rate and alloy composition. Deposition rate of NiFe(B) alloy increases with increasing bath temperature and the concentration of reducing agent. Solution agitation has an adverse effect on deposition rate. Lower temperature and higher concentration of the reducing agent yield FeNi(B) deposit with a higher Fe content. A deposition rate of 3 micron per hour was achieved and maintained for the duration of five weeks. A stable Fe content around 30% in NiFe deposit was achieved in this timeframe. It was found that the crystal structure of NiFe(B) deposit was correlated to Fe content in the deposit. As the Fe content goes up, NiFe(B) deposit becomes more crystalline. IMC spalling was observed at the interface of Pb-free SnAgCu solder and electroless NiFe(B) after solder reflow, which is believed to be related to a high stress in NiFe(B) deposit. The present NiFe(B) process can be applied for PCB as a surface finish.

Acknowledgements

The authors would like to thank Clare J Mccarthy (IBM STG, Hopewell Junction) for SEM/EDS work, Dr. Yu Zhu

(IBM Research, Yorktown) for TEM work, and Paul Lauro for supporting cross-section sample preparation. The authors also acknowledge the collaboration with Dr. Jae-ho Lee (Hongik Univ, Korea) through the IBM-Hongik-Univ joint study program.

References

1. K.N. Subramanian (ed), "Lead-free Solders; Materials Reliability for Electronics," Wiley Series in Materials for Electronic & Optoelec. Appl, (2012).
2. S.K. Kang, et al (eds), "Special Issue on Pb-free Solders," J. Electronic Mat's, Vol.38, No.12, Dec. (2009).
3. K.N. Subramanian (ed), "Special Issue on Lead-Free Electronic Solders," J. Materials Sci., Vol.18, No.1-3, Mar. (2007).
4. S.K. Kang, D.Y. Shih, W.E. Bernier, "Flip-Chip Interconnections – Past, Present & Future,"in *Advanced Flip-Chip Packaging* (eds,H.Tong, Y. Lai, C. Wong), Chap.5, p.85, Springer, April, (2013).
5. S.K. Kang, R.S. Rai, S. Purushothaman, "Interfacial Reactions During Soldering with Lead-Tin Eutectic and Pb-Free, Tin-Rich Solders," J. Electron. Mater., Vol. 25, No.7 p.1113, (1996).
6. S.K. Kang, D.Y. Shih, K. Fogel, P. Lauro, M.J. Yim, G. Advocate, M. Griffin, C. Goldsmith, D.W. Henderson, T. Gosselin, D. King, J. Konrad, A. Sarkhel, K.J. Puttlitz, "Interfacial Reaction Studies on Lead (Pb)-Free Solder Alloys," IEEE Trans. Electron. Packag. Manuf., Vol. 25, No.3, p.155, (2002).
7. S.K. Kang, W.K. Choi, D.Y. Shih, P. Lauro, D.W. Henderson, T. Gosselin, D.N. Leonard, "Interfacial Reactions, Microstructure and Mechanical Properties of Pb-Free Solder Joints in PBGA Laminates," Proc 52nd ECTC, San Diego, p. 147, (2002).
8. K.W. Moon, W.J. Boettinger, U.R. Kattner, F.S. Biancaniello, C.A. Handwerker, "Experimental and Thermodynamic Assessment of Sn-Ag-Cu Solder Alloys", J. Electron. Mater., Vol. 29, No. 10, p.1122, (2000).
9. G. Ghosh, "Thermodynamic Modeling of the Nickel-Lead-Tin System," Metall. Mater. Trans. A, 30A, p.1481, (1999).
10. http://en.wikipedia.org/wiki/Electroless_nickel_immersion_gold
11. Y.C. Sohn, J.Yu, S.K. Kang, D.Y. Shih, and T.Y. Lee, "Spalling of Intermetallic Compounds During the Reaction Between Lead-free Solders and Electroless Ni-P Metallization," J. Materials Research, Vol.19, No.8, p.2428, (2004).
12. C.E. Ho, S.C. Yang, C.R. Kao, "Interfacial Reaction Issues for Lead-free Electronic Solders," J. Materials Sci., Vol.18, No.1-3, p. 155, Mar. (2007).
13. K. Zeng, R. Stierman, D. Abbott, M. Murtuza, "The Root Cause of Black Pad Failure of Solder Joints with Electroless Ni/Immersion Gold Plating," JOM, p.75, June, (2006).

14. P.C. Andricacos, M. Datta, W.J. Horkans, S.K. Kang, K.T. Kwietniak, G.S. Mathad, S. Purushothaman, L. Shi, H-M Tong, "Flip-Chip Interconnections Using Lead-free Solders," US Patent No.6,224,690 B1, May 1, (2001).

15. Y-T Cheng, S.R. Chiras, D.W. Henderson, S.K. Kang, S.J. Kilpatrick, H.A. Nye, C.J. Sambucetti, D-Y Shih, "Ball Limiting Metallurgy, Interconnection Structure Including the same, and Method of Forming an Interconnection Structure," US Patent No.7,273,803, Sep.25, (2007).

16. K.E. Fogel, B. Ghosal, S.K. Kang, S. Kilpatrick, P.A. Lauro, H.A. Nye, D-Y Shih, D.S. Zupanski-Nielsen, "Interconnections for Flip-Chip using Lead-free Solders and Having Reaction Barrier Layers," US Patent No.7,410,833 B2, Aug.12, (2008).

17. S.K. Kang, J.Horkans, P.Andricacos, R.Crruthers, J.Cotte, M.Datta, P.Gruber, J.Harper, K.Kwietniak, C.Sambucetti, L.Shi, G.Brouillette and D.Danovitch, "Pb-Free Solder Alloys for Flip Chip Applications", Proc.49th ECTC, San Diego, CA, p.283, June, (1999).

18. J. Guo, L. Zhang, A. Xian, J.K. Shang, "Solderability of Electrodeposited Fe-Ni Alloys with Eutectic SnAgCu Solders," J. Mater. Sci. Tech., Vol.23, No.6, p.811, (2007).

19. B. Dang, S. Wright, J. Maria, C. Tsang, P. Andry, L. Wiggins, and J. Knickerbocker "NiFe-based BLM for Microbumps at 50 μm Pitch in 3D Chip Stalks," 63rd ECTC, p.1595, (2013).

20. S.K. Kang, P. Lauro, M. Lu, D.Y. Shih, "Electromigration-resistant Under-bump Metallizaiton of Nickel-Iron Alloys for Sn-rich Soder Bumps of Pb-Free Flip-Chip Applications," US Patent No.2011/0156256 A1, Jun. 30, (2011).

21. D. W. Hall, J. A. Linholm, L.T. Romankiw,, A.F. Schmeckenbecher "Process for Electrolessly Plating Magnetic Thin Films," US Patent No. 3,702,263, Nov.7, (1972).

22. H. Zhou, J. Guo, Q. Zhu, J.K Shang, "Application of Electroless Fe-42Ni(P) Film for Under-bump Metallization on Solder Joint," J. Mater. Sci. Technol., Vol.29 (1), p.7, (2013).

23. M.W. Jung, S.K. Kang, J-H. Lee, "Effects of Sodium Citrate Concentration on the Electroless Ni-Fe Bath Stability and Deposits," J. Electronic Materials, Vol.43, No.1, p.290, Jan. (2014).

Novel Conductive Paste Using Hybrid Silver Sintering Technology for High Reliability Power Semiconductor Packaging

Howard(HWA IL) Jin, Senthil Kanagavel, Wai Foo Chin
Alpha Advanced Materials, an Alent plc Company
3950 Johns Creek Court, Suite 300, Suwanee GA 30041, USA
E-mail : hjin@alent.com

Abstract

A number of industries are developing new product-lines with innovative electronics packages that require low thermal resistance and high temperature stability, including hybrid electric vehicles, concentrator photovoltaic and wide bandgap RF amplifiers. One of the largest obstacles in the design and manufacture is the thermal management of the devices. Localized heat generation is the characteristic of the semiconductor chips used in these devices. For high power applications the thermal impedance of the die attach layer can play significant role in the thermal management and the operating temperature. Therefore, one would like to use the highest thermal conductivity and lowest thermal resistance die attach material that is capable for high volume manufacturing.

In a typical electronic packaging process, chips are attached to substrates and electrically connected before they are encapsulated or sealed for protection. The attachment and electrical interconnections provide the chip with an infrastructure for the flow of electrical signals, mechanical support and heat removal. The die attach materials[1] used in the packaging of high performance power semiconductors are required to have high thermal conductivity. Lead solders, eutectic gold-tin, transient liquid phase sintering (TLPS) pastes[2] and nano-silver sintering technology[3] are typical materials used for the die attachment of power semiconductor. This paper will introduce a new die attach material using Hybrid Silver Sintering Technology (HSST).

Introduction

Power semiconductor packaging engineers are looking for Pb-free alternative to traditional high Pb solder die attach paste and wire . Lead solders have respectable thermal conductivity of 30-50W/mk, but have known process difficulties in high volume mass production such as voiding, bond line control, and the requirement inert gas(Nitrogen or Forming gas) environment. Lead is now categorized as hazardous substance to human body and environment and products containing are scheduled to be banned. Silver epoxy paste used in standard semiconductor packaging is another die attach technology but its thermal conductivity is not high enough for Power devices. Silver sintering materials have became an attractive alternative for power devices because they possess high thermal conductivity (150~200W/MK) achieved through solid state diffusion or "silver sintering". But most often high bonding temperature and pressure are required to achieve a high reliability joint when silver sintering technology is employed. Similar to soldering, silver

sintering technology requires backside metallization due to slow diffusion into bare silicon. HSST is composed of micron size silver powder and an organic phase. This technology overcomes the limitations of conventional silver epoxy and silver sintering products by using the unique design of polymer composition. The unique organic and polymer composition facilitates silver sintering at relatively low temperature compared with sintering silver and enables up to 150W/MK thermal conductivity without pressure during cure. The polymer enables adhesion to a variety of surfaces, bare silicon, gold and silver metalized die and silver and copper metal surfaces.. The viscosity of hybrid silver sintering product is similar to standard epoxy paste and its application process is as easy as conventional silver epoxy die attach paste. HSST paste can easily drop into existing commercial die bonders and serve as a replacement for soft solder die attach.

Experimental Procedure

A. Sample Preparation

Five silver paste samples were prepared. Sample A-1, A-2, A-3, C-1 have 7 wt% thermoset resin and B-1 and D-1 have 13wt% organic resin. The thermoset resin has more than one reaction group in a molecule and reacts above 150°C. All the samples have organic enhancers except C-1 and D-1. E-1 has silver, diluents and organic enhancers but doesn't have any thermoset resin. Three different diluents are used for the sample preparation of A-1, A-2 and A-3 and the property of each diluent is in Table2. A-1 has the highest boiling temperature diluent and A-3 has the lowest boiling temperature diluent. The detail compositions are listed in Table 1. The silver used in the formulation is micron size flake and the average size is 5um. The SEM image of the silver flake is in Figure 1.

Table 1. Silver paste composition

	Silver wt%	Thermoset resin	Diluent	Organic Enhancer
A-1	80%	7%	Type1, 13%	Yes
A-2	80%	7%	Type2, 13%	Yes
A-3	80%	7%	Type3, 13%	Yes
B-1	80%	13%	Type1, 7%	Yes
C-1	80%	7%	Type1,13%	No
D-1	80%	13%	Type1, 7%	No
E-1	80%	0%	Type1, 20%	Yes

Table 2. Diluents

	Type 1	Type 2	Type 3
Boiling temperature	230 °C	190 °C	171 °C
Vapor pressure at 20°C	0.04 mmHg	0.3 mmHg	0.85 mmHg

Figure1. SEM image of micron size silver flake

B. Measurement and evaluation result

B-1. Dry-out and void

A-1, A-2 and A-3 were tested on dry-out and void. The dry-out test is to test if paste can achieve uniform bond line thickness up to one hour staging time. Each paste was dispensed on lead frame by needle using an automated dispenser and die was placed on the paste immediately after dispensing and then subsequently after one hour staging. The samples were cured by box oven following the curing profile of Figure 3. The adhesive thicknesses of 0 minute and 60 minutes staged samples were measured using a micrometer. Less than 25% change in thickness after the one hour staging is considered a pass for the dry-out test.

Figure2. Dry-out Test process

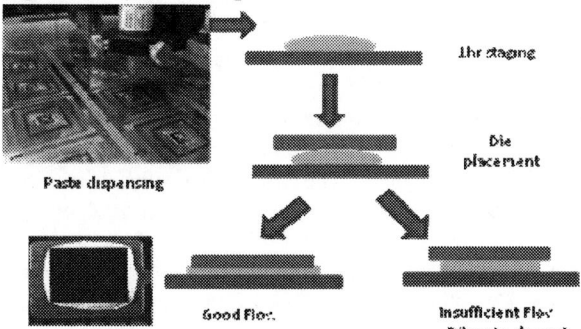

To determine the level of voiding in the bond line, the cured sample of A-1, A-2 and A-3 were tested by X-ray. A 3mm x 3mm gold plated silicon die was used along with a silver plated copper lead frame. The die attached samples

were prepared without staging and cured in box oven with Nitrogen environment used during the curing process.

1. Ramp 30 min from 23°C to 150°C
2. Hold 30 min at 150°C
3. Ramp 40 min from 150C to 250°C
4. Hold 60 min at 250°C
5. Cool down from 250°C to room temperature

Figure3. Curing profile with box oven

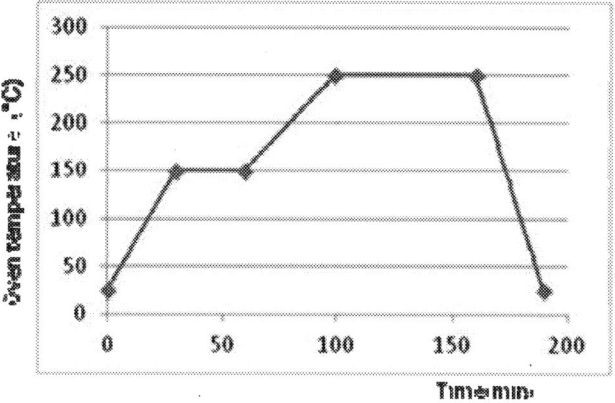

B-2.Die shear strength

The die shear strength of A-1, B-1 and C-1 were tested using a Dage 4000 bondtester equipped with heat block was used for the evaluation. Figure 4 showed the measurement set up. The paste was first dispensed on a lead frame and die was placed on the dispensed paste with appropriate bonding force to achieve ~300 micron filet around the die. Silver plated copper lead frames and gold plated or bare silicon dies were used for this experiment. The die size is 3mm x 3mm and the thickness is 250 microns. The die attached samples were cured in box oven and the final dry thickness is 25 microns +/- 3 microns. The curing profile is shown in Figure 3 and the die shear strength was measured on heated block as shown in Figure 4.

Figure4. Schematic illustration of die shear test

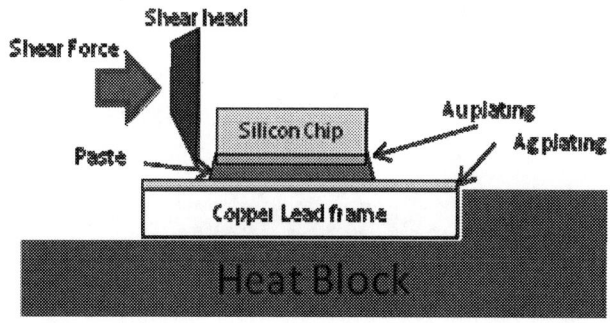

Figure 4 shows the die shear strength of A-1, B-1, C-1 and D-1 without heating the heat block. Figure 5 shows the die shear strength with 260°C heat block temperature.

B-3.Thermal conductivity measurement: Bulk & K_{eff}

The bulk thermal conductivity of A-1, B-1, C-1 and D-1 were measured by Netzsch LFA 447N Nanoflash instrument. The samples were cured with the same cure profile as the die shear sample preparation of Figure 3. The sample size is 8mm by 8 mm with 1.5mm thickness.

In many cases, the bulk thermal conductivity data of die attach paste doesn't have correlation with the thermal resistance in the package level because it only represents bulk conductivity and doesn't consider interfacial resistance between die and adhesive layer. To obtain more information about the contact or interfacial resistance, we developed a test method in which we determine the "effective" conductivity, termed K_{eff}, which accounts for the loss of thermal conductivity due to die attach layer to substrate and die attach to die resistance. The K_{eff} sample is prepared by making sandwich with die attach paste and two gold metalized silicon die as shown in Figure 5. The thickness of the adhesive layer is 23-27 micron thickness.

Figure5. K_{eff} sample

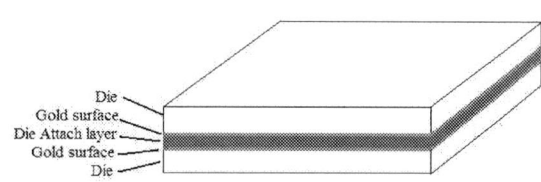

Netzch LFA 447 Nanoflash instrument was used to measure K_eff .

B-4.Adhesive bond line thickness measurement during cure

A 50 micron thick samples of A-1, B-1 and E-1 were prepared by printing on glass slide. In order to print 50 micron thickness of die attach paste, 50 micron PSA (pressure sensitive adhesive) tapes were used as a spacer and taped on the left and right sides of glass slide.

Figure 6. 50 micron thickness sample preparation of A-1, B-1 and E-1.

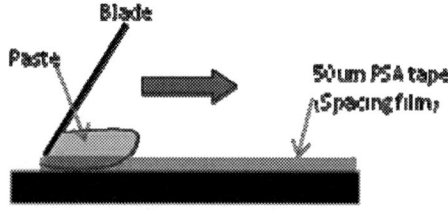

The curing profile of the printed samples is shown in Figure 3 and the thickness was measured during the curing process. The measurement was conducted at three points of below.

①Before cure
②After 150°C for 30min
③After full cure

Figure7. Bond line thickness measurement during cure

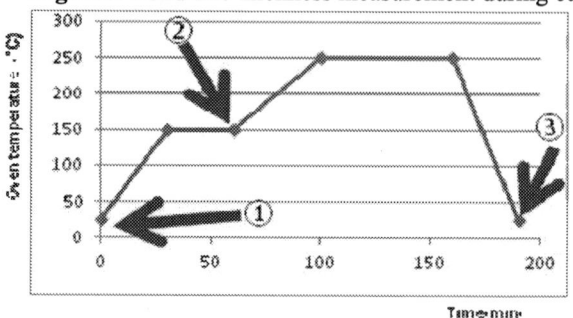

B-5. Viscosity measurement and needle dispensing

Viscosity and thixotropic index are the material properties to indicate the dispensing performance of die attach material. The viscosity of A-1 and B-1 were measured by Brookfield HBDVIII+ viscometer and compared with a silver epoxy, which has good dispensing performance with needle.

Thixotropic index was calculated from the viscosity numbers with 0.5rpm and 5rpm spindle speed.

Thixotropic index = (0.5rpm viscosity)/(5rpm viscosity)

Result and Discussion

A. Diluents test

The properties of the diluent are very critical for making consistent bond line thickness, controlling flow and being void free after the curing process. If the diluent in die attach paste dries prematurely at room environment, the die attach adhesive will thicken in short time period after dispensing and not be able to flow well enough to cover the die. The result is a thicker bond line than targeted and insufficient die coverage. Die attach paste needs to flow consistently and uniformly to achieve good die coverage at least one hr after dispensing. Thus, the diluents in die attach paste should not evaporate for at least one hour under ambient conditions. The results are summarized in Table 3. A-2 and A-3 dried too quickly and failed the one hour staging test and resulted in too thick of a bond line. Thus, we concluded that the diluents type 2 and type 3 in the formulations are not appropriate diluents for die attach application.

On the other hand, if the diluents do not evaporate during oven cure process, they will be captured inside of die attach layer. The capture diluents would become voids and negatively affect thermal conductivity, adhesion strength and reliability performance. The void test results were shown in Table 4 and the three samples showed no void or minimal voids. Thus, we concluded the diluents of three formulations were adequately removed during the curing profile described in Figure3. Type 1 diluent was found to be the best diluent with acceptable staging time and minimal voiding.

Table 3. Dry-out result

	A-1	A-2	A-3
Bond line thickness: No staging	35um	37um	38um
Bond line thickness after 1hr staging	38um	49um	73um
Bond line thickness change	7.9%	32.4%	92.1%
Dry-out test result	Pass	Fail	Fail

Table4. X-ray image of void after cure. Die size is 3mm x 3mm

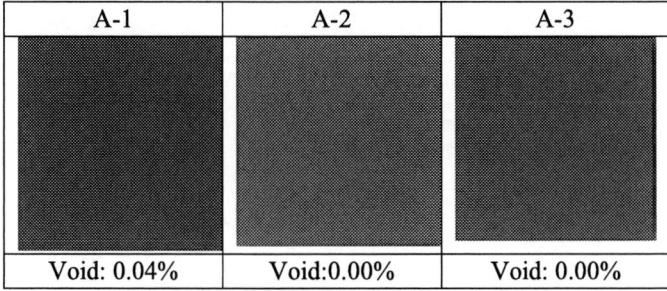

A-1	A-2	A-3
Void: 0.04%	Void:0.00%	Void: 0.00%

B. Die shear strength

The die shear strength of A-1, B-1, C-1 and D-1 were measured at room temperature and 260°C(heat block temperature) and the results are shown in Figure 8 and 9. A-1 and B-1have organic enhancers and higher die shear strength than C-1and D-1, which don't have organic enhancers in the formulation. We believe the organic enhancers play a role in the die shear strength because they seem to enhance the sintering of silver particles. The silver sintering can increase cohesive strength of die attach layer and also adhesion strength to metal substrate and metalized die. If comparing the die shear strength of A-1 and B-1 to gold plated and bare silicon dies, the die shear strength to bare silicon die is as good as to gold plated die. Other die attach materials such as solder, gold-tin and nano-silver sintering products requires metal plated surface to have respectable adhesion strength. They cannot have good adhesion to bare silicon die because they cannot diffuse into silicon surface. But HSST die attach paste shows good adhesion with bare silicon die because it has polymer component to enable the paste to bond to the silicon surface.

Figure8. Box plot of the die shear strength at room temperature

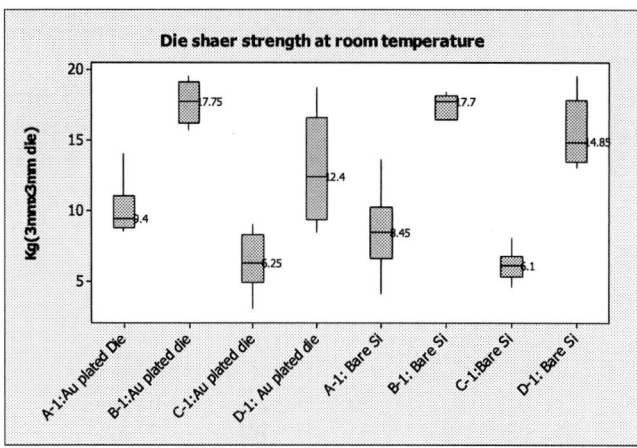

If comparing the die shear strength of A-1 and B-1, A-1 has lower adhesion at room temperature than B-1 but higher adhesion at 260°C shown in Figure9. A-1 has the higher silver loading and can have more sintering than B-1. The die shear strength of A-1 at 260°C is good and not much different from the room temperature because the cohesive strength of the sintered silver is maintained through 260°C as long as it is below the melting temperature of silver. Formulation B-1 has relatively low silver and high resin ratio, so the adhesion is affected by die shear temperature. The die shear strength at room temperature is relatively high due to the thermoset resin but drops dramatically at 260°C because the thermoset resin has the glass transition temperature at 150°C and becomes soft at the 260°C die shear test condition.

Figure9. Box plot of die shear test with 260°C heat block temperature

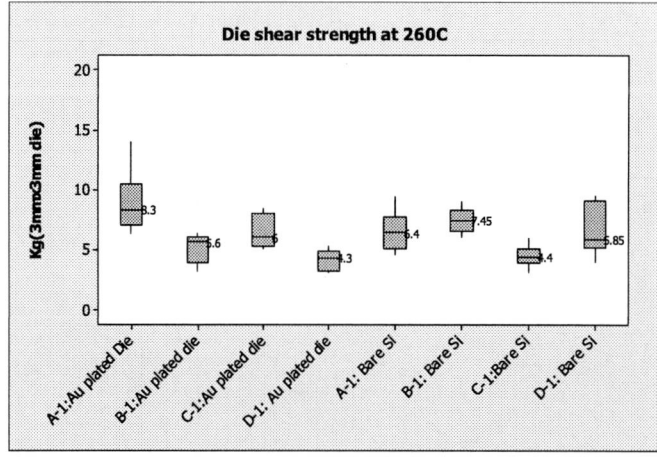

C. Bulk thermal conductivity and K_eff

Figure10 showed the bulk thermal conductivity and K_{eff} of A-1, B-1, C-1 and D-1 after cured by the curing profile of Figure3. A-1 has the highest thermal conductivity and K_{eff} among them. B-1 has higher thermal conductivity than D-1. The organic enhancer is doing significant role for bulk thermal conductivity and K_{eff}. A-1 has higher thermal conductivity than B-1 because it has higher silver ratio than the other after curing.

K_{eff} is measured of the sandwiched sample by two dies(Figure 5) and represents bulk and interface conductivity. K_{eff} number is more relevant than bulk thermal conductivity to the die attach thermal resistance in package level.

Pb solder has the bulk conductivity in the range of 30 to 50 W/MK and tin-gold has 58 W/MK. A-1 and B-1have comparable thermal conductivity with tin-gold and Pb-solder.

Figure10. Bulk thermal conductivity and K_{eff} of A-1, B-1, C-1 and D-1

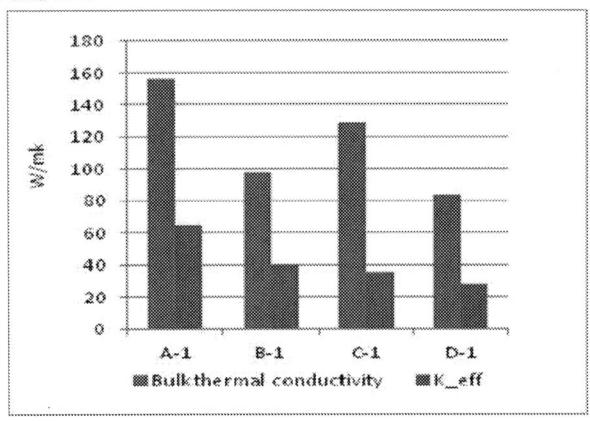

D. Bond line thickness change during cure

The wet paste thickness shrinks during cure due to the evaporation of the diluents and the shrinkage from the thermoset resin cross linking. The paste samples of A-1, B-1 and E-1 were cured by the curing profile shown in Figure 3 and the bond line thickness was measured before and after cure and during cure.

Figure11. Bond line thickness change during cure : A-1, B-1 and E-1

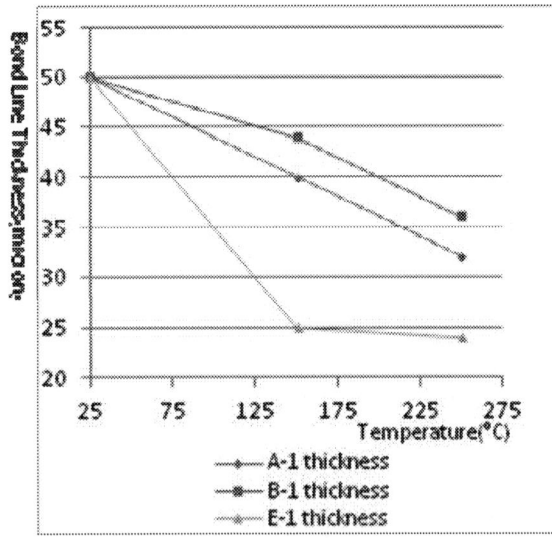

E-1, which has no thermoset resin and 20 wt% of diluent in the formulation, showed significant shrinkage from room temperature to 150°C but little shrinkage from 150°C to

250°C. However, A-1 and B-1 showed significant shrinkage from 150°C to 250°C as well as room temperature to 150°C. The weight loss of A-1, B-1 and E-1 measured by TGA(thermal gravimetric analysis). The temperature profile is the same as the curing profile of Figure 3. According to the TGA graph shown in Figure 12, the diluent of E-1 was fully removed by the end of 150°C. The diluents of A-1 and B-1 were also removed more than 98% from room temperature to 150°C. But A-1 and B-1 shrank 20% and 18% respectively from 150°C to 250°C. The shrinkage of E-1 is mostly driven by the diluent evaporation. However, the shrinkage of A-1 and B-1 were induced not only by diluent removal but also by resin shrinkage. It is known that the epoxy cure reaction can induce the resin shrinkage more than 6% by volume[4],[5].

Figure12. TGA(thermal gravimetric analysis) weight loss of A-1, B-1 and C-1 by the temperature profile of Figure3.

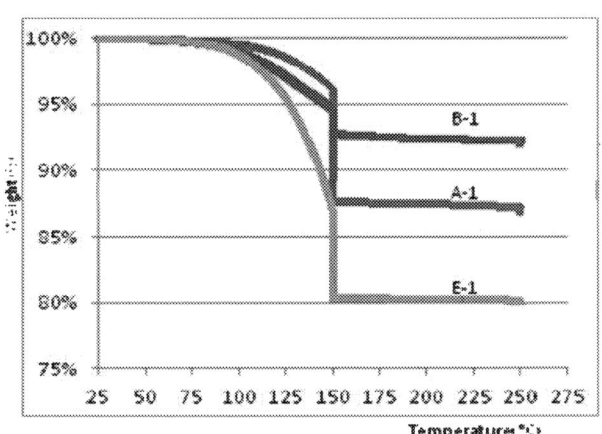

E. Viscosity and needle dispensing

Viscosity and thixotropic index are important material properties related to dispensing performance. The viscosity and thixotropic index of A-1 and B-1 are not much different from silver epoxy used in mass production today as shown in Table 5. If viscosity is higher than 25K cps, the paste cannot flow through small needle such as 23gage needle which will result in poor dispensing. If thixotropic index would be lower than 3.5, the dispensed pattern will not be uniform.

Table5. Viscosity of A-1 and B-1 comparing with Ag Epoxy

	A-1	B-1	Ag Epoxy Atrox™ 5582A)
Viscosity	10500 cps	13800 cps	9000 cps
Thixotropic Index	4.9	5.7	4.2

The dispensing test was conducted to validate the performance of A-1 and B-1. A Camalot 1818 time pressure dispenser was used for the evaluation. The needle used for the test was 23 gage size and the pressure was 20 psi. Both A-1 and B-1 showed uniform dispensed patterns without tailing.

Figure13. Dispensed pattern of A-1 and B-1

Paste Manufacturing and Application

Figure 14 illustrates the basic procedure of making hybrid silver paste and processing it.

Silver flakes are dispersed in the mixture of thermoset resin and diluents. The mixed paste samples are cured in box oven following the two step curing profile shown in figure 3. The diluents evaporate in the 1st step cure temperature (150°C). The thermoset resin starts cross-linkage from 150°C to 250°C. The resin shrinks by the cross-linking reactions and the resin shrinkage forces silver flakes get closer each other. The silver flakes eventually sinter at or below 250°C due to their close proximity

Figure14. Process diagram for making hybrid silver sintering formulation and silver sintering by resin cross-linking

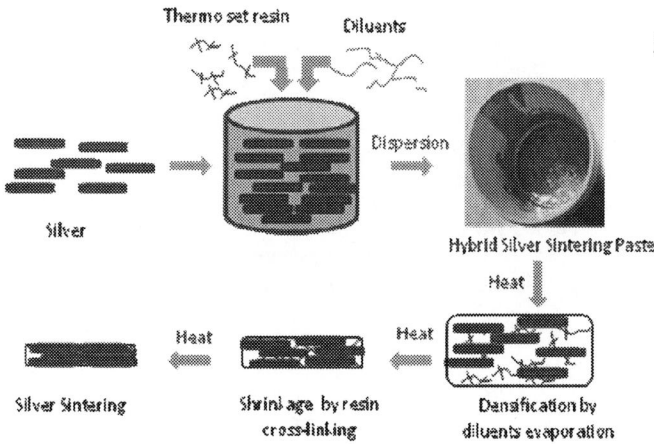

Summary and Conclusion

Hybrid silver sintering technology (HSST) is the mixture of micron silver flakes, thermoset resin and diluents. HSST described in this paper takes advantage of the shrinkage force of the selected thermoset resin to bring particles closer together. We have demonstrated that with the correct formulation approach micron silver can effectively sinter at reasonable process temperatures without pressure during the curing process.

The formulation approach to use thermosetting resin enables hybrid silver sintering products to have excellent adhesion to bare silicon die. This is a primary advantage over Pb solder, eutectic gold-tin, transient liquid phase sintering (TLPS) and nano-silver sintering materials that require back side metallization

The formulations presented in this paper are capable of achieving very high bulk, high effective conductivities, low thermal resistance and can serve as a solder replacement. The HSST formulations presented also have the correct rheology and dispense characteristics for high volume manufacturing on commercial available die bonders

References

1. Yi Li, Daniel Lu and C.P. Wong, Electrical Conductive Adhesives with Nanotechnologies, Springer, New York, 2009, pp. 343–345.
2. Matt Wrosch and Arsenia Soriano "Sintered Conductive Adhesives for High Temperature Packaging". *IEEE Electronic Components and Technol. Conf. (ECTC)*, Las Vegas, NV, June 1–4, 2010, pp. 973–978.
3. Cyril Buttay, Amandine Masson, Jianfeng Li and Mark Johnson, "Die Attach of Power Device Uisng Silver Sintering-Bonding Process Optimization and Characterization," HiTEN, Oxford, United Kingdom, hal-00672619, version 1, 21Feb., 2012
4. Mauro Zarrelli, Alexandros ASkordos and Ivana K Partridge, "Investigation of cure induced shrinkage in unreinforced epoxy resin", Plastics, Rubber and Composites Processing and Applications 31, 377-384, 2002
5. B. Yates, B. A. McCalla, L. N. Philips, D. M. Kingston-Lee and K. F. Rogers, J Mater Sci, 14, 1207-1217, 1979

Novel Low Temperature Curable Photo-sensitive Insulator

Kenji OKAMOTO, Hikaru MIZUNO, Tomohiko SAKURAI, and Katsumi INOMATA
Device Integration Materials Laboratory, Fine Electronic Materials Research Laboratories
Yokkaichi Research Center, JSR Corporation
100, Kawajiri-cho, Yokkaichi, Mie, 510-8552, Japan

Abstract

Currently we achived to develop new lower modulus (<1.8GPa) materials with a small film shrinkage (<15%), lower residual stress (<20Mpa), and good chemical resistance at low curing temperature around 200C. These new materials provide good lithography performince at verious film thikness from 5um to 30um without deterioration of physical properties. The whole performance have been designed with new polymer concept which is containing flexible unit and crosslinking unit in polymers, and are also well-balanced to satisfy various reliability performance at thermal cycle test (TCT), PCT and high accelated stress test (HAST) conditions. For this new deisgned polymers, we hamonized our own lithography knowledge for semiconductior and pertochemical technology for polymerization technique.

Introduction

In recent years, packaging structure for semiconductors became more diversified and complicated to satisfy various requirements for 3D-IC and WL packaging technologies [1-5] such as 2.5D interposers, 3D-TSV, FO-WLP, WL-CSP,and Flip-chip wafer bumping. For these structures, photo-sensitive organic materials used for buffer coating, passivation layers, and insulator for re-distribution, are required for a low cure temperature, low modulus, low residual stress, high impact resistance, high resolution, and etc. Specially, low curing temperature for thinner and larger wafer is required to keep higher reliability without any errors such as delaminations, crackings and etc. On the other hand, one of the most famous and conventional heat resistance insulator materials is polyimide [6] [7]. And originally photo-sensitive polyimide type insulators were negative-tone and were developed with organic solvents [8] [9]. In these days, various type of alkaline developable matrials [10] [11] [12] are developping due to reduce the running cost and the enviromental burden. And also, the other potential issue of these materials is higher curing temperature over 300°C and it causes wafer bow and possible damage to devise wafers.

Our current products of photo-sensitive insulator are containing the phenolic resin as the main polymer to perform good lithgrapy and cured film properties. Generally the phenolic resins have known well for the print circuit bords (PCB) because of good insulation, good heat resistance, and so on. And the other components for our photo-sensitive insulator are epoxy compounds and melamine compounds for the cross-linker and naphthoquinone diazide (DNQ) type photo-sensitive compounds for having positive tone lithography performance. And also our current design contains the cross-linking rubber particles to satisfy the required reliabilities such as TCT, PCT, HAST, and so on, as the insulator. This rubber partciles are mainly composed with styrene and butadiene, and are applied hydrophilic treatment with several monomers to have alkali-solubility and compatibility with the other components. Figure 1 shows the picture of TEM observation and the design concept of our cross-linking rubber particles. With TEM observation results, the rubber particles are uniformly dispersed in the cured film, and the particle size are around 70nm. This uniformed dispersion of the cross-linking rubber particles in the cured films can release the residual stress to achive good realibility performance. And the current products including the cross-linking rubber particles are curable at 200°C with less film shrinkage, and satisfy the physical properties, chemical resistances, adhesion to the several material substrates, specially the residual stress is around 20MPa. And additionally, the lithograpy performance margin such as exposure dose, and development time are wide enough to proceed stable process. This current products are used for FO-WLP, WL-CSP, and Flip-chip at low temperature cure process.

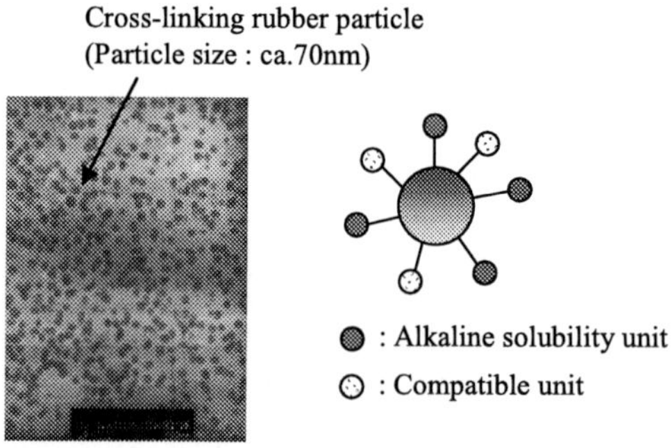

Cross-linking rubber particle
(Particle size : ca.70nm)

● : Alkaline solubility unit
○ : Compatible unit

Figure 1. TEM observation picture and design concept of cross-linking rubber particle.

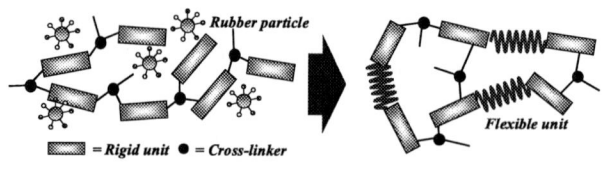

= Rigid unit ● = Cross-linker

Rigid structure   **Flexible cross-linking structure**

Figure 2. Design concept of flexible cross-linking structure

Moreover, to support the current market requirement such as down-sizing, multi-functionalization, weight-reduction of the electronic devices, semiconductor packages are getting smaller and thinner. Under this situation, we newly developed the phenolic resins containing the monomers with flexible unit to improve the flexibility of cured film, comparing the current design containing the dispersed cross-linking rubber particles.

2014 Electronic Components & Technology Conference

Physical performances of new designed polymers

<u>Polymer synthesis</u>: The new phenolic resins are polymerized p-tert-butoxy styrene (PTBST), the monomer with flexible unit, and so on. After polymerization, those resins are purified with re-precipitation or liquid-liquid separation to remove the remaining monomers. Then, PTBST in resins is hydrolyzed with HCl or H_2SO_4 to p-hydroxy styrene. After washing with water and solvent swapping, finally the new phenolic resins containing flexible unit are prepared. Mw is from 3,000 to 12,000. And the monomer ratio is analyzed with NMR.

<u>Composition of photo-sensitive insulator</u>: Our positive-tone insulators are composed with the phenolic resins as main polymer, cross-linker such as epoxide compounds and melamine compounds [13], photo-initiators and additives. And at this new study, the new designed photo-sensitive insulator contained the new phenolic resins containing the monomer with flexible unit. In the case of the current products contained the cross-linking rubber particles. Regarding lithography performance, DNQ compounds contained as photo-initiators work for positive-tone resists and decomposed with UV light as the following reaction scheme as shown Figure 3. After the exposure with UV light, DNQ unit is decomposed to indene carboxylic acid [14], and then the exposed area is developable with TMAH solutions.

And also, DNQ compounds are well-known to perform higher alkaline solubility contrast between exposed area and unexposed area as Figure 4, because DNQ compounds have the electro-static interaction with phenolic resins [15]. As the results, DNQ system can be expected to generate higher remaining film thickness, good pattern profile, and better resolution due to that high contrast.

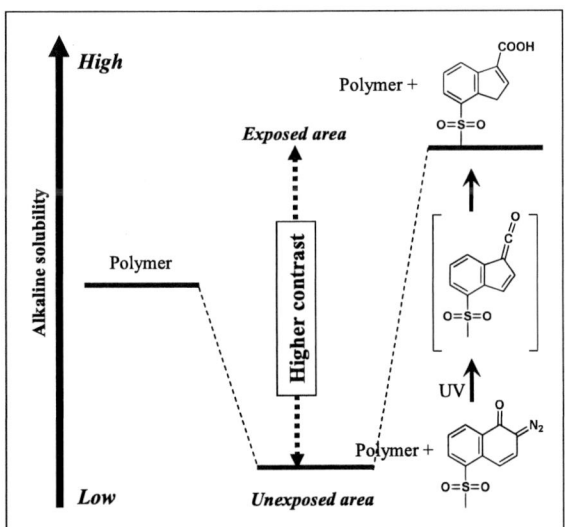

Figure 3. Reaction scheme of DNQ compounds and conceptual scheme of positive tone resist dissolution contrast

<u>DOE for polymer screening</u>: Table 1 shows the DOE for polymer screening to find out the optimal Mw and monomer ratio of monomer with the flexible unit. The mixture of the normal phenolic resin, which is p-hydroxy styrene and styrene, and the cross-linking rubber particle were as reference for this comparison of the various performances such as physical properties, chemical resistance with PR stripper, and basic lithography performance. In this study, the variation of Mw as from around 3000 to around 11000, and the variation of monomer ratio of the flexible unit was from 30mol% to 70mol%.

These samples except for the reference on Table 1 were prepared with the same chemical compositions which was discribed as above and with the predetermined preparation procedure to compare the performances properly.

Sample	Ref.	Sample 1	Sample 2	Sample 3	Sample 4	Sample 5
Polymer	Current system *1)	Polymer 1	Polymer 2	Polymer 3	Polymer 4	Polymer 5
Monomer ratio with flexible unit	0 mol%	50 mol%	30 mol%	30 mol%	30 mol%	70 mol%
Mw	10000	11000	3300	6800	9000	10000
Cured film properties @ 200°C*60min						
Tg	210°C	> 250°C	210°C	> 250°C	> 250°C	> 250°C
Elastic Modulus	2.5 Gpa	1.8 Gpa	1.7 GPa	2.0 GPa	2.1 GPa	1.6 GPa
Chemical resistance (TMAH/DMSO) *2)	OK	OK (better)	OK (worse)	OK	OK	OK (worse)
Lithography performance	OK	OK	OK	OK	OK	NG

*1) Phenolic resin and cross-linking rubber particle, *2) RT*10min

Table 1. DOE for polymer screening and physical properties of cured film

Table 2 shows the standard sample conditions for sample evaluations such as lithgraphy performances and physical properties. For the cure film preparation conditions, which was considered the acutal process as shown Table 2. First, each resist was coated on Bare-Si wafer to target the optimimal coating film thickness to achive the predetermined cured film thickness for each evaluation. Then, those were baked at 110°C for 5mins on the hot plate. In this study, even though all the samles were the positive tone resists, we applied the development process for the cured film preparations to dupulicate the actual process as much as we can. Finally, these sample films after development and rinse were cured at our standard curing process with several step bake and the final maximum temperature was 200°C.

And Table 3 shows the evaluation methods of physical properties. Tg and CTE were analyzed with TMA method, and Tensile strength, Elastic modulus. For these tests, the films were prepared with patterning and curing at the pre-determined condition as discribed above. And Residual stress was measured the bow of coated wafer before and after cure.

Process	Condition	
Film thickness	For lithography 10um after cure	For physical properties 15um after cure
Soft-bake	105°C*5min	
Exposure (i-line stepper)	600mJ/cm²	1200mJ/cm²
Development	60sec x 2times / 2.38wt% TMAH	
Rinse	60sec / DI water	
Post cure (3°C/min)	120°C*30min -> 150°C*30min -> 200°C*60min	

Table 2. Standard process conditions for sample evaluation

Item	Testing method
Tg	TMA method
CTE (-65 <--> 150°C)	TMA method
Tensile Strength	Tension test Cured film thickness: 15um
Elastic Modulus	
Residual Stress	Cured film thickness: 10um On 200mm Si wafer
Chemical resistance (TMAH/DMSO)	Test wafers prepared with condition at Table 2

Table 3. Evaluation methods for physical properties

Figure 4 shows the comparison data of elastic modulus with the samples on Table 1. In this DOE, the samples containing the monomer with the flexible unit showed lower elastic modulus than the current design as reference which was the normal phenolic resin and the cross-linking rubber particle, as we expected. The current design was that the small amount of the cross-linking rubber particle was dispersed in the cure film due to the limited solubility to the resist compounds. But this new polymer containing the monomer with the flexible unit had good solubility and compatibility with the other regents in resist. Based on this new polymer design, the amount of polymer with the flexible unit instead of the cross-linking rubber particle was increased effectively.

Regarding the Mw effect study from Sample 2 to Sample 4 with 30mol% flexible unit in polymer, there was the tendency that lower Mw had lower elastic modulus with less chemical resistance to the stripper, TMAH/DMSO, at RT for 10mins. Sample 2 showed the lowest elastic modulus at Figure 4, but less chemical resistance. On the other hand, as easily expected higher Mw showed better chemical resistance with higher elastic modulus and in this study Sample 3 and 4, those Mw were over 6800, showed almost similar performances with the cured films. But this time, we didn't evaluate the cured film with the higher Mw than 9000 at this monomer ratio.

Regarding the flexible amount effect study among Sample 1, 2, and 5, as we expected, this is also that higher amount of flexible unit in polymer was lower elastic modulus because the large amount of the flexible units included in cross-linked cure films work more effectively than the limited amount of dispersed the cross-linking rubber particles. But higher

amount of the flexible unit was less chemical resistance with the deterioration of lithography performances. Especially higher amount of the flexible unit as Sample 5 in this resist composition, the lithography performance was not good.

In these studies about Mw and flexible unit amount, Sample 1 showed the best balanced performances about elastic modulus, chemical resistance, and lithography performance. We think there are still rooms to optimize Mw and flexible unit amount in polymer, and also to apply another type of flexible unit. But based on this study, we also proceeded with the additional performance evaluation of Polymer 1 which was for Sample 1

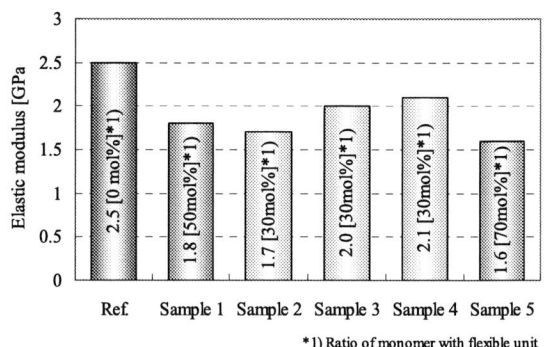

*1) Ratio of monomer with flexible unit

Figure 4. Elastic modulus of cured films

<u>Additional study of physical properties with Sample 1</u>:Based on the results of polymer comparison as Table 1 and Figure 4, we evaluated the process effects, such as curing temperature and curing time for the physical properties, with Sample 1 as shown at Table 4 and Table 5. Regarding the curing temperature effect, when the curing temperature was over 200°C, the physical properties were almost constant and the chemical resistance was strong enough to treat with the typical stripper such as TMAH/DMSO. And also, regarding the curing time, over 60mins at 200°C is enough to have the constant physical performances and the chemical resistance.

In these results, under this curing condition such as at over 200°C for over 60mins, the remarkable physical performances of Sample 1 were low modulus, around 1.8GPa, and low residual stress, around 20MPa. Comparing this residual stress of our material with that of the typical Polyimide (PI) film, it was drastically lower than PI. Normally the residual stress of PI is from 30MPa to 40MPa. The physical performance of our new material is potentially satisfied at the requirement for thinning device wafers under the current technology trends such as down-sizing and multi-functionalizing of the semiconductor devices.

Item	Unit	Curing temperature (@60min)			
		180°C	200°C	220°C	240°C
Tg	°C	210	>250	>250	>250
CTE	ppm	69	65	65	65
Tensile Strength	MPa	81	85	86	86
Elastic Modulus	GPa	1.8	1.8	1.8	1.8
Residual stress	MPa	17	17	17	17
Chemical resistance	-	NG	OK	OK	OK

Table 4. Curing temperature effect for physical properties

Item	Unit	Curing time (@200°C)		
		30min	60min	90min
Tg	°C	>250	>250	>250
CTE	ppm	66	65	65
Tensile Strength	MPa	85	85	86
Elastic Modulus	GPa	1.8	1.8	1.8
Residual stress	MPa	18	17	17
Chemical resistance	-	NG	OK	OK

Table 5. Curing time effect for physical properties at 200°C

Chemical resistances of Sample 1: The films were prepared with the standard conditions from coating to curing as shown Table 2, and then the cured film on wafer was dipped into the various process materials such as organic solvent, stripper, acid and so on at RT or 40°C for the predetermined time. After dipping, the film was treated with water and dried up. The swelling ratio was calucurated with the film thickness difference before and after dipping, and also we inspected the appearance abnormalities such as cracking, swelling, wrinking, and discoloring [16]. As shown Table 6, the cured films of Sample 1 had better chemical resistances compared with Ref.

Chemicals	Conditions	Ref.	Sample-1
PGME/PGMEA	r.t./10min	OK	OK
PGMEA	40C/10min	NG(crack)	OK
NMP	r.t./10min	OK	OK
	40C/10min	OK	OK
DMSO	r.t./10min	OK	OK
PR stripper (TMAH/DMSO)	r.t./10min	OK	OK
	40C/10min	NG(104%)	OK
IPA	r.t./10min	OK	OK
	40C/10min	OK	OK
Hydrofluoric acid aq. (0.5%)	r.t./5min	OK	OK
Sulfuric acid aq. (10%)	70C/5min	OK	OK
Sodium hydroxide aq. (7%)	r.t./30min	OK	OK
TMAH (2.38%)	r.t./30min	OK	OK

Definition of NG: Swelling (thickness change is over 3%) and/or cracking

Table 6. Chemical resistance to various process materials

Adhesion properties of Sample 1: Ths adhesion on various substrates are one of the most important properties. Figure 5 shows the test method for adhesion properties. First, the patterned film was cured at the standatard condition, 200°C *

60mins, on the various substrates, and then the adhesive force of these cured films were measured before and after PCT (Pressure Cooker Test) with share bond tester. PCT condition was 121°C/85% /24hrs. As shown Table 7, we compared the current product, reference (Ref.), containing the phenolic resin and the cross-linking rubber particle with Sample 1 which was newly designed to use the polymer containing the flexible unit. The adhesive force of Ref. and Sample 1 on the various substrates were good enough to keep the same adhesive properties even after PCT without significant deterioaration. Specially, Sample 1 was better adhesive properties than Ref.

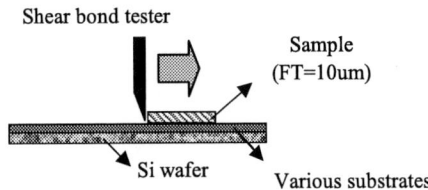

Figure 5. Adhesion test using Share bond tester

Surface layer (kg/cm2)	Ref.		Sample-1	
	Initial	After PCT	Initial	After PCT
SiN	140	150	180	185
SiO₂	150	155	185	180
Al	120	120	165	175
Cu	140	155	170	175

PCT condition : 130C/85%RH/168hr

Table 7. Adhesion properties of Sample 1

Results: Based on these results, Polymer 1, the new polymer containing the monomer with the flexible unit, performed well for lower modulus, lower residual stress, good adhesion properties on various substrates, and good chemical resistance to various process materials at low curing temperature. This novel polymer concept is expected to apply to our photo-sensitive insulator for WL-CSP and so on.

Lithography performances

Lithography performance of the photo-sensitive insulator, Sample 1, containing Polymer 1 was as follow. First, this new polymer is mixed with cross-linkers, photo-sensitizers, additives and solvents at RT for the predetermined time, and filtrated with the fine filter to remove small particles.

Film thickness change through process: As shown Figure 6, the remining film thickness after development was about 9.9um with slight filmloss, and that after cure was 9.2um compared with 10um right after coating and soft-baking. The ratio of film thickness loss is around 10% which is much better than that of PI (40% to 50%). This is the pottential advantage to have good lithography performance such as focus margin, development margin, resolution and etc, because this higher remaining film thickness after cure comparing with that after coating can support to proceed the lithography procee with thiner coated film.

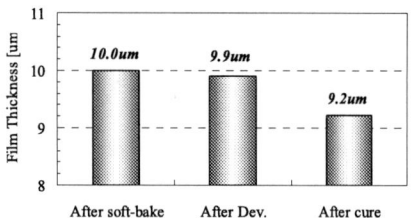

Figure 6. Film thickness change through process

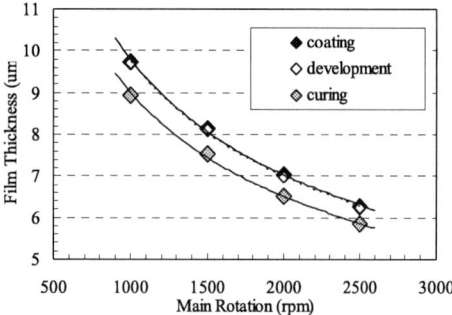

Figure 7. Film thickness change through process

And Figure 7 shows the coating curve of Sample 1 at each steps. Sample 1 can support the cured film from 6um to 9um with less film shrinkage after the curing process, moreover the target film thickness after cure are adjustable with the solid content. This data indicates our material design can provide the insulator performance with less coating amount.

Dissolution curve: Figure 8 shows the correlation between the remaing thickness and exposure dose, and these data points of remaining thickness ratio were collected with various esposure steps. And the other conditions such as coating, soft-baking, and developping are optimized for Sample 1. The remaining film thickness of Sample 1 was almost linearly decreased, and it was almost zero at over 500mJ/cm^2. As shown Figure 6, there were alomost no film loss at the unexposed area compared with the exposed area over 500mJ/cm^2. This development contrast can support the good pattern profile and lithography margins.

Exposure latitude: Figure 9 shows the exposure latitude of Sample 1. And the gentle slope means that the exposure latitude was wide enough to provide the prcess window. Based on this data, for example, if the dose variation is +/- 50 mJ/cm2, the pattern CD variation would be around +/- 0.25um which is roughly 1% CD variation at 20um CD target. Figure 10 is the cross-section SEM observation around the optimum expose dose (Eop). The Eop of was 560mJ/cm2, and the patterns around Eop was good lectangle profiles with no deterioration. These data shows that Sample 1 has wide exposure margin for the pattern CD and profiles.

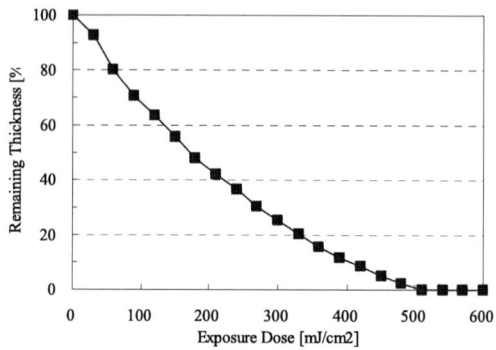

Figure 8. Correlation between remaining film thickness and exposure dose

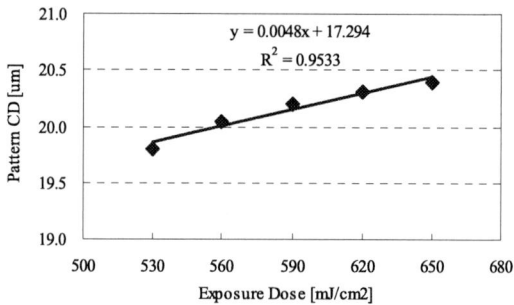

Figure 9. Exposure latitude of Sample 1

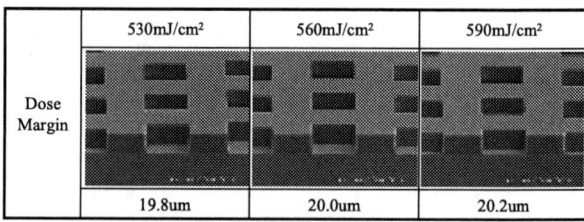

Figure 10. Cross-section SEM observation around optimum expose dose of Sample 1

Development process margin: Figure 11 shows the development process margin of the Sample 1. In this study, the various development time from 50 seconds to 70 seconds with the fixed frequency, 2 times, are applied to evaluate the development process margin. As the results of cross-section SEM as shown Figure 11, Sample 1 has wide process margin with less CD change and without pattern profile deterioraton.

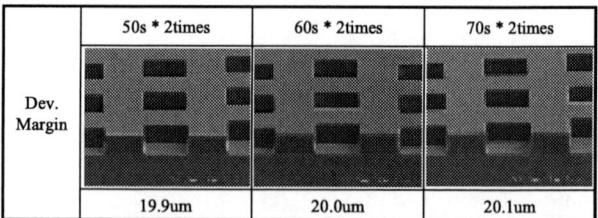

Figure 11. Cross-section SEM observation of development margin of Sample 1

978-1-4799-2408-0/14 $31.00 © 2014 IEEE

Patterning resolution: Figure 12 shows the patterning resolutuon of Sample 1 at 10um film thickness. As the result of this evaluation, Sample 1 achieved the higher aspect ratio, over 3, and good linearity from 10um-C/H to 3um-C/H. In addition, after the curing process at 200°C * 60mins with our standard temperatuture profile, the pattern CDs were still constant without drastical thermal flow. This data shows the possibilities to apply Sample 1 for the small and high aspect bumps.

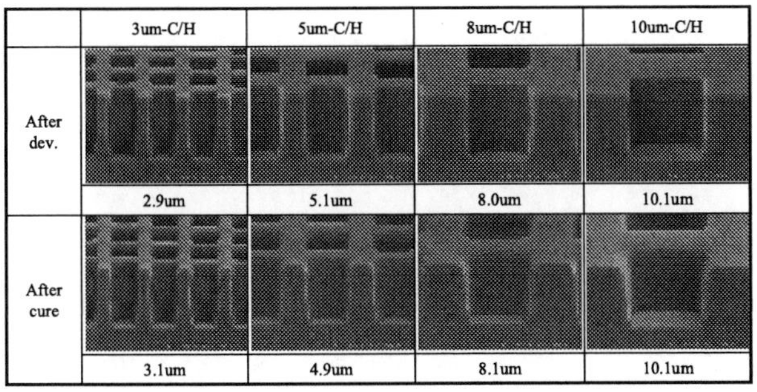

	3um-C/H	5um-C/H	8um-C/H	10um-C/H
After dev.				
	2.9um	5.1um	8.0um	10.1um
After cure				
	3.1um	4.9um	8.1um	10.1um

Figure 12. Cross-section SEM observation of patterning resolution of Sample 1

Curing temperature program effect: Figure 13 shows the effect of curing temperature program for Sample 1. Genrrally holding at 120°C is the purpose for decreasing the residual solvents, and holding at 150°C is the purpose for accelerating the thermal decomposition of DNQ compounds. As the result of this study, there are no significant deferences about the cured pattern profile and the pattern CD, and also the surface roughness due to no holding at 120°C and/or 150°C did not detected. Sample 1 has wide curing process margin.

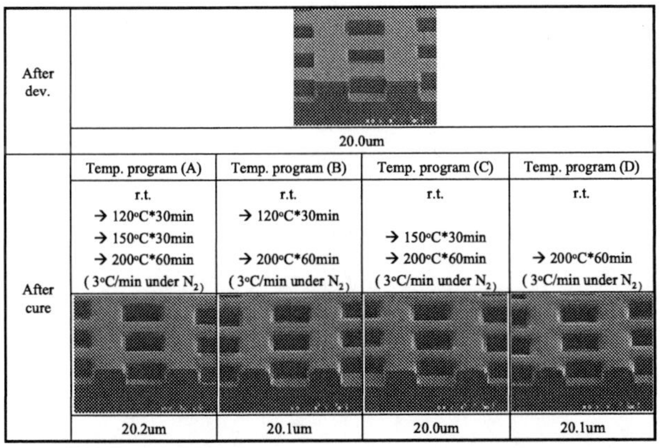

After dev.				
		20.0um		
	Temp. program (A)	Temp. program (B)	Temp. program (C)	Temp. program (D)
After cure	r.t. → 120°C*30min → 150°C*30min → 200°C*60min (3°C/min under N₂)	r.t. → 120°C*30min → 200°C*60min (3°C/min under N₂)	r.t. → 150°C*30min → 200°C*60min (3°C/min under N₂)	r.t. → 200°C*60min (3°C/min under N₂)
	20.2um	20.1um	20.0um	20.1um

Figure 13. Cross-section SEM observation at various curing program

Applicable film thickness: Figure 14 shows the pattern profile before and after cure at various cured film thickness from 6um to 14um. The patterns after development were rectangular profile, and those after development were little bit tapered profile. This results shows that Sample 1 had wide film thickness magin.

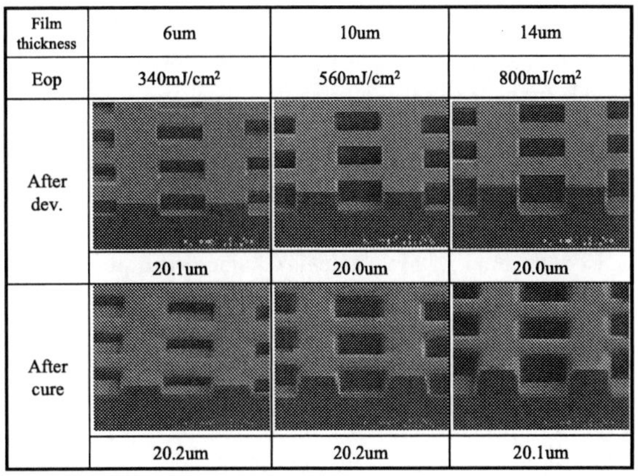

Film thickness	6um	10um	14um
Eop	340mJ/cm²	560mJ/cm²	800mJ/cm²
After dev.			
	20.1um	20.0um	20.0um
After cure			
	20.2um	20.2um	20.1um

Figure 14. Cross-section SEM observation at various film thickness before and after cure

Results: Based on this lithography evaluation of Sample 1 with the new polymer 1 containing the flexible unit, this new concept for cross-linking structure after cure worked effectively to balance elastic modulus, chemical resistance, and adhesion properties with any deterioration of lithography performances such as expose dose, development margin, resolution, etc. And also, those physical properties were achieved at lower curing temperature at around 200°C. Additionally, Sample 1 had wider margin for the applicable film thickness.

Conclusions

In this paper we reported that our new polymer design to achieve the lower elastic modulus at lower curing temperature around 200°C without any deterioration of the other performances which are basically required for the photo-sensitive insulator. In this study, we found out that the implementation of the monomer with the flexible unit to polymer reduced the elastic modulus of the cross-linking film after cure. And we optimized the other insulator performances for this new polymer concept. The futures of our new photo-sensitive insulator are as follow.

1. Low modulus (<1.8GPa) at lower curing temperature (at 200°C)
2. Good chemical resistance with various process reagents
3. Good adhesion properties with various substrates
4. Wider expose latitude with rectangular profiles
5. Wider development margin without deterioration of pattern profiles
6. Higher resolution; 3um-C/H at 10um cured film thickness
7. Higher remaining film thickness
8. Wider curing program margin for pattern profiles

Finally, this new polymer design was ready to implement to our current development for the new photo-sensitive insulators.

References

1. G. J. Jung, "Structure and Process Development of Wafer Level Embedded SiP (System in package) for Mobile Applications", 2009 11th EPTC, p.191

2. Yoichiro Kurita, "A 3D Stacked Memory Integrated on a Logic Device Using SMAFTI Technology", 2007 ECTC, p.821

3. S. Yoon, et.al., "Mechanical Characterization of Next Generation eWLB (embedded Wafer Level BGA) Packaging", ECTC, pp441-446, 2011

4. M. Santarini, "Stacked and Loaded: Xilinx SSI, 28-Gbps I/O Yield Amazing FPGAs", Xcell Journal, No.74, pp.8-13, 2011

5. M. Murugesan, et.al., "Wafer Thinning, Bonding, and Interconnects Induced Local Strain/Stress in 3D-LSIs with Fine-Pitch High-Density Microbumps and Through-Si Vias", IEDM, pp.2.3.1-2.3.4, December 2010

6. Y. K. Lee and J. D. Craig, "Polyimide Coatings for Microelectronic Applications", Polymer Materials for Electronic Applications, ACS Symposioum Series, Washington, 1982

7. R. Wayne Johnson, et.al., "Effects of Enviromental Exposure on Under-Fill Materials Behavior", Proceedings of the International Microelectronics Conference, October 2006

8. R. Rubner, H. Ahne, E. Kuhn, and G. Koloddieg, "Photogr. Sci. Eng.", 23, pp304, 1979

9. N. Yoda, and H. Hiramoto, "J. Macromol. Sci.," A21, p1641, 1984

10. S. Hayase, K. Takano, Y. Mikogami, and Y. Nakano, "J. Electrochem.Soc.", 138, pp3625, 1991

11. H. Seino, A. Mochizuki, O. Haba, and M. Ueda, "J. Polym. Sci. Technol.", 10, pp55, 1997

12. T. Fukushima, K. Hosokawa, T. Oyama, T. Iijima, M. Tomoi, and H. Itatani, "J. Polym. Sci.", Part A: Polym. Cem., 39, pp934, 2001

13. W. E. Feely, J. C. Imhof, and C. M. Stein, "Poly. Eng. Sci.", 26, pp1101, 1986

14. F. Arndt, and B. Eistert, "Ber.", 68, pp200, 1935.

15. M. Hanabata, Y. Uetani, and A. Furuta, "Design of PACs for High-Performance Photoresists (II)", SPIE, vol 1925, pp227-234, 1993

16. J. Yota, et.al., "Variable Frequency Microwave and Furnace Curing of Polybenzoxazole Buffer Layer for GaAs HBT Technology", IEEE Transactions on Semiconductor Manufacturing, Vol.20, No.3, 2007

3D and 2.5D Packaging Assembly with Highly Silica filled One Step Chip Attach Materials for both Thermal Compression Bonding and Mass Reflow Processes

Daniel Duffy[1], Chris Gregory[3], Christopher Breach[2] and Alan Huffman[3]

Kester Inc., [1] 800 West Thorndale Ave, Itasca, IL 60143, [2] 500 Chai Chee Lane, Singapore 469024

[3] RTI International, Research Triangle Park, NC 27709

[1] dduffy@kester.com

Abstract

One Step Chip Attach (OSCA) materials are chemically engineered organic fluids that function both as underfills and fluxes for flip chip assembly. These emerging materials chemically remove solder oxides and surface finishes to facilitate soldering during reflow while simultaneously curing to form a rigid reinforcing polymeric underfill that improve the reliability of the finished product. The key advantage of OSCA is a reduction in the number of steps in the assembly process, which naturally results in faster cycle times and more efficient flip chip assembly. With suitable chemistry, fillers and the correct filler loading, OSCA materials are dispensable and capable of achieving high yield and reliability. OSCA materials may be designed for use in thermo-compression bonding (OSCA-T) and in more standard mass or gang reflow (OSCA-R), the choice of process depending on the package functionality and design requirements.

This paper shares information on the properties and processing techniques of OSCA materials and explores the assembly data concerning control of solder wetting, over collapse, voiding, and HAST behavior. Data will be presented from testing at the material level and from testing of micro bumped lead free Silicon to Silicon vehicles designed to simulate chip to wafer or chip to chip bonding. Challenging loadings of up to 65% silica with distributions as low as 0.5 micron average particle size will be shown at both the assembled and material level.

1.0 Introduction

The need for continued increase in interconnect density in advanced microelectronics packaging, like mobile computing applications for example, is driving the need for new assembly process and materials. [1][2] Traditional flip chip assembly fluxes bumps and reflows ICs, which is then followed by a capillary underfill process [3] as shown schematically in Figure 1. However, demand for smaller and thinner ICs with reduced power consumption and greater functionality requires smaller bumps and chip standoff heights and finer bump pitches. With smaller bumps and standoffs and high bump density, getting enough flux onto bumps to ensure all bumps are soldered is important. However, overfluxing to ensure all bumps are fluxed can result in flux residues that interfere with the underfill flow and adhesion with no clean processes. With water soluble fluxes fine bump pitches and low standoff heights can make cleaning very difficult at best, increasing the cleaning process costs and/or cycle time and at worst it may be impossible to fully remove all of the flux residue prior to underfilling.

One Step Chip Attach Materials (OSCA) have been proposed as a solution to these issues by combining the functionality of a flux prior to curing and an underfill after

curing. These materials have the advantage of combing two materials into one and removing the problems of flux interference or flux cleaning. The OSCA materials need to be tuned for the desired assembly process: the newer Thermo-Compression Bonding (TCB) or the more traditional mass reflow. The process flow of the OSCA materials is shown schematically in Figure 2 below. The OSCA-T path shows the process for TCB, while the OSCA-R path shows the process for mass reflow.

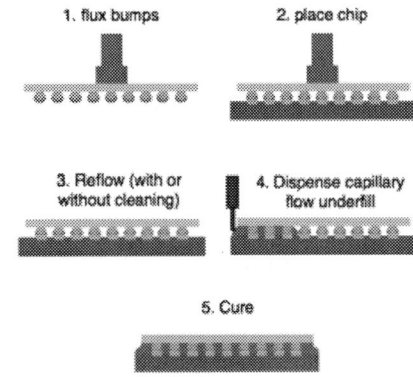

Conventional capillary flow process

Figure 1: Traditional Flip Chip Assembly Process

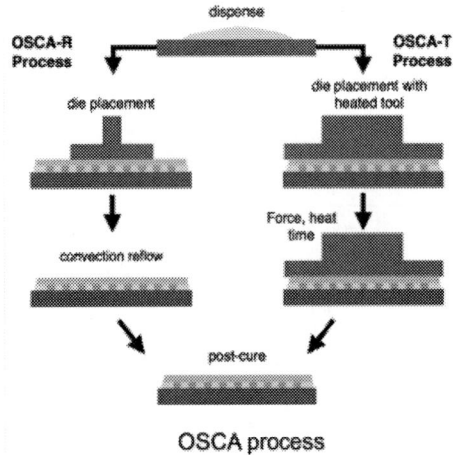

OSCA process

Figure 2: New OSCA Material Flip Chip Assembly Processes

TCB can be used for thin, small pitch ICs with solder bumps or copper pillars for which traditional chip attach processes suffer from die misalignment. The TCB process can be used for various types of assembly including Die to Wafer

(D2W), 2.5D Packaging, 3D Packaging, and die to substrate (D2S) which all require different thermal profiles. Materials properties such as curing kinetics, fluxing activity, and viscosity all need to be optimized for the intended process and tool capabilities to give good results. The materials described in prior works for focused on mass reflow processing [4][5][6] do not work well with the rapid TCB processes.

This paper explores the assembly of solder bumped test chips with epoxy-based materials specifically designed for either TCB or mass reflow. Two Kester materials will be presented for TCB processes in this paper. One for processes requiring potentially long pre-staging time, long working life, and wider process windows, and another for operations more focused on shorter cycle times and maximizing throughput. In the following sections, details of the test chips, materials, assembly and reliability tests are given followed by a description of the main test results and their interpretation.

2.0 Experimental Details

2.1 Mechanical Test Chip for TCB and Mass Reflow

A solder bumped mechanical test vehicle was used for TCB and mass reflow assembly experiments. Specifics of the mechanical test vehicles are contained in Table 1.

Table 1: Mechanical Test Chip Characteristics

Property	Test Vehicles		Unit
	Vehicle 1	Vehicle 2	
Die Material	Silicon	Silicon	NA
Substrate Material	Silicon	FR4	
Die Passivation	Silicon Oxide	Silicon Oxide	
Die Solder	SnAg3	SnAg3	
Substrate Pad Metallurgy	Copper	SnAg3	
Die Bump Height	40	90	microns
Die Bump Diameter	60	80	
Die Bump Pitch	100	250	
Die Thickness	750	750	
Die Dimensions	10x10	6x6	mm

2.2 TCB Materials and Assembly Process

Uncured material properties are shown in the Table 2. Material A is the wider process window material and material B is the material designed for faster throughput and for improved HAST performance and will be described in more detail later in the paper. Mechanical builds of test vehicle 1 were made with OSCA Material A. The material was dispensed with a 21-gauge stainless steel needle with an Auger style pump. A SET FC150 Bonder was used to assemble chips using the bonding profile in Figure 3. The following process was programmed to prevent over collapse of the solder bumps. After bonding, assembled chips were post cured at 180°C.

Table 2: Uncured properties of OSCA Materials for TCB

Property	Material A Value	Material B Value	Units	Method
Viscosity at 75C	25	40	Poise	Brookfield, 5 RPM
Complex Viscosity at 25C	29	38	Pa-sec	Oscillatory Rheometer
Dispense Temp Range	20 -70	20-30	C	
Filler Content	65	65	wt %	TGA
Avg. Filler Size	0.5	0.5	micron	
Max Filler Size	2	2	micron	
Density	1.7	1.7	g/cm^3	Liquid pycnometry
Pot life at 25C	>8	5	hours	Dispensing R&R
Pot life at 25C	>8	2	hours	+25% in viscosity
Appearance	White	White		

Table 3: Cured Properties of OSCA Materials for TCB

Property	Material A Value	Material B Value	Units	Method
Tg	117	129	C	TMA inflection
CTE1	29	28	ppm/C	TMA
CTE2	91	92	ppm/C	TMA
Modulus	5.813	Pending	GPa	ASTM D790
Stress at Break	91	Pending	MPa	ASTM D790
Strain to Break	2.3	Pending	%	ASTM D790
Fracture Toughness K1C	2.7	Pending	MPA(m)$^{0.5}$	ASTM D5045
Density	1.7	1.7	g/cm^3	Gas pycnometry
Cure Shrinkage	0.9	0.9	%	
Thermal Conductivity	0.52	Pending	W/m-K	ASTM 1461
Die Shear	50	80	kg-f	J-STD
Die Post HAST 50 hr	35	76	kg-f	J-STD

978-1-4799-2408-0/14 $31.00 © 2014 IEEE

The properties of the uncured OSCA material are shown in Table 2. Material A is the wider process window material and Material B is the material designed for faster throughput and for improved HAST performance and will be described in more detail later in the paper. The properties of the cured OSCA materials for use in TCB processing are shown in Table 3.

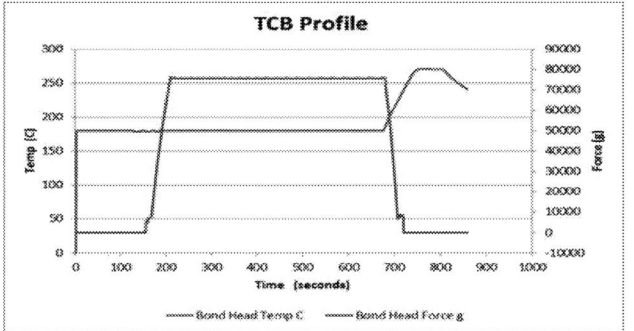

Figure 3: TCB Process for Vehicle 1 builds with Material A.

2.3   Mass Reflow Materials and Assembly Process

OSCA materials for mass reflow are of interest to develop for end markets in need of high device throughput. However, restrictions on using chemicals such as organometallics, along with the drive for smaller geometries have prompted a need to develop new materials that improve upon the previous generation of materials. [7] The uncured properties of the OSCA material for mass reflow are listed in Table 4 and the cured properties are listed Table 5. The material was dispensed with a 21-gauge stainless steel needle with an Auger style pump. A SET FC150 Bonder was used to pick and place chips. Test vehicle 1 was used. After placement chips were reflowed as shown in Figure 4. After bonding, assembled chips were post cured at 180°C.

Table 4: Un-Cured Properties of the OSCA Materials for Mass Reflow

Property	Material C Value	Units	Method
Viscosity at 25C	0.9	Pa-s	Brookfield, 5 RPM
Dispense Temp Range	18-25	C	
Filler Content	40	wt%	TGA
Avg. Filler Size	0.5	micron	
Max Filler Size	2	micron	
Density	1.2	g/cm^3	Liquid pycnometry
Pot life at 25C	8	hours	+25% in viscosity

Table 5: Cured Properties of the Cured OSCA Material for Mass Reflow

Property	Material C Value	Units	Method
Tg	141	C	TMA inflection
CTE1	77	ppm/C	TMA
CTE2	217	ppm/C	TMA
Modulus	488	MPa	ASTM D638
Tensile Strength	23	MPa	ASTM D638

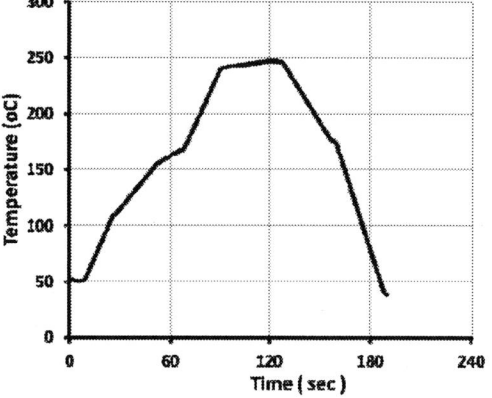

Figure 4: Mass Reflow Profile used in the Assembly Process

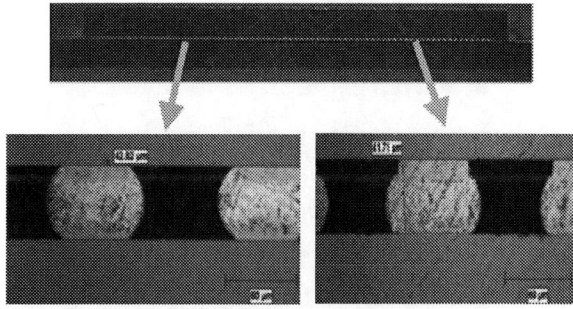

Figure 5. Optical images of a bonded cross-sectioned chip with close up views of individual bumps. Bonded with Material A.

2.4   Scanning Acoustic Microscopy (CSAM)

After assembly a Sonoscan Gen5 innovative acoustic imaging system was used to look for assembly defects in the assembly.

2.5   Die Shear Testing

The die shear was performed 6mm square blank die were bond to silicon substrates with a 50-micron spacer. The shear test was performed using a Dage 4000 with a DS100 kg cartridge operated at a test speed of 25 mils per second.

3.0   Results and Discussion

3.1   TCB Materials

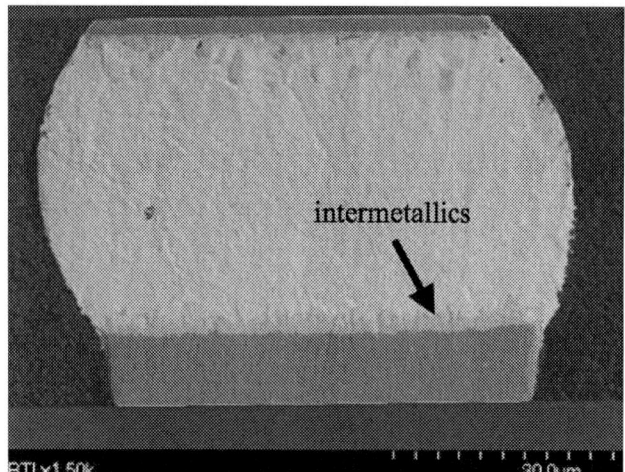

Figure 6. An SEM image of a typical solder bump after the TCB process. Bonded with Material A.

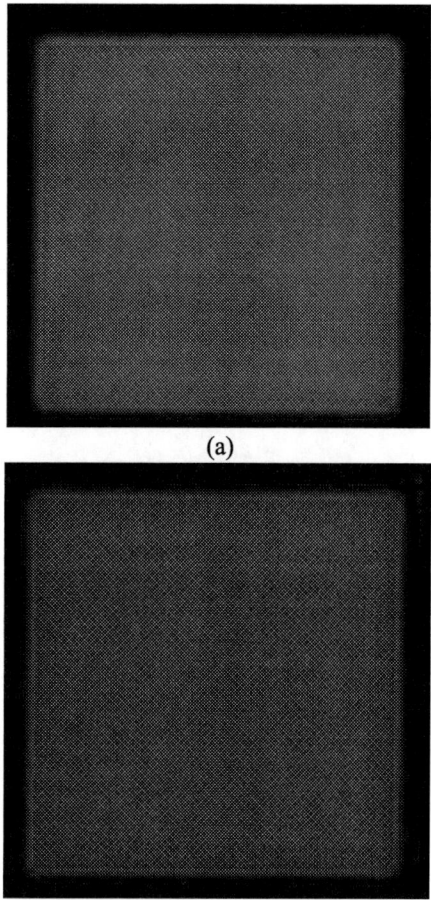

Figure 7. CSAM images of Test Vehicle 1 assembled with Material A after post cure.

A representative example of a test chip bonded using the TCB process and Material A in Figure 5 shows that the solder balls appeared exhibited the typical barrel shape of a well interconnected ball-pad joint. There was some variation in collapse height. Different collapse heights were observed due to the difficulty in evenly removing pressure to prevent over collapse during the ramp to liquidus, although this is not expected to be a problem with a solder capped copper pillar bump because the pillar height controls the standoff height. An SEM image in Figure 6 shows a relatively uniform intermetallic layer was formed after the TCB process. Assembled modules were analyzed by CSAM for voiding after assembly and post cure. Typical examples in Figure 7 show that neither voids nor filler or organic entrapment or filler separation were seen within the solder bump or interfaces.

After confirming void free assemblies after post curing, further chips were assembled with Material A and subjected to J-STD HAST testing at 121°C/2 atm/100% RH and were examined using CSAM. The CSAM images in Figure 8 show some voids at time zero due to air entrapped during dispensing of the material. After 264 hours, the CSAM images show some voiding as well as the "X" shape corresponding to the geometry on the bond head. The voiding after 264 hours of HAST does show the moisture in the HAST chamber is penetrating into the assembly and degrading some of the chemical bonds. The glass transition temperature of Material A is 117°C (Table 3) which is below the IPC HAST chamber meaning that the polymer network was in an expanded state, making penetration of moisture into the assembly easier.

Figure 8. CSAM images of Vehicle 1 at (a) 0 and (b) 264 hours of IPC HAST conditions. Bonded with Material A.

Die shear test results in Figure 9 show that the die shear strength started to decline at 75 hours and continued to decrease until 240 hours. This is consistent with the penetration of moisture into the package, which most probably attacks the chemical bonds between Material A and the substrate and/or die surfaces.

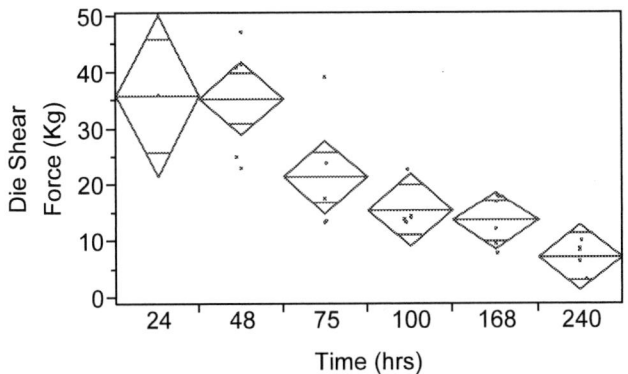

Figure 9. HAST performance of Material A in Die Shear

Material B was developed with two ideas in mind: minimizing the time in the bonder and improving performance in the HAST chamber. A four minute b-staging process at 110°C was done after dispensing but before placement prior to bonding, for 4 seconds at 250°C. The bonding profile is shown in Figure 10.

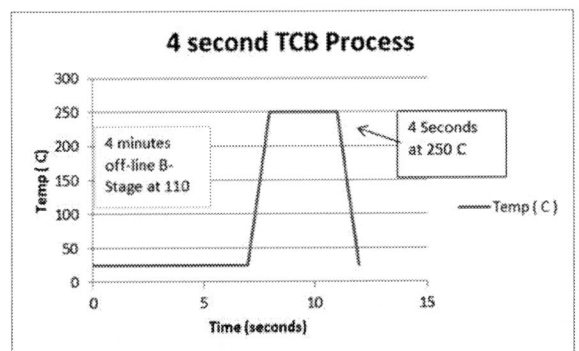

Figure 10. TCB profile for decreased cycle time of Material B.

In addition to the mechanical builds in shown above, die shear samples were also made with material A to quantify the adhesion strength. The die shear testing was done with a 5x5mm square die to a blank silicon substrate. A standoff height of 25 microns was obtained through the use of glass spacer beads. The CSAM images of the die shear samples of Material A and Material B are shown in Figure 11. It is clear from the images that Material B has held up better than Material A after 264 hours.

The die shear force of Material B after HAST in Figure 12 also degraded with time. Figure 13 shows that the die shear strength of Material B degraded slightly less than Material A.

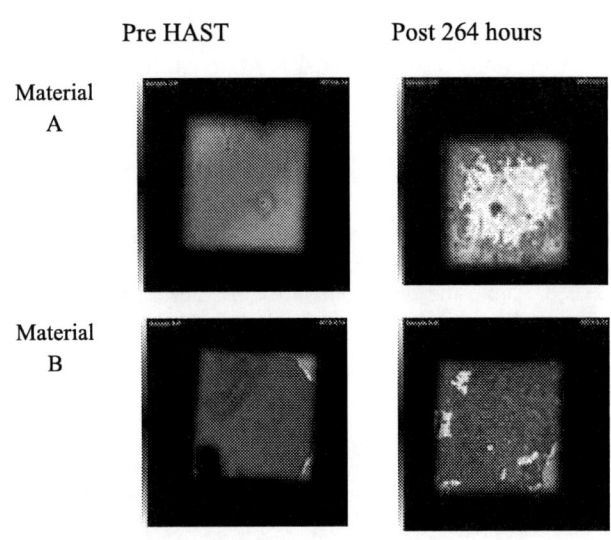

Figure 11. CSAM Image of the Die Shear Assemblies of Material A and B pre and post HAST.

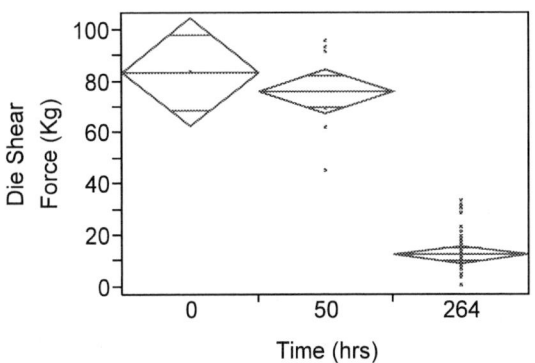

Figure 12. HAST performance of Material B in Die Shear

Figure 13 shows a comparison of the 48 and 50 hour test HAST die shear results for A and B respectively, and shows that Material B did increase the HAST performance by increasing the glass transition temperature. A comparison of the process times for Material A and B is shown in Table 6.

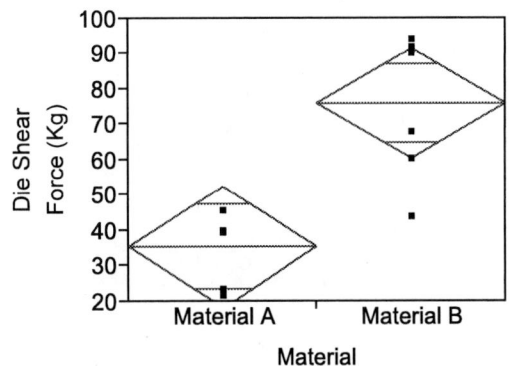

Figure 13. HAST Die Shear Comparison of Material A and B.

978-1-4799-2408-0/14 $31.00 © 2014 IEEE

Table 2. Process Comparison for Material A and B.

	Material A	Material B	Unit
Time in the TCB Bond Step	120	4	seconds

Good solder ability was observed after cross-sections and is shown in the images below for Test Vehicle A in images 6, 7, and 8. Note the dark spots are not filler entrapment, nor voids, but diamond grit from the cross-sectioning.

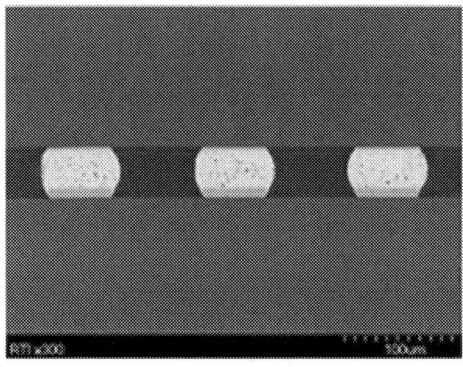

Figure 14. Cross Section of Vehicle 1 with Material C.

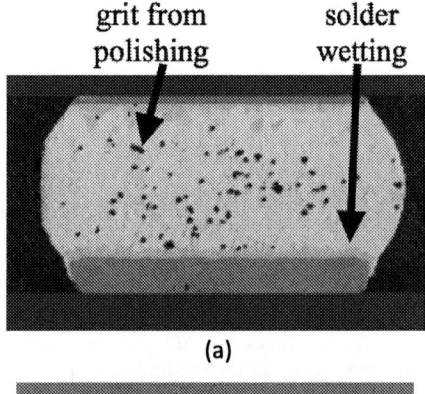

(a)

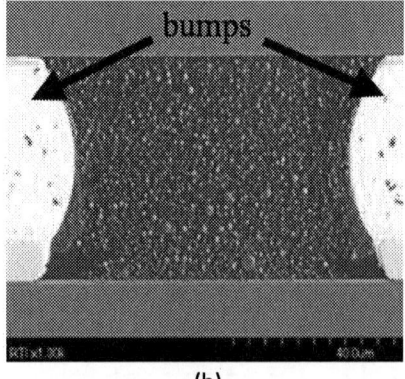

(b)

Figure 15. Cross Sections of Vehicle 1 with Material C.

3.2 Mass Reflow Materials

Low and high magnification SEM images of cross-sectioned chips are shown in Figure 14 and Figure 15

respectively. Good Intermetallic formation is seen to the copper substrate pad and even wetting down the side of the copper pad showing that the OSCA material had not gelled during this process. The dark masses in the solder joint are not voids, but rather artifacts of the cross-section sample preparation. A typical CSAM image in Figure 16 shows the test assemblies did not shown any voids.

Figure 16. CSAM of Test Vehicle 1 built with Material C.

The same process was then applied to the Mechanical Test Vehicle 2 and its FR4 based substrate. A cross section is shown below in image 10 and 11. Good wetting of the solder to solder is observed, (Note the dark spots are not voids or entrapment but diamond grit from the polish process). Excellent solder to solder wetting of the solder to solder is observed, (again note the dark regions are diamond grit artifacts from the cross section polishing.

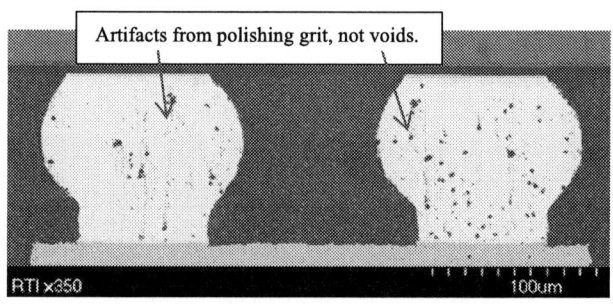

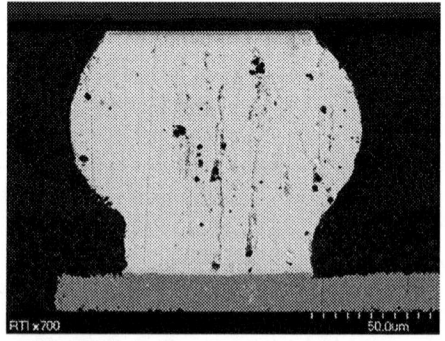

Figure 17. Cross Sections of Test Vehicle 2 with Material C.

These FR4 substrates were outgassed by preheating to drive out any volatile components prior to processes. This pre heating was done in Nitrogen to minimize any oxidation of the metal surfaces of the substrate. A reduction in volatiles was observed by processing the substrate through a 150°C bake out and the entire reflow profile, however it was not eliminated. A comparison of the outgassing is illustrated below of optical images of the substrate when processed with a clear glass cover slide over the OSCA material.

No Substrate Degassing Substrate Degassing

 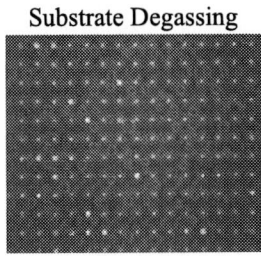

Figure 18. Vehicle 2 Substrate Outgassing Images.

This outgassing of the substrates clearly helped to reduce the number of voids observed, however it is did not eliminate them. With this conclusion, that the outgassing of the substrate could not be completely eliminated, further work will be conducted on non-FR4 substrates such at BT, or silicon interposer. Reliability testing such as HAST was not performed on these samples due to substrate outgassing and the detrimental effect it would have on the assemblies.

Conclusions

OSCA materials for both TCB and mass reflow processes have been introduced. Solderability has in a TCB processes have been shown with two different materials tuned for processes at the extremes of 120 to 4 second bonds. CSAM images have shown void free assemblies with both of these processes with these tuned materials to the process with filler loading up to 65%. HAST data up to 264 hours was presented, along with CSAM images after testing. Material B with a Tg greater than the test condition had a better HAST performance than the Material A whose Tg was 117 C.

Mass reflow solderability has been shown with 40% loading. Void free assemblies were shown on Silicon substrates, however outgassing from organic FR-4 based substrates were entrained in the OSCA-R material.

Acknowledgments
The Authors would like to thank:
  Jean Liu, Lin Xin, Kal Choski, Maulik Shah, David Eichstadt of Kester
  Sonoscan, Elk Grove Village, IL
  Illinois Tool Works Tech Center, Glenview, IL:

References

1. R. Huemoeller, "Market Demand Readiness for 2.5/3D TSV Products" *IMAPS 2012 Device Packaging Conference,* Scottsdale, AZ. March 5-2012.

2. S. Arkalgud, "2.5 and 3D- Scaling Walls", *IMAPS 2013 Device Packaging Conference,* Scottsdale, AZ March 12, 2013.

3. J. Lau. Low Cost Flip Chip Technologies: for DCA,WLSCP, and PBGA assemblies. New York: McGraw and Hill. 2000. Chapter 6.

4. R. Thorpe, McGovern, L.P., Baldwin, D.F., "Analysis of Process Yield - in low cost Flip Chip on Board Assembly Process." Thermo-mechanical Characterization of Evolving Packaging materials and structure, *1998 ASME International Mechanical Congress and Exposition*, Anaheim, CA, Nov. 1998

5. R.Thorpe, D. Baldwin, G.W.Woodruff, L.P. McGovern, "High Throughput Flip Chip Processing and Reliability Analysis using No-Flow Underfills", *Proc. 49th ECTC*, 1999, San Diego

6. C.P.Wong, S. Shi, "Study of the Fluxing Effects on the properties of No-Flow Underfill Materials For Flip Chip Applications", *Proc. 48th ECTC*, Seatle, WA. 1998

7. J. Liu, R. Kraszewski, X. Lin, L. Wong, SH Goh, J. Allen. "New Developments in Single Pass Reflow Encapsulant for Flip Chip Application", *Proc. 51 ECTC*, Orlando, FL 2002

Process Compatibility of Conventional and Low-Temperature Curable Organic Insulation Materials for 2.5D and 3D IC Packaging – A User's Perspective

Guilian Gao, Bong-Sub Lee, Andrew Cao, and Ellis Chau
Invensas Corporation
3025 Orchard Parkway, San Jose, CA 95134
ggao@invensas.com

Abstract

For 2.5D and 3D IC packaging, wafer back side processing often has temperature limits substantially lower than curing temperature of conventional polyimide (PI). New low temperature curable insulation materials possess very different properties and require thorough evaluation to ensure process compatibility and product reliability. In this work, we present a set of evaluation methods that can be used to compare materials and curing processes against bench mark materials and against each other, to facilitate optimal materials selection, minimize cost, and shorten development cycle time. We also evaluated conventional PI cured in a Variable Frequency Microwave (VFM) oven, which achieved comparable properties at lower temperature compared with conventional curing.

Introduction

Commonly used via first and via middle through-silicon-via (TSV) fabrication processes use a temporary bonding material with a silicon or glass carrier for handling of thin wafers[1]. Today, despite of more than 10 suppliers of temporary bonding materials on the market, temporary bonding/deboning remains a major challenge for high volume manufacturing[2]. Among other concerns, thermal stability of the temporary bonding adhesives is still the limiting factor for TSV backside processing temperature. Most temporary bonding adhesives currently on the market have thermal budgets that allow curing of organic passivation materials around 200°C. If backside processing temperature exceeds process window of the adhesive, it either outgases and causes delamination of the thin wafer, or reacts with the passivation material on the wafer, resulting in permanent bonding of the thin wafer to the adhesive or the carrier.

Polyimide has been used extensively for surface passivation and protection for thin and ultra-thin silicon devices. It is very process friendly. Spin coating produces well controlled and uniform thickness. The cured coating has excellent chemical resistance, is stable at solder reflow and underfill cure temperatures, and acts not only as dielectric layer but also stress buffer. However, conventional polyimide material requires curing temperature of 280°C to 400°C, much higher than the temperature that most temporary bonding adhesive can tolerate.

In response to this challenge, several material suppliers have introduced new passivation coatings that can be cured at lower temperatures [3]. Equipment suppliers have also introduced Variable Frequency Microwave (VFM) curing ovens to lower curing temperatures of conventional polyimides [4], [5]. However, the new low temperature curable passivation materials may have very different properties from those of conventional polyimides, which can have significant impact on process compatibility and product reliability. Same is true of conventional polyimide materials cured at much lower temperature in a VFM oven.

Material suppliers usually provide basic properties of their materials. These include, but are not limited to: curing temperature; mechanical properties, including tensile strength, elongation, Young's modulus, coefficient of thermal expansion (CTE), and glass transition temperature (Tg); thermal stability, typically given as the 5% weight loss temperature; chemical resistance, typically including resistance to acids, alkaline and solvents commonly encountered in wafer processing. While such information is very helpful for initial literature review for material selection, our experience showed that they are inadequate.

For example, the 5% weight loss temperature is commonly interpreted as the upper limit of application temperature. However, such interpretation is misguided. The 5% weight loss temperature is most likely derived from a thermo-gravimetric analysis (TGA) curve as illustrated in Figure 1, which is typically generated at a temperature ramp rate of 5-40°C/min. Since weight loss is a function of temperature and time, such a curve tends to overestimate the upper limit temperature for most processes due to the high ramp rate during the data collection.

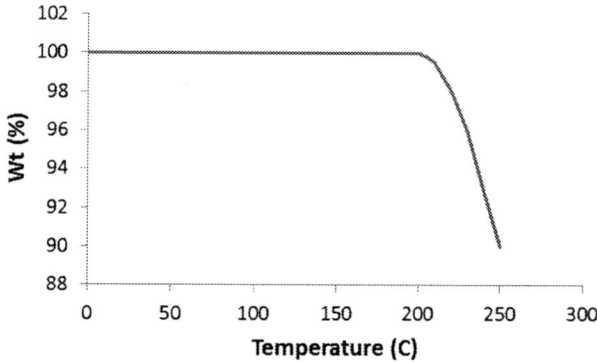

Figure 1: Illustrated TGA thermal curve for determining 5% weight loss temperature

Similarly, chemical resistance data from the material supplier also tend to be general and limited in scope. It is impossible for the material supplier to know all details of each customer's process and provides comprehensive process compatibility data. Moreover, there is no standard for testing and grading, making it difficult to compare data from different suppliers. In addition, it does not provide information on cumulative effect of exposure to multiple chemicals and process steps. Therefore, thorough evaluation by the customer is critical for successful process integration. Since cost and cycle time of changing a material escalates exponentially with progression of a project, such evaluation should be done at the earliest stage possible, starting with blank test coupons or test samples with minimum features required to simulate the actual process.

Test Methodology

In this study, we present our approaches to achieve optimal material selection. Based on our silicon interposer fabrication process, we developed a battery of tests for material evaluation and applied them to several low temperature curable passivation materials (with curing temperatures ranging from 170°C to 200°C) and a conventional polyimide that were cured in a conventional oven or a VFM oven. The degree of imidization was measured by Fourier-transform infrared (FTIR), which compared the absorption peaks with a fully cured material. Details are shown in previous reports [5].

Table 1: Information of five resins evaluated in this study

Resin	Photo sensitive?	Recommended minimum curing Temperature (°C)	Curing time (min)
A	Yes	170	60
B	Yes	200	60
C	Yes	200	60
D	No	350	30
E	Yes	320	60

Table 2: Sample information (resin information given in Table 1)

Sample ID	Resin	Curing conditions
A	A	170°C, 60 min in air
B	B	200°C, 60 min in air
C	C	200°C, 60 min in air
C-2	C	200°C VFM oven, 60 min in air
D	D	350°C, 30 min in low pressure N$_2$ (bench mark)
D-2	D	225°C, 60 min in air
D-3	D	250°C, 60 min in air
D-4	D	200°C VFM, 60 min in air
D-5	D	225°C VFM, 60 min in air
D-6	D	250°C VFM, 60 min in air
E	E	320°C, 60 min in low pressure N$_2$ (Bench mark)

We evaluated three low temperature curable photo sensitive passivation materials (Materials A, B and C in Table 1) at test coupon level. Two conventional PI coatings were also included in the study. Material D is not photo sensitive and was included for VFM curing evaluation. Material E is the most relevant bench mark for the three low temperature curable materials because it is photo sensitive. The materials were spin coated onto blank wafers and cured according to the process guides from respective material suppliers to a final thickness of 5μm-10μm. Test coupons were cut from the wafers for evaluation of thermal stability, moisture absorption, tensile strength, chemical resistance, and adhesion to underfill. Test protocols were developed based on thorough analysis of downstream processes from back side passivation and served to simulate the worst case conditions that the coating would encounter. Detailed information on curing conditions of the samples is given in Table 2.

Passivation materials cured at low temperature tend to outgas at temperatures above the curing temperature. Outgassing can cause failures such as delamination between the PI and underfill. As we discussed earlier, it is dangerous to assume the 5% weight loss temperature on the material supplier's datasheet as the upper limit process temperature. From a user's perspective, it is important to ensure no excessive outgassing at any process step downstream. Since the lead free solder reflow process, which requires highest process temperature after the passivation coating curing, can reach a peak temperature of 260°C, we chose baking at 275°C for 10 minutes to simulate the worst case thermal excursion. Silicon coupons coated with 5-10μm thick coating were weighed before and after baking using a Mettler Toledo XP-6U micro balance with resolution of 0.1μg. Coating weight was calculated using surface area, thickness and density data provided by material suppliers. Baking was carried out in a TDM-Compact by Insidix with 1°C/s temperature ramp-up and ramp-down to simulate solder reflow. Weight loss measured was attributed 100% to the change in the passivation coating since silicon is very stable at 275°C.

For moisture absorption test, samples were first baked at 150°C for 10 minutes to dry out any moisture absorbed at ambient condition. The samples were then exposed to 85°C/85% relative humidity (RH) condition for 68 hours. After humidity exposure, the samples were left at ambient condition for 2 hours to dry out any surface moisture before weight measurement. The % weight gain was measured and calculated using the same procedure as for thermal stability evaluation.

For tensile strength evaluation, the coated wafers were cleaved into strips of approximately 10mm x 50mm. The strips were then immersed in HF solution to separate the passivation film from silicon. The free-standing film was pulled in an Instron 5565 Tensile Tester at a tensile speed of 5mm/min until fracture. Due to limitation of our test instrument and the non-ideal cleavage lines of the samples, data generated showed large scatter. Tensile rate of 5mm/min was chosen because it gave higher data consistency than other speeds we tried.

For chemical resistance, the test coupons were first coated with a thin layer of rosin flux typically used for wafer level solder reflow. They were then run through a typical lead free reflow profile with 245°C peak temperature. After the reflow process, the coupon was exposed to AZ 300T photoresist stripper solution for 5 minutes at 50°C, followed by DI water rinse. Samples with good chemical resistance showed no cracking during the test, while samples with poor chemical resistance showed massive cracking in the film.

Test samples in form of rectangular die approximately 4mm x 8mm were used for testing film adhesion to underfill. A commercial underfill qualified for fine pitch flip chip packaging application at Invensas was used to attach the die to a PCB substrate. As shown in Figure 2, a 75 µm thick polytetrafluoroethylene (PTFE) film with a 2mm wide slot was used as a spacer to control the underfill thickness and contact area. After the sample was cured according to the underfill supplier's process guide, the die was sheared off using a Dage 4000 Multi-purpose Bond Tester and examined under optical microscope for failure mode.

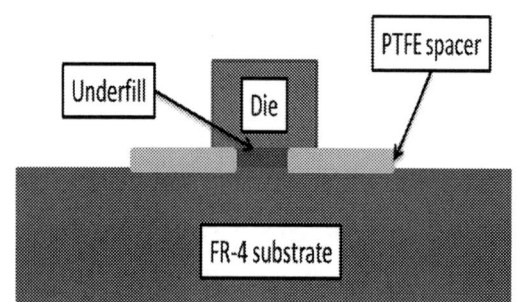

Figure 2: Illustration of test sample construction for underfill adhesion test.

Results and Discussion

Thermal stability

Figure 3 shows average weight losses after 10 minute baking at 275°C. The non-photo-sensitive material (sample D) cured at 350°C shows very low weight loss. This is expected since the curing temperature is 75°C above the baking temperature. The same resin VFM cured at 200°C (sample D-4) showed weight loss of 3.5%. Clearly the low-temperature VFM curing process left more volatile compound behind, which can potentially outgas and cause delamination during solder reflow. More detailed discussion of on the effect of VFM curing will be given later. Although sample E was cured at 320°C, which is 45°C above the baking temperature, weight loss measured was still >2%. Volatilization of residual photo pack is a plausible explanation for the higher weight loss.

The three low temperature curable passivation materials showed weight loss ranging from 3-6%. Thus delamination of the coating at the coating/silicon interface or coating/ underfill interface may occur during lead free reflow. More process evaluation should be carried out to assess such risk prior to committing the material to design.

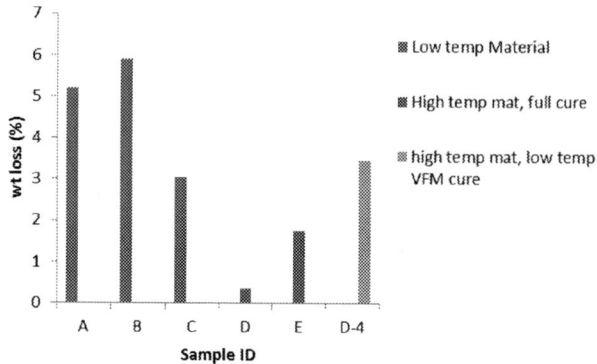

Figure 3: Weight loss of cured polyimide coatings on silicon after 10 minutes bake at 275°C.

Moisture Absorption

In addition to the chemical residues from curing that can volatilize during downstream process, moisture absorption is also well known for causing delamination during solder reflow. Hence most material data sheets provide moisture absorption data. The data sheet does not normally specify test conditions used to obtain the data. Some suppliers use room temperature immersion test. Such test simulates water absorption during wafer processing in aqueous solutions. We supplemented supplier data with data from accelerated high humidity exposure (85C/85% RH) since it better simulates moisture absorption in storage before assembly.

As shown in Figure 4, the three low temperature curable materials showed very different moisture absorption behavior. The worst performing sample B showed weight gain of 9.7%. Sample C, on the other hand, showed <2% absorption, comparable to that measured on the bench mark sample E cured at 320°C. Sample D showed lowest absorption among all samples. Sample D-4, which is resin D VFM cured at 200°C, showed slightly higher absorption, but still lower than all of the photo sensitive materials.

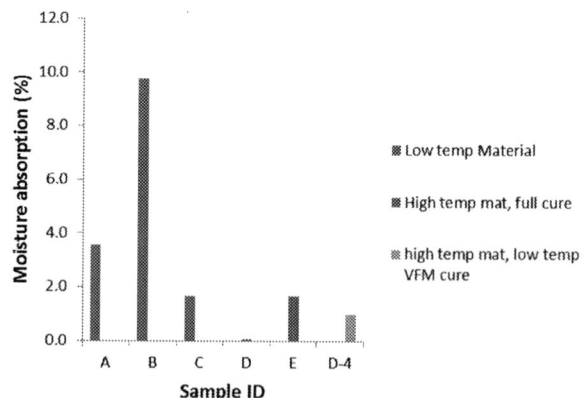

Figure 4: Moisture absorption of different coatings after 68 hours of exposure at 85°C/ 85%RH.

Chemical Resistance

For our process, we found that the sequence of flux coating, thermal excursion using lead-free solder reflow profile, and exposure to photoresist strip provides a simple and fast way to weed out materials with poor chemical resistance. As shown in Figure 5, the material (sample A) with poor chemical resistance showed massive cracking in the coating after test, while as the material with better chemical resistance (Sample C) showed no cracking. On Sample C, only some dust particles and flux residue are visible (the experiment and observation were carried out outside a clean room).

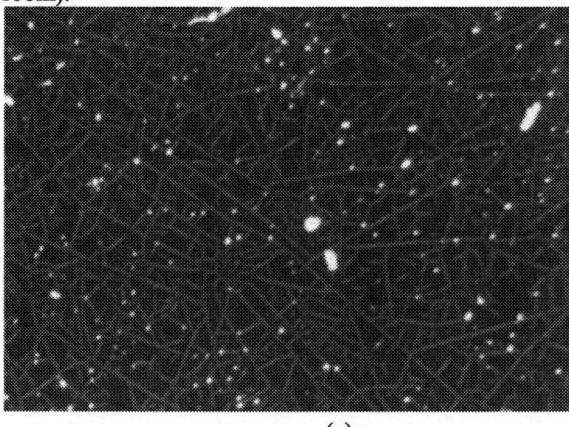

(a)

(b)

Figure 5: Optical image of samples after flux coating, lead free reflow and photoresist striper exposure. (a) Cracking in the sample A; (b) Surface of sample C showing no crack.

A summary of chemical resistance test results is given in Table 3. Among the three low temperature curable mateirals, sample C is the only one showing good chemical resistance. By contrast, both comventional PIs showed good chemical resistance. Sample D-4 (VFM cure at 200°C) also passed the test.

Table 3: Chemical resistance test results. Fail: Film cracked; Pass: No film cracking.

Sample ID	Resin	Chemical resistance
A	A	Fail
B	B	Fail
C	C	Pass
C-2	C	Pass
D	D	Pass
D-2	D	Pass
D-3	D	Pass
D-4	D	Pass
D-5	D	Pass
D-6	D	Pass
E	E	Pass

Adhesion to Underfill

Table 4 summarizes the results of the adhesion test. Both low temperature curable materials tested (samples A & C) showed very good adhesion to the underfill, with die fracture as the failure mode (Figure 6a). By comparison, adhesion of resin D to the underfill is not optimal. Although sample D cured at 350°C showed high adhesion strength, failure mode was clear separation of the PI from underfill (Figure 6b). Sample D-4, which was VFM cured at 200°C, showed further deterioration in adhesion strength.

Table 4: Test results of adhesion to an underfill

Sample ID	Resin	Adhesion strength	Failure mode
A	A	High	Die fracture
C	C	High	Die fracture
D	D	High	Interface separation
D-4	D	Low	Interface separation
D-5	D	High	Interface separation

(a)   (b)

Figure 6: Pictures of shear test failure surface. (a) Die fracture (Sample A); (b) Underfill/polyimide coating interface separation (sample D).

Tensile Strength and Elongation

The Instron tensile tester used in this study is not ideal for pulling thin films. As shown in Figure 7, elongation at fracture showed large data scatter. We believe that the scatter resulted from non-uniform stress distribution in the sample, which induced variation in crack initiation. Such data scatter makes the test less useful for comparing small differences between samples. However, it can still detect very large differences.

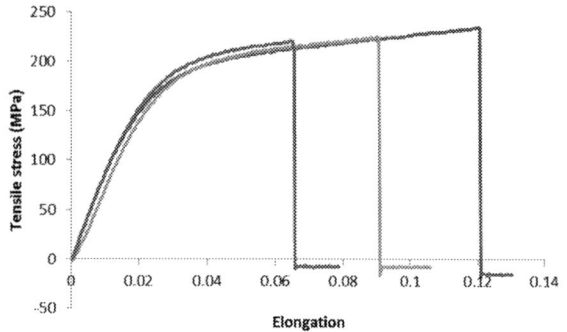

Figure 7: Tensile stress vs elongation curves for three replica of sample D-5, showing consistent tensile strength value but considerable scatter in elongation at failure.

Mechanical property data provided by material supplier show that tensile strength and elongation of low temperature curable passivation material are lower than fully cured conventional PI. As shown in Figure 8, our test data agree with this trend. Film made of sample C was too fragile to be tested with the fixture. All 5 replica fractured during sample mount. Sample A yielded a Young's modulus of 3.02 GPa and elongation at fracture of 9.1%, compared to sample D of 8.2GPa modulus and 40% elongation at fracture.

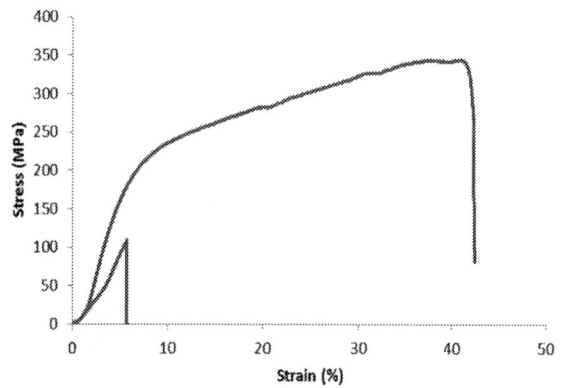

Figure 8: Stress-strain curves measured on Sample A (purple) and Sample D (red).

Table 5 shows overall ranking of the materials tested. Sample B ranked last for weight loss, moisture absorption and chemical resistance test. It was therefore eliminated from adhesion and tensile testing. Comparing remaining two low temperature curable materials, sample A has better tensile properties and sample C has better chemical resistance. Final

material choice would depend on specific needs of the customer.

Table 5: Overall ranking of materials tested. Low number denotes better property

ID	Wt. loss	H₂O absorption	Chem. Resistance	Adhesion	Tensile elongation
A	5	5	Fail	1	2
B	6	6	Fail	/	/
C	3	3	Pass	1	4 (0%)
D	1	1	Pass	2	1
D-4	4	2	Pass	3	3
E	2	4	Pass	/	/

VFM Curing Impact

While conventional heating causes vibration of molecules, microwave energy from a VFM oven rotates the polarizable dipoles in uncured resins. This mechanism can effectively promote reaction and curing in a polymer material, even if the microwave-induced temperature is relatively low. When we applied VFM to resin D, similar properties were obtained at lower temperatures than from a conventional oven. The degree of imidization became close to 100% after it was cured by VFM at 200°C for 1 hour, while it required 225 – 250°C by conventional curing (Figure 9). The thermal stability of a VFM-cured sample was also significantly better than that of a sample oven-cured at the same temperature (Figure 10). However, although ~100% curing was achieved by VFM at 200°C (resin D-4), it was not equivalent to 350°C oven curing that was recommended by the manufacturer. Resin D-4 showed much less elongation (Table 5) and somewhat inferior thermal stability (Figure 10). VFM was also tested for the low-temperature curable resin C, but there was no noticeable difference between VFM and conventional curing.

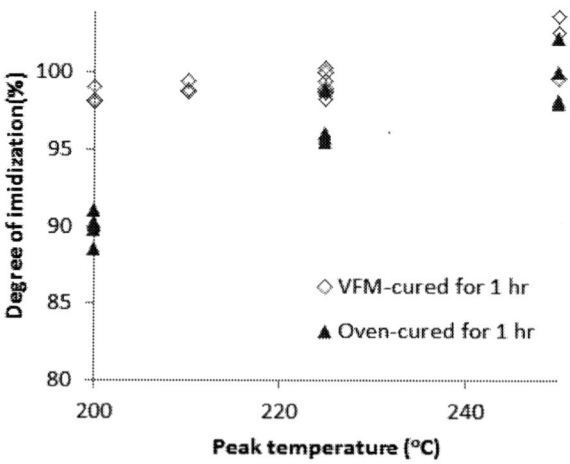

Figure 9: Curing of resin D by VFM and conventional oven heating. At the same temperature (microwave-induced temperature in case of VFM), VFM achieved higher degree of imidization.

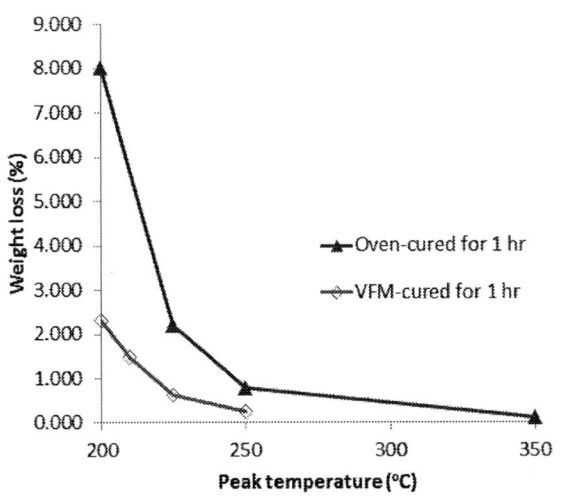

Figure 10: Weight loss from cured resin D samples by a 10 minute baking at 275°C. VFM curing was more effective in reducing outgassing than conventional oven curing, when the same temperature was used.

Conclusions

A battery of tests has been developed for comprehensive evaluation of organic passivation materials for 3D IC backside application. These tests realistically simulate process conditions encountered in wafer processing and package assembly. All three low temperature curable passivation materials tested are inferior to the bench mark conventional photo sensitive polyimide fully cured at 320°C in at least one of following properties: thermal stability, moisture absorption, or chemical resistance. They are also much inferior in tensile properties to the non-photo sensitive polyimide material cured at 350°C.

The high temperature polyimide resin (D) can be cured in the VFM oven at reduced temperature. When cured at 200°C in a VFM oven, the material shows thermal stability, moisture absorption and tensile properties comparable to the materials cured at 225°C or higher in a conventional oven. However, the low temperature curable polyimide material C does not benefit from curing in VFM oven. Careful consideration should be given when choosing a low-temperature curable material to ensure that it can satisfy the actual application environment.

Acknowledgments

The authors like to thank all the material manufactures that provided materials to Invensas for test/ evaluation.

References

1. J. Hermanowski, " Thin Wafer Handling –Study of Temporary Wafer Bonding Materials and Processes", http://www.suss.com/fileadmin/user_upload/technical_pub lications/thin_wafer_handling_temporary_wafer_bonding. pdf
2. S. Arkalgud, A. Cao, E. Chau, G. Gao, H. Katske, L. Wang, "Reaching A Low Cost, Manufacturable Interposer", in *3D Architectures For Semiconductor Integration And packaging*, December 11-13, 2014, Burlingame, CA
3. M. Töpper, T. Fischer, T. Baumgartner, H. Reichl, "A Comparison of Thin Film Polymers for Wafer Level Packaging", in *Proc. Electronic Components and Technol. Conf. (ECTC)* , Las Vegas, NV, June 1-4, 2010, pp769-776.
4. R. L. Hubbard, I. Ahmad, K. Hicks, "Low Temperature Curing of Polymer Films for Wafer Level Packaging", in *Proc. of the iMAPS Device Packaging Conf.*, Scottsdale, AZ, March, 2005
5. R. L. Hubbard and B.-S. Lee, "Low Warpage and Improved 2.5/3DIC Process Temperature Capability with a Low Stress Polyimide Dielectric", in *Proc. of Internation Wafer Level Packaging Conf.*, Santa Clara, CA, November 2013.

Optimization of CMP Process for TSV Reveal in Consideration of Critical Defect

DongHoon Lee[1,2], DoHyeong Kim[1], SeungChul Han[1], JooHyun Kim[1], JungSoo Park[1], BoRa Jang[1], YoungSuk Chung[1],
SeongMin Seo[1], YongSang Kim[2], and ChoonHeung Lee[1]

[1] 151, Dongil-ro, SeongDong-gu, Amkor Technology Korea, Inc , Seoul, Korea

[2] Schools of Electronic and Electrical Engineering, Sungkyunkwan University, Suwon 440-746, Republic of Korea

Donghoon.Ree@amkor.co.kr

Abstract

In this paper, we discuss the optimization of CMP process for mid-end-of-line (MEOL). TSV breaking and blister are the two types of critical defects during CMP. TSV breaking occurs due to high principle stress at the bottom of TSV and blister occurs because of adhesive deformation. Therefore, we suggest a new process schematic for CMP and the use of thermally stable adhesive material to prevent these critical defects. Later we optimized CMP process in consideration of via height, throughput and non-uniformity without any critical defects. TSV was successfully revealed without TSV breaking. The result shows less 10 % WIWNU, WIDNU and less 10 nm TSV dishing.

Introduction

To meet the increasing demand for longer, faster, cheaper and smaller electronic products; semiconductor packaging industry needs more innovative and emerging packaging technologies [1]. Three dimensional integration (3D) technologies has been a promising method for future package. It can meet demand as low power consumption from lower RC delay, fast device, and short-interconnection with small space. For these reasons, 3D technologies were researched by many companies and research center [2],[3].

Through silicon via (TSV) is one of the strong candidates of 3D technologies. TSV can be fabricated by via-first, via-middle and via-last with backside process. Among these, via-first process faces problem due to thermally unstable copper during CMOS process and in via-last process the cost of die for via-filling is too high. For these reasons, via-middle process has been widely used [4],[5].

Therefore, in this paper, we focused on the backside process with via-middle schematics. Commonly, the backside processes comprehend backside bumping process and Mid-End-Of-Line (MEOL) process. Fig. 1 shows step of backside process which consists of a wafer support system (WSS), wafer back grinding (WBG), etch, plasma enhanced chemical vapor deposition (PECVD), chemical mechanical polishing (CMP) and wafer bumping [6],[7]. Among these processes, CMP process is erratic and the key source of defects.

CMP process has been used to increase depth of focus (DOF) margin in semiconductor industrial areas. It can be classified into shallow trench isolation (STI), W plug, interlayer dielectric (ILD), Cu damascene and utilize as applications of poly, metal and oxide CMP by polished material according to some purpose. These applications of the CMP process focus on how to remove each layer without defects as particle, delamination and scratch. On the other

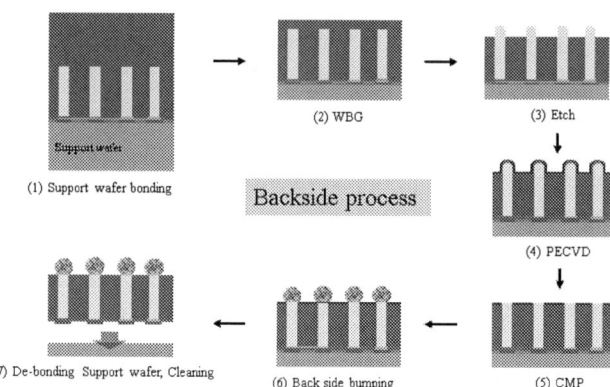

Figure 1. Backside Process.

hand, TSV is a new application which has many challenges as it involves polishing of two kinds of materials. Therefore, this study focuses on how to optimize the CMP process for the application of TSV backside reveal.

Critical defects after CMP process

Types of defect after the CMP process of MEOL process is different from common defects of CMP process. Therefore, we discuss the two types of critical defects i.e. TSV breaking and blister before the discussion of process optimization.

1. TSV breaking

Fig. 2 shows SEM image of TSV breaking (a) and cracking (b) which are one of the critical defect in MEOL process. Bending of TSV is the typical defect during initial stage of CMP. In general, TSV height is protruded over 3 um after etch process due to total thickness variation (TTV) of adhesive from WSS process and TSV depth variation. TSV breaking occurs at highly protruded TSV over a few microns by high stress during CMP process. At this time, stress analysis of CMP process can be explained using shear and principle stress. They are maximized at different points as shown in Fig. 3. The upper side of TSV faces the maximum shear stress. On the other hands the lower edge faces the maximum principal stress (MPS) [8]. Among these stresses, MPS is the more effective reason to break TSV because of MPS position is same with breaking and cracking position.

TSV breaking could induce interconnection failure from an irregular pad of bumping process and bad reliability from NCP trap during the chip attachment process. Stress analysis with passivation type, via height and varied parameter of CMP process, as pressure and spin speed, can be a good tool for research to understand TSV breaking mechanism.

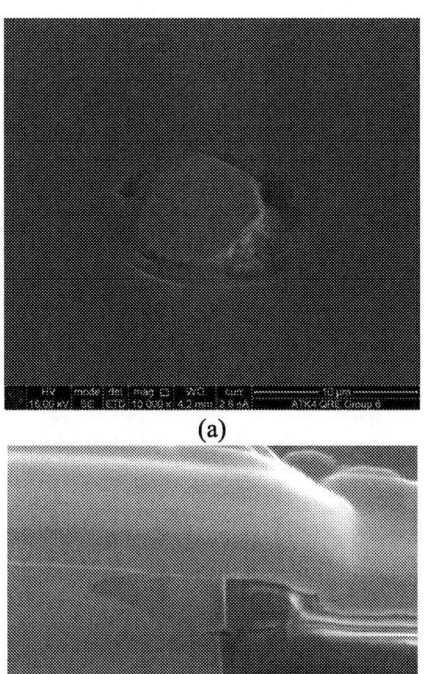

(a)

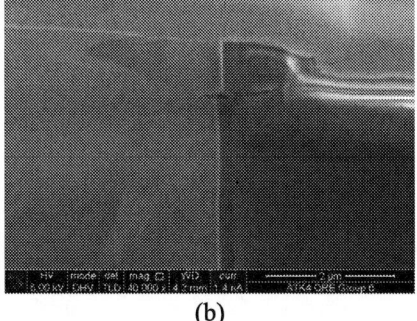

(b)

Figure 2. SEM Image of (a) TSV breaking & (b) Cracking

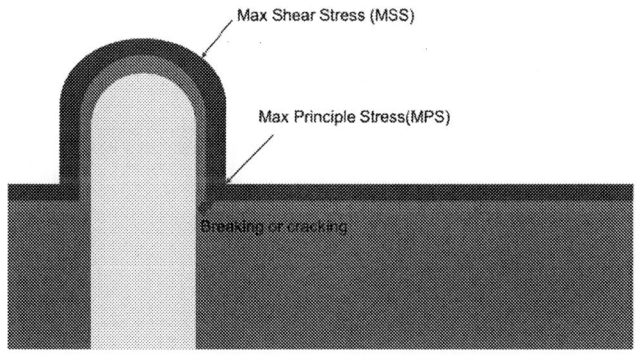

Figure 3. TSV breaking mechanism

2. Blister & dimple

Fig. 4 shows blister and dimple defects which are also critical after CMP process. The wafer was locally over-polished, exposing the silicon area. Scanning Acoustic Tomography (SAT) was used to find the reason of this defect. As shown Fig. 4, voids in adhesive layer were detected. The position of blister defect matched with that of voids in SAT image. The void was checked before and after PECVD using the SAT as shown Fig. 5. Small voids of 100 um in size occurred after the wafer bonding process, and they were expanded after PECVD.

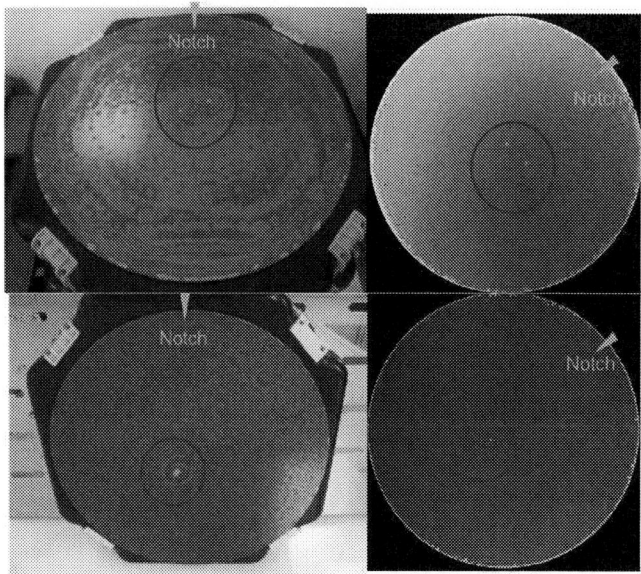

Figure 4. SAT analysis result of blister defect

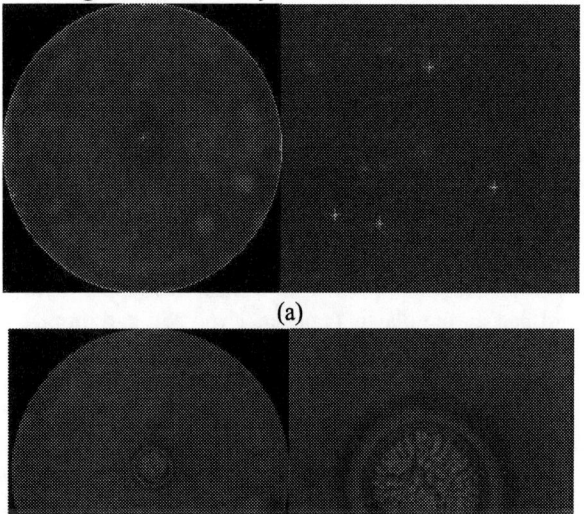

(a)

(b)

Figure 5. SAT image of adhesive deformation
(a) before PECVD (b) after PECVD

The high temperature (200 °C) and pressure (2000 mTorr) of PECVD process caused the adhesive deformation near the void which induced blister. On the other hand, dimple which is an unpolished area with unrevealed TSV also occurred nearby blister as shown in Fig. 6 (a). Fig. 6 shows a cross section image of blister and dimple area that shows adhesive deformation. Adhesive thickness at normal area (Fig. 6 (b)) is 50 um, but blister area (Fig. 6 (c)) shows over 70 um thick adhesive while the adhesive thickness of the dimple area (Fig. 6 (d)) is less than 30 um. Through exposing silicon surface, unexpected electrical leakage might be induced, whereas dimple area can cause high resistance and interconnection failure

978-1-4799-2408-0/14 $31.00 © 2014 IEEE   1817

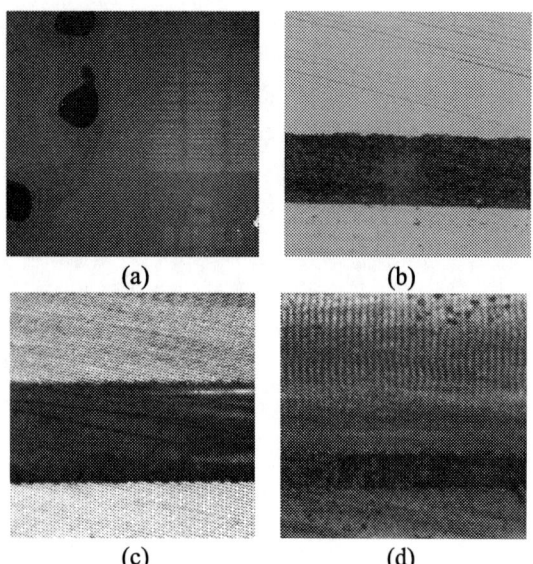

Figure 6. Optical images of defect die (a), adhesive of normal area (b), adhesive of blister area (c), and adhesive of dimple area (d)

Discussion

Here, we suggest process schematic to solve above mentioned critical defects and discuss CMP process optimization.

1. Process step

The process step of CMP to damascene copper layer uses a hard pad on the platen 1, platen 2 and soft pad on platen 3. Platen 1 rapidly removes the copper layer with high pressure and then platen 2 controls the end point to reach a target thickness. Finally a soft pad with platen 3 polishes wafer to remove particle and barrier layer [9]. However, this schematic is not recommended for MEOL process due to the TSV breaking issue. Therefore, we propose new process schematic as shown in fig. 7 to prevent TSV breaking.

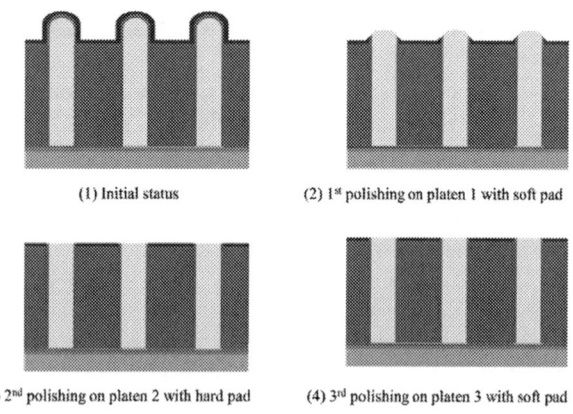

(1) Initial status
(2) 1st polishing on platen 1 with soft pad
(3) 2nd polishing on platen 2 with hard pad
(4) 3rd polishing on platen 3 with soft pad

Figure 7. New process schematic

In the new process schematic, platen 1 used soft pad to reduce via height. CMP process using the soft pad reduced the stress compared to that with hard pads. Soft pad showed higher compressibility and so the via could get buried under the soft pad. At the second step, platen 2 with hard pad and high pressure was used to remove passivation layer to the target thickness. Lower aspect ratio of TSV can endure higher load during polishing. In order to decide the target height at platen 1, pressure at platen 2 should be considered. Finally, platen 3 used the soft pad for planarization of via, passivation and cleaning particles. This schematic of the CMP process can polish backside of the TSV wafer without TSV breaking issue.

2. Adhesive material

The adhesive material is one of the critical factors for MEOL process and affects the CMP process the most. Adhesive layer changes randomly due to high pressure and temperature during the PECVD process which causes blister defect. Therefore, the stability of adhesive should be considered before optimization of CMP process. Fig. 8 shows the non - uniformity result of the remaining passivation thickness after CMP process with the varied PECVD condition according to two kinds of adhesive material. Table. 1 shows the physical properties of adhesives, where material A has a higher viscosity before bonding. But after bonding, viscosity of material A and B changed to 4220 cP and 5790 cP, respectively. Lower viscosity tends to deform under high pressure and temperature. For this reason, uniformity data of material A shows instable performance with large variation and material B shows very stable performance.

As a result, zone pressure control of polish head can make up for adhesive bonding TTV and small adhesive deformation after considering the temperature of PECVD and choice of adhesive material.

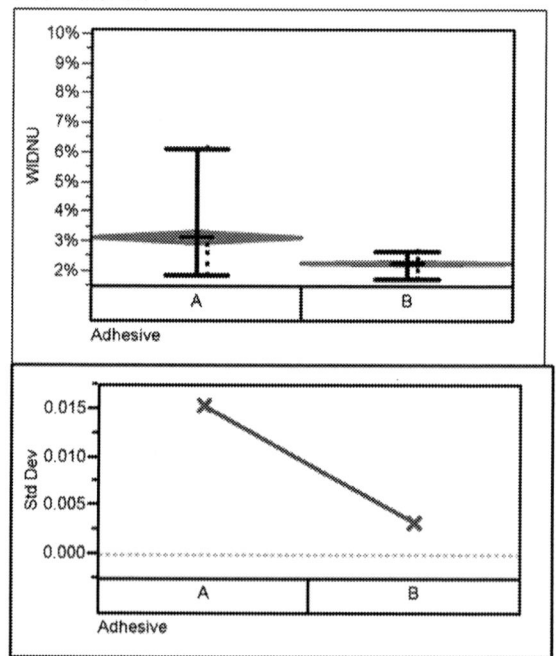

Figure 8. Uniformity data with 2 types of adhesive materials

Table 1. Adhesive property

Item	A	B
Viscosity at 215 °C during bonding	204 cP	1270 cP
TGA	> 300 °C	> 300 °C
Viscosity at 220 °C after bonding	4220 cP	5790 cP

3. Process optimization

CMP process was evaluated using passivation thickness, uniformity, TSV height and throughput. Removal rate profile, slurry selectivity, removal rate and die design should be considered to optimize CMP process with thermal stable adhesive material.

First, uniformity of passivation thickness is evaluated using wafer in die non uniformity (WIDNU), wafer in wafer non uniformity (WIWNU), wafer to wafer non uniformity (WTWNU). WIWNU is related to removal profile of the CMP process with zone pressure control. Fig. 9 shows removal profile of bonded pair. Red line is before optimization of zone pressure control and blue line is after the zone pressure optimization with 3mm edge exclusion. Edge area of red line was over polished due to cumulative TTV through MEOL process. Non uniformity of the red line is over 20% of the edge area. Non uniformity decreased within 10 % after zone pressure was changed to match TSV wafer profile after PECVD.

TSV pattern of die is affected to WIDNU. As shown fig. 9, remained passivation thickness at high density areas of TSV is thicker than the at low density areas. It shows difference of almost 7000A which was related to density of TSV and polishing time. The difference of passivation thickness might give some potential to affect die warpage and bump connection at assembly process. Polishing time of platen 1 is one of important factor to optimize WIDNU because passivation on surface is also polished when TSV is polished. Therefore, reduction of polishing time of platen 1 and increase of polishing time of platen 2 can improve WIDNU. Additionally, suitable pad with compressibility and hardness and slurry with considered copper selectivity and removal rate can also improve WIDNU.

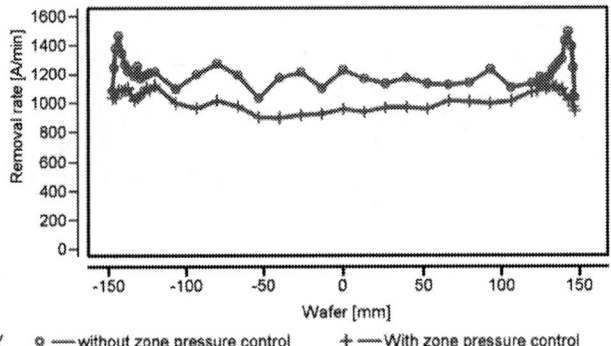

Figure 9. Removal profile on bonded wafer

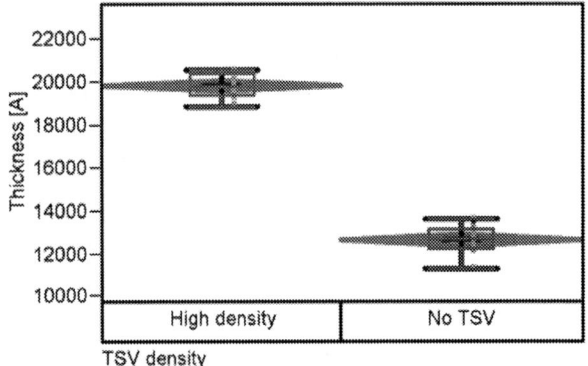

Figure 10. Difference of passivation thickness according to TSV density

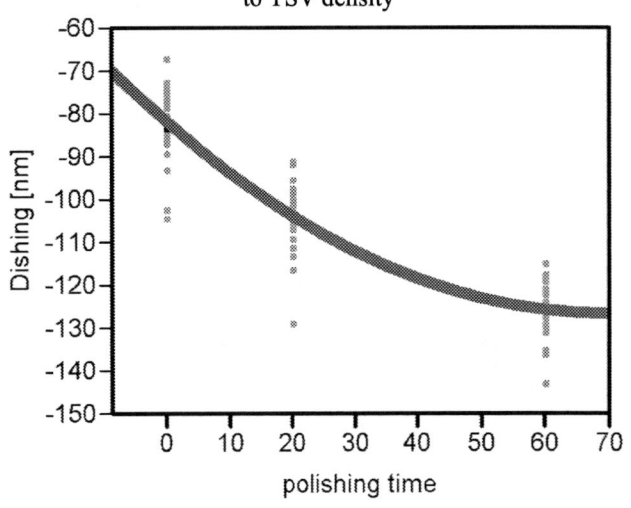

Figure 11. TSV height change with polishing time

The second evaluation point is the TSV height, which can change the exposed Cu tip to dishing or protruding because of the copper selectivity of slurry. High copper selective slurry can result in dishing and low copper selectivity slurry in protrusion. Fig. 11 shows dishing trend according to polishing time at platen 3. Initial dishing condition is 60 nm and is increased to 80 nm by polishing for about 20 s. Dishing variation of each step was reduced because the flat surface blocked the contact TSV with pad. Hence, minimum polishing time of platen 3 should be optimized to reduce dishing level without particle and scratch issues.

Final evaluation point is the throughput which can determine cost of CMP process. Throughput of CMP process depends on maximum polishing time of each platen. In the suggested process schematic, platen 1 shows the longest polishing time. Therefore, in order to increase throughput, polishing time of platen 1 should be reduced. Slurry A shows high oxide removal rate with lower copper removal rate and slurry B has a higher copper removal rate as shown in table. 2. Fig. 12 shows removal rate of TSV using two types of slurries. TSV was polished on a soft pad with low pressure to prevent TSV breaking. Slurry A shows TSV removal

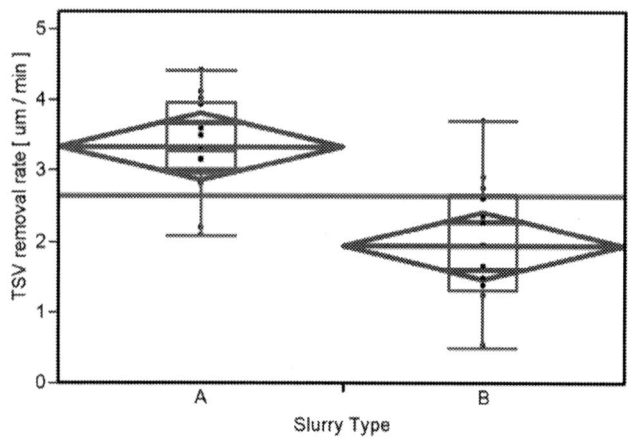

Figure 12. TSV removal rate, according to selectivity

Table 2. Slurry performance with blanket wafer on soft pad

Material	Slurry A	Slurry B
SiO_2	1113 Å	721 Å
Copper	75 Å	3070 Å

rate of around 3.5 um/min and slurry B shows around 2 um/min. Slurry A with high SiO_2 removal rate shows higher removal rate of TSV height than other slurry because TSV is surrounded by SiO_2 passivation which can interrupt to reduce TSV height using a slurry A. According to this result, the SiO_2 removal rate is a more effective factor to planarize backside of TSV wafer and higher removal rate of passivation layer can show better throughput.

Optimized CMP process

Fig. 13-a shows the wafer image without critical issue such as blister and fig13-b shows a FIB SEM image of TSV after CMP. Optimized process resulted in good planarization and prevented breaking of TSV. Table. 3 summarizes the result of TSV bonded pair after the CMP process. Non-uniformities lie within 10 % and via height was within 100 nm. And then, throughput of CMP process can be maximized through the optimization.

Table 3. Measurement result after optimization of CMP

Measurement Item	Result
TSV height	- 89 nm
WIWNU	8.41 %
WIDNU	7.09 %

Conclusions

We optimized the CMP process of TSV bonded pair without critical defect. To prevent issues such as blister and TSV breaking, we analyzed problems and found the major root cause. The summarized consideration and recommendation to optimize CMP process are as below.

i) The height of TSV should be reduced using a soft pad on the platen 1, because high aspect ratio of TSV can induce breaking by stress during CMP process.

ii) Temporary bonding adhesive is critically affected to the CMP process. When choosing this kind of adhesive, its thermal stability should be considered.

iii) Consumable and process recipe of CMP could be optimized considering non-uniformity, TSV height, and throughput. Finally, backside of the TSV wafer after CMP process optimization showed less 10 % WIWNU and WIDNU and less 100 nm TSV dishing with competitive throughput. Which are sufficient results for TSV backside reveal application.

(a)

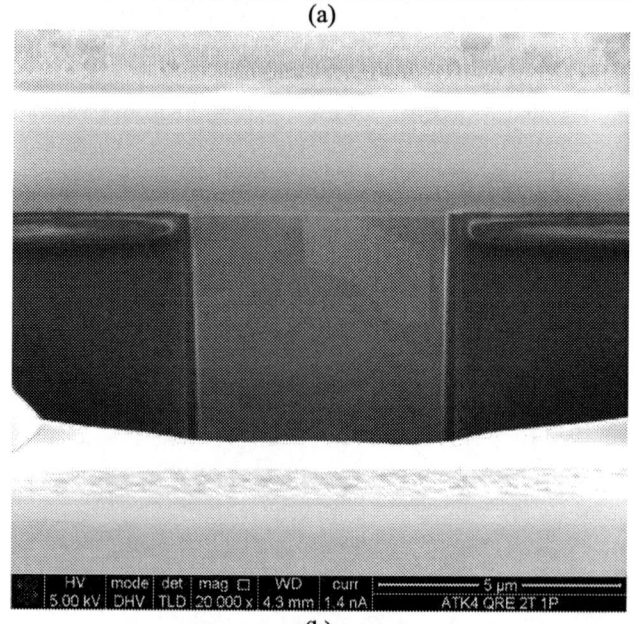

(b)

Figure 13. CMP process result
(a) wafer image (b) TSV FIB SEM image

Acknowledgments

Authors would like to thank engineers, technicians and operator who develop TSV wafer process in Amkor Technology Korea.

References

1. S.W Yoon, "TSV MEOL (Mid-End-Of-Line) and its Assembly/packaging Technology for 3D/2.5D Solutions," ICEP-IAAC 2012, TA1-1.

2. B. Wu, A. Kumar, S. Ramaswami, "3D IC stacking technology," Mc Graw Hill Book Publication, 2011.

3. J.C Chen, "Challenges of Cu CMP of TSVs and RDLs Fabricated from the Backside of a Thin Wafer," 3D Systems Integration Conference (3DIC), 2013 IEEE International.

4. Dong Wook Kim, "Development of 3D Through Silicon Stack (TSS) Assembly for Wide IO Memory to Logic Devices Integration", , IEEE ECTC 2013, pp.77-80.

5. Stephen Olson, "TSV Reveal Etch for 3D intergration," 3D Systems Integration Conference (3DIC), 2011 IEEE International.

6. B.K Huang, "Integration Challenges of TSV Backside Via Reveal Process," 2013 Electronic Components & Technology Conference. 2013, pp.915-917.

7. T. Mourier, "3D Integration challenges today : from technological toolbox to industrial prototypes", IEEE ECTC 2013.

8. C Liao, "Stress analysis of Cu/low-k interconnect structure during whole Cu-CMP process using finite element method", Microelectronics Reliability 53 (2013) 767–773.

9. J.C Chen, "Impact of Slurry in Cu CMP (Chemical Mechanical Polishing) on Cu Topography of Through Silicon Vias (TSVs), Re-distribution Layers, and Cu Exposure, IEEE ECTC 2011.

High Throughput Roller Type Nano-pattern Transfer Technique on Both Rigid Flexible Substrates and Mold Deformation Analysis under Atmospheric Imprint Environment

Yinsheng ZHONG, Matthew M. F. YUEN
Department of Mechanical and Aerospace Engineering
The Hong Kong University of Science and Technology
Clear Water Bay, Kowloon, Hong Kong
rinto@ust.hk, meymf@ust.hk
Phone: (852) 2358 8814, Fax: (852) 2358 8357

Abstract

This paper provides a roller type nano-pattern transfer process, which is based on UV nanoimprint lithography, for nano-patterning on both rigid and flexible substrates. By using UV-curable polymer resist, room temperature and low pressure process, which is possible to apply to flexible films, is introduced. PDMS and PVA soft molds that can have conformal contact with large area non-flat substrate surface are fabricated. The nano-scale mold patterns are duplicated from Si wafer after E-beam lithography and DRIE dry etching. A homemade roller type printing machine with an elastic buffer layer is designed and fabricated to achieve uniform printing results over entire working area. High-throughput printing process with printing speed up to 50 mm/s is demonstrated. Glass, Si wafer and PET film were tested as printing substrates. Micro and nano-patterns were transferred clearly to the polymer layer on substrates. Images from optical microscope, SEM and AFM were used to evaluate the transferred nano-patterns. Detail process flow, including mold fabrication, is described. The governing parameters in the process are discussed. Simulation and theoretical analysis on the mold deformation under atmosphere imprint environment is studied. The effect of surface property during the pattern transfer process is identified.

Introduction

The consumer electronics market pursues higher performance, smaller size and also energy saving in the products. In order to catch up with the trend, the industry is pushing high-precision low cost nano-scale patterning process to manufacturing. Flexible electronics is also attractive to many researchers in this area. The current photolithography process is not able to apply to non-flat substrate surface. Nano-patterning technology which can be applied to both Si wafer and flexible substrates is wanted.

Nanoimprint lithography (NIL) is a promising process to produce high resolution nano-patterns [1]. NIL is based on the mechanical embossing principle. A mold with nano-patterns is imprinted against a thin polymer layer which covers the substrate. Then the low viscosity liquid phase polymer goes through a curing process and changes to solid phase. After de-molding, opponent pattern is transferred to the polymer layer. The NIL process can be applied to non-flat surface, as long as the mold can have conformal contact with substrate. And it is not as sensitive as photolithography process to the environment, which can reduce the cost of facility. By using UV curable resist, room temperature UV-NIL process is

developed. It makes the application of elastic mold become possible [2]. The high productivity of nanoimprint can be achieved by having large imprint area. However, the plane type nanoimprint requires huge imprint force, which is proportional to the imprint area, to reduce the residual polymer layer thickness. Then roller type nanoimprint (RNIL) was introduced by researchers. Cylindrical molds are generally used in RNIL [3], [4]. Schematic diagrams of UV type RNIL process for flexible and rigid substrates are shown in Fig.1. By rotating and pressing the cylinder mold into the polymer thin film on substrate, the feature on the mold can be transferred to the polymer layer. Compare to the early imprint process, it reduces the total imprint force and allows the resist flow more freely inside the nano-pattern. Then the phase of polymer layer is changed from gel or liquid to solid during this contact by UV light. After further dry etching process, the nanopattern is defined on the substrate.

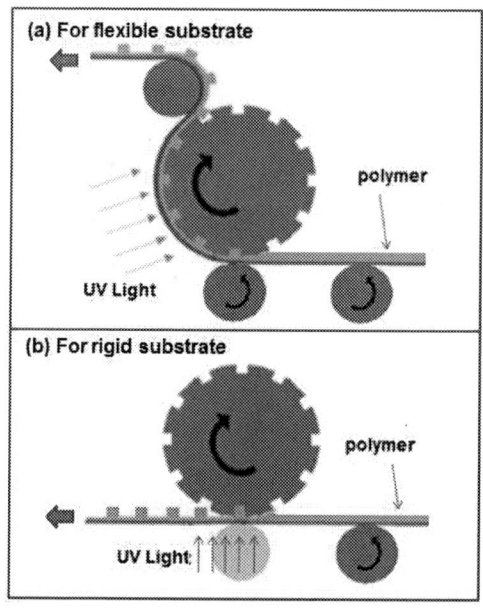

Figure 1. Schematic of the RNIL process: (a) for flexible substrate, (b) for rigid substrate.

Although RNIL is a continuous process, the rolling speed is still limited by the curing process. To make sure getting the fine pattern duplication, substrate must keep tightly conformal contact with the mold until the polymer layer is fully cured. If the rolling speed is too fast, substrate will separate from the mold when the polymer layer is still in gel phase. The cylindrical mold is not suitable for the RNIL process on rigid substrate, because of the small contact area between mold and substrate. Printing speed on the flexible substrate is faster. Bending the substrate provides larger contact area.

Another problem for current RNIL system is the relative motion between the mold and the substrate. Ideally, there is only vertical motion that the mold is pressed to the substrate. At the beginning of the imprint process, when the polymer still has low viscosity, the motion between the mold and substrate helps the polymer to fill into the mold pattern. However, when the polymer is partially cured, the shear motion will damage the pattern. It is important to keep the mold and substrate at a steady state during the phase transfer process. But there is an unavoidable shear force acting on the mold in traditional RNIL.

In this paper, a high throughput roller type soft mold nanoimprint process, which avoids the shear motion during polymer curing, is introduced. Instead of the cylindrical mold, a patterned flexible thin film is used as the imprint mold. And a smooth roller presses the mold against the substrate subsequently to gradually apply imprint pressure across the whole working area. High fidelity nano-pattern can be transferred to polymer layer after simply UV exposure and de-molding. The basic process flow is shown in Fig.2 below. Detail of the experiment is described in the later chapter. Similar process has been successfully demonstrates to produce micro-scale patterns with feature size of 50 µm. [5]

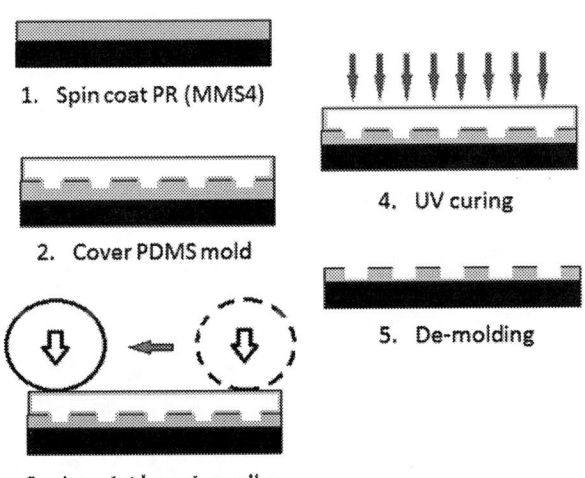

Figure 2. Process flow of the roller type UV nanoimprint.

Mold Fabrication

The original patterns on Si wafer were made by using E-beam lithography. The patterns included feature sizes from 300 nm to 5 µm. And the flexible molds were duplicated from the Si hard mold by casting polymer materials, likes PDMS and PVA.

PDMS was used by many research groups, due to its good mechanical properties and chemical stability [6], [7]. PDMS (Sylgard 184, Dow Corning), blending with 10 wt.% curing agents, was put in ultrasound 15min for mixing and degassing. Then the mixture was poured onto the patterned Si

wafers and degassed in a vacuum chamber for 60 min. Hydrophobic coating material Trichloro(1H,1H,2H,2H-perfluorooctyl)silane was coated on Si mold before PDMS casting by vapor deposition to prevent sticking. The Si wafers with PDMS were then placed into an oven and made the PDMS cured at 90 ºC for 1 hour. The cured PMDS film can be easily peeled off from the wafer without breaking.

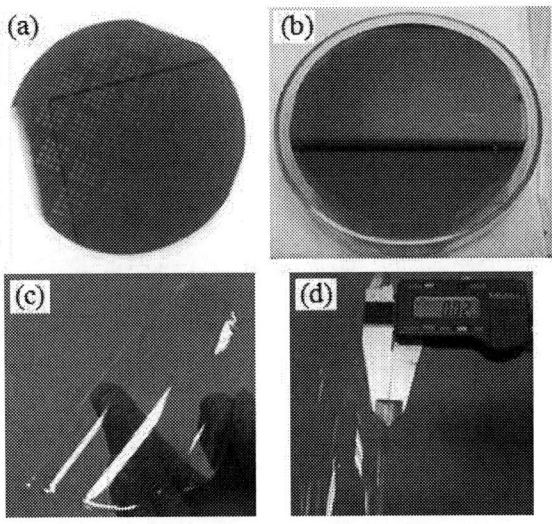

Figure 3. Picture of the flexible mold fabrication: (a) Si mold, (b) PVA/PDMS casting, (c) flexible mold (d) 20 µm thick PVA mold.

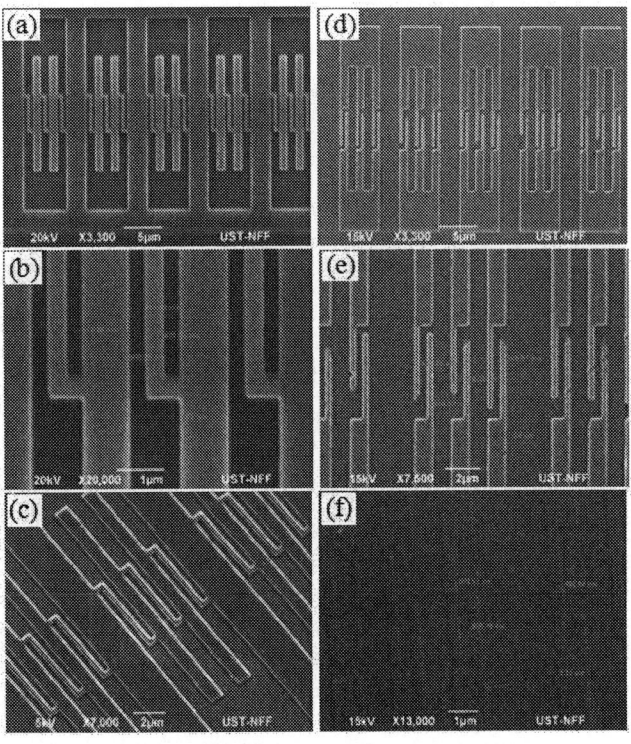

Figure 4. SEM of the nanopattern on the mold: (a-c) original nanopattern on Si mold, (d-e) nanopattern on PVA, (f) nanopattern on PDMS.

PVA mold was obtained by the similar casting process. But the hydrophobic coating on Si mold surface needs to be changed to a plasma treatment. The plasma treatment was used to form thin layer of PVA film on Si mold. The PVA aqueous solution had higher surface tension then PDMS mixture. On bare Si surface, the aqueous solution tended to form droplet rather than liquid thin film. If the PVA costing was too thick, the surface tension of water caused wrinkles during the film drying step. 10 wt.% PVA powder was dissolved in water at 80 °C. Then the solution was spin coated on Si mold and dry at 60 °C. Highly transparent PVA film is shown in Fig.3. The film thickness is about 20 μm.

Fig.4 shows the SEM images of the Si and soft molds. Patterns in the PDMS and PVA were the opponent of Si mold. And the final pattern which was transferred to substrate kept the same as original Si pattern. As the PDMS had low elastic module, the mold need to be thicker to reduce the strain during the process. The poor electric conductivity of the thick mold caused that the image contrast of PDMS mold in SEM is not as high as PVA mold. The dark line in Fig.4b is about 310 nm. The corresponding line pattern in PVA and PDMS are 320 nm and 340 nm respectively from Fig.4e, f.

Roller Type Nanoimprint Process

After the mold fabrication, the roller type nanoimprint process was done under atmospheric imprint environment by using a homemade roller imprint machine. In the machine, a metal roller with rolling speed up to 50 mm/s was used to provide the imprint force to the mold. The working area for one imprint cycle was 220 mm x 300 mm. The maximum imprint force was 110 N. To make sure the uniform contact between the roller and the mold, an elastic buffer layer was added in the machine. Fig.5a and b are the optical images of imprinted substrate from optical microscope. The color differences in the images show the uniformity of imprinted polymer layer is greatly improved by adding the buffer layer.

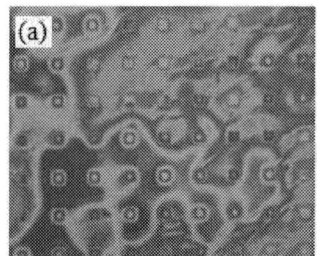

Figure 5. Optical images of imprinted substrate: (a) without buffer layer, (b) with buffer layer.

UV-curable AMONIL MMS4 was used as the pattern transfer polymer layer. Low viscosity (50mPa.s) of polymer was suitable to low imprint pressure. 200 nm thick polymer thin film covered the substrate by spin coating. The coating speed was 3000 rpm. Si wafer, glass and PET film were tested as substrate materials. Then the PDMS or PVA mold was placed on top of the polymer layer. And the roller pressed the mold against the substrate subsequently on the back side of the mold. The moving speed of the substrate holder kept the

same as the rolling speed to minimize the shear motion. After rolling, there was no extra force acting on the mold or substrate. The mold and the substrate were in a steady state. No shear motion between the mold and substrate during the next curing process. The polymer flow inside the nanopattern was eliminated. Then the UV light irradiation with exposure dose 2J/cm^2 cured the polymer within 3 min. After curing and de-molding, patterns from Si mold were transferred to the polymer layer. One process cycle was finished within 10 minutes. And the curing process was independent to the imprint process. Therefore, second imprint process could be started, while the first cycle was undergoing the UV curing.

Imprinted Results

SEM and AFM were used to evaluate the quality of transferred pattern. Fig.6 shows the photo of the imprinted Si wafer and the SEM images of the transferred nanopattern. The width of dark line in Fig.6d is 330 nm, while the corresponding pattern in original Si mold was 310 nm. Similar imprint results were achieved on both PET and glass substrates. Fig. 7 (a) is the result by using PVA mold, while Fig.(b) is came from PDMS mold. AFM images were used to measure the depth of the imprint patterns. Fig.8a shows that the original pattern depth on the Si mold was 108 nm. And the pattern depths were about 95 nm (Fig.7b) and 97 nm (Fig.7c) on PET and glass substrate respectively.

High fidelity nanopattern was transferred to the substrate. 300 nm wide features were successfully achieved. Height contrast was about 100 nm. The differences between the mold and transferred patterns were within 20 nm. The measurement results prove that the proposed roller type UV nanoimprint process can be applied to both rigid and flexible substrate in nano-patterning application. Nano-scale patterns were transferred clearly to the substrate. The process shows good uniformity over the whole imprint area.

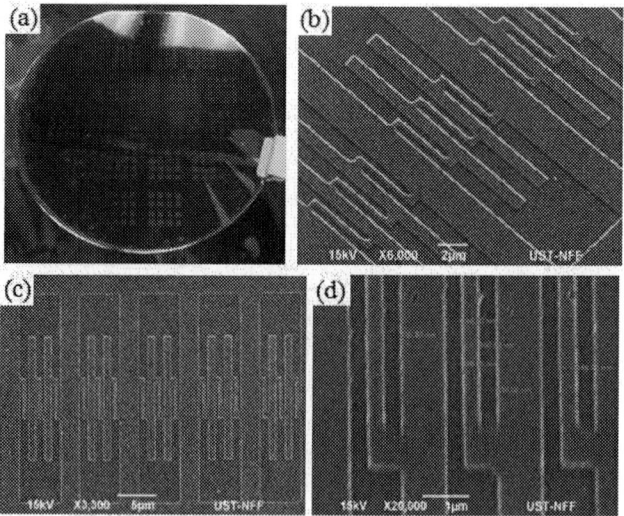

Figure 6. Pictures of the imprinted pattern on Si wafer: (a) patterned 4-inch wafer, (b) 45° tilt angle SEM image, (c) micro and nanopattern, (d) 300 nm features.

978-1-4799-2408-0/14 $31.00 © 2014 IEEE

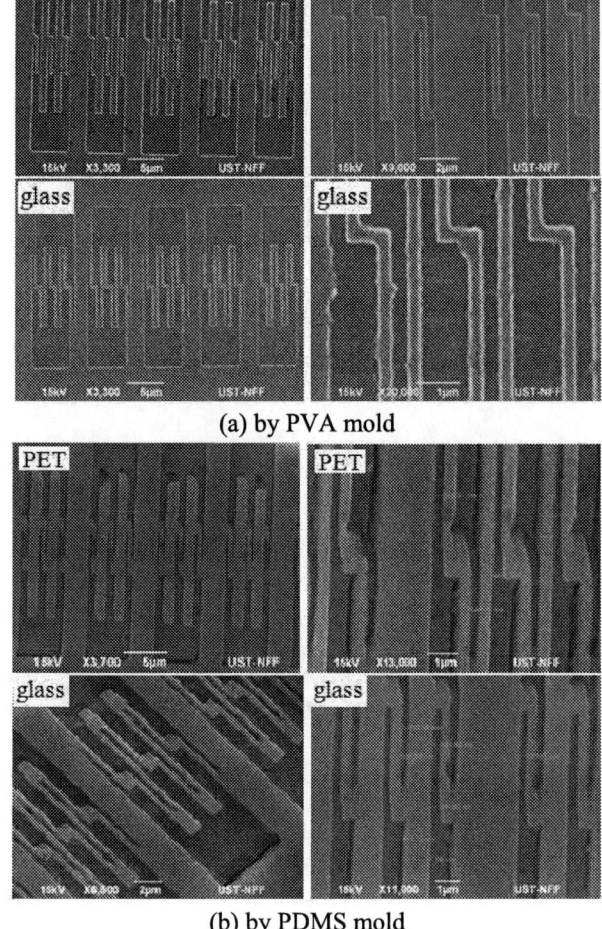

(a) by PVA mold

(b) by PDMS mold

Figure 7. SEM pictures of the patterns on non-flat substrates

Effect of Surface Property

During the nanoimprint process, every steps of pattern transfer were related to mechanical contact. Surface properties of the materials had great influence to the imprint results.

At the first step of imprint, the polymer was preferred that it can fully fill into the nano-patterns spontaneously. When the mold surface had high surface energy, the liquid polymer tended to wet the mold. It was necessary for fidelity pattern replication and also reduces the imprint force. If the polymer got a large contact angle on mold surface, additional pressure was required to press the polymer in the patterns. And this pressure increased as the pattern size decreasing. The wetting minimized the forming of bubbles. While the nanoimprint process was carried out in the atmosphere, air bubbles were likely to be trapped between the mold and substrate. Lower down the rolling speed helped to squeeze out the bubbles and extra polymer during the rolling.

But in the following de-molding process, mold surface should have low surface energy for easily separation. Hydrophobic coating was usually applied to the mold to perform this property. And the UV-curable polymer also had anti-sticking feature. These two situations formed a contradiction. Same problem happened at the flexible mold fabrication.

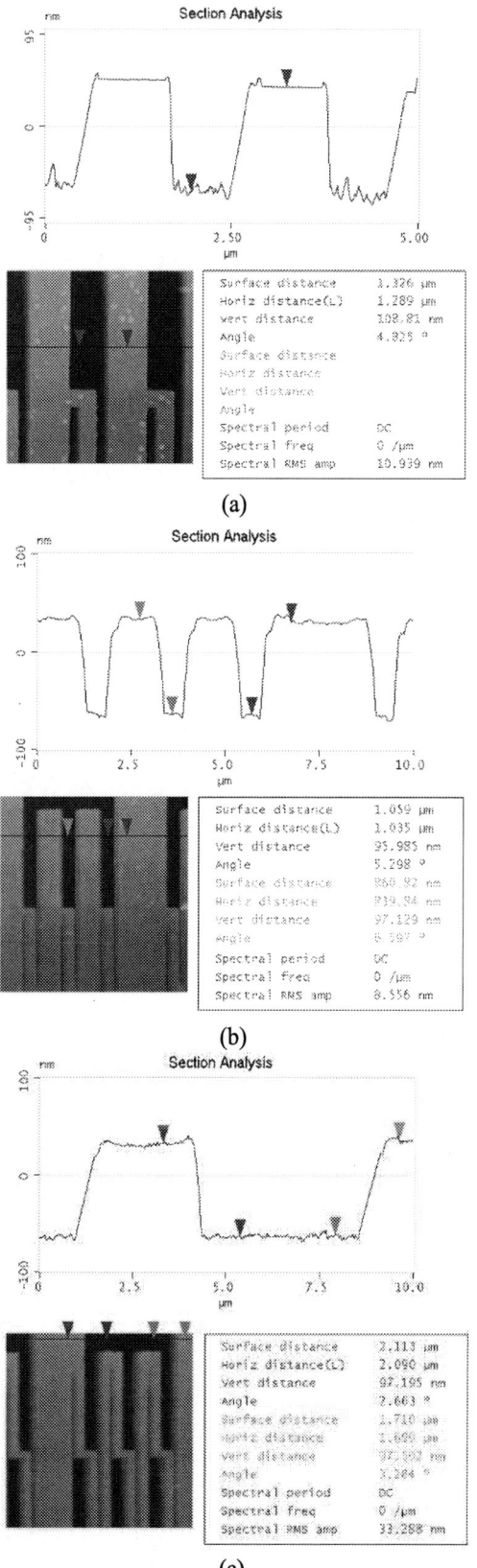

(a)

(b)

(c)

Figure 8. Cross-section analysis from AFM measurements: (a) pattern on Si mold, (b) pattern on PET substrate, (c) pattern on glass substrate.

Fig.9a shows the residual material stick on the mold. And Fig. 9b is the clean mold surface with silane coating. The coating also increased the durability of the mold. The flexible mold could be re-used up to 10 times without extra cleaning process. It saved the cost of mold fabrication.

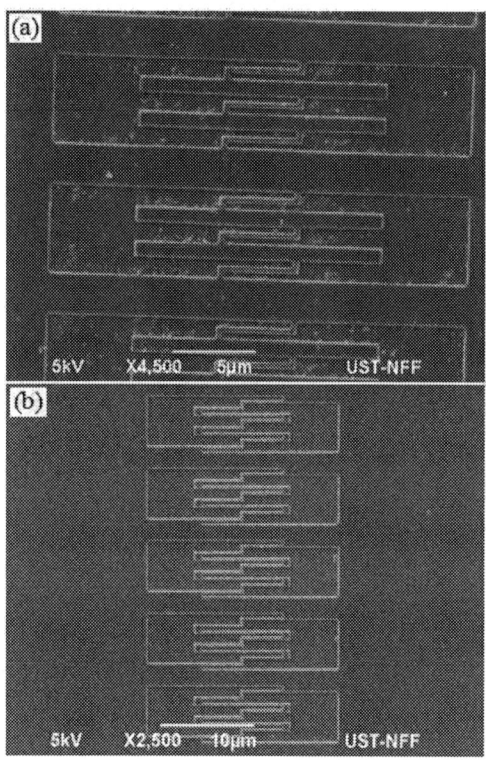

Figure 9. SEM of the mold after 10 printing cycles: (a) mold without silane coating, (b) mold with silane coating.

A good property of PVA was that it can be dissolved in the water. So the mechanical de-molding can be replaced by water soaking. On the other hand, PDMS was very hard to be removed by chemical process.

Deformation of Flexible Mold

The shape of transferred pattern was mainly dependent by the mold. Si or quartz hard mold can provide higher resolution pattern transfer than the flexible soft mold in NIL process. One of the reasons was that the low elastic modules of flexible molds. They were easily deformed under the vertical imprint force, even though the process minimized the shear force. The distortion of flexible molds is a critical issue to be solved for high fidelity nanopattern fabrication in the imprint process. In this study, PVA mold had high elastic modules than the PDMS mold. And the imprint results supported that the PVA mold provides higher quality in pattern transfer.

For the same PDMS mixture, thickness of the mold determined the deformation. From the imprint test, when the thickness of PDMS mold was less than 0.5 mm, wrinkles were observed in the mold after rolling. The thin PDMS mold could be easily stretched under small force when it was covered on the substrate. The large strain of thin mold caused

that only 60% of the mold patterns were transferred to the substrate. When the thickness of PDMS mold was larger than 0.8 mm, over 90% of the patterns were transferred to the substrate. Compare to PDMS, PVA mold could maintain the transfer efficiency even the mold thickness was less than 40 µm. Simulation results (Fig.10) support that deformation of the nanopattern in thin mold (0.3 mm) is larger (over 4 times) than the thick mold (1 mm) under the same imprint pressure.

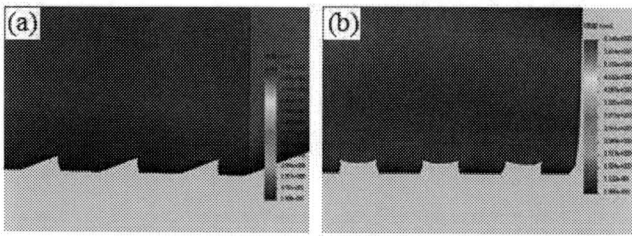

Figure 10. Simulation results of PDMS mold deformationin in different thickness: (a) 1 mm, (b) 0.3 mm.

N. Koo suggested adding one step which releases the imprint pressure [8]. It helps to control of the flexible mold deformation and the local variation of residual layer thickness in soft mold imprint process. In the proposed process, because the curing process was carried out in the atmosphere, 1 bar of the pressure was acting to the soft mold after rolling. Put the sample into vacuum environment for the step of UV curing can minimize the deformation of the soft mold.

Aspect ratio of the mold pattern was another parameter that affects the mold deformation. Either too high or too low would make the nanopattern become unstable. For the nano-patterning in elastic material, the aspect ratio was suggested to stay between 0.2 and 2 in order to obtain defect-free mold [9].

Conclusions

By using the roller type UV nanoimprint process, 300 nm patterns were successfully transferred to both rigid and flexible substrates in an area of 220 mm x 300 mm. The process was carried out in room temperature and under atmosphere. PVA mold provided a better imprint quality than PDMS mold. High elastic module of PVA had smaller pattern distortion. And the de-molding could be replaced by water soaking.

High-throughput process with roller printing speed up to 50 mm/s was provided. One patterning cycle could be finished within 10 minutes. Shear motion between mold and substrate, which damaged the imprint patterns, was minimized in the proposed process. The temperature variation and mechanical vibration from the environment did not show obvious affect to the imprint results.

References

1. Stephen Y. Chou, Peter R. Krauss, and Preston J. Renstrom, "Nanoimprint lithography", *J. Vac. Sci. Technol. B*. 14(6), Nov/Dec 1996.
2. Heon Lee, Sunghoon Hong, Kiyeon Yang, Kyungwoo Choi, "Fabrication of 100 nm metal lines on flexible plastic substrate using ultraviolet curing nanoimprint lithography", *Appl. Phys. Lett.*, Vol. 88 (2006), 143112.

3. Se Hyun Ahn, L. Jay Guo, "High-speed Roll-to-Roll Nanoimprint Lithography on Flexible Substrate and Mold-separation Analysis", *Proc. of SPIE*, Vol. 7205, 72050U

4. Shuhuai Lan, Hyejin Lee, Jun Ni, Soohun Lee, Moongu Lee, "Survey on Roller-type Nanoimprint Lithography (RNIL) Process", *International Conference on Smart Manufacturing Application*, Gyeonggi-do, Korea, April. 9-11, 2008, pp. 371-376.

5. Y. Zhong, Matthew M. F. Yuen, "Roller Type Soft Imprint Process on Rigid and Flexible Substrates", in *Proc. 15th International Conference on Electronic Materials and Packaging (EMAP)*, Seoul, Korea, Oct 6-9, 2013.

6. D. Y. Khang, H. Kang, T. Kim, H. H. Lee, "Low-Pressure Nanoimprint Lithography", *Nano Lett.*, Vol. 4 (2004), 633.

7. X. Cheng, L. J. Guo, P. F. Fu, "Room-Temperature, Low-Pressure Nanoimprinting Based on Cationic Photopolymerization of Novel Epoxysilicone Monomers", *Adv. Mater.* Vol. 17 (2005), 1419.

8. N. Koo, M. Otto, J. W. Kim, J. H. Jeong, H. Kurz, "Press and release imprint: Control of the flexible mold deformation and the local variation of residual layer thickness in soft UV-NIL", *Microelectronic Engineering*, 88 (2011) 1033–1036

9. Delamarche E, Schmid H, Biebuyck HA, Michel B., "Stability of molded polydimethylsiloxane microstructures", *Adv. Mater.*, Vol. 9 (1997), pp. 741–746.

Capacitive Deionization of Water Coolant Using Hybrid Carbon Electrodes for High Power Electronic Applications

Ziyin Lin,[1] Zhuo Li,[1] Kyoung-sik Moon[1] and Ching-Ping Wong[1,2]*

[1]School of Materials Science and Engineering, Georgia Institute of Technology
771 Ferst Drive, Atlanta, GA, USA 30332
[2] Department of Electronic Engineering, the Chinese University of Hong Kong, Hong Kong
[*]Phone: (404) 894-8391 Fax: (404) 894-9140   Email: cp.wong@mse.gatech.edu

Abstract

Water is promising for liquid cooling of high power electronics due to its high heat capacity and thermal conductivity. However, the ionic impurities and contaminants cause corrosions and threaten the device reliability. We explored capacitive deionization (CDI) of water coolant using hybrid carbon electrode consisting of carbon fiber paper (CFP) and vertically aligned carbon nanotubes (VACNTs). A facile method was developed to transfer VACNTs to CFP by using a polymeric adhesive, which achieved high bonding strength and good electrical conductivity. Moreover, we fabricated a miniaturized test vehicle for CDI measurement to evaluate the performance for such carbon hybrids and demonstrate the effective CDI effect.

1. Introduction

Liquid cooling has become more attractive for the thermal management of high power electronics due to the excellent heat transfer coefficient as well as the capability of hot-spot driven thermal management. Water is one of the best liquid coolants due to its superior heat capacity (4 times larger than perfluorocarbons and hydro-fluoro-ethers) and thermal conductivity (10 times larger than perfluorocarbons). In addition, water is by far the most economic and environmental friendly liquid. Although tap water meets the requirements for some liquid cooling applications, the ionic impurities in tap water impose threats to the reliability of electronic devices. For example, the minerals may precipitate and block the coolant flow, which is especially problematic for micro channel based cooling systems. Another risk is the electrical arcing due to static charge built up from the circulating coolant, which can seriously damage the electronics being cooled. Therefore, de-ionized water is a critical requirement for effective thermal management.

Incorporating a water de-ionizer into the water cooling loops is a convenient way to provide and maintain the high purity of de-ionized water coolant. However, traditional de-ionization methods on the basis of the ion exchange, electrodialysis or reverse osmosis need bulky components such as ion exchange resin columns or high pressure pumps. As a result, they cannot be incorporated into small devices with micro channels. Capacitive deionization (CDI, also called electrosorption) is a novel technology that has been proposed to remove ions from water by forcing the ions into oppositely polarized electrodes.[1] The structure of a CDI is very similar to energy storage devices such as ultracapacitors consisting of two electrodes immersed in electrolytes. When an electric potential is applied on the electrodes, the electrode will attract and adsorb oppositely charged ions at the surface, forming an electrical double layer. When the electric potential is reversed, the ions will be released to regenerate the electrodes. Therefore, CDI offers many advantages including low energy consumption, facile regeneration, high sorption capacitance (8 times higher than pure chemical adsorption), fast adsorption rate (~60% higher than pure chemical sorption) and most importantly, the ability for miniaturization.[2-4] Nevertheless, the ion sorption capacity (i.e. how many ions can be adsorbed in total), deionization rate and reliability of a CDI are strongly dependent on the material properties of electrodes such as their electrical conductivity, surface area, wettability to water, pore size distribution, and mechanical robustness.

We investigated a miniaturized built-in capacitive deionizer using vertically aligned carbon nanotubes (VACNTs) as the electrodes, which can be incorporated into cooling loops used as the de-ionized water source. The use of VACNT can potentially enhance the electrosorption capacity by 4~10 times, [2, 3, 5] and the adsorption rate can be significantly enhanced compared with random CNTs due to the short and straight path of ion diffusion in VACNT structure. We have developed a facile method to transfer VACNTs onto a carbon fiber paper (CFP) substrate using a polymeric adhesion layer. The resulting VACNT–CFP carbon hybrid structure displays a very low interface resistance VACNT–CFP interface. We further demonstrated that the VACNT–CFP can be used as CDI electrodes with excellent rate performance and cycling stability, which could be attributed to the vertically alignment and chemical stability of CNTs in the hybrid structure. Moreover, VACNT–CFP bonding structure showed good resistance against the impact of flowing water, and the electrochemical performance was not affected, suggesting the superior mechanical robustness of the hybrid structure.

2. Experimental Section

2.1. Growth of VACNTs

VACNT arrays were synthesized in a 1.5-inch diameter aluminum tube furnace. The catalysts were prepared by a sequential deposition of Al (10 nm) and Fe (2.0 nm) using an electron-beam evaporator on Si wafers with a 300 nm thermal SiO_2 layer. CNT growth was carried out at 750 °C with 350 sccm Ar, 150 sccm C_2H_4, and 300 sccm H_2. The water vapor was introduced into the furnace by bubbling Ar gas through water.

2.2. Transfer of VACNTs

In a typical process, the silicone resin (Dow Corning HIPEC Q1-4939, Part A: Part B = 1:1) was carefully coated on the as-received carbon fiber paper (CFP, 2050-A from Fuel Cell Store). The loading of silicone resin is ~20 mg/cm^2.

VACNTs of ~ 1 × 1 cm² was flipped on the CFP. Then the silicone was pre-cured at 110 °C for 1 h, during which 20 g weight was applied to improve the contact between VACNTs and CFP. After that, the Si substrate was removed and the VACNTs were left on CFP. Finally, the silicone resin was fully cured at 150 °C for 1 h.

2.3. Characterizations

Scanning electron microscopy (SEM, LEO 1530) was used to characterize the morphology of samples using an accelerating voltage of 4 kV. The curing behavior of silicone was studied by differential scanning calorimetry (DSC, Q-600 TA Instruments) and discovery hybrid rheometer-2 (HR2, TA Instruments). Nitrogen-adsorption/desorption measurements were conducted on an Autosorb-1 analyzer (Quantachrome Instruments, Boynton Beach, FL, USA). The electrical resistance of VACNT–CFP was measured using a Keithley 2000 multimeter and a Hewlett-Packard 6553A DC power supply. The electrochemical measurements, including cyclic voltammetry (CV) and galvanostatic charge/discharge, were measured on a Versastat 2-channel system (Princeton Applied Research). The electrochemical measurements were carried out in a beaker-type cell using a two-electrode configuration. Each VACNT–CFP electrode was clipped to the end of a CFP strip and the other end of the CFP strip was connected to metal clips. The electrolyte was 1 M NaCl aqueous solution. Only the VACNT–CFP and CFP strip was immersed in the electrolyte to avoid the corrosion of metal clips. The distance between VACNT–CFP electrodes is ~ 5 mm. Before the electrochemical measurements, VACNT–CFP was treated by UV–ozone at room temperature for 20 min in order to improve the wetting of VACNTs with aqueous electrolyte.

3. Results and Discussion

The VACNTs were grown using chemical vapor deposition method in a tube furnace. The morphology of as-gown VACNTs is shown in Figure 1, which suggests obvious alignment of CNTs with diameters of ~ 20 nm.

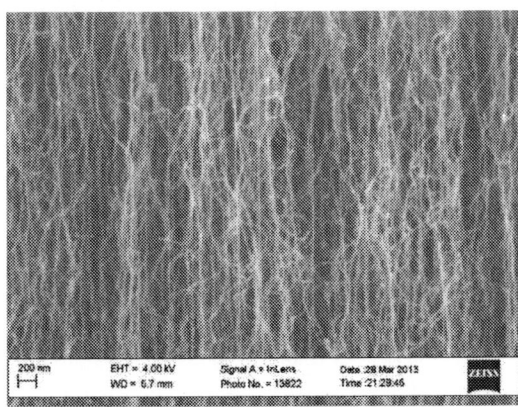

Figure 1. SEM images of VACNTs.

Figure 2 schematically depicts the fabrication process of VACNT–CFP by transferring VACNT onto the CFP substrate using the silicone resin as an adhesive. The CFP is specially chosen because of its low cost, good electrical conductivity and chemical inertness. The silicone elastomer adhesive is chosen due to its chemically stability, good

resistance to water, and excellent flexibility that will preserve the mechanical property of substrate. The curing of silicone resin is an important step in the transfer process. We used DSC and rheological measurements to monitor its curing reaction and optimize the curing profile, as shown in Figure 3. From the DSC curve, the curing onset temperature and peak temperature are 103 and 114 °C, respectively. However, the initial curing of silicone does not lead to change of its viscosity. The viscosity starts to increase rapidly at ~ 120 °C and stabilize after 150 °C. The viscosity of silicone resin plays an important role in the transfer. Low viscosity is desirable in order for the resin to spread and cover the surface of CFP and VACNT, and for VACNT to penetrate the resin and make contact with CFP. On the other hand, the high viscosity is desirable for stronger adhesion and better transfer. A two-step curing was therefore used, which contains a pre-cure at 110 °C for 1 hour and a post cure at 150 °C for 1 hour. After the pre-cure, the Si growth substrate was peeled off, leaving CNT transferred on CFP. The post cure completes the fabrications process.

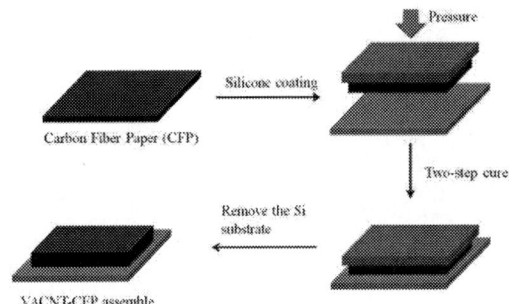

Figure 2. Schematics of the transfer of VACNTs onto CFP substrate using a silicone adhesive.

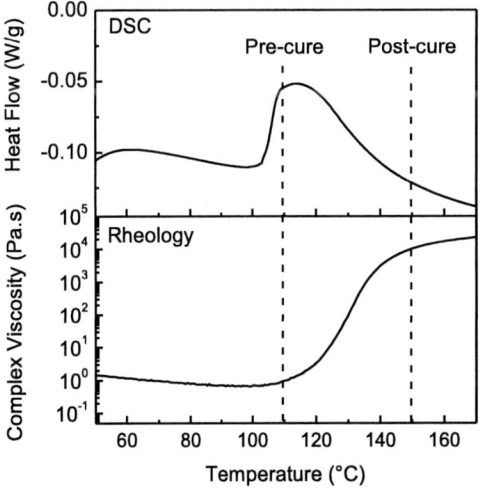

Figure 3. The curing behavior of the silicone adhesive.

The SEM images in Figure 4 show the morphology of VACNT-CFP. The VACNTs are well anchored on the CFP substrate and the alignment is preserved (Figure 4a). The silicone adhesive impregnates the CFP substrate and part (~ 5–20 μm) of VACNT tips with the help of the capillary force.

Figure 4b shows the top surface of VACNT-CFP, which corresponds to the roots of as-grown VACNT. The bright spots on the tips of VACNTs are the iron catalyst nanoparticles.

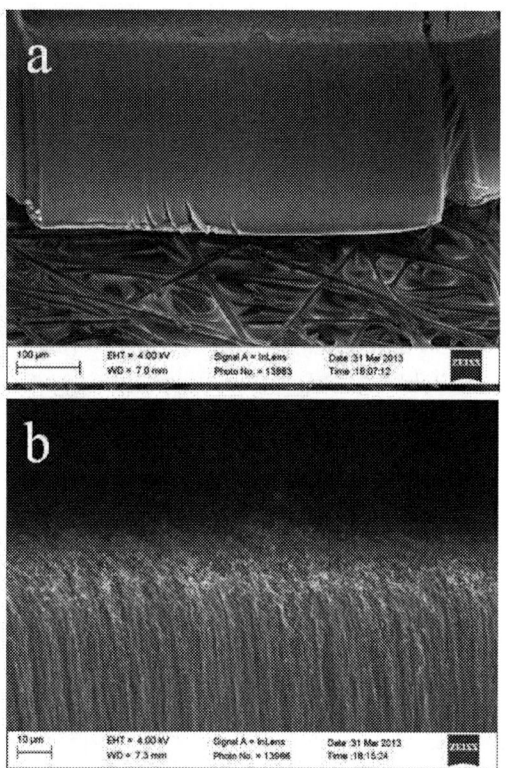

Figure 4. SEM images of transferred (a) VACNT-CFP and (b) the top surface of transferred CNT.

We measured the surface area of CFP, VACNT and VACNT-CFP, as summarized in Table 1. The BET surface of CPF is only 0.9 m^2/g, because it consists of large carbon fibers with diameters of ~ 8 μm. In contrast, the BET surface area of VACNTs is 474.3 m^2/g, which is a typical value for multi-walled carbon nanotubes. After combining VACNT with CFP, the hybrid shows a BET surface of 5.8 m^2/g.

Table 1. BET surface area measurement results

Sample	BET surface area (m^2/g)
CFP	0.9
VACNT	474.3
CFP-VACNT	5.8

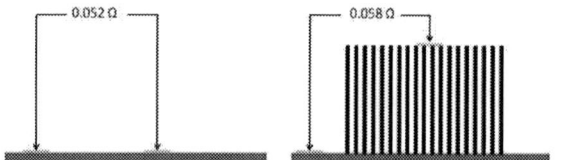

Figure 5. The Illustration of the measurement of electrical resistance of the VACNT-CFP interface.

To measure the electrical resistance of VACNT-CFP interface, we deposited Ti/Au (30/150 nm) electrodes with a diameter of 500 μm the top surface of VACNT-CFP and CFP substrate by electron-beam evaporation. The metallization layer ensures good ohmic contact with carbon surfaces. The resistance measured on CFP substrate is 0.052 Ω, whereas the resistance between VACNT top surface and CFP is 0.058 Ω. The slight different between these values suggests that the interface resistance is as low as 0.006 Ω, which is considered sufficient for CDI applications. The nature of VACNT and CFP interaction at the interface is the Van der Waals force that is responsible for the electrical resistivity of graphite in c-axis (~0.005 Ω·cm).[6, 7] The large contact area and entangled CNT tips also contributes to the low electrical resistance.

The electrochemical performance of VACNT-CFP was measured using a two electrode system with 1 M NaCl aqueous solution as the electrolyte. The CV curves of VACNT-CFP and CFP are shown in Figure 5. The CV area of VACNT–CFP is much larger than that of CFP, owing to the large surface area of VACNTs. The areal capacitance could be calculated from the CV curves using the equation,

$$C = \frac{\int I dV}{S V r}$$, in which C, I, V, S, and r are the areal capacitance, measured current, voltage, a single electrode area, and the scan rate, respectively. The VACNT-CFP has an areal specific capacitance of 6.1 mF/cm^2. In addition, the charge-discharge curve was collected at a current density of 0.1 mA/cm^2, as shown Figure 5b. The corresponding areal capacitance is 6.7 mF/cm^2, calculated by the

equation $C = \frac{It}{SV}$, where t is the discharge time.

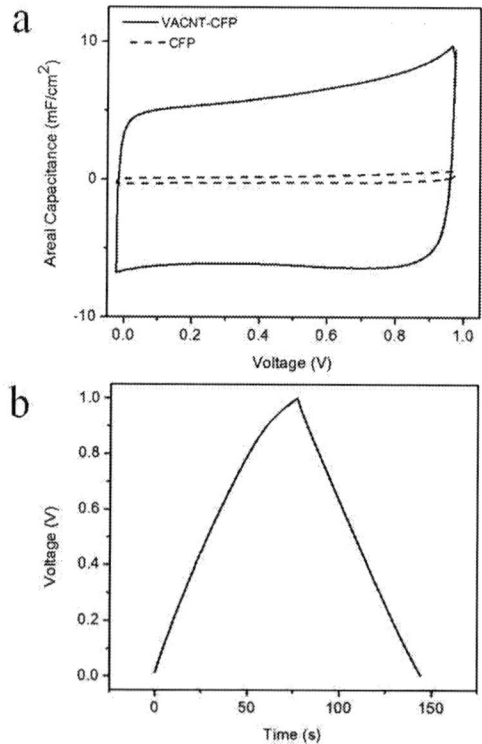

Figure 5. (a) The CV curves of VACNT-CFP and CFP at a scan rate of 100 mV/s; (b) the charge-discharge curve of VACNT-CFP at a current density of 0.1 mA/cm^2.

For CDI, the good adhesion of VACNT on CFP substrate is critical in order for the VACNT to survive from the water flow. To evaluate the mechanical robustness, the CV curves were also recorded in flowing electrolyte at a flow rate of 2 L/min. As shown in Figure 6, the CV curve maintains a rectangle shape, suggesting good mechanical stability. In addition, electrochemical cycling was carried out to evaluate the chemical robustness. After 10,000 cycles, the areal capacitance shows negligible degradation. [8]

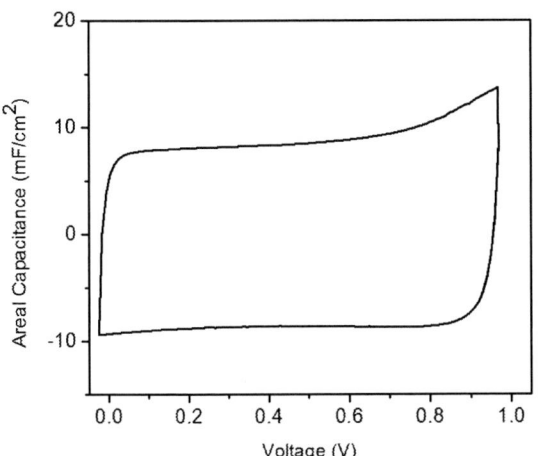

Figure 6. The CV curve of VACNT-CFP at 100 mV/s and a water flowing rate of 2L/min.

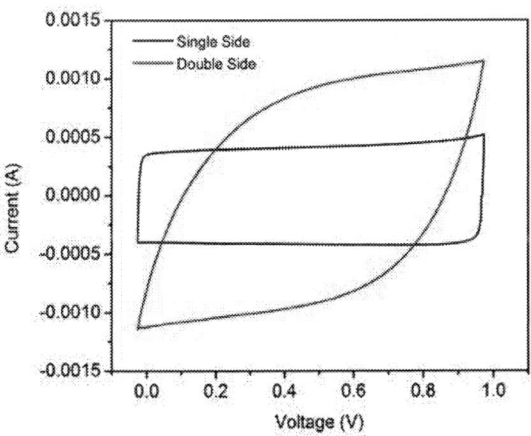

Figure 7. CV curves of single-side and double-side transferred VACNTs at a scan rate of 100 mV/s.

An effective way to improve the areal capacitance is to use double side transferred VACNTs on CFP to increase the actual loading of VACNTs. As shown in Figure 7, the CV area of double side VACNT-CFP is roughly twice of that of single side VACNT-CFP. However, the rate performance is compensated, which is evidenced by the distorted CV curve for double side transferred VACNTs. This result is likely due to the increased ionic transport pathway for double side VACNT-CFP.

To investigate the material performance, we built a CDI setup with a continuously flowing system as schematically shown in Figure 8a. In each experiment, the aqueous solution was continuously pumped with a peristaltic pump into the cell and the effluent returned to the cell. A constant flow rate around 20 ml/min and a DC voltage of 2 V were applied. 100 ppm aqueous NaCl solution was used as the electrolyte. The electrical conductivity of the NaCl solution was continuously monitored at the outlet of the CDI cell using an ion conductivity meter.

The structure of CDI cell is shown in Figure 8b. The assembly of one half of the unit cell is in the order: poly(methyl methacrylate) (PMMA) retaining plate/PDMS rubber gasket/electrode/polydimethylsilane (PDMS) rubber spacer/filtration paper spacer. The spacing of 1 mm between the electrodes was maintained by the PDMS rubber spacer. The cell was fixed by putting 6 screws through all the layers and the two PDMS rubber gaskets help to prevent water leakage from the cell.

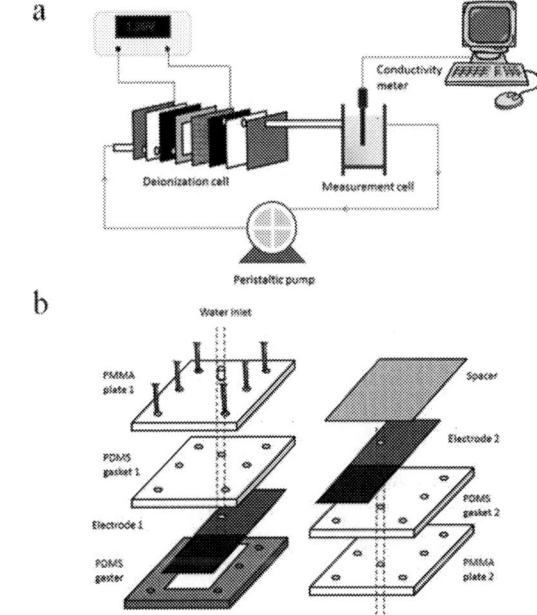

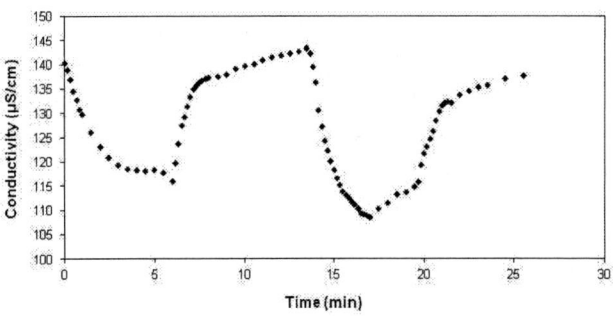

Figure 8. Schematics of (a) the experimental setup for CDI measurement and (b) the CDI cell.

Figure 9. CDI performance for random CNT-CFP.

We tested the CDI performance of random CNT-CFP to verify the applicability of the test setup. The CNT used is multi-walled CNT that was treated by HNO_3 to improve the hydrophilicity. A paste of 90 wt% CNT and 10 wt% PTFE was applied to CFP to make the electrode. As shown in Figure 9, the electrical conductivity of NaCl solution decreased rapidly when a 2 V bias was applied on the electrodes. The electrosorption process is very fast at beginning and slows down after several minutes. Within 3.5 minutes, the cell removed ~ 24 % NaCl. Moreover, we studied the regeneration of the electrode by removing the bias. After removing the bias, the ions desorbed from the CNT electrodes and within 10 minutes, the electrical conductivity was back to the original value, indicating a complete regeneration of the CNT electrodes. When re-applying the voltage, the electrical conductivity dropped again, and the deionization efficiency is slightly improved and the deionization rate is higher compared to the first cycle, which may result from the activation of the electrodes from the first cycle. These results tell the excellent applicability of our CDI test setup, and we will further test the CDI performance of VACNT-CFP.

Conclusions

We developed a facile method to fabricate VACNT-CFP hybrids using a silicone adhesive, which achieved high bonding strength and good electrical conductivity. The VACNT-CFP can be used as high performance electrodes for CDI with excellent rate performance and mechanical robustness. Moreover, we fabricated a CDI test setup and demonstrate the CDI effect of CNT materials.

Acknowledgments

The authors would like to acknowledge the Defense Advanced Research Projects Agency (DARPA, W31P4Q-13-1-0009) for financial support.

References

1. S. Porada, R. Zhao, A. van der Wal, V. Presser, and P. M. Biesheuvel, "Review on the science and technology of water desalination by capacitive deionization" *Prog Mater Sci* vol. 58 pp. 1388-1442 Oct 2013.
2. S. Wang, D. Z. Wang, L. J. Ji, Q. M. Gong, Y. F. Zhu, and J. Liang, "Equilibrium and kinetic studies on the removal of NaCl from aqueous solutions by electrosorption on carbon nanotube electrodes" *Sep Purif Technol* vol. 58 pp. 12-16 Dec 1 2007.
3. X. Z. Wang, M. G. Li, Y. W. Chen, R. M. Cheng, S. M. Huang, L. K. Pan, and Z. Sun, "Electrosorption of ions from aqueous solutions with carbon nanotubes and nanofibers composite film electrodes" *Appl Phys Lett* vol. 89 Jul 31 2006.
4. J. Yang, L. D. Zou, H. H. Song, and Z. P. Hao, "Development of novel MnO2/nanoporous carbon composite electrodes in capacitive deionization technology" *Desalination* vol. 276 pp. 199-206 Aug 2 2011.
5. L. Wang, M. Wang, Z. H. Huang, T. X. Cui, X. C. Gui, F. Y. Kang, K. L. Wang, and D. H. Wu, "Capacitive deionization of NaCl solutions using carbon nanotube sponge electrodes" *J Mater Chem* vol. 21 pp. 18295-18299 2011.
6. D. Z. Tsang and M. S. Dresselhaus, "C-Axis Electrical-Conductivity of Kish Graphite" *Carbon* vol. 14 pp. 43-46 1976.
7. W. Primak, "C-Axis Electrical Conductivity of Graphite" *Phys Rev* vol. 103 pp. 544-546 1956.
8. Z. Y. Lin, Z. Li, K. S. Moon, Y. N. Fang, Y. G. Yao, L. Y. Li, and C. P. Wong, "Robust vertically aligned carbon nanotube-carbon fiber paper hybrid as versatile electrodes for supercapacitors and capacitive deionization" *Carbon* vol. 63 pp. 547-553 Nov 2013.

Gap in pagination due to withheld paper.

Pages 1833-1837

RF Energy Harvesting

Parvizso Aminov and Jai P. Agrawal
Purdue University Calumet
2200 169th Street, POTT 227, Hammond, IN 46307
agrawajp@purduecal.edu

Abstract

This paper presents the results of a project in RF Energy Harvesting for scavenging energy from the ubiquitous radio-frequency (RF) electromagnetic waves. Such a device can be very useful to charge mobile phone in jungles and in remote areas or where the electric utility is not available or not reliable. In comparison to other methods of energy harvesting, RF has the smallest energy density and therefore poses big challenges. Based on the experiments, we find that the most efficient range of operation lies in the medium wave frequency band: 531–1,611 kHz. The experiment uses an antenna, LC tuning circuit, 5-stage of Villard voltage multiplier circuit and super-capacitor as energy storage. The experiment could harvest a RF signal from 1-mile distant transmitter that generates a field-strength of 103.724 dBu at the location of the receiver. The maximum charge on storage capacitor achieved was recorded 2.8V. A limitation of using this band of frequency is the large size of antenna that limits its portability.

Introduction

Emergence of portable devices in today's world creates technology that is largely dependent on the battery power. This creates a problem of having the battery constantly charged. RF Energy Harvesting System is aimed at providing solution to this problem. The RF energy is ubiquitous. TV, radio, cellphones, and a lot of other electronic devices transmit RF energy into the air continuously. Since the RF energy is available almost at any location, the project is aimed at harvesting that energy and use stored energy to power an electric load. One of the examples of RF energy harvesting is in the RFID technology where wireless tags receive energy from the reader and use this energy to send information back to reader. Most of the RF energy harvesting is achieved using stronger near field RF energy, whereas this project is aimed at harvesting weaker far field RF energy.

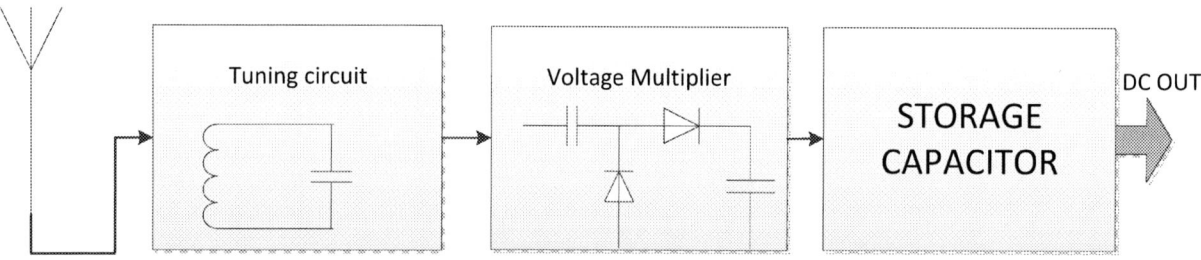

Fig. 1.   Block Diagram: Ambient RF energy harvesting

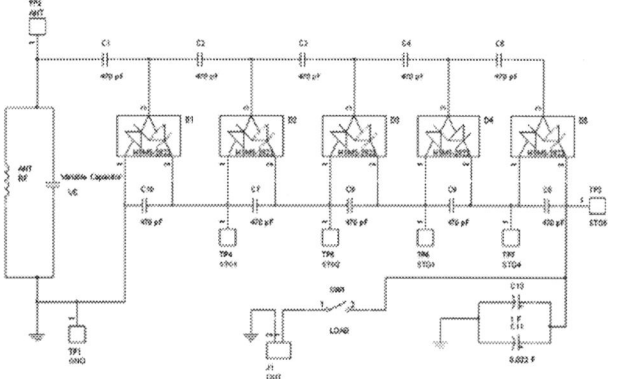

Fig. 2 Schematic diagram

System Design

The design of the system is based on a simple concept, capture the ambient RF energy, increase received voltage, and use acquired energy to charge a device. Based on the concept the design of the system consists of several parts: antenna, tuning circuit, voltage multiplier circuit, and Storage capacitor, as shown in Figure 1.

Antenna

Antennas are metallic structures designed for radiating and receiving electromagnetic energy [1]. The system is harvesting electromagnetic energy from radio frequencies. Therefore, antenna design is extremely important. Several versions of antennas were used in this project, this includes monopole antenna, microstrip patch antenna, and loop antenna.

Design with Monopole antenna:

The monopole antenna was found to be not efficient. The circuit was tuned to 101.1 MHz signal with transmitted power of 5.7 kW 22.4 miles away from the circuit. The outcome of the system with monopole antenna created a charge of storage capacitor by 10mV in 4 days.

Design with Microstrip patch antenna:

The three essential parameters for the design of the microstrip patch antenna: frequency of operation (f_o), dielectric constant of the substrate (ε_r), height of dielectric substrate (h) were

978-1-4799-2408-0/14 $31.00 © 2014 IEEE   1838   2014 Electronic Components & Technology Conference

$f_o = 1.8\ GHz$
$\varepsilon_r = 4.34\ \ (dielectric\ constant\ of\ fiberglass)$
$h = 1.524\ mm$

The values above were used to calculate the width and length of the patch and the ground plane. All of the design equations where taken from reference [1].

Equation (1) is used to calculate the width (W) of the patch:

$$W = \frac{c}{2f_o\sqrt{\frac{\varepsilon_r+1}{2}}} \qquad (1)$$

Equation (2) is used to calculate effective dielectric (ε_{reff}) constant:

$$\varepsilon_{reff} = \frac{\varepsilon_r+1}{2} + \frac{\varepsilon_r-1}{2}\left[1 + 12\frac{h}{W}\right]^{-1/2} \qquad (2)$$

Equation (3) is used to calculate effective length (L_{eff}) of the patch:

$$L_{eff} = \frac{c}{2f_o\sqrt{\varepsilon_{reff}}} \qquad (3)$$

Equation (4) is used to calculate length extension (ΔL) of the patch:

$$\Delta L = 0.412\ h\ \frac{(\varepsilon_{reff}+0.3)\left(\frac{W}{h}+0.264\right)}{(\varepsilon_{reff}-0.258)\left(\frac{W}{h}+0.8\right)} \qquad (4)$$

Equation (5) is used to calculate actual length of patch (L):

$$L = L_{eff} - 2\Delta L \qquad (5)$$

Equations (6) and (7) are used to calculate the length (L_g) and width (W_g) of the ground plane of the antenna:

$$L_g = 6h + L \qquad (6)$$

$$W_g = 6h + W \qquad (7)$$

Based on the equations (1) through (7), and the values of frequency of operation (f_o), dielectric constant of the substrate (ε_r), height of dielectric substrate (h), the values were calculated as below:

$W = 50.57\ mm$
$\varepsilon_{reff} = 4.1$
$L_{eff} = 41.2\ mm$
$\Delta L = 7.078\ e\text{-}4\ mm$
$L = 39.8\ mm$
$L_g = 48.9\ mm$
$W_g = 59.7\ mm$

The circuit with microstrip patch antenna has been more promising, however the storage capacitor would demonstrate faster charge rate only when near field RF was applied to patch antenna.

With attachment of patch antenna the circuit did not display any improvements. However, when a cellphone was placed directly onto the antenna, the circuit displayed

significant charge within several hours. The result of this test is below:

Initial state of capacitor:	1.50 V
2 hours of operation:	2.46 V

Capacitor was discharged to 1.88V with use of a resistive load
Capacitor was discharged to 1.19 V within 28 hours
Capacitor was further discharged to 1.131 V within 13.5 hours

50 minutes of transmission:	1.16 V
75 minutes of transmission:	1.2 V

Fig. 2.   Microstrip Patch Antenna

The circuit has demonstrated improvements with the Patch antenna. However the input RF was near field, whereas the project was aimed at harvesting the far field RF energy.

Loop antenna:
The strongest RF signal in the Hammond, IN area was determined to be AM 1230 radio station, located in Hammond, with signal strength of 115.713 dBu. Loop antenna was implemented in the circuit in order to tune to AM frequency, and retain the portability of the device. The designed antenna is square loop antenna with a side of 64.77 cm and 10 turns of 24 AWG enameled wire, with inductance of the coil of approximately 188 µH. Circuit with Loop antenna has demonstrated a constant charge of the storage capacitor.

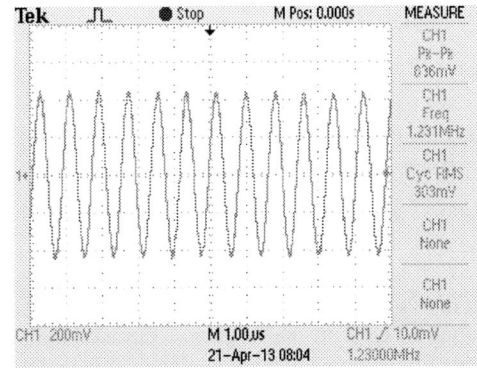

Fig. 3. Source of energy harvesting - AM 1230

The circuit with loop antenna was the most efficient. With use of variable capacitor, the circuit was tuned to 1230 AM station. Figure 3 shows the signal from the radio station. The measurement is taken from the tuning circuit connected to the oscilloscope. The waveform measurement shows a signal of 1.23MHz and amplitude of 896 mV$_{\text{P-P}}$. The peak to peak voltage varies, since it's an amplitude modulated signal.

The test results are shown in Table 1 below.

Elapsed Time	Voltage
0:00:00	1 V
0:00:20	1.25 V
0:00:55	1.27 V
0:01:00	1.275 V
0:02:20	1.281 V
0:03:30	1.284 V
0:04:30	1.286 V
0:07:00	1.288 V
0:08:40	1.29 V
0:25:00	1.299 V
0:40:00	1.304 V
1:00:00	1.31 V
1:30:00	1.315 V

Table 1. Test results of the circuit tuned to AM 1230

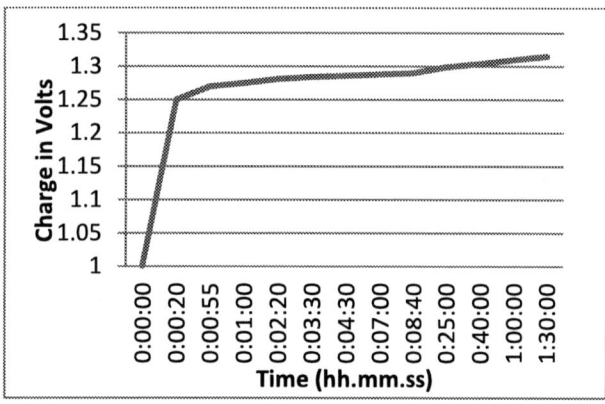

Graph 1. Super capacitor charge vs. hours of reception of RF energy

The results above are demonstrated in Graph 1. From the graph it can be observed that the voltage rises exponentially. This is true since step response calculated by the equation (8):

$$V_c(s) = \frac{V_0(s)}{s} - \frac{V_0(s)}{s + \frac{1}{\tau}} \qquad (8)$$

where $V_c(s)$ is the voltage on capacitor in frequency domain, $V_0(s)$ is the input voltage on the frequency domain, and τ is the time constant.

Applying inverse Laplace to equation (8) we will get the equation (9):

$$V_c(t) = V_0(s)(1 - e^{-\frac{t}{\tau}}) \qquad (9)$$

From the graph it can be observed that super capacitor discharges exponentially. This is true since the discharge voltage on the capacitor is calculated by equation (10)

$$V_c(t) = V_0(s)(e^{-\frac{t}{\tau}}) \qquad (10)$$

The maximum voltage that was stored into the capacitor from the RF energy is 2.8 volts. Over time the super capacitor discharges. The leakage of super capacitor is displayed in Graph 2.

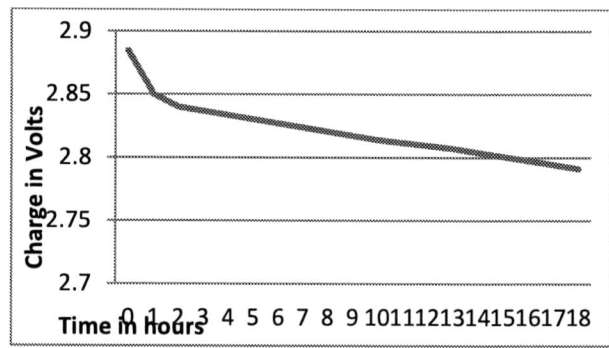

Graph 2. Super capacitor leakage vs. time

Tuning circuit

Tuning circuit is a band-pass filter designed with resonant frequency of a desired signal. Selection of matching circuit depended on the frequency of the harvesting signal. The circuit with monopole antenna harvested energy from FM Radio Signal - 101.1MHz with tuning inductor and capacitor in parallel with voltage doubler circuit. For the inductance 124 nH were chosen and 20 pF for the capacitance.

In order to broaden the bandwidth of the energy the circuit with Microstrip Patch antenna did not use any tuning circuit, the antenna itself was designed for 1.8 GHz frequency.

Lastly, the tuning circuit for 1230 MHz AM signal, was the antenna itself with variable capacitor in parallel, which allowed tuning to different AM frequencies.

Voltage Multiplier and Storage Capacitor

Voltage multiplier circuit is used to raise the voltage collected from RF signal. Based on researched material, Villard voltage doubler was chosen to be used in the Voltage Multiplier circuit. The advantage of Villard voltage doubler is that it doubles the voltage and rectifies the AC signal. Figure 3 shows an example of Villard voltage doubler.

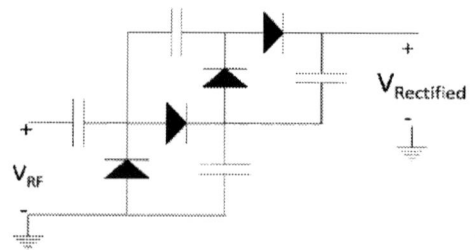

Fig. 3.   Villard voltage doubler

978-1-4799-2408-0/14 $31.00 © 2014 IEEE

In order to increase the voltage to the required amount, several stages of voltage doubler should be used. Nintanavongsa writes "the output voltage is directly proportional to the number of stages in the energy harvesting circuit. However, practical constraints force a limit on the number of permissible stages, and in turn, the output voltage." [2]. The voltage gain decreases as number of stages increases due to parasitic effect of the component capacitance of each stage, and finally it becomes negligible [2]. Based on Nintanavongsa's experiments, 5 stage systems should be most efficient for wider range of RF signals.

The choice of components was made based on Harrist's experiments. His analysis shows that the best output of the system is made with the 5-6 stages of voltage doubler, stage capacitors with value of 0.47 nF, and storage capacitor of 1.5 nF [3]. The multiplier circuit uses Schottky barrier diodes to get a higher output voltage at weak RF signals [4]. The proposed Schottky diodes are HSMS-2822 [3][4].

The stage capacitor values where modified once the source of the RF energy was modified from FM frequencies to the AM frequencies. With smaller frequencies, the 470 pF capacitors are not very efficient, therefore the 2200 pF capacitors were substituted.

Conclusions

This paper demonstrates a system that is capable of harvesting RF energy. As described above, the system consists of an antenna, matching circuit, 5 stages of voltage doubling and a storage capacitor. Based on conducted research and testing it could be concluded that far field RF energy harvesting is possible when the center frequency of the circuit is within medium wave frequency band: 531–1,611 kHz. There are limitations of the amount of energy that could be harvested by the RF Energy Harvesting System which could be caused by the strength of RF field, implemented antenna, and the multiplier circuit. The system can be used to capture the RF energy from a local AM radio station by aiming the loop antenna towards the location of the station. The system was able to harvest enough energy to charge a super capacitor to 2.8 V and sustain the voltage while no load connected to the circuit. This charge is sufficient to power a 1kΩ load for approximately 1 hour.

Reference

[1] Nakar, P. S. (2004) *Design of a Compact Microstrip Patch Antenna for use in Wireless/Cellular Devices,* The Florida State University, Tallahassee.

[2] Nintanavongsa, P., Muncuk, U., Lewis, D. R., Chowdhury, K. R. (2012). Design Optimization and Implementation for RF energy Harvesting Circuits. *IEEE Journal on Emerging and Selected Topics in Circuits and Systems, 2*(1), 24-33.

[3] Harrist, W. D. (2004). *Wireless Battery Charging System Using Radio Frequency Energy Harvesting,* University of Pittsburgh, Pittsburgh.

[4] Kitazawa, S., Ban, H., Kobayashi, K. (2012). Energy Harvesting from Ambient RF Sources. *IMWS-IWPT2012 Proceedings,* 39-42.

Localized Metal Plating on Aluminum Back Side PV Cells

M.Balucani[1,2,*], K. Kholostov[1], L. Serenelli[3], M.Izzi[3], D. Bernardi[4], M.Tucci[3]

[1]Sapienza University of Rome-DIET, Via Eudossiana, 18 - 00184 Roma ITALY

[2]Rise Technology S.r.l. Lungomare P. Toscnelli, 170 - 00122 Roma ITALY

[3]ENEA Casaccia Research Centre Rome, Via Anguillarese 301 – 00123 Roma ITALY

[4]2BG S.r.l. Via Monte Bianco 18 – 35018 Padova ITALY

* balucani@diet.uniroma1.it

Abstract

In this work we demonstrate a new selective metallization technique to perform localized plating on the screen-printed Al contact using the innovative approach based on Dynamic Liquid Drop/Meniscus that is able to touch the cell back contact in specific defined positions and show that it is possible to produce suitable electrical and mechanical contact with Al-Si and thus to replace the silver from the back contact in the cell manufacturing process reducing the solar cell cost. A fast pre-treatment process was developed to clean and prepare the surface of the aluminum on the back side of PV cells allowing direct plating with good electrical contact. Several commercial aluminum screen printable pastes have been experimented also having different distribution of sphere particles dimensions.

We have used high resolution Scanning Electron Microscopy (SEM) and compositional microanalysis with Energy Dispersive X-Ray microanalysis (EDX) to evaluate the metal dispersion within aluminum-silicon inter-diffused region and Transfer Length Method and current-voltage measurements to estimate the specific contact resistivity of the metal contact and series resistance of the overall solar cell device. We have found that the interconnection ribbon soldered on tin contacts plated on screen printed aluminum back contact shows adhesion higher (> 1N/mm) than that verified on screen printed silver over silicon. The main difference between a tin pad and a nickel-tin pad will be shown. Efficiency increase and fill factor are compared respect standard Al-Ag back contact PV cell.

Introduction

In standard manufacturing process of p-type doped crystalline silicon based solar cell, the back contact is made with screen-printed aluminum and silver pastes. The former is able to form deep and effective Back Surface Field (BSF) providing surface recombination velocity values in the range of few hundreds cm/s and performing a backside segregation gettering of undesired metal content within silicon network [1]. The latter, in shape of tabs or stripes, is helpful for interconnection ribbon soldering. The screen printable silver paste has several drawbacks indeed cannot be directly printed on Al contact, does not form either BSF, or ohmic contact on p-type doped c-Si and its cost is growing rapidly despite the general tendency of reducing the solar cell market price.

In this evidence an attractive idea is to reduce the amount of silver or totally eliminate it by developing the process of the direct local metal plating on the Al layer thus keeping uniform BSF and preserving high efficiency all over the surface. The main requirements for the deposited layer are low contact resistance and high adhesion and ability of soldering.

An innovative technique based on dynamic liquid meniscus (DLM) that is able to touch the silicon solar cell back contact in specific defined positions and perform wet processing described in [2,3] was used. A dynamic liquid meniscus (DLM) represents a liquid bridge between two surfaces where a continuous liquid flow exists (Figure 1).

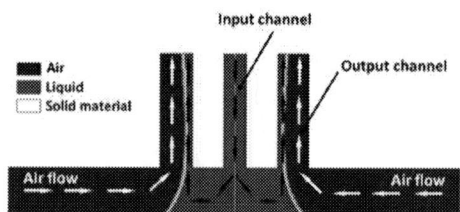

Figure 1. A schematic view of the DLM obtained in contact with the substrate.

The system in a principle implementation is composed by an internal jetting outlet where a liquid flow is forced, and an external recalling inlet throughout due to depression (obtained by a vacuum pump) the liquid is recalled back into the system. The input channel, that could be of any shape (e.g. circular, rectangular, etc.), confined by rigid wall (i.e. solid material), pumps a constant liquid flux that, depending on the input nozzle dimension, defines the velocity of the liquid exiting the input channel. Due to a lower pressure in the surrounding of output channel, the airflow sustains the liquid forming a DLM. The dynamic characteristics are due to the constantly refreshment of the liquid, during time, inside the drop. The contact angle depends on the wettability of the surface.

By placing an electrode in the liquid flow and applying a voltage to DLM respect to the substrate various electrochemical processing (i.e. anodization or metal electroplating) can be performed depending on the electrolyte used. In this work as the first step a fast chemical pre-treatment of the Al on the backside of PV cell was made and then metal electroplating in predefined places directly on Al was carried out using a mechanical contact to the substrate as the second electrode.

For characterization of samples we used high resolution Scanning Electron Microscopy (SEM) and compositional microanalysis with Energy Dispersive X-Ray microanalysis (EDX) using FESEM Auriga 405 equipped with Quantax EDX detector to evaluate the metal dispersion within aluminum-silicon interdiffused region and Transfer Length Method and current-voltage measurements to estimate the specific contact resistivity of the metal contact and series resistance of the overall solar cell device. Several commercial aluminum screen printable pastes have been experimented also having different distribution of sphere particles dimensions. We have investigated the dependence of Sn dispersion in Al and then the film adhesion, on the Al sphere particles dimensions. Finally we have performed the deposition on the screen-printed aluminum contact of standard size solar cell then adhesion test soldering metal ribbon on metal film plated.

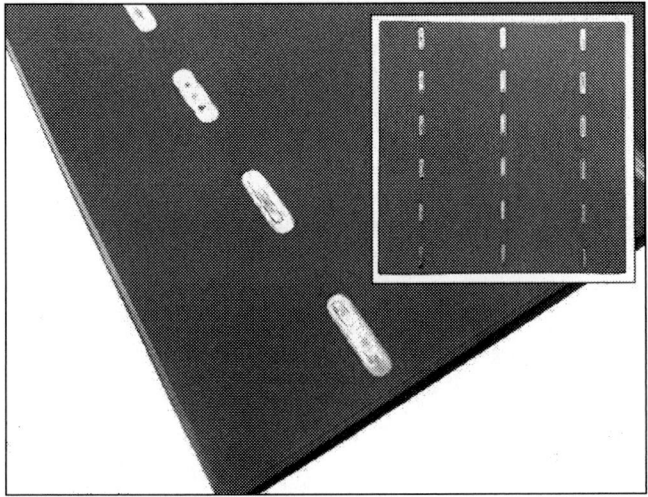

Figure 2. Results of the localized metal plating on Al on the backside of the PV Cell.

Experimental

In order to see the influence of the backside surface properties on the characteristics of the cells and metal coating two different Al screen printable pastes have been tested, differing one from each other in the particle size distribution and composition. The two pastes were produced as experimental lots by R&D labs of Chimet S.p.A. thick film division, according to the following labels reported in table I. Different particles size is also expected to give different surface roughness and thus influence the adhesion of the deposited metal film.

Table I. Aluminum paste distribution: the value Y = d(0.X) represents the X·10% probability that Al particles diameter are smaller than Y

Description	Ø (min), um	d(0.1), μm	d(0.5), μm	d(0.9), μm	Glass frit
S29 (small particle size)	0.955	1.537	2.733	4.746	2% Pb free
S30 (large particle size)	2.512	3.928	5.849	8.703	2% Pb free

As the both powders were supplied from the same dealer they were produced with the same process and same surface characteristics. Each paste has been tested by measuring: a) the specific contact resistivity with the Si wafer, evaluated with the Transfer Length Method (TLM) technique; b) the conductivity, evaluated with 4 points probe sheet resistance measurement. For specific contact resistivity measurements of the several samples have been produced by screen printing 8 patterns for TLM measurements of each paste on 5 Ohm·cm p-type doped CZ wafers. All pastes have been printed, dried and fired according to the following process parameters:

(i) Screen printing: squeegee hardness 70 – 75 shore, force 7 Kg, speed 100 mm/s, snap off 0.5 mm.

(ii) Screen: 250 mesh stencil screen, wires 36 microns.

(iii) Drying: 4 zones IR belt Aurel furnace at 250 °C, total duration time – 2 minutes.

(iv) Firing: 3 zones IR belt RTC furnace, settled temperatures 580 – 640 – 910 °C speed 50 ipm.

To estimate the fired paste conductivity a 4x2 cm^2 area has been fully printed on a similar substrate and using the same parameters as described above. The conductivity of the layers has been measured by 4-point probe method, mapping the whole area at 6 different points and extracting an average. The bulk conductivities were calculated by considering the layer thicknesses.

After that solar cells have been fabricated on p-type doped 1 Ohm·cm, 200 μm thick CZ alkaline textured wafers. The front side emitter has been diffused up to 70 Ohm/□. The backside electrode has been made by screen-printed Al full coverage and the front side has been ensured by screen-printed Ag grid. A co-firing process has been performed in a 3 zones IR belt RTC furnace with settled temperatures of 580 – 640 – 950 °C respectively and belt speed of 80 ipm. For avoiding the need for edge isolation the wafer has been cut into smaller samples. The solar cells have been measured by testing the open circuit voltage (V_{OC}) and Fill Factor (FF) under AM1.5G class A sunlight simulator conditions. Internal Quantum Efficiency (IQE) evaluation has been performed in the spectral range between 900 nm and 1200 nm to evaluate the BSF depth. Hall profile measurements have been performed on the backside of the cell with Al removed to evaluate the active doping concentration in the BSF region of the cell.

Concentration profiles were measured by the EDX at different magnification and with two acceleration voltage settings to modify the electron range (i.e. about 1.0 μm @ 10 kV and 0.3 μm @ 5 kV) and consequently the pear-shape of the volume analyzed. In table II the amount of Al, Si and O is listed for each of the samples as deduced from quantitative microanalysis performed at different magnifications and energies.

For the selective plating of Sn by DLM technique [2] a standard industrial solution Dow Chemical Solderon ST-200 was used. In order to reduce the influence of native oxide, pretreatment of the fired Al was performed in acid and alkaline solution by means of DLM technique to remove the alumina oxide and the silicon precipitates that are around the Al balls [4].

Nickel electroplating was made from a sulfamate solution with composition close to standard, supplied by Enthone®.

Results and Discussions

In table II the results of the microanalysis composition depending on the energy is presented.

Table II. Microanalysis composition of the samples.

Magnification and Energy	Materials	Sample S29		Sample S30	
		at.%	1σ	at.%	1σ
10000, 10KV	Al	68.67	3.29	53.87	2.53
	Si	15.68	0.73	29.31	1.32
	O	10.04	0.81	10.84	0.85
2500, 10kV	Al	68.20	3.29	60.87	2.82
	Si	15.25	0.72	22.16	0.99
	O	10.34	0.83	10.93	0.85
1000, 5kV	Al	58.98	2.55	51.27	2.24
	Si	19.93	0.90	24.61	1.12
	O	17.42	1.20	19.38	1.34

The electrical characteristics of the different samples are given in table III, with the specific contact resistivity (ρ_c) of the Al-Si contact, the Al paste resistivity (ρ after firing, the open circuit voltage (V_{oc}) and the BSF thickness.

Table III. Electrical properties of the samples.

Sample	ρ_c, mOhm·cm^2	ρ, µOhm·cm	V_{OC}, mV	BSF, µm
S29	53.8	28.4	625	5.9
S30	58.7	23.8	628	4.9

In figure 3 it is shown the typical view of the Al film obtained after the firing process. The structure of the film, consisting of the particles (balls) of different dimensions can be clearly seen. From the composition analysis one can observe that there is a high percentage of oxygen coming from the natural oxide of the Al. In order to perform controllable and uniform electroplating on top of such layer a pretreatment should be done to chemically remove oxide layer.

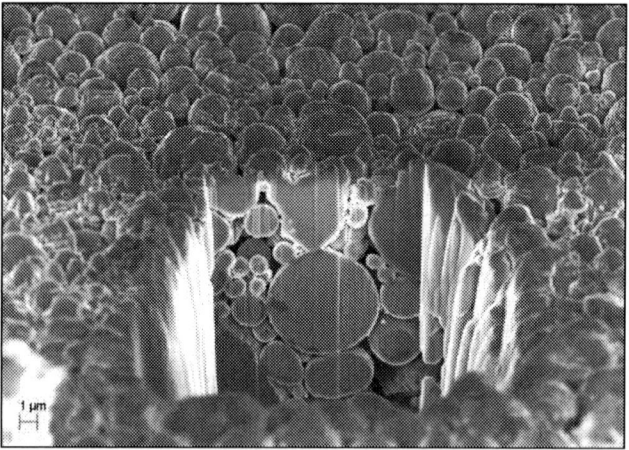

Figure 3. SEM image of the backside Al layer after the firing process.

In figure 4 is shown a SEM image of the Al ball before the pretreatments. Observing the figure is possible to see that the alumina thickness of the as-fired Al balls is about 140nm and a surface is smooth.

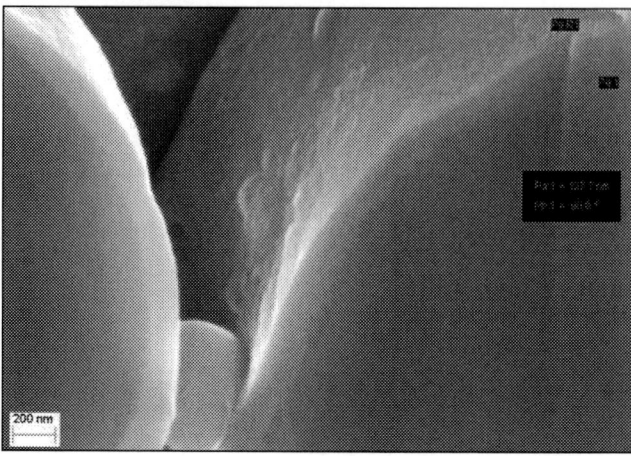

Figure 4. SEM image of magnified part of aluminum ball of backside silicon solar cell with measured alumina thickness.

In figure 5 is shown the Al ball after the acid and alkaline pretreatments showing how the alumina thickness is reduced to 25 – 40 nm and the surface roughness in increased.

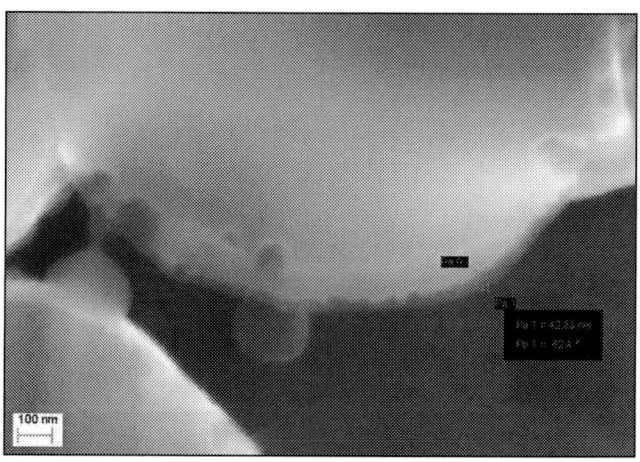

Figure 5. SEM image of magnified part of Aluminum ball after the acid and alkaline pretreatments.

In figure 6 is shown a SEM image after the tin electroplating process performed by DLM technique where is possible to see a cross-section of the Al ball obtained by a Focused Ion Beam (FIB) cut. In the image there is the EDX color map showing respectively in red Al, in yellow Si and in blue the deposited Sn.

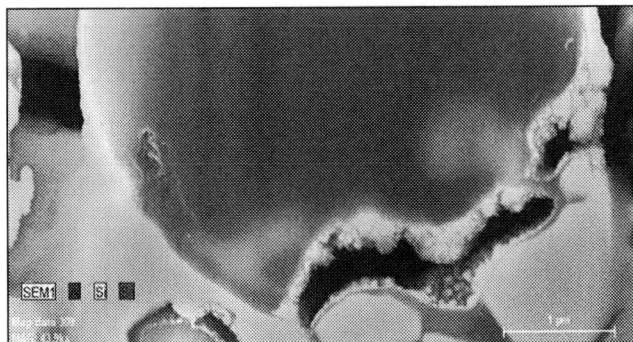

Figure 6. Aluminum Ball after Tin Plating.

In table IV the measured efficiencies of the different solar cells made using the two different Al paste are reported. The solar cell of 4x4 cm^2 with back Ag localized contact and were compared with Sn localized contact and direct Al contact. Fill Factor and efficiency are normalized respect to the cell with the direct Al contact.

Table IV. The measured efficiency of the samples.

Sample	V_{oc}, mV	J_{sc}, mA·cm^{-2}	FF	Eff.	R_s, Ohm·cm^2
S29 – Al contact	625	36.3	1.000	1.000	2.3
S29 – Ag contact	586	34.6	0.782	0.700	3.0
S29 – Sn contact	624	36.1	0.992	0.992	2.3
S30 – Al contact	628	35.4	1.000	1.000	2.9
S30 – Ag contact	590	34.5	0.825	0.756	3.0
S30 – Sn contact	628	34.7	0.998	0.980	2.8

Analyzing the table IV it is possible to see the beneficial influence of a uniform Al contact possessing the BSF in contrast to the Ag contact where the BSF is missing. For both kinds of initial pastes we were able to reach almost the same efficiency using direct Sn electroplating.

Sn layer electroplated on Al allows direct bonding of the cell to the ribbon without using any flux. No noticeable difference in adhesion was seen for two types of Al backside. In the figure 4 a typical load-deformation curve is shown obtained by the peel test for a bonded 2 mm-wide ribbon.

Currently industry looks for peel strengths higher 1.5 N on a 2 mm wide ribbon [5]. Considering the value reported in figure 4 such technique is expected to allow obtaining not less then 2 N for a 2 mm wide ribbon. But still the obtained value is close to the limit and is much lower to the values of the tensile strength of typical Sn joint. Considering this the detailed investigation of the failure reasons has been performed revealing that the main issue with the tin pad contact on Al backside solar cell is due to the reflow process during bonding.

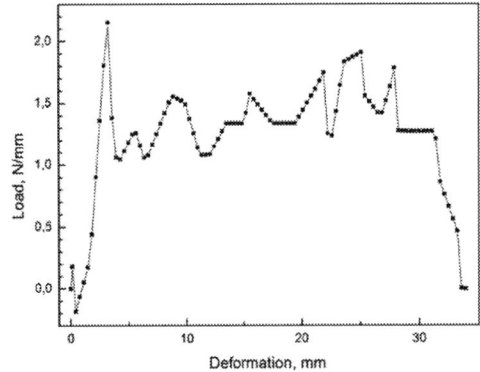

Figure 7. Peel test of the tin pad plated by DLM technique on the screen-printed Al contact.

If we have a look on the figure 8 where a SEM and EDX color map image are shown it is possible to see that Sn is penetrating inside Al balls and fills the voids between them. But during the bonding process Sn melts and tends to move and adhere more to the ribbon that is more wettable than oxidized Al balls.

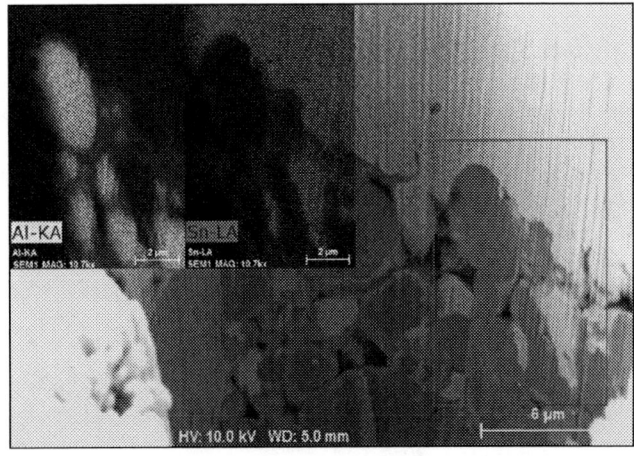

Figure 8. SEM and EDX color map of Sn deposited on Al balls before bonding.

This effect is evident from figure 9 that shows the place where the tin layer has been delaminated (red line) and that some tin is still inside the Al balls but voids are formed between Sn and Al.

Such effect makes the peel strength dependent from the tin pad dimension and thickness. Moreover, we also observed during peel test that quite often ribbon removes the Al balls up from the Al-Si interface.

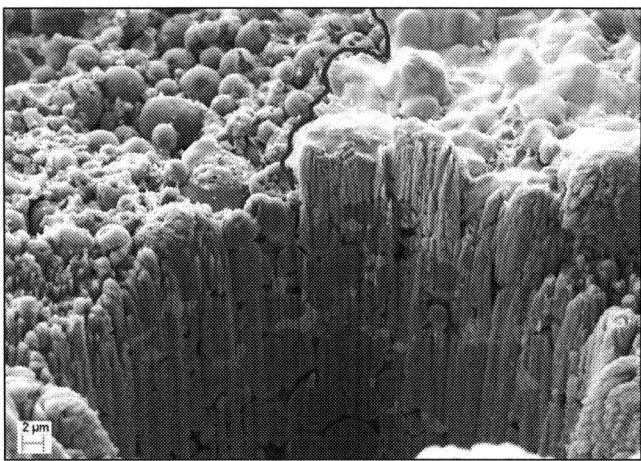

Figure 9. SEM image of tin pad on Al balls after peel test (the red line shows the line where the tin pad broke during peel test).

In order to gain a control over the adhesion and increase reproducibility, the plating process has been modified and a double-step metal electroplating was performed after the pretreatment step. On the first step nickel is electroplated and only then a thin layer of tin is plated on top. As a result, the voids between Al balls are totally filled with compact nickel and during the reflow process this structure remains intact. A cross-section of the structure representing Ni electroplated on top of pretreated Al balls obtained by FIB is shown on figure 10.

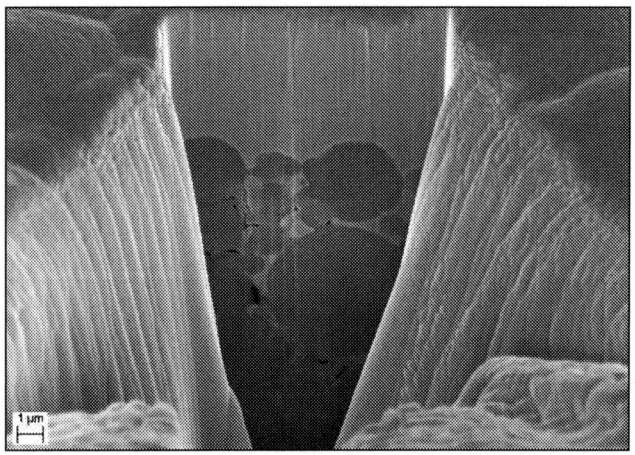

Figure 10. SEM image of Ni pad showing Ni penetrating inside Al balls.

A thin (i.e. 3 μm) layer of Sn was deposited on the Ni afterwards to ensure ribbon bonding without flux. The results of the peel tests on the full-size PV Cell are presented on the figure 11. The number of peaks corresponds to the number of metal pads created locally on the backside in the places corresponding to the ribbon bonding points.

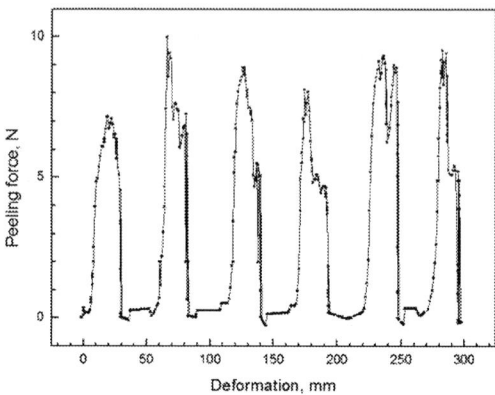

Figure 11. Peel test of the nickel-tin pad plated by DLM technique on the screen-printed Al contact.

As can be seen from the figure 11 the peel test showed a significant increase of adhesion (in comparison with that one without nickel) – to more than 7N for a 2 mm wide ribbon for all of the six Ni pads on the Al back side of the silicon solar cell.

Conclusions

A new selective processing technique based on a confined dynamic liquid drop/meniscus (DLM) was used to perform localized metal (nickel and tin) plating on Al back PV cell. In order to warranty the necessary adhesion pretreatment process was developed using acid an alkaline solutions. Different Al pastes were used to compare the uniform Al contact with Ag contact and Al-Sn contact. The results clearly show the beneficial influence of a uniform BSF. It was found that the main issue with the direct Sn plating on Al for a real application in solar cell is the reliability of the soldering. It was proved that the main failure is due to tin reflow and escape with formation of voids between Sn and Al. As a solution for improving adhesion a two-step process laying in the consecutive electroplating of Ni and Sn was proposed. As a result of the process voids in the Al layer are filled with compact nickel which stays untouched during the Sn reflow. The proposed solution allowed reaching the adhesion force values of more than 7 N for a 2 mm wide ribbon.

The DLM technique adopted to form plated Nickel and/or Tin film on screen printed Al back contact of solar cell manufacturing ensures localized contact formation and low quantities of chemical waste.

Fabrication of the high adhesion nickel-tin pads demonstrated on screen printed Al contact allow to replace the screen printable silver paste from the back contact of the cell thus reducing solar cell cost. Removing the silver tabs or stripes from the back side of the cell allows three main advantages in solar cell device: homogeneous BSF formation, simplified design of the back electrode and soldering of the ribbon without flux.

References

1. Narasinha, S. Rohatgi, A. "Optimized aluminum back surface field techniques for silicon solar cells",

Photovoltaic Specialists Conference, 1997., Conference Record of the Twenty-Sixth IEEE, pp. 63-66

2. Balucani, M.; Ciarniello, D.; Nenzi, P.; Bernardi, D.; Crescenzi, R.; Kholostov, K. "New selective wet processing" Electronic Components and Technology Conference (ECTC), 2013 IEEE 63rd, 247-254

3. M. Balucani, K Kholostov, P Nenzi, R Crescenzi, D Ciarniello, D Bernardi, L Serenelli, M Izzi, M Tucci; "New Selective Processing Technique for Solar Cells", Energy Procedia (2013) 43, 54-65

4. Balucani M., Serenelli L., Kholostov K., Nenzi P., Miliciani M., Mura F., Izzi M., Tucci M., " Aluminum-silicon interdiffusion in screen printed metal contacts for silicon based solar cells applications", Energy Procedia (2013) 43, 100-110

5. L. Hamann, R. Zapf Gottwick, M. Haas, W. Wille, J. Mattheis "30% Silver Reduction in Rear Side Busbar Pastes" 4th Workshop on Contacting Silicon Solar Cells, Constance, Germany May 6th-8th, 2013

Wet Etching of Deep Trenches on Silicon with Three-dimensional (3D) Controllability

Liyi Li[1], Ching-Ping Wong[1,2]*

1. School of Materials Science and Engineering, Georgia Institute of Technology, Atlanta, GA, USA
2. Department of Electronic Engineering, The Chinese University of Hong Kong, HongKong
*Corresponding author: cpwong@cuhk.edu.hk

Abstract

Trenches on silicon have found important applications in microelectromechanic system, microfluidic devices, photonic devices, capacitor memory devices and etc. Etching trenches with controllability of 3D geometry receives growing interests from academia as well as industry. In this paper we introduce a novel wet etching method, named metal assisted chemical etching, as a promising trench etching technology with 3D geometry variation. Both vertical and tapered etching results are presented. Slanted trenches from few-micron scale to sub-micron scale are also demonstrated with complex 3D features. Etchant composition, temperature and catalyst type are identified as key parameters in tuning 3D geometry of trenches by MaCE. Compared to currently available etching technology such as wet etching and reactive ion etching, the presented data in this paper demonstrate the merit of flexible 3D geometry capability, high-aspect ratio capability and low cost, which uniquely belongs to MaCE.

Introduction

Trenches on silicon are key components in electronic industry. In terms of the geometry in three dimensional (3D) spaces, the "simplest" fashion of trenches is the vertical trench. As illustrated in Figure 1 (a), in a vertical trench, the central axis aligns perpendicular to the top surface of the substrate. Ideally, the sidewall of vertical trenches should also be perpendicular to the top surface (referred as surface normal direction in the following discussion). Thus, the geometry of vertical trenches is majorly defined by their width (w), length (l, not shown in the cross-sectional schematic image) and depth (d). In practice, reactive ion etching (RIE) technology has successfully fabricated vertical trenches with a wide range of dimensions from hundreds of micron to tens of nanometers [1-3]. These vertical trenches found enormous applications in microelectromechanical system (MEMS) [4-6], microfluidic device [7], and capacitors for information storage.

Besides the vertical trenches, deep trenches with geometric variation through 3D spaces have received intensive academic interest and are spurring novel application. Here we briefly discuss two types of 3D geometric variation derived from the vertical trenches. As shown in Figure 1 (b), if the width of the trenches is non-uniformly distributed at different depth, then the cross section of the trenches will evolve from a rectangle to a trapezoid. Now the width of the trench opening (w_T) does not equal to that of the trench bottom (w_B). Thus the sidewalls of the trench are tilted from surface normal. This type of sidewalls is often named as tapered trenches. Here we defined the angle between surface normal and the sidewall normal as tapering angle (θ_T). A tapered trench evolves to a vertical trench if the tapering angle approaches 90°. In a more general sense, if w_B shrinks to zero, then the trapezoidal cross section further evolve into a

triangle. Due to their overall geometry, this type of trenches is sometimes referred as V-grooves. The tapered trenches with triangular and trapezoidal cross sections have been fabricated by basic wet etching method for decades. Various basic reagents, such as KOH [8, 9], NaOH [10], tetramethylammonium hydroxide (TMAH) [11, 12] and ethylenediamine pyrocatechol (EPW) [13], have been used for tapered trenches with different etching rate, surface roughness and other properties. In basic wet etching, the controllability of tapering angle originates from the different etch rate of crystalline plane in single crystal silicon substrate. Tapering angle of 90°, 54.7° and 45° have been achieved from various combinations of etchant, additives [14] and silicon substrate. The trenches etched by basic wet etching have been widely used for MEMS [15], microfluidic system[16] and optical devices[17, 18].

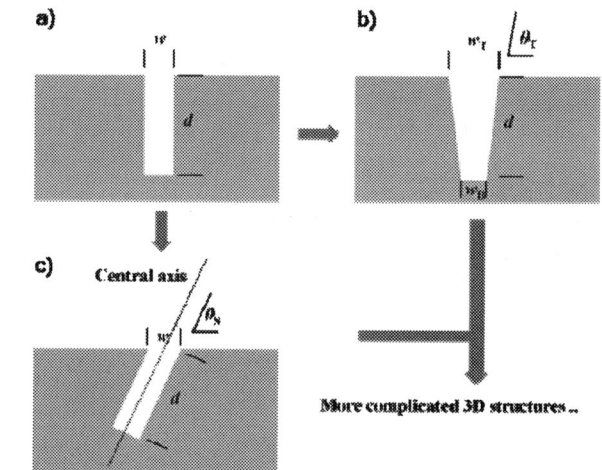

Figure 1 Schematics of (a) vertical trench, (b) tapered vertical trench and (c) slanted trench.

It should be noted that the sidewalls of the tapered trench are not necessary to be flat surfaces. Actually, it has long been known that silicon substrate can be isotropically etched in acid etching solution. For example, a mixture of nitric acid and hydrofluoric acid solution is reported to etch trenches with semi-circular cross sections. In RIE, stepwise tapering of trenches has been achieved [19-22]. The tapered features are proposed to be beneficial to the related electronic packaging process.

On the other hand, if the central axis of trenches rather the sidewall is tilted, they can be named as slanted trenches (Figure 1 (c)). The geometry of the slanted trenches is determined by the opening width (w), depth along the central axis (d) and slanting angle (θ_S). The slanting angle is defined as the angle between central axis and the top surface plane. Recently, submicron-wide slanting trenches have been successfully etched by focus ion beam (FIB) technology, which shows high coupling efficiency as grating couplers in silicon waveguides [23]. Also, slanted etching of silicon has been realized in submicron scale by RIE technology using Faraday cage [24] or ion-shealth control plate [25]. The novel slanted etching technologies are spurring novel applications [26] that cannot be achieved by traditional vertical etching technology.

Although various trenches with 3D geometric complexity have been fabricated, several serious challenges still exist in both wet etching and dry etching:

1. For wet etching, including basic etching and acidic etching, fabricating trenches with width less than 1μm is extremely hard. Since the tapering angle of basic etching depends on the crystalline orientation of the substrate, vertical trenches can only be achieved on (110)-oriented substrate;

2. For dry etching, including RIE and FIB, the cost is prohibitively high. Due to the nature of discontinuous etching in RIE, the sidewall of trenches often possesses scalloped roughness, which may harm the performance of devices thus produced.

In order to overcome the challenges, a novel etching method with the following merits is highly desirable:

1. Low cost and compatible with current micro-processing technology;

2. Controllability of 3D dimensions of trenches in a wide range with good uniformity;

3. High-aspect ratio etching capability.

Recently, a novel silicon etching technology, named metal-assisted chemical etching (MaCE), has attracted significant attention in academia as well as industry. In MaCE, silicon substrate is loaded with noble metal as catalyst on its top surface. The substrate is etched by solution composed of hydrofluoric acid (HF) and hydrogen peroxide (H_2O_2):

$$6HF+2H_2O_2+Si \xrightarrow{\text{Metal}} H_2SiF_6+4H_2O \quad (1)$$

At the initial stage, silicon adjacent to metal catalyst is etched and then metal catalyst sinks into the etched cavity to assist further etching. High-aspect ratio silicon nanowires [27], pillars [28], pores [29] and 3D-complicated cavity [30] have been successfully fabricated by MaCE in the past decade. Due to the nature of wet etching, MaCE shows promising result in deep trenches etching that may fulfill the merits mentioned above[31]. In this paper, we show that MaCE is not only capable of etching perfect vertical deep trenches, but also manage to obtain trenches with tunable tapering angle. On the other hand, the etching direction in MaCE is found to be influence comprehensively by multiples parameters, such as etchant composition [32], lateral geometry of catalyst [33]. By tuning these parameter, both vertical and slanted etching have been observed using MaCE. However, the slanted trench etching using MaCE is still uninvestigated. In this paper, we demonstrated that high aspect ratio slanted etching can also be achieved by MaCE.

Discussion

In order to fabricate deep trenches by MaCE, the silicon substrate is first patterned by photolithography (PL) or electron beam lithography (EBL). In this paper, (100)-type

single crystalline silicon substrates are used. We use an array of long stripe-shape patterns for trench etching in the following steps. After lithographic patterning, metal catalyst is deposited on the patterned silicon substrate by electron beam evaporation (EBE). The etching is conducted in HF-H_2O_2 etchant solution for a certain time. After etching, the sample is washed copiously by deionized water and dry under nitrogen gas. The geometry and morphology of etched trenches are obtained by scanning electron microscope (SEM). Figure 2 shows the successful etching of paralleled trenches. In the top view image, only the region covered by catalyst is etched; other region keeps intact. The width of the trenches is 2.0 μm, the same as the initial width of catalyst. The depth of trenches on the same substrate falls within the range of 6.3-6.8 μm, indicating a uniform etching rate between each trench. The result demonstrates that uniform vertical trenches can be achieved by MaCE. The width of the trenches can be defined by the standard lithographic methods that are ready available in common microfabrication lab.

a)

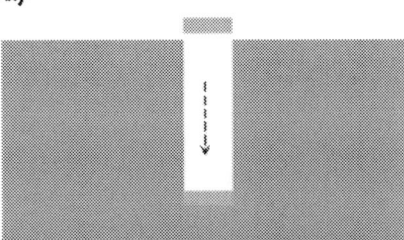

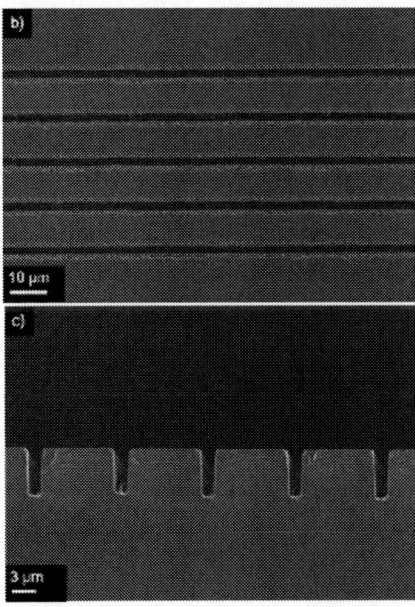

Figure 2 Paralleled vertical trenches etched by MaCE. (a) schematic etching process; top view (b) and cross-sectional (c) SEM image of paralleled vertical trenches.

Figure 3 shows the trenches etched at a temperature higher than that used in the vertical trench etching experiment. Significant tapering of sidewalls can be observed in Figure 3 (b). The tapering phenomenon can be explained from the etching mechanism of MaCE. In MaCE, the overall etching reaction can be treated as the combination of two half reactions:

$$H_2O_2 + 2H^+ \xrightarrow{\text{Metal}} 2h^+ + 2H_2O \qquad (2)$$

$$6HF + 4h^+ + Si \xrightarrow{\text{Metal}} H_2SiF_6 + 4H^+ \qquad (3)$$

where h^+ refers to electronic holes. At the elevated temperature, the rate constant k of both half reactions tends to increase. The tapering phenomenon indicates that k of reaction (2) probably increases more than that of reaction (3). Under this condition, h^+ generated from reaction (2) exceeds that can be consumed by reaction (3). The excessive h^+ will diffuse from catalyst towards the sidewalls and cause the extra etching, which finally renders the sidewall as tapered.

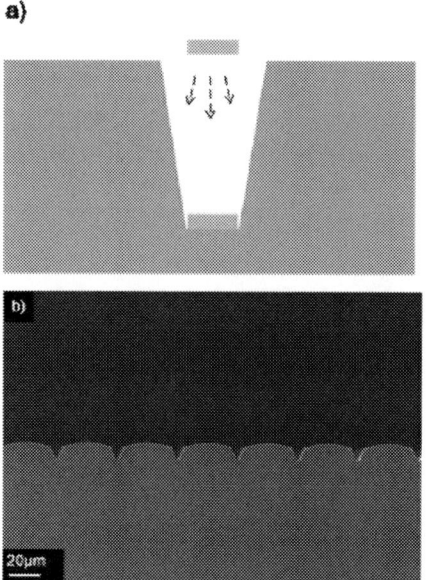

Figure 3 Tapered trenches etched by MaCE. (a) schematic etching process; (b) cross-sectional SEM image.

In Figure 3 (b), the tapering angle of the trenches is measured to be 65°. Compared to the tapered trenches etched by traditional basic wet etching, the unique 65° tapering angle shown in Figure 3 (b) indicates that the sidewalls of the tapered trenches do not belong to any low-index crystalline plane. In other words, the tapering phenomenon is not caused by the difference of etching rate on various crystalline planes. Thus, MaCE may be able to fabricate trenches with arbitrary tapering angle. Also, the sidewall of the trenches appears smooth and flat, which is different from the stepwise tapered trenches made by RIE. Thus, MaCE is potentially advantageous over both traditional basic wet etching as well as RIE in that the sidewalls of trenches are smooth and flat with flexible range of tapering angle.

It has been reported that addition of non-aqueous liquid induced bending of silicon nanowires made by MaCE [34]. In this regards, we repeat the vertical etching experiment shown in Figure 2 with the addition of ethanol (EtOH). Tapered trenches can also be found, as shown in Figure 4. Compared to the result of vertical etching, the vertical etching rate drops from 0.7 µm/min to 0.1 µm/min, and the opening width of trenches increases to 2 µm to 14 µm. The tapering angle drops to 28°. Trenches with such a low tapering angle are hardly reported by literature. Again, the tapering angle of 28° cannot be related to any low index crystalline plane of silicon, which further supports the point that trenches with arbitrary tapering

angle can be obtained by MaCE. Interestingly, upon addition of EtOH, the sidewall and the adjacent region shows nanoporous morphology, indicating the h^+ density might be essentially different from the vertical etching conditions.

Figure 4 Cross-sectional SEM image of a tapered trench fabricated by MaCE with addition of EtOH during etching.

Further, we change the type of catalyst and find the etched trenches possessing semi-circular cross section (Figure 5), which highly resembles the result of acidic wet etching [35]. The istotropic etching profile may be caused by the dissolution and re-deposition of catalyst during etching.

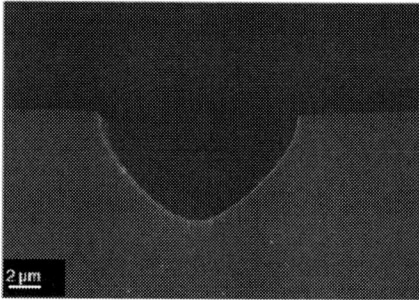

Figure 5 Cross sectional of a trench with semi-circular cross sections by MaCE.

Besides the tapered trenches shown above, we investigate the capability of MaCE in slanted trench etching. Figure 6 shows a slanted trench etched with a higher HF concentration compared to that of vertical trenches. After etching, the top surface keeps intact except the trench opening; some not fully-etched silicon "grass" remains in the trench (Figure 6 (b)). The catalyst strip breaks into two parts after etching into the substrate for a short time, and the two subunits proceed into two different directions (Figure 6 (c)). On another case where catalyst strips are patterned with submicron width, a different slanted profile is observed. Figure 7 shows the etching of 600 nm-wide catalyst stripes in the same etchant as that of vertical etching. From top view SEM image, a lateral "swinging" of the catalyst strip can be observed (Figure 7 (b)). The cross-sectional image further illustrates that the stripe moves up and down inside silicon substrate and forms slanted trenches with 3D complexity (Figure 7 (c)). The slanted trenches mentioned above possess high aspect ratio and 3D complexity that cannot be achieved by currently available etching methods. Based on these findings, more research is under way to achieve higher controllability of etching profile in MaCE.

978-1-4799-2408-0/14 $31.00 © 2014 IEEE

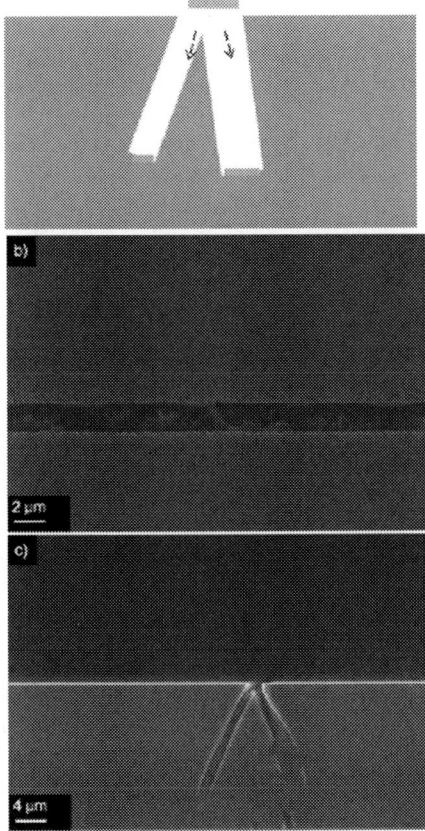

Figure 6 Slanted trenches with branches fabricated by MaCE. (a) schematic etching process; (b) top-view SEM and (c) cross-sectional SEM image.

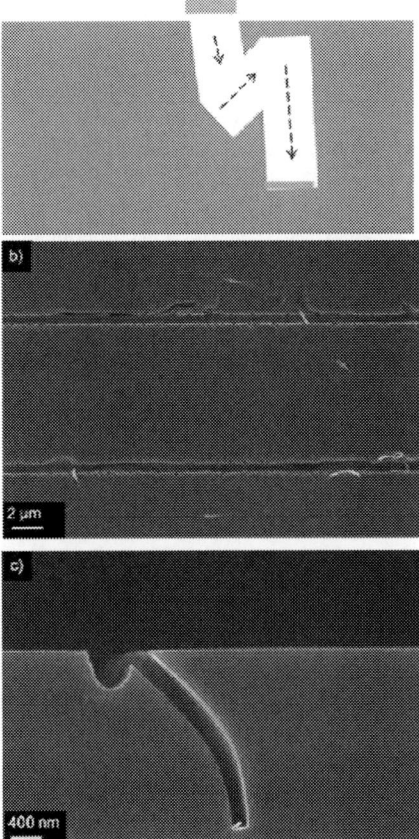

Figure 7 Slanted trenches with 3D complexity fabricated by MaCE. (a) schematic etching process; (b) top-view SEM and (c) cross-sectional SEM image.

Conclusion

In conclusion, we have shown that a novel wet etching technology, MaCE, can be employed in trench etching on silicon with controllability in their 3D geometry. Vertical etching by MaCE is demonstrated, which is comparable to the result of RIE. Trenches with tapering angle of 28° and 65° are successfully fabricated with smooth sidewalls; isotropic trench etching is also realized by MaCE. Slanted trenches with 3D complexity are also demonstrated. Several key experimental parameters, including etchant composition, temperature, type and lateral dimension of catalyst, are found to effectively tune the geometry. The presented result in this paper indicates the capability of MaCE in etching trenches on silicon with high 3D geometric controllability, high aspect ratio and low cost, which is promising for application such as silicon based MEMS, photonics, electronic packaging while spurring novel technology.

Acknowledgments

The authors thank IEN at Georgia Tech for cleanroom facility training and usage. The authors thank CNC at Georgia Tech for SEM facility. The authors thank funding support from National Science Foundation (NSF CMMI #1130876).

Reference

1. H. Jansen, M. Deboer, R. Legtenberg, and M. Elwenspoek, "The Black Silicon Method - a Universal Method for Determining the Parameter Setting of a Fluorine-Based Reactive Ion Etcher in Deep Silicon Trench Etching with Profile Control," *J. Micromech. Microeng.,* vol. 5, pp. 115-120, Jun 1995.

2. F. Marty, L. Rousseau, B. Saadany, B. Mercier, O. Francais, Y. Mita, and T. Bourouina, "Advanced etching of silicon based on deep reactive ion etching for silicon high aspect ratio microstructures and three-dimensional micro- and nanostructures," *Microelectron. J.,* vol. 36, pp. 673-677, Jul 2005.

3. J. Parasuraman, A. Summanwar, F. Marty, P. Basset, D. E. Angelescu, and T. Bourouina, "Deep reactive ion etching of sub-micrometer trenches with ultra high aspect ratio," *Microelectron. Eng.,* vol. 113, pp. 35-39, Jan 2014.

4. F. Ayazi and K. Najafi, "High aspect-ratio combined poly and single-crystal silicon (HARPSS) MEMS technology," *J. Microelectromech. Syst.,* vol. 9, pp. 288-294, 2000.

5. F. Ayazi and K. Najafi, "A HARPSS polysilicon vibrating ring gyroscope," *J. Microelectromech. Syst.,* vol. 10, pp. 169-179, Jun 2001.

6. J. W. Weigold, W. H. Juan, and S. W. Pang, "Etching and boron diffusion of high aspect ratio Si trenches for

released resonators," *J. Vac. Sci. Tech. B,* vol. 15, pp. 267-272, Mar-Apr 1997.

7. C. Wu, F. Bendriaa, F. Brunelle, and V. Senez, "Fabrication of AD/DA microfluidic converter using deep reactive ion etching of silicon and low temperature wafer bonding," *Microelectron. Eng.,* vol. 88, pp. 1878-1883, Aug 2011.

8. K. Sato, M. Shikida, Y. Matsushima, T. Yamashiro, K. Asaumi, Y. Iriye, and M. Yamamoto, "Characterization of orientation-dependent etching properties of single-crystal silicon: effects of KOH concentration," *Sensors and Actuators a-Physical,* vol. 64, pp. 87-93, Jan 1 1998.

9. E. D. Palik, O. J. Glembocki, I. Heard, P. S. Burno, and L. Tenerz, "Etching Roughness for (100) Silicon Surfaces in Aqueous Koh," *J. Appl. Phys.,* vol. 70, pp. 3291-3300, Sep 15 1991.

10. P. Allongue, V. Costakieling, and H. Gerischer, "Etching of Silicon in Naoh Solutions .1. Insitu Scanning Tunneling Microscopic Investigation of N-Si(111)," *J. Electrochem. Soc.,* vol. 140, pp. 1009-1018, Apr 1993.

11. M. Shikida, K. Sato, K. Tokoro, and D. Uchikawa, "Differences in anisotropic etching properties of KOH and TMAH solutions," *Sensors and Actuators a-Physical,* vol. 80, pp. 179-188, Mar 10 2000.

12. K. Sato, M. Shikida, T. Yamashiro, K. Asaumi, Y. Iriye, and M. Yamamoto, "Anisotropic etching rates of single-crystal silicon for TMAH water solution as a function of crystallographic orientation," *Sensors and Actuators a-Physical,* vol. 73, pp. 131-137, Mar 9 1999.

13. A. Reisman, M. Berkenblit, S. A. Chan, F. B. Kaufman, and D. C. Green, "Controlled Etching of Silicon in Catalyzed Ethylenediamine-Pyrocatechol-Water Solutions," *J. Electrochem. Soc,* vol. 126, pp. 1406-1415, 1979.

14. K. P. Rola and I. Zubel, "Triton Surfactant as an Additive to KOH Silicon Etchant," *J. Microelectromech. Syst.,* vol. 22, pp. 1373-1382, Dec 2013.

15. K. Biswas and S. Kal, "Etch characteristics of KOH, TMAH and dual doped TMAH for bulk micromachining of silicon," *Microelectron. J.,* vol. 37, pp. 519-525, Jun 2006.

16. P. Pal and K. Sato, "Various shapes of silicon freestanding microfluidic channels and microstructures in one-step lithography," *J. Micromech. Microeng.,* vol. 19, May 2009.

17. C. T. Chen, P. K. Shen, C. C. Chang, H. L. Hsiao, J. Y. Li, K. Liang, T. Y. Huang, R. H. Chen, G. F. Lu, and M. L. Wu, "45 degrees-Mirror Terminated Polymer Waveguides on Silicon Substrates," *IEEE Photonics Technol. Lett.,* vol. 25, pp. 151-154, Jan 15 2013.

18. C. Strandman, L. Rosengren, H. G. A. Elderstig, and Y. Backlund, "Fabrication of 45 degrees mirrors together with well-defined V-grooves using wet anisotropic etching of silicon," *J. Microelectromech. Syst.,* vol. 4, pp. 213-219, Dec 1995.

19. N. Ranganathan, D. Y. Lee, L. Ebin, N. Balasubramanian, K. Prasad, and K. L. Pey, "The development of a tapered silicon micro-micromachining process for 3D microsystems packaging," *J. Micromech. Microeng.,* vol. 18, Nov 2008.

20. S. Hamaguchi and M. Dalvie, "Intrinsic and Passivation - Induced Trench Tapering during Plasma Etching," *J. Electrochem. Soc.,* vol. 141, pp. 1964-1972, July 1, 1994 1994.

21. N. Roxhed, P. Griss, and G. Stemme, "A method for tapered deep reactive ion etching using a modified Bosch process,"

J. Micromech. Microeng., vol. 17, pp. 1087-1092, May 2007.

22. M. Sato and Y. Arita, "Etched Shape Control of Single - Crystal Silicon in Reactive Ion Etching Using Chlorine," *J. Electrochem. Soc.,* vol. 134, pp. 2856-2862, November 1, 1987 1987.

23. J. Schrauwen, F. Van Laere, D. Van Thourhout, and R. Baets, "Focused-ion-beam fabrication of slanted grating couplers in silicon-on-insulator waveguides," *IEEE Photonics Technol. Lett.,* vol. 19, pp. 816-818, May-Jun 2007.

24. J.-K. Lee, S.-H. Lee, J.-H. Min, I.-Y. Jang, C.-K. Kim, and S. H. Moon, "Oblique-Directional Plasma Etching of Si Using a Faraday Cage," *J. Electrochem. Soc,* vol. 156, pp. D222-D225, July 1, 2009 2009.

25. S. Takahashi, K. Suzuki, M. Okano, M. Imada, T. Nakamori, Y. Ota, K. Ishizaki, and S. Noda, "Direct creation of three-dimensional photonic crystals by a top-down approach," *Nat Mater,* vol. 8, pp. 721-725, 2009.

26. Z. Xu, J. Jiang, M. R. Gartia, and G. L. Liu, "Monolithic Integrations of Slanted Silicon Nanostructures on 3D Microstructures and Their Application to Surface-Enhanced Raman Spectroscopy," *J. Phys. Chem. C,* vol. 116, pp. 24161-24170, 2012/11/15 2012.

27. K. Peng, Y. Xu, Y. Wu, Y. Yan, S.-T. Lee, and J. Zhu, "Aligned Single-Crystalline Si Nanowire Arrays for Photovoltaic Applications," *Small,* vol. 1, pp. 1062-1067, 2005.

28. J. Yeom, D. Ratchford, C. R. Field, T. H. Brintlinger, and P. E. Pehrsson, "Decoupling Diameter and Pitch in Silicon Nanowire Arrays Made by Metal-Assisted Chemical Etching," *Adv. Funct. Mater.,* vol. 24, pp. 106-116, 2014.

29. L. Li and C. P. Wong, "High aspect ratio sub-100 nm silicon vias (SVs) by metal-assisted chemical etching (MaCE) and copper filling," in *Electronic Components and Technology Conference (ECTC), 2013 IEEE 63rd,* 2013, pp. 2326-2331.

30. O. J. Hildreth, A. G. Fedorov, and C. P. Wong, "3D Spirals with Controlled Chirality Fabricated Using Metal-Assisted Chemical Etching of Silicon," *ACS Nano,* vol. 6, pp. 10004-10012, 2012.

31. L. Li, Y. Liu, X. Zhao, Z. Lin, and C.-P. Wong, "Uniform Vertical Trench Etching on Silicon with High Aspect Ratio by Metal-Assisted Chemical Etching Using Nanoporous Catalysts," *ACS Appl. Mater. Interfaces,* vol. 6, pp. 575-584, 2014/01/08 2014.

32. M.-L. Zhang, K.-Q. Peng, X. Fan, J.-S. Jie, R.-Q. Zhang, S.-T. Lee, and N.-B. Wong, "Preparation of Large-Area Uniform Silicon Nanowires Arrays through Metal-Assisted Chemical Etching," *The Journal of Physical Chemistry C,* vol. 112, pp. 4444-4450, 2008/03/01 2008.

33. Z. Huang, T. Shimizu, S. Senz, Z. Zhang, X. Zhang, W. Lee, N. Geyer, and U. Gösele, "Ordered Arrays of Vertically Aligned [110] Silicon Nanowires by Suppressing the Crystallographically Preferred <100> Etching Directions," *Nano Lett.,* vol. 9, pp. 2519-2525, 2009.

34. Y. Kim, A. Tsao, D. H. Lee, and R. Maboudian, "Solvent-induced formation of unidirectionally curved and tilted Si nanowires during metal-assisted chemical etching," *J. Mater. Chem. C,* vol. 1, pp. 220-224, 2013.

35. B. Hanrahan, C. M. Waits, and R. Ghodssi, "Isotropic etching technique for three-dimensional microball-bearing raceways," *J. Micromech. Microeng.,* vol. 24, Jan 2014.

978-1-4799-2408-0/14 $31.00 © 2014 IEEE

An Innovative Bumpless Stacking with Through Silicon Via
for 3D Wafer-On-Wafer (WOW) Integration

*Sue-Chen Liao[1], Erh-Hao Chen[1], Chien-Chou Chen[1], Shang-Chun Chen[1], Jui-Chin Chen[1], Po-Chih Chang[1],
Yiu-Hsiang Chang[1], Cha-Hsin Lin[1], Tzu-Kun Ku[1], Ming-Jer Kao[1], Young Suk Kim[2], Nobuhide Maeda[2],
Shoichi Kodama[2], Hideki Kitada[2], Koji Fujimoto[2], Takayuki Ohba[2]
[1]Electronics and Optoelectronics Research Labs, Industrial Technology Research Institute
Rm.256, Bldg.17, No.195, Sec. 4, Chung Hsing Road, Chutung, Hsinchu 310, Taiwan
*Correspondent, scl@itri.org.tw, +886-3-591-3750
[2]ICE Cube Center, WOW Alliance, Tokyo Institute of Technology (Tokyo Tech)
4259 Nagatsuda, Midori-ku, Yokohama 226-8503, Japan

Abstract

An adequate sequential etching though dielectrics, silicon and permanent adhesive material was successfully developed for the damascene interconnects in the face-to-back bumpless TSV Wafer on Wafer (WOW) processes. The induced bowing taken place at the etching of permanent adhesive was optimized and no void Cu metallization was achieved. According to those TSV technology, the upper and lower stacked wafers was electrically connected without bump electrodes. The improved process such as chemical mechanical planarization (CMP) of Cu re-distribution layer (RDL) is also developed successfully to provide uniform and straight line resistance distribution and reduce the loading of TSV over-etching to avoid the interconnect open issue.

Introduction

In the modern digital society, files are becoming bigger and bigger due to the high resolution photos and videos are embedded. It is really a challenge to process effectively such big data in today's semiconductor industry. Obviously, the three dimension integrated circuit (3DIC) technology is a promising way to provide the solution [1-2]. Chips with TSVs adopted can provide shorter path to connect stacked chips to enhance the performance and reduce the power consumption at the same time. Moreover, as the emergence of guides and standards such as hybrid memory cube (HMC) and wide input/output (I/O) spec, the 3DIC stacked architecture is clearly a cost effective solution for handling such big data.

Currently, the main stream approaches to achieve the three dimension (3D) stacked chips are through the chip on chip (COC) or chip on wafer (COW) stacking. But these ways would meet a significant bottleneck when the chips are keeping scaling down (i.e. more chips in a wafer). In COW, the stacking throughput is inversely proportional to the chip quantities in a wafer. In other words, the more chips in a wafer, the more time is needed to finish all the COW stacking in a wafer. Only the WOW approach can keep the same throughput no matter the chip quantities in a wafer. Therefore, WOW is a preferred solution with high chip stacking throughput and low manufacturing cost to be adopted when the chip sizes are becoming smaller and with a higher stacking alignment accuracy requirement.

In this work, the innovative 300mm WOW stacking approach with permanent adhesive material coated between bonding wafers is investigated to achieve a low cost 3DIC stacked chips. Also the related key modules such as TSV etching, CMP and oxygen plasma are introduced to show the challenges of this study.

Results and Discussion

Figure 1 shows the bumpless WOW process flow [3-5]. For the top wafer (shown as MC wafer), after the definition of RDL, this MC wafer is thinned to 20μm with the help of temporary glass carrier as shown in Figure 1(a). The RDL is also defined in the bottom wafer (shown as MB wafer) and capped with silicon nitride and permanent adhesive material after the CMP process as shown in Figure 1(b). These two wafers MC and MB are stacked together and then electrically connected using the bumpless TSV.

Obviously, the TSV process is a key module to realize this bumpless WOW integration scheme [6-7]. In order to interconnect the MC to MB wafers, the TSV etching should etch through the silicon oxide of RDL, the silicon nitride and silicon oxide, the silicon of MC wafer, the permanent bonding adhesive material, the oxide liner layer and the capped silicon nitride layer on MB wafer. The thickness of photoresist (PR) should be controlled carefully to ensure its thickness is enough for TSV etching to etch all these film schemes and do not leave too much remained PR to increase the loading of subsequent removal and cleaning processes. In this work, the PR is removed after the etching of permanent bonding adhesive material and then deposit the oxide liner layer. Next, break through the TSV bottom oxide liner layer and also the capped silicon nitride layer by using self-aligned etching back process.

Another key module is CMP process. The CMP process should be controlled precisely to keep the uniform and straight line resistance distribution [8]. And also the improved lower Cu dishing after CMP can reduce the loading of TSV over-etching and minimize the possibility of interconnect open issue. The 10 μm Cu line dishing after CMP can be improved to less than 10nm in the whole 300mm wafer by adopting the silicon oxide etching back and CMP buffing processes as shown in Figure 2.

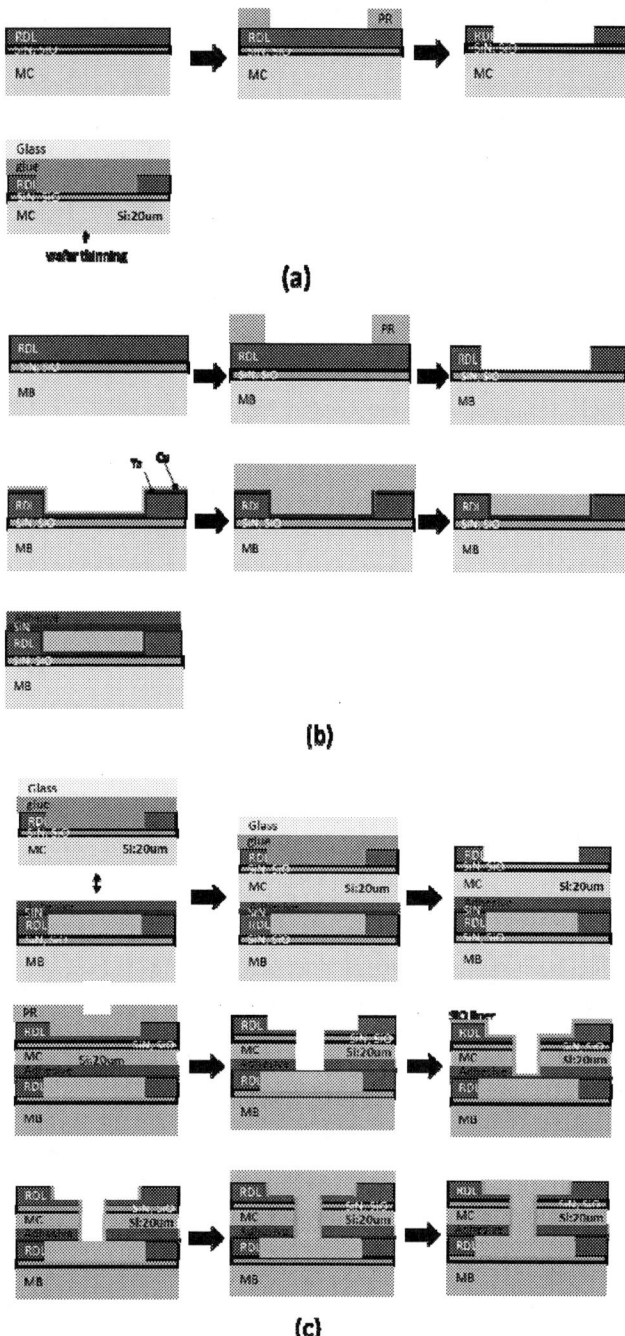

Figure 1. Brief process flows for (a) MC wafer formation (b) MB wafer formation and (c) the formation of 2-layer stacked bumpless WOW

After the oxide liner etching back process, the wafer surface is inspected by the X-ray photoelectron spectroscopy (XPS) and an abnormal large amount of C-F element is detected. This C-F element is originated from the residue of etching back process and would cause the adhesion problem in the subsequent film deposition processes. In order to integrate the WOW processes successfully, an effective way to remove such kind of residue is needed. Therefore, an oxygen plasma process is tried to be included in the flow to

test the capability of C-F residue removal. As shown in Figure 3(a), the oxide liner film is deposited first on silicon wafer and only the C-C and C-H peaks (blue line) can be detected. But the C-F peak (red line) is appeared after the oxide liner etching back process. Moreover, this C-F residue is still existed after the solvent cleaning process (black line). As expected, this extent of C-F residue can be reduced significantly by performing one-minute oxygen plasma process (purple line). From Figure 3(b), it is clear to see that oxygen plasma process is an feasible way to reduce the C-F residue but cannot remove it completely. More oxygen plasma time (from 1 minute to 3 minutes) has no effect in further removing the C-F residue. Currently, a tuning of etching recipe with lower C-F residue after etching process and an evaluation of including another sputtering process are on-going to deliver a successful WOW integration flow.

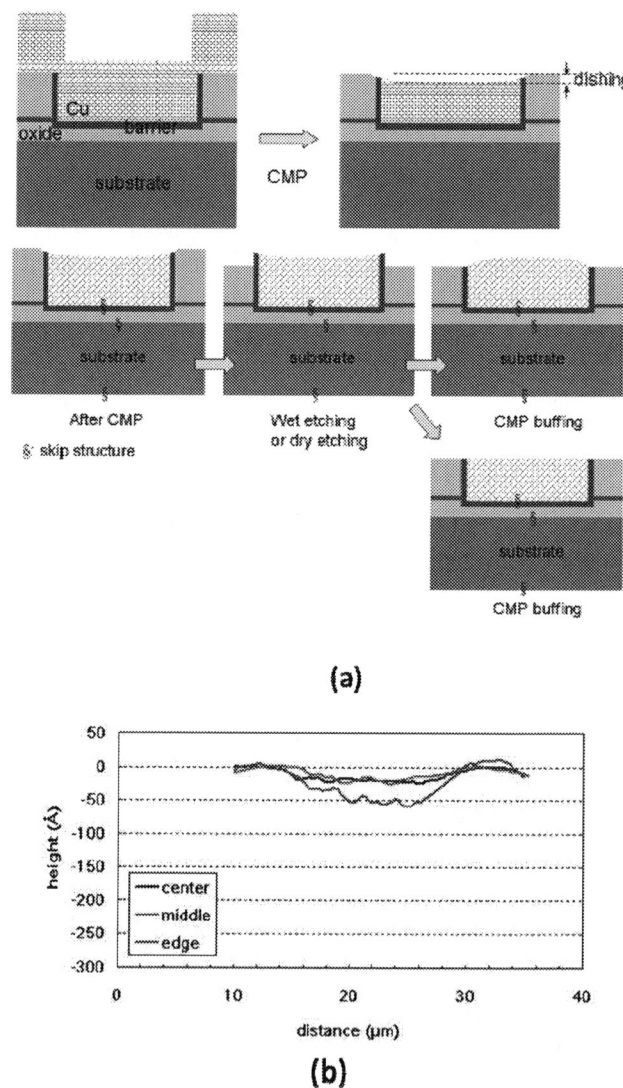

Figure 2. (a) Using oxide etch back and CMP buffing to improve the Cu dishing. (b) The Cu dishing extent of 10μm Cu line after improvement.

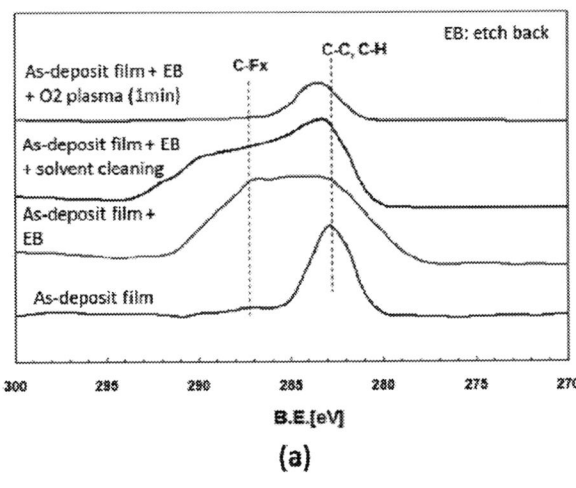

(a)

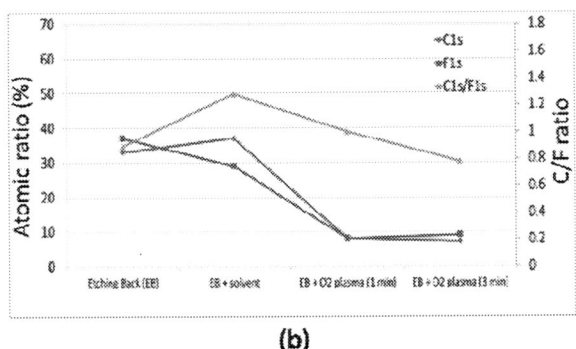

(b)

Figure 3. (a) The XPS spectrum showing the detected elements at different process stages (b) The variation of atomic ratios of C1s and F1s and the relative C/F ratio at different process stages

Two-layer stacked WOW structure is achieved successfully as shown in Figure 4. According to etching profile, a little bit bowling of TSV profile is observed and would be improved by reducing the thermal budget of the whole WOW processes. The daisy chain with 216 TSVs is designed in this WOW structure to connect the RDLs of MC and MB wafers together. This resistance distribution of TSV daisy chain is plotted in Figure 5. This straight resistance distribution (over 90%) reveals the stability of WOW process and shows the nearly 18.5ohm average daisy chain resistance.

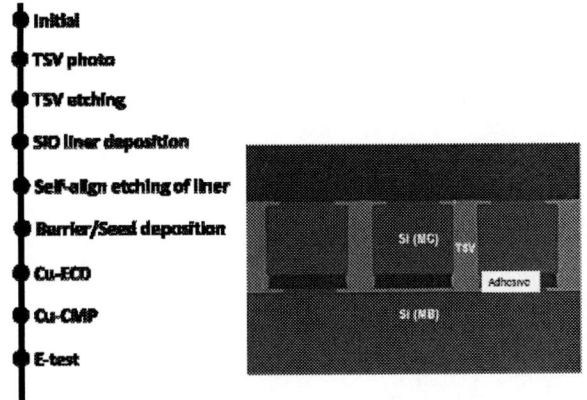

Figure 4. 2-layer WOW stacking

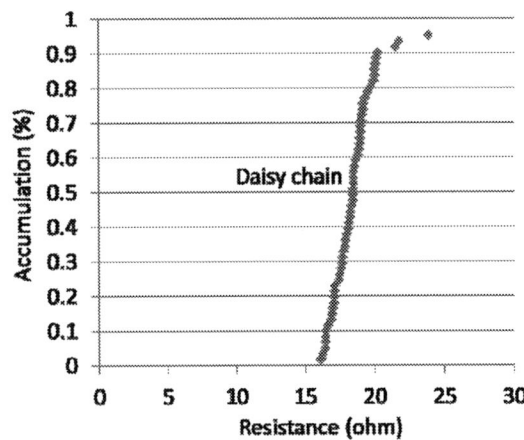

Figure 5. The resistance distribution of TSV daisy chain in 2-layer WOW structure

Except the resistance measurement of TSV daisy chains, the line resistances of RDLs with 40μm width in MC and MB wafers are also measured as shown in Figure 6. Due to the RDL is suspected to be over-polished in the top MC wafers, their resistance distributions are not as straight as that of MB wafers. And also the MC wafer reveals the larger RDL resistance value (4.7ohm) as compared with that of MB wafer (3.1ohm).

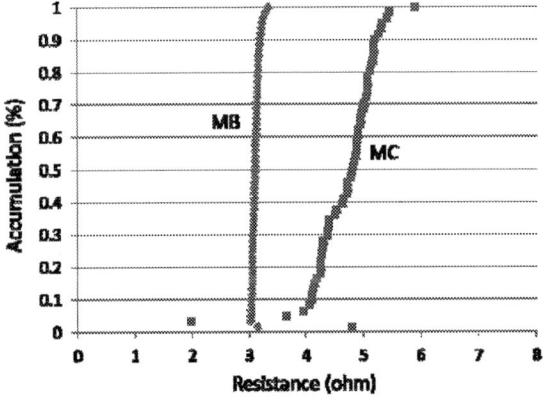

Figure 6. The line resistance distribution of 10μm and 40μm RDLs in MC and MB wafers

Conclusions

An optimized bumpless integration process including TSV etching, CMP, oxygen plasma process, etc is employed to successfully achieve the 2-layer WOW structure. The electrical measurements of TSV daisy chains and RDLs of MC and MB wafers can provide quantitative characteristics to monitor the maturity of this innovative three dimension integration process.

References

1. Pei-Jer Tzeng, et al., " Key Enabling Technologies of 300mm 3DIC Process Integration", VLSI-TSA 2012, pp. 1-2.

2. Erh-Hao Chen, et al., "Permanent and Bumpless Wafer to Wafer Bonding Technologies Development for Low Cost 3DIC Wide I/O Memory Cube", IITC 2014, in submitting.

3. N. Maeda et al., Proc. AMC 2008, Eds. M. Naik, R. Shaviv, T. Yoda, and K. Ueno, Mat. Res. Soc., p. 501, 2009.

4. Y. S. Kim et al., IEEE IEDM Tech. Dig., p. 365, 2009.

5. T. Ohba et al., Microelectronic Eng., Elsevier, **87**, pp. 485-490, 2010.

6. Yu-Chen Hsin, et al., " Effects of Etch Rate on Scallop of Through-Silicon Vias (TSVs) in 200mm and 300mm Wafers", IEEE/ECTC Proceedings, Lake Buena Vista, FL, May 2011, pp. 1130-1135.

3D Integration and Assembly of Wireless Sensor Nodes for 'Green' Sensor Networks

Jian Lu[1,2*], Hironao Okada[1,2], Toshihiro Itoh[1,2], Takeshi Harada[2], Ryutaro Maeda[1,2]

1. Research Center for Ubiquitous MEMS and Micro Engineering (UMEMSME),
National Institute of Advanced Industrial Science and Technology (AIST)
Namiki 1-2-1, Tsukuba, Ibaraki, JAPAN
2. NMEMS Technology Research Organization,
Namiki 1-2-1, Tsukuba, Ibaraki, JAPAN
*Corresponding author, E-mail: jian-lu@aist.go.jp, Phone: +81-(0)29-849-1180

Abstract

Integration and assembly of wireless sensor nodes within limited size and volume may not only reduce production cost, but also enable layout-free distribution of autonomous sensor nodes for environmental monitoring, energy consumption management, remote monitoring of civil infrastructure, as well as for biological and human health monitoring. Pursuit of the lowest power consumption of the wireless sensor nodes is another critical issue, which may allow us to use solar cell or other energy harvesting device as the power source, and then reduces the system installation difficulties and maintenance cost. Therefore, in this work, we have engaged in developing a practical applicable approach for flexible assembly of wireless sensor nodes with lowest possible size and lowest power consumption form both system block and physical interconnection points-of-view. A customized RF-transmitter IC (operation frequency: 315MHz) with universal interface to both digital sensors and analog sensors was developed to manage the power consumption of those sensors as well as to reduce the power consumption of data transmission. Buried bump interconnection technology (B2it[TM]) with internal cavities for bare die IC chips and passive components was introduced for 3D integration and assembly of MEMS sensors, customized RF-transmitter IC, crystal oscillator, resistors, capacitors and other passive components. One of the world smallest wireless sensor nodes, 3.9 mm (W) × 3.9 mm (D) × 3.5 mm (T) (except for power source and antenna), was demonstrated for humidity and temperature monitoring. Power consumption of above sensor nodes was evaluated and compared with our previous prototype, in which a commercial available RF-transmitter IC was integrated for data transmission. The developed approach in this work is believed practically valuable as platform technologies for mass production of wireless sensor nodes as well as for ubiquitous applications of those sensor nodes in wireless sensor networks.

Introduction

In past few decades, sensing technologies have been well developed and commercialized by semiconductor industries. It becomes an emerging technology and continues to grow quickly across various industry sectors in recent years due to the advances of microelectromechanical systems (MEMS) to reduce production cost and the utilization of wireless communication technologies to improve sensing performance and sensing capabilities [1][2]. Because remarkable technological breakthroughs have been achieved, wireless sensor network (WSN) [3], which consists of spatially distributed autonomous sensor nodes and cooperatively pass their data through the network to a main location, has been applied for environmental monitoring, energy consumption management, remote monitoring of civil infrastructure, as well as for biological and human health monitoring [4][5][6]. At present, WSN is more essential than ever before, since natural disasters are increasing around the world as the climate changes, costing lives and millions of civil infrastructures in damage.

'Green Sensor Network System' (GSN) project , which is supported by New Energy and Industrial Technology Development Organization (NEDO), aims at developing revolutionary new MEMS sensor nodes that incorporate wireless communication functions, stand-alone power source, and ultra-low power consumption functions for applications in low-cost, layout-free and maintenance-free 'green' sensor networks to reduce energy consumption though visibility and optimization. Details of GSN project can be found elsewhere [7]. In this project, not only revolutionary MEMS sensors, stand-alone power source, networking technologies, data storage and processing technologies, but also integration approaches for flexible and low cost assembly of above independent modules will be developed through collaboration between 20+ industry and academic members. The practical evaluation of 'green' sensor networks is also scheduled in final stage of the project to demonstrate its potential applications, such as power consumption management in smart factory and smart office, environmental monitoring and control in smart cleanroom and smart convenient store, etc. Fig. 1 shows concept and structure of the 'green' wireless sensor node and its potential applications in GSN.

In this work, we engaged in developing a practical applicable approach for flexible assembly of 'green' wireless sensor nodes with lowest possible volume and lowest power consumption form both system block and physical interconnection points-of-view. Reducing the size and volume of the wireless sensor nodes enables layout-free distribution and low cost mass production, while the pursuing of low power consumption allows us to use solar cell or other energy harvesting device as the power source to reduce the system installation difficulties and maintenance cost. The developed technologies are also expected to be used as platform for integration of many other MEMS devices, remote sensing systems, and consumer electronics products.

978-1-4799-2408-0/14 $31.00 © 2014 IEEE

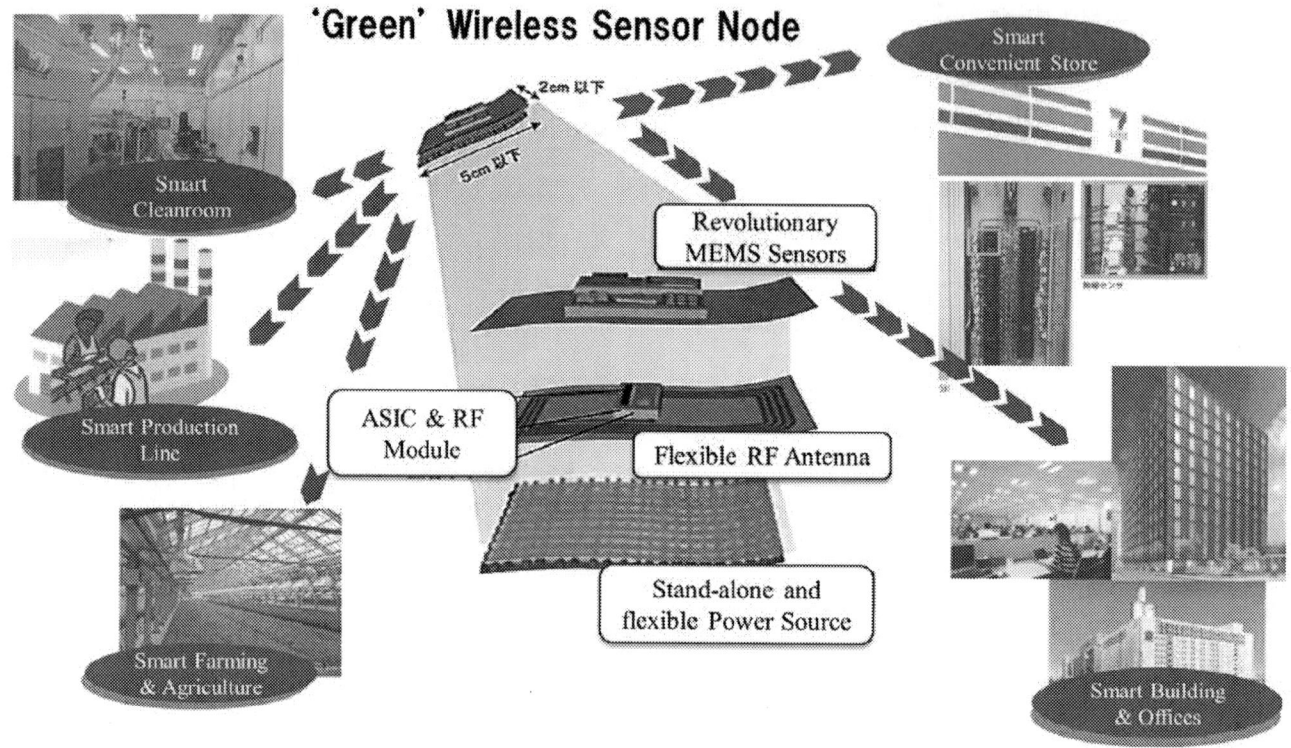

Figure 1. Concept and structure of our 'green' wireless sensor node (center figure), which includes MEMS sensors, RF-IC, crystal oscillator, resistors, capacitors, and other passive components. The sensor node will be assembled onto a flexible antenna and stand-alone power source. The left and right insets show its potential applications in 'green' sensor networks (GSN).

State-of-the-art technologies for sensor node assembly

Compared to MEMS or other electronic devices, wireless sensor node is rather complicated since it includes MEMS sensors, ASIC, RF-IC, power management unit, crystal oscillator, capacitors, resistors, and many other passive components. Therefore, most of the wireless sensor nodes in market were assembled by using traditional printed circuit board (PCB) technology, and the size of those sensor nodes is usually from a few cm to ten cm.

Wafer scale integration of MEMS with IC is an inevitable trend to industry for the pursuit of infinite new applications with smallest volume, dramatically improved performance, and reasonably high integration efficiency. However, for realizing above wireless sensor nodes, a silicon interposer wafer with through silicon via (TSV), trench capacitors, and other passive components is required in between a MEMS processed wafer and an IC processed wafer. Although tremendous progress and encouraging results have been achieved both in industries and in academies [8][9], this approach still needs our great efforts to reduce process difficulties and process cost. Besides, the yields after wafer to wafer direct bonding needs to be further improved for practical applications.

In our previous work, a 2-step approach has been developed too for size-free MEMS-IC integration, which has high process flexibility and reasonably high efficiency [10]. In that approach, bare die chips were self-aligned and temporarily bonded onto carrier wafer in first step. In second step, those chips are de-bonded from carrier wafer and then transferred onto target interposer wafer or IC processed wafer simultaneously by wafer level low temperature or room temperature permanent bonding. Further works are undergoing to improve bonding accuracy and transfer yields.

Therefore, a practical applicable approach for flexible assembly of 'green' wireless sensor nodes with lowest possible volume and lowest power consumption is essentially important to 'Green Sensor Network System' and other wireless sensor networks as well.

3D assembly by B2it™ and the process flow

In this work, buried bump interconnection technology (B2it™, Dai Nippon Printing Co., Ltd.) was introduced for the first time for the assembly of 'green' wireless sensor nodes. A 3D structure with internal cavities between PCB substrates was designed to integrate MEMS sensors, RF-IC, crystal oscillator, resistors, capacitors and other passive components within limited spaces.

978-1-4799-2408-0/14 $31.00 © 2014 IEEE

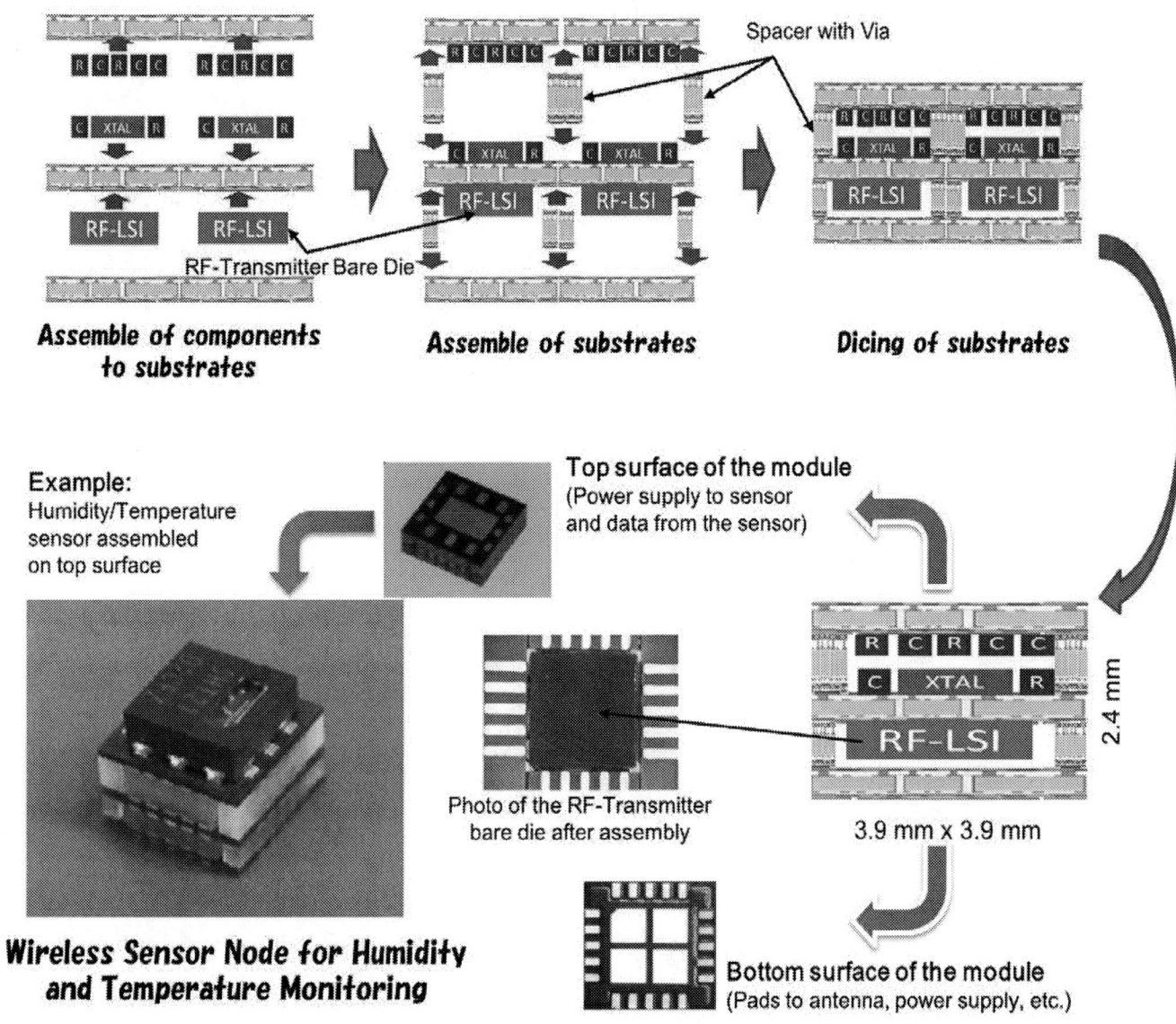

Figure 2. Schematic view of the process for 3D assembly of wireless sensor node by using buried bump interconnection technology (B2it™, Dai Nippon Printing Co., Ltd.): (1) assembly of components to each substrate; (2) assembly of substrates by using spacer with via; and (3) dicing of substrates. As an example, one of the world smallest wireless sensor nodes, 3.9 mm (W)×3.9 mm (D)×3.5 mm (T) (except for power source and antenna), was successfully obtained for humidity and temperature monitoring.

Fig. 2 shows detailed process for 3D assembly of wireless sensor node and the structure of the wireless sensor node. As can be seen from fig.2, two cavities were formed for customized RF-transmitter bare die chip and for passive components (crystal oscillator, resistor, capacitor), respectively. The interconnection between each PCB substrate was realized by using substrates with via, which was also used as the spacer for cavities. In the customized RF-transmitter LSI, both digital interface (I2C compatible) and analog interface (100X amplifier; 8 bits ADC) were included, so it is no need to integrate ASIC chip for data processing. The count of passive components was >25, and the size of

RF-transmitter bare die was 2.54 mm × 2.54 mm. The minimum module size that we achieved was 3.9 mm (W)× 3.9 mm (D)×2.4 mm (T).

Pads on top surface of the module were used as interconnection to sensor, including power supply to sensor and data acquisition from the sensor (both digital interface and analog interface). Therefore, a digital sensor or an analog sensor can be directly assembled onto top surface of the module if size of the sensor is small enough.

Figure 3. The wireless sensor node was assembled onto a flexible antenna substrate for evaluation. A humidity and temperature sensor was assembled on top surface of the module. Other sensors can be connected to the module by using sockets on flexible substrate.

Pads on bottom surface of the module were designed for inter-connection with flexible antenna substrate and power source. The interconnection to sensor was also included in bottom surface of the module, if sensor needs to be assembled separately, i.e. on flexible antenna substrate. Besides, several pads on bottom surface of the module were used as the monitor to check the operation of the customized RF-transmitter LSI as well as the communications between the customized RF-transmitter LSI and the sensors.

A commercial available digital humidity and temperature sensor (SHT21: interface: I2C; package: 3 mm×3 mm 6-pins DFN) was assembled on top surface of the module to evaluate performance of the module and the customized RF-transmitter LSI. Size of the module with sensor, as small as 3.9 mm (W) ×3.9 mm (D)×3.5 mm (T), was successfully achieved in this work. It was believed as one of the world smallest wireless sensor nodes up-to-date.

Above wireless sensor node was then assembled onto a flexible antenna substrate (2 cm×5 cm) for evaluation, as shown in Fig. 3. Besides low power consumption during data transmission, the customized RF-transmitter LSI has the capability of controlling power supply to sensors through analog switches. This unique function enables zero power consumption of the sensors when sensor node operates in standby mode. This function also enables the assembly of another sensor through bottom surface of the module by using sockets on flexible substrate. Carrier frequency of the customized RF-transmitter LSI was set to 315 MHz, which is the certificated license-free frequency range in Japan. Detailed specifications of the customized RF-transmitter LSI and its circuit block will be presented elsewhere.

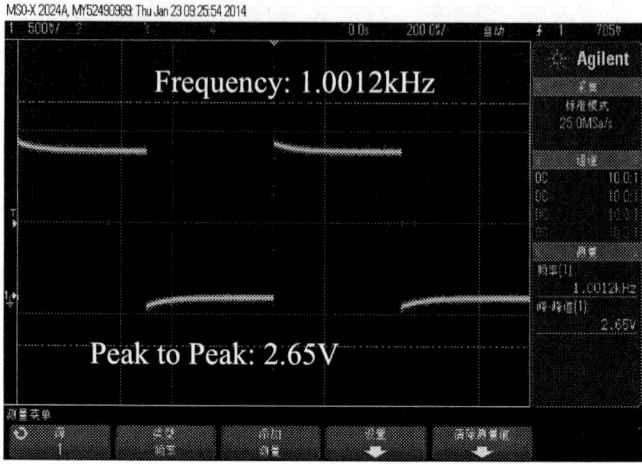

Figure 4. Measured internal clock (1.0012 kHz) of the customized RF-Transmitter LSI in wireless sensor node. The designed frequency of the internal clock was 1 kHz.

Evaluation of the sensor node and the RF-Transmitter

Fig. 4 shows measured internal clock (frequency: 1.0012 kHz) of the customized RF-transmitter LSI, which is in agreement with designed specifications of the LSI (designed frequency: 1 kHz). Moreover, it was found that the data from both analog sensor and digital sensor can be successfully acquired by customized RF-transmitter LSI, indicating that the interconnection between the humidity and temperature sensor SHT21, customized RF-transmitter LSI, passive components, power supply can be successfully established inside the module without failure.

Electrical evaluation indicated that the customized RF-transmitter LSI works under ultra-low power consumption when compared with that of our previous prototype, in which a commercial available RF-transmitter IC (Si4010) was integrated for data transmission [11]. The electrical performance of customized RF-transmitter LSI is summarized in Table 1 for comparisons. The integration of MCU8051 in Si4010 offers powerful performances, while at the cost of additional power consumption.

Table I Electric performance of the customized RF-Transmitter LSI during one cycle of data acquisition from the sensor SHT21 and data transmission.

Communication between Sensor and RF-LSI	Current	15uA
	Duration	300ms
	Pull-up Resistor	100kΩ
	Clock	1kHz
Data Transmission	Current	10mA
	Duration	4ms

Since a novel algorithm between the customized RF-transmitter LSI and the RF receiver was designed to save energy by shorten the length of the data frame during transmission, a customized RF-receiver is required to evaluate

978-1-4799-2408-0/14 $31.00 © 2014 IEEE   1860

RF performance of above sensor node. However, this work is still undergoing. RF performance of the module will be evaluated after the customized receiver is developed. The results will be presented soon in our future publications.

Conclusions and future works

This paper presented the successful integration and assembly of the world smallest 'green' MEMS sensor nodes by using buried bump interconnection technology (B2it™). Experimental results indicated that this approach is practically valuable as a platform for mass production of various wireless sensor nodes in GSN project as well as for ubiquitous applications of wireless sensor networks. Those developed technologies are also believed useful for scaling down of other electronic products.

Further optimization, system integration, and evaluation with revolutionary MEMS sensors, flexible stand-alone power source, etc. are on progress to practically extend the application of our wireless sensor nodes not only for humidity and temperature monitoring but also for unattended monitoring of wide variety of environments, infrastructures, and human health.

Acknowledgments

This work is supported by New Energy and Industrial Technology Development Organization (NEDO) through Green Sensor Network System Project.

The authors also would like to thank Dai Nippon Printing Co., Ltd. (DNP) for their technical support on assembly of the sensor node.

References

1. B.A.Warneke, K.S.J.Pister, "MEMS for distributed wireless sensor networks", in *Proc. 9th Int. Conf. on Electronics, Circuits and Systems*, vol.1, pp.291-294, Dubrovnik, Croatia, Sept.15-18, 2002

2. C.Y.Chong, S.P.Kumar, "Sensor networks: evolution, opportunities, and challenges", in *Proc. IEEE*, vol.91, no.8, pp.1247-1256, Aug. 2003

3. P.Garg, K.Saroha, R.Lochab, "Review of wireless sensor networks – architecture and applications", *IJCSMS Internationa Journal of Computer Science & Management Studies*, vol.11, no.1, pp.34-38, May 2011

4. T.Itoh, T.Masuda, K.Tsukamoto, "Development of a sensor system for animal watching to keep human health and food safety", *Synthesiology*, vol.3, no.3, pp.231-240, 2010.

5. Santoshkumar, C.Kelvin, C.Chavhan, "Development of wireless sensor node to monitor poultry farm", *Communications in Computer and Information Science*, vol.296, pp.27-32, 2013.

6. B.F.Spencer Jr, M.E.Ruiz-Sandoval, N.Kurata, "Smart sensing technology: opportunities and challenges", *Structure Control and Health Monitoring*, vol.11, no.4, pp.349-368, Oct. 2004

7. http://www.nmems.or.jp/gsnpj/index.html

8. S.Yoshimi, K.Fujimoto, M.Akazawa, H.Matsumoto, H.Mawatari, K.Suzuki, T.Itoh, R.Maeda, "Development of 300 mm TSV interposer with redistribution layers on both sides using MEMS process", in *Proc. 63rd Electronic Components and Technology Conference (ECTC2013)*, Las Vegas, USA, May 2013, pp.2168-2172.

9. Y.Morikawa, T.Murayama, Y.Nakamuta, T.Sakuishi, A.Suzuki, and K.Suu, "Total cost effective scallop free Si etching for 2.5D & 3D TSV fabrication technologies in 300mm wafer", in *Proc. 63rd Electronic Components and Technology Conference (ECTC2013)*, Las Vegas, USA, May 2013, pp.605-607.

10. J.Lu, H.Takagi, Y.Nakano, R.Maeda, "Size-free MEMS-IC high efficient integration by using carrier wafer with self-assembled monolayer (SAM) fine pattern", in *Proc. 63rd Electronic Components and Technology Conference (ECTC2013)*, Las Vegas, USA, May 2013, pp.1508-1513.

11. J.Lu, H.Okada, T.Itoh, T.Harada, R.Maeda, "Assembly of super compact wireless sensor nodes for environmental monitoring applications", in *Proc. Symp. Design, Test, Integration & Packaging of MEMS/MOEMS 2013*, Barcelona, Spain, April 16-18, 2013. pp.294-297.

New Demultiplexer Component For Optical Polymer Fiber Communication Systems

S. Höll*, M. Haupt*, U.H.P. Fischer*

Harz University of Applied Sciences

Friedrichstr. 57-59, 38855 Wernigerode, Germany

Phone: +49-(0)3943-659 825 Fax: +49-(0)3943-659 5825, Email: shoell@hs-harz.de

Abstract

Data communication over Polymer Optical Fibers (POF) is limited to only one channel for data transmission. Therefore the bandwidth is strongly restricted. By using more than one channel, it is possible to break through the limit. This technique is called Wavelength Division Multiplexing (WDM). It uses different wavelengths in the visible spectrum to transmit data parallel over one fiber. Two components are essential for this technology: A multiplexer (MUX) and a demultiplexer (DEMUX). The multiplexer collects the light of the different sources to one fiber and the demultiplexer separates the light at the end of the fiber into the different fiber output ports.

To separate the channels at the output ports, one interesting option for high multimode transmission systems is to use an optical grating. Here, the optical grating is placed on an aspheric mirror, which focuses the monochromatic parts of light into the outgoing fibers.

In order to keep the advantage of cost-effective POFs it is necessary to mass-produce the MUX and DEMUX component at reasonable prices. For polymers, injection molding is the only technology, which offers high potential to achieves this goal.

Before starting the production of the mold insert, a demonstrator of the DEMUX is fabricated by directly machining it in the PMMA material by means of diamond turning technique. Thus, the same diamond-turning technology is used for the manufacture of the mold insert. This step is done due to validate the simulation results with the produced component. Several measurements are required to validate the demonstrator for example to locate the exact position of the focus points of the separated wavelength.

The paper discusses the results of the different development steps, the measurements done with the first demonstrator and the challenges related to the injection molding process.

Introduction

Polymer Optical Fibers (POF) are used in various fields of applications. The core material consists of PMMA (polymethylmethacrylate), while the cover is made of fluorinated PMMA (fig.1). The whole fiber has a diameter of 1 mm. POFs are used for optical data transmission based on the same principle as glass fiber. As a communication medium they offer a couple of advantages related to other data communication systems such as copper cables, glass fibers and wireless systems, and have great potential to replace them in different applications.

Namely, in comparison with glass fibres (GOF), POFs have the advantage of easy and economical processing and are more flexible for optical connections [1]. However, one advantage of using glass fibres is their low attenuation, which is below 0.2 dB/km in the infrared range. The larger core diameter of POFs leads to higher mode dispersion and thus to higher attenuation across the electromagnetic spectrum. This increased attenuation leaves only one remaining transmission window, namely the visible spectrum of light (400 – 700 nm). Hence, POFs are best suited for the use in short distance data communication.

Here, POFs can outperform the current standard of copper cable as communication medium. On the one hand, they feature lower weight and space. On the other hand, POFs are not susceptible to electromagnetic interference. [2-3].

Also in comparison to wireless communication POFs offer two solutions to problems in the wireless domain. Firstly, like any radio frequency transmission, wireless networking signals are subject to a wide variety of interference, as well as complex propagation effects, which are beyond the control of the network administrator. Secondly, another main disadvantage of wireless communication is the susceptibility to unwanted access to the transmitted data by third parties. Hence, it is not usable for secure transmission of sensitive information.

For these reasons, POFs are already used in various application domains, for example in the automotive sector and for in-house communication.

In the area of automotive multimedia applications in the passenger compartment POFs have successfully replaced copper as transmission medium, mainly because of the huge weight reduction. It was first introduced by BMW in the 7 Series in 2001. Since then not only high-class cars but also models targeted at the mass-market were equipped with POFs [4].

As of now, one limitation to the use of POFs in cars remains. The POF has a maximum operating temperature of 85 °C, which makes it impossible to use the fibres near the engine compartment [4]. However, this problem might be solved in the foreseeable future, which would allow the use of POFs as sensors for various in-car pressures or forces.

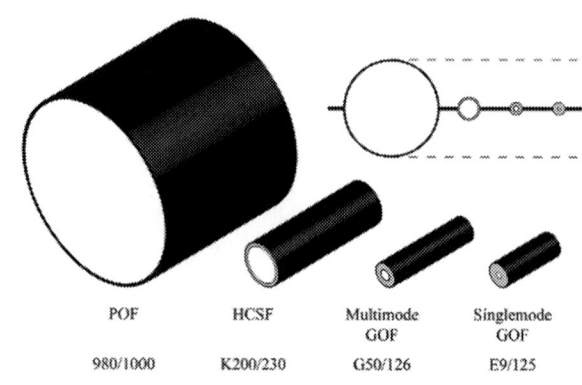

POF	HCSF	Multimode GOF	Singlemode GOF
980/1000	K200/230	G50/126	E9/125

978-1-4799-2408-0/14 $31.00 © 2014 IEEE   1862   **2014 Electronic Components & Technology Conference**

Figure 1. POF overview related to other fibers.

Another sector where POFs replace the traditional communication medium is the in-house communication [5]. Various home appliances can be connected and controlled through the use of POFs (see fig. 2). But the possibilities of application are not confined to the inside of the house itself. Today, copper cables are the most significant bottleneck for high-speed internet. In the future, POF will most likely replace copper cables for the so-called last mile between the last distribution box of the telecommunication company and the end-consumer.

The so-called "Triple Play" (a combination of VoIP, IPTV and classical Internet) is being introduced to the market with force, therefore high-speed connections are essential. It is highly expensive to realize any VDSL system using copper components, thus the future will be FTTH (Fibre to the Home).

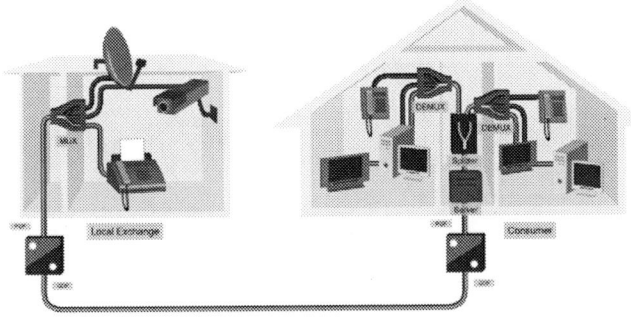

Figure 2. In-house communication with POF

WDM over POF

Currently, the transmission over standard POF is limited to one wavelength [1-2]. Making more than one wavelength per fiber available for data transmission would significantly increase the bandwidth of POFs. This can be achieved through wavelength division multiplexing (WDM). In glass fibre technology, the use of the WDM in the infrared range at about 1550 nm has long been established [6, 7]. This principle of WDM is now transferred to the visible spectrum for POF communication. Figure 3 illustrates the data transmission of WDM over POF.

WDM has widely expanded the overall transmission bit rate in glass fiber long-range systems. The reason is the easily expandable system approach. By adding a new source with a different transmission wavelength in combination with a MUX/DEMUX-element, it is possible to expand the usable transmission rate directly by this source.

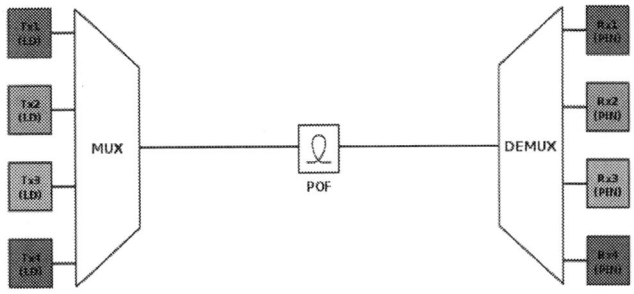

Figure 3. Schematic of the WDM over POF

When implementing WDM technology in POF-systems, the same key elements are needed: the multiplexer and the demultiplexer. Due to the different target spectrum it is necessary to completely redesign these components. The multiplexer combines rays with different wavelengths to one fiber. At the end of the fiber the demultiplexer separates the light back into the different wavelengths.

This separation can be realized via different techniques. One possibility is to apply an optical phased array to change the phases of every different channel and, therefore, divide the light in different channels [4]. Another possible technique is interference filters, which are well known in the infrared range but also available for the visible spectrum [1, 8].

However, applying these approaches to the visible spectrum of light results in costly solutions. This would hinder the use of POFs in mass-market applications. Therefore, this paper investigates a third alternative, namely demultiplexing the light ray with a diffractive grating. This grating can eventually be produced using injection molding, which greatly reduces production costs. In the following, this paper will address various challenges in the design of this production process.

Components for WDM

For WDM two essential components are needed: a multiplexer and a demultiplexer. To create a functional demultiplexer for POF, several preconditions must be fulfilled. Firstly, a mirror must focus the divergent light beam coming form the POF. The shape of this mirror cannot be spherical because of the appearing spherical aberration. Instead, a toric shape of the mirror prevents spherical aberration.

The second function is the separation of the different transmitted wavelengths, which can be achieved by a diffraction grating. This principle is illustrated in figure 4. The light is split into different orders of diffraction. The first order is the important one to regain all information. There, the outgoing POFs can be placed.

978-1-4799-2408-0/14 $31.00 © 2014 IEEE 1863

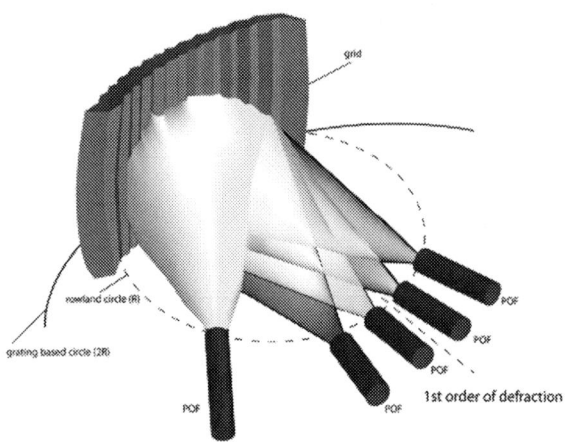

Figure 4. Schematic of a Rowland spectrometer

Both functions are combined, so that the grid is directly located on the toric mirror. Hence, the light is not afflicted by any aberrations or attenuations of a focusing lens or other diffractive or refractive elements, which would be necessary for any other setup [9, 10].

In case of a DEMUX for glass fiber, the Rowland grating can be designed planar. But POF has a large numerical aperture, which leads to a higher opening angle of the emitted light. A planar setup would result in high losses, therefore a three dimensional design is necessary.

Design of the first Demonstrator

Now these elements must be manufactured in injection molding technology. The reason for this technology is the ability for cost-effective mass-production of plastic components. Today, it is the standard manufacturing method for high-volume plastic parts. Through developments in the past injection molding is used to produce high-quality optical components. One of these developments is the injection compression molding, which has the ability to produce dimensional stable and relatively stress free moldings. Especially the reduction of the internal mechanical stress makes this technique ideal for optical molded lens and media applications [11]. With a cost-effective producing the components for WDM over POF are available for a broader market. Another reduction of the production cost is the elimination of expensive adjustment of the optical parts. That's why the different functions are unified in one molded component.

From a technological point of view, the DEMUX is more difficult to realize and will thus be discussed in more detail.

In the first step a demonstrator will be produced. The demonstrator is used to verify the concept of the demultiplexer and to compare the simulation results with the real setup. Therefore a special design was created. Fig. 5 illustrates the new design, which includes a hemisphere at the output of the DEMUX. The radius of the hemisphere is equal to the radius of the Rowland circle, which can be seen in the cross-section view in figure 6. The outgoing light is focused on this radius. Therefore the light is coupled into the center of the hemisphere and the separated wavelength can be detected on the right surface of the hemisphere as shown in figure 6. A

scan of the surface will be done to locate the positions of the outgoing separated light for each wavelength.

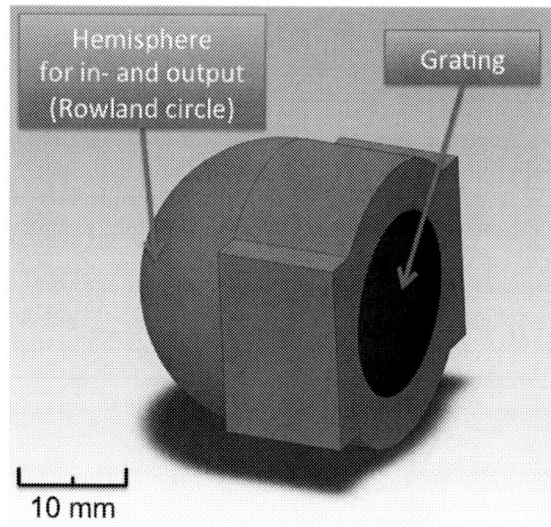

Figure 5. 3D-model of the DEMUX demonstrator.

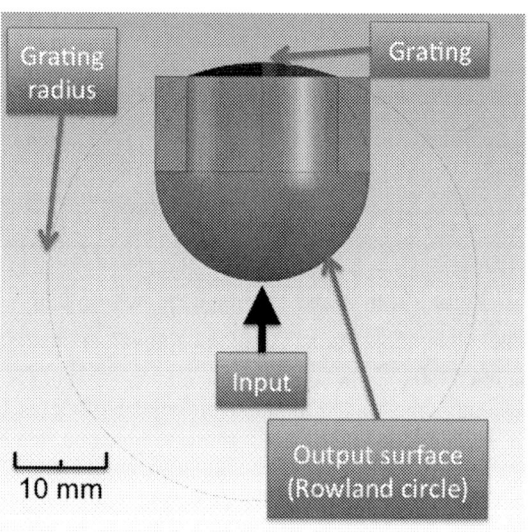

Figure 6. Cross-section view of the DEMUX demonstrator

Materials and Methods

In front of the manufacturing of the DEMUX some preliminary investigation have taken place. The goal was to find the best suitable material for the demultiplexer. Therefore the processability of the material as well as the optical parameters has to be considered.

The processability is tested with a thick-walled mold tool, which has the same shape like the DEMUX except of the grating. The tests were done with the Babyplast 6-10P because of its ability to injection mold small-volume parts. . Table 1 lists all materials for the investigation and compares their respective Melt Volume Rate (MVR) and light transmission (according to the manufacturer's specification). The test is also used to find the optimized injection molding process parameters for the material.

978-1-4799-2408-0/14 $31.00 © 2014 IEEE   1864

Table 1 Injection molding materials for
MUX/DEMUX-element

Name	Type	MVR [cm³/10min]	Transmission [%]
Plexiglas 6N	PMMA	12	92
Plexiglas POQ62	PMMA	21	92
Topas 5013L-10	COC	48	91,4
Topas 6013M-07	COC	14	91
ZEONEX F52R	COP	22	92
ZEONEX 350R	COP	26	92
Makrolon LED2245	PC	35	90

Within the optimization phase some defects occurred relating to the injection molding of the test structures. These are the so-called diesel effect, burrs, and voids. The diesel effect occurs when the molten polymer encapsulates air and the high pressure during the injection phase causes high temperatures. The result is that the polymer decomposes and grey clouds appear in the polymer body (Fig. 7). These grey clouds strongly reduce the transmission of light through the molded part. Among other things an increase of the backpressure leads to clear parts without grey clouds.

Another defect are burrs on the injection-molded parts through over molding. This can be seen in Fig. 8. Burrs appear during the injection molding process and are located at the parting line and the spacing between the ejector bolt and the ejector-guiding hole. An additional mechanical processing is necessary to remove the burrs. The adjustment of the process parameters clamping force, melt temperature in the cylinder, and injection speed was needed. After the optimization of these three parameters the burrs do not occur anymore.

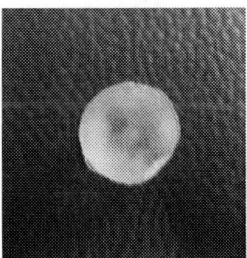 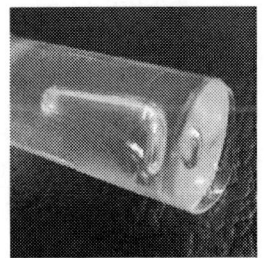

Figure 7. Diesel effect   **Figure 8.** Voids and over molding

The last defect, namely the voids, is seen in Fig. 8. The optimization of the injection speed and the adjustment of the backpressure reduce the dimensions of the voids inside the polymer body (Fig. 9).

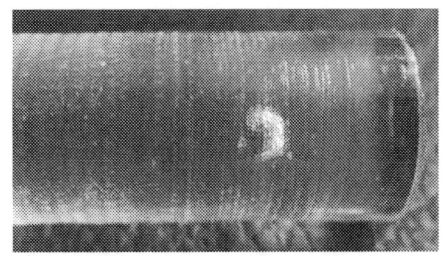

Figure 9. After the first step of process optimization

Additionally, the optical quality of the polymer materials must be investigated. Therefore a molding tool was designed to injection mold test-plates. The test-plates are shown in figure 10 and have a thickness of 2 mm. The molding tool is used to produce samples of every material listed in Table 1. The DIN EN ISO 13468-2 standard describes the measurement of the optical transmission of polymer materials. Therefore the test-plates are designed to conform to this standard.

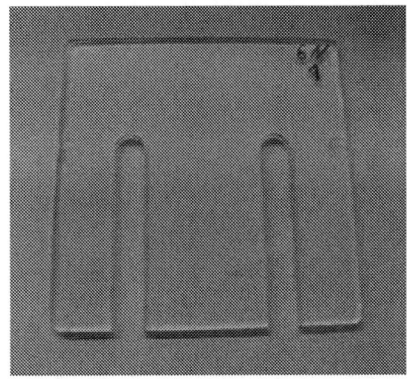

Figure 10. Injection molded test plate

With all test-plates transmission measurements were done. The results are shown for 405 nm in figure 11. It can be seen that both ZEONEX types and PMMA POQ62 show the highest value for light transmission. PMMA POQ62 is a polymer type with high purity of the polymer granulates. The measurement is done at the wavelength of 405 nm because it is one of the used wavelengths for the WDM system.

The transmission measurements of our own injection-molded test-plates were compared to samples directly from the manufacturer. Additional, measurements were done by our project partners at the University of Braunschweig. The result for one example is depicted in figure 12. The transmission for the type TOPAS 5013L-10 across the whole visible spectrum are shown. For example, the comparison of 405 nm wavelength results in an average transmission of 89,50 % for the manufacturer plate and 89,68 % for our samples.

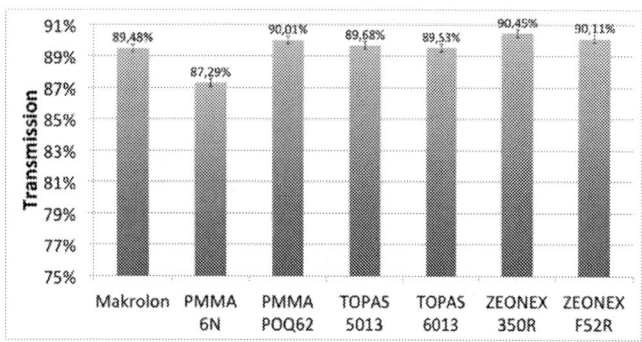

Figure 11. Transmission of different material at 405 nm

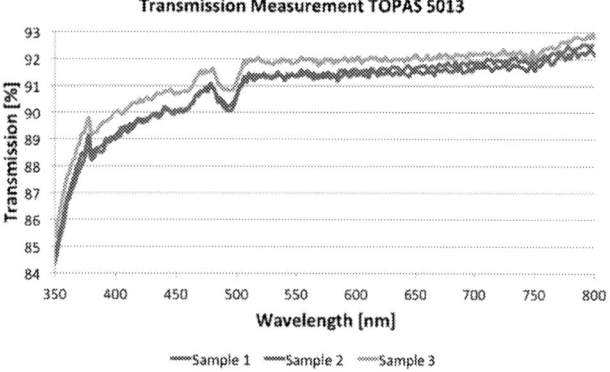

Figure 12. Transmission of TOPAS 5013 over the visible spectrum

Manufacturing of the Demonstrator

By using the injection molding process, the manufacturing of the mold insert is the most important factor. Due to the three-dimensional toric structure of the grating planar manufacturing methods like lithography, especially LIGA (German acronym for **L**ithographie, **G**alvanoformung, **A**bformung - Lithography, Electroplating, and Molding) cannot be used. LIGA is used to manufacture planar spectrometers based on the glass fiber technology [12]. But in our case, the three-dimensional grating needs another machining method. Especially the microstructure of the grating and the exact curve shape of the toric surface require high precision. The microstructure has the shape of a saw tooth with a pitch between the teeth of 2.5 µm. Figure 13 shows an enlarged 3D-Model of the grating. After investigate several machining methods only the diamond turning meets the high demands of the micro structured grating.

Diamond turning is a special machining process, which uses a single-crystal diamond-cutting tool. It is possible to achieve a surface of optical grade also to the edge of the optical component. It offers the following advantages:

- True three-dimensional contour generation
- Accuracy of one part in 10^6 with absolute accuracy of 1 part in 10^8 on a single axis for ideal conditions
- Surface finish of 5 nm R_a for a range of materials and as good as 1 nm R_z

- Ability to generate surfaces with variable aspect ratios, and
- Feature sizes that exceed the limits of optical microscopy. [13, 14]

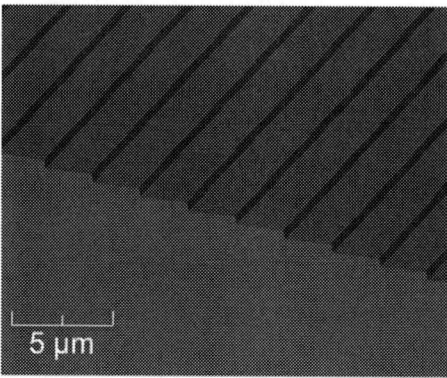

Figure 13. Grating of the demultiplexer

Before the first mold tool is manufactured a demonstrator is produced. This will be done by directly machining of a PMMA preform instead of the negative mold tool. It is used to verify the concept and to make sure the manufacturing method produce the structures needed for the functionality of the DEMUX. All manufacturing steps regarding to the diamond turning process are done by an external partner. The first part is depicted in figure 14.

Figure 14. First sample of the DEMUX demonstrator

To analyze the surface of the grating a metallization process was applied. The surface was sputtered with a thin aluminum layer. Now it is possible to measure the shape of the surface with a white-light interferometer and to examine the grating structure under the raster electron microscope (REM). The metallized surface of the grating is shown in figure 15. It can be seen that the structure has a dull and mat surface on the left side instead of the shiny rest of the surface. That is a first indicator that the surface roughness is higher in this part and does not meet the requirements. The first impression was confirmed by the analysis under the REM. Figure 16 shows the rough structure of the surface. The analysis shows that the quality of the grating structures depends on the position on the surface. On one half of the surface the milled structures are exact (figure 17). However, on the other half the quality degrade from the middle to the edge. Due to the degradation of the grating quality is directed parallel to the lines an abrasion of the diamond tip can be excluded.

Figure 15. Metallized surface of the grating

The source of the deterioration of the grating quality is the change in strain when the milling tool passes the highest point in the center of the surface. It changes the way the force is applied to the surface from a pushing to a pulling movement. That leads to a rough structure on the other half of the surface.

From the measurement of the structure size in figure 17 can be determine a width of 2.55 µm, which is within the tolerances of the reference of 2.5 µm.

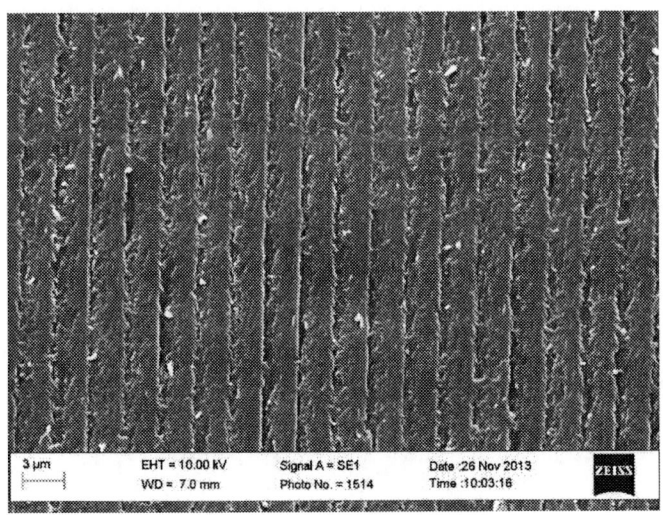

Figure 16. Rough structure of the grating

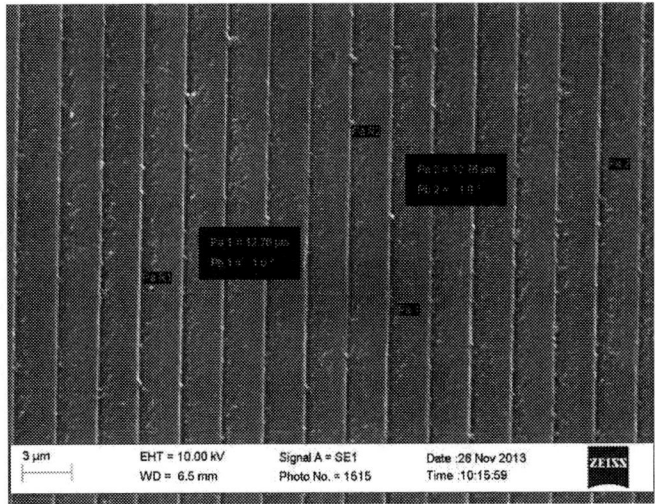

Figure 17. High quality structures of the grating

Besides of structure quality, the dimensions of the surface are also important for the functionality of the DEMUX and have to be considered. The shape of the radius of the toric surface is designed to focus the separated wavelength to the Rowland circle. Therefore it was analyzed by using a white light interferometer (FRT MicroProf). In figure 18 a cross section of the toric surface is depicted. The black graph shows the measurement of the surface and the red graph is the ideal shape of the curve. The measurement exposes that the dimensions of the surface meet the tolerances of the DEMUX, except of the diameter in x-axis, which is slightly outside the tolerance (table 2).

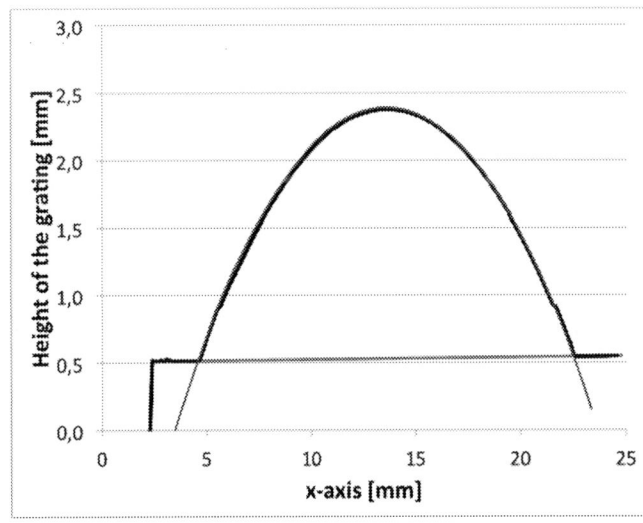

Figure 18. Measured contour of the toric grating surface (black graph are the measured points; red graph is the ideal shape)

978-1-4799-2408-0/14 $31.00 © 2014 IEEE   1867

Table 2 Measurement results of the DEMUX dimensions

Dimension	Measurement [mm]	Reference [mm]
Diameter x-axis	15.869	> 16.000 ± 0,1
Diameter y-axis	15.887	> 15.170 ± 0,1
Height of grating	1.862	1.872 ± 0,05

To solve the defects of the manufacturing several parameters have to be optimize. This is done in iterations in close cooperation with the manufacturer. For example the adjustment of the applied force on the surface by the diamond tip. The next part with optimized parameters is now in production and will be analyzed in the same way to verify the adjustments of the parameters.

Optical measurements

In order to measure the position of the focal points of the different separated wavelength a special measurement setup was build. It uses a parallel-kinematics precision alignment system to align a fiber at the surface of the hemisphere. A fixed input-fiber is used to couple white light into the DEMUX as shown in figure 19. In this figure can be seen that the separated wavelength are focused at a ring on the hemisphere. This ring is scanned by the fiber on the alignment system. The light of the scanning fiber is analyzed by a spectrometer.

Figure 19. Measurement setup with white light coupled into the DEMUX

Out of the spectra along the ring the location of the peaks of the wavelengths are determined. For the first part the whole measurement was done and compared to the simulation results. Therefore four different wavelengths are used: 405 nm, 450 nm, 520 nm and 650 nm. The simulation results of the focal points positions are shown in figure 20, where the positions are related to the central point on the grating sphere expressed by the angle between the input axis and the output axis.

The positions of the wavelengths measured by the setup are depicted in figure 21. In comparison to the simulation a shift of the positions can be recognized. Nevertheless, the separation of the wavelengths was measured and confirms the functionality of the demultiplexer.

The derivations to the simulation could be caused through following reasons:

- Derivation of the blaze angle of the saw tooth grating
- Inhomogeneous structure of the grating
- Manufacturing tolerances

These highly depend on the precision of the manufacturing process. As mentioned before the production of such complex structures on a toric surface is a high challenge. Therefore the process parameters must be improved and optimized to meet the optical requirements of the demultiplexer component.

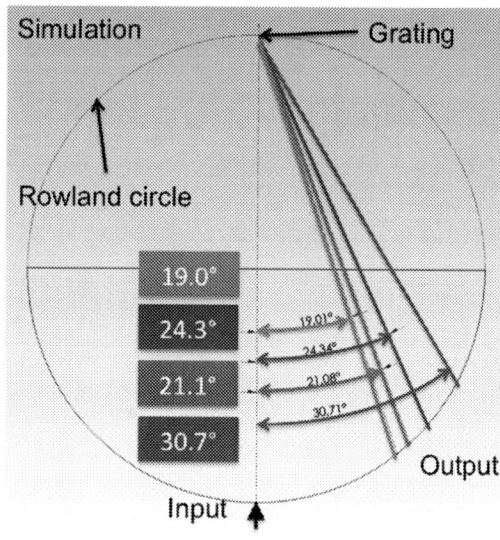

Figure 20. Simulation results of the focal points for different wavelengths (405 nm violet ray; 450 nm blue ray; 520 nm green ray; 650 nm red ray)

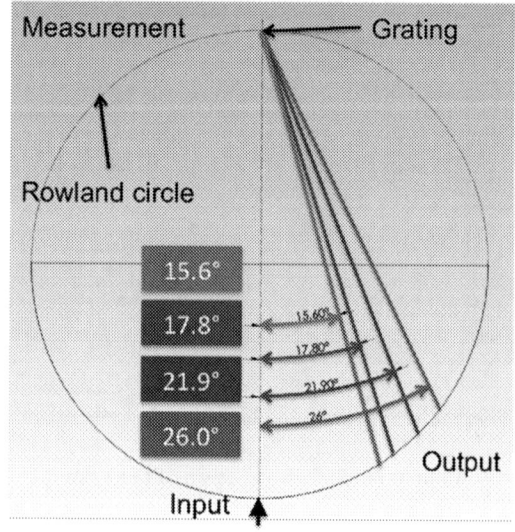

Figure 21. Measurement results of the focal points for different wavelengths (405 nm violet ray; 450 nm blue ray; 520 nm green ray; 650 nm red ray)

Conclusions

In summary, injection molding the DEMUX element for POFs poses several challenges, especially the microstructure of the grating on a three-dimensional surface. It is shown that the current manufacturing process is able to produce the structure size and the exact radius needed for the DEMUX. The high challenge of the grating lead to some defects regarding to the milling process, which has to be eliminated. This is done by the optimization of the process parameter. The next parts will be produced with the optimized parameters and then analyzed.

The first optical measurements to prove the principle were done. These confirm the separation of the wavelength in the visible spectrum. In comparison with the simulation a shift of the position of different wavelengths can be determined. This can be traced back to the inhomogeneous grating structure, which was detected during the analysis.

This demultiplexer is not limited to the application of WDM over POF communication. It can be applied as a spectrometer for several analysis methods e.g. in medical devices. This expands the range of application extremely.

The next steps are the optimization of the diamond turning process parameters to produce a component, which meet the requirements. The optical measurement of the DEMUX examines the location of the focal points of the different wavelength. This is used to design the final component with fixed input and output fibers. Then a mold insert can be manufactured using the diamond turning method and finally first parts can be injection molded.

Acknowledgments

The project on which this report is based was funded by the German Ministry of Education and Research (BMBF) under grant number 16V0009 (HS Harz) /16V0010 (TU BS).

All injection molded parts are done with the support of the Institute of Micro and Sensor Systems at the Otto-von-Guericke University Magdeburg and Prof. Bertram Schmidt

References

1. W. Daum, J. Krauser, P. E. Zamzow, O. Ziemann, *POF Handbook: Optical Short Range Transmission Systems*, Springer-Verlag, 2008
2. H. S. Nalwa (Ed.), *Polymer Optical Fibres*, American Scientific Publishers, California 2004
3. Club des Fibres Optiques Plastiques (CFOP) France, *Plastic Optical Fibres – Practical Applications*, J. Marcou, John Wiley & Sons, Masson, 1997
4. J. Brandrup, E. H. Immergut, E. A. Grulke, *Polymer Handbook*, 4th Edition, Wiley-Interscience, 1999
5. R. T. Chen and G. F. Lipscomb, Eds, "WDM and Photonic Switching Devices for Network Applications", in *Proceedings of SPIE*, vol. 3949, 2000
6. Colachino, J., "Mux/DeMux Optical Specifications and Measurements", *Lightchip Inc. white paper*, Lightreading, 2001
7. A. H. Gnauck, A. R. Chraplyvy, R. W. Tkach, J. L. Zyskind, J. W. Sulhoff, A.J. Lucero, et. al., "One terabit/s transmission experiment", *Proceedings OFC'96*, 1996
8. U. H. P. Fischer, *Optoelectronic Packaging*, VDE-Verlag, 2002
9. U. H. P. Fischer, M. Haupt, "WDM over POF: the inexpensive way to breakthrough the limitation of bandwidth of standard POF communication", *SPIE Symposium on Integrated Optoelectronic Devices*, Photonics West San Jose, 2007
10. U. H. P. Fischer, M.Haupt, "Integrated WDM System for POF Communication with Low Cost Injection Moulded Key Components", *Access Networks and In-house Communications*, 2010
11. M. Stricker, G. Pillwein, J. Giessauf, "Focus on Precision - Injection Molding Optical Components" in *Kunststoffe international*, vol. 4, pp. 15-19, 2009
12. J. P. Ferguson, S. Schoenfelder, "Micromoulded spectrometers produced by the Liga Process," Searching for Information: Artificial Intelligence and Information Retrieval Approaches, IEE Two-day Seminar (Ref. No. 1999/199), pp.11/1-11/4, 1999
13. M. A. Davies ; C. J. Evans ; R. R. Vohra ; B. C. Bergner and S. R. Patterson, "Application of precision diamond machining to the manufacture of microphotonics components", *Proc. SPIE 5183, Lithographic and Micromachining Techniques for Optical Component Fabrication II*, 94, November 2003
14. D. Dornfeld, S. Min, Y. Takeuchi,"Recent Advances in Mechanical Micromachining", *CIRP Annals - Manufacturing Technology*, vol. 55, Issue 2, 2006, p. 745-768

Nanofiller Based Spin-On Materials for Negligible Reflection of Silicon Photonic External Coupling

Yoichi Taira[1], Ryuma Mizusawa[2], Rie Matsumoto[2], Kuniaki Sueoka[1], and Hidetoshi Numata[1]

[1] IBM Research – Tokyo
NANOBIC 7-7 Shinkawasaki, Saiwai-ku, Kawasaki 212-0032 JAPAN
e-mail: taira@jp.ibm.com
[2] Tokyo Ohka Kogyo Co. Ltd.
1590 Tabata, Samukawa-machi, Koza-gun, Kanagawa 253-0114 JAPAN

Abstract

Optical out coupling is one of the key research items of silicon photonic packaging. It is necessary to realize robust, high-bandwidth, low-power optical data communication system in an economical way. Because material interfaces are always involved for the coupling, the reflection loss at the interface often occupies a major part of loss. Here we report on an ideal anti-reflection coating method for use with silicon nanophotonics devices. Because the refractive index of silicon is large at about 3.5, the optical reflection at the silicon waveguide boundary is large. We describe a set of nanoparticle-based spin-on materials having indices of 1.8 and 1.2. The former material is suitable for a single layer anti-reflection coating on a silicon surface and realizes a first order thin reflection suppression which allows large bandwidth and large angle tolerance. The latter material is for coating of the fiber or polymer interface. We studied the feasibility of uniform coating over microlens structures. High index material is also photo patternable, which is desirable for silicon chip integration.

Introduction

Significant progress has been made in silicon nanophotonics technology. New opportunities are being brought to datacom, telecom, and high performance computing by this technology [1-10]. However, the large scale commercialization of silicon photonics still requires more cost effective optical inputs and output methods. Although there are several potential methods to realize these couplings, the methods have to satisfy the condition of spectral wide band, optical low loss, mechanical reliability, and cost efficiency.

There are several approaches that enable the optical coupling between optical waveguides on a silicon photonics chip and an external single mode (SM) optical cable. Current approaches to interfacing silicon nanophotonic waveguides to standard single-mode fibers often use specialized equipment and show significant cost as well as scalability questions [11-15].

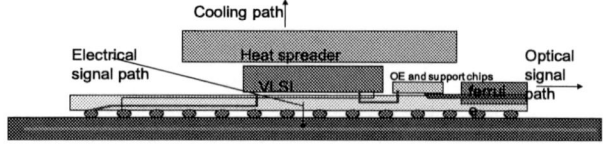

Fig. 1 Silicon photonics based VLSI packaging

Figure 1 shows a concept of a silicon photonics based VLSI, or CPU, module, where the electrical signals are connected at the bottom surface of the carrier and the optical signals are connected at the side. In this module there two levels of electric packaging – chip to carrier, and carrier to PCB. Similar hierarchy is necessary for optical packaging – photonic chip to intermediate waveguides, and waveguides to external fibers.

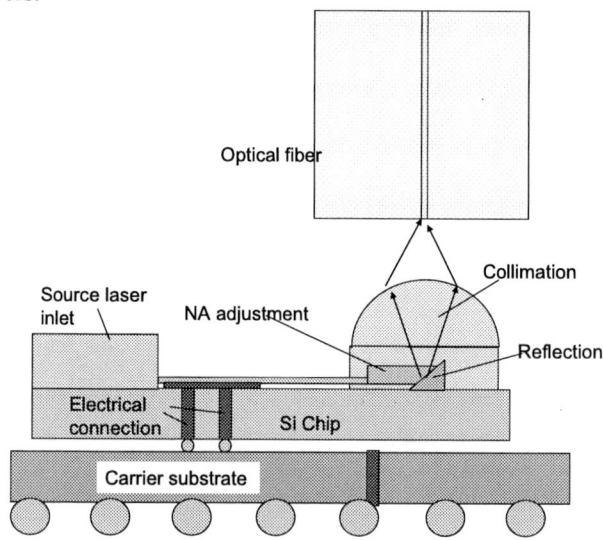

Fig. 2 An example of optical transceivers showing the necessary functions of external interfacing

Since the refractive index of silicon at the working wavelength is about 3.5, there is a large difference of the numerical aperture (NA) of the optical mode between the silicon waveguide on a photonic chip and that of the optical fiber. Therefore the optical coupling also requires some form of NA conversion, which is equivalent to the spot size conversion (SSC).

Figure 2 shows a possible configuration of a silicon photonics based optical transceiver. In this configuration, NA is adjusted by using a spot size convertor on a silicon photonic chip. Some amount of reflection occurs at each interface along with the optical path,

The NA conversion can be achieved either at the interface of the two optical media, silicon waveguides and glass fibers, or on a silicon chip. For the on-chip NA conversion, optical mode transition between two difference waveguide materials is conducted, such as by using an adiabatic mode conversion from a tapered silicon waveguide to silicon nitride waveguide on silicon chip. Then the light can be butt coupled to an optical fiber with a good coupling efficiency.

For the NA conversion at the interface, a grating or lens system can be used. The grating method always changes the direction of the optical path, which makes placement of optical fiber interface rather easier together with the process simplicity. The disadvantage of the silicon grating approach, however, is that coupling optical path has two symmetric directions unless one of them is restricted. To enable a single direction coupling, some additional structure is required such as a metal reflector. The grating approach works by the optical interference, which is wavelength dependent or sensitive. Therefore the direction of the re-directed optical path is a function of the wavelength, which limits the allowable bandwidth of the grating coupler. This may give the limitation of the applicability of the grating coupling when wavelength domain multiplexing is incorporated in the future. The interface to the next optical medium is often either silicon or medium index silicon oxide/nitride. In either case, reflection has to be suppressed to avid loss and undesirable back reflection.

The lens coupling is more straightforward, the optical beam size is adjusted by lens imaging. By using a lens or a set of lens, one can easily manipulate the numerical aperture or beam spot size as shown in Fig. 3.

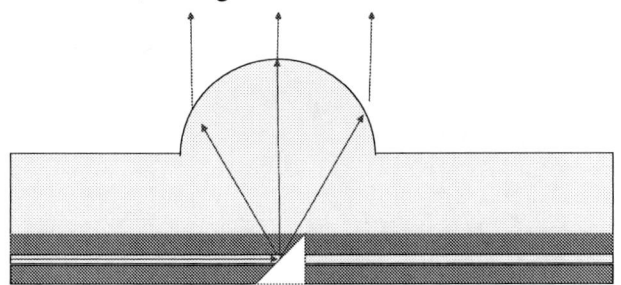

Fig. 3 A possible structure allowing lens coupling of silicon waveguide

However, one may have to consider the special situation related to the high refractive index of silicon about 3.5. When the light exits from the silicon surface, there is a strong Fresnel reflection. For example, the reflectivity of normal incident light at the silicon air interface is (3.5-1)2/(3.5+1)2=0.3 or 30% of the light power is reflected back at the surface. Light inside silicon also undergoes the total internal reflection at the air-silicon interface when the incident angle is larger than 16 degrees. Therefore one has to pay a special attention to every silicon interface, where light goes through, so that the power is not lost unexpectedly.

Although the same caution has to be paid at the polymer or glass interface, the reflectivity due to the Fresnel reflection of a regular polymer having a refractive index of 1.5 is 4%. Even though the each reflection is not large, the accumulated amount reflection is non-negligible when multiple polymer/glass interfaces are involved as in the most of the cases. To reduce the reflection, one- or few-layer thin-film anti-reflection coatings is used. A quarter wavelength thick single thin layer coating of an appropriate refractive index material is

often sufficient. The refractive index for polymer is around 1.2-1.3 and 1.7-1.8 for silicon.

Depending on the actual interface structure, the process option need to be considered. Although spin coating option often allows a simpler economical process, vacuum evaporation is used because of the material limitation. To coat on a photonic device, patterning is often necessary. Patterning of the evaporated layer requires several process steps including photolithography or a masked deposition is used. Photo definable spin-on process is suitable for this purpose.

Because the interface surface is often non-planar as in the case of a lens, it is also necessary that such surfaces can be coated with a controlled thickness.

Packaging Levels

There are two levels of the light coupling of silicon photonics devices. The first level deals with a direct coupling of a silicon waveguide to and an intermediate medium, whereas the second level handles optical connection of the intermediate medium and external fiber [16]. The first level also needs to realize adjustment of NA difference between the silicon waveguides and the intermediate medium. The second level supports a repeatable connection to external fibers. Both levels also need to satisfy low loss and low reflection to avoid performance degradation due to back reflection. We recently proposed a use of compliant polymer waveguides between the first and second level couplings. This allows mechanical isolation when the external fiber is connected. When the polymer waveguides are used, there are several possible options of first level coupling, which include adiabatic, grating, butt, and lens couplings. In these coupling methods there is a silicon interface normal to the light propagation direction. There is often a non-negligible amount of light and reflection at this interface, which causes problematic back reflection and signal loss. This reflection should be minimized for the proper operation of the optical transmitter and receiver devices.

Spin-On Materials

We evaluated a set of "hybrid" spin-on materials having refractive indices of about 1.7 and 1.2. The materials contains nanometer sized fillers in a curable polymer. Although the contents of the nanofillers are not disclosed, it is generally known that inclusion of high index metal oxide such as titanium oxide gives a high index and inclusion of nanometer sized silica sphere gives a low index beyond what the ordinary bulk material can achieve.

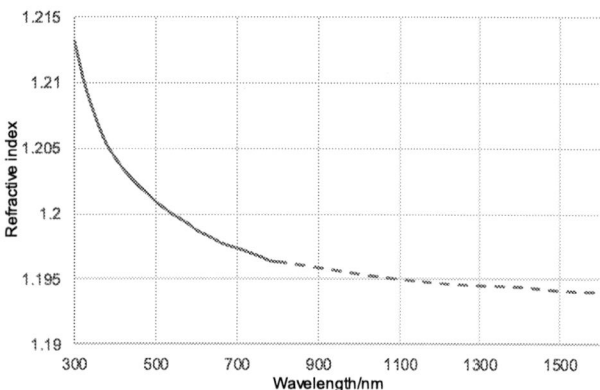

Fig. 4 Refractive index of the low index material. Dotted line represents simulation result assuming the normal dispersion.

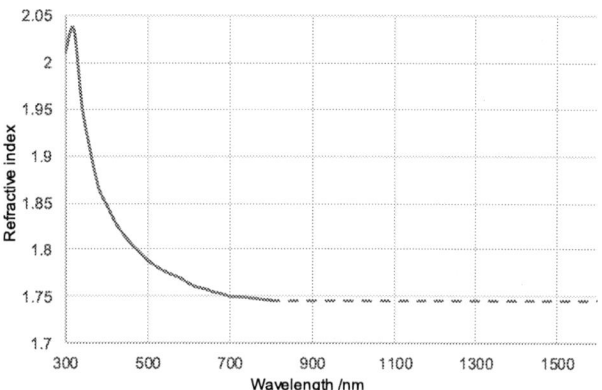

Fig. 5 Refractive index of the high-index spin-on material. Dotted line represents simulation result assuming the normal dispersion.

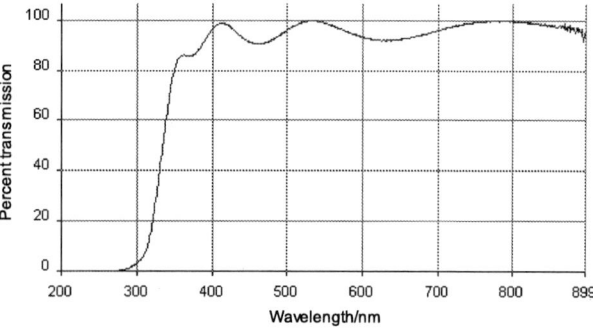

Fig. 6 Transmission of 443 nm thick high index film on a sheet of glass. The transmission is normalized relative to a reference bare glass.

The high index material is UV curable and photo-patternable. The coating process flow is 1) spin coat on a substrate 2) prebake, 3) UV exposure, and 4) post bake. The low index material under study is high temperature cure type and is not photo-patternable. The step 3) above can be skipped for this material.

Both of them are transparent in IR region up to 1.7 μm. without any significant absorption peak. Figure 4 and 5 show measured index of spin-on materials on glass.

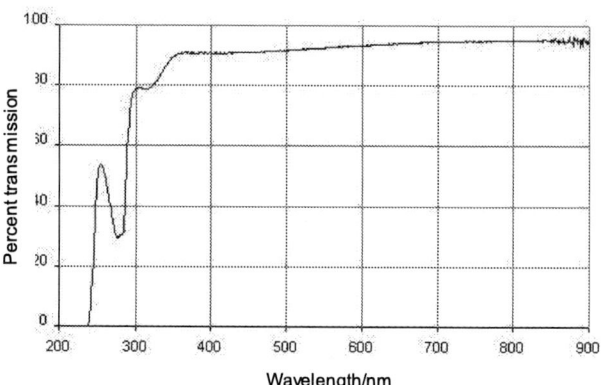

Fig. 7 Transmission of 200 nm thick low index film on a sheet of glass. Absolute transmission is shown.

Fig. 8 SEM image of the embossed microlens array. We cut the sample to observe the cross section of the lens hemisphere.

Figures 6 and 7 show transmission spectra of spin-on samples on a planar sheet of glass. In Fig. 6 maxima and minima are due to interference. Note that the transmission maximum occurs when the thickness equals to an even integer multiple of a quarter wave coating thickness for this unusual high index film coating case. Transmission maxima at 790, 530, 410 nm represent 2nd, 3rd, and 4th order interference peaks. The transmission maximum around 850 nm of the low index coating shown in Fig. 7 represents the 1st order interference peak.

Preliminary thermal resistance test was also carried out for the both films. After raising the temperature to 260 C for 5 minutes, about 1% or less thickness reduction was observed in the both cases. We also observed no refractive index change.

Coating on a curved surface

For the optical coupling, one may have to coat the material on a curved or convex lens surface. Although the thickness uniformity on a planar glass sheet was confirmed by interference measurement at various spots, the thickness uniformity is rather hard to measure. The test method was to coat the material on microlens array structure. The base structure we used was 250 mm pitch 100 mm diameter 4x12 microlens array. The base structure was made by using a hot embossing of concave lens array mold made by using plating and electroforming [16]. Each microlens forms a hemisphere.

Then the high and low index spin on materials were coated on the polymer film with microlens array structures.

After coating of the materials, we measured the coating material thickness in various method. The most straightforward method was to observe the cross-section by using SEM. We applied reworkable epoxy resin on top of the coated sample to avoid delamination of the coated film, then we cut the sample. As shown in Figs 9 and 10, the coating is uniform over the lens surface in the case of high index coating.

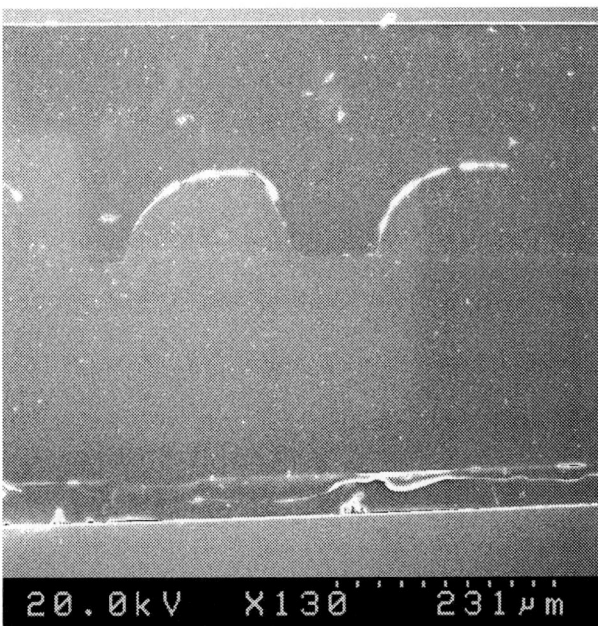

Fig. 9 SEM image of a cross section of the high index coated sample

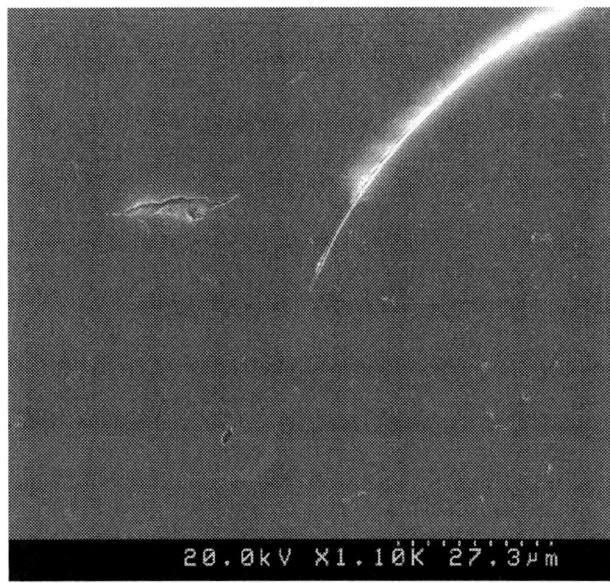

Fig. 10 Blow up SEM image of the high index sample.

The surface profile was also measured by using a confocal microscope (CFM). (Fig 11) This method allows us to determine the depth with a few nm accuracy although the high angle part cannot be measured due to the low reflection from the high tilt part. (Fig. 12)

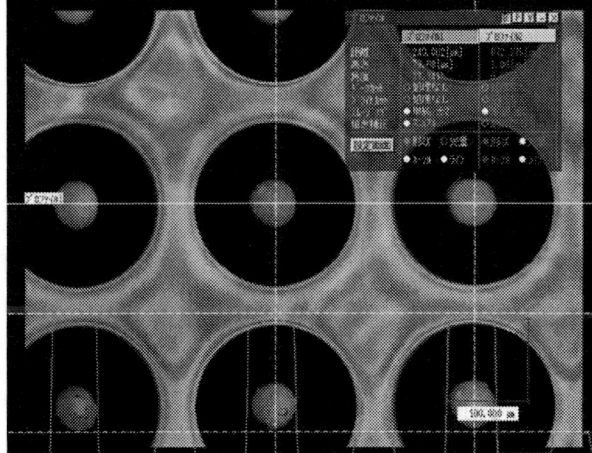

Fig. 11 Surface profile of the high index coated sample by using confocal microscopy

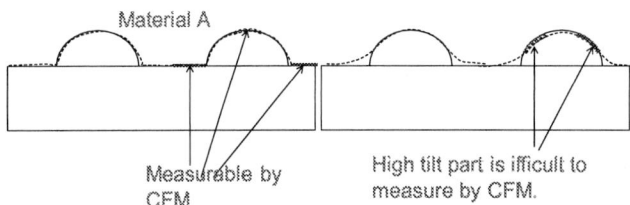

Fig. 12 Part of the sample measurable by confocal microscopy (CFM). Due to low reflection high tilt part is not measurable.

Achieving uniformity was more difficult in the coating of the low index material. The material contains high volume content of nanospheres, which makes the material more viscous. For the sample preparation the coating material was diluted to achieve the best result.

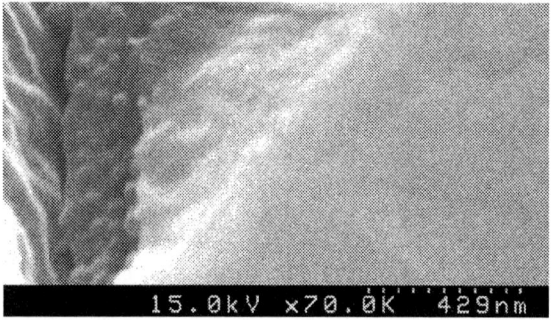

Fig. 13 SEM image of the coating surface cross section. Spheres with a diameter about 40-50 nm are observed.

The coating uniformity was confirmed by using confocal microscopy although there is some thick area where nano fillers gather at the ridge of the lens hemisphere. This phenomenon however, is not serious for the application because only the top part with less incline is used for the lens and the high angle part is not used in the real case.

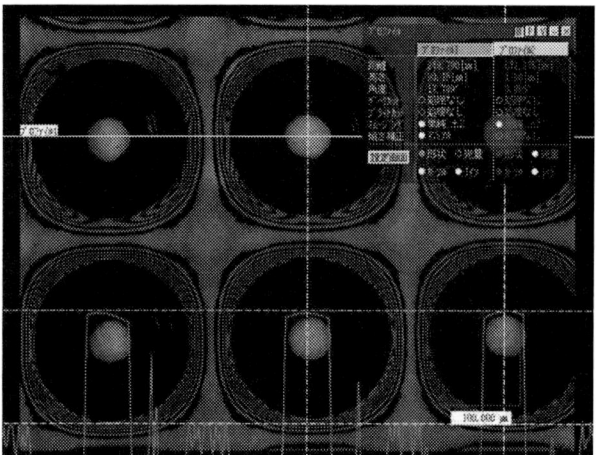

Fig. 14 Surface profile of the low index coating measured by confocal microscopy

To confirm the optical performance, we then used fused silica based microlens as a base substrate and coated it by the low index material with a target thickness of 200 nm. The transmission spectrum has a similar shape which was effective to determine the thickness. After confirming the thickness, we tried to coat the real microlens samples for optical characterization. (Fig. 15).

The set of materials with high and low refractive indices will be the basis of optical integration for the various configurations of silicon photonics optical packaging. The detailed design and hardware evaluation are under way.

Fig. 15 Surface profile of the low index coating measured by confocal microscopy

Conclusions

We report on an ideal anti-reflection coating method for use with a polymer waveguide based optical coupler to be placed between the silicon nanophotonic chip and the external parallel glass fibers, where the optical reflection at the silicon waveguide boundary is minimized. The set of materials to be used are nanoparticle based spin-on materials having indices of 1.8 and 1.2. The former material is suitable for a single layer anti-reflection coating on a silicon surface and realizes a first order thin reflection suppression which allows large bandwidth and large angle tolerance. The latter material is suitable for a single layer anti-reflection coating on a polymer or glass surface where the similar effect is expected. The material set can be used for the coupling of silicon waveguides and polymer waveguides where the interface is normal to the light propagation direction. This situation includes grating, butt and lens coupling cases. We studied the optical performance of a spin coated film in terms transmission, reflection, and scattering. A good thermal stability has been also shown. We can use this material set for the compliant polymer based silicon nanophotonics parallel coupler that bridges the silicon waveguides and the external fibers effectively. Although the current coupler uses the MT standard physical-contact parallel fiber connection to the fiber where the interface is angled, the low index material reported here opens a new possibility of lens coupled connector method of single mode fibers with a negligible back-reflection and further cost effectiveness.

Acknowledgments

This work is supported by NEDO under the Low Power Innovative Technology Research and Development Program. We thank JSR and JSR Microtech for realization of the polymer microlens array.

References

1. http://www.top500.org
2. http://www.green500.org
3. R. Berridge, et al. : "IBM POWER6 microprocessor physical design and design methodology", IBM Journal of Research and Development, 51, No. 6, Page 658 (2007)

4. http://www-06.ibm.com/systems/jp/saiteki/green/vdc.html

5. J. F. Fulzacchelli, et al.: "A 10-Gb/s 5-Tap DEF/4-Tap FFE Transceiver in 90-nm CMOS Technology", IEEE J. Solid State Circuits, 41, No. 12 2885-2900 (2008).

6. http://www.lanl.gov/orgs/hpc/roadrunner/rrfullimages3/index.htm.

7. D. Kuchta, et al.: "Optical Interconnects for Servers," Japan. J. Appl. Phys., 47 pp. 6642-6645 (2008)

8. Y. Taira, et al., "OE Device Integration for Optically Enabled MCM" Proceedings of 57th Electronics Components & Technology Conference 2007, IEEE (2007) p.1262.

9. S. Nakagawa, et al.: "High-Bandwidth, Chip-Based Optical Interconnects on Waveguide-Integrated SLC for Optical Off-Chip I/O," Electronic Components and Technology Conference , 2009

10. F. E. Doany, C. L. Schow, et al. "Gb/s 24-Channel Bidirectional Si Carrier Transceiver Optochip for Board-Level Interconnects" 58th Electronic Components & Technology Conference 2008, IEEE (2008)   Section 6 paper no.1

11. P. W. Coteus, J. U. Knickerbocker, C. H. Lam, and Y. A. Vlasov, "Technology for exascale system" IBM J. Research and Dev. 55, 14:1-14:12 (2011)

12. Y. A. Vlasov: "Silicon CMOS-integrated nano photonics for computer and data communications beyond 100G," IEEE Comm. Mag. (2012)

13. Y. A. Vlasov, "On-chip Si nanophotonics," Tutorial in European Conf. Optical Comm (2008)

14. Y. Taira and H. Numata, "High Channel Count Optical Interconnection for Servers," Electronic Components and Technology Conference, 2010

15. Y. Taira and H. Numata, "Optical Package Design for Silicon Photonic Chips," OSA Frontiers in Optics/Laser Science, 2010.

16. Y. Taira, H. Numata, and F. Yamada: "Optical Packaging of Silicon Photonic Devices for External Connection of Parallel Optical Signals," ECTC 2013 paper 747, Las Vegas, 2013.

Effect of Patterned Substrate on Light Extraction Efficiency of Chip-on-Board Packaging LEDs

Huai Zheng[1], Zhili Zhao[3], Yiman Wang[1], Lang Li[1], Sheng Liu[2], and Xiaobing Luo[1*]

[1] School of Energy and Power Engineering, Huazhong University of Science & Technology,
Wuhan, Hubei, 430074, China

[2]School of Mechanical Science and Engineering, Huazhong University of Science and Technology,
Wuhan, Hubei, 430074, China

[3]School of Optical and Electronics Information, Huazhong University of Science and Technology,
Wuhan, Hubei, 430074, China

*Corresponding author: Telephone: 86-13971460283, Fax number: 86-27-87557074,
Email: Luoxb@hust.edu.cn

Abstract

We studied the effect of substrate with patterned structure on the light extraction efficiency (LEE) of chip-on-board （CoB） packaging light-emitting diodes (LEDs) by optical simulations based on Monte Carlo ray-tracing method. Two kinds of typical patterned substrates were analyzed which were with conical and spherical structure, respectively. Results show that when the encapsulation layer is without phosphor particles, patterned structures strongly enhance the LEE compared with the flat substrate. The maximum LEE enhancement reaches 41.1% at 1/2 inclination angle α of 30° for the conical structure and 39.5% at 1/2 center angle β of 50° for the spherical structure. However, with the phosphor concentration in the encapsulation layer increasing, the LEE enhancement effect of patterned structures significantly reduces. Even, the enhancement effect almost disappears at the high phosphor concentrations which result in low color correlated temperatures (CCT) of 4000-6000K.

Introduction

Light-emitting diodes (LEDs) are green light sources and regarded as the most promising future illumination methods owing to their high efficiency, hence lower power consumption and exceptional reliability [1-3]. With LEDs rapidly developing in recent years, they have emerged into many illumination areas, such as large size smooth backlighting, street lighting, vehicle forward lamp, museum illumination and residential illumination [1, 4]. However, in order for LEDs to replace traditional light sources and become a main player in lighting, many technical challenges still need to be overcome. One of major challenges is the improvement of light extraction efficiency (LEE).

Currently, chip-on-board (CoB) packaging LEDs integrating multi chips have been more and more popular in a lot of illumination applications thanks to their better thermal performance and low manufacture cost [5-7]. Nevertheless, compared with the traditional single lead-frame packaging LEDs, its LEE is still at a low level because of total internal reflection (TIR) phenomenon at the interface between the flat encapsulation layer and the air. In order to enhance the LEE of CoB packaging LEDs, many LEE enhancement methods have been introduced. They consist of dome-shape lens, microstructure array encapsulation layer and patterned substrate [8-11]. Due to the large substrate size, the solution by dome-shape lens causes some disadvantages, such as high encapsulation material consumption and non compact packaging structure. The microstructure array encapsulation layer approach brings an additional imprint process to the current LED packaging. So, it increases the total manufacture cost. In order to overcome problems of above approaches, the patterned substrate method have been introduced recently. Owing to its advantages in process simplicity and cost, this method has attracted much interesting in industries. But, there are few literatures which systematically studies this method.

In this study, we systematically studied the effect of patterned substrate on the LEE enhancement of CoB packaging LEDs by Monte Carlo ray-tracing method. Two kinds of typical pattern structures, conical and spherical structure which are easily fabricated were analyzed. Results show that the LEE enhancement effect of patterned substrates is significantly influenced by the detail patterned structure and phosphor coating condition.

Optical Model and Simulation

Fig. 1 schematically presents the optical model of CoB packaging LED with patterned substrate. The optical model comprises four pieces of blue LED chips, the encapsulation layer and the patterned substrate. A kind of the common vertical injection structure chip was applied in simulations. The chip size is 1.2 mm×1.2 mm and the P-N junction area is 1.05mm×1.05mm. The thicknesses of the layers of the blue LED chip are N-GaN 3 μm, MQW 0.1μm, P-GaN 0.3 μm, and mental alloy substrate 140 μm, respectively. Their absorption coefficients and refractive indexes are listed in the table 1 [12]. The pitch between LED chips is 3mm. The area of patterned substrate is 9×9 mm^2. As shown in Fig. 1(b), the diameters of the bottom circle of conical structure and spherical structure keep 0.8 mm and their center axis separates with 0.6mm. Thus adjacent patterns are overlapped each other for increasing the number of pattern surfaces on the substrate. The specular and diffused reflection coefficients of substrate surfaces are assumed to be 85% and 5%, respectively [13].

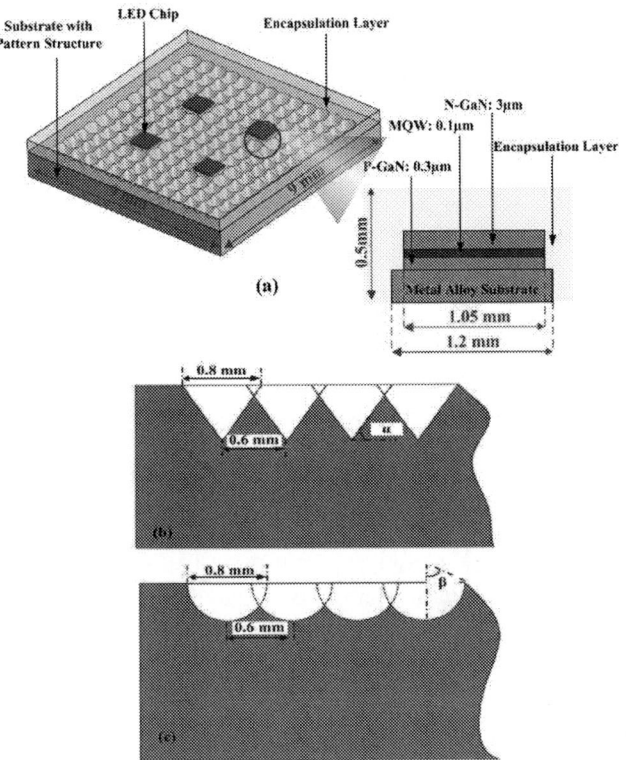

Fig. 1. Schematic illustration of optical model. (a) CoB packaging LED module with patterned substrate. (b) and (c) Patterned substrates with conical structure and spherical structure, respectively.

TABEL Ⅰ. Optical properties of blue LED chip

Symbol	Wavelength	454 nm	571nm
N-GaN	Refractive index	2.43	2.36
	Absorption efficient	2 mm^{-1}	1.5 mm^{-1}
P-GaN	Refractive index	2.43	2.36
	Absorption efficient	2 mm^{-1}	1.5 mm^{-1}
MQW	Refractive index	2.51	2.39
	Absorption efficient	120 mm^{-1}	8 mm^{-1}

The encapsulation layer is also with 9×9 mm² area and 0.6 mm thickness. The encapsulation layer material is set as the silicone or composite of silicone and phosphor. The refractive index of the silicone is 1.54. The optical properties of the composite of the silicone and phosphor are calculated by the Mie theory. Here, the phosphor particle size was approximated by a Gaussian distribution with a mean diameter of 8 μm and a standard deviation of 3.2 μm. In order to simplify the calculation, specific wavelengths of 454 and 571 nm were used in the calculation to represent blue and yellow light, respectively. The blue light is assumed to be emitted from the MQW with isotropic pattern. The yellow light is remitted from the encapsulation layer. After ray-tracing, escaped rays are collected by a detector and the optical

information such as the LEE and color correlated temperature can be gained.

Results and Discussion

Fig. 2 shows the relationship between the LEE and patterned structures when the encapsulation layer only includes the silicone. It can be seen that patterned structures can significantly enhance the LEE, compared with the flat substrate. It can be also seen that the LEE enhancement trend presents first increasing, then reducing with both the 1/2 inclination angle α and 1/2 center angle β increasing for the conical structure and the spherical structure. The maximum LEE enhancements reach up to 41.1% and 39.5% by the conical structure with α of 30° and spherical structure with β of 50°.

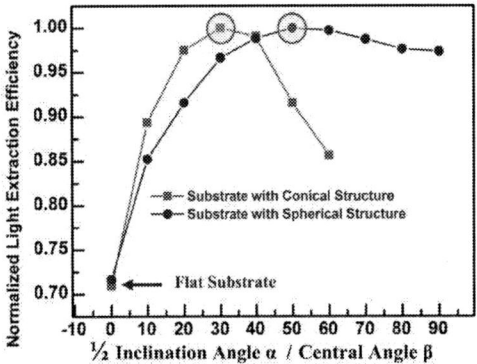

Fig. 2. Normalized LEE at different inclination angles and central angles when encapsulation layer only consists of silicone.

The reason why patterned structures can improve the LEE is due to reducing TIR phenomenon at the interface between encapsulation layer and the air by patterned structures. For the flat substrate packaging structure, lights which are reflected by the interface between the encapsulation layer and the air are specularly reflected again by flat substrate. Thus, their entry angles at encapsulation-air interface are the same with those of the last TIR. So these lights are still total internally reflected and can't escape. The cycle propagation proceeds until these lights were finally absorbed by the materials. While by the patterned substrate surface, directions of these reflected lights by patterned substrate are changed. Entry angles of some of reflected lights get small, thus they can escape from the encapsulation-air interface. Thus, patterned substrates present higher efficiency that flat substrate. However, the pattern parameters in terms of the inclination angle and the central angle strongly influence the LEE enhancement effect. The patterned structure with small inclination angles or center angles leads to multiple reflections before the escaping. And the larger inclination angle or center angle results in light trapping into patterns. They both result in large energy loss. So, the maximum LEE occurs at the middle inclination angle and central angle, as shown in Fig. 2.

Fig. 3 presents the LEE variations of the flat structure and patterned structures with the phosphor concentration. It can be found that for patterned substrates, the LEE reduces with the phosphor concentration increasing. While for flat substrates, the LEE firstly rises, then drops with the phosphor concentration increasing. In finally, the LEE difference between the flat substrate and patterned substrates obviously turns to be small as the phosphor concentration increases. The LEE difference reduced up to less than 2%, when the phosphor concentration is larger than 0.8 g/cm³ which leads to CCTs of 4000-6000K.

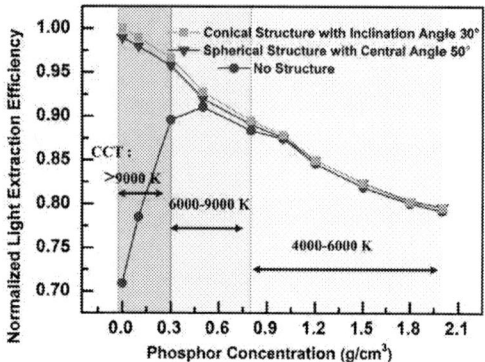

Fig. 3. Effect of phosphor concentration on LEE.

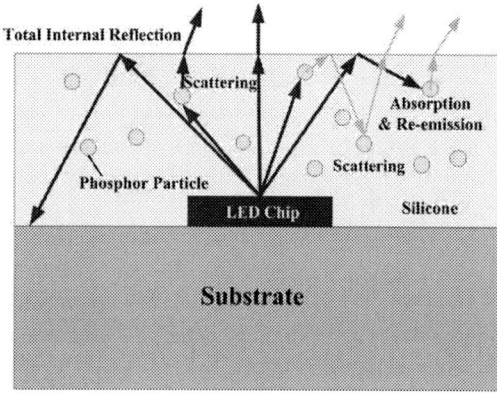

Fig. 4. Ray dynamics in encapsulation layer with phosphor particle.

Results shown in Fig. 3 can be explained as the following. Fig. 4 schematically describes the ray dynamics in the encapsulation layer with phosphor particles. It can be clearly see that the blue light is absorbed and scattered by phosphor particles and the yellow light is re-emitted from phosphor particles by downconverting the blue light. For CoB LEDs with patterned substrates, the LEE reduction as the phosphor concentration increases is owing to more energy lose of the blue light downconverting to yellow light. While for CoB LEDs with the flat substrate, the light scattering of phosphor particles change light directions and it benefits light escaping. Therefore, the LEE rises with adding the phosphor in spite of the energy lose with light downconverting. With the phosphor concentration further increasing, the energy loss with light downconverting becomes dominant, the LEE begins to get down at the flat substrate. At the high phosphor concentration, on the one hand, most of lights can escape from the

encapsulation layer by light scattering of phosphor particles. On the other hand, the reflected lights by encapsulation-air interface is absorbed and scattered by phosphor particles during transmitting backward and less of them reaches the substrate. These two reasons lead to that the patterned substrates have less effect on LEE enhancement. Thus, patterned substrates and flat substrate present the almost same LEE.

Conclusions

The effect of patterned substrates on LEE enhancement for CoB packaging LEDs was researched by optical simulations. Results clearly indicate that the LEE enhancement effect of patterned substrates strongly affected by the coating phosphor concentration in the encapsulation layer. Under the condition of the encapsulation layer without the phosphor, the optimal patterned substrates could significantly improve the LEE, compared with the flat substrate. However, with the high phosphor concentration, the LEE enhancement effect of patterned substrates almost is faint due to the strong absorption and scattering of phosphor particles. Based on the study, we can concluded that besides the pattern optimizing in such LEE enhancement method by patterned substrates, the application condition should be carefully considered, especially for the condition with the high phosphor concentration packaging.

Acknowledgments

The authors would like to acknowledge the financial support in part from 973 Project of The Ministry of Science and Technology of China (2011CB013105) and in part by National Science Foundation of China (51376070).

References

1. S. Liu and X. B. Luo, LED Packaging for Lighting Applications: Design, Manufacturing and Testing (John Wiley & Sons, USA, 2011).
2. A. Zukauskas, M. S. Shur, and R. Caska, Introduction to Solid-state Lighting (John Wiley & Sons, New York, USA, 2002).
3. B. L. Wu, X. B. Luo, H. Zheng, and S. Liu, "Effect of gold wire bonding process on angular correlated color temperature uniformity of white light-emitting diode, " Opt. Express, vol. 19, pp. 24115-24121, 2011.
4. K. Wang, D. Wu, F. Chen, Z. Y. Liu, X. B. Luo, and S. Liu, "Angular color uniformity enhancement of white light-emitting diodes integrated with freeform lenses," Opt. Lett. , vol. 35, pp. 1860-1862, 2010.
5. M. C. Hyun and J. K. Hyeong, "Metal core printed circuit board with alumina layer by aerosol deposition process", IEEE Electron Dev Lett, vol. 29, pp. 991–993, 2008.
6. P. Hartmann, F. P. Wenzl, and C. Sommer, et al., "White LEDs and modules in chip-on-board technology for general lighting", Proc SPIE 6337, pp. 63370I-1–63370I-7, 2006.
7. H. H. Wu, K. H. Lin, and S. T. Lin, "A study on the heat dissipation of high power multi-chip COB LEDs," J. Microelectron., vol. 43, no. 4, pp. 280–287, 2012.
8. K. Wang, F. Chen, Z. Y. Liu, X. B. Luo and S. Liu, "Design of compact freeform lens for application specific

light-emitting diode packaging", Optics Express, vol. 18, pp. 413-425, 2010.

9. D. Wu, K. Wang, and S. Liu, "Enhancement of Light Extraction Efficiency of Multi-Chips Light-Emitting Diode Array Packaging with Various Microstructure Arrays", 61st Electronic Components and Technology Conference, USA, pp. 242-245, 2011.

10. H. Luo, J. K. Kim, E. F. Schubert, J. Cho, C. Sone, and Y. Park, "Analysis of high-power packages for phosphor-based white-light-emitting diodes," Appl. Phys. Lett., vol. 86, no. 24, pp. 243505-1–243505-3, 2005.

11. Z. Li, Q.Wang, Y. Tang, C. Li, X. Ding, and Z. He, "Light Extraction Improvement for LED COB Devices by Introducing a Patterned Leadframe Substrate Configuration", IEEE Trans Electron Dev., vol. 60, pp.1397-1403, 2013.

12. Z. Y. Liu, C. Li, B. H. Yu, Y. H. Wang, and H. B. Niu, "Uniform White Emission of WLEDs Realized by Multilayer Phosphor With Pyramidal Shape and Inversed Concentration Distribution," IEEE Photon. Technol. Lett., vol. 24, pp. 1558-1560, 2012.

13. Z. Y. Liu, S. Liu, K. Wang, and X. B. Luo, "Optical Analysis of Color Distribution in White LEDs With Various Packaging Methods," EEE Photon. Technol. Lett., vol. 20, pp. 2027-2029, 2008.

Transferrable Fine Pitch Probe Technology

Y. Liu, S.L. Wright, B. Dang, P. Andry, R. Polastre, J. Knickerbocker
IBM T.J. Watson Research Center
1101 Kitchawan Rd, Yorktown Heights, NY 10598
E-mail: yangliu@us.ibm.com, phone: 914-945-2142

Abstract

Two vertical probe technologies were explored to evaluate wafer-level and 3D stack-level die test at 50 μm pitch. In one approach, Cu probe tips were serially built on a 3D silicon wafer using lithography and wet chemical etching. In a second approach, metal probe tips were fabricated by filling a silicon mold made with anisotropic etching and transferred to a silicon die. Each approach had advantages which were demonstrated through test vehicles. The second approach appears promising to offer a robust, scalable probe technology, highly suited for high speed test of 3D die. As compared to traditional probe technologies, this approach also offers an opportunity toward low cost test at fine pitch.

Introduction

Through-Silicon-Via (TSV)-based 3D integration has received a lot of interest due to inherent features such as high density, high speed, low power consumption, and small height [1], [2], [3], [4]. Both known-good die (KGD) and known-good stack (KGS) issues can strongly impact product yield. Wafer level test, die- and stack-level test, are major challenges to achieve low-cost [5]. Probe technology is regarded as a major requirement in 3D testing due to challenges including: fine pitch, low force, high probe accuracy, co-planarity, low probe damage to contacts, number of contacts, low parasitics, high current capacity, and scalability. In particular, high-power and high-speed test of integrated components is desired.

Historically, buckling-type cobra probes or thin-film vertical probes have been used to test bumped wafers at I/O pitch in the range 130 to 200 μm. A high degree of compliance is required to overcome probe card non-uniformities and non-planarity of die bump interconnections. Cantilever MEMS probes have been widely researched for fine pitch, good probe accuracy, low contact force, and low probe damage to contacts. B. Kim et al. reported an area-arrayed MEMS probe card with cantilevers of 1600 μm long, 80 μm wide, and 70 μm tall with a co-planarity of ±10 and X-Y alignment of ±10 μm, probe tip surface vary from 6 x 6 μm to 13 x 13 μm, the probe mark is in the range of 10x15μm [6]. F. Wang et al. reported a hoe-shaped area-arrayed MEMS probe of 90 μm x 196 μm with probe co-planarity ±4 μm range in 10x10 probe array with hundreds of mΩ probing resistance, the probe mark is in the range of 1 x 8μm [7]. Past probe technologies have been passive, with active circuits for test far removed from the active devices in the die under test. Recently Namburi reported an area-arrayed MEMS of 50 μm x 50 μm on active silicon stack. Scrub length is 10~15 μm and the contact resistance is 600 mΩ [8]. While MEMS technology can create probes that are compliant, any lateral probe movement scrubs the bumps, creating both damage and debris. In general, challenges in scalability, number of contact, co-planarity, accuracy, probe damage to contacts, contact force, current capacity, mechanical reliability, and cost have inhibited effective probe solutions for finer pitch.

In this work, two rigid vertical probe technologies were explored for fine-pitch applications. Since fine pitch bumps have inherently smaller non-uniformity than coarse-pitch bumps, a highly-planar rigid probe precisely built on stable silicon substrates can provide good contact with minimal damage. In one approach, Cu probe tips with tip dimensions in the range 1-10 μm were serially built up on a 3D silicon wafer using copper damascene, lithography and wet chemical etching. In a second approach, metal probe tips were fabricated by filling a silicon mold made with anisotropic etching. The probe tips fabricated with this technique are precise 4-sided pyramids with face angles of 70.5 degrees, and are precisely planar. The probe tip structure can also contain micro-pillars. Using a low-temperature process, the probe tips were transferred to a silicon space transformer which contains through-silicon-vias. Fine pitch test probes were used to probe 50 μm pitch bumped die containing 45,400 bumps, 12,644 of which could be electrically tested. Low contact resistances of 20-30 mΩ were obtained at forces of 0.05 to 0.5 grams per probe. Preliminary results suggest that a DC current approaching 1A can be applied without damage. This approach appears promising to offer scalable fine pitch test probes, with low cost, and low maintenance in comparison to traditional probe technologies used for coarse pitch die.

50 μm Probe Claw

Probe claw structures (Fig. 1) were fabricated at 50 μm pitch to obtain preliminary data on fine pitch probing of micro-C4. The structure was designed with a center probe pin and four peripheral points which act as "claws" to tolerate mechanical misalignment and thermal CTE mismatch. Copper was deposited on a silicon wafer by a damascene process, patterned, and then wet-etched to create a claw shape. An environmentally-friendly proprietary etchant was developed to control a uniform center pin diameter across an 8-inch wafer within ±0.5 μm. Finally, the probe is finished with a layer of hard coating to enhance mechanical strength and improve electrical performance and reliability. Probe co-planarity and accuracy are ensured by mature back end of line (BEOL) technologies and process control.

Mechanical test was performed by pressing the probe chip and a bumped silicon chip in a flip-chip bonder, which is calibrated for parallelism and accuracy. The silicon probe chip was 5x6 mm. As shown in Fig. 2, at 0.5 g/probe, uniform probe mark was achieved on all 11,192 lead-free micro-C4s. The center pin and one or more of the peripheral points penetrated respective bump surfaces. A cross-sectional image of an indentation is shown in Figure 3. Aged micro-bump samples were chosen to provide thick surface oxides for easy identification of surface oxide breakage by focused ion beam analysis. At 0.5 g/probe low force, the probes created 2.6 μm indentation on lead-free bumps, which is sufficient to break surface oxide. It was also interesting that the probe pressure induced grain boundary change in the solder bump.

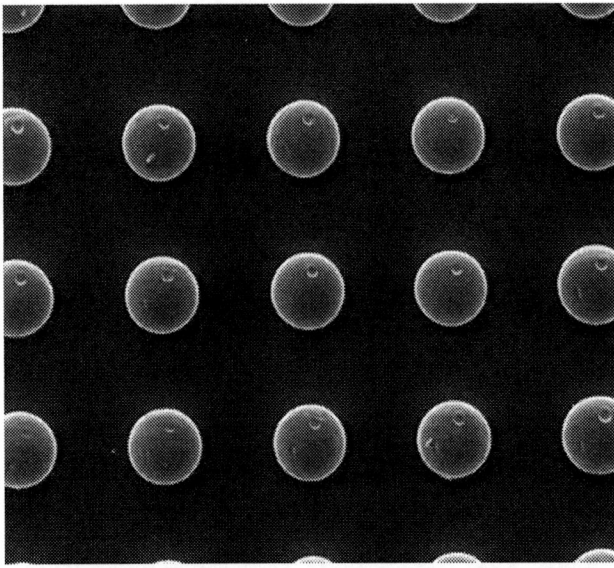

Figure 2. Uniform probe mark of 50 μm probing claw.

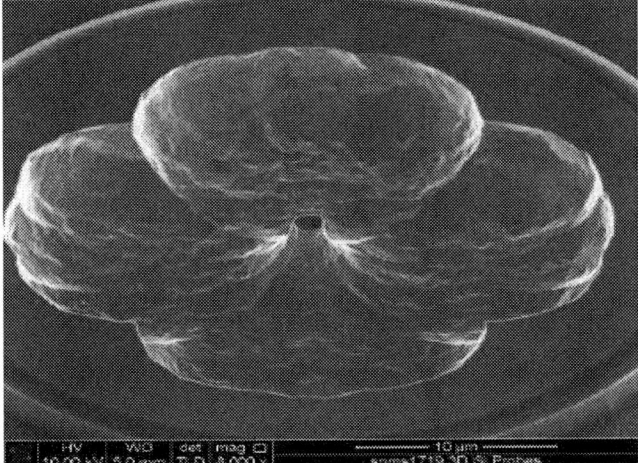

(a)

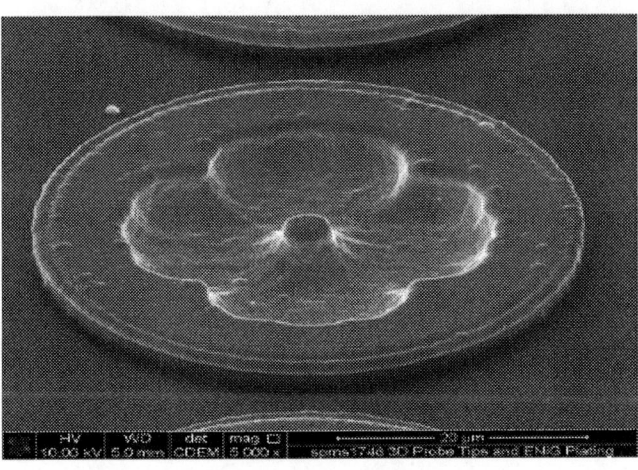

(b)

Figure 1. (a) 50 μm probing claw after copper etch
(b) 50 μm probing claw after hard coating.

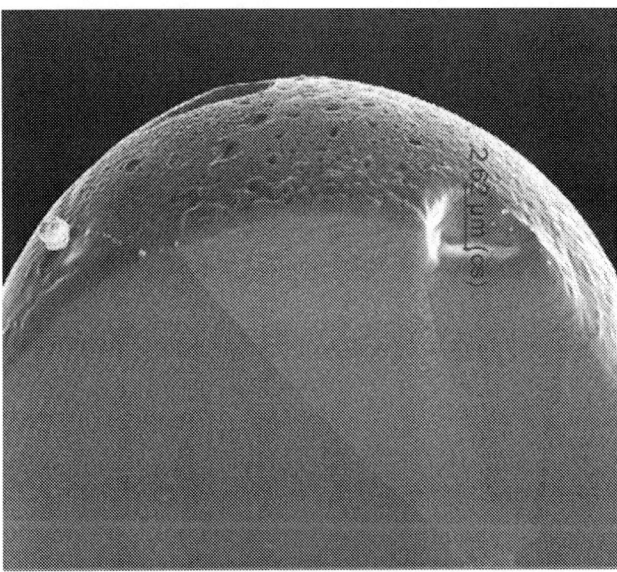

Figure 3. Lead-free micro-C4 surface oxide broken by vertical indentation.

Transferable Probe Tip

While the mechanical test result for probe claw structures were a serial build-up approach, transferable probe tips were designed and developed, offering an improved process and structure with further scalability and versatility.

Silicon (100) wafers were anisotropically etched to form pyramidal molds. Because the etch rate in <111> directions are 1/300 the etch rate in <100> directions and 1/600 the rate in <110> directions, only {111} planes are left in the etch pits and the mold pit reaches a sharp vertex [9]. This allows a large process window and good controllability. Several layers of metal were then deposited to fill the mold and form a micro-pillar. Next, various low-temperature processes were developed to transfer the probes to the surface of a substrate. The substrates may include 2D and 3D silicon chips or wafers, glass chips or wafers, chip stack, multi-chip module (MCM), ceramic, and organic substrates. The probes could also be transferred to MEMS cantilevers, membranes, or springs for additional compliance. After joining, the probes were mechanically released with low force. The released silicon molds could be reused for filling and transfer.

The silicon crystal structure guarantees precise tip sharpness as well as submicron co-planarity of the tips. The relationship between the lithographically defined base of the pit (L) and the depth of the pit (H) is given by:

$$H = L \times \tan(54.7°)/2 = 0.707\ L,$$

Where L is determined by mask design and lithography. Since modern state-of-art photolithography can routinely reproduce critical dimensions and accuracy far below micron level, the probe tip co-planarity and accuracy are well-controlled. Also because of {111} planes act as good etch stop, scalability is good to a pitch of a few microns. The good probe co-planarity, accuracy, and scalability make it possible for direct probing TSVs.

The bonding method and materials can be designed to accommodate some degree of non-planarity of the substrate. Lead-free solders, such as SnAu, SnAg, SnAgCu, are valid candidates for different applications. Thermal compression or a reflow process could be chosen to join the probes and the substrates.

Transferable Probe Tip Fabrication and Mechanical Test

Probe tips, shown in Fig. 4, were fabricated at 50 μm pitch using previously-described process and transferred to a 10x12mm 760 μm thick silicon chip with 45,400 interconnects at high yield. They were also transferred to a 10x12mm 50 μm thin 3D silicon chip with TSVs as shown in Fig. 5 to confirm that the probes can be successfully released on thin and fragile 3D silicon.

The full thickness probe chip was pressed into a bumped silicon chip in a flip chip bonder. Probe mark depth was measured against a 3x3 matrix of probe force and temperature. Uniform probe marks were found over the chip as shown in Fig. 6, with a typical indentation cross section shown in Fig. 7. A stress-induced grain boundary change was also observed. The probe mark at room temperature formed by 0.5 grams per probe is 5x5 μm size and 3.5 μm deep. Probe mark depth vs temperature and force is shown in Fig 8. As expected, lower force is needed to at higher temperature due to reduced solder hardness. By controlling the pressure and temperature, the probe mark can be controlled on a certain type of bump material. Recovery of micro bump shape may be important for a subsequent bonding step. Figure 9 shows complete solder ball recovery after reflow at 250 °C in formic acid atmosphere. Vertical indentation eliminates scrubbing material off the bump, and thus reduced solder volume loss and contamination.

Figure 4. Transferable Probe Tip.

Figure 5. Transferable Probe Tip on 50 μm 3D silicon chip.

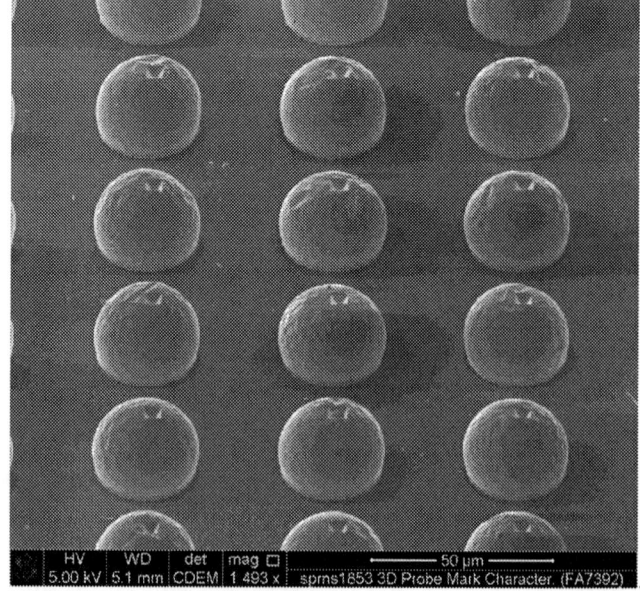

Figure 6. Uniform Probe Mark formed by Transferable Probe Tips.

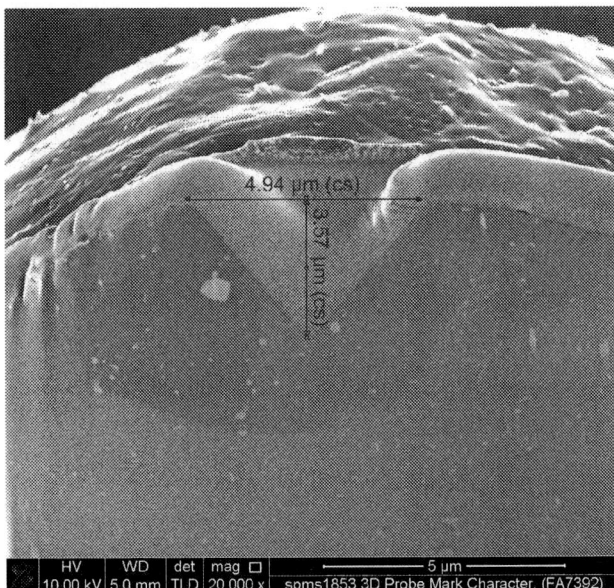

Figure 7. Probe Mark formed at 0.5g/probe room temperature.

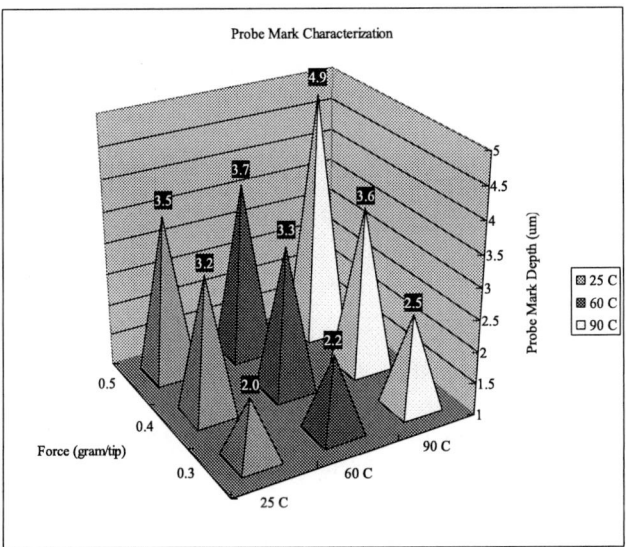

Figure 8. Probe Mark Depth vs Force and Temperature.

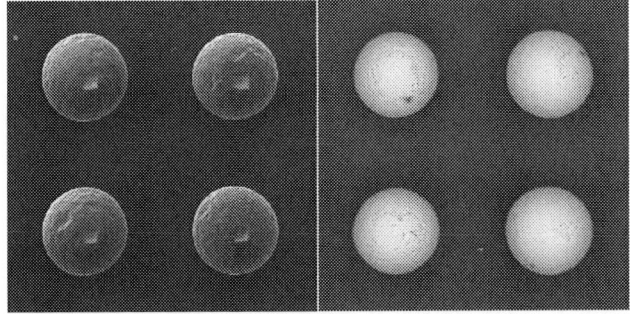

Figure 9. Complete solder recovery at 250°C in formic acid atmosphere.

Electrical Test

A first series of electrical tests were performed on a set of 2-layer samples in which a flip-chip bonder was used to tack down the top die with a force of 0.5 g/bump applied at temperatures of 25°C, 90°C, or 150°C. Following mechanical tack down; the samples were unloaded and gently placed in a conventional probe station, since the stiction forces holding the two die together were small. The die stack was then tested using a conventional probe card with tungsten cantilever probes contacting peripheral pads on the bottom die, outside the perimeter of the top die. The samples contain many chains of micro bumps connected in series, along with other test sites specifically designed to measure contact resistance, using a 4-point transmission line method. Each sample contained 45,400 physical micro bumps at 50 μm pitch, 12,644 of which could be electrically tested.

For tack down at room temperature, very tentative contact was established, with only a few functioning chains and an estimated contact resistance of about 0.5 Ω. Some of the working chains became open upon subsequent test, possibly due to the vibration of stage movement during the tests. This result was consistent with the observation that the chips could be very easily separated.

For tack down at 90°C, about 50% of the chains were functioning. From the distribution of chain lengths in the structure, this corresponds to an individual bump contact yield of about 99%.

For tack down at 150°C, about 90% of the chains worked, corresponding to a bump contact yield of 99.99%. The total bump contact resistance was about 40 mΩ, the sum of resistance on both sides of the bump. On the substrate side of the bump, the current is crowded due to electron flow into the thin Cu pad and trace. The current is also crowded in the region where the tip indents the bump. For comparison, typical contact resistances we have obtained are in the range 10-20 mΩ for the same structures without tips, and which have been joined in a flip chip bonder by bump melting. From this, we can conclude that the resistance of the tip portion of the interconnection is in the range 20-30 mΩ. The individual link resistances of several 20-link chains were also measured. The resistance standard deviation was within a few percent of the average, indicating uniform contact resistance. All electrical tests were repeated on the samples after storage for 2 days, with no change in results.

A second series of electrical tests was performed using a probe station equipped several cameras to allow alignment without the need to view through the probe card. For those tests, the top die contained probe tips, and was attached to a probe card backed by a stiffener plate. Electrical wirebond connections were made between pads on the top die and traces on the probe card. On top of the stiffener plate, four load cells were used to measure the forces on the plate as the bottom die was brought into contact with the top probe die. Indications of electrical contact were found at forces as low as 0.05 g/bump, with contact resistances above 100 mΩ. As the force was increased further, the contact resistance decreased. In some cases, the resistance dropped abruptly as a threshold force was applied, and in others there was a gradual decrease in total contact resistance down to about 40 mΩ for forces between about 0.1 g/bump to 0.4 g/bump. These results were obtained on a single test site requiring the functionality of a small number of bumps. In other tests, larger forces of about 0.5

g/bump or larger were required to get uniform contact over all the bumps in a die. This variability is most likely due to bump height variation within the die, and/or the presence of foreign particles inhibiting uniform probe contact.

Another single electrical test was performed with a different configuration for touchdown, in which manual tungsten probes were used to contact peripheral pads connected to the bump chains. After contact touchdown was established at high force, 4-point resistance measurements were made on a 2-bump chain at high DC current levels to assess the current-carrying capability of these probes. As the current was increased, the chain resistance increased due to Joule heating. At currents of 0.5 A and 1.0 A, the temperature rise was about 5°C and 25°C, respectively. At a current of 1.25 A, the current was held steady with a dwell of several minutes, during which the temperature exceeded 100°C. At a current of 1.5 A, the test site resistance began to show instability, along with signs of Joule heating in the external circuit. A maximum current of 2 A was applied without opening the chain. Following the test, the chips were separated and the chain site examined. Although electrical connectivity was maintained throughout the current stress, as well as post-stress, there were clear indications of bump melting and oxidation. The currents used in this stress test with rigid vertical probes are well beyond the maximum current which can be applied with conventional compliant probe technology, including MEMS. In this case, at a current of 1.25 A, the power dissipation in the test chain exceeded 100 W/mm², and limitations due to Cu trace wiring are likely to dominate over the probe tips.

Conclusions

A transferrable probe tip technology was successfully demonstrated. The probes were precisely formed from a silicon mold and transferred to 2D and 3D silicon chips. Electrical contacts were formed by probe vertical indentation with probe marks well-controlled in shape and depth. The technology can deliver very low probe force and high current capacity. Furthermore, the parallel process and low-force release enables the probes to be built on a 3D structure without causing further complication in material and process compatibility, yield and cost, and supply chain. By using this technology, test circuitry can be placed close to the device under test (DUT) for high-power and high-speed functional test. Transferrable probe tip technology is promising to offer scalable fine pitch test probes, with low cost, low maintenance, and high performance for 3D fine-pitch test. Prior patents and new patent applications are in progress which support the probes and test technology.

Next steps for this test and test probe technology include further wafer level test hardware build and characterization evaluations in research and manufacturing test equipment evaluations for fine pitch interconnection wafers.

References

1. P. Garrou, C. Bower, P. Ramm, *Handbook of 3D Integration*, Wiley-VCH, Germany, 2008, pp. 13–23.
2. J.U. Knickerbocker; P.S. Andry; E. Colgan; B. Dang; T. Dickson; X Gu.; C. Haymes; C. Jahnes; Y. Liu; J. Maria; R.J. Polastre; C. K. Tsang; L. Turlapati; B.C. Webb; L. Wiggins; S.L. Wright, "2.5D and 3D technology challenges and test vehicle demonstrations," *IEEE Electronic Components and Technol. Conf. (ECTC)*, Las Vegas, NV, May 29-June 1 2012, pp.1068-1076
3. J. Maria; B. Dang; S.L. Wright; C.K. Tsang; P. Andry; R. Polastre; Y. Liu; L. Wiggins; J.U. Knickerbocker, "3D Chip stacking with 50 μm pitch lead-free micro-c4 interconnections," *IEEE Electronic Components and Technol. Conf. (ECTC)*, Lake Buena Vista, FL, May 31-Jun 3, 2011, pp.268-273
4. L. Schaper; S. Burkett; M. Gordon; L. Cai; Y. Liu; G. Jampana; I.U. Abhulimen, "Integrated system development for 3-D VLSI," *IEEE Electronic Components and Technol. Conf. (ECTC)*, Reno, NV, May 29- Jun 1, 2007, pp. 853-857
5. E. J. Marinissen, "Challenges in testing TSV-based 3D stacked ICs: Test flows, test contents, and test access," *IEEE Asia Pacific Conference on Circuits and Systems (APCCAS)*, Kuala Lumpur, Malaysia, Dec 6- 9, 2010, pp. 544-547
6. B. Kim, J. Kim, "Design and fabrication of a highly manufacturable MEMS probe card for high speed testing," *Journal of Micromechanics and Microengineering*, vol.18, no. 7, Jun.2008
7. F. Wang, R. Cheng, X. Li, "MEMS Vertical Probe Cards With Ultra Densely Arrayed Metal Probes for Wafer-Level IC Testing," *Journal Of Microelectromechanical Systems*, vol. 18, no. 4, pp. 933-941, Jul. 2009
8. L.Namburi, G. Maier, F. Cros, Y. Hong, T. Hu, R. Smith, V. T. Truong, H. C. Liu, "A Fine Pitch MEMS Probe Card with Built in Active Device for 3D IC Test," *Semiconductor Wafer Test Workshop (SWTW)*, San Diego, CA, Jun 9-12, 2013
9. Microchemicals GmbH, "Wet-Chemical Etching of Silicon", http://www.microchemicals.eu/technical_information /silicon_etching.pdf.

Improvement of the Crystallinity of Electroplated Copper Thin Films for Highly Reliable 3D Interconnections

Chuanhong Fan, Osamu Asai, Ryosuke Furuya, Ken Suzuki, and Hideo Miura

Fracture and Reliability Research Institute, Graduate School of Engineering, Tohoku University

6-6-11-712 Aobayama, Aoba-ku, Sendai, Miyagi 9808579, Japan

E-mail: hmiura@rift.mech.tohoku.ac.jp

Abstract

The degradation process of the crystallographic quality of copper thin films, which are used for interconnections and micro bumps for 3D integration, during electromigration and stress-induced migration tests is dominated by the diffusion along grain boundaries and the diffusion constant of copper varies drastically depending on the crystallinity of the films. The degradation process was visualized clearly by applying an electron back-scatter diffraction method. The copper atoms in the electroplated copper thin films migrated mainly in the area with low crystallinity, in other words, the area with high defect density. Since the crystallinity of the films was found to be dominated by the lattice mismatch between copper and the seed layer material used for electroplating, the integrity of the interface structure was improved by minimizing the lattice mismatch. It was validated that the introducing the thin layer with fine grains and random orientation is effective for minimizing the lattice mismatch and thus, improving the crystallographic quality of the electroplated copper thin-film interconnections.

Introduction

Electroplated copper thin films have started to be applied to thin film interconnections and TSV (Through Silicon Via) in semiconductor devices because of its low electric resistivity and high thermal conductivity. However, it has been reported that both electrical and mechanical properties of the electroplated-copper thin-film interconnections show wide variation or fluctuation comparing with those properties of bulk copper [1-6]. The main reason for the variation was found to be the fluctuation of the crystallinity of gains and grain boundaries in the films. In particular, the density of the porous or sparse grain boundaries with high defect density causes the very high resistivity and brittle fracture of the films [7, 8]. These porous grain boundaries also decreases the thermal conductivity of the thin-film interconnections, and some porous grain boundaries with high electronic resistivity cause local high Joule heating and thus, open failures due to fusion [9]. The high porosity causes the shrinkage of the film during its densification, and large voids often appear in the interconnections after high temperature annealing [10].

The authors have reported that the mechanical properties of the electroplated copper thin films easily change from brittle ones to ductile ones depending on their electroplating conditions and annealing condition after the electroplating. [11, 12] Not only the yielding stress and tensile strength of the films, but also their Young's modulus were found to vary drastically. Brittle fatigue fracture appears even under high cycle fatigue loads and it shortens the fatigue life of the films. Such unstable fracture mode deteriorates the reliability of electronic products. Therefore, it is very important to control

and improve the crystallinity of the electroplated copper thin films.

The authors have developed a new method for visualizing the crystallinity of polycrystalline thin film materials quantitatively by applying the conventional EBSD (Electron Back-Scatter Diffraction) method [13]. In this analysis, an electron beam with 50 nm in diameter is scanned on the surface of a thin film and the crystallinity of grains and grain boundaries is evaluated by analyzing the average sharpness of Kikuchi lines observed by the EBSD analysis quantitatively. The degradation process of the electroplated copper thin-film interconnections can be observed in-situ during both an electromigration and stress-induced migration tests. In this study, the degradation process of the crystallinity of the interconnections during stress-induced migration tests was observed in detail by applying this new EBSD method. In addition, the effect of the crystallographic structure of the interface between the electroplated copper layer and the seed layer on the crystallinity of the electroplated copper thin film was validated quantitatively. The introduction of the intermediate layer with fine grains and random crystallographic orientation was effective for minimizing the lattice mismatch between them and thus, for improving their crystallinity, electronic properties, and long-term reliability.

Sample Preparation

The test thin-film interconnections were made by damascene process for stress-induced migration (SM) and electromigration (EM) tests as shown in Fig. 1. First, a 300-nm thick SiO_2 layer was deposited on a 280-μm thick Si substrate by chemical vapor deposition (CVD). The SiO_2 layer was locally etched off by BHF (Buffered Hydrogen Fluoride) to make a mask-pattern, and the Si substrate was etched off to make 1-μm deep trenches for modeling a TSV interconnection structure by TMAH (Trimethyl-phenyl-ammonium hydroxide). The width of the trenches was 5-μm. After the SiO_2 layer was removed, a 50-nm thick copper diffusion barrier layer (Ta) and a 150-nm thick copper seed layer (Cu) or ruthenium (Ru) layer were deposited by physical vapor deposition (PVD). Then, the trenches were filled by electroplated copper film. Diluting 80-g of CuO powder and 186-g of H2SO4 into 1000 ml of purified water controlled the composition of a plating bath used for the electroplating. Finally, the overburdened copper was mechanically polished to make isolated interconnections. Test samples were electroplated under the constant current density from 10 to 50 mA/cm^2 at 30°C. In addition, some interconnections were annealed after the electroplating in pure argon gas at 200°C and 400°C for 30 minutes.

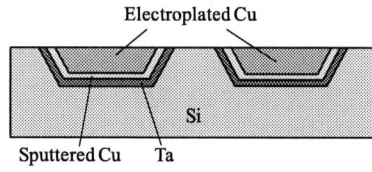

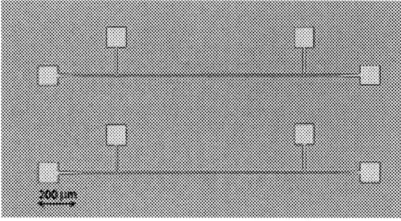

Fig. 1 Schematic of the electroplated copper thin film interconnections

The reason for the introduction of the ruthenium layer was that the lattice constants of Ru(0001) (2.71 Å) is very close to that of Cu(111) (2.55 Å). It is possible to decrease the lattice mismatch from about 18% which exists between copper and tantalum to about 5% by changing the seed layer material from tantalum to ruthenium [12].

Figure 2 shows the change of the residual stress in the copper thin films during the annealing process at 400°C as a function of the current density during electroplating. The residual stress was measured by detecting the deflection of the silicon substrate on which the electroplated copper thin film was deposited. During the initial heating process, compressive stress increased in all the films because of thermal stress. It started to decrease, however, when the temperature exceeded at about 100°C. This decrease in the residual stress was caused by the densification and recrystallization in the annealed films as shown in Fig. 3. Initially, the electroplated films consist of fine columnar grains and grain boundaries with a lot of defects such as vacancies. It was found that the both the densification and grain coarsening of the films started at about 100°C, and grain coarsening was clearly observed at 200°C (Fig. 3(b)). This recrystallization should cause shrinkage of the films and thus, tensile stress due to the constraint of the deformation of the films by silicon substrates. The recrystallization continued up to 400°C as shown in Fig. 3(c). The average grain size of the films annealed at 400°C was about 20 times higher than that of the as-electroplated films. During the cooling process, tensile stress increased almost linearly with the decrease in temperature from 400°C to room temperature. Finally, tensile stress higher than their initial residual stress remained in all the annealed films. In particular, the residual stress in the film electroplated at the current density of 50 mA/cm2 exceeded its yielding stress of about 80 MPa which was measured by a simple tensile test of the stand-alone film. This high tensile residual stress was caused by the constraint of the shrinkage of the film by a rigid silicon substrate. When the residual stress in the electroplated copper thin films is higher than their yielding stress, stress-induced migration should be activated in the films, and both void and hillock formations should be accelerated in the thin-film interconnections.

In addition, the gradient of their thermal stress during the

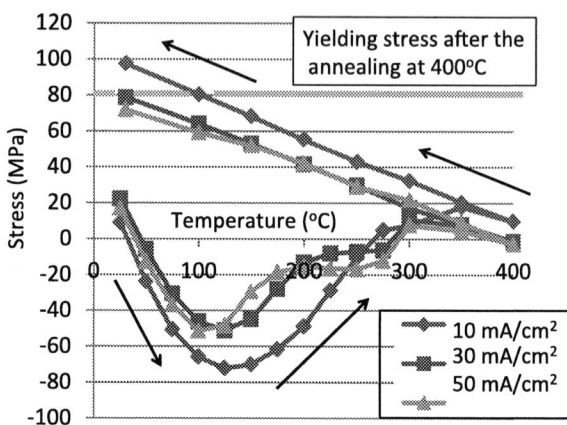

Fig. 2 Change of the residual stress in the electroplated copper thin film during annealing at 400°C

(a) Initial   (b) At 200°C   (c) At 400°C

Fig. 3 Change of the micro texture of the electroplated copper thin film as a function of annealing temperature

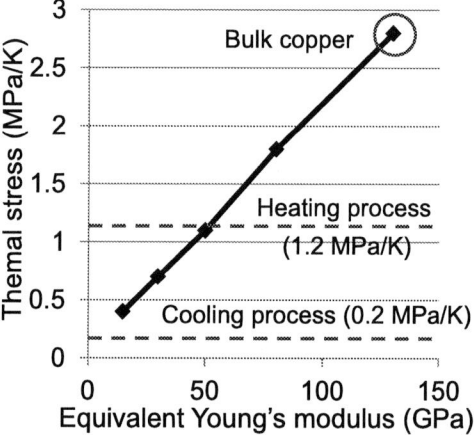

Fig. 4 Comparison of the measured thermal stress with the estimated thermal stress as a function of equivalent Young's modulus of the electroplated copper thin film

cooling process was clearly different fro that during the heating process. It was about 1.2 MPa/K in the heating process up to 100°C, while it was about 0.2 MPa/K in the cooling process. Since this gradient was mainly caused by the difference in the coefficient of thermal expansion between the copper thin film and the silicon substrate, this change of the gradient indicates that mechanical properties of the films annealed at 400°C were different from those of the as-electroplated films. Figure 4 shows the comparison of the measured thermal stress with the estimated thermal stress calculated by using a finite element method. The estimated thermals stress using material properties of bulk copper was 2.8 MPa/K, and it was much higher than the measured result

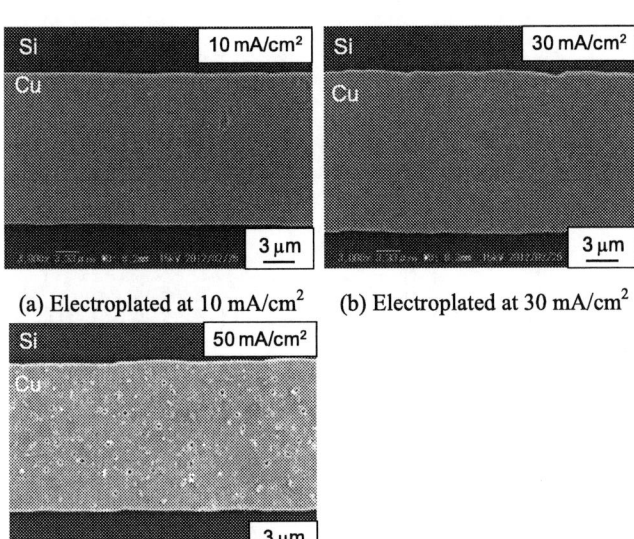

(a) Electroplated at 10 mA/cm²　　(b) Electroplated at 30 mA/cm²

(c) Electroplated at 50 mA/cm²

Fig. 5 Change of the surface morphology of the interconnections 15 days after the annealing at 400°C

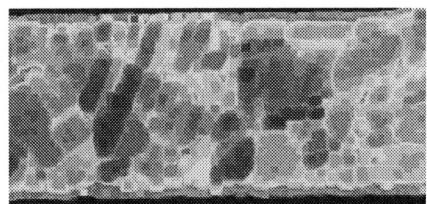

(a) IQ map of as-annealed film

(b) IQ map of the film after 15 days

Fig. 6 Change of the crystallinity of grains in the annealed film caused by stress-induced migration

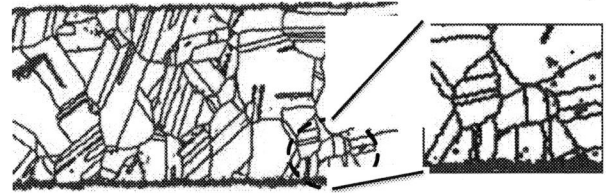

(a) CI map of as-annealed film

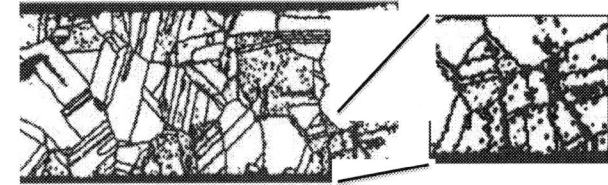

(b) CI map of the film after 15 days

Fig. 7 Change of the crystallinity of grain boundaries in the annealed film caused by stress-induced migration

of 1.2 MPa/K The estimated value agreed well with the measured value when the assumed Young's modulus of the electroplate copper thin film was about 50 GPa. This value agreed well with the measured result obtained from the simple tensile test of the s-electroplated copper thin film [13]. The thermal stress measured during the cooling process agreed well with the estimated one when the equivalent Young's modulus of the annealed film was assumed to be 10 GPa. These values are clearly different from the Young's modulus of bulk copper. Thus, this measured results indicated that the mechanical properties of the electroplated copper thin films vary drastically depending on their electroplating conditions and thermal history after the electroplating.

Long-Term Reliability of the Interconnections

Figure 5 shows the change of the surface morphology of the interconnections 15 days after the annealing at 400°C as a function of the current density during electroplating. After the annealing at 400°C, some small voids appeared on the surface of all the interconnections during its cooling process. But the surface morphology of the films annealed at temperatures lower than 200°C did not change even after 15 days after the annealing. However, a lot of voids and hillocks appeared on the interconnection electroplated at 50 mA/cm² as shown in Fig. 5(c). This surface morphology change was caused by the stress-induced migration in the interconnection, and the stress-induced migration was activated by the residuals stress higher than its yielding stress as shown in Fig. 2. Since this high tensile stress occurred due to the constraint of the shrinkage of the porous copper thin film by the surrounded rigid substrate, it is indispensable for evaluating the crystallinity of the electroplated copper thin films just after electroplating to assure the long-term durability of the interconnection.

The change of the crystallographic quality of the interconnections caused by the stress-induced migration was evaluated by using the EBSD method. This method can evaluate the quality of grain boundaries quantitatively by using two parameters obtained from the signal processing of the measured Kikuchi pattern. First parameter is Image Quality (IQ value), which indicates the periodic regularity of the atomic alignment in the observed area. The second parameter is Confidence Index (CI value), which shows the reliability of the analyzed crystallographic orientation of the observed area [14]. The CI value was used for the determination of the position of a grain boundary, and the IQ value was average quality of the grain boundary. Since the stress-induced migration accelerates the diffusion of copper atoms along grain boundaries, the appearance of hillocks on the surface of the interconnections indicates that there are a lot of vacancies in the interconnection. The accumulation of the vacancies degrades the quality of the grain boundary, and it should cause the high Joule heating and thus, local fusion at that grain boundary. Figure 6 shows the change of the IQ value map in the interconnection electroplated at the current density of 50 mA/cm². In these color maps, the red color indicates the area with high crystallinity and the blue color shows the area with low crystallinity. It is clear that the crystallinity of most grains in this film degraded due to the

978-1-4799-2408-0/14 $31.00 © 2014 IEEE　　　1887

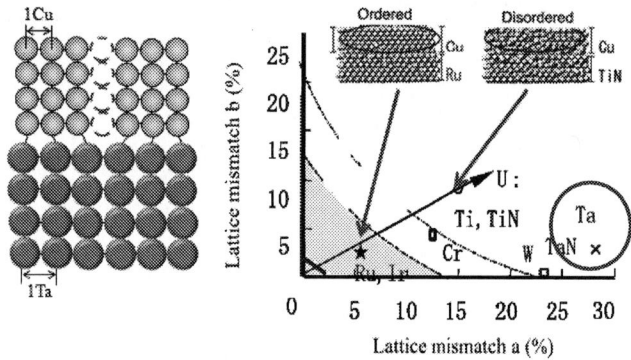

Fig. 8 Lattice mismatch between copper and other materials used for the thin-film interconnections[14]

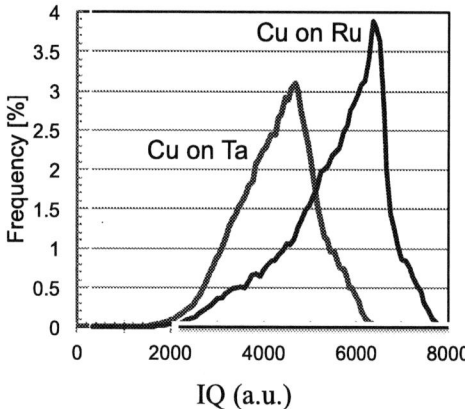

Fig. 9 Difference in the crystallinity of the copper thin film electroplated on tantalum and ruthenium

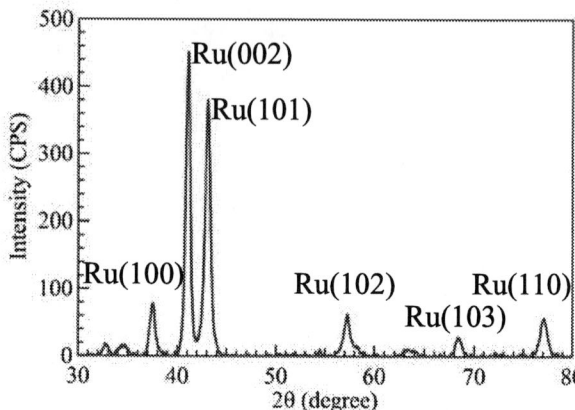

Fig. 10 X-ray diffraction spectrum obtained from the thin ruthenium layer

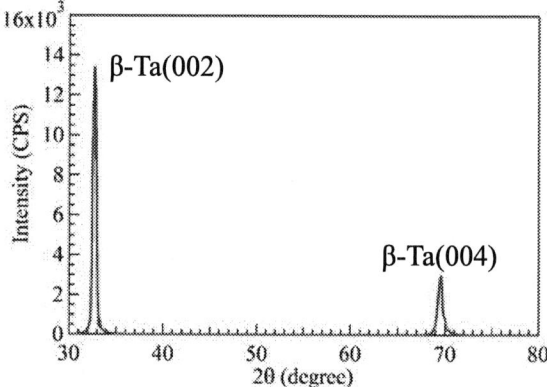

Fig. 11 X-ray diffraction spectrum obtained from the thin tantalum layer

stress-induced migration. But the size of the grains did not change. The change of the crystallinity of the grain boundaries is summarized in Fig. 7. In these maps, the blue lines indicate the matched grain boundaries and the red ones are random grain boundaries. Some grain boundaries were heavily damaged, while others were not. It is clear that most matched gran boundaries did not change so much, while most random grain boundaries showed the expansion of their area. These results indicate that the stress-induced migration proceeded mainly around random grain boundaries. In addition, the crystallinity of the grains surrounded by random grain boundaries was degraded seriously.

The main reason for the degradation was the acceleration of the stress-induced migration of copper atoms and this acceleration was caused by the low crystallinity of grain boundaries of the interconnection films, in other words, the high density of vacancies around the grain boundaries. This high density of defects in the interconnection occurs in the interconnection due to the high mismatch in the lattice constant between copper and tantalum, which is used for a diffusion barrier layer against copper, as shown in Fig. 8. Since there is the lattice mismatch of about 18% between them, the atomic alignment of copper formed on tantalum should be disordered seriously and the quality of the electroplated copper should be low. In addition, the porous grain boundaries with high density of vacancies should grow

in the electroplated copper thin films in order to reduce the high lattice mismatch. This low crystallinity enhances the diffusion of copper atoms in the films. Therefore, it was concluded that the crystallinity of the grain boundaries of the electroplated copper thin films varies the long-term reliability of the film. It is very important, therefore, to improve the quality of their crystallinity in order to assure the long-term reliability of electronic products.

Ruthenium is one of the effective materials, which can minimize the lattice mismatch with copper. The expected mismatch is about 5% when the ruthenium film consists of strong (001) crystallographic orientation as shown in Fig. 8. Figure 9 shows the effect of the introduction of the ruthenium thin layer as a seed layer on the crystallinity of the electroplated copper thin film. It is clear that the quality of the copper thin film electroplated on the ruthenium layer is higher than that on the tantalum layer. In addition, the full at half maximum width of the film on ruthenium was much smaller than that on tantalum. This result indicates that the ruthenium layer also improved the homogeneity of the quality of the film. However, there still remained the grains with low crystallinity in the film on the ruthenium layer. The reason for this fluctuation was the random crystallographic orientation of the ruthenium layer as shown in Fig. 10. Not only grains with (001) crystallographic orientation, but also grains with other

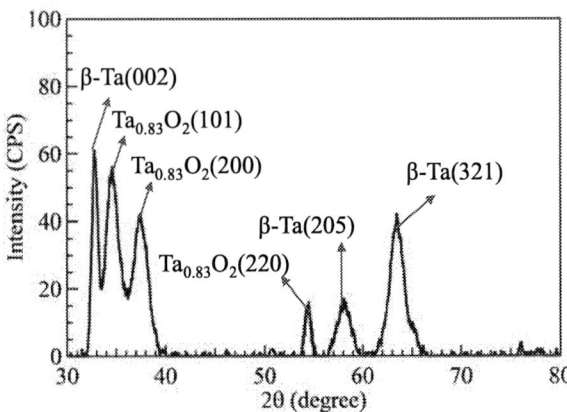

Fig. 12 X-ray diffraction spectrum obtained from the thin tantalum-oxide layer

orientation grew in the film. It was difficult to form the strong (001)-oriented ruthenium thin film on the tantalum layer.Thus, another method for improving the crystallinity of the electroplated copper thin film was discussed.

Improvement of the crystallinity of the electroplated copper thin film by surface oxidation of tantalum layer

Figure 11 shows the x-ray diffraction spectrum obtained from the diffusion barrier layer of tantalum. This layer consisted of the strong (001)-orientated film. This was the main reason for the degradation of the crystallinity of the electroplated copper formed on this layer. Because the lattice mismatch of other crystallographic plane of ruthenium is higher than that of (001) plane. One of the other effective methods for minimizing the lattice mismatch between copper and tantalum is destruction of this highly (001)-oriented surface structure of tantalum. It is expected that an amorphous layer under the electroplated copper thin film can reduce the lattice mismatch with copper. In order to destroy the strong surface orientation of the tantalum layer, the slight surface oxidation of the layer was added after the deposition of the tantalum layer. Figure 12 shows the x-ray diffraction spectrum obtained from the surface oxide layer. Since the measured spectrum indicates low intensity and broad full at half maximum width, the crystallinity of the grown tantalum oxide was quite low with random orientation.

Figure 13 summarizes the effect of the surface material before electroplating on the crystallinity of the electroplated copper thin films. There are three IQ maps in this figure. The red color region indicates the grains with high quality and the blue color region shows the grains with low crystallinity. It is clear that the crystallinity of the electroplated copper thin film on the ruthenium layer was improved drastically comparing with that on the tantalum layer. Furthermore, the crystallinity of the film formed on the tantalum oxide layer was highest among the three films as was expected. This result clearly indicates that it is very important to control the crystallinity of the surface layer before electroplating in order to improve the crystallographic quality of the electroplated copper thin films.

Figure 14 shows the effect of the improvement of the crystallinity of the electroplated copper thin-film interconnections on their long-term reliability under high

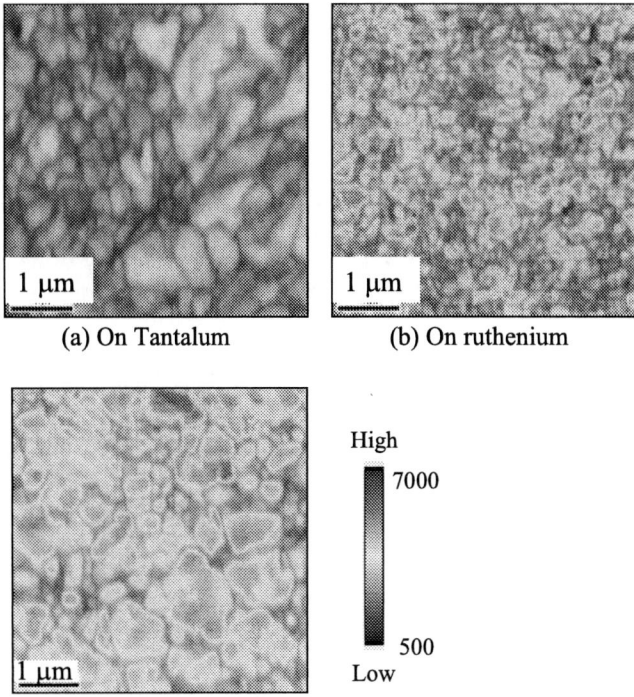

(a) On Tantalum   (b) On ruthenium

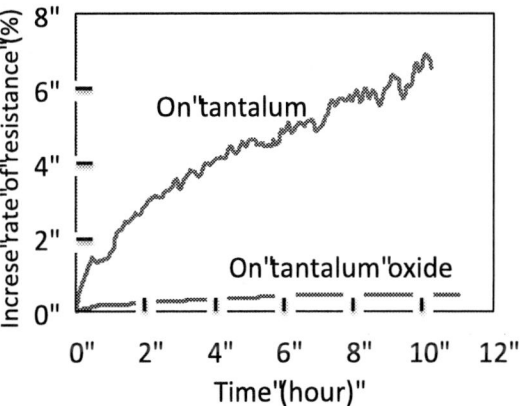

(c) On tantalum oxide

Fig. 13 IQ map of the copper thin film electroplated on various seed layer

Fig. 14 Change of the resistance of the copper interconnections electroplated on the tantalum and tantalum oxide layers during the electromigration test at 10 MA/cm^2

current density. The resistance of the interconnection formed on the tantalum layer increased quickly under the application of the current density of 10 MA/cm^2. This increase was caused by the degradation of the crystallinity of the interconnection due to electromigration. However, the resistance of the interconnection formed on the tantalum oxide layer did not change under the same condition. This result validates that the crystallinity of the electroplated copper thin film dominates the long-term reliability of the interconnection. Even when the electroplating condition is fixed, the crystallinity of the electroplated copper thin film varies significantly depending on the surface crystallographic structure of the seed layer used for electroplating. It is, therefore, very important to form the thin layer on the

978-1-4799-2408-0/14 $31.00 © 2014 IEEE

tantalum diffusion barrier layer, which can minimize the lattice mismatch between copper and tantalum in order to assure the long-term reliability of the electroplated copper thin-film interconnections.

Conclusions

The effect of the crystallinity of the electroplated copper thin-film interconnections was investigated by applying a conventionally used EBSD method. The crystallinity of both grains and grain boundaries in the electroplated copper thin films was evaluated by analyzing the quality of Kikuchi patterns obtained from the observed area. It was validated that the periodic regularity of the atomic alignment in the films varied drastically depending on their electroplating conditions such as the seed layer material and the current density during electroplating. Both the mechanical and electrical properties of the copper films changed significantly depending on their crystallinity. In addition, the long-term reliability of the interconnections varies drastically. The degradation process of the crystallinity of the interconnections during both electromigration and stress-induced migration tests was dominated by the grain boundary diffusion of copper atoms and it was dominated by the crystallinity of the interconnections.

The initial quality of the as-electroplated copper thin film was dominated by the lattice mismatch between copper and the seed layer material used for electroplating. It is effective to introduce a thin ruthenium layer on the tantalum layer for improving the crystallinity of the copper thin films electroplated on it. Furthermore, the slight surface oxidation of the tantalum layer improves the crystallinity of the copper thin film electroplated on it, and thus, the long-term reliability of the interconnection. It is, therefore, very important to form the thin layer on the tantalum diffusion barrier layer, which can minimize the lattice mismatch between copper and tantalum in order to assure and improve the long-term reliability of electronic products.

Acknowledgments

This research was partially supported by the Ministry of Education, Sports, Science and Technology: Special Coordination Funds for Promoting Science and Technology. The authors would also like to thank Prof. Mitsumasa Koyanagi and Prof. Masayoshi Esashi of Tohoku University for their kind support and helpful discussions.

References

1. S-C. Chang, J-M. Shieh, K-C. Lin, B-T. Dai, T-C. Wang, C-F. Chen, M-S. Feng, and C-P. Lu, " Investigation of effects of bias polarization and chemical parameters on morphology and filling capability of 130 nm damascene electroplated copper", *J. of Vacuum Science and Technology B*, vol. 19, (2001), pp. 767-773.

2. Rui, H., Robl, W., Ceri, H., and Dehm, G, "Stress, Sheet Resistance, and Microstructure Evolution of Electroplated Cu Films During Self-Annealing, *IEEE Trans. on Device and Materials Reliability*, vol. 10, (2010), pp. 47-54.

3. L. Lu, M-L. Sui, and K. Lu, "Superplastic extensibility of nanocrystalline copper at room temperature", *Science*, vol. 287, (2000), pp. 1463-1466.

4. C. T. Lin, and K. L. Lin, "Effects of current density and deposition time on electrical resistivity of electroplated Cu layers", *Journal of Materials Science; Materials in Electronics*, vol. 15, (2004), pp.757-762.

5. A. V. Vairagar, S. G. Mhaisalkar, and Ahila Krishnamoorthy, " Electromigration behaviour of dual-damascene Cu interconnections-Structure, width, and length dependences", *Microelectronics Reliability*, vol. 44 (2004), pp. 747-754.

6. H. Lee, S. S. Wong, and S. D. Lopatin, "Correlation of stress and texture evolution during self- and thermal annealing of electroplated Cu films", *Journal of Applied Physics*, col.93 (2003), pp. 3796-3804

7. Naoki Saito, Naoakzu Murata, Kinji Tamakawa, Ken Suzuki, and Hideo Miura, "Evaluation of the Crystallinity of Grain Boundaries of Electronic Copper Thin Films for Highly Reliable Interconnections", *Proc. of IEEE ECTC 2012*, No. 173, (2012), pp.1153-1158.

8. Ryosuke Furuya, Chuanhong Fan, Osamu Asai, Ken Suzuki, and Hideo Miura, "Improvement of the Reliability of TSV Interconnections by Controlling the Crystallinity of Electroplated Copper Thin Films", *Proc. of IEEE ECTC 2013*, No. 572, (2013), pp.1153-1158.

9. Rittinon Pornvitoo, Osamu Asai, Ken Suzuki and Hideo Miura, "Effect of Microtexture in Electroplated Copper Thin Films on Their Thermal Conductivity", *Proc. of EMAP2013*, No.0084, (2013), pp. 1-4. (CDROM)

10. Ryosuke Furuya, Chuanhong Fan, Osamu Asai, Ken Suzuki, and Hideo Miura, "Evaluation of the Crystallographic Quality of Electroplated Cpper Thin-Film Interconnection Embedded in a Silicon Substrate", *Proc. of the ASME 2013 International Technical Conference and Exhibition on Packaging and Integration of Electronic and Photonic Microsystems*, No. 73148, (2013), pp. 1-6. (CDROM)

11. Miura, H., Sakutani, K., and Tamakawa, K., "Fluctuation Mechanism of Mechanical Properties of Electroplated-Copper Thin Films Used for Three Dimensional Electronic Modules", *Key Engineering Materials*, vol. 353-358, (2007), pp. 2954-2957.

12. Tamakawa, K., Sakutani, K., and Miura, "Effect of Micro Texture of Electroplated Copper Thin Films on Their Mechanical Properties", H., *Journal of The Society of Materials Science Japan*, vol. 56, (2007), pp. 907-912.

13. N. Murata, N. Saito, K. Suzuki, and H. Miura, "Effect of the crystallinity of electroplated copper thin filmson their mechanical and electrical reliability", *Proc. of JIEP ICEP 2012*, (2012), pp. 1-6. (CDROM)

14. Tomio Iwasaki and Hideo Miura, Molecular dynamics analysis of adhesion strength of interfaces between thin films, Journal of Materials Research, Vol. 16, No. 6, (2001.6) pp. 1789-1794.

978-1-4799-2408-0/14 $31.00 © 2014 IEEE

Process, Assembly and Electromigration Characteristics of Glass Interposer for 3D Integration

Chun-Hsien Chien[1], Ching-Kuan Lee[1], Chun-Te Lin[1], Yu-Min Lin[1], Chau-Jie Zhan[1], Hsiang-Hung Chang [1], Chao-Kai Hsu [1], Huan-Chun Fu[1], Wen-Wei Shen [1], Yu-Wei Huang [1], Cheng-Ta Ko[1], Wei-Chung Lo[1], Yung Jean (Rachel) Lu[2]

[1]Electronics and Optoelectronics Research Laboratories (EOL), Industrial Technology Research Institute (ITRI),
Hsinchu, Taiwan
[2]Corning Incorporated
[#]Tel: 886-3-5917854, Fax: 886-3-5917357, Email: ch_chien@itri.org.tw

Abstract

Glass interposer is proposed as a superior alternative to organic and silicon-based interposers for 3DIC packaging in the near future. Because glass is an excellent dielectric material and could be fabricated with large size, it provides several attractive advantages such as excellent electrical isolation, better RF performance, better feasibility with CTE and most importantly low cost solution.

In this paper, we investigated the EM performance of Cu RDL line with glass substrate. Three different physical properties of glass materials were used for studying the EM performance of Cu RDL line. The used testing conditions are under 150~170 ℃ and 300~500mA. The glass type material with best performance was applied for glass interposer process integration and assembly investigation. Therefore, a wafer-level 300mm glass interposer scheme with topside RDLs, Cu TGVs, bottom side RDLs, Cu/Sn micro-bump and PBO passivation has been successfully developed and demonstrated in the study. The chip stack modules with glass interposer were assembled to evaluate their electrical characteristics. Pre-conditioning test was performed on the chip stacking module with the glass interposer to assess the reliability of the heterogeneous 3D integration scheme. All the results indicate that the glass interposer with polymer passivation can be successfully integrated with lower cost processes and assembly has been successfully developed and demonstrated in the study.

Introduction

Primary approach of 3DIC packaging usually adopts organic substrates or silicon interposer as the inter-medium between multi-integrated circuits (ICs) and printed circuit board. Base on Moore's law and the market demands of high data rate and speed, high I/O densities and fine pitch are especially addressed for the applications of server, mobile and communication systems.

While manufacturabilities of TSVs and organic substrates continues to improve, there still are difficult challenges around cost and electrical performance that are prompting consideration of alternative solutions for interposer applications. Glass as an interposer material provides a number of merits of the intrinsic electrical property as an insulator, excellent insertion loss in high frequency application, better feasibility with CTE, process cost reduction and large size availability.[1-4]

Therefore, the main focus of this paper is on (a) EM performance of Cu RDL line with different types of glass substrate and characterization, (b) wafer level integration in TGV formation, two RDL on the front-side, one RDL on the backside and polymer-based PBO for the passivation, copper interconnects designed in glass substrate also eliminates the need for an oxide barrier layer on both via and backside surface, which provides the benefit of cost reduction and reduced complexity (c) assembly process of silicon chip stack on the glass interposer with daisy chain resistance measurement. The glass interposer was assessed to have successfully integrated and is potentially to be applied for 3D product applications.

Experimental

Electromigration (EM) evaluation of copper redistribution line on glass substrate

Electromigration (EM) is one of the primary reliability failure mechanisms in the advanced package. EM induced voiding due to the diffusion of metal atoms in the direction of electron flow. The voids occurred bring the RDL line open, and then circuit operation fail. In addition, to avoid the EM phenomenon copper will instead of aluminum due to it have easy processes and good electrical properties. In this investigation, electromigration phenomenon of Cu line was studied to determine the mechanisms in the different glass substrate.

Three different physical properties of glass substrates used for electromigration (EM) evaluation. The material properties of three glass substrates listed in table 1. Then, metal formation applied on the glass substrate. Copper metal was adopted and intended of aluminum line for RDL, due to copper has higher thermal conductivity and lower electric resistivity, thermal expansion as well as temperature coefficient of electric resistivity. The process flow and wafer level test vehicle were shown in the Figure. 1.

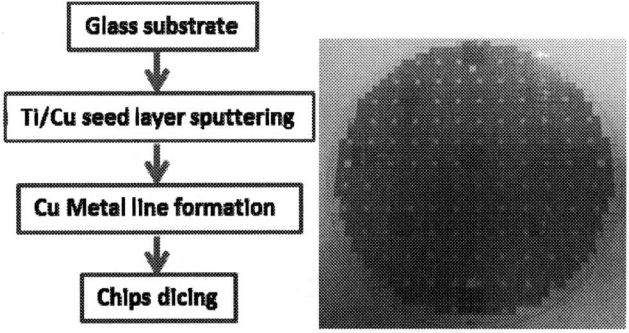

Figure 1. process flow and wafer level test vehicle

Figure 2 showed the test vehicle package structure and EM evaluation pattern. The test vehicle consisted of a 1000 x 40 x 6 um Cu RDL line. Test conditions were shown in Table 2. The Cu line was applied a constant electric DC current of

300mA and 400mA as well as the test vehicle including packaging structure put into the oven under constant temperature of 125^{0}C. The current density was calculated by electric DC current divide the cross-section area of Cu RDL line which the current density is 125 kA/cm^2.

Table 1. Glass Material Properties

	Glass A	Glass B	Glass C
Young's module (Gpa)	73.6	64.0	71.5
Poisson's	0.23	0.2	0.21
Density (g/cm^3)	2.38	2.2	2.42
CTE (ppm/°C)	31.7	32.5	8
Dielectric constant	5.3	4.6	7.24
Thermal conductivity (W/m-k)	1	1.2	1.35

(a)

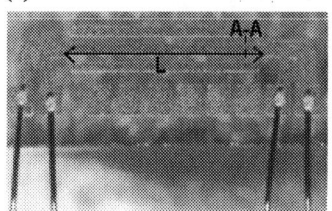 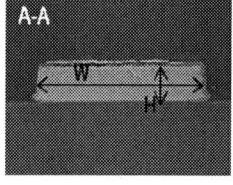

(b) LxWxH=1000x40x6 um

Figure 2. (a) package structure; (b) EM evaluation pattern

Results and discussion
Electromigration (EM) evaluation of copper redistribution line on glass substrate

Figure 3 showed the resistance versus evaluation time for EM test of Cu RDL. The failure was define the circuit was functionally increasing or open. The criterion of resistant was increasing of 10% or open circuit. According to the Fig. 3, the resistance of test vehicle including substrate A, B and C are stable. It means the electromigration didn't occurred alone the Cu RDL line under evaluation condition of the current density is 125 kA/cm^2 and temperature is 125 ^{0}C after 500 hours. Therefore, all of three glass substrate can be selected for interposer used. . In addition, EM can be explanted from the Black's equation [5][6]:

$$N_f = \frac{C}{J^n} e^{\frac{E_a}{RT}}$$

where J is the current density, E_a is activation energy, R is the Boltzmann constants, T is the kelvin temperature, C and n are the experimental constant. According to this equation, the dominate of failure life time including temperature and current density. In the future, the higher current density will be applied for three type of glass substrate. In this paper, the Cu atom will along the RDL line moving if EM phenomenon occurred. Furthermore, another EM path will be also investigated, shown in Fig. 4. According to the Fig.4, we will investigate another EM phenomenon which the Cu atom will transferred along glass substrate surface.

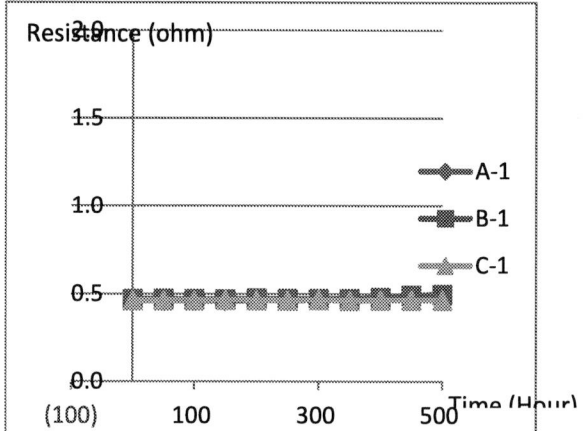

Figure 3. The resistance versus evaluation time

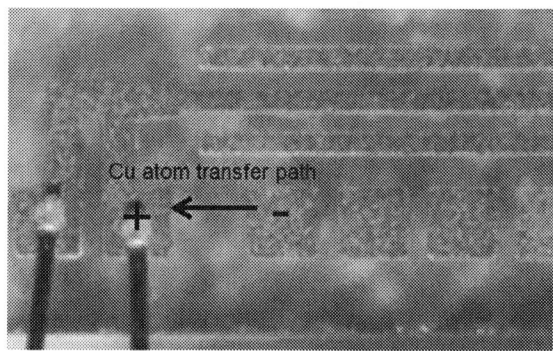

Figure 4. Different EM mechanism

300mm glass interposer evaluation

The interposer of test vehicle is 300mm glass wafer. The process for glass interposer with 100um thickness was fabricated by present machine for Si interposer. The interposer includes a glass substrate with two metal layers (RDL) and one metal layer which were disposed on the opposite sides of the substrate. The processes in detail are shown in figure 5 and described in below.

978-1-4799-2408-0/14 $31.00 © 2014 IEEE

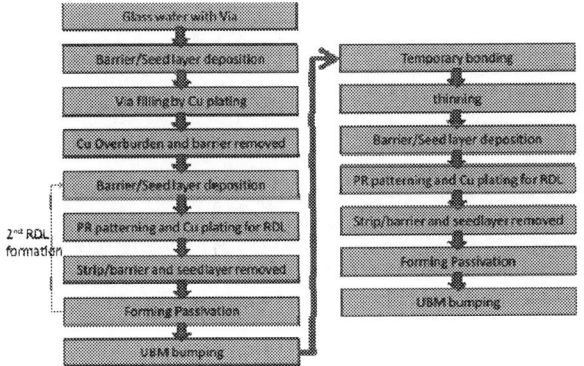

Figure 5. The process flow for 300mm Glass interposer

The initial 300mm glass wafer with via is provided from Corning Co. After Ti barrier and Cu seed-layer process on wafer, Cu plating with a bottom-up mechanism is applied to minimum the overburden on the wafer surface. Cu overburden and Ti barrier are removed by wet etching process. The via formation and overburden removed are major differences between Glass interposer and Si interposer, the cost of process for glass interposer will be simply easier and cheaper than dry etching for through Si Via. The uniformity of Cu wet etching is very important. The result is shown in figure 6. The deviation between Cu and Glass is below 1um.

For top RDL (line-width = 20μm), Cu plating process with seed layer (Ti/Cu) wet-etching process is applied. The polymer-based PBO material is used for isolation and passivation, the equipment and process for proceeded PBO is cheaper than oxide or nitride inorganic-based isolation. This process temperature is below 200°C. Because of the concern for glue of temporary bonding, the low temperature process for passivation formation is needed. Top UBM (Cu/Sn) is formed with a top passivation opening and then do temporary bonding process. For temporary bonding process, the glass substrate shows no stopper here. By using C shape or round shape glass substrate with standard notch, the temporary bonder could do the fully automatic process without any issues. The bonding quality could be easily inspected by eyes after bonding. Since the CTE of the glass substrate could be adjusted, the warp could be controlled after bonding. Table 2 shows the temporary bonding result of 5 pairs of 300 mm glass-Si bonding. The thickness of the glass is 700 μm and glue thickness is around 110 μm. The average TTV is 6.5 μm and the average warp is 55.4 μm.

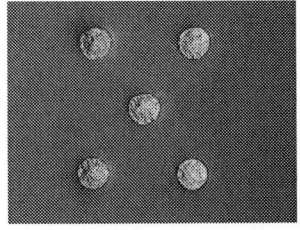

Figure 6. After wet etching of Cu overburden

Table 2. TTV and warp data after bonding (Unit: μm)

Wafer No.	Si Average Thickness	Si TTV	glass + glue	glass + glue TTV	Total Thickness	Total TTV	Warp
1	777.7	0.2	813.1	6.4	1590.8	6.4	48.7
2	776.0	0.3	817.7	5.8	1593.7	5.7	33.0
3	775.6	0.1	807.7	6.5	1583.3	6.5	57.8
4	773.8	0.5	806.0	7.3	1579.8	7.2	49.7
5	776.2	0.5	810.4	6.5	1586.6	6.3	87.6

ITRI collaborated with Ad-STAC member Disco for glass thinning and TSV reveal process. Grinding/CMP is used for the Cu revealing (Fig. 7). There are three steps for thinning process, containing two grinding steps and one polish step. The glass thickness is 100um.

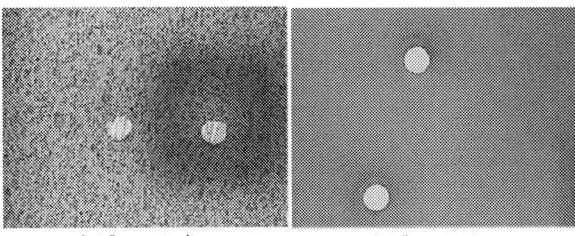

Figure 7. process for backside grinding, a)after grinding; b)after CMP

We used the different wheel to test grinding process. The result is displayed in Figure 8 and Figure 9. The spindle current of wheel A(6.4 A) is lower than Wheel B(8.4 A). The wheel wear of Wheel A is almost half compared to Wheel B.

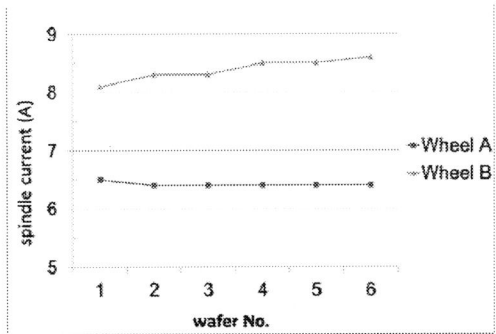

Figure 8. grinding test: spindle current for different wheel

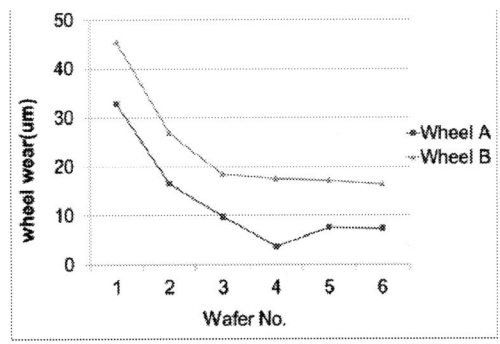

Figure 9. grinding test: wheel wear for different wheel

After grinding/CMP the glass surface roughness is shown in figure 10. The roughness can meet our criteria.

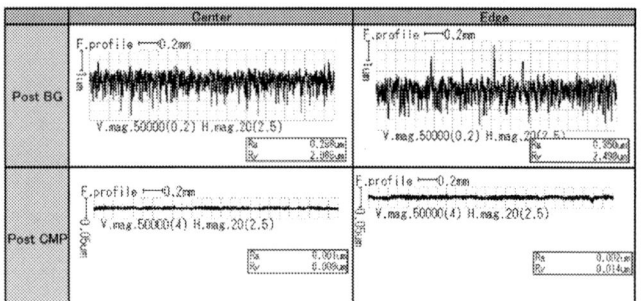

Figure 10. Glass wafer surface roughness after BG/CMP process.

And then, backside RDL is formed by Cu plating after lithography patterning. Bottom UBM is formed with a PBO passivation opening of 85~120μm.

The photo of interposer from optical microscope is shown on Figure 11&12, the backside pad of glass interposer is clearly seen through from the view of front side. The test vehicle for 300mm Glass interposer is successfully processed and the photo is shown in Figure 11&12.

Assembly and Reliability

After the fabrication process of glass interposer, the assembly of chip-on-glass interposer stack was proceeding. The fluxless chip-to-chip bonding scheme was selected and thermo compression bonding process with gap-height control was adopted for the chip stack. Before the bonding process, both the surfaces of chip and glass interposer was pretreated by Ar/H$_2$ mixed plasma for 10 minutes to remove the tin oxidation layer upon the surface of solder micro bumps. After several trial-run boning tests, the optimized bonding condition of 275°C for 10 seconds dwell time under the bonding force of 15N with appropriate gap-height setting was determined, and the joined-well solder micro bump interconnections could be obtained.

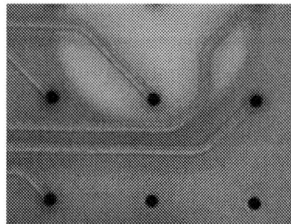

Figure 11. optical microscopic photo of interposer

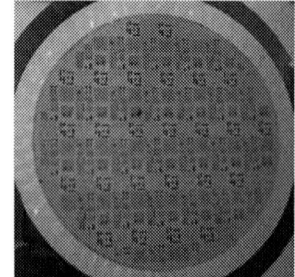

Figure 12. 300mm Glass interposer after debonding

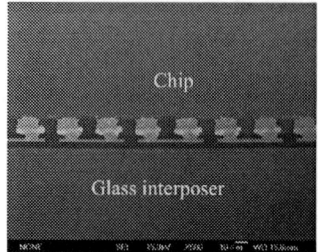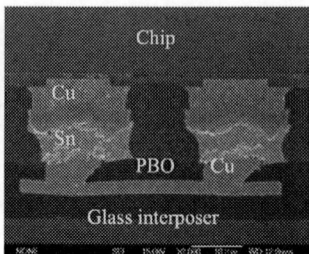

Figure 13. The cross-sectioned microstructure of daisy-chain solder micro bump joints within the chip-on-glass interposer stack.

Figure 13 showed the cross-sectioned microstructure of daisy-chain solder micro bump joints and the pillar-like solder micro joints could be found. By the bonding method with gap-height control, the joint shape could be well controlled to avoid the occurrence of solder squeezing during bonding when compared to the normal thermocompression bonding mode. Under such bonding method, more tin solder volume still could be remained within the solder micro joints, as seen in Fig. 13. The thickness of the remained tin layer was about 4μm while the gap height between chip and glass interposer was around 15μm. In addition, Fig. 13 presented that the Ni$_3$Sn$_4$ and Cu$_6$Sn$_5$ IMCs with a thickness of about 2μm formed upon the UBMs of chip and glass interposer after bonding process. After chip bonding process, underfill dispensing was conducted to fill the gap between chip and glass interposer. C-SAM inspection results indicated that the gap could be fully filled without any voids.

After assembly process, reliability test of pre-conditioning was selected to evaluate the reliability performance of chip-on-glass interposer stack. The test conditions were that firstly the samples were baked at the temperature of 125°C for 24 hours and followed by soaking under the environment of 30°C/60%RH for 192 hours. Finally, the samples were reflowed 3 times under the peak temperature of 260°C. Before reliability test, the daisy chain resistances of 134 micro joints measured from those tested samples were in the range of 19.4 ~ 22.0 Ω. However, no any daisy chain resistances could be measured again after pre-conditioning test, which meant that all the samples were failed after test.

To find out the reason for electrical failure after reliability test, the cross-sectioned microstructure analysis was conducted. Figure 14 showed the representative microstructure of micro joints between chip and glass interposer after reliability test. The image displayed that a crack propagated along the interface of underfill/ PBO passivation layer and intersected all the micro interconnections as well. The propagating crack either crossed the interface of IMCs/tin solder or passed the interface of Cu UBM/RDL in glass interposer side. This circumstance, therefore, caused the totally electrical failure of micro joints after reliability test.

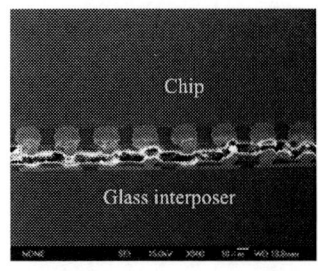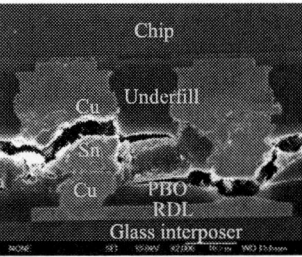

Figure 14. The representative microstructure of micro joints between chip and glass interposer after reliability test.

From the results of microstructure analysis, it was difficult to indicate the origin of the propagating crack during reliability test though the interfaces mentioned above were probable locations for the start of crack. However, delamination of underfill/PBO interface was an unusual failure phenomenon when compared to the case of Si chip-to-Si chip stack where the passivation layer upon the chip was Si_3N_4 or SiO_2. We considered that the material compatibility of the underfill used in this investigation and PBO passivation layer needed to be evaluated in detail especially under environmental testing. All in all, we had successfully demonstrated the process, assembly and electrical characteristics of glass interposer in this study, and the glass interposer showed the highly potential feasibility for 3DIC integration.

Conclusions

A wafer-level 300mm glass interposer scheme includes 1) Electromigration evaluation with different glass substrates, 2). Wafer-level glass interposer fabrication with two topside RDLs, Cu TGVs, one bottom side RDL, Cu/Sn micro-bump and PBO passivation, 3). characterization of glass interposer assembly has been successfully developed and demonstrated in the study.

1. Electromigration phenomenon didn't happened under 125 kA/cm2 and 125^{0}C after 500 hours for three different glass substrate. Therefore, all of glass substrate can be selected for interposer used. Different electromigration path and high current density will be investigated for the different glass substrates in the further.

2. The glass interposer has been demonstrated with potential low cost in glass via process (TGV 30um is formed by Corning), no CMP process (Cu overburden removal by wet process) and polymer-based PBO process is used for passivation. The structure is analyzed and demonstrated by SEM analysis.

3. Well-connected solder micro bump joints could be obtained in glass interposer module under the peak bonding temperature of 275°C with a dwell time of 10 seconds and the bonding force of 15N.

4. By measuring the daisy-chain solder micro interconnections, the daisy-chain resistance of 134 micro joints with a diameter of 18μm ranged between 19.4 to 22.0 Ω. That a crack propagated along the interface of underfill/ PBO passivation

layer and intersected all the micro interconnections was the reason for electrical failure after pre-conditioning test. We considered that the material compatibility of the underfill used in this investigation and PBO passivation layer needed to be evaluated in the near future.

Acknowledgments

This research is supported by the ITRI-Corning collaboration project and the Ministry of Economic Affairs (MOEA), Taiwan, ROC. Authors greatly thank to the Ad-STAC members, AMAT, SUSS, DISCO, and Brewserscience for the experiment assistance

References

1. Phillip Garrou "Qualcomm's Nowak: 3-D Faces Cost Issues" Semiconductor International, October/6/2009

2. Yannou, J.-M. Neuilly, F. Moreno, J.-A. Pommier, M. Bellenger, S. Biermans, P "NXP System-in-Package vision and latest 3D technology developments"ADVANCING MICROELECTRONICS, VOL 34;NUMB 6, pages 16-21 2007

3. M. Young, The Technical Writer's Handbook. Mill Valley, CA: University Science, 1989.

4. Sukumaran, V.; Chen, Q.; Fuhan Liu; Kumbhat, N.; Bandyopadhyay, T.; Chan, H.; Min, S.; Nopper, C.; Sundaram, V.; Tummala, R., "Through-package-via formation and metallization of glass interposers," Electronic Components and Technology Conference (ECTC), 2010 Proceedings 60th , vol., no., pp.557,563, 1-4 June 2010

5. J. R. Black, "Electromigration failure modes in aluminum metallization for semiconductor devices", Proceedings of the IEEE, Vol. 57, no. 9, pp. 1587–1594, 1969.

6. Y. S. Lai, C. L. Kao, Y. T. Chiu and B. K. Appelt, "Electromigration reliability of redistribution lines in wafer-level chip-scale packages", Electronic components and technology conference, pp. 326-331, 2011.

Improved PCB Via Pattern to Reduce Crosstalk at Package BGA Region for High Speed Serial Interface

Yujeong Shim and Dan Oh

Altera Corporation
101 Innovation DR, San Jose, CA, 95051, USA
Dr.yujeong.shim@ieee.org / yshim@altera.com
+1-408-544-7000

Abstract

Crosstalk is one of the root causes to violate jitter compliance in high speed serial link, as the data rate increase and signal density goes up. To minimize cancellation or crosstalk impact, many techniques have been investigated. However, unlike horizontal crosstalk, crosstalk at vertical interconnection has limitation to eliminate. In particular, coupling occurs at the interface of chip and PCB is one of the unavoidable sources of crosstalk. Among three coupling types, TX to TX, RX to RX and TX to RX, TX to RX coupling is detrimental compared to others since the aggressor TX swing is large with high slew rate and the victim RX signal has low swing and slow edge. RX jitter performance is more sensitive to noise coupling. Therefore, it is important to investigate crosstalk impact on RX performance on the system level including IOs, package and PCB. In this paper, the impact of transmitter to receiver crosstalk induced at package balls and PCB vias is characterized. Critical factors causing this near-end crosstalk are identified and analyzed. Additionally, the PCB via pattern is proposed to minimize field interference between TX and RX as simply rotating TX via pair perpendicularly to RX via pair. It is demonstrated that the proposed structure reduces crosstalk significantly.

Keywords—crosstalk, near-end crosstalk, transceiver crosstalk, ball crosstalk, via crosstalk, crosstalk cancellation

I. Introduction

Crosstalk is one of the important factors determining the performance of the system as data rate increases and signal density goes up. Even though it is nature that coupling becomes more according to frequency increase, timing budget allocated for crosstalk gets tighter and tighter due to less timing margin. Crosstalk is induced between all kinds of interconnections on chip, package and PCB. There have been a lot of papers and efforts to cancel crosstalk or minimize its impact on system performance. However, more analysis has been done for horizontal interconnections rather than vertical interconnections such as ball grid arrays (BGAs), socket pins and vias [1]-[4]. In spite of much shorter length of vertical interconnections than horizontal interconnections, vertical coupling issue is not negligible due to difficulty of control. Package BGA are not only source of vertical coupling but also socket pins and PCB vias connecting BGAs and PCB. Location of PCB vias and pads is dedicated depending on package ball assignment (ball map). In order to reduce coupling, further distance between two conductors helps. However, package ball pitch is fixed so that distance between signals is not controllable for vertical interconnection. Another way is adding ground balls surrounding important signals so that PCB also has shielding vias around signals, accordingly. Nevertheless, it is difficult to achieve perfect shielding .

Transmitter (TX) to receiver (RX) coupling is the most significant among TX to TX, RX to RX, TX to RX and RX to TX, since TX to RX coupling impact is more significant than other cases since RX signals lose high frequency contents through lossy channels, while TX slew rate is still high at the package ball area. Therefore, TX channels' high frequency signals are couple more to RX signal than the other way around. Especially, crosstalk impact could be even worse for long reach application that many of high speed serial link protocols support. In case of FPGA, it is common that every TX and RX run individually with different protocols although many high speed serial channels don't run TX and RX simultaneously. TX to RX coupling is studied and its impact on receiver performance is characterized in [5].

In this paper, TX to RX coupling impact on RX performance is characterized by the bath-tub curve measurement using an internal bit error checker. From a sort of characterization, it has been demonstrated that near-end crosstalk from transmitter to receiver impact is not ignorable. As well as cross talk impact characterization, the PCB escaping pattern is proposed and implemented to minimize cross talk through vertical interconnections. Even though BGA pins of TX and RX are laterally placed, TX PCB via pair is able to be located perpendicularly to RX via pair with short neck lines. The proposed dog bone PCB via structure enables to minimize field interference as changing field direction orthogonally and putting more distance between TX and RX vias. In order to verify this proposal, the test vehicle has been manufactured to model package BGAs and PCB vias. Both the typical PCB via structure and the proposed dog bone structure are implemented or comparison. Frequency domain measurement shows that the reduction is about 10dB to 15dB in the fundamental frequency region. Improvement is also observed by time domain noise measurement and the internal eye monitoring circuit.

II. Signal Coupling at Vertical Interconnection and its characteristics

There have been intensive study and many design techniques to cancel horizontal coupling. However, most of these techniques are not able to be applied to vertical coupling. For high speed serial link design, horizontal coupling is well isolated while vertical coupling becomes significant along with impedance discontinuity. The package ball grid arrays (BGAs) on the bottom layer of the package For high speed differential channels, each pair of channels is surrounded by six ground balls for shielding. It is shown in Fig. 1 that one signal of a differential pair faces diagonally another signal because, typically, they are not perfectly shielded by ten ground balls due to cost increase and routing issue. Therefore, there is field interference between receiver and transmitter BGAs and PCB vias, which induces a common mode noise not to be cancelled by differential scheme. In many cases, these high speed signals are routed with strip lines on inner layers in PCB and package so that coupling impact is minimized. Length of these vertical interconnections is short compared to package or PCB traces. However, coupling is not negligible depending on signal condition of victims and aggressors since horizontal coupling is relatively better controlled than vertical coupling.

According to ball placement in Fig. 1 , there are three kinds of coupling, transmitter to transmitter, receiver to receiver and transmitter to receiver as depicted in Fig. 2. Since the balls are evenly placed, coupling ratio determined by physical dimension is exactly same for all three cases. Even though the coupling ratio is identical for each case, impact on performance is totally different. It is because amount of crosstalk is not determined by only physical dimension and dielectric material but also by frequency contents of aggressors. Moreover, crosstalk impact is not only depending on coupling but also on sensitivity of victims. Since cross talk impact on the system performance is more important than amount of crosstalk itself, we have to consider all of coupling ratio, aggressor's frequency contents/strength and victim's sensitivity. Definitely, transmitter is a more significant crosstalk source than receiver because the transmitter's signal edge is still sharp at the package substrate and the received signal is slowed down as passing through a lossy channel.

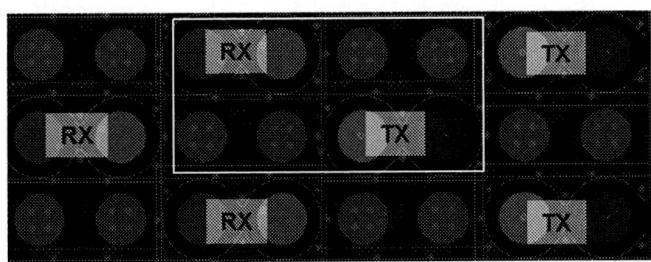

Fig. 1. The package BGAs on the bottom layer. One pair of channels are surrounded by 6 ground BGAs. P signal of one pair and N signal of another pair face diagonally so that there is field interference.

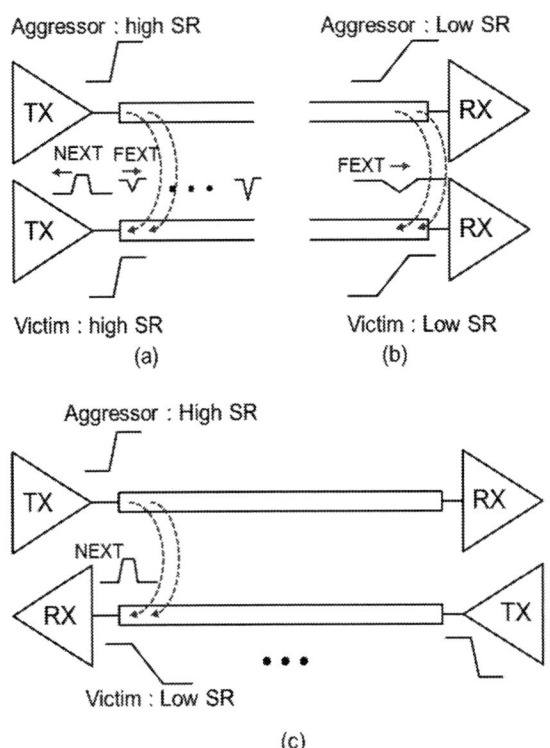

Fig. 2. There are three types of vertical coupling [5]. (a) transmitter to transmitter (b) receiver to receiver (c) transmitter to receiver

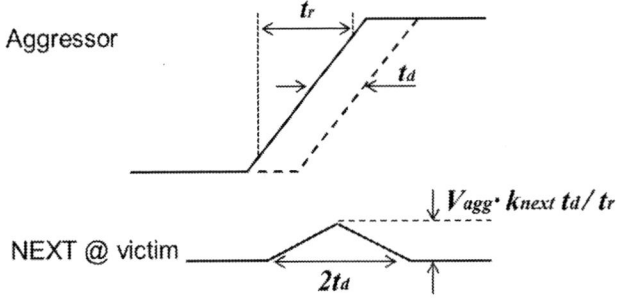

Fig. 3. Near-end crosstalk of short interconnection. The shape of NEXT is a triangle whose amplitude is inverse proportional to t_r (or t_f), when t_r/t_f is longer than t_d

For TX to TX coupling case in Fig. 2, TX slew rate is high to contain high frequency contents which are coupled easily to other adjacent channels. However, the sensitivity of victim to FEXT is not significant. Therefore, impact on the victim's jitter performance is not significant as well. In case of RX to RX coupling depicted in Fig. 2 (b) is not serious in spite of high sensitivity of receiver because aggressor's high frequency contents are filtered out by lossy channels. Crosstalk type is far-end crosstalk (FEXT) as both aggressor and victim move forward in the same direction for RX-RX and TX-TX coupling cases.

Unlike RX to RX or TX to TX coupling, crosstalk type between TX and RX is near-end crosstalk (NEXT). The slowed down receiver signal is sensitive to any noise coupling and TX has high frequency contents which have high coupling ratio. RX and TX signals' edges are not aligned timing wise. Therefore, the coupling impact is shown in voltage and also timing margin loss depending on victim / aggressor's slew rate, victim's voltage swing and dimension of vertical interconnections.

Typically, near-end crosstalk has a rectangular shape whose width is two times of flight time (td) and doesn't depend on tr/tf when the aggressor is a step function. However, if length of vertical interconnection is too short to reach the maximum voltage (knext x Vagg), NEXT has a triangle shape as shown in Fig. 3. The triangle's height and width are determined by rising (tr) / falling (tf) time and flight time (td) relating to length of interconnection. The height of the triangle is derived in [5] and noted in Fig. 3. Therefore, crosstalk impact on RX performance is bigger when the received signal has less slew rate and less voltage/timing margin, or the interconnection is longer as shown in **Error! Reference source not found.**. **Error! Reference source not found.** shows the coupling depending on the interconnection length. **Error! Reference source not found.**(a) is the measured crosstalk without a socket and PCB vias, while **Error! Reference source not found.**(b) is the one with a socket and 3mm PCB vias. The coupling is only 15 mV without long vertical interconnections (The oscilloscope's base noise is 9 mV). But the measured crosstalk is 61 mV with 1.5 mm socket pins and 3 mm PCB vias when there are two aggressors whose swing is 1V.

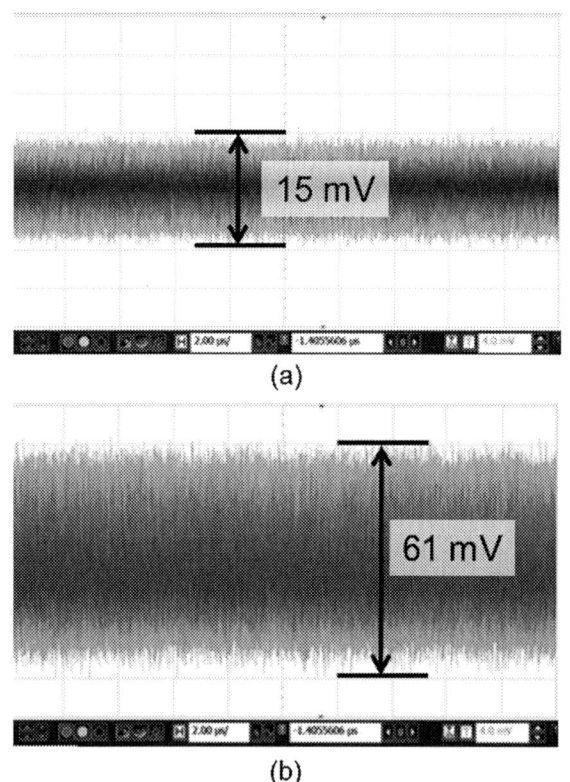

(a)

(b)

Fig. 4. Coupling ratio depending on interconnection legnth. 3mm PCB vias case has much more coupilng than the case without PCB vias. (a) measured crosstalk without a socket and PCB via is 15mV. (b) measured cross talk with a socket and 3mm PCB via is 61 mV.

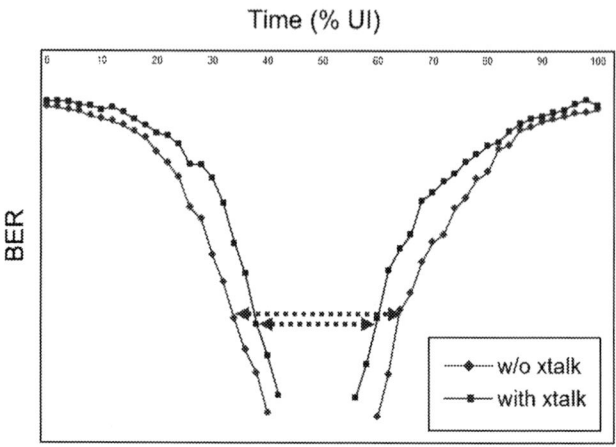

Fig. 5. Internal bath tub with and without TX to RX coupling. The red one is the bathtub without coupling and the blue one is with coupling. RX input swing is 120mV while two TX aggressor's swing is 1.0 V.

Fig. 5 shows internal bathtub measurement with and without TX to RX coupling when the RX input signal swing is 120 mV and TX output swing is 1V. The measurement is taken with CDR bypass (fixed clock phase) and sweeping input data phase. The red curve is the one without crosstalk and the blue one has two TX aggressors with 3mm PCB vias (**Error! Reference source not found.**(b)). The crosstalk impact on RX eye-opening is about 0.06 UI with the given RX input condition. Crosstalk impact depending on RX input condition is studied in [5]. For ultra high speed serial interface, this impact is not ignorable. Therefore, vertical interconnection design on PCB level is very important.

3. Proposal of PCB Via Patterns To Minimize Vertical Coupling and Its Impact on Receiver Performance

A. PCB Via Patterns to Minimize Vertical Coupling

In order to reduce TX to RX coupling impact on RX eye margin loss, there are two techniques. One is to cancel by using decision feedback equalizer (DFE). DFE improved sensitivity to crosstalk noise as recovering original signal. Another way is reducing coupling ratio by shielding or coding. To isolate important differential signals, two signal balls should be shielded by 10 ground balls.

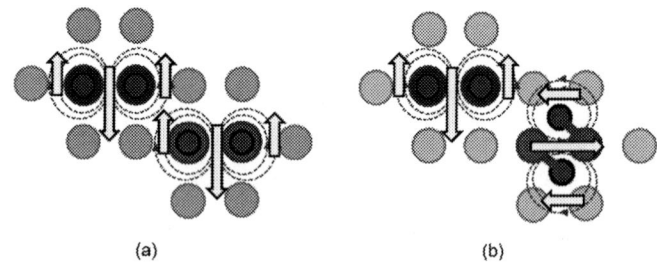

(a)   (b)

Fig. 6. (a) typical via location. There is field interference between TX and RX (b) Dog Bone structure. Field interference is minimal since their direction is orthogonal.

This shielding method costs too much as ground balls occupies signal placing/routing area. As shown in Fig. 1, 6 ground balls surrounding one differential pair are typical ball placement. Indeed, the coupling induced at BGA level is less than -50 dB which is not significant. However, since PCB via location is also dedicated depending ball map, long PCB via can induce severe coupling. The best way to route the channels on PCB is that RX or TX channels are on top level. However, due to lack of routing place limitation, this design can't be done always. Therefore, in this paper, the PCB via routing method to reduce coupling is proposed. To reduce coupling without cost increase, it is feasible to change signal assignment or via locations so that parallel field interference is minimized by as changing field to orthogonal direction. Fig. 6 (a) is the typical PCB via location and BGAs along with H-filed direction. It shows that the magnetic fields of RX and TX have interference. Fig. 6 (b) is the proposed dog bone structure that TX vias go straight down from BGAs all the way to the signal layer with the same manner in Fig. 6 (a), while RX vias are connected with short trace necks to be located diagonal to BGAs. Fig. 7 shows simulation results with 3 mm PCB vias in frequency domain. Fig. 7(a) is near-end crosstalk and (b) is far-end crosstalk. The red graph is coupling ratio of the original via structure, while the blue one is the coupling ratio of the dog bone via structure. In both near-end and far-end, coupling is reduced by about 10 dB.

There are two reasons that the proposed via structure reduces crosstalk. One is that there is minimal H field interference since TX and RX's field directions are orthogonal. Another reason is that the distance between TX and RX vias gets further to induce loose coupling and the short distance with ground vias increase shielding impact. Meanwhile, the distance between signal via and ground get closer by 0.3 times of ball pitch so that shielding is improved.

B. Measurement Results

The test coupon of the proposed dog bone via structure is implemented. For both the original via structure and dog bone structure, TX channels are routed on the bottom layer and RX channels are routed on the deep inner layer (16th). The via to link the top layer and bottom layer is 3 mm and the via from the top layer to 16th layer is 2.6 mm. The layout is exactly same with Fig. 6. There are ground and signal pads on the top layer according to BGA location. For the original structure, vias go down to inner/bottom layer. For the dog bone structure, TX vias go straight down, while the RX vias are located diagonally to the pads with very short traces. The measured coupling ratio in frequency domain is depicted in Fig. 8. The red curve shows the near-end crosstalk of the original structure while the blue curve is the NEXT of the dog bone via structure. Since this measurement includes PCB traces and connectors, Fig. 8 has difference with simulated NEXT curves. As shown in the simulated NEXT, about 10 ~ 15 dB reduction is achieved in Fig. 8.

Time domain measurement is also performed in Fig. 9. The aggressor TX's data rate is 12.5Gbps with 1V swing and 15 ps rising/falling time. There is one implemented aggressor. Fig. 9 (a) is the crosstalk of the original via structure. The measured peak to peak crosstalk is 32 mV with one TX aggressor. The below graph is the spectrum of the crosstalk. The center frequency is 6.25 GHz.

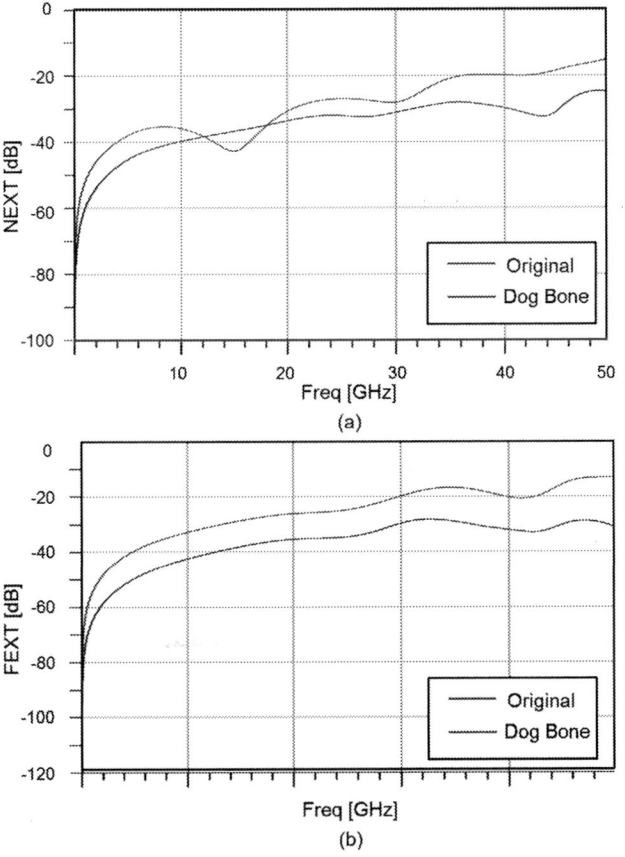

Fig. 7. Simulated crosstalk of the via structure. The red curve is the original structure and the blue one is the dog bone structure. (a) NEXT (b) FEXT

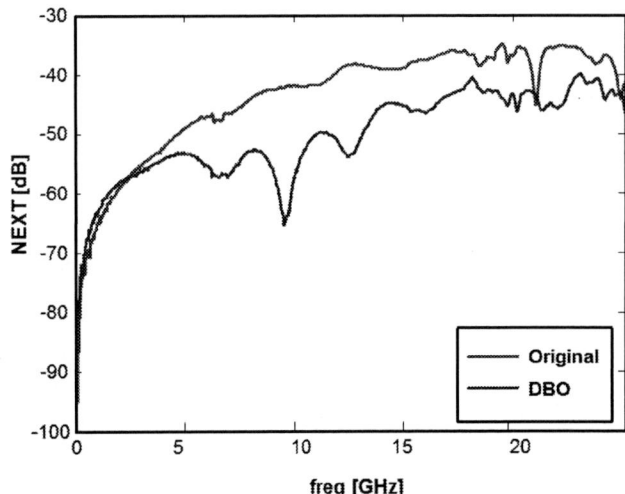

Fig. 8. Measured near-end crosstalk of typical via structure and dog bone structure including PCB traces. The original structure is appered in the red graph and the dog bone structure is shown in the blue curve.

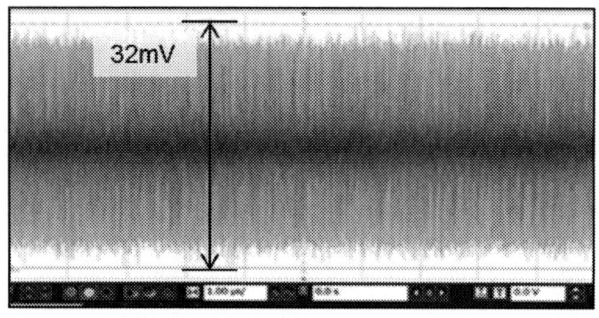

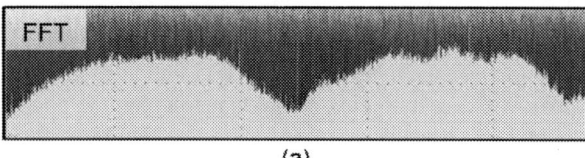

(a)

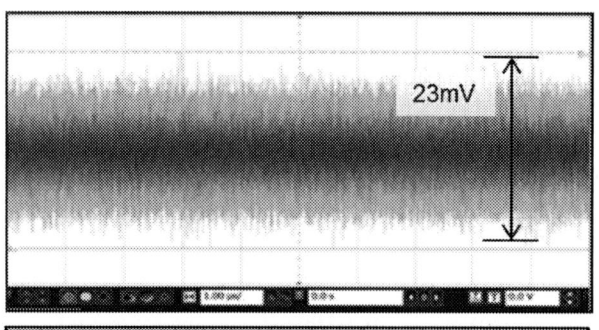

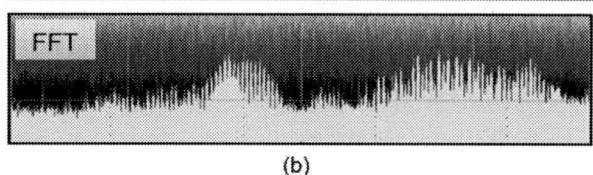

(b)

Fig. 9. Measured crosstalk profile in time domain and spectrum in requency domain (below) (a) original. The coupled noise is 32mV. (b) dog bone structure. The coupled noise is 23 mV, which is 70% of the original structure.

Fig. 9 (b) is the crosstalk of the dog bone via structure. The measured peak to peak crosstalk is 23 mV, which is only 70 % of the coupled noise of the original structure. As shown in the spectral diagram, remarkable coupling reduction is also observed.

The internal eye-monitoring circuit enables to measure received eye [6]. The horizontal unit of the phase interpolator is 1/32UI and the vertical step of the voltage slicer is 10mV. There is error term since the horizontal and vertical units are coarse. The aggressor signal is generated by 12.5Gbps TX channels whose output swing is controllable from 0V to 1.2V. TX swing is set with 1V and data pattern is clock pattern in this measurement condition. For the quiet condition case (no aggressor) as a reference, TX output drivers' swing is set with 0. The 120mV output (12.5 Gbps, PRBS9) of the pattern generator goes into RX through a 60cm SMA cable and the test coupon. As mentioned, the test coupon's via length is 2.6 mm. TABLE I. summarizes the eye opening (width and height) depending on aggressors' condition. The eye width is 0.5 UI and eye height is 50mV without TX coupling to RX in the given input condition. The original structure reduces the eye opening by 0.03 UI and 20 mV as this structure induced

32 mV of near-end crosstalk. Meanwhile, the proposed dog bone structure shrinks eye-margin by 0.02 UI and 10 mV. Therefore, the proposed structure improves eye-opening by 0.01 UI and 10mV. In case of signaling in the one direction, crosstalk affects timing directly. The crosstalk affects voltage margin loss as described in II. However, it affects timing as well when the received input signal is under stress condition, for example, high lossy RX channel, small swing, long tr/tf etc.

TABLE I.   TX TO RX COUPLING IMPACT

	Eye Width	Eye Height
No aggressor	0.50 UI	50mV
Original structure	0.47 UI	30mV
Dog bone structure	0.49 UI	40mV

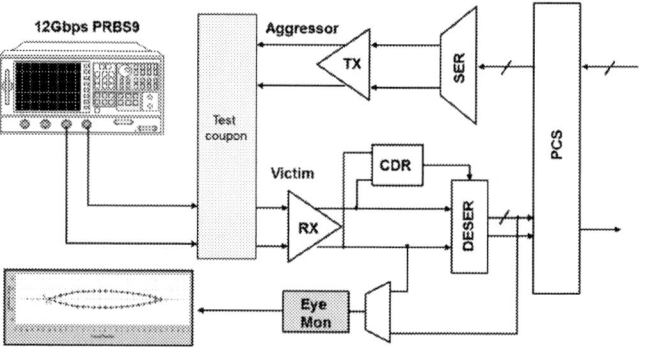

Fig. 10. Measurement setup for characterization of TX to RX coupling impact on RX eye performance.

4. Conclusion

Crosstalk cancellation techniques for vertical coupling have not been intensively studied unlike horizontal coupling. Therefore, it is important to understand vertical crosstalk impact on system performance and improve package/PCB design. It is shown that near-end crosstalk from transmitter to receiver impact is significant compared to TX to TX or RX to RX coupling. Even though, TX and RX are not aligned, coupling can affect RX performance depending on RX input condition, which could be easily under stress condition due to channels / back planes. In this paper, TX to RX crosstalk impact on receiver eye performance is addressed and characterized. Then, the new dog bone via structure is proposed to reduce TX to RX coupling at PCB vias. In order to verify the prosed PCB via structure, the test coupon has been manufactured. The proposed structure reduces crosstalk at fundamental frequency by 10 ~ 15dB. It is also proven by time domain measurement and internal eye-monitoring circuit. This proposal works well when the high speed differential channels are routed in deep PCB layers.

Acknowledgments

The authors would like to appreciate Alex Razmadze, Jianmin Zhang, Shishuang Sun, Yuri Tretiakov and Janani Chandrasekhar for supporting.

978-1-4799-2408-0/14 $31.00 © 2014 IEEE

References

1. X.Chen, J.E.Schutt-Aine, and A.C.Cangellaris, "A strategy to optimize modal signaling over microstip lines with spacing variablity," *Signal and Power Integiry (SPI), 17th worksghop on*, France, May.2013, p 1-4

2. X.Ye, "Intentional and un-intentional far end crosstalk cancellation in high speed differential link," *Electromagnetic Compatibility, IEEE international symposium on*, US, Aug.2011, pp791-796

3. Chung, S. Sudhakaran, V.Satagopan, S.Hwang, "Rigorous breakdown of crosstalk in single-ended high speed memory interface," *DesignCon 2012,* Santa Clara, US, Feb. 2012

4. S.H.Hall and H.L.Heck, "Advanced signal integrity for high speed digital designs," Wiley, 2009, pp177-193

5. Y.Shim, D.Oh, S.Sun, J.Zhang, J.Jiang, C. Nguyen, J.Chandrasekhar, and Yuri Tretiakov, "Characterization and Analysis of Vertical Coupling Impact on Receiver Perormance in High Speed Serial Interface," *IEEE EPEPS*, US, Oct.2013

A Wafer Level Through-Stack-Via Integration Process with One-time Bottom-up Copper Filling

Yunhui Zhu[1], Shenglin Ma[2, 1], Xin Sun[1], Runiu Fang[1], Xiao Zhong[1], Yuan Bian[1], Yong Guan[1],
Jing Chen[1*], Min Miao[1, 3], Yufeng Jin[1]
1. National Key Laboratory of Science and Technology on Micro/Nano Fabrication,
Peking University, Beijing, 100871, China.
2. Department of Mechanical & Electrical Engineering, Xiamen University, Xiamen, Fujian, 361005, China.
3. Inst. of Information Microsystem, Beijing Information Science and Technology University, Beijing, 100101, China.
*E-mail: j.chen@pku.edu.cn, phone: 86-10-62766595

Abstract

We reported a wafer level through-stack-via (TSV) integration approach for stacked memory module using one-time bottom-up copper filling. This bumpless TSV integration approach simplified the fabrication process and provided better reliability compared with solder based technologies. Silicon wafer with blind vias was first bonded to a carrier wafer face to face with pre-patterned BCB, and then thinned from backside to reveal the TSVs. The carrier wafer was coated with a release layer and a seed layer, which provided a uniform seed layer for bottom-up TSV filling and was easy to be debonded. A layer of copper RDL was pre-deposited on the silicon wafer before bonding, which enhanced the wettability of the sidewall of TSVs during bottom-up copper filling. More silicon wafers could be bonded and thinned in the same way. At last, one-time bottom-up TSV filling was performed and the carrier wafer was released. A 4-layer wafer stacking with TSVs of 173μm × 52μm has been successfully demonstrated with the thinnest wafer of 22μm. The electrical test results shown that this process had a significant yield improvement. The lowest resistance measured was 7.6mΩ with the yield of over 84% on the 4-inch wafer. This proposed TSV integration process was ready for stacked memory application.

Introduction

3D integration using through silicon via (TSV) has many advantages, such as high density, small form factor, high bandwidth and low power due to the short connection lengths [1-3]. Wafer to wafer (W2W) approach is attractive for 3D integration compared to chip to chip (C2C) or chip to wafer (C2W), in that W2W can provide high throughput and reduce the process cost [4]. In this paper, a wafer level TSV integration approach using one-time bottom-up copper filling was proposed. By swapping the procedure of TSV filling and wafer bonding, this bump-less TSV integration process reduces the process complexity and is supposed to have better reliability. Comparing with other 3D-TSV integration approaches, the proposed TSV integration approach has no need of solder bumping and underfill filling. This all-copper interconnect integration approach is expected to have better electrical and mechanical properties, as well as better scalability and reliability compared with soder-based bonding [5, 6]. A four-layer wafer level integration sample has been demonstrated in this paper using the proposed TSV integration approach. The electrical characteristics of the four-layer TSV integration sample has also been investigated.

Process Flow

The process flow of wafer level through-stack-via integration approach with one-time bottom-up copper filling was shown in fig. 1. This is a via-last process in which through multi-layer TSVs were filled after wafer bonding. With the coming wafers, TSVs were first etched from the front side with DRIE (Deep Reactive Ion Etching) process. A layer of silicon oxide was deposited with low temperature ICP-CVD (Inductive Coupled Plasma Chemical Vapor Deposition) process. Then, a layer of RDL was pre-deposited using double-layer spin coating technique without any residuals left in TSVs, as shown in fig. 1(a). A carrier wafer with seed layer on top of a release layer was prepared and bonded with the device wafer using wafer level pre-patterned BCB bonding, as shown in fig. 1(b). Then, the device wafer was thinned from backside until exposing the TSVs, as shown in fig. 1(c). After that, as shown in fig. 1(d), multiple wafers were bonded and thinned by repeating steps from fig 1(a) to fig. 1(c). Fig. 1(e) showed the bottom-up copper electro-chemical deposition (ECD) to fill the through-stack-via at one time. At last, the carrier wafer was debonded and the wafer level one-time bottom-up TSV filling process was finished.

Comparing with traditional TSV-enabled 3D integration approaches, the proposed process has no need of solder bumping and underfill filling, which removes the problems of bump contact, for instance, bump height uniformity, surface roughness, oxidation, or voids. Besides, a carrier wafer with an additional release layer was applied, which provided a high quality seed layer for bottom-up TSV ECD and was easy to be debonded.

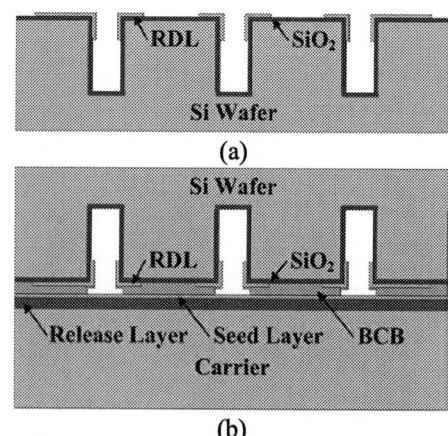

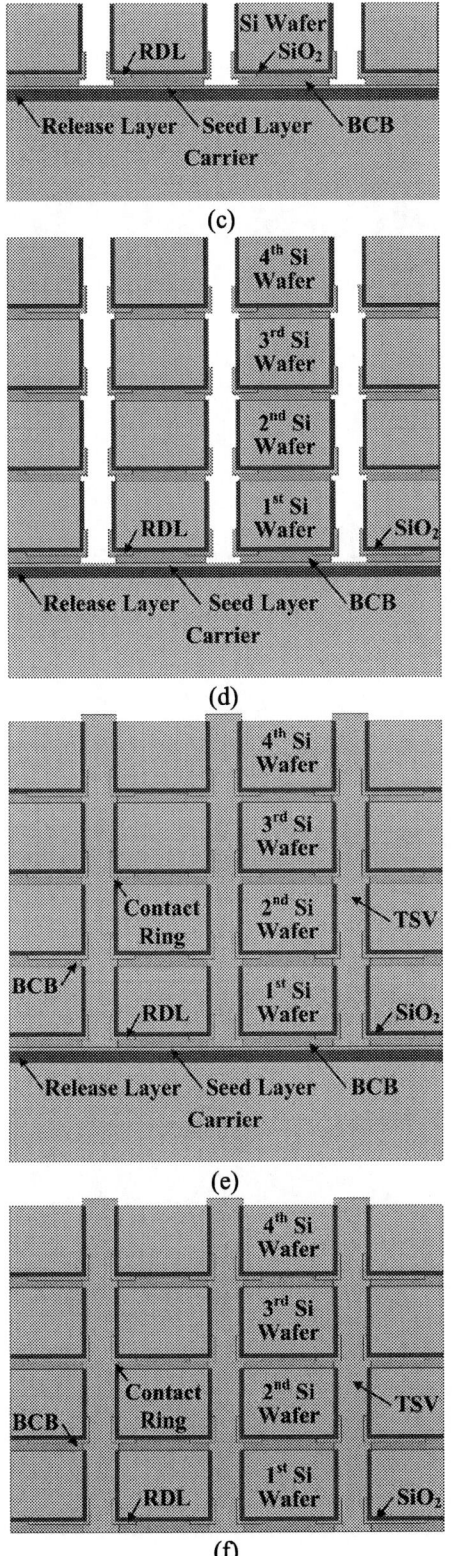

(c)

(d)

(e)

(f)

Figure 1. Process flow of the proposed wafer level TSV integration approach with one-time bottom-up copper filling. (a) TSV etching and RDL pre-deposition; (b) Wafer level pre-patterned BCB bonding with carrier wafer; (c) Backside thinning and TSV exposure; (d) Multi-layer wafer stacking; (e) One-time bottom-up TSV filling; (f) Release from carrier wafer.

Process Evaluation

The process evaluation was carried out on 4-inch wafers. Firstly, TSVs were etched with Bosch process to form blind vias with diameter of 50μm. A layer of SiO_2 insulation layer was deposited with low temperature ICP-CVD process at 75°C. Then a layer of RDL was deposited before TSV filling. To address the challenge of RDL pre-deposition on the nonplanar substrate surface in presence of blind vias, a double-layer photoresist spin coating technique was employed. The lower layer of photoresist with high viscosity was spin coated first to form a suspended film over TSVs, while the upper layer was a normal positive photoresist, as shown in fig. 2(a). This can prevent photoresist residuals left inside TSVs. After development, undercut was formed at the edge of the opening, which was helpful for the subsequent lift-off process. Titanium-tungsten and copper was sequentially sputtered and lifted-off to form the pre-deposited RDL, as shown in fig. 2(b). This double-layer RDL lift-off process was able to form electrical interconnection before via filling process.

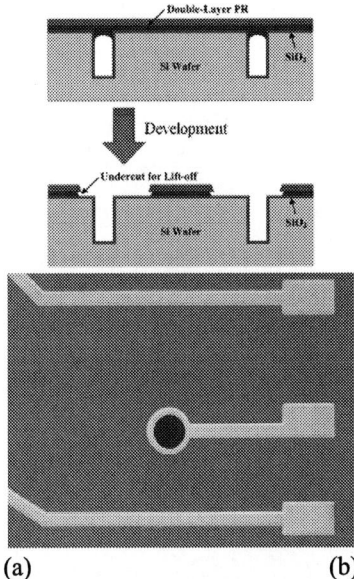

(a)   (b)

Figure 2. Double-layer spin coating technique for RDL pre-deposition. (a) Schematic diagram of double-layer spin coating; (b) RDL pre-deposition over TSV.

In the bottom-up filling scheme, two approaches are usually employed to provide the seed layer for copper electroplating. One method is direct metal seed layer deposition on one side of the wafer and electroplating sealing the TSV form this side [7, 8]. This method is for thicker wafers which can be processed in fabrication equipment directly. Another method is attaching a carrier wafer with seed layer to hold the tinned wafer [9, 10]. This method has the capability to handle ultra thin wafers and provides a uniform seed layer with good quality. In this paper, we employed the carrier wafer with a release layer process. Thus, the carrier wafer can be easily removed after bottom-up TSV filling.

Wafer level pre-patterned adhesive bonding is a critical process step in this 3D integration approach. A layer of photosensitive BCB was spin coated and patterned after RDL pre-deposition. Considering the large number of blind TSVs on the wafer surface, recipes for spin coating, pre-curing, exposure and development need to be fine tuned to prevent BCB residues left in TSVs. The BCB opening was larger than TSVs, to expose the contact ring, which connected the RDL with TSV when performing bottom-up TSV filling. Fig. 3 shows the microscope image of RDL pre-deposition and BCB pre-patterning with the optimized recipe. Fig. 4 shows the cross-section of wafer level pre-patterned BCB bonding and no residues of BCB is observed inside TSVs.

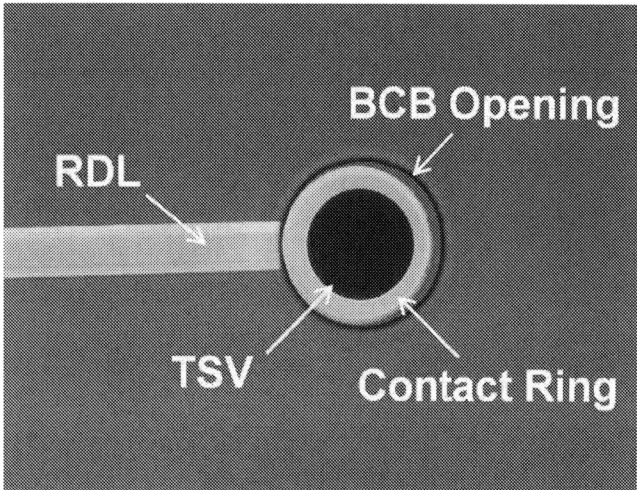

Figure 3. RDL pre-deposition and BCB pre-patterning.

Figure 4. Wafer level pre-patterned BCB bonding, no residues of BCB is observed inside TSV

The carrier wafer preparation including a release layer coating and a copper seed layer deposition. The silicon wafer and the carrier wafer were bonded in a SUSS CB6L wafer bonder. The bonding temperature was lower to 210°C in order to be compatible with the release layer. At the lower temperature, the bonding strength was enough to go through the subsequent backside thinning process. After bonding, ultrasonic inspection was performed to examine the bonding quality, as shown in fig. 5. Thickness of the bonded wafer was also measured, as shown in fig. 6. The thickness of the

carrier wafer and the silicon wafer was 500μm and 400μm respectively. That suggested the thickness of the add-on layers, including the release layer, the seed layer, the BCB bonding layer and the pre-deposited RDL, added up to be 21μm, with the bonding TTV of 6.1μm.

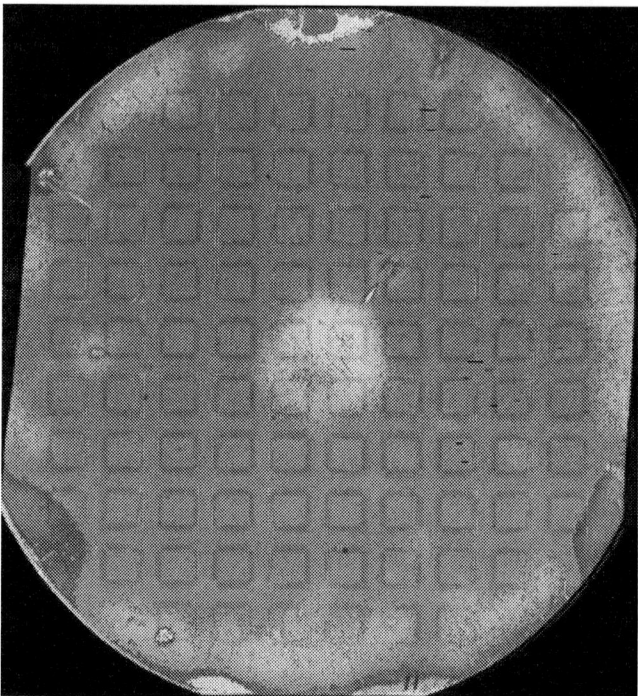

Figure 5. Ultrasonic inspection of wafer with blind TSVs bonded with the carrier wafer.

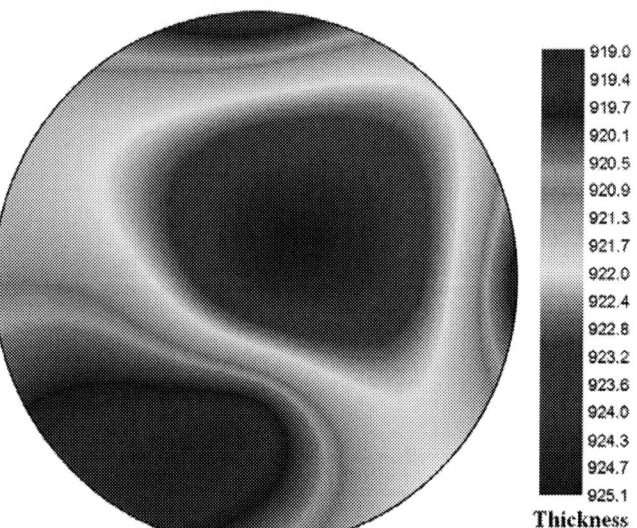

Figure 6. Wafer thickness mapping of wafer with blind TSVs bonded with the carrier wafer. (Mean=921.3μm, TTV=6.1μm)

After wafer bonding, backside thinning of the silicon wafer was performed to expose TSVs. The bonded wafer pair went through a standard 3-step backgrinding process including coarse grinding, fine grinding and polishing. The silicon wafer was thinned down to 50μm to 150μm, which showed the compatibility of the low temperature BCB pre-patterned wafer bonding with ultrathin wafer handling process. Fig. 7 shows the typical experimental results with the optimized thinning recipe. No obvious mechanical damage can be found at the opening of TSVs.

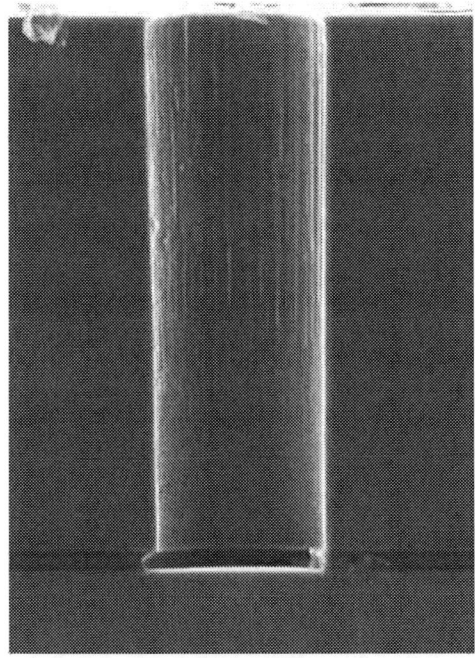

Figure 7. Backside thinning and TSV exposure after wafer bonding.

After backside thinning, more wafers with blind vias and pre-deposited RDL were bonded and thinned in the same way, as shown in fig. 8. No obvious misalignment can be observed in the cross section SEM image. In this paper, a 4-layer wafer stacking with TSV was demonstrated.

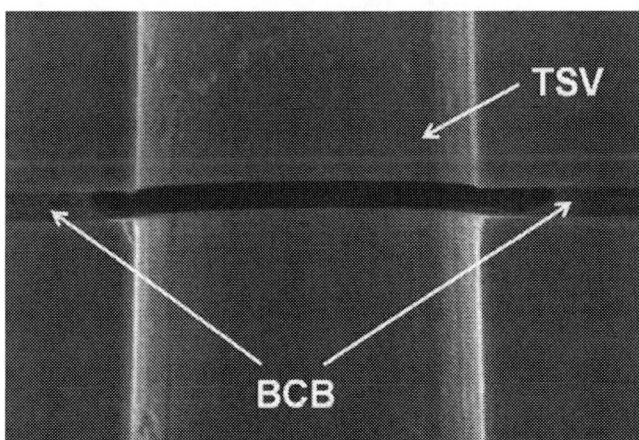

Figure 8. Multiple wafers with blind vias stacked together.

At the last step, one-time bottom-up filling of through-stack-via was carried out. As for bottom-up filling of TSVs, sidewall wettability is critical to the filling quality. Various methods are introduced to enhance the sidewall wettability, including ultrasonic treatment, plasma treatment, and SiN_x sidewall insulation [10]. In this paper, the pre-deposited RDL has form a thin layer of seed layer on the sidewall of TSV, which can enhance the sidewall wettability. The RDL between different layer of wafers are unconnected, which eliminates the concern of pinch off at the opening during TSV

bottom-up filling. Fig. 9 showed the X-Ray image of void free four-layer bottom-up TSV filling. The traces in light gray are copper RDL pre-deposited before TSV filling. Finally, the carrier wafer was removed, as shown in fig. 10. The SEM cross section image confirmed the quality and uniformity of one-time bottom-up filling of four-layer through-stack-via. The 4-layer bottom-up filling TSV has a diameter of 52μm with an aspect ratio of 3.3. The total thickness of four-layer stacked wafers was 173μm with the thinnest wafer thickness of 22μm.

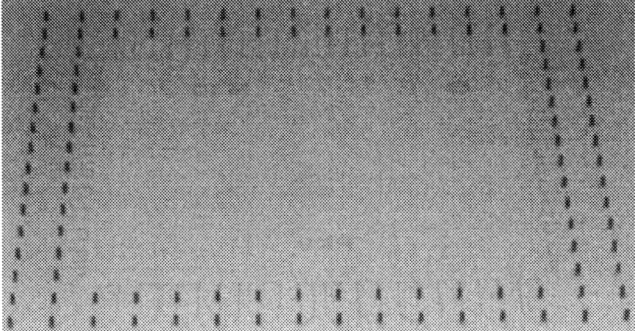

Figure 9. X-Ray image of void free four-layer bottom-up TSV filling.

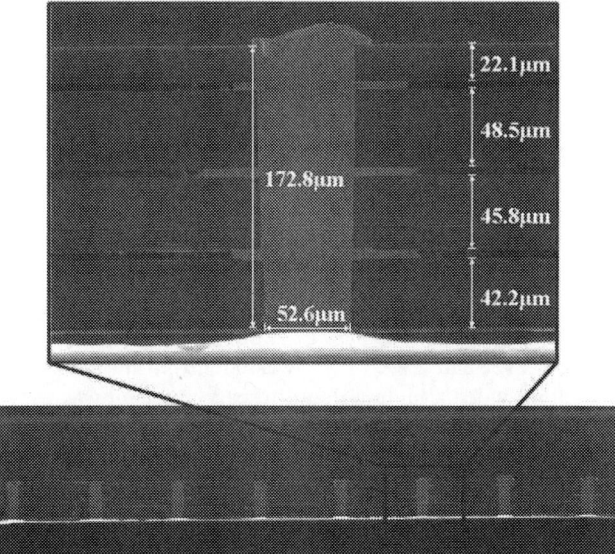

Figure 10. Four-layer wafer stacking with one-time bottom-up TSV filling.

Electrical Test Results

After bottom-up TSV filling, the top side of the stacked wafer was patterned with photoresist, followed by copper electro-chemical deposition. The copper trace on the top side acted as electrodes for the Kelvin test structure, as shown in fig. 11. Electrical measurement of single TSV resistance was carried out with HP 4156B semiconductor parameter analyzer and SUSS PM8 manual probe station. The current applied was from -25 to 25mA. Fig. 12(a) showed the measured resistance of single TSV under current stressing. The measured resistance of single TSV mostly matches the theoretical values. Fig. 12(b) showed the resistance distribution of single TSV Kelvin structure on the 4-inch wafer. The average resistance result of single TSV was

9.0mΩ, with the lowest and highest measured value being 7.6mΩ and 12.1mΩ respectively. The overall yield of the four-layer wafer level bottom-up filled TSV integration was calculated to be more than 84%.

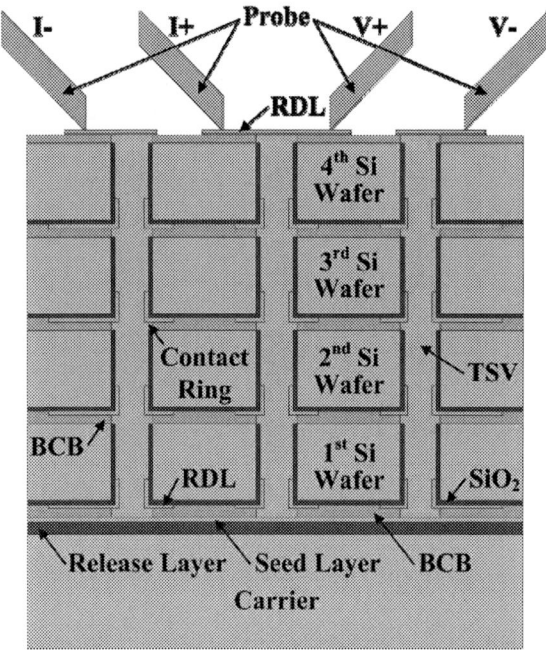

Figure 11. Schematic of four-point TSV Kelvin test structure.

Conclusions

In this study, a wafer level through-stack-via integration process with one-time bottom-up copper filling was developed. A four-layer wafer level bottom-up filled TSV integration sample has been successfully demonstrated. The total thickness of four-layer stacked wafers was 173μm with the thinnest wafer thickness of 22μm. A carrier wafer with a release layer and a seed layer was employed in this process, which provided a high uniformity seed layer and was easy to be debonded. The cross section and X-Ray inspection had confirmed the feasibility and quality of the proposed TSV integration process. The electrical test results showed that this process had a significant yield improvement. The lowest resistance measured was 7.6mΩ with the yield of over 84% on the 4-inch wafer. This proposed TSV integration process was ready for stacked memory application.

Acknowledgments

The work presented is funded by National Science and Technology Major Project of China (Project No. 2009ZX02038-02).

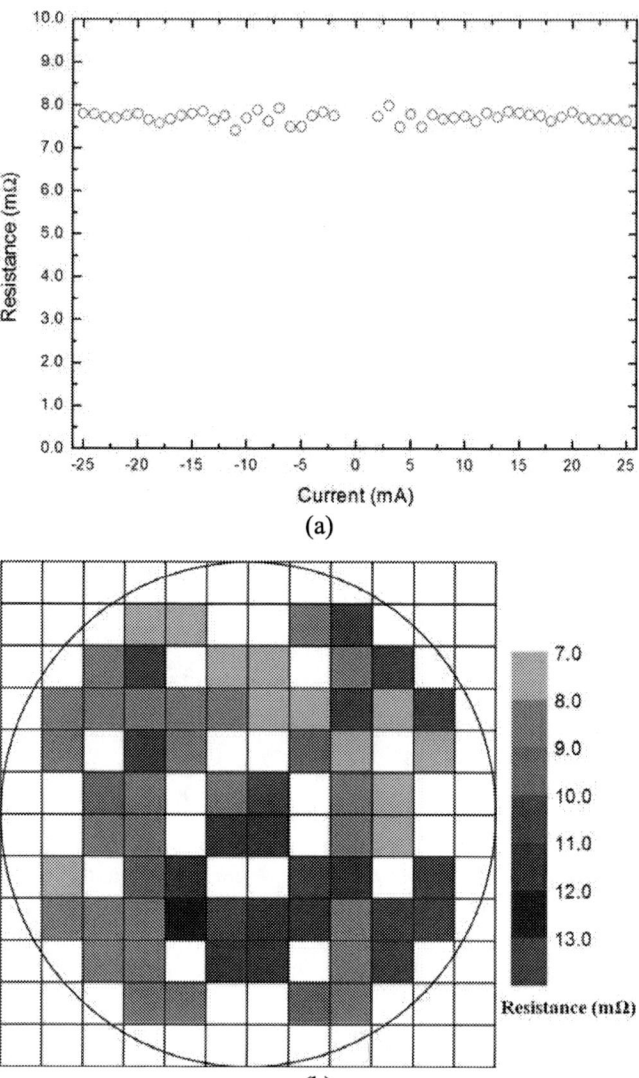

(b)

Figure 12. Test results of four-layer wafer level bottom-up filled TSV resistance with four-point Kelvin structure. (a) Resistance of single TSV under current stressing; (b) Wafer scale mapping of single TSV resistance.

References

1. G. Pares, "Full integration of a 3D demonstrator with TSV first interposer, ultra thin die stacking and wafer level packaging," in *Proc. IEEE Electronic Components and Technol. Conf. (ECTC)*, Las Vegas, NV, May 28-31, 2013, pp. 305-306.

2. B. Dang, "Three-Dimensional Chip Stack With Integrated Decoupling Capacitors and Thru-Si Via Interconnects," *IEEE Electron Device Letters*, vol. 31, no. 12, Dec. 2010, pp. 1461-1463.

3. J. U. Knickerbocker, "Three-dimensional silicon integration," *IBM Journal of Research and Development*, vol. 52, no. 6, Nov. 2008, pp. 553-569.

4. K. Fujimoto, "TSV (through silicon via) interconnection on wafer-on-a-wafer (WOW) with MEMS technology," in *International Solid-State Sensors, Actuators and Microsystems Conference, (TRANSDUCERS)*, Denvor, CO, June 21-25, 2009, pp. 1877-1880.

5. K. Hummler, "TSV and Cu-Cu direct bond wafer and package-level reliability," in *Proc. IEEE Electronic Components and Technol. Conf. (ECTC)*, Las Vegas, NV, May 28-31, 2013, pp. 41-48.

6. H. Son, "Reliability studies on micro-bumps for 3-D TSV integration," in *Proc. IEEE Electronic Components and Technol. Conf. (ECTC)*, Las Vegas, NV, May 28-31, 2013, pp. 29-34.

7. N. T. Nguyen, "Through-wafer copper electroplating for three-dimensional interconnects," *Journal of Micromechanics and Microengineering*, vol. 4, no.12, Jul. 2002, pp. 395-399.

8. Z. Wang, "Silicon micromachining of high aspect ratio, high-density through-wafer electrical interconnects for 3-D multichip packaging," *IEEE Trans. on Advanced Packaging*, vol. 29, no. 3, Aug. 2006, pp. 615 -622.

9. H. H. Chang, "TSV Process Using Bottom-up Cu Electroplating and its Reliability Test," in *Electronics Systemintegration Technology Conference (ESTC)*, London, UK, Sep. 1-4, 2008, pp. 645-650.

10. H. H. Chang, "3D stacked chip technology using bottom-up electroplated TSVs," in *Proc. IEEE Electronic Components and Technol. Conf. (ECTC)*, San Diego, CA, May 26-29, 2009, pp. 1177-1184.

978-1-4799-2408-0/14 $31.00 © 2014 IEEE

Effect of Joint Shape Controlled by Thermocompression Bonding on the Reliability Performance of 60μm-pitch Solder Micro Bump Interconnections

Yu-Wei Huang[1], Chau-Jie Zhan[1*], Jing-Ye Juang[1], Lin Yu-Min[1], Shin-Yi Huang[1], Su-Mei Chen[1], Chia-Wen Fan[1], Ren-Shin Cheng[1], Shu-Han Chao[2], Wan-Lin Hsieh[2] and Chih Chen[2], John H. Lau[1]

[1]Electronics and Optoelectronics Research Laboratories, Industrial Technology Research Institute
Rm. 266, Bldg. 17, 195, Sec.4, Chung Hsing Rd., Chutung, Hsinchu, 31040, Taiwan, R.O.C
[2]Department of Materials Science and Engineering, National Chiao Tung University
1001, University Rd., Hsinchu, Taiwan 30010, Taiwan, R.O.C
*itri960087@itri.org.tw, +886 3 591 7107

Abstract

Three dimensional integration circuits technology has received much attention recently since the demands of functionality and performance in microelectronic packaging for electronic products are rapidly increasing. For high-performance 3D chip stacking, high density interconnections are essential. In the current types of interconnects, solder micro bumps have been widely used and thermocompression bonding process are well adopted to form the connection between bumps. However, the prefect joint contour is difficult to obtain and control by such kind of bonding process in solder micro bump joints. For fine-pitch solder micro bump interconnections, the effect of joint shape on the reliability performances of the solder micro bump joints is not concluded yet till now and needs to be clarified. In this study, the effect of joint shape controlled by thermocompression bonding on the reliability performance of solder micro bump interconnections with a pitch of 60 μm was discussed.

The chip-to-chip test vehicle having more than 4000 solder micro bump interconnections with a bump pitch of 60 μm was used in this study. A solder micro bump structure of Cu/SnAg having a thickness of 7 μm/10 μm was fabricated in both the silicon chip and substrate. To evaluate the effect of joint shape, four types of joint shape were controlled and made. The first type had a conventional shape of micro joint. Compared to the first one joint structure, the second type of joint structure showed the compressed shape. The third type of joint structure was the pillar-like micro joint while the fourth type of joint structure presented a neck shape having the highest joint height among all the joint structures tested. We used the fluxless thermocompression bonding process to form these four types of micro joints. After bonding process, the chip stack was assembled by capillary-type underfill. Reliability tests of temperature cycling test (TCT), high temperature storage (HTS) and electromigration test (EM) were selected to assess the effect of joint shape on the reliability properties of those four types of solder micro bump interconnections.

The reliability results presented that all the types of joint structures could pass TCT of 1000 cycles and HTS of 1000 hours but high variation of daisy chain resistance more than 15% would happen in the neck-shape micro joint after TCT. For the neck-shape micro joint, the high variation of daisy chain resistance after TCT resulted from the cracking propagated along the interface of Cu UBM/Cu_6Sn_5 IMC and across the tin solder. The cracking situation was more serious as compared to the other three tested micro joints. The results of HTS revealed that resistance variation mainly depended on the micro structural evolution within micro joints tested. Electromigration test was conducted under the testing condition of 0.56 A/150°C. A daisy chain structure was adopted. For both the pillar-shape and neck-shape micro joints, Cu UBM consumption and formation of large void were the major microstructure evolutions within the micro interconnections during EM testing. The conpressed-shape showed the longer electromigration lifetime among all the types of micro joints tested.

Introduction

Three-dimensional integration circuit (3DIC) has become a very promising technology in the semiconductor industry, recently [1-5]. By introducing the structure of through silicon via (TSV) inside the die, vertically multiple-die stack would be practicable. Traditional wire bonding used for current chip module could be replaced by the TSV interconnection technology and the introduction of TSV structure could bring the advantages of increasing communication bandwidth, reducing form factor and low power consumption in multi-die stack. Additionally, the most attractive advantage of 3DIC technology is highly heterogeneous integration among different types of functional chips. By using 3DIC technology, IC designer could develop a stacked-die system that has more design flexibilities when compared to system on chip (SoC) and higher electrical performance in contrast with system in package (SiP). Therefore, 3DIC would be a highly potential packaging technology to meet the demands of high performance electronics.

Within high performance chip stack module, high I/Os interconnections are required and solder micro bumps have been widely used due to its low material cost and well-developed fabrication process. In the case of solder joints within flip chip package, due to relatively large solder volume and the effect of surface tension, perfect joint shape would be easy to form and obtain by reflow process. The effect of joint shape on the reliability performance of solder joints has been discussed in literatures [6-8]. In the case of solder micro bump joints, thermocompression bonding process is often adopted to form the fine-pitch interconnections between stacked dies instead of reflow process. However, the shape of solder micro bump joints produced by thermocompression bonding is difficult to control especially in fine-pitch micro interconnections. In comparison with solder joints, the influence of joint shape on the reliability response of solder micro bump interconnections is not clear. Therefore, from the view points of process limitation and reliability concern, the effect of joint shape on the reliability performance of solder micro bump joints needs to be clarified in detailed.

In this study, we designed and fabricated four types of 60μm-pitch solder micro bump joints to evaluate the effect of joint contour on the reliability response. Under thermo-mechanical and electrical reliability tests, the influence of joint shape was discussed.

Experimental Procedure

Test Vehicle

The chip-to-chip test vehicle was adopted to evaluate the effect of joint shape on the reliability performance of 60μm-pitch solder micro bump interconnections in this study. The size of Si chip was 6 mm x 6 mm and the Si substrate showed a dimension of 16 mm x 16 mm. The pattern of electroplating solder micro bumps upon the test chip was designed as a nearly full array type. There were 4290 I/Os within a chip-to-chip stack after chip bonding. The diameter of micro bumps on both the chip and substrate was 30 μm. The Cu/SnAg solder micro bump with a thickness of 7 μm/10 μm was fabricated upon both the silicon chip and substrate. To evaluate the effect of joint shape, four types of joint shape were controlled and made. They were conventional shape, compressed shape, pillar shape and neck shape, respectively. Fig. 1 presented the schematic structures of the four types of joint shape tested.

Formation of shape-control solder micro bump interconnections

The fluxless chip-to-chip bonding scheme was used in chip bonding process. Plasma treatment was applied on both the chip and substrate just before bonding process to achieve the purpose of fluxless bonding. The plasma with Ar/H$_2$ mixed gas flow was used to remove the tin oxidation layer upon the surface of solder micro bumps. Chip-to-chip bonding was conducted by Toray FC-3000WS bonder. To control the joint shape, we adopted thermocompression bonding mode plus Z-axis control function. Different bonding parameters including temperature, force, time and gap height were chosen and determined to obtain the joined-well micro joints. The underfill dispensing process was then conducted to fill the gap between chips and followed by C-SAM inspection to check no voids existed within the chip gap.

Reliability Assessments

After assembly of chip stack module, temperature cycling test (TCT), high temperature storage (HTS) and electromigration (EM) test were performed to evaluate the effect of joint shape on the reliability response of solder micro bump interconnections. Firstly, all assembled samples were experienced pre-conditioning test (JESD22-A113D, LV3) to screen out the early-failed samples. Subsequently, the samples passed pre-conditioning test were tested by TCT, HTS and EM. The failure criterion in TCT and HTS was considered as the variation of contact resistance over 15%. All the testing conditions of reliability assessments were listed in Table 1. The microstructure evolution of micro joints after reliability test was observed by cross-sectioned microstructural analysis and the chemical composition of the phases formed was identified by an energy dispersive spectrometer (EDS).

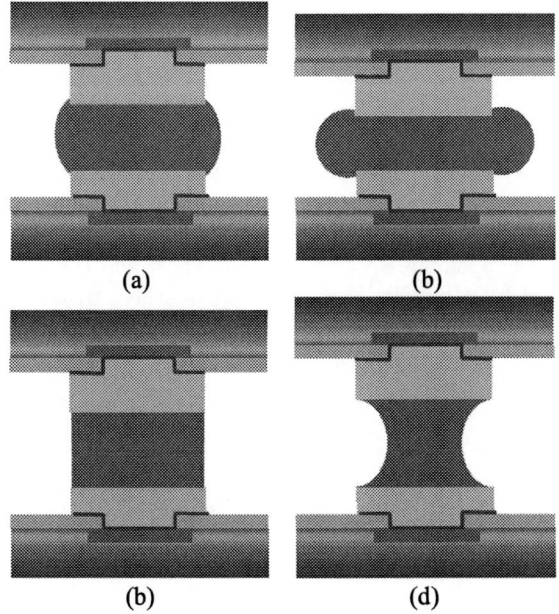

Fig.1 Schematic solder micro joints of (a) conventional shape, (b) compressed shape, (c) pillar shape and (d) neck shape.

Table 1 Testing conditions of reliability assessment

Item	Condition
Pre-conditioning	Baking (125°C, 24 hours) → Soaking (30°C /60%RH, 192 hours) → Reflow (260°C, 3 times)
TCT	-55°C ~ 125°C, 1000 cycles, Dwell time = 5 min, Ramp rate =15°C / min
HTS	150°C, 1000 hours
EM	0.56 A/150°C

Results and Discussions

Microstructure Analysis of Solder Micro Bump Joints

Fig. 2 showed the cross-sectioned microstructures of the four types of solder micro bump interconnections, which displayed by secondary electron images. The bump heights of the conventional-shape joint, compressed-shape joint, pillar-shape joint and neck-shape joint were 29 μm, 17 μm, 37 μm and 46 µm, respectively. Though the conventional-shape joint were preferred contour of solder micro bump micro joints, the compressed-shaped joints were mostly obtained after thermocompression bonding process. In this study, we could well control the bump shape by thermocompression bonding with Z-axis control function.

In the bonding profile, through the adjustment of two sets of bump height values, the control of joint shape could be achieved. Two sets of bump height value were used to control the joint shape. The first set of bump height value was related the deformation of micro joints during peak bonding temperature while the second set of value was associated with the shape control just at the beginning of cooling step. We used the first set of bump height value to counteract the thermal expansion of micro joint, which could prevent the happening of squeezed solder. Subsequently, the second set of bump height value was used to alter the joint shape by pressing or pulling the joint. Finally, after the cooling step, the deformed micro joint would be kept.

Peak bonding temperature ranged from 250°C to 300°C for few seconds was selected to connect the solder micro bumps. At the temperature of 250°C, the joining between micro bumps was not good and interface-like defects were easy to observe. However, growth of Cu_xSn_y IMC would be prominent within micro joints when bonded at the temperature of 300°C. After several times of trial boning, optimized bonding conditions of 275°C/10 seconds with propriety bump height values were chosen and determined and then the join-well micro joints with specific shape could be achieved, as seen in Fig. 2.

upon the Cu UBMs after bonding. The Cu_6S_5 IMC phase with a chemical composition of 57.6 at% Cu and 42.4 at% Sn was identified by EDS. In addition, the thickness of Cu_6S_5 IMC formed upon the Cu UBM of the substrate side was thicker than that of the chip side, irrespective of what the joint shape was. This situation of asymmetrical IMC growth upon both Cu UBM sides could be ascribed to the thermomigration of Cu during thermocompression bonding process.

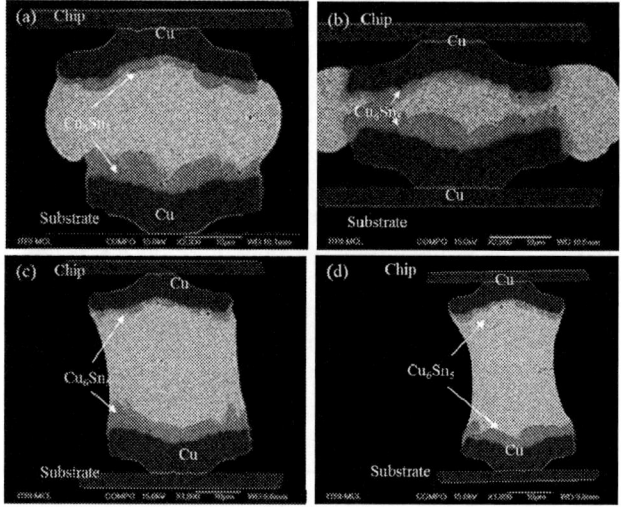

Fig. 3 Back-scattered electron images of (a) conventional-shape, (b) compressed-shape, (c) pillar-shape and (d) neck-shape micro joints.

Guo et al. have reported that an asymmetrical growth of Cu_6Sn_5 IMC on the two interfaces of Cu/SnAg/Cu solder joints during reflow at 260°C on a hot plate is observed [9]. In their experiment design, the thickness of the solder layer is only 30 µm but the thermal gradient of approximately 51°C/cm across the solder layer could be obtained. Under this situation, thermomigration of Cu would be very pronounced, resulting in the thick IMC layer formed upon the cold end. In this investigation, the solder micro bump interconnections were formed by thermocompression bonding process. During bonding process, the bonded chip was placed on a hot stage with a temperature of 150°C and the heat and pressure was applied by a bonding head. The applied heat from bonding head was often higher than 350°C to make sure of the joining between solder micro bumps. By such bonding method, thermal gradient across the solder layer would be anticipated and caused thermomigration of Cu within bonded chip. As the thickness of solder layer decreased, thermomigation of Cu became more serious. Therefore, the thick Cu_6S_5 IMC layer formed upon the Cu UBM of the substrate side was reasonable, as seen in Fig. 3.

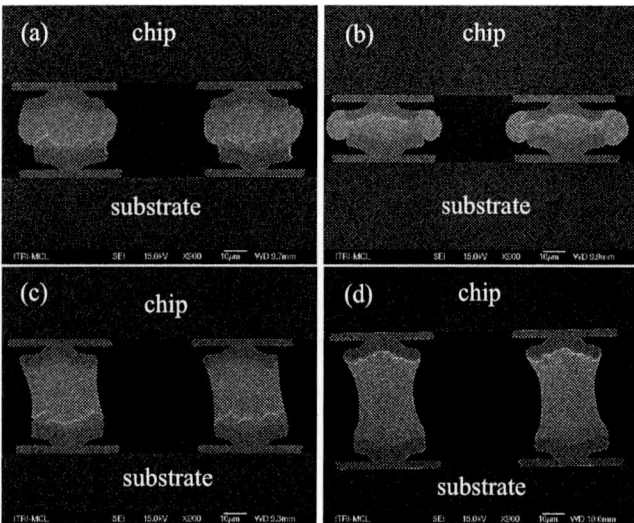

Fig. 2 Cross-sectioned microstructures of (a) conventional-shape, (b) compressed-shape, (c) pillar-shape and (d) neck-shape micro joints.

Fig. 3 showed the cross-sectioned microstructures of the four types of solder micro bump interconnections, which presented by back-scattered electron images. The images clearly exhibited that a continuous layer of Cu_6S_5 IMC formed

Reliability Assessment
Temperature Cycling Test

After assembly process of the chip stack and before reliability test, the assembled chip modules firstly underwent pre-conditioning test to remove the early-failed samples and followed by TCT, HTS and EM. The test results of the four types of solder micro joints in the testing items of TCT and HTS were summarized in Table 2. The results indicated that only neck-shape micro joints failed but the other three types of micro joints passed in TCT. In addition, all the types of micro joints passed in HTS. It should be noted that the failure criterion in TCT and HTS was considered as the variation of daisy chain resistance higher than 15%.

Table 2 Results of TCT and HTS

Joint shape	Conventional	Compressed	Pillar	Neck
TCT	pass	pass	pass	Fail[*]
HTS	pass	pass	pass	pass

*fail was judged as the resistance variation over than 15%.

Fig. 4 was the plot of the variation of daisy chain resistance versus testing cycle during 1000 cycles of TCT. The plot displayed that the resistance variation was gradually increased with increasing testing cycles for all the types of micro joints tested except for the neck-shape micro joint. After TCT of 750 cycles, the resistance variation of neck-shape micro joints suddenly ramped up and was over than 15% in TCT of 1000 cycles. Therefore, they were judged as electrical failed. Additionally, the resistance variations for conventional-shape, compressed-shape and pillar-shape micro joints after TCT of 1000 cycles were 11.5%, 5.8% and 2.6%, respectively.

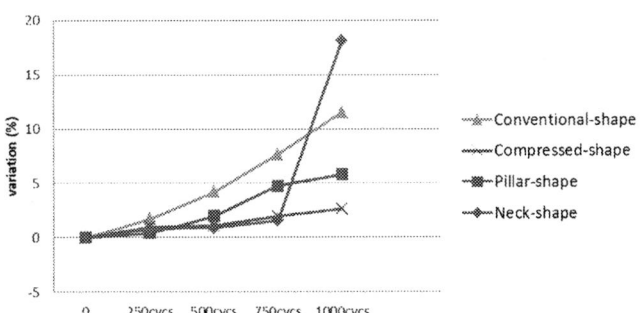

Fig. 4 Plot of resistance variation versus cycles.

Fig. 5 was the plot of the variation of daisy chain resistance versus testing time during 1000 hours of HTS. The plot presented that the variation of daisy chain resistance was all gradually increased with increasing testing time for all the types of tested micro joints. No any electrical failures happened after HTS. The resistance variations for conventional-shape, compressed-shape, pillar-shape and neck-shape micro joints after HTS of 1000 hours were 9.5%, 10.2%, 8.1% and 8.3%, respectively.

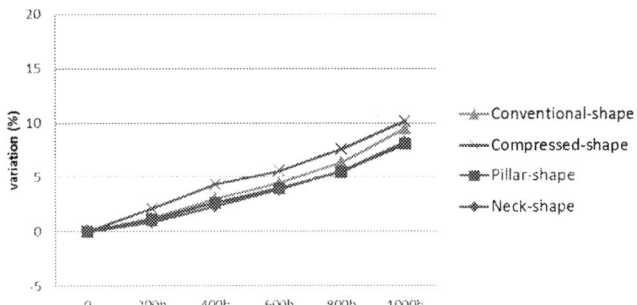

Fig. 5 Plot of resistance variation versus time.

To verify the root cause for electrical failure happened in TCT and microstructure evolution within the micro joint after reliability tests, cross-sectioned microstructural analysis was carried out. Fig. 6 showed the cross-sectioned microstructures of the four types of solder micro bump interconnections after TCT of 1000 cycles, which displayed by back-scattered electron images. As seen from Fig. 6, few cracks formed at the interface of Cu_6Sn_5 IMC/tin solder after TCT. Compared to the pillar-shape micro joints, the situation of cracking within the conventional-shape and neck-shape micro joints was more serious and cracks even propagated across the tin solder, as seen in Figs. 6(a) and 6(d). Furthermore, no obvious cracking happened within the compressed-shape micro joints, as showed in Fig. 6(b).

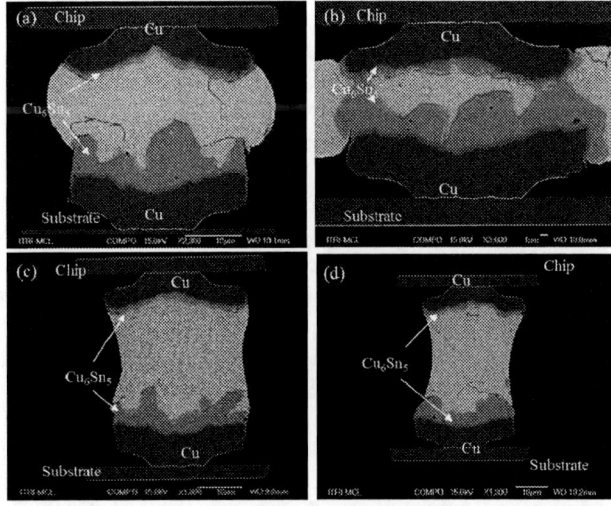

Fig. 6 Cross-sectioned microstructures of (a) conventional-shape, (b) compressed-shape (c) pillar-shape and (d) neck-shape micro joints after TCT of 1000 cycles.

The extent of cracking within the micro joints exactly reflected the circumstance of resistance variations among the four types of solder micro bump interconnections after TCT.

The neck-shape micro joints showed the serious cracking inside the tin solder, resulted in the highest resistance variation among the four types of micro joints tested. Though formation of cracking within the micro joints might cause the high resistance variation during TCT, clear failure mode after testing was not observed.

The effect of elongated-shape solder joints on the thermo-mechanical reliability has been reported. The elongated-shape solder joint is identical to the neck-shape micro joint in this study. Rajoo et al. have shown that the stretched solder interconnections under mechanical and thermal cycling reliabilities are found to be significantly better than those of conventional solder joints [10]. Yang et al. also have shown that elongated solder joints with an optimized shape demonstrate far superior board-level thermo-mechanical reliability to that of the conventional BGA bumps [11]. Their finite element modeling reveals that the elongated joints could disperse the thermal strain throughout the whole length of the joint and result in longer lifetime. In addition, due to the dispersion of stain, multiple cracks are constantly observed in the elongated joints, indicating the elongated joint has better compliance to absorb stresses. Hence, better reliability response of the elongated solder joints is expected during reliability test.

Compared to the testing results in this investigation, whether the neck-shape micro joints showed the same effect could not be confirmed, since we did not observe any clear fracture mode after TCT of 1000 cycles. On the other hand, based on the microstructure analysis, it was difficult to interpret the differences of reliability response among the four types of micro joints tested in TCT results. More works needed to be done, such as numerical simulation, to clarify the reason for the difference of reliability performance.

High Temperature Storage

Fig. 7 revealed the cross-sectioned microstructures of the four types of solder micro bump interconnections after HTS of 1000 hours, which showed by back-scattered electron images. Compared to the samples after TCT, apparent microstructure evolution could be seen in those samples testing by HTS. As seen in Fig. 7, Cu_xSn_y IMC layer became thick on both Cu UBM sides and new IMC phase formed between the Cu UBM and Cu_6Sn_5 IMC. Even in the case of compressed-shape micro joints, the residual solder almost fully transformed into Cu_xSn_y IMCs. According to EDS results, this new phase showed a chemical composition of 78.6 at% Cu and 21.4 at% Sn and could be identified as the Cu_3Sn IMC. It was well known that the growth of Cu_3Sn IMC was related to the diffusion reaction of Cu-Cu_6Sn_5 during annealing and usually accompanied Kirkendall voids [12], as seen from Fig. 7. Additionally, gray granular phases distributed within the tin solder and IMCs also could be found. They were with a chemical composition of 74.8 at% Ag and 25.2 at% Sn and could be recognized as the Ag_3Sn IMC.

From the results of HTS and microstructure observation, the resistance variation mainly depended on the microstructural evolution within micro joints could be confirmed. Although no any electrical failure happened during HTS, consumption of Cu UBM by tin solder would be a reliability concern especially during long-term reliability test.

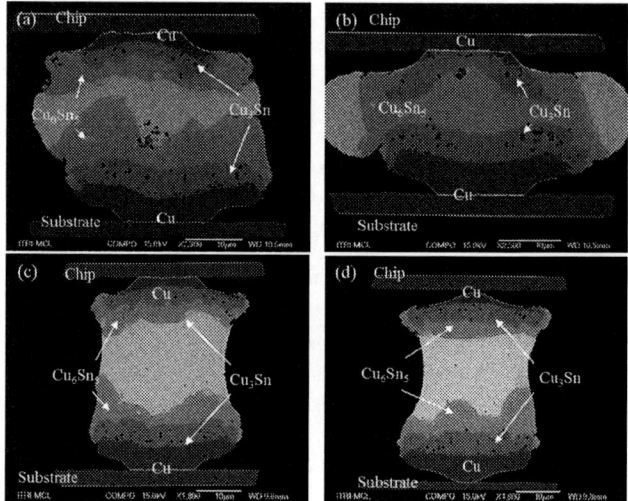

Fig. 7 Cross-sectioned microstructures of (a) conventional-shape, (b) compressed-shape (c) pillar-shape and (d) neck-shape micro joints after HTS of 1000 hours.

Electromigration Test

With intent to investigate the mechanisms of electromigration in micro joints with different solder shape, three types of micro joints were selected and tested. They were compressed-shape, pillar-shape and neck-shape micro joints. A daisy chain structure was adopted. There were 400 micro joints in the daisy chain structure. Solder joints were stressed with 0.56 A on a hot plate of 150°C. The average current density was 8×10^4 A/cm^2, calculated based on the UBM opening.

The resistance increase of neck-shape micro joints mostly came from the dissolution of UBM and voids. The resistance increased 20% of its initial value after 10.4 hours as shown in Figs. 8(a) and 8(b). Fig. 8(a) showed the microstructure changed with electron flow upward. A layer type Cu_3Sn occurred in both chip and substrate sides uniformly. There were many Kirkendall voids inside Cu_3Sn. The Cu_6Sn_5 almost bridged together across the micro joint. Fig. 8 (b) showed electron flowed downward. A void was observed in the chip side. There was also layer-type Cu_3Sn formation in the chip and substrate side. The resistance increased 100% of its initial value after 12.8 hours as shown in Figs. 8(c) and 8(d). Fig. 8(c) showed electron flowed upward. Cu UBM dissolved severely in substrate side. Fig. 8(d) showed electron flowed downward. The Cu UBM in chip side was consumed completely, which caused the resistance increment.

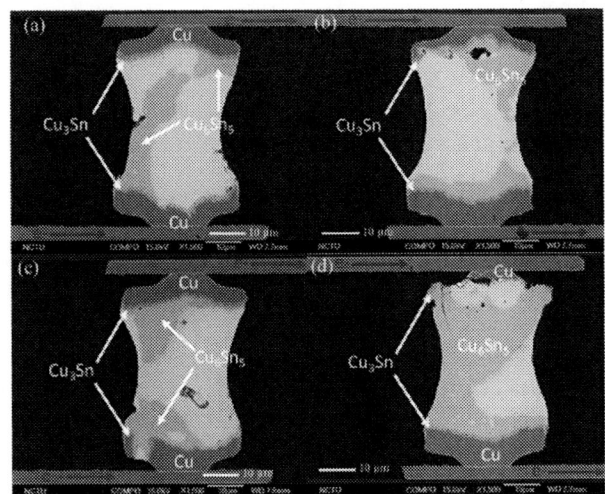

Fig. 8 Cross-sectioned microstructures for the neck-shape micro joints after current stressing; (a), (b) with resistance increased 20% after 10.4 hours and (c), (d) with resistance increased 100% after 12.8 hours.

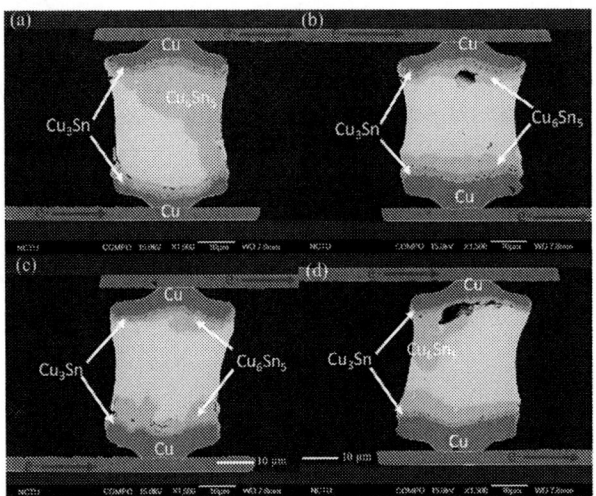

Fig.9 Cross-sectioned microstructures for the pillar-shape micro joints after current stressing; (a), (b) with resistance increased 30% after 21.3 hours and (c), (d) with resistance increased 100% after 40.7 hours.

The failure mechanisms in the neck-shape and pillar-shape micro joints were quite similar. The resistance increased 30% of its initial value after 21.3 hours as shown in Figs. 9(a) and 9(b). Fig. 9(a) showed microstructure of the micro joints with an upward electron flow. Cu_3Sn IMC grew uniformly in layer type in both chip and substrate sides. There were many Kirkendall voids inside Cu_3Sn IMC. The Cu UBM in the substrate side was also severely consumed. Cu_6Sn_5 bridged together across the solder joint. The reason that IMC formed on the right side might because of Sn grain orientation [13]. Fig. 9(b) showed electron flowed downward. A large void was observed in chip side and Cu_6Sn_5 IMC did not bridge together. The resistance increased 100% of its initial value after 40.7 hour as shown in Figs. 9(c) and (d). Fig. 9(c) showed electron flowed upward. Fig. 9(d) showed electron flowed downward. There was a large void formed in chip side.

The compressed-shape micro joints was stressed with 0.56 A on a hot plate of 150°C for more than 200 hours, but its resistance was just slightly increased. Based on the above experimental results, we could conclude that electromigration lifetime of compressed-shape micro joints was longer than those of neck shape and pillar shape micro joints. More works needed to be done to understand the reason for the better electro migration resistance of the compressed-shape micro joints.

Conclusions

The effect of joint shape on the thermo-mechanical and electrical reliability performances of 60μm-pitch solder micro bump interconnections was investigated and discussed. Emphases were placed on: (1) the shape control of micro joints by thermocompression with Z-axis control function; (2) the reliability characterizations of the shape-control micro joints in terms of temperature cycling test, high temperature storage and electromigration test; and (3) the failure analyses of the solder micro joints tested. Some important results were summarized in the following.

➤ Four types of shape-control 60μm-pitch solder micro interconnections could be formed by thermocompression plus Z-axis control function under the bonding conditions of 275°C for 10 seconds. They were conventional shape, compressed shape, pillar shape and neck shape micro joints, respectively.

➤ After TCT of 1000 cycles, only neck-shape micro joints failed but the other three types of micro joints passed. The daisy-chain resistance variation of neck-shape micro joint was higher than 15% after TCT. The situation of crack propagation along the interface of Cu UBM/IMC within this type of micro joint was more manifest than that within the other three types of micro joints, but no clear failure mode was observed.

➤ After HTS of 1000 hours, all the types of micro joints passed. The resistance variation during test mainly depended on the microstructural evolution within the micro joints.

➤ During EM testing, Cu UBM consumption and formation of large void were the major microstructure evolutions within the pillar-shape and neck-shape micro joints, while the compressed-shape micro joint showed the longest EM lifetime.

References

1. H. Jihwan et al., "Fine Pitch Chip Interconnection Technology for 3D Integration ," *Electronic Components and Technology Conference*, 2010 , pp. 1399 – 1403.

2. T. Fukushima et al., "New Three-Dimensional Integration Technology Based on Reconfigured Wafer-on-Wafer Bonding Technique," *International Electron Devices Meeting*, 2007, pp. 985-988.

3. J. U. Knickerbocker et al., "Three-Dimensional Silicon Integration," *IBM Journal of Research and Development*, 2008, Vol. 52, pp. 553-569.

4. K. Sakuma et al., "Characterization of Stacked Die using Die-to-Wafer Integration for High Yield and Throughput," *Electronic Components and Technology Conference*, 2008, pp. 18-23.

5. E. Beyne et al., "3D Interconnection and Packaging: Impending Reality or Still a Dream?" *International Solid-State Circuits Conference*, 2004, pp. 138-145.

6. X. Liu et al., "Effects of Solder Joint Shape and Height on Thermal Fatigue Lifetime," *Transactions on Components and Packaging Technologies*, 2003, pp. 455-465.

7. X. Liu et al., "Stacked Solder Bumping Technology for Improved Solder Joint Reliability," *Microelectronics Reliability*, 2001, Vol. 41, pp. 1979-1992.

8. T. H. Ho et al., "Linear Finite Element Stress Simulation of Solder Joints on 225 I/O Plastic BGA Packages Under Thermal Cycling," *Electronic Components and Technology Conference*, 1995, pp. 930-936.

9. M. Y. Guo et al., "Asymmetrical Growth of Cu_6Sn_5 Intermetallic Compounds due to Rapid Thermomigration of Cu in molten SnAg solder joints," *Intermetallics*, 2012, Vol. 29, pp. 155-158.

10. R. Rajoo et al., "Super Stretched Solder Interconnects for Wafer Level Packaging," *Electronic Components and Technology Conference*, 2013, pp. 1227-1232.

11. S. C. Yang et al., "Optimization of Solder Heitht and Shape to Improve the Thermo-Mechanical Reliability of Wafer-Level Chip Scale Packages," *Electronic Components and Technology Conference*, 2013, pp. 1210-1218.

12. D. Q. Yu et al., "Electromigration Study of 50 μm Pitch Micro Solder Bumps using Four-Point Kelvin Structure," *Electronic Components and Technology Conference*, 2009, pp. 930-935.

13. D. C. Yeh et al., "Extreme Fast-Diffusion System: Nickel in Single-Crystal Tin," *Physical Review Letters*, 1984, Vol. 53, pp. 1469-1472.

Development of Micro Bump Joints Fabrication Process
Using Cone Shape Au bumps for 3D LSI Chip Stacking

Fumito Imura, Naoya Watanabe, Shunsuke Nemoto, Wei Feng, Katsuya Kikuchi,
Hiroshi Nakagawa, and Masashiro Aoyagi
Nanoelectronics Research Institute (NeRI),
National Institute of Advanced Industrial Science and Technology (AIST)
Tsukuba Central 2, 1-1-1 Umezono, Tsukuba, Ibaraki, 305-8568 Japan
E-mail: m-aoyagi@aist.go.jp, Phone: +81-29-861-5529

Abstract

3D LSI chip stacking technology have been developed using cone shape Au micro bumps fabricated by nanoparticle deposition method. The cone shape bumps with less than 10 μm diameter are suitable for a thermocompression bump joint process with low temperature and low load force. High yield micro bump joints can be obtained. In this study, the property evaluation of the cone shape bumps, and the cone shape bump joints were investigated in details. The collapsed bump height and the electrical resistance can be controlled by compression force. The low resistance (average 8.6 mΩ) bump joint with a 10 μm diameter cone shape Au bump was successfully achieved.

Introduction

3D system integration technologies are playing an important role in system miniaturization and performance improvement of future LSI system technology. 3D LSI chip stacking technology have been developed in AIST using cone shape Au micro bumps fabricated by nanoparticle deposition method [1-2]. The cone shape bumps are suitable for a thermocompression bump joint process with low temperature and low load force, where the bumps are easy to collapse with loading due to the pointed structure. High yield micro bump joints can be obtained [3-4].

The cone shape bumps can be fabricated by lift-off process using thick photo resist with 10 μm diameter bump pattern. The cross section of the bump pattern must have vertical straight side wall.

During the nanoparticle deposition, sample cooling is very important to assure the lift-off process. The target sample should moves continuously during the deposition in order to avoid the heat concentration.

A set of a plane pad and a cone shape bump is used for the micro bump joint formation. The cone shape bump can smoothly collapse by thermocompression method without buckling. The collapsed bump height and the electrical resistance can be controlled by compression force. The temperature range of the bonding process is from RT to 200 °C. The bump hardness decreases with increasing the process temperature.

In this paper, we report about the nanoparticle deposition system, the cone shape bump formation process, the property evaluation of the cone shape bumps, the bump joint formation process, and the property evaluation of the cone shape bump joints in details.

Nanoparticle Deposition System

The diagram of a nanoparticle deposition system for fine bump formation is shown in Fig. 1, where a nanoparticle generation chamber and a film deposition chamber are connected through a gas transfer tube. A prepared sample is placed on a sample stage in the film deposition chamber, and the nanoparticle deposition is performed from a nozzle placed at the end of the gas transfer tube. We use the revised nanoparticle deposition system NP150H (Mikuni Kogyo) [5, 6]. It was significantly improved about the gas flow rate (1/3), the power consumption (1/3), the operational cycle time (1/2), the variation of the deposition rate (±5 %), the large (larger than 3 μm) size particle generation (eliminated), etc.

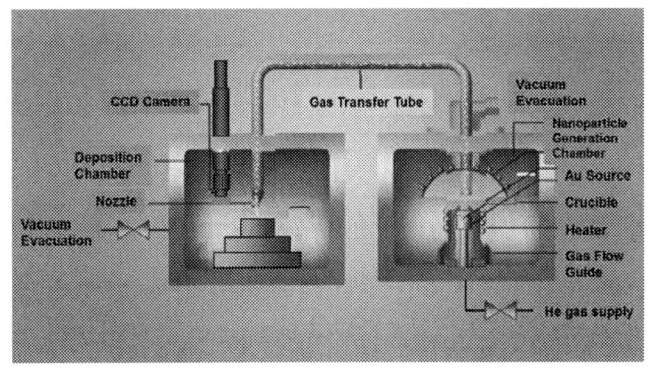

Figure 1. Diagram of nanoparticle deposition system for fine bump formation.

At the nanoparticle generation chamber, where it was filled with He gas, the crucible with Au evaporation source was heated up under the He gas pressure of 82 kPa. Au vapor is quickly cooled by He gas and Au nanoparticles are formed.

At the film deposition chamber, where it was in vacuum of 700 Pa, the generated nanoparticles were ejected on the target sample from the nozzle of the 0.5 mm diameter with He gas through the transfer tube. The flow rate of He gas was 3 L/min.

The Au evaporation source was heated by DC resistance heating to the evaporation temperature of 1600 °C. The sample holder which has large heat capacity is cooled by water flow to suppress the temperature increase of the target sample.

Smooth Au film can be obtained on the targeted area by the scanning motion of the precision X-Y stage holding the target sample, ejected Au nanoparticles with He gas from the nozzle. By the scanning motion, local heating can be avoided, and the heat damage of the sample can sufficiently be suppressed.

Cone Shape Au Bump Formation Process Using Nanoparticle Deposition System

A process flow of fine Au bump formation by the nanoparticle deposition method is shown in Fig. 2. It is explained that Au nanoparticle condensation film is grown, and cone shape bumps are formed by self-organized method.

First, the resist hole pattern for cone bump formation is formed on a substrate with photolithography process using a thick coating type photoresist and a g-line 1/2.5 reduction stepper PrA II-SS (Nikon Engineering) with deep focus depth of 30 μm (NA 0.125). Au nanoparticles with He gas flow are deposited within the bump area on the substrate. The stagnation of He gas flow is made at the top edge of the resist hole. Au nanoparticle accumulates also at the resist hole top edge, and the hole opening becomes gradually narrow. The bump shape in the hole becomes tapering off gradually. When the hole opening is closed completely, the bump formation will be finished. Finally, the resist film is removed in the solvent, and cone shape Au bumps can be obtained.

By such a self-organized deposition method, high uniformity about the height and the aspect ratio of cone shape bumps can be achieved. We can control precisely these bump parameters even for less than 10 μm size. This is a potential merit compared to the other bumping method.

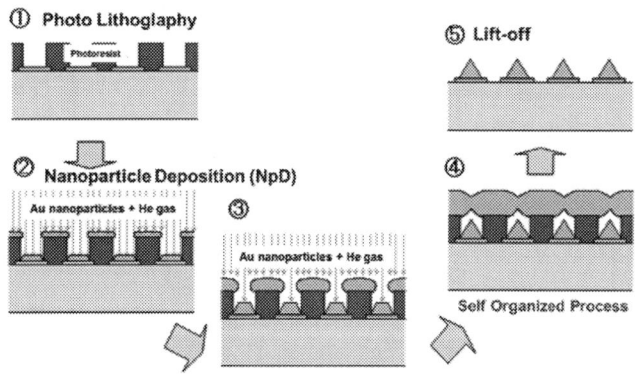

Figure 2. Cone-shape Au bump formation process flow.

About the resist hole pattern for cone shape bump formation, resist selection was carried out in order to obtain the vertical side wall shape. The relationships between ln (exposure dose) and development depth for thick coating type photoresists AZP6130, AZP4620, and AZP4930 (AZ Electronic Materials) are shown in Fig. 3, where development depth was measured by laser scanning microscope after development. The development condition was as follows; pre-bake (temperature: 100 ℃, time: 5 min), development (solution: TMAH 2.38%, time: 3 min), post-bake (temperature: 100 ℃, time: 10 min). There is a linear relationship between ln (exposure dose) and development depth. Slope of the straight lines express development speed. The required exposure dose for AZP6130 is small. Since the

development speed is low, resist hole patterns with taper side walls are formed. On the other hand, the required exposure dose for AZP4620 and AZP4930 is large. Since the development speed is high, resist hole patterns with vertical side walls are formed. From these experimental results, we decided to use AZP4620 with high development speed in order to secure the perpendicular side wall.

About thick coating type photoresist AZP4620, the top diameter of the resist hole was measured with changing the exposure dose with 100-1000 mJ/cm². The variation of the top diameter was evaluated. The measurement of the hole size were performed using a conventional optical microscope.

The optical microscope photograph of circular and square resist hole patterns is shown in Fig. 4. It turns out that the perfect resist hole without any resist residual can be obtained by exposure dose larger than 500 mJ/cm². The relationship between the size of the circular or square resist hole and the exposure dose is shown in Fig. 5, where the error bar indicate the standard deviation of the hole size within 20 samples. At the exposure dose of 600 mJ/cm², the top diameter of the circular resist hole was close to the designed value of 10 μm. Furthermore, the variations in the top diameter of the circular resist hole were ±0.12 μm. The optimum exposure dose was determined as 600 mJ/cm².

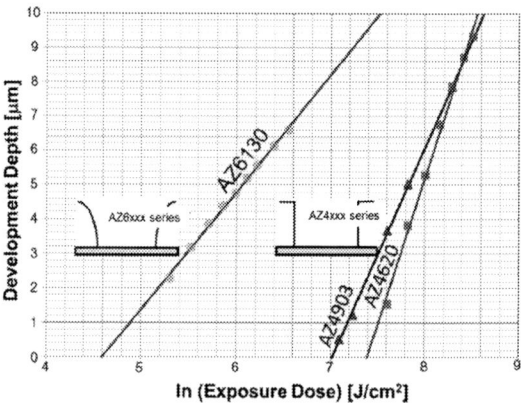

Figure 3. Dependency of development depth with exposed dose value for thick coating type photoresists AZP6130, AZP4620, and AZP4930.

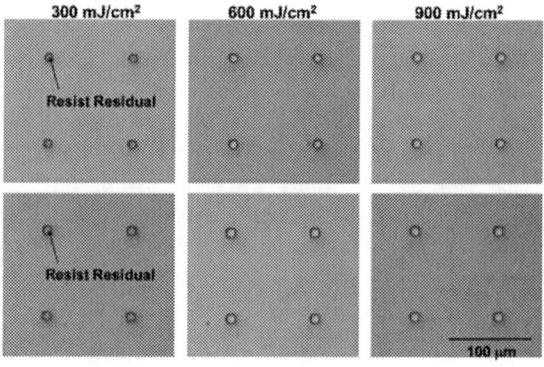

Figure 4. Photographs of circular and square photoresist patterns with different exposure dose value

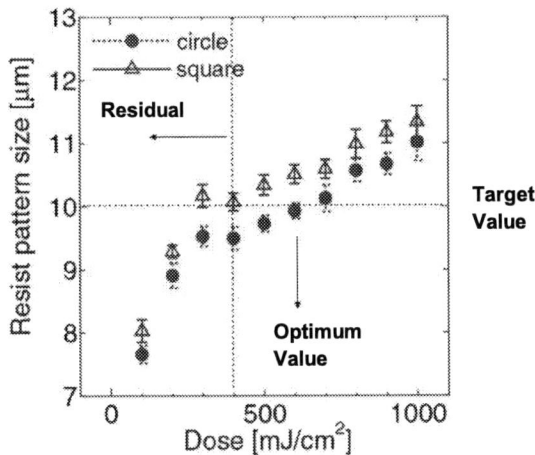

Figure 5. Dependency of circular and square resist pattern size with exposed dose value.

Properties Evaluation of Cone Shape Au Bumps Formed Using Nanoparticle Deposition System

Using circular photoresist AZP4620 hole pattern exposed with 600 mJ/cm^2, cone shape Au bumps were formed using nanoparticle deposition system. A SEM photograph of 10000 cone shape bumps (diameter: 10 μm, height: 12 μm, pitch: 20 μm) is shown in Fig. 6, where magnified view (x50) is also shown. Then, the bump height was measured using a scanning laser microscope VK9700 (KEYENCE). 40 bumps were measured, and the height variation was evaluated.

The bump height histogram of the 40 cone shape bumps is shown in Fig. 7. About typical measured cone shape bump height, the average value μ was 12.6 μm, standard deviation σ was 0.19 μm, and variation 3 σ was 4.5 %. We can achieve uniform cone shape bump formation.

We are developing a new optical 3D shape measurement system for cone shape bumps under the collaboration with Soft Works. We will able to measure all 10000 cone shape bumps soon [7, 8].

In order to evaluate mechanical strength properties of a cone shape bump, the compression test for the cone shape bump with 10 μm diameter was done at room temperature using a microhardness measurement instrument H100C (Fischer Instruments) with a flat diamond press tool.

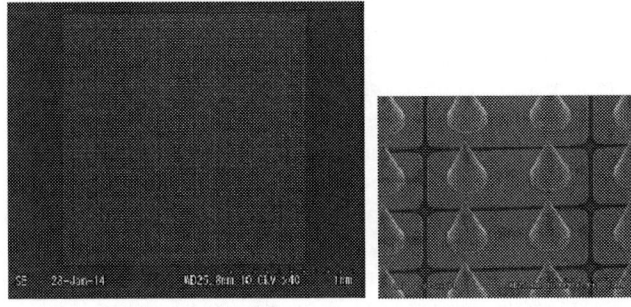

Figure 6. SEM photograph of 10000 bumps and the magnified view (x50) of some bumps.

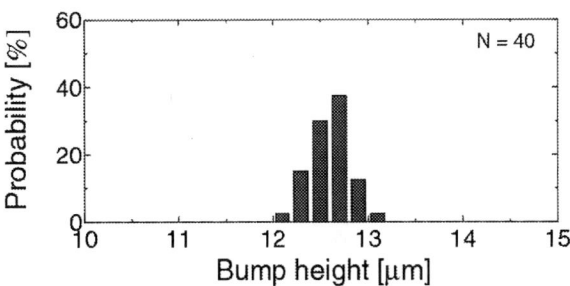

Figure 7. Measured height histogram of 40 cone shape Au bumps.

The compression curves of the single cone shape bump for 5 samples are shown in Fig. 8, where the maximum load was 20 mN. The average bump height reduction was 3.8 μm with the load of 10 mN and 5.6 μm with the load of 20 mN. The single cone shape bump was collapsed smoothly without buckling with increasing load. The variation of the 5 compression curves was very small, where the variation of the collapsed bump height was ±0.08 μm (±2.1 %) with the load of 10 mN and ±0.15 μm (±2.7 %) with the load of 20 mN.

We have tried to measure the compression curves of the single cone shape bumps with a heating sample stage varying the temperature from RT to 200 ºC. However, initial contact area was very small, and the press tool temperature increase very slow. So, it is difficult to keep the constant temperature of the measured bump. The heating press tool is required for the precise measurement.

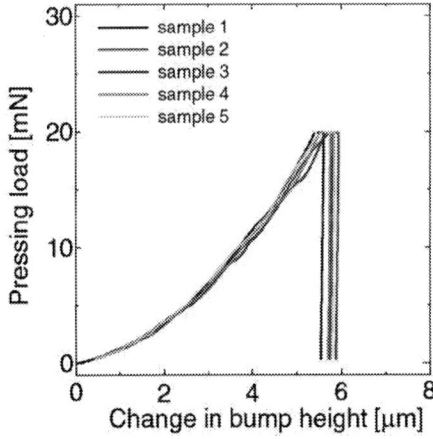

Figure 8. Compression test curve for a single cone shape bump under the press condition with the maximum load of 20 mN.

Bump Joint Formation with Cone Shape Au Bumps

The photograph of two test chips for process evaluation of bump joint formation by thermocompression method is shown in Fig. 9. About a bump test chip with the size of 10 mm square, fine cone shape bumps (diameter: 10 μm, height: 12 μm, pitch: 20 μm, total number: 10012) were formed on it. About a measurement test chip with the size of 14 mm square, Ti-Au counter electrodes (width: 16 μm, thickness: 1 μm, pitch: 20 μm, number 10012) were formed on it. The wirings for daisy chain measurement of 10000 bump joints and 4 terminal measurement of a single bump joint were prepared on these chips.

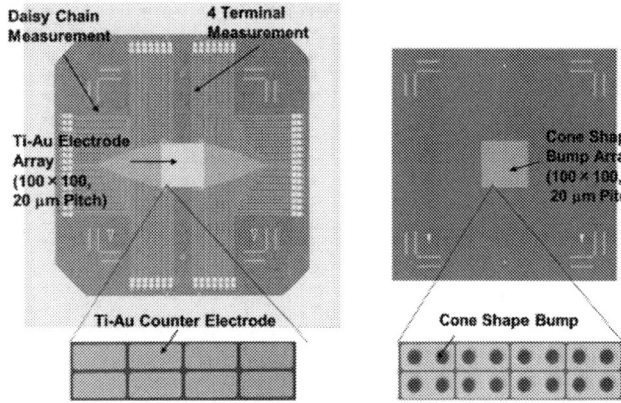

Figure 9. Photographs about two kinds of test chips with bumps and measurement wirings.

For the measurement test chip:
(a) 300 nm thick SiO_2 film was deposited on a Si wafer by TEOS-CVD.
(b) The resist patterns for the wirings and the counter electrodes are made by photolithography process using photoresist ZPN1150-90 (ZEON).
(c) Ti-Au deposition was done on the resist patterns by evaporation method. (Ti: 20 nm thick, Au: 200 nm thick)
(d) The resist patterns were removed by solvent. Ti-Au wirings and electrodes were obtained.

For the bump test chip:
(a) 1 μm thick SiO_2 film was formed on a Si wafer by thermal oxidation.
(b) The resist patterns for the wirings and the counter electrodes are made by photolithography process using photoresist ZPN1150-90.
(c) Ti-Au deposition was done on the resist patterns by evaporation method. (Ti: 20 nm thick, Au: 200 nm thick)
(d) The resist patterns were removed by solvent. Ti-Au electrodes were obtained.
(e) The resist patterns for the bumps with the diameter of 10 μm are made by photolithography process using photoresist AZP4620. The thickness of the photoresist was about 14 μm.
(f) Au nanoparticles were deposited on the resist patterns.
(g) The resist patterns were removed by solvent. The cone shape bumps were obtained.

In consideration of the alignment tolerance between a fine bump and a counter electrode, it is necessary to carry out the bump joint formation with the alignment error less than ±2 μm for the cone shape bump with 10 μm diameter. For the

alignment process and the thermal compression bonding process, the flip-chip bonding machine CA300SS (PMT) with the infrared transparent alignment module was used, where the alignment module can achieve high precision optical alignment of ±0.4 μm. The total bonding process is as follows.

1) The bump test chip and the measurement test chip were treated by plasma cleaning with Ar gas.
2) The bump test chip was chucked on the press tool head, and the measurement test chip was placed on the sample stage. The press tool head was moved down to have a 40 μm gap between two chips.
3) The alignment marks on both chips were observed in the same view area using an infrared camera. The infrared light was introduced through the backside of the measurement test chip from the light source equipped in the sample stage. The sample stage can be precisely moved by a piezodriven actuator and a step motor.
4) Finally, the press tool head was down with the load of 100 N, where the load value for a single cone shape bump was 10 mN. It was heated at 200 °C for 30 sec.

Parallelism between the press tool head surface and the sample stage surface is a critical issue to obtain a uniform gap between bonded two chips. We measured the parallelism using 10 mm square test chips with 40 μm tall 40 μm diameter 200 μm pitch 204 stud bumps on the peripheral region. After pressing the test chip with 100 N on a glass substrate, the bump heights were measured. The parallelism was precisely adjusted to retain the value less than 1.0 μm (best case 0.6 μm) in the 10 mm square area.

In the mechanical bearing cylinder of the press tool, horizontal slipping movement slightly occurs during thermocompression bonding process. In the case of the cone shape bump, this movement is larger than that of the pillar shape bump due to less friction force. The two step compression with a realignment process is one of the solutions. The use of an air bearing cylinder is another solution.

Properties Evaluation of Cone Shape Bump Joints

The cross sectional structure of a single cone shape bump joint with the diameter of 10 μm was observed by a scanning ion microscope (SIM) as shown in Fig. 10. The bump height was reduced from 12.6 μm to 7.1 μm by 5.5 μm (44 %) during the bonding process. The collapsed height at 200 °C was 40 % larger than that of room temperature. The collapsed Au metal was spread out around the bump tip.

Four terminal measurement of a single bump joint was carried out through two bump joints used for I and V ports. The circuit configuration of 4 terminal resistance measurement is shown in Fig. 11. When the current I was added from the current source, potential difference VH-VL between the bump test chip and the measurement test chip was measured using the high precision semiconductor parameter analyzer 4156C (Agilent). The resistance value of the single bump joint was calculated from V_H-V_L/I, where the influence of wiring resistance and probe contact resistance was totally eliminated.

978-1-4799-2408-0/14 $31.00 © 2014 IEEE

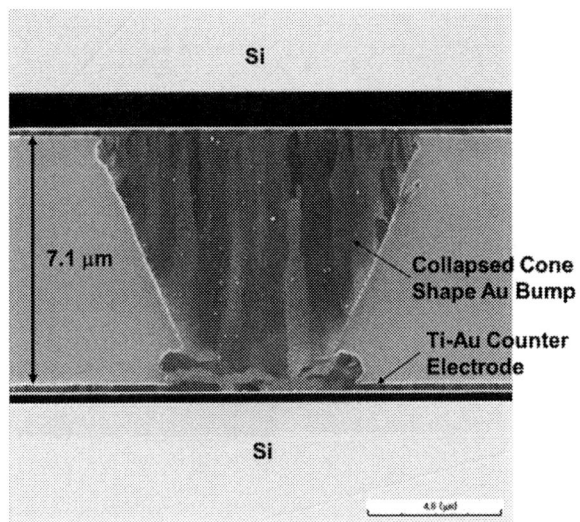

Figure 10. Cross sectional SIM photograph of a single cone shape bump joint.

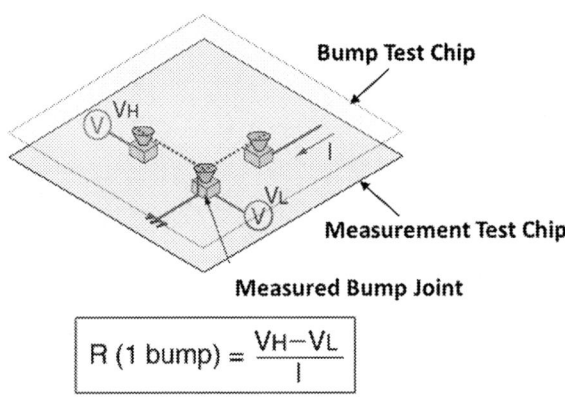

$$R\ (1\ bump) = \frac{V_H - V_L}{I}$$

Figure 11. Electrical resistance measurement circuit configuration for a single bump joint by 4 terminal method.

Figure 12 shows the plot of I and V_H-V_L. The resistance value of the single bump joint was obtained to be 8.8 mΩ from Fig. 12. From the other 3 circuits of single bump joint resistance measurement, the resistance values of 7.6 mΩ, 8.1 mΩ and 10 mΩ were obtained. The average value was 8.6 mΩ, where the resistance variation caused by the collapsed bump height variation should be considered.

The typical resistance model of the cone shape bump joint is considered as follows, based on the cross sectional SIM observation in Fig. 10. The model consists of combination of a truncated cone and a disk, where upper diameter was 4.5 μm, bottom diameter was 10.1 μm, height was 6.0 μm, disk diameter was 6.1 μm, and disk thickness was 1.2 μm. The resistance values are calculated 4.3 mΩ for truncated cone, 0.5 mΩ for disk, 4.8 mΩ for total.

The electrical resistance R of the truncated cone bump is calculated using a formulation $R = \rho h/\pi r r_0$, where ρ is resistivity, h is bump height, r is radius of upper plane, r_0 is

radius of bottom plane. Resistivity of Au nanoparticle thick (6 μm) film was measured to be 2.6 x 10^{-8} Ωm [9].

From the comparison between the measurement and the calculation, the resistance difference 8.6 – 4.8 = 3.8 mΩ is slightly large. The reason should be considered about the collapsed bump height variation.

The experimental evaluation about the daisy chain of 10000 bump joints is in progress. The results of the electrical and mechanical evaluations will be reported soon.

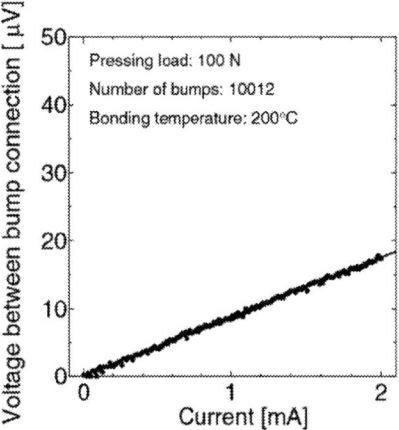

Figure 12. Electrical resistance measurement result of a single bump joint by 4 terminal method.

Conclusions

The cone shape bumps are suitable for a low temperature low load force thermocompression micro bump joint process required for 3D LSI chip stacking technology. The cone shape bump formation process using nanoparticle deposition system, the property evaluation of the cone shape bumps, the bump joint formation process, the property evaluation of the cone shape bump joints were investigated in details. The collapsed bump height and the electrical resistance can be controlled by compression force. The low resistance (average 8.6 mΩ) bump joints with a 10 μm diameter cone shape Au bump were successfully achieved.

Acknowledgments

The authors wish to thank Mr. Gomi, Mr. Saito, and Mr. Hasegawa (Mikuni Kogyo) about Au cone bump processing. This work was done in NEDO Research and Development Program for Innovative Energy Efficiency Technology and METI Strategic Key Technology Advancement Support Projects for Small and Medium Sized Enterprises.

References

1. F. Imura, H. Nakagawa, K. Kikuchi, Y. Yamaji, T. Yokoshima, M. Aoyagi, S. Baba, and J. Akedo: "Micro-cone-shaped Au-bump by gas deposition method for high-density interconnection of LSI chips", Proceedings of IMAPS/ACerS 4th Inter. Conf. on Ceramic Interconnect and Ceramic Microsystems Tech. (CICMT2008), pp.269-271, April 2008.

2. F. Imura, S. Liu, S. Nemoto, F. Kato, K. Kikuchi, M. Suzuki, H. Nakagawa, M. Aoyagi, Y. Gomi, I. Saito, and H. Hasegawa: "Fabrication of Au Cone Bump with Nanoparticle Deposition Technology", 25th Proceedings of JIEP Annual Meeting, pp. 229-232, March 2011 [in Japanese].

3. M. Aoyagi, F. Imura, S. Nemoto, N. Watanabe, F. Kato, K. Kikuchi, H. Nakagawa, M. Hagimoto, H. Uchida and Y. Matsumoto: "Wide Bus Chip-to-Chip Interconnection Technology Using Fine Pitch Bump Joint Array for 3D LSI Chip Stacking", Proceedings of IEEE CPMT Symposium Japan, pp. 183-186, December 2012.

4. M. Aoyagi, F. Imura, S. Melamed, S. Nemoto, N. Watanabe, K. Kikuchi, H. Nakagawa, M. Hagimoto, and Y. Matsumoto: "Development of Testing Technology for Wide Bus Chip-to-Chip Interconnection in 3D LSI Chip Stacking System", Digests of IEEE International Workshop on Testing Three-Dimensional Stacked Integrated Circuits, September 2013.

5. Y. Gomi, H. Hasegawa, I. Saito: "Nanoparticle Deposition System NP150H", Proceedings of 25th JIEP Annual Meeting, pp. 437-440, March 2011 [in Japanese].

6. http://www.mixnus.jp/index_E.html

7. M. Aoyagi, N. Watanabe, M. Suzuki, K. Kikuchi, S. Nemoto, N. Arima, M. Ishizuka, K. Suzuki, and T. Shiomi: "New Optical Three Dimensional Structure Measurement Method of Cone Shape Micro Bumps Used for 3D LSI Chip Stacking", Proceedings of IEEE International 3D Systems Integration Conference, October 2013.

8. https://www.softworks.co.jp/index.html

9. S. Nemoto, N. Watanabe,. T. Bui, W. Feng, S. Melamed, F. Kato, K. Kikuchi, H. Nakagawa, and M. Aoyagi: "Electrical Evaluation of Micro-Cone Au Bump Joints Using Nanoparticle Deposition Method", Proceedings of 28th JIEP Annual Meeting, March 2014 [in Japanese].

Effect of Polymer Liners in CNT based Through Silicon Vias

Archana Kumari, M. K. Majumder, B. K. Kaushik and S. K. Manhas

Microelectronics and VLSI Group, Department of Electronics and Communication Engineering

Indian Institute of Technology Roorkee, Roorkee – 247667, INDIA

E-mail: {karchpec, m1985dec, bkk23fec, samanfec}@iitr.ac.in

Abstract

Polymer liners have provided potentially attractive solutions over conventional silicon dioxide (SiO_2) in single-walled carbon nanotube (SWCNT) bundle based through silicon vias (TSVs). Using SiO_2 and different polymer liners (such as BCB, PPC and polyimide), this paper analyzes the delay with and without crosstalk for different TSV heights and radius. It is observed that the crosstalk coupling is prominently influenced by the dielectric constant of the materials lying between the TSV and the Si substrate. For a SWCNT bundled TSV with higher aspect ratio, the crosstalk delay is reduced by 43.8%, 36.5% and 19.3% for BCB, PPC and polyimide, respectively, in comparison to conventional SiO_2 liner.

Introduction

Three-dimensional (3D) integration is an emerging technology that forms multi-functioning high performance integrated circuits (ICs) by providing significant benefits with improved performance, packing density, power consumption and heterogeneous technology integration capabilities. Traditionally, different technologies such as wire bonding, flip-chip bonding and micro-bumps were used for the electrical interconnection of stacked dies [1, 2]. In comparison to the traditional interconnection methods, through silicon via (TSV) based 3D integration is more preferred technology that passes completely through a silicon wafer or die. The TSV based 3D integration provides improved performance, smaller size and lower power consumption in comparison to the conventional 2D technology [2].

The development of a reliable 3D integrated system is largely dependent on the choice of filler materials used in TSVs. During recent past, carbon nanotubes (CNTs) have emerged as an interesting choice for filler material due to their higher current carrying capability, long ballistic transport length, higher thermal conductivity and mechanical strength compared to Cu based TSV [3, 4]. These unique properties of CNTs substantially reduce the electromigration induced hillocks and void formation that results in improved crosstalk and delay performance. An improved crosstalk performance can be obtained for a lower value of oxide capacitance that exists between two single-walled CNT (SWCNT) bundle based TSVs. This capacitance is primarily dependent on the dielectric constant of the liner material. SiO_2 is the most commonly used liner material for TSVs and provides a capacitance of 50fF-1pF with large relative dielectric constant and small thickness (100nm-1µm) [5, 6]. Several researchers [7-14] reported different electrical and

mechanical reliability problems of SiO_2 liners. Recently, Hsin *et al.* [7] reported the problem of deposition of uniform thin SiO_2 layer on the sidewalls of DRIE (deep reactive ion etching) etched vias. Due to the inherent scallops on the sidewalls of TSVs, stress concentration at the scallop ridges results in failure of ultra thin barrier layers and seed layers, current leakage, degradation of electrical performance, etc [8, 9]. Few mechanical reliability problems [10, 11] such as die cracks or breakage [12] and interfacial layer contamination [13] comes into play because of the large differences in the thermal expansion coefficients between the SiO_2 and the CNT bundle. It largely affects the performance and tolerance of the 3D ICs [14]. Therefore, in order to improve the reliability, the capacitive coupling of TSVs can be reduced by replacing SiO_2 with polymers liners. Certain features like large and uniform thickness and low elastic modulus allow the polymers to be a suitable material for TSV liners that results in an improved electrical performance [15]. Current leakage can be avoided by eliminating the discontinuity of barriers using large and uniform thickness of polymer liners. In addition, due to the low elastic modulus, polymer liner can act as a buffer layer between silicon substrate and metal plugs to avoid the thermal stresses [16].

Crosstalk in coupled lines is broadly classified in two categories: (1) functional and (2) dynamic crosstalk. Under functional crosstalk category, victim line experiences a voltage spike when the aggressor line switches. On the other hand, dynamic crosstalk is observed when adjacent line (aggressor and victim) switches either in same direction (in-phase) or in opposite direction (out-phase). This paper presents an analysis of delay with and without crosstalk for different liner materials such as SiO_2, PPC, BCB and polyimide. Using a pair of SWCNT bundled TSV, the impact of the liner materials on delay is analyzed for different via heights and radius.

Physical configuration and equivalent electrical model

The embedded TSVs on Si substrate primarily have a CNT bundle as filler material. The physical configuration of a pair of CNT bundle based TSV is shown in Fig. 1. The TSV is surrounded by a dielectric for *dc* isolation. Furthermore, the isolation dielectric is surrounded by a depletion region (Fig. 1). The thickness of this depletion region is mainly dependent on the applied bias voltage, interface charge density, material properties, etc. Depending on the physical configuration of Fig. 1, the equivalent electrical model of SWCNT bundle based TSV is shown in Fig. 2.

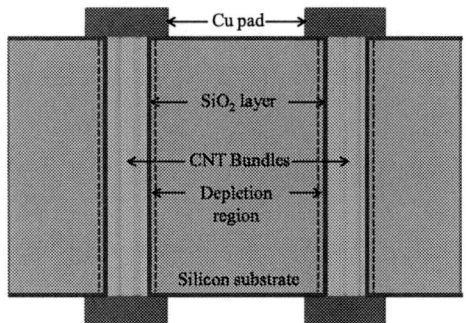

Figure 1. Physical configuration of a pair of SWCNT bundle based TSVs

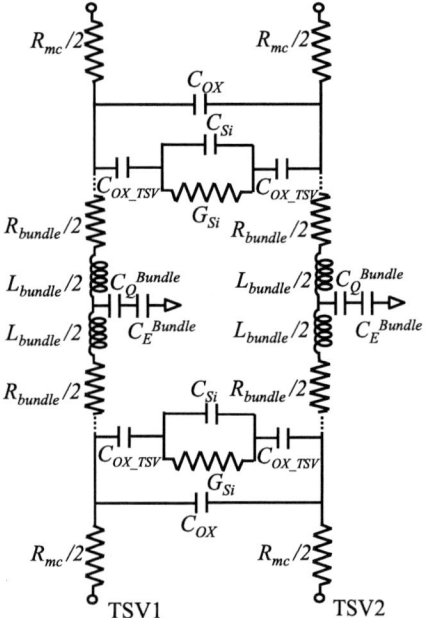

Figure 2. Equivalent electrical model of a pair of SWCNT bundle based TSVs

The equivalent via resistance primarily consists of (1) quantum or intrinsic resistance that is due to the quantum confinement of electrons in a nano-wire [17], (2) imperfect metal-nanotube contact resistance (R_{mc}) that primarily depends on the fabrication process [18] and (3) scattering resistance that arises due to the static impurity scattering, defects, line edge roughness scattering, acoustic phonon scattering, etc. The equivalent scattering resistance (R_{bundle}) of the bundled TSV can be expressed as [19, 20]

$$R_{bundle} = \frac{h}{2e^2 \lambda_{mfp} N_{total}} \quad (1)$$

where h and e represent the Planck's constant and charge of an electron, respectively. N_{total} is the total number of conducting channels and can be expressed as

$$N_{total} = N_{channel} \times N_{CNT} \quad (2)$$

where $N_{channel}$ is the number of conducting channels of each SWCNT in a bundle [17]. N_{CNT} represents the total numbers of SWCNTs in a bundled TSV and can be expressed as [20]

$$N_{CNT} = \frac{2\pi r_{via}^2}{\sqrt{3}(D + \delta)} \quad (3)$$

where r_{via} and $\delta(\approx 0.34nm)$ are the via radius and the distance between neighboring SWCNTs in the bundle, respectively. The conduction mechanism in CNT is ballistic or dissipative due to the long mean free path (mfp) in the range of micrometers. The diameter dependent mfp can be expressed as

$$\lambda_{mfp} = \frac{1000D}{(T/T_i) - 2} \quad (4)$$

where T_i and T represents the temperature that is equal to 100K and 300K, respectively [19]. D is the diameter of each SWCNT in the bundled TSV. The equivalent bundle inductance (L_{bundle}) consists of (1) kinetic inductance (L_K) that originates from the kinetic energy of the electrons in each conducting channel and (2) magnetic inductance (L_M) that represents the magnetic field induced by the current flowing through a nanotube [19]. The equivalent L_{bundle} can be expressed as

$$L_{bundle} = \frac{L_K}{2N_{total}} + L_M \quad (5)$$

where $\quad L_K = \dfrac{h}{2e^2 v_F}$ and $L_M = \dfrac{\mu}{2\pi} \ln\left(\dfrac{y}{D}\right)$ (6)

y and v_F represents the distance of CNT bundle from the ground plane and the Fermi velocity of CNT ($\approx 8 \times 10^5$m/s), respectively. The equivalent via self capacitance primarily comprises of (1) quantum capacitance (C_Q^{Bundle}) and (2) parallel plate electrostatic capacitance (C_E^{Bundle}). The C_Q^{Bundle} and C_E^{Bundle} can be expressed as [19, 21]

$$C_Q^{Bundle} = C_{Q0} \times 2N_{total} \; ; \text{ where } \; C_{Q0} = \frac{2e^2}{hv_F} \quad (7)$$

$$C_E^{Bundle} = 2C_{En} + \frac{(n_w - 2)}{2} C_{Ef} + \frac{3(n_H - 2)}{5} C_{En} \quad (8)$$

where C_{En} and C_{Ef} are the parallel plate capacitances of isolated SWCNT with respect to near and far neighboring interconnects, respectively [21], and can be expressed as

$$C_{En} = \frac{2\pi \varepsilon_0 \varepsilon_r}{\ln(2r_{via} / D)} \quad \text{and} \quad C_{Ef} = \frac{2\pi \varepsilon_0 \varepsilon_r}{\ln(4r_{via} / D)} \quad (9)$$

For a circular TSV, $n_w = n_H = \sqrt{N_{CNT}}$. The MOS capacitance primarily consists of (1) C_{OX} that represents the oxide capacitance between two TSVs and (2) C_{OX_TSV} that appears between the via and the Si substrate [21]. These two capacitances can be expressed as

$$C_{OX\_TSV} = \frac{4\varepsilon_0 \varepsilon_r H_{TSV}(r_{via} - t_{ox})}{t_{ox}} \quad (10)$$

$$C_{OX} = \left(\frac{2}{C_{OX\_TSV}} + \left(\frac{\varepsilon_0 \varepsilon_r A}{d_{pitch}}\right)^{-1}\right)^{-1} ; \text{ where } A = \pi r_{via} H_{TSV} \quad (11)$$

where H_{TSV}, d_{pitch} and t_{ox} are the via height, center-to-center distance between two TSVs and the oxide thickness, respectively. The capacitance and conductance of the lossy Si substrate are represented as C_{Si} and G_{Si}, respectively. These components are primarily dependent on the via radius and d_{pitch} and can be expressed as [21]

$$C_{Si} = \frac{\varepsilon_0 \varepsilon_r A}{d_{pitch}} \tag{12}$$

$$G_{Si} = \pi\sigma / \ln\left(\frac{d_{pitch}}{2r_{via}} + \sqrt{\left(\frac{d_{pitch}}{2r_{via}}\right)^2 - 1} \right) \tag{13}$$

where $\sigma = 0.1(\Omega.cm)^{-1}$ represents the conductivity of the silicon substrate. The quantitative values of abovementioned via self parasitics are summarized in Table 1. For different bundle aspect ratios (AR) of 1.2:1 and 5:1, the parasitic values are obtained for 45000, 37522, 18761 and 100 number of SWCNTs in a CNT bundle based TSV.

Table 1. Via parasitics for SWCNT bundle based TSVs

Via parasitics	Parasitic values for different bundle aspect ratios (AR) of			
	AR=1.2:1			AR=5:1
N_{CNT}	45000	37522	100	18761
N_{total}	30015	25027	67	12513
R_{bundle} (Ω)	8.06	9.67	3628.1	3.22
L_{bundle} (fH)	4.61	5.52	2072.3	1.73
C_Q^{Bundle} (nF)	0.39	0.33	0.0008	0.03
C_E^{Bundle} (pF)	0.39	0.36	0.02	0.06
C_{Si} (fF)	10.96	10.96	10.96	0.07
G_{Si} (mho)	39.51	39.51	39.51	6.64

Effect of polymer liners

This section analyzes the delay with and without crosstalk for SiO$_2$ and different polymer liners. Using a SWCNT bundled TSV with $AR = 1.2:1$, Figs. 3(a), 3(b) and 3(c) present the out-phase (OP), propagation (PD) and in-phase delay (IP), respectively. It is observed that the out-phase delay is more in comparison to the in-phase and propagation delay. The reason behind this is the Miller capacitive effect that leads to almost doubling of coupling parasitics. Under out-of-phase transitions, the Miller Coupling Factor (MCF) tends to a value of 2. Additionally, it is observed that the delay substantially reduces for higher number of SWCNTs in a bundle for a fixed via height and radius. It is due to the lower parasitic values that mainly depend on the number of conducting channels (N_{total}) of each SWCNT in the bundle. The higher N_{total} substantially reduces the overall resistance and inductance with a small increase in capacitance (Table 1). It results in reduced delay for bundled TSV having 45000 SWCNTs.

Using SiO$_2$ and different polymer liners, the propagation delay and crosstalk induced delay of bundled TSV with AR=1.2:1 (N_{CNT}=45000) is compared with a TSV bundle

having AR=5:1 (N_{CNT}=18761). Table 2 presents the percentage reduction in delay with and without crosstalk for different polymer liners. In comparison to SiO$_2$ liner, the overall delay of bundled TSV with AR=1.2:1 is reduced by 30.2%, 24.3% and 10.9% for BCB, PPC and polyimide, respectively; similarly, the overall reduction in delay for bundled TSV with AR=5:1 is 43.8%, 36.2% and 18%, respectively. The primary reason behind this reduction is the smaller quantitative values of oxide capacitances (Table 3). The oxide capacitances are primarily dependent on the dielectric constant of the liner material between two TSVs. The lower dielectric constant of BCB substantially reduces the capacitance values that results in lesser crosstalk induced delay in comparison to SiO$_2$ and other polymer liners with relatively higher dielectric constant.

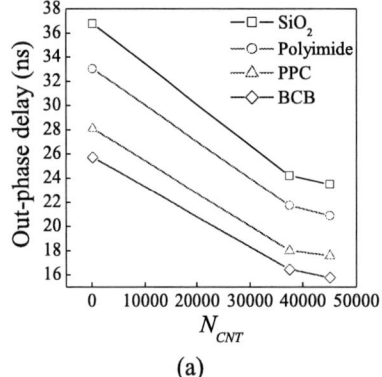

(a)

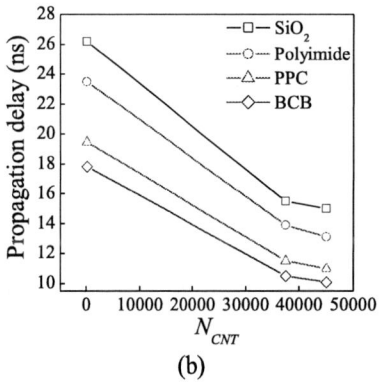

(b)

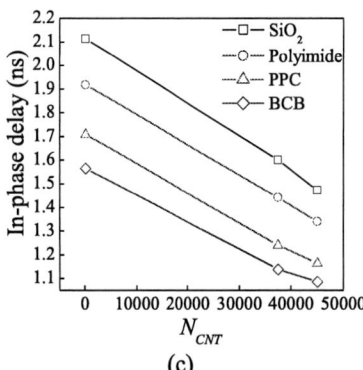

(c)

Figure 3. (a) Out-phase (b) propagation and (c) in-phase delay for SWCNT bundled TSV with AR=1.2:1

Table 2: Percentage reduction in delay using polymer liners with respect to SiO_2 liner

Bundle AR	Delay	% reduction in delay for different polymer liners		
		Polyimide	PPC	BCB
1.2:1	OP	12.49	26.68	32.73
	PD	11.17	25.17	31.75
	IP	9.05	21.01	26.24
5:1	OP	27.95	39.84	46.20
	PD	15.32	35.72	43.63
	IP	10.73	33.09	41.44

Table 3: Oxide capacitances for different liner materials

Liner material	Bundle AR	C_{OX} (fF)	C_{OX_TSV} (pF)
SiO_2	1.2:1	10.936	4.649
	5:1	0.073	0.029
Polyimide	1.2:1	10.933	4.172
	5:1	0.072	0.026
PPC	1.2:1	10.927	3.457
	5:1	0.071	0.022
BCB	1.2:1	10.924	3.159
	5:1	0.070	0.019

Conclusions

This research paper presented a comparative analysis of delay with and without crosstalk using SWCNT bundled TSV with SiO_2 and different polymer liners. Using BCB as liner material, the oxide capacitance between two TSVs is reduced upto 34.48% in comparison to the conventional SiO_2 liner. It results in reduced delay with and without crosstalk for higher bundle aspect ratio. For a SWCNT bundled TSV with higher aspect ratio, the overall delay is reduced by 36.2% for PPC whereas the reduction becomes 43.8% and 18% for BCB and polyimide, respectively, compared to the SiO_2 liner. The above mentioned results are obtained by considering only the effect of dielectric constant that demonstrates BCB as the most suitable liner material. However, PPC can be preferred over BCB and polyimide because of its advantage in obtaining the smooth surface, readily adjustable viscosity of precursors, good fluidity and adhesion. These outstanding properties are the primary reasons that make PPC coating process CMOS compatible.

References

1. J. Q. Lu, "3-D hyperintegration and packaging technologies for micro-nano systems," *Proc. of IEEE*, vol. 97, no. 1, pp. 18-30, Jan. 2009.

2. W. R. Davis, J. Wilson, S. Mick, J. Xu, H. Hua, C. Mineo, A. M. Sule, M. Steer, P. D. Franzon, "Demystifying 3D ICs: The pros and cons of going vertical," *IEEE Des. & Test of Comput.*, vol. 22, no. 6, pp. 498-510, Nov.-Dec. 2010.

3. H. Jiang, B. Liu, Y. Huang, and K. C. Hwang, "Thermal expansion of single wall carbon nanotubes: Special section on mechanics and mechanical properties of carbon nanotubes," *J. Eng. Mater. Technol.*, vol. 126, no. 3, pp. 265-270, Jul. 2004.

4. H. Li, C. Xu, N. Srivastava, and K. Banerjee, "Carbon nanomaterials for next-generation interconnects and passives: Physics, status and prospects," *IEEE Trans. Electron Devices*, vol. 56, no. 9, pp. 1799-1821, Sep. 2009.

5. G. VanderPlas, P. Limaye, I. Loi, A. Mercha, H. Oprins, C. Torregiani, S. Thijs, D. Linten, M. Stucchi, G. Katti, D. Velenis, V. Cherman, B. Vandevelde, V. Simons, I. DeWolf, R. Labie, D. Perry, S. Bronckers, N. Minas, M. Cupac, W. Ruythooren, J. V. Olmen, A. Phommahaxay, M. de Potter, A. Opdebeeck, M. Rakowski, B. De Wachter, M. Dehan, M. Nelis, R. Agarwal, A. Pullini, F. Angiolini, L. Benini, W. Dehaene, Y. Travaly, E. Beyne, and P. Marchal, "Design issues and considerations for low-ccst 3D TSV IC technology," *IEEE J. of Solid State Circuits*, vol. 46, no. 1, pp. 293–307, Dec. 2010.

6. L. Zhang, X. Gu, and J. Guo, "Clinical observation on dental caries treatment using nanometer composite resin," in Proc. *IEEE 2011 Int. Conf. Human Health and Biomedical Engineering (HHBE 2011)*, Jilin, FL, Aug. 19–22, 2011, pp. 668–670.

7. Y. -C. Hsin, C. -C. Chen, J. H. Lau, P. -J. Tzeng, S. -H. Shen, Y. -F. Hsu, S. -C. Chen, C. -Y. Wn, J. -C. Chen, T. -K. Ku, M. -J. Kao, "Effects of etch rate on scallop of through-silicon vias (TSVs) in 200mm and 300mm wafers," in Proc. *IEEE Electronic Components and Technol. Conf. (ECTC 2011)*, Lake Buena Vista, FL, May 31 - Jun. 3, 2011, pp. 1130-1135.

8. N. Ranganathan, D. Y. Lee, L. Youhe, G. -Q. Lo, K. Prasad, and K. L. Pey, "Influence of bosch etch process on electrical isolation of TSV structures," *IEEE Trans. Components, Packaging and Manufacturing Technol.*, vol. 46, no. 1, pp. 293–307, Dec. 2010.

9. J. Bea, K. Lee, T. Fukushima, T. Tanaka, and M. Koyanagi, "Evaluation of Cu diffusion from Cu through-silicon via (TSV) in three-dimensional LSI by transient capacitance measurement," *IEEE Electron Device Letts.*, vol. 32, no. 7, pp. 940-942, Jul. 2011.

10. X. Liu, Q. Chen, P. Dixit, R. Chatterjee, R. R. Tummala, S. K. Sitaraman, "Failure mechanisms and optimum design for electroplated copper through-silicon vias (TSV)," in Proc. *IEEE Electronic Components and Technol. Conf. (ECTC 2009)*, San diego, CA, May 26-29, 2009, pp. 624-629.

11. C. S. Selvanayagam, J. H. Lau, X. Zhang, S. K. W. Seah, K. Vaidyanathan, and T. C. Chai, "Nonlinear thermal stress/strain analyses of copper filled TSV (through silicon via) and their flip-chip microbumps," *IEEE Trans. Advanced Packaging*, vol. 32, no. 4, pp. 720-728, Nov. 2009.

978-1-4799-2408-0/14 $31.00 © 2014 IEEE

12. L. C. Shen, C. W. Chien, H. C. Cheng, and C. T. Lin, "Development of three-dimensional chip stacking technology using a clamped through-silicon via interconnection," *Microelectron. Reliab.*, vol. 50, no. 4, pp. 489-497, Apr. 2010.

13. S. -K. Ryu, K. -H. Lu, X. Zhang, J. -H. Im, P. S. Ho, and R. Huang, "Impact of near-surface thermal stresses on interfacial reliability of through-silicon vias for 3-D interconnects" *IEEE Trans. Device and Material Relaib.*, vol. 11, no. 1, pp. 35–43, Mar. 2011.

14. S. E. Thompson, G. Sun, Y. S. Choi, and T. Nishida, "Uniaxial-process-induced strained-Si: Extending the CMOS roadmap," *IEEE Trans. Electron Devices*, vol. 53, no. 5, pp. 1010-1020, May 2006.

15. Z. Chen, X. Song, and S. Liu, "Thermo-mechanical characterization of copper filled and polymer filled TSVs considering nonlinear material behaviors," in Proc. *IEEE Electronic Components and Technol. Conf. (ECTC 2009)*, San Diego, CA, May 26-29, 2011, pp. 1374–1380.

16. P. J. Burke, "Lüttinger liquid theory as a model of the gigahertz electrical properties of carbon nanotubes," *IEEE Trans. Nanotechnol.*, vol. 1, no. 3, pp. 129-144, Sep. 2002.

17. A. Srivastava, Y. Xu, and A. K. Sharma, "Carbon nanotubes for next generation very large scale interconnects," *J. of Nanophotonics*, vol. 4, no. 041690, pp. 1-26, Jan. 2010.

18. M. S. Sarto and A. Tamburrano, "Single-conductor transmission-line model of multiwall carbon nanotubes," *IEEE Trans. Nanotechnol.*, vol. 9, no. 1, pp. 82–92, Jan. 2010.

19. A. Naeemi and J. D. Meindl, "Performance modeling for single- and multiwall carbon nanotubes as signal and power interconnects in gigascale systems," *IEEE Trans. Electron Devices*, vol. 55, no. 10, pp. 2574-2582, Oct. 2008.

20. W. –S. Zhao, W. –Y. Yin and Y. –X. Guo, "Electromagnetic compatibility-oriented study on through silicon single-walled carbon nanotube bundle via (TS-SWCNTBV) arrays," *IEEE Trans. Electromag. Compat.*, vol. 54, no. 1, pp. 149-157, Feb. 2012.

21. S. Kannan, A. Gupta, B. C. Kim, F. Mohammed, and B. Ahn, "Analysis of carbon naotube based through silicon vias," in Proc. *IEEE 60th Electronic Components and Technology Conference (ECTC 2010)*, Las Vegas, USA, pp. 51-57, Jun. 1-4, 2010.

Investigation of Low-temperature Deposition High-uniformity Coverage Parylene-HT as a Dielectric Layer for 3D Interconnection

Bui Thanh Tung[1], Xiaojin Cheng[1,2], Naoya Watanabe[1], Fumiki Kato[1], Katsuya Kikuchi[1], and Masahiro Aoyagi[1]

[1]National Institute of Advanced Industrial Science and Technology (AIST), Japan
[2]Loughborough University, United Kingdom
tung.bui@aist.go.jp, Tel. +81-29-862-6510

Abstract

Polymer low-k materials have been considered in literature to meet the requirements of lowering the dielectric constant of the dielectric layer to decrease the problem of signal delay, lower power consumption, and reduce cross-talk between the neighboring paths, as well as, lower the fabrication temperature budget. In this paper, the feasibility of using Parylene-HT as a low-temperature deposition intelever dielectric in 3D interconnection is investigated and the results are presented. In particular, the diffusivities of Cu in room temperature deposited high-uniformity coverage Parylene-HT at 250 °C and 350 °C are evaluated to be 5.7E-18 cm^2/s and 1.3E-16 cm^2/s respectively, by dynamic secondary ion mass spectrometry (D-SIMS) technique. In addition, the capability of embedding Parylene-HT in through-Si-via (TSV) fabrication process through the demonstration of 36-μm-diameter 100-μm-depth copper-filled TSVs using Parylene-HT as a liner, are reported.

Introduction

Driven by "More than Moore", various 3D interconnection technologies are emerging and playing important roles in furthering electronics miniaturization. Beside fine-pitch high precision inter-chip interconnection [1, 2], through-Si-via (TSV) technology has become a promising candidate for 3D integration in many prospective applications. Owning to its low electrical resistivity, high stress migration resistance, and high melting point, copper-filled TSVs (Figure 1) are the most common, as well as, cost effective mass producible TSVs [3]. However, with the aggressively scaling down of the interconnect system for higher integrity and better performance, the requirement of lowering the dielectric constant of the dielectric layer to decrease the problem of signal delay, lower power consumption, and reduce cross-talk between neighboring paths, is given. Besides, in emerging applications where the device's fabrication thermal budget is limited, it is critical to find appropriate materials as the insulation layer between copper and Si, which not only offer excellent dielectric and mechanical properties but are also able to withstand various thermal conditions during fabrication. Polymer low-k materials have been considered to meet these desires. Moreover, by using a polymer buffer layer, the thermal stress in the TSV structure can be considerably reduced [4], and the electrical performance can be improved by reducing the capacitive coupling. Parylene has several distinct advantages over other polymer materials, Kapton, and polyimide materials [5–7]. Parylene has a low dielectric constant, excellent electrical insulation, high dielectric strength, high mechanical durability and can be deposited on many substrates such as silicon, glass, metal, plastics and ceramic, in the form of conformal thin film. It is deposited under vacuum at room temperature by means of vapor phase. As the entire process can be implemented at room temperature, stresses generated from different thermal expansion coefficient between the temperature of cure and room temperature, as would be the case in some of other cured polymers, are avoidable. Besides, parylene has been known to facilitate a good conformal coverage of sidewall and process stability of the deposition process, not only on surface of the wafers but also inside the vias of different dimensions. Additionally, it can be patterned by dry techniques such as plasma etching, reactive ion beam etching, reactive ion etching or high-density plasma etching [8–12]. Hence, it has been

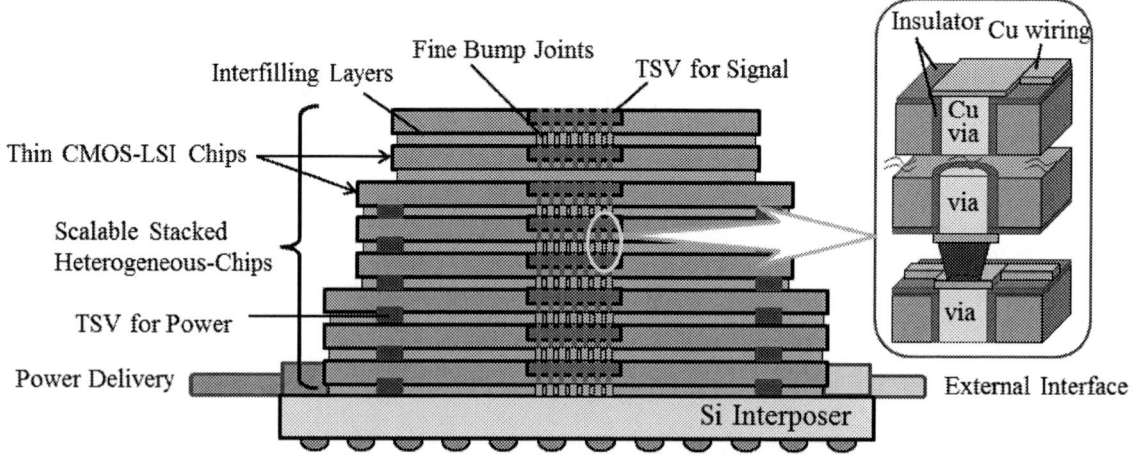

Figure 1. Schematic of the 3D integrated LSI system.

Table 1. Mechanical, thermal and electrical properties of Parylene HT in compared with conventional liners.

Parameter	Parylene-HT	Parylene-N	Parylene-C	Parylene-D	SiO_2	Si_3N_4
Density (g/cm^3)	1.32	1.10 - 1.12	1.289	1.418	2.2	3.1
Young's modulus (GPa)	2.55	2.41	2.76	2.62	69	270
Melting point (°C)	>500	420	290	380	1600	1900
Thermal expansion (ppm/°C)	36	69	35	38	0.5	3.2
Dielectric constant	1.17 - 2.21	2.65	2.95 - 3.15	2.80 - 2.84	3.9	7.5
Resistivity (Ω·cm)	2×10^{17}	1.4×10^{17}	8.8×10^{16}	1.2×10^{17}	1×10^{14} - 1×10^{16}	~1E14

investigated to be used for 3D interconnect applications. Parylene-N has been investigated at IMEC as an insulation layer in the TSV structures [13]. Parylene-C also has been used in the process of TSV copper electroplating filling, as sidewall protection to ensure bottom-up filling of blind TSV [14].

Parylene-HT, the newest parylene material with advantages of the lowest dielectric constant (1.17 to 2.21) and highest temperature withstanding (short term up to 450°C) among the commercially available variant of Parylene-N, -C, -D and –HT, would be a promising candidate by means of vapor phase polymerization at low temperatures (i.e., 25°C) [15]. In this context, this work is aimed to investigate the feasibility of using Parylene-HT for 3D interconnection. In particular, fundamental investigations of the diffusion of Cu in Parylene-HT, and the capability of embedding Parylene-HT into back-end-of-line (BEOL) TSVs will be reported in this paper.

Parylene Deposition Process

Chemical structures of the common types of parylene (Parylene -N, -C, and -D) and Parylene-HT is illustrated in Figure 2. Parylene-HT, same as other parylenes (poly-p-xylylene) is polymeric film. It is chemical vapor deposited by the process as schematically shown in Figure 3. Parylene dimer (di(poly-p-xylylene)) is vaporized at about 150 °C, pyrolized to the monomer poly-p-xylylene at about 680 °C, and then deposited at room temperature. The mechanism of deposition begins with condensation of parylene monomer on and diffusion to the surface, follows by chain initiation and propagation, in which the monomer molecules form long polymer chains [16]. Since polymerization is preceded by condensation, the parylenes are highly transparent, conformal, pinhole-free coatings over large surface areas. Some mechanical, thermal and electrical properties of Parylene HT compared with other common parylenes and the two dielectrics, which have been the workhorse of the semiconductor industry, are listed in Table 1.

In this work, parylene-HT with goal thickness of 1 μm was coated on Si substrates by a parylene deposition system (PDS 2060, Specialty Coating Systems, Inc.). The uniformity of parylene-HT on the surface of the wafers after the deposition was measured using a surface profilometer. The roughness of parylene surface was inspected by a scanning probe microscopy (SPM). Pin-hole free, 1.05 μm-thick (standard deviation of and 20 nm, i.e., about 2%) Parylene-HT layers deposited on 3-inch Si wafers at 25°C, with average roughness value of Ra = 4.402 nm (Figure 4), were achieved. In the next

part, the diffusion of copper in these parylene films at different temperatures will be presented.

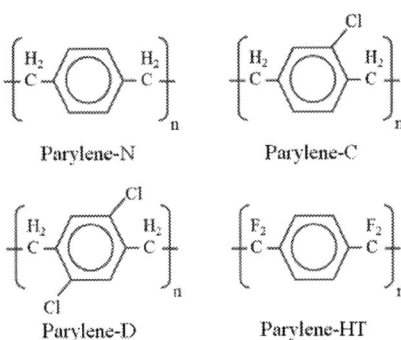

Figure 2. Chemical structures for different types of parylene.

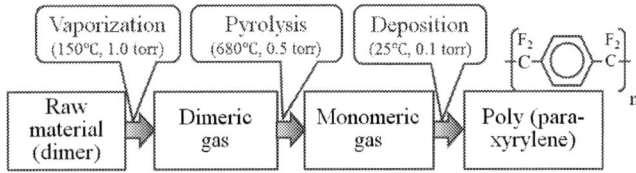

Figure 3. Parylene deposition process.

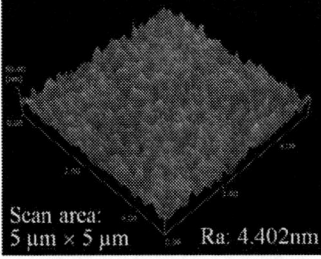

Figure 4. Surface roughness of deposited parylene thin film on Si substrate.

Diffusion of Copper into Parylene-HT

Copper is known as quite easy to diffuse into another material. It can diffuse into Si and SiO_2 to form Cu silicides compounds at temperatures of around 200°C [17–20]. For that reason, barrier materials such as TaN, Ta , Ti, TiN, W and so on, have to be used to prevent Cu atoms from diffusing into either dielectric materials or into Si substrates [21–23]. There-

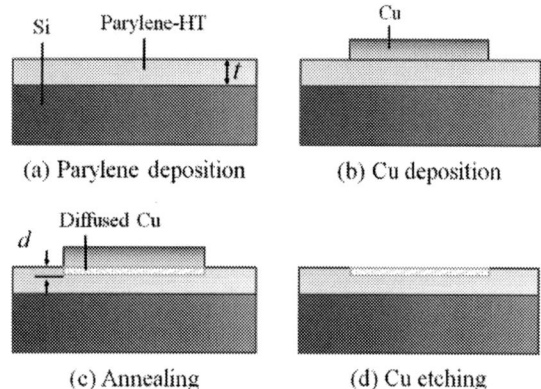

Figure 5. Sample preparation procedure for SIMS analysis.

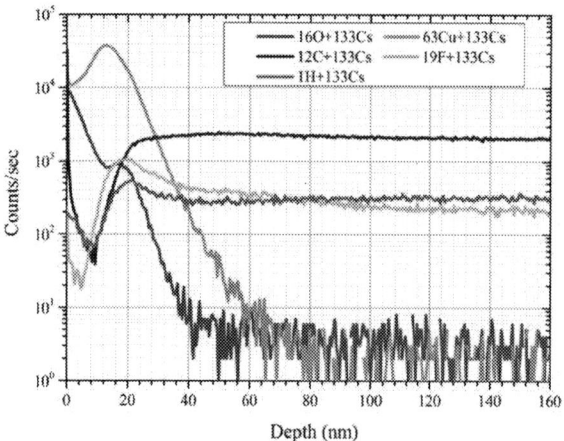

Figure 6. Depth profile obtained from SIMS analysis. The sample was annealed at 250 °C in 48 hours.

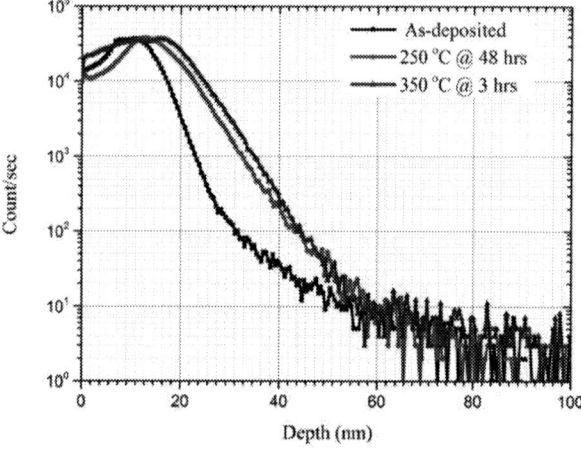

Figure 7. Diffusion of copper at different annealing conditions.

fore, diffusion of Cu in Parylene-HT needs to be examined to estimate the possibility of using this parylene in microelectronics.

In this work, the diffusivity of copper in Parylene-HT is examined using depth-profiling technique. Samples with metal-insulator-silicon (MIS) structure were prepared for investigating the diffusion of Cu in Parylene-HT. The preparation steps are summarized in Figure 5. First, on the substrate coated with 1 μm of Parylene-HT (Figure 5 (a)), 500 nm copper was deposited by sputtering (Figure 5 (b)). The samples then were annealed at temperatures up to 350°C, i.e., the thermal budget of most BEOL processes, for different periods. After that, the top copper layer was wet-etched to expose the parylene surface to the analyzing beam for copper diffusion inspection through depth profile obtaining by secondary ion mass spectrometry (SIMS).

The diffusion depth of Cu was examined using ADEPT-1010 (LVAC-PHI, Inc.,) with Cs primary ion bombardment (acceleration voltage of 1kV, system vacuum of 1E-11 Pa). The sputter crater depth was measured by a profilometer to convert the time axis into depth. We could confirm that Parylene-HT did not degas under the high vacuums conditions during the SIMS sputtering process. Figure 6 shows the depth profile of a sample annealed at 250 °C for 48 hours. A crater depth of 160 nm was measured. The profiles showed few nanometers of a copper oxide layer, followed by the detection of copper. The copper diffusion depth can be determined by observing the concurrence of Parylene-HT, i.e., the appearance of C, H and F elements. From the profile, we could also confirm that the atomic ratio of C:H:F is approximately 2:1:1, corresponding to the traces of elements of Parylene-HT (i.e., $[-F_2C-C_6H_4-CF_2-]_n$).

The depth profile of a non-annealing sample was used as the reference and copper diffusion depths were determined through comparison of the Cu depth profile of non-annealing and annealed samples at different annealing conditions. The Cu atom profiles of non-annealed samples, samples annealed at 250 °C for 24 hours and at 350 °C for 3 hours are drawn in Figure 7. It can be seen from this figure that the profile becomes broader showing the diffusion of Cu. However, the results reveal that Cu atoms are difficult to be diffused into the Parylene-HT film at 250 °C. Event at higher annealing temperatures, i.e., up to 350 °C, the diffusion is still not significant. The diffusion depths are estimated to be 10 nm and 12 nm, which correspond to annealing conditions of 250 °C for 24 hours and 350 °C for 3 hours respectively.

The diffusion coefficient at these given temperatures can be extrapolated from the experimental results. As a 500-nm-thick copper is deposited on top of parylene film, a constant-surface-concentration boundary can be applied for Flick's diffusion equation for having [24]:

$$C(x,t) = C_s erfc\left\{\frac{x}{2\sqrt{Dt}}\right\} \qquad (1)$$

where *erfc* is the complementary error function, x is the coordinated axis in the direction of flow, C_s is the constant-surface-concentration. D and t are the diffusion coefficient and diffusion time, respectively. Diffusion depth d can be expressed as:

$$d = \sqrt{Dt} \qquad (2)$$

Using (2) the diffusivities corresponding to 250 °C and 350 °C are calculated to be 5.7E-18 cm²/s and 1.3E-16 cm²/s, respectively. The diffusion of copper in Parylene-HT was evaluated through transmission electron microscopy (TEM)

978-1-4799-2408-0/14 $31.00 © 2014 IEEE

analyses. The result confirming the low diffusivity of copper will be presented in the next part.

As is well known, the integration of copper wiring with silicon dioxide requires barrier encapsulation since the diffusion coefficient of Cu atoms is quite high in silicon or silicon oxide. The diffusivities of Cu in Parylene-HT were noticed to be low, showing the possibility of using Parylene-HT as a liner layer for TSV without any barrier layer, i.e., advanced in integration complexity. In the next part, TSVs with Parylene-HT as a liner will be demonstrated.

Demonstration of Copper Filling Parylene-HT Liner TSV

CMOS compatible bottom-up copper filling TSVs with Parylene-HT as a liner have been demonstrated on 100-μm-thick silicon wafer. The fabrication flow is demonstrated in Figure 8. First, the Si substrate was temporary bonded to a carrier wafer (Figure 8(a)). Then, vias were etched by BOSCH process (Figure 8(b)). After de-bonding and cleaning steps (Figure 8(c)), Parylene-HT was deposited on the substrate (Figure 8(d)). Next, dry film and copper foil were consequently laminated on the backside of the substrate (Figure 8(e, f)). Dry film was then selective etched and parylene surface was treated with oxy plasma (Figure 8(g)). Finally, copper was bottom-up filled by electroplating (Figure 8(h)). The fabrication results are shown in Figure 9 and Figure 10, respectively.

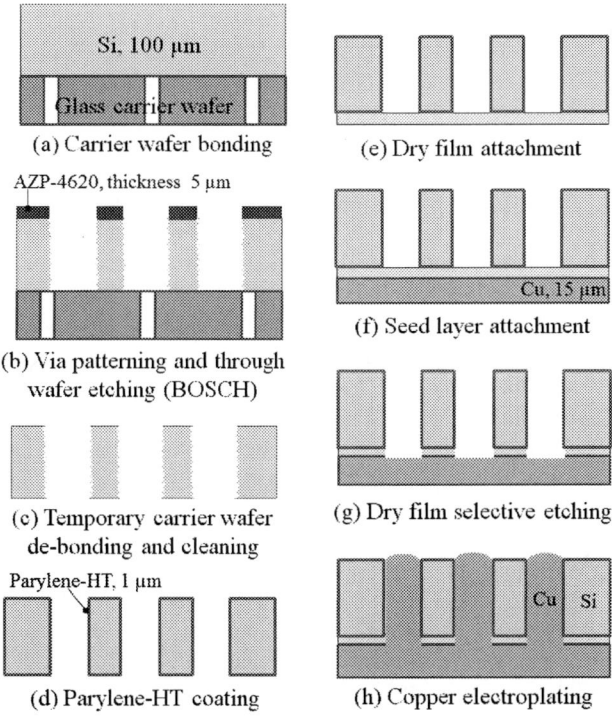

Figure 8. Copper filling TSV fabrication process.

TSV after the parylene deposition process are depicted in Figure 9. We could confirm pin-hole free, thickness high-uniformity Parylene-HT layers were deposited on 3-inch Si wafers. The thickness and standard deviation of parylene layer are 1.05 μm and 20 nm (~2%), respectively. Moreover, SEM inspections revealed that Parylene-HT can also be well deposited on the side wall of vias, with standard deviation of deposited thickness less than 5%, regardless the tiny and high aspect ratio of vias (diameter [μm] : depth [μm] = 7 : 100). Besides, as the entire parylene deposition and Cu electroplating processes are implemented at room temperature, stresses generating from thermal expansion coefficient mismatch between the materials are avoided.

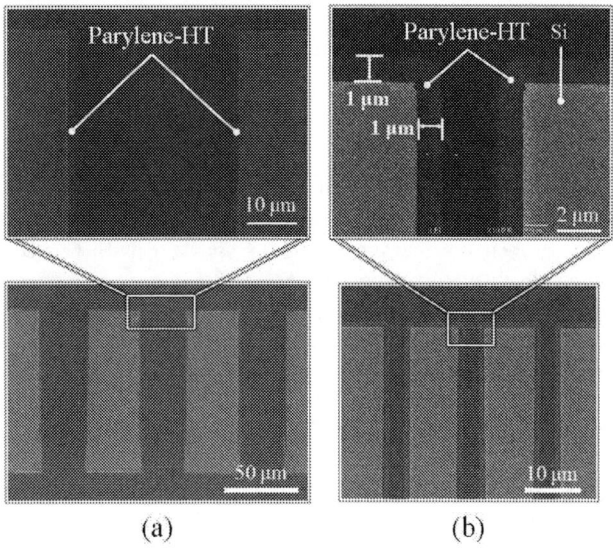

Figure 9. TSVs with pin-hole free, high uniformity Parylene-HT coating. (a) 36 μm diameter vias. (b) 7 μm diameter vias.

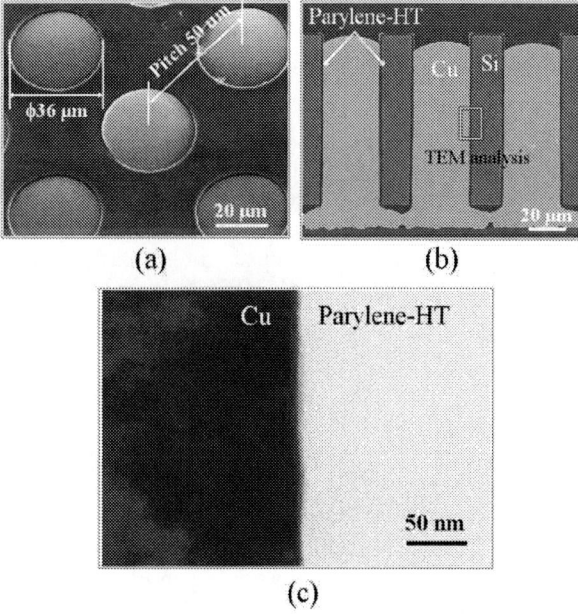

Figure 10. Bottom-up copper filled TSV with Parylene-HT liner. (a) Top view (tilt angle of 30°). (b) Cross sectional-view. (c) Cross section TEM analysis.

Figure 10 displays the images of copper filled TSV with Parylene-HT coated vias. The top and cross sectional-views are presented in Figure 10 (a) and (b), clarifying the success-

978-1-4799-2408-0/14 $31.00 © 2014 IEEE   1929

ful of embedding Parylene-HT for 3D interconnection process. Figure 10 (c) shows a cross section high-resolution transmission electron microscopy (HR-TEM) image at the interface between Cu and parylene of the sample annealed at 250°C, confirmed the low diffusivity of Cu in Parylene-HT.

While the integration of copper wiring with silicon dioxide requires a barrier encapsulation as the diffusion coefficient of Cu atoms is quite high in silicon or silicon oxide, the diffusivities of Cu in Parylene-HT were noticed to be low, showing the possibility of using Parylene-HT as a liner layer for TSV without any barrier layer, i.e., enabling more advanced integration applications.

Conclusion

The applicability of using Parylene-HT for 3D interconnections was investigated. The diffusion of copper at various temperatures was measured on MIS structure with parylene-HT as the insulator, using D-SIMS technique. Low diffusivities, i.e., 5.7E-18 cm^2/s and 1.3E-16 cm^2/s at 250°C and 350°C respectively, of room temperature deposited high-uniformity coverage Parylene-HT reveal the possibility of using Parylene-HT without any barrier layer for the applications with the thermal budget of 350°C and below. Furthermore, the capability of embedding Parylene-HT in through-Si-via (TSV) fabrication process was demonstrated. 36-μm-diameter 100-μm-depth bottom-up copper-filling TSVs with Parylene-HT as a liner was successfully realized.

In addition to a low deposition temperature, high-uniformity coverage, excellent dielectric properties, good thermal stability properties, as well as CMOS compatibility, the results from this study imply that the use of Parylene-HT as an insulating liner layer for TSV and also as an interlayer dielectric in 3D interconnection continues to look highly promising.

Acknowledgements

The authors would like to thank Mr. N. Igawa for his support in experiments. A part of this work was conducted at the Nano-Processing Facility, supported by NPF, AIST and Nanotechnology Platform Project (NIMS Nanofabrication Platform) sponsored by the Ministry of Education, Culture, Sports, Science and Technology (MEXT), Japan.

References

1. B. T. Tung, F. Kato, N. Watanabe, S. Nemoto, K. Kikuchi, and M. Aoyagi, "15-μm-pitch Cu/Au Interconnections Relied on Self-aligned Low-temperature Thermosonic Flip-chip Bonding Technique for A dvanced Chip Stacking Applications," Jpn. J. Appl. Phys., vol. 53, no. 4, 2014 (in press).
2. B. T. Tung, M. Suzuki, F. Kato, S. Nemoto, N. Watanabe, and M. Aoyagi, "Sub-Micron-Accuracy Gold-to-Gold Interconnection Flip-Chip Bonding Approach for Electronics–Optics Heterogeneous Integration," Jpn. J. Appl. Phys., vol. 52, no. 4, p. 04CB08, 2013.
3. J. H. Lau, "Overview and outlook of through-silicon via (TSV) and 3D integrations," Microelectron. Int., vol. 28, no. 2, pp. 8–22, May 2011.
4. K. H. Lu, X. Zhang, S.-K. Ryu, J. Im, R. Huang, and P. S. Ho, "Thermo-mechanical reliability of 3-D ICs

containing through silicon vias," in Electronic Components and Technology Conference, 2009. ECTC 2009. 59th, 2009, pp. 630–634.
5. J. Jakabovič, J. Kováč, M. Weis, D. Haško, R. Srnánek, P. Valent, and R. Resel, "Preparation and properties of thin parylene layers as the gate dielectrics for organic field effect transistors," Microelectron. J., vol. 40, no. 3, pp. 595–597, Mar. 2009.
6. Y. Temiz, M. Zervas, C. Guiducci, and Y. Leblebici, "A CMOS-compatible chip-to-chip 3D integration platform," in Electronic Components and Technology Conference (ECTC), 2012 IEEE 62nd, 2012, pp. 555–560.
7. M. Santos, S. Soo, and H. Petridis, "The effect of Parylene coating on the surface roughness of PMMA after brushing," J. Dent., vol. 41, no. 9, pp. 802–808, Sep. 2013.
8. E. Meng, P.-Y. Li, and Y.-C. Tai, "Plasma removal of Parylene C," J. Micromechanics Microengineering, vol. 18, no. 4, p. 045004, Apr. 2008.
9. T. E. F. M. Standaert, P. J. Matsuo, X. Li, G. S. Oehrlein, T.-M. Lu, R. Gutmann, C. T. Rosenmayer, J. W. Bartz, J. G. Langan, and W. R. Entley, "High-density plasma patterning of low dielectric constant polymers: A comparison between polytetrafluoroethylene, parylene-N, and poly(arylene ether)," J. Vac. Sci. Technol. A, vol. 19, no. 2, pp. 435–446, Mar. 2001.
10. B. Ratier, Y. S. Jeong, A. Moliton, and P. Audebert, "Vapor deposition polymerization and reactive ion beam etching of poly(p-xylylene) films for waveguide applications," Opt. Mater., vol. 12, no. 2–3, pp. 229–233, Jun. 1999.
11. B. P. Levy, S. L. Campbell, and T. L. Rose, "Definition of the Geometric Area of a Microelectrode Tip by Plasma Etching of Parylene," IEEE Trans. Biomed. Eng., vol. BME-33, no. 11, pp. 1046–1049, Nov. 1986.
12. J. T. C. Yeh and K. R. Grebe, "Patterning of poly-para-xylylenes by reactive ion etching," J. Vac. Sci. Technol. A, vol. 1, no. 2, pp. 604–608, Apr. 1983.
13. B. Majeed, N. P. Pham, D. S. Tezcan, and E. Beyne, "Parylene N as a dielectric material for through silicon vias," in Electronic Components and Technology Conference, 2008. ECTC 2008. 58th, 2008, pp. 1556–1561.
14. M. Miao, Y. Zhu, M. Ji, S. Ma, X. Sun, and Y. Jin, "Bottom-up filling of Through Silicon Via (TSV) with Parylene as sidewall protection layer," in Electronics Packaging Technology Conference, 2009. EPTC '09. 11th, 2009, pp. 442–446.
15. "Parylene Properties http://scscoatings.com/what_is_parylene/parylene_properties.aspx." [Online]. Available: http://scscoatings.com/what_is_parylene/parylene_properties.aspx. [Accessed: 20-Feb-2014].
16. M. Gazicki-Lipman, "Vapor Deposition Polymerization of para-Xylylene Derivatives —Mechanism and Applications," Shinku, vol. 50, no. 10, pp. 601–608, 2007.
17. S. W. Loh, D. H. Zhang, C. Y. Li, R. Liu, and A. T. S. Wee, "Study of copper diffusion into Ta and TaN barrier

978-1-4799-2408-0/14 $31.00 © 2014 IEEE

materials for MOS devices," *Thin Solid Films*, vol. 462–463, pp. 240–244, Sep. 2004.

18. Y. Shacham-Diamand, A. Dedhia, D. Hoffstetter, and W. G. Oldham, "Copper Transport in Thermal SiO2," *J. Electrochem. Soc.*, vol. 140, no. 8, pp. 2427–2432, Aug. 1993.

19. R. Chan, T. N. Arunagiri, Y. Zhang, O. Chyan, R. M. Wallace, M. J. Kim, and T. Q. Hurd, "Diffusion Studies of Copper on Ruthenium Thin Film A Plateable Copper Diffusion Barrier," *Electrochem. Solid-State Lett.*, vol. 7, no. 8, pp. G154–G157, Aug. 2004.

20. Eicke R. Weber, "Transition metals in silicon," *Appl. Phys. A*, vol. 30, no. 1, pp. 1–22, 1983.

21. M. T. Wang, Y. C. Lin, and M. C. Chen, "Barrier Properties of Very Thin Ta and TaN Layers Against Copper Diffusion," *J. Electrochem. Soc.*, vol. 145, no. 7, pp. 2538–2545, Jul. 1998.

22. J. O. Olowolafe, C. J. Mogab, R. B. Gregory, and M. Kottke, "Interdiffusions in Cu/reactive-ion-sputtered TiN, Cu/chemical-vapor-deposited TiN, Cu/TaN, and TaN/Cu/TaN thin-film structures: Low temperature diffusion analyses," *J. Appl. Phys.*, vol. 72, no. 9, pp. 4099–4103, Nov. 1992.

23. Y.-H. Shin and Y. Shimogaki, "Diffusion barrier property of TiN and TiN/Al/TiN films deposited with FMCVD for Cu interconnection in ULSI," *Sci. Technol. Adv. Mater.*, vol. 5, no. 4, p. 399, Jul. 2004.

24. H. Mehrer, *Diffusion in Solids: Fundamentals, Methods, Materials, Diffusion-Controlled Processes*. Springer, 2007.

Arrays of Millimeter-Wave Silicon Waveguides for Interchip Communication on Glass Interposer

Qidong Wang[1,2*], Daniel Guidotti[1,2], Liqiang Cao[1,2], Delong Qiu[1], Daquan Yu[1,2], Shuling Wang[1,2], Xugang Wang[1,2], Tianchun Ye[1,2] Lixi Wan[1]

[1]Microsystem Packaging Research Center, Institute of Microelectronics of Chinese Academy of Sciences, No. 3, Bei Tu Cheng West Road, Beijing, China 100029
[2]National Center for Advanced Packaging (NCAP China), Wuxi, Jiangsu, China, 214135
*wangqidong@ime.ac.cn

Abstract

High capacity, guided, millimeter wave communication between integrated circuits is investigated with emphasis on a glass interposer integrated architecture. Silicon rectangular waveguides with integrated internal probes are designed both from the point of view of circuit theory and antenna theory in the presence of a sub cutoff wavelength backwall taper. Simulation results are compared with vector network analyzer results up to 110 GHz.

Background

Terabit per second aggregate interchip data communication is a desirable goal [1] over distances between 2 cm to 20 cm, for example, provided that the power efficiency is of the order of 1 pJ/bit.

Optical interconnects based on discrete component integration, for example VCSELs, has limited single channel scalability and low channel density due to the size of discrete lasers, photo-detectors (PDs) and their drivers/amplifiers. Wide variations are reported (5 - 25 pJ/bit) for power efficiency, and it is difficult to scale much above 15 Gb/s without seriously compromising the MTBF of the VCSEL [2, 3].

Still at an early research stage, as of this writing, Si photonics is limited by 1) Ge PD dark current requiring high signal levels [4], 2) Ge PD BW ~ 10 GHz [5], 3) M-Z modulator BW ~ 15 GHz [6], 4) the high insertion losses of these components and silicon waveguides, and 5) the lack of silicon integrated lasers in the wavelength range 1.3 - 1.5 um. 6) In addition, no fully integrated pump laser source or amplifiers are available in this wavelength range. Descrete edge emitting lasers (EELs) require discrete component integration and silicon-bench type assembly and hermetic packaging for long MTBF as well as single mode optical fiber delivery I/Os to/from linked chips. Even with 25 Gb/s optical I/Os the number of required optical I/Os numbers in the thousands [1]. Means of connecting 1000 single mode optical fibers to a 2 cm x 2 cm chip are unknown to the author, as of this writing, at least in the case when chips are separated by distances between 2 cm and 20 cm.

Clearly, to extend single-channel data rates much beyond 25 Gb/s, innovative solution are required. It is noted that both VCSEL optical data communication in data centers and supercomputers, as well as most copper connections are based on ON-OFF or ASK modulation. Exceptions are some copper coaxial cables and some back-plane communication that may adopt bipolar signaling, an example of which is duobinary. Duobinary may also be used in optical signaling using zero-biased Mach-Zenhnder modulators [7]. Additional multiple voltage level signaling used in copper channel communication are: pulse amplitude modulation (PAM) [8] or high order quadrature amplitude modulation (QAM).

Research in 100 G and 400 G telecommunication is exploring high order QAM and orthogonal frequency domain multiplexing (OFDM). These signaling methods are aimed at increasing the spectral efficiency of the optical channel. The price is the inevitable digital processing overhead, power overhead, increased form factor and the loss of "real time" data transmission, where "real time" means delay by "a few" processor clock cycles.

There are few known options for "real time", data communication between ICs that are spatially separated by 2 cm to 20 cm and that can provide sufficient channel density for an escape bandwidth of THz [1].

An option that dates back to the 1970s is re-visited in the light of the present day need for efficient, high BW communication with a much reduced form factor and greatly increased power efficiency.

We re-visit guided millimeter wave communication. In the late 1940s and up to the mid 1970s, prior to the advent of commercially viable optical fiber communication, guided millimeter waves were envisioned to provide cost-effective, metropolitan area network (MAN) trunk-level telecommunication [9]. The basic idea was simple. Provide a hollow millimeter wave channel that can support a high carrier frequency bandwidth, partition the carrier frequency bandwidth into many millimeter wave channels with guard bands, send coded data at one end and decode data at another end. The two ends being separated by several miles, for example [10]. Historically, improved optical fiber performance and long lived semiconductor lasers carried the day.

As good ideas often return in different forms, one envisioned application of smaller mm waveguides may be described as follows: In a von Neumann computer architecture, a processor derives data and instruction stored in separate memory, cache memory that may be available on the processor die is small and, without loss of generality, is ignored for the purpose of the summary considerations that follow.

Over the evolution of general purpose digital computing, the data processing efficiency of the chip multiprocessor (CMP) has far exceeded the efficiency with which data and instructions can be fetched or stored in main memory, which may be shared and updated by other CMPs. These circumstances lead to a highly unwelcomed latency that is attributable, in part, to the limits with which memory cells can be addressed to store or fetch data in the ubiquitous synchronous dynamic random access memory (SDRAM). Another aspect of the latency is the communication channel between SDRAM and CMP, based on copper traces formed on a frequency dispersive and generally high insertion loss polymer substrate. A third aspect of the latency is attributable to the communication path in which data crosses disparate wiring hierarchies that require re-synchronization and buffers, for example, serializer/deserializer for chip to board communications. Additional delays can be due to queuing of pending memory operations, and network delays due to contention and arbitration.

The problem of memory access latency may be improved significantly in principally three ways: 1) provide sufficient volume and sufficiently fast cache memory and pre-fetching, 2) increase the aggregate bandwidth of the main memory bus, preferably by increasing the lane capacity, 3) decrease data transfer overhead such as arbitration and 4) establish a direct path between SDRAM I/O buffers and the processor I/Os. In part, this has been previously discussed [11, 12, 13].

The guided mm wave solution can be summarized as follows [11, 12, 13]: Arrays of rectangular waveguides (RWGs) filled with high resistivity silicon are first formed; each waveguide in the array being quipped with an integrated pair of simplified mm wave transmitter (Tx) and receiver (Rx) and an integrated internal probe antenna at each end. Each RWG is substantially enclosed in a conductor except where electrical isolation is necessary. The RWG array may be "flip-chipped" as a whole onto a substrate to provide data communication between stacks of SDRAM dies and a CMP. See Fig. 1. The substrate may be ceramic or glass. The concept of a self contained compute socket that incorporates sufficient SDRAM for balanced system operation was previously advanced [11]. The entire compute socket may be internally cooled by a flowing refrigerant as discussed in [13].

It should be noted that the millimeter carrier wave, which may be in the range 60 GHz to 120 GHz, (accessible by CMOS or Si-Ge compatible CMOS, for example) is confined to the interior of the RWG except for minor leakage near the internal antenna feed from the integrated mm wave Tx/Rx. The local oscillator (LO) signal for each Tx/Rx is confined to the sub-mm wavelength range and multiplied within the integrated Tx/Rx. External baseband data enters the Tx and is retrieved at the Rx to be sent to its destination, as discussed in [11]. Also discussed in Refs.11 and 14 is the concept of the single source optical LO, whose signal is distributed to all Tx/Rx in a system over optical fiber drops. Each fiber drop constitutes an electrical-to-optical conversion point that may serve an entire mm waveguide array, for example. After several optical drops it becomes necessary, the optical LO signal can be amplified by the use of a common Er-doped fiber amplifier.

The theory of rectangular waveguides predicts a large free spectral frequency bandwidth between the dominant TE_{10} mode and the first higher order modes TE_{01}, TE_{20}, degenerate when the wall width ratios are 2:1. Most of this frequency band is usable except close the cutoff frequency where phase dispersion is largest. As a result, each RWG can, in principle, supports a sufficiently wide frequency spectrum to accommodate four 10 GHz wide mm wave channels, plus guard bands, or two simultaneous, fully duplex mm wave channels with guard bands.

Because each RWG in the array is shielded from its neighbors, measured cross talk is as low as -50 dB, as is shown in Fig.3. Consequently, frequency re-use is not limited.

Millimeter carrier waves confined to a rectangular waveguide are disposed to behave like free space mm waves and the same high order coding protocols, 16 QAM, for example, can be directly adopted to increase spectral efficiency. In addition, a closed system having a fixed number of transmitters and receivers that are, in addition, stationary, has certain simplifying advantages over a mobile system.

1) All local oscillators can be physically linked to a common LO source. Consequently, frequency and phase are self tracking and there is no need for a phase lock loop (PLL) or a voltage controlled oscillator (VCO) in the RF IC.

2) All crystal oscillators and frequency synthesizers can also be removed from the RF IC and replaced by a 3X frequency multiplier, for example.

3) As the Poynting vector is confined to the RWG and the insertion loss for a RWG filled with high resistivity Si is about -1 dB/cm in the range 60-120 GHz, the required amplification of the Tx and Rx signals is minimal even up to a 20 cm long waveguide.

4) A sinusoidal varying optical LO up to 40 GHz can be conveniently generated by a Mach-Zehnder phase modulator and tuned for minimum distortion. An optical phase noise of about -150 dBc/Hz at 1 kHz offset is obtainable optically [15].

The simplified RF IC resulting from a closed, stationary system and the use of an optical LO is offset by an increase in power usage of the Mach-Zehnder modulator and the photo-detector amplifier used at each optical drop per array. In a small system having low number of mm waver Tx/Rx arrays, power efficiency may be negated. However, in a large system such as a supercomputer or data center, having many thousands of mm waver Rx/Tx arrays, the simplified Tx/Rx structure clearly provides advantages in power, cost and RF IC fab yield. Where the cross-over point is, remains to be calculated.

It is clear, however, that if a single RF channel is defined by a 10 GHz carrier frequency bandwidth and data is encoded according to 16 QAM protocol (4 bits/symbol) its capacity at a S/N of about 25 to 30 is about 40 Gb/s with a an error probability of about 10^{-12} [16]. All else being equal, each waveguide can be operated in double duplex, giving 160 Gb/s as the waveguide data carrying capacity.

Compute socket architecture
In Fig.1, is shown a portion of the cross section of the proposed compute node socket, also discussed in Refs. 11 and 13. The essential features are 1) a glass or silicon substrate of about 8 cm on a side with through connections for power, ground and baseband signaling. 2) stacked memory, 3) mm waveguide arrays and 4) a supporting ceramic base with baseband wiring networks and power/ground feeds. The fine pitch of the baseband through substrate I/Os is mitigated somewhat by an architecture that places the highest speed data communications in arrays of mm waveguides at the top of the substrate. The large area of the socket that is necessary to accommodate 256 GBytes of SDRAM main memory also

mitigates the fine pitch requirement. The mm waveguides have conductive grounded walls and may be in physical contact without significant cross-talk, maximizing frequency re-use and minimizing form factor. The dimensions of the compute node with 256 GB of stacked SDRAM and heat exchanger should measure about 8 cm on a side, making it only slightly larger than existing processor sockets that house a single processor with heat exchanger [17] but hold no memory.

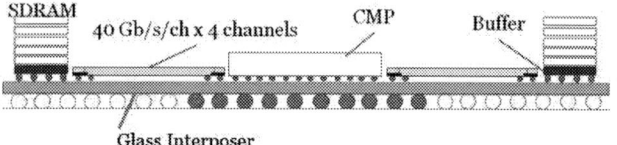

Fig.1 Cross section of the proposed compute node architecture with CMP, DRAM and arrays of mm waveguide. The structure will mount on a rigid substrate. The buffer die in the SDRAM stack allows for free data transit or data pass-through. A single mm waveguide has a theoretical data capacity of 160 Gb/s. With proper enclosure, this structure is compatible with two phases, immersion fluid flow.

It should be emphasized that all high speed signals and carrier waves are confined to the mm waveguides. The operational length of waveguides depends on the required symbol BW and can be greater than 10 cm. BW is limited primarily by waveguide dispersion, aggravated by high electrical permittivity fillers, notably silicon or Al_2O_3.

A top view of the proposed compute node architecture is shown in Fig.2. An 8 cm by 8 cm platform can house one 1.9 cm x 1.9 cm processor slot and 64 DRAM dies slots, each roughly measuring 8 mm by 8 mm. If each DRAM slot accommodates a stack of about four dies and each die holds 1 GB. The total on-socket memory is 256 GB and matches the memory requirements of contemporary 256 GFLOPS CMPs.

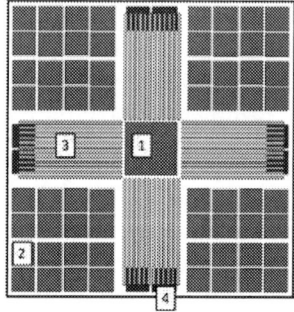

Fig.2 A representation of the proposed compute node socket. 1 CMP ~ 1.9 cm on a side. 2 DRAM dies stacked 4 deep 3 Millimeter waveguide array about 3 cm long. 4 Memory controller and buffer die with pass-through. The structure may have pin plugs in the back.

When single DRAM dies with 2 GB capacity become available, or higher DRAM stack heights are developed, then the total on-interposer memory will be 512 GB or greater, and the platform is scalable to future CMPs. It can be pointed out here that one of the advantages of the compute module and two-phase cooling is that larger DRAM stack heights need not be limited by concerns of heat removal and cooling fluid easily flows all around the stack preferentially cooling hot spots.

Each millimeter waveguide will be about 3 - 10 cm long and have cross section 0.6 mm wide and 0.3 mm deep and operate at 40 Gb/s per channel (simplex) using 16-QAM digital modulation in about a 10 GHz carrier frequency band

centered at about 90-120 GHz. Each waveguide can support 4 10 GHz wide channels for a capacity of 160 Gb/s. Arrays of 32 waveguides can arranged on four sides for a total of 128 WGs for a theoretical cumulative bandwidth between DRAM and CMP of 2.5 TB/s. An array of 32 WGs with integrated Tx/Rx fits on each side of the CMP and draws a width of about 19 mm. Thus the envisioned compute node will contain one ~ 256 GFLOPS CMP, and 512 GB DRAM, for a Byte-to-FLOPS ratio of 2.

Each high resistivity (HR) silicon or Al_2O_3 waveguide mm waveguide has a large free spectral range between the dominant mode and the next highest order mode, as discussed in Ref. 11. This spectral range can be used to accommodate 4 millimeter wave channels, well above the cutoff frequency, each having 10 GHz of usable symbol BW. For example, 80-90; 90-100; 100-110; 110-120 GHz. 1 GHz guard bands can be inserted to diminish cross talk. Four channels per waveguide, each having a capacity of 40 Gb/s amounts to an aggregate capacity of 160 Gb/s per channel. The entire compute node can have an internal BW of 2.5 TB/s. The aggravating factor limiting BW will be waveguide dispersion which needs to be considered.

A globally distributed optical clock signal originating from a single source semiconductor laser provides coherent self tracking and local oscillator (LO) frequency to all Tx/Rx channels on the compute node. The optical local oscillator is also expected to provide a phase noise of less than -150 dBc/Hz @ 1 MHz. The globally distributed optical clock signal is discussed elsewhere [14, 18] and greatly simplifies the Tx/Rx architecture by eliminating the need for un-necessary components as PLLs, VCOs, X'tal oscillators and frequency synthesizers. Further, the optical LO can provide the low phase noise necessary for low bit error rate (BER) in M-QAM symbol modulation/demodulation [19].

The above performance and form factor depend on being able to operate silicon filled rectangular waveguides over a distance of at least 6 cm at a symbol rate of 10 to 15 GHz, without equalization, and with BER less than 10^{-12} s^{-1}. To achieve I/O pitch competitiveness it is necessary to choose rectangular mm waveguides filled with a high electrical permittivity, low loss material. Intrinsic or compensated silicon and aluminum oxide, meet that requirement at reasonable base cost, but only intrinsic silicon can also be used to homogeneously integrate mm wave, CMOS RF IC transceivers up to 120 GHz. The price for I/O pitch competitiveness is high phase dispersion that limits channel bandwidth over distance.

Propagation constant dispersion is predictive and causes different frequencies to experience different rates of phase shift as propagation distance. This is a form of phase velocity dispersion and will cause inter symbol interference (ISI) by time skewing and broadening digital pulses.

The characteristics of a low loss, high relative permittivity-filled rectangular waveguide are discussed elsewhere [14].

Silicon filled rectangular waveguides design considerations

Ceramic filled rectangular waveguides and their fabrication are discussed in [12].

In the case of dielectric waveguides with embedded antennas, such as a folded ¼ wave quasi-Yagi antenna, the waveguide itself plays no significant role in the design of the transmitter (Tx) and receiver (Rx) circuit [20]. In this case the role of the dielectric guide is similar to that of an optical fiber guiding the electromagnetic wave by virtue of its real dielectric permittivity being greater than that of its surroundings.

In the case of a rectangular (or cylindrical) hollow or filled waveguide, the conducting walls and the relative placement and length of an internal probe (or monopole antenna) with respect to the walls strongly influences the performance of the waveguide channel [21, 22].

A rectangular waveguide with conducting walls acts as a high pass filters. A large free spectral range is one of the main reasons for choosing millimeter waveguides for high bit rate transmission. A second reason is the absence of cross talk as the walls of each waveguide is a grounded conductor. A third reason for the choice is form factor. The width of a silicon-filled waveguide designed to have a cutoff frequency of 120 GHz and aspect ratio a/b=2 is approximately 500um and the thickness is about 300um. It is projected that each waveguide can carry four mm wave channels, each 10 GHz wide, in addition to sidebands. The channel pitch is therefore equivalent to the bare optical fiber pitch, 125 μm. A silicon filled rectangular waveguide array is easily defined by conventional deep trench Si etching. Furthermore, as there is no cross talk the array pitch of is 500um (125 μm per RF channel) is competitive with the pitch of balanced, differential copper transmission lines. A sketch of a rectangular waveguide with conducting wall of aspect ratio a/b is shown in Fig. 3(a), along with a sketch of the electric field of the dominant mode at an instant in time. HFSS simulation of the dominant mode propagation is shown in Fig. 3(b).

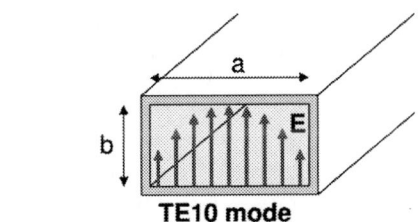

Fig. 3(a). Crosstalk without taper structure.

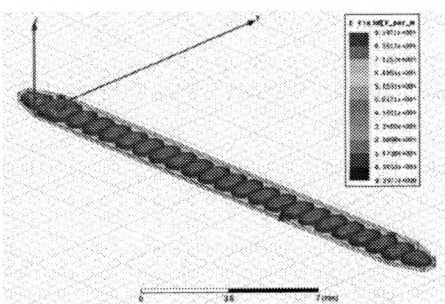

Fig. 3(b). EM wave propagating along the waveguide.

Clearly, if the waveguide array is fabricated directly on the wafer without appropriate end termination, the signal will leak into the wafer and into neighboring wavegides, as depicted in Fig. 4.

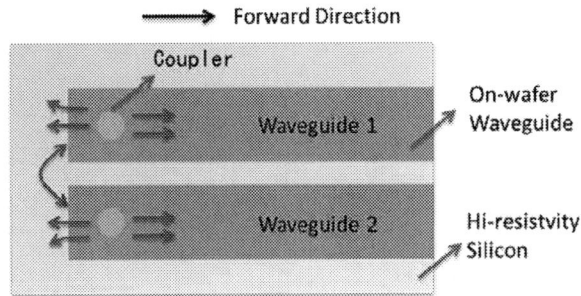

Fig.4. Crosstalk without taper structure.

Termination of the ends of a rectangular waveguide strongly influences return loss and usable carrier frequency bandwidth [23]. Given the size or our waveguides and the need to minimize round-trip bit reflection interference or bit echo, we have chosen to terminate the backwall of our Si waveguides in a short sub cutoff wavelength taper that is designed to prevent cross talk with adjacent waveguides. A sketch of the H-field direction taper is shown in Fig.5.

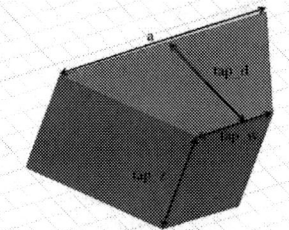

Fig.5. The terminating H-field taper structure.

In Fig.6 is shown the behaviour of the TE01 guided mode in the taper zone, propagating along the tap_d direction, or along the axis of the waveguide. It is to be noted that as the cross section of the taper shrinks, the waveguide impedance changes and part of the wave is reflected back into the waveguide. Fig. 6 (left) shows the E-field contours in the taper, while Fig. 6 (right) shows the E-field at the two opposite faces of the taper. It is clear that the coherent E-field at the smaller end-face has dissipated, as predicted by fundamental guided wave propagation.

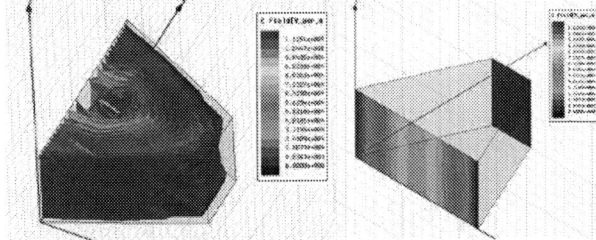

Fig.6. Field display in the H-field taper structure.

One of the effects of the taper structure is to introduce dispersion due to the fact that EM waves at higher frequencies penetrate deeper into the taper. Thus high frequency waves display longer delays upon reflection with consequences for intersymbol interference. Thus, optimization of the taper should minimize reflection and dispersion but not be too long as to compromise form factor. Fig.7 shows the simulated effects of the taper length on the propagation loss in the taper region as a function of carrier wave frequency.

978-1-4799-2408-0/14 $31.00 © 2014 IEEE   1935

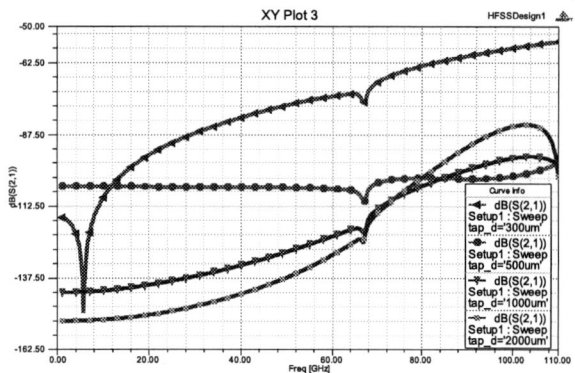

Fig.7. Propagation loss in the taper region as a function of taper length and frequency.

When analyzed in terms of circuit theory, the probe can be regarded as being inductively or capacitively coupled to the wall, depending on its length, and is generally characterized by a superposition [21, 22].

When analyzed in terms of antenna theory, the probe can be regarded as being, approximately a ¼ wave monopole radiator.

The real impedance of the monopole or probe is estimated to be about 200 Ohms and a balun is required. Given our constraints, a stripline-to-microstrip tansistion balun [24] may be suitable but requires further simulation. Our actual antenna feed remains to be optimized and presently consist of a short stripline landing pad designed to capture the probe via at one end, and capture a 50 Ohm contact probe (Cascaed Microtech Infinity Probe) at the other end.

Silicon filled rectangular waveguides method of construction

Float zone silicon wafers having bulk resistivity greater than 10,000 Ω-cm are used as the starting material for the fabrication of test arrays of silicon filled rectangular waveguides. The (001) oriented wafers are 4-inches in diameter, 300 μm thick and are polished on both sides. The waveguides are defined and separated by 100 μm wide trenches through the thickness of the wafer. Through waveguide walls and blind probe antenna vias are formed simultaneously during a single Bosch reactive ion etching step by taking advantage of the dependence of the etch rate on feature size and pattern loading. After silicon etching, a 1 μm thick insulating SiO_2 layer is deposited everywhere by using a low temperature TEOS process. This is followed by magnetron sputtering deposition of a Ti barrier/adhesion layer about 60 nm thick, followed by a three micron layer of copper on both sides of the wafer, or the rough equivalent of three skin depths at 75 GHz. It is surmised that the blind via should have at least a 2 μm Cu wall coverage, to be verified by cross section.

Silicon waveguides were designed having widths of 0.55, 0.7 and 0.8 mm and lengths of 1, 2, 4 and 8 cm. Waveguides are terminated at the end by a narrowing width or H-taper designed to prevent EM field from back leaking into the wafer thus causing interference between adjacent waveguides.

Simple probe antennas serve as transmitters and receivers and are placed at the onset of the waveguide taper. The probe antenna is formed by a blind via having a diameter of 60 μm.

Arrays of waveguides of unequal length and width are constructed to monitor process consistency. A stripline is used to feed signal to or extract signal from the probe antennas. A circular area around the probe via is metalized and Ground-signal-ground test pads were formed to provide an interface with external probes and measurement instrumentation.

Float zone silicon wafers having bulk resistivity greater than 10,000 Ω-cm are used as the starting material for the fabrication of test arrays of silicon filled rectangular waveguides. The (001) oriented wafers are 4-inches in diameter, 300 μm thick and are polished on both sides. The waveguides are defined and separated by 100 μm wide trenches through the thickness of the wafer. Through waveguide walls and blind probe antenna vias are formed simultaneously during a single Bosch reactive ion etching step by taking advantage of the dependence of the etch rate on feature size and pattern loading. After silicon etching, a 1 μm thick insulating SiO_2 layer is deposited everywhere by using a low temperature TEOS process. This is followed by magnetron sputtering deposition of a Ti barrier/adhesion layer about 60 nm thick, followed by a three micron layer of copper on both sides of the wafer, or the rough equivalent of three skin depths at 75 GHz. It is surmised that the blind via should have at least a 2 μm Cu wall coverage, to be verified by cross section.

The antenna and pad feed structure is defined by copper etch back in ($FeCl_3$ + HCl + H_2O) spray etch. Because of the existing wafer topography, dry film photoresist processing is used with consistently good process control.

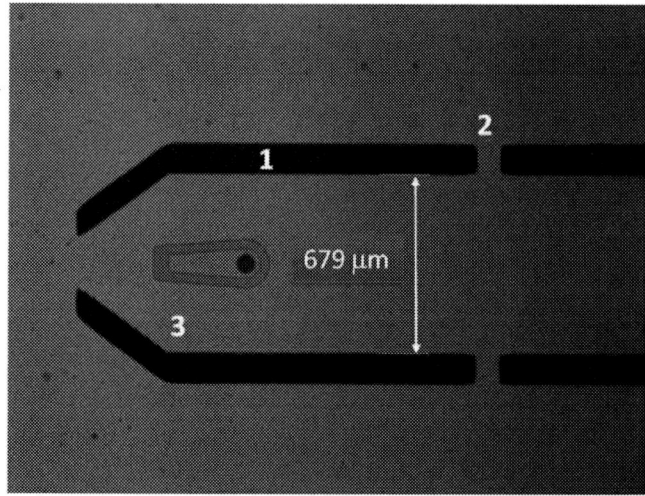

Fig. 8. Portion of a top view of one silicon waveguide covered with copper which extends into the trench wall and into the probe via. 1) through trench separation, 2) bridge stabilizer, 3) taper termination.

A single waveguide in an array is shown in Fig. 8. Clearly visible is the end taper at one end of the waveguide and 50 μm wide bridges (2) that span the 100 μm wide trench wall (1) at 1 mm intervals. These bridges provide mechanical support for the waveguide. Also visible are the antenna coplanar microstrip feed, feed isolation, and the 60 μm diameter probe via hole. Details of the probe feed are shown in Fig. 9. The width of the feed isolation is 36 μm and was defined by dry film lithography and spray etching of a solution of $FeCl_3$ and HCl.

978-1-4799-2408-0/14 $31.00 © 2014 IEEE

There is a particular impedance mismatch between the microstrip feed which has an impedance of about 35 Ohms where it meets the probe, and the radiation resistance of the probe itself which is estimated to be about 200 Ohms.

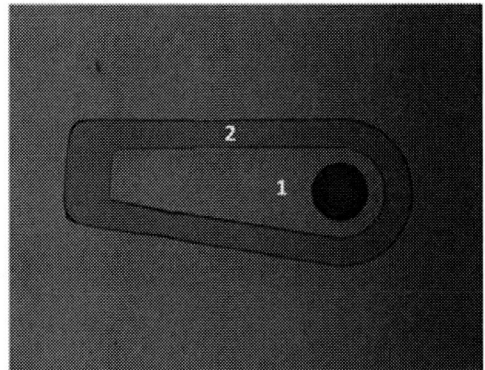

Fig. 9. Details of the microstrip connection to the probe via. The isolation gap is about 36 μm. 1) antenna probe feed, 2) isolation.

Silicon filled rectangular waveguide channel characterization

The full electromagnetic wave analysis package HFSS, now commercialized by Ansys was used to optimize various aspects of the mm wave channel, especially the end tapers and the probe location and depth. Optimization criterions were insertion loss to be less than -1 dB/cm and return loss to be greater than -10 dB over a free spectral range of at least 40 GHz in the W band. The optimization process is continuing. Simulation results to date are shown in Fig. 10 which summarizes the waveguide in the E band and having a cutoff frequency of 63 GHz.

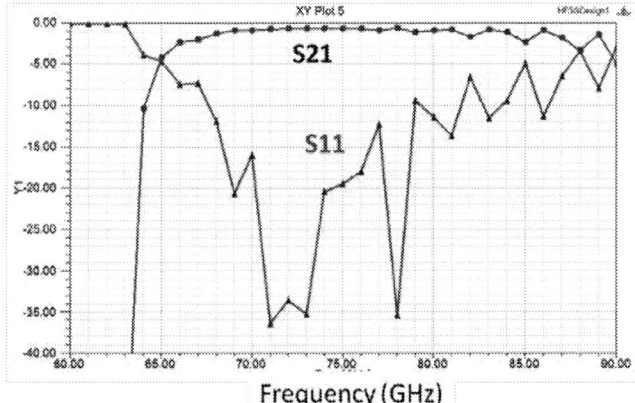

Fig. 10. Measured insertion loss and return loss in a waveguide with a cutoff frequency of about 63 GHz.

The usable bandwidth of this waveguide design is between 68 GHz and 80 GHz. It is believed that the lack of a balun in the antenna feed line contributes substantially to the high return loss at high frequencies.

Four inch diameter wafers containing arrays of waveguides having different lengths and widths were characterized from 75 GHz to 110 GHz. Throughput and return loss for each waveguide was measured over the W band. The fabrication process showed excellent consistency. The probe station with a visible copper coated wafer can be seen in Fig. 11.

Fig. 11. W band probe station and 4-inch wafer containing waveguide arrays under test.

Characterization of a 2 cm long representative channel having cutoff frequency above 81 GHz is shown in Fig. 12. The return loss is high as the waveguide was originally optimized for 60-90GHz while the testing on this band is not carried out yet.

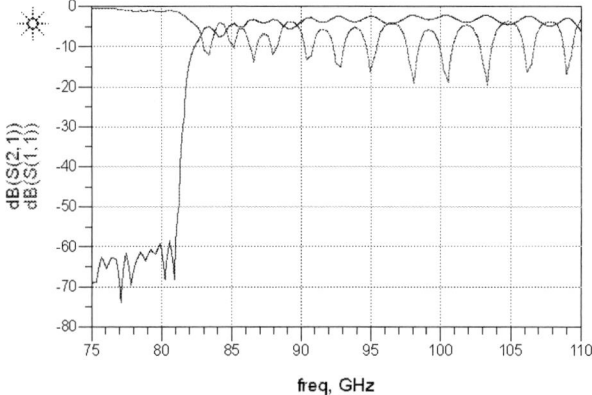

Fig. 12. W band response of a 2 cm long silicon waveguide with feed structure shown in Fig. 9. The cutoff frequency is about 81 GHz.

Since individual waveguides are mechanically re-enforced by the silicon bridge structure shown in Fig. 8 and adjacent waveguides in an array share a common set of bridges spaced 1 mm apart, the extent of cross talk has to be investigated. The near end and far end cross talk was measured for a number of adjacent waveguides. Representative results are shown in Fig. 13 for two adjacent waveguides, 2 cm long, measured in the near end configuration.

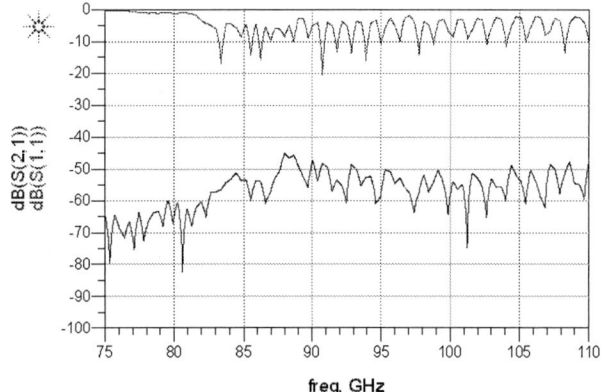

Fig. 13. Throughput and near end cross talk of two adjacent waveguides. The cross talk level being about -50 dB is acceptable.

Silicon filled rectangular waveguide integration on glass interposer

Glass has a much lower loss tangent than silicon which is advantageous both minimize insertion losses for vias and transmission lines. In one application, short, balanced transmission lines on glass may be used to couple signals from a CMP to mm waveguide arrays at one end, and from a DRAM buffer to the mm waveguide array at the other end. We therefore demonstrated two schemes for integrating mm waveguides onto a substrate carrier. The first is direct bonding, the second is surface mounting to short microstrips.

For the direct bonding scheme, the functional chip is mounted directly on top of the waveguide wafer, the modulated carrier wave is completely contained within the waveguide while the baseband signal propagates through solder bumps and wiring re-distribution layers (RDL) microstrip or stripline as shown in Figs. 14(a) and 14(b).

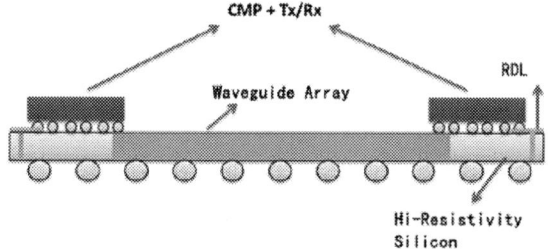

Fig.14(a). Scheme for direct bonding integration.

Fig.14(b). Cross section of a flip-chip mounted die on pads on high-resistivity silicon.

Surface mounting of the mm waveguide array is represented in Fig. 15(a). The waveguide array is diced, gold-stud-bumped and "flip chip" mounted onto the substrate glass carrier as a unit. Other chips are similarly "flip chip" attached to RDL pads. In Fig. 15(b) are shown the flip chip tool, a glass wafer with RDL, and six, 1x4 mm waveguide arrays and six test dies. In Fig. 15(c) are shown a mm waveguide with ground and antenna probe bumps (left) and gold stud bumping to RDL pads on a glass interposer or substrate. In Fig. 15(d) is shown a top-down view, clockwise from lower left, of an unused array of RDL traces on glass, a flip chip die connected to a flip chip 1x4 array of mm waveguides by RDL traces.

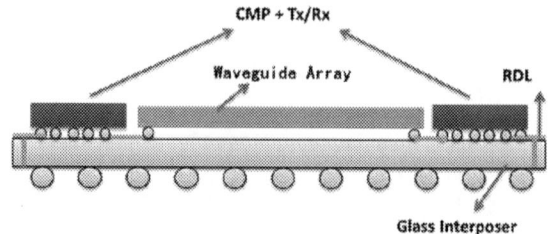

Fig.15(a). Scheme of surface mounting integration.

Fig.15(b). Surface mounting process.

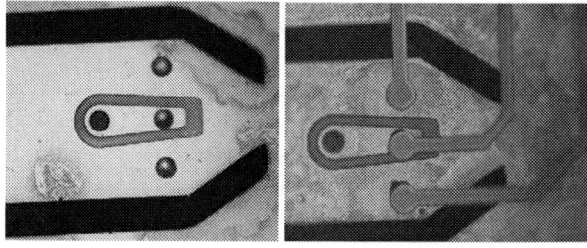

Fig.15(c). Schemes for surface mounting. Waveguide with gold stud-bumps (left) and waveguide surface mounted to glass interposer RDL (right).

Fig.15(d) surface mounting demonstration, backside copper peel off due to improper dicing.

Conclusion

A novel chip-to-chip interconnect system is reported, based on guided millimeter waves. Rectangular millimeter waveguides offer wide free spectral bandwidth that can support multiple RF channels per waveguide. When coupled with high order modulation RF protocols, extraordinarily high data rates per waveguide can be obtained. While work is continuing toward the design of an efficient antenna probe feed line, the measured characteristics of actual data transport will have the final word. An integration scheme is also discussed in the paper. Work is continuing to identify and characterize channel impediments by measuring the bit error rate as a function of waveguide length, data rate and cross talk.

Acknowledgments

Partial supported by the Opening Project of Key Laboratory of Microelectronics Devices & Integrated Technology, Institute of Microelectronics, Chinese Academy of Sciences. In addition, Daniel Guidotti gratefully acknowledges support from the Chinese Academy of Sciences and the Foreign Expert Program.

References

1. M.B. Ritter, Y. Vlasov, J.A. Kash, and A. Benner, "Optical technologies for data communication in large parallel systems," Topical workshop on electronics for particle physics 2010, Aachen, Germany, p. 1, 2010.
2. J.E. Cunningham, D. Beckman, D. McElfresh, C. Forrest, D. Cohen and A.V. Krishnaoorthy, "Scaling VCSEL reliability up to 250 Terabits/s of sysstem bandwidth," OSA/IP, paper IThA3 (2005).
3. Neinyi Li, Chuan Xie, Wenlin Luo, Chris J. Helms, Li Wang, Chiyu Liu, Qi Sun, Shenghong Huang, Chun Lei, K. P. Jackson, and Rich F. Carson, "Emcore's 1 Gb/s to 25 Gb/s VCSELs," Proc. of SPIE Vol. 8276, p. 827603-1, (2012).
4. Huapu Pan,Solomon Assefa, William M.J. Green, Daniel M. Kuchta, Clint L. Schow, Alexander V. Rylyakov, Benjamin G. Lee, Christian W. Baks Steven M. Shank, and Yurii A. Vlasov, "High-speed receiver based on waveguide germanium photodetector wire-bonded to 90nm SOI CMOS amplifier," Optics Express, Vol. 20, p. 18145 (2012).
5. Solomon Assefa, Steven Shank, William Green, Marwan Khater, Edward Kiewra, Carol Reinholm, Swetha Kamlapurkar, Alexander Rylyakov, Clint Schow, Folkert Horst, Huapu Pan, Teya Topuria, Philip Rice, Douglas M. Gill, Jessie Rosenberg, Tymon Barwicz, Min Yang, Jonathan Proesel, Jens Hofrichter, Bert Offrein, Xiaoxiong Gu, Wilfried Haensch, John Ellis-Monaghan, and Yurii Vlasov, "A 90nm CMOS Integrated Nano-Photonics Technology for 25Gbps WDM Optical Communications Applications," IEEE International Electron Devices Meeting (IEDM), postdeadline session 33.8 (2012).
6. Hui-Wen Chen, Ying-hao Kuo, John E. Bowers, "High speed silicon modulators," paper ThG1, IEEE (2009).
7. Mahmud, R.R., Khan, M.A.G., Razzak, S.M.A., "Design of a Duobinary encoder and decoder circuits for communication systems," Electrical and Computer Engineering (ICECE), 2010 International Conference on, pp 49-52, (2010).
8. Jri Lee, Ming-Shuan Chen, and Huai-De Wang, "Design and Comparison of Three 20-Gb/s Backplane Transceivers for Duobinary, PAM4, and NRZ Data," IEEE Journal of Solid-State Circuits, Vol. 43, p. 2120 (2008).
9. Thomas A. Abele, D.A. Alsberg and P.T. Hutchison, "A High-capacity Communication System Using TE01 Transmission in Circular Waveguides," IEEE Trans. on Microwave Theory and Techniquesm, Vol. MTT-23, p. 326, 1975.
10. S.E. Miller, "Waveguide as a Communication Medium," The Bell System Technical Journal, p. 1209 (1954).
11. Qidong Wang, Guidotti, D. Fujiang Lin, Guang Zhu, Jie Cui, Qian Wang, Liqiang Cao, Tianchun Ye, Lixi Wan, "Low latency high throughput memory-processor interface," Electronic Components and Technology Conference (ECTC), p.2098-2105, (2012).
12. Qidong Wang, Guidotti, D., Jie Cui, Liqiang Cao, Tianchun Ye, Lixi Wan,"A highly integratable millimeter-wave silicon waveguide array for Terabit application," Electronic Components and Technology Conference (ECTC), p.140-145, (2013).
13. Qidong Wang, Guidotti, D., Lixi Wan, Liqiang Cao, Jie Cui, Fujiang Lin, Guang Zhu, Qian Wang, Tianchun Ye, "Low latency compute node architecture cooled by a two phase fluid flow,"Low latency compute node architecture cooled by a two phase fluid flow," International Conference on Electronic Packaging Technology and High Density Packaging (ICEPT-HDP), p. 74-81, (2012).
14. Daniel Guidotti, Arshad Chowdhury, Hung-Chang Chien, Shu-Hao Fan and Gee-Kung Chang, "Toward a 60 GHz wireless, low power, high throughput memory access system," Microwave and Optical Technology Letters, Vol. 51, pp. 2969 (2009).
15. W. Shieh and L. Maleki, "Phase Noise Characterization by Carrier Suppression Techniques in RF Photonic Systems," IEEE Photonics Technology Letters, Vol. 17, p 474, (2005).
16. P.V. Vitthaladevuni and Mohamed-Slim Alouini, "BER computation of 4/M-QAM hierarchical constellations," IEEE Trans. on Broadcasting, Vol. 47, p. 228 (2001).
17. Jackson Braz Marcinichen, John Richard Thome, Bruno Michel, "Cooling of microprocessors with micro-evaporation: A novel two-phase cooling cycle," Intl. J. of Refrigeration, Vol. 33, p.1264-1276 (2010).
18. Guang Zhu, Guidotti, D., Fujiang Lin, Qidong Wang, Jie Cui, Qian Wang, Liqiang Cao, Tianchun Ye, Lixi Wan, "Millimeter wave interchip communication," Proceedings of the 2012 5th Global Symposium on Millimeter Waves (GSMM 2012), p. 471-6 (2012).
19. Pavan K. Vitthaladevuni and Mohamed-Slim Alouini, "BER Computation of 4/M-QAM Hierarchical Constellations," IEEE Trans. on Broadcasting, Vol. 47, p. 228 (2001).
20. Satoshi Fukuda, Yasufumi Hino, Sho Ohashi, Takahiro Takeda, Hiroyuki Yamagishi, Satoru Shinke, Kenji Komori, Masahiro Uno, Yoshiyuki Akiyama, Kenichi Kawasaki, and Ali Hajimiri, "A 12.5 + 12.5 Gb/s Full-Duplex Plastic Waveguide Interconnect," IEEE J. of Solid-state Circuits, Vol. 46, p. 3113 (2011).
21. Islam A. Eshrah, Ahmed A. Kishk, Alexander B. Yakovlev and Allen W. Glisson, "Equivalent Circuit Model for a Waveguide Probe With Application to DRA Excitation," IEEE Trans. ON Antenns and Propagation, Vol. 54, p.1433 (2006).
22. Roger F. Harrington, "TIME-HARMONIC ELECTROMAGNETIC FIELDS," Copyright 2001 by the Institute of Electrical and Electronics Engineers, Inc., IEEE Press. Chapter 8, Sections 8-10 through 8-14.
23. C. C. H. Tang, "Microwave delay equalization by tapered cutoff wavegiodes," presented at the 1964 ICMCI, Tokyo, Japan. See also Eugene N. Torgow, "Equalization of Waveguide Delay Distortion," IEEE Trans. on Microwave Theory and Techniques, Vol. MTT-13, p. 756 (1965).
24. Y.-G. Kim, D.-S. Woo, K. W. Kim, Y.-K. Cho, "A New Ultra-wideband Microstrip-to-CPS Transition", IEEE/MTT-S International Microwave Symposium, pp. 1563-1566 (2007).

Effect of Ag and Cu content in Sn based Pb-free solder on Electromigration

Minhua Lu, Charles Goldsmith*, Thomas Wassick*, Eric Perfecto*, Charles Arvin*

IBM T. J. Watson Research Center
Yorktown Heights, NY 10598
*IBM System and Technology Group
Hopewell Junction, NY 12533

Abstract

The effect of Ag and Cu concentrations on the electromigration of Pb-free solder was investigated. A nine cell experiment with Ag concentrations ranging from 1.2% to 2.2% and Cu concentrations ranging from 0.2 to 2.7% was conducted. Although the EM performance was found to be insensitive to the Ag and Cu concentrations in the range of the study, solder with 1.7% Cu showed the least Ag content dependence. EM failures occurred primarily on the substrate side. An effect of Sn grain texture with Ag and Cu concentrations was also observed.

Introduction

Sn, Ag, and Cu alloys are widely used in Pb-free interconnects. Alloy additions of Ag and Cu in a Sn based solder can modify both the Sn texture and the solder mechanical properties. The higher the Ag content, the more stable the Sn texture [1-3]. The stiffer mechanical properties due to higher Ag content may have a severe impact to the mechanical integrity of chips built with fragile ultra- low k dielectric layers. The addition of Cu can improve solder joint yield and produces less hardening than with Ag addition. However, Cu is a fast diffuser in Sn and the diffusion of Cu is detrimental to electromigration in Sn based solder. Attempts have been made to dope Sn solder with elements such as Ni, Al, Co, Fe, Zn, Sb, Ce, and Nd, as well as others, in a desire to improve both mechanical and electromigration reliability [4-9]. Nevertheless, investigations of alloy doping is still limited to academic studies, since co-plating processes for doping elements are difficult to achieve. The mainstream industrial solder composition is typically limited to Sn, Ag, and Cu alloys. In this paper, a quantitative study of the effect of Ag and Cu on Pb-free electromigration was conducted to understand the alloy impact and define the processing window.

Experiment

To control the Cu content, a chip with a Ni UBM (under bump metallurgy) and NiP/Au laminate surface finished substrates were used. The Cu was introduced into the joint by plating a Cu layer of varying thicknesses on top of a 9 um thick Ni UBM, which was then completely reacted with solder after reflow. The thicknesses of the Cu layers were 0, 1.5 um and 3 um, respectively. The Ag content was controlled by plating a layer of Sn followed by a layer of SnAg that upon reflow mixed together to produce samples which were 0.57%, 1.54% and 2%, respectively.

The following plating baths were used for this investigation. A sulfamate based Ni bath with 75 g/L Ni, 25 g/L boric acid, operated at 50°C and 35 mA/cm² was used to produce the Ni structures. A sulfate based Cu bath with 28 g/L copper, 175 g/L sulfuric acid, operated at 25C and 40 mA/cm² was used to produce the Cu structures. Ultra low alpha MSA based plating baths (< 2cts / cm² / 1000 hrs) operated at 25C were used to produce the solder structures. For the Sn bath, steps of 30, 60, 90 and 120 mA/cm² were used. For the SnAg bath, a single step of 40 mA/cm² was used.

The NiPAu substrates were supplied with SAC305 presolder, and the volume of the SAC305 was about 1/3 of the plated solder. The final Cu and Ag concentrations in the experiment were the combination of the Cu from plated Cu layer on Ni UBM and the Cu from the SAC305 presolder from the substrate. Table 1 shows a summary of the Ag and Cu concentrations in the nine experimental cells. The column in yellow shows the Cu thickness on the UBM and the row in blue shows the Ag content in the plated solder; the 3x3 table in white denotes the final Cu and Ag concentrations in the tested solder joints.

Plated	Sn0.57 Ag	Sn1.54Ag	Sn2Ag
0 Cu	Sn1.2Ag0.2Cu	Sn2Ag0.2Cu	Sn2.2Ag0.2Cu
1.5um Cu	Sn1.2Ag1.7Cu	Sn2Ag1.7Cu	Sn2.2Ag1.7Cu
3um Cu	Sn1.2Ag2.7Cu	Sn2Ag2.7Cu	Sn2.2Ag2.7Cu

Table 1. List of the experimental cells.

Four modules were used in each of the nine cells and each module contained 22 EM test sites, providing 88 test samples for each Ag and Cu concentration. The details of the EM test vehicle and testing setup were the same as described in a previous paper [10]. The C4's tested had a diameter of 110 um, joined to a laminate through a solder mask opening of 100 microns. Of the 22 EM bumps tested within a module, ten were wired so that the electrons entered from substrate side and the remaining twelve are wired such that electrons entered from the chip side. The resistance of every bump was monitored by a four point measurement. The samples were all tested at 150°C and 7400A/cm² (700 mA per bump) and the modules were stressed up to 4500 hours. The samples were cross sectioned to EM sites and analyzed by SEM and optical microscope for EM damage and grain texture. The EM time to failure data was analyzed and compared for the various cells.

Results: Electrical

Figure 1 shows the plot of the cumulative failure data for the nine cells. Overall the EM performance was found to be insensitive to Cu and Ag concentrations in the range of the experiment. Two cells, 2.2%Ag with 2.7%Cu and 2.2% Ag with 0.2%Cu had a large sigma due to a number of early fails. Figure 2 is a plot of the value of the distribution sigma for the different Cu concentrations. The cells with 1.7% Cu had the lowest sigma. The cell with 0.2% Cu and 2.7% Cu were similar, although not as good as the cells with 1.7% Cu, suggesting the existence of an optimum Cu concentration.

Among the nine Ag and Cu combinations, the samples with 1.2% Ag and 1.7% Cu have the best EM performance. Given the Ag content variations available in solder plating, 1.7% Cu is a preferred choice.

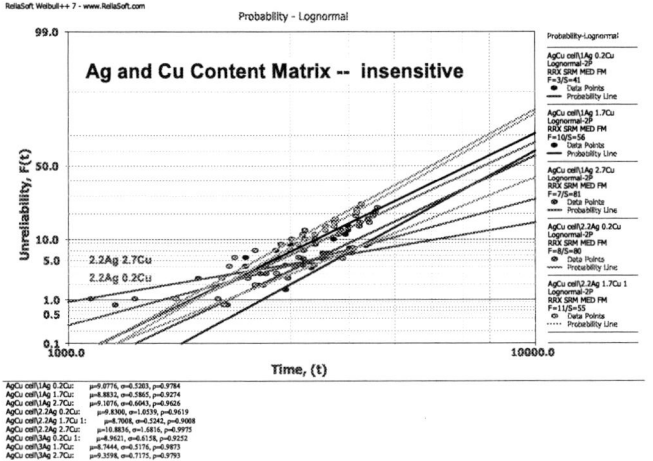

Figure 1. Lognormal plot of cumulated failure vs. stress time for the Ag and Cu experiment.

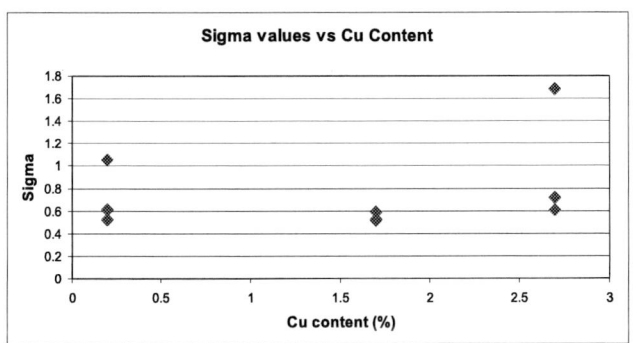

Figure 2. Plot of the value of the distribution sigma for the different Cu concentrations.

Results: Failure Analysis

Figure 3 shows SEM images of the solder to chip and solder to substrate interfaces for a typical as-joined module. There are many voids in the intermetallic compound (IMC) on the substrate side, while the voiding in the IMC is minimal on the chip side for all three Cu contents, including the one without Cu on Ni UBM. The reason for the difference in voiding is not clear. It is suspected that it might be related to the chemistry and morphology of the electroless NiP/Au surface, flux in the presolder, and/or the bond and assembly process.

The electrical stress data showed that 86% of the EM fails were associated with the substrate side, and included both early and late fails. The chip side EM fails were fewer and occurred later in time.

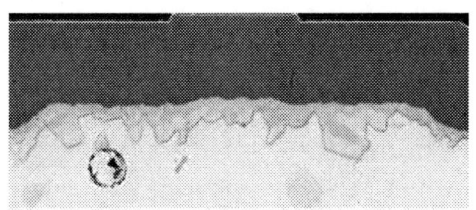

Figure 3 SEM images of the solder to chip and solder to substrate interfaces for a typical as-joined module.

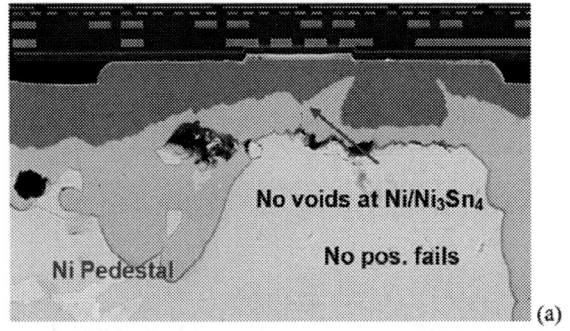

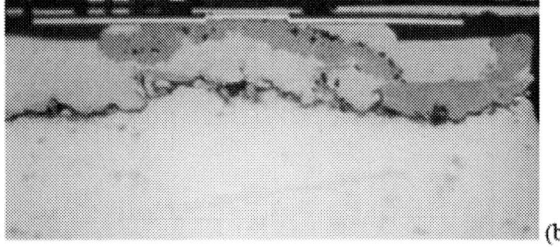

Figure 4 Chip side EM fail comparison between (a) Ni UBM and (b) Cu UBM

Figure 4(a) shows a chip side EM failure on a Ni UBM and Figure 4(b) is a chip side EM fail with a Cu UBM. For the Ni UBM, the voids only occur at the solder and IMC interface. For Cu UBM, voids can be found at the solder Cu/IMC interface and inside the IMC, in addition to being present at the solder/IMC interface. The differences in the UBM voiding might contribute to the lack of UBM failure for a Ni UBM and more UBM failures for a Cu UBM.

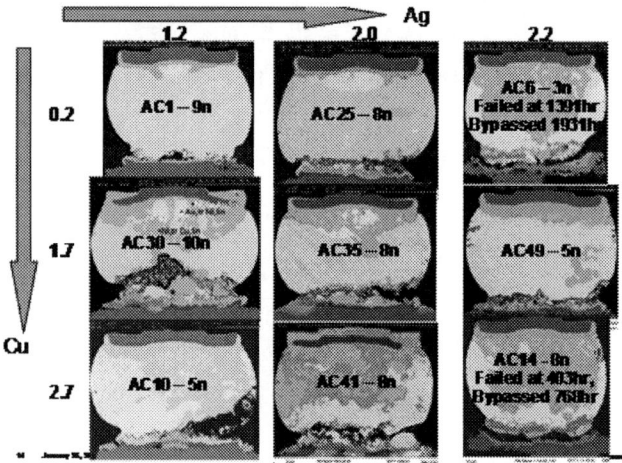

Figure 5. SEM images from early substrate fails in the 9 cells. Electron flow direction is from bottom up. Columns from left to right are for Ag concentration 1.2%, 2.0%, and 2.2% respectively. Rows from top to bottom are for Cu concentration from 0.2%, 0.7% and 2.7% respectively.

Figure 7 Cross polarized microscope images of the nine cells in the study. Cu and Ag concentrations are marked on the images. Columns from left to right are for Ag concentration 1.2%, 2.0%, and 2.2% respectively. Rows from top to bottom are for Cu concentration from 0.2%, 0.7% and 2.7% respectively.

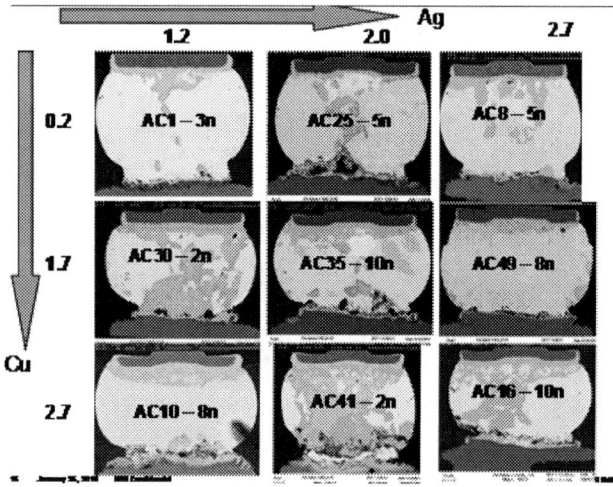

Figure 6. SEM images from late substrate fails in the 9 cells. Electron flow direction is from bottom up. Columns from left to right are for Ag concentration 1.2%, 2.0%, and 2.2% respectively. Rows from top to bottom are for Cu concentration from 0.2%, 0.7% and 2.7% respectively.

Figure 5 and Figure 6 show SEM images from early and late substrate fails in the 9 cells, respectively. Columns from left to right show the Ag concentrations of 1.2%, 2% and 2.2% respectively. The rows from top to bottom identify the Cu concentrations of 0.2%, 1.7% and 2.3%, respectively. The early EM fails were associated with severe NiP/Au damage. They are likely defect fails. The low and high Cu concentration EM cells showed more damage than the middle Cu concentration cell. The late fails, as shown in Figure 5, exhibited less substrate damage. There is no obvious Cu and Ag concentration dependence on the late fails. Improvement in the bond and assembly process could help to reduce defect induced failures and extend EM lifetimes.

The dependence of the Sn texture on Ag and Cu concentration was also investigated. Figure 7 shows cross polarizer microscope images of the nine cells in the study. The number of grains decreases as the Cu concentration increases and the number of grains increases as the Ag concentration increases. This result is consistent with a study published a few years ago [4,11]. The diffusion in Sn is highly anisotropic, as are the electromigration failures. The grain orientation effect is more pronounced at higher current densities, such as at current concentration points resulting from current crowding. A thick UBM greatly reduces the current density and therefore reduced the sensitivity of the EM to grain orientation. An unfavorable grain orientation can still exist in solder joints and induce earlier EM failures. A higher Ag concentration in Sn based solders has a certain advantage in reducing grain orientation dependence and improving the EM reliability, although Ag concentration effect on EM is not significant in the range used in this study.

Conclusions

The effect of Ag and Cu on Sn based, Pb-free solder electromigration was studied with Ni UBM on chip and NiP/Au finishing on substrates. The electromigration performance was found to be insensitive to both Ag and Cu concentrations in the range of the study and the EM failures were dominated by failures on the substrate side.

Acknowledgments

The authors would like to acknowledge IBM Research and STG for support, in particular J. Rosa, F. Dicesare and M. Jones for the sample preparation and SEM failure analysis.

References

1. Bieler, T. R., et al., "Influence of Sn Grain Sixe and Orientation on the Thermomechanical Response and Reliability of Pb-free Solder Joints," *Proc 56th Electronic*

Components and Technology Conf, San Diego, CA, May. 2006, pp. 1462-1467.

2. Sylvestre, J., and Blander, A., "Large-scale correlation in ther orientaiton of grains for lead- free solder bumps," to be published in J. Electronic Materials.

3. M. Lu, D.-Y. Shih, C. Goldsmith, T. Wassick, "Comparison of electromigration behaviors of SnAg and SnCu solders," *Proc International Syposium of Reliability Physics*, Montreal, Canada, April. 2009, p.150.

4. M. Lu, D.-Y. Shih, P. Lauro, Sung Kang, C. Goldsmith, and S. K. Seo, "The Effects of Ag, Cu Compositions and Zn Doping on the Electromigration Performance of Pb-Free Solders," *Proc. 59th Electronic Components and Technology Conf.*, 2009, p.922.

5. S. K. Kang, J. Horkans, P. Andricacos, R. Crruthers, J. Cotte, M.Datta, P. Gruber, J. Harper, K. Kwietniak, C. Sambucetti, L. Shi, G. Brouillette and D. Danovitch, . "Pb-Free Solder Alloys for Flip Chip Applications", Proc.49th Elec. Comp. Tech. Conf., San Diego, CA, June 1999, p.283-88.

6. M. Lu, D.-Y. Shih, Sung K. Kang, C. Goldsmith, P. Flaitz, "Effects of Zn Doping on SnAg Solder Microstructure and Electromigration Stability", J. Appl. Phys. **106**, 053509 (2009)

7. M. Lu, P. Lauro, C. Goldsmith, "Study of Interfacial reaction and electromigration reliability of Pb-free solders with NickEL iron barrier layer", Proc. of Interpack 2011, Portland OR, 2011

8. Y. W. Wang, Y. W. Lin, C.CT.Tu, C.CR. Kao, "Effects of monor Fe, Co, and Ni addition on the reaction between SnAgCu and Cu", J. Alloy and Compounds, **478**, (2009) p121

9. T. H. Chuang, H.-F. Wu, "Effects of Ce Addtion on the microstructure and Mechanical Properties of Sn-58Bi Solder, J. Electronic Materials, **40**, (2011), p71

10. M. Lu, S. L. Wright, G. McVicker, S. M. Sri-Jayantha, ECTC 2012

11. M. Lu, "Effects of Microstructure on Electromigration in Pb-free Solder Interconnect", Stress-Induced Phenomena in metallization, Editors, E. Zschech, P. Ho, S. Ogawa, AIP Conf. Proceedings **1300**, (2010), p229

Low Loss Transmission Lines on Flexible COP Substrate by Standard Lamination Process

Chang-Ho Liou[1], Hsin-Chia Lu[2], Yi-Fan Lin[2], Shih-Keng Chuang[2],
Wen-Ching Ko[1] and Je-Ping Hu[1]
[1]Electronics and Optoelectronics Research Laboratories,
Industrial Technology Research Institute, Chutung, Hsinchu, Taiwan, 310, Republic of China.
[2]Graduate Institute of Electronics Engineering, Graduate Institute of Communication Engineering and
Department of Electrical Engineering,
National Taiwan University, Taipei, Taiwan, 106, Republic of China.
E-mail:leonardo@ntu.edu.tw

Abstract

Wearable electronics are gaining more attention in the market. They usually demand a non-planar structure to conform the shape of human body. Flexible substrate provides a good basis to design electronics systems that can be non-planar and conformal. On the other hand, to support high speed data or high quality video transmission, the loss of transmission lines at microwave, even at millimeter wave band becomes an important performance index.

Typical low loss flexible substrate such as liquid crystal polymer (LCP) requires processing temperature higher than 250ºC, this presents a challenge for standard flexible substrate processing lines. A low dielectric constant and low loss tangent (tanδ) material called cyclo-olefin polymers (COP) developed at ZEON Corporation, Japan, requires processing temperature under 200ºC while maintaining similar low loss characteristics as LCP materials that require high processing temperature. In the future, the insulation layer will be made of COP for the 2 layer flexible copper clad laminate (2L FCCL) structure. Coating process of the copper foil on COP may also be possible in the future. The coating process will be more cost-competitive than the current high temperature lamination process. As the material cost and processing cost are both reduced for this promising material, we therefore need a more thoroughly study of the high frequency characteristics of this material.

In this study, both microstrip lines (MSL) and grounded co-planar waveguide (GCPW) are designed and fabricated on COP. The dielectric constant and loss tangent of COP is 3.0 and 0.005. To protect the copper traces, PI cover layer with thickness of 38μm is also applied on top of COP. The dielectric constant and loss tangent of PI cover layer is 3.5 and 0.02. Simulation results show that with 40μm of COP and 12 μm thickness of copper trace, the loss at 60 GHz is about 1.0~1.2dB/cm for both types of transmission lines under 50Ω of characteristic impedance. Measurement results shows about 2.5dB/cm for MSL and 3.3dB/cm for GCPW. The extra loss may due to that some solvents are not fully evaporated during laminating process.

Material and fabrication process description

The comparison of material parameters for popular substrate is given in Table 1. COP shows similar dielectric constant and fair loss tangent. For fabrication of test patterns, we adopted a F C C L process as shown in Fig. 1. The test setup for measuring the peeling strength fabricated sample is shown in Fig. 2. The peeling strength results of sample1 are given in Table 2. In order to investigate high frequency propagation properties, 50Ω microstrip line and grounded coplanar waveguide are fabricated on COP.

Table1. Material parameters of popular high frequency substrates

	Rogers PTFE/ Ceramic	Polyimide (normal)	Kuraray (LCP)	Zeon (COP)
Dk (10GHz)	3.0	3.87	2.9	3.0
Df (10GHz)	0.0013	0.012	0.002	0.005
Water absorption (%,24hr)	<0.1	1.58	0.04	<0.1
CTE(ppm/ºC)	<25	<30	18	24

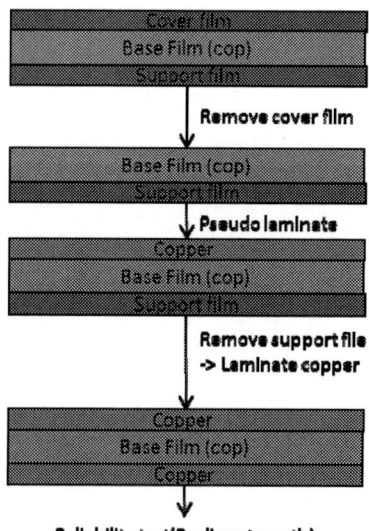

Fig. 1. FCCL lamination process

Sample 1
(W:2.4mm;L:200mm)

Fig. 2. FCCL peeling strength test setup.

Table 2. The peeling strength result of sample1

Peel strength test (Method: TM-650)			
Unit	Peeling force (Average Load) -Unit:(gf)	Peeling force (Average Load) -Unit:(KN)	Peeling force -Unit:(KN/m)
Sample 1	183.580	0.002	0.562

Transmission line design

As microstrip line and grounded coplanar waveguide are the most popular planar type transmission lines, we will design these two types of transmission lines on this COP substrate. The thickness of copper and COP are 12μm and 40μm respectively. At 10GHz the dielectric constant ε_r of the substrate is 3.0, and dielectric loss (tanδ) is 0.005. The cross section view of microstrip line and grounded coplanar waveguide are shown in Fig. 3. The dimension parameters of both transmission lines are shown in Table 3. To prevent oxidation of copper traces, PI cover layers are added on top surface of both transmission lines. The design process will be given in the following paragraphs.

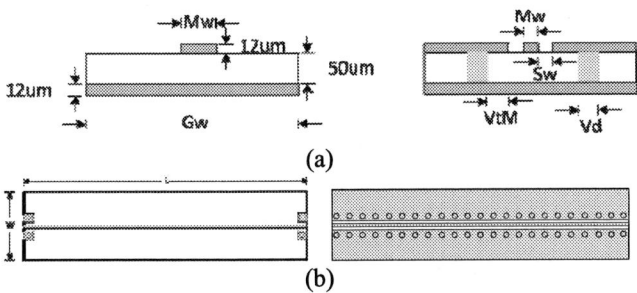

Fig. 3. (a)The cross section and (b) top view of microstrip line and grounded coplanar waveguide

Table 3 Design parameters of transmission lines

Spec.	Mw	Gw	Sw	VtM	Vd
Unit. (μm)	120	10000	100	40	150

With known dielectric constant and substrate thickness, we can use a transmission line calculator like TXLine from AWR to find the dimensions of both microstrip line and grounded CPW for 50Ω characteristic impedance. The results are shown in Fig. 4.

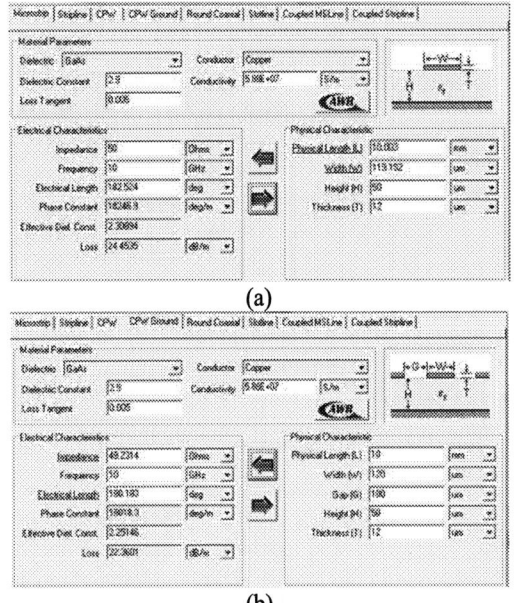

Fig. 4. Dimension synthesis results for (a) microstrip line and (b) ground CPW.

The cross sectional dimension of microstrip line is imported into Ansys HFSS. Wave ports are used to find impedance and propagation constant (γ=α+jβ) of this microstirp line. The results are shown in Fig. 5. We can see the the port impedance is close to 50 Ω. This results only need the cross section dimension of microstrip line.

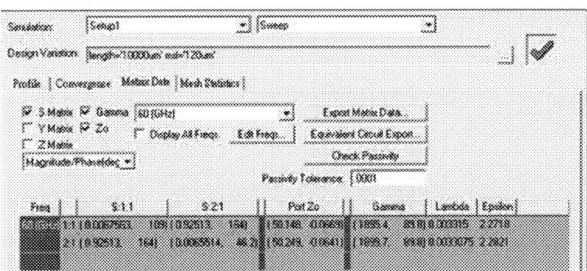

Fig. 5. The simulation result of a microstrip line.

In addition, we can also simulate the two port scattering parameters of the 'thru' calibrator, which is a microstrip line with length of 1cm. The simulated scattering parameters are shown in Fig. 6. The reflection coefficient is below -30dB in whole frequency range, this indicates the impedance of this transmission line is close to 50Ω. The simulated transmission coefficient shows a negative slope vesus frequency.

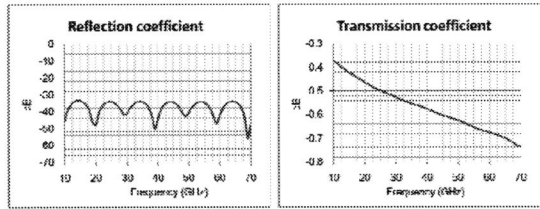

Fig. 6. HFSS simulation results of the thru calibrator.

As the length of this transmission line is 1cm, its S21 is identical to loss per cm. We can compare the loss per cm between results from cross sectional dimension and two-port

978-1-4799-2408-0/14 $31.00 © 2014 IEEE   1945

simulation of this microstrip line. We can see in Fig. 7, both results agree well.

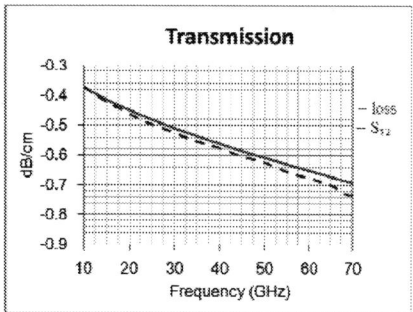

Fig. 7. Loss calculated from cross sectional dimension (loss) and two-port simulation (S12).

In the above simulation, we assumed an ideal smooth metal surface. However, the copper foil has some surface roughness. This will cause some additional loss in transmission lines. Table 4 shows the measured surface roughness of metal in current fabrication process.

Table 4. Measured surface roughness

	Cu (50μm*50μm)
Area Ra	~0.236μm
Area RMS	~0.299μm
Max. Range	~3.472μm

The loss due to surface roughness can be calculated by using the following equations from [4]:

$$R_{s1} = R_s \left\{ 1 + \frac{2}{\pi} tan^{-1} \left[1.4 \left(\frac{\Delta}{\delta_s} \right)^2 \right] \right\}$$

$$R_s = \sqrt{\frac{\pi \mu f}{\sigma}}$$

$$\delta_s = \sqrt{\frac{1}{\pi \mu \sigma f}}$$

R_s is the surface resistance due to finite metal conductivity. δ_s is the skin depth of metal. R_{s1} is the revised surface resistance that includes the effect of roughness. Δ is the surface roughness of metal. The surface resistance with smooth metal surface and rough metal surface are shown in Fig. 8. We can see both surface resistances are increased as frequency increases. And the rough metal surface will give higher surface resistance than smooth surface.

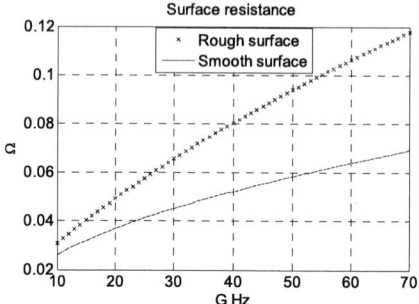

Fig. 8. Surface resistance for smooth and rough metal surface.

We can convert the increased surface resistance into its effective metal conductivity. The results are shown in Fig. 9. We can see the conductivity of copper is σ=5.8x10^7 S/m for smooth surface, while effective conductivity for rough surface decrease over frequency and is only 2.2x10^7 S/m at 60 GHz.

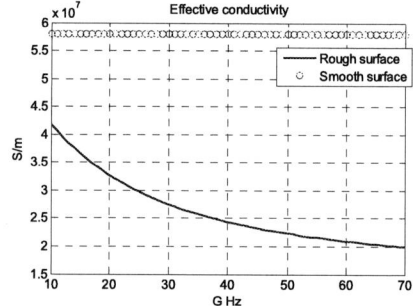

Fig. 9. Effective conductivity of smooth and rough metal surface.

We then use HFSS to compare the loss due to metal roughness. Fig. 10 shows the loss per cm with smooth and rough metal surface. We can clearly see the loss is increased by about 0.33dB/cm due to roughness.

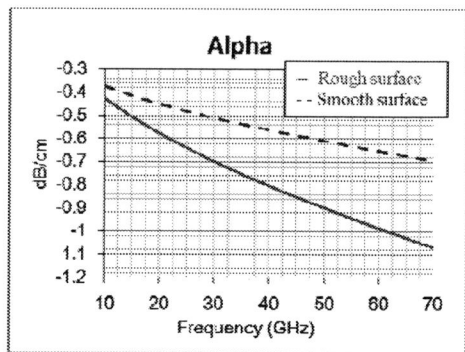

Fig. 10. Comparison of loss between smooth and rough metal surface for microstrip line.

Fig. 11 shows the simulated scattering parameters of microstrip lines with or without roughness. The length of thru is 1cm, while the length of line1 is 2cm. We can see that reflection coefficients are not changed too much among these transmission lines, while rough surface increased the loss of thru by about 03.dB. As the length of line1 is double that of thru, its loss is also doubled.

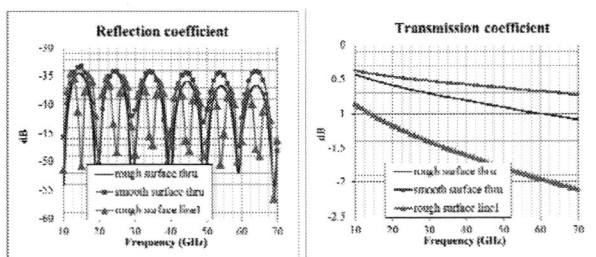

Fig. 11. Scattering parameters with smooth and rough metal surface for microstrip line.

We also use the same design process for grounded CPW. The port impedance and propagation constant from cross sectional dimension is given in Fig. 12.

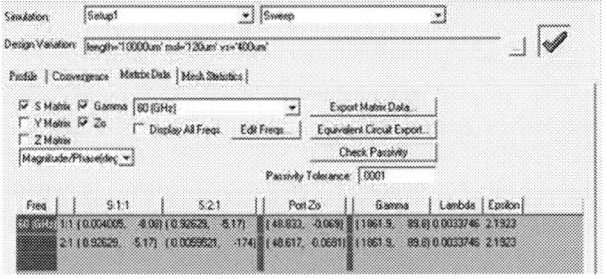

Fig. 12. The simulation result of a GCPW.

The loss per cm with smooth and rough surface for GCPW is shown in Fig. 13. Similar increase in loss as microstrip line is observed.

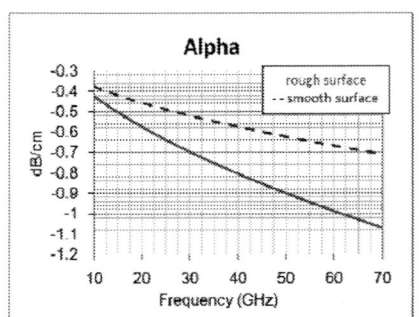

Fig. 13. Comparison of loss between smooth and rough metal surface for GCPW.

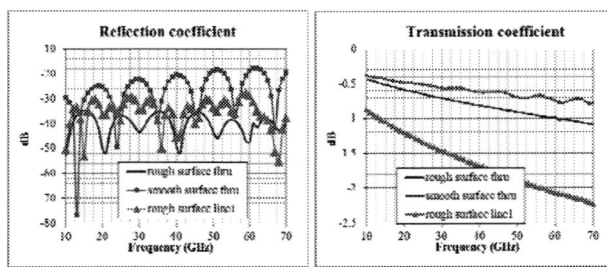

Fig. 14. Scattering parameters with smooth and rough metal surface for GCPW.

Fig. 14 shows the simulated scattering parameters of transmission lines with or without roughness. The length of thru is 1cm, while the length of line1 is 2cm. We can see that reflection coefficients are not changed too much among these transmission lines, while rough surface increased the loss of

thru by about 0.3dB. As the length of line1 is double that of thru, its loss is also doubled.

TRL calibration method and layout design

As GSG probe pads are used to provide connection between probes and transmission lines, they will add some additional phase and reflection in the measurement. Thru-reflection-line (TRL) calibration method [8] is selected to remove the effects of probing pads based on its simple structure and no need for accurate resistor calibration standard. As the PI covert layer has to be removed in area around probing pads, longer thru calibrator length allows easier fabrication of PI cover layer opening. We choose 1cm as the length of thru calibrator. The length of line calibrator is 1000μm longer than thru standard. This length is about the quarter wavelength at 60 GHz. 3 transmission lines with different lengths are also designed for loss measurement. The length of Line1~Line3 are 2cm, 4cm and 6cm respectively. After TRL calibration, their S21 will give loss of line length at 1cm, 3cm and 5cm respectively.

Fabricated transmission lines and measurement results

Fig. 15 shows the SEM of the cross section of a microstrip line sample. The thickness of COP is 48μm which is larger than the thickness used in simulation. This is larger than the thickness used in simulation. We are tuning process parameters to achieve 40μm thickness right now. The total metal thickness (including laminated copper, plated copper, Ni and Au) is 23μm.

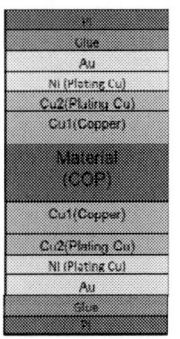
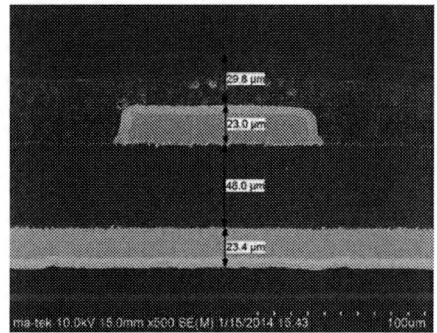

Fig. 15. SEM photo of the cross section cut of a microstrip line.

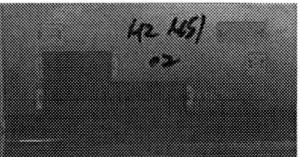

Fig. 16a. Photo of fabricated microstrip line samples.

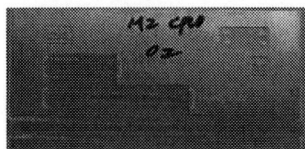

Fig. 16b. Photo of fabricated coplanar waveguide samples.

Photos of fabricated microstrip line (MSL) and grounded coplanar waveguide (GCPW) are shown in Fig. 16a and 16b respectively. We can see the cover PI layer in brown above all

978-1-4799-2408-0/14 $31.00 © 2014 IEEE

transmission lines. These transmission lines are measured by Agilent network analyzer E8361A with measurement frequency range from 10MHz to 67GHz. We use GSG 200μm pitch probes from Allstron with calibration substrate AC-2 to calibrate the reference planes to the tips of probes. The measurement setup is shown in Fig. 17.

Fig. 17. Measurement setup

TRL calibration method is used to remove the effect of probing pads. The measured normalzed S21 of microstrip lines after calibration for Line1~Line3 are shown in Fig. 18a. Normalized S21 with respect to length show similar loss in about 2.5 dB/cm for all three lines at 60 GHz except for Line1. Some dips at 55GHz are also observed for all three transmission lines. This may due to the self resonant of the probing pads.

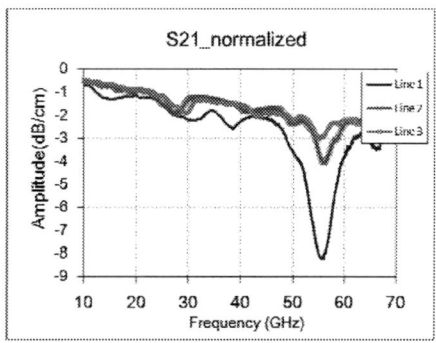

Fig.18a. The measurement results of microstrip lines with cover layer and without cover layer.

The measured scattering parameters for GCPW after calibration for Line1~Line3 are shown in Fig. 18b. Normalized S21 show similar loss in 3.3dB/cm for all three lines. Some dips at 55GHz similar to microstrip lines are also observed for all three GCPW lines. This may also due to the self resonant of the probing pads.

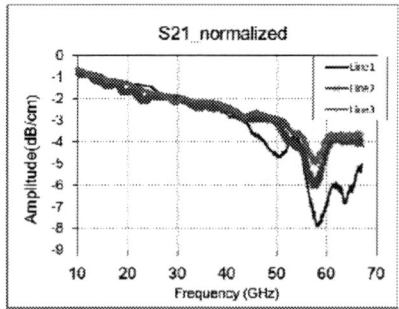

Fig. 18b. The measured results of GCPW with cover layer and without cover layer.

Conclusions

In this study, both microstrip lines (MSL) and grounded co-planar waveguide (GCPW) are designed and fabricated on COP. The dielectric constant and loss tangent of COP is 3.0 and 0.005. To protect the copper traces, PI cover layer with thickness of 38μm is also applied on top of COP. The dielectric constant and loss tangent of PI cover layer is 3.5 and 0.02. Simulation results show that with 40μm of COP and 12μm thickness of copper trace, the loss at 60 GHz is about 1.0~1.2dB/cm for both types of transmission lines with 50Ω of characteristic impedance. Measurement results show larger loss. At 60GHz, the loss is about 2.5dB/cm for MSL and 3.3dB/cm for GCPW. The extra loss may due to that some solvents are not fully evaporated during laminating process. We will continue tuning the fabrication process parameters to achieve lower losses of transmission lines.

Acknowledgments

The authors would like to thank the research team members of Zeon Corporation for providing the material and the technical support of this work. This work is partially supported by National Science Council of Taiwan under Grant NSC 102-2219-E-002 -020 and Bureau of Energy, Ministry of Economic Affair, Taiwan, under project 103B51033.

References

1. D. C. Thompson, O. Tantot, H. Jallageas, G. E. Ponchak, M. M. Tentzeris, and J. Papapolymerou, "Characterization of liquid crystal polymer (LCP) material and transmission lines on LCP substrates from 30 to 110 GHz," *IEEE Trans. Microw. Theory Tech.,* vol. 52, no. 4, pp. 1343-1352, Apr. 2004.

2. L. N. Lewis and D. Katsamberis, "Ultraviolet- curable, abrasion - resistant, and weatherable coatings with improved adhesion," *Journal of Applied Polymer Science,* vol. 42, no.6, pp. 1551-1556, Mar. 1991.

3. J. M. Liu, T. M. Lee, C. H. Wen, and C. M. Leu, "High-performance organic- inorganic hybrid plastic substrate for flexible displays and electronics," *Journal of the Society for Information Display,* vol. 19, no.1, pp. 63-69, Jan. 2011.

4. G. Zou, H. Gronqvist, J. P. Starski, and J. Liu, "Characterization of liquid crystal polymer for high frequency system-in-a-package applications," *IEEE Trans. Adv. Packag.,* vol. 25, no. 4, pp. 503-508, Nov. 2002.

5. A. Tsuchiya et al, "Low-loss and high-speed transmission flexible printed circuits based on liquid crystal polymer films," *Electrics Letters,* Vol. 48, No. 19, 2012.

6. J. Izadian et al, "Novel transmission line for 40 GHz PCB applications," *Design Con,* 2011.

7. X. Zhang, "Design, fabrication and characterization of microwave passive devices on liquid crystal polymer substrates," Chalmers University of Technology, Doctoral Thesis, Sweden, 2009.

8. G. F. Engen and C. A. Hoer, "Thru-Reflect-Line": An improved technique for calibrating the dual six-port automatic network analyzer," *IEEE Trans. Microwave Theory and Tech.,* vol. MTT-27, pp. 987~993. Dec. 1979

FBEOL No-Aluminum Pad Integration in Pb-Free C4 Products for Environmental, Cost and Reliability Benefits

E. Misra, T. Daubenspeck, T. Wassick, K. Tunga, D. Questad

IBM Microelectronics Division

2070 Route 52, Hopewell Junction, NY 12533

(845)892-3185 , emisra@us.ibm.com

Abstract

Integrated circuits with Pb-free C4's have two major reliability concerns: thermal-mechanical stress induced mechanical fails within the Si chip and C4 electromigration (EM). Decreasing feature sizes, and increasing power and performance requirements have exacerbated these concerns and necessitated the development of innovative solutions to address the reliability issues and support the building of more robust and reliable packaged parts. Aluminum (Al) pads have been historically used in the Far Back End of Line (FBEOL) levels of the Si chip primarily for mechanical and Chip Package Interactions (CPI) benefits. Previous studies have shown that the Al pads used in legacy FBEOL process integration can serve as a stress redistribution layer to dissipate the detrimental thermal-mechanical stresses from reaching the underlying weaker BEOL ULK/Low-K levels. Aluminum processing in the wafer fab however typically uses carcinogenic chemicals such as hexavalent chromium for passivation and for corrosion inhibition. The current work evaluates non-Al FBEOL structures for the obvious environmental reasons but also as means for reducing processing costs and fab cycle times. Mechanical finite element analysis have been performed to determine the effect of the FBEOL structural changes on the stresses in the ULK/Low-k BEOL levels. White "C4" bump and C4 EM data comparing the Al pad structure to the non-Al pad structures will also be reviewed and some of the key process integration challenges with the non-Al structures will be discussed.

Introduction

The continuously shrinking device dimensions and the increasing power and electrical performance requirements in integrated circuits has led to the replacement of Aluminum (Al, typically Al(Cu)) by Cu as the primary interconnect metallization to reduce resistance, replacement of SiO2 by mechanically weaker low-K and ultra low-K (ULK) inter layer dielectrics to reduce parasitic capacitance between metal lines, and use of a more stressful under bump metallurgy (UBM) /Pb-free solder bumps and introduction of organic laminates [1-3]. All these material and structural changes have led to a weaker and highly stressed packaged part susceptible to white C4 bumps during the assembly or chip-join processing and/or during reliability stressing of the parts. The common thermo-mechanical induced failure mechanisms are solder fatigue or delaminations and cracking either in the underfill or in the low-k or ultra low-K BEOL levels [4].

The thermal mechanical stresses that are generated in the packaged part during the chip-join reflow process are due to the mismatch of the coefficient of thermal expansion of the Si, the interconnect metallization, low-K/ultra low-K ILD's, lead-free solder, and the organic flip-chip substrate, which leads to

thermally induced shear stresses being developed in the UBM, solder bump and the weaker ILD's within the Si Chip. These thermal stresses induce a rotational moment on the solder bumps leading to tensile stress on the solder bump edge away from the chip center and compressive stress on the opposite edge as demonstrated in Figure 1. These thermally induced tensile stresses are typically worse at the chip edge than the chip center [2, 5]. The tensile stress pulls the bump away from the chip surface and transfers stress to the final polymeric passivation via (FV) and the underlying weak BEOL low-k and ultra low-k levels leading to delaminations and cracks in those levels [6].

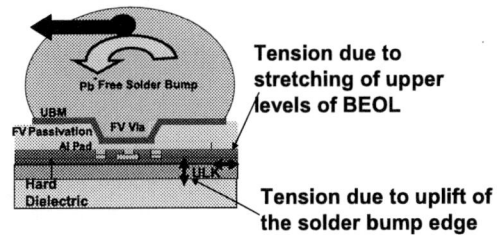

Figure 1. Schematic Depicting the Thermally Induced Rotational Moment Leading to Tensile Stresses on the Side of C4 Bumps Away From the Chip Center [6]

Although the interconnect metallization in the BEOL levels have changed to Cu to help in the reduction of RC delay for high speed device applications, Al pads are still used in the FBEOL for C4 and wirebond applications. One of the reasons for the continued usage of Al pads in FBEOL is the lower cost of the sputtered Al process compared to the damascene process used for Cu. Also a prior study by the authors [6] has shown that the Al pads could act as a stress redistribution layer to dissipate the thermal-mechanical tensile stresses from damaging (delaminating or cracking) the weaker BEOL low-k or ULK levels.

Aluminum processing in the FBEOL traditionally involves the following process steps - metal deposition, photo patterning, plasma etch and solvent clean to passivate the Al pads and prevent corrosion during downstream processing. The processing cost for the Al pads is typically about 20-30% of total FBEOL process cost. Hence for the non-Al pad structure there will be a 20-30% FBEOL process cost savings in comparison to the FBEOL structure with Al pads. However if Al is replaced by Cu in the FBEOL then the cost benefits expected with the non-Al pad structures would go away. In fact the FBEOL process cost of the Cu structure would be comparable or higher than the Al structure.

978-1-4799-2408-0/14 $31.00 © 2014 IEEE

2014 Electronic Components & Technology Conference

Aluminum pads can undergo corrosion due to the plasma etch byproduct left on the surface of the Aluminum post etch or if the etch chemistry used is Fluorine or Chlorine based. These etch by-products can cause corrosion of the Al pads leading to pits on the surface which could lead to adhesion concerns for the UBM. There are several wet clean solvents that could be used to remove the plasma etch residues/byproducts and passivate the surface of the Al pad. Some examples of such wet chemical solvents are hexavalent Chromium [7], EKC-265, Aleg310 and NE-111 [8], all could be used to inhibit corrosion in the Al pads. However some of these solvents are carcinogens and would require careful and special handling. Because of these processing concerns related to Al and to further improve the mechanical and electrical performance of the chip, the non-Al pad structures are of interest.

In this paper the three FBEOL structures that are being evaluated for mechanical robustness and C4 EM are (a) legacy Al pads (b) Al pad removed (c) Al pad replaced by Cu.

Three dimensional mechanical modeling is used to predict stresses in the ULK and oxide levels for all three structures under consideration. Prior work by the authors has shown that removal of the Al pad leads to high stresses in the ULK levels and possible white C4 bump fails [6]. Mechanical modeling is therefore done to determine the impact of increasing the final passivation via (FV) thickness on the ULK stresses for the no Al pad structure.

Mechanical white C4 bump analysis was performed to determine the impact of the Al removal and the Cu FBEOL structure on the BEOL stresses and determine any potential fails related to these structures when subjected to accelerated reflow conditions during chip-join. In addition C4 electromigration (EM) analysis was done to compare the EM performance of the Cu FBEOL structure to the legacy Al FBEOL structure.

Predictive Mechanical Modeling

In this section, results from the three dimensional Finite Element Analysis (FEA) done to predict the mechanical stresses in the Back-End-of-the-Line (BEOL) levels due to the FBEOL process integration and metallization changes will be discussed.

Figure 2 depicts the global/local (Macro/Micro) modeling approach used for this purpose. The global model consisted of the chip, substrate and individual lead-free C4 joints. The chip size assumed in the model was consistent with the size of a typical large die and the substrate was a multi-layered organic laminate. The local model included the ULK material and all other BEOL features present within the silicon chip design.

The global model was subjected to a typical reflow temperature loading condition involving cool-down from the solder melting temperature to room temperature. The displacements resulting from solving the global model were applied to the local model using cut-boundary interpolation. The local model was then used to obtain the peeling (adhesive) stress and the principal (cracking) stress within the ULK layer. The typical contour plots (peeling and principal stress) at the BEOL level of interest obtained from mechanical modeling are also shown in Figure 2. The contour plots depict

that the stresses are highest at the UBM edge away from the center of the chip. This region of increased stress within the weaker BEOL ULK level is more susceptible to adhesive and cohesive fails.

Mechanical modeling was done to determine the stresses in the ULK layers as a function of C4 pitch for the cases of (1) FBEOL structure with Aluminum Pad and (2) FBEOL with Cu replacing Aluminum Pad. The variation of ULK Principal stress for both the FBEOL structures is plotted in Figure 3. The FBEOL hard dielectric thickness, final passivation (FV) via diameter and the UBM size were held constant for all cases modeled.

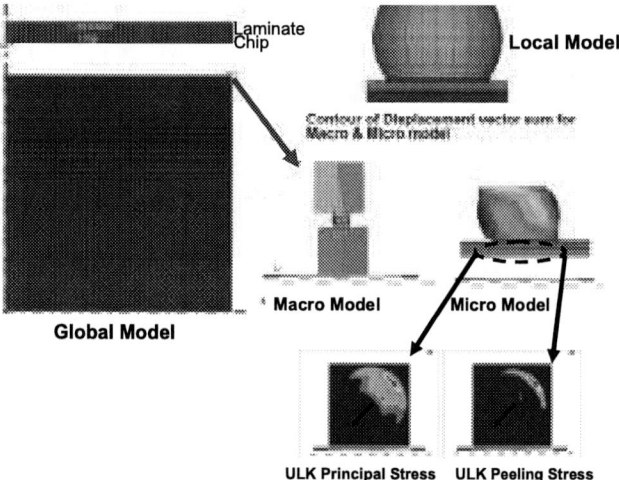

Figure 2. Global/Local Modeling Approach for ULK Stress Prediction [9]

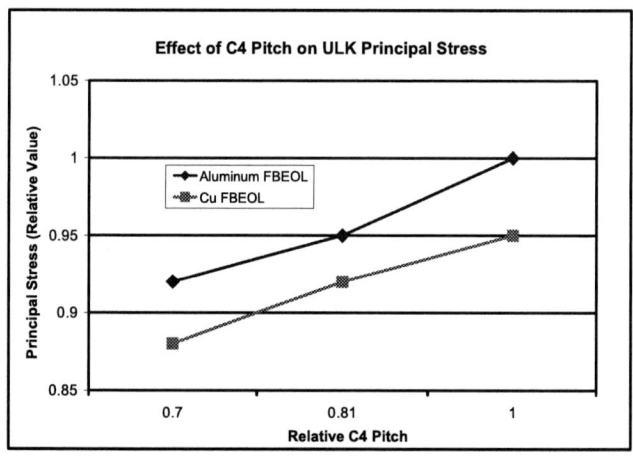

Figure 3. ULK Principal Stress as a Function of Pitch for (1) FBEOL with Aluminum & (2) FBEOL with Cu

It can be seen from the plots in Figure 3 that the principal stress in the ULK layer decreases with reduction in pitch. When the C4 pitch is reduced by 19% and 40%, the ULK principal stress for the Al FBEOL structure reduces by 5% and 9% respectively and the reduction in the C4 pitch by 19% and 40% causes the ULK stresses in the FBEOL Cu structure to go down by 4% and 8% respectively. The number of C4 bumps for a medium die size product is expected to increase

978-1-4799-2408-0/14 $31.00 © 2014 IEEE

by approximately 49% when the pitch is reduced by 19% and by 97% when the pitch is reduced by 40%. Therefore the reduction in the ULK Principal Stresses observed with the reduction in C4 pitch in both the Al and Cu structures (Figure 3) could be attributed to the stresses being more evenly distributed over a larger number of C4's. The stress in the Cu FBEOL structure is 5%, 8% and 12% lower compared to the Al FBEOL structure at relative C4 pitch of 1.00, 0.81 and 0.7 respectively. The lower stresses in the Cu FBEOL structure could be attributed to the lower coefficient of thermal expansion (CTE) of Cu (16.9 ppm/C)[5] compared to Aluminum. The CTE mismatch is lower between the FBEOL Cu and the surrounding oxide/UBM materials in comparison to FBEOL Aluminum and this in turn reduces the stresses associated with the FBEOL Cu structure. Also since Cu has a higher Young's modulus (129 GPa)[5] therefore it's stiffer and thus restrains the movement of the solder bump sitting over it during thermal cycling and thus reducing the tensile stresses generated at the UBM edge.

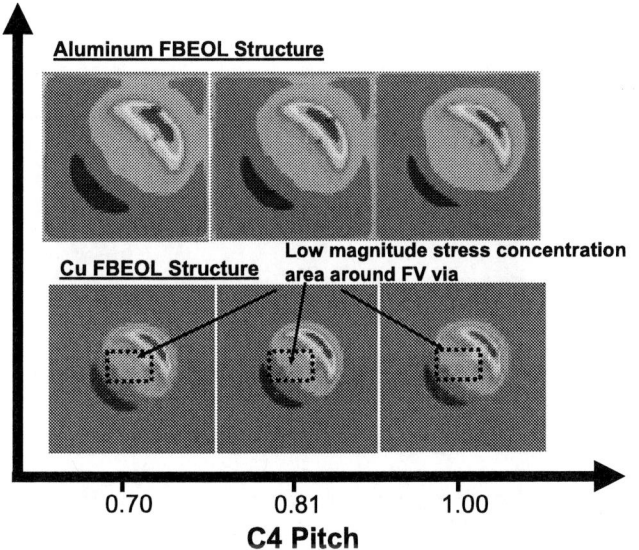

Figure 4. Contour Plots of ULK Principal Stress as a Function of Pitch for (1) FBEOL with Aluminum Pad & (2) FBEOL with Cu

Figures 4 and 5 are the contour plots of ULK principal and peeling stress for both the FBEOL structures (FBEOL structure with Al Pad and the Cu structure) under consideration as a function of C4 pitch. In both the FBEOL structures and for all 3 C4 pitches the max stress is underneath the UBM edge away from the center of the chip. The area of the peak stress underneath the UBM edge in the contour plots seems to be smaller for the Cu structure in comparison to the Al structure. This indicates that the high modulus Cu in the FBEOL helps prevent and dissipates the tensile mechanical stresses developed during the chip-join reflow process at the UBM edge from propagating into the weaker BEOL ULK levels. Also in Figures 4 and 5 for the case of the Cu FBEOL structure there seems to be a region of low magnitude stress concentration underneath the FV via edge which is not seen in the Al structures. This stress concentration around the FV via

could be attributed to the stiffness of the Cu within the final passivation via and also the CTE mismatch between the polymeric passivation and Cu during the thermal chip-join process.

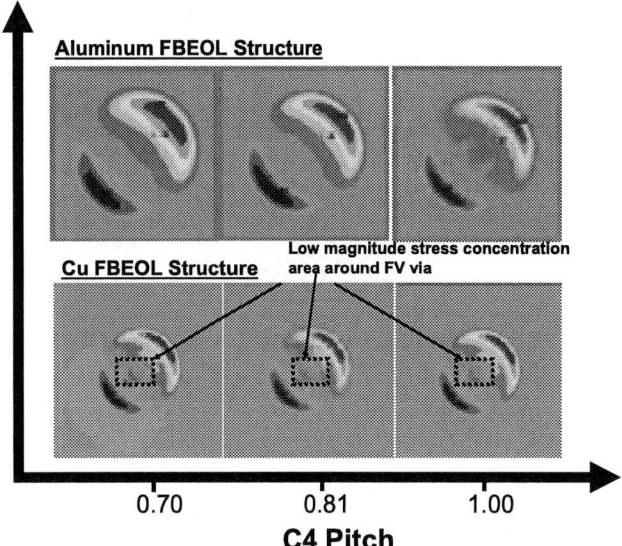

Figure 5 Contour Plots of ULK Peeling Stress as a Function of Pitch for (1) FBEOL with Aluminum Pad & (2) FBEOL with Cu

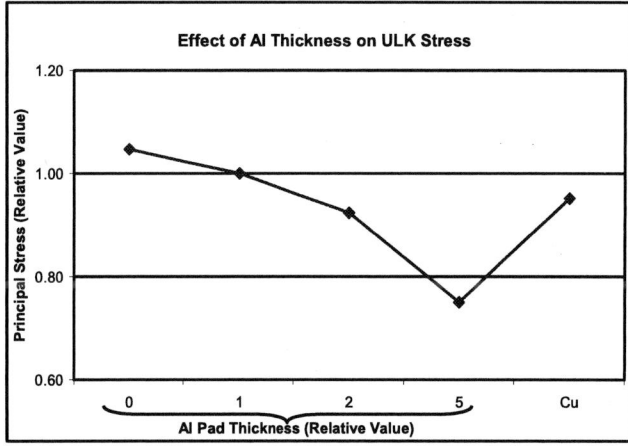

Figure 6. Effect of Al Pad thickness on ULK Principal Stress

Figure 6 is a plot of the relative value of the principal stress in the BEOL ULK level for the FBEOL structure with Al pad of varying thicknesses and the Cu FBEOL structure. The relative principal stress in the weak ULK levels is highest when Al pad thickness =0 and it reduces by 5% and 29% when the relative Al Pad thickness is increased to 1 and 5 respectively. Also the principal stress in the ULK level of the Cu FBEOL structure appears to be close to the stress seen in the Al structure with relative pad thickness of 2 and about 9% lower in comparison to the no-Al Pad case.

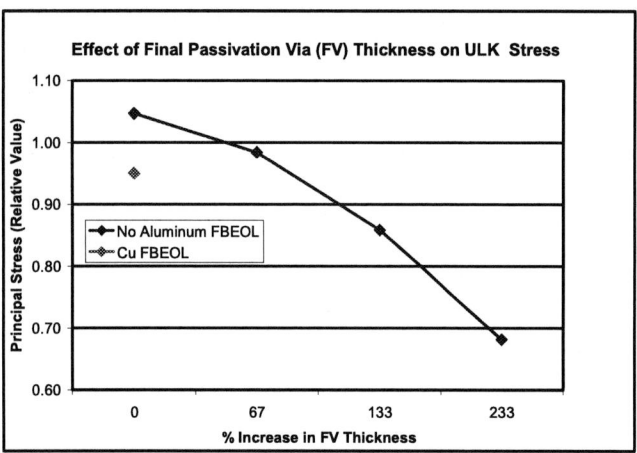

Figure 7. ULK Principal (or Cracking) Stress as a Function of Final Passivation Via (FV) Thickness For Al Thickness = 0 and Cu FBEOL Structure

Prior studies have shown that increasing the final polymeric passivation via (FV) thickness reduces stress in the BEOL ULK levels [2]. As shown in Figure 5 stresses in the ULK levels increase by about 5% for the no-Al pad structure w.r.t structure with relative Al Pad thickness of 1. Thus the ULK level stresses could be reduced for the No-Al pad FBEOL structure by increasing the FV thickness as depicted in Figure 7. The stresses in the ULK level reduce by 6%, 18% and 35% as the FV thickness is increased by 67%, 133% and 233% respectively. Also the ULK stress in the Cu FBEOL structure is about 3% lower than the stress in the No-Al Pad structure with 67% thicker FV.

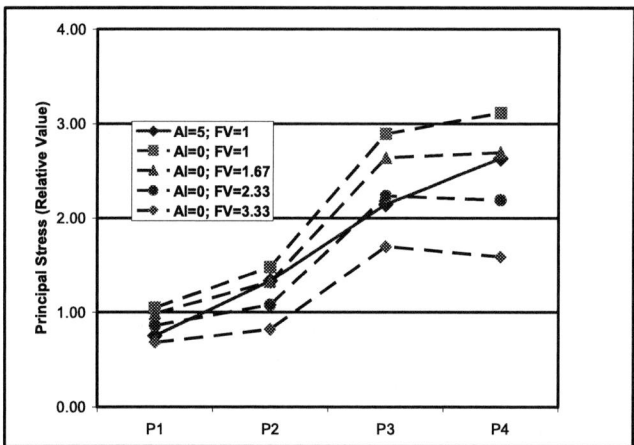

Figure 8. ULK Principal Stress for Al Thickness=0 & 5 (Relative Value) and Varying FV Thicknesses at Various Locations in the BEOL Stack (P1 – is furthest away from the C4 bump and P4 is closer to the C4 bump in the BEOL stack)

Figure 8 shows the plot of the principal stress at various points on the BEOL stack. For all the cases modeled the principal stress increases as one moves up the BEOL stack. For the no-Al pad and the Al Pad (relative thickness =5) structure with relative FV thickness of 1 the stresses increase by 200-250% as one moves up the BEOL stack from the ULK level (P1 – farthest away from the C4 bump) to the FBEOL

hard dielectric (P5 – closest to the C4 bump). However as the FV thickness is increased it is observed that the stresses do not increase as drastically as one moves up the BEOL stack.

Thus, the plots in Figure 8 show that the stresses in the weaker ULK levels could be dissipated by building a tall BEOL stack comprising of both thicker hard dielectric and final polymeric passivation such that the ULK level is further away from the stress inducing UBM and solder bump.

Mechanical White C4 Bump Analysis

Mechanical white bump experimentation was also done to determine the effect of the Al pad thickness and the replacement of Al by Cu in the FBEOL on the CPI stresses.

The experiments were done using a medium die size test vehicle and with ULK material in at least one BEOL level. The hardware was built with even more stressful solder bumps consisting of a thick Cu UBM and lead-free Sn-Ag C4 bumps. After the modules are subjected to an accelerated reflow condition Scanning Acoustical Analysis (CSAM) was used to image the parts after module build to detect white C4 bumps.

The white bump analysis was done for three different cases of the FBEOL structure (1) with No Al Pad (Thickness=0) (2) with Al Pad of relative thickness =1, and (3) Cu structure

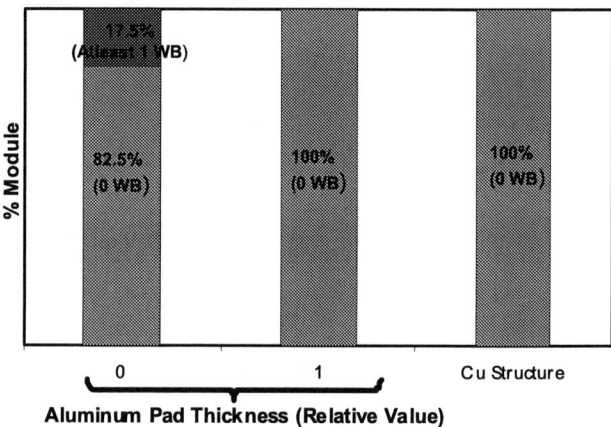

Figure 9. Percentage of Modules with White C4 Bumps (WB) for Modules Built (a) With Al Pads (b) Without Aluminum Pads and (c) Cu Structure.

Figure 9 shows the plot of percentage Modules with white bumps for all 3 structures that were inspected with the CSAM. The No Al Pad structure had 17.5% of the modules having atleast one white bump, while for the cases of relative Al Pad thickness of 1 and the Cu structure there were no white C4 bumps detected during the CSAM analysis.

The mechanical modeling done in the previous section indicated that the ULK stress increased by about 5% when the Al pad was eliminated (Figure 6). However as depicted in Figure 8 the stresses increase by 78%-251% higher up in the BEOL stack away from the ULK level and are highest closer to the UBM/solder bump interface for the No Al Pad and FV thickness =1 case. Thus in this particular case of white bump analysis the fails detected in the no Al Pad structure are expected higher up in the stack close to the FBEOL levels if there are any inherent defects or weakness in those levels. In

978-1-4799-2408-0/14 $31.00 © 2014 IEEE   1952

the Al pad and Cu FBEOL structures the Al and Cu metallizations act as buffer to any crack inducing stresses from reaching the weaker BEOL levels. The non-POR BEOL stack was built in order to make the modules more sensitive to white C4 bump fails during the accelerated chip-join reflow process.

The white bump CSAM analysis was repeated on a similar medium die size test vehicle with the same stressful solder bumps as described above. However in this case the BEOL stack constituting at least one ULK level is comparatively less robust due to elimination of atleast one hard dielectric level underneath the FBEOL stack. This eliminated level typically acts as buffer against the detrimental tensile stresses generated underneath the UBM edge from reaching the weaker BEOL ULK levels.

Figure 10 shows the percentage of Modules with white bumps as a function of the Aluminum pad thickness and a Cu FBEOL for the stack described above. The percentage of modules having white bumps increases slightly for the no Al pad case because of the inherent weakness of the BEOL stack.

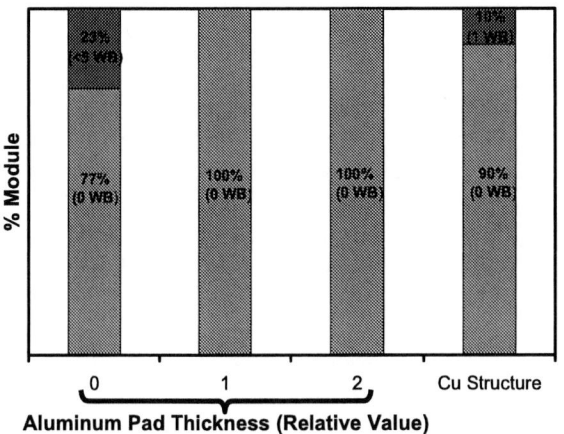

Figure 10. Percentage Modules With White C4 Bumps (WB) for Modules Built Varying Relative Al Pad Thickness (=0, 1, and 2) and Cu FBEOL Structure.

In this study 10% of the modules built with the Cu structure have at least 1 white C4 bump. While the modules built with Al pad of relative thickness 1 and 5 respectively have no white C4 bumps. Mechanical modeling data in Figure 5 indicated that the stresses in the ULK level for the Cu structure are equivalent to the Al pad structure with the relative thickness of 2. However in this white C4 bump analysis the fail is not expected to be in the ULK level. In fact as in the previous analysis shown in Figure 9 the fail is expected higher up in the BEOL stack close to the FBEOL levels for both the Cu and no Al pad structures. This could be due to the random defects that are potentially generated in the hard dielectric underneath the FBEOL level due to the non-POR processing that was done to make the structure more sensitive. This explains the increase in the number of white bump fails for the modules built with the no Al pad and the low level of white bumps seen in the modules built with Cu.

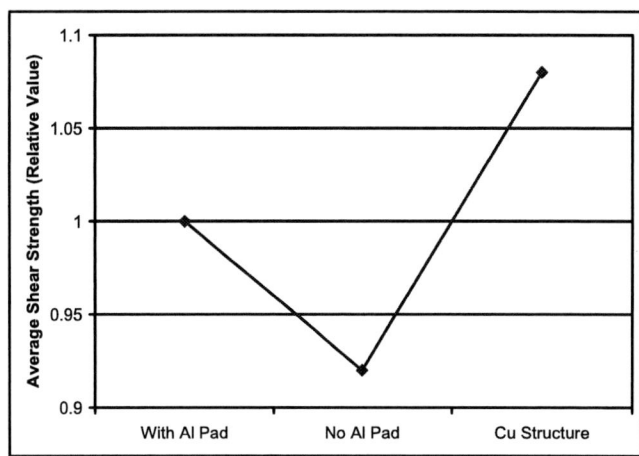

Figure 11. Average Single Bump Shear Strength (Relative Value) of FBEOL Structures (a) With Out Al Pad (b) With Al Pad (Relative Thickness =1) (c) Cu Structure

The robustness of the FBEOL stack with no Al pad and the Cu structure was assessed in comparison to the FBEOL stack with the Al pad of relative thickness = 1 using the single bump shear test under constant shearing speed and stand-off height. This test is used to evaluate the integrity of the FBEOL, UBM and solder bump interfaces for each solder bump subjected to a shearing force. Figure 11 is the plot of the relative value of the average single bump shear strength for the three FBEOL structures under consideration. The average shear strength of the hardware built without the Al pad was found to be 8% lower in comparison to the hardware built with the Al pad. The hardware with Cu FBEOL had an 8% increase in average shear strength in comparison to the hardware built with Al structure. This slight improvement in the average shear strength for the Cu structure could be attributed to the less compliant Cu underneath the UBM and solder bump preventing any thermal stress induced movement of the solder bump thus preventing the generation of the high tensile stress region on the UBM edge away from the die center. This data is in agreement with the mechanical modeling and the white C4 bump analysis.

C4 Electromigration

C4 electromigration (EM) study was done for two different FBEOL structure (a) with Al Pad of relative thickness =1 and (b) Cu structure. The same UBM structure and solder bump composition was used for both the structures. The Cu FBEOL structure seems to have about 10% improvement in EM performance in comparison to the Al FBEOL structure under the standard EM stress conditions and time (Figure 12). The improvement in the EM performance could be attributed to the lower resistivity of the Cu in comparison to Al and thus more uniform distribution of the current density through the UBM and solder bump. The uniform current density at the critical UBM and solder bump interface helps eliminate any current crowding and thus localized joule heating which could lead to early failures.

978-1-4799-2408-0/14 $31.00 © 2014 IEEE   1953

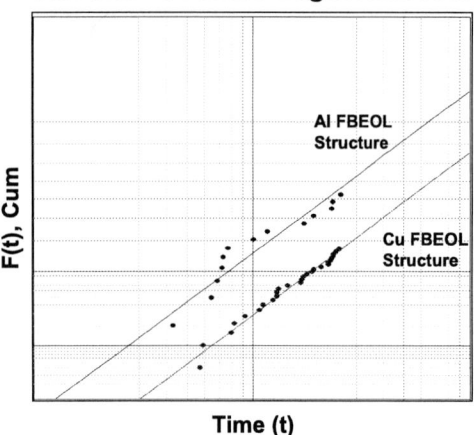

Figure 12. Comparing C4 Electromigration performance of Modules built with Al Pad and with Cu FBEOL Structure

Conclusion

In this paper 3D mechanical modeling was done to study the effect of FBEOL structures (a) with Aluminum pad (varying pad thickness), (b) no Al Pad, and (c) with Cu on the stresses in the BEOL ULK levels. Modeling showed that the ULK stresses were always higher when there was no Aluminum pad. The Al pad structure and Cu structure had comparatively lower ULK stress thus indicating that the Al pad and Cu might be acting as a buffer and helping dissipate potentially damaging tensile stresses from reaching the weaker BEOL levels. In addition the model also showed that for the Cu and Al structures reducing the C4 pitch helped further reduce the ULK stresses due to the distribution of the stress over a larger number of bumps. Mechanical modeling also showed the benefits of increasing the FV thickness for the no Al pad case in reduction of the ULK stresses.

White C4 bump analysis was also done to show that the stresses are higher for the no Al pad case thus leading to white C4 bump fails. The Cu FBEOL structure had no white C4 bump fails for a POR BEOL stack but for a non-POR weaker stack there were low level of white C4's detected.

C4 EM analysis was performed to compare the Al pad structure and Cu pad structure under standard EM test conditions. The Cu FBEOL structure had better EM performance in comparison to the Al structure due to the lower resistance and better current distribution.

Acknowledgements

The Authors are grateful to Judith Wright and Tammy Bardwell for their help with the logistics for executing the experiments. The authors are also grateful to Kurt Smith for the help with the Single Bump Shear Tests used in this paper. The authors are thankful to Ian Melville, Troy Graves-Abe, John Cincotta, Randy Werner, Gordon Osborne Jr, Richard Bisson, Eric Giguere and David Turnbull for their support with the project.

References

1. Yoon, S. W. et al "150um Pitch Pb-Free FlipChip Packaging with Cu/Low-k Interconnects", 2005 Electronic Components and Technology Conference, pp. 100-106.
2. Landers, W. et al "Chip-to-Package Interaction for a 90nm Cu/PECVD Low-k Technology", Proceedings of the IEEE 2004 Internations Interconnect Technology Conference, June 2004, pp. 108-110
3. Van Driel, W.D. "Reliability Consequences of the Chip-Package Interactions" 11th Electronics Packaging Technology Conference, 2009, pp. 406-410.
4. Banijamali, B. et al "Reliability of Fine-Pitch Flip-Chip Packages", 2009 Electronic Components and Technology Conference, pp. 293-300.
5. Lee, M.W. et al "Below 45nm Low-k Layer Stress Minimization Guide for High –Performance Flip-Chip Packages with Copper Pillar Bumping", 2010 Electronic Components and Technology Conference, pp. 1623-1630.
6. Misra, E. et al, "Role of FBEOL Al pads and hard dielectric for improved mechanical performance in lead-free C4 products", 2013 Electronic Components and Technology Conference, pp. 2208-2213.
7. Kendig, M.W. et al, "Corrosion Inhibition of Aluminum and Aluminum Alloys by Soluble Chromates, Chromate Coatings, and Chromate-Free Coatings", Critical Review of Corrosion Science and Engg, May 2003, pp. 379-400
8. Hua, Y. et al, Method to Prevent Corrosion of Bond Pads, U.S. Patent application number: 20100184285, Publication Date 2010-07-22.
9. Misra, E. et al, "Novel design and integration enhancements in the final polymeric passivation for improved mechanical performance and C4 electromigration in lead-free C4 products", 2012 Electronic Components and Technology Conference, pp. 571-576.

Preparing 25Gbps Electrical I/O for Exascale Computing Systems

*Lei Shan, Young Kwark, Renato Rimolo-Donadio, Christian Baks, Michael Gaynes, Timothy Chainer
IBM T J Watson Research Center, Yorktown Heights, NY 10598. leis@us.ibm.com
**Manabu Hoshino, Masakazu Hashimoto, Toshihiko Jimbo, Junji Kodemura, Ikkei Matsuura, Zeon Corporation, Tokyo,
Japan

Abstract

This paper summarizes the exploratory work conducted at IBM Research and Zeon Corporation, which seeks to expand bandwidth and reduce power consumption of electrical I/O's for future exascale computing systems. The development of novel low-loss dielectric materials was coupled with high-speed and power scalable circuit designs to achieve 25Gbps per channel data-rate and to investigate power limits as low as 1pJ/bit at 10Gbps for aggregated parallel links. Results from test vehicles confirmed over 20% reduction in channel loss at 20GHz when compared to currently leading commercial materials, which will permit the extension of electrical link performance to meet future system requirements.

Introduction

The next milestone of super computing systems is to achieve a target of 10^{18} FLOPS (exascale) performance by 2018. In addition to physical, mechanical, and thermal challenges, I/O bandwidth and power consumption will also require inventive solutions [1]. As IC dimensions and C4 pitches are approaching physical limits, the growth in I/O bandwidth will be largely dictated by the scaling of signal speed, i.e. increasing from the current 5Gbps to 25Gbps. Such significant increases in I/O data rates often require channel equalization, which in turn, elevates driver/receiver power consumption and may push exascale system power well beyond 20 megawatts, three times more than the current IBM Blue Gene/Q system.

In order to alleviate this power challenge, a new low-loss and low-cost dielectric material was developed, which when used in printed circuit boards (PCB), offers a 20% reduction in channel loss compared to the best currently available commercial alternatives. To take full advantage of this new material, power scalable I/O circuits were also developed, so that I/O power may be accurately tailored to suit individual channel loss characteristics, thereby minimizing total I/O power consumption. Ultra-low power I/O links running at 1 pJ/bit, 4X lower than in current systems, was targeted while supporting on-board channels up to 10 inches in length.

In addition, a 25Gbps high speed serial test bed was designed to evaluate the new material's real-time impact on performance, distance, and power savings in a complex transmission environment including two PCBs assembled with backplane connectors to mimic long-reach I/O channels in high-end computing and communication systems. PCBs made with various material sets can be readily swapped, with eye-diagrams and bit-error-rates displayed on a PC control interface. With the new low-loss material, more than a 2X increase in channel length was achieved compared to a standard-loss material (FR4 equivalent). These innovative I/O solutions were developed to meet the stringent requirements of bandwidth, distance, and power consumption in future exascale computing systems.

Channel Loss Reduction

In addition to channel impedance and current return path optimization, the insertion loss of an electrical channel is mostly dominated by conductor surface roughness as well as dielectric loss. Therefore, together with the development of low-loss dielectric materials, lamination adhesion mechanisms were also investigated in order to take the advantage of ultra-low roughness copper foils.

Dielectric materials used in PCB construction are mostly compounds of polymer matrices, fillers, and glass cloths. Since fillers and glass cloths are common for all PCB materials, the polymer matrices play the most critical role in determining electric performance and reliability. The newly developed ZEONIF^{TM} XL-series dielectric materials are based on a unique Cyclo-Olefin Polymer (COP) matrix (by Zeon Corporation), which is categorized as a hydro-carbon material with a cyclic structure in the backbone unlike linear or branched polyethylene and polypropylene polymers. As the COP does not contain any polar groups, it exhibits extremely low dielectric constant, dielectric loss, and water absorption. For the same reason (elimination of polar groups), these XL-series materials show much higher chemical resistance than existing epoxy or PPO based materials. In addition, by attaching cross-linking functional groups to the COP chains, ultra-smooth copper foil (R_a=0.1um) may be cladded by pure chemical bonding with excellent heat resistance and peeling strength. Figure 1 shows the peeling strength passing the IPC/JEDEC J-STD-020D standard and remaining high even after 10 reflow cycles. The unique capability of using ultra-smooth copper assists further channel loss reduction.

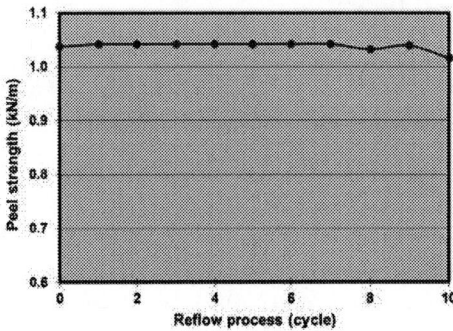

Figure 1. The peeling strength of XL-series material

Table 1 includes the key properties of the cost-effective ZEONIF^{TM} XL-Series material. Besides the above mentioned ultra-low dielectric/copper loss and superior reliability (high peeling strength and low water absorption), it also exhibits low dielectric constant (low cross-talk), high T_g, and low CTE. Most importantly, the XL-series materials have demonstrated wider process windows than conventional epoxy-based and PPO-based materials, which can be expected to significantly enhance manufacturing yield.

Table 1. Properties of COP-based XL-series material

Dielectric constant (Dk) (10GHz)	3.3
Dielectric loss tangent (Df) (10GHz)	0.002
Tg (deg. C)	195
CTE, ppm/deg. C(X-Y 30-150C)	17
Surface roughness, Ra(um)	0.1
Peel strength (kN/m)	1.0
Water absorption (%)	0.06

To validate the loss improvement of the newly developed ZEONIF[TM] XL-Series over commercially available materials, test vehicles were designed and fabricated as shown in figure 2. The panel measures 220x280mm with a relatively simple layer stackup, i.e. tri-layer stripline on top and bottom with a thick (50mil) supporting layer in the middle. In addition to circular disks for basic dielectric property extraction, the majority of the test structures are transmission lines, including single-ended traces and differential pairs. Most lines are equal in length and arranged at various angles to test dielectric uniformity (especially glass weave) across the panel. Other lines at various lengths were also included for calibration and precise dielectric property extraction purposes.

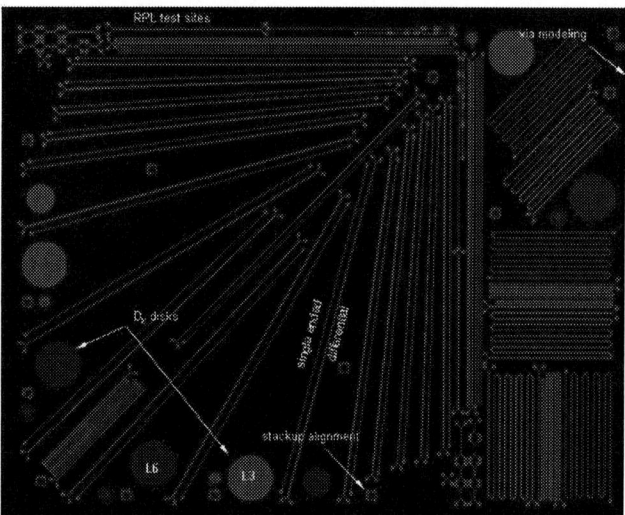

Figure 2. Layout of material evaluation test vehicles

Three materials were evaluated with the test vehicles, i.e. the new ZEONIF[TM] XL-Series material, an epoxy-based "standard-loss" material (FR4 equivalent), and a currently leading commercial material (PPO-based) referred to as "low-loss" with properties comparable to data published by Glenn Oliver et al. [2] In all three cases, glass cloths were matched to permit direct comparisons. As mentioned earlier, XL-series materials may form chemical bonds to ultra-smooth copper foils, while the other two materials require relatively rougher copper surfaces. To improve measurement accuracy, an

Agilent[TM] E8364C 4-port 50GHz network analyzer was used combined with an innovative method of accessing the transmission lines called the "recessed probe launch" to minimize launch parasitics [3]. The measured results on 20cm transmission lines (100um in width) of the three materials are shown in Figure 3. Greater than 50% and 20% loss reductions were observed on the new XL-series when compared to "standard-loss" and "Low-loss" materials respectively at 20GHz, which is significant and also essential for high-speed and low-power links. Measurements on various orientations were also performed with little variation that indicates excellent dielectric uniformity of the XL-series materials.

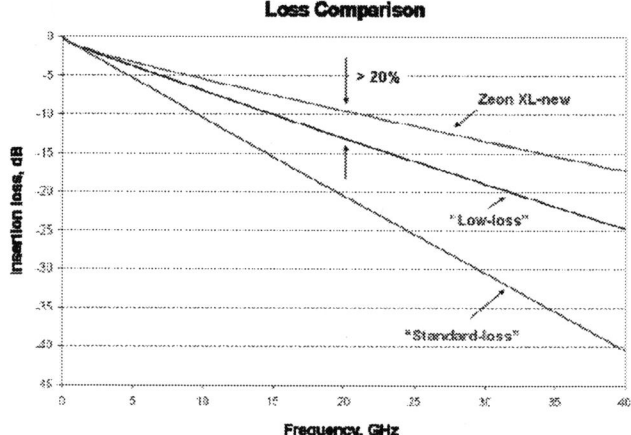

Figure 3. Various dielectric comparisons over 20cm striplines of 100um in width.

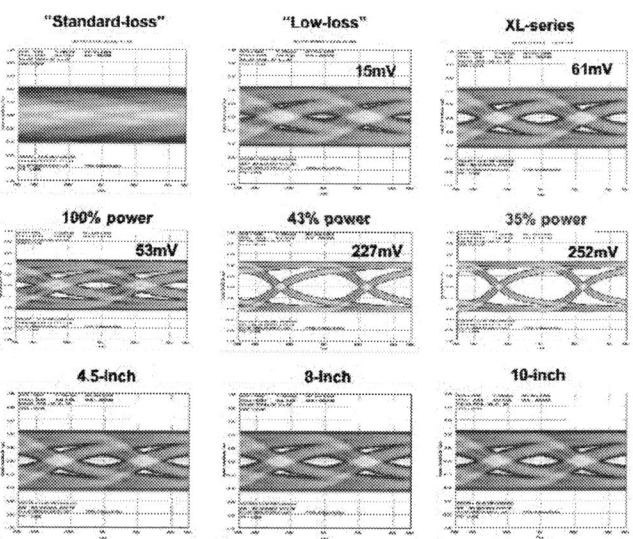

Figure 4. Time domain simulations on board-level links

The loss advantage of the XL-series materials may be translated into to I/O performance improvement, power reduction, and distance extension as shown in Figure 4. The time domain (eye-diagram) simulations were based on measured S-parameter results without signal conditioning such as equalizations. A 20Gbps random data sequence was simulated with +/-500mV differential driving amplitude and 15ps rise/fall slew rate. The three columns are for three different materials, i.e. "standard-loss", "low-loss", and XL-series. The three charts in top row show that, for a 25cm channel, both "standard-loss" and "low-loss" materials failed, while only the XL-series material passed the commonly adopted 40mV requirement. The second row demonstrates potential power saving on a relatively short channel (12.5cm). When "standard-loss" material is used as a reference, the use of "low-loss" material may save 57% power, and the new XL-series material may help reduce power by 65%. The last row shows that, for the same vertical "eye" opening, the use of XL-series materials may permit links more than twice the length of those allowed by "standard-loss" materials.

Active Material Comparisons

A more sophisticated test vehicle was designed to allow real-time material comparisons at 25Gbps data-rate. The complete setup is shown in Figure 5. The core used is an IBM High-Speed-Serial (HSS) device that contains eight TX/RX channels with data-rates adjustable from 1 Gbps to 28Gbps. The HSS chip was assembled onto a PCB using a socket, so that the same chip can be applied to multiple PCB's made of different materials ("standard-loss", "low-loss", and XL-series) for direct performance comparisons. The PCB design is shown in Figure 6. It consists of 26 metal layers and is built with multiple materials by combining very low-cost materials with XL-series materials or other comparable materials for cost reduction - the so called "hybrid" construction. Two high dielectric constant material layers are also incorporated on top and bottom to improve power supply decoupling.

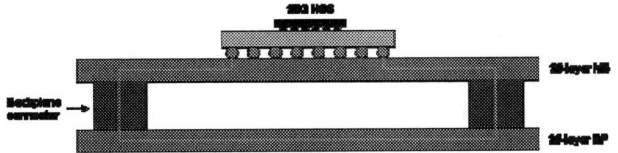

Figure 5. Setup for real-time material comparisons

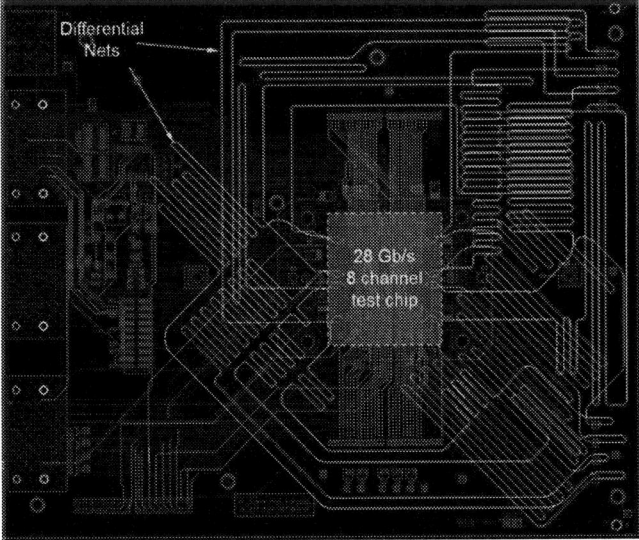

Figure 6. 26-layer hybrid-material PCB design

Eight Channels were used in four 2-channel groups of different lengths, including 26-inch (66cm), 35-inch (89cm), 44-inch (111cm), and 52.5-inch (133cm). Before assembling the HSS package, passive measurements were performed with the aforementioned 4-port VNA, and the results are shown in Figure 7. Compared to the current commercial "Low-loss" material, the new XL-series material outperforms by over 20% (at 20GHz) for all four channel lengths.

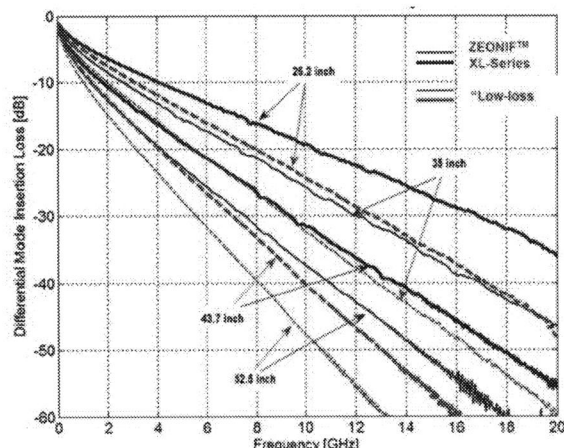

Figure 7. Insertion loss vs. net length

Figure 8 shows the assembled test vehicle connected to a control board that allows the use of a proprietary interface for link parameter setup and performance monitoring. The active-link was running on-board with a PRBS7 data pattern, and the test results are shown in Figure 9. The Y-axis is the vertical eye-opening at the receiving end normalized to 100% of the IBM HSS standard, with "pass" rule set at 20%. Figure 9(a) shows the test results measured on ZEONIF™ XL-Series test board, and Figure 9(b) for the board built with commercial "low-loss" material. For all channel lengths, the XL-series material yields greater eye-opening than the "Low-loss" material, which correlates with the passive results shown in Figure 7. At 25Gbps data-rate, the board built with "Low-loss" material failed for all channel lengths. By replacing the "Low-loss" with the XL-series material, the HSS link may operate properly up to 35-inch (89cm) on-board channel distance. This significant improvement will offer the customer more design room and flexibility when applying IBM HSS devices to their individual systems.

Figure 8. Assembled test vehicle in operation

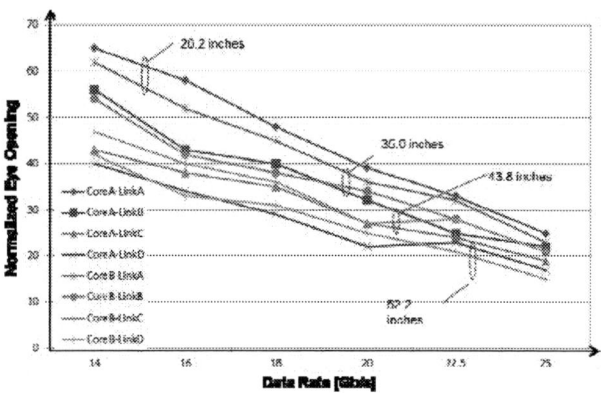

(a) ZEONIF™ XL-Series material

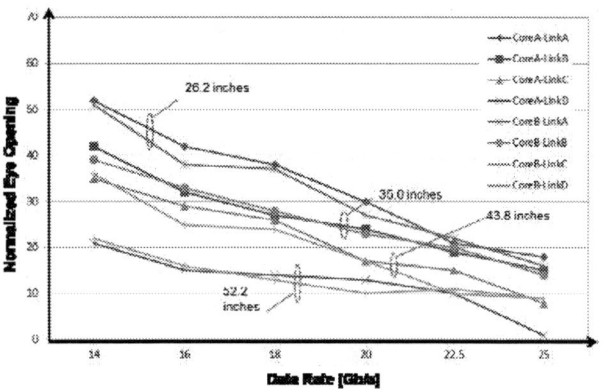

(b) Commercial "Low-loss" material

Figure 9. Normalized eye-opening vs. data-rate

For more complex channels with backplane connectors, "backplane" boards were also designed, fabricated, and mated to the main board through Molex Impact backplane connectors. The "backplane" boards were built with the three materials to be compared, i.e. "standard-loss", "low-loss", and XL-series. The active-link test results on "standard-loss" and XL-series are included in Table 2. At 12.5Gbps data-rate, a 2.5X channel length extension was observed by replacing the "standard-loss" material with the new XL-series for similar receiver "eye" openings. The board built with XL-series material may pass 12.5Gbps data over 50-inch (125cm) total channel distance with 40% normalized vertical eye-opening, which will meet the most challenging application needs.

Table 2. Results: complex channels with backplane connectors

Material	Channel Length (Inches)	Data Rate (Gbps)	VO (%)	HO (% UI)
XL-series	25"	12.5	50	19
XL-series	50"	12.5	39	16
"Standard-loss"	10"	12.5	49	20
"Standard-loss"	25"	12.5	30	18

Conclusions

A new Cyclo-Olefin Polymer (COP) based ultra-low loss dielectric material was developed for printed circuit boards. As the COP does not contain polar groups, it shows extremely low dielectric constant, dielectric loss, and water absorption.

By adding cross-linking functional groups to the COP chains, ultra-smooth copper foil (R_a=0.1um) may be clad relying only on chemical bonding for further loss reduction. Together with proven processability, it is a promising candidate to enable high data bandwidth and low I/O power required by future exascale computing and communication systems. Test vehicles were designed, fabricated, and assembled to characterize this new ZEONIF™ XL-series low-loss material and compare it with other commercial materials, including epoxy based "standard-loss" and PPO based "low-loss" materials. The transmission-line test results showed significant loss reduction (>20% at 20GHz) when compared to leading commercial low-loss material). By incorporating this new material with power-scalable TX/RX circuits, ultra-low power I/O may be realized. Active-link tests confirmed the loss advantage of the XL-series material. Comparing with commercial "standard-loss" and "low-loss" materials, the board built with XL-series dielectric may help elevate electrical I/O bandwidth as well as extend channel distance by 2.5X and 35% respectively.

Acknowledgments

The authors would like to thank to the collaboration of IBM system design team, and we gratefully acknowledge the support of Zeon Corporation on the joint development efforts.

References

1. J. Shalf, S. Dosanjh, and J. Morrison, "Exascale Computing Technology Challenges", VECPAR, Berlin, Heidelberg, Springer-Verlag, 2010, pp1-25

2. G. Oliver, J. Nadolny, and D. Nair, "Comprehensive Analysis of Flexible Circuit Materials Performance in Frequency and Time Domains", DesignCon 2012

3. Y. Kwark, C. Schuster, L. Shan, C. Baks, and J. Trewhella, "The Recessed Probe Launch – A New Signal Launch for High Frequency Characterization of Board Level Packaging", DesignCon 2005

Large Low-CTE Glass Package-to-PCB Interconnections with Solder Strain-Relief Using Polymer Collars

Gary Menezes, Vanessa Smet, Makoto Kobayashi[+], Venky Sundaram, Pulugurtha Markondeya Raj, and Rao Tummala

3D Systems Packaging Research Center, Georgia Institute of Technology, Atlanta, USA
+ Namics Corporation, Niigata, Japan

Abstract

This paper reports the use of circumferential polymer collars as a strain-relief mechanism to improve the fatigue life of low-CTE package-to-PCB solder interconnections, while preserving SMT-compatibility and reworkability. Acting as a partial underfill, the polymer- collar serves to block shear deformation at the solder-package interface, and redistributes the load to reduce the overall plastic strain concentration in the solders. It also suppresses failure initiation from defective surface sites and,thus further enhances reliability.

Ultra-thin glass 100µm interposers were fabricated in 18.4 mm x 18.4 mm size to model, design and demonstrate the reliability enhancement with the polymer-collar approach. The detailed interposer design and fabrication process with laminated dielectric and metallization layers on both sides is presented. A new class of epoxies with low modulus, without the incorporation of silica fillers, was used to act as the polymer collars. The polymer collars are formed by spin-coating with an optimized thickness to provide the best compromise between the effective strain relief and reworkability. Board-level assembly was performed using standard SMT processes for glass interposers with and without polymer collars. Thermal cycling reliability testing (-40°C to 125°C) of interposers, assembled on PCBs with and without polymer collars for various thicknesses of the collar was performed.

Introduction

Large, low-CTE packages are emerging as major need to interconnect multiple chips with fine-pitch chip-level interconnections with advanced RDL ground rules, both for high-performance 2.5D packages and ultrathin consumer products. These low-CTE packages are also needed to minimize stress on the ultra-low k on-chip dielectrics. This low-CTE package approach, however, creates CTE-mismatch related stress issues for package-to-board-level interconnections, resulting in degradation of board-level reliability. Innovative, low-cost, interconnection technologies at finer pitches are therefore highly sought after for direct assembly to the board, thus eliminating the need for additional interposer packaging level. For wider acceptance of low-CTE packages in high-volume production, it is also critical to assemble them using standard SMT with high degree of automation, reduced labor costs and higher production rates.

A number of unique stress-relief interconnection approaches have been reported in the literature, originally developed for WLP, but then extending them to solve the board-level interconnection reliability challenges. The G-helix [1] and stretched-solder interconnections [2] are based on compliant metal interconnection structures to provide stress-relief and improve the reliability. The double solder ball wafer-level technique by IZM [3] is another such approach to improve the interconnection compliance. Ball-on-polymer [4] and ELAStec WLP [5] place the joints on a layer of compliant polymer to reduce the strain in the joint itself. Wide Area Vertical Expansion (WAVE) [6] technology introduces a compliant low-modulus layer between the die and the interconnection layer. These approaches compromise the electrical performance, increase process complexity, or create additional challenges by increasing the package height, or use non-standard assembly processes.

GT-PRC is pioneering novel low-cost stress-relief approaches with minimal additional process steps using standard package- to board SMT processes by extending them to large package sizes at fine pitch. Past work on such glass package-to-PCB interconnections at GT-PRC used a dielectric strain buffer on either side of the glass to enhance reliability of 7.2 mm x 7.2 mm packages. Two variations of glass CTE were studied, 3.8 ppm/K and 9.8 ppm/K. Both variations of glass were shown to give better reliability than silicon interposer at board level, undergoing up to 1500 thermal cycles before the first occurrence of joint failure [8].

This paper reports the use of circumferential polymer collar as a strain-relief mechanism to improve the fatigue life of low CTE package-to-PCB solder interconnections, while preserving SMT-compatibility and reworkability. Acting as a partial underfill, the polymer collar serves to block shear deformation at the solder-package interface, and redistributes the load to reduce the overall plastic strain concentration in the solders. It starts with thermo-mechanical modeling to predict the solder joint strains ,leading to design of the polymer-collar structure for meeting the reliability requirements, followed by process development to achieve the polymer-collar structure design, large glass package test-vehicle fabrication and surface mount assembly process for high-volume manufacturing, and finally ends with reliability characterization and model-to-hardware correlation.

Finite Element Modeling

Two-dimensional half-symmetry models of glass interposers were built along the diagonal of the package to simulate the plastic strain in the furthest solder joint. The left boundary of the package was given symmetry boundary conditions with the bottom corner pinned.

The modeled assembly was first subjected to a drop in temperature from 260°C to 25°C to simulate the cool-down phase of the SMT reflow process. Five thermal cycles between -40°C and 125°C (following the JESD22-A106B thermal shock standard) were applied. The ramp-up and ramp-down times were 1 minute and the dwell time at the temperature extremes was 5 minutes. Equivalent plastic strain range was extracted after cycling.

Lead-free SAC105 solder material was used during modeling and test-vehicle fabrication due to better performance under drop test [9]. Solder was modeled as a visco-plastic material using parameters derived from Anand's model. Copper was modeled as a bilinear elastic-plastic material while all other materials were considered elastic (Table 1). The viscosity of polymer materials is neglected for the temperature range in consideration.

Table 1: Material properties used for FEA

Material	E (GPa)	CTE (ppm/°C)	Poisson's ratio	Material model
Glass (High CTE)	74	9.8	0.23	Elastic
Glass (Low CTE)	77	3.8	0.22	Elastic
Silicon	130	2.7	0.28	Elastic
FR-4	24	16	0.3	Elastic
Solder	Temp. dep.	22	0.34	Visco-plastic
Copper	121	17.3	0.3	Elastic-plastic

Finite element modeling results and analysis

FEA was used to analyze the strain relief mechanism behind the polymer collar approach. Five variations in polymer height on the solder ball were simulated so as to provide guidelines for experimental design (Figure 1, Table 2).

Figure 1: Variations in polymer collar height

Table 2: Variations in polymer collar height

Collar type	Height between solder balls
UF	Collar material as underfill
T0	~170 μm
T1	~110 μm
T2	~80 μm
T3	~55 μm

Strain relief mechanism

A high-CTE glass package assembly at 18.4 mm x 18.4 mm size was simulated with each of the collar types. Plastic strain was extracted at the four critical locations on the outer most solder ball (Figure 2).

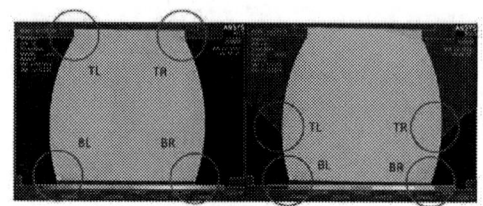

Figure 2: Critical locations for maximum solder plastic strain with (left) and without (right) polymer collar

The underfilled structure shows the same plastic strain at all four points on the joint (Figure 3). With the addition of the polymer collar, the region of high plastic strain was transferred from the package interface to that between the joint and collar, decreasing the package side strain as compared to the UF case. Presence of the collar reduces plastic strain range at all four critical points, though maximum strain range is still seen on the board side. The collar cannot be applied on the board side as this would hinder reworkability.

A trend of decreasing strain range with increasing collar thickness is seen from the results of these simulations. The thickest collar variation (T0) provides the largest buffer to shear deformation resulting in the lowest strain range at the bottom left of the joint. Reduction in strain range for this case is 16.7%. The polymer collar therefore acts as a partial under-fill.

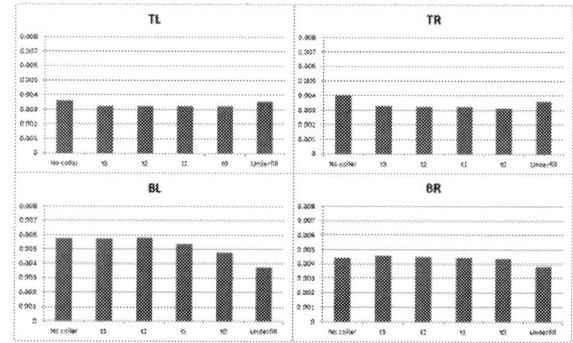

Figure 3: Plastic strain range at the 4 critical locations on the joint

Glass panel fabrication

Glass-interposer test vehicles were designed at a body size of 18.4 mm x 18.4 mm. High CTE, 100 μm thick interposers were fabricated with two dielectric build-up layers RXP-4M (20 μm) and ZS-100 (22.5 μm), laminated on either side of the 6" x 6" bare CF-XX glass panels provided by Asahi Glass Company. The dielectric materials were procured from Rogers Corporation and Zeon Corporation respectively. The panels were then metalized to provide patterning for test daisy-chain structures. Two variations in package pads were introduced with non-solder mask defined (NSMD) pads created with dry-film photosensitive solder resist. The second variation with resin mask defined (RMD) pads were created by laminating an additional layer of ZS-100 over the patterning and drilling holes in the polymer to create openings.

ENEPIG was used as the surface finish on the copper pads (carried out at Atotech, Germany). The interposer fabrication flow is depicted in Figure 4. The panels were then sent to Nanium for attachment of 250 μm solder BGAs by ball drop and interposer dicing.

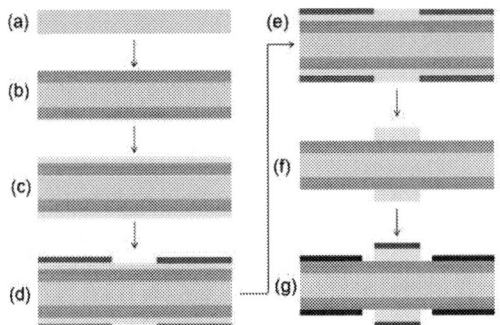

Figure 4: Panel-level fabrication process flow for glass interposers. (a)-(g): Substrate cleaning, dielectric lamination, seed copper, lithography, electroplating, seed etching, surface finish and solder resist processing.

Polymer collar process

Epoxy materials without fillers were spin-coated onto the BGA side of the diced-glass interposers to form the circumferential collars. The interposers were then oven dried at 70 °C for 1 hour to allow the solvent content to evaporate (Figure 5).

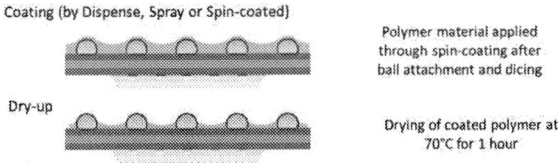

Figure 5: Polymer collar application and drying process

When a square substrate is coated with polymer collar, its corners experience high air friction resulting in a higher rate of evaporation at these locations. The material at the corners dries up and impedes flow of material, being driven radially outward by the centrifugal forces. This behavior leads to material build-up at the corners of the substrate.

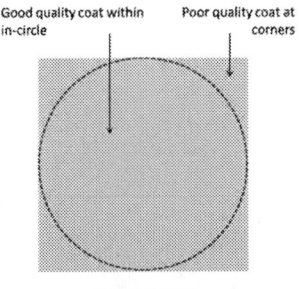

Figure 6: Material build-up at substrate corners during spin coating

The shape of the larger 18.4 mm interposers deviates even more from an ideal circular substrate. Trials using a simple ramp-up and dwell process resulted in highly non-uniform deposition (Figure 7).

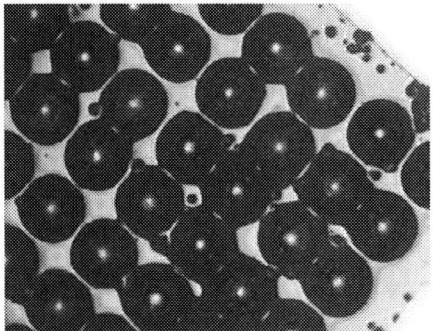

Figure 7: Non-uniform coating on 18.4mm interposer

Two enhancements were therefore used to improve spin-coating uniformity. Firstly, dummy interposers were placed around the test interposer to simulate a larger substrate size and eliminate corner effects (Figure 8).

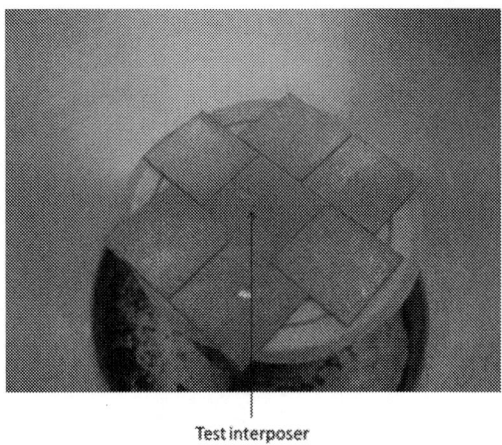

Figure 8: Enhancement to make region around test substrate radially uniform

In addition, the coating profile was changed to include spreading and uniformity phases to allow the material to deposit uniformly over the surface of the test interposer (Figure 9, 10). The final height of coated material depends on the spin coating rate applied in the uniformity phase. It is therefore possible to control the height of the polymer collar using this parameter. The maximum variation in coated height after these enhancements was about 20% (Figures 11,12).

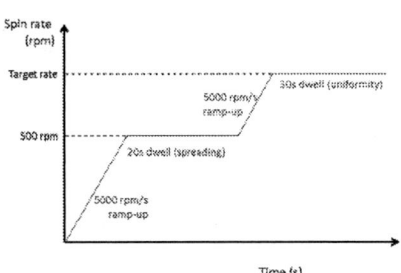

Figure 9: Coating profile for 18.4 mm interposers

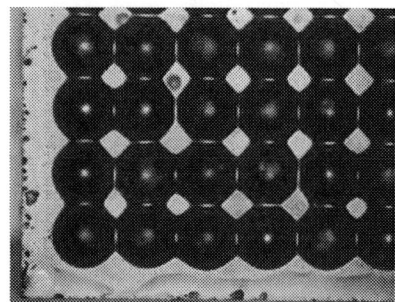

Figure 10: Uniform coating on 18.4 mm interposers

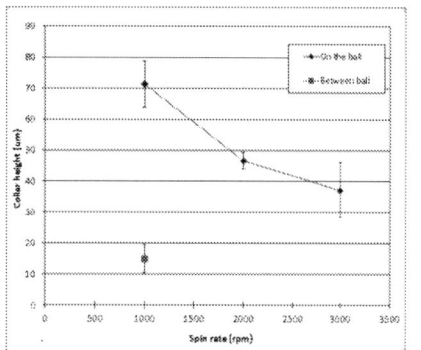

Figure 11: Measured thickness of deposited collar material on 18.4 mm interposers

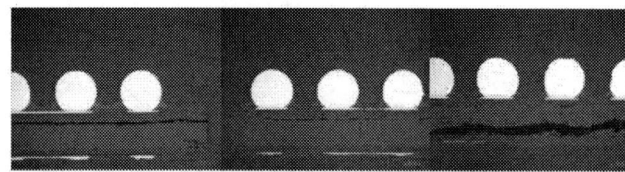

Figure 12: Cross sections of uniformly deposited 18.4 mm interposers

Polymer collar residue removal

SEM imaging was used to confirm the presence of a polymer residue on the ball surface after the application process (Figure 13). Although the material is designed to flow from the bonding area, this residue could be perceived as contamination of the solder balls by automated defects detection systems used in industry. Samples were assembled in order to check for contamination, however, no difference in the surface topography and chemical composition was seen at the solder-pad interface after careful analysis of the intermetallic compounds by SEM/EDX. The next generation of collar material may contain fillers, making residue removal a possible challenge.

Ball surface before coating   Ball surface after coating

Figure 13: Coating with polymer collar leaves a residue on the ball surface

With the objective of manufacturability, three options for residue removal were explored.

1. *Plasma etching*

The parameters used in the plasma etching approach are listed in Table 4. It was observed that while removing residue from the surface of the ball, etching also attacks the circumferential collar (Figure 14). In addition, longer etching times can change the microstructure of the ball surface, which is not acceptable by industry standards.

Table 3: Plasma etching conditions

Parameter	Condition
Composition	CHF_3 (10 sccm) / O_2 (50 sccm)
Pressure	10 mtorr
Power	300 W
Time	15 and 30 minutes

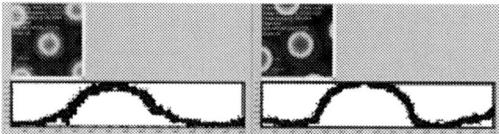

Figure 14: Side profile of solder ball before (left) and after (right) 15 minutes of plasma etching

2. *Sandblasting*

Sand blasting of the ball surface was explored as a possible approach (Table 5). With 250 μm diameter pumicite, the solder surface and, in some samples, the interposer suffered serious damage. Trials done with 35 μm diameter pumicite suffered no visible damage up to 10 seconds of processing. Pumicite was found embedded in the collar material for all variations in process time.

Table 4: Sand blasting conditions

Medium / grain size: pumicite / 250 μm	
Pressure	30 psi
Time (sec)	10, 20, 30
Medium / grain size: pumicite / 35-75μm	
Pressure	25 psi
Time (sec)	5, 10, 30

While this process removes the residue from the ball surface, it is very hard to optimize. Damage to the ball and embedded particles that compromise the integrity of the collar prevent it from being used as a manufacturable solution.

Solvent cleaning

Acetone was used to wipe the ball surface right after the polymer drying stage. As the residue is not cured at this stage, it is easily removed (Figure 15). The material forming the collar remains intact as the solvent comes in contact with only the top surface of the ball (Figure 16).

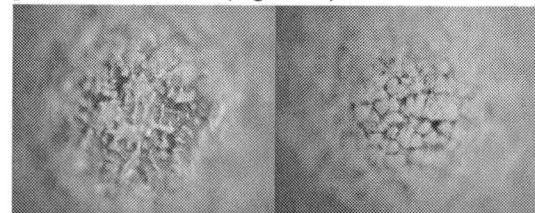

Figure 15: Close-up of ball surface before (left) and after (right) solvent wipe cleaning

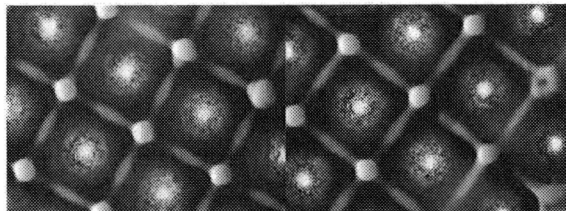

Figure 16: The polymer collar remains intact after cleaning

A comparison of the results from the three approaches shows that the solvent wipe method is most effective in removing the polymer-collar residue with negligible damage to the collar. The process can possibly be implemented by dipping the balls in acetone and then IPA using infrastructure similar to that used for flux dipping prior to assembly.

Assembly for reliability

Tacky flux was first applied to the PCB, followed by a pick-and-place step using a Finetech Matrix flip-chip bonder with a 20 mm x 20 mm pre-leveled vacuum-locked spring gimbal tool. Reflow conditions were optimized to match the recommended profile by the flux supplier. Less than 25% voiding was achieved on assembled samples as required by industry standards.

Assembly process evaluation

In samples assembled with collar, leakage of polymer material onto the PCB was observed. This occurs because the melting point of the collar material is lower than the reflow temperature. The most severe occurrence of board-side leakage is seen in the case of thick collar. A compromise has therefore to be made between buffering capability of the thicker collar and board-side leakage which hinders reworkability.

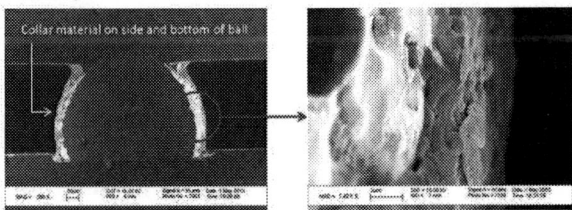

Figure 17: Leakage of material onto ball and PCB after assembly with 90μm thick collar

Self-alignment of solder joints

An important consideration for assembly with circumferential collars is the ability of the solder joints to self-align during reflow. Samples with thickest collar, serving as the worst possible scenario, were used to evaluate self-alignment capability. Increasing degrees of misalignment were introduced during the pick-and-place process. Samples were x-ray imaged before and after reflow to document the introduced misalignment and evaluate alignment after reflow.

The experiment showed that introducing a thick collar around the solder balls does not prevent self-alignment even with the interposer is off by 3/4 of the PCB pad (~170

μm) unless the ball is placed in contact with multiple pads. This verifies that the polymer collar is compatible with the SMT process.

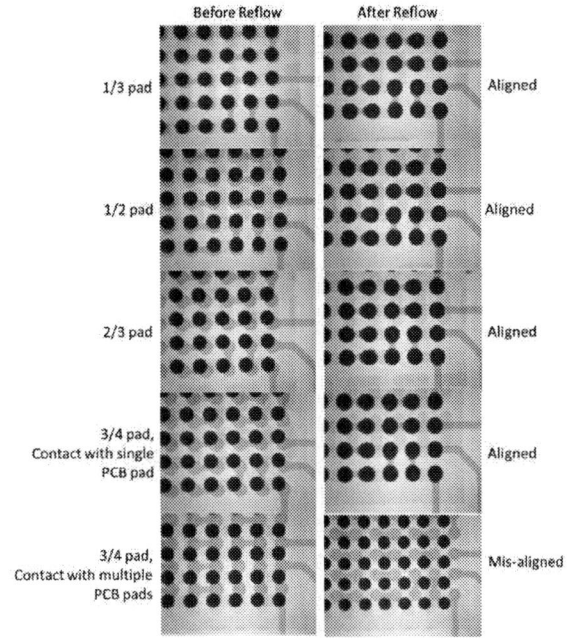

Preliminary Reliability assessment

Reliability testing was performed using JESD22-A104D standards for thermal cycling with assembly, following a pre-conditioning step comprising of 2 additional reflows at 260 °C. Thermal cycling was then performed between -40 °C and 125 °C with a 15 minute dwell time and a rate of 1 cycle/hour. A dummy Si die (12mm x 12mm and 100μm thickness) was underfilled on the interposer to see the impact of die-attach on board-level reliability. The daisy-chain resistances were measured after every 50 cycles. No failures were detected during the testing. The reliability characterization data indicates that polymer collar was successful in preventing solder-joint reliability failures.

Conclusion

Solder-pad interfaces on the package side play a critical role in determining the board-level interconnection reliability, particularly with low CTE packages. This paper develops and demonstrates a polymer-collar approach to alleviate the failures at such interfaces. The polymer collar, which forms a circumferential layer around the solder joint on the package side, blocks the shear deformation of the joint at this interface, transferring the high plastic strain to the lower side of the ball which has lower propensity for failure. The collar therefore acts as a partial under-fill and maintains the reworkability of the assembly. This circumferential polymer collar is deposited around the solder ball on the package side by spin-coating. SMT-compatible assembly processes for large 18.4 mm x 18.4 mm glass interposers with polymer collars were demonstrated. Results demonstrate that the reliability of low-CTE glass interposers on organic packages or baords is significantly enhanced with this polymer collar approach.

Acknowledgments

This study was supported by the "Interconnection and Assembly" focused program at the Georgia Tech Packaging Research Center. The authors would like to thank industry mentors, especially Dr. Jaesik Lee from Qualcomm, for their active guidance, suggestions and support. The authors would also like to thank Asahi Glass Company for providing glass substrates, Namics Corporation for their polymer collar material, Atotech for surface finish processing, Nanium for BGA attachment and interposer singulation, and Zeon Corporation and Rogers Corporation for providing dielectric materials. The authors also thank the Toshitake Seki from NTK/NGK, Yoichiro Sato from Asahi Glass Company and Akira Mieno from Atotech for their support and advice on fabrication, and Anna Stumpf and Anne Matting for their experimental contribution.

References

1. Zhu, K., Ma, L., Sitaraman, S., "Development of G-Helix Structure as Off-Chip Interconnect," Journal of Electronic Packaging, June 2004.

2. Lim, S.S., Rajoo, R., Wong, E.H., Hnin, W.Y., "Reliability Performance of Stretch Solder Interconnections," International Electronic Manufacturing Technolgy, 2006.

3. Topper, M., et. al., "Wafer Level Package using Double Balls," International Symposium on Advanced Packaging Materials, 2000.

4. Varia, B., Fan, X., Han, Q., "Effects of Design, Structure and Material on Thermal-Mechanical Reliability of Large Array Wafer Level Packages," ICEPT-HDP, 2009.

5. Dudek, R., et. al., "Thermo-mechanical Design of Resilient Contact Systems for Wafer Level Packaging," 7th International Conference on Thermal, Mechanical and Multiphysics Simulation and Experiments in Micro-Electronics and Micro-Systems, 2006.

6. Li, D., et. al., "A Wide Area Vertical Expansion (WAVE) Package Process Development," IEEE Electronic Components and Technology Conference, 2001..

7. Bakir, M., et. al., "Sea of Leads Compliant I/O interconnect Process Integration for the Ultimate Enabling of Chips with Low-k Interlayer Dielectrics," IEEE Transacations on Advanced Packaging, 2005.

8. Qin, X., Kumbhat, N., Sundaram, V., Tummala, R., "Highly-Reliable Silicon and Glass Interposers-to-Printed Wiring Boaring SMT Interconnections: Modeling, Design, Fabrication and Reliability," 62nd IEEE Electronic Components and Technology Conference, 2012.

9. Pandher, R., et.al., "Drop Shock Reliability of Lead Free Alloys – Effect of Micro-Additives," ECTC, 2007.

The Study of Bare-Die FCBGA Die Damage in Response to Applied Mechanical Stress During Heat Sink Assembly

Heidi S.Y. Ho[1], Daijiao Wang[1], Michael Johnson[2], C.J. Berry[2]

[1]Broadcom Corporation,
5300 California Ave, Irvine, CA92617, USA
heidi.ho@broadcom.com

[2]Amkor Technology, Inc.
1900 South Price Rd., Chandler, AZ 85286, USA

Abstract

Die chipping or cracking in bare-die FCBGA packages is occasionally but consistently reported by OEMs during system-level heat sink installation. Since heat sink assembly is often a manual process, it can be difficult to determine if the damage was the result of a latent component defect or was induced during heat sink assembly. Most heat sinks require some type of spring force to maintain sufficient mechanical contact with the target electrical component, such as spring-loaded push-pins or Z-clips. When installed properly, these spring systems will not damage the electrical component. However, if the heat sink was installed improperly (i.e., tilted, with excessive force, etc.), or if the full spring deflection is maxed out, die damage may occur. Published studies on this topic are limited, so the following analysis was conducted. In this study, a test method was developed to exert a repeatable, controllable force on the top of a bare-die BGA component. The force was varied in magnitude, load/impact angle, and load rate. In addition, criteria for systematically characterizing component damage response were developed. Based on the data collected, the authors were able to characterize and compare the propensity to create damage by different types of assembly forces, such as forces perpendicular to the component, off-angle, excessive, slow, etc. Detailed quantitative results are presented. The test methods and empirical data generated during this study are useful in identifying certain limits in procedures used for system-level heat sink assembly that should result in less component damage and higher yields.

Introduction

Over the last decade, increases in integrated circuit (IC) process technology have led to increased IC functionality and shrinking die sizes. These changes have led to continually increasing IC power density. As a result, more and more IC components require heat sinks as a part of the system thermal solution to optimize performance. Heat sink application is commonly seen associated with high-power flip-chip ball grid array (FCBGA) packages, since due to the package structure heat sinks may be required for heat dissipation.

During the heat sink assembly process, the heat sink is clamped with a specified force. This force is necessary to ensure mechanical stability of the system during assembly and to minimize thermal impedance of the thermal interface material (TIM) that lies between the die and heat sink [1]. Lopez et al. [2] used Monte Carlo analysis to characterize

push-pin heat sink assembly and found that the compressive load asserted on the assembly is typically between 224.0 and 279.5 lbs.

Studies have shown that heat sink installation may cause a static and/or dynamic strain on the IC package and result in interconnection issues [3]. Zhu et al. [4] investigated flip-chip ball grid array (FCBGA) solder joint reliability under preloaded heat sink conditions. Roberts et al. [5] looked into die stress due to heat sink clamping, and reported that the typical heat sink clamping force is around 1000 N in high-performance application, which results in −60 MPa stress at the center of the die, and −40 MPa at the corner of the die.

However, most of the studies focus on the static stress present in clamped heat sink applications, and only limited discussion could be found on die damage caused by the compression force process step and its associated variables during heat sink installation. This study considers various scenarios that could occur during heat sink installation. The variables of compression force, load rate, and tilting angle/direction, all of which can come into play during the heat sink clamping/installation process, are investigated for their impact on die damage

Method and Test Setup

Compression testing on bare-die FCBGA packages was performed using an Instron 5965 Materials Testing System with a 5K N load cell. The general structure of an FCBGA package is shown in Figure 1.

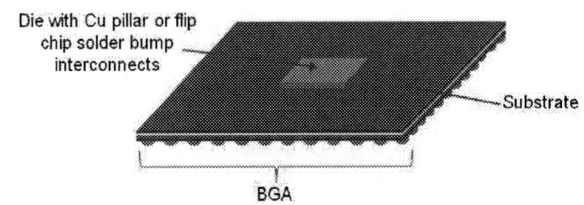

Figure 1. FCBGA

For all tests, Honeywell PTM3180 thermal interface material (TIM) was used to cover the Si die to simulate the actual manufacturing setup. See Table 1.

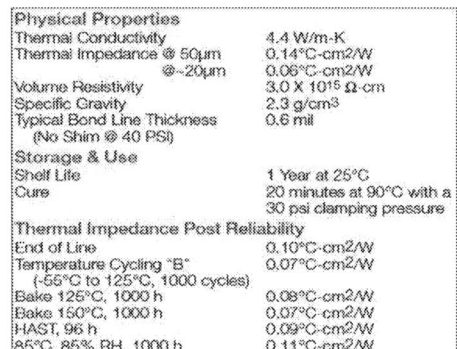

Physical Properties	
Thermal Conductivity	4.4 W/m-K
Thermal Impedance @ 50μm	0.14°C-cm2/W
@~20μm	0.06°C-cm2/W
Volume Resistivity	3.0 X 10¹⁵ Ω-cm
Specific Gravity	2.3 g/cm3
Typical Bond Line Thickness (No Shim @ 40 PSI)	0.6 mil
Storage & Use	
Shelf Life	1 Year at 25°C
Cure	20 minutes at 90°C with a 30 psi clamping pressure
Thermal Impedance Post Reliability	
End of Line	0.10°C-cm2/W
Temperature Cycling "B" (-55°C to 125°C, 1000 cycles)	0.07°C-cm2/W
Bake 125°C, 1000 h	0.08°C-cm2/W
Bake 150°C, 1000 h	0.07°C-cm2/W
HAST, 96 h	0.09°C-cm2/W
85°C, 85% RH, 1000 h	0.11°C-cm2/W

Table 1. Thermal interface material property – Honeywell PTM3180 [6]

Three types of compression tests were performed. The first set of tests involved applying a normal force over the entire flat Si die surface (Fig.2). For the normal force testing, test vehicles were 35 mm x 35 mm Bare-Die FCBGA packages having a die size of 10.7 mm x 10.4 mm x 787 μm. Die were of the 28nm technology node variety with Cu-Pillar interconnections to the laminate substrate and standard capillary underfill was applied. All samples were exposed to the same maximum load and load rate. Testing was also conducted on packages of the same size and type, but having 40nm technology node die that were solder-bumped and slightly smaller (10.2 mm x 10.5 mm x 795 μm) than the aforementioned 28nm, Cu-Pillar case. These solder bumped die were also underfilled using the standard capillary process. Results showed very similar mechanical behavior between these two types of test vehicles.

The second set of tests (one-side tilt) involved applying a force over the elevated edge of a Si die that had been tilted/raised to a specific angle (Fig. 3). For one-side tilt testing, the 40nm solder-bump test vehicle was used. Load rate, maximum load, and tilt angle were all varied.

The third set of tests (two-side tilt) involved applying a force over the elevated corner of a Si die that had been tilted/raised to specific angles on two sides (Fig.4). For the two-side tilt testing, the test vehicle was the same as used for the one-side tilt testing. Load rate and maximum load were varied. Tilt angle was held constant.

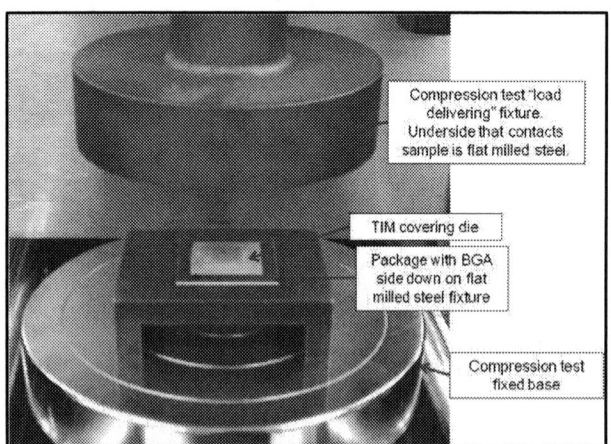

Figure 2. Normal force test setup

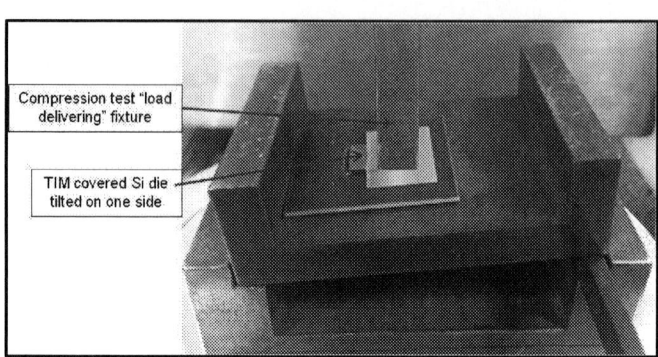

Figure 3. One-Side Tilt Test Setup

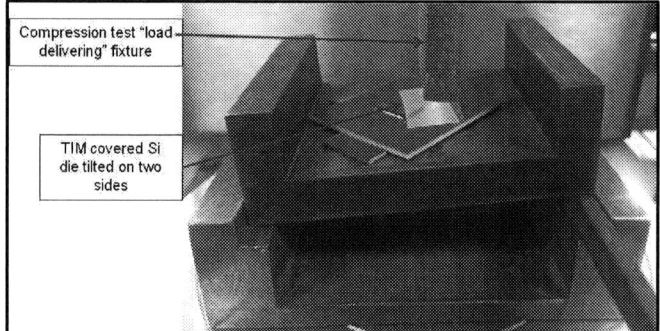

Figure 4. Two-side tilt test setup

Prior to testing, the system load versus displacement was measured to determine the displacement at increasing loads that should be attributed to the load train itself and TIM. The data was plotted and fitted to a regression line which was subsequently subtracted from the actual compression tests. This is commonly referred to as eliminating the 'load train compliance' and ensures that displacement data for any test reflects only the displacement of the sample and does not include displacements due to the 'system.' See Fig. 5 for an example plot.

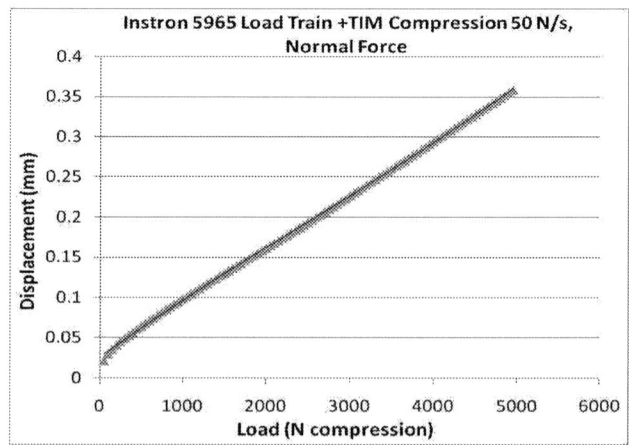

Figure 5. System load vs. displacement

Post testing, the normal force test group was thoroughly inspected for external and internal damage using optical inspection of the die and package, laser profilometry, CSAM, and cross-section. The one-side and two-side tilt groups were analyzed for Si damage only using optical inspection.

Result and Discussion

Normal Force Testing

For normal force compression testing, loads up to 4950N were applied at a rate of 50N/s. The 4950N load was near, and limited by, the maximum capacity of the available load cell.

A representative load versus displacement plot for one of the samples tested is shown in Fig.6. From 0N to 2000N the load induced a mostly linear style deformation in the package which displayed a stiffness of about 6500N/mm. After 2000N the package became stiffer due to BGA ball deformation and the associated increase in load-bearing surface.

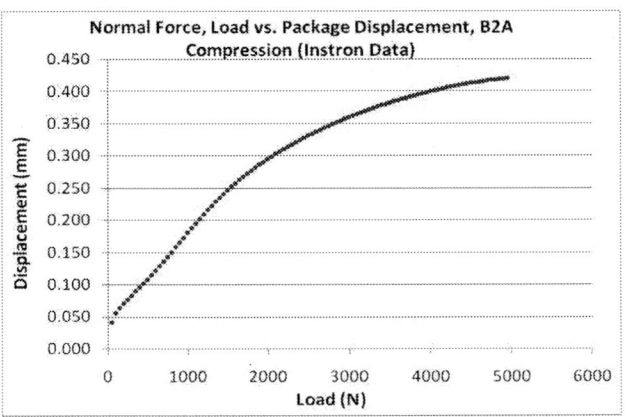

Figure 6. Normal force load vs. package displacement

Representative pre and post-test laser profilometer data (Fig.7) shows an uneven permanent deformation of the BGAs; center BGAs (those directly under the die) are significantly deformed, whereas BGAs outside the die shadow are only slightly coined. This uneven deformation is the result of the natural package deflection shape during testing (Fig 8). Table 2 summarizes permanent deformation data for select BGA along the profilometer scan path as well as overall package height reduction (due to BGA deformation). Specifically and for example, the average height change of BGA 9 which is outside the die shadow is 86um whereas the average height change of BGA 17 which is under the die shadow is 180um. This clearly illustrates the aforementioned trend.

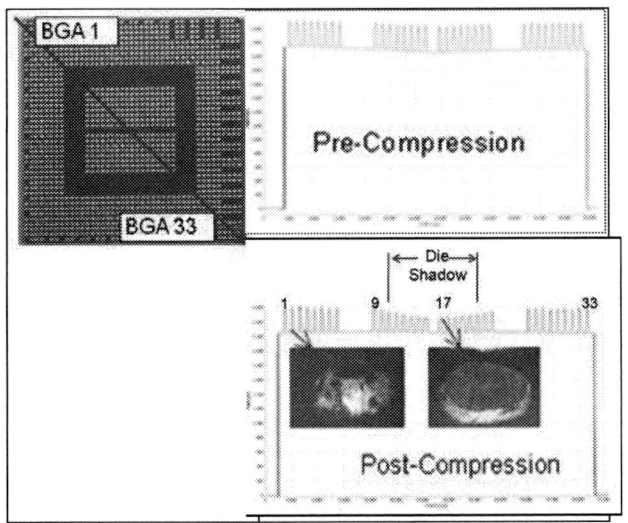

Figure 7. BGA laser profilometer scan

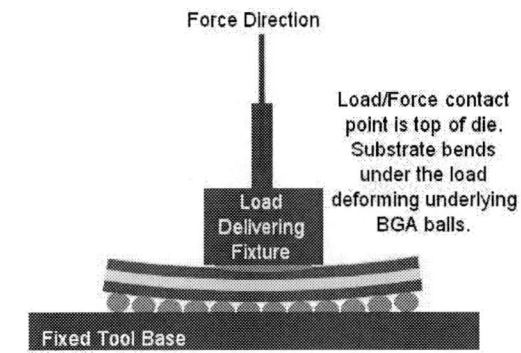

Figure 8. FCBGA package flex during normal force compression testing

Pre/Post Compression Package Height Change (um)						
Note:" To top of Die" data was measured from profilometer stage table to top of die with BGA side down on table. "To top of BGA" data was measured from profilometer stage table to top of selected BGA.						
Sample		To Top of Die	To Top of BGA			
			BGA 1	BGA 9	BGA 17	BGA 33
B-2A	Height Change	50	140	100	200	100
B-3A	Height Change	60	70	90	180	130
B-4A	Height Change	60	90	70	160	110

Table 2. Normal force package/BGA deformation data

All normal force samples were inspected post-test using CSAM and cross section (Figs.9, 10) and compared to pre-test inspection. No delamination, cracking, or other damage could be found anywhere in the solder, die, dielectric layers, underfill, PCB, or package in general as a result of the normal force compression testing. Sample Cu pillar joint heights for the pre-compression sample B-1A and post-compression samples B-2A, B-3A, and B-4A were also measured. Comparison of the pre-test vs. post-test joint heights shows no variation, indicating the internal Cu pillar joints were not "deformed" as a result of compression testing (Table 3).

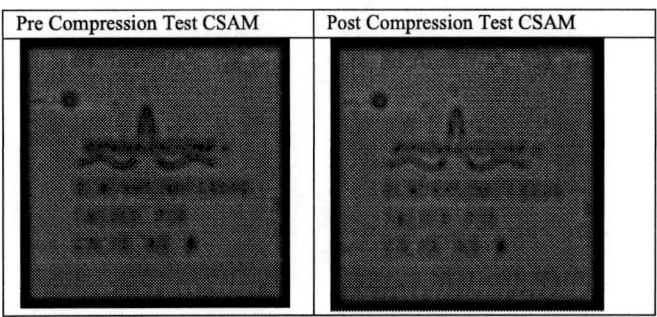

Figure 9. CSAM inspection

Bump Sample Cross-Section	
Pre Compression Test	Post Compression Test– sample 1
Post Compression Test– sample 2	Post Compression Test– sample 3

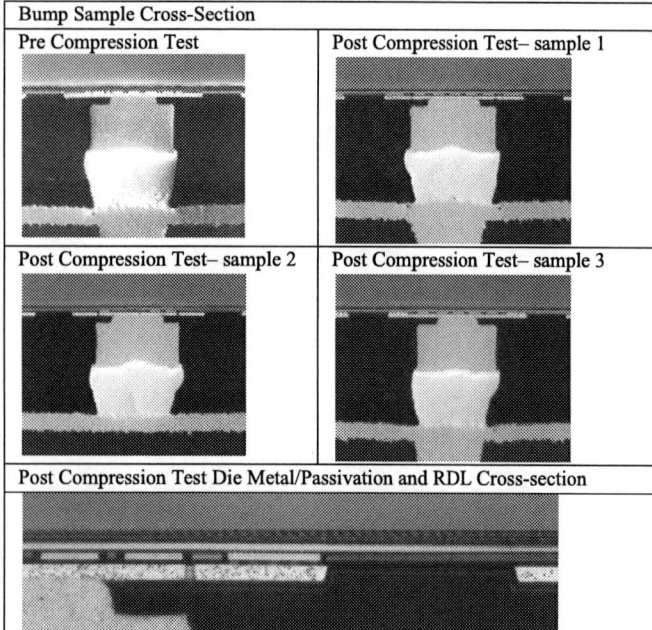

Post Compression Test Die Metal/Passivation and RDL Cross-section

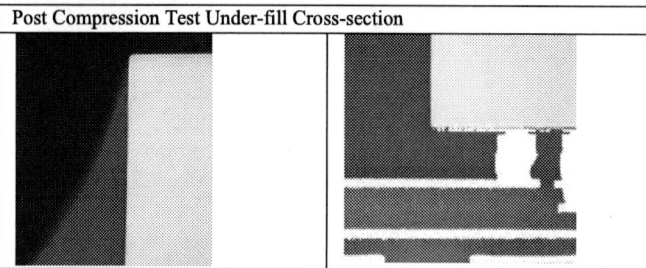

Post Compression Test Under-fill Cross-section

Figure 10. Die/package cross-section

Joint Standoff Height (microns)				
	Sample			
Joint	Pre Compression B-1A	Post Compression B-2A	Post Compression B-3A	Post Compression B-4A
1	93.1	93.5	93.9	92.6
10	91.5	90.7	92.4	91.3
20	94.4	92.4	93.6	92.8
30	93.7	91.3	92.2	91.1
40	93.2	94.1	93.8	93.9
Avg	93.2	92.4	93.2	92.3
Stdev	1.1	1.4	0.8	1.2
Min	91.5	90.7	92.2	91.1
Max	94.4	94.1	93.9	93.9

Table 3. Pre-/post-compression test joint heights

One-Side Tilt Testing

For the one-side tilt compression testing, uneven load force conditions were investigated by tilting the package on one-side to create a one-side force contact arrangement. A representative load vs. displacement plot is shown in Fig.11.

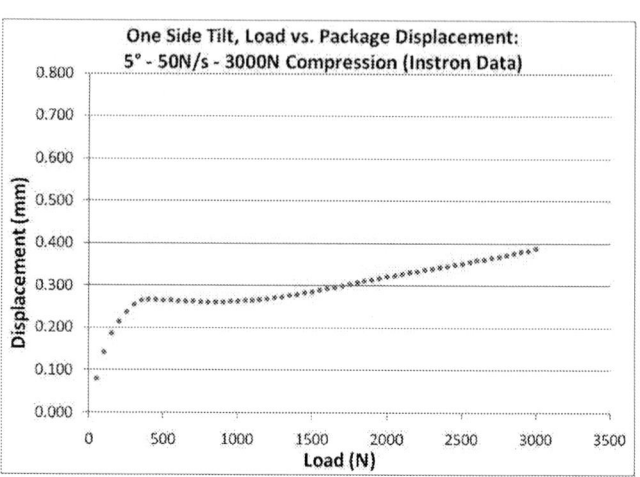

Figure 11. One-side-tilt load vs. displacement

Multiple tilting angles (5°, 10°, 20°, 30°), loading speeds (50N/S,500N/S) and load magnitudes (250 - 3000N) were tested. Die damage response criteria were developed using visual inspection to evaluate the severity of silicon cracking (Table 4, Fig 12).

Die/Si Cracking Severity Scale	
No Damage	0%
Slight edge / corner chipping	1%
Corner Crack	5%
Small near edge crack	10%
Large near edge crack	15%
Severe cracking on ½ of die	50%
Sever cracking throughout die	80%
Die Shattered	100%

Table 4. One-side-tilt Si cracking severity scale

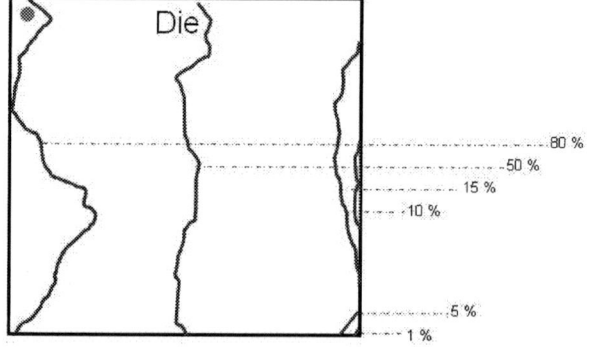

Figure 12. Definition of severity of silicon cracking in one-side-tilt testing

Representative images of Si die cracking from one-side tilt testing are shown in Fig.13.

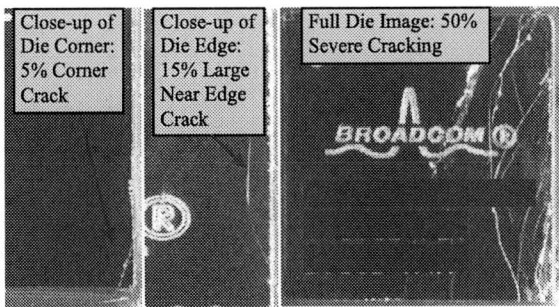

Figure 13. One-side tilt test die cracking examples

The test configurations and silicon damage responses are shown in Table 5 and Fig. 14. In contrast to normal force loading which showed no defects at loads up to ~5000N, significant die cracking was seen from 1000N onward.

A similar silicon damage response pattern as a function of load was observed throughout the tilt-angle-groups (Fig. 14). This pattern indicates that tilting angle is an on-and-off factor leading to silicon damage, but that once the tilt is present, the angle is not a major factor in determining the severity of silicon damage.

A similar silicon damage response was also seen at both 50N/s and 500N/s loading rates. From this finding, it appears that for a one-side-tilt condition the loading rate may not be a major factor either in contributing to silicon damage.

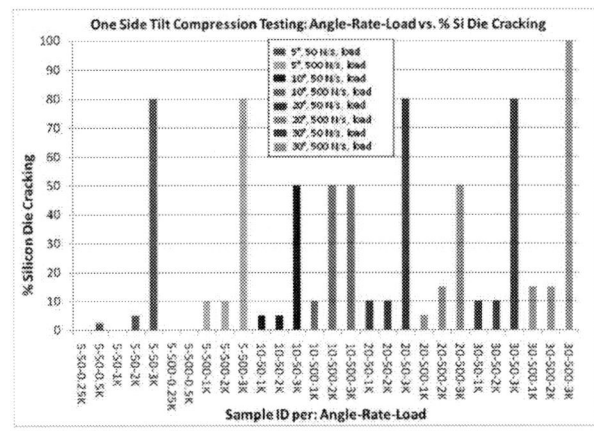

Figure 14. One-side-tilt testing results

The conclusion that applied load is the key factor contributing to the extent of Si damage is further illustrated by the plot in Fig. 15 which shows maximum load vs. % Si cracking for all samples. Similar scatter plots (not shown here) were made for tilt angle vs. % Si cracking and load rate vs. % Si cracking. Those data showed only slight trends towards increased die cracking with increased load rate and therefore, within the scope of this study, were not significant enough to warrant being labeled as key factors.

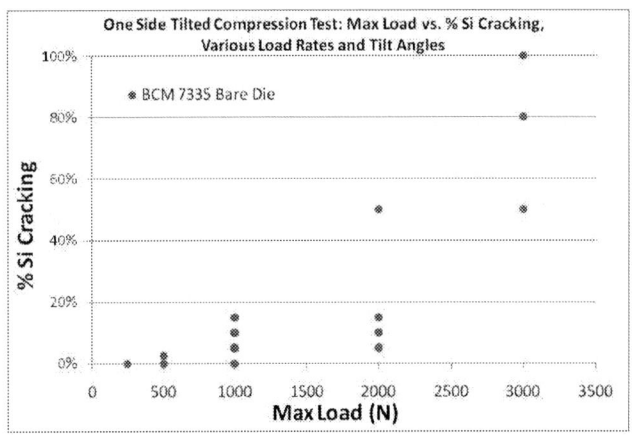

Figure15. One-side tilt testing, maximum load vs. % Si cracking

Two-Side Tilt Testing

For the two-side tilt compression testing, corner-point load force conditions, as may occur during the push-pin heat sink assembly process, were investigated. A representative load versus displacement plot is shown in Fig.16.

Tilting Angle (°)	Load Rate(N/S)	Max Load (N)	% Si Crack
5	50	250	0
		500	1
		1000	0
		2000	5
		3000	80
	500	250	0
		500	0
		1000	10
		2000	10
		3000	80
10	50	1000	5
		2000	5
		3000	50
	500	1000	10
		2000	50
		3000	50
20	50	1000	10
		2000	10
		3000	80
	500	1000	5
		2000	15
		3000	50
30	50	1000	10
		2000	10
		3000	80
	500	1000	15
		2000	15
		3000	100

Table 5. One-side-tilted testing configuration and silicon damage responses

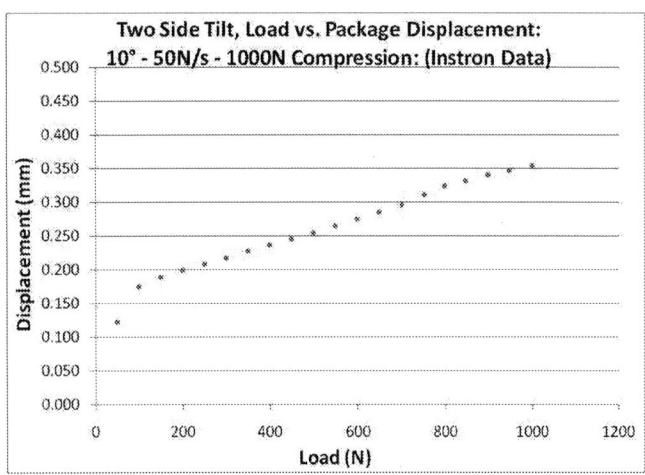

Figure 16. Two-side-tilt load vs. displacement

Based on results from the previous "one-side tilt" configuration, a 10 degree tilt angle on both the x and y axis was chosen to represent the corner-point load condition. Loading rate and applied force were varied. As for the one-side tilt tests, a die damage response criterion for two-side tilt testing was established using visual inspection to evaluate the severity of silicon cracking (Table 6, Fig. 17). Fig. 18 shows representative images of Si die cracking from two-side tilt testing.

Severity of Silicon Die Cracking (X = 1/2 diagonal length of the silicon, see Figure 17)	
No cracking / chipping	0%
Slight corner chipping	1%
Si cracking up to ~0.07X into the die	5%
Si cracking up to ~0.2X into the die	15%
Si cracking up to ~0.3X into the die	25%
Si cracking up to ~0.5X or greater into the die	45%

Table 6. Two-side-tilt Si cracking severity scale

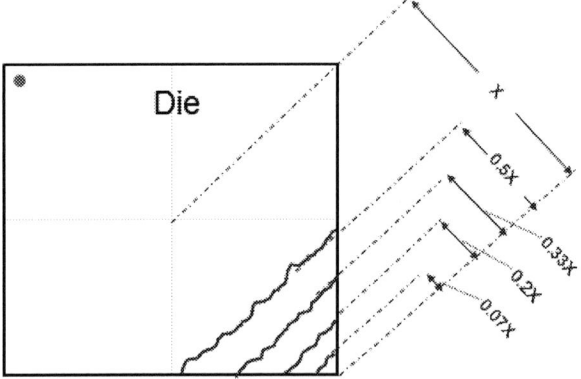

Figure 17. Definition of severity of silicon cracking in two-side-tilt testing

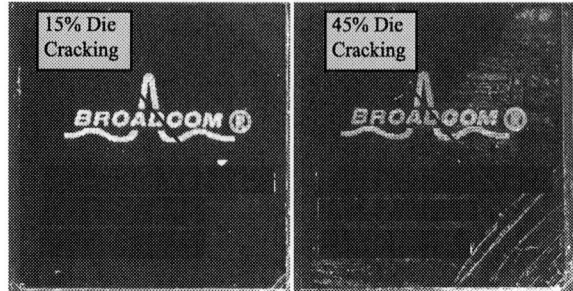

Figure 18. Two-side tilt test die cracking examples

Test results are shown in Table 7 and Fig. 19. Two-side-tilt testing resulted in die damage becoming significant (15% Si cracking) at a load of 150N, which is less than 1/6th of the one-side-tilt force load tolerance. Damage did occur at lower loads, however all damage at lower loads was 1% or 5% Si cracking, and while this level of cracking may be significant in some instances, for purposes of this study it was considered to be cosmetic in nature.

As in the one-side-tilt tests, when the same magnitude of maximum loading force was used, a similar pattern of silicon damage was seen throughout different load rate groups. This pattern suggests that the magnitude of the force is the dominant factor related to silicon damage when a two-side-tilt condition is present.

Max Load (N)	Load Speed (N/S)	% Silicon Crack
10	3000	0
	5000	0
25	50	0
	500	0
	1000	0
	3000	0
	5000	5
50	50	0
	500	0
	1000	1
	3000	5
	5000	5
150	50	1
	500	1
	1000	5
	3000	5
	5000	15
250	50	15
	500	15
	1000	5
	3000	5
	5000	15
500	50	25
	500	5
	1000	15
	3000	15
	5000	25
1000	50	45
	500	45

Table 7. Two-side-tilt testing configuration and silicon damage responses

978-1-4799-2408-0/14 $31.00 © 2014 IEEE

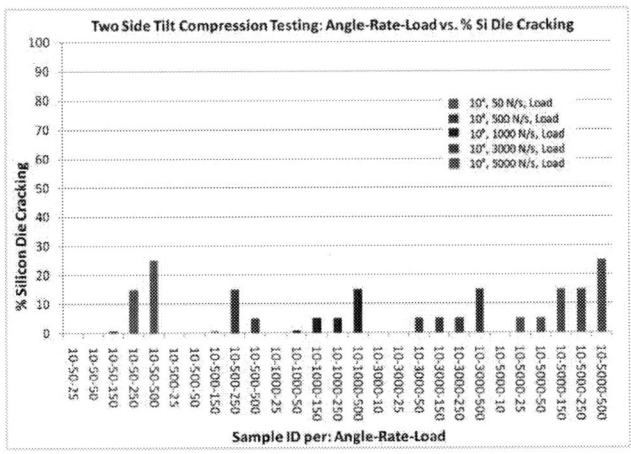

Figure 19. Two-side-tilt testing results

The conclusion that the applied load is the key factor contributing to the extent of Si damage in two-side tilt testing is further illustrated by the plot in Fig. 20, which shows maximum load vs. % Si cracking for all samples. A similar scatter plot (not shown here) was made for load rate vs. % Si cracking. No trend relating to load rate and % Si cracking was noted.

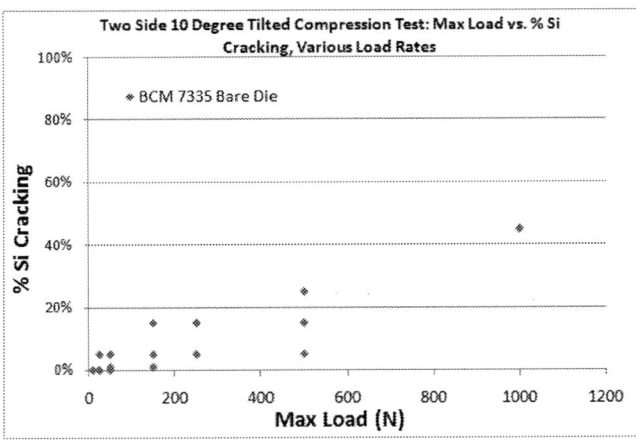

Figure 20. Two-side tilt testing, maximum load vs.% Si Cracking

Conclusions

- Typical "normal force" compression on TIM covered bare-die package assemblies, such as is present during heat sink assembly processes does not pose a damage risk to the Si die; the die/package assemblies used in this study were able to withstand very high forces of up to 4950N without sustaining damage other than BGA flattening.
- Tilting a TIM covered bare-die/package assembly on one-side significantly increases the probability that damage will occur during heat sink assembly. One-side tilt resulted in up to 15% Si die damage occurring at 1000N, which is a significantly lower force than in the "normal" force compression case (4950N) which resulted in virtually no damage.
- Tilting a TIM covered bare-die/package assembly on two sides increases even further the probability that

damage will occur during heat sink assembly. Two-side tilt resulted in significant Si die damage occurring at 150N; a 97% reduction in the maximum force (tested) in "normal" force compression without significant damage.

- For TIM covered bare-die/package assemblies that are tilted during heat sink assembly, the main factor leading to Si die damage is the magnitude of the load.
- For the one-side tilt angles tested in this study (5 to 30 degrees), tilt is an on/off factor leading to silicon damage, but the value of the angle (within the range tested) is not believed to be a key factor in determining the severity of damage.
- Load rate was not found to be a significant factor in creating die damage on TIM covered bare-die package assemblies.
- Typical TIM materials such as those used in this study do not protect the Si die from damage once the package assembly is tilted.

Future Work

A future study evaluating tilt angles less than the 5 degree minimum tested here would be valuable in further understanding the impact of package tilt on Si die damage during heat sink assembly. It may also help to define tolerances on the planarity of any fixtures or equipment used in heat sink attach.

Typical die saw/dicing processes are known to result in varying degrees of edge chipping. Therefore, testing a population of die having very clean cut edges vs. a population with rougher/chipped (but within specification) edges would provide further insight into how other factors, such as die edge condition, may influence damage induced during system level assembly processes which impart compressive loads on packages.

Acknowledgments

The authors would like to thank Quan Pham for his CSAM and x-section work on this project.

References

[1] Chen, C.-I., Ni, C.-Y., Lee, C.-C., Pan, H.-Y. and Yuan T.-D., 2007, "Mechanical Characterization and performance Optimization for GPU Fan-Sink Cooling Module Assembly," IEEE Trans. Electron. Packg. Manuf., 30, pp.173-181.

[2] Lopez, L.D., Nathan, S. and Santos S., 2004, "Preparation of Loading Information for Reliability Simulation," IEEE Tran. Compon.Packag. Technol., 27, pp. 732-735.

[3] Quinones, H. and Babiarz, A., 2000, "Chip Scale Packaging Reliability," Proceedings of the International Symposium on Electronic Materials and Packaging (EMAP), pp. 298-405.

[4] Zhu, L., Summers, M., Uppalapati, R. and Clyne, K., 2006, "Thermal Fatigue Reliability Modeling and Analysis of BGF Socket Assembly in System Board with Preloaded Use Condition," Proceedings of 2006 Electronic Components and Technology Conference, pp. 748-755.

[5] Roberts, J.C., Motalab, M., Hussain, S., Suhling, J.C., Jaeger, R.C. and Lall, P., 2012, "Characterization of

Compressive Die Stresses in CBGA Microprocessor Packaging Due to Component Assembly and Heat Sink Clamping," ASME Journal of Electronic Packaging, Vol. 134, pp. 031005-1 – 031005-17.

[6] Honeywell "Technical Data Sheet," https://www.honeywell-pmt.com/sm/em/common/documents/PTM3180_Phase_Change_Thermal_Interface_Material_Application_Note.pdf

Prognostication of Copper-Aluminum Wirebond Reliability under High Temperature Storage and Temperature-Humidity

Pradeep Lall [1], Shantanu Deshpande [1], Luu Nguyen [2], Masood Murtuza [2]

[1]Auburn University
NSF-CAVE3 Electronic Research Center
Department of Mechanical Engineering
Auburn, AL. 36849
[2]Texas Instruments, Santa Clara, CA 95052
Tele: (334) 844-3424
E-mail: lall@auburn.edu

Abstract

Gold wire bonding has been widely used as first-level interconnect in semiconductor packaging. The increase in the gold price has motivated the industry search for alternative to the gold wire used in wire bonding and the transition to copper wire bonding technology. Potential advantages of transition to Cu-Al wire bond system includes low cost of copper wire, lower thermal resistivity, lower electrical resistivity, higher deformation strength, damage during ultrasonic squeeze, and stability compared to gold wire. However, the transition to the copper wire brings along some trade-offs including poor corrosion resistance, narrow process window, higher hardness, and potential for cratering. Formation of excessive Cu-Al intermetallics may increase electrical resistance and reduce the mechanical bonding strength. Current state-of-art for studying the Cu-Al system focuses on accumulation of statistically significant number of failures under accelerated testing. In this paper, a new approach has been developed to identify the occurrence of impending apparently-random defect fall-outs and pre-mature failures observed in the Cu-Al wirebond system. The use of intermetallic thickness, composition and corrosion as a leading indicator of failure for assessment of remaining useful life for Cu-al wirebond interconnects has been studied under exposure to high temperature and temperature-humidity. Damage in wire bonds has been studied using x-ray Micro-CT. Microstructure evolution was studied under isothermal aging conditions of 150°C, 175°C, and 200°C till failure. Activation energy was calculated using growth rate of intermetallic at different temperatures. Effect of temperature and humidity on Cu-Al wirebond system was studied using Parr Bomb technique at different elevated temperature and humidity conditions (110°C/100%RH, 120°C/100%RH, 130°C/100%RH) and failure mechanism was developed. The present methodology uses evolution of the IMC thickness, composition in conjunction with the Levenberg-Marquardt algorithm to identify accrued damage in wire bond subjected to thermal aging. The proposed method can be used for quick assessment of Cu-Al parts to ensure manufactured part consistency through sampling.

Introduction

Gold wire has been widely used in the electronic industry for first-level interconnects because of its high electrical conductivity, low hardness and resistance to oxidation. The recent steep rise in the gold prices has forced the electronic industry to look for other low cost alternatives. Copper and palladium coated copper are leading candidates to replace the incumbent gold wire [Breach 2010]. The transition to copper wire bond has been encouraged by not only the lower cost of copper wire but also desirable properties such as higher thermal conductivity and electrical conductivity compared with gold. In addition, copper wire also has higher mechanical strength and a low reaction rate with aluminum, which bodes well for long term reliability of the copper-aluminum system [Boettcher 2010; Kim 2003]. However, introduction of copper in to wire bonding process have also created new set of challenges. Copper is harder than gold, and thus requires more ultrasonic energy and mechanical force during the mechanical bonding process. Higher energy during the bonding process requires a narrow bonding window, increasing the potential for chip cratering beneath bonding surface. Further, copper and copper intermetallics tend to oxidize at high temperatures reducing bond strength. Substantial amount of work has been done on developing and optimizing thermosonic wire bonding process for copper and significant improvements are noted in literature [Breach 2010; Qin 2011; Shah 2009; Chen 2004; Schmitz 2011].

In spite of process advancements, reliability is still one of the major concerns for copper wire bonding. Past research has shown that formation of excessive intermetallics at Cu-Al interface is a major contributor to failure of the wire bond accompanied by resistance increase and fracture or bond lifts [Boettcher 2011; England 2011]. Past researchers have studied the IMC growth under isothermal aging. IMC has been studied using various techniques, including the crystal structure, effect of annealing time and effect of temperature on IMC growth [Na 2011]. Composition of IMC has been studied using TEM and XRD, and reported to be $CuAl_2$ and Cu_9Al_4 [Wieczorek-Ciurowa 2005; Laik 2008; Lee 2005]. Tian, et.al. [2011] reported IMC growth behavior of Cu-Al wirebond system when subjected to multiple thermal stresses. Cu-Al IMCs grow much faster in thermal aging compared to thermal shock and that rate of growth is higher for higher aging temperatures. Three predominant phases of Cu-Al IMCs include $CuAl_2$, Cu_9Al_4, CuAl. However, the timing of appearance of each phases is still not clear. It has also been found that IMC growth is diffusion driven [Chen 2011; Hand 2008]. Activation energy of Cu-Al IMCs reported in literature have wide range from 2.5l kcal/mol to 23.9 kcal/mol [Lee 2005; Goh 2013; Simonovic 2009; Levenson 1989]. Prior data indicates that cracking in the vicinity of the Cu-Al intermetallics is a predominant location of failure. The initial cracks may form in the copper-aluminum bond because of ultrasonic squeeze during the bond process, and that the cracks may propagate with aging time [Boettcher 2010]. The Cu-Al IMCs grow at a slower pace compared to the Au-Al system [Goh 2013]. Oxidation of Cu-Al intermetallics during operation at high temperature and high humidity may cause oxidation of IMC, followed by crack initiation, and eventual failure. The process of corrosion is accelerated in the presence of ionic species, such as halide, hydroxyl ions, and elevated temperature. Prior data indicates that under 85%RH/85°C conditions, the aluminum pad gets corroded faster and causes accelerated failure compared to the Au-Al wirebond system [Breach 2010]. Different corrosion reactions have been proposed for Cu rich and Al rich phases of IMC. Copper rich interface (Cu_9Al_4) undergoes preferential corrosion compared to the aluminum rich phase. Reaction rate is greatly affected by concentration of hydroxyl, halide ions in surrounding. Corrosion of IMC causes sudden increase in resistance and reduction in ball shear strength [Osenbach 2013; Liu 2011; Su 2013; Zeng 2013]. Understanding this abnormal behavior is important step in troubleshooting reliability issues related to Cu-Al wire bonding.

Electronic components operating in harsh environment are often subjected to high temperature and/or humidity conditions. It is often not feasible to capture the continuous time temperature operational history over the use life of the product. Current health management systems provide nearly zero visibility into health of electronics and packaging for prediction of impending failures [McCann 2005; Marko 1996; Schauz 1996; Shiroishi 1997]. The built-in-self tests (BIST) are generally used to give electronic assemblies the ability to test and diagnose themselves with minimal interaction from external test equipment. BIST has a built-in circuit capable of providing error detection and correction [Zorian 1994; Chandramouli 1996; Drees 2004; Hassan 1992]. However, several studies conducted [Drees 2004; Gao 2002; Daniel 1990] have shown that BIST can be prone to false alarms and can result in unnecessary costly replacement, requalification, delayed shipping, and loss of system availability. There is need to develop scientific methodology to predict damage in copper-aluminum electronic packages, without knowing there prior aging history. Prognostic health management (PHM) is a method for assuring the reliability of a system by monitoring the system in real time as it is used in the field. PHM system if implemented for copper-aluminum wire bond interconnects could provide information of the electronic system's state and the possibility of impending failure. Leading-information on near failure systems could allow enough time for repair or replacement of the damaged modules. In addition, it is envisioned that the leading indicators for the copper-aluminum system could be used for defect detection.

Previously, leading indicators based prognostic and health management methodologies for residual life computation of electronic solder joints subjected to single, multiple and superimposed thermal environments comprising of isothermal aging and thermal cycling have been developed [Lall 2008[a,b]; Lall 2007[c], 2009[c]]. Examples of second-level interconnect damage precursors include micro-structural evolution, inter-metallic compound growth, stress and stress gradients. Previously, damage pre-cursors have been developed for

various lead-free alloy compositions on a variety of area-array architectures. PHM methodology for second-level interconnects was successfully developed for microstructure evolution of damage based leading indicator for estimating prior accrued damage [Lall 2012[a,c]]. Researchers have also implemented PHM technique for assessment of damage and remaining useful life in electronic system, which are subjected to thermo-mechanical load, sequential multiple thermal environment [Lall 2012[b]; 2010[b]]. In this paper; PHM framework has been introduced using Levenberg-Marquardt (LM) algorithm based on physics based state-vectors for assessment of accrued damage and future degradation prediction, using IMC layer development when Cu-Al system is subjected to thermal aging, as a leading indicator.

Test Vehicle

The test vehicle used for the study is the 32-pin chip scale package. The package is 4.5 mm in length, 5.5 mm in width, 0.7 mm in height. Each pin has a length of 0.45 mm and width of 0.3 mm. The package interconnects has an I/O pitch of 0.5 mm. The package has 30μm diameter copper wires and aluminum pads. The packages used for the study were not daisy chained. Package dimensions are listed in Table 1.

Table 1: Dimensions of Test Vehicle

Parameter	Dimensions (mm)
Width of Package	4.5
Length of Package	5.5
Height of Package	0.7
Length of Pin	0.45
Width of Pin	0.3
Pitch	0.5

Figure 1 shows the optical microscopic images of the package top and bottom. Figure 2 shows X-ray images of the package taken using the YXLON Cougar μCT System. Figure 3 shows the 3D μ-CT reconstruction showing copper-aluminum wirebonds. The chip and the electronic mold compound have been deselected for better visibility of the copper-aluminum interconnects. Figure 4 shows magnified view of single wire-bond.

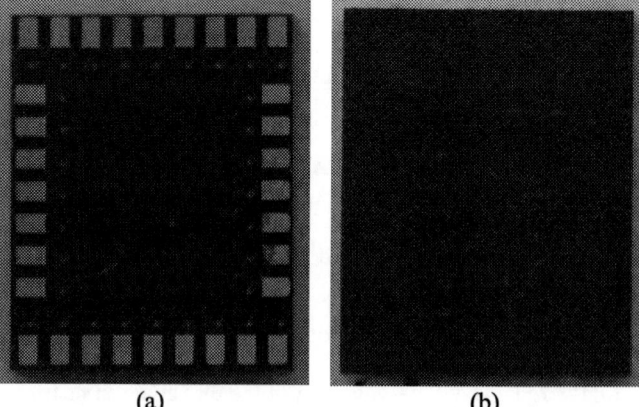

(a)   (b)

Figure 1 – Optical images (a) bottom view, (b) top view.

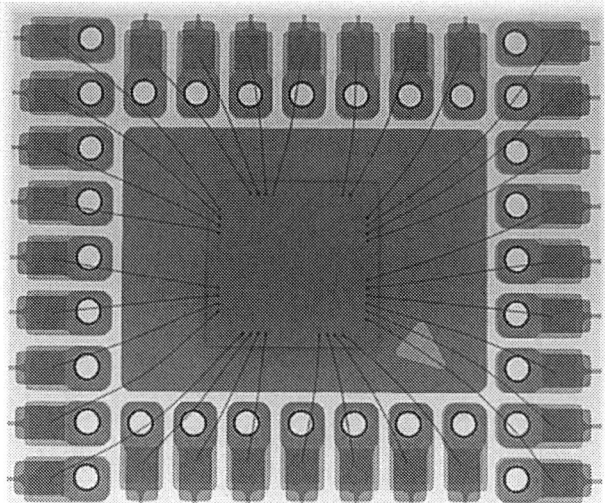

Figure 2 - X-ray image (Top View)

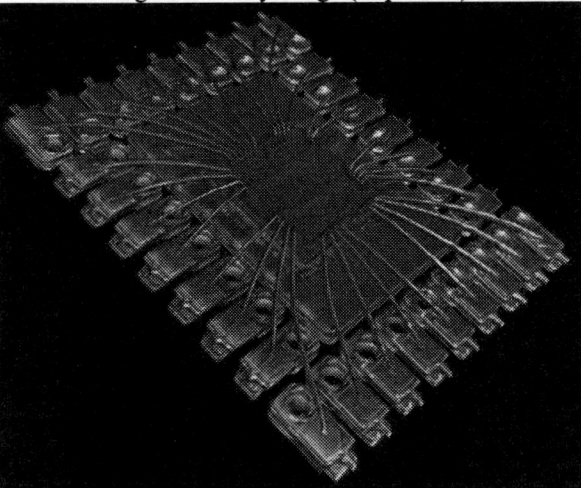

Figure 3– 3D µCT reconstruction of package

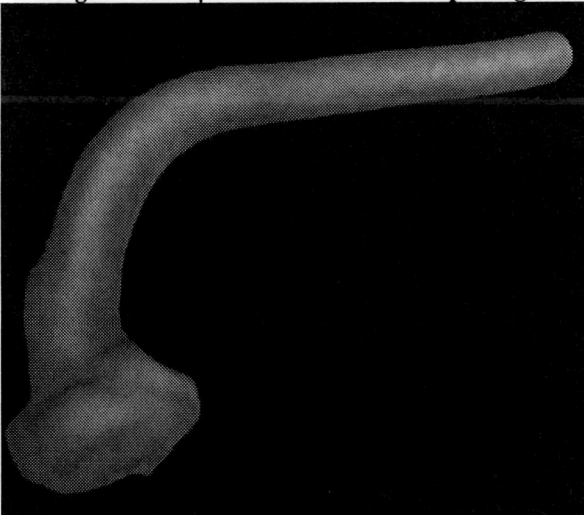

Figure 4– Close-up of copper wirebond

Test Matrix

Test packages were subjected to 150°C, 175°C, and 200°C isothermal aging environment. For aging at 150°C and 175°C, IMC growth was monitored at the interval of 1 week. For aging temperature of 200°C, reading interval was 2 days.

To study effect of temperature-humidity bias, set of packages were subjected to three different conditions, 130°C/100%RH, 120°C/100%RH, 110°C/100%RH. Reading interval for these tests was 24 hours. Packages were potted into resin, polished, and then polished surface was sputter coated with gold, at 25µA, for 45 seconds. IMC growth was observed using scanning electron microscopy. IMC thickness at multiple locations was recorded in each image, and mean value of readings was considered as final IMC thickness. This ensured accuracy in measurement. Different modes of EDS scans (e.g. line scan, point scans) were performed for material characterization.

Experimental Results and Analysis
High Temperature Thermal Aging

Figure 5 shows development of IMC layer when packages were subjected to 150°C isothermal aging. IMC growth over period of temperature can be observed. Similarly, images of IMC growth have been captured at thermal aging conditions of 175°C and 200°C as shown in Figure 6 and Figure 7.

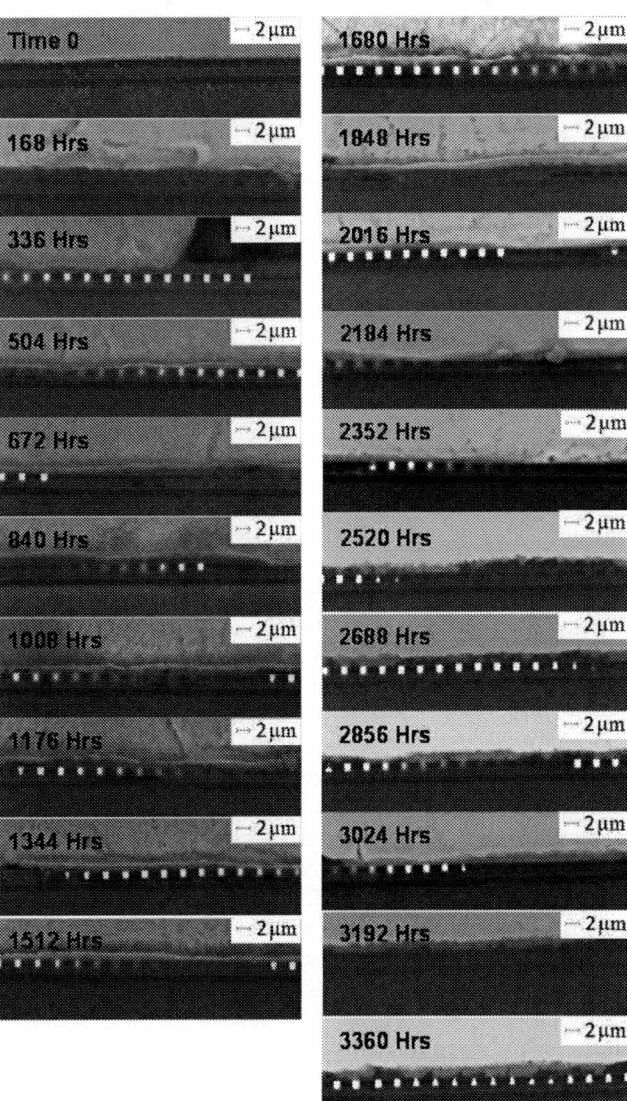

Figure 5 – IMC Development at 150°C

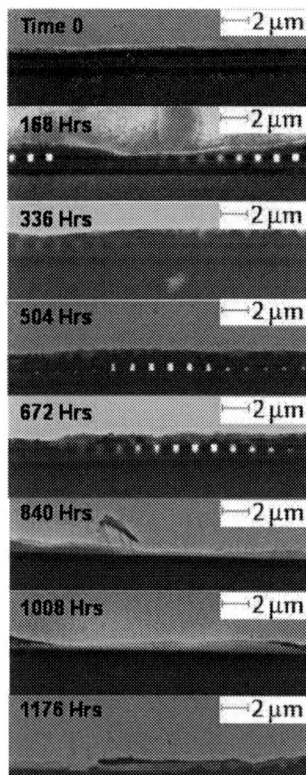

Figure 6 - IMC Development
at 175°C

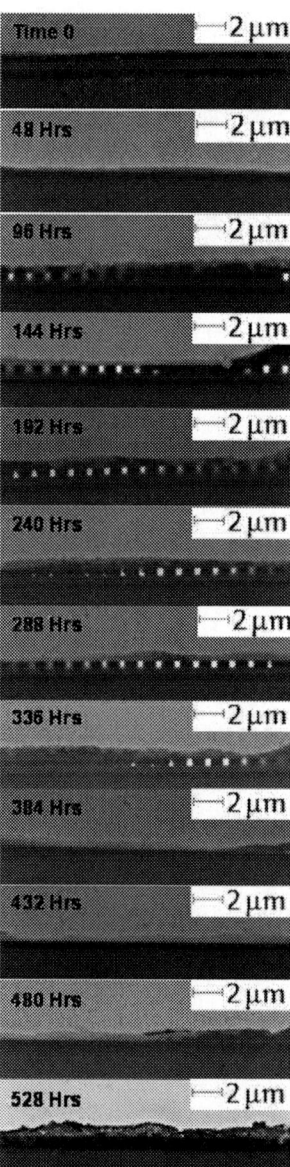

Figure 7 - IMC Development
at 200°C

Their IMC growth is plotted in Figure 13 (b) and (c) respectively. Even after aging for 20 weeks, at 150°C, IMC corrosion was not found. IMC growth is very slow. Initially only Cu-rich phase of IMC was visible. However, after aging for 840 hours, Al-rich phase was distinctly visible in the Cu-Al intermetallics. In case of thermal aging at 175°C, the Al-rich phase was detected after aging after only 336 hours of thermal aging, much sooner than the 840 hours required at 150°C. Initial cracking of wirebond was observed after 1008 hours of aging at 175°C. Initial cracks were observed at edges of ball bond, in copper rich interface, and then they propagated towards center, resulting into complete cracking in most of the wirebonds after 1176. Even after complete cracking it was observed that the IMC growth continued resulting in formation of singular IMC phase below the crack. The aluminum from the Al-pad continued to diffuse into the IMC, but diffusion of copper from ball bond got restricted due to crack. IMC growth after crack formation resulted in the development of a continuous layer of aluminum rich phase

below the crack. Similar trend was observed in third thermal aging condition at 200°C. IMC growth at 200°C was the fastest of the three thermal conditions measured in this study. Crack initiation and propagation was faster at higher thermal aging temperature with the crack initiation observed after 432 hours and near complete cracking in majority of the wirebonds after 528 hours of 200°C aging. Crack initiation and propagation is shown in Figure 8.

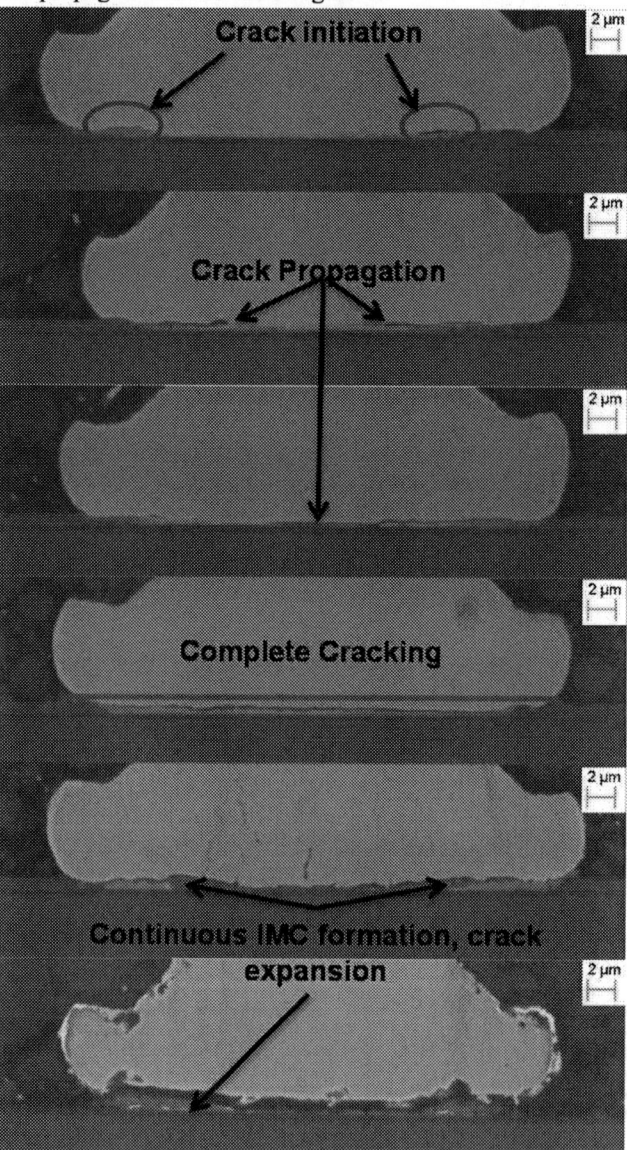

Figure 8 – Crack initiation and propagation

EDX analyses were performed on IMC phases, as well as region of cracking. Point scans and line scanning techniques were used. EDX analysis on IMC phases to determine its composition is shown in Figure 10. It was found that two phases found are Cu-rich and Al-rich phases. They preliminary consist Cu_9Al_4 (Spectrum-4; Figure 10) and $CuAl_2$ (Spectrum-3; Figure 10) respectively, which is in good agreement with previous reports [Wieczorek-Ciurowa 2005; Liak 2008; Lee 2005]. EDX spot scan also shows that cracking takes place in copper rich interface as shown in Figure 11. Location of the crack has also been confirmed using the EDX elemental line scan in Figure 12. IMC thickness at each reading interval was measured using a RMS average of the IC thickness at 20-points. The average value of the IMC thickness was plotted as shown in Figure 9.

Figure 9 – Method for measurement of the IMC growth.

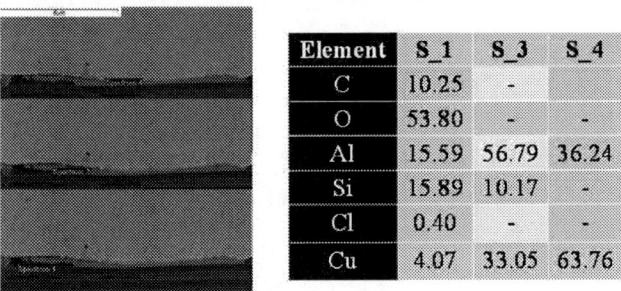

Element	S_1	S_3	S_4
C	10.25	-	
O	53.80	-	-
Al	15.59	56.79	36.24
Si	15.89	10.17	-
Cl	0.40	-	-
Cu	4.07	33.05	63.76

Figure 10 – EDX sport analysis to determine IMC Composition

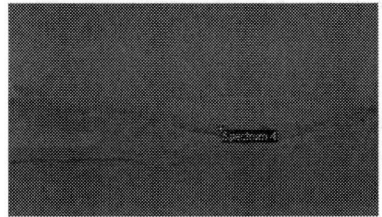

Element	Atomic%
Al	22.85
Cu	73.20
Au	3.96

Figure 11 – EDX Spot analysis of crack.

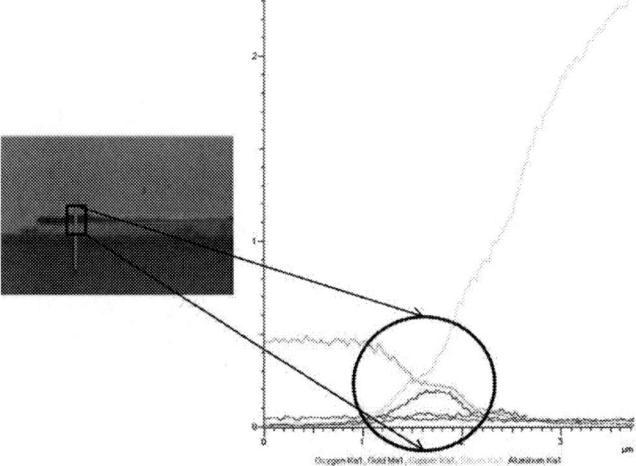

Figure 12 – EDX Line scan through crack.

In line scan image shown in Figure 12, the area of crack and IMC is marked by black box, projected as circle in scan result. Image analysis shows that Cu rich zone has undergone oxidation, and cracking. Source of this oxygen is hypothesized to be outgassing from the molding compound under thermal aging, or broken polymer chain in molding compound.

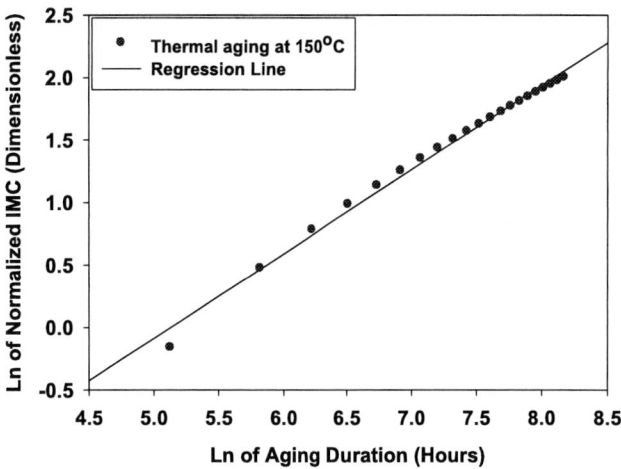

Figure 13 – IMC growth for thermal aging at 150°C

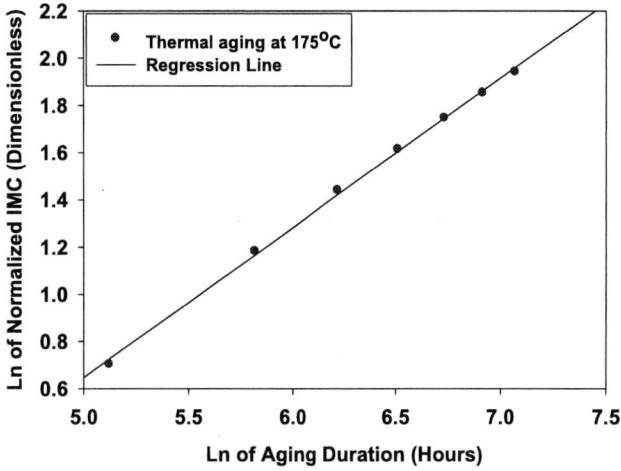

Figure 14 – IMC growth for thermal aging at 175°C

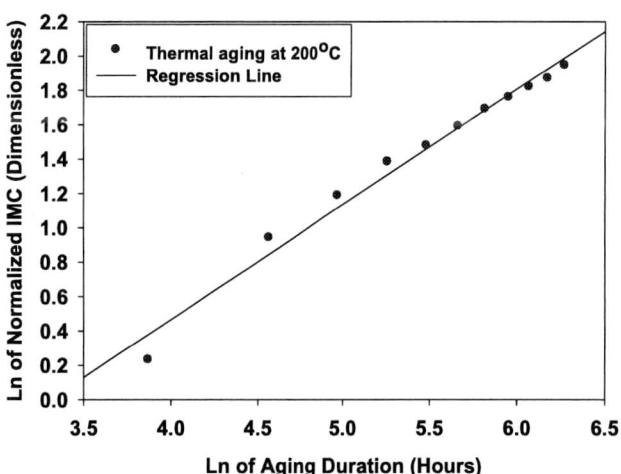

Figure 15 – IMC growth for thermal aging at 200°C

Accelerated Test for Corrosion-Susceptibility of Cu-Al

In this test, two sets of 10-packages each were subjected to three different conditions of 110°C/100%RH, 120°C/100%RH, and 130°C/100%RH. To achieve the test conditions of 110°C/100%RH, 120°C/100%RH and 130°C/100%RH, a Parr bomb test apparatus was used. Parr bomb test apparatus shown in Figure **16** is a closed pressure vessel in which 100%RH can be maintained at high ambient temperatures. PTFE cup inside the Parr bomb was filled with water, and packages were placed in cup such that they will not be immersed into water but rather only in contact with the water vapors.

Figure 16 – Parr Bomb (Courtesy of Parr Instruments Company)

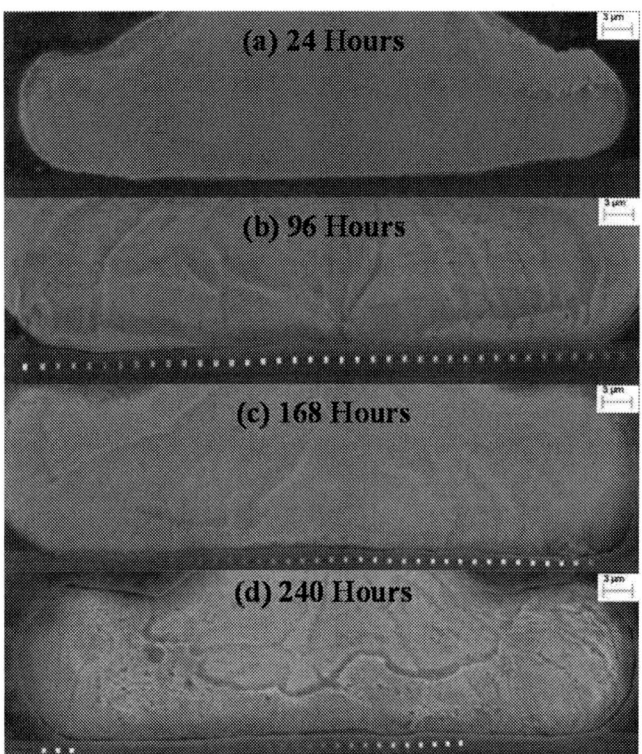

Figure 17 - Crack Development and Propagation at 110°C/100%RH

Microstructure analysis using scanning electron microscope was performed on parts subjected to the highly-accelerated stress test. EDS line scan, point scan and area scanning techniques were used. Figure 20 shows the point scan analysis at the Cu-Al wirebond interface. Analysis indicates that oxygen was present at the Cu-Al interface. Percentage atomic weight of oxygen was higher at edges, showing that initial corrosion took place at edges, and then it propagated towards center. For all three cases, oxidation was predominantly found in the Cu-rich phase.

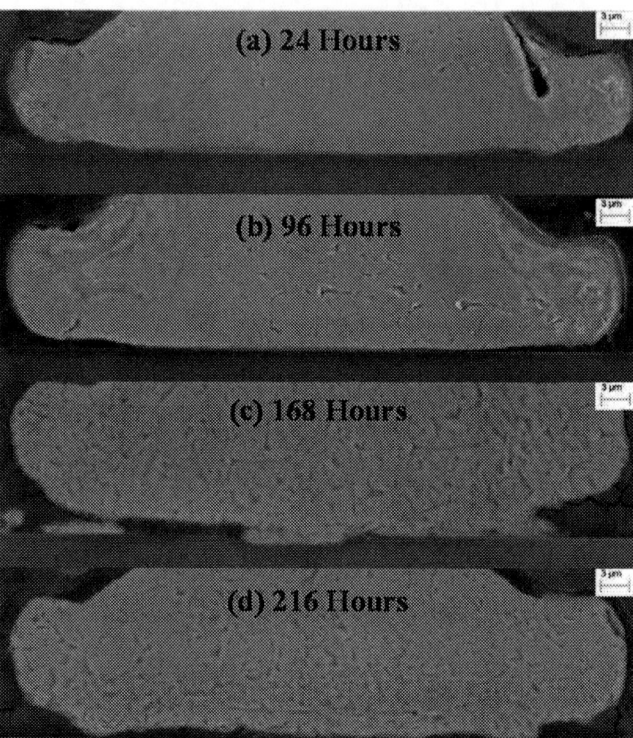

Figure 18 – Crack Development and Propagation at 120°C/100%RH

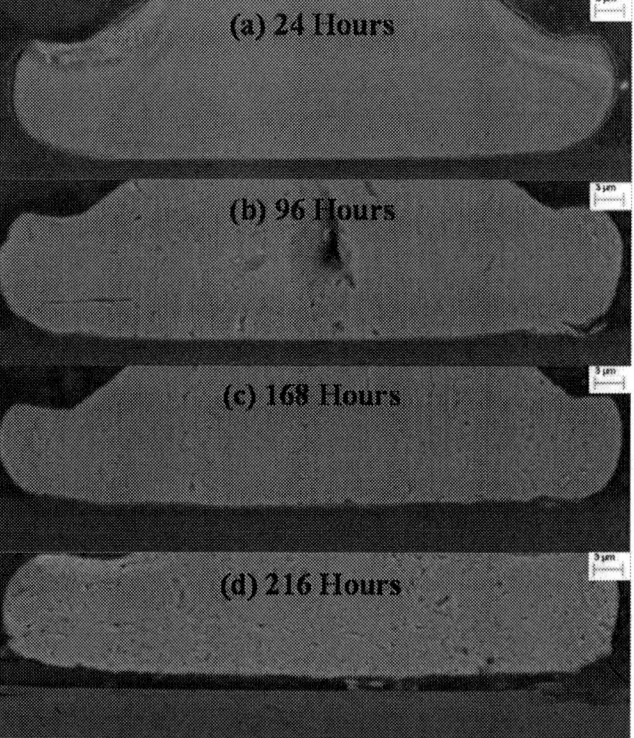

Figure 19 - Crack Development and Propagation at 130°C/100%RH

Figure 21 shows EDS line scan from Cu ball bond through Al pad to the Si-Chip. Analysis results indicate that cracking of wirebond took place in Cu-rich zone, and the Al-rich zone was still intact. In order to ascertain that the cracking is due to corrosion; but a visual polishing artifact, area mapping was done. Figure 22 shows area mapping performed to find oxygen content of selected area. Intensity of red dots indicates concentration of oxygen at the location of the red dot. Figure 22 shows that dark red spots were spotted in the region of cracking, and area near it. Density of the oxygen rich region indicated by the red-dots shows that the cracks shown are due to oxidation.

| (a) | (b) | (c) |

Figure 20 – EDS spot analysis (a) Spectrum 1, (b) Spectrum 2, (c) Spectrum 3

Table 2 –Percentage atomic Weight of Elements.

Elements	Spectrum 1	Spectrum 2	Spectrum 3
O	33.36	17.37	21.56
Al	3.08	2.74	2.63
Si	13.00	5.25	7.80
Cl	0.19	0.03	0.03
Cu	50.37	74.61	67.98

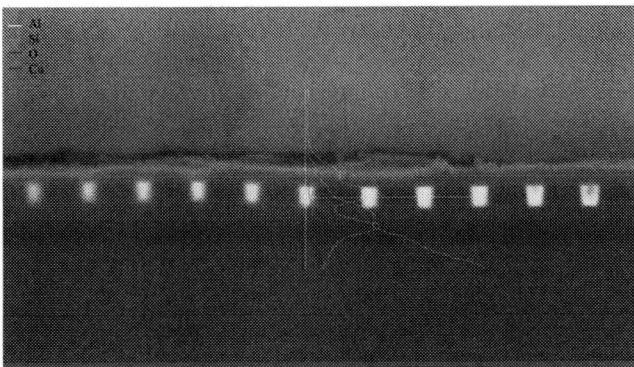

Figure 21 – EDS Line Scan of crack in wirebond

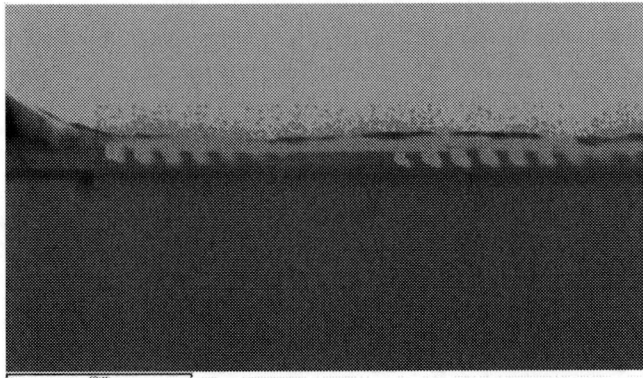

Figure 22 – Area Mapping for Oxygen in the region of cracking.

The corrosion mechanism has been analyzed using the Pourbaix Diagram or the potential/pH diagram which maps out the potential stable phases of the aqueous electrochemical system. The vertical axis is labeled for the voltage potential with respect to a standard hydrogen electrode (SHE) as calculated by the Nernst Equation. The lines in the Pourbaix diagram show the equilibrium condition where the activities for each of the species are equal on either side of the line. On either side of the line one of the ionic species is said to be predominant. The Pourbaix Diagram has been used to identify the regions of immunity, corrosion, and passivity for both copper and aluminum. The vertical lines indicate the species that are in acid-alkali equilibrium. The non-vertical lines separate the species at redox equilibrium. Specifically, the horizontal lines separate redox equilibrium species not involving hydrogen or hydroxide ions. The diagonal lines separate redox equilibrium species involving hydrogen or hydroxide ions. The dashed lines enclose the practical region of stability of the aqueous solvent to oxidation or reduction and thus the region of interest in aqueous systems. Outside the dashed region, water breaks down and not the metal.

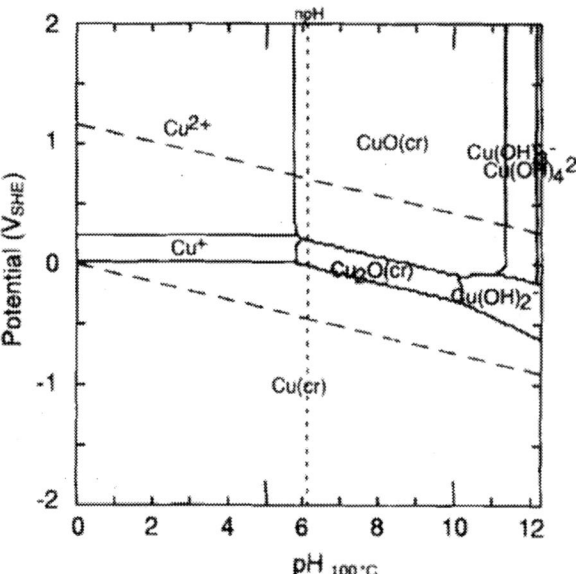

Figure 23 – Pourbaix Diagram of Copper at 100°C [Beverskog 1997]

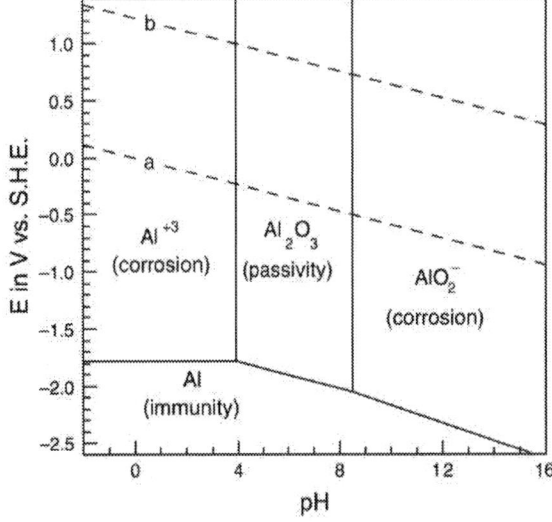

Figure 24 - Pourbaix Diagram of Aluminum [McCafferty 2010, Pg. 96]

In general, the metal is not attacked and forms stable unreacted metal species in the region of immunity, metal forms a stable oxide or stable hydroxide in the region of passivity, and metal is susceptible to corrosion in the region labeled as corrosion. Low E (or pE) values represent a reducing environment. High E values represent an oxidizing environment. Electrochemical potential represents the driving force for oxidation or reduction. The electrochemical potential is generally cited as standard reduction potential and quantifies the tendency of the chemical species to be reduced. More positive values of reduction potential indicate higher ease with which that the chemical species will be reduced. Corrosion of metal usually occurs at the anode. Standard electrochemical potential of pure Cu is +0.34V. Electrochemical potential of aluminum is -1.67V. Thus in the copper-aluminum system, copper is easier to reduce and aluminum is easier to oxidize. Exact value of electrochemical potential of IMC components, i.e. Cu_9Al_4 and $CuAl_2$ is unknown and it reported that it should be in between potential values of pure Cu and Al [Benedeitit 1995]. So theoretically based on electrochemical potential values, being most active metal in the system, Al would act as sacrificial anode, and undergo following reaction –

$$Al + 3OH^- \rightarrow Al(OH)_3 + 3e^- \qquad (1)$$

Figure 24 shows Pourbaix diagram for Al, for $1e^{-6}$ concentration, 373°K. In regions where Al^{+++} is stable, corrosion is possible. In region where aluminum oxide is stable, resistance or passivity is possible. If the pH is between 4 and 8.3, Al_2O_3 is stable and thus protects the aluminum. Aluminum hydroxide forms stable passivation layer around bare Al pad. The pH values of most of the commercial electronic molding compounds falls in this range of 4 to 8.3. Passivation layer stability window might get narrow at in the presence of ionic contaminations, such as halide ions. With the industry migration to green molding compound, which are halide ion free or having very less halide concentration the possibility of narrowing of the stability window and the possibility corrosion of Al pad are lower.

During testing of samples in the current study, corrosion of wirebond was found during highly accelerated stress test in the Parr Bomb apparatus. Two main phases found in the IMC development study were Cu_9Al_4 and $CuAl_2$. Cu_9Al_4 is on copper ball side, while $CuAl_2$ was found on Al pad side of IMC. A higher fraction of Aluminum in Al rich phase forms strong passivation layer which stops further reaction, even at low pH values [Birbilis 2008, Osenbach 2013]. On the other hand, Cu_9Al_4 which is Cu rich IMC layer has lower Al fraction – which means it will have weaker passivation layer than $CuAl_2$ and Al pad. This layer can be easily attacked by moisture and even small concentration of ionic contamination can trigger the corrosion reaction, making it corrosion prone. Detailed microstructure analysis of the aged samples revealed that IMC formed during bonding process was corroded. IMC layer formed near copper ball region (Cu rich interface) found to be corrosion prone than other IMC layer. Line scans and point scans showed that higher oxygen content i.e. corrosion followed by cracking was found in Cu-rich interface (Figure 21, Figure 22). There are different possibilities in which reaction may occur:

$$Cu_9Al_4 + Cl_2 \rightarrow 4AlCl_3 + 9Cu \qquad (2)$$

$$Cu_9Al_4 + 6H_2O \rightarrow 2Al_2O_3 + 9Cu + 6H_2 \qquad (3)$$
$$2AlCl_3 + 3H_2O \rightarrow Al_2O_3 + 6HCl \qquad (4)$$

Main byproducts of reaction are Al_2O_3 and HCl. Al_2O_3 is brittle oxide and can cause cracks in oxidized region. Hydrochloric acid further gets dissolved into moisture and breaks into chlorine ions. These chlorine ions again attack Cu_9Al_4 and continue corrosion process. The process continues and can eat away Cu-rich IMC phase, making it mechanically and electrically unstable, causing excessive stresses, and mostly results into crack formation, and propagation. The observations noted in this experiment are in good agreement with previous literature [Boettcher 2010; Breach 2010; Osenbach 2013; Liu 2011; Zeng 2013; Gan 2013]. Prior studies were limited to maximum 120 hours of aging. However, for 120°C and 130°C Parr bomb testing after 168+ hours of aging, it was found that along with Cu rich IMC, Al rich IMC and Al pad were also oxidized. This might be due to extreme conditions, and prolonged exposure to such environment, which might have caused degradation of molding compound, releasing some byproduct which can make reaction more aggressive. Initially Al will react with moisture to produce passivation layer of $Al(OH)_3$. This can be attacked by chlorine ions and broken down as follows.

$$2Al(OH)_3 + Cl^- \rightarrow 2Al(OH)_2Cl + OH^- \qquad (5)$$

$$Al + 4Cl^- \rightarrow AlCl_4^- + 3e \qquad (6)$$

$$2\,AlCl_4^- + 3H_2O \rightarrow 2Al(OH)_3 + 8Cl^- \qquad (7)$$

One of the products of reaction is chlorine ion. So once the reaction initiates, it will keep on going and Al pad will undergo pitting corrosion. Typical corrosion reaction of $CuAl_2$ can be described as following reaction [Osenbach 2013]

$$CuAl_2 + 6Cl^- \rightarrow 2AlCl_3 + Cu \qquad (8)$$

Levenberg-Marquardt Algorithm

The relationship between the IMC growth parameter and time is nonlinear. Inverse solution for interrogation of system-state is challenging for damage evolution in such systems. Levenberg-Marquardt (LM) algorithm n iterative technique that computes the minimum of a non-linear function in multidimensional variable space has been used for identifying the solution in the prognostication neighborhood [Madsen 2004, Lourakis 2005, Nielsen 1999]. Let "f" be an assumed functional relation between a measurement vector referred to as prior damage and the damage parameter vector, p, referred to as predictor variables. The measurement vector is the current values of the leading-indicator of failure and the parameter vector includes the prior system state, and accumulated damage and the damage evolution parameters. An initial parameter estimate p0 and a measured vector x are provided and it is desired to find the parameter vector p, that best satisfies the functional relation "f" i.e. minimizes the squared distance or squared-error, $\varepsilon^t\varepsilon$. The minimize parameter vector p, given by

$$F(p) = \frac{1}{2}\sum_{i=1}^{m}(g_i(p)^2)) = \frac{1}{2}g(p)^T g(p) \qquad (9)$$

$$F'(p) = J(p)^T g(p) \qquad (10)$$

$$F''(p) = J(p)^T J(p) + \sum_{i=1}^{m} g_i(x)g_i''(x) \quad (11)$$

Where $F(p)$ represents the objective function for the squared error term $\varepsilon^t\varepsilon$, $J(p)$ is the Jacobian, and $F'(p)$ is the gradient, and $F''(p)$ is the Hessian. The variation of an F-value starting at "p" and with direction "h" is expressed as a Taylor expansion, as follows:

$$F(p + \alpha h) = F(p) + \alpha h^T F'(p) + o\alpha^2 \quad (12)$$

Where α is the step-length from point "p" in the descent direction, "h". Mathematically, "h" is the descent direction of $F(p)$ if $h^T F'(p) < 0$. If no such "h" exists, then $F'(p) = 0$, showing that in this case the function is stationary. Since the condition for the stationary value of the objective function is that the gradient is zero, i.e.

$$F'(p + h) = L'(h) = 0 \quad (13)$$

The descent direction can be computed from the equation,

$$(J^T J + \mu I)h = -J^T g \quad (14)$$

The term μ is called as the damping parameter, $\mu > 0$ ensures that coefficient matrix is positive definite, and this ensures descent direction. When the value of μ is very small, then the step size for LM and Gauss-Newton are identical. Algorithm has been modified to take the equations of inter-metallic growth under isothermal aging to calculate the unknowns.

Calculation of Activation Energy for Cu-Al System
Consistent with the acquired data on the Cu-Al IMC system, a square dependence of the IMC thickness on the aging duration has been assumed to calculate the activation energy.

$$X^2 = Kt + C \quad (15)$$

Where, X is the IMC thickness (μm), t is the aging time (s), K is the reaction rate of IMC formation (μm²/s), C is the constant related to initial IMC thickness (μm²) and, the reaction rate is represented as function of temperature:

$$K = K_0 e^{\frac{-\Delta Q}{RT}} \quad (16)$$

Where, K_0 is the Multiplication Factor (μm²/s), R is the Gas constant (1.99 cal/mol K), T is the Aging Temperature (K), ΔQ is the Activation Energy (Kcal/mol). Using equation (15), data on square of IMC thickness for the temperatures of 150°C, 175°C and 200°C has been fit versus versus aging time in seconds (Figure 25). Three values of growth rate i.e. k were obtained, and plotted against 1/T as shown in Figure 26. Activation energy calculated from the plot is 12.893 Kcal/mol, i.e. 0.559eV. Table 3 and Table 4 shows comparison of energy of activation and IMC growth rate with data previously published in literature. Activation energy obtained in current experiment is in good agreement of activation energy reported in [Goh 2013; Na 2011]. In all the papers, growth rate of IMC because of temperature bias is in

the range of 10^{-6} μm²/s to 10^{-8} μm²/s, which is also in good agreement with current study (10^{-7} μm²/s).

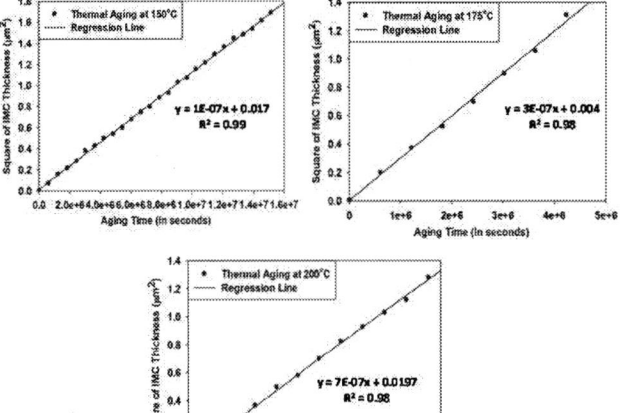

Figure 25 – Intermetallic growth rates at different temperatures.

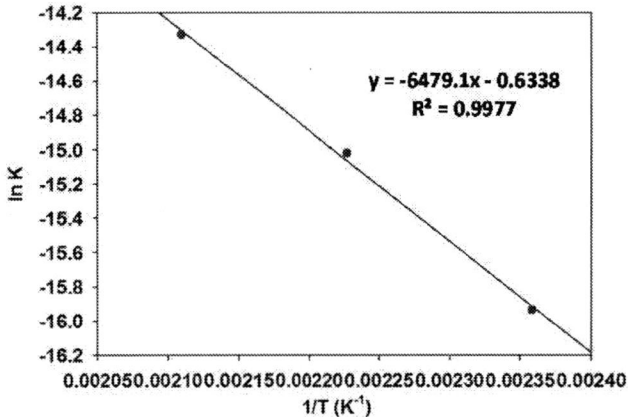

Figure 26 – ln(k) vs 1/T

Table 3 – Comparison of Activation Energies for Cu-Al

WB Type	Reference	Test Temperatures (°C)	ΔQ (Kcal/mol)
Copper	[Goh 2013]	175, 200, 225	6.1
	[Kim 2003]	150, 250, 300	26
	[Na 2011]	150, 200, 250	10.71
	Current Study	150, 175, 200	12.89

Table 4 – Comparison of IMC Growth Rate for Cu-Al System

WB Type	Reference Temperature	Temperature (°C)	Growth Rate (μm²/s)
Copper	150°C	[Kim 2003]	$1.88*10^{-8}$
		[Na 2011]	$2.15*10^{-8}$
		Current Study	$1.21*10^{-7}$
	175°C	[Goh 2013]	$3.57*10^{-7}$
		Current Study	$3.25*10^{-7}$
	200°C	[Na 2011]	$2.56*10^{-8}$
		[Goh 2013]	$6.26*10^{-7}$
		Current Study	$7.02*10^{-7}$

Prognostication Approach for HAST

In order to assess the accrued damage in the Cu-Al package, IMC growth has used as leading indicator of failure for interrogation of state of system and remaining useful life calculations. Measurements of IMC thickness growth have been fit into equation (17).

$$\frac{y_1 - y_0}{y_0} = kt_1^{\,n} \tag{17}$$

Prior damage in each case has been prognosticated based on the IMC evolution. Prognostication involves withdrawal of three samples at three periodic intervals. The samples were then cross-sectioned, potted and polished to measure intermetallic thickness. Prior damage accrued was prognosticated. IMC growth parameter was used to compute life consumed due to exposure to thermal aging:

$$\frac{y_1 - y_0}{y_0} = kt_1^{\,n} \tag{18}$$

$$\frac{y_2 - y_0}{y_0} = k(t_1 + \Delta t)^n \tag{19}$$

$$\frac{y_3 - y_0}{y_0} = k(t_1 + 2\Delta t) \tag{20}$$

Where, y_1 is IMC thickness at time t_1, Δt is the time interval at which future IMC thickness i.e. y_2, y_3. The parameter y_0 is IMC thickness before initiation of thermal aging. The solution requires three equation and three unknowns. Levenberg- Marquardt (LM) algorithm was used to solve these three nonlinear equations (18), (19), and (20) and optimization of three unknowns. Figure 27 gives an overview of the methodology used for prognostication of Cu-Al wire bond for prior accrued damage using IMC layer as leading indicator of failure. Consider a field deployed electronic package, which has been used for certain time, say t_1, which is unknown. For prognostication, samples will be taken from field at uniform-interval of time Δt. IMC layer of all packages can be measured and then using equation (18), (19) and (20) providing measurements of y_1, y_2, y_3. The LM algorithm is then used to solve for prior unknown damage.

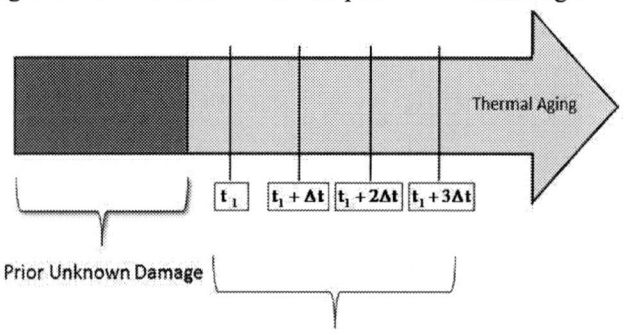

Figure 27 - Prognostication of thermally aged samples

In case of isothermal aging, in which IMC development is primarily driven by diffusion, value of "n" from equation (17)

is theoretically known to be 0.5 so, equation has been updated to,

$$\frac{y_1 - y_0}{y_0} = kt_1^{\,0.5} \tag{21}$$

$$\frac{y_2 - y_0}{y_0} = k(t_1 + 168)^{0.5} \tag{22}$$

$$\frac{y_3 - y_0}{y_0} = k(t_1 + 336)^{0.5} \tag{23}$$

Equation (17) was used to plot graph of normalized IMC against square root of aging time, for all three cases, shown in Figure 28. Linear plot implies that value of n in equation (17) is 0.5. Levenberg- Marquardt (LM) algorithm was used to solve these three nonlinear equations, and prediction of remaining useful life.

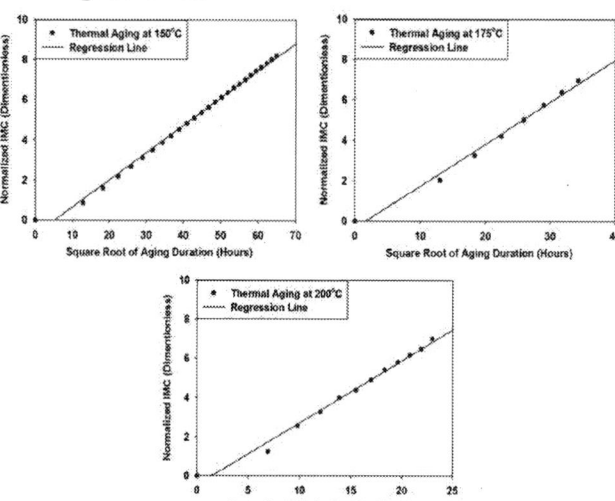

Figure 28 – Normalized IMC vs Square root of aging time

Table 5 : Damage accrual Relationship Using IMC as Leading Indicators

Aging Condition	150°C	175°C	200°C
Equation	$K_n = 0.023(t)^{0.482}$	$K_n = 0.427(t)^{0.498}$	$K_n = 0.045(t)^{0.513}$

Table 6 - Comparison of experimental and prognosticated Results; for isothermal aging at 150°C

	Experimental (Hours)	Prognosticated (Hours)	% Error
t_1	1176	1286	8.5%
t_2	1344	1298	3.5%
t_3	1512	1578	4.1%

Levenberg-Marquardt algorithm was developed based on equations (21), (22) and (23) for prediction of remaining useful life, when packages were subjected to isothermal aging conditions. Table **5** represents damage accrual relationships, where K_n is IMC thickness at any time t. The equations were derived from experimental data. Exponent of time "t" in all three cases is in the vicinity of 0.5.

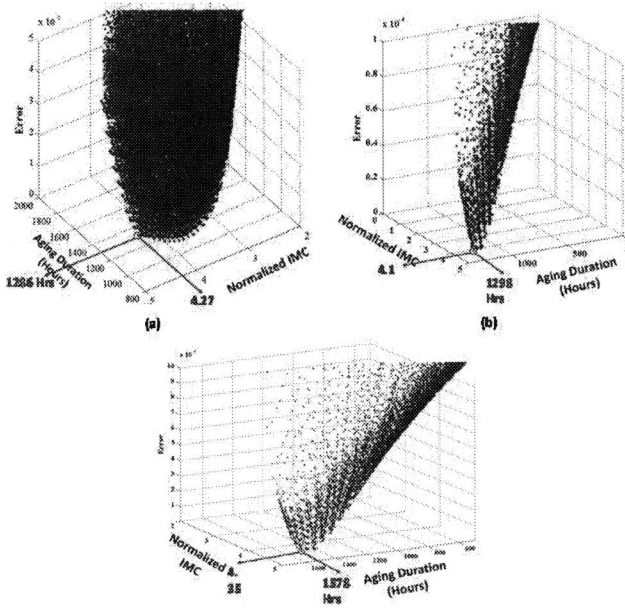

Figure 29 - 3D plot of error versus aging duration in hours versus normalized IMC thickness, for isothermal aging at 150°C, for time duration (a) 1176 hours, (b) 1344 hours, (c) 1512 hours.

Table 7 - Comparison of experimental and prognosticated results; for isothermal aging at 175°C.

	Experimental (Hours)	Prognosticated (Hours)	% Error
t_1	504	539	6.5%
t_2	672	656	2.5%
t_3	840	897	6.3%

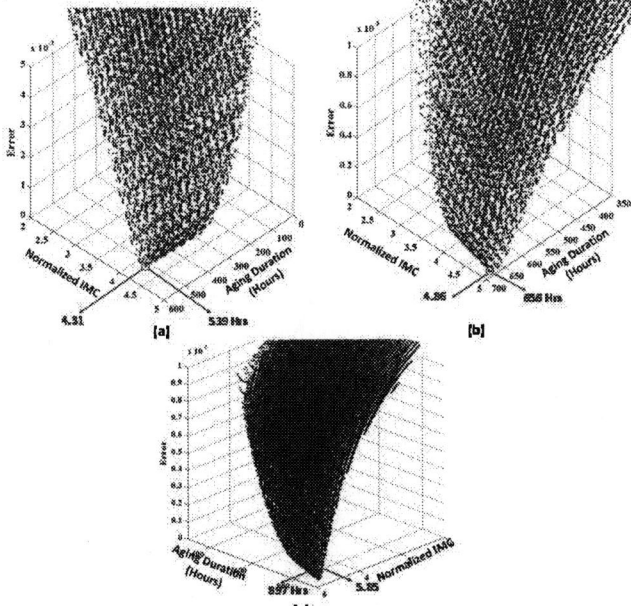

Figure 30 - 3D plot of error versus aging duration in hours versus normalized IMC thickness, for isothermal aging at 175°C, for time duration (a) 504 hours, (b) 672 hours, (c) 840 hours.

It shows that IMC growth in Cu-Al wire bond system is driven by Fickian diffusion. For thermal aging at 150°C, prognostication was done at 1176, 1344, 1512 hours of thermal aging. Results of the prognostication are shown in Table 6. 3D plots of algorithm output are shown in Figure 29. Percentage error in between experimental aging duration and prognosticated aging duration was calculated. The error bound was in the range of 3.5% to 8.5%. The solution in Figure 29 corresponds to the point at minimum error. The minimum error point represents prognosticated aging duration. Similar approach was used to prognosticate prior accrued damage when packages were subjected to thermal aging of 175°C. Results are tabulated in Table 7 and 3D plot of output are shown in Figure 30. Percentage error bound for this model was in between 2.5% to 6.5%. The approach was extended to packages subjected to thermal aging at 200°C. Results are shown in Table 8 and 3D plots are shown in Figure 31. Percentage error in models prediction was in between 2.8% to 8.7%.

Table 8 - Comparison of experimental and prognosticated Results; for isothermal aging at 200°C.

	Experimental (Hours)	Prognosticated (Hours)	% Error
t_1	288	265	8.7%
t_2	336	361	7.0%
t_3	384	395	2.8%

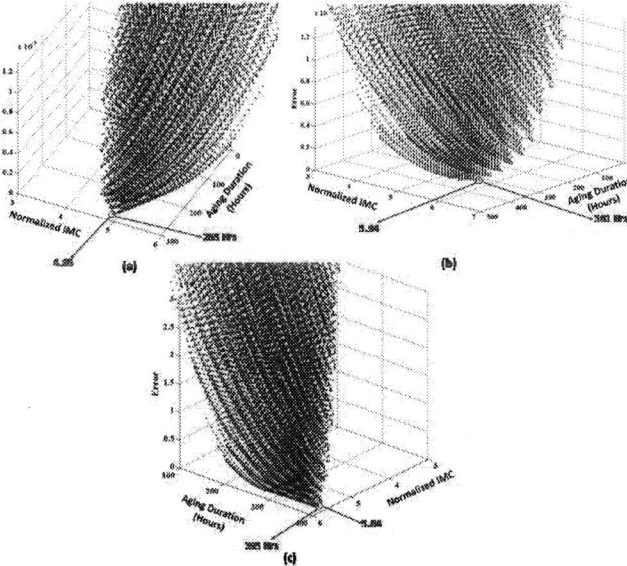

Figure 31 - 3D plot of error versus aging duration in hours versus normalized IMC thickness, for isothermal aging at 200°C, for time duration (a) 288 hours, (b) 336 hours, (c) 384 hours.

Conclusions

Microstructure evolution of Cu-Al intermetallics when subjected to thermal aging has been studied in this paper. Two distinct phases were found, namely Cu-rich interface (Cu_9Al_4) and Al rich interface ($CuAl_2$). Complete cracking of wirebond was found after prolonged thermal aging at 175°C and 200°C.

Crack initiates from edges of ball bond, and propagates towards center. Cu-rich IMCs were found to be corrosion at the Cu-Al interface. A method has been developed for prognostication of accrued prior damage and remaining useful life after exposure to thermal aging. The presented approach uses the Levenberg-Marquardt Algorithm in conjunction with development of damage based leading indicator for estimating prior accrued damage. Specific damage proxies examined is the intermetallic thickness in Cu-Al wire bond. The viability of the approach has been demonstrated 32 pin chip-scale package without any prior knowledge aging duration, subjected to thermal aging at 150°C, 175°C, and 200°C. The prognosticated values have been validated versus experimental data. Correlation between the prognosticated damage and the actual accrued damage demonstrates that the proposed approach can be used to assess prior damage accrued because of aging.

Acknowledgments

The work presented here in this paper has been supported by a research grant from the Semiconductor Research Corporation (SRC), Research ID 2284.

References

Appelt, Bernd K., Andy Tsenga, Chun-Hsiung Chenb, Yi-Shao Lai. "Fine pitch copper wire bonding in high volume production", Microelectronics Reliability 51, pp.13- 20, 2011

Benedeitit, A.V., P.T.A. Sumodjo, K. Nobe, P. L. Cabot and W. G. Proud, "Electrochemical Studies of Copper, Copper-Aluminium and Copper-Aluminium-Silver Alloys: Impedance in 0.5M NaCl", Electrochimica Acta 1995. Vol. 40, pp. 2657-2668.

Birbilis, N., and R.G. Buchheit, "Investigation and Discussion of Characteristics for Intermetallic Phases Common to Aluminum Alloys as a Function of Solution pH", Journal of The Electrochemical Society 155, C117- C126 (2008).

Boettcher, T., Michael Rother Stefan Liedtke, Mandy Ullrich, Marc Bollmann, Andreas Pinkernelle, Daniel Gruber, Hans-Juergen Funke, Michael Kaiser, Kan Lee, Martin Li, Karina Leung, Tina Li, Mark Luke Farrugia, Orla O'Halloran, Matthias Petzold, Benjamin März, Robert Klengel "On the Intermetallic Corrosion of Cu-Al wire bonds", Proceedings of 12th Electronics Packaging Technology Conference, pp 585-590 2010.

Breach, C. D., Ng Hun Shen ; Tee Wai Mun ; Teck Kheng Lee. "Effects of Moisture on Reliability of Gold and Copper Ball Bonds", Proceedings of 12th Electronics Packaging Technology Conference, pp 44-51, 2010.

Breach, C. D., What is the future of bonding wire? Will copper entirely replace gold?, Gold Bulletin, Volume 43, No 3, pp- 150-167, 2010.

Chandramouli, R., Pateras, S., "Testing Systems on a Chip", IEEE Spectrum, Vol. 33, No. 11, pp. 42-47, Nov. 1996.

Chen, J., Dominiek Degryse, Petar Ratchev, Ingrid De Wolf, "Mechanical Issues of Cu-to-Cu Wire Bonding", IEEE Transactions on Components and Packaging Technologies, Vol. 27, No. 3, pp.539-545, September 2004.

Chen, Jiunn, Yi-Shao Lai, Yi-Wun Wang, C.R. Kao. "Investigation of growth behavior of Al–Cu intermetallic

compounds in Cu wire bonding", Journal of Microelectronics Reliability, Vol 51, pp 125–129, 2011.

Chong L.G.., Francis Classe, Bak Lee chan, Hasjhim Uda, "Extended Reliability of gold and copper ball bonds in microelectronic packaging", gold Bulletin, June 2013, Volume 46, Issue 2, pp 103-115.

Drees, R., and Young, N., Role of BIT in Support System Maintenance and Availability, IEEE A&E Systems Magazine, pp. 3-7, August 2004.

England, L. Siew Tze Eng,, Chris Liew, Hock Heng Lim. "Cu wire bond parameter optimization on various bond pad metallization, and barrier layer material schemes", Microelectronics Reliability, Volume 51, No.1, pp.81-87, 2011

Facility for the Analysis of Chemical Thermodynamics. http://www.crct.polymtl.ca/factweb.php

Gao, R. X., Suryavanshi, A., BIT for Intelligent System Design and Condition Monitoring, IEEE Transactions on Instrumentation and Measurement, Vol. 51, Issue: 5, pp.1061-1067, October 2002

Goh, Chwee Sim, Wee Ling Eddy Chong, Teck Kheng Lee and Christopher Breach. "Corrosion Study and Intermetallics Formation in Gold and Copper Wire Bonding in Microelectronics Packaging", Crystals Journal, Vol 3, pp 391-404, 2013.

Hang, C.J., C.Q. Wang, M. Mayer, Y.H. Tian, Y. Zhou, H.H. Wang "Growth behavior of Cu/Al intermetallic compounds and cracks in copper ball bonds during isothermal aging", Journal of Microelectronics Reliability, Vol 48, pp 416–424 2008.

Hassan, A., Agarwal, V. K., Nadeau-Dostie, B., Rajski, J., BST of PCB Interconnects Using Boundary- Scan Architecture, IEEE Transactions on Computer-Aided Design, Vol. 11, No. 10, pp. 1278-1288, October 1992

Kim, Hyoung-Joon, Joo Yeon Lee, Kyung-Wook Paik, Kwang-Won Koh, Jinhee Won, Sihyun Choe, Jin Lee, Jung-Tak Moon, and Yong-Jin Park "Effects of Cu/Al Intermetallic Compound (IMC) on Copper Wire and Aluminum Pad Bondability", IEEE Transactions On Components And Packaging Technologies, Vol. 26, No. 2, June 2003

Laik, A., K. Bhanumurthy, G.B Kale. "Diffusion in Cu(Al) Solid Solution", Defect and Diffusion Forum Vol. 279, pp 63-69, 2008.

Lall, P., Bhat, C., Hande, M., More, V., Vaidya, R, Pandher, R., Suhling, J., Goebel, K., Interrogation of System State for Damage Assessment in Lead-free Electronics Subjected to Thermo-Mechanical Loads, Proceedings of 58th ECTC conference pp. 918-929, 2008.

Lall, P., Hande, M., Bhat, C., More, V., Vaidya, R., Suhling, J., Algorithms for Prognostication of Prior Damage and Residual Life in Lead-Free Electronics Subjected to Thermo-Mechanical Loads, Proceedings of the 10th ITherm, pp. 638-651, May 28-31, 2008.

Lall, P., M. Hande, C. Bhat, J. Suhling, Jay Lee, Prognostics Health Monitoring (PHM) for Prior-Damage Assessment in Electronics Equipment under Thermo-Mechanical Loads,Proceedings of 57th ECTC conference, pp. 1097-1111, 2007c.

Lall, P., Mahendra Harsha, Jeff Suhling, Kai Goebel. "Sustained damage and remaining useful life assessment in lead-free electronics subjected to sequential multiple thermal environments", electronics components and technology conference 2012, pp. 1-14, 2012a

Lall, P., Mahendra Harsha, Kai Goebel, "Level of Damage and Remaining Useful Life Assessment in Lead-free Electronics Subjected to Multiple Thermo-mechanical Environments", IEEE International Conference on Prognostics and Health Management, pp.1-14, 2012b

Lall, P., Mahendra Harsha, Kai Goebel, Jim Jones. "Interrogation of Thermo-Mechanical Damage in Field-Deployed Electronics", IEEE International Conference on Thermal, Mechanical and Multi-Physics Simulation and Experiments in Microelectronics and Microsystems, pp.1-16, 2012c

Lall. P., More, V., Vaidya, R., Goebel, K., "Assessment of Residual Damage in Lead-free Electronics Subjected to Multiple Thermal Environments of Thermal Aging and Thermal Cycling", Proceedings of 60th ECTC, pp. 206-218,June 2-5, 2010b.

Lall. P., More, V., Vaidya, R., Goebel, K., "Prognostication of Latent Damage and Residual Life in Lead-free Electronics Subjected to Multiple Thermal-Environments", 59th ECTC,pp. 1381-1392, 2009c

Lee, Won-Bae, Kuek-Saeng Bang, Seung-Boo Jung. "Effects of intermetallic compound on the electrical and mechanical properties of friction welded Cu/Al bimetallic joints during annealing", Journal of Alloys and Compounds, Vol 390, pp 212–219 2005.

Levenson, L. L., "Grain boundary diffusion activation energy derived from surface roughness measurements of aluminum thin films", Applied Physics Letters, Vol 55, pp 2617-2619, 1989.

Liu, Hai, Zhenqing Zhao, Qiang Chen, Jianwei Zhou, Maohua Du, Senyun Kim, Jonghyun Chae, Myungkee Chung "Reliability of Copper Wire Bonding in Humidity Environment", Proceedings of 13th Electronics Packaging Technology Conference, pp 53-58, 2011.

Marko, K.A., J.V. James, T.M. Feldkamp, C.V. Puskorius, J.A. Feldkamp, and D. Roller, Applications of Neural Networks to the Construction of "Virtual" Sensors and Model-Based Diagnostics, Proceedings of ISATA 29th International Symposium on Automotive Technology and Automation, pp.133-138, June 3-6, 1996.

McCann, R. S., L. Spirkovska, "Human Factors of Integrated Systems Health Management on Next-Generation Spacecraft, First International Forum on Integrated System Health Engineering and Management in Aerospace", Napa, CA, pp. 1-18, November 7-10, 2005.

Na, Seok Ho, TaeKyeong Hwang, JungSoo Park, JinYoung Kim, HeeYeoul Yoo and ChoonHeung Lee "Characterization of Intermetallic Compound (IMC) growth in Cu wire ball bonding on Al pad metallization", Proceedings of 2011 IEEE Electronic Components and Technology Conference, pp 1740-1745, 2011.

Osenbach, John, Wang, B.Q. ; Emerich, S. ; DeLucca, J. "Corrosion of the Cu/Al Interface in Cu-Wire-Bonded Integrated Circuits", Proceedings of IEE Electronic Components & Technology Conference, pp 1574-1586, 2013.

Qin, I., A., Shah, C. Huynh M. Meyer, M. Mayer, Y. Zhou. "Role of process parameters on bond ability and pad damage indicators in copper ball bonding", Microelectronics Reliability 51, pp.60-66, 2011.

Rosenthal, Daniel, Brian C. Wadell, "Predicting and Eliminating Built-In Test False Alarms", IEEE Transactions On Reliability, Vol. 39, No. 4, pp 500-505, 1990.

Schauz, J. R., Wavelet Neural Networks for EEG Modeling and Classification, PhD Thesis, Georgia Institute of Technology, 1996.

Schmitz, S., M. Schneider-Ramelow, S. Schröder, "Influence of bonding process parameters on chip cratering and phase formation of Cu ball bonds on AlSiCu during storage at 200C", Microelectronics Reliability 51, pp.107-112, 2011

Shah A, Mayer M, Zhou Y, Hong SJ, Moon JT. "Low-stress thermo sonic copper ball bonding", IEEE Trans Electronics Packaging Manufacturing (32) 2009.

Shiroishi, J., Y. Li, S. Liang, T. Kurfess, and S. Danyluk, "Bearing Condition Diagnostics via Vibration and Acoustic Emission Measurements", Mechanical Systems and Signal Processing, Vol.11, No.5, pp.693-705, Sept. 1997.

Simonovic, Darko, Marcel H. F. Sluiter, "Impurity diffusion activation energies in Al from first principles", Journal of The American Physical Society- Physics Review, pp 1-12, 2009.

Su, Peng, Hidetoshi Seki, Chen Ping, Shingo Itoh Louie Huang, Nicholas Liao, Bill Liu, Curtis Chen, Winnie Tai, and Andy Tseng. "Effects of Reliability Testing Methods on Microstructure and Strength At the Cu Wire-Al Pad Interface", Proceedings of Electronic Components & Technology Conference, pp 179-185, 2013.

Tian, Y.H., C.J. Hang, C.Q. Wanga, G.Q. Ouyanga, D.S.Yanga, J.P. Zhao. "Reliability and failure analysis of fine copper wire bonds encapsulated with commercial epoxy molding compound", Microelectronics Reliability, volume 51, pp 157–165, 2011.

Wieczorek-Ciurowa, K., K. Gamrat, Z. Sawlowicz "Characteristics Of $Cual_2$–Cu_9al_4/Al_2O_3 Nanocomposites Synthesized By Mechanical Treatment", Journal of Thermal Analysis and Calorimetry, Vol. 80, pp 619–623 2005.

Zeng, Yingzhi, Kewu Bai, Hongmei Jin "Thermodynamic study on the corrosion mechanism of copper wire bonding", Microelectron Reliab (2013), http://dx.doi.org/10.1016/j.microrel.2013.03.006.

Zorian, Y., Hakim Bederr, A Structured Testability Approach for Multi Chip Boards Based on BIST and Boundary Scan, IEEE Transactions on Components, Packaging, and Manufacturing Technology-Part B, Vol. 17, No. 3, pp. 283- 290, August 1994.

Low-frequency Testing of Through Silicon Vias for Defect Diagnosis in Three-dimensional Integration Circuit Stacking Technology

Yichao Xu[1, 3], Min Miao[2, 1], Runiu Fang[1], Xin Sun[1], Yunhui Zhu[1], Minggang Sun[2], Guanjiang Wang[1, 3], Yufeng Jin[1, 3*]
1. National Key Laboratory of Science and Technology on Micro/Nano Fabrication, Peking University, Beijing, 100871, China.
2. Information Microsystem Institute, Beijing Information Science & Technology University, Beijing, 100101, China.
3. Peking University Shenzhen Graduate School, Shenzhen, 518055, China.
*Email: yfjin@pku.edu.cn , Phone: 86-10-62752591

Abstract

In order to qualify through silicon via (TSV) structures during manufacturing effectively and efficiently, two low-frequency testing methods were proposed here for electrical property and defect diagnosis of TSV samples. The first method (Method I) based on four-point probe test was adopted to measure via resistance and contact resistance, while the second method (Method II) based on two-point probe test was used to verify insulation integrity of TSVs. We adopted Simulated Method of Moments to illustrate the testing principles, carried out self-designed tests, and prepared samples on an industrial TSV packaging line as experimental validation. The effectiveness of the tests methods on four common types of defects were demonstrated by the simulation and test data analysis. The test methods proposed in this paper introduce a simple and low-cost solution for improving production efficiency in 3D integration circuit stacking technology, and they are being integrated into practical techniques for the industrial TSV packaging line.

Introduction

Through silicon via (TSV) is one of the vital supporting technologies for three-dimensional integration circuit (3D IC) and system-in-package (3D SIP). It was used especially for shortening electrical and thermal interconnections between multiple stacked up chips, offering high performance, low power consumption and low packaging profiles. Despite the aforementioned advantages brought by TSVs, fabrication of TSVs are sometimes not ideal so that there may exist conduction defects, such as defects inside vias, between TSV and the re-distribution layer (RDL), and within isolation layers. The yield losses experienced during the fabrication process especially limits the commercialization of 3D stacked IC or SIP products. Thus developing highly reliable TSVs is critical for enabling practical 3D IC manufacturing and commercialization.

Up to date, various measurement techniques focused on characterization and qualification of the TSVs have been proposed, such as Radio-frequency (RF) measurements [1], inductance and capacitance extraction from TSV chains [2], time dependent dielectric breakdown (TDDB) methodology [3], and Fourier transform infrared spectroscopy (FTIR) analysis [4]. However, most of these measurement methods cannot function effectively during the production process and also not easily distinguish different defects types.

The low-frequency testing proposed for TSV defects diagnosis includes using four point probe methods to measure via resistance and contact resistance (referred to as Method I below), and using a two point probe method to assess the leakage current and verify insulation integrity of TSVs (referred to as Method II below). Both of these are easy to carry out and cause little damage to TSV samples. According to data analysis and comparison, specific defect types could be diagnosed. Particularly, this methodology can be used to determine quality of TSVs in-line during manufacturing and has potential to be a significant improvement of production efficiency.

Theory and Principle

When the measured resistance value is less than a few ohms, resistance of test leads and contact resistance between the resistance-measuring probe and test point are not negligible compared to the measured value. In this situation, using a two-wire test methods would lead to test error; therefore the four point Kelvin resistor configuration as shown in Figure 1 is adopted to achieve an accurate TSV resistance measurement of the device under test (DUT) via [5].

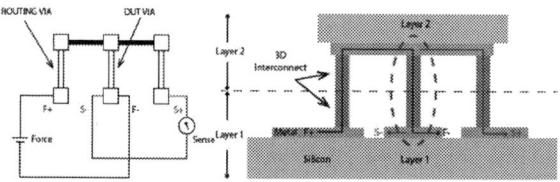

Figure 1. Schematic view of four point Kelvin resistor measurement configurations.

Also, the isolation layer would result in leakage current between TSVs when DC voltages were applied. This current could be measured by the two point probe method [6, 7].

Process Description

In this study, various TSV structures were built for testing. Figure 2 shows the layout of the sample chip, where the green ovals on the right side indicate the Kelvin structure and the orange ones on the left side mark the leakage current structure.

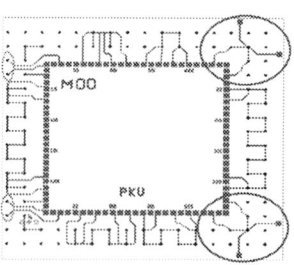

Figure 2. Layout of the sample chip for testing.

2014 Electronic Components & Technology Conference

Figure 3 shows schematic setup of the wafer test, in which the TSV is cylinder with 40μm in diameter and 100μm in height. An 1μm thick oxide layer is used as insulating material. The re-distribution layer (RDL) connecting two vias is 400μm in length, 20μm in width, and 0.4μm in thickness.

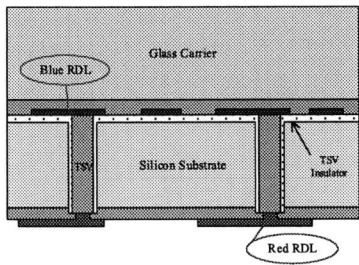

Figure 3. Schematic cross section of TSV samples.

In Method I, a constant current (I) is applied to the circuit through two probes, while the voltage (V) is measured from the other two probes. There are two setting configurations of electric current excitation and response (Method IA and Method IB).

Method IA, as shown in Figure 4, measures the summation (R_1) of via resistance (R_{via}) and contact resistance (R_c). The resistance is given by

$$R_1 = V_1/I_1 = R_{via} + R_c \qquad \text{Eq. 1}$$

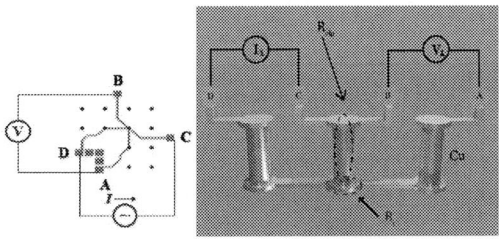

Figure 4. Schematic view of Method IA.

Figure 5 describes Method IB, where the contact resistance alone (R_2) is measured [8]. The resistance is given by

$$R_2 = V_2/I_2 = R_c \qquad \text{Eq. 2}$$

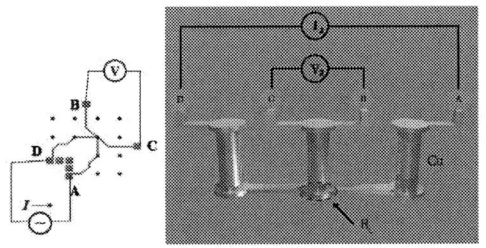

Figure 5. Schematic view of Method IB.

In addition, R_{via} can be calculated by the formula below.

$$R_{via} = R_1 - R_2 \qquad \text{Eq. 3}$$

The Method II test of leakage current is to probe upper end of two or more neighboring TSVs with excitation voltage as shown in Figure 6, then the leakage current can be read from the analyzer.

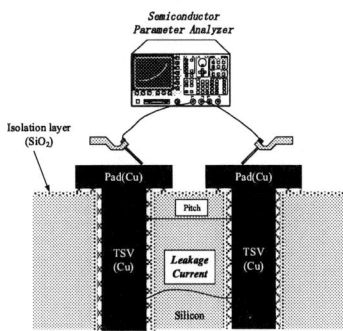

Figure 6. Schematic view of Method II measurement configuration.

Simulation for Measurement Principles

Simulated Method of Moments (MOM) is performed using a 3D electromagnetic quasi-static extraction tool (Q3D™). TSV structures are modeled the same as practical sample chips designed for testing. Since the influence of some extra parts, such as the pads and RDLs connecting pads and vias, have been offset by the test method, the model in simulation is slightly simplified, omitting these extra parts.

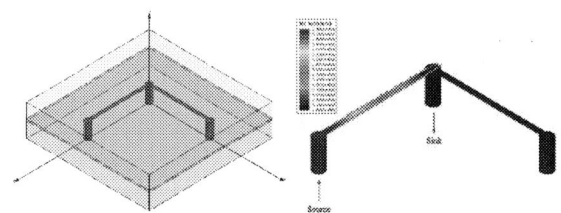

Figure 7. Diagram of test structures on Ansoft Q3D extractor and stimulation results of Method IA.

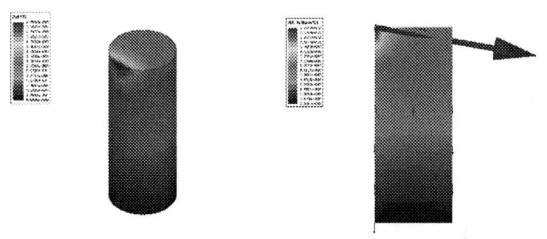

Figure 8. Detail stimulation results of the electric field and current lines of tested TSV in Method IA.

In Method IA simulation, the current line and potential difference mainly focus on TSVs and RDLs under excitation.

Figure 7 shows the model structure and simulation of Method IA on Ansoft Q3D extractor, where a constant current source is added from the left via to the middle one. It could be found out that current lines mainly flowed from source to sink. Since voltage response is set between the middle via and the right one, the voltage result turns out to be that of the middle TSV. The simulated R_1 is 0.026Ω. Details of the tested via, the middle one, could be found in Figure 8. Spiral current flows downwards to the sink side, and current line is relatively obvious around the sidewall.

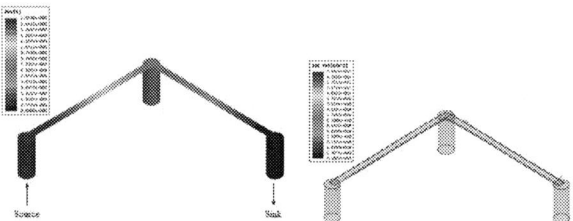

Figure 9. Diagram of Method IB and stimulation results of the current distribution.

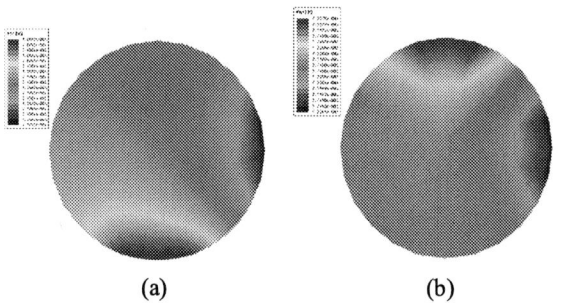

(a)   (b)

Figure 10. Up (a) and down (b) surfaces of tested TSV in Method IB.

Electric potential and current line distribution simulation results of Method IB are shown in Figure 9, where the constant current source is added from the left via to the right one. The voltage response of the middle TSV can be measured. Simulated R_2 is 0.05Ω. Figure 10 shows the cross-sectional electric potential of the middle via. Upper and lower surfaces display the similar distribution and value.

As fabricated TSVs could not be ideal for every lot or even for each wafer, structures with the defect inside via, at the connection between via and RDL, or in the RDL are modeled. The response of current line and electric potential turns out to change significantly due to the existence of these defects.

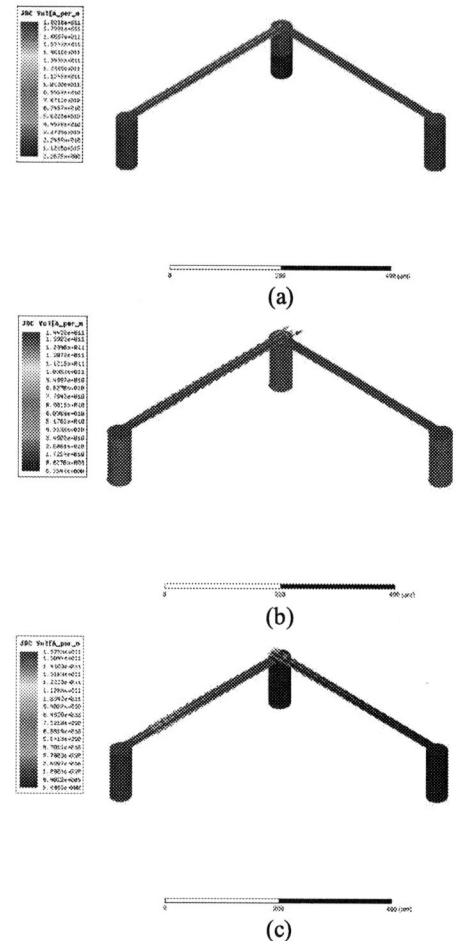

(a)

(b)

(c)

Figure 11. Stimulation result of the electric field and current distribution of tested TSVs with: (a) defect in the middle of TSV, (b) defect in the contact position, (c) defect in the RDL. (Method IA).

Figure 11 (a), (b), and (c) represent simulated results of samples with defect in the middle of TSV, in the contact position, and in the RDL. Take the open circuit situation for example, current line is interrupted in the breakage, or it just bypasses the defect. In situation (a) and (b), R_1 is simulated to be around 60Ω, and R_2 is around 1.5Ω. In situation (c), the simulated R_1 is 0.05Ω, while R_2 is more than 70Ω. According to these results, it could be predicted that R_1 and R_2 would change in different ways depending on the defects circumstances.

According to the simulation of Method II model, the existence of leakage current is proved. In order to further investigate the mechanism, TSVs with the defect on isolation layer are modeled as in Figure 12 [9], whilst the stimulating leakage current increases significantly and current lines becomes more intense around the defects than the non-defects model before.

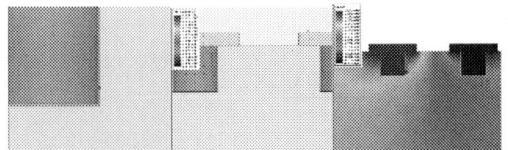

Figure 12. Diagram of test structures on Ansoft Q3D extractor and stimulation results of Method II.

To sum up, the simulation results display the electric potential and current line distribution in the ideal circumstance as well as the situation with defects in the model, which provide the basis for the experiment. All of the simulation results verify the feasibility of test methods.

Experiment Setup and Procedure

Figure 13 presents the photographs of the test apparatus including analytical probe station and the semiconductor parameter analyzer.

Figure. 13 Photographs of micromanipulator probe station designed and setup by the authors and HP-4156B semiconductor parameter analyzer.

During test verifications, a HP-4156B semiconductor parameter analyzer was used as source of electric current and voltage. Then it exported the corresponding measured voltage and current data results.

In Method I, a constant current ranging from -0.005A to 0.005A was applied to the circuit, while the corresponding voltage was measured by HP-4156B. We limited the voltage from -5V to 5V to protect these sample chips. Sampling time was set to be medium, since the output was more accurate in this case. After data analysis, V-I characteristic curve would be drawn and resistance value would be calculated.

In Method II, operating voltages ranging from 0V to 16V was used as source, and then the leakage current measurement was performed by the semiconductor parameter analyzer. After test, I-V characteristic curve would be plotted using the data obtained.

Test Results and Defect Diagnosis

Summation of via and contact resistance (R_1) as well as the contact resistance (R_2) are obtained from test results of Method IA and Method IB. The typical V-I characteristic curves of these two methods have good linearity as shown in Figure 14 (a), while R_1 is 2.63Ω and R_2 is 1.84Ω. R_{via} is 0.79Ω according to Eq. 3. The measured resistances are reasonably larger than the theoretical value. However, tested R_c is nearly twice of R_{via}, which is consistent with the simulation results. Hence V-I

curves in Figure 14 (a) could mean that via, contact position and RDL are of high integrity.

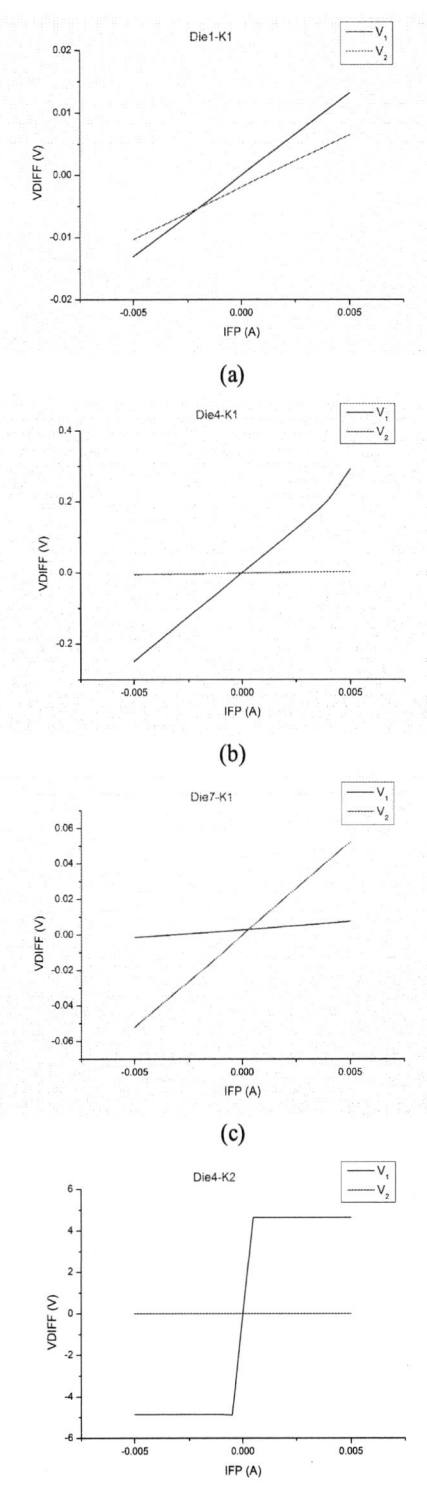

Figure 14. Representative V-I characteristic curves from Method IA and Method IB.

There are three other situations. In Figure 14 (b), V-I curves of Method IA and IB are both linear, but R_1 is much larger than R_2. This consists with the simulations where defects exit inside of the tested via or in the contact position. Figure 14 (c) also displays linear V-I curves, however, R_1 is smaller than R_2. This phenomenon shows similarity with simulation in which RDL has the defect. In Figure 14 (c), V-I curve of Method 1 exhibits a big step and reaches the limit voltage while curve of Method 2 seems like one parallel to horizontal axis, which indicates tested TSV is broken down. Such result also matches with the simulation of Q3D™ above.

Table 1 summaries these typical test results and corresponding defect types for Method I.

Table 1. Summary of representative test results from Method IA and Method IB.

Test results	Situation (a)	Situation (b)	Situation (c)	Situation (d)
R_1	2.63Ω	50Ω	1.5Ω	--
R_2	1.84Ω	1Ω	10.5Ω	0
Linearity	Linear	Linear	Linear	Step
Conclusion	Qualified	Defect in via or contact	Defect in RDL	Breakage in TSV

Besides, it is worth mentioning that there is also one special situation where curve of Method IA appears to have the characteristics similar to Schottky Barrier Diode. As shown in Figure 15, it has a voltage barrier of 0.02V. Also, TSV structure has the metal-semiconductor junction, which conforms to Schottky Diode rule. This means sometimes the TSV could have Schottky contact rather than Ohmic contact.

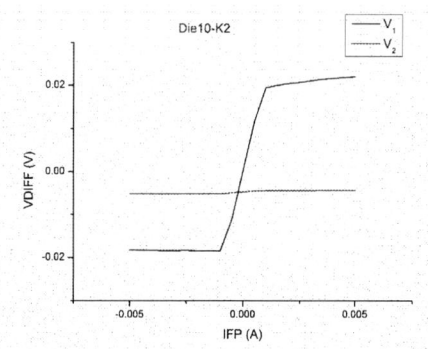

Figure 15. Curve of tested TSV with characteristic like Schottky Barrier.

For the Method II of leakage current, as shown in Figure 16, TSVs can be determined to have qualified insulation integrity when I-V curve rises linearly around 10V to 15V while the leakage current is dozens of pA. If there are abrupt rises of leakage current appearing at voltages equaling to or below 10V, it can be concluded that one or both of tested TSVs have defects in isolation layers.

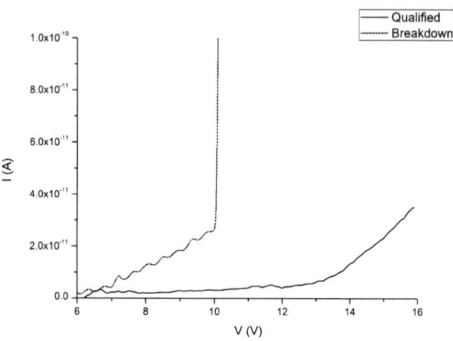

Figure 16. I-V characteristic plots of leakage current test.

Cross-sectional electron microscopy image of TSV with qualified testing curves shows good appearance in Figure 17, which consists with the conclusions above.

Figure 17. Cross section of TSV with good testing results.

The low-frequency electrical testing methodology has been shown as potential qualification techniques, for manufacturing in semiconductor and packaging industry. Here a designed test on sample wafers, which is obtained on the packaging lines of Q-Tech.Ltd (Kunshan, Jiangsu Province, China), is done for further verification. As shown in Figure 18, dry etched annulus (and partially filled) TSVs are of 40μm diameter and 100μm height. Thickness of sidewall copper is 3μm, while that of bottom is the same. Insulating layer is 1.5μm to 3.5μm silicon oxide.

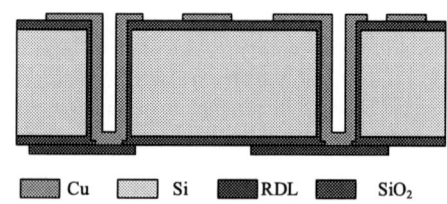

Figure 18. Schematic cross section of annulus TSVs for test.

From the test results of Kelvin Method IA and Method IB, it could be found in Figure 19 and 20 that R_1 was between 2Ω to 12.5Ω, while the R_2 was almost zero.

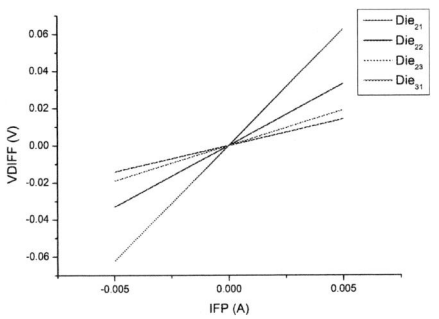

Figure 19. V-I characteristic curves of annulus TSVs from Method IA.

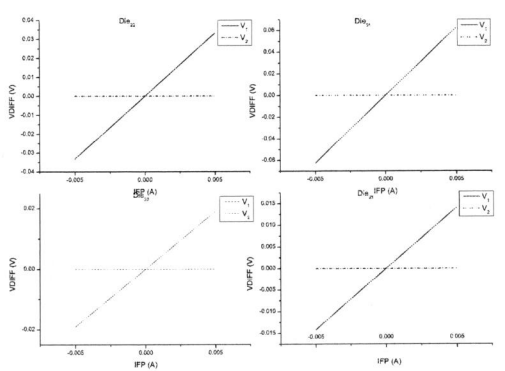

Figure 20. V-I characteristic curves of annulus TSVs from Kelvin Method IA and Method IB.

This consists with the situation where defects exit inside of the tested via. It is reasonable because annulus TSV just has a cavity inside via, which brings effect similar to the defect.

Conclusions

We proposed a low-frequency testing methodology of TSV, and also discussed defect types according to the test results. Characteristics of TSV resistance and leakage current were obtained by the four-point probe test (Kelvin Method IA and Method IB) and two-point probe test (Method II). Through data analysis, qualified TSVs could be picked up and enter the next round of production, whilst TSV with the four types of TSV failures could be detected and screened out, i.e. defects exiting inside of the tested via or in the contact position, defect in the RDL, breakage of TSV, as well as defects in isolation layers. Additionally, sometimes TSVs would seem to have a Schottky Barrier Diode characteristic, which can be seen as one defect. TSVs with defects should be fed back to the fabricators in order to enhance the production efficiency.

A verification test for samples prepared on an industrial TSV packaging line has been carried out, using a method designed by the authors. The test has proved the rationality and feasibility of these test methods. Thus, these method could lead the way to an in-line test set for reduction in overall cost of 3D IC and 3D SIP products in manufacturing and an improvement of production efficiency.

Acknowledgments

The work presented is co-funded by National Science and Technology Major Project of China (Project No. 2009ZX02038), National Natural Science Foundation of China (Project. No. 61176102 and 60976083), Funding Project for Academic Human Resources Development in Institutions of Higher Learning Under the Jurisdiction of Beijing Municipality (Project. No. PHR201108257, "Research on technical basics for three-dimensional system-in-package"). The authors wish to thank Q-Tech.Ltd (Kunshan, Jiangsu Province, China) for providing sample wafers on the packaging lines.

References

1. Yann P. R. Lamy, "RF Characterization and Analytical Modelling of Through Silicon Vias and Coplanar Waveguides for 3D Integration," *IEEE Transactions on Advanced Packaging*, vol. 33, no.4, pp. 1072–1079, Nov. 2010.
2. F. Liu, X. Gu, "Electrical Characterization of 3D Through-Silicon-Vias," in *Proc. IEEE Electronic Components and Technology Conference (ECTC)*, Las Vegas, NV, Jun. 1–4, 2010, pp. 1100–1105.
3. Y.-L. Li, "Electrical Characterization Method to Study Barrier Integrity in 3D Through-Silicon Vias," in *Proc. IEEE Electronic Components and Technol. Conf. (ECTC)*, San Diego, CA, May 29–Jun. 1, 2012, pp. 304–308.
4. Songfang Zhao, "Investigation on the properties and processability of polymeric insulation layers for through silicon via," in *Proc. IEEE Electronic Components and Technol. Conf. (ECTC)*, Las Vegas, NV, May 28–31, 2013, pp. 81–85.
5. M. Stucchi, "Test Structures for Characterization of Through-Silicon Vias," *IEEE International Conference on Microelectronic Test Structures*, Hiroshima, Japan, Mar. 22–25, 2010, pp. 130–134.
6. Jui-Feng Hung, "Electrical Testing of Blind Through-Silicon Via (TSV) for 3D IC Integration," in *Proc. IEEE Electronic Components and Technol. Conf. (ECTC)*, San Diego, CA, May 29–Jun. 1, 2012, pp. 564–570.
7. Min Miao, "An In-line Test Method for the TSV Insulation Integrity," *Journal of Test And Measurement Technology*, no. 6, pp. 1–7, Dec. 2012.
8. Tzu-Ying Kuo, "Reliability tests for a three dimensional chip stacking structure with through silicon via connections and low cost," in *Proc. IEEE Electronic Components and Technol. Conf. (ECTC)*, Lake Buena Vista, FL, May 27–30, 2008, pp. 853–858.
9. Yichao Xu, "In-Line Testing of Blind TSVs for 3D IC Integration And M/NEMS Packaging," *IEEE International Conference on Nano/Micro Engineered and Molecular Systems (NEMS)*, Suzhou, China, Apr. 7–10, 2013, pp. 233–236.

Fast Estimation of LED's Accelerated Lifetime By Online Test Method

Qi Chen, Quan Chen, Xiaobing Luo*
School of Energy and Power Engineering, Huazhong University of Science and Technology,
Wuhan, Hubei, 430074, China
*Corresponding author: Telephone: 86-13971460283, Fax number: 86-27-87557074,
Email: luoxb@hust.edu.cn

Abstract

High power light emitting diodes (HPLEDs) based semiconductor solid-state lighting has been hailed as the new energy-saving lighting source that can replace the conventional lighting sources due to its high efficiency, long life, environmental protection and compact size. So far, there are a lot of different LED products on the market, including LED bulb, LED spot light, LED landscape lamp, etc. All these LED products need to take a series of parameters testing before pushed to the market, among which the lifetime is an important one.

However, to measure LED product's lifetime by conventional process costs a lot due to long test time. Thus elevated temperature accelerated reliability test is usually used to obtain the accelerated lifetime, which is essential for projecting the product's life under rated conditions. Most of the present accelerated test methods can be identified as "offline" methods, which mean the processes need to be stopped now and then for data acquisition. This will produce problems like small sample data capacity and further, disability to reflect small changes of the product performance within short time. So it is difficult to accurately predict the LED product life since the error is significant.

In this paper, we provided an "online" test method. The accelerated temperature and driving current were set to 125°C and 350mA, respectively. An optical cable was utilized to transmit the LED-emitted light from high test temperature to room temperature so that the optical property parameters can be measured and acquired synchronously without throwing the detection equipment into the risk of overheat.

During the whole process, the LED module was lighted. The degradation details were very clear. After around 420 hours, the output power declines about nine percent. According to Arrhenius model, the projected medium life L50 is calculated to be 3912 hours with high accuracy r2=0.95, which is much better than those offline results. Therefore, in terms of lifetime projection, the operated time of accelerated life test could be shortened effectively on the guarantee of precision by this online method.

0 Introduction

High power light emitting diodes (HPLEDs) based semiconductor solid-state lighting has been hailed as the new energy-saving lighting source that can replace the conventional lighting sources due to its high efficiency, long life, environmental protection and compact size [1-3]. Among these advantages, it's worth mentioning that LED's theoretical lifetime can reach as long as more than 50,000 hours [4]. However, LED products from different companies don't share the same long life because of different materials used, various

manufacturing techniques and so on. When a new product is released, a series of product parameters will be published along, which include the lifetime one. As mentioned above, the lifetimes of LED products from different manufacturers can be different, while still many small companies are using the theoretical one which is apparently not appropriate. Thus, the real one needs to be measured or projected.

There are different ways of measuring or projecting a product's lifetime. For high power LED product, to measure its lifetime by conventional process costs a lot of time due to its long lifetime. That may make an obstacle for those small companies to do such tests rather than using the theoretical life directly. Therefore, highly accelerated reliability test is used to project the product's life [5]. Most of the present test methods can be identified as "offline" methods, which means the specimens must be taken out from the experimental environment for parameter measuring and put back afterwards. In these methods, the testing environment change will destroy the test continuity and may lead to other problems like getting the specimens scratched or stained. To decrease the impact, the space of every two measuring time points is designed to be large. That is to say, within a certain time only a relative small amount of data can be obtained using the offline process. Apart from that, the small changes of the LED product performance will not be reflected. Therefore, in accelerated reliability test, how to capture as much data as possible in a certain time without changing the testing environment becomes a big interest.

In this paper, we provided an "online" test method to deal with those problems. A huge amount of the LED light output data was acquired in an elevated temperature accelerated life test using this method. The time sampling interval was short enough to capture the output modification of the LED module. Based on the data, the LED's accelerated life was projected. The results showed that the projected medium life was 3912 hours using the data of a 424 hours span. And the coefficient of association of the fitting was up to 0.95.

1 Online Test Method

The main reason why the specimens need to be taken away from experimental environment for parameter measuring is because that the measuring devices can't endure the harsh experimental conditions such as high temperature, temperature cycle, thermal shock and so on. For HPLEDs, the mostly concerned parameter is the light output. Therefore, by transmitting the LED's emitted light from harsh conditions to a normal one, the light output parameter can be monitored and acquired online.

The principle of the online test method we provided is shown in Fig.1. The LED emitted light is transmitted outside to normal conditions through a heat-resistant optical cable, and then monitored and measured by a photometer. The light output data is stored in a computer. The coaxality of the LED module, optical cable's entrance and exit is guaranteed by using a manual fixture. The fixture is placed in a manual box, which is also designed to prevent other light sources from effecting the transmission and detection. There are two kinds of boxes used in the experiment, one of which is heat-resistant for the testing environment, the other is common for measuring environment. In the internal corners of the box, four small holes are designed to insure the environment consistency inside and outside of the box on the basis of minimizing the influence of light from outside the box.

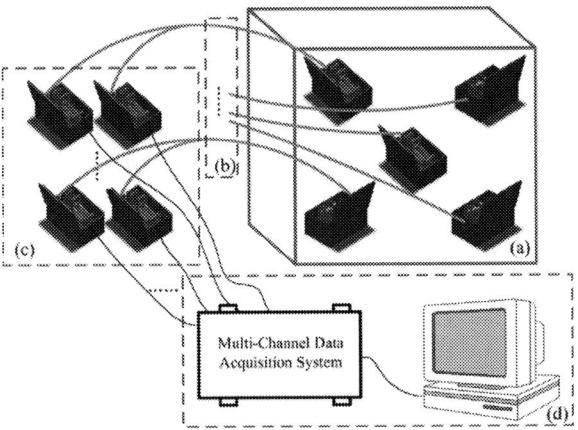

Figure 1. Configuration of experimental equipment for measuring LEDs optical degradation
(a) Light transmission environment enclosure (b) Heat-resistant armored cable
(c) Sensing environment enclosure (d) Photometric and colorimetric measurement system

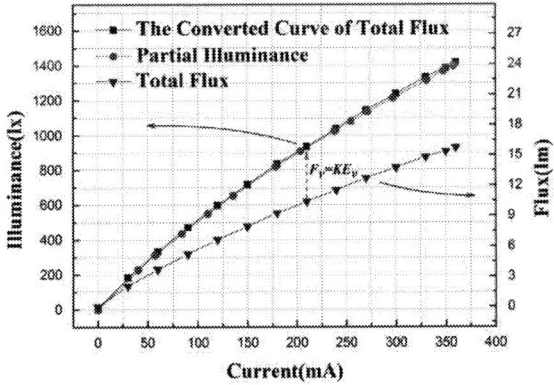

Figure 2. The relationship between measured partial illuminance and the total flux

A comparison test was conducted to contrast the data obtained by using the online test method with that obtained by using the integrating sphere under a variety of currents to verify this online method. Fig.2 reports the results. When the current is 350mA, the illuminance obtained by online test method is 1385.22lx, while the light flux obtained by integrating sphere is 15.39lm. The relationship between the two parameters can be represented as:

$$K = \frac{F_v}{E_v} = \frac{F_v \cdot S}{F_v'}$$

Where F_v is the total light flux, E_v is the measured illuminance, F_v' is the relative flux, and S is the aperture area of detector. According to the experimental data, the transfer coefficient K is a constant which value is 0.0111. As seen from the figure, there is a good linear relationship between the measured illuminance and the total flux. The maximum deviation arises under the current of 330mA, where the relative error is 1.4%, and the standard deviation is around 0.93%. This means that the real changes of the light output can be well reflected by using the online test method.

2 Experiment

The treated device was 1 W high power GaN based blue LED module. A cross-section diagram is shown in Fig.3. Unlike normal blue LEDs, the treated sample doesn't has silicone gel above the chip. In the online test method, equipments like stabilized current supply, calorstat oven, heat-resistant armored cable, manual fixture, multichannel photometer and so on were used. They were connected with each other in the way shown in Fig.1. The experimental temperature was set to 125°C, the drive current was maintained at 350mA. The PC software was set to capture experimental data every 30 seconds for 500 hours.

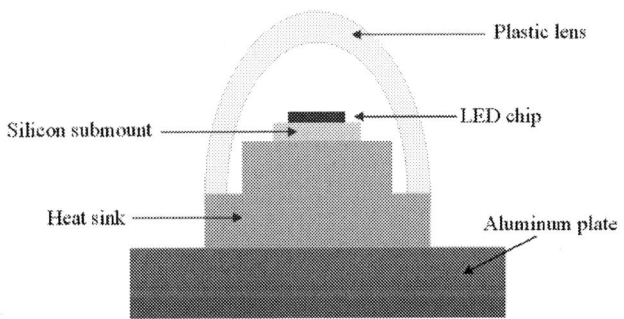

Figure 3. Schematic diagram of LED structure

3 Results and Discussion

Fig.4 plots the relative light output decay of first few hours as a function of time under the experimental temperature 125°C. The normalized light output goes up to about 103% in the short run before slowly declines down. This is supposed to be caused by the annealing influence [6]. For semiconductor lighting device like LED, the annealing effect of encapsulation can lead to a fluctuation at the start. As time moves on, the influence will decrease and disappear after the saturation.

As displayed in Fig.5, a huge amount of experimental data was obtained by using the online test method. After an aging time of more than 400 hours, the relative light output decay reached 9% and continued to go higher. The sample data error reflected as relative light output fluctuation may be attributed to the uncertainty of the photometer probe.

978-1-4799-2408-0/14 $31.00 © 2014 IEEE   1993

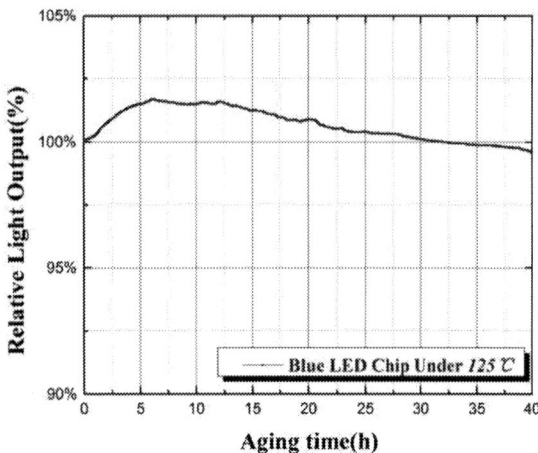

Figure 4. Relative light output decay of first few hours

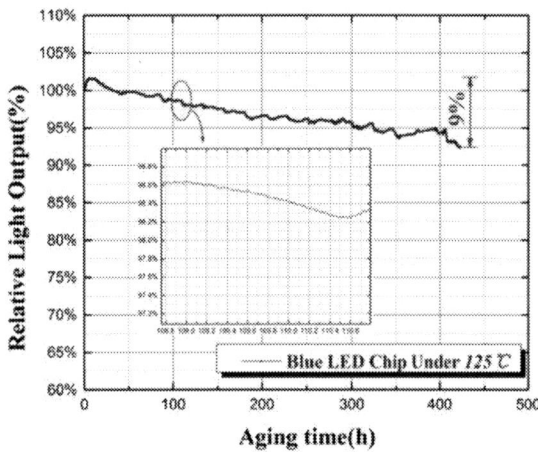

Figure 5. Light output degradation as a function of aging time under 125°C

It is known from the IES TM-21 Standard [7] that the influence of temperature on LED's life can be represented by an exponential decay model, that is

$$\Phi(t) = Be^{-at}$$

Where t is the operated time, $\Phi(t)$ is the relative light output, B is the initial constant and α is the degradation rate. As for the accelerated life projection of the LED module, due to the annealing influence, the data of first few hours must be abandoned. Thus only data from hour 12 to 424 was utilized to predict the lifetime. And for calculation, least square method was adopted to get the parameters. The parameter B is inferred to be 0.98926, α is 1.74404×10^{-4}. Fig.6 plots both the sample data and the fitting curve deduced by the exponential decay model. As seen from Fig.6, the projected medium life L_{50} is 3912 hours. And the correlation coefficient is up to 0.95354, which means that the exponential decay model is appropriate. In the reference papers [8-10], within an experimental time of about 500 hours, only a few data points about four or five are displayed. Many models such as linear model, exponential model and so on can be fitted well if there

are only a few data points. Thus, to specify the most appropriate model, more data points need to be acquired, which means more experimental time is needed. In comparison, employing the provided online test method, a huge amount of data can be acquired in a short time and high correlative coefficient can be achieved. Thus the time needed is shortened.

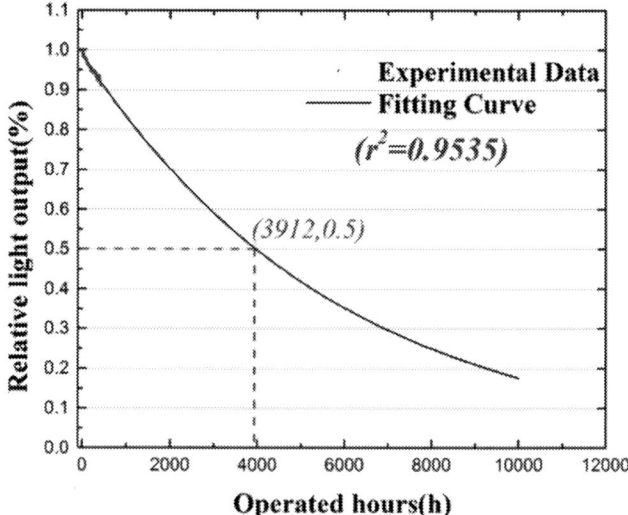

Figure 6. The schematic diagram of the accelerated life projecting

4 Conclusions

We provided an online test method to quickly project LED's lifetime under accelerated life test, which is distinguished from offline method. A contrast experiment was performed to verify the feasibility of the online method. An accelerated reliability life test was conducted using this online method. And the accelerated medium lifetime was projected by an exponential decay model with high correlative coefficient. The experimental time needed to specify the most appropriate model was shortened effectively. However, further study is necessary to better understand the relationship between the accelerated life and the normal one.

Acknowledgment

The authors would like to acknowledge the financial support in part by Natural Science Foundation of China (51376070), and in part by 973 Project of the Ministry of Science and Technology of China (2011CB013106).

References

1. Krames M R, Shchekin O B, Mueller-Mach R, et al. Status and future of high-power light-emitting diodes for solid-state lighting[J]. Display Technology, Journal of, 2007, 3(2): 160-175.
2. Steranka F M, Bhat J, Collins D, et al. High power LEDs–Technology status and market applications[J]. physica status solidi (a), 2002, 194(2): 380-388.

3. Schubert E F, Kim J K, Luo H, et al. Solid-state lighting—a benevolent technology [J]. Reports on Progress in Physics, 2006, 69(12): 3069.

4. Koh S, Van Driel W, Zhang G Q. Degradation of light emitting diodes: a proposed methodology[J]. Journal of Semiconductors, 2011, 32(1): 014004.

5. Silverman M. Summary of HALT and HASS results at an accelerated reliability test center[C]//Reliability and Maintainability Symposium, 1998. Proceedings., Annual. IEEE, 1998: 30-36.

6. Lee T W, Park O O. The effect of different heat treatments on the luminescence efficiency of polymer light-emitting diodes[J]. Advanced materials, 2000, 12(11): 801-804.

7. Illuminating Engineering Society, "Projecting Long Term Lumen Maintenance of LED Light Sources", ISBN:978-0-87995-259-4,2011.

8. Ishizaki S, Kimura H, Sugimoto M. Lifetime estimation of high power white LEDs[J]. Journal of Light & Visual Environment, 2007, 31(1): 11-18.

9. Yang S C, Lin P, Wang C P, et al. Failure and degradation mechanisms of high-power white light emitting diodes[J]. Microelectronics Reliability, 2010, 50(7): 959-964.

10. Kang J M, Kim J W, Choi J H, et al. Degradation characteristics of blue GaN-LED chip related to packages[J]. physica status solidi (c), 2010, 7(7-8): 2205-2207.

Methodology and Apparatus for Rapid Power Cycle Accumulation and In-Situ Incipient Failure Monitoring for Power Electronic Modules

Roy I. Davis and Daniel J. Sprenger
Fairchild Semiconductor Corporation
640 Avis Drive, Suite 300, Ann Arbor, MI, USA, 48108
roy.davis@fairchildsemi.com, dan.sprenger@fairchildsemi.com

Abstract

Power cycling lifetime is a topic of continued interest in the power electronic module industry. For inverter power modules consisting of IGBT and diode power devices, wear-out mechanisms usually encountered are wirebond lift and solder joint degradation. Wirebond failures typically manifest in an upward shift in $V_{ce(sat)}$ well before an open circuit condition occurs. Solder joint degradation usually manifests as an increase in die temperature that may be detected by a calculation of thermal impedance. Design and manufacturing processes have matured so that highly reliable, high quality modules are available from numerous manufacturers. However, the burden still remains on manufacturers to demonstrate the power cycling lifetime of their products. Given the range of application conditions to which power modules may be exposed, it is beneficial to customers for manufacturers to provide power cycling lifetime data in terms of number of cycles to failure for various levels of delta-T_j and various starting temperatures, T_0. It is also beneficial, during design cycles, for module designers and manufacturing engineers to obtain quantified failure data on prototype designs regarding onset of the failure mechanisms mentioned. In this paper, we describe and demonstrate a method for the rapid accumulation of delta-T_j cycles under various conditions and develop *in-situ* monitoring methods for $V_{ce(sat)}$ and V_f which are of sufficient accuracy to detect incipient failures in the module, and which do not require the cessation of power cycling to perform a readout.

In particular, a three phase inverter module consisting of six IGBTs and six anti-parallel diodes is tested in a multi-module, multi-operating point test stand. Three phase inductive loading with typical switching frequency and various amplitude sinusoidal current profiles is carried out under controlled starting temperatures for each module, while monitoring $V_{ce(sat)}$, diode V_f, and module substrate temperature. The $V_{ce(sat)}$ and V_f are sampled at instants of known phase current so as to correlate with nominal device behavior in order to determine health of the wirebond and solder interfaces. Module control, data acquisition, load determination, failure detection, and data logging are described. Simulation techniques used to target the operating conditions necessary to achieve a particular delta-T_j are shown, along with experimental data used to validate these simulations. Experimental results from the power cycling to date are shown.

Introduction

Reliability of power modules used in various applications is a key concern of manufacturers and users alike, with the well-known wirebond and die attach degradation due to thermo-mechanical stresses induced by power cycling being of clear interest [3]. Numerous authors have addressed the various modeling approaches to predict power module lifetime [1][2][5]. These techniques continue to be improved and adjusted to take into account various design features and aspects of the applications, improving the accuracy of the models. Such analytical approaches are valuable in providing guidance to designers, but also must be confirmed during design validation or prototype stages with reliability testing.

One drawback of testing is the time duration required to accumulate enough power cycles on a sufficient sample size to provide meaningful data. Depending on the application, power pulses may be various durations with different junction temperature slew rates [4][8]. Typically, one or more power modules will be connected in series and subjected to externally controlled current profiles to induce the desired delta-T_j. The device under test (DUT), be it an IGBT or MOSFET, operates in the fully on state to simplify control hardware and software. Such tests are interrupted at regular intervals to make measurements of the device characteristics to determine if degradation has occurred, necessitating a partial disassembly of the test apparatus. Some authors have reported systems that allow the test apparatus to remain undisturbed, but use alternative operating modes to make the health measurements [6][7][9][10].

This paper describes a test stand that has been developed to assess power module degradation *in-situ* while the module is being operated in essentially real-world application conditions. The DUT is a power module intended for automotive applications such as electric vehicle air conditioning, consisting of a three phase bridge of IGBTs and anti-parallel free-wheeling diodes (FRDs). Details of a similar industrial module are found in [11]. During the power cycling test, three modules drive three independent three phase inductive loads under PWM control so that all power die are exposed to conduction and switching losses comparable to the end application. Sinusoidal phase currents are generated at a low fundamental frequency, allowing the necessary delta-T_j excursion to be achieved during each cycle. $V_{ce(sat)}$ and V_f are monitored for each die at the peak of their sinewave. Embedded within these sampled data is confirmation of the achieved peak junction temperature and the integrity of the wirebond and die attach.

We propose the novel features of this approach include: (a) loading the module in a manner that is similar to field applications with an inductive load that introduces losses into all die in the power module including the diodes, as they would be in real usage, (b) achieving the rapid accumulation of delta-T_j cycles, and (c) making the assessment of degradation in real-time *in-situ* without interrupting the test.

978-1-4799-2408-0/14 $31.00 © 2014 IEEE

Figure 1 is a block diagram of the test apparatus. A water-cooled baseplate provides a common mounting surface for the power modules and the inductive load. The coolant temperature can be varied to achieve various starting temperatures, while the current amplitude and fundamental frequency can be varied to achieve the desired delta-T_j. As shown, the apparatus is configured for three power modules and loads, but can be extended to include additional modules running in parallel if necessary.

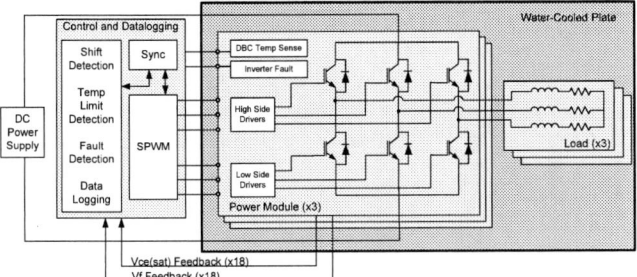

Figure 1. Test Apparatus Block Diagram

Simulation Used to Determine Cycle Conditions

The cycle conditions are determined by selecting a realistic load current which will accomplish the desired delta-T_j in each die. Non-intrusive direct measurement of junction temperature on the DUT is almost impossible with transfer molded power modules. Therefore, it is calculated via simulation and then validated in a separate test, and again estimated *in-situ* based on V-I characteristics of the device. The simulation starts with datasheet information to determine individual device losses, such as shown in Figure 2. Measurement data is then used to refine the simulation to reflect the performance of the specific power modules employed for the power cycle test. The losses excite the thermal model of the module package to calculate the junction temperature over time.

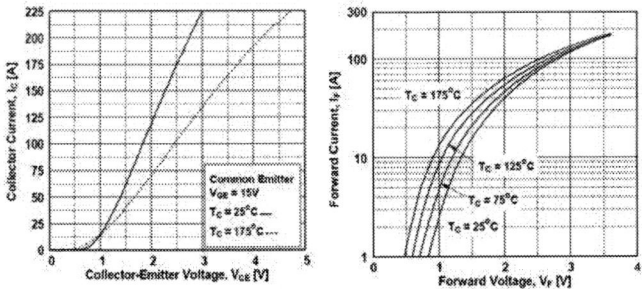

Figure 2. IGBT $V_{ce(sat)}$ and V_f vs. Current

Total loss measurements were made using a Yokogawa WT1600 digital power meter with DanFysik current transformers. In addition, switching losses were measured over several current levels for both the upper and lower IGBTs in two-pulse tests. These measurements were used to make minor adjustments to the magnitude of the datasheet curves which are used in the simulation routine. The power meter data and simulation data were then compared over several current levels and case temperature values. Figure 3 shows a comparison at a 95°C case temperature and a 5 kHz switching frequency.

The loss characteristics shown in Figure 3 reflect a range of application operating conditions over which the module will operate. These data were collected during application testing of the module and have been used to validate the accuracy of the simulation projections.

The simulation thermal model is generated using a finite element analysis with FloTherm. The analysis produces a thermal impedance curve for the IGBT and diode as shown in Figure 4.

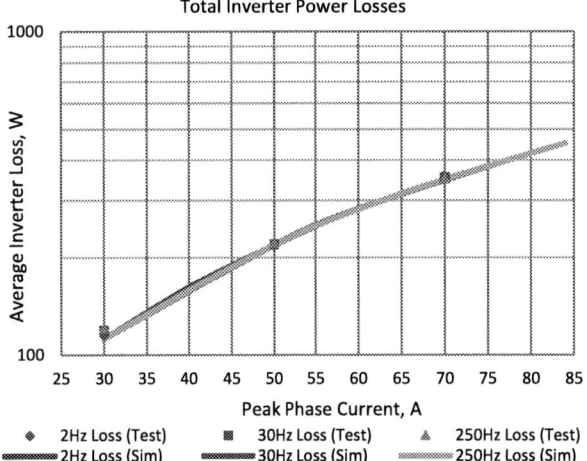

Figure 3. Simulation vs. Test Loss Measurements

Individual die exhibit slightly different junction-case thermal impedances, varying by perhaps 10%, depending on the layout of the module. The curves of Figure 4 are those of the die with the highest thermal impedance. This model was verified by decapsulating a module and installing a thermocouple onto the worst-case IGBT and diode die. The module was employed to measure the $R_{\theta j-c}$ of the IGBT at 1.026 °C/W, which is also plotted in Figure 4. As shown, the measured value correlates nicely with the model.

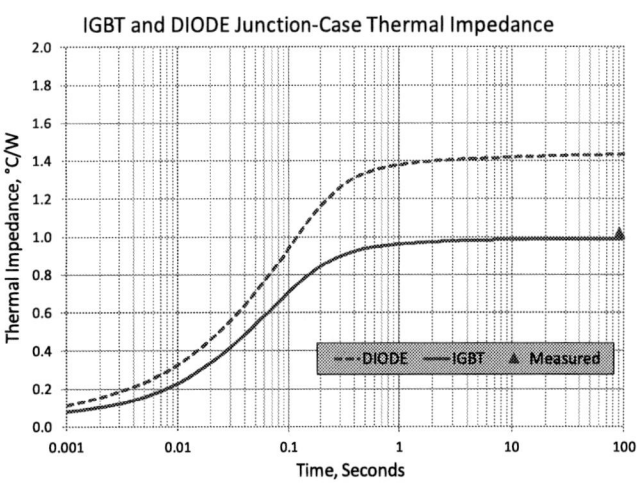

Figure 4. Device Thermal Response Curves

With the loss and thermal impedance models validated, simulations were carried out to determine operating conditions that would introduce the desired delta-T_j from a desired starting T_0. Table I shows a range of achieved values, and demonstrates clearly that the typically targeted delta-T_j of 100°C to 125°C is easily achievable for this module.

Freq, Hz	Ic, A_{rms}	T_0, C	T_f, C	delta-T_j, C	Peak Loss, W	1M Cycles, Hrs
0.5	28	60	142	82	98	555.5
0.5	42	60	222	162	195	
2.0	28	63	123	60	98	138.9
2.0	42	67	182	115	195	
3.0	28	66	114	48	98	92.6
3.0	42	72	168	96	195	

Table I. Range of Simulation Results

In Table I it should be noted that the initial temperature T_0 varies due to the rise of the heatsink temperature, which was not controlled.

In order to exercise the test apparatus to demonstrate the desired degradation in the wirebonds or die attach layer as quickly as possible, delta-T_j of 125°C was targeted first. After some experimentation, it was found that with three modules and their inductive loads mounted on the heatsink, 30-35°C was the minimum T_0 that could be maintained with the available recirculating chiller. Hence the simulation was run to determine the necessary current and fundamental frequency to achieve delta-T_j on the IGBTs of approximately 115°C. Figure 5 shows the resulting 65.5A_{pk} / 46.3A_{rms} collector current needed.

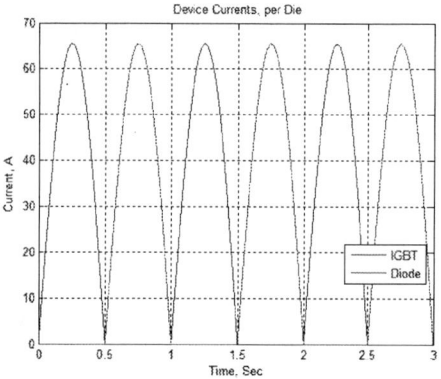

Figure 5. Simulated Device Currents for Target delta-T_j

Figures 6 and 7 show the losses and junction temperatures for, respectively, the IGBTs and diodes that are introduced into the module with the current of Figure 5. These include the conduction, switching, and reverse recovery losses of the devices while operating at 2.5 kHz switching frequency, and 1.0 Hz fundamental, with an inductive load with power factor = 0.999.

In-Situ Data Acquisition and Test Apparatus Description

An accurate method of measuring on-state voltage is essential for determining device health for this test methodology. This is more difficult with high-voltage devices. Most accurate voltage sensors, such as the FPGA analog I/O module used in this test, can only handle low voltage. Thus, the high off-state voltage must be isolated

from the measurement equipment. For this test it was desired to measure *in-situ*, which necessitates a method to take an isolated voltage measurement accurately.

Linear analog opto-couplers enable accurate measurement, while galvanically isolating the measurement from the device voltage. The opto-coupler must have both high steady-state accuracy and high bandwidth for measuring within the device on-time. The opto-coupler used for this test has ±0.5% accuracy with unity gain and a 100kHz nominal bandwidth.

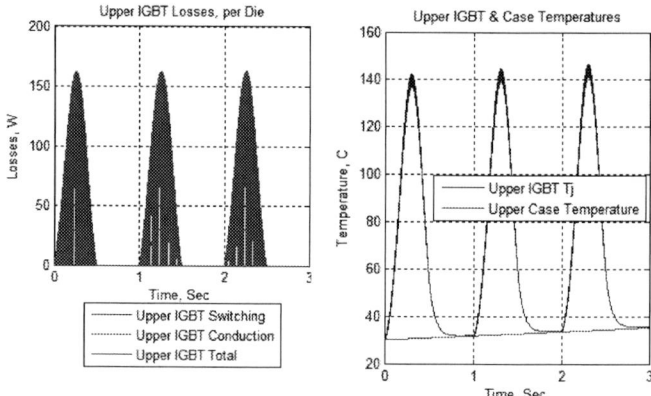

Figure 6. Simulated IGBT Losses and Junction Temperature

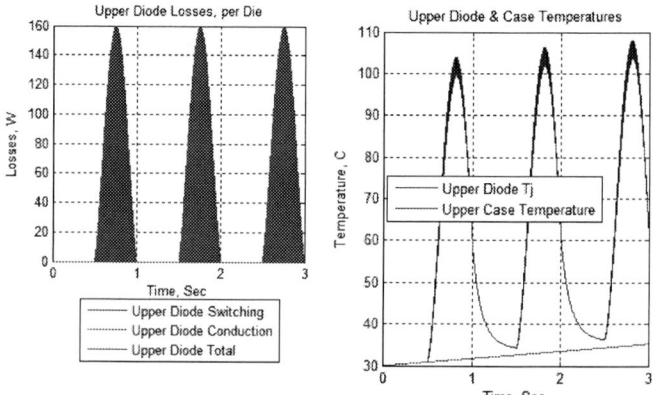

Figure 7. Simulated Diode Losses and Junction Temperature

The selected IC also has 1GΩ input impedance, which is important for the clamping circuit which must be used between the input and the opto-coupler IC. The circuit is displayed in Figure **8**. The IC has an acceptable accuracy range of 0-2V and a maximum safe voltage of 5.5V. During off-state, the clamp circuit limits the input voltage to 10V. Excess voltage is dissipated across R1, the current-limiting resistor. The resistive divider then further limits the input voltage to 5V, which keeps the IC within a safe range. The on-state characteristics are more important, since this is when the device is measured. Design considerations include:

- The clamp circuit should not allow any leakage current through the diode path, or this will dissipate extra voltage across R1.
- R1, R2, and R3 should be highly accurate resistors (0.1% tolerance) so that they don't detract from the

opto-coupler accuracy. Furthermore, R1 should stay cool to maintain its accuracy as much as possible.

- R1-3 should be high enough to not limit the bandwidth, but R3 should be several orders of magnitude lower than the IC input impedance.
- The buffer amplifier should produce a single-ended output, allowing a common reference point for all measurements.

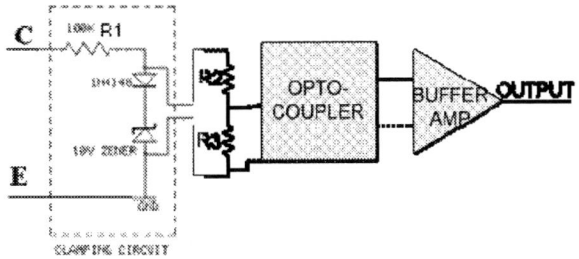

Figure 8. Saturation Voltage Detection Circuit

Multiple tests verify the performance of the detection circuit. Figure 9 shows the accuracy of the circuit. This test compared a precision DC reference voltage to the amplified output. After compensating out a small offset voltage (-6.5mV), the measured value is within 0.1% of the input. Next, the bandwidth of the circuit was checked by measuring on a switching inverter. Sample data from this test can be seen in Figure 10. The figure shows a voltage measurement at about 70A phase current. The measurement settling time after gate turn-on is about 100µs. This sets an upper limit on the switching frequency and duty cycle, so that the measurements take place within the circuit bandwidth.

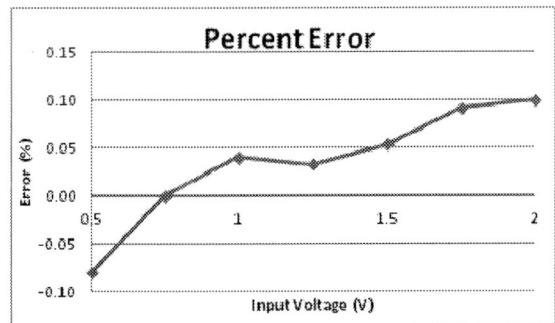

Figure 9. V_{sat} Detection Circuit Accuracy

The full test uses LabView to program an FPGA. The code uses open-loop sine-triangle PWM control, for which the basic block diagram is shown in Figure 11. The test operator inputs a fundamental frequency (f_1) and duty-cycle variation (D), which sets the sinusoidal voltage applied to the load. The sinusoid and the switching frequency (f_{sw}) set the on-times for the PWM control signals (IN1-6). The saturation voltage detection circuit feeds back the six $V_{ce(sat)}$ and six V_f signals per module. Each module provides the NTC-measured case temperature and over-current/over-temperature fault flag.

Additional programming is needed to correctly measure and process the on-state voltages. The signals are measured with a delay of 110µs after device turn-on in order to allow for the signals to settle within the circuit's bandwidth. To eliminate the inherent noise that is involved with measuring in a high-voltage switching environment, there is some signal processing. The signals are measured at the peak of the current wave to maintain a consistent measurement reference. By choosing the peak of the sine wave, the current remains essentially constant across a relatively wide range of angle. Choosing a measurement range in which the sinewave varies by only +/- 1% allows time to take over 100 samples per fundamental (f_1) period. The samples are then averaged to smooth out the individual measurements. By averaging this way, the standard deviation in $V_{ce(sat)}$ measurements over one full hour on 18 IGBTs was only 0.17%.

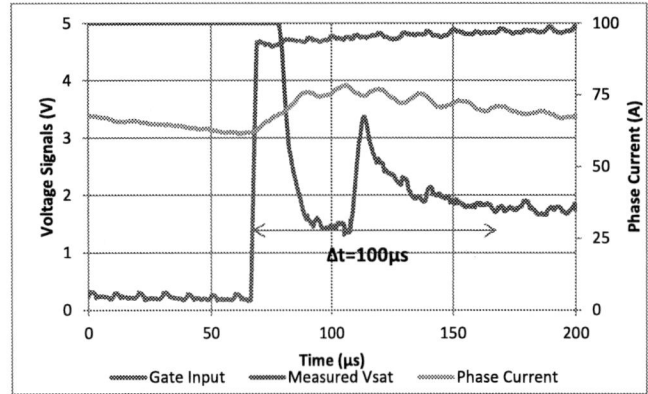

Figure 10. V_{sat} Detection Circuit Bandwidth
(Ch1 = V_{sat}/4, Ch2 = Gate Input, Ch4= Phase Current)

The four feedback signals are sampled once per 10ms to check for device failure or test error. A device failure is defined as a sustained 20% difference from initial on-state value [7]. Additionally, a temperature variation of greater than 5°C will cause an error. These signals are then logged in a *csv* file once per second to show the performance over time. A new *csv* file is generated every hour to avoid too large file size. Finally, all measurements are averaged over each full hour to display the test results charts (see Figures 13-15).

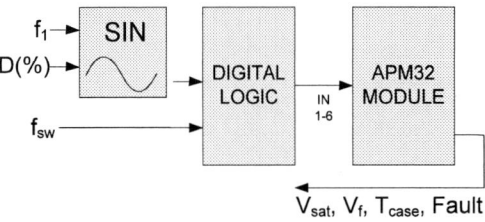

Figure 11. Test Control and Monitoring Program Structure

Figure 12 is a photo of the coldplate assembly and the controller in the foreground. The visible surface of the coldplate has the three power modules and interface boards mounted to it, and on the reverse side the three phase inductive load for each module is mounted with thermal epoxy to give good thermal interface to the coldplate.

The controller is a National Instruments Compact-RIO chassis containing a real-time processor, FPGA, and configurable I/O modules (digital outputs for PWM signals, digital inputs for fault signals, analog inputs for NTC reading and $V_{ce(sat)}$ and V_f feedback).

Cooled process water is recirculated through the coldplate at 10°C to maintain a case temperature of ~30°C. Not shown is the DC supply that provides approximately 50A dc to source the 46A$_{rms}$ phase currents.

Figure 12. Photo of Test Apparatus

Test Results

The charts in Figures 13 – 15 show the results for the first test condition ($T_{case}=T_0=30°C$, $T_j=145°C$). The device degradations are clearly seen with the $V_{ce(sat)}$ and V_f shits, and suspected to be due to wirebond cracking, starting between 800k – 1M cycles. The shift in voltage observable in these charts is smaller than expected, but clearly has the character of the expected shift [6] [7]. This illustrates that the saturation detection circuit worked well to identify incipient device failures, and the sampling and averaging algorithm described above is able to achieve stable and clean signals. One published general rule [7] regarding the shift in Vce that indicates an incipient failure is

$$\text{delta-}V_{ce(sat)} \, (\%) = 1500/I_{rated}$$

For this module, I_{rated} is 75A, hence the shift was expected to be 20%. The detection circuit was designed with much higher accuracy, and proved to be needed to capture the much smaller shift seen here of about 5%. As of publication date, the modules have not yet been subjected to DPA to determine exact failure mode.

Two types of failure signatures were noted. The first was a concurrent IGBT/diode voltage shift, which happened in Module 1-Switch 1 and in Module 2-Switches 2 and 5. This likely manifested as a wirebond crack or a wirebond lift at the diode die. This bondwire joint carries current to both the diode and the IGBT due to the wirebond pattern. The power module DUT has a construction similar to that shown in Figure 16. Thus, an increase in resistance due to degradation near the diode would show up in the voltage drop measured for both the diode and IGBT. Figure 17 shows the second-by-second change in V_f and $V_{ce(sat)}$ for Module 1 that occurred over a roughly 5 minute period. This type of concurrent shift in voltage drop for the IGBT and diode is suspected to be indicative that a wirebond carrying current to both devices has degraded. This will be confirmed by failure analysis. Failure analysis results are expected prior to the delivery of this paper at the ECTC conference.

The second failure signature happens only in the IGBT and is seen in Module 1-Switch 1 and in Module 3-Switch 1. This bond wire only carries current to the IGBT; however the IGBT junction will reach a higher temperature than the diode. Post-stress test physical analysis will be performed to determine the exact nature of the degradation.

To the extent possible with open loop control on the load current, these modules were stressed at the same operating conditions (2.5 kHz swiching frequency, 1 Hz fundamental frequency, 46 Arms, and sine triangle-based duty cycle peak of yy%). Slight variations in load impedance phase to phase and module to module may exist which allow differences in phase current to develop. Also, variation in the thermal interface material (grease) thickness may contribute to different die heating conditions. These small variations, over hundreds of thousands of cycles may be responsible for differences seen in the number of cycles to failure among die locations and modules. This is not entirely understood at this point, but may be determined by destructive physical analysis that will be carried out.

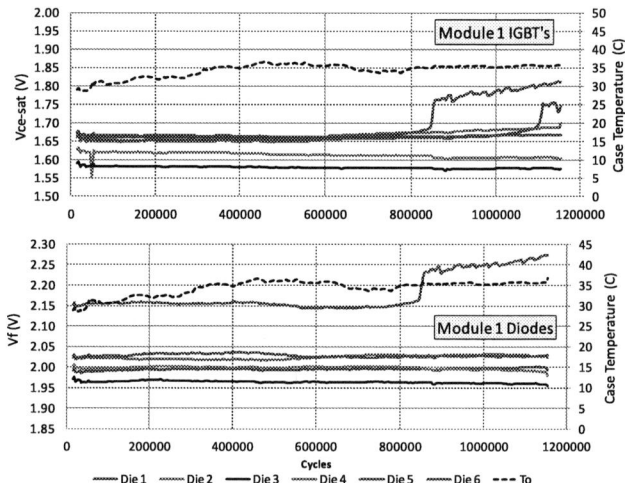

Figure 13. Evolution of Voltages on Module 1

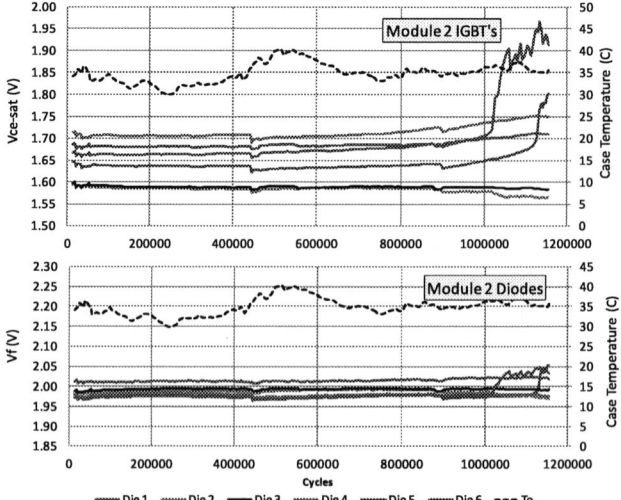

Figure 14. Evolution of Voltages on Module 2

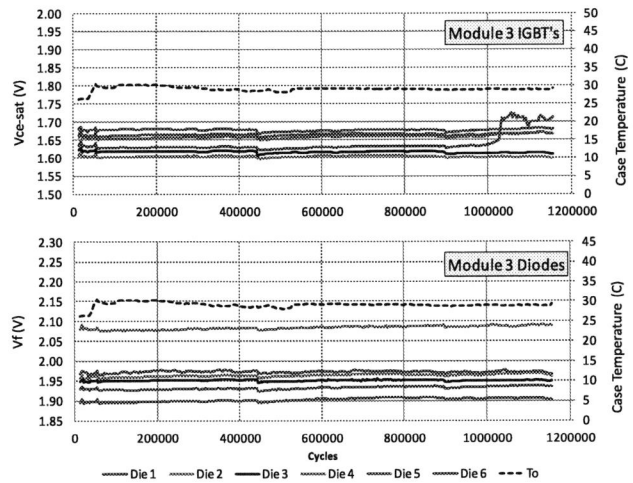

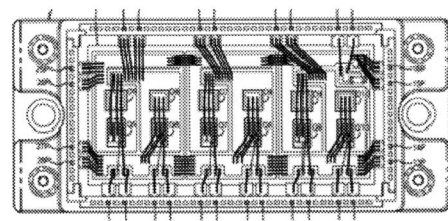

Figure 15. Evolution of Voltages on Module 3

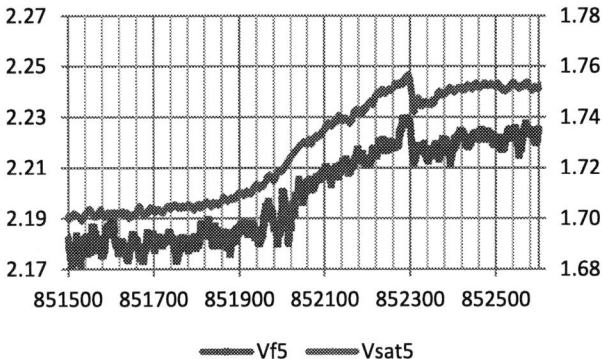

Figure 16. Typical 3-phase IGBT module wirebond layout [ref: US Patent Application US20130248883 A1 (Cree)]

due to variations in the module thermal interface layer of grease and in differences in load current caused by open loop load current control. These differences may be improved by closed loop control and by utilization of a thermal interface material whose properties are more easily controlled during laboratory assembly of the test apparatus. This is an improvement that can be made in future work.

Figure 17. Detail View of IGBT & DIODE Voltage Shift, Module 1

Acknowledgments

The authors would like to acknowledge our Fairchild colleagues, Darshan Gandhi, Tom Kopley, Fabio Necco, Byoungok Lee, John Grabowski, Ikgyoo Song, Eric Beaty, Michael Tighe, and Yuna Im for invaluable consultations and technical support.

Conclusions

A method and apparatus to accumulate power cycles resulting in pre-determined delta-T_j swings has been conceived, developed, and tested for three phase IGBT power modules. The technique makes use of detailed modeling and characterization test results normally carried out during product development to define the load current amplitude, fundamental frequency, switching frequency, and starting case temperature necessary to achieve the desired delta-T_j cycle.

The method uses inexpensive, off the shelf control hardware and software with inductive load components to establish the test facility, and provides a means of accumulating millions of cycles within a few weeks of elapsed time.

The implementation has shown excellent sensitivity to voltage shifts across the power devices in the module.

The technique stresses the entire module as it would be stressed in the intended motor control applications by driving the inductive load using sine-triangle or space vector PWM patterns ubiquitous in the motor drive market.

Future work will include destructive physical analysis to determine precise failure mechanisms for the die which have been seen to have a voltage shift. This work is expected to be completed prior to the ECTC conference and presented in the live presentation.

Variations in first failures at the 46Arms operating condition with T_o = 30C and delta-Tj = 115C may have been

References

1. Mainka, Krzysztof; Thoben, Markus; and Schilling, Oliver, "Lifetime Calculation for Power Modules, Application and Theory of Models and Counting Methods," Proceedings of the 14th European Conference on Power Electronics and Applications, 2011.
2. Lutz, Josef, "IGBT-Modules: Design for Reliability," Proceedings of the 13th European Conference on Power Electronics and Applications, 2009.
3. Bredtmann, R.; Olesen, K.; Osterwald, F.; Eisele, R.; "Options for Electric Power Steering Modules: A Reliability Challenge," Automotive Power Electronics, Paris, Sept. 2007.
4. Musallam, Mahera; Johnson, Mark C.; Yin, Chunyan; Lu, Hua; and Bailey, Chris, "Real-Time Comparison of Power Module Failure Modes Under In-Service Conditions," Proceedings of the 13th European Conference on Power Electronics and Applications, 2009.
5. Pollar, T., D'Arco, S., Hernes, M., and Lutz, J., "Influence of thermal cross-couplings on power cycling lifetime of IGBT power modules," Conference on Integrated Power Electronics Systems, Paper 05.4, March 2012.
6. Smet, Vanessa; Forest, Francois; Huselstein, Jean-Jacques; Rashed, Amgad; and Richardeau, Frederic, "Evaluation of Vce Monitoring as a Real-Time Method

978-1-4799-2408-0/14 $31.00 © 2014 IEEE

the Estimate Aging of Bond Wire-IGBT Modules Stesssed by Power Cycling," IEEE Transactions on Industrial Electronics, Vol. 60, No. 7, July 2013, pp. 2760-2770.

7. Nielsen, Rasmus; Due, Jens; and Munk-Nielsen, Stig, "Innovative Measuring System for Wear-out Indication of High Power IGBT Modules," Proceedings of the 2011 Energy Conversion Congress and Exposition, IEEE, pp. 1785-1790, 2011.

8. Stupar, A; Bortis, D; Drofenik, U.; and Kolar, J.W., "Advanced Setup for Thermal Cycling of Power Modules following Definable Temperature Profiles," Proceedings of the 2010 International Power Electronics Conference, IEEE, pp. 962-969, 2010.

9. Hartmann, Samuel; Bayer, Martin; Schneider, Daniel; and Feller, Lydia, "Observation of Chip Solder Degradation by Electrical Measurements During Power Cycling," Proceedings of the 8th International Conference on Integrated Power Electronics Systems, Nuremburg, Germany, 2014.

10. Wagenitz, Dennis; Westerholz, Andre; Erdmann, Eike; Hambrecht, Andreas; and Dieckerhoff, Sibylle, "Power Cycling Test Bench for IGBT Power Modules Used in Wind Applications," Proceedings of the 14th European Conference on Power Electronics and Applications, 2011.

11. Fairchild Semiconductor, "Smart Power Module Motion SPM Device in DIP (SPM2 V1) User's Guide," Application Note AN-9043, 9 July 2012.

Fine-Pitch Probing on TSVs and Microbumps
Using a Chip Prober Having a Transparent Membrane Probe Card

Naoya Watanabe[1*], Michiyuki Eto[2], Kenji Kawano[2], and Masahiro Aoyagi[1]

[1]Nanoelectronics Research Institute, National Institute of Advanced Industrial Science and Technology
AIST Tsukuba Central 2, 1-1-1 Umezono, Tsukuba-shi, Ibaraki 305-8568, Japan
[2]STK Technology Co., Ltd.
2468-10 Misa, Oita-shi, Oita 870-0108, Japan
*Phone: +81-29-849-1463, Fax: +81-29-862-6511, E-mail: naoya-watanabe@aist.go.jp

Abstract

For the chip-level pre-bond testing of a three-dimensional integrated circuit, we fabricated a transparent membrane probe card which had 20-μm-pitch probes on a polyethylene naphthalate film. It was fabricated by photolithography, Ti-Au sputtering, Au electroplating, Ni electroplating, electroless Au plating, and wet etching of the seed layer. Using the transparent membrane probe card and the chip prober, we could successfully conduct direct probing of 20-μm-pitch Cu-SnAg microbumps and through-silicon vias even when the tested chip was very thin (thickness: 30–50 μm).

Introduction

A three-dimensional integrated circuit (3D-IC) containing through-silicon vias (TSVs) is a key technology that can lead to a paradigm shift in the semiconductor industry because of its many benefits such as improved performance, low power consumption, and reduced footprints. Recently, 3D imagers [1-4], 3D processors [5,6], 3D memories [7,8], and 3D (or 2.5D) field-programmable gate arrays (FPGAs) [9,10] have been proposed and developed. However, some defects occur during 3D-IC fabrication processes such as wafer thinning, TSV [and redistribution layer (RDL)] formation, microbump formation, thinned wafer dicing, and chip (wafer) stacking processes. For a high-yield 3D-IC fabrication, dies with no defects must be obtained by pre-bond testing, that is, testing before stacking.

Pre-bond testing methods can be classified into two categories: nonprobing and probing methods. In nonprobing method, pre-bond testing is performed using on-die design-for-testability (DfT) circuits [11-13]. However, DfT circuits cannot detect functional faults. Moreover, DfT circuits require a relatively large area when the number of TSVs is large. On the other hand, in probing method, pre-bond testing is performed by direct probing on fine-pitch microbumps or TSVs. This method can detect most defects in 3D-ICs; however, it has two problems (Fig. 1). The first problem concerns wafer-level probing [14]. The parallelism between the wafer prober and the wafer is not good, causing probe misalignment and damage to microbumps and TSVs in the overstressed part of the wafer [15]. Moreover, damaged chips generated by the debonding of the support substrate and dicing cannot be removed before chip stacking because a thinned wafer with a support substrate is tested. The second problem involves the probe card. The probe card for pre-bond testing must meet the following conditions: (1) fine pitch, (2) low-force contact (typically less than 10 mN/pin), (3) high alignment accuracy, and (4) low cost. However, conventional probe cards [16-18] do not meet the pitch requirement.

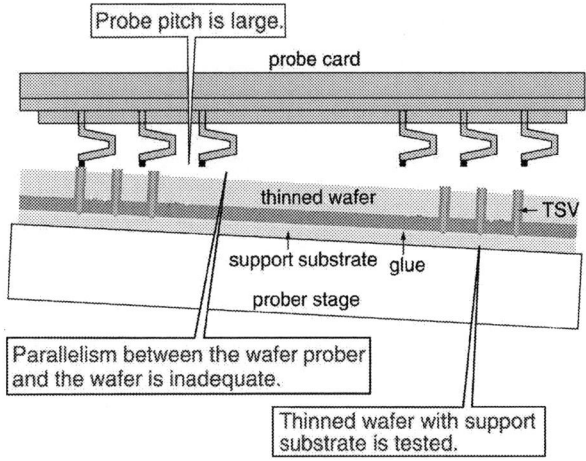

Figure 1. Problems of pre-bond testing using a wafer prober.

To overcome these problems of probing method, we proposed chip-level 3D-IC testing (Fig. 2) [19]. By performing 3D-IC testing at the chip level, parallelism adjustment becomes simple, and it is possible to perform pre-bond testing just before chip stacking. On the other hand, damageless probing of very thin chips must be performed. For chip-level 3D-IC testing, we developed a chip prober (Figs. 3 and 4) with a transparent membrane probe card using a polyethylene naphthalate (PEN) film. The chip prober has the following features: (1) it exhibits uniform probing enabled by a parallelism-adjustment system with a three-point support, (2) it allows small-gap alignment (between the probe card and the tested chip) and realizes high alignment/probing accuracy because of the precise control of the XYZθ stage and the transparency of the probe card, and (3) it allows low-load probing enabled by the compliancy of the transparent membrane probe card. Using this chip prober, we performed probing on 76-μm-pitch Al electrodes of very thin (30–50-μm-thick) static random access memory (SRAM)/flash memory chips and successfully carried out pre-bond testing of the memory chips without causing any damage [19, 20].

In this study, we fabricate a 20-μm-pitch probe card using a PEN film and perform fine-pitch probing on microbumps and TSVs. It is demonstrated that 20-μm-pitch probing on microbumps and TSVs is possible.

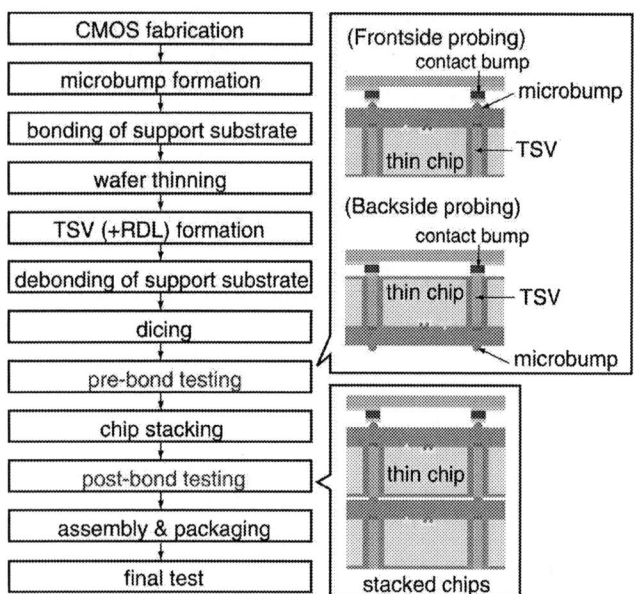

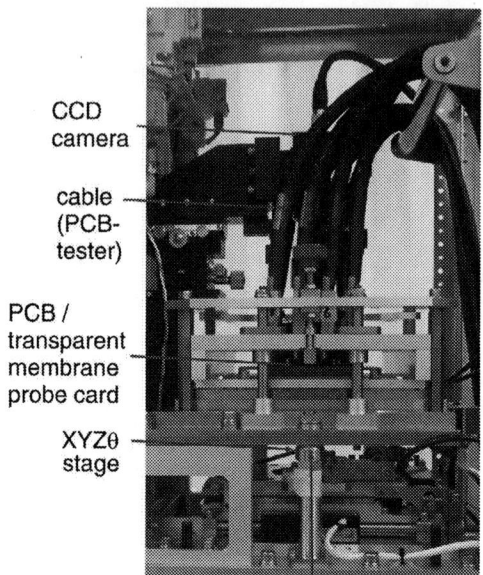

parallelism-adjustment system with three-point support

Figure 2. Flow of chip-level 3D-IC testing. In this test flow, it is possible to perform pre-bond testing just before chip stacking, and damaged chips generated by debonding of the support substrate and dicing can be removed before chip stacking.

Figure 4. Photograph of alignment/probing system in chip prober. By introducing an image processing system using a CCD camera and electrically driven XYZθ stage, alignment and probing are carried out automatically.

Fabrication of 20-μm-pitch probe card using PEN film

Figure 5 shows the fabrication process of the 20-μm-pitch probe card using a PEN film. First, 300-μmφ through-holes were formed by a UV-YAG laser beam. Next, photolithography was performed. Ar plasma irradiation was then performed on the backside of the PEN film to improve the adhesion between the PEN film and the Ti-Au layer. Following this, the backside electrodes were formed by Ti-Au sputtering and photoresist removal with acetone. After Ar plasma irradiation on the front surface, a seed layer was formed by Ti-Au sputtering. Photolithography was followed by Au electroplating under the following conditions: a MICROFAB Au100 bath (Electroplating Engineers of Japan Ltd.), current density of 0.2 A/dm^2, TiPt anode, and temperature of 62 °C. The seed layer (Au) was removed using KI + I$_2$ solution, and the photoresist was removed using acetone. Subsequently, Ni-Au bumps were formed by photolithography on both surfaces, Ni electroplating, and electroless Au plating. The Ni electroplating conditions were as follows: a MICROFAB Ni100 bath (Electroplating Engineers of Japan Ltd.), current density of 2 A/dm^2, Ni anode, and temperature of 52 °C. The electroless Au plating conditions were as follows: a bath solution containing Muden Noble AU-1, Muden Noble AU-2, Muden Noble AU-3 (Okuno Chemical Industries Co., Ltd.), gold(I) trisodium disulphite solution, and a temperature of 60 °C. Finally, the photoresist was removed using acetone, and the seed layer (Ti) was removed by KOH + H$_2$O$_2$ solution.

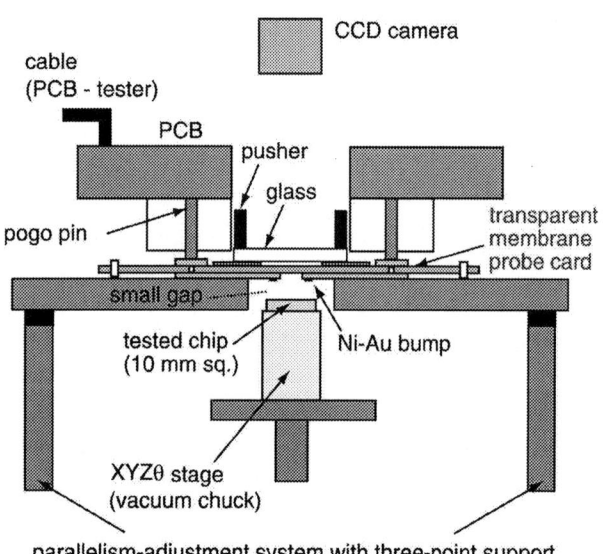

parallelism-adjustment system with three-point support

Figure 3. Schematic illustration of the chip prober. The chip prober has the following features: (1) it exhibits uniform probing enabled by a parallelism-adjustment system with a three-point support, (2) it allows small-gap alignment (between the probe card and the tested chip) and realizes high alignment/probing accuracy because of the precise control of the XYZθ stage and the transparency of the probe card, and (3) it allows low-load probing enabled by the compliancy of the transparent membrane probe card.

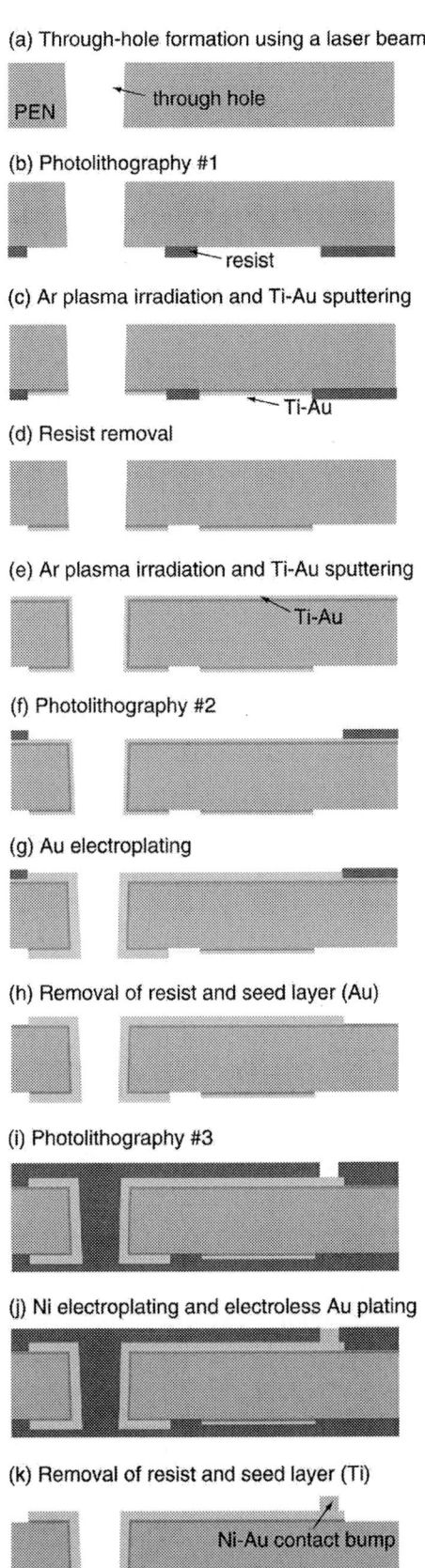

Figure 5. Fabrication process of 20 μm-pitch probe card using PEN film.

(a) Through-hole formation using a laser beam

(b) Photolithography #1

(c) Ar plasma irradiation and Ti-Au sputtering

(d) Resist removal

(e) Ar plasma irradiation and Ti-Au sputtering

(f) Photolithography #2

(g) Au electroplating

(h) Removal of resist and seed layer (Au)

(i) Photolithography #3

(j) Ni electroplating and electroless Au plating

(k) Removal of resist and seed layer (Ti)

Figure 6 shows photographs and the scanning electron microscope (SEM) image of the 20-μm-pitch probe card. The diameter of the probe card was 76 mm. The thickness of the PEN film was 225 μm. 20-μm-pitch probes consisting of Ni-Au contact bumps and Ti-Au wirings were formed on the frontside surface. The diameter of each Ni-Au contact bump was approximately 10 μm. The thicknesses of the Ti-Au wiring and the Ni-Au contact bumps were approximately 3 μm and 4 μm, respectively. 2-mmφ backside electrodes and a ground plane were formed on the backside surface. They were connected to the tester by Pogo pins and a printed circuit board (PCB).

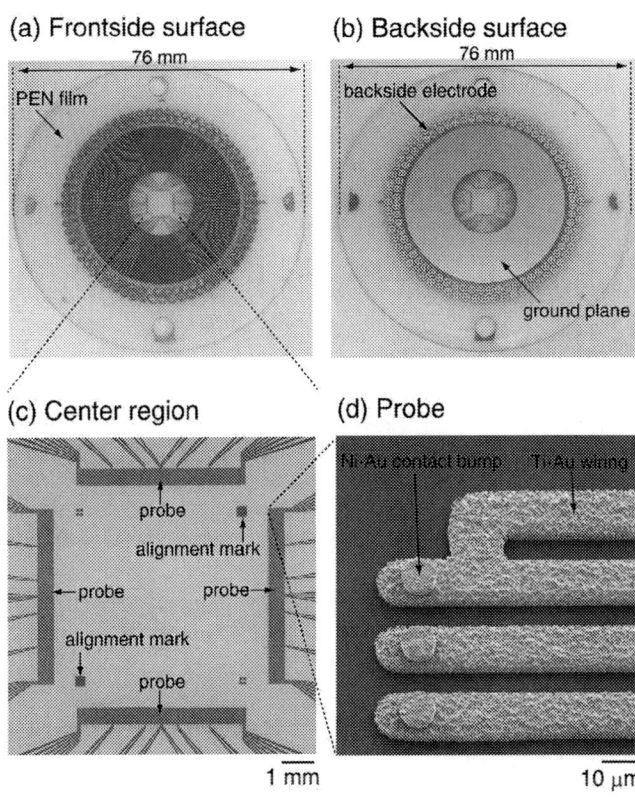

Figure 6. Photographs and SEM image of 20-μm-pitch probe card. (a) Entire image of frontside surface. (b) Entire image of backside surface. (c) Extended image of center region. (d) SEM image of 20-μm-pitch probes for four-terminal meausuremt.

Figure 7 shows the relationship between the displacement and the applied pressing load. The measurement was performed using a microhardness tester (FISCHERSCOPE H-100CXYp, Fischer Instruments). The applied pressing load per Ni-Au contact bump was varied from 1.0 mN to 10 mN. The displacement increased with the load; it was approximately 0.8 μm with 1.0 mN of load and 4.3 μm with 10 mN of load. These values are relatively low, but direct and uniform probing on fine-pitch TSVs and microbumps is possible.

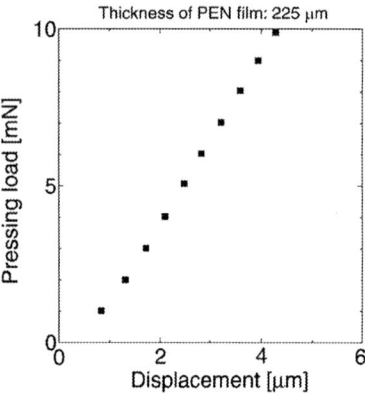

Figure 7. Relationship between displacement and applied pressing load.

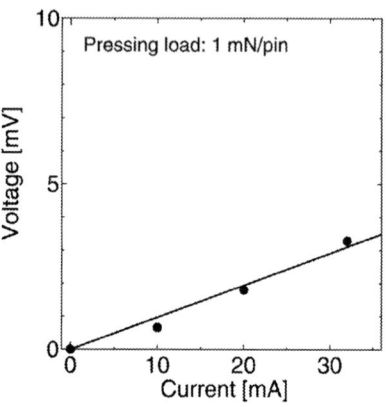

Figure 9. Result of four-terminal measurement of Cu-SnAg bump.

Probing on microbumps and TSVs

Using the chip prober with 20-μm-pitch probe card, we performed probing on 20-μm-pitch Cu-SnAg bumps on a 50-μm-thick Si chip (Fig. 8). The pressing load applied was 1 mN/pin. Figure 9 shows the result of the four-terminal measurement of the Cu-SnAg bump. The sum of resistances of the Cu-SnAg bump and the frontside contact (between the Cu-SnAg bump and the probe) was approximately 104 mΩ.

We also performed probing on 20-μm-pitch TSVs in a 30-μm-thick Si chip (Fig. 10). The contact to the Cu pad was performed using the transparent membrane probe card whereas the contact to the exposed region of TSVs was performed using the metal (Al) stage of the chip prober as a common short circuit. Figure 11 shows the result of the four-terminal measurement of TSVs. The applied pressing load was 5 mN/pin. The sum of the resistances of the TSV and the frontside contact (between the Ti-Cu pad and the probe) and the resistance of the backside contact (between TSV and the metal stage) was relatively high and approximately 518 mΩ. This is because the resistance of the backside contact is high. The metal stage will need to be optimized to reduce the resistance of the backside contact.

These results demonstrate that the chip prober with the transparent membrane probe card is effective for probing on microbumps and TSVs in a very thin chip.

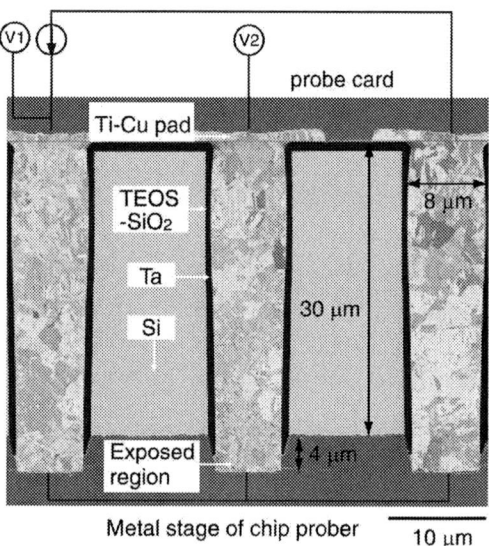

Figure 10. Scanning ion microscope (SIM) image of cross section of 20-μm-pitch TSVs in 30-μm-thick Si chip. The diameter of the TSV is approximately 8 μm. The TSVs are exposed from the backside surface of the chip, and the height of the exposed region is 4 μm. The Ti-Cu pads are formed on the frontside surface of the chip.

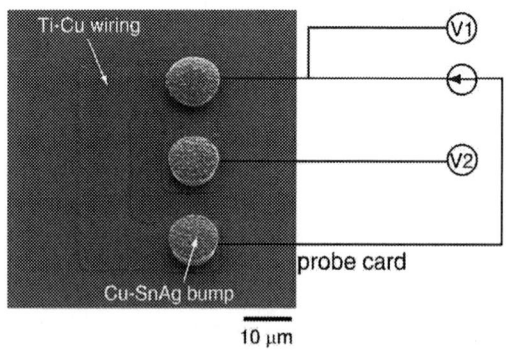

Figure 8. SEM image of 20-μm-pitch Cu-SnAg bumps on 50-μm-thick Si chip. The diameter and height of the Cu-SnAg bumps are approximately 10 μm and 5 μm, respectively.

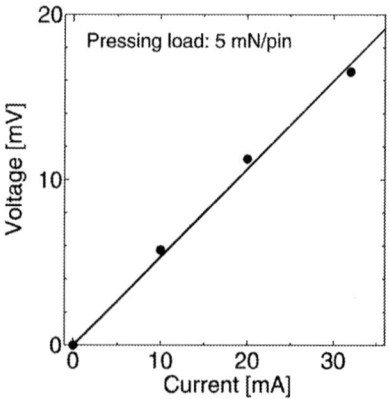

Figure 11. Result of four-terminal measurement of 20 μm-pitch TSVs.

Conclusions

We fabricated a transparent membrane probe card which had 20-μm-pitch probes on a PEN film for chip-level pre-bond testing of 3D-ICs. Using the transparent membrane probe and the chip prober, we successfully conducted direct probing on 20-μm-pitch Cu-SnAg microbumps and TSVs even when the tested chip was very thin.

Acknowledgments

This work was partially supported by the supporting-industry project of the Ministry of Economy, Trade and Industry (METI). The authors greatly appreciate Dr. S. Hara and Mr. N. Igawa at the National Institute of Advanced Industrial Science and Technology for their help. The authors also appreciate Mr. K. Himeno at STK Technology Co., Ltd. for his assistance with the measurements. The authors would like to thank Teijin DuPont Films Japan Limited for preparing the PEN films.

References

1. M. Koyanagi, Y. Nakagawa, K. W. Lee, T. Nakamura, Y.Yamada, K. Inamura, K. T. Park, and H. Kurino, "Neuromorphic Vision Chip Fabricated Using Three-Dimensional Integration Technology," ISSCC Dig. Tech. Papers, 2001, p. 270.

2. K. Kameyama, Y. Okayama, M. Umemoto, A. Suzuki, H. Terao, M. Hoshino, and K. Takahashi, "Application of High Reliable Silicon Thru-Via to Image Sensor CSP," Ext. Abstr. Solid State Devices and Materials, 2004, p. 276.

3. J. A. Burns, B. F. Aull, C. K. Chen, C. L. Chen, C. L. Keast, J. M. Knecht, V. Suntharalingam, K. Warner, P. W. Wyatt, and D. R. W. Yost, "A Wafer-Scale 3-D Circuit Integration Technology," IEEE Trans. Electron Devices vol. 53 (2006) pp. 2507–2516.

4. N. Watanabe, I. Tsunoda, T. Takao, K. Tanaka, and T. Asano, "Fabrication of Back-Side Illuminated Complementary Metal Oxide Semiconductor Image Sensor Using Compliant Bump," Jpn. J. Appl. Phys. vol. 49 (2010) pp. 04DB01-1–8.

5. P. Morrow, B. Black, M. J. Kobrinsky, S. Muthukumar, D. Nelson, C. M. Park, and C. Webb, "Design and Fabrication of 3D Microprocessors," MRS Proc. vol. 970 (2007) pp. 0970-Y03-02-1–12.

6. D. H. Kim, K. Athikulwongse, M. Healy, M. Hossain, M. Jung, I. Khorosh, G. Kumar, Y. J. Lee, D. Lewis, T. W. Lin, C. Liu, S. Panth, M. Pathak, M. Ren, G. Shen, T. Song, D. H. Woo, X. Zhao, J. Kim, H. Choi, G. Loh, H. H. Lee, and S. K. Lim, "3D-MAPS: 3D Massively Parallel Processor with Stacked Memory," ISSCC Dig. Tech. Papers, 2012, p. 188.

7. Y. Kurita, S. Matsui, N. Takahashi, K. Soejima, M. Komuro, M. Itou, and M. Kawano, "Vertical Integration of Stacked DRAM and High-Speed Logic Device Using SMAFTI Technology," IEEE Trans. Adv. Packag. vol. 32 (2009) pp. 657–665.

8. U. Kang, H. J. Chung, S. Heo, S. H. Ahn, H. Lee, S. H. Cha, J. Ahn, D. Kwon, J. H. Kim, J. W. Lee, H. S. Joo, W. S. Kim, H. K. Kim, E. M. Lee, S. R. Kim, K. H. Ma, D. H. Jang, N. S. Kim, M. S. Choi, S. J. Oh, J. B. Lee, T. K. Jung, J. H. Yoo, and C. Kim, "8Gb 3D DDR3 DRAM Using Through-Silicon-Via Technology," ISSCC Dig. Tech. Papers, 2009, p. 130.

9. R. S. Patti, "Three-Dimensional Integrated Circuits and the Future of System-on-Chip Designs," Proc. IEEE vol. 94 (2006) pp. 1214–1224.

10. M. Santarini, "Stacked & Loaded: Xilinx SSI & 28-Gbps I/O Yield Amazing FPGA," Xcell journal vol. 74 (2011) pp. 8–13.

11. M. Tsai, A. Klooz, A. Leonard, J. Appel, and P. Franzon, "Through Silicon Via(TSV) Defect/Pinhole Self Test Circuit for 3D-IC," Proc. IEEE International Conference on 3D System Integration, 2009, p. 1 (Paper 32).

12. M. Cho, C. Liu, D. Kim, S. Lim, and S. Mukhopadhyay, "Design Method and Test Structure to Characterize and Repair TSV Defect Induced Signal Degradation in 3D System," Proc. IEEE/ACM International Conference on Computer-Aided Design, 2010, p. 694.

13. P. Y. Chen, C. W. Wu, and D. M. Kwai, "On-Chip Testing of Blind and Open-Sleeve TSVs for 3D IC before Bonding," Proc. IEEE VLSI Test Symposium, 2010, p. 263.

14. E. J. Marinissen and Y. Zorian, "Testing 3D Chips Containing Through-Silicon Vias," Proc. IEEE International Test Conference, 2009, p. 1 (Paper ET1.1).

15. W. R. Bottoms, "Test Challenges for 3D Integration," Proc. IEEE Custom Integrated Circuits Conference (2011) p. 1 (Paper 6-3).

16. The homepage of MicroProbe Inc. (http://www.microprobe.com/products/technology/cantilever/)

17. K. Smith, P. Hanaway, M. Jolley, R. Gleason, E. Strid, T. Daenen, L. Dupas, B. Knuts, E. J. Marinissen, M. V. Dievel, "Evaluation of TSV and Micro-bump Probing for Wide I/O Testing," Proc. IEEE International Test Conference, 2011, p. 1 (Paper 17.2).

18. L. Namburi, F. Cros, Y. Hong, T. Hu, R. Smith, G. Maier, V. T. Truong, H. Liu, and K. Sakuma, "A Fine Pitch MEMS Probe Card with Built in Active Device for 3D IC Test," presented at IEEE SW Test Workshop, 2013.

19. N. Watanabe, M. Suzuki, M. Eto, K. Kawano, and M. Aoyagi, "Development of a Chip Prober for Pre-Bond Testing of a 3D-IC," Proc. IEEE CPMT Symposium Japan, 2013, p. 43.

20. N. Watanabe, M. Eto, K. Kawano, and M. Aoyagi, "3D IC Testing Using a Chip Prober and a Transparent Membrane Probe Card," Proc. IEEE International Conference on Microelectronic Test Structures, 2014, in press.

Thermal Management of 3D RF PoP Based on Ceramic Substrate

Fengze Hou*[1,2], Fengman Liu[1,2], Yi He[2], Xiaomeng Wu[2], Xia Zhang[1,2], Liqiang Cao[1,2], Yuan Lu[1,2], Dongkai Shangguan[1,2]

[1]National Center for Advanced Packaging (NCAP China), Wuxi, Jiangsu, China, 214135
[2]Microsystem Packaging Research Center, Institute of Microelectronics of Chinese Academy of Sciences,
No. 3, Bei Tu Cheng West Road, Beijing, China 100029
*houfengze@ime.ac.cn

Abstract

In this paper, a new high performance three dimensional radio frequency package on package (3D RF PoP) based on ceramic substrate is designed for micro base station which is able to cover the multicasting of critical messages to as many mobile users as possible even under communications network failure events, such as the failure of macro base stations. The RF PoP integrates receiver (RX) module, transmitter (TX) and digital predistortion (DPD) module, and analog-to-digital/digital-to-analog (AD/DA) and clock (CLK) module vertically, has better signal integrity and faster data-rate transfer due to shorter signal paths among the three modules. Additionally, the ceramic substrate has higher thermo-mechanical reliability and better heat dissipation performance compared with organic substrate. The paper firstly studies the thermal performance of the RF PoP without external heat sinks using commercial software ANSYS Icepak. At the ambient temperature of 25 °C, the highest junction temperature of the RF PoP is 239.8 °C, which is above the acceptable baseline for a silicon chip. Secondly, in order to improve the heat dissipation capability of the RF PoP, a large copper bottom heat dissipation plate is employed. The impact of bottom heat dissipation plate on the thermal performance of the RF PoP is investigated. It is found that a bottom heat dissipation plate of $200 \times 200 \times 2$ mm^3 is reasonable, and the highest junction temperature is about 47.0 °C. Thirdly, we study the thermal performance of the entire and internal package structure of the RF PoP mounted on the bottom heat dissipation plate. The temperature distributions of the top and middle substrates are almost uniform. Heat generated from active devices on the top and middle packages is mainly transferred to the edges, conducted to the bottom heat dissipation plate through the edges and bottom substrate and heat slug, heat generated from active devices on the bottom package is conducted to the bottom heat dissipation plate through the bottom substrate and heat slug, and then dissipated into the ambient by natural convection, so that the temperature at the edge of the bottom substrate is higher than the other parts. Lastly, the effect of different ambient temperatures on the thermal performance of the RF PoP is investigated. When the bottom heat dissipation plate size is $200 \times 200 \times 2$ mm^3 and the ambient temperature reaches 100 °C, the highest junction temperature is about 121 °C. In order to further improve heat dissipation capability of the RF PoP, a copper top heat spreader is employed. The simulation result shows that the junction temperature drops to about 115.6 °C.

Introduction

In recent years, some innovative 3D packages that have emerged, mainly include PoP, 3D chip integration with through silicon vias (TSVs), system in package (SiP) with embedded components, wafer level stacking, 3D package based on flexible substrate, etc. [1-6]. The application space for 3D packages is increasing because of several advantages: higher performance, smaller volume, lighter weight, shorter time-to-market, lower cost, multi-functionality, etc.

One of the important applications of PoP is the vertical integration of DRAM memory package and Logic or Application Processor (AP) packages [7-11]. The thermal performance of the PoP is a challenge due to its vertical stack. Under continuous operation, the PoP can generate a lot of heat. If the heat is not dissipated in time, the operational lifetime and reliability of the PoP may be significantly reduced. Many scholars have investigated the thermal performance of the PoP. For example, Yang investigated the thermal characteristics of the PoP, especially on the thermal interactions between the top and bottom packages [7]. Yen et al. studied the effects of power dissipation (in both packages), mold compound thermal conductivity, number of through mold via (TMV), number of solder ball, size of an inter-package heat slug, thermal conductivity of a top heat spreader on the thermal performance of the PoP [8]. Han analyzed the impact of the thermal properties of underfill, passivation layer and mold compound, the percentage of copper in each redistribution layer (RDL), the geometries of mold compound, passivation layer and copper layer in RDL on the thermal performance of the PoP. A top heat spreader, thermal via array, bottom heat dissipation plate and two types of top thermal cases were employed to enhance the heat dissipation capability [9]. Hung et al. investigated electrical, thermal and warpage of three high bandwidth PoPs [10].

However, there are few reports on PoP for RF SiP application. In this paper, a new and high performance 3D RF PoP based on ceramic substrate is designed for micro base station, which is able to cover the multicasting of critical messages to as many mobile users as possible even under communications network failure events, such as the failure of macro base stations [12]. The RF PoP integrates three different RF modules vertically, has better signal integrity and faster data-rate transfer due to shorter signal paths among the three modules. Additionally, the ceramic substrate has higher thermo-mechanical reliability and better heat dissipation performance compared with organic substrate. The paper firstly studies the thermal performance of the RF PoP without external heat sinks using commercial software ANSYS Icepak. Secondly, in order to improve the heat dissipation capability of the RF PoP, a large copper bottom heat dissipation plate is employed. The impact of the bottom heat dissipation plate on the thermal performance of the RF PoP is investigated. Thirdly, we study the entire and internal package

2014 Electronic Components & Technology Conference

structure of the RF PoP mounted on the bottom heat dissipation plate. Lastly, the effect of different ambient temperatures on the thermal performance of the RF PoP is investigated. At higher ambient temperatures, in order to further improve the heat dissipation capability of the RF PoP, a copper top heat spreader is employed.

Package structure and material

A. Package structure

Figure 1 shows a cross sectional view of a 3D RF PoP based on ceramic substrate. The 3D RF PoP includes top, middle, bottom packages, metal kovar wall, metal kovar cover plate, heat slug, leads, etc. In this case, the top, middle, and bottom packages are the RX module, TX and DPD module, and AD/DA and CLK module, respectively. It is noted that the top and middle substrates are both cavity structure, so that the active and passive devices can be assembled on the middle and bottom substrates. The top package is covered by metal kovar cover plate, so that the RX module doesn't affect other electronic devices, or is not affected by other electronic devices. Active devices on the three packages are all quad flat no lead (QFN) devices. The size and heat power of the active devices are shown in Table 1. The sizes of the top, middle, and bottom substrates are $50{\times}50{\times}5.1$ mm³, $51{\times}51{\times}5.1$ mm³, and $56{\times}56{\times}3$ mm³, respectively. The size of the entire RF PoP is $80{\times}80{\times}16.9$ mm³. The heat powers of the top, middle, and bottom are 6.715 W, 7.354 W, and 4.711 W, respectively. The total heat power of the RF PoP is 18.78 W.

In order to improve the heat dissipation capability, the 3D RF PoP is mounted on a copper bottom heat dissipation plate and its leads are soldered on a cavity printed circuit board (PCB) test board, as shown in Figure 2. The length and width of the PCB test board and bottom heat dissipation plate are the same. The test PCB board is a four-layer substrate of thickness 1.6 mm.

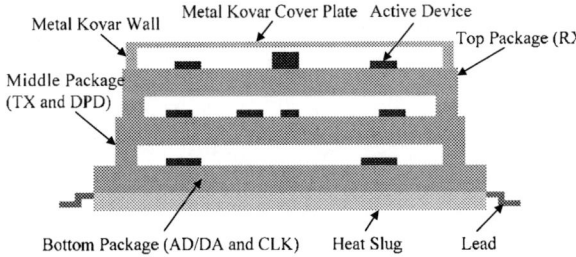

Figure 1. Cross sectional view of a 3D RF PoP based on ceramic substrate

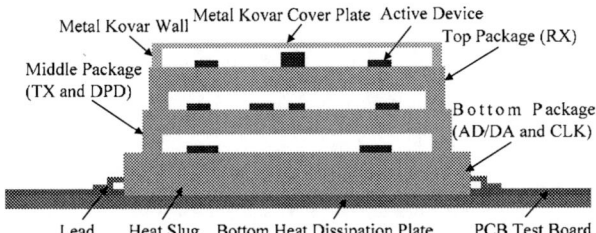

Figure 2. Cross sectional view of the RF PoP mounted on a bottom heat dissipation plate

Table.1 Size and heat power of the active devices on the three packages

Active devices	Size (L×W×T) （mm³）	Heat Power (W)
U1	10×10×0.85	0.891
U2	8×8×0.85	0.69
U3	12×12×1	1.63
U4	10×10×0.85	1.5
Bottom package	--	**4.711**
2×U5	6×6×0.85	2×1.25
2×U6	4×4×0.9	2×0.35
U7	5×5×0.75	1.1
U8	5×5×1	1.354
U9	4×4×0.85	1.5
U10	2×1.25×0.9	0.2
Middle package	--	**7.354**
U5	6×6×0.85	1.25
U6	4×4×0.9	0.35
U10	2×1.25×0.9	0.2
U11	6.2×5.3×2	0.5
U12	4×4×0.85	1.5
U13	4×4×0.85	1.37
U14	4×4×0.85	1.11
U15	5×5×0.85	0.435
Top package	--	**6.715**
Total	--	**18.78**

B. Package material

The steady-state thermal conduction equation with heat source, shown in Eq. (1), is used to obtain the temperature distribution. Thermal convection can be expressed as Eq. (2).

$$\frac{\partial^2 T}{\partial x^2} + \frac{\partial^2 T}{\partial y^2} + \frac{\partial^2 T}{\partial z^2} + \frac{Q_v}{\lambda} = 0 \tag{1}$$

$$-\lambda \frac{\partial T}{\partial n}\bigg|_{\Gamma} = h(T_w - T_f) \tag{2}$$

Where T is the temperature in a steady-state condition, x, y, and z are the spatial coordinates, Q_v is the heat generation rate, λ is the thermal conductivity, Γ is the wall surface of one component, n is the normal direction of the surface, h is the convective heat transfer coefficient, T_w is the wall surface temperature and T_f is the bulk temperature of the fluid [13]. From the Eq. (1) and Eq. (2), it can be seen that the thermal conduction and convection are both relative to thermal conductivities of the package materials, as shown in Table 2.

Table 2 Thermal conductivities of the package materials for the 3D RF PoP

Components	Materials	Thermal conductivity (W/m·°C)
Active device	Silicon	148
Ceramic substrate	Al2O3	17.4
Heat slug	W85Cu15	248
Lead	4J29	16.7
Metal kovar wall	4J34	17
Metal kovar cover plate	4J42	14.7
Heat dissipation plate	Copper	385
PCB	FR-4	0.35
Thermal interface material	--	2.5

Thermal simulation

In the process of thermal simulation, in order to reduce the element size, improve the mesh quality, accelerate the convergence speed, save the computation time, the following simplification and assumptions are listed under the premise of simulation precision:

- For the boundary condition, natural convection is assumed.
- The initial temperature is ambient temperature.
- Gravity is in the −z direction.
- All active devices are assumed to be silicon blocks having uniform heat power.
- The multilayer top, middle, and bottom ceramic substrates are all simplified as blocks. The effect of electrical traces of the ceramic substrates is ignored.
- Perfect adhesion is assumed at interfaces.

Figure 3 shows the thermal simulation model of the 3D RF PoP mounted on a copper bottom heat dissipation plate in a large cabinet filled with air.

Figure 3. Thermal simulation model of the 3D RF PoP mounted on a heat dissipation plate

A. Thermal performance of the 3D RF PoP without external heat sinks

As a baseline, the heat dissipation capability of the 3D RF PoP based on the ceramic substrate without external heat sinks is firstly investigated. The RF PoP is soldered on a cavity PCB test board. When the PCB test board is 200×200×1.6 mm^3 and the ambient temperature is 25 °C, the temperature distribution of the RF PoP without external heat sinks is shown in Figure 4. The highest junction temperature of the RF PoP is 239.8 °C, which is above the acceptable baseline for a silicon chip. Therefore, in order to dissipate the heat generated from the RF PoP, effective heat dissipation method must be used. In the following, a large copper bottom heat dissipation plate is employed, and the effect of the bottom heat dissipation plate on the thermal performance of the 3D RF PoP is studied.

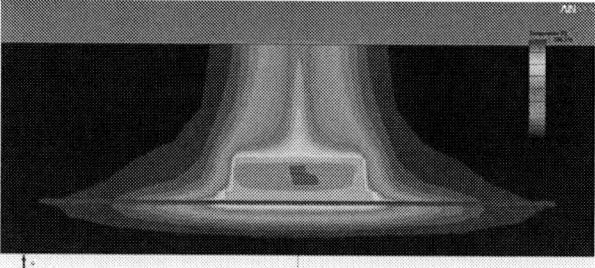

Figure 4. Temperature distribution of the RF PoP without external heat sinks

B. Effect of the copper bottom heat dissipation plate on the thermal performance of the 3D RF PoP

Figure 5 shows the highest junction temperature of the RF PoP when the bottom heat dissipation plate size increases from 150×150×2 mm^3 to 250×250×2 mm^3 at the ambient temperature of 25 °C. From the figure, it can be seen that when the bottom heat dissipation plate size increases from 150×150×2 mm^3 to 175×175×2 mm^3, the highest junction temperature of the RF PoP is higher. However, when the bottom heat dissipation plate size increases from 175×175×2 mm^3 to 200×200×2 mm^3, the highest junction temperature of the RF PoP drops abruptly from 141.6 °C to 47.0 °C at about a heat dissipation plate size of 200×200×2 mm^3. When the heat dissipation plate size increases from 200×200×2 mm^3 to 250×250×2 mm^3, the highest junction temperature remains almost the same, and it is about 46 °C. Therefore, for the thermal performance of the 3D RF PoP, choosing a heat dissipation plate size of 200×200×2 mm^3 is reasonable.

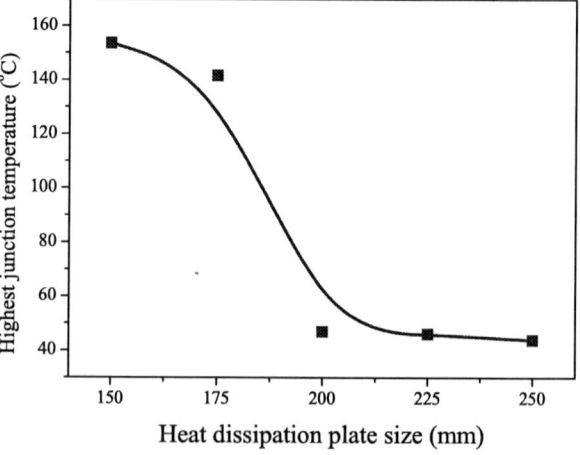

Figure 5. Effect of increasing the bottom heat dissipation plate size on the thermal performance of the 3D RF PoP

C. Thermal performance of the entire and internal package structure of the 3D RF PoP mounted on the copper bottom heat dissipation plate

In this section, the thermal performance of the entire and internal package structure of RF PoP mounted on the bottom heat dissipation plate is studied. Figure 6 illustrates the temperate distribution of the entire RF PoP mounted on the bottom heat dissipation plate. It can be seen that the highest and lowest temperature are about 47.0 °C and 30.4 °C, respectively, and that the lowest temperature occurs at the bottom heat dissipation plate. The temperature at the surface of metal kovar cover plate is about 38.7 °C. The temperature gradient from the maximum heat flux active device to the bottom heat dissipation plate is about 17 °C. Figure 7, Figure 8, and Figure 9 show the internal temperature distributions of the RF PoP based on ceramic substrate. The temperature distribution of each active device can be seen from the three figures. The highest temperature occurs in U9 and U12, which have the same size and heat power, show the maximum heat generation of 0.11×10^9 W/m^3 and have relatively cool neighbors. In addition, the temperature distributions of the top and middle substrate are almost uniform. Heat generated from active devices on the top and middle packages is mainly transferred to the edges, conducted to the bottom heat dissipation plate through the edges and bottom substrate and heat slug, heat generated from active devices on the bottom package is conducted to the bottom heat dissipation plate through the bottom substrate and heat slug, and then dissipated into the ambient by natural convection, so that the temperature at the edge of the bottom substrate is higher than the other parts except for active devices.

Figure 6. Temperate distribution of the RF PoP mounted on the bottom heat dissipation plate

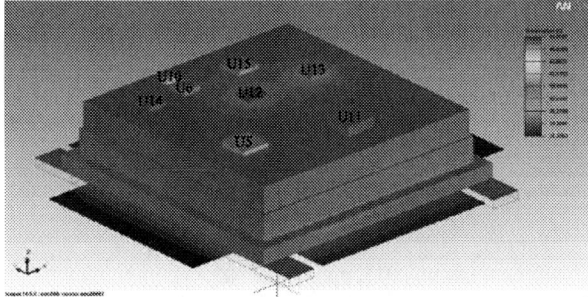

Figure 7. Temperate distribution of the RF PoP without metal kovar wall and cover plate

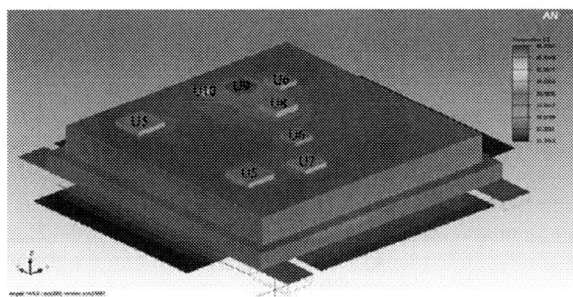

Figure 8. Temperate distribution of the middle, bottom package, heat slug, and leads

Figure 9. Temperate distribution of the bottom package, heat slug and leads

D. Effect of different ambient temperatures on the thermal performance of the 3D RF PoP

When the 3D RF PoP operates at full load, the ambient temperature of may be in the range of 0~100 °C, the effect of different ambient temperatures on thermal performance of the 3D RF PoP is investigated in the following manner.

When the heat dissipation plate size keeps 200×200×2 mm^3, Figure 10 represents the simulated highest junction temperature of the 3D RF PoP in the ambient temperature range of 0~100 °C with an increasing step of 25 °C. From the figure, it is found that there is an almost linear relation between the highest junction temperature of the 3D RF PoP and the ambient temperature. The highest junction temperature is higher about 21 °C. When the ambient temperature reaches 100 °C, the highest temperature is about 121.2 °C.

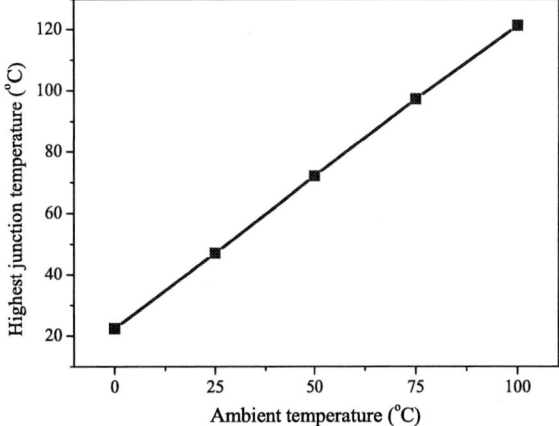

Figure 10. Effect of rising the ambient temperature on the thermal performance of the 3D RF PoP

E. Effect of a top heat spreader on the thermal performance of the 3D RF PoP

At high ambient temperatures, in order to further improve the heat dissipation capability of the RF PoP, a copper top heat spreader is employed as illustrated in Figure 11. In order to prove heat generated from active devices on the top package can be conducted to the metal kovar cover plate, a high thermal conductivity thermal interface material1 (TIM1) is used to fill the gap between the active devices and metal kovar cover plate. Simultaneously, in order to reduce the contact thermal resistance, a layer of TIM2 is coated between the metal kovar cover plate and top heat spreader.

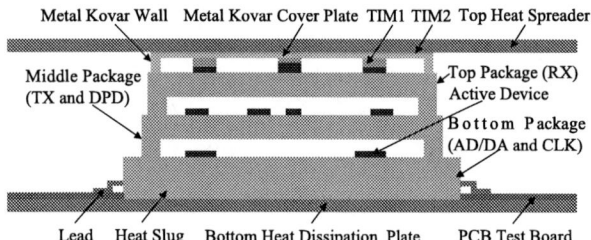

Figure 11. Cross sectional view of the RF PoP cooled by a top heat spreader and a bottom heat dissipation plate

When the size of top heat spreader and bottom heat dissipation plate are both 200×200×2 mm^3 and the ambient temperature is 100 °C, the temperature distribution of the RF PoP is shown as in Figure 12. The highest junction temperature of the RF PoP is about 115.6 °C, and is lower by 5.6 °C than the junction temperature without top heat spreader.

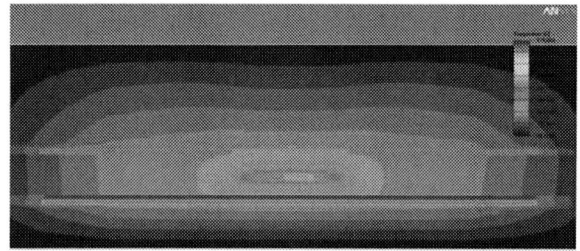

Figure 12. Temperature distribution of the RF PoP cooled by top heat spreader and bottom heat dissipation plate

Conclusions

In this study, a new and high performance 3D RF PoP based on ceramic substrate is designed for micro base station. The thermal performance of the 3D RF PoP is investigated. Some thermal management conclusions can be drawn:

(1) When the RF PoP is soldered on a PCB test board of 200×200×1.6 mm^3 and the ambient temperature is 25 °C, the highest junction temperature of the RF PoP without external heat sinks is 239.8 °C.

(2) In order to improve the heat dissipation capability of the RF PoP, a large copper bottom heat dissipation plate is employed. The impact of bottom heat dissipation plate on the thermal performance of the RF PoP is investigated. It is found that a bottom heat dissipation plate of 200×200×2 mm^3 is reasonable, and the highest junction temperature of the RF PoP is about 47.0 °C.

(3) The temperature distributions of the top and middle substrates are almost uniform. Heat generated from active devices on the top and middle packages is mainly transferred to the edges, conducted to the bottom heat dissipation plate through the edges and bottom substrate and heat slug, heat generated from active devices on the bottom package is conducted to the bottom heat dissipation plate through the bottom substrate and heat slug, and then dissipated into the ambient by natural convection, so that the temperature on the edge of the bottom substrate is higher than the other parts except for active devices.

(4) When the bottom heat dissipation plate size is 200×200×2 mm^3 and the ambient temperature reaches 100 °C, the highest junction temperature is about 121 °C. In this case, in order to further improve the heat dissipation capability of the RF PoP, a copper top heat spreader is employed. The simulation result shows that the junction temperature drops to about 115.6 °C, and is lower by 5.6 °C than the junction temperature without top heat spreader.

Acknowledgments

The authors acknowledge the support of the National Science and Technology Major Project (Project No. 2013ZX02503003-005 and Project No. 2013ZX02501004). The authors also would like to thank Professor Daniel from NCAP and IMECAS to review this paper.

References

1. H. Eslampour, M. Joshi, S. W. Park, et al. "Advancements in Package-on-Package (PoP) Technology, Delivering Performance, Form Factor & Cost Benefits in Next Generation Smartphone Processors," in *Proc. IEEE Electronic Components and Technol. Conf. (ECTC)*, Las Vegas, NV, 2013, pp. 1823-1828.

2. Y. Kim and S. B. Park. "Optimization of Underfill Material for Better Reliability and Thermal Behavior of 3D Packages with TSVs," in *Proc. IEEE Electronic Components and Technol. Conf. (ECTC)*, Las Vegas, NV, 2013, pp. 2310-2318.

3. J. H. Lau and T. G. Yue. "Thermal Management of 3D IC Integration with TSV (Through Silicon Via)," in *Proc. IEEE Electronic Components and Technol. Conf. (ECTC)*, San Diego, CA, 2009, pp. 635-640.

4. N. Khan, V. S. Rao, S. Lim, et al. "Development of 3-D Silicon Module With TSV for System in Packaging," *IEEE Transactions on Components and Packaging Technologies*, vol. 33, no.1 (2010), pp.3-9.

5. C. W. Chien, L. C. Shen, T. C. Chang. "Chip Embedded Wafer Level Packaging Technology for Stacked RF-SiP Application," in *Proc. IEEE Electronic Components and Technol. Conf. (ECTC)*, Sparks, NV, 2007, pp.305-310.

6. F. Z. Hou, X. Zhang, X. P. Guo, et al. "Thermo-mechanical reliability study for 3D package module based on flexible substrate," in *Proc. IEEE Int. Conf. Electron. Packag. Technol. High Density Packag. (ICEPT-HDP)*, Dalian, China, 2013, pp. 1296-1300.

7. C. Yang. "Thermal Characterization of Package-on-Package (PoP) Configuration through Modeling," *in Proc. IEEE Electronics Packag. Technol. Conf. (EPTC)*, Singapore, 2008, pp. 731-736.

978-1-4799-2408-0/14 $31.00 © 2014 IEEE

8. Y. Y. Hoe, C. S. Choong, V. S. Rao, et al. "Thermal Modeling and Simulation of a Package-on-Package Embedded Micro Wafer Level Package (EMWLP) Structure at the Package and System-level," *in Proc. IEEE Electronics Packag. Technol. Conf. (EPTC)*, Singapore, 2010, pp. 285-291.

9. Y. Han, B. Y. Zhang, C. S. Choong, et al. "Package-Level Thermal Management of a 3D Embedded Wafer Level Package," *in Proc. IEEE Electronics Packag. Technol. Conf. (EPTC)*, Singapore, 2013, pp. 83-87.

10. M. Hung, C. C. Lee, H. H. Cheng, et al. "Electrical, Thermal and Warpage Investigation on High Bandwidth PoP", in *Proc. IEEE Electronics Packag. Technol. Conf. (EPTC)*, Singapore, 2013, pp. 535-538.

11. S. C. Chong, D. H. S. Wee, V. S. Rao, et al. "Development of Package-on-Package Using Embedded Wafer-Level Package Approach," *IEEE Trans. CPMT*, vol. 3, no. 10, pp. 1654–1662, Oct. 2013.

12. I. Rubin, H. B. Chang, and R. Cohen. "Robust Multicasting in Micro Base Station Aided Wireless Cellular Networks", in *Proc. Wireless Networking Symposium*, Anaheim, CA, 2012, pp. 5464-5469.

13. J. P. Holman, *Heat Transfer*, China Machine Press, Beijing, 2005, pp. 1-5.

Bump Pattern Optimization and Stress Comparison Study for DCA Packages

Akash Agrawal[1], Owen Fay[2], Mark Johnson[2]
Micron Technology Inc.
8000 S Federal Way, Boise ID - 83716
Email: aagrawal@micron.com

Abstract

With the increasing demand for reduced package size and enhanced package performance, the semiconductor industry has intensified focus on the development of alternate technologies, resulting in significant interest in flip chip or direct chip attach (DCA) packages using copper pillar bump interconnects. In DCA packages, the coefficient of thermal expansion (CTE) mismatch between different materials causes warpage/reliability issues during the reflow process and thermal cycling. These CTE mismatch issues interact with various copper pillar patterns and affect package reliability. This study reports on simulation work to optimize the 1st level copper pillar patterns on DCA packages. Finite element models have been established to understand the effects of various copper pillar patterns on stress conditions in solder regions. In addition to material factors like low CTE core and different underfill compounds, other factors including solder joint thickness, a portion of the copper pillar bumps and die thickness are also simulated to determine the impact on solder stress in relation to the die and substrate interface.

Keywords: Direct chip attach, finite element, modeling, copper pillar pattern, solder stress

Introduction

The emergence of compact products has suddenly increased the necessity of small, light weight, low cost, and more reliable products. Direct chip attach (DCA) technology is a new and reliable method of attaching small chips to polymeric substrates. DCA packages are basically described as flip chip packages where chips are connected to the substrate using copper pillar bumps with solder caps. The space between substrate and chip is encapsulated by underfill material to enhance the reliability and performance of the package. During reflow, thermal stresses and strains occur in the package due to the mismatch of coefficient of thermal expansion (CTE) of various materials, causing warpage and reliability issues in chips and solder joints. It is important to understand how a package performs in thermal conditions and what changes need to be made to the package to avoid any mechanical failures.

An earlier study by Gall, et al. [4] explained the generation of thermo-mechanical stresses in DCA packages while Schubert et. al., [5] performed thermo-mechanical reliability finite element simulation to capture stresses in solder bumps during thermal cycling. Lau et. al., [6] also ran a finite element study of creep analysis of solder bumps in DCA packages. It is important to understand the reliability of copper pillar bumps and how to arrange copper pillar bumps in DCA packages to get higher bump life. In this paper, 3-dimensional finite element models of DCA packages with different outrigger copper pillar bump patterns were developed, and thermo-mechanical analysis was conducted at different temperatures. A comparison study of stress generated in copper pillar bumps solder joints and die warpage was performed. ANSYS 14.0 was used to conduct the finite element simulation.

Package Configuration

A typical stack up of flip chip DCA packages is shown in Figure1. In DCA packages, the I/O bond pads (Cu pillars) on the die are solder bumped with lead-free solder alloy. Once the flip chip has been aligned and placed on the substrate with active face down, the whole assembly goes into a mass reflow process.

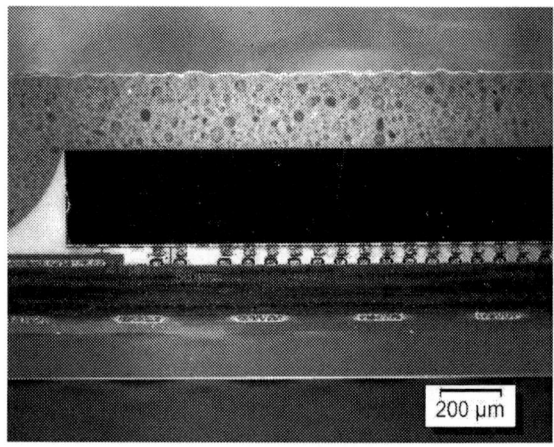

Figure 1. Direct Chip Attach Package

The die in DCA packages is connected to the substrate by copper pillar bumps with solder caps as shown in Figure 2.

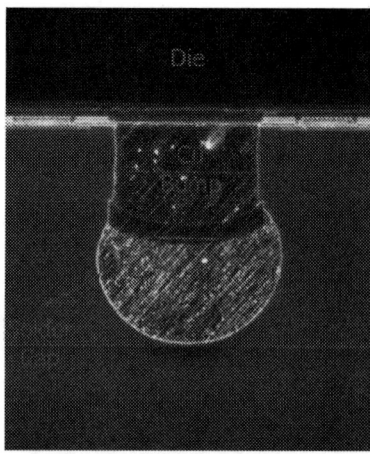

Figure 2. Copper Pillar Bump with Solder Cap

In mass reflow process, solder reflows and wets to the substrate traces and forms interconnects between the die and package substrate. The mismatch of thermal expansion between different regions of the package induces stress and strain in the package during reflow process. The package warpage and stress/strain in the die will highly depend on the copper pillar bump pattern. It is critical to avoid both warpage in die and stresses in solder joints of copper pillar bumps to minimize reliability and performance issues with the package. If warpage is too high, solder opening or die cracking may occur. High stress in solder joints may also cause solder failure, not only at room temperature, but also at reflow temperature.

DCA packages face many challenges during manufacturing, such as high die warpage during reflow and intact connection between outrigger copper pillar bumps due to high die warpage, solder wetting, die misalignment during reflow, etc. Figure 3 shows an example of solder wetting and an intact copper pillar bump condition due to high die warpage.

This paper focuses on the effect of different copper pillar bump patterns. Therefore, all the copper pillar bumps are assumed to be symmetrical and uniform in shape and die is assumed to be attached to the copper pillar bumps.

Finite Element Modeling

Use of finite element analysis provides a design tool in the product development process for numerical stress screening. All the models of DCA packages in this study were developed using ANSYS 14.0. Figure 4 shows the FEA model with detailed active and outrigger copper pillar bumps. The model was developed at two levels: The 1st level model includes only die and copper pillar bumps on the substrate (no underfill or mold compound), to capture the effect of copper pillar bump patterns. The 2nd level model has mold compound (550µm thick) and underfill to capture the effect of different mold compounds and underfill.

The active pillars are modeled as rectangular pillars (50x70µm) while outrigger bumps are modeled as cylindrical pillars (40µm diameter). An outrigger copper pillar bump pitch of 100µm is used for different outrigger copper pillar bump patterns, as shown in Figure 5. All the pillars are modeled with detailed information of copper pillars on the die and substrate side. Lead-free solder joints are modeled as a portion of copper pillars.

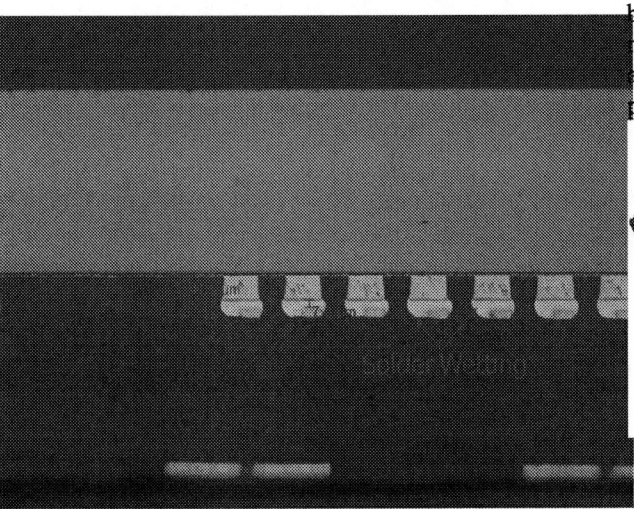

Figure 3. (a) Solder Wetting

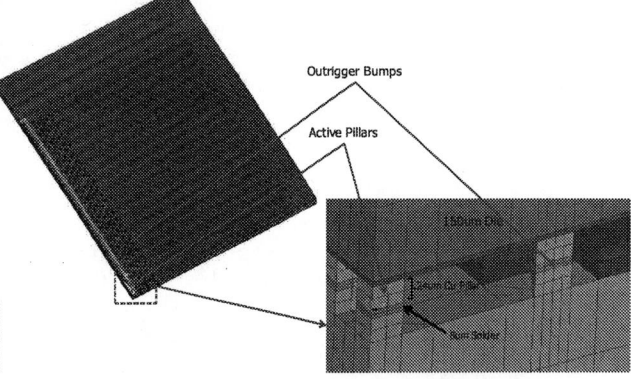

Figure 4. (a) 1st Level Quarter Symmetry FEA Model

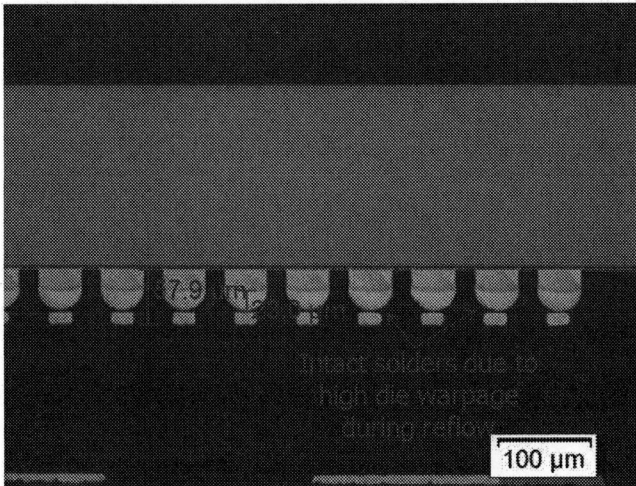

Figure 3. (b) Intact Copper Pillar Bumps Condition (High Die Warpage)

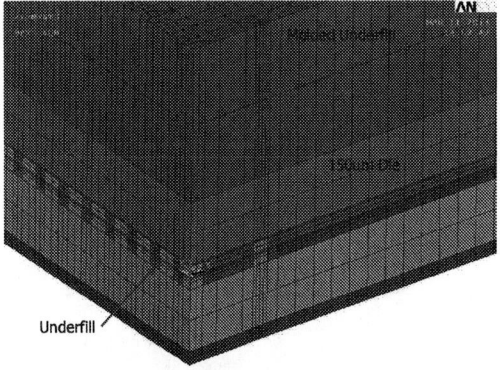

Figure 4. (b) 2nd Level Quarter Symmetry FEA Model

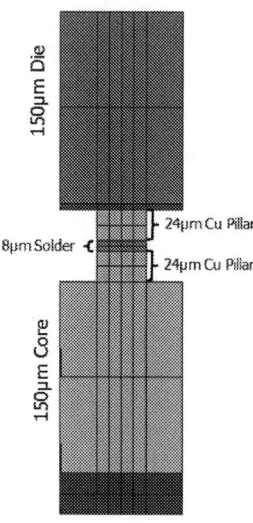

Figure 5. Outrigger Copper Pillar Bump

Figure 6 shows different bump layout patterns (quarter symmetry models) for 1st level DCA packages. Pattern1 has the least number of outriggers, while Pattern2 to Pattern7 have many outrigger copper pillar bumps. The main differences between Pattern2 to Pattern5 are fewer outriggers near the die center. Patten6 has extra copper pillar bumps near the die edge and die center, while Pattern7 has no copper pillar bumps near die center. All the copper pillar bumps for Pattern7 are near the die edges.

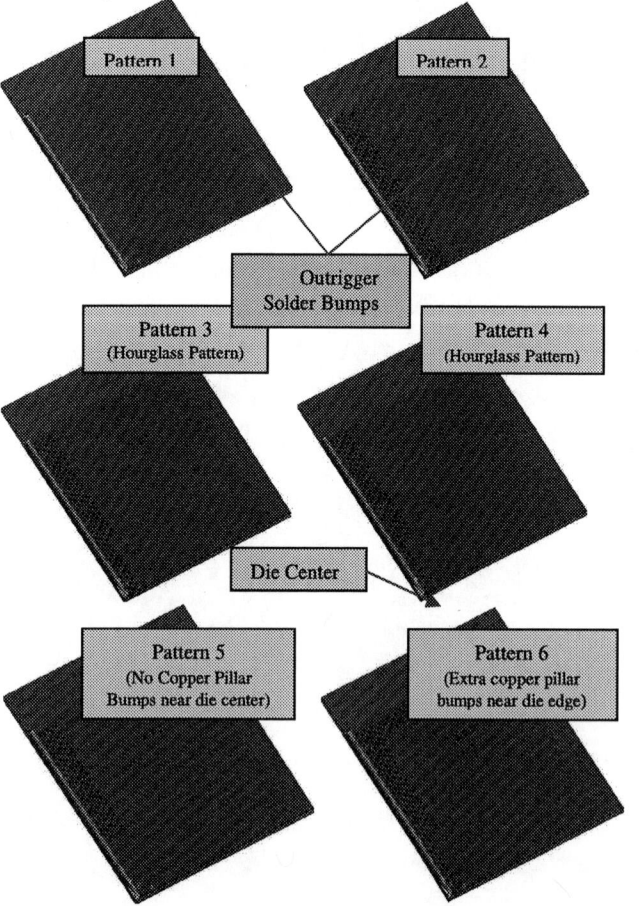

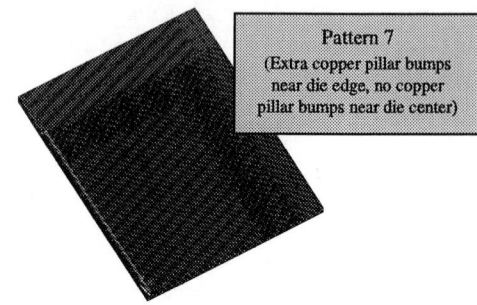

Figure 6. Outrigger Copper Pillar Bump Patterns

Elasto-Plastic Solder Material Model

The lead-free solder in this study is assumed to exhibit elastic, bilinear kinematic hardening plastic behavior after yield. Elasto-plastic material properties in ANSYS can be defined with BKIN option. This option includes the Bauschinger effect and assumes that the total stress range is equal to twice the yield stress.

The initial slope of the curve is taken as the elastic modulus of the material and the tangent modulus of material is calculated after yield stress. The constants (C_1 and C_2) used for the solder bilinear material model are given in the Figure 7.

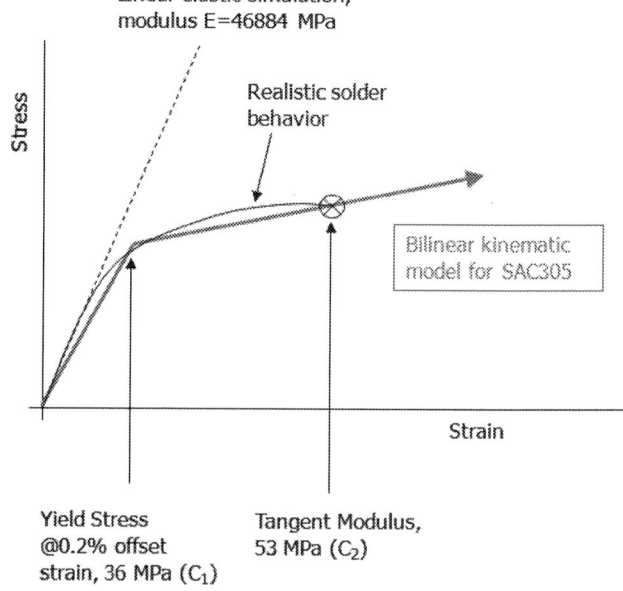

Figure 7. Lead Free Solder Bilinear Material Model

Results and Discussion

For simplicity, all models are simulated at different static temperatures, and results are collected only at room temperature (maximum stress in solder bumps and maximum die warpage is observed at low temperature).

A. Die Warpage

Figure 8 shows the warpage results in the die for 1st level models at room temperature. All the die warpage results are

978-1-4799-2408-0/14 $31.00 © 2014 IEEE

calculated for quarter symmetry models. Maximum die warpage is calculated for Pattern7.

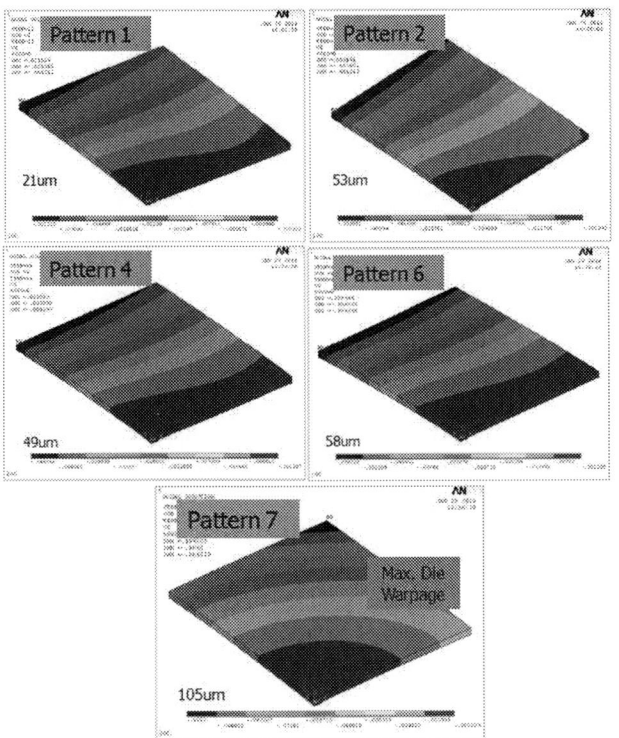

Figure 8. Die Warpage at Room Temperature

B. *Solder Stress*

Figure 9 shows solder stress in copper pillar bumps at room temperature for 1st level models. Maximum active solder stress is calculated for Pattern1.

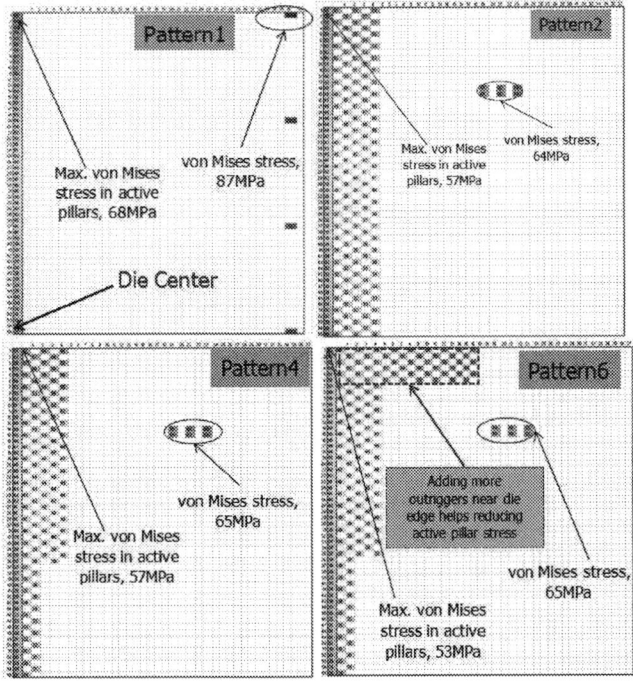

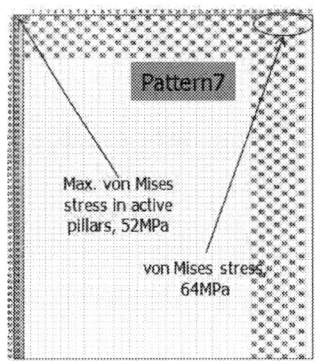

Figure 9. Copper Pillar Bump Solder Joint Stresses at Room Temperature

Table 1 shows maximum von Mises stresses in solder for the active and outrigger copper pillar bumps. Stresses in solder joints for active copper pillar bumps are up to 25% lower than stresses in outrigger copper pillar bump solder joints. The outrigger copper pillars in Pattern 1 show the maximum solder stress because there are a small number of outrigger copper pillars to support the die, but it also has low die warpage due to almost no restraint when die is trying to warp or expand at higher temperatures. Pattern2 to Pattern5 have comparatively higher warpage and lower solder joint stress in active and outrigger arrays than Pattern1. Pattern6 has few outrigger copper pillars near the die edge and comparatively lower solder joint stresses for active pillars bumps.

Stresses in both active and outrigger copper pillar solder joints are significant, but outrigger pillars are basically formed to distribute stresses in the package and lower the stresses on active pillars. Outrigger pillars may be sacrificed to lower the stresses on active pillars.

Pattern7 has no copper pillars near the die center but does have some near the die edge. This causes comparatively low stress on solder joints but high die warpage due to more constraints for the die. Although solder joint stress and die warpage does not vary much from outrigger Pattern2 to Pattern7, outrigger Pattern7 shows the lowest solder joint stress and highest die warpage.

TABLE I
MAX. VON MISES STRESS IN SOLDERS

Outrigger Bump Patterns	Max. von Mises Stress (MPa)	
	Active Solder Joints	Outrigger Solder Joints
Outrigger pattern 1	68	87
Outrigger pattern 2	57	64
Outrigger pattern 3	57	65
Outrigger pattern 4	57	65
Outrigger pattern 5	58	65
Outrigger pattern 6	53	62
Outrigger pattern 7	52	64

Results in Table 1 show that adding or taking out copper pillars near the die center does not affect the solder joint stress and die warpage significantly (Pattern2 to Pattern5), while adding supplementary copper pillars near the die edge may dramatically reduce stresses in the active array (Pattern6 and Pattern7).

C. Effect of Low-CTE Substrate Core

The substrate core's coefficient of thermal expansion is one of the major parameters affecting package warpage and stresses in die-level interconnect. CTE mismatch between die, substrate and mold material causes warpage and leads to die cracking and other reliability issues. Lowering substrate CTE closer to die CTE should reduce the CTE mismatch and reduce die warpage. Simulations were run with two different lower CTE core materials. Pattern6 was selected for this core CTE comparison study. Results are shown in Table 2.

TABLE II
LOW CTE CORE WARPAGE AND SOLDER STRESS

Core Material (60µm)	Die Warpage (µm)	Von Mises Stress (MPa)	
		Active Solders	Outrigger Solders
NXA core (18ppm/C)	58	53	62
MGC NS(FLCA) (3ppm/C)	42	38	46
MGC NS(LCM) (8ppm/C)	48	47	55

Changing NXA core to a lower CTE core material reduced the stresses in the solder joints significantly (20 - 40%). Die warpage was also reduced (15 - 40%). Thus, low CTE cores help packages with warpage control.

D. Effect of Mold Compound and Underfill Material Property

Another major parameter which may affect solder joint stress in copper pillars is mold compound and underfill material. In this paper, two different underfill compounds, molded underfill (MUF) and capillary underfill (CUF), were used to run the simulation studies to determine the effect on solder joint stresses. The differences in material properties in these two materials are shown in Figure 10.

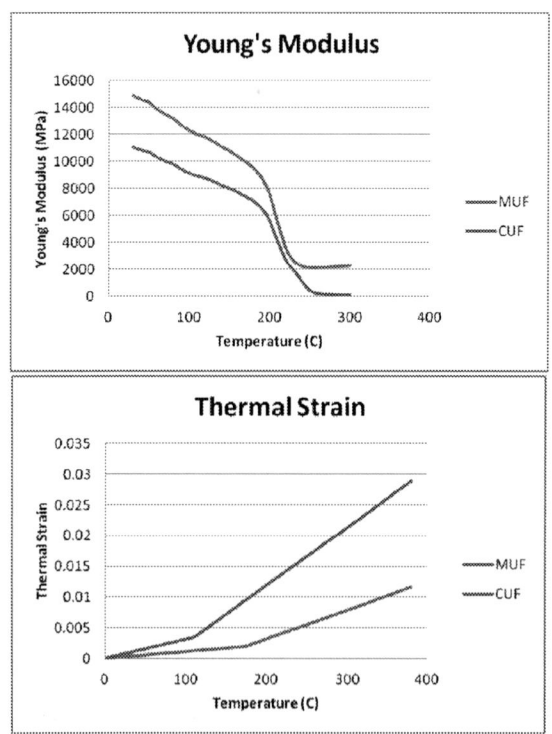

Figure 10. Molded Underfill vs. Capillary Underfill Mold Compound

Simulations were run on FEA models, as shown in Figure 3(b). The results are shown in Table 3.

TABLE III
MUF VS. CUF MOLD COMPOUND MATERIAL

Mold Compound (550µm)	Von Mises Stress (MPa)	
	Active Solder Joints	Outrigger Solder Joints
MUF	30	30
CUF	29	31

Solder stresses are comparable with both MUF and CUF mold compounds.

E. Effect of Increased Solder Joint Thickness

Solder joint thicknesses of active and outrigger copper pillar bumps were increased to determine the effect on solder joint stress and die warpage. To keep the package height the same, solder joint heights were increased from 8µm to 16µm and copper bump thicknesses were reduced from 24µm to 16µm as shown in Figure 11.

With increased solder joint thickness and reduced copper bump thickness, von Mises stresses in solder joints were reduced by 20 – 30%, die warpage was reduced by 10 – 15%.

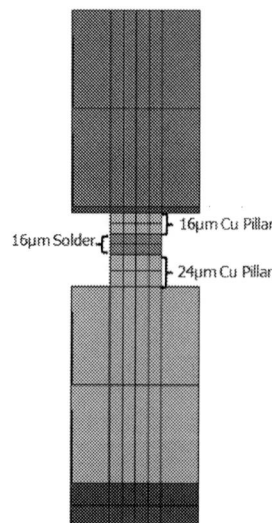

Figure 11. Outrigger Copper Pillar with Increased Solder Joint Thickness

F. Effect of Die Thickness

Die thickness is the main parameter in calculating stresses in copper pillar bump solder joints. The ratio of die thickness to package thickness is very critical. Keeping the package thickness the same, a comparison study was performed to see the effect of die thickness. Simulations were run for two different die thicknesses: 100 microns and 150 microns. Pattern7 was used to run simulations for both 150µm and 100µm die thickness to compare solder bump stresses. Results are shown in Table 4.

TABLE IV
EFFECT OF DIE THICKNESS

Die Thickness	Die Warpage (µm)	Von Mises Stress (MPa)	
		Active Solder Joints	Outrigger Solder Joints
150µm	105	52	64
100µm	160	47	57

Simulation results for 150 micron die thickness show 10-15% higher copper pillar solder joint stresses than with 100 micron die thickness. Increasing die thickness reduces the mold thickness, thus reducing the effective CTE of the package and increasing resultant stress in solder joints.

Conclusions

This paper provides a preliminary study to optimize copper pillar patterns for direct chip attach packages using copper pillar interconnects. FEA models were developed to predict the behavior of various copper pillar patterns and a parametric study was performed to determine the effect on copper pillar bump solder joint stresses and die warpage. The model with low CTE core and molded underfill material show low stresses in solder joints for copper pillars compared to solder joint stresses with high CTE core material and non-underfill packages. Also, thicker solder and die help reduce the stresses in solder joints. Comparing different copper pillar patterns suggests that including more copper pillars near the die edge helps to reduce solder joint stress, but there is trade-off in die warpage.

In this study, lead-free solder is modeled using bilinear kinematic material model. A detailed model can be developed with more advanced time/temperature-dependent material properties, and a more comprehensive study on copper pillar pattern optimization can be performed.

References

1. H. L J. Pang, T.I. Tan, G.Y. Lim, C.L. Wong, "Thermal stress analysis of direct chip attach electronic packaging assembly," Electronic Packaging Technology Conference, 1997, pp. 170-176, 8-10 Oct 1997.

2. D.E.H. Popps, T. Koschmieder, A. Mawer, "Optimization of Direct Chip Attach Variables for Improved Board Level Reliability," International Microelectronics Assembly and Packaging Society Workshop - Flip Chip, 2003.

3. Wei Lin and Min Woo Lee, "PoP/CSP warpage evaluation and viscoelastic modeling," Electronic Components and Technology Conference, 2008, pp.1576-1581, 27-30 May 2008.

4. C. A. Le Gall, J. Qu, D. L. McDowell "Thermomechanical Stresses In An Underfilled Flip Chip DCA". 47th Electronic Components and Technology Conference 1997, Braselton, GA, USA, Mar 1997, pp. 128-129.

5. A. Schubert, R. Dudek, D. Vogel, B. Michel, H. Reichl "Thermo-Mechanical Reliability of Flip Chip Structures Used in DCA and CSP". 48th Electronic Components and Technology Conference 1998, Braselton, GA, USA, Mar 1998, pp. 153-160.

6. J. H. Lau, S. H. Pan, C. Chang "Creep Analysis of Solder Bumped Direct Chip Attach (DCA) on Microvia Build-up Printed Circuit Board with Underfill". 50th Electronic Components and Technology Conference 2000, Hong Kong, pp. 127-135.

Characterization of In-Plane Stress in TSV Array – A Unit Model Approach

Cheng-fu Chen

Department of Mechanical Engineering, University of Alaska Fairbanks
PO BOX 755905, Fairbanks, AK 99775-5905, USA
cf.chen@alaska.edu, (907) 474-7265

Abstract

This paper presents an analytical model to describe the in-plane thermomechanical stresses of a unit TSV model. The goal is to use this model to evaluate the influence of key design parameters in TSV interposers prior to rigorous investigations. The key parameters considered herein are via pitch and array size. This analytical model is based on a circular unit TSV structure, the size of which (in terms of its radius R) determines both the via's volume fraction and the behavior of the stress. The value of R can be $\sqrt{(A/\pi)}$, where A is the area of a unit polygonal pattern in a fabricated TSV array. With a superposition of many such unit TSV structures at a given pitch, we illustrate the stress-interplay issue that becomes apparent in densely populated TSVs. We also use this simple strategy to highlight the stress concentration at the Cu/Si interface, which is critical to thermomechanical reliability design of TSV interposers.

Introduction

Building packaging into the z-direction provides an economic avenue to reduction in form factors and costs for higher pin counts and better performance. It allows an even higher packaging density without a significant increase in the footprint [1]. 3D packaging can be traced back to the early 1990s [2,3], and evolves to the stacked package-on-package in the early 2000s [4]. In the recent year, 3D IC integration is emerging [5] for packaging Moore's law chips in various sectors of consumer electronics. This new 3D technique enables homogeneous and heterogeneous integration [2] because of the key component, the thru-silicon via (TSV) technology [6]. Vertical interconnection and smaller form factors with high pin-counts in 3D IC integration bypasses the limitations in traditional 2D packaging for more efficiency, and performance with less power consumption and costs.

The TSV interposer is one key component for 3D IC. Very fine-pitch and high pin-count can be accomplished through vias on the silicon interposer [7]. However, the TSV is vulnerable to thermal loading, mainly due to the thermal mismatch between the vias and substrate. For copper vias in silicon substrates, the thermomechanical stress is a concern. Although the silicon dioxide passivation liner, landing pad, and the interdielectric layer all contribute and complicate the analysis of thermomechanical stress [8], the reliability issue is roots at the great thermal mismatch between the via and substrate. Often, characterizing the thermomechanical stresses for design-for-reliability is expensive, either numerically [8-11] or experimentally [12]. Such an endeavor is indeed necessary. Yet, it becomes rigorous and not trivial, in particular for systems such as 3D IC with TSV interposers that exhibit a wide span of feature dimensions across a few orders of magnitude.

In this paper, we provide a tactics allowing designers to conduct first-principle evaluation of a TSV design prior to rigorous investigations. An analytical model is developed to serve for this purpose. This method, for simplicity, uses a circular unit TSV model to describe the elastic in-plane thermomechanical stresses. For illustration, we use this model to address a few important features of the thermomechanical stress in a TSV array. These features include the stress interplay that exists in densely populated TSVs, and the stress concentration at the Cu/Si interface. We also discuss the condition under which a single unit model can be representative of the entire interposer for stress analysis.

Description of Stresses and Displacements in TSV

Consider a 2D single unit TSV model in Figure 1, which is a circular two-phase Cu/Si material loaded by the field temperature ΔT and surface traction σ^∞. Here we have neglected other features of small dimensions such as the SiO2 passivation liner, for simplicity. It is noted that a TSV model without the SiO2 passivation liner can be served as a worst-case scenario study, because the liner effectively absorbs thermo-mechanical stress at the TSV/liner interface [13].

This circular model literally is less practical, as most TSV arrays are fabricated in a regular pattern of polygons (e.g., rectangle, square, or hexagon). For this concern, we propose that the radius R of this circular unit TSV model is determined by $\sqrt{(A/\pi)}$, where A is the area of a unit from any polygonal pattern of a TSV array. Nevertheless, this proposal needs more work for validation.

The stresses in the circular model can be described by the equilibrium equations:

$$\frac{\partial \sigma_r}{\partial r} + \frac{1}{r}\frac{\partial \tau_{r\theta}}{\partial \theta} + \frac{\sigma_r - \sigma_\theta}{r} = 0 \tag{1}$$

$$\frac{1}{r}\frac{\partial \sigma_\theta}{\partial \theta} + \frac{\partial \tau_{r\theta}}{\partial r} + \frac{2\tau_{r\theta}}{r} = 0 \tag{2}$$

Axisymmetry is assumed in this model, which yields all stress variables independent of θ. Therefore, no shearing stress exists in this model due to this assumption. The compatibility relation are $\varepsilon_r = \partial u_r / \partial r$, $\varepsilon_\theta = u_r / r$, and $\gamma_{r\theta} = 0$. Together with the use of the plane-stress constitutive equation, the governing equation can be expressed in terms of the radial normal stress σ_r as:

$$r^2 \frac{d^2 \sigma_r}{dr^2} + 3r \frac{d\sigma_r}{dr} = \alpha r \frac{d(\Delta T)}{dr} \tag{3}$$

where α is the coefficient of thermal expansion. (We will shortly address the plane-strain framework.) Here ΔT can be a function of the radial location in the above equation. Below

2014 Electronic Components & Technology Conference

we present the solution for a constant distribution of ΔT for Eq. (3):

$$\sigma_r = C_1 + \frac{C_2}{r^2} \qquad (4)$$

$$\sigma_\theta = C_1 - \frac{C_2}{r^2} \qquad (5)$$

The radial displacement is formulated per the constitutive law:

$$\frac{u_r}{r} = \varepsilon_\theta = \frac{1}{E}\left(\sigma_\theta - \upsilon\sigma_r\right) + \alpha\Delta T \qquad (6)$$

For plane strain problems, it simply replaces the Young's modulus E by $E/1-\upsilon^2$ and the Poisson's ratio υ by $\upsilon/1-\upsilon$ in all the equations addressed above. These equations serve as the platform for solving the stresses and displacements in each of the Cu and Si phases that will be described in the following.

The unknown constants C_1 and C_2 in Eq.(4)-(5) for each material phase are different and can be determined by the boundary conditions of each material phase that will be shortly addressed. To facilitate the following derivation, here we associate labels C_1 and C_2 with the Cu phase, while D_1 and D_2 with the Si phase.

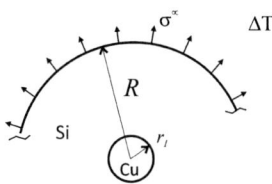

Figure 1. Unit TSV model (without SiO2) under surface traction σ^∞ and a change in the field temperature ΔT.

For the Cu phase, the stresses in terms of Eq. (4)-(5) are determined by the boundary conditions: (1) $u_r = 0$ at $r = 0$, which gives $C_2 = 0$; and (2) $\sigma_r^{Cu} = \sigma^T$ at $r = r_I$. It gives:

$$C_1 = \sigma^T \qquad (7)$$

where σ^T is the thermomechanical stress at the Cu/Si interface and will be determined shortly in Eq. (14). By introducing C_1 and C_2 into Eq. (4)-(5), the stress field in the copper via is:

$$\sigma_r^{Cu} = \sigma_\theta^{Cu} = \sigma^T, \ \tau_{r\theta}^{Cu} = 0 \qquad (8)$$

The displacement, per Eq. (6), is:

$$u_r^{Cu} = \left(\frac{1-\upsilon^{cu}}{E^{cu}}\sigma^T + \alpha^{cu}\Delta T\right)r \qquad (9)$$

The stresses in Si are:

$$\sigma_r^{Si} = D_1 + \frac{D_2}{r^2} \qquad (10)$$

$$\sigma_\theta^{Si} = D_1 - \frac{D_2}{r^2} \qquad (11)$$

Unknown constants D_1 and D_2 are determined by the boundary conditions: (1) $\sigma_r^{Si} = \sigma^T$ at $r = r_I$ and (2) $\sigma_r^{Si} = \sigma^\infty$ at $r = R$. It gives:

$$D_1 = \sigma^\infty\frac{R^2}{R^2 - r_I^2} - \sigma^T\frac{r_I^2}{R^2 - r_I^2}$$
$$D_2 = \left(\sigma^T - \sigma^\infty\right)\frac{r_I^2 R^2}{R^2 - r_I^2} \qquad (12)$$

The displacement field in Si, per Eq. (6), is:

$$u_r^{Si} = \left(\frac{1-\upsilon^{Si}}{E^{Si}}D_1 - \frac{1+\upsilon^{Si}}{E^{Si}}\frac{D_2}{r^2} + \alpha^{Si}\Delta T\right)r \qquad (13)$$

The interfacial stress σ^T at the Cu/Si interface can be determined by requiring that $u_r^{Cu} = u_r^{Si}$ at $r = r_I$:

$$\sigma^T = \frac{-\dfrac{2E^{Cu}}{\left(r_I/R\right)^2 - 1}\sigma^\infty - E^{Si}E^{Cu}\left(\alpha^{Cu} - \alpha^{Si}\right)\Delta T}{E^{Cu}\left(\left(\dfrac{\left(R/r_I\right)^2 + 1}{\left(R/r_I\right)^2 - 1}\right) + \upsilon^{Si}\right) + E^{Si}\left(1-\upsilon^{Cu}\right)} \qquad (14)$$

where σ^T and σ^∞ are positive for tension. Because σ^T is by all means unequal to σ^∞, the constant D_2 is not zero in any condition, which features a decreasing stresses in Si by the order $1/r^2$.

The use of either the plane-stress model or the plain-strain model should be determined by the thickness-to-via aspect ratio. Noted that the aspect ratio is usually in the range [1, 20] for various 3D TSV technologies [14].

Verification of the Theoretical Descriptions

In this section, we use known solutions to verify the derived formulas. For example, the distribution of stress components in Cu and Si has an analytical expression for a uniform traction σ^∞ and $\Delta T = 0$ [15] (p.70). Compared to that, our formulated results show a good agreement (Figure 2). (In this comparison $R = 4r_I$ is used.) Verification of our results by FE simulations, which was conducted in COMSOL, also shows in Figure 2. The normal stresses are proportional to $1/r^2$ in the Si phase. This feature agrees with the published data [8,10,11,16,17].

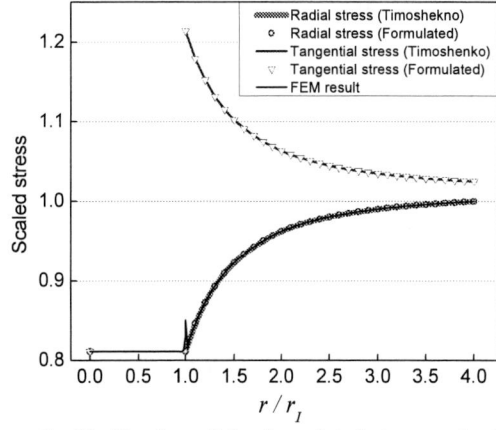

Figure 2. Verification of the formulated stress to the known solution given in [15] for $\Delta T = 0$. Stress scaled by σ^∞.

978-1-4799-2408-0/14 $31.00 © 2014 IEEE

Discussions

The features of the stress in the unit TSV model are characterized first in this section. In the remaining contents we demonstrate how to use this analytical model to study the distribution and interplay of stresses in a TSV array converged value, which corresponds to the limit $R/r_I \rightarrow \infty$ for the unit TSV model. The converged value of each stress is calculated from Eq. (8), (10)-(12) and Eq. (14) as $R/r_I \rightarrow \infty$:

$$\sigma_{r,\text{inf}}^{Cu} = \sigma_{\theta,\text{inf}}^{Cu} = \frac{2E^{Cu}\sigma^{\infty} - E^{Si}E^{Cu}\left(\alpha^{Cu} - \alpha^{Si}\right)\Delta T}{E^{Cu}\left(1 + v^{Si}\right) + E^{Si}\left(1 - v^{Cu}\right)} \quad (15)$$

$$\sigma_{r,\text{inf}}^{Si} = \left(1 - \left(\frac{r_I}{r}\right)^2\right)\sigma^{\infty} + \sigma_{r,\text{inf}}^{Cu}\left(\frac{r_I}{r}\right)^2 \quad (16)$$

$$\sigma_{\theta,\text{inf}}^{Si} = \left(1 + \left(\frac{r_I}{r}\right)^2\right)\sigma^{\infty} - \sigma_{\theta,\text{inf}}^{Cu}\left(\frac{r_I}{r}\right)^2 \quad (17)$$

where the subscript "inf" labels the converged value. In practice, $R/r_I \rightarrow \infty$ ensembles a very dilute TSV population in silicon. Here, the unit TSV structure is stressed-up by two external loads: surface traction σ^{∞} and thermal loading ΔT. For $\Delta T \neq 0$ and $\sigma^{\infty} = 0$ which represents a field-temperature scenario for 3D IC integration, Figure 3 shows a decreasing stress of Eq. (16) and Eq. (17) with an increasing r at an order $\sim 1/r^2$. The tendency agrees with the published results [11].

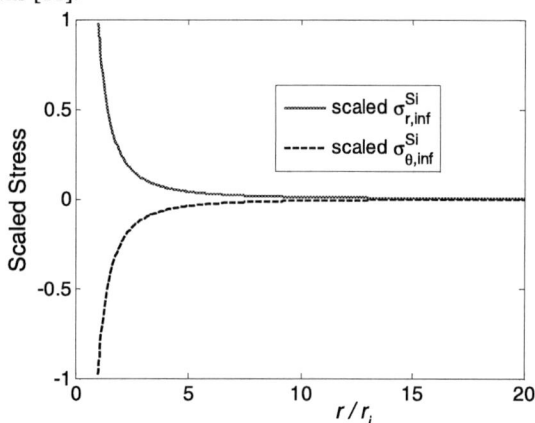

Figure 3. Converged value of the stress in Si for $\Delta T \neq 0$ and $\sigma^{\infty} = 0$. Stress scaled by $\sigma_{r,\text{inf}}^{Cu}$.

I. Stress Induced by Surface Traction σ^{∞}

$\sigma^{\infty} \neq 0, \Delta T = 0$ is not the thermomechanical loading condition seen in packaging. However, with the introduction of σ^{∞} it allows our TSV modeling work to include the SiO$_2$ layer in the future. Figure 4 shows the magnitude of the scaled stresses $\sigma_r^{Cu}/\sigma_{r,\text{inf}}^{Cu}$, $\sigma_{\theta}^{Cu}/\sigma_{\theta,\text{inf}}^{Cu}$, $\sigma_r^{Si}/\sigma_{r,\text{inf}}^{Si}$, and $\sigma_{\theta}^{Si}/\sigma_{\theta,\text{inf}}^{Si}$ when $\Delta T = 0$ is used in Eq. (14)-(17). The stress components in silicon are calculated at five different radial locations: $r = r_I + 0.1dR$, $r_I + 0.25dR$, $r_I + 0.5dR$, $r_I + 0.8dR$, and $r_I + 0.99dR$, individually, where $dR = R - r_I$.

From Figure 4 we observe that: (1) as $R > 5r_I$, which resembles a dilute population of TSVs, the stress in each phase of the TSV model can be well approximated by its converged value; and (2) the scaled value of each stress component in either material phase is universal, because these parametric curves are identical and thus they all overlap among another into a single curve shown in Figure 4.

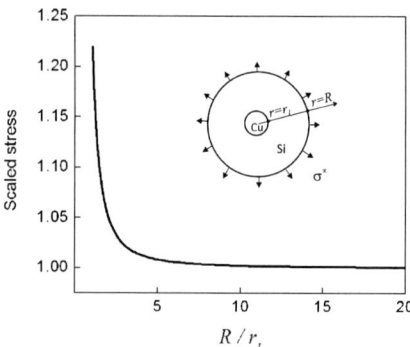

Figure 4. Magnitude of normal stresses in a single unit TAV model of varying size R under a uniform traction with $\Delta T = 0$. Stress scaled by its converged value Eq. (15)-(17).

II. Stress Induced by Thermal Loading ΔT

The loading condition $\sigma^{\infty} = 0, \Delta T \neq 0$ corresponds to the thermal mismatch problem. These radial and tangential stresses are calculated against the parameter R/r_I and scaled per $\sigma_r^{Si}/\sigma_{r,\text{inf}}^{Si}$ and $\sigma_{\theta}^{Si}/\sigma_{\theta,\text{inf}}^{Si}$, as shown in Figure 5. Similar to the results in Figure 4, each stress can be well approximated by its converged value when $R > 5r_I$. Figure 5 shows that a smaller radial stress occurs in the outer region of the silicon matrix, where the tangential normal stress dominates the loading status.

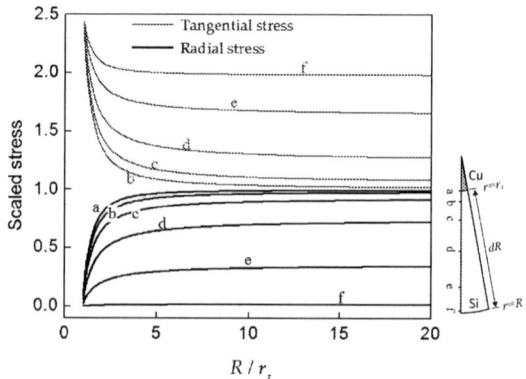

Figure 5. Thermomechanical stresses in the unit TSV model under a non-zero ΔT at various locations: (inset, labels a-f). Each stress is scaled by its converged value. Labels locations—a r_I (at the Cu/Si interface). b $r_I + 0.10dR$. c $r_I + 0.25dR$. d $r_I + 0.50dR$. e $r_I + 0.80dR$. f $r_I + 0.99dR$.

III. Stress Interplay in TSV Array

Stress interplay is the interaction of stresses in TSVs, which has been studied in [8,18] by the finite element simulation, and in [11] by an analytical description of the strain field. The interaction of stress, when happens in the elastic regime, is the superposition (overlay) of the thermomechanical stress induced by the thermal mismatch in one unit TSV over the stress in another in the same substrate continuum. Here we illustrate how our analytical unit TSV model can be used to characterize the stress interplay among TSVs.

By superposition, we place a number of TSV structures in lattice and overlay the individual stress field, each of which was described in Eq. (8), (10), and (11). The extent of stress interplay is evaluated by observing any distortion in the distribution of the thermomechanical stresses around one via in an array of TSVs at various pitches and array size.

Figure 6 shows the contour plot of von Mises stress in a 2×2 TSV array at $3r_I$ pitch. The contours are at an evenly divided interval. Stress concentration is evident by the denser contours, which occur in between vias.

Figure 7 shows a polar plot of the von Mises stress around each via at a radial distance $r = 1.01r_I$, $1.05r_I$ and $1.10r_I$, respectively, in the 2×2 array of Figure 6. The circumferential expression of the von Mises stress concentrates at $90°$ apart, nearly at a polar symmetry (by $90°$) around a via. This exhibition agrees with the result by finite element modeling [13]. The four branches outline the stress concentration regions, where the "keep-away" zone [19] should be placed for better electron mobility [20].

Noted that the polar plot curve at $r = 1.10r_I$ changes more than at $r = 1.01r_I$. Also noted that the larger "contour branches" in between vias break the polar symmetry, and thus may be served as an indicator to the extent of stress interplay, which may be calculated as follow. For the polar plot of TSV1 in Figure 7, for example, the von Mises stress at $r = 1.01r_I$ in the $0°$ and $90°$ directions is about $0.2\,\sigma_{r,\inf}^{Cu}$ larger than in the $180°$ and $270°$ directions. This value $0.2\,\sigma_{r,\inf}^{Cu}$ may be used as an indicator to the stress interplay for this case.

As compared to the case of pitch $3r_I$, Figure 8 show the contour of von Mises stress in a 2×2 TSV array at a larger pitch ($5r_I$). It shows that at a larger via pitch, there is less stress interplay and less stress concentration, as evidenced by comparing Figure 6 (at a smaller pitch) and Figure 8 (at a larger pitch). There is barely any stress interplay when TSVs are at pitch $5r_I$, as shown in Figure 9 with a better polar symmetry than in Figure 7.

The thermomechanical stress is repeated in pattern in a large TSV array. It can be seen, for instance, in Figure 10 of a 7×7 TSV array at $3r_I$ pitch. Because the thermomechanical stress decreases by $1/r^2$ at a radial distance r, the thermomechanical stress from far-field vias will be insignificant. Therefore, the pattern of distribution of the thermomechanical stresses becomes repeatable and predictable from a smaller array.

Again, in the silicon substrate a decreasing stress follows the order $1/r^2$ at a radial distance r. Therefore, the added stress from the thermal mismatch of far-field TSVs imposes barely anything to the stress at the near site of concern. An increasing array size would barely load up the thermomechanical stress in the TSV array, as evidenced by Figure 11. Therefore, it expects that the thermomechanical stress in a TSV interposer should reach a converged value as the array size increases. This converged value should be a useful indicator to an initial evaluation of a TSV design.

The distribution of thermomechanical stress along various y-levels across a TSV array (Figure 11) sheds insight into two important features of the stress in TSV interposers. First, concentration of thermomechanical stress occurs at the Si/Cu interface. Stress concentration is less in a smaller TSV array. The second feature observed from Figure 11 is that the thermomechanical stress in Cu via is much higher than in the Si substrate. It implies that failure modes, if any, will be mainly associated with Cu vias and their boundary annex to SiO2 and Si.

This theoretical model enables a quick evaluation of the thermomechanical stress in TSV interposers, by varying the pitch size and pitch-to-via-diameter ratio for meeting a satisfactory design.

Conclusions

In this paper, an analytical description of the stress was given for a unit circular TSV model subject to thermal loading ΔT and uniform traction. This 2D analytical model was established by assuming axisymmetry and elasticity, and suitable for either plane-stress or plane-strain problems. The determination of which framework should be used, however, is beyond the scope of this paper and will be explored. The features of stresses in the unit TSV model were characterized against the parameter R/r_I, which is the ratio of radii of the unit model and the via. We have proposed an approach to relate the radius R of the unit TSV model to the size of any polygonal unit pattern in fabricated TSV interposers. In the second part of this paper, a few case studies were used to demonstrate the effects of pitch and array size on the distribution and magnitude of the thermomechanical stress in an TSV array. The stress interplay is apparent in between vias that are densely distributed (in our study, pitch at $3r_I$ is a dense distribution). The maximal thermomechanical stress occurs at the Cu/Si interface, and is bigger in a 7×7 array than in a smaller array. The stress in Cu vias is larger than in Si, which means vias and their annex layer are more vulnerable to failure. The distribution of the thermomechanical stress in a large TSV array is repeated, and may be predicted by the distribution of the thermomechanical stress in a smaller TSV array patched from its mother counterpart.

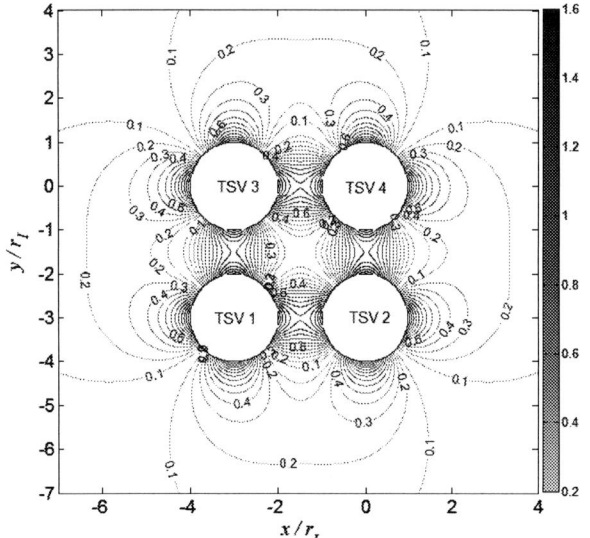

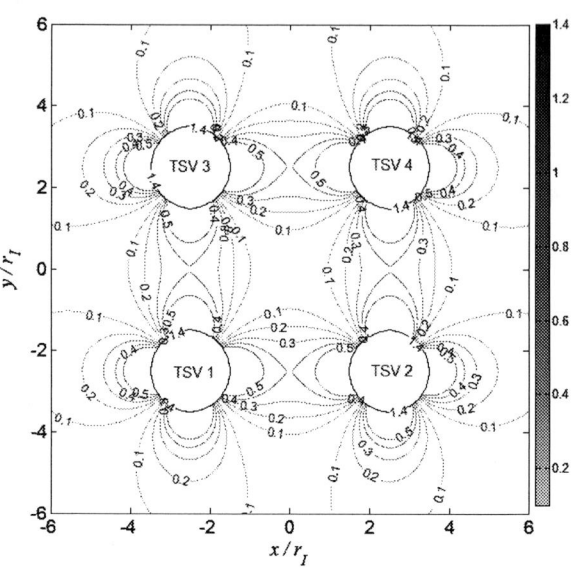

Figure 6. Contour of von Mises stress around a 2×2 TSV array at $3r_I$ pitch. Vias centered at (-3,-3), (0, -3), (-3,0), and (0,3) in the scaled coordinate system. Dimension scaled by via's radius r_I. Stress scaled by $\sigma_{r,\text{inf}}^{Cu}$.

Figure 8. Contour of von Mises stress around a 2×2 TSV array at $5r_I$ pitch. Vias centered at (-2.5, 2.5), (2.5, 2.5), (2.5, -2.5), and (-2.5, -2.5). Dimension scaled by via's radius r_I. Stress scaled by $\sigma_{r,\text{inf}}^{Cu}$.

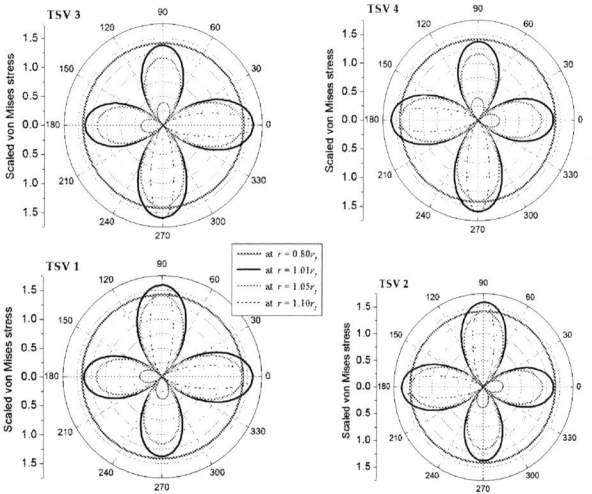

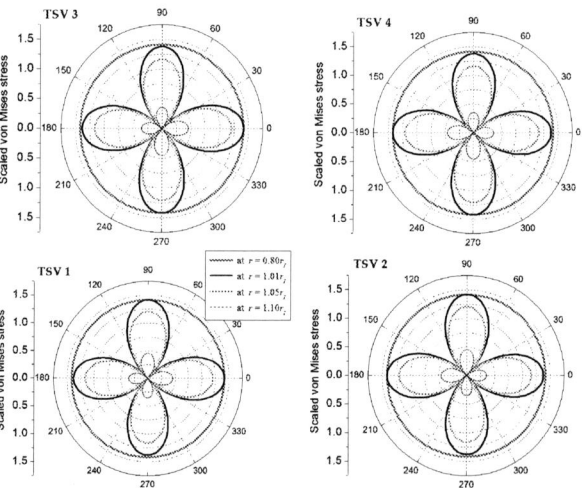

Figure 7. Polar plot of the von Mises stress around the center of each TSV in 2×2 TSV array at $3r_I$ pitch. (Refer to Figure 6 for the location of each via.)

Figure 9. Polar plot of the von Mises stress around the center of each TSV in 2×2 TSV array at $5r_I$ pitch. (Refer to Figure 8 for the location of each via.)

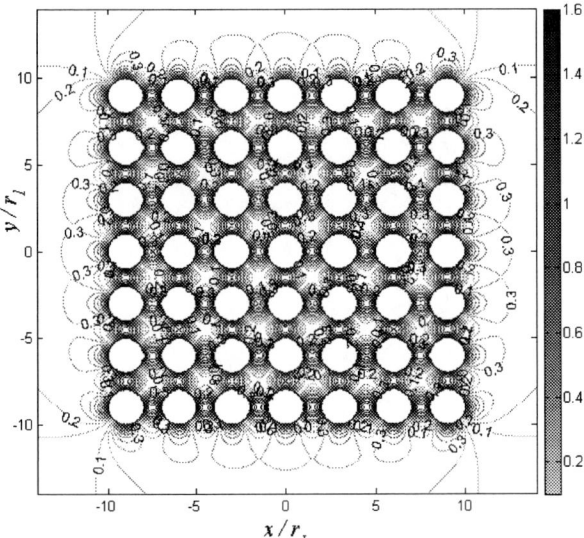

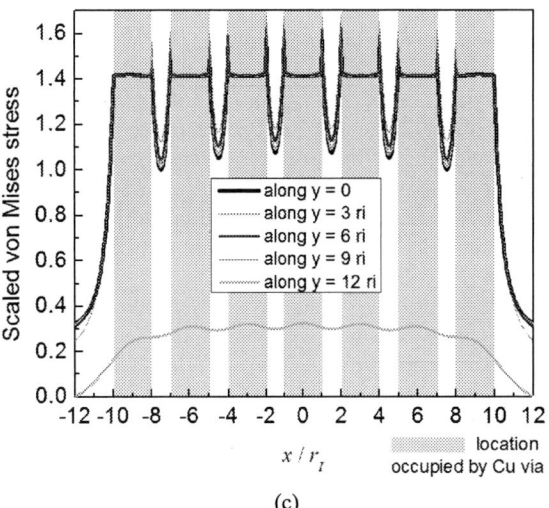

Figure 10. Contour of von Mises stress around a 7×7 TSV array at $3r_I$ pitch. Dimension scaled by via's radius r_I. Stress scaled by $\sigma_{r,\mathrm{inf}}^{Cu}$.

Figure 11. Comparison of von Mises stress at various y-levels of square TSV array under thermal loading. (a) 3×3 array at $4r_I$ pitch. (b) 3×3 array at $3r_I$ pitch. (c) 7×7 array at $3r_I$ pitch (c.f. Figure 10). Dimension scaled by via's radius r_I. Stress scaled by $\sigma_{r,\mathrm{inf}}^{Cu}$.

References

1. T. S. Cale, J.-Q. Lu., and R. J. Gutmann, "Three-Dimensional Integration in Microelectronics: Motivation, Processing, and Thermomechanical Modeling," *Chem. Eng. Commun.,* vol. 195, no. 8 pp. 847-888, 2008.

2. J. U. Knickerbocker, P. S. Andry, D. Dang *et al.*, "Three-Dimensional Silicon Integration," *IBM J. Res. & Dev.,* vol. 32, no. 6 pp. 553-569, 2008.

3. A. Papanikolaou, D. Soudris, and R. Radojcic, Three Dimensional System Integration IC Stacking Process and Design, Springer, 2011, pp. 13-32.

4. Smith, L., Achieving the 3rd generation from 3D packaging to 3D IC architectures. (2010), *Future Fab International Issue,* Chap. 34, p. 7 p.

5. J. H. Lau, "Start-of-the-Art and Trends in 3D IC/Si Integrations and WLP," *Short Course Development in Electronic Manufacturing Technology Symposium (IEMT)*, pp. 1-3.

6. J. H. Lau, "Key Enabling Technologies for 3D IC Integrations," *Professional Development Course, IEEE Electronic $ Components Technology Conferences.*

7. C. Selvanayagam, J. Lau, X. Zhang *et al.*, "Nonlinear Thermal Stress/Strain Analyses of Copper Filled TSV (Through Silicon Via) and Their Flip-Chip Microbumps," *IEEE Trans. Adv. Packag.,* vol. 32, no. 4 pp. 720-728, 2009.

8. M. Jung, J. Mitra, D. Z. Pan *et al.*, "TSV Stress-Aware Full-Chip Mechanical Reliability Analysis and Optimization for 3D IC," *Design Automation Conference (DAC)*, pp. 188-193.

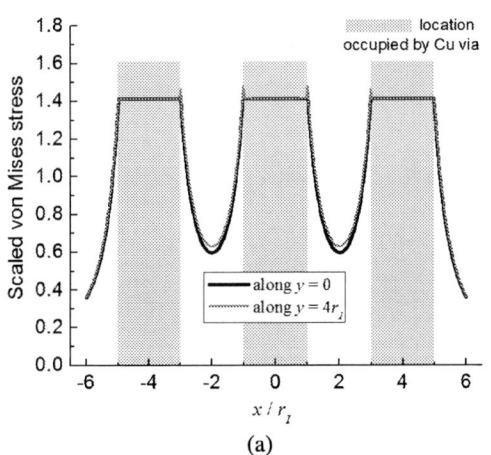

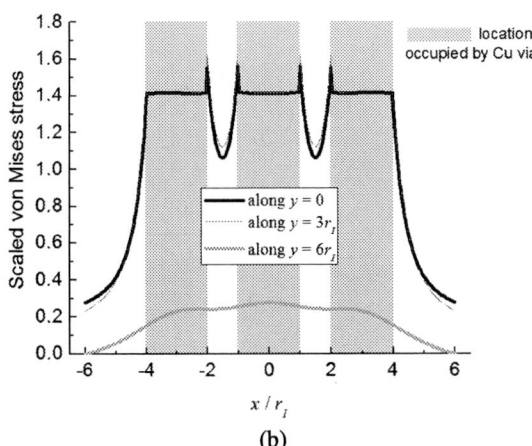

9. S. Ryu, K. Lu, J. Im *et al.*, "Stress Induced Delamination of Through Silicon Via Structures," *AIP Conf. Proc.*, pp. 153-167.

10. S. Ryu, K. Lu, X. Zhang *et al.*, "Thermomechanical Reliability Challenges for 3D Interconnects With Throughsilicon Vias," *AIP Conf. Proc.*, pp. 189-202.

11. S.-R. Jan, T.-P. Chou, C.-Y. Yeh *et al.*, "A Compact Analytic Model of the Strain Field Induced by Through Silicon Vias," *IEEE Trans. Electron Devices,* vol. 59, no. 3 pp. 777-782, 2012.

12. J. A. Wakil, P. W. Dehaven, N. R. Klymko *et al.*, "Thermo-Mechanical Response of Thru-Silicon Vias Under Local Thermal Transients Using Experimentally Validated Finite Element Models," *J. Electronic Packaging,* vol. 133, p. 031001, 2011.

13. J. Mitra, M. Jung, S.-K. Ryu *et al.*, "A Fast Simulation Framework for Full Chip Thermo-Mechanical Stress and Reliability Analysis of Through-Silicon Via Based 3D ICs," *Electronic Components and Technilogy Conference*, Lake Buena Vista, FL, USA, pp. 746-753.

14. G. Van der Plas, P. Limaye, I. Loi *et al.*, "Design Issues and Considerations for Low-Cost 3-D TSV IC Technology," *IEEE J. Solid-State Circuits,* vol. 46, no. 1 pp. 293-307, 2011.

15. S. Timoshenko *et al.*, *Theory of Elasticity,* 2nd ed., McGraw-Hill, New York, 1970,

16. T. C. Lu, J. Yang, Z. Suo *et al.*, "Matrix Cracking in Intermetallic Composites Caused by Thermal Expansion Mismatch," *Acta Metall. Mater.,* vol. 39, no. 8 pp. 1883-1890, 1991.

17. K. Lu, X. Zhang, S.-K. Ryu *et al.*, "Thermo-Mechanical Reliability of 3-D ICs Containing Through Silicon Vias," *59th Electronic Components and Technology Conference*, San Diego, pp. 630-634.

18. L. J. Ladani, "Numerical Analysis of Thermo-Mechanical Reliability of Through Silicon Vias (TSVs) and Solder Interconnects in 3-Dimensional Integrated Circuits," *Microelectron. Eng.,* vol. 87, pp. 208-215, 2012.

19. S.-H. Hwang, B.-J. Kim, H.-Y. Lee *et al.*, "Electrical and Mechanical Properties of Through-Silicon Vias and Bonding Layers in Stacked Wafers for 3D Integrated Circuits," *J. Electronic Materials,* vol. 41, no. 2 pp. 232-240, 2012.

20. S. E. Thompson, G. Y. Sun, Y. S. Choi *et al.*, "Uniaxial-Process-Induced Strained-Si: Extending the CMOS Roadmap," *IEEE Trans Electronic Devices,* vol. 53, pp. 1010-1020, 2006.

Electrical-Thermal Characterization of Wires in Packages

Kai Liu, Robert Frye*, HyunTai Kim, YongTaek Lee, Gwang Kim, Susan Park, and Billy Ahn

STATS ChipPAC

1711 W. Greentree Dr., Suite #117, Tempe, Arizona, 850284

kai.liu@statschippac.com, 480-222-1722

*RF Design Consulting, LLC

Abstract

The electrical-thermal co-simulation approaches, for wires-in-air and wires-in-package, are developed by the coupling between their electrical and thermal properties, using ADS (Agilent Design System) Symbolically-Defined Devices (SDD) models for multiple wire segments. Key parameters for these simulation models are then derived from experimental results. These experimentally validated (or assisted) simulation models can be used to predict electrical-thermal behavior of bond wires in situations of interest, and to develop design guidelines for reliable operation.

Test boards with wires-in-air were made, and fusing currents on wires of different materials (Au, Cu, Ag), lengths, and diameters were measured and compared with published data. Some QFN package testers having wires in different materials(Au, Cu, Ag), lengths, and different diameters, with mold material were also made to characterize the wires in real package environment. Simulation and experiment data, as well as some failure-analysis (FA) data through X-ray and SEM methods, are presented in the paper.

Introduction

Trends in package technology have led to the adoption of new materials and the use of smaller conductor geometries in package structures. Copper bond wires, for example, are used increasingly in place of gold wires. To accommodate increased numbers of die signal pads in smaller die sizes, chip designs use smaller pad-dimensions, requiring the use of smaller bond wire diameters in wire-bonded packages. These same trends have led to the adoption of smaller bump sizes in flip-chip designs and smaller line-width and space dimensions in package substrates.

On the other hand, trends in CMOS IC technology have also been toward smaller dimensions. To accomplish these changes in the physical dimensions of the transistors, it has been necessary (and desirable) to scale down the operating voltage of digital circuits. The general approach that has been adopted by the semiconductor industry is referred to as *constant power scaling*, in which the decreased operating voltages are accompanied by a proportional increase in overall operating current levels.

Together, these trends have led to appreciable increases in current density levels in the interconnecting conductors used in packages. Therefore, it is imperative that new design guide lines should be established to reflect these trends.

Many of the reliability concerns for package interconnection structures at high currents arise from the elevated temperatures that are developed as a result of resistive heating in the conductors. In contrast to electrical parameters, which can often be measured with great precision, thermal effects are much more difficult to precisely measure. This is especially true for very small structures like wire bonds or package traces. Because of their small thermal mass, temperature probes such as thermocouples or thermistors often perturb the result by conducting heat away from the Device Under Test (DUT). However, with reasonable simulation models, the DUT's thermal state can in many cases be inferred from its electrical behavior.

Thermal Circuit

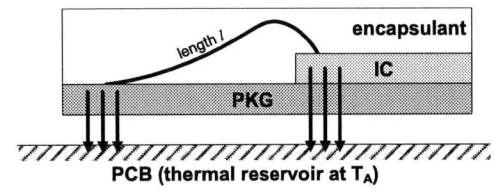

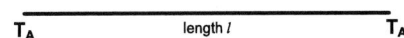

Figure 1. Typical bond wire package configuration and idealized representation.

Figure 1A shows a typical configuration for a bond wire in a package. The bond wire connects pads on the top surface of the IC to the underlying package. Heat is generated by the electrical resistance of the wire, the IC and the package, and flows into the underlying PCB, mainly through the electrical contacts, where it spreads out and is dissipated into the surrounding ambient. Typically, the package and PCB are very massive compared with the bond wire itself, and the resistance to thermal flow into the PCB is small. The dielectric encapsulation material (usually molding compound) surrounding the bond wire is a poor thermal conductor compared with the metal pathways, including the wire itself. So, especially near the ends of the wire, a key pathway for heat dissipation is via conduction along the length of the wire and through the package into the PCB. For analysis, the PCB itself is usually considered to be an infinite thermal reservoir at the ambient temperature, T_A. However, the overall thermal resistance of this conductive pathway grows with wire length. So, especially for longer wires, in the regions farther from the ends direct thermal conduction through the molding compound may be the dominant mechanism.

The heat flow equation for the case of an electrically-heated conductor in which the heat predominantly flows through the conductor itself is

$$\frac{\partial T}{\partial t} = \frac{k}{c_p \rho} \nabla^2 T + \frac{1}{c_p \rho} \vec{J} \cdot \vec{E} \qquad (1)$$

The left-hand side of this equation is the time rate of change in the temperature, T. The first term on the right-hand side represents heat flow by conduction and the second represents electrical heat generation. In this equation, c_p is the heat capacity at constant pressure, ρ is the density and k is the thermal conductivity. The electrical power dissipated in the material is given by the scalar product of the current density, J, and the electric field, E. More generally, a third term on the right-hand side would be included to account for radiated power, but this is assumed to be negligible in this case.

In an electrically conducting material the current density and electric field are related by Ohm's law

$$\vec{J} = \sigma \vec{E} \qquad (2)$$

where σ is the electrical conductivity. In this case, Equation (1) becomes

$$\frac{\partial T}{\partial t} = \frac{k}{c_p \rho} \nabla^2 T + \frac{1}{c_p \rho} \frac{J^2}{\sigma}. \qquad (3)$$

Figure 1B shows a highly idealized representation of the bond wire that is used for analysis. It is assumed that the thermal resistances of the pathways from the wire ends into the PCB are low compared with the resistance of the wire itself, and can be neglected. In this case, the ends of the wire

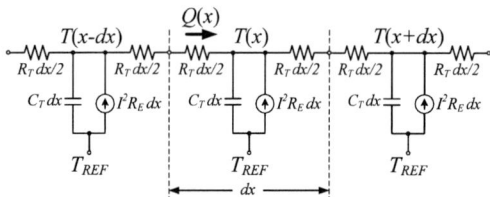

Figure 3. Distributed thermal circuit model.

can be assumed to be at the ambient temperature T_A. Furthermore, additional dissipation by conduction into the surrounding encapsulation material is assumed to be negligible. In this case, the only heat flow is along the length of the wire, and the heat flow equation is one-dimensional:

$$\frac{\partial T}{\partial t} = \frac{k}{c_p \rho} \frac{\partial^2 T}{\partial x^2} + \frac{1}{c_p \rho} \frac{J^2}{\sigma} \qquad (4)$$

For uniform current density through a wire of cross-section area A, Equation (4) becomes

$$\frac{\partial T}{\partial t} = \frac{k}{c_p \rho} \frac{\partial^2 T}{\partial x^2} + \frac{1}{c_p \rho} \frac{I(t)^2}{A^2 \sigma} \qquad (5)$$

where $I(t)$ is the total current flowing through the wire. It is assumed that the current is only a function of time, and is uniform along the wire's length.

Solutions of the heat-flow equation describe wave-type behavior for the temperature distribution. It has long been recognized that flow of heat in response to gradients in the temperature is directly analogous to the flow of charges (electrical current) in response to gradients in the potential (electrical voltage). Furthermore, heat flow can be modeled using circuit element concepts [1]. For this one-dimensional heat flow problem, we can define several circuit-element analogs.

We define the thermal capacitance per unit length to be

$$C_T \equiv c_p \rho A \qquad (6)$$

The thermal resistance per unit length is defined as

$$R_T = \frac{1}{kA} \qquad (7)$$

and the electrical resistance per unit length as

$$R_E \equiv \frac{1}{\sigma A} \qquad (8)$$

With these definitions, Equation (5) becomes

$$C_T \frac{\partial T}{\partial t} - \frac{1}{R_T} \frac{\partial^2 T}{\partial x^2} = I^2 R_E \qquad (9)$$

The left side of this equation is identical in form to the electrical equations describing voltage propagation in a distributed RC delay line. In this equation, the temperature T is analogous to the voltage in an electrical circuit. At any point along the distributed line, the first term represents a "current" (actually heat flow) flowing out of that point through a shunt capacitance. The second term represents the net heat flow out of that point through the series resistance.

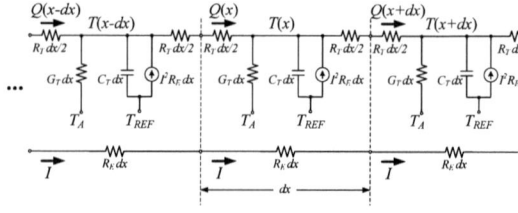

Figure 2. Coupled thermal-electrical circuit model.

The right-hand side of the equation is a heat-flow current that is injected into the region (i.e. a source) representing the electrical joule heating.

Figure 2 shows a distributed circuit representation of the heat flow equation. In this circuit, the node voltages represent temperatures and the branch currents represent heat flows. The length of the wire bond is subdivided into incremental sections of length dx, and the lumped circuit elements are calculated from the per-unit-length parameters described in Equations (7)-(9). The reference temperature T_{REF} in this circuit is analogous to the ground potential in an electrical circuit.

The one-dimensional heat flow equation (Equation (9)) assumes that the only appreciable mechanism of heat flow is along the wire. In practical situations, however, some of the heat is removed into the material surrounding the wire, typically molding compound or, in the case of some of the experimental results, air. This mechanism becomes more important for longer wires. The resistance to heat flow into the surrounding material is the same at all points along the length of the wire, but the resistance to heat flow out the ends of the wire grows with distance from the ends. The modified version of Equation (9) that accounts for this is

$$G_T(T - T_A) + C_T \frac{\partial T}{\partial t} - \frac{1}{R_T}\frac{\partial^2 T}{\partial x^2} = I^2 R_E \qquad (10)$$

where G_T is a thermal conductance per unit length. For a cylindrical wire of diameter d, the thermal conductance for heat transfer into an infinite surrounding medium is typically represented by

$$G_T = \pi d h \qquad (11)$$

where h is the heat-transfer coefficient. In this model it is assumed that the heat transfer is proportional to the outer surface area of the wire and the local temperature difference between the wire and the ambient. Unlike the other parameters, the heat-transfer coefficient is not an intrinsic property of the wire material. Instead, it depends mainly on the nature of the surrounding medium.

Figure 3 shows the coupled electrical-thermal simulation model. In this circuit model, the lower part represents the electrical pathway through the wire, and the upper part represents the thermal pathway. The main coupling between the two is via the dependent sources that drive "current" (i.e. heat) into the thermal part of the circuit that is proportional to the electrical power generated in each section. Further coupling occurs that is not shown, since the electrical resistance R_E is dependent on the temperature. This is one of several nonlinear effects that must be included in the model.

Nonlinear Effects

Equation (9) is a linear partial-differential equation, and can be solved analytically for the simple boundary conditions described in Figure 1. However, a careful examination of the material properties of the metals used in bond wires shows that the key thermal and electrical parameters are, themselves, temperature-dependent. This makes the equations nonlinear.

The electrical conductivity, for example, shows a large variation with temperature. Over the range from room temperature to the melting temperature, the electrical resistivity of metals is reasonably well described by a first-order behavior. This dependence is given by

$$\sigma = \frac{\sigma_0}{1 + \alpha(T - T_{REF})} \qquad (12)$$

where σ_0 is the conductivity at temperature T_{REF} and α is the temperature coefficient of resistance. Typically, these parameters are defined at a reference temperature of 20C. For the calculation of fusing currents, the nonlinear variation in electrical conductivity is a significant effect. At its melting temperature, the conductivity of the metal is typically only 25-30% of its value at room temperature.

The thermal conductivity also shows variation with temperature, and over the range from room temperature to the melting point it can also be represented by a first order dependence

$$k = k_0[1 + \beta(T - T_{REF})] \qquad (13)$$

where k_0 is the thermal conductivity measured at temperature T_{REF} and β is the first-order temperature coefficient of the thermal conductivity. Like the electrical conductivity, the thermal conductivity of bond wire metals decreases with increasing temperature, so β is negative. However, the magnitude of the effect is less. Typically the thermal conductivity at the melting temperature is 5-10% less than at room temperature.

Finally, the heat capacity of the metal shows a similar dependence

$$c_p = c_{p0}[1 + \gamma(T - T_{REF})] \qquad (14)$$

Heat capacity also increases slightly with increasing temperature.

It is also the case that the density is temperature dependent. This is related to the coefficient of thermal expansion. But this effect is relatively minor compared with the above effects and can be ignored in this analysis. Table 1

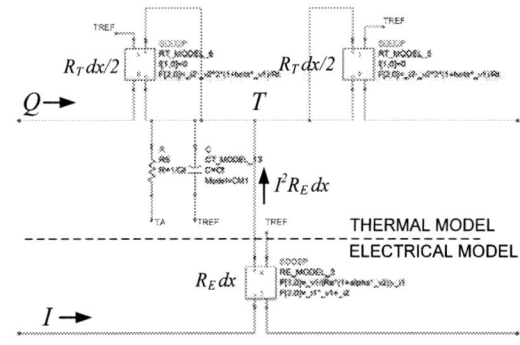

Figure 4. Agilent ADS implementation.

lists values for the nonlinear effects described above in pure metals, derived from a variety of sources [2]-[4].

Table 1. Material parameters for pure metals at T_{REF}=20C.

Parameter	Units	Ag	Au	Cu
σ_0	Sm^{-1}	$6.30 \cdot 10^7$	$4.10 \cdot 10^7$	$5.95 \cdot 10^7$
α	K^{-1}	$4.31 \cdot 10^{-3}$	$3.73 \cdot 10^{-3}$	$3.93 \cdot 10^{-3}$
k_0	Wm^{-1}K^{-1}	427	315	398
β	K^{-1}	$-1.82 \cdot 10^{-4}$	$-1.87 \cdot 10^{-4}$	$-1.56 \cdot 10^{-4}$
c_0	Jg^{-1}K^{-1}	0.230	0.126	0.386
γ	K^{-1}	$1.93 \cdot 10^{-4}$	$2.39 \cdot 10^{-4}$	$2.49 \cdot 10^{-4}$
ρ	gm^{-3}	$1.05 \cdot 10^7$	$1.89 \cdot 10^7$	$8.94 \cdot 10^6$

When these more accurate temperature-dependent material parameters are used, Equation (5) is nonlinear and cannot be solved analytically. To circumvent this difficulty, in an analysis of fusing time and current Loh [5] effectively used a more approximate form in which these parameters were replaced by their average value over the temperature range of interest, such as,

$$\frac{I^2}{\overline{\sigma} A} = -A\overline{k}\frac{d^2 T}{dx^2} + A\rho\overline{c}_p \frac{dT}{dt} \quad (15)$$

In addition, Loh's analysis focused on the calculation of the fusing time. In the more general case, we are interested in a detailed analysis of the spatial distribution of the temperature along the wire at lower temperatures. This makes it possible to determine the changes in the observed wire resistance, which can be used to relate experimental observations with the peak temperatures in the wire.

An advantage of the thermal circuit model is that it can be conveniently implemented in an electrical circuit simulator to find the transient behavior of the temperature, taking into account the nonlinear effects. An additional advantage, as will be seen, is that the thermal circuit model is easily extended to include the effects of non-ideal boundary conditions, which are difficult to achieve in practical measurements.

Simulation Methodology

The coupled flow shown in the model in Figure 3 can be implemented in a transient circuit simulator such as Agilent ADS. The basic circuit implementation for the incremental length of line described above is shown in Figure 4. In this circuit, the two-port block elements are Symbolically-Defined Devices (SDD) from the "Eqn Based-Nonlinear" palette of ADS. The upper part of the circuit models the heat flow path along the wire, and the bottom half models the electrical charge flow path.

In the ADS implementation, the incremental, temperature-dependent electrical resistance, R_E, and thermal resistance, R_T are represented by the SDD blocks. Furthermore, the block representing R_E generates an output "current" at port-2, representing the heat flow into the thermal part of the circuit.

In the SDD block representing R_E, the net voltage on port-2 is

$$\_v2 = T - T_{REF} \quad (16)$$

The "current" (i.e. heat flow) from port 2 of the R_E block is equal to the power dissipated in the electrical part (port-1) of the circuit:

$$\_i2 = -\_v1 * \_i1 \quad (17)$$

The minus sign in the right-hand side of Equation (17) takes into account the fact that positive power in port-1 should cause current to flow *out* from the positive terminal of port-2. In the ADS convention, this corresponds to negative current. Equation (17) is implemented in the ADS SDD component using the implicit relationship

$$0 = \_i2 + \_v1 * \_i1 = F(2,0) \quad (18)$$

Port-1 of this device represents the temperature-dependent electrical resistance of the line segment, which is given by

$$R_E = \frac{dx}{\sigma_0 A}\left[1 + \alpha(T - T_{REF})\right] \quad (19)$$

In the SDD device, using Equation (10), the equation defining the electrical characteristics of port-1 is

$$\_i1 = \_v1/R = \_v1/\left[Re*(1 + alpha * \_v2)\right] \quad (20)$$

where

$$Re = \frac{dx}{\sigma_0 A} \quad (21)$$

The implicit relationship for port-1 is

$$0 = \_v1/\left[Re*(1 + alpha * \_v2)\right] - \_i1 = F(1,0) \quad (22)$$

The two SDD blocks in the upper part of the circuit represent the two nonlinear thermal resistances in Figure 3. Port-1 of these devices only senses temperature. The voltage across this port is given by

$$\_v1 = T - T_{REF} \quad (23)$$

No current flows in this port, so the explicit equation describing its I-V characteristics is

$$I(1,0) = 0 \quad (24)$$

Port-2 represents the thermal resistance. Its voltage-current relationship is given by

$$\_i2 = \_v2/(R_T/2) = 2*\_v2*(1 + beta*\_v1)/Rt \quad (25)$$

where

$$Rt = \frac{dx}{k_0 A} \quad (26)$$

978-1-4799-2408-0/14 $31.00 © 2014 IEEE

Table 2: Measured parameters of the wires at low current.

Metal	d_{NOM} (mil)	σ_{NOM} (S/m)	R_E (Ω/m)	d_{EFF} (mil)	σ_{EFF} (S/m)	R_{PKG} (mΩ)
Au-99	1.0	3.42e7	82.7	0.83	2.46e7	6.07
	0.6		206.2	0.53	2.75e7	14.14
Pd-coated Cu	1.0	5.57e7	38.3	0.96	5.32e7	4.32
	0.6		112.5	0.56	5.03e7	8.37
Ag-88	1.0	2.00e7	104.8	0.97	1.94e7	7.94
	0.8		167.5	0.77	1.90e7	9.09
Ag-96	1.0	3.57e7	65.8	0.92	3.10e7	6.89
	0.8		106.1	0.72	3.00e7	7.19

The final nonlinear element in the co-simulation model is the thermal capacitance. In ADS this is most conveniently implemented as a simple nonlinear capacitance. The "voltage" (i.e. temperature) across the capacitor is $T\text{-}T_{REF}$. Consequently, the linear temperature coefficient of the heat capacity, γ, is equivalent to the linear voltage coefficient of capacitance in the electrical analog. This is implemented by the capacitor model element CM1 in the simulation.

The thermal conductance per unit length, G_T, is a linear element in the model. It is implemented by a simple resistance connecting the thermal node, at temperature T, to the ambient at temperature T_A. Similar to the reference temperature, the ambient temperature is set by a dc source in the simulation.

The model element shown in Figure 4 represents a short segment of the bond wire. The overall model is built up from a cascade of these elements (typically ten segments). These parts of the simulation represent the bond wire. Additional model elements are needed to account for the thermal and electrical boundary conditions of the physical structure. For example, the experimental sample may also introduce additional contact resistances in the electrical path of the model and additional thermal resistance in the thermal path. A key advantage of the electrical-thermal co-simulation methodology described above is that these boundary conditions can easily be added or modified to suit the circumstances of each particular experiment. The details of these boundary conditions are discussed in the following sections describing the experiments.

QFN-Mounted Samples

Figure 5 shows the bonding diagram for the QFN packages. Each package contained 8 wires bonded from one of the QFN pins to the central die paddle of the packages. As shown in the diagram, the wires had a nominal length of 1, 2, 3 and 4mm. These packages were assembled using automated wire bond equipment. The loop height and end-points of the wires were adjusted so that the actual wire length closely matched the nominal length. The one exception to this was the 4mm wire which, because of equipment limitations, had an actual length of 3.9mm.

Eight different wire types were measured in these tests: Au-99 wire of 1-mil and 0.6-mil diameters, Pd-coated Cu wire of 1mil and 0.6mil diameters, Ag-88 wire of 1-mil and 0.8-mil diameters and Ag-96 wire of 1-mil and 0.8mil diameters. The material and model parameters used in the simulation comparisons are listed in Table 2.

The package pins of the QFN-mounted samples are large, allowing the use of 4-point probe contact. The QFN package itself was mounted upside-down on the chuck of a probe station. A metal block that served as a heat sink was clamped onto the package such that it covered most of the exposed center pad of the package. Two high-current probes were used to supply the test current to the samples, and the voltage was measured through a second pair of finer probes connected to the package terminals.

The four-point probe eliminates the probe electrical resistance from the measurement, so there was no need to characterize the fixture. This has the further advantage of eliminating the variable contact resistance. The metal block that forms the heat sink has a very large thermal mass, so the paddle of the QFN package can be assumed to be held at ambient temperature.

Because the wire lengths were well-controlled in these samples, it was possible to very accurately determine the resistance per unit of the wires. This was done by finding a linear least-squared error fit to the measure resistance at low

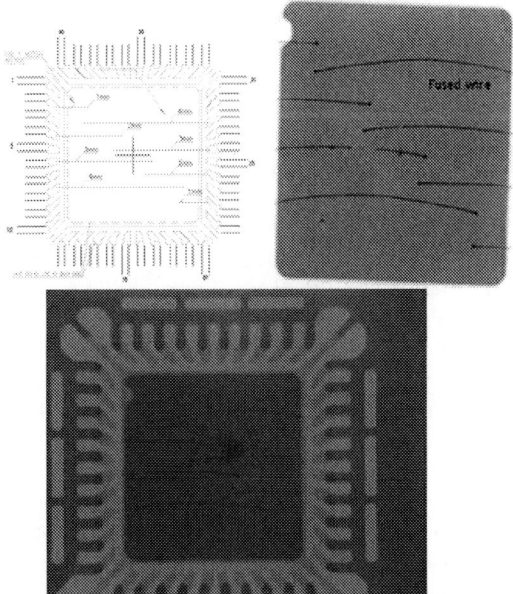

Figure 5. Bonding diagram for the QFN-mounted samples (top left), and X-ray picture after fusing test on 3mm long wire (top right). Tester QFN package with wires for measurement (bottom).

Table 3. Fitted thermal conductance values.

Metal	Diameter (mil)	G_T, QFN (WK^{-1}m^{-1})	G_T, air (WK^{-1}m^{-1})
Au-99	1	1.5	0.090
	0.6	1.5	0.054
Pd-coated Cu	1	1.2	0.090
	0.6	1.2	0.054
Ag-88	1	1.5	0.090
	0.8	1.5	0.072
Ag-96	1	1.7	0.090
	0.8	1.7	0.072

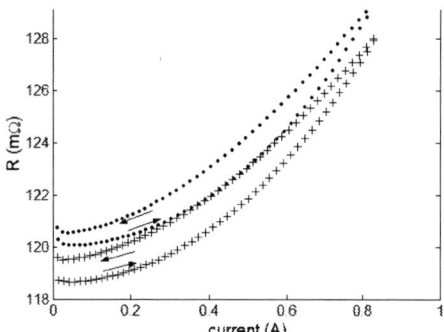

Figure 6. Example resistance versus current result from a QFN-mounted sample (3mm-long 1-mil Pd-coated Cu wire).

current as a function of the wire length. The slope of this fit gives the resistance per unit length and the intercept gives the parasitic resistance of the package leads. These measured values are listed in Table 2.

Table 2 shows the nominal metal conductivity, σ_{NOM}, and nominal wire diameter, d_{NOM}. In all cases we observed that the measured wire resistance per unit length, R_E, is slightly higher than expected based on the manufacturer's specifications for the metal resistivity. We speculate that this is likely a result of the wire drawing process, which alters to some extent the grain structure of the metal compared with its original bulk form. (The resistivity is characterized by measurement of the bulk material, not the wire.) As mentioned above, some of this variation may also be a result of variation in the actual wire diameter, which is specified to a tolerance of ±1μm. The observed higher resistance may result from either reduced effective conductivity (σ_{EFF}) or reduced effective wire diameter (d_{EFF}) or a combination of these effects. But since in all cases the resistance was higher than expected, we believe that the more likely explanation is that the conductivity is reduced. This effect is especially pronounced in the gold wires. However, the simulation result is equivalent using either a reduced effective diameter or electrical conductivity.

A simple ramp-up, ramp-down sequence of applied current was used in these measurements. The maximum current used in the tests was based on the previous results from PCB-mounted samples, and was intended to result in a maximum wire of about 200C. Figure 6 shows a typical

result. In all cases the resistance versus current showed a small hysteresis, most likely arising from some temperature increase in the heat-sink block. Since the overall resistance in these samples is dominated by the wire itself, from the relative difference between the ramp-up and ramp-down curves we can estimate that this hysteresis corresponds to about a 2-degree difference in the wire temperature, which is negligible. In addition, in all of the measurements the resistance versus current curves have negative slope at low current. This is almost certainly an experimental artifact, perhaps resulting from a small temperature dependence in the external current sensing resistance. In the worst case, it results in a relative drop of 0.5% in the low current resistance.

One reason that these effects are so apparent in the data is that the overall temperature rise in the wires is much smaller than expected. In the example shown in Figure 10, the relative change in resistance is about 8%. This indicates that the average wire temperature increases by about 20°C above ambient at a current of 800mA.

The coupled thermal-electrical model shown in Figure 3 can be applied to QFN-mounted samples. As discussed above (Equation (10)), in the steady state, the time rate of change is zero. In this condition the heat balance, Equation (10) becomes

$$I_{MAX} = \sqrt{\frac{G_T}{R_E} \frac{(T_{MAX} - T_A)}{[1 + \alpha(T_{MAX} - T_{REF})]}} \qquad (27)$$

Max current depends only on the electrical resistance per unit length and the thermal conductance per unit length from the wire to the ambient. In the case of a wire in air (as in the standard fusing current test) the thermal conductance per unit length is determined by convection. In that case, heat is transferred to air molecules where the physical processes of conduction and diffusion transport the heat away from the wire. In the case of the QFN packages, heat is transported from the wire mainly into the nearby package paddle, which is near the ambient temperature. The molding compound, which is much more thermally conductive than air, acts as the medium for the heat conduction.

Using the electrical resistance per unit length derived from the low-current limits of the measurements (Table 2) and the same value of probe thermal resistance as in the PCB-mounted measurements (120°K/W) previously done, the only remaining parameter to be fitted is the thermal conductance per unit length, G_T. For convective cooling in air, G_T is typically assumed to be proportional to the wire's diameter (Equation (11)) but in the case of the molded package, we would expect that the conductance depends on proximity to the cool paddle, so its dependence on diameter is not so obvious. The approach used in interpreting these measurements is to determine for each wire type the value of G_T that best fits simulation to measurement. In addition, it can be seen in the package diagram in Figure 5 that for part of its length the wire does not lie immediately above the package paddle. In the model this region is assumed to have negligible conductance.

The experimental results, along with simulations using fitted values of G_T from this simulation model are shown in

Figure 7. The fitted values of G_T are shown in Table 3. Because the overall temperature rise in these samples was low and the resistance change is small. Consequently, these data can be reasonably well modeled by a range of values of G_T. In fusing experiments, the data spans a much wider range of temperature and resistance change, and in those measurements the value of G_T is much less ambiguous. For this reason, the values of G_T used in the simulations are derived from the fusing experiments, and the method used for this derivation will be described in a subsequent section. These values are listed in Table 3. For comparison, the values obtained in the PCB samples for another project are also listed.

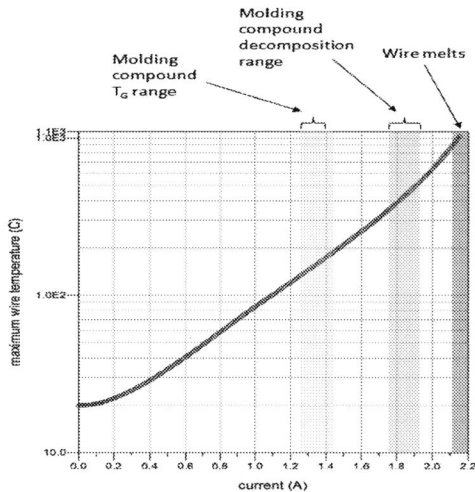

Figure 8. Simulated maximum wire temperature versus current for a 3mm long, 1-mil diameter Au-99 wire in a QFN package.

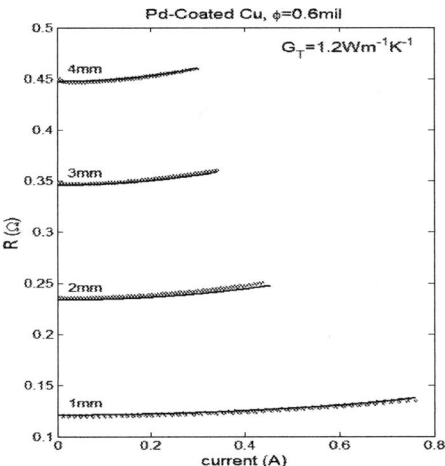

Figure 7. Resistance versus current relationship for QFN-mounted Pd-coated Cu wires. Green lines from measurement, and black lines from simulation.

The differences in the magnitude of the thermal conductance values are apparent in this table. Thermal conductance in molded packages is more than an order of magnitude greater than in air. There are other interesting differences as well. Whereas thermal conductance via convection in air is proportional to the wire diameter, it appears on the basis of these measurements that in the molded packages the thermal conductance is independent or very weakly dependent, on diameter. In air, the thermal conductance does not depend on the wire material, but it appears that in the case of the molded packages it does.

A key consideration in the fitted values of G_T is the distance between the wire and the package paddle. We would expect that as this distance decreases the thermal conductance should increase. The model assumes a single value for the entire length of the wire when, in fact, the distance tapers to zero at the point of the wire attachment to the paddle. Nonetheless, it appears that the observed behavior is well-explained by these conductance values, which may be considered to be an effective, average conductance over the wire length.

Current Capacity Guidelines for Wires in QFN Packages

Using the model values derived from these samples, we can run simulations to find the maximum internal wire temperature as a function of applied current, and extrapolate these results to higher currents. An example result is shown in Figure 8. It can be seen in this figure that the maximum wire temperature increase is approximately exponential with applied current. Figure 9 shows the wire temperature as a function of distance along the wire, from the thermal-electrical simulation.

The temperature profiles in this plot show that the point of maximum wire temperature is located near the mid-point of the wire length, but slightly closer to the pin connection, which is slightly hotter than the package paddle. The three current levels shown in this plot correspond, approximately, to the lower limit of the three temperature ranges identified in Figure 9. Especially at the lower current levels, the temperature profile near the wire midpoint is very flat. Because there is little temperature gradient along the wire, this indicates that there is little heat flow along the wire's axis. The main mechanism of cooling in this case is thermal conduction through the molding compound into the package paddle, and the temperature in this section is approximately described by the long-wire limit.

From failure analysis lately done, even before the wire was driven at fusing current, the molding material was decomposed (Figure. 10). In other words, the wire will not fail before the molding material does. Therefore we use temperature below Tg of the molding material as the maximum temperature to safely provide a guideline.

The results listed in Tables 2 and 3, in particular the values of R_E and G_T, can be used to predict the maximum temperature rise in long wires. These may be considered to be worst-case estimates, since shorter wires will generally be somewhat cooler than longer ones for a given current. The maximum current in long bond wires as a function of the maximum allowable wire temperature is given by Equation (27).

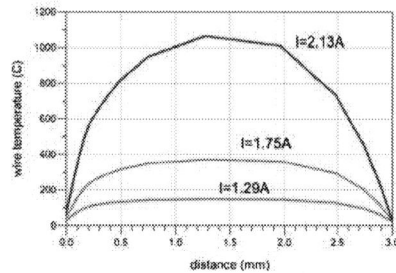

Figure 9. Example simulated wire temperature as a function of distance along the wire for a 3mm-long 1-mil diameter Au-99 wire bond in a QFN package. In this plot the wire connection to the package pin is on the left.

The experimental results in this study were obtained at ambient temperature of about 20C, which is also the reference temperature for the resistance. For the specification of maximum current guidelines, it is necessary to consider worst-case operating conditions of higher ambient temperature. For industrial applications it is typically 85C. Table 4 lists values of maximum operating current for various values of maximum wire temperature in the long wire limit for typical industrial applications.

Conclusions

The thermal conductance per unit length is much higher than in air. This is to be expected, since the molding compound is generally much more thermally conductive. The unexpected result, however, is that the conductance is independent of the wire diameter, at least within the limits of the accuracy to which we are able to resolve it. The other unexpected result is that the thermal conductance appears to depend on the wire material(Au, Cu, Ag). This result suggests that interfacial characteristics between the bond-wire metal and the molding compound may play a role in this heat transfer. It can be seen in Table 3 that values of G_T for palladium-coated copper wire are significantly different than those for gold wire of the same diameter.

Table 4. Maximum current capacity guidelines for ambient temperature of 85C (industrial).

Metal	Diameter (mil)	Maximum Current (A)		
		T_{MAX}=100 C	T_{MAX}=125 C	T_{MAX}=150 C
Au-99	1	0.47	0.74	0.92
	0.6	0.30	0.47	0.58
Pd-coated Cu	1	0.60	0.95	1.17
	0.6	0.35	0.55	0.68
Ag-88	1	0.44	0.71	0.89
	0.8	0.35	0.56	0.70
Ag-96	1	0.57	0.91	1.13
	0.8	0.45	0.71	0.89

The experiment-assisted simulation method simulates very fast (in few seconds) and predicts the wires' thermal and electrical behaviors reasonably well. When wires are in

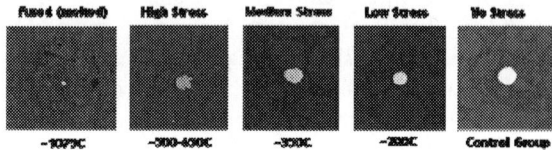

Figure 10. Damage signatures of wires in a QFN package driven at different currents.

molding material of a package, they will not fail before the molding material does, as Tg temperature and decomposition temperature of molding material are much lower than melting temperature of a metal wire. The maximum current capacity guidelines listed in Tables 6 are the main significant result of this investigation, and designing within these guidelines is necessary for long-term reliability.

Acknowledgments

The current-driven tests for this project was done by Robert Melville at Emecon, LLC. The authors are grateful for the help of Linda Chua at STATS ChipPAC, Singapore in the preparation of experimental samples.

References

1. K. W. Awkward, "Model for determining thermal profiles of bond wires using 'PSICE' analysis", *Proc. IEEE SEMI-THERM* VII, 86-90, Feb 1991, pp.12-14.
2. E. Loh, "Physical Analysis of Data on Fused-Open Bond Wires," *IEEE Trans. Components, Hybrids and Manufacturing Technology*, **CHMT-6**, No. 2, JUne 1983, pp. 209-217.
3. R. W. Powell, C. Y. Ho and P. E. Liley, "*The Thermal Conductivity of Selected Materials*," National Standard Reference Series, NSRDS-NBS 8, 1966.
4. S. I. Abu-Eishah, Y. Haddad, A. Solieman and A. Bajbouj, "A New Correlation for the Specific Heat of Metals, Metal Oxides and Metal Fluorides as a Function of Temperature," *Latin American Applied Research*, **34**, 2004, pp. 257-265.
5. R. C. Dorf, ed. *The Electrical Engineering Handbook*, CRC Press, ISBN 0-8493-0185-8, 1993.
6. A. Teverovsky, "Effect of environments on degradation of molding compound and wire bonds in PEMs", *Proc. 56th Electronic Components and Technology Conference*, 2006.

Computational Investigation of Failure in Anodized Aluminum

Sabrina Ball[1], Ibrahim Guven[1*], Pankaj Sinha[2], Rajiv Rastogi[2] and Brian McCarson[2]

[1]The University of Arizona, Tucson, AZ

[2]Intel Corporation, Chandler, AZ

[*] guven@email.arizona.edu

Abstract

This study concerns a common wafer manufacturing problem due to fracturing of anodized layer typically used to protect aluminum components from corrosive environment in plasma chamber. A computational approach is introduced for modeling of mechanical behavior of such structures. Predicted fracture morphology is consistent with the experimentally observed phenomena. Suggestions for mitigation of anodized aluminum fracturing related wafer failure are made.

Introduction

Plasma etching is an important step of wafer manufacturing in integrated circuit industry. The process aims to precisely engineer the surface features of the wafer. It involves generation of plasma, which requires a sealed chamber. The plasma is exposed to the target (wafer) and any other material surface inside the chamber including the plasma chamber walls. Anodized aluminum is a common plasma chamber and component material in the IC industry; an oxide coating (alumina) is produced on the aluminum surface leading to strong anodic polarization, significantly reducing reactivity. A common problem is the landing of foreign particles on the wafer while it is being etched rendering part of the wafer with zero yielding die. These particles are believed to originate from the plasma chamber components surrounding the wafers. One of the models is mechanical failure in the form of fracturing of the coating and flaking off. There are two main reasons for the coating failure [1]: (i) the difference between the coefficients of thermal expansion of the coating and the underlying metal, and (ii) change of surface chemistry due to plasma environment. These two mechanisms may or may not work together. Very few studies have been conducted to address the problems associated with integrity of anodized aluminum parts used in plasma chambers; most focus on cleaning and conditioning of the surface [2, 3]. A physics based understanding of the failure mechanisms at play is essential for improving the surface integrity of anodized aluminum parts.

Prediction of fracture initiation and propagation through computational simulations has been an important research area. Most of such studies involve finite element analysis (FEA), which is a well-established and robust numerical method. However, it is based on the solution of partial differential equations (PDEs) of classical continuum mechanics, which suffers from the inherent limitation that the spatial derivatives do not exist at geometric discontinuities such as crack tips, crack surfaces or along material interfaces. Recently a new theory that does not require spatial derivatives was introduced [4, 5]: Peridynamic (PD) theory. It allows crack initiation and propagation at multiple locations in the body with arbitrary crack paths. Further, the material interfaces have their own mechanical properties.

This study aims to investigate the underlying physics of the failure modes observed in anodized aluminum parts used in plasma chambers. Experimental investigation involves mechanical property characterization and clear identification of failure modes examination through SEM. Simulations of aluminum substrate with alumina coating under expansion loading conditions are performed. Computational investigation also explores the potential effect of an additional thin layer between the coating and the substrate.

Formulation

The computational methodology utilized here was first introduced by Silling [4]. It is a nonlocal theory in that the deformation state of a material point in a body is influenced by not only the nearest material points, but also those that are a finite distance away. This concept is schematically illustrated in Fig. 1. where a material point $\mathbf{x}$ interacts with other material points $\mathbf{x}'$ in its neighborhood $\mathcal{R}$ defined by a sphere (circle in 2-D) of a radius δ. The radius δ is commonly referred to as "horizon."

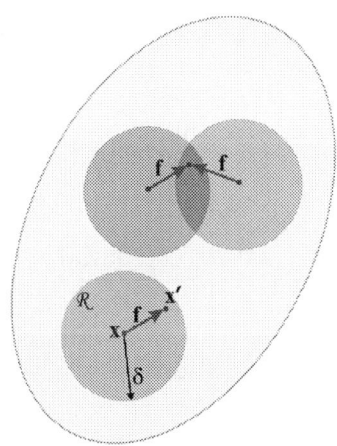

Figure 1. Peridynamic material points and neighborhood.

The interactions between material points are governed by a force-deformation response function, $\mathbf{f}$, in the following form

$$\mathbf{f}(\mathbf{u}, \mathbf{u}', \mathbf{x}, \mathbf{x}', t) = \frac{\boldsymbol{\xi}+\boldsymbol{\eta}}{|\boldsymbol{\xi}+\boldsymbol{\eta}|} \mu c s \qquad (1)$$

where $\mathbf{u}$ and $\mathbf{u}'$ are the displacement vectors of material points $\mathbf{x}$ and $\mathbf{x}'$, respectively. The relative position of material point $\mathbf{x}'$ with respect to material point $\mathbf{x}$ at times $t = 0$ and $t = t$ are $\boldsymbol{\xi}$ and $\boldsymbol{\eta}$, respectively. The material response is represented by the material parameter c, which is derived in terms of the elastic property of the material and the radius of the sphere defining the material point's neighborhood, δ as

$$c = \frac{18\,K}{\pi\delta^4} \qquad (2)$$

in which K is the bulk modulus of the material [6]. Based on the definitions of relative positions at initial and current times, stretch between two material points can be defined as [4]

$$s = \frac{|\xi - \eta| - |\xi|}{|\xi|}. \tag{3}$$

An equation of motion at time t at a material point in the body is be written as follows

$$\rho \ddot{\mathbf{u}} = \int_{\mathcal{R}} \mathbf{f}(\mathbf{u}, \mathbf{u}', \mathbf{x}, \mathbf{x}', t) dV \tag{4}$$

in the absence of body loads. In Eq.(4), ρ denotes mass density and the integral is taken over the spherical volume defining the neighborhood of material point $\mathbf{x}$. This integral represents the total force acting on the material point at a specific time value, which is evaluated numerically. Once the right hand side is known, the acceleration of the material point is calculated; subsequent calculation of velocity and displacement components round up the computations at the time step. The body under consideration is discretized into smaller volumes with a material point placed at the center of each small volume. The calculations are performed for each material point at every time step. As the body deforms, some material points move away from each other, leading to increased stretch. Material failure is introduced by defining a critical value for the stretch, s_0; if the stretch (defined in Eq.(3)) between two material points exceeds the critical stretch, the interaction between the material points is terminated permanently, and the force that was carried by this interaction is redistributed. The relationship between force and stretch along with failure and critical stretch are shown in Fig. 2. In light of the definition of critical stretch, the binary function μ in Eq.(1) controls the interaction between two material points: Its value is 1 as long as the stretch is less than the critical stretch and 0 if the stretch exceeded the critical stretch.

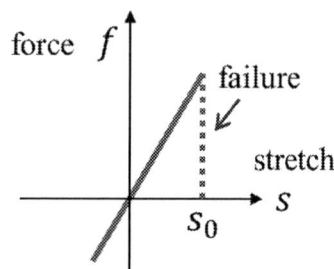

Figure 2. Critical stretch and failure.

The value of critical stretch for a homogeneous isotropic material is derived in [6] in terms of bulk modulus, K, horizon δ, and energy release rate G_f as follows:

$$s_0 = \sqrt{\frac{5 G_f}{9 K \delta}}. \tag{5}$$

Finally, a measure of "damage" at a material point is defined by the ratio of number of terminated interactions to the initial number of interactions. As it can easily be deduced, the damage parameter varies between 0 and 1.

Fracture Observations

Ground electrodes encase wafers processed in a plasma chamber. These electrodes are exposed to plasma and corrosive gases used during the etching process. The chamber temperature varies a few tens of degrees from room temperature during the etch process. Commercially available ground electrode after use was cross sectioned. The images were taken under bright field conditions at 15kV. The interior and exterior surfaces both showed similar sized crazing (Fig. 4). The mean crack-free area for the inside surface was 3952 μ^2 while that for the exterior was 3521 μ^2. The thicknesses of exterior and interior coatings were approximately 22 μm and 19 μm, respectively (Fig. 5). EDS mapping of the interior surface showed the same elements common in the top-down spectra but without the F signal. No morphological or compositional layering was evident.

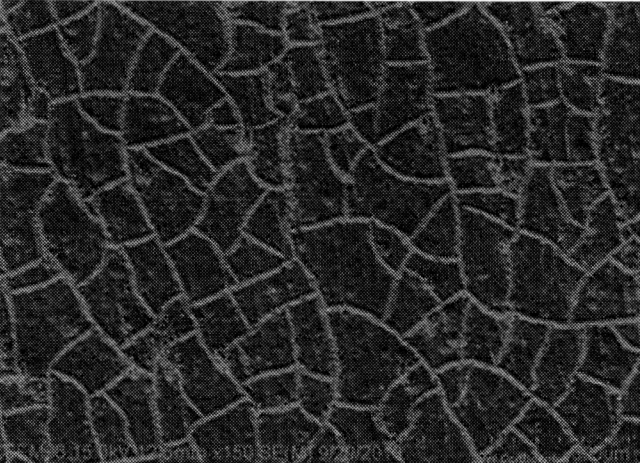

Figure 4. Fracture pattern observed along the inside surface of the plasma chamber.

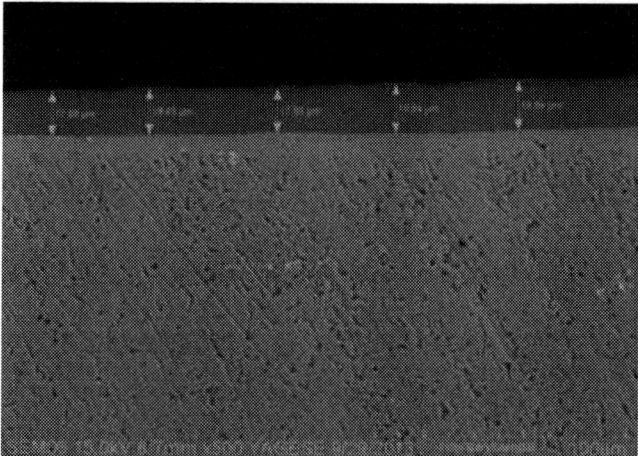

Figure 5. Cross-sectional view of the alumina coating and aluminum substrate. Fracturing of alumina coating is observed.

Computational Analysis and Results

The fracture due to coefficient of thermal expansion mismatch between the alumina coating and the aluminum substrate is simulated using the peridynamic theory. Results for the material system with a single alumina layer are presented first, followed by the investigation of effect of adding a compliant layer between the aluminum and the alumina.

The geometry of the material system (two-layer system) is shown in Fig. 6. The problem domain is a rectangular prism with length $l = 400$ μm, width $w = 400$ μm, and a total height of $h_t = 220$ μm. The thickness of the alumina film is taken as $h_f = 20$ μm. The grid spacing in the model is taken as 3.33 μm leading to approximately 950,000 grid points.

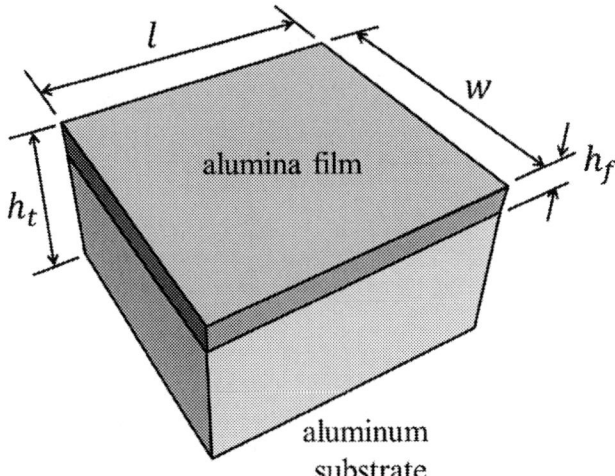

Figure 6. Geometry of the anodized aluminum material system.

The material properties for the aluminum substrate are: elastic modulus of 68 GPa and density of 2,700 kg/m³. Similarly, for the alumina film, the elastic modulus is taken as 370 GPa while its density is kg/m³. The critical stretch for the thin alumina coating is calculated to be around 5% while aluminum is about 20% owing to its ductile nature.

In order to simulate the deformation field due to the mismatch between coefficients of thermal expansion values of the constituent materials, a large portion of the substrate in the depth direction (80%) was subjected to isotropic expansion in the planar directions (length and width as defined earlier). The expansion is then transferred to the film through deformation. The loading is applied in a ramped fashion gradually so as to prevent premature fracture near the boundary regions.

Figure 7 shows the top view of the damage progression along the surface of the coating. Four different time steps are shown, with time increasing from top to bottom. Fracture starts at one edge and propagates as the expansion of the substrate continues. Due to the nature of the isotropic extension, a number of branches emerge leading to the final fracture configuration, shown in the bottom segment of Fig. 7. However, there is further fracture beneath the surface that is of interest. In order to examine the fracture morphology inside the coating, damage contours are plotted such that only those material points with damage values ranging between 0.2 and 1.0 are shown in Fig. 8 from top view. Same set of results are shown from an oblique angle in Fig. 9 in order to clarify the extent and geometric distribution of the damage in 3-D. Examination of Figs. 8 and 9 and their comparison to Fig. 7 reveal that a considerable damage is accumulated along the interface between the aluminum substrate and the alumina coating. This is consistent with the failure mode observed in experiments, where part of the crazed coating peels/flakes off, suggesting interface delamination. Peridynamic simulations appear to capture the correct failure modes observed in experimental setup.

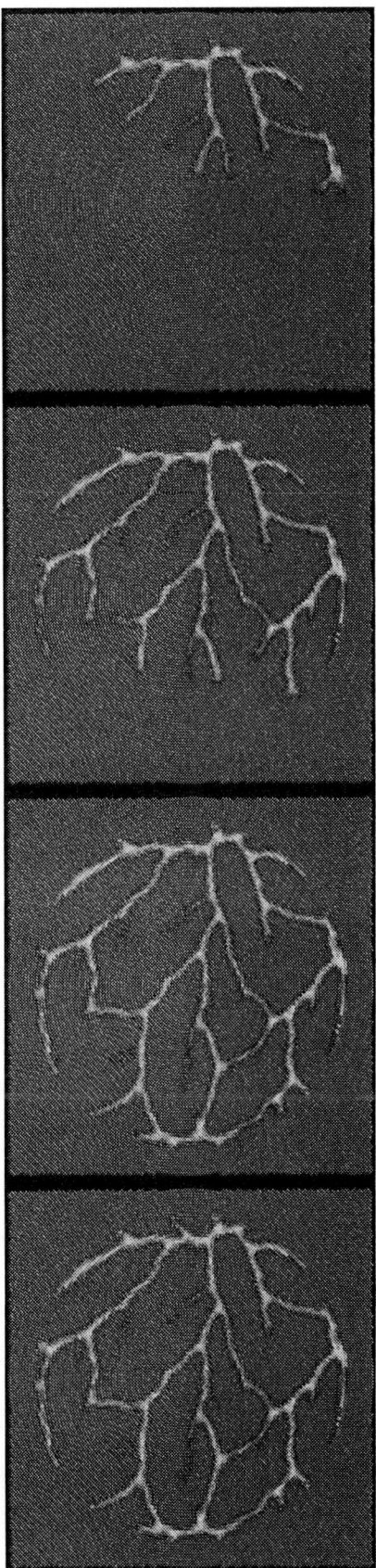

Figure 7. Damage progression along the top surface of the alumina layer predicted by peridynamic theory for the two-layer system.

978-1-4799-2408-0/14 $31.00 © 2014 IEEE   2037

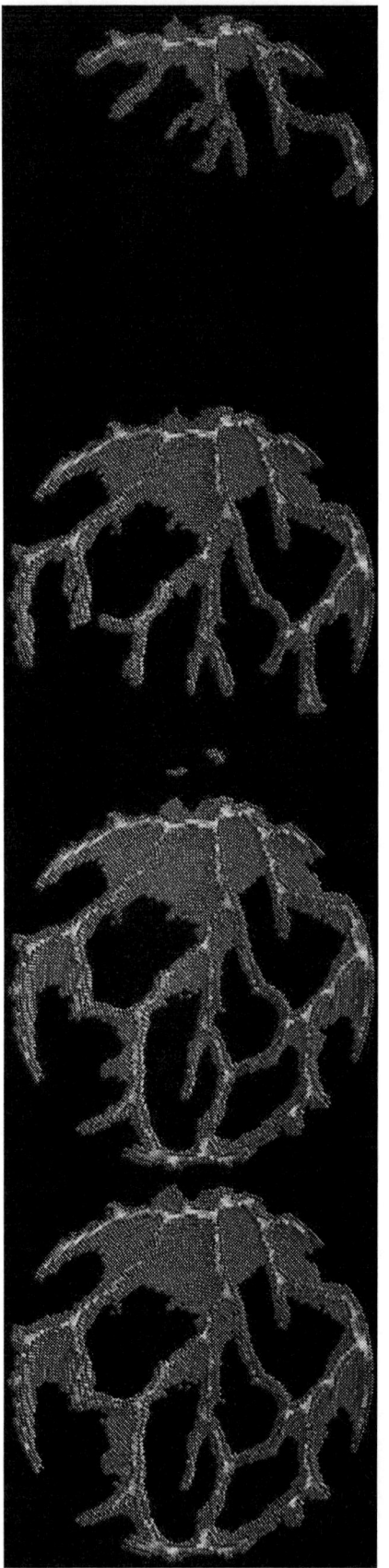

Figure 8. Damage progression in the alumina layer through the thickness from top view predicted by peridynamic theory for the two-layer system.

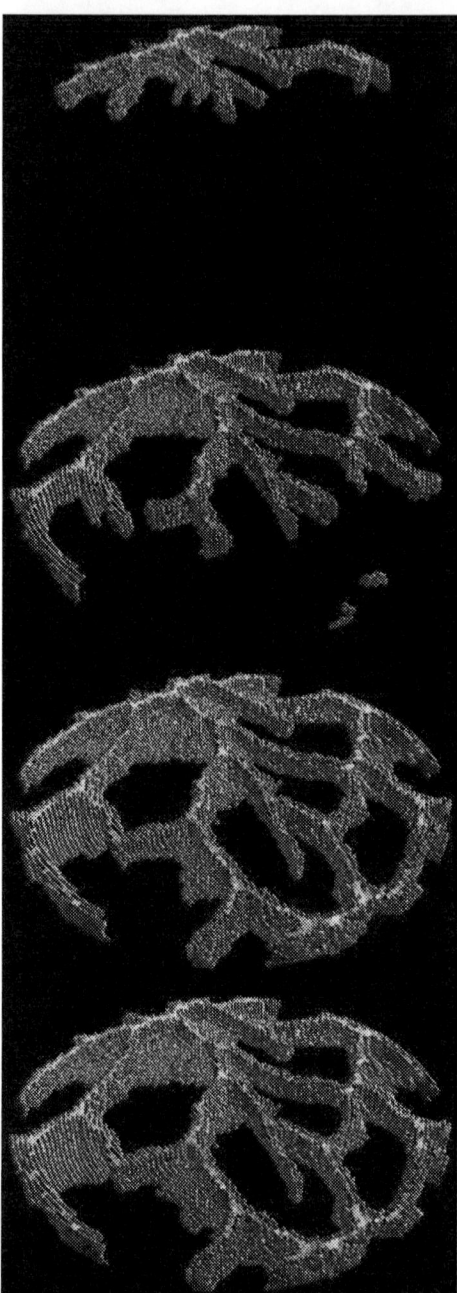

Figure 9. Damage progression in the alumina layer through the thickness from oblique view predicted by peridynamic theory for the two-layer system.

Additionally, the top surface fracture pattern predicted by peridynamic theory is compared to the micrographs of the crazed coating inside the plasma chamber in Fig. 10. A scale bar of 300 μm is included; it applies to both the SEM micrographs (left) and the peridynamic results (right). The size and shape of the surface fracture predicted by peridynamic theory closely resemble those observed experimentally. The "cellular" characteristic of the fracture is prominent on both sides. Peridynamic theory is able to predict the shape and size of the fractured-coating pieces with satisfactory accuracy.

Figure 10. Comparison of experimentally observed fracture patterns (left) against those predicted by peridynamic theory for the two-layer system.

Modification of the anodization process may allow a layered coating system with intention of designing a layer sequence to alleviate or eliminate cracking in the exposed surfaces. The anodized coatings are inherently porous; depending on the durations and concentrations of acid bath process, different porosity values are attained. Therefore, there may be configurations of layered anodized coatings that might lead to better fracture performance of the chamber interior coating. With this hypothesis, the same problem is considered with an added thin layer between the alumina coating and the aluminum substrate (leasing to a three-material system) as sketched in Fig. 11. This intermediate layer has the same thickness as the alumina coating but with different material properties. The relationship between the mechanical properties and porosity is well-documented for alumina [7-9]. As the porosity increases, elastic modulus decreases, which makes the material more compliant allowing the material to deform more before fracturing. By making the intermediate layer compliant, the effect of the mismatch between the substrate and the original coating will be lessened, leading to less cracks and flaking off. This added layer will serve as a buffer layer.

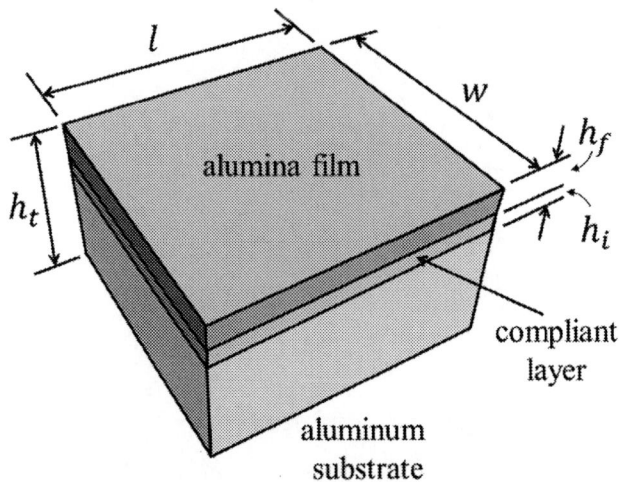

Figure 11. Geometry of the anodized aluminum material system with the compliant intermediate layer.

Therefore, in this hypothetical test case, the intermediate layer elastic modulus was decreased to half of the original value while the critical stretch was increased to about 7.5% (compared to 5%). The remaining parameters of the model is identical to the previous problem.

The damage progression along the top surface of the exposed coating is shown in Fig. 12. The amount and distribution of the damage in the three-material system are less than those of the two-layer system. Further, similar to the previous case, the damage contours for the current configuration are plotted for values between 0.2 and 1.0 in order to examine the damages in the thickness direction as shown in Fig. 13. In order to make the comparison easier, side-by-side comparison of damage patterns for the two-layer and three-layer systems are shown in Fig. 14. The two-layer system results are shown in the left column while the right column is shows the three-layer system damages. Top row in Fig. 14 shows only the top surface while bottom row shows damages greater than 0.2 through the thickness. It is clear that the interface delamination problem is significantly reduced; the predicted cracks have vertical faces. This prediction suggests that the peeling/flaking off phenomena could be significantly reduced or eliminated by carefully engineering an anodization process leading to a desired layered system.

Conclusions

In this study, a common problem in IC manufacturing, which involves peeling/flaking off of fractured coating of anodized aluminum plasma chamber walls, is described. A computational approach, peridynamic theory, is used to simulate the surface fracturing due to uniform expansion is demonstrated. It was shown that the peridynamic theory captures the correct failure modes observed in experiments. Also, it is able to predict the shape and size of the fractured-coating pieces with satisfactory accuracy.

In order to reduce or eliminate the peeling/flaking off problem, incorporation of an intermediate layer between the coating and substrate is considered. Peridynamic simulation of this hypothetical structure under the identical expansion conditions suggests that a compliant intermediate layer between the coating and substrate has potential to reduce the problem through eliminating interface fracture (delamination) and reducing the surface fracture.

Further study would explore the effects a range of elastic properties and critical stretch values as well as geometry parameters (e.g. thickness).

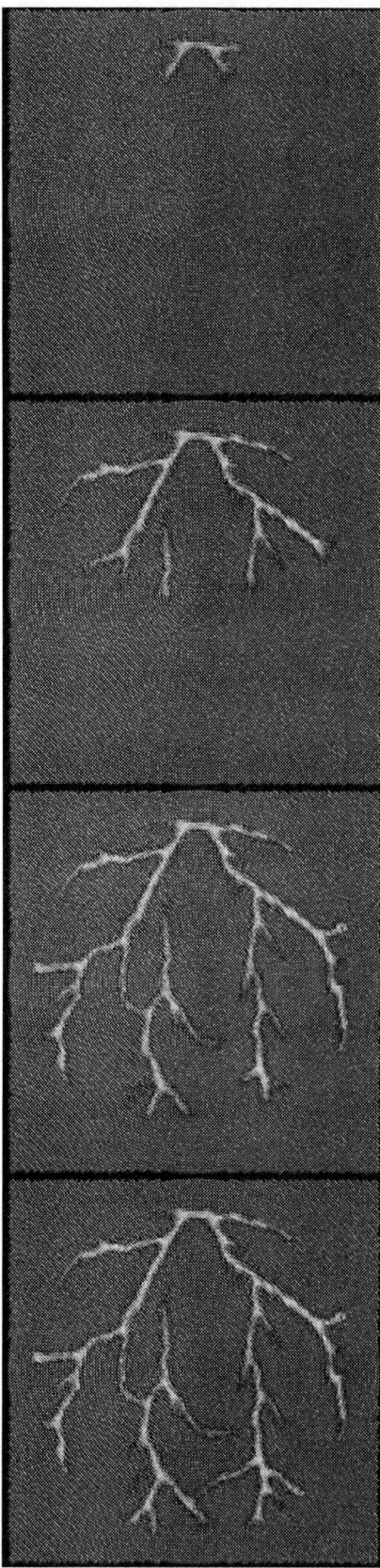

Figure 12. Fracture progression along the top surface of the alumina layer predicted by peridynamic theory for the three-layer system.

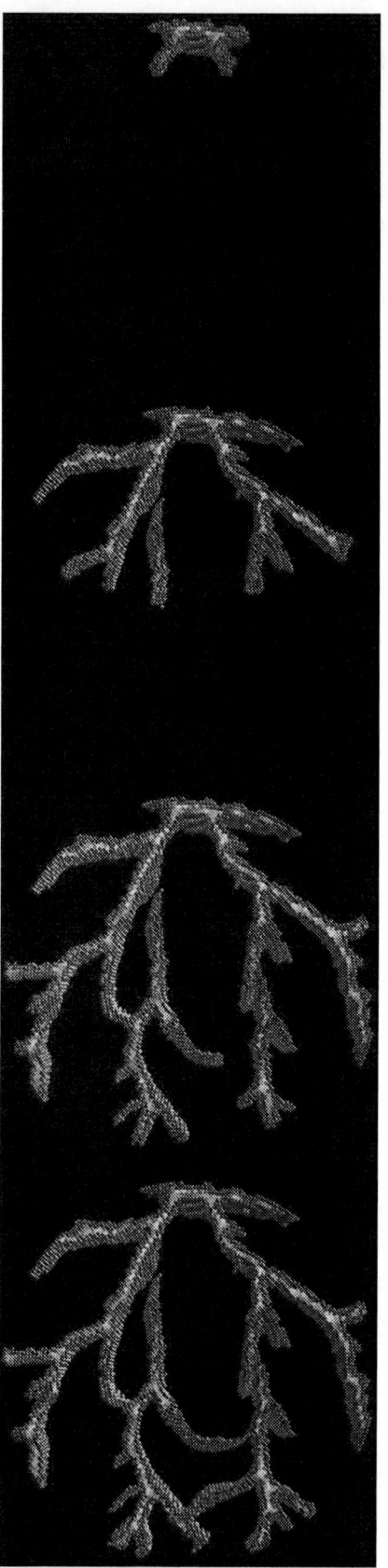

Figure 13. Fracture progression in the alumina layer through the thickness from top view predicted by peridynamic theory for the three-layer system.

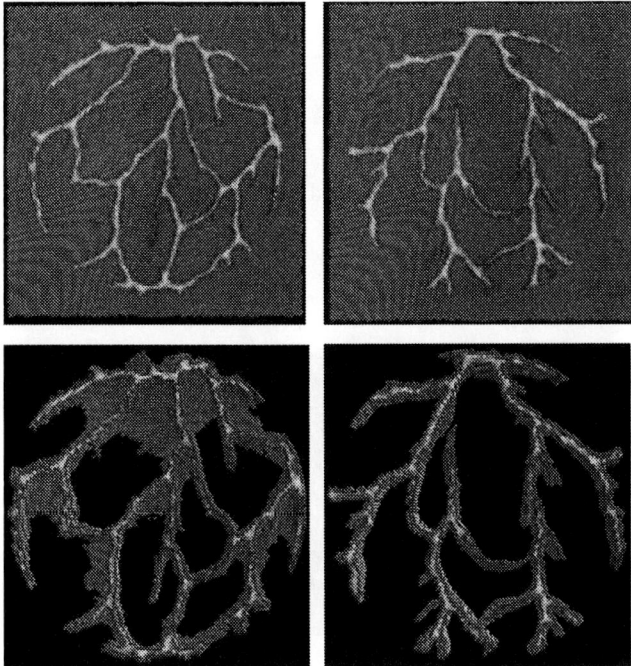

Figure 14. Side-by-side comparison of damage patterns for the two-layer (left) and three-layer (right) systems. Top row shows only the top surface while bottom row shows damages greater than 0.2 through the thickness.

References

1. S. Ponnekanti, "Failure mechanisms of anodized aluminum parts used in chemical vapor deposition chambers," *J. Vac. Sci. Technol. A Vacuum, Surfaces, Film.*, vol. 14, no. 3, pp. 1127, May 1996.

2. G. Cunge, B. Pelissier, O. Joubert, R. Ramos, and C. Maurice, "New chamber walls conditioning and cleaning strategies to improve the stability of plasma processes," *Plasma Sources Sci. Technol.*, vol. 14, no. 3, pp. 599–609, Aug. 2005.

3. R. Ramos, G. Cunge, B. Pelissier, and O. Joubert, "Cleaning aluminum fluoride coatings from plasma reactor walls in SiCl 4 /Cl 2 plasmas," *Plasma Sources Sci. Technol.*, vol. 16, no. 4, pp. 711–715, Nov. 2007.

4. S. A. Silling, M. Epton, O. Weckner, J. Xu, and E. Askari, "Peridynamic States and Constitutive Modeling," *Journal of Elasticity*, vol. 88, no. 2., , pp. 151–184, 2007.

5. S. A. Silling, "Reformulation of elasticity theory for discontinuities and long-range forces," *J. Mech. Phys. Solids*, vol. 48, no. 1, pp. 175–209, Jan. 2000.

6. Silling, S. A., and Askari, E., "A Meshfree Method Based on the Peridynamic Model of Solid Mechanics," *Computers and Structures*, vol. 83, pp. 1526-1535, 2005.

7. M. Asmani, C. Kermel, A. Leriche and M. Ourak, "Influence of porosity on Young's modulus and Poisson's ratio in alumina ceramics," *Journal of the European Ceramic Society*, vol. 21, pp.1081–1086, 2001.

8. W. Pabst, E. Gregorov and G. Tich, "Elasticity of porous ceramics—A critical study of modulus–porosity relations," *Journal of the European Ceramic Society*, vol. 26, pp.1085–1097, 2006.

9. J. Kovacik, "Correlation between Young's modulus and porosity in porous materials," *Journal of Materials Science Letters*, vol. 18, pp.1007–1010, 1999.

Study on Prediction about Residual Position of Void Generated by Resin Flow

Masayuki Mino[1], Naoya Suzuki[2], Hiroshi Takahashi[2], Tsutomu Kono[1]
[1]Hitachi, Ltd. Yokohama Research Laboratory
Yoshida-cho 292, Totsuka-ku, Yokohama, 244-0817 Japan
[2]Hitachi Chemical Co., Ltd. Advanced Performance Materials Operational Headquarters
Wadai 48, Tsukuba-shi Ibaraki, 300-4247, Japan
E-Mail: masayuki.mino.rh@hitachi.com

Abstract

Liquid epoxy resin is used in various types of mobile information equipment and devices and expected to be a key material for flip chip packaging [1-2].

The flip chip connection process is a thermo-compression molding process. In this process, air voids may remain behind due to the bump shape and resin properties; therefore, an appropriate molding condition is necessary.

We developed a computer-aided engineering (CAE) technique to calculate the generation of air voids and flow of resin and evaluated it in terms of prediction about the position of residual voids. This technique involves global resin-and-particle coupled moving and local air-resin two-phase flow analyses.

In this paper, we give an example of prediction about the position of residual voids generated by resin flow by using our CAE technique. By clarifying the molding phenomena through analysis, we clarified the material properties suitable for a molding process.

1. Introduction

The flip chip connection process is a thermo-compression molding process. Figure 1 shows the structure of a flip chip package, which is comprised of a liquid sealant applied between the substrate and chip. Figure 2 shows the assembly process for a flip chip package. First, liquid epoxy resin is applied to the top of the substrate. Second, a tool head attached to the chip is lowered with the bumps above aligned with the lands below. This compresses the resin, which is also being heated by heat transmitted from the tool head through the chip. As a result, the resin flows outward along the top of the substrate. The speed of the descending tool head is controlled and the movement is terminated when the bumps and the lands come into contact. At this time, air voids are left in the resist openings surrounding the lands. Finally, the temperature of the tool head is raised and lowered to cure the resin while also melting and setting the solder between the bumps and the lands. In short, this connection process enables interconnections between the bumps and the lands and seals the space between the chip and substrate using the resin.

The voids remaining in the resin result in defective reliability assessment. Therefore, for an adequate thermo-compression molding process of the liquid epoxy resin, we developed a computer-aided engineering (CAE) technique for prediction about the entrainment of air in the resin flow and void movement in the resin flow.

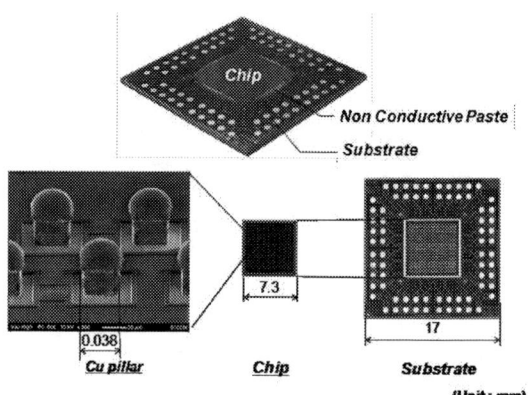

Figure 1: Structure of flip chip package

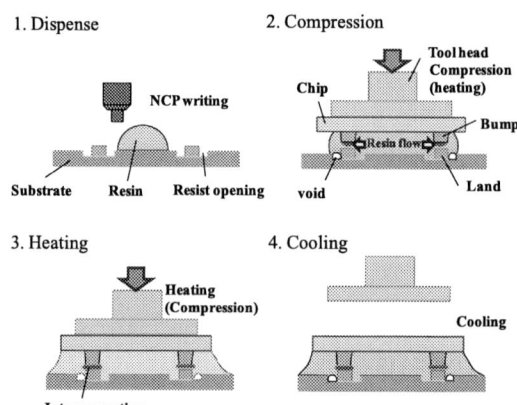

Figure 2: Assembly process for flip chip package

2. Method of prediction about position of residual voids
2.1 Modeling of resin property

Thermosetting resin is used as a liquid sealant. This resin goes through phase conversion from liquid to solid by producing a hardening reaction to change into a three-dimensional molecular structure. Therefore, taking into account heat generation and viscosity change by the chemical reaction is necessary.

We used the following formulas to describe the heat generation of resin [3-5].

$$d\alpha/dt = (K1 + K2\alpha^M)(1-\alpha)^N \qquad (1)$$
$$K1 = Kaexp(-Ea/T) \qquad (2)$$
$$K2 = Kbexp(-Eb/T) \qquad (3)$$
$$\alpha = Q/Qo \qquad (4)$$
$$dQ/dt = Qo(K1 + K2\alpha^M)(1-\alpha)^N , \qquad (5)$$

where α is curing degree, t is time, T is temperature, N, M, Ka, Kb, Ea, and Eb are material-specific constants, Q is the heat generation quantity, and Qo is the total heat generation at the end of the reaction.

Figure 3 compares the measured and calculated results of heat flow profiles, and Table 1 lists the heat generation parameters.

We used the following formulas to describe the viscosity of resin.

$$\eta=\eta_0[(1+\alpha/\alpha_{gel})/(1-\alpha/\alpha_{gel})]^C \qquad (6)$$
$$\eta_0=a \cdot exp(b/T) \qquad (7)$$
$$C=f/T-g \quad , \qquad (8)$$

where η is viscosity, α is curing degree, η_0 is the initial viscosity, α_{gel} is the curing degree at the gelation, T is temperature, and a, b, f, and g are material-specific coefficients.

Figure 4 compares the measured and calculated results of viscosity, and Table 2 lists the viscosity parameters.

For different temperature speeds, the measured and calculated values were in good agreement, as shown in figures 3 and 4.

Table 1: Heat generation parameters

Ka	Kb	Ea	Eb	N	M	Qo
1.5e+6	1.1e+6	8600	7000	1.9	1.5	139844

Table 2: Viscosity parameters

α_{gel}	a	b	f	g
0.8	0.005	2530	1500	1.8

2.2 Resin flow simulation

The flow analysis for thermosetting resin is conducted by substituting Eqs. (1) through (8) for the conservation of momentum, conservation of mass, and conservation of energy equations in the 3D flow simulation program FLOW3D®(FLOW SCIENCE Inc.).

Figure 5 shows a flowchart of prediction about the position at residual voids generated by resin flow. This CAE flow involves global resin-and-particle coupled moving and local air-resin two-phase flow analyses. We first input the bump and resist-opening shapes as product-shape parameters, as well as resin properties and process conditions. Then, in Step 1, we conduct global analysis of resin flow and calculate the resin flow-speed field. Next, we select a local location for evaluation, and based on the resin flow-speed field we calculated in Step 1, we use the flow speed for that location as a boundary condition and perform Step 2 calculations. Specifically, we conduct local air-resin two-phase flow analysis to calculate the volume of a void trapped near the bump and the land. Next, in Step 3, we input the void volume calculated in Step 2, and treating a void as a particle, we conduct another global resin-and-particle coupled moving analysis and calculate the position of the residual voids after electrode connection is made. Therefore, we conduct multi-scale analysis combining global and local analyses to investigate any position of residual voids by using a computer. If voids remain, we review the resin properties and process conditions.

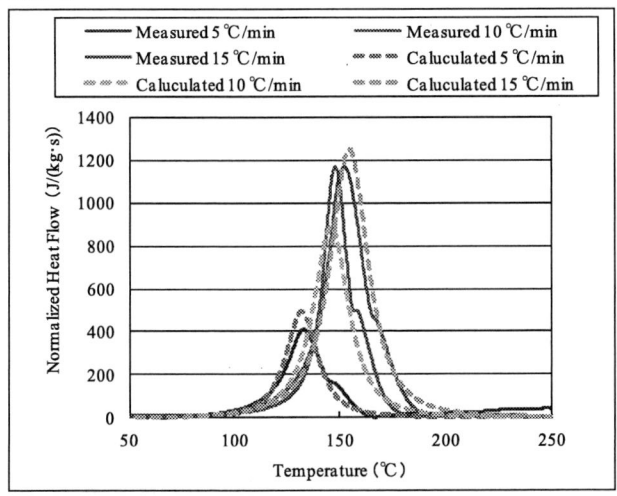

Figure 3: Comparison of measured and calculated results of heat flow profiles

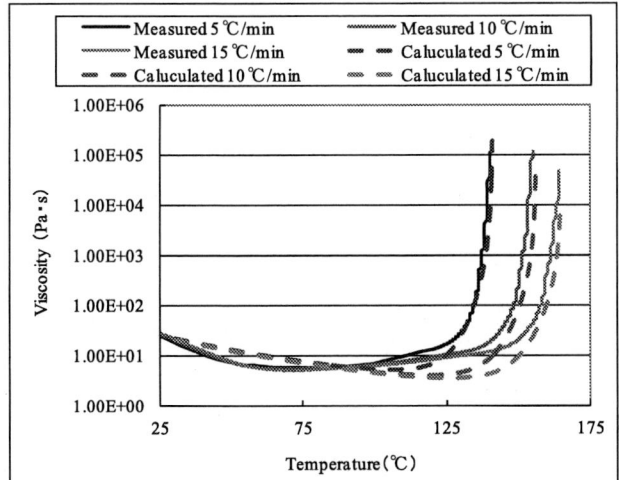

Figure 4: Comparison of measured and calculated results of viscosity

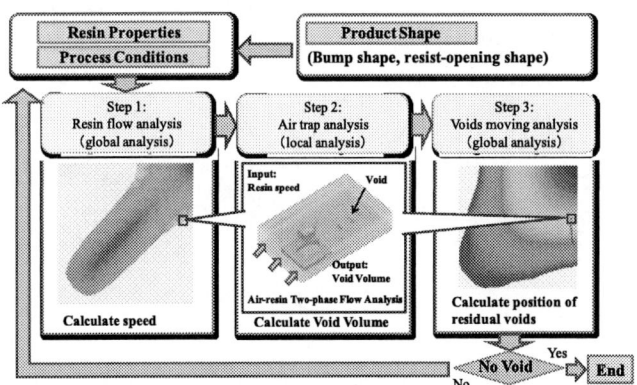

Figure 5: Flowchart of prediction about position of residual voids generated by resin flow

3. Calculated Results and Discussion

3.1 Resin flow analysis (Global analysis)

The flowchart shown in Figure 5 was calculated to examine the residual void position of the flip chip package. A general resin flow analysis model used for the first calculation is shown in Figure 6. The thickness of the initial application domain of the resin was assumed to be 159 μm at 674 μm in width on the diagonal from the chip center. A bump formed on a chip with 300 μm pitch. The analysis model was assumed to be a 1/4 model in terms of symmetry. The properties input the constant parameters listed in Tables 1 and 2 and measurements listed in Table 3.

Figure 7 shows the analysis results. The time when the chip came into contact with the top surface of the resin of 159 μm in height from the board side was with 0 s. The tool head compress to drop at 200μm/s. Bumps came into contact with lands in 0.8 s, and the resin flow finished. In addition, the resin speed reached maximum just before connection. The speed of the resin at the time of application was slow, and that when spreading out was fast.

Figure 8 shows the flow front of the resin flow at each time. The interval between the lines from 0.6 s to 0.75 s greatly increased. The representative points (a) and (b) where speed increased were extracted. The local speed distribution is shown in Figure 9. At point (a), resin flowed at 0.002 m/s around a bump from the 45-degree direction. At point (b), resin flowed at 0.004 m/s around a bump from the 90-degree direction.

The volume of air voids was examined using the second calculation for points (a) and (b).

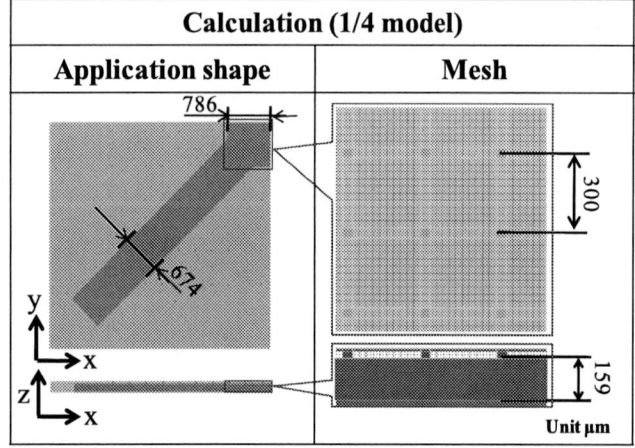

Figure 6: Analysis model shape

Table 3: Resin Properties

Density (kg/m³)	1610
Thermal coefficient (W/m·K)	0.512
Specific heat (J/kg·K)	1190

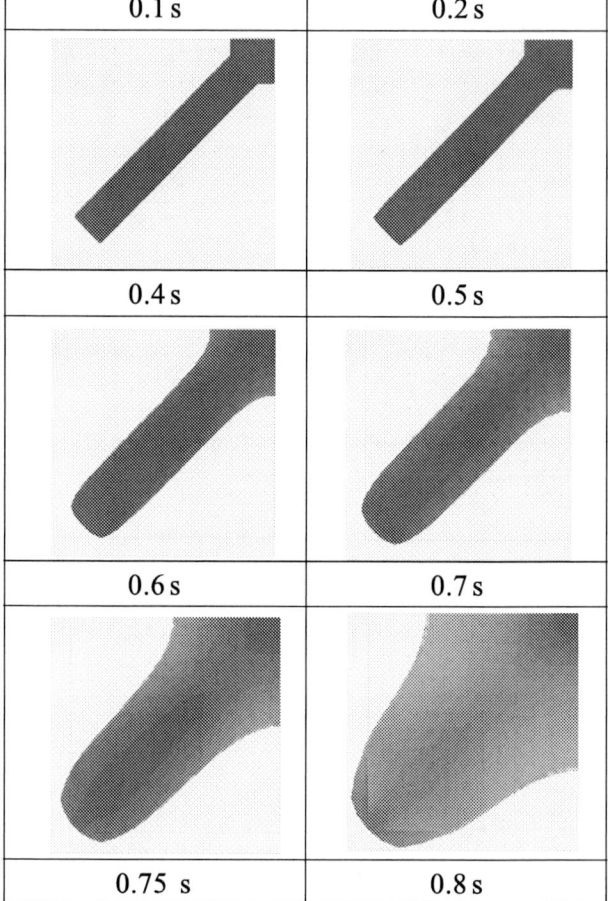

Figure 7: Results of resin flow analysis

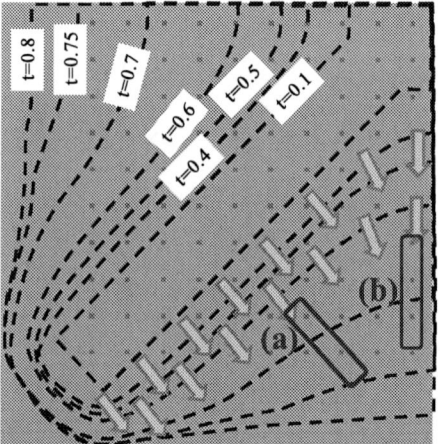

Figure 8: Results of the flow front of the resin flow

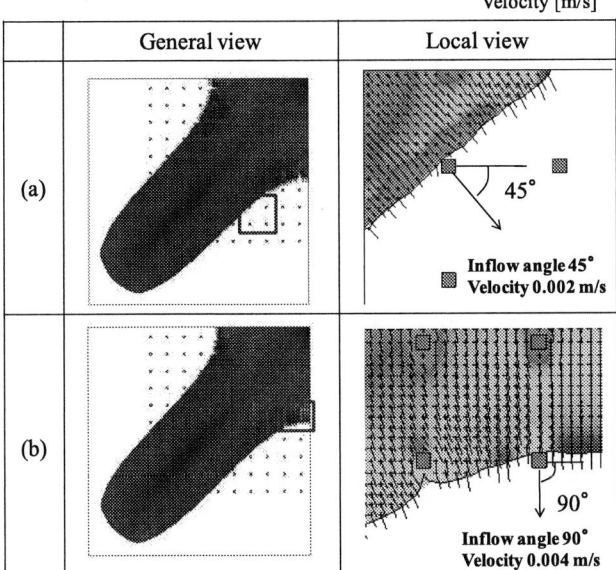

Figure 9: Local flow at representative point

3.2 Air trap analysis (Local analysis)

The air void volume around the bump was calculated using air-resin two-phase flow analysis for points (a) and (b).

The model shape and boundary conditions used for analysis are shown in Figure 10. The speed of the output at each representative point of the first calculation and an inflow angle input as the boundary condition of the second calculation.

The analysis results in three dimensions and each section at point (a) are shown in Figure 11 to Figure 14. The resin that flowed in from the 45-degree direction surrounded the land. The resin confluent in the back of the land, and air voids remained behind.

The analysis results in three dimensions and each section at point (b) are shown in Figure 15 to Figure 18. The resin that flowed in from the 90-degree direction also surrounded the land. The resin confluent in the back of the land, and some of the air voids remained behind and some were discharged into the resin flow.

The residual void volume decreased due to resin pressure, while discharged voids were carried away by resin flow.

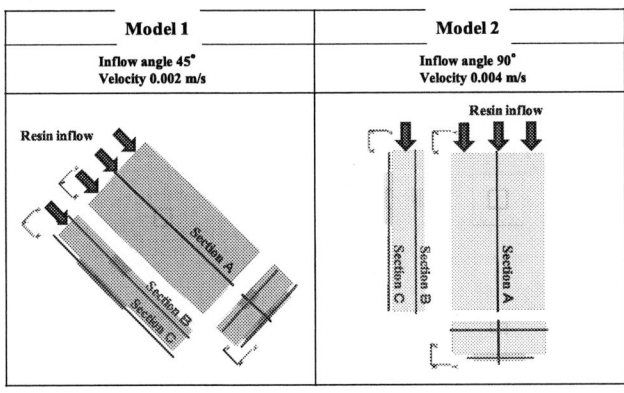

Figure 10: Two-phase flow analysis model around representative point

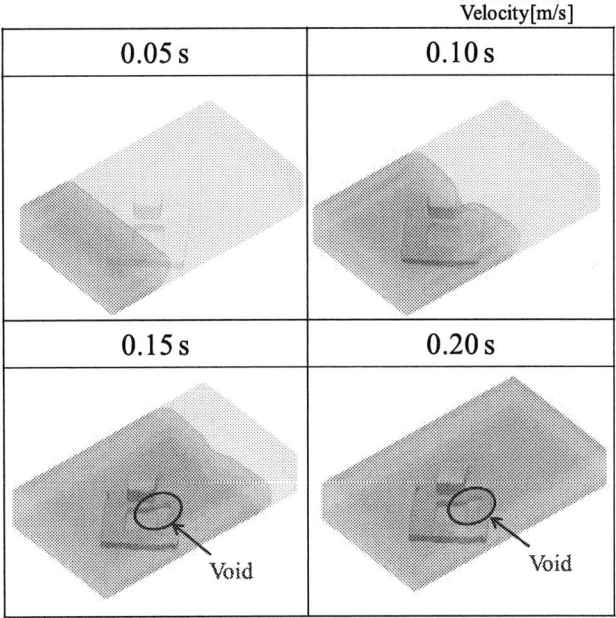

Figure 11: Results of two-phase flow analysis in 3D of model 1

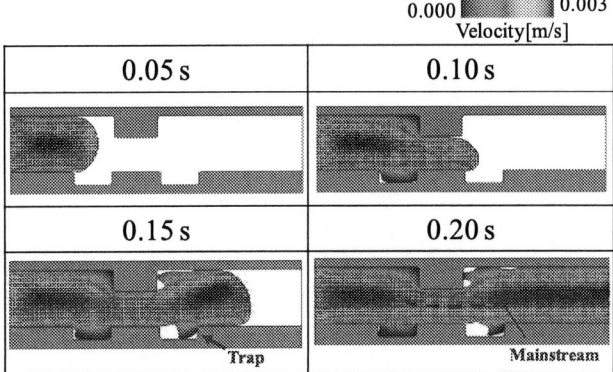

Figure 12: Results of two-phase flow analysis in section A of model 1

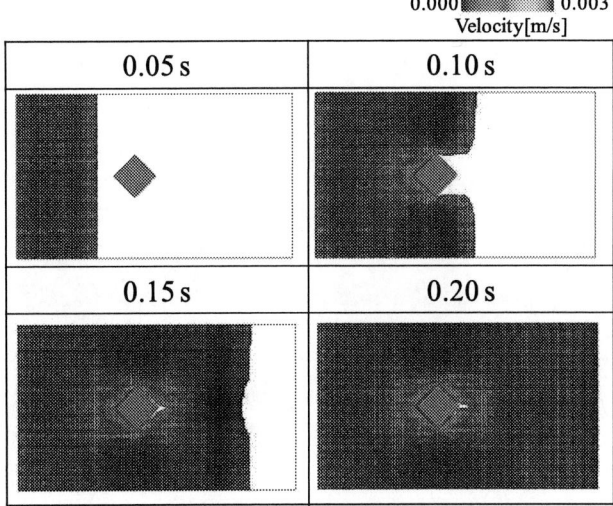

Figure 13: Results of two-phase flow analysis in section B of model 1

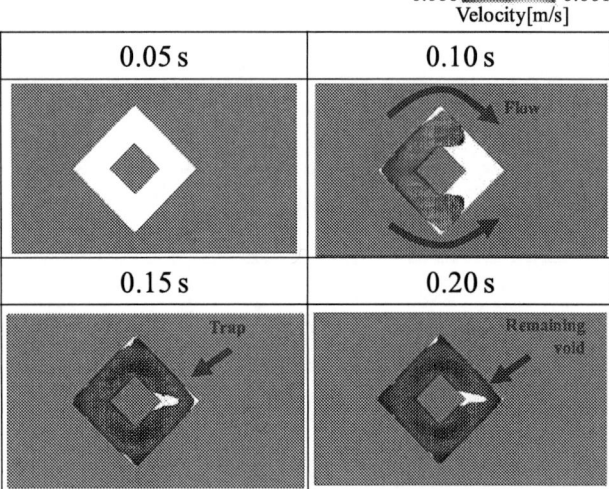

Figure 14: Results of two-phase flow analysis in section C of model 1

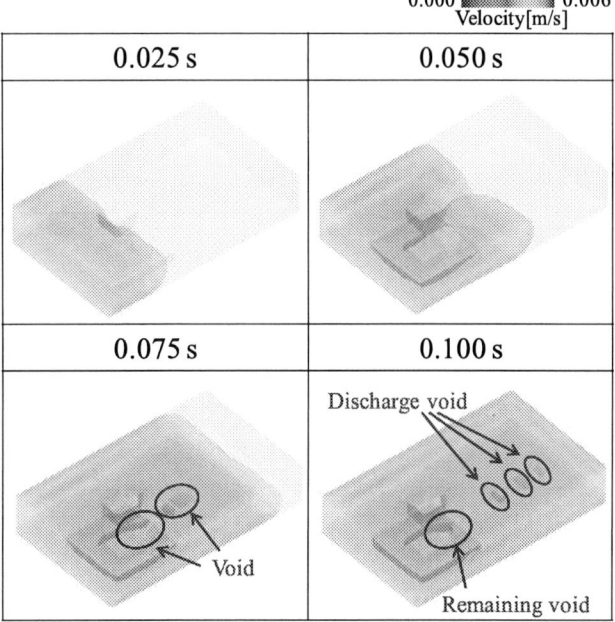

Figure 15: Results of two-phase flow analysis in 3D of model 2

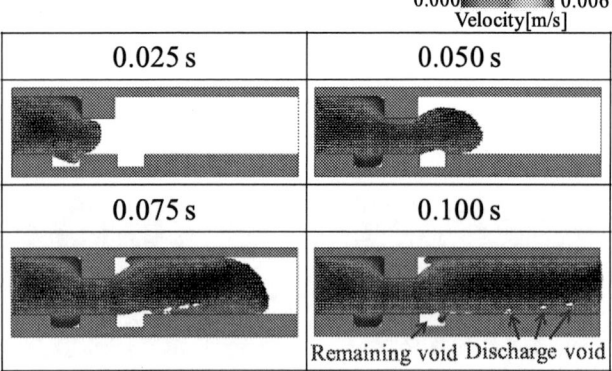

Figure 16: Results of two-phase flow analysis In section A of model 2

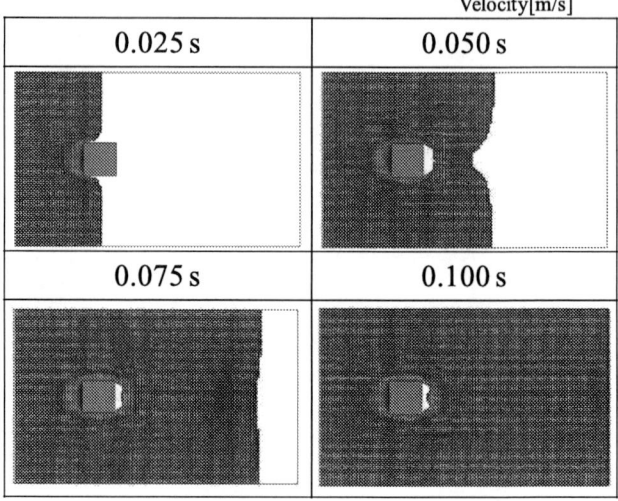

Figure 17: Results of two-phase flow analysis in section B of model 2

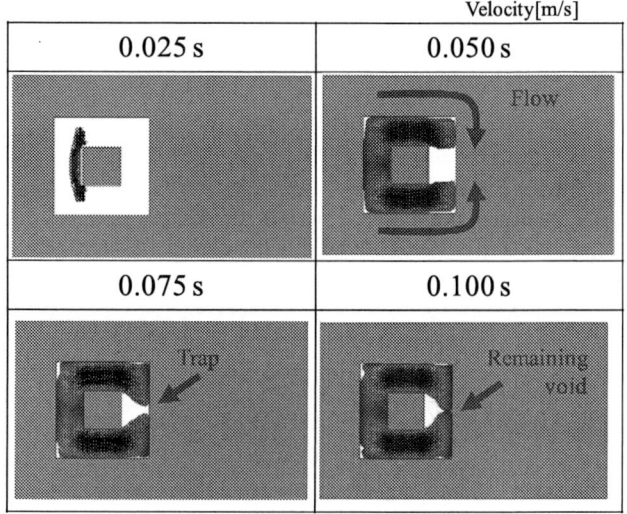

Figure 18: Results of two-phase flow analysis in section C of model 2

3.3 Void moving analysis (Global analysis)

For the discharge voids detected from air-resin two-phase flow analysis, the location where air voids remained behind after connection was calculated using resin-and-particle coupled moving analysis.

The model shape and particle outbreak position used for analysis are shown in Figure 19. Particle outbreak was conducted at point (b), where discharged voids were calculated using air-resin two-phase flow analysis. The particle diameter was assumed to be 25 μm that was the output of the two-phase analysis. Air was assumed to be present, and the density was set to 1.293 kg/m³.

The results of resin-and-particle coupled moving analysis are shown in Figure 20. The voids discharged at point (b) were carried into the resin flow.

978-1-4799-2408-0/14 $31.00 © 2014 IEEE   2046

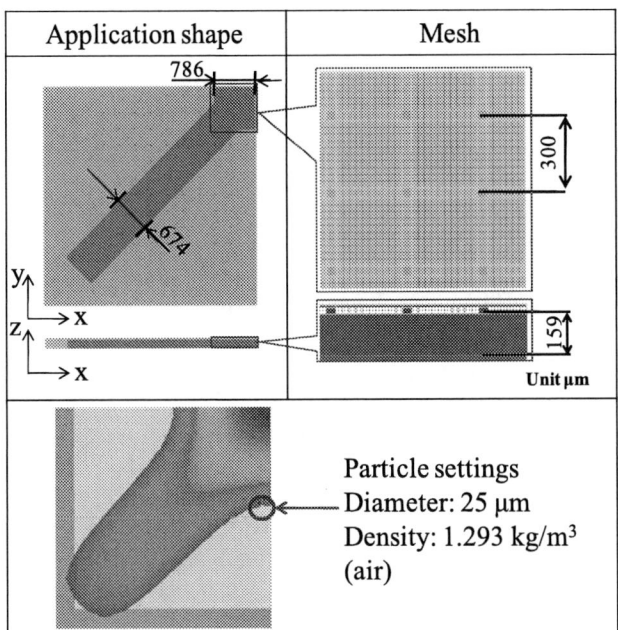

Figure 19: Particle movement analysis model

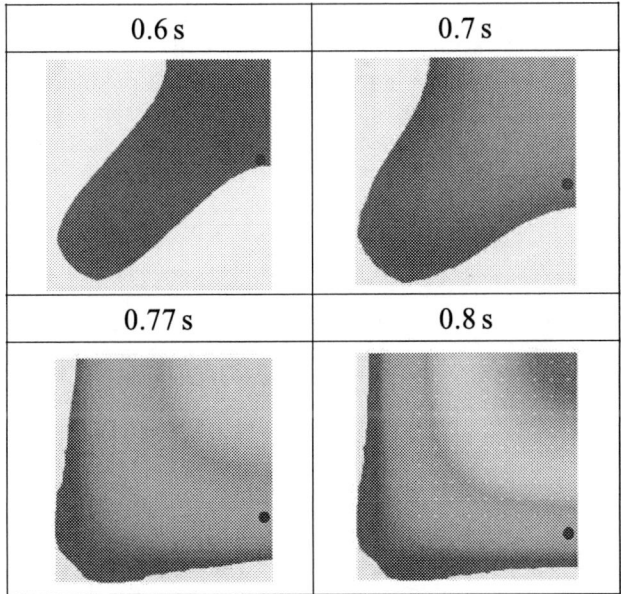

Figure 20: Results of Particle movement analysis model

the resin at the time of application was slow, and the speed of the resin when spreading out was fast. From the results of air trap analysis, the resin confluent in the back of the land and some of the air voids remained behind and some were discharged into the resin flow. This phenomenon depends on the position of the land, resin speed, and inflow angle of the resin. From the results of void-moving analysis, the discharged voids were carried into the resin flow.

We gave an example of prediction about the position of residual voids generated by resin flow. By clarifying the molding phenomena through analyses, we could clarify the material properties suitable for a molding process.

References

1. Minjae Lee, Min Yoo, Jihee Cho, Seungki Lee, Jaedong Kim, Choonheung Lee,Daebyoung Kang, Curtis Zwenger, Robert Lanzone, "Study of Interconnection Process for Fine Pitch Flip Chip" ECTC2009, pp. 720-723, May. 2009.

2. Min Woo Lee, Woon Kab Jung, Eun Sook Sohn, Joon Yeob Lee, Chan Ha Hwang, Choon Heung Lee, "A Study on the Rheological Characterization and Flow Modeling of Molded Underfill (MUF) for Optimized Void Elimination Design" ECTC2008, pp. 382-388, May 2008.

3. Junichi Saeki: "CAE on Macroscopic System in Polymer Processing-Foundation and Application for Thermosets", Seikei-Kakou, 2006, Vol. 18, No. 4, pp. 280-289 (In Japanese).

4. Junichi Saeki, Aizou Kaneda: "Flow Analysis of an Epoxy Compound for Low-Pressure Transfer Molding in a Circular Cross-Sectional Channel", JSME international journal. Ser.2, 1990, Vol. 33, No. 3, pp. 486-493.

5. Junichi Saeki, Isamu Yoshida: "Analysis of Heat Generation of Eplxy Conpounds for Encapsulation of Semiconductor Devices", Seikei-Kakou, 2001, Vol. 13, No. 2, pp. 118-124 (In Japanese).

Conclusions

We developed a CAE technique for evaluating the positions of residual voids generated by resin flow using global void-moving and local air-resin two-phase flow analyses.

Thermosetting resin goes through phase conversion from liquid to solid by producing a hardening reaction. Therefore, the viscosity and reaction profile of the thermosetting resin were formulated by measuring the heat-generation rate.

From the results of resin flow analysis, the resin speed reached maximum just before connection end. The speed of

Modeling and Analysis of Temperature Effect on MEMS Gyroscope

Ming Wen[1], Weihui Wang[1], Zhang Luo[1], Yong Xu[2,3], Xin Wu[4], Fei Hou[5], Sheng Liu[2]*

1. Huazhong University of Science & Technology, Wuhan, 430074, China
2. Cross-disciplinary Institute of Engineering Sciences, School of Power and Mechanical Engineering, Wuhan University, Wuhan, Hubei, 430072, China
3. Department of Electrical and Computer Engineering, Wayne State University, Detroit, MI 48202 USA
4. Department of Mechanical Engineering, Wayne State University, Detroit, MI 48202 USA
5. Dongfeng Automobile Electronics Co., Ltd
* Corresponding Author: Sheng Liu, Fax: 86-27-87557074, Email: victor_liu63@126.com

Abstract

It is well known that temperature variation affects a MEMS device's performance. In this paper, the effect of temperature on the whole MEMS based gyroscope is observed by recording and analysis the zero rate output (ZRO). The effect of temperature on the ZRO comes from material parameter changing and electronic parameter changing.

With temperature changing, the ZRO of the gyroscope suffers from changes of stiffness coefficient of beams, the damping ratio and other parameters [1]. A matlab Simulink model is built to research how much influence the material parameter has on the ZRO. The simulation result shows that material parameter changing induced effects have influence on the ZRO for 0.2% in the range of around room temperature.

Application Specific Integrated Circuit (ASIC) is an important part of a gyroscope. As most ASICs are made by semiconductors like silicon, the features of ASIC components (like output voltage, gain factors and resistance of doped silicon) are temperature sensitive by nature. A model based on a simplified ASIC component of capacitance to voltage converter circuit (C/V circuit) with temperature sensitive parasitic resistances is built to explore how much influence the electronic parameter has on the ZRO. The simulation result shows that electronic parameter changing induced effects have influence on the ZRO for less than 3ppm in temperature range from 273 K to 318 K.

The experiment is conducted by exposing a MEMS gyroscope into a thermal chamber, and the temperature range is from 273 K to 318 K. The experiment result indicates the temperature fluctuation has influence of about 5% on ZRO when the reference is the ZRO at 300 K. Material parameter changing induced effect has the greatest impact on ZRO, and that effect needs to be compensated to improve the ZRO stability of a MEMS gyroscope.

I. Introduction

MEMS gyroscope is a kind of widely used component for obtaining angular velocity in applications like inertial measurement unit (IMU) and cellphones, etc. [2]. MEMS gyros can be divided into many kinds according to their working principle or sensing method, etc. A linear vibration MEMS gyroscope is used in our modeling and experiment.

There have been many articles studying on temperature effects on gyroscopes. Many past studies have characterized the relationship of gyroscope frequency with temperature, and the frequency change is believed to be one of the main reasons to affect ZRO. Temperature variation induced

frequency change and phase change can also be used as a thermometer to compensate ZRO [1].

Parasitic resistance and capacitance have great influence on gyroscopes. However, the effect of temperature dependent characteristic of the parasitic resistance on the output of gyroscopes has never been considered [9].

In the following article, the working principles of the MEMS gyroscopes are discussed. Next, material parameters changing induced effects (MPCIE) and electronic parameters changing induced effects (EPCIE) are discussed. At last, the experiment result is compared with the simulation result.

II. Working Principles of the MEMS Gyroscopes

The operation principle of the majority of all existing commercial MEMS vibratory gyroscopes relies on the Coriolis force shown in formula 1[3], which is proportional to the result of the linear velocity of the proof-mass multiplies by the orthogonal angular-rate input. The proof mass and its flexible supporting beams above the substrate are fabricated in silicon-on-insulator (SOI)-based process [4].

$$F_C = -2m \cdot \vec{\Omega} \times \vec{v} \qquad (1)$$

Where F_C is the Coriolis force applied to the proof-mass, and m is the mass of the proof-mass, Ω is the angular velocity of the proof-mass, and v is the velocity of the proof-mass.

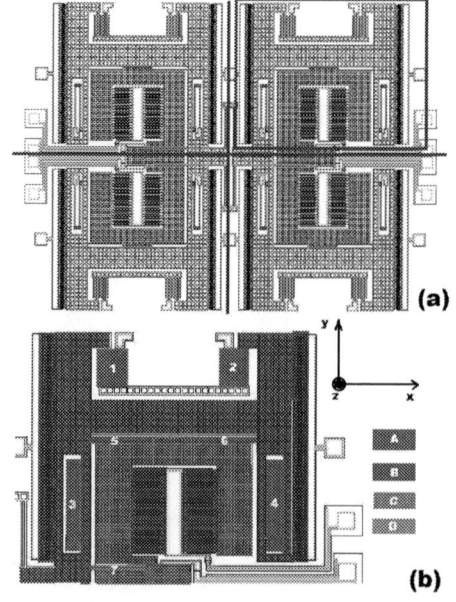

Figure 1. Sketches of the MEMS Part of a Gyroscope.

978-1-4799-2408-0/14 $31.00 © 2014 IEEE

2014 Electronic Components & Technology Conference

The sketch of the MEMS part in our experiment gyroscope is shown in Figure1. Fig.1 (b) is an enlarged drawing of the top right part (surrounded by a red rectangle) of Fig.1 (a). It's obvious that Fig.1 (a) is symmetric along the two blue lines (one is vertical, and the other is horizontal), so we could take Fig.1 (b) to illustrate the whole gyroscope.

Fig.1 (b) consists of four main segments (A, B, C and D) emphasized in different colors. Segment A and C are proof-masses that can move in x direction, while segment C can move in both x and y directions. Segment D1, D2, D3 and D4 are flexible beams to support segment A and C above the substrate, and segment D5, D6 and D7 are flexible beams to connect segment C to segment A. Segment B is comb structure capacitance for driving and sensing.

As an inertial component cannot avoid working in a certain degree of vibration environment that will bring small undesired Coriolis response, it's necessary to use an anti-phase system, also known as tuning fork gyroscopes (TFG) shown in Fig.1[5].

There are two identical masses in TFG architecture, which are working in opposite direction and in the same frequency. When the response of two masses is added, the common-mode response of TFG is canceled out. The lumped model TFG is shown below in Fig.2.

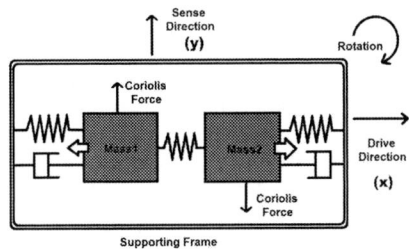

Figure 2. Lumped Model of TFG.

The drive-mode oscillator is comprised of mass1, mass2, suspension beams to allow masses move in x direction, actuation electrodes and feedback electrodes. The sense-mode oscillator is formed by masses, supporting beams and sense-mode detection electrodes. The driving forces applied to the masses are usually sinusoidal generated by a closed-loop drive system shown in Fig.3 below.

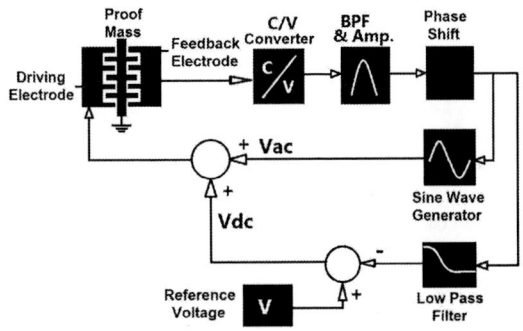

Figure 3. Closed-loop Drive-Mode Oscillator System with AGC.

The overall dynamical system is typically a two degrees-of-freedom (2-DOF) mass-spring-damper system. When the gyroscope is subjected to an extern angular rotation around axis z, a sinusoidal Coriolis force at the same frequency of driving frequency is induced in the direction orthogonal to the drive-mode oscillation.

Even though designing the drive and sense resonant frequency to match is helpful to attain the maximum sensitivity, it's more common that the sense mode is designed to be slightly shifted from the drive-mode to improve robustness.

III. Material Parameter Changing Induced Effects

There has been many articles discussing thermal characteristic of MEMS linear vibration gyroscopes, and most of these articles focus on the change of the size configuration and elastomeric modulus due to the change of the temperature [6]. From the simulation result of Sun, we can see that the change of the amplitude of the driving and detecting mode due to the change of the size of the configuration is very small (within 0.8%)[6]. So we can ignore the effects of configuration's changing, and focus on the effects of elastic modulus's variation.

Temperature variation may influence the elastic coefficient of the supporting beams, thereby the resonance frequency and the amplitude of drive mode.

The change of the silicon's elastic modulus with temperature can be described as follows [7]:

$$E(T) = E_0[1 - k(T - T_0)] \qquad (2)$$

where $E(T)$ and E_0 are the elastic modulus of silicon at temperature of T and 300K respectively. T_0=300K, and k=70ppm.

The supporting's stiffness coefficient is proportional to the elastic modulus, so the relationship of the stiffness coefficient and temperature is shown as equation (3).

$$K(T) = K_0[1 - k(T - T_0)] \qquad (3)$$

where $K(T)$ and K_0 are the stiffness coefficient of silicon at temperature of T and 300K respectively. T_0=300K, and k=70ppm.

The drive mode oscillators are two coupled 1-DOF resonators, and each one can be modeled as a mass-spring-damper system. With a sinusoidal drive-mode excitation force, the motion equations along the x-axis are

$$m_1 \ddot{x}_1 + c_d \dot{x}_1 + k_c(x_1 - x_2) + k_d x_1 = F_d \qquad (4)$$

$$m_2 \ddot{x}_2 + c_d \dot{x}_2 - k_c(x_1 - x_2) + k_d x_2 = -F_d \qquad (5)$$

$$F_d = \sin(\omega_d t) \qquad (6)$$

where m_1 and m_2 are the mass of mass1 and mass2 in Fig.2 respectively, and x_1 and x_2 are the displacement of each mass. k_c is the coupling spring's stiffness coefficient, and k_d is the supporting beams' stiffness coefficient. F_d is a sinusoidal drive force, and it's sign reversing in equation (4) and equation (5), which indicate the motion of two masses are anti-phase. There are modules like AGC and PLL in our experiment gyroscope to enable drive mode resonant stable in both amplitude and phase, so we can believe that temperature has little impact on the phase.

The Q of gyroscope and pressure has a relationship: if the pressure increases for four orders of magnitude, the Q decreases for one order of magnitude. The pressure is proportional to the temperature in a MEMS package. Q is defined as

$$Q = \frac{1}{2\xi} = \frac{\sqrt{km}}{c} \qquad (7)$$

where k is the stiffness coefficient, c is the damping coefficient, and ξ is the damping ratio. Generally, the variation of k vs. temperature is smaller than the variation of c vs. temperature. The relationship of damping ratio vs. temperature is

$$c(T) = c_0 \cdot 10^{\frac{\ln(T/300\,K)}{4\ln(10)}} \approx c_0 \cdot 1.28^{\ln(T/300\,K)} \qquad (8)$$

where c_0 is the damping coefficient at 300 K, and T is temperature in degree of Kelvin.

To solve these equations at different temperature, we use matlab Simulink. The model we build is shown below in Fig.4.

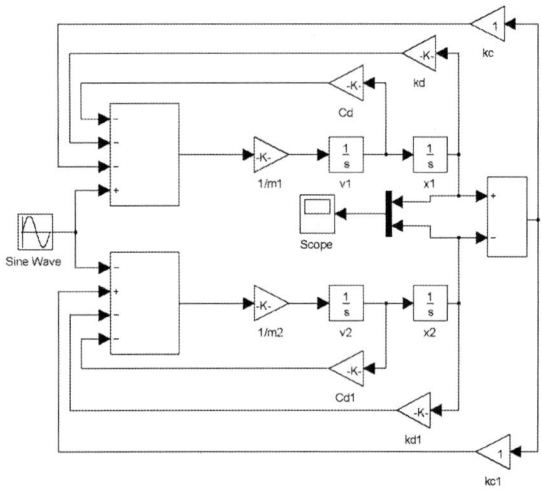

Figure 4. The Simulink Model of Drive-Mode Oscillator.

One single gyroscope cannot work, and gyroscopes need suitable circuits and packaging to work well. The packaging enables the temperature of gyroscopes to vary in a range of nearby room temperature. The temperature range in our simulation is set to be from 0 ℃ to 45 ℃ with the interval being 5 ℃.

The elastic module at 300 K is set as a reference, so the value can be set to be one, and likewise the value of damping coefficient at 300 K can be set to be one as a reference. These values can be calculated before simulation, and the results are shown below in Tab.1.

Table 1. Relative values of elastic module and damping coefficient nearby room temperature

Temp. (K)	273	278	283	288
Elastic module	1.0019	1.0015	1.0012	1.0008
Damping coefficient	0.9770	0.9814	0.9857	0.9900
Temp. (K)	293	298	303	308
Elastic module	1.0005	1.0001	0.9998	0.9994
Damping coefficient	0.9942	0.9984	1.0025	1.0065
Temp. (K)	313	318		
Elastic module	0.9991	0.9987		
Damping coefficient	1.0105	1.0145		

The parameters used in Simulink take value as below.

$$m = 8.5 \times 10^{-8}\, kg \qquad (9)$$

$$k_{d0} = 215 \quad N/m \qquad (10)$$

$$k_{c0} = 20 \quad N/m \qquad (11)$$

$$c_{d0} = 6 \times 10^{-6} \quad N \cdot s/m \qquad (12)$$

where m is the mass of the drive-mode proof mass. k_{d0} and k_{c0} are the elastic module of supporting beams and coupling beams at 300 K, respectively. c_{d0} is the damping coefficient at 300 K.

The Simulink simulation result (Fig.5) shows that the amplitude of drive-mode oscillator is 4.0071 μm at 300 K. The amplitude would change with temperature, and the relationship is almost linear. In the range of nearby room temperature, the amplitude variation ratio changes from -0.2% at 273 K to 0.14% at 313 K. In Fig.5, the scatter plot is the simulation result, and the red solid line is the fitting curve with a slope of 0.2918 nm/K.

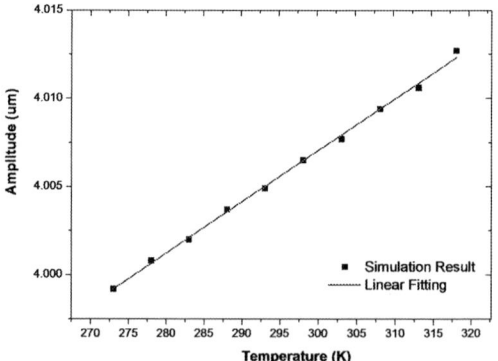

Figure 5. Amplitude vs. Temperature.

IV. Electronic Parameter Changing Induced Effects

There are many comb structures in the MEMS part of the gyroscope, and these structures are capacitances for three different usages. Three usages for capacitances are driving, feedback and sensing. Driving capacitance is designed to generate the electrostatic force to drive proof masses to vibrate. Feedback capacitance is designed to obtain the displacement of driving electrodes, which is helpful for

978-1-4799-2408-0/14 $31.00 © 2014 IEEE

closed-loop control of the driving oscillator. Sensing capacitance is designed to detect the displacement of proof masses in Coriolis force direction.

However, the electrodes of capacitances in most gyroscopes are not connected to ASIC pads directly. The components to connect capacitance electrodes and ASIC pads are doped silicon and gold wires usually. The resistance of gold wire is quite smaller than that of doped silicon, so that the resistance of gold wire is ignored in most articles. It is also common that the resistance of doped silicon is ignored to simplify analysis.

The pads of the gyroscope in our experiment are placed along one edge of the MEMS part, and this kind of design lengthens the doped silicon conductor trace line, and finally results in larger resistance.

C/V converter circuit is a kind of circuit that converts capacitance variation into voltage variation. A typical C/V circuit schematic is shown below in Fig.6.

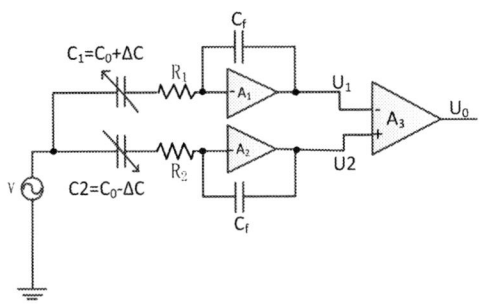

Figure 6. C/V Circuit with Parasitic Resistance

where V is a sinusoidal voltage source, C_1 and C_2 are sensing capacitances of MEMS, C_f is a feedback capacitance in ASIC, and A_1 A_2 A_3 are ideal amplifiers. R_1 and R_2 are parasitic resistances in MEMS structure, and they are zero in ideal situation.

C_1 and C_2 vary in anti-phase to reject the common mode signal and increase the output signal. The output voltage U_0 is proportional to ΔC (the variation of feedback capacitance of MEMS) as shown in Equ.13.

$$U_0 = \frac{2\Delta C}{C_f} \cdot V \cdot \frac{1}{\sqrt{1+(2\pi f C_0 R)^2}} \qquad (13)$$

In equation (13), f is the frequency of carrier wave, which is a few times the driving mode frequency (8000Hz in our experiment gyroscope), and in our simulation f is 50000 Hz. C_0 is the capacitance of sensing capacitance at static state, and its value is usually in the level of pF and in our simulation we can regard it as 10pF. The equation also indicates that in ideal situation (R=0), the circuit gets its largest output voltage.

The resistivity of highly doped silicon is around 0.02 $\Omega\cdot$cm [8]. The length of highly doped silicon conductor-trace-line is around 3 mm and the width is around 5 μm. The aspect ratio of MEMS gyroscope is about 10, so the thickness is about 50 μm. Based on the data provided, we can calculate that the parasitic resistance is about 2400 Ω.

The mismatch ratio is defined to be the real output voltage to the ideal output voltage ratio, so we get equation (14).

$$h = \frac{1}{\sqrt{1+(2\pi f C_0 R)^2}} \qquad (14)$$

From equation (14) we can see that the real output voltage is always smaller than that of ideal.

The resistivity of highly doped silicon changes with temperature and the temperature coefficient is up to 0.2%/℃ [9]. The simulation result is shown in Fig. 7. The scatter plot is the simulation result, and the red solid line is the fitting curve.

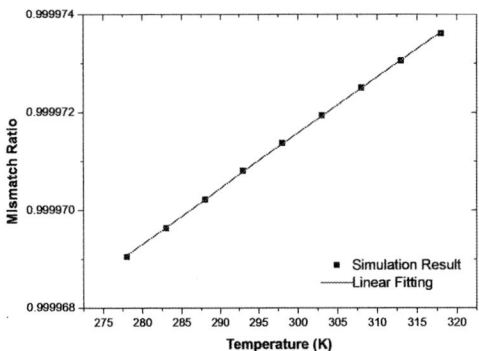

Figure 7. Mismatch Ratio vs. Temperature.

Mismatch ratio changes with temperature slightly, and their relationship is linearity with a slop of about 0.146 ppm/K. The mismatch ratio changes by from 0.0003% at 273 K to -0.0002% at 318 K, relative to the mismatch ratio at 300 K.

V. Experiment Results

The output of the gyroscope in out experiment is analog voltage within the range from 0.5 V to 4.5 V. The ideal ZRO is 2.5 V at room temperature (300 K). The gyroscope PCB board is placed in thermal cycling chamber, and the data processing circuit connected to the gyroscope by wires is placed outside the chamber.

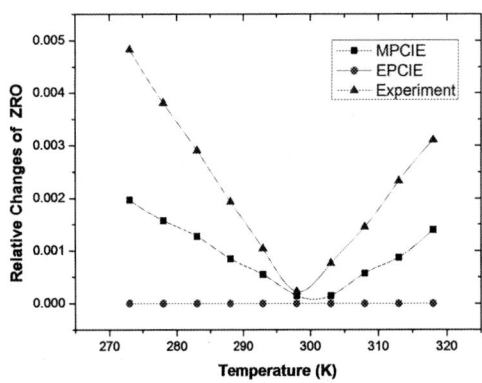

Figure 8. Comparison between Simulation and Experiment.

The results are shown in absolute value of relative changes of ZRO, and the value at 300 K is selected as the reference ZRO to calculate ZRO drift. The trends of MPCIE and EPCIE match the experiment result well.

VI. Conclusions

In this article, two models are built to analysis ZRO drift with temperature variation. The simulation result shows that material parameter changing induced ZRO drift is within 0.2% and is the main reason of ZRO drift nearby room temperature. Electronic parameter changing induced effects has very little influence on ZRO drift of about 0.0003% which is three orders magnitude smaller than that of MPCIE. In the range of room temperature (273 K ~ 318 K), the relationship of ZRO and temperature is linear, so are MPCIE and EPCIE.

VII. Acknowledgments

This work is supported by National High Tech Program (863) of Ministry of Science and Technology with contract number of 2012AA040501 and MEMS-volume Manufacture Technology of National Basic Research Project (973) under Minister of Science and Technology with No. 2011CB309504.

VIII. References

1. D. Liu, X. Chi, et al., "Research on temperature dependent characteristics and compensation methods for digital gyroscope", in *3rd International Conference on Sensing Technology*, Tainan, Taiwan, Nov. 30-Dec. 3, 2008, pp. 273-277.

2. Sheng Liu, Yong Liu, *Modeling and Simulation for Microelectronic Packaging Assembly, Manufacturing, Reliability and Testing*, Chemical Industry Press and John Wiley & Sons, 2011.

3. W. A. Clark, R. T. Howe, and R. Horowitz, "Surface micromachined z-axis vibratory rate gyroscope", *Tech. Dig. Solid-State Sensor and Actuator Workshop*, pp.283-287, 1996.

4. Z. Luo, S. Liu, G. Cao, and X. Chen, "Modeling and simulation of the comb structure in the presence of imperfections," *Electronic Components and Technology Conference (ECTC), 2013 IEEE 63rd*. pp. 2214-2217.

5. Cenk Acar, Andrei Shkel, *MEMS Vibratory Gyroscopes: Structural Approaches to Improve Robustness*, Springer 2006, pp. 80-81.

6. F. Sun, Q. Guo, et al., "Research on thermal characteristic in slow-small temperature changing for MEMS linear vibration gyroscope", *Proc. IEEE Int. Conf. Mechatronics Autom.*, pp. 25-28, 2006.

7. Y. Y. Tan, H. Yu, Q. A. Huang, T. Q. Liu, "Effect of temperature on the young's modulus of silicon Nano-filmsA", *Chinese Journal of Electrics Devices*, vol. 30, pp. 755-758, 2007.

8. H. T. Ding, X. S. Liu, J. Cui, X. Z. Chi, Z. C. Yang and G. Z. Yan, "A bulk micromachined z-axis single crystal silicon gyroscope for commercial applications", *Proc. IEEE NEMS*, pp. 1039-1042, 2008.

9. Z. Hou, D. Xiao, X. Wu, et al., "Effect of Parasitic Resistance on a MEMS Vibratory Gyroscopes due to Temperature Fluctuations", 6th IEEE *International Conference on Nano/Micro Engineered and Molecular Systems,* 2011, pp. 299-302.

Life Prediction and Classification of Failure Modes in Solid State Luminaires Using Bayesian Probabilistic Models

Pradeep Lall, Junchao Wei, Peter Sakalaukus
Auburn University
NSF-CAVE3 Electronic Research Center
Department of Mechanical Engineering
Auburn, AL. 36849
Tele: (334) 844-3424
E-mail: lall@auburn.edu

Abstract

A new method has been developed for assessment of the onset of degradation in solid state luminaires to classify failure mechanisms by using metrics beyond lumen degradation that are currently used for identification of failure. Luminous Flux output, Correlated Color Temperature Data on Philips LED Lamps has been gathered under 85°C/85%RH till lamp failure. The acquired data has been used in conjunction with Bayesian Probabilistic Models to identify luminaires with onset of degradation much prior to failure through identification of decision boundaries between lamps with accrued damage and lamps beyond the failure threshold in the feature space. In addition luminaires with different failure modes have been classified separately from healthy pristine luminaires. It is expected that, the new test technique will allow the development of failure distributions without testing till L70 life for the manifestation of failure.

Introduction

The lighting industry is undergoing a change from the incandescent lamps and compact fluorescent lamps (CFL) to light emitting diodes (LED). Mercury widely used in CFLs has the potential of contaminating large amounts of drinking water to beyond drinkable levels even in trace amounts. Transition to LEDs can impact energy efficiency tremendously because nearly 17% of the annual energy consumption is used for lighting. LEDs are being used in a wide variety of applications including automotive lighting, LED displays, street and home lighting. Traditional methods of failure-detection often used for identification of failure in incandescent lamps may not be applicable to LEDs.

Traditional light sources "burn out" at end-of-life. For an incandescent bulb, the lamp life is defined by B50 life. However, the LEDs have no filament to "burn". The LEDs continually degrade and the light output decreases eventually below useful levels causing failure. LED failure is characterized by L70 life or 70% degradation of the lumen output. Currently, it is not possible to qualify SSL lifetime of 10-years and beyond often necessary of high reliability applications, primarily because of lack of accelerated test techniques and comprehensive life prediction models. SSL comprises of several length scales with different failure modes at each level. Interactions between optics, drive electronics, controls and thermal design. Accelerated testing for one sub-system may be too harsh for another sub-system. New methods are needed for predicting SSL reliability for new and unknown failure modes. Presently, there is scarcity of life distributions for LEDs and SSLs which are needed to assess the promised lifetimes. Several cities are experimenting with large scale deployment of luminaries. In order to keep high availability of the system, it is essential that the onset of damage in form of color shift, luminous output degradation, and change in CCT be detected early.

Previous researchers have studied the failure modes of luminaires. Junction temperature of a luminaire plays a substantial role on its lifetime. The degradation rate of the plastic encapsulation material (PEM) on the diode is predominately affected by junction temperature causing the attenuation of the light output [Narendran 2004, Baillot 2010]. Excessive temperatures inside the LED package or the ingress of moisture can produce thermal-mechanical and hydro-mechanical stresses between the various material layers of LED packages causing delamination [Lumileds 2006, Luo 2010]. Elevated temperatures and humidity can produce delamination between the die and silicone encapsulant [Lumileds 2006] and between the encapsulant and packaging lead frame [Luo 2010]. The stresses can also produce a hairline cracks known as lens cracking, which occurs due to thermal expansion at various operating temperatures [Lumileds 2006, Hsu 2008], as well as when a long-term exposure to moisture [Hewlett Packard 1997]. In this paper, wet high temperature operating life environmental conditions of 85°C and 85%RH have been used to understand the reliability of solid state luminaires. SSL failure is quantified by the deterioration of luminous flux output and correlated color temperature (CCT) with respect to the time during accelerated testing. The Illuminating Engineering Society test standards LM-80-08 and IES TM-21-11 define the lifetime of an LED for lighting as the degradation to 70-percent of the original luminous flux output at room temperature [IES LM-80-08; IES TM-21-11]. Bayesian generative models have been used for classification of damaged assemblies and Bayesian regression models have been used to model the damage progression in solid state luminaires. The luminous flux, CCT, and the color shift have been used as input variables for identification of the onset of damage and separation of the healthy SSLs from those with significant accrued damage. Discriminant functions have been used to identify the class boundaries and classify SSLs significantly prior to the development of complete failure distributions. The models have been used to estimate the remaining useful life for each sample under test and the model predictions validated versus experimental data. It is expected that, the new test technique will allow early identification of failure distributions.

Test Vehicle

One of the original off-the-shelf 60W LED Lamps has been used as the test vehicle (Figure 1). The lamp has a total of 9 LUXEON Royal-Blue LEDs which are divided into three systematic lamp housings with a yellow cerium doped yttrium aluminium garnet phosphor shell.

Figure 1 Ambient LED 60W Lamp

The lamp produces white light through the color mixing of the blue LEDs and the yellow phosphor. The luminaires have been subjected to temperature-humidity at 85°C/85RH in an accelerated test chamber. The luminaires were non-functional during the accelerated test and placed upright inside a lamp holder to prevent movement inside the test chamber. Each of the lamps were extracted on a weekly basis to exam the spectral data for luminous maintenance, chromaticity shift, and correlated color temperature. All of the lamps in the test set were aged in the temperature-humidity condition for a total duration of 2537 hours.

Experimental Set-Up

The LED lamp measurement has been accomplished using an integrating sphere. Typically, the lamp measurement system contains parts: (1) Light Emitting Device. (2) Light Gathering System. (3) Light Transmitting System. (4) Light Analyzing System. The light emitting device provides the AC voltage power connection to the LED lamp that is producing continuous measureable light. The Light Gathering System, in this case the integrating sphere redistributes and collects the entire light beam emitting from the LED lamp. The integrating sphere is an optical component that uniformly scatters the light, which has a special coating on its surface of inside sphere. With a small exit ports on the side of sphere, the LED lights can be transmitted through the cosine diffuser, which is a detector, filtering and transferring the distributed light to the cable optical fiber. Then, the light is carried into the Labsphere 'USB4000' Spectrometer. Data on the Lumen Flux and CCT is collected using SpecraSuite Software. The total spectral radiant flux, $\Phi_{test}(\lambda)$, of the LED lamps under test was obtained by comparison of the total spectral radiant flux of the test lamp, $\Phi_{TEST}(\lambda)$, to the total spectral radiant flux of a reference standard, $\Phi_{REF}(\lambda)$. The following equation was used to compute the total spectral radiant flux:

$$\Phi_{TEST}(\lambda) = \Phi_{REF}(\lambda) * \frac{y_{TEST}(\lambda)}{y_{REF}(\lambda)} * \frac{1}{a(\lambda)} \quad (1)$$

Where $y_{TEST}(\lambda)$ is the are the spectrometer readings for the lamp under test, $y_{REF}(\lambda)$ is the spectrometer readings for the reference-lamp, respectively, and $a(\lambda)$ is the self-absorption

factor measured using an auxiliary lamp as described in LM-79. From the measured total spectral radiant flux $\Phi_{TEST}(\lambda)$ [W/nm], the total luminous flux $\Phi_{TEST}(\lambda)$[lm] is obtained by

$$\Phi_{TEST} = K_m \int_\lambda \Phi_{TEST}(\lambda) * V(\lambda) d(\lambda)$$
$$K_m = 683 lm/W \quad (2)$$

Where $V(\lambda)$ is the photopic sensitivity as a function of the wavelength. Self-absorption is the effect, in which the response of the sphere system is affected due to the absorption of light by the lamp itself in the sphere. Errors can also occur if the size and shape of the test light source are significantly different from those of the standard light source. The self-absorption factor is given by,

$$a(\lambda) = \frac{y_{aux, TEST}(\lambda)}{y_{aux, REF}(\lambda)} \quad (3)$$

Where $y_{aux,TEST}(\lambda)$ is the spectrometer readings for the auxiliary lamp with the LED lamp in the sphere, and $y_{aux,REF}(\lambda)$ is the spectrometer readings for the auxiliary lamp with the reference standard in the sphere.

Computation of Decay Rate and Failure Threshold

The decay rate of the luminous flux and the correlated color temperature in the LED lamps and the LUXEON LEDs inside the LED lamps has been calculated from the exponential model,

$$\Phi = \beta \cdot e^{\alpha \cdot t} \quad (4)$$

Where, β is the pre-decay factor, α is the decay rate, t is the test time, and Φ is either the luminous flux output or the correlated color temperature depending on the decay rate being calculated. The decay rate is a function of temperature and represented by:

$$\alpha = A \cdot e^{-\left(\frac{E_A}{k_B \cdot T}\right)} \quad (5)$$

Where T is the temperature in kelvin, K_B is the Boltzmann constant, and E_A is the activation energy. The Method of Least Square (LS) has been used to compute the decay rate for both CCT and Lumen Maintenance. The data for the LED lamps has been taken from accelerated test data under 85°C/85%RH. The LUXEON LEDs data from [DR05-1-LM80; Philips 2012] is under conditions of 55°C at 1A. Measured values of both the luminous flux output and the correlated color temperature have been normalized with respect to the measured value at time zero. LED lamp data shows the degradation of CCT to 96% from the initial value of 100% after 2500 hours. Similarly, the Lumen Maintenance shows degradation to 68% after 2000 hours of accelerating test.

Typically, L70 (70% Lumen Maintenance) life has been treated as the failure threshold for the luminous flux output of the solid state luminaire. Further, the 7-step MacAdam ellipse states that the target 'Duv' and its tolerance is ±0.006, and the corresponding target CCT and tolerance is 3000±175K for a nominal 3000K lamp [ANSI C78.377-2008 Specification]. One can therefore conclude that variation of CCT of greater than 94.17% of the original CCT values are deemed as unacceptable. The 94.17% value for a 3000K lamp is 2825K for the LED lamp. The lumen decay is more significant than the CCT decay. For the purpose of computing the remaining useful life of the luminaire and LEDs, the degradation of lumen maintenance was used. Figure 2 and Figure 3 show the decay rates for luminous flux output and correlated color temperature.

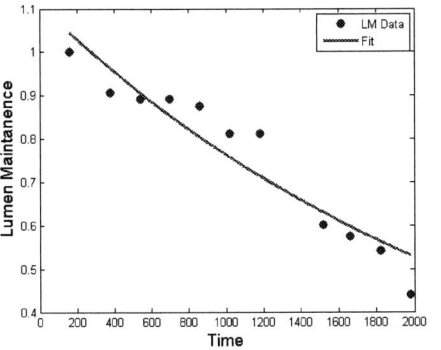

Figure 2: Lumen Maintenance Evolution for the LED Lamp in 85°C/85RH

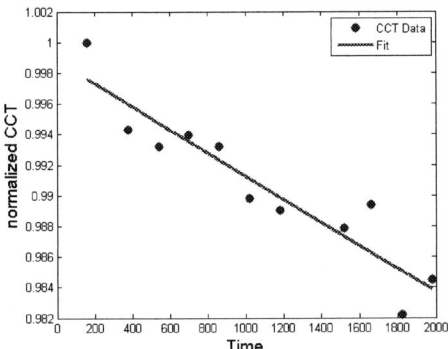

Figure 3: Correlated Color Temperature Evolution for the LED Lamp in 85°C/85RH

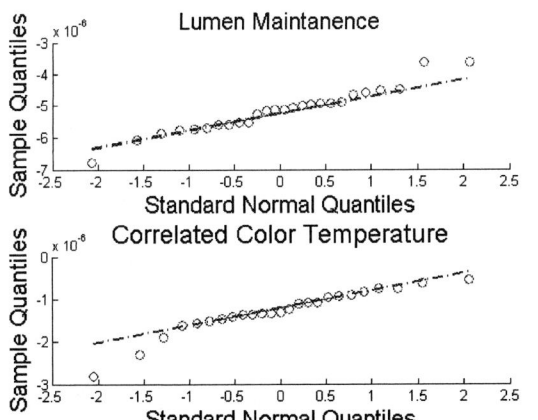

Figure 4: QQ plot for normalized luminous flux output and normalized correlated color temperature for LUXEON LEDs

The mean decay rate of normalized luminous flux output for the LUXEON LEDs has been calculated to be -5.14e-06 per hours, and mean decay rate of the normalized CCT for the LUXEON LEDs has been calculated to be -1.28e-06 per hours. Normality of the luminous flux output and the correlated color temperature distributions has been checked using the QQ-plot (Figure 4). The blue line shows the quantile of normal distribution, the red dots show the normalized luminous flux output and correlated color

temperature data. Analysis results indicate that the data is normally distributed with only two or three outliers.

The decay rates for the normalized luminous flux output and the correlated color temperature of the lamp has been similarly computed to be -4.05e-04 per hours, and -6.30e-06 per hour respectively. The Arrhenius model has been used to calculate the activation energy for the normalized luminous flux output and the normalized correlated color temperature. The computed activation energy has been used to evaluate the effect of temperature on the luminous flux output and the correlated color temperature.

$$
E_A = -\left(\frac{\ln \dfrac{\alpha_1}{\alpha_2}}{\left(\dfrac{1}{T_1} - \dfrac{1}{T_2} \right) \cdot K_B} \right)
\tag{6}
$$

The activation energy for the normalized lumen degradation is 1.47 eV and the activation energy for the normalized correlated color temperature is 0.53 eV. The failure threshold for the normalized luminous flux and the normalized correlated color temperature was identified by computing the 95% confidence bounds. Data that fell below the failure threshold at any time during the life test was deemed as a failure. Remaining useful life predictions were done for samples that did not fall below the failure threshold. The failure criterion is the curve of the maximum normalized decay rate, which will envelop all the degradation lines in the tested sample-set. For the LUXEON LEDs, the maximum decay rate ($\alpha_{max,LM}$) for the normalized luminous flux output is -6.77e-06 hour^{-1} and maximum decay rate ($\alpha_{max,CCT}$) for the normalized CCT is -2.8e-06 hour^{-1}. The 95% Confidence Interval for the maximum decay rate has been used to compute the highest possible decay rate, i.e. the lower boundary, for formulating the failure criterion.

$$
\alpha_{LM}^{failure} = \alpha_{max,LM} - 1.96 \frac{\sigma}{\sqrt{N}}
\tag{7}
$$
$$
= -7.04 \times 10^{-6} \, hour^{-1}
$$

$$
\alpha_{CCT}^{failure} = \alpha_{max,CCT} - 1.96 \frac{\sigma}{\sqrt{N}}
\tag{8}
$$
$$
= -3.01 \times 10^{-6} \, hour^{-1}
$$

Where $\alpha_{LM}^{failure}$ is the failure threshold of the decay rate for the normalized luminous flux output, and $\alpha_{CCT}^{failure}$ is the failure threshold of the decay rate for the normalized correlated color temperature. The decay rate values from the LUXEON LEDs have been used to compute the LED related decay rate for the lamp. The failure threshold decay rate has been calculated using an Arrhenius model:

$$
L\alpha_{LM}^{failure} = \alpha_{LM}^{failure} \cdot e^{-\frac{E_A}{K_B} \left(\frac{1}{T_1} - \frac{1}{T_2} \right)}
\tag{9}
$$
$$
= -5.55 \times 10^{-4} \, hour^{-1}
$$

$$L\alpha_{CCT}^{failure} = \alpha_{CCT}^{failure} \cdot e^{\dfrac{E_{act}}{k_B}\left(\dfrac{1}{T_1}-\dfrac{1}{T_2}\right)} \tag{10}$$

$$= -1.48 \times 10^{-5} \text{ hours}^{-1}$$

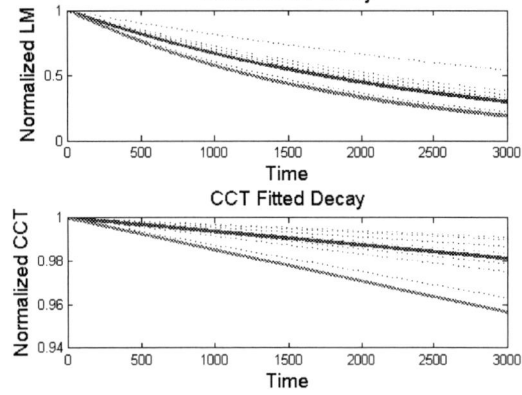

Figure 5: Characterized Decay Rate Curve for the LED Lamp

The time dependent decay curve for the LED Lamp is shown in Figure 5a where the bold red line is the mean decay rate. The lower-bound is depicted as a green line in Figure 5a. The CCT decays almost linearly. However, the Lumen Maintenance exhibits much more exponential pattern. The green dash line in Figure 5b shows the failure criterion for the LED Lamp. Figure 5b shows that all of CCT degradation lines are higher than the failure threshold. Only one of the tested samples lies on the failure threshold for Lumen Maintenance.

Bayesian Probabilistic Model

In this paper, Bayesian Probabilistic Generative Models [Bishop 2006] have been used to classify and separate damaged solid state luminaire assemblies from healthy assemblies. The goal of classification is to analyze input vector, x consisting of CCT, Color-Shift and Luminous Flux Output and to assign it to one of the classes, C_k. There are two possible classes including damaged or healthy. The classes are taken to be disjoint, so that each input is assigned to only one class. The input space is divided decision regions whose boundaries are called the decision boundaries. The target variable has been represented as a binary variable such that t=1 represents class C_1 and t=0 represents class C_2. The value of 't' is the probability that the class is C_1 with the values of probability taking only extreme values of 0 and 1.

The conditional probability distribution, $p(C_k \mid X)$, has been modeled in the inference stage and then the distribution has subsequently been used to make optimal decisions of classification. A generative approach has been adopted for computing the conditional probability distribution $p(C_k \mid X)$. In this procedure, the class conditional probabilities, $p(X \mid C_k)$ have been modeled as well as the class priors, $p(C_k)$, and then used to compute posterior probabilities through Bayes Theorem. For the purpose of the analysis, it was assumed that the class conditional probability density function is Gaussian, represented by:

$$P(X \mid C_k) = \frac{1}{(2\pi)^{d/2}} \cdot \frac{1}{|\Sigma_k|} \cdot \tag{11}$$

$$\exp(-\frac{1}{2}(X-\mu_k)^T \Sigma_k^{-1}(X-\mu_k))$$

Where the input vector, x, is a d-component column vector, μ is the d-component mean vector, Σ is the d-by-d covariance matrix. The class-prior $p(C_k)$ utilizes the weighted group form, for classifying two groups, where the probabilities of each group are given by:

$$p(C_1) = \frac{N_1}{N_1 + N_2}; p(C_2) = \frac{N_2}{N_1 + N_2} \tag{12}$$

$$p(C_1)/p(C_2) = N_1/N_2 \tag{13}$$

Where, N_l is the number of samples in the first group, and N is the total numbers of samples. The conditional probability distribution for classifying the 'k' group has been normalized, based on the weighted value between its posterior and the sum of posteriors from all the groups. Minimum error rate classification has been achieved through the use of discriminant functions, $p(C_k \mid X)$:

$$g_i(x) = p(C_k \mid X) = \frac{p(X \mid C_k)p(C_k)}{\sum p(X \mid C_j)p(C_j)} \tag{14}$$

Where, $g_i(x)$ is a discriminant function which is used as a classifier. Thus, the discriminant function for the multiple-class classification is defined as:

$$g_i(x) = p(X \mid C_k)p(C_k) \tag{15}$$

Alternatively, the discriminant may be represented in log-form as,

$$g_i(x) = \ln p(X \mid C_k) + \ln p(C_k) \tag{16}$$

In a general multivariate normal case, the covariance matrices are different for each category. The discriminant function can be computed by substituting Equation (11) for the class conditional probability density function into Equation (16) for the long-form of the discriminant function as follows:

$$g_i(x) = -\frac{1}{2}(x-\mu_i)^T \Sigma_i^{-1}(x-\mu_i) \tag{17}$$

$$-\frac{d}{2}\ln 2\pi - \frac{1}{2}\ln|\Sigma_i| + \ln p(C_i)$$

Where the input vector, x, is a d-component column vector, μ is the d-component mean vector, Σ is the d-by-d covariance matrix. The resulting discriminant terms are inherently quadratic:

$$g_i(x) = X^T.W_i \cdot X + w_i^T \cdot X + \omega_{i0} \tag{18}$$

Where the quadratic coefficients are solved as:

$$W_i = -\frac{1}{2}\left(\Sigma_i^{-1}\right) \tag{19}$$

$$w_i = \left(\Sigma_i^{-1} \cdot \mu_i\right) \tag{20}$$

$$w_{i0} = -\frac{1}{2}\left(\mu_i^T \Sigma_i^{-1} \mu_i\right) - \frac{1}{2}\left(\ln|\Sigma_i|\right) + \ln p(C_k) \tag{21}$$

978-1-4799-2408-0/14 $31.00 © 2014 IEEE

The discriminant functions have been computed for all the samples and the samples assigned to the class corresponding to the highest discriminant. The decision boundaries have been computed by setting

$$g_1(x) = g_2(x) \qquad (22)$$

Feature Space Creation
A two dimensional feature space has been created for classification of the test data. The two dimensions include the normalized luminous flux output and the correlated color temperature. The decay rate failure thresholds for the solid state luminaires which have been computed previously (Equations (9) and (10)) are used to construct the failure boundary for luminous flux output and a second boundary for the correlated color temperature. Lamps could fail because they breach the failure boundary for either the luminous flux output, correlated color temperature or both. The time at which the lamp breaches either boundary is termed as the failure time and represented by T_{CF}. The luminous flux output and the correlated color temperature at the failure boundary for failure time (T_{CF}) has been computed based on the previously calculated maximum decay rate.

$$\Phi = \beta \cdot e^{\alpha \cdot t} \qquad (23)$$

$$LM_{\text{max failure}} = (100) \cdot e^{L\alpha_{LM}^{\text{failure}} \cdot T_{CF}} \qquad (24)$$

$$CCT_{\text{max failure}} = (100) \cdot e^{L\alpha_{CCT}^{\text{failure}} \cdot T_{CF}} \qquad (25)$$

Where the multiplier of '100' in Equations (24) and (25) was used to change the computed normalized values into percentages. The computed locations in the feature space allow the location of the failure threshold versus the current state of the lamp in the feature space.

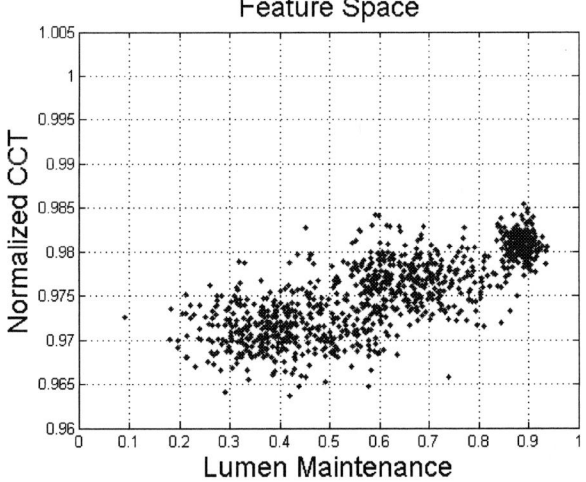

Figure 6: Feature space containing data of lamp's current state, lamp's failure threshold, and pristine healthy distribution of lamps prior to classification.

The mean and variance of the failure threshold and location at the failure time has been computed for all the devices under test. The classification of the healthy lamps versus the damaged lamps was accomplished using a decision boundary computed based on the discriminant function (Equation (18)). The test lamps have been classified as belonging to the failure

threshold distribution or the healthy distribution. The correlation between the luminous flux output and the correlated color temperature has been removed by computing the principal directions of the variance to yield uncorrelated x-axis and y-axis variances. The covariance matrix for the 85C/85%RH dataset is presented by,

$$\Sigma = \begin{pmatrix} \Sigma_{aa} & \Sigma_{ab} \\ \Sigma_{ba} & \Sigma_{bb} \end{pmatrix} \qquad (26)$$

The covariance matrix has been decorrelated by computing the principal components, thus rendering the correlation matrix in the form,

$$\Sigma = \begin{pmatrix} \Sigma_{11} & 0 \\ 0 & \Sigma_{22} \end{pmatrix} \qquad (27)$$

Where the subscripts '1' and '2' indicate the principal directions. The distributions of the lamp-state and the lamp's failure threshold have been transformed into the decorrelated principal component feature space for the purpose of classification. The data groups plotted in Figure 6 include the lamp's current state, lamp's failure threshold, and the pristine healthy distribution of lamps prior to classification.

Bayesian Regression Model
The response variables of luminous flux output, and CCT are the target variables (t) for the Bayesian regression models. Input parameters (w) include weights for the input parameters of time. The posterior probability has been computed based on the conditional probability:

$$p(t \mid w) = \frac{p(w \mid t)p(t)}{P(w)} \qquad (28)$$

Where, $p(t \mid w)$ is the normalized conditional posterior of the target variables, and $p(t)$ is the prior distribution of the target variables. The Bayesian conjugate prior Gaussian probability is represented as follows:

$$P(W) = N(W \mid M_0, S_0) = N(W \mid 0, \alpha^{-1}I) \qquad (29)$$

Where, α is the precision parameter of the weight distribution. The real-valued input variable column vector is:

$$X = [x_1 \quad x_2 \quad \dots \quad x_n]^T \qquad (30)$$

The real-valued predict target column vector is represented as,

$$T = [t_1 \quad t_2 \quad \dots \quad t_n]^T \qquad (31)$$

Candidate basis functions used may include polynomial functions (ϕ) with weights (w). The basis function 1×M matrix can be shown as the following:

$$\Phi(x) = [1 \quad x \quad x^2 \quad \dots \quad x^M]^T \qquad (32)$$

The weights M×1 matrix is represented as:

$$W = \begin{bmatrix} w_0 \\ w_1 \\ w_2 \\ \dots \\ w_M \end{bmatrix} \qquad (33)$$

The future degradation of the luminaire can be calculated from the estimation matrix as follows:

$$t_i = W^T \cdot \Phi(x_i) \tag{34}$$

The likelihood function will be represented with a Gaussian probability distribution as follows,

$$P(t \mid x, W, \beta) = N(t \mid W^T \cdot \Phi(x_i), \beta^{-1}I) \tag{35}$$

Where W is the weight vector and β is the precision of the target variable distribution, t. The n- set of observations $t_1, \ldots, t_N$, have been combined into a matrix T of size $N \times K$ such that the nth row is given by t_{T_n}. Similarly, we can combine the input vectors $x_1, \ldots, x_N$ into a matrix X. The log-likelihood of the data-set is given by:

$$\ln P(T \mid X, W, \beta) = \sum_{n=1}^{N} \ln N(t \mid W^T \cdot \Phi(x_i), \beta^{-1}I) \tag{36}$$

The likelihood represented by Equation (36) that the target, t corresponds to the input variable sets being considered is maximized with respect to β. The target parameter's variance is represented by:

$$\frac{1}{\beta_{ML}} = \frac{1}{N} \sum_{i=1}^{N} (y_i - t_i)^2 = \frac{1}{N} \cdot (Y - T) \cdot (Y - T)^T \tag{37}$$

The variance computed from equation (37) corresponds to the maximum value of the likelihood function. We can substitute the β_{ML} into Equation (36) for $P(T \mid X, W, \beta)$, which gives:

$$P(T \mid X, W, \beta) = \prod_{i=1}^{n} N(t \mid W^T \cdot \Phi(x_i), \beta_{ML}^{-1}) \tag{38}$$

The weight vector will be updated using the Bayesian posterior conditional probability represented as follows:

$$P(W \mid X, T, \alpha, \beta) \propto P(T \mid X, W, \beta) \cdot P(W \mid \alpha) \tag{39}$$

$$= N(W \mid M_N, S_N)$$

$$M_N = S_N(S_0^{-1}M_0 + \beta\Phi^T T)$$

$$S_N^{-1} = S_0^{-1} + \beta\Phi^T\Phi$$

Where, M_N is the mean and S_N is the covariance of the Bayesian posterior conditional probability

$$M_N = S_N(S_0^{-1}M_0 + \beta\Phi^T T) \tag{40}$$

$$S_N^{-1} = S_0^{-1} + \beta\Phi^T\Phi$$

The prediction of the target vector at the next time step is represented as:

$$P(t \mid T, \alpha, \beta) = \int P(t \mid W, \beta) \cdot P(W \mid T, \alpha, \beta)dW \tag{41}$$

The condition distribution $P(t \mid T, \alpha, \beta)$ has been calculated out as the distribution and probability with its mean and variance depending on the variable 'x'; Therefore, we can finally predict each output 't' including luminous flux and correlated color temperature from each time series input 'x', such as:

$$P(t \mid x, T, \alpha, \beta) = \tag{42}$$

$$N(t \mid M_N^T\Phi(x), \beta^{-1} + \Phi^T(x)S_N\Phi(x))$$

Failure Analysis Results

Once the Bayesian classifier has finished the training process, the data mapped onto the feature space is classified. The discriminant function has been used to classify the samples in the feature space and formulate a decision boundary between the lamps with accrued damage and pristine samples. The lamps migrate in the feature space from the top right to the

bottom left with the increase in the amount of accrued damage. In Figure 7, the red data points are the healthy samples and the green data points are the samples with accrued damage. The red dash line shows the failure threshold between the healthy lamps and lamps with accrued damage. The decision boundary has been calculated such that the discriminant for the two classes has an equal value along the boundary.

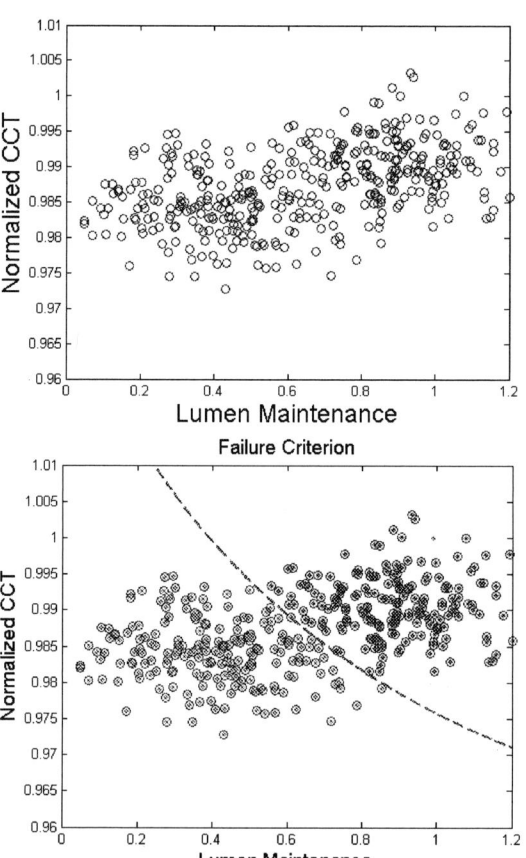

Figure 7: End of Life distribution and pristine for LED Lamp

The values of the coefficients of the polynomial that describes the decision boundary have been calculated using the following equations,

$$F(x, y) = I^T W_2 I + W_1^T I + W_0 \tag{43}$$

$$I = \begin{bmatrix} x \\ y \end{bmatrix} \tag{44}$$

$$W_2 = -\frac{1}{2}\Sigma_i^{-1} \tag{45}$$

$$W_1 = \Sigma_i^{-1}\mu_i$$

$$W_0 = -\frac{1}{2}\mu_i^T\Sigma_i^{-1}\mu_i - \frac{1}{2}\log|\Sigma_i| + \log(p(C_i))$$

The matrix expression of $F(x, y)$ has been expanded as the quadratic area function:

$$F(x, y) = W_2(1,1)x^2 + 2W_2(1,2)xy + W_2(2,2)y^2 \tag{46}$$

$$+ W_1(1,1)x + W_1(2,1)y + W_0$$

The classification decision boundary for the failure threshold can be calculated from previous analysis as equating the PDFs for the classes on either side of the decision boundary:

978-1-4799-2408-0/14 $31.00 © 2014 IEEE

$$G(x, y) = F_1(x, y) - F_2(x, y) = 0 \qquad (47)$$
$$= 3.72 x^2 - 157.74 xy + 131.11 x +$$
$$1823.78 y^2 - 3984.71 y + 2170.75$$

From the classification, the calculated coefficients of the polynomial are:

$$W0 = -2.7019 \cdot 10^4 \qquad (48)$$

$$W1 = \begin{bmatrix} 0.0305 \\ 5.4802 \end{bmatrix} \cdot 10^4 \qquad (49)$$

$$W2 = \begin{bmatrix} -0.0016 & -0.0149 \\ -0.0149 & -2.7783 \end{bmatrix} \cdot 10^4 \qquad (50)$$

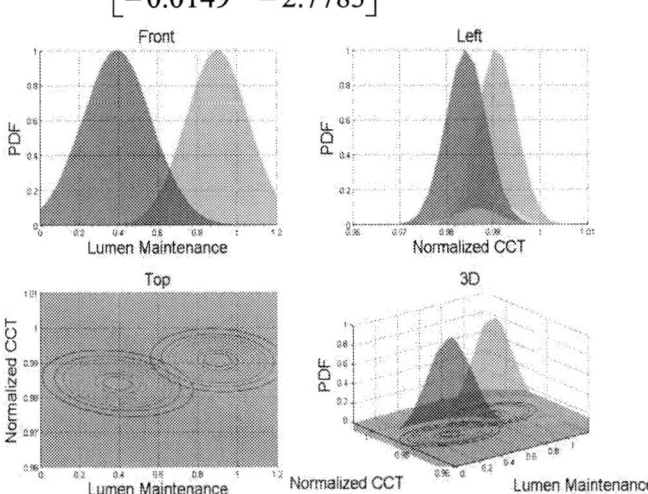

Figure 8: Classification of the lamps with accrued damage and pristine lamp PDFs.

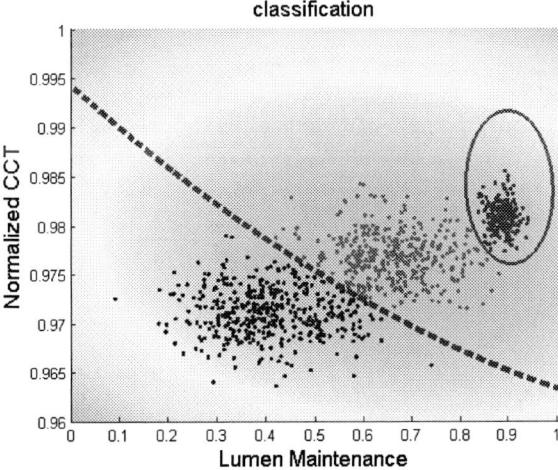

Figure 9: L70 Time Decision Boundary

Overall, the Bayesian unsupervised classifier is powerful classification tool. Even though two groups have been classified, the technique presented is applicable to multiple groups. The distributions corresponding to the healthy group with significant accrued damage has been plotted. The red PDF is corresponding to the lamps with accrued damage while the green PDF are the healthy lamps. The overlapping area displays the transition failure area between the healthy and the lamps with accrued damage. Typically, we want this

overlapping PDF region to be as small as possible. The decision boundary has been updated as more data becomes available for the different classes. Figure 9 shows the three groups classification, which we assign the initial parametric distributions for the (a) failure threshold (b) the pristine LED Lamp group and (c) damaged LED Lamp group. The testing data has been grouped, and Bayesian Classifier calculates the mean and variance numerically.

Figure 9 shows the migration of decision boundary. The decision boundary between pristine lamp group and failure lamp group is shown with a solid magenta ellipse, and the decision boundary between the failure threshold and the lamps with accrued damage is shown with a dashed red line. The Figure 10 demonstrates the three PDFs for the pristine LED Lamp, damaged lamp group as well as lamp group beyond the failure threshold. The decision boundary between the damaged lamp group and the group beyond the failure threshold has been termed as the critical failure boundary, which should not be breached to avoid failure.

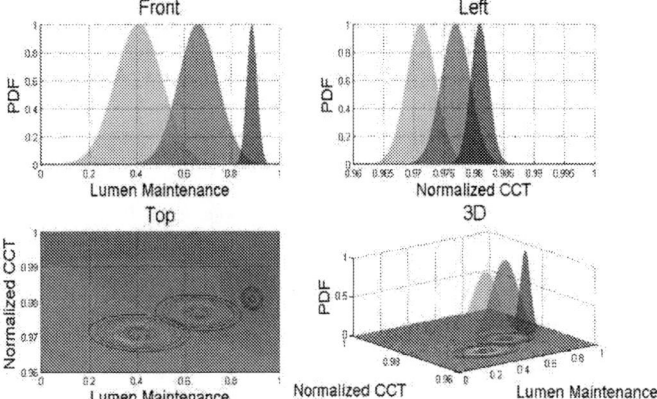

Figure 10: Critical Failure, LED Lamp Failure and Pristine Lamp PDF Distribution

Bayesian Regression for the System Remaining Useful Life
Bayesian regression method has been used to determine the Remaining Useful Life (RUL) for every test lamp. Lumen Maintenance (LM) degradation has been used as the main indicator of system decay, by fitting the Lumen Maintenance degradation curve (Figure 11).

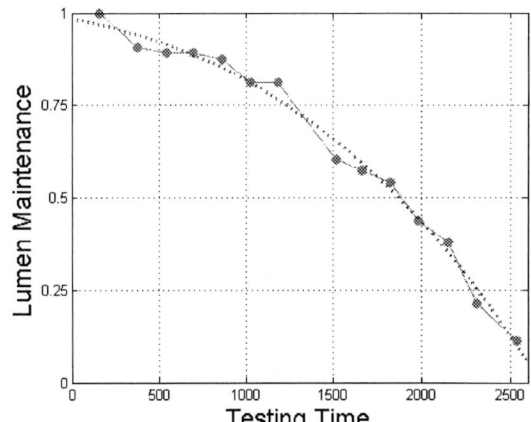

Figure 11: Lumen Maintenance Regression for LED Lamp

Figure 12 shows the training of the Bayesian regression model through the maximum likelihood function and prediction of the posterior distributions. The process discussed previously in the Bayesian regression section, has been used for the future state prediction of the lamp's luminous flux and the remaining useful life. Figure 13 shows the Bayesian linear regression for the third order polynomial model with four weights. The green dots are the measured data points, and the red dots show the predicting decay curve. The testing length is up to 2537 hours. The Remaining Useful Life (RUL) has been calculated by predicting the future luminous flux output state till the L70 threshold.

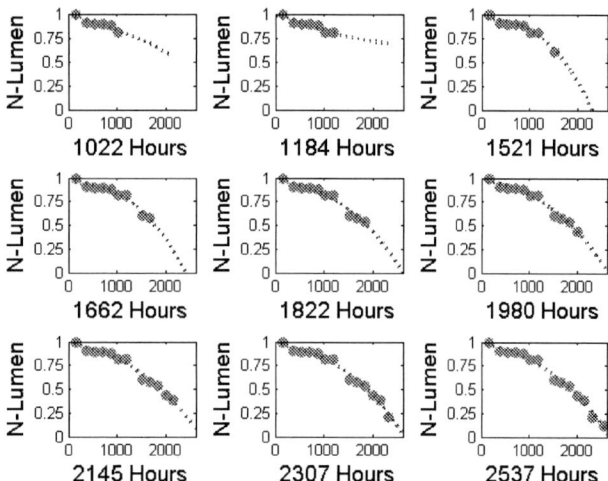

Figure 12: Bayesian Regression Learning Process

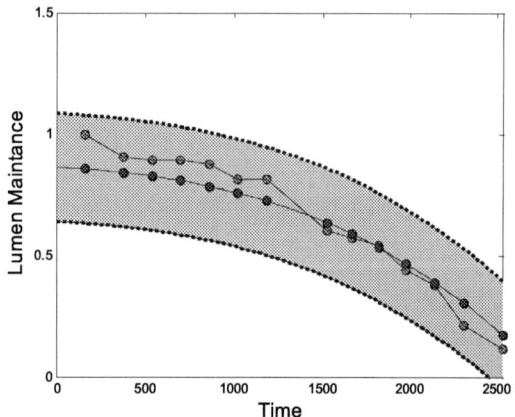

Figure 13: Bayesian Regression with Confidence Interval

The predicted RUL is known as the predicted End of Life (EoL) minus the sampling time, represented by:

$$T_{predict} = P_{EoL} - T_{sample} \qquad (51)$$

The real RUL is known as the actual EoL minus the sampling time. So the algebra equation presents as following:

$$T_{actual} = A_{EoL} - T_{sample} \qquad (52)$$

Figure 14 shows the α-λ prediction to assess the accuracy of the prediction results. The red line is the actual Remaining Useful Life of lamp. The blue dash line is the prognosticated remaining useful life. The green dash line is the ±20% confidence interval. While the Bayesian prediction is outside

the 20% confidence interval initially, the model converges to the true health of the system fairly quickly and tracks the degradation well.

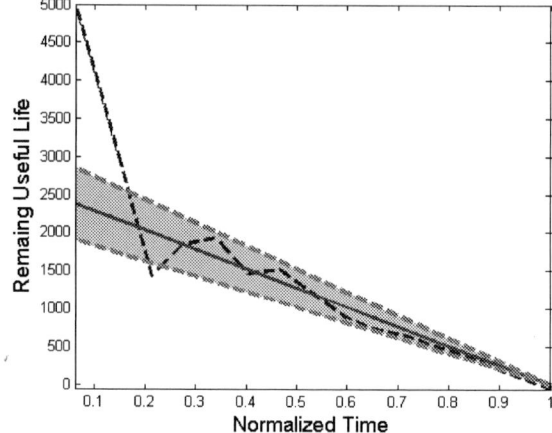

Figure 14: α-λ Plot of RUL Prediction

A two parameter weibull distribution has been used to model the lamp failures. The probability density function for the two parameter Weibull distribution has the following form:

$$f(t) = \frac{\beta}{\eta}\left(\frac{t}{\eta}\right)^{\beta-1} e^{-\left(\frac{t}{\eta}\right)^{\beta}} \qquad (53)$$

Where β is the shape parameter, η is the characteristic life. The estimated shape parameter and the characteristic life are: β = 7.1 and η = 1790 hours. Since the β > 1.0, it indicates that the failures are wear out failures. The Weibull cumulative distribution, the population fraction failing by time t is given as following CDF:

$$F(t) = 1 - e^{-(t/\eta)^{\beta}} = 1 - e^{-(t/1790.1)^{7.1}} \qquad (54)$$

The reliability function is thereby given by the 1-F(t), shown in the Figure 15. The CDF shows that once time reached 1700.2 hours the LED lamp reliability dropped to 50%. The characteristic life (B63.2 life) is 1790.1 hours, which says 63.2% LED lamps have failed at this time in the accelerated test condition of 85°C/85%RH.

$$R(t) = 1 - F(t) = e^{-(t/\eta)^{\beta}} = e^{-(t/1790.1)^{7.1}} \qquad (55)$$

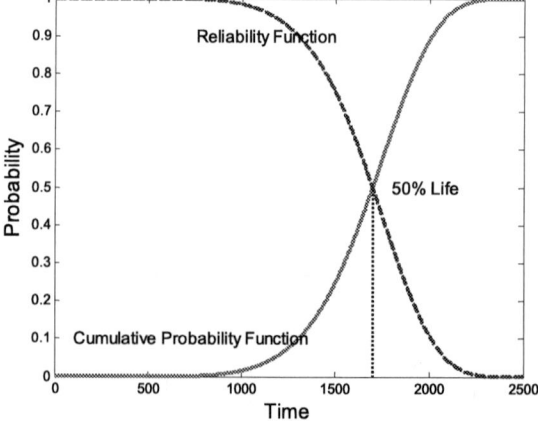

Figure 15 System Reliability Function and Cumulative Probability Function

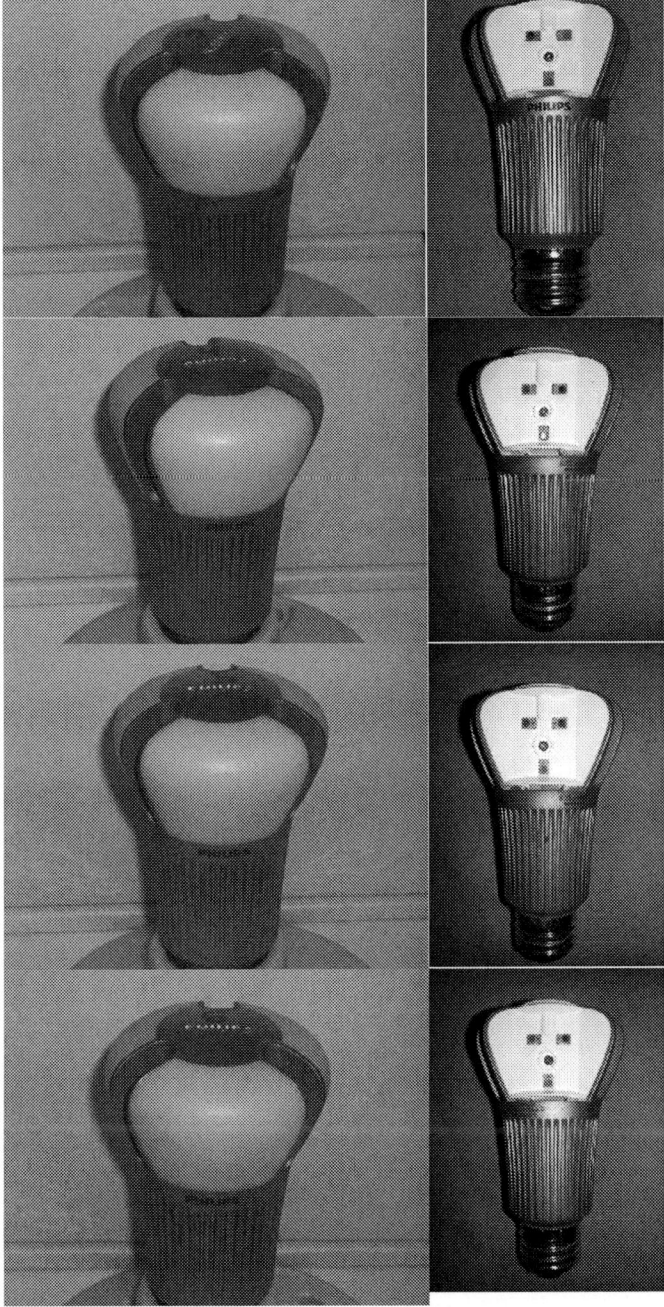

Figure 16 Representative Samples of the Failed Lamps with and without the Lens.

Figure 16 shows the representative samples of the failed lamps with and without the lens. Note that encapsulant of several of the LEDs in the failed lamps shows distinct discoloration. It is hypothesized that the discoloration of the encapsulant was a major contributor to the degradation in the luminous flux output and the color shift during the 85C/85%RH accelerated test.

Summary and Conclusion

The 60W LED lamps have been studied under the accelerated test conditions of 85°C/85%RH for both luminous flux output and the correlated color temperature. A Bayesian framework for early classification of the failed lamps in the luminous flux and correlated color temperature feature space has been formulated and demonstrated on the test-population of the lamps. Failures have been identified because of problems of luminous flux degradation or color shift or both. In addition the Bayesian regression model has been developed to predict the luminous flux degradation till the L70 threshold widely used as definition of failure for the solid state luminaires. The proposed methodology allows the early identification of the onset of failure much prior to development of complete failure distributions and can be used for assessing the damage state of SSLs in fairly large deployments. The α-λ plots have been used to evaluate the robustness of the proposed methodology. Results show that the predicted degradation for the lamps tracks the true degradation observed during 85°C/85%RH during accelerated life test fairly closely within the ±20% confidence bounds. Failure modes of the test population of the lamps have been studied to understand the failure mechanisms in 85°C/85%RH accelerated test. Results indicate that the dominant failure mechanism is the discoloration of the LED encapsulant inside the lamps which is the likely cause for the luminous flux degradation and the color shift.

Acknowledgments

The work presented here in this paper has been supported by a research grant from the Department of Energy under Award Number DE-EE0005124.

References

Baillot, R., Deshayes, Y., Bechou, L., Buffeteau, T., Pianet, L., Armand, C., Voillot, F., Sorieul, S., Ousten, Y. "Effects of Silicone Coating Degradation on GaN MQW LEDs Performances Using Physical and Chemical Analysis." *Microelectronics Reliability* 50(2010): 1568-1573.

Bishop, C. M., Pattern Recognition and Machine Learning, Springer Science-Business Media, LLC, 2006.

Cree. "Cree Xlamp XR Family LED Reliability, CLD-AP06 Rev. 7." Cree Inc. 2009.

Hewlett Packard. "Reliability of Precision Optical Performance AllnGaP LED Lamps in Traffic Signals and Variable Message Signs." *Application Brief I-004* (1997)

Hsu, Y., Lin, Y., Chen, M., Tsai, C., Kuang, J., Huang, S., Hu, H., Su, Y. and Cheng, W. "Failure Mechanisms Associated with Lens Shape of High-Power LED Modules in Aging Test." *IEEE Trans. on Electron Devices* 55(2008): 689-694.

IES (Illuminating Engineering Society), IES LM-80-08. "Approved Method: Measuring Lumen Maintenance of LED Light Sources." 2008

IES (Illuminating Engineering Society), IES TM-21-11. "Projecting Long Term Lumen Maintenance of LED Light Sources" 2008

Lall, P., Sakalaukus, P., Davis, L., Prognostics of Damage Accrual in SSL Luminaires and Drivers Subjected to HTSL Accelerated Aging, Proceedings of the ASME 2013 International Technical Conference & Exposition on Packaging and Integration of Electronic and Photonic Microsystems, InterPACK2013-73250, pp. 1-8, Burlingame, CA, 2013.

Lall, P., Wei, J., & Davis, L., Prediction of L70 lumen maintenance and chromaticity for LEDs using extended Kalman filter models, SPIE Optical Engineering+ Applications (pp. 88350M-88350M-19), San Diego, CA, 2013.

Lall, P., Wei, J., Davis, L., L70 Life Prediction for Solid State Lighting Using Kalman Filter and Extended Kalman Filter Based Models, Electronic Components and Technology Conference, ECTC, 63rd, pp.1452-1465, 2013.

Lall, P., Wei, J., Davis, L., Solid State Lighting Life Prediction Using Extended Kalman Filter, Proceedings of the ASME 2013 International Technical Conference & Exposition on Packaging and Integration of Electronic and Photonic Microsystems, InterPACK2013-73288, pp. 1-11, Burlingame, CA, 2013.

Lall, P., Zhang, H., Davis, L., Assessment of Lumen Degradation and Remaining Life of LEDs using Particle Filter, Proceedings of the ASME 2013 International Technical Conference & Exposition on Packaging and Integration of Electronic and Photonic Microsystems, InterPACK2013-73305, pp. 1-13, Burlingame, CA, 2013.

Lumileds. "Luxeon Reliability." *Philips Lumileds Reliability Datasheet RD25* (2006).

Luo, X., Wu, B. and Liu, S. "Effects of Moist Environments on LED Module Reliability." *IEEE Trans. on Device and Materials Reliability* 10(2010): 182-186.

Meneghini, M., Trevisanello, L., Meneghesso, G. and Zanoni, E. "A Review on the Reliability of GaN-Based LEDs." *IEEE Trans. on Device and Materials Reliability* 8(2008): 323-331.

Narendran, N., Gu, Y., Freyssinier, J.P., Yu, H. and Deng, L. "Solid-State Lighting: Failure Analysis of White LEDs." *Journal of Crystal Growth* 268(2004): 449-456.

Nichia. "Specifications for Nichia Chip Type White LED Model: NCSW119T-H3. Nichia STS-DA1-0990A." Nichia Corp. 2009

Saxena, A., Celaya, J., Saha, B., On Applying the Prognostic Performance Metrics, IEEE Annual Conference of the Prognostics and Health Management Society, 2009.

Modeling for Reliability of Ultra Thin Chips in a System in Package

Richard Qian and Yong Liu
Fairchild Semiconductor
South Portland, Maine 04106, USA

Abstract

In this paper, ultrathin silicon chip failure mechanism and reliability performance of a system in package (SIP) package are investigated. Advanced finite element analysis (FEA) modeling is carried out to simulate die stress and delamination in the temperature cycling test. Parametric models with different ultrathin die thicknesses, different delamination areas and different bond wire positions are simulated. Temperature cycling modeling results have shown that the thinner die has lower tensile stress than the thicker die. Delamination increases die tensile stress in temperature cycling test. Larger delamination areas in the package induce higher die tensile stress. Comparing with three different delamination models, the model without delamination in the whole package induces lowest die tensile stress, the model with the largest delamination areas at die top surface, aluminum bond wire and lead frame DAP induces the highest die tensile stress. Finally, the modeling results disclose that the bond wire at different position has very limited impact on die stress during temperature cycling test.

Introduction

The SIP package, which is used for power management has been developed with multiple chips, including power IGBT Mosfet, diode and IC controllers [1-3]. To maximize the product performance, the power chips have been made in ultra thin thickness. Ultra thin die minimize the Rds(on), maximize thermal performance, and minimize the board standoff height by allowing the package to be thinner. Therefore, it is critical to understand the impact of thinning die to the reliability performance of the product [1]. The ultra thin die could be a potential risk for die cracking if it is done without careful evaluation [4-5]. In this package, a ceramic based heat sink is attached to the lead frame pad through silicone elastomer. The dimension of ceramic is relative large since all of the power chips are attached to the ceramic layer through lead frame pad. The ceramic layer has much lower thermal expansion coefficient than the mold compound. The big gap of CTE between ceramic and mold compound induces possible delamination at silicon die top surface; aluminum bond wire and lead frame DAP of the package. This could be potential risk for speeding up the die cracking. Therefore, it is very critical to fully investigate the die stress and the delamination in reliability test.

In this paper, a 3D delamination model for a power SiP, which considers the initial defects between bond wire and EMC, die and EMC, and DAP and EMC, is built with contact pairs. Parametric models with different ultrathin die thickness, different delamination percentage and different bond wire position are simulated in the temperature cycling test. Three

delamination models are studied in the simulation: The first model has no delamination area at all. The mold compound glues everything perfectly. The second delamination model simulates delamination at ultrathin die top surface and aluminum bond wire. The third model simulate the delamination at the ultrathin die top surface, aluminum bond wire and lead frame DAP. All of above 3 delamination models consider the impact of two different IGBT thicknesses, and to simplify the model, above models consider one bond wire at edge only due to it has largest distance to the package center. Finally, the delamination model with different bond wire position is studied.

Comparison of IGBT die thickness and delamination areas

Figure 1 shows the internal layout of the power SIP package we will investigate. There are 5 analog IC chips, 6 IGBT chips and 6 diodes. Each IGBT chip is bonded with heavy aluminum bond wire to lead frame through diode. This model only considers one bond wire at position 1 to simplify the model.

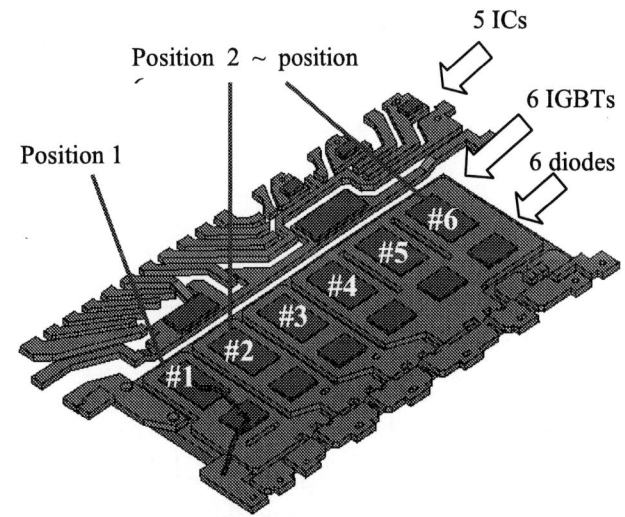

Figure 1: Internal structure of the package.

Figure 2 shows the local view of the silicon dice and the bond wire. Fig. 2 (a) shows the finite element model of the internal structure at position 1 (hide the mold compound). Fig. 2 (b) shows the side view of the internal structure. The IGBT die thickness is made to be ultra thin to improve its product performance. Diode will keep the same thickness. These dice are bonded to lead frame DAP through Pb free solder. The lead frame is attached to a ceramic based heat sink through silicone elastomer.

978-1-4799-2408-0/14 $31.00 © 2014 IEEE   2063   2014 Electronic Components & Technology Conference

Table 1 and Table 2 list the material properties of the SIP package for simulation.

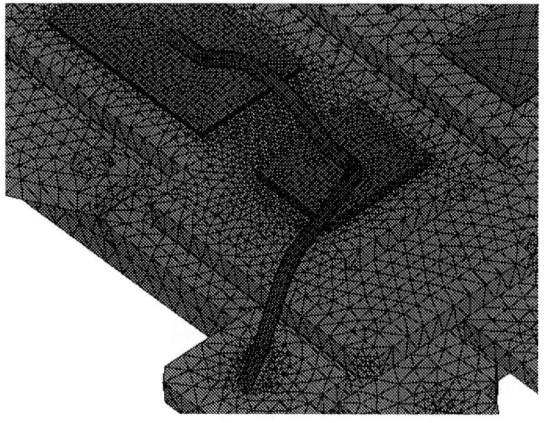

(a) Finite element model of bond wire at position 1.

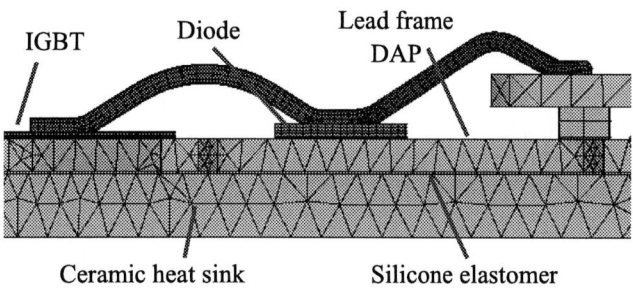

(b) Left view of bond wire at position 1.

Figure 2: Local finite element model of internal structure.

Table 1: Material properties of each material.

Material	Elastic modulus (Gpa)	Poisson ratio	CTE (ppm)
ceramic	340	0.22	6.8
EMC	24 @ 25C 1.44 @ 175C	0.3	8.6 (<140C) 36 (>140C)
solder	50	0.36	21.9
Chip	161	0.26	2.6
epoxy	4.65 @ 25C 1.39 @ 150C	0.4	60 (<73C) 135 (>73C)
Al wire	68.9	0.33	20
lead frame	125.53	0.3	17.5
ceramic adhesive	19.75@ 25C 0.402 @ 175C	0.4	17 (<80C) 76 (>80C)

Table 2: Yield stress and tangent modulus of solder, aluminum bond wire and lead frame.

Material	Yield stress (Mpa)	Tangent Modulus (Mpa)
solder	41	410
Al wire	200 @ 25C 164.7 @ 125C	300 @ 25C 150 @ 125C
lead frame	345	3450

Table 3: Modeling DOE of die thickness and delamination areas.

Model	IGBT thickness	Delamination
1	75μm	No delamination
2	97μm	No delamination
3	75μm	@ die top & wire
4	97μm	@ die top & wire
5	75μm	@ die top, DAP & wire
6	97μm	@ die top, DAP & wire

Table 3 lists the simulation DOE. There are 6 models with two different IGBT thickness and three different delamination areas. The IGBT is designed with thinner die. Model 1, Model 3 and Model 5 simulates 75μm thick IGBT. Model 2, Model 4 and Model 6 simulates 97μm thick IGBT. Model 1 and Model 2 simulate the first model without the delamination Model 3 and Model 4 simulates the delamination, in which the delamination locates at die top surface and bond wire. Two contact pairs are defined at the delamination areas. One contact pair is defined between mold compound and die top surface. The other contact pair is defined between mold compound and aluminum bond wire. Model 5 and Model 6 simulates the delamination. In which the delamination locates at die top surface, bond wire and lead frame DAP. Three contact pairs are defined at delamination areas of die top surface; aluminum bond wire and lead frame DAP.

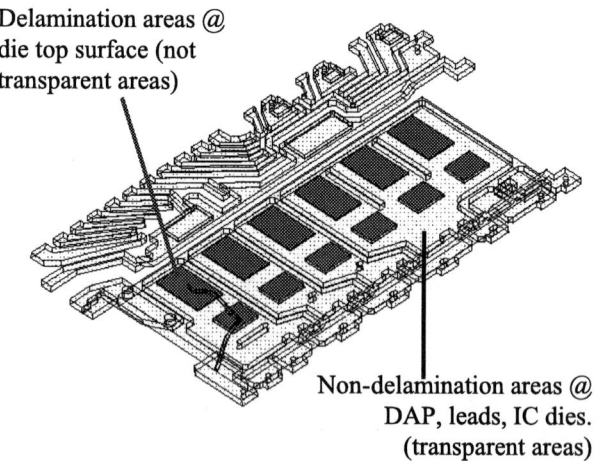

(a) Delamination areas at die top.

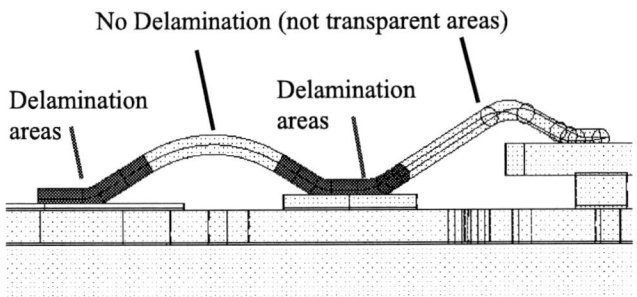

b) Delamination areas at bond wire.

Figure 3: Delamination at die top surface and bond wire.

Figure 3 (a) shows the internal structure of the model with delaminations at die top surface and bond wire only. Figure 3 (b) shows the side view of the internal structure. The bond wire's surface was divided to two parts. The surfaces near the IGBT and diode have delamination with mold compound. The other surfaces will not have the delamination.

Figure 4 (a) shows the internal structure of the model with delaminations at die top surface, bond wire and lead frame DAP. Figure 4 (b) shows the side view of the internal structure.

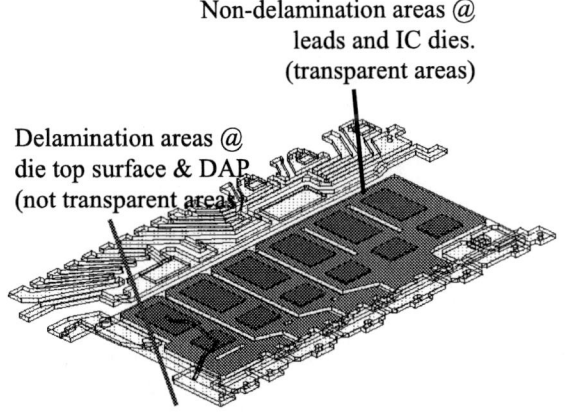

(a) Delamination areas at die top surface, bond wire and lead frame DAP.

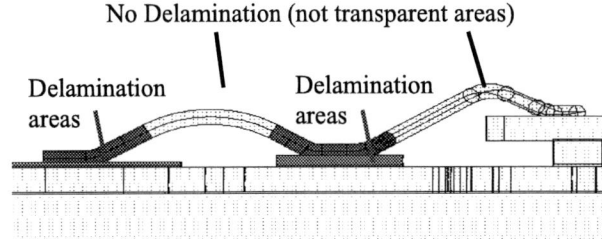

b) Side view of delamination model.

Figure 4: Delamination at upper surface of DAP, die top surface and bond wire.

Figures 5~10 show die's first principle stress S1 in temperature cycling test. Figure 5 shows S1 in die for model 1 with IGBT thickness 75μm and no delamination. The max S1 in IGBT is 154.1 Mpa. The max S1 in diode is 147.8 Mpa. Figure 6 shows S1 in die for model 2, with IGBT thickness 97μm and no delamination. The max S1 in IGBT is 164.6 Mpa. The max S1 in diode is 147.5 Mpa.

Figure 7 shows die first principal stress S1 for model 3 with IGBT thickness 75μm and delaminations at die top surface and bond wire. The max S1 in IGBT is 184.9 Mpa. The max S1 in diode is 199.2 Mpa. Figure 8 shows the die first principal stress S1 for model 4, with IGBT thickness 97μm and delamination at die top surface and bond wire. The max S1 in IGBT is 181.5 Mpa. The max S1 in diode is 198.9 Mpa.

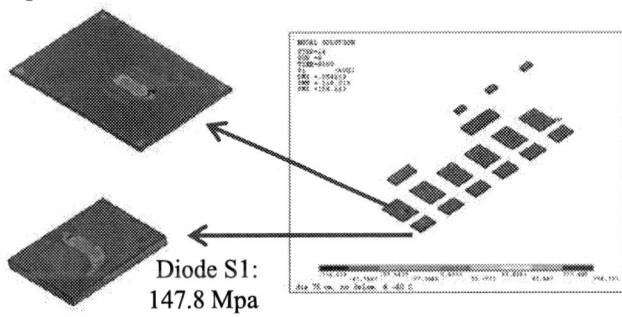

Figure 5: First principle stress S1 of IGBT and diode. (Model 1, IGBT thickness: 75μm, no delamination).

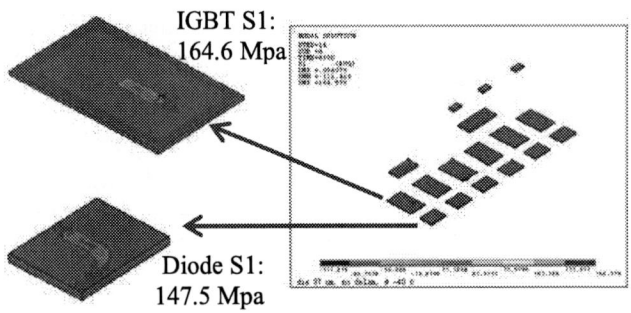

Figure 6: First principle stress S1 of IGBT and diode. (Model 2, IGBT thickness: 97μm, no delamination).

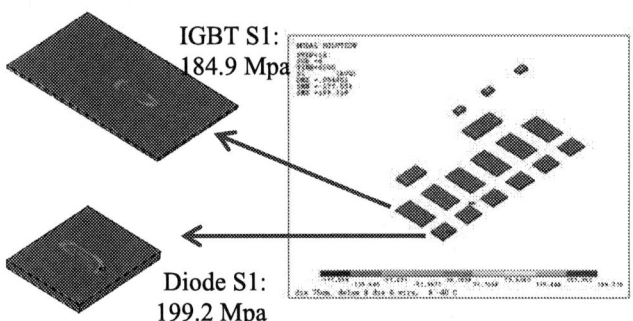

Figure 7: First principle stress S1 of IGBT and diode. (Model 3, IGBT thickness: 75µm, delamination at die & wire).

Figure 9 shows die S1 for model 5, with IGBT thickness 75µm and delaminations at lead frame upper surface of DAP, die top surface and bond wire. The max S1 in IGBT is 217.6 Mpa. The max S1 in diode is 258.6 Mpa. Figure 10 shows die S1 for model 6, with IGBT thickness 97µm and delaminations at lead frame upper surface of DAP, die top surface and bond wire. The max S1 in IGBT is 224.3 Mpa. The max S1 in diode is 258.0 Mpa.

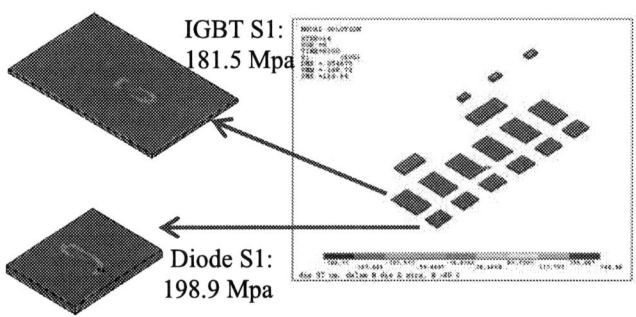

Figure 8: First principle stress S1 of IGBT and diode. (Model 4, IGBT thickness: 97µm, delamination at die & wire).

Figure 11 shows the comparison of first principal stress S1 of IGBT die in temperature cycling test. Figure 12 shows the diode first principle stress S1 comparison in temperature cycling test. The thinner IGBT slightly reduces S1. There is no stress change in diode. The model without delamination shows lowest tensile stress S1 in IGBT and diode. The models with severe delamination areas at lead frame DAP, die top surface and bond wire induce the highest tensile stress S1 in IGBT and diode.

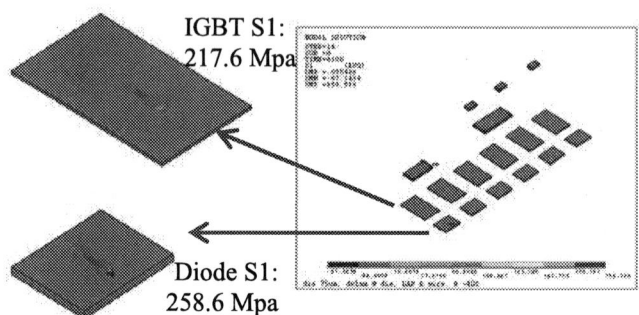

Figure 9: First principle stress S1 of IGBT and diode. (Model 5, IGBT thickness: 75µm, delamination at DAP, die & wire).

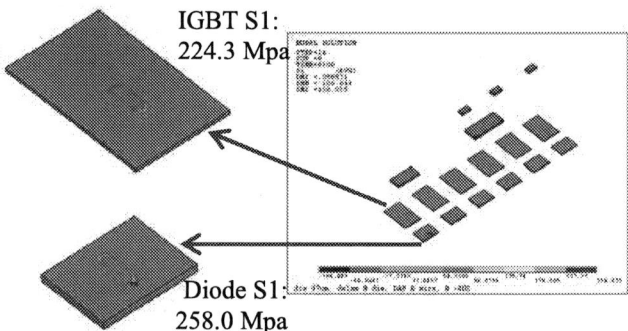

Figure 10: First principle stress S1 of IGBT and diode. (Model 6, IGBT thickness: 97µm, delamination at DAP, die & wire).

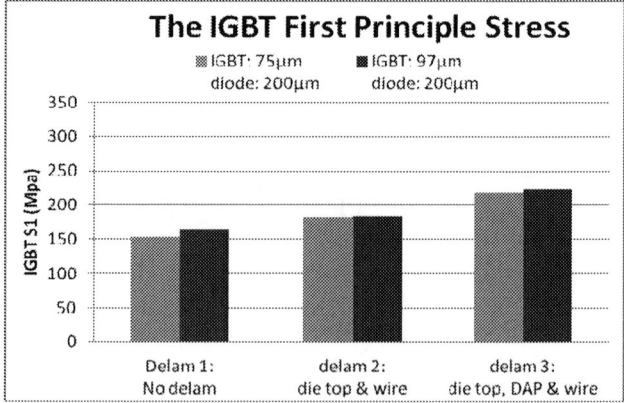

Figure 11: Max first principle stress S1 of IGBT with different die thickness and delamination area.

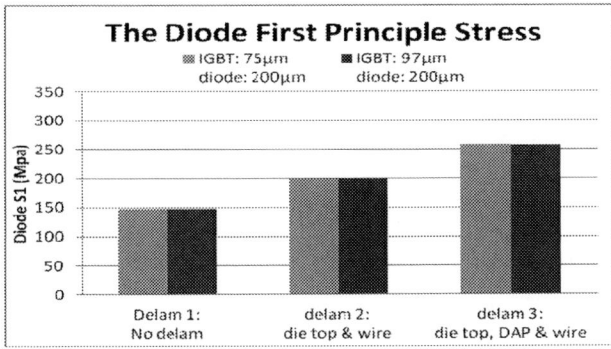

Figure 12: Max first principle stress S1 of diode with different IGBT thickness and delamination area.

Comparison of different bond wire position

Figure 13 shows the internal structure of the SIP package with 2 bond wires at different positions. There are 4 DAP in the lead frame. The left small DAP at position 1 attached with one IGBT and one diode. The right big DAP attached three IGBT dice and three diodes at position 4~6. The bond wire at position 1 and at position 5 are checked in the temperature cycling test. The delamination area locates at lead frame DAP, die surface and part of bond wire surface.

Figure 14 shows the max tensile stress S1 comparison in IGBT die and diode. Figure 15 shows the max shear stress Syz comparison in IGBT die and diode. Both figures show the same trends. The tensile stress and shear stress in IGBT position 1 and position 5 are almost the same. The tensile stress and shear stress in diode at position 5 is slightly higher than at position 1. The tensile stress and shear stress in diode is higher than the stress in IGBT due to diode is higher than the IGBT die.

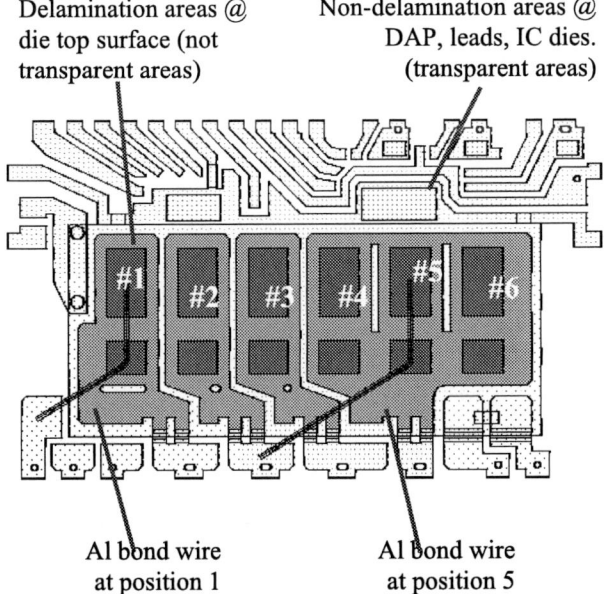

Figure 13: Internal structure of the package with wire bonding at different position.

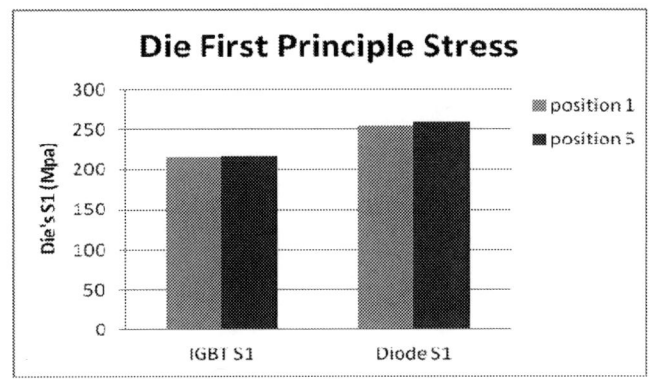

Figure 14: Max first principle stress S1 of IGBT and diode at different position (with delamination at DAP, die and wire).

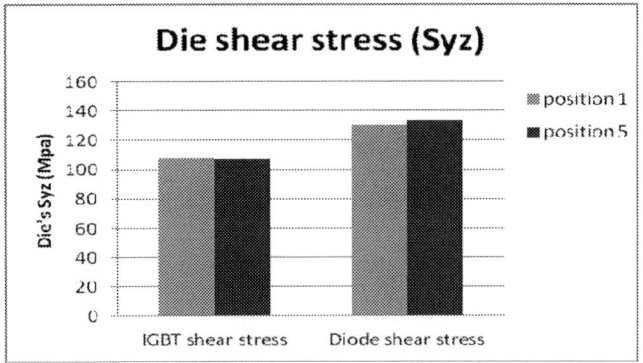

Figure 15: Max shear stress Syx of IGBT and diode at different position (with delamination at DAP, die and wire).

Conclusions

A complicated modeling with consideration of delamination is investigated to check the die stress in temperature cycling test. Parametric models with different ultrathin die thickness, different delamination areas and different bond wire position are simulated. The simulation results have shown that:

(1) The max first principal stress (tensile) in 75μm thick IGBT is slightly lower than that in 97μm thick IGBT in temperature cycling test.

(2) The delamination has increased die tensile stress significantly in temperature cycling test. Both models with 75μm thick IGBT and 97μm thick IGBT show the same trends. The model without delamination shows lower tensile stress in IGBT and diode than the model with delamination. The model with severe and larger delamination area induces higher tensile stress in IGBT and diode than the model with smaller delamination area.

(3) The bonding wire position slightly impacts the stress in diode. The tensile stress and shear stress in diode at position 5 is slightly greater than position 1.

Acknowledgments

The authors wish to thank the support from Package Development Fairchild Semiconductor.

References

1. Y. Liu., *Power electronic Packaging,* Springer, 2012,

2. Y. Liu, Y. Liu, J. Yang, Q. Qian, S. Irving, "Reliability Study of Ceramic Substrate in a SIP Type Package," *ICEPT*, Shenzhen, Aug., 2005

3. Y. Liu, S. Irving, D. Desbiens, T. Luk, Q. Qian, "Simulation and Analysis for Typical Package Assembly Manufacture," *Thermal, Mechanical and Multi-Physics simulation and Experiments in Microeletronics and Microsystems (EuroSimE)*, 2006.

4. Reiche, M. and Wagner, G., "Wafer Thinning: Techniques for Ultra-thin Wafers," *Advanced Packaging*, March 2003.

5. Wetz, L., White, J., and Keser, B., "Improvement in WLCSP Reliability by Wafer Thinning," in Proc. *IEEE Electronic Components and Technology Conf. (ECTC)*, New Orleans, may 2003.

Development of Effective Thermal Characterization on Handheld Devices by Matrix Method

Tai-Yu Chen, Chung-Fa Lee
MediaTek Inc.
No 1., Dusing 1st Rd., Hsinchu Science Park, Hsinchu City 30078, Taiwan, R.O.C.
E-mail: kidd.chen@mediatek.com

Abstract

The thermal coupling effect between multiple chips cannot be ignored in handheld devices as they were not only all inside a compact system but also consumed comparable power simultaneously in practical scenarios. To shorten the time needed for thermal analysis, besides thermal simulation, an efficient analysis method by matrix method based on the concept of thermal network was developed in this study. It involves two matrixes as one was for the prediction of chipsets junction temperature to eliminate reliability or performance issue for electronics components, and the other was used for device skin temperature.

In order to validate this proposal, in the first part, the thermal model of a handheld device was validated through a series of package and mock-up phone thermal experiments. The experiment includes not only one heat source in the standard JEDEC package level test but multiple heat sources on the phone level. Compared with experimental data, a range of +/-6.5% and +/-9% error was achieved for package and system level thermal model respectively.

In the second part, two matrixes of thermal characteristic resistance derived from a simulate solution space were proven to be applicable in the concerned temperature-rising range of consumer electronics devices. It's indicated by the experimental data that the two matrixes method can serve as an effective thermal characterization tool with +/-4% and +/-2% average error on all chips junction temperature and device skin temperature in all scenarios respectively.

With the proposed 2-matrix-based method, the analysis time can be significantly reduced and the benefits of various thermal designs were discussed to deliver some thermal management guidelines from the thermal network's point of view.

Introduction

As a result of the relentless pursuit of new and more functionalities for handheld devices, total system power consumption of handheld devices is higher than ever and could exceed the thermal capacity under its form factor. Therefore, thermal characterization for handheld devices was becoming increasingly critical.

There were numerous researches on the thermal analysis for package used in a handheld device [1], [2], [3]. Some studies were based on detail thermal simulation while other works combined both simulation and experiment [4], [5]. In the recent years, a variety of studies had been done to provide insightful system level thermal modeling result and valuable experimental data for handheld devices such as phones or tablets [6], [7], [8].

Although previous works offered different approaches for detail thermal characterization of handheld devices, a comprehensive discussion on multiple heat sources in a close enclosure was still missing. There was no simple but efficient method to estimate critical temperature for multiple objects in a phone device as it could be done for a package by using standard theta_JA. The challenge to establish a simple but efficient method was to consider not only multiple heat sources in a device and thermal coupling effect between each other, but also the skin temperature of device had to be considered for the perspective of ergonomics (the illustration of a phone with major chipsets is in Figure 1.).

In this work, inspired by the matrix method which was frequently adopted for thermal characterization of power module and package-on-package [9], [10], two matrixes were proposed to fully encompass the concerned scope of handheld device. The matrix was deemed as thermal characteristic resistance and could be derived from the data of a basic set of simulations which covered the power range of interest. The calculation from matrix method was proved to have good correlation with experiment data as long as thermal model is with high accuracy.

Experiment

In order to study the thermal characterization of handheld devices, allocating multiply heat source in a system is a must to produce a system level thermal condition similar to real usage mode of the phone (refer to Figure 1.). Therefore, a mock-up phone was designed to encompass three ball-grid-array packages and one QFN package of thermal test vehicles (TTV), and all were deployed on the same thermal test board (TTB). (Refer to Figure 2 and 3 for TTB and Figure 4 for the mock-up phone). With the assistance of thermal test vehicle, the power consumption of each chip can be easily controlled and its junction temperature can be recorded simultaneously [11]. As for other system level thermal information, such as local air temperature around chip in the phone (T_local), surface temperature of the device (T_skin), were recorded by thermocouple directly. All TTV package assembly conditions were summarized in Table 1, and the size of phone mock-up is 128x65x11(unit: mm).

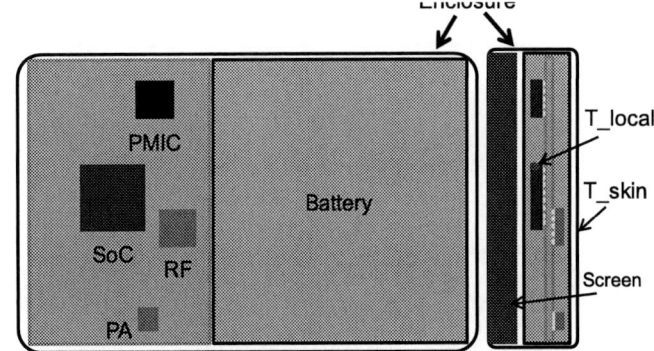

Figure 1. Schematic diagram of a mainstream phone design with multiple heat sources inside.

Table 1. Dimension of thermal test vehicles

TTV Package (unit: mm)	TTV1	TTV2	TTV3	TTV4
Type	BGA	BGA		QFN
Size	12.2x12.2	7.1x7.1		3x3
Height	1.2 (max)	1.2 (max)		0.85
Die thickness	0.125	0.125		0.2
Die attach thickness	0.03	0.03		0.025
Mold cap thickness	0.7	0.65		0.83
Substrate thickness	0.26	0.26		n/a
Solder ball diameter	0.25	0.3		n/a
Solder ball pitch	0.4	0.5		n/a
Exposed pad size	n/a	n/a		1.6

Standard JEDEC package tests were firstly conducted for each kind of TTV [12], and standard package thermal resistances of all TTV were verified. Detail test data were listed in Table 2.

Table 2. Standard JEDEC package thermal test result

Device	Measurement Power (W)	Tj (°C)	Ta (°C)	θ_{JA} (°C/W)	Avg_θ_{JA} (°C/W)
TTV1	1.822	89.6	23.3	36.37	36.44
	1.816	89.5	23.2	36.50	
TTV2	1.004	92.8	23.5	69.10	69.15
	1.003	92.8	23.5	69.19	
TTV4	0.8072	73.0	23.5	61.18	61.28
	0.8088	73.2	23.5	61.37	

Secondly, the TTB was assembled into the mock-up phone to execute the multiple heat sources tests. For phone level thermal validation, a series of tests was done with various power combinations as listed in Table 3, and the test result was summarized in Figure 5.

Table 3. Platform power setting of experiments

Power Loading (unit: W)	TTV1	TTV2	TTV3	TTV4	Platform Total Power
Load 1	0.4	0.5	0.3	0	1.2
Load 2	0.8	0.5	0.3	0	1.6
Load 3	1.2	0.5	0.3	0	2
Load 4	1.6	0.5	0.3	0	2.4
Load 5	1.6	0.5	0.3	0.7	3.1
Load 6	1.6	0.8	0.8	0.7	3.9

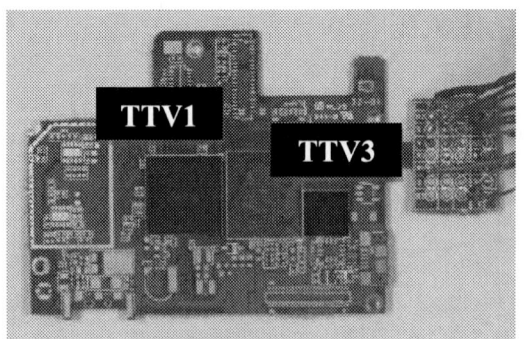

Figure 2. Top view of thermal test board. (Facing phone screen)

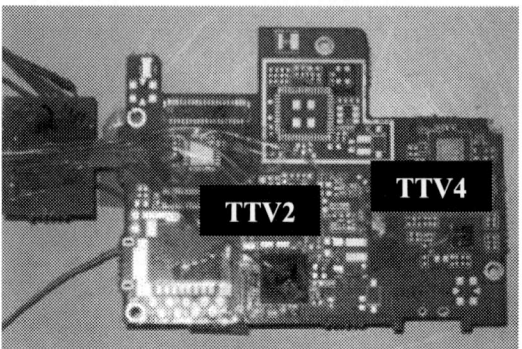

Figure 3. Bottom view of thermal test board. (Facing phone back cover)

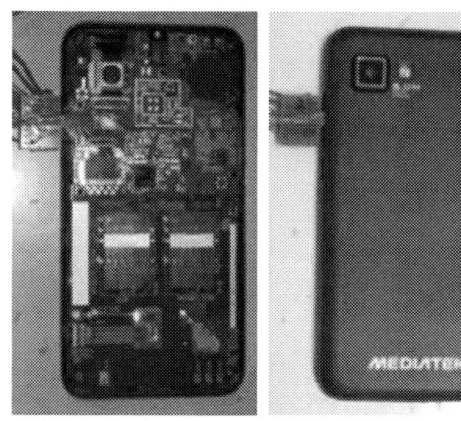

Figure 4. Back side view of phone mock-up test device. (TTB inside)

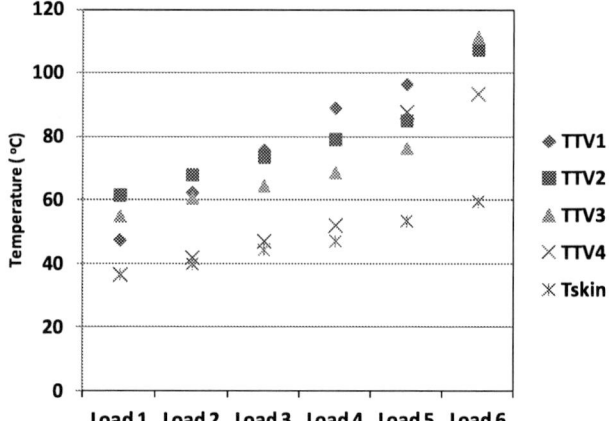

Figure 5. Phone thermal test result of 6 power loadings.

Simulation

Computational fluid dynamic tool of ANSYS Icepak was used in this study to perform thermal simulation. The validation of simulation was divided into two parts. The first part of the thermal model validation was to verify the TTV package model by comparing with the package test result.

The thermal models of 2 kinds of BGA TTV and 1 kind of QFN TTV were built in detail according to the package design file, and using standard bill of material shown in Table 4. The thermal test board was set up in a similar fashion to insure the quality of detail models. Detail modeling method of the

978-1-4799-2408-0/14 $31.00 © 2014 IEEE   2070

package thermal model can be referred to [13]. The accuracy of three TTV package models were achieved within a range of +/-6.5% error as summarized in Table 5.

Table 4. Thermal property of material

Component in package	Thermal Conductivity (w/m-K)
Die	148
Die attach adhesive (BGA)	1.2
Mold compound (BGA)	1.0
Solder ball	50.7
BT core of substrate	0.35
Metal of substrate	387
Die attach adhesive (QFN)	1.0
Mold compound (QFN)	0.9
Lead Frame	268

The second part of the validation was to confirm the accuracy of phone model which had included the validated package model inside. Unlike single package thermal test which usually can be done with only one power setting as the result was quite consistent in the reasonable power range. In order to make sure the thermal coupling effect among several chips in the phone was accurately modeled, phone level thermal correlation was compared with a series of power combination setting for 4 TTV listed before. Both phone and TTV's experimental data were collected, as shown in Figure 6 and 7. The correlation result showed a consistent accuracy between simulation and experiment as overall +/-9% error was met for the entire phone model in all six loading conditions (Figure 8).

Table 5. Package thermal model correlation

Device	Measurement θ_{JA} (^{o}C/W)	Simulation θ_{JA} (^{o}C/W)	Error (%)
TTV1	36.44	37.6	3.18
TTV2	69.15	66.8	-3.4
TTV4	61.28	65.1	6.23

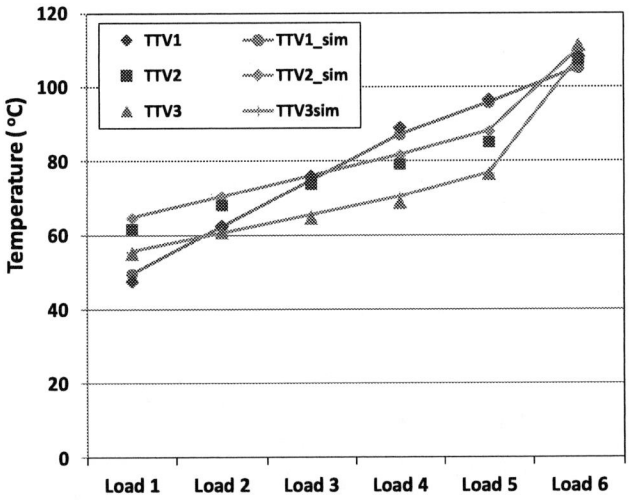

Figure 6. Simulation result of TTV1, TTV2 and TTV3 in 6 sets of power loading.

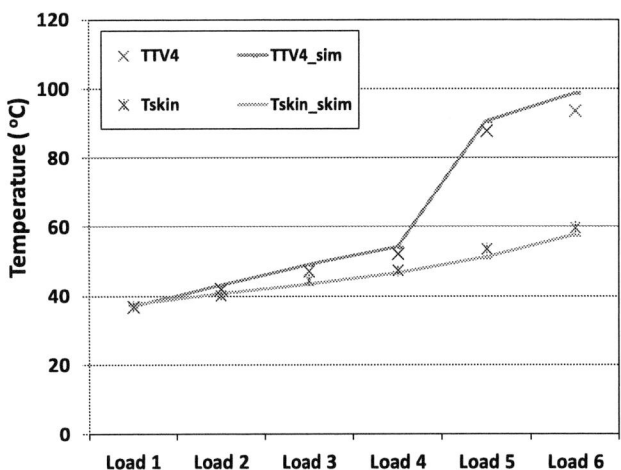

Figure 7. Simulation result of TTV4 and Tskin in 6 sets of power loading.

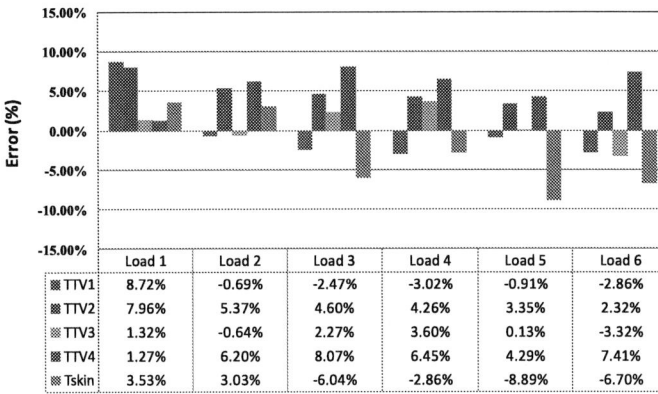

	Load 1	Load 2	Load 3	Load 4	Load 5	Load 6
TTV1	8.72%	-0.69%	-2.47%	-3.02%	-0.91%	-2.86%
TTV2	7.96%	5.37%	4.60%	4.26%	3.35%	2.32%
TTV3	1.32%	-0.64%	2.27%	3.60%	0.13%	-3.32%
TTV4	1.27%	6.20%	8.07%	6.45%	4.29%	7.41%
Tskin	3.53%	3.03%	-6.04%	-2.86%	-8.89%	-6.70%

Figure 8. Error chart of phone and chipsets simulation.

Effect Thermal Characterization Method

Abundant thermal analysis can be gained by simulation to provide insight of the phone thermal result. Although simulation can deliver the most accurate result, however, it's usually not time-efficient, especially with the huge increment on mesh quantities required for detail system modeling.

It's learnt from the thermal network concept and the CFD modeling result, the overall thermal coupling effect in-between all heat sources seems reasonably to be represented in a thermal matrix form and we could use it to calculate all the temperature increment contributed from different heat sources on each concerned object. The thermal matrix concept could be expressed in a matrix form as shown in equation 1.

$$[\Delta T] = [P] \cdot [\Psi]$$ Eq. (1)

where the element $[\Delta T]$ represents the increment of temperature of each object; the element $[P]$ represents the power of each object; the element $[\Psi]$ represents the thermal characteristic resistance between each object.

It's required to have a reference temperature to come out the actual temperature of each object with the calculated increment of temperature as expressed in equation 2.

$$[T] = [\Delta T] + [T_{ref}]$$ Eq. (2)

978-1-4799-2408-0/14 $31.00 © 2014 IEEE

where [T] represents the actual temperature of object; [T_{ref}] represents the reference point of system.

It had been used to take the local air temperature of chip as the reference point for thermal evaluation in the conventional way. But from the trend of local air shown in Figure 9, it could be observed that for a compact and close system such as phone, the local air temperature was strongly affected by the power dissipate inside the system, and therefore it couldn't be a constant value. Instead of it, ambient temperature should be the best reference point because it could be treated as a constant as an infinite heat sink. Once the rise of temperature can be calculated by equation 2, the actual temperature of concerned objects can be derived.

An example of N heat source system could be expressed in the matrix form:

$$
[\Delta T_1 \quad \Delta T_2 \quad ... \quad \Delta T_i] = [P_1 \quad P_2 \quad ... \quad P_i] \cdot
\begin{bmatrix}
\Psi_{11} & \Psi_{12} & ... & \Psi_{1j} \\
\Psi_{21} & ... & ... & ... \\
... & ... & ... & ... \\
\Psi_{i1} & ... & ... & \Psi_{ij}
\end{bmatrix}
\quad \text{Eq. (3)}
$$

where the element of [ΔT_i] represents the increment of temperature of object where i=1 to N; the element of [P_i] represents power of object i, where i=1 to N; the element of [Ψ_{ij}] represents the matrix of thermal characteristic resistance from object i to object j, where i, j = 1 to N.

Furthermore, as skin of devices was actually not a heat source at all, it was just the "victim" of heat sources within the system. Another matrix was utilized to calculate its result:

$$
[\Delta T_{skin}] = [P_1 \quad P_2 \quad ... \quad P_i] \cdot
\begin{bmatrix}
\Psi_{1S} \\
\Psi_{2S} \\
\Psi_{3S} \\
\Psi_{4S}
\end{bmatrix}
\quad \text{Eq. (4)}
$$

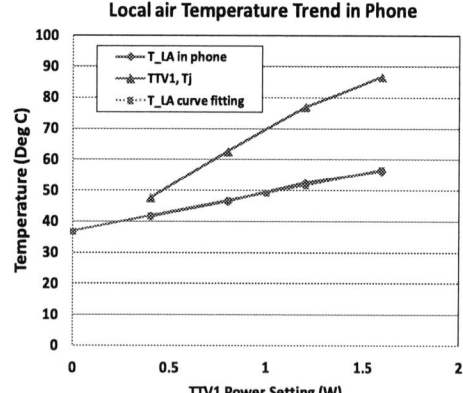

Figure 9. Local air temperature vs. power dissipation (load1~load4)

In order to have the proper thermal characteristic matrix to represent a handheld device, it's a necessary to derive the matrix from a solution space of simulation which covered the possible power range of all heat sources (refer to Table 6.). Taking this mock-up phone which had four heat sources for concept validation, total five scenarios of power combination were simulated with the validated model to provide a solution space and the thermal characteristic matrix from equation 3 and 4 can be derived.

Table 6. Power setting for various phone usage scenarios

Item	Component	TTV1	TTV2	TTV3	TTV4	Skin
Phone power scenario (W)	scenario 1	0.8	0.5	0.3	0	n/a
	scenario 2	1.2	0.25	0.6	0.1	n/a
	scenario 3	1.6	0.5	0.3	0	n/a
	scenario 4	1.6	0.5	0.3	0.7	n/a
	scenario 5	1.6	0.8	0.8	0.7	n/a
Thermal result (°C)	scenario 1	62.34	70.52	60.67	43.16	41.66
	scenario 2	75.88	66.42	79.24	56.68	42.55
	scenario 3	87.26	81.72	70.48	54.06	46.76
	scenario 4	95.82	88.16	76.8	90.6	51.15
	scenario 5	105.2	110.1	107.8	98.79	57.56

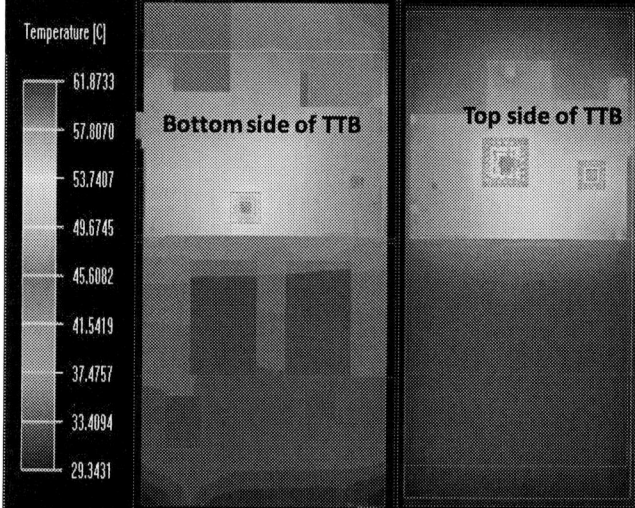

Figure10. PCB thermal contour of scenario 1.

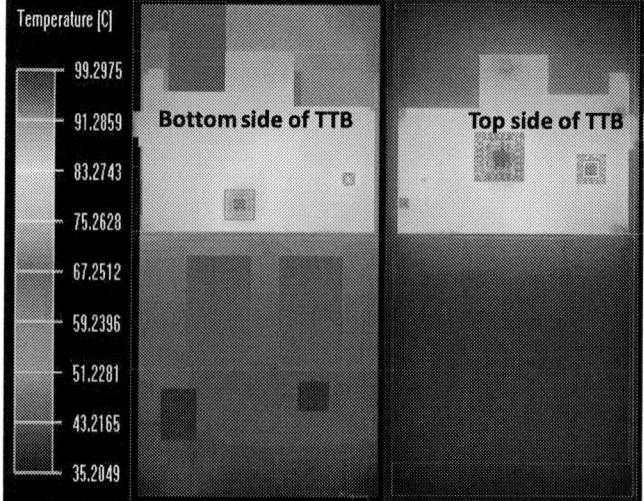

Figure 11. PCB thermal contour of scenario 5.

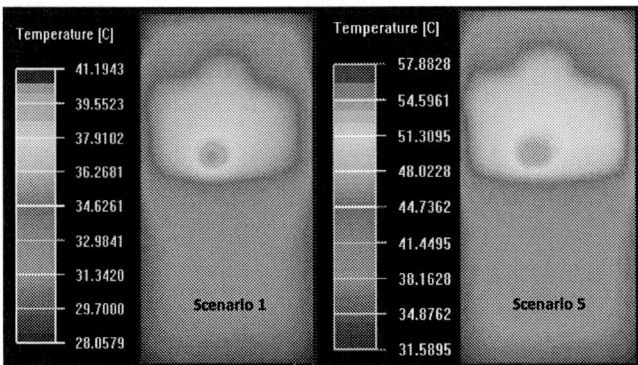

Figure 12. Phone skin temperature from simulation

Result

As a matter of fact that superposition principle was not theoretically perfect for thermal application when the heat transfer was dominated by nature convection. In order to confirm the applicability of this approach, the calculation result from 2-matrix-based method was examined with the experiment data in term of chipset junction temperature and phone skin temperature.

The thermal characteristic matrixes obtained from the data listed in Table 6 were presented in equation 5 and 6. With these 2 matrixes, we could calculate the junction temperate from TTV1 to TTV4 and phone skin temperature, and then compare those with previous experimental data.

$$[\Psi_{chipset}] = \begin{bmatrix} 32.48 & 15.84 & 13.82 & 13.94 \\ 14.65 & 57.02 & 15.02 & 5.79 \\ 11.43 & 11.69 & 54.7 & 13.25 \\ 11.07 & 7.59 & 7.67 & 51.93 \end{bmatrix} \quad \text{Eq. (5)}$$

$$[\Psi_{skin}] = \begin{bmatrix} \Psi_{1S} \\ \Psi_{2S} \\ \Psi_{3S} \\ \Psi_{4S} \end{bmatrix} = \begin{bmatrix} 8.05 \\ 15.82 \\ 5.17 \\ 4.81 \end{bmatrix} \quad \text{Eq. (6)}$$

Table 7. Temperature-rise comparison table.

Experiment data of junction temperature increment				
Condition	TTV1	TTV2	TTV3	TTV4
Load 1	22.7	36.8	30.2	11.8
Load 2	37.6	43.2	35.9	17.1
Load 3	51.1	48.9	39.7	22.3
Load 4	64.2	54.4	43.9	27.3
Load 5	71.7	60.3	51.7	62.9
Load 6	83.3	82.6	86.5	68.7
Calculated result of junction temperature increment				
Condition	TTV1	TTV2	TTV3	TTV4
Load 1	23.75	38.35	29.44	12.44
Load 2	36.74	44.69	34.97	18.02
Load 3	49.73	51.03	40.50	23.59
Load 4	62.73	57.36	46.02	29.17
Load 5	70.47	62.68	51.40	65.52
Load 6	80.59	85.63	83.25	73.88

It's indicated by this examination that good correlation by using 2-matrix-based method could be maintained in the concerned range of temperature rise which was less than 75 deg C with using 25 deg C as the ambient temperature. The error of chip junction temperature was about +/-7.5% among all TTV, and an average of +/-3.7% was met under every platform power loading condition. Although the range of phone skin temperature error was +6.55% ~ -10.43% which actually only meant 1~3 degree C, and the average of error to skin temperature for all test conditions was -1.3%.

Table 8. Summary of error for the rise of junction temperature of each TTV.

Condition	TTV1	TTV2	TTV3	TTV4	Average
Load 1	4.62%	4.22%	-2.50%	5.46%	**2.95%**
Load 2	-2.29%	3.45%	-2.59%	5.38%	**0.99%**
Load 3	-2.67%	4.35%	2.01%	5.80%	**2.37%**
Load 4	-2.30%	5.45%	4.84%	6.85%	**3.71%**
Load 5	-1.71%	3.95%	-0.59%	4.16%	**1.45%**
Load 6	-3.26%	3.67%	-3.76%	7.54%	**1.05%**

Table 9. Summary of error for the rise of skin temperature.

Condition	Experiment data	Calculated data	Error
Load 1	11.9	12.68	6.55%
Load 2	15.2	15.90	4.60%
Load 3	19.7	19.12	-2.94%
Load 4	22.4	22.34	-0.27%
Load 5	28.7	25.71	-10.43%
Load 6	34.9	33.04	-5.34%

Conclusions

The correlation between the matrix-method and the experiment was acceptable within the concerned temperature-rising range of consumer handheld devices. The proposed 2-matrix-based method was proved to be a great alternative for efficient thermal characterization which can provide the temperature of critical object without doing simulation for every single scenario. Nevertheless, the two matrixes shall firstly be derived from the result of a series of system level simulation which was recommended to cover possible range of the platform power. Once the two matrixes were derived, one can complete the sensitivity study on power of each heat source and understand the corresponding thermal impact efficiently.

One benefit of the 2-matrix-based method was it can be considered as a system level thermal power budgeting tool. It can be utilized to allocate the adequate power budget to each chip with the consideration of thermal performance to minimize thermal issues on system or package level in the early design phase.

Another advantage of this method is to help decide the strategy of thermal management. Based on the thermal characterization matrix, we could find out the weighting of each heat source regarding the concerned object. Then the effectiveness of power throttling on various chips can be calculated to trade off power for a suitable thermal management guideline. Such as it's obvious that from the matrix of skin thermal characteristic resistance, the weighting of TTV2 was the most dominant one as it faces back cover directly and was the hot spot on PCB on the bottom side as shown in Figure 11. This could be easily observed from simulation result in Figure 12 as well, but with thermal matrix, the improvement of TTV2's power reduction can be estimated without running simulation again.

Acknowledgments

The authors would like to acknowledge the support from MediaTek PCB team for thermal test board design, and express the sincere appreciation to SPIL Laboratory team and Miaowen for the tremendous assistance in thermal test vehicle design and assembly as well as executing the experiments.

References

1. Reimer, C. J., Smy, D. J. Walkey, et al., "A simulation study of IC layout effects on thermal management of die attached GaAs ICs ", Components and Packaging Technologies, IEEE Transactions on Vol. 23, 2000, pp.341-351.
2. Chun-Kai Liu, Yi-Hsiang Cheng, "Thermal analysis of poewr amplifier package in cellular phones," International Symposium on Electronic Materials and Packaging 2002, pp.415-421.
3. Victor Adrain Chiriac, Tien-Yu Tom Lee, "Thermal performance comparison between various RF transceiver modules", in Proc. 10th ITHERM Conference, 2006, pp. 274-280.
4. Tetsuro Nozu, "Thermal chacacterization of power amplifers for CDMA cellular phone application", 21st IEEE SEMI-THERM symposium.
5. Siva P. Gurrum, Darvin R. Edwards, et al., "Generic thermal analysis for phone and tablet systems", in *Proc. IEEE Electronic Components and Technol. Conference*, 2012, pp. 1488–1492.
6. Yen Yi Germaine Hoe, Chong Ser Choong, et al., "Thermal modeling and simulation of a package-on-package embedded micro wafer level package (EMWLP) structure at the package and system level", 12th Electronics Packagine Technology Conference 2010, pp.285-291.
7. Jaecho Lee, David W. Gerlach, et al., "Parametric thermal modeling of heat transfer in handheld electroni devices", in Proc. 11th ITHERM Conference, 2008, pp. 604-609.
8. Sung-Won Moon, Prstic, S., Chia-pin Chiu, "Thermal management of a stacked-die package in a handhled electronic device using passive solutions", Components and Packaging Technologies, IEEE Tansactions on Vol. 31, issue 1, 2008, pp. 204-210.
9. Morris Bowers, Yeon Lee, Bennett Joiner et al., "Thermal characterization of Package-on-Package (PoP)" 25th IEEE SEMI-THERM Symposium 2009, pp.309-316.
10. Manu Mital, Ying-Feng Pang and Elaine P. Scott, "Evaluation of thermal resistance matrix method for an embedded power electronic module", Components and Packaging Technologies, IEEE Tansactions on Vol. 31, issue 2, 2008, pp. 382-387.
11. Bernie Siegal, J. Galloway, "Thermal test chip design and performance considerations", 24th IEEE SEMI-THERM Symposium, 2008, pp.59-62.
12. EIA/JEDEC Standard JESD15 Methodology for the thermal measurement of component packages, Electroni Industries Association, JC15 committee, 1995.
13. Chen, K., Hsu, I., et al., "Chip-package-PCB thermal co-design for hot spot analysis in SoC", Electrical Design of Advanced Packaging and System Symposium IEEE, 2012.

Comprehensive Design Optimization for 2.133 Gbps LPDDR3 Extension for Mobile Platform System

Chanmin Jo, Jaemin Shin, BaekKyu Choi, Sangmin Lee, Seongjae Moon, Sungjoo Kim, and Woong Hwan Ryu
Samsung Electronics
1-1, Samsungjeonja-ro, Hwaseong-si, Gyeonggi-do, South Korea
chanmin.jo@samsung.com, Tel: 82-31-8037-3374

Abstract

Recent fast-evolving mobile system demands high bandwidth and low power consumption, necessitating extension of LPDDR3 beyond 1.6 Gbps. This demand, however, brings significant technical challenges from the perspective of signal and power integrity. A simple way of mitigating signal integrity issue at the challenging speed is to use ODT (On-Die Termination). Since ODT significantly increase power consumption, the use of ODT is not an attractive solution in a mobile system which considers low power consumption as primary metric. Thus, comprehensive design optimization with given design constraint is a practical solution which can avoid penalty of power consumption in the mobile system.

In this paper, we have demonstrated design success of stretched LPDDR3 up to 2.133Gbps by applying rigorous design optimization based on comprehensive PI-aware SI analysis. It includes signal quality improvement and power delivery optimization. The Both metrics are organically considered during optimization process to achieve successful system operation up to 2.133Gbps without additional power consumption by ODT use.

Introduction

As a mobile system evolves fast to satisfy end-user appetite with high graphic resolution and performance, it requires high operation speed for mobile application process (AP) core and memory interface. Even with increase in system speed, a mobile device should be designed to consume less power to lead competitive mobile market. Figure 1 shows graphic resolution and memory interface trends. In this figure,

it can be inferred that fast increase in graphical resolution stimulates memory bandwidth requirement in a mobile AP platform.

A conventional mobile memory interface such as LPDDR3 does not employ ODT option to lower power consumption, which can easily bring significant signal reflection to impact timing margin and overshoot/undershoot. Also, the memory interface operates in single-ended source synchronous timing parallel topology susceptible to cross-talk and simultaneous switching noise (SSN). Thus, in order to mitigate the noises for proper operation at the extended LPDDR3, comprehensive design optimization needs to be implemented carefully based on PI-aware SI analysis and solid design guidance.

In this paper, we have analyzed key design limiters and enablers to extend standard LPDDR3 (1.6 Gbps) up to 2.133Gbps in a mobile platform. The following technical areas have been tackled to achieve the stretched LPDDR3 operation.

- Reduction of input capacitance
- Reduction of channel length
- PDN design optimization
- Reduction of clock jitter

The step-by-step design optimization has been applied to enable LPDDR3 extension at 2.133 Gbps. Finally, we have demonstrated design success of extended LPDDR3 up to 2.133 Gbps without power consumption penalty.

1. Design Optimization

1.1. Reduction of input capacitance of IO

Input capacitance of the IO (Cio) is one of important factors affecting electrical performances such as voltage swing and time constant. As input capacitance gets reduced, we can make full voltage swing with longer channel. In other words, effective electrical length to make full voltage swing can be shortened by lowering Cio. In general, it is hard for an AP developer to lead to lower memory Cio and also to reduce driver size. Instead, ESD (ElectroStatic Discharge) can be compromised to decrease input capacitance of AP for proper operation at high data rate. In this paper, we simulated full link memory interface to find proper Cio level for LPDDR3 extension at 2.133Gbps. It was turned out that Cio of the platform that works at 1.6 Gbps can be reduced by 40% for proper operation at 2.133Gbps. Figure 3 illustrates eye difference due to Cio reduction. The result tells us that Cio reduction is more dominant in the case of short length where Cio is not overshadowed by channel loss. More precisely,

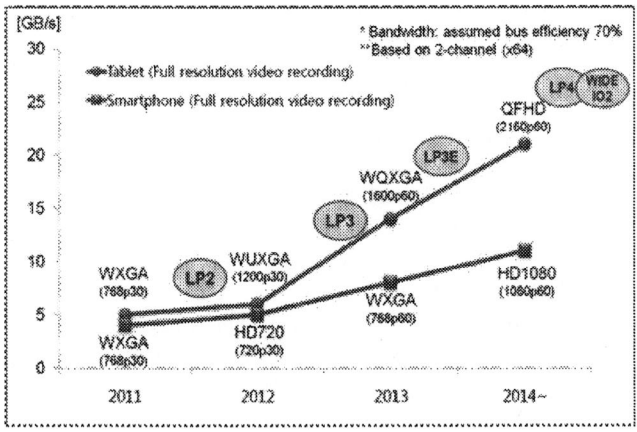

Figure 1. Memory bandwidth trend in a mobile AP platform.

40% reduction of Cio can improve timing margin about 40ps, which is ~ 8% UI at 2.133Gbps data rate.

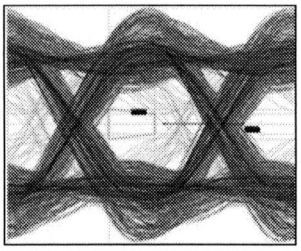 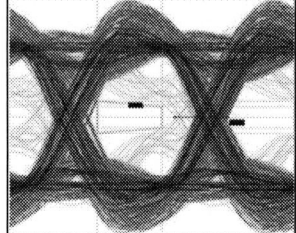

[Before reduction of Cio]   [After reduction of Cio]

(a) Long channel.

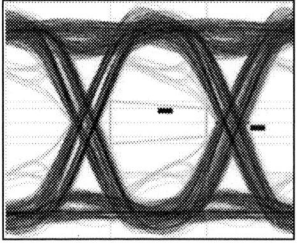 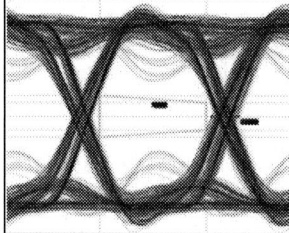

[Before reduction of Cio]   [After reduction of Cio]

(b) Short channel.

Figure 3. Eye difference due to Cio reduction.

1.2. Reduction of physical channel length

A typical LPDDR3 interface between AP and memory in package-on-package is about 10~30 mm. The interface has open termination to reduce power consumption as shown in Figure 2.

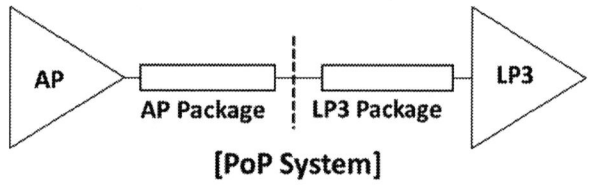

Figure 2. LPDDR3 interface in PoP.

Since open termination causes significant signal reflection, effective electrical length of channel with respect to data rate needs to be reduced to minimize non-ideal transmission line effect. With the given mobile design constraint, the channel acts as a resonator with quarter wavelength over 1GHz and non-idea signal/return path effect gets more significant as the data rate goes higher [1]. Thus, it is very important that effective electrical channel length does not match around quarter wavelength of operating frequency and its odd harmonic frequency range.

As mentioned above, one of effective ways of enabling LPDDR3 at 2.133Gbps is to reduce physical channel length. Practically it is very difficult to shorten physical channel length with the pre-defined jointed ball by JEDEC standard because package area in AP and memory is very limited in a PoP design. Thus, we has customized jointed ball map based on optimal routability. This jointed ballmap optimization is inspired by the previous LPDDR3 design practice [2-3] in the sense that signal quality of memory interface was limited by the longest byte group. With JEDEC standard joint ball map, the longest byte group was over 20mm in channel length [4]. To reduce the channel length of each byte, we optimized joint ball map and reduced the channel length by about 30~40%. Moreover, the location of CA and DQ groups has been determined by considering location of those groups in memory. This effective placement can help to reduce physical routing lengths by 40%. Shortening channel length significantly improves eye-diagram as shown in Figure 4.

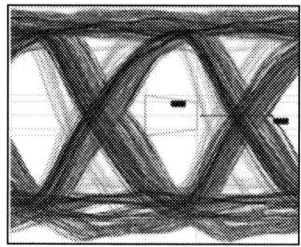 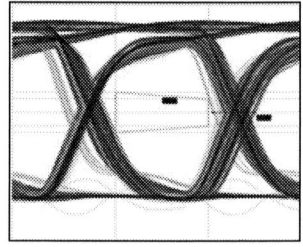

(a) Long channel   (b) Short channel

Figure 4. Eye improvement due to physical channel length reduction

1.3 PDN Design Optimization

Area-limited mobile platform is lacking of return path and power/ground planes, resulting in poor power distribution network (PDN). Since a mobile platform having low voltage supply and high current demand is very vulnerable to PDN noise such as simultaneously switching noise (SSN), careful PDN optimization by lowering PDN impedance is a very crucial to achieve LPDDR3 operation at 2.133Gbps [5]. Subsection describes how to optimize PDN on- and off- chip, respectively.

On-chip decoupling capacitance can reduce impedance at high frequencies to suppress high frequency PDN noise into the memory driver. On-chip intrinsic parasitic capacitance of signal and power to ground inherently plays a role as a part of on-chip decoupling capacitance. In the most of cases, the intrinsic parasitic capacitance is not enough to suppress PDN noise, resulting in degradation of voltage margin and timing margin. It is a common design practice to add intentionally substantial amount of on-chip decoupling capacitance to protect the system from PDN noise. However, it is a trade-off between die area and amount of decoupling capacitance. Finding an optimal point to save area is very important.

In general, a mobile AP platform has a small form factor but the design routing is very complicated. In addition, its area is very limited due to a number of components to be placed on outer layer of a board. Because of this area limitation and low cost approach, minimum number of on-board decoupling capacitors is allowable at certain position. This design constraint has an impact on PDN design optimization. In the case that on-board decoupling capacitors is not sufficient to reduce impedance, on-package decoupling capacitors can be utilized to compensate for this limitation.

978-1-4799-2408-0/14 $31.00 © 2014 IEEE   2076

Figure 5 shows self-impedance (Z11) curves of memory IO PDN with on-chip decoupling capacitor or/and on-package decoupling capacitor. The red curves represent Z11 of original LPDDR3 platform (1.6 Gbps) and the blue curves represent Z11 of optimized LPDDR3 extension platform (2.133Gbps). Figure 5(a) is comparison graph by insertion of on-chip decoupling capacitor. In this figure, the peak impedance of optimized LPDDR3 extension platform is ~10% lower than that of original LPDDR3 platform and the resonance frequency shifts. Since the peak impedance can be determined by combination of on-chip decoupling capacitance and on-package/board decoupling capacitors. On-package/board impedance also needs to be optimized. Figure 5(b) shows impedance optimized by only on-package decoupling capacitor. Also, Figure 5 (c) illustrates PDN design optimization by insertion of both on-chip capacitance and on-package capacitor.

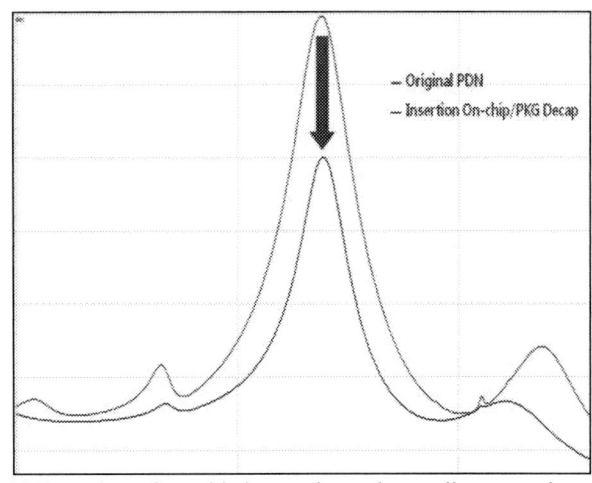

(c) Insertion of on-chip/on-package decoupling capacitors.

Figure 5. Self-impedance by decoupling capacitance.

Finally, Figure 6 shows eye-diagram improvement by PDN optimization by insertion of both on-die capacitance and on-package capacitors. In this figure, the insertion of on-chip capacitance and package decoupling capacitor can improve timing margin by over 20% unit interval (UI).

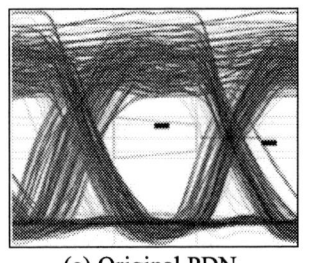 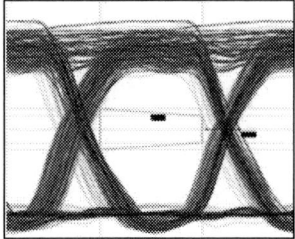

(a) Original PDN   (b) Optimization Decap.

Figure 6. Timing margin by decoupling capacitance.

Figure 7 shows an example of time-domain measurement results of DQ and DQS impacted by PDN noise due to poor power design. In this case, memory operation fails due to high voltage droop when all DQ signals start switching simultaneously. This fail can be resolved by putting on-package decoupling capacitor in IO power domain. This on-package capacitor can effectively suppress high peak impedance, as shown in Figure 8. Figure 9 shows the curves of voltage droops with and without on-package decoupling capacitors. Accordingly, insertion of on-package embedded decoupling capacitor can lessen power fluctuation about 55%. This reduced power fluctuation is directly conducive to improvement of signal timing margin. As a result, the eye is almost closed due to strong SSN as shown in Figure 10 (a). On the other hands, the closed eye can be improved by inserting on-package decoupling capacitor, as shown in Figure 10 (b). Thus, in the design described here, it can be concluded that insertion of decoupling capacitor at package can improve poor board PDN design and make the mobile platform design robust.

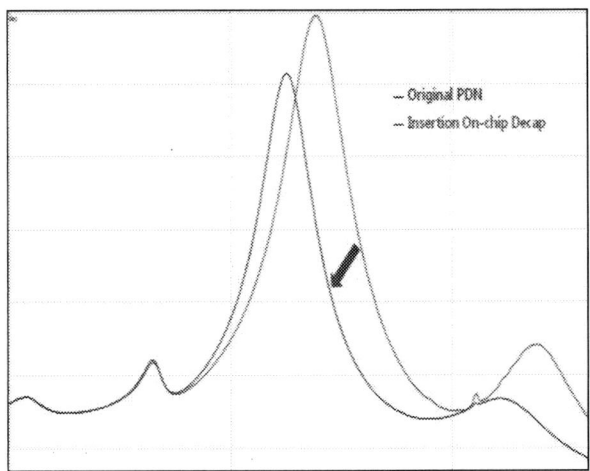

(a) Insertion of on-chip decoupling capacitor.

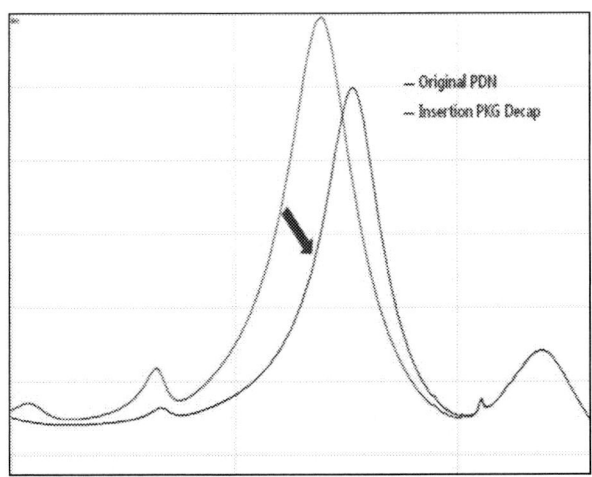

(b) Insertion of on-package decoupling capacitor.

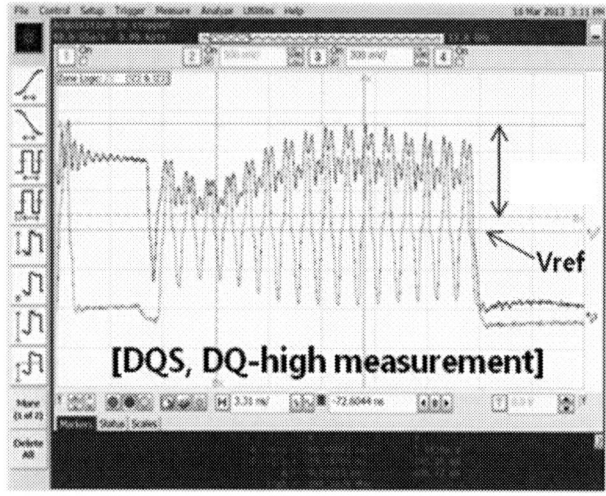

Figure 7. Time-domain measurement of DQ and DQS in poor PDN board design.

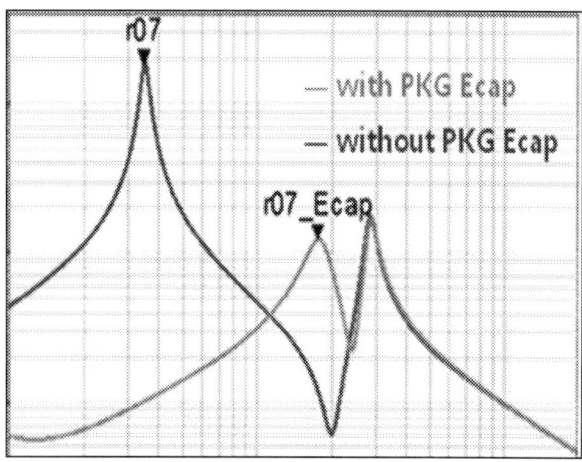

Figure 8. Self-impedance with and with on-package decoupling capacitor.

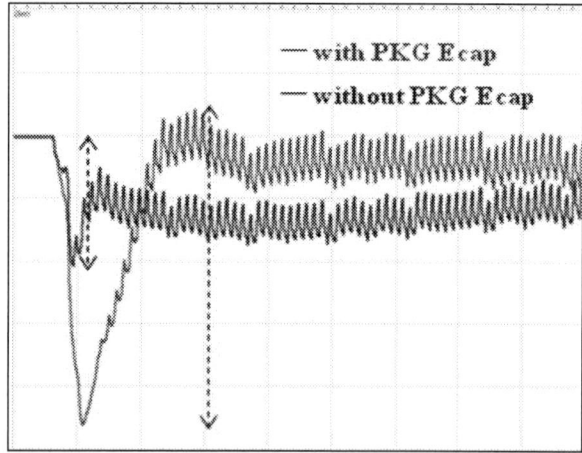

Figure 9. Voltage droops with and without on-package decoupling capacitor.

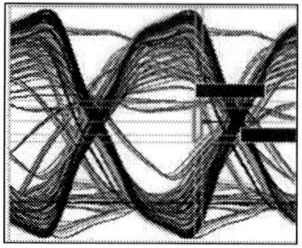 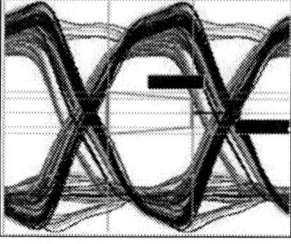

(a) without on-PKG decap   (b) with on-PKG decap

Figure 10. Eyes with and without on-package decoupling capacitor.

1.4. Reduction of Clock Jitter

Another performance challenge for LPDDR3 and beyond is clock jitter increase because its requirement in specification gets stringent and becomes one of key performance metrics. In this paper, we analyzed clock jitter trend with respect to data rates and found that power noise is a main factor impacting clock jitter performance. Particularly, since clock IO block sits near other IO drivers like DQ/CA signals and shares power domain, PDN noise from other drivers has a significant impact on clock jitter. The analysis result shows clock jitter was reduced by order of tens pico-second by isolation of DQ and clock power domains. Thus, it is proven that power isolation of the clock from DQ driver is a very effective way to minimize clock jitter. In spite of benefit of power isolation, there is a practical limitation to use separate LDO or buck for DQ and clock. Clock and DQ power planes in general are merged in a board.

In this paper, we have simulated two different types of boards to understand design optimization for clock jitter reduction. Figure 11(a) shows that the PDN between CA and DQ is routed with wide plane. In this case, even if there is no actual power connection between CA and DQ on package, power noise stimulated by DQ/CA switching can be conveyed to clock jitter. In order to reduce the transfer of power noise from DQ IOs, we optimized board PDN by utilizing narrow trace between DQ and CA as shown in Figure 11 (b).

 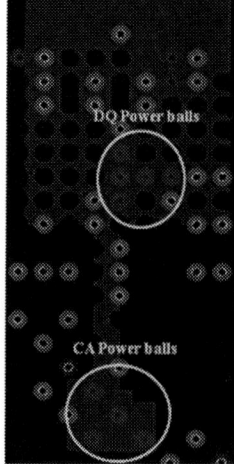

(a) Wide trace   (b) Narrow trace

Figure 11. Board IO PDN between CA and DQ

Reducing power trace width increases transfer-impedance between CA and DQ, which blocks noise transfer to clock. As shown in Figure 11, clock jitter of wide power trace is ~ 150% of JEDEC spec. but clock jitter of narrow power trace is ~ 70% of JEDEC spec. The board design optimization can reduce 40% of clock jitter.

2. Summary of the results

With all the efforts of design optimization described in this paper, we successfully have enabled the operation of LPDDR3 extension at 2.133 Gbp. Among those efforts, shortening signal channel length is straightforward and effective. However, the other optimizations need to go thru comprehensive analysis using simulation tool and theoretical understanding. Figure12 shows overall eye opening difference between before and after design optimization. Also, Table 1 quantitatively summarizes their performance difference.

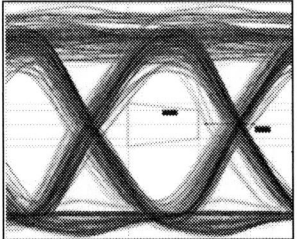 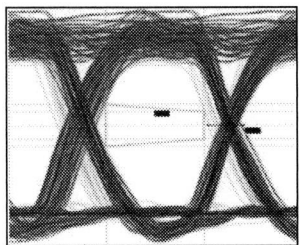

[Before optimization]   [After optimization]

(a) Eye openings for DQ

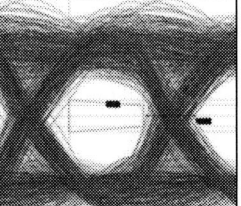 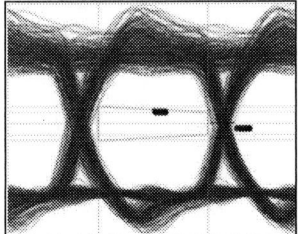

[Before optimization]   [After optimization]

(b) Eye openings for CA

Figure 12. Eye Openings before and after design optimizations.

Table 1. Eye Openings and clock jitter before and after design optimizations.

Timing parameter		Before optimization	After optimization
Eye opening [% of UI]	DQ	49%	68%
	CA	51.8%	77%
Clock Jitter [% of JEDEC spec.]		152%	70%

Conclusions

In this paper, we have demonstrated design success of stretched LPDDR3 up to 2.133 Gbps, which is the first industry achievement in mobile market, by applying rigorous design optimization based on comprehensive PI-aware SI analysis. Design optimization includes signal quality improvement and power delivery optimization. Both the metrics are organically considered during optimization process to achieve successful system operation up to 2.133 Gbps without power consumption sacrifice.

In order to operate the system at high data rate, one of efficient ways is to reduce effective electrical channel length to stay away from the channel resonance. We performed power-signal combined analysis using IO transistor-level model. The first design optimization targeted Cio (IO input capacitance) reduction based on eye-diagram simulation. Next step focus on shortening physical channel length to expand both the voltage and timing margin for 2.133 Gbps operation. Power delivery network was also optimized to get better performance margin in both on-chip and off-chip areas. On-chip power delivery network needs more on-die decoupling capacitance to support high frequency energy into a driver. Increase in on-die decoupling capacitance leads to timing margin increase. Since a design challenge in a mobile platform is high design complexity even in small form factor, passive components such as on-package and on-board decoupling capacitors in a mobile platform should be effectively allocated in terms of number, location and performance. We analyzed effectiveness of decoupling capacitor both on-package and on-board to improve off-chip power delivery. This analysis indicates this optimization process reduce voltage noise substantially (by ~ order of tens of mv). Another performance challenge at high data rate is clock jitter increase because its requirement in specification gets stringent and becomes a key performance criterion. We analyzed clock jitter trend at various data rate and found power noise is a main factor to impact clock jitter performance. Particularly, in a compact on-chip design, a clock IO block is located near the other IO drivers like DQ and CA signals. SSN (simultaneously switching noise) from DQ IO blocks has a significant impact on clock jitter because the clock driver shares power domain with DQ drivers. The study concludes the clock power isolation from DQ driver power is very effective by showing the result that clock jitter was reduced by order of tens ps by splitting DQ-Clock power domain.

Therefore, this paper has demonstrated design success of LPDDR3 extension from 1.6 Gbps up to 2.133 Gbps in a mobile application without the use of ODT by comprehensive PI-aware SI design optimization. Eventually, this achievement is a great practical example to prove the importance of power/signal integrity design optimization to maximize the performance of existing industry standard.

Acknowledgement

The authors would like to thank Samsung Electronics Corporation for their full support for this works.

References

1. Hall, S. H. & Heck, H. L. Advanced Signal Integrity for High-Speed Digital Designs, IEEE: Wiley, 2008

2. C. Jo, B. Choi, S. Lee, S. Moon, S. Lee, M. Seo and Y. Kim, "World's First LPDDR3 for Enabling Mobile Application Processor Systems", UBM DesignCon Conference, Santa Clara, CA, USA, Jan. 28-31, 2013.

3. K. Koo, W. Ryu, S. Lee, B. Choi, "Robust I/O circuit scheme for world's first over 1.6Gbps LPDDR3", UBM DesignCon Conference, Santa Clara, CA, USA, Jan. 28-31, 2013.

4. JEDEC Standard JESD209-3A, Low Power Double Data Rate 3 (LPDDR3) SDRAM, 2013

5. Swaminathan, M. & Eugin, A. E. Power Integrity Modeling and Design for Semiconductors and Systems: Prentice hall, 2007, pp. 37.

Estimation of Mode Conversion and Crosstalk Impact from a Single-Ended Aggressor to a Differential Victim using Statistical BER analysis

Arun Reddy Chada*, Jun Fan*, James L. Drewniak*, Bhyrav Mutnury**

Missouri S&T EMC Laboratory*, DELL Inc**

Rolla, Missouri*, Roundrock, Tx**

ac253@mst.edu*,drewniak@mst.edu*, bhyrav_mutnury@dell.com**

Abstract

Increase in the cost of printed circuit board (PCB) with the increase in layer count has led to the design of PCB stack-ups that have broadside coupled signals. The coupling that occurs between the single-ended signal trace coupled to a differential signal traces at various angles and at multiple instances creates a imbalance in the differential signal line thereby generating common-mode signal at the victim receiver. The periodic routing of single-ended trace also generate Floquet modes where the near end crosstalk (NEXT) peaks impacting the signal at the victim receiver. In this paper, various design parameters that impact the eye opening such as number of coupled sections (which dictates the periodicity), horizontal spacing of victim and aggressor, and signaling speeds are included. Different styles of routing of single-ended trace are studied to quantify the impact of differential-common mode conversion and NEXT at the receiver using statistical bit error rate (BER) analysis. Guidelines for best routing practices are developed for the designer to maintain signal integrity.

Keywords-broadside, Floquet modes, Statistical analysis, bit error rate, Near end crosstalk

Introduction

Increase in the cost of printed circuit board (PCB) as the layer count increases has led to denser signal routing forcing high speed signal traces to route closely for long lengths. Broadside coupling of signals in adjacent layers observed in real boards that is difficult to model is shown in Figure 1.

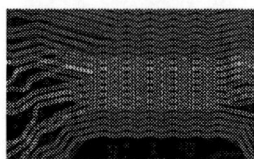

Figure 1. Broadside coupled traces in real design

Previous work to model the coupling between the single-ended traces and differential traces is studied in [1-2]. However the coupling study did not have design guidelines based on bit error rate (BER) eye contour metric using statistical analysis. Lot of analytical work to model crosstalk for transmission lines and lines crossing at angle is done in [1-13].

Designers tend to route a single-ended trace close to differential trace as the rise times of the single-ended signals are slow. The single-ended trace impacts the balance of the differential trace which influences the differential-common mode conversion. Some percentage of differential input at the victim source is manifested as common-mode signal at the receiver. There is also crosstalk from the singled-ended trace impacting the victim at the receiver. The common-mode signal creates a radiation problem in the system.

A topology of broadside coupled traces in a real design can be classified as shown in Figure 2. Designers tend to route signals sometimes in a zig-zag fashion to negate the fiber-weave effect. The broadside coupling between the victim and aggressor traces in adjacent layers is assumed to be periodic in nature to simplify the analysis. Real layouts have certain sections of uncoupled traces followed by coupled sections where the coupling occurs at an angle.

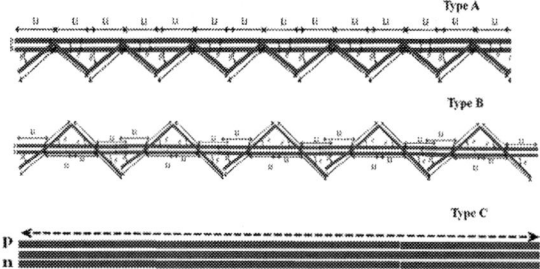

Figure 2. Routing topologies to be studied for single to differential crosstalk

The focus of the paper is to quantify the effect of broadside coupling in terms of BER for various routing topologies commonly observed in practical PCB design. The study would help electrical designers to come up with set of design and routing guidelines that can save PCB cost and at the same time maintain electrical integrity. Various test cases are studied in this paper to demonstrate the near end crosstalk (NEXT) impact and differential-common mode conversion effect on high speed serial links. The impact of all these examples is studied on the horizontal eye opening at the receiver in terms of BER for victim link running at 8Gbps, 12Gbps and 16Gbps. Aggressor is a DDR4 trace running at 3.2Gbps. Further, the aggressor is coupled in parallel along the entire length of the victim. The horizontal spacing between the center of the aggressor trace and differential trace is varied to find the best location of minimum NEXT impact. Impact on the horizontal eye opening from various design parameters such as the horizontal spacing between the traces, number of coupled sections, and crossover angle of the aggressor over the victim, and routing type are investigated. Section III summarizes the paper.

Crosstalk Impact on victim eye opening

The impact of broadside coupling on horizontal eye opening at the victim receiver is observed using three different broadside routing topologies typically used in the complex high speed board design layouts. A quantitative comparison of the NEXT impact on the horizontal eye opening and differential-common mode conversion between different testcases are presented.

Routing Style Unitcell A : Test structures for Unitcell A with 200 and 2000 mils periodicity with 30 degree routing angle as shown in Figure 3 are constructed by cascading unitcell 50 times and 5 times to generate a 10 inch broadside coupled model using standard FR-4 as the dielectric material with dielectric constant 3.7 and loss tangent of 0.025. 2000 mils periodicity case is studied as designers tend to route in such a fashion to negate fiber-weave effect.

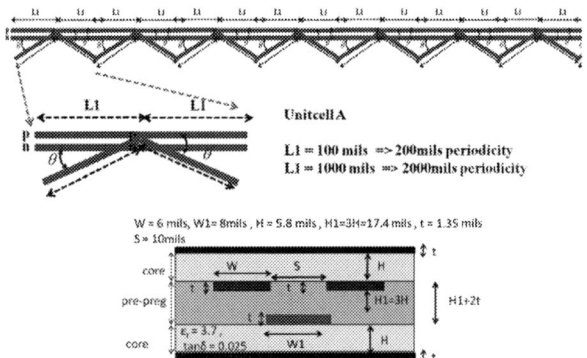

Figure 3. 10 inch broadside trace Unitcell A routing

An input of 500mV pk-pk with a rise/fall time of 35ps at data rate of 5Gbps, rise/fall time of 22ps at data rate of 12Gbps and rise/fall time of 15ps at data rate of 16Gbps is used on victim. An input of 250mV pk-pk with a rise/fall time of 80ps at data rate of 3.2Gbps is used at aggressor input. It is assumed that broadside coupling occurs not at the transmitter(Tx) but somewhere around center of the channel. The eye width and eye height of models for Unitcell A for 200 and 2000mils periodicities are compared at 8Gbps, 12Gbps and 10Gbps in Tables 1, 2 and 3. The impact of broadside coupling from single-ended aggressor on the differential victim can be seen on eye height and eye width at different periodicities at 16Gbps in Figures 4-5.

Table 1. Eye diagram results for victim at 8Gbps

Unitcell A periodicity	EW (ps)	EH (mV)	BER
200mils	103	240	1E-15
200mils - No crosstalk	104	245	1E-15
2000mils	103	240	1E-15
2000mils - No crosstalk	105	245	1E-15

Table 2. Eye diagram results for victim at 12Gbps

Unitcell A periodicity	EW (ps)	EH (mV)	BER
200mils	54.17	134	1E-15
200mils - No crosstalk	55.67	139	1E-15
2000mils	54.17	137	1E-15
2000mils - No crosstalk	56	141	1E-15

Table 3. Eye diagram results for victim at 16Gbps

Unitcell A periodicity	EW (ps)	EH (mV)	BER
200mils	25.3	47	1E-15
200mils - No crosstalk	26.4	51	1E-15
2000mils	24	46	1E-15
2000mils - No crosstalk	27	55	1E-15

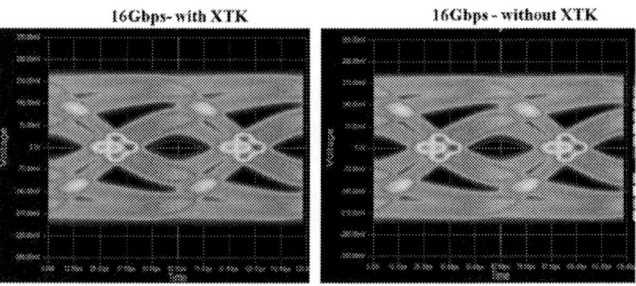

Figure 4. Eye Diagram of Unitcell A 200mils periodicity at 16Gbps

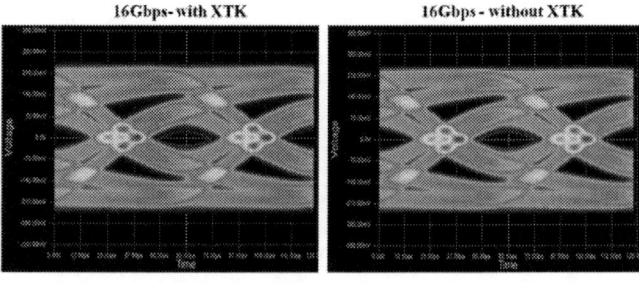

Figure 5. Eye Diagram of Unitcell A 2000mils periodicity at 16Gbps

Unitcell A has 1.2% impact on eye width @ 8Gbps, 2.6% impact on eye width @ 12Gbps and 4.3 % impact on eye width @ 16Gbps for 200 mils periodicity as seen from Tables 1-3. Unitcell A has 1.4% impact on eye width @ 8Gbps, 2.8% impact on eye width @ 12Gbps and 11.5 % impact on eye width @ 16Gbps for 200 mils periodicity as seen from Tables 1-3. The single-ended aggressor shifts the Vref from 0v for victim differential trace to either positive or negative voltage. The voltage offset decreases the eye width measured by the receiver with decision threshold set at 0. 2000 mils periodicity case has more crosstalk impact above 12Gbps with aggressor running at 3.2Gbps as seen in Figures 4,5 and 7. Most of crosstalk energy of aggressor is within 5GHz which is well above the bandwidth of the aggressor signal as shown in Figure 6. 2000 mils periodicity model has more crosstalk compared to 200mils periodicity model within the 5GHz bandwidth as shown in Figure 6.

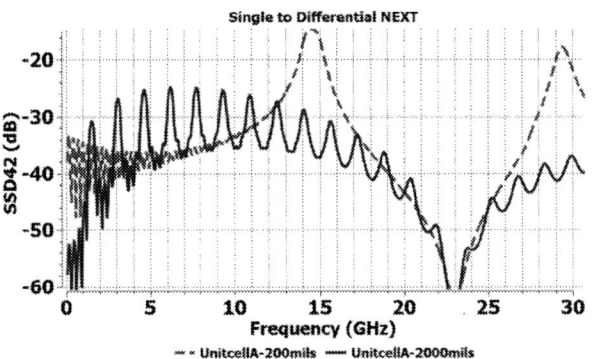

Figure 6. Single to differential NEXT for Unitcell A

The crosstalk impact on the horizontal eye opening at different data rates of the victim is shown in Figure 7.

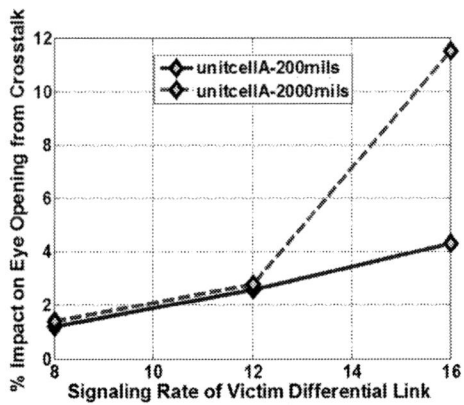

Figure 7. Crosstalk impact at different data rates of victim with aggressor at 3.2 Gbps for Unitcell A

Routing Style Unitcell B : Test structures for Unitcell B with 200 and 2000 mils periodicity with 30 degree routing angle as shown in Figure 8 are constructed by cascading unitcell 50 times and 5 times to generate a 10 inch broadside coupled model using standard FR-4 as the dielectric material with dielectric constant 3.7 and loss tangent of 0.025. 2000 mils periodicity case is studied as designers tend to route in such a fashion to negate fiber-weave effect.

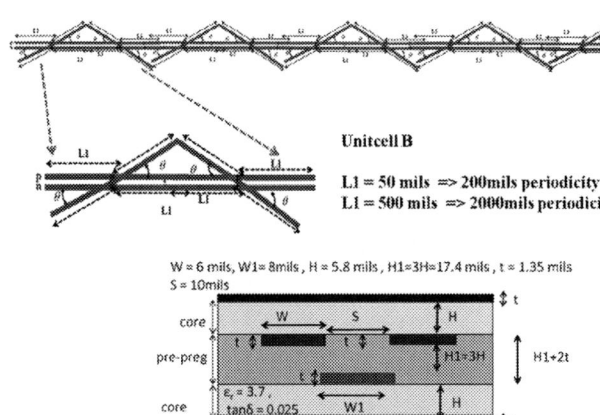

Figure 8. 10 inch broadside trace Unitcell B routing

An input of 500mV pk-pk with a rise/fall time of 35ps at data rate of 5Gbps, rise/fall time of 22ps at data rate of 12Gbps and rise/fall time of 15ps at data rate of 16Gbps is used on victim. An input of 250mV pk-pk with a rise/fall time of 80ps at data rate of 3.2Gbps is used at aggressor input. It is assumed that broadside coupling occurs not at the transmitter(Tx) but somewhere around center of the channel. The eye width and eye height of models for Unitcell B for 200 and 2000mils periodicities are compared at 8Gbps, 12Gbps and 10Gbps in Tables 4, 5 and 6. The impact of broadside coupling from single-ended aggressor on the differential victim can be seen on eye height and eye width at 2000 mils periodicity at 16Gbps in Figures 9 an 10.

Table 4. Eye diagram results for victim at 8Gbps

Unitcell B periodicity	EW (ps)	EH (mV)	BER
200mils	104.5	240	1E-15
200mils - No crosstalk	104.5	240	1E-15
2000mils	104.1	245	1E-15
2000mils - No crosstalk	104.1	245	1E-15

Table 5. Eye diagram results for victim at 12Gbps

Unitcell B periodicity	EW (ps)	EH (mV)	BER
200mils	55	141	1E-15
200mils - No crosstalk	55	142	1E-15
2000mils	55.8	141	1E-15
2000mils - No crosstalk	55.8	141	1E-15

Table 6. Eye diagram results for victim at 16Gbps

Unitcell B periodicity	EW (ps)	EH (mV)	BER
200mils	26.8	52	1E-15
200mils - No crosstalk	26.8	52	1E-15
2000mils	27.3	54	1E-15
2000mils - No crosstalk	27.3	55	1E-15

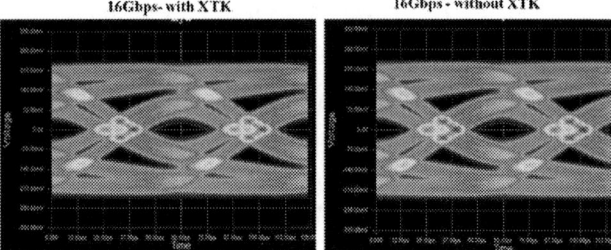

Figure 9. Eye Diagram of Unitcell B 200mils periodicity at 16Gbps

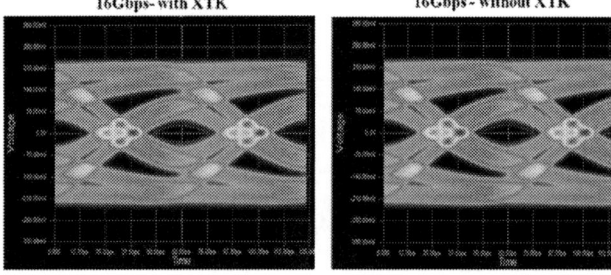

Figure 10. Eye Diagram of Unitcell B 2000mils periodicity at 16Gbps

Unitcell B has 0% impact on eye width @ 8Gbps, @ 12Gbps and @ 16Gbps for 200 and 2000 mils periodicity as seen from Tables 4-6. Unitcell B routing topology has no impact from crosstalk due to aggressor running at 3.2Gbps as seen in Figures 9 and 10. Most of crosstalk energy of aggressor is within 5GHz which is well above the bandwidth of the aggressor signal as shown in Figure 11. The crosstalk level is well below -50dB which has negligible impact on the eye diagram. Due to symmetry of routing of aggressor around the victim negates the differential crosstalk at the receiver for aggressor signals within 5GHz bandwidth. For single-ended aggressor running at higher data rate crosstalk becomes a issue, but speeds are typically below 5Gbps.

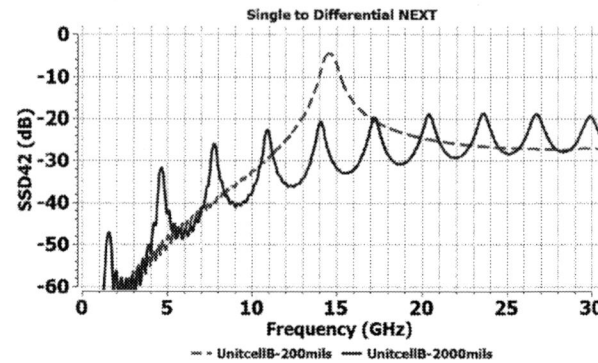

Figure 11. Single to differential NEXT for Unitcell B

The crosstalk impact on the horizontal eye opening at different data rates of the victim is shown in Figure 12.

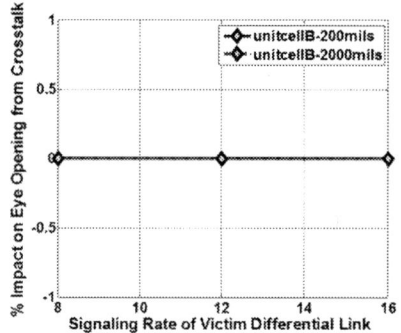

Figure 12. Crosstalk impact at different data rates of victim with aggressor at 3.2 Gbps for Unitcell B

Routing Style Unitcell C : Test structures for Unitcell C : CENTER with aggressor parallel to the victim pair and located in the middle of victim pair is shown in Figure 13. Unitcell C : Offset P has aggressor located right under the positive net of the victim pair is also shown in Figure 13. 10 inch broadside coupled lines are modeled using standard FR-4 as the dielectric material with dielectric constant 3.7 and loss tangent of 0.025 using Unitcell C : center and Unitcell C : offset P routing styles.

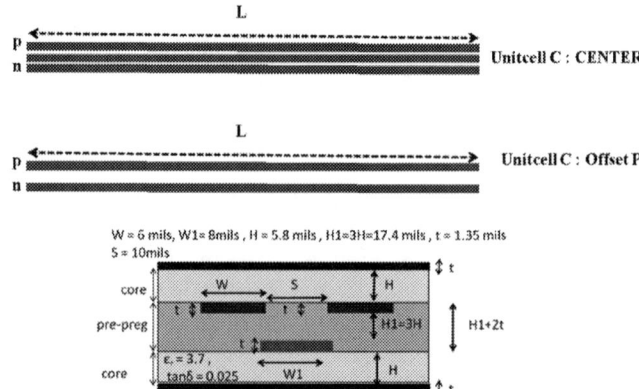

Figure 13. 10 inch broadside trace Unitcell C routing

An input of 500mV pk-pk with a rise/fall time of 35ps at data rate of 5Gbps, rise/fall time of 22ps at data rate of

12Gbps and rise/fall time of 15ps at data rate of 16Gbps is used on victim. An input of 250mV pk-pk with a rise/fall time of 80ps at data rate of 3.2Gbps is used at aggressor input. It is assumed that broadside coupling occurs not at the transmitter(Tx) but somewhere around center of the channel. The eye width and eye height of models for Unitcell C : CENTER and Unitcell C : Offset P are compared at 8Gbps, 12Gbps and 10Gbps in Tables 7, 8 and 9. The impact of broadside coupling from single-ended aggressor on the differential victim can be seen on eye height and eye width for offset-P routing at 16Gbps in Figures 14 and 15.

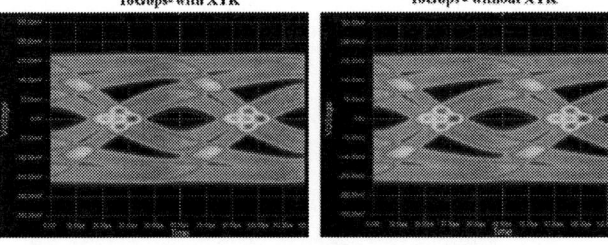

Figure 14. Eye Diagram of Unitcell C : CENTER at 16Gbps

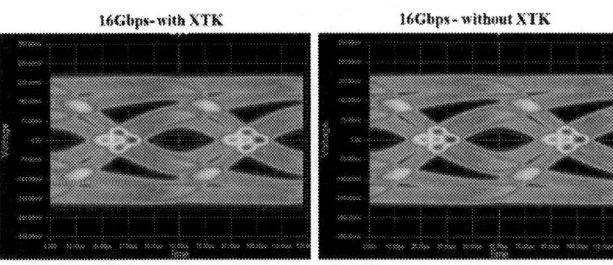

Figure 15. Eye Diagram of Unitcell C : Offset P at 16Gbps

Unitcell C : CENTER has 0% impact on eye width @ 8Gbps, @ 12Gbps and @ 16Gbps as seen from Tables 7-9. Unitcell C : CENTER routing topology has no impact from crosstalk due to aggressor running at 3.2Gbps as seen in Figures 14. Due to symmetry of routing for Unitcell C : CENTER, the crosstalk is minimal. Unitcell C : Offset P has 2.6% impact on eye width @ 8Gbps, 3.5% impact on eye width @ 12Gbps and 13% impact on eye width @ 16Gbps as seen from Tables 7-9. Unitcell C : CENTER routing can be achieved in real board layouts as aggressor lies within the space bounds of the victim pair. Unitcell C : Offset P routing has more crosstalk compared to Unitcell C : CENTER routing which correlates directly with the geometry as shown in Figure 16. By increasing the horizontal separation between the traces crosstalk goes down. This rule is difficult to implement in real layouts due to spacing constraints. Unitcell C : CENTER is best option for minimal crosstalk for any aggressor speeds.

Table 7. Eye diagram results for victim at 8Gbps

Unitcell C Position	EW (ps)	EH (mV)	BER
Center	105.1	250	1E-15
Center- No crosstalk	105.1	250	1E-15
Offset-P	104.1	238	1E-15
Offset P - No crosstalk	104.1	249	1E-15

Table 8. Eye diagram results for victim at 12Gbps

Unitcell C Position	EW (ps)	EH (mV)	BER
Center	56.7	144	1E-15
Center- No crosstalk	56.7	144	1E-15
Offset-P	54.8	136	1E-15
Offset P - No crosstalk	56.7	144	1E-15

Table 9. Eye diagram results for victim at 16Gbps

Unitcell C Position	EW (ps)	EH (mV)	BER
Center	28.1	58	1E-15
Center- No crosstalk	28.1	58	1E-15
Offset-P	24.5	47	1E-15
Offset P - No crosstalk	28.1	58	1E-15

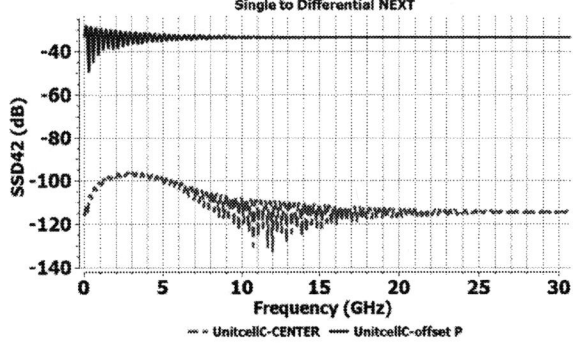

Figure 16. Single to differential NEXT profile for Unitcell C : center and Unitcell C : Offset P

The crosstalk impact on the horizontal eye opening at different data rates of the victim is shown in Figure 17.

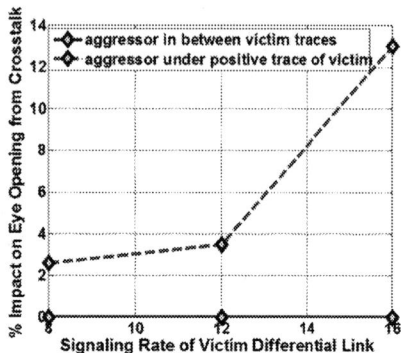

Figure 17. Crosstalk impact at different data rates of victim with aggressor at 3.2 Gbps for Unitcell C : center and Unitcell C : Offset P

Mode conversion due to imbalance of victim pair

The differential-common mode conversion occurs when routing single-ended trace underneath differential trace. Depending on the imbalance created by the single-ended trace the common mode radiation is generated. Higher the imbalance , higher the common mode radiation from the victim pair. Common mode radiation can be a issue for EMC compliance. The mode conversion plot is shown in Figure 18 for different routing styles and periodicities. An input of 1V pk-pk step input with a rise time of 15ps is used on victim. The common mode response observed at the victim receiver due to mode conversion is shown in Figure 19.

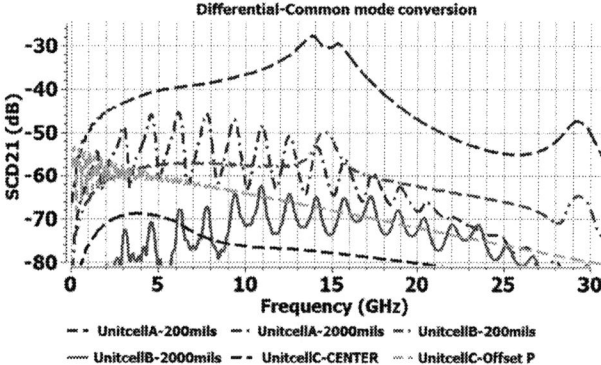

Figure 18. Differential-to-Common mode conversion of routing styles Unitcells A,B and C

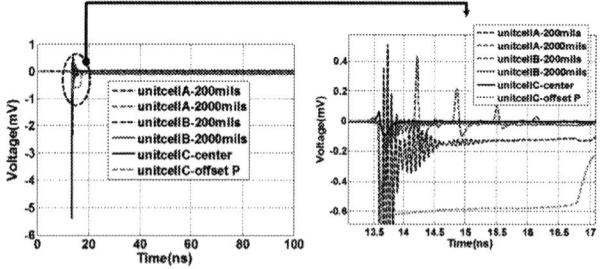

Figure 19. Common-mode voltage disturbance observed at the victim receiver due to step input of Unitcells A,B and C

It is observed that Unitcells A and B routing topology for smaller periodicitiy create more differential-common mode conversion compared to longer periodicity. Unitcell A with 200 mils periodicity has maximum differential-common mode conversion impact compared to all other routing styles due to asymmetry. From crosstalk point of view, smaller periodicity is better compared to longer periodicity. Unitcell C : CENTER routing style has minimal crosstalk and minimal common-mode radiation. Unitcell C : CENTER and Unitcell B with 2000 mils periodicity have similar common-mode performance.

The Floquet mode location[13] observed in the NEXT frequency response of periodic routing types A and B is given by

$$f_p = \frac{n(3 \times 10^8)}{2\sqrt{\varepsilon_r}\,p} \quad n = 1,2,.. \tag{1}$$

where p is periodicity of the unitcell, ε_r is the dielectric constant of the medium.

The frequency at which the first peak occurs for unit cell of length 200 mils as in Unitcells A and B with 200mils periodicity is

$$p = 200 mils \tag{2}$$
$$f_{p1} = \frac{3 \times 10^8}{2\sqrt{3.7}(200mils)} \approx 15.3 GHz$$

The frequency at which the first peak occurs for unit cell of length 2000 mils as in Unitcells A and B with 2000mils periodicity is

$$p = 2000 mils \tag{3}$$
$$f_{p2} = \frac{3 \times 10^8}{2\sqrt{3.7}(2000mils)} \approx 1.53 GHz$$

Unitcells A and B with 200 and 2000mils periodicities have peaks at the locations determined by equations 2-3 in the NEXT frequency profile at multiples of 15.3 GHz and 1.53GHz as shown in Figure 5 and 9. These peaks in the NEXT frequency profile impact the horizontal eye opening depending the aggressor signal bandwidth. Unitcell A has more impact compared to Unitcell B even though both are periodic and have floquet modes. Unitcell B is symmetric and hence the crosstalk impact is negligible similiar to Unitcell C : CENTER routing for speeds below 5Gbps. Unitcell A with 2000 mils periodicity is worse compared 200 mils periodicity for any data rate.

Floquet mode resonances are also observed at same frequency locations in the differential-common mode conversion frequency profile as shown in Figure 18. The ringing is observed in the common mode output at the victim as seen in Figure 19. The ringing frequency is same as resonance frequency of Floquet modes. These resonances can be the trouble makers for engineers trying to achieve EMC compliance. If vertical spacing between the adjacent signal layers decreases then the crosstalk and common-mode radiation will increase for Unitcell A, B and C : Offset P routing styles.

Conclusions

The focus of the paper is to quantify the effect of broadside coupling in terms of BER for various routing topologies commonly observed in practical PCB design. Unitcell C : CENTER has minimal impact from crosstalk among all the routing styles. Unitcell C : Offset P is worst routing style compared to all routing styles. Unitcell A with longer periodicity is exposed to crosstalk compared to smaller periodicity. Unitcell B routing styles has also negligible impact from crosstalk for aggressor speeds lower than 5Gbps. Smaller periodicity of Unitcells A and B have more common-mode radiation compared to longer periodicity. Unitcell C : CENTER has least common-mode radiation problem among all routing styles. It is observed from Unitcells A and B that periodicities with low crosstalk has more common-mode radiation and vice versa. The study would help electrical designers to come up with set of routing guidelines that can maintain electrical integrity and EMC compliance.

Acknowledgements

This material is based on work supported by the National Science Foundation under Grant No. 0855878.

References

1. F. Xiao, R. Hashimoto, K. Murano, and Y. Kami, "Analysis of crosstalk between single-ended and Differential lines," PIERS., vol. 3, No.1, pp. 16-20, 2007.

2. F. Xiao and Y. Kami, "Modeling and analysis of crosstalk between Differential lines in high-speed interconnects," PIERS., vol. 3, No.8, pp. 1293-1297, 2007.

3. F. Xiao, W. Liu and Y. Kami, "Analysis of crosstalk between finite-length microstrip lines: FDTD approach and circuit-concept modeling," IEEE Trans. on EMC., vol. 43, No.4, pp. 573-578, Nov. 2001.

4. K. Araki, F. Xiao and Y. Kami, "Simplified interference coupling model for two orthogonal striplines on adjacent layers," IEEE Trans. Commun., vol. E91-B, pp. 3983-3989, Dec. 2008.

5. K. Araki, F. Xiao, Y. Kami, H. Bishnoi and J. Drewniak, "Modeling interference coupling between two orthogonal strip lines on adjacent layers," International symposium on EMC Europe, pp. 1-6, 2008.

6. Y. Kami and W. Liu, "Analysis of coupling between transmission lines in arbitrary directions," IEEE International Symposium on EMC, vol. 2, pp. 952-957, 1998.

7. S. Wu, "Modeling of Crosstalk between non-parallel striplines on adjacent layers," PhD Dissertation, Missouri S&T ,2011

8. C. R. Paul, "Modeling electromagnetic interference properties of printed circuit boards," IBM J. Res. Dev., vol.33, no.1, pp. 33-50, 1989.

9. F. M. Romeo and M. M. Santomauro, "Time-domain simulation on n-coupled transmissionline network, " IEEE Trans. Microwave Theory Tech., Vol. MTT-35, pp. 131-137, Feb. 1987.

10. C. Wei, R. F. Harrington, J. R. Mautz and T. K. Sarkar, "Multiconductor transmission lines in multilayered dielectric media," IEEE Trans. Microwave Theory Tech., Vol. MTT-32, pp. 439-450, Apr. 1984.

11. C. R. Paul, Analysis of multiconductor transmission lines, Hoboken, NJ: John Wiley & Sons, Inc., 2008.

12. A.R Chada , S. Wu, J. Fan, J. L. Drewniak, B. Mutnury, and D.N. de Araujo, "Modeling broadside coupled traces using Eq PUL RLGC model," IEEE EPEP Conference, 2012.

13. A.R Chada , S. Wu, J. Fan, J. L. Drewniak, B. Mutnury, and D.N. de Araujo, "Efficient Complex Broadside Coupled Trace Modeling and Estimation of Crosstalk Impact using Statistical BER analysis for High Volume, High Performance Printed Circuit Board Designs," IEEE ECTC Conference, 2013.

Power Distribution Network Worst-Case Power Noise and an Efficient Estimation Method

Jiangyuan Qian [1*] and Shiji Pan [2]

[1] Broadcom Corporation, 5300 California Ave, Irvine, CA 92617

[2] University of California, Irvine, CA 92617

*Email: chan.qian@broadcom.com

Abstract

In this study, we examine a novel insight into the worst-case power noise in power integrity analysis for the core power of an ASIC. It is found that the traditional target impedance method resorting minimizing the peak impedance cannot guarantee a minimal worst-case power noise. The reason is because the worst-case power noise may not occur when the ASIC switching current is modulated at the peak impedance frequency of the power delivery network, contrary to what has been suggested in other studies. This paper discusses the fundamental mechanism behind this phenomenon. Additionally, we propose a new guideline to assess the worst-case power noise, i.e., characterizing the relationship between the power noise and the modulation frequency of the switching current. An efficient method to quickly characterize the relationship is provided, which is capable of greatly reducing the simulation time to reveal the 'true' worst-case power noise.

Introduction

One of the outcomes in the continual effort to downsize integrated circuits is that the core voltage of the CMOS process keeps dropping to a lower value, and the circuit becomes more sensitive to power supply noise due to the increasing ratio of the noise over the core voltage [1]. The power noise is the effect of the on-chip switching current interacting with the power distribution network (PDN) system. As a detrimental factor, power noise determines the minimal operating voltage of the chip and therefore affects the chip yield. To minimize the power noise, an accurate analysis of power integrity (PI) is critical for a successful system design.

It is important to understand how the PDN and switching current interact with each other, as well as the conditions under which the worst-case power noise occurs. In [2]–[3], it is claimed that the worst-case power noise occurs when the switching current pattern is modulated with a square wave at the PDN resonant frequency, where the peak PDN impedance is observed.

In this study, we show that the observation in [2]–[3] may not be always true, and the largest power noise could occur when the switching current is modulated with a square wave at a *lower* frequency than the PDN resonant frequency. We also discuss the basic mechanism behind this phenomenon.

Additionally, we propose an estimation method to characterize the relationship between the power noise and the frequency of the square wave, which could be used to determine the worst-case power noise in a very efficient manner.

Worst-Case Switching Current

PDN power noise is the response to the ASIC current load on the entire PDN system. A typical equivalent circuit model of an entire PDN system is shown in Fig. 1. To obtain the worst-case power noise, one needs to have the profile of the worst-case switching current load. Based on this worst-case switching current, a transient analysis should be performed to monitor the voltage noise on the core power rail. The voltage noise is the most critical factor in determining the quality of a PDN design.

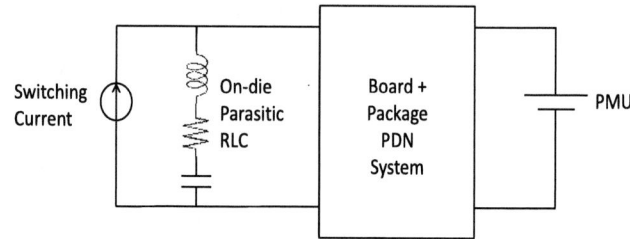

Fig. 1. An equivalent circuit model of a typical PDN system.

According to [2], the chip power model (CPM), which was generated using RedHawk simulation tools, captures the switching and leakage current in the scenario of highest operating power, together with the on-chip parasitics of the ASIC. The simulation involves creating an accurate model representing the actual die operating behavior such that the switching current profile matches the ASIC's real functional activities. A period of the current waveform in one CPM and its related frequency spectrum are shown in Fig. 2. The CPM was extracted while the ASIC was operating at 1.4 GHz. Fig. 2(b) shows that the peak frequency component of switching current occurs at exactly 1.4 GHz, which is the operating frequency of the chip.

In order to effectively evaluate a PDN design, the worst-case power noise should be used as a general rule of thumb. However, using the switching current profile taken directly from the CPM in the PI analysis cannot guarantee a true worst-case power noise on the power rail. In [3], it was reported that a worst-case noise occurs when the switching activity has a major frequency component that coincides with the resonant frequency of the PDN. For a typical PDN with adequate on-die decoupling, its resonant frequency is in the range of 10 MHz to 100 MHz, which rarely coincides with the operating frequency of the chip. In the case being discussed here, the PDN resonant frequency is much lower than the ASIC operating frequency.

To obtain the worst-case power noise, as reported in [4], ANSYS has provided an improved version of their CPM, in which a higher percentage of power is concentrated in the

978-1-4799-2408-0/14 $31.00 © 2014 IEEE   2088   2014 Electronic Components & Technology Conference

resonant frequency of the PDN system. The PDN resonant frequency is used as a control factor to generate the CPM current profile. In practice, the new CPM model could result in a worse perturbation on the power rail. However, this method would result in a much more complex simulation flow for PCB designers (i.e., for each PCB design, one needs to generate its corresponding CPM), and therefore may not be a practical solution.

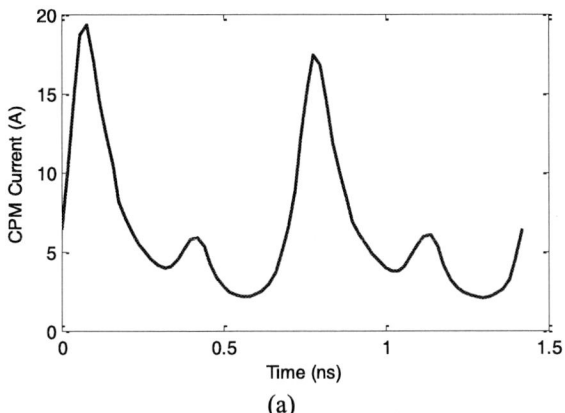

(a)

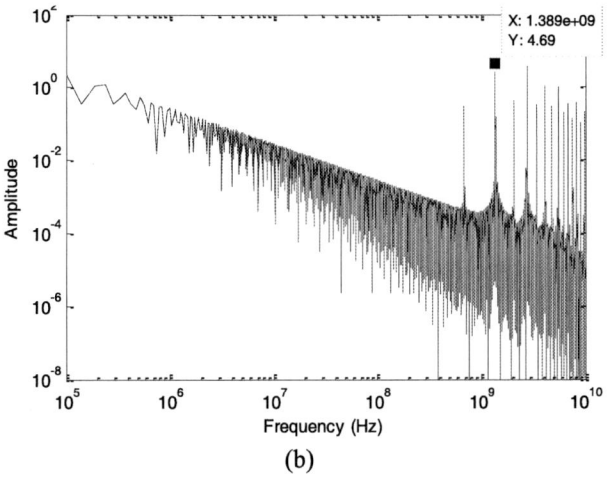

(b)

Fig. 2. (a) A period of the current waveform in the CPM, and (b) the frequency spectrum of the CPM in (a).

A better approach, as described in [3], is to add modulation to the switching current profile in the PI transient simulation. According to [3], a burst of square wave is fed into the AISC current profile to mimic the realistic logic burst activity such that the circuit starts toggling after some period in an idle state. In other words, the ASIC current profile is postprocessed by modulating with a square function at a certain frequency (i.e., the power virus frequency). It has been reported that a stimulus with a periodic burst at the PDN resonant frequency, where the PDN shows its highest impedance, would result in the worst-case power noise on the PDN power rail. In addition, the same report stated that the worst power noise also depends on the PDN's quality factor around the resonant frequency. It is worth mentioning that, although a perfect periodic burst would probably never occur during actual ASIC operation, this particular stimulus

provides an effective way to approach the worst-case scenario of the power noise.

If one considers the effect of modulation in the frequency domain, a periodic burst at the PDN resonant frequency would strongly highlight the frequency component of the CPM at the PDN resonant frequency. Because the power rail voltage is the interaction of the ASIC switching current and the PDN, it is natural to assume that an 'enhanced' peak power noise could be obtained at the frequency where both the switching current and PDN impedance show their peaks.

Regardless, there is no proof to confirm that this local peak power noise would end up being the worst-case power noise. In this paper, we will prove that the 'true' worst-case power noise may not occur at the resonant frequency of the PDN (that is, where the peak impedance is located).

Fig. 3(a) shows a case of the CPM current with periodic burst modulation at 8 MHz (i.e., the power virus frequency is 8 MHz) with a 50% duty cycle. To better model the current drawn by the ASIC, instead of using constant '0' as the current value to define the idle state of the inverters, we consider a static low current at the ASIC idle state. For a digital CMOS inverter, the total current drawn from V_{dd} can be separated into dynamic current (mainly the switching current) and the static current (including the leakage current and the internal current). When the inverter transistors are OFF, the transistors still draw static current from V_{dd}. In this case, the idle current is around 3.44A according to the chip-level current extraction produced by RedHawk software.

Fig. 3(b) shows a comparison of the fourier transform of the CPM current with (in red) and without (in blue) adding the effect of power virus. Without periodic burst modulation, the frequency profile of the CPM current shows a stronger frequency component at a lower frequency and decays at a rate of –10 dB per decade over the frequency. As expected, with periodic burst modulation at 8 MHz in the CPM current, the frequency component around 8 MHz is greatly enhanced (almost 100 times more than without modulatioin at 8 MHz).

Moreover, the frequency component at the 3rd order harmonics (24 MHz), 5th order harmonics (40 MHz), and other higher-order odd harmonics are also elevated to a much higher amplitude level compared with other frequencies. Indeed, if one considers the Fourier series of a time-domain square function with frequency f_0 and a 50% duty cycle, it has major components at its central frequency (f_0) and its odd order harmonics ($3 \cdot f_0$, $5 \cdot f_0$, $\cdots$) as shown below

$$square(t) = \frac{2}{\pi}[\sin(2\pi f_0 t) + \frac{1}{3}\sin(6\pi f_0 t) + \frac{1}{5}\sin(10\pi f_0 t) + \cdots\cdots]. \quad (1)$$

Because the frequency components at the high-order odd harmonics are much higher than those of non-harmonic frequencies, it implies that these frequency components also strongly contribute to the power noise, as could be perceived from the frequency domain perspective. When the frequency of higher-order harmonics coincides with the resonant frequency of the PDN, the power noise contribution from the higher-order odd harmonics will be particularly strong.

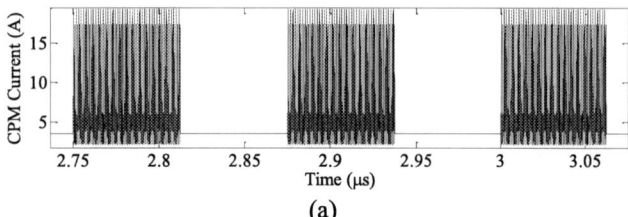

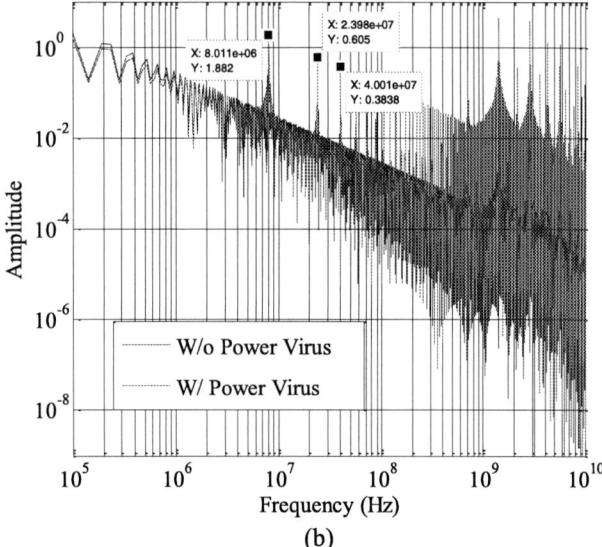

Fig. 3. (a) CPM current with power virus at 8 MHz. (b) Comparison of the Fourier transform of the current in (a) and the current without power virus.

As an example, Fig. 4 shows the comparison of the PDN impedance over frequency (in blue dashed line) and its peak-to-peak power noise over the power virus frequency for one prototyped system (in red solid line). Each point in the red curve requires a transient simulation to obtain the peak-to-peak power noise. The two curves have similar shapes. The PDN peak impedance occurs around 40 MHz, where a local peak power noise can be observed. Meanwhile, the global peak power noise occurs when the power virus frequency is 8 MHz, where the local peak impedance can be observed. Based on the discussion above, this phenomenon is due to the fact that (i) when the power virus frequency is 8 MHz, its 5th order harmonics (40 MHz) coincide with highest PDN impedance such that it generates a large contribution to the overall power noise, and (ii) the local peak impedance at 8MHz also results in a relatively large power noise by itself, considering that the frequency response of the switching current typically contains a larger magnitude component at low frequencies (i.e., 8 MHz) than high frequencies (i.e., 40 MHz). The sum of the two effects described above results in a larger power noise at 8MHz.

More importantly, this phenomenon prompts us to reassess the traditional target impedance method [5] that is widely used in PDN design evaluation. For the case shown in Fig. 4, if one ignores the local peak impedance around 8 MHz and only attempts to reduce the peak impedance around 40 MHz,

the design may not provide the lowest power noise. Instead, creating a design with the lowest possible power noise requires minimizing the impedance for all frequencies at which the local and global impedance peak occurs (for the case in Fig. 4, this is mainly 8MHz and 40 MHz).

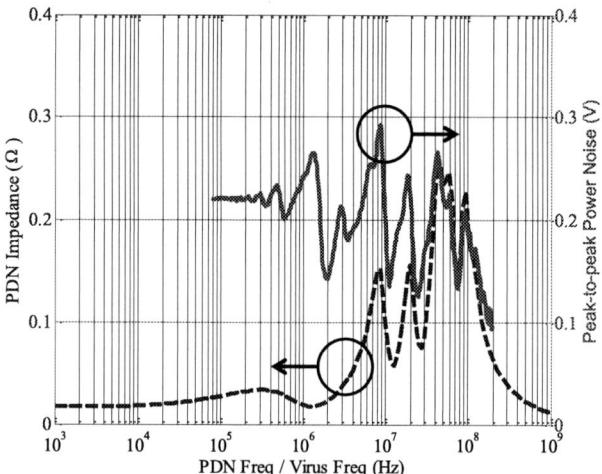

Fig. 4. Comparing PDN impedance over frequency and the peak-to-peak power noise with the power virus frequency.

In the previous discussion, we only considered modulation using a square wave (50% duty cycle). If the switching current is modulated by a pulse wave (i.e., the duty cycle is not 50%), enhancement in the switching current frequency spectrum would appear at all possible higher-order harmonics (including both odd and even harmonics) of the modulation frequency. Therefore, a more definitive conclusion is that all the local impedance peaks should be minimized with carefully selected decoupling capacitors (de-caps), especially when the local impedance peak occurs at the sub-harmonics of the global peak impedance frequency.

In addition, it would be very convenient if one could characterize the relationship between the power virus frequency and the power noise as shown in Fig. 4, because the curve explicitly reveals the location and magnitude of the worst-case power noise. However, it is very time-consuming to retrieve such a curve using the traditional method because each point requires a transient simulation. In the next section, we propose an efficient estimation method to provide this curve based on postprocessing instead of transient simulation.

A Fast Approach to Estimate the Worst Power Noise
The amount of power noise produced is highly dependent on the power virus frequency. Therefore, it is critical to characterize the relationship between the peak-to-peak power noise and the power virus frequency.

The traditional simulation flow used to obtain power noise is shown in Fig. 5(a). For each virus frequency point, one needs to run one transient simulation for the whole PDN system using the switching current. Since the frequency span of interest could vary between 1 MHz and 100 MHz (and sometimes even larger ranges), it is a tremendous amount of

978-1-4799-2408-0/14 $31.00 © 2014 IEEE

work to characterize the whole curve with adequate frequency points covering the feature frequency range.

We propose a novel method that could significantly reduce the simulation time required to obtain this curve and reveal the relationship between power noise and the power virus frequency. The flow of the proposed approach is shown in Fig. 5(b). The diagram can be divided into two separate stages. In the first stage EDA tools are used to extract the PDN model and the switching current. In the second stage the data is postprocessed using numerical calculation (e.g., with MATLAB).

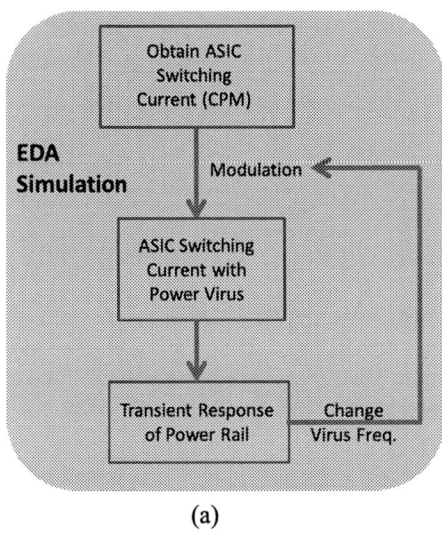

(a)

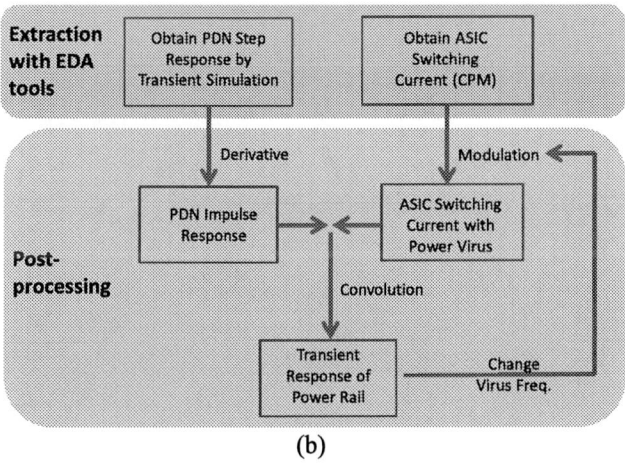

(b)

Fig. 5. Simulation flow used to obtain the power noise through (a) the traditional approach and (b) the proposed approach.

During the first stage, a transient simulation using a circuit simulator (e.g., HSPICE) must be performed to obtain the step response of the PDN circuit. Also, the switching current profile of the ASIC must be extracted using EDA tools (e.g., RedHawk). The transient simulation of the PDN should include the decoupling and parasitic effects on all levels, including on-die, in-package, and onboard. The obtained step response is used later as a representation of the entire PDN system. Note that the rise time of step function should be set

to zero to avoid any additional filtering effect in the system response. In fact, the PDN, comprising all de-caps and on-die parasitics, is a passive linear time invariant (LTI) system with two ports—one at the die node and the other at the Power Management Unit (PMU) end (where we assume perfect clean power). Any LTI system could be represented as an impulse response between its input and output. That is, the PDN could be regarded as the voltage response excited by an impulse current. The impulse response could be calculated by applying a derivative operation on the step response obtained from the circuit simulation.

The second stage involves all the postprocessing based on the model extracted in the first stage. First, we need to get the impulse response of the PDN ($h(t)$) from the step response by applying a derivative operation. Next, based on the raw switching current profile, we add the power virus into the current profile by applying square function modulation at a certain frequency. Here, if one ignores the voltage dependency of the current drawn by the ASIC, the transient response of the power rail voltage $V_{dd}(t)$ is obtained by the convolution of the modulated current load $i(t)$ with the impulse response of the PDN network $h(t)$. The calculation can be expressed as

$$v_{dd}(t) = i(t) \otimes h(t), \qquad (2)$$

in which $\otimes$ is the convolution symbol.

The most obvious advantage of the proposed approach is that no transient simulation tool is required in the second stage, whereas in the traditional approach a transient simulation must be run for each virus frequency. Moreover, only one transient simulation is required to characterize the PDN system, and the power rail voltage response can be obtained through postprocessing using convolutions, which could be hundreds of times faster than using circuit transient simulations (the comparison between the two methods in terms of calculation complexity is shown in Table I). In addition, the calculation time can be reduced by using Fast Fourier Transform (FFT) and Inverse Fast Fourier Transform (IFFT).

Table I Comparison of calculation complexity
(N indicates the number of power virus frequency points)

Method	Calculation Complexity
Traditional	N transient simulations
Proposed	Only 1 transient simulation + N convolution calculations

Nevertheless, although the proposed approach provides a significantly faster simulation time, it is unrealistic to expect that it can produce the exact voltage waveform as traditional methods. The example in Fig. 6 shows that the peak-to-peak power noise obtained by the traditional approach (in blue line with cross symbol) and the proposed approach (in black solid line) could deviate significantly; the difference is around 0.1V in the case shown. The main reason for this is that, while the chip is operating, the real-time current acquired by the ASIC does not exactly follow the current profile of the CPM because it also has a dependency on the real-time voltage at the chip node. However, since the proposed approach presumes the PDN is an LTI system, this dependency effect is removed.

To validate whether the results are reasonably close, we purposely remove this voltage-dependency term when simulating the switching current profile using the traditional approach in order to mimic the effect of our presumption. Fig. 6 shows the comparison of the transient response between using the proposed approach (in black solid line) and using the traditional approach but removing the voltage-dependency of the current load (in red dashed line). The two power rail curves match very well with each other, thereby proving our interpretation above.

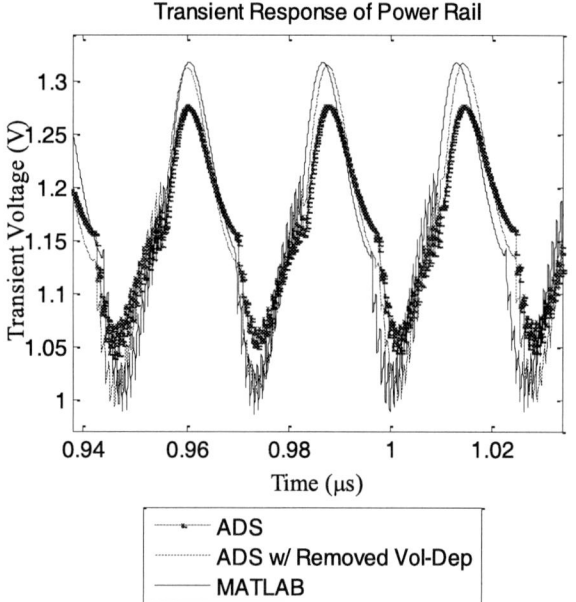

Fig. 6. Comparison of the transient response of the on-die power rail using (i) the traditional approach, (ii) the traditional approach with real-time voltage-dependency in the ASIC current load removed, and (iii) the proposed approach.

Application: Determining a Better PDN Design

With a limited number of de-caps, a PI engineer always faces the issue of determining which de-cap schemes are best suited for the PDN. Although the PDN target impedance method is widely used for this purpose, in some cases it is hard to make the decision based only on the impedance. For example, Fig. 7 shows that one de-cap scheme (in blue dashed line) has a better (smaller) peak impedance over the whole frequency range, and the other scheme (in red solid line) exhibits smaller impedance in the lower frequency range.

Given that power noise is the most explicit factor in determining PDN quality, the proposed method can be used to quickly characterize the peak-to-peak power noise versus the power virus frequency to determine the true worst-case power noise. For example, in the power noise comparison shown in Fig. 8, it can be observed that, although the blue scheme shows an overall lower peak impedance compared to the red scheme, the highest power noise would occur at 8 MHz for

the blue scheme, and is much more severe than the power noise at other frequencies. Based on Fig. 8, it is clear that the red scheme is better.

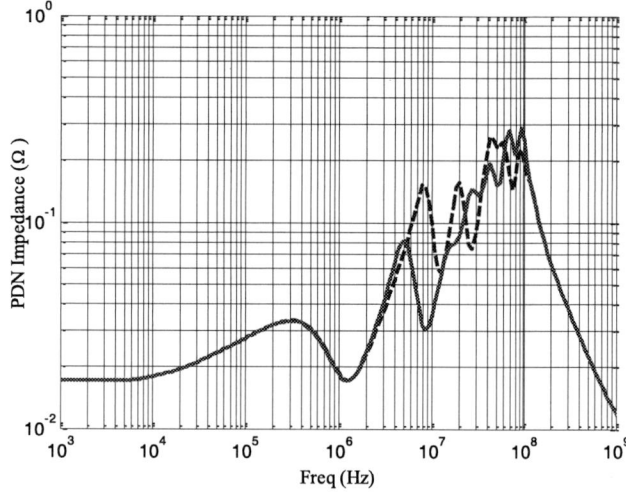

Fig. 7. Comparison of two PDN designs based on the same ASIC, but with different de-cap schemes. The red curve shows higher peak impedance, whereas the blue curve shows higher local peak impedance below 30 MHz.

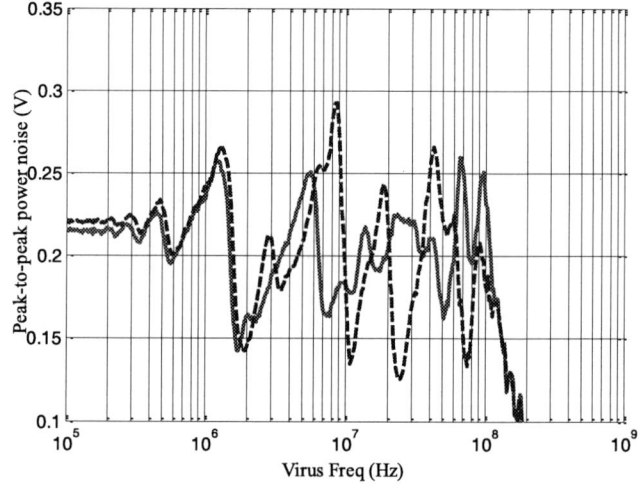

Fig. 8. Comparison of peak-to-peak power noise and virus frequency for two PDN schemes (the PDN impedance is the same as in Fig. 7).

Conclusions

In this study, we demonstrate that the worst-case power noise may not occur when the switching current is modulated by a square wave at the frequency where the peak impedance is located. Instead, it may occur at a lower frequency than the peak impedance frequency. The basic mechanism behind this phenomenon is the combined effect of (i) because the frequency response of the switching current typically contains a larger magnitude component at low frequencies, a 'local peak' impedance located at a low frequency could possibly generate a stronger power noise, and (ii) the high-order harmonics of the modulated switching current coincides with the PDN in the resonant frequency region. Therefore, we conclude that for a guaranteed design, instead of only reducing the peak impedance, one must to minimize all the local and peak impedances over the entire frequency spectrum.

Also, in order to better evaluate a PDN design, one needs to characterize the curve between the power noise and the modulation frequencies (i.e., the power virus frequency). Used as part of the criteria to determine PDN design quality, this relationship is more suitable than PDN target impedance. However, the characterization process requires a significant amount of time because, for each modulation frequency, one transient simulation with a reasonable time span must be performed.

To reduce the simulation time, we propose an efficient method to quickly characterize the relationship between the power noise and the modulation frequency. Based on a linear time-invariant assumption of the PDN system, the power noise is calculated with the impulse response of the PDN system and the pattern of the switching current. The calculation results show good correlation with the ones obtained by transient simulation. The proposed method not only reduces the simulation time significantly, but also automates the analysis flow.

References

1. M, Swaminathan,K. Joungho;I. Novak, J.P. Libous, "Power distribution networks for system-on-package: status and challenges," *IEEE Transactions on Advanced Packaging*, vol.27, no.2, pp.286,300, May 2004
2. W. Cheng, A. Sarkar, S. Lin, and J. Zheng, "Worst-Case Switching Pattern for Core Noise Analysis", presented at DesignCon 2009, Santa Clara, CA.
3. S. Sun, L. D. Smith, and P. Boyle, "On-chip PDN Noise Characterization and Modeling," presented at the DesignCon 2010, Santa Clara, CA.
4. Apache, "Advanced Modeling Technologies for Chip Package System Co-analysis and Co-optimization" (White paper).
5. L. Smith, D. Becke, S. Weir and I. Novak, "Comparison of Power Distribution Network Methods: Bypass Capacitor Selection Based on Time Domain and Frequency Domain Performances", presented at the DesignCon 2006, Santa Clara, CA.

Fast Calculation of Electromagnetic Interference by Through-Silicon Vias

Aosheng Rong*, Andreas C. Cangellaris*, and Feng Ling**

(*) ECE Department, University of Illinois at Urbana-Champaign, 1406 W. Green St. Urbana, IL 61801, U.S.A.
(**) School of Electronic and Optical Engineering, Nanjing University of Science & Technology, Nanjing 210094, China
rongas@gmail.com, cangella@illinois.edu, fengling@ieee.org

Abstract

A methodology is presented for the expedient calculation of radiated electromagnetic interference by dense arrays of through-silicon vias in layered substrates. The proposed methodology builds upon the efficiency of the Foldy-Lax scheme for the fast calculation of the electromagnetic interactions between multiple vias in planar substrates, by exploiting translation invariant expressions of the governing equations that lend themselves to fast calculation of the convolution integrals for the radiated fields using Fast Fourier Transform. Applications of the developed model and algorithm involving the investigation of radiated emissions from through-silicon via clusters are used to demonstrate the efficiency of the proposed algorithm.

Introduction

Electromagnetic radiated emissions by Through-Silicon Vias (TSVs) and their interference with other interconnect circuitry and functional blocks, is a pressing signal integrity issue in 3D IC integration. In particular, the complexity of the three-dimensional topography of 3D ICs impedes the routine application of state-of-the-art electromagnetic CAD (EM-CAD) tools to assess such electromagnetic interference (EMI) effects, quantify their impact, and guide the development of design guidelines for the mitigation and/or suppression of TSV-generated EMI.

A methodology that overcomes this shortcoming is presented in this paper. Key to the proposed methodology is the fast calculation of the primary electromagnetic field radiated by the TSVs. Following commonly used terminology in electromagnetic field radiation and scattering, the primary electromagnetic field is defined to be the field radiated by the TSVs in the absence of the victim device, whose possible functionality degradation due to interference by the TSVs is of interest. With the primary field available over the volume occupied by the victim device, the subsequent calculation of the induced noise in the device is obtained through the solution of an electromagnetic boundary value problem that is amenable to solution using commercially-available EM field solvers. For example, a common case of particular interest is the radiated noise coupling to the interconnect wiring exposed to the primary field, the solution of which is well documented in the literature [1].

The proposed approach relies upon the extension of the methodology in [2] for the full-wave electromagnetic analysis of dense arrays of TSVs in layered substrates. The paper is organized as follows. In Section 2, the mathematical formulation used for the calculation of the electromagnetic fields generated by the TSVs is reviewed and the proposed methodology to expedite their calculation is presented. This is followed by the demonstration of the resulting computer

algorithm to the modeling of EMI by TSV clusters in Section 3. The paper concludes with a summary of the attributes of the proposed methodology and an outlook of on-going and future work.

Mathematical Formulation

Figure 1 below depicts a cross-sectional view of a 3×3 TSV array, with the vias arranged in a uniform pattern. The array traverses a layered, planar semiconductor substrate of the type encountered in integrated circuits [2]. For the purposes of this paper, the material composition of each one of the cylindrical vias assumes one of the following two types. The first type consists of a copper (Cu) core with a SiO_2 outer coating and will be referred to as a Cu-SiO2 TSV. The second type consists of a copper core followed by a sequence of three layers, the innermost being a titanium coating, followed by a SiO_2 layer and an outermost, depletion layer. It will be referred to as a multi-layered TSV.

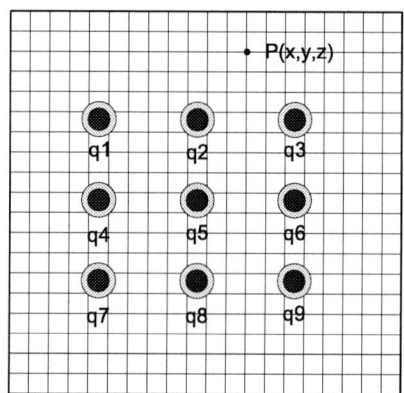

Figure 1. Cross-sectional view of a 3×3 via cluster.

In the following, the index q, $q = 1, 2, ..., NT_s$, will be used for the numbering of the vias in the cluster. Following [2] and the mathematical formulation and convention used in it and associated papers [3] − [6], the electric field and the magnetic field, respectively, inside the substrate and exterior to the vias are written in terms of a superposition of cylindrical harmonics according to the Foldy-Lax equations as follows,

$$
\mathbf{E}_t^{(e)}(x, y, z) = \sum_{q=1}^{NT_s} \sum_{l=0}^{L} \sum_{n=-\infty}^{+\infty} w_{q,ln} T_{q,ln} \frac{\left(k_\rho^{(i)}\right)^2}{k^{(i)}} \times
$$
$$
H_n^{(2)}\left(k_\rho^{(i)} \left|\vec{\rho} - \vec{\rho}_q\right|\right) \exp\left(-jn\phi_q\right) \cos(k_i z)\hat{z} \tag{1}
$$

$$\mathbf{H}_t^{(e)}(x,y,z) = \sum_{q=1}^{NT_z}\sum_{l=0}^{L}\sum_{n=-\infty}^{+\infty} w_{q,ln} T_{q,ln} \frac{-j\omega\varepsilon^{(i)}}{k^{(i)}}\left(\frac{jn}{\left|\vec{\rho}-\vec{\rho}_q\right|}\right)\times$$

$$H_n^{(2)}\left(k_\rho^{(i)}\left|\vec{\rho}-\vec{\rho}_q\right|\right)\exp\left(-jn\phi_q\right)\cos(k_l z)\hat{\rho}_q \quad (2)$$

$$+\sum_{q=1}^{NT_z}\sum_{l=0}^{L}\sum_{n=-\infty}^{+\infty} w_{q,ln} T_{q,ln}\frac{-j\omega\varepsilon^{(i)}}{k^{(i)}}\times$$

$$k_\rho^{(i)} H_n^{(2)\prime}\left(k_\rho^{(i)}\left|\vec{\rho}-\vec{\rho}_q\right|\right)\exp\left(-jn\phi_q\right)\cos(k_l z)\hat{\varphi}_q$$

In the above expressions, $j=\sqrt{-1}$, ω is the angular frequency, $k^{(i)}$ denotes the wave number in medium i, while $k_\rho^{(i)}$ is the radial wavenumber in medium i with electric permittivity $\varepsilon^{(i)}$. The subscript q in the radial and polar angle and associate unit vectors indicate the use of a local reference coordinate system with its z axis along the axis of the qth via. Finally, the coefficients $w_{q,ln}, T_{q,ln}$, are, respectively, the excitation coefficients and the reflection coefficients that need to be computed subject to the boundary and excitation conditions for the vias. Expressions similar to (1) and (2) must be written for the electric and magnetic fields inside each one of the layers in the composite vias [2]. Enforcement of the boundary conditions at all material interfaces along with the excitation conditions yields the linear system of equations that needs to be solved numerically for the calculation of the unknown coefficients in the mathematical statements of the electromagnetic fields. The numerical solution of these equations, which has already been discussed in [2], can be expedited through the application of an adaptive integral method. Such an approach will be presented in a forthcoming paper. In the following, we focus on the development of a fast methodology for the fast calculation of the expressions (1) and (2) after the unknown coefficients have been obtained.

Equations (1) and (2) are of the convolution type. However, except for the case $n=0$, their kernels do not exhibit translation invariance due to the existence of the factor $\exp\left(-jn\phi_q\right)$. This means that the direct application of well-known Fast Fourier Transform (FFT) means to expedite their numerical calculation is not possible in their form shown in (1) and (2). This hurdle can be overcome by means of proper decompositions of the kernels as shown next.

Considering first the expression in (1) for the calculation of the electric field, the kernel that controls the convolution is

$$G_n^{(E_z)} = H_n^{(2)}\left(k_\rho^{(i)}\left|\rho-\rho_q\right|\right)\exp\left(-jn\phi_q\right)$$

$$=\begin{cases} G_0^{(E_z)}, & n=0 \\ G_{e,k}^{(E_z)(+)}, & n=2k \\ G_{e,k}^{(E_z)(-)}, & n=-(2k) \\ G_{o,k}^{(E_z)(+)}, & n=2k+1 \\ G_{o,k}^{(E_z)(-)}, & n=-(2k+1) \end{cases} \quad (3)$$

Where the index k is a positive integer and, it is,

$$G_0^{(E_z)} = H_0^{(2)},$$

$$G_{e,k}^{(E_z)(+)} = G_{e1,k}^{(E_z)} - jG_{e2,k}^{(E_z)}, \quad G_{e,k}^{(E_z)(-)} = G_{e1,k}^{(E_z)} + jG_{e2,k}^{(E_z)},$$

$$G_{e1,k}^{(E_z)} = H_{2k}^{(2)}\cos\left((2k)\phi_q\right), \quad G_{e2,k}^{(E_z)} = H_{2k}^{(2)}\sin\left((2k)\phi_q\right), \quad (4)$$

$$G_{o,k}^{(E_z)(+)} = G_{o1}^{(E_z)} - jG_{o2}^{(E_z)}, \quad G_{o,k}^{(E_z)(-)} = -\left(G_{o1}^{(E_z)} + jG_{o2}^{(E_z)}\right),$$

$$G_{o1,k}^{(E_z)} = H_{2k-1}^{(2)}\cos\left((2k-1)\phi_q\right), \quad G_{o2,k}^{(E_z)} = H_{2k-1}^{(2)}\sin\left((2k-1)\phi_q\right)$$

With the decompositions above, it is immediately evident that different terms behave differently with regards to their translation invariance on the $x-y$ plane. More specifically, $G_0^{(E_z)}$ is translation invariant because of its independence of the polar angle ϕ. On the other hand, $G_{ei,k}^{(E_z)}$, $i=1,2$, and $G_{oi,k}^{(E_z)}$, $i=1,2$, exhibit a quadrant-dependent translation invariance that is easily determined from straightforward application of trigonometric formulas. Once these translation invariant properties are identified the application of well-known FFT algorithms for the expedient calculation of (1) is straightforward [7], [8].

A similar decomposition is required for the components of the magnetic field vector in (2). Starting with the radial component and, once again, focusing on the kernel that controls the convolution we obtain the following decomposition,

$$G_n^{(H_\rho)} = \left(\frac{jn}{\left|\vec{\rho}-\vec{\rho}_q\right|}\right) H_n^{(2)}\left(k_\rho^{(i)}\left|\vec{\rho}-\vec{\rho}_q\right|\right) e^{-jn\phi_q}\hat{\rho}_q$$

$$=\begin{cases} G_{ex,k}^{(H_\rho)(+)}\hat{u}_x + G_{ey,k}^{(H_\rho)(+)}\hat{u}_y, & n=2k \\ G_{ex,k}^{(H_\rho)(-)}\hat{u}_x + G_{ey,k}^{(H_\rho)(-)}\hat{u}_y, & n=-(2k) \\ G_{ox,k}^{(H_\rho)(+)}\hat{u}_x + G_{oy,k}^{(H_\rho)(+)}\hat{u}_y, & n=2k+1 \\ G_{ox,k}^{(H_\rho)(-)}\hat{u}_x + G_{oy,k}^{(H_\rho)(-)}\hat{u}_y, & n=-(2k+1) \end{cases} \quad (5)$$

Where for the case of even values of the index n, $n=2k$, it is,

$$G_{ex,k}^{(H_\rho)(+)} = jG_{e1,k}^{(H_\rho)} + G_{e2,k}^{(H_\rho)}, \quad G_{ex,k}^{(H_\rho)(-)} = -jG_{e1,k}^{(H_\rho)} + G_{e2,k}^{(H_\rho)},$$

$$G_{e1,k}^{(H_\rho)} = \frac{1}{\left|\vec{\rho}-\vec{\rho}_q\right|}(2k)H_{2k}^{(2)}\cos\left(2k\phi_q\right)\cos\left(\phi_q\right),$$

$$G_{e2,k}^{(H_\rho)} = \frac{1}{\left|\vec{\rho}-\vec{\rho}_q\right|}(2k)H_{2k}^{(2)}\sin\left(2k\phi_q\right)\cos\left(\phi_q\right),$$

$$G_{ey,k}^{(H_\rho)(+)} = jG_{e3,k}^{(H_\rho)} + G_{e4,k}^{(H_\rho)}, \quad G_{ey,k}^{(H_\rho)(-)} = -jG_{e3,k}^{(H_\rho)} + G_{e4,k}^{(H_\rho)}, \quad (6)$$

$$G_{e3,k}^{(H_\rho)} = \frac{1}{\left|\vec{\rho}-\vec{\rho}_q\right|}(2k)H_{2k}^{(2)}\cos\left(2k\phi_q\right)\sin\left(\phi_q\right),$$

$$G_{e4,k}^{(H_\rho)} = \frac{1}{\left|\vec{\rho}-\vec{\rho}_q\right|}(2k)H_{2k}^{(2)}\sin\left(2k\phi_q\right)\sin\left(\phi_q\right),$$

While for the case of odd values of the index n it is,

$$G_{ox,k}^{(H_\rho)(+)} = jG_{o1,k}^{(H_\rho)} + G_{o2,k}^{(H_\rho)}, \quad G_{ox,k}^{(H_\rho)(-)} = jG_{o1,k}^{(H_\rho)} - G_{o2,k}^{(H_\rho)},$$

$$G_{o1,k}^{(H_\rho)} = \frac{1}{|\vec{\rho} - \vec{\rho}_q|}(2k-1)H_{2k-1}^{(2)}\cos\big((2k-1)\phi_q\big)\cos\big(\phi_q\big),$$

$$G_{o2,k}^{(H_\rho)} = \frac{1}{|\vec{\rho} - \vec{\rho}_q|}(2k-1)H_{2k-1}^{(2)}\sin\big((2k-1)\phi_q\big)\cos\big(\phi_q\big),$$

$$G_{oy,k}^{(H_\rho)(+)} = jG_{o3,k}^{(H_\rho)} + G_{o4,k}^{(H_\rho)}, \quad G_{oy,k}^{(H_\rho)(-)} = jG_{o3,k}^{(H_\rho)} - G_{o4,k}^{(H_\rho)}, \tag{7}$$

$$G_{o3,k}^{(H_\rho)} = \frac{1}{|\vec{\rho} - \vec{\rho}_q|}(2k-1)H_{2k-1}^{(2)}\cos\big((2k-1)\phi_q\big)\sin\big(\phi_q\big),$$

$$G_{o4,k}^{(H_\rho)} = \frac{1}{|\vec{\rho} - \vec{\rho}_q|}(2k-1)H_{2k-1}^{(2)}\sin\big((2k-1)\phi_q\big)\sin\big(\phi_q\big).$$

A similar process is followed for the decomposition of the kernel that controls the convolution attribute of the ϕ component of the magnetic field in (2). More specifically, we have,

$$G_n^{(H_\phi)} = H_n^{(2)\prime}\left(k_\rho^{(i)}|\vec{\rho} - \vec{\rho}_q|\right)\exp\left(-jn\phi_q\right)\hat{\varphi}_q$$

$$= \begin{cases} G_{ex,k}^{(H_\phi)(+)}\hat{u}_x + G_{ey,k}^{(H_\phi)(+)}\hat{u}_y, & n = 2k \\ G_{ex,k}^{(H_\phi)(-)}\hat{u}_x + G_{ey,k}^{(H_\phi)(-)}\hat{u}_y, & n = -(2k) \\ G_{ox,k}^{(H_\phi)(+)}\hat{u}_x + G_{oy,k}^{(H_\phi)(+)}\hat{u}_y, & n = 2k+1 \\ G_{ox,k}^{(H_\phi)(-)}\hat{u}_x + G\big(H_\phi\big)_{oy,k}^{(-)}\hat{u}_y, & n = -(2k+1) \end{cases} \tag{8}$$

The terms for n even are of the form,

$$G_{ex,k}^{(H_\phi)(+)} = G_{e1,k}^{(H_\phi)} - jG_{e2,k}^{(H_\phi)}, \quad G_{ex,k}^{(H_\phi)(-)} = G_{e1,k}^{(H_\phi)} + jG_{e2,k}^{(H_\phi)},$$

$$G_{e1,k}^{(H_\phi)} = H_{2k}^{(2)\prime}\cos\big(2k\phi_q\big)\big(-\sin\big(\phi_q\big)\big),$$

$$G_{e2,k}^{(H_\phi)} = H_{2k}^{(2)\prime}\sin\big(2k\phi_q\big)\big(-\sin\big(\phi_q\big)\big),$$

$$G_{ey,k}^{(H_\phi)(+)} = G_{e3,k}^{(H_\phi)} - jG_{e4,k}^{(H_\phi)}, \quad G_{ey,k}^{(H_\phi)(-)} = G_{e3,k}^{(H_\phi)} + jG_{e4,k}^{(H_\phi)}, \tag{9}$$

$$G_{e3,k}^{(H_\phi)} = H_{2k}^{(2)\prime}\cos\big(2k\phi_q\big)\cos\big(\phi_q\big),$$

$$G_{e4,k}^{(H_\phi)} = H_{2k}^{(2)\prime}\sin\big(2k\phi_q\big)\cos\big(\phi_q\big)$$

The ones for n odd are,

$$G_{ox,k}^{(H_\phi)(+)} = G_{o1,k}^{(H_\phi)} - jG_{o2,k}^{(H_\phi)}, \quad G_{ox,k}^{(H_\phi)(-)} = -\big(G_{o1,k}^{(H_\phi)} + jG_{o2,k}^{(H_\phi)}\big),$$

$$G_{o1,k}^{(H_\phi)} = H_{2k-1}^{(2)\prime}\cos\big((2k-1)\phi_q\big)\big(-\sin\big(\phi_q\big)\big),$$

$$G_{o2,k}^{(H_\phi)} = H_{2k-1}^{(2)\prime}\sin\big((2k-1)\phi_q\big)\big(-\sin\big(\phi_q\big)\big),$$

$$G_{oy,k}^{(H_\phi)(+)} = G_{o3,k}^{(H_\phi)} - jG_{o4,k}^{(H_\phi)}, \quad G_{oy,k}^{(H_\phi)(-)} = -\big(G_{o3,k}^{(H_\phi)} + jG_{o4,k}^{(H_\phi)}\big), \tag{10}$$

$$G_{o3,k}^{(H_\phi)} = H_{2k-1}^{(2)\prime}\cos\big((2k-1)\phi_q\big)\cos\big(\phi_q\big),$$

$$G_{o4,k}^{(H_\phi)} = H_{2k-1}^{(2)\prime}\sin\big((2k-1)\phi_q\big)\cos\big(\phi_q\big)$$

In a manner similar to the case for the electric field, the decompositions above, allow us to exploit the quadrant-dependent translation invariance for each term, determined

easily through a straightforward application of trigonometric formulas. Once these translation invariant properties are identified the application of well-known FFT algorithms for the expedient calculation of (2) is straightforward [7], [8].

Numerical Demonstrations

A computer tool was developed to implement the FFT-accelerated computation of (1) and (2) making use of the aforementioned decomposition schemes. This tool was used to calculate radiated fields by TSV clusters under different excitation conditions.

The first case is the 4×4 TSV uniform array with Cu-SiO2 vias and material, geometric, and substrate properties identical to the ones in the example in [1]. Depicted in Fig. 2 below is the excitation pattern for the vias in the cluster.

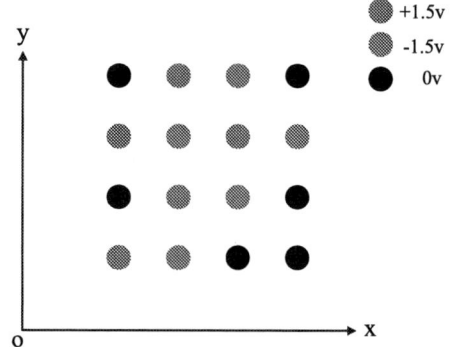

Figure 2. Cross-sectional view of a 4×4 TSV.

Field intensity plots for the calculated electric and magnetic fields at 10 GHz are shown in Figs. 3 and 4, respectively.

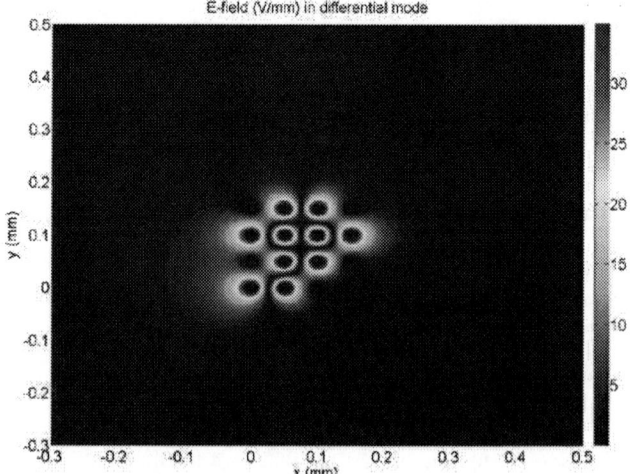

Figure 3. Magnitude of radiated electric field for the excitation of Fig. 2 at 10 GHz.

The next case considered is that of a dense 16×16 TSV cluster. Each via in the cluster is of the multilayered type. As depicted in Fig. 5, the cluster is uniform with substrate properties and via to via pitch the same with the previous example. As far as the properties of the different layers in each via, those were as follows. The copper core was of diameter of 20 μm and electric conductivity of 5.8×10^4 S/mm. The thickness of the titanium layer was 1 μm and electric conductivity of 2.0×10^5 S/mm. The thickness of the SiO$_2$ layer was 1 μm and relative permittivity of 4. Finally, the thickness of the depletion layer was 1 μm and relative permittivity of 6.

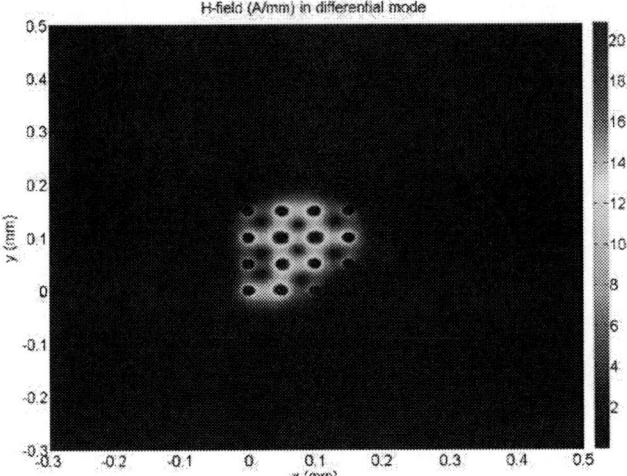

Figure 4. Magnitude of radiated magnetic field for the excitation of Fig. 2 at 10 GHz.

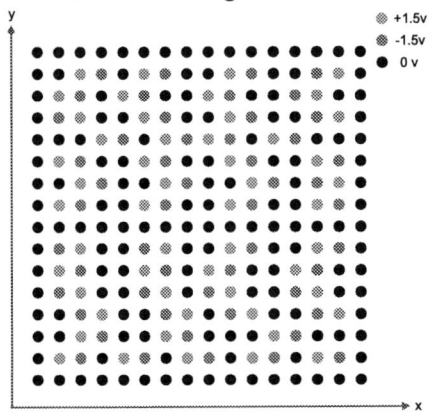

Figure 5. Excitation pattern for a 16×16 TSV cluster with the outermost vias grounded.

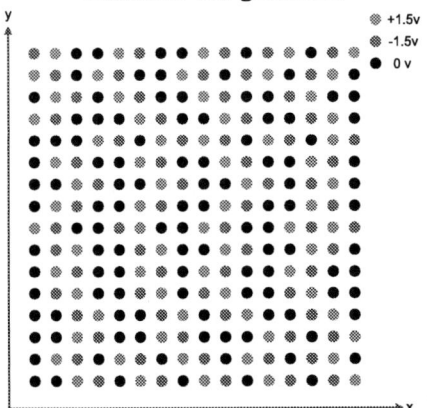

Figure 6. Excitation pattern for a 16×16 TSV cluster with only a portion of outermost vias grounded.

For this dense TSV array, two different excitations at 10 GHz are considered, depicted in Figs. 5 and 6. For the one in Fig. 5, all the outermost vias are grounded, thus providing a continuous shield for the interior ones. For the one in Fig. 6, only a subset of the outermost vias are grounded; thus, the shield is imperfect. Figures 7 and 8 contrast the calculated electric field intensity plots for the two cases. The superior effectiveness of the continuous shield is apparent.

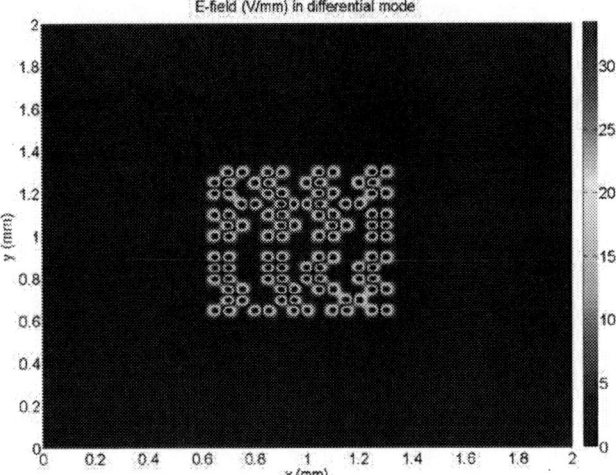

Figure 7. Magnitude of radiated electric field for the excitation of Fig. 5 at 10 GHz.

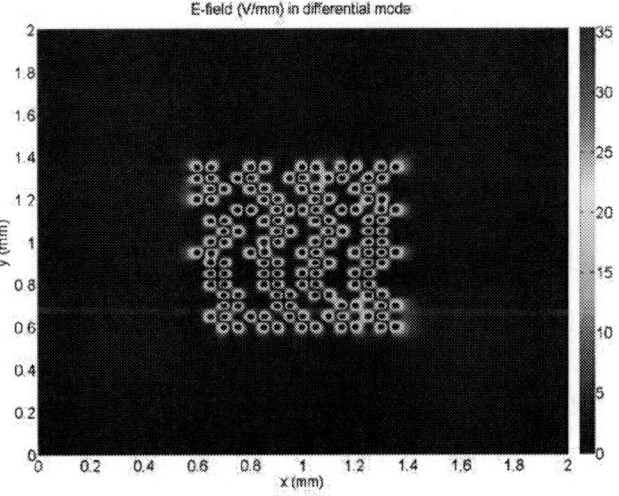

Figure 8. Magnitude of radiated electric field for the excitation of Fig. 6 at 10 GHz.

Concluding Remarks

The FFT-based methodology presented in this paper for the fast computation of electromagnetic field radiation by TSV clusters has a computational cost that is, for all practical purposes, independent of the number of vias in the cluster. This is to be contrasted to the cost of the direct numerical computation of the electromagnetic fields, which, as evident by the governing equations stated in the paper, grows linearly with the number of vias in the cluster. The computational efficiency of the proposed algorithm results from the proper decomposition of the original kernel functions in the governing equations in terms of component functions with quadrant-dependent translation invariance. It is the translation invariance of the terms that allows the use of FFT algorithms for the efficient calculation of the convolution expressions of the field distributions.

A computer tool based on this methodology provides for the efficient investigation of the effectiveness of various options for mitigating and/or suppressing electromagnetic interference between vias in the TSV cluster as well as the shielding of radiated emissions from the cluster by proper assignments of grounding vias. In the latter case, the fast calculation of radiated emissions from the vias anywhere in the substrate lends itself to the efficient prediction of TSV-radiated noise coupling to other interconnect wiring and sensitive electronic devices in the substrate.

References

1. C. R. Paul, *Analysis of Multiconductor Transmission Lines*, 2nd ed., Wiley, 2007.
2. X. Gu, B. Wu, M. Ritter and L. Tsang, "Efficient Full-wave Modeling of High Density TSVs for 3D Integration", Proceedings of the 2010 Electronic Components and Technology Conference, pp. 663-666, 2010.
3. Joohee Kim et al, "High-frequency scalable electrical model and analysis of a Through Silicon Via (TSV)", *IEEE Transactions on Components, Packaging and Manufacturing Technology*, vol.1, pp. 181-195, 2011.
4. Joohee Kim et al, "Through Silicon Via (TSV) equalizer", Proceedings of the IEEE Electrical Performance of Electronic Packaging, pp. 13-16, 2009.
5. K. Han et al, "Electromagnetic modeling of Through-Silicon Via (TSV) interconnections using cylindrical modal basis functions", *IEEE Trans. Advanced Packaging*, vol. 33, no. 4, pp. 804-817, 2010.
6. L. Tsang, H. Chen, C.-C. Huang, and V. Jandhyala, "Modeling of multiple scattering among vias in planar waveguides using Foldy-Lax equations", *Microwave & Optical Technology Letters*, vol. 31, pp. 201–208, 2001.
7. J. C. Bowman and M. Roberts, "Efficient dealiased convolutions without padding", *SIAM Journal on Scientific Computing*, vol. 33, no. 1, pp. 386-406, 2011.
8. H. J. Nussbaumer, *Fast Fourier Transform and Convolution Algorithms*, New York: Springer-Verlag, 1982.

Electrical Simulation and Analysis of Si Interposer for 3D IC Integration

Xin SUN[1], Min MIAO[2]*, Yunhui ZHU[1], Runiu FANG[1], Guanjiang WANG[1,3], Wengao Lu[1], Jing CHEN[1], Yufeng JIN[1,3]

1. National Key Laboratory of Science and Technology on Micro/Nano Fabrication, Peking University, Beijing, 100871, China.

2. Information Microsystem Institute, Beijing Information Science & Technology University, Beijing, 100101, China.

3. Shenzhen Graduate School, Peking University, Shenzhen, China.

*Email: miaomin@ime.pku.edu.cn, miaomin@bistu.edu.cn

Abstract

This paper focuses on the electrical simulation and analysis of silicon interposer. Basic interconnect elements such as TSV and RDL are simulated and verified with measurement results, electrical parameters are extracted and analyzed. Segmentation is used in long signal path modeling to improve simulation efficiency with a fine accuracy. Silicon interposer is segmented into many interconnect pieces. Parasitic RLCGs of each segment are extracted and interconnected to form the whole circuit model of Si interposer. A Silicon interposer for a 4-SRAM module integration, proposed and implemented as a leading demo for a Logic+Memory high-speed digital signal processing module, was used as a test vehicle for the modeling and analysis methodologies mentioned.

Background

3D integrations with through silicon via (TSV) and Si interposer have attracted substantial attention and been on the way to commercialization. TSV provides vertical interconnection through silicon substrate, which greatly helps shorten interconnect length and improve signal transmission performance. Si interposer has been considering as one of the most important TSV-enabled technologies for its low cost, high yield and great flexibility. Redistribution layers (RDL) on Si interposer help to reroute signal lines to provide chip-to-chip and chip-to-substrate connection and power delivery. It compensates the mismatch between chip I/O pad and TSV landing, while keeping IC layout design unchanged. Multilayered RDL on interposer can meet the needs of high density and large numbers of electrical interconnects. [1-3]

TSV and RDLs form the interconnect network in Si interposer [4, 5]. As design gets larger and routing complexity increases, the impact of wire resistance, capacitance and inductance parasitics becomes significant, which also give rise to a whole set of signal integrity issues. Large-scaled 3D interconnect network also makes 3D electromagnetic field modeling impractical for the electrical analysis, due to the long running time and occupation of many computational resources. Circuit model based analysis is an effective way to estimate the electrical performance of interconnects on Si interposer. In this paper, we focus on the electrical simulation and analysis of silicon interposer. Basic interconnect elements TSV and RDL are simulated using 3D electromagnetic field solver and verified with measurement results. Long signal path on interposer will be segmented into small parts and reconnected in circuit model to improve simulation efficiency. A Silicon interposer design for a 4-SRAM module integration,

which was proposed and implemented as a leading demo for a Logic+Memory high-speed digital signal processing module development, is introduced. The design is segmented into many interconnect pieces. Parasitic RLCGs of each segment are extracted and connected to form the whole circuit model of Si interposer. Some of the measurements are compared with modeling and analysis methodologies presented above.

TSV and RDL Analysis

There are TSVs, RDLs, micro bumps etc. on Si interposer. To start the electrical analysis of interposer, all structures need to be characterized. Electrical model of TSV has been widely studied these years with many circuit models coming forth [6-11], available in different designs and applications. The characteristics of TSV are dependent on its geometrical parameters like diameter, pitch, sidewall liner thickness, and material parameters, such as liner permittivity, and silicon substrate resistivity. Fig.1 shows the transmission performance of ground-signal TSV by 3D electromagnetic filed solver, as TSV diameter (18, 20, 22μm) and pitch (30, 40, 50μm) vary. TSV height is set as 80μm, and SiO$_2$ liner 0.5μm thick. It's shown that, as TSV diameter increases, the loss |S21| increase, probably because the consequently increased capacitive and inductive coupling of TSV. As for TSV pitch, opposite variation tendency shows, as coupling between TSV decreases when TSV pitch increases.

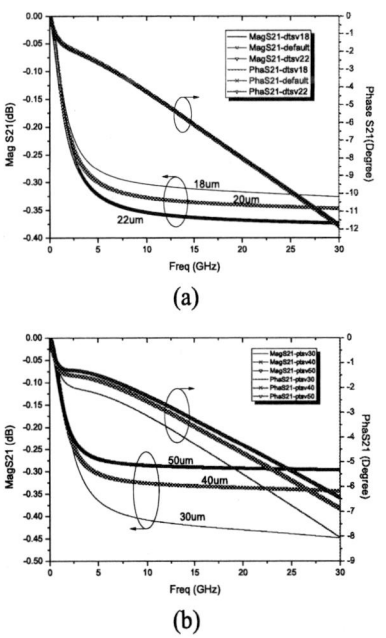

Fig.1 parametric sweep of TSV S21 parameter, (a)TSV diameter, (b) TSV pitch.

Substrate loss is also an important factor in TSV electrical performance. TSV-RDL structure as in Fig.2 is used to simulate the impact of Si conductivity, and to evaluate the performance of TSV-RDL signal path, as depicted in Fig.3. The TSV here has the same configuration as in Fig.1, test samples are fabricated on a Si wafer with a conductivity of ~25S/m. As can be seen from Fig.3, high resistive Si is a good choice for high-quality signal transmission. The measured results fit well with simulation above 8GHz. Lower frequency measurements are still in progress with a better de-embedding solution. Resistance and inductance of TSV can also be extracted based on test structures shown in Fig.4, which shows frequency-dependent features. DC resistance of TSV (80μm deep, 20μm in diameter) is about 30 mΩ, and inductance of 48 pH.

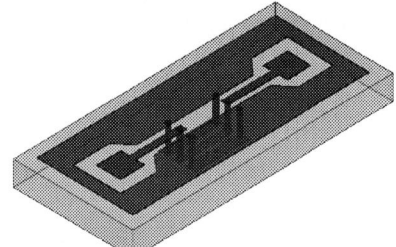

Fig.2 GSG-TSV-RDL structure

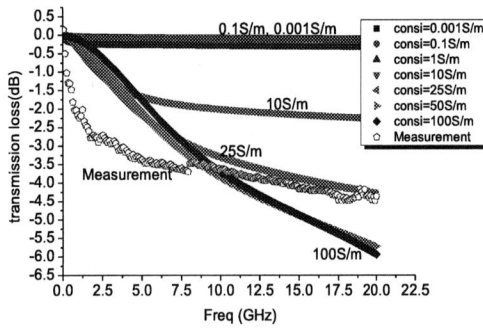

Fig.3 TSV S21 curve, with different Si substrate conductivity, and measurement results.

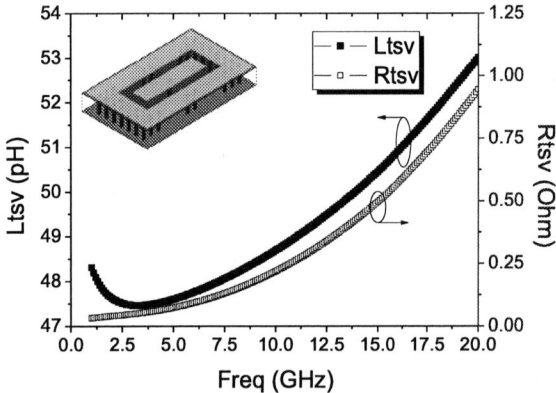

Fig.4 Extracted R/L parameters of TSV derived from measurement results.

RDLs (Redistribution layers) provide horizontal interconnect on interposer and pitch match to different chips or chip-substrate combination. As routing complexity and planar interconnect layers increases, the scale of RDLs also increases and will be of great significance on Si interposer's performance.

Fig.5 shows the copper CPW (coplanar waveguide) simulated and measured for RDL analysis. It's 1mm long, with signal width/ space 30μm/21μm for impedance match. The Si substrate is of low resistivity. The modeling shows a good agreement with the measurement result, and the structure is suitable for broad-range frequency applications. Electrical field distribution of CPW can be captured by 3D EM field simulation, as in Fig.6. Large part of the electrical field is confined between substrate and CPW signal line, where the parasitics mainly come from. Distributed resistance, inductance, capacitance, and conductance (RLCG) parameter model for CPW is derived from S parameters, as shown in Fig.7, which can be used in circuit modeling.

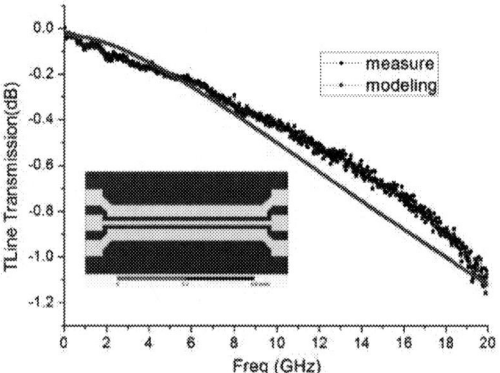

Fig.5 transmission loss of CPW

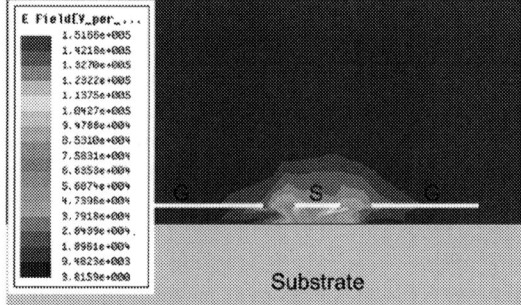

Fig.6 Simulated Electrical field distribution of CPW

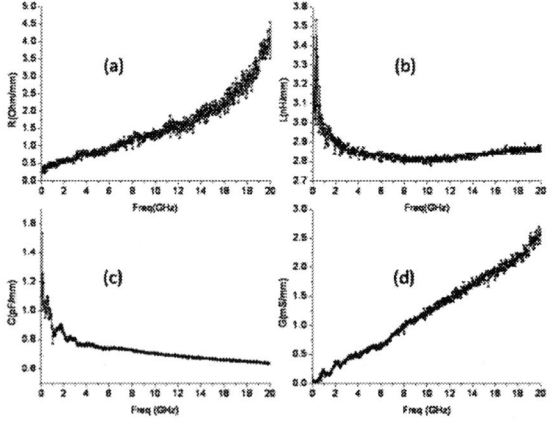

Fig.7 Distributed RLCG parameters of CPW derived from measurement results,(a)Resistance, (b)Inductance, (c)Capacitance,(d)conductance

Si interposer Architecture and Modeling

Taking TSV, RDL, micro bump into account, the physical dimension of interposer ranges from μm to cm. Using 3D EM field solver to simulate the interposer will need large mesh and computational resources. To evaluate the electrical performance of Si interposer, we decompose the interposer interconnects into small parts, and simulate each part, and recombine them in circuit model. [12-17]

Fig.8 is a schematic of interconnect path on Si interposer. It's segmented into three part, board CPW, TSV & microbump, and RDL on interposer. Each part is simulated in ANSYS HFSS, and SPICE netlist is exported to perform circuit model simulation by ADS, as shown in Fig.9. From Fig.10(a), we can see that, long RDL on interposer dominates the electrical performance of the signal path. Comparison of 3D EM field simulation and circuit model segmentation is depicted in Fig.10(b), which proves effective. Improvements are still needed to increase the accuracy of simulation results.

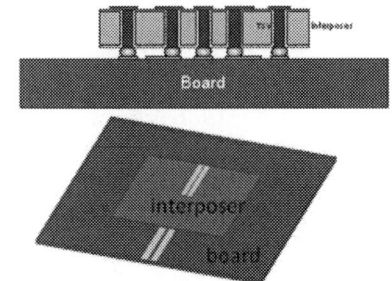

Fig.8 Schematic of Signal path on Si interposer

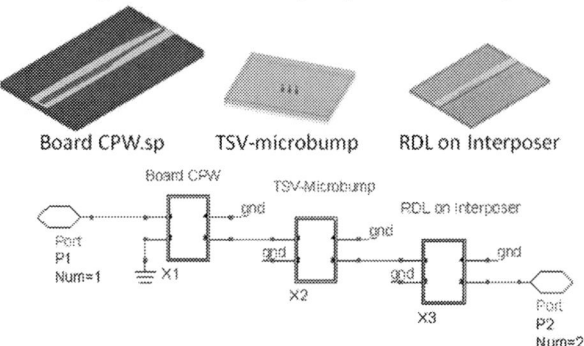

Fig.9 cascaded circuit model of signal path on Si interposer

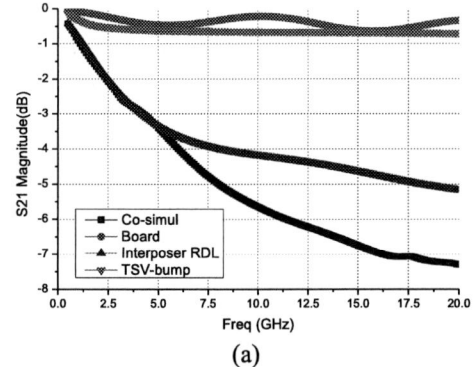

(a)

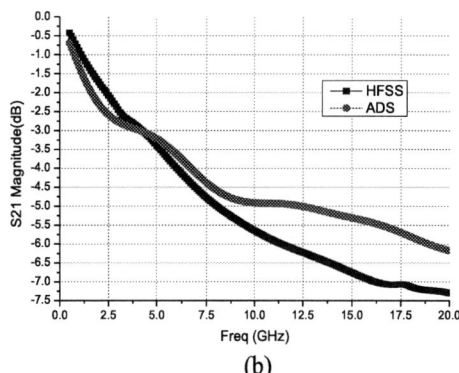

(b)

Fig.10 (a)Simulated S21 of Board CPW,TSV-microbump, and interposer RDL, (b)simulation comparison of 3D field solver and SPICE modeling

In an actual interposer design, complicated interconnect is as shown in Fig.11. The Si interposer has a footprint of 17mm ×17mm×0.2mm. Four chips can be placed on the front side with TSVs arranged on the peripheral edges. In the design, 4 SRAM chips are arranged on the Si interposer to extend SRAM capacity and provide a chip module. There are two metal layers on the top and one on the bottom, as shown in Fig.12. With proper structural parameters, the layout is directly translated into 3D pattern for a better understanding, as shown in Fig.13.

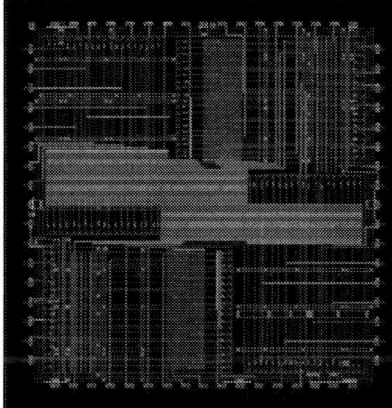

Fig.11 Schematic of interconnect on Si interposer

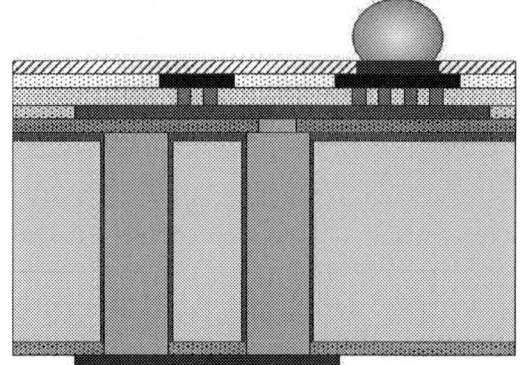

Fig.12 Cross-section of Si interposer

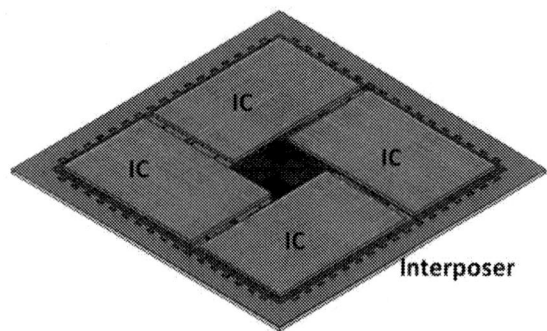

Fig.13 3D model of Si interposer

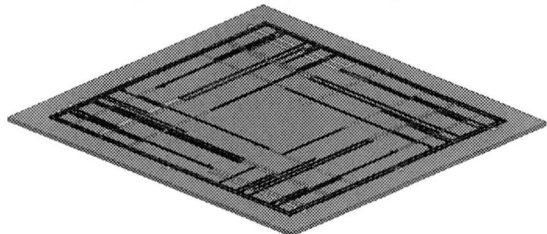

Fig.14 Power delivery network of Si interposer

The power delivery network of this Si interposer sample is extracted, as shown in Fig.14. Parasitic capacitance can be calculated between power and ground line, which is about 30pF in total, the coupling between power/ground and substrate is even large, up to 170pF. Interconnects will be segmented into different pieces, such as TSV, the first RDL topRDL0, the second RDL topRDL1, overlap between topRDL0 and topRDL1, etc. Detailed results will be presented at the conference, when the final Logic + Memory sample will come up.

Conclusions

In this study, TSV and RDL are simulated and analyzed. Segmentation is used in long signal path simulation to improve efficiency. A Silicon interposer design for a 4-SRAM module integration, which was proposed and implemented as a leading demo for a Logic+Memory high-speed digital signal processing module development, is introduced. The design is segmented into many interconnect pieces. Parasitic RLCGs of each segment are extracted and connected to form the whole circuit model of Si interposer. Some of the measurements are compared with modeling and analysis methodologies presented above. More detailed results will be presented at the conference.

Acknowledgments

The work presented is co-funded by National Science and Technology Major Project of China (Project No. 2009ZX02038), National Natural Science Foundation of China (Project No. 61176102 and 60976083), Funding Project for Academic Human Resources Development in Institutions of Higher Learning Under the Jurisdiction of Beijing Municipality (Project No. PHR201108257, "Research on technical basics for three-dimensional system-in-package").

References

1. N. Vodrahalli,"Silicon TSV interposers with embedded capacitors for high performance VLSI packaging", IEEE CPMT Symposium Japan, 2010, pp. 1-4.
2. E. J. Vardaman,"3D IC infrastructure status and issues", Electronic System-Integration Technology Conference(ESTC), 2010, pp. 1-6.
3. J. U. Knickerbocker, P. S. Andry, E. Colgan, B. Dang, T. Dickson, X. Gu, C. Haymes, C. Jahnes, Y. Liu, J. Maria, R. J. Polastre, C. K. Tsang, L. Turlapati, B. C. Webb, L. Wiggins, S. L. Wright,"2.5D and 3D Technology Challenges and Test Vehicle Demonstrations", ECTC, 2012, pp. 1068-1076.
4. Namhoon Kim, Daniel Wu, Jack Carrel, Joong-Ho Kim, Paul Wu,"Channel Design Methodology for 28Gb/s SerDes FPGA Applications with Stacked Silicon Interconnect Technology", ECTC, 2012, pp. 1786-1794.
5. Kim Namhoon, D. Wu, Kim Dongwook, A. Rahman, P. Wu,"Interposer design optimization for high frequency signal transmission in passive and active interposer using through silicon via (TSV)", Electronic Components and Technology Conference(ECTC), 2011, pp. 1160-1167.
6. Liang Yuanjun, Li Ye,"Closed-Form Expressions for the Resistance and the Inductance of Different Profiles of Through-Silicon Vias", IEEE Electron Device Letters, Vol. 32, No. 3, pp.393-395, 2011.
7. I. Savidis, E. G. Friedman,"Closed-Form Expressions of 3-D Via Resistance, Inductance, and Capacitance", IEEE Transactions on Electron Devices, Vol. 56, No. 9, pp.1873-1881, 2009.
8. Xu Chuan, Li Hong, R. Suaya, K. Banerjee,"Compact AC Modeling and Performance Analysis of Through-Silicon Vias in 3-D ICs", IEEE Transactions on Electron Devices, Vol. 57, No. 12, pp.3405-3417, 2010.
9. K. Salah, A. El Rouby, H. Ragai, K. Amin, Y. Ismail,"Compact lumped element model for TSV in 3D-ICs", IEEE International Symposium on Circuits and Systems(ISCAS), 2011, pp. 2321-2324.
10. R. Weerasekera, M. Grange, D. Pamunuwa, H. Tenhunen, Zheng Li-Rong,"Compact modelling of Through-Silicon Vias (TSVs) in three-dimensional (3-D) integrated circuits", IEEE International Conference on 3D System Integration(3DIC), 2009, pp. 1-8.
11. Yu Le, Yang Haigang, T. T. Jing, Xu Min, R. Geer, Wang Wei,"Electrical characterization of RF TSV for 3D multi-core and heterogeneous ICs", IEEE/ACM International Conference on Computer-Aided Design (ICCAD), 2010, pp. 686-693.
12. G. Charles, P. D. Franzon, Kim Jaemin, A. Levin,"Analysis and approach of TSV-based hierarchical power distribution networks for estimating 1st-Droop and resonant noise in 3DIC", Electrical Performance of Electronic Packaging and Systems (EPEPS), 2011, pp. 267-270.
13. Kim Kiyeong, Lee Woojin, Kim Jaemin, Song Taigon, Kim Joohee, So Pak Jun, Kim Joungho, Lee Hyungdong, Kwon Yongkee, Park Kunwoo,"Analysis of power distribution network in TSV-based 3D-IC", Electrical

Performance of Electronic Packaging and Systems(EPEPS), 2010, pp. 177-180.

14. Kim Jaemin, Lee Woojin, Shim Yujeong, Shim Jongjoo, Kim Kiyeong, So Pak Jun, Kim Joungho,"Chip-Package Hierarchical Power Distribution Network Modeling and Analysis Based on a Segmentation Method", IEEE Transactions on Advanced Packaging, Vol. 33, No. 3, pp.647-659, 2010.

15. Kim Kiyeong, Min Yook Jong, Kim Junchul, Kim Heegon, Lee Junho, Park Kunwoo, Kim Joungho,"Interposer Power Distribution Network (PDN) Modeling Using a Segmentation Method for 3-D ICs With TSVs", IEEE Transactions on Components, Packaging and Manufacturing Technology, Vol. 3, No. 11, pp.1891-1906, 2013.

16. Zheng Xu, Xiaoxiong Gu, Michael Scheuermann, Kenneth Rose, Buckwell C. Webb, John U. Knickerbocker, Jian-Qiang Lu,"Modeling of Power Delivery into 3D Chips on Silicon Interposer", Electronic Components and Technology Conference(ECTC), 2012, pp. 683-689.

17. Zheng Xu, Xiaoxiong Gu, Michael Scheuermann, Kenneth Rose, Buckwell C. Webb, John U. Knickerbocker, Jian-Qiang Lu,"Multiple Voltage-Supplies in TSV-Based Three-Dimensional (3D) Power Distribution Networks", Electronic Components and Technology Conference(ECTC), 2012, pp. 1819-1825.

A SPICE Model of Multi-Mode Optical Fiber in Mid-Channel Link for Package System SI Transient Simulations

Zhaoqing Chen
IBM Corporation
2455 South Rd, B002, Poughkeepsie, NY 12601
zhaoqing@us.ibm.com, (845)435-5595

Abstract

In high-performance computer system design, optical links are applied to many system channels to replace traditional copper cables as mid-channel links. Multi-mode optical fiber is an important component in the mid-channel link. An accurate multi-mode fiber SPICE model is a key issue in the system signal integrity transient simulation as well as in generating IBIS-AMI optical link model. In this paper, a multi-mode fiber SPICE modeling method is proposed for graded-index core multi-mode fibers to include the signal integrity effects by modal dispersion, chromatic dispersion, attenuation, and even the total time flight. The proposed model can be used together with the VCSEL, photodiode, and trans-impedance amplifier SPICE models to simulate the mid-channel optical link using a general-purpose SPICE or SPICE-like simulator which supports the I/O device transistor level models and other interconnect, packaging component SPICE models. Test cases on the 25Gb/s system eye diagram signal integrity simulation are shown using a graded-index multi-mode fiber.

1. Introduction

In high-performance computer system design, optical links are applied more and more widely in many system channels for replacing traditional copper cables for mid-channel links [1] as the system bit rate increases and the cost of optical components decreases. The multi-mode fiber (MMF) with Vertical Cavity Surface Emitting Laser (VCSEL), photodiode, and trans-impedance amplifier (TIA) is one of most commonly accepted low-price optical solutions for the high-performance computer design.

The MMF has been applied to the multi-gigabit network for many years [2]-[4]. Accurate modeling and simulation of the MMF is a very important issue in the optical mid-channel signal integrity (SI) design and verification. Several literatures have been published [3][5][6] in the MMF modeling. Recently, the IBIS-AMI model have been extended to support the mid-channel MMF links [7][8]. However, the IBIS-AMI models usually do not support general transient simulations and are not suitable for the optical component parameter tolerance and sensitivity study. In addition, the IBIS-AMI needs accurate MMF model for the transient simulation in the modeling procedure. However, most of the published MMF modeling methods and corresponding models depend on either mixed tool simulation (like mixed SPICE and MATLAB simulations) or special simulation tools like Agilent ADS[7] or OptiSPICE[9][10]. Using existing general-purpose SPICE and SPICE-like simulation tool has a very high priority for the high-performance computer system designers since the existing packaging component models and

sometimes I/O device transistor level models need to be supported. In addition, the mid-channel optical link is simpler compared to most other optical systems, so we just need to include very limited number of the optical components like MMF, VCSEL with direct intensity modulation, photodiode, etc. The modeling effects in this approach will be smaller than other approaches. In this paper, an MMF SPICE model is developed to support any general-purpose SPICE or SPICE-like transient simulation tools.

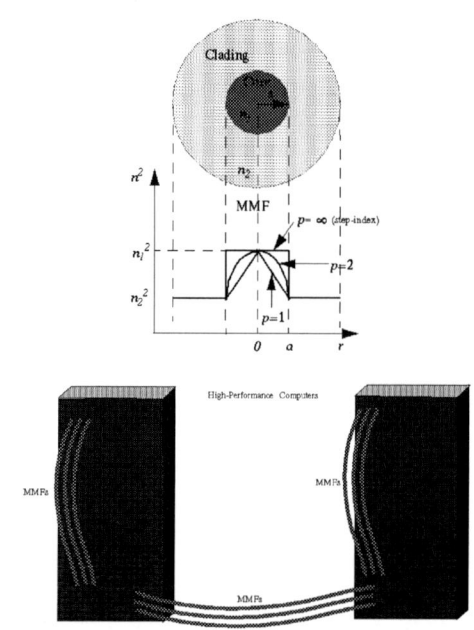

Fig.1 Multi-mode fiber (MMF) and its application in the high-performance computer system as mid-channel links

2. Methodology

The major difficulty to include the mid-channel optic link in a SPICE simulation is that the regular signal voltage and current do not exist in the optical part. In the optical part, the electromagnetic waves are laser light waves. The original high-speed electrical signal becomes an envelop of the directly intensity-modulated lightwave.

To simplify the procedure in this paper, the modulated light power is represented by the node voltage in the SPICE model and simulation. In this case, the node voltage represents high-speed signals in the measure of the light power in the optical part. Meanwhile, in the electrical part, the node voltage is just as usual (regular electrical signal). In this paper, all reflected light waves are absorbed and affect the attenuation or transform coefficient only. They will not travel back towards the laser source. In this approach, ideal transmission lines with all terminations equal to the characteristic impedance Z_0 are just the behavioral models to represent the unidirectional propagation of the modulated signal in the optical part. In other words, in the case the lossless fiber and long time period all modulated light energy generated at the laser source like a VCSEL optical output will arrive at the photodiode input. However, because of the dispersion effect there is no one-to-one signal time mapping between two ends of the optical fiber.

A. MMF Chromatic Dispersion and Attenuation SPICE Model

The chromatic dispersion exists in both single-mode fiber and multi-mode fiber and can be expressed as [6],[11]-[13]

$$S_{21} = Ae^{j(\beta_2 L\omega^2/2 - \beta_1\omega L_1)} \tag{1}$$

$$S_{11} = S_{22} = 0 \tag{2}$$

$$S_{12} = 0 \text{ (unidirectional case)} \tag{3}$$

$$\beta_2 = -D(\lambda)\,\lambda_0^2/(2\pi c) \tag{4}$$

$$\beta_1 = 1/c \tag{5}$$

$$c = c_0/n_1 \tag{6}$$

$$D(\lambda) = S_0(\lambda - \lambda_0^4/\lambda^3)/4 \tag{7}$$

$$A = 10^{-Att\,L/10.0} \tag{8}$$

Where L is the MMF length, L_1 is a short compensate length if $D(\lambda) < 0$, Att is the fiber attenuation in dB/m, λ_0 is the fiber zero dispersion wavelength, λ is the laser operating wavelength, S_0 is the MMF zero dispersion slope, , c_0 is speed of the light in vacuum, the definition of n_1 is shown in Fig.1.

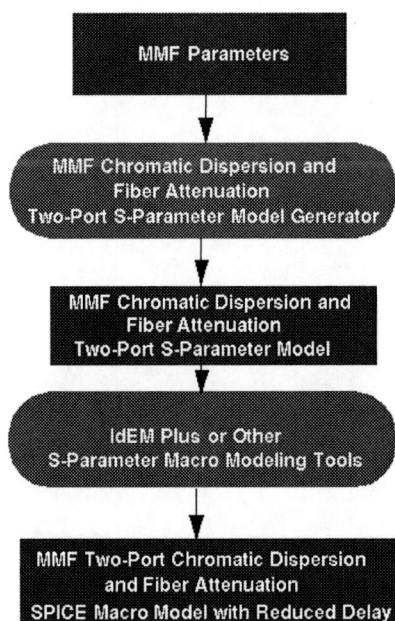

Fig.2 Flow Chart of the SPICE Modeling of the MMF Chromatic Dispersion and Fiber Attenuation

Unlike the SPICE modeling procedures in [11] and [13], a simpler approach is proposed here by using a S-parameter modeling tool, IdEM Plus [14], to export the SPICE circuit model after the curve-fitting for the chromatic dispersion and attenuation of the fiber though the unidirectional two-port S-parameter model. Fig.2 shows the flow chart of the methodology. In this approach, the chromatic dispersion associated with β_2 is counted for the whole fiber length L while the signal group delay associated with β_1 is eliminated except for a very short length L_1 in the case of a negative $D(\lambda)$. The original time flight of the fiber can be recovered later as shown in Fig.7 just in case some applications need it.

B. MMF Modal Dispersion SPICE Model

Unlike the single mode optical fiber (SMF) in which only one mode propagates, there are hundreds even thousands modes propagating in the MMF [2]-[6]. The time difference at the MMF output between the propagating modes causes the modal dispersion which results in signal integrity problems and limits the length of the MMF to a range much shorter than the SMF. Fortunately, the working length of the MMF (1-1000m) still satisfies the requirement of the mid-channel link in most high-performance computer systems if using state-of-the-art MMF technologies like the optimal graded-index fiber [15] instead of the step-index fiber.

The graded-index MMF has much smaller modal dispersion than the step-index fiber. The equations to calculate the MMF dispersion parameters are listed as the following Equations (9)-(16), most of them are from References [5] and [16].

$$n^2(r) = n_1^2(1 - 2(r/a)^p\,\Delta) \qquad r \leqslant a \tag{9}$$

$$\Delta = (n_1^2 - n_2^2)/(2n_1^2) \tag{10}$$

$$NA = (n_1^2 - n_2^2)^{1/2} \tag{11}$$

$$V = 2\pi(a/\lambda)NA \tag{12}$$

$$\tau_q = L/v_q \tag{13}$$

$$N_g = 2\pi n_1 a(p\Delta/(p+2))^{1/2}/\lambda \tag{14}$$

for $p \neq 2$

$$v_q = (c_0/n_1)(1 - (p-2)(q/N_g)^{2p/(p+2)}\Delta/(p+2)) \quad q = 1,2,...,N_g \tag{15}$$

for $p = 2$

$$v_q = (c_0/n_1)(1 - q^2\Delta^2/(2N_g^2)) \qquad q = 1,2,...,N_g \tag{16}$$

In Equations (9)-(16), $n(r)$ is the fractional reflective-index profile, p is the graded profile parameter, Δ is the fractional-index change, NA is the numerical aperture of the fiber, V is the fiber parameter or V parameter, τ_q is the group delay (distinguish from the phase delay) of every mode in Mode Group q, v_q is the group velocity (distinguish from the phase velocity) of every mode in Mode Group q, L is the fiber length, N_g is the total number of modes, c_0 is speed of the light in vacuum, the definitions of n_1, n_2, and a are shown in Fig.1, λ is the laser operating wavelength.

Based on the mode groups, the MMF SPICE model can be generated as described in Fig.3. In Fig.3, the parameter Z_0 is used as a constant impedance for all ideal transmission lines and terminators in the optical part of the model to guarantee a non-reflection optical system. To make it easier, we can just use $Z_0 = 50\Omega$. This is just a behavioral model to represent the light power propagation in the MMF. The time flight of each ideal transmission line τ_q $(q=1,2,...N_g)$ in Fig.2 can be replaced by a new value $(\tau_q - \tau_{min} + \tau_0)$ where τ_{min} is the minimum time flight of all N_g groups, τ_0, is a short time flight to avoid a zero length ideal transmission line and keep the SPICE simulation stable. In Fig.3, the principle mode index is defined as $q = l + 2m - 1$, l and m are the indexes of Mode LP_{lm} where LP stands for Linear Polarization.

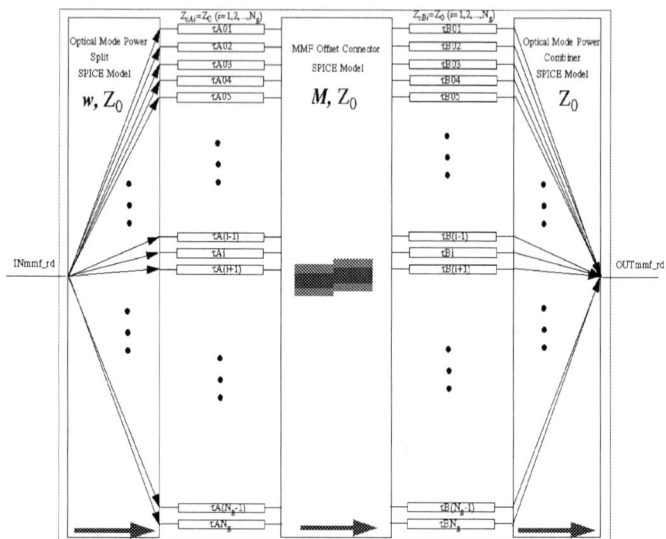

Fig.4 MMF SPICE model including an offset fiber connector in the middle of the fiber

The split, combiner, and connector models in Figs. 3 and 4 are all SPICE models consisting of controlled sources and terminators. For some advance SPICE simulators, the math equations can be used in the controlled source parameters. Figs. 5-7 show the SPICE circuit schematic of those models. For those SPICE simulators not supporting the math equations in the controlled source parameters, multiple single controlled sources can be used instead. To save the computation efforts, using the Norton equivalent by the current sources will be better.

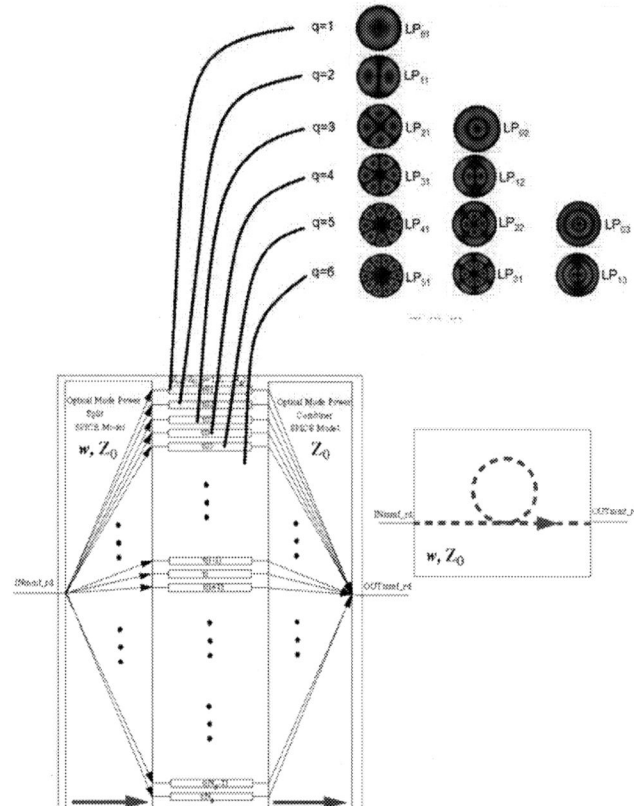

Fig.3 the MMF SPICE model in which each ideal transmission line represents a principle mode group, Mode Group q which may include several modes (the mode distribution color maps are derived from a step-index case and for illustration purpose only)

In the SPICE model, the parameter w represents the mode power distribution determined by the laser-to-MMF coupling and, in some cases, higher order mode filtering device at the beginning of the MMF [17][18].

In the case there is a MMF offset connector, mode transferring will occur at the offset transition described by a mode transfer matrix M as in Reference [3]. An MMF offset connector SPICE model will be inserted into the connector location as shown in Fig.4.

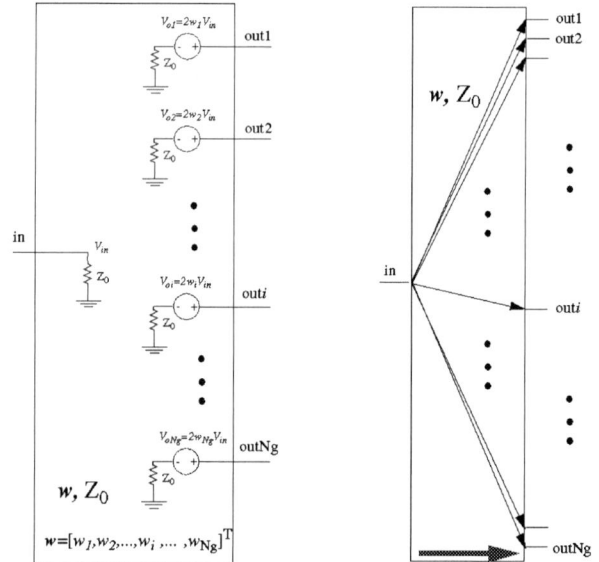

Fig.5 Schematic and symbol of the mode split SPICE model

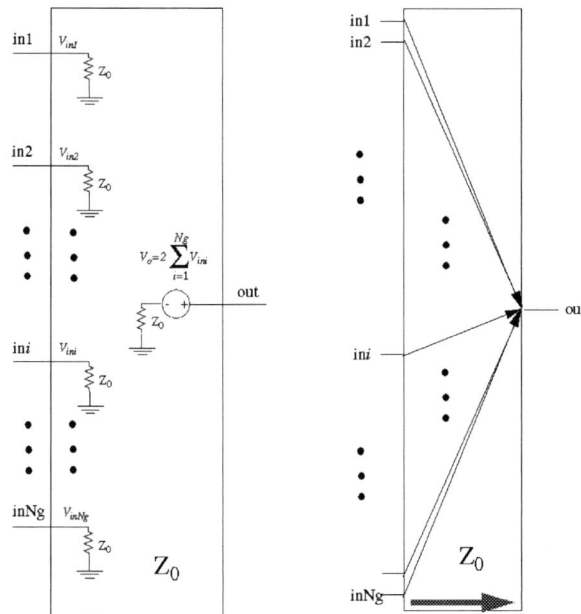

Fig.6 Schematic and symbol of the mode combiner SPICE model

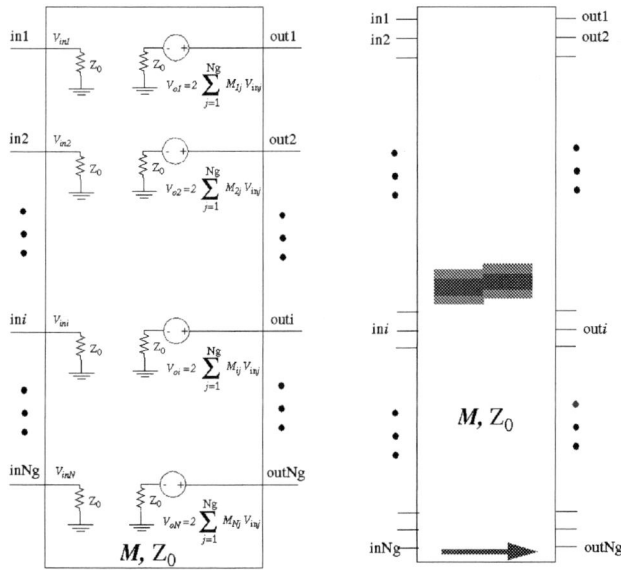

Fig.7 Schematic and symbol of the MMF offset connector SPICE model

C. Complete MMF SPICE Model Including Chromatic and Modal Dispersions, Attenuation, and the Full Fiber Delay

The models shown in Figs. 3 and 4 are on MMF modal dispersion only. The chromatic dispersion, fiber loss, and fiber delay are not included. It is easy to combine the MMF chromatic and attenuation SPICE model with the modal dispersion model by simply cascading the two models since they have the same Z_0. In addition, an ideal transmission line model with full MMF delay less the small ones in the above two models can be cascaded to the complete MMF SPICE

model in case the user needs the full fiber delay. The fiber delay should be calculated by the signal first order group velocity which should very close the lightwave phase velocity. The complete MMF SPICE model is shown in Fig. 8. The corresponding flaw chart is shown in Fig.9.

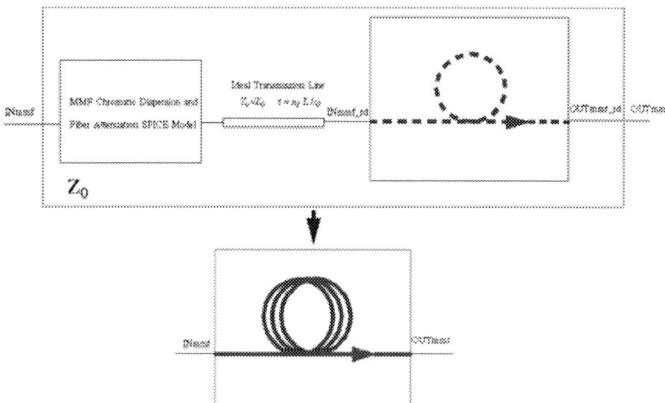

Fig.8 The MFF Complete SPICE model including chromatic and modal dispersions, fiber attenuation, and full time delay

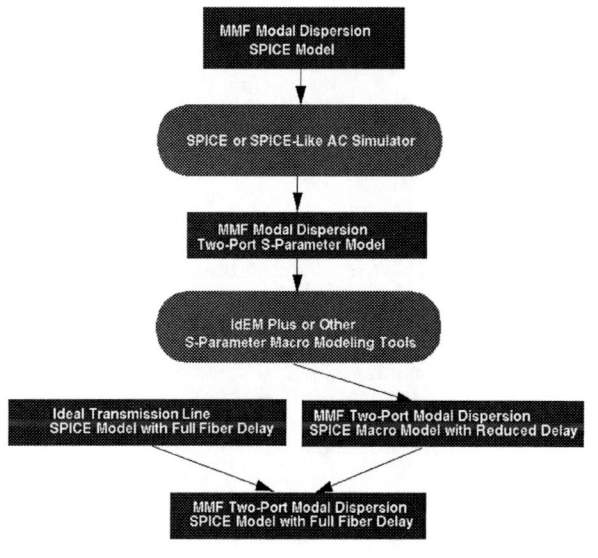

Fig.9 Flaw Char for the MMF complete SPICE modeling

D. The VCSEL, Photodiode, and Trans-Impedance Amplifier SPICE Models

The VCSEL SPICE model based on the rate equations [7][8],[19]-[21] is used in this paper for testing the MMF SPICE model. Equations (17)-(19) are the VCSEL Rate Equations. Using two arbitrary constants z_n and δ (in test cases of this paper, $z_n=10^8$, $\delta=0.01$) to help the convergence of the circuit simulations, the transformation equations are Equations (20)-(22) for linking the VCSEL Rate Equations to the SPICE circuit model.

978-1-4799-2408-0/14 $31.00 © 2014 IEEE

$$\frac{dN}{dt} = \frac{\eta_i(I - I_{off}(T))}{q} - \frac{N}{\tau_n} - \frac{G_0(N-N_0)S}{1+\varepsilon S} \qquad (17)$$

$$\frac{dS}{dt} = -\frac{S}{\tau_p} + \frac{\beta N}{\tau_n} + \frac{G_0(N-N_0)S}{1+\varepsilon S} \qquad (18)$$

$$T = T_0 + (IV - P_0)R_{th} - \tau_{th}\frac{dT}{dt} \qquad (19)$$

$$N = z_n v_n \qquad (20)$$

$$S = (v_m + \delta)/k \qquad (21)$$

$$P_o = kS = v_m + \delta \qquad (22)$$

In Equations (17)-(22)

$R_n = \eta_i \tau_n /(qz_n)$

$C_n = qz_n/\eta_i$

$I_{stn} = qG_0(z_n v_n - N_0)(v_m + \delta)/(\eta_i(k + \varepsilon(v_m + \delta)))$

$I_{off}(T) = a_0 + a_1 T + a_2 T^2 + a_3 T^3 + a_4 T^4 + ...$

$C_{th} = \tau_{th}/R_{th}$

$I_{th} = T_0/R_{th} = (I_{tot}V - P_o)$

$C_{ph} = \tau_p/R_{ph}$

$I_{sp} = \tau_p \beta k z_n v_n /(\tau_n(v_m + \delta))$

$I_{stm} = G_0\tau_p(z_n v_n - N_0)(v_m + \delta)/(1 + \varepsilon(v_m + \delta)/k)$

S is the photon number,
N is the carrier number,
T is the transient temperature,
τ_{th} is a thermal time constant,
R_{th} is the VCSEL's thermal impedance,
T_0 is the ambient temperature,
V is the VCSEL input voltage,
I is the VCSEL input current excluding the parasitic leakage,
η_i is the injection efficiency,
τ_n is the carrier recombination life time,
G_0 is the gain coefficient,
N_o is the carrier transparency number,
τ_p is the photon lifetime,
β is the spontaneous emission coupling coefficient,
ε is the gain-compression factor, and
k is a scaling factor accounting for the output coupling efficiency od the VCSEL.

The VCSEL SPICE circuit model is shown in Fig.10 where the circuit block with Node *n* representing Equation (17), the circuit block with Node *m* represents Equation (18), and the circuit block with Node *td* represents Equation (19). The controlled voltage source E_{p0} represents the output optical power with a value of the $2P_0$ because the terminated ideal transmission line is used to represent the non-reflection light power transmission. By the factor of 2 to the light power we can exactly deliver P_0 to the photodiode if the fiber attenuation and dispersion are zero.

The photodiode small signal model [6] is used because the small signal assumption is usually satisfied at this stage (Fig.11). The rate equation based photodiode model [22] can also be used to capture more state-of-the-art properties.

In the VCSEL and photodiode SPICE models, the termination Z_0 has been added to match the non-reflection termination condition of the MMF SPICE model.

The trams-impedance amplifier small signal model [12] is used for simplicity (Fig.12). In the test cases of this paper, the VCSEL driver [23] before VCSEL and the electrical driver after the trans-impedance amplifier, and any filter devices like FFE, voltage level limiter, CTLE, and DEF, etc., are not included. In most cases, they are based on the transistor level models depending on each semiconductor manufacturer's technology. In case they are available, they can be included in the system signal integrity simulations.

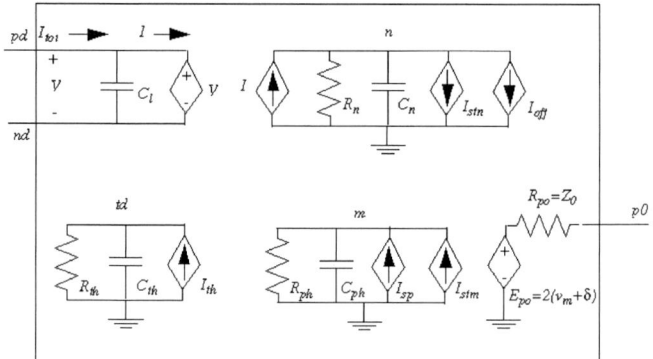

Fig.10 VCSEL SPICE model based on rate equations

The VCSEL SPICE circuit model parameters for the test case in this paper are listed as the following

$V = 1.721 + 275.000I - 2.439\times10^4 I^2 + 1.338\times10^6 I^3$
$\quad -4.154\times10^7 I^4 + 6.683\times10^8 I^5 - 4.296\times10^9 I^6$

$\eta_i = 1$, $\beta = 10^{-6}$, $\tau_n = 5$ns, $k = 2.6\times10^{-8}$W, $G_0 = 1.6\times10^4$s^{-1}, $N_0 = 1.94\times10^7$
$\tau_p = 2.28$ps, $R_{th} = 2.6 \times 10^3$ Ω (in place of K/W), $\tau_{th} = 1\mu$s, $R_{ph} = 1\Omega$,
$a_0 = 1.246\times10^{-3}$A, $a_1 = -2.545\times10^{-5}$A/K, $a_2 = 2.908\times10^{-7}$A/K^2,
$a_3 = -2.531\times10^{-10}$A/K^3, $a_4 = 1.022\times10^{-12}$A/K^4, $C_l = 0.1$fF

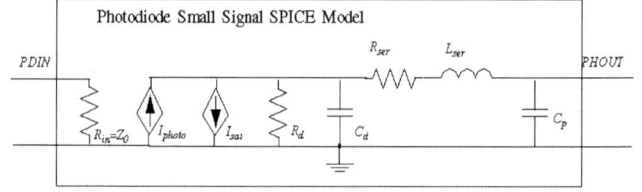

Photo Current:	$I_{photo} = q\eta_{pd}V_{pdin}/(hv)$
Electronic Charge:	$q = 1.6021\times10^{-19}$C
Quantum Efficiency:	$\eta_{pd} = 0.8$
Plank's Constant:	$h = 6.6261\times10^{-34}$ J s
Optical Frequency:	$v = 3.53\times10^{14}$ Hz
Reverse Satuation Current:	$I_{sat} = 0.5$nA
Dark Resistance:	$R_d = 5$MΩ
Junction Capacitance:	$C_d = 17.57$fF (175.7fF)
Series Diode Resistance:	$R_{ser} = 1.06\Omega$ (10.6Ω)
Series Diode Inductance:	$L_{ser} = 0.004$nH (0.04nH)
Parasitic Capacitance:	$C_p = 4.06$fF (40.6fF)

Fig.11 Photodiode small signal SPICE model including interconnect parasitic elements, in the parentheses are the original parameters [6] for the10Gb/s application.

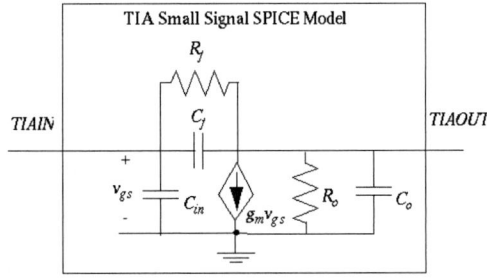

R_f=50.0Ω, C_f=0.02pF, g_m=0.0035 S, R_o=1kΩ, C_o=10fF

Fig.12 Trans-impedance amplifier small signal amplifier SPICE model

3. Test Case Modeling and Simulations

For test purpose, an MMF with the core diameter of 50mm is used. Other MMF parameters include n_1=1.562, n_2=1.540, operating wavelength λ=850nm, zero chromatic dispersion wavelength λ_0=1310nm, S_0=0.101ps/(nm^2.km), fiber attenuation Att=2.4dB/km=0.0024dB/m.

Equations (1)-(8) are applied for deriving the S-parameter model with the fiber length L=30m for calculation the dispersion and a shorter group delay length L_1=5mm for adding a small group delay to compensate the negative chromatic dispersion value $D(\lambda)$<0 to make IdEM Plus [14] curve fitting work. The short group delay by L_1 is not needed if the chromatic dispersion factor $D(\lambda)$>0. One of the advantages of including the whole fiber attenuation in the chromatic dispersion model is to avoid a potential passivity violation introduced by the curve-fitting. As we can see from Fig. 13, the ripple in the S_{21} magnitude curve-fitting may not cause $|S_{21}|$ >1 if some amount of fiber attenuation is included.

The SPICE model generated by the proposed method can be used in transient simulations by any SPICE or SPICE-like circuit simulation tools. In test cases of this paper, PowerSPICE, an IBM version of SPICE-like circuit simulation tool, is used. Fig.14 shows the single-bit response by 25Gb/s signal. Actually, the effect of chromatic dispersion by this 30m fiber is very small and can be negligible in some cases.

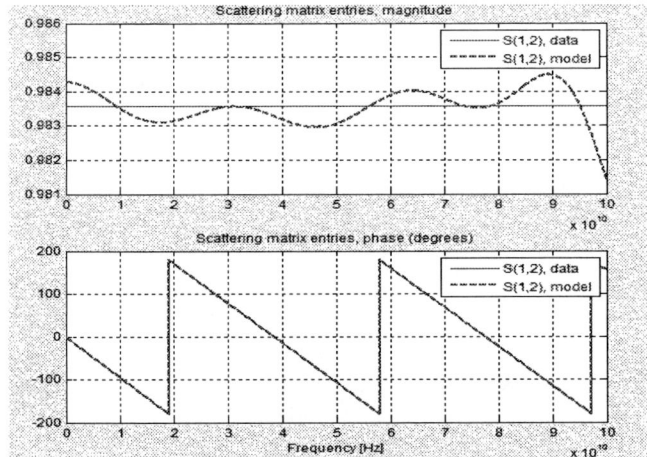

Fig.13 Curve-fitting S_{21} of the 30m MMF chromatic dispersion and fiber attenuation model by IdEM Plus compared with the input data by Equations (1)-(3)

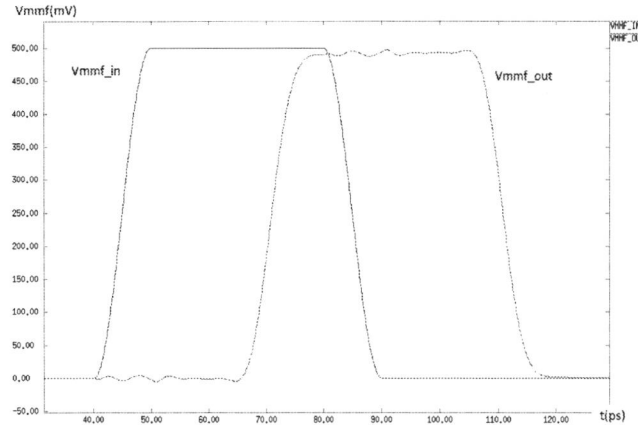

Fig.14 Single-bit input and output voltage (representing the light power) waveforms by chromatic dispersion and fiber attenuation model at ends of a 30m MMF.

The SPICE model for the modal dispersion of MMF can be obtained by the proposed method by assuming the mode power distribution as shown in Fig.15 which is similar to a figure in Reference [17]. In Fig.15 the mode powers are normalized so that $w_1+w_2+...+w_{Ng}=1$

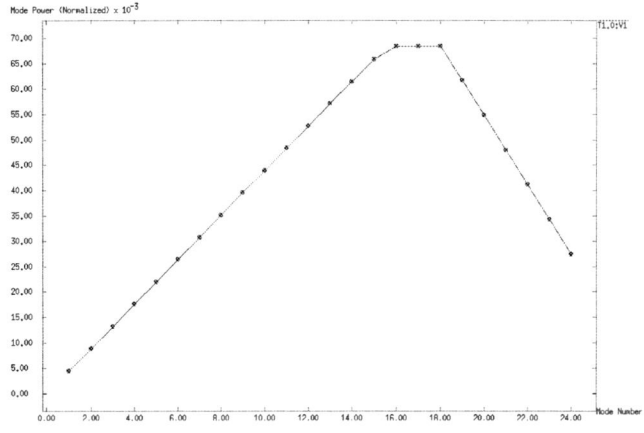

Fig.15 Mode power distribution used in the test case

The magnitudes of S_{21} up to 80GHz for four different fiber core graded-indexes p=1.9, 2.0, 2.04, and 2.1 are shown in Fig.16. Please note in this paper, the MMF mode analysis is based on the simplified equations 9-16 for illustration of the SPICE model. In real MMF products the core dielectric distribution may be more complex than the ideal graded-index exponential distribution. The 90m MMF 25Gb/s single-bit output responses for the four different p values are shown in Fig. 17.

978-1-4799-2408-0/14 $31.00 © 2014 IEEE   2109

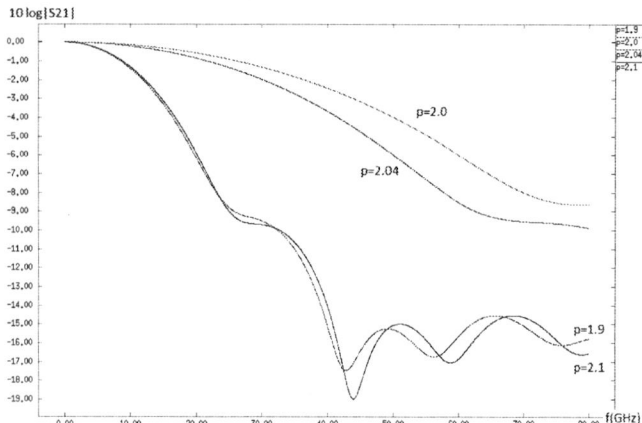

Fig.16 $|S_{21}|$ of a 30m graded-index MMF with 3 different p-values p=1.9, 2.0 (optimal index), 2.04, and 2.1

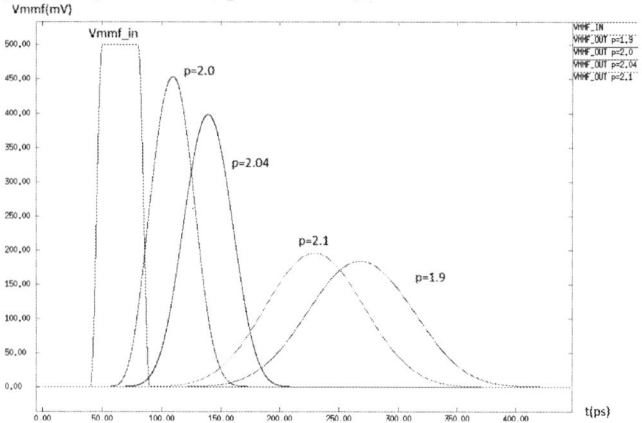

Fig.17 Single-bit impulse response of a 90m graded-index MMF in 25Gb/s application with 4 different p-values p=1.9, 2.0 (optimal index), 2.04, and 2.1

A 25Gb/s mid-channel optical link eye diagram by transient simulation is illustrated in Fig.18 by using the proposed MMF SPICE model for a fiber length of 180m, the modified VCSEL model, photodiode model, and trans-impedance amplifier model.

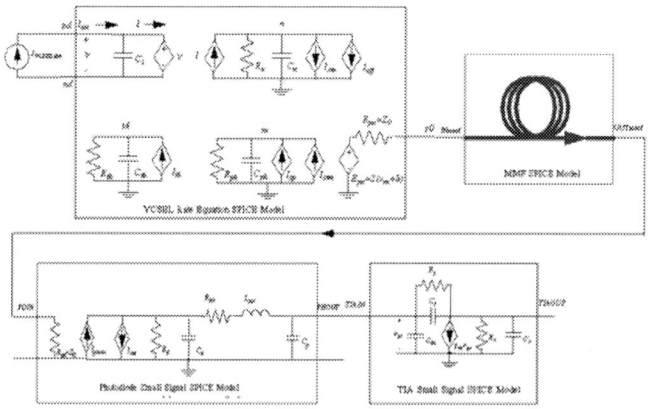

Fig.18 An optical link schematic from the VCSEL input to the TIA output for the eye diagram transient simulation

The SPICE simulated eye diagram by using the original photodiode SPICE model [6] for 10Gb/s applications is not so good in this 25Gb/s case as shown in Fig.19 because of the

larger parasitic interconnect parameters in the original photodiode SPICE model which was for 10Gb/s applications. After reducing the parasitic parameters at the photodiode output part to smaller values, we get much better system eye diagram as shown in Fig.20. This test demonstrates the ability and convenience of the proposed modeling and simulation method in testing the sensitivity of individual component in the optical link by the SPICE or SPICE-like simulation tools.

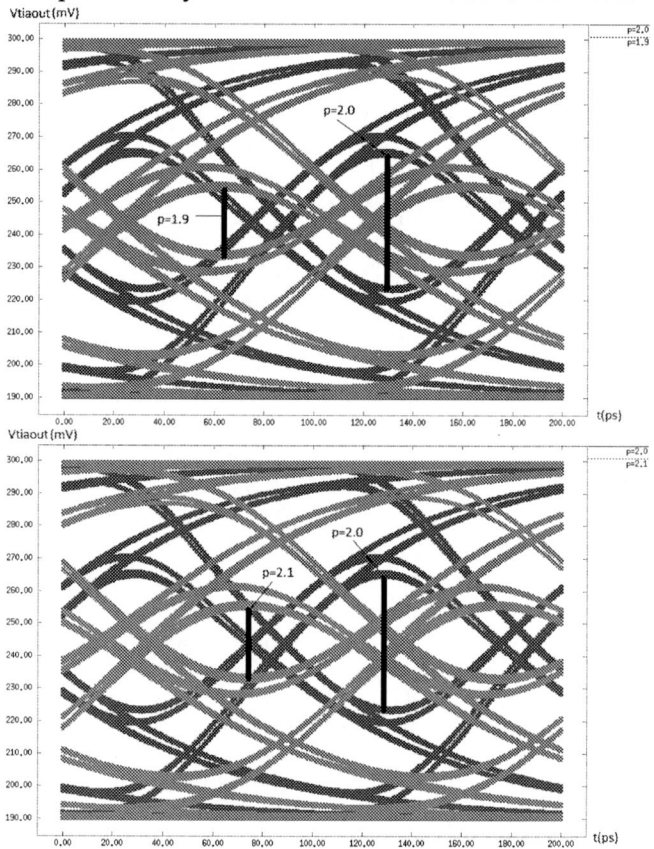

Fig.19 The 25Gb/s eye diagrams at the TIA output of the 180m graded-index MMF link with 3 different values of fiber core graded-index parameter p (p=1.9, 2.0, and 2.1) and larger photodiode package parasitic parameters.

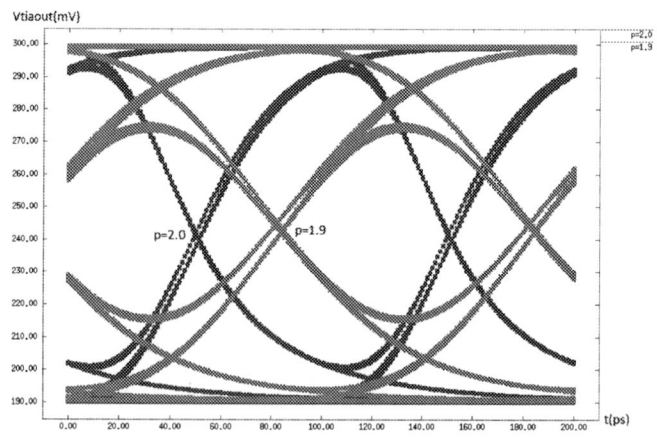

978-1-4799-2408-0/14 $31.00 © 2014 IEEE

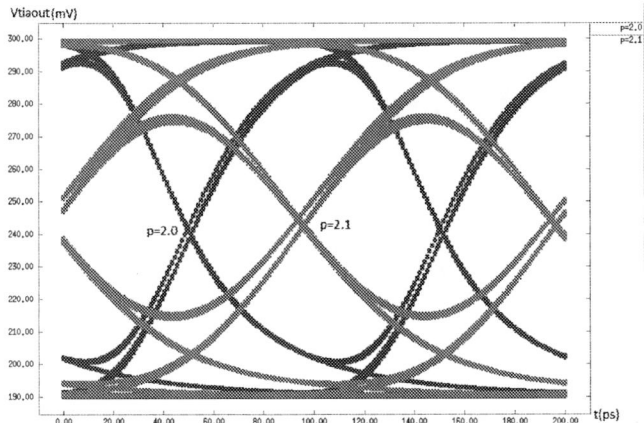

Fig.20 The 25Gb/s eye diagrams at the TIA output of the 180m graded-index MMF link with 3 different values of fiber core graded-index parameter p (p=1.9, 2.0, and 2.1) and smaller photodiode package parasitic parameters.

4. Conclusions

In this paper a multi-mode fiber SPICE modeling method has been proposed for both step-index and graded-index multi-mode fibers to include the SI effects by modal dispersion, chromatic dispersion, fiber attenuation, and even the total delay. The proposed model can be used together with the VCSEL, photodiode, and trans-impedance amplifier SPICE models to simulate the mid-channel optical link in any general-purpose SPICE simulator. Test cases on the system eye diagram signal integrity simulation are shown with 25Gb/s signal and 180m graded-index multi-mode fiber. The proposed modeling method is convenient for electric package system designers to include the signal integrity effect of the mid-channel optical link using their familiar and favorite SPICE-like simulation platform.

Acknowledgement

The author would like to thank Dale Becker and Petar Pepeljugoski of IBM for helpful discussions and suggestions on this paper.

References

[1] Marc A. Taubenblatt, "Optical interconnects for high-performance computing," *J. of Lightwave Technol*ogy, vol.30, no.4, 2012, pp.448-458.

[2] Stefano Bottacchi, *Multi-Gigabit Transmission Over Multimode Optical Fibre*, John Willy & Sons, Ltd, West Sussex, England, 2006.

[3] Petar Pepeljugoski, Steven E. Golowich, A.John Ritger, Paul Kolesar and Aleksandar Risteski, "Modeling and simulation of next-generation multimode fiber links," *J. Lightwave Technology*. vol.21, no.5, 2003, pp.1242-1245.

[4] Petar Pepeljugoski, Michael J. Hackert, John S Abbott, Steven E. Swanson, Steven, E. Golowich, A. John Ritger, Ye. C. Chen, and Peter Pleunis, "Development of system specification for laser-optimized 50-mm multimode fiber for multigigabit short-wavelength LANs, *J. Lightwave Technology*. vol.21, no.5, 2003, pp.1256-1275.

[5] Bahha E.A. Saleh, Malvin Carl Teich, *Fundamentals of Photonics*, Chapter 8, John Wiley & Sons, Inc., 1991.

[6] Ashok Prabhu Masilamani, *Radio Over Multimode Fiber Using VCSELS*, Thesis for the Degree of Master of Engineering, National University of Singapore, 2005.

[7] Sanjeev Gupta, Fangyi Rao, Jing-Tao Liu, and Amolak Badesha, "Efficient end-to-end simulations of 25G optical links," *DesignCon IBIS Summit*, February, 2012, Santa Clara, CA.

[8] Zhaokai Yuan, "Modeling, extraction and verification of VCSEL model for IBIS AMI," *Asian IBIS Summit*, November 2013, Taipei, Taiwan.

[9] Optiwave, *OptiSPICE, the First Opto-Electronic Circuit Design Software*, www.optiwave.com.

[10] Pavan Gunupudi, Tom Smy, Jackson Klein, and Z. Jan Jakubczyk, "Self-consistent simulation of opto-electronic circuits using a modified nodal analysis formulation," *IEEE Trans. on Advanced Packaging*, vol. 33, no. 4, 2010, pp.979-993.

[11] M.K.Jackson and G.S.Burley, "Modeling dispersive optical fibres with a circuit simulator," *Electron. Lett.* vol.30, no.15, 1994, pp.1245-1246.

[12] James J. Morikuni and Sung-Mo Kang, *Computer-Aided Design of Optoelectronic Integrated Circuits and Systems*, Pretice Hall PTR, Upper Saddle Revire, NJ, 1997.

[13] Benjamin Pen-Cheng Tsou, *Equivalent-Circuit Modeling of Quantum-Well Lasers and Fibre-Optic Communications Channels*, PhD Dissertation, the University of British Columbia, Vancouver, Canada, 2000.

[14] IdemWorks, *IdEM*, http://idemworks.com/products/idem/

[15] Tony Irujo, *OM4-The Next Generation of Multimode Fiber*, Furukawa Electric North America, www.ofsoptics.com/ofs-fiber, 2011.

[16] Andrew G. Hallam, *Mode Control in Multimode Optical Fibre and Its Applications*, PhD Dissertation, Aston University, Nov. 2007.

[17] FOA, *Modal Effects on Multimode Fiber Loss Measurement*, FOA Reference for Fiber Optics, 2011, www.thefoa.org/tech/ref/testing/test/MPD.html

[18] Yahei Koyamada and Katsuya Yamashita, "Lauching condition dependence of graded-index multimode fiber loss and bandwidth," *J. Lightwave Technology*. vol. 6, no.12, 1988, pp.1256-1275.

[19] Pablo V. Mena, Sung-Mo Kang, and Thomas A DeTemple, "Rate-equation-based laser models with a single solution regime," *J. of Lightwave Technology*, vol. 15, no.4, 1997, pp.717-730.

[20] P.V.Mena, J.J.Morikuni, S.-M. Kang, A.V.Harton, and K.W.Wyatt, "A simple rate-equation-based themal VCSEL model," *J. of Lightwave Technology* , vol.17, no.5, 1999, pp.865-872.

[21] Houssam Kanj and Michael Steer, Vertical Cavity Self-Emitting Laser Diode (Mena et al. Model), 2003, http://www.freeda.org/doc/elements/vcsel.pdf.

[22] Weiyou Chen and Shiyong Liu, "PIN Avalance photodiodes model for circuit simulation", *IEEE J. of Quantum Electronics*, vol. 32, pp.2105-2111, 1996.

[23] Jianjun Gao, *Optoelcetronic Integrated Circuit Design*, Higher Education Press, Beijing, 2011.

Next Generation Package-on-Package Solution to Support Wide IO and High Bandwidth Interface

Hung-Hsiang Cheng, Chang-Chi Lee, Ming-Feng Chung, Po-Chih Pan, Ping-Feng Yang, Chi-Tsung Chiu, Chih-Pin Hung, and [1]Chen-Chao Wang

Advanced Semiconductor Engineering (ASE) Inc., Kaohsiung, Taiwan

[1]email: alexcc_wang@aseglobal.com

Abstract

A potential technology by substrate interposer enables high bandwidth and low power application processing devices of the future, because the demand of smart mobile products are driving for higher logic-to-memory bandwidth (BW) over 30 GB/s with lower power consumption and ultra-capacity of memory. This paper presents a new High Bandwidth Package-on-Package (HB-PoP) structure with substrate interposer to support 2-channel 64-bit LPDDR3 memory and demonstrate electrical performances including signal integrity (SI) and power integrity (PI). In addition, the thermal performance and warpage behavior of HB-PoPs structures are also better than conventional PoP. Utilizing the structures proposed by Advance Semiconductor Engineering (ASE) Inc., the package of Wide-IO memory can be realized by them as well.

Introduction

The mobile system as smart phone and tablet plays an important role on driving the trend of semiconductor industry. From past to present, the marketing demands of mobile system continue toward miniaturization, higher performance, more functionalities, and lower cost to drive semiconductor technical development. Conventional PoP structure as aMAP PoP is widely used on mobile devices which memory module is directly stacked to the top side of the application processor. The advantages for this kind of structure are smaller area occupied with multiple functions and good electrical behavior due to shorter interconnection distance between application processor and memory module. As the market demands from smart phone and tablet high speed signal transmission and more bandwidth, the requests of memory module are moving to 2 channels LPDDR3 with 32-bits [2]-[3] or even 2 channels one with 64-bits. In addition, the trend of display on mobile system, the high display resolution accompanies with high bandwidth requirement [1]. For small display size, to achieve retina resolution is easier than large one, but large display size is today's major requirement from end-users for smart phone and tablet. It will have big problem on the requirement of memory bandwidth by using present package structure as conventional PoP due to not enough ball counts supported on top package. For increasing memory bandwidth, there are two strategies. One is increasing the speed of I/O signal, and another is by adding more than more data lines or data busses. For the first solution, more design challenges will be encountered on package electrical design owing to the requirements of electrical performance [4]-[5]. Choosing the second solution will increase the difficulty on package manufacturing because the ball counts can't be added unlimitedly.

For balancing long term requirement of bandwidth and cost benefit, ASE proposes new PoP structures named high

bandwidth PoP (HB-PoP) showed as Fig. 1. There are three kinds of high bandwidth PoP (HB-PoP). That is, HB-PoPI, HB-PoPII and HB-PoPIII. HB-PoPI can support up to 384 top interconnects for application processor to stack with memory module. HB-PoPII provides 512 top interconnects, while HB-PoPIII supports over 1000 top interconnects. Base on the structures, more channels for LPDDR IO can be realized to increase system bandwidth with cost benefit than 2.5D-IC or 3D-IC. In addition, there are three options for interconnect materials depending on pitches. The interconnect can be made by either pure solder or Cu to solder or Cu to Cu bumps. The proposed options are able to provide the scalability of interconnect pitches from 0.27, 0.20 and even down to 0.13mm. The latter one allows 1200 interconnects on a 15x15mm HB-PoP. Three HB-PoP structures are proposed to meet application processor to memory interconnection requirements:

1. SeLI (Solder embedded Laminated Interconnect): supports HB PoP-I and HB PoP-II

2. CuPiS (Cu Pillar in Solder): supports HB PoP-I and HB PoP-II

3. CuPI (Cu Pillar Interconnect): supports HB PoP-I, HB PoP-II and HB PoP-III

In this paper, the performance comparisons of thermal and warpage behavior have been investigated and the electrical behavior for signal integrity and power integrity with 1.6Gbps LPDDR3 has been demonstrated by using HB-PoP package. Compared with conventional PoP, the study shows that the proposed HB-PoPs have superior warpage performance, and comparable electrical and thermal performance with conventional PoP.

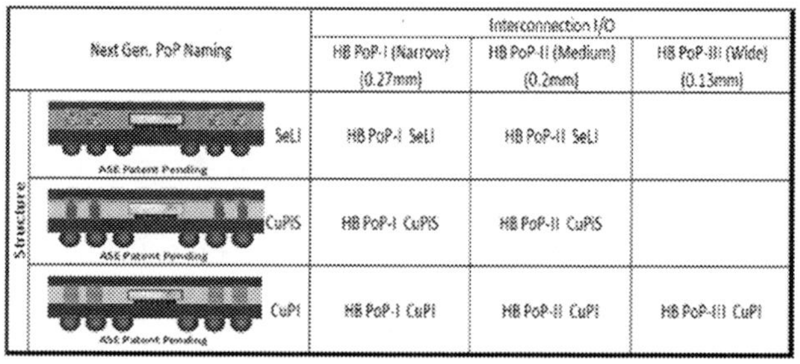

Figure 1. The high bandwidth HB-PoP structures.

Test Vehicle Design

In order to evaluate the impact of various PoP solutions on thermal and thermo-mechanical performance, the test vehicles of aMAP (Through Mold Via), SeLI, CuPiS, and CuPI PoPs are designed. In this study, package size and top ball pitch of bottom package are kept the same for each PoP. The bottom package size is 14 x 14 mm^2 with 11 x 11 mm^2 die, and the top ball pitch is 0.4 mm.

For electrical performance evaluation, a real package substrate is applied on various PoP solutions including aMAP and SeLI PoPs. A memory module with substrate interposer is attached on bottom package of SeLI PoP. The substrate interposer is 2-layer substrate. The bottom package is 4-layer substrate and the size is 14 x 14 mm^2 with 11 x 11 mm^2 die as well, and ball pitch on top side is 0.4mm for aMAP PoP, and 0.27mm for SeLI PoP.

To demonstrate the feasibility of wide IOs applied on SeLI PoP, two package substrates are designed. One is for 2-channel 32-bits LPDDR3solution, the other is for 2-channel 64-bits LPDDR3 interface. The substrate drawings are showed in Fig. 2 and Fig. 3.

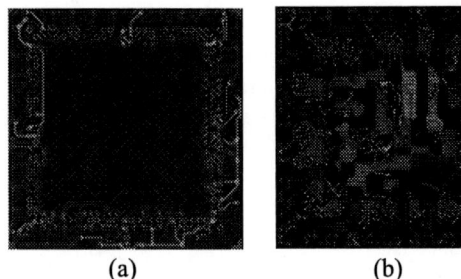

(a)   (b)

Figure 2. (a) The top view of substrate interposer of 2 channel 32-bits SeLI PoP. (b) The top view of substrate interposer of 2 channel 32-bits SeLI PoP

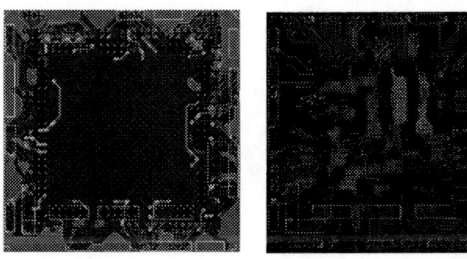

Figure 3. (a) The top view of substrate interposer of 2 channel 64-bits SeLI PoP. (b) The top view of substrate interposer of 2 channel 64-bits SeLI PoP

Thermal Modeling and Analysis

Fig. 4 shows the temperature contour for SeLI PoP. Three-dimensional finite volume analysis is performed based on commercial software FloTHERM. One-watt power dissipation is applied to the bottom surface of the die. The boundary conditions for environment are set up following the JEDEC EIA/JESD51-2 standard [6] with 4-layer PCB at 55°C ambient temperature. The material properties for thermal simulation are provided in Table I. The thermal simulation results are shown in Fig. 5. The results show that the thermal performance difference among the four PoP structures is less than 0.7%. It is because the major heat flow path is downward to PCB in JEDEC standard environment.

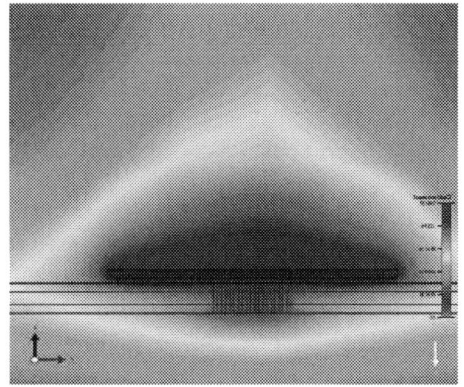

Figure 4. Temperature contour of SeLI PoP.

Table I: Material properties for thermal and thermo-mechanical analysis

Material	E (GPa)	α (ppm/C)	Tg (C)	ν	K (W/mK)
Silicon	131	2.8	-	0.23	149 @ 25C 107 @ 125C
Compound	23.0 @ 25C 0.7 @ 260C	8 / 35	145	0.30	0.9
Adhesive	13.0 @ 25C 0.26 @ 260C	12 / 34	156	0.30	0.628
Copper	130	17	-	0.30	389
Soldermask	6.8 @ 25C 0.24 @ 260C	42.5 / 127.5	135	0.30	0.20
Core for bottom substrate	34.0 @ 25C 26.5 @ 260C	3.0 / 3.0	270	0.20	0.55
Core for top substrate	27.0 @ 25C 13.24 @ 260C	X, Y: 10 / 3 Z: 22 / 150	230	0.20	0.55
Underfill	5.817 @ 25C 1.03 @ 260C	31.1 / 63.5	88	0.30	0.64
SAC	41.0 @ 5C 27.0 @ 215℃	20	-	0.30	58
Sn/Ag	57.57 @ 25C 9.918 @ 220C	27.9	-	0.30	58

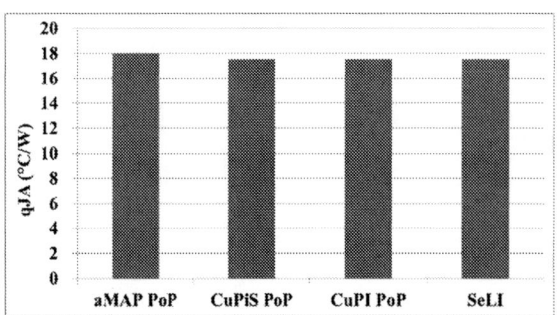

Figure 5. Thermal simulation results for different PoP structures.

Thermo-mechanical Modeling and Analysis

PoPs are composed of various materials. Base on the Bernoulli-Euler beam theory and the compatibility on the material interface, analytical solution for the curvature of beam is dependent on each layer of component thickness, Young's modulus and coefficients of thermal expansion (CTE) [7]-[8]. Fig. 6 shows quarter finite element model of SeLI PoP for the bottom package only. The analysis is carried out using ANSYS v. 13.0. The material properties for finite element analysis are provided in Table I. Warpage comparison between simulation and shadow moire measurement for SeLI PoP is shown in Table II. From the comparison, it indicates that the model is well correlated with the measurement data at both 25°C and 260°C. The total warpage simulation results with different package structures are shown in Fig. 7. From the results, it is noted that HB-PoP shows much lower package warpage compared with aMAP PoP. In other words, the balanced structure of HB PoP reduces the warpage. Therefore, it ensures good memory stacking yield.

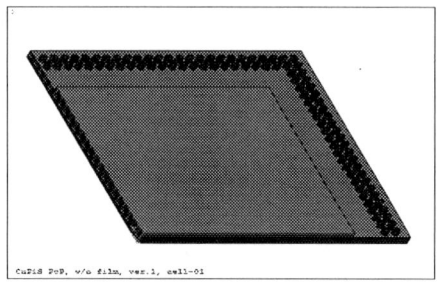

Figure 6. The quarter finite element model of SeLI PoP.

Table II: Warpage simulation vs. shadow moire measurement
(warpage: "-": concave; "+": convex)

SeLI structure	Warpage @ 25°C	Warpage @ 260°C
Simulation result (package top view)	-54 um	+43 um
Shadow moire result (package bottom view)	-44 um	+53 um

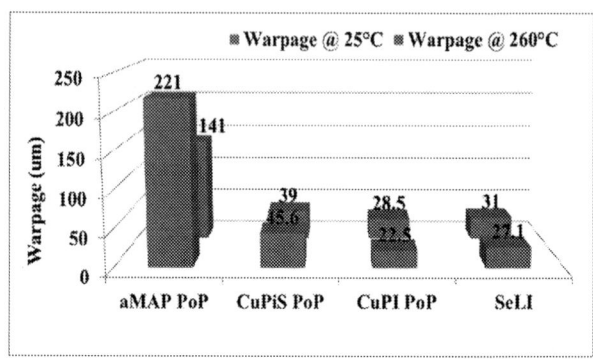

Figure 7. Warpage simulation results for different PoPs.

Electrical Characteristic and Modeling

Due to SeLI PoP are composed of top package for memory module with substrate interposer and bottom package for application processor, a commercial electromagnetic (EM) solver, Ansys/Apache PSI, is applied on model extraction of stacking package structure. The 3D simulating model is showed as Fig 8. Fig. 9 shows the S-parameter results of LPDDR3 signals including aMAP and SeLI PoPs. In Fig. 9, the results of insertion loss are around -1.5dB at 10GHz between them. For coupling noise showed in Fig. 10, there is no obvious difference for the coupling effect for aMAP PoP and SeLI PoP between DQ6 and DQ7 signals.

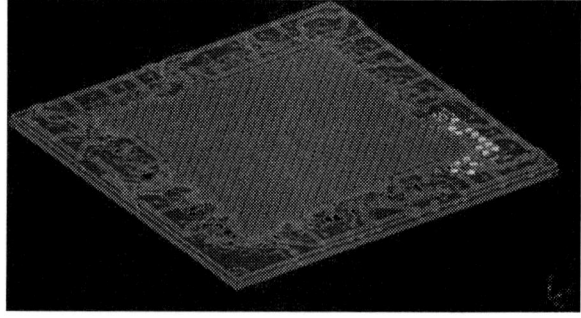

Figure 8. 3D electromagnetic model.

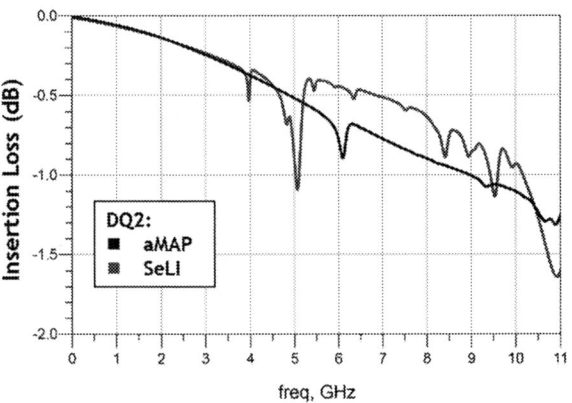

Figure 9. The S-parameter results of LPDDR3 signal for DQ0 and DQ6 including aMAP and SeLI PoPs.

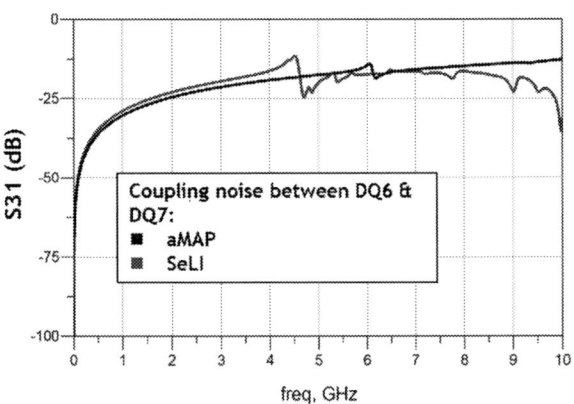

Figure 10. The coupling effect between DQ6 and DQ7 for aMAP and SeLI PoPs,.

The power delivery networking system of LPDDR3 IOs is designed independently for bottom and top package substrates. Fig. 11 shows the inductance values of power/ground planes of aMAP and SeLI PoPs on top and bottom packages for LPDDR3 IO portion. It is very close between the two PoP structures for bottom package. For top package, the inductance value of power/ground planes of aMAP PoP is better than SeLI. This is because there is addition substrate interposer on SeLI PoP.

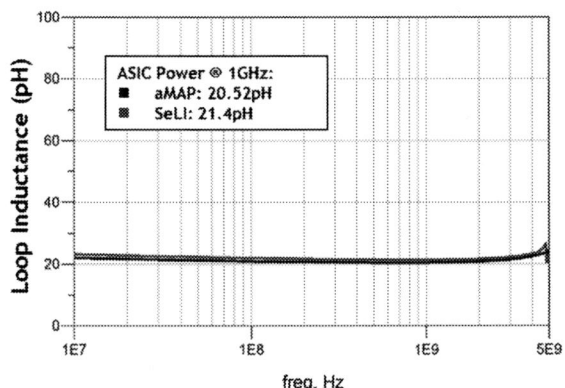

(a)

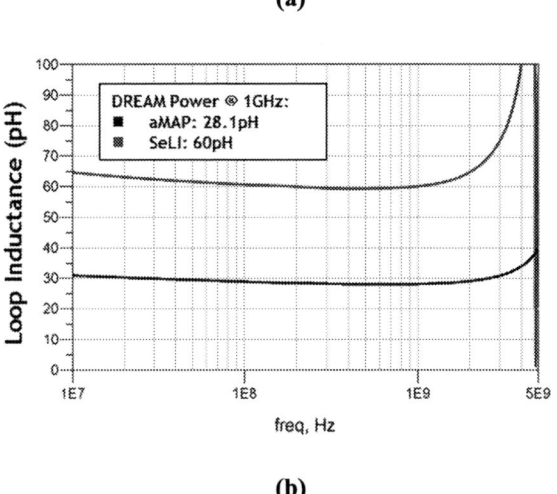

(b)

Figure 11. (a) The inductance values of power/ground planes of aMAP and SeLI PoPs on bottom package. (b) The inductance values of power/ground planes of aMAP and SeLI PoPs on top package.

Electrical Performance Analysis

Next, a systematization analysis has been demonstrated the feasibility applied on 2-channel 32-bits and 64-bits 1.6Gbps LPDDR3 IOs. The LPDDR3 IO model on application processor is from IP vendor, and receiver model on memory module is from Micron Technology Inc.. All these models are IBIS 5.0 version. The system modeling extraction including top package, bottom package, and system board uses Ansys/Apache PSI to generate S-parameter macro model. The Fig. 12 displays simulation architecture for signal integrity and power integrity analysis. The on-die decoupling capacitance is 500pF for one byte lane, so the total capacitance for on-die decoupling capacitance is 2nF..

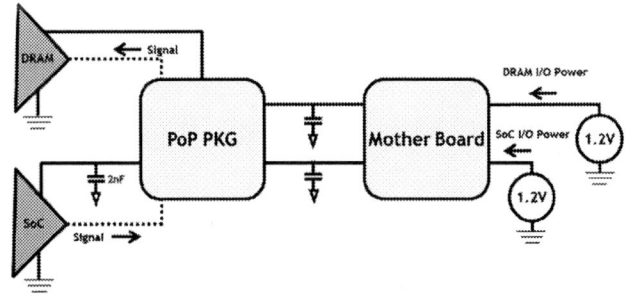

Figure 12. Simulation architecture for electrical analysis.

Fig. 13 and Fig.14 show the simulation waveforms of eye-diagram for aMAP and SeLI PoPs with 2-channel 32-bits. From the results, the performance between aMAP and SeLI PoPs is comparable. A summary data as Table III shows the electrical performance on both PoP structures meets JDEC LPDDE3 criterions. For dynamic power noise analysis, the waveforms show on Fig. 15 and Fig. 16, and the results summarize on Table IV. The addition substrate interposer

978-1-4799-2408-0/14 $31.00 © 2014 IEEE   2115

induces extra inductance on SeLI PoP, so the peak-to-peak noise of SeLI PoP is a little high than aMAP PoP. In addition, when the on-die-termination (ODT) is enabled, the peak-to-peak noise is better than disable situation. From Table IV, the peak-to-peak noises of both structures meet the specification of voltage swing.

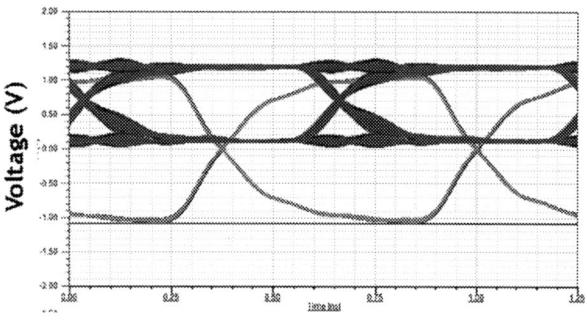

(a)

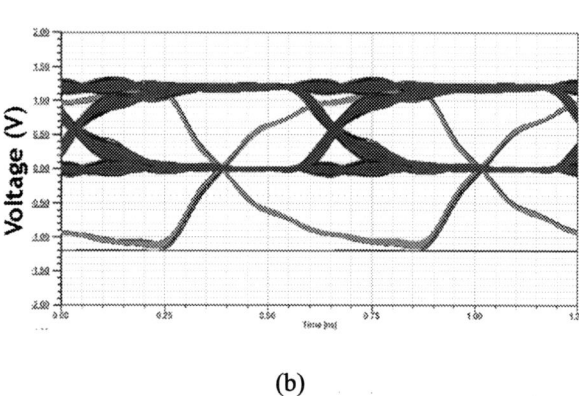

(b)

Figure 13. The simulation waveforms of eye-diagram for aMAP PoP with 2-channel 32-bits. (a) With ODT. (b) Without ODT.

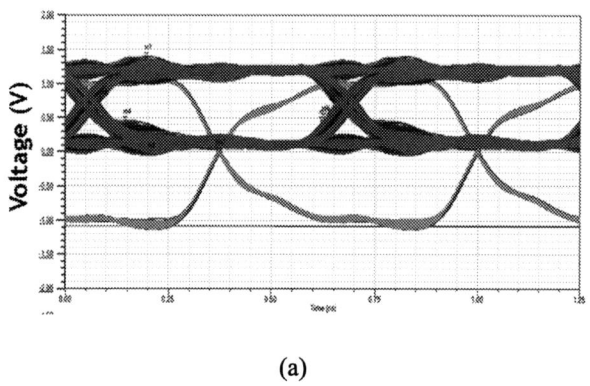

(a)

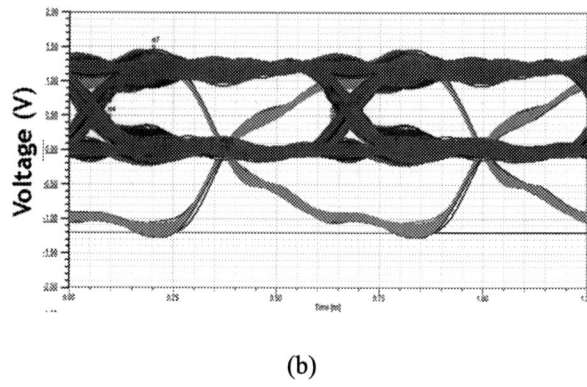

(b)

Figure 14. The simulation waveforms of eye-diagram for SeLI PoPs with 2-channel 32-bits. (a) With ODT. (b) Without ODT.

Table III: Summary data for signal integrity analysis

Analysis Item	ODT enabled		ODT disabled		Spec
	aMAP	SeLI	aMAP	SeLI	
ODW (ps)	480	482	528	529	N/A
Setup time tDS (ps)	232	223	288	266	113
Hold time tDH (ps)	240	250	227	249	125
tVAC (ps)	592/473	556/476	530/537	524/537	48
Overshoot (V)	1.308	1.394	1.3464	1.456	1.55
Undershoot (V)	0.034	-0.049	-0.113	-0.233	-0.35

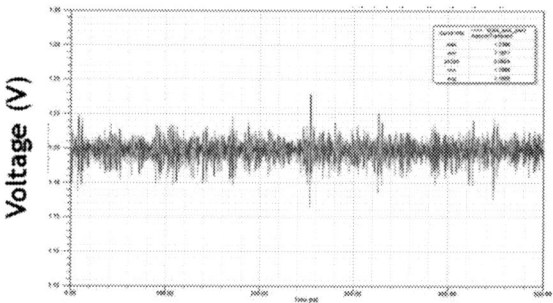

(a)

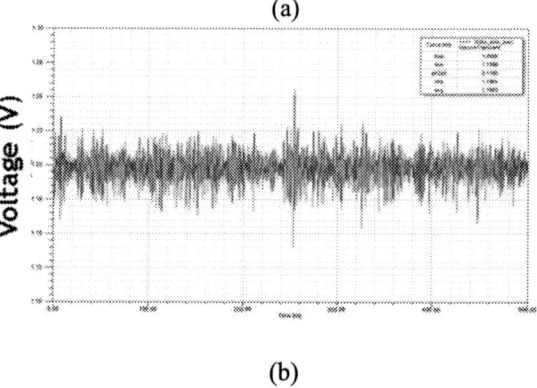

(b)

Figure 15. The dynamic power noise waveform of application processor for aMAP PoP. (a) With ODT. (b) Without ODT.

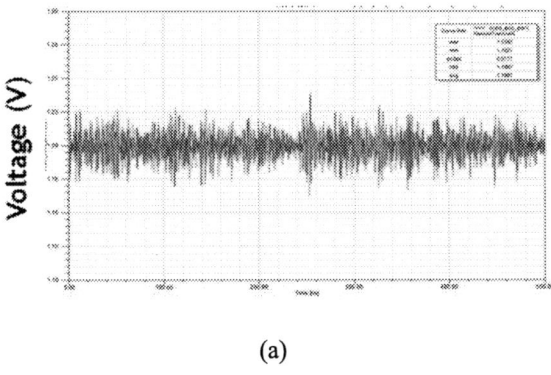

(a)

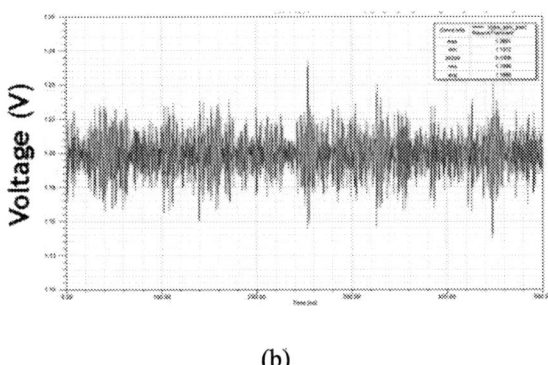

(b)

Figure 16. The dynamic power noise waveform of application processor for SeLI PoP. (a) With ODT. (b) Without ODT.

Table IV: Dynamic power noise

ODT Type	ASIC IO Power		DRAM IO Power	
	aMAP	SeLI	aMAP	SeLI
Enabled ODT	82	77	61	119
Disenabled ODT	116	131	21	39

For 2-channel 64-bits situation, Fig. 17 shows the simulation waveforms of eye-diagram for SeLI PoPs without ODT. Due to ball pitch over 0.4mm, aMAP PoP can't support wide IOs application, especially the package size is below 15mm by 15mm area. The timing waveform meets JDEC LPDDR3 specification as well. The transient waveform for dynamic power noise is as Fig. 18. Around 80mV peak-to-peak voltage swing is for SeLI PoP. From SI and PI results, the electrical performance of SeLI PoP is able to be comparable with conventional PoP (aMAP PoP), and all testing items pass LPDDR3 JDEC criterions on 2 channels 32-bits and 64-bits conditions.

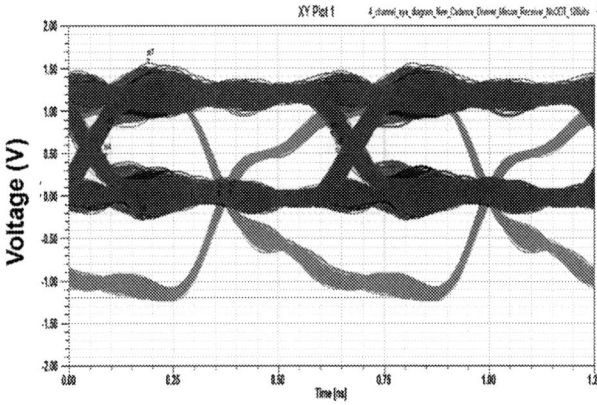

Figure 17. The simulation waveforms of eye-diagram for SeLI PoP with 2-channel 64-bits without ODT.

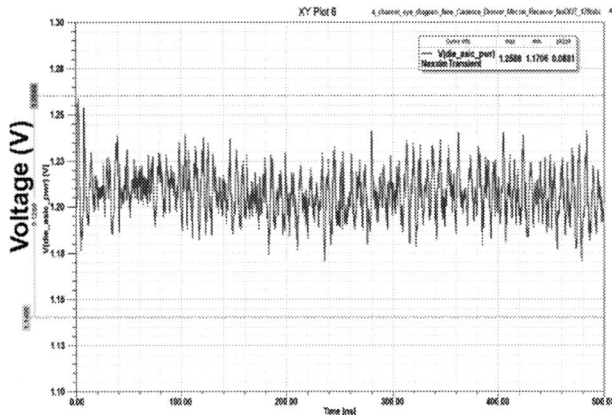

Figure 18. The dynamic power noise waveform of application processor for SeLI PoP without switching on ODT.

Conclusions

Thermal and warpage modeling are carried to investigate the performance difference among conventional and three high bandwidth PoPs. As for thermal performance, the difference among all PoPs is negligible. It is because the major heat flow path is downward to PCB in JEDEC standard environment. Regarding thermo-mechanical performance, all high bandwidth PoPs exhibit much lower warpage than conventional PoP thanks to balanced package structure. Therefore, good memory stacking yield can be expected.

For electrical performance, a real product has been designed on 2 channels 64-bits situation and the S-parameter indicates that the insertion loss difference is acceptable between conventional and high bandwidth PoP. In terms of eye diagram and dynamic power noise, SeLI PoP is comparable with aMAP PoP, and it corresponds to JEDEC specifications with large timing margin.

978-1-4799-2408-0/14 $31.00 © 2014 IEEE   2117

References

1. J. J. Lee, et al, "2-Channel 2-Layer Inner-Stack Memory-module Design for LPDDR2/3 DRAM," *Signal and Power Integrity (SPI) workshop*, 2012, pp. 13-15.

2. C. Jo, et al, "World's first LPDDR3 Enabling for Mobile Application Processors System," DesignCon 2013

3. S. H. Hall, et al, "Advanced Signal Integrity for High-Speed Digital Designs," IEEE: Wiley, 2008.

4. N. Matt, *3D Packaging Magazine*, No. 19, Nov, 2010, pp. 2.

5. J. M. Yannou, et al, "Evolution of the market drivers and development of the infrastructure," SemiconWest, No.11, July, 2011, pp. 35.

6. "Integrated Circuits Thermal Test Method Environment Conditions" EIA/JEDEC JESD51-2, December, 1995.

7. S. Wiese, et al, "Time-independent Plastic Behaviour of Solders and its Effect on FEM Simulations for Electronics Packages," ISAPM, 2002, pp. 104-111.

8. J. H. Lau, "Modeling and Analysis of 96.5Sn–3.5Ag Lead-Free Solder Joints of Wafer Level Chip Scale Package on Buildup Microvia Printed Circuit Board," *IEEE Trans-EPM*, Vol. 25, No. 1 (2002), pp. 51-58.

Package- Level Electromagnetic Interference Analysis

Namhoon Kim, Leo Hongyu Li, Sam Karikalan, Reza Sharifi and Henry Kim
Broadcom Corporation
5300 California Avenue
namhoon@broadcom.com

Abstract

Electromagnetic Interference (EMI) is becoming more problematic as both the number of components in a system and operating speed increase. This paper introduces possible EMI problems at the package level, and various package stack-up structures and templates are compared. In addition, die shielding, package edge shielding, and package lid shielding effects on EMI are discussed. Primarily, the package lid is used for heat spreading and warpage control. However, the impact of EMI is significant, and properly grounding the lid is a very important.

A major contributor of EMI in the package is cavity resonance. The package structure, along with the lid, can create a rectangular resonant cavity, where the resonant frequency of its TMzxx0 modes can be calculated using equations based on the substrate material and package geometry. The resonant frequency is related to the dielectric material properties and substrate size. This phenomenon was verified through simulation using a full 3M EM field solver. Furthermore, a test vehicle was fabricated and compared to the simulation results.

These analyses help identify possible weak spots in the package and device and determine the most effective solution for improving package-level EMI problems.

Introduction

High-speed serial channel design has become more important as operating frequencies increase. Most signal integrity engineers focus on various types of signal integrity problems in packages including impedance discontinuity, reflection, attenuation, undershoot/overshoot, and crosstalk [1][2][3]. However, these challenges are very EMI related. A design that has better crosstalk, less reflection, and less discontinuity in the return current path has a better chance of having lower EMI.

EMI comprises good Signal Integrity (SI) and Power Integrity (PI). However, reducing EMI is a complicated problem because it is a system-level issue that includes the active die, package, and PCB [4][5]. EMI is related to circuit performance that generates the source of energy spectrum through entire frequency range. And it is also related to package and PCB structure that can radiate those frequency components into the air. Thus, frequency-dependent package and PCB characteristic analysis is required to characterize the behavior of EMI. The package structure is relatively smaller than PCB structure, but it needs to be carefully considered as the source energy spectrum tends to go into the higher frequency region. Current package sizes are no longer ignorable for multi-gigabit channels in terms of EMI. Therefore, package-level EMI analysis is the focus of this paper. It would be useful to know the most important factors and what can be done at the package level to minimize EMI problems in the system.

EMI Fundamentals

EMI is defined as a process by which disruptive electromagnetic energy is transmitted from one electronic device to another via radiated and/or conducted paths. Commonly, the term refers to RF signals, but EMI can occur in the frequency range "from dc to daylight" [6].

EMI behavior can be explained by the energy (noise) source and radiating structure (propagation path). In this case, the radiating structure is the package. Depending on the operating conditions, the energy source model can be extracted and converted to the frequency domain spectrum by switching the current of circuits. The main frequency components and their harmonics should be captured at this stage. Figure 1 (a) shows an example of a frequency domain energy spectrum. There are several specific frequency points that have strong energy.

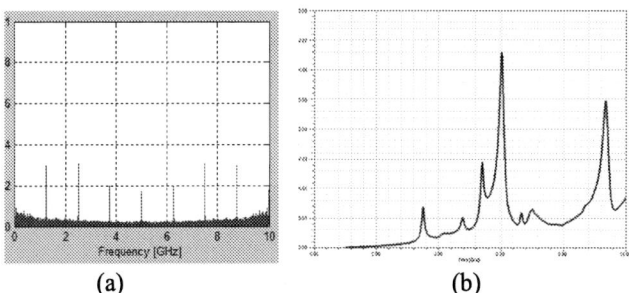

Figure1. Examples of frequency domain (a) Source spectrum (b) Radiation efficiency of victim (Antenna) structure.

EMI behavior can be simply described as the multiplication of the source energy spectrum and radiation efficiency of the victim structure, which is shown in Figure 1 (b). If one of these high radiating frequencies hits the strong source energy source frequency, it will be amplified. At times, certain noise source frequencies are ignored because only the main operating frequency and harmonic frequencies (e.g., first, third, and fifth harmonics) tend to be considered. But, second and fourth harmonic frequency components may exist, which are important if there is a rise/fall time mismatch. Therefore, it is important to understand where the noise comes from and which frequency component propagates best into the air.

978-1-4799-2408-0/14 $31.00 © 2014 IEEE   2119   2014 Electronic Components & Technology Conference

Package Break- Down

What is the major radiation source in a FCBGA package (which is the most common package for high-speed SerDes products)? For simplification, a package structure can be segmented into multiple pieces:

I)   Bumps between the die and the substrate

II)   Trace, via, and reference planes interconnecting the die to the BGA.

III)   BGA solder balls for package to PCB interconnection.

IV)   Lid (provides die protection, heat spreading, and warpage control).

Each part of the package was compared and analyzed to determine and characterize the most dominant EMI contributors.

Microstrip line trace and stripline trace comparison is an interesting topic, as it is a common knowledge that a stripline trace has better EMI performance than a microstrip line trace because of solid ground references. But microstrip line structures for most high speed SerDes packages have coplanar grounds even though the distance between the trace and same plane grounds is much longer than dielectric thickness. In this case, the difference may not be important for typical package trace lengths if we consider all other package elements such as bumps, vias, planes, and BGA balls at the same time. To verify this, the microstrip line and stripline traces were routed with the same amount of impedance. All traces in microstrip line and stripline structures have coplanar grounds on same plane. The only difference between the two designs is that the microstrip line structure has traces on the first layer and the stripline structure has traces on the second layer. Both designs used same size bump, plane, via, and BGA balls. 3D FEM full wave simulations were performed to compare the two cases.

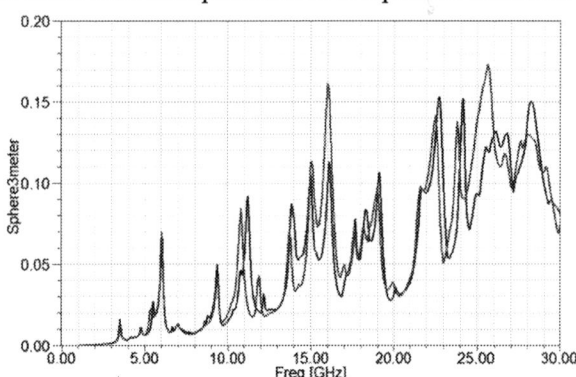

Figure 2. Radiation [V/m] at 3m sphere vs. frequency [GHz]. Solid Red – Stripline package structure; Solid Blue – Microstrip line package structure.

Interestingly, the measured radiation pattern of a 3-meter sphere shows very similar behavior between the microstrip line and stripline structures over the frequency spectrum of interest. To determine the most radiated location in the package, the electric field was plotted. Figure 3 shows the vertical direction of the electric field plot of microstrip line and stripline structures. Figure 4 shows the horizontal direction of the electric field plot in the core plane area of the microstrip line and stripline structures. Where there are traces,

there are strong radiating fields from the package edges, even though the vertical distance is much shorter than the horizontal distance.

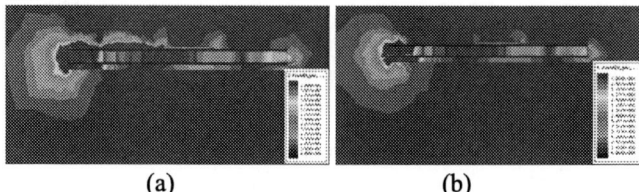

(a)   (b)

Figure 3. Vertical direction Electric Field Plot [V/m] of (a) Microstrip line structure (b) Stripline structure.

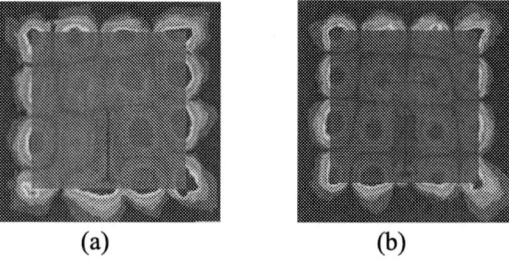

(a)   (b)

Figure 4. Horizontal direction Electric Field Plot [V/m] in between core ground planes of (a) Microstrip line structure (b) Stripline structure.

From these results, it can be concluded that where the traces are located, substrate edge side radiation is larger than horizontal radiation. This behavior is observed wherever a parallel plate structure exists. Having a parallel plate structure by those planes is one of the biggest contributors of the radiation in the package. Other parallel plate structures also exist in the system, such as die plane to package plane, package plane to PCB plane, and lid to package plane. The package lid is commonly used to protect the die, to control warpage, and to provide for good thermal characteristics by enabling heat spreading. The package lid should be carefully considered because it can be one of the biggest contributors of EMI radiation since it is located close to the noise source (die). To verify this, a package lid was added to the simulation. Figure 5 shows the lid effect. The radiation was greatly increased at specific frequency points (e.g., 7.5 GHz, 17.5 GHz).

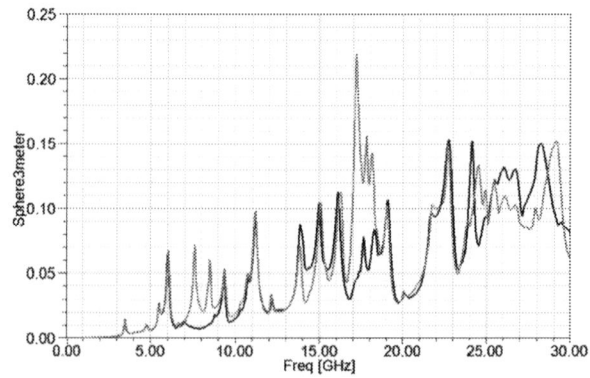

Figure 5. Radiation [V/m] at 3m sphere vs. frequency [GHz]. Solid Green – Microstrip line package structure with package lid; Solid Blue – Microstrip line package structure without package lid.

978-1-4799-2408-0/14 $31.00 © 2014 IEEE

A package lid can also be a good radiation structure since it can create parallel plane resonance with the package plane. In addition, lids are not normally grounded for assembly and reliability reasons. However, since the size and shape of the package lid could determine plane resonant frequencies, the lid should be engineered and designed and not randomly selected.

To understand the nature of each radiating frequency in the package, each parallel plate structure (around the core planes, die, and lid) should be shielded at the edges one at a time. The package model and lid was chosen to identify the frequency components in the package structure. Figure 6 shows the amount of radiation from each location. In this figure, the solid green line represents the microstrip line package structure and package lid used in the Figure 5. First, the core plane and build-up plane edges were shielded using a perfect conductor. Many of the frequency components disappeared, as represented by the dotted purple line in Figure 6. Radiation at those frequencies came from the package planes. The constant difference of the dielectric material between the core material and the build-up material makes various parallel plate resonant frequencies. After shielding the die edges, except for the microstrip line trace area, most of radiating frequencies disappeared, as represented by the solid red line in Figure 6. In this case, all vertical noise sources (bump, build-up via, and core via) were shielded, but they still have strong horizontal noise sources (microstrip line) and lid openings.

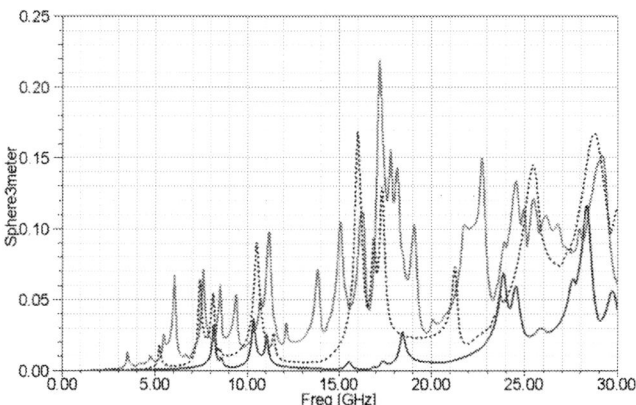

Figure 6. Radiation [V/m] at 3m sphere vs. frequency [GHz]. Solid Green – Microstrip line package structure with package lid; Dotted Purple – Shielded core planes; Solid Red – Shielded die edges to the package plane.

Figure 7 shows the electrical field plot of the horizontal direction in the package lid area. The only difference between plot (a) and plot (b) is the die boundary shielding. Both have strong horizontal E-fields from the microstrip line traces, but because of the shielding, (b) does not have vertical direction excitation. It can be concluded that the horizontal trace is not a good source of exciting parallel plane resonance in a package structure.

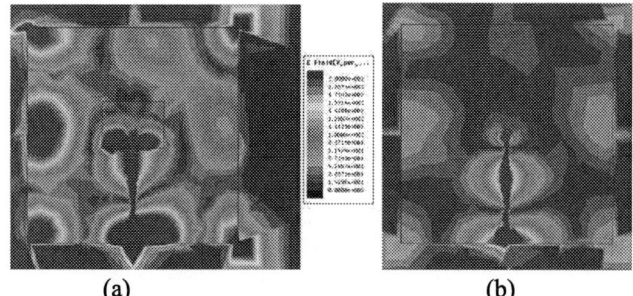

(a)   (b)

Figure 7. Horizontal direction Electric Field Plot [V/m] of microstrip line package structure in package lid area (a) without die shielding (b) with die boundary shielding.

The resonant frequency due to the package lid can be controlled in several ways. All other parallel plane structures in the package, except the lid, were shielded using a perfect conductor. If the size of the lid is reduced, the parallel plane resonant frequency shifts, as shown by the dotted purple and solid light blue lines in Figure 8. The solid light blue line represents radiation using 0.6 mm smaller size lid. A smaller size lid tends to make the resonant frequency higher. When four corner ground posts were added to the lid, the resonant frequency shifted as shown by the solid orange line in Figure 8. But, at the same time, it can be verified that all radiation peaks do not disappear by adding the four corner ground posts. When more ground posts with a 2.2 mm pitch on all sides of the lid were added, most radiation peaks below 20 GHz disappeared. As stated earlier, package lid size and the proper grounding are very important to avoid possible EMI issues.

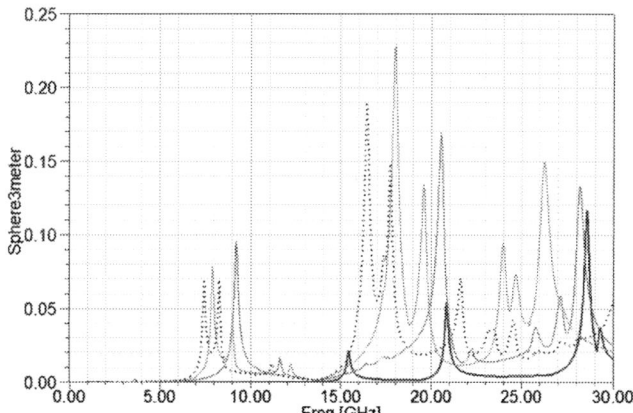

Figure 8. Radiation [V/m] at 3m sphere vs. frequency [GHz]. Package lid radiation only. Dotted Purple – Original size lid; Solid Light Blue – Smaller size lid; Solid Orange – Smaller size lid with 4x corner ground post; Solid Red- Smaller size lid with more ground posts on each side of the lid.

The following section describes parallel plane resonance in greater detail. The basic theory will be proved by actual measurement using test vehicle.

Parallel Plane Resonance

A package has several plane cavity structures and the resonant frequency of its TMzxx0 modes can become significant contributor of EMI. The resonant frequency can be calculated using the following equation:

$$f_{Zmn0} = \frac{c}{2\sqrt{\varepsilon}} \sqrt{\left(\frac{m}{a}\right)^2 + \left(\frac{n}{b}\right)^2} \qquad (1)$$

Where f is the resonance frequency, c is the speed of light in free space, ε is the permittivity of the dielectric material, a and b are the length and width of the plane, and m and n are integers.

Experiments

A test chip, as illustrated in Figure 9, was built to verify the argument that plane resonance dominates the direct emission from the IC, which is a 17 mm by 17 mm substrate. The relative permittivity of the dielectric material is 4.77 at 20 GHz. A 10.3Gb/s SerDes circuit is implemented on the die to excite the substrate.

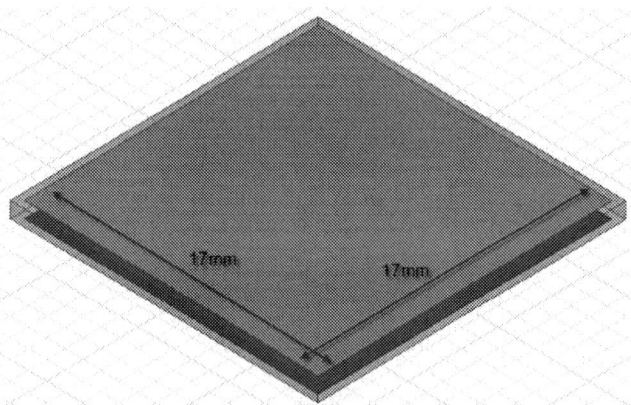

Figure 9. Test chip illustration.

The resonance frequencies of different modes calculated with (1) are shown in Figure 10.

Figure 10. Resonance frequencies of different modes.

From the calculation, the resonance frequency of TM_{Z150} and TM_{Z510} is overlapping with the fourth harmonic (20.6 GHz) of the fundamental frequency of the signal (5.15 GHz).

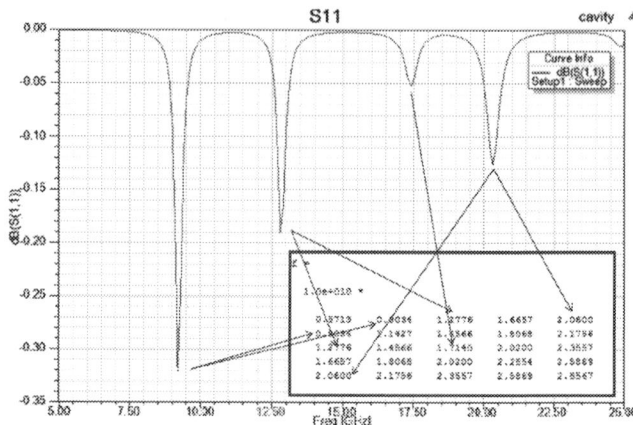

Figure 11. 3D FEM full-wave simulation results.

A 3D FEM full-wave simulation was performed to verify the calculation. The results plotted in Figure 11) show the plane structure resonating around 20.6 GHz. Not all the modes can be observed in the simulation results because they are not excited due to the positions of the excitation on the test chip.

To further verify that plane resonance dominates the direction of emissions from the substrate, the test chip was modified by cutting off 0.6 mm on each side as illustrated in Figure 12. The small cut ensures that no critical connections are cut off, and that the test chip is still functional.

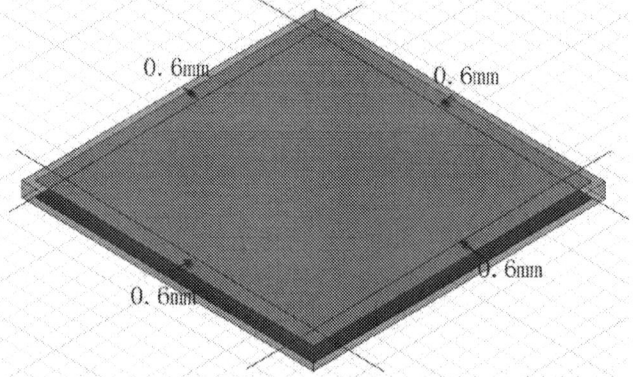

Figure 12. Modified test chip.

The modified test chip is exactly the same as the original one, except with smaller planes. The smaller planes shift the resonances to a higher frequency and lower the emissions because of the 20.6 GHz plane resonance, as shown in Figure 13 and Figure 14. If the plane resonance dominates direct emission from the substrate, the reduction should be seen in the overall emission test since everything is the same and only the planes are smaller.

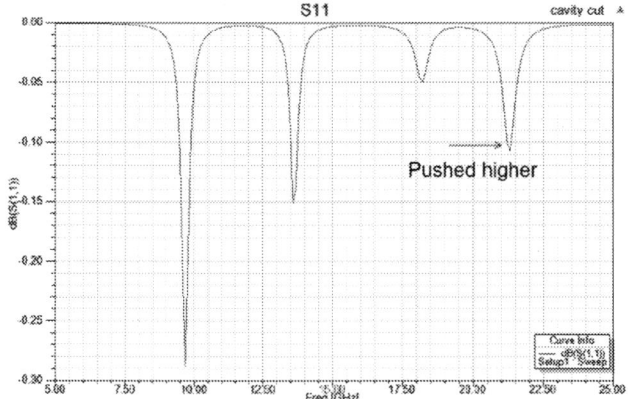

Figure 13. The resonance is pushed to higher frequency.

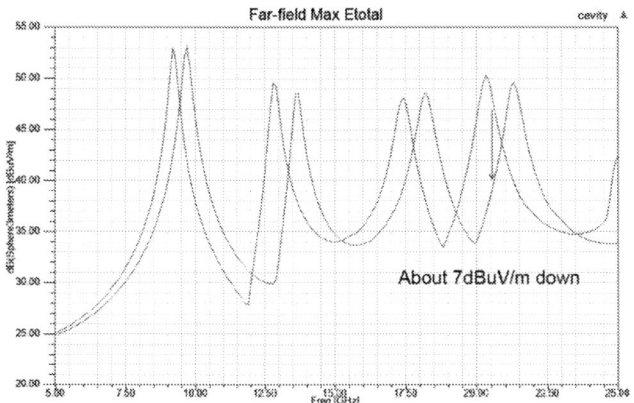

Figure 14. Emission at 20.6 GHz is reduced by 7 dBuV/m assuming the same excitation.

The direct emission from the test chips was measured in anechoic chamber with a carefully designed test board for which there is no other emission sources. It was observed that the direct emission from the modified test chip is 4 to 5 dBuV/m lower than the original test chip. These results confirm that the plane resonance dominates the direct emission from the test chip, which should be true in most cases.

Conclusion

Possible EMI weak spots at the package level were discussed in this paper. Microstrip line and stripline structures in a package were compared from an EMI point of view. In addition, die, package edge, and package lid shielding effects on EMI were shown. A major EMI contributor in the package is a cavity resonance. The package structure and lid could be a rectangular resonant cavity and could become one of the biggest contributors of EMI radiation. This phenomenon was verified through simulation using a 3D FEM full- wave solver. Furthermore, a test vehicle was fabricated and compared to the simulation results. Proper grounding of the package lid is very important. All these analyses help to provide a better understanding of possible weak spots in the package and find the most effective solution for improving package-level EMI problems.

Acknowledgments

The authors would like to thank to Dong Cho, Qui Nguyen and Argha Nandy in system testing group for valuable comments and making available the data.

References

1. Ivan Ndip, Werner John and Herbert Reichl, "Effects of Discontinuities and Technological Fluctuations on the RF Performance of BGA Packages", *Electronic Components and Technology Conference,* Lake Buena Vista, FL, May 31-Jun 3, 2005, pp. 1769-1775 Vol. 2.
2. Namhoon Kim, Hongsik Ahn, Chris Wyland, Ray Anderson and Paul Wu, "Spiral via structure in a BGA package to mitigate discontinuities in multi-gigabit SERDES system", *IEEE Electronic Components and Technol. Conf. (ECTC)*, Las Vegas, NV, Jun 01–04, 2010, pp. 1474-1478.
3. Guang-Hwa Shiue, Jia-Hung Shiu and Po-Wei Chiu, "Analysis and Design of Crosstalk Noise Reduction for Coupled Striplines Inserted Guard Trace with an Open-Stub on Time-Domain in High-Speed Digital Circuits", Components, *Packaging and Manufacturing Technology, IEEE Transactions*, Vol.1, No.10, October 2011.
4. Toshio Sudo, Hideki Sasaki, Norio Masuda and James L. Drewniak, "Electromagnetic Interference (EMI) of System-on-Package (SOP)", *IEEE Transactions on Advanced Packaging*, Vol.27, No.2, May 2004
5. Nozad Karim, Jingkun Mao and Jun Fan, "Improving Electromagnetic Compatibility Performance of Packages and SiP Modules Using a Conformal Shielding Solution", *Asia-Pacific International Sympoisum on EMC*, Arpil 12-16, 2010.
6. Mark I. Montrose, EMC and The Printed Circuit Board, IEEE PRESS, pp. 2–3.

A Path Finding Based SI Design Methodology for 3D Integration

Bill Martin[*], KiJin Han[&] and Madhavan Swaminathan[+]

[*]E-System Design, Atlanta, GA, USA

[&]School of ECE, UNIST, S. Korea

[+]School of ECE, Georgia Institute of Technology, Atlanta, GA, USA

Abstract: 3D integration is being touted as the next semiconductor revolution by industry. 3D integration involves the use of various interconnects that include balls, pillars, bond wires, through silicon vias (TSV) and redistribution layers (RDL) for enabling chip stacking, interposer and printed circuit board (PCB) based technologies. More recently 2.5D integration using silicon interposers has gained momentum as a viable solution for 3D integration. For such new integration schemes to be viable, mixing and matching of technologies is required to evaluate system performance early in the design cycle. The role of path finding is therefore to enable early exploration (planning) prior to costly implementation. Path finding must be based on an efficient electromagnetic analysis (EM) methodology which offers a good balance between speed and accuracy. In this work a hybrid solver is used which combines the Method of Moments (MoM) technique with specialized basis functions and the Partial Element Equivalent Circuit (PEEC) method to accurately provide design guidance.

Three types of test cases are used to explore key implementation areas. The impact of local interconnection density and routing topology for a 2 or 3 layer low cost Silicon interposer technology is investigated. Eleven signal lines are routed within a PWR/GND mesh grid for this example where the line width and spacing is varied to determine the variation in performance. A 49 TSV array is implemented in order to analyze near-end crosstalk (NEXT) between various TSVs in the array. The TSV array is varied to determine the crosstalk impact to determine where signals can be assigned. Wirebonds in a PoP (Package on Package) structure are designed to analyze the effect of design variations on performance. It is predicted that classical PCB designs for consumer electronic devices will continue to shrink as PoP implementations prove more advantageous for speed, area, power and weight related issues. The interconnect length and other parameters of the bond wires is varied to determine the impact on SI performance metrics. Various configurations of the wirebond structure have been demonstrated.

While several variations for each of the test cases described above is analyzed to determine the impact on signal integrity and performance, a larger parameter set can be explored using a Design of Experiments (DoE) methodology, which is not covered in this paper. From these findings, a set of rules can be created for detailed implementation. The examples covered show the attractiveness of using an exploratory tool early in the design cycle.

Keywords — path finder, insertion loss, cross talk, TSV, package on package, bond wire

I. INTRODUCTION

As the semiconductor and packaging industry moves towards 3D integration, the impact of vertical interconnections is becoming very important. This coupled with high density redistribution lines (RDL) in the layers of the interposer is allowing for high integration density. Three embodiments of vertical integration is shown in Figure 1 namely, a) chip stacking using wirebonds, b) package on package and c) chip integration using through silicon vias (TSV) and interposers. With technologies rapidly changing, modifications of the three embodiments shown in Figure 1 are necessary to meet both the cost and performance targets. Hence, systems of the future will contain a mixture of these technologies, where as an example, ICs can be assembled on to a substrate using a combination of wire bonds and micro bumps with high density redistribution layers on a silicon interposer containing through silicon vias providing the necessary conduit for communication, with a host of other materials and structures providing for an inhomogeneous interconnection environment.

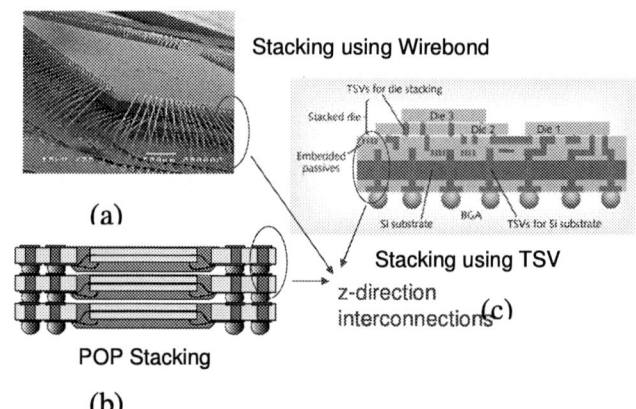

Figure 1: (a) Stacking of ICs using wirebond, (b) Package on Package stacking and (c) 3D ICs on silicon interposer with TSV [1]

Integration of such disparate technologies for a system architect or designer can be challenging since doubts often exist as to whether the combination of such technologies can meet the performance and cost targets required. These doubts can continue even after a combination of technologies is chosen since the structures used in the design needs careful evaluation as to whether they meet the performance targets. In addition process variations and other electrical interactions can make the technologies difficult to implement. The role of path finding is therefore to enable early analysis prior to implementation to minimize expensive modifications to either the technologies chosen or the structures being implemented. This is illustrated in Figure 2 showing an improved design flow where path finding as an important exploratory tool is shown which can help reduce overall cost of the design cycle.

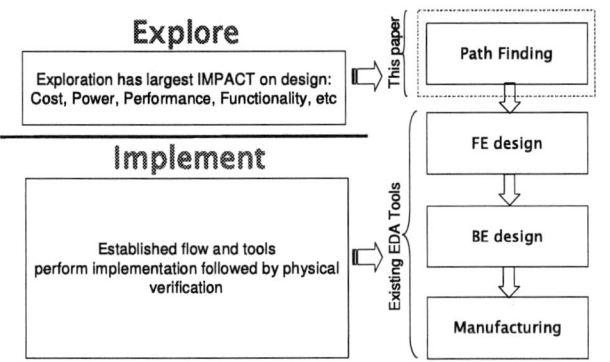

Figure 2: Exploration Vs Implementation and Design Flow [2]

In [2], a path finding methodology for 3D integration was introduced. In this paper details on the model development and hybrid solver is discussed along with examples that elaborate on the application of path finding in the context of vertical interconnections that include wire bonds and TSVs along with redistribution layers in a silicon interposer. This paper is organized as follows: In section II the 3D Path Finding methodology is briefly discussed for completeness along with a description of model development in section III followed by the solver in section IV. Three examples in the context of path finding are discussed in section V with the conclusions summarized in section VI.

II. 3D PATH FINDER (3DPF) METHODOLOGY

A discussion on the 3D path finder methodology is available in [2] which is briefly repeated here for completeness. Consider as an example ICs that are assembled on to a package using wire bonds with the lines routed in the top and the bottom package to create a connection between the ICs. The packages connect to each other through solder bumps. The objective of path finding is to develop a methodology for analyzing the interconnection path by changing the dimensions of the structures (on the fly) so that issues related to signal and power integrity can be assessed early in the design cycle. This may require modifying the diameter of the wire bonds, dimensions of the RDL in the package, diameter of the solder bumps, signal-to-ground ratio, material properties of the package, underfill material between the two packages, to name a few. Hence, the number of alterations to the design is many and therefore a methodology is required whereby a) the structures and test cases can be generated with ease, b) the structures can be analyzed in a reasonable time with manageable memory and c) rules can be developed based on this analysis that can be used for design once the technology options are finalized (not covered in this paper). A lego block concept is used in this paper to create the structures where all the critical parameters

are parameterized to generate test cases. Details of model development are described in the next section. It is to be noted that most vertical structures such as wire bonds, vias, TSVs and pillars have cylindrical cross section while the redistribution layer (RDL) has rectangular cross section. In addition, spherical structures such as solder balls can be represented using cylindrical objects. Since the vertical interconnections are critical for 3D integration and are often times difficult to discretize due to the large number of mesh elements required, specialized basis functions are used in a Method of Moments (MOM) formulation to minimize memory and CPU time for analyzing these structures. The RDLs are then meshed using an MOM formulation where the Partial Element Equivalent Circuit (PEEC) approach is used to connect the lateral and vertical interconnection structures together. Details of the solution procedure are discussed in this paper.

III. MODEL DEVELOPMENT

Path Finding requires an evaluation of a large set of alternative solutions to determine which one or ones are suitable for a specific design implementation. This requires any path finding tool to be 1) fast to construct a large set of test cases, 2) support a rigid but flexible methodology; and 3) provide a reasonably fast but accurate analysis.

3DPF, an EDA tool described in this paper, has been developed using the concept of LEGO™ blocks, a toy franchise that has existed for decades allowing kids of any age to imagine and build their creations. LEGO™ blocks allow stacking various sized and shaped blocks requiring minimal instructions or effort. Each block can contain specific via/metal structures and blocks can be stacked to form larger 2.5/3D structures using balls, pillars and/or bond wires. All 3D interconnects are defined as unique profiles and are contained in libraries. The various libraries and profiles can be modified to meet specific manufacturing process electrical and physical requirements. The basic idea behind 3DPF is therefore to allow users to create their own building blocks and then use (and reuse) these blocks to build more complex 2.5/3D structures. Once the various path finding test cases are constructed, each test case's performance can be evaluated against design requirements.

Since 3DPF was specifically developed for 3D integration, three design types are supported namely, Through Silicon Via (TSV), Column Grid Array (CGA) and 3D System. The GUI operations are divided into 3 groups based on frequency of use. The difference between the TSV and CGA based designs is that in the former, the basic building block is a semiconductor while in the latter it is a dielectric, with both supporting through vias. These two design approaches can be connected together in the 3D System using blocks and by adding redistribution layers, wirebonds, solder balls and micro-bumps. Figure 3 shows the methodology used for model development.

In Figure 3, the user first constructs or acquires a library of blocks that can be used to construct 3D systems, where each block has its own electrical and physical properties and can be analyzed separately before its integration into a larger system. These can be thought of as passive ICs that are constructed with via and metal interconnect layers (without transistors). Blocks can be created in GUI or an ASCII file can be created from various scripting languages such as PERL, TCL, Excel spreadsheets or any Word/Notepad application and imported. In either case, only ports need to be placed prior to analysis. Each block can be used to evaluate the effect of different electrical properties, via profiles/topologies or can be used to connect to other blocks.

To construct the 3D system, all building blocks are imported into the workspace and connected to each other. This could be in the form of stacks (as in 3D integration) or placed side by side (as in 2D or 2.5D integration). The base block is the original 'parent' of the structure. This parent base could be thought as a package, within which all other blocks are arranged using several interconnection schemes.

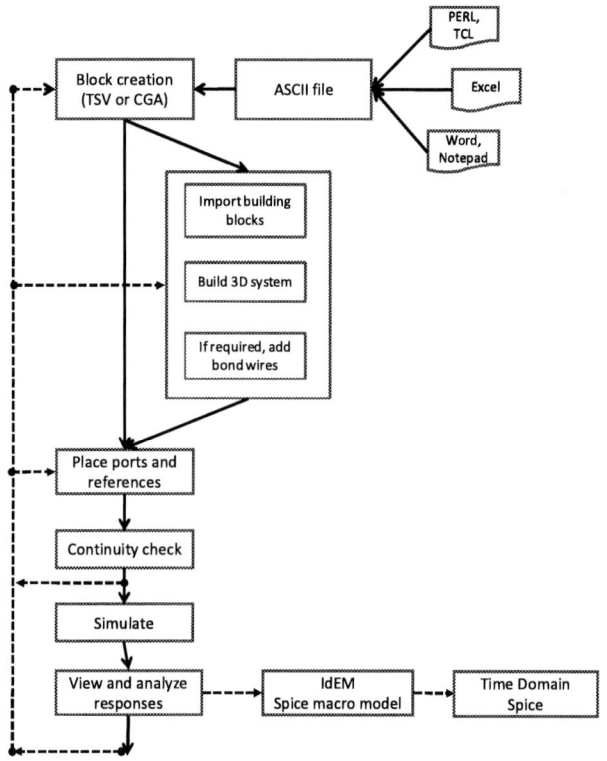

Figure 3: Model Development

Any blocks added become children of this parent. If additional blocks are stacked on these children, they become grandchildren of the base block. Each time a block is added, balls, bumps or pillars can be added at the interface or the interface can be left empty (example is two packages brick-walled to each other). The stack-up design tree can be used to define the parent/child relationship for the individual blocks that form the 3D system. The design tree only affects the Z dimension where the XY dimensions can be suitably aligned with other blocks in the parent/child hierarchy.

As an example consider Figure 4, where the design consists of four unique blocks namely, CGABase, CGA3x3, MemTier and MemCube [2]. Each block contains an array of vias. The blocks are placed on each other with CGABase as the parent and MemCube, MemTier and CGA3x3 as the

children. In addition, CGA3x3 is placed as the child to CGA3x3. This parent/child relationship is shown in Figure 4. Also, it is important to note that block CGA3x3 is repeated to create three instances of CGA3x3. Each block before being placed in the system is attached to micro-bumps, and the blocks are then aligned to each other.

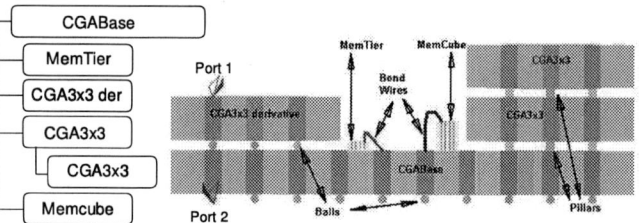

Figure 4: 3D Design showing parent/child relationship

The user can continue to add blocks until all blocks have been added and vertically interconnected. If errors are found during the stacking process, the user can perform *Edit in Place* to correct the errors. When saving individual block edits, the user can either replicate the edits to all blocks with the same name OR create a derivative block (CGA3x3 derivative in Figure 4). If a derivative block is created, only that instance will reflect the changes while all other instances of this block will remain as originally designed. Once all the blocks are correctly placed, bond wires can be added to the design. As with balls and pillars, a library of bond wires can be used for placement which includes custom structures as well.

Two final steps are required before analysis can begin which includes port placement and a continuity check. Ports represent positions where the electrical response can be computed in the form of insertion loss, return loss and cross talk. Continuity checks are similar to DRC (design rule checker), where the physical continuity of the signal path is verified along with the associated references. This provides an elegant method to check for opens or shorts in the design, especially when complicated structures are created.

A design that has passed continuity checks can be simulated to obtain the frequency domain response, which can then be reviewed to ensure if they meet design specifications over the frequency band of interest. Another possibility is to convert the frequency domain response in the form of touchstone files into spice netlists through Idem [3]. Spice netlists can then be used for time domain analysis as well in standard circuit simulators.

IV. SOLVER

After the 3D structure is subdivided into several blocks as described in the previous section, each block can be modeled as a multi-port network by using numerical procedures based on the mixed-potential integral equation. This section summarizes the numerical method, which is composed of three separate procedures, namely 1) modeling dielectric based structures with through vias; 2) modeling of semiconductor based structures with through vias and 3) connectivity between structures to obtain the overall electrical response, which have been deployed in the 3D Path Finder.

978-1-4799-2408-0/14 $31.00 © 2014 IEEE   2126

Further details on the numerical solution techniques can be found in [1]. Since the most important interconnections in 3D integration are the z-directed wires, special attention is given to the vias during model extraction.

A. Modeling of Package Interconnections and other Planar Structures in dielectrics and insulators (CGA Block)

Any block including wirebonds, package vias, column grid arrays (CGA), balls, and planar structures can be modeled by using the mixed-potential integral equation, as in the Partial Element Equivalent Circuit (PEEC) approach. The main difference between the method implemented in 3DPF and conventional PEEC method is that it removes the discretization process for structures with cylindrical cross section by using specialized modal basis functions for capturing current and charge density distributions [5]. Since the required number of modal basis functions is less than seven for wirebonds and vias, the resultant equivalent circuit model becomes simplified, and the modeling of a large number of vertical interconnections in 3DPF [4] becomes possible. As an example, modeling bond wires using the cylindrical modal basis functions are illustrated in Figure 5.

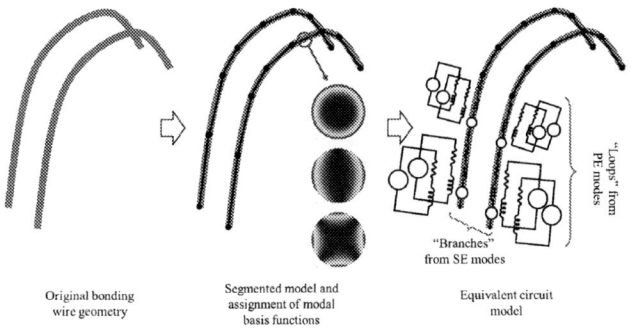

Figure 5: Procedure for modeling bond wires

Planar structures such as microstrip lines, strip lines, pads, and ground planes are modeled using the conventional PEEC method, where the exact interconnection model capturing the thickness effect can be generated. Also, to reduce the computational time, thin-metal approximation that only includes the planar coupling effects can be used (though both the thin and thick metal extraction has been implemented).

The coupling between cylindrical and planar structures requires integration involving piecewise constant basis functions and cylindrical modal basis functions. For considering solid and large ground planes, the image method has been used instead of the PEEC plane model.

B. Modeling of Semiconductor Substrate with TSVs (TSV Block)

If part of a 3D structure contains TSV interconnections, it is considered as a separate building block, and the associated TSVs are modeled by using another modeling method, with details available in [5], [6] and [1].

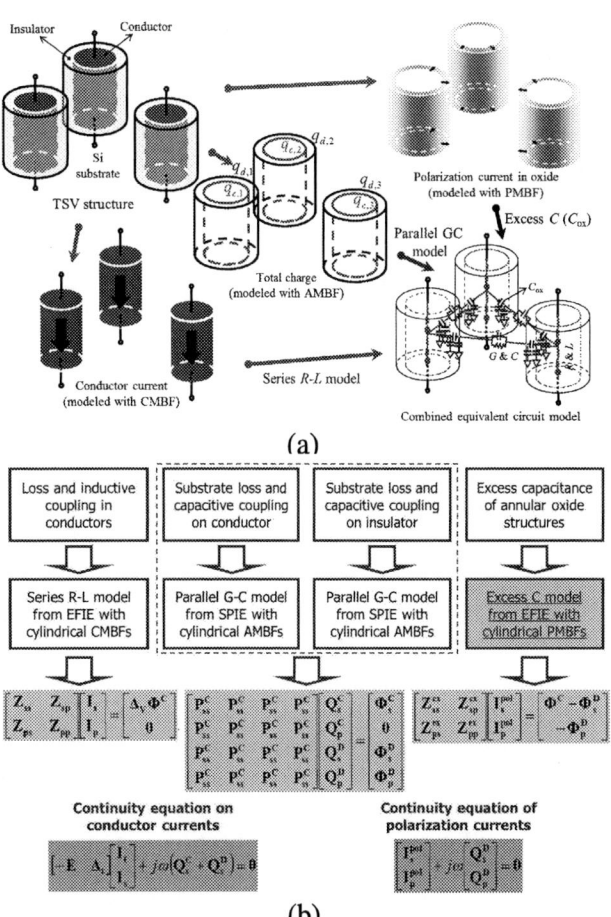

Figure 6: (a) TSV modeling using specialized basis functions and (b) Formulation for modeling TSVs

The TSV modeling method is an extension of the integral-equation-based method in the previous sub-section, where the effects of lossy silicon substrate and oxide liner around each TSV conductor are captured by adding excess capacitors generated with the polarization mode basis functions. The entire TSV model is composed of series resistances/inductances and parallel capacitances/conductances, as described in Figure 6 [1]. Although the generated model is similar to the analytical circuit models based on physics or intuition based methods, the method used in 3DPF is robust and can address the electrical coupling from arbitrary configurations of TSVs. The effect of temperature of the bulk silicon substrate is also considered by including the temperature-dependent conductivity of silicon.

Extension of this method for modeling the depletion effect due to DC biasing is also possible, as discussed in [7]. In addition, tapered vias can be modeled by discretizing the length of the TSVs, which is useful for process optimization.

C. Node Merging and Port Assignment

To generate the network parameter model of the entire 3D structure, submodels obtained from the modeling methods (discussed in the previous subsections) for separate blocks should be combined, and ports should be assigned for the overall structure.

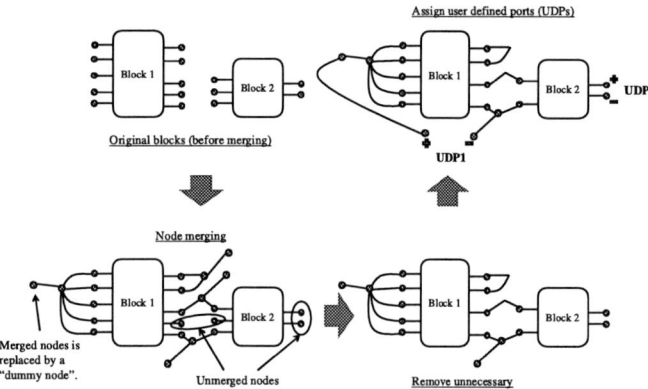

Figure 7: Node merging and port assignment procedure

The process of combining blocks is internally performed through a node merging process, as described in Figure 7. Firstly, some nodes of the original network for individual blocks are merged into a single node. As shown in Figure 7, the nodes to be merged can be in the same block or they can belong to different blocks. After merging, several pairs of reduced nodes are assigned as user defined ports (UDP). The remaining nodes that are not used for UDP are left as open nodes.

V. EXAMPLES

We present three examples in this section using the 3D Path Finder to demonstrate its functionality in the early exploration phase.

Example 1: Comparison of wire bond topologies

In this example, various 1 mil diameter wire bond structures are simulated up to 20GHz to examine impact of: materials (Al vs. Au vs. Cu), wire length (40 mils to 100 mils), wire bond pitch (2.36 mils to 4.72 mils) and impact of return path (location and diameter of PDN wires). The return path is defined as one of the wire bonds contained in Figure 8. As expected, the choice of material had insignificant affect on insertion loss for 40 mil long wires and approached -1dB at 20GHz. All other wire bond simulations were performed using only aluminum (Al) wires.

The insertion loss for the wirebonds is shown in Figure 9a when the length is varied from 40 mils to 100 mils with the frequency response plotted over a frequency band of 20GHz. In Figure 9b, the location of the bond wire and its diameter is varied to understand its impact while in Figure 9c, the effect of wire pitch has been modeled where a tighter pitch results in a lower insertion loss due to the close proximity of the return path. The path finding activity in this example is therefore to understand the impact of various wire bond parameters on its performance.

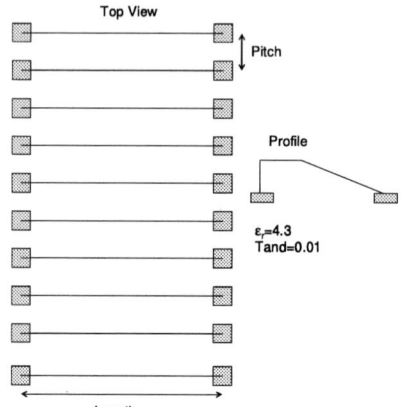

Figure 8: Wirebond parameters

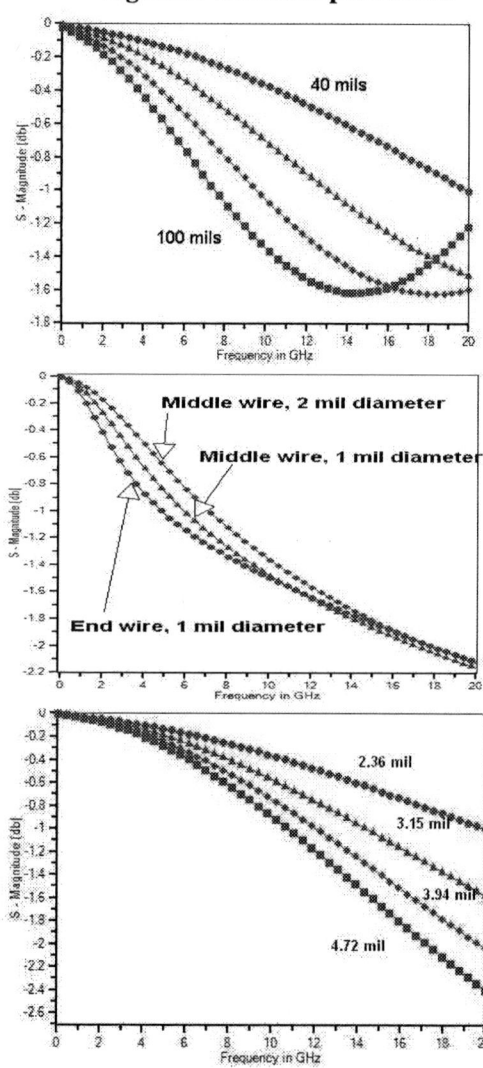

Figure 9: Insertion Loss for Wirebond (a) Top: Function of length, (b) Middle: Function of location/diameter and (c) Bottom: Function of pitch

Example 2: Signal Lines (RDL) and referencing

Consider a group of 11 L-shaped signal lines as shown in Figure 10 with Vss lines beneath it and Vdd lines on the same layer as the signal lines. The bottom layer is a solid ground plane. The dimensions of the lines and cross sectional dimensions of the structure (3um and 30um thick dielecric) are shown in Figure 10. The dielectric used has a relative permittivity of 4.0. The objective is to compute the insertion loss for the various pitches: 1) Width and space of 3um with unique return paths, 2) Width and spacing of 3um with grouped return paths, 3) Width of 3um and spacing of 7um and 4) Width and spacing of 5um. A more detailed analysis of this structure is provided here as compared to [2].

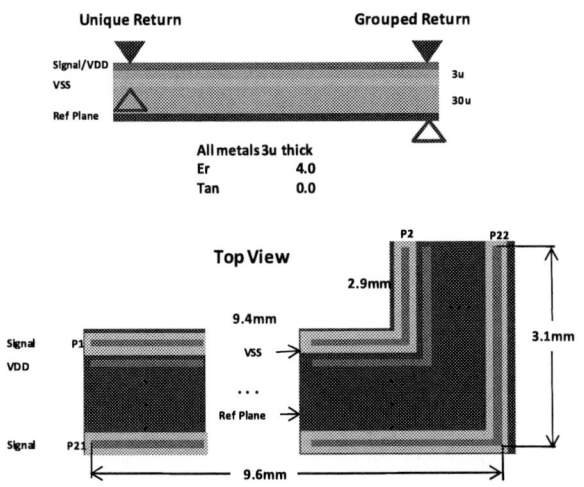

Figure 10: L-Shaped Lines with Vss and Ground reference

The insertion loss for the eleven lines (Cases 1-4) is shown in Figure 11 up to 20GHz. There are several noteworthy effects that can be seen from the figure as follows: For all four cases, each test case has a unique insertion loss profile and significantly degrades in the 4-8GHz range; Unique vs. grouped return paths have dramatic effect on signal line insertion loss due to the return path's lower resistance when grouped; and for this structure, it appears that 3um width and 7um spacing has the best overall insertion loss.

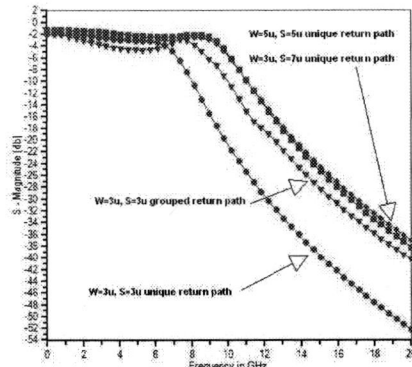

Figure 11: Insertion Loss for the L-Shaped Lines

However insertion loss is only part of the analysis since cross talk needs to be analyzed as well. As shown in Figure 12, Cases 1 and 2 show the effect of grouped return paths on

near end cross talk (NEXT). Cases 1, 3 and 4 also show the impact of varying width/spacing combinations. As the separation between signal lines increases NEXT improves as well.

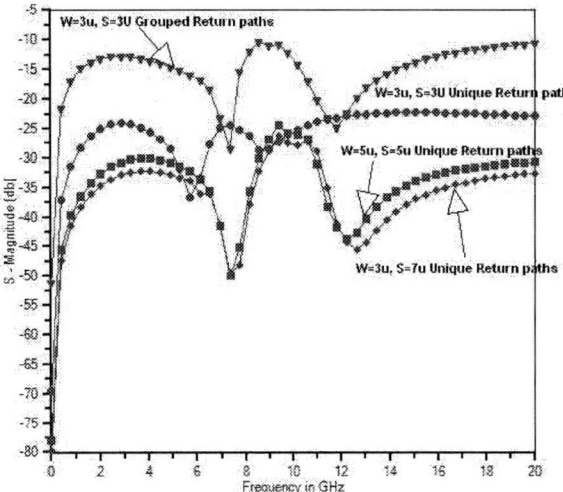

Figure 12: NEXT for L-Shaped Lines

Example 3: TSV Array topology and PDN assignments

This relates to a problem consisting of a TSV array shown in Figure 13a with all physical and electrical parameters provided [2].

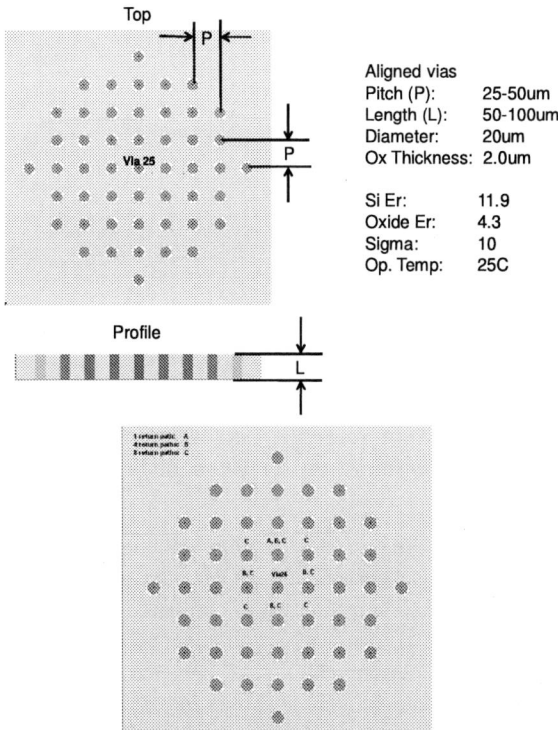

Figure 13: (a) Top: TSV array and parameters and (b) Bottom: Center Via (25) with return path cases A, B and C

The objective is to calculate the coupling between TSVs in the array with uniform cross section with varying via length/pitch and return path configurations (1 vs. 4 vs. 8

return paths). Figure 13b shows Via25 and the three return path test cases A, B and C. Each TSV has ports on top and bottom with infinite references. The first aspect investigated was to determine the effect of length and pitch (maintaining constant via diameter) on cross talk performance. Figure 14a shows insertion loss and Figure 14b shows near end crosstalk (NEXT), respectively.

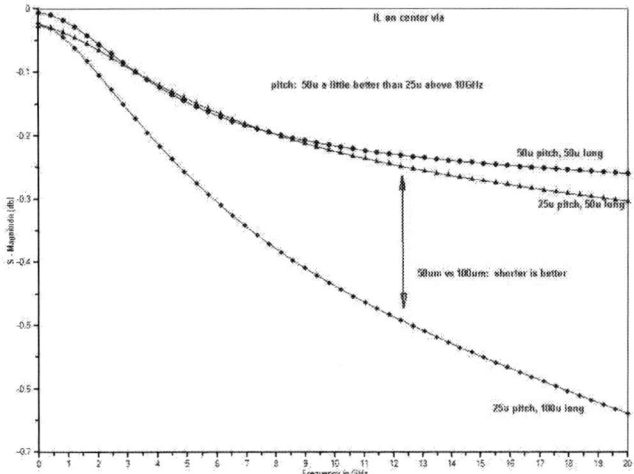

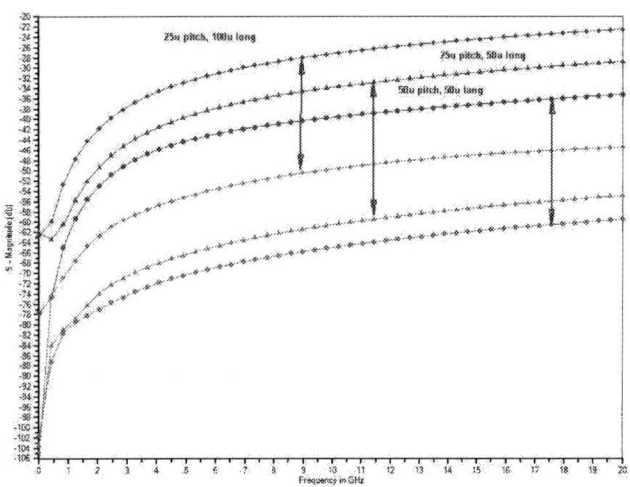

Figure 14: (a) Top: Insertion Loss and (b) Bottom: Near End Cross Talk for 25um Vs 50um pitch and 50um Vs 100um length

In Figure 14a the impact of varying via length with a 25um pitch is shown. Approaching 20GHz, the insertion loss is ~2X worse for a 100um long via as compared to a 50um long via. A thinner silicon wafer will always lead to a better design implementation. In Figure 14b, the worst and best near end cross talk (NEXT) is shown for each configuration. This shows that a wider pitch coupled with a shorter via length reduces NEXT by ~15dB.

Another aspect investigated was the three return path test cases and their impact on NEXT. In Figure 15, the NEXT responses for each test case are shown. The NEXT is measured from the center via to a via next to the return path via. As can be seen, Case A with a single return path has the

maximum cross talk on a nearby via whereas Case C with eight return paths has the least. Through path finding analysis, a design's specific signal and PDN assignment can be defined to minimize the performance impact.

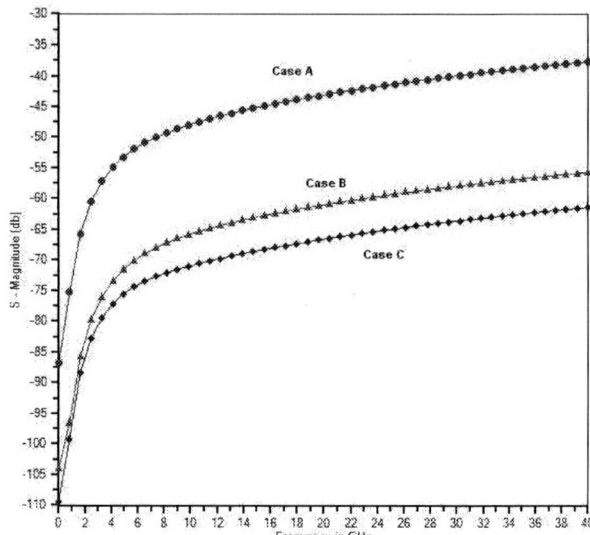

Figure 15: NEXT for different return path cases

VI. CONCLUSION

In this paper, a path finding methodology has been introduced for assessing electrical performance in the context of 3D integration early in the design cycle. The model development is based on a lego block concept where all the critical variables are parameterized for generating test cases easily. This is then used for analysis using an electromagnetic solver specifically developed for 3D path finding. Several 3D structures have been analyzed to demonstrate the value of path finding for exploration prior to design implementation.

VII. REFERENCES

[1] M. Swaminathan, K. J. Han, *"Design and Modeling for 3D ICs and Interposers"*, WSPC, 2013.
[2] M. Swaminathan, K. J. Han and B. Martin, "3D Path Finder Methodology for the Design of 3D ICs and Interposers", pp. 21-24, EDAPS, Nara, Japan, Dec. 2013.
[3] Idem, www.idemworks.com, 2013.
[4] 3D Path Finder (3DPF), www.e-systemdesign.com, 2013.
[5] K. J. Han, M. Swaminathan, "Inductance and Resistance Calculations in Three-Dimensional Packaging Using Cylindrical Conduction Mode Basis Functions," IEEE Transactions on Computer-Aided Design of Integrated Circuits and Systems, vol. 28, no. 6, pp. 846-859, Jun. 2009.
[6] K. J. Han, M. Swaminathan, T. Bandyopadhyay, "Electromagnetic Modeling of Through-Silicon Via (TSV) Interconnections Using Cylindrical Modal Basis Functions," IEEE Trans. on Advanced Packaging, vol. 33, no. 4, pp. 804-817, Nov. 2010.
[7] K. J. Han, M. Swaminathan, "Consideration of MOS capacitance effect in TSV modeling based on cylindrical modal basis functions," EDAPS, pp. 42-45, Taipei, Taiwan, Dec. 2012.

Gap in pagination due to withheld paper.

Pages 2131-2136

Dielectric Lens Optimization for Conical Helix THz Antennas

Paolo Nenzi[1], Volha Varlamava[2], Frank Silvio Marzano[2], Fabrizio Palma[2], Marco Balucani[2]

[1]ENEA Frascati Research Center – UTAPRAD-SOR, Via E. Fermi, 45 - 00044 Frascati (RM), Italy

[2]University of Roma "La Sapienza" – DIET, Via Eudossiana, 18 - 00184 Roma, Italy

Email: paolo.nenzi@enea.it, varlamava@diet.uniroma1.it

Abstract

In this work the design and characteristics of an extended hemispherical dielectric lens applied to a non-planar antenna for the detection of THz radiation is presented. Antenna and lens behavior is numerically investigated in a large frequency range around the central frequency of 1 THz. Common printable materials widely characterized at THz range are used for the lens realization. Numerical results of the full-wave electromagnetic analysis of some proposed structures are reported and discussed.

1. Introduction

In last years the interest addressed to the THz region (the frequency range between 0.1 and 10 THz) has grown because of the congestion of the electromagnetic spectrum at lower frequencies and of the need for higher data rate in digital wireless communications [1,2]. Whereas the communication industry is the strongest driver in the exploitation of the THz range, spectroscopy applications [3] and non-destructive testing [4] are today's major applications for THz radiation.

The successful utilization of THz technology is subject to the availability of compact, stable, room temperature and low cost sources and detectors. In particular, detectors with high sensitivity are required because of the very high attenuation of THz radiation in the atmosphere and because of the difficulty of producing high-power compact sources.

The integrated circuit technology has been identified as the most promising to manufacture reliable and economical detectors for the THz region [5,6]. In particular, in [6] a prototype of 1kpixel THz camera fabricated in commercial 65nm CMOS process has been presented. The camera is feed by USB power, operates at room temperature and has a bandwidth going from 600 GHz to 1 THz. In order to enhance the coupling between its 1024 pixels focal plane array and the outer environment, a silicon lens is places above the camera.

Most common materials used for the fabrication of dielectric lenses at THz region are high resistivity silicon and PTFE (Teflon). Usually, a single lens focuses the incident radiation onto a focal plane array. The approach adopting the high resistivity silicon is expensive, thus, low-loss polymers (as PTFE) should be preferred for low cost applications. In a recent work [7], 3D printable and injection moldable materials have been characterized at THz frequencies, as their attractive characteristics in terms of the reduction of lens cost and the possibility to apply wafer level process (3D printing or molding) to produce reproducible arrays of micro-lenses. This configuration contemplates a dielectric micro-lens over every single pixel (detector) constituting the focal plane array [8,9]. Dielectric lenses have been thoroughly studied for millimeter wave and sub-millimeter wave imaging systems to increase the gain of the planar antennas coupled to detectors (slot, bowtie, spiral, log-periodic, etc).

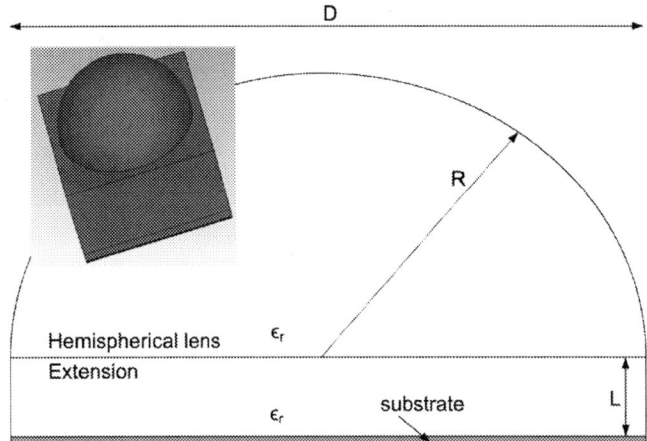

Figure 1. Cross sectional view of the extended hemispherical lens with principal parameters defined: the extension length L, the lens radius R (and the corresponding diameter D), and the permittivity ϵ_r. A tridimensional view of the extended lens antenna is shown in the inset on the top left of the picture.

The application of dielectric lens is an efficient technique to increases the gain and the radiation efficiency by the enhancement of the coupling between the electromagnetic wave impinging on the detector surface and by preventing the excitation of surface modes that take place when the substrate thickness is comparable with the wavelength of the exciting wave. This situation is particularly detrimental in THz systems, as the wafer thickness (usually between 50 μm and 800 μm) is comparable with the wavelength of radiation in the THz range (both in vacuum and in commonly used dielectrics).

An in-depth analysis of effects of a dielectric hemispherical lens has been reported in [10] where a power gain increase of 27dB was obtained for a double slot antenna coupled to an extended hemispherical lens at 94 GHz operational frequency. Authors also report an increase in the radiation efficiency in the lens coupled antenna: 98% for a dielectric with permittivity $\epsilon_r = 4$ and a lens radius $R>0.25\cdot\lambda_d$, with λ_d the wavelength in the dielectric. The verification of the reported condition provides the power gain to be independent from the lens radius.

Different lens configurations have been studied in literature: hemispherical, hyper/hypo-hemispherical and extended hemispherical lenses [5,8,9,11]. The extended hemispherical lens is the most versatile solution as it permits to synthesize the other types by varying the extension length. An example of hemispherical lens is shown in Fig. 1, with its three-dimensional model in the inset.

In this work we present the results of the coupling of a non-planar antenna to an hemispherical dielectric lens to verify the effectiveness of the described technique for non-

978-1-4799-2408-0/14 $31.00 © 2014 IEEE

2137

2014 Electronic Components & Technology Conference

planar antennas. The u-helix antenna developed by authors [12, 13] is used as embedded antenna because of its interesting characteristics in the THz range.

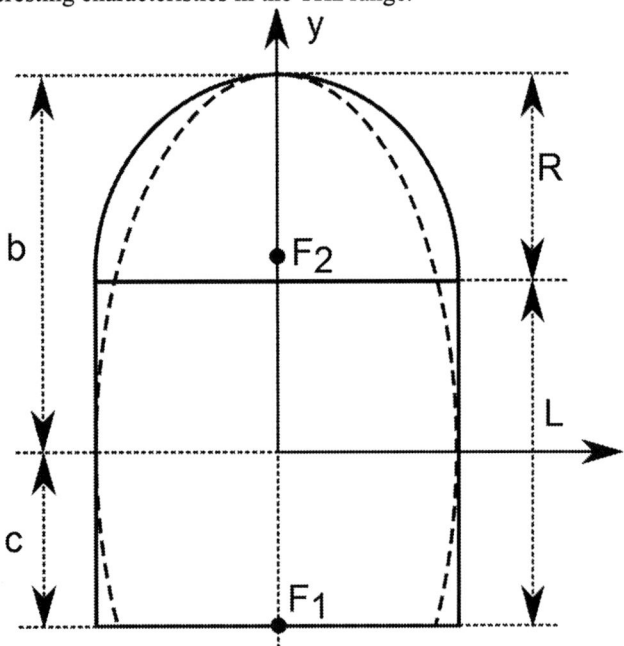

Figure 2. Extended hemispherical lens superposed to its elliptical approximation.

2. Lens design

As the condition of the rotational symmetry is satisfied, we perform our analysis for a two-dimensional case (see Fig. 2). In xy-plane ,the defining equation for an ellipse with foci at $c = \pm\sqrt{b^2 - a^2}$ is

$$\left(\frac{x}{a}\right)^2 + \left(\frac{y}{b}\right)^2 = 1, \qquad (1)$$

with a and b, called ellipse half-axis, both positive. The basic optics provides the condition that permits the geometric and the optical focus to match:

$$e = \frac{\sqrt{b^2 - a^2}}{b} = \frac{1}{n}, \qquad (2)$$

where e is the ellipse eccentricity and n is the refraction index of the material that constitutes the lens. Combining (1) and (2), after some algebra simple relations between the ellipse geometry and the refraction index are obtained:

$$a = b\sqrt{1 - \left(\frac{1}{n}\right)^2}, \qquad (3)$$

$$c = \frac{b}{n} \qquad (4)$$

In order to superimpose the synthetized lens with an ellipse, we make some assumption on relations between ellipse (a, b, c) and design (L, R) parameters . Firstly, we suppose the total extension of the hemispherical lens is equal to the distance from the tip of the ellipse and its more distant focus. Moveover, we suppose the a is a good approximation for the radius of the synthetized lens R. Thus, to derive:

$$\begin{cases} L + R = b + c \\ R = a \end{cases} \qquad (5)$$

Further transformations lead to the design equation of an extended hemispheric lens studied in the following:

$$L = b\left(1 + \frac{1}{n}\right) = \frac{a}{\sqrt{1 - \left(\frac{1}{n}\right)^2}}\left(1 + \frac{1}{n}\right) = R\sqrt{\frac{n+1}{n-1}} \cdot \qquad (6)$$

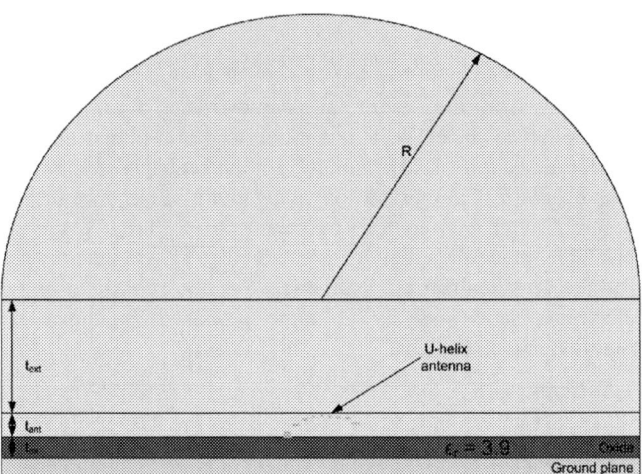

Figure 3. Cross sectional view of the model simulated with CST.

3. The simulation model

The model of the antenna embedded into the dielectric lens has been simulated with CST electromagnetic tool in order to numerically derive the directivity and the far field pattern of the structure.

The cross sectional view of the simulated model is depicted in Fig. 3. The u-helix antenna embedded into the dielectric lens is designed to operate at 1THz in free space, and its main characteristics was presented in [12] and are reported in Table 1 for convenience.

Table 1. Geometric parameters of the u-helix antenna used in the electromagnetic simulation (from [12]).

f [Hz]	R_{ext} [µm]	$W_{pad(ext)}$ [µm]	$W_{pad(int)}$ [µm]	T_{metal} [µm]	H_{tot} [µm]
1e12	55	9	3	3	48

In Table 1, f is the center design frequency, R_{ext} is the radius of the external branch of the spiral (see Fig. 4(a)), $W_{pad(ext)}$ is the width of the metal at the anchor point (see Fig. 4(a)) and $W_{pad(int)}$ is the width of the metal at the center of the spiral. The thickness of the metal is denoted as T_{metal} and the total height of the antenna, once pulled and embedded into the polymer, is H_{tot}. The antenna has been imported into CST and simulated as composed of PEC (Perfect Electric Conductor) material.

The antenna has been placed over ground plane and isolated from it with a 6µm thick (t_{ox} in Fig. 3) oxide layer. The oxide has the permittivity of SiO_2 (i.e. $\epsilon_r = 3.9$) to simulate an antenna placed over the topmost metal layer of an integrated circuit (the oxide layer simulates the total thickness of inter-metal dielectrics stack). The antenna is embedded into a polymer with the thickness of t_{ant} (see Fig. 3) and equal to H_{tot}. This layer is composed of the same material that the extended dielectric lens and is necessary to confer a

978-1-4799-2408-0/14 $31.00 © 2014 IEEE

mechanical stability to the structure. The thickness of this layer must be taken into account when computing the length of the lens extension. The extended hemispherical dielectric lens is built on top of this layer and consists of a square extension t_{ext} thick and a spherical dome of radius R. All the polymeric materials have been modeled in CST with the real and imaginary part of the permittivity or the dielectric constant and loss tangent. Data for the two polymers used: ABS and HDPE were taken from previously published works [7,14]. In particular, data for ABS were taken from [7] because they were extracted for a material actually used in 3D printing. Data for HDPE were taken from [14].

Our CST simulations were carried out using the time domain solver. This choice permits to achieve a good accuracy of results and to reasonably reduce the simulation time.

4. Simulation results

Six different configuration of the dielectric lens have been simulated as shown in Table 2, and the results, compared with the simulations of the basic antenna without the lens.

Table 2. Geometric parameters of the simulated dielectric lens configurations. Each configuration is assigned a code name composed by three fields separated by the underscore character. The first field identifies the material, the second filed the lens radius and the third one, the extension length.

Code	R [μm]	t_{ext}[μm]
HDPE_400_80	400 $2.0\lambda_d$	80
HDPE_400_472	400 $2.0\lambda_d$	472
HDPE_1250_250	1250 $6.4\lambda_d$	250
HDPE_1250_1500	1250 $6.4\lambda_d$	1500
ABS_400_400	400 $2.2\lambda_d$	400
ABS_1250_1200	1250 $7\lambda_d$	1200

Simulation results for the basic antenna (i.e. without the coupled lens) have been already presented in [12]. The simulated directivity of 7.6 dBi was calculated with a FWHM (Full Width Half Maximum) greater than 30°.

The values of the maximum gain of the simulated configurations, versus frequency, are collected in Fig. 4.

The single antenna operates in the axial mode, i.e. the maximum gain is expected to be directed along antenna vertical axis. The same behavior is observed with the dielectric lens.

The plot in fig. 4 clearly shows the focusing effect of the lens on the gain: the two configurations with higher maximum gain values are the ones with larger lenses (HDPE_1250_1500 and ABS_1250_1200).

In particular, maximum gain values of larger antennas follow the same trend (with ABS_1250_1200 being higher)

up to 700GHz where the gain of HDPE_1250_1500 becomes higher, up to 1100GHz.

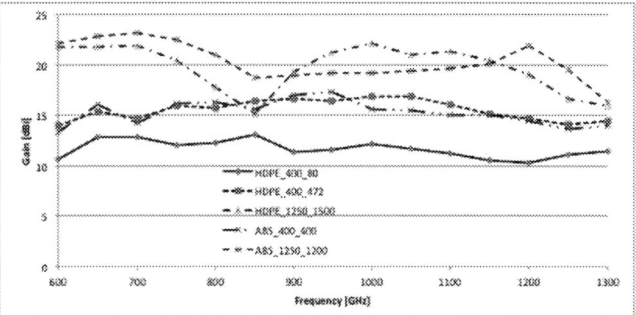

Figure 4. Maximum gain versus frequency of the simulated configuration of the dielectric lens antenna.

This behavior is attributed to the influence of the lens extension that changes the position of foci of the lens with respect to the antenna. The smaller lens, HDPE_400_80, with an extension thickness of 80μm shows the lowest gain of the lenses with 400μm radius. This confirms that, for a given lens radius, the extension length t_{ext} strongly influences the maximum gain value because it varies the position of the internal focus of the lens. The maximum gain is obtained when the u-helix antenna is placed between the two foci of the lens.

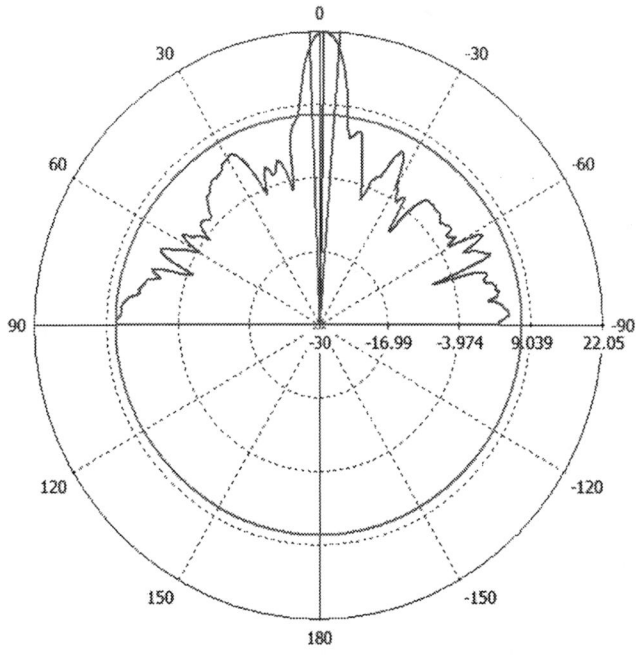

Figure 5. Farfield plot of HDPE_1250_1500 at 1THz, plane phi = 0. Maximum gain is 22.5 dB, an angular width (FWHM) is 6.6 degrees, and the side-lobe level is -14.8 dB.

Fig. 5 and Fig. 6 show the farfield plots of the HDPE_1250_1500 antenna for two different phi angle values (90° apart), simulated at 1THz. From the picture it is evident the effect of multiple reflections inside the lens that generate side-lobes with level as high as -14.8dB.

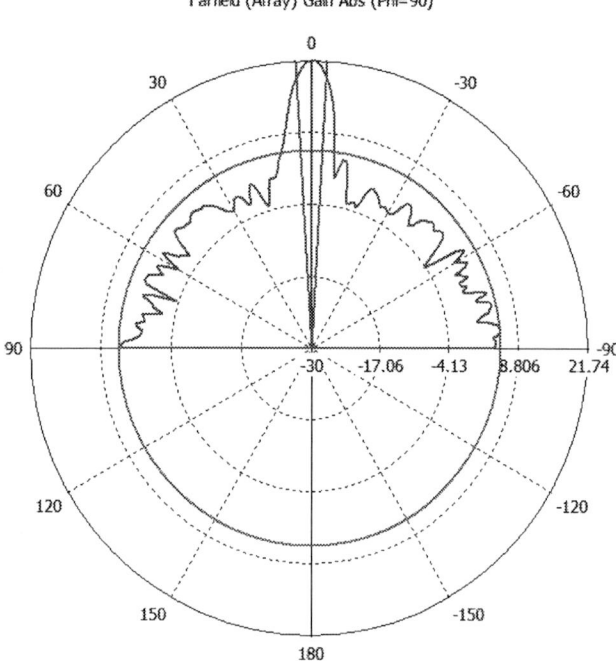

Figure 6. Farfield plot of HDPE_1250_1500 at 1THz, plane phi = 90. Maximum gain is 21.74 dB, an angular width (FWHM) is 6.5 degrees, and the side-lobe level is -16.2 dB.

The maximum gain direction presents a tilt angle of 1 degree and this is common for the u-helix as this structure is fed from one edge and not from the center. The feed position introduces a slight tilt in the position of the gain maximum. The tilt is directed along the phi=0 direction. The FWHM angular width of the antenna is of 6.6 degrees (with 0.1 degree of difference between orthogonal phi directions), corresponding to the 22% of the original antenna FWHM.

The maximum directivities for this configuration are 22.5 (phi=0) and 22.2 dBi (phi=90). The effect of the lens in directivity an increase of 14.9dB. The three dimensional view of the farfield plot for this configuration is shown in fig. 7, with the lens structure below.

As comparison, fig. 8 and 9 show the farfield gain plot for the HDPE_400_472 configuration. In this configuration, lower gain (16.8dB) and wider FWHM are obtained (20.3 degrees). Side lobe level is comparable (-14.4 dB) and the effect of the asymmetry in the u-helix antenna (the tilt angle of the direction of maximum gain), is more evident (2 degrees). The directivities, for this configuration are: 16.9 along phi=0 and 16.8 dBi along phi=90. The three-dimensional view of the far field of this configuration is shown in fig. 10.

The last configuration presented is the HDPE_400_80 (see figs. 11 and 12). In this configuration the lens only distorts the u-helix antenna pattern (see [12]). There is a slight increase in directivity (4.3dB) with respect to the original u-helix, but a strong distortion of the radiation pattern. The direction of the maximum gain is off-axis by 9 degrees.

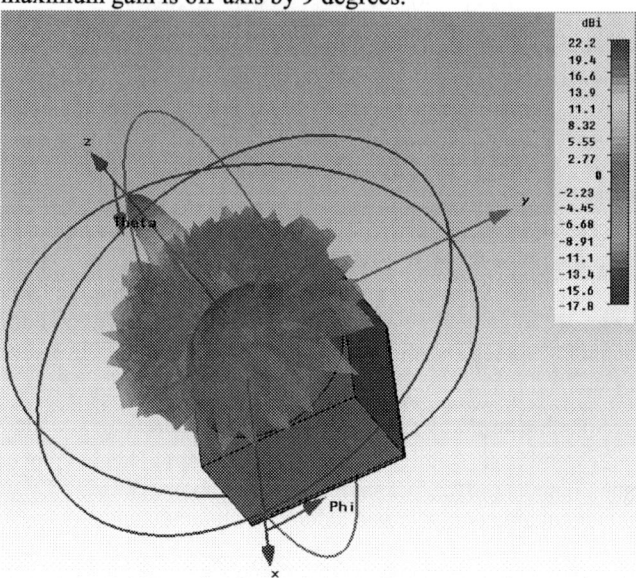

Figure 7. Farfield of HDPE_1250_1500 at 1THz, 3D view.

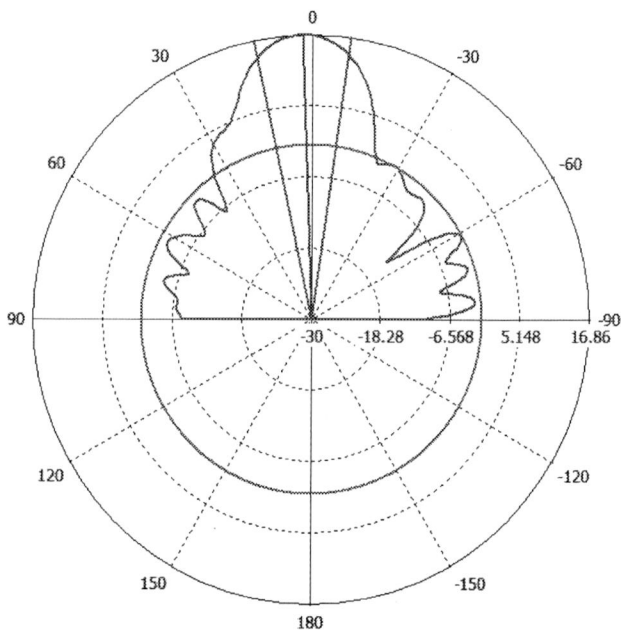

Figure 8. Farfield plot of HDPE_400_472 at 1THz, plane phi = 0. Maximum gain is 16.9 dB, an angular width (FWHM) is 20.3 degrees, and the side-lobe level is -18.1 dB.

The side-lobe level is higher than the other configurations (-11.6dB). This configuration, together with the other two presented, demonstrate the role of the extension length for the extended hemispherical lens. The optimum found in this work is placing the antenna at the middle between the two foci (see fig. 2). The radiation patters for larger configurations, with lens radius of 1250 µm, show analogous results.

978-1-4799-2408-0/14 $31.00 © 2014 IEEE

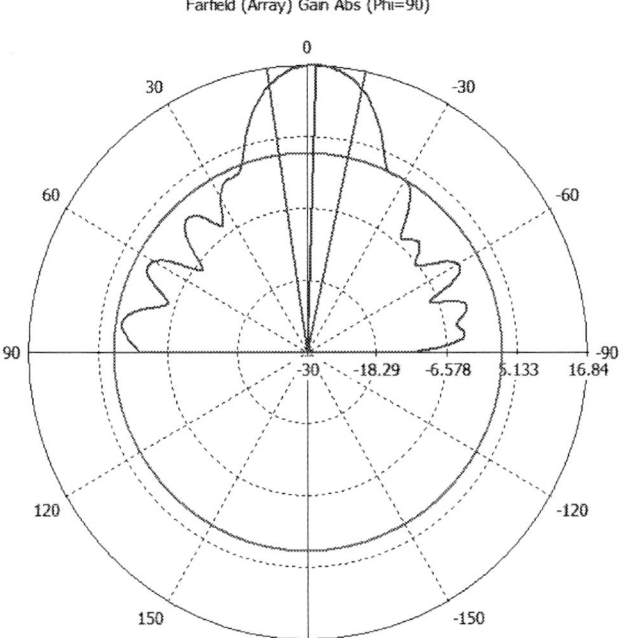

Figure 9. Farfield plot of HDPE_400_472 at 1THz, plane phi = 90. Maximum gain is 16.8 dB, an angular width (FWHM) is 20.4 degrees, and the side-lobe level is -14.4 dB.

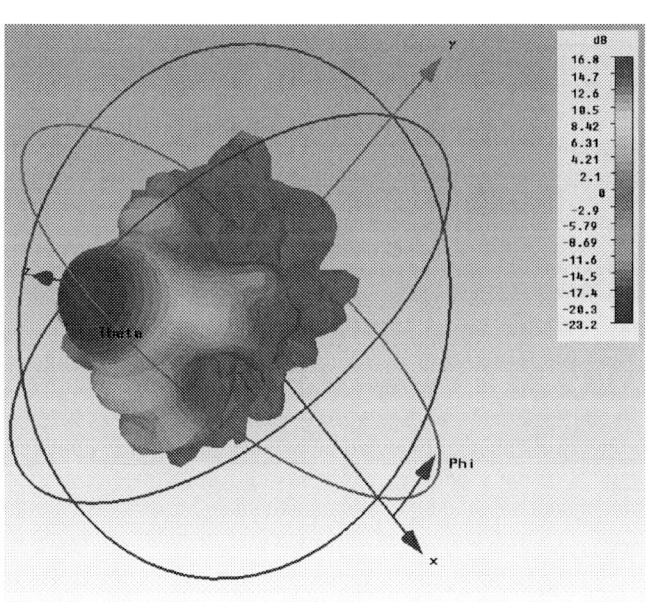

Figure 10. Farfield of HDPE_400_472 at 1THz, 3D view.

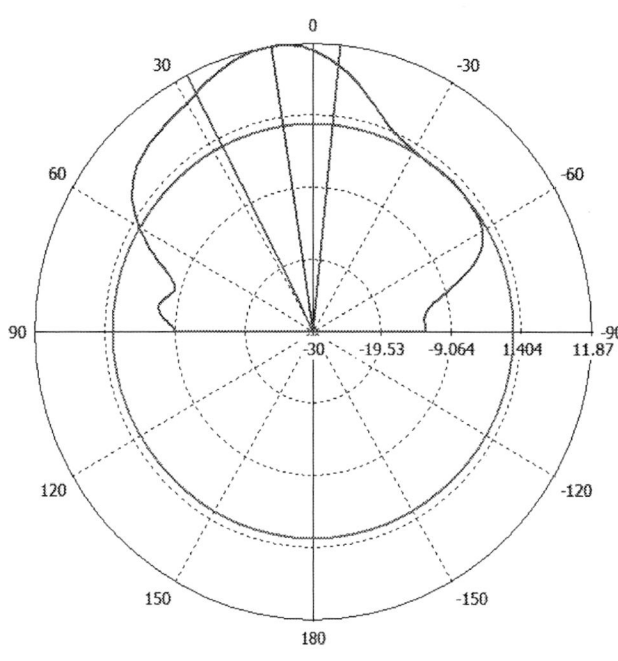

Figure 11. Farfield plot of HDPE_400_80 at 1THz, plane phi = 0. Maximum gain is 11.9 dB, an angular width (FWHM) is 32.9 degrees, and the side-lobe level is -11.6 dB.

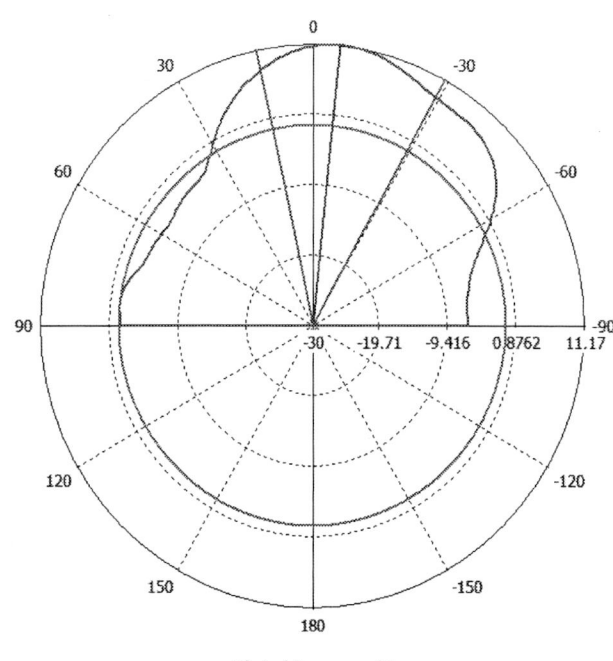

Figure 12. Farfield plot of HDPE_400_80 at 1THz, plane phi = 90. Maximum gain is 11.2 dB, an angular width (FWHM) is 41.5 degrees, and the side-lobe level is -11.8 dB.

978-1-4799-2408-0/14 $31.00 © 2014 IEEE   2141

5. Lens construction

The methodology followed to design the U-Helix antenna has been presented in [12] where an in-depth analysis of the mechanical design has been presented.

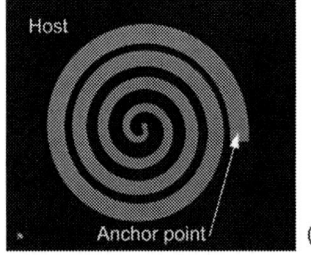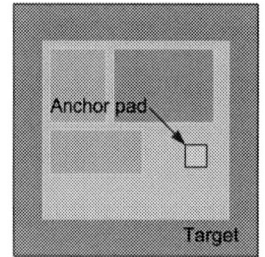

Figure 13. Top view of the antenna on the host substrate (a) and of a target substrate (b), where the antenna will be deployed. Anchor point and anchor pad are bonded together.

The antenna is initially patterned onto a host substrate with common lithographic techniques as shown in Fig. 13(a). A second substrate (so called target substrate, Fig. 13(b)), has a pad that is bonded to a specific point of the antenna (anchor point).

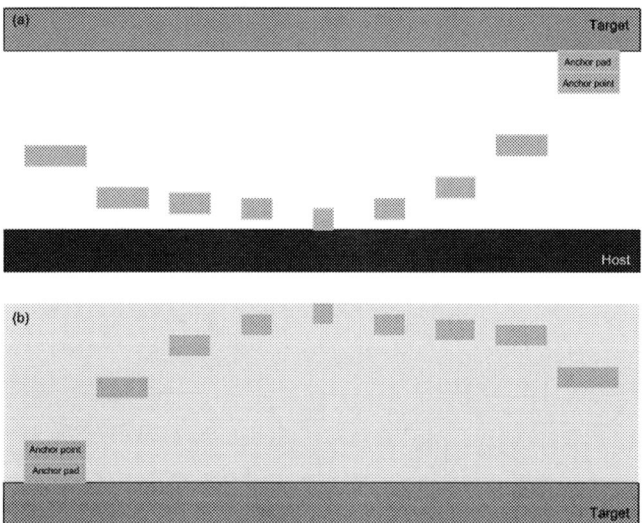

Figure 14. Cross sectional view of the antenna deployed between the host and the target substrates (a), and of the antenna embedded into a polymer over the target substrate with the host removed (b).

The complete procedure to deploy the antenna is described in [12]. The two panels in fig. 8 show the antenna standing between the host and the substrate wafer (Fig. 14(a)) and antenna inside a polymer with the host substrate removed (Fig. 14(b)). The dielectric lens antenna is built on the top of the polymer using the same polymer that includes the antenna to avoid reflections due to a discontinuity of the refraction index.

The two main technologies that can be used to apply the lens are 3D printing or transfer molding. 3D printing, is a particularly interesting technology due to the recent availability of low-cost printers and its feasibility has been already demonstrated in [7].

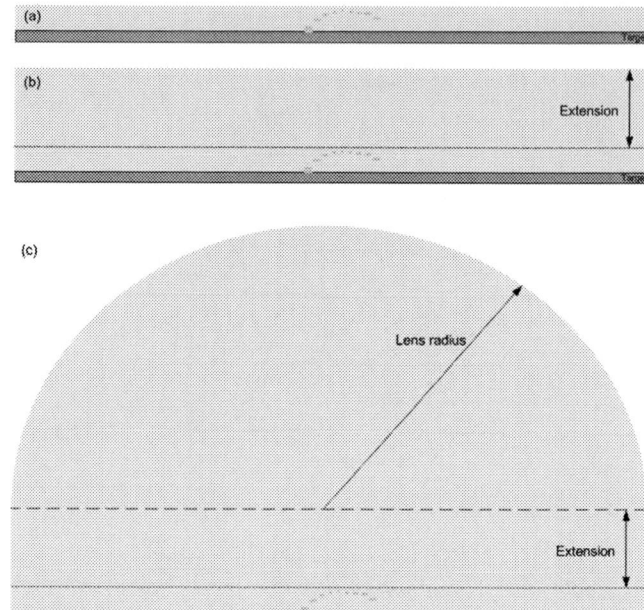

Figure 15. Cross sectional view of the extended lens realization. This is a two-step process. In the first step a planar extension (a) is added to the embedded antenna (a) and in the second step (c), the lens is added by molding or 3D printing.

The lens is added into two distinct steps, in the first on (Fig. 15(b)) a uniform layer of polymer is dispensed on top of the antenna (Fig. 15(a)). The thickness of this polymer is defined by the electromagnetic simulations and is necessary to move the second focus of the extended hemispherical lens at the apex of the antenna wire. In the second step (Fig. 15(c)) a hemispherical lens is added on top.

Conclusions

In this work we presented the results obtained from simulation of the embedding of a three-dimensional antenna into an extended hemispherical dielectric lens as the first step for the design of a high gain focal plane array employing u-helix antennas for THz radiation detection. The extension length of the lens has been proved to be the most critical factor in determining the gain of the antenna, and its radiation pattern characteristics (main lobe direction, FWHM). The most promising results have been obtained by placing the antenna apex at the midpoint between lens's two foci. It has also been found that the focusing property of the lens helps in correcting the axial deviation of maximum gain direction, typical of the u-helix antenna, arising from the asymmetrical feed.

References

1. J. Federici, L. Moeller, "Review of terahertz and subterahertz wireless communications," *J. Appl. Phys.*, vol. 107, no. 11, 111101, Jun. 2010.
2. S. Koenig, D. Lopez-Diaz, J. Antes, F. Boes, R. Henneberger, A. Leuther, A. Tessmann, R. Schmogrow, D. Hillerkuss, R. Palmer, T. Zwick, C. Koos, W. Freude, O. Ambacher, J. Leuthold, I. Kallfass, "Wireless sub-THz communication system with high data rate", *Nature Photonics,* 7(12), pp. 977-981, Oct. 2013.

3. P.U. Jepsen, D.G Cooke, M. Koch, "Terahertz spectroscopy and imaging – Modern techniques and applications", *Laser and Photonics Reviews*, vol. 5, no. 1, pp. 124-166, Jan. 2011.

4. G.P. Gallerano, A. Doria, E. Giovenale, G. Messina, A. Petralia, I. Spassovsky, K. Fukunaga, I. Hosako, "THz-ARTE: non-invasive terahertz diagnostics for art conservation," in *Proc. Infrared, Millimeter and Terahertz Waves Conf. (IRMMW-THz)*, Sept. 15-19, 2008, pp. 1-2.

5. G.M. Rebeiz, "Millimeter-wave and terahertz integrated circui antennas", *Proceedings of the IEEE*, vol. 80, no. 11, pp.1748-1770, Nov. 1992.

6. H. Sherry, J. Grzyb, R. Al Hadi, A. Cathelin, A. Kaiser, U. Pfeiffer, "A 1kpixel CMOS camera chip for 25fps real-time terahertz imaging applications," in *Proc. IEEE Solid-State Circuits Conference (ISSCC)*, Feb. 19-23, 2012, pp. 252-254.

7. K. Y. Park, N. Wiwatcharagoses, P. Chahal, "Wafer-level integration of micro-lens for THz focal plane array application," in *Proc. IEEE Electronic Components and Technol. Conf. (ECTC)*, May 28-31, pp.1912-1919.

8. D.F. Filippovich, S.S. Gearhart, G.M. Rebeiz, "Double-Slot Antennas on Extended Hemispherical and Elliptical Silicon Dielectric Lenses," *IEEE Trans. MTT*, vol. 41, no. 10, Oct. 1993, pp. 1738-1749.

9. D.F. Filippovich, G.P. Gauthier, S. Raman, G.M. Rebeiz, "Off-Axis Properties of Silicon and Quartz Dielectric Lens Antennas," *IEEE Trans. Ant. Prop.*, vol. 45, no. 5, May 1997, pp. 760-766.

10. H. Kobayashi, Y. Yasuoka, "Receiving Properties of Extended Hemispherical Lens Coupled Slot Antennas for 94-GHz Millimeter Wave Radiation", *Electronics and Communications in Japan Part 1*, vol. 84, no. 6, Feb. 2001, pp. 32-40.

11. W.B. Dou, G. Zeng, Z.L. Sun, "Pattern prediction of extended hemispherical-sens/objective-lens antenna system at millimetre wavelenghts," IEE Proc. Microw. Antennas Propag., vol. 145, no. 4, pp. 295-298, Aug. 1998.

12. P. Nenzi, F. Tripaldi, V. Varlamava, F. Palma, M. Balucani, "On-chip THz 3D antennas," in *Proc. IEEE Electronic Components and Technol. Conf. (ECTC)*, San Diego, CA, May 29-June 1, 2012, pp. 102–108.

13. P. Nenzi, V. Varlamava, F.S. Marzano, F. Palma, M. Balucani, "U-Helix: On-chip short conical antenna", in *Proc. European Conference on Antennas and Propagation (EuCAP)*, Apr. 8-12, 2013, pp. 1289-1293.

14. Y.-S. Jin, G.-J. Kim, S.-G. Jeon, "Terahertz Dielectric Properties of Polymers," J. Kor. Phys. Soc., vol. 49, no. 2, pp. 513-517, Aug. 2006.

978-1-4799-2408-0/14 $31.00 © 2014 IEEE

Embedded Diodes for Microwave and Millimeter Wave Circuits

Xianbo Yang, Amanpreet Kaur, and Premjeet Chahal
Terahertz System Lab (TeSLa)
Department of Electrical and Computer Engineering
Michigan State University
chahal@egr.msu.edu   yangxian@msu.edu

Abstract

This paper presents a novel embedded active fabrication process for high frequency circuit applications. Multiple active devices with high placement accuracy can be embedded on a common substrate at low processing temperature, and the process is compatible with large area fabrication. For the design of a millimeter and terahertz (THz) imaging array and other circuits, multiple high cutoff-frequency GaAs Schottky diodes are embedded on a polymer substrate. DC and microwave characterizations are carried out to test the fabrication repeatability, reliability and the performance of high frequency circuits. Circuits measured in this work include rectifier, multiplier and mixer. The measurement results show that the process is reliable and can be readily utilized in the manufacture of circuits operating at millimeter wave and well into the Terahertz frequency regime.

Introduction

Low cost, high frequency (microwave, millimeter wave, and terahertz) circuits and their applications are currently one of the most active research topics. High frequency devices enable faster data transfer and broadband communications [1 - 2]. The rapid development in this area has also widened applications ranging from mobile devices, such as 3G and 4G communications, biomedical imaging to adverse weather landing and driving [3 - 7]. The demands of high performance RF integrated circuits have already pushed the operating frequency to as high as the terahertz (THz) region, where the parasitics associated with active and passive elements begin to dominate. Not only the parasitics associated with active and passive elements need to be small to achieve high cut-off frequency, but also the devices are required to be integrated in close proximity to each other as the wavelengths decrease. Furthermore, for the design of efficient and highly functional high frequency systems, a host of devices made with different semiconductor technologies are desired to form on a single circuit. For example, a simple THz imaging array requires high cut-off frequency diodes (e.g., InP based diodes) for rectification and CMOS circuits for data reading out.

III-V semiconductors such as GaAs and InP are generally required as the building blocks to fabricate millimeter wave (MMW) and THz integrated circuits and devices like imaging detectors as well as frequency multipliers and mixers. The cost to manufacture these circuits is relatively high since less than 1% of the entire III-V substrate is occupied by the active elements, while the rest of the area is filled with passive devices where the semiconductor substrate is not a necessity. The compatibility of III-V epitaxial fabrication to existing commercialized CMOS technology is poor due to large lattice mismatch. Thus, it is desired that fabrication approaches allow seamless integration of different semiconductor technologies on a common substrate at low process temperatures, and have

high compatibility with large-area manufacturing. An embedded active process is an attractive approach to meet this challenge. The embedded integration of electronic components for commercial CMOS integrated circuits designs have been widely studied [8 - 10]. Major benefits include reducing the manufacturing cost, decreasing the form factor for multilayer and 3D integration, increasing the packaging reliability and maintaining the electrical performances.

Embedded integration packaging technology can also offer advantages for high frequency circuits. The short interconnects enhance the electrical performances, and the homogeneous material surrounding the devices provides a uniform electrical and mechanical environment thus improving the system reliability [10]. Several monolithic microwave integrated circuits designs (MMIC) and chips such as a 77GHz SiGe Mixer [11] and high frequency power amplifier ICs [12], have been reported using embedded packaging techniques to achieve excellent performance. Also, embedded actives can be utilized in the design of novel 3D circuit structures through stacking of active devices or chips.

The aim of this research is to push the boundary of embedded actives applications into the THz frequency regime, to demonstrate integration of multiple active devices on a common substrate, and to study a host of high frequency circuits that can be designed using embedded active process. Processing at low temperatures (below 120°C) is preferred as this allows the use of host of low-temperature substrates having low glass transition temperature (T_g). Key challenge to designing high frequency circuits using embedded actives is to maintain the low parasitics associated with the device close to their original value. One of the key elements that significantly affect the device performance is the post processed via connecting the peripheral circuits to the pads of the embedded devices, and these should be optimally designed to keep minimum disturbance to the high frequency performances. Here the process is optimized to be able to design high frequency circuits such as rectifier, multiplier and mixer. The following discussion will start from the selection of the active devices to be utilized as embedded cores in this research.

Active Device Selection and Circuits Design

For many applications such as frequency multiplier, mixer and rectifier (detector), strong nonlinear devices are indispensable components. Diodes and transistors are the most commonly used active devices to realize such functions [13 - 15]. Furthermore, diodes have only two terminals and are easy to operate, which simplifies the design work and fabrication procedure. In this work, GaAs Schottky diodes are selected to design and then embedded to from variables of microwave, MMW and THz circuits, which also reduce the complexity of the embedded active fabrication process. One of the

978-1-4799-2408-0/14 $31.00 © 2014 IEEE

fundamental components among these circuits is a detector. The need of accurate diode parameters can be well understood based on a detector circuit design.

Analytical calculations for the NEP as well as sensitivity of a Schottky diode based detector have been studied systematically in [15 - 17]. From the conclusions of their analysis, thermal noise V_n and flicker noise V_{1f} should be decreased as much as possible, while maintaining high voltage sensitivity, β_v, in order to achieve low NEP values. Moreover, there is strong relationship between the NEP and the resistance parameters of the devices. The total resistance (R_s+R_d) must be maintained small enough to reduce thermal noise and to achieve good impedance matching to the antenna element. Another criteria is that the Schottky diode should be able operable well into the THz region in order to produce high resolution imaging. This requires the junction as well as parasitic capacitance of the diode to be small enough to obtain high cut-off frequencies. In brief, the parasitic of a diode are critical in the design of high frequency circuits. Thus, the intrinsic cutoff frequency of the diodes should be high and it should not degrade after the diode has been embedded in a dielectric layer and interconnected with other elements.

Based on the simple analysis above, a commercial flip-chip GaAs Schottky diode DBES105a, from United Monolithic Semiconductor (UMS), is chosen for the design of circuits using embedded integration. The equivalent circuit of this diode is shown in Figure 1. The parameters of the equivalent circuit model for single diode at zero bias are shown in Table 1. The chip has dual Schottky diodes connected in series. This Schottky diode flip-chip has been used in many applications, such as frequency multiplication and mixing above 200GHz [19 - 21].

Table 1 Specs of the DBES105a dual Schottky diode flip-chip

R_s (Series resistance)	C_{j0} (Junction capacitance)	C_{par} (Parasitic capacitance)	f_{co} (Cut-off frequency)
4.4Ω	9.5fF	5.8fF	2.4THz

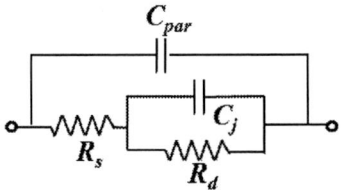

Figure 1. Small signal circuits of a single diode on DBES105a Schottky diode flip-chip

Antenna based circuits are designed according to this diode specs in embedded format. The antenna coupled diode can be applied in the design of high frequency detectors, mixers and multipliers. Broadband planar antenna (log-periodic bow-tie antenna, LPA) is used here in order to characterize the detector over a wide frequency range. The junction resistance R_d of the diode is usually at hundreds Ohms at the strongest nonlinear position. To be able to achieve good impedance matching over a wide band, complimentary LPA geometry is chosen. Its impedance

reaches its highest value at 188.5Ω, and the antenna is relatively frequency independent. Figure 4 (b) shows the geometry of the designed LPA. Details of the design technique for the LPA have been explained in [22 - 24]. Apart from diode coupled to antenna elements, diodes coupled to a coplanar transmission line are also fabricated in order to measure the S-parameters for deriving an equivalent circuit model of an embedded diode.

Circuit Fabrication

The concept of the fabrication process applied here was first reported in [25]. The process steps are as follows: 1) A 100μm thick Rogers 3003 substrate with double-side copper cladding is used as the starting material to form "housing" layer for the diode. Cu was removed from one side and the other side is patterned to outline the cavity structures. 2) A CO_2 laser engraving is then applied on the patterned copper side to remove exposed Rogers polymer-ceramic material, the patterned copper acts as the protection mask, after the cavities have been etched open, this Cu layer is then removed by wet etching. 3) The "blank" Rogers thin film with opened diode cavities was then laminated on Zeonor plastic substrate, which is an extremely low loss material from the GHz to THz regime [26]. 4) The diode is then placed into the cavity and a layer of 100μm SU-8 is spun on the whole substrate, and patterned to open via on the diode pads. 5) A thin layer of copper was sputtered on the sample so that the various elements can be patterned to realize functional systems. In this work, however, multiple diodes are required to be embedded on the same wafer through one single fabrication processes. Figure 2 shows the schematic cross-section view of the finished sample. One of the key challenges to this process is the ability to integrate multiple diodes on the substrate and their relative placement to each other. To be able to achieve good tolerance, tight cavities with vertical side walls are fabricated.

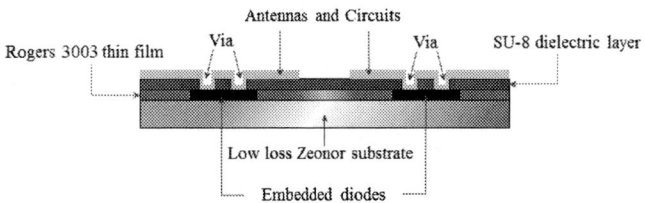

Figure 2. Schematic cross-section view of a finished sample with two diodes embedded.

There are several problems raised when performing multi diode embedded integration. During the patterning of the outline of the cavity on the copper, the non-uniformity of the copper wet etching causes the sizes of the outlines not to be identical. This results in different size cavities across the wafer, and leads to a significant gap between the diode and cavity side walls so that the diodes misaligned relative to each other. This also leads to further misalignment of via to the pad structures. To solve this, the wet etching procedure was improved by fine controlling the bath temperature and the incorporating periodic agitation of the liquid to achieve uniform temperature distribution as well as to uniformly replenish the etchant, respectively. This fine-tuned etching improves the alignment of via in the follow on process steps. Achieving uniform coating of SU-8 layer is critical to create fine-line patterns. Uniform filling of spacing between the diode and the pocket walls is critical. This was attained by multiple coating of thin SU-8 layers and also longer pre-bake time on a leveled hot plate to allow SU-8 reflow and planarization. Overall thickness of this SU-8 layer is made thin (approx. $10 - 40$ µm) so that the via connecting top metal to the diode pads have low parasitics. Figure 3 shows one diode that has been placed into a cavity. The spacing between the cavity wall and the diode wall is made less than 15 µm on each side providing a tight fit. In this work, a 3 by 3 array is designed on a single substrate, and for preliminary experiments, only four diodes are placed into the cavities next to each other to form a 2 by 2 imaging array. Figure 4 (a) shows the photo of a fabricated wafer and Figure 4 (b) shows close up view of an antenna element coupled to an embedded diode.

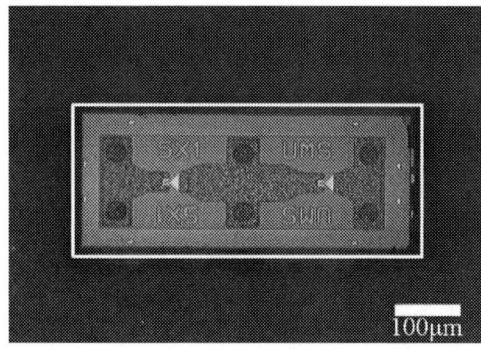

Figure 3. One diode places into the cavity on Rogers 3003 thin film, white window represents the outline of the cavity.

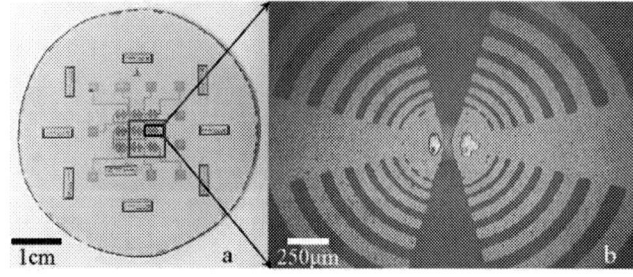

Figure 4. Fabricated sample with four diodes embedded to form 2 x 2 array highlighted by a red square box (a), and zoom in on one single device (b).

Measured Results

Measurements were first carried out using a semiconductor parameter analyzer (SPA) to acquire the I-V characteristics of the four embedded diodes after fabrication. Figure 5 shows the I-V characteristics of an embedded diode. A Fitted curve for the I-V measurement indicates that the saturation current is equal to 3.5×10^{-14}A, and the ideality factor is approximately 1.2, which matches well with the values provided in the diode datasheet by the manufacturer. Figure 6 shows the measured I-V curves of the four diodes from 0.65-0.75V, at which bias the diodes start to turn on. Overlapped curves show that the diodes in one fabrication batch have close I-V characteristics. This shows that this

embedded fabrication process is suitable for heterogeneous integration of multi-functional high frequency integrated circuits. However, the series resistance R_s is approximately 27Ω, which is larger than the 4.4Ω stated on the datasheet. This may be due to the polymer residue from the SU-8 layer present on the diode pads, and it can be cleaned in RIE prior to the deposition of the top metal layer.

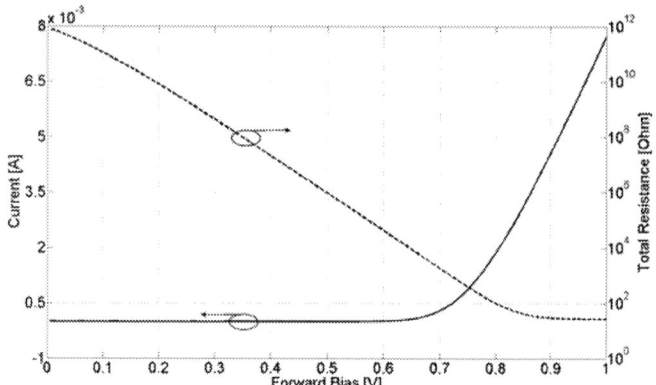

Figure 5. Example I-V characteristics and total resistance (R_s+R_d) of one of the embedded diodes.

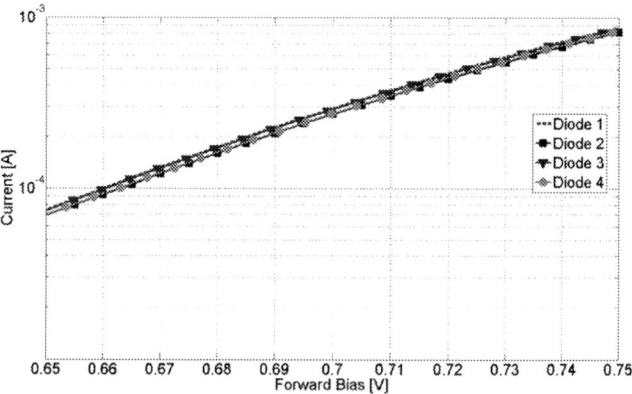

Figure 6. I-V characteristics within 0.65-0.75V forward bias for 4 embedded diodes on the same fabricated substrate.

In order to acquire the equivalent circuit model for the embedded diode structure, same process was used to fabricate another sample with embedded diodes which were coupled to coplanar measurement structures. The S-parameters between 10 and 20GHz of the sample were measured by probing the device using a 40GHz ground-signal-ground (GSG) probe. For comparison, a bare diode was also measured. The measured S-parameters were used to derive the equivalent circuit model of the diodes (bare and embedded), which is shown in Figure 1. Fitting was carried out using Agilent ADS simulator. Table 2 outlines the equivalent circuit values for each of the diode elements at a bias of 0.7V. This voltage provides the strongest non-linearity in the I-V characteristics of the diodes, and it will be applied to all of the circuit applications presented later. Figure 7 shows the measured and modeled S-parameters of bare and embedded diodes on the Smith chart.

The derived cutoff frequency for the embedded diode is similar to that of bare diode, which indicates the diodes' high frequency responses have not been compromised. There is an increase in the series inductance which is largely due to the line length of the CPW structure used in the characterization. Also, diode total resistance (R_s+R_d) is different and this may be attributed to inherent difference in diode resistance across the wafer. Overall, this embedded fabrication process is reliable and can be used in the design and fabrication of MMW and THz circuits. Note that RIE process has been applied to this sample to remove the polymer residue in the via prior to post processing, which results in smaller series resistance, and is comparable to that of pristine bare diode value.

Table 2 Value for each element of the equivalent circuit for bare and embedded diodes.

	R_s	R_d	C_{par}	C_j	L_s	f_{co}
Bare diode	6Ω	42.5Ω	3fF	21fF	130pH	0.44 THz
Embedded diode	5Ω	35.7Ω	3fF	21fF	400pH	0.48 THz

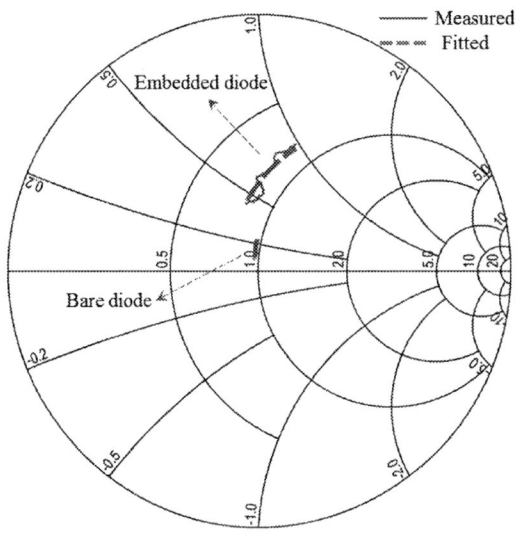

Figure 7. Measured and model fitted S-parameters for embedded and bare diodes at 0.7V.

For the first RF measurements, a 40 GHz ground-signal (GS) probe is used to probe the devices and carry out preliminary measurements in the low frequency range. For some the circuits, the probe is brought into contact to one of the antenna arrays connecting the diodes. Several circuit functions were tested including rectification, multiplication and mixing.

For rectification tests, 18GHz excitation with a power of -6.33dBm from the RF source is fed in through the probe to the embedded diode. Figure 8 shows the output rectified voltage as a function of forwards bias applied to the diode. The applied bias is maintained below 0.8V to prevent any damage to the device. The diode has the strongest nonlinearity around 0.7V, where the measured output rectified voltage is 4.56mV. This matches the I-V characteristic where around 0.7V is the turn-on voltage. Note that due to the impedance mismatch between the probe and the diode, RF power provided by the

source was not completely absorbed by the device, further measurement shows that at 0.7V, only about -23.33dBm had been fed in through the probe to the diode element.

The rectified voltage was also measured as a function of input RF power and the measured results are shown in Figure 9. Here, the applied bias is fixed at 0.7V, where the diode has the strongest nonlinear behavior.

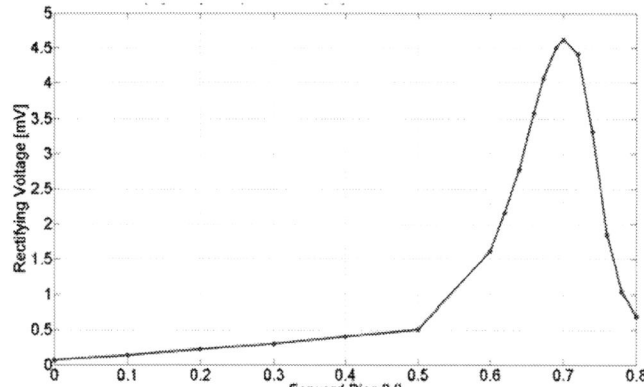

Figure 8. Rectifying voltage changing respected to sweeping forward bias.

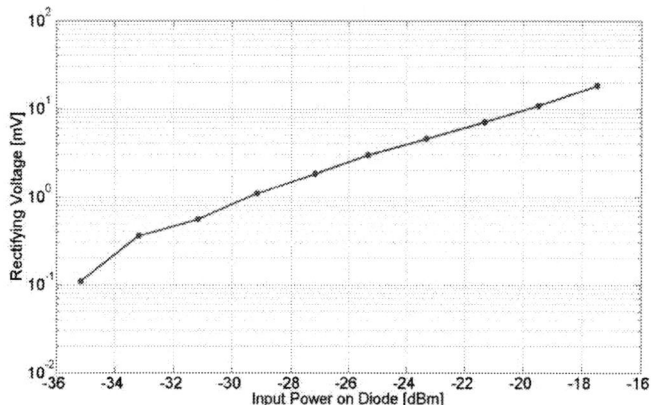

Figure 9. Output rectified voltage as a function of input RF power.

The voltage sensitivity β and the NEP at 18GHz can be derived based on the diode equivalent circuit as well as the measured equivalent model parameters of the embedded diode. The calculated values are plotted as a function of applied voltage in Figure 10. The voltage sensitivity at 0.7V is approximately 0.9mV/µW, which matches close to the measurement results. The lowest NEP in Figure 10 is around 4.5pW/Hz$^{0.5}$, which can be further improved by impedance matching as shown in Figure 11. Both optical and electrical NEPs (under matched condition) are shown in Figure 11 for comparison. With good impedance matching NEP as low as 2pW/Hz$^{0.5}$ can be achieved.

NEP and voltage sensitivity are also characterized for frequencies higher than 18GHz. Assuming the designed antenna is used instead of the GS probe, Figure 12 shows the NEP and voltage sensitivity β spanning within 10GHz to 1THz frequency range. As the frequency increases above 100GHz, the voltage sensitivity degrades and the NEP increases dramatically. At 1THz, the sensitivity is about 0.04mV/µW, and NEP is around 150pW/Hz$^{0.5}$, this is due to the strong impedance mismatch between the antenna and the embedded diode. This needs to be improved for applications in the design of THz imagers and detectors. As discussed earlier, to overcome this, small series resistance and capacitance diodes with smaller parasitics are desired. Further simulations indicate that under matched condition, the NEP can be as low as 15pW/Hz$^{0.5}$ at 1THz, and the sensitivity can be maintained around 0.4mV/µW at the same frequency, which is 10 times better than those under the mismatched condition.

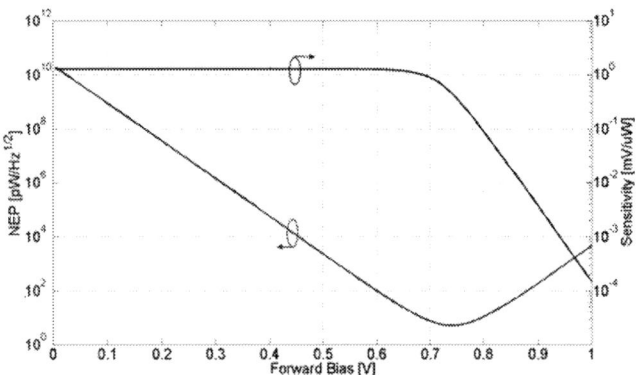

Figure 10. Voltage sensitivity β and the optical NEP at 18GHz under impedance mismatch condition.

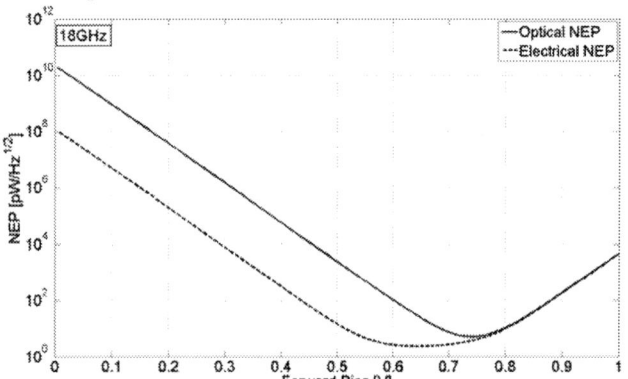

Figure 11. Optical (mismatched) and Electrical (matched) NEP at 18GHz under different bias conditions.

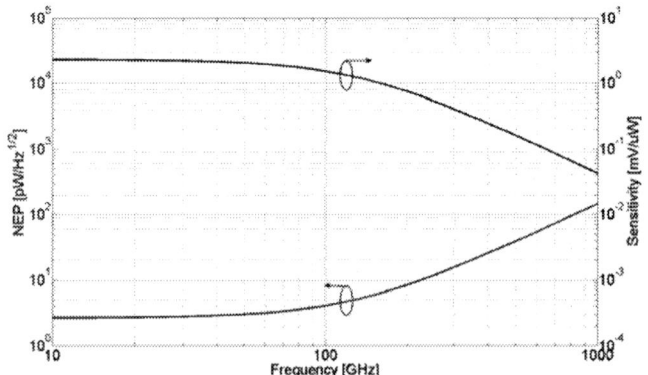

Figure 12. Optical NEP and voltage sensitivity under mismatched condition as a function of frequency.

Another application of the fabricated embedded diode structures is in frequency multiplication. The experimental setup is the same as used above for the voltage rectification experiments. The GS probe is used to both pump the diode with the fundamental frequency and also to pick up the harmonics generated by the nonlinear characteristics of the diode. Figure 13 shows the 3rd harmonics of the fundamental frequencies for 4GHz and 5GHz. The input power on x axis represents the power that is absorbed by the embedded diode, and it is swept by controlling the RF source power level. The bias is again fixed at 0.7V where the best nonlinear behavior is achieved from the diode.

Figure 14 shows the output power of the frequency multiplier at different fundamental frequencies, when the input power (absorbed by diode) is fixed at around -13dBm. The output powers for both of the 2nd and 3rd harmonics decreases when the fundamental pump frequency increases. For higher frequency multiplication, more power needs to be fed into the device to achieve good conversion efficiency.

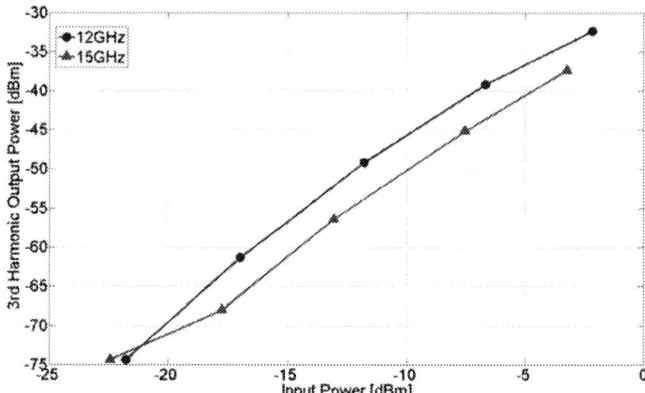

Figure 13. 3rd harmonics output power of 4GHz and 5GHz fundamental input frequencies.

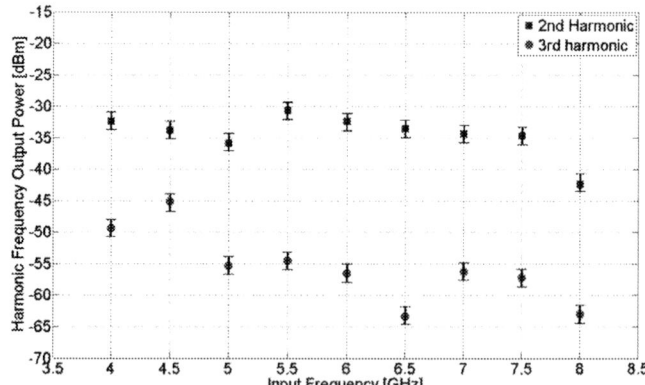

Figure 14. Output power of 2nd ($2f_o$) and 3rd ($3f_o$) harmonics at different fundamental frequencies. The input power at f_o is approximately -13dBm.

A frequency mixer is also tested using this embedded diode structure. Again, the GS probe is used for feeding in the RF signal and also to measure the output signal. A separate RF source was used as a local oscillator (LO), whose frequency is fixed at 18GHz. The RF input frequency is fixed at 19GHz, and its power is being swept in order to see the variation of IF output power at 1GHz for different input power values. Figure 15 shows the results of output power for 1GHz at different LO power while sweeping the RF input power. Here again, the conversion efficiency improves with increase in the LO power.

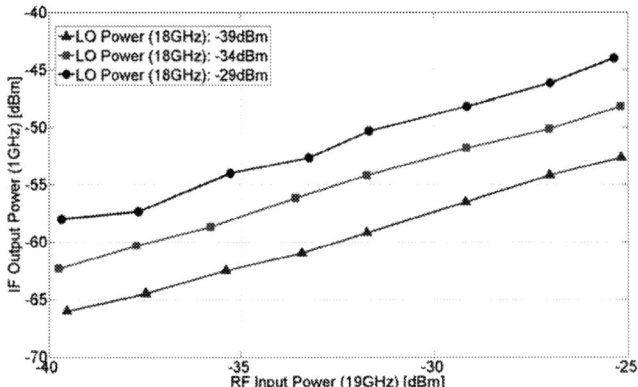

Figure 15. Output power of IF (1GHz) at different LO power and RF power.

In summary, similar DC characteristics of embedded diodes on the same substrate indicate the embedded fabrication procedure is reliable for multiple diodes assembling on a single wafer. Strong RF responses of the embedded diodes allow designing large variety of MMW and THz integrated circuits based on the embedded integration. Further measurements in the W-band and higher frequencies will be carried out and presented at the conference. However the results presented here indicate that an embedded diode process is compatible with the design of high frequency circuits.

Conclusions

This work demonstrates a novel embedded fabrication process that can be used to realize low cost, large area heterogeneous integration of active devices for MMW and THz integrated circuits. Overlapped I-V curves of the tested embedded diodes show that the new process preserves the strong nonlinear DC characteristics of the embedded active devices and is repeatable across the wafer. Preliminary measurements of microwave circuits (rectifier, mixer and multiplier) show that the fabricated circuits are functional and reliable for high frequency circuit applications. Further optimization of the process will improve the cut-off frequency of the embedded diodes through reduced series resistance.

Acknowledgments

The authors would like to thank all of the TeSLa group members for their help with this project. The authors would also like to thank Brian Wright and Karl Dersch for assisting with circuit fabrication. This work is supported in part by DARPA YFA program (Grant Number: N66001-12-1-4238).

References

1. A. Hirata, T. Kosugi, H. Takahashi, R. Yamaguchi, F. Nakajima, T. Furuta, H. Ito, H. Sugahara, Y. Sato and T. Nagatsuma, "120-GHz-band millimeter-wave photonic wireless link for 10-Gb/s data transmission," *Microwave Theory and Techniques, IEEE Transactions on* , vol.54, no.5, pp.1937,1944, May 2006.

2. S. Montusclat, F. Gianesello and D. Gloria, "Silicon full integrated LNA, filter and antenna system beyond 40 GHz for MMW wireless communication links in advanced CMOS technologies," *Radio Frequency Integrated Circuits (RFIC) Symposium,* pp.80, 11-13 June, 2006.

3. K. Humphreys, J.P. Loughran, M. Gradziel, W. Lanigan, T. Ward, J.A. Murphy, and C. O'Sullivan, "Medical applications of terahertz imaging: a review of current technology and potential applications in biomedical engineering," *Proc. 26th Annual International Conference of Engineering in Medicine and Biology Society*, vol.1, pp. 1302-1305, Sep, 2004.

4. Z.D. Taylor, R.S. Singh, D.B. Bennett, P. Tewari, C.P. Kealey, N. Bajwa, M.O. Culjat, A. Stojadinovic, H. Lee, J. Hubschman, E.R. Brown, and W.S. Grundfest, "THz Medical Imaging: in vivo Hydration Sensing," *Terahertz Science and Technology*, vol.1, no.1, pp. 201-219, Sep, 2011.

5. S. Clark and D.W. Hugh, "Autonomous land vehicle navigation using millimeter wave radar." *Proc.* of *International Conference of Robotics and Automation,* vol. 4, pp. 3697-3702, 1998.

6. K. Sarabandi, M. Vahidpour, M. Moallem, and J. East, "Compact beam scanning 240GHz radar for navigation and collision avoidance." In *SPIE Defense, Security, and Sensing*, pp. 803113-803113. International Society for Optics and Photonics, 2011.

7. E. Öjefors, U. R. Pfeiffer, A. Lisauskas, and H. G. Roskos, "A 0.65 THz Focal-Plane Array in a Quarter-Micron CMOS Process Technology," *Journal of Solid-State Circuits*, vol. 44, no. 7,pp. 1968-1976, Jul., 2009.

8. C.Ko, S. Chen, C. Chiang, T. Kuo, Y.C. Shih, T.H. Chen, "Embedded active device packaging technology for next-generation chip-in-substrate package, CiSP," Proc. on 56th *Electronic Components and Technology Conference,*pp. 322-329.8 May 30th –June 2nd, 2006, San Diego, CA.

9. Y.P. Hung, T.C. Chang, C.K. Lee, Y.C. Lee, J.Y. Chang, S.Y. Huang, C.K. Hsu, S.M. Li, J.H. Huang, F.J. Leu, R.S. Cheng, Y.W. Huang, and T.H. Chen, "Processing characteristics and reliability of embedded DDR2 memory chips," *5th International Microsystems Packaging Assembly and Circuits Technology Conference (IMPACT), 2010*, pp.1,4, 20-22 Oct. 2010.

10. L. Boettcher, D. Manessis, A. Ostmann, S. Karaszkiewicz, H. Reichl, "Embedding of Chips for System in Package realization - Technology and Applications," *3rd International Microsystems, Packaging, Assembly & Circuits Technology Conference, 2008. IMPACT 2008.*,pp.383-386, 22-24 Oct. 2008.

11. M. Wojnowski, M. Engl, B. Dehlink, G. Sommer, M. Brunnbauer, K. Pressel, R. Weigel, "A 77 GHz SiGe mixer in an embedded wafer level BGA package," Proc. on 58th *Electronic Components and Technology Conference.*

ECTC 2008, pp. 290-296, 27-30 May 2008, Lake Buena Vista, FL.

12. M. Itoh, S. Hoshi and H. Linaga, "Technology of Embedded Ultra-High Frequency Power Amplifier ICs in the Print Circuit Boards", *Oki Technical Review*, Issue 216 Vol.77, No.1, April 2010.

13. M. Sakhno, F. Sizov, and A. Golenkov, "Uncooled THz/sub-THz Rectifying Detectors: FET vs. SBD", *Journal of Infrared, Millimeter, and Terahertz Waves*, Springer, Sep. 2013, DOI: 10.1007/s10762-013-0023-2.

14. W. Knap, M. Dyakonov, D. Coquillat, F. Teppe, N. Dyakonova, J. Łusakowski, K. Karpierz, M. Sakowicz, G. Valusis, D. Seliuta, I. Kasalynas, A.E. Fatimy, Y. M. Meziani, T. Otsuji, "Field Effect Transistors for Terahertz Detection: Physics and First Imaging Applications", *Journal of Infrared, Millimeter, and Terahertz Waves*, Vol. 30, Issue 12, pp. 1319-1337, December 2009.

15. P. Chahal, F. Morris, and G. Frazier, "Zero Bias Resonant Tunnel Schottky Contact Diode for Wide-Band Direct Detection," *Electron Device Letters*. vol. 26, no. 12, pp. 894-896, 2005.

16. Xianbo Yang, A. Kaur, P. Chahal, "Implementation of semiconducting nanowires for the design of THz detectors," *Proc. 63rd Electronic Components and Technology Conference (ECTC)*, pp.2375,2380, 28-31 May 2013, Las Vegas, NV.

17. X. Yang, and P. Chahal, "Large-area Low-cost Substrate Compatible CNT Schottky Diode For Thz Detection", *Proc 61st Electronic Components and Technology Conference (ECTC)*, Lake Buena Vista, FL, May 31st-June 1st, pp. 2158-2164, 2011.

18. J.L. Choi, V. Mitin, R. Ramaswamy, V. A. Pogrebnyak, M. P. Pakmehr, A. Muravjov, M.S. Shur, J. Gill, I. Mehdi, B.S. Karasik, A.V. Sergeev, "THz Hot-Electron Micro-Bolometer Based on Low-Mobility 2-DEG in GaN Heterostructure," *Sensors Journal*, vol.13, no.1, pp.80,88, Jan. 2013

19. B. Zhang;, Y. Fan, Zhang, S. X., X. Yang, F. Zhong and Z. Chen, "110GHz high performanced varistor tripler," *2012 International Workshop Microwave and Millimeter Wave Circuits and System Technology (MMWCST)* , pp. 19-20 April 2012.

20. D. Schneiderbanger, C. Kneuer, M. Sterns, R. Rehner, S. Martius, L. Schmidt, "A 75–110 GHz seventh-harmonic balanced diode mixer in a novel circuit configuration," *European Conference of Wireless Technology, EuWiT 2008.*, pp.158,161, 27-28 Oct. 2008.

21. D. Schneiderbanger, A. Cichy, R. Rehner, M. Sterns, S. Martius, L. Schmidt, "A hybrid broadband millimeter-wave diode ring mixer with advanced IF extraction technique," *European Microwave Conference, 2007.*,pp.656,659, 9-12 Oct. 2007.

22. M. M. Gitin, F. W. Wise, G.Arjavalingam, Y. Pastol and R. C. Compton, "Broad-Band Characterization of Millimeter-Wave Log-Periodic Antennas by Photoconductive Sampling", *Transactions on Antennas and Propagation*, vol. 42, no. 3, Mar. 1994.

23. A. Scheuring, A. Stockhausen, S. Wuensch, K. Ilin, and M. Siegel, "A new analytical Model for log-periodic Terahertz Antennas, " *In Proc. of 4th European Conference on Antennas and Propagation (EuCAP)*,April, 12-16th, 2010, Barcelona, Spain.

24. Y. Mushiake, "Self-complementary antennas," *Antennas and Propagation Magazine*, vol. 34, Issue 6, pp. 23-29, Dec. 1992.

25. Xianbo Yang; Chahal, P., "Embedded actives for terahertz circuit applications: Imaging array," *Proc. 62nd Electronic Components and Technology Conference (ECTC)*, pp.2082-2086, May 29 2012-June 1 2012.

26. J. A. Hejase, P. R. Paladhi, and P. Chahal, "Terahertz Characterization of Dielectric Substrates for Component Design and Nondestructive Evaluation of Packages", *Tran on Components, Packaging and Manufacturing Technology*, Vol. 1, No. 11, pp. 1685-1694, 2011.

PCIe Gen3 Link Design and Tuning in Server Systems with End Devices from Multiple IP Suppliers

Si T. Win, Daniel Rodriguez and Nanju Na

IBM Systems and Technology Group
11400 Burnet Road, Austin, TX 78758
stwin@us.ibm.com, dirodrig@us.ibm.com, nananju@us.ibm.com

Abstract

This paper discusses link routing budget considerations for PCIe Gen3 designs in server systems. Special attention will be given to channel discontinuities and their effect on eye opening. Link training complications will be discussed with respect to equalization and tuning behavior when accommodating multiple transmitters and receivers from different vendor sources. Insertion loss plots and eye simulation data will be analyzed and interpreted. Hardware data will show that optimally simulated equalization cases may not necessarily occur.

Introduction

As the most commonly employed IO bus, PCIe links operate on a wide variety of system configurations. Channel topology can range from short routing on a single planar to server system topologies that often require longer PCB routing across multiple cards, thicker stackups, and various connector types. For past high speed bus designs, routing budgets were easily arrived at based on insertion loss at a reference frequency defined by standards or rules of thumb. This worked well for PCIe Gen1 and Gen2 designs on typical topologies. As the signaling rate increases to 8Gbps for Gen3, extra complication is added to the overall design process. The non-linearities in insertion loss profile due to discontinuities such as via stubs more severely degrade eye openings at the higher signaling rate than at Gen1 and Gen2 speeds. Because of this, the routing length budget can vary widely based on interconnect configuration.

Eye opening requirements for Gen1 and Gen2 were specified at the receiver chip pins; signal quality measurements with an oscilloscope were easily done. However, the Gen3 eye opening requirement is defined at a chip-internal node after equalization. Oscilloscope probing at this node is impossible. Only vendor provided chip diagnostic tools or built-in error indicator registers can give some clue about performance. While some vendors do provide diagnostic tools and error registers, their outputs are sometimes ambiguous or open to interpretation.

Budgeting decisions for a design are further complicated when systems are required to accommodate devices from various suppliers. While the PCIe specification defines a particular phase of training where the TX FFE coefficients will be tuned, it does not specify exactly how the optimal FFE coefficients should be found. Different vendors do in fact implement different FFE tuning algorithms. In addition, the great leeway provided by the specification in RX equalization procedure adds further complexity.

Simulations can be optimistic when they assume that the equalization tuning procedure is predictable and will find only the optimal equalization settings. This can lead to excessively aggressive channels being seen as feasible, when in reality, ICs from suppliers are black boxes with unpredictable equalization behavior and results.

This paper will study the loss behavior of two channel topologies and expand the discussion with statistical eye simulations and lab digital eye measurements on operating links with supplier devices given the complications described above.

PCB channel insertion loss behavior and resulting statistical eye simulations

In the past, loss budgeting at a single frequency point (typically half the baud rate) has been used to determine maximum allowed PCB trace lengths. However, loss behavior is not so neatly linear over the frequency ranges of interest when discontinuities due to layer transitions and connectors are factored in.

Figure 1a and Figure 2a show two example topologies for PCIe links in order to examine loss behavior as a simulation case study. These topologies represent possible server channels from root complex (RC) to endpoint (EP) where via discontinuities are particularly severe due to the thicker stackups.

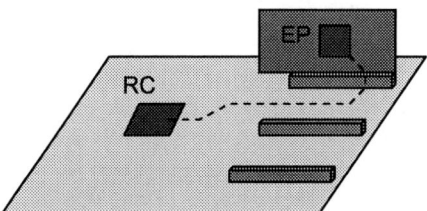

Figure 1a. Channel A, a one-connector PCB channel configuration with SMT and PTH connector variations.

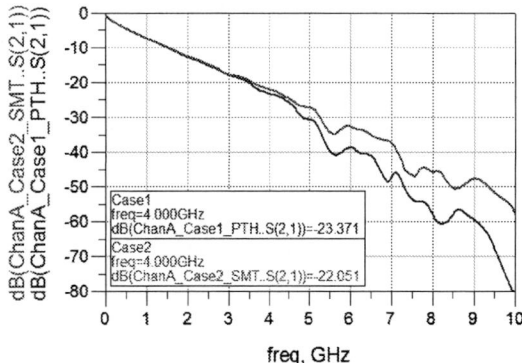

Figure 1b. Differential insertion loss plots for Channel A with SMT and PTH connector variations, 23in total length.

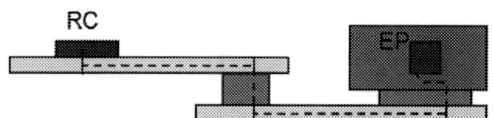

Figure 2a. Channel B, a two-connector PCB channel configuration with SMT and PTH connector variations.

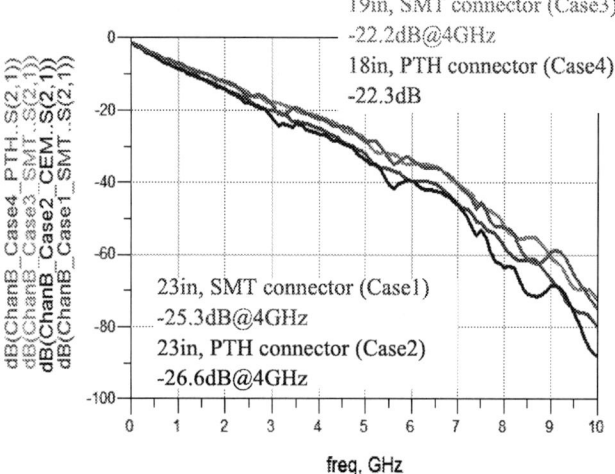

Figure 2b. Differential insertion loss plots for Channel B with varying connector types and trace lengths.

For Channel A, as depicted in Figure 1a, two connector technologies are considered: a surface mount (SMT) type and a pin-through-hole (PTH) type. The insertion loss plots in Figure 1b show that using the PTH connector adds more than 1dB of loss. This is due to the fact that PTH connectors require large vias with finished hole sizes of 28mils. SMT connectors can use much smaller 10mil vias if routing on internal layers is desired. The larger via for PTH connectors creates a greater impedance discontinuity and unwanted capacitance than the smaller via used with SMT connectors.

In Figure 2a, a two-connector channel (named Channel B) is depicted with similar routing length as Channel A. The insertion loss plots shown in Figure 2b indicate that loss at 4GHz in Channel B is higher than that of Channel A by more than 3dB. This 3dB difference holds true whether the PCIe connector is SMT or PTH.

Figure 2b also shows that in order for Channel B to stay within the 22dB loss envelope seen with Channel A, routing length must be limited to 19in with the SMT connector and 18in with the PTH connector.

While insertion loss at the fundamental frequency can provide a first order insight on channel performance, it is not a direct translation to eye opening size. In multiple connector channels with higher order discontinuity characteristics such as in Channel B, differential insertion loss plots display more bumpiness and non-linearity. Equalization using CTLE (continuous time linear equalizer) to recover the signal from high frequency PCB loss will not be as efficient on those

bumpy channels as on channels with more linear frequency response.

The effects of multiple discontinuities can be more fully investigated via eye simulation. Using IBM's statistical eye simulator tool "HSSCDR", behavioral models of the TX and RX devices were created. As shown in Figure 3, the models incorporated worst case jitter and minimum required characteristics such as drive amplitude and equalization. These models did not include die capacitance in the simulation study; worse results are to be expected when including die capacitance.

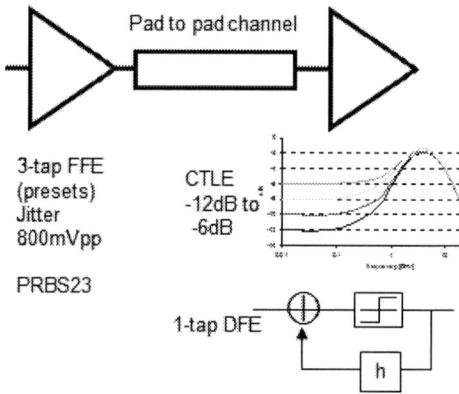

Figure 3. Behavioral model setup for eye opening simulation on PCIe Gen3 channels.

Eye opening simulations were performed for the following cases: Channel A with SMT connector, Channel A with PTH connector, the 19in Channel B with SMT connector, and the 18in Channel B with PTH connector. The two Channel A cases were chosen to compare the effect of the connectors, while Channel B cases were chosen because they matched Channel A SMT in terms of insertion loss at 4GHz, but had different overall frequency profile.

Results are mapped according to equalization level with color grading representing the horizontal and vertical eye measurement. Figure 4 describes how the results map is arranged and the requirements for red, yellow, and green shading. Figures 5 through 8 show the results for the four channel cases previously described.

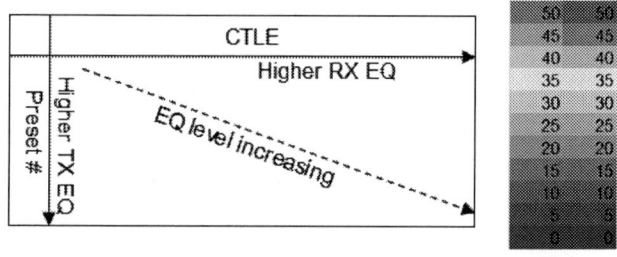

Figure 4. Simulation results format for TX FFE Preset and RX CTLE Adc sweep and color grading criteria for eye opening.

The data in Figures 5 and 6 show that greater eye opening is achieved across all equalization cases for Channel A when

978-1-4799-2408-0/14 $31.00 © 2014 IEEE   2152

the SMT connector is used. These results were expected based on the 1dB difference in insertion loss and greater discontinuity at the connector.

Note however, that when comparing channels with similar insertion loss at 4GHz, there is noticeable difference in eye opening. It is observed that the worst results occurred on the channel with the most discontinuities despite having the shortest trace length.

Figure 9 summarizes the equalization amount that yielded the single best eye opening result for each of the four cases. Despite the Channel A cases having more overall trace length than the Channel B cases, less equalization was needed. Maximum allowed trace length is not solely a function of insertion loss; effects of discontinuities can significantly reduce overall allowed length.

Channel A, 23in, SMT connector, Heye (%UI) @10^{-12}BER

TX Preset # \ CTLE	0 dB	-1 dB	-2 dB	-3 dB	-4 dB	-5 dB	-6 dB	-7 dB	-8 dB	-9 dB	-10 dB	-11 dB	-12 dB
4	9.1	31.5	37.7	42.3	43.9	44.0	49.2	50.1	37.7	42.3	43.9	44.0	49.2
5	16.0	39.9	43.1	46.5	48.5	51.3	52.9	49.7	43.1	46.5	48.5	51.3	52.9
3	27.8	40.8	46.1	46.3	46.4	49.0	47.1	45.7	46.1	46.3	46.4	49.0	47.1
6	15.2	37.4	43.5	40.5	44.0	49.0	48.3	49.2	43.5	40.5	44.0	49.0	48.3
1	31.5	48.2	44.0	46.1	48.8	46.0	45.2	40.1	44.0	46.1	48.8	46.0	45.2
9	6.7	37.7	37.4	42.8	45.7	46.0	46.1	46.7	37.4	42.8	45.7	46.0	46.1
2	34.6	45.7	46.0	46.6	43.9	42.9	39.7	39.6	46.0	46.6	43.9	42.9	39.7
0	40.8	46.2	42.6	44.5	39.2	37.2	34.2	33.0	42.6	44.5	39.2	37.2	34.2
8	28.4	48.7	46.7	49.7	48.3	40.6	40.9	43.7	46.7	49.7	48.3	40.6	40.9
7	39.1	48.6	46.9	43.2	43.5	39.3	37.7	33.9	46.9	43.2	43.5	39.3	37.7

Channel A, 23in, SMT connector, Veye (mVpp) @10^{-12}BER

TX Preset # \ CTLE	0 dB	-1 dB	-2 dB	-3 dB	-4 dB	-5 dB	-6 dB	-7 dB	-8 dB	-9 dB	-10 dB	-11 dB	-12 dB
4	18.3	44.0	47.8	52.2	53.4	55.6	57.4	58.2	47.8	52.2	53.4	55.6	57.4
5	31.4	50.8	52.8	55.8	57.2	60.6	61.8	60.8	52.8	55.8	57.2	60.6	61.8
3	41.4	47.2	49.8	46.8	49.0	51.2	52.2	50.2	49.8	46.8	49.0	51.2	52.2
6	28.4	50.8	54.0	53.4	56.4	58.0	58.6	60.2	54.0	53.4	56.4	58.0	58.6
1	46.4	44.8	46.8	47.4	49.2	49.0	47.2	44.6	46.8	47.4	49.2	49.0	47.2
9	14.6	48.2	49.2	53.8	54.8	52.8	54.6	54.0	49.2	53.8	54.8	52.8	54.6
2	50.0	46.2	48.2	47.2	46.6	44.4	42.2	41.6	48.2	47.2	46.6	44.4	42.2
0	53.0	45.6	43.8	44.8	40.0	40.0	36.2	34.4	43.8	44.8	40.0	40.0	36.2
8	48.2	51.2	51.2	52.0	51.2	50.2	46.8	46.2	51.2	52.0	51.2	50.2	46.8
7	55.8	48.6	48.8	45.2	45.4	41.2	40.6	36.4	48.8	45.2	45.4	41.2	40.6

Figure 5. Eye opening at 10^{-12}BER of Channel A 23in with SMT connector (Case 1).

Channel A, 23in, PTH connector, Heye (%UI) @10^{-12}BER

TX Preset # \ CTLE	0 dB	-1 dB	-2 dB	-3 dB	-4 dB	-5 dB	-6 dB	-7 dB	-8 dB	-9 dB	-10 dB	-11 dB	-12 dB
4	8.8	30.8	35.8	37.3	40.8	45.3	49.0	46.4	37.3	35.8	40.8	45.3	49.0
5	9.7	34.6	37.8	45.8	43.9	44.0	49.7	50.1	45.8	37.8	43.9	44.0	49.7
3	18.7	40.2	40.2	43.9	43.9	46.4	45.4	43.0	43.9	40.2	43.9	46.4	45.4
6	8.7	34.6	40.1	39.8	42.8	45.9	49.0	46.4	39.8	40.1	42.8	45.9	49.0
1	24.5	40.8	40.5	43.3	40.8	43.5	41.8	37.7	43.3	40.5	40.8	43.5	41.8
9	9.0	34.6	37.7	37.8	40.3	43.6	49.2	45.2	37.8	37.7	40.3	43.6	49.2
2	29.9	40.3	37.8	40.2	41.8	37.0	36.9	35.7	40.2	37.8	41.8	37.0	36.9
0	34.6	37.4	39.0	37.1	33.9	31.5	30.2	25.0	37.1	39.0	33.9	31.5	30.2
8	25.3	43.9	43.8	46.3	46.4	42.5	44.7	39.2	46.3	43.8	46.4	42.5	44.7
7	40.2	46.1	40.3	40.4	39.6	37.0	31.2	31.0	40.4	40.3	39.6	37.0	31.2

Channel A, 23in, PTH connector, Veye (mVpp) @10^{-12}BER

TX Preset # \ CTLE	0 dB	-1 dB	-2 dB	-3 dB	-4 dB	-5 dB	-6 dB	-7 dB	-8 dB	-9 dB	-10 dB	-11 dB	-12 dB
4	6.1	39.2	43.6	45.0	48.6	50.8	53.4	53.4	45.0	43.6	48.6	50.8	53.4
5	20.8	42.8	45.6	50.2	50.0	51.2	54.8	54.8	50.2	45.6	50.0	51.2	54.8
3	30.8	41.2	42.6	43.2	45.6	46.4	46.0	43.8	43.2	42.6	45.6	46.4	46.0
6	14.5	42.8	46.4	45.4	48.0	51.0	53.6	51.4	45.4	46.4	48.0	51.0	53.6
1	36.2	40.4	41.8	41.4	42.4	42.8	40.8	38.4	41.4	41.8	42.4	42.8	40.8
9	14.7	41.8	45.2	44.8	46.0	49.0	50.8	50.8	44.8	45.2	46.0	49.0	50.8
2	41.6	39.8	39.2	38.8	40.2	37.0	36.8	34.8	38.8	39.2	40.2	37.0	36.8
0	45.8	37.8	38.2	36.0	34.0	31.8	29.6	25.6	36.0	38.2	34.0	31.8	29.6
8	38.6	45.2	43.0	45.4	46.0	42.4	43.6	39.6	45.4	43.0	46.0	42.4	43.6
7	51.8	43.4	39.6	39.4	38.2	36.4	31.6	30.6	39.4	39.6	38.2	36.4	31.6

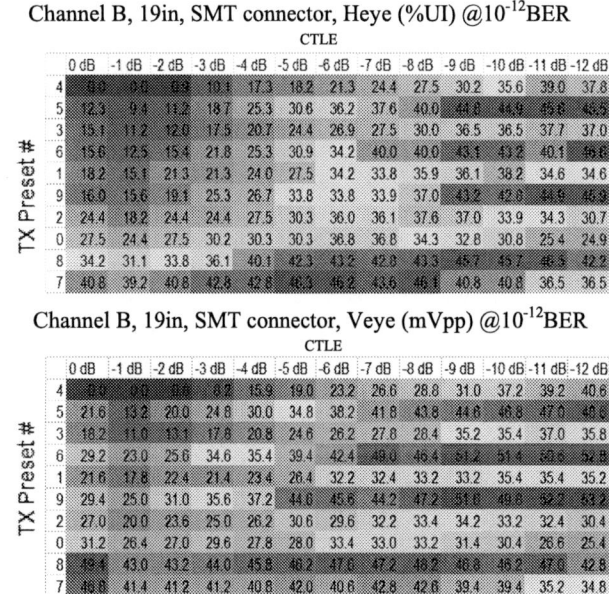

Figure 6. Eye opening at 10^{-12}BER of Channel A 23in with PTH connector (Case 2).

Channel B, 19in, SMT connector, Heye (%UI) @10^{-12}BER

TX Preset # \ CTLE	0 dB	-1 dB	-2 dB	-3 dB	-4 dB	-5 dB	-6 dB	-7 dB	-8 dB	-9 dB	-10 dB	-11 dB	-12 dB
4	0.0	0.0	0.0	10.1	17.3	18.2	21.3	24.4	27.5	30.2	35.6	39.0	37.8
5	12.3	9.4	11.2	18.7	25.3	30.6	36.2	37.6	40.0	44.8	44.9	45.6	45.5
3	15.1	11.2	12.0	17.5	20.7	24.4	26.9	27.5	30.0	36.5	36.5	37.7	37.0
6	15.6	12.5	15.4	21.8	25.3	30.9	34.2	40.0	40.0	43.1	43.2	40.1	46.6
1	18.2	15.1	21.3	21.3	24.0	27.5	34.2	33.8	35.9	36.1	38.2	34.6	34.6
9	16.0	15.6	19.1	25.3	26.7	33.8	33.8	33.9	37.0	43.2	42.0	44.9	45.9
2	24.4	18.2	24.4	24.4	27.5	30.3	36.0	36.1	37.6	37.0	33.9	34.3	30.7
0	27.5	24.4	27.5	30.2	30.3	30.3	36.8	36.8	34.3	32.8	30.8	25.4	24.9
8	34.2	31.1	33.8	36.1	40.1	42.3	43.2	42.8	43.3	45.7	45.7	46.5	42.2
7	40.8	39.2	40.8	42.8	42.8	46.3	46.2	43.6	46.1	40.8	40.8	36.5	36.5

Channel B, 19in, SMT connector, Veye (mVpp) @10^{-12}BER

TX Preset # \ CTLE	0 dB	-1 dB	-2 dB	-3 dB	-4 dB	-5 dB	-6 dB	-7 dB	-8 dB	-9 dB	-10 dB	-11 dB	-12 dB
4	0.0	0.0	0.0	8.2	15.9	19.0	23.2	26.6	28.8	31.0	37.2	39.2	40.6
5	21.6	13.2	20.0	24.8	30.0	34.8	38.2	41.8	43.8	44.6	46.8	47.0	48.6
3	18.2	11.0	13.1	17.8	20.8	24.6	26.2	27.8	28.4	35.2	35.4	37.0	35.8
6	29.2	23.0	25.6	34.6	35.4	39.4	42.4	49.0	46.4	51.2	51.4	50.6	52.8
1	21.6	17.8	22.4	21.4	23.4	26.4	32.2	32.4	33.2	33.2	35.4	35.4	35.2
9	29.4	25.0	31.0	35.6	37.2	44.6	45.6	44.2	47.2	51.0	49.8	52.2	53.2
2	27.0	20.0	23.6	25.0	26.2	30.6	29.6	32.2	33.4	34.2	33.2	32.4	30.4
0	31.2	26.4	27.0	29.6	27.8	28.0	33.4	33.0	33.2	31.4	30.4	26.6	25.4
8	49.4	43.0	43.2	44.0	45.8	46.2	47.6	47.2	48.2	46.8	46.2	47.0	42.8
7	46.8	41.4	41.2	41.2	40.8	42.0	40.6	42.8	42.6	39.4	39.4	35.2	34.8

Figure 7. Eye opening at 10^{-12}BER of Channel B 19in with SMT connector (Case 3).

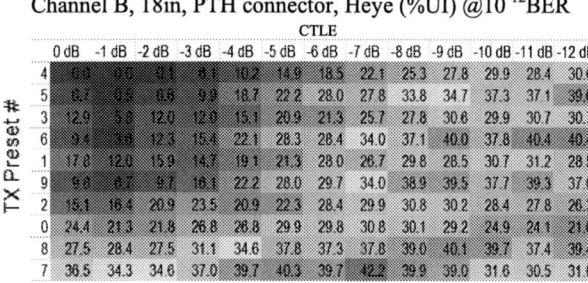

Channel B, 18in, PTH connector, Heye (%UI) @10⁻¹²BER

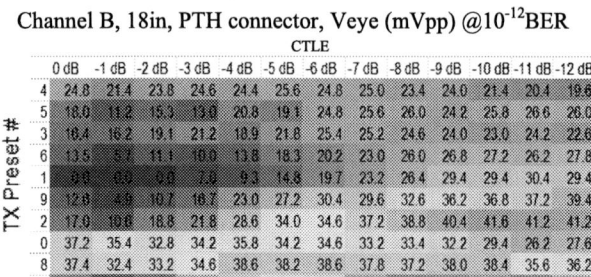

Channel B, 18in, PTH connector, Veye (mVpp) @10⁻¹²BER

Figure 8. Eye opening at 10^{-12}BER of Channel B 18in with PTH connector (Case 4).

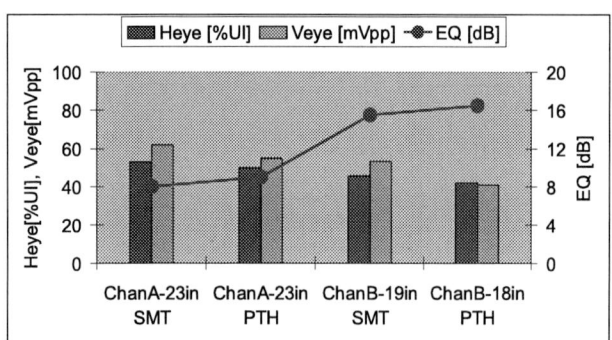

Figure 9. Maximum eye opening and associated equalization on channels under test.

Channel	TX EQ			RX CTLE	EQ sum
	Preset#	Pre	Post		
Chan A-23in/SMT	P5	2dB	dB	-6dB	8dB
Chan A-23in/PTH	P5	2dB	0dB	-7dB	9dB
Chan B-19in/SMT	P9	3.5dB	0dB	-12dB	15.5dB
Chan B-18in/PTH	P7	3.5dB	-6dB	-7dB	16.5dB

Variations in implementation of PCIe Gen3 adaptive equalization among vendor devices

For PCIe Gen3, the TX FFE setting is determined through a negotiation process during link training as defined by the standard. In the first step of the process, the root complex sets both its own TX and the endpoint TX to an initial FFE preset. In the next phase, the endpoint should evaluate the incoming waveform and request different root complex TX FFE settings if the waveform is unacceptable. Finally, the root complex requests the endpoint TX to modify its FFE settings as it deems necessary.

Implementation of the above process varies widely among vendors. Firstly, each RX must essentially measure its own eye diagram to determine if the signal will lead to a low bit error rate. How this is to be accomplished is not explicitly defined by the specification. It is possible that identical waveforms will be judged differently by different vendor devices.

It has also been observed that some devices allow the root complex to retain the initial FFE preset without requesting different settings. In a situation such as this, the endpoint RX equalization must be carefully tuned to account for all possible incoming waveforms.

The format of exchanging FFE settings can also vary. Some devices give preset numbers when requesting new settings. Other devices request specific coefficient values. Problems arise when these different implementations attempt to negotiate FFE settings with each other.

Even in cases where FFE is indeed chosen adaptively, the method to determine the next FFE setting to try is not universal. Some devices sweep through a list of predefined presets. Others change coefficients on a finer granularity and try even those FFE settings not listed as a specification defined preset.

Due to all of these variations, the same channel but with different root complexes and endpoints can have different resulting eye opening. This is an added dimension of complexity not covered by simulation. Because of this, the optimal equalization combination found in simulation is unlikely to be used by hardware.

Equalization training variation over link training events

Hardware prototype PCIe links were continually retrained to gather equalization parameters and eye opening data to judge variance of these statistics. Devices are labeled as RC#n and EP#n throughout this section to indicate different supplier sources without revealing suppliers by name. Two channels were studied which are shown in Figures 10 and 11.

Figure 10 and Figure 11 also show link equalization training results on two PCIe links over 10 successive link training tests. Trained equalization values such as TX FFE, RX CTLE, and VGA were captured for all lanes of the x8 link.

The vertical axis for the CTLE and voltage gain plots represents index numbers and not actual dB gains. The vertical axis for the TX equalization plots in Figures 10 and 11 is the sum of preshoot and de-emphasis in dB.

The data show that successive trains can lead to variation in equalization results despite channel configuration remaining the same. More surprising is the wide variation in equalization results among lanes, despite having similar routing length. This is most clearly noticed in the CTLE plot of Figure 11.

Because of the inconsistent nature of adaptive equalization, some extra margin should be added to designs to account for device unpredictability.

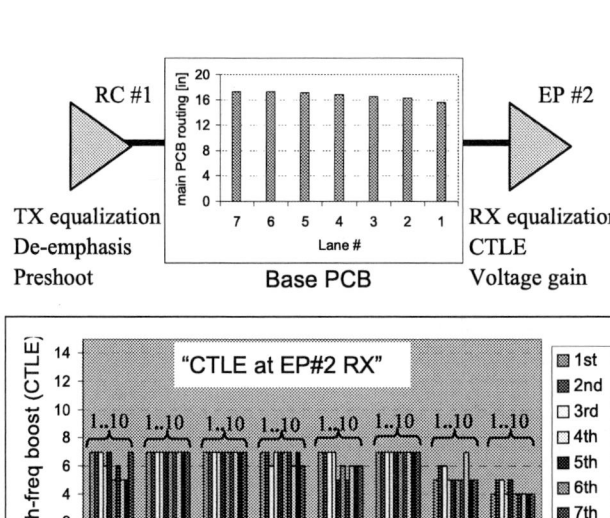

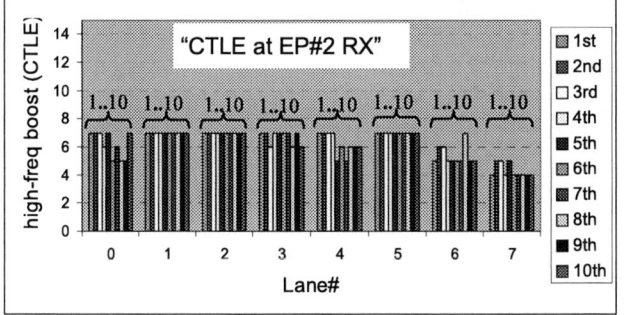

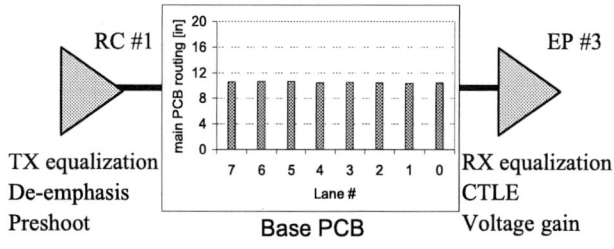

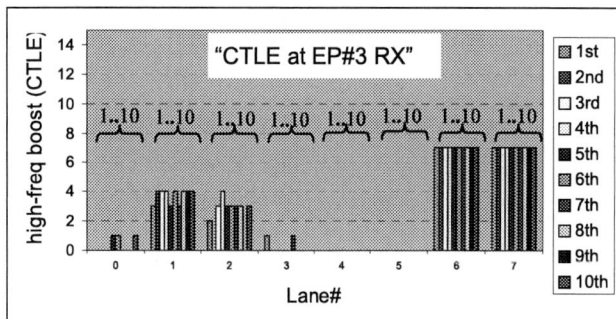

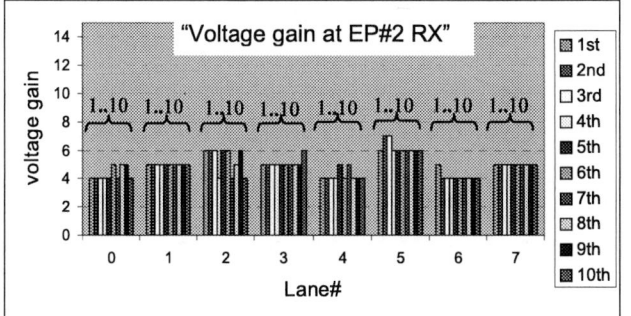

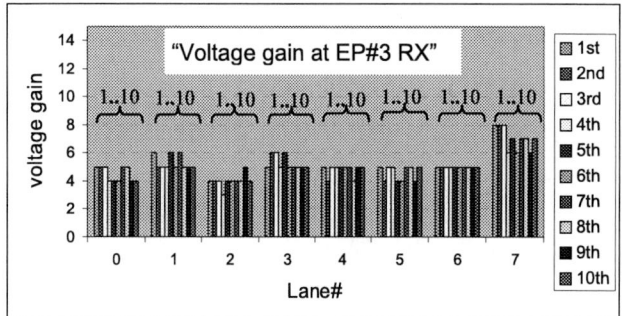

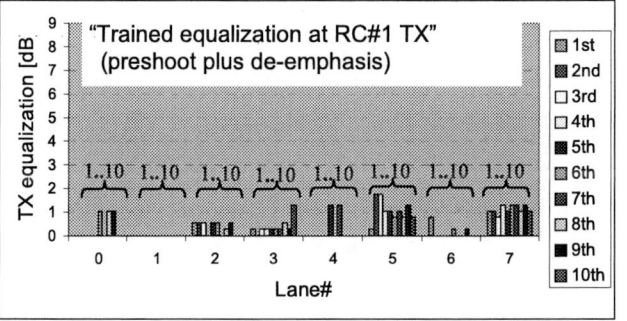

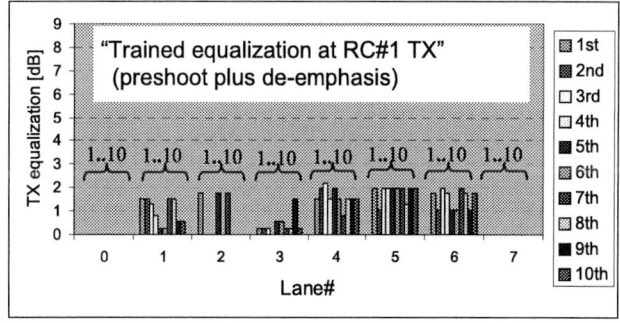

Figure 10. Equalization training results over iterated link training events on RC#1 to EP#2 with approximately 17in total channel length.

Figure 11. Equalization training results over iterated link training events on RC#1 to EP#3 with approximately 11in total channel length.

Adaptive RX equalization and eye opening under different TX FFE conditions

During phase 2 of normal Gen3 equalization training, TX FFE coefficients for the root complex are determined by the endpoint. Though the root complex TX FFE starts at the initially advertised preset, it typically ends up at different coefficients unless equalization phase bypass mode is enabled.

Adapters from different suppliers were plugged into PCIe slots of different systems. Equalization behavior on the links was examined, and eye opening data was captured from internal chip registers. Since the Gen3 RX eye opening requirement is defined at a node internal to the chip, vendor supplied tools to gather this data must be used. It was observed that some of these endpoint devices behaved as if phase 2 of equalization training was set to bypass. Only the initial FFE preset advertised by the root complex was used.

Figure 12 describes a PCIe link with 14in of trace routing to a standard PCIe slot. One of the above described adapters that use only the initial preset was tested.

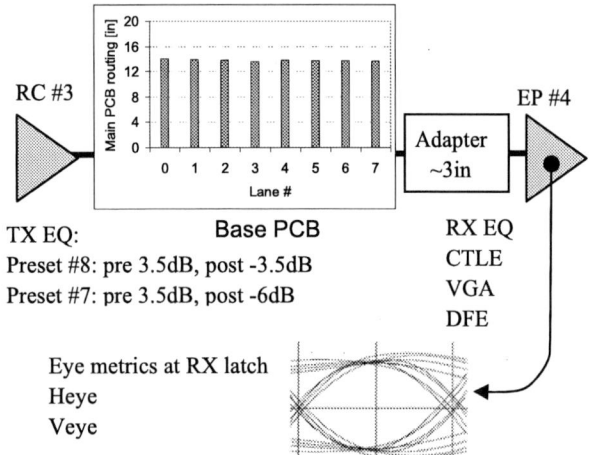

Figure 12. Diagram of channel and digital eye data measurement location after equalization.

Figure 13 shows eye opening and equalization data for the link in Figure 12 in which the TX FFE was set to preset 8 (preshoot 3.5dB, de-emphasis -3.5dB). Again, this is the situation where TX FFE does not get modified. The results reflect data for the root complex FFE set to preset 8. Figure 14 shows eye opening and equalization results with TX FFE preset 7 (preshoot 3.5dB, de-emphasis -6dB).

As a side note, the eye opening data reported was gathered from vendor tools, not an oscilloscope. This data is captured over a very short time and as such would not represent a long term persistent eye like an oscilloscope would show, though some vendor tools do statistically extrapolate to 10^{-12}BER. The eye opening data presented in Figure 13 and Figure 14 is based on eye opening extrapolation to 10^{-12}BER. Actual criteria for good reliability may be different from supplier to supplier due to differences in chip circuitry.

It is observed in Figure 13 that TX preset 8 resulted in relatively uniform behavior over all lanes in RX equalization levels and eye opening. Note that DFE coefficients and CTLE

levels are given as integers. Normally, DFE coefficients range between 0 and 1, and CTLE is described as a dB peaking amount. This level of detail was not provided by the vendor tool.

On the other hand, Figure 14 shows that TX preset 7 resulted in smaller eye opening on many of the lanes despite higher TX FFE equalization; this likely indicates over-equalization. Variation among the lanes in eye opening and equalization is also larger.

Another interesting observation for the preset 7 case is that the adapted CTLE levels ended on a much lower setting while VGA levels ended up higher. The need for high frequency boosting from CTLE is less since preset 7 already includes greater de-emphasis. The overall signal swing envelope would also decrease with greater de-emphasis, requiring greater levels of flatband gain.

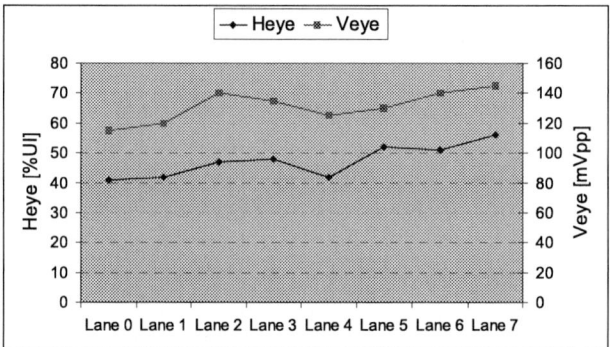

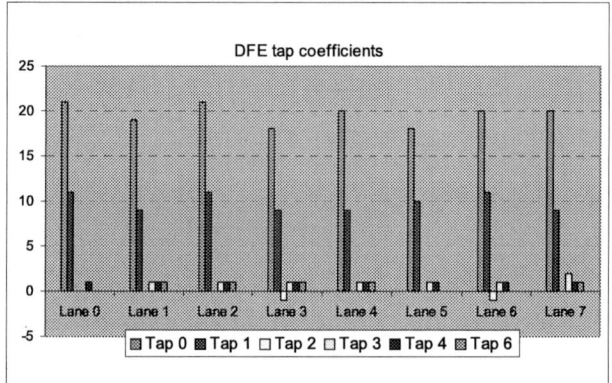

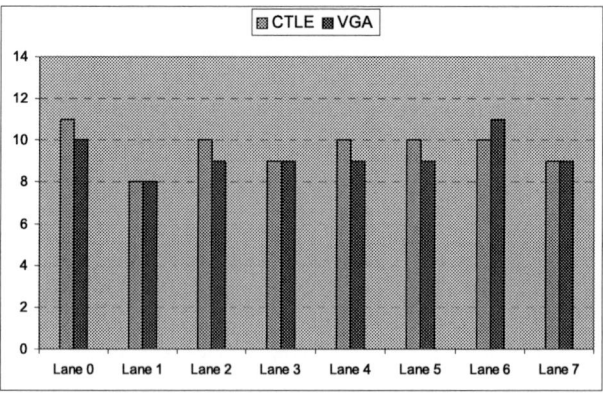

Figure 13. Eye opening and RX equalization on endpoint RX with root complex TX preset 8.

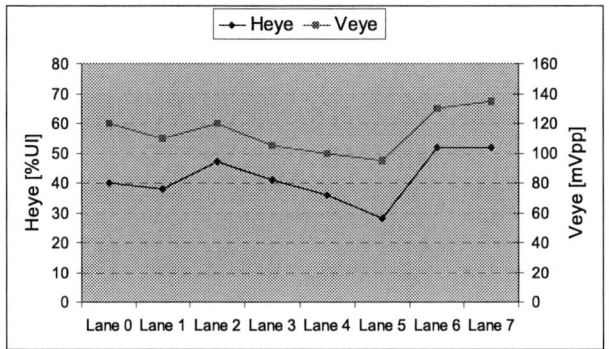

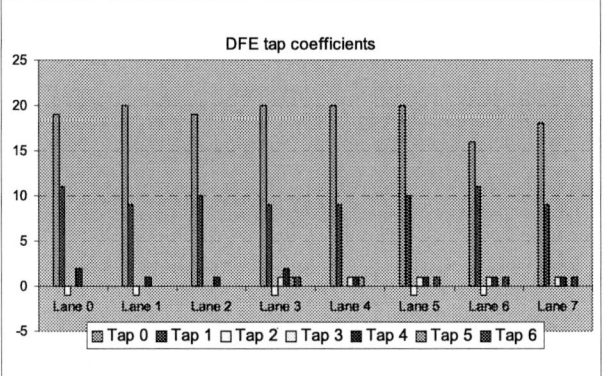

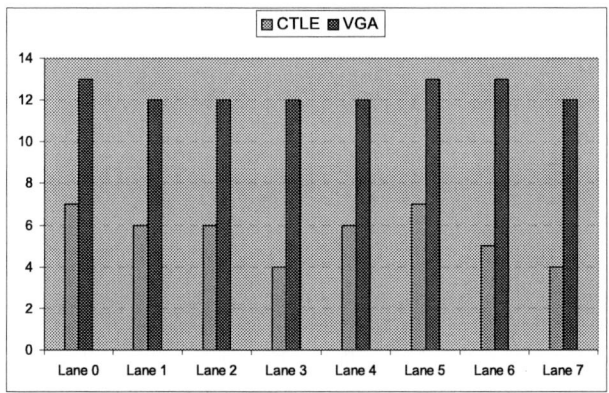

Figure 14. Eye opening and RX equalization on endpoint RX with root complex TX preset 7.

RX tuning efficiency in supplier device with TX preset sweep

Two identical adapters with different RX equalization tuning (set by firmware) were tested on similar system channels. All spec defined TX FFE presets were tested and eye opening data was gathered. Figure 15 and Figure 16 show the gathered data for both firmware versions. Trace routing lengths for the two channels are similar, as also shown in the figures.

The differences seen in eye opening results would mostly be due to the RX equalization tuning. These settings are coded through firmware programming. Details of which are not available to system designers to protect vendor proprietary design information.

Whereas both firmware versions achieved the best eye openings with presets 4, 5, and 6, it is observed that results of earlier firmware version A show significantly smaller horizontal opening compared to the later firmware version B.

It appears that the built in RX equalization tuning is effective only for specific channel configurations. Retuning the equalization may help improve eye opening. Unfortunately, this option is not always readily available to system designers.

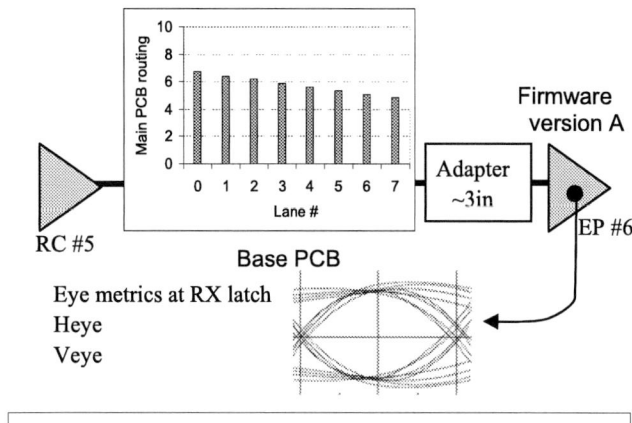

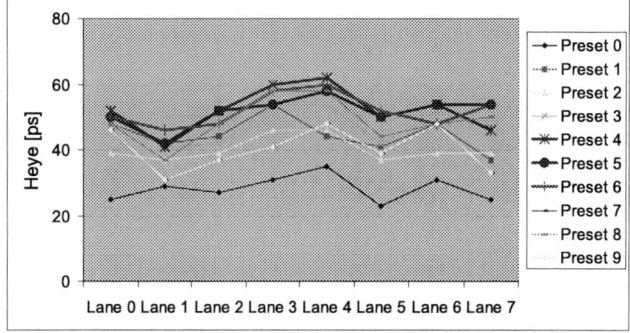

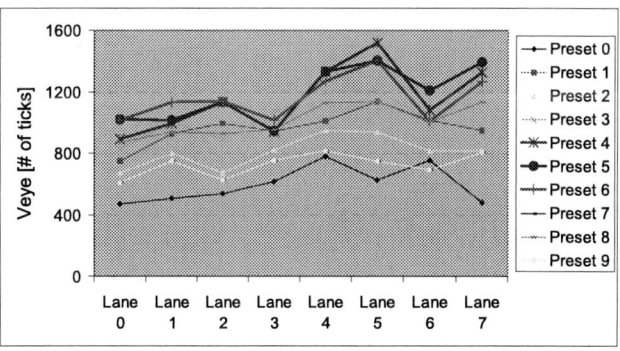

Figure 15. Eye opening variation with TX preset sweep with firmware version A at EP#6.

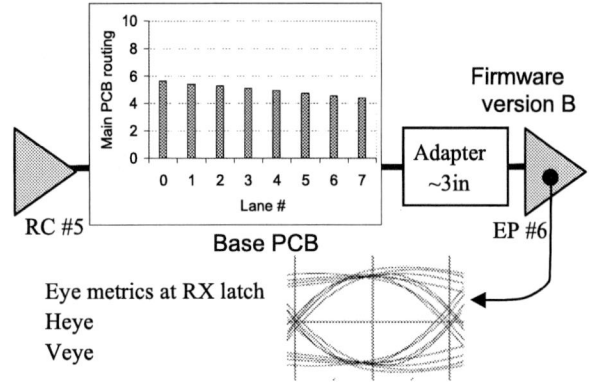

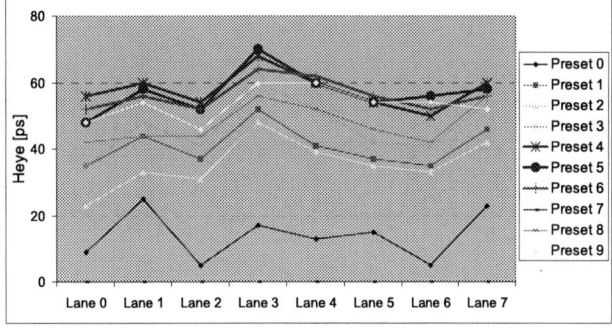

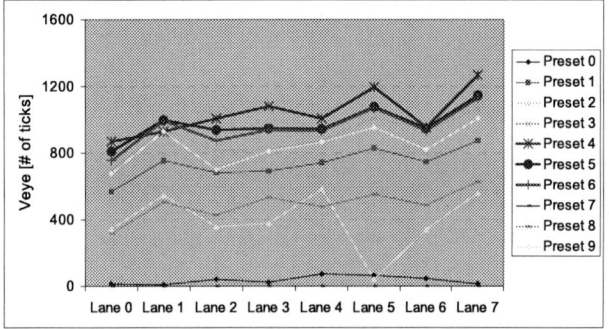

Figure 16. Eye opening variation with TX preset sweep with firmware version B at EP#6.

Conclusions

When defining a trace routing budget for high speed designs such as PCIe Gen3, discontinuities must be carefully designed for in addition to overall routing length. Eye simulations presented in this paper show that discontinuities can significantly decrease PCB reach.

Even eye simulations do not always represent the non-ideal situations that can arise in hardware. PCIe devices have shown great variability due to the loose interpretation of adaptive equalization allowed by the specification.

The best design procedure under such circumstances is to thoroughly test all devices intended to be supported by hardware. Monitoring of error registers, gathering digital eye and equalization data, and training success tests are all advised.

Good signal integrity work can no longer be confined to the design of good PCB channels. With the introduction of PCIe Gen3, there is need for consideration of root complex and endpoint idiosyncrasies.

References

1. PCI Express Base Specification Revision 3.0
2. PCI Express Card Electromechanical Revision 3.0 Version 0.9

A Low-Cost PCB Fabrication Process

[1]Jack Ou, [1]Alberto Maldonado, [1]Chio Saephan, [1]Farid Farahmand, [2]Michael Caggiano
[1]Engineering Science, Sonoma State University, Rohnert Park, California, United States
[2]Electrical and Computer Engineering, Rutgers University, Piscataway, New Jersey, United States
Tel: (707) 664-3462, Fax: (707) 664-2361, Email: jack.ou@sonoma.edu

Abstract

This paper investigates the resolution of a low-cost printed circuit board (PCB) fabrication process. A set of frequently used footprints is fabricated on a PCB and examined under a digital microscope. The results indicate that using the process described in this paper, a thin wire with a 0.38 mm (14.96 mils) width can be fabricated. The process described in this paper is useful for educators who wish to fabricate a fine structure on a PCB, but do not have access to a milling machine. It is also be useful for researchers who wish to quickly build an inexpensive prototype before sending out the final design to a commercial venue.

1. Introduction

The need for fabricating a prototype on a PCB arises frequently in electrical engineering, particularly in areas such as antenna and radio frequency circuits. Even though access to a high quality PCB process is widely available for commercial companies and research institutions, access to an inexpensive yet accurate PCB process with a quick turn-around time remains non-existent for educators who teach mid-size classes.

Simple do-it-yourself (DIY) techniques such as using an iron to transfer ink printed on a transparency to a PCB can be useful for through-hole components, but lack sufficient accuracy and consistency for surface mount devices (SMD). In this paper, we investigate the resolution of a low-cost PCB fabrication process that utilizes a photopolymer film [1].

This paper is organized as follows: The materials and the procedure for fabricating a PCB with a photopolymer film are described in Section 2. The results of our study are discussed in section 3. We present our conclusion in Section 4.

2. Fabrication Process

2.1 Materials

Materials used in this paper are documented below:

- Transparency Film (3M, PP2500, water-based transparency coating.)
- Laser Printer (HP Laserjet, 4050N model)
- Printed Circuit Board (double-sided, 1.4 mm in thickeness, 6 in. x 6 in., $3.38 from Amazon.)
- Photopolymer dry film (30 cm x 200 cm, $16.00 from Amazon.)
- Heat sealing tool (TF Top Flite Monokote)
- Photopolymer developer (sodium carbonate, 0.85 wt%)
- Etchant (Ferric Chloride)
- UV Box (converted from a used Astra 1220 UMAX scanner.)

2.2 Process

We provide a description for fabricating a PCB with a photopolymer film in this section. Figure 1 shows the overall fabrication process in four steps. We begin by preparing the transparency, the photosensitive film, and the PCB board for fabrication. Next, we carefully place the photosensitive film on the PCB and position the photosensitive film above the layout printed on a transparency. The photosensitive film is then subjected to a brief exposure of ultra violet light. The exposed film is rinsed with a developer solution. Excess film is removed with sodium carbonate. The remaining film on the PCB corresponds to the pattern printed on the transparency. Finally, the copper unprotected by the photosensitive film is removed with an etchant. The details of each step are described below.

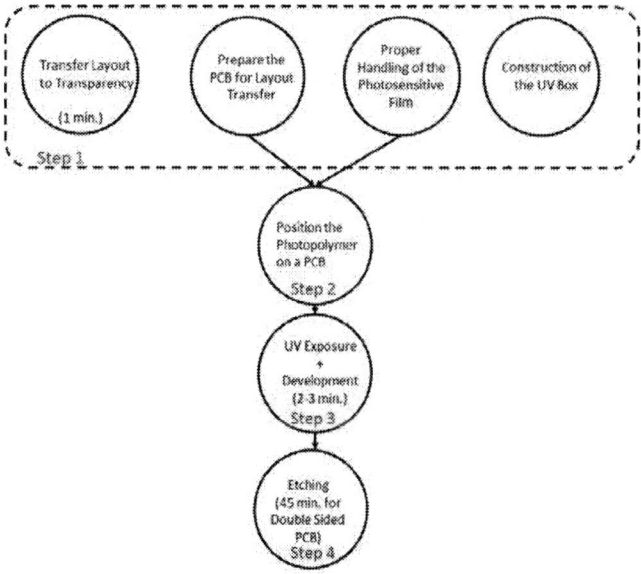

Figure 1. PCB fabrication with a photopolymer dry film.

Step 1: Preparation

The *first* step deals with proper handling of materials used in the fabrication as well as the construction of an ultra violet (UV) exposure box. The details are described below:

A. Layout Transfer

The layout is printed on a transparency in black ink. To improve the quality, the layout is printed twice and touched up with a Sharpie®.

B. Prepare the PCB for Layout Transfer

A PCB with dimensions matching the layout is selected and cut with a hacksaw. The surface of the PCB is sanded with a 400-grit sand paper and rinsed under water. As an alternative to sand paper, Emory Cloth with appropriate grain size can also be used. The copper should be clean and shiny after this step.

C. Proper Handling of the Photosensitive Film

A piece of photosensitive film with dimensions matching the layout is selected and cut. The film should be large enough to cover the board. Transparent tapes are placed on both sides of the film so that one of the protective films covering the photosensitive film can be peeled off.

D. Construction of the UV Exposure Box

The UV exposure box (Figure 2) was constructed from a used flatbed scanner. The original electronics and the mechanical parts were removed. A high density polyethylene (HDPE) sheet with appropriate dimensions was used to cover the entire glass area. Six 12-inch fluorescent tubes powered by a ballast transformer were mounted on the HDPE. The fluorescent light bulbs were chosen to compliment the photopolymer film which has a peak resist response from 350 nm to 380 nm.

Figure 2. An UV exposure box converted from a flatbed scanner.

Step 2: Proper Placement of the Photosensitive Film on the PCB

One of the transparent protective films on the photosensitive film is peeled off. The exposed side photosensitive film is placed against the copper. Air bubbles are eliminated by applying pressure from center out. Next, we separate copper with a piece of paper and apply heat gently with an iron through the paper so that the photosensitive film attach properly to the copper. The iron should be set to a temperature between 105 degree Celsius and 120 degree Celsius.

Step 3: UV Exposure and Board Development

The transparency with a layout printed in black ink is placed on the exposure box as shown in Figure 3. The PCB is attached by the photopolymer film. The photopolymer film is separated from the transparency by a protective film. The ink side of the transparency is placed against the PCB. The exposure time depends on the quality of the laser printer. A high quality printer, capable of producing a consistent, dark and crisp layout, produces an image that can protect the photopolymer film from a longer UV exposure, therefore, leading to a better layout pattern transfer. We typically expose the board to 10s UV light for a light pattern on the transparency and 30s to 45s for a dark pattern on the transparency. The protective film on the photosensitive film is peeled after UV exposure. The protective film should be brittle and hard after exposure.

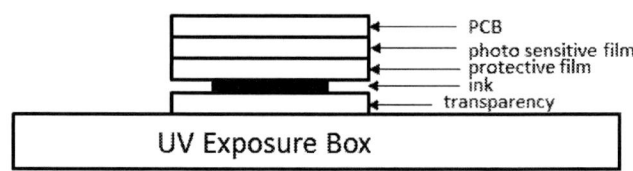

Figure 3. A PCB positioned on the UV exposure box.

After UV exposure, the board is developed with sodium carbonate. Other developers such as potassium carbonate can also be used [1]. The board is rinsed with the developer solution to bring out the layout pattern on the board. Excess photosensitive film is removed by rubbing the board with sodium carbonate with a finger for 2-3 minutes. A darker layout pattern will emerge on the PCB if it is exposed to the UV light for a few more seconds. The layout can be touched up with a Sharpie® at this point as necessary.

Step 4: Etching

Finally, the board is etched with 30% Ferric Chlorid. For a double sided 3" x 3" PCB, the etching usually takes approximately 45 minutes. The PCB is rinsed with water and cleaned with acetone to complete the process.

3. Results

We present two sets of experimental results. In the first experiment, we investigate the resolution of the fabrication process with a set of frequently used footprints on a PCB. In the second experiment, we demonstrate the application of the fabrication process in the context of a radio frequency filter design.

3.1 Investigation of Process Resolution Using Frequently Used Footprints

In this experiment, we fabricate frequently used footprints on a PCB in order to evaluate the resolution of the fabrication process. The layout was drawn in a PCB layout editor, printed onto a transparency (Figure 4), and transferred to a PCB. A photograph of the fabricated board is shown in Figure 5.

Nine groups of footprints are fabricated on the PCB. The footprints in group A represent the footprints commonly used by SMD components. The footprints in group B represent a variety of wires with widths ranging from 0.1 mm to 1.06 mm. The footprint in group C corresponds to that of a 44-pin LQFP. A spiral inductor is shown in group D. An SO32 footprint is shown in group E. A Micro-8 footprint is shown in group F; a TO72 footprint is shown in group G; an SOT footprint is shown in group H; an 18-pin DIP footprint is shown in group I. The dimensions of the shapes are measured as follows: *first*, the layout on the transparency is examined under a Celestron 44308 digital microscope with a maximum of 200x magnification. The dimensions are measured with a Neiko 6" digital caliper with a resolution of 0.01 mm and an accuracy of 0.02 mm. Next, the same geometry implemented on the PCB is measured. The measurement results of selected groups are discussed next.

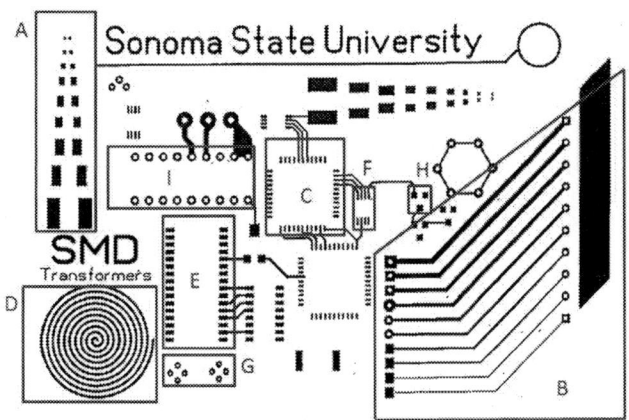

Figure 4. Frequently used footprints printed on a transparency.

Figure 5. PCB layout.

Group A: SMD Footprints

Footprints in group A correspond to the pad dimensions of the 0201, 0402, 0603, 0805, 1008, 1206, 1210, 1812, and 2220 SMD components. They are numbered from 1 to 9 in Figure 6. The length (l), the width (w), and the spacing (s) of the pads are defined in Figure 6. Δh is calculated by subtracting the h of the pad on the transparency from the h of the pad on the PCB, similarly for Δs and Δw. The average of Δh and the average of Δw are 0.025 mm and 0.086 mm respectively, indicating that the h and the w of the pads on PCB are larger than those found on the transparency. The average of Δs is -0.22, indicating that s the pads on the PCB is on smaller than the s of the pads on the transparency. The negative Δs is expected because Δw is positive. The 0402

SMD footprint is the smallest SMD footprint we can fabricate reliably with this process.

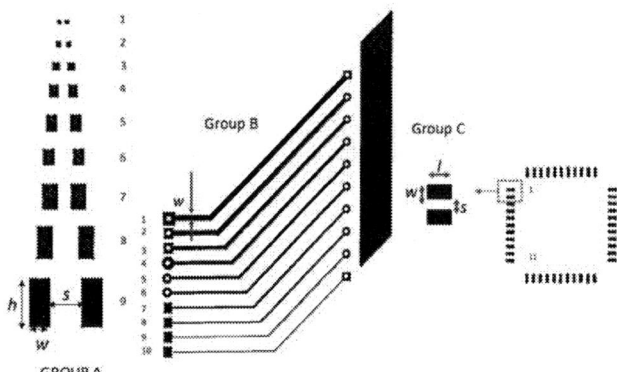

Figure 6. Footprints of group A, B and C.

Unit	Transparency			PCB			Comparison		
(mm)	h	w	S	h	w	s	Δh	Δw	Δs
A1	0.43	0.4	0.44	0.53	0.58	0.29	0.1	0.18	-0.15
A2	0.65	0.59	0.58	0.62	0.75	0.27	-0.03	0.16	-0.31
A3	0.92	0.84	0.72	0.94	0.95	0.51	0.02	0.11	-0.21
A4	1.42	1.07	0.97	1.55	1.22	0.7	0.13	0.15	-0.27
A5	1.75	1.25	1.36	1.69	1.4	1.03	-0.06	0.15	-0.33
A6	1.79	1.44	1.76	1.92	1.39	1.65	0.13	-0.05	-0.11
A7	2.97	1.91	1.34	2.68	1.71	1.19	-0.29	-0.2	-0.15
A8	3.61	1.79	2.58	3.82	2.03	2.49	0.21	0.24	-0.09
A9	5.46	2.52	3.43	5.48	2.56	3.05	0.02	0.04	-0.38

Table 1. Dimensions of SMD footprints.

Group B: Wire Width

We examine wire width in this experiment. The width (w) of the wire is measured as shown in Figure 6. The deviation in w (i.e., Δw) is equal to PCB width (w_{PCB}) minus the transparency width ($w_{transparency}$). The thinnest wire (W10), which has a transparency width of 0.1 mm, was *broken* after UV exposure, and had to be repaired with a Sharpie. The PCB width of W10 was not measured because its width is not uniform throughout the wire.

Unit (mm)	$w_{transparency}$	w_{PCB}	Δw
W1	1.06	1.33	0.27
W2	0.94	1.07	0.13
W3	0.75	0.98	0.23
W4	0.7	0.75	0.05
W5	0.54	0.7	0.16
W6	0.59	0.65	0.06
W7	0.45	0.6	0.15
W8	0.3	0.49	0.19
W9	0.22	0.38	0.16
W10	0.1	N/A	N/A

Table 2. Widths of wires in group B.

Group C: 44-pin LQFP

The footprint in group C corresponds to the pads of a 44-pin LQFP. l, w and s are defined in Figure 6. We only measure the l and w of the pad in the upper left hand corner (i.e. pad number 1) because the pads are extremely *well matched*. As a result, we were not able to measure the differences in l and w accurately. $l_{transparency}$ and l_{PAD} are 1.12 mm and 1.21 mm respectively. $w_{transparency}$ and w_{PAD} are 0.49

978-1-4799-2408-0/14 $31.00 © 2014 IEEE   2161

mm and 0.51 mm respectively. We defined $S_{i\text{-}i+1}$ as the spacing of pad i to pad $i+1$. As indicated by Table 3, $\Delta S_{i\text{-}i+1}$ is relatively *uniform* for pads in a 44-pin LQFP; and the average space is 0.29 mm.

	Transparency	PCB	$\Delta S_{i\text{-}i+1}$
$S_{1\text{-}2}$	0.41	0.34	-0.07
$S_{2\text{-}3}$	0.42	0.25	-0.17
$S_{3\text{-}4}$	0.42	0.3	-0.12
$S_{4\text{-}5}$	0.42	0.28	-0.14
$S_{5\text{-}6}$	0.42	0.28	-0.14

Table 3. Spacing measurement for pads in a 44-pin LQFP.

Group D: PCB Inductor

The layout of a spiral inductor is shown in Figure 7. The inner diameter (d_{in}), outer diameter (d_{out}), the width (w) and the spacing (s) are defined in the figure. The deviation in geometry is shown in Table 4.

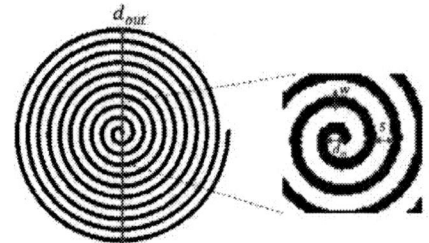

Figure 7. PCB layout.

	Transparency	PCB	Deviation
d_{in}	0.64	0.54	-0.1
d_{out}	20.46	20.81	0.35
s	0.87	0.44	-0.43
w	0.53	0.67	0.14

Table 4. Deviation in the geometry of a spiral inductor.

3.2 A 3.5 GHz Coupled Bandpass *Filter*

To illustrate the integration of this fabrication process in a classroom environment, a 3.5 GHz band-pass filter (BPF) was constructed on a PCB. The BPF was designed and optimized using Agilent's Advanced Design System (ADS). The filter geometry was first analyzed in Momentum, and then again in EMPro. Once the final geometry was determined, the geometry was drawn in Eagle, fabricated on a double-sided PCB, and measured on a network analyzer. A photograph of the coupled-band pass filter is shown in Figure 8. S_{21}, the gain of the filter is measured using a network analyzer and compared to the S_{21} obtained using ADS, Momentum and EMPro. The center frequency (f_o) and the 3dB bandwidth (f_{bw}) of the filter are summarized in Table 5.

Figure 8. Coupled Band-Pass Filter.

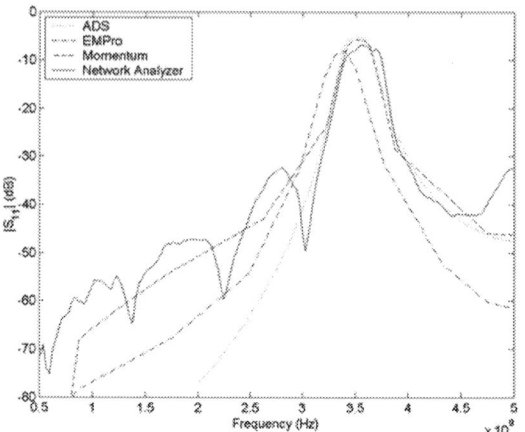

Figure 9. Gain of the Filter.

	f_o (GHz)	S21 at f_o (dB)	f_{bw} (MHz)
ADS	3.5	-5	280
Momentum	3.5	-5.4	250
EMPro	3.4	-7.8	210
Network Analyzer	3.57	-6.8	330

Table 5. Center frequency and bandwidth of the filter.

3.3 Cost Analysis

A. One Time Cost

The UV exposure box was converted from a used scanner. The total cost of constructing the exposure box is sixty dollars. The total cost includes three electronic ballast ($15.57), universal power cable ($4.01), six 12-inch fuorescent light bulbs ($28.99), one HDPE sheet ($6.45), and miscellaneous items such as bolts, washers, zip tiles, and zip ties.

B. Cost for fabricating a 3"x"3 PCB

The cost of developing a 3" x 3" double-sided PCB is as follows: $0.85 for the PCB, $1.65 for the etchant and $0.20 for the photosensitive film. The total cost for 3" x 3" board is $2.85 if we account for other consumable expenses such as plastic gloves, paper towels. If a pre-sensitized PCB is used to eliminate the bubbles between the photopolymer film and the PCB, the total cost for a double-sided PCB increases to $5.07.

4. Conclusions

In this paper, we investigate the resolution of a low-cost PCB fabrication process that utilizes a photopolymer dry film. The thinnest wire we can fabricate with the process has a width of 0.38 mm. The smallest SMD footprint that we can fabricate is that of a 0402 footprint. Pads that are fabricated in a uniform footprint have the minimum pad-to-pad variation and can achieve an average spacing of 0.29 mm. The resolution reported in this paper is useful to educators and researchers who wish to fabricate a low-cost PCB accurately with a quick turn-around time of 1.5 to 2 hours.

References

1. DuPont, "DuPont Riston MultiMaster Photopolymer Dry Film," MM500 datasheet, 2011.

Gap in pagination due to withheld paper.

Pages 2163-2167

Study of Microwave Circuits Based on Metal-Insulator-Metal (MIM) Diodes on Flex Substrates

Amanpreet Kaur, Xianbo Yang, and Premjeet Chahal

Department of Electrical and Computer Engineering, Michigan State University

kaurama1@msu.edu, yangxian@msu.edu, chahal@egr.msu.edu

Abstract

This paper demonstrates fabrication and characterization of thin-film Ti-TiO$_2$-Pd based Metal-Insulator-Metal (MIM) diodes on flexible substrates. MIM diodes with contact areas of 9 μm^2 and 48 μm^2 were fabricated, and a comparison is made for their DC and RF performances. The current-voltage characteristics of the fabricated diodes show strong non-linearity. The diodes are also tested for microwave circuit applications such as detection, frequency multiplication and mixing over a frequency range of 1 – 18 GHz. The devices show strong second harmonic frequency multiplication for fundamental frequencies of 1-10 GHz. Details of DC characteristics, RF rectification, mixing and multiplication using MIM diodes are presented.

Introduction

The microwave circuits on flex substrates enable various applications in wireless communication, identification systems, portable/wearable communication devices, wireless power transfer, security and bio-sensing. Flexible electronics have potential advantages over conventional system as they are light weight, large area compatible, enables roll to roll fabrication, low temperature processing and thus very important for low cost manufacturing [1-4]. Over the last decade, significant work has been carried out on the design and fabrication of RF passive devices on flexible substrates [5, 6]. However, direct integration of active components such as diodes and transistors on flex substrates is still a challenge. For applications that require low cost but high-performance RF/Microwave circuitry capable of operating up to several gigahertzs, flexible substrate is a good choice. Many RF applications like portable communications systems require microwave wave antenna arrays integrated with active electronic devices. The cost of assembling passive and active device modules is high as the cost of the individual III-V electronic circuitry is very high. This problem can be solved by integration of passive devices like antenna arrays and high-speed electronics onto large-area, flexible substrates for applications like imaging system. The metal-insulator-metal (MIM) diodes are made entirely of thin film materials i.e. metals and insulators and provide non-linear characteristics dominated by tunneling effects. They can be easily fabricated on wide variety of flexible substrate or even on top of existing CMOS circuitry. Thus RF and microwave devices based on MIM diodes is a promising area of research, as they can operated at very high frequencies, they are simple to implement and process, and a host of metal dielectric combinations can be used to achieve desired diode characteristics. Thus, this allows fabrication of RF circuits onto low cost plastic substrates, which in-turn may be attached to objects of interest like automobile, aircraft or on buildings for sensing and trans-receiving applications. The MIM diodes can also be easily coupled with other thin film components like MIM capacitors, metal inductors, and thin film resistors leading to formation of complex circuits. The high speed and frequency response in comparison to III-V Schottky diodes as well as the possibility to choose flexible substrate due to thin film fabrication makes MIM diodes a good choice for RF flex devices.

The MIM diodes can operate at high frequencies with high switching speed and faster response time due to tunneling effect. MIM diodes are also preferred due to their temperature insensitive characteristics [7]. High frequency MIM diodes coupled with antennas/waveguides are increasingly explored for application in microwave circuits as they provide good device scalability for Microwave/mm wave detectors [8, 9]. In the past, point-contact MIM diodes also known as whiskers were used for millimeter wavelength detection and mixing, but they lack reproducibility and stability [10].Therefore thin film MIM diodes are utilized more often as they are more reliable and stable. In a MIM diode, the electrons flow between top and bottom metal electrodes through a very thin insulator layer. Also depending on the thickness of insulator and the barrier height, i.e., difference in work function between the two metals, either quantum tunneling or thermionic emission may dominate [11]. Generally MIM diodes with dissimilar metals electrodes show significant non-linearity. The asymmetry is more pronounced when the work function between the two metal electrodes is large. Various combination of dissimilar metals had been investigated in past such as Ni-NiO-Au [8], Ni-NiO-Ni [12], Ti-TiO$_2$-Al [13], Ni-Nio-Cr/Au [14], Al-AlO$_x$-Pt [15]

Although MIM devices have been studied in great detail for a range of applications, to date their fabrication and implementation in the design of RF circuits on flex substrates has not been investigated. In this paper, we present the design, fabrication and characterization of MIM based RF/Microwave devices on flex substrate, Polyetheretherketone (PEEK). PEEK is a low loss substrate which is also compatible with standard micro-fabrication process (temperature and chemicals). For high frequency operation the contact area should be smaller or thickness of the dielectric layer should be high to keep the capacitance low. But increasing the oxide thickness reduces the tunneling current and thus decreasing the contact area is more attractive. In this paper, we report the fabrication of thin film MIM diodes with two different contact areas of 9 μm^2 (Diode A) and 48 μm^2 (Diode B) using Ti-TiO$_2$-Pd. Ti has a work function of 4.3 eV and Pd has work function of 5.2 eV, and thus creating a work function difference of 0.9 eV. This difference in work function has exhibited a higher degree of non-linearity and has been verified by studying the Current–Voltage (I–V) characteristics of the diode. This paper also presents microwave circuits

978-1-4799-2408-0/14 $31.00 © 2014 IEEE

2014 Electronic Components & Technology Conference

measurements over a frequency range of 1 – 18 GHz using these diodes.

Experiment

Deign of coplanar waveguide coupled diode

In this paper coplanar waveguide (CPW) coupled MIM diodes were used to characterize its function as detector, mixer and multiplier. CPW consists of a center (signal) conductor and a pair of ground planes on each side of the center conductor (GSG). The CPW structures were designed using Linecalc tool form ADS (Advance Design System). The CPW's are designed for 50Ω characteristics impedance for substrate with a dielectric constant of 3.3 and thickness of 250 µm. Along with CPW structures for embedding diodes, calibration circuits are also designed for de-embedding the characteristics of MIM diodes using S-parameters. The selected dimensions for GSG are shown in Figure 1(a). The diode was designed to fit within the CPW design as shown in Figure 1(c). The overlap area of narrow region of center conductor and ground conductors defines the MIM diode area.

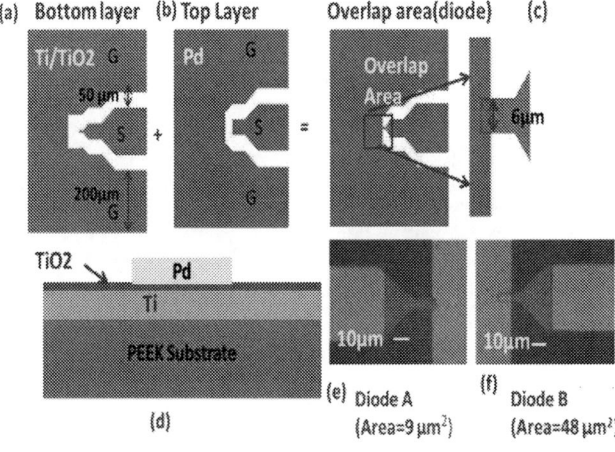

Figure 1. Top view of CPW structures (a) Bottom layer of Ti/TiO2. (b) Top layer of Pd. (c) Small overlap area of two layers defining the diode. (d) Schematic of cross-section of fabricated diodes. (e, f) Optical Micrograph of Diode A with area of 9 µm² and Diode B with area 48µm².

Microelectrodes Fabrication

The devices were fabricated on flexible substrate. The key selection criteria's for polymer substrates are high glass transition temperature, chemical compatibility and low loss. Commercially available thin polymer films like Polyetheretheretherketone (PEEK), Polyimide (PI) and polyethyleneterephthalate (PET) are compatible with chemicals used in the micro fabrication processes and also have low dielectric loss (loss tangent) over a wide range of frequency [16]. Devices in this paper are built on PEEK which has a glass transition temperature of 143 °C and a thermal expansion co-efficient of $2.6 \times 10^{-5} K^{-1}$.

The electrodes were fabricated using standard optical lithography to achieve two different contact area for diodes i.e. Diode A (Figure 1(e)) and Diode B (Figure 1(f). The PEEK substrate wafers were first cleaned using ultrasonic cleaning. For the fabrication of diodes, two lithographic mask

layers were utilized to realize the structure. The first mask layer was used to define structure on Ti/TiO2 layer (Figure 1(a)) and second mask is used to define the Pd layer (Figure 1(b)). The bottom electrode of Ti (150 nm) was deposited using standard e-beam evaporation and the insulator layer (TiO2) was formed during the Ti deposition by introducing a trace amount of oxygen in the chamber. This is followed by a lift off process to define the CPW test structures.

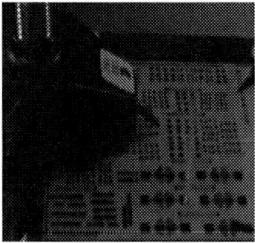

Figure 2: Fabricated structures on flexible PEEK substrate.

The top layer of Pd (150 nm) was deposited by e- beam evaporation .The patterning of top metal layer was carried out by selective etching of Pd using FeCl3. The small overlapping area between Ti/TiO2-Pd defines the diode as shown in Figure 1(c). The fabricated circuits on flexible PEEK substrate are shown in Figure 2.

Results

I-V Measurements

The Current -Voltage (I-V) measurements of the fabricated MIM diodes were carried out at room temperature. The I-V characteristics of diode A was obtained over a voltage range of -0.4V to +0.4 V; while Diode B is studied over a range of -0.3 V to +0.3 V. Figure 3(a) and (b) shows the I-V characteristics of a diode A and diode B respectively before and after removing the effect of series resistance. The measurements show series resistance of approximately 100Ω , which is largely due to thin Ti layer and contact resistance. In spite of the series resistance, the diodes reported here exhibits a significant degree of non-linearity required for microwave circuits of interest. The diodes exhibited current in the range of 1mA at ± 0.2 V of applied bias. For the given contact area, the observed range of current is higher than previously reported diodes using Ni-NiO-Cr/Au and with area of 1µm² [17], this can be contributed to barrier height and thickness of insulator layer [18]. To obtain rectification and multiplication at higher frequencies, MIM diodes with strong non-linearity are required [19].

RF/Microwave Measurements

All of the RF measurements on MIM diode were carried out using a high frequency 50 Ω coplanar GSG probe (Infinity probes) with a pitch size of 150 µm. All measurements were performed at room temperature. Losses at higher frequencies were measured in order to estimate the actual power delivered to the diode. The results mentioned in the following sections are after correction/calibration determined from the reflected signal and the source signal.

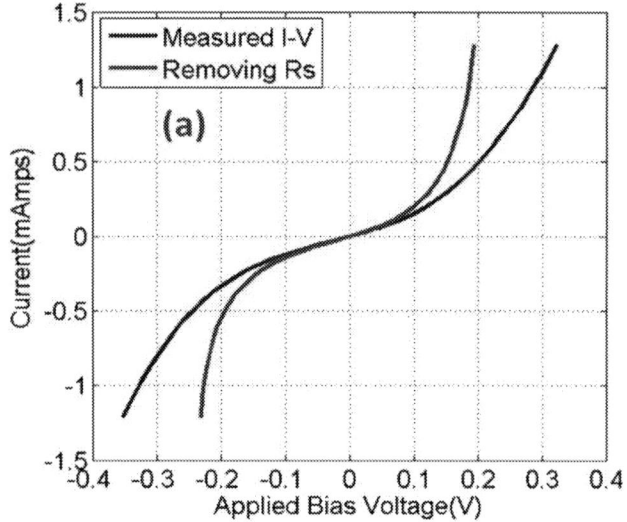

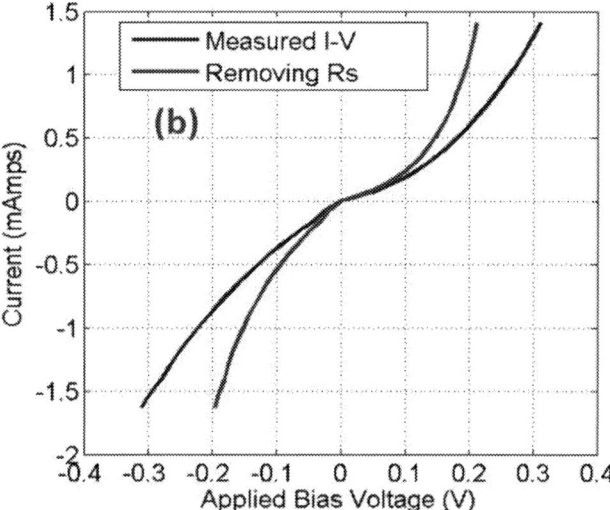

Figure 3. Measured I-V characteristics of MIM diodes before and after removing the effect of series resistance (a) Diode A (b) Diode B.

S-Parameter Measurements and Equivalent Diode Model

One port on-wafer characterization of the MIM devices was carried out using a CPW probe connected to a network analyzer over a frequency range of 1 – 20 GHz. Biasing of the device was carried out using a T-bias and S-parameters were measured at different bias points (0, 0.1 and 0.2V). Equivalent RLC circuit was developed using ADS. An equivalent circuit representation is shown in Figure 4. The circuit includes diode resistance (R_d), diode Capacitance (C_d), series resistance (Rs) and series inductance (Ls). Figure 5 shows the measured and fitted results on a Smith Chart for Diode A and B under bias condition of 0.2V. The measured and equivalent model matched closely. The values of the extracted components for Diode A and B at two different biases is shown in Table 1. The diode resistance and capacitance depends on area of diode which determines the RC time constant and thus the cut off frequency also. The diode resistance and capacitance

scales according to the diode contact area. The series resistance is large and is largely due to the high resistance associated with the first metal layer (Ti) which is very thin. The series resistance can be reduced by increasing the thickness of this layer. The equivalent model shows that the cut-off frequency can be increased significantly by decreasing the diode area.

Table 1. Equivalent model values derived from S-parameters

Diode	Bias	R_{series}	R_{diode}	C_{diode}	L_{series}
Diode A	0 V	114 Ω	490 Ω	0.6 pF	120 pH
	0.2V	114 Ω	140 Ω	0.6 pF	120 pH
Diode B	0 V	114 Ω	110 Ω	3.0 pF	70pH
	0.2V	114 Ω	49Ω	3.1pF	70pH

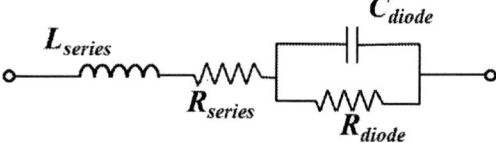

Figure 4. Equivalent circuit model for MIM diode

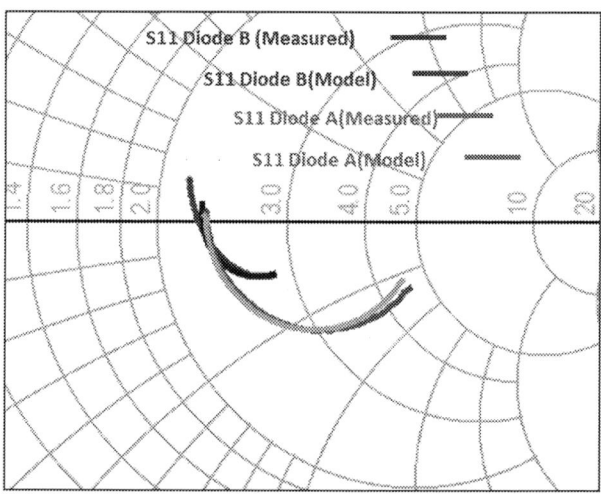

Figure 5. S parameter results for measurement and equivalent model for Diode A and B within frequency range of 1-20 GHz.

MIM diode Rectification

RF rectification measurements were carried out at 10 and 18 GHz for Diode A, while Diode B shows rectification at comparatively lower frequency range (1 – 6 GHz). Rectification for both of the diodes was carried out at a fixed bias of 0.2 V, where the diode shows strongest non-linearity. For experimental set-up, the RF signal from a signal generator was supplied to the diode through a directional coupler (HP87300B) and T-Bias. The T-bias was used to bias the diode and also to separate low and high frequency signals. The power absorbed by the diode was measured using a spectrum analyzer and the detected voltage was checked

978-1-4799-2408-0/14 $31.00 © 2014 IEEE   2170

through a nano-voltmeter. When a RF signal is applied it is rectified by the diode and is measured using a DC nano-voltmeter. Figure 6 (a) and (b) shows measured rectified voltage signal as a function of power absorbed by the Diode A and B, respectively, for a fixed bias of 0.2 V. The measured results show that device has good sensitivity and it can significantly be improved by reducing the series resistance of the diodes and through good impedance matching. Detected voltage for Diode B at frequencies above 3 GHz is much smaller in comparison to Diode A.

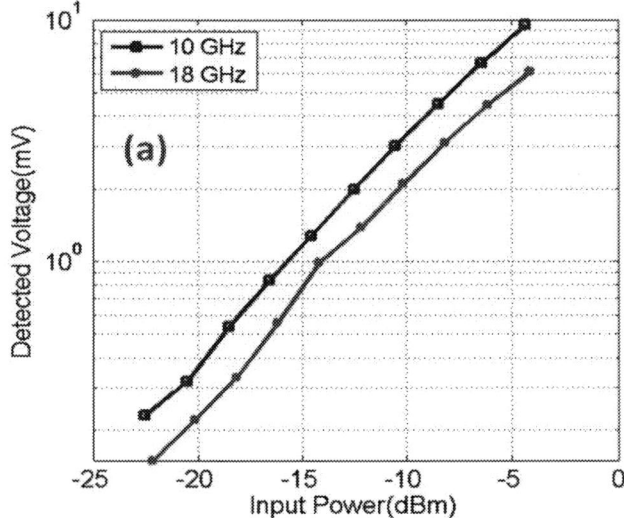

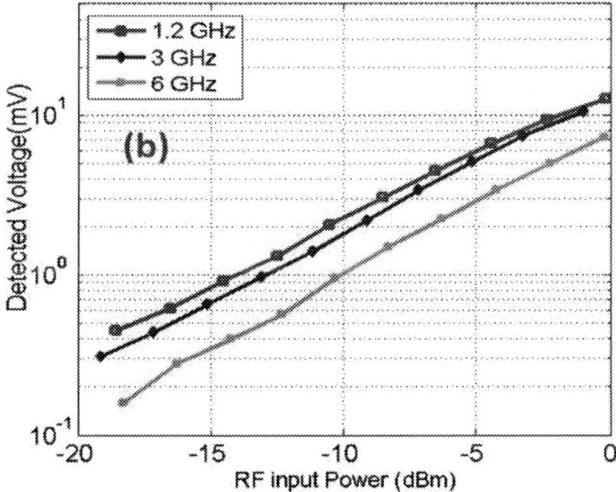

Figure 6. Detected voltage as a function of input power (a) input frequency of 10 and 18 GHz for Diode A. (b) Input frequency of 1.2, 3 and 6 GHz for diode B.

The voltage detection was also tested at different bias voltages for a fixed RF power and frequency. Figure 7 (a) and (b) shows the detected voltage for a input power of -13.6 dBm (18 GHz) for diode A and -16 dBm (3 GHz) for Diode B, respectively. The highest rectification is observed close to the strongest non-linear region of the diode.

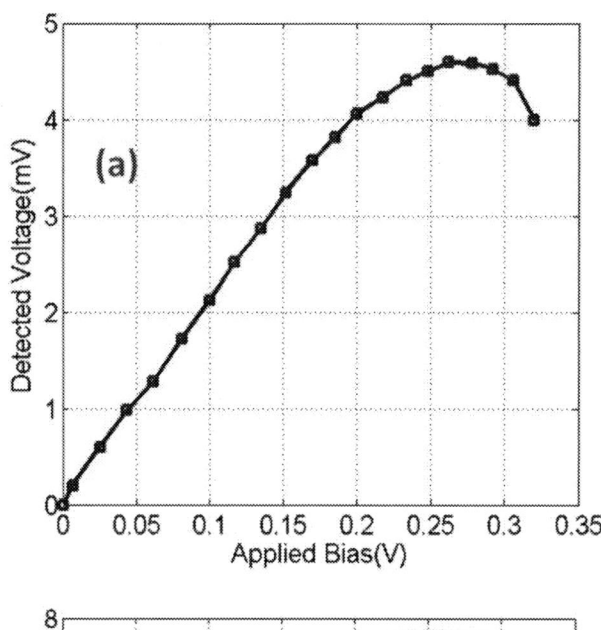

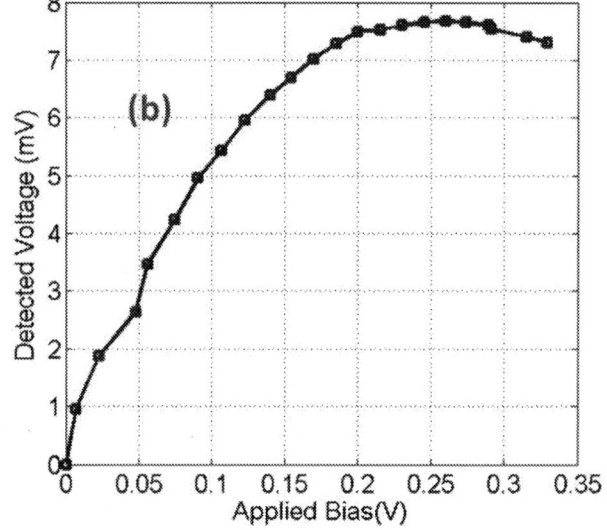

Figure 7. Measured output voltage as a function of applied bias voltage at a fixed power (a) -13.6 dBm (18 GHz) Diode A (b) -16 dBm (3GHz) Diode B.

Frequency multipliers

Use of MIM diodes for frequency multiplication is analyzed by probing the same CPW diode structures. Frequency multiplication is an important part of RF communication and can be realized using non-linear devices like diodes and FETs. Diodes based multipliers have higher bandwidth but the conversion efficiency decreases at the higher harmonics. Frequency multiplier generates signal with a frequency that is a multiple of the input or fundamental frequency (f_0). In this work, 2nd ($2 \times f_o$) and 3rd ($3 \times f_o$) order frequency multiplication was observed for both of the diodes. The power of 2nd order harmonic changes with a small applied bias, while the power of 3rd harmonic does not change much with bias. A fixed bias of 0.2 V is applied through T-Bias for these measurements.

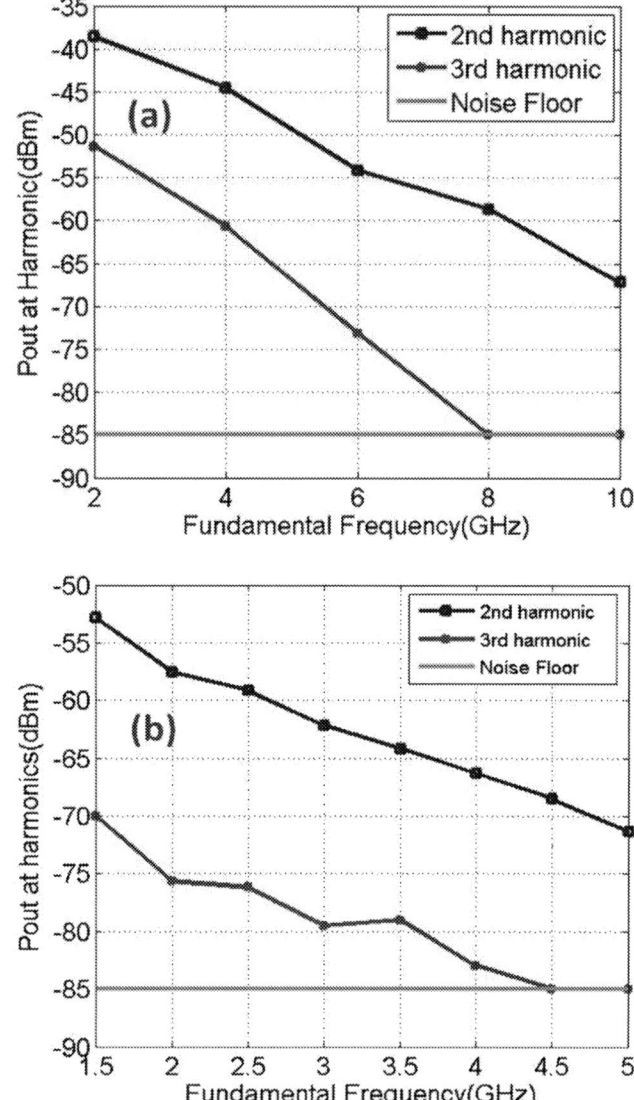

The fabricated diodes exhibited multiplication over a wide range of frequency. Comparing the performance of these two diodes for a $f_0 = 3$ GHz and for approximately the same input power, Diode A has much high output (~ -43 dBm) in comparison to Diode B (~ -70 dBm) for 2nd harmonic. The device with much smaller contact area can perform at even higher frequencies. To further increase the operational frequencies of the diode, scaling in area is necessary as discussed earlier under the S-parameter measurement section.

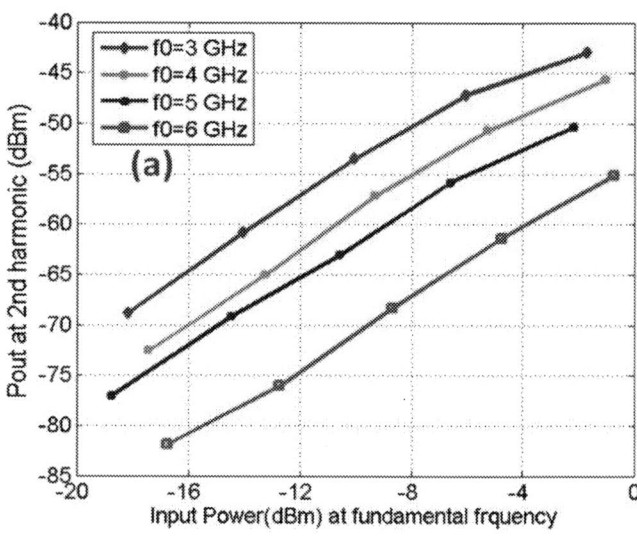

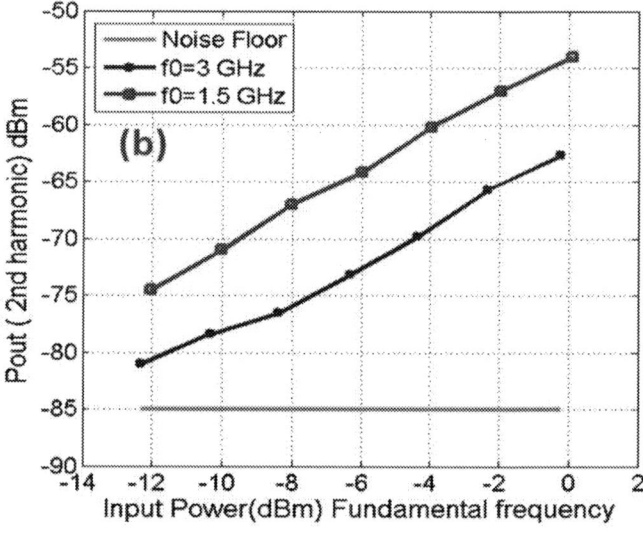

Figure 8. Measured output power of second and third harmonic for (a) $f_0 = 2, 4, 6, 8$ snd 10 GHz for Diode A. (b) $f_0 = 1.5, 2, 2.5, 3, 3.5, 4$ and 4.5 GHz for Diode B

For Diode A, 2nd harmonic was measured for input frequencies of 2, 4, 6, 8 and 10 GHz and third harmonic was observed for input frequencies of 2, 4 and 6 GHz. For diode B, 2nd harmonic was observed for lower frequencies i.e. from 1.5- 5 GHz while 3rd harmonic is observed for 1.5-4 GHz with much smaller output signal. Figure 8 (a) and (b) shows the output power of 2nd and 3rd harmonic for Diode A and B for a range of fundamental frequencies. In the measured result, the conversion efficiency decreases as the fundamental frequency increases. The output power of 2nd harmonic was also measured as a function of input power of the fundamental frequency (f_o) at a fixed bias of 0.2 V. For Diode A, measurements were carried out at $f_0 = 3, 4, 5$ and 6 GHz, while Diode B was measured for $f_0 = 1.5$ and 3 GHz. The measured results are shown in Figure 9 (a) and (b) for Diode A and B, respectively. A linear increase in output power with increase in input power was observed as expected.

Figure 9. Measured output power (2nd harmonic) as a function of the input power for fundamental frequencies (a) $f_0 = 3, 4, 5$ and 6 GHz for Diode A. (b) $f_0 = 1.5$ and 3 GHz for Diode B.

Frequency mixers

The non-linearity of MIM diodes can also be utilized for RF frequency mixing to generate the frequency difference. Here, we demonstrate the use of these diodes for mixing applications. For the experimental set up, the Local oscillator(LO) signal and RF input signal(RF) were fed using a power splitter (11667B, DC-26.5 GHz) which was used here as a power combiner and also to provide isolation between LO and RF signal. The output of the power combiner was connected to a directional coupler (HP 873008) which in turn was connected to CPW probe through T-Bias. The output of coupler is connected to the spectrum analyzer to measure the output intermediate frequency (IF) signal.

Diode A was measured for (f_{RF}=4 GHz and f_{LO}=3 GHz) and (f_{RF}=16 GHz and f_{LO}=15 GHz). The IF output signal was measured as a function of input RF power. All measurements were carried out at fixed bias of 0.2 V and at room temperature. Figure 10 (a) and (b) shows the output power for IF signal for two different set of frequencies for Diode A. The IF signal is also measured at three different LO powers, and the results show close to a linear behavior as expected. Diode A performs well even at f_{RF}=16 GHz and f_{LO}=15 GHz with a strong down-conversion of ~ -63 dBm at a LO power of -14.67 dBm. Down conversion efficiency can further be improved by increasing the LO power.

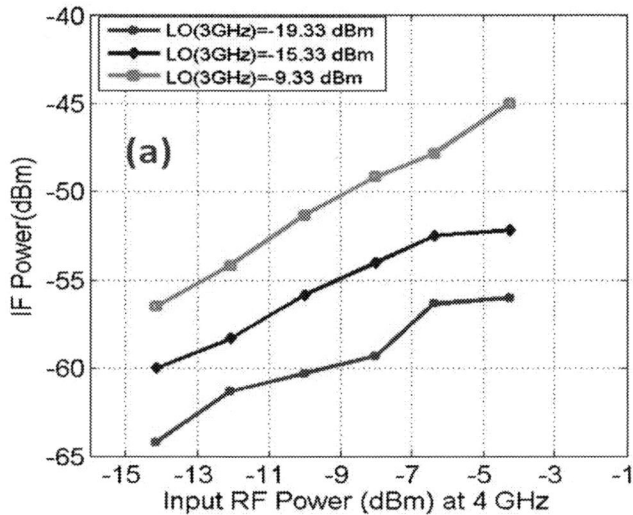

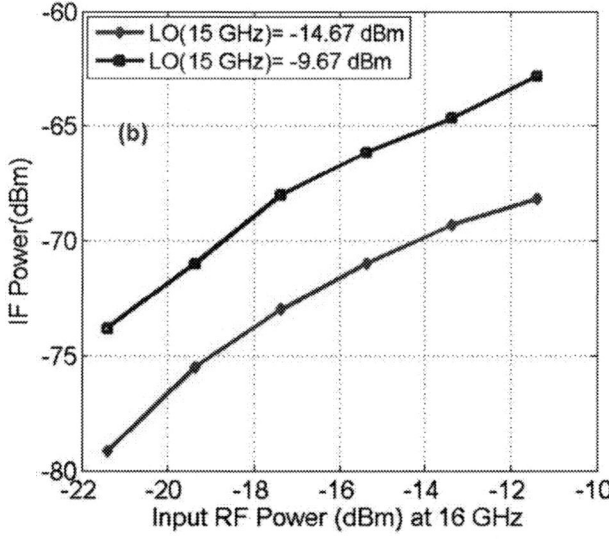

Figure 10. Measured IF signal power versus the input RF power for Diode A (a) f_{RF} = 4 GHz and f_{LO} = 3GHz. (b) f_{RF} = 16 GHz and f_{LO} = 15 GHz.

Diode B was also tested for two different sets of

frequencies, i.e. for (f_{RF}=1.5 GHz and f_{LO}=1 GHz) and (f_{RF}=4 GHz, f_{LO}=3GHz). Figure 11 (a) and (b) shows the output power for IF signals for Diode B at two different set of frequencies. The results show good linear behavior. The performance of Diode A at f_{RF}=4 GHz and f_{LO}=3 GHz is much better than Diode B for approximately same RF and LO input power. The Diode B due to its large contact area and thus smaller cut off frequency doesn't show down conversion at frequencies above 4 GHz.

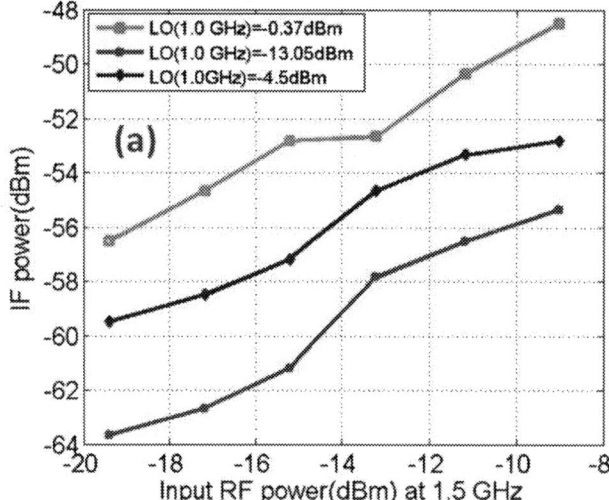

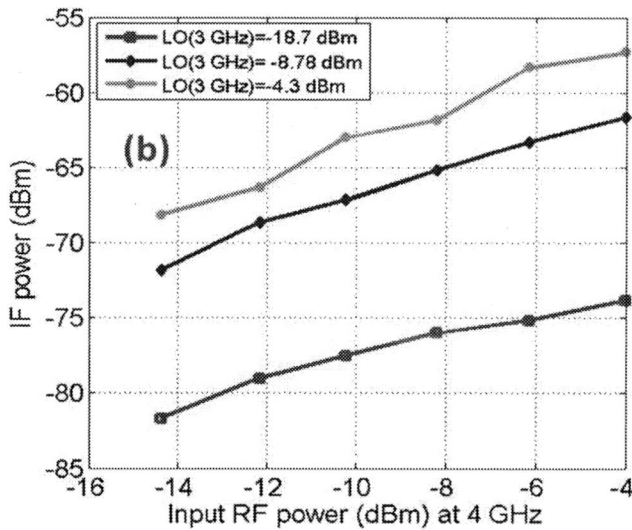

Figure 11. Measured IF signal power versus the input RF power for Diode B (a) f_{RF} = 1.5 GHz and f_{LO} = 1GHz. (b) f_{RF} = 4 GHz and f_{LO} = 3 GHz.

Conclusion

This paper demonstrates the characterization of Ti-TiO$_2$-Pd MIM diodes on flexible substrate for Microwave applications. All of the devices were fabricated using a low temperature, low cost and scalable process. The MIM diodes with contact area of 9 μm^2 and 48 μm^2 were presented and their DC and RF performance was compared. The diodes have strong non-linear characteristics. However, the series resistance of the fabricated devices was large and this can be reduced by increasing the thickness of the first metal layer. A set of RF circuits were studied including: rectifiers, multiplier and mixers. For direct detection (rectifiers), diode with the smaller contact area shows good detection up to 18 GHz. The paper also demonstrated that MIM diodes are able used for frequency multiplication and diodes here were used up to an input frequency of 10 GHz. The fabricated diodes also show strong down conversion for f_{RF} = 16 GHz for smaller diode and f_{RF} =4 GHz for diodes with larger contact area. In the future diodes can be scaled down further to work at much higher frequencies and their effective series resistance will be decreased.

Acknowledgments

The authors would like to thank members of TeSLa research lab for helpful discussions. This work was supported in part by the DARPA YFA program (Grant Number: N66001-12-1-4238).

References

1. H. Kudo, T.Sawada, E. Kazawa, H. Yoshida, Y. Iwasaki and K. Mitsubayashi, "A flexible and wearable glucose sensor based on functional polymers with Soft-MEMS techniques," *Biosensors and Bioelectronics*, Vol. 22, Issue 4, , pp. 558-562, 15 October 2006.

2. Y. Chen, J. Au, P. Kazlas, A. Ritenour, H. Gates and M. McCreary," Electronic paper: Flexible active-matrix electronic ink display", *Nature 423*, pp.136-137, 8 May 2003.

3. H.C. Yuan and Z. Ma, "Microwave thin-film transistors using Si nanomembranes on flexible polymer substrate", *Appl. Phys. Lett.*, Issue 89, No.212105, 2006.

4. L. Sun, G. Qin, H. Huang, H. Zhou, N. Behdad, W. Zhou and Z. Ma, "Flexible high-frequency microwave inductors and capacitors integrated on a polyethylene terephthalate substrate", *Appl. Phys. Lett.* Issue 96, No. 013509, 2010.

5. J.S. Meena, M.C. Chu, S.W. Kuo, F.C. Chang and F.H. Ko, "Improved reliability from a plasma-assisted metal-insulator-metal capacitor comprising a high-k HfO2 film on a flexible polyimide substrate", *Chem. Phys.*, vol. 12, pp. 2582-2589, 2010.

6. M. K. Hota, M. K. Bera and C. K. Maiti, "Flexible metal–insulator–metal capacitors on polyethylene terephthalate plastic substrates", *Semicond. Sci. Technol.*, vol. 27 no.105001, 2012.

7. F. J. González, B. Ilic, J. Alda, and G. D. Boreman, "Antenna-Coupled Infrared Detectors for Imaging Applications", *Quantum Electronics*, Vol. 11, No. 1, , pp.117-120, January/February 2005.

8. A. B. Hoofring, V. J. Kapoor and W. Krawczonek, "Submicron nickel-oxide-gold tunnel diode detectors for rectennas", *J. Appl. Phys.*, Vol. 66, pp. 430, 1989.

9. S. Krishnan, S. Bhansali, E. Stefanakos, Y. Goswami," Thin Film Metal-Insulator-Metal Junction for Millimeter Wave", Volume 1, Issue 1, *Proc Eurosensors XXIII conference*, pp. 409–412, Sep.2009.

10. J. W. Dees, "Detection and harmonic generation in the sub-millimeter wavelength region," *J. Microw.*, vol. 9, pp. 48–55, Sep. 1966.

11. J. G. Simmons, "Electric Tunnel Effect between Dissimilar Electrodes Separated by a Thin Insulating Film", *J. Appl. Phys.*, vol. 34, pp. 2581, 1963.

12. I. Wilke, Y. Oppliger, W. Herrmann, F. K. Kneubühl, "Nanometer thin-film Ni-NiO-Ni diodes for 30 THz radiation," *Applied Physics A*, Vol.58, Issue 4, pp.329-341. April 1994.

13. Y. Rawal, S. Ganguly, and M. S. Baghini, "Fabrication and Characterization of New Ti-TiO$_2$-Al and Ti-TiO$_2$-Pt Tunnel Diodes," *Active and Passive Electronic Components* Vol.2012, Article ID 694105, 2012

14. S. Krishnan, H. La Rosa, E. Stefanakos, S. Bhansali, K. Buckle, "Design and development of batch fabricatable metal–insulator–metal diode and microstrip slot antenna as rectenna elements," *Sensors and Actuators A: Physical* Vol. 142, Issue 1, pp. 40–47, 10 March 2008.

15. J.A. Bean, A. Weeks, G. D. Boreman, , "Performance Optimization of Antenna-Coupled Al-AlO$_x$-Pt Tunnel Diode Infrared Detectors," *Quantum Electronics*, vol.47, no.1, pp.126-135, Jan. 2011.

16. J. A. Hejase, P. R. Paladhi, and P. Chahal, "Terahertz Characterization of Dielectric Substrates for Component Design and Nondestructive Evaluation of Packages," *Components, Packaging and Manufacturing Technology, IEEE Transactions on*, vol. 1, pp. 1685-1694, 2011S. Krishnan, E.

17. Subramanian Krishnan a,b, Elias Stefanakos a, Shekhar Bhansali, "Effects of dielectric thickness and contact area on current–voltage characteristics of thin film metal–insulator–metal diodes," Thin Solid Films 516 (2008) 2244–2250

18. P. Periasamy, H. L. Guthrey, A. I. Abdulagatov, P. F. Ndione, J. J. Berry, D. S. Ginley, S. M. George, P. A. Parilla, and R. P. O'Hayre, "Metal–Insulator–Metal Diodes: Role of the Insulator Layer on the Rectification Performance", *Adv. Mater.*, vol. 25, pp. 1301–1308, 2013.

19. D. M. Pozar, *Microwave engineering*: John Wiley & Sons, 2009

Nanocomposite Pastes for Thermal and Mechanical Bonding

Tingting Zhang, Bahgat Sammakia, Howard Wang*
Institute for Materials Research and Department of Mechanical Engineering
Binghamton University, State University of New York, Binghamton, NY 13902
E-mail: wangh@binghamton.edu, Phone: 607-768-4801

Abstract

Heat dissipation is a major challenge in high performance electronic devices. Current thermal interface materials (TIMs) have either low conductivity, such as conventional thermal greases, or high costs, such as solder materials and indium metals. [1-2] We address TIM challenges by integrating silver nanoparticles (AgNPs) and copper micropowders (CuMPs) in a resin-free TIM paste. The nanocomposite TIMs optimize both the bulk and interfacial thermal performances: CuMPs with a particle size of 1-10 μm offer the promise of high bulk thermal conductivity, while AgNPs with a diameter of 3-8 nm provide the flexibility in interfacial engineering. The assembling temperature can be varied from 125 to 200 °C, due to the low sintering temperature of AgNPs. Fused AgNPs in the TIMs can form strong metallic bonds with CuMPs and the substrates, resulting in low interfacial thermal resistance and high mechanical bond strength. Morphological and compositional analyses have shown percolated network structures in the sintered TIM pastes, which are responsible for high thermomechanical reliability. Hybrid TIMs bonded between two metal substrates, such as a copper foil and an invar foil, with their linear coefficients of thermal expansion (CTEs) difference exceeding 15 ppm/ °C, have gone through more than 1000 cycles of thermal shocks between -50 and 150 °C without failure. The studies have shown promises of the hybrid TIMs for broader applications of thermal, electrical and mechanical performance in electronics packaging.

Keywords: thermal interface materials, silver nanoparticles, low temperature sintering process, themomechanical

Introduction

In the microelectronics field, the power densities in electronic devices have increased from 10 W/cm^2 to more than 100 W/cm^2 in the last decade. As the trend continues, heat dissipation has become a major challenge in high performance electronics applications. To solve this problem, TIM is introduced to fill the air gap and increase the actual contact area between the central processor unit (CPU) and the heat sink, thus minimizing the thermal resistance. [2] Currently, there are mainly four types of TIMs, including thermal greases, filled polymer matrices (elastomers), carbon based materials and phase change materials. [3] The thermal greases usually have high thermal conductivity, but because of their flow property, the thickness is difficult to control and excess greases can contaminate the other components. What's more, the thermal grease will easily dry-out over time if used at high temperature range, which causes low reliability. The filled polymers have a lower thermal conductivity than greases, but

they are easy to use and have good dielectric property and higher reliability. The disadvantages are higher cost and the need for permanent clamping to maintain the joints. The third category is carbon based materials, such as carbon fibers, graphene and carbon nanotubes, which have been the focuses of research community recently. It is reported that the theoretical thermal conductivity of individual multiwalled carbon nanotube (MWNT) in the aixial direction can be as high as 3000 W/mK. [4] However, the discrete distribution nature of MWNTs in TIMs results in much lower overall thermal conductivity to 15 W/mK or lower. [5] It is believed that, the widespread use of nanocarbon-based TIMs still has a long way to go. Phase change materials include thermal pads, low-melting-temperature alloys (LMTAs), shape memory alloys, exfoliated clay and fusible/ non-fusible fillers, among which LMTAs are attractive because of their high thermal conductivity, low operating temperature, ease of flow to fill voids, and ease of disassembly.

In this paper, we report the development of a new resin-free hybrid paste that consists of AgNPs (3-8 nm) and CuMPs (1-10 μm). The assembling temperature could be as low as 125 °C due to the small size of AgNPs. At the sintering temperature, the organics in the TIM paste evaporate, and AgNPs fuse together to form a firm metallic network at the same time, resulting in good thermal and mechanical performance. In our previous studies, TIMs consisting of silver flakes ranging from 1 to 10 μm and AgNPs of 3-8 nm show thermal conductivity of 20-100 W/mK, higher than the best commercial thermal grease in the market. In our current study, we use CuMPs instead of Ag flakes in order to reduce the cost. The resin-free pastes are expected to have good thermal and mechanical performance, because both matrix and fillers are good thermal conductors, and they can sinter together to form a strong metallic network.

Experiment

AgNPs were synthesized by two-phase method, [6] in which silver acetate was used as metal source and $NaNH_4$ as reduction agent. As the synthesized AgNPs are stabilized in oil phase with alkane amine surfactant, they were precipitated and purified using methanol/ acetone mixture. The final products were dried under vacuum and stored in the refrigerator for later use.

Commercial CuMPs were soaked in the HOAc water solution to remove oxide layers on the surface. After rinsing with DI water, CuMPs were dried in the vacuum oven and stored in the refrigerator. In this study, two different average particle sizes of CuMPs were used, 10 μm and 3 μm, respectively.

978-1-4799-2408-0/14 $31.00 © 2014 IEEE

AgNPs and CuMPs with a mass ratio of 3:7 (used through this paper) were mixed in organic solvent with the help of surfactant to form a homogeneous paste as TIMs. TIMs were coated on both surfaces of two joining metal substrates. Upon evaporation of most organic solvent, they were hot pressed to form a sandwich of TIMs between. The hot press temperature ranged from 125 °C to 200 °C, while the press time varied from 5 min to 30 min.

The morphology of the AgNPs and composite TIMs was investigated using scanning electron microscope (SEM), while the macrostructures of TIMs were examined in situ using scanning acoustic microscope (SAM). The possible phase change of the metal alloy was measured using x-ray diffraction (XRD). Mechanical tests of the TIM pastes were carried out on a Dage Plus 4000 setup, and the thermal shock using an Espec TSE-11-A instrument. The purpose of the study is to explore the mechanism of the sintering process of the TIM pastes, and help to control the preparation process and improve their thermal, mechanical and thermomechanical performances.

Results and Discussion

SEM and SEM characterizations of as synthesized AgNPs are shown in figure 1. SEM micrograph shows that AgNPs are fairly uniform in size and can pack to form hexagonal orders locally. Statistics on particle sizes from TEM micrograph show AgNPs have a size range of 3-8 nm, centered at around 6 nm with a narrow distribution of ± 2 nm. AgNPs have reduced melting temperature comparing to the bulk silver, and can be sintered at low temperatures.

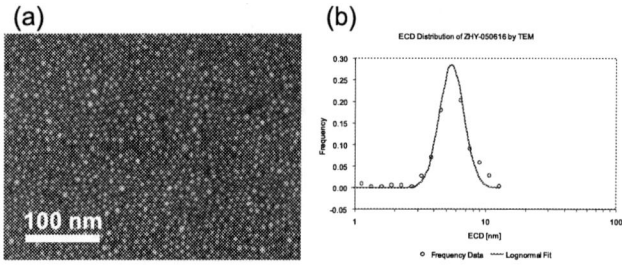

Figure 1. (a) SEM and (b) TEM characterizations of AgNPs.

Common sintering methods include laser sintering [7], microwave [8], and the conventional radiation-conduction-convection heating, or thermal sintering in short. In our studies, thermal sintering method has been used due to its easy operation, low-cost setup and high efficiency. The sintering process of metal nanoparticles protected by surfactant molecules is illustrated in figure 2. [9] Upon heating, solvent and surfactant molecules surrounding nanoparticles evaporate, resulting in surface metal atoms diffusion to join adjacent particles. As particles continue to grow in order to reduce the overall free energy of the system, isolated nanoparticles merge to form a contiguous network of metallic bonding. The connectivity via metallic bonding among AgNPs, CuMPs and the substrates is the basis for high mechanical strength, as well as electrical and thermal conductivity.

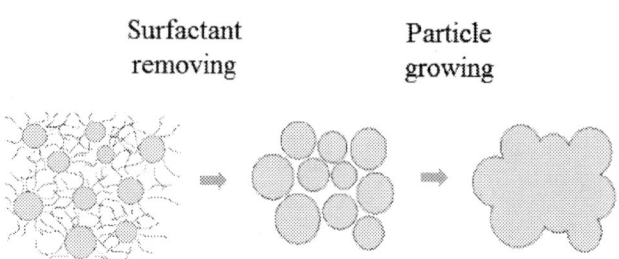

Figure 2. Schematic illustration of metal nanoparticle sintering.

To understand the application conditions of TIMs, the sintering process of AgNPs has been studied first using XRD. Neat AgNP films were cast from paste solution on the kapton substrate, which is stable during the sintering temperature range while considered relatively clean in XRD spectrum. The temperature of the heating stage was preset to 120 °C, and the AgNP films were quickly transferred to the stage and stayed for 10 s, 30 s, 1 min, 2 min, 5 min, 10 min, and 30 min, respectively, before quenching to room temperature. The XRD spectra of as cast and sintered specimen are shown in figure 3.

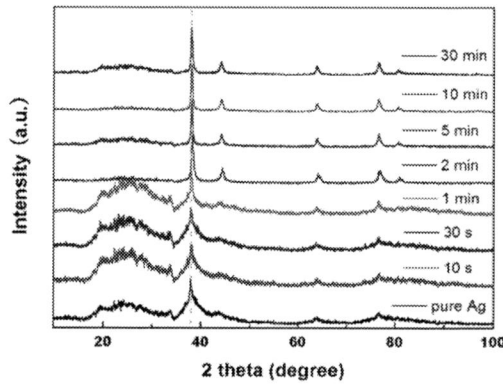

Figure 3. XRD spectra of neat and sintered AgNPs at 120 °C for different times.

The peaks in XRD spectra are from crystalline silver, while the broad signals at angles lower than 35° come from kapton substrate. The first XRD peak for Ag at 38.1° is (111) reflection and analyzed for evaluating the size evolution of AgNPs in films, according to the Scherrer equation, $L = K \lambda / \beta \cos\theta$, [10] where L is the mean size of the crystalline domains, K is the shape factor ($K = 0.89$), λ is the x-ray wavelength (1.54 Å), β is the line width at half of the maximum intensity in radians, and θ is the Bragg angle. For the Ag (111) peak in Figure 3, K, λ, and θ are all constants. As the peaks become sharper with increasing sintering time, the size of Ag particles grow from 10 nm as synthesized to about 22 nm after 1 min sintering and 65 nm after 5 min. It is consist with the fact that AgNPs sinter to form larger particles in order to minimize the overall free energy of the system by reducing surface areas.

For TIMs consisting of AgNPs and CuMPs, fused AgNPs could form strong metallic bonding with CuMPs and the substrates upon sintering, resulting in a strong network. Figure 4 shows the SEM images of TIMs sintered at 125 °C (figure

4a), 150 °C (figure 4b) and 175 °C (figure 4c) for 20 min and at 200 °C for 5 min (figure 4d), 10 min (figure 4e) and 20 min (figure 4f). They all show metallic networks formed through AgNPs sintering. Close examination of the network morphology in TIMs through higher magnification SEM micrographs is shown in figure 5. Sintered AgNPs can be seen at the surface of and among bigger CuMPs, with a size range from tens to hundreds of nanometers, together they form a continuous metallic network responsible for good thermal, mechanical and electrical performances.

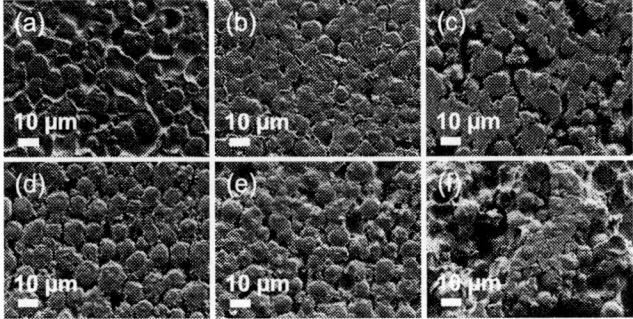

Figure 4. SEM images of hybrid TIM pastes of AgNPs and 10 μm CuMPs sintered at (a) 125 °C, (b) 150 °C, (c) 175 °C for 20 min; and at 200 °C for (d) 5 min, (e) 10 min, and (f) 20 min.

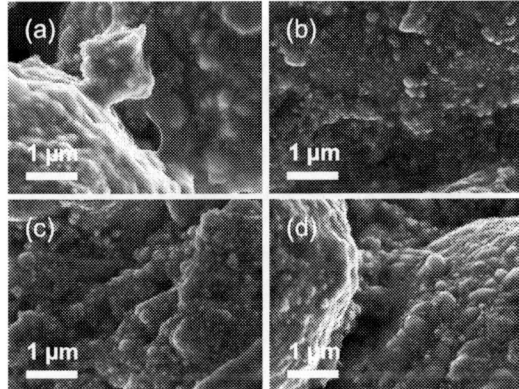

Figure 5. SEM images in higher magnification of the TIM pastes of AgNPs and 10 μm CuMPs sintered at (a) 125 °C, (b) 150 °C, (c) 175 °C, and (d) 200 °C for 20 min.

Figure 6 shows XRD spectra for neat CuMPs and AgNPs, and their composite TIM pastes with the Ag-Cu mass ratio of 3:7 sintered at 200 °C for 20 min. The TIM spectrum composes of both Cu and Ag peaks in their neat form, and there is no appearance of new peaks, implying no formation of other significant phases. It is expected interaction between AgNPs and CuMPs at or near their corresponding particle boundaries. The existence of Ag-Cu interphases, if they do exist upon sintering of AgNPs, is beyond the detection limit in this study.

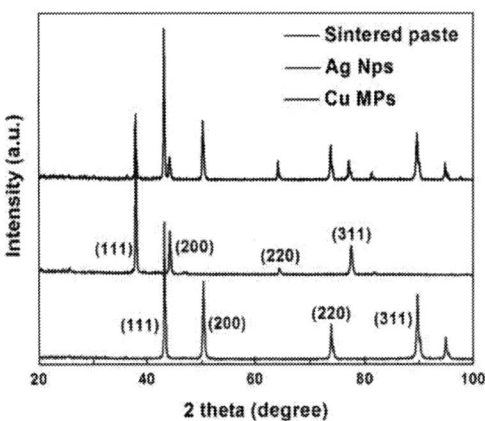

Figure 6. XRD patterns for neat CuMPs and AgNPs, and their TIM paste sintered at 200 °C for 20 min.

We also investigated the thermo and mechanical performance of TIMs, which is critical for reliable TIM applications. The bond strength of TIMs prepared at different conditions, as well as after thermal shocks up to 2000 cycles has been studied.

Figure 7a illustrates a TIM assembly with two identical 100 μm Cu foils as the substrates, and the size of a paste is about 3 mm* 10 mm. A required typical shear stress-strain curve of TIM layers shown in figure 7b was obtained from an elongational stress-strain test on the Dage Plus 4000 setup for wire pull mode. The TIM paste in figure 7b was 3:7 by mass for AgNPs and 10 μm CuMPs, and sintered at 200 °C for 20 min. For assessing the quality of the TIM assemblies and their reliability, a maximum load of 5 kg in pull test was used, and the sample could survive the load of 5 kg. In a semi-quantitative measurement, we use the integral area under the entire stress-strain curve, i.e., the energy density (toughness) of the TIMs, to indicate the sample characteristics and compare the effects of sintering conditions and cyclic thermal shocks.

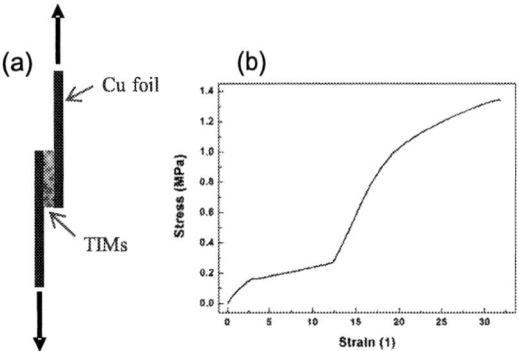

Figure 7. An example for (a) the assembled TIM paste and (b) a shear stress-strain curve obtained from the pull test.

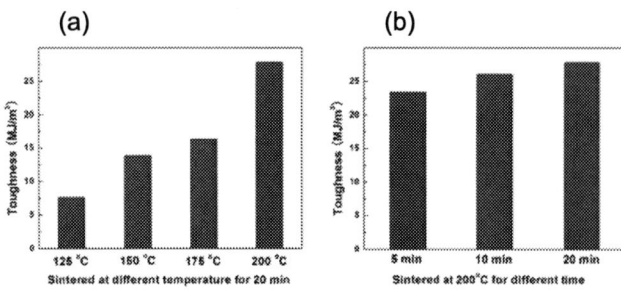

Figure 8. Comparison of the toughness of the TIM pastes of AgNPs and 10 μm CuMPs sintered (a) at different temperatures for 20 min, and (b) at 200 °C for different times.

Figure 8 shows the effects of sintering temperature and time. Data shows that the partial toughness ranges from below 10 MJ/m³ to above 25 MJ/m³. It increases with both sintering temperature and time. The effect of temperature is more prominent in this study, the toughness of TIMs after 20 min annealing increases from 8 MJ/m³ at 125 °C to 28 MJ/m³ at 200 °C, respectively. On the other hand, sintering time from 5 min to 20 min does not alter the toughness result much.

Considering the constraints of process temperature and duration, we choose specimens sintered at 200 °C for 10 min for thermomechanical performance study. Typical profiles of 50 thermal cycles between -50 to 150 °C in Espec TSE-11-A thermal shock chamber are shown in figure 9. The duration of each cycle is 20 min. After a certain number of thermal cycling, TIM assemblies are tested with the same mechanical test carried out on the Dage machine. The change of the mechanical behavior with increasing thermal cycles indicates the thermomechanical reliability of the TIMs. According to the result in figure 10, the toughness value reduces with the ever-increasing number of thermal cycles, demonstrating the TIMs stiffening over the thermal cycling.

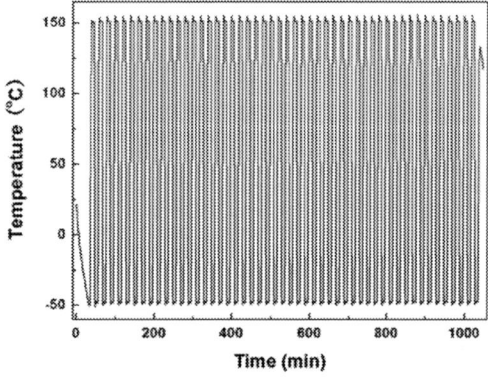

Figure 9. Thermal profiles of 50 thermal cycles in the thermal shock chamber.

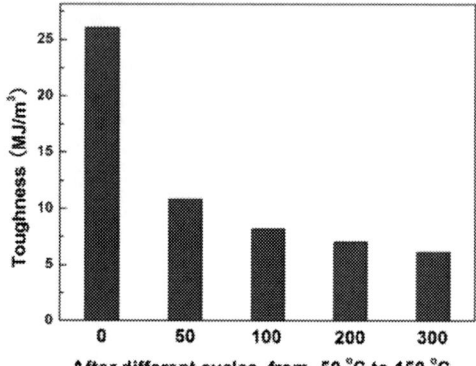

Figure 10. Comparison of the toughness of the TIM pastes containing 10 μm CuMPs (sintered at 200 °C for 10 min) after different thermal shock cycles.

As most previous studies were carried out on TIMs containing 10 μm CuMPs, comparative studies have been carried out on TIMs containing 3 μm CuMPs. Similar mechanical and thermomechanical behaviors were found. Figure 11 shows a series of shear stress-strain curves and toughness comparison of a TIM paste containing 3 μm CuMPs sintered at 200 °C for 20 min. The TIM assembly could survive 5 kg load even after 2000 thermal shock cycles between -50 to 150 °C. The stress-strain curves in figure 11a show clearly the stiffening of TIMs upon cycling, resulting in shortened strain in a stress-limiting test. The toughness value reduces from the initial value, but it appears to be only during the first 500 cycle as shown in Figure 10, after which the partial roughness remains mostly constant. Combing with the result in figure 10, the toughness seems to decreases more quickly at the beginning of the thermal cycling.

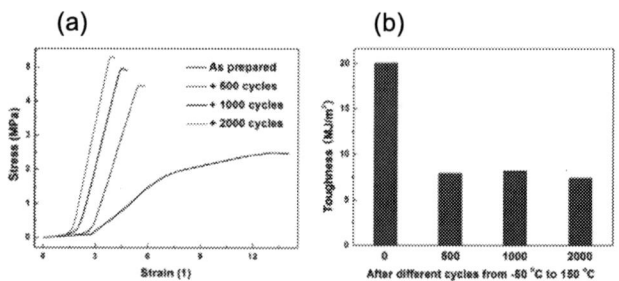

Figure 11. The effects of thermal cycles on the TIMs containing 3 μm CuMPs sintered at 200 °C for 10 min reflected in the change of (a) the shear stress-strain curves, and (b) the partial toughness.

Scanning acoustic microscope (SAM) was used to image the structure of TIMs in situ without breaking the assembly in order to gain better insights in the extraordinary thermomechanical performance of hybrid composite TIMs in this study. SAM is a non-destructive diagnosis technique that is based on differences in the propagation and reflection of ultrasonic sound waves through different medium [10]. In a typical test, most contrast is obtained from the existence of voids in a condensed medium, which makes SAM particularly suitable for probing the uniformity and defects in TIM assemblies. Different imaging modes are possible in SAM

978-1-4799-2408-0/14 $31.00 © 2014 IEEE

investigation. We show the most commonly used amplitude mode (Amp) images to indicate the macrostructure changes in TIM assembly upon thermal cycling.

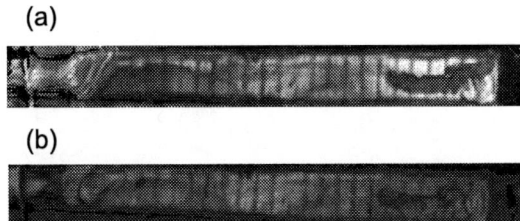

Figure 12. SAM images for the TIM paste containing 10 μm CuMPs sintered at 200 °C for 20 min (a) as prepared, and (b) after 50 thermal cycles.

Figure 12 shows SAM images of a TIM assembly containing 10 μm CuMPs sintered at 200 °C for 20 min, as prepared (figure 12a) and after 50 thermal shock cycles (figure 12b). The main body of the sample is uniform, while defects and voids appear mostly near two ends and at boundaries, which are due mostly to the manual coating processes. After 50 thermal cycles, some texture features appear to be smoothened, and some voids disappear, which may be related to the increased stiffness upon thermal cycling.

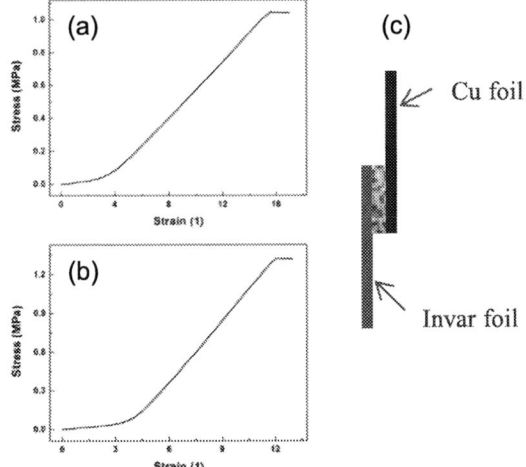

Figure 13. The shear stress-strain curve for TIMs of AgNPs and 10 μm CuMPs sintered at 200 °C for 20 min (a) as prepared, (b) after 500 thermal shock cycles, and (c) schematics of the assembled mismatched TIM paste.

As for the themomechanical reliability of electronic devices, the differences in coefficients of thermal expansion (CTEs) between bonding solids play a critical role. Most of previous studies focus on TIMs bonding two identical Cu foils as substrates, which have no CTE mismatch. To confirm the good thermomechanical performance of the composite TIMs, we used invar foils as the other substrate bonding Cu foils. The linear CTE of invar is 1.2 ppm/°C, which is similar to that of silicon, while the linear CTE of the Cu foil is 17 ppm/°C. The schematics of a TIM assembly with mismatched substrates and its shear stress-strain curves are shown in figure 13. The TIMs consist of 30% AgNPs and 70% 10 μm CuMPs by mass and sintered at 200 °C for 20 min in bonding the invar

and Cu substrate. A comparison of shear stress-strain curves in figure 13a and b indicates the stiffening of TIM assembly upon 500 thermal shock cycles, similar to TIMs of identical substrates. The assembly has gone through more than 1000 thermal cycles without failure, and survived further pulling tests up to 5 kg of load without broken. We have demonstrated high thermomechanical reliability of the resin-free hybrid nanocomposite TIMs.

Summary

We have investigated a new resin-free nanocomposite TIM paste, which contains AgNPs of 3-8 nm and CuMPs with particle sizes of 1-10 μm. We have tested sandwich TIM assemblies with different substrates, and prepared under various conditions (sintering temperature, time, pressure, etc). The assembling temperature could be as low as 125 °C due to the low sintering temperature of AgNPs. During the sintering process, AgNPs fuse together to form strong metallic bonding with CuMPs and the substrates. The assembled TIM pastes remain intact after more than 2000 thermal shocks between -50 °C and 150 °C.

SEM micrographs of TIMs prepared under various conditions confirm percolated network structures that are responsible to extraordinary thermomechanical performance. XRD studies indicate growth of AgNPs upon thermal annealing, which help to bond the surface of CuMPs and substrates without changing phase structures. SAM images show the existence of voids in as prepared TIMs and ones after thermal cycling. The evolution of void structures during the thermal cycling may be partially responsible for stiffening of the TIM assemblies. Stiffening occurs mostly during the first several hundred cycles, after which TIM assemblies appear stabilized.

Further demonstration of the excellent thermomechanical performance of hybrid TIM pastes is achieved in sandwich assemblies using two substrates, invar and copper foils, with a linear CTEs mismatch exceeding 15 ppm/°C, which have gone through more than 1000 cycles of thermal shocks between -50 and 150 °C without failure. This study has shown the promises of the hybrid TIMs for broader applications of thermal, electrical and mechanical performance in electronics packaging.

Acknowledgments

Financial support from the Integrated Electronics Engineering Center (IEEC) at the Binghamton University is acknowledged.

References

1. R. L. Webb, et al., "Low melting point thermal interface material", *Inter Society Conf. on Thermal Phenomena*, Itherm, May 30-June 01, 2002, pp. 671-676.
2. J. P. Gwinn, R. L. Webb, "Performance and testing of thermal interface materials," *Microelectronics Journal*, vol. 34, pp. 215-222, 2003.
3. F. Sarvar, et al., "Thermal interface materials - A review of the state of the art," *Electronics Systemintegration Technol. Conf.*, Dresden, Sep. 1st, 2006, pp. 1292-1302.

4. P. Kim, et al., "Thermal Transport Measurements of Individual Multiwalled Nanotubes", *Phys. Rev. Lett.*, vol. 87, no. 21, pp. 215502-1-215502-4, Oct. 2001.

5. D. J. Yang, et al., "Thermal conductivity of multiwalled carbon nanotubes", *Physical Review B*, vol. 66, no. 16, pp. i.d. 165440, 2002.

6. D. V. Goia, et al., "Preparation and formation mechanisms of uniform metallic particles in homogeneous solutions," *J. Matter. Chem.,* vol. 14, pp. 451-458, 2004.

7. S. H. Ko, et al., "Air stable high resolution organic transistors by selective laser sintering of ink-jet printed metal nanoparticles," *Appl. Phys. Lett.,* vol. 14, no. 90, pp. 141103, Apr. 2007.

8. T. Someya, et al., "Direct inkjet printing of silver electrodes on organic semiconductors for thin-film transistors with top contact geometry," *Appl. Phys. Lett.,* vol. 93, no. 4, pp. 043303-1-043303-3, July 2008.

9. L. W. Huang, *Sintering Metal Nanoparticle Films,* Doctoral dissertation, Binghamton University, New York, 2012, pp. 2-9.

10. U. Holzwarth, et al., "The Scherrer equation versus the Debye-Scherrer equation," *Nat. Nanotechnol.,* vol. 6, no. 9, pp. 534, Aug. 2011.

11. T. M. Nelson, et al., "Scanning acoustic microscopy," *Adv. Mater. & Processes,* vol. 162, no. 12, pp. 29-32, Dec. 2004.

Assembly and Packaging Technologies for High-Temperature and High-Power GaN HEMTs

A. A. Bajwa[1], Y. Qin[1], J. Wilde[1], R. Reiner[2], P. Waltereit[2], R. Quay[2],

[1] Laboratory for Assembly and Packaging Technology, IMTEK, University of Freiburg, Germany

[2] Fraunhofer Institute for Applied Solid State Physics, Freiburg, Germany
adeel.bajwa@imtek.uni-freiburg.de

Abstract

In this work, assembly and packaging technologies for high-temperature high-power GaN high electron mobility transistors (HEMTs) are presented. GaN HEMTs with epitaxial growth on Silicon substrates were used during these experiments. Both die-attachment and interconnection techniques were investigated and a performance comparison is given before and after the assembly process. State-of-the-art silver sintering and transient liquid phase bonding were used as die-attachment methods [2], [3]. For the die-attach material, various characterizations such as shear strength, Energy Dispersive X-ray (EDX) spectroscopy and Differential Scanning Calorimetery (DSC) were performed to characterize the operation up to 500 °C. An estimation of the thermal behavior of the sintered and TLP-bonded GaN HEMTs is performed. For interconnection, gold- and palladium-based materials were investigated for wire-bonding. The complete bonding process was characterized. Estimations about the current carrying capabilities are made for both materials. Passive temperature cycling from -40 to +150 °C was performed as an indication of initial reliability for both die-attachments and interconnections. A systematic electrical characterization of HEMTs is performed starting from the on-wafer measurements up to the final assembly process. The influence of thermal effects on the electrical properties, such as on-state resistance at higher power levels, i.e., 350 W were studied before and after the assembly process. A combination of sintered device with the gold wire bonds is considered as the optimum packaging of GaN HEMTs.

Motivation

GaN material is well suited for high-temperature and high-power applications due to its wide bandgap, high breakdown field, and high current density. GaN based HEMTs are used for power-switching applications. In these devices high breakdown voltages can be achieved in the off-state while low on-state resistances and high on-state currents are possible due to high sheet carrier density at the AlGaN/GaN hetero-interface and high electron mobility in transistor channels [6]. In order to take full advantage of these characteristics, an assembly process is needed which minimizes the thermal and thermo-mechanical degradation of the device performance. For instance, the thermal resistance of the die-attachment material should be minimized in order to dissipate the heat produced in the active area of the device. Furthermore thermal heating causes a reduction of the electron mobility in the channel which consequently leads to an undesired increase of the on-state resistance. Moreover, the interconnection material must withstand high temperatures while offering the minimum contribution to the on-state resistance. In order to reduce influences originating from the assembly process, an efficient design is required that helps to minimize the parasitic influences.

Die-attachment technologies

Die-attachment as the first level packaging provides a thermal dissipation path, an electrical conduction path and mechanical support between the semiconductor and the substrate. The thermal resistance of the die-attachment material must be low to dissipate the heat produced in the active area of the chip. During the die-attachment process, thermo-mechanical stresses are induced. This is one of the main causes leading to the failure of bonded dies.

The optional set of the materials and processes is reduced significantly for high temperature applications. Several key factors have to be taken into account while selecting the die-attachment technology. Various process parameters, such as temperature, heating rate, pressure, time and environment have to be carefully selected in a way that they must not deteriorate the device characteristics during the die-attachment process. In addition the materials aspects such as thermal conductivity and coefficient of thermal expansion (CTE) matching of attachment, die and substrate have to be taken into account. An optimum combination of die-bonding process and associated material properties is required to ensure a stable contact. Two die-attachment methods were used in this work, silver sintering [2] and transient liquid phase bonding [3].

Figure 1: RF transistor (IXYS) attached onto an AlN DCB using transient liquid phase die-attachment [3].

Silver Sintering

Silver sintering was selected as one of the die-attachment techniques for GaN HEMTs. The main advantages associated with this technique are low bonding temperature i.e. 230 °C, good electrical and thermal properties and a melting point near to the bulk silver value, i.e. 961 °C [4] A process optimization and characterization were performed for this method. A novel micro-scaled Ag-paste LTS 275 3P2, obtained from Heraeus GmbH was used as sintering material. The sintering paste was applied by screen printing or needle dispensing.

For better adhesion of the chip on the substrate, a noble metal finish such as Au or Ag was electroplated both on the substrate and also on the chip backside.

Transient Liquid Phase Bonding

A foil-based transient liquid phase (TLP) die-attachment method was tested as an alternative to the silver sintering for performance comparison [3]. In the past, this method has been used for the contacting of solar cells and also for the electronic chips [5]. The main advantages associated with this technique are the formation of high-melting binary alloys, void-free interfaces, and flux-less joining [1, 5, 7].

The silver-tin binary system is investigated as the metallurgical basis for die bonding. Tin has a low melting temperature of 232 °C. With temperature, external pressure and under appropriate environmental conditions, tin first melts and subsequently diffuses into the silver resulting in the formation of a high-temperature stable binary alloy. The transient liquid phase bonding process can be summarized in the following steps [6, 8].

- Preparation of interlayer material in the form of multilayer foils using electroplating.
- Heating of the assembly above the melting point of tin or alternatively indium to produce liquid layer.
- Keeping the assembly at bonding temperature until the isothermal solidification has been accomplished.
- Optional annealing step after bonding.

The scheme of the bonding process is shown in figure 1 in the case of the silver-tin binary system.

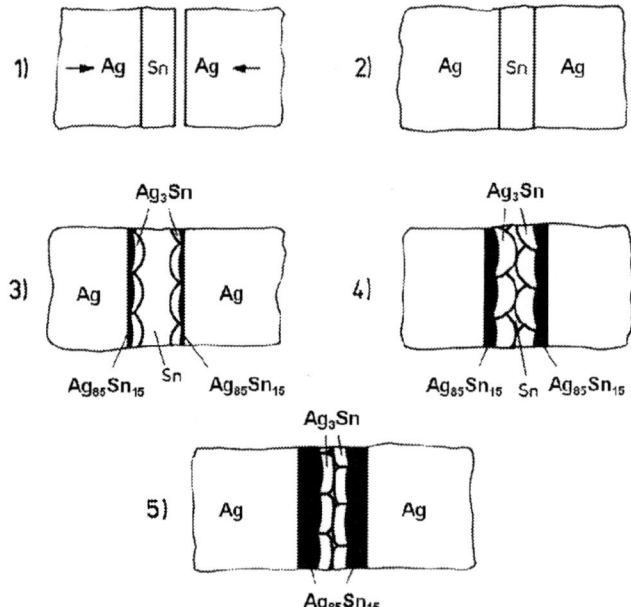

Figure 2: Principle of transient liquid phase bonding

The following reactions can take place during the formation of high-temperature stable binary alloys in case of Sn-Ag system.

$$3Ag + Sn \rightarrow Ag_3Sn$$

Where, Ag_3Sn (ε) is the high temperature stable phase of the Sn-Ag binary system. The table below shows the desired high temperature stable binary phases along with their melting temperatures.

Binary system	Phase	Composition (wt. % of Sn)	Melting temperature [°C]
Ag-Sn	Ag	0 - 12.5	961
	ε (Ag_3Sn)	25	480
	ζ ($Ag_{85}Sn_{15}$)	12.8 - 24.58	> 600

Table 1: Binary systems with high-temperature stable phases

Preform Foils for the TLP Process

The foils were produced in this work as multilayer material for TLP bonding using the electroplating method. A multilayer Sn-Ag-Sn based foil was used during the experiments. The thickness of the Sn, and Ag layers was 4 μm and 20 μm, respectively. The thickness of Sn- and Ag-layers in the foil were designed to maintain an overall composition of 78 wt.% Ag and 22 wt.% Sn. The data were selected based on the phase diagrams in order to keep the melting points of all resulting phases above 500 °C [3].

Process parameters for die-attachment methods

The table below summarizes the key process parameters that were used during the die-attachment process. The bonding process can lead to a generation of packaging stresses due to the mismatch of the coefficient of thermal expansion (CTE). Moreover, electrical properties of the device must not be deteriorated during the assembly process.

Process parameters		
Die-attachment method	Silver sintering	Transient liquid phase bonding
Material	Silver paste, LTS 275 3P2	Sn-Ag-Sn foil (In-house production)
Temperature, T	240 °C	235 °C
Pressure, P	10 MPa	5 MPa
Bonding time, t	5 min	30 min
Heating rate, dT/dt	10 °C/min	34 °C/min
Optional annealing	-	48 h at 200 °C

Table 2: Used process parameters for the sintering and TLP bonding process [2, 3].

Metallurgical Aspects on Test Chips and Substrates

The test chips were made out of silicon and diced to various chip sizes, i.e., 3 mm × 4 mm and 4 mm × 4 mm . Direct bonded copper (DBC) substrates from Curamik GmbH were used as base materials for the mounting. Inclusion of Ni as diffusion barrier layer for the chip backside metallization was necessary for the TLP bonding process.

point of the resultant phases is definitely beyond 480 °C.

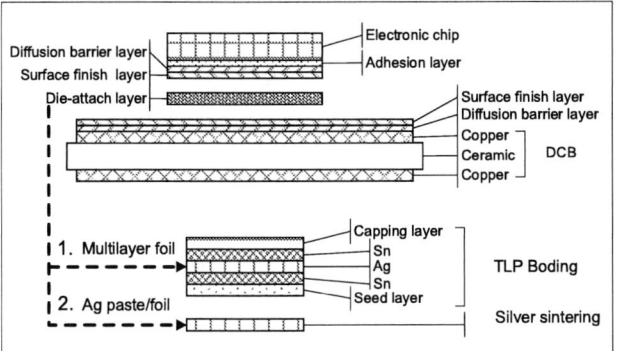

Figure 3: Assembly concept for sintering and transient liquid phase bonding.

	Material	Backside metallization	Thickness
Chip	Si	Ti/Au or Ti/Ag	50/200 nm
		Ti/Ni/Au or Ti/Ni/Ag	50/100/200 nm
Substrate	AlN DCB	Ni/Au	5000/500 nm

Table 3: Metallurgical aspects for chip and substrate [2]

Characterization of Die-attachments

Both silver sintering and TLP bonding were aimed at high-temperature operation up to 500 °C. The following characterizations were made for each die-attachment method.

1. SEM/EDX analysis.
2. Differential Scanning Calorimetery (DSC).
3. Shear strength.

Energy Dispersive X-ray Spectroscopy

Scanning Electron Microcopy (SEM) along with Energy Dispersive X-ray Spectroscopy (EDX) were performed to analyze the local microstructure and composition of the binary alloys formed in the joint region during TLP bonding. Silicon chips with a backside metallization of Ti/Ni/Ag were bonded to silicon substrates using TLP method for the characterization of the composition of the joint. The TLP-bonded joints were optionally annealed at 200 °C for 48 h. The joint compositions for the annealed and unannealed samples are shown in the figures 5 and 6, respectively. A homogeneous composition of Sn and Ag was found across the joint from the top Si/TLP bond interface (0 μm) to the bottom TLP bond/Si interface (30 μm) after the annealing step, as shown in figure 7. The resultant phases in case of Sn-Ag joint were found to be Ag and ζ, as indicated in the phase diagram in figure 4.

Differential Scanning Calorimetery

Differential Scanning Calorimetery (DSC) measurements were performed to measure the melting points of the resultant intermetallic phases. The measurements confirm the presence of identified phases in Sn-Ag binary system. The melting peak in the unannealed TLP bonded sample corresponds to the melting of the Ag_3Sn phase. A small portion of the same phase was also found in the annealed samples as well. It can be stated that within the accuracy limits of the equipment, the melting

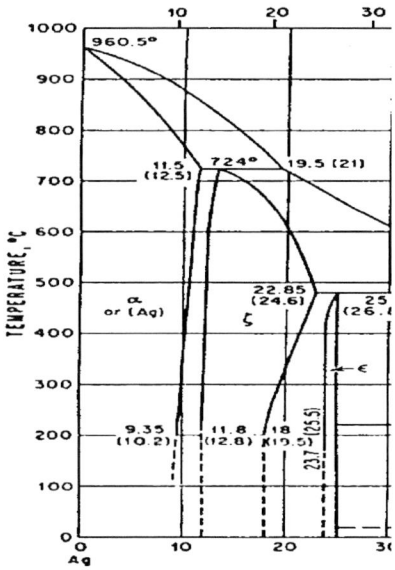

Figure 4: Phase diagram of Ag-Sn [16]

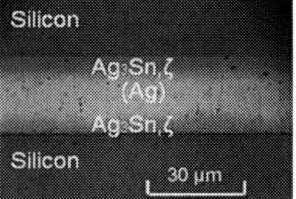

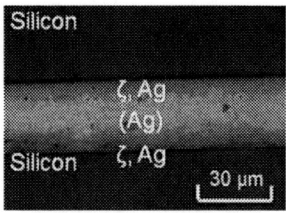

Figure 5: Unannealed Sn-Ag-Sn joint bonded at 235 °C 30min 5 MPa

Figure 6: Annealed Sn-Ag-Sn joint at 200 °C for 48 h

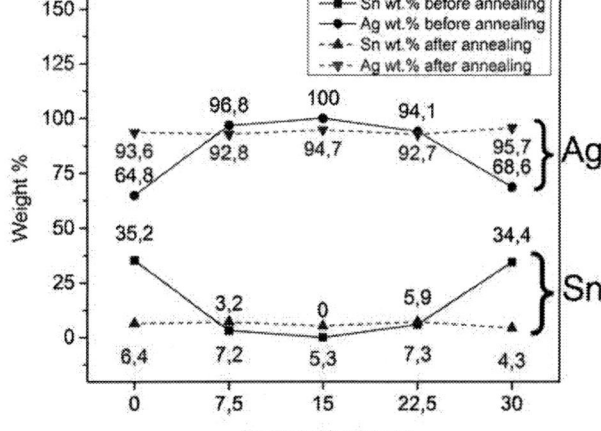

Figure 7: Joint composition measured by Energy Dispersive X-ray Spectroscopy

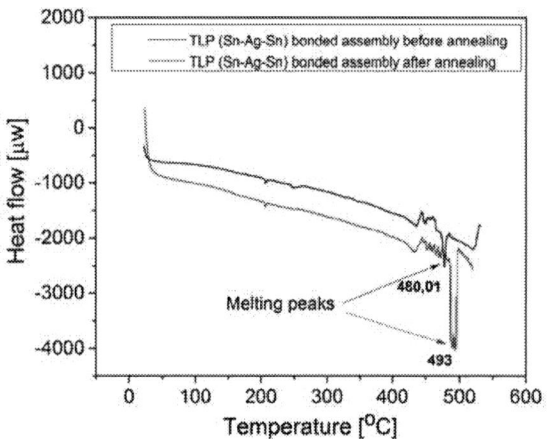

Figure 8: Measurement of melting point using DSC measurement for annealed and annealed bonded Sn-Ag-Sn system [3].

Bonding Stability

Shear tests were performed to check the stability of the die-attachment for both TLP bonded and silver sintered dies. The samples were also subjected to passive temperature cycling from -40 to 150 °C for 300 cycles. The shock cycles were performed as initial reliability test. Both die bonding methods led to shear strength values above 30 MPa.

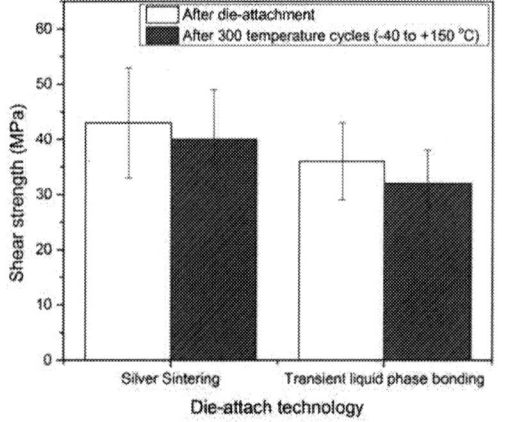

Figure 9 Shear strength before and after passive temperature cycling

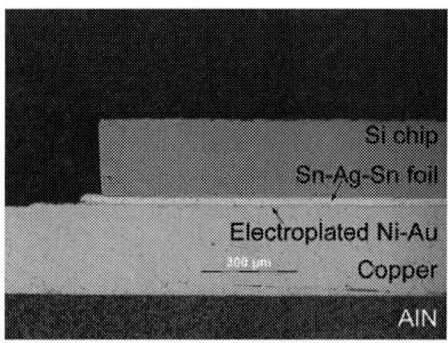

Figure 10: Cross section of a Si chip TLP bonded on to an AlN DCB

Interconnection Technologies

Wire bonding has been reported as the interconnection method for devices such as GaN HEMTs. There are still challenges associated with the materials used for wire bonding. These must withstand high temperatures of great than 250 °C. They must be mechanically stable. Metallurgical aspects are to be explored for the bond pads on chip and substrate, in order to avoid the formation of intermetallic phases and to reduce interface diffusion. Good electrical properties of the bonding materials are required, which means they must be capable of high current densities even at high temperatures. Some materials oxidize in the heat, which can reduce the effective cross sectional area of the conductor. Thermo-mechanical stresses at the bonding sites need to be minimized due to local and global CTE mismatch in the assembly.

Gold and Palladium Wire Bonding

Gold and palladium were investigated as materials for high-power high-temperature devices. Thick gold wires were already demonstrated for high-temperature high-power interconnections [9]. However gold wires aging at high temperatures $\geq$ 300 °C show a decline of mechanical properties like the tensile strength [10]. Palladium has been chosen due to its superior melting point and mechanical stability. It is also very resistant to corrosion because of its chemical inertness. This material has already been demonstrated as the interconnection material for SiC sensors, which are deployed in high-temperature environments [11]. Another important aspect is the matching of the coefficient of thermal expansion (CTE) of the wire material to the chip and substrate metallization. This is necessary to achieve a reliable electrical interconnection especially during the active power cycling of the devices when the interconnections are subjected to high thermal loads as well. Hence, the bond pad metallization of the chip and substrate side must be of the same material to reduce the local interfacial stresses. Another aspect is the electrical resistivity of the material. At high power levels the resistance of the wires may become similar to the on-resistance of the devices as in case of GaN HEMTs. This reduces the on-state current during the device operation. The table below gives a comparison of the material properties [12].

Property	Au	Pd
α in ppm/K [0 to 300 °C]	13.4	11.5
Melting point [°C]	1063	1552
ρ at 20 °C [μΩ cm]	2.2	10.8

Table 4: Material properties of gold and palladium

Thermosonic Wire Bonding Process

Gold and palladium wires with a diameter of 50 μm were used for the bonding process. The bonding wires were obtained from Heraeus GmbH. A Delvotec 5430 ultrasonic wire bonder was used. Palladium has a significantly higher hardness than gold at room temperature, table 5 [12]. Therefore it was necessary to heat the sample during the ultrasonic wire bonding process in the range from 220 to 230 °C. This range did not exceed that of the die-bonding process in order to avoid additional thermo-mechanical stresses. A further optimization of the bonding process was performed. The parameters that influenced the bond quality for the interconnections were bonding force, ultrasonic power, ultrasonic-time and -temperature. The optimization was performed by varying one parameter while keeping the others constant. Subsequently the bonding strengths were measured by the pull strength.

Material	Gold		Palladium	
Hardness in HV at 25°C	25		55	
Bond	1st*	2nd*	1st*	2nd*
Bondpad metallization	Au	Au	Au	Au
Bonding force [g]	30	25	35	30
Ultrasonic power [W]	0.78	0.63	0.79	0.64
Ultrasonic time [ms]	100	85	100	95
Temperature [°C]	220	220	230	230

Table 5: Used bonding parameters after optimization, 1st *: substrate side bond, 2nd *: chip side bond

Characterization of Interconnects

Wire-pull Strength

The stability of the wire bonds was carried out using the wire-bond pull test. The tests were performed using a Dage 4000 tester. A load cartridge DS 100 was used during the pull test, which can apply forces up to 1 N. The wire bonds were subjected to temperature cycling from -40 °C to 150 °C as initial reliability tests. Both Au and Pd yielded pull strength of over 300 mN and would have passed the standard MIL-STD-883G 2011.7.

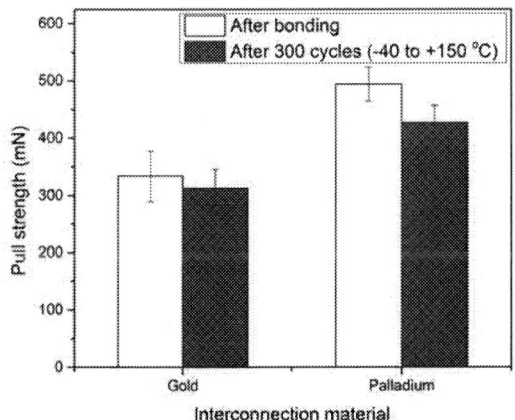

Figure 11: Pull strength of 50 μm gold and palladium wires bonds

GaN HEMTs

GaN-based transistors were used for power switching applications. These are lateral devices which are capable of delivering high output currents of over 70 A at high breakdown voltages of 900 V. GaN structures are grown by epitactic metal-organic chemical vapor deposition (MOCVD) on high resistivity Si(111) substrates, which are 675 μm thick. The grown structures consist of an AlN-nucleation layer, a thick GaN buffer (3.2 μm), an AlGaN barrier layer (25 nm) and a thin GaN cap (3 nm). The transistor yields a Schottky gate and thus the device has a negative threshold voltage of $V_{TH} = -3V$. Due to a lateral interdigital finger layout structure it has a total gate width of $W_g = 219$ mm. The drain to gate

distance is $L_{GD} = 20$ μm, while the gate length is $L_G = 2$ μm, and the gate to source distance amounts to $L_{GS} = 2$ μm. The device has a gate and a source field plate to reduce electric field peaks and dynamic dispersion. The table below shows the metallurgical aspects of the chip for the assembly process.

GaN HEMT	Metallization	Thickness [nm]	Comments
Backside	Ti/Au	10-400	Used for silver sintering
	Ti/Ni/Ag	20/100/200	Used for transient liquid phase bonding
Frontside	Au	7000	Used for gold and palladium wire bonding

Table 6: Metallurgical aspects of GaN HEMTs

Figure 12: GaN transistor assembled on AlN DCB using TLP bonding and 50 μm gold wire bonding

Assembly process for GaN HEMTs

The lateral GaN devices were mounted on AlN DCBs. Both silver sintering and transient liquid phase bonding were used for the die-attachment. The die-bond layer thickness in case of sintered transistors was on average 20 μm, whereas, for TLP bonded samples, it was 28 μm. For electrical interconnections, gold and palladium thin wires with diameter of 50 μm were used. Drain and source were connected with 16 wire bonds each and the gate was connected with a single wire bond. The size of the used GaN devices was 4 mm × 3 mm.

Electrical characterizations

A systematic analysis of the devices' electrical characteristics was performed before and after the assembly process. First on-wafer measurements for thick and thin device were performed. Later on the wafer was thinned and measurements were performed again. The devices were then diced, assembled and again electrically characterized. The influence of the thermal effects on the electrical properties after the assembly process was studied. The self heating effects in the transistor cause a reduction of the electron mobility in the channel, which can lead to an undesired increase of the on-state resistance. The forward characteristics show that the self heating of the devices can be avoided by thinning the Si substrate from 675 µm to 150 µm. A further increase of forward currents for the assembled device shows an improved thermal dissipation behavior. For the sake of comparison, the on-state resistance was measured for the sintered and TLP bonded transistor with gold wire bonds at $V_{gs} = 1V$ and $V_{ds} = 6V$. The measurements were performed in pulse mode with a drain current pulse width of $t_{PLS} = 100$ µs, and a continuous voltage signal at the gate. A 4-point measurement setup was used to measure the $R_{ds,on}$. The on-state resistance for both sintered and TLP bonded joints decreased as compared to on-wafer measurement. The assembly effects are explained in the table 8.

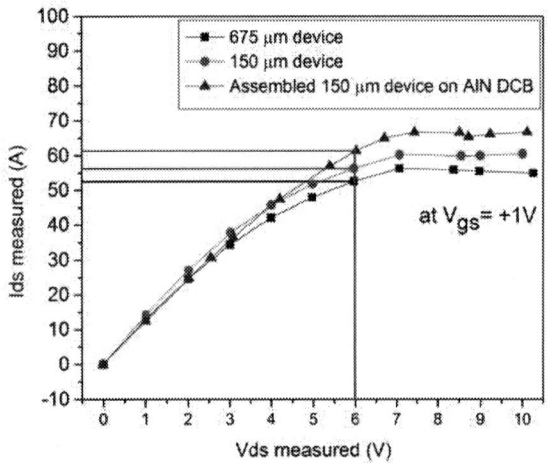

Figure 13: Comparison of forward characteristics for on-wafer measurement of thick (675 µm), thin (150 µm) GaN devices and assembled on AlN DCB with gold wire bonds and TLP die-attachment.

Effect of Interconnect Resistivity on $R_{ds,on}$

In addition to the thermal effects, the electrical resistivity of the interconnection material also effect strongly the $R_{ds,on}$. The resistance offered by the wire bonds, is added in series to the $R_{ds,on}$ of the device. The primary reason for selecting Pd was that it can endure high temperatures, i.e., 500 °C. However, the electrical resistivity of palladium is approximately five times higher than that of gold. In order to see the effect of electrical resistivity of the interconnect materials; two 675 µm GaN devices were mounted on AlN DCB with TLP bonding. One device was bonded with 50 µm gold wires and other was bonded with 50 µm palladium wires. The table below summarizes the analytical calculation of the additional resistance by the interconnection material.

Material	Resistivity ρ at 20 °C [µΩ cm]	Chip side	No of wire bonds	Bond length [mm]	Resistance [mΩ]
Au	2.2	Drain	16	4	2.8
		Source	16	6	4.2
Pd	10.8	Drain	16	4	13.7
		Source	16	6	20.6

Table 7: Calculation of resistances for Au and Pd bonding wires (diameter 50 µm)

The forward characteristics were measured for the both assembled devices attached with gold and palladium wires. A

decrease of maximum drain-source current was observed in case of Pd wire bonding as shown in table 8.

Device	$R_{ds,on}$ at $V_{gs}=1V$ and $V_{ds}=6V$ On-state resistance (mΩ)	$I_{dsat,max}$ at $V_{gs}=1V$ and $V_{ds}=10V$ Max. Saturation Current (A)
HEMT 675 µm	114	55
HEMT 150 µm	106	61
HEMT, 150 µm, TLP, Au wire	98	67
HEMT, 675 µm, TLP, Au wire	102	59
HEMT, 675 µm, TLP, Pd wire	112	54

Table 8: Effects of assembly on electrical characteristics

Thermal Resistance of assembled GaN devices

The investigations were carried out to estimate the thermal resistance of the bonded GaN HEMTs on AlN substrates. The temperature on the surface of the chip is measured using an infrared camera, QFI MWIR-512 from Infrascope™. The AlN DCB, with the assembled GaN device, was mounted onto a chuck with constant temperature of 40 °C. The DCB was held onto the chuck with a pressure contact. The thermal resistance was measured using the following formula [13].

$$R_{th} = \frac{\Delta T}{P} = \frac{T_{top(chip\ surface)} - T_{bottom(DCB\ bottom)}}{V.I}$$

where, T_{top} is the temperature measured on the chip surface, T_{bottom} is the fixed temperature at the bottom of the DCB, V is the applied voltage and I is the applied current.

Using above equation the thermal resistance of the assembly was calculated at 100 Watts for both TLP and sintered dies. The layers thicknesses for the chip, die-attachment and AlN DCB substrate were 150 µm, 30 µm, and 1.27 mm respectively

The thermal resistances in both cases are in the same range. The thermal conductivity of the sintered silver layers is described in the literature to range from 100 to 200 W/mK [15]. The thermal conductivity for TLP bonded layer can be estimated from the following formula.

$$\frac{R_{th,sintered}^{**}}{R_{th,TLP}} = \frac{\lambda_{TLP}}{\lambda_{sintered}^{**}}$$

$$\lambda_{sintered}^{**} = 200\ \text{W/mK}$$

$$R_{th,sintered}^{**} = \frac{d}{\lambda_{sintered}^{**} \times A} = 0.0125\ \text{K/W}$$

$$\Delta R_{th} = R_{th,TLP} - R_{th,sintered}^{**}$$
$$= 0.03\ \text{K/W}$$

$$\lambda_{TLP} = 60\ \text{W/mK}$$

978-1-4799-2408-0/14 $31.00 © 2014 IEEE   2186

	Sintered	TLP bonded
Thermal resistance, R_{th}	1.40	1.43
Thermal conductivity, λ	200*	60**
Thermal conductivity, λ	100*	46**

Table 9: Comparison of measured thermal resistances of assembled GaN HEMTs with different die-attachments, *literature value, **own calculated value

Both die-attachment types seem to be promising. The real power dissipation in the active area of the device occurs between the gate finger and the drain, where drain to gate distance is $L_{GD} = 20$ µm. In addition, the field plates cover the area above. Therefore, the temperature readings include reflections from the metal lines and field plates, which can be near to the junction temperatures. Therefore more accurate measurements of the junction temperature must be performed using micro-Raman spectroscopy for the same devices without any field plates [14].

Conclusions

This works reports on a systematic investigation of die-attachment and interconnection technologies for GaN devices. Both silver sintering and transient liquid phase bonding turned out to be promising die-attachment techniques. GaN devices survived after the assembly and packaging process. With the used parameters, such as temperature, pressure, bonding times etc. both processes did not deteriorate the electrical properties of GaN devices after the die-attachment. The melting point of the TLP bonding was found to be beyond 480°C, whereas for the sintered layer it is 961°C. Shear tests yielded high strength before and after 300 temperature cycles from -40 to 150°C. For the mounted devices on AlN DCBs, the thermal resistance of the sintered devices (1.41 K/W) is slightly less than the TLP bonded devices (1.43 K/W). The systematic investigation of electrical characteristics such as IV (drain current Vs drain voltage) curves for GaN devices from the on-wafer measurements till the final assembled devices was performed. On-state resistance at high power levels (350 Watts) i.e. Vgs=1V, and Vds=6V were the comparison criteria. It was found out that the on-state resistance of the thinned devices (106 mΩ) decreases as compared to the thick devices (114 mΩ). This shows a decrease of self heating effects in the GaN devices.

A good thermal-dissipation behavior was found for the devices after the assembly process. The self-heating effects of the GaN devices were further reduced after the assembly process as the on-state resistance was further reduced to 99 mΩ for sintered device using gold wire bonding. With the good thermal dissipation behavior it is possible to increase the power levels without deteriorating the electrical characteristics of the GaN devices at high temperatures. The table below evaluates the die-attachment and interconnection technologies on the basis of this investigation.

Die-attachment technology	
Silver sintering	++
Transient Liquid Phase bonding	+
Interconnection technology	
Gold wire bonding	++
Palladium wire bonding	O

Table 10: Comparison of the technologies

It was found out that silver sintering and gold wire bonding is the best combination for the die-attachment process. This combination offers better thermal and electrical properties for the GaN devices. Power levels up to 130 Watts were achieved using the suggested assembly, where the average surface temperature of the devices was 260 °C.

Acknowledgments

The authors would like to thank Dr. Herbert Walcher and Heiko Czap for their valuable contributions to the experiments. Moreover, the authors would also acknowledge Fritz Hüttinger Foundation for their support of the research work.

References

1. G. O. Cook III, C. D. Sorensen, "Overview of transient liquid phase and partial transient liquid phase bonding", *J. Mater Sci*, vol. 46, pp. 5305-5323, 2011.

2. A.A. Bajwa, R. Zeiser, J. Wilde, "Process optimization and characterization of a novel micro-scaled silver sintering paste as a die-attach material for high temperature high power semiconductor devices"; *36th International Spring Seminar on Electronics Technology (ISSE)*, Alba Iulia, Romania, May 8-12, 2013, pp. 53-58.

3. A.A. Bajwa, Y.Y. Qin, R. Zeiser, J. Wilde, "Foil based transient liquid phase bonding as a die-attachment method for high temperature devices", *8th International Conference on Integrated Power Electronic Systems (CIPS)*, Feburary 25-27, 2014.

4. C. Buttay, A. Masson, J. Li, M. Johnson, M. Lazar, C. Raynaud, H. Morel, "Die attach of power devices using silver sintering- bonding process optimization and characterization", *Proc High Temperature Electronics Network Conf* (HiTEN 2011), Oxford, United Kingdom, July 18- 20.

5. J. Wilde, N. Pchalek, "Kontaktierung von Solarzellen durch Isotherme Erstarrung", *Verbindungstechnik in der Elektronik* 1993, 5(4), pp 172-179.

6. P. Waltereit, R. Reiner, H. Czap, D. Peschel, S. Müller, R. Quay, M. Mikulla, and O. Ambacher, " GaN-based high voltage transistors for efficient power switching ", *Phys. Status Solidi C 10*, No. 5, 2013, pp. 831– 834.

7. H. A. Mustain, W. D. Brown, Simon S, Ang, " Transient liquid phase die attach for high temperature silicon carbide power devices", *IEEE Trans. Components and Packaging Technologies*, Vol. 33, No. 3, September 2010, pp.563-570.

8. W. D. MacDonald, T. W. Eager, "Transient liquid phase bonding", *Ann. Rev. Mat. Sci.* 1992, vol. 22, pp. 23-46.

9. R. W. Johnson, C. Wang, Y. Liu, J. D, Scofield, "Power device packaging technologies for exterme environments",

IEEE Transactions on Electronics Packaging Manufacturing, Vol. 30, No. 3, (2007), pp. 182-193.

10. R. K. Burla, C. Li, C. A. Zorman, M. Mehregany, "Development of nickel wire bonding for hightemperature packaging of SiC devices," *IEEE Transactions on Advanced Packaging,* Vol. 32, No. 2, (2009), pp. 564-574.

11. R. Zeiser, P. Wagner, J. Wilde, "Investigation of ultrasonic platinum and palladium wire bonding as interconnection technology for high temperature SiC MEMs", *Porc 4th Electronic System-Integration Technology Conf*, Amsterdam, Netherlands, Sep. 17 – 20, 2012, pp. 1- 6.

12. DEGUSSA, Handbook for noble metals (german book), DEGUSSA press (Frankfurt am Main, 1967), pp. 34-36.

13. E. Rastjagaev, J. Wilde, B. Wielage, T. Grund, S. Kuemmel, "Foil based transient liquid phase bonding as a die-attachment method for high temperature devices", *7th International Confercenc on Integrated Power Electronic Systems (CIPS)*, March 6-8, 2012, pp. 1-6.

14. M. Kuball, S. Rajasingam, A. Sarua, M. J. Uren, T. Martin, B. T. Hughes, K. P. Hilton, R. S. Balmer, "Measurement of temperature distribution in multifinger AlGaN/GaN heterostructure field-effect transistors using micro-Raman spectroscopy", *Applied physics letters*, Vol. 82, No. 1, Jan 2003, pp. 124-126.

15. J. G. Bai, Z. Z. Zhang, J. N. Calata, G. Q. Lu, "Characterization of low temperature sintered nano-sclaed silver paste for attaching semiconductor devices", *Conference on High Density Microsystem Design and Packaging and Component Failure Analysis*, Shanghai, June 27-29, 2005, pp.1-5.

16. ASM Handbook, Alloy phase diagrams, Volume 3. ASM international, 1992.

Flip-Chip on Glass (FCOG) Package for Low Warpage

Scott R. McCann[1,2], Venkatesh Sundaram[1,3], Rao R. Tummala[1,3,4], and Suresh K. Sitaraman[1,2,*]

[1]Packaging Research Center
Georgia Institute of Technology
Atlanta, GA 30332

[2]The George W. Woodruff School of Mechanical Engineering
Georgia Institute of Technology
Atlanta, GA 30332
*contact email: suresh.sitaraman@me.gatech.edu

[3]School of Electrical and Computer Engineering
Georgia Institute of Technology
Atlanta, GA 30332

[4]School of Materials Science and Engineering
Georgia Institute of Technology
Atlanta, GA 30332

Abstract

As microelectronic industry moves toward stacking of dies to achieve greater performance in smaller footprint, there are several reliability concerns when assembling the stacked dies on current organic substrates. These concerns include excessive warpage, interconnect cracking, die cracking, and others. Silicon interposers are being developed to assemble the stacked dies, and then to assemble the silicon interposers onto organic substrates. Although such an approach could address stacked-die to interposer reliability concerns, there are still reliability concerns between the silicon interposer and the organic substrate. The ongoing work at the Packaging Research Center is exploring the use of glass substrates as a superior alternative to organics in I/Os and to silicon in electrical performance. In addition, glass provides intermediate and tunable coefficient of thermal expansion between silicon and organic, good mechanical rigidity, large-area panel processing for low cost, planarity, and better electrical properties. However, glass is brittle and low in thermal conductivity, and there is very little work in existing literature to examine glass as a potential substrate material.

In this paper, we examine large glass panels as substrates for microelectronic packages through experiments and simulation. Starting with a 150 x 150 mm glass panel with a thickness in the range of 100 to 300 um μm, we have built alternating layers of dielectric and copper on both sides of the panel. The panels go through typical cleanroom processes such as lithography, electroplating, etc. Upon fabrication, the panels are diced into individual substrates of 25 x 25 mm, and a 10 mm x 10 mm Si die with a peripheral staggered bump pitch of 80/40 um μm is then assembled on the glass substrate by thermocompression bonding with a pre-applied no-flow underfill. The warpage of the flip-chip assembly is measured. In parallel to the experiments, numerical models have been developed. These models account for temperature-dependent properties of the dielectric as well as viscoplastic behavior of the solder. The models also mimic material addition and etching through element "birth-and-death" approach. The warpage from the models has been compared against experimental measurements for glass substrates with flip- chip assembly. It is seen that the glass substrates provide significantly lower warpage compared to organic substrates, and thus could be a potential candidate for future 3D and 2.5D systems.

1.0 Introduction

As integrated circuits (ICs) have scaled according to Moore's Law [1], microelectronic packages have also continued to scale over the last several decades with higher interconnect density. As the technologies available have ranged from 2D wire-bonded packages through area-array flip-chip and, more recently, 2.5D and 3D, the capabilities of packaging have exponentially increased to provide more I/Os. In doing so, the interconnect pitch has decreased proportionally.

Packaging today is commonly done with an organic substrate such as FR4, which has been the case since transitioning from ceramics in the 1990s. As packaging continues to scale, the limits of organic substrates are being approached. The demand for thinner packages, primarily from a mobile perspective, reduces the mechanical support and rigidity an organic substrate can provide. Organics face larger fine line and space limits than ceramics. As size increases, organic packages have limited dimensional stability. The higher CTE of the organic substrate creates assembly yield issues due to the differential lateral displacement between the substrate and the die. Also, the large CTE mismatch creates die-to-substrate reliability concerns as well. On the other hand, although ceramics can address most of these issues associated with organic substrates, cost is a major impediment.

Glass has the potential to combine some of the benefits of ceramics and organics. For example, glass is rigid. The CTE of the glass can be tailored to meet silicon or organic CTE. In other words, glass can function as an interposer with an intermediate CTE between silicon and organic board. Glass is smooth and planar, and therefore, amenable to fabricating fine lines and spaces. Glass is available in large panels, and therefore, will facilitate large-area processing. Glass is inexpensive compared to ceramic. However, glass has other challenges that need to be addressed and studied. Fabrication and assembly processes have been well established for silicon, ceramic, and organic materials, while process development for glass is still in its infancy. For example, metallization of glass, lamination on glass, fine via drilling and metallization, and other processes need to be extensively studied and characterized. The thermal, mechanical, and electrical properties and performance of glass have not been adequately studied in literature for microelectronic packaging applications. The Packaging Research Center is actively pursuing glass as a substrate over the past few years [2][3] [4].

The objective of this work is to study the processing of glass substrates through numerical models and experiments, and to compare the results against organic and silicon substrates. In particular, this paper focuses on warpage induced in glass substrates during thin-film processing as well as during die assembly. This work employs finite element birth-and-death approach to simulate thin-film processing as well as die assembly process. The simulation includes the thermal history associated with such processes as well as material and geometry parameters at various stages of the processes. This work compares the predicted warpage results against experimental data. Also, this work compares the warpage behavior of glass substrate against silicon and organic substrates, and develops design guidelines for glass substrates to minimize warpage. In this paper, the terms substrate and interposer are used interchangeably. This is because this work has made warpage comparison against flip chip on glass, organic, and silicon with different intended functionalities.

Section 2 presents details on sequential fabrication of glass interposer with appropriate process details. Section 3 presents the finite element modeling and design approach. Section 4 presents and discusses results of this work. Section 5 highlights the conclusions.

2.0 Sequential Fabrication of Glass Interposer

Fabrication begins with a bare thin glass panel, shown at the top. The panel is a low-CTE EN-A1 glass, measuring 150 x 150 mm and will produce a six by six array of 25 x 25 mm interposers. Although the processing is carried out on a panel, for the sake of clarity, the process steps are described using one interposer, as illustrated in Figure 1. Figure 1a depicts the cross section of a 25 x 25 mm interposer. The panel is cleaned and a 17.5 µm polymer layer of ZEONIF™ ZS-100 is vacuum laminated on, then hot pressed and cured, shown in Figure 1b. A copper seed layer is then deposited, a dry photoresist film is laminated, exposed, and developed. Copper is then electroplated and the photoresist is then stripped. The resulting copper is approximately 10 µm thick. An identical pattern is used on both sides of the two metal layer ("2ML") interposers. The result is shown as a cross section in Figure 1c. A palladium finish is applied to prevent oxidation of the copper. Although two metal layers are present on top and bottom of the glass interposer, no vias are fabricated in this work. Via fabrication through glass interposers is being pursued by other researchers within the Packaging Research Center at Georgia Tech, and will be presented and discussed elsewhere.

A 10 x 10mm die is then assembled, as shown in Figure 1d. To assemble the die to the interposer, flux is first applied to the interposer. The die is thermocompression bonded with a peak temperature of 250 °C for five seconds with a pressure of

1.5 MPa. Next, a capillary underfill is dispensed and cured at 165 °C for 90 minutes.

Figure 1: Cross section schematic of fabrication process steps. From the top, (a) bare glass; (b) polymer laminated glass; (c) interposer with trace pattern; (d) after die assembly (with partially hidden underfill).

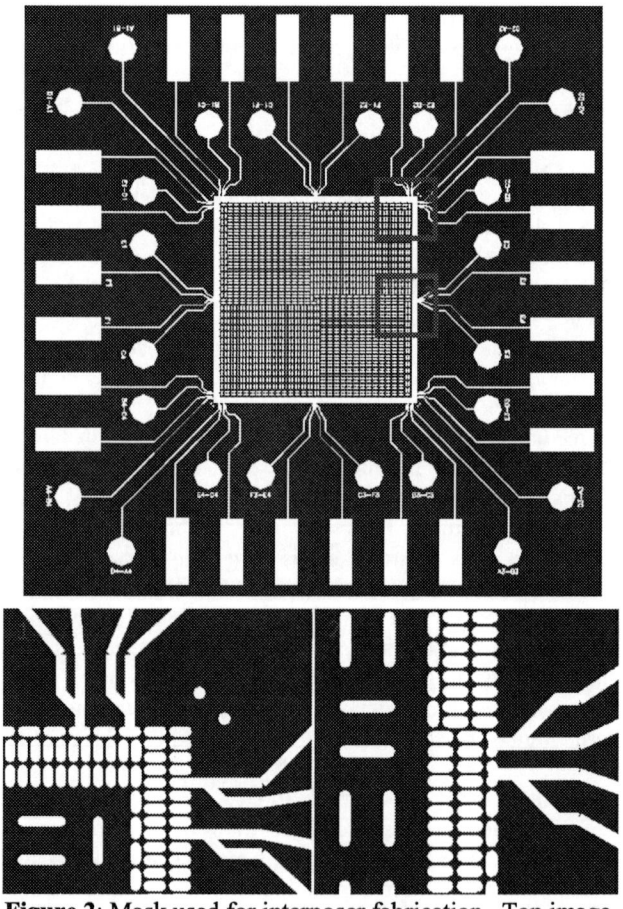

Figure 2: Mask used for interposer fabrication. Top image is full interposer layout with die region in center. Regions 1 and 2 are expanded below. The large pads surrounding the die are for test purposes.

Figure 2 shows the mask on the substrate, which is a daisy chain pattern with test pads around the outside. The two insets, labeled (1) and (2), show detailed regions of the mask. The pitch of the pads is 80 µm at the edge of the die and 150 µm in the central region for a total I/O count of approximately 7700. Each quadrant of the die is connected as an independent daisy-chain loop. At the end of assembly, the daisy chains are probed to ensure assembly yield.

3.0 Modeling and Design

In parallel to experiments, finite-element models are created to understand the role of substrate properties on warpage and to develop design guidelines. The modeling is done parametrically in ANSYS™, using plane-strain approximation. Although 2.5D or 3D models are desirable to account for property and geometry variations in the third dimension, 2D models are appropriate for comparison against different cases. Figure 3 shows a schematic of the plane-strain model for a 2ML substrate. As shown, the model captures the glass substrate, the polymer layers, copper redistribution layers, solder interconnects, underfill, and silicon die. The model is half symmetric, with symmetry boundary conditions at the left side. One node at left bottom is fixed in y direction to prevent rigid body motion. The fixed node is within the glass, as the glass is present from the beginning of fabrication. Figure 4a zoomed in view of the mesh of an interconnect. Further away from critical regions, such as in the silicon and glass, the mesh is less fine.

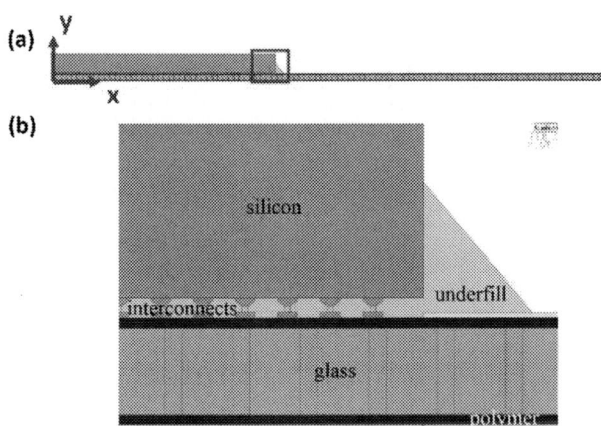

Figure 3: (a) Plane-strain model showing all modeled components (b) Zoomed-in image of the boxed area from (a).

Material properties used in the models are given in Table 1, Table 2, and Table 3. Table 1 shows the thermo-mechanical material properties and includes the stress-free temperature for various materials, based on fabrication process steps. Table 2 shows the temperature-dependent material properties for electroplated copper, which has a lower modulus than bulk copper [5]. Table 3 presents Anand's viscoplastic model parameters for 96.5 Sn-3.5 Ag solder [6].

In addition to the plane-strain models, 2.5D models are created. The mesh of the 2.5D model is shown in Figure 6. Underfill is not shown in the figure so the interconnects can be seen. The 2.5D model is one pitch wide, allowing it to capture properties and geometry variations in the third dimension. Unlike a full 3D model, a 2.5D model takes less computational time. Similar to the plane-strain model, half symmetry is employed, and symmetry boundary conditions are applied on the left. Again, one node is constrained in y direction to prevent rigid body motion. All nodes on the front and the rear face (as

seen in Figure 6) are coupled in z direction so that the z displacements are uniform for all nodes on those faces. The meshes shown in Figure 4 and Figure 5 are example meshes, and multiple mesh densities were used to ensure that the results converged.

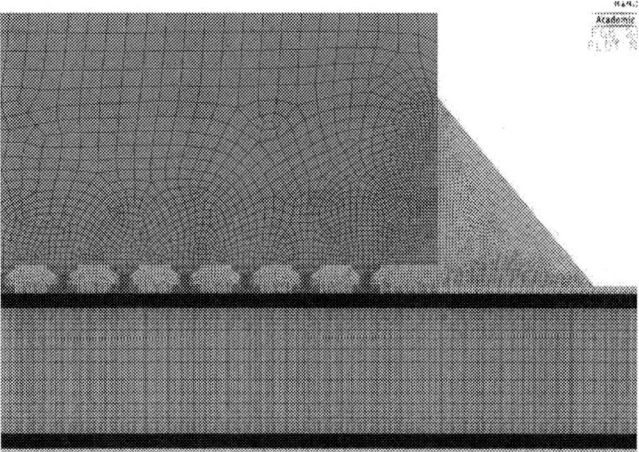

Figure 4: Mesh of plane-strain model from Figure 3b.

Process modeling is done using manual birth and death. This is accomplished by creating additional materials that are "killed" (modulus is reduced by a factor of 10^6). Materials are given a stress-free temperature based on when they are added to the package during fabrication.

Table 1: Isotropic, temperature-independent material properties.

Material	E [GPa]	CTE [ppm/°C]	ν	Stress Free Temp. [C]
Silicon chip	$E_{x,z}$ = 169 E_y = 130	2.6	.28	220
96.5Sn-3.5Ag Solder	58	24	.4	220
Underfill	3	50	.4	140
Low CTE Glass	77	3.8	.22	160
Copper	--	17	.33	40
Polymer	6.9	31	.3	160
Organic substrate	20	20	.14	160
Silicon interposer	$E_{x,z}$ = 169 E_y = 130	2.6	.28	160

Table 2: Temperature dependence of copper modulus [5].

Temp. [C]]	0	100	140	150	160
E [GPa]	80	72	68.8	68	67.2
Temp. [C]	170	180		220	260
E [GPa]	66.4	65.6		62.4	59.2

Table 3: Anand's viscoplastic model for tin silver solder [6].

Coefficient	Value	$\hat{s}$ (MPa)	73.81
A (s⁻¹)	2.23(10⁴)	n	0.018
Q/R (K)	8900	h_0 (MPa)	3321.15
ξ	6	a	1.82
m	0.182	s_0 (MPa)	39.09

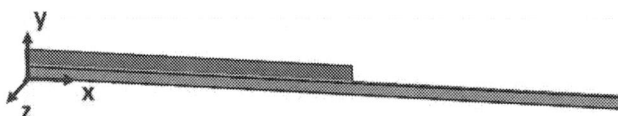

Figure 5: 2.5D model.

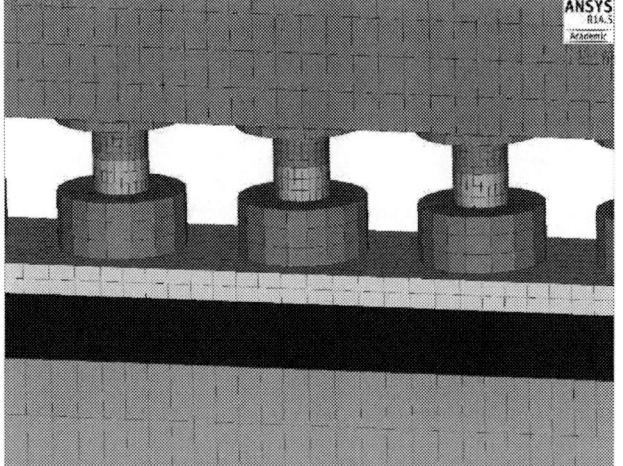

Figure 6: Example mesh from 2.5D model.

4.0 Results and Discussion

The following sections discuss experimental measurement as well as theoretical modeling of glass interposer warpage over a range of processing temperatures.

4.1 Warpage Measurement

Warpage is measured using the shadow moiré technique. Measurement can be done at any point during fabrication or assembly. This paper reports warpage measurements after interposer fabrication and die assembly. Measurements are taken using an akrometrix TherMoiré PS400™, which has an accuracy of ±2.5 μm and a lateral resolution of 640 x 480 pixels. Warpage profiles are taken at different temperatures over a temperature range starting at 20 °C and ramping to 260 °C Measurements are taken from the top of the die.

Shadow moiré results for a 10 mm, 200 μm-thick die on 150 μm thick glass are obtained and shown in Figure 7. Each 3D plot corresponds to the data obtained for a given temperature. Warpage is determined as the vertical deflection of the assembly between the center and the middle of an edge of the die. As the die and the substrate are quarter-symmetric, the plotted warpage values in Figure 7 are the average of the warpage values for the four edges of the sample for a given temperature. At room temperature, the die exhibits positive warpage ("dome" like) of 17.5 μm. This is because when a silicon is die is assembled on a higher-CTE glass interposer and cooled to room temperature, the assembly will have a dome shape. As the temperature in the shadow moiré is increased from room temperature, the warpage decreases and the sample becomes nearly flat. This is because the assembly approaches the stress free temperature of approximately 150 °C, which is the glass transition temperature for the polymer and underfill. When the temperature is further increased, the sample changes the shape to a bowl shape. Above the glass transition temperature for underfill, the package loses rigidity and

becomes less tightly coupled, warping less for a given temperature change.

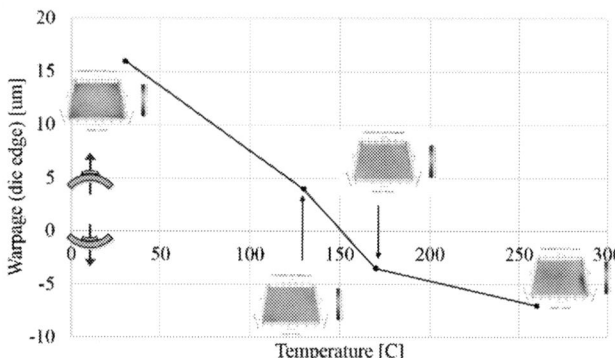

Figure 7: Shadow moiré data of a 2ML, 18.4 mm, 150 μm-thick glass interposer with 10 mm die (courtesy of Namics).

4.2 Modeling Validation

An example of this validation is seen in Figure 8, which shows temperature model comparison with the shadow moiré data presented in Figure 8. In the figure, the shadow moiré data is plotted in black, the plane-strain model is plotted in gray, and the 2.5D model is plotted in orange. As seen, the simulated results track the experimental data. Also, as different materials have different stress-free temperatures, the assembly appears flat near the underfill cure temperature where two of the most dissimilar CTE materials (interposer and die) are fully bonded. Although one experimental validation is presented here, the measured data is used to validate the simulation results for a variety of samples with die thicknesses of 200, 300, and 400 μm, glass thicknesses of 100, 150, and 300 μm, and interposer sizes of 18.4 and 25 mm.

4.3 Role of Substrate Material and Thickness on Package Warpage

The primary objective of this work is to reduce warpage through the use of glass interposers, and therefore, the warpage results from glass interposer are compared against organic and silicon interposers, as shown in Figure 9. The organic substrate has the highest warpage of 97.4 μm compared to the warpage of 17.5 μm for the low CTE glass interposer and 6.12 μm of Si interposer. This is because the organic substrate has the largest CTE mismatch compared to the silicon die, followed by low-CTE glass substrate and silicon substrate, in that order.

Although Si substrate has a matched CTE compared to the die, the assembly still has some warpage, albeit small. This is because of the presence of polymer and copper materials that are present on the silicon substrate and thus, would increase its effective CTE compared to the Si die. The board level assembly will pose a greater challenge for a silicon interposer as the CTE mismatch between a silicon interposer and organic printed circuit board will be high. Glass provides an intermediate CTE between a silicon die and an organic system board.

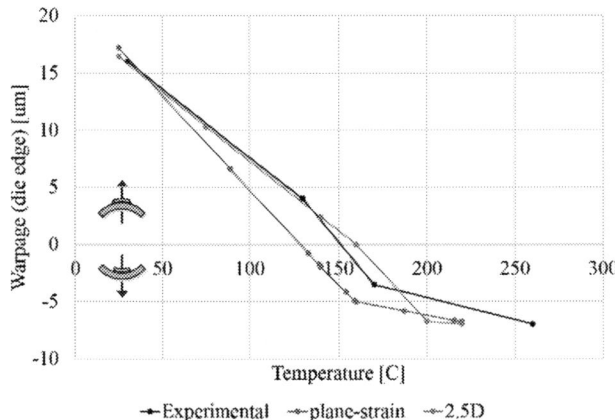

Figure 8: Comparison of plane-strain model (gray), 2.5D model (orange), and shadow moiré (black) from Figure 7.

The warpage depends on both the die and substrate thickness as seen in Figure 10. Die warpage is plotted as a normalized value to highlight the importance of the ratio of die thickness to substrate thickness. For a glass package, the worst ratio is when the die is 1.2 to 1.5 times as thick as the glass substrate. This ratio will vary based on substrate materials and build-up, but the idea will still apply. Moving either the die or substrate to be thicker or thinner will reduce the overall warpage.

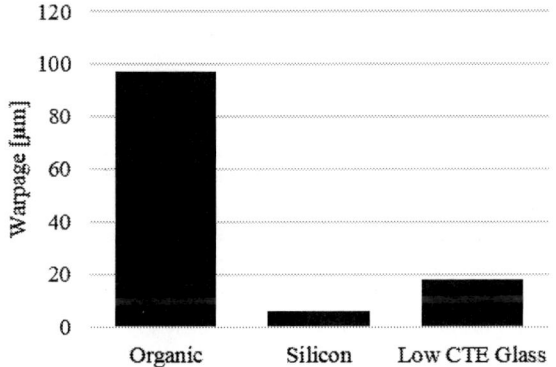

Figure 9: Warpage of 10 mm, 200 μm-thick die on 18.4 mm, 150 μm-thick substrate as a function of substrate core material.

5.0 Conclusions

For comparable glass and organic substrates, glass has significantly less warpage than organics substrate. With glass interposers, the die warpage is worst when the die is 1.2 to 1.5 times as thick as the interposer (including build-up layers), and thus, a suitable die to substrate thickness ratio needs to be selected for reducing the package warpage. Glass substrates, due to their intermediate CTE, can effectively serve as interposers between silicon die and an organic printed circuit board. Thus, glass interposers will be appropriate for enhancing the reliability of first as well as second-level interconnects. In the ongoing efforts at Georgia Tech, such first-level and second-level interconnection reliability are being studied and will be presented in future publications.

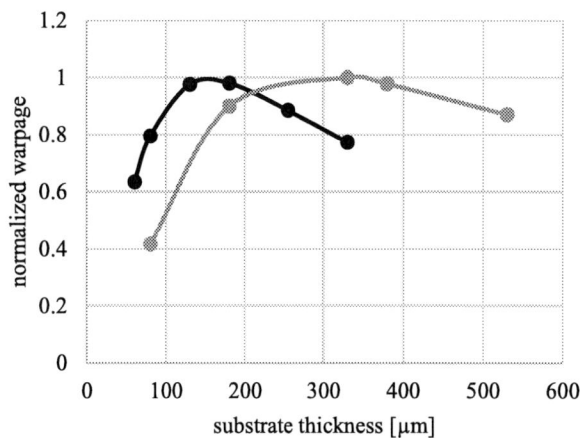

Figure 10: Normalized warpage as a function of substrate thickness for a glass package. The black data shows a 200 μm die and the gray data shows a 400 μm die.

Acknowledgments

The authors would like to thank Sathyanarayanan Raghavan, Dr. Vanessa Smet, Makoto Kobayashi, Tao Wang, and Yoichiro Sato for their valuable help in modeling and experimental studies. The authors would like to thank Namics Corporation for making shadow moiré measurements, Asahi Glass Co., Ltd. for providing glass panels, and Zeon Corporation for providing ZEONIF™ ZS-100.

References

1. G. E. Moore, "Cramming More Components onto Integrated Circuits," *Proc. IEEE,* vol. 86, no. 1, pp. 82-85, 1998.

2. V. Sukumaran, et al. "Low-Cost Thin Glass Interposers as a Superior Alternative to Silicon and Organic Interposers for Packaging of 3-D ICs," *CPMT,* vol. 2, no. 9, pp. 1426-1433, 2012.

3. X. Qin, et al. "Finite Element Analysis and Experimental Validation of Highly Reliable Silicon and Glass Interposers-to-Printed Wiring Board SMT Interconnections," *CPMT,* vol. PP, no. 99, 2014.

4. B. Chou, et al. "Modeling, Design, and Fabrication of Ultra-high Bandwidth 3D Glass Photonics (3DGP) in Glass Interposers," ECTC 2013.

5. R. S. Li and J. Jiao, "The Effects of Temperature and Aging on Young's Moduli of Polymeric Based Flexible Substrates," *Int. J. Microcircuits & Electronic Packaging,* vol. 23, no. 4, pp. 456-461, 2000.

6. N. Bai, X. Chen and H. Gao, "Simulation of Uniaxial Tensile Properties for Lead-free Solders with Modified Anand Model," *Materials and Design,* vol. 30, pp. 122-128, 2009.

Laser-based Conductive Film Forming with Gold Nanoparticles for Electrical Contacts

Mitsugu YAMAGUCHI*, Shinji ARAGA**, Mamoru MITA***, Kazuhiko YAMASAKI*, Katsuhiro MAEKAWA*
* Ibaraki University, Hitachi, Ibaraki 316-8511, Japan, 13nd205h@hcs.ibaraki.ac.jp, +81-80-6605-4780
** Ibaraki Giken Ltd., Kitaibaraki, Ibaraki 319-1541, Japan, araga@iba-giken.co.jp, +81-293-43-0193
*** M&M Research Laboratory, Hitachi, Ibaraki 316-8511, Japan

Abstract

The present study discusses the formation of a conductive film from noble metal nanoparticles as an alternative to conventional electroplating for electrical components, such as connectors, switches, and memory cards. The proposed method consists of inkjet printing with nanoparticle paste followed by laser sintering. The aims are fourfold: to establish sintering technology for gold nanoparticles placed on a nickel-electroplated-phosphor bronze substrate, to characterize the laser-sintered film, to discuss the laser sintering mechanism, and to examine applicability to industry. The major results obtained are as follows: the laser sintering formed a gold film with a diameter of 0.3-0.8 mm and a thickness of 0.3-0.5 μm on the nickel-electroplated phosphor-bronze substrate; a laser with a wavelength of 915 nm enabled instantaneous sintering within one second in an atmosphere; the laser-sintered gold nanoparticle film had such a high adhesion to the substrate that no separation occurred after 90°-0.5R bend-peel tests; the high adhesion was attributed to interdiffusion of gold and nickel in the course of sintering; optical properties of the gold nanoparticle paste depend on preheat conditions. A relatively high-preheat temperature around 523 K produced a paste surface with a suitable absorbance of the infrared laser; and a primary sintering of the preheated gold nanoparticles with a small amount of solvents, followed by an auxiliary sintering from the substrate side made possible an efficient sintering of the nanoparticles as well as high adhesion to the substrate with a high thermal conductivity.

Introduction

Electroplating with noble metals is conventionally used in the electronics industry, including wiring and the formation of electrical contacts because of its high reliability with respect to corrosion resistance and electrical inter-engagement. Recently, the rising prices of noble metals necessitate resource-saving and low-cost processing for electronic applications: for example, the reduction of processing areas by local plating, and of plating thickness. When electroplating is localized in a small region, it requires such additional photochemical processes as making masks for resist coating, UV development, and resist cleaning [1]. In the case of gold plating on a copper substrate, nickel plating is usually added to the substrate as a barrier layer for the purpose of preventing the diffusion of copper to the gold surface. Nickel plating requires such complicated processes as strike plating and acid pickling [2]. These treatments use large amounts of chemicals and water, for which it is necessary to set up plants for purifying waste disposal in order to meet environment-friendliness standards.

Unlike the above-mentioned wet processes, dry processes based on inkjet printing of metal nanoparticle paste followed by sintering have been developed for fine wiring and extremely small functional-film formation [3]. It is well known that the use of metal nanoparticles with a high specific surface area depresses melting point, which enables low-temperature sintering around 500 K [4]. It takes more than 30 min in conventional furnace sintering to obtain a good electrical conductivity similar to that of the bulk material; in this case, moreover, the substrate suffers from thermal damage to some extent. Therefore, much attention has been paid to the use of a laser to sinter metal nanoparticles locally in a short time. Researches on laser sintering with silver nanoparticles have been reported extensively [5, 6], and a few reports have been published on laser sintering of gold nanoparticles coated on semiconductor substrates or metal substrates for the purpose of replacing electroplating [7-9]. In the area of application to electrical-contact materials such as nickel-plated phosphor-bronze, however, there are few researches on the mechanism of film formation.

The present study investigates the potential of laser-sintering technologies with gold nanoparticles to replace conventional electroplating in the formation of electrical contacts to be used for connectors, switches, and memory cards. The dry process has a potential for solving most of the problems related to electroplating, as described above. The paper first describes the experimental procedures of the dry process designed to form a gold film on a spring material made of phosphor-bronze. Second, adhesion between the sintered film and the substrate is tested with a view to its practical applications. Finally, the discussion moves to the mechanism of the laser-based conductive film formation with reference to Auger electron spectroscopy and spectroscopic analysis.

Experimentation

(a) Material and procedures

Figure 1 illustrates the experimental procedures based on the inkjet printing of dispersed metal nanoparticles and laser sintering. First, (a) the dispersed nanoparticles is atmospherically dispensed onto the substrate to make a specimen; (b) the specimen is heated by a hotplate, causing evaporation of organic solvents in the paste, which prevents the laser heating explosively dispersing the metal nanoparticles; (c) the laser is irradiated onto the dispensed area in an atmosphere to form a thin film. The use of these processes obviates the need of additional treatments such as acid cleaning of the specimen and unmasking.

Table 1 lists the materials and laser parameters used. The substrate is a thin sheet of phosphor-bronze (JIS C5210, 25 $\times 20 \times 0.25$ mm^3), which is widely used as a spring material

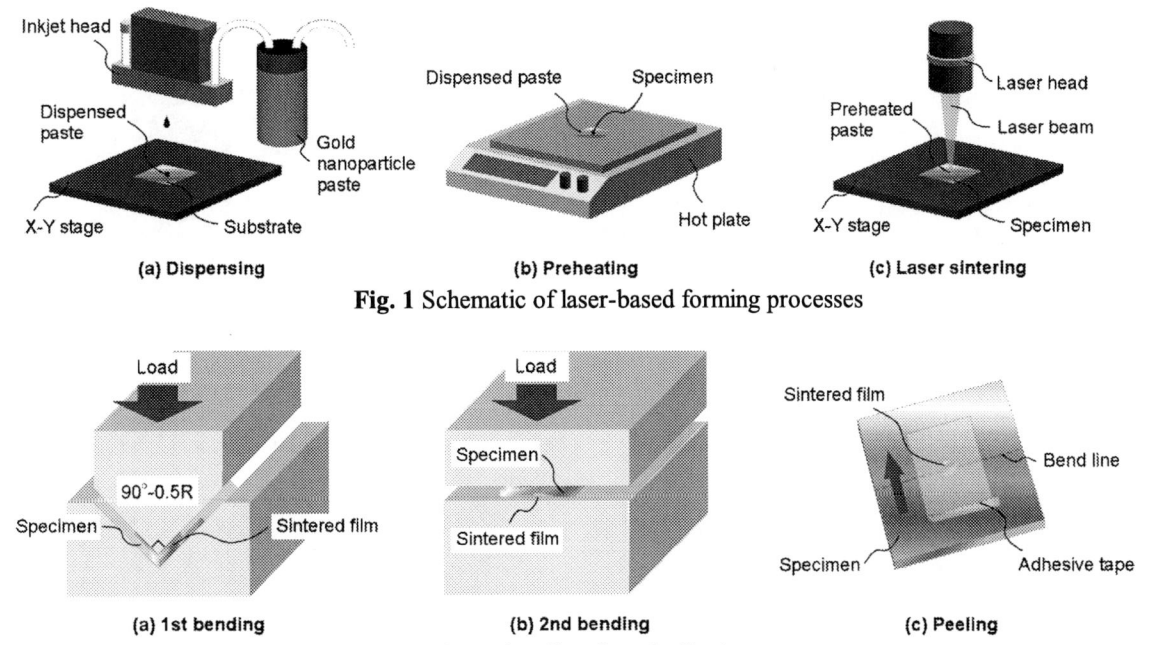

Fig. 1 Schematic of laser-based forming processes

(a) Dispensing **(b) Preheating** **(c) Laser sintering**

(a) 1st bending **(b) 2nd bending** **(c) Peeling**

Fig. 2 Schematic of bend-peel adhesion tests

Table 1 Substrate, paste and laser conditions

Substrate	
Material	Phosphor-bronze (JIS C5210)
Composition	0.01 Zn, 7.71 Sn, 0.12 P,
(mass %)	0.001 Fe, 0.001 Pb,
	99.95 Cu+Sn+Pb
Dimension	$25 \times 20 \times 0.25$ mm^3
Thickness of nickel plating	0.8 - 1.5 μm
Paste	
Nanoparticle	Gold
Mean particle size	7 nm
Concentration	57 mass%
Solvent	Naphthene-based solvent
Dispense amount	2200 pL
Laser	
Type	Laser diode
Wavelength	915 nm
Power	100 W
Irradiation time	0.1 s
Beam diameter	φ1.2 mm

for electrical contacts. Its surface has been electroplated with nickel (0.8-1.5 μm in thickness) as a barrier to diffusion between the foundation material and the contact film. The dispersed gold nanoparticles, provided by Harima Chemical Inc., have a mean diameter of 7 nm, a metal content of 57 mass%, and a viscosity of 7.5 mPa·s. The viscosity has been adjusted with naphthenic-base solvent.

The inkjet printer, assembled by PMT Inc. （IJ-AXIS-MICROBE-C1）, consists of a piezo-type head, an X-Y stage, an ink reservoir, and a system controller. The inkjet head has 128 micro-nozzles, and discharges a minimum volume of 11 pL. The laser used is a fiber coupling laser diode with a wavelength of 915 nm, a beam diameter of 1.2 mm, and a maximum output of 100 W.

The whole process was conducted in an atmosphere. First, the nickel-electroplated phosphor-bronze substrate was ultrasonically cleaned with ethanol, and then partially coated with the gold nanoparticle paste of 2200 pL. Second, the hotplate provisionally heated the specimen at 373-573 K for 60 s. Finally, the specimen was irradiated using the laser diode at 100 W for 0.1 s, yielding a gold film around 0.3-0.8 mm in diameter and about 0.3-0.5 μm in thickness.

(b) Experimental evaluation

The thickness of the sintered film was measured with a stylus-type surface roughness tester (Mitsutoyo Corp., FORMTRACER, SV-C624). Bend-peel tests were carried out to evaluate the adhesive strength of the sintered film. As shown in Fig. 2, the specimen was bent into a right angle by a punch with a sharpness of 0.5R, and then bent back and reflattened; this was followed by peeling by an adhesive tape (adhesive peel strength: 3.93 N/ 10 mm) at a low speed of 10 mm/s. The interface between the sintered film and the substrate was analyzed by Auger electron spectroscopy (ULVAC-PHI Inc., FE-SAM Model 670). The cross-section of the specimen was machined stepwise by a focused beam of argon, and then surface analysis was carried out to provide AES element maps. On the other hand, sputtering and analysis were carried out alternately at increasing depths to provide the AES depth profile. Acceleration voltage of the primary electron gun, sample current, beam diameter, and area of analysis were set at 10 kV, 10 nA, about 30 nm, and □2.5 μm, respectively.

Visible and near-infrared spectroscopic analysis (Hitachi Spectrophotometer, U-4100) of the gold nanoparticle paste was conducted as follows: a nanoparticle paste of 20 μL was dispensed onto a borosilicate glass substrate (□22 mm, 0.12-0.17 mm in thickness), followed by forming a thin uniform film 0.3 μm thick using a spin coater at 2000 rpm × 60 s, and then the specimen was dried by a hotplate at 373-573 K for 60 s before spectroscopic analysis.

978-1-4799-2408-0/14 $31.00 © 2014 IEEE

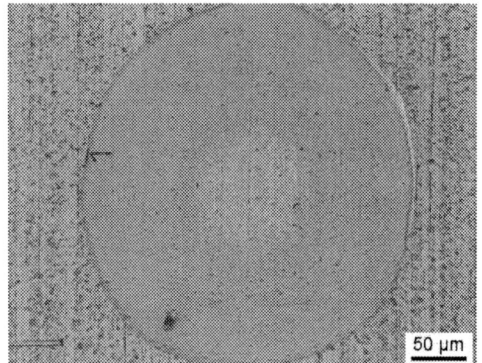

Fig. 3 Optical micrograph of sintered gold nanoparticle film

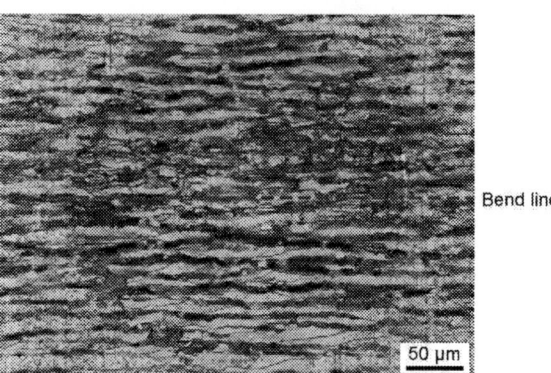

Fig. 6 Optical micrograph of sintered gold nanoparticle film after peel-bend test

Fig. 4 SEM image of sintered gold nanoparticle film after preheat at 523 K for 60 s

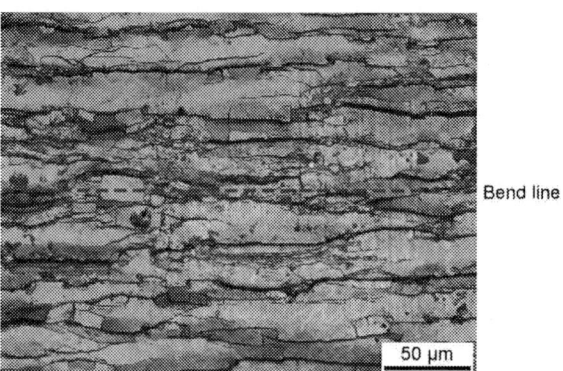

Fig. 7 Enlarged view of sintered gold nanoparticle film after peel-bend test

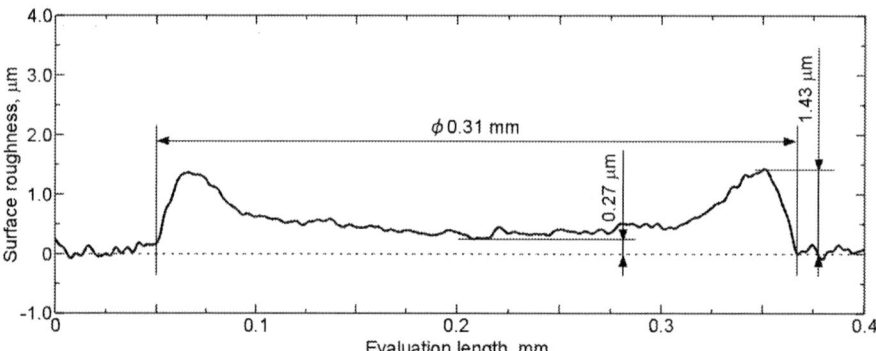

Fig. 5 Cross-sectional profile of sintered gold nanoparticle film

Experimental Results and Discussion

(a) Formation of laser-sintered gold film

Figures 3 and 4 show the optical micrograph and the SEM image of the laser-sintered gold film, respectively, under the following forming conditions: preheat at 523 K for 60 s, and laser irradiation with a laser power of 100 W, an irradiation time of 0.1 s, and a spot diameter of 1.2 mm. It can be observed that a good surface integrity without thermal damage and a dense film without any voids are obtained. It is known that metal nanoparticles start bonding together with the vaporization of solvent and dispersant [3]. This short sintering time is sufficient in the case of the gold nanoparticles placed on the nickel-electroplated phosphor-bronze substrate.

Figure 5 depicts the cross-sectional profile in the middle of the laser-sintered film, showing that a pad with a diameter of 0.3 mm and a thickness of 0.3 μm is formed, except at the circumference. The Marangoni effect takes place on the periphery, where the film thickness is increased [10, 11]. This phenomenon is based on convection caused by non-uniformity of surface tension due to temperature gradients and stemming from the heat of vaporization and concentration gradient when the paste is dried.

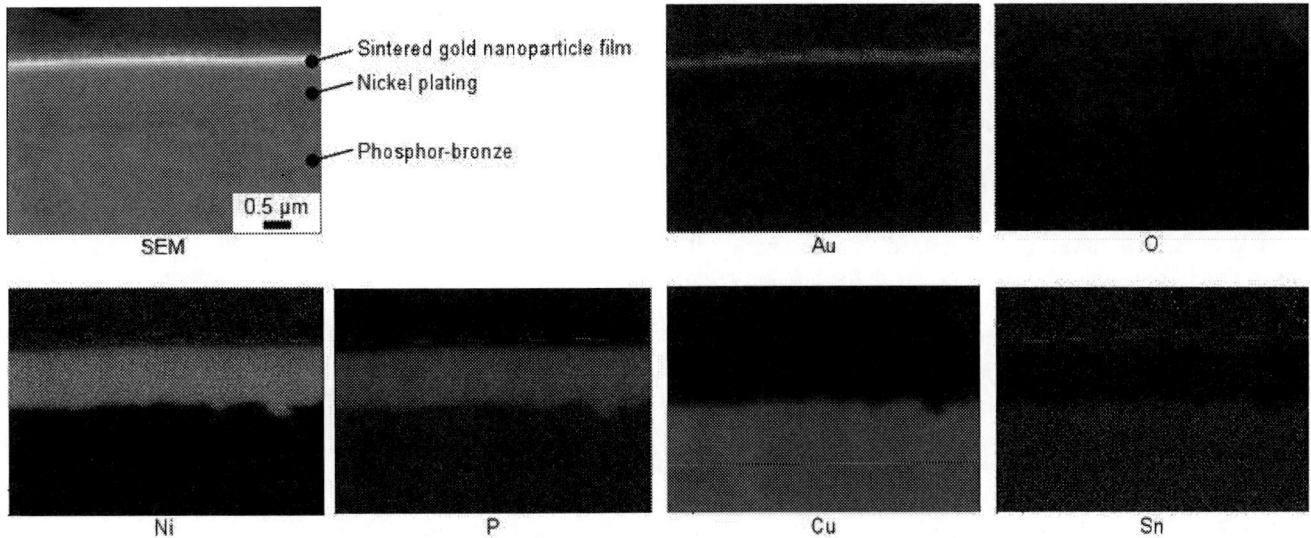

Fig. 8 AES element maps of sintered gold nanoparticle film and substrate interface

Figure 6 shows an optical image of the surface after the bend-peel test; Fig. 7 is an enlarged view. The broken line indicates the place where the bend and bendback was carried out. The specimen is the one illustrated in Fig. 3. No separation of the film can be seen, showing that a good adhesion to the nickel base film has occurred. It is difficult to achieve this result in the conventional electroplating of gold on a barrier nickel without sophisticated pretreatments. This point is a feature of the current laser sintering method.

(b) AES analysis of laser-sintered gold film

Figure 8 shows the AES surface analysis results with respect to the interface between the laser-sintered gold film and the nickel-plated phosphor-bronze: the SEM image and the mappings of Au, O, Ni, P, Cu, and Sn are presented. It can be seen that the surface is almost 100% gold, and the other elements are not interdiffused. In contrast, Fig. 9 shows the AES depth profile of the specimen. At the start of sputtering 40 at% gold, 40 at% carbon, 10 at% oxygen, 10 at% nickel were observed. The presence of nickel and oxygen means that nickel has diffused into the gold nanoparticles, forming an oxide layer on the surface. The carbon is considered to be one of the contaminant elements adhering to the surface. It can be seen that 80 at% gold content at the 6 min sputtering time decreases from 10 min sputtering time onwards. At the same region, nickel and phosphor contents increase.

It seems that interdiffusion at the interface contributes to good adhesion between the gold and nickel. Besides, nickel diffusion into the gold may lead to surface hardening. In conventional gold electroplating for electrical contacts, some amounts of impurities, such as 0.2 mass% nickel or cobalt, are added to increase hardness and wear resistance [12]. These points should be studied further in the case of laser-sintered gold film.

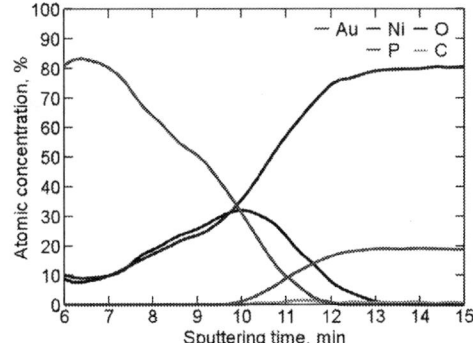

Fig. 9 AES depth profile of sintered gold nanoparticle film

(c) Laser sintering mechanism of gold nanoparticle paste

It has been reported that characteristics of absorption and transmittance of the metal nanoparticle paste in the laser wavelength region to be irradiated are important factors affecting the sintering mechanism [5, 6]. The visible and near-infrared spectroscopic analysis revealed that the optical spectrum of the gold nanoparticle paste varies with preheat temperature and time. In this study, before reporting sintering experiments conducted with gold nanoparticle paste having different spectra with varying preheat conditions, we discuss the sintering mechanism in the case of using an infrared laser of 915 nm wavelength.

Figure 10 shows the transmittance spectra of the gold nanoparticle paste at different preheat temperatures and a fixed heating time of 60 s. The spectra monotonically increase with increasing wavelength. In the infrared region, at 915 nm wavelength, the transmittance of the gold nanoparticle paste is approximately 40% at 373 K, but is only about 1% at 523 K.

Figure 11 shows the reflection spectra of the gold nanoparticle paste at different preheat temperatures. All three spectra sharply increase from 500 nm wavelength, and become almost constant above 600 nm. In the infrared region, at 915 nm wavelength, the reflectance of the gold nanoparticle paste is about 20% at 373 K, and about 50% at both 523 K and 573 K. Since the boiling point of the naphthenic base solvent in the gold nanoparticle paste is 555 K, it is considered that almost all the solvents were removed by preheating over 523 K, yielding a similar level of reflectance. Consequently, the absorption of the gold nanoparticle paste in the infrared region at 915 nm can be estimated to be about 40% at the preheat temperature of 373 K, and about 50% at 523 K.

Figure 12 shows an SEM image of the gold nanoparticle film when it was sintered under the forming conditions of 373 K for 60 s preheating, a laser power of 100 W, and an irradiation time of 0.1 s. It can be seen that at the preheat temperature of 373 K laser sintering is insufficient, as fine cracks have been generated on the sintered film surface. Besides, the sintered part was easily wiped off with a cloth. In contrast, an almost perfect sintered film was obtained at the preheat temperature of 523 K, as was shown in Fig. 3. This difference can be explained by the below laser sintering characteristics mentioned.

Figure 13 is a schematic of the sintering situation with (a) low-temperature preheat and (b) high-temperature preheat. In case (a) the laser energy reaches the substrate through the gold nanoparticle paste because the transmittance of the paste is high. The laser energy transmitted to the substrate surface generates a high temperature there, being conducted by the gold nanoparticle paste. As a result, sintering proceeds from the substrate side, as was reported elsewhere [5]. Moreover, as phosphor-bronze has a high thermal conductivity of 65 Wm⁻¹K⁻¹, the heat generated at the interface quickly diffuses into the base metal. Additional energy for vaporizing the solvent remaining in the paste, as well as this heat loss, probably leads to an insufficiently laser-sintered film, as shown in Fig. 12. Further, rapid temperature rise in the substrate causes the generation of gases in the course of sintering, so that the film surface becomes rough.

As mentioned above, in case (b) the absorption of the gold nanoparticle paste in the infrared region at 915 nm is about 50% when the preheat temperature is 523 K. The laser energy transmitted to the substrate is 10%; this is low compared to the Figure for the preheat temperature of 373 K. Although the energy is reflected from the half-dried surface to some extent, a large amount of energy is converted to heat on the surface, being conducted through the paste into the substrate without large heat loss. Simultaneously, the small amount of the laser energy transmitted to the substrate starts sintering at the interface. On top of this advantage, a small amount of residual solvents is present in the paste, the selection of a high preheat temperature prevents explosive vaporization; the fine texture of the sintered film can be attributed to this. Furthermore, the use of a large-diameter laser beam of around 1.2 mm which covers the whole sintering area of 0.3 mm diameter leads to increase in sinterability.

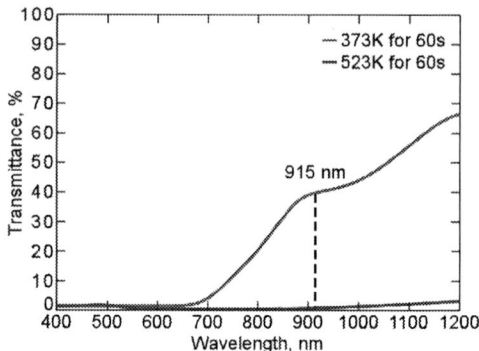

Fig. 10 Transmission spectra of preheated gold nanoparticle paste

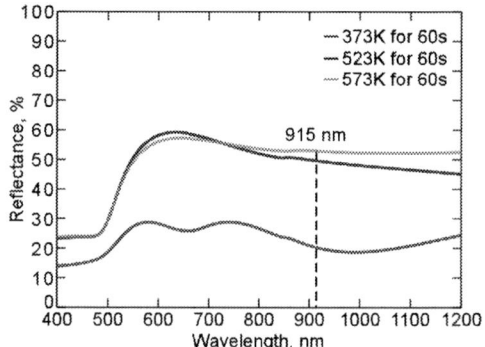

Fig. 11 Reflection spectra of preheated gold nanoparticle paste

Fig. 12 SEM image of sintered gold nanoparticle film after preheat at 373 K for 60 s

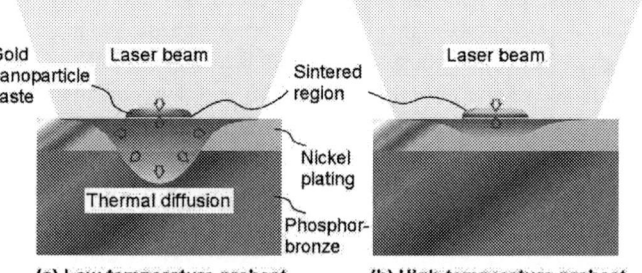

Fig. 13 Schematic of laser sintering mechanism with different preheat conditions

It has been reported that the use of a shorter wavelength of around 532 nm efficiently sinters gold nanoparticles [7]. An attempt has been made to confirm this by using a green laser of 532 nm wavelength with preheat conditions of 373 K and 60 s. The result was unfavorable to yield the gold film with thickness of more than 0.1 μm. The proper selection of wavelength and preheat conditions is thus key to the laser sintering of gold nanoparticle paste. A laser having a wavelength in the infrared region enables a primary sintering of the preheated gold nanoparticles with a small amount of solvents, and an auxiliary sintering from the substrate side, as well. As a result, efficient sintering of metal nanoparticles as well as high adhesion to the substrate with a high thermal conductivity has been achieved.

Conclusions

The present study focused on forming a conductive film with noble metal nanoparticles as an alternative to the conventional electroplating method for electrical contacts. The proposed method consists of inkjet printing with metal nanoparticle paste followed by laser sintering. The major results obtained are as follows:

(1) The laser sintering formed a gold film with a diameter of 0.3-0.8 mm and a thickness of 0.3-0.5 μm on the nickel-electroplated phosphor-bronze substrate.

(2) A laser with a wavelength of 915 nm enabled instantaneous sintering within one second in an atmosphere.

(3) The laser-sintered gold nanoparticle film had such a high adhesion to the substrate that no separation occurred after 90°-0.5R bend-peel tests.

(4) The high adhesion was attributed to interdiffusion of gold and nickel in the course of sintering.

(5) Optical properties of the gold nanoparticle paste depend on preheat conditions. A relatively high-preheat temperature around 523 K produced a paste surface with a suitable absorbance of the infrared laser.

(6) A primary sintering of the preheated gold nanoparticles with a small amount of solvents, followed by an auxiliary sintering from the substrate side made possible an efficient sintering of the nanoparticles as well as high adhesion to the substrate with a high thermal conductivity.

The electrical-contact forming method combining inkjet printing using a noble metal nanoparticle paste with laser sintering has various advantages, including an on-demand process with a minimum consumption of resources, a short-time atmospheric process, no need of pretreatments except preheat of the paste, and, last but not least, an alternative to electroplating or chemical plating on a nickel-plated substrate.

Acknowledgments

The authors would like to thank Mr. A. Ando at the Industry Technology Center of Ibaraki Prefecture for his contribution to the spectroscopic analysis of the gold nanoparticle paste.

References

1. Electroplating Research Society, *Electroplating manual*, NIKKAN KOGYO SHIMBUN Ltd. (1986), p. 198 (in Japanese).

2. T. Sasabe, M. Akiyama and K. Kato, *Plating technique of really useful printed circuit boards for business*, NIKKAN KOGYO SHIMBUN Ltd. (2012), p. 280 (in Japanese).

3. N. Terada, "Fine Pattern Formation by Ink-Jet Printing Using Conductive NanoPaste and Its Application", *Journal of Japan Institute of Electronics Packaging*, Vol. 11, No. 4 (2008), pp. 300‑306.

4. Ph. Buffat and J-P. Borel, "Size effect on the melting temperature of gold particles", *Physical Review A*, Vol. 13, No.6 (1976), pp. 2287-2298.

5. K. Maekawa, K. Yamasaki, T. Niizeki, M. Mita, Y. Matsuba, N. Terada and H. Saito, "Drop-on-Demand Laser Sintering with Silver Nanoparticles for Electronics Packaging", *IEEE Trans. CPMT*, Vol. 2, No. 5 (2012), pp. 868-877.

6. K. Maekawa, K. Yamasaki, T. Niizeki, M. Mita, Y. Matsuba, N. Terada and H. Saito, "Laser Sintering Technology for Printed Electronics: Minute Wiring and Functional Coating with Ag-Nanoparticle Paste", *Journal of Japan Institute of Electronics Packaging*, Vol. 15, No. 1 (2012), pp. 96-105.

7. Y.-T. Cheng, R.-H. Uang, Y.-M. Wang, K.-C. Chiou and T.-M. Lee "Laser annealing of gold nanoparticles thin film using photothermal effect", *Microelectronics Engineering*, Vol. 86, No. 4-6 (2009), pp. 865-867.

8. Z. Cai, K. C. Yung and X. Zeng, "Fabrication and adhesion performance of gold conductive patterns on silicon substrate by laser sintering", *Applied Surface Science*, Vol. 258, No. 1 (2011), pp. 478-481.

9. A. Watanabe, "Laser Sintering of Nanoparticle Film", *Journal of Photopolymer Science and Technology*, Vol. 26, No. 2 (2013), pp. 199-205.

10. N. R. Bieri, J. Chung, D. Poulikakos and C. P. Grigoropoulos, "Manufacturing of nanoscale thickness gold lines by laser curing of a discretely deposited nanoparticle suspension", *Superlattices and Microstructures*, Vol. 35, No. 3-6 (2004), pp. 437-444.

11. N. R. Bieri, J. Chung, D. Poulikakos and C. P. Grigoropoulos, "An experimental investigation of microresistor laser printing with gold nanoparticle-laden inks", *Applied Physics A*, Vol. 80, No. 7 (2005), pp. 1485-1495.

12. A. Sugiyama, T. Yokoshima, T. Hachisu, Y. Okinaka and T. Osaka, "Gold Plating Films for Electronic Components", *Journal of the Surface Finishing Society of Japan*, Vol. 62, No. 12 (2011), pp. 635-641.

Analysis of Modes Effect on Signal/Power Integrity in Finite Cavity for Chip and Die Level Packaging Based on A Hybrid Full Wave Method

Xin Chang and Leung Tsang
Electrical Engineering Department of University of Washington - Seattle
185 Stevens Way, Paul Allen Center – Room AE100R, Seattle, WA, 98105
changx@uw.edu, tsang1@uw.edu

Abstract

Most of the via modeling methods assumed that only the TEM mode which is the fundamental mode can propagate due to the thickness of substrate being electrically small, and all the higher order waveguide modes and anisotropic modes decay rapidly along the radial direction thus they can be ignored. However, for high speed chip and die level packaging system, due to the interested frequency range is higher and higher, the high order modes can be excited. At the same time, since the size of via holes can be comparable to the size of cavity, the excited modes for the via-plane pair structures may also be different from the cavity modes. These modes effects do contribute to the interactions among adjacent vias and cavity. Ignoring the modes effects may lead to inaccurate network parameters prediction for signal and power integrity analysis. In this paper, we first applied a hybrid full wave method based on Foldy-Lax multiple scattering equations method to model dense via array and multiple vias sharing same antipad, e.g. the case of differential via pair, in finite cavity. Then we investigate the geometric parameters and parasitic modes effects on signal/power integrity. Numerical results comparisons with commercial full wave solver show the hybrid full wave method has great accuracy and efficiency for modeling multiple vertical vias in finite cavity. We also explain the physical roots of the modes effect in order to improve signal link path bandwidth due to the modes effects.

Discussion 1 Introduction

Complex interconnects system result in that timings now are interconnects driven. With the increase of high density 3D integrations, via size is decreasing while the number of vias is increasing now, which leads to a higher and higher density of vias. Also, 3D packaging saves space by stacking separate chips in a single package; its enabling technology adopts die stacking and die to die vias, the components in package is from nanometer to micron in size. These features which allow signal propagate in the high frequency range and hence may excite higher order waveguide modes and axially anisotropic modes will definitely introduce signal integrity issue, power and clock distribution issue and thermal issue, with the ever-rising clock rate and edge rate of chip-package-board systems.

In the previous work [5-12], we use a semi-analytical technique of Foldy-Lax equations and MoM to compute the full wave solution of Maxwell equations that includes multiple scattering among cylindrical vias in both infinite large planar waveguides and finite cavity. For finite ground planes, the propagating modes of the waveguide will be reflected from the walls of the ground plane. In [10], we discussed the modes effects for the coupling among vias in infinite waveguide, for closely spaced vias in dense vias array.

In this paper, we will discuss the modes effects for the coupling effects among vias and cavity walls with finite power/ground(P/G) plane, especially for chip and die level packaging system, based on the hybrid method of generalized Foldy-Lax equations and 1D MoM, as proposed in [9], up to 100GHz.

Discussion 2 Foldy-Lax Scattering Equations Method

In this section, I briefly describe the 1D MoM for the problem of finite cavity of irregular shape we derived in [9]. The walls of the cavity will be treated as perfect magnetic conductors (PMC). In the multiple scattering using Foldy-Lax equations, we treat the TM modes with

$$k_{zl} = \frac{l\pi}{d}; k^2 - k_{zl}^2 = k_{\rho l}^2 \tag{1}$$

where $l = 0,1,2.....$, d is the separation between the two plates, and k is the wavenumber. All the notations are same as shown in [9].

We use the t coordinate to describe the line contour of the caivty boundary. In the MoM formulation, 1-dimensional discretization of the boundary wall is used. Let there be N_t segments. We use pulse basis functions and point matching for the t_v coordinates with $v = 1,2...N_t$ and the length of segment v is Δt_v. Then the impedance matrix $\bar{\bar{Z}}$ is of dimension $N_t \times N_t$. The impedance matrix elements are

$$Z_{\mu v} = \begin{cases} [\hat{n}' \cdot \nabla_t' g(\bar{\rho}, \bar{\rho}')\Delta t_v]_{\bar{\rho}'=\bar{\rho}(t_\mu),\bar{\rho}'=\bar{\rho}(t_v)} & \mu \neq v \\ \frac{1}{2} & \mu = v \end{cases} \tag{2}$$

for $\mu, v = 1,2...N_t$, and $g(\bar{\rho}) = \frac{1}{4j}H_0^{(2)}(k\rho)$.

Consider multiple vias, $q = 1,2,3,....N$, the Foldy-Lax equations are

$$\bar{w}^q = \bar{a}^{q,inc} + \sum_{\substack{p=1 \\ p \neq q}}^{N}[\bar{\bar{\alpha}}_{qp}^+ + \bar{\bar{X}}^{Wqp}]\bar{A}^p \tag{3}$$

In general, we include harmonics of $m = 0, \pm 1, ...\pm M$. Harmonics of $m \neq 0$ give rise to anisotropic effects.

In the above equation, $\bar{w}^q$ and $\bar{A}^q$ are of dimensions $(2M + 1) \times 1$, and contain exciting field coefficients $[\bar{w}^q]_m = w_m^q$ and scattered field coefficients $[\bar{A}^q]_m = A_m^q$, $m = 0, \pm 1, ...\pm M$; $\bar{a}^{inc}$ is the incident field coefficients, $\bar{\bar{\alpha}}_{qp}^+$ matrix is the translation matrix for cylindrical waves and is of dimension $(2M + 1) \times (2M + 1)$

$$[\bar{\bar{\alpha}}_{qp}^+]_{nm} = H_{n-m}^{(2)}\big(k|\overline{\rho_p} - \overline{\rho_q}|\big)e^{j(n-m)\emptyset_{\overline{\rho_p}\overline{\rho_q}}} \qquad (4)$$

and

$$\bar{\bar{X}}^{Wqp} = \bar{\bar{\gamma}}^{(qW)}\bar{\bar{Z}}^{-1}\bar{\bar{Q}}^{(Wp)} \qquad (5)$$

In equation (5), the $\bar{\bar{\gamma}}^{(qW)}$ matrix which has dimension $(2M + 1) \times N_t$ where W stands for wall. It is of mixed dimension as it represents discretized MoM points coupled to wave harmonics, $m = 0, \pm 1, \ldots \pm M$. $\bar{\bar{Q}}^{(Wq)}$ is with dimension $N_t \times (2M + 1)$. It represents coupling from harmonic to discretized point on the boundary. $\bar{\bar{Z}}^{-1}$ is the inverse of the MoM impedance matrix. Their expressions can be found in [9]. Note that the waveguide modes for $l \geq 1$ are included in this paper, as the distance between vias and cavity side walls for the test vehicles shown in the numerical results can be compared with the substrate thickness.

Using the $\bar{\bar{\tau}}^{(p)}$, the T-matrix of single via including wall effects, we have

$$\overline{w}^q = \bar{a}^{q,inc} + \sum_{\substack{p=1 \\ p \neq q}}^{N}[\bar{\bar{\alpha}}_{qp}^+ + \bar{\bar{X}}^{Wqp}]\bar{\bar{\tau}}^{(p)}\overline{w}^p \qquad (6)$$

where

$$\bar{\bar{\tau}}^{(p)} = [\bar{\bar{I}} - \bar{\bar{T}}^p\bar{\bar{X}}^{Wpp}]^{-1}\bar{\bar{T}}^p \qquad (7)$$

After the Foldy-Lax equations are solved, $\overline{w}^q, q = 1,2\ldots N$ are obained. Then the magnetic field in the $\hat{\phi}$ direction can be obtained. The surface currents J_s^q as shown in [9] is in the z direction and is calculated from the magnetic field on the surface of the via.

Once the currents are determined, the calculation of the admittance parameter matrix of multiple vias inside finite P/G planes $\bar{\bar{Y}}_{P/G}$ can be obtained, and the S parameters are then can be also calculated.

Discussion 3 Numerical Results and Discussions

As similarly shown in [10], the indices l and n stand for the mode order for waveguide modes and azimuthal modes for vertical vias respectively. We discuss the high order modes effects for single through hole dense via array and differential signaling dense via array in this paper. For excitation in the antipad, only TEM or TM_{z00} mode is considered, hence only TM_z modes are excited in the waveguide. Then the modes can be classified as

- TEM or TM_{z00} mode, with $l = n = 0$, it is propagating and isotropic.
- TM_{zn0} modes, with $l = 0$ and $n \neq 0$, they are propagating and anisotropic.
- TM_{z0l} modes, with $l > 0$ and $n = 0$, they are nonpropagating and isotropic.
- TM_{znl} modes, with $l > 0$ and $n \neq 0$, they are nonpropagating and anisotropic.

We illustrate all modes effects for two cases. L_{max} and N_{max} are the truncation mode order numbers of l and n.

All single-ended S-parameters provided here are referenced to 50 Ω, and all mixed-mode S-parameters are referenced to 100 Ω for differential mode. The configurations of PC used in the simulations are: Intel(R) Core(TM)2 Duo E7300 2.66GHz processor, 3GB RAM and Windows Vista 32-bit operating system. The average CPU times per frequency for simulation of single-ended S parameters by using hybrid Foldy-Lax/1D MoM method and HFSS are compared and shown in Table I. More details about the computational resources for the hybrid method can be found in [9]. In the case specifications, d stands for waveguide thickness, and t stands for the plane thickness in the specifications of both cases.

A. 5 X 10 Dense Vias Array for Each Via Going through One Antipad

Fig. 1 shows top view of structure about 5x10 dense vias geometry. Each signal via goes through one antipad. The specifications are: $\varepsilon_r = 3.84$, $R_{via} = 7\ mil$, $R_{antipad} = 15\ mil$, $tan\delta = 0.033$, $p = 40\ mil$, $d = 30\ mil$, $t = 1\ mil$. The cavity length is $300\ mil$ in x (vertical) direction and $500\ mil$ in y (horizontal) direction. The first via which is the left top comer signal via locate at $(70,70)\ mil$. The via order increases along y (horizontal) direction first and then increases by along x (vertical) direction.

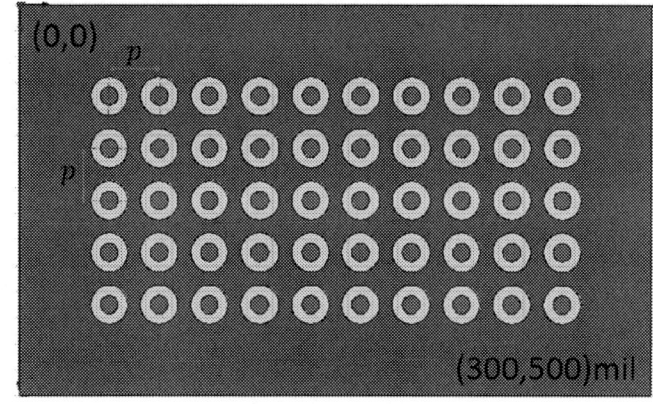

Figure 1. Top view of 5x8 dense via array for case A.

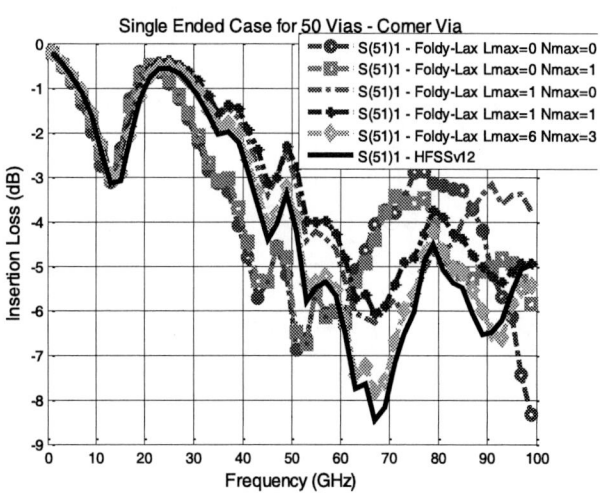

Figure 2. Insertion loss for corner via.

978-1-4799-2408-0/14 $31.00 © 2014 IEEE

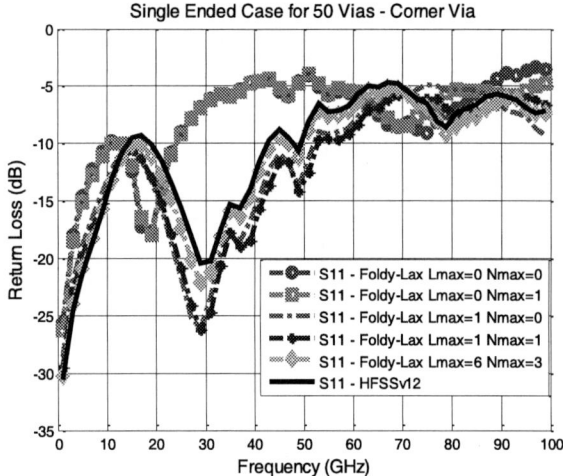

Figure 3. Return loss for corner via.

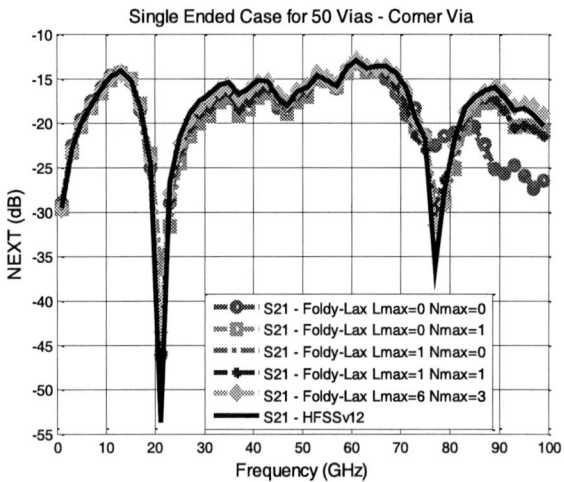

Figure 4. NEXT for corner via.

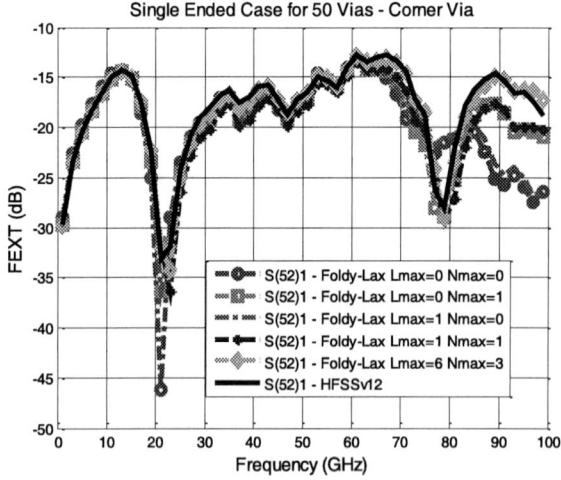

Figure 5. FEXT for corner via.

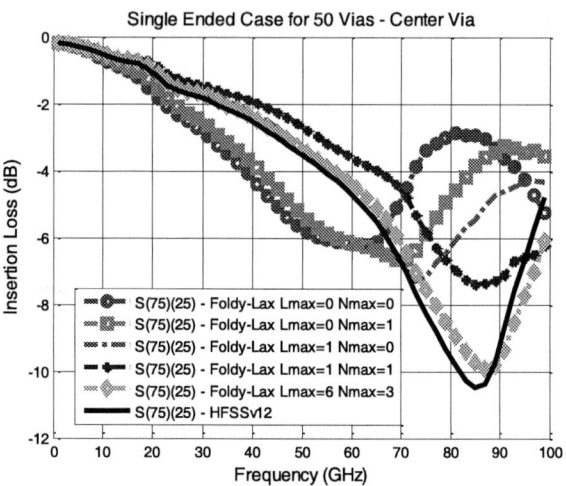

Figure 6. Insertion loss for center via.

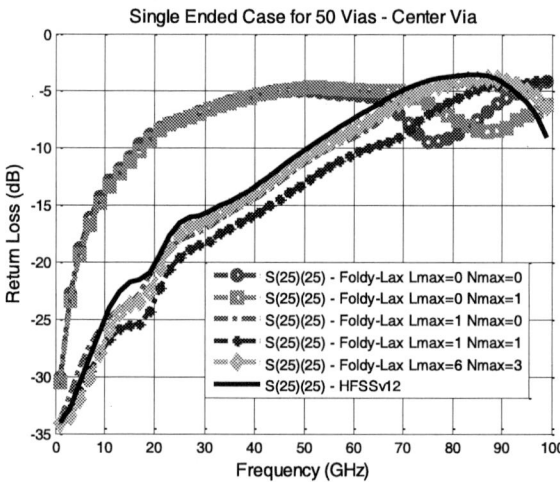

Figure 7. Return loss for center via.

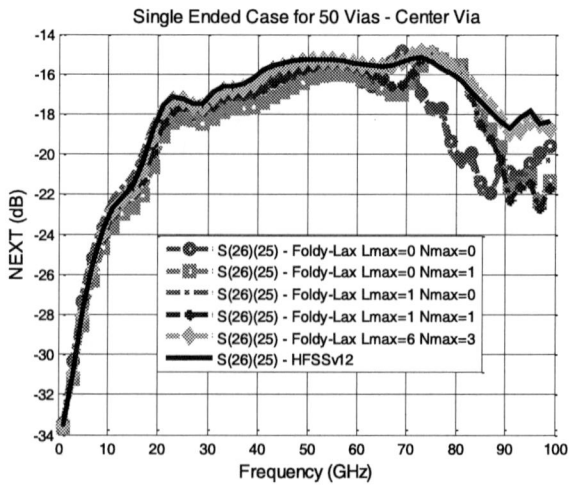

Figure 8. NEXT for center via.

978-1-4799-2408-0/14 $31.00 © 2014 IEEE

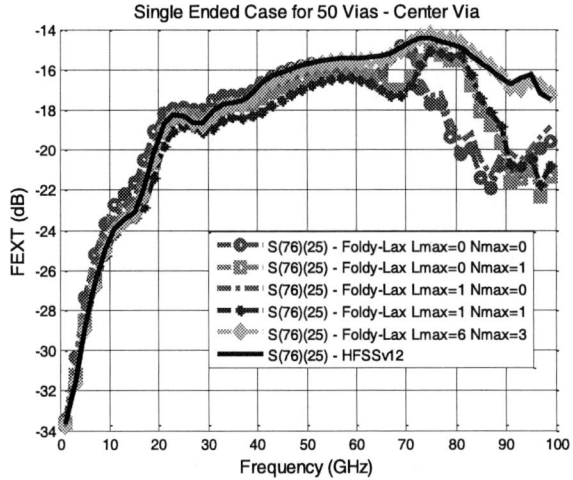

Figure 9. FEXT for corner via.

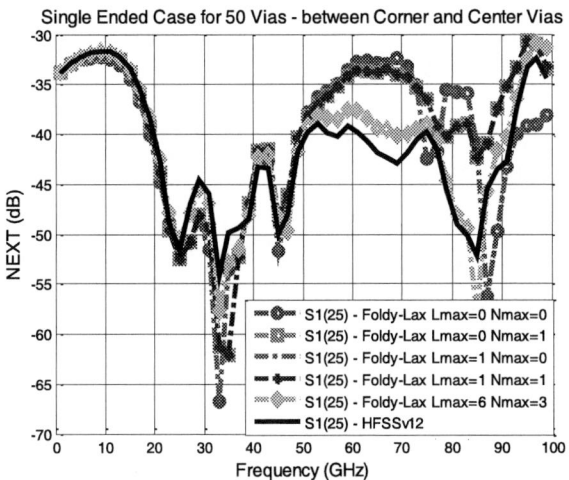

Figure 10. NEXT for center via.

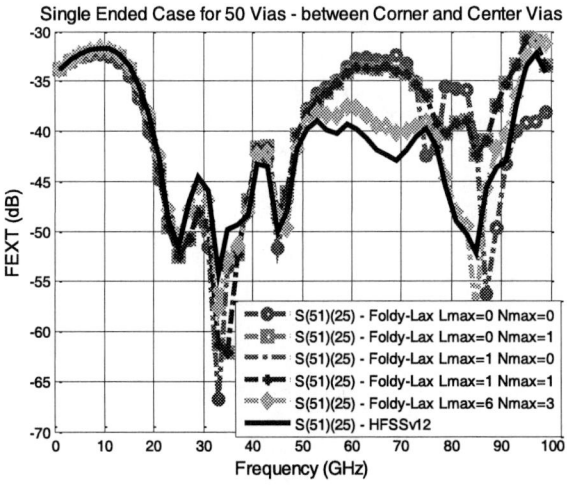

Figure 11. FEXT for corner via.

Fig. 2-5 show the insertion loss, return loss, NEXT and FEXT for corner via which is the left top corner via. Fig. 6-9 show the insertion loss, return loss, NEXT and FEXT for center via which is the left middle center via. Fig. 10-11 show the NEXT and FEXT between the corner and center vias.

From the Fig 2-11, we can see that for dense via array in small cavity, for return loss, when the interested frequency is above 10GHz (for corner via) or 20GHz (for center via), the high order modes effects need to be included for accurate design and modeling. This is because the corner via is placed closer to the cavity side walls than the center via is. For insertion loss, high order modes effects need to be included at as low frequency as possible, as we can observe their strong contributions for the return loss. However, the higher order modes effects make weak contributions for the NEXT and FEXT, for both corner and center vias case. Their effects need to be considered when the interested frequency is above 50GHz.

B. 3X 10 Dense Vias Array for Differential Singling Pairs

Fig. 12 shows top view of structure about 3x10 dense vias geometry. Each two signal via sharing same antipad. The specifications are: $\varepsilon_r = 4.4$, $R_{via} = 10\ mil$, $R_{antipad} = 20\ mil$, $tan\delta = 0.02$, $p = 35\ mil$, $p_x = p_y = 50\ mil$, $d = 50\ mil$, $t = 1\ mil$. The cavity is square with length of $350\ mil$ in both x (vertical) direction and y (horizontal) direction. The first via which is the left top comer signal via locate at $(75,75)\ mil$. The differential signaling pair order increases along y (horizontal) direction first and then increases by along x (vertical) direction.

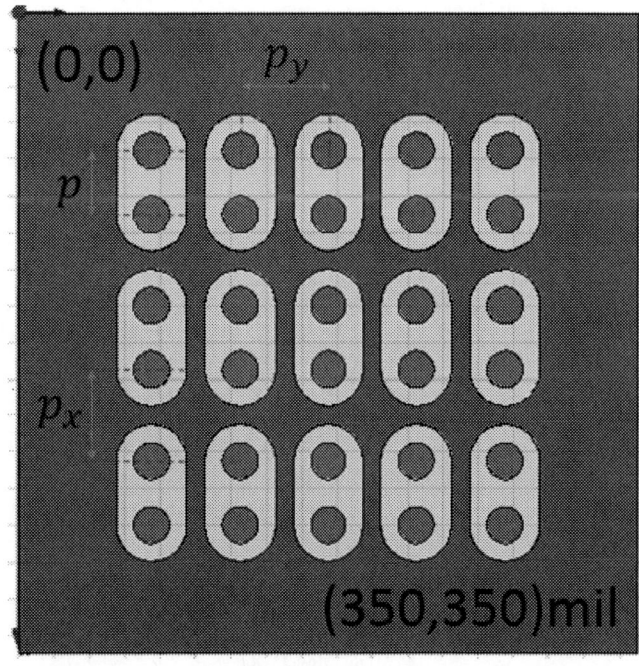

Figure 12. Top view of 3x10 dense via array for case B.

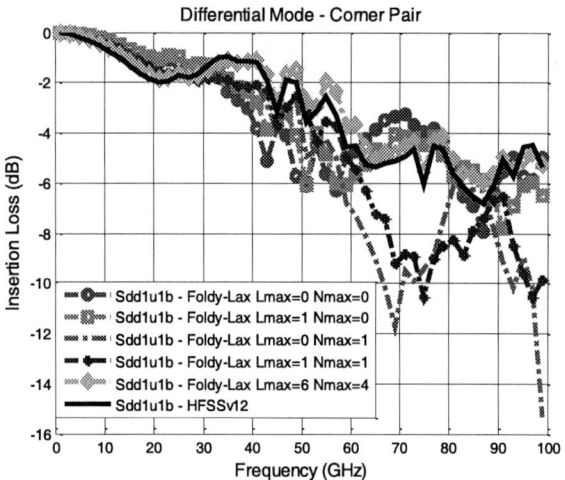

Figure 13. Insertion loss of differential mode for corner pair.

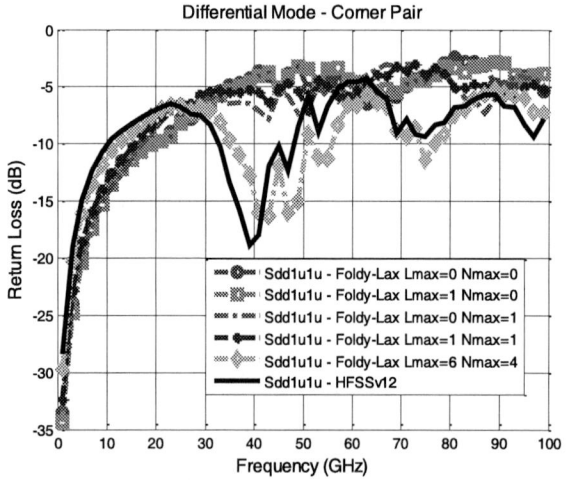

Figure 14. Return loss of differential mode for corner pair.

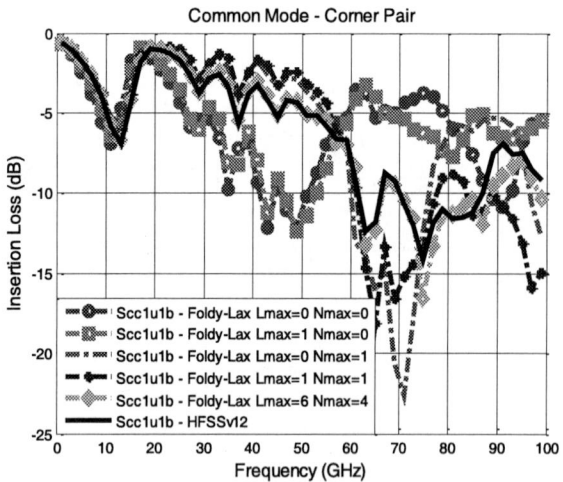

Figure 15. Insertion loss of common mode for corner pair.

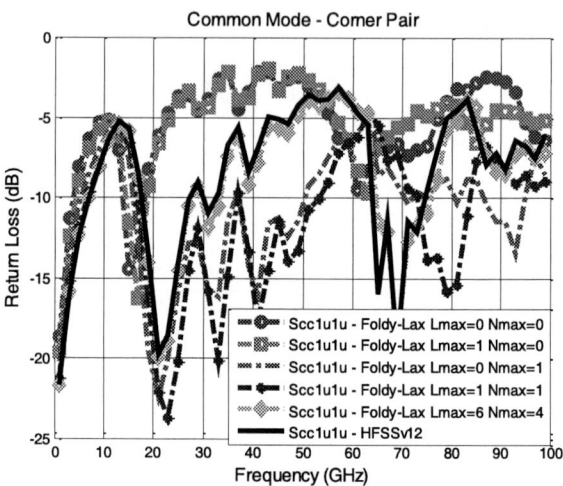

Figure 16. Return loss of common mode for corner pair.

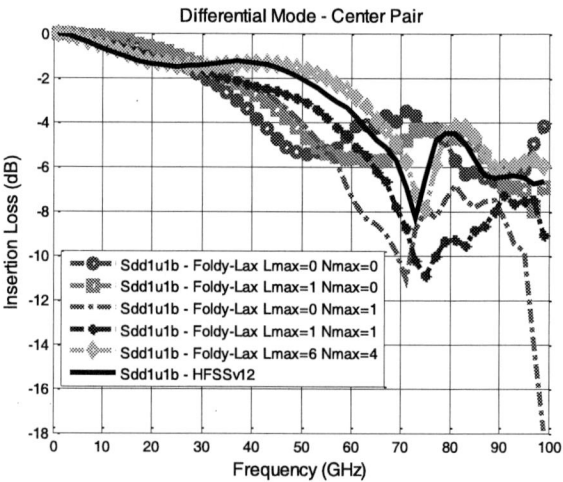

Figure 17. Insertion loss of differential mode for center pair.

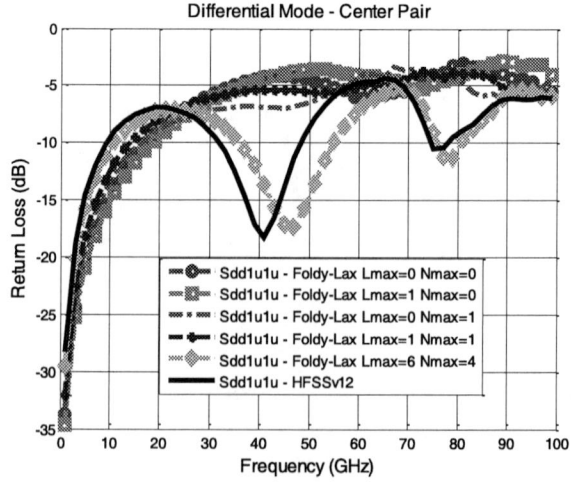

Figure 18. Return loss of differential mode for center pair.

978-1-4799-2408-0/14 $31.00 © 2014 IEEE

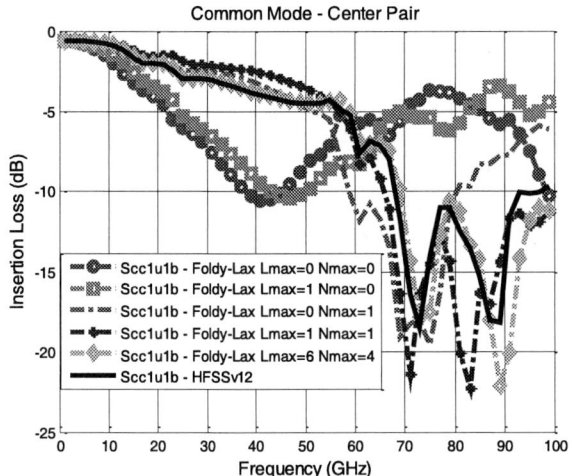

Figure 19. Insertion loss of common mode for center pair.

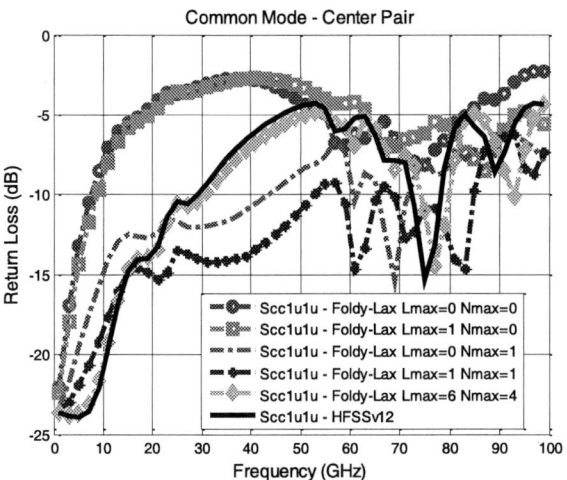

Figure 20. Return loss of common mode for center pair.

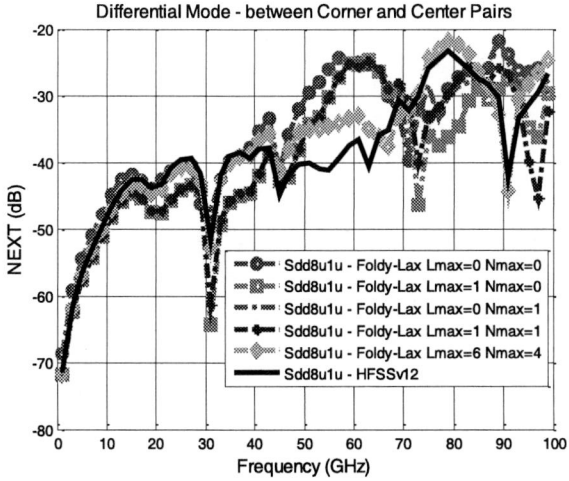

Figure 21. NEXT differential mode for center/corner pairs.

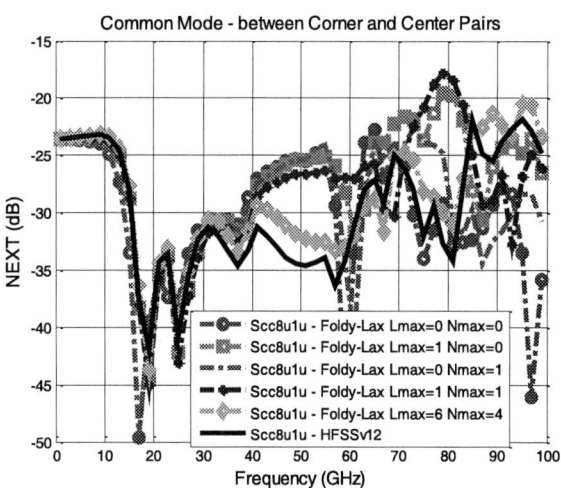

Figure 22. NEXT common mode for center/corner pairs.

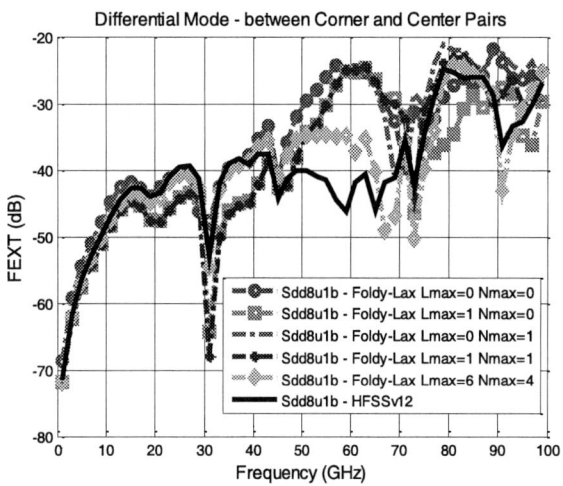

Figure 23. FEXT differential mode for center/corner pairs.

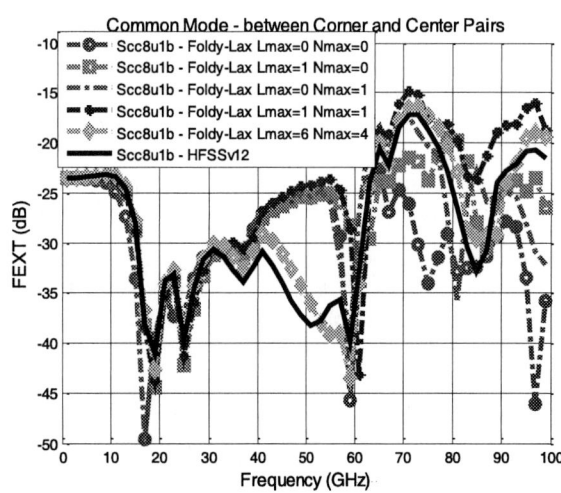

Figure 24. FEXT common mode for center/corner pairs.

978-1-4799-2408-0/14 $31.00 © 2014 IEEE

TABLE I. CPU Run Time Per Frequency Comparisons For Case A-B

	Foldy-Lax/1D MoM	HFSS v12
A	*25 sec*	*720 sec*
B	*10 sec*	*196 sec*

Fig. 13-16 show the insertion loss and return loss, for differential mode and common mode respectively, for corner pair which is the left top via pair. Fig. 17-20 show the insertion loss and return loss, for differential mode and common mode respectively, for center pair which is the middle pair. Fig. 21-24 show the NEXT and FEXT, for differential mode and common mode respectively, between the corner and center vias,

From the Fig 13-24, we can see that for dense via array for differential singling in small cavity, for both insertion loss and return loss, the higher order modes effects make stronger contributions for the corner pair than the center pair, for both differential mode and common mode. This is due to the physical closer locations to the cavity side walls for the corner pair than the center pair is. Also, without considering higher order modes effects may make the simulation show the resonances at lower frequency than they should be, which then limit the real design requirements.

For NEXT and FEXT between the corner pair and center pair, unlike the single through hole via case as shown in case A, the higher order mode effects show up in lower frequency as 10GHz. This means their effects are more important for the differential signaling vias than the case of single through hole via case, due to two vias sharing same antipad and strong near field coupling in high frequency.

Based on the simulations shown above, we suggest that the vias should be placed not close to the cavity side walls in order to avoid the reflections from the walls. These modes reflections may lead to resonances at undesired frequency. More vias placement also help shift the cavity modes to higher frequency.

Conclusions

Two dense via array cases are simulated based on an efficient hybrid full wave method for investigation of modes effects for chip and die level packaging system. The hybrid method is much faster and also are keeping much smaller memory than HFSSv12. From the simulation results, higher order modes effects are suggest to be included at as low frequency as possible, for accurate signal/power integrity modeling and design.

References

1. X. Chang, B. Archambeault, M. Cocchini, F. De Paulis, V. Sivarajan, et al., "Return via connections for extending signal link path bandwidth of via transitions," in *Proc. Int. Symp. Electromagn. Compat. Eur.*, Hamburg, Germany, pp. 1-6, Sep. 2008.
2. J. Kim, L. Ren, and J. Fan, "Physics-based inductance extraction for via arrays in parallel planes for power distribution network design," *IEEE Trans. Microw. Theory Tech.*, vol. 58, no. 9, pp. 2434–2447, Sep. 2010.
3. S. Wu, X. Chang, C. Schuster, X. Gu, and J. Fan, "Eliminating Via-Plane Coupling Using Ground Vias for High-Speed Signal Transitions", *2008 Electrical Performance of Electronic Packaging International Symposium*, Santa Rose, CA, USA, 2008.
4. Y. Zhang, G. Feng, and J. Fan, "A novel impedance definition of a parallel-plate pair for an intrinsic via circuit model," *IEEE Trans. Microwave Theory Tech.*, vol. 58, no. 12, pp. 3780-3789, Aug. 2010.
5. L. Tsang, H. Chen, C.-C. Huang, and V. Jandhyala, "Modeling of multiple scattering among vias in planar waveguides using Foldy-Lax equations," Microwave Optical Technol. Lett. vol. 31, pp. 201-208, Nov. 2001.
6. L. Tsang and D. Miller, "Coupling of vias in electronic packaging and printed circuit board structures with finite ground plane", *IEEE Trans. Advanced Packaging*, vol. 26, pp. 375-384, Nov. 2003.
7. C.J. Ong, et al, "Application of the Foldy-Lax multiple scattering method to the analysis of vias in ball grid arrays and interior layers of printed circuit boards," *Microw. Opt. Tech. Lett.*, vol. 49, no. 1, pp. 225-31, Jan. 2007.
8. L. Tsang and X. Chang, "Modeling of vias sharing the same antipad in planar waveguide with boundary integral equation and group T matrix method," *IEEE Trans. Comp.Packag. Manuf. Technol.*, vol. 3, pp. 315–327, Feb. 2013.
9. X. Chang and L. Tsang, "Fast and Broadband Modeling Method for Multiple Vias with Irregular Antipad in Arbitrarily Shaped Power/Ground Planes in 3-D IC and Packaging Based on Generalized Foldy-Lax Equations", in press, *IEEE Trans. Compon. Packag. Manuf.*, 2014.
10. X. Chang and L. Tsang, "A new efficient method for modeling dense via arrays with 1D discretization in 2D method of moment and group T matrix", *2012 Electrical Performance of Electronic Packaging International Symposium*, pp. 163-166, Tempe, AZ, USA, Oct. 2012.
11. X. Chang and L. Tsang, "Modeling multiple scattering among vertical interconnects for SIW structures and 3D ICs in arbitrarily shaped waveguide", *2013 IEEE International Symposium on Antennas and Propagation and USNC-URSI National Radio Science Meeting*, Orlando, FL, USA, July, 2013.
12. X. Chang and L. Tsang, "A generalized modeling method for signal/power integrity analysis of 3D coupled interconnects in finite cavity based on 1D technology", *2013 Electrical Performance of Electronic Packaging International Symposium*, Santa Clara, CA, USA, October 2013.
13. R. Rimolo-Donadio, H.-D. Brüns, and C. Schuster, "Including stripline connections into network parameter based via models for fast simulation of interconnects," in *Proc. 20th Int. Electromagn. Compat. Symp.*, Zurich, Switzerland, Jan. 2009, pp. 345–348.
14. R. Rimolo-Donadio, et al, "Physics-based via and trace models for efficient link simulation on multilayer structure up to 40 GHz," *IEEE Trans. Microw. Theory Tech.*, vol. 57, no. 8, pp. 2072–2083, Aug. 2009.
15. E.-P. Li, et al, "Progress review of electromagnetic compatibility analysis technologies for packages, printed circuit boards, and novel interconnects," *IEEE Trans. Electromagn. Compat.*, vol. 52, no. 2, pp. 248-265, May, 2010.

Directed Self-Assembly of Mesoscopic Dies using Magnetic Force and Shape Recognition

Anton Tkachenko, Robert F. Karlicek, Jr., and James J.-Q. Lu
NSF Smart Lighting Engineering Research Center,
Rensselaer Polytechnic Institute, Troy, NY 12180
E-mail: tkacha@rpi.edu

Abstract

This paper reports on a directed self-assembly approach suitable for assembly of mesoscopic dies, such as LEDs, in a parallel fashion. This approach utilizes magnetic force, shape recognition, and capillary force of the solder to assemble the dies on the receiving substrate in a correct orientation. Assembly takes place in liquid medium. Effects of the die geometry, assembly setup and magnet configuration on the assembly results are presented along with the simulation of the solder-assisted self-alignment.

1. Introduction

Currently, the majority of LED light fixtures imitate the shape of incandescent bulbs or fluorescent tubes [1]. Such design is widely used because it allows fitting the new LED luminaires into the already existing infrastructure. However, LED bulbs suffer from several disadvantages, such as difficult heat extraction and difficult integration of additional *Smart Lighting* functionality. Sparse assembly of LED dies on a flat large-area substrate eliminates these disadvantages. Such large-area luminaires can also utilize flexible substrates such as PET-copper laminate, which can provide sufficient heat dissipation and interconnection reliability during flexing [2] for a flexible LED luminaire to be viable.

However, cost-effective fabrication of large LED panels is challenging because a large number of components have to be assembled. The sub-millimeter (mesoscopic) size of typical LEDs makes assembly even more challenging due to different scaling of the forces involved in the assembly process with the component size [3], as shown in Figure 1. In particular, higher surface-to-volume ratio for LED dies means that surface forces such as van der Waals force become stronger compared to gravity force as the die size gets smaller. As a result the dies may remain stuck to the head of a pick-and-place robot instead of remaining on the receiving substrate.

Directed self-assembly (DSA) presents an alternative way of assembling a large amount of small components sparsely on a large substrate [4, 5]. DSA avoids the limitations of sequential pick-and-place process by assembling the dies in a parallel fashion. It relies on a combination of at least two forces. A strong and short-range *bonding force* attracts the dies to a desired location on the substrate, while a weaker long-range *mixing force* moves the dies in the assembly medium until they get into the range of the bonding force. Gravity [6,7], capillary [8,9], mechanical [10,11] as well as electrostatic [12] and magnetic forces [13,14,15] are used for assembly of the mesoscopic objects. Chemical gradients and steric forces are also used in nanoscale and microscale regions [16, 17, 18]. Air, liquid or the interface between two different phases can be used as an assembly medium. Assembly in liquid medium can utilize weaker forces to move the dies as compared to the assembly on the air/solid interface.

Directed assembly of electronic components, such as LEDs, requires not only placing them on a correct spot on the receiving substrate, but also forming the permanent electrical and mechanical connections to the substrate. Using solder interconnects allows to achieve this goal, while providing an opportunity to utilize solder-assisted self-alignment to improve placement accuracy [9].

This paper describes a directed self-assembly approach based on shape-recognition and magnetic pick-up of dies in fluidic medium, followed by self-alignment of dies using the capillary force of the solder.

2. Experimental

Silicon test dies, with sizes representative of typical vertical LEDs, are used in this study. Additional die preparation steps are necessary before the dies can be used in this directed self-assembly process.

2.1 Die Preparation

Thin silicon dies with a thickness of 150 μm are used in this work. Several metallization layers are deposited on the bottom side of the dies in order to provide:

- a magnetic layer so that dies can be attracted by a strong rare-earth magnet;
- a solder-wettable layer.

A nickel layer (1.3 μm, from 99.999% pure source) is used as the magnetic layer. Copper is used as the solder-wettable layer. Ti is used to improve the adhesion between Si and Ni. All metal layers are deposited using e-beam evaporation.

Dies are made buoyant by depositing syntactic foam consisting of hollow glass microspheres (H20/1000 Glass

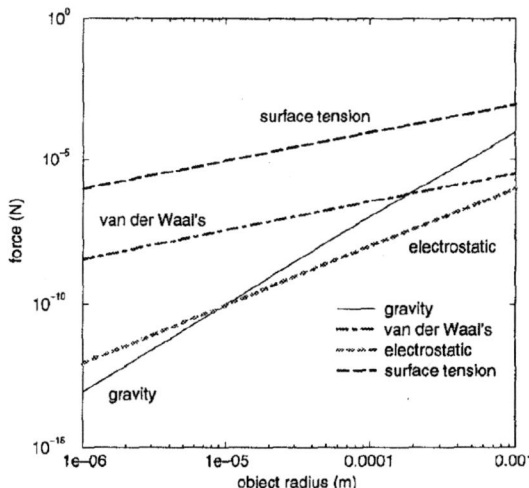

Figure 1. Scaling of gravitational, van der Waals, electrostatic and surface tension forces between a silicon sphere and a conductive plane [3].

Bubbles from 3M) in a 2-part epoxy matrix (Dow Epoxy Resin 332 + Triethylenetetramine). Benzocyclobutene (BCB, Cyclotene 3022-35) is used to improve adhesion between the syntactic foam and the silicon die. Both syntactic foam and BCB are deposited by spin-coating using the recipe shown in Figure 2. The process is repeated until the thickness of syntactic foam is sufficient to make the dies buoyant in the deionized water.

After the syntactic foam deposition, the samples are diced and the dies are chemically released from the dicing tape as described in [4]. The final die structure is shown in Table 1.

2.2 Setup for Directed Self-Assembly

The dies are assembled on 50x75 mm substrates. Each substrate contains a rectangular 7x5 array of pads with a 10 mm pitch. Substrates with the following thicknesses are used: 127 μm for polyimide film (Kapton 500 HPPST, from American Durafilm), 1 mm for microscope glass slide and 1.59 mm for FR4 PCB.

Dies are attracted to the substrate using permanent NdFeB magnets in one of the following configurations: an array of 1.59 mm cube magnets (grade N42), a single 25.4 mm cube magnet (grade N52), a single cylindrical magnet with a thickness of 25.4 mm and a diameter of 6.35 mm (grade N42, axially magnetized). Multiple magnets are arranged in an array configuration, while a single large magnet is moved below the substrate as shown in Figure 3:

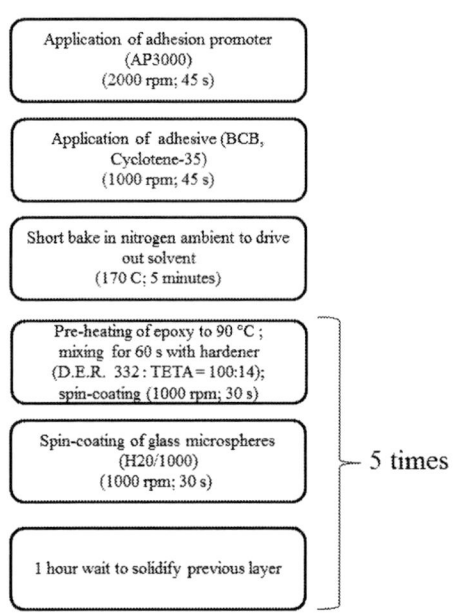

Figure 2. Recipe for spin-coating the syntactic foam buoyant layer.

Table 1. Structure of dies used used in DSA process.

Layer	Thickness	Function
Syntactic foam	~850 μm	Buoyancy
Silicon	150 μm	Substrate ("LED")
Ti	60 nm	Adhesion
Ni	1.3 μm	Magnetic
Cu	60 nm	Solder-wettable layer

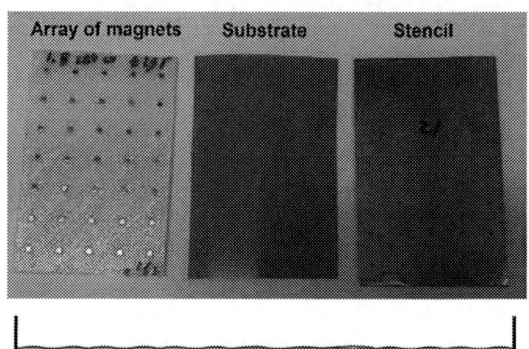

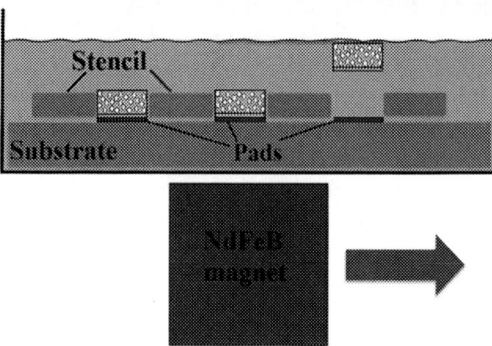

Figure 3. Magnetic die pick-up using: a) an array of small magnets b) a large movable magnet.

The substrates are covered with soldering flux to hold the dies in place after the assembly. Water-soluble flux (Superior No. 30, in gel form) and water-insoluble flux (rosin flux; composition: colophony, isopropanol) are used for this purpose.

The assembly setup with magnets in an array configuration utilizes vibration of the liquid to get the dies within the range of the magnets. The assembly setup with a single large movable magnet utilizes a motorized stage holding a container with the substrate, stencil, and dies in the carrier liquid. Both setups are shown in Figure 4.

2.3 Simulation of Solder-Assisted Self-Alignment

After the assembly, the die placement precision is improved by solder-assisted self-alignment. It utilizes the capillary force of the molten solder joint. The capillary force of the solder moves the die placed with some error until the solder joint shape reaches a geometry with minimal surface area, which corresponds to the situation when the die and the pad on the substrate are aligned. This geometry has the lowest total energy.

The solder-assisted self-alignment process is modelled using Surface Evolver software to estimate the magnitude of self-aligning forces acting on dies and their dependence on placement offsets and rotational misalignment.

The simulation is performed under the following conditions:

- Both the die pad and the substrate pad are of equal size, and completely wetted with the solder.
- The area surrounding the pad is not wettable by the solder.
- Initial form of the liquid solder is a parallelepiped with the top surface on the die pad and bottom surface on the substrate pad.

978-1-4799-2408-0/14 $31.00 © 2014 IEEE   2208

The following dependencies are modeled:

- Lateral alignment force as a function of 1D offset and solder volume;
- Solder joint height as a function of solder volume;
- Restoring torque as a function of rotational misalignment of the die.

3. Results and Discussion

3.1 Results of Directed Self-Assembly

Dies are fabricated using the recipe shown in Figure 2 and diced into 0.71x0.71 mm² and 1.61x1.61 mm² dies. Figure 5 shows the dies after the dicing and chemical release from the dicing tape.

The dies are assembled on a thin polyimide substrate using an array of 1.59 mm magnets. The range of the magnetic interaction between the dies and the permanent magnets is not

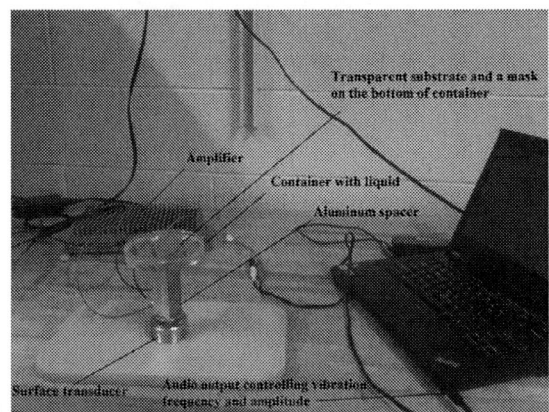

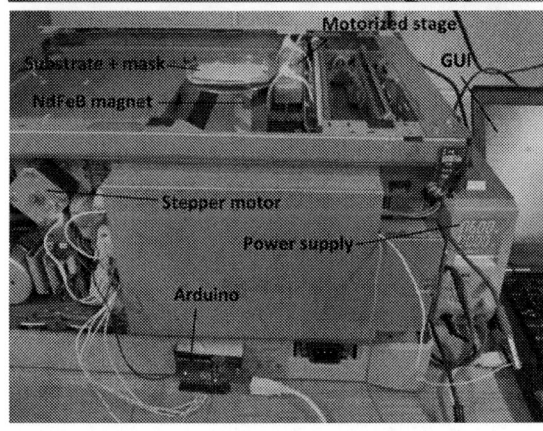

Figure 4. Setups for DSA with: a) magnets in an array configuration and a vibrating assembly medium; b) one movable magnet.

Table 2. Die parameters used in the simulation.

Parameter	Die 1	Die 2
Lateral size, mm	1.61	0.71
Die mass*, mg	2.235	0.387
H_COM**, mm	0.325	0.304

*Average value for 100 (for die 1) and for 300 (for die 2) dies (incl. syntactic foam layer).

**H_COM is the height of center of mass of the die above the bottom pad.

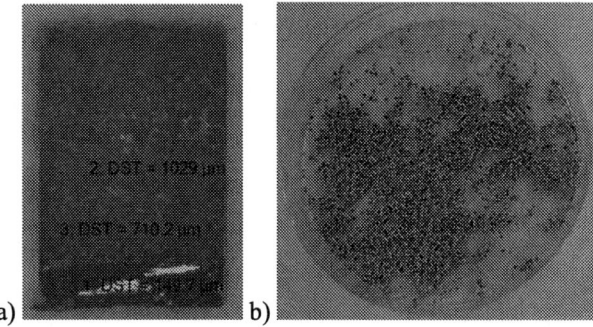

Figure 5. 0.71 mm dies with a syntactic foam layer on top.

sufficient for reliable die assembly on thicker substrates of 1 mm and 1.59 mm.

In order to assemble the dies onto the receiving substrate using the magnetic force, the following conditions need to be satisfied:

$$d_{mag} > t_{sub} + t_{stencil} + h_{liq} - t_{die}, \quad (1)$$

$$h_{liq} > t_{die}, \quad (2)$$

where d_{mag} is the range of the magnetic attraction between the die and the permanent magnet; t_{sub}, $t_{stencil}$, and t_{die} are the thicknesses of the substrate, stencil and the die (incl. syntactic foam), respectively; and h_{liq} is the height of the liquid above the top of the stencil.

If d_{mag} is too small, the dies do not respond to the field of the permanent magnets. Additionally, if h_{liq} is smaller than the die thickness, the dies may be stuck on the surface of the stencil even if they are within the range of the magnets.

For the simplest case (the die is located directly above the magnet), the field of a permanent block magnet along the axis of magnetization can be calculated as:

$$B_x = \frac{B_r}{\pi} \left(arctan\left(\frac{lw}{2x\sqrt{4x^2+l^2+w^2}} \right) - arctan\left(\frac{lw}{2(t+x)\sqrt{4(t+x)^2+l^2+w^2}} \right) \right), \quad (3)$$

where B_x is the field along the axis of magnetization; B_r is the residual flux density; l, w and t are the length, width and thickness of the magnet, respectively; and x is the distance from the surface of the magnet (along the axis of magnetization). Using this formula we can estimate the magnetic field along the axis of magnetization for cube ($l=w=t$) NdFeB magnets (grade N42, B_r = 13200 Gauss) with the size from 0.159 cm to 2.54 cm.

As it can be seen in Figure 6, small (millimeter-sized) magnets are of limited use for assembly on thick substrates (e.g. PCB) as their field drops down to several hundred Gauss over a distance of less than 2 mm. Considering t_{die} of 1 mm and a stencil of several hundred micrometers thick, substrate thickness (t_{sub}) should be very small to keep the magnetic layer on the dies within less than 2 mm from the surface of the magnets. The field of a large 25.4 mm cube magnet remains strong (several thousand Gauss) over the distance of more than 1 cm, making assembly possible on the substrates of any thickness. Its range also allows avoiding exposing the magnet

978-1-4799-2408-0/14 $31.00 © 2014 IEEE   2209

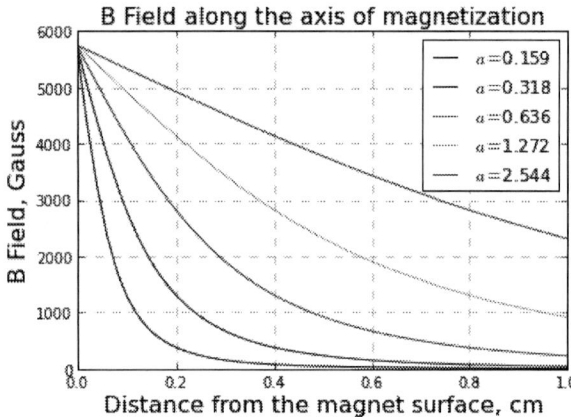

Figure 6. B-field as a function of the distance and the magnet size (a = 0.16 to 2.54 cm) for cuboid NdFeB magnets (grade N42) along the axis of magnetization.

Table 3. Comparison of magnet configurations.

	Array of small magnets	**Strong movable magnet**
Role of magnetic force	Attracting and holding the dies	Only attracting the dies
No. of masks	2	1
Die aspect ratio	AR>1 is preferable but not required	AR>1 is required
Range of magnetic interaction	Very Limited: Less than 2 mm	Long: up to 1 cm

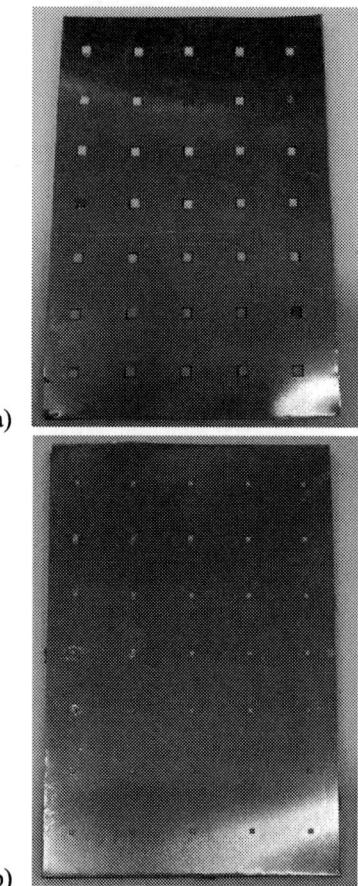

Figure 7. Results of die assembly with: a) AR<1 (1.61 mm dies, contains dies rotated by 90 degrees) and b) AR>1 (0.71 mm dies, error-free).

to the carrier liquid by placing it below the container where the DSA takes place. A brief comparison of both magnet configurations is shown in Table 3.

Assembly yield strongly depends on the geometry of the die. During DSA with an array of weak magnets, the top/bottom orientation of the die is preserved. The dies float with syntactic foam layer on top and a magnetic contact layer on bottom, and can be assembled on the substrate in the same orientation. When the dies are assembled with a single strong magnet, its field is strong enough to sink the dies and drag them along the surface of the stencil as the magnet (or the substrate) moves. The top/bottom orientation of the dies in this case may not be preserved. While assembly in "upside down" position (with dies rotated by 180°) was not observed, assembly with dies rotated by 90° is possible. This assembly defect is eliminated if the dies with assembly ratio (AR) > 1 are used. Here, AR is defined as the ratio between the thickness of the die with buoyant layer and its lateral dimension. AR = 1000 μm/700 μm = 1.43 for the 0.71 mm dies, such as the one shown in Figure 5. AR < 1 for the 1.6 mm dies with the same thickness.

For the dies with the AR > 1, thickness of the die is the largest dimension; the aperture size for the stencil can be selected in such a way that it is larger than the lateral dimension of the die, but smaller than the die thickness.

Examples of assembled dies with AR > 1 and < 1 are shown in Figure 7.

Laser-cut stainless steel (grade 304) stencils with a 380 μm thickness are used with the following aperture sizes:

- 770x770 μm, 840x840 μm, 910x910 μm for 0.71 mm dies;
- 1760x1760 μm, 1920x1920 μm, 2080x2080 μm dies for 1.61 mm dies.

If L is the lateral size of the die, then the stencil aperture size A is selected so that:

$$L < A < \sqrt{2} \cdot L.$$

In this case the dies are assembled with sub-45° misalignment angle.

As long as the AR > 1, all dies are assembled in a correct orientation with 100% yield (before forming permanent interconnects) and the size of the stencil apertures influences only the assembly speed and die assembly precision. Larger aperture size allows faster assembly, but lowers the assembly precision by increasing the maximal possible offset and rotational misalignment of the die.

Both Superior No. 30 and rosin fluxes provide sufficient tack to prevent the dies from leaving the substrate and returning to the surface of the liquid due to the buoyant force. Superior No. 30 gel flux slowly dissolves in water, which allows reducing the amount of flux after the DSA and before the solder reflow process.

3.2 Solder-Assisted Self-Alignment

In the simulations to determine the parameters affecting the solder-assisted self-alignment (as described in 2.3), the following forces are taken into account: the surface tension of the solder, the gravity force acting on the die, and the gravity force acting on the solder.

Figure 8 shows the dependence of equilibrium solder joint height on the solder volume for 0.71 mm dies. Figure 9 shows the dependence of the restoring torque on the misalgniment angle (yaw only, pitch and roll angles are set to zero) for a 0.71 mm die. Maximal total energy and zero restoring torque are observed at 45° misalignment. Maximal torque is observed between 10-20° and 70-80° misalignment, which are identical due to the 4-fold symmetry of the square dies. Zero restoring torque at 45° misalignment is undesirable, but irrelevant for this process as the dies assembled with a stencil do not allow such a large rotational misalignment.

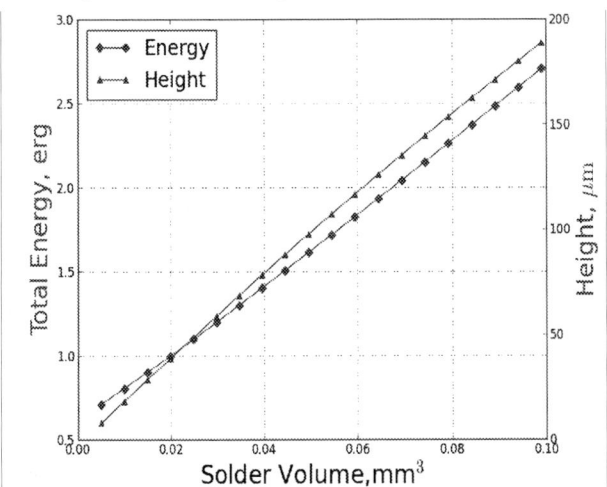

Figure 8. Total energy and the solder joint height as a function of the solder volume for 0.71 mm die aligned to the substrate (zero offset and misalignment angle).

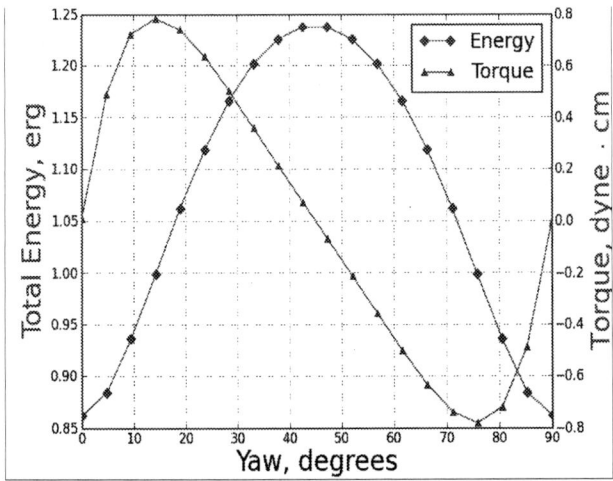

Figure 9. Total energy and the restoring torque as a function of the misalgnment angle (zero offset, V_{solder} = 0.0318 mm³).

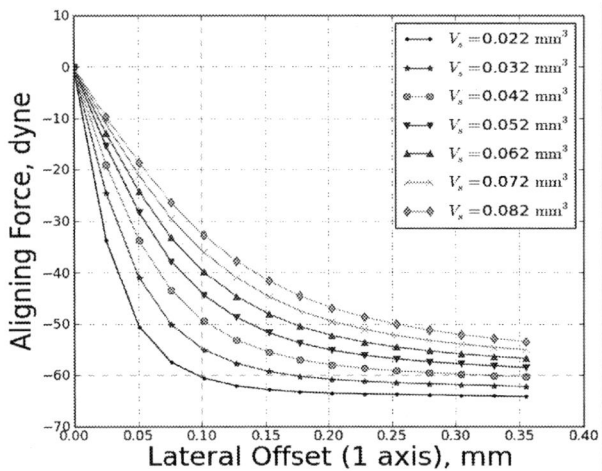

Figure 10. Lateral aligning force as a function of the die offset and solder volume for 0.71 mm die (yaw = 0°).

Dependence of the lateral aligning force on the die offset and solder volume is depicted in Figure 10. Typical force acting on a die in this scenario is on the order of several tens of dynes (1 dyne = 10^{-5} N), while the gravity force acting on a 0.71 mm die is only 0.380 dyne.

Die self-alignment is empirically tested with a SAC305 solder. A solder ribbon with a thickness of 50 μm (96.5Sn 3.0Ag 0.5Cu, from Indium Corp.) is diced into 1.61x1.61 mm and 0.71x0.71 mm preforms. The substrate is covered with liquid rosin flux before the solder reflow step. Both rotational and lateral self-aligning are observed. Final die offset and rotational misalignment are limited by the imperfect wettability of the die contact pads. Dies with the well-wetted pads are aligned to the substrate with less than 3 degree misalignment, which is also affected by the difference between the die pad size and the size of the pads on the substrate. The variation of pad size from substrate to substrate is observed to be close to 25 μm. A typical result of the substrate with the dies assembled is shown in Figure 11.

4. Summary and Conclusions

A directed self-assembly approach is demonstrated, which is suitable for assembly of mesoscopic dies in a parallel fashion. The magnetic force and shape-recognition are used

Figure 11. A substrate with 35 assembled dies (with syntactic layer still on top).

for attracting the dies to the receiving substrate and placing them in a correct orientation with a relatively low precision. The solder-assisted self-alignment is used to improve the precision of die placement after the assembly. Directed assembly in liquid medium reduces the forces necessary to move the dies as compared to the assembly on the air/solid interface. High aspect ratio of the die dimensions is required for achieving high assembly yield. The buoyant force returns the dies to their initial position (surface of the liquid) if the dies do not fill the stencil apertures, so they are ready to be assembled on the next substrate. This approach can be scaled up for assembly of LED dies on large substrates and, if a flexible stencil is used, a roll-to-roll implementation is also possible.

Acknowledgments

This work was supported primarily by the Engineering Research Centers Program (ERC) of the National Science Foundation under NSF Cooperative Agreement No. EEC-0812056 and in part by New York State under NYSTAR contract C090145.

References

1. T. Taguchi, "Present status of energy saving Technologies and future prospect in white LED lighting", *IEEJ Transactions on Electrical and Electronic Engineering*, Vol. 3, Issue 1, 2008, pp. 21-26.

2. D.A. van den Ende, R.H.L. Kusters, M. Cauwe, A. van der Waal, J. van den Brand, "Large area flexible lighting foils using distributed bare LED dies on polyester substrates", *Microelectronics Reliability*, Vol. 53, Issue 12, 2013, pp. 1907-1915.

3. S. Fearing, "Survey of sticking effects for micro parts handling", Proceedings of 1995 IEEE/RSJ International Conference on Intelligent Robots and Systems, Human Robot Interaction and Cooperative Robots, Vol. 2, 1995, pp. 212-217.

4. A. Tkachenko, R. F. Karlicek, Jr., and J. J.-Q. Lu "Evaluation of directed self-assembly process for LED assembly on flexible substrates", 63rd IEEE Electronic Components and Technology Conference, Las Vegas, NV, May 2013.

5. C. J. Morris, "Self-Assembly for microscale and nanoscale packaging: steps toward self-packaging", *IEEE Transactions on Advanced Packaging*, Vol. 28, Issue 4, 2005, pp. 600-611.

6. G. S. W. Craig, E. J. Snyder, and J. K.-J. Tu, "Apparatus relating to block configurations and fluidic self-assembly processes", U.S. Patent No. 6,657,289. 2 Dec. 2003.

7. J. Jacobsen et al, "Plastic-film displays with NanoBlock IC drivers integrated by a fluidic self-assembly process", *Journal of the Society for Information Display*, Vol. 11, Issue 1, 2002, pp. 726-729.

8. P.A. Kralchevsky and K. Nagayama, "Capillary interactions between particles bound to interfaces, liquid films and biomembranes", *Advances in Colloid and Interface Science*, Vol. 85, 2000, pp. 145-192.

9. W. Zheng, P. Buhlmann, and H. O. Jacobs, "Sequential shape-and-solder-directed self-assembly of functional microsystems", *Proceedings of National Academy of Science (PNAS)*, Vol. 101, No. 35, 2004, pp. 12814-12817.

10. L.S. Penrose and R. Penrose "A self-reproducing Analogue", *Nature* 179, 1183, 08 June 1957.

11. S. A. Stauth and B. A. Parviz, "Self-assembled single-crystal silicon circuits on plastic", *Proceedings of National Academy of Science (PNAS)*, Vol. 103 no. 38, 2006, pp. 13922–13927.

12. J. Tien, A. Terfort, and G. M. Whitesides, "Microfabrication through electrostatic self-assembly", *Langmuir*, Vol. 13, No. 20, 1997, pp. 5349-5355.

13. J. D. Lohn, G. L. Haith, and S. P. Colombano, "Two electromechanical self-assembling systems", 6th Foresight Conference on Molecular Nanotechnology, 1998.

14. S. Shet, V. R. Mehta, A. T. Fiory, M. P. Lepselter, and N.M. Ravindra, "The magnetic field-assisted assembly of nanoscale semiconductor devices: a new technique", *JOM*, Volume 56, Issue 10, October 2004, pp 32-34.

15. Q. Ramadan, Y. S. Uk, and K. Vaidyanathan, "Large scale microcomponents assembly using an external magnetic array", *Applied Physics Letters*, Vol. 90, Issue 17, 2007, pp. 172502 - 172502-3.

16. M.-P. Valignat, O. Theodoly, J. C. Crocker, W. B. Russel, and P. M. Chaikin, "Reversible self-assembly and directed assembly of DNA-linked micrometer-sized colloids", Proceedings of National Academy of Science (PNAS), Vol. 102, No. 12, 2005, pp. 4225–4229.

17. I. Ziemecka, G. J. M. Koper, A. G. L. Olivea and J. H. van Esch "Chemical-gradient directed self-assembly of hydrogel fibers", *Soft Matter*, Vol. 9, Issue 5, 2013, pp. 1556-1561.

18. Y. Ke, L. L. Ong, W. M. Shih, P. Yin, "Three-dimensional structures self-assembled from DNA bricks", *Science*, Vol. 338, No. 6111, 2012, pp. 1177-1183.

Controlled Silicon IC Thinning on Individual Die Level for Active Implant Integration Using a Purely Mechanical Process

Vasiliki Giagka[1], Nooshin Saeidi[2], Andreas Demosthenous[1] and Nick Donaldson[2]

[1]Dept. of Electronic and Electrical Engineering, University College London, UK
[2]Dept. of Medical Physics and Bioengineering, University College London, UK
{vasiliki.giagka.10; n.saeidi}@ucl.ac.uk

Abstract

We are developing an electrode array for epidural spinal cord stimulation and a thin integrated circuit (IC) is to be embedded in it. This paper focuses on the development and characterization of a manual process for thinning individual IC die and discusses the issues associated with thinning small dice by a manual process. The procedure allows easy and controlled post-separation thinning of small (about 1 mm^2) silicon chips by grinding. A systematic approach was followed to characterize the technique and repeatability of the results. With the setup we introduced we were able to control the final thickness of the IC with a standard deviation of 9.2 μm. Although no chemical processing is used, a small grit size film can create smooth surfaces, with roughness comparable to reported values after etching, acting as the so-called "stress-relief" step. Electrical tests performed on a thinned stimulator output stage IC indicated that no die damage was caused by the procedure. Some issues regarding the integration of thinned ICs on flexible substrates and the reliability of gold ball rivet bonds on the ICs' aluminium pads are also discussed.

Introduction

One of the main motivations for this work is the integration of ICs on a small and thin implant. We are developing an electrode array for chronic epidural spinal cord stimulation [1]. We are using laser-cut platinum foil for the conductive part of the array and silicone rubber for insulation. An IC is to be embedded in it [2]. Due to volume restrictions related to the implantation site, the total thickness of the implant needs to be no more than 300 μm. On this basis, a conventional 500 μm thick IC would not be suitable for this application.

The total thickness of the active layers of an IC may comprise a total thickness of 5 – 10 μm, depending on the complexity of the design. Nevertheless, the typical thickness of fabricated chips is in the order of 500 μm – 1 mm. This extra material, which mainly serves as mechanical support, can be removed after processing to create thinner devices. Chips with functional structures thinner than 25 μm have been reported [3].

Several techniques for silicon back-thinning have been developed from a combination of different mechanical and chemical processes. Most commonly, grinding, lapping or polishing are used for the removal of the major part of the material, usually followed by a stress-relief step, where dry etching, wet etching or laser chemical etching are employed for the removal of the last 10 – 100 μm of silicon, to reduce the backside damage caused by the previous coarse step and ensure a smooth surface finish ([4] – [6]).

In nearly all the efforts described above, the problem of silicon thinning is dealt at a wafer level. This approach has several advantages when mass production is required. It allows the processing of all the ICs on a wafer simultaneously, saving huge amounts of time and effort, and consequently, money. On the other hand, there are cases where this approach is not suitable. This includes multi-project wafers where different devices share the same wafer but only some of them need thinning, or applications where devices of different thicknesses are required. Also, in some cases, existing ICs might need to be thinned after they have been separated from the wafer and post-processed. In these cases, a procedure that deals with the thinning of individual die is necessary—and could also be cheaper.

Although thinning of individual die is not a new concept there does not seem to be an established procedure based on quantitative results. In most of the reported cases, the thinning procedure has been treated as the means towards another goal, therefore, it has not been reported in detail together with the difficulties one faces along the way. The small dimensions of individual dice, pose some quite different challenges compared to wafer level thinning. Specifically, the mounting and handling requirements of individual dice are radically different compared to wafer level thinning and accurately predetermining the final thickness is not straightforward. Some of these issues are discussed in [7], where mechanical thinning of Gallium Arsenide dice is demonstrated.

In wafer level processes, several methods have been developed to control the final thickness. In [4], the dicing-by-thinning concept was introduced, where the depth of dicing grooves, which have been dry-etched at the front side of the wafer, defines the final silicon thickness, after the chips have been separated. In [6], the authors tried to control the thinning locally, using the signal obtained from an optical beam induced current, which varies depending on the remaining silicon.

In this work we aimed to establish and characterize a procedure that allows easy and controlled post-separation thinning of individual dice featuring areas as small as about 1 mm^2. The procedure is based on mechanical thinning only. No extra automated equipment is used other than the main device, a commercial grinder-polisher. Quantitative results of the thickness and surface roughness of the ICs were obtained and a systematic approach was followed in order to achieve repeatable results with this purely manual method. In this paper, we present these results while discussing the issues one faces when small devices are thinned without access to more elaborate equipment.

The next section of the paper describes the materials and methods used in this work and results are presented and discussed in the section after that. A series of experiments were run to test the repeatability of the procedure and, by

---

This work is part of the European Research project NEUWalk, funded by the European Community's Seventh Framework Programme [FP7/2007-2013] under grant agreement no. 258654.

Figure 1. The polishing device. A glass platen is used to ensure a flat surface. Diamond lapping films are held on the glass platen by water tension. Samples are mounted on a tri-point polisher kit, which is manually held against the abrasive surface of the film. Polishing is performed while the glass platen rotates at a constant speed, set by the user. Water is used to remove residues throughout the process.

introducing our own custom-made setup we have managed to achieve results with a standard deviation of 9.2 μm. Surface roughness values of 1.75 nm were achieved. ICs were found electrically functional after thinning, verifying that no die damage was caused by the procedure. Some issues regarding the integration of thin dice on flexible substrates and the reliability of the gold ball rivet bonds on the IC's aluminium pads are discussed in a separate section. A summary is given in the conclusions section followed by acknowledgements.

Materials and Methods

Integrated circuits (1324 μm x 1324 μm) were laser-cut from a wafer which was back-lapped by the foundry to an initial thickness of roughly 500 μm.

High precision polyvinylidene fluoride carbon fibre reinforced tweezers (Ideal-Tek SA, Balerna, Switzerland) were used for handling in order to prevent damaging the ICs. The material can be used at a constant temperature of up to 150 °C, it is electrostatic discharge (ESD) safe and suitable to use with very scratch-sensitive components.

Each IC to be thinned is attached on the setup using Crystalbond 509 (SPI Supplies / Structure Probe, Inc). Crystalbond 509 is a reversible mounting adhesive that melts at 121 °C, with a viscosity at flow point of 6000 cps, and is soluble to acetone. A nozzle head in combination with a vacuum system were transformed to a custom manual pick and place setup, to handle the thinned dice during bonding and de-bonding.

A commercial grinder-polisher is used for the thinning (EcoMet 250, Buehler, Düsseldorf, Germany) (Fig. 1). The device accepts diamond lapping films (diamond particles resin bonded to polyester film) (8 inch, UltraPrep, Buehler, Düsseldorf, Germany), that come in eight different abrasive sizes for coarse to fine grinding. For this work, films with a grit size of 6 μm, 3 μm and 1 μm were used.

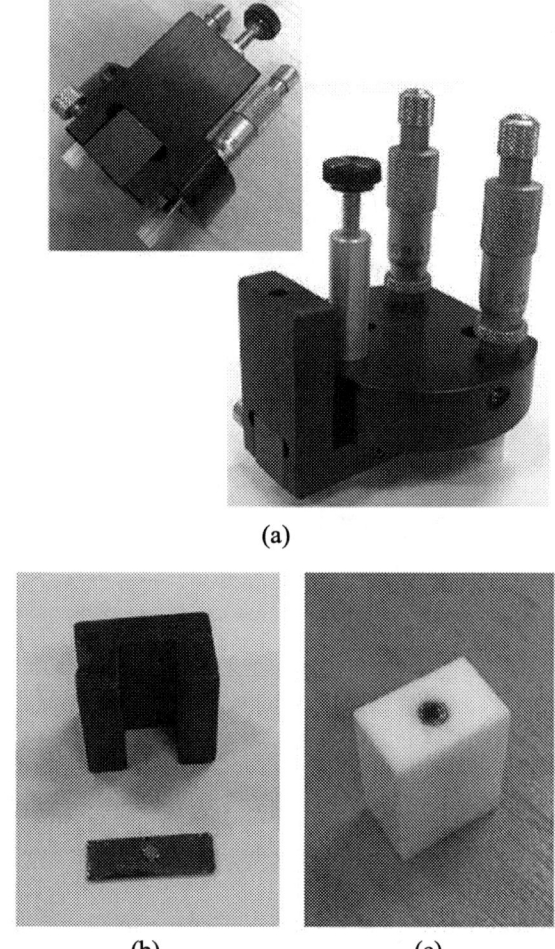

(a)

(b)   (c)

Figure 2. Tri-point polisher kit (holder) (a), mount that comes with the polisher (b), and custom made PTFE mount (c). The mount is inserted in the holder with the backside of the IC facing down for polishing. The two adjustable spacers on the holder are used to keep the sample flat against the surface of the lapping film. The small ICs are visible in both (b) and (c).

These abrasive films are held in place on the polisher by water tension. After the lapping film is firmly attached on the glass surface of the polisher, it is carefully cleaned with a wet tissue and water is used to keep the film wet throughout the procedure. This helps remove residues created during the process and leads to a more uniform result. The sample is attached on a mount, which, in turn, is securely fixed on a tri-point polisher kit (holder). The holder (Fig. 2(a)), has two adjustable spacers, adjusted to keep the sample flat against the surface of the lapping film whatever its thickness. These spacers are manually adjusted with a resolution of 50 μm. Samples are mounted on the holder, which is held by hand against the abrasive surface of the film. Polishing is performed while the glass platen rotates at a constant speed, automatically set by the user.

We performed experiments with two different setups. In the first one, the blue mount, supplied by the polisher's

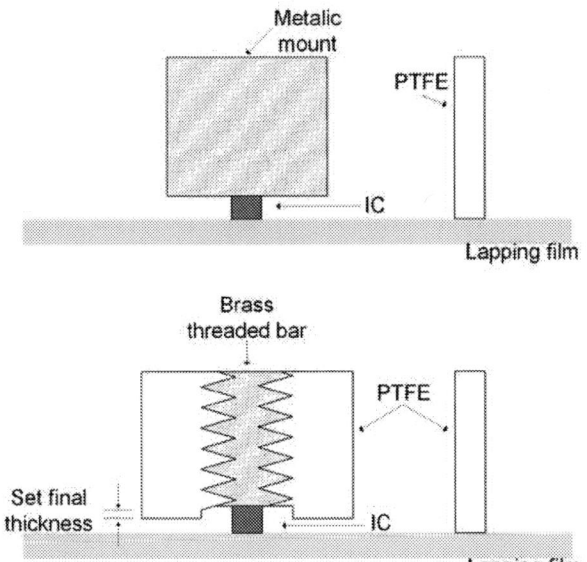

Figure 3. The kinematics of the thinning in the first (top) and second (bottom) setup. By adjusting the z-position of the brass threaded bar in the second setup, the final thickness of the IC can be better controlled.

manufacturer (illustrated in Fig. 2(b)), is used. The IC is mounted on a silicon substrate (Fig. 2(b)), which serves as a carrier and is, in turn, attached on the mount.

Attempts to control the final thickness of the IC using this setup presented a high variability (results are presented in the next section). To address this we used our own custom made mount of Fig. 2(c). This is made of polytetrafluroethylene (PTFE), else known as Teflon, with a brass threaded bar in its center where the chip is directly mounted. The screw has a pitch of 200 μm and allows adjustment of its z-position to control the amount of silicon that protrudes outside the cavity and will be removed. This can be done under a microscope during sample preparation. When the silicon surface has reached the same level as the mount, thinning slows down, as PTFE is also polished but at a considerably slower rate. Nevertheless, so far, after our experiments, no signs of worn PTFE are yet visible.

A comparison of the kinematics related to the first and second setup is graphically illustrated in Fig. 3.

Results and Discussion

A. Thinning Rates

In the first set of tests, to evaluate the thinning rate of each lapping film 8 – 10 samples were polished with each film at a constant speed (30 revolutions per minute (rpm)) using the setup of Fig. 2(b). The thickness of the ICs was measured before and after the polishing.

Fig. 4 shows the variability of the measured thinning rates for the all the thinned samples. It should be noted that, although placed in a common graph, these samples are not the same for each film, so direct comparison among them would not be sensible. As illustrated here, finer films exhibit lower thinning rates and a much smaller variation. The extreme low value observed in case of sample 6_8 (sample 8 polished with

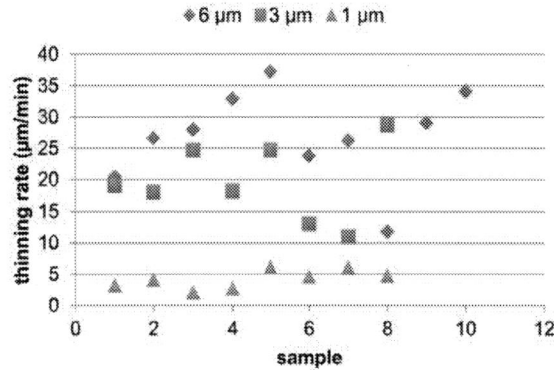

Figure 4. Measured thinning rates for lapping films of 6, 3 and 1 μm grit. Coarser films exhibit a larger variation.

the 6 μm-film) indicates that the abrasive ability of the lapping film has deteriorated a lot, so the film is close to its lifetime limit. For samples 6_9 and 6_10, a brand new, previously unused film was used, and the measured value was inside the range we had already defined from samples polished using the previous film. Unfortunately, it has not been possible to accurately measure the amount of silicon that can be removed with each lapping film before there is a noticeable change in its abrasive rate, but it is estimated that coarser films could be used for more than 4500 revolutions. The average thinning rates for each lapping film are summarized in Table I (together with the lowest thinning rates we recorded for each film, used later for calculations in Fig. 5).

TABLE I.   AVERAGE THINNING RATES

Lapping film	Average thinning rate (μm/min)	Lowest thinning rate (used in Fig. 5) (μm/min)
6 μm	27	11.8
3 μm	19.6	5.4
1 μm	4.2	2.1

B. Control of the final thickness

After the polishing rates for each film were determined, more samples were prepared using the same setup (Fig. 2(b)) and polished using a predetermined set of parameters (3 consecutive steps with descending grit size films for a predetermined time and speed) in order to verify the results in practice.

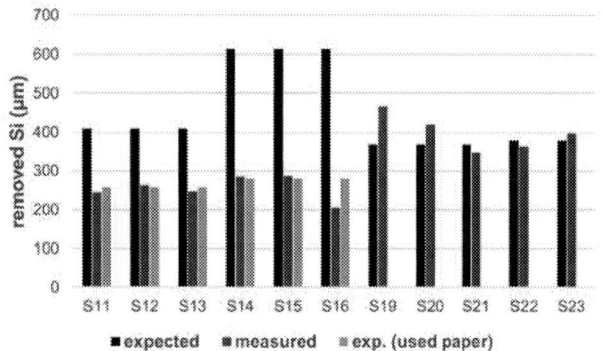

Figure 5. Comparison between measured (red) and expected amount of removed silicon as calculated from the average thinning rates for each lapping film (black) (using the first setup). Samples 11 – 16 were polished on old, used films. The third, grey, column that has been included for these samples is an estimation of the expected amount of removed silicon when the lowest recorded thinning rates for each film are used in the calculations. From sample 19 onwards a new, unused set of films was introduced.

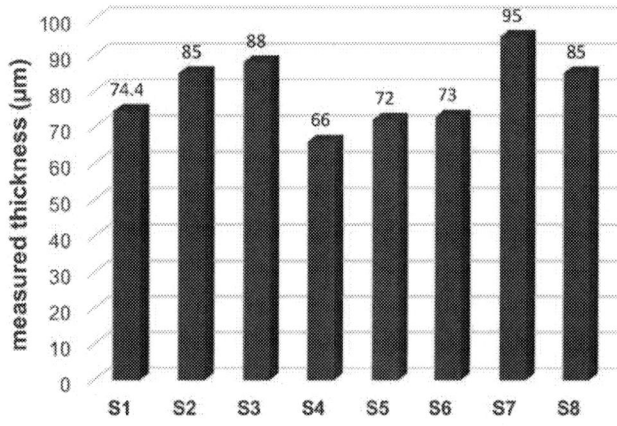

Figure 6. Measured thickness of each of the 8 samples thinned using our own custom-made mount of Fig. 2(c). Results have a mean value of 79.8 μm, with a standard deviation of 9.2 μm.

We calculated the anticipated amount of removed material at the end of the processing based on the selected parameters. Fig. 5 shows how the measured results deviate from the expected ones. For each sample, the first (black) column indicates the expected amount of removed silicon, as calculated using the average removal rate for each film times the polishing time. The second (red) column illustrates the measured results. Samples 11 – 16 were polished on the old, used films, therefore, the difference between the expected and measured results is quite large. Actually, when the anticipated amount of removed silicon is calculated using the lowest recorded thinning rates for each film, (Fig. 5, third (grey) column), the results match. The lowest recorded thinning rates for each film used in these calculations are those in the

third column of Table I. This is an indication that the film has reached its lifetime limit and needs to be replaced. A new set of films was introduced, starting from sample 19.

During this manual process, several factors, such as the force that the user applies on the holder and the amount of water used during the procedure, cannot, unfortunately, be accurately controlled and, thus, affect the repeatability of the results. Fig. 5 suggests that the final thickness can be predicted with an accuracy in the range of about 30 – 85 μm. This variability is quite high, when targeted thicknesses could be in the range of 70 – 100 μm.

We therefore performed a new set of experiments using our own, custom made setup of Fig. 2(c), described in the previous section.

A set of 8 samples was thinned using this setup to a targeted thickness of 80 μm. Again, a sequence of 3 steps was used (6 μm film for 10 minutes, 3 μm film for 5 minutes, and 1 μm film for > 7.5 minutes until the end of the procedure, all at 30 rpm). Measured results of the final thickness of each sample are graphically illustrated in Fig. 6. The mean thickness was 79.8 μm, with a standard deviation of 9.2 μm. An example of an individually thinned IC after it was de-bonded from its substrate is illustrated in Fig. 7.

As the results suggest, using this second setup we have managed to better control the final thickness of the chips. This method though, does not allow precise control over the amount of material that is going to be thinned by each individual step, unless the z-position of the threaded bar is readjusted under the microscope between steps.

C. Total Thickness Variation

One of the main challenges of a manual thinning process is the limited control over the total thickness variation that appears across each thinned IC. The two adjustable spacers of the holder in Fig. 2(a) that need to be manually adjusted to keep the sample flat against the surface of the lapping film, introduce an uncertainty to the procedure. Gradients that appeared across the chips mainly depended on aligning the 3 points of contact of the holder at the beginning, or in between the procedure steps. This problem was more prominent in the first setup, mainly due to the third point of contact being the very small IC (1 mm^2). Visual inspection was not sufficient to achieve a total thickness variation in the range of less than 30 μm. Fig. 8(a) and 8(b) show an example of the profile of an IC measured before and after the thinning using the first

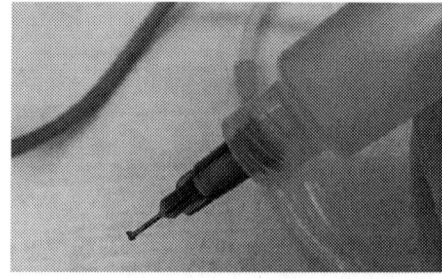

Figure 7. A 1.3 mm^2 IC, individually thinned down to 80 μm with our setup.

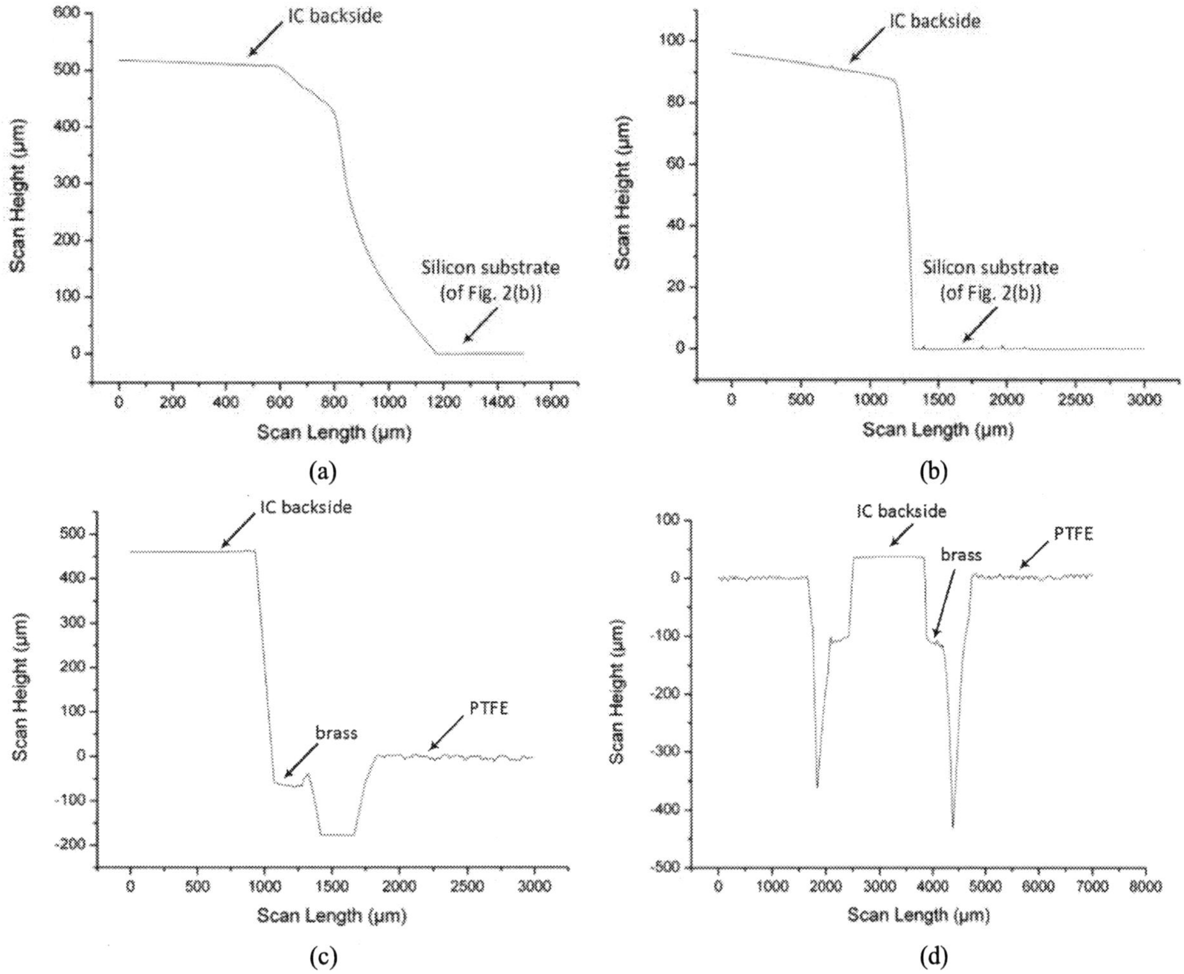

Figure 8. Recorded surface profiles; chip on substrate before (a) and after (b) thinning using the first setup; and before (c) and after (d) thinning using our custom made setup.

setup. The profile in this case is between the IC and the substrate and was recorded using a stylus profilometer (DekTak XT).

In the second setup we have tackled this problem by using the PTFE mount as the third point of contact, which, due to its larger area, made the alignment easier. Fig. 8(c) and 8(d) are profile recordings of the ICs on the PTFE mount before and after the thinning and illustrate that a flatter profile can be achieved.

Nevertheless, even in this case, the measured average gradients, after the chips were de-bonded, were roughly 10 μm. That is mainly due to the uncertainty that the Crystalbond layer adds to the procedure; although the Crystalbond is liquid at high temperatures, its viscosity, when used in undiluted form, is higher than one would want for this application. In fact, when the material melts and the IC is placed on top of it, even if force is applied to push the IC against the silicon substrate during bonding, the thickness of the adhesive layer can vary from one edge of the IC to the other. This can lead to uneven polishing. Similar issues have been reported in [7]. We have found that Crystalbond diluted in acetone can provide sufficient retention levels for this application, while

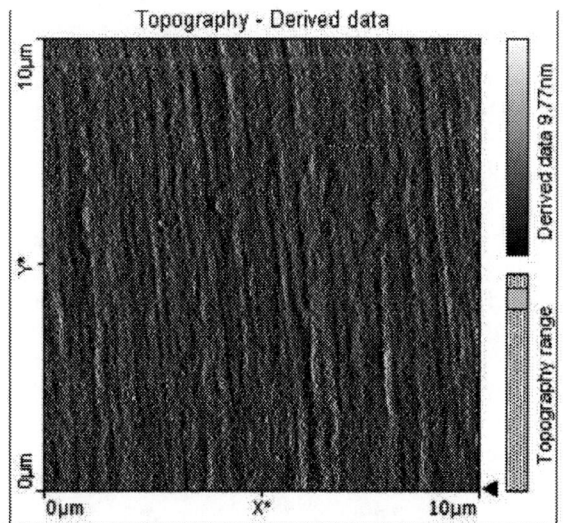

Figure 9. AFM image of the surface topology of a 10 μm x 10 μm area of a typical sample, polished with a sequence of 6, 3, 1 and 0.1 μm lapping films. (Surface roughness: 1.75 nm).

978-1-4799-2408-0/14 $31.00 © 2014 IEEE

being much less viscous. Another potential solution would be the use of a uniform thickness double-sided adhesive film, but preliminary experiments held by the authors using the 1.5 mil thick X4 retention level DGL film from Gel-Pak, USA, indicated that a higher retention level would be necessary as the adhesion of the gel to the substrate was not enough for this application.

D. Surface Roughness

Typical measured surface roughness values after treatment with different polishing foils were in the order of a hundred nm down to below ten nm, for films of 6, 3 and 1 μm. Using finer films even smoother surfaces can be achieved. Fig. 9 shows an atomic force microscopy (AFM) image of the surface topology of a sample polished with a sequence of 6, 3, 1 and 0.1 μm lapping films. It reveals a surface roughness of 1.75 nm.

Although mechanical tests (ball-ring, 3-point and 4-point bending tests) are needed for proper characterization of mechanical stability of the ICs, it has been extensively reported in the literature that there is a strong correlation between the surface topology of an IC and its mechanical strength, with smoother surfaces presenting greater strength ([3], [4] and [8] – [10]). These results indicate that purely mechanical polishing with a small grit size film can act as a "stress-relief" step, as very smooth surfaces can be achieved. This suggestion eliminates the need for potentially slow and expensive etching processes, and all the difficulties that these include in terms of handling and protecting the active area, when thinning is done on IC level.

E. Electrical Testing

To verify that no damage is caused by the procedure, a stimulator output stage IC [2], thinned down to 100 μm, was found fully operational in electrical tests.

Integration on a flexible substrate

Very smooth surfaces, as these produced after polishing, can be more difficult to handle. When thinned ICs are thinned for implant integration on a flexible substrate, the targeted surface roughness should be a trade-off between the desired mechanical stability and the adhesion of the polished dice to the substrate.

Regarding the electrical interconnections, Beutel at al. [11] have developed a procedure to interconnect integrated circuits onto flexible implants. In the Microflex technique, conductive tracks are thermosonically bonded on a substrate using gold ball studs through via holes, as micro-rivets. Mechanical tests that recorded the maximum stress before failure for interconnections between thick platinum foil structures and printed gold tracks on an alumina substrate have been investigated in [12]. Similar tests we run on thin chips indicated that they can be reliably bonded with this technique, provided that the dice are properly clamped and supported during the procedure. Using a 12.5 μm thick stainless steel foil, we have performed a series of bonding tests using different gold ball and hole sizes on standard aluminium pads and recorded the maximum stress the joints could be subjected to before failure occurred. Our joints were double gold ball studs, rather than single studs, vertically aligned one on top of the other. Our results indicated that the average maximum strength of the bonds depends on the relative size of the ball stud to the hole. We have recorded average values (of 10 bonds) as low as 9.6 cN, and, as high as 60 cN, for different parameter combinations. Larger holes, provide larger contact areas with the substrate and generally result in stronger bonds, but the right combination of ball and hole sizes, could lead to strong bonds even with smaller holes.

Conclusions

We have developed and characterized a manual method for controlled thinning of small dice. We have addressed the problem of accurate control of the final die thickness by introducing a custom made setup, tailored to the specific needs of small chips. We were able to control the final thickness with a standard deviation of 9.2 μm, and achieve surface roughness values of 1.75 nm, all with a purely mechanical and manual process. Compared to the standard setup, we have improved the total thickness variation across chips after the thinning, but some gradients are still present, mostly due to the liquid adhesive used throughout the procedure. Electrical tests have verified the post-thinning operation of real ICs. Finally, some issues regarding the integration of thin ICs on flexible implants and the reliability of the gold ball rivet bonds on the ICs' aluminium pads have also been discussed.

Acknowledgments

The authors would like to thank Dr. Oleg Mitrofanov for allowing us access to his equipment and Mr. Joe Evans for making the custom made mount used in this work.

References

1. V. Giagka, A. Vanhoestenberghe, N. Wenger, P. Musienko, N. Donaldson, and A. Demosthenous, "Flexible platinum electrode arrays for epidural spinal cord stimulation in paralyzed rats: An *in vivo* and *in vitro* evaluation," in *Proc. 3rd Annual Conf. IFESSUKI 2012*, Birmigham, UK, Apr. 2012, pp. 52-53.
2. V. Giagka, C. Eder, V. Valente, A. Vanhoestenberghe, N. Donaldson, and A. Demosthenous, "A dedicated electrode driving ASIC for epidural spinal cord stimulation in rats," in *Proc. ICECS 2013*, Abu Dhabi, UAE, Dec. 2013.
3. J. Burghartz, W. Appel, C. Harendt, H. Rempp and H. Richter, 'Ultra-thin chip technology and applicattions, a new paradigm in silicon technology,' *Solid State Electron.*, vol. 54, pp. 818-829, 2010.
4. G. Hawkins, H. Berg, M. Mahalingam, G. Lewis and L. Lofgran, 'Measurements of silicon strength as affected by wafer back processing,' in *Proc. IEEE International Reliability Physics Symposium (IRPS)*, San Diego, CA, Apr. 7-9, 1987, pp. 216-223.
5. M. Feil, C. Adler, G. Klink, M. König, C. Landesberger, S. Scherbaum, G. Scwinn and H. Spöhrle, 'Ultra thin ICs and MEMS elements: techniques for wafer thinning, stress-free separation, assembly and interconnection,' *Microsyst. Technol.*, vol. 9, pp. 176-182, 2003.
6. R. Goruganthu, M. Bruce, J. Birdsley, V. Bruce, G. Gilfeather and R. Ring, 'Controlled Silicon Thinning for Design Debug of C4 Packaged ICs,' in *Proc. IEEE International Reliability Physics Symposium (IRPS)*, San Diego, CA, Mar. 23-25, 1999, pp. 327-332.

7. E. Bosman, J. Missinne, B. Van Hoe, G. Van Steenberge, S. Kalathimekkad, J. Van Erps, I. Milenkov, K. Panajotov, T. Van Gijseghem, P. Dubruel, H. Thienpont and P. Van Daele, 'Ultrathin Optoelectronic Device Packaging in Flexible Carriers,' *IEEE J. Sel. Top. Quant.*, vol. 17, no. 3, pp. 617-628, May-Jun. 2011.

8. G. Omar, N. Tamaldin, M. R. Muhamad and T. C. Hock, 'Correlation of Silicon Wafer Strength to the Surface Morphology,' in *Proc. IEEE International Conference on Semiconductor Electronics (ICSE)*, Guoman Port Dickson Resort, Malaysia, Nov. 13-15, 2000, pp. 147-151.

9. W. Kröninger and F. Mariani, 'Thinning and Singulation of Silicon: Root Causes of the Damage in Thin Chips,' in *Proc. IEEE Electronic Components and Technol. Conf. (ECTC)*, San Diego, CA, May 30-Jun. 2, 2006, pp. 1317-1322.

10. S. Schönfelder, J. Bagdahn, M. Ebert, M. Petzold, K. Bock and C. Landesberger, 'Investigations of Strength Properties of Ultra-Thin Silicon,' in *Proc. Intern. Conf. Thermal, Mechanical and Multi-Physics Simulation and Experiments in Micro-Electronics and Micro-Systems (EuroSimE)*, Apr. 18-20, 2005, pp. 105-111.

11. H. Beutel, T. Stieglitz, and J. U. Meyer, "Versatile 'microflex'-based interconnection technique," in *Proc. SPIE 3328, Smart Structures and Materials 1998: Smart Electronics and MEMS*, San Diego, CA, Mar. 1998, pp. 174 – 182.

12. M. Schuettler, C. Henle, J. S Ordonez, W. Meier, T. Guenther, and T. Stieglitz, "Interconnection technologies for laser-patterned electrode arrays," in *Proc. IEEE EMBC*, Vancouver, Canada, Aug. 2008, pp. 3212 – 3215.

Gap in pagination due to withheld paper.

Pages 2220-2226

Study of Extreme Low Temperature and Load
Solid-Phase Sn-Ag System Bonding Mechanism for 3D ICs

Kiyoto Yoneta[*1], Ryohei Sato[*1], Yoshiharu Iwata[*1], Koichiro Atsumi[*2], Kazuya Okamoto[*2], Yukihiro Satio
and Takumi Shigemoto[*1]
[*1]Osaka University, [*2]Osaka Univ. Office for University-Industry
2-1 Yamadaoka, Suita Osaka, 565-0871
Tel.: +81-6-6879-4191, Contact email: yonetakiyoto@mapse.eng.osaka-u.ac.jp, satohr@mapse.eng.osaka-u.ac.jp

Abstract

The objective of this study is i) to optimize a new nano solid phase, Ag-Sn thin-film bonding system for wafer-level 3D-stacking for 3D ICs, and ii) to clarify its bonding mechanism.

As reported in our previous study, we achieved bonding at a much lower temperature (180°C), with lower load (0.4MPa) and much shorter time (5 min) compared to Cu-Cu direct bonding. Moreover, the bonded interface had high heat resistance (> 480°C) when we deposit the main bonding material of Sn (low melting point metal) and Ag (formation compound with Sn) as a multi-layer film.

By performing detailed analysis using TEM, we find that formation of Sn-Ag type intermetallic compound occurs and that the bonding volume contracts, i.e., Sn+Ag ===> Sn-Ag IMC: about 6%. This contraction results in extremely high pressure being applied to an un-bonding process caused by micron roughness and an impurity layer, e.g., oxide, at the interface. A bonding process we call "self-compression bonding" then sequentially proceeds at unbonded interface regions. We hypothesize good bonding is achieved over the entire interface in a short time period due to self-compression bonding.

1. Introduction

3D-ICs using TSV (Through-Silicon Via) as one possible breakthrough method that can overcome semiconductor scaling limits is desired strongly [1]. To realize this breakthrough, the bonding technology between chips (Fig. 1) is a key technology. However, TSV stack bonding is so complex that mass production of 3D-ICs has been problematic. Of interest, joinable bonding systems using low temperature, low pressure (low load to the low-k insulation material), low vacuum, short time (productivity drive), and narrow pitch (10 μm) have yet to be realized [2], [3], especially for the bonding of Cu electrodes.

As reported in our previous work, we achieved bonding at a much lower temperature (180°C), with a low load (0.4MPa) and a much shorter time (5min) compared to Cu-Cu direct bonding [4], [5]. Moreover, the bonded interface has high heat resistance (more than 480°C) when we deposit the main bonding material of Sn (low melting point metal) and Ag (formation compound with Sn) as a multi-layer film.

However, the phenomenon at the bonding interface is still unclear. The objective of this study is to clarify the mechanism of the bonding, and establish the validity of the bonding system.

2. Experiment Method

The following bonding components in this experiment are used. On a Si board (3mm×3mm), a Cu film (2μm electrode pad) is deposited with electron beam vapor deposition. Cr film (50nm) is deposited under Cu to strengthen adhesion of Cu to Si. Next, to prevent Cu-Sn reaction, wafer curvature, diffusion time, etc…, an Ag (2μm) film is deposited as a film on the Cu side followed by deposition of a Sn film (1μm) on the surface side to form a 3μm Ag₃Sn after bonding.

Finally, an Ag film (0nm, 10nm) is deposited on top of the Sn film to prevent Sn oxidation. A bonding experiment was conducted using two multilayer thin film attached face to face (Fig. 2).

A bonding experiment was conducted under a bonding pressure 0.4MPa and a 1 Pa vacuum. Furthermore, samples were produced using bonding temperature of 180°C and bonding times of 0sec, 10sec, 100sec, 300sec, and 600sec.

Analysis of the bonding interface state was based on observations using FE-SEM (Field Emission Scanning Electron Microscopy) and TEM (Transmission Electron Microscope) after processing with a flat milling system and FIB (focused ion beam). Atomic identification was performed by EDX (Energy Dispersive X-ray Spectroscopy). Furthermore, elemental analysis including the check of the oxidation state of Sn near the surface was performed by ESCA (Electron Spectroscopy for Chemical Analysis). In this case, element ratios of Ag, Sn, and Sn oxide in the arbitrary depth were analyzed by quantitative analysis of the depth with Ar etching.

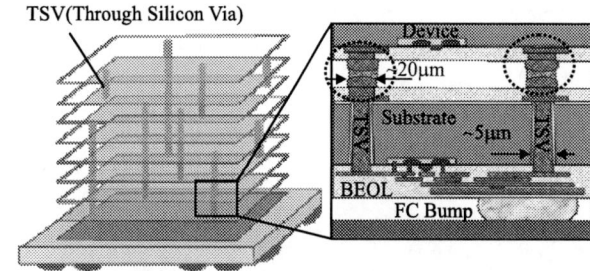

Fig. 1 3D-SiP image with TSV.

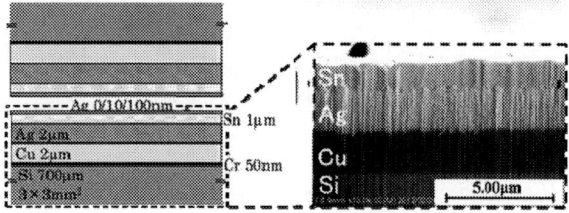

Fig. 2 Our composition of bonding materials.

3. Result and discussion

3.1 Influence on the bonding characteristic of surface Ag film

The bonding experiment was conducted using two multilayer thin films attached face to face. Furthermore, we observed their cross section using FE-SEM after polishing using FIB. The result is shown in Fig. 3. Due to space limitations, we show only the result of bonding samples joined in the bonding time of 600s. One result of the bonding experiment is that bonding was achieved for both surface Ag thicknesses (0nm, 10nm). However when the film of 10nm Ag was formed in surface, we clearly found that better bonding was achieved when there was no interface. Although the effects of nm order Ag films are understandable, conventional wisdom does not explain why superior bonding is even formed on the predicted Sn oxide surface created from the oxidation of the Sn surface on 0nm Ag. Consequently, we carried out an analysis and new study of the surface and its interface to explain the bonding mechanism.

3.2 Analysis of the bonding interface using TEM

To deduce the bonding mechanism details, we analyzed the interface condition of the bonding interface. We used TEM to observe a bonding sample with 10nm surface Ag which was formed in 600sec. The result is shown in Fig. 4. The bonding interface is not observed in the vicinity of central position of the multilayer thin films after bonding. This suggests that the bonding is very good. Additionally, Fig. 4 shows that several intermetallic compounds (IMC) are formed and solid state diffusion bonding is achieved at low temperature.

Next, we performed compound identification analysis using EDX and TEM diffraction patterns TEM (Fig. 4). From these results, it was found that two kind of Ag-Sn type IMC (Ag4Sn or Ag3Sn) are created at the bonding interface. Futhermore, Cu-Sn type IMC (Cu6Sn5 or Cu3Sn) were also created on the Cu side. These IMC's have a high melting point so it is possible to satisfy the stacking temperature hierarchy. However, since we did not initially assume the Cu-Sn type IMC would be generated, it is believed that a lower temperature and shorter bonding time is possible by optimizing the process condition. This is left as future work.

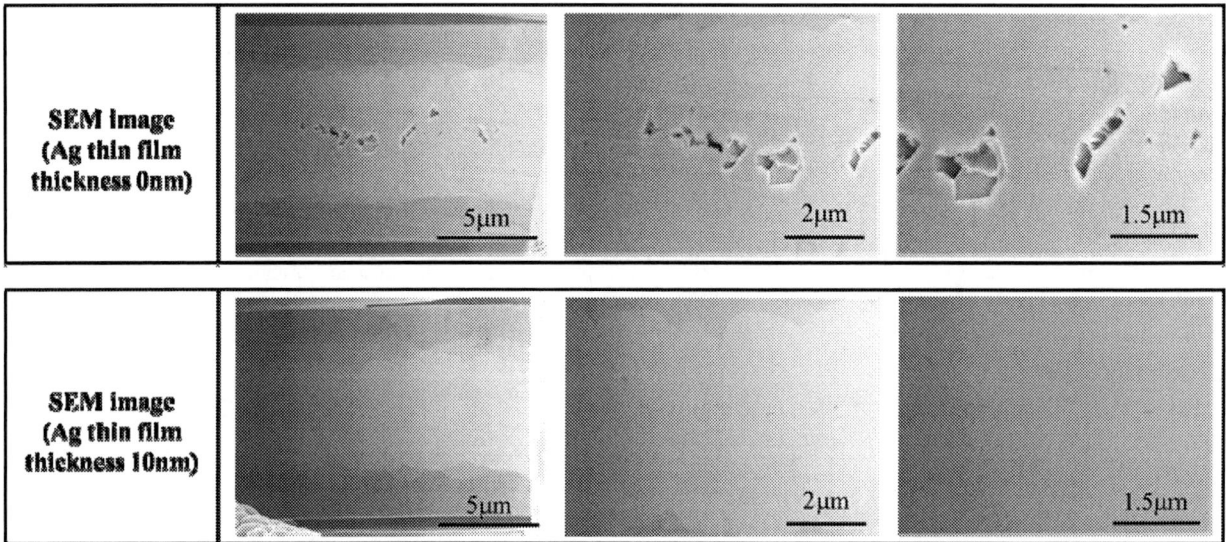

Fig. 3 Cross-sectional image of bonding interface by FE-SEM after milling by FIB.

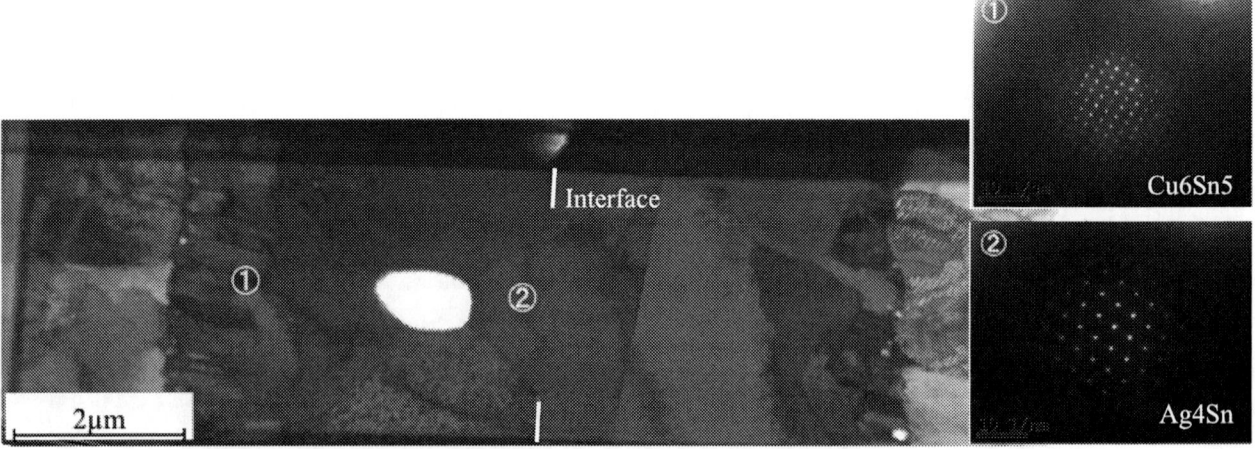

Fig. 4 Cross-sectional image of bonding interface by TEM and diffraction pattern.

3.3 Influence of an anti-oxidization film (Ag)

After attaining low temperature and short time bonding, we consider the influence of the surface Sn oxide on bonding.

First, we confirmed the effect of the surface Ag to anti-oxidation using vacuum film deposition (around 180°C) with the vacuum level lowered to ~10^4Pa. The TEM observations of cross section before bonding are shown in Fig. 5. It shows that a thin Sn oxide is uniformly formed on the surface for both 0nm and 10nm surface Ag thickness cases. The oxide thicknesses are 20 ~ 25nm and 40 ~ 50nm, respectively. The thickness in the case of Ag10 nm is thinner, so the suppression effect of surface Ag is confirmed. However, we assume that preferential oxidation of Sn proceeds, and oxide film is formed since surface Ag cannot cover the Sn surface completely. In this way, the samples, whose surface have uniform Sn oxide cannot be bonded under low temperature (about 180°C), low pressure (0.4MPa), deformation of the surface irregularities only occur as shown in Fig. 6 Through the mechanical removal of the surface Sn oxide layer, bonding was achieved over almost the entire surface. The cross section of bonded part (Fig. 7) shows that solid state bonding at low temperature and low pressure is achieved.

3.4 Bonding mechanism hypothesis

In this study, we confirmed that bonding cannot be generated when creating multilayer films using low vacuum since even for the Ag 10nm surface film case, a 20-25nm Sn surface film is formed. This result suggests that the bonding is strongly dependent on the thickness of the surface oxide film.

Based on these results, we hypothesize a mechanism of low-temperature and low load bonding.

We find from Fig. 3 that good bonding can be achieved even though there is a thin Sn oxide film. In order to break this Sn oxide, high pressure is required locally. We expect this pressure to come from the IMC which is partially formed. We show a schematic diagram of the mechanism in Fig. 8. In this mechanism, the initial pressure is concentrated on the convex portion in contact to the surface. Plastic deformation of Sn also occurs and Sn oxide is locally destroyed by slip deformation. Finally, the newly formed surfaces of Sn reach the solid phase diffusion. With heating time, the bonding part grows to IMC Ag3Sn. The volume of Ag3Sn shrinks about 6% compared to the volume of Sn and Ag. With the growth of the IMC, high compression forces are arise at the extreme ends of the unbonded Sn oxide. With the destruction of Sn oxide, there is solid state diffusion of a newly formed surface of Sn where bonding progresses sequentially. Bonding of the entire structure is achieved with the progress of Ag3Sn growth. We name this bonding "Self-Compression Bonding" (SCB) due to the fact the bonding generates a compression force on itself with the growth of the compound in order to bond.

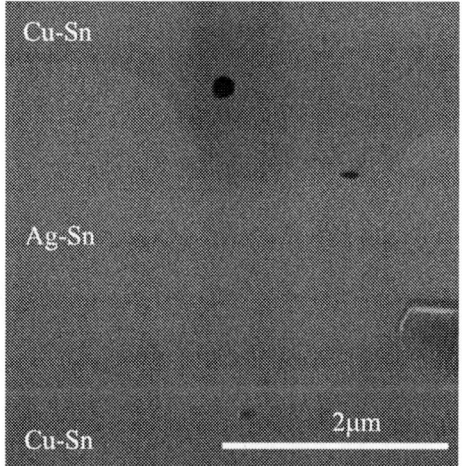

Fig. 7 Bonding part without interface.

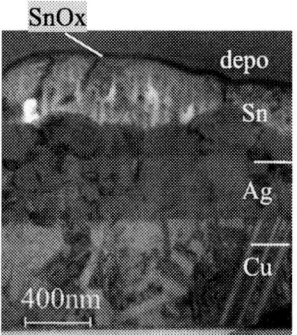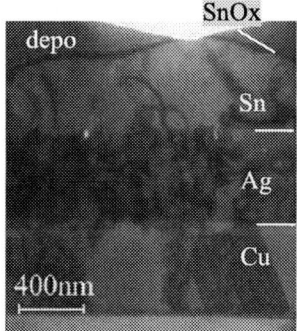

(a) Surface Ag is 0nm.   (b) Surface Ag is 10nm.

Fig. 5 Cross-sectional image of multilayer before bonding by TEM

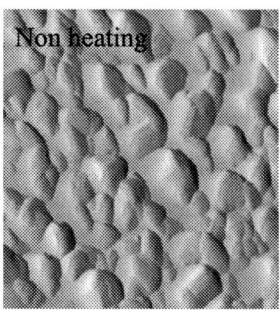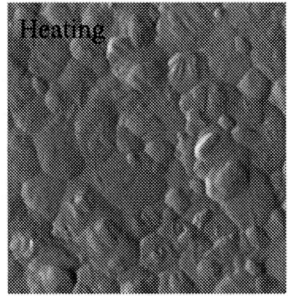

Fig. 6 3D images of AFM analysis for heating and non heating samples.

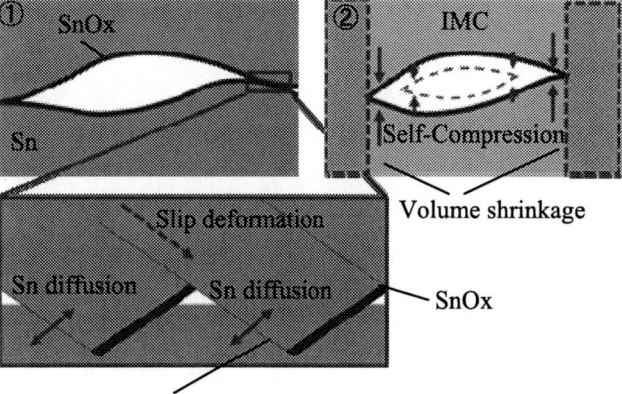

Fig. 8 Self-Compression Bonding (SCB) mechanism.

We need to control the surface oxide film in order to generate effectively SCB. In the future, we consider design of surface Ag film thickness and optimization of film depositing conditions.

Conclusions

We have studied a low temperature, low load, solid-phase bonding associated with the formation of high heat resistant Ag-Sn type IMC for wafer-level 3D stacking for 3D ICs. We found the following

(1) We showed that Ag-Sn system IMC occurred at the bonding interface under a bonding pressure 0.4MPa, a 1 Pa vacuum, 180°C bonding temperature and 600sec bonding time.

(2) We found bonding is formed even for samples with surface oxide layers and where we predict 0nm Ag surfaces.

(3) In the case of low vacuum and heat deposition, we found that oxide layers of 20-50nm are formed and that bonds cannot be formed.

(4) From (1)-(3), we have shown the possibility of Self-compression Bonding arising from volume contraction due to formation of Ag-Sn type IMC.

From these results, we found that it is possible to create bonds using low temperature and low pressure even if there is some degree of Sn surface oxidation. In the future, we will consider control of the Sn surface oxidation.

References

1. Seiichi Denda, *Sanjigen jisso no tameno TSV gijutsu*, Japanese Industrial Standards Committee, 2009, pp. 70–78.

2. Akitsu Shigetou, Naoe Hosoda, Toshihiro Itoh and Tadatomo Suga, "Room-Temperature Direct Bonding of CMP-Cu film for Bumpless Interconnection," in *IEEE Electronic Components and Technology Conf. (ECTC)*, Orlando, May 29–June 1, 2001, pp. 755–760.

3. Katsumi Taniguchi, Tomoaki Goto, Kiyokazu Yasuda, Kozo Fujimoto, "Interfacial Microstructure and Joint Properties of Copper Direct Bond Inserted by the Thin Film of Indium", in *Solid State Devices and Materials conf. (SSDM)*, Tokyo, September 15– 17, 2004, pp. 282–283.

4. B. Swinnen, W. Ruythooren, P. De Moor, L. Bogaerts, L. Carbonell, K. De Munck, B. Eyckens S. Stoukatch, D. Sabuncuoglu Tezcan, Z. Tőkei, J. Vaes, J. Van Aelst, E. Beyne, "3D integration by Cu-Cu thermo-compression bonding of extremely thinned bulk-Si die containing 10 μm pitch through-Si vias", in *IEEE International Electron Device Meeting Tech. Digest (IEDM)*, December 11–13, 2006, pp. 371–374.

5. A. Fan, A. Rahman, and R. Reif, "Copper Wafer Bonding", *Electrochemical and Solid-State Letters (ESL)*, 1999, Vol. 2, Issue 10, pp. 534-536.

Self-Patterning, Pre-Applied Underfilling Technology for Stack-Die Packaging

Chia-Chi Tuan[1], Ziyin Lin[1], Yan Liu[1], Kyoung-Sik Moon[1], and Ching-Ping Wong[1,2*]

[1]School of Materials Science and Engineering, Georgia Institute of Technology,
771 Ferst Drive, Atlanta, GA 30332
[2]Department of Electronic Engineering, Chinese University of Hong Kong, Hong Kong
[*]Corresponding author: 404-894-8391, cp.wong@mse.gatech.edu

Abstract

Die stacking is one of the next-generation 3D IC packaging methods, but its stringent material requirements are unlikely to be met by traditional underfills. Moreover, filler trapping is becoming an increasingly serious issue in no-flow and wafer-level underfills. In this report, we demonstrate a novel underfilling technology for the reduction of filler trapping in fine-pitch interconnects. In our method, we fabricate superhydrophobic bond pads, and control the flow of the underfill material by the surface energy difference between the bond pads and the Si_3N_4 substrate. The superhydrophobic bond pads are shown to have no effect on the bonding of soldering materials to the pads.

Introduction

Underfilling is a key step in electronic packaging that greatly affects the production yield and device reliability. Current packaging technologies are rapidly progressing towards low profile packaging and 3D packaging, giving rise to new challenges to underfilling process and materials due to fine pitch and low stand-off heights. Traditional underfills are unlikely to satisfy the material requirements for this type of new packaging solution. Capillary underfills meet difficulties when highly filled underfills need to flow into gaps smaller than 80 μm. On the other hand, it is possible to use no-flow and wafer–level underfills in such packages, but the freedom in underfill formulation is constrained by material property requirements. For example, filler trapping at the bond pads is a serious issue in no-flow and wafer-level underfills with filler loading higher than 30~40 wt.%. The trapped filler particles interfere with solder bonding, leading to defects in interconnections and issues in reliability. As a result, many of the no-flow and wafer-level underfills on the market have low filler contents in order to circumvent the filler trapping issue. But a low filler loading in underfills translates to large mismatch in coefficient of thermal expansion (CTE) between interconnects and underfill, leading to thermomechanical issues and compromised package reliability.

Researches have been conducted on controlling fluids by substrate surface engineering. Microfluidic channels and oil-water separation are a few examples [1], [2]. Kobayashi et al. have shown that the combination of hydrophilic-hydrophobic surfaces can lead to surface wettability differences and liquid manipulation. They successfully controlled droplet formation and split the fluid into 14.1 nL in a microfluidic chip [3].

The preparation of superhydrophobic Cu bond pads is critical in this underfilling technology. The state of superhydrophobicity is defined as having water contact angle greater than 150° and hysteresis smaller than 10° [4]. Wenzel and Cassie states are usually observed on rough superhydrophobic surfaces, which refer to complete wetting and partial wetting, respectively [5]. Additionally, the Wenzel equation correlates the surface roughness to the contact angle; as surface roughness increases, so does the contact angle [6]. This leads to the first requirement of superhydrophobic surfaces- the surface requires a rough morphology. The second requirement is low surface energy. This is often done via chemical modification using low surface energy molecules.

The preparation of rough surfaces on Cu substrates has been reported. Long copper oxide microwires of high aspect ratios can be produced by chemical etching methods using basic solutions [7]–[10].

In this paper, we demonstrate a mask-less underfill patterning process by controlling the surface wettability (Figure 1). Neat epoxy resin and filled underfill materials are shown to selectively wet the passivation surfaces, leaving the bonding pads clean for high-quality interconnect bonding. We also show that the surface modification performed on the bond pads do not hinder the formation of interconnects. Compared to conventional capillary, no-flow, and wafer-level underfilling processes, our process not only provides a new route to the reduction of filler trapping at bond pads for fine-pitch, low profile packaging, but also has potential for fluxless metal-to-metal bonding [11].

Figure 1. Schematic of our mask-less, self-patterning underfilling process.

Methods

Test vehicle design

We designed test vehicles with circular Cu bond pads on the Si_3N_4 substrate. Two levels of pad feature sizes were used in this study: (1) 1 cm diameter and 1.5 cm pitch, and (2) 100 μm diameter and 150 μm pitch. The Cu pads were 20-30 μm in height. The test vehicle was fabricated via standard photolithography and Cu electroplating processes using commercially available photoresist and electroplating solution (Figure 2). The photoresist was exposed using 365 nm wavelength light.

Surface treatments

Plasma treatments are widely used for surface property modifications. Here, the Si_3N_4 surface was treated hydrophilic by oxygen plasma using a Novascan UV Ozone Cleaner at elevated temperatures. A second photolithography process was carried out in order to coat a photoresist layer on the test vehicle (Figure 3). The photoresist was exposed by 365 nm light.

The Cu bond pads on the photoresist-patterned test vehicle were treated superhydrophobic by an ammonia solution etching and subsequent surface chemical modification with superhydrophobic molecules. After the superhydrophobic treatment, the photoresist coating on the Si_3N_4 surface was removed.

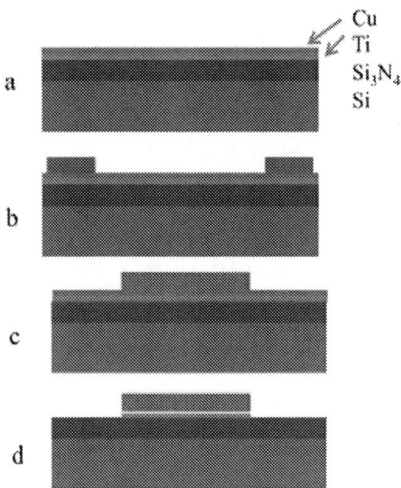

Figure 2. Schematic of the test vehicle fabrication process. (a) Si substrate undergoes nitride growth, followed by subsequent Ti and Cu depositions. (b) Photoresist application and development after exposure for pattern formation. (c) Bond pad deposition by Cu electroplating (d) Removal of Cu seed layer.

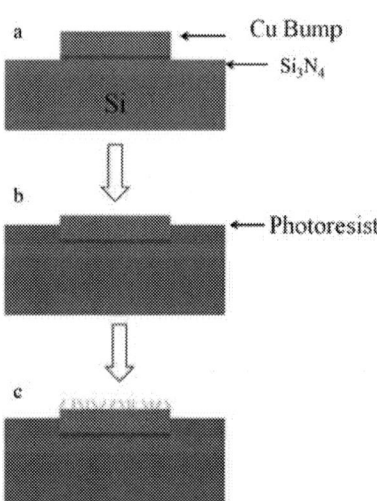

Figure 3. Second lithography process and surface roughness treatment: (a) test vehicle with Cu bond pad, (b) photoresist application and development, and (c) surface roughness produced by wet chemical etching.

Characterization

A contact angle measurement system equipped with a Rame-Hart goniometer and a built-in CCD camera was used to evaluate the water contact angles of test vehicle surfaces.

Scanning electron microscopy (SEM) images were obtained using a Ziess LEO 1550 thermally assisted field emission (TFE) SEM using an acceleration voltage of 4-5 kV. X-ray diffraction (XRD) analysis was performed using Cu Kα radiation (45 kV and 40 mA) on a Philips X-pert alpha-diffactometer. Thermo K-alpha x-ray photoelectron spectroscopy (XPS) was used to analyze the presence of chemical species and bonding on the sample surfaces.

Results and Discussion
Surface treatments

The morphology of the etched Cu bond pads was characterized using SEM. Microscale surface roughness was produced by ammonia etching, and Figure 4 shows the morphology evolution as etching time increases. The length of the micro-needles grows with increasing reaction time, as well as the overall surface roughness. From an XRD pattern, the surface species were determined to be copper hydroxide, and XPS spectrum also indicated that the surface Cu has a bivalent charge (Figure 5).

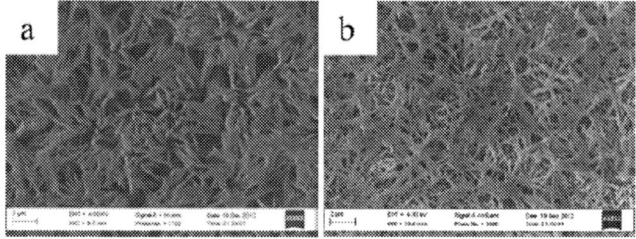

Figure 4. SEM images of Cu pads etched for (a) 12 hours, and (b) 24 hours.

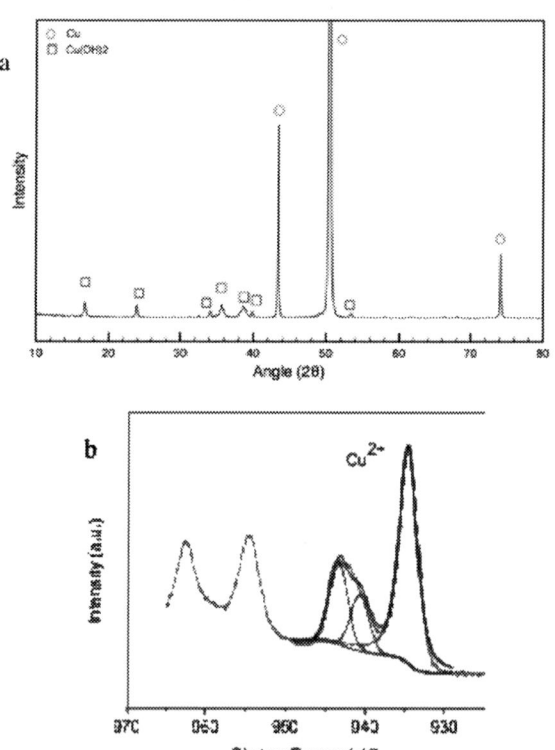

Figure 5. (a) The XRD pattern shows surface $Cu(OH)_2$ and underlying metallic bulk Cu. (b) The XPS spectrum indicates bivalently-charged Cu.

The surface properties of the treated Cu bond pads and Si_3N_4 substrate were characterized by XPS and the water contact angle measurement system. A Cu C1s XPS spectrum was obtained after the superhydrophobic fluorocarbon molecule treatment, which suggested bonding of the fluorocarbon molecule to the pad (Figure 6). As shown in Figure 7, the contact angle on the hydrophilic Si_3N_4 surface is below 20º, and the contact angle on superhydrophobic Cu pads is above 160º. The water contact angle results of Cu samples etched for different durations indicate that the surface roughness is essential in the preparation of the superhydrophobic surface, as the surface morphology affects both the static water contact angle and the contact angle hysteresis (Table 1).

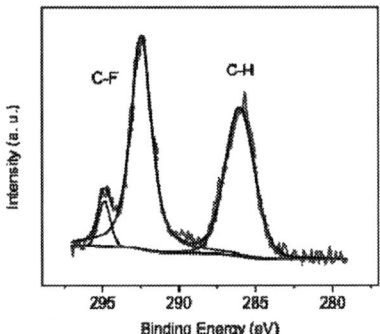

Figure 6. C1s XPS spectrum of the superhydrophobic Cu bond pad.

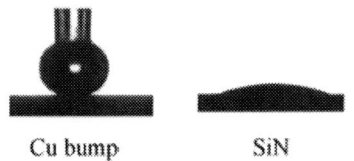

Figure 7. Side views of water contact angle tests.

Table 1. The effect of etching duration and surface morphology on water contact angle.

Etching time (hr)	12	24
Contact angle (º)	160	165
Hysteresis (º)	5	3

Solder bonding

Commercially available SnAg- and SnPb-based solder pastes were used to bond to the surface-treated superhydrophobic Cu bond pads. The solder pastes were applied directly onto the Cu pads, and the substrate temperature was raised to above the respective melting temperatures of the solder materials. It was observed that the flux contained in the pastes reduced the surface species to metallic Cu, and the solders wetted the pads to form bonding. In cross-sectional images obtained by SEM, the typical scallop-like intermetallic compound (Sn_6Cu_5) can be clearly observed at the solder-Cu interface (Figure 8). The soldering material was preferentially etched by an acid-isopropanol solution to better show the intermetallic compounds in the image.

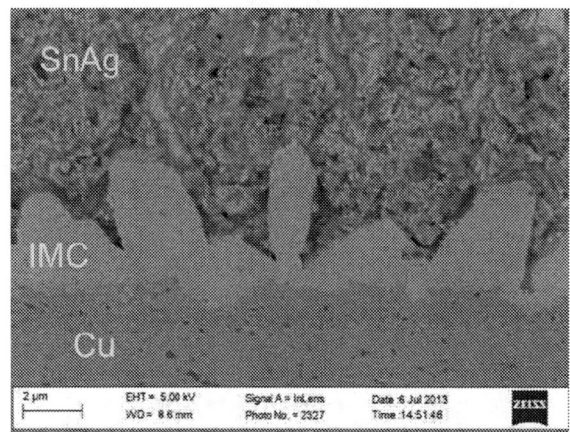

Figure 8. The SEM image of the SnAg solder and Cu pad interface after reflow. The formation of the intermetallic compound shows good bonding.

Selective wetting on test vehicles

Liquid deposition on the test vehicles was performed with fluids of various viscosities and surface energies: water, neat epoxy resin, low-viscosity underfill, and commercial filled underfill systems. The fluids were deposited onto the test vehicles under a combination of conditions, including spin coating, tilting, and heated substrate.

On the large feature size (1 cm diameter) test vehicles, self-patterning of underfilling materials were performed using neat epoxy, in-house prepared SiO_2-filled underfill, and commercial low-viscosity underfill (Underfill A). Some of the relevant properties of the materials are listed in Tables 2 and 3. Figure 9 shows the wetting behavior of Underfill A on the test vehicle at room temperature, and Figure 10 shows those of Underfills B and C at elevated temperatures.

Table 2. Material properties of fluids used in selective wetting demonstrations.

Fluid	Water	Epoxy resin
Surface energy (mJ/m^2)	72	42-46
Viscosity (Pa·s at 25ºC)	$8.9 \cdot 10^{-4}$	1.8

Table 3. Material properties of commercially available underfills used in selective wetting demonstrations.

Underfill	A	B	C
Filler content (wt%)	30-60	65	70-75
Viscosity (Pa·s at 25ºC)	4.5	45	25

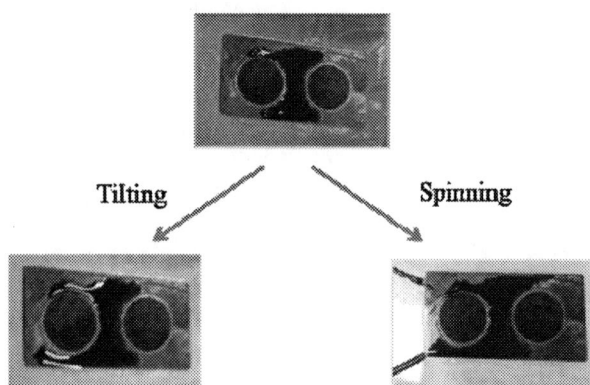

Figure 9. Selective wetting of Underfill A: as deposited (top), and after different wetting conditions (as indicated).

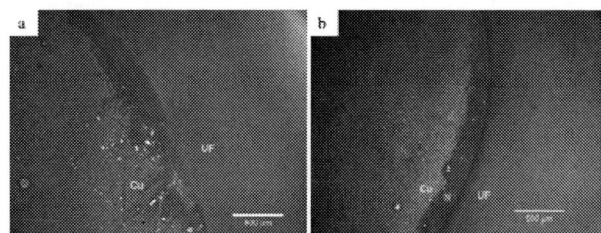

Figure 10. (a) Underfills B and (b) C deposited to show selective wetting on the large feature size test vehicles.

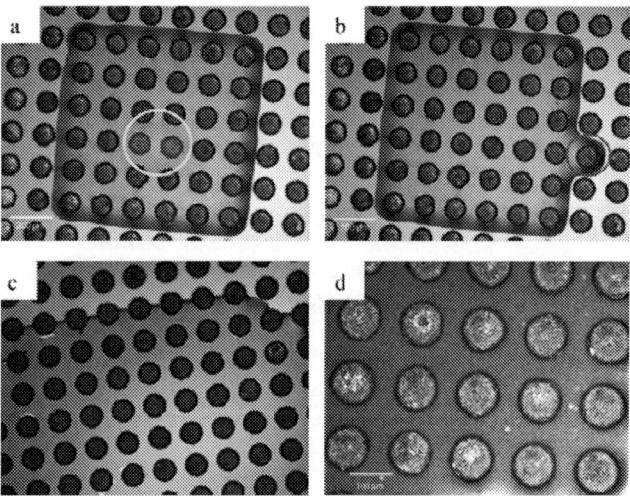

Figure 11. Self-patterning demonstrations with epoxy resin (a-b), water (c), and Underfill C.

Similar tests were carried out on the fine feature size test vehicles (100 μm diameter) using water, epoxy resin, and commercially available, filled underfill materials (Underfills B & C). Due to the finer pitch of the test vehicles, the deposition of underfilling materials posed more condition requirements, such as substrate temperature at deposition. Self-patterning can be achieved under controlled temperature and volume conditions, as shown in Figure 11. Figures 11 (a)~(b) show the top views of self-patterning of epoxy resin; in 11a, two Cu pads were covered by the epoxy resin (yellow circle), and in 11b, the epoxy has flew off the Cu pad to surround a new bond pad (red circle). In Figure 12, the SEM cross-sectional image of the test vehicle shows that Underfill

C fully filled the gaps between the Cu bonding pads, leaving the Cu pads clean on the surface, and that no gaps are observed at the underfill-pad interface.

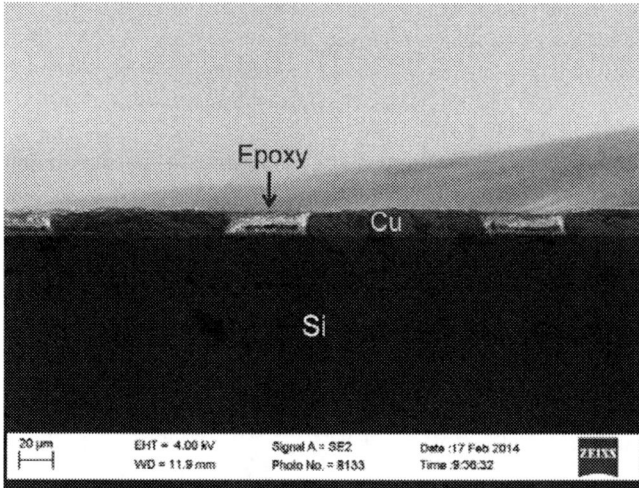

Figure 12. SEM image of the cross-section of test vehicle wetted with Underfill C.

Conclusions

We fabricated superhydrophobic Cu bond pads on hydrophilic Si_3N_4 substrate via a combined photolithography, electroplating, wet etching, and surface treatment process. Furthermore, we demonstrated the self-patterning of underfill materials on the test vehicle, wetting only the Si_3N_4 substrate but not the bond pads. It is shown that the surface engineering on the bond pads does not interfere with solder bonding for interconnect formation. Compared to conventional underfilling processes, this technology can reduce filler trapping at the bond pads, thus improving the reliability of interconnects in the package.

Acknowledgments

The authors would like to thank Dr. Yonghao Xiu for helpful discussions. This work was supported by Semiconductor Research Corporation (SRC) under Research ID 2287.

References

[1] Z. Cheng, H. Lai, Y. Du, K. Fu, R. Hou, N. Zhang, and K. Sun, "Underwater superoleophilic to superoleophobic wetting control on the nanostructured copper substrates.," *ACS Appl. Mater. Interfaces*, vol. 5, no. 21, pp. 11363–70, Nov. 2013.

[2] Z. Cheng, H. Lai, Y. Du, K. Fu, R. Hou, C. Li, N. Zhang, and K. Sun, "pH-Induced Reversible Wetting Transition between the Underwater Superoleophilicity and Superoleophobicity.," *ACS Appl. Mater. Interfaces*, Dec. 2013.

[3] T. Kobayashi, K. Shimizu, Y. Kaizuma, and S. Konishi, "Novel combination of hydrophilic/hydrophobic surface for large wettability difference and its application to liquid manipulation.," *Lab Chip*, vol. 11, no. 4, pp. 639–44, Feb. 2011.

[4] A. Lafuma and D. Quéré, "Superhydrophobic states," *Nat. Mater.*, vol. 2, no. 7, pp. 457–60, Jul. 2003.

[5] Y. Liu, Z. Lin, K. Moon, and C. P. Wong, "Robust, novel, and low cost superhydrophobic nanocomposites coating for reliability improvement of microelectronics," in *2012 IEEE 62nd Electronic Components and Technology Conference*, 2012, pp. 2135–2139.

[6] R. N. Wenzel, "Resistance of Solid Surface to Wetting by Water," *Ind. Eng. Chem.*, vol. 28, no. 8, pp. 988–994, Aug. 1936.

[7] Z. Cheng, R. Hou, Y. Du, H. Lai, K. Fu, N. Zhang, and K. Sun, "Designing heterogeneous chemical composition on hierarchical structured copper substrates for the fabrication of superhydrophobic surfaces with controlled adhesion.," *ACS Appl. Mater. Interfaces*, vol. 5, no. 17, pp. 8753–60, Sep. 2013.

[8] J. Ou, W. Hu, S. Liu, M. Xue, F. Wang, and W. Li, "Superoleophobic textured copper surfaces fabricated by chemical etching/oxidation and surface fluorination.," *ACS Appl. Mater. Interfaces*, vol. 5, no. 20, pp. 10035–41, Oct. 2013.

[9] D.-D. La, T. A. Nguyen, S. Lee, J. W. Kim, and Y. S. Kim, "A stable superhydrophobic and superoleophilic Cu mesh based on copper hydroxide nanoneedle arrays," *Appl. Surf. Sci.*, vol. 257, no. 13, pp. 5705–5710, Apr. 2011.

[10] F. Zhang, W. Bin Zhang, Z. Shi, D. Wang, J. Jin, and L. Jiang, "Nanowire-haired inorganic membranes with superhydrophilicity and underwater ultralow adhesive superoleophobicity for high-efficiency oil/water separation.," *Adv. Mater.*, vol. 25, no. 30, pp. 4192–8, Aug. 2013.

[11] R. Zhang, K.-S. Moon, W. Lin, Y. Liu, and C. P. Wong, "Fluxless Metal-Metal Bonding for 3D IC Stacking," GTRC #5382, 2010.

Study of High CRI White Light-emitting Diode Devices with Multi-chromatic Phosphor

Min Zheng,[1] Wen Ding,[1,2] Feng Yun,[1,2,*] Deyang Xia[1], Yaping Huang[1], Yukun Zhao[1], Weihan Zhang[1],
Minyan Zhang[1], Maofeng Guo[2], and Ye Zhang[2].

[1]Key Laboratory for Physical Electronics and Devices of the Ministry of Education & Shaanxi Key Laboratory
of Information Photonic Technique, School of Electronics and Information Engineering, Xi'an Jiaotong
University, Xianning West Road 28, Xi'an 710049, China

[2]Solid State Lighting Engineering Research Center, School of Electronics and Information Engineering,
Xi'an Jiaotong University, Xianning West Road 28, Xi'an 710049, China

Abstract

In this paper, we investigated the connection between the employment of multi-chromatic phosphor and CRI value for WLED devices by simulation, using the excitation and emission spectra of phosphor and surface source property of LED chips acquired from experiments.

CRI value was examined in the range of 56.8-76.0 due to the change of phosphor concentration, for traditional blue-pumped yellow phosphor. When red phosphor was added, it was found that as red contents increased, red shift occurred in CIE, and CRI value was enhanced only within a limited range. The highest enhancement in such case was 13.8% for blue-pumped yellow phosphor, and when more multi-chromatic phosphor such as red, yellow, green was mixed, the value of CRI enhancement was 19.2% higher than that of dichromatic LED. Ray tracing simulation revealed that multi-chromatic phosphor also had an impact on luminous efficacy and color temperature for high-power WLED devices.

It was also showed in our simulation that CRI value increased with the increase of total phosphor concentration, up until the point of an optimum concentration where CRI value started to decrease. Other parameters such as quantum efficiency and molar absorbance index also contributed to white-LED devices performance. Such simulation results are useful to design the optimum phosphor mixture concentration and are helpful to fabricate high CRI blue-pumped or ultraviolet-pumped WLED devices with the best multi-chromatic phosphor proportion.

Introduction

So far, white light-emitting diode (WLED) devices have mainly been manufactured based on blue chips with the coating of YAG:Ce^{3+}, a representative kind of conventional yellow phosphor [1-3]. However, the realization of high color rendering index (CRI) in such dichromatic LEDs appeared to be difficult. To break this bottleneck, multi-chromatic phosphor method had been attempted and showed great improvements.

Some efforts have been devoted to addressing this issue and providing high CRI WLEDs by searching for new phosphor powder material. Based on the conventional system of YAG:Ce, R. Marin tried to make Pr^{3+} doped, adjusted the doping concentration by controlling the annealing temperature and annealing time, and successfully broke the bottleneck (CRI≤80) [4]. Besides, multi-chromatic phosphor method, which was done by mixing different colors of phosphor, both for blue-pumped and ultraviolet-pumped WLED had been

attempted and showed great improvements [5-8]. It had been proved that CRI displayed various values within different concentration for the same phosphor, while in multi-chromatic phosphor system, concentration ratio as a factor should be taken into account. To achieve the best CRI value, a large number of experiments had been used which cost a lot.

In this paper, we investigated the connection between the employment of multi-chromatic phosphor and CRI value for WLED devices by software TracePro based on Monte Carlo method. Tracepro had been used by many researchers to analyze optical consistency [10], irradiance [11], luminous flux, color temperature, color coordinates, CRI [12-13] and other optical properties of WLEDs. With ray tracing simulation, optical properties of WLEDs could be easily achieved. Finally, data processing was carried out and a comparison is made between yellow phosphor and multi-chromatic phosphor with the same blue chip.

Theory and Model

Monte Carlo method is used to analyze and simulate the optical processes. In the process of transmission in the phosphor layer, photons would be reflected or refracted by boundaries, be scattered or absorbed by colloidal substances, be absorbed by phosphor, emit new photons etc. These events above could be regarded as function of photon transmission occurred randomly in the phosphor layer, the values of which were the factors of continued transmission and frequency shift of photons.

In this simulation, we assumed that the angle distribution of surface source depicted a Lambertian surface [9]:
$$I(\theta) = I_0 \cos \theta \qquad (1)$$
Emitted from light source into phosphor layer surface, photon was absorbed by fluorescent powder after it had transferred the random distribution of free path, and then, made contribution to fluorescence excitation with a chance. Fluorescence photon was emitted as a grid source. In this WLED simulation process, two types of light (blue light emitted by chip and light with another color converted by phosphor) were traced in different steps.

In this paper, a vertical structure white LED model was strictly designed as the experiment used in our laboratory, which was set up in the commercial software TracePro.

As shown in Fig. 1, the white LED model includes a ceramic substrate, a blue chip which is placed on the surface of the board, a phosphor layer which is formed as a conventional proximate phosphor-in-cup packaging, and a hemispherical lens with a diameter of 5.6mm.

---

*)Author to whom correspondence should be addressed. Electronic mail:
*syzm_2005@qq.com，Tel.: t86 029 82668015.

978-1-4799-2408-0/14 $31.00 © 2014 IEEE

2014 Electronic Components & Technology Conference

Figure 1. Optical models of white LED.

High power blue LED chip with vertical structure has a dimension of 45mil×45mil (1.143mm×1.143mm), the model of which includes four parts as depicted in Fig 2.

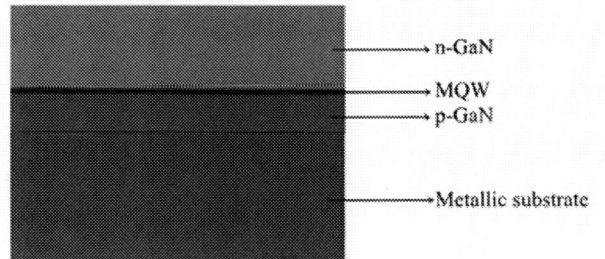

Figure 2. Blue LED chip model.

The thickness of substrate is 200μm with the material of copper-tungsten. Multiple quantum well (MQW) thickness has an impact on emission wavelength，and for GaN-based blue LED，with an undoped InGaN/GaN MQW active layer and an emission wavelength of 450nm, the active layer consists of ten periods of 3-nm-thick InGaN well and 7-nm-thick GaN barrier [16]. Different layers and their thickness comprising the LED chip is given in Table I, along with their refractive indexes and absorption coefficients.

Table I
Parameters of LED chip

Layer	Thickness (μm)	Refractive Index	Absorption coeff. (mm⁻¹)
n-GaN	5	2.43	2
MQW	0.1	2.40	36
p-GaN	0.9	2.43	2

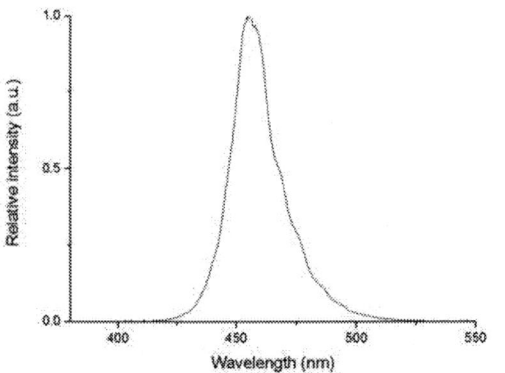

Figure 3. Emission spectra of LED chip.

Light source of LED is considered to be the upper surface of the chip, emission spectrum of which is acquired from experiment (EL measurement) showed as Fig. 3.

The hemispherical lens in Fig. 3 is transparent quartz glass with the diameter of 5.6mm, the refractive index of which is 1.54. The surface reflectance of the reflective cup (both for the wall and the bottom) is considered to be 95%.

Phosphor parameters were consistent with the oxynitride yellow-emitting phosphor and nitride red-emitting phosphor which were used in our experiment with perfect chemical stability and thermal stability. These two phosphors, emission peak of which were 570nm and 640nm respectively, illustrated as Fig. 4, could emit stable white light with blue chip after being well mixed. In this model, refractive index and thickness of phosphor are 1.8 and 0.92mm respectively which will not vary with the change of molar concentration or other parameters.

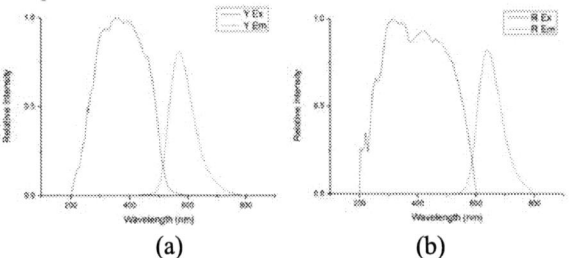

(a)	(b)

Figure 4. Excitation and emission spectra for (a) yellow phosphor and (b) red phosphor.

In the simulation, we assumed that yellow and red phosphor could be well-distributed mixed with concentration ratio of x:y and the fluorescence spectrum intensity of I_1 and I_2 in the wavelength of λ. Then, the fluorescence spectrum intensity of phosphor after being mixed I_m could be expressed as follows:

$$I_m = x * I_1 + y * I_2 \qquad (2)$$

According to the formula (2) and excitation and emission spectra for monochromatic phosphor, the fluorescence spectrum of multi-chromatic phosphor for different ratio could be easily fitted.

Results and Discussion

According to the spatial color mixing principle [14]，if the wavelength of mixed light might cover the entire visible region (390nm-760nm), the formation of white light would be well created.

Firstly, CRI value for various concentrations was measured. In order to reflect the increase of CRI with the utilization of multi-chromatic phosphor, the situation of blue-pumped yellow phosphor was simulated as shown in Fig.5. The CRI was examined in the range of 56.8-76.0 due to the change of phosphor concentration. At low concentration, the luminous flux of emitted light was dominated by the chip, and CRI value would increase with the increase of concentration up until 4.8mole where phosphor light started to be dominant, and CRI value started to decrease.

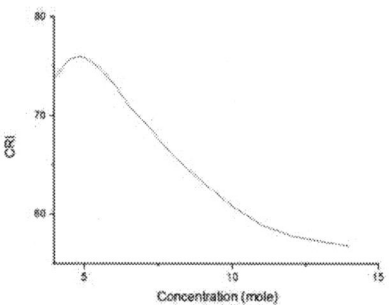

Figure 5. CRI of blue-pumped yellow phosphor package for different concentrations.

When red phosphor was added, CRI value was improved as the Fig.6 shown, as concentration was varied from 2mole to 28mole with seven different concentration ratio for yellow phosphor and red phosphor (yellow: red = 2:8; 3:7; 4:6; 5:5; 6:4; 7:3; 8:2). The best concentration ratio was Y:R=3:7 for its highest CRI value was 86.5, 13.8% higher than that of blue-pumped yellow phosphor.

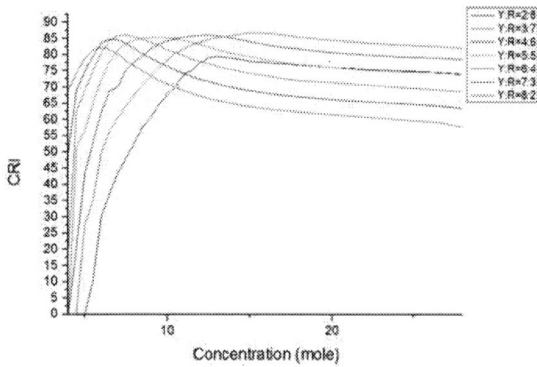

Figure 6. CRI of blue-pumped different ratio phosphors for different concentrations.

These simulation results had proved that the mix of red phosphor into yellow helped make CRI value increase, here, the optimal concentration ratio of which was Y:R=3:7.

When green phosphor with the emission peak of 500nm was added (yellow: red: green = 3:7:1), as illustrated in Fig.7, simulation had shown the highest CRI value was 90.6, 4.8% and 19.2% higher than that of blue-pumped bicolor phosphor and that of blue-pumped yellow phosphor respectively.

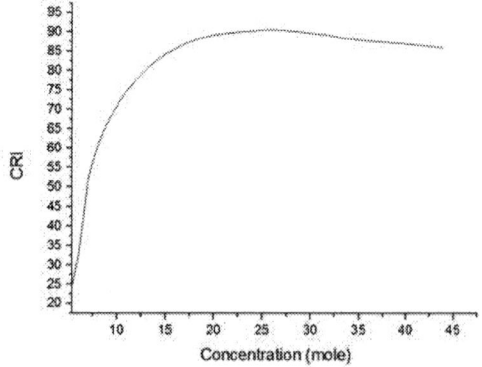

Figure 7. CRI of blue-pumped trichromatic phosphor for different concentrations.

Simulation had also showed when red phosphor was added, if phosphor concentration remained constant, red shift would occur in CIE and color temperature would decline from 4682K to 2803K as shown in Fig.8. What was more, it was found that as red contents increased, red shift performed more obviously.

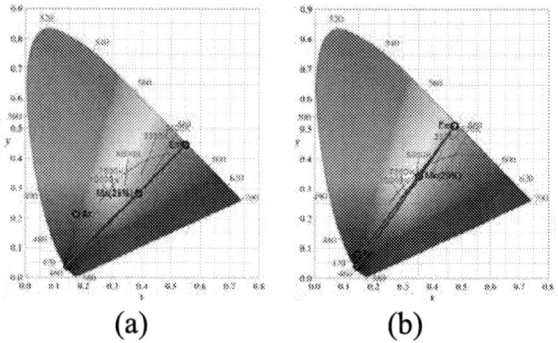

(a)   (b)

Figure 8. Chromaticity coordinates of WLEDs with (a) multi-chromatic (red and yellow) phosphor and (b) yellow phosphor when concentration keeps 8mole.

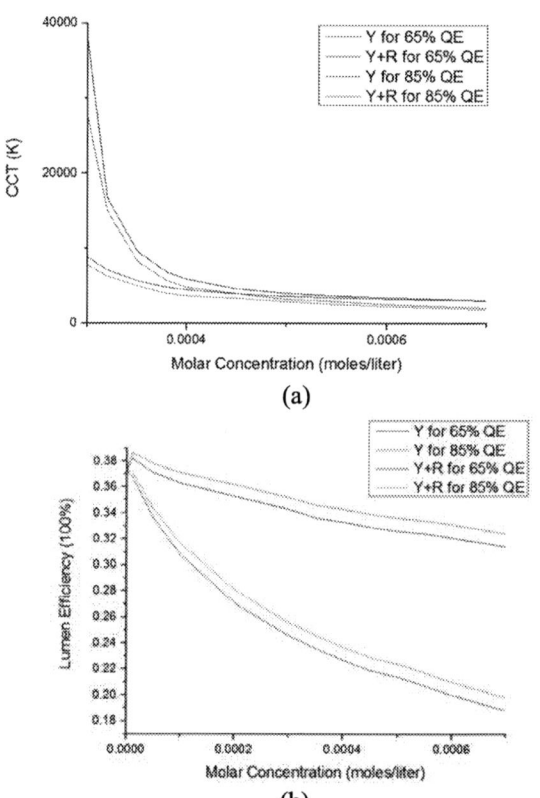

Figure 9. Dependences of (a) correlated color temperature (b) lumen efficiency on phosphor molar concentration at two kinds of phosphor and two quantum efficiency when peak molar extinction keeps to be 25000 liter/(mole*cm)

Our ray tracing simulation also revealed that multi-chromatic phosphor also had an impact on luminous efficacy and correlated color temperature (CCT) for high-power WLED devices. In this work, luminous efficacy is characterized by the ratio of luminous flux escaping from the package over the luminous flux emitted from the active layer of LED chip. Correlated color temperature (CCT) described the color of light source was achieved as simulation result in colorimetric distribution of white LEDs based on the colorimetry [17]. CCT varied with the exit angle [15], and in this simulation we only recorded the maximum CCT.

Molar concentration of the phosphor was varied from 0.0001 mole/liter to 0.0007 mole/liter with two quantum efficiency (65% and 85%) and the constant absorbance index of 25000 liter/(mole*cm). The lumen efficiency and color temperature of LED with yellow phosphor and multi-chromatic phosphor (Y: R=3:7) uniformity for various molar concentration and quantum efficiency was compared and analyzed.

The CCT of WLEDs decreased as phosphor concentration increased and quantum efficiency increased for fluorescence components increased presented in Fig.9 (a). When red phosphor was added, luminous efficacy would decline overall, while color temperature would be closer to warm white. As Fig.9 (b) showed, lumen efficiency increased with the increase of molar concentration, up until the point of an optimum concentration where lumen efficiency started to decrease.

The results can be discussed that at low concentration, phosphor would transform some blue light emitted by the chip into fluorescence with higher luminous flux, what was more, the components of light source contained parts of non-visible light that would be absorbed and transformed into visible light, both of which made contribution to the enhancement of lumen efficiency. If red phosphor was added, this loss of energy conversion would be even more obvious due to the unavoidable Stokes loss.

To receive desired white light, moderate elevation of phosphor concentration was needed accompanying with the decrease of lumen efficiency for the energy loss in photoluminescence. Simulation results in this paper would help produce high CRI value WLEDs with various light extraction efficiency and color temperature under different needs of production.

Conclusions

In this work, the connection between the employment of multi-chromatic phosphor and CRI value for WLED devices was investigated by simulation. For traditional blue-pumped yellow phosphor, the highest CRI value was examined to be 76.0 due to the change of phosphor concentration. When red phosphor was added, it was found that as red contents increased, red shift occurred in CIE, and the highest CRI value was enhanced to be 86.5, which was 13.8% higher than that of blue-pumped yellow phosphor. When more multi-chromatic phosphor such as red, yellow, green was mixed, the value of CRI enhancement was 19.2% higher than that of dichromatic LED. Such simulation results are useful to design the optimum phosphor mixture concentration and are helpful to fabricate high CRI blue-pumped or ultraviolet-pumped WLED devices with the best multi-chromatic phosphor proportion.

Acknowledgments

This work is supported by the National High Technology Research and Development Program of China (Project No. 2012AA041004), the National High Technology Research and Development Program of China (Project No. 2011AA03A111), the Shaanxi Technology Committee Industorial Public Relation Project (Project No. 2011K08-32) and the Xi'an City Technology Bureau (Project No. CX1261（4）).

References

1. Schubert E F, Gessmann T, Kim J K. *Light emitting diodes*. John Wiley & Sons, Inc. 2005.
2. Xingren, Liu. "Phosphors for white LED solid state lighting." *Chin. J. Lumin.* 28.3 (2007): 291-301.
3. Nakamura, Shuji, Takashi Mukai, and Masayuki Senoh. "Candela-class high-brightness InGaN/AlGaN double-heterostructure blue-light-emitting diodes." *Applied Physics Letters* 64 (1994): 1687.
4. Marin, R., et al. "Photoluminescence properties of YAG: Ce^{3+}, Pr^{3+} phosphors synthesized via the Pechini method for white LEDs." *Journal of Nanoparticle Research* 14.6 (2012): 1-13.
5. Korthout, Katleen, P. F. Smet, and Dirk Poelman. "Rare earth doped core-shell particles as phosphor for warm-white light-emitting diodes." *Applied Physics Letters* 98.26 (2011): 261919-261919
6. Kuo, Te-Wen, Wei-Ren Liu, and Teng-Ming Chen. "High color rendering white light-emitting-diode illuminator using the red-emitting Eu^{2+}-activated CaZnOS phosphors excited by blue LED." *Opt. Express* 18.8 (2010): 8187-8192.
7. Kuznetsov, A. S., et al. "Ultraviolet-driven white light generation from oxyfluoride glass co-doped with Tm^{3+}-Tb^{3+}-Eu^{3+}." *Applied Physics Letters* 102 (2013): 161916.
8. Lin, Chun Che, and Ru-Shi Liu. "Advances in phosphors for light-emitting diodes." *The Journal of Physical Chemistry Letters* 2.11 (2011): 1268-1277.
9. Da-wei, T. U., et al. "Effect of optical structure on output light intensity distribution in LED package." *Optics and Precision Engineering* 5.16 (2008): 832-838.
10. Liu, Zong-Yuan, et al. "Studies on optical consistency of white LEDs affected by phosphor thickness and concentration using optical simulation." *Components and Packaging Technologies, IEEE Transactions on* 33.4 (2010): 680-687.
11. Sommer, Christian, et al. "The impact of light scattering on the radiant flux of phosphor-converted high power white light-emitting diodes." *Lightwave Technology, Journal of* 29.15 (2011): 2285-2291.
12. Tran, Nguyen T., and Frank G. Shi. "Studies of phosphor concentration and thickness for phosphor-based white light-emitting-diodes." *Journal of lightwave technology* 26.21 (2008): 3556-3559.
13. Liu, Mu-Qing, et al. "Study on methodology of LED's luminous flux measurement with integrating sphere." *Journal of Physics D: Applied Physics* 41.14 (2008): 144012.
14. Muthu, Subramanian, Frank J. Schuurmans, and Michael D. Pashley. "Red, green, and blue LED based white light generation: issues and control." *Industry Applications Conference, 2002. 37th IAS Annual Meeting. Conference Record of the.* Vol. 1. IEEE, 2002.

15. Li, Shuiming, et al. "Realization of high-quality light output based on a novel LED packaging." *Electronic Components and Technology Conference (ECTC), 2013 IEEE 63rd.* IEEE, 2013.

16. Hsin-Ying Lee, Member, IEEE, Yu-Chang Lin, I-Hsing Chen, and Chia-Hsin Chao, "Effective Color Conversion of GaN-Based LEDs via Coated Phosphor Layers," *IEEE Photon. Technol. Lett.*, vol. 25, no. 8, pp. 1041-1135, Apr. 15, 2013.

17. Commission Internationale de I'Eclairage (CIE), Colorimetry, CIE 15:2004.

The Effects of Self-Fluxing Additives in Solder Anisotropic Conductive Films (ACFs) on Solder Wettability and Joint Reliability of Flex-On-Board (FOB) Assemblies

Seung-Ho Kim, Yongwon Choi, Yoosun Kim and Kyung-Wook Paik

Nano Packaging and Interconnect Lab. (NPIL)

Department of Materials Science and Engineering

Korea Advanced Institute of Science and Technology (KAIST)

291 Daehak-ro, Yuseong-gu, Daejeon, 305-701, Korea

phone: +82-42-350-3375, fax:+82-42-350-8124

e-mail: dedoop1@kaist.ac.kr

Abstract

In this study, self-fluxing additives were added in solder anisotropic conductive films (ACFs) in order to eliminate solder oxide resulting in an excellent solder wetting on metal pads during ACF bonding. Since the solder oxide causes poor wettability of solder particles, solder oxide was chemically removed by self-fluxing additives in solder ACFs.

The test boards were 25 μm-thick polyimide based FPCs and 1 mm-thick FR-4 organic rigid PCBs which have 400 μm pitch Cu patterns with electroless nickel and immersion gold (ENIG) surface finish. Newly formulated solder ACFs were acrylic based adhesives film which can be fully cured above 150°C containing self-fluxing additives. The film contained 25 μm diameter Sn58Bi particles which has 138°C melting point and 8 μm diameter Ni particles as a spacer to maintain the gap between metal electrodes.

According to the experimental results, the addition of self-fluxing additive caused significant improvement of solder wettability and reliability of solder ACF joints. Therefore, using a flux function added solder ACFs can be used for various applications such as FOB and FOF assemblies, and can provide an alternative interconnection method for high power and fine pitch assemblies for many other applications.

Introduction

The Anisotropic Conductive Film (ACF) which consists of thermo-setting polymer matrix and conductive balls is commonly used in display manufacturing to make the electrical and mechanical connections between two electrical parts such as chips, glasses, FPCs, and PCB boards [1]. The ACF has advantages of low bonding temperature, elimination of lead, underfill, and flux, compared to traditional soldering technology. However, it has a shortcoming in the power handling capability caused by narrow conduction paths. Nevertheless, there have been many needs of alternative ACFs which have high power handling properties and improved reliabilities because the applications of ACFs bonding have been extended to module interconnection requiring higher power and finer pitch assembly in portable electronic and flexible electronic devices [2]. Therefore, solder ACFs which consist of a thermosetting adhesive matrix and solder particles were introduced from our group.

Using the solder ACFs, soldering on metal electrodes in a thermosetting adhesive matrix was expected. However, unstable soldering occurred due to solder native oxides in

conventional thermo-compression bonding method. In the first approach to remove the oxides, we adopted ultrasonic bonding method. In this method, excellent soldering on electrodes was obtained because a mechanical vibration broke the oxides well above the solder melting temperature [3]. Nevertheless, the other method which can remove the oxides without mechanical vibration have been required in packaging industry for mechanically sensitive applications [4]. Therefore, we have introduced the epoxy based solder ACF bonding technology with self-fluxing additive which can be applied in 250°C bonding processes before [5]. Even though stable soldering could be obtained in the bonding processes, there was needs of a minimized bonding temperature to prevent a thermal damage of substrates due to too high hot bar temperature (over 300°C). Thus, we have investigated ways of lowering bonding temperature using low temperature curable adhesives and Sn58Bi particles which has 138°C melting point.

Experiment

1. Material preparation

Figure 1 shows the test vehicles. The test substrates consists of a 25 μm thick polyimide based FPC and a 1 mm thick FR-4 PCB. The test vehicles contain 300 μm pitch Cu patterns with ENIG surface finish. Newly formulated solder ACFs were 50 μm thick acrylic based adhesive films containing 2 wt% of Thermal Acid Generator (TAG) as a self-fluxing additive. The films contain 25-32 μm diameter Sn58Bi solder particles which act as conductive particles and 8 μm diameter Ni particles which act as a spacer to maintain constant gap between two electrodes. The curing temperature of the adhesive films and melting temperature of solder particles were measured by Differential Scanning Calorimeter (DSC).

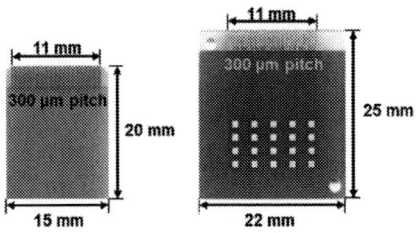

Figure 1. Test vehicles; polyimide based FPC (left) and FR-4 PCB (right) (left), FR-4 organic rigid boards (right)

978-1-4799-2408-0/14 $31.00 © 2014 IEEE

2241

2014 Electronic Components & Technology Conference

2. Thermo-compression bonding method

In order to investigate the effects of bonding temperature, 150°C and 180°C bonding were conducted with 2 MPa bonding pressure for 10 s. To measure the in-situ temperature of the internal ACF layer, a K-type thermocouple was used with 0.1 s sampling rate. Figure 2 shows the in-situ ACF temperature measurement structure.

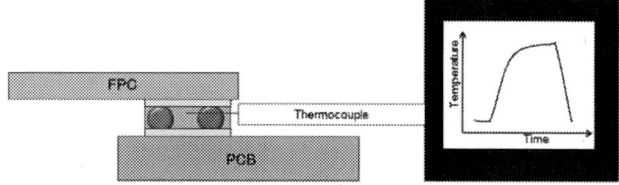

Figure 2. Schematic diagram of in-situ ACF temperature measurement

3. Wettability observation method

In order to observe the solder wettability in solder ACFs, the following method in Figure 3 was used.

1. Solder ACF bonding using thermo-compression bonding
2. Peeling away the FPC from the PCB
3. Cleaning the ACF resin using an acetone
4. Residual solders on the electrode indicated wetted solder.

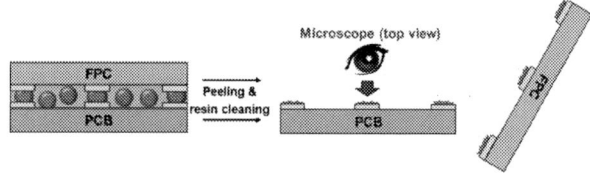

Figure 3. Schematic diagram of wetting area observation method

4. Reliability evaluation of solder ACF joints

For reliability evaluation, an unbiased autoclave test (121°C, 2 atm, 100%RH) was performed in order to compare contact resistances during the test. To measure the contact resistances, 4-terminal sensing method described at Figure 4 was used.

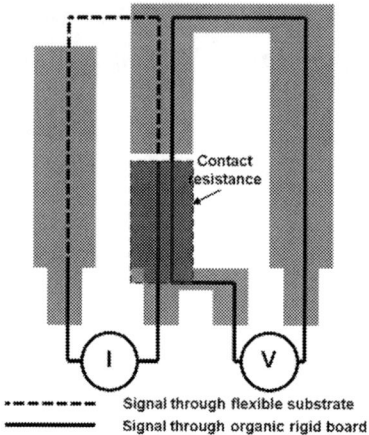

Signal through flexible substrate
Signal through organic rigid board

Figure 4. Schematic diagram of 4-terminal sensing method

Results and discussion

1. In-situ ACF temperature during thermo-compression bonding

As shown in Figure 5, the bonding temperatures were measured. Because the saturated ACF temperatures (150°C and 180°C) were higher than melting temperature of Sn58Bi solder shown in Figure 6 as well as activation temperature of self-fluxing additive (120°C), sufficient solder wetting was expected.

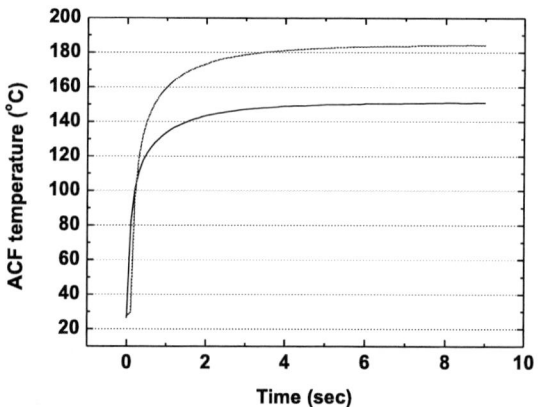

Figure 5. In-situ ACF temperature during thermo-compression bonding

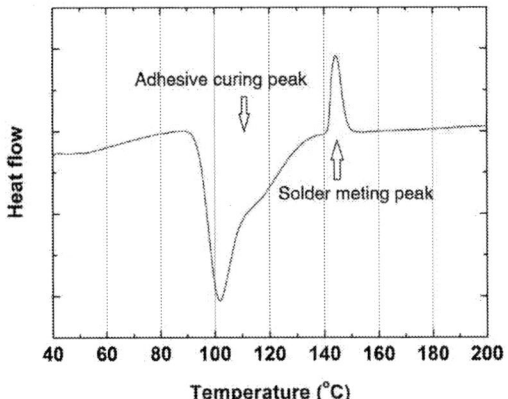

Figure 6. DSC dynamic scan results of the solder ACF.

2. Solder wettability on PCB electrodes

The solder residues on PCB electrodes are shown in Figure 7. In the case of the 150°C bonding sample without an additive solder ACF, the center parts of solder residues were empty. It meant that insufficient solder wetting occurred due to solder oxides. On the other hand, solder wettability increased as an amount of the additive and bonding temperature increased due to self-fluxing additive.

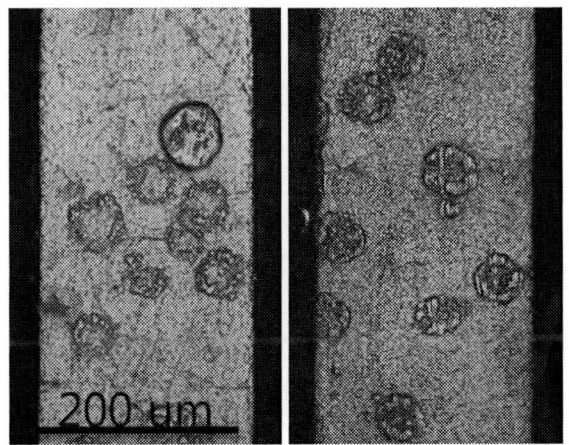

a) Solder residues on PCB electrode of 150°C bonding sample; no additive (left) and 2 wt% additive (right)

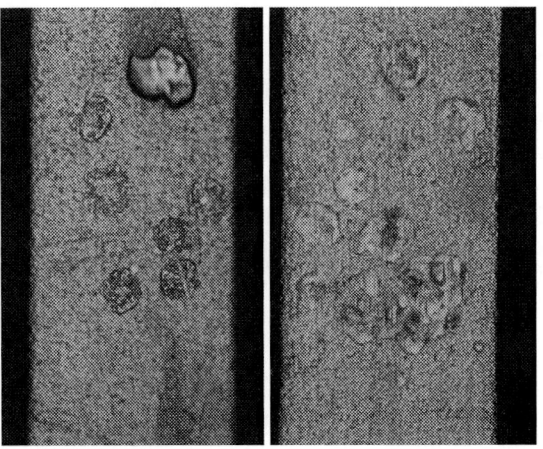

b) Solder residues on PCB electrode of 180°C bonding sample; no additive (left) and 2 wt% additive (right)

Figure 7. Top view images of wetted solder on the PCB electrodes after FPCs were peeled away

The solder wettability observed with not only solder residues on electrodes but also cross-sectional images of solder joints as shown in Figure 8. The interface between a solder and a PCB electrode in the 150°C bonding sample with no additive, showed poor wettability. However, the solder wettability improved as the amount of the additive and bonding temperature increased because rates of the additive activation and the oxide reduction increased.

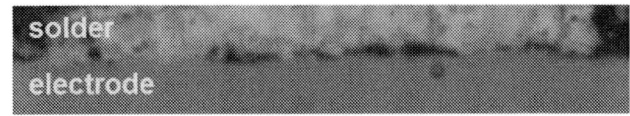

a) 150°C bonding, no additive

b) 150°C bonding, 2 wt% additive

c) 180°C bonding, no additive

d) 180°C bonding, 2 wt% additive

Figure 8. Cross-sectional images of interfaces between solders and PCB electrodes

3. Effects of the self-fluxing additive on reliability of solder ACF joints

After 72 h of the unbiased autoclave test (121°C, 2 atm, 100%RH), solder ACF joints without self-fluxing additive showed unstable contact resistances as shown in Figure 9. However, the contact resistances of solder ACF joints with 2 wt% additive was quiet stable. Moreover, the bonding temperature also affected contact resistances of solder ACF joints. The results well correlated with the wettability results above-mentioned because solder ACF joints which have good wettability must show the stable reliability.

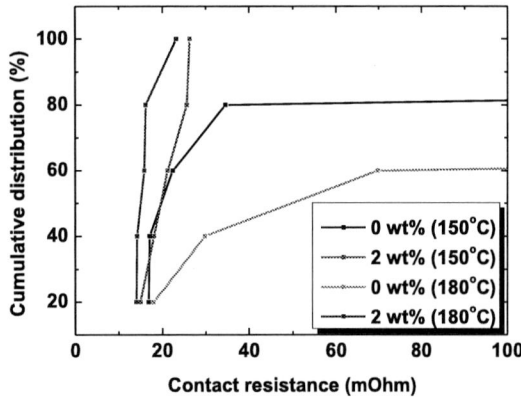

Figure 9. Contact resistance of solder ACF joints with various contents of the self-fluxing additive and bonding temperatures after 72 h auto clave test (121°C, 2 atm, and 100%RH)

Conclusions

In this study, in order to realize a highly reliable FOB module assembly, flux function added solder ACF was successfully demonstrated using a conventional thermo-compression bonding method.

According to the experimental results, the addition of self-fluxing additive caused significant improvement in the solder wettability and reliability of solder ACF joints. After the thermo-compression bonding, solder ACF joints containing 2 wt% of the additive showed stable solder metallurgical joints with metal electrodes. At the same time, solder ACF joints containing 2 wt% of the additive showed enhanced reliability in an unbiased autoclave test (121°C, 2 atm, and 100%RH) compared to the solder ACF joints containing no additive. Moreover, the bonding temperature was the one of factor which affected above properties due to the correlation with a reaction rate of the additive.

The significance of this result is that stable solder ACF joints can be obtained by normal 180°C thermo-compression bonding. Therefore, the flux function added solder ACF can be used for various FOB and FOF applications which provides an alternative interconnection for high power and fine pitch assemblies.

References

1. N. SHIOZAWA, K. ISAKA, and T. OHTA, "ELECTRIC PROPERTIES OF CONNECTIONS BY ANISOTROPIC CONDUCTIVE FILM", Journal of Electronics Manufacturing, Vol.05, Issue 01, pp33-37, March 1995
2. S.H. FAN and Y.C. CHAN, "Current-Carrying Capacity of Anisotropic-Conductive Film Joints for the Flip Chip on Flex Applications", Journal of Electronic Materials, Vol. 32, No. 2, 2003
3. Kiwon Lee, Saarinen Ilkka J, Pykari Lasse, and Kyung-Wook Paik, "Micro-solder/adhesive hybrid joints for high-density, high-power, high-reliability, and reworkable module interconnection in mobile phones", In proceedings of Electronic Components and Technology Conference, 412-415, 2012
4. Yoosun Kim, Kiwon Lee, and Kyung-Wook Paik, "Ultrasonic-assisted thermo-compression bonding method for high-performance solder Anisotropic Conductive Film (ACF) joints", In proceedings of Electronic Components and Technology Conference, 465-468, 2012
5. Seung-Ho Kim, Yongwon choi, Yoosun Yim, and Kyung-Wook Paik, "Flux Function added Solder Anisotropic Conductive Films (ACFs) for High Power and Fine Pitch Assemblies", In proceedings of Electronic Components and Technology Conference, 1713-1716, 2013

Modeling and Analysis of Frequency Shift of MEMS Gyroscope Subjected to Temperature Change

Weihui Wang[1], Sheng Liu[2]*, Zhang Luo[1], Ming Wen[1], Qiang Dan[1], Man Yu[1], Yong Xu[2,3], Xin Wu[4]

[1]School of Mechanical Science and Engineering, Huazhong University of Sci & Tech, Wuhan, Hubei, 430074, China
[2]Cross-disciplinary Institute of Engineering Sciences, School of Power and Mechanical Engineering, Wuhan University, Wuhan, Hubei, 430072, China
[3]Department of Electrical and Computer Engineering, Wayne State University, Detroit, MI 48202 USA
[4]Department of Mechanical Engineering, Wayne State University, Detroit, MI 48202 USA
* victor_liu63@126.com

Abstract

The mechanism of resonant frequency shift of MEMS gyroscope subjected to environment temperature change was investigated comprehensively in this paper, and its computational model was set up based on certain assumptions. Taking the structure of MEMS gyroscope as an example using the finite element method, firstly, the authors derived the simplified M-C-K dynamic model. Then, thermal stress and the thermal deformation of the model were calculated within temperature range from -40℃to 120℃, and the regions of maximum value of thermal stress and thermal deformation were selected in the regions between anchors and driving spring beams. Finally, the analysis of modes was implemented under varying temperature ranging from -40℃ to 120℃, and the relation between resonant frequency shift and temperature change was given at the reference temperature of 20℃. The results showed that the driving resonant frequency decreased while the resonant sensing frequency fluctuated in certain value when the environment temperature increased, which indicated that frequency shift was greatly dependent on thermal stress of the driving spring beams that suspended the driving comb structure.

Introduction

MEMS gyroscope is a typical sensor which can measure the input angular rate information with respect to an inertial reference frame based on a transfer of energy between two modes of vibration caused by the Coriolis effect. In recent years, MEMS gyroscopes with advantages on their miniaturization, light weight, low-power, low-cost and the ability to mass fabricate, have a wide range applications, such as in consumer electronics, automotive industry and navigation system [1]. The necessity for stable high performance inertial sensors has always been the force of driving development of MEMS gyroscopes. Although, there are some critical factors like structure design, material inhomogeneity, manufacturing process and environment temperature variation which may degrade its performance and even induce errors, and among these disturbances, environment temperature change and its negative effects on the performance are of great interest [2]. Temperature change could cause the resonant frequency shift of driving and sensing mode, which directly affects the sensitivity, the quality factor, the bandwidth, the bias instability, etc., so that it is necessary to make sense of the relation between resonant frequency shift and temperature change. Ferguson et al. [3]

indicated that the drive and sense resonant frequency have a linear relationship with temperature over a range. Sun et al. [4] investigated the influence of changing temperature upon the amplitude and phase of the drive and sense modes. The deviations of the frequencies and dynamic characteristics are calculated accurately in [5], which provides a reference for the design of temperature compensation method. The origin of thermal elastic nonlinearities in contour mode of AlN resonators was analyzed, and the thermal nonlinear coefficients were extracted experimentally in [6]. In this paper, the mechanism of resonant frequency shift due to temperature change was investigated comprehensively, and the reminders of this article is organized as follows: Section 2 introduces the structure and principle of MEMS gyroscope taking a double mass gyroscope as an example; thermal stress and thermal deformation were analyzed in Section 3; the simulation results of frequency shift due to temperature change from -40℃ to 120℃ come in Section 4, followed by discussion and conclusions in Section 5.

Structure and Principle of MEMS Gyroscope

One linear vibrating MEMS gyroscope with symmetrical and coupled structure is used to explain its working principle in this paper, and its schematic diagram is shown as Figure 1.

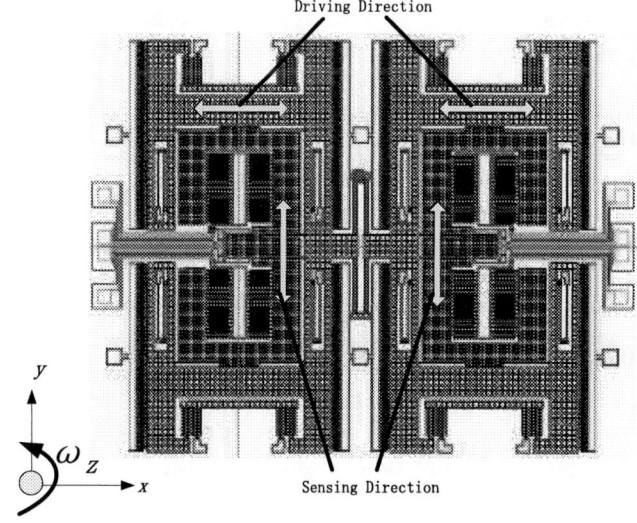

Figure 1. Schematic diagram of an MEMS gyroscope.

We can see that the presented gyroscope has two proof masses, and use differential capacitance method to detect the input angular velocity ω_z. The structure is driven to oscillate in the x direction, and the proof mass would also oscillate in the y direction by the Coriolis force when there is input angular velocity perpendicular to the vibration plane. The input angular velocity would be modulated with capacitance detection in the direction. The model could be described as the simplified "*Mass-Damping-Spring*" dynamic system shown in Figure 2.

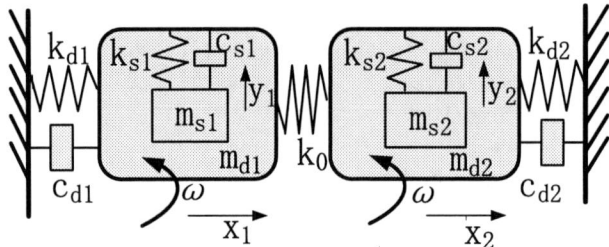

Figure 2. Model of the MEMS gyroscope system.

The vibratory gyroscope is modeled by the second order vector differential equation

$$M\ddot{q} + (C + 2MG\omega)\dot{q} +$$
$$(K + MG^2\omega^2 + MG\dot{\omega})q = F + Ma \tag{1}$$

where M is the symmetric mass damping matrix, C is the symmetric damping matrix, K is the symmetric stiffness matrix, q is the four dimensional displaces, F is the forcing vector, a is the external acceleration vector, G is skew symmetric gyroscope matrix, and ω is the input angular velocity. Eq.(1) is the full governing equation, and there are many simplified styles with some assumptions in practice.

Influence on Resonant Frequency Shift Due to Temperature Change

The modal equation of the dynamic system is expressed as

$$M\ddot{q} + C\dot{q} + Kq = 0 \tag{2}$$

The effect of temperature change on MEMS gyroscope consists of the three different changes in the Young's modulus, thermal expansion and thermally induced stresses [5] which would induce change of the stiffness matrix K, therefor, the resonant frequencies would shift with temperature variations.

The model of MEMS gyroscope is constructed and then sufficiently meshed by the FEM software ABAQUS. The geometry feature of MEMS gyroscope is shown as Figure1, and its material parameters with temperature change such as Young's modulus and thermal expansion coefficient come from references [7,8,9]. As the third important factor affecting the resonant frequencies, the thermal stress simulation [10] would be done next.

In the process, we take "μm-N-s" dimension system, so that the units of deformation, stress, and frequency would be "μm", "N/(μm)^2" and "Hz".

The thermal stress from -40℃ to 120℃ with respect to the reference temperature of 20℃ is calculated, and the maximum value of von Mises stress are presented in Figure 3, and also von Mises stress contours of the structure at the temperature of 120℃ is shown in Figure 4. From Figure 4, we can find that there is large thermal stress in the regions of driving spring

beams between anchors and structure while nearly no thermal stress in the regions of sensing spring beams, which would have different effects on resonant frequency shift of driving and sensing when environment temperature changes. From Figure 3, the maximum value of von Mises stress would come to be larger when the temperature difference is larger with respect to room temperature.

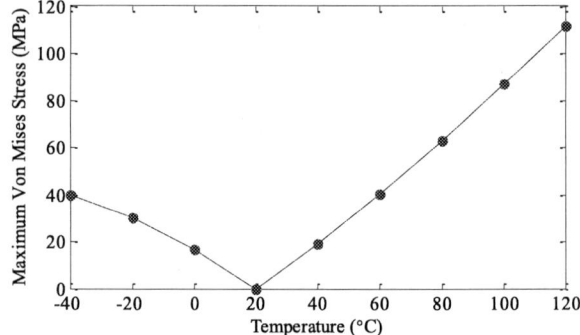

Figure 3. Maximum von Mises stress with temperature change.

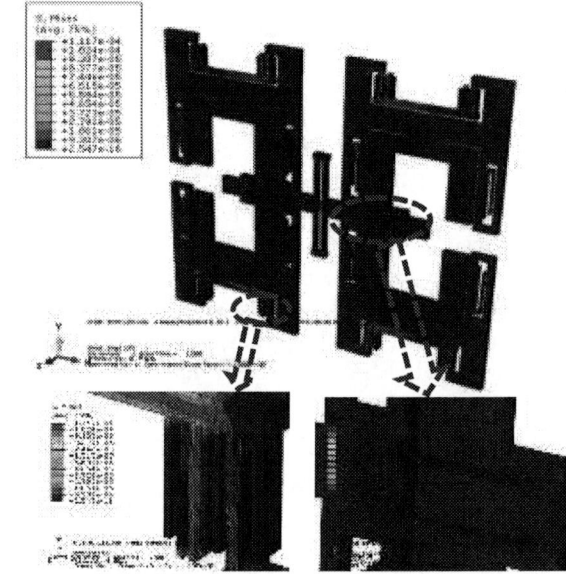

Figure 4. Global and local von Mises stress contours of MEMS gyroscope structure at 120℃.

At the same time, the thermal deformation of the structure in the temperature range is also calculated, and he max value of thermal deformation ranging from -40℃ to 120℃ are presented in Figure 5, then, its contours at 120℃ are shown in Figure 6.

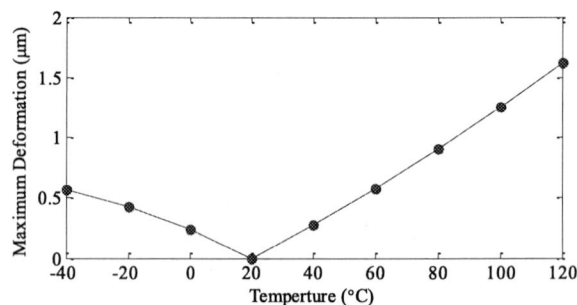

Figure 5. Maximum thermal deformation with temperature change.

978-1-4799-2408-0/14 $31.00 © 2014 IEEE

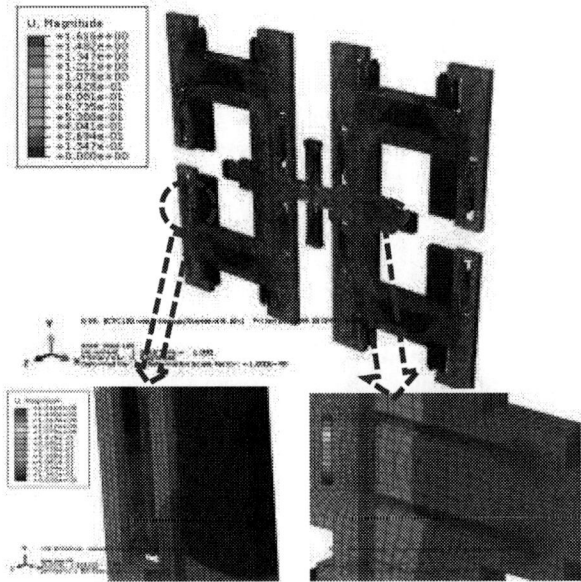

Figure 6. Global and local thermal deformation contours of MEMS gyroscope structure at 120℃.

From Figure 5 and Figure 6, we find that there is only little thermal deformation comparing to the structure dimension in the regions near the driving spring beams, which means such thermal deformation almost have no effect on structure stiffness, therefore, no effect on the resonant frequency shift in the temperature range.

Above, we have obtained the three crucial factors which have effects on the resonant frequency shift of MEMS gyroscope when environment temperature changes, Young's modulus, thermal expansion from literatures, and thermal stress from calculation with the FEM software ABAQUS. Considering the three factors above, we can do model analysis of MEMS gyroscope structure in the temperature range from - 40 ℃ to 120 ℃ , and the resonant frequencies at room temperature 20℃ are chosen as the referenced state. The first four modes of the MEMS gyroscope structure at the temperature of 20℃ are shown in Figure 7.

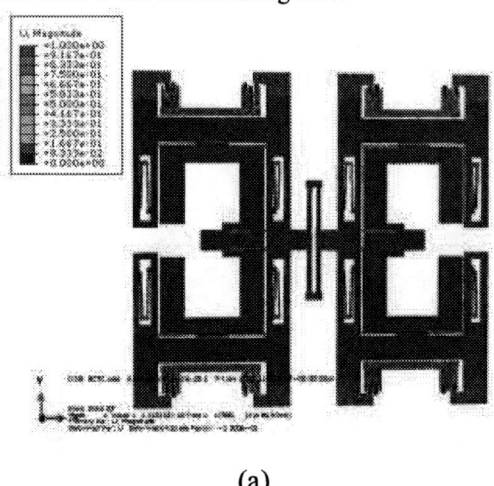

(a)

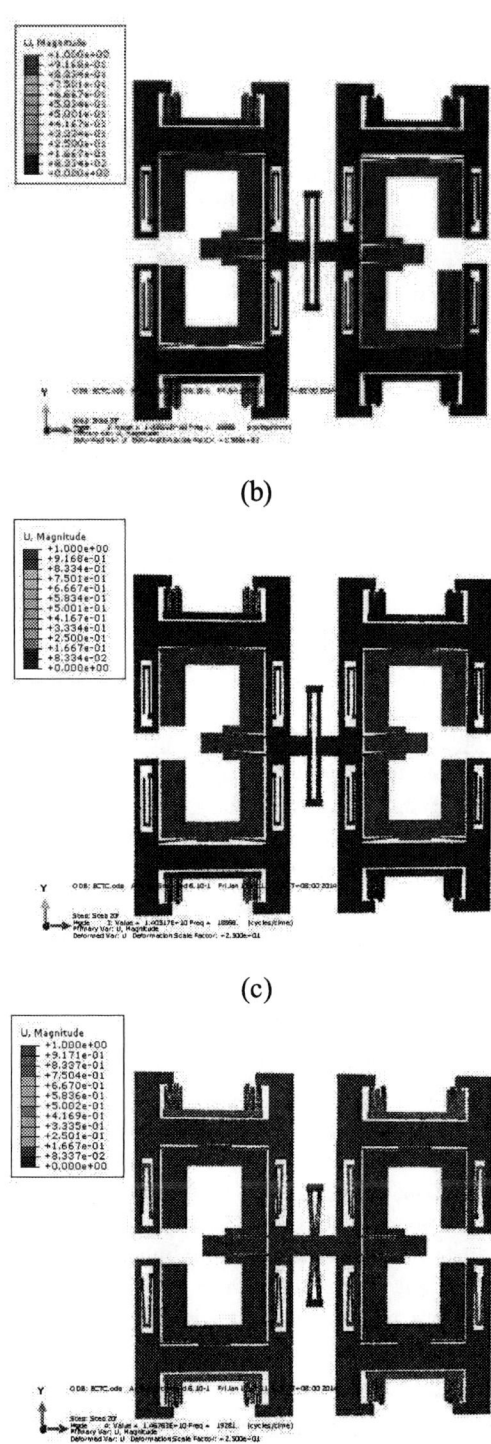

(b)

(c)

(d)

Figure 7. Schematics of modes analysis: (a) Drive in-phase mode (17566 Hz); (b) Sense reverse-phase mode (18866 Hz); (c) Sense in-phase mode (18866 Hz); (d) Drive reverse-phase mode (19281 Hz).

From Figure 7, there are two drive modes, the drive in-phase mode (a) and the drive reverse-phase mode (d), which both vibrate in the x direction. And also, there are two sense mode, the in-phase mode (c) and the reverse-phase mode (b), which both vibrate in the y direction. The two sense frequencies are between the two drive frequencies because of the modulation effect of center drive spring beams, otherwise, the drive and sense resonant frequencies would match well. Theoretically, the smaller the stiffness of the center drive spring beams is, the better the two resonant frequencies match. In practice, drive reverse-phase mode (d) and sense reverse-phase mode (c) are often used, which can take the difference detection to resist external vibration noise.

The first four resonant frequencies are respectively presented in Table 1 with the environment temperature changing from -40℃ to 120℃.

Table 1. Frequency Shift due to Temperature Change

Frequency(Hz) / Temp(℃)	f_1	f_2	f_3	f_4
-40	17605	18883	18884	19324
-20	17590	18873	18873	19308
0	17575	18865	18866	19291
20	17566	18866	18666	19281
40	17565	18877	18878	19280
60	17556	18882	18883	19269
80	17538	18880	18880	19250
100	17528	18888	18888	19238
120	17506	18886	18886	19215

* f_1: drive in-phase frequency;
f_2: sense reverse-phase frequency;
f_3: sense in-phase frequency;
f_4: drive reverse-phase frequency.

According to Table 1, we could find that the resonant frequencies change with temperature change; the drive resonant frequencies (f_1 and f_4) descend while the temperature increases, and their variable magnitude is within 110 Hz from -40℃ to 120℃; at the same time, the sense resonant frequencies (f_2 and f_3) fluctuate in the certain value with variable magnitude of 23 Hz when environment temperature increases.

All results above indicated that change of Young's modulus, thermal expansion and thermal stress give contribute to the resonant frequency shift because they together influence the stiffness of MEMS gyroscope structure. Furthermore, thermal stress contributes more to resonant frequency shift than the two others due to large thermal stress in the regions between anchors and drive spring beams while almost no thermal stress near the sense spring beams.

Discussion

The performance of MEMS gyroscope is greatly affected by environment temperature change since it may cause the resonant frequency shift of drive mode and sense mode. In order to make sense of the relationship, the mechanism of resonant frequency shift due to temperature change was comprehensively investigated in this paper. The authors set up a finite element model of MEMS gyroscope to illustrate these issues. We find that there are three critical factors causing the resonant frequency shift of MEMS gyroscope structure, change of Young's modulus, thermal expansion and thermal stress with temperature change. What's more, thermal stress plays more important role than the two others based on the fact that drive resonant frequency descends while sense

resonant frequency fluctuate in certain value, that is because there is large thermal stress in the regions of drive spring beams while almost no thermal stress in the regions of sense spring beams. All of results above could give suggestions to improve temperature robustness of MEMS gyroscope structure, such as designing perfect structure to decrease thermal stress.

Acknowledgment

This work is supported by National High Tech Program (863) of Ministry of Science and Technology with No. 2012AA040501 and MEMS-volume Manufacture Technology of National Basic Research Project (973) under Minister of Science and Technology with No. 2011CB309504.

References

1. Fei, Juntao, et al, "Robust adaptive neural sliding mode approach for tracking control of a MEMS triaxial gyroscope," *Int J Adv Robotic Sy*, vol. 9, no.24, pp.1-8, 2012.

2. Jiancheng F, Jianli L, "Integrated model and compensation of thermal errors of silicon microelectromechanical gyroscope," *Instrumentation and Measurement, IEEE Transactions on*, vol.58, no.9, pp. 2923-2930, 2009.

3. Ferguson MI, Keymeulen D, Peay C, Yee K, Li DL (2005) Effect of temperature on MEMS vibratory rate gyroscope. *Aerospace Conference IEEE*, pp.1-6

4. Sun F, Guo QF, Ge YS, Li JS, "Research on thermal characteristic in slow-small temperature changing for MEMS linear vibration gyroscope," *Proceedings of the 2006 IEEE International Conference on Mechatronics and Automation*, Luoyang, China, 2006, pp.475-479

5. Liu, Guangjun, et al, "Effects of environmental temperature on the performance of a micromachined gyroscope," *Microsystem Technologies*, vol.14, no.2, pp.199-204, 2008.

6. Segovia-Fernandez, Jeronimo, and Gianluca Piazza, "Thermal nonlinearities in contour mode AlN resonators," *Journal of Micromechanics and Microengineering*, vol.22, no.4, pp.976-985, 2013.

7. Painter C C, Shkel A M, "Structural and thermal modeling of a z-axis rate integrating gyroscope," *Journal of Micromechanics and Microengineering*, 2003, vol.13, pp. 229-237.

8. Hopcroft M A, Nix W D, Kenny T W, "What is the Young's Modulus of Silicon?" *Journal of Microelectromechanical Systems*, vol.19, no.2, pp.229-238, 2010.

9. Cho C H, "Characterization of Young's modulus of silicon versus temperature using a "beam deflection" method with a four-point bending fixture," *Current Applied Physics*, vol.9, no.2, pp.538-545, 2009.

10. S. Liu, Y. Liu, *Modeling and Simulation for Microelectronic Packaging Assembly: Manufacturing, Reliability and Testing*, John Wiley & Sons, 2011.

Interaction Effect between Electromigration and Microstructure Evolution

in Cu/Sn-58Bi/Cu Solder Interconnect

Hong-Bo Qin[a], Bin Li[b], Wu Yue[a], Chang-Bo Ke[a], Min-Bo Zhou[a], Xin-Ping Zhang[a],*

[a] School of Materials Science and Engineering, South China University of Technology, Guangzhou 510640, China

[b] Southern Methodist University, Dallas, TX,75205,USA

* Email: mexzhang@scut.edu.cn; Tel: +86-20-22236396

Abstract

The interaction effect between electromigration (EM) and microstructure evolution in Cu/Sn-58Bi/Cu solder interconnect with asymmetric configuration along current direction was studied by cellular automaton (CA) modeling embedded with finite element (FE) simulation, with a comparison to in-situ scanning electron microscope (SEM) observation and focus ion beam (FIB) microanalysis. The CA method was proposed to simulate the eutectic structure of Sn-58Bi solder alloy. Results show that the accumulation of Bi-rich phase tends to occur near the bottom corner position of the interconnect owing to the smallest local electrical resistance and severest current crowding effect. At the bottom corner of the interconnect, the current density in Sn-rich phase is much higher than that in Bi-rich phase. Bi atoms in Sn-rich phase are more prone to migrate to the anode first, rather than migrating directly along the interface between Bi-rich phase and Sn-rich phase. By employing the criterion of EM induced atomic flux of Bi in FE analysis under CA rules, simulation results of Bi-rich phase segregation are consistent with the experimental observation under current stressing.

1. Introduction

In electronic products and devices, solder interconnects provide effective electrical, thermal and mechanical connections among different components and various circuits, including chips. With increasing miniaturization of electronic devices and systems, the dimension of solder interconnects and pitches has been continuously scaling down. Accordingly, current density in solder interconnects has increased dramatically, resulting in severe electromigration (EM) effect which leads to severe electromigration-related structural problems and defects such as microstructure coarsening, intermetallic compound (IMC) polarity or abnormal polarity growth, whiskers, hillocks and voids [1-3]. These structural problems and defects may deteriorate the reliability of solder interconnects which are generally regarded as the weakest parts in packaging systems and electronic assemblies [4]. EM has now become one of the most predominated reliability issues in interconnects of microelectronic devices. Among all lead-free solders, eutectic Sn-Bi alloy is a promising candidate for low-temperature applications due to the low melting point and competitive price. Previous studies found that, in eutectic Sn-58Bi of solder interconnects, EM has significant influence on the formation of IMC, phase coarsening and accumulation of Bi atoms [5, 6]. Although eutectic structure has been confirmed to be highly inhomogeneous in the solder matrix of microscale interconnects by experimental observation, in most experimental and numerical simulation studies, Sn-58Bi alloy

was treated as a homogeneous composition [5, 7]. Due to the lack of both numerical analysis and experimental validation, the interaction effect between EM and inhomogeneity of microstructure in the solder matrix of microscale solder interconnects is still ambiguous.

In numerical simulation, there are two methods to be used to investigate the initiation and evolution of eutectic structure and morphology, i.e., phase-field (PF) method and cellular automaton (CA) method. The PF method is based on the diffusion interface theory, and the interface is treated as a region with an extremely small thickness. Thus, the grid numbers in PF model have to be set large enough to ensure the calculation accuracy. This makes the simulation complicated and time-consuming, and strongly limits both the size and geometry of PF model in applications. For the CA method, the state of each cell is determined by the previous physical state of the cell and the physical states of cells in their neighborhood according to governing mathematical function (which may have information such as undercooling, solute concentration and surface energy, etc). For the cell (or grid) of CA model can be in arbitrary dimension and treated as "element" in finite element (FE) simulation, it is more appropriate to apply CA model in a wide range of dimension and geometry, which can combine with FE method (CA-FE) for further multiple physical analysis, such as stress, thermal and electric fields.

In this study, EM behavior of the microscale Cu/Sn-58Bi/Cu right-angle type solder interconnect under a direct current (DC) density was investigated by CA-FE modeling and simulation with a comparison to the experimental observation and microanalysis.

2. Experiments and simulation method

The solder interconnect samples were assembled using the eutectic Sn-58Bi solder on a specially designed fixture by modeling reflow soldering process at the temperature 190 °C with a dwelling time of 60 s and then cooled in air. After soldering, the Cu/Sn-58Bi/Cu interconnect samples were embedded into the epoxy; finally, the embedded interconnect samples were carefully machined and polished to the final geometrical dimension as shown in Fig. 1(a). The procedure was detailed in our previous work [8]. The electromigration tests were conducted at a constant temperature (25°C) and humidity (60%) in the cleaning room, the direct current density is 1.5×10^8 A/m^2. In order to observe the locally entire segregation of Bi-rich and Sn-rich phases, the stressing time was prolonged to 288 h in this study. The whole observation section (including copper wires) of the solder interconnect was exposed to the air, the interconnect was cooled well and the joule heating could be alleviated effectively and timely. The microstructural evolution of the solder interconnects at different current stressing times was characterized by a scanning electron microscope (SEM, Supra 55) equipped with an energy dispersive X-ray spectroscopy (EDX, Oxford-7659). The phase segregation at the different inner microregions in the solder interconnect was investigated through cross-sectional analysis of the interconnects using focus ion beam (FIB, SMI 3050MS2) and SEM.

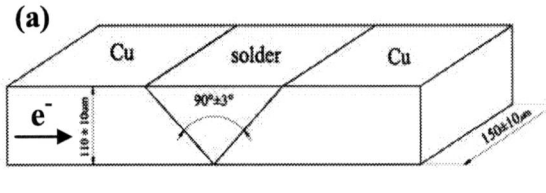

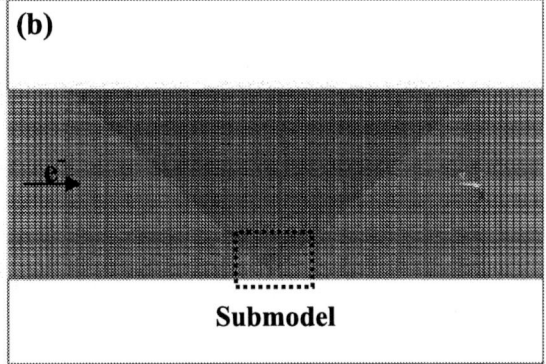

Submodel

Fig. 1. The right-angle type solder interconnect used in this study: (a) Geometry of the interconnect; and (b) FE model of the interconnect.

Due to the inhomogeneity of microstructure in the extremely small region (or volume) of the solder matrix and the influence of interconnect geometry on the current density distribution, it is very difficult to characterize the magnitude of current density and its vector in the solder matrix of interconnects by experimental methods. To understand the current crowding effect and electromigration behavior in the right-angle type solder interconnect as shown in Fig.1(a), FE model based on ANSYS 11.0 was performed to simulate the current density and EM induced atomic flux, in which the thermal-electric element type PLANE67 was employed and the material properties are given in Table 1. Considering that the IMC layers at both interfaces of the solder interconnect are very thin and the Joule heating can be dissipated timely, the influence of the IMC layers on the current density distribution can be neglected. The finite element model is shown in Fig. 1 (b), in which each element is a square with a length of 1 μm. Boundary constraints and current loads are applied according to the experimental condition.

Table 1 Material properties of Cu and Sn-58Bi solder [10-12].

Materials	Density, g/cm^3	CTE, °C	Resistivity, $\times 10^{-8}\,\Omega\cdot m$
Cu	8.90	17.0×10^{-6}	1.68
Sn-58Bi	8.70	15.0×10^{-6}	38.3

Clearly, the mesh density of the FE model shown in Fig. 1(b) is not fine enough to evaluate the phases of eutectic structure such as Sn-rich and Bi-rich phases. Therefore, combined with CA program, the submodel approach is employed in FE analysis according to the Saint-Venant's Principle [13]. CA programming is a very intuitive way to model physical systems and allows the programmer to ultimate control over the behavior of each cell (or each element). A cellular automaton evolves in discrete time steps, with the value of the variable at one cell being affected by the values of variables at its neighborhood cells on the previous time [14]. In this study, the nine-neighbor square (or square of Moore neighbor type) is employed to simulate the eutectic microstructure of Sn-58Bi alloy, see Fig. 2, in which cell e (center cell) is determined by itself and its nearest neighbor cells a to i. With the control of outer rule F, physical state of cell e at time $t+1$ (i.e., $e(t+1)$) is updated according to its previous value $e(t)$ and the values of neighbor cells at time t, as expressed in Eq. (1)

$$e(t+1) = F\{a(t),b(t),c(t),d(t),e(t),f(t),g(t),h(t),i(t)\} \quad (1)$$

Previous studies [15, 16] argued that in the Cu/Sn-58Bi/Cu system the Bi atoms would mainly migrate along the interface between Sn-rich and Bi-rich phases and congregate in front of the anode under current stressing. Accordingly, the physic state of Bi-rich phase is critical in the CA program under current stressing.

In the submodel, there are 3600 cells (or elements, with an identical length of 0.2 μm). Above all, the phase composition should be assigned to each cell in the program. Assuming that phase composition of cell e is phase A (either Sn-rich phase or Bi-rich phase), cell e is controlled by the following rules after each cellular automata step (CAS):

Rule 1: If all of its neighbors are cells of phase A at time t, then it will remain unchanged in the next CAS.

Rule 2: Considering the influence of solute concentration and surface energy of neighbor cells on the physic sate of cell e, if more than four of its neighbors are cells of phase B (i.e., a different phase composition), then the cell e will change to a cell of phase B in the next CAS. Else, the Cell e remains the same.

Rule 3: If one or more than one of its neighbors is a cell of phase B, then cell e will be defined as a boundary cell. If the cell e is a Bi-rich phase cell as well as a boundary cell, then it may change to a Sn-rich phase cell under current stressing with a probability P, which is defined by Eq. (2),

$$P = \begin{cases} 0 & \text{if } J_{em} < J_{emc} \\ 0.5 & \text{if } J_{em} = J_{emc} \\ 1 & \text{if } J_{em} > J_{emc} \end{cases} \quad (2)$$

where J_{em} is the EM induced atomic flux of Bi, and J_{emc} is the critical value of J_{em}, both of them will be discussed in Section 3.3.

Rule 4: Considering the mass conservation, if one of the boundary cell e changes from Bi-rich phase to Sn-rich phase according to Rule 3, then a boundary cell of Sn-rich phase nearest to the anode will transform into Bi-rich phase in the solder matrix. If there are many eligible boundary cells nearest to the anode, then select one of them randomly.

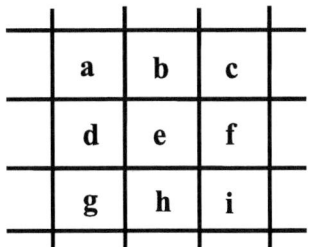

Fig. 2. Moore type square.

3. Results and discussion

3.1 CA-FE modeling of eutectic structure

For the ratio of Sn-rich phase area to Bi-rich phase area in Sn-58Bi eutectic microstructure is 1:1.03 [9], the ratio of generating probability of Sn-rich phase cell to that of Bi-rich phase cell is set to 1:1.03 using CA method. Fig. 3 (a) shows the generated initial state of Sn-58Bi alloy before the formation of primary phases, where cells of Sn-rich phase (dark contrast) and cells of Bi-rich phase (bright contrast) distribute randomly and uniformly in the solder matrix. After the first CAS, a eutectic structure is generated randomly according to Rule 1 and 2 in Section 2, see Fig.3 (b). By evolving the eutectic structure continuously (i.e., minimize the influence of solute concentration and surface energy on the microstructure), a stable state of eutectic structure is reached, as shown in Fig. 3(d). For the cell of CA model can be treated as an "element" in FE method, then FE model is created according to the CA simulation result for further analysis.

Fig. 4(a) presents the submodel of eutectic structure of Sn-58Bi alloy in the bottom corner of the interconnect, which is generated randomly by the CA-FE method. The location of the submodel is marked in Fig. 1(b). Fig. 4(b) shows the SEM image of the bottom corner of the right-angle type solder interconnect after reflow soldering. Clearly, both the Sn-rich phase (dark contrast) and Bi-rich phase (bright contrast) distribute randomly and inhomogeneously in the extremely small region of the solder matrix.

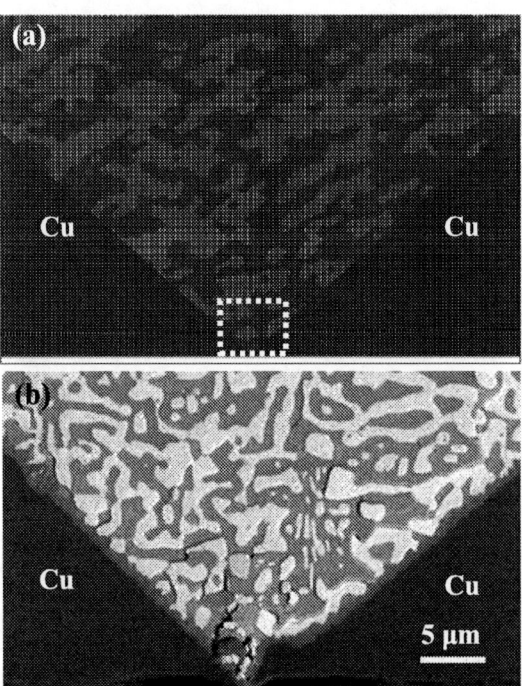

Fig. 4. Eutectic structure generated randomly in the submodel by CA-FE method (a) and the in-situ SEM image of eutectic structure of the solder matrix in the as-reflowed state (b).

3.2 Influence of eutectic structure on the current density

The thicknesses of both the congregated Bi-rich phase layer at the anode and Sn-rich phase layer at the cathode do not distribute uniformly along the two interfaces. After current stressing for 288 h, the segregation between Bi-rich and Sn-rich phases occurs significantly in the matrix of the solder at the bottom corner of the right-angle type interconnect (i.e., position A) by analyzing the microstructures in position A to I, as shown in Fig. 5 (a) and (b). That is, there is an obvious current crowding effect at the bottom corner. Fig. 5(c) shows the simulation result of current density distribution in the right-angle type solder interconnect. Clearly, the maximum value of current density value appears at the bottom corner of the interconnect, which is 3.30×10^8 A/m^2, being eighty five times as much as that at the top side with a value of 3.89×10^6 A/m^2. This can be explained by the shortest distance in the solder matrix along the current direction in the bottom corner. In other words, the local electrical resistance at the bottom corner is much smaller compared with that in the top side, so the smaller electrical resistance at the bottom corner of the interconnect leads to the local current crowding there. The simulation results are in good coincidence with the experimental observation.

Furthermore, the current crowding region at the bottom corner is extremely small and may be just several micrometers. However, the thicknesses of lamellar phases (or

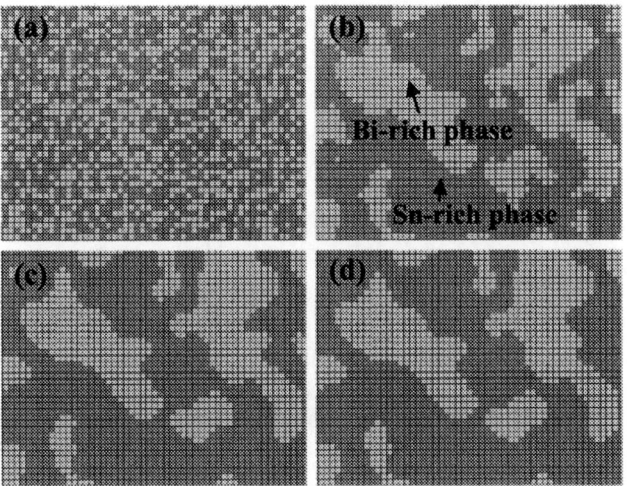

Fig. 3. Schematic of the evolution of eutectic Sn-58Bi solder alloy in the submodel at different values of CAS: (a) 0 CAS (initial state); (b) 1 CAS; (c) 4 CAS; and (d) 6 CAS.

the diameters of rod-like phases) are in the range of 0.5~2 μm (see Fig. 4). Therefore, it seems like that the inhomogeneous eutectic structure in the current crowding region may have an influence on the distribution of current density, which plays a crucial role in electromigration.

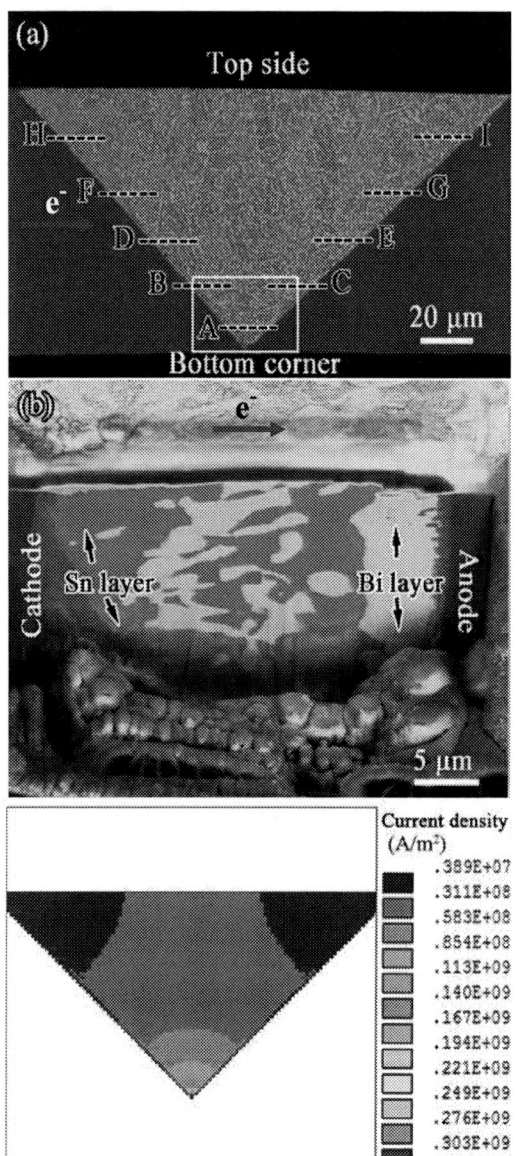

Fig. 5. Illustration of the cross-sectional position of the right-angle interconnect after current stressing (a), cross-sectional FIB-SEM image at position A undergoing current stressing for 288 h (b) and simulation result of distribution of current density in the solder interconnect (c).

Fig. 6 (a) shows the magnitude of current density in Sn-rich phase, which is much higher than that in Bi-rich phase in the zone shown in Fig. 6 (b), and the vector plot indicates that Bi-rich phases look like some "islands" in the "river" of current. It can be explained by the difference in resistivity between Bi and Sn, which is $1.1 \times 10^{-6}\,\Omega\cdot m$ and $1.1 \times 10^{-7}\,\Omega\cdot m$, respectively. Thus, the high current density is dominantly distributed in the Sn-rich phase. Accordingly, the Bi atoms

may not migrate directly along the interface between Sn-rich and Bi-rich phases under current stressing, as also mentioned in previous studies [15, 16]; in contrast, they migrate from the Sn-rich phase to the anode. Then, Bi atoms will diffuse from Bi-rich phase to Sn-rich phase, which can be explained by Fick's first law of diffusion, as given in Eq. (3),

$$ J = \lambda \frac{dc}{dx} \qquad (3) $$

where J is the atom flux, λ is the diffusion coefficient, and dc/dx is the concentration gradient. The decrease of Bi concentration in Sn-rich phase increases dc/dx between Sn-rich and Bi-rich phases, this will generate a force of concentration gradient. Then, Bi atoms along the interface between Sn-rich and Bi-rich phases diffuse from Bi-rich phase to Sn-rich phase to compensate the loss of Bi atoms in the Sn-rich phase under EM effect. If the diffused Bi atoms can not offset the loss of Bi atoms in the Sn-rich phase, then the structural damage in the Sn-rich phases would occur.

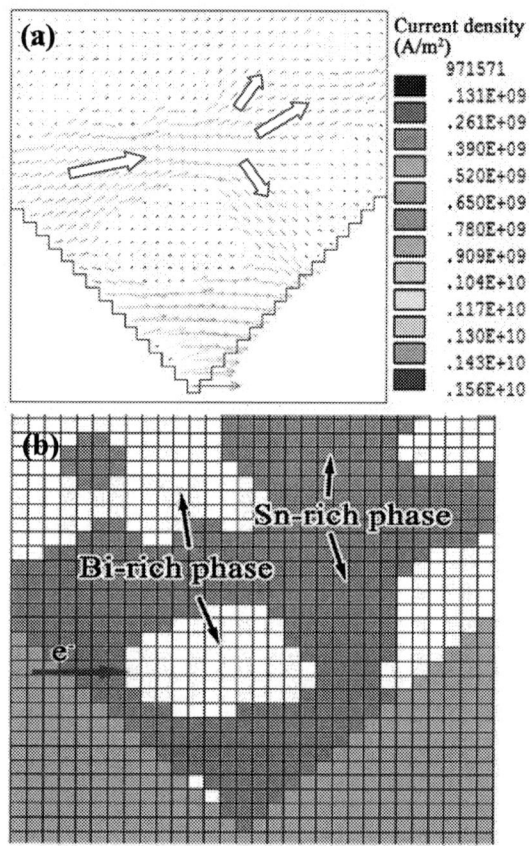

Fig. 6. Current density in the eutectic structure at the bottom corner of the interconnect: (a) The simulation result of current density and current direction in the bottom corner marked in Fig. 4 (a); and (b) Magnification of the region marked in Fig. 4(a).

3.3 Electromigration induced microstructure evolution

As mentioned above, Bi is the main diffusing species in the Cu/Sn-58Bi/Cu interconnect system. To calculate the electromigration induced atomic flux of Bi, the following Eq.

978-1-4799-2408-0/14 $31.00 © 2014 IEEE   2252

(4) is utilized [17]:

$$J_{em} = C\frac{D}{kT}Z^*e\rho j \tag{4}$$

where C, D, and Z^* are the concentration, diffusivity and effective charge number of the diffusing species Bi in the eutectic SnBi solder respectively, ρ is the resistivity of the eutectic Sn-58Bi solder, k is the Boltzman's constant (1.38×10^{-23} J/K), T is the temperature (298 K), e is the electron charge (1.6×10^{-19} C), and j is the current density, which can be calculated by FE method. According to a previous study, the concentration of Bi in the eutectic Sn-Bi solder (C) is 1.41×10^{22} atoms/cm^3, and the product of $D \times Z^*$ was calculated to be 4.72×10^{-10} [6]. Then, Eq. (4) can be expressed as:

$$J_{em} = 9.92 \times 10^9 j \tag{5}$$

By referring to the evolution of eutectic structure in the experimental observation, the critical current density in Bi-rich phase (j_c) used in simulation is 1.7×10^4 A/cm^2, and the value of J_{emc} is 16.9×10^{13} atoms/cm^2·sec according to Eq. (5). By evolving the boundary cell of Bi-rich phase according to Rule 3 and 4 in Section 2, EM behavior of Bi atoms in the cross-sectional region shown in Fig. 5(b) can be simulated. Fig. 7(a) shows the initial eutectic structure of Sn-58Bi alloy generated randomly by CA-FE method. After evolving 160 CAS under current stressing, Bi-rich phase congregates near the anode interface obviously, while the amount of Bi-rich phase near the cathode decreases and disappears gradually, see Fig. 7(c) and (d). The Bi-rich phase segregation or the accumulation of Bi atoms at the anode destroys the eutectic structure and induces the damage in the Sn-rich phase. Consequently, this will deteriorate the integrity and reliability of the solder interconnect.

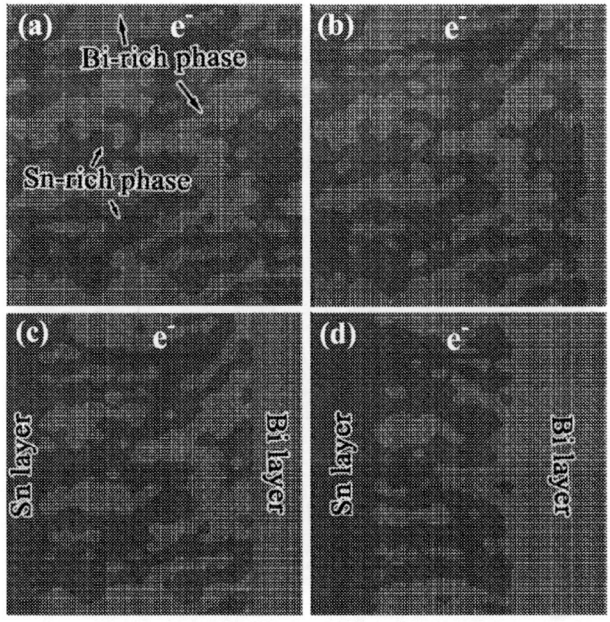

Fig. 7. Simulation of electromigration induced microstructure evolution under current stressing at different times: (a) 0 CAS; (b) 80 CAS; (c) 160 CAS; and (d) 240 CAS.

4. Conclusions

The interaction effect between electromigration and microstructure evolution in Cu/Sn-58Bi/Cu solder interconnect is studied by CA-FE simulation and experimental characterization. Owing to the smallest local electrical resistance and severest current crowding effect, the accumulation of Bi atoms easily occurs near the bottom corner position of the right-angle type solder interconnect, where the current density in the Sn-rich phase is much higher than that in the Bi-rich phase. Bi atoms in the Sn-rich phase are more inclined to migrate to the anode first, rather than migrating directly along the interface of Bi-rich and Sn-rich phases. The accumulation of Bi atoms at the anode destroys the eutectic structure and induces the structural damage in the Sn-rich phase, and hereby severely impairs the integrity and reliability of the solder interconnect. Simulation results using CA-FE method proposed in this study are in good coincidence with the experimental observation of Bi-rich phase segregation induced by EM.

Acknowledgments

This research was supported by the National Natural Science Foundation of China under grant Nos. 51275178 and 51205135, the Research Fund for the Doctoral Program of Higher Education of China under grant No. 20110172110003 and Fundamental Research Fund for the Central Universities (SCUT-2013ZM0026).

References

1. K. Zeng, K. N. Tu, "Six cases of reliability study of Pb-free solder joints in electronic packaging technology," *Materials Science & Engineering R-Reports*, vol. 38, no. 2, pp. 55-105, 2002.
2. K. N. Tu, "Recent advances on electromigration in very-large-scale-integration of interconnects," *Journal of Applied Physics*, vol. 94, no. 9, pp. 5451-5473, 2003.
3. Y. C. Chan, D. Yang, "Failure mechanisms of solder interconnects under current stressing in advanced electronic packages," *Progress in Materials Science*, vol. 55, no. 5, pp. 428-475, 2010.
4. A. Sharif, Y.C. Chan, R.A. Islam, "Effect of volume in interfacial reaction between eutectic Sn-b solder and Cu metallization in microelectronic packaging," *Materials Science and Engineering: B*, vol. 106, no. 2, pp. 120-125, 2004.
5. X. Gu, Y.C. Chan, "Electromigration in line-type Cu/Sn-Bi/Cu solder joints," *Journal of Electronic Materials*, vol. 37, no. 11, pp. 1721-1726, 2008.
6. L. T. Chen, C. M. Chen, "Electromigration study in the eutectic SnBi solder joint on the Ni/Au metallization," *Journal of Materials Research*, vol. 21, no. 4, pp. 962-969, 2011.
7. W. Mu, W. Zhou, B. Li, P. Wu, "Janus-faced Cu-core periphery formation and Bi phase redistribution under current stressing in Cu-cored Sn58Bi solder joints," *Journal of Alloys and Compounds*, vol. 584, pp. 483-486, 2014.
8. W. Yue, M. B. Zhou, H. B. Qin, X. P. Zhang, "A comparative investigation of the electromigration behavior

between wedge-type and line-type Cu/Sn3. 0Ag0. 5Cu/Cu interconnects," *in Proc. International Conference on Electronic Packaging Technology and High Density Packaging (ICEPT-HDP)*, Shanghai, China, 2011, pp. 971-975.

9. W. Yue, H. B. Qin, M. B. Zhou, X. Ma, X. P. Zhang, "Influence of solder joint configuration on electromigration behavior and microstructural evolution of Cu/Sn-58Bi/Cu microscale joints," *Acta Metallurgica Sinica*, vol. 48, no. 6, pp. 678-686, 2012.

10. J. M. E. Harper, C. Cabral, P. C. Andricacos, L. Gignac, I. C. Noyan, K. P. Rodbell, C. K. Hu, "Mechanisms for microstructure evolution in electroplated copper thin films near room temperature," *Journal of Applied Physics*, vol. 86, no. 5, 2516-2525, 1999.

11. K. Yoshida, H. Morigami, "Thermal properties of diamond/copper composite material," *Microelectronics Reliability*, vol. 44, no. 2, pp. 303-308, 2004.

12. J. Sun, G. Xu, F. Guo, Z. D. Xia. Y. P. Lei, Y. W. Shi, X. Y. Li, X. T. Wang, "Effects of electromigration on resistance changes in eutectic SnBi solder joints," *Journal of Materials Science*, vol. 46, no. 10, pp. 3544-3549, 2011.

13. J. Swanson, *Release 10.0 Documentation for ANSYS*, ANSYS Company, Pennsylvanian, 2006.

14. Y. M. Wei, S. J. Ying, Y. Fan, B. H. Wang, "The cellular automaton model of investment behavior in the stock market," *Physica A: Statistical Mechanics and its Applications*, vol. 325, no. 3-4, pp. 507-516, 2003.

15. C. Liu, C. Chen, K. Tu, "Electromigration in Sn–Pb solder strips as a function of alloy composition," *Journal of Applied Physics*, vol. 88, no. 10, pp. 5703-5709, 2000.

16. Q. Yang, J. Shang, "Interfacial segregation of Bi during current stressing of Sn-Bi/Cu solder interconnect," *Journal of electronic materials*, vol. 34, no. 11, pp. 1363-1367, 2005.

17. K. N. Tu, "Electromigration in stressed thin films," *Physical Review B*, vol. 45, no. 3, pp. 1409-1413, 1992.

Effects of Alignment of Graphene Flakes on Water Permeability of Graphene-epoxy Composite Film

Seong-Yoon Jung and Kyung-Wook Paik
Nano Packaging and Interconnection Lab. (NPIL)
Department of Materials and Science Engineering
Korea Advanced Institute of Science and Technology (KAIST)
291 Daehak-ro, Yuseong-gu, Daejeon, 305-701, Korea
Phone: +82-42-350-3386, Fax: +82-42-350-8124
E-mail : Jung_seungyoon@kaist.ac.kr

Abstract

In this study, the electric field-assisted alignment of graphene flakes in B-stage graphene-epoxy composite film was demonstrated. Compared with conventional electric field-induced alignment method which involves curing liquid phase polymer matrix, the electric field application on B-stage composite film with controlled viscosity can achieve graphene flake alignment without any curing reactions, thereby enabling further processes such as film application, assembly and curing.

B-stage graphene-epoxy composite was prepared by solution-mixing method, and electric field was applied parallel to the film with various temperature and electric field strengths. Cross-sectional SEM images revealed that graphene flakes were horizontally aligned better as film viscosity decreased and electric field strength increased. However, curing reaction at high temperature should be considered because high temperature was needed to achieve low viscosity. The aligned graphene-epoxy composite film showed lower water vapor transmission rate (WVTR).

1. Introduction

Graphene, which is a layer of carbon atoms in a hexagonal lattice, has attracted scientists and engineers due to its excellent mechanical, electrical, and thermal properties [1-3]. In addition, it was theoretically reported that graphene had good gas barrier properties due to its electronic structure. [4]. One of the application of graphene is graphene-polymer composite, where graphene flakes are incorporated to polymeric matrix. It was reported that small amount of graphene flakes can significantly improve electrical, thermal, and gas permeability properties of polymeric materials [5]. Therefore, graphene-polymer composite has potential applications in electronics such as thermal management, electrical interconnection, and gas barrier materials for sealing.

In general, the property of composites can be tailored by arranging or orienting fillers in the matrix. Thus, there has been many attempts to fabricate oriented polymer composites, and one of the ways to orient rod-like or planar fillers is to apply electric field to polymer composite [6-9]. Several authors have reported oriented composites including CNT- and graphene-polymer composite by electric field application, however in most studies; the polymeric matrices were bulk liquid phase since low viscosity was desired for better flake alignment. Moreover, the electric field was applied while the composite was cured. However, this process can limit further

assembly and curing processing. On the other hand, B-stage, or partially-cured adhesive film has processing advantages over liquid phase paste, due to its ability to precisely control residues and its compatibility to different kinds of substrates such as polymeric film, glass and metals. This B-stage adhesive film is in solid state at ambient temperature, but the viscosity can be lowered by applying heat, which graphene flakes can easily interact with external electric field.

In this study, graphene flakes were incorporated to B-stage epoxy adhesive film, and aligned structure of graphene flakes was demonstrated by applying external electric field. The applied electric field direction was parallel to the film direction, and flake orientation distribution was observed depending on the electric field strength, film viscosity. In order to observe the effects of horizontal alignment on properties of graphene-epoxy composite film, water vapor transmission rate (WVTR) was also measured.

2. Experimental
2.1. Preparation of graphene-epoxy composite film

Graphene flakes used in this study were commercially available exfoliated graphene nanoplatelet. The mean size of flakes was 5 μm and the thickness was 6-8 nm according to the manufacturer. Also, commercially available bisphenol-A epoxy, thermoplastic resin, and a curing agent (dicy; dicyandiamide) were used as polymer materials. In addition, methyl-ethyl ketone (MEK) was used as an organic solvent.

Procedure of preparing graphene-epoxy composite film is schematically illustrated in figure 1. Since as-received graphene flakes were granular powders, mechanical exfoliation of powders was necessary. The powder was first mixed with MEK in a bottle and sonicated for 24 hours in the sonication bath in order to exfoliate powders into individual graphene flakes. Epoxy resin and curing agent were then added to the solution, and mixed by using the ball-milling technique for 24 hours. After that, the remaining organic solvent was dried out resulting graphene-epoxy composites. After the graphene-epoxy composites were prepared, film was coated on the releasing film using comma-roll film coater (Fig. 2).

978-1-4799-2408-0/14 $31.00 © 2014 IEEE

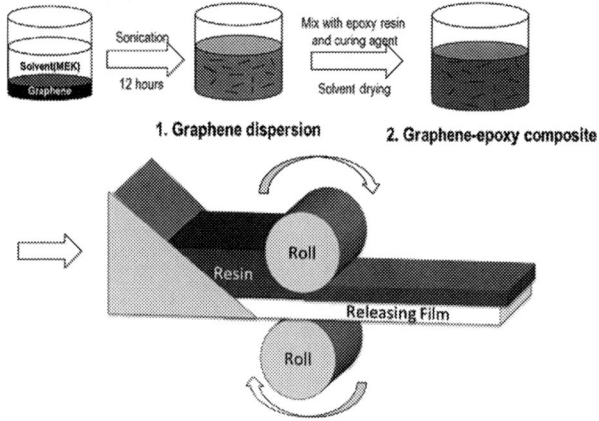

Figure 1 Illustration of fabrication of graphene-epoxy composite film

Figure 2 Image of graphene-epxy composite film coated on releasing film (3 wt%, 32 μm)

2.2. Electric field application for graphene flake alignment

Experimental setup for electric field application is shown in figure 2. Graphene-epoxy composite film with 3 wt% graphene content was coated on the releasing film by using comma-roll film coater, and the thickness was fixed to be 32μm. Two needles were connected to the DC power supply, where the electric field direction was parallel to the film.

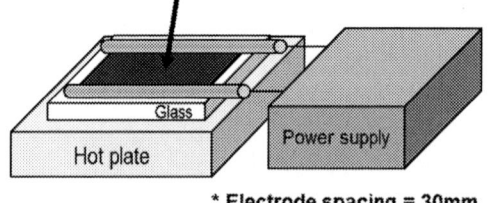

Figure 3 Illustration of electric field application setup

After the electric field was applied, the processed film on releasing film was immersed in liquid nitrogen and cut so that cross-section would be revealed. Cross-sectional scanning

electronic microscopy (SEM) images of processed film were obtained. Graphene flakes seen in the SEM images were analyzed by measuring the flake orientation using image processing software. In order to compare the degree of alignment of graphene flakes, the portion of graphene flakes tilted less than 15 degrees was compared for each experimental condition.

There were two main variables in this study; temperature and electric field strength, which were tabled in table 1. In all electric field application, the time was fixed to be 60 min. Electric field strength was controlled by adjusting electrical voltage, and film viscosity was measured as a function of temperature. In order to compare the effects of film viscosity, electric field strength of 30 V/mm was applied to the film at 30, 80, 100 °C. In order to investigate the effects of electric field strength, 30 and 100 V/mm electric field was applied to the film at 80 °C

Table 1 Summary of temperature, corresponding viscosity and electric field strength used in the experiment

Temperature (°C)	Viscosity (Pa · s)	Electric field strength (V/mm)
30	90440	30
80	4840	10, 30, 100, 200
100	1294	30

2.3. Effects of graphene flake alignment on properties of graphene-epoxy composite film

In order to investigate the effect of horizontal alignment on the water permeation property of graphene-epoxy composite film, water vapor transmission rate (WVTR) was measured by the method according to ASTM standard F-1249. Graphene content was fixed to be 3 wt%. The thickness of composite film was 20 μm and the area was 50 mm x 50 mm. Graphene-epoxy composite film was first laminated on polyimide (PI, 30 μm) film and then cured at 120 °C for 40 min (Fig. 3). In order to prepare the electric field applied sample, electric field was applied on laminated composite film. The electric field strength was 100 V/mm and the field application was done at 120 °C for 20 min and the temperature was hold for 20 min for full cure. For reference, WVTR of epoxy film with no graphene was measured.

3. Results and Discussion

3.1. Graphene flake alignment by electric field application

Theoretical approach on rotation of ellipsoidal particles by electric field application has been well studied [9-11]. Especially, rotation behavior of a particle in polymer matrix under electric field was theoretically interpreted in terms of electrostatic torque by polarized particle and hydrodynamic torque by viscous polymer matrix. The response time of the particle under electric field was proportional to viscosity and inversely proportional of the square of electric field strength [11]. In addition, computational calculations on graphene under electric field have demonstrated that graphene in free space tend to be aligned to the direction of external electric field by polarization [12]. Therefore, the response time of graphene flakes in polymer matrix was a function of electric field strength and polymer viscosity mentioned above.

Figure 4 shows the cross-sectional SEM images of as-coated and electric field applied graphene-epoxy composite film. As shown in the figure, graphene flakes in the composite film appeared as bright lines. Graphene flakes in as-coated composite film were in a linear shape, which indicates that the flakes were somewhat oriented in the three-dimensional space. This might be explained that the coating process caused flakes to be oriented toward the coating direction. However, the flake orientation was quite random for as-coated composite film. After electric field was applied, more flakes were aligned toward the electric field direction, which demonstrated that graphene flakes could be aligned in a solid composite film.

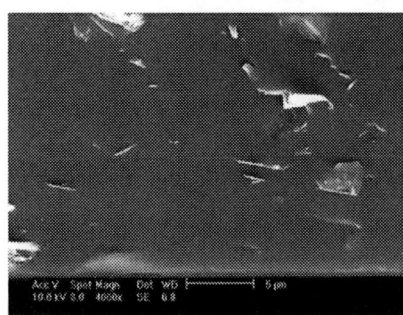

(a) As-coated (3wt%)

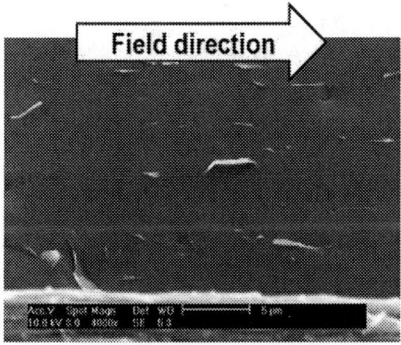

(b) 30V/mm, 60min

Figure 4 Cross-sectional SEM images of (a) as-coated composite film and (b) electric-field applied film sample. The electric field was applied at room temperature.

In order to compare the degree of alignment, the angle relative to the electric field direction was statistically analyzed. Figure 5 compares the portion of graphene flakes under 15 degrees, with different electric field strengths at 80 °C for 60 min. As indicated in the figure, more graphene flakes were horizontally aligned when applying higher electric field strength at same field application time of 60 min. The portion drastically increased even by weak electric field strength, and the maximum portion of 84.5 % was obtained by applying 100 V/mm for 60 min at 80°C.

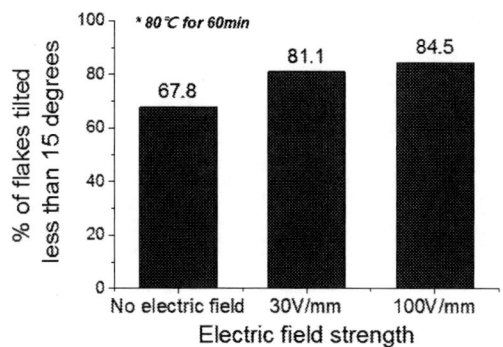

Figure 5 Degree of alignment of 200 graphene flakes with different electric field strength applied.

However, the effect of temperature was quite different. Figure 6 and 7 shows the film viscosity versus temperature and the effect of temperature on degree of graphene flake alignment respectively. As shown in figure 5, film viscosity decreased with temperature until the minimum viscosity around 120 Pa•s at 140 °C, then drastically increased because of the curing reaction of epoxy film. Comparing film viscosity and the results of electric field application at 80 °C and 100 °C in figure 6, the degree of alignment was higher at 80 °C than 100 °C even though the corresponding viscosity was 4 times lower at 100 °C. This phenomenon could occur at 100 °C since the temperature was maintained for a long period of time (electric field application time). In order to examine whether the film at 100 °C has been cured or not, attenuated reflectance mode of Fourier transform infrared spectroscopy (ATR-FTIR) was used to measure the degree of curing (Fig. 8). Comparing the spectra between as-coated film and fully cured film, the peak of 912 cm^{-1} in the spectrum decreased, indicating that degree of curing of epoxy can be measured by this peak. As seen in figure 8, the composite film has partially cured by heating at 100 °C for 60 min while at the same time, the film at 80 °C has not been cured.

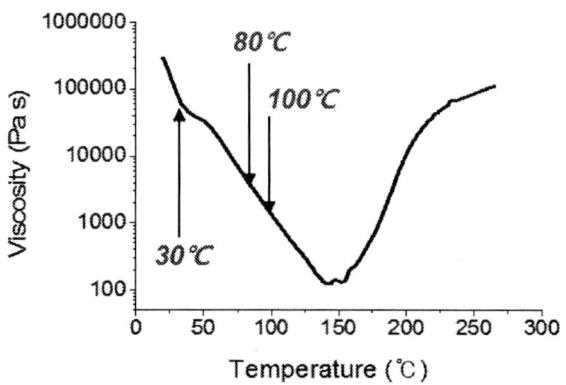

Figure 6 Temperature vs. viscosity of 3 wt% graphene-epoxy composite film.

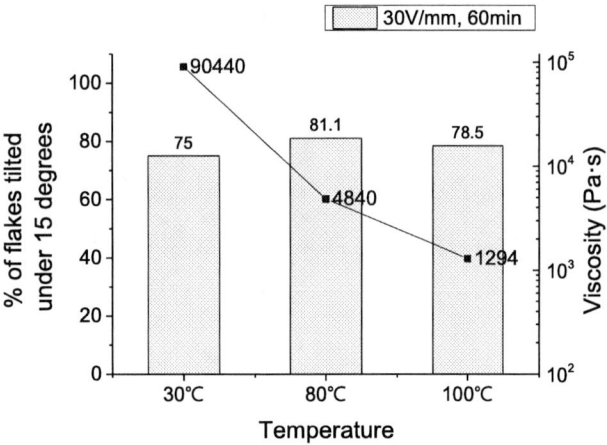

Figure 7 Degree of alignment of graphene flakes after electric field application with different temperature. The corresponding viscosity was also indicated.

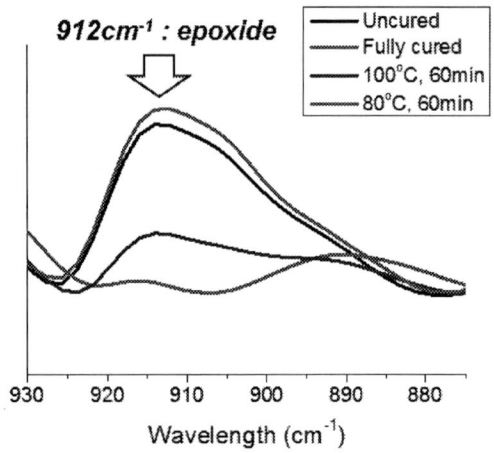

Figure 8 FTIR spectra of graphene-epoxy composite films

3.3. Effects of graphene flake alignment on water permeation property of graphene-epoxy composite film

In order to investigate the effects of horizontal alignment on properties of graphene-epoxy composite film, Water vapor transmission rate (WVTR) was measured. In preparing the samples for WVTRs of the films were shown in table 2. As demonstrated in table 2, the aligned graphene-epoxy composite film showed the lowest WVTR of 22.85 g/m^2/day. Several authors have reported improved gas barrier property of graphene-polymer composites [13-14]. Theoretically, the gas barrier effect maximizes when the flakes were aligned to the direction perpendicular to gas permeating direction by diffusion length increase [14]. Therefore, the lower WVTR of graphene-epoxy composite film was achieved by horizontally aligning graphene flakes.

Table 2 WVTR of graphene-epoxy composite film laminated on PI film. Film thickness of polyimide substrate was 30 μm, while the composite film was 20 μm

Sample description	WVTR (g/m^2/day)	In detail
Polyimide (PI)	96.90	Reference
Epoxy (No graphene)	48.26	No graphene
3 wt% graphene	33.25	
3 wt% graphene (aligned)	22.85	100V/mm 20min

4. Conclusion

In this study, the electric field-assisted alignment of graphene flakes in B-stage composite film was demonstrated. This alignment process has advantages over conventional flake alignment in liquid polymer composite, since further processing such as assembly and curing can be possible. Better alignment was achieved by applying high electric field strength at low viscosity, and the low viscosity was obtained at high temperature. Also, electric field application at high temperature for long time caused curing reaction, which in turn, caused poor alignment by rapid viscosity increase. . During the curing reaction, viscosity drastically decreases to a certain degree, in a short time period. By optimizing the electric field strength and field application time in terms of curing behavior, B-stage graphene-polymer composite film with totally aligned graphene flakes can be achieved.

Furthermore, the effect of horizontal alignment on gas permeation property was investigated. The horizontally aligned composite film showed lower sheet resistance and water permeation property. Since graphene has good barrier properties, aligned structure of graphene flakes could be used in various applications. Gas barrier membrane, for example, is one of the possible applications. Graphene has the ability to prevent gas molecules to be passed through therefore, aligned graphene-polymer gas barrier membranes could drastically reduce the gas permeability compared to as-prepared membranes.

Acknowledgments

This work was financially supported by Graphene Materials and Components Development Program of MOTIE/KEIT (10044412, Development of basic and applied technologies for OLEDs with graphene)

References

1. A. Geim and K. Novoselov, "The rise of graphene," *Nature Materials*, vol. 6, no.3, pp. 183-191, 2007.
2. A. Balandin, "Thermal properties of graphene and nanostructured carbon materials," *Nature Materials*, vol. 10, no.8, pp. 569-581, 2011.

3. C. Lee et al., "Measurement of the Elastic Properties and Intrinsic Strength of Mono- layer Graphene," *Science*, vol. 321, no.5887, pp. 385-388, 2008.

4. B. M. Yoo, H. J. Shin, H. W. Yoon, and H. B. Park, "Graphene and graphene oxide and their uses in barrier polymers," *Journal of Applied Polymer Science*, vol. 131, no.1, pp. 39268, 2014.

5. J.R. Potts et al., "Graphene-based polymer nanocomposites," *Polymer*, vol.52, no.1, pp. 5-25, 2011.

6. R. Verdejo et al., "Graphene filled polymer nanocomposites," *Journal of Materials Chemistry*, vol. 21, no.10, pp. 3301-3310, 2011.

7. X.Q. Chen et al., "Aligning single-wall carbon nanotubes with an alternating-current electric field," *Applied Physics Letters*, vol.78, no. 23, pp. 3714-3716, 2001.

8. G. Chen, "Fabrication of highly ordered polymer/graphite flake composite with eminent anisotropic electrical property," *Polymers for Advanced Technologies*, vol. 19, pp. 1113-1117, 2008.

9. G. Kim and Y. M. Shkel, "Polymeric Composites Tailored by Electric Field," *Journal of Materials Research*, vol.19, no. 4, pp. 1164–1174, 2011.

10. A. Okagawa, R. Cox, and S. Mason, "Particle behavior in shear and electric fields. VI. The microrheology of rigid spheroids," *Journal of Colloid and Interface Science*, vol. 47, no.2, pp. 536–567, 1974.

11. T. Z. Kosc, K. L. Marshall, S. D. Jacobs, and J. C. Lambropoulos, "Polymer cholesteric liquid-crystal flake reorientation in an alternating-current electric field," *Journal of Applied Physics*, vol. 98, no. 1, pp. 013509, 2005.

12. Z. Wang, "Alignment of graphene nanoribbons by an electric field," *Carbon*, vol. 47, no. 13, pp. 3050-3053, 2009.

13. H. Kim et al., "Graphene/Polyurethane Nanocomposites for Improved Gas Barrier and Electrical Conductivity," *Chemistry of Materials*, vol. 22, no. 11, pp. 3441-3450, 2010.

14. O. Compton et al., "Crumpled Graphene Nanosheets as Highly Effective Barrier Property Enhancers," *Advanced Materials*, vol. 22, no. 42, pp. 4759-4763, 2010.

Characterization of Alternate Power Distribution Methods for 3D Integration

David C. Zhang, Madhavan Swaminathan, David Keezer and Satyanarayana Telikepalli

School of Electrical and Computer Engineering,

Georgia Institute of Technology

Atlanta, GA 30332, USA.

dzhang43@gatech.edu, madhavan@ece.gatech.edu

Phone: 404.894.3340, Fax: 404.894.9959

Abstract

Signal return path discontinuities, parasitic inductance and impedance mismatch within interconnects are major factors that contribute to degraded high-speed signal quality in three-dimensional (3D) integrated circuits and systems. In this paper, we apply an alternate power delivery method and a novel I/O signaling scheme to a 3D system to address these issues. Two test vehicles made of stacked PCBs that resemble 3D integrated systems will be presented. One test vehicle is designed based on our proposed approach while the other is based on the conventional power delivery network design. The signal integrity and power supply noise performance will be shown in both simulated environment and actual test measurement. At data rates up to 3Gbps, our proposed design produces higher signal quality than the conventional design with better eye height, lower timing jitter, and lower power supply noise.

Index Terms—**Power delivery network (PDN), power transmission line (PTL), simultaneous switching noise (SSN), power supply noise (PSN), return path discontinuity (RPD).**

I. INTRODUCTION

In the current and future high-speed digital systems, the I/O speed is being pushed to tens or even hundreds of gigabits per second with transition time precision in femtoseconds. Additionally, the supply voltages in an integrated chip (IC) also trends down to sub-volt range to prevent current leakage and gate breakdown due to shrinking transistor size. As a result, noise tolerance in a high-speed digital system decreases as the circuits switch faster. Therefore good signal integrity (SI) and power integrity (PI) are vital for the success of any high-speed systems.

A 3D IC system such as the one shown in Fig. 1, it normally consists of a PCB and a stack of heterogeneous IC dies. Maintaining high SI and PI in such a complex structure has its own challenges. The first challenge is return path discontinuities (RPD). The modern printed circuit board (PCB) designs have become more challenging due to increased number of layers, complex power delivery network and high routing density. As a result, maintaining uninterrupted return paths and avoiding impedance mismatch for high-speed signals in the face of current power delivery network design is very difficult. Many times additional ground layers have to be added for the purpose of providing a complete reference for fast switching signals. This approach significantly increases the manufacturing cost. Return path discontinuities occur when transmission lines carrying fast-

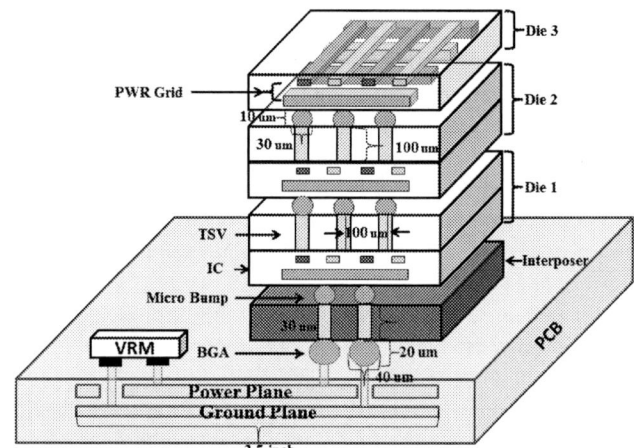

Fig. 1 A conventional 3D model for simulation

edged signals make via transitions without proper ground reference or are routed over splits of planes underneath. Return path discontinuities cause increased loop inductance formed by the forward and return signals, excite cavity resonance within the PCB, and facilitate unwanted electromagnetic coupling which can then lead to degraded signal quality and increased simultaneous switching noise [1][2]. Fig. 2 shows an illustration of how RPD can occur in a PCB. The return path for the forward signal is interrupted by the middle plane 1 layer which forces return current to jump between plane 1 and 2. This return path interruption can lead to increased effective loop inductance and undesired signal and PDN coupling.

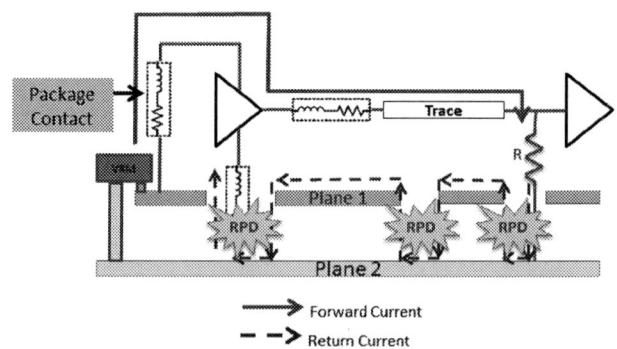

Fig. 2 Illustration of RPD cause-effect in a PCB

The second issue is the increasing parasitic inductance as the height of the IC stack grows. As illustrated in [3] and [4], due to increasing parasitic inductance from interconnects among the stacked dies, the impedance of the PDN of a higher die is more than that of a lower die in the inductive region. Consequently this increasing parasitic inductance contributes to higher simultaneous switching noise (SSN) as one moves up the stack.

The third issue is impedance mismatch among the various interconnect for signal, power and ground. As one can see from Fig. 1 that there are various bonding structures such as ball grid array (BGA), soldering micro bump and through silicon via (TSV) along a power or ground rail interconnect that runs from bottom to the top. It would be impractical to expect their impedances would match among themselves. The mismatch among interconnects causes voltage and current reflection as well as unwanted coupling among the interconnects.

The aggregate effect of the aforementioned three major problems with 3D ICs systems causes increased system level noise and reduced SI and PI. To reduce the noise, the most common method used is by using on-chip and/or off-chip decoupling capacitors. Therefore, many studies are devoted to the characterization, selection and placement of decoupling capacitors. The works done in [5]-[7] focus on relatively large two dimensional systems such as a computer server in which space may not be a critical constrain. However for many modern "smart" consumer electronics such smartphones and 3D integrated systems, on-board or on-chip real estate is a precious commodity. There may not be enough space to place the required number of off-chip decoupling capacitors in order to meet the impedance target. In addition, adding decoupling capacitors whether on-chip or off-chip does not always help to meet the impedance target or design requirement in ICs due to current leakage and parasitic inductance. For on-chip decoupling capacitors, the effectiveness of the capacitors also heavily depends on the distance between the capacitors and where noise needs to be controlled. Researchers and designers also propose to put voltage regulators on chip. However, the effectiveness of these regulators also heavily depends on the placement [8]. Power efficiency of the on-chip regulars is another important factor to determine whether it is a worthwhile approach.

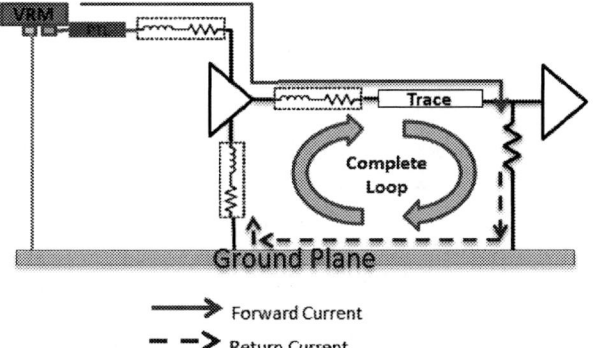

Fig. 3 Power transmission line based PDN in a PCB

The design method we propose in this paper uses few or no decoupling capacitors to suppress noise. Instead, we tackle

the problem in a proactive way by redesigning the power delivery network in a 3D system working in tandem with a new signaling scheme. We propose to use power transmission lines (PTL) to replace power planes in conventional PDN designs in PCBs so that the forward signal and its return path will be able to form a complete loop and hence eliminate the RPD effect as shown in Fig. 3 and [1]. We further improve the design by introducing a novel high-speed signal scheme that would add a current compensation mechanism in order to reduce current transient on the power supply rail [1]-[2]. The design approach is named constant current power transmission line or CCPTL. The CCPTL proof of concept was demonstrated in [1].

The PTL design concept was then applied to 3D models and produced optimistic results in the previously published work in [4] and [9]. The simulated result showed the SI and PI performances of our proposed method clearly exceed that of the traditional design in a 3D system. However, there was no measurement that had been done in an actual 3D test vehicle with our proposed design until now. This paper will report the design of two 3D test vehicles based on conventional PDN design and the PTL design respectively. The high-speed signal eye-height, peak to peak jitter and power supply noise will be shown and compared between the two designs. We will show that with minimum decoupling capacitors, our proposed PTL-based design can produce better SI and PI than conventional design in a 3D environment even with parasitic inductance of several nanohenries at a data rate of 3Gbps PRBS.

This paper is arranged as follows: in section II, a brief summary of CCPTL design is introduced. The manufacturing details of the 3D test vehicles are reported in section III. Measurement results of two test cases are shown to demonstrate proof of CCPTL concept in a 3D test vehicle. Finally, we conclude this paper through a summary in section V.

II. Proposed PTL-Based design

The fundamental difference in PTL based design is the use of transmission lines as the medium to carry power from the supply source to local digital logics instead of the traditional use of power planes [1]. By removing power planes, only ground planes exist in a PCB design. Regardless of the complexity of the PCB, continuous reference for the return path of a forward signals is made easily accessible. Therefore, a complete loop of forward and return path can be formed without inducing any RPDs as shown in Fig. 3. As a result, both SI and PI can be significantly improved [1] and [2]. A PTL based 3D IC system model is shown in Fig. 4.

By using PTL as a conduit to channel power from the power supply to local active devices, dynamic data dependent DC drop will occur. In addition, impedance of the PTL needs to be matched with that of the on-resistance of the connected drivers to minimize reflection. This is a difficult task for practical designs. Therefore, a signaling method was proposed in [1] named constant current PTL or CCPTL which can address both dynamic DC drop and impedance mismatch issues by providing a current compensating mechanism,

978-1-4799-2408-0/14 $31.00 © 2014 IEEE   2261

namely the dummy path, based on the input data as shown in Fig. 5. This mechanism ensures that regardless of the state of the input bit, the current draw from the power supply source through the PTL remains the same. Therefore current fluctuation or di/dt is minimized [1] and [2]. With a minimum di/dt, even we have significant parasitic inductance along the path of PDN in a 3D structure; noise can be kept at a minimum [4].

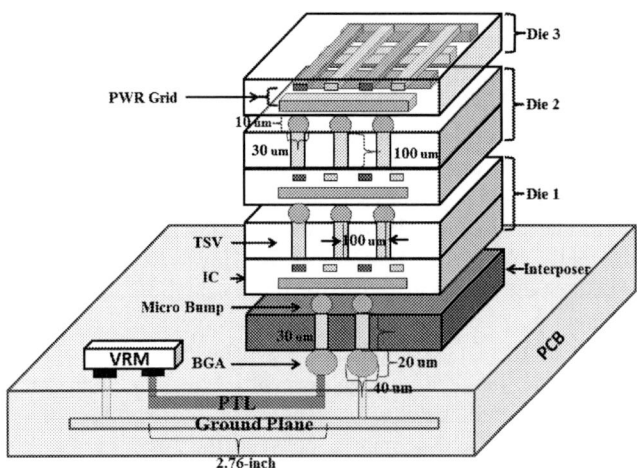

Fig. 4 A PTL-based 3D model for simulation

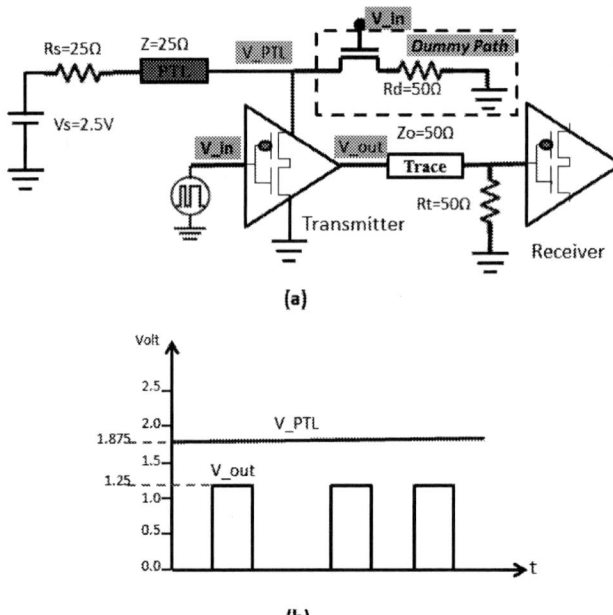

Fig. 5 CCPTL. (a) schematic. (b) simulated waveform at nodes V_PTL and V_out

III. 3D PCB test vehicle information

In order to model a 3D IC system that consists of a stack of IC dies and a PCB at the bottom as shown in Fig. 1 and 4, two versions 3D test vehicles made of PCBs were designed and fabricated. The first version was designed based on PTL power delivery network and the second, conventional PDN.

For each version of the TVs, the designed PCBs are stacked together to create a 3D structure as illustrated in Fig. 8. The bottom PCB is the motherboard. On top of the motherboard are three stacked PCBs that are named daughter cards. Each daughter card has multiple high speed I/O drivers. The PCBs are joined together by board-board connectors. The board-board connector pins can be treated as interconnects of the 3D model as described in section I. The inductance of each connection within the board-board connector is estimated to be 7.77nH between adjacent PCBs. The parasitic inductance between adjacent stacked dies in our previously presented simulation model [4] was about 66.9pH. Therefore, the inductance between two stacked PCBs in the TVs is approximately 116 times more than the parasitic inductance from the simulation model presented in [4]. We will show through measurement that even with this large parasitic inductance we can still maintain good SI and PI with our proposed PTL based design with minimum number of decoupling capacitors.

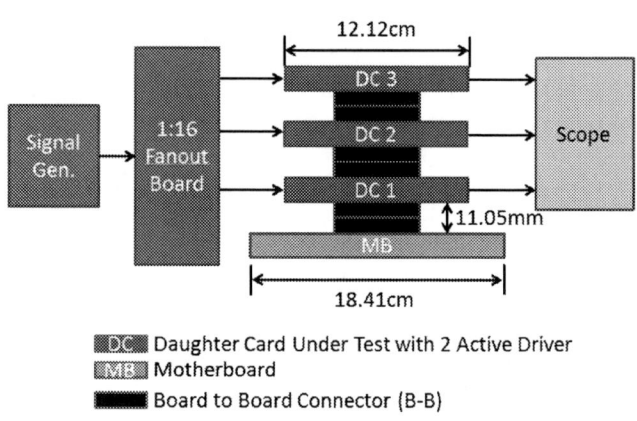

Fig. 6 3D PCB structure illustration and test setup

Both motherboard and daughter cards are 4-layer FR-4 PCBs whose stack-up is shown as shown in Fig 7. The top and bottom layers are used for routing purpose. The middle two layers are ground layers for PTL based TV as the power plane is replaced by PTLs which are routed on the bottom layer. For conventional design, layer-2 is set as VDD power layer and layer-3 is assigned to ground. The dimension of the motherboard is 18.41cm by 10.16cm. The daughter-card is approximately 12.12cm by 35.56cm. Mated height between adjacent PCB is 11.05mm. In the conventional design TV, the top daughter card has eighteen 0402 and eighteen 0603 decoupling capacitors strategically placed around the board to suppress the power supply noise. On the PTL based TV; however, only two 0402 decoupling capacitors are placed for every active driver. The ratio of the placed decoupling capacitors on the convention design to that on the PTL design at the time of measurement was approximately 9:1.

The test setup is illustrated in Fig. 6. The signal generator is an Agilent 81133A 3.35GHz Pulse/Pattern Generator. The oscilloscope is an Agilent DCA-X 86100D with HP 54752A Two-Channel 50GHz Module probe. The IC drivers are On-Semiconductor NBSG16 12Gbps SiGe drivers. Since the signal generator can only generate one pair of differential

978-1-4799-2408-0/14 $31.00 © 2014 IEEE

signals, a 1:16 fan-out board is used to drive more drivers on the TVs. The lab-bench setup is shown in Fig. 8.

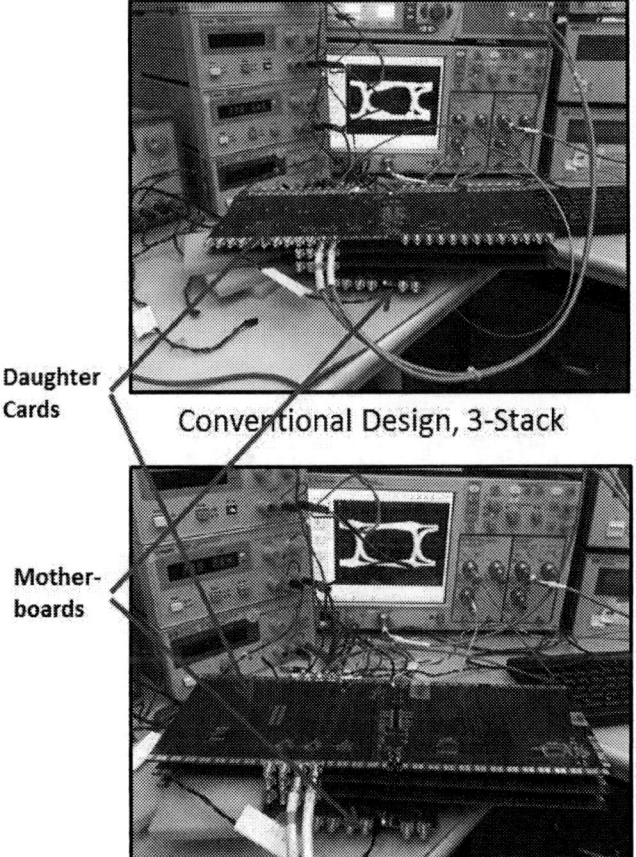

Fig. 7 PCB stackup for the 3D test vehicles

Fig 8 Lab-bench setup for 3D test vehicle measurement

IV. 3D PCB Test Vehicle Measured Results

As with the 3D IC system model which was previously shown in [7], the impedance profile of the conventional 3D test vehicle is measured with a VNA with SMA connectors mounted directly onto the test points as shown in Fig. 9.

Due to the larger scale of the TV the impedance is about eight times higher than that in [7] for the simulated 3D IC system model; nonetheless, the general trend of higher impedance for higher stacked PCB can be seen from Fig. 10.

The intermittent crossover among the impedance curves is due to board to board process variation.

Fig. 9 Top view of the daughter card with mounted SMA connectors

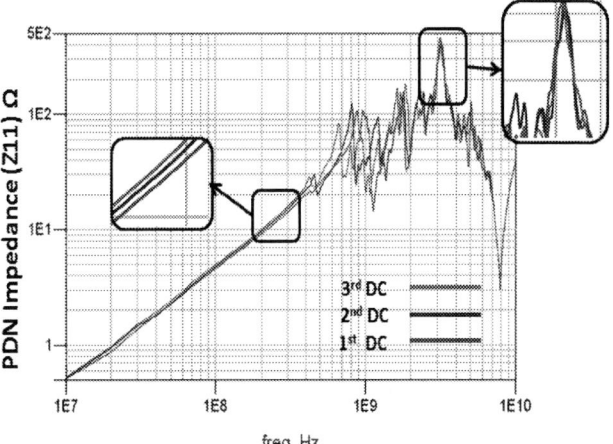

Fig. 10 Measured impedance profile of the test vehicle based on conventionally PDN design

Next, we excite a total of six drivers with a 3Gbps pseudo-random bit stream, two drivers on each daughter card, to compare the SI and PI performance. The eye diagrams and performance data plot of the conventional and PTL based designs are shown in Fig. 15. From Fig. 15 and Fig. 10, both simulated and actual lab measurement, the proposed CCPTL design performance exceed that of the conventional design in term of eye height and especially in p-p jitter and power supply noise. For example, at the 3rd stack the eye height of the CCPTL design is 104% higher, p-p jitter is 53.3% lower and power supply noise is 72.9% lower than the conventional design. Both the power supply noise and p-p jitter at each daughter card of the PTL-based TV varies very little with a standard deviation of only 5.5 and 4.4 respectively while those in the conventional design varies by as much as almost 26 and 13 respectively. Table 1 presents a summary of measured results and percentage of improvement.

978-1-4799-2408-0/14 $31.00 © 2014 IEEE   2263

Table 1 Performance comparison between PTL based and conventional designs

	Eye Height (mV)			P-P Jitter (PS)			PSN (mV)		
	CNV*	CCPTL	% Δ	CNV*	CCPTL	% Δ	CNV*	CCPTL	% Δ
DC3	203	414	104	196	91.7	-53.3	140	37.9	-72.9
DC2	258	429	66.3	185	85.4	-53.7	190	27.0	-85.8
DC1	251	463	84.5	210	83.2	-60.5	177	31.2	-82.4

*CNV=Conventional Based Design

We have also compared the SI performances of the two TVs at different speeds. Fig. 12 shows the eye diagrams measured at the 3rd stacks of the two TVs at 1Gbps and 3Gbps respectively with one active driver on each TV. The performance of CCPTL design again exceeds that of the conventional design. The peak to peak jitter improves by over 40%. While the eye heights between the two designs are similar at 1Gbps, the eye of the conventional design is significantly reduced at a higher speed of 3Gbps. Table 2 summarizes the measured data.

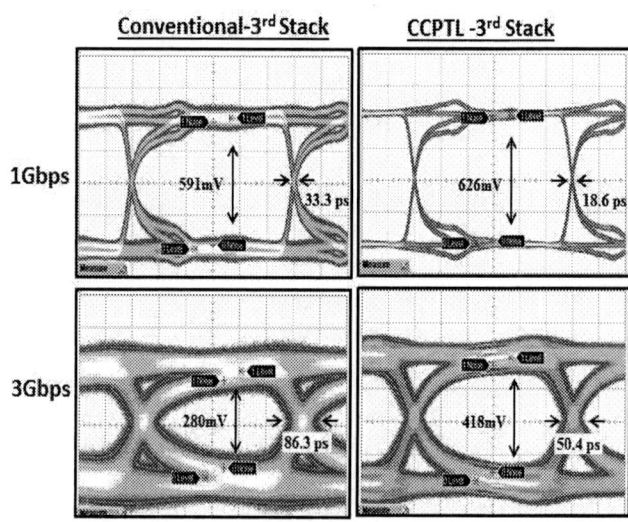

Fig. 12 Measured Eye Diagrams at the 3rd stacks of the Two TV's with 1Gbps and 3Gbps Data rates

Table 2 Performance Comparison between CCPTL Based and Conventional Designs at 1Gbps and 3Gbps at the 3rd Stack

	Eye Height (mV)			P-P Jitter (PS)		
	CNV*	CCPTL	% Δ	CNV*	CCPTL	% Δ
1Gbps	591	626	5.92	33.3	18.6	-44.2
3Gbps	280	418	49.3	86.3	50.4	-41.6

*CNV=Conventional Based Design

V. Conclusion

In this paper, we present an alternate power delivery network design by proposing to replace conventional power planes with power transmission lines. The advantage of this approach is to eliminate return path discontinuities in a PCB. We further offered a high speed signaling scheme to work in tandem with the PTL design in order to solve impedance mismatch and parasitic inductance issues along the power and ground rail interconnects in a 3D system. To show proof of concept and advantage of our design approach, two 3D test

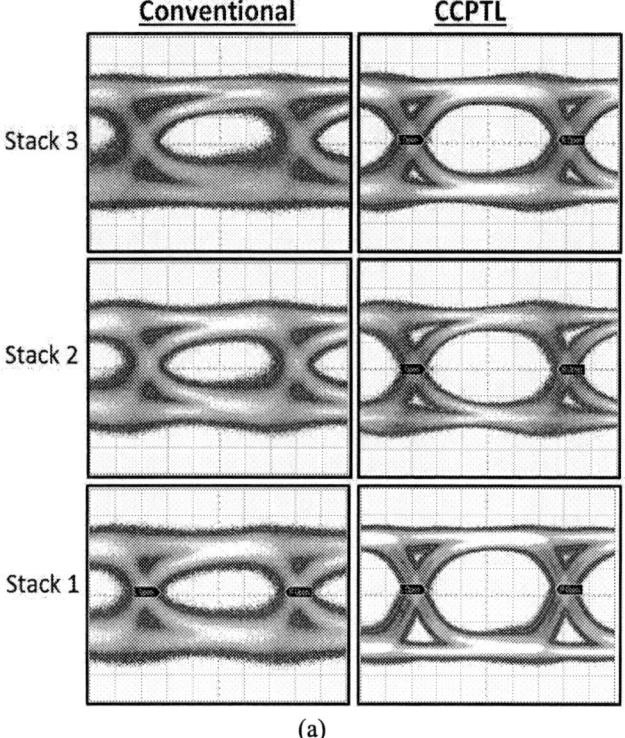

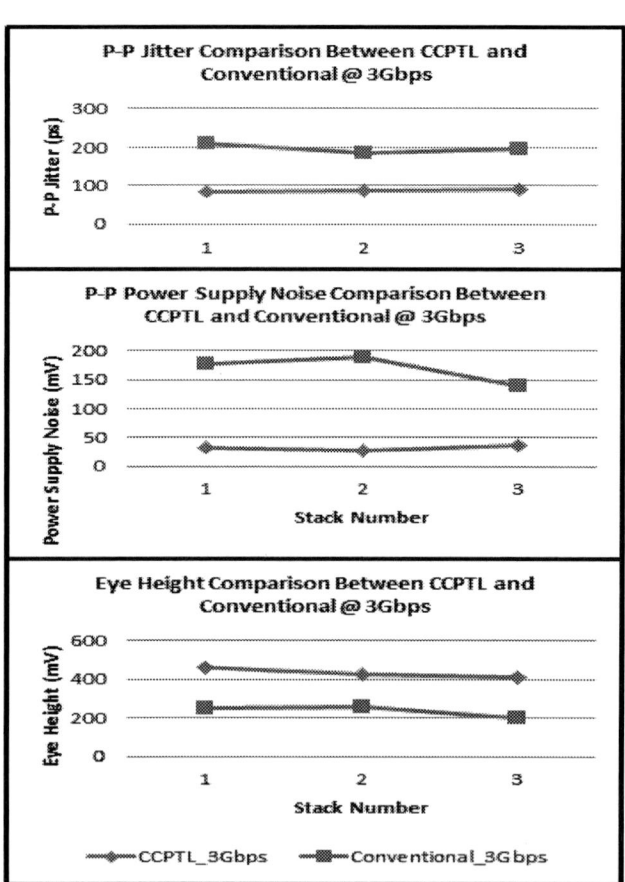

Fig. 11 (a) Eye diagrams measured at each level of the TVs (b) Plots of the SI and PI Performance Comparison of the PTL-base and Conventional Design TVs

978-1-4799-2408-0/14 $31.00 © 2014 IEEE

vehicles made of stacked PCBs were made and subsequently tested and measured. The stack consists of three daughter cards stacked on a mother board with sockets in between them, which mimics a 3D IC stack with approximately 160 times more inductance between the stacks.

We successfully demonstrated for the first time the measured SI and PI results on a complex 3D stacked test vehicle that implements our proposed alternate PDN design and signaling method, namely constant current power transmission line. The communication between the stacked boards using this new comprehensive method is made possible due to better and more consistent signal and power integrities as compared to the conventional designs. Eye height, jitter and power supply noise measurements for channels operating at 1Gbps and 3Gbps using the alternate methods were shown and compared with the more traditional methods used today. Even with far more decoupling capacitors placed on top of the conventional test vehicle, the SI and PI performance of the PTL-based test vehicle still outperforms the former by significant margins. As an example, the eye height has improved by 104%; p-p jitter and power supply noise have reduced by 53.3% and 72.9% respectively at the 3rd stacked daughter card in the CCPTL-based test vehicle as compared to the conventionally designed test vehicle when all daughter cards are active. This PTL design concept was also applied to silicon interposer based 3D IC systems with advantages quantified through simulations [4] and [9].

ACKNOLWEDGEMENT

This research was supported by NSF under the reward number ECCS-0967134. The main author would like to thank Professor Swaminathan for his guidance, confidence and support throughout the development of the test vehicles. He would also like to thank Professor Keezer for his valuable feedback during the design and layout of the test vehicles. Professor Keezer also generously provided the access to the many of his lab equipment.

REFERENCE

[1] S. Huh, M. Swaminathan and D. Keezer, "Constant current power transmission line based power delivery network for single-ended signaling," IEEE Transactions on Electromagnetic Compatibility, Vol. 53, Issue: 4, pp: 1050 - 1064, 2011.

[2] S. Huh, D. Chung and M. Swaminathan, "Achieving near zero SSN power delivery networks by eliminating power planes and using constant current power transmission lines," in Proceedings of EPEPS, pp. 17–20, 2009

[3] K. Kim, W. Lee, J. Kim, T. Song, J. Kim, J. S. Pak, J. Kim, H. Lee, Y. Kwon and K. Park, "Analysis of Power Distribution Network in TSV-based 3D-IC," Electrical Performance of Electronic Packaging and Systems (EPEPS), 2010.

[4] Zhang, D.C. ; Swaminathan, M. ; Huh, S., "New power delivery scheme for 3D ICs to minimize simultaneous switching noise for high speed I/Os," 2012 IEEE 21st Conference on Electrical Performance of Electronic Packaging and Systems (EPEPS).

[5] Novak, I., Miller, J., "Frequency-dependent characterization of bulk and ceramic bypass capacitors," Electrical Performance of Electronic Packing (EPEP), pp 101-104, 2013.

[6] Novak, I., "Power Distribution Network Design Methodologies," IEC publications, December, 2008.

[7] Larry D. Smith, Raymond E. Anderson, Douglas W. Forehand, Thomas J. Pelc, and Tanmoy Roy, "Power Distribution System Design Methodology and Capacitor Selection for Modern CMOS Technology", IEEE Trans. on AP, vol. 22, no. 3, pp. 284-291, 1999.

[8] Kose, S, Friedman, E.G., "Distributed power network co-design with on-chip power supplies and decoupling capacitors," 2011 13th International Workshop on System Level Interconnect Prediction (SLIP), pp. 1-5, June 2011.

[9] Telikepalli, S.; Zhang, D.C.; Swaminathan, M.; Keezer, D., "Constant Voltage-Based Power Delivery Scheme for 3-D ICs and Interposers," Components, Packaging and Manufacturing Technology, IEEE Transactions on , vol.PP, no.99, 2013

Adhesion and Reliability of Direct Cu Metallization of Through-Package Vias in Glass Interposers

Timothy Huang, Venky Sundaram, P. Markondeya Raj, Himani Sharma, and Rao Tummala
3D Systems Packaging Research Center
813 Ferst Drive NW
Atlanta, GA 30332-0250
tim.huang@gatech.edu

Abstract

Direct metallization of bare glass with copper is required to reach the full potential low-cost benefit of glass interposers. However, this poses a fundamental materials challenge associated with copper-to-glass adhesion. Intermediate polymer liners on glass have been used by others, adding an extra material and processing step. In this paper, three approaches to direct metallization of copper to glass interposers are explored and reported. Electroless plating, sputtering followed by electrolytic plating, and sol-gel were investigated as Cu deposition methods with an emphasis on adhesion and reliability of copper to bare glass. The adhesion and reliability performance of films were characterized by tape-testing, peel-strength measurements, and thermal-shock testing. Based on these results, individual assessments are made for each approach and compared with others to assess future directions.

Introduction

The need for high logic-memory bandwidth in both mobile and high-performance applications with emerging 2.5D and 3D package architectures has been driving new advances in interposer and substrate materials. The semiconductor industry is primarily focusing on Si as the next-generation interposer material to address the challenges associated with dimensional instability and coefficient of thermal expansion (CTE) mismatch with organic packages. Based on recent pioneering technical advances by Georgia Tech, glass has been demonstrated as the best next-generation interposer and package material to meet the high-I/O, high-performance, and low-cost requirements for advanced packaging [1]. In addition, because of glass's matched coefficient of thermal expansion (CTE) (few ppm/°C) with that of Si IC (2.5 ppm/°C), thermomechanical reliability is expected to be superior when compared to organic substrates of much higher CTE (18-22 ppm/°C), where CTE mismatch causes serious reliability concerns. Compared to organic packages, glass can be manufactured with a higher density of through-vias and with higher wiring density and I/Os to the chip because of its Si-like dimensional stability, and in addition, demonstrating better electrical performance and lower loss while scalable to large panels with high manufacturing through-put.

One of the major challenges with glass, however, is its metallization with copper conductors with sufficient adhesion. Due to the differences in their chemical structures, most metallic materials such as Cu do not bond strongly to oxide networks of silicate glasses. This results in insufficient adhesion and failure to meet reliability criteria. One approach that has been demonstrated to circumvent this is to coat both sides of glass with a polymer film before the glass through package vias (TPVs) are made. After the creation of vias by

excimer laser ablation, the polymer surface and the roughened-glass via sidewalls can be metallized by standard electroless processes [2]. The major disadvantage of this approach is its limited compatibility with via-formation methods. Glass manufacturers have recently developed proprietary TPV formation processes, some of which are not fully compatible with polymer-coated glass. Direct metallization of copper on smooth glass surfaces without polymer also provides more opportunities for lower cost, miniaturization as well as reliability and electrical performance improvements.

In general, covalent-bonded oxide networks such as glass bond directly to materials composed of oxides or oxide interfaces that are stable. Since Cu is composed of metallic-bonded atoms, some surface modification that results in an oxide formation can be expected to result in enhanced bonding. Surface modifications can be physical (e.g., mechanical interlocking) or chemical, with the latter being ideal. For good chemical adhesion between bare copper metal and glass, fundamental adhesion theory states that it is important to have an intermediate metal oxide interface for strong and lasting adhesion. Based on this fundamental principle, a process that enhances chemical bonding through gradual, continuous material-property changes in the transition from glass to metal oxide to metal is preferred over mechanical-based methods to achieve strong adhesion. For example, it is well known that certain, reactive metals with high oxygen affinity (e.g., Cr, Ti, Zr, Mg) exhibit strong bonding "directly" to glass by forming thin oxide layers at the interfaces [3,4]. These oxides are very stable even at high temperatures and in reducing gas atmospheres. While standard vacuum deposition techniques benefit from this approach by depositing an adhesion layer such as Ti between Cu and glass, new innovations [5,6] are required to achieve this intermediate "adherence oxide" layer with wet metallization techniques as described next.

Electroless copper deposition is a standard wet metallization process typically used for organic substrates. It is a relatively low-cost, wet processing method involving a series of steps which includes chemical surface treatments, deposition of catalyst (typically Pd/Sn) particles, and electroless copper reduction on the catalyst particles. This creates three separate, ionic-bonded interfaces between the glass (negatively charged OH^- surface), conditioner (positively charged polymer electrolyte), and catalyst (negatively charged surface) (Fig. 1). Except for the bonding between the Pd catalyst and the electroless-deposited Cu, each of the interfaces between glass and catalyst rely on ionic bonding. To improve adhesion, surface etching and roughening is a typical part of electroless plating for organic substrates. Adhesion is improved through increased surface area and mechanical interlocking with the roughened surface. However, increasing the surface roughness of the glass-metal interface will result in higher conductor losses, especially at higher frequencies due to the skin-effect [7]. Additionally, glass roughening processes are difficult to control precisely and can introduce flaws or cracks, resulting in increased risk of mechanical failure. Since chemical, covalent bonds are not directly made between the Cu and the glass oxide network, glass surface roughness is mandatory for achieving sufficient adhesion. To maximize electrical performance and mechanical integrity, it is therefore of interest to find the minimum roughness required to achieve sufficient adhesion through roughening for electroless plating. Companies such as Atotech, Inc. are pioneering novel surface treatments and nanoparticle-based copper deposition approaches to achieve metal-glass adhesion with minimal roughness [5,6].

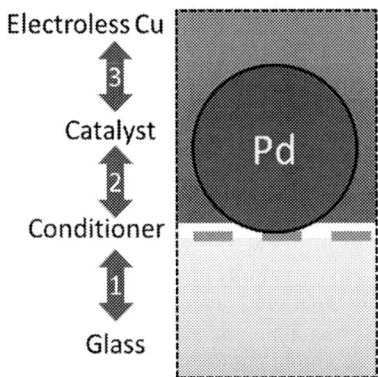

Figure 1. Schematic of chemical interfaces between electroless Cu and glass (not to scale).

In the MCM-D (MultiChip Module – Deposition) and flat-panel display industries, sputtering is a standard metallization process for ceramic substrates and large panels of bare glass. High-performance packaging requires thicker metal lines than what sputtering alone can achieve cost-effectively. The approach in this paper uses sputtered adhesion (Ti) and seed (Cu) layers, which are subsequently electroplated at high deposition rates to reach the necessary Cu thickness. Since glass/Ti and Ti/Cu are known to be strong interfaces individually, the glass/Ti/Cu composite structure is also expected to be strong.

Sol-gel is a common approach to deposit thin films due to its versatility with available chemistries and its relative simplicity as a process. In this wet deposition process, a metal-organic precursor solution (sol) is deposited onto the substrate. Because of the low viscosity of the solutions, the substrate can be dip-coated, spin-coated, or sprayed with controlled thickness to form a three-dimensional network (gel). The film is then subjected to a temperature high enough to pyrolyze the organic components, resulting in a three-dimensional network of the metal oxide. Finally, it is sintered at a higher temperature to crystallize the film, creating a glass/CuO_x/Cu structure.

This paper investigates three approaches to metallize glass directly, without relying on organic materials that may hinder subsequent processing: electroless Cu deposition, sputtering with electrolytic plating, and sol-gel synthesis. For sputtering and sol-gel, the resulting glass-metal oxide-metal structure is expected to demonstrate strong adhesion. As the electroless Cu does not introduce an oxide interface, weaker adhesion is expected. Based on the results from each of the approaches described above, individual assessments are made in terms of potential feasibility and future directions.

Experimental Methods

The three approaches to metallize glass are briefly described here. The key characterization techniques employed in this work are also outlined. Smooth and roughened glass-substrates were provided by Corning, Inc. and Life BioScience, Inc. Roughened samples were characterized by atomic force microscopy (Veeco AFM) prior to processing to obtain roughness (R_a), and ranged from < 1 nm to 0.559 µm.

Electroless processes were used to deposit electroless copper. The electroless-plating steps are substrate cleaning, conditioning (deposition of polymer electrolyte), catalyst deposition, catalyst activation, and finally electroless copper deposition. Triple rinses in water were performed between each step. Electroless deposition was performed until 0.2 µm of Cu was deposited. Finally, samples were annealed at 155°C for 30 minutes in air. Tape-tests were then performed according to IPC-TM-650 test standards. For samples which failed the tape-test, the substrate and peeled interfaces were analyzed by X-ray photoelectron spectroscopy (XPS) (Thermo K-Alpha XPS) for the presence of Cu, Si, and C [8].

Sputtering was performed by Tango Systems, Inc. Ti (100 nm) was first deposited as an oxide-forming adhesion layer, followed by Cu deposition to various thicknesses. Samples were then electroplated with a current density of 2.0 ASD (amperes per square decimeter) until the desired thickness was reached, as measured by a Cu gauge. This was followed by annealing at 155°C for 30 minutes in air. Lines and spaces of 1.0 and 0.5 cm were patterned by photolithography and subtractive etching (Transene Company, Inc.) for peel-testing. Peel-testing was performed at a pull-rate of 12 in/min. Thermal shock testing was performed according to JEDEC Standard JESD22-A106B Condition C.

Cu sol-gel solution was prepared by dissolving Cu-ethoxide in 2-methoxyethanol and acetic acid. The solution was then spin-coated on glass and baked at 300°C. This was repeated until a sufficient thickness was deposited. It was then placed in a rapid thermal processing (RTP) chamber in a forming gas atmosphere at 560°C for 10 minutes for Cu metal crystallization in the bulk, and simultaneous copper oxide interface formation with glass. Characterization of the metal and glass-metal interfaces was performed with X-ray diffraction (XRD) and Scanning Electron Microscopy (SEM). This paper shows preliminary results of using the sol-gel route to deposit a copper-oxide film directly on glass, followed by thermally converting the oxide film to metallic copper.

Results and Discussion

Electroless Cu Approach: Results of the tape-test of electroless Cu on glass are shown in Fig. 2, in the order of increasing adhesion. It can be seen that there is a regime of surface roughness values (R_a) where the structure of peeled copper changes. In the samples with worst adhesion, Cu outside of the tape dimensions were peeled, as can be seen by the jagged peels in Fig. 2a,b. As the adhesion increases, the copper film does not peel beyond the edge of the tape (Fig. 2c), and then smaller amounts of Cu peel from the glass (Fig. 2d-f).

978-1-4799-2408-0/14 $31.00 © 2014 IEEE

Finally, the glass chemically roughened to 0.559 µm passed the tape test, exhibiting no peeled Cu (Fig. 2g).

Figure 2. Tape tests for electroless copper on various glasses: (a) Corning SGW8.5 (R_a = 0.5 nm) (b) Corning SGW3 (R_a = 1.2 nm) (c) Corning SGW3 (R_a = 0.9 nm) (d) Corning SGW8.5 (R_a = 1.3 nm) (e) Corning SGW8.5 (R_a = 19 nm) (f) Corning SGW8.5 (R_a = 46 nm) (g) Life BioScience APEX (R_a = 0.559 µm).

The large range in tape-test results for similar roughness values (Fig. 2a-d) can be attributed to several factors. Roughness values were measured with AFM and could vary depending on the selected area because the topography was generally not homogenous in the areas inspected (10x10 µm and 30x30 µm). The values in this range are within the instrumental error. In addition to surface roughness, adhesion to glass depends on the glass chemical composition. High CTE glasses (SGW8.5) are composed of more glass network-modifying ions than low CTE glasses (SGW3); as these ions are positively charged, they can assist with adhesion to the Pd catalysts.

Representative XPS elemental Pd and survey scans of the peeled Cu and the exposed glass are shown in Fig. 3. It is apparent that Pd was detected only on the peeled Cu, while C was detected only on the glass surface. Assuming that the only source of carbon in the system is the polymer electrolyte conditioner, this suggests that the weakest interface in the electroless copper-glass is that between the conditioner and the Pd catalyst.

With the current electroless process, a surface roughness on the order of hundreds of nanometers is necessary to achieve sufficient adhesion to pass the tape test. Based on the skin-effect, Brist et al. [7] modeled the effects of surface roughness on electrical loss and introduces an additional electrical loss factor K_{sr}, where a value of 1 represents the factor for smooth surfaces (i.e., no additional loss). At an electrical frequency of 10 GHz, surface roughnesses of 0.2 µm and 0.6 µm result in K_{sr} values of ~1.1 and 1.55, respectively [7]; i.e., Cu with 0.6 µm roughness will have 55% more loss (dB/length) than smooth Cu.

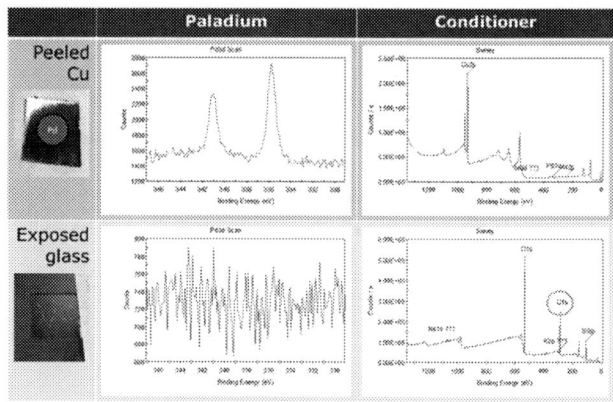

Figure 3. XPS elemental Pd and survey scans of the peeled Cu and the exposed glass.

Sputtering Approach: To investigate the effects of the sputtered Cu seed-layer thickness and total thickness of the Cu layer, three different seed layer thicknesses (200, 500, and 800 nm) were sputtered on unroughened glass. Each film was electroplated to a total Cu thickness of 5 µm and 10 µm, for a total of six different samples. Peel-tests of these samples showed no significant correlation with the seed-layer thickness for the ranges investigated.

Peel-testing for smooth and roughened (R_a = 10, 50, 100, 200, 300, 400, 500, 1000 nm) glass electroplated to 10 µm Cu was performed before and after thermal shock testing (5 µm thick Cu strips tore during peel testing, and could not be measured). Prior to thermal shock testing, smooth glass exhibited peel strength values over 0.2 kg/cm (Fig. 4). On the other hand, the Cu lines on all roughened glass samples could not be peeled from the glass and therefore exceeded the strength limits that the technique can measure. After 1000 thermal shock cycles, the smooth glass was again the only peelable sample, and did not show any significant change in peel strength when compared to those from pre-thermal shock testing.

In the microelectronics packaging industry, a typical benchmark for the peel strength is 0.7 kg/cm for Cu thickness of 30 µm on organic substrates. The peel strength W_o of a metal strip as a function of film thickness t_s has been modeled by Bikerman to follow the relationship [9]:

$$W_o \propto t_s^{3/4} \qquad \text{Equation (1)}$$

Using this relationship, a comparable peel strength value for Cu plated to 10 µm thickness was calculated to be 0.18 kg/cm. As seen in Fig. 4, the peel strengths before and after thermal shock testing were over 0.25 kg/cm, well-above this scaled benchmark value.

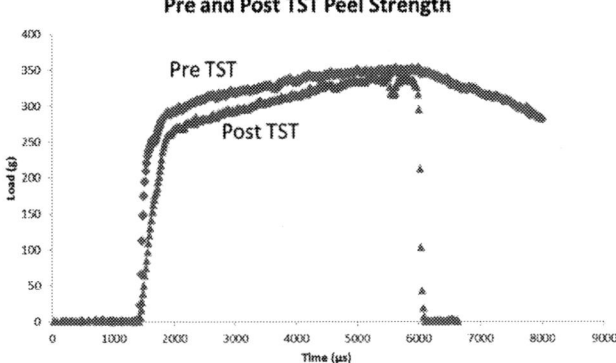

Figure 4. Peel strength of 10 µm Cu on smooth glass before and after 1000 thermal shock cycles.

Sol-gel Approach: The steps of Cu film deposition by sol-gel are shown in Fig. 5. The resulting Cu film is continuous, and the adhesion was sufficient to pass the tape test. The film was confirmed to be reduced to the metallic state by XRD (Fig. 6).

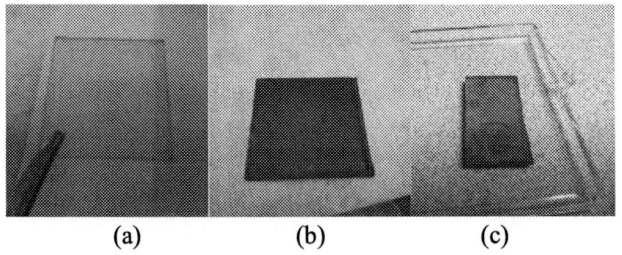

Figure 5. Cu deposited on glass by sol-gel after (a) spin-coating, (b) baking, and (c) rapid thermal processing.

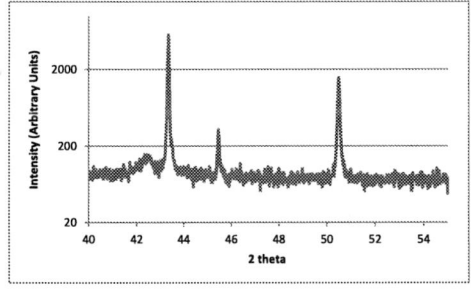

Figure 6. X-ray diffraction scan of sol-gel Cu film.

While the film appeared to be continuous across the glass surface, SEM images show that the Cu film thickness was not uniform (Fig. 7). The non-uniform film thickness is partly due to insufficient wetting of the Cu solution onto the glass surface during spin-coating, which can lead to minor islanding effects. Additionally, the Cu solution showed some cloudiness, indicating that some Cu hydroxyl-complexes precipitated as particles in the solution, adding to the non-uniformity of the resulting film. Further optimization of the sol-gel process to increase glass-wettability and Cu solution quality should result in more uniform and higher-quality Cu films.

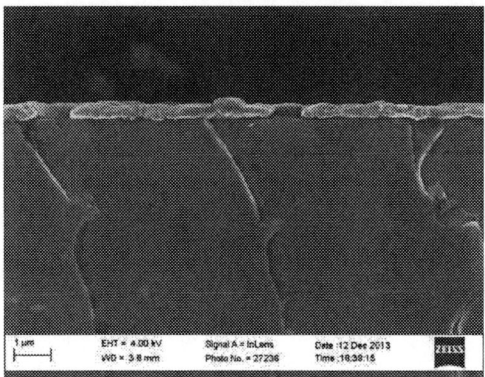

Figure 7. SEM image of cross section of Cu on glass deposited by sol-gel.

Conclusion

Three approaches for direct copper metallization of glass substrates were investigated for the resulting metal-to-copper adhesion.

With electroless deposition, the effect of glass surface roughness on adhesion strength was analyzed with tape-test results. The required roughness for mechanical adhesion of copper on glass is of the order of hundreds of nanometers, which will negatively affect the electrical performance at high frequencies. Analysis of the electroless Cu/glass interface through XPS suggests that the weakest interface is the ionic bond between the polymer electrolyte and the Pd catalyst particles.

With sputtered Ti/Cu, the seed-layer thickness had no measurable effect on the resulting peel strength. Sputtering on smooth glass, as known previously, met the peel-strength performance targets before and after thermal shock testing. Ti/Cu sputtered on roughened-glass exceeded the peel-strength measurement capability of the technique even after thermal shock testing. Therefore, reliability has been demonstrated for sputtering on smooth and roughened glass.

Cu oxide deposited with sol-gel and reduced to metallic copper passed the tape test, but exhibited non-uniform Cu films. Enhancements to sol-gel processing can improve the quality of Cu films.

As expected from fundamental glass-to-metal bonding theory with stable oxides at the interface, Cu deposited by sputtering and sol-gel showed significantly greater adhesion to smooth glass than electroless plating because of the presence of an intermediate "adherence oxide". The stability of this Cu oxide is the subject of future work.

Acknowledgments

The authors would like to acknowledge Jason Bishop and Chris White for their guidance and support in sample fabrication, as well as Dibyajat Mishra for XPS data acquisition and Yuya Suzuki for assistance in peel strength analysis.

References

1. V. Sukumaran, "Low-cost thin glass interposers as a superior alternative to silicon and organic interposers for packaging of 3-D ICs," *IEEE Trans. CPMT,* vol. 2, no. 9, pp. 1426-1433, Sept. 2012.

2. V. Sukumaran, "Through-package-via formation and metallization of glass interposers," in *Proc. IEEE Electronic Components and Technol. Conf. (ECTC)*, Las Vegas, NV, Jun. 1-4, 2010, pp. 557-563.

3. P. Benjamin, "The adhesion of evaporated metal films on glass," *Proc. R. Soc. Lond. A*, vol. 261, no. 1307, pp 516-531, May 1961.

4. N. Jiang, "Observations of reaction zones at chromium/oxide glass interfaces," *J. Appl. Phys.*, vol.87, no. 8, pp. 3768-3776, Apr. 2000.

5. S. Bamberg, "Novel wet chemical metallization for glass interposers", 3D Materials and Processes Session, *International Conference on Device Packaging*, Scottsdale/Fountain Hills, AZ, Mar. 7-10, 2011.

6. R. Taylor, "Challenges of Adhesion Promotion for Metallization of Glass Interposers"in *Global Interposer Technology Workshop (GIT)*, Atlanta, GA, Nov. 17-20, 2013.

7. G. Brist, "Non-classical conductor losses due to copper foil roughness and treatment," in *Proc. Electron. Circuits World Conv. (ECWC)*, Anaheim, CA, Feb. 22-24, 2005, pp. 22-24.

8. X. Cui, "Copper deposition and patterning for glass substrate manufacture," in *IEEE Trans. Electronics Packaging Technol. Conf. (EPTC)*, Singapore, Dec. 10-12, 2007, pp. 37-42.

9. C. Jouwersma, "On the theory of peeling," *Journal of Polymer Science*, vol. 45, issue 145, pp. 253-255, July 1960.

High-Frequency Characterization of Through Package Vias Formed by Focused Electrical-Discharge in Thin Glass Interposers

Jialing Tong[*], Yoichiro Sato[+], Shintaro Takahashi[+], Nobuhiko Imajyo[+], Andrew F Peterson[++], Venky Sundaram, and Rao Tummala

3D Systems Packaging Research Center
813 Ferst Drive NW, Georgia Institute of Technology, Atlanta, GA, USA
[+]Asahi Glass Company, Tokyo, Japan
[++]School of Electrical & Computer Engineering, Georgia Institute of Technology
[*]jtong@gatech.edu, (404) 358-3588

Abstract

This paper presents the modeling, design, fabrication and characterization, up to 30 GHz, of low loss and high aspect-ratio 55 μm diameter through package vias (TPVs) in 300 μm thick glass interposers. These TPVs were fabricated using a novel, high-throughput, focused electrical discharge method and low cost panel-based double-side metallization processes. Such a glass interposer is targeted at two emerging applications, (a) large 30 mm to 60 mm body size 2.5D interposers to achieve 28.8 Gbps logic-memory bandwidth and (b) 3D interposers for mm wave applications at 28 GHz local multipoint distribution service (LMDS) for future 5G networks. Accurate measurement of the electrical performance of fine pitch metallized through vias in glass up to 30 GHz and beyond is critical for both these high performance interposer applications. In this paper, two novel characterization methods are applied: 1) the short-circuit-and-open-circuit method and 2) the dual-via-chain method. The resistance and the inductance of a single via are extracted by using a short-circuit structure along with an open-circuit structure. At 10 GHz, the values for the series resistance and inductance have average values of 0.1 Ω and 160 pH respectively. Long dual-via chains were designed to evaluate their performance in insertion loss, delay and eye diagram. The insertion loss achieved with the longest dual-via chain was found to be less than 1 dB/cm up to 30 GHz with only a 6.2 ps delay in the TPVs, and the simulations indicate a wide open eye.

I. Introduction

Two major applications are driving the need for low loss and high frequency interposers, (a) high bandwidth in 2.5D and 3D logic-memory and multi-memory architectures to achieve 28.8 Gbps and 56 Gbps high speed channels, and (b) mm-wave modules operating at 28 GHz, 39 GHz and other spectrum bands for future 5G mobile networks. Both these applications require ultra-low electrical loss and high precision circuits with high speed signal propagation and high density interconnections. 2.5D and 3D Silicon interposers, using through silicon vias (TSVs), face severe technical challenges including high electrical loss in large body sizes, and high cost, thus limiting their widespread commercial use. Glass has been proposed to be a superior alternative to silicon because of its excellent electrical property and the scalability to large panel sizes leading to lower cost [1].

The two key building block technologies required for glass interposers are fine pitch through package vias (TPVs) and re-distribution layers (RDLs). This paper presents detailed electrical modeling, design and characterization of TPVs in 300 μm thick 3D glass interposers, using a new high-throughput drilling technique, namely the focused electrical discharge method, capable of high throughputs, greater than 1000 vias per second [2]. The focused electrical discharge method is viewed to be a major enabler for high-performance glass interposers, since it is capable of forming high aspect-ratio TPVs at fine pitch with smooth side wall surfaces, unlike other approaches. There is limited literature about the high frequency electrical behavior of such TPVs in glass, and this paper represents the first comprehensive modeling and characterization study of high aspect-ratio TPVs in glass using the electrical discharge method for forming vias.

Five parameters are necessary to comprehensively evaluate the transmission performance with such TPVs: series resistance, series inductance, insertion loss, delay and eye diagrams. While the electrical performance of glass TPVs, formed by excimer laser ablation has been characterized [3], TPVs drilled by the focused electrical-discharge method have yet to be evaluated. Though the short-circuit-and-open-circuit method is not accurate, in the case of through-silicon-vias (TSVs) due to the electrical lossiness of silicon [4], it is the preferred method to extract the series resistance and inductance of TPVs in glass due to its simplicity and miniaturization, compared to other two-port measurement techniques requiring a large-area embedded capacitor [5]. The dual-via-chain method is the common method of choice to evaluate the insertion loss, but has only been applied so far up to 10 GHz [3, 6], which is insufficient for high-speed applications. Furthermore, the delay and the eye diagram of TPVs formed by the focused electrical-discharge method have not been investigated so far.

In this paper, two specific TPV configurations were designed and implemented in 300 μm thin glass interposers to fully assess the electrical performance of 55 μm diameter TPVs drilled by the focused electrical-discharge method up to 30 GHz. The first configuration was based on the short-circuit-and-open-circuit method, by which the series resistance and inductance of a single via can be retrieved from the Z-parameter that is calculated using the measured S-parameter. The second configuration was the commonly-used dual-via chain with long 50 Ω lines for insertion loss, delay and eye diagram. The insertion loss was measured up to 30

GHz by a Vector Network Analyzer (VNA) with excellent correlation of measured results to those from the 3D electromagnetic (EM) solver-Computer Simulation Technology (CST), while the delay and the eye diagram were generated using the simulated data in CST.

In Section II, the detailed high-frequency modeling and design are described. Then, the characterization data on the fabricated test vehicles is presented in Section III. Finally, the analysis and discussion of the results and the conclusions are given in Section IV and Section V, respectively.

II. High-Frequency Modeling and Design

The through package vias in the glass interposer can either be used for power and ground connections or signal transmission. The short-circuit-and-open-circuit method is an accurate technique to extract the series resistance and inductance for the power and ground TPVs, and the dual-via-chain method is widely used to characterize signal transmission TPVs. In this section, the modeling and the design for both methods are presented.

A. Short-Circuit-and-Open-Circuit Method

The short-circuit-and-open-circuit method was initially proposed in [4] to characterize the series resistance and inductance of TSV. The structures based on this methodology are shown in Fig. 1, including one structure that has a TPV shorting the signal to the backside ground and another structure that has identical dimensions but no shorting TPV for de-embedding the parasitic capacitance and conductance. The backside ground is constructed to be a plane rather than a trace, in that a ground plane has negligible inductance for perfect shorting. On the other hand, the probing pad is minimized to reduce the pad-introduced inductance.

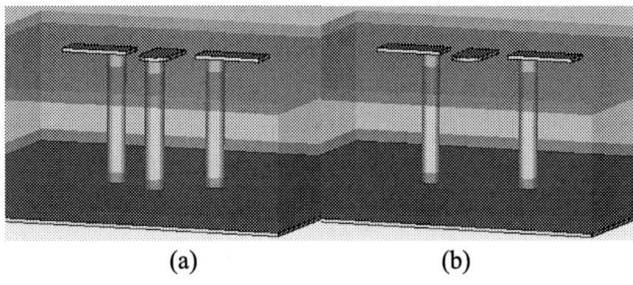

(a)   (b)

Figure 1. (a) Short-circuit structure shorted at the backside for retrieving resistance and inductance. (b) Open-circuit structure for de-embedding parasitic capacitance and conductance.

Based on this physical structure of Fig. 1(a), an equivalent circuit model can be derived, which is shown in Fig. 2. There are a total of three TPVs in this scenario, namely two ground TPVs and one signal TPV, where each TPV is modelled as a resistor in series with an inductor. For the Ground–Signal–Ground (GSG) probe, one signal pad and two ground pads are directly on the three TPVs, and parasitic capacitance and conductance respectively denoted by C and G in Fig. 2 exist between them, which can be de-embedded by the open-circuit structure.

Due to the symmetry of the GSG configuration, the current flowing through each of the two ground TPVs is estimated to be half of that in signal TPV. In other words, it is reasonable to assume that

$$L = 2 \cdot L^*$$
$$R = 2 \cdot R^* \tag{1}$$

where L and R are the inductance and the resistance of the signal TPV respectively, while L^* and R^* are the inductance and the resistance of one ground TPV respectively.

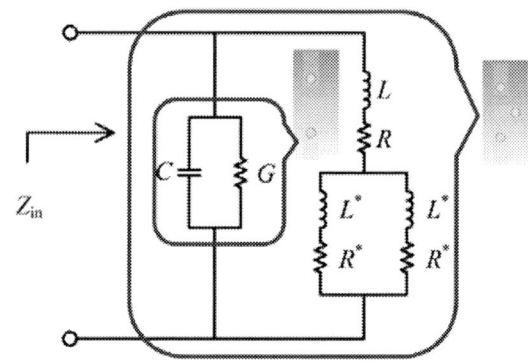

Figure 2. Equivalent circuit model of the short-circuit structure with the open-circuit structure included.

The input impedance of the short-circuit structure can be derived from the equivalent circuit model, which is

$$\frac{1}{Z_{in}} = j\omega C + G + \cfrac{1}{j\omega L + R + \cfrac{j\omega L^* + R^*}{2}} \tag{2}$$

where $\omega = 2\pi f$ is the radian frequency.

B. Dual-Via-Chain Method

A dual via chain is widely used to study the transmission performance of TPVs, which consists of one set of TPVs transiting from top to bottom and another set of TPVs transiting from bottom back to top, as shown in Fig. 3.

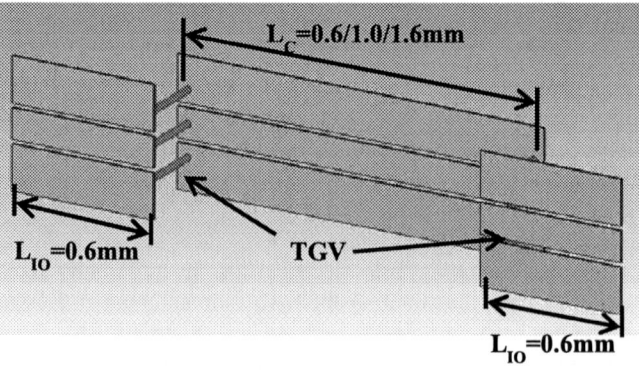

Figure 3. Dual via chain structure for studying the transmission performance of TPVs.

To support the GSG probes, co-planar waveguide (CPW) lines were applied and designed for 50 Ω matching. According to the stack-up shown in Fig. 4 and the electrical property of EN-A1 glass from Asahi Glass Company (AGC) [2] and ZEONIF™ ZS-100 polymer from Zeon Cooperation (Zeon) [7] listed in Table I, a 50 Ω CPW line was designed with the center conductor width of 141.025 µm and the gap between conductors of 20 µm. The length of the bottom connection lines was varied as L_C=0.6 mm, 1 mm and 1.6 mm, and the top CPW lines were fixed to L_{IO}=0.6 mm.

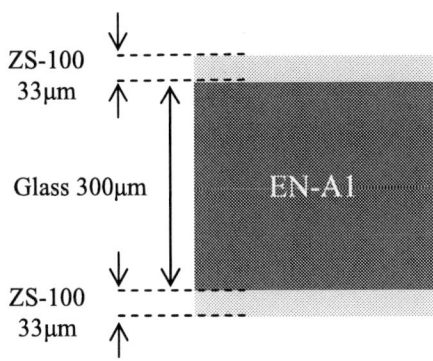

Figure 4. Cross-section view of the stack-up.

TABLE I
MATERIAL PROPERTY

Electrical Property	AGC EN-A1	Zeon ZS-100
Dielectric Constant at 10 GHz	5.46	3.0
Loss Tangent at 10 GHz	0.0056	0.005

The delay of the structure shown in Fig. 3 has three segments: the delay τ_{IO} from the top L_{IO} line, the delay τ_{TPV} from the TPV buried in the glass and the delay τ_C from the bottom L_C line. Then, the total delay of the whole structure can be expressed as

$$\tau = 2 \cdot \tau_{IO} + 2 \cdot \tau_{TPV} + \tau_C \qquad (3)$$

Thus, the delay caused by the TPV can be captured by deducting the delay τ_{IO} and the delay τ_C from the total delay τ.

III. Characterization Results

Based on the modeling and design structures presented in the previous section, a test vehicle was fabricated using 300 µm AGC EN-A1 glass and 33 µm Zeon ZS-100 polymer, which is shown in Fig. 5. The TPVs were formed by the electrical-discharge method developed by Asahi Glass Company, with a 55 µm via diameter and 150 µm minimum center-to-center via pitch.

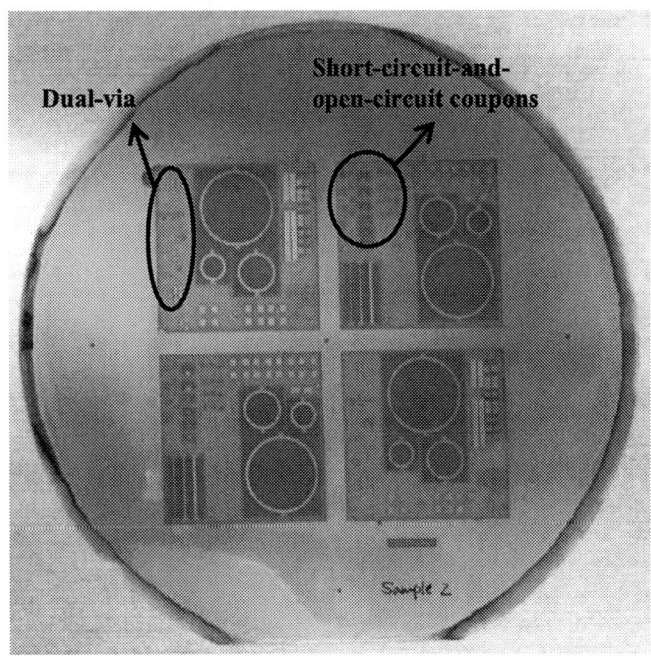

Figure 5. Top view of the fabricated test vehicle.

An on-wafer probe was used to measure the S-Parameters from 100 MHz to 30 GHz with an Agilent 8510C VNA and Cascade Microtech Ground–Signal–Ground (GSG) probes.

A. Results by Short-Circuit-and-Open-Circuit Method

TPVs should have no resistance or inductance in an ideal case. From the perspective of the impedance Smith Chart shown in Fig. 6, the ideal TPV is supposed to be located in the leftmost point of the impedance Smith Chart. Unfortunately, there is series resistance with each TPV because of the finite copper conductivity. Also, due to the skin effect, the input impedance moves from the outermost circle into the inside of the impedance Smith Chart. More importantly, the inductance effect introduced by the physical length of the TPV plays a critical role for high-speed digital applications, which might cause current overshooting. This inductance effect makes the input impedance travel along the outermost circle. Thus, the skin effect and the inductance effect of the TPV result in the impedance moving along the impedance Smith Chart to the inside.

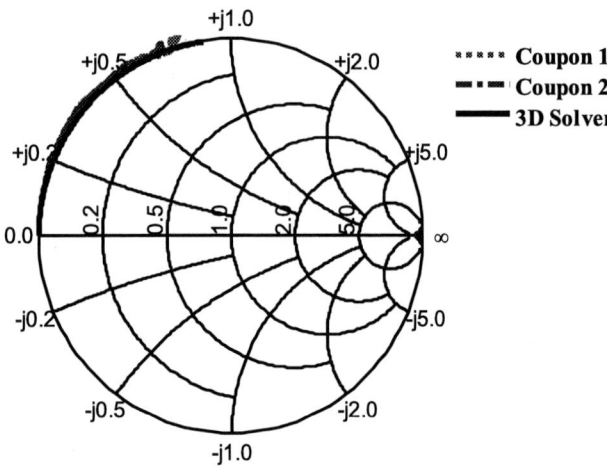

Figure 6. Measured and simulated results shown in the Impedance Smith Chart.

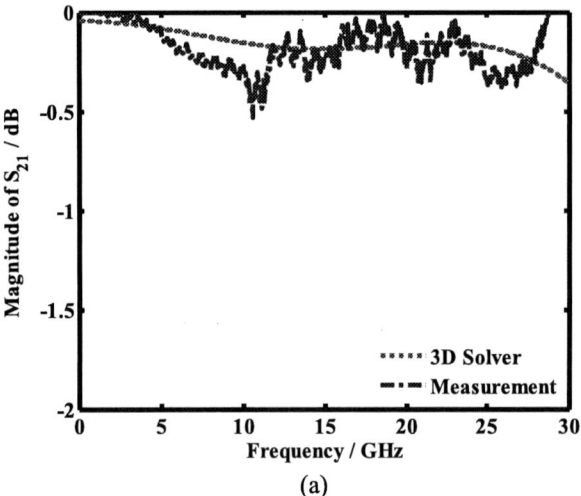

(a)

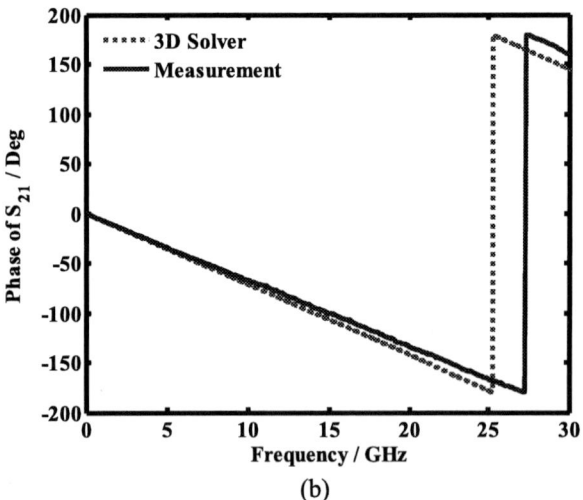

(b)

Figure 7. Comparison of the measured and simulated results: (a) the magnitude of S_{21} in dB; (b) the phase of S_{21} in Degree.

Two coupons with the same via location were measured up to 30 GHz, and the measured results are depicted in Fig.6 with the simulated results for comparison. It can be seen that the simulated results agree well with the measured results and more detailed interpretation of the data will be presented in the next section.

B. Results by Dual-Via-Chain Method

Three dual via chains were designed and implemented with different lengths, and the longest one had a total length of 2.8 mm excluding the length of the TPV. All these three lines were characterized up to 30 GHz. It is well known that the insertion loss increases with line length. Thus, the simulated and measured results of the longest chain are presented in Fig. 7, which shows excellent match between the simulated and measured results. The measurements demonstrate the superior electrical transmission of the TPV in the glass interposer, compared to TSVs. More analysis of the data will be provided in the next section to discuss the delay and the eye diagram for high-speed digital applications.

IV. Analysis and Discussion

In this section, detailed analysis of the results by the short-circuit-and-open-circuit method will be presented first. The extracted resistance and the inductance over a wide frequency range will be given, and the skin effect and the proximity effect will be discussed. Then, the delay profile as frequency and the eye diagram were generated using 3D EM solver – CST, based on the dual-via-chain method.

A. Results by Short-Circuit-and-Open-Circuit Method

The S-Parameters were measured using Agilent 8510C VNA and Cascade GSG probes. Once the S-parameters were obtained, they were converted into Z-parameters. Since these structures are a one-port network, the obtained S-parameter is essentially the return loss (Γ), which is related to the Z-parameters by the following equation [8]

$$Z_{in} = \frac{1}{Y_{in}} = Z_0 \frac{1+\Gamma}{1-\Gamma} \tag{4}$$

Then, according to formula (2), the resistance and the inductance of each TPV were calculated, shown in Fig. 8 and Fig. 9, respectively.

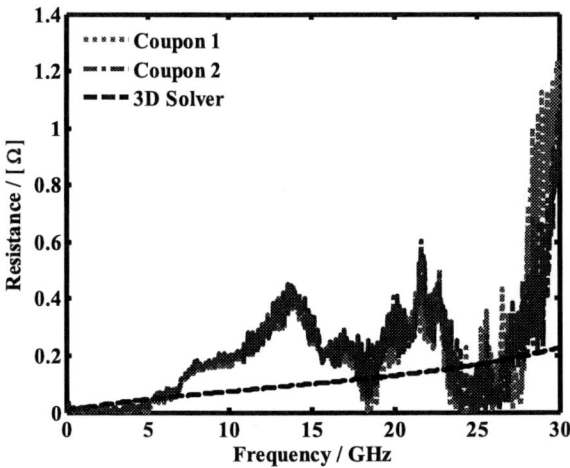

Figure 8. The resistance retrieved from the simulated and measured results.

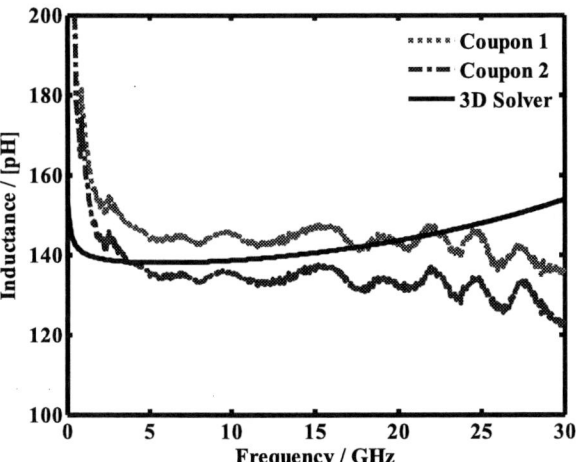

Figure 9. The inductance retrieved from the simulated and measured results.

In Fig. 8, it can be seen that the resistance of a single TPV is small and very challenging to measure. At 10 GHz, the simulated resistance was around 100 mΩ, which is close to the measured resistances with an average value of around 185 mΩ. Due to the skin effect, the resistance increases as frequency increases. In addition, because the skin depth follows the square root of frequency, the resistance also has the same square root relation with the frequency.

The extracted inductance is plotted in Fig. 9, and the measured results match well with the simulations. The inductance of a single via was estimated to be 140 pH which is much lower than the typical values for wire-bond interconnects, generally in the nH order of magnitude. At low frequencies below 5 GHz, the inductance decreases with frequency, because the skin effect causes the TPV to lose its internal inductance. Then, the inductance remains almost constant till the parasitic capacitance comes into play and results in LC resonance.

There are minor mismatches between the measured results and the simulated results at the lowest and at the highest frequencies. The minor mismatch at the lowest frequency is because the inductance effect is not pronounced, which makes the measurement rather challenging; while the minor mismatch at the highest frequency is due to a capacitance shift caused by the fabrication. In addition to these, the proximity effect also increases the mismatch between modeling and measurement. The magnitude of the magnetic field around the TPV is shown in Fig. 10. According to the boundary condition of the magnetic field shown below

$$\vec{J}_s = \hat{n} \times \vec{H} \qquad (5)$$

the current flowing in the signal via is almost uniformly distributed along the via circumference. However, it is not the case for the two ground vias, as shown in Fig. 10. The magnetic field is almost crowded in half of the ground via while the other half is not contributing much to the current conduction. In other words, the assumption made in (1) is good but not very precise, considering the proximity effect.

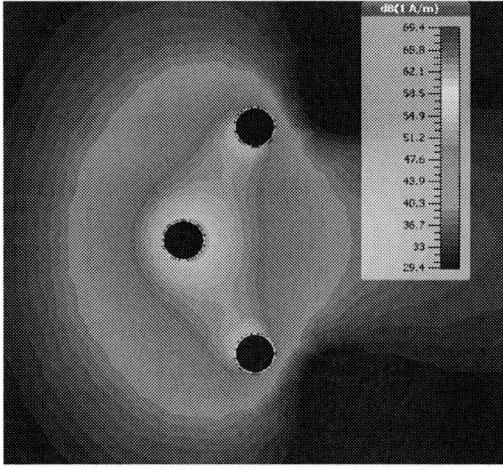

Figure 10. The magnitude of the magnetic field surrounding the TPV array.

B. Analysis of Dual-Via-Chain Results

Signal delay is a very important parameter for high speed digital applications, because it is critical for correct timing. Using the simulated results and the formula (3), all the delay profiles including that of the TPV were computed, and presented in Fig. 11. It can be seen that the total 6.2 ps delay of the TPV in glass is longer than that of the 0.6 mm long CPW line and it is close to the delay of the 1.6 mm long CPW line, though the physical length of the TPV is very short. This is due to the higher effective permittivity in the glass than in the CPW line. However, the delay of a typical TSV in silicon will be larger than that of TPV in glass, because the electrical permittivity of silicon – 11.9 is larger than that of glass – 5.46. Thus, TPVs in the glass interposer can support faster transistor speed than TSVs.

978-1-4799-2408-0/14 $31.00 © 2014 IEEE   2275

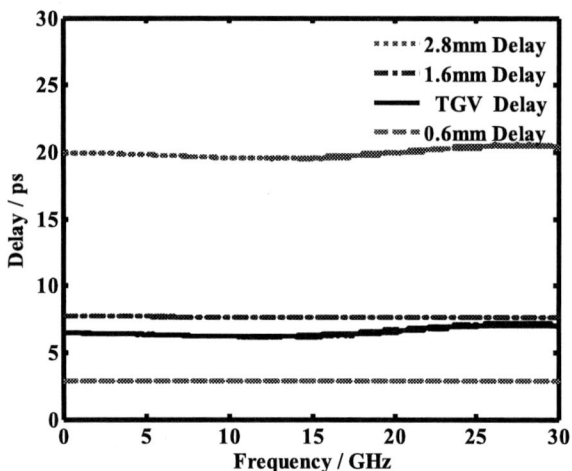

Figure 11. All the delay profiles generated by the simulated results from CST.

Fig. 12 shows the eye diagram obtained in 3D EM Solver - CST, for a 20 Gbps pseudo-random bit stream (PRBS) transmitted through a dual via chain (shown in Fig. 3) with the bottom CPW line length set at 1.6 mm. It can be seen that the eye is clearly open with 1.43 mV jitter, 50 ps eye width and 0.91 V eye height.

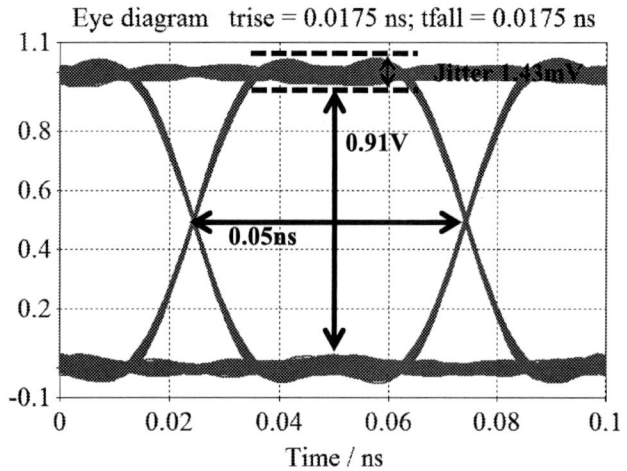

Figure 12. The simulated eye diagram at 20 Gbps for 2.8 mm glass CPW channel.

V. Conclusions

A detailed electrical modeling, design and high frequency characterization, up to 30 GHz, was presented for high aspect-ratio 55 μm diameter TPVs in 300 μm thin glass, formed by a novel focused electrical discharge method that is capable of greater than 1000 vias per second throughput. Such a glass interposer is ideal for 2.5D and 3D package integrations for two major applications: high performance digital systems with high logic-memory bandwidth, and mm-wave modules at 28 GHz and higher frequencies for future 5G mobile networks. The high-aspect-ratio and smooth-side-wall TPVs drilled by the focused electrical discharge method, produce exceptionally low resistance and inductance, and achieved high-quality signal transmission enabled by low loss

and delay, even at high frequencies. These results establish the superior electrical performance of glass interposers with such TPVs, thus making glass the ideal package material for high-speed digital and 5G mobile systems.

Acknowledgments

The authors would like to thank the full-member and supply-chain companies of the Low-cost Glass Interposer and Package (LGIP) consortium at the 3D Systems Packaging Research Center (PRC), Georgia Institute of Technology, Atlanta, for their fabrication support.

References

1. Sukumaran, V.; Bandyopadhyay, T.; Sundaram, V.; Tummala, R., "Low-Cost Thin Glass Interposers as a Superior Alternative to Silicon and Organic Interposers for Packaging of 3-D ICs," *IEEE Transactions on Components, Packaging and Manufacturing Technology*, vol.2, no.9, pp.1426,1433, Sept. 2012.

2. Takahashi, S.; Horiuchi, K.; Tatsukoshi, K.; Ono, M.; Imajo, N.; Mobely, T., "Development of Through Glass Via (TGV) formation technology using electrical discharging for 2.5/3D integrated packaging," in *Proceeding of 2013 IEEE 63rd Electronic Components and Technology Conference (ECTC)*, vol., no., pp.348,352, 28-31 May 2013.

3. Sukumaran, V.; Bandyopadhyay, T.; Chen, Q.; Kumbhat, N.; Fuhan Liu; Pucha, R.; Sato, Y.; Watanabe, M.; Kitaoka, Kenji; Ono, M.; Suzuki, Y.; Karoui, C.; Nopper, C.; Swaminathan, M.; Sundaram, V.; Tummala, R., "Design, Fabrication and Characterization of Low-Cost Glass Interposers with Fine-Pitch Through-Package-Vias," in *Proceeding of 2011 IEEE 61st Electronic Components and Technology Conference (ECTC)*, pp. 583-588, May 31-June 3 2011.

4. Leung, L.L.W.; Chen, K.J., "Microwave Characterization and Modeling of High Aspect Ratio Through-Wafer Interconnect Vias in Silicon Substrates," *IEEE Transactions on Microwave Theory Technology*, Vol. 53, No. 8, pp. 2472-2480, Aug. 2005.

5. Poh, C.H.J.; Bhattacharya, S.K.; Ferguson, J.; Cressler, J.D.; Papapolymerou, J., "Extraction of a Lumped Element, Equivalent Circuit Model for Via Interconnections in 3-D Packages Using a Single Via Structure with Embedded Capacitors," in *Proceeding of 2010 IEEE 60th Electronic Components and Technology Conference (ECTC)*, pp. 1783-1788, June 1-4 2010.

6. Shorey, A.; Keech, J.; Piech, G.; Bor-Kai Wang; Tsai, L., "Glass Substrates for Carrier and Interposer Applications and Associated Metrology Solutions," in *Proceeding of 2013 24th Annual SEMI Advanced Semiconductor Manufacturing Conference (ASMC)*, pp. 142-147, May 14-16 2013.

7. ZEON COOPERATION, "ZEONIF™: Insulation Materials for Printed Circuit Board," http://www.zeon.co.jp/business_e/enterprise/imagelec/zeonif.html.

8. D. M. Pozar, Microwave Engineering (4th Edition), Wiley Global Education, 2011, pp. 56–78.

Interfacial Reactions between Cu and Sn, Sn-Ag, Sn-Bi, Sn-Zn Solder under Space Confinement for 3D IC Micro Joint Applications

T. L. Yang*, W. L. Shih, J. J. Yu, and C. R. Kao

Department of Materials Science & Engineering, National Taiwan University

No. 1, Sec. 4, Roosevelt Road, Taipei, 10617 Taiwan (R.O.C)

*E-mail: d99527005@ntu.edu.tw

Abstract

Recently, three-dimensional integrated circuit (3D IC) integration technology consisting of TSV and micro joints has been viewed as one of the very promising solutions to go beyond Moore's law. Typical solder joints for 3D IC stacking technique under development today has a solder volume about $1000\ \mu m^3$, which is roughly 1/500 that of a conventional flip-chip solder joint with usually an 100 μm in diameter. One well-perceived effect of such a small solder volume is the solder joint would have very high possibility to end in a relatively large fraction of intermetallic compounds (IMCs) after common reflowing or thermal compression bonding process. Under this circumstance, instead of solder, the mechanical properties of IMCs and other reaction-induced microstructural changes and characteristics will undoubtedly play dominant roles on the reliability of the micro joints.

In this study, solid-state interfacial reactions under a very confined space between Cu and Sn, Sn-Ag, Sn-Bi, Sn-Zn solders are systematically investigated. The objective of this paper is not merely to propose and identify some critical reliability concerns arising from interfacial reactions but to provide essential data for microelectronic packaging industry to the design of sturdy enough micro solder joints for multi-chip stacking applications. Sandwich structures of Cu/Solder/Cu are fabricated through chip-to-chip thermal compression bonding process. The thickness of the solder layer in this study is well-controlled at 10 μm. High temperature storage tests are conducted by isothermal aging at 120 °C, 150 °C, 180 °C, and 200 °C for different time periods, respectively.

Issues to be discussed in this paper include (a) Impingement behavior of IMC grains after solid-state isothermal aging, (b) Rise of concentrations of minor inert alloying constituents in low solder volume joint, (c) Promising approach to reduce the growth rate of Cu-Sn reactants within limited solder volume joint. These critical issues will be proposed and discussed through experimental evidence, and the implications based on these finding will be discussed as well.

Introduction

Interfacial reactions between Sn-based solders and two common substrate metallization layers, typically Cu and Ni, have been widely researched and reviewed in both microelectronic packaging industry and academic research activities worldwide for several decades [1-4]. With the coming of novel integration technology for electronic packaging arena; however, some new concerns resulted from interfacial reactions are still worthy to keep an eye on from the viewpoint of a reliable solder joint. For example, recently, three-dimensional integrated circuit integration (3D IC) through implementing of TSV and micro-joints has been considered to be a very promising solution to go beyond Moore's law. In comparison with conventional 3D die stacking technique, such as Wire Bonding (WB) or Package on Package (PoP), it provides a more direct way and better performance to integrate several heterogeneous functional components, such as Memory, Processor, Logic, and Flash, etc. together in a vertically stacking way [5-6].

Unfortunately, there are only limited studies regarding solid-state interfacial reactions in micro joints. Most of prior investigations reported in literatures [7-12] mainly focused on discussing the applications of solid liquid inter-diffusion (SLID) bonding method for micro-joints due to low solder volume in essence, where a layer of low melting solder materials, such as Sn, is taken as an intermediate layer to interconnect two higher melting point substrate materials, typically Cu. Under appropriate temperature and reflowing time frame, a configuration of Cu/IMCs/Cu sandwich can be obtained as Sn had been completely reacted with Cu to form IMCs.

However, it worth noting that it is still uncertain up to the present time whether a micro-joint with full IMCs is better than a joint with remaining solder after assembly particularly as one take the entirety reliability of a 3D IC package (typically Chip Package Interaction (CPI) issue) into consideration. Therefore, despite its controversy, in terms of academic research views, it pays to carry out more comprehensive research activities to have a better knowledge of interfacial reactions aspects in micro-joints for oncoming 3D IC integration applications.

Experimental

Four kinds of lead-free solder including pure Sn, Sn-Ag, Sn-Bi and Sn-Zn solders are used in the current experiments. The metallization layer at both chip and substrate sides are electroplated Cu (30 μm) onto silicon wafers with a pre-sputtered seed layer of Cr (300 Å) Cu (3000 Å). The reason for choosing such Cu/solder combination system is twofold. One is because Cu is one of the most popular choices for UBM and surface finish material, and the other is that Cu is known to have fast reaction kinetics with not merely liquid solder but solid solder as well. The thickness of the solder layer is controlled at 10μm ±1μm after chip to chip thermal compression bonding process at 250°C for 1 minute. Figure 1 illustrates the dimension of the sandwich structure as well as the fabrication process of sandwich sample used in this study. The unnecessary solder would be completely squeezed out during bonding process. Under such a configuration, the lateral inter-diffusion can be considered as infinite, so the atomic flux is merely along the vertical direction in this study.

It worth mentioning that reason for utilizing such a sandwich scheme, instead of a real micro-joint, for this work is because the most important objective in this study is to control the thickness of the joints so that the key features resulted from interfacial reactions under space confinement can be spotlighted accurately. A homemade sandwich sample with a well-controlled spacer in Z direction is viewed as the simplest way to achieve such a destination. Also, the 1x1 mm lateral dimension of sandwich structure allows one to have a relatively wide region for observation and make the experimental results more meaningful and statistical.

The sandwiches are aged at 120 °C, 150 °C, 180 °C, and 200 °C for different periods, respectively. Then the sandwiches are subjected to the sequential polishing process for further cross-cut bservation. Scanning electron microscopy (SEM) and an energy dispersive X-ray spectrometer (EDX) are used to examine the microstructure and identify the chemical compositions of the intermetallic compounds (IMCs). Electron backscattered diffraction analysis (EBSD-EDAX, OIM6.1-TSL) equipped on the FE-SEM (JSM-7001, JEOL) is applied to investigate the crystallographic orientation of IMCs within the sandwich sample.

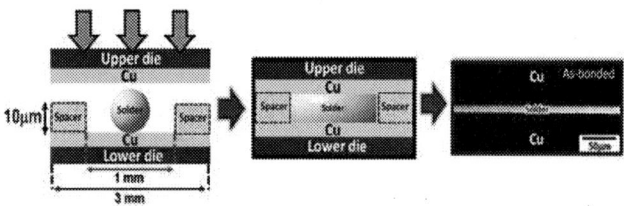

Figure 1. Dimension of the sandwich structure and the fabrication process of sample used in this study.

Results and Discussion

Impingement behavior of IMC grains after solid-state isothermal aging

It is well-recognized that the remaining solder in micro joints will continue to react with metallization layer to form more IMCs during solid-state aging process. According to our previously results [13], even for Ni, a common surface finish layer that is known to have the slowest reaction kinetics with both liquid and solid solder, the solder layer could be easily transformed into Ni_3Sn_4 completely after a few hundreds hours of aging at a reasonable temperature range. Compared with Ni, Cu is often thought to have fast reaction kinetics with most of solder materials. Therefore, one can expect that a large portion or even the entirety of the solder layer within micro-joint might be occupied by Cu-Sn IMCs during solid-state isothermal aging. Figure 2 shows the growth kinetics of IMCs in Cu/Sn/Cu sandwiches at different temperature with 10 μm Sn solder layer in thickness. It should be noted that the data points in Fig. 2 show the averaged thickness of total IMCs layer (i.e. $Cu_6Sn_5+Cu_3Sn$) per interface. The results indicate that the growth of $Cu_6Sn_5+Cu_3Sn$ IMCs still follows parabolic kinetics during solid-state reactions. These results,

to some extent, clearly imply that as the thickness of solder layer in Cu/Sn/Cu sandwich structure is reduced to a few μm, the micro-joints would be totally occupied by Cu-Sn reactants within a very short time even the temperature is as low as 120 °C or 150 °C, as shown in Fig. 2.

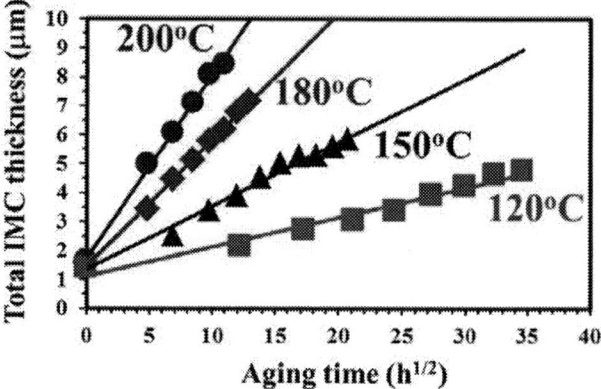

Figure 2. Growth kinetics of total IMCs layer (i.e. $Cu_6Sn_5+Cu_3Sn$) per interface in Cu/Sn 10 μm/Cu sandwiches at 120 °C, 150 °C, 180 °C, 200 °C, for different time periods, respectively.

In addition, when the distance between two opposite interfaces in a micro joint is only a few μm, IMCs growing from these two interfaces will gradually approach each other as the interfacial reactions proceeds. Figure 3(a)-(f) show the cross-sectional image of Cu/Sn (10 μm)/Cu sandwiches aged at 150 °C for 0, 48, 96, 144, 192, and 432 h, respectively. During aging, two layers of Cu_6Sn_5 grew toward each other from the opposite interfaces. The original Sn layer transformed to Cu_6Sn_5 and Cu_3Sn gradually, which is well consistent with literature results, and the rationale has been reported elsewhere [1]. In addition, the growth rate of Cu_3Sn is much faster than Cu_6Sn_5, and the ratio of Cu_3Sn/Cu_6Sn_5 had nearly exceeded 1/2. Some micro voids or Kirkendall voids can be observed within the Cu_3Sn layer. It pays to notice that the morphology of the Cu_6Sn_5 is not so flat during aging, which may be safe for flip chip solder joint or larger solder ball. For micro joints, however, it may carry a critical reliability concern as it is easy to make those Cu_6Sn_5 growing from the opposite interfaces impinge on each other during reasonable time frame, as clearly shown in Fig. 3(f) emphasized by a white arrow with red highlighter.

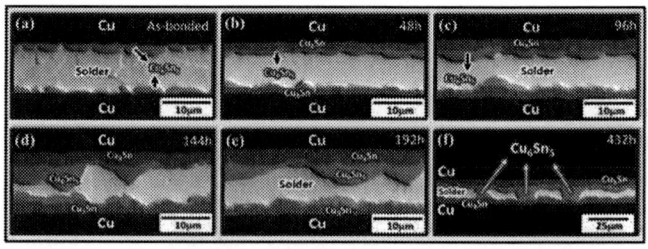

Figure 3. Micrograph showing the Cu/Sn (10μm)/Cu sandwiches after reaction at 150 °C for (a) As-bonded, (b) 48, (c) 96, (d) 144, (e) 192, and (f) 432 h.

978-1-4799-2408-0/14 $31.00 © 2014 IEEE

According to the literature, the morphology of Cu_6Sn_5 layer would usually start to transform from scallop shape to planar shape in sequence during aging process in the solid-state interfacial reactions between Cu and Sn solder on the bulk scale [1-2]. However, the morphology of Cu_6Sn_5 in this study is more faceted, instead of being planar, during solid-state thermal aging, as shown in Fig. 3 (d) and (e). Such roof-top shape morphology is not new for Cu_6Sn_5. Tu et al. [14] and Suh et al. [15] reported the effects of solder composition (pure Sn and Sn-Pb solder alloy) on the morphology of Cu_6Sn_5 scallops in wetting reaction between molten solder and copper foil. The results showed that the morphology of Cu_6Sn_5 scallops was found to be highly faceted as the solder alloy was pure Sn, and round when the solder was eutectic Sn-Pb composition. Zou et al. [16] systematically investigated the morphologies and orientation relationships of Cu_6Sn_5 formed between Sn and (001), (011), (111), and (123) Cu single crystals under liquid- and solid-state condition. They found the regular prism-type Cu_6Sn_5 grains, which is also regarded as faceted morphology of Cu_6Sn_5 grains in this study, will form on (001) and (111) Cu single crystals. In contrast, mainly scallop-type with only a few regular prism-type Cu_6Sn_5 grains will form on (011) and (123) Cu single crystal surfaces after liquid-state soldering reactions. Most importantly, the results indicated that the regular prism-type Cu_6Sn_5 grains will change into scallop-type grains after long time reflowing or solid-state aging. Therefore, it is certainly amazing to observe such kind of roof-top shape Cu_6Sn_5 grains under solid-state aging reactions in the present work particularly when the Cu metallization layer was made by electroplating, which usually resulted in poly-crystalline Cu grains. More fundamental studies and experiments are needed to dig out other underlying causes for this occurrence.

Despite the fact that the root cause for forming intriguing roof-top shape Cu_6Sn_5 grains during solid-state aging is still unclear, such a faceted morphology of Cu_6Sn_5 may carry a critical reliability concern especially for micro-joints because it tends to make those Cu_6Sn_5 grains growing from the opposite interfaces impinge on each other quickly during reasonable time frame, as shown in Fig. 3 (f). Also, a new reference value of "Time-to-impingement of IMCs" for micro joints is going to be established first time under solid-state interfacial reactions, which will be discussed in-depth later.

On the other hand, it is interesting to explore whether those grains will further coarsen into larger grains during sequential isothermal aging. It can be anticipated that if the situation is still two individual grains contacted with each other, a continuous grain boundary will undoubtedly run across the entire micro joint in a direction nearly parallel to the two interfaces, which will become a potential weakening spot as the micro joint subject to an unexpected shear force. Figure 4 (a) is a micrograph of Cu/Sn (10 μm)/Cu sandwich aged at 180 °C for 192 h. In order to have a better understanding on those Cu_6Sn_5 grains that had contacted with each other, Fig. 4 (b) shows the corresponding orientation image map (Image Quality + Inverse Pole Figure) of Cu_6Sn_5, Cu_3Sn, and Sn grains with ND directions. As vividly shown in Fig. 4 (b), the result shows some of the Cu_6Sn_5 grains growing from the opposite directions have merged into a large

single Cu_6Sn_5 grain (i.e. grain C or grain G); however, other Cu_6Sn_5 grains growing from opposite direction are still independent grains with a clear grain boundary between themselves (i.e. interface between grain A/B, D/F, and E/F). For Cu/Sn/Cu structure, it had been reported that, as long as the reflowing time is long enough, all the Cu_6Sn_5 grains would merge into several large grains even though those grains were from the opposite directions and with its own crystallographic orientation [9-10, 12-13]. Compared with the results of liquid-state soldering reactions, whether these unmerged Cu_6Sn_5 grains, as shown in Fig. 4 (b), will completely transform into several large Cu_6Sn_5 grain spanned two opposite interfaces eventually under solid-state reactions is still unknown. If it would happen, what kinds of responsible mechanism for ruling such two Cu_6Sn_5 grains with its different crystallographic orientation to merge into a large one with only one orientation is the greatest area of interests, and more detail investigations are ongoing now.

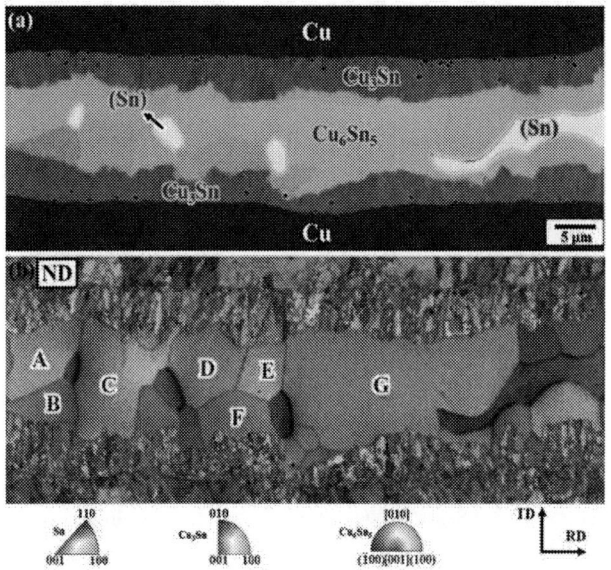

Figure 4. (a) Cross-sectional micrograph of Cu/Sn (10 μm)/Cu sandwich aged at 180 °C for 192 h. (b) The corresponding orientation image map (Image Quality + Inverse Pole Figure) of Cu_6Sn_5, Cu_3Sn, and Sn grains with ND directions.

Rise of concentrations of minor inert alloying constituents in low solder volume joint

It is well known that minor alloying constituent additions (such as Ag, Co, Ni, Bi, Ti, Fe, Co, Ni, and Zn etc.) into solder have pronounced effects on the interfacial reactions between solder and metallization layer. These alloying elements might substantially change the reaction rate and/or the morphology of the reaction products. Among these additives, Ag and Bi are two typical additives that are inert to the soldering reaction. What would happen when one adding these inert alloying constituents in the space confinement micro joint is the area of interest in this part?

978-1-4799-2408-0/14 $31.00 © 2014 IEEE

Ag is a key alloy additive often presented in many commercial lead-free solders, such as Sn-Ag or Sn-Ag-Cu solder because of its benefits in improving the wettability [17] as well as mechanical properties [18]. However, it is reported that Ag atoms have negligible solubility in Cu-Sn reactants [19], so Ag atoms will be undoubtedly rejected into the remaining solder, forming Ag_3Sn, an inert phase, as the interfacial reaction proceeds.

Figure 5 (a)-(f) show the cross-sectional micrograph of Cu/Sn3.5Ag (10μm)/Cu sandwiches aged at 150 °C for 0, 48, 96, 144, 288, and 432 h, respectively. After thermal compression bonding process, Ag_3Sn particles disperse uniformly in the solder matrix. During solid state aging, the thickness of Cu_6Sn_5 and Cu_3Sn increase with the aging time. Some micro voids or Kirkendall voids is observed within the Cu_3Sn layer. In addition, the dispersed Ag_3Sn particles coarsen together and some of them are detected at the interface between Cu_6Sn_5 and solder with the aging process.

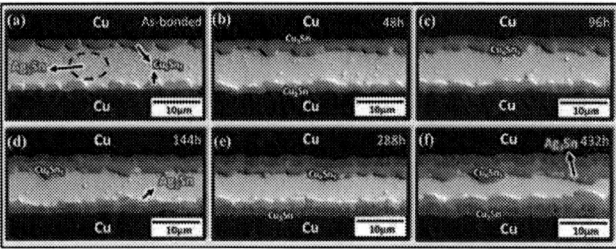

Figure 5. Micrograph showing the Cu/Sn3.5Ag (10μm)/Cu sandwiches after reaction at 150 °C for (a) As-bonded, (b) 48, (c) 96, (d) 144, (e) 288, and (f) 432 h.

Compared with the results of Cu/Sn/Cu sandwiches under solid-state aging, in short, there are two important novel benefits with regard to Ag additions into solder for micro-joints applications. One is relating to the growth kinetics of Cu-Sn reactants. The results show the growth rate of IMCs in Cu/Sn3.5Ag (10μm)/Cu system seems to be much slower than that in Cu/Sn(10μm)/Cu system at 150 °C. Figure 6 (a)-(f) show the growth kinetics of $Cu_6Sn_5+Cu_3Sn$ as well as Cu_3Sn IMCs, respectively, in both Cu/Sn/Cu and Cu/Sn3.5Ag/Cu system at 120 °C 150 °C, 180 °C, and 200 °C for different aging periods. The results indicate that adding Ag into solder can effectively reduce the growth rates of not merely total IMCs (i.e. $Cu_6Sn_5+Cu_3Sn$) but also Cu_3Sn, particularly at the lower temperature condition.

The other important advantage of adding Ag is that the morphology of Cu_6Sn_5 will become much smooth and flat, as shown in Fig. 6, which can effectively prolong the time for IMCs impingement we discussed earlier. Based on above results, the new reference value of "Time-to-impingement of IMCs" between Cu/Sn/Cu and Cu/Sn3.5Ag/Cu system can be established during solid/solid reactions at different temperature, as listed in Table 1.

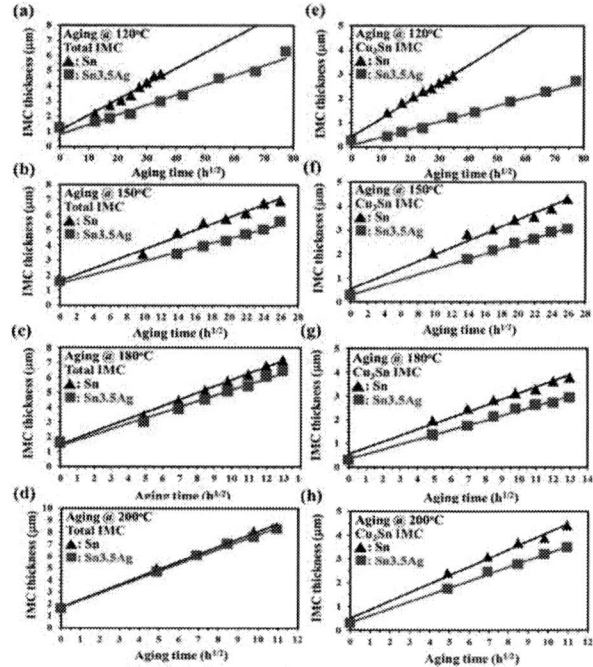

Figure 6. Growth kinetics of IMCs in Cu/Sn (10 μm)/Cu and Cu/Sn3.5Ag (10 μm)/Cu sandwiches at 120 °C, 150 °C 180 °C, and 200 °C for different aging periods. (a)-(d) averaged thickness of $Cu_6Sn_5+Cu_3Sn$ per interface, (e)-(h) averaged thickness of Cu_3Sn per interface. Solid symbols (▲, ■) denote data points from Cu/Sn (10 μm)/Cu and Cu/Sn3.5Ag (10 μm)/Cu sandwiches, respectively.

Table 1

	Time-to-Impingement of IMCs (hr)	
Temperature	Cu/Sn(10μm)/Cu	Cu/Sn3.5Ag(10μm)/Cu
120°C	450h ~ 600h	>5000h
150°C	144h ~ 192h	432h ~ 480h
180°C	48h ~ 72h	144h ~ 168h
200°C	72h~ 96h	72h ~ 96h

Here, a very intriguing question is noteworthy for Cu/Sn-Ag/Cu system in micro-joints. Generally, it is well-recognized that some impurity molecules or atoms are insoluble in IMCs, and they will be rejected and pushed away from the IMC/solder interface to the center of micro solder joint during interfacial reactions proceeds. However, our results here show that most of Ag_3Sn, an inert phase, are not always pushed by Cu_6Sn_5 grains to the middle of the sandwich, forming a continuously layer of Ag_3Sn. Some of them are inclined to be entrapped or wrapped in the Cu_6Sn_5 grains during solid-state aging, as illustrated in Fig. 7. Such an intriguing behavior of Ag_3Sn might be strongly related to the interfacial free energy for the interface between Cu_6Sn_5 and Ag_3Sn. A more detailed analysis for verifying such a preliminary deduction is in progress.

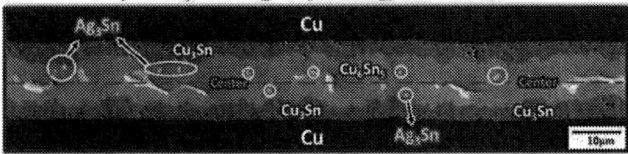

Figure 7. Micrograph showing the Cu/Sn3.5Ag (10μm)/Cu sandwiches after reaction at 180 °C for 216 h.

Another typical example of inert alloying constituent is Bi, which is also a common alloying constituent in many commercial solder systems. Similar to Ag, Bi is not inclined to incorporate itself into Cu-Sn reactants, and it will be rejected into the remaining solder as interface IMCs grow. In addition, there is one thing difference between Ag and Bi, which should be paid attention to: Ag will react with Sn to form Ag_3Sn but Bi does not. Figure 8 (a)-(c) show the Cu/Sn10Bi (10μm)/Cu sandwiches aged at 120 °C for 0, 600, and 1050 h, respectively. It can be obviously seen that Bi particles are rejected into the remaining solder as the interfacial reactions proceeds. At 1050 h, as shown in Fig. 8 (c), all the solder had been completely consumed, and the space was replaced by Cu_6Sn_5, Cu_3Sn, and Bi-rich phase. One should notice that if a higher Bi-bearing solder is chosen for the purpose of decreasing the melting point of the solder, such as eutectic Sn58Bi solder alloy. Finally, the Bi-rich phase might become continuous Bi-rich layer and run across the entire micro joint along the center of the joint, which might pose an adverse impact on the micro joint reliability.

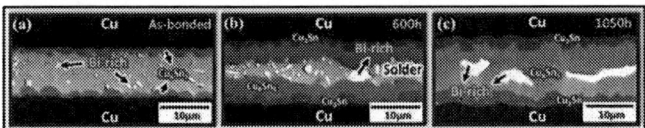

Figure 8. Micrograph showing the Cu/Sn10Bi (10μm)/Cu sandwiches after reaction at 120 °C for (a) As-bonded, (b) 600, (c) 1050 h.

Promising approach to reduce the growth rate of Cu-Sn reactants within limited solder volume joint

Up to the present time, it is worth mentioning that it is still controversial about whether a micro joint occupied by fully IMCs is better than a joint with remaining solder. In other words, if full of IMCs in joint is detrimental to the micro joint's reliability, it seems better to find other approaches to reduce the growth rates of IMCs growing from the opposite directions and further suppress the formation of micro voids or Kirkendall voids in thick Cu_3Sn layer. The conterminous third issue is relating to the effects of adding highly reactive constituent additions (typically Zn) on the interfacial reactions within limited solder volume joint in order to reduce the growth rates of Cu-Sn reactants. Two kinds of solder concentrations, i.e. Sn0.4Zn, and Sn1.0Zn, are taken into investigation in this part.

Figure 9 shows the micrographs of Cu/Sn (10μm)/Cu, and Cu/Sn-xZn (10μm)/Cu (x=0.4, and1.0 w.t.%) sandwiches

aged at 180 °C for for 0, 48, and 96 h, respectively. The results clearly show that minor Zn added in solder matrix can effectively suppress the growth rates of Cu-Sn reactants, especially for the growth of Cu_3Sn layer. Micro voids or Kirkendall voids are markedly suppressed by adding Zn in solder. In addition, the results also indicate the ability to suppress the growth of Cu_3Sn has a strong dependence on Zn concentration. In this study, the thickness of Cu_3Sn layer is much thinner in Sn1.0Zn case than that in Sn0.4Zn case during the same aging period, indicating that both solder volume effect and solder concentration effect are important particularly for micro solder joints. Industry should take these two critical effects into consideration simultaneously and carefully so that the optimum level of Zn addition for micro joints in 3D IC integration applications can be determined.

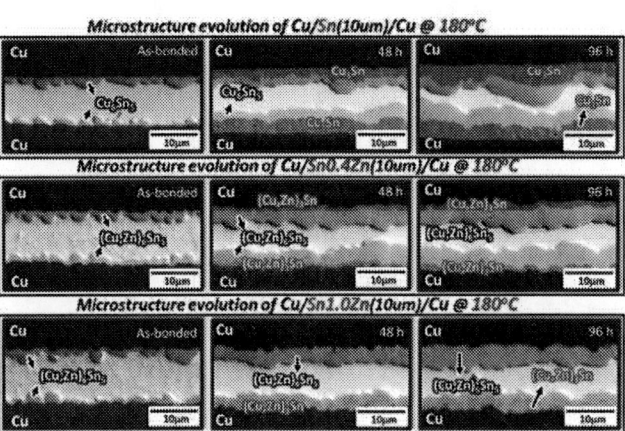

Figure 9. Cross-sectional micrograph showing the Cu/Sn (10μm)/Cu, Cu/Sn0.4Zn (10μm)/Cu, and Cu/Sn1.0Zn (10μm)/Cu sandwiches aged at 180 °C for 0, 48, and 96 h, respectively.

Conclusions

In summary, three critical issues or concerns relating to interfacial reactions for 3D IC integration applications is unequivocally presented and investigated in this present work. The key common factor that makes these issues unique is that the interfacial reactions occur under a very sever space confinement solder joint. In Cu/Sn/Cu sandwiches, the morphology of Cu_6Sn_5 is found to be roof-top shape during solid-state aging. Such a faceted Cu_6Sn_5 morphology would tend to make those Cu_6Sn_5 grains growing from the opposite interfaces impinge on each other very quickly, carrying a potential reliability concern for micro-joints.

In contrast, the most noteworthy observation results in Cu/Sn3.5Ag/Cu sandwiches is the morphology of Cu_6Sn_5 grains change to be more planar shape when one adds Ag into the solder alloy, which could effectively prolonged the time for IMCs impingement behavior. In addition, the growth kinetics of Cu-Sn reaction products (i.e. Cu_6Sn_5 and Cu_3Sn) could be effectively suppressed by means of Ag addition particularly at lower temperature. Also, a new reference value of "Time-to-impingement of IMCs" for Cu/Sn/Cu and Cu/Sn3.5Ag/Cu sandwiches in micro-joint with 10 μm solder

978-1-4799-2408-0/14 $31.00 © 2014 IEEE

layer in thickness is established first time during solid-state interfacial reactions at different temperature.

In Cu/Sn10Bi/Cu sandwiches, the results show that the concentrations of minor inert alloying constituents will continue rising as the interfacial reactions proceeds. Such a result, to some extent, can represent the influence of impurities from electroplating in micro bumping industry. At last, if a full IMCs joint is detrimental to the entire chip's reliability, the best alternation is to reduce the growth rate of interfacial reactants by adding some highly reactive element. In this work, the results show that minor Zn addition can effectively suppress the growth rates of Cu-Sn reactants, especially for Cu_3Sn IMCs. However, the results also indicate that both solder volume effect and solder concentration effect will influence the ability of Zn inhibition for Cu-Sn reactants, which should be taken into consideration carefully.

Acknowledgments

This work is financially supported by National Science Council of Taiwan through grant 101-2221-E-002-162-MY3

References

1. K. N. Tu, K. Zeng, Mater. Sci. Eng. **R34**, 1 (2001).
2. T. Laurila, V. Vuorinen, J. K. Kivilahti, Mater. Sci. Eng. **R49**, 1 (2005).
3. C. E. Ho, S. C. Yang, C. R. Kao, J. Mater. Sci., **18**, 155 (2007).
4. T. Laurila, V. Vuorinen, Materials, **2**, 1796 (2009).
5. K. N. Tu, Microelectron. Reliab,, **51**, 517 (2011).
6. K. Sakuma, P. S. Andry, C. K. Tsang, S. L. Wright, B. Dang, C. S. Patel, B. C. Webb, J. Maria, E. J. Sprogis, S. K. Kang, R. J. Polastre, R. R. Horton, and J. U. Knickerbrocker, Y. W. Lin, IBM J. Res. & Dev., **52**, 611 (2008).
7. Y. H. Cao, W. G. Ning, L. Luo, IEEE Trans. Electron. Packag. Manuf. **32**, 125 (2009).
8. L. L. Yan, C. K. Lee, D. Q. Yu, A. B. Yu, W. K .Choi, J. H. Lau, S. U. Yoon, J. Electron. Mater. **38**, 200 (2008).
9. C. Hang, Y. Tian, R. Zhang, D. Yang, J. Mater. Sci. **24**, 3905 (2013).
10. R. Zhang, Y. Tian, C. Hang, B. Liu, C. Wang, Mater. Lett., **110**, 137 (2013).
11. B. Lee, J. Park, S. J. Jeon, K. W. Kwon, and H. J. Lee, J. Electrochem. Soc., **157**, 420 (2010).
12. J. F. Li, P. A. Agyakwa, and C. M. Johnson, Acta Mater., **59**, 1198 (2011).
13. H. Y. Chuang, T. L. Yang, M. S. Kuo, Y. J. Chen, J. J. Yu, C. C. Li, and C. R. Kao, IEEE Trans. Device Mater. Reliab., **12**, 233 (2012).
14. K. N. Tu, A. M. Gusak, M. Li, J. Appl. Phys., **93**, 1335 (2003).
15. J. O. Suh, K. N. Tu, G. V. Lutsenko, A. M. Gusak, Acta Mater., **56**, 1075 (2008).
16. H. F. Zou, H. J. Yang, Z. F. Zhang, Acta Mater., **56**, 2649 (2008).
17. C. Kanchanomai, Y. Mutoh, Mater. Sci. Eng. A **381**, 113 (2004).
18. K. S. Kim, S. H. Huh, and K. Suganuma, Mater. Sci. Eng. A **333**, 106 (2002).
19. T. Laurila, V. Vuorinen, and M.P. Paulasto-Krockel, Mater. Sci. Eng. **R68**, 1 (2010).

Simulation and Optimization of a Micro Flow Sensor

Xing Guo[1], Chunlin Xu[1], Shengzhi Zhang[1], Yong Xu[3], Xin Wu[4], and Sheng Liu[2].*

[1]School of Mechanical & Engineering, Huazhong University of Science & Technology, Wuhan, 430074, China

[2]Cross-disciplinary Institute of Engineering Sciences, School of Power and Mechanical Engineering, Wuhan University, Wuhan, Hubei, 430072, China

[3]Department of Electrical and Computer Engineering, Wayne State University, Detroit, MI 48202 USA

[4]Department of Mechanical Engineering, Wayne State University, Detroit, MI 48202 USA

* victor_liu63@126.com

Abstract

In this paper, simulation and optimization of a micro flow sensor is presented. Modeling of the micro flow sensor using ANSYS Fluent for temperature distribution is studied. The detailed structure of the micro flow sensor chip is considered in the simulation model. What's more, the temperature dependencies of physical properties of air are defined to improve the model precision. As a key parameter, optimized position of thermistor is obtained. The effects of mass flow rate, input power and depth of flow channel on the performance of the micro flow sensor are investigated in detail, thus optimized parameters are obtained. The effect of sensor chip installation error in radial direction of flow channel on temperature distribution of the micro flow sensor is also investigated. At last, comparison of simulation model with and without considering sensor chip structure is conducted.

Introduction

Gas flow measurement plays a very important role in many fields such as industrial process control, automobile, aeronautics and astronautics, and so on. Micro thermal flow sensors rely on the flow induced cooling of a heated region and there are mainly three types. Hot-wire sensors which measure the effect of the flowing air on a heater. Calorimetric sensors which measure the asymmetry of temperature profile around the heater which is modulated by the air flow. Time-of-flight sensors which measure the passage of time of a heat pulse over a known distance [1]. Compared with traditional volumetric air flow sensor, micro thermal flow sensors develop very fast due to their advantages on accuracy of measurement, response time and reliability [2, 3].

With gas flow measurement becoming more prominent, modeling of flow sensor is becoming important [4-6]. The materials of sensing resistors were not included in the numerical simulation model performed by N. Sabat. [3]. The temperature dependencies of physical properties of fluid have been ignored in finite element model of previous study [6]. Due to the wide use of micro flow sensors, an accurate model and simulation offer the designer many advantages. These include reduced cycle time for development and the possibility for device optimization through software instead of iterations on physical devices. These in turn help to reduce the development cost of the device. The micro flow sensor based on thermal calorimetric principle considered in this study is developed by Institute of Microsystems [7]. It is based on silicon as substrate material. The micro flow sensor consists of a heater and two thermistors embedded in a silicon nitride membrane which acts as isolation layer and masking layer.

Thermistors symmetrically situated on both sides of the heater measure the shift of temperature distribution due to forced convection which can be interpreted as flow. The heater and thermistors are made of platinum.

The purpose of this paper is to present simulation model of a micro flow sensor. The whole sensor structure is considered in the model. Furthermore, the temperature dependencies of viscosity, density, specific heat capacity and thermal conductivity of the air have been considered. The effects of distance between heater and thermistors, mass flow rate, input power, and depth of flow channel on the performance of the micro flow sensor are investigated in detail. In addition, the effect of installation error and comparison of model with sensor chip structure or not are also studied.

Method

The simulations were modeled in a two-dimensional environment, as cross-section of a flow channel with an embedded sensor chip. The sensor chip is flush-mounted with frame of the flow channel. The two-dimensional model of micro flow sensor geometry with flow channel is shown in Figure 1.

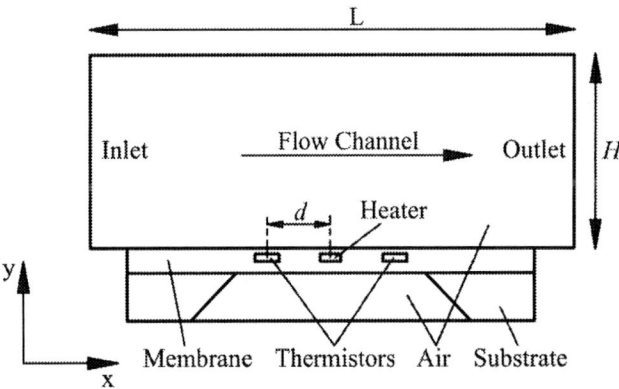

Figure 1. Micro flow sensor geometry with flow channel.

The cross-section of the flow channel has a rectangular geometry, and its dimensions are 10mm (length) × 1mm (width) × 1mm (height). The model incorporates the whole sensor chip structure and two air regions. The whole sensor chip structure includes substrate and membrane. The dimensions of the silicon substrate are 1.5mm (length) × 1mm (width) × 50μm (height).The 1.4μm thick silicon nitride membrane has an area of 1.5 × 1mm². The fine structures of heater and thermistors meander in the membrane have been also integrated into the model. The heater width is 13μm (two lines 3μm wide separated 7μm) and the thermistors width are 53μm (six lines 3μm wide separated 7μm). The thicknesses of heater and thermistors are 0.1μm. The two air regions are divided by the membrane plane. The region underneath contains air in no flow condition while the region above the membrane is flowing in the flow channel.

Physical properties of thermal conductivity λ, specific heat capacity c_p and density ρ of materials of the sensor chip are shown in Table 1 [7]. For each region of heater and thermistors, equivalent thermal conductivity, specific heat capacity and density are calculated as the weighted averages of physical properties of the stacked layers composing each structure. Moreover, thermal conductivity of these regions is also calculated for the normal and the in-plane directions. Thus, thermal conductivity of normal direction is 25W/(m·K) and that of in-plane direction is 7W/(m·K). Weighted averages of specific heat and density are 537 (J/(kg·K)) and 8699 kg/m³, respectively. Temperature dependencies of density ρ, viscosity μ, thermal conductivity λ and specific heat capacity c_p of air have been defined by the material library.

Table 1. Physical properties of sensor chip materials.

Materials	λ (W/(m·K))	c_p (J/(kg·K))	ρ (kg/m³)
Platinum	71.5	133	21450
Silicon	150	700	2340
Silicon Nitride	5	710	3234

Based on the Navier-Stokes equations, steady fields of airflow velocity are calculated approximately. The heat transport model used includes conduction and convection, whereas radiative heat transfer is neglected. Natural convection caused by density variation due to temperature gradients is also considered. Coupled heat transfer boundary condition is defined between different materials. Heating power is applied to the heater of the micro flow sensor and uniform velocity profile is applied to the inlet of the flow channel to investigate the temperature difference between downstream and upstream thermistors. Various mass flow rates from 6SCCM (standard-state cubic centimeter per minute) to 600SCCM (i.e. velocity from 0.1m/s to 10m/s), various distance between thermistors and heater from 80μm to 200μm, various input powers from 6mW to 12mW, various flow channel depths from 1mm to 2mm and various installation errors of one percent to minus one percent are used to simulate the temperature distribution of the micro flow sensor. The results are shown as follows.

Result and Discussion

Figure 2 shows the relationship between the temperature of downstream and upstream thermistors and mass flow rate. As shown in Figure 2, the temperature in the upstream section of the membrane decreases and the temperature in the downstream section also decreases with increasing mass flow rate. If mass flow rate is zero, the temperature of downstream and upstream thermistors is equivalent as the thermistors are positioned symmetrically on both sides of the heater. Temperature distribution profile is distorted to downstream if mass flow rate exists. In the case with flow, the upstream part of the membrane is cooled by air so that the temperature of this part becomes less than the case without flow. The energy supplied to heater transfer to the air as it passes above the heater. Heated air lowers the cooling rate on the downstream part. Therefore, the temperature of the downstream thermistor decreases slowly than that of the upstream thermistor. As a result, the temperature difference between the downstream

and upstream thermistors increases with increasing mass flow rate for constant heating powers. The results are different from previous results by T. H. Kim [2], which shows that the temperature of the downstream thermistor increases first and then decreases. But the variation tendency of the temperature difference with variation of mass flow rate is the same.

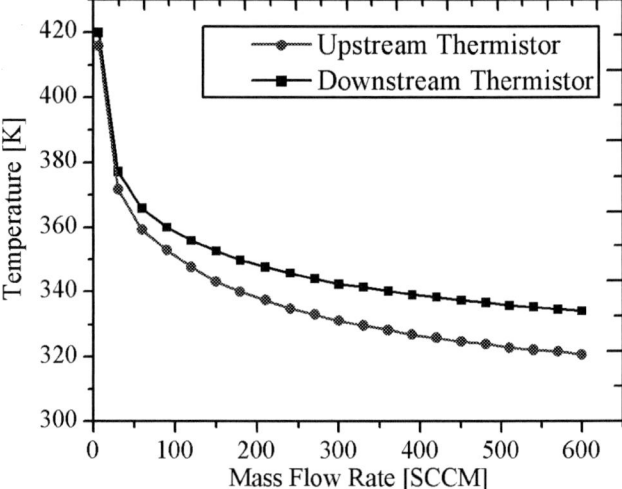

Figure 2. Temperature response of the thermistors as a function of the mass flow rate.

Figure 3 shows the relationship between the temperature difference of the downstream and upstream thermistors and mass flow rate. As mentioned above, temperature difference increases with the increase of mass flow rate. As shown in Figure 3, the temperature difference varies linearly at first, and the relationship becomes nonlinearly when mass flow rate is larger than a certain value. At larger mass flow rate, the temperature difference increases slowly and tends to be saturated.

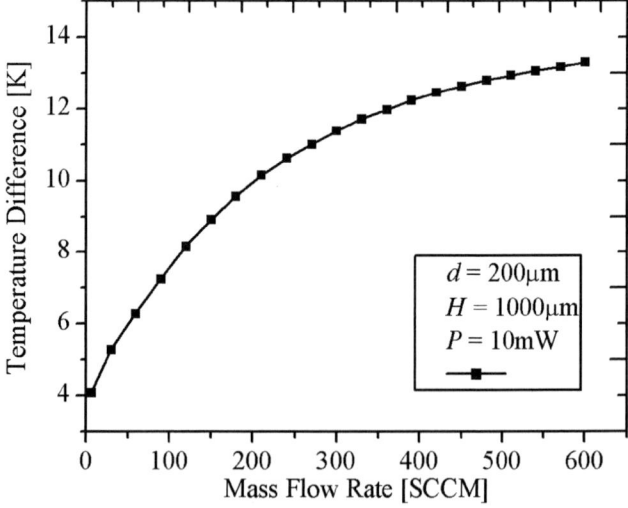

Figure 3. Temperature difference of the micro flow sensor with variation of mass flow rate.

Figure 4 indicates the relationship between temperature difference of the downstream and upstream thermistors and mass flow rate for various positions of thermistors when the input power is constant. In Figure 4, d is the distance from center of heater to center of thermistors. As the value of d increases from 80μm to 160μm, the temperature difference increases obviously for a fixed mass flow rate. As the value of

d varies from 160μm to 200μm, the temperature difference decreases first and then increases for a fixed mass flow rate. The temperature difference variation is slightly and the turning point is at 240SCCM. The value of d affects the sensitivity of the micro flow sensor, and sensitivity variation is non-monotonic with increasing value of d. If the thermistors are positioned near the heater, the temperature difference is very small. On the other hand, if the thermistors are positioned quite far away from the heater, the temperature difference should be also close to zero because their temperatures are equal to the ambient temperature. This implies that the temperature difference increases first and then decreases as the value of d increases. Thus there is an optimized distance for maximum temperature difference. It is important to place thermistors at appropriate location to achieve excellent performance.

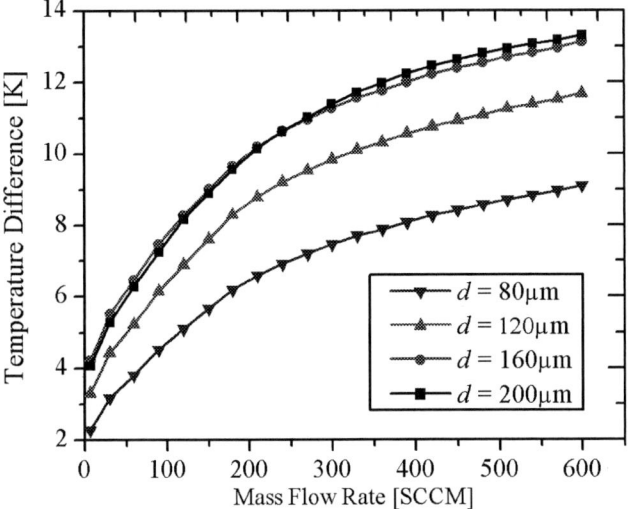

Figure 4. Effect of position of thermistors on the relationship between the temperature difference and the mass flow rate.

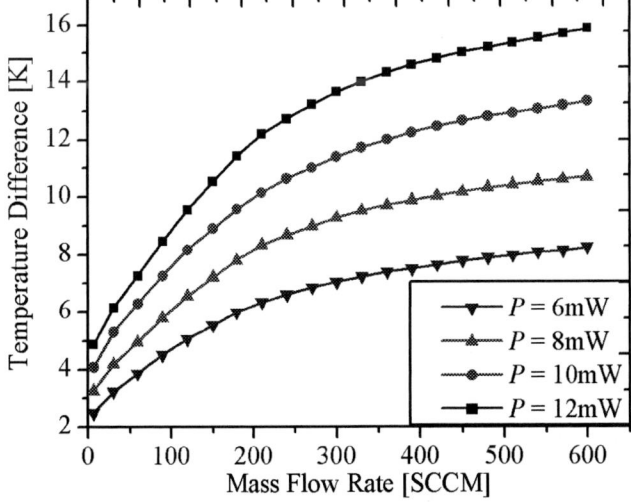

Figure 5. Effect of input power on the relationship between the temperature difference and the mass flow rate.

Figure 5 shows the relationship between temperature difference between the downstream and upstream thermistors and mass flow rate in terms of the input power variation. In Figure 5, P is the power supplied to heater. The larger the value of P is, the larger the temperature difference for a fixed

mass flow rate. The variation tendency is almost the same and the input power P doesn't affect the linearly measurement range. As sensitivity of the micro flow sensor is proportional to temperature difference, it can be improved by increasing the value of P. However, a larger P will increase the energy consumption, and thermal stress caused by coefficient of thermal mismatch of sensor chip materials will be larger, which will reduce the reliability of the micro flow sensor.

The effect of depth of flow channel on performance of the micro flow sensor is investigated. In Figure 6, H is the depth of flow channel. As H increases, the temperature difference between the downstream and upstream thermistors decreases for a fixed mass flow rate. For a smaller H, saturated tendency is occurring at lower mass flow rate, thus measurement range is smaller. The flow area decreases as the values of H decreases. For a fixed mass flow rate, smaller H corresponds to a larger velocity. According to heat transfer theory, forced convection depends on velocity distribution. As mentioned above, temperature difference increases with the increase of mass flow rate (i.e. increase of velocity when flow area is constant).

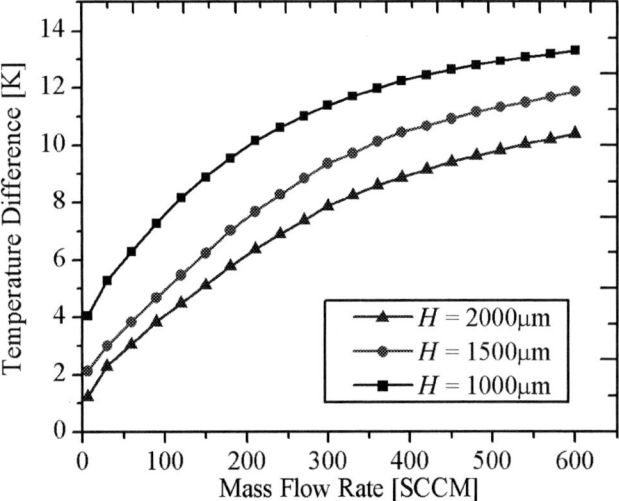

Figure 6. Effect of depth on the relationship between the temperature difference and the mass flow rate.

As installation error is inevitable during the assembly process, the effect of sensor chip installation error on performance of the micro flow sensor is also investigated. ΔH is variation of H in radial direction of flow channel induced by installation error. In Figure 7, ε is the ratio of ΔH to H. As the value of ε varies, temperature difference between the downstream and upstream thermistors varies for a fixed mass flow rate. As ε varies from -1% to +1%, temperature difference decreases for a fixed mass flow rate smaller than 360SCCM, but the temperature difference increases first and then decreases when mass flow rate is larger than 360SCCM. For negative installation error, the sensor positions in a higher flow velocity of the air (close to the center of the channel) and the heat is transferred along the channel in a better manner. For positive installation error, relative flow channel depth increases and the flow velocity decreases, which restrict the axial heat transfer. The saturation point occurs at smaller mass flow rate for the situation with installation error. The suddenly change of flow channel depth induced by installation error will lead to disturbance of flow field and lead to the saturation of temperature difference at lower mass flow rate.

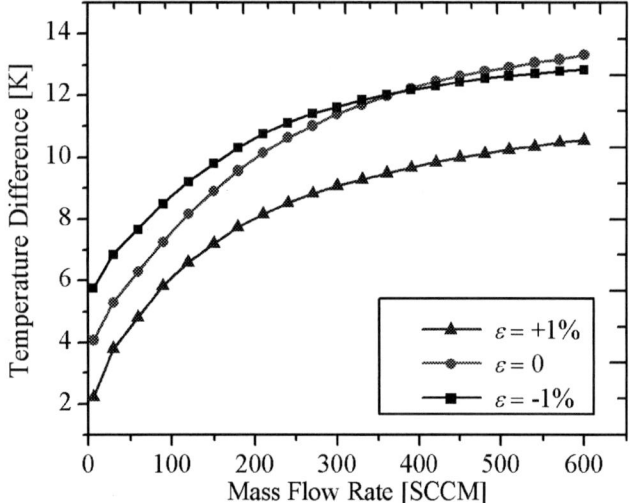

Figure 7. Effect of installation error on the relationship between the temperature difference and the mass flow rate.

The comparison of simulation model considering sensor chip structure or not is conducted and the result is shown in Figure 8. For a fixed mass flow rate, temperature difference between downstream and upstream thermistors (right y axis label) without sensor chip structure is larger than that of considering sensor chip structure (left y axis label). As the mass flow rate increases, temperature difference without sensor chip structure increases and reaches its maximum at mass flow rate of 90SCCM and then decreases.

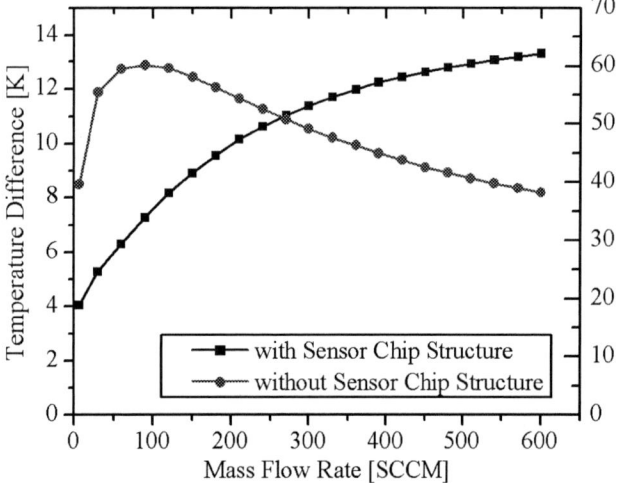

Figure 8. Temperature difference of the micro flow sensor with variation of mass flow rate considering sensor chip structure or not.

For the model considering sensor chip structure, heat supplied by the heater is transferred not only by forced convection of the air flow in flow channel but also by conduction of membrane and substrate and natural convection in no flow air region. The heat transfer to upstream by conduction and natural convection reduce the cooling effect of upstream air and thus the temperature difference of downstream and upstream thermistors is smaller. For the model without sensor chip structure, the temperature distribution is dominated by forced convection. When mass

flow rate increases to a certain value, upstream thermistor temperature approaches to ambient temperature while downstream thermistor temperature decreases due to global cooling effect, which leads to the decrease of temperature difference.

Conclusions

This paper presents simulation and optimization of a micro flow sensor using ANSYS Fluent. Temperature difference of the downstream and upstream thermistors increases with the increase of mass flow rate, and the temperature difference increases rapid first and then tends to be saturated. Temperature difference increases rapidly as distance between thermistor and heater increases for a fixed mass flow rate, and this tendency slows down when the value of distance exceeds 160µm. Sensibility of the micro flow sensor increases as the input power increases but the linearly measurement range is constant. As the flow channel depth increases, the temperature difference between the downstream and upstream thermistors decreases. Thus the optimized input power, position of thermistors and depth of flow channel are obtained for the micro flow sensor. Linear measurement range decreases when installation error exists. At last, the comparison of simulation model considering sensor chip structure or not shows that the model with fine structure of the sensor chip is more reasonable.

Acknowledgments

The support of Hightech Program (863) with a contract number of 2012AA040501 and MEMS-volume Manufacture Technology of Basic Research (973) with a contact number of 2011CB309504 of Ministry of Science and Technology of PR China is highly appreciated.

References

1. N. T. Nguyen, "Micromachined flow sensorsA review", *Flow Meas. Instrum.*, vol. 8, pp.7 -16, 1997.
2. T. H. Kim and S. J. Kim, "Development of a micro-thermal flow sensor with thin-film thermocouples", *Journal of Micromechanics and Microengineering*, vol. 16, pp. 2502, 2006.
3. N. Sabat, J. Santander, L. Fonseca, I. Grcia and C. Can, "Multi-range silicon micromachined flow sensor", *Sens. Actuators: A. Phys.*, vol. 110, no. 13, pp.282 -288, 2004.
4. Sheng Liu, Yong Liu, *Modeling and Simulation for Microelectronic Packaging Assembly, Manufacturing, Reliability and Testing*, Chemical Industry Press and John Wiley & Sons, 2011.
5. Dillner, U., et al, "Thermal simulation of a micromachined thermopile-based thin-film gas flow sensor, " *Microelectronics journal*, vol. 29, pp.291-297, 1998.
6. Franz Kohl, Roman Beigelbeck, Patrick Loschmidt, Jochen Kuntner, and Artur Jachimowicz, "FEM-based analysis of micromachined calorimetric flow sensors", *Sensors, 2006. 5th IEEE Conference* on, page1215-1218, 2006.
7. Chunlin Xu, Research on air mass flow sensor chip with high accuracy. Unpublished dissertation. Huazhong Universityof Science & Technology, 2013.

978-1-4799-2408-0/14 $31.00 © 2014 IEEE

Minimizing Coupling of Power Supply Noise Between Digital and RF Circuit Blocks in Mixed Signal Systems

Satyanarayana Telikepalli, Madhavan Swaminathan, David Keezer
Department of Electrical & Computer Engineering
Georgia Institute of Technology
266 Ferst Dr., Atlanta, GA, USA 30327
stelikepalli@ece.gatech.edu, madhavan.swaminathan@ece.gatech.edu, david.keezer@ece.gatech.edu

Abstract

Isolation of supply noise between disparate circuit blocks is crucial. When powered by the same voltage supply, the switching noise created at the supply node of the digital devices can couple into the power path of the RF circuitry and cause significant performance degradation. Electromagnetic bandgap (EBG) structures, ferrite beads, and split planes are all commonly used to mitigate this problem, but each have drawbacks which can be detrimental to signal and power integrity. Furthermore, previous works in [1] and [2] have shown that by utilizing a power transmission line (PTL) in place of a power plane, one can significantly reduce the effect of switching noise in high speed digital I/Os by preventing the occurrence of return path discontinuities.

The method proposed here extends the concept of the PTL to mitigate the effect of supply noise coupling between a set of digital I/O buffers and an RF low noise amplifier (LNA). In this work, the approach is to place a notch filter with a bandstop frequency corresponding to center frequency of the LNA in the power supply path of the LNA. Therefore, any frequency content of the switching noise close to operating frequency of the LNA is prevented from entering into its supply node. A board-level test vehicle was built to demonstrate this concept with off-the-shelf components. Through theory, simulation, and lab measurements, is has been shown that utilizing this method can reduce the amount of the switching noise that couples into the output of the LNA by 84%.

Keywords—simultaneous switching noise; noise isolation; electromagnetic band gap; power transmission line

I. INTRODUCTION

In complex mixed signal systems, the isolation of supply noise between disparate circuit blocks is very critical. Due to the varying voltage swings of different circuit blocks as well as each circuit's sensitivity to power supply noise, it is crucial to isolate digital and RF devices. Digital devices can have relatively large voltage swings as compared to RF signals. For example, the output voltage swing for Low Voltage Differential Signaling (LVDS) and Positive Emitter Coupled Logic (PECL) devices is 400mV and 800mV, respectively. Integrated RF devices, however, can have very small voltages (< -20dBm or approximately 25mV for a 50Ω system). When powered by the same source voltage, the supply noise created at the power supply node of the digital devices can couple into the RF circuitry. If, for example, the digital devices are powered by a 2.5V supply, the conventional design methodology is to design the power distribution network (PDN) such that the maximum supply noise is within 5 to 10% of the supply voltage. Therefore, the maximum tolerated noise magnitude can be upwards of 125 to 250mV. If even 10% of this noise is coupled into the supply node of a low

noise amplifier, then 25mV of noise voltage can be injected into the LNA. Consequently, a large amount of switching noise that is injected into the supply rail by the digital circuitry can couple into sensitive RF devices and cause significant performance degradation in the form of reduced gain and linearity, and can even produce a false signal at the output.

A common method for mitigating noise coupling is to use an electromagnetic bandgap (EBG) structure. EBGs are repeating etched metal patterns on the power or ground plane that can be optimized to give a specific frequency response. EBG structures can be very effective in noise isolation, commonly achieving over 30dB of isolation. However, these structures can cause signal degradation, due to the presence of many split planes and via transitions, as discussed in [3] and [4].

Consequently, a method is proposed here that aims to mitigate this problem. The objective of this method is to power both digital and RF devices with a single power source while preventing switching noise from coupling from the digital path into the RF signal path.

II. PROPOSED METHOD

In this proposed configuration, by adding a filter in the power supply path of the LNA, it is possible to reduce the amount of noise that is coupled between the two circuit blocks. This configuration is shown in detail in Figure 1.

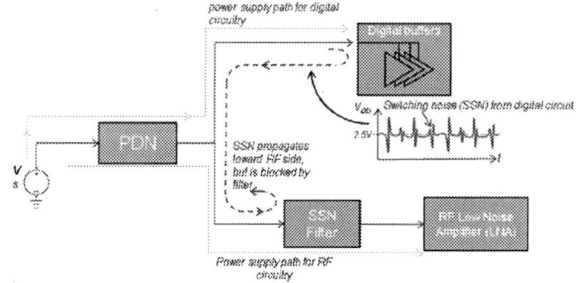

Figure 1.   Mixed signal noise isolation

Suppose the digital buffers are operating at a certain data rate, f_d. The LNA has a desired center frequency of f_{LNA}. If the lower-order harmonics of the simultaneous switching noise (SSN) generated by the buffers happen to fall close to or exactly at f_{LNA}, then the noise will be coupled into the LNA and will be expressed at the output. A notch filter is designed to have a bandstop frequency at f_{LNA}, and is used as the SSN filter. With the SSN filter is connected between the PDN and the LNA circuitry, any power supply noise generated by the digital circuit will be attenuated before coupling into the LNA.

The power delivery network used for this circuit is a power transmission line. The power transmission line concept is described in detail in [1] and [2]. If the PTL is placed on the same layer are the signal traces, then the signal network and the PDN will share the same ground reference, such that a continuous current loop is established and therefore removes return path discontinuity (RPD) effects[1][2].

III. SIMULATIONS

Simulations were performed to demonstrate this idea in Agilent Advanced Design Systems (ADS). The LNA used in the simulation is a simple common-gate configuration operating at a center frequency of 2.4GHz with $V_{DD} = 2.5V$ [5]. At the operating frequency, the gain of the amplifier is approximately 11.95dB.

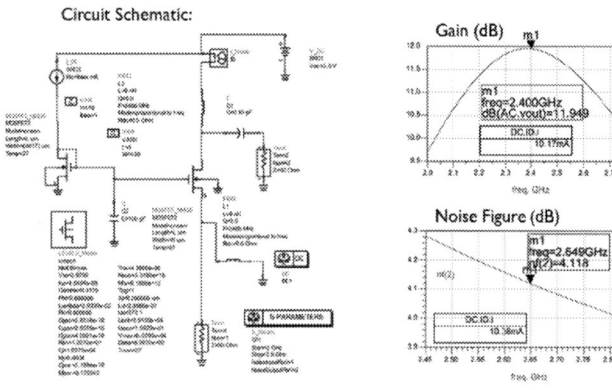

Figure 2. RF LNA used in simulations

A 10Ω power transmission line is used to serve as the PDN and four 2.5V CMOS buffers are operating with a 1.2Gbps PRBS-8 input. Figure 3 shows the power spectrum of the noise generated at the V_{DD} node of the buffers.

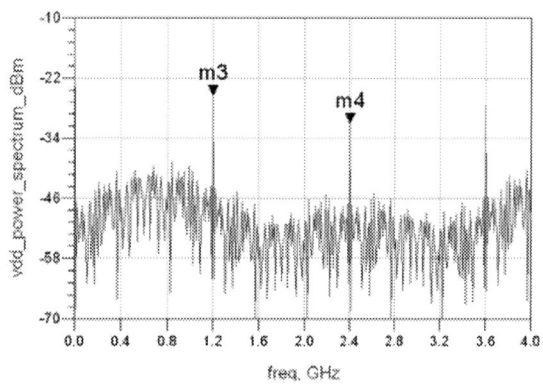

Figure 3. Spectrum of noise due to SSN

Most of the switching noise generated by the digital buffers will be at the switching frequency, 1.2GHz. However, there will also be noise generated at the 1st harmonic, 2.4GHz. This corresponds to the operating frequency of the LNA, so any coupling between the two devices will cause noise to show up at the output of the LNA. In this simulation, the SSN filter is a 3rd order transmission line stub filter that is optimized to give a stop-band response at 2.4GHz [6] with very high insertion loss. This will prevent noise from the 1st harmonic of the SSN to couple into the RF output. The circuit is simulated under various conditions, which are summarized below.

_____

Identify applicable sponsor/s here. If no sponsors, delete this text box (*sponsors*).

A. Case 1: Digital buffers are OFF, RF LNA is ON

This is the control case, in which no input signal is applied to the digital buffers and there is a -50dBm signal at the input to the LNA. As expected, there is a -41dBm signal at the desired frequency at the output of the LNA due to the ~10dB gain.

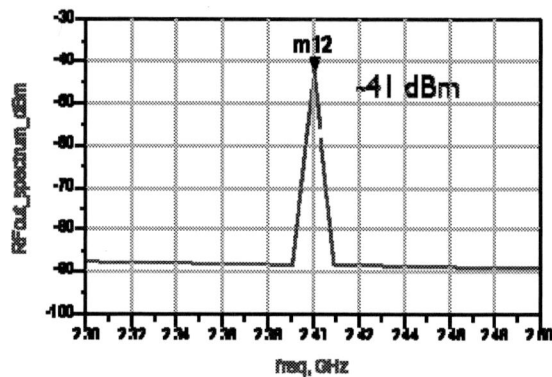

Figure 4. Frequency spectrum of LNA output

B. Case 2: Digital buffers are ON, RF LNA is OFF; no SSN filter

Figure 5a and Figure 5b show the noise voltage at the supply node of the buffers and the LNA, respectively. In this case, even without an input signal to the LNA, there is an output signal at the desired frequency band, as shown in Figure 5c. This is due to the switching noise from the digital circuits that is coupling into the supply node of the LNA.

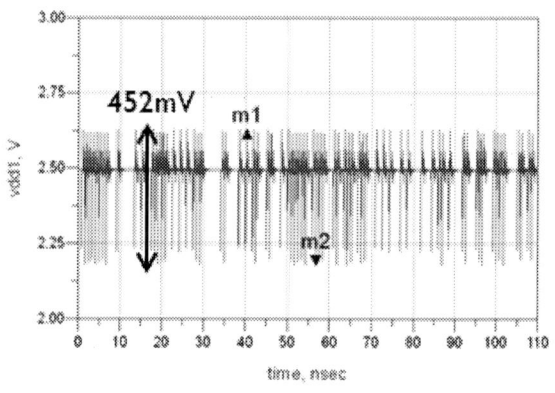

(a)

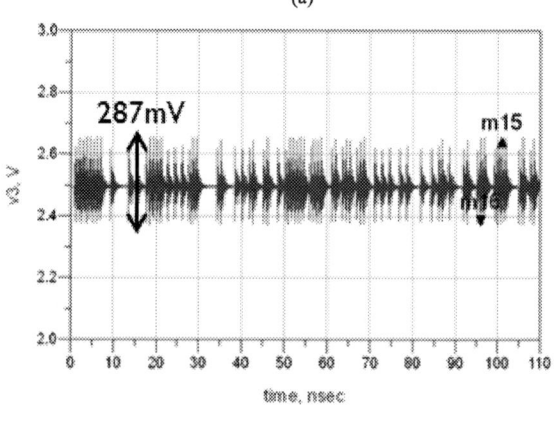

(b)

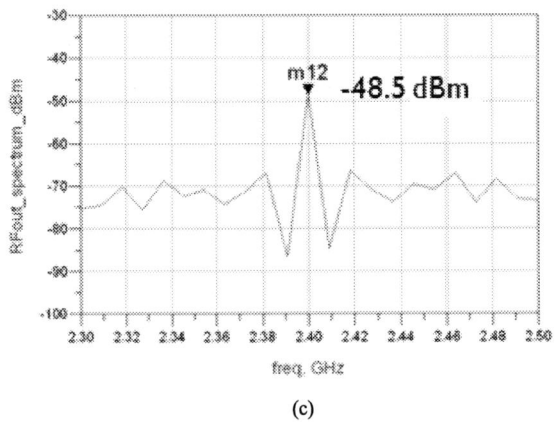

(c)

Figure 5. (a) Supply noise on digital size (b) supply noise on RF side (c) LNA output spectrum at 2.4 GHz

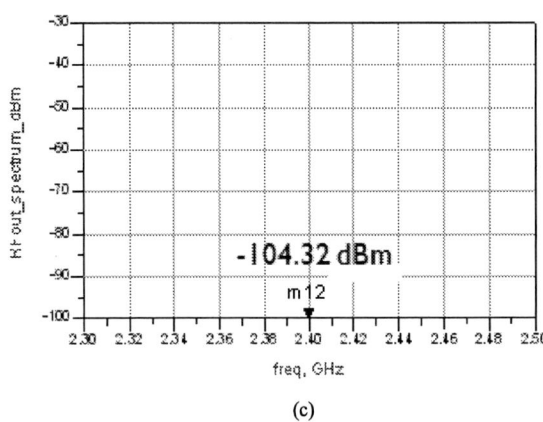

(c)

Figure 6. (a) Supply noise on digital size (b) supply noise on RF side (c) LNA output spectrum

C. Case 3: Digital buffers are ON, RF LNA is OFF with SSN filter

In this case, the transmission line notch filter placed in the supply path of the LNA. The filter has a very high insertion loss at 2.4GHz, and consequently any supply noise generated by the digital circuitry at the bandstop frequency is highly attenuated and does not propagate to the output of the LNA, as shown in Figure 6c. Notice that the peak-to-peak supply noise that is generated by the digital circuits is approximately the same with and without the presence of the notch filter (Figure 5a and Figure 6a, respectively). Consequently, the filter only helps provide isolation between the two circuits, and does not contribute more noise. In addition, although the supply noise from other frequencies may couple into the LNA, these signals can generally be filtered out further down the receiver chain in a real system.

D. Case 4: Digital buffers are ON, RF LNA is ON; no SSN filter

In this case, both the digital and RF sections are operating simultaneously. Consequently, there is significant noise at the supply node of the LNA and the in-band noise is being coupled into the output of the LNA, as shown by the larger than expected output in Figure 7c.

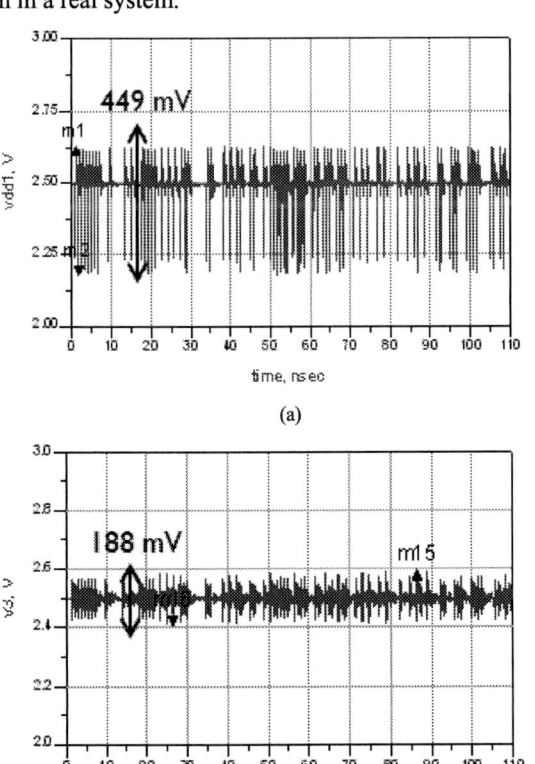

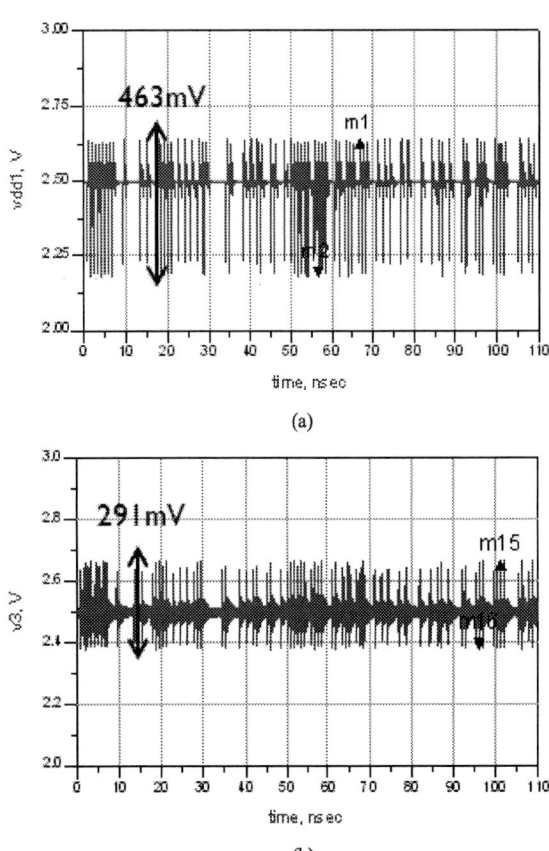

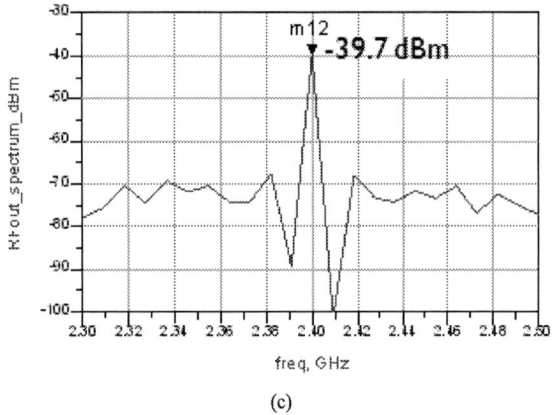

(c)

Figure 7. (a) Supply noise on digital size (b) supply noise on RF side (c) LNA output spectrum

E. Case 5: Digtial buffers are ON, RF LNA is ON; with SSN filter

When Case 4 is repeated with the presence of the SSN filter, the noise at the output of the LNA is significantly reduced and the output spectrum of the LNA is nearly identical to that of Case 1, which is desired.

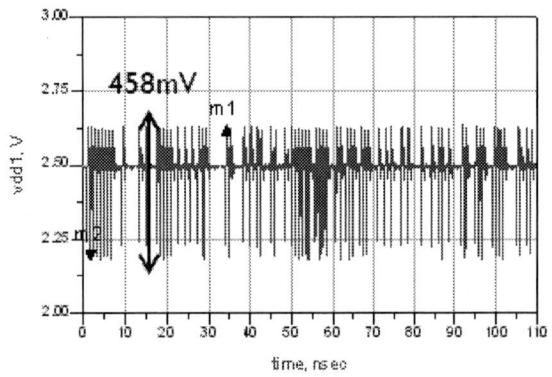

(a)

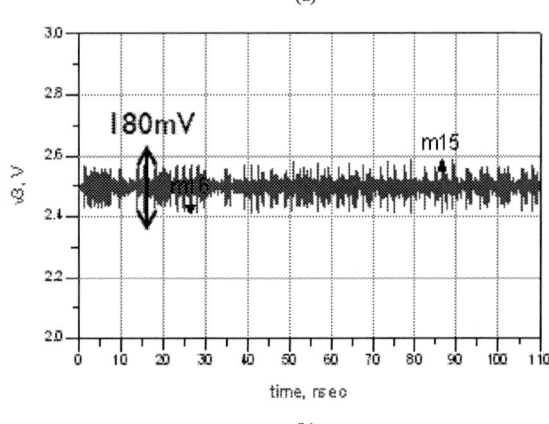

(b)

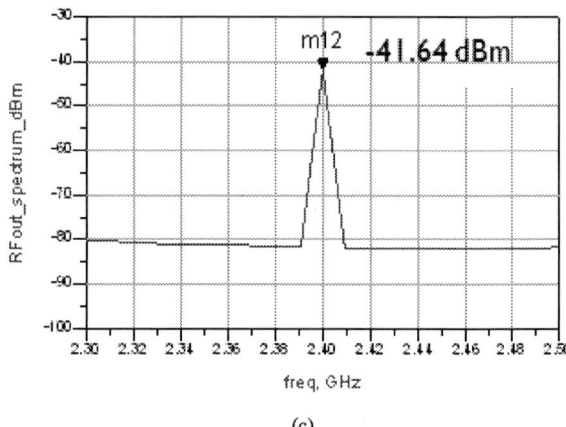

(c)

Figure 8. (a) Supply noise on digital size (b) supply noise on RF side (c) LNA output spectrum

These simulation results show the effectiveness of this strategy. By carefully designing the power delivery network and the SSN filter, it is possible to isolate the power supply noise between digital and RF circuit blocks in mixed signal systems. In order to further demonstrate this concept, several test vehicles were designed and measured.

IV. TEST VEHICLE DESIGN

Test vehicles were designed and fabricated to demonstrate the effectiveness of this noise isolation strategy, shown in Figure 9 below. The test vehicles utilize an unterminated 25Ω power transmission line as the PDN.

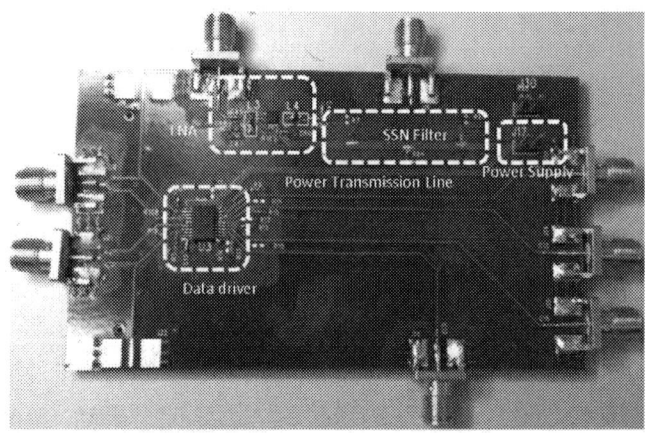

Figure 9. Noise isolation test vehicles with unterminated PTL

The low noise amplifier used in all the test vehicles is an off-the-shelf component from Skyworks with a frequency range from 1.5-3.0GHz and a bias voltage of 2.5V. The operating frequency of the LNA was chosen to be 1.80GHz. Four digital PECL buffers with a supply voltage of 2.5V and 50Ω loads comprise the digital section. The SSN filter was designed using the transmission line stub matching technique as described in [6] and has a Chebyshev band-stop response with a center frequency of 1.8GHz. The filter was designed and simulated using a 3D electromagnetic solver, and the physical layout is shown in Figure 10. By collapsing the stubs into a serpentine pattern, the overall area of the filter was reduced for easier layout onto the PCB.

978-1-4799-2408-0/14 $31.00 © 2014 IEEE

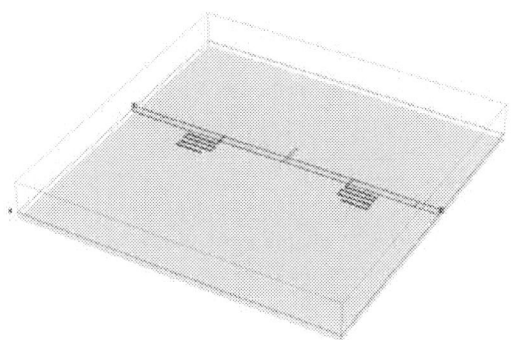

Figure 10. Microstrip line stub filter layout

Figure 11 shows the frequency response of the SSN filter. Due to imperfections in the PCB material, the resulting band-stop frequency occurs at approximately 1.88 GHz.

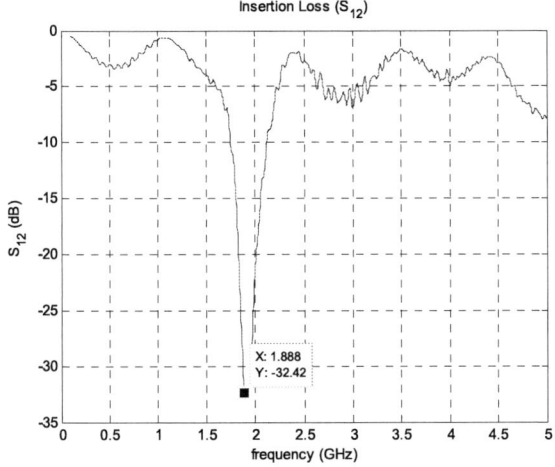

Figure 11. Noise isolation test vehicles with unterminated PTL

In addition, an identical test vehicle was made in which the transmission line stubs of the filter were removed, resulting in a straight 25Ω transmission line connecting the supply voltage to the LNA circuitry. Without the stubs, the standard transmission line does not exhibit a bandstop response and does not isolate noise between the digital and RF sections of the board. Other than this change, this test vehicle is identical to Figure 9.

V. PTL NOISE ISOLATION MEASUREMENTS

The noise isolation test vehicles were tested for the four cases presented previously. For both test vehicles, the input data for the digital buffers was provided by a separate FPGA board which was programmed to provide a 4-bit pseudo-random bit sequence (PRBS-8). An Agilent E8257D signal generator was used to provide a -15dBm input signal for the LNA at 1.88GHz. A 6dB power splitter is used to split the RF signal so that one signal can be used as the trigger source for the oscilloscope to properly view the waveform. Therefore, the amplitude of the signal applied to the input of the LNA is -21dBm.

When the -21dBm input is applied to the LNA, and the digital drivers are disabled, the output of the LNA for each test vehicle is shown in Figure 12. One can see that with and without the filter in the PDN, the output waveforms are very similar, with a peak of approximately -41dBV and -43dBV, respectively. This shows that the presence of the filer does not adversely affect the performance of the LNA.

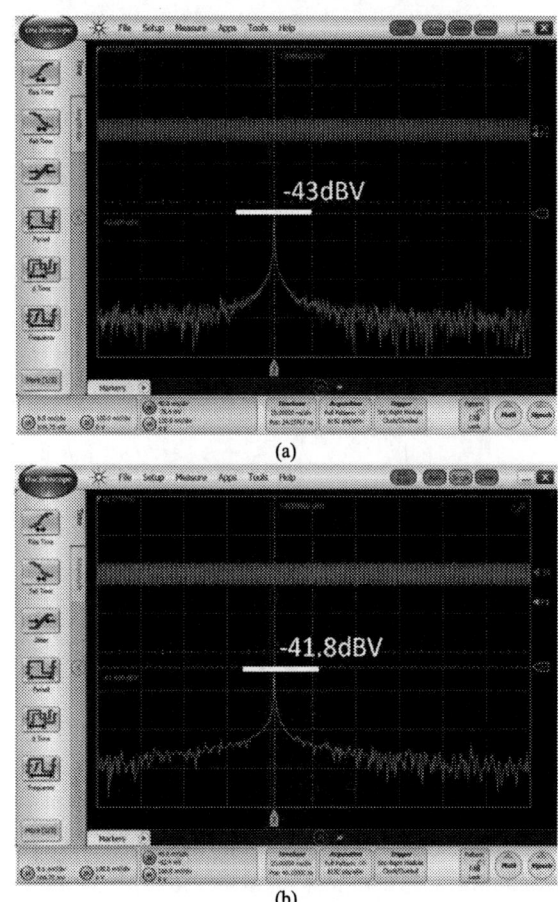

Figure 12. (a) LNA output without filter (b) LNA output with SSN filter

Figure 13 shows the frequency spectrum of the output of the LNA when there is no RF input signal but the digital buffers are switching at 610MHz. At this data rate, the switching behavior of the drivers generates a harmonic at 1.83GHz, which is within the bandstop of the SSN filter. Without the SSN filter, the peak in the spectrum at 1.83GHz has a magnitude of approximately -64dBV. When the filter is present, the peak is at -80dBV. Consequently, the presence of the filter reduces the amount of switching noise that is coupled to the output of the LNA by 16dB. Since the LNA is terminated with a 50Ω load, this translates to an 84% decrease in the noise voltage that is produced at the output terminal of the LNA.

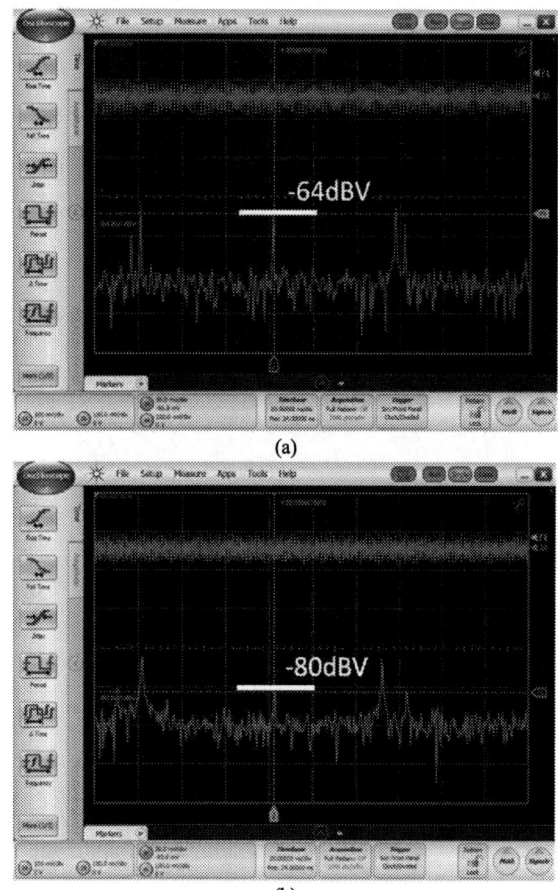

(a)

(b)

Figure 13.   LNA output without filter (b) LNA output with SSN filter

CONCLUSION

The proposed method presented here extends the concept of the power transmission line to mitigate the effect of supply noise coupling between digital and RF circuit blocks. The devices under test are a set of digital buffers and an RF low noise amplifier. If both the digital and RF blocks of a mixed signal design are operating simultaneously and powered by the same source, and a lower-order harmonic of the digital switching noise falls within the operating bandwidth of the LNA, then the switching noise can affect the LNA output. Once present, this noise is very difficult to remove since it is within the operating bandwidth of the LNA. In this work, a method is proposed in which a notch filter with a bandstop frequency at f_{LNA} is placed in the power supply path of the LNA. Therefore, any frequency content of the switching noise close to f_{LNA} is blocked from entering into the supply node of the LNA. A board-level test vehicle was built to demonstrate this concept with off-the-shelf components, with the digital drivers operating at 610Mbps and the LNA having a center frequency at 1.83GHz. Through theory, simulation, and lab measurements, is has been shown that utilizing this method can reduce the amount of the switching noise that couples into the output of the LNA by 84%.

REFERENCES

[1] A. Ege Engin and M. Swaminathan, "Power transmission lines: A new interconnect design to eliminate simultaneous switching noise," in *Electronic Components and Technology Conference, 2008. ECTC 2008. 58th*, 2008, pp. 1139-1143.

[2] S. Telikepalli, M. Swaminathan, and D. Keezer, "Minimizing simultaneous switching noise at reduced power with constant-voltage power transmission lines for high-speed signaling," in *Quality Electronic Design (ISQED), 2013 14th International Symposium on*, 2013, pp. 714-718.

[3] Q. Jie, O. M. Ramahi, and V. Granatstein, "Novel Planar Electromagnetic Bandgap Structures for Mitigation of Switching Noise and EMI Reduction in High-Speed Circuits," *Electromagnetic Compatibility, IEEE Transactions on,* vol. 49, pp. 661-669, 2007.

[4] A. C. Scogna, A. Orlandi, and V. Ricciuti, "Signal and Power Integrity Performances of Striplines in Presence of 2D EBG planes," in *Signal Propagation on Interconnects, 2008. SPI 2008. 12th IEEE Workshop on*, 2008, pp. 1-4.

[5] M. Egels, J. Gaubert, P. Pannier, and S. Bourdel, "Design method for fully integrated CMOS RF LNA," *Electronics Letters,* vol. 40, pp. 1513-1514, 2004.

[6] D. M. Pozar, *Microwave Engineering*, 4 ed.: Wiley, 2011.

A Feasibility Study of Flip-Chip Packaged Gallium Nitride HEMTs on Organic Substrates for Wideband RF Amplifier Applications

Spyridon Pavlidis[1], A. Cagri Ulusoy[1], Wasif T. Khan[1], Outmane Lemtiri Chlieh[1], Edward Gebara[2],
John Papapolymerou[1]

[1]School of Electrical and Computer Engineering, Georgia Institute of Technology, Atlanta, USA 30332
[2]I2R Nanowave Inc., Atlanta, USA 30308
spavlidis@gatech.edu and ioannis.papapolymerou@ece.gatech.edu

Abstract

Gallium nitride (GaN) technology has emerged as a frontrunner for high power electronics applications. By performing a survey of wire-bond and flip-chip- packaged GaN HEMTs on either AlN (a ceramic with high thermal conductivity) or LCP (an organic polymer with low thermal conductivity), the thermal and electrical limits of each package are established. Flip-chip packaging has the benefit of improving the bandwidth of a hybrid PA. Dies that were wire-bonded on AlN showed best performance, and were able to dissipate more than 6 W of power while remaining below the maximum operating junction temperature. On the other hand, flip-chipped devices on LCP were severely limited by thermal effects, even at a 10% duty cycle. This study motivates the need for advanced packaging techniques, such as integrated microfluidics or backside heat-sinking, in order to make LCP a viable material for high-power applications.

Introduction

Gallium nitride (GaN) technology has garnered much attention in recent years for use in solid-state radio frequency (RF) power amplifiers (PA) due to its high power density and high carrier mobility. For state-of-the-art performance, monolithic microwave integrated circuits (MMICs) on silicon carbide (SiC) substrates have been preferred, since they offer low parasitics, which enable wide bandwidth designs, high scalability and excellent thermal management due to SiC's high thermal conductivity [1, 2]. The drawback of MMICs, however, is that they are costly and require more time to manufacture compared to hybrid systems. If a hybrid approach is taken, the GaN HEMTs are usually wire-bonded on brittle ceramic substrate materials (e.g., alumina), or bulky carriers made from copper (Cu) or other copper-based alloys [3, 4]. Though this approach yields optimum thermal performance by leveraging the high thermal conductivity of these carriers, the use of wire-bonds introduces high interconnect parasitics (i.e., inductance) which greatly limit the achievable bandwidth of PAs assembled in this manner.

An alternative packaging method to wire-bonding is the flip-chip technique, which minimizes the interconnect length by replacing wire-bonds with short metallic bumps and, in turn, improves the bandwidth. Flip-chipped wideband GaN PAs have already been demonstrated on aluminum nitride (AlN) due to its high thermal conductivity [5, 6], but the limitations of this packaging scheme have not been studied in detail and reported. Despite its advantages, AlN poses a relatively high cost and mechanical fragility, which might limit its suitability for rugged or portable applications.

To address these issues, low-cost organic packaging materials have been widely investigated in recent years for high-frequency electronics. A prominent example is liquid crystal polymer (LCP), which has now been characterized up to 170 GHz [7], showing a low dielectric constant (ε_r = 2.95) and low loss tangent (tanδ = 0.0025). Moreover, it has a coefficient of thermal expansion (CTE) that matches that of copper and a low melting temperature that enables multilayer and embedded systems-on-package (SOP). Thus far, CMOS [8], silicon-germanium (SiGe) [9] and gallium arsenide (GaAs) amplifiers [10] have all been successfully flip-chipped on LCP. Though one study presented a wire-bonded GaN transmitter module on LCP [11], a flip-chipped amplifier on LCP has yet to be demonstrated.

The purpose of this work, therefore, is to evaluate the feasibility of flip-chip bonding high power GaN high electron mobility transistors (HEMTs) on organic substrates, particularly LCP, and to identify performance limits given LCP's poor thermal conductivity. Since maximum power is dissipated at DC, pulsed IV curves at different duty cycles will be used to evaluate the thermal impact of this packaging on the electrical performance of the GaN HEMTs. This technique is coupled with the use of infrared (IR) spectroscopy to quantify the maximum temperature on the die during operation. In order to provide context for this study, three further packaging schemes are also compared: 1) wire-bonded on AlN (WB-AlN), 2) flip-chipped on AlN (FC-AlN) and 3) wire-bonded on LCP (WB-AlN).

Design

Cree Inc. GaN Die (CGHV1J006D)

To ensure a fair comparison across all packages, the same commercially-available GaN HEMT from Cree Inc. is employed: the CGHV1J006D [12]. This 800 μm x 840 μm bare die (Figure 1), which is fabricated using Cree's GaN-on-SiC process with a 0.25 μm gate length and 1.2 mm gate periphery, is characterized by a typical saturation power (P_{SAT}) of 6 W and drain bias of 40 V. Absolute ratings include a maximum operating junction temperature ($T_{J,max}$) of 225 °C, maximum drain current of of 800 mA and maximum drain voltage of 100 V at 25 °C. Through-SiC vias provide a backside gold-plated ground contact, thus eliminating the need for source wire-bonds. For reference, if the die is attached to a 40 mil thick copper-molybdenum-copper (CuMoCu) carrier using 80/20 gold-tin (Au-Sn) solder, the maximum dissipated power (P_{diss}) is 7.2 W.

Figure 1. Top-view of the CGHV1J006D die [12].

Package Layouts

In order to facilitate testing, it was important to make the packages directly probable. Therefore, coplanar waveguide (CPW) lines were chosen for compatibility with high-frequency ground-signal-ground (GSG) probes. The signal line width and gap were limited by the width of the gate and drain contact pads (200 μm) as well as and the pitch from gate/drain to source pads (approx. 290 μm). For the wire-bonded package (Figure 2), the two ground planes were connected to form a contact pad to which the die's bottom-side source was attached. To ensure alignment between the die's pads and the flip-chip package's pads for bonding, particular attention was given to the location of the contact pads, which do not sit at the edge of the die. Vias were also dispersed throughout the ground planes in order to ensure a common ground for the top and bottom metallization.

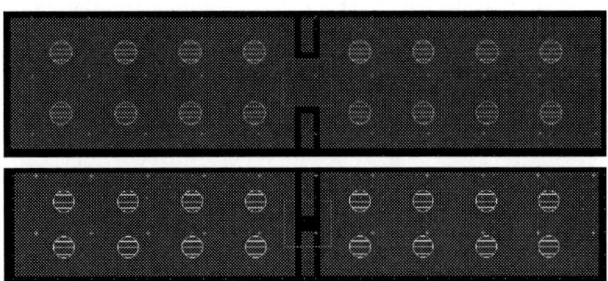

Figure 2. (Top) Layout of the wire-bond package. (Bottom) Layout of the flip-chip package. For each package, the outline of the die is shown in blue.

Fabrication and Assembly

Aluminum Nitride (AlN) Board Fabrication

The ground vias were first drilled through a bare 15 mil thick AlN board using a CO_2 laser. A 5 μm thick Au film was deposited onto the top and bottom of the board through sputtering, which also coated the via sidewalls. The top side was then patterned through a photolithographic process.

Liquid Crystal Polymer (LCP) Board Fabrication

A 4 mil thick LCP substrate from Rogers Corporation was used. The double-clad copper was stripped from the top side only using nitric acid etchant. A CO_2 laser was then used to drill 4 mil diameter vias. Subsequently, a 5 μm thick film of copper was redeposited onto the top side using a DC sputterer, which also served to metallize the via sidewalls. In order to protect the metalized vias adequately, a Suss AltaSpray Spray Coater was used to deposit photoresist for lithography.

Wire-bond Packaging

For the wire-bond on AlN (WB-AlN) package, the die was attached to the gold pad using a 97/3 indium-silver (InAg) ribbon solder from Indium Corporation whose liquidus temperature is 143 °C. Since the WB-LCP sample was metalized with copper, a eutectic bond was not possible. Therefore the die was attached using silver epoxy, a method which has commonly been used thus far for LCP-based packages. Following the die attach step, the gate and drain pads were wire-bonded using a wedge tip and 1-mil diameter gold wire. A cross-section of the wire-bond package is shown at the top of Figure 3. Figures 4a) and 4b) depict the fully packaged wire-bonded devices.

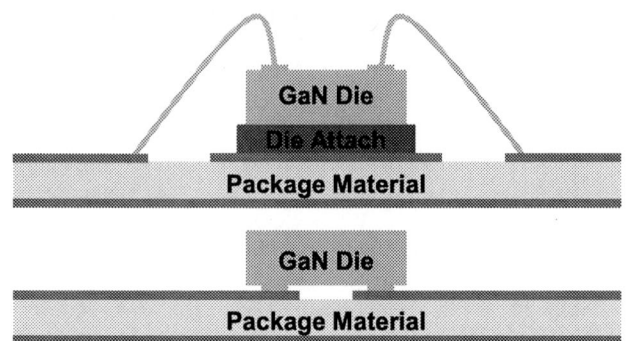

Figure 3. (Top) Cross-section of the wire-bond package. (Bottom) Cross-section of the flip-chip package.

Flip-Chip Packaging

In order to minimize parasitics and facilitate heat transfer, two Au bumps (40 μm tall) were applied to each pad on the Cree die (see Figure 4c). This was done using an Au wire ball bumper that combines thermocompression and ultrasonic energy. A Finetech Submicron Flipchip Bonder was used to align and thermally compress the chip to either the AlN or LCP substrate. To help reinforce the bond between the gold bumps and the package, a controlled dose of silver epoxy was also used. The completed flip-chip packaged devices are shown in Figures 4d) and 4e).

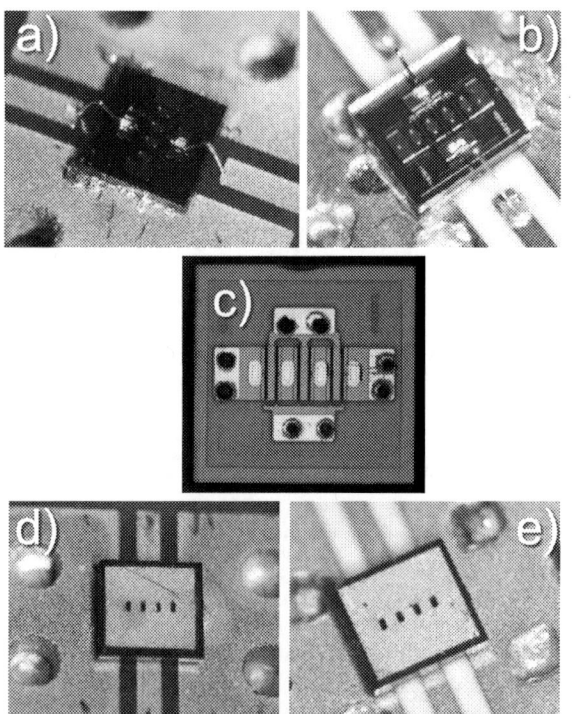

Figure 4. (a) Wire-bond on AlN package, b) Wire-bond on LCP package, c) Double stud-bumped GaN die, d) Flip-chip on AlN package, e) Flip-Chip on LCP package.

Overview of Materials

Table 1 provides an overview of the thermal conductivity of the various materials that were used during the fabrication and packaging process. The materials are categorized depending on their role in the system. It is important to note that both the package substrate material (LCP vs. AlN) as well as the die attach material (silver epoxy vs. InAg) play a vital role in influencing the thermal management of the package.

Table 1. Thermal conductivity of the materials used

Role	*Material*	*Thermal Conductivity [W/m/°K]*
Die Materials	SiC	370
	GaN	130
	AlGaN	19
Package Materials	LCP	0.2
	AlN	180
Package Metallization	Cu	400
	Au	310
Die Attach	Silver Epoxy	2.5-29

Experimentation

Both pulsed IV curves and IR thermal images were obtained on a QFI InfraScope II system (Figure 5). 500 um pitch GSG probes from GGB Picroprobe were used to probe the devices. Short fixture cables on both the input (Gate) and output (Drain) sides connected the probes to Agilent Bias Tees (Model No. 11612A with High Current 001 option). In order to ensure device stability throughout testing, 50 Ohm terminations were also connected to the RF input ports of the bias tees. An AMCAD Pulsed IV system operated through Maury Microwave's IVCAD software was used to bias the devices. This system is capable of sourcing up to 20 A at 250 V and can provide pulse widths as short as 200 ns. For these tests, a 50 µs period was chosen along with three different duty cycles (δ): 10% (5 µs drain pulse width), 20% (10 µs drain pulse width) and 50% (25 µs drain pulse width).

It should be noted that the stage of the QFI system was maintained at 60 °C throughout all testing. To provide a good thermal contact from the stage to the package, thermal paste was applied.

Results

Thermal IR Imaging

The goal of using IR imaging was to determine the maximum temperature (T_{max}) on the die under different operating conditions and, in turn, identify the range of bias points (or dissipated powers) that would allow the device to function below the abovementioned maximum operational junctional temperature of 225 °C given by Cree Inc. Each package was tested under different duty cycles and a fixed drain bias (V_{DS}) of 20 V. The dissipated power (Pd_{iss}) was adjusted through changes to the drain current (I_D), which itself was controlled by the gate voltage (V_{GS}) given the fixed V_{DS}. In this experiment, two values for P_{diss} were calculated: peak dissipated power ($P_{diss,peak}$) which is defined as $I_D \times V_{DS}$ during the pulse, and average dissipated power, which is defined as the peak dissipated power multiplied by the duty cycle ($P_{diss,avg} = P_{diss,peak} \times \delta/100$).

Figure 5. Experimental setup used for both pulsed IV and IR thermal characterization of the packages.

Examples of the captured IR images are shown in Figure 6. In this case, the WB-AlN device was biased at $V_{DS} = 20$ V, $I_D = 329$ mA at $\delta = 50\%$, corresponding to $P_{diss,peak} = 6.56$ W and $P_{diss,avg} = 3.28$ W. A peak temperature of 104.2 °C was registered, as expected, in the middle of the die where the fingers of the HEMT are located. Out of the four vertical source vias, it can be seen that the central ones, which are located within the hotspot, are the hottest. Due to the die's orientation in the FC-AlN package, it was only possible to obtain thermal information about the backside of the device. The image of the FC-AlN package in Figure 6 was taken for a bias condition of $V_{DS} = 20$ V and $I_D = 103$ mA at $\delta = 50\%$. At $P_{diss,peak} = 2.06$ W and $P_{diss,avg} = 1.03$ W, the T_{max} was 93.8 °C. Interestingly, T_{max} was recorded within the four vias, signifying that they indeed play a critical role in dissipating the heat away from the active region of the device located on the reverse side of the die.

IR images were collected for each of the packages over a range of dissipated powers and duty cycles in order to

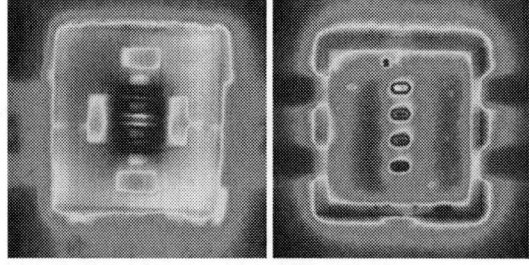

Figure 6. IR images of the WB-AlN package at $P_{diss,avg} = 3.3$ W (left) and FC-AlN package at $P_{diss,avg} = 1$W (right).

978-1-4799-2408-0/14 $31.00 © 2014 IEEE

determine what range of bias points would cause the device to approach temperatures that could be detrimental to its function (approximately 200 °C). Figure 7a) contains a plot of $P_{diss,avg}$ vs. T_{max} for the two WB packages at different values of δ. Given the vastly superior thermal conductivity of AlN over LCP, it came as no surprise to see that the WB-LCP package reached these higher temperatures at significantly lower dissipated powers compared to the WB-AlN package. In fact, at δ = 10% and 20%, the performance of the WB-AlN package was not limited by T_{max}, rather it was the peak current (800 mA) that determined the upper limit of testing. The WB-LCP package reached T_{max}>180 °C below $P_{diss,avg}$ = 1 W. To quantify this trend, the total thermal resistance (R_{TH}) was determined for each package by drawing a line through each set of data and extracting the slope.

The analogous comparison for the flip-chip packages was made in Figure 7b, demonstrating, once again, the superior thermal performance of AlN. Due to rapid heating in the FC-LCP package, it was only tested at δ = 10%, where T_{max} > 180 °C was reached at $P_{diss,avg}$ < 500 mW, thus greatly limiting this type of package for real-life circuit implementation. In addition to a dependence on the package materials, the data show that, for all packages, R_{TH} is also a function of the duty cycle. Finally, it should be noted that each line intersects with the y-axis at 60 °C, which relates to the base plate temperature that was maintained through the tests.

Table 2 summarizes the performance limitations of each package. Given the 6 W rating for this device, it can be concluded that, when operated under the pulsed conditions that were considered here, all of the AlN-based packages are capable of reaching the necessary peak power levels. On the other hand, the LCP packages are very limited in use, such that only the WB-LCP package operated at δ = 10% reaches >6 W peak power. It should be noted that with proper matching for these packages in order to achieve maximum power added efficiency (PAE), the dissipated power can be reduced, thus widening the potential for the LCP packages. All in all, however, the FC-LCP package is severely limited in its use, with extremely high temperatures being recorded even at $P_{diss,peak}$ = 2.72 W at δ = 10%.

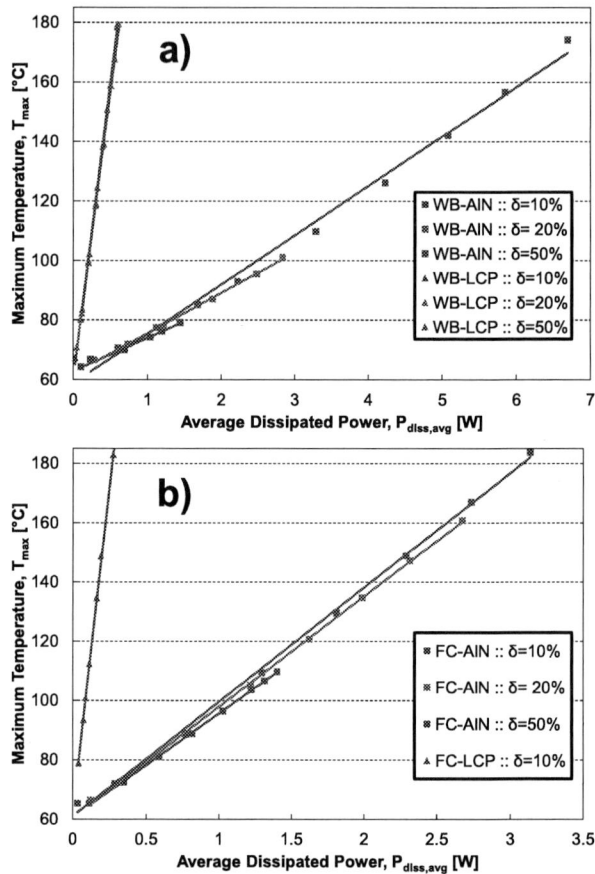

Figure 7. a) Comparison of IR measurements done on the wire-bond packages (WB-AlN vs. WB-LCP) at different δ. b) Comparison of IR measurements done on the flip-chip packages (FC-AlN vs. FC-LCP) at different δ.

Pulsed IV Curves

From the IR measurements, maximum power compliance levels (see Table 2) were extracted which could be defined in IVCAD to safely perform pulsed IV sweeps. This is similar to defining a load-line. In Figure 8, the output characteristic (V_{DS}

Table 2. Overview of the thermal performance for the various packages

Package	Thermal Resistance, R_{TH} [°C/W]	$P_{diss,avg}$ at T_{max} = 180 °C [W]	$P_{diss,peak}$ at T_{max} = 180 °C [W]
WB-AlN			
δ = 10%	11.19	10.73[*]	100.72[*]
δ = 20%	13.61	8.81[*]	44.05[*]
δ = 50%	18.16	6.61	13.22
WB-LCP			
δ = 10%	193.41	0.62	6.2
δ = 20%	199.75	0.60	3.0
δ = 50%	204.98	0.59	1.17
FC-AlN			
δ = 10%	34.87	3.44[*]	34.4[*]
δ = 20%	37.31	3.22[*]	16.1[*]
δ = 50%	38.57	3.11	6.22
FC-LCP			
δ = 10%	440.47	0.27	2.72

[*]*Extrapolated from measured data.*

vs. I_D) for the WB-AlN package is plotted under pulsed conditions for various duty cycles and values of V_{GS}. It is observed that as the duty cycle rises, heat effects come into play, which lead to a drop in the saturation current. This phenomenon is especially prominent at higher current levels; at low values of V_{GS} the difference from $\delta = 10\%$ to 50% is not significant. In Figure 9, similar IV curves are plotted for the FC-AlN package. Here, as expected, the more rapid increase in temperature due to the higher R_{TH} of this package leads to more noticeable falls in current. When compared to the WB-AlN package (Figure 10), it is clear that the WB-AlN package is superior to the FC-AlN package. At $\delta = 10\%$, the two curves overlap, showing little difference in performance between the two packages. However, as δ is increased to 50%, the measured saturation current drops off faster in the FC-AlN package than for the WB-AlN package.

Figure 11 shows the IV curves for the WB-LCP package. Due to the temperature and power limitations identified with the IR camera the characteristics were only measured at lower current levels to avoid damaging the device. Despite this lower range of operation, the effects of heating are clear to be seen. Measurements on the FC-AlN package were only conducted for $\delta = 10\%$ (Figure 12). Even under these conditions, there are clear aberrations in the IV curves which confirm that this package would not be suitable for high-power applications, which are typically well-suited to GaN. A comparison between the two LCP packages in Figure 13, shows the extent to which the performance of the device degrades when it is flip-chipped on LCP: not only does the current fall at this low duty cycle of operation, but the knee current is also lower than in its wire-bonded counterpart.

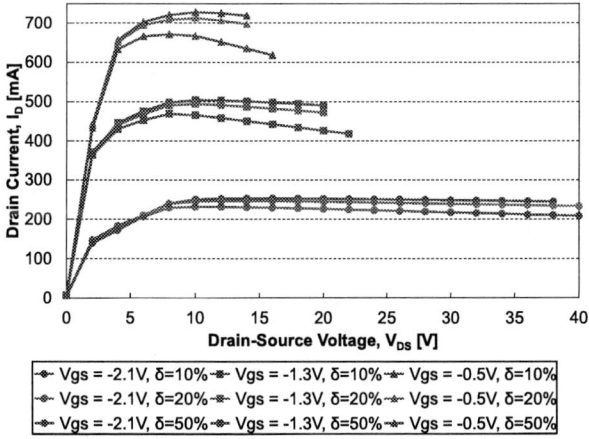

Figure 8. Pulsed IV characteristics of the WB-AlN package.

Conclusions

This work has studied the thermal and electrical performance of GaN HEMTs packaged in AlN and LCP, with either wire-bonding or flip-chipping methods. Devices that were wire-bonded on AlN fulfilled their 6 W rating, thus demonstrating that this package does not impose thermal restrictions on the device's performance. Devices wire-bonded on LCP, however, showed significant degradations in performance, such that 6 W was only achieved at $\delta = 10\%$. Flip-chipped devices on AlN showed favorable performance,

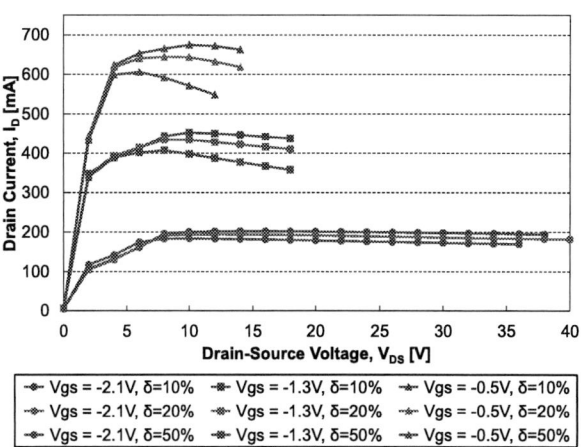

Figure 9. Pulsed IV characteristics of the FC-AlN package.

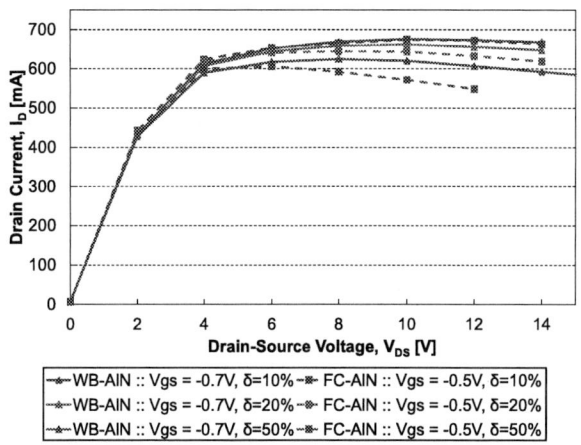

Figure 10. Comparison of pulsed IV curves of the AlN-based packages (WB-AlN vs. FC-AlN).

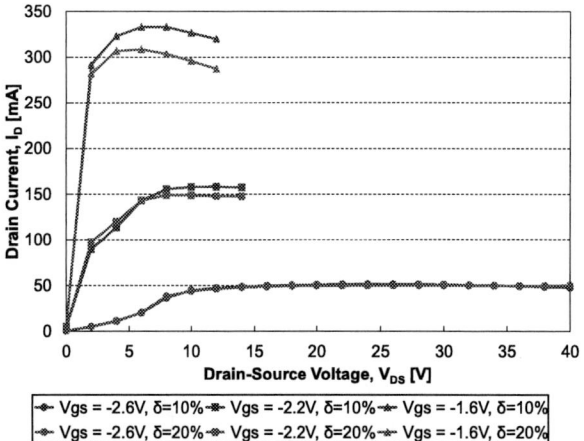

Figure 11. Pulsed IV characteristics of the WB-LCP package.

978-1-4799-2408-0/14 $31.00 © 2014 IEEE

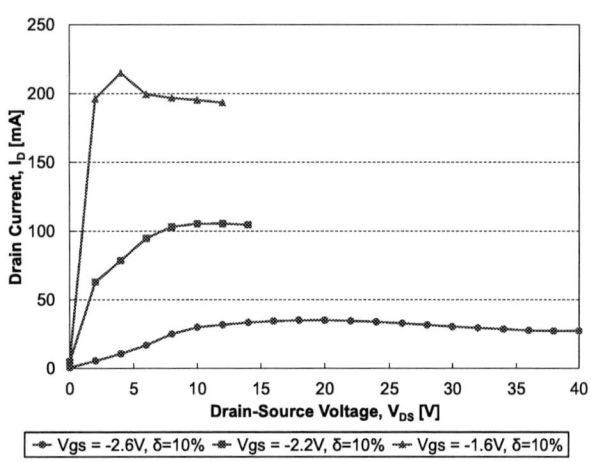

Figure 12. Pulsed IV characteristics of the FC-LCP package.

with more than 6 W of power capable of being dissipated, even at δ= 50%. Finally, when the die was flip-chipped on LCP, severe performance limitations were identified, whereby a maximum average dissipated power of only 2.72 W could be handled without surpassing the maximum junction temperature of 225 °C specified for the device.

In conclusion, though flip-chip packaged devices make it possible to extend the bandwidth of hybrid GaN PAs, the thermal implications of not providing a heat sink to the bottom side of the die cannot be ignored. Therefore, in order for flip-chipping – particularly on organics -- to become feasible for GaN modules, further studies should be carried out. Possible avenues include: 1) the development of a multilayer package incorporating a high thermal conductivity heat sink for the otherwise floating bottom side 2) integrated microfluidics [13] or 3) optimized use of thermal vias [14].

Acknowledgments

The authors would like to acknowledge I2R Nanowave Inc. for supporting this research. In addition, the authors would like to thank the cleanroom staff at the Georgia Tech Institute for Electronics and Nanotechnology (IEN) for their help with fabrication.

References

[1] C. Campbell, L. Taehun, V. Williams, K. Ming-Yih, T. Hua-Quen, P. Saunier, *et al.*, "A Wideband Power Amplifier MMIC Utilizing GaN on SiC HEMT Technology," *Solid-State Circuits, IEEE Journal of*, vol. 44, pp. 2640-2647, 2009.

[2] S. Masuda, M. Yamada, Y. Kamada, T. Ohki, K. Makiyama, N. Okamoto, *et al.*, "GaN single-chip transceiver frontend MMIC for X-band applications," in *Microwave Symposium Digest (MTT), 2012 IEEE MTT-S International*, 2012, pp. 1-3.

[3] R. S. Pengelly, S. M. Wood, J. W. Milligan, S. T. Sheppard, and W. L. Pribble, "A Review of GaN on SiC High Electron-Mobility Power Transistors and MMICs," *IEEE Transactions on Microwave Theory and Techniques*, vol. 60, pp. 1764-1783, Jun 2012.

[4] A. Margomenos, M. Micovic, A. Kurdoghlian, K. Shinohara, D. F. Brown, C. Butler, *et al.*, "X band highly efficient GaN power amplifier utilizing built-in

electroformed heat sinks for advanced thermal management," in *2013 IEEE MTT-S International Microwave Symposium Digest*, 2013, pp. 1-4.

[5] J. J. Xu, S. Keller, G. Parish, S. Heikman, U. K. Mishra, and R. A. York, "A 3-10-GHz GaN-based flip-chip integrated broad-band power amplifier," *IEEE Transactions on Microwave Theory and Techniques*, vol. 48, pp. 2573-2578, 2000.

[6] S. Piotrowicz, R. Aubry, E. Chartier, O. Jardel, J. C. Jacquet, E. Morvan, *et al.*, "Broadband hybrid flip-chip 6-18 GHz AlGaN/GaN HEMT amplifiers," in *2008 IEEE MTT-S International Microwave Symposium Digest - MTT 2008, 15-20 June 2008*, Piscataway, NJ, USA, 2008, pp. 1131-4.

[7] W. T. Khan, C. A. Donado, A. C. Ulusoy, and J. Papapolymerou, "Characterization of Liquid Crystal Polymer (LCP) from 110- GHz to 170 GHz," in *IEEE Radio and Wireless Symposium (RWS)*, Newport Beach, CA, USA, January 2014.

[8] C. E. Patterson, D. Dawn, and J. Papapolymerou, "A W-band CMOS PA encapsulated in an organic flip-chip package," in *Microwave Symposium Digest (MTT), 2012 IEEE MTT-S International*, Montreal, QC, Canada 2012, pp. 1-3.

[9] C. A. Donado Morcillo, C. E. Patterson, B. Lacroix, C. Coen, C. H. J. Poh, J. D. Cressler, *et al.*, "An Ultra-Thin, High-Power, and Multilayer Organic Antenna Array With T/R Functionality in the X-Band," *IEEE Transactions on Microwave Theory and Techniques*, vol. 60, pp. 3856-3867, 2012.

[10] C. E. Patterson, W. T. Khan, G. E. Ponchak, G. S. May, and J. Papapolymerou, "A 60-GHz Active Receiving Switched-Beam Antenna Array With Integrated Butler Matrix and GaAs Amplifiers," *IEEE Transactions on Microwave Theory and Techniques*, vol. 60, pp. 3599-3607, 2012.

[11] S. Pavlidis, C. A. D. Morcillo, P. Song, W. T. Khan, R. Fitch, J. Gillespie, *et al.*, "A Hybrid GaN/Organic X-Band Transmitter Module," in *Radio and Wireless Symposium (RWS), 2013 IEEE*, Austin, Texas, USA, 2013, pp. 241-243.

[12] Cree Inc. (2013, Oct.). CGHV1J006D Data Sheet. *Rev. 03*. Available: www.cree.com

[13] O. L. Chlieh, C. A. D. Morcillo, S. Pavlidis, W. T. Khan, and J. Papapolymerou, "Integrated microfluidic cooling for GaN devices on multilayer organic LCP substrate," in *Microwave Symposium Digest (IMS), 2013 IEEE MTT-S International*, 2013, pp. 1-4.

[14] J. Coonrod and A. F. Horn III. (2010) High Frequency Circuit Materials With Increased Thermal Conductivity. *High Frequency Electronics*. 40-49.

A Novel Molding Process for Wafer Level LED Packaging Using Uniform Micro Glass Bubble Arrays

Yu Zou[1], Jintang Shang[1*], Yu Ji[1], Li Zhang[2], Chiming Lai[2], Dong Chen[2], Kim-HuiChen[2], Ching-Ping Wong[3]

1 Key Laboratory of MEMS of Ministry of Education, Southeast University, Nanjing, CHINA, 210096
2 Jiangyin Changdian Advanced Packaging Co. Ltd, Jiangyin, CHINA
3 Chinese University of Hong Kong, Hong Kong, CHINA
*Corresponding author,+86 13913869603,Email: jshang@seu.edu.cn

Abstract

Light Emitting Diodes (LEDs) as an emerging light source has been rapidly developed due to its considerable advantages including high energy efficiency, extremely long life and environmentally friendly. Since packaging accounts for the major part of the total cost, a novel molding process for wafer level LED packaging will be presented in this paper, which could decrease the total cost by the wafer level process. The packaging was performed using micro glass bubble arrays combined with silicon substrate. The micro glass bubble arrays wereprepared using Pyrex 7740 glass wafer by a WLP chemical foaming process which was suitable for volume production of micro glass bubbles. In addition, the uniformity of the packaged micro glass bubbles were characterized by SEM, which shows a good uniformity of the micro glass bubbles. The experimental results showed a brilliant application prospect in wafer level LED packaging as well as other semiconductor devices using the uniform wafer level micro glass bubble arrays.

Introduction

Light emitting diode (LED)had a history over 50 years since its invention in 1962.However, LED was not widely used as industrial products from the beginning.Its application wasmainly focused on traffic light, decoration and simple display. [1-2] During the past 20 years, the application of LEDs has been rapidly extended, which resulted in a rapid progress in LED technologies. Fig.1 shows the rapid growth in global LED packaging market.[3]

Country	2005	2006	2007*	2008*	2009*	2006**	2009**
China	425	534	695	853	1,089	25.6%	27.7%
Taiwan	1,164	1,313	1,445	1,589	1,732	12.8%	9.0%
Korea	613	750	931	1,111	1,282	22.3%	15.4%
Europe	666	1,106	1,381	1,350	1,546	66.1%	14.5%
US	845	709	851	1,002	1,226	(16.1%)	22.4%
Japan	2,987	3,337	3,637	3,964	4,281	11.7%	8.0%
Total	6,700	7,749	8,740	9,869	11,156	15.7%	13.0%

Source: PIDA, compiled by Digitimes, April 2007
*: forecast **: year to year growth rate

Fig.1. Global LED chip packaging trend by region (US $,million)

Compared with traditional light sources, many considerable advantages such as long life, low energy consumption, small form factor, highly energy efficiency, environmentally friendly and fast response time that made LED light sources an ideal alternative. [3-6]It has an ultra-long lifetime over 20000 hours and its power is commonly below 0.1 W. The chip size of LED is usually 0.2~1.0 mm square. Such a small form factor improves its design flexibility. Besides, mercury is a toxic substance which wouldcause environment pollution. Environment pollution is still one of the most important global issues at present. LED products have no mercury which is good for environment.

Nowadays, LEDs play more and more important role in our daily life including mobile phone, automobiles and billboards. However, it still has a long way to go to completely replace the general lighting by LEDs due to the high cost.It is recognized that packaging and testing is a major part of the total cost of the electronic products. [7-8] WLP has many advantages including low cost, high reliability and small form factor.Wafer-level LED packaging not only decreases the total cost, but also improves its reliability as well as the yield. [9-10]

However, the packaging materials for LED have special requirements, such as optically transparency.Besides, good UV and weathering resistance is also demanded for special applications. Epoxy resins are commonly used for LED packaging by molding process due to its low cost. A large amount of heat will be generated in the local place when they work especially for high power LED chips. As a result, the optical transparency and the optical transmission performances of the commercial lenses made of epoxy or other polymer will be decreased. [11]Compared with polymers, glass has its excellent thermal and physical properties including low CTE, smooth surface, optical transparence, thermal stability.

In this paper, a novel molding process for wafer level LED packaging using uniform wafer level micro glass bubble arrays will be presented. The wafer level micro glass bubble arrays were prepared by a chemical foaming process which could be used for volume production of glass bubbles in a cost effectiveway. The details were described in the corresponding author's previous research. [12-14]Fig.2 shows the wafer level micro glass bubble arrays prepared by the CFP. The presented process has the advantages bothin cost and in reliability.Silicon substrate was selected forits advantages in thermal conductivity and the well matched coefficient of thermal expansion with Pyrex 7740 glass.

Molding process for LED packaging

Fig.3 shows the novel molding process for wafer level LED packaging schematically which will be described in detail as follows:

Before packaging, a 4-inch or larger uniform glass bubble array wafer was prepared as well as a patterned Si wafer in same size which was sputtered with metal films. The glass bubbles interconnected with micro channelsserve as housing or a cap to hold the LED chips. Metal films on silicon

978-1-4799-2408-0/14 $31.00 © 2014 IEEE

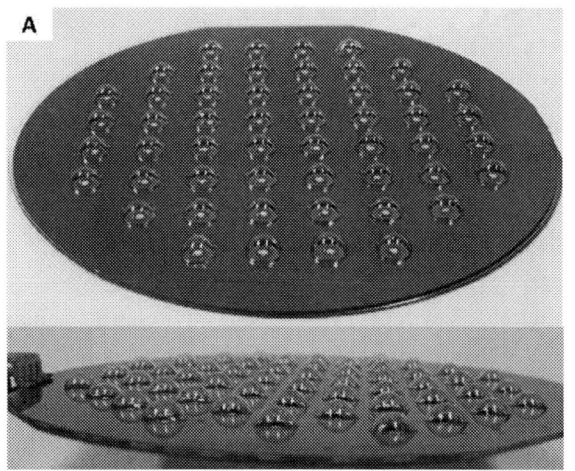
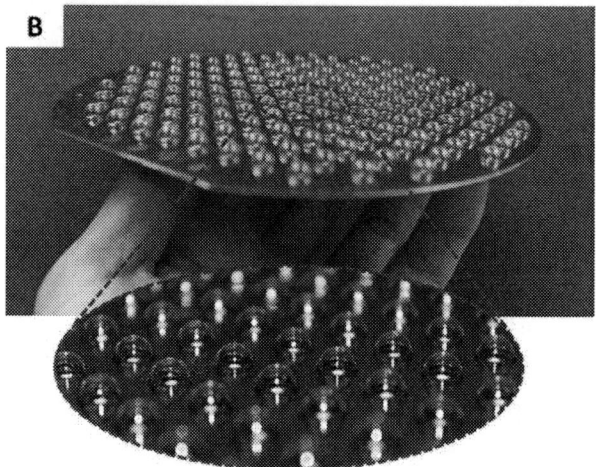

Fig.2 Uniform micro glass bubble arrays prepared by the CFP: (A) 4-inch uniform wafer level micro glass bubble array; (B) 6-inch uniform wafer level micro glass bubble array.

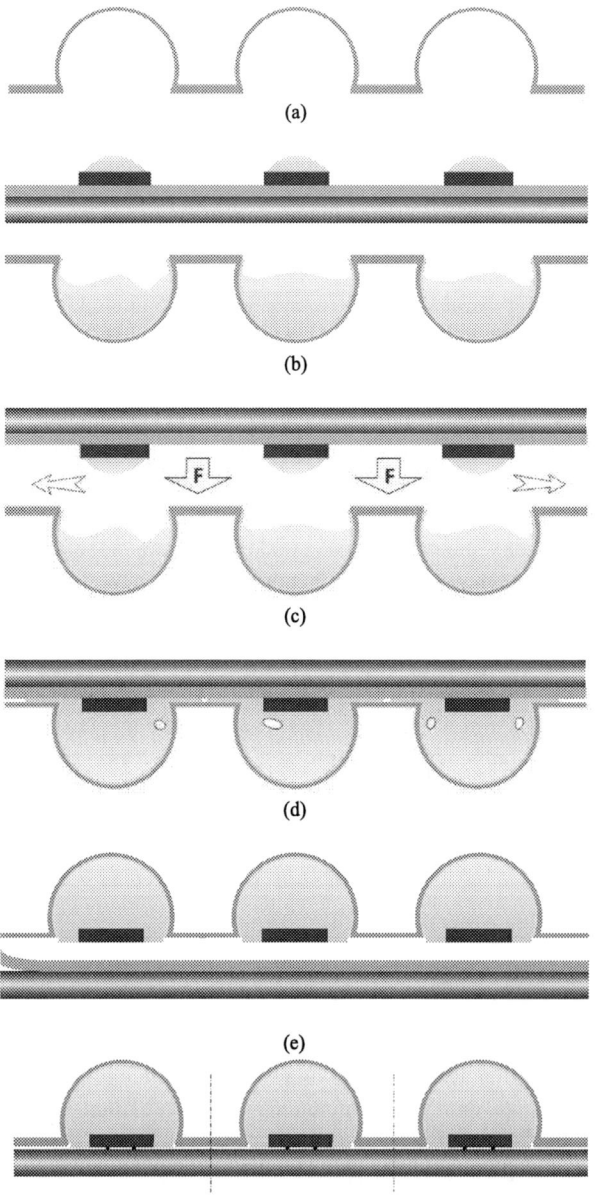

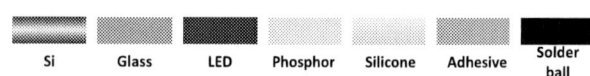

Fig.3. Schematic illustration of the package process.

substrate provide electrical routings for LED chips.As illustrated in Fig.3 (a), the process started witha uniform glass bubble array wafer. The glass bubble is 5mm in diameter and 2.5mm high which could be changed flexiblyaccording to different application. Next, a siliconcarrier wafer was spin coated with temporary bonding adhesives and LED chips were flip-chip glued on the proper surface of the adhesives. In addition, a small amount of silicone was dispensed on chip surface. Meanwhile, the inner surface of micro glass bubble array was coated with phosphor and filled with silicone partially in sequence shown in Fig.3 (b). Silicone is used to fill the glass bubbles to protect the LED and electrical routing from mechanical vibration.Then, the two wafers were bonded together under vacuum shown in Fig.3 (c). The air between the two wafer surfaces was evacuated. On the other hand, the excessive silicone will be squeezed into the interconnected micro channels. Voids might be trapped into the silicone, as depicted in Fig.3 (d). The silicone in interconnected micro channels will flow into micro glass bubble to fill the voids in atmosphere. In addition, when the silicone is solidified, a certain degree of shrinkage of silicone might happen which also would result in formation of voids, the silicone in micro channel could fill the generated voids as well. As a result, its performance was improved. And next, silicone cure was performed. Then the two wafers were debonded and the temporary bonding adhesive was stripped from the silicon carrier wafer by heating (see Fig.3 (e)). Finally, an electrical routed silicon wafer was bonded to the micro glass bubble array wafer and dicing was performed subsequently, as illustrated in Fig.3 (f).

The LED packaging concept using uniform wafer level micro glass bubble array provides a wide and symmetry view angle for LED chip which is larger than most of the commercial LED packages. [4]Besides, the package shows

978-1-4799-2408-0/14 $31.00 © 2014 IEEE   2300

stronger heat dissipation potential due to the high thermal   conductivity of silicon substrate.

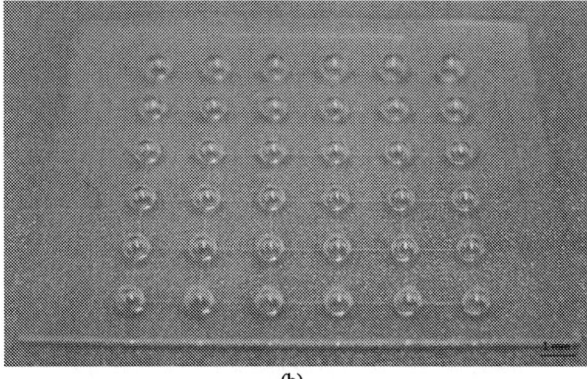

(a)   (b)

Fig.4. Small arrays of uniform micro glass bubbles: (a) micro glass bubble array with diameter of 2mm. (b) micro glass bubble array with diameter of 0.9mm. [16]

Enabling technologies for the proposed LED WLP

In the proposed packaging process, the key technology is how to fabricate uniform wafer level micro glass bubble arrays. Compared with methods such as free inflation, wet etching, dry etching, micro electrochemical discharge machining (ECDM) and laser drilling, CFP has the advantages of low cost, simple process, smooth surface, size controllable.Micro glass bubble arrays in different size could be prepared, as was shown in Fig.2 and Fig.4. The uniformity of micro glass bubbles was characterized by SEM, which was shown in Fig.5. Several glass bubbles of the 6-inch wafer in different region were chosen. The figures showed a good uniformity of micro glass bubbles. And the size of glass bubble could be preciselyestimated by controlling the mass of foaming agents. [14]Different size of micro glass bubbles could be fabricated for various applications.The average surface roughness of the inner and outer surfaces of the micro glass bubble is 0.404 nm and 0.868 nm respectively, which was measured by atomic force microscopy using the tapping mode.Such smooth surface indicated that it could be employed for optical lens.

Besides, phosphor could be remotely coated on the inner surface of the micro glass bubble rather than conformal coated on the chip surface. And in this way, the extraction efficiency would be improved because of reducing the reabsorption probability of wavelength-converted light.[15]

In the previous research [13], a single blue LED chip was packaged on a silicon substrate using a micro glass bubble prepared by this way. Meanwhile the sample was testedat a condition that the temperature is 29 °C and humidity is 45% as well as a commercial packaged LED chip.The ideal color temperature of white light, Tc, is 6000K, its coordinate is (0.33, 0.33). And the result was shown in Fig.6, which shows a brilliant prospect of LED packaging using uniform wafer level micro glass bubble on silicon substrate.

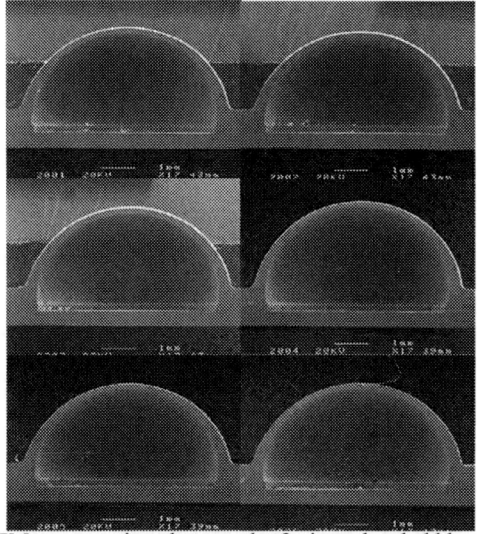

Fig.5. SEM cross-section photograph of micro glass bubbles

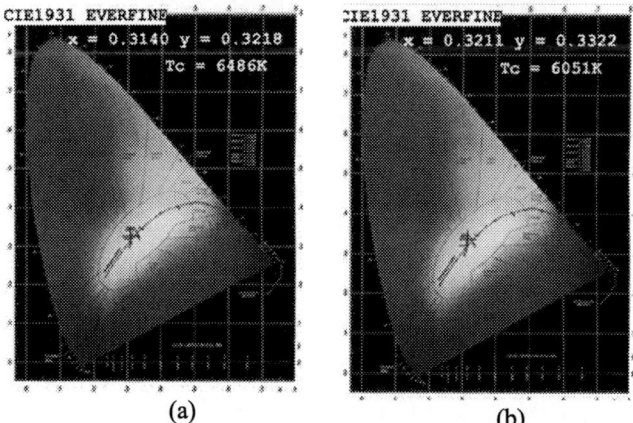

(a)   (b)

Fig.6. The chromaticity diagram of two packaged LED samples: (a) commercial packaged LED sample; (b) experimental packaged LED sample.

Conclusion

In this paper, we have presented a novel molding process for LED WLP using uniform wafer level micro glass bubble arrays. And the characteristics of micro glass bubble showed that micro glass cavity was a good alternative for LED packaging. In addition, the results of packaged LED chip also indicated a significant potential for cost reduction in LED WLPas well as other semiconductor devices (e.g. MEMS) using uniform wafer level micro glass bubble arrays.

The wafer level packaging of LED with this process has not been finished and will be demonstrated in future.

Acknowledgments

This work is financially supported by the National Science Foundation of China (No.51275091).

References

1. Rong Zhang and S. W. Ricky Lee, "Wafer Level LED Packaging with Integrated DRIE Trenches for Encapsulation,"*Proceedigns of International Conference on Electronic Packaging Technology & High Density Packaging (ICEPT-HDP)*, Shanghai,CHINA, Jul 2008, pp.1-6.

2. S. W. Ricky Lee, "Advanced LED Wafer Level Packaging Technologies,"*Proceedings of 6th Microsystems, Packaging, Assembly and Circuits Technology Conference (IMPACT)*, Taipei, Oct. 2011, pp. 71-74.

3. Daniel Lu and C.P. Wong, *"Materials for Advanced Packaging,"* 2009, pp.630-636.

4. Chien-Lin Chang-Chie et al," 'Flip glass substrate' package technologyfor LED yield and performanceenhancement,"*J. Micromech. Microeng*, 22, 2012, pp. 1-10.

5. Chun Kai Liu et al,"High efficiency silicon-based high power LED package intergrated with micro-thermoelectric device,"*Proceedings of Microsystems, Packaging, Assembly and Circuits Technology Conference (IMPACT)*, Taipei, Oct. 2007, pp. 29-33.

6. WonKyu Jeung et al, "Silicon-Based, Multi-Chip LED Package,"*Proceedings of 57thElectronic Components and Technology Conference*, Reno, NV, USA, Jun 2007, pp. 722-727.

7. Chingfu Tsou and Yu-Sheng Huang, "Silicon-Based Packaging Platformfor Light-Emitting Diode,"*IEEE Trans. Adv. Packag.*, 29, Aug. 2006, pp. 607-614.

8. Thomas Uhrmann, Thorsten Matthias and Paul Lindner,"Silicon-Based Wafer-Level Packaging for Cost Reduction of High Brightness LEDs,"*Proceedings of 61th Electronic Components and Technology Conference*, Lake buena Vista, FL, USA, Jun 2011, pp. 1622-1625.

9. Rao R. Tummala, *"Fundamentals of Microsystems Packaging,"* 2004, pp. 400-405.

10. Jyh-Rong Lin et al,"Wafer-level LED-SiP based mobile flash module and characterization,"*Microelectronics Reliability*, 52, 2012, pp.916–921.

11. Joon-Soo Kim, SeungCheol Yang, and Byeong-Soo Bae, "Thermally Stable Transparent Sol-Gel Based Siloxane Hybrid Materialwith High Refractive Index for Light Emitting Diode (LED)Encapsulation,"*Chem. Mater.*, 22, 2010, pp. 3549-3555.

12. Jintang Shang et al,"Preparation of wafer-level glass cavities by a low-cost chemical foamingprocess", *Lab Chip*, 11, 2011, pp. 1532–1540.

13. Yu Zou, Jintang Shang et al,"Preparation of Wafer-Level LED Packaging Used Uniform Micro Glass Cavities by an Improved Chemical Foaming Process (CFP),"*Proceedings of 15th Electronics Packaging Technology Conference (EPTC)*, Singapore, Dec. 2013.

14. Yu Zouet al,"Study on Volume Production of Uniform Wafer-Level Micro Glass Cavities by A Chemical Foaming Process (CFP),"*Proceedigns of International Conference on Electronic Packaging Technology & High Density Packaging (ICEPT-HDP)*, Dalian,CHINA, Aug. 2013.

15. Chien-Lin Chang-Chien et al,"Development of a 'flip glass substrate' LED package technology for color bin yield and view angle enhancement," *Proceedings of 16th Solid-State Sensors, Actuators and Microsystems Conference*, Beijing, CHINA, Jun 2011, pp. 1919-1922.

16. Qin S J et al, "Fabrication of Micro-Polymer Lenses With Spacers Using Low-Cost Wafer-Level Glass-Silicon Molds,"*IEEE Trans. Compon. Packag. Manufact. Technol.* 3, 2013,pp. 2006-2013.

Analysis of Room-Temperature Bonded Compliant Bump with Ultrasonic Bonding

Keiichiro Iwanabe, Takanori Shuto, and Tanemasa Asano
Graduate School of Information Science and Electrical Engineering, Kyushu University
744 Motooka, Nishi-ku, Fukuoka 819-0394, Japan
E-mail: iwanabe@fed.ed.kyushu-u.ac.jp, phone: +81-92-802-3727

Abstract

Room temperature microjoining of Au or Cu bumps in the air ambient has been achieved by bonding cone-shaped microbumps with ultrasonic application. This technology has been applied to fabrication of near infrared (NIR) image sensor of q-VGA (quarter video graphic array) resolution, where InGaAs/InP phtosensor array is joined with CMOS read out in the pixel level and, therefore, low temperature bonding is strongly required to address problems caused by mismatch in thermal expansion of the two materials. In this work, we investigate bonding mechanism of the cone shaped microbump using bumps made of Au. Die shear tests shows that shear strength of the bonded chips is proportional to the bonded contact area of the bump and that room temperature bonding gives sufficient bonding strength for applications while bonding at elevated temperature results in higher bonding strength. Analysis of change in bump height shows that "softening" of Au bump takes place under the application of ultrasonic vibration. Transmission electron microscopy shows that crystal grains at the bonded interface transform to small crystallites.

1. Introduction

Integration of heterogeneous materials with Si CMOS LSI is attracting a great deal of attention to create new electronic function. Flip chip bonding of compound semiconductor, where an array of photosensors was fabricated, to CMOS read-out circuit can provide us image sensors for invisible wavelength. Near infrared (NIR) image sensors offer wide applications ranging from security to medicals [1-3]. These applications demand high resolution. To meet this requirement, flip chip bonding should be performed at temperatures as low as possible. This is because that the difference in thermal expansion coefficient between the two materials gives constraints in pixel pitch. NIR image sensors have been fabricated by using mostly solder bumps made of indium (In). The In bump technology provides flip chip bonding at temperatures less than 200°C. Even at this temperature, however, the difference in thermal expansion becomes as 4 mm as large for a stacking of an InP chip and a Si chip of 10 mm-square.

It has been shown that cone-shaped microbump is effective to realize low temperature microjoining owing to its mechanical compliance [4]. Room temperature bonding of Au/Au and Cu/Cu bumps has been demonstrated realized by introducing a hole in the counter electrodes [5, 6]. This technology provides high alignment accuracy. However, the formation of a hole in the counter electrode requires additional photolithography process which increases the fabrication cost.

In the previous works, ultrasonic bonding of the cone-shaped microbumps (Fig. 1) has been shown to provide room temperature bonding of Au/Au [7] and Cu/Cu [8] microjoining. Moreover, the bonding time is very short, on the order of several hundred milliseconds, which increase productivity. The ultrasonic bonding process is, therefore, a practical solution to solve the issues of the difference in thermal expansion coefficient between materials. This room temperature bonding technology has been applied to fabricate InGaAs/Si-CMOS NIR image sensor having q-VGA (320 x 256) resolution [9]. Figure 2 shows an example of image taken by the NIR image sensor. Extension to VGA (640x512) resolution is in progress [10].

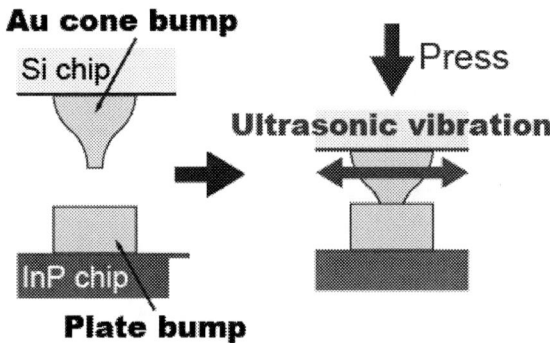

Figure 1. Schematic illustration of ultrasonic bonding of the Au cone-shaped microbump.

To further extend this technology to mega-pixel range, understanding of the room temperature bonding mechanism is highly demanded. In this paper, we report results of investigation of bonding characteristics of the cone bump with ultrasonic vibration and analysis carried out on bonded interface.

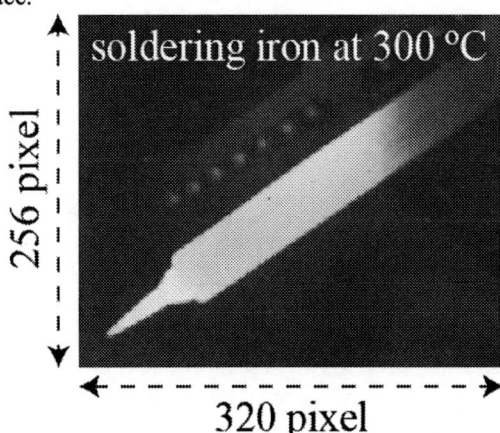

Figure 2. NIR image of the soldering iron heated to 300 °C. This image was taken by a q-VGA NIR image sensor fabricated by using room-temperature ultrasonic bonding of cone bumps.

978-1-4799-2408-0/14 $31.00 © 2014 IEEE
2303
2014 Electronic Components & Technology Conference

2. Experimental

In this study, cone-shaped microbumps and counter electrodes were made of Au. Figure 2 shows the fabrication process of the Au cone-shaped microbumps. Details of the fabrication process have been described elsewhere [7]. The following is the process flow. First, the seed metal layer composed of a 250-nm-thick TiW film and an 80-nm-thick Au film was deposited on a oxidized Si wafer by sputtering. [Fig. 3(a)]. Next, a photoresist pattern having undercut holes was formed by photolithography [Fig. 3(b)]. A negative tone photoresist of chemical amplification type was used. The undercut profile was controlled by adjusting the process parameters. Then, Au electroplating was applied to fill the undercut holes in the resist film [Fig. 3(c)]. The photoresist was then removed by immersion in a solvent [Fig. 3(d)]. The seed layer was removed by reactive ion etching using a mixture of CF_4 and O_2 gases [Fig. 3(e)]. Finally, cone bumps were annealed at 300°C for 15 min in order to improve the adhesiveness between the cone bumps and seed layer [Fig. 3(f)].For the counter electrode, planar Au electrodes were fabricated using the conventional technique that combined standard photolithography with Au electroplating.

Figure 4 shows optical micrographs and SEM images of the test chips. The 7-mm-square chip [Fig. 4(a)] contains Au cone-shaped 12,100 microbumps of 20 μm-pitch. The 5-mm-square chip [Fig. 4(b)] contains Au planar electrodes as the counter electrodes of the cone-shaped bumps. Prior to bonding, in order to remove contaminants on the Au cone-shaped microbumps and counter electrodes, surface cleaning by using Ar plasma was performed. The cleaning conditions were as follows: the radio frequency power was 400 W, the chamber pressure was 1.8-2.2 Pa, and the cleaning time was 1.5 min.

Bonding of the test chips was carried out using a flip chip bonder equipped with an ultrasonic bonding tool made by Adwelds Inc.. The bonding was performed at room temperature in ambient air. Bonding conditions are described in the next section.

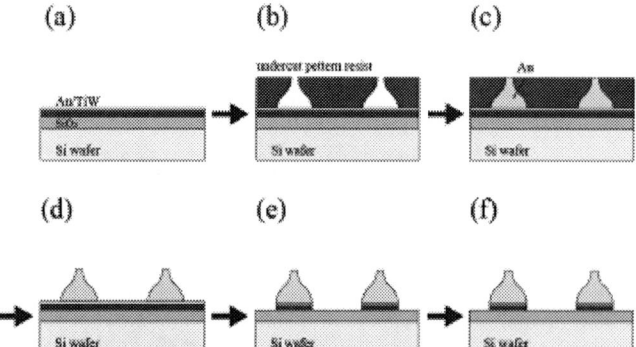

Figure 3. Process flow of cone-shaped microbump formation: (a) formation of seed layer, (b) photolithography, (c) electroplating, (d) removal of photoresist, (e) removal of seed layer, and (f) annealing.

Change in bump height with bonding parameters and area of the bonded interface of a bump were measured using scanning electron microscope (SEM). The observation was carried out after de-bonding the bonded chips using a die-

share test tool. In addition, transmission electron microscope (TEM) observation was carried out to analyze the texture of the bonded metal near the interface.

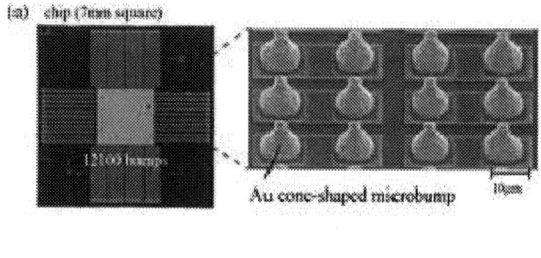

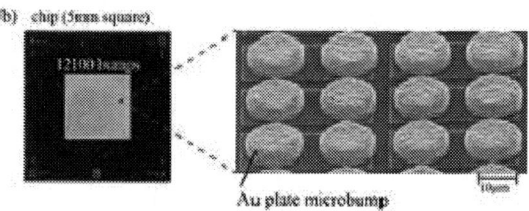

Figure 4. Optical micrographs and scanning electron microscope (SEM) images of test chips: (a) Chip photo with SEM of Au cone-shaped microbumps. (b) Chip photo with planar counter electrodes made of Au.

3. Results and discussion

3.1 Effect of ultrasonic on bump deformation

First, we investigate effect of the application of ultrasonic vibration on change in magnitude of plastic deformation of the cone bump. To examine effect of ultrasonic application, ultrasonic bonding and thermo compression bonding were compared. Table I shows bonding parameters employed to carry out this investigation. Figure 5 shows SEM images of bump before bonding (Fig. 5(a)) and after bonding using either compression without heat (Fig, 5(b)), thermo compression (Fig. 5(c)), or ultrasonic bonding (Fig. 5(d)). All bondings were carried out under the same bonding force, 0.75 gf/bump. SEM observations for thermo compression bonding and ultrasonic bonding were carried out after de-bonding the bonded chips using a die share tester, while chips bonded by compression bonding without heat resulted in boding failure, i. e; chip release without force. These SEM images clearly show that the bump height change is the largest in the case of ultrasonic bonding and is the smallest in the case of compression without heat.

Similar observations were carried out by changing boding force. Figure 6 plots the observed bump height change with bonding force for the cases of compression without heat, thermo compression and room-temperature ultrasonic bonding. These results clearly demonstrate that compression deformation is significantly enhanced by the application of ultrasonic vibration. The deformation enhancement by the ultrasonic application is larger than that by the heat application for elevation to 250 °C.

It has been reported that application of ultrasonic to aluminum wire results in "softening" of aluminum [11]. The softening was found for tension test of aluminum wires. The enhanced change in bump height by the application of ultrasonic to Au cone bump can be attributed to the softening of metal. The results shown in Fig. 5 indicates that the application of ultrasonic with the amplitude of 0.9 μm gives larger softening effect than heat up to 250 °C.

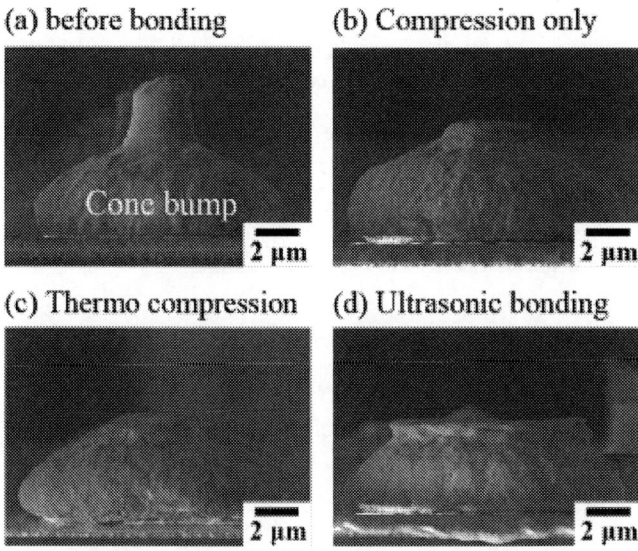

(a) before bonding (b) Compression only

(c) Thermo compression (d) Ultrasonic bonding

Figure 5. SEM images of bumps deformed as the result of bonding process. (a) before bonding, (b) bonded by compression without heat, (c) thermo compression at 250 °C (no ultrasonic), (d) ultrasonic bonding at room temperature. The bonding force applied was 0.75 gf/bump for the cases (b) – (d).

From the applied force per bump shown in Fig. 6 and the area bonded interface measured by SEM observation after de-bonding, we are able to estimate effective yield strength of the bump and its change with application of ultrasonic or heat during bonding. The effective yield strength thus determined were approximately 630, 170, and 130 MPa for bonding using compression at room temperature, thermo compression at 250 °C, ultrasonic at room temperature, respectively. Although the absolute values of these results needs further investigation, the results suggest that the effective yield strength is reduced to approximately 1/4 by the application of ultrasonic of 0.9 μm amplitude at 48.5 kHz.

Table I. Bonding parameters employed to investigate effect of ultrasonic application on deformation of bump.

	Ultrasonic bonding	Thermo compression	Compression only
Bonding temperature	RT	250 [°C]	RT
Vibration amplitude	0.9 [μm]	-	-
Vibration frequency	48.5 [kHz]	-	-
Bonding time	0.5 [s]	30 [s]	
Pressing load	10.1 - 90.9 [N] (0.10 - 0.75 [gf/bump])	30.3 - 90.9 [N] (0.25 - 0.75 [gf/bump])	

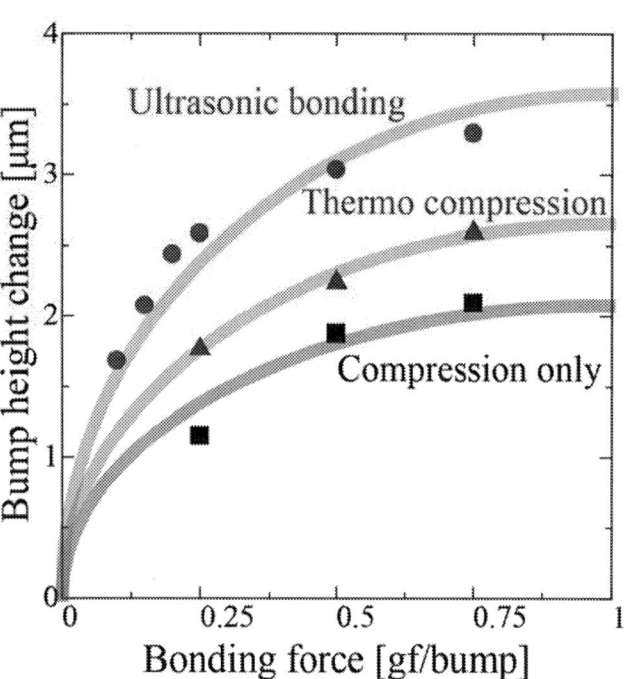

Figure 6. Bump height change with bonding force for the bonding cases, compression binding without heat, thermo compression bonding, and ultrasonic bonding at room temperature. Bump height change was measured with SEM by observing bumps before bonding and after bonding and de-bonding.

3.2 Bonding strength

Next, bonding strength tests using die share test were carried out. We compared ultrasonic bonding and thermo compression bonding in terms of relationship between bonding temperature, bump deformation, and bonding strength. Table II shows bonding parameters employed for this study. Figure 7 plots the results of the die share test. The horizontal axis represents the area of the bonded interface, i.e. contact area, for a bump measured by observing the bonded interface with an optical microscope with metering function after die share tests. The vertical axis is the die share strength normalized by the number of the bump. The plots can be divided into two groups; one is the plots of 250°C bonding and the other is the plots of room temperature bonding. We also find the followings: (1) Die share strength is almost proportional to the bonded interface area. (2) Thermo compression bonding at 250°C gives higher bonding strength than the bonding at room temperature when the bonded interface area is same. This result suggests that thermally enhanced interdiffusion of constituent atoms at the interface effectively increases. (3) Ultrasonic bonding can increase the bonded interface area and, therefore, bonding strength can be maximized even at room temperature. (4) Increasing vibration amplitude can increase bonded interface area, and, therefore, bonding strength. From these result, mechanical bonding strength is higher in bonding at elevated temperature (250°C) than in the room temperature bonding. But the difference in mechanical strength can be easily compensated by changing, vibration amplitude and pressing load during bonding.

Table II. Bonding conditions employed to investigate bonding strength for the cases of ultrasonic bonding and thermo compression bonding.

	Ultrasonic bonding	Thermo compression
Bonding temperature	RT	250 [°C]
Vibration amplitude	0.9, 1.5 [μm]	-
Vibration frequency	48.5 [kHz]	-
Bonding time	0.5 [s]	30 [s]
Pressing load	30.3 - 90.9 [N] (0.25 - 0.75 [gf/bump])	

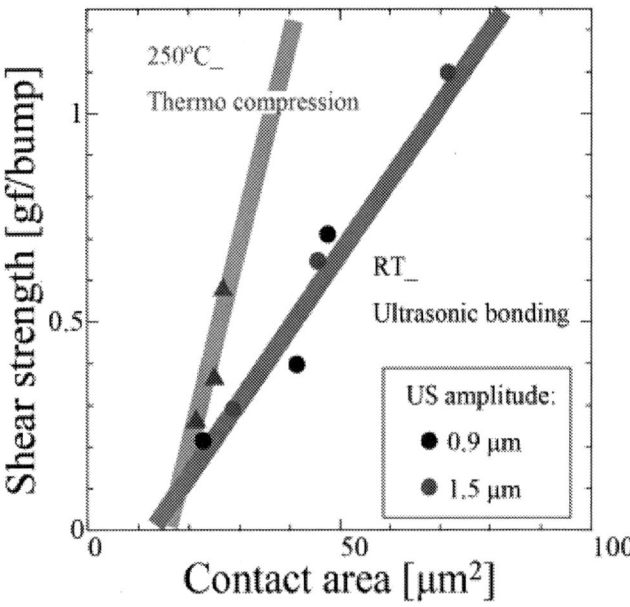

Figure 7. Results of die shear tests. Contact area is the area of the bonded interface of a bump evaluated by observing the surface of bumps after de-bonding as the result of die share test. Shear strength was normalized by the number of bumps on a chip.

3.3 Crystal texture at bonded interface

The above results showed that the bonding strength is almost proportional to the bonded interface area for the cases of ultrasonic bonding and thermo compression bonding. Then, a hypothesis may be formulated, that is, a large deformation produced by simply applying a large bonding force at room temperature would produce bonded interface. We investigated this hypothesis by carrying out room temperature bonding at a large bonding force. The result is shown in Fig. 8(b), where an SEM image indicates that the height of the bump thus deformed is almost same as that of the ultrasonic bonded bump shown in Fig. 8(a). However, the bonding without ultrasonic was failed, i. e., bonded chips were released each other without application of a force. This result demonstrates that the removal and/or dispersion of contaminants from the interface are necessary to produce bonded interface.

In order to investigate how the ultrasonic boding produce bonded interface at room temperature, crystal texture at and near the interface was analyzed using transmission electron microscopy (TEM). Figure 9 shows cross-sectional TEM images of a bump bonded by ultrasonic bonding at room-temperature. From Fig. 9(a), we find that crystal grains of Au transform to small crystallites near the interface. The fine crystallites are probably formed by breaking original large crystal grains as a result of ultrasonic vibration and deformation. This transformation would break the original contaminated interface and disperse contaminants from the interface, which results in metallic bonding at room temperature. As is observed in Fig. 9(b), joining at atomic level is performed by the ultrasonic bonding.

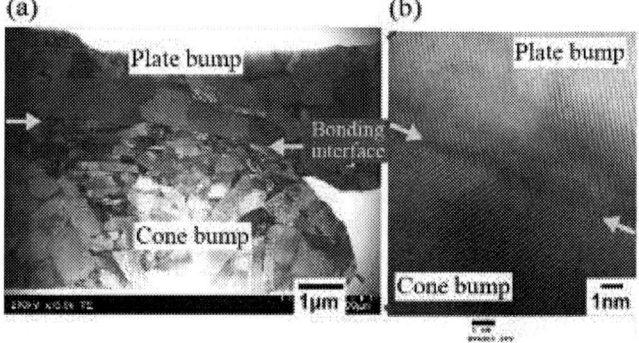

Figure 8. SEM images of bump deformation by bonding at room-temperature. (a) bonding with ultrasonic (b) bonding without ultrasonic and without heat (compression only)

Figure 9. Cross-sectional TEM images of a bump bonded by ultrasonic bonding at room-temperature. The ultrasonic frequency was 48.5 kHz with amplitude of 0.9 μm. Bonding time was 0.5 s. Pressing load during bonding was 0.75gf/bump.(a) Overview of the bonded bump. (b) High resolution image of the interface

Conclusions

We investigated the room temperature bonding mechanism of ultrasonic bonding of the cone-shaped microbumps. We clarify the following points with clear experimental data.

1. Owing to its geometrical property, the cone-shaped bump plastically deforms under small compression force. Besides, the geometrical effect, the application of ultrasonic vibration induces "softening" of the bump.

978-1-4799-2408-0/14 $31.00 © 2014 IEEE

While elevation of bonding temperature, i. e. thermo-compression, also provides softening phenomenon, the application of ultrasonic vibration with 0.9 um amplitude gives much larger degree of softening than heating up to 250 degrees.

2. Die share strength is proportional to the bonded bump area, in other word, the interface area after deformation. The ultrasonic bonding at elevated temperature gives larger die share strength than the room temperature bonding. However, room temperature bonding provides a large enough strength for applications by changing, vibration amplitude and pressing load during bonding.

3. Recrystallization of grains takes place near the interface to transform to fine crystallites. This transformation may break the original interface to produce fresh interface and produces metallic bonds.

These results will provide useful knowledge for developing very-high-resolution arrayed sensors composed of heterogeneous materials.

Acknowledgments

The authors are grateful to Dr. Akihiro Ikeda, Dr. Ryo Takigawa and Mr. Takayuki Takao for their discussion and technical help. They are also thankful to Dr. Mutsuo Ogura and Dr. Katsuhiko Nishida of IR Spec, Inc. for their discussion. This work was partly supported by Adaptable and Seamless Technology transfer Program (A-STEP) of JST and the Supporting Industry Program of METI, Japan.

References

1. M. Fendler, C. Davoine, F. Marion, D. S. Patrice, R. Fortunier, and H. Ribot, "A Fluxless and Low-Temperature Flip Chip Process Based on Insertion Technique," IEEE Trans. Components Packag. Technol. Vol. 32, No. 1, (2009), pp. 207-215.

2. S. D. Gunapala, S. V. Bandara, J. K. Liu, J. M. Mumolo, D. Z. Ting, C. J. Hill, J. Nguyen, B. Simolon, J. Woolaway, S. C. Wang, W. Li, P. D. LeVan, and M. Z. Tidrow, "Demonstration of Megapixel Dual-Band QWIP Focal Plane Array," IEEE Journal of Quantum Electronics, Vol. 46, No. 2, (2010), pp. 285-293.

3. S. Wakiyama, H. Ozaki, Y. Nabe, T. Kume, T. Ezaki, and T. Ogawa, "Novel Low-Temperature CoC Interconnection Technology for Multichip LSI (MCL)," Proc. ECTC, 2007, pp. 610-615.

4. N. Watanabe, T. Kojima, and T. Asano "Wafer-level Compliant Bump for Three-Dimensional LSI with High-Density Area Bump Connections," International Electron Devices Meeting Technical Digest (2005) pp. 687-690.

5. N. Watanabe, I. Tsunoda, T. Takao, K. Tanaka, and T. Asano "Fabrication of Back-Side Illuminated Complementary Metal Oxide Semiconductor Image Sensor Using Compliant Bump" Jpn. J. Appl. Phys 49, 04DB01 (2010)

6. N. Watanabe and T. Asano, "Room-temperature Cu–Cu bonding in ambient air achieved by using cone bump," Appl. Phys. Express, vol. 4, 016501 (2011).

7. K. Iwanabe, T. Shuto, K. Noda, S. Nakai and T. Asano, "Room-temperature microjoining using ultrasonic bonding of compliant bump"Proc. LTB-3D, 2012, pp. 167 - 170.

8. L. Qiu, A. Ikeda, K. Noda, S. Nakai, and T. Asano, "Room-temperature Cu microjoining with ultrasonic bonding of cone-shaped bump", Jpn. J. Appl. Phys., Vol. 52, 04CB10 (2013).

9. T. Shuto, K. Iwanabe, L. Qiu, K. Noda, S. Nakai, and T. Asano, "Room-Temperature High-Density Interconnection using Ultrasonic Bonding of Cone Bump for Heterogeneous Integration," Proc. ECTC, 2013, pp. 1141.

10. T. Shuto, K. Iwanabe, M. Ogura, K. Nishida and T. Asano, "Room-Temperature bonding of heterogeneous materials for near-infrared image sensor", Jpn. J. Appl. Phys.Vol. 53, 04EB01 (2014).

11. B. Langenecker, "Effects of Ultrasound on Deformation Characteristics of Metals" IEEE Transactions on Sonics and Ultrasonics, Vol. SU-13,1 (1966).

AUTHOR INDEX

A

Aasmundtveit, Knut	139, 498
Abe, Hidenori	1523
Agarwal, Rahul	590
Agrawal, Akash	2014
Agrawal, Jai P.	1838
Ahn, Billy	2027
Ahn, Byungki	1616
Ahn, Yesul	1361
Aizawa, Mitsuhiro	1166
Akaike, Masakate	1143
Akin, Meriem Ben-Salah	1595
Alatorre, Roseann	215
Alcira, Cecille	47
Allen, Aileen M.	1433
Allen, Craig	878
Almeida, Rodrigo	935
Althoff, Simon	1549
Alvanos, Tyson	452
Alvarado, Rey	100, 925, 1173
Amara, Karima	1198
Aminov, Parvizso	1838
Amirkhany, Amir	560
Anderson, R.	1641
Andreassen, Erik	139
Andry, P.	576, 883, 1372, 1880
Angyal, Matthew	647
Anselm, Martin	119
Anzai, Nobuhiro	829
Aoki, Issei	1776
Aoyagi, Masahiro	62, 1915, 1926, 2003
Araga, Shinji	2194
Arai, Yoshiyuki	913
Arakere, Guruprasad	395
Arfaei, Babak	425, 655, 1769
Arik, Mehmet	209
Arkalgud, Sitaram	862
Armutlulu, Andac	1098
Arnold, Kim	894
Arvin, Charles	1940

Aryasomayajula, Lavanya 348
Asahi, Noboru 913
Asai, Kosuke 186
Asai, Osamu 1885
Asai, Satoshi 1028
Asano, Tanemasa 2303
Aschenbrenner, R. 940
Atsumi, Koichiro 2227
Audoin, M. 1714
Auerswald, E. 1134
Augustine, Anne E. 528
Aung, Kyaw Oo 596
Awatsuji, Yasuhiro 186

B

Baba, Mikio 763
Baba, Shunji 68
Bader, V. 940
Bae, Byeong-Hyun 1635
Bae, Byung-Hyun 1735
Bae, Hyun-Cheol 1154
Bae, In-Tae 1562
Bae, J.-C. 304
Bae, Jangyong 973
Baek, Hyunho 554
Bailey, Chris 1342
Bailey, Susan 894
Baillin, X. 1714
Bajwa, A.A. 1707, 2181
Bakir, Muhannad S. 13
Baks, Christian 1016, 1272, 1955
Ball, Sabrina 2035
Baloglu, Bora 1231, 1401
Balucani, M. 194, 1571, 1842, 2137
Bandarenka, H. 194
Bao, Andy 47
Barros, Isabel 935
Barth, Holly D. 1457
Barwicz, Tymon 179
Bauer, J. 940
Bchir, Omar 1396
Bea, JiChel 636
Becker, K.-F. 940
Beer, Gottfried 1183
Belardini, A. 194
Beleran, John D. 490

Benedetti, A. ... 194
Benedetto, Elizabeth ... 425, 1433
Benjamin, Shuki ... 1021
Bennett, N.S. ... 1517
Berg, J. ... 1071
Berger, Daniel ... 647
Berger, François ... 478
Berghuvud, A. ... 2220
Bernardi, D. ... 1571, 1842
Berry, C.J. ... 1965
Beyene, Wendemagegnehu ... 730
Beyer, G. ... 33, 309, 572, 850, 894
Beyne, E. ... 26, 33, 309, 572, 613, 850, 894
Bezuk, Steve ... 47, 925, 1173, 1396
Bian, Yuan ... 1902
Bieler, Thomas R. ... 697
Björnängen, T. ... 2220
Blair, Justin ... 609
Böck, J. ... 956
Bock, K. ... 1482
Bohm, Johannes ... 1509
Bollmann, D. ... 1482
Bondarenko, V. ... 194
Borgesen, P. ... 371
Bösch, Wolfgang ... 1183
Böttcher, M. ... 625
Bowen, Terry ... 1054
Boyer, Nicolas ... 179
Boyne, Dan ... 1539
Bozack, Michael J. ... 379
Bozack, Mike ... 242
Brady, D. ... 74
Brand, Sebastian ... 850
Brandt, Lutz ... 279, 1652
Braun, T. ... 940
Bravin, Julian ... 888
Breach, Christopher ... 1803
Brofman, Peter ... 255, 1308
Brusberg, Lars ... 1033, 1498, 1759
Budd, R. ... 883
Burggraf, Jürgen ... 888

C

Caffey, Kevin ... 1173
Caggiano, Michael ... 2159
Cai, Jian ... 1378

Calvert, Jeffrey..342
Campos, José..935
Cangellaris, Andreas C..717, 2094
Cao, Andrew..862, 1810
Cao, Liqiang...1116, 1932, 2008, 2131
Cao, Zhihua..464
Caplet, S...1714
Capuz, G...572
Cardoso, Paulo..935
Castagné, Laetitia...1198
Castro, José..935
Cate, S..74, 952
Cha, Seungyong..354
Chada, Arun Reddy..2081
Chahal, Premjeet...775, 2144, 2168
Chainer, Timothy...1955
Chakraborti, Parthasarathi......................................541, 1492
Chan, M...952
Chan, Yan Cheong..1342
Chang, Chih-Wei...512
Chang, David..947
Chang, Gee-Kung...1054
Chang, Hong-Da..947
Chang, Hsiang-Hung..1891
Chang, Hung-Hsein..868, 1628
Chang, Jenny...1562
Chang, Nistec...81
Chang, Pai-Cheng...290
Chang, Po-Chih...1853
Chang, Shih-Chieh..316
Chang, Xin...2200
Chang, Yiu-Hsiang..1853
Chao, Chun-Chieh...868
Chao, Shu-Han..1908
Chao, Yu-Lin..290
Chau, Ellis..1810
Chen, Agassi...1750
Chen, Cheng-Fu...963, 2020
Chen, Chien-Chou...1853
Chen, Chih...1908
Chen, Chunwei...26
Chen, Dong...2299
Chen, Eason...81
Chen, Erh-Hao..1853
Chen, Gang...1080
Chen, George..470

Chen, Guang 566, 748
Chen, Hsien-Wen 1
Chen, Jing.......................................1902, 2099
Chen, Jui-Chin 1853
Chen, K.M. 297
Chen, Kim-Hui 2299
Chen, Kuan-Neng 512
Chen, Kuo-Hua 512
Chen, M.T. 572
Chen, Qi 1992
Chen, Qianwen 1372
Chen, Quan 1992
Chen, Scott 1562
Chen, Shang-Chun 1853
Chen, Shi-Ching 868
Chen, Shih-Ching 1628
Chen, Stephen 1
Chen, Su-Mei 1908
Chen, Tai-Yu 2069
Chen, Xu 717, 1080
Chen, Yan-Heng 1750
Chen, Yan-Siang 1584
Chen, Yen-Chi 1258
Chen, Yi 1584
Chen, Yi-Ling 1529
Chen, Yin-Fa 419
Chen, Youpeng 1488
Chen, Yu 1378
Chen, Yu Hua 360, 1590
Chen, Zhaoqing 723, 2104
Cheng, Hung-Hsiang 2112
Cheng, Lianxi 1335
Cheng, M.D. 572
Cheng, Ren-Shin 290, 1908
Cheng, Xiaojin 1926
Cheng, Yu-Mei 290
Chéramy, S. 906
Cherman, V. 309, 1517
Chia, Pierre 1122
Chiang, Tzu-Hsing 419
Chien, Chun-Hsien 290, 1891
Chien, F.L. 56
Chien, Heng-Chieh 290, 963
Chin, Wai Foo 1790
Chiou, Jin-Chern 512
Chiu, Chi-Tsung 512, 2112

Chiu, Steve 1, 947
Chiu, Ying-Ta 419
Chlieh, Outmane Lemtiri 2293
Cho, Byoungwoo 1361
Cho, Jounghyun 541
Cho, Junghyun 1328
Cho, Sangbeom 1247
Choi, Alan 925
Choi, BaekKyu 2075
Choi, Eun-Kuk 1735
Choi, Hoi Wai 919
Choi, Hyeseon 1004
Choi, Joonyoung 836
Choi, Kwang-Seong 1154
Choi, Won Kyung 596
Choi, Yongwon 1128, 1661, 1755, 2241
Choi, Youjoung 836
Choi, Young Won 1765
Choki, Koji 1148
Chong, Chan Kai 490
Choo, Joung-Hoon 1735
Chou, Bruce C. 1054
Chou, Lei-Chun 512
Chua, S.L. 324
Chuang, Ching-Te 512
Chuang, Shih-Keng 1944
Chung, Ming-Feng 2112
Chung, YoungSuk 582, 1816
Chylak, Bob 1523
Ciarniello, Danilo 1571
Clauberg, Horst 1523
Co, Rey 215, 1389
Cochet, Philippe 20, 523
Collado, Ana 796
Collins, Sian 878
Conti, Fosca 1464
Cook, Jeffery 684
Cordes, Steven A. 1782
Cotts, Eric 655, 690, 1769
Coutier, C. 1714
Cowley, A. 1517
Coyle, Richard 425, 655
Cremaldi, Joseph 255
Crescenzi, Rocco 1571
Croes, Kristof 613
Crouthamel, D. 74

Cui, Tong 100
Cunningham, G. 8
Czurratis, Peter 850

D

Daerhan, Daerhan 759
Dai, Ming-Ji 290
Daily, Derek 1308
Daily, R. 165, 309, 572
Dal Molin, Renzo 1198
Dalal, Mitul 145
Dan, Qiang 1010, 1690, 2245
Dang, B. 576, 883, 1372, 1880
Dang, J. 74
Danilewsky, A.N. 1517
Darveaux, Robert 703
Daubenspeck, T. 1949
Davis, Roy I. 1996
Davis, Taryn 236
De Messemaeker, J. 33, 613
De Vos, J. 309
De Wolf, I. 309, 613, 850, 1517
Defay, E. 1296
Dehag, G. 1714
Dej, Sebastian 590
Demir, Kaya 1098
Demosthenous, Andreas 2213
Derix, Robert 1464
Deshpande, Shantanu 242, 1449, 1973
Desmaris, V. 1071
Detalle, Mikael 33
D'hiver, Philippe 1198
Dieng, K. 1296
Dinan, Thomas 862
Ding, L. 1212
Ding, Wen 2236
Djuric, Tatjana 850
Doany, Fuad E. 1016
Dobashi, Masahiro 763
Dobritz, Stephan 873
Donaldson, Nick 2213
Dong, Jianwei 342
Dong, Mingzhi 1192
Dornala, Kalyan 85
Dragoi, Viorel 888
Dressler, C. 1714

Drewniak, James L. 2081
Drost, A. 1482
Du, Ellen 748
Duffy, Daniel 1803
Durfee, Loren 236
Dzarnoski, John 157

E

Eastep, Brian 279
Eggen, Trym 498
Ehrhardt, Christian 1321
Eichstadt, David 1803
Eisenstadt, William R. 554
Elger, Gordon 1464
Ellis, Charles D. 1086
Elmer, John W. 1457
Elsherbini, Adel A. 1610
Emerich, S. 74, 952
Endo, Yoshinori 1523
Engelmann, Sebastian 179
Eom, Yong-Sung 1154
Esfahani, Zahra Kolahdouz 1556
Eto, Michiyuki 2003

F

Fan, Chia-Wen 1908
Fan, Chuanhong 1885
Fan, Jun 2081
Fan, Xuejun 967
Fang, Runiu 641, 1902, 1986, 2099
Fanget, S. 1714
Farahmand, Farid 2159
Farcy, A. 1296
Farrugia, Mark-Luke 411
Fay, Owen 2014
Feng, June 748
Feng, Wei 1915
Fiedler, C. 625
Fischer, Thorsten 1421
Fischer, U.H.P. 1862
Flack, Warren 26
Fortier, Paul 179
Franiatte, R. 906
Frank, Torrey W. 1610
Frye, Robert 1284, 2027
Fu, Hailuo 1652

Fu, Huan-Chun **290, 1891**

Fu, Shancan **1080**

Fu, Xianzhu **464**

Fu, Xingming **170, 1189**

Fu, Yifeng **459**

Fujii, Atsushi **829**

Fujimaru, Koichi **913**

Fujimoto, Koji **1853**

Fujino, Masahisa **1504**

Fujita, Mitsuru **829**

Fukuoka, Yoshitaka **1673**

Fukushima, T. **304, 636, 856, 1110, 1148**

Fukuzono, Kenji **68**

Furuya, Ryosuke **1885**

G

Gaherty, Lee **279**

Gandhi, Saumya **541, 1492**

Gao, Guilian **862, 1810**

Garant, John **452**

Gardner, Donald S. **1290**

Garnier, A. **906**

Gaynes, Michael **255, 1372, 1955**

Ge, Y.P. **1212**

Ge, Yun **684**

Gebara, Edward **2293**

Gelorme, Jeffrey **1308**

Georgakopoulos, Stavros V. **759**

Georgiadis, Apostolos **796**

Ghaffari, Roozbeh **145**

Gharaibeh, Mohammad **119**

Giagka, Vasiliki **2213**

Gieser, H. **1482**

Gilham, David **504**

Gill, Harpreet **354**

Gissila, T. **2220**

Glodde, M. **883**

Goldsmith, Charles **1940**

Golick, L. **952**

Gonzalez, M. **33, 309, 1517**

Goodwin, S. **8**

Gottfried, Knut **1218**

Goumans, Leon **411**

Goyal, Deepak **1457**

Grafe, Juergen **873**

Grams, A. **625**

Graves-Abe, Troy 647
Greco, F. 1714
Gregory, C. 8
Greve, Hannes 1314
Grosinger, Jasmin 1183
Grymyr, Ole Johannes 139
Gu, Ping 464
Gu, Sam 609
Gu, Xiaoxiong 548, 1272
Gu, Yingke 1378
Guan, Yong 1902
Guerin, Luc 647
Guerrero, Alice 894
Guevara, Gabe 215
Guidotti, Daniel 1932
Guiller, O. 1296
Gundurao, Anil 354
Günther, Wolfgang 1218
Guo, F.M. 1212
Guo, Maofeng 2236
Guo, W. 309
Guo, Xing 1010, 2283
Gurrum, Siva P. 821
Guthrie, William 647
Guven, Ibrahim 2035

H

Hagelauer, A. 956
Hahm, Yeon-Chang 730
Hale, Cassandra 236
Halonen, Eerik 151
Halvorsen, Per Steinar 139
Hamasha, S. 371
Hamilton, Michael C. 441, 1086
Han, C.J. 1641
Han, Gyuwan 1361
Han, KiJin 2124
Han, Kyu 782
Han, Michael 47
Han, Minghui 560
Han, SeungChul 582, 1816
Han, Sungwon 973
Hanna, Carlton 1278
Hao, Jifa 1241
Harada, Takeshi 1857
Harel, Stephane 179

Hartner, W.	956
Hasharoni, Kobi	1021
Hashiguchi, H.	856, 1110
Hashimoto, H.	304, 636
Hashimoto, Masakazu	1955
Hashimoto, Tomoaki	596
Hasnine, Mohammad	379
Haupt, M.	1862
Hau-Riege, Christine	1173
He, Hongwen	1116
He, Huanyu	548
He, Yi	2008, 2131
Heinrich, T.	108
Hell, W.	1482
Hemsel, Tobias	1549
Henriques, Vitor	935
Herbold, Christian	203
Herbst, Christian	1033
Hernandez, George A.	1086
Heuer, Henning	1509
Heumann, Wolfgang	1595
Higashi, Mitsutoshi	1166, 1366
Higgins, Leo	1539
Hill, Michael J.	528
Hilton, A.	8
Hiner, Dave	590
Hirai, Hiroyuki	1729
Ho, Heidi S.Y.	1965
Ho, Lung-Hua	316
Ho, Paul S.	1122
Hoff, Lars	139
Hoffrogge, Peter	850
Hoivik, Nils	139
Hölck, O.	625, 1134
Höll, S.	1862
Holmes, Pat	47
Holweg, Gerald	1183
Hong, J.B.	1533
Hong, Shengping	1622
Honrao, Chinmay	1160
Horibe, A.	803
Hoshino, Manabu	1955
Hoshiyama, M.	803
Hou, Fei	1690, 2048
Hou, Fengze	2008
Howell, Keith	425

Hsiao, Zhi-Cheng	290
Hsieh, Robert	26
Hsieh, Wan-Lin	1908
Hsu, Chao-Kai	1891
Hsu, Chih-Chung	1258
Hsu, H.S.	947
Hsu, Yung-Yu	145
Hu, Dyi-Chung	360, 1590
Hu, Hao	759
Hu, Je-Ping	1944
Hu, Y.H.	572
Hu, Yating	1189
Huang, Chen-Yu	1
Huang, Cui	1722
Huang, Hsiao-Chun	868, 947
Huang, Joseph	1533
Huang, Kuo-Hsin	620
Huang, Louie	419
Huang, Po-Tsang	512
Huang, Rui	1122
Huang, Shin-Yi	1908
Huang, Timothy	2266
Huang, Ting-Chia	1160
Huang, Yaping	2236
Huang, Yu-Wei	290, 1891, 1908
Huffman, A.	8, 20
Hui, Ho-Yee	1492
Hummler, Klaus	1648
Hunegnaw, Sara	1652
Hung, Chih-Pin	2112
Hung, Yin-Po	360
Huppert, Gil	145
Hutter, Matthias	1321
Hutzler, Aaron	1700
Huylenbroeck, S.V.	572
Hwang, Lih-Tyng	1303
Hwang, Seung Min	74
Hwang, Wei	512
Hwang, Yuchul	973

I

Iijima, Yu	452
Ikari, Gary	1366
Im, Jay	1122
Imai, H.	899
Imajyo, Nobuhiko	2271

Imanari, Masaaki 342
Imenes, Kristin 139, 498
Imura, Fumito 1915
Indyk, Richard 452
Inomata, Katsumi 1796
Iruvanti, Sushumna 236, 1253
Ishida, Hiroya 404
Ishigure, Takaaki 1042
Ishiguro, Kazuya 1308
Ishino, H. 1179
Islam, Md. R. 1272
Islam, Nokibul 50, 836
Ito, Yuka 1148
Itoh, Toshihiro 1857
Ivankovic, A. 309, 1517
Iwanabe, Keiichiro 2303
Iwata, Yoshiharu 2227
Iyer, Subramanian 647
Izzi, M. 194, 1842

J

Jang, BoRa 582, 1816
Jang, Myong-Gi 447
Jao, Pitfee 789
Jeong, Min-Su 1735
Jeong, Yonghyuk 836
Jeong, Youchul 753
Ji, Jason 1750
Ji, Liang 1116
Ji, Yu 1488, 2299
Jiang, D.S. 1533
Jiang, Feng 1740
Jiang, Hanqing 470
Jiang, Tengfei 1122
Jimbo, Toshihiko 1955
Jin, Howard (Hwa II) 1790
Jin, Yufeng 641, 1902, 1986, 2099
Jin, Zhenrong 535
Jing, Xiangmeng 1116, 1740
Jinka, Oblesh 1782
Jo, Chanmin 2075
Joblot, S. 1296
Johansson, A. 1071
Johansson, Susie 157
John, P. 625
Johnson, Christopher 1577

Johnson, Mark 2014
Johnson, Michael 1965
Jomaa, Houssam 1396
Jonah, Olutola 759
Joshi, Gaurang 119
Joshi, Yogendra 1247
Jourdain, Anne 894
Jouve, A. 906
Juang, Jing-Ye 1908
Jung, Dae Young 1562
Jung, Seong-Yoon 2255
Jürgensen, Nils 1498
Juskey, Frank 1264

K

Kabir, M.S. 1071
Kahle, R. 940
Kaletta, K. 1204
Kamgaing, Telesphor 1264, 1610
Kamlapurkar, Swetha 179
Kanagavel, Senthil 1790
Kandaswamy, Shri Vishnu 1464
Kang, Dongchul 1523
Kang, Kuiwon 1396
Kang, Sung K. 1782
Kang, SungGeun 590
Kang, Un Byung 1128, 1755
Kannan, Sukeshwar 590
Kao, C.R. 2277
Kao, Ming-Jer 290, 1853
Kao, Nicholas 1750
Karikalan, Sam 2119
Karlicek, Jr., Robert F. 2207
Karsli, Kivanc 209
Kashyap, Anirudh 279
Kata, Keiichirou 596
Katkar, Rajesh 1389
Kato, Fumiki 62, 1926
Katoh, Y. 1179
Kaur, Amanpreet 2144, 2168
Kaushik, B.K. 1091, 1921
Kawamoto, S. 803
Kawanami, Satoshi 186
Kawano, Kenji 2003
Kaynak, M. 1204
Ke, Chang-Bo 2249

Keech, John...20

Keezer, David 2260, 2287

Kenyon, Gareth...26

Keser, Beth 100, 925, 1173

Khan, Wasif T. 2293

Khim, JooHyun 582

Kholostov, K. 194, 1571, 1842

Ki, Wing-Hung 919

Kida, Tsuyoshi 596

Kijkanjanapaiboon, Kasemsak..............................967

Kikuchi, Katsuya 62, 1915, 1926

Kim, Cheolbok 1103

Kim, Cheolgyu 1004

Kim, Chimyung 1616

Kim, Chin Kwan 1396

Kim, Choong-Un 133, 697

Kim, Chunho 263

Kim, DoHyeong...............................582, 590, 1816

Kim, Dong Wook 609

Kim, Dongsu 41

Kim, Ga Won 354

Kim, Gwang 50, 2027

Kim, Hakgu 1616

Kim, Henry 2119

Kim, Hui Joong 712

Kim, HyunTai 2027

Kim, Hyup Jong 1103

Kim, Il...1661

Kim, Jaemin 753

Kim, Jinseong 1361

Kim, JooHyun 1816

Kim, Joungho 541

Kim, Jun Chul 41

Kim, June-Bum 1635

Kim, Kwonil 279

Kim, KyungOe...50

Kim, Min Sung 1004

Kim, Namhoon 2119

Kim, Nam-Seog 1122, 1635

Kim, Samuel 1471

Kim, Seung-Ho 841, 2241

Kim, Seung-Hyun...1635

Kim, Sung Jin..1264, 1384

Kim, Sungjoo 2075

Kim, Tae Wan 1060

Kim, Taek-Soo 1004

Kim, Taewan 271, 1616

Kim, YongSang 1816

Kim, Yoosun 841, 2241

Kim, Young Soon 1128, 1661, 1755, 1765

Kim, Young Suk 1853

Kim, Younghoon 354

Kimura, Kazushi 230

Kimura, Michitaka 596

King, A. 601

Kino, H. 856, 1110

Kintaka, Kenji 186

Kitada, Hideki 1853

Klink, G. 1482

Klyshko, A. 194, 1571

Knickerbocker, J. 576, 647, 883, 1372, 1880

Ko, Cheng-Ta 1891

Ko, Wen-Ching 1944

Kobayashi, Makoto 284, 365, 484, 742, 1160, 1384, 1959

Kobayashi, Yuta 913

Kodama, Shoichi 1853

Kodani, K. 1077

Kodemura, Junji 1955

Kohara, S. 647, 803

Koide, Masateru 68

Koiwa, Sumio 1673

Kono, Tsutomu 2042

Kotake, Tomohiko 1407

Koyama, Toshinori 348

Koyanagi, M. 304, 636, 856, 1110, 1148

Kraft, Silke 1700

Krieger, William E.R. 983

Krüger, Michael 114

Ku, Tzu-Kun 1853

Kubo, A. 899

Kuchta, Daniel M. 1016

Kumar, Gokul 541

Kumar, Santosh 712

Kumar, Vobulapuram Ramesh 1091

Kumari, Archana 1921

Kumpulainen, Tero 1685

Kunimoto, Yuji 348

Kuo, An-Yu 1303

Kuo, C.L. 297

Kuo, Chih-Ming 316

Kuo, H.J. 572

Kuo, Kuei Hsiao (Frank) 56

Kuramochi, Satoru 1673
Kurihara, Takashi. 348
Kurokawa, Tsuyoshi 1776
Kutlu, Zafer 348
Kutter, C. 1482
Kwark, Young 1955
Kwatra, Abhishek 983

L

La Manna, A. 33, 309
Lachner, R. 956
Laflamme, Simon 179
Lagae, Liesbet 165
Lai, Chiming 2299
Lai, J.Y. 1
Lai, Yi-Che 947, 1750
Lall, Pradeep 85, 242, 379, 666, 990, 1449, 1973, 2053
LaManna, A. 572
Lambert, William J. 528
Lamy, Y. 1296
Lan, Albert 81, 1533
Lan, Jia-Shen 1235
Landesberger, C. 1482
Lang, K.-D. 114, 625, 873, 940, 1033, 1204, 1218, 1321, 1421, 1498, 1759
Langlois, Richard 236, 647
Lannon, J. 8, 1641
Larson, Lyndon 236
Lau, John 56, 290, 1908
Lau, Kei May 919
Lauro, Paul 1308
LaVoie, Annique 236
Law, Edward 518
Lazerand, Thierry 1577
Le, Fuliang 919
Le, Taoran 769
Lee, Bong-Sub 862, 1810
Lee, Byeong Rok 1635, 1735
Lee, Chang-Chi 2112
Lee, Chin C. 1335, 1529
Lee, Ching-Kuan 290, 1891
Lee, ChoonHeung 582, 1361, 1816
Lee, Chu-Chung (Stephen) 1539
Lee, Chung-Fa 2069
Lee, DongHoon 1816
Lee, DongHun 582
Lee, Fred C. 504

Lee, Haksun	1154
Lee, Heng	815
Lee, Inho	342
Lee, Jae Hong	712
Lee, Jason	56
Lee, K.W.	304, 856
Lee, Kangwook	636, 1110, 1148
Lee, Kenny	47
Lee, KiWook	590
Lee, Minkyu	1616
Lee, Ning-Cheng	655
Lee, Rick	56
Lee, S.W. Ricky	919
Lee, Sang Hoon	271, 1060
Lee, Sangmin	2075
Lee, Seungbae	354
Lee, Seungyong	1616
Lee, Shih-Wei	512
Lee, Taeik	1004
Lee, Tae-Kyu	133, 697
Lee, Yil-Hak	342
Lee, YongTaek	2027
Lee, Young Woo	712
Lee, Yuan-Chang	290
Lee, Yuh-Lang	1584
Levine, Lyle E.	1648
Lewandowski, Eric	255
Leyrer, Benjamin	203
Li, Bin	2249
Li, C.	1641
Li, Guangfeng	1231
Li, Heng	1411
Li, Jun	2131
Li, K.H.	324
Li, Lang	1876
Li, Leo Hongyu	2119
Li, Li	1122, 1366
Li, Liyi	631, 1492, 1745, 1848
Li, Long	1690
Li, Menglu	609
Li, Qiang	504
Li, Shidong	1253
Li, Xin	1080
Li, Y.Q.	1212
Li, Yan	1457
Li, Yuan	338

Li, Yuefa...170
Li, Zhe...338
Li, Zhuo...1745, 1828
Liang, Hanshuang...470
Liang, Jiang-Long...1584
Liao, Anmou...2131
Liao, Jash...1750
Liao, Li-Ling...290
Liao, Mark...81
Liao, Sue-Chen...1853
Libsch, Frank...1016
Lii, M.J....572
Liimatta, Toni...151
Lin, C.F....297
Lin, Cha-Hsin...1853
Lin, Chun-Tang...947
Lin, Chun-Te...1891
Lin, Chun-Ting...1590
Lin, Edward...748
Lin, Frank M.-S....620
Lin, Kung-An...316
Lin, Kwang-Lung...419, 1584
Lin, M.J....297
Lin, Puru...1590
Lin, Wei...647, 1401
Lin, Y.C....297
Lin, Y.J....952
Lin, Yang-Kai...1258
Lin, Yaojian...931
Lin, Yi-Fan...1944
Lin, Yu-Min...1891, 1908
Lin, Ziyin...447, 769, 1828, 2231
Lindner, Paul...888
Ling, Feng...2094
Liou, Chang-Ho...1944
Liu, C.S....572
Liu, Changqing...1348
Liu, Chaojun...1010, 1189
Liu, Cheng-Hsiang...868, 1628
Liu, Chengxun...165
Liu, Duixian...1272
Liu, Fengman...2008, 2131
Liu, Fuhan...1384
Liu, Hsichang...647
Liu, Hung-Wen...1750
Liu, Johan...459

Liu, Kai 1284, 2027
Liu, Kenny 947
Liu, Li 1348
Liu, Sheng 170, 1010, 1189, 1690, 1876, 2048, 2245, 2283
Liu, X.Y. 1212
Liu, Y. 1880
Liu, Yan 447, 2231
Liu, Yingxia 609
Liu, Yi-Wei 1750
Liu, Yong 808, 1241, 1601, 2063
Liu, Yumin 808, 1601
Liu, Zhiming 1652
Lo, Wei-Chung 290, 360, 1891
Lofrano, M. 309
Loh, Chooi Ian 748
Longgood, Stuart 425
Longworth, Hai 236
Lopez-Montesinos, Pedro 342
Lu, Chang-Lun 868, 947, 1628
Lu, Guo-Quan 1080
Lu, Hao 742, 1416
Lu, Hsin-Chia 1944
Lu, Hua 1342
Lu, James J.-Q. 2207
Lu, Jian 1857
Lu, Jian-Qiang 548
Lu, Minhua 690, 1940
Lu, PingHung 26
Lu, Terren 1
Lu, Wengao 2099
Lu, Yongqiang 878
Lu, Yuan 1354, 2008
Lu, Yung Jean (Rachel) 1891
Luckenbach, Thomas 1606
Lueck, M. 8, 20
Luesebrink, Helge 1021
Luo, L. 815, 1212, 1411
Luo, Xiaobing 170, 1189, 1876, 1992
Luo, Yihua 242
Luo, Yu 1782
Luo, Zhang 1690, 2048, 2245
Lv, Cheng 470, 1290
Lwo, Ben-Je 620

M

Ma, Shenglin 1902
Ma, Teng 470
Mackenzie, Kenneth D. 1577
Madenci, Erdogan 973
Maeda, Nobuhide 1853
Maeda, Ryutaro 1857
Maekawa, Katsuhiro 2194
Magaya, Tafadzwa 1652
Maikowske, Stefan 203
Majeed, Bivragh 165
Majumder, M.K. 1921
Maldonado, Alberto 2159
Mallampati, Sandeep 1328
Malta, D. 8
Maman, Avi 1021
Manhas, S.K. 1921
Maniatty, Antoinette 1241
Manier, C.-A. 1204
Mäntysalo, Matti 151, 1685
Mao, Cindy 56
Marcoux, Phil 1071
Maria, Joana 1372
Mariappan, Murugesan 636
Martin, Bill 2124
Martinez, Linnell 1577
Marzano, Frank Silvio 2137
Mason, Maurice 1782
Mathewson, Alan 1064
Matsubara, Takahiro 1028
Matsumae, Takashi 1504
Matsumoto, Rie 1870
Matsushita, Kiyoto 404
Matsuura, Ikkei 1955
Matthias, Thorsten 888
Mauer, Laura B. 878
Mavinkurve, Amar 411
Mawer, Andrew 1539
Mayer, Michael 1471
McCann, Scott R. 2189
McCarson, Brian 2035
McCleary, Roger 523
McCluskey, F. Patrick 1314
Mclaughlin, Kevin 878
McLeod, Mark 1308

McMullen, Tom 126
McNally, P.J. 1517
Mehr, M. Yazdan 1477
Mehta, Gaurav 490
Mei, Yunhui 1080
Meindl, Manfred 1183
Meinecke, Christoph 1218
Melville, Robert 1284
Mendoza, Nestor 1523
Menezes, Gary 1959
Mesh, Michael 1021
Miao, Min 641, 1902, 1986, 2099
Middendorf, Andreas 114
Milanes II, Ninoy 490
Miller, A. 33
Miller, Allen 1033
Miller, Andy 26, 894
Milton, Basil 1523
Min, Max (Sungwan) 354
Minami, Shigenobu 1729
Mino, Masayuki 2042
Mirza, Kazi 990
Misra, E. 1949
Mita, Mamoru 2194
Mitachi, Seiko 230
Miura, Hideo 1885
Miura, Testunosuke 186
Miyamoto, Yoshinori 913
Miyatake, Masato 1407
Miyazaki, Tomokazu 165
Mizuno, Hikaru 1796
Mizusawa, Ryuma 1870
Mizutani, Daisuke 68
Moeini, S. Ali 1314
Mok, Philip K.T. 919
Möller, Eike 1707
Moon, Jeong Tak 712
Moon, Kyoung-Sik 447, 1828, 2231
Moon, Seongjae 2075
Morales, Jorge Mario Herrera 478
Morey, Briana 145
Mori, H. 803
Morikawa, Y. 846, 1697
Morris, Jeffrey 221
Mu, Mingkai 504
Mukai, Kenichiroh 279

Mullen, Don 730
Murai, Hikari 1407
Murayama, Kei 1166
Murayama, T. 846, 1697
Murtuza, Masood 242, 1973
Murugesan, M. 304, 1148
Mustafa, Muhannad 666
Mutnury, Bhyrav 2081
Mutuku, Francis 425, 655, 1769

N

Na, Duk Ju 596
Na, Nanju 2151
Nad, Suddhasattwa 684
Nagao, S. 1077, 1179
Nah, Jae-Woong 1308, 1372
Nair, Vijay 782, 1264, 1278
Nakagawa, Hiroshi 1915
Nakamoto, Mark 1226
Nakamura, Shinya 1523
Nakamuta, Y. 1697
Nakazuru, Kazumi 1028
Nam, Kiyoung 1616
Natarajan, Arun 1272
Nauchi, Takashi 1308
Ndip, Ivan 1498
Neitz, Marcel 1759
Nemoto, Shunsuke 1915
Nenzi, Paolo 1571, 2137
Neo, Chong-Wei 518
Neuhaus, Jan 1549
Newman, Keith 1433
Nguyen, Anh Tuan Thai 139
Nguyen, Hoa 470
Nguyen, Hoang-Vu 498
Nguyen, Luu 242, 1973
Ni, Chih-Hsien 316
Ni, Jiamin 1241
Nicholls, Lou 1361
Nicolas, S. 1714
Niittynen, Juha 151
Niizeki, Shoichi 913
Nimura, Masatsugu 913
Ning, Wenguo 815, 1411
Niotaki, Kyriaki 796
Nishio, Kenzo 186

Nishizono, Shinji 763
Niwa, Hiroyuki 913
Nolmans, P. 33
Nonaka, Toshihisa 913
Noriki, A. 856
Numata, Hidetoshi 179, 1870
Nuss, M. 625

O

Obeng, Yaw 1648
Ochiai, Toshihiko 596
Ochoa, Juan S. 717
O'Connell, Barry 1241
Oh, Dan 566, 748, 1896
Oh, Tac-Keun 1635
O'Halloran, G.M. 411
Ohba, Takayuki 1853
Oi, Kiyoshi 348
Okada, Hironao 1857
Okamoto, Kazuya 2227
Okamoto, Keishiro 68
Okamoto, Kenji 1796
Okoro, Chukwudi 1648
Onishi, M. 304
Ooshima, Nobuhiro 1729
Oppermann, H. 1204, 1321
Orii, Y. 803, 1308
Osaka, Masahiko 1523
Ose, Masahisa 1407
Osenbach, J. 74, 952
Oster, Sasha N. 1610
Ostrowicki, Gregory T. 821
Otaka, S. 348, 899
Oterkus, Erkan 973
Oterkus, Selda 973
O'Toole, Eoin 935
Ou, Jack 2159
Ouyang, Eric 836
Oya, S. 899

P

Pacheco, Mario 1457
Pachler, Walther 1183
Paek, JongSik 590
Pagliucci, Roberto 1571
Paik, Kyung-Wook 271, 841, 1060, 1128, 1661, 1755, 1765, 2241, 2255

Pakbaz, Faraydon ... 535
Palma, Fabrizio ... 2137
Pan, Jie ... 2163
Pan, Po-Chih ... 2112
Pan, Shiji ... 2088
Pan, Yi ... 332, 1253
Pang, Cheng ... 2163
Papakyrikos, Cole ... 145
Papapolymerou, John ... 2293
Parat, Guy ... 1198
Park, Dongjoo ... 1361
Park, John ... 1667
Park, JungSoo ... 582, 1816
Park, Kyoung Youl ... 775
Park, S.W ... 1077, 1179
Park, Susan ... 2027
Park, Yongsun ... 1616
Park, Young-Bae ... 1635, 1735
Parker, Ben ... 1272
Parker, Richard ... 425
Parkinson, Dilworth Y. ... 609
Parkinson, Dula ... 1457
Parks, Gregory ... 690
Patnaik, Amalendu ... 1091
Paul, Jens ... 590
Pavlidis, Spyridon ... 2293
Pays-Volard, David ... 1577
Pedreira, Olalla Varela ... 613
Pei, Min ... 395, 684
Peng, Jr-Wei ... 1584
Peng, Shih-Liang ... 1
Perfecto, Eric ... 647, 690, 1308, 1606, 1940
Pesika, Noshir ... 255
Peterson, Andrew F. ... 2271
Peterson, K. ... 1641
Pham, Nam ... 535
Philipsen, Harold ... 613, 850
Phommahaxay, Alain ... 850, 894
Pinho, Nelson ... 935
Pitarresi, James ... 119
Pitwon, Richard ... 1033
Plante, David ... 236
Pleyer, Wolfgang ... 1021
Polastre, R. ... 883, 1880
Pollard, Scott ... 20
Prange, Jonathan ... 342

Pressel, Klaus 1183
Prewitz, T. 625
Privett, Mark 894
Prorok, Barton C. 379
Pucha, Raghuram 1098
Pufall, R. 1134

Q

Qasaimeh, A. 371
Qian, Jiangyuan 2088
Qian, Richard 2063
Qin, Hong-Bo 2249
Qin, Ivy 1523
Qin, Y. 342, 2181
Qin, Zheng 2163
Qiu, Delong 1932
Qiu, Jinghao 1690
Qu, Shichun 808
Quay, R. 2181
Queisser, Marco 1759
Questad, D. 332, 1253, 1949

R

Radhakrishnan, Kaladhar 528
Radojcic, Riko 1226
Raghavan, Sathyanarayanan 983
Rahimi, Arian 736, 789
Raj, Milan 145
Raj, P. Markondeya 284, 484, 541, 782, 1160, 1264, 1384, 1492, 1959, 2266
Rajendran, H. 1134
Rajoo, Ranjan 490
Ralph, W. Carter 1433
Ramachandran, Koushik 647
Ramachandran, Vidhya 1226
Ramm, P. 1134
Ranjan, Manish 26
Rao, Valluri R. 1610
Rastjagaev, Eugen 1707
Rastogi, Rajiv 2035
Ratel, David 478
Razdan, Sandeep 1054
Razeeb, Kafil M. 1064
Razzaq, A. 324
Rebibis, K.J. 572
Rebibis, Kenneth 894
Reiner, R. 2181

Ren, Xiaoli 2163
Reynolds, Scott K. 1272
Rimolo-Donadio, Renato 1955
Rissing, Lutz 1595
Roa, Fernando 1656
Robertazzi, Raphael 1372
Roberts, Jordan C. 666
Röder, Julia 1033
Rodriguez, Daniel 2151
Rogoff, Rich 523
Romankiw, Lubomyr T. 1782
Rong, Aosheng 2094
Rongen, René 411
Rosenthal, Christopher 452
Rouhana, Layal 1396
Roy, Rajiv 523
Ruhmer, Klaus 20, 523
Rylyakov, Alexander V. 1016
Ryu, Woonghwan 354, 2075

S

Sabuncuoglu, Deniz 165
Saeidi, Nooshin 2213
Saephan, Chio 2159
Sakai, Taiji 742
Sakalaukus, Peter 2053
Sakamoto, S. 1077
Sakuishi, T. 846, 1697
Sakuma, Katsuyuki 647, 1372
Sakurai, Tomohiko 1796
Sakuyama, Seiki 68
Saleem, A.M. 1071
Sammakia, Bahgat 2175
Samoilov, Arkadii 1679
Sandhu, Javed 518
Santagata, Fabio 1192
Sasaki, Hideki 763
Satio, Yukihiro 2227
Sato, Osamu 452
Sato, Ryohei 2227
Sato, Y. 304
Sato, Yasuhiro 1729
Sato, Yoichiro 365, 1247, 2271
Sato, Yutaka 636
Sawyer, Brett 742, 1416
Scheuermann, Michael 1372

Schletz, Andreas 1700
Schmitz, D. 371
Schmitz, Stefan 114
Schneider, Marc 203
Schoeller, Harry 1328
Schow, Clint L. 1016
Schröder, Henning 1033, 1759
Schultz, Mark 1016
Schutt-Ainé, José E. 717
Schwarz, Mark 100, 925
Secker, Dave 730
Seki, Toshitake 365
Seler, E. 956
Senior, David E. 789, 1103
Seo, Hochul 1616
Seo, SeongMin 582, 1816
Seo, Sun Kyung 1128, 1755
Seo, YoungChul 582
Serenelli, L. 1842
Sextro, Walter 1549
Shaba, Hala 215
Shaddock, David 1328
Shafiee, S. 1071
Shah, Milind 1396
Shan, Lei 1955
Shang, Jintang 1488, 1833, 2299
Shang, Wenya 2163
Shangguan, Dongkai 1116, 1740, 2008, 2131, 2163
Sharifi, Reza 2119
Sharma, Himani 782, 1492, 2266
Shen, Hong 862
Shen, J.H. 1212
Shen, Wen-Wei 1891
Shi, Shawn 263
Shibuya, Hiroki 763
Shigemoto, Takumi 2227
Shih, W.L. 2277
Shim, Yujeong 1896
Shimizu, Kozo 68
Shimizu, Noriyoshi 348
Shimizu, Tadashi 763
Shin, Jaemin 2075
Shin, Jiwon 841, 1128, 1661, 1755, 1765
Shiraishi, Takashi 1048
Shirangi, M.H. 108
Shirazi, S. 371

Shlepnev, Yuriy	730
Shorey, Aric	20, 1103
Shoukrun, Aki	1606
Shuto, Takanori	2303
Sibilia, C.	194
Sikka, Kamal K.	332
Sillanpää, Hannu	151
Simon, Gilles	478, 1198
Simons, V.	309
Sin, Johnny K.O.	919
Sinha, Pankaj	2035
Sitaraman, Srikrishna	1264
Sitaraman, Suresh K.	983, 2189
Skidmore, G.	1641
Skordas, Spyridon	647
Slabbekoorn, John	26
Sleeper, Scott	263
Smet, Vanessa	284, 365, 484, 742, 1054, 1160, 1384, 1959
Smith, Dan	590
Snyder, S.	601
Soler, Vilmarie	1606
Soltani, Ayat	1685
Soma, Kazutomo	1042
Son, Ho-Young	1122, 1635
Son, Yong	590
Sone, H.	803
Song, Chongshen	1116
Sorce, Peter	1308
Souchon, F.	1714
Souriau, Jean-Charles	478, 1198
Spreitzer, Ronald	1278
Sprenger, Daniel J.	1996
Steffek, Roland	1021
Steller, Wolfram	1218
Sten-Nilsen, Bjørnar	498
Stepanov, Stanislav	1021
Stewart, Aaron	119
Stömmer, Christian	1021
Stratton, Ken	1401
Strothmann, Tom	931
Struyf, Herbert	850
Su, Meiying	1116
Su, Peng	1122
Su, Quang	119
Su, Yipeng	504
Suaoke, Kuniaki	1870, 647

Suga, Tadatomo 1143, 1504

Sugahara, T. 1077, 1179

Suganuma, Daisuke 1042

Suganuma, K. 1077, 1179

Sugase, Naoki 452

Sugawara, Mariko 1048

Sugawara, Yohei 1110

Sugiura, K. 1179

Suh, Min-Suk 1122, 1635

Suhling, Jeff 379, 666, 990, 1449

Suk, Kyung-Lim 1661

Sun, Jibin 1054

Sun, Minggang 641, 1986

Sun, Rong 464

Sun, Xiaofeng 1354, 1378

Sun, Xin 641, 1902, 1986, 2099

Sun, Yangyang 47

Sun, Zhuowen 862

Sundaram, Venkatesh 1098, 2189

Sundaram, Venky 284, 365, 541, 742, 1054, 1247, 1264
.... 1384, 1416, 1427, 1959, 2266, 2271

Sung, Baegin 753

Suthau, Eike 173

Suthiwongshunthorn, Nathapong 490

Suu, K. 846, 1697

Suzuki, A. 1697

Suzuki, Hiroko 763

Suzuki, Ken 1885

Suzuki, Kousuke 1673

Suzuki, Naoya 2042

Suzuki, Ryoichi 1308

Suzuki, Takuji 1729

Suzuki, Yuya 742, 1264, 1416, 1427

Swaminathan, Madhavan 782, 2124, 2260, 2287

Swan, Johanna 1278

Sweatman, Keith 425, 655

Syed, Ahmer 100, 1173

Sze, Henry 518

T

Taddei, John 878

Tai, Rui-Feng 947

Tain, Ra-Min 290, 360

Taira, Yoichi 179, 1870

Takagi, Yutaka 365, 742, 1416, 1427

Takaguchi, Akira 1308

Takahashi, Hiroshi	2042
Takahashi, Naoki	1028
Takahashi, Shintaro	2271
Takanezawa, Shin	1407
Takano, Akihito	1366
Takasugi, Hiroshi	1776
Takegami, Toshifumi	913
Takekoshi, Masaaki	1407
Takenobu, Shotaro	179
Takizawa, Hideo	452
Tamura, K.	899
Tan, C.S.	324
Tan, Keith	518
Tanaka, Kazuhiro	1048
Tanaka, Masato	348
Tanaka, T.	304, 636, 856, 1110, 1148
Taner, Ozgur	967
Tang, Rui	470
Tang, Yin	236
Tanikawa, Seiya	1110
Tao, Jing	1064
Taylor, Robin	1416
Telikepalli, Satyanarayana	2260, 2287
Temple, D.	8, 1641
Tentzeris, M.M.	759, 769
Terajima, Katsushi	763
Teraki, Shin	1776
Thangaraju, Sara	590
Thomas, T.	940
Thomas, Windsor	1103
Thompson, J.	601
Tian, Shurong	1016
Tillack, B.	1204
Tjulkins, Fjodors	139
Tkachenko, Anton	2207
Toepper, Michael	1421
Tokarski, Adam	1700
Tong, Ho-Ming	512
Tong, Jialing	1098, 2271
Töpper, Michael	1498
Toriyama, Kazushige	1308
Toukhy, Medhat	26
Trampert, Stefan	114
Tran, TuAnh	1539
Tsai, Jensen	1533
Tsai, Jyun-Ling	868, 1628

Tsai, Wen-Li..290
Tsang, C. 576, 883, 1372
Tsang, Leung...2200
Tsebo, Simo G...108
Tseng, Stephen 1
Tsukuda, Tatsuaki..763
Tsunoda, Masatoshi...1028
Tsuruta, K. 1179
Tu, Chia-Jung..316
Tu, K.N..609
Tuan, Chia-Chi..447, 2231
Tucci, M. 194, 1842
Tummala, Rao 284, 365, 484, 541, 742, 782, 1054, 1098, 1160, 1247,
...............................1264, 1384, 1416, 1427, 1492, 1959, 2189, 2266, 2271
Tung, Bui Thanh...62, 1926
Tunga, K. 1949
Tyberg, Christy..1372
Tyler, P. 601

U

Uhrmann, Thomas...888
Ulusoy, A. Cagri..2293
Unnikrishnan, R. 108
Ura, Shogo..186
Uwataki, R. 1179
Uzoh, Cyprian...862

V

Valdes-Garcia, Alberto..1272
Van Acker, Lut..165
Van der Donck, Tom...613
Van der Plas, G. 309
Van Driel, W.D. 1477
van Zeijl, Henk..1556
Vandevelde, B. 33, 1517
Vanstreels, K. 309
Vardakas, John...796
Varlamava, Volha..2137
Velenis, D. 572
Verbinnen, Greet..894
Vick, E. 8, 1641
Vlasov, Yurii...179
Vogel, D. 1134
Voges, S. 940
Voitsekhivska, Tetiana..173
von Kouwen, Maarten...1464
Vujosevic, Milena...395, 684

W

Wagner, Rebecca **236**

Wahrmund, Wieland **873**

Walls, Lloyd **535**

Walter, H. **625, 1421**

Waltereit, P. **2181**

Walters, E. **601**

Wan, Lixi **1354, 1378, 1932**

Wang, Chen-Chao **2112**

Wang, Daijiao **1965**

Wang, Guanjiang **641, 1986, 2099**

Wang, Hanguo **221**

Wang, Howard **2175**

Wang, Huijuan **2163**

Wang, Liang **215, 862**

Wang, M.J. **1212**

Wang, Nan **459**

Wang, Qian **1378**

Wang, Qibin **1740**

Wang, Qidong **1932**

Wang, Shiqiang **1086**

Wang, Shuling **1932**

Wang, T. **309, 572**

Wang, Tao **284, 484**

Wang, Tzu-Chang **1258**

Wang, W. **1212, 2048, 2245**

Wang, Xianyan **145**

Wang, Xugang **1932**

Wang, Yiman **1876**

Wang, Z. **108**

Wang, Zheyao **1722**

Wang, Zhihua **1378**

Wassick, T. **1940, 1949**

Watanabe, Manabu **68**

Watanabe, Naoya **62, 1915, 1926, 2003**

Watanabe, Shoji **348**

Weatherspoon, M.R. **601**

Webb, Bucknell **576**

Weber, Daniel **1033**

Wee, K.H. **324**

Wegner, M. **1204**

Wei, Frank **452**

Wei, Jia **1192**

Wei, Junchao **1449, 2053**

Wei, Pinghung **145**

Weigel, R.	956
Wen, Ming	1010, 2048, 2245
Wen, Shengmin	1231, 1361
Wentlent, L.	371
Werner, Randall J.	1253
Westerman, Russell	1577
Whalley, Simon	1033
White, Christopher	1264
Wietstruck, M.	1204
Wilde, Jürgen	1707, 2181
Wilke, M.	1204
Wimplinger, Markus	888
Win, Si T.	2151
Winstel, Kevin	647
Wittler, O.	625
Woertink, Julia	342
Wöhrmann, Markus	1421, 1498
Wojewoda, Leigh	528
Wojnowski, M.	956
Woldt, Gregor	1218
Wolf, H.	1482
Wolf, M.J.	625, 873, 1218
Wolter, Klaus-Juergen	173, 1509
Wong, C.P.	447, 464, 631, 769, 1488, 1492, 1745, 1828, 1833, 1848, 2231, 2299
Wong, C.S.	1517
Wong, K.	952
Woo, Min	126
Wood, James	1606
Wright, S.L.	1880
Wu, C.Y.	297
Wu, Chenglin	1122
Wu, Chuan-Yu	316
Wu, Chung-Hsi	512
Wu, Dong	1722
Wu, Fei-Jain	316
Wu, Hao	470, 1290
Wu, Li Wei	1533
Wu, Mei-Ling	1235
Wu, Peng	2131
Wu, Shang-Lin	512
Wu, Sheng-Tsai	290, 963
Wu, Wei-Hsin	316
Wu, Xiaomeng	2008, 2131
Wu, Xin	1010, 1189, 2048, 2245, 2283
Wu, Yuan-Yun	1529

Wu, Yung Shen...316
Wu, Zhongming...126
Wu, Zihan...1384
Wunderle, B...1134

X

Xia, Deyang..2236
Xie, Dongji..85, 126
Xie, John...338
Xie, Weidong...697
Xie, Xiang...1378
Xiong, Wei...560
Xu, Cheng...1116
Xu, Chunlin...2283
Xu, Gaowei...815, 1411
Xu, Hui..1523
Xu, Huili...133
Xu, Jiafeng...1833
Xu, P..1086, 1212
Xu, Ruqing..1648
Xu, Sha...1342
Xu, Steven...100
Xu, Yichao...641, 1986
Xu, Yong.................................170, 1010, 1189, 2048, 2245, 2283
Xue, Jie...1366
Xue, Kai..1116, 1740

Y

Yacoub-George, E..1482
Yagisawa, Takatoshi..1048
Yamada, Hiroshi..1729
Yamaguchi, Mitsugu...2194
Yamamoto, Tsuyoshi...68
Yamasaki, Kazuhiko...2194
Yan, Xiaolong...1622
Yang, Hyung Suk..13
Yang, Melinda (Ling)...354
Yang, Ming-Hsien...1
Yang, Ping-Feng...419, 2112
Yang, T.L...2277
Yang, Wenhua..1143
Yang, Xianbo...2144, 2168
Yau, YouWen...1173
Yazdani, Farhang..1667
Ye, J.T...1212
Ye, Jiaotuo..815

Ye, Lilei 459
Ye, Tiachun 1932
Yeap, Geoffrey 47
Yeh, C.T. 297
Yeung, Tak-Sang 518
Yin, L. 371, 1328
Yoneta, Kiyoto 2227
Yong, Andy Chang Bum 596
Yoo, Sehoon 447
Yook, Jong-Min 41
Yoon, Jong-Sun 1735
Yoon, Juhoon 1361
Yoon, S.W. 596, 931, 952
Yoon, Yong-Kyu 736, 789, 1103
Yoshikawa, Tomoyasu 1308
Yoshioka, T. 899
You, Eileen 354
You, Se-Ho 354
Youssef, Ramey 878
Yu, Daquan 1116, 1740, 1932, 2163
Yu, Doug C.H. 572
Yu, H. 324, 470, 1290
Yu, J.J. 2277
Yu, Man 2245
Yuan, Cadmus 1192
Yue, C. Patrick 919
Yue, Wu 2249
Yuen, Matthew M.F. 464, 1822
Yun, Feng 2236

Z

Zandén, Carl 459
Zhan, Chau-Jie 290, 1891, 1908
Zhang, Andy 85, 1441
Zhang, Chaoqi 13
Zhang, David C. 2260
Zhang, Di 85
Zhang, G.Q. 1477, 1556
Zhang, Gaugping 464
Zhang, Guoqi 1192
Zhang, Haipeng 1054
Zhang, Jinshen 170
Zhang, Kai 464
Zhang, Li 2299
Zhang, Mingchuan 2163
Zhang, Minyan 2236

Zhang, Pengtu	459
Zhang, Ron	215
Zhang, S.H.	1212
Zhang, Shengzhi	1010, 1690, 2283
Zhang, Tingting	2175
Zhang, Tonglong	1622
Zhang, Weihan	2236
Zhang, Weihong	1622
Zhang, Wenli	504
Zhang, Wenqi	1116
Zhang, Xia	2008
Zhang, Xin-Ping	2249
Zhang, Xuefeng	47
Zhang, Ye	2236
Zhao, Lily	47
Zhao, Wei	1226
Zhao, Xueying	1745
Zhao, Yukun	2236
Zhao, Zhili	1876
Zheng, H.Z.	1212
Zheng, Huai	1876
Zheng, Min	2236
Zhong, Jie	1086
Zhong, Xiao	1902
Zhong, Yinsheng	1822
Zhou, Bite	684
Zhou, Eric	221
Zhou, Feng	1622
Zhou, Longzao	1348
Zhou, Min-Bo	2249
Zhou, Tiao	1679
Zhu, Chunsheng	815, 1411
Zhu, Xiaoxin	1342
Zhu, Xunxun	1378
Zhu, Yunhui	641, 1902, 1986, 2099
Zilch, Christian	1183
Zischler, Sigrid	1700
Zitz, Jeffrey	332, 647
Zohni, Wael	1389
Zoschke, K.	1204
Zou, Simin	441
Zou, Yu	2299
Zou, Zhihua	1290
Zschenderlein, U.	1134